Manufacturing Processes for Engineering Materials

To grandson

Bennett Matthew Kalpakjian

and

Shelly Petronis

Manufacturing Processes for Engineering Materials

FOURTH EDITION

Serope Kalpakjian
Illinois Institute of Technology, Chicago, Illinois

Steven R. Schmid
University of Notre Dame, Notre Dame, Indiana

Pearson Education, Inc.
Upper Saddle River, New Jersey 07458

Library of Congress Cataloging-in-Publication Data

Kalpakjian, Serope, 1928–
Manufacturing processes for engineering materials/Serope Kalpakjian, Steven R. Schmid.—4th ed.
p. cm.
Includes bibliographical references and index.
ISBN 0-13-040871-9
1. Manufacturing processes. I. Schmid, Steven R. II. Title
TS183.K34 2002
670.42—dc21

2002070096

Vice President and Editorial Director, ECS: *Marcia J. Horton*
Acquisitions Editor: *Dorothy Marrero*
Editorial Assistant: *Brian Hoehl*
Vice President and Director of Production and Manufacturing, ESM: *David W. Riccardi*
Executive Managing Editor: *Vince O'Brien*
Managing Editor: *David A. George*
Production Editor: *Tamar Savir*
Director of Creative Services: *Paul Belfanti*
Creative Director: *Carole Anson*
Art Editor: *Xiaohong Zhu*
Cover Designer: *Geoffrey Cassar*
Manufacturing Manager: *Trudy Pisciotti*
Manufacturing Buyer: *Lisa McDowell*
Marketing Manager: *Holly Stark*

Pearson Education, Inc.
Upper Saddle River, NJ 07458

Printed in the United States of America.

10 9 8 7 6 5 4 3 2 1

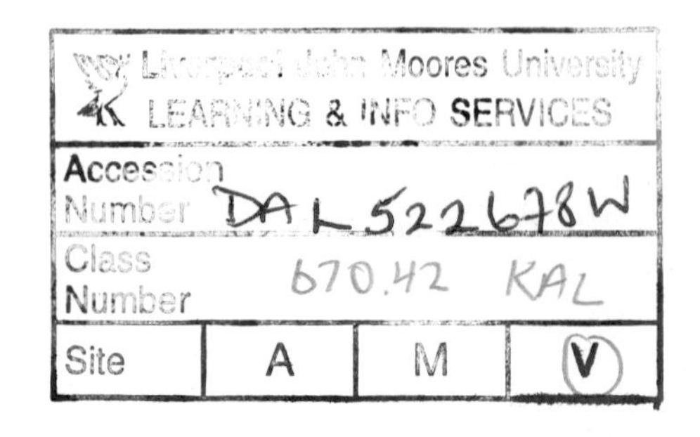

ISBN 0-13-040871-9

Pearson Education Ltd., *London*
Pearson Education Australia Pty. Ltd., *Sydney*
Pearson Education Singapore, Pte. Ltd.
Pearson Education North Asia Ltd., *Hong Kong*
Pearson Education Canada, Inc., *Toronto*
Pearson Educación de Mexico, S.A. de C.V.
Pearson Education—Japan, *Tokyo*
Pearson Education Malaysia, Pte. Ltd.
Pearson Education, Inc., *Upper Saddle River, New Jersey*

CONTENTS

3 Structure and Manufacturing Properties of Metals 81

4 Surfaces, Tribology, Dimensional Characteristics, Inspection, and Product Quality Assurance 128

5 Metal-Casting Processes and Equipment; Heat Treatment 186

11 Properties and Processing of Metal Powders, Ceramics, Glasses, Composites, and Superconductors 630

12 Joining and Fastening Processes 688

15 Computer-Integrated Manufacturing Systems 852

16 Product Design and Competitive Aspects of Manufacturing 893

Index 928

PREFACE

The fourth edition of this text, like the previous editions, continues to make an attempt at presenting a comprehensive, balanced, and up-to-date coverage of the relevant fundamentals and real-world applications of engineering materials and manufacturing processes and systems and the analytical approaches involved. The topics covered throughout the 16 chapters reflect the rapid and significant advances that have taken place in various areas in manufacturing, and they are organized and described in such a manner to draw the interest of students. The case studies, mostly from industry, make the subject of manufacturing science and engineering less abstract by showing students the practical aspects of process applications.

Integration of product design and manufacturing has justly resulted in greater recognition and prestige for these disciplines, and this edition therefore puts greater emphasis and better focus on this increasingly important subject. While studying this text, students should be able to assess the capabilities, limitations, and potential of production processes, including, particularly, the processes' economic and competitive aspects. The discussions throughout the chapters are aimed at motivating and challenging students to explore technically and economically viable solutions to a wide variety of important questions and problems in product design and manufacturing.

This book has been written mainly for undergraduate students in mechanical, industrial, and metallurgical and materials engineering programs; it is also useful for graduate courses in manufacturing science and engineering. The text, as well as the numerous examples and case studies in each chapter, clearly shows that manufacturing engineering is an interdisciplinary and complex subject and that it is as important, exciting, and challenging as any other engineering discipline.

What's new in this edition

- A new chapter has been added that covers the fabrication of microelectronic and micromechanical devices.
- The discussion of design considerations in each chapter have been expanded.
- Several new examples and case studies have been added throughout all chapters.
- A total of 1230 questions and problems has been added—30% more than in the third edition.
- Questions and problems now include a total of 140 design problems.
- Figures have been improved for better graphical impact.
- More cross-references to sections and chapters have been added throughout the text.
- All chapter bibliographies have been thoroughly updated.

New or expanded topics for this edition include the following:

- Automated guided vehicles
- Biodegradable plastics
- Cryogenic machining and grinding
- Cryogenic treatment of cutting tools
- Design considerations for casting
- Design considerations for powder metallurgy

- Design considerations for sheet-metal forming
- Electrically conducting adhesives
- Enterprise resource planning
- Flexible fixturing
- ISO 9000 and 14000 standards
- Life-cycle assessment
- LIGA process
- Metal foams
- Microelectromechanical systems fabrication
- Microelectronics device fabrication
- Micromachining
- Modeling of casting
- Nanomaterials
- Plasma and wet etching
- Polymer processing
- Printed circuit boards
- Rapid protoyping and rapid tooling
- Rotary ultrasonic machining
- Silicon microstructure
- Six-sigma quality
- Solid free-form fabrication
- Stick–slip in extrusion
- Superconductor processing
- Surface-mount technology
- Value assessment
- Vibration and chatter in rolling
- Water-jet peening

Acknowledgments

We gratefully acknowledge the following colleagues and associates for their contributions to the various sections of this book:

K. Anderson (Baxter Healthcare Corp.), P. J. Courtney (Loctite Corp.), D. Furrer (Ladish Corp.), K. L. Graham (Guidant Corp.), M. Hawkins (Zimmer, Inc.), K. M. Kalpakjian (Micron Technology, Inc., Mr. Kalpakjian is principal author of the sections on fabrication of microelectronic devices in Chapter 13), R. Kassing (University of Kassel), K. M. Kulkarni (Advanced Manufacturing Practices, Inc.), M. Madou (Nanogen, Inc.), A. Marsan (Ford Motor Co.), S. Paolucci (University of Notre Dame), C. Petronis (ST Microelectronics, Inc.), S. Petronis (Zimmer Holdings, Inc.), M. Pradheeradhi (Concurrent Technologies Corp.), Y. Rong (Worcester Polytechnic Institute), P. Saha (Boeing Co.), S. Shepel (University of Notre Dame), M. T. Siniawski (Northwestern University), P. Stewart (Ford Motor Co.), S. Vaze (Concurrent Technologies Corp.), J. E. Wang (Texas A&M University), K. J. Weinmann (Michigan Technological University), and K. R. Williams (Agilent, Inc.).

Many thanks to our colleagues at various institutions for their help, their detailed reviews, comments contributions, and many constructive suggestions for this edition:

D. D. Arola (University of Maryland), S. Mantell (University of Minnesota), M. H. Miller (Michigan Technological University), J. Moller (Miami University), D. J. Morrison (Clarkson University), U. Pal (Boston University), B. S. Thakkar (Lucent Technologies), and A. Tseng (Arizona State University), Dr. Rajiv Shivpuri (Ohio State University), Dr. Shaochen Chen (Iowa State University), Dr. Nicholas Zabaras (Cornell University), Dr. Mark Tuttle (University of Washington), Dr. Donald W. Radford (Colorado State University), Dr. Mica Grujicic (Clemson University).

We would also like to acknowledge the dedication and continued help of our editor, Laura Fischer, and the editorial staff at Prentice Hall, including Tamar Savir, David George, and Xiaohong Zhu. We appreciate the help of the many organizations that supplied us with various case studies and numerous illustrations. And, finally, many thanks to Jean Kalpakjian for her help in the final preparation of this book.

SEROPE KALPAKJIAN
Chicago, Illinois

STEVEN R. SCHMID
Notre Dame, Indiana

ABOUT THE AUTHORS

Serope Kalpakjian taught and conducted research at the Illinois Institute of Technology for 38 years prior to his retirement in 2001 as professor emeritus of mechanical and materials engineering. After graduating from Robert College (high honors), Harvard University, and the Massachusetts Institute of Technology, he joined Cincinnati Milacron, where he was a research supervisor in advanced metal-forming processes. He is the author of numerous technical papers and several articles in handbooks and encyclopedias and has edited various conference proceedings. In addition, Professor Kalpakjian has served on the editorial boards of several journals and the *Encyclopedia Americana* and is the coauthor of *Lubricants and Lubrication in Metalworking Operations*. Both first editions of his textbooks, *Manufacturing Processes for Engineering Materials* and *Manufacturing Engineering and Technology*, have received the M. Eugene Merchant Manufacturing Textbook Award. He is a Life Fellow of the ASME, Fellow of the SME, Fellow and Life Member of ASM International, emeritus member of CIRP (International Institution for Production Engineering Research), and a founding member and past president of NAMRI/SME.

Among the awards Professor Kalpakjian has received are the Forging Industry Educational and Research Foundation Best Paper Award (1966), an Excellence in Teaching Award from IIT (1970), the Centennial Medallion from the ASME (1980), the International Education Award from the SME (1989), a Person of the Millennium Award from IIT (1999), and the Albert Easton White Distinguished Teacher Award from ASM International (2000). SME named the Outstanding Young Manufacturing Engineer Award after Professor Kalpakjian for the year 2002.

Steven R. Schmid is an associate professor with the Department of Aerospace and Mechanical Engineering, University of Notre Dame, where he teaches and performs research in the general areas of manufacturing, machine design, and tribology. As the director of the Manufacturing Tribology Laboratory at the university, he oversees industry- and government-funded research on a variety of manufacturing topics, including tribological issues in rolling, forging and sheet-metal forming, polymer processing, medical-device design and manufacture, and nanomechanics.

He received the B.S. degree in mechanical engineering from the Illinois Institute of Technology (with honors) and M.S. and Ph.D. degrees, both in mechanical engineering, from Northwestern University. He has received numerous awards, including the John T. Parsons Award from the SME (2000), the Newkirk Award from the ASME (2000), and the Kaneb Center Teaching Award (2000). He is the recipient of a National Science Foundation CAREERS Award (1996) and an ALCOA Foundation Award (1994). Dr. Schmid is the author of over 50 technical papers; has edited three conference proceedings; has coauthored two books, *Fundamentals of Machine Elements* and *Manufacturing Engineering and Technology*; and has contributed two chapters to the *CRC Handbook of Modern Tribology*. He serves on the Tribology Division Executive Committee of the ASME, is an associate editor of the *Journal of Manufacturing Science and Engineering*, and is a registered professional engineer and certified manufacturing engineer.

CHAPTER

1 Introduction

1.1 What Is Manufacturing?

As you read this Introduction, take a few moments to inspect the different objects around you: your watch, chair, stapler, pencil, calculator, telephone, and light fixtures. You will soon realize that all these objects have been transformed from various raw materials into individual parts and assembled into specific products. Some objects, such as nails, bolts, and paper clips, are made of one material; however, most objects, such as light bulbs, toasters, bicycles, computers, all types of instruments and machinery, and automotive engines (Fig. 1.1), and countless other products, are made of numerous parts from a wide variety of materials. A ballpoint pen, for example, consists of about a dozen parts, a lawnmower about 300 parts, a grand piano

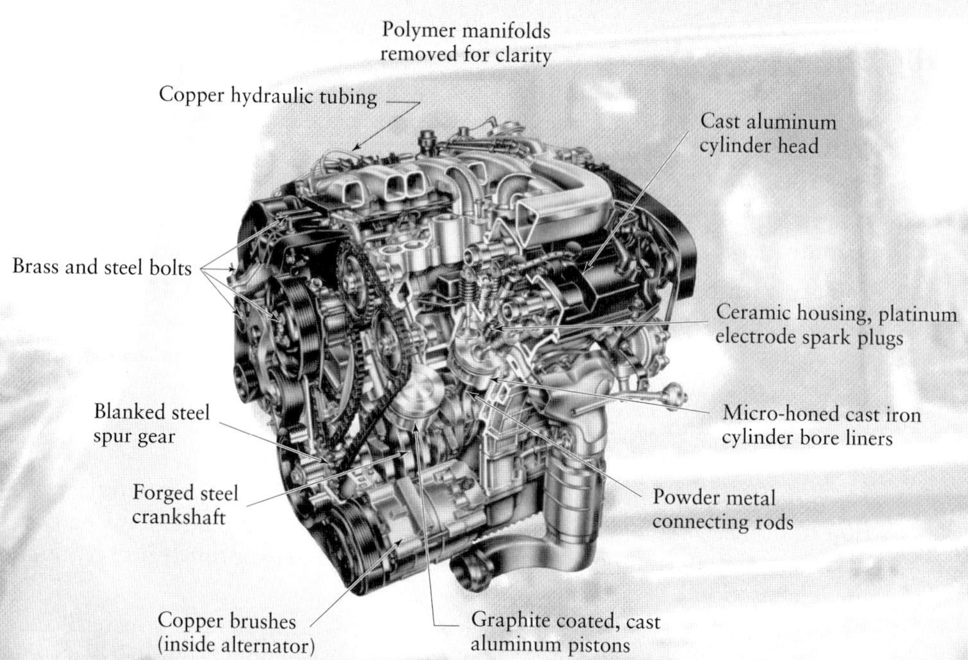

FIGURE 1.1 Section of an automotive engine—the Duravec V-6—showing various components and the materials used in making them. *Source*: Courtesy of Ford Motor Company. Illustration by David Kimball.

12,000 parts, a typical automobile 15,000 parts, a C-5A transport plane more than 4 million parts, and a Boeing 747-400 about 6 million parts; all are produced by a combination of various processes called *manufacturing*.

Manufacturing, in its broadest sense, is the process of converting raw materials into products; it encompasses the design and fabrication of goods by means of various production methods and techniques. Manufacturing began around 5000 to 4000 BC with the production of various articles of wood, ceramic, stone, and metal (Table 1.1). The word *manufacturing* is derived from the Latin *manu factus*, meaning made by hand; the word *manufacture* first appeared in AD 1567 and the word *manufacturing* in 1683. The word **production** is also used interchangeably with the word *manufacturing*. Manufacturing represents approximately 20% to 30% of the value of all goods and services produced in industrialized countries. Generally, the higher the level of manufacturing in a country, the higher the standard of living of its people. Manufactured products are also used to make other products; examples include large presses to form sheet metal for car bodies, metalworking machinery used to make parts for other products, and sewing machines for making clothing.

Manufacturing may produce *discrete products*, meaning individual parts or pieces of parts, or it may produce continuous products. Nails, gears, steel balls, beverage cans, and engine blocks are examples of discrete products. Wire, metal or plastic sheet, hose, and pipe are *continuous products* that may be cut into individual pieces and thereby become discrete products.

Because a manufactured item has undergone a number of changes during which raw material has become a useful product, it has **added value**, defined as monetary worth in terms of price. For example, clay has a certain value when mined. When the clay is used to make a ceramic dinner plate, cutting tool, or electrical insulator, value is added to the clay; similarly, a wire coat hanger or a nail has added value over and above the cost of a piece of wire.

Manufacturing is generally a complex activity involving people who have a broad range of disciplines and skills, together with a wide variety of machinery, equipment, and tools with various levels of automation, including computers, robots, and material-handling equipment. Manufacturing activities must be responsive to several demands and trends:

1. A product must fully meet **design requirements** and **specifications** and **standards.**
2. A product must be manufactured by the most **economical** and **environmentally friendly** methods.
3. **Quality** must be built into the product at each stage, from design to assembly, rather than relying on quality testing after the product is made.
4. In a highly competitive environment, production methods must be sufficiently **flexible** to respond to changing market demands, types of products, production rates, and production quantities and to provide on-time delivery to the customer.
5. New developments in **materials, production methods**, and **computer integration** of both technological and managerial activities in a manufacturing organization must constantly be evaluated with respect to their timely and economic implementation.
6. Manufacturing activities must be viewed as a large **system** in which all individual components are interrelated. Such systems can now be modeled in order to study the effects of various factors, such as changes in market demands, product design, materials, costs, and production methods, on product quality and cost.

TABLE 1.1

Historical Development of Materials and Manufacturing Processes

Period	Dates	Metals and casting	Various materials and composites	Forming and shaping	Joining	Tools, machining, and manufacturing systems
Egypt: ~3100 BC to ~300 BC Greece: ~1100 BC to ~146 BC Roman Empire: ~500 BC to AD 476 Middle Ages: ~476 to 1492 Renaissance: 14th to 16th centuries	Before 4000 BC	Gold, copper, meteoric iron	Earthenware, glazing, natural fibers	Hammering		Tools of stone, flint, wood, bone, ivory, composite tools
	4000–3000 BC	Copper casting, stone and metal molds, lost-wax process, silver, lead, tin, bronze		Stamping, jewelry	Soldering (Cu–Au, Cu–Pb, Pb–Sn)	Corundum (alumina, emery)
	3000–2000 BC	Bronze casting and drawing, gold leaf	Glass beads, potter's wheel, glass vessels	Wire by slitting sheet metal	Riveting, brazing	Hoe making, hammered axes, tools for ironmaking and carpentry
	2000–1000 BC	Wrought iron, brass				
	1000–1 BC	Cast iron, cast steel	Glass pressing and blowing	Stamping of coins	Forge welding of iron and steel, gluing	Improved chisels, saws, files, woodworking lathes
	AD 1–1000	Zinc, steel	Venetian glass	Armor, coining, forging, steel swords		Etching of armor
	1000–1500	Blast furnace, type metals, casting of bells, pewter	Crystal glass	Wire drawing, gold- and silversmith work		Sandpaper, windmill-driven saw
	1500–1600	Cast-iron cannon, tinplate	Cast plate glass, flint glass	Water power for metalworking, rolling mill for coinage strips		Hand lathe for wood
	1600–1700	Permanent-mold casting, brass from copper and metallic zinc	Porcelain	Rolling (lead, gold, silver), shape rolling (lead)		Boring, turning, screw-cutting lathe, drill press

(continues on next page)

Historical Development of Materials and Manufacturing Processes (cont.)

Period	Dates	Metals and casting	Various materials and composites	Forming and shaping	Joining	Tools, machining, and manufacturing systems
Industrial Revolution: ~1750 to 1850	1700–1800	Malleable cast iron, crucible steel (iron bars and rods)		Extrusion (lead pipe), deep drawing, rolling		
	1800–1900	Centrifugal casting, Bessemer process, electrolytic aluminum, nickel steels, babbitt, galvanized steel, powder metallurgy, open-hearth steel	Window glass from slit cylinder, light bulb, vulcanization, rubber processing, polyester, styrene, celluloid, rubber extrusion, molding	Steam hammer, steel rolling, seamless tube, steel-rail rolling, continuous rolling, electroplating		Shaping, milling, copying lathe for gunstocks, turret lathe, universal milling machine, vitrified grinding wheel
	1900–1920		Automatic bottle making, bakelite, borosilicate glass	Tube rolling, hot extrusion	Oxyacetylene; arc, electrical-resistance, and thermit welding	Geared lathe, automatic screw machine, hobbing, high-speed-steel tools, aluminum oxide and silicon carbide (synthetic)
WW I	1920–1940	Die casting	Development of plastics, casting, molding, polyvinyl chloride, cellulose acetate, polyethylene, glass fibers	Tungsten wire from metal powder	Coated electrodes	Tungsten carbide, mass production, transfer machines
WW II	1940–1950	Lost-wax process for engineering parts	Acrylics, synthetic rubber, epoxies, photosensitive glass	Extrusion (steel), swaging, powder metals for engineering parts	Submerged arc welding	Phosphate conversion coatings, total quality control
	1950–1960	Ceramic mold, nodular iron, semiconductors, continuous casting	Acrylonitrile-butadiene-styrene, silicones, fluorocarbons, polyurethane, float glass, tempered glass, glass ceramics	Cold extrusion (steel), explosive forming, thermomechanical processing	Gas metal arc, gas tungsten arc, and electroslag welding; explosive welding	Electrical and chemical machining, automatic control

Historical Development of Materials and Manufacturing Processes (cont.)

Period	Dates	Metals and casting	Various materials and composites	Forming and shaping	Joining	Tools, machining, and manufacturing systems
Space Age	1960–1970	Squeeze casting, single-crystal turbine blades	Acetals, polycarbonate, cold forming of plastics, reinforced plastics, filament winding	Hydroforming, hydrostatic extrusion, electroforming	Plasma-arc and electron-beam welding, adhesive bonding	Titanium carbide, synthetic diamond, numerical control, integrated circuit chip
	1970–1990	Compacted graphite, vacuum casting, organically bonded sand, automation of molding and pouring, rapid solidification, metal-matrix composites, semisolid metalworking, amorphous metals, shape-memory alloys (smart materials), computer simulation	Adhesives, composite materials, semiconductors, optical fibers, structural ceramics, ceramic-matrix composites, biodegradable plastics, electrically conducting polymers	Precision forging, isothermal forging, superplastic forming, dies made by computer-aided design and manufacturing, net-shape forging and forming, computer simulation	Laser beam, diffusion bonding (also combined with superplastic forming), surface-mount soldering	Cubic boron nitride, coated tools, diamond turning, ultraprecision machining, computer-integrated manufacturing, industrial robots, machining and turning centers, flexible-manufacturing systems, sensor technology, automated inspection, expert systems, artificial intelligence, computer simulation and optimization
Information Age	1990–2000s	Rheocasting, computer-aided design of molds and dies, rapid tooling	Nanophase materials, metal foams, advanced coatings, high-temperature superconductors. machinable ceramics, diamondlike carbon	Rapid prototyping, rapid tooling, environmentally friendly metalworking fluids	Friction stir welding, lead-free solders, laser butt-welded (tailored) sheet-metal blanks, electrically conducting adhesives	Micro- and nano-fabrication, LIGA (a German acronym for a process involving lithography, electroplating, and molding), dry etching, linear motor drives, artificial neural networks, six sigma

Source: J.A. Schey, C.S. Smith, R.F. Tylecote, T.K. Derry, T.I Williams, S.R. Schmid, and S. Kalpakjian.

7. The manufacturer must work with the customer to get timely feedback for **continuous product improvement.**
8. The manufacturing organization must constantly strive for higher **productivity,** defined as the optimum use of all its resources: materials, machines, energy, capital, labor, and technology. Output per employee per hour in all phases must be maximized.

1.2 Product Design and Concurrent Engineering

Product design is a critical activity: It has been estimated that 70% to 80% of the cost of product development and manufacture is determined at the initial design stages. The design process for a product first requires a clear understanding of the functions and the performance expected of the product. The product may be new, or it may be a new and improved model of an existing product. The market for a product and the product's anticipated uses must be defined clearly, with the assistance of sales personnel, market analysts, and others in the organization.

Traditionally, design and manufacturing activities have taken place sequentially rather than concurrently or simultaneously (Fig. 1.2a). Designers would spend considerable effort and time in analyzing components and preparing detailed drawings of parts; these drawings would then be forwarded to other departments in the organization, such as materials departments, where, for example, particular alloys and vendor sources would be identified. The specifications would then be sent to the manufacturing department, where the detailed drawings would be reviewed and processes selected for efficient production. While this approach seems logical and straightforward in theory, it has been found in practice to be extremely wasteful of resources.

In theory, a product can flow from one department in an organization to another and then directly to the marketplace, but in practice difficulties are usually encountered. For example, a manufacturing engineer may wish to taper the flange on a part to improve its castability or may decide that a different alloy is desirable; such changes necessitate a repeat of the design analysis stage in order to ensure that the product will still function satisfactorily. These iterations, also shown in Fig. 1.2a, certainly waste resources, but, more importantly, they waste time.

A more advanced approach to product development is shown in Fig. 1.2b. While it still has a general product flow from market analysis to design to manufacturing, it contains deliberate iterations. The main difference between it and the older approach is that all disciplines are now involved in the early design stages and the stages progress concurrently, so that the iterations (which occur by nature) result in less wasted effort and less lost time. A key to the new approach is the recognition of the importance of communication among and within disciplines. That is, while there must be communication between engineering, marketing, and service functions, so, too, must there be avenues of interaction between engineering subdisciplines—for example, design for manufacture, design for recyclability, and design for safety.

Concurrent engineering (CE), also called *simultaneous engineering*, is a systematic approach that integrates the design and manufacture of products with the view of optimizing all elements involved in the life cycle of the product. **Life cycle** means that all aspects of a product, such as design, development, production, distribution, use, its ultimate recycling and disposal and its environmental impact are

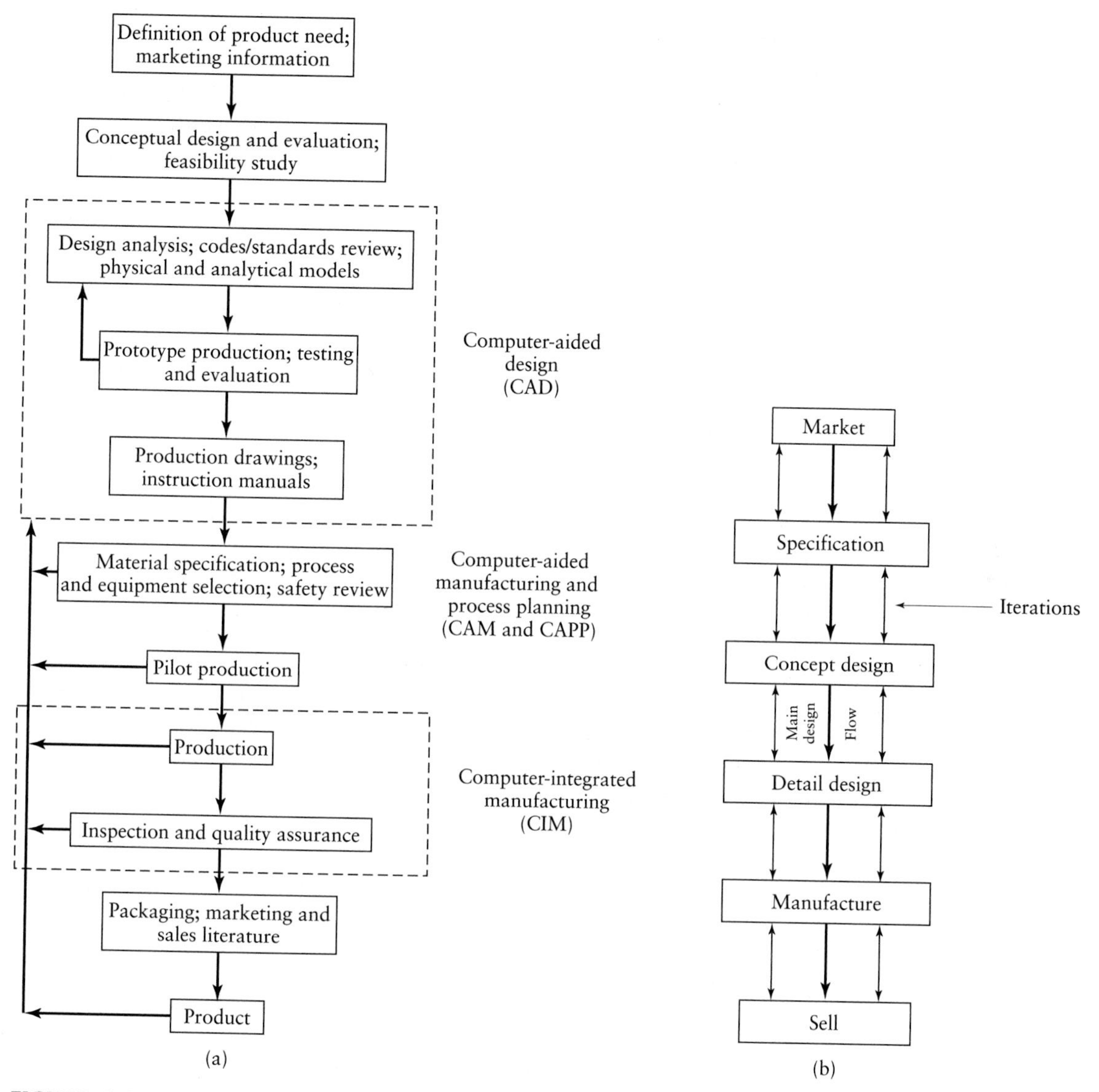

FIGURE 1.2 (a) Chart showing the various steps involved in designing and manufacturing a product. Depending on the complexity of the product and the type of materials used, the time span between the original concept and the marketing of a product may range from a few months to many years. (b) Chart showing general product flow, from market analysis to selling the product, and depicting concurrent engineering. *Source*: After S. Pugh, *Total Design*, Addison-Wesley, 1991.

considered simultaneously. The basic goal of concurrent engineering is to minimize product design and engineering changes and the time and costs involved in taking the product from design concept to production to introduction into the marketplace. An extension of concurrent engineering, called **direct engineering**, has also been proposed. It uses a database representing the engineering logic applied during the design of each component of a product. If a design modification is made on a part, direct engineering will determine the manufacturing consequences of that change.

Although the concept of concurrent engineering appears to be logical and efficient, its implementation can take considerable time and effort when those using it either do not work as a team or fail to appreciate its real benefits. It is apparent that for concurrent engineering to succeed, it must (1) have the full support of top management, (2) have a multifunctional and interactive work team, including support groups, and (3) use all available technologies.

For both large and small companies, product design often involves preparing analytical and physical models of the product as an aid for analyzing factors such as forces, stresses, deflections, and optimal part shape. The necessity for such models depends on product complexity; today, however, the process of constructing and studying analytical models is simplified through the use of **computer-aided design, engineering,** and **manufacturing** techniques. On the basis of these models, the product designer selects and specifies the final shape and dimensions of the product, its dimensional tolerances and surface finish, and the materials to be used. The selection of materials is generally made with the advice and cooperation of materials engineers, unless the design engineer is sufficiently experienced and qualified in this area. An important design consideration is how a particular component is to be assembled into the final product. To understand the concept of assembly, take apart a ballpoint pen or a toaster, or lift the hood of a car and observe how hundreds of components are put together in a confined space.

A powerful and effective tool, particularly for complex production systems, is **computer simulation,** which can be used to evaluate the performance of the product and plan the manufacturing system to produce it. Computer simulation also helps in early detection of design flaws, identification of possible problems in a particular production system, and optimization of manufacturing lines for minimum product cost. Several computer simulation languages that use animated graphics and have various capabilities have been developed.

The next step in the production process is to make and test a **prototype,** that is, an original working model of the product. An important development is **rapid prototyping** (Section 10.12), which relies on computer-aided design/computer-aided manufacturing (CAD/CAM) and various manufacturing techniques (typically using polymers or metal powders) to rapidly produce prototypes in the form of a solid physical model of a part. Rapid prototyping can significantly cut costs as well as development times. These techniques are currently being advanced further to such an extent that they can now be used for low-volume economical production of parts.

Tests on prototypes must be designed to simulate as closely as possible the conditions under which the product is to be used. Such factors include environmental conditions such as temperature and humidity, as well as the effects of vibration and repeated use and misuse of the product. During this stage, modifications of the original design, materials selected, or production methods may be necessary. After this phase has been completed, appropriate process plans, manufacturing methods (Table 1.2), equipment, and tooling are selected with the cooperation of manufacturing engineers, process planners, and all other involved in production.

1.3 Design for Manufacture, Assembly, Disassembly, and Service

It is apparent that design and manufacturing must be closely interrelated; they should never be viewed as separate disciplines or activities. Each part or component of a product must be designed so that it not only meets design requirements and specifications, but also can be manufactured economically and with relative ease.

TABLE 1.2

Shapes and Some Common Methods of Production

Shape or feature	Production method
Flat surfaces	Rolling, planing, broaching, milling, shaping, grinding
Parts with cavities	End milling, electrical-discharge machining, electrochemical machining, ultrasonic machining, casting
Parts with sharp features	Permanent-mold casting, machining, grinding, fabricating, powder metallurgy
Thin hollow shapes	Slush casting, electroforming, fabricating
Tubular shapes	Extrusion, drawing, roll forming, spinning, centrifugal casting
Tubular parts	Rubber forming, expanding with hydraulic pressure, explosive forming, spinning
Curvature on thin sheets	Stretch forming, peen forming, fabricating
Openings in thin sheets	Blanking, chemical blanking, photochemical blanking, laser
Square edges	Fine blanking, machining, shaving, grinding
Small holes	Laser, electrical-discharge machining, electrochemical machining, micromachining
Surface textures	Knurling, wire brushing, grinding, belt grinding, shot blasting, etching, deposition, laser texturing
Detailed surface features	Coining, investment casting, permanent-mold casting, machining
Threaded parts	Thread cutting, thread rolling, thread grinding, chasing
Very large parts	Casting, forging, fabricating
Very small parts	Investment casting, machining, etching, powder metallurgy, nanofabrication, LIGA, micromachining

This approach improves productivity and allows a manufacturer to remain competitive.

This broad concept is now recognized as the area of **design for manufacture** (DFM): a comprehensive approach to the production of goods that integrates the product design process with selection of materials, consideration of manufacturing methods, process planning, assembly, testing, and quality assurance. Effective implementation of design for manufacture requires that designers acquire a fundamental understanding of the characteristics, capabilities, and limitations of materials, production methods, and related operations, machinery, and equipment. This knowledge includes an understanding of characteristics such as variability in machine performance, dimensional accuracy and surface finish of the workpiece, processing time, and the effect of processing method on part quality.

Designers and product engineers must be able to assess the impact of design modifications on manufacturing-process selection, tools and dies, assembly, inspection, and product cost. Establishment of quantitative relationships is essential in order to optimize the design for ease of manufacturing and assembly at *minimum cost* (also called *producibility*). Computer-aided design, engineering, manufacturing, and process-planning techniques, using powerful computer programs, have become indispensable to those conducting such analysis; such computerized tools include **expert systems**, which are computer programs with optimization capabilities, that expedite the traditional iterative process in design optimization.

After individual parts have been manufactured, they are assembled into a product. **Assembly** is an important phase of the overall manufacturing operation and requires

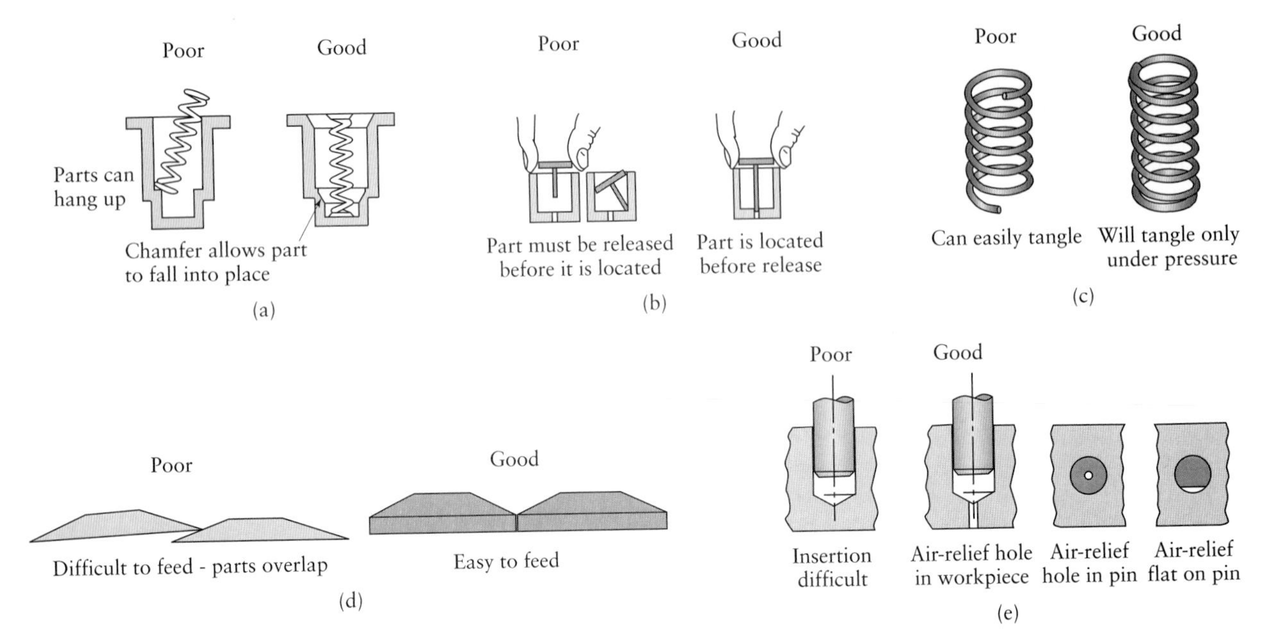

FIGURE 1.3 Redesign of parts to facilitate automated assembly. *Source*: Reprinted from G. Boothroyd and P. Dewhurst, *Product Design for Assembly*, 1989, by courtesy of Marcel Dekker, Inc.

consideration of the ease, speed, and cost of putting parts together (Fig. 1.3). Also, products must be designed so that **disassembly** is possible with relative ease and less time involved, enabling the products to be taken apart for maintenance, servicing, or recycling of their components. Because assembly operations can contribute significantly to product cost, **design for assembly** (DFA) and **design for disassembly** are now recognized as important aspects of manufacturing; typically, a product that is easy to assemble is also easy to disassemble. The trend now also includes **design for service**, ensuring that individual parts in a product are easy to reach and service. In addition, current methods combine design for manufacture and design for assembly into the more comprehensive **design for manufacture and assembly** (DFMA), which recognizes the inherent and important interrelationships among design, manufacturing, and assembly.

Design principles for economic production may be summarized as follows:

- The design should be as simple as possible to manufacture, assemble, disassemble, service, and recycle.
- Materials should be chosen for their appropriate manufacturing characteristics as well as for their service life.
- Specified dimensional accuracy and surface finish should be as broad as permissible.
- Because they can add significantly to cost, secondary and finishing operations should be avoided or minimized.

1.4 Environmentally Conscious Design and Manufacturing

In the United States alone, more than five billion kilograms of plastic products are discarded each year. Every three months, U.S. industries and consumers discard enough aluminum to rebuild the country's commercial air fleet. Globally, countless tons of automobiles, television sets, appliances, and computer equipment are discarded each year. Furthermore, metalworking fluids such as lubricants and coolants are often used in machining, grinding, and forming operations, and various fluids and solvents are used in cleaning manufactured products; some of these fluids can pollute the air and waters if discarded in an untreated form. Likewise, many byproducts from manufacturing plants have, for years, been discarded: sand with additives used in metal-casting processes; water, oil, and other fluids from heat-treating facilities and plating operations; slag from foundries and welding operations; and a wide variety of metallic and nonmetallic scrap produced in operations such as sheet forming, casting, and molding. Consider also the effects of water and air pollution, acid rain, ozone depletion, the greenhouse effect, hazardous wastes, landfill seepage, and global warming.

The present and potential future adverse effects of these activities, their damage to our environment and to the earth's ecosystem, and, ultimately, their effect on the quality of human life are now well recognized by the public as well as by local and federal governments. In response, a wide range of laws and regulations have been promulgated by local, state, and federal governments as well as professional organizations in the United States and other industrial countries. These regulations are generally stringent, and their implementation can have a major impact on the economic operation and financial health of industrial organizations. Regardless, and notwithstanding arguments by some that environmental damage from these activities is overestimated, manufacturing engineers and the management of companies have a major responsibility to plan and implement environmentally safe materials and manufacturing operations.

Major developments also have been taking place regarding **design for recycling** (DFR), indicating universal awareness of the aforementioned problems, and waste has become unacceptable. A major emphasis is on **design for the environment** (DFE), or **green design** ("green" meaning environmentally safe and friendly). This approach anticipates the possible negative environmental impact of materials, products, and processes so that it can be considered at the earliest stages of design and production.

It is apparent that it is essential to conduct a thorough analysis of the product, materials used, and the manufacturing processes and practices employed in making the product. Certain guidelines can be followed in this regard:

- Reduce waste of materials at their source by implementing refinements in product design and reducing the amount of materials used.
- Reduce the use of hazardous materials in products and processes.
- Ensure proper handling and disposal of all waste.
- Make improvements in recycling, waste treatment, and reuse of materials.

1.5 Selecting Materials

An ever-increasing variety of materials is now available, each having its own characteristics, applications, advantages, and limitations (Fig. 1.4). The following are the general types of materials used in manufacturing today, either individually or in combination:

1. **Ferrous metals:** carbon, alloy, stainless, and tool and die steels (Chapter 3);
2. **Nonferrous metals and alloys:** aluminum, magnesium, copper, nickel, titanium, superalloys, refractory metals, beryllium, zirconium, low-melting alloys, and precious metals (Chapter 3);

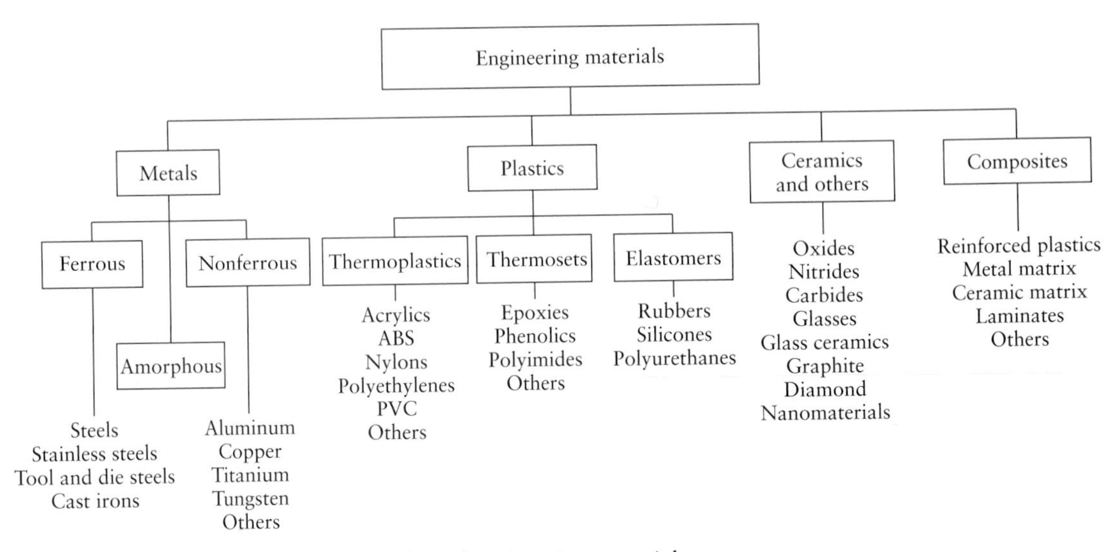

FIGURE 1.4 An outline of engineering materials.

3. **Plastics:** thermoplastics, thermosets, and elastomers (Chapter 10);
4. **Ceramics:** glass ceramics, glasses, graphite, and diamond (Chapter 11);
5. **Composite materials:** reinforced plastics, metal-matrix and ceramic-matrix composites, and honeycomb structures—these materials are also known as *engineered materials* (Chapters 10 and 11);
6. **New materials:** nanomaterials, shape-memory alloys, amorphous alloys, superconductors, and various other materials with unique properties (Chapter 3).

As new materials are developed, important trends arise with respect to their selection and application. Aerospace structures as well as products such as sporting goods especially have been at the forefront of new-material usage. Because of the vested interest of the producers of different types of metallic and nonmetallic materials, there are constantly shifting trends in the usage of these materials, driven principally by economics. For example, by demonstrating steel's technical and economic feasibility, steel producers are countering the increased use of plastics in cars and aluminum in beverage cans. Likewise, aluminum producers are countering the use of various materials in cars (Fig. 1.5). Obviously, making materials-selection choices that facilitate recycling should be a fundamental part of these considerations.

Some examples of the use or substitution of materials in common products are the following: steel vs. plastic paper clips, plastic vs. sheet-metal light-switch plates, wood vs. metal handles for hammers, glass vs. metal water pitchers, plastic vs. leather car seats, sheet-metal vs. reinforced-plastic chairs, galvanized-steel vs. copper nails, and aluminum vs. cast-iron frying pans.

Properties of materials. When selecting materials for products, the first consideration generally involves **mechanical properties** (Chapter 2), which typically are strength, toughness, ductility, hardness, elasticity, and fatigue and creep resistance. The mechanical properties of materials can be significantly modified by various heat treatment methods (Chapter 5). The strength-to-weight and stiffness-to-weight ratios of materials are also important, particularly for aerospace and automotive applications.

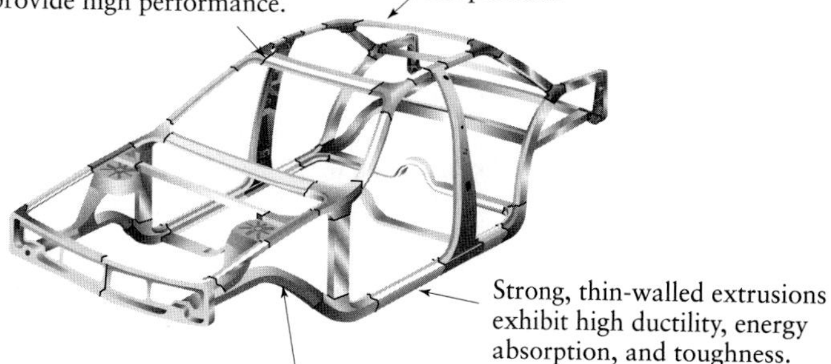

(a) (b)

FIGURE 1.5 (a) The Audi A8 automobile, an example of advanced materials construction. (b) The aluminum body structure, showing various components made by extrusion, sheet forming, and casting processes. *Source*: Courtesy of ALCOA, Inc.

Aluminum, titanium, and reinforced plastics, for example, have higher strength-to-weight ratios than those of steels and cast irons. The mechanical properties specified for a product and its components should, of course, be appropriate for the conditions under which the product is expected to function.

Further aspects to consider are the **physical properties** (Chapter 3) of the materials, such as density, specific heat, thermal expansion and conductivity, melting point, and electrical and magnetic properties. **Chemical properties** can play a significant role in hostile as well as normal environments. Oxidation, corrosion, general degradation of properties, and flammability of materials are among the important factors to consider, as is toxicity. (Note, for example, the development of lead-free solders; see Section 12.12.3). Both the physical and the chemical properties of materials are important in advanced machining processes (Chapter 9). Finally, the **manufacturing properties** of materials determine whether they can be cast, formed, shaped, machined, welded, or heat treated with relative ease. An important consideration is that the methods used to process materials to the desired shapes should not adversely affect the product's final properties, service life, and cost.

Cost and availability. The economic aspects of material selection are as important as the technological considerations of properties and characteristics of materials. The cost and availability of raw and processed materials and manufactured components are major concerns in manufacturing. If raw or processed materials or manufactured components are not commercially available in the desired shapes, dimensions, and quantities, substitutes or additional processing will be required, which can contribute significantly to product cost (Chapter 16). For example, if we need a round bar of a certain diameter and it is not available from the supplier, then we will have to purchase a larger rod and reduce its diameter by a process such as machining, drawing though a die, or grinding.

Reliability of both supply and demand affects cost. Most countries import numerous raw materials that are essential for production. The United States, for example, imports the majority of such raw materials as natural rubber, diamond, cobalt, titanium, chromium, aluminum, and nickel from other countries. The broad political implications of such reliance on other countries are self-evident. In addition, various costs are involved in processing materials by different methods. Some methods require expensive machinery, others require extensive labor, and still others require personnel with special skills, a high level of formal education, or specialized training.

Service life and recycling. Time- and service-dependent phenomena such as wear, fatigue, creep, and dimensional stability are important considerations, as they can significantly affect a product's performance and, if not controlled, can lead to failure of the product. Likewise, compatibility of materials used in a product is important; for example, galvanic action between mating parts made of dissimilar metals causes them to corrode. Recycling or proper disposal of components in a product at the end of its useful life has become increasingly important as we become more and more aware of the importance of conserving materials and energy and living in a clean and healthy environment. The proper treatment and disposal of toxic wastes is also a crucial consideration.

1.6 Selecting Manufacturing Processes

As can be seen in Table 1.2, a wide range of manufacturing processes is used to produce a variety of parts, shapes, and sizes. Also note that there is usually more than one method for manufacturing a part from a given material, and, as expected, each of these processes has its own advantages, limitations, production rates, and cost (Fig. 1.6). The broad categories of processing methods for materials are as follows:

- **Casting:** expendable mold and permanent mold (Chapter 5);
- **Forming and shaping:** rolling, forging, extrusion, drawing, sheet forming, powder metallurgy, and molding (Chapters 6, 7, 10, 11);
- **Machining:** turning; boring; drilling; milling; planing; shaping; broaching; grinding; ultrasonic machining; chemical, electrical, and electrochemical machining; and high-energy beam machining (Chapters 8 and 9);
- **Joining:** welding, brazing, soldering, diffusion bonding, adhesive bonding, and mechanical joining (Chapter 12);
- **Finishing:** honing, lapping, polishing, burnishing, deburring, surface treating, coating, and plating (Chapter 9).

Selection of a particular manufacturing process or a series of processes depends not only on the component or part shape to be produced, but also on many other factors. Brittle and hard materials, for example, cannot be shaped easily, whereas they

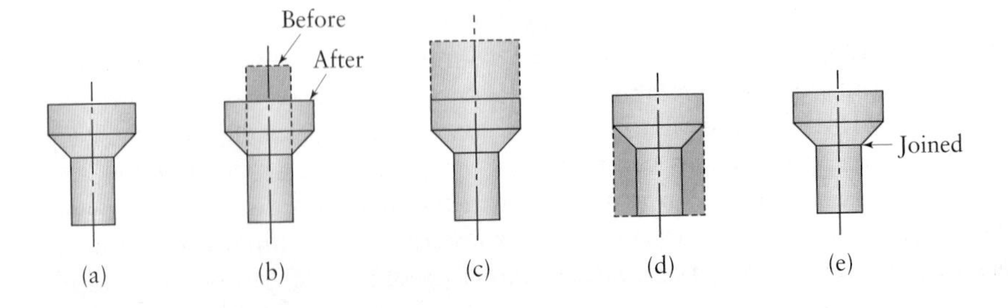

FIGURE 1.6 Various methods of making a simple part: (a) casting or powder metallurgy, (b) forging or upsetting, (c) extrusion, (d) machining, (e) joining two pieces.

can be cast or machined by several methods. The manufacturing process usually alters the properties of materials; metals that are formed at room temperature, for example, become stronger, harder, and less ductile than they were before processing. Thus, it is important to study characteristics such as *castability*, *formability*, *machinability*, and *weldability* of materials. Some examples of manufacturing processes used or substituted for common products are the following: forging vs. casting of crankshafts, sheet-metal vs. cast hubcaps, casting vs. stamping sheet metal for frying pans, machined vs. powder-metallurgy gears, thread rolling vs. machining of bolts, and casting vs. welding of machine structures.

Manufacturing engineers are constantly challenged to find new solutions to production problems as well as to find means for significant cost reduction. For a long time, for example, sheet-metal parts were cut and made by traditional tools such as punches and dies. Although they are still widely used, these operations are now being replaced by laser cutting techniques. With advances in computer controls, path of the laser can be automatically controlled, thus producing a wide variety of shapes accurately, repeatedly, and economically without the use of expensive tooling.

Size and dimensional accuracy. Obviously, size, shape complexity, and thickness of a part have a major influence on the process selected to produce the part. Complex parts, for example, cannot be formed easily and economically, whereas they may be cast, injection molded, or fabricated from individual pieces or by powder-metallurgy techniques; likewise, flat parts with thin cross-sections cannot be cast properly. Dimensional tolerances and surface finish (Chapter 4) obtained in hot-working operations cannot be as fine as those obtained in cold-working operations, because dimensional changes, warping, and surface oxidation occur during processing at elevated temperatures. Also, some casting processes produce a better surface finish than others, because of the different types of mold materials used. The appearance of materials after they have been manufactured into products influences their appeal to the consumer; color, feel, and surface texture are characteristics that we all consider when making a purchasing decision.

The size and shape of manufactured products vary widely. For example, the main landing gear for the twin-engine, 400-passenger Boeing 777 jetliner is 4.3 m (14 ft) high, with three axles and six wheels, made by forging and machining processes (Chapters 6, 8, and 9). At the other extreme is the manufacturing of microscopic parts and mechanisms (Fig. 1.7). Such components are produced through surface micromachining operations, referred to as **nanotechnology** and **nanofabrication**, typically using electron-beam, laser-beam, and etching techniques on materials such as silicon. Such mechanisms, called **microelectromechanical systems** (MEMS; see Chapter 13), can potentially power microrobots to repair human cells, and produce microknives for surgery and camera shutters for precise photography, and some are now widely used in sensors, ink-jet printing mechanisms, and magnetic storage devices. The most recent trend is the development of **nanoelectromechanical systems** (NEMS), which operate on the same scale as biological molecules.

Ultraprecision manufacturing techniques and machinery are now being developed and are coming into more common use. For machining mirrorlike surfaces, for example, the cutting tool is a very sharp diamond tip, and the equipment has very high stiffness and must be operated in a room where the temperature is controlled within less than one degree. Highly sophisticated techniques such as molecular-beam epitaxy and scanning-tunneling engineering are being implemented to reach dimensional accuracies on the order of the atomic lattice (nanometer).

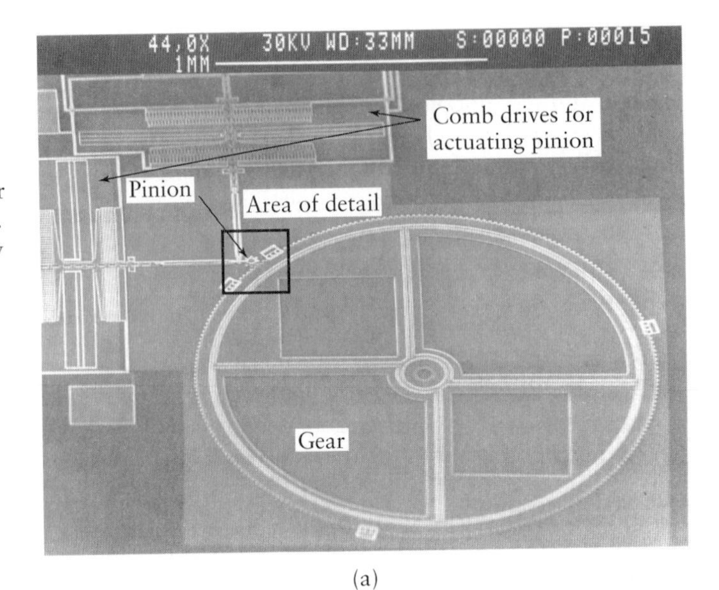

(a)

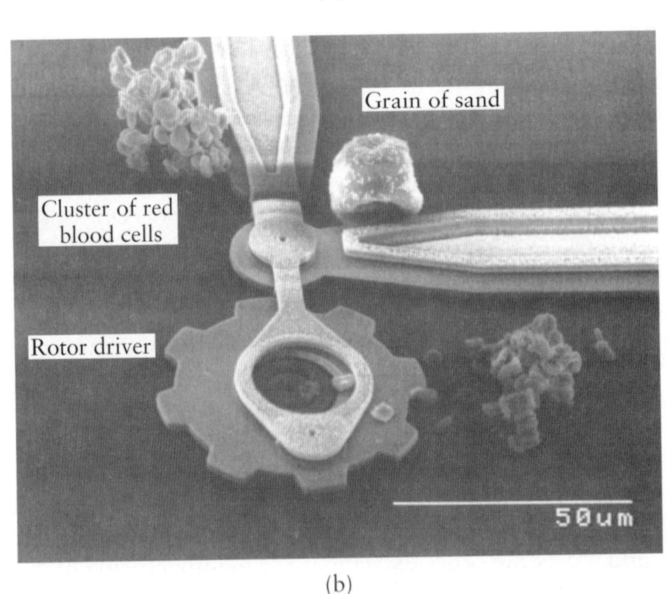

(b)

FIGURE 1.7 An example of nanofabrication: gear assembly driven by resonant combdrives. (a) A view of the entire assembly. (b) Details of the rotor-driver portion of the gear assembly. *Source*: R. Muller, University of California at Berkeley.

Manufacturing and operational cost considerations. The design and cost of tooling, the lead time required to begin production, and the effect of workpiece materials on tool and die life are major considerations in manufacturing (Chapter 16). The cost of tooling can be substantial, depending on tool size, design, and expected life; for example, a set of steel dies for stamping sheet-metal fenders for automobiles may cost about $2 million. For parts made from expensive materials, the lower the scrap rate, the more economical the production process will be. Because machining takes longer time and produces chips, it may not be as economical as forming operations, all other factors being equal.

The number of parts or products required and the desired production rate (in pieces per hour) help determine the processes to be used and the economics of production. Beverage cans and transistors, for example, are consumed in numbers and at rates much higher than telescopes and propellers for ships. The availability of machines and equipment, operating experience, and economic considerations within the manufacturing facility are also important cost factors. If certain parts are not produced internally, they must be manufactured by outside firms; automobile manufacturers, for example, purchase numerous parts from outside vendors or have the parts made by outside firms, according to their specifications.

The operation of machinery has significant environmental and safety implications. Depending on the type of operation and the machinery involved, some processes adversely affect the environment. Unless properly controlled, such processes may cause air, water, and noise pollution. The safe use of machinery is another important consideration, requiring precautions to eliminate hazards in the workplace.

Net-shape manufacturing. Because not all manufacturing operations produce finished parts or products, additional processes may be necessary. For example, a forged part may not have the desired dimensional accuracy or surface finish; thus, additional operations such as machining or grinding may be necessary. Likewise, it may be difficult, impossible, or economically undesirable to produce a part with holes in it by using just one manufacturing process. The holes produced by a particular process may not have the proper roundness, dimensional accuracy, or surface finish, thus necessitating additional operations such as boring, drilling, and reaming.

These additional operations can contribute significantly to the cost of a product. Consequently, there is currently an important trend, called *net-shape* or *near-net-shape manufacturing*, in which the part is made as close to the final desired dimensions, tolerances, and specifications as possible. Typical examples of such manufacturing methods are near-net-shape forging and casting of parts, powder-metallurgy techniques, metal injection molding of metal powders, and injection molding of plastics (Chapters 5, 6, 10, and 11).

1.7 | Computer-Integrated Manufacturing

The major goals of automation in manufacturing facilities (Chapters 14 and 15) are to integrate various operations so as to improve productivity, increase product quality and uniformity, minimize cycle times, and reduce labor costs. Beginning in the 1940s, automation has accelerated because of rapid advances in control systems for machines and in computer technology.

Few developments in the history of manufacturing have had a more significant impact than computers. Computers are now used in a very broad range of applications, including control and optimization of manufacturing processes, material handling, assembly, automated inspection and testing of products (Section 4.8), inventory control, and numerous management activities. Beginning with computer graphics and computer-aided design and manufacturing, the use of computers has been extended to *computer-integrated manufacturing* (CIM).

Computer-integrated manufacturing is particularly effective because of its capability for making the following possible: (1) responsiveness to rapid changes in market demand and product modification; (2) better use of materials, machinery, and personnel, and reduction in inventory; (3) better control of production and management of the total manufacturing operation; and (4) manufacture of high-quality products at low cost.

The following is an outline of the major applications of computers in manufacturing (Chapters 14 and 15):

a) **Computer numerical control** (CNC). This is a method of controlling the movements of machine components by direct insertion of coded instructions in the

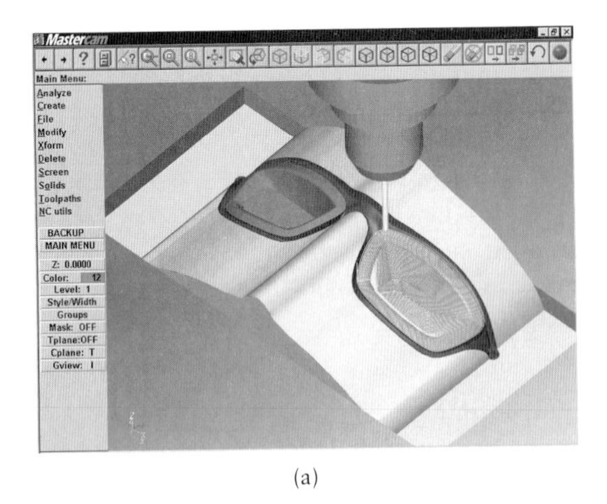

(a)

(b)

(c)

FIGURE 1.8 Machining a mold cavity for making sunglasses. (a) Computer model of the sunglasses as designed and viewed on the monitor. (b) Machining the die cavity using a computerized numerical-control milling machine. (c) Final product produced from the mold. *Source*: Courtesy of Mastercam/CNC Software, Inc.

form of numerical data (Fig. 1.8). Numerical control was first implemented in the early 1950s and was a major advance in the automation of machines.

b) **Adaptive control** (AC). The parameters in a manufacturing process are adjusted automatically to optimize production rate and product quality and to minimize cost. Parameters such as forces, temperatures, surface finish, and dimensions of the part are monitored constantly. If they move outside the acceptable range, the system adjusts the process variables until the parameters again fall within the acceptable range.

c) **Industrial robots**. Introduced in the early 1960s, industrial robots have been replacing humans in operations that are repetitive, boring, and dangerous, thus reducing the possibility of human error, decreasing variability in product quality, and improving productivity. Robots with sensory-perception capabilities (intelligent robots) are being developed, with movements that simulate those of humans (Section 14.7).

d) **Automated handling.** Computers have made possible highly efficient handling of materials and products in various stages of completion (i.e., work in progress, WIP), such as when the materials and products are being moved from storage to machines or from machine to machine and at the points of inspection, inventory, and shipment.

e) **Automated and robotic assembly systems.** These systems are replacing costly assembly by human operators. Products are being designed or redesigned so that they can be assembled more easily by machine.

f) **Computer-aided process planning** (CAPP). This approach is capable of improving productivity in a plant by optimizing process plans, reducing planning costs, and improving the consistency of product quality and reliability. Functions such as estimation of cost and monitoring of work standards (i.e., the time required to perform a certain operation) can also be incorporated into the system.

g) **Group technology** (GT). The concept of group technology is that parts can be grouped and produced by classifying them into families, according to similarities in design and similarities in the manufacturing processes employed to produce each part. In this way, part designs and process plans can be standardized, and families of like parts can be produced efficiently and economically.

h) **Just-in-time production** (JIT). The principal of JIT is that supplies are delivered just in time to be used, parts are produced just in time to be made into subassemblies and assemblies, and products are finished just in time to be delivered to the customer. In this way, inventory-carrying costs are low, part defects are detected right away, productivity is increased, and high-quality products are made at low cost.

i) **Cellular manufacturing.** Cellular manufacturing involves workstations, called *manufacturing cells*, usually containing several machines controlled by a central robot, each machine performing a different operation on the part.

j) **Flexible manufacturing systems** (FMS). This methodology integrates manufacturing cells into a large unit, all interfaced with a central computer. Flexible manufacturing systems have the highest level of efficiency, sophistication, and productivity of all manufacturing systems (Fig. 1.9). Although very costly, they are capable of efficiently producing parts in small runs and of changing manufacturing sequences on different parts quickly. This flexibility enables them to meet rapid changes in market demand for various types of products.

k) **Expert systems.** These systems are basically complex computer programs. They are rapidly developing the capability to perform tasks and solve difficult real-life problems much as human experts would.

l) **Artificial intelligence** (AI). This important field involves the use of machines and computers to replace human intelligence. Computer-controlled systems are becoming capable of learning from experience and of making decisions that optimize operations and minimize costs. Artificial neural networks, which are designed to simulate the thought processes of the human brain, have the capability of modeling and simulating production facilities, monitoring and controlling manufacturing processes, diagnosing problems in machine performance, conducting financial planning, and managing a company's manufacturing strategy.

Although large corporations can implement modern technology and afford to take risks, smaller companies, with their limited resources, personnel, and capital, generally have difficulty in doing so. More recently, the concept of *shared manufacturing* has been proposed. This system would consist of a regional or nationwide network of manufacturing facilities with state-of-the-art equipment for training, prototype

FIGURE 1.9 General view of a flexible manufacturing system, showing several machines (machining centers) and an automated guided vehicle (AGV) moving along the aisle. *Source*: Courtesy of Cincinnati Machine, a UNOVA Company.

development, and small-scale production runs, and it would be available to help small companies develop products that compete in the global marketplace.

In view of the aforementioned advances, some experts have envisioned *the factory of the future*. Although highly controversial, and viewed as unrealistic by some, this is a system in which production will take place with little or no direct human intervention. The human role is expected to be confined to the supervision, maintenance, and upgrading of machines, computers, and software.

The implementation of some of the modern technologies requires significant technical and economic expertise, time, and capital investment. Some of the advanced technologies can be applied improperly, or they can be implemented on too large or ambitious a scale, involving major expenditures with questionable return on investment. Consequently, it is essential to perform a comprehensive analysis and assessment of the real and specific needs of a company and of the market for its products, as well as of whether there is good communication among the parties involved, such as the vendors.

1.8 Quality Assurance and Total Quality Management

Product quality has always been one of the most important concerns in manufacturing, as it directly influences the marketability of a product and customer satisfaction. Traditionally, quality assurance has been obtained by inspecting parts after they have been manufactured; parts are inspected to ensure that they conform to a detailed set of specifications and standards, such as dimensional tolerances, surface finish, and mechanical and physical properties.

However, quality cannot be inspected into a product after it is made. The practice of inspecting products after they are made has now been replaced by the broader view that *quality must be built into a product*, from the design stage through all subsequent stages of manufacture and assembly. Because products typically are made by using several manufacturing processes that can have significant variations in their performance, even within a brief period of time, the control of processes is a critical factor in product quality.

Thus, *we control processes, and not products*. Producing defective products can be very costly to the manufacturer, creating difficulties in assembly operations, necessitating repairs in the field, and resulting in customer dissatisfaction. Contrary to general public opinion, low-quality products do not necessarily cost less than high-quality products. Although it can be described in various ways, **product integrity** is a term that can be used to define the degree to which a product (1) is suitable for its intended purpose, (2) fills a real market need, (3) functions reliably during its life expectancy, and (4) can be maintained with relative ease.

Total quality management (TQM) and *quality assurance* are the responsibility of everyone involved in designing and manufacturing a product. Awareness of the technological and economic importance of built-in product quality has been heightened by pioneers in quality control, primarily Deming, Taguchi, and Juran. They have pointed out the importance of management's commitment to product quality, pride of workmanship at all levels of production, and the use of powerful techniques such as **statistical process control** (SPC) and *control charts* for on-line monitoring of part production and rapid identification of the sources of quality problems (Section 4.9). Ultimately, the major goal is to prevent defects from occurring rather than to identify defective products; for example, computer chips can now be produced in such a way that only a few parts out of a million may be defective. The importance of this trend to the reliability of products and customer satisfaction is self-evident.

Important developments in quality assurance include the implementation of **design of experiments**, a technique in which the factors involved in a manufacturing process and their interactions are studied simultaneously. Thus, for example, variables affecting dimensional accuracy or surface finish in a machining operation can readily be identified, allowing appropriate actions to be taken. The use of computers has greatly enhanced our capability to rapidly utilize such techniques.

The major trend toward global competitiveness has created a need for international conformity in the use of (and for consensus regarding the establishment of) quality control methods. This need has resulted in the International Organization for Standardization ISO 9000 series (Quality Management and Quality Assurance Standards), as well as QS 9000. A company's registration for this standard, which is a quality process certification, not a product certification, means that the company conforms to consistent practices as specified by its own quality system. These two standards have permanently influenced the manner in which companies conduct business in world trade, and they have become the world standard for quality.

1.9 Product Liability

We are all familiar with the consequences of using a product that has malfunctioned, causing bodily injury, or even death, and the financial loss to an employee as well as the organization manufacturing that product. This important topic is referred to as **product liability**. Because of the related technical and legal aspects, for which laws can vary from state to state and from country to country, this subject is complex and can have a major economic impact on all the parties involved.

Designing and manufacturing safe products is an important and integral part of a manufacturer's responsibilities. All those involved with product design, manufacture, and marketing must be fully cognizant of the consequences of product failure, including failures occurring during possible misuse of the product. Numerous examples of products that could involve liability can be cited: (1) a grinding wheel that shatters and blinds a worker; (2) a cable that snaps, allowing a platform to drop; (3) brakes that become inoperative due to the failure of one component; (4) machines with no guards or inappropriate guards; and (5) electric or pneumatic tools without proper warning labels.

Human-factors engineering and **ergonomics** (human–machine interactions) are important aspects of the design and manufacture of safe products. Examples of products for which these factors are faulty include (1) an uncomfortable or unstable workbench or chair; (2) a mechanism that is difficult to operate manually, causing back injury; and (3) a poorly designed computer keyboard, causing pain to the user's hands and arms as a result of repetitive use.

1.10 Manufacturing Costs, Lean Production, and Agile Manufacturing

The cost of a product is often the overriding consideration in its marketability and general customer satisfaction. Typically, manufacturing costs represent about 40% of a product's selling price. The total cost of manufacturing a product consists of costs of materials, tooling, and labor, as well as fixed and capital costs, and several factors are involved in each cost category. Manufacturing costs can be minimized by analyzing the product's design to determine whether part size and shape are optimal and the materials selected are the least costly, while still possessing the desired properties and characteristics. The possibility of substituting materials is an important consideration in minimizing costs (Chapter 16).

While economic considerations in manufacturing have always been a major factor, they have become even more so as international competition (**global competitiveness**) for high-quality products and low prices becomes entrenched in worldwide markets. The markets have become multinational and dynamic, product lines have become extensive and technically complex, and demands by customers for quality have become commonplace (**world-class manufacturing**). To respond to these needs while keeping costs low is a constant challenge to manufacturing companies and is crucial to their very survival. These approaches require that manufacturers **benchmark** their operations; this concept means understanding the competitive position of other manufacturers with respect to one's own position and setting realistic goals for the future. Benchmarking is thus a reference from which various measurements can be made and compared.

The trends just briefly outlined have led to the concept of **lean production,** also called *lean manufacturing*. Not a novel concept, it basically involves a major assessment of each activity of a company: the efficiency and effectiveness of its various operations, the necessity of retaining some of its operations and managers, the efficiency of the machinery and equipment used in the operation while maintaining and improving quality, the number of personnel involved in a particular operation, and a thorough analysis to reduce the cost of each activity, including both productive and nonproductive labor. This concept requires a fundamental change in corporate culture as well as necessitates cooperation and teamwork between management and the workforce. It does not necessarily mean cutting back resources, but aims at *continuously improving* efficiency and profitability of the company by removing all kinds of waste from the operations (*zero-base waste*) and dealing with problems right away.

Agile manufacturing is a term that has been coined to indicate the use of the principles of lean production on a broader scale. The principle behind agile manufacturing is ensuring flexibility (agility) in the manufacturing enterprise so that it can quickly respond to changes in product variety and demand and customer needs. This agility is to be achieved through machines and equipment with built-in flexibility (**reconfigurable machines**), using modular components that can be arranged and rearranged in different ways, advanced computer hardware and software, reduced changeover time, and advanced communications systems. It has been predicted, for example, that in the future, the automotive industry will be able to configure and build a car in three days and that, eventually, the traditional assembly line will be replaced by a system in which a nearly custom-made car will be produced by connecting modules.

1.11 General Trends in Manufacturing

With advances in all aspects of materials, processes, and production control, there have been certain important trends in manufacturing, briefly outlined as follows:

Materials. The trend is for better control of material compositions, purity, and defects (e.g. impurities, inclusions, flaws) in order to enhance the material's overall properties, manufacturing characteristics, reliability, and service life while keeping costs low. Developments are continuing on superconductors, semiconductors, nanomaterials and nanopowders, amorphous alloys, shape-memory alloys (*smart materials*), coatings, and various other engineered metallic and nonmetallic materials. Testing methods and equipment are being improved, through methods including the use of advanced computers and software, particularly for materials such as ceramics, carbides, and various composites.

Concerns over energy and material savings are leading to better recyclability and higher strength- and stiffness-to-weight ratios. Thermal treatment of materials is being conducted under better control of relevant variables for more predictable and reliable results, and surface treatment methods are being advanced rapidly. Included in these developments are advances in tool, die, and mold materials, with better resistance to a wide variety of process variables, thus improving the efficiency and economics of manufacturing processes. As a result of these developments, production of goods has become more efficient, with high-quality products and at low cost.

Processes, equipment, and systems. Continuing developments in computers, controls, industrial robots, automated inspection, handling and assembly, and sensor technology are having a major impact on the efficiency and reliability of all manufacturing processes and equipment. Advances in computer hardware and software, communications systems, adaptive control, expert systems, and artificial intelligence and neural networks have all helped enable the effective implementation of concepts such as group technology, cellular manufacturing, and flexible manufacturing systems, as well as modern practices in the efficient administration of manufacturing organizations.

Computer simulation and modeling are becoming widely used in design and manufacturing, resulting in the optimization of processes and production systems, and better prediction of the effects of relevant variables on product integrity. As a result of such efforts, the speed and efficiency of product design and manufacturing are improving greatly, which also affects the overall economics of production and reduces product costs in an increasingly competitive marketplace.

1.12 Responsibilities of Manufacturing Engineers

The various manufacturing activities and functions described in this chapter must be organized and managed efficiently and effectively in order to maximize productivity and minimize costs while maintaining high quality standards. Because of the complex interactions among the various factors involved in manufacturing (materials, machines, people, information, power, and capital), proper coordination and administration of diverse functions and responsibilities are essential.

Manufacturing engineers traditionally have had several major responsibilities:

1. Plan the manufacture of a product, and select the processes to be used. This function requires a thorough knowledge of the product, its expected performance, and standards and specifications.
2. Identify the machines, equipment, tooling, and personnel needed to carry out the plan. This function requires evaluation of the capabilities of machines, tools, and workers so that proper functions and responsibilities can be assigned.
3. Interact with design and materials engineers to optimize productivity and minimize production costs.
4. Cooperate with **industrial engineers** when planning plant-floor activities such as plant layout, machine arrangement, material-handling equipment, time-and-motion study, production methods analysis, production planning and scheduling, and maintenance. Some of these activities are carried out under the name **plant engineering,** and some are interchangeably performed by both manufacturing and industrial engineers.

Manufacturing engineers, in cooperation with industrial engineers, are also responsible for evaluating new technologies, their applications, and how they can be implemented. In view of the rapidly growing amount of global technical information, this task in itself can present a major challenge. Gaining a broad perspective of computer capabilities, applications, and integration in all phases of manufacturing operations is crucial. This knowledge is particularly essential for long-range production-facility planning, because of the wide variety of products available, competition, and constantly changing global markets.

SUMMARY

- Manufacturing is the process of converting raw materials into products by means of a variety of processes and methods (Section 1.1).
- Product design is an integral part of manufacturing, as evidenced by trends in concurrent engineering, design for manufacture, design for assembly, disassembly, and service (Sections 1.2, 1.3).
- Designing and manufacturing safe and environmentally friendly products are important and integral parts of a manufacturer's responsibilities (Section 1.4).
- A key task is to select appropriate materials and an optimal manufacturing method from several possible alternatives, given product design goals, process capabilities, and cost considerations (Sections 1.5, 1.6).
- Computer-integrated manufacturing technologies use computers to automate a wide range of design, analysis, manufacturing, and quality-control tasks (Section 1.7).
- Ensuring product quality is now a concurrent engineering process rather than a last step in the manufacture of a product. Total quality management and statistical

process control techniques have increased our ability to build quality into a product at every step of the design and manufacturing process (Section 1.8).

- Human-factors engineering, ergonomics, and product liability are important considerations in design and manufacturing (Section 1.9).
- Lean production and agile manufacturing are approaches that focus on the efficiency and flexibility of the entire organization in order to help manufacturers respond to global competitiveness and economic challenges (Section 1.10).
- There are certain general trends that have an important bearing on the future of manufacturing processes and materials, and systems, including better control of materials, development of new materials improved recyclability, and increased availability and application of computer hardware and software (Section 1.11).
- Manufacturing engineers, in cooperation with other professionals in a variety of disciplines, have major responsibilities regarding all aspects outlined previously (Section 1.12).

HAPTER

2 Fundamentals of the Mechanical Behavior of Materials

2.1 Introduction

Chapter 1 outlined the manufacturing methods and systems by which materials can be shaped into useful products. In manufacturing, one of the most important groups of processes is **plastic deformation** (also known as **deformation processing**), namely, shaping materials by applying forces in various ways. It includes *bulk deformation* (forging, rolling, extrusion, and rod and wire drawing) and *sheet-forming processes* (bending, drawing, spinning, and general pressworking). This chapter describes the fundamental aspects of the mechanical behavior of materials during deformation. Individual topics described are deformation modes, stresses, forces, work of deformation, effects of rate of deformation and temperature, hardness, residual stresses, and yield criteria.

In stretching a piece of metal to make an object such as an automobile fender or a length of wire, the material is subjected to *tension*. A solid cylindrical piece of metal is forged in the making of a turbine disk, thus subjecting the material to *compression*. Sheet metal undergoes shearing stresses when, for example, a hole is punched through its cross-section. A piece of plastic tubing is expanded by internal pressure to make a bottle, subjecting the material to tension in various directions.

In all these processes, the material is subjected to one or more of the basic modes of deformation, namely, tension, compression, and shear, as shown in Fig. 2.1. The degree of deformation to which the material is subjected is defined as **strain.** For tension or compression, the **engineering strain,** or **nominal strain,** is defined as

$$e = \frac{\ell - \ell_o}{\ell_o}. \tag{2.1}$$

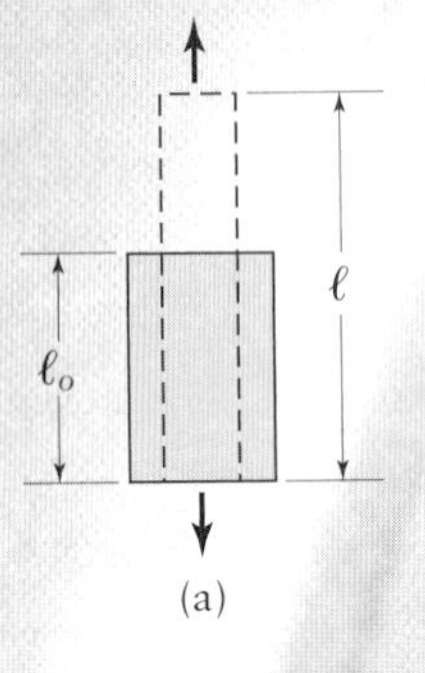

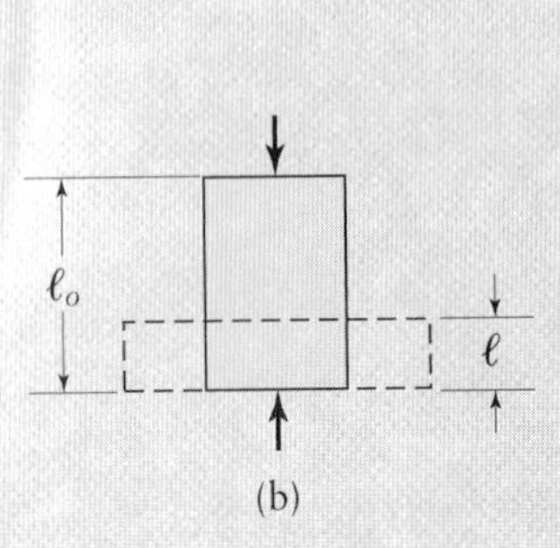

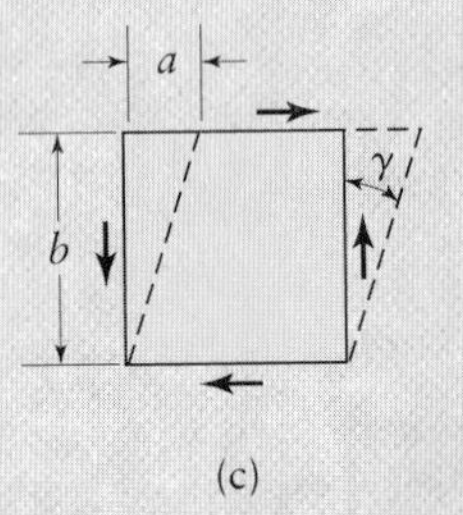

FIGURE 2.1 Types of strain. (a) Tensile, (b) compressive, and (c) shear. All deformation processes in manufacturing involve strains of these types. Tensile strains are involved in stretching sheet metal to make car bodies, compressive strains in forging metals to make turbine disks, and shear strains in making holes by punching.

Note that in tension the strain is positive, and in compression it is negative. The **shear strain** in Fig. 2.1c is defined as

$$\gamma = \frac{a}{b}. \tag{2.2}$$

In order to change the geometry of the elements or bodies shown in Fig. 2.1, *forces* must be applied to them, shown by the arrows. The determination of these forces as a function of strain is very important in manufacturing processes. We have to know the forces in order to design the proper equipment, select the tool and die materials for proper strength, and determine whether a specific metalworking operation can be accomplished on certain equipment. Thus, the relation between a force and the deformation it produces is an essential parameter in manufacturing.

2.2 Tension

Because of its relative simplicity, the *tension test* is the most common test for determining the *strength-deformation characteristics* of materials. It involves the preparation of a test specimen (according to ASTM specifications) and testing it under tension on any of a variety of available testing equipment.

The specimen has an original length ℓ_o and an original cross-sectional area A_o (Fig. 2.2a). Although most specimens are solid and round, flat-sheet and tubular specimens are also tested under tension. The original length is the distance between **gage marks** on the specimen and is generally 50 mm (2 in). Longer lengths may be used for larger specimens, such as structural members, and shorter lengths can be used as well.

Figure 2.2b shows typical results from a tension test. The **engineering stress**, or **nominal stress**, is defined as the ratio of the applied load to the original area,

$$\sigma = \frac{P}{A_o}, \tag{2.3}$$

and the engineering strain is given by Eq. (2.1).

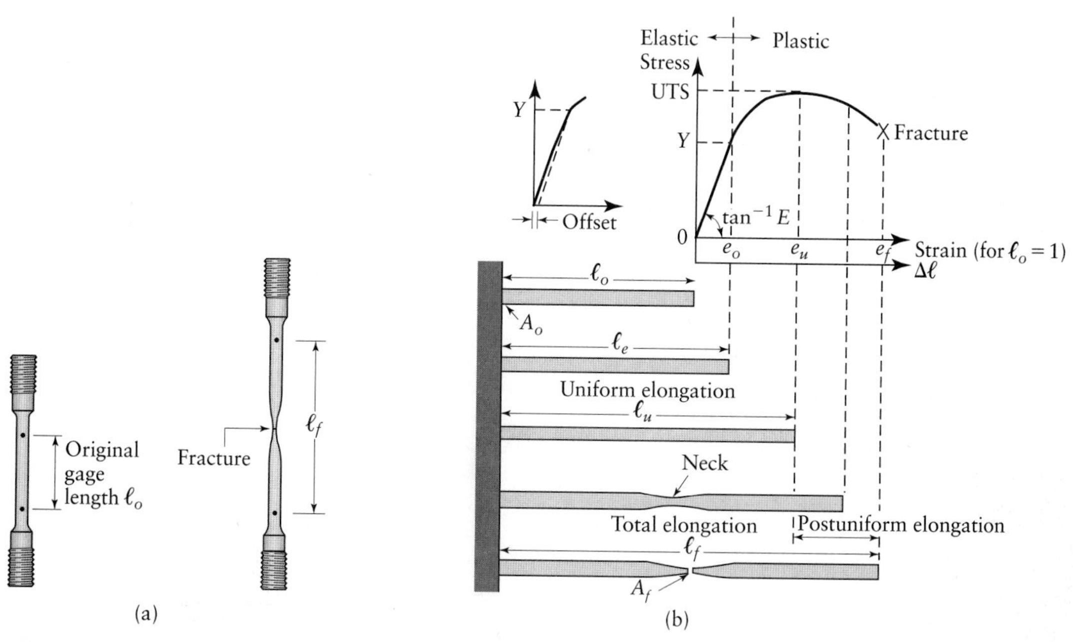

FIGURE 2.2 (a) Original and final shape of a standard tensile-test specimen. (b) Outline of a tensile-test sequence showing different stages in the elongation of the specimen.

When the load is first applied, the specimen elongates proportionately to the load up to the **proportional limit**; this is known as **linear elastic behavior.** The material will continue to deform elastically, although not strictly linearly, up to the **yield point** Y. If the load is removed before the yield point is reached, the specimen will return to its original length. The **modulus of elasticity,** or **Young's modulus,** E, is defined as follows:

$$E = \frac{\sigma}{e}. \tag{2.4}$$

This linear relationship between stress and strain is known as **Hooke's law,** the more generalized forms of which are given in Section 2.11. The elongation of the specimen is accompanied by a contraction of its lateral dimensions. The absolute value of the ratio of the lateral strain to the longitudinal strain is known as **Poisson's ratio,** ν. Table 2.1 gives typical values for E and ν for various materials.

The area under the stress–strain curve up to the yield point Y of a material is known as the **modulus of resilience:**

$$\text{Modulus of resilience} = \frac{Ye_o}{2} = \frac{Y^2}{2E}. \tag{2.5}$$

TABLE 2.1

Typical Mechanical Properties of Various Materials at Room Temperature

	E (GPa)	Y (MPa)	UTS (MPa)	Elongation in 50 mm (%)	Poisson's Ratio (ν)
METALS (WROUGHT)					
Aluminum and its alloys	69–79	35–550	90–600	45–5	0.31–0.34
Copper and its alloys	105–150	76–1100	140–1310	65–3	0.33–0.35
Lead and its alloys	14	14	20–55	50–9	0.43
Magnesium and its alloys	41–45	130–305	240–380	21–5	0.29–0.35
Molybdenum and its alloys	330–360	80–2070	90–2340	40–30	0.32
Nickel and its alloys	180–214	105–1200	345–1450	60–5	0.31
Steels	190–200	205–1725	415–1750	65–2	0.28–0.33
Stainless steels	190–200	240–480	480–760	60–20	0.28–0.30
Titanium and its alloys	80–130	344–1380	415–1450	25–7	0.31–0.34
Tungsten and its alloys	350–400	550–690	620–760	0	0.27
NONMETALLIC MATERIALS					
Ceramics	70–1000	–	140–2600	0	0.2
Diamond	820–1050	–	–	–	–
Glass and porcelain	70–80	–	140	0	0.24
Rubbers	0.01–0.1	–	–	–	0.5
Thermoplastics	1.4–3.4	–	7–80	1000–5	0.32–0.40
Thermoplastics, reinforced	2–50	–	20–120	10–1	–
Thermosets	3.5–17	–	35–170	0	0.34
Boron fibers	380	–	3500	0	–
Carbon fibers	275–415	–	2000–5300	1–2	–
Glass fibers (S, E)	73–85	–	3500–4600	5	–
Kevlar fibers (29, 49, 129)	70–113	–	3000–3400	3–4	–
Spectra fibers (900, 1000)	73–100	–	2400–2800	3	–

Note: In the upper table, the lowest values for E, Y, and UTS and the highest values for elongation are for the pure metals. Multiply GPa by 145,000 to obtain psi, and MPa by 145 to obtain psi. For example, 100 GPa = 14,500 ksi, and 100 MPa = 14,500 psi.

This area has the units of **energy per unit volume** and indicates the **specific energy** that the material can store elastically. Typical values for the modulus of resilience are 2.1×10^4 N-m/m^3 (3 in-lb/in^3) for annealed copper, 1.9×10^5 (28) for annealed medium-carbon steel, and 2.7×10^6 (385) for spring steel.

With increasing load, the specimen begins to yield; that is, it begins to undergo **plastic (permanent) deformation**, and the relationship between stress and strain is no longer linear. The rate of change in the slope of the stress–strain curve beyond the yield point is very small for most materials, so the determination of Y can be difficult. The usual practice is to define the yield stress as the point on the curve that is *offset* by a strain of (usually) 0.2%, or 0.002 (Fig 2.2b); other offset strains may also be used and should be specified when reporting the yield stress of a material.

It is important to note that yielding does not necessarily mean failure. In the design of structures and load-bearing members, yielding is not acceptable, since it leads to permanent deformation. However, yielding is essential in metalworking processes, such as bending, forging, rolling, and sheet-metal forming, where materials have to be subjected to permanent deformation.

As the specimen continues to elongate under increasing load beyond Y, its cross-sectional area decreases permanently and uniformly throughout its gage length. If the specimen is unloaded from a stress level higher than Y, the curve follows a straight line downward and parallel to the original elastic slope (Fig. 2.3). As the load, and hence the engineering stress, is further increased, it eventually reaches a maximum and then begins to decrease. The maximum stress is known as the **tensile strength**, or **ultimate tensile strength** (UTS), of the material (Table 2.1). Ultimate tensile strength is a practical measure of the overall strength of a material.

When the specimen is loaded beyond its UTS, it begins to *neck* (Fig. 2.2a), and the elongation is no longer uniform. That is, the change in the cross-sectional area of the specimen is no longer uniform between the gage marks of the specimen, but is concentrated locally in a "neck" formed in the specimen (known as **necking**, or *necking down*). As the test progresses, the engineering stress drops further, and the specimen finally fractures in the necked region. The final stress level (marked by an × in Fig. 2.2b) is known as the **breaking**, or **fracture, stress.**

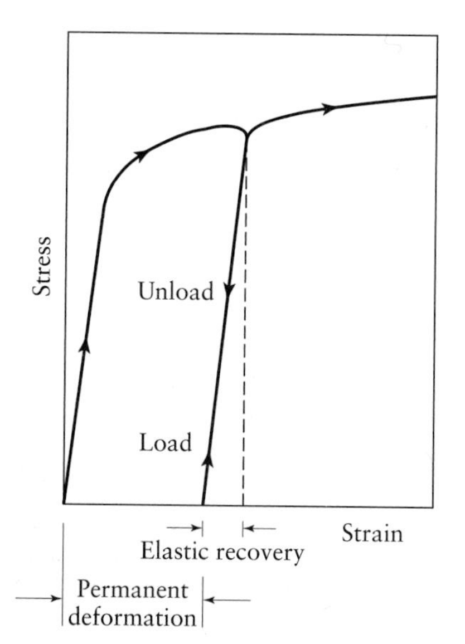

FIGURE 2.3 Schematic illustration of loading and unloading of a tensile-test specimen. Note that during unloading, the curve follows a path parallel to the original elastic slope.

2.2.1 Ductility

The strain at fracture is a measure of *ductility*, that is, how large a strain a material withstands before failure. Note from Fig. 2.2b that until the UTS is reached, elongation is *uniform*; the strain up to the UTS is known as **uniform strain.** The elongation at fracture is known as the **total elongation.** It is measured between the original gage marks after the two pieces of the broken specimen are placed together.

Two quantities commonly used to define ductility in a tension test are *elongation* and *reduction of area*. **Elongation** is defined as

$$\text{Elongation} = \frac{\ell_f - \ell_o}{\ell_o} \times 100 \tag{2.6}$$

and is based on the total elongation (Table 2.1).

Necking is a *local* phenomenon. If we put a series of gage marks at different points on the specimen, pull and break the specimen under tension, and then calculate the percent elongation for each pair of gage marks, we will find that with decreasing gage length, the percent elongation increases (Fig. 2.4). The closest pair of gage marks undergoes the largest elongation, because these gage marks are closest to the necked region. However, the curves do not approach zero elongation with increasing gage length, because the specimens have all undergone a finite permanent elongation before fracture. It is therefore important to report gage length in conjunction with elongation data; other tensile properties, however, are generally independent of gage length.

A second measure of ductility is **reduction of area,** defined as

$$\text{Reduction of area} = \frac{A_o - A_f}{A_o} \times 100. \tag{2.7}$$

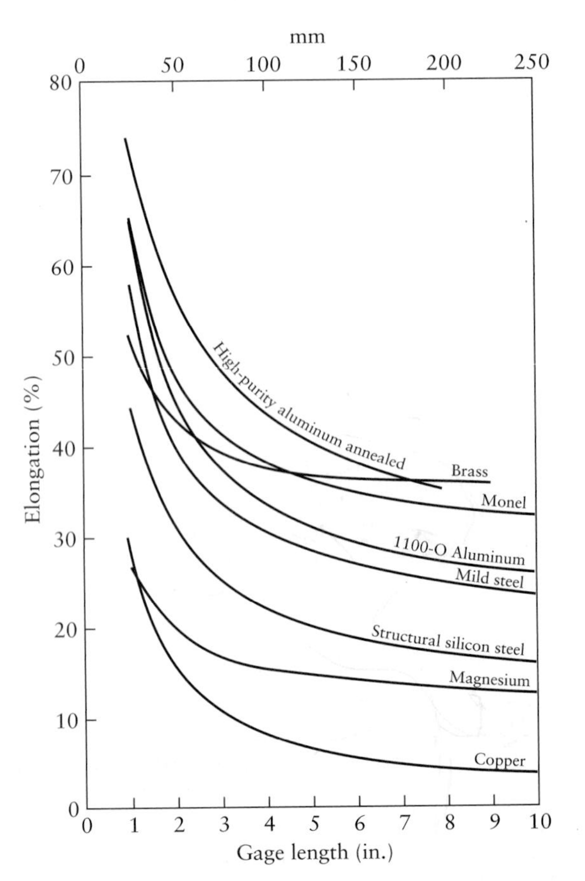

FIGURE 2.4 Total elongation of a specimen in a tension test as a function of original gage length for various metals. Because necking is a local phenomenon, elongation decreases with gage length. Standard gage length is usually 50 mm (2 in.), although shorter lengths can be used if larger specimens are not available.

Thus, a material that necks down to a point at fracture, such as a glass rod at elevated temperature, has 100% reduction of area.

The elongation and reduction of area are generally related to each other for many engineering metals and alloys. Elongation ranges approximately between 10% and 60%, and values between 20% and 90% are typical for reduction of area for most materials. *Thermoplastics* (Chapter 10) and *superplastic* materials (Section 2.2.7) exhibit much higher ductility. Brittle materials, by definition, have little or no ductility. Examples of brittle materials include glass at room temperature, a piece of chalk, gray cast iron, and ceramics.

2.2.2 True stress and true strain

Because stress is defined as the ratio of force to area, *true stress* is defined as

$$\sigma = \frac{P}{A}, \tag{2.8}$$

where A is the actual (hence true) or instantaneous area supporting the load.

Likewise, the complete tension test may be regarded as a series of incremental tension tests where, for each succeeding increment, the original specimen is a little longer than the previous one. Thus, *true strain* (or *natural or logarithmic strain*), ε, can be defined as

$$\varepsilon = \int_{\ell_o}^{\ell} \frac{d\ell}{\ell} = \ln\left(\frac{\ell}{\ell_o}\right). \tag{2.9}$$

Note that, for small values of engineering strain, $e = \varepsilon$, since $\ln(1 + e) = \varepsilon$. For larger strains, however, the values diverge rapidly, as shown in Table 2.2.

The volume of a metal specimen in the plastic region of the test remains constant (*volume constancy*, Section 2.11.5). Hence, the true strain within the uniform elongation range can be expressed as

$$\varepsilon = \ln\left(\frac{\ell}{\ell_o}\right) = \ln\left(\frac{A_o}{A}\right) = \ln\left(\frac{D_o}{D}\right)^2 = 2\ln\left(\frac{D_o}{D}\right). \tag{2.10}$$

Once necking begins, the true strain at any point in the specimen can be calculated from the change in cross-sectional area at that point; thus, by definition, the largest strain is at the narrowest region of the neck.

We have seen that, for small strains, the engineering and true strains are very close, and therefore either one can be used in calculations. However, for the large strains encountered in metalworking, the true strain should be used. This value is the true measure of the strain and can be illustrated by the following two examples: (a) Assume that a tension specimen is elongated to twice its original length. This deformation is equivalent to compressing a specimen to one half its original height. Using the subscripts t and c for tension and compression, respectively, we find that $\varepsilon_t = 0.69$ and $\varepsilon_c = -0.69$, whereas $e_t = 1$ and $e_c = -0.5$. Hence, true strain is a correct measure of strain. (b) As another example, assume that a specimen 10 mm in

TABLE 2.2

Comparison of Engineering and True Strains in Tension

e	0.01	0.05	0.1	0.2	0.5	1	2	5	10
ε	0.01	0.049	0.095	0.18	0.4	0.69	1.1	1.8	2.4

height is compressed to a final thickness of zero; thus, $\varepsilon_c = -\infty$, whereas $e_c = -1$. We have deformed the specimen infinitely, which is exactly what the value of the true strain indicates. From these examples, it is obvious that true strain is consistent with the actual physical phenomenon, which is not the case for engineering strain.

2.2.3 True-stress–true-strain curves

The relation between engineering and true values for stress and strain can now be used to construct *true-stress–true-strain curves* from a curve such as that in Fig. 2.2b. The procedure for this construction is given in Example 2.1. Figure 2.5a shows typical true-stress–true-strain curve. For convenience, such a curve is typically approximated by the equation

$$\sigma = K\varepsilon^n. \tag{2.11}$$

Note that Eq. 2.11 indicates neither the elastic region nor the yield point Y of the material. These quantities, however, are readily available from the engineering stress–strain curve. (Since the strains at the yield point are very small, the difference between the true and engineering yield stress is negligible for metals. The reason is that the difference in the cross-sectional areas A_o and A at yielding is very small.)

We can rewrite Eq. (2.11) as

$$\log \sigma = \log K + n \log \varepsilon.$$

If we now plot the true-stress–true-strain curve on a log–log scale, we obtain the graph in Fig. 2.5b. The slope n is known as the **strain-hardening exponent**, and K is known

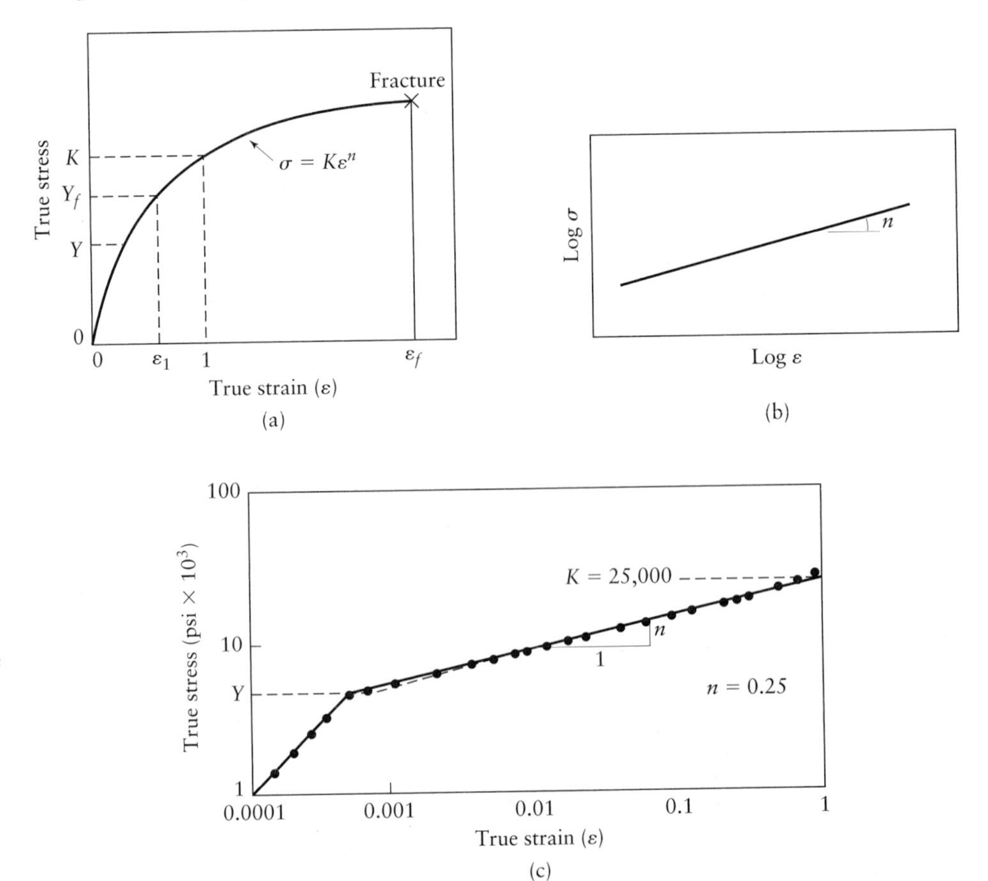

FIGURE 2.5 (a) True-stress–true-strain curve in tension. Note that, unlike in an engineering stress–strain curve, the slope is always positive, and the slope decreases with increasing strain. Although stress and strain are proportional in the elastic range, the total curve can be approximated by the power expression shown. On this curve, Y is the yield stress and Y_f is the flow stress. (b) True-stress–true-strain curve plotted on a log–log scale. (c) True-stress–true-strain curve in tension for 1100-O aluminum plotted on a log–log scale. Note the large difference in the slopes in the *elastic* and *plastic* ranges. *Source*: After R. M. Caddell and R. Sowerby.

TABLE 2.3

Typical Values for *K* and *n* in Eq. (2.11) at Room Temperature

Material	K (MPa)	n
Aluminum, 1100-O	180	0.20
2024-T4	690	0.16
5052-O	210	0.13
6061-O	205	0.20
6061-T6	410	0.05
7075-O	400	0.17
Brass, 70-30, annealed	895	0.49
85-15, cold rolled	580	0.34
Bronze (phosphor), annealed	720	0.46
Cobalt-base alloy, heat treated	2070	0.50
Copper, annealed	315	0.54
Molybdenum, annealed	725	0.13
Steel, low carbon, annealed	530	0.26
1045 hot rolled	965	0.14
1112 annealed	760	0.19
1112 cold rolled	760	0.08
4135 annealed	1015	0.17
4135 cold rolled	1100	0.14
4340 annealed	640	0.15
17–4 P–H, annealed	1200	0.05
52100 annealed	1450	0.07
304 stainless, annealed	1275	0.45
410 stainless, annealed	960	0.10

Note: 100 MPa = 14,500 psi.

as the **strength coefficient**. Note that K is the true stress at a true strain of unity. Table 2.3 lists values of K and n for a variety of engineering materials. The true-stress–true-strain curves for some materials are given in Fig. 2.6. Some differences between the values listed in Table 2.3 and these curves exist because of different sources of data and test conditions.

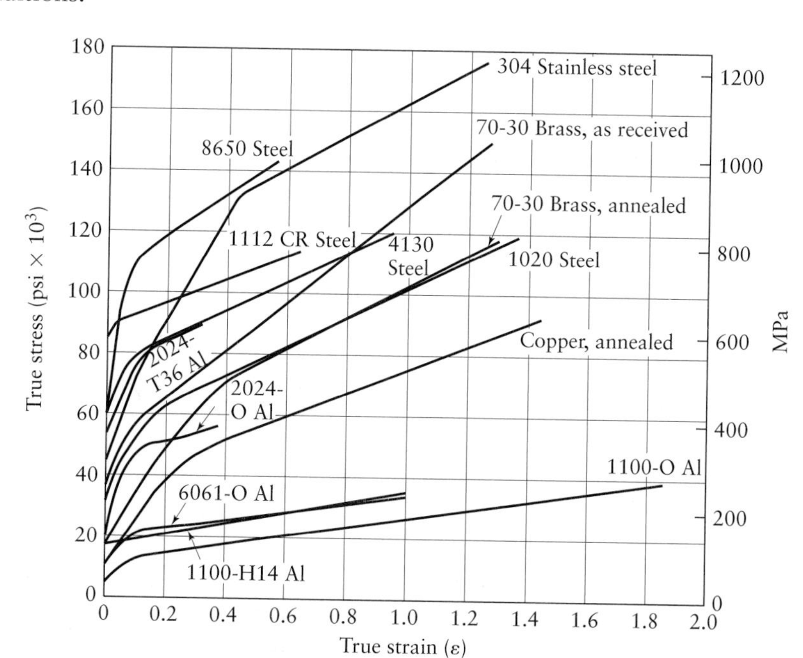

FIGURE 2.6 True-stress–true-strain curves in tension at room temperature for various metals. The point of intersection of each curve at the ordinate is the yield stress Y; thus, the elastic portions of the curves are not indicated. When the K and n values are determined from these curves, they may not agree with those given in Table 2.3, because of the different sources from which they were collected. *Source:* S. Kalpakjian.

In Fig. 2.5a, Y_f is the **flow stress**, defined as the true stress required to continue plastic deformation at a particular true strain ε_1. Thus, for strain-hardening materials, the flow stress increases with increasing strain. Note from Fig. 2.5c that the elastic strains are much smaller than the plastic strains. Consequently, and although both effects exist, we will ignore elastic strains in our calculations for forming processes throughout the rest of this text; the plastic strain will be the total strain.

The area under the true-stress–true-strain curve is known as the **toughness** and can be expressed as

$$\text{Toughness} = \int_o^{\varepsilon_f} \sigma \, d\varepsilon, \tag{2.12}$$

where ε_f is the true strain at fracture. Toughness is hence defined as the energy per unit volume (specific energy) that has been dissipated up to the point of fracture. It is important to remember that this specific energy pertains only to that volume of material at the narrowest region of the neck; any volume of material away from the neck has undergone less strain and thus has dissipated less energy.

As used here, toughness is different than the concept of fracture toughness as treated in textbooks on fracture mechanics. Fracture mechanics (the study of the initiation and propagation of cracks in a solid medium) is beyond the scope of this text. It is of limited relevance to metalworking processes, except in die design and die life.

EXAMPLE 2.1 Construction of a true-stress–true-strain curve

The following data are taken from a stainless-steel tension-test specimen:

$A_o = 0.056 \text{ in}^2$

$A_f = 0.016 \text{ in}^2$

$\ell_o = 2$ in.

Load P (lb)	Extension $\Delta\ell$ (in)
1600	0
2500	0.02
3000	0.08
3600	0.20
4200	0.40
4500	0.60
4600 (max.)	0.86
3300 (fracture)	0.98

Draw the true-stress–true-strain curve for the material.

Solution. In order to determine the true stress and true strain, we use the following relationships:

$$\text{True stress } \sigma = \frac{P}{A};$$

$$\text{True strain } \varepsilon = \ln\left(\frac{\ell}{\ell_o}\right) \text{ up to necking;}$$

$$\varepsilon_f = \ln\left(\frac{A_o}{A_f}\right) \text{ at fracture;}$$

$$\ell = \ell_o + \Delta\ell \text{ up to necking.}$$

Assuming that the volume of the specimen remains constant, we have

$$A_o\ell_o = A\ell,$$

or

$$A = \frac{A_o \ell_o}{\ell} = \frac{(0.056)(2)}{\ell} = \frac{0.112}{\ell}\,\text{in}^2.$$

Using these relationships, we obtain the following data:

$\Delta\ell$ (in)	ℓ (in)	ε	A (in^2)	True stress (psi)
0	2.00	0	0.056	28,600
0.02	2.02	0.01	0.055	45,000
0.08	2.08	0.039	0.054	55,800
0.20	2.20	0.095	0.051	70,800
0.40	2.40	0.182	0.047	90,000
0.60	2.60	0.262	0.043	104,500
0.86	2.86	0.357	0.039	117,300
0.98	2.98	1.253	0.016	206,000

The true stress and true strain are plotted (solid line) in Fig. 2.7. The point at necking is connected to the point at fracture by a straight line, because there are no data on the instantaneous areas after necking begins. The reason for the correction on this curve (broken line) is explained in Example 2.7.

2.2.4 Instability in simple tension

As noted previously, the onset of necking in a tension test corresponds to the ultimate tensile strength, UTS, of the material. The slope of the load–elongation curve at this point is zero (or $dP = 0$), and it is here that *instability* begins; that is, the specimen begins to neck and cannot support the load, because the neck is becoming smaller in cross-sectional area.

Using the relationships

$$\varepsilon = \ln\left(\frac{A_o}{A}\right), \qquad A = A_o e^{-\varepsilon}, \quad \text{and} \quad P = \sigma A = \sigma A_o e^{-\varepsilon},$$

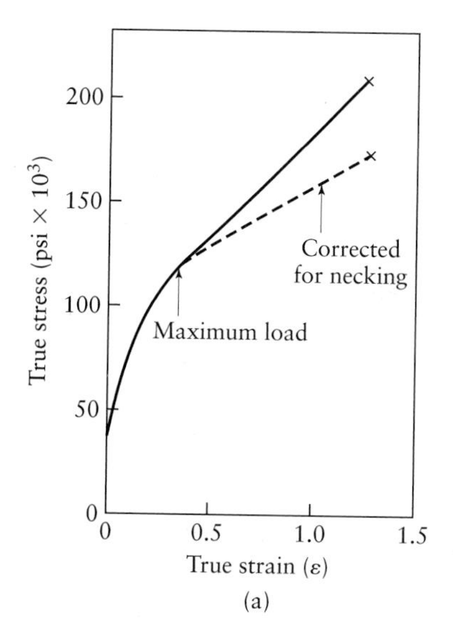

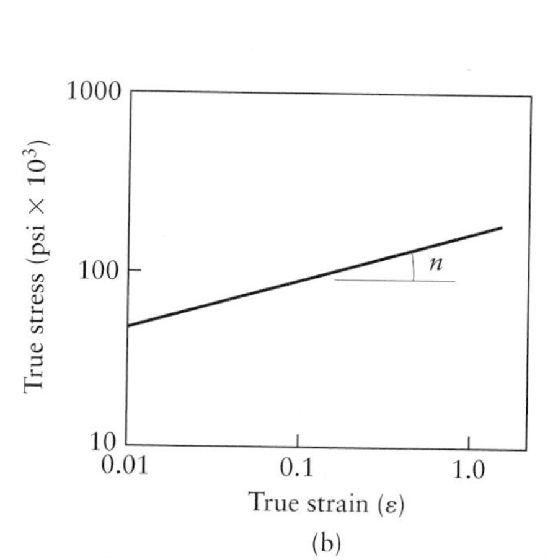

FIGURE 2.7 (a) True-stress–true-strain curve developed from the given data. Note that this curve has a positive slope, indicating that the material is becoming stronger as it is strained. (b) True-stress–true-strain curve plotted on a log–log scale and based on the corrected curve. The steeper the slope n, the higher the capacity of the material to become stronger and harder as it is worked.

we can determine $dP/d\varepsilon$:

$$\frac{dP}{d\varepsilon} = A_o\left(\frac{d\sigma}{d\varepsilon}e^{-\varepsilon} - \sigma e^{-\varepsilon}\right).$$

Because $dP = 0$ at the UTS where necking begins, we set this expression equal to zero, obtaining

$$\frac{d\sigma}{d\varepsilon} = \sigma.$$

However,

$$\sigma = K\varepsilon^n,$$

and consequently,

$$nK\varepsilon^{n-1} = K\varepsilon^n,$$

or

$$\varepsilon = n. \tag{2.13}$$

Thus, the true strain at the onset of necking (i.e., the termination of uniform elongation) is numerically equal to the strain-hardening exponent n. Hence, the higher the value of n, the greater the strain to which a piece of material can be stretched before necking begins. Table 2.3 shows that metals such as annealed copper, brass, and stainless steel can be stretched uniformly to a greater extent than the other materials listed. These observations are covered in greater detail in Chapters 7 and 10, because of their relevance to sheet-forming processes.

Instability in a tension test can be viewed as a phenomenon in which two competing processes are taking place simultaneously. As the load on the specimen is increased, its cross-sectional area becomes smaller; this behavior is more pronounced in the region where necking begins. With increasing strain, however, the material becomes stronger, due to strain hardening. Because the load on the specimen is the product of area and strength, instability sets in when the rate of decrease of cross-sectional area is greater than the rate of increase of strength; this condition is known as **geometric softening.**

EXAMPLE 2.2 Calculation of ultimate tensile strength

A material has a true-stress–true-strain curve given by

$$\sigma = 100{,}000\varepsilon^{0.5} \text{ psi}.$$

Calculate the true UTS and the engineering UTS of this material.

Solution. Since the necking strain corresponds to the maximum load, and the necking strain for this material is given as

$$\varepsilon = n = 0.5,$$

then from

$$\sigma = Kn^n,$$

we have

$$\text{UTS}_{\text{true}} = 100{,}000(0.5)^{0.5} = 70{,}710 \text{ psi}.$$

The cross-sectional area at the onset of necking is obtained from

$$\ln\left(\frac{A_o}{A_{neck}}\right) = n = 0.5.$$

Consequently,

$$A_{neck} = A_o e^{-0.5},$$

and the maximum load P is expressed as

$$P = \sigma A = \sigma A_o e^{-0.5},$$

where σ is the true UTS. Hence,

$$P = (70{,}710)(0.606)(A_o) = 42{,}850\ A_o\ \text{lb}.$$

since the engineering UTS $= P/A_o$, we have

$$\text{UTS} = 42{,}850\ \text{psi}.$$

2.2.5 Types of stress–strain curves

Each material has a differently shaped stress–strain curve; its shape depends on its composition and many other factors, to be treated in detail later. Some of the major types of curves are shown in Fig. 2.8 with their associated stress–strain equations and have the following characteristics:

1. A **perfectly elastic** material behaves like a spring with stiffness E. The behavior of brittle materials, such as glass, ceramics, and some cast irons, may be represented by such a curve (Fig. 2.8a). There is a limit to the stress the material can sustain, after which it breaks. Permanent deformation, if any, is negligible.
2. A **rigid, perfectly plastic** material has, by definition, an infinite value of E. Once the stress reaches the yield stress Y, it continues to undergo deformation at the same stress level. When the load is released, the material has undergone permanent deformation, with no elastic recovery (Fig. 2.8b).
3. An **elastic, perfectly plastic** material is a combination of the first two types of material: it has a finite elastic modulus, and it undergoes elastic recovery when the load is released (Fig. 2.8c).
4. A **rigid, linearly strain-hardening** material requires an increasing stress level to undergo further strain; thus, its **flow stress** (the magnitude of the stress required to maintain plastic deformation at a given strain; see Fig. 2.5a) increases with increasing strain. It has no elastic recovery upon unloading (Fig. 2.8d).

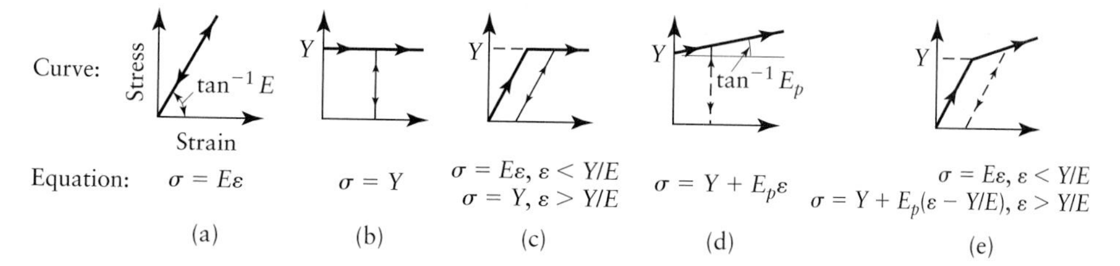

FIGURE 2.8 Schematic illustration of various types of idealized stress–strain curves. (a) Perfectly elastic. (b) Rigid, perfectly plastic. (c) Elastic, perfectly plastic. (d) Rigid, linearly strain hardening. *Ep* is referred to as the *plastic modulus* and is the slope of the stress–strain curve after yiedling. (e) Elastic, linearly strain hardening. The broken lines and arrows indicate unloading and reloading, respectively, during the test. Most engineering metals exhibit a behavior similar to that shown in curve (e).

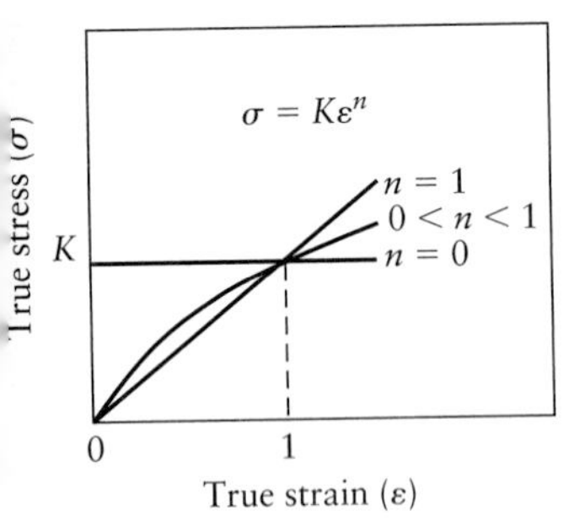

FIGURE 2.9 The effect of the strain-hardening exponent, n, on the shape of true-stress–true-strain curves. When $n = 1$, the material is elastic, and when $n = 0$, it is rigid and perfectly plastic.

5. An **elastic, linearly strain-hardening** curve (Fig. 2.8e) is an approximation of the behavior of most engineering materials, with the modification that the plastic portion of the curve has a decreasing slope with increasing strain (see Fig. 2.5a).

Note that some of these curves can be expressed by Eq. 2.11 by changing the value of n (Fig. 2.9), or by other equations of a similar nature.

2.2.6 Effects of temperature

The various factors that have an influence on the shape of stress–strain curves will be discussed in this and subsequent sections. The first factor is temperature. Although it is somewhat difficult to generalize, increasing temperature usually increases ductility and toughness and lowers the modulus of elasticity, yield stress, and ultimate tensile strength (Fig. 2.10). Temperature also affects the strain-hardening exponent, n, of most metals, in that n decreases with increasing temperature. Depending on the type of material and its composition and level of impurities, elevated temperatures can have other significant effects as well, as detailed in Chapter 3. However, the influence of temperature is best discussed in conjunction with strain rate, for the reasons explained in the next section.

2.2.7 Effects of strain rate

Depending on the particular manufacturing operation and equipment, a piece of material may be formed at low or high speeds; thus, in performing a tension test, the specimen can be strained at different rates to simulate the actual deformation process. Whereas the **deformation rate** may be defined as the *speed* (in m/s, for instance) at which a tension test is being carried out, the strain rate is a function of the geometry of the specimen.

The **engineering strain rate**, $\dot{e}$, is defined as

$$\dot{e} = \frac{de}{dt} = \frac{d\left(\frac{\ell - \ell_o}{\ell_o}\right)}{dt} = \frac{1}{\ell_o}\frac{d\ell}{dt} = \frac{v}{\ell_o} \tag{2.14}$$

and the **true strain rate**, $\dot{\varepsilon}$ as

$$\dot{\varepsilon} = \frac{d\varepsilon}{dt} = \frac{d\left[\ln\left(\frac{\ell}{\ell_o}\right)\right]}{dt} = \frac{1}{\ell}\frac{d\ell}{dt} = \frac{v}{\ell}, \tag{2.15}$$

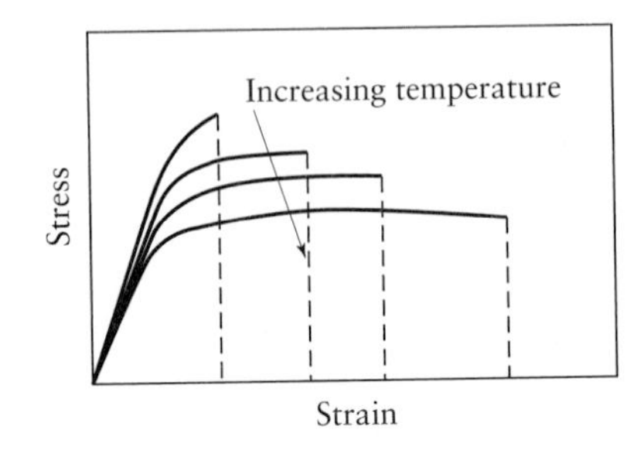

FIGURE 2.10 Typical effects of temperature on engineering stress–strain curves. Temperature affects the modulus of elasticity, yield stress, ultimate tensile strength, and toughness of materials.

TABLE 2.4

Typical Ranges of Strain, Deformation Speed, and Strain Rates in Metalworking Processes

Process	True Strain	Deformation Speed (m/s)	Strain Rate (s^{-1})
Cold working			
Forging, rolling	0.1–0.5	0.1–100	$1–10^3$
Wire and tube drawing	0.05–0.5	0.1–100	$1–10^4$
Explosive forming	0.05–0.2	10–100	$10–10^5$
Hot working and warm working			
Forging, rolling	0.1–0.5	0.1–30	$1–10^3$
Extrusion	2–5	0.1–1	$10^{-1}–10^2$
Machining	1–10	0.1–100	$10^3–10^6$
Sheet-metal forming	0.1–0.5	0.05–2	$1–10^2$
Superplastic forming	0.2–3	$10^{-4}–10^{-2}$	$10^{-4}–10^{-2}$

where v is the speed or rate of deformation (e.g., the speed of the jaws of the testing machine in which the specimen is clamped).

Although the deformation rate v and the engineering strain rate, $\dot{e}$, are proportional, the true strain rate, $\dot{\varepsilon}$, is not proportional to v. Thus, in a tension test with constant v, the true strain rate decreases as the specimen becomes longer. Therefore, in order to maintain a constant $\dot{\varepsilon}$, the speed must be increased accordingly; however, for small changes in length of the specimen during a test, this difference is not significant.

Table 2.4 lists typical deformation speeds employed in various metalworking processes, as well as the strain rates involved. Note that there are considerable differences in the magnitudes. Because of this wide range, strain rates are quoted in orders of magnitudes, such as $10^2\ s^{-1}$, $10^4\ s^{-1}$, etc.

The general effects of temperature and strain rate on the strength of metals are shown in Fig. 2.11. The figure clearly indicates that increasing strain rate increases strength and that the sensitivity of strength to the strain rate increases with

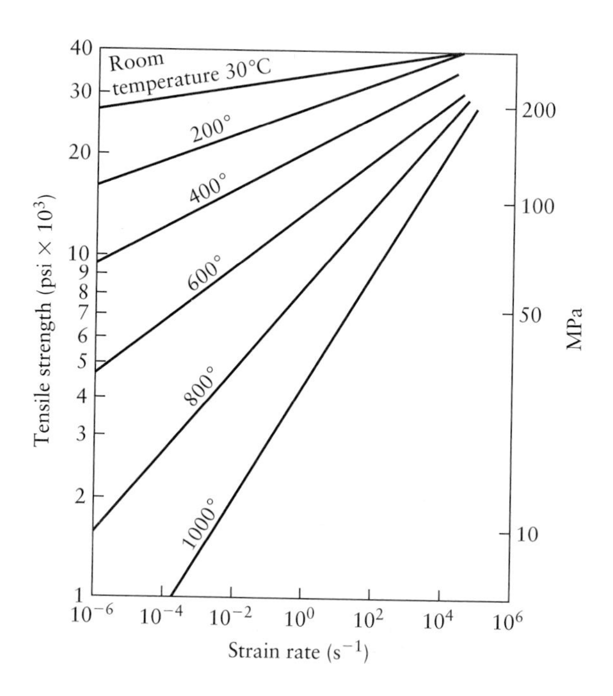

FIGURE 2.11 The effect of strain rate on the ultimate tensile strength of aluminum. Note that as temperature increases, the slope increases. Thus, tensile strength becomes more sensitive to strain rate as temperature increases. *Source*: After J. H. Hollomon.

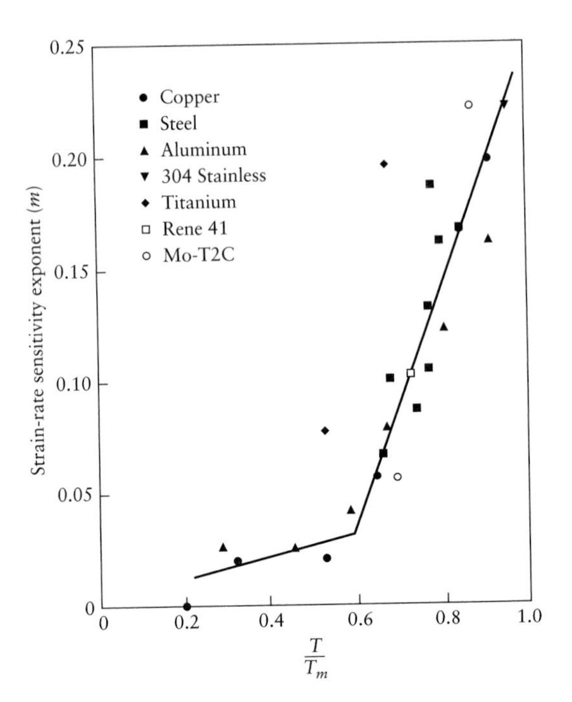

FIGURE 2.12 Dependence of the strain-rate sensitivity exponent m on the homologous temperature T/T_m for various materials. T is the testing temperature, and T_m is the melting point of the metal, both on the absolute scale. The transition in the slopes of the curve occurs at about the recrystallization temperature of the metals. *Source*: After F. W. Boulger.

temperature, as shown in Fig. 2.12. Note, however, that this effect is relatively small at room temperature. Figure 2.11 shows that the same strength can be obtained either at low temperature and low strain rate or at high temperature and high strain rate. These relationships are important in estimating the resistance of materials to deformation when processing them at various strain rates and temperatures.

The effect of strain rate on strength also depends on the particular level of strain: it increases with strain. The strain rate also affects the strain-hardening exponent, n, because n decreases as the strain rate increases. The effect of strain rate on the strength of materials is generally expressed as

$$\sigma = C\dot{\varepsilon}^m, \tag{2.16}$$

where C is the **strength coefficient**, similar to K in Eq. (2.11), and m is the **strain-rate sensitivity exponent** of the material. A general range of values for m is as follows: up to 0.05 for cold working, 0.05 to 0.4 for hot working, and 0.3 to 0.85 for superplastic materials. For a *Newtonian fluid*, for which the shear stress increases linearly with the rate of shear, the value of m is unity. Table 2.5 lists some specific values for C and m. (See also Section 10.3.)

The term **superplastic** refers to the capability of some materials to undergo large uniform elongation prior to failure. The elongation may be on the order of a few hundred percent to as much as 2000%. An extreme example is hot glass; other materials exhibiting superplastic behavior include polymers at elevated temperatures, very fine-grain alloys of zinc–aluminum, and titanium alloys.

The magnitude of m has a significant effect on necking in a tension test. Experimental observations show that, with high m values, the material stretches to a greater length before it fails; this behavior is an indication that necking is delayed with increasing m. When necking is about to begin, the region's strength with respect to the rest of the specimen increases, due to strain hardening. However, the strain rate in the neck region is also higher than in the rest of the specimen, because the material is elongating faster there. Since the material in the necked region becomes stronger as it is strained at a higher rate, the region exhibits a greater resistance to necking. This increased resistance to necking thus depends on the magnitude of m. As the test progresses, necking becomes more *diffuse*, and the specimen becomes longer before fracture; hence, total elongation increases with increasing values of m (Fig. 2.13). As expected, the elongation after necking (*postuniform elongation*) also increases with increasing m. It has been observed that the value of m decreases with metals of increasing strength.

TABLE 2.5

Approximate Range of Values for C and *m* in Eq. (2.16) for Various Annealed Metals at True Strains Ranging from 0.2 to 1.0

Material	Temperature, °C	C		m
		psi × 10^3	MPa	
Aluminum	200–500	12–2	82–14	0.07–0.23
Aluminum alloys	200–500	45–5	310–35	0–0.20
Copper	300–900	35–3	240–20	0.06–0.17
Copper alloys (brasses)	200–800	60–2	415–14	0.02–0.3
Lead	100–300	1.6–0.3	11–2	0.1–0.2
Magnesium	200–400	20–2	140–14	0.07–0.43
Steel				
Low carbon	900–1200	24–7	165–48	0.08–0.22
Medium carbon	900–1200	23–7	160–48	0.07–0.24
Stainless	600–1200	60–5	415–35	0.02–0.4
Titanium	200–1000	135–2	930–14	0.04–0.3
Titanium alloys	200–1000	130–5	900–35	0.02–0.3
Ti-6Al-4V*	815–930	9.5–1.6	65–11	0.50–0.80
Zirconium	200–1000	120–4	830–27	0.04–0.4

* at a strain rate of $2 \times 10^{-4}\ s^{-1}$.
Note: As temperature increases, C decreases and m increases. As strain increases, C increases and m may increase or decrease, or it may become negative within certain ranges of temperature and strain.
Source: After T. Altan and F.W. Boulger.

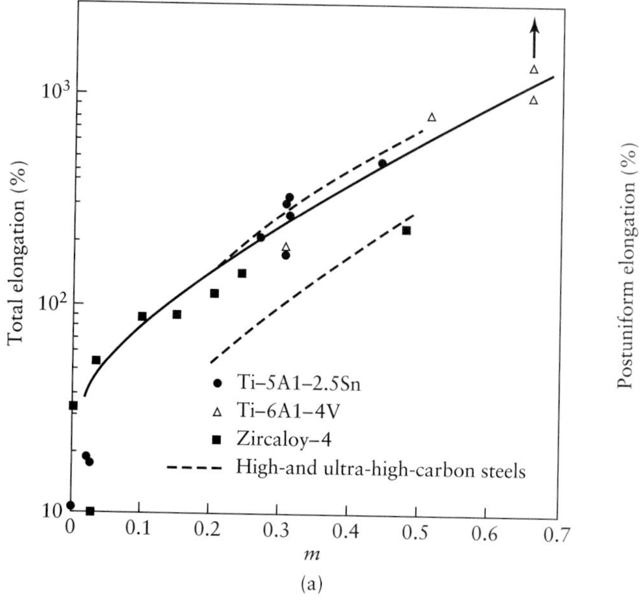

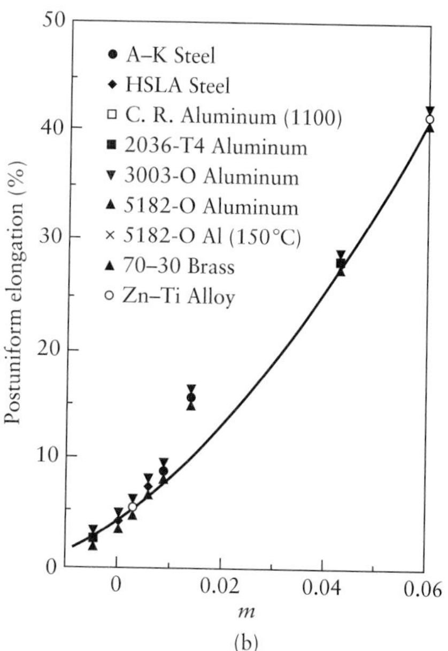

FIGURE 2.13 (a) The effect of the strain-rate sensitivity exponent m on the total elongation for various metals. Note that elongation at high values of m approaches 1000%. *Source*: After D. Lee and W. A. Backofen. (b) The effect of the strain-rate sensitivity exponent on the postuniform (after necking) elongation for various metals. *Source*: After A. K. Ghosh.

Because the formability of materials depends largely on their ductility, it is important to recognize the effect of temperature and strain rate on ductility. Generally, higher strain rates have an adverse effect on the ductility of materials. (The increase in ductility due to the strain-rate sensitivity of materials has been exploited in superplastic forming of metals, as described in Section 7.10.)

2.2.8 Effects of hydrostatic pressure

Although most tests are generally carried out at ambient pressure, tests also have been performed under hydrostatic conditions, with pressures ranging up to around 10^3 MPa (10^5 psi). Three important observations have been made concerning the effects of high hydrostatic pressure on the behavior of materials: (1) It substantially increases the strain at fracture (Fig. 2.14); (2) it has little or no effect on the shape of the true-stress–true-strain curve, but only extends it; and (3) it has no effect on the strain or the maximum load at which necking begins. Experiments have shown that, generally, the mechanical properties of metals are not altered after the material has been subjected to hydrostatic pressure.

The increase in ductility due to hydrostatic pressure has also been observed in other tests, such as compression and torsion tests. This increase in ductility has been observed not only with ductile metals, but also with brittle metals and nonmetallic materials. Materials such as cast iron, marble, and various rocks acquire some ductility (or an increase in ductility) and thus deform plastically when subjected to hydrostatic pressure. The level of pressure required to enhance ductility depends on the material.

2.2.9 Effects of radiation

In view of the various nuclear applications of many metals and alloys, studies have been conducted on the effects of radiation on material properties. Typical changes in the mechanical properties of steels and other metals exposed to high-energy radiation are (1) increased yield stress, (2) increased tensile strength and hardness, and (3) decreased ductility and toughness. The magnitudes of these changes depend on the material and its condition, temperature, and level of radiation.

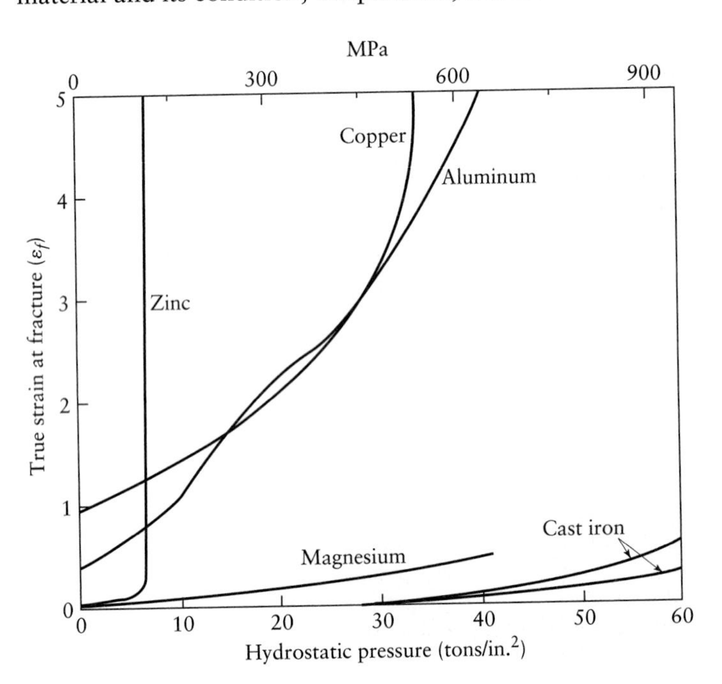

FIGURE 2.14 The effect of hydrostatic pressure on true strain at fracture in tension for various metals. Note that even cast iron becomes ductile under high pressure. *Source*: After H. L. D. Pugh and D. Green.

2.3 Compression

Many operations in metalworking, such as forging, rolling, and extrusion, are performed with the workpieces under compressive forces. The compression test, in which the specimen is subjected to a compressive load, as shown in Fig. 2.1b, can give useful information for these processes, such as stresses required and the behavior of the material under compression. The deformation shown in Fig. 2.1b is ideal. The compression test is usually carried out by compressing a solid cylindrical specimen between two flat platens; the friction between the specimen and the dies is an important factor. Friction causes **barreling** (Fig. 2.15). In other words, friction prevents the top and bottom surfaces from expanding freely.

This situation makes it difficult to obtain relevant data and to construct properly the compressive stress–strain curve, because (1) the cross-sectional area of the specimen changes along its height; and (2) friction dissipates energy, and this energy is supplied through an increased compressive force. Therefore, it is difficult to obtain results that are truly indicative of the properties of the material. With effective lubrication or other means, it is possible, of course, to minimize friction (and hence barreling) to obtain a reasonably constant cross-sectional area during this test.

The *engineering strain rate*, $\dot{e}$, in compression is given by

$$\dot{e} = -\frac{v}{h_o}, \tag{2.17}$$

where v is the speed of the die and h_o is the original height of the specimen. The *true strain rate*, $\dot{\varepsilon}$, is given by

$$\dot{\varepsilon} = -\frac{v}{h}, \tag{2.18}$$

where h is the instantaneous height of the specimen. If v is constant, the true strain rate increases as the test progresses. In order to conduct this test at a constant true strain rate, a **cam plastometer** has been designed that, through a cam action, reduces v proportionately as h decreases during the test.

FIGURE 2.15 Barreling in compression of a round solid cylindrical specimen (7075-O aluminum) between flat dies. Barreling is caused by friction at the die–specimen interfaces, which retards the free flow of the material. See also Figs. 6.1 and 6.2. *Source*: K. M. Kulkarni and S. Kalpakjian.

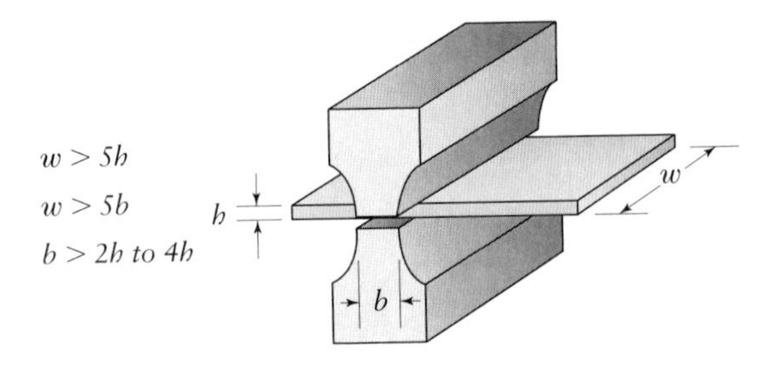

FIGURE 2.16 Schematic illustration of the plane-strain compression test. The dimensional relationships shown should be satisfied for this test to be useful and reproducible. This test gives the yield stress of the material in plane strain, Y'. *Source*: After A. Nadai and H. Ford.

The compression test can also be used to determine the ductility of a metal by observing the cracks that form on the barreled cylindrical surfaces of the specimen. Hydrostatic pressure has a beneficial effect in delaying the formation of these cracks. With a sufficiently ductile material and effective lubrication, compression tests can be carried out uniformly to large strains. This behavior is unlike that in the tension test, where, even for very ductile materials, necking sets in after relatively little elongation of the specimen.

2.3.1 Plane-strain compression test

Another useful test is the plane-strain compression test (Fig. 2.16), which simulates processes such as forging and rolling (Chapter 6). In this test, the die and workpiece geometries are such that the width of the specimen does not undergo any significant change during compression; that is, the material under the dies is in the condition of *plane strain*. (See Section 2.11.3.) The yield stress of a material in plane strain, Y', is given by

$$Y' = \frac{2}{\sqrt{3}}Y = 1.15Y, \tag{2.19}$$

according to the distortion-energy yield criterion. (See Section 2.11.2.) As the geometric relationships in Fig. 2.16 show, the test parameters must be chosen properly to make the results meaningful. Furthermore, caution should be exercised in test procedures, such as preparing the die surfaces and aligning the dies, lubricating the surfaces, and accurately measuring the load.

For *ductile* metals, when the results of tension and compression tests on the same material are compared, the true-stress–true-strain curves for both tests coincide (Fig. 2.17); however, this is not the case for brittle materials, particularly in regard to ductility. (See also Section 3.8.)

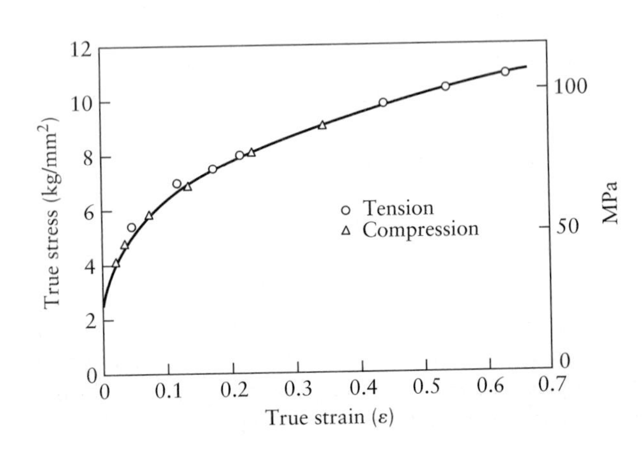

FIGURE 2.17 True-stress–true-strain curve in tension and compression for aluminum. For ductile metals, the curves for tension and compression are identical. *Source*: Courtesy of Alan Cottrell.

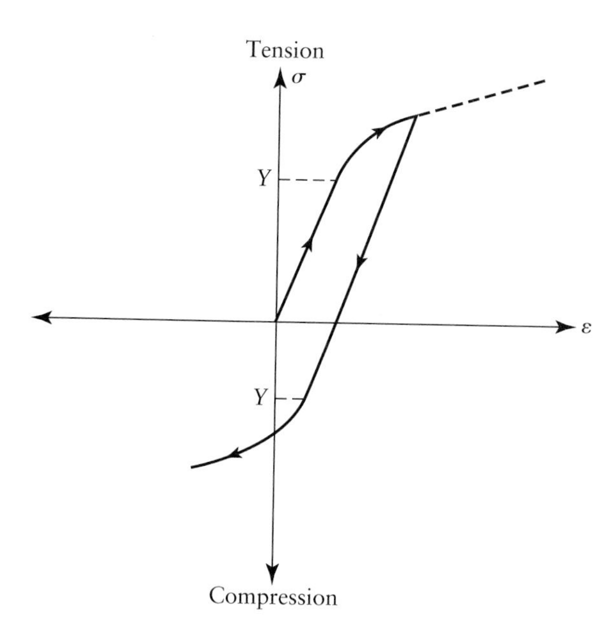

FIGURE 2.18 Schematic illustration of the Bauschinger effect. The arrows show loading and unloading paths. Note the decrease in the yield stress in compression after the specimen has been subjected to tension. The same result is obtained if compression is applied first, followed by tension, whereby the yield stress in tension decreases.

2.3.2 Bauschinger effect

In deformation processing of materials, a workpiece is sometimes first subjected to tension and then to compression, or vice versa. Examples of such situations are bending and unbending, roller leveling (see Fig. 6.41), and reverse drawing (see Fig. 7.62). When a metal with a tensile yield stress, Y, is subjected to tension into the plastic range and then the load is released and applied in compression, the yield stress in compression is lower than that in tension (Fig. 2.18). This phenomenon is known as the *Bauschinger effect* and is exhibited in varying degrees by all metals and alloys. This effect is also observed when the loading path is reversed, i.e., compression followed by tension. Because of the lowered yield stress in the reverse direction of load application, this phenomenon is also called **strain softening**, or **work softening**. This phenomenon is also observed in torsion (Section 2.4).

2.3.3 The disk test

For brittle materials, such as ceramics and glasses, a *disk test* has been developed in which the disk is subjected to compression between two hardened flat platens (Fig. 2.19). When the disk is loaded as shown in Fig. 2.19, tensile stresses develop perpendicular to the vertical centerline along the disk, fracture begins, and the disk splits vertically in half. (See also *rotary tube piercing* in Section 6.3.5.)

The tensile stress, σ, in the disk is uniform along the centerline and can be calculated from the formula

$$\sigma = \frac{2P}{\pi d t}, \tag{2.20}$$

where P is the load at fracture, d is the diameter of the disk and t is its thickness. In order to avoid premature failure at the two contact points, thin strips of soft metal are placed between the disk and the platens. These strips also protect the platens from being damaged during the test.

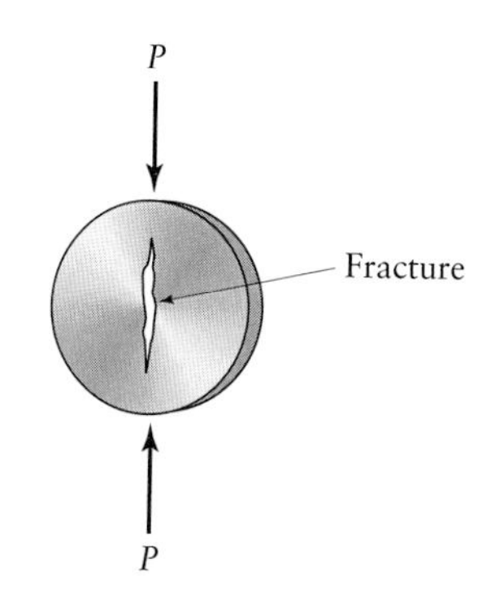

FIGURE 2.19 Disk test on a brittle material, showing the direction of loading and the fracture path. This test is useful for brittle materials, such as ceramics, carbides, and round specimens from grinding wheels.

2.4 Torsion

Another method of determining properties of materials is the *torsion test*. In order to obtain an approximately uniform stress and strain distribution along the cross-section, this test is generally carried out on a tubular specimen (Fig. 2.20). The *shear stress*, τ, can be determined from the equation

$$\tau = \frac{T}{2\pi r^2 t}, \tag{2.21}$$

where T is the torque applied, r is the mean radius, and t is its thickness of the reduced section of the tube. The *shear strain*, γ, is determined from the equation

$$\gamma = \frac{r\phi}{\ell}, \tag{2.22}$$

where ℓ is the length of the reduced section and ϕ is the angle of twist, in radians. With the shear stress and shear strain thus obtained from torsion tests, we can construct the shear-stress–shear-strain curve of the material. (See also Section 2.11.7.)

In the elastic range, the ratio of the shear stress to shear strain is known as the **shear modulus,** or the **modulus of rigidity,** G:

$$G = \frac{\tau}{\gamma}. \tag{2.23}$$

The shear modulus and the modulus of elasticity are related by the formula

$$G = \frac{E}{2(1 + \nu)}, \tag{2.24}$$

which is based on a comparison of **simple-shear** and **pure-shear** strains (Fig. 2.21). Note from Fig. 2.21 that the difference between the two strains is that simple shear is equivalent to pure shear plus a rotation of $\gamma/2$ degrees. (See also Section 2.11.7 for the state of stress shown in Fig. 2.21 for pure shear.)

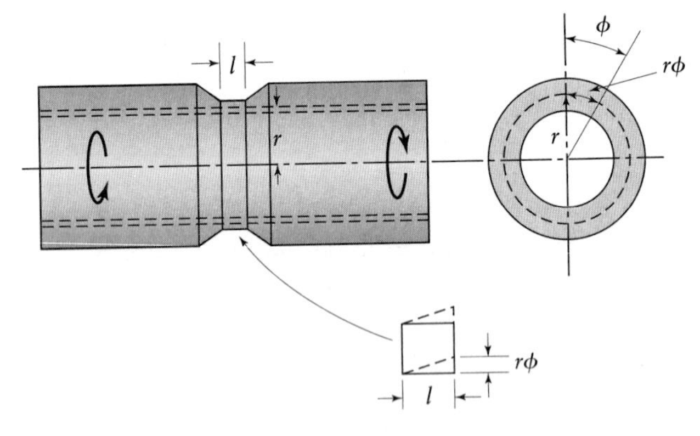

FIGURE 2.20 A typical torsion-test specimen. It is mounted between the two heads of a machine and is twisted. Note the shear deformation of an element in the reduced section of the tubular specimen.

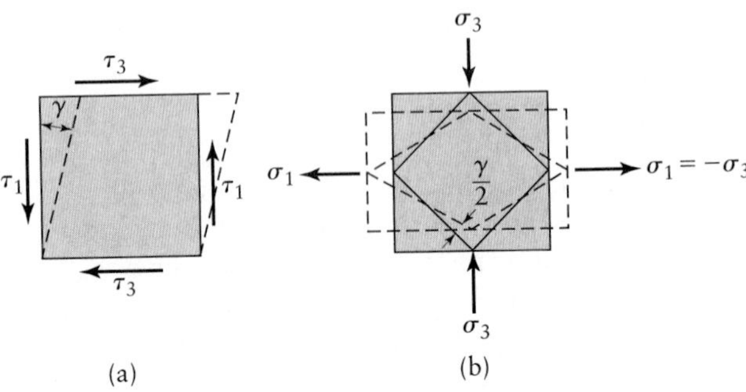

FIGURE 2.21 Comparison of (a) simple shear and (b) pure shear. Note that simple shear is equivalent to pure shear plus a rotation.

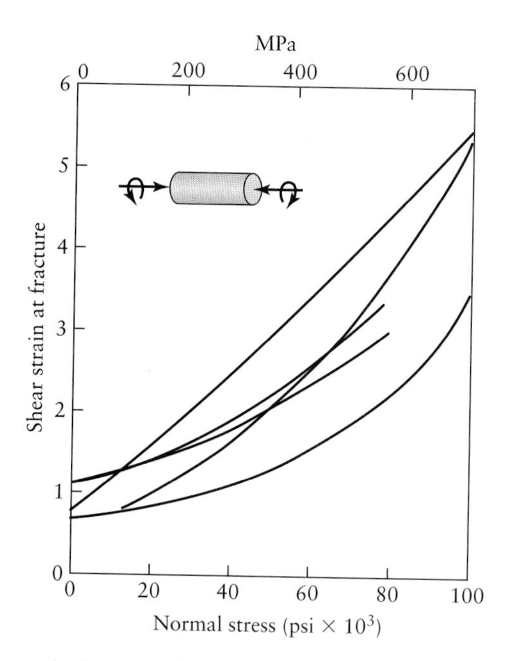

FIGURE 2.22 The effect of axial compressive stress on the shear strain at fracture in torsion for various steels. Note that the effect on ductility is similar to that of hydrostatic pressure, shown in Fig. 2.14. *Source*: Based on data in P. W. Bridgman, *Large Plastic Flow and Fracture*, McGraw-Hill, 1952.

In Example 2.3, we show that a thin-walled tube does not neck in torsion, unlike the tension-test specimens described in Section 2.2.4; therefore, we do not have to be concerned with changes in the cross-sectional area of the specimen in torsion testing. The shear-stress–shear-strain curves obtained from torsion tests increase monotonically just as do in true-stress–true-strain curves.

Torsion tests are performed on solid round bars at elevated temperatures in order to estimate the *forgeability* of metals. (See Section 6.2.6.) The greater the number of twists prior to failure, the greater the forgeability of the material. Torsion tests have also been conducted on round bars that are compressed axially (Fig. 2.22). The maximum shear strain at fracture is measured as a function of the compressive stress. The shear strain at fracture increases substantially as the compressive stress is increased. This observation again indicates the beneficial effect of a compressive environment on the ductility of materials. Other experiments show that, with increasing normal *tensile* stresses, the curves in Fig. 2.22 follow a trend downward and to the left, signifying reduced ductility.

The effect of compressive stresses in increasing the maximum shear strain at fracture has also been observed in metal cutting. (See Section 8.2.) The normal compressive stress has no effect on the magnitude of shear stresses required to cause yielding or to continue the deformation, just as hydrostatic pressure has no effect on the general shape of the stress–strain curve.

EXAMPLE 2.3 Instability in torsion of a thin-walled tube

Show that necking cannot take place in the torsion of a thin-walled tube made of a material whose true-stress–true-strain curve is given by $\sigma = K\varepsilon^n$.

Solution. According to Eq. (2.21), the expression for torque, T, is

$$T = 2\pi r^2 t\tau,$$

where for this case the *shear stress*, τ, can be related to the *normal stress*, σ, using Eq. (2.56) for the distortion-energy criterion, as $\tau = \sigma/\sqrt{3}$. (See Section 2.11.7 for details.) The criterion for instability in torsion is

$$\frac{dT}{d\varepsilon} = 0.$$

Because r and t are constant, we have

$$\frac{dT}{d\varepsilon} = \frac{2}{\sqrt{3}} \frac{\pi r^2 \tau \, d\sigma}{d\varepsilon}.$$

For a material represented by $\sigma = K\varepsilon^n$,

$$\frac{d\sigma}{d\varepsilon} = nK\varepsilon^{n-1}.$$

Therefore,

$$\frac{dT}{d\varepsilon} = \frac{2}{\sqrt{3}} \pi r^2 \tau n K \varepsilon^{n-1}.$$

Since none of the quantities in this expression are zero, $dT/d\varepsilon$ cannot be zero. Consequently, a tube in torsion does not undergo necking.

2.5 Bending

Preparing specimens from brittle materials, such as ceramics and carbides, is difficult, because of the problems involved in shaping and machining the materials to proper dimensions. Furthermore, because of their sensitivity to surface defects and notches, clamping brittle specimens for testing is difficult. Also, improper alignment of the test specimen may result in nonuniform stress distribution along the cross-section of the specimen.

A commonly used test method for brittle materials is the *bend*, or *flexure*, *test*, usually involving a specimen that has a rectangular cross-section and is supported at its ends (Fig. 2.23). The load is applied vertically, either at one or two points; hence, these tests are referred to as **three-point** and **four-point bending**, respectively. The stresses in these specimens are tensile at their lower surfaces and compressive at their upper surfaces, and they can be calculated by using simple beam equations, described in texts on the mechanics of solids. The stress at fracture in bending is known as the **modulus of rupture**, or **transverse rupture strength**, and is obtained from the formula

$$\sigma = \frac{Mc}{I}, \tag{2.25}$$

where M is the bending moment, c is one half of the specimen depth, and I is the moment of inertia of the cross-section.

Note that there is a basic difference between the two loading conditions in Fig. 2.23. In three-point bending, the maximum stress is at the center of the beam,

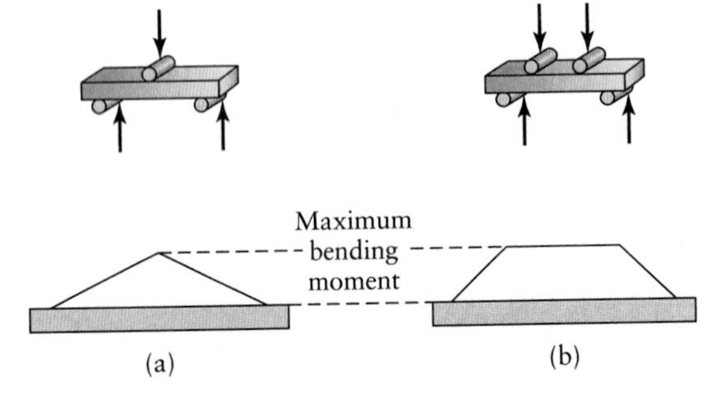

FIGURE 2.23 Two bend-test methods for brittle materials: (a) three-point bending; (b) four-point bending. The lower sketches represent the bending-moment diagrams, described in texts on the mechanics of solids. Note the region of constant maximum bending moment in diagram (b), whereas the maximum bending moment occurs only at the center of the specimen in diagram (a).

whereas in the four-point test, the maximum stress is constant between the two loading points. The stress is the same in both cases when all other parameters are maintained. However, there is a greater probability for defects and imperfections to be present in the volume of material between the loading points in the four-point test than in the much smaller volume under the single load in the three-point test. This difference means that the four-point test is likely to result in a lower modulus of rupture than that of the three-point test, as also verified by experiments. Furthermore, the results of the four-point test show less scatter than those of the three-point test. The reason is that the strength of a small volume of brittle material (such as under the single load in the three-point test) is subject to greater variations than is the strength of the weakest point in a larger volume (such as in the four-point test).

2.6 Hardness

One of the most common tests for assessing the mechanical properties of materials is the hardness test. The *hardness* of a material is generally defined as its resistance to permanent indentation. Less commonly, hardness may also be defined as resistance to scratching or to wear. Various techniques have been developed to measure the hardness of materials, using different indenter geometries and materials. Because resistance to indentation depends on the shape of the indenter and the load applied, hardness is not a fundamental property. Among the most common standardized hardness tests are the Brinell, Rockwell, Vickers, Knoop (Fig. 2.24), and scleroscope tests.

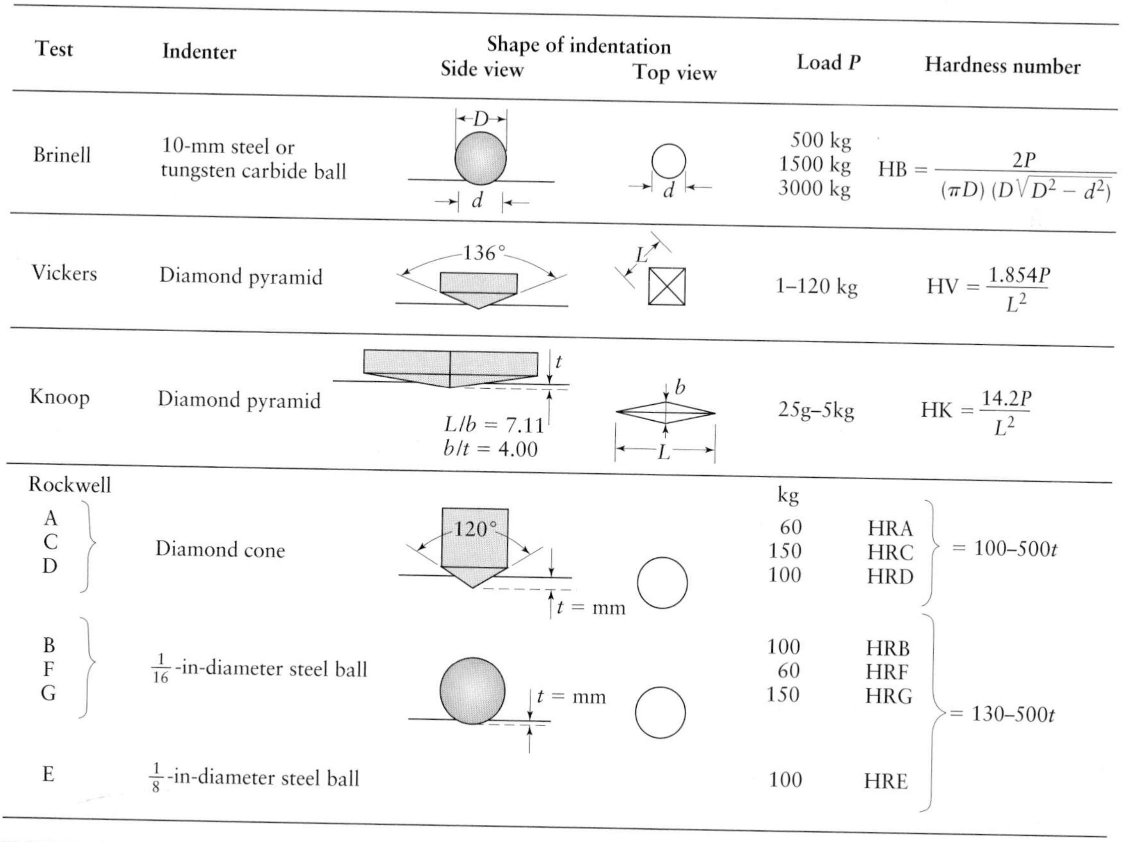

Test	Indenter	Shape of indentation: Side view	Shape of indentation: Top view	Load P	Hardness number
Brinell	10-mm steel or tungsten carbide ball	D, d	d	500 kg 1500 kg 3000 kg	$HB = \frac{2P}{(\pi D)(D - \sqrt{D^2 - d^2})}$
Vickers	Diamond pyramid	136°	L	1–120 kg	$HV = \frac{1.854P}{L^2}$
Knoop	Diamond pyramid	t $L/b = 7.11$ $b/t = 4.00$	b, L	25g–5kg	$HK = \frac{14.2P}{L^2}$
Rockwell				kg	
A, C, D	Diamond cone	120°, t = mm		60 150 100	HRA, HRC, HRD $= 100–500t$
B, F, G	$\frac{1}{16}$-in-diameter steel ball	t = mm		100 60 150	HRB, HRF, HRG $= 130–500t$
E	$\frac{1}{8}$-in-diameter steel ball			100	HRE $= 130–500t$

FIGURE 2.24 General characteristics of hardness testing methods. The Knoop test is known as a *microhardness* test, because of the light load and small impressions. *Source*: Courtesy of Dr. William G. Moffatt.

2.6.1 Brinell Test

In the *Brinell test,* a steel or tungsten carbide ball 10 mm in diameter is pressed against a surface with a load of 500, 1500, or 3000 kg. The *Brinell hardness number* (HB) is defined as the ratio of the load P to the curved area of indentation, or

$$\text{HB} = \frac{2P}{(\pi D)\left(D - \sqrt{D^2 - d^2}\right)} \text{ kg/mm}^2, \tag{2.26}$$

where D is the diameter of the ball and d is the diameter of the impression, in mm.

Depending on the condition of the material tested, different types of impressions are obtained on the surface after a Brinell hardness test has been performed. Annealed materials generally have a rounded profile, whereas cold-worked (strain-hardened) materials have a sharp profile (Fig. 2.25). The correct method of measuring the indentation diameter d for both cases is shown in Fig. 2.25.

Because the indenter (with a finite elastic modulus) also undergoes elastic deformation under the applied load P, measurements of hardness may not be as accurate as expected. One method of minimizing this effect is to use tungsten carbide balls, which, because of their high modulus of elasticity, deform less than steel balls. Also, since harder workpiece materials produce very small impressions, a 1500-kg or 3000-kg load is recommended in order to obtain impressions that are sufficiently large for accurate measurement. Tungsten carbide balls are generally recommended for Brinell hardness numbers higher than 500. In reporting the test results for these high hardnesses, the type of ball used should be cited.

Because the impressions made by the same indenter at different loads are not geometrically similar, the Brinell hardness number depends on the load used. Consequently, the load employed should also be cited with the test results. The Brinell test is generally suitable for materials of low to medium hardness.

Brinelling is a term used to describe permanent indentations on a surface between contacting bodies—for example, a component with a hemispherical protrusion or a ball bearing resting on a flat surface. Under fluctuating loads or vibrations, such as during transportation or vibrating foundations, a permanent indentation may be produced on the flat surface by dynamic loading.

2.6.2 Rockwell test

In the *Rockwell test,* the depth of penetration is measured. The indenter is pressed on the surface, first with a minor load and then with a major one. The difference in the depth of penetration under the loads is a measure of the hardness. There are several Rockwell hardness test scales that use different loads, indenter materials, and indenter geometries. Some of the more common hardness scales and the indenters used are listed in Fig. 2.24. The Rockwell hardness number, which is read directly from a dial on the testing machine, is expressed as follows: if the hardness number is 55 using the C scale, then it is written as 55 HRC. *Rockwell*

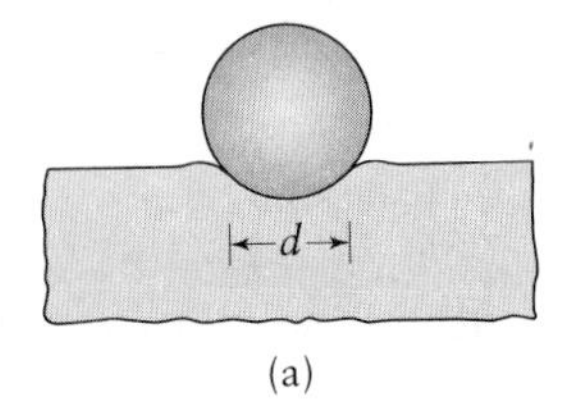

(a)

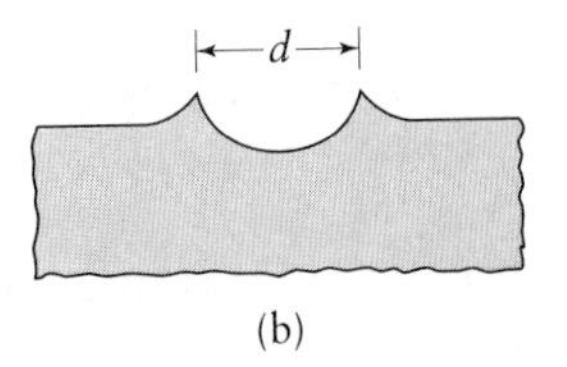

(b)

FIGURE 2.25 Indentation geometry for Brinell hardness testing: (a) annealed metal; (b) work-hardened metal. Note the difference in metal flow at the periphery of the impressions.

superficial hardness tests, that use lighter loads and the same type of indenters, are also available.

2.6.3 Vickers test

The *Vickers test*, also known as the *diamond pyramid hardness test*, uses a pyramid-shaped diamond indenter (see Fig. 2.24) with loads ranging from 1 to 120 kg. The Vickers hardness number (HV) is given by the formula

$$\mathrm{HV} = \frac{1.854P}{L^2} \tag{2.27}$$

The impressions are typically less than 0.5 mm on the diagonal. The Vickers test gives essentially the same hardness number regardless of the load. It is suitable for testing materials with a wide range of hardnesses, including very hard steels.

2.6.4 Knoop test

The *Knoop test* uses a diamond indenter in the shape of an elongated pyramid (see Fig. 2.24) with loads ranging generally from 25 g to 5 kg. The Knoop hardness number (HK) is given by the formula

$$\mathrm{HK} = \frac{14.2P}{L^2}. \tag{2.28}$$

The Knoop test is referred to as a *microhardness test* because of the light loads used; hence, it is suitable for very small or thin specimens and for hard and brittle materials, such as gemstones, carbides, and glass. This test is also used for measuring the hardness of individual grains in a metal. The size of the indentation is generally in the range of 0.01 to 0.10 mm; therefore, surface preparation is very important. Because the hardness number obtained depends on the applied load, test results should always cite the load applied.

2.6.5 Scleroscope

The *scleroscope* is an instrument in which a diamond-tipped indenter (hammer), enclosed in a glass tube, is dropped on the specimen from a certain height. The hardness is determined by the *rebound* of the indenter; the higher the rebound, the harder the specimen. Indentation is slight, and because the instrument is portable, it is useful for measuring the hardness of large objects.

2.6.6 Mohs test

The *Mohs test* is based on the capability of one material to *scratch* another. Mohs hardness is on a scale of 1 to 10, with 1 for talc and 10 for diamond (the hardest substance known); thus, a material with a higher Mohs hardness can scratch materials with a lower Mohs hardness. The Mohs scale generally is used by mineralogists and geologists. However, some of the materials tested are of interest in manufacturing, as discussed in Chapters 8, 9, and 11. Although the Mohs scale is qualitative, good correlation is obtained with Knoop hardness. Soft metals have a Mohs hardness of 2 to 3, hardened steels about 6, and aluminum oxide (used in grinding wheels) 9.

2.6.7 Durometer

The hardness of elastomers, rubbers, plastics, and similar soft and elastic materials is generally measured with an instrument called a *durometer*. The durometer provides an empirical test in which an indenter is pressed against the surface, with a constant load applied rapidly. The depth of penetration is measured after one second; the hardness is inversely related to the penetration. There are two commonly used scales for this test. Type A has a blunt indenter, a load of 1 kg, and is used for softer materials. Type D has a sharper indenter, a load of 5 kg, and is used for harder materials. The hardness numbers in these tests range up to 100; there are other scales, the lowest of which is 000 for measuring the hardness of soft foams.

2.6.8 Hardness and strength

Since hardness is the resistance to permanent indentation, testing for it is equivalent to performing a compression test on a small portion of a material's surface. Therefore, we would expect some correlation between hardness and yield stress, Y, in the form of

$$\text{Hardness} = cY, \tag{2.29}$$

where c is a proportionality constant.

Theoretical studies, based on plane-strain slip-line analysis with a smooth flat punch indenting the surface of a semi-infinite body (see Fig. 6.12), show that for a perfectly plastic material of yield stress, Y, the value of c is about 3. This value is in reasonably good agreement with experimental data (Fig. 2.26). Note that cold-worked materials (which are close to being perfectly plastic in their behavior) show better agreement than annealed ones. The higher c value for the annealed materials is explained by the fact that, due to strain hardening, the average yield stress they exhibit during indentation is higher than their initial yield stress.

The reason that hardness, as a compression test, gives higher values than the uniaxial yield stress, Y, of the material can be seen in the following analysis: if the volume under the indenter were a column of material, it would exhibit a uniaxial compressive yield stress, Y. However, the volume being deformed under the indenter

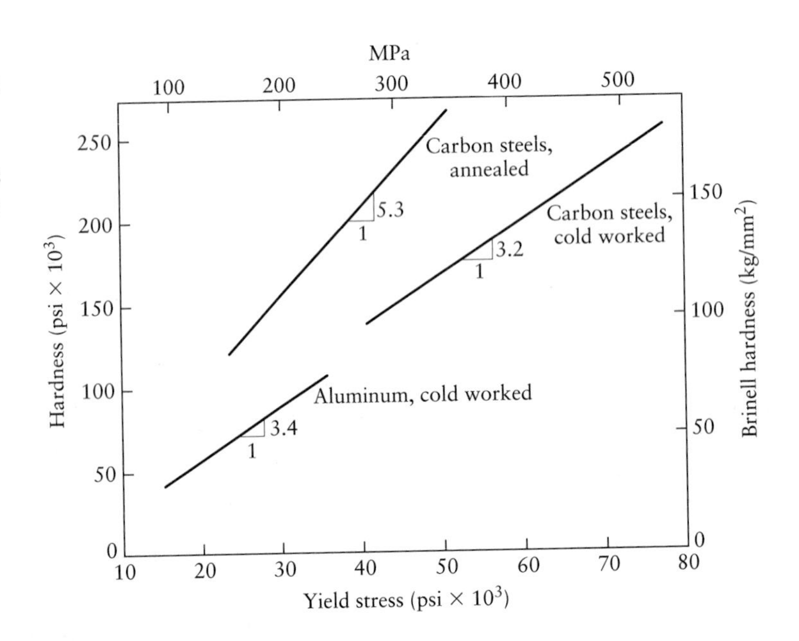

FIGURE 2.26 Relation between Brinell hardness and yield stress for aluminum and steels. For comparison, the Brinell hardness (which is always measured in kg/mm^2) is converted to psi units in the scale on the left.

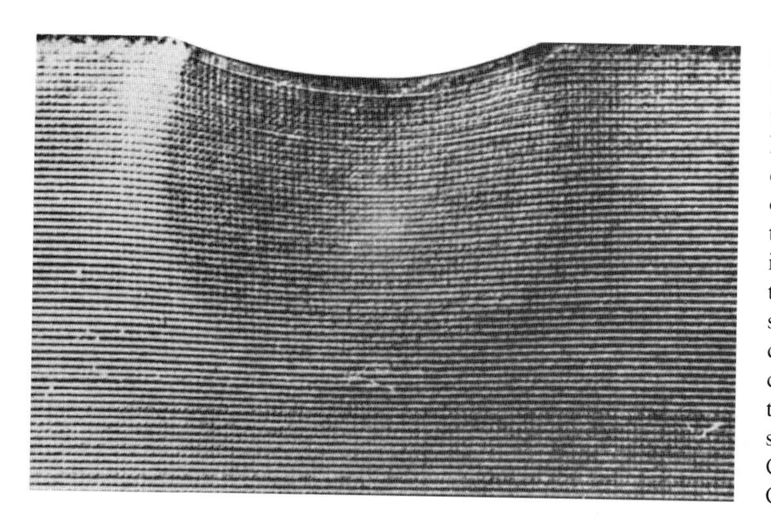

FIGURE 2.27 Bulk deformation in mild steel under a spherical indenter. Note that the depth of the deformed zone is about one order of magnitude larger than the depth of indentation. For a hardness test to be valid, the material should be allowed to fully develop this zone. This difference in depth is why thinner specimens require smaller indentations. *Source*: Courtesy of M. C. Shaw and C. T. Yang.

is, in reality, surrounded by a rigid mass (Fig. 2.27). The surrounding mass prevents this volume of material from deforming freely. In fact, this volume is under *triaxial compression*. As shown in Section 2.11, on yield criteria, this material requires a normal compressive yield stress that is higher than the uniaxial yield stress of the material.

More practically, a relationship also has been observed between the ultimate tensile strength (UTS) and Brinell hardness number (HB) for steels,

$$\mathrm{UTS} = 500(\mathrm{HB}), \tag{2.30}$$

where UTS is in psi and HB is in $\mathrm{kg/mm^2}$ as measured with a load of 3000 kg. In SI units, the relationship is given by

$$\mathrm{UTS} = 3.5(\mathrm{HB}), \tag{2.31}$$

where UTS is in MPa.

Hot hardness tests can also be carried out using conventional testers with certain modifications, such as enclosing the specimen and indenter in a small electric furnace. The hot hardness of materials is important in applications where the materials are subjected to elevated temperatures, such as in cutting tools in machining and dies for metalworking (See Fig. 8.31).

EXAMPLE 2.4 Calculation of the modulus of resilience

A piece of steel is highly deformed at room temperature. Its hardness is found to be 300 HB. Estimate the modulus of resilience for this material.

Solution. Since the steel has been subjected to large strains at room temperature, it may be assumed that its stress–strain curve has flattened considerably, thus approaching the shape of a perfectly plastic curve. According to Eq. (2.29) and using a value of $c = 3$, we obtain

$$Y = \frac{300}{3} = 100\ \mathrm{kg/mm^2} = 142{,}250\ \mathrm{psi}.$$

The modulus of resilience is defined by Eq. (2.5):

$$\text{Modulus of resilience} = \frac{Y^2}{2E}.$$

For steel, $E = 30 \times 10^6$ psi. Hence,

$$\text{Modulus of resilience} = \frac{(142{,}250)^2}{(2 \times 30 \times 10^6)} = 337 \text{ in} \cdot \text{lb/in}^3.$$

2.7 Fatigue

Various structures and components in manufacturing operations, such as tools, dies, gears, cams, shafts, and springs, are subjected to rapidly fluctuating (cyclic or periodic) loads, as well as static loads. Cyclic stresses may be caused by fluctuating mechanical loads, such as gear teeth, or by thermal stresses, such as a cool die coming into repeated contact with hot workpieces. Under these conditions, the part fails at a stress level below which failure would occur under static loading. This phenomenon is known as **fatigue failure** and is responsible for the majority of failures in mechanical components.

Fatigue test methods involve testing specimens under various states of stress, usually in a combination of tension and compression, or torsion. The test is carried out at various stress amplitudes, S, and the number of cycles, N, to cause total failure of the specimen or part is recorded. Stress amplitude is the maximum stress, in tension and compression, to which the specimen is subjected.

A typical plot, known as an *S–N curve*, is shown in Fig. 2.28. S–N curves are based on complete reversal of the stress–that is, maximum tension, maximum compression, maximum tension, and so on—such as that obtained by bending an eraser or piece of wire alternately in one direction, then the other. The test can also be performed on a rotating shaft with a constant downward load. The maximum stress to which the material can be subjected without fatigue failure, regardless of the number of cycles, is known as the **endurance limit,** or **fatigue limit.**

The endurance limit for metals is related to their ultimate tensile strength (Fig. 2.29). For steels, the endurance limit is about one half their tensile strength. Although many metals, especially steels, have a definite endurance limit, aluminum alloys do not, and their S–N curve continues its downward trend. For metals exhibiting such behavior (most face-centered cubic metals; see Fig. 3.2b), the fatigue strength is specified at a certain number of cycles, such as 10^7. In this way, the useful service life of the component can be specified.

FIGURE 2.28 Typical S–N curves for two metals. Note that, unlike steel, aluminum does not have an endurance limit.

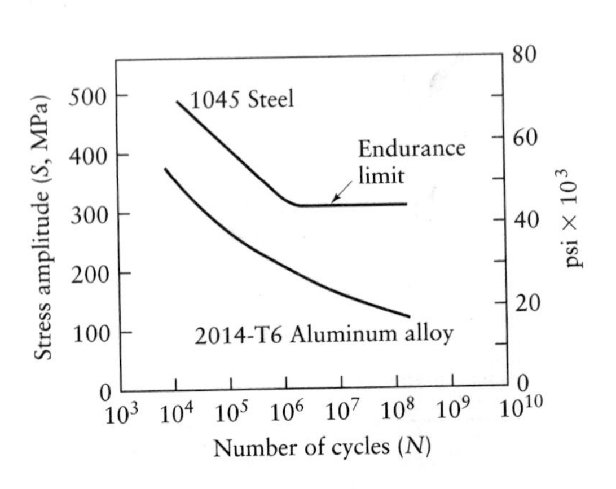

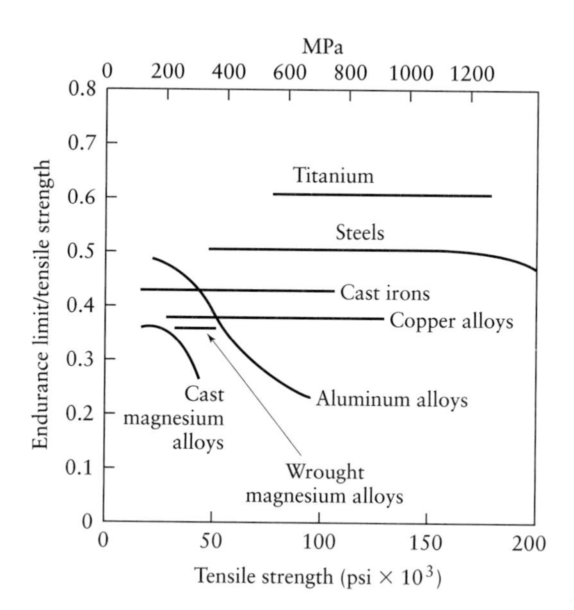

FIGURE 2.29 Ratio of endurance limit to tensile strength for various metals, as a function of tensile strength. The fatigue strength of a metal can thus be estimated from its tensile strength.

2.8 Creep

Creep is the permanent elongation of a component under a static load maintained for a period of time. It is a phenomenon of metals and certain nonmetallic materials, such as thermoplastics and rubbers, and can occur at any temperature. Lead, for example, creeps under a constant tensile load at room temperature, and the thickness of window glass in old houses has been found to be greater at the bottom than at the top of windows, the glass having undergone creep by its own weight over many years.

For metals and their alloys, creep of any significance occurs at elevated temperatures, beginning at about 200°C (400°F) for aluminum alloys, and up to about 1500°C (2800°F) for refractory alloys. The mechanism of creep at elevated temperatures in metals is generally attributed to *grain-boundary sliding*. Creep is especially important in high-temperature applications, such as gas-turbine blades and similar components in jet engines and rocket motors. High-pressure steam lines and nuclear-fuel elements also are subject to creep. Creep deformation also can occur in tools and dies that are subjected to high stresses at elevated temperatures during hot-working operations such as forging and extrusion.

A creep test typically consists of subjecting a specimen to a constant tensile load, and hence a constant engineering stress, at a certain temperature and measuring the change in length over a period of time. A typical creep curve (Fig. 2.30)

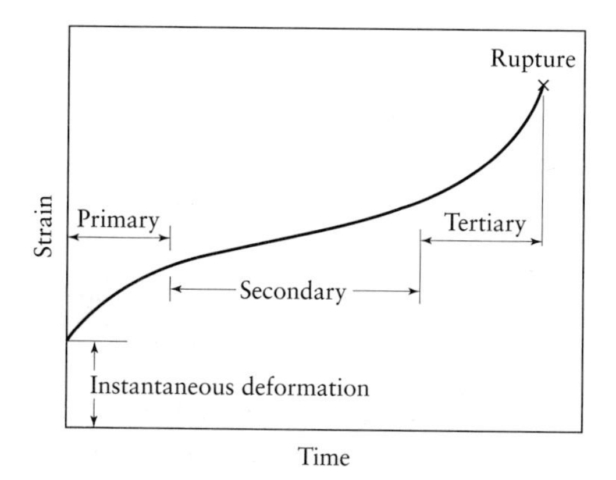

FIGURE 2.30 Schematic illustration of a typical creep curve. The linear segment of the curve (constant slope) is useful in designing components for a specific creep life.

usually consists of primary, secondary, and tertiary stages. The specimen eventually fails by necking and fracture, as in a tension test; this failure is called **rupture** or **creep rupture**. As expected, the creep rate increases with temperature and the applied load.

Design against creep usually involves a knowledge of the secondary (linear) range and its slope, because the creep rate can be determined reliably when the curve has a constant slope. Generally, resistance to creep increases with the melting temperature of a material; this trend serves as a general guideline for design purposes. Stainless steels, superalloys, and refractory metals and alloys are commonly used in applications where creep resistance is required.

Stress relaxation is closely related to creep. In stress relaxation, the stresses resulting from external loading of a structural component decrease in magnitude over a period of time, even though the dimensions of the component remain constant. Examples include rivets, bolts, guy wires, and similar parts under tension, compression, or bending. Stress relaxation is particularly common and important in thermoplastics. (See Section 10.3.)

2.9 Impact

In many manufacturing operations, as well as during the service life of components, materials are subjected to *impact* (or *dynamic*) *loading*, as in high-speed metalworking operations. Tests have been developed to determine the impact toughness of metallic and nonmetallic materials under impact loading. A typical impact test consists of placing a *notched* specimen in an impact tester and breaking it with a swinging pendulum. In the **Charpy test**, the specimen is supported at both ends (Fig. 2.31a), whereas in the **Izod test**, it is supported at one end, like a cantilever beam (Fig. 2.31b). From the amount of swing of the pendulum, the *energy* dissipated in breaking the specimen is obtained; this energy is the **impact toughness** of the material.

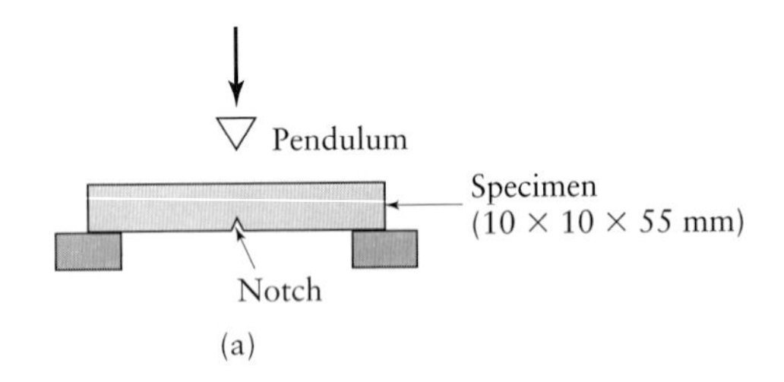

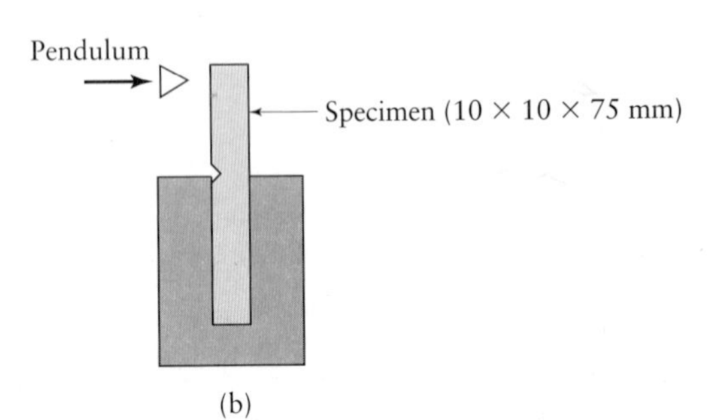

FIGURE 2.31 Impact tests: (a) Charpy; (b) Izod.

Impact tests are particularly useful in determining the ductile–brittle *transition temperature* of materials. (See Fig. 3.26.) Materials that have high impact resistance are generally those that have high strength and high ductility, and hence high toughness. Sensitivity to surface defects (**notch sensitivity**) is important, as it lowers impact toughness.

2.10 Residual Stresses

In this section, we show that inhomogeneous deformation leads to *residual stresses*, which are stresses that remain within a part after it has been deformed and all external forces have been removed. A typical example of inhomogeneous deformation is the bending of a beam (Fig. 2.32). The bending moment first produces a linear elastic stress distribution. As the moment is increased, the outer fibers begin to yield, and, for a typical strain-hardening material, the stress distribution shown in Fig. 2.32b is eventually obtained. After the part is bent (permanently, since it has undergone plastic deformation), the moment is removed by unloading. This unloading is equivalent to applying an equal and opposite moment to the beam.

As previously shown in Fig. 2.3, all recovery is elastic; thus, in Fig. 2.32c, the moments of the areas *oab* and *oac* about the neutral axis must be equal. (For purposes of this treatment, let's assume that the neutral axis does not shift.) The difference between the two stress distributions produces the residual-stress pattern shown within the beam in Fig. 2.32d. Note that there are compressive residual stresses in layers *ad* and *oe*, and tensile residual stresses in layers *do* and *ef*. With no external forces, the residual stresses in the beam must be in static equilibrium. Although this example involves stresses in one direction only, in most situations in deformation processing, the residual stresses are three dimensional.

The equilibrium of residual stresses may be disturbed by altering the shape of the beam, such as by removing a layer of material by machining. The beam will then acquire a new radius of curvature in order to balance the internal forces. Another example of this effect is the drilling of round holes on surfaces with residual stresses. It

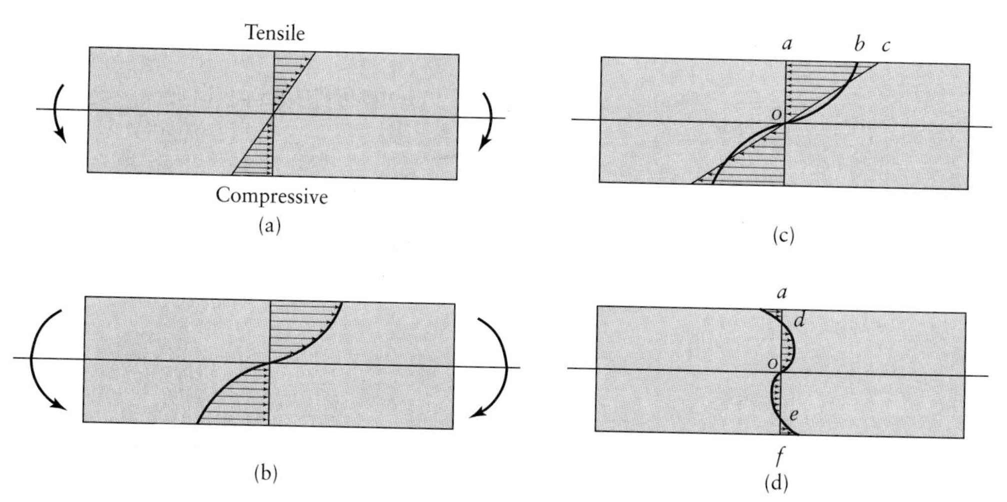

FIGURE 2.32 Residual stresses developed in bending a beam made of an elastic, strain-hardening material. Note that unloading is equivalent to applying an equal and opposite moment to the bent part, as shown in diagram (c). Because of nonuniform deformation, most parts made by plastic-deformation processes contain residual stresses. Note that for equilibrium, the forces and moments due to residual stresses must be internally balanced.

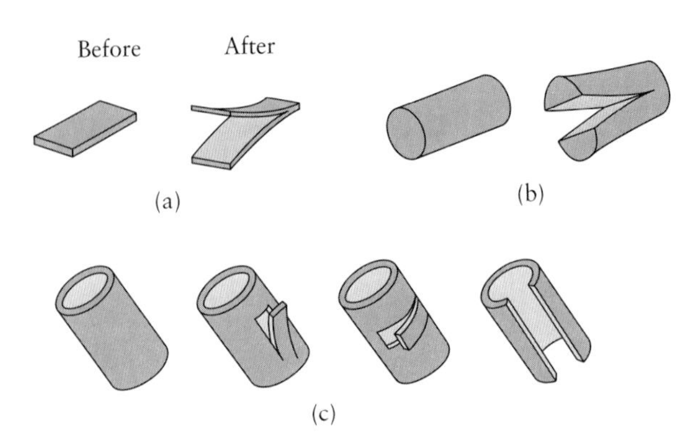

FIGURE 2.33 Distortion of parts with residual stresses after cutting or slitting: (a) rolled sheet or plate; (b) drawn rod; (c) thin-walled tubing. Due to the presence of residual stresses on the surfaces of parts, a round drill may produce an oval-shaped hole, because of relaxation of stresses when a portion is removed.

may be found that, as a result of removing the material, the equilibrium of the residual stresses is disturbed, and the hole becomes oval. Such disturbances of residual stresses lead to *warping*, some simple examples of which are shown in Fig. 2.33.

The equilibrium of residual stresses may also be disturbed by *relaxation* of these stresses over a period of time, which results in instability of the dimensions and shape of the component. These dimensional changes can be an important consideration for precision machinery and measuring equipment.

Residual stresses may also be caused by *phase changes* in metals during or after processing, because of density differences between phases (such as between ferrite and martensite in steels). Phase changes cause microscopic volumetric changes and result in residual stresses. This phenomenon is important in warm and hot working and in heat treatment following cold working.

Residual stresses can also be caused by *temperature gradients* within a body, as during the cooling cycle of a casting (Chapter 5), when applying brakes to a railroad wheel, or in a grinding operation (Section 9.4). The expansion and contraction due to temperature gradients are analogous to nonuniform plastic deformation.

2.10.1 Effects of residual stresses

Tensile residual stresses on the surface of a part are generally undesirable, because they lower the fatigue life and fracture strength of the part. A surface with tensile residual stresses will sustain lower additional tensile stresses due to external loading than can a surface that is free from residual stresses. This is particularly true for relatively brittle materials, where fracture takes place with little or no plastic deformation. Tensile residual stresses in manufactured products can also lead to **stress cracking**, or **stress-corrosion cracking**, (see Section 3.8.2) over a period of time.

Conversely, compressive residual stresses on a surface are generally desirable. In fact, in order to increase the fatigue life of components, compressive residual stresses are imparted on surfaces by techniques such as **shot peening** and **surface rolling**. (See Section 4.5.1.)

2.10.2 Reduction of residual stresses

Residual stresses may be reduced or eliminated either by **stress-relief annealing** or by further *plastic deformation*. Given sufficient time, residual stresses may also be diminished at room temperature by *relaxation*; the time required can be greatly reduced by increasing the temperature of the component. Relaxation of residual stresses by stress-relief annealing is generally accompanied by warpage of the part; hence, a "machining allowance" is commonly provided to compensate for dimensional changes during stress relieving.

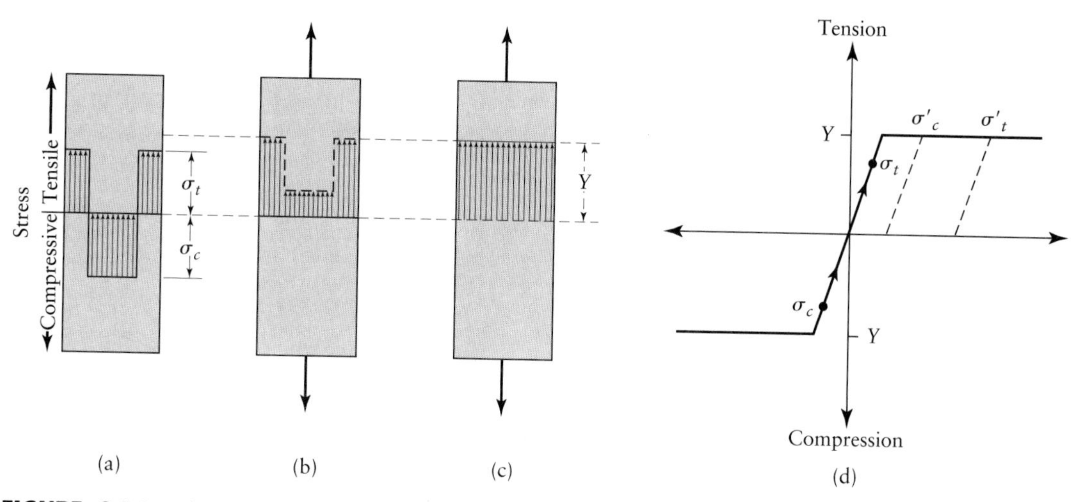

FIGURE 2.34 Elimination of residual stresses by stretching. Residual stresses can be also reduced or eliminated by thermal treatments, such as stress relieving or annealing.

The mechanism of residual-stress reduction or elimination by plastic deformation is as follows: assume that a piece of metal has the residual stresses shown in Fig. 2.34a, namely, tensile on the outside and compressive on the inside; these stresses are in equilibrium. Also assume that the material is elastic and perfectly plastic, as shown in Fig. 2.34d. The levels of the residual stresses are shown on the stress–strain diagram; both are below the yield stress, Y. If a uniformly distributed tension is now applied to this specimen, points σ_c and σ_t in the diagram move up on the stress–strain curve, as shown by the arrows. The maximum level that these stresses can reach is the tensile yield stress, Y. With sufficiently high loading, the stress distribution eventually becomes uniform throughout the part, as shown in Fig. 2.34c. If the load is then removed, the stresses recover elastically, and the part has no residual stresses. Note that very little stretching is required to relieve these residual stresses. The reason is that the elastic portions of the stress–strain curves for metals are very steep; hence the elastic stresses can be raised to the yield stress with very little strain.

The technique for reducing or relieving residual stresses by plastic deformation, such as by stretching as described, requires sufficient straining to establish a uniformly distributed stress in the part. A material such as the elastic, linearly strain-hardening type (Fig. 2.8e) therefore can never reach this condition, since the compressive stress, σ'_c, will always lag behind σ'_t. If the slope of the stress–strain curve in the plastic region is small, the difference between σ'_c and σ'_t will be rather small, and little residual stress will be left in the part after unloading.

EXAMPLE 2.5 Elimination of residual stresses by tension

Refer to Fig. 2.34a, and assume that $\sigma_t = 140$ MPa and $\sigma_c = -140$ MPa. The material is aluminum, and the length of the specimen is 0.25 m. Calculate the length to which this specimen should be stretched so that, when unloaded, it will be free from residual stresses. Assume that the yield stress of the material is 150 MPa.

Solution. Stretching should be to the extent that σ_c reaches the yield stress in tension, Y. Therefore, the total strain should be equal to the sum of the strain required to bring the compressive residual stress to zero and the strain required to bring it to the tensile yield stress. Hence,

$$\varepsilon_{\text{total}} = \frac{\sigma_c}{E} + \frac{Y}{E}. \tag{2.32}$$

For aluminum, let $E = 70$ GPa. Thus,

$$\varepsilon_{\text{total}} = \frac{140}{70 \times 10^3} + \frac{150}{70 \times 10^3} = 0.00414.$$

Hence, the stretched length should be

$$\ln\left(\frac{\ell_f}{0.25}\right) = 0.00414, \quad \text{or} \quad \ell_f = 0.2510 \text{ m}.$$

As the strains are very small, we may use engineering strains in these calculations. Hence,

$$\text{strain} = \frac{\ell_f - 0.25}{0.25} = 0.00414, \quad \text{or} \quad \ell_f = 0.2510 \text{ m}.$$

2.11 Triaxial Stresses and Yield Criteria

In most manufacturing operations involving deformation processing, the material, unlike that in a simple tension or compression test, is generally subjected to triaxial stresses. For example, in the expansion of a thin-walled spherical shell under internal pressure, an element in the shell is subjected to equal biaxial tensile stresses (Fig. 2.35a). In drawing a rod or wire through a conical die, an element in the deformation zone is subjected to tension in its length direction and to compression on its conical surface (Fig. 2.35b). An element in the flange in deep drawing of sheet metal is subjected to a tensile radial stress and compressive stresses on its surface and in the circumferential direction (Fig. 2.35c). As shown in subsequent chapters, many

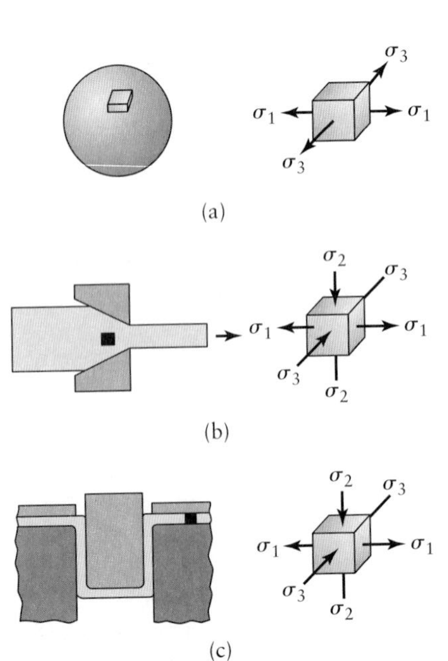

FIGURE 2.35 The state of stress in various metalworking processes. (a) Expansion of a thin-walled spherical shell under internal pressure. (b) Drawing of round rod or wire through a conical die to reduce its diameter; see Section 6.5 (c) Deep drawing of sheet metal with a punch and die to make a metal cup; see Section 7.12.

similar examples can be given in which the material is subjected to various normal and shear stresses during processing. The three-dimensional stresses shown in Fig. 2.35 cause distortion of the elements.

In the *elastic range*, the strains are represented by the *generalized Hooke's law* equations:

$$\varepsilon_1 = \frac{1}{E}[\sigma_1 - \nu(\sigma_2 + \sigma_3)]; \quad (2.33a)$$

$$\varepsilon_2 = \frac{1}{E}[\sigma_2 - \nu(\sigma_1 + \sigma_3)]; \quad (2.33b)$$

$$\varepsilon_3 = \frac{1}{E}[\sigma_3 - \nu(\sigma_1 + \sigma_2)]. \quad (2.33c)$$

Therefore, for simple tension where $\sigma_2 = \sigma_3 = 0$,

$$\varepsilon_1 = \frac{\sigma_1}{E},$$

and

$$\varepsilon_2 = \varepsilon_3 = -\frac{\nu\sigma_1}{E}.$$

The negative sign indicates a contraction of the element in the 2 and 3 directions, respectively.

In a simple tension or simple compression test, when the applied stress reaches the uniaxial yield stress Y, the material will deform *plastically*. However, if the material is subjected to a more complex state of stress, relationships between the stresses will predict yielding. These relationships are known as *yield criteria*, the most common ones being the maximum-shear-stress criterion and the distortion-energy criterion.

2.11.1 Maximum-shear-stress criterion

The maximum-shear-stress criterion, also known as the **Tresca criterion**, states that yielding occurs when the *maximum shear stress* within an element is equal to or exceeds a critical value. As shown in Section 3.3, this critical value of the shear stress is a *material property* and is called **shear yield stress**, k; hence, for yielding to occur, the following condition must be met:

$$\tau_{max} \geq k. \quad (2.34)$$

The most convenient way of determining the stresses acting on an element is by use of **Mohr's circles for stresses**, described extensively in textbooks on mechanics of solids. **Principal stresses** and their directions can be determined easily from Mohr's circles or from stress-transformation equations.

If the maximum shear stress, as determined from the Mohr's circles or from appropriate equations, is equal to or exceeds k, then yielding will occur. It can be seen that there are many combinations of stresses (or *states of stress*) that can produce the same maximum shear stress. From the simple tension test, we note that

$$k = \frac{Y}{2}, \quad (2.35)$$

where Y is the uniaxial yield stress of the material. If, for some reason, we are unable to increase the stresses on the element in order to cause yielding, the solution is simply to lower Y by raising the temperature of the material; this solution is the basis of and one major reason for hot working of materials. We can now write the maximum-shear-stress criterion as

$$\sigma_{max} - \sigma_{min} = Y. \tag{2.36}$$

This relationship means that the maximum and minimum normal stresses produce the largest circle and hence the *largest* shear stress. Consequently, *the intermediate stress has no effect on yielding*. It should be emphasized that the left-hand side of Eq. (2.36) represents the *applied stresses* and that the right-hand side is a *material property*.

We have assumed here that the material is *continuous*, *homogeneous*, and *isotropic* (i.e., it has the same properties in all directions) and that tensile stresses are positive, compressive stresses negative, and the yield stress in tension and that in compression are equal (see the discussion of the Bauschinger effect in Section 2.3.2). All of these assumptions are important.

2.11.2 Distortion-energy criterion

The distortion-energy criterion, also called the **von Mises criterion,** states that yielding occurs when the relationship between the principal applied stresses and uniaxial yield stress Y of the material is

$$(\sigma_1 - \sigma_2)^2 + (\sigma_2 - \sigma_3)^2 + (\sigma_3 - \sigma_1)^2 = 2Y^2. \tag{2.37}$$

Note that, unlike the case for the maximum-shear-stress criterion, the intermediate principal stress is included in this equation. Here again, the left-hand side of the equation represents the applied stresses and the right-hand side a material property.

EXAMPLE 2.6 Yielding of a thin-walled shell

A thin-walled spherical shell is under internal pressure, p. The shell has a diameter of 20 in. and is 0.1 in. thick. It is made of a perfectly plastic material with a yield stress of 20,000 psi. Calculate the pressure required to cause yielding of the shell according to both yield criteria.

Solution. For this shell under internal pressure, the membrane stresses are given by

$$\sigma_1 = \sigma_2 = \frac{pr}{2t}, \tag{2.38}$$

where $r = 10$ in. and $t = 0.1$ in. The stress in the thickness direction, σ_3, is negligible, because of the high r/t ratio of the shell. Thus, according to the maximum-shear-stress criterion,

$$\sigma_{max} - \sigma_{min} = Y,$$

or

$$\sigma_1 - 0 = Y$$

and

$$\sigma_2 - 0 = Y$$

Hence, $\sigma_1 = \sigma_2 = 20{,}000$ psi. The pressure required is then

$$p = \frac{2tY}{r} = \frac{(2)(0.1)(20{,}000)}{10} = 400 \text{ psi}.$$

According to the distortion-energy criterion,

$$(\sigma_1 - \sigma_2)^2 + (\sigma_2 - \sigma_3)^2 + (\sigma_3 - \sigma_1)^2 = 2Y^2,$$

or

$$0 + \sigma_2^2 + \sigma_1^2 = 2Y^2.$$

Hence, $\sigma_1 = \sigma_2 = Y$. Therefore, the answer is the same: $p = 400$ psi.

EXAMPLE 2.7 Correction factor for true-stress–true-strain curves

Explain why a correction factor has to be applied in the construction of a true-stress–true-strain curve, shown in Fig. 2.7a based on tensile-test data.

Solution. The neck of a specimen is subjected to a triaxial state of stress, as shown in Fig. 2.36. The reason for this stress state is that each element in the region has a different cross-sectional area; the smaller the area, the greater the tensile stress on the element. Hence, element 1 will contract laterally more than element 2, and so on; however, element 1 is restrained from contracting freely by element 2, and element 2 is restrained by element 3, and so on. This restraint causes radial and circumferential tensile stresses in the necked region, resulting in an axial tensile stress distribution, as shown in Fig. 2.36.

The *true* uniaxial stress in tension is σ, whereas the *calculated* value of true stress at fracture is the *average* stress; hence, a correction has to be made.

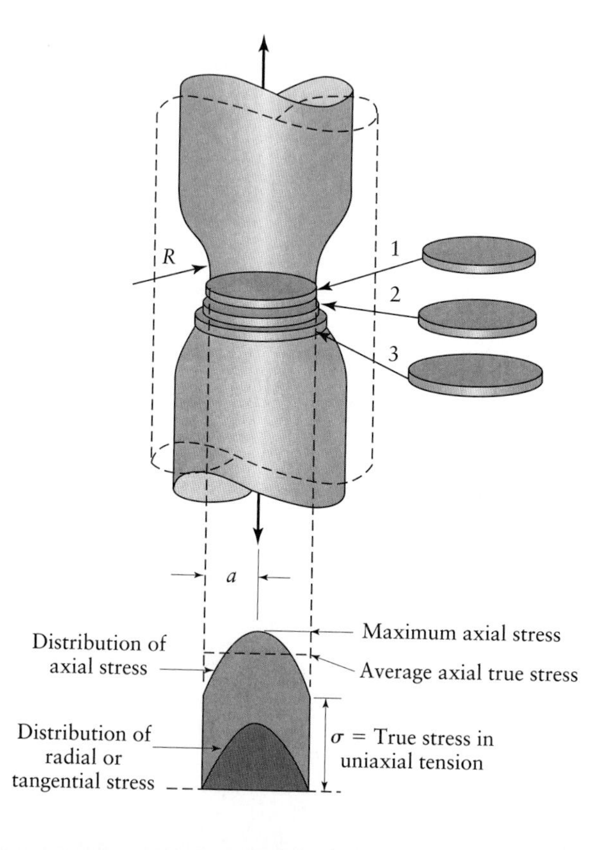

FIGURE 2.36 Stress distribution in the necked region of a tensile-test specimen.

A mathematical analysis by P. W. Bridgman gives the ratio of true stress to average stress as

$$\frac{\sigma}{\sigma_{av}} = \frac{1}{\left(1 + \frac{2R}{a}\right)\left[\ln\left(1 + \frac{a}{2R}\right)\right]}, \tag{2.39}$$

where R is the radius of curvature of the neck and a is the radius of the specimen at the neck. However, because R is difficult to measure during a test, an empirical relation has been established between a/R and the true strain at the neck. The corrected true-stress–true-strain curve is shown in Fig. 2.7a in Example 2.1.

2.11.3 Plane stress and plane strain

Plane stress and plane strain are important in the application of yield criteria. *Plane stress* is the state of stress in which one or two of the pairs of faces on an elemental cube are free from stress. An example is the torsion of a thin-walled tube (See Fig. 2.20). There are no stresses normal to the inside or outside surface of the tube; hence the state of the stress of the tube is one of plane stress. Other examples are shown in Fig. 2.37.

The state of stress where one of the pairs of faces on an element undergoes zero strain is known as *plane strain* (e.g., Figs. 2.37c and d). Another example is the **plane-strain compression** test shown in Fig. 2.16. There, by proper choice of specimen dimensions, the width of the specimen essentially is kept constant. (Note that an element does not have to be physically constrained on the pair of faces for plane-strain conditions to exist.) A third example is the torsion of a thin-walled tube in which the wall thickness remains constant. (See Section 2.11.7.)

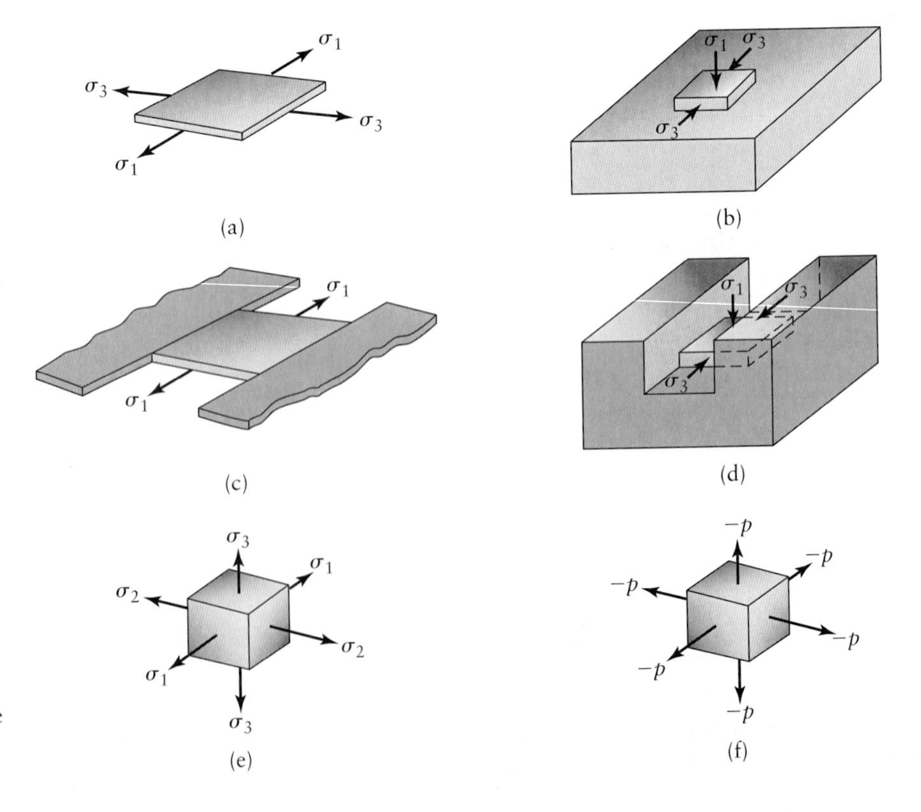

FIGURE 2.37 Examples of states of stress: (a) Plane stress in sheet stretching; there are no stresses acting on the surfaces of the sheet. (b) Plane stress in compression; there are no stresses acting on the sides of the specimen being compressed. (c) Plane strain in tension; the width of the sheet remains constant while being stretched. (d) Plane strain in compression (see also Fig. 2.16); the width of the specimen remains constant, due to the restraint by the groove. (e) Triaxial tensile stresses acting on an element. (f) Hydrostatic compression of an element. Note also that an element on the cylindrical portion of a thin-walled tube in torsion is in the condition of both plane stress and plane strain. (See also Section 2.11.7.)

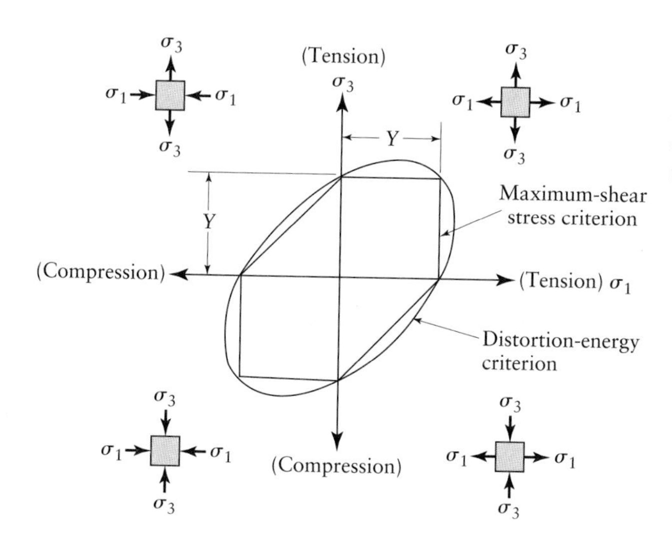

FIGURE 2.38 Plane-stress diagrams for maximum-shear-stress and distortion-energy criteria; note that $\sigma_2 = 0$.

A review of the two yield criteria previously outlined indicates that the *plane-stress* condition (where $\sigma_2 = 0$) can be represented by the diagram in Fig. 2.38. The maximum-shear-stress criterion gives an envelope of straight lines. In the first quadrant, where $\sigma_1 > 0$ and $\sigma_3 > 0$, and σ_2 is always zero for plane stress, Eq. (2.36) reduces to $\sigma_{\max} = Y$. Therefore, the maximum value that either σ_1 or σ_3 can acquire is Y—hence the straight lines in the diagram. In the third quadrant, the same situation exists, because σ_1 and σ_3 are both compressive. In the second and fourth quadrants, σ_2 (which is zero for the plane-stress condition) is the intermediate stress. Thus, for the second quadrant, Eq. (2.36) reduces to

$$\sigma_3 - \sigma_1 = Y, \tag{2.40}$$

and for the fourth quadrant, it reduces to

$$\sigma_1 - \sigma_3 = Y. \tag{2.41}$$

Equations (2.40) and (2.41) represent the 45° lines in Fig. 2.38.

The distortion-energy criterion for plane stress reduces to

$$\sigma_1^2 + \sigma_3^2 - \sigma_1\sigma_3 = Y^2 \tag{2.42}$$

and is shown graphically in Fig. 2.38. Whenever a point (with its coordinates representing the two principal stresses) falls on these boundaries, the element will yield.

The three-dimensional *elastic* stress–strain relationships are shown in Eqs. (2.33a)–(2.33c). When the stresses are sufficiently high to cause *plastic deformation*, the stress–strain relationships are obtained from **flow rules** (*Lévy–Mises equations*), described in detail in texts on plasticity. These relationships are in strain increments in the following equations:

$$d\varepsilon_1 = \frac{d\bar{\varepsilon}}{\bar{\sigma}}\left[\sigma_1 - \frac{1}{2}(\sigma_2 + \sigma_3)\right]; \tag{2.43a}$$

$$d\varepsilon_2 = \frac{d\bar{\varepsilon}}{\bar{\sigma}}\left[\sigma_2 - \frac{1}{2}(\sigma_1 + \sigma_3)\right]; \tag{2.43b}$$

$$d\varepsilon_3 = \frac{d\bar{\varepsilon}}{\bar{\sigma}}\left[\sigma_3 - \frac{1}{2}(\sigma_1 + \sigma_2)\right]. \tag{2.43c}$$

Note that these expressions are similar to those given by the elastic stress–strain relationships of the generalized Hooke's law.

For the plane-strain condition shown in Figs. 2.37c and d, we have $\varepsilon_2 = 0$. Therefore,

$$\sigma_2 = \frac{\sigma_1 + \sigma_3}{2}. \tag{2.44}$$

Note that σ_2 is now an intermediate stress.

For the *plane-strain compression* of Figs. 2.16 and 2.37d, the distortion-energy criterion (which includes the intermediate stress) reduces to

$$\sigma_1 - \sigma_3 = \frac{2}{\sqrt{3}}Y = 1.15Y = Y'. \tag{2.45}$$

Note that, whereas for the maximum-shear-stress criterion $k = Y/2$, we have $k = Y/\sqrt{3}$ for the distortion-energy criterion for the plane-strain condition.

2.11.4 Experimental verification of yield criteria

The yield criteria described thus far have been tested experimentally. A suitable specimen commonly used is a thin-walled tube under internal pressure and/or torsion. Under such loading, it is possible to generate different states of plane stress. Various experiments, with a variety of *ductile* materials, have shown that the distortion-energy criterion agrees better with the experimental data than does the maximum-shear-stress criterion.

These findings suggest use of the distortion-energy criterion for the analysis of metalworking processes (Chapters 6 and 7), which generally make use of ductile materials. The simpler maximum-shear-stress criterion, however, can also be used, particularly by designers. The difference between the two criteria is negligible for most practical applications.

2.11.5 Volume strain

By summing the three equations of the generalized Hooke's law [Eq. (2.33)], we obtain

$$\varepsilon_1 + \varepsilon_2 + \varepsilon_3 = \frac{1 - 2\nu}{E}(\sigma_1 + \sigma_2 + \sigma_3), \tag{2.46}$$

where it can be shown that the left-hand side of the equation is the *volume strain,* or **dilatation,** Δ. Thus,

$$\Delta = \frac{\text{volume change}}{\text{original volume}} = \frac{1 - 2\nu}{E}(\sigma_1 + \sigma_2 + \sigma_3) \tag{2.47}$$

In the plastic range, where $\nu = 0.5$, the volume change is zero. Hence, in plastic working of metals,

$$\varepsilon_1 + \varepsilon_2 + \varepsilon_3 = 0, \tag{2.48}$$

which is a convenient means of determining a third strain if two strains are known.

The **bulk modulus** is defined as

$$\text{Bulk modulus} = \frac{\sigma_m}{\Delta} = \frac{E}{3(1 - 2\nu)}, \tag{2.49}$$

where σ_m is the *mean stress,* defined as

$$\sigma_m = \frac{1}{3}(\sigma_1 + \sigma_2 + \sigma_3). \tag{2.50}$$

From Eq. (2.47), in the *elastic* range, where $0 < \nu < 0.5$, it can be seen that the volume of a tension-test specimen increases, and that of a compression-test specimen decreases during the test.

2.11.6 Effective stress and effective strain

A convenient means of expressing the state of stress on an element is by the *effective* (*equivalent* or *representative*) *stress*, $\bar{\sigma}$, and the *effective strain*, $\bar{\varepsilon}$. For the maximum-shear-stress criterion, the effective stress is

$$\bar{\sigma} = \sigma_1 - \sigma_3, \tag{2.51}$$

and for the distortion-energy criterion, it is

$$\bar{\sigma} = \frac{1}{\sqrt{2}}[(\sigma_1 - \sigma_2)^2 + (\sigma_2 - \sigma_3)^2 + (\sigma_3 - \sigma_1)^2]^{1/2}. \tag{2.52}$$

The factor $1/\sqrt{2}$ is chosen so that, for simple tension, the effective stress is equal to the uniaxial yield stress, Y.

The strains are likewise related to the *effective strain*. For the maximum-shear-stress criterion, the effective strain is

$$\bar{\varepsilon} = \frac{2}{3}(\varepsilon_1 - \varepsilon_3), \tag{2.53}$$

and for the distortion-energy criterion, it is

$$\bar{\varepsilon} = \frac{\sqrt{2}}{3}[(\varepsilon_1 - \varepsilon_2)^2 + (\varepsilon_2 - \varepsilon_3)^2 + (\varepsilon_3 - \varepsilon_1)^2]^{1/2}. \tag{2.54}$$

Again, the factors $2/3$ and $\sqrt{2}/3$ are chosen so that for simple tension, the effective strain is equal to the uniaxial tensile strain. It is apparent that stress–strain curves may also be called effective-stress–effective-strain curves. (See also Example 2.8.)

2.11.7 Comparison of normal stress/normal strain and shear stress/shear strain

Stress–strain curves in tension and torsion for the same material are, of course, comparable. Also, it is possible to construct one curve from the other, since the material is the same; the procedure for doing so is outlined next.

The following observations are made with regard to tension and torsional states of stress:

- In the tension test, the uniaxial stress σ_1 is also the effective stress and the principal stress;
- in the torsion test, the principal stresses occur on planes whose normals are at 45° to the longitudinal axis; the principal stresses σ_1 and σ_3 are equal in magnitude, but opposite in sign;
- the magnitude of the principal stress in torsion is the same as that of the maximum shear stress.

We now have the following relationships:

$$\sigma_1 = -\sigma_3, \quad \sigma_2 = 0, \quad \text{and} \quad \sigma_1 = \tau_1.$$

Substituting these stresses into Eqs. (2.51) and (2.52) for effective stress, the following relationships are obtained: for the maximum-shear-stress criterion,

$$\bar{\sigma} = \sigma_1 - \sigma_3 = \sigma_1 + \sigma_1 = 2\sigma_1 = 2\tau_1, \tag{2.55}$$

and for the distortion-energy criterion,

$$\bar{\sigma} = \frac{1}{\sqrt{2}}[(\sigma_1 - 0)^2 + (0 + \sigma_1)^2 + (-\sigma_1 - \sigma_1)^2]^{1/2} = \sqrt{3}\sigma_1 = \sqrt{3}\tau_1. \quad (2.56)$$

With regard to strains, the following observations can be made:

- In the tension test, $\varepsilon_2 = \varepsilon_3 = -\varepsilon_1/2$;
- in the torsion test, $\varepsilon_1 = -\varepsilon_3 = \gamma/2$;
- the strain in the thickness direction of the tube is zero, i.e., $\varepsilon_2 = 0$.

The third observation is true because the thinning caused by the principal tensile stress is countered by the thickening effect under the principal compressive stress of the same magnitude. Hence, $\varepsilon_2 = 0$. Since σ_2 is also zero, a thin-walled tube under torsion is both a plane-stress and a plane-strain situation.

Substituting these strains into Eqs. (2.53) and (2.54) for effective strain, the following relationships can be obtained: for the maximum-shear-stress criterion,

$$\bar{\varepsilon} = \frac{2}{3}(\varepsilon_1 - \varepsilon_3) = \frac{2}{3}(\varepsilon_1 + \varepsilon_1) = \frac{4}{3}\varepsilon_1 = \frac{2}{3}\gamma, \quad (2.57)$$

and for the distortion-energy criterion,

$$\bar{\varepsilon} = \frac{\sqrt{2}}{3}[(\varepsilon_1 - 0)^2 + (0 + \varepsilon_1)^2 + (\varepsilon_1 - \varepsilon_1)^2]^{1/2} = \frac{2}{\sqrt{3}}\varepsilon_1 = \frac{1}{\sqrt{3}}\gamma. \quad (2.58)$$

This set of equations provides a means by which tensile-test data can be converted to torsion-test data, and vice versa.

2.11.8 Instability in biaxial tension

In Section 2.2.4, we showed that in simple tension and for a material whose behavior can be represented by $\sigma = K\varepsilon^n$, instability occurs at a true strain of $\varepsilon = n$. Using the equations obtained from flow rules, one can derive the strain at instability for various conditions.

2.12 Work of Deformation

Work is defined as the product of collinear force and distance, so a quantity equivalent to work per unit volume is the product of stress and strain. Because the relation between stress and strain in the plastic range depends on the particular stress–strain curve, this work is best calculated by referring to Fig. 2.39.

Note that the area under the true-stress–true-strain curve for any strain ε is the **energy per unit volume** u **(specific energy)** *of the material deformed*. It is expressed as

$$u = \int_0^{\varepsilon_1} \sigma \, d\varepsilon. \quad (2.59)$$

As seen in Section 2.2.3, true-stress–true-strain curves can be represented by the expression

$$\sigma = K\varepsilon^n.$$

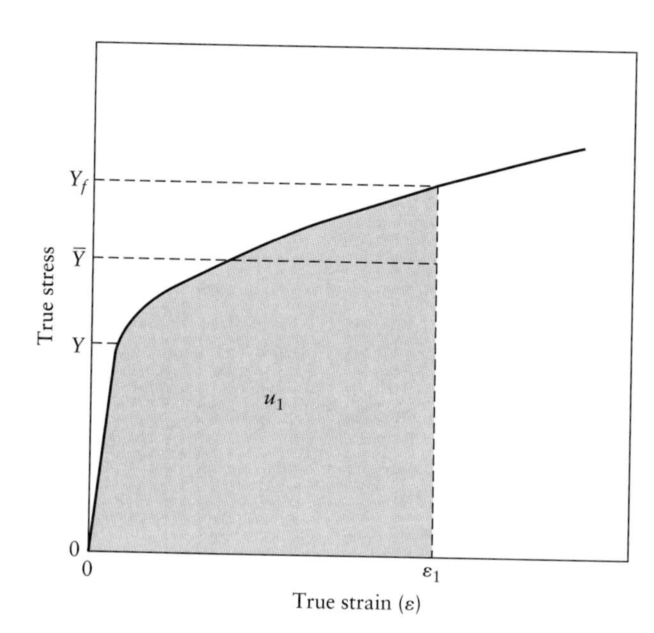

FIGURE 2.39 Schematic illustration of a true-stress–true-strain curve showing the yield stress Y, the average flow stress $\overline{Y}$, the specific energy u_1, and the flow stress Y_f.

Hence, Eq. (2.59) can be written as

$$u = K \int_0^{\varepsilon_1} \varepsilon^n \, d\varepsilon,$$

or

$$u = \frac{K\varepsilon_1^{n+1}}{n+1} = \overline{Y}\varepsilon_1, \tag{2.60}$$

where $\overline{Y}$ is the **average flow stress** of the material.

This energy represents the work dissipated in uniaxial deformation. For triaxial states of stress, a general expression is given by

$$du = \sigma_1 \, d\varepsilon_1 + \sigma_2 \, d\varepsilon_2 + \sigma_3 \, d\varepsilon_3.$$

For an application of this equation, see Example 2.8. For a more general condition, the effective stress and effective strain can be used. The energy per unit volume is then

$$u = \int_0^{\varepsilon} \bar{\sigma} \, d\bar{\varepsilon}. \tag{2.61}$$

To obtain the work expended, we multiply u by the volume of the material deformed:

$$\text{Work} = (u)(\text{volume}). \tag{2.62}$$

The energy represented by Eq. (2.62) is the *minimum energy,* or the *ideal energy,* required for uniform (homogeneous) deformation. The energy required for actual deformation involves two additional factors. One factor is the energy required to overcome *friction* at the die–workpiece interfaces; the other is the **redundant work** of deformation, which is described as follows.

In Fig. 2.40a, a block of material is being deformed into shape by forging, extrusion, or drawing through a die, as described in Chapter 6. As shown in Fig. 2.40b, this deformation is uniform, or homogeneous. In reality, however, the material more often than not deforms as in Fig. 2.40c, from the effects of friction and die geometry. The difference between b and c is that c has undergone additional shearing along horizontal planes.

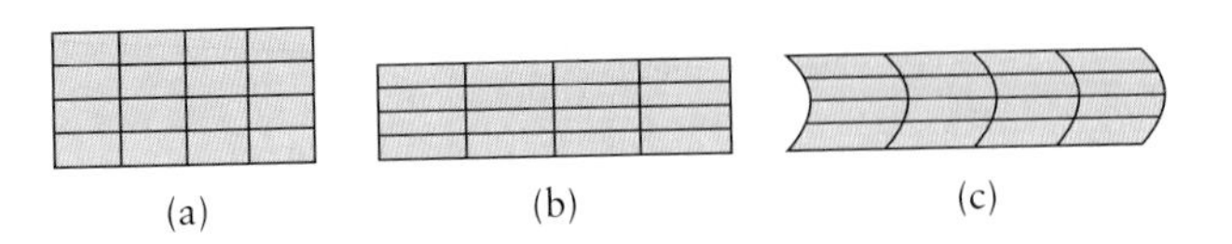

FIGURE 2.40 Deformation of grid patterns in a workpiece: (a) the original pattern; (b) after ideal deformation; (c) after inhomogeneous deformation, requiring redundant work of deformation. Note that grid pattern (c) is basically the same as grid pattern (b), but with additional shearing, especially at the outer layers. Thus, grid pattern (c) requires higher work of deformation than grid pattern (b). See also Figs. 6.2 and 6.51.

This shearing requires expenditure of energy, because additional plastic work has to be done in subjecting the various layers to shear strains. This additional work is known as *redundant work*; the word "redundant" reflects the fact that this work does not contribute to the shape change of the material. (Note that grid patterns (b) and (c) in Fig. 2.40 have the same overall shape and dimensions.)

The *total specific energy* required can now be written as

$$u_{\text{total}} = u_{\text{ideal}} + u_{\text{friction}} + u_{\text{redundant}}. \tag{2.63}$$

The *efficiency*, η, of a process is defined as

$$\eta = \frac{u_{\text{ideal}}}{u_{\text{total}}}. \tag{2.64}$$

The magnitude of this efficiency varies widely, depending on the particular process, frictional conditions, die geometry, and other process parameters. Typical values are estimated to be 30% to 60% for extrusion and 75% to 95% for rolling.

EXAMPLE 2.8 Expansion of a thin-walled spherical shell

A thin-walled spherical shell made of a perfectly plastic material of yield stress, Y, original radius, r_o, and thickness, t_o, is being expanded by internal pressure. (a) Calculate the work done in expanding this shell to a radius of r_f. (b) If the diameter expands at a constant rate, what changes take place in the power consumed as the radius increases?

Solution. The membrane stresses are given by

$$\sigma_1 = \sigma_2 = Y$$

(from Example 2.7), where r and t are instantaneous dimensions. The true strains in the membrane are given by

$$\varepsilon_1 = \varepsilon_2 = \ln\left(\frac{2\pi r_f}{2\pi r_o}\right) = \ln\left(\frac{r_f}{r_o}\right).$$

Because an element in this shell is subjected to equal biaxial stretching, the specific energy is

$$u = \int_0^{\varepsilon_1} \sigma_1\, d\varepsilon_1 + \int_0^{\varepsilon_2} \sigma_2\, d\varepsilon_2 = 2\sigma_1\varepsilon_1 = 2Y\ln\left(\frac{r_f}{r_o}\right)$$

Since the volume of the shell's material is $4\pi r_o^2 t_o$, the work W done is

$$W = (u)(\text{volume}) = 8\pi Y r_o^2 t_o \ln\left(\frac{r_f}{r_o}\right).$$

The specific energy can also be calculated from the effective stresses and strains. Thus, according to the distortion-energy criterion,

$$\bar{\sigma} = \frac{1}{\sqrt{2}}[(0)^2 + (\sigma_2)^2 + (-\sigma_1)^2]^{1/2} = \sigma_1 = \sigma_2$$

and

$$\bar{\varepsilon} = \frac{\sqrt{2}}{3}[(0)^2 + (\varepsilon_2 + 2\varepsilon_2)^2 + (-2\varepsilon_2 - \varepsilon_2)^2]^{1/2} = 2\varepsilon_2 = 2\varepsilon_1.$$

(The thickness strain $\varepsilon_3 = -2\varepsilon_2 = -2\varepsilon_1$, because of volume constancy in plastic deformation, where $\varepsilon_1 + \varepsilon_2 + \varepsilon_3 = 0$.) Hence,

$$u = \int_0^{\bar{\varepsilon}} \bar{\sigma}\, d\bar{\varepsilon} = \int_0^{2\varepsilon_1} \sigma_1\, d\varepsilon_1 = 2\sigma_1\varepsilon_1.$$

Thus, the answer is the same.

Power is defined as work per unit time; thus,

$$\text{Power} = \frac{dW}{dt}.$$

Since all other factors in the expression are constant, the expression for work can be written as being proportional to strain:

$$W \propto \ln\left(\frac{r}{r_o}\right) \propto (\ln r - \ln r_o).$$

Therefore,

$$\text{Power} \propto \frac{1}{r}\frac{dr}{dr}.$$

Because the shell is expanding at a constant rate, dr/dt = constant. Hence, the power is related to the instantaneous radius r by the following expression:

$$\text{Power} \propto \frac{1}{r}.$$

2.12.1 Work, heat, and temperature rise

Almost all the mechanical work of deformation in plastic working is converted into heat. This conversion is not 100% because a small portion of energy is stored within the deformed material as elastic energy. This is known as *stored energy* (see Section 3.6); it is generally 5% to 10% of the total energy input, but it may be as high as 30% in some alloys.

In a simple frictionless process, and assuming that work is completely converted into heat, the *temperature rise*, ΔT, is given by

$$\Delta T = \frac{u_{total}}{\rho c}, \tag{2.65}$$

where u_{total} is the specific energy from Eq. (2.63), ρ is the density, and c is the specific heat of the material. Thus, higher temperatures are associated with large areas under the stress–strain curve and smaller values of specific heat. The theoretical temperature rise for a true strain of 1 (such as a 27-mm-high specimen compressed down to 10 mm) has been calculated to be as follows: aluminum, 75°C (165°F); copper, 140°C (285°F); low-carbon steel, 280°C (535°F); and titanium, 570°C (1060°F).

The temperature rise given by Eq. (2.65) is for an ideal situation, where there is no heat loss. In actual operations, heat is lost to the environment, to tools and dies, and to any lubricants or coolants used. If the process is performed very rapidly, these losses are relatively small. Under extreme conditions, an *adiabatic* state is approached, with a very high temperature rise, leading to **incipient melting**. The rise in temperature can be calculated if the stress–strain curve used is at the appropriate strain-rate level (see Section 2.2.7); however, if the process is carried out slowly, the actual temperature rise will be a small portion of the calculated value. It should also be noted that properties such as specific heat and thermal conductivity also depend on temperature, and they should be taken into account in the calculations.

EXAMPLE 2.9 Temperature rise in compression

A cylindrical specimen 1 in. diameter and 1 in. high is being compressed by dropping a weight of 100 lb on it from a certain height. The material has the following properties: $K = 15{,}000$ psi; $n = 0.5$; density $= 0.1\ \text{lb/in}^3$ and specific heat $= 0.3\text{Btu/lb-°F}$; Assuming no heat loss and no friction, calculate the final height of the specimen if the temperature rise is 100°F.

Solution. The expression for heat is given by

$$\text{Heat} = (c_p)(\rho)(\text{volume})(\Delta T),$$

or

$$\text{Heat} = (0.3)(0.10)[(\pi)(1^2)/4](100)(778),$$

where 778 ft-lb = 1BTU. Thus,

$$\text{Heat} = 1830\ \text{ft-lb} = 22{,}000\ \text{in.-lb}.$$

Also, ideally,

$$\text{Heat} = \text{Work} = (\text{volume})(u) = \frac{\pi}{4}\,\frac{15{,}000\varepsilon^{1.5}}{1.5}$$

and

$$\varepsilon^{1.5} = \frac{(22{,}000)(1.5)(4)}{(\pi)(15{,}000)} = 2.8;$$

hence,

$$\varepsilon = 1.99.$$

Using absolute values, we have

$$\ln\left(\frac{h_o}{h_f}\right) = \ln\left(\frac{1}{h_f}\right) = 1.99,$$

and therefore, $h_f = 0.137$in.

SUMMARY

- Many manufacturing processes involve shaping materials by plastic deformation; consequently, mechanical properties such as strength, elasticity, ductility, hardness, toughness, and the energy required for plastic deformation are important factors. The behavior of materials, in turn, depends on the particular material and its condition, as well as other variables, particularly temperature, strain rate, and the state of stress. (Section 2.1)

- Mechanical properties measured by tension tests include the modulus of elasticity (E), yield stress (Y), ultimate tensile strength (UTS), and Poisson's ratio (ν). Ductility, as measured by elongation and reduction of area, can also be determined by tension tests. True-stress–true-strain curves are important in determining such mechanical properties as strength coefficient (K), strain-hardening exponent (n), strain-rate sensitivity exponent (m), and toughness. (Section 2.2)
- Compression tests model manufacturing processes such as forging, rolling, and extrusion. Properties measured by compression tests are subject to inaccuracy, due to the presence of friction and barreling. (Section 2.3)
- Torsion tests are typically conducted on tubular specimens that are subjected to twisting. These tests model such manufacturing processes as shearing, cutting, and various machining processes. (Section 2.4)
- Bend, or flexure, tests are commonly used for brittle materials; the outer fiber stress at fracture in bending is known as the modulus of rupture, or transverse rupture strength. The forces applied in bend tests model those incurred in such manufacturing processes as forming of sheet and plate, as well as testing of tool and die materials. (Section 2.5)
- A variety of hardness tests is available to test the resistance of a material to permanent indentation. Hardness is related to strength and wear resistance, but is itself not a fundamental property of a material. (Section 2.6)
- Fatigue tests model manufacturing processes whereby components are subjected to rapidly fluctuating loads. The quantity measured is the endurance limit, or fatigue limit—that is, the maximum stress to which a material can be subjected without fatigue failure, regardless of the number of cycles. (Section 2.7)
- Creep is the permanent elongation of a component under a static load maintained for a period of time. A creep test typically consists of subjecting a specimen to a constant tensile load at a certain temperature and measuring its change in length over a period of time. The specimen eventually fails by necking and rupture. (Section 2.8)
- Impact tests model some high-rate manufacturing operations as well as the service conditions in which materials are subjected to impact (or dynamic) loading, such as drop forging or the behavior of tool and die materials in interrupted cutting or high-rate deformation. Impact tests determine the energy required to fracture the specimen, known as impact toughness. Impact tests are also useful in determining the ductile–brittle transition temperature of materials. (Section 2.9)
- Residual stresses are those stresses that remain in a part after it has been deformed and all external forces have been removed. The nature and level of residual stresses depend on the manner in which the part has been plastically deformed. Residual stresses may be reduced or eliminated by stress-relief annealing, by further plastic deformation, or by relaxation. (Section 2.10)
- In metalworking operations, the workpiece material is generally subjected to three-dimensional stresses through various tools and dies. Yield criteria establish relationships between the uniaxial yield stress of the material and the stresses applied; the two most widely used yield criteria are the maximum-shear-stress (Tresca) and the distortion-energy (von Mises) criteria. (Section 2.11)
- Because energy is required to deform materials, the work of deformation per unit volume of material, u, is an important parameter and is composed of ideal, frictional, and redundant work components. In addition to supplying information on force and energy requirements, the work of deformation also indicates the amount of heat developed and, hence, the temperature rise in the workpiece during plastic deformation. (Section 2.12)

SUMMARY OF EQUATIONS

Engineering strain, $e = \frac{\ell - \ell_o}{\ell_o}$

Engineering strain rate, $\dot{e} = \frac{v}{\ell_o}$

Engineering stress, $\sigma = \frac{P}{A_o}$

True strain, $\varepsilon = \ln\left(\frac{\ell}{\ell_o}\right)$

True strain rate, $\dot{\varepsilon} = \frac{v}{\ell}$

True stress, $\sigma = \frac{P}{A}$

Modulus of elasticity, $E = \frac{\sigma}{e}$

Shear modulus, $G = \frac{E}{2(1 + \nu)}$

Modulus of resilience $= \frac{Y^2}{2E}$

Elongation $= \frac{\ell_f - \ell_o}{\ell_o} \times 100$

Reduction of area $= \frac{A_o - A_f}{A_o} \times 100$

Shear strain in torsion, $\gamma = \frac{r\phi}{\ell}$

Hooke's law, $\varepsilon_1 = \frac{1}{E}[\sigma_1 - \nu(\sigma_2 + \sigma_3)]$, etc.

Effective strain (Tresca), $\bar{\varepsilon} = \frac{2}{3}(\varepsilon_1 - \varepsilon_3)$

Effective strain (von Mises), $\bar{\varepsilon} = \frac{\sqrt{2}}{3}[(\varepsilon_1 - \varepsilon_2)^2 + (\varepsilon_2 - \varepsilon_3)^2 + (\varepsilon_3 - \varepsilon_1)^2]^{1/2}$

Effective stress (Tresca), $\bar{\sigma} = \sigma_1 - \sigma_3$

Effective stress (von Mises), $\bar{\sigma} = \frac{1}{\sqrt{2}}[(\sigma_1 - \sigma_2)^2 + (\sigma_2 - \sigma_3)^2 + (\sigma_3 - \sigma_1)^2]^{1/2}$

True-stress–true-strain relationship (power law), $\sigma = K\varepsilon^n$

True-stress–true-strain rate relationship, $\sigma = C\dot{\varepsilon}^m$

Flow rules, $d\varepsilon_1 = \frac{d\bar{\varepsilon}}{\bar{\sigma}}\left[\sigma_1 - \frac{1}{2}(\sigma_2 + \sigma_3)\right]$, etc.

Maximum-shear-stress criterion (Tresca), $\sigma_{max} - \sigma_{min} = Y$

Distortion-energy criterion (von Mises), $(\sigma_1 - \sigma_2)^2 + (\sigma_2 - \sigma_3)^2 + (\sigma_3 - \sigma_1)^2 = 2Y^2$

Shear yield stress, $k = Y/2$ for Tresca and $k = Y/\sqrt{3}$ for von Mises (plane strain)

Volume strain (dilatation), $\Delta = \frac{1 - 2\nu}{E}(\sigma_1 + \sigma_2 + \sigma_3)$

Bulk modulus $= \frac{E}{3(1 - 2\nu)}$

BIBLIOGRAPHY

ASM Handbook, Vol. 8: *Mechanical Testing*, ASM International, 2000.

ASM Handbook, Vol. 10: *Materials Characterization*, ASM International, 1986.

Boyer, H. E. (ed.), *Atlas of Creep and Stress-Rupture Curves*, ASM International, 1986.

________, *Atlas of Fatigue Curves*, ASM International, 1986.

________, *Atlas of Stress–Strain Curves*, ASM International, 1986.

Budinski, K. G., *Engineering Materials: Properties and Selection*, 6th ed., Prentice Hall, 1998.

Cheremisinoff, N. P., and P. N. Cheremisinoff, *Handbook of Advanced Materials Testing*, Dekker, 1994.

Courtney, T. H., *Mechanical Behavior of Materials*, 2d ed., McGraw-Hill, 1999.

Davis, J. R. (ed.), *Metals Handbook: Desk Edition*, ASM International, 1998.

Dieter, G. E., *Mechanical Metallurgy*, 3d ed., McGraw-Hill, 1986.

Dowling, N. E., *Mechanical Behavior of Materials: Engineering Methods for Deformation, Fracture, and Fatigue*, 2d ed., Prentice Hall, 1998.

Han, P. (ed.), *Tensile Testing*, ASM International, 1992.

Hsu, T. H., *Stress and Strain Data Handbook*, Gulf Publishing Co., 1986.

Pohlandt, K., *Material Testing for the Metal Forming Industry*, Springer, 1989.

QUESTIONS

2.1 Can you calculate the percent elongation of materials based only on the information given in Fig. 2.6? Explain.

2.2 Describe your observations regarding the contents of Table 2.1.

2.3 Explain if it is possible for the curves in Fig. 2.4 to reach 0% elongation as the gage length is increased further.

2.4 Explain why the difference between engineering strain and true strain becomes larger as strain increases. Is this phenomenon true for both tensile and compressive strains? Explain.

2.5 Using the same scale for stress, we note that the tensile true-stress–true-strain curve is higher than the engineering stress–strain curve. Explain whether this condition also holds for a compression test.

2.6 Which of the two tests, tension or compression, requires a higher capacity testing machine than the other? Explain.

2.7 Explain how the modulus of resilience of a material changes, if at all, as it is strained: (1) for an elastic, perfectly plastic material, and (2) for an elastic, linearly strain-hardening material.

2.8 Take an element under the anvils in the deformation zone in the plane-strain compression test shown in Fig. 2.16. Is the element subjected to simple shear or pure shear during deformation? Explain. (*Hint*: View this element from the sides, in the width direction, and compare it with Fig. 2.21.)

2.9 If you pull and break a tensile-test specimen rapidly, where would the temperature be the highest? Explain why.

2.10 Comment on the temperature distribution if the specimen in Question 2.9 is pulled very slowly.

2.11 In a tension test, the area under the true-stress–true-strain curve is the work done per unit volume (the specific work). We also know that the area under the load–elongation curve represents the work done on the specimen. If you divide this latter work by the volume of the specimen between the gage marks, you will determine the work done per unit volume (assuming that all deformation is confined between the gage marks). Will this specific work be the same as the area under the true-stress–true-strain curve? Explain. Will your answer be the same for any value of strain? Explain.

2.12 The note at the bottom of Table 2.5 states that as temperature increases, C decreases and m increases. Explain why.

2.13 You are given the K and n values of two different materials, respectively. Is this information sufficient to determine which material is tougher? If not, what additional information do you need, and why?

2.14 Modify the curves in Fig. 2.8 to indicate the effects of temperature. Explain your changes.

2.15 Using a specific example, show why the deformation rate, say in m/s, and the true strain rate are not the same.

2.16 Referring to Fig. 2.3, explain how the magnitude of elastic recovery will change as an elastic, linearly strain-hardening material is strained.

2.17 It has been stated that the higher the value of m, the more diffuse the neck is, and likewise, the lower the value of m, the more localized the neck is. Explain the reason for this behavior.

2.18 Explain why materials with high m values, such as hot glass and silly putty, when stretched slowly, undergo large elongations before failure. Consider events taking place in the necked region of the specimen.

2.19 Assume that you are running four-point bending tests on a number of identical specimens of the same length and cross-section, but with increasing distance between the upper points of loading. (See Fig. 2.23b.) What changes, if any, would you expect in the test results? Explain.

2.20 Would Eq. (2.10) hold true in the elastic range? Explain.

2.21 Why have different types of hardness tests been developed? How would you measure the hardness of a very large object?

2.22 Which hardness tests and scales would you use for very thin strips of material, such as aluminum foil? Why?

2.23 Would the answer to Example 2.2 double if the strength coefficient of the material is doubled? Explain.

2.24 List and explain the factors that you would consider in selecting an appropriate hardness test and scale for a particular application.

2.25 Describe the difference between creep and stress relaxation, giving two examples for each as they relate to engineering applications.

2.26 Referring to the two impact tests shown in Fig. 2.31, explain how different the results would be if the specimens were impacted from the opposite directions.

2.27 If you remove the layer *ad* from the part shown in Fig. 2.32d, such as by machining or grinding, which way will the specimen curve? (*Hint*: Assume that the part in diagram (d) is composed of four horizontal springs held at the ends. Thus, from the top down, we have compression, tension, compression, and tension springs.)

2.28 Is it possible to completely remove residual stresses in a piece of material by the technique described in Fig. 2.34 if the material is elastic, linearly strain hardening? Explain.

2.29 Referring to Fig. 2.34, would it be possible to eliminate residual stresses by compression? Assume that the piece of material will not buckle under the uniaxial compressive force.

2.30 List and explain the desirable mechanical properties for (1) an elevator cable, (2) a paper clip, (3) a leaf spring, (4) a bracket for a bookshelf, (5) a piano wire, (6) a coat hanger, (7) a gas-turbine blade, and (8) a staple.

2.31 Using the generalized Hooke's law [Eq. (2.33)], show whether or not a thin-walled tube undergoes any change in thickness when subjected to torsion in the elastic range.

2.32 Make a sketch showing the nature and distribution of the residual stresses in Figs. 2.33a and b before the parts were cut. Assume that the split parts are free from any stresses. (*Hint*: Force these parts back to the shape they were in before they were cut.)

2.33 It is possible to calculate the work of plastic deformation by measuring the temperature rise in a workpiece, assuming that there is no heat loss and that the temperature distribution is uniform throughout. If the specific heat of the material decreases with increasing temperature, will the work of deformation calculated using the specific heat at room temperature be higher or lower than the actual work done? Explain.

2.34 Explain whether or not the volume of a homogeneous, isotropic specimen changes when the specimen is subjected to a state of (a) uniaxial compressive stress and (b) uniaxial tensile stress, all in the elastic range.

2.35 How would you explain the general behavior of a metal as shown in Fig. 2.11 to someone over the telephone?

2.36 We know that it is relatively easy to subject a specimen to hydrostatic compression, such as by using a chamber filled with a liquid. Devise a means whereby the specimen (say, in the shape of a cube or a round disk) can be subjected to hydrostatic tension, or one approaching this state of stress. (Note that a thin-walled, internally pressurized spherical shell is not a correct answer, because it is subjected only to a state of plane stress.)

2.37 Referring to Fig. 2.20, make sketches of the state of stress for an element in the reduced section of the tube when it is subjected to (1) torsion only, (2) torsion while the tube is internally pressurized, and (3) torsion while the tube is externally pressurized. Assume that the tube is a closed-end tube.

2.38 How would you conduct the test shown in Fig. 2.22? Make a sketch of the setup you would use.

2.39 A penny-shaped piece of soft metal of a certain diameter is brazed to the ends of two flat, round steel rods of the same diameter as the piece. The assembly is then subjected to uniaxial tension. What is the state of stress to which the soft metal is subjected? Explain.

2.40 A circular disk of soft metal is being compressed between two flat, hardened circular steel punches of having the same diameter as the disk. Assume that the disk material is perfectly plastic and that there is no friction or any temperature effects. Explain the change, if any, in the magnitude of the punch force as the disk is being compressed plastically to, say, a fraction of its original thickness.

2.41 A perfectly plastic metal is yielding under the stress state $\sigma_1, \sigma_2, \sigma_3$, where $\sigma_1 > \sigma_2 > \sigma_3$. Explain what happens if σ_1 is increased.

2.42 What is the dilatation of a material with a Poisson's ratio of 0.5? Is it possible for a material to have a Poisson's ratio of 0.7? Give a rationale for your answer.

2.43 Can a material have a negative Poisson's ratio? Give a rationale for your answer.

2.44 As clearly as possible, define plane stress and plane strain.

2.45 What test would you use to evaluate the hardness of a coating on a metal surface? Would it matter if the coating was harder or softer than the substrate? Explain.

2.46 Explain clearly the difference between resilience and toughness. Give an application for which each is important.

2.47 List the advantages and limitations of the stress–strain relationships given in Fig. 2.8.

2.48 Plot the data in Table 2.1 on a bar chart, showing the range of values, and comment on the results.

2.49 A hardness test is conducted on as-received metal as a quality check. The results show indicate that the hardness is too high, indicating that the material may not have sufficient ductility for the intended application. The supplier is reluctant to accept the return of the material, instead claiming that the diamond cone used in the Rockwell testing was worn and blunt, and hence the test needed to be recalibrated. Is this explanation plausible? Explain.

2.50 Explain why a 0.2% offset is used to obtain the yield strength in a tension test.

2.51 Referring to Question 2.50, would the offset method be necessary for a highly strained hardened material? Explain.

PROBLEMS

2.52 A strip of metal is originally 1.5 m long. It is stretched in three steps: first to a length of 2.0 mm, then to 2.5 m, and finally to 3.0 m. Show that the total true strain is the sum of the true strains in each step, that is, that the strains are additive. Show that, using engineering strains, the strain for each step cannot be added to obtain the total strain.

2.53 A paper clip is made of wire 1.25 mm in diameter. If the original material from which the wire is made is a rod 15 mm in diameter, calculate the longitudinal and diametrical engineering and true strains that the wire has undergone during processing.

2.54 A material has the following properties: UTS = 50,000 psi and $n = 0.3$. Calculate its strength coefficient K.

2.55 Based on the information given in Fig. 2.6, calculate the ultimate tensile strength of annealed copper.

2.56 Identify the two materials in Fig. 2.6 that have the lowest and highest uniform elongations, respectively. Calculate these quantities as percentages of the original gage lengths.

2.57 Calculate the ultimate tensile strength (engineering) of a material whose strength coefficient is 800 MPa and of a tensile-test specimen that necks at a true strain of 0.30.

2.58 A cable is made of four parallel strands of different materials, all behaving according to the equation $\sigma = K\varepsilon^n$, where $n = 0.5$. The materials, strength coefficients and cross-sections are as follows:

Material A: $K = 500$ MPa, $A_o = 7\ \text{mm}^2$;
Material B: $K = 800$ MPa, $A_o = 2\ \text{mm}^2$;
Material C: $K = 400$ MPa, $A_o = 3\ \text{mm}^2$;
Material D: $K = 600$ MPa, $A_o = 2\ \text{mm}^2$.

(a) Calculate the maximum tensile force that this cable can withstand prior to necking.
(b) Explain how you would arrive at an answer if the n values of the three strands were different from each other.

2.59 Using only Fig. 2.6, calculate the maximum load in tension testing of a 304 stainless-steel specimen with an original diameter of 0.5 in.

2.60 Using the data given in Table 2.1, calculate the values of the shear modulus G for the metals listed in the table.

2.61 Derive an expression for the toughness of a material represented by the equation $\sigma = K(\varepsilon + 0.2)^n$ and whose fracture strain is denoted as ε_f.

2.62 A cylindrical specimen made of a brittle material 1 in. high and with a diameter of 1 in. is subjected to a compressive force along its axis. It is found that fracture takes place at an angle of 45° under a load of 40,000 lb. Calculate the shear stress and the normal stress, respectively, acting on the fracture surface.

2.63 What is the modulus of resilience of a highly cold-worked piece of steel with a hardness of 350 HB? Of a piece of highly cold-worked copper with a hardness of 125 HB?

2.64 Calculate the work done in frictionless compression of a solid cylinder 40 mm high and 20 mm in diameter to a reduction in height of 50% for the following materials: (1) 1100-O aluminum, (2) annealed copper, (3) annealed 302 stainless steel, and (4) 70–30 brass, annealed.

2.65 Calculate the toughness of the material in Example 2.1 by using the area under the stress–strain curve. Then determine the K and n values for this material and obtain the toughness by integration. Compare the two answers.

2.66 A material has a strength coefficient $K = 125{,}000$ psi and $n = 0.2$. Assuming that a tensile-test specimen made from this material begins to neck at a true strain of 0.2, show that the ultimate tensile strength of this material is 74,200 psi.

2.67 A tensile-test specimen is made of a material represented by the equation $\sigma = K(\varepsilon + n)^n$. (a) Determine the true strain at which necking will begin. (b) Show that it is possible for an engineering material to exhibit this behavior.

2.68 Take two solid cylindrical specimens of equal diameter, but different heights. Assume that both specimens are compressed (frictionless) by the same percent reduction, say 50%. Prove that the final diameters will be the same.

2.69 A horizontal rigid bar *c–c* is subjecting specimen *a* to tension and specimen *b* to frictionless compression such that the bar remains horizontal. (See the accompanying figure.) The force *F* is located at a distance ratio of 2:1. Both specimens have an original cross-sectional area of 1 in^2, and the original lengths are $a = 10$ in and $b = 5$ in The material for specimen *a* has a true-stress–true-strain curve of $\sigma = 100{,}000\varepsilon^{0.5}$ psi. Plot the true-stress–true-strain curve that the material for specimen *b* should have for the bar to remain horizontal.

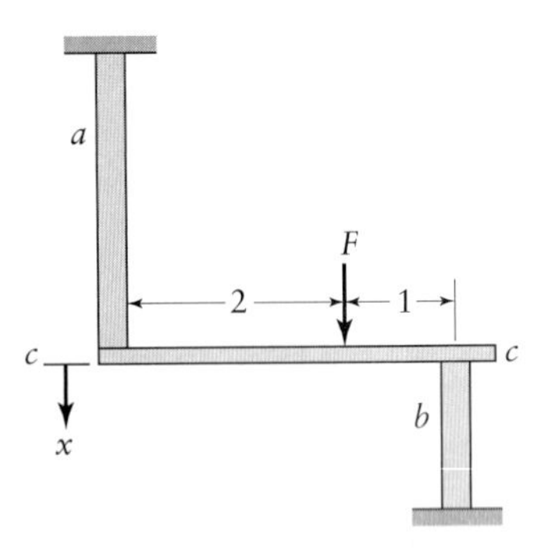

2.70 Inspect the curve that you obtained in Problem 2.69. Does a typical strain-hardening material behave in that manner? Explain.

2.71 In a disk test performed on a specimen 30 mm in diameter and 5 mm thick, the specimen fractures at a stress of 500 MPa. What was the load on it at fracture?

2.72 In Fig. 2.34a, let the tensile and compressive residual stresses both be 30,000 psi, and the modulus of elasticity of the material be 30×10^6 psi, with a modulus of resilience of 30 in.-lb/in^3. If the original length in diagram (a) is 40 in., what should be the stretched length in diagram (b) so that, when unloaded, the strip will be free of residual stresses?

2.73 Show that you can take a bent bar made of an elastic, perfectly plastic material and straighten it by stretching it into the plastic range. (*Hint*: Observe the events shown in Fig. 2.34.)

2.74 A bar 0.5 m long is bent and then stress relieved. The radius of curvature to the neutral axis is 0.30 m. The bar is 30 mm thick and is made of an elastic, perfectly plastic material with $Y = 600$ MPa and $E = 200$ GPa. Calculate the length to which this bar should be stretched so that, after unloading, it will become and remain straight.

2.75 Derive the three generalized Hooke's law equations given in Eq. (2.33).

2.76 A thin cylindrical sleeve of thickness *t*, length *L*, and inside radius *r* is placed on a rigid shaft with no clearance. (See the accompanying figure.) The sleeve is then stretched elastically by tension σ_a in the axial direction. Let the coefficient of friction at the sleeve–shaft interface be denoted as μ. Derive an expression for the force necessary to strip the sleeve from the shaft. Assume that σ_a does not affect the stripping force.

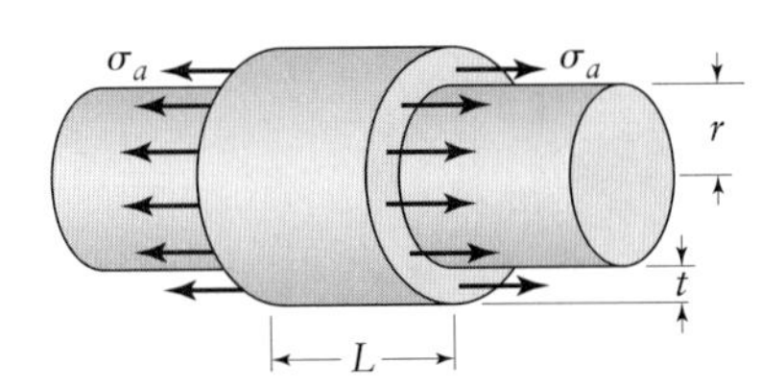

2.77 Assume that a material with a uniaxial yield stress *Y* yields under a stress state of principal stresses $\sigma_1, \sigma_2, \sigma_3$, where $\sigma_1 > \sigma_2 > \sigma_3$. Show that the superposition of a hydrostatic stress *p* on this system (such as placing the specimen in a chamber pressurized with a liquid) does not affect yielding. In other words, the material will still yield according to yield criteria.

2.78 Give two different and specific examples in which the maximum-shear-stress and the distortion-energy criteria give the same answer.

2.79 A thin-walled spherical shell with a yield stress *Y* is subjected to an internal pressure *p*. With appropriate equations, show whether or not the pressure required to yield this shell depends on the particular yield criterion used.

2.80 Show that, according to the distortion-energy criterion, the yield stress in plane strain is 1.15*Y*, where *Y* is the uniaxial yield stress of the material.

2.81 What would be the answer to Problem 2.80 if the maximum shear stress criterion were used?

2.82 A closed-end, thin-walled cylinder of original length ℓ, thickness t, and internal radius r is subjected to an internal pressure p. Using the generalized Hooke's law quations, show the change, if any, that occurs in the length of this cylinder when it is pressurized. Let $\nu = 0.33$.

2.83 A round, thin-walled tube is subjected to tension in the elastic range. Show that both the thickness and the diameter decrease as tension increases.

2.84 In Problem 2.83, assume that the tube is subjected to compression. Explain what changes, if any, take place in the thickness and diameter of the tube. Assume that the tube does not buckle.

2.85 Take a long cylindrical balloon and, with a thin felt-tip pen, mark a small square on it. What will be the shape of this square after you blow up the balloon, (1) a larger square, (2) a rectangle with its long axis in the circumferential directions, (3) a rectangle with its long axis in the longitudinal direction, or (4) an ellipse? Perform this experiment, and, based on your observations, explain the results, using appropriate equations. Assume that the material the balloon is made up of is perfectly elastic and isotropic and that this situation represents a thin-walled closed-end cylinder under internal pressure.

2.86 Take a cubic piece of metal a side length ℓ_o and deform it plastically to the shape of a rectangular parallelepiped of dimensions ℓ_1, ℓ_2, and ℓ_3. Assuming that the material is rigid and perfectly plastic, show that volume constancy requires that the following expression be satisfied: $\varepsilon_1 + \varepsilon_2 + \varepsilon_3 = 0$.

2.87 What is the diameter of an originally 20-mm-diameter solid steel ball when the ball is subjected to a hydrostatic pressure of 10 GPa?

2.88 Determine the effective stress and effective strain in plane-strain compression according to the distortion-energy criterion.

2.89 It can be shown that a thin-walled tube under longitudinal tension (in the elastic range) decreases in diameter. It is, of course, possible to internally pressurize the tube so that its diameter remains constant while being stretched. Assuming that the tube is open ended, derive an expression for the pressure required to keep the diameter constant as the tube is stretched.

2.90 What is the magnitude of compressive stress for the plane-strain compression test shown in Fig. 2.16, according to the maximum-shear-stress criterion?

2.91 (a) Calculate the work done in expanding a 1-mm-thick spherical shell from a diameter of 100 mm to 140 mm, where the shell is made of a material for which $\sigma = 200 + 50\varepsilon^{0.5}$ MPa. (b) Does your answer depend on the particular yield criterion used? Explain.

2.92 A cylindrical slug that has a diameter of 1 in. and is 1 in. high is placed at the center of a 2-in.-diameter cavity in a rigid die. (See the accompanying figure.) The slug is surrounded by a compressible matrix, the pressure of which is given by the relation

$$p_m = 40{,}000\, \Delta V / V_{om} \text{ psi},$$

where m denotes matrix and V_{om} is the original volume of the compressible matrix. Both the slug and the matrix are being compressed by a piston and without any friction. The initial pressure of the matrix is zero, and the slug material has the true-stress–true-strain curve of $\sigma = 25{,}000\varepsilon^{0.5}$. Obtain an expression for the force F versus piston travel d up to $d = 0.5$ in.

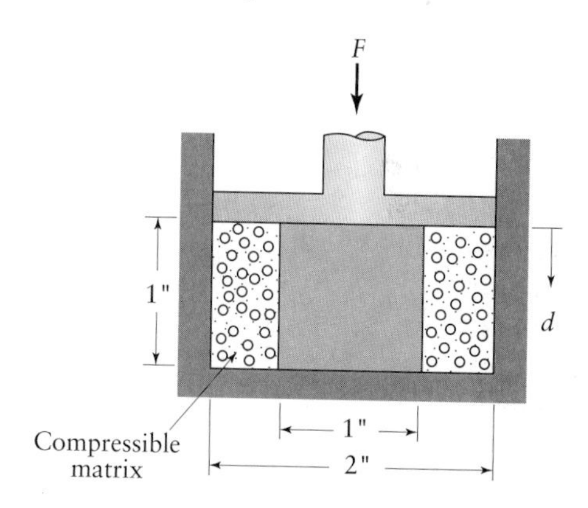

2.93 A specimen in the shape of a cube 10 mm on each side is being compressed without friction in a die cavity, as shown in Fig. 2.37d, where the width of the groove is 15 mm. Assume that the linearly strain-hardening material has the true-stress–true-strain curve given by $\sigma = 75 + 20\varepsilon$ MPa. Calculate the compressive force required when the height of the specimen is 3 mm, according to both yield criteria.

2.94 Assume that you are asked to give a quiz to students on the contents of this chapter. Prepare five quantitative problems, and supply the answers.

2.95 Obtain expressions for the specific energy for a material for each of the stress–strain curves shown in Fig. 2.8, similar to those shown in Section 2.12.

2.96 A material with a yield stress of 100 MPa is subjected to principal (normal) stresses of $\sigma_1, \sigma_2 = 0$, and $\sigma_3 = -\sigma_1/2$. What is the value of σ_1 when the metal yields according to the von Mises criterion? What if $\sigma_2 = \sigma_1/3$?

2.97 A steel plate is 100 mm × 100 mm × 5 mm thick. It is subjected to biaxial tension $\sigma_1 = \sigma_2$, with the stress in the thickness direction of $\sigma_3 = 0$. What is the largest possible change in volume at yield, using the von Mises criterion? What would this change in volume be if the plate were made of copper?

2.98 A 25-mm-wide, 1-mm-thick strip is rolled to a final thickness of 0.6 mm. It is noted that the strip has increased in width to 26 mm. What is the strain in the rolling direction?

2.99 An aluminum alloy yields at a stress of 50 MPa in uniaxial tension. If this material is subjected to the stresses $\sigma_1 = 25$ MPa, $\sigma_2 = 15$ MPa, and $\sigma_3 = -26$ MPa, will it yield? Explain.

2.100 A cylindrical specimen 1 in. in diameter and 1 in. high is being compressed by dropping a weight of 200 lb on it from a certain height. After deformation, it is found that the temperature rise in the specimen is 300°F. Assuming no heat loss and no friction, calculate the final height of the specimen, using the following data for the material: $K = 30{,}000$ psi, $n = 0.5$, density $= 0.1$ lb/in^3, and specific heat $= 0.3$ BTU/lb · °F.

CHAPTE

3

Structure and Manufacturing Properties of Metals

3.1 Introduction

The structure of metals greatly influences their *behavior* and *properties*. A knowledge of structures guides us in controlling and predicting the behavior and performance of metals in various manufacturing processes. Understanding the structure of metals also allows us to predict and evaluate their properties. This then helps us make appropriate selections for specific applications under particular force, temperature, and environmental conditions. For example, we will come to appreciate the reasons for the development of single-crystal turbine blades (Fig. 3.1) for use in jet engines; these blades have better properties than those made conventionally by casting or forging.

This chapter begins with a review of the crystal structure of metals and the role of various imperfections as they affect plastic deformation and work-hardening characteristics. We will also describe the importance of grain size, grain boundaries, and environment. We will then review the failure and fracture of metals, both ductile and brittle, together with factors that influence fracture, such as state of stress, temperature, strain rate, and external and internal defects in metals. The rest of the

FIGURE 3.1 Turbine blades for jet engines, manufactured by three different methods: (a) conventionally cast; (b) directionally solidified, with columnar grains, as can be seen from the vertical streaks; and (c) single crystal. Although more expensive, single-crystal blades have properties at high temperatures that are superior to those of other blades. *Source*: Courtesy of United Technologies Pratt and Whitney.

chapter is devoted to a brief discussion of physical and other properties of ferrous and nonferrous metals and alloys relevant to their use as engineering materials. Some data on properties are also given as a guide in design considerations and material selection.

3.2 The Crystal Structure of Metals

When metals solidify from a molten state, the atoms arrange themselves into various orderly configurations, called **crystals**. The arrangement of the atoms in the crystal is called **crystalline structure**. A crystalline structure is like metal scaffolding in front of a building under construction, with uniformly repetitive horizontal and vertical metal pipes and braces.

The smallest group of atoms showing the characteristic **lattice structure** of a particular metal is known as a **unit cell**. It is the building block of a crystal, and a single crystal can have many unit cells. Think of each brick in a wall as a unit cell. The wall has a crystalline structure, that is, it consists of an orderly arrangement of bricks; thus, the wall is like a single crystal, consisting of many unit cells.

The three basic patterns of atomic arrangement found in most metals are as follows (Figs. 3.2a–c):

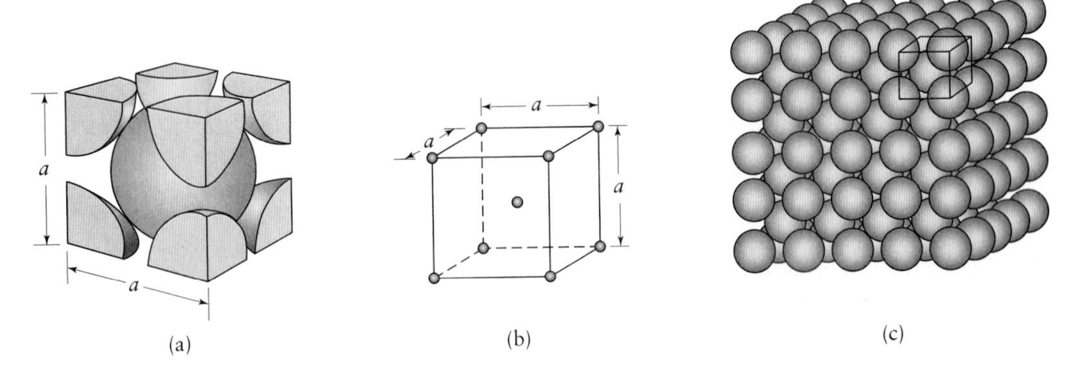

FIGURE 3.2a The body-centered cubic (bcc) crystal structure: (a) hard-ball model; (b) unit cell; and (c) single crystal with many unit cells. *Source*: Courtesy of Dr. William G. Moffatt.

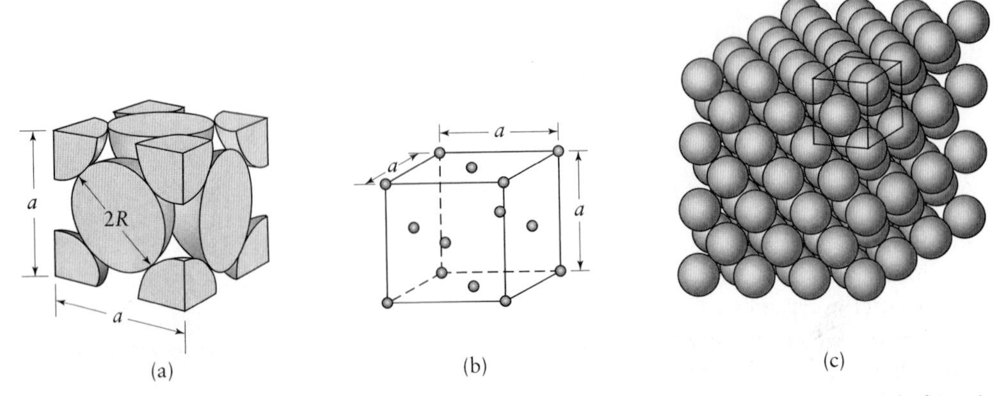

FIGURE 3.2b The face-centered cubic (fcc) crystal structure: (a) hard-ball model; (b) unit cell; and (c) single crystal with many unit cells. *Source*: Courtesy of Dr. William G. Moffatt.

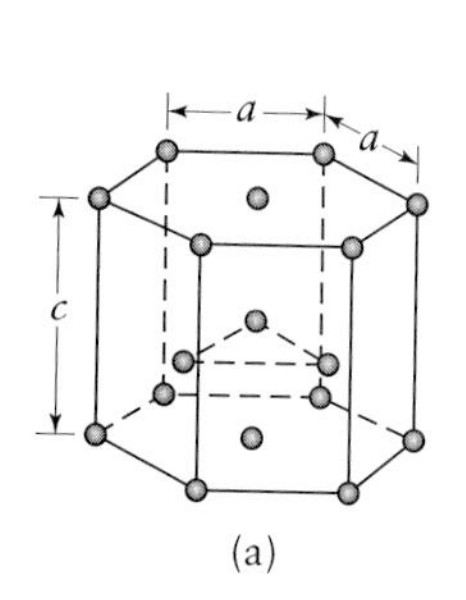

(a)

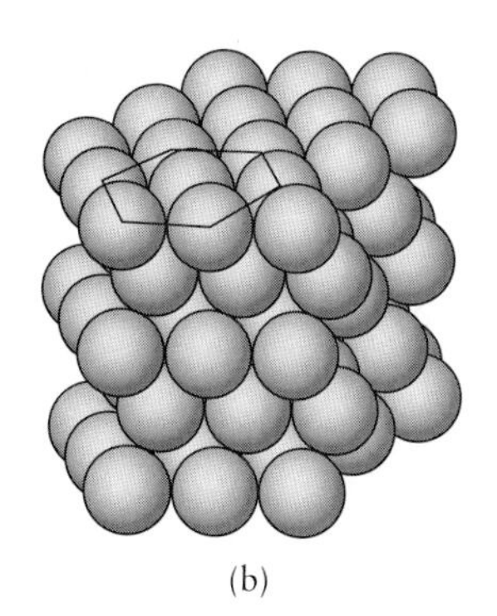
(b)

FIGURE 3.2c The hexagonal close-packed (hcp) crystal structure: (a) unit cell; and (b) single crystal with many unit cells. *Source*: Courtesy of Dr. William G. Moffatt.

a) **body-centered cubic** (bcc),
b) **face-centered cubic** (fcc), and
c) **hexagonal close packed** (hcp).

The order of magnitude of the distance between the atoms in these crystal structures is 0.1 nm (10^{-8} in.). The manner in which the atoms are arranged has a significant influence on the properties of a particular metal. These arrangements can be modified by adding atoms of some other metal or metals, known as **alloying**, and often improves the properties of the metal. (See Section 5.2.)

In the three structures illustrated in Figs. 3.2a–c, the fcc and hcp crystals have the most densely packed configurations. In the hcp structure, the top and bottom planes are called **basal planes**. The different crystal structures form due to differences in the energy required to form these structures. Thus, for example, tungsten forms a bcc structure because it requires less energy than other structures; similarly, aluminum forms an fcc structure. However, at different temperatures, the same metal may form different structures, because of differences in the amount of energy required at each temperature. For example, iron forms a bcc structure below 912°C (1674°F)) and above 1394°C (2541°F), but it forms an fcc structure between 912°C and 1394°C. The appearance of more than one type of crystal structure is known as **allotropism**, or **polymorphism**; as described in Section 5.2.5, these structural changes are an important aspect of heat treatment of metals and alloys.

3.3 Deformation and Strength of Single Crystals

When crystals are subjected to an external force, they first undergo **elastic deformation**; that is, they return to their original shape when the force is removed. (See Section 2.2.) An analogy to this type of behavior is a helical spring that stretches when loaded and returns to its original shape when the load is removed. However, if the force on the crystal structure is increased sufficiently, the crystal undergoes **plastic deformation**, or *permanent deformation*; that is, it does not return to its original shape when the force is removed.

There are two basic mechanisms by which plastic deformation may take place in crystal structures:

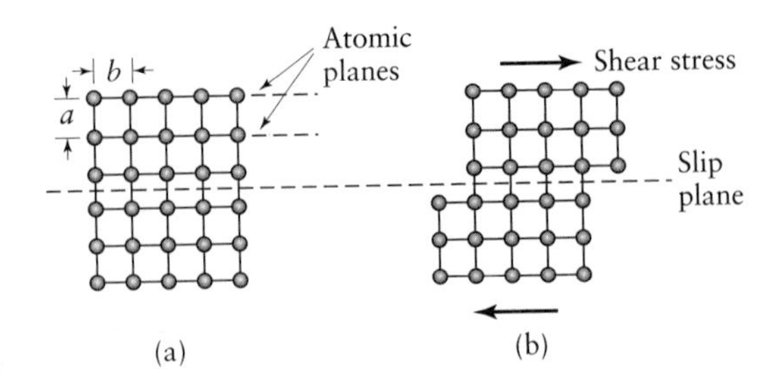

FIGURE 3.3 Permanent deformation, also called plastic deformation, of a single crystal subjected to a shear stress: (a) structure before deformation; and (b) deformation by slip. The b/a ratio influences the magnitude of the shear stress required to cause slip.

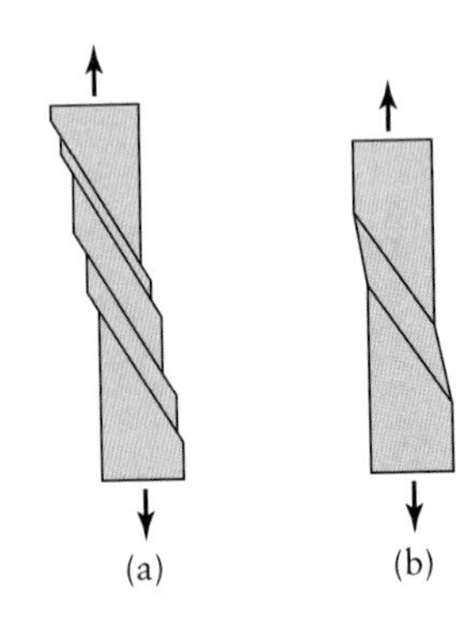

FIGURE 3.4 (a) Permanent deformation of a single crystal under a tensile load. Note that the slip planes tend to align themselves in the direction of pulling. This behavior can be simulated using a deck of cards with a rubber band around them. (b) Twinning in tension.

1. **Slip.** The first deformation mechanism is the slipping of one plane of atoms over an adjacent plane (the **slip plane**) under shear stress (Fig. 3.3). The deformation of a single-crystal specimen by slip is shown schematically in Fig. 3.4; this situation is much like sliding playing cards against each other. Just as it takes a certain amount of force to slide playing cards against each other, so a crystal requires a certain amount of shear stress (**critical shear stress**) to undergo permanent deformation. Thus, there must be a shear stress of sufficient magnitude within a crystal for plastic deformation to take place.

   The **maximum theoretical shear stress,** τ_{max}, to cause permanent deformation in a perfect crystal is obtained as follows: when there is no stress, the atoms in the crystal are in equilibrium (Fig. 3.5). Under shear stress, the upper row of atoms is moved to the right, where the position of an atom is denoted as x. Thus, when $x = 0$ or $x = b$, the shear stress is zero. Each atom of the upper row is attracted to the nearest atom of the lower row, resulting in nonequilibrium at positions 2 and 4, where the stresses are maximum, but opposite in sign. At position 3, the shear stress is again zero, since this position is symmetric. In Fig. 3.5, we assume that, as a first approximation, the shear stress varies sinusoidally. Hence, the shear stress at a displacement x is

$$\tau = \tau_{max} \sin \frac{2\pi x}{b}, \tag{3.1}$$

   which, for small values of x/b, can be written as

$$\tau = \tau_{max} \frac{2\pi x}{b}.$$

   From Hooke's law, we have

$$\tau = G\gamma = G\left(\frac{x}{a}\right).$$

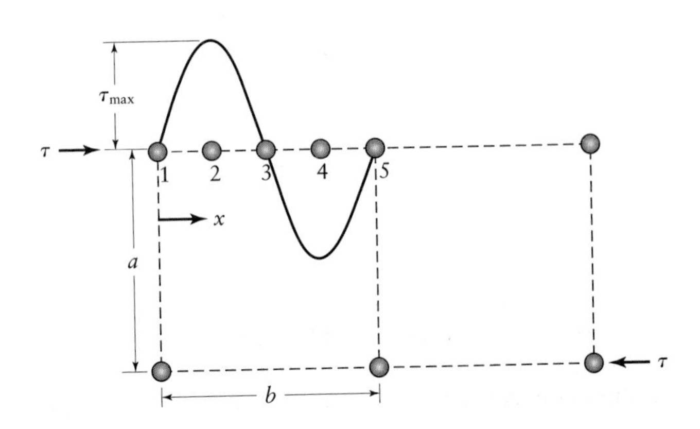

FIGURE 3.5 Variation of shear stress in moving a plane of atoms over another plane.

Hence,

$$\tau_{max} = \frac{Gb}{2\pi a}. \tag{3.2}$$

If we assume that b is approximately equal to a, then

$$\tau_{max} = \frac{G}{2\pi}. \tag{3.3}$$

More refined calculations have set the value of the maximum theoretical shear stress as between $G/10$ and $G/30$.

The shear stress required to cause slip in single crystals is directly proportional to the ratio b/a in Fig. 3.5. It can therefore be stated that slip in a crystal takes place along planes of maximum atomic density, or that slip takes place in closely packed planes and in closely packed directions. Because the b/a ratio is different for different directions within the crystal, a single crystal has different properties when tested in different directions; thus, a crystal is anisotropic. A common example of **anisotropy** is woven cloth, which stretches differently when we pull it in different directions, or plywood, which is much stronger in the planar direction than along its thickness direction (where it splits easily).

2. **Twinning.** The second mechanism of plastic deformation is *twinning*, in which a portion of the crystal forms structurally a mirror image of itself across the *plane of twinning* (Fig. 3.4b). Twins form abruptly and are the cause of the creaking sound (*tin cry*) when a tin or zinc rod is bent at room temperature. Twinning usually occurs in hcp and bcc metals by plastic deformation and in fcc metals by annealing. (See Section 5.11.4.)

3.3.1 Slip systems

The combination of a slip plane and its direction of slip is known as a *slip system*. In general, metals with five or more slip systems are ductile, whereas those with less than five are not ductile. Each pattern of atomic arrangement has a different number of potential slip systems:

1. In body-centered cubic crystals, there are 48 possible slip systems; thus, the probability is high that an externally applied shear stress will operate on one of the systems and cause slip. However, because of the relatively high b/a ratio, the required shear stress is high. Metals with bcc structures (such as titanium, molybdenum, and tungsten) have good strength and moderate ductility.
2. In face-centered cubic crystals, there are 12 slip systems. The probability of slip is moderate, and the required shear stress is low. Metals with fcc structures (such as aluminum, copper, gold, and silver) have moderate strength and good ductility.
3. The hexagonal close-packed crystal has three slip systems and thus has a low probability of slip. However, more systems become active at elevated temperatures. Metals with hcp structures (such as beryllium, magnesium, and zinc) are generally brittle at room temperature.

Note in Fig. 3.4a that the portions of the single crystal that have slipped have rotated from their original angular position toward the direction of the tensile force. Note also that slip has taken place along certain planes only. With the use of electron microscopy, it has been shown that what appears to be a single slip plane is actually a **slip band**, consisting of a number of slip planes (Fig. 3.6).

3.3.2 Theoretical tensile strength of metals

In addition to obtaining the maximum theoretical shear strength, we can also calculate the theoretical tensile strength of metals. In the bar shown in Fig. 3.7, the interatomic distance is a when no stress is applied. In order to increase this distance, we have to apply a force to overcome the cohesive force between the atoms, the cohesive force being zero in the unstrained equilibrium condition. When the tensile stress reaches σ_{max}, the atomic bonds between two neighboring atomic planes break; this σ_{max} is known as the *ideal tensile strength*.

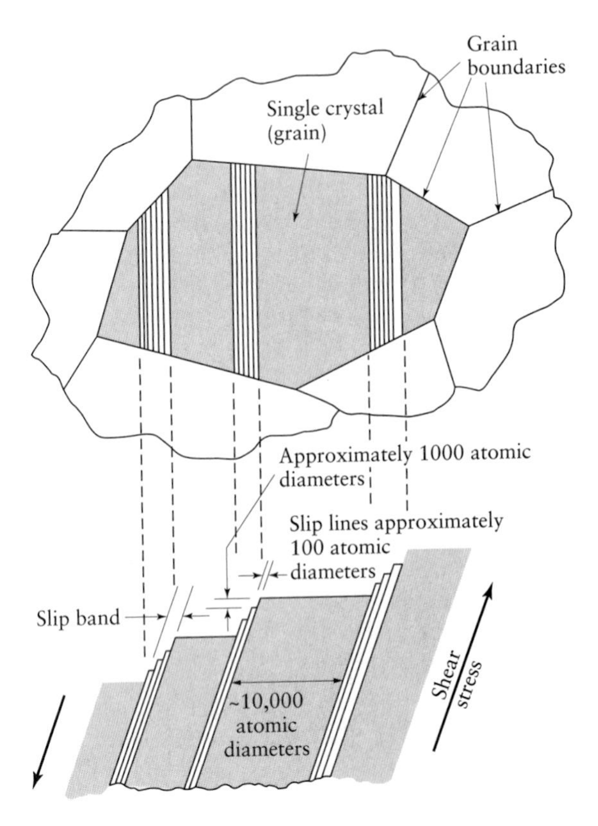

FIGURE 3.6 Schematic illustration of slip lines and slip bands in a single crystal subjected to a shear stress. A slip band consists of a number of slip planes. The crystal at the center of the upper drawing is an individual grain surrounded by other grains.

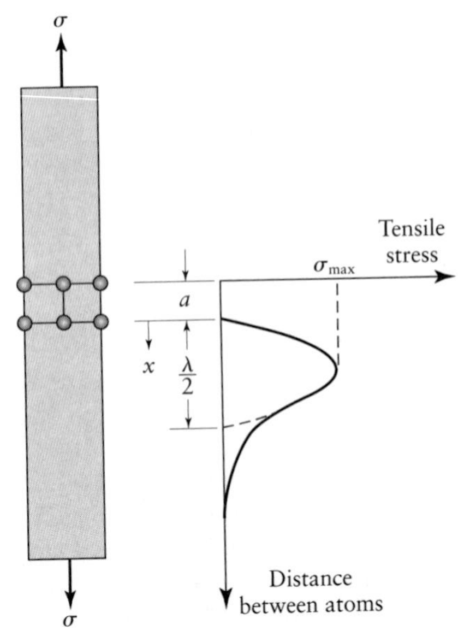

FIGURE 3.7 Variation of cohesive stress as a function of distance between a row of atoms.

The ideal tensile strength can be derived by the following approach: Referring to Fig. 3.7, it can be shown that

$$\sigma_{\max} = \frac{E\lambda}{2\pi a} \tag{3.4a}$$

and that the work done per unit area in breaking the bond between the two atomic planes (i.e., the area under the cohesive-force curve) is given by

$$\text{Work} = \frac{\sigma_{\max}\lambda}{\pi}. \tag{3.4b}$$

This work is dissipated in creating two new fracture surfaces, thus involving surface energy of the material, γ. The total surface energy then is 2γ. Combining these equations, we find that

$$\sigma_{\max} = \sqrt{\frac{E\gamma}{a}}. \tag{3.4c}$$

When appropriate values are substituted into this equation, we have

$$\sigma_{\max} \simeq \frac{E}{10}. \tag{3.5}$$

Thus the theoretical strength of steel, for example, would be on the order of 20 GPa (3×10^6 psi).

3.3.3 Imperfections

The actual strength of metals is approximately one to two orders of magnitude lower than the strength levels obtained from the foregoing theoretical calculations. This discrepancy has been explained in terms of *imperfections* in the crystal structure. Unlike the idealized models described previously, actual metal crystals contain a large number of imperfections and defects, which are categorized as follows:

1. *Point defects*, such as a **vacancy** (missing atom), an **interstitial atom** (extra atom in the lattice), or an **impurity atom** (foreign atom that has replaced an atom of the pure metal) (Fig. 3.8);
2. *Linear*, or *one-dimensional, defects*, called **dislocations** (Fig. 3.9);
3. *Planar*, or *two-dimensional, imperfections*, such as **grain boundaries** and **phase boundaries**;
4. *Volume*, or *bulk, imperfections*, such as **voids, inclusions** (nonmetallic elements such as oxides, sulfides, and silicates), other **phases**, or **cracks**.

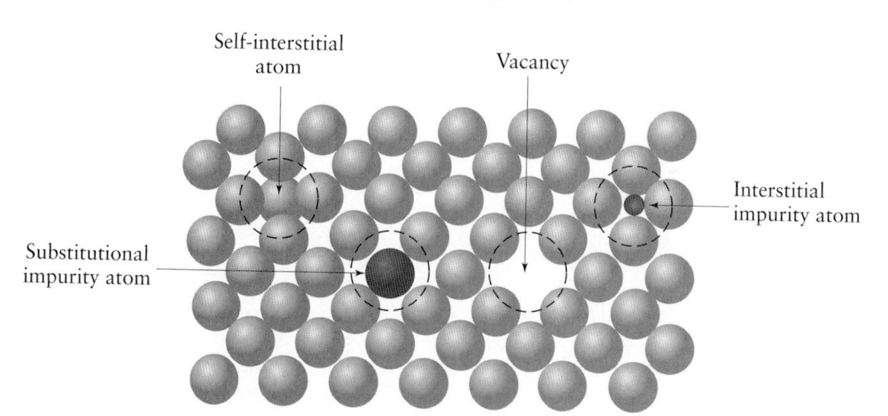

FIGURE 3.8 Various defects in a single-crystal lattice. *Source*: Courtesy of Dr. William G. Moffatt.

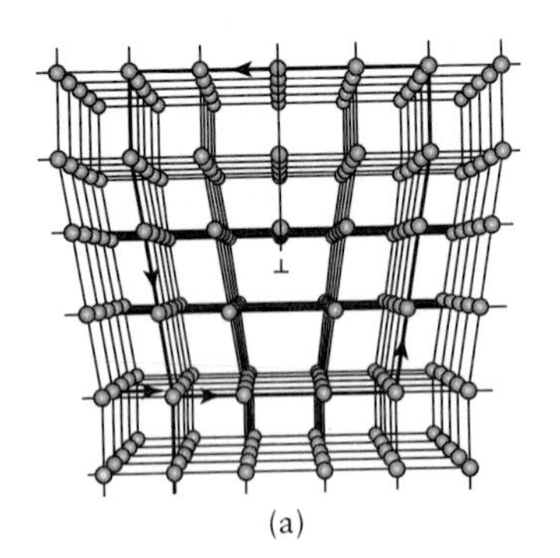

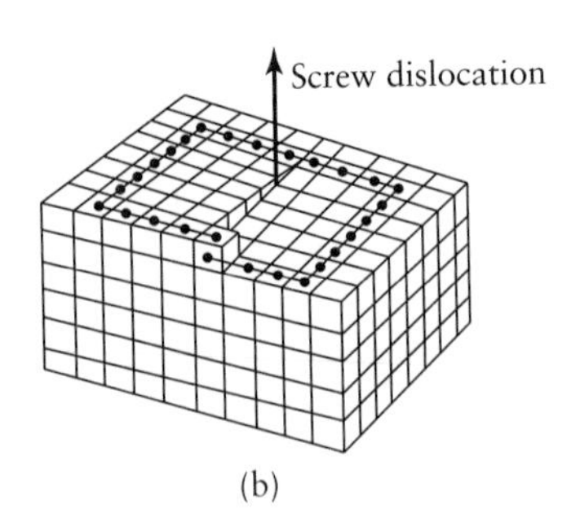

(a) (b)

FIGURE 3.9 (a) Edge dislocation, a linear defect at the edge of an extra plane of atoms. (b) Screw dislocation, a helical defect in a three–dimentional lattice of atoms.

A slip plane containing a dislocation requires lower shear stress to cause slip than does a plane in a perfect lattice (Fig. 3.10). One analogy used to describe the movement of an edge dislocation is the earthworm, which moves forward through a hump that starts at the tail and moves toward the head. A second analogy is that of moving a large carpet by forming a hump at one end and moving the hump forward toward the other end. The force required to move a carpet in this way is much less than that required to slide the whole carpet along the floor. *Screw dislocations* are so named because the atomic planes form a spiral ramp.

The **density of dislocations** (i.e., the total length of dislocation line per unit volume, $mm/mm^3 = mm^{-2}$) increases with increasing plastic deformation, by as much as 10^6 mm^{-2} at room temperature. The dislocation densities for some conditions are as follows:

a) very pure single crystals: 0 to 10^3 mm^{-2};
b) annealed single crystals: 10^5 to 10^6;
c) annealed polycrystals: 10^7 to 10^8;
d) highly cold-worked metals: 10^{11} to 10^{12}.

The mechanical properties of metals, such as yield and fracture strength, as well as electrical conductivity, are affected by lattice defects; these properties are known as **structure-sensitive properties.** On the other hand, physical properties, such as melting point, specific heat, coefficient of thermal expansion, and elastic constants are not sensitive to these defects and are thus known as **structure-insensitive properties.**

3.3.4 Strain hardening (work hardening)

Although the presence of a dislocation lowers the shear stress required to cause slip, dislocations can (1) become entangled and interfere with each other, and (2) be impeded by barriers, such as grain boundaries and impurities and inclusions in the material. Entanglement and impediments increase the shear stress required for slip.

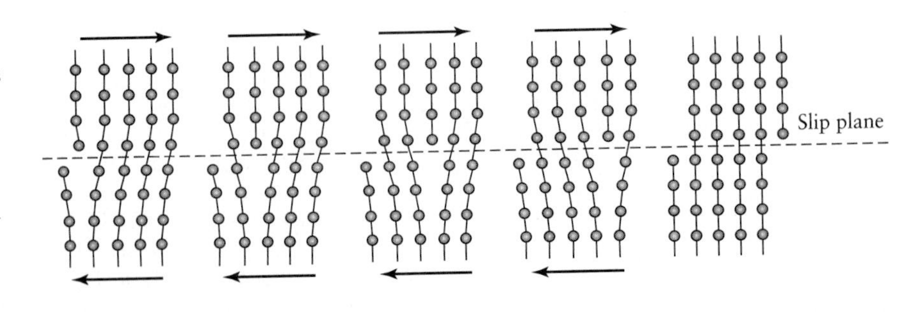

FIGURE 3.10 Movement of an edge dislocation across the crystal lattice under a shear stress. Dislocations help explain why the actual strength of metals is much lower than that predicted by theory.

The increase in the shear stress required, and hence the increase in the overall strength of the metal, is known as *strain hardening*, or *work hardening*. (See Section 2.2.3.) The greater the deformation, the more the entanglements, which increase the metal's strength. Work hardening is used extensively to strengthen metals in metalworking processes at ambient temperature. Typical examples are strengthening wire by drawing it through a die to reduce its cross-section (Section 6.5), producing the head on a bolt by forging it (Section 6.2.4), and producing sheet metal for automobile bodies and aircraft fuselages by rolling (Section 6.3). As shown by Eq. (2.11), the degree of strain hardening is indicated by the strain-hardening exponent, n. Among the three crystal structures, hcp has the lowest n value, followed by bcc and then fcc, which has the highest.

3.4 | Grains and Grain Boundaries

Metals commonly used for manufacturing various products are composed of many individual, randomly oriented crystals (*grains*). We are thus dealing with metal structures that are not single crystals, but **polycrystals**, that is, many crystals.

When a mass of molten metal begins to solidify, crystals begin to form independently of each other, with random orientations and at various locations within the liquid mass (Fig. 3.11). Each of the crystals grows into a crystalline structure, or grain. The number and size of the grains developed in a unit volume of the metal depend on the rate at which **nucleation** (the initial stage of formation of crystals) takes place. The number of different sites in which individual crystals begin to form (seven in Fig. 3.11a) and the rate at which these crystals grow affect the size of grains developed. Generally, rapid cooling produces smaller grains, whereas slow cooling produces larger grains. (See also Section 5.3.)

If the crystal nucleation rate is high, the number of grains in a unit volume of metal will be greater, and consequently grain size will be small. Conversely, if the rate of growth of the crystals is high compared with their nucleation rate, there will be fewer, but larger, grains per unit volume.

Note in Fig. 3.11 how the grains eventually interfere with and impinge upon one another. The surfaces that separate the individual grains are called *grain boundaries*.

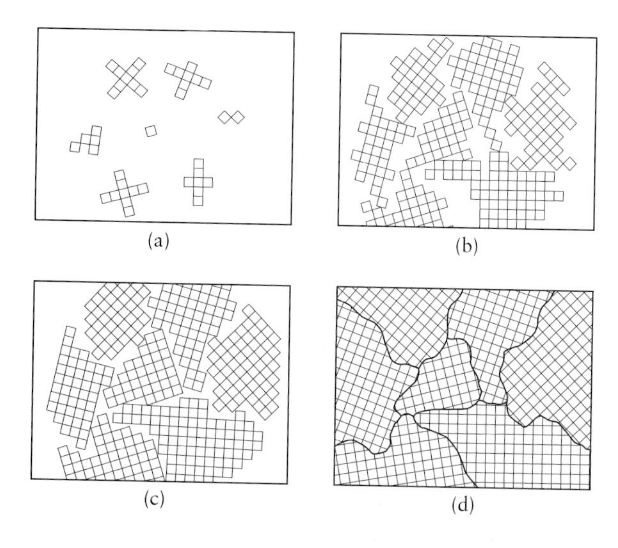

FIGURE 3.11 Schematic illustration of the various stages during solidification of molten metal. Each small square represents a unit cell. (a) Nucleation of crystals at random sites in the molten metal. Note that the crystallographic orientation of each site is different. (b) and (c) Growth of crystals as solidification continues. (d) Solidified metal, showing individual grains and grain boundaries. Note the different angles at which neighboring grains meet each other. *Source*: W. Rosenhain.

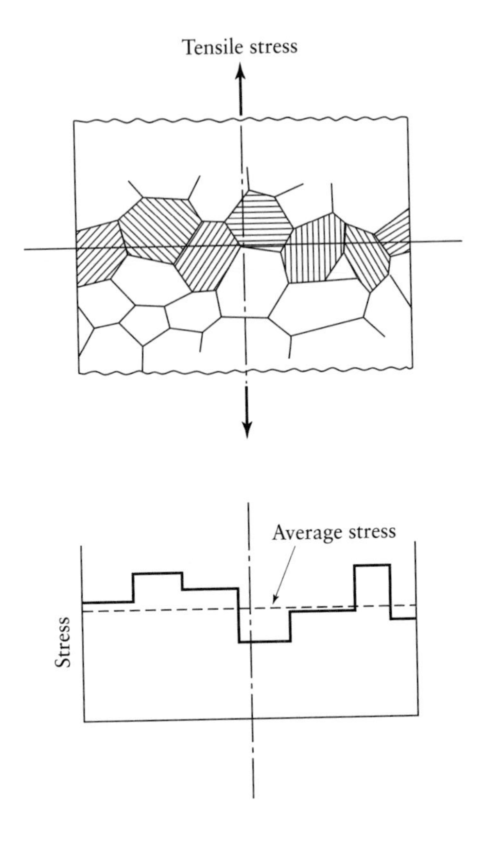

FIGURE 3.12 Variation of tensile stress across a plane of polycrystalline metal specimen subjected to tension. Note that the strength exhibited by each grain depends on its orientation.

Each grain consists of either a single crystal (for pure metals) or a polycrystalline aggregate (for alloys). Note that the crystallographic orientation changes abruptly from one grain to the next across the grain boundaries. Recall from Section 3.3 that the behavior of a single crystal or single grain is *anisotropic*. The ideal behavior of a piece of polycrystalline metal (Fig. 3.12) is *isotropic*, because the grains have random crystallographic orientations; thus, the metal's properties do not vary with the direction of testing. In practice, however, this situation rarely exists.

3.4.1 Grain size

Grain size significantly influences the mechanical properties of metals; large grain size is generally associated with low strength, low hardness, and high ductility. Large grains, particularly in sheet metals, also have a rough surface appearance after being stretched. (See Section 3.6.) The yield strength, Y, is the most sensitive property and is related to grain size by the empirical formula (*Hall-Petch equation*):

$$Y = Y_i + kd^{-1/2}, \tag{3.6}$$

where Y_i is a basic yield stress that can be regarded as the stress opposing the motion of dislocations, k is a constant indicating the extent to which dislocations are piled up at barriers (such as grain boundaries), and d is the grain diameter. Equation (3.6) is valid below the recrystallization temperature of the material.

Grain size is usually measured by counting the number of grains in a given area or the number of grains that intersect a given length of a line randomly drawn on an enlarged photograph of the grains on a polished and etched specimen taken under a microscope. Grain size may also be determined by referring to a standard chart. The American Society for Testing and Materials (ASTM) grain-size number, n, is related to the number of grains, N, per square inch at a magnification of 100× (equal to 0.0645 mm^2 of actual area) by the expression

$$N = 2^{n-1}. \tag{3.7}$$

Because grains generally are extremely small, many grains occupy a unit volume of metal. Grains of sizes between 5 and 8 are generally considered fine grains. A grain size of 7 generally is acceptable for sheet metals used to make car bodies, appliances, and kitchen utensils. Grains can also be large enough to be visible to the naked eye, such as grains of zinc on the surface of galvanized sheet steel.

3.4.2 Influence of grain boundaries

Grain boundaries have an important influence on the strength and ductility of metals. Also, because they interfere with the movement of dislocations, grain boundaries also influence strain hardening. These effects depend on temperature, rate of deformation, and the type and amount of impurities present along the grain boundaries. Grain boundaries are more reactive than the grains themselves, because the atoms along the grain boundaries are packed less efficiently and are more disordered and hence have a higher energy than the atoms in the orderly arrangement within the grains.

At elevated temperatures and in materials whose properties depend on the rate of deformation, plastic deformation also takes place by means of **grain-boundary sliding.** The **creep mechanism** (elongation under stress over a period of time, usually at elevated temperatures; see Section 2.8) results from grain-boundary sliding.

When brought into close atomic contact with certain low-melting-point metals, a normally ductile and strong metal can crack under very low stresses, a phenomenon referred to as **grain boundary embrittlement** (Fig. 3.13). Examples include aluminum wetted with a mercury–zinc amalgam, and liquid gallium and copper at elevated temperature wetted with lead or bismuth; these elements weaken the grain boundaries of the metal by *embrittlement.* The term **liquid-metal embrittlement** is used to describe such phenomena, because the embrittling element is in a liquid state. However, embrittlement can also occur at temperatures well below the melting point of the embrittling element, a phenomenon known as **solid-metal embrittlement.**

Hot shortness is caused by local melting of a constituent or an impurity in the grain boundary at a temperature below the melting point of the metal itself. When such a metal is subjected to plastic deformation at elevated temperatures (*hot working*; see Section 3.7), the piece of metal crumbles and disintegrates along its grain

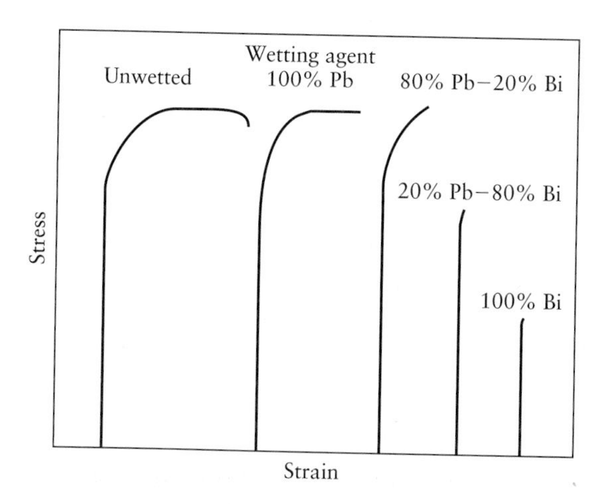

FIGURE 3.13 Embrittlement of copper by lead and bismuth at 350°C (660°F). Embrittlement has important effects on the strength, ductility, and toughness of materials. *Source*: After W. Rostoker.

boundaries; examples include antimony in copper, and leaded steels and brass. To avoid hot shortness, the metal is usually worked at a lower temperature, which prevents softening and melting along the grain boundaries. Another form of embrittlement is **temper embrittlement** in alloy steels, which is caused by segregation (movement) of impurities to the grain boundaries.

3.5 Plastic Deformation of Polycrystalline Metals

If a piece of polycrystalline metal with uniform equiaxed grains (i.e., having equal dimensions in all directions, as shown in the model in Fig. 3.14) is subjected to plastic deformation at room temperature (cold working), the grains become permanently deformed and elongated. The deformation process may be carried out either by compressing the metal (as in forging) or by subjecting it to tension (as in stretching sheet metal). The deformation within each grain takes place by the mechanisms described in Section 3.3 for a single crystal.

During plastic deformation, the grain boundaries remain intact, and mass continuity is maintained. The deformed metal exhibits higher strength, because of the entanglement of dislocations with grain boundaries. The increase in strength depends on the amount of deformation (strain) to which the metal is subjected; the greater the deformation, the stronger the metal becomes. Furthermore, the increase in strength is greater for metals with smaller grains, because they have a larger grain-boundary surface area per unit volume of metal.

Anisotropy (texture). Figure 3.14 shows that, as a result of plastic deformation, the grains in a piece of metal become elongated in one direction and contracted in the other. Consequently, the metal has become *anisotropic*, and thus its properties in the vertical direction are different from those in the horizontal direction. Many products develop anisotropy of mechanical properties after they have been processed by metalworking techniques. The degree of anisotropy depends on how uniformly the metal is deformed. Note from the direction of the crack in Fig. 3.15, for example, that the ductility of the cold-rolled sheet in the vertical (transverse) direction is lower than in its longitudinal direction.

Anisotropy influences both mechanical and physical properties of metals. For example, sheet steel for electrical transformers is rolled in such a way that the resulting deformation imparts anisotropic magnetic properties to the sheet, thus reducing magnetic-hysteresis losses and improving the efficiency of transformers. (See

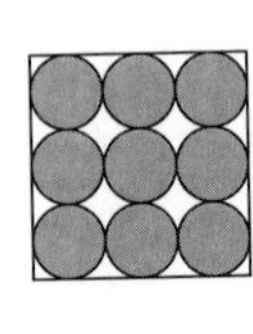

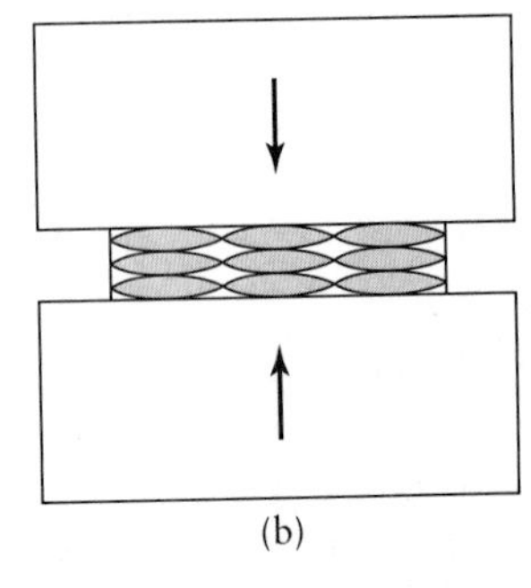

FIGURE 3.14 Plastic deformation of idealized (equiaxed) grains in a specimen subjected to compression, such as is done in rolling or forging of metals: (a) before deformation; and (b) after deformation. Note the alignment of grain boundaries along a horizontal direction.

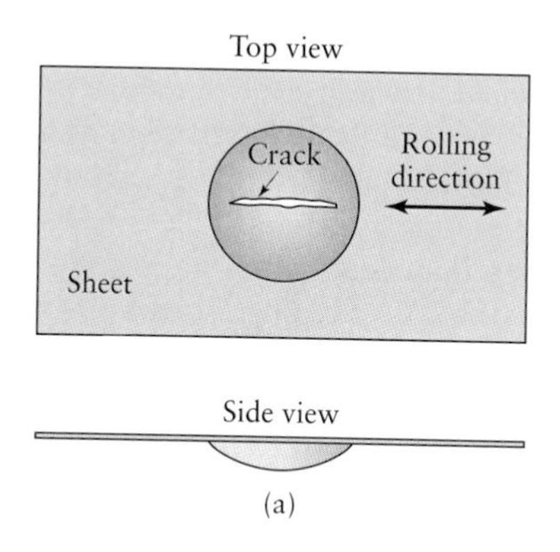

(a)

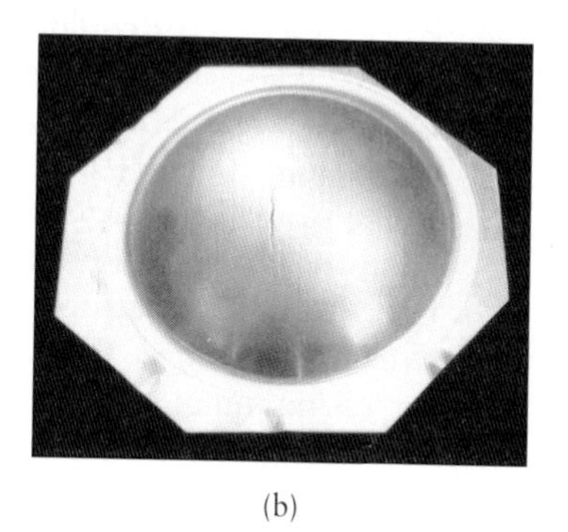
(b)

FIGURE 3.15 (a) Illustration of a crack in sheet metal subjected to bulging, such as by pushing a steel ball against the sheet. Note the orientation of the crack with respect to the rolling direction of the sheet. This material is anisotropic. (b) Aluminum sheet with a crack (vertical dark line at the center) developed in a bulge test. *Source*: Courtesy of J. S. Kallend, Illinois Institute of Technology.

also the discussion of amorphous alloys in Section 3.11.9.) There are two general types of anisotropy in metals: *preferred orientation* and *mechanical fibering*.

Preferred orientation. Also called **crystallographic anisotropy**, preferred orientation can be best described by reference to Fig. 3.4. When a metal crystal is subjected to tension, the sliding blocks rotate toward the direction of pulling; thus, slip planes and slip bands tend to align themselves with the direction of deformation. Similarly, for a polycrystalline aggregate, with grains in various orientations (Fig. 3.12), all slip directions tend to align themselves with the direction of pulling. Conversely, under compression, the slip planes tend to align themselves in a direction perpendicular to the direction of compression.

Mechanical fibering. Mechanical fibering results from the alignment of impurities, inclusions (stringers), and voids in the metal during deformation. Note that if the spherical grains in Fig. 3.14 were coated with impurities, these impurities would align themselves generally in a horizontal direction after deformation. Since impurities weaken the grain boundaries, this piece of metal would be weak and less ductile when tested in the vertical direction than in the horizontal direction. An analogy to this case would be plywood, which is strong in tension along its planar directions, but peels off easily when tested in tension in its thickness direction.

3.6 Recovery, Recrystallization, and Grain Growth

We have shown that plastic deformation at room temperature results in the deformation of grains and grain boundaries, a general increase in strength, and a decrease in ductility, as well as causing anisotropic behavior. These effects can be reversed and the properties of the metal brought back to their original levels by heating the metal to within a specific temperature range for a period of time. The temperature range and amount of time depend on the material and several other factors. Three events take place consecutively during the heating process:

1. **Recovery.** During recovery, which occurs at a certain temperature range below the **recrystallization temperature** of the metal, the stresses in the highly deformed regions are relieved and the number of mobile dislocations is reduced. Subgrain boundaries begin to form (**polygonization**) with no appreciable change in mechanical properties, such as hardness and strength, but with some increase in ductility (Fig. 3.16).

2. **Recrystallization.** The process in which, at a certain temperature range, new equiaxed and strain-free grains are formed, replacing the older grains, is called **recrystallization**. The temperature for recrystallization ranges approximately between $0.3T_m$ and $0.5T_m$, where T_m is the melting point of the metal on the absolute scale. The recrystallization temperature is generally defined as the temperature at which complete recrystallization occurs within approximately one hour. Recrystallization decreases the density of dislocations and lowers the strength, but raises the ductility, of the metal (Fig. 3.16). Metals such as lead, tin, cadmium, and zinc recrystallize at about room temperature.

Recrystallization depends on the degree of prior cold work (work hardening); the higher the amount of cold work, the lower the temperature required for recrystallization to occur. The reason for this relationship is that as the amount of cold work increases, the number of dislocations and the amount of energy stored in the dislocations (*stored energy*) also increase. The stored energy supplies the work required for recrystallization. Recrystallization is a function of time, because it involves *diffusion*, that is, movement and exchange of atoms across grain boundaries.

The effects on recrystallization of temperature, time, and reduction in the thickness or height of the workpiece by cold working are as follows (Fig. 3.17):

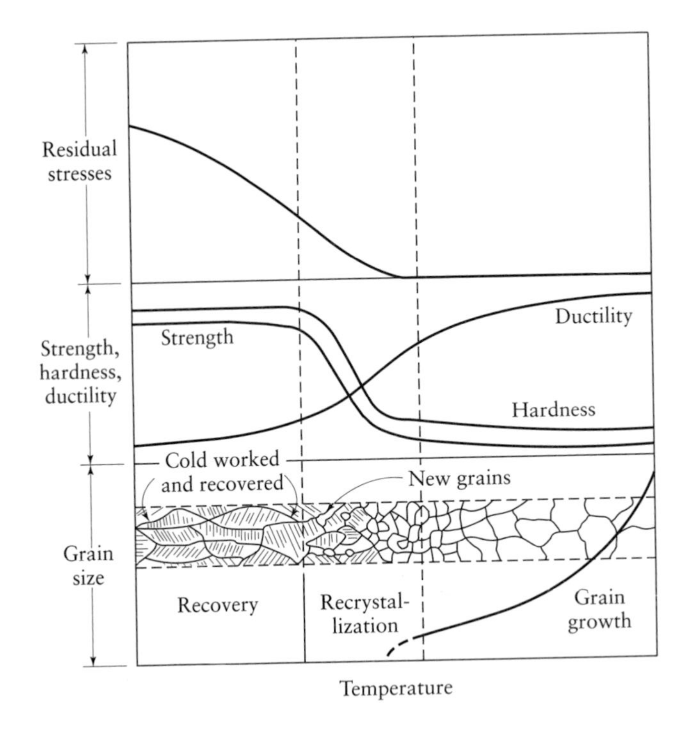

FIGURE 3.16 Schematic illustration of the effects of recovery, recrystallization, and grain growth on mechanical properties and shape and size of grains. Note the formation of small new grains during recrystallization. *Source*: G. Sachs.

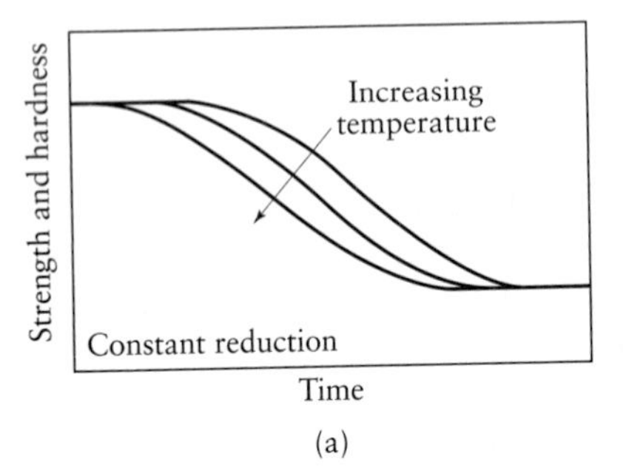

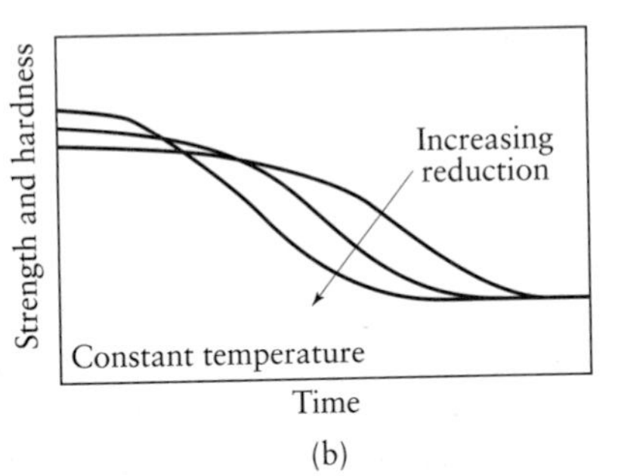

FIGURE 3.17 Variation of strength and hardness with recrystallization temperature, time, and prior cold work. Note that the more a metal is cold worked, the less time it takes to recrystallize, because of the higher stored energy from cold working due to increased dislocation density.

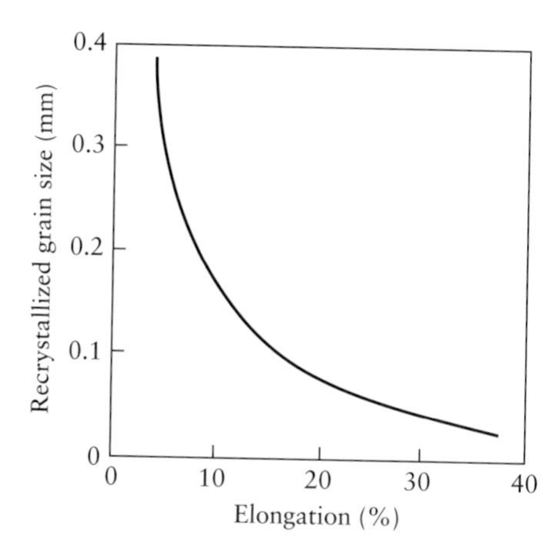

FIGURE 3.18 The effect of prior cold work on the recrystallized grain size of alpha brass. Below a critical elongation (strain), typically 5%, no recrystallization occurs.

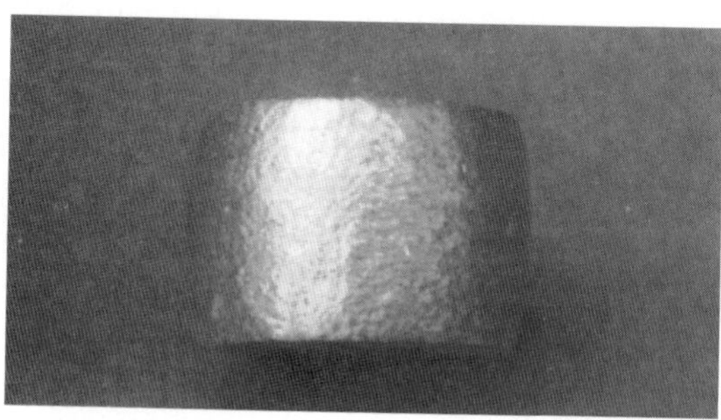

FIGURE 3.19 Surface roughness on the cylindrical surface of an aluminum specimen subjected to compression. *Source*: A. Mulc and S. Kalpakjian.

- For a constant amount of deformation by cold working, the time required for recrystallization decreases with increasing temperature.
- The more prior cold work done, the lower the temperature required for recrystallization.
- The higher the amount of deformation, the smaller the grain size during recrystallization (Fig. 3.18). Subjecting a metal to deformation is a common method of converting a coarse-grained structure to one of fine grain, with improved properties.
- Anisotropy due to preferred orientation usually persists after recrystallization. To restore isotropy, a temperature higher than that required for recrystallization may be necessary.

3. **Grain growth.** If we continue to raise the temperature of the metal, the grains begin to grow, and their size may eventually exceed the original grain size. This phenomenon is known as *grain growth*, and it has a slightly adverse effect on mechanical properties (Fig. 3.16). However, large grains produce a rough surface appearance on sheet metals (**orange-peel effect**) when they are stretched to form a part or when a piece of metal is subjected to compression (Fig. 3.19), such as in forging operations.

3.7 Cold, Warm, and Hot Working

When plastic deformation is carried out above or below the recrystallization temperature, it is called *hot working* or *cold working*, respectively. As the name implies, *warm working* is carried out at an intermediate temperature; thus, warm working is a compromise between cold and hot working. The temperature ranges for these three

TABLE 3.1

Homologous Temperature Ranges for Various Processes

Process	T/T_m
Cold working	<0.3
Warm working	0.3 to 0.5
Hot working	>0.6

categories of plastic deformation are given in Table 3.1 in terms of a ratio, where T is the working temperature and T_m is the melting point of the metal, both on the absolute scale. Although it is a dimensionless quantity, this ratio is known as the **homologous temperature.**

There are important technological differences in products that are processed by cold, warm, or hot working. For example, compared with cold-worked products, hot-worked products generally have less dimensional accuracy, because of uneven thermal expansion and contraction during processing, and a rougher surface appearance and finish, because of the oxide layer that usually develops during heating. Other important manufacturing characteristics, such as formability, machinability, and weldability, are also affected by cold, warm, and hot working to different degrees.

3.8 Failure and Fracture

Failure is one of the most important aspects of a material's behavior, because it directly influences the selection of a material for a certain application, the methods of manufacturing, and the service life of the component. Because of the many factors involved, failure and fracture of materials is a complex area of study. In this section, we consider only those aspects of failure that are of particular significance to selecting and processing materials.

There are two general types of failure: (1) **fracture** and separation of the material, through either internal or external cracking; and (2) **buckling** (Fig. 3.20). Fracture is further divided into two general categories: ductile and brittle (Fig. 3.21). Although failure of materials is generally regarded as undesirable, certain products are designed to fail. Typical examples include (a) food and beverage containers with tabs or entire tops that are removed by tearing the sheet metal along a prescribed path, and (b) screw caps for bottles.

3.8.1 Ductile fracture

Ductile fracture is characterized by plastic deformation, which precedes failure of the part. In a tension test, highly ductile materials, such as gold and lead, may neck down to a point and then fail (Fig. 3.21d). Most metals and alloys, however, neck

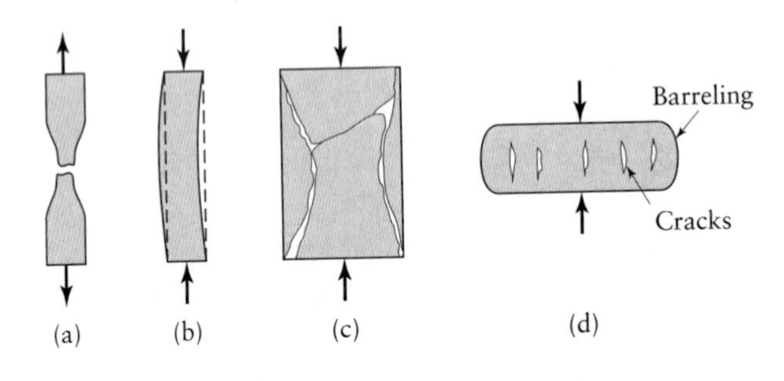

FIGURE 3.20 Schematic illustration of types of failure in materials: (a) necking and fracture of ductile materials; (b) buckling of ductile materials under a compressive load; (c) fracture of brittle materials in compression; (d) cracking on the barreled surface of ductile materials in compression. (See also Fig. 6.1b)

down to a finite area and then fail. Ductile fracture generally takes place along planes on which the *shear stress is a maximum*. In torsion, for example, a ductile metal fractures along a plane perpendicular to the axis of twist, that is, the plane on which the shear stress is a maximum. Fracture in shear is a result of extensive slip along slip planes within the grains.

Upon close examination of the surface of a ductile fracture (Fig. 3.22), we see a *fibrous* pattern with *dimples*, as if a number of very small tension tests have been carried out over the fracture surface. Failure is initiated with the formation of tiny voids, usually around small inclusions or preexisting voids, which then grow and coalesce, developing cracks that grow in size and lead to fracture. In a tension-test specimen, fracture begins at the center of the necked region from the growth and coalescence of cavities (Fig. 3.23). The central region becomes one large crack and propagates to the periphery of this necked region. Because of its appearance, the fracture surface of a tension-test specimen is called a **cup-and-cone fracture.**

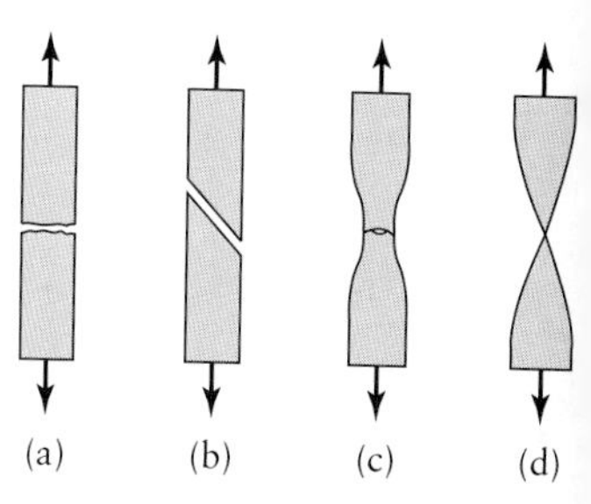

FIGURE 3.21 Schematic illustration of the types of fracture in tension: (a) brittle fracture in polycrystalline metals; (b) shear fracture in ductile single crystals (see also Fig. 3.4a); (c) ductile cup-and-cone fracture in polycrystalline metals (see also Fig. 2.2); (d) complete ductile fracture in polycrystalline metals, with 100% reduction of area.

Effects of inclusions. Because they are nucleation sites for voids, *inclusions* have an important influence on ductile fracture and thus on the formability of materials (Fig. 3.24). Inclusions may consist of impurities of various kinds and second-phase particles such as oxides, carbides, and sulfides. The extent of their influence depends on factors such as their shape, hardness, distribution, and volume fraction. The greater the volume fraction of inclusions, the lower will be the ductility of the material. Voids and porosity developed during processing, such as from casting, reduce the ductility of a material. (See Section 5.4.6.)

Two factors affect void formation. The first is the strength of the bond at the interface of an inclusion and the matrix; if the bond is strong, there is less tendency for void formation during plastic deformation. The second factor is the hardness of

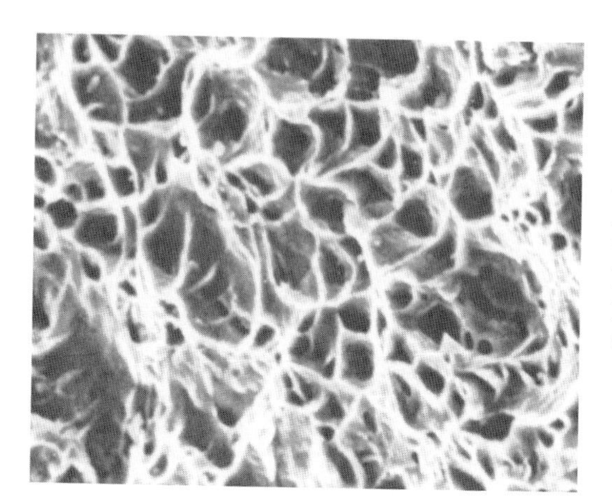

FIGURE 3.22 Surface of ductile fracture in low-carbon steel, showing dimples. Fracture is usually initiated at impurities, inclusions, or preexisting voids in the metal. *Source*: K.-H. Habig and D. Klaffke. Photo by BAM, Berlin, Germany.

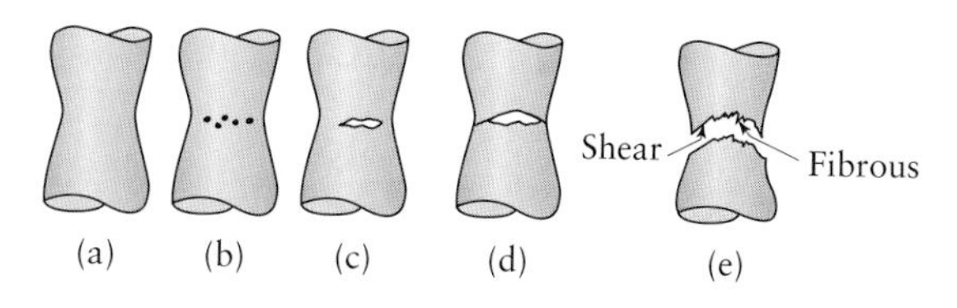

FIGURE 3.23 Sequence of events in necking and fracture of a tensile-test specimen: (a) early stage of necking; (b) small voids begin to form within the necked region; (c) voids coalesce, producing an internal crack; (d) rest of cross-section begins to fail at the periphery by shearing; (e) final fracture surfaces, known as cup-(top fracture surface) and-cone (bottom surface) fracture.

the inclusion. If the inclusion is soft, such as manganese sulfide, it will conform to the overall change in shape of the specimen or workpiece during plastic deformation. If it is hard, such as a carbide or oxide, it could lead to void formation (Fig. 3.25). Hard inclusions may also break up into smaller particles during deformation, because of their brittle nature. The alignment of inclusions during plastic deformation leads to **mechanical fibering.** Subsequent processing of such a material must therefore involve considerations of the proper direction of working for maximum ductility and strength.

Transition temperature. Many metals undergo a sharp change in ductility and toughness across a narrow temperature range called the *transition temperature* (Fig. 3.26). This phenomenon occurs in body-centered cubic and some hexagonal close-packed metals; it is rarely exhibited by face-centered cubic metals. The transi-

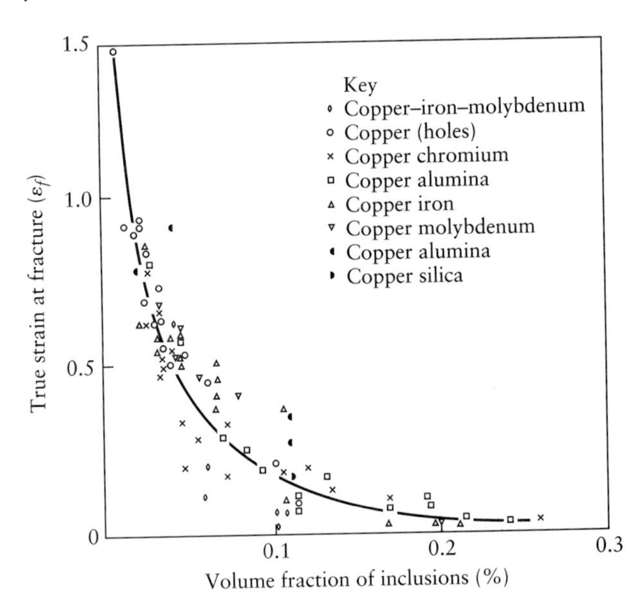

FIGURE 3.24 The effect of volume fraction of various second-phase particles on the true strain at fracture in a tensile test for copper. *Source*: After B. I. Edelson and W. M. Baldwin.

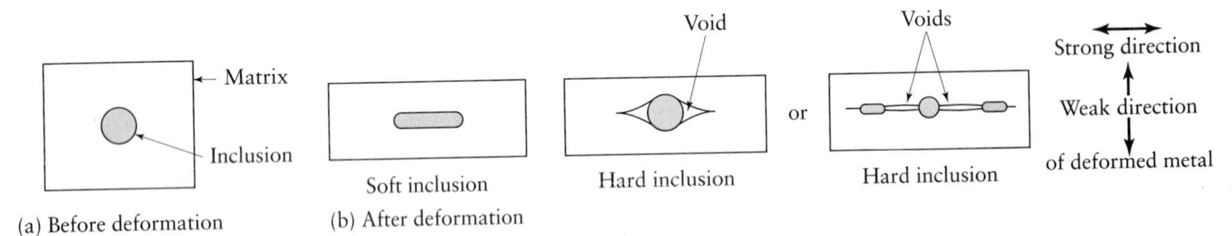

FIGURE 3.25 Schematic illustration of the deformation of soft and hard inclusions and their effect on void formation in plastic deformation. Note that hard inclusions, because they do not comply with the overall deformation of the ductile matrix, can cause voids.

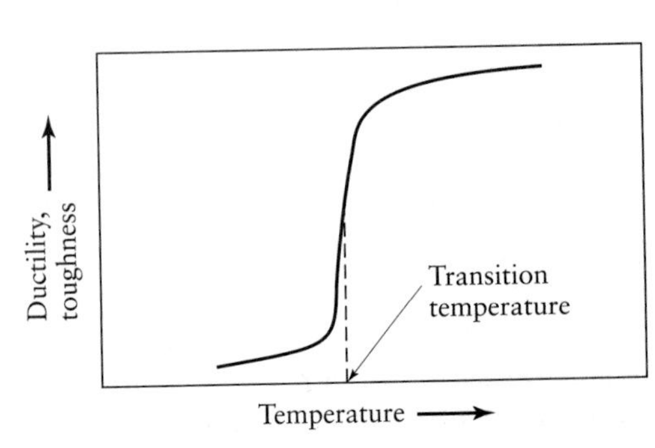

FIGURE 3.26 Schematic illustration of transition temperature. Note the narrow temperature range across which the behavior of the metal undergoes a major transition.

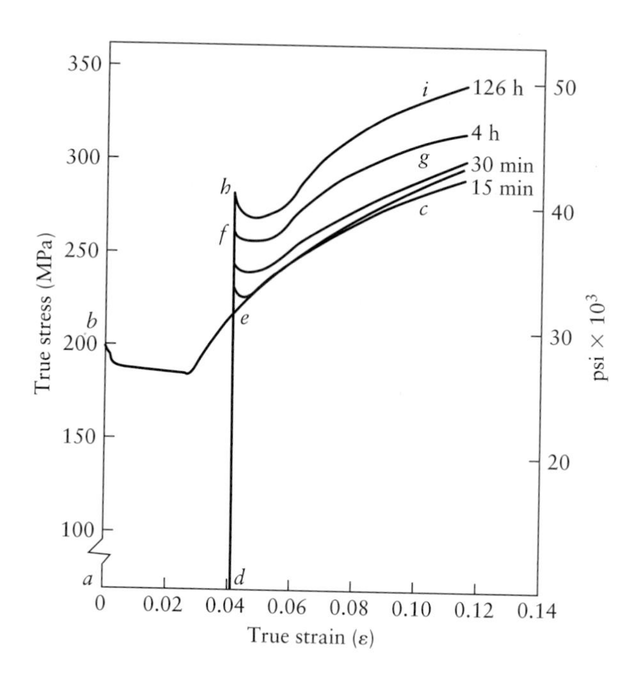

FIGURE 3.27 Strain aging and its effect on the shape of the true-stress–true-strain curve for 0.03% C rimmed steel at 60°C (140°F). *Source*: A. S. Keh and W. C. Leslie.

tion temperature depends on factors such as composition, microstructure, grain size, surface finish and shape of the specimen, and rate of deformation. High rates, abrupt changes in shape, and surface notches raise the transition temperature.

Strain aging. *Strain aging* is a phenomenon in which carbon atoms in steels segregate to dislocations, thereby pinning them and thus increasing the steel's resistance to dislocation movement. The result is increased strength and reduced ductility. Figure 3.27 shows the effects of strain aging on the shape of the stress–strain curve for low-carbon steel at room temperature. Curve *abc* is the original curve, with upper and lower yield points typical of these steels. If the tension test is stopped at point *e* and the specimen is unloaded and tested again, curve *dec* is obtained. Note that the upper and lower yield points have disappeared. However, if four hours pass before the specimen is strained again, curve *dfg* is obtained; if 126 hours pass, curve *dhi* is obtained. Instead of taking place over several days at room temperature, this same phenomenon can occur in just a few hours at a higher temperature; it is called **accelerated strain aging.** For steels, the phenomenon is called **blue brittleness.**

3.8.2 Brittle fracture

Brittle fracture occurs with little or no gross plastic deformation preceding the separation of the material into two or more pieces. In tension, fracture takes place along a crystallographic plane, called a **cleavage plane,** on which the normal tensile stress is a maximum. Face-centered cubic metals usually do not fail in brittle fracture, whereas body-centered cubic and some hexagonal close-packed metals fail by cleavage. In general, low temperature and high rates of deformation promote brittle fracture.

In a polycrystalline metal under tension, the fracture surface has a bright granular appearance, because of the changes in the direction of the cleavage planes as the crack propagates from one grain to another. Figure 3.28 shows an example of the surface of brittle fracture. Brittle fracture of a specimen in compression is more complex and may follow a path that is theoretically 45° to the direction of the applied force.

Examples of fracture along a cleavage plane include the splitting of rock salt and the peeling of layers of mica. Tensile stresses normal to the cleavage plane, caused by

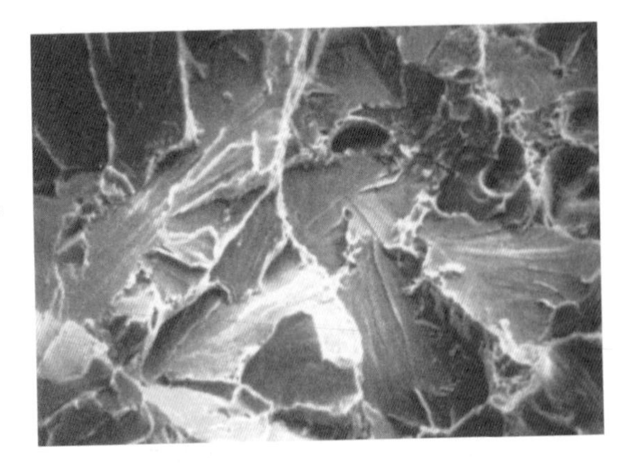

FIGURE 3.28 Typical fracture surface of steel that has failed in a brittle manner. The fracture path is transgranular (through the grains). Compare this surface with the ductile fracture surface shown in Fig. 3.22. Magnification: 200×. *Source*: Courtesy of Packer Engineering.

pulling, initiate and control the propagation of fracture. Another example is the behavior of brittle materials, such as chalk, gray cast iron, and concrete. In tension, they typically fail in the manner shown in Fig. 3.21a. In torsion, however, they fail along a plane at 45° to the axis of twist, that is, along a plane on which the tensile stress is a maximum.

Defects. An important factor in fracture is the presence of *defects* such as scratches, flaws, and external or internal cracks. Under tension, the tip of a crack is subjected to high tensile stresses, which propagate the crack rapidly, because the material has little capacity to dissipate energy. The tensile strength of a specimen with a crack perpendicular to the direction of pulling is related to the length of the crack as follows:

$$\sigma \propto \frac{1}{\sqrt{\text{crack length}}}. \tag{3.8}$$

The presence of defects is essential in explaining why brittle materials are so weak in tension compared with their strength in compression. Under tensile stresses, cracks propagate rapidly, causing what is known as *catastrophic failure*. With polycrystalline metals, the fracture paths most commonly observed are **transgranular** (*transcrystalline*, or *intragranular*), meaning that the crack propagates *through* the grain. **Intergranular** fracture, where the crack propagates *along* the grain boundaries (Fig. 3.29), generally occurs when the grain boundaries are soft, contain a brittle phase, or have been weakened by liquid-or solid-metal embrittlement (Section 3.4.2).

The maximum crack velocity in a brittle material is about 62% of the elastic wave-propagation (or acoustic) velocity of the material; this velocity is given by the formula $\sqrt{E/\rho}$, where E is the elastic modulus and ρ is the mass density. Thus, for steel, the maximum crack velocity is 2000 m/s (6600 ft/s).

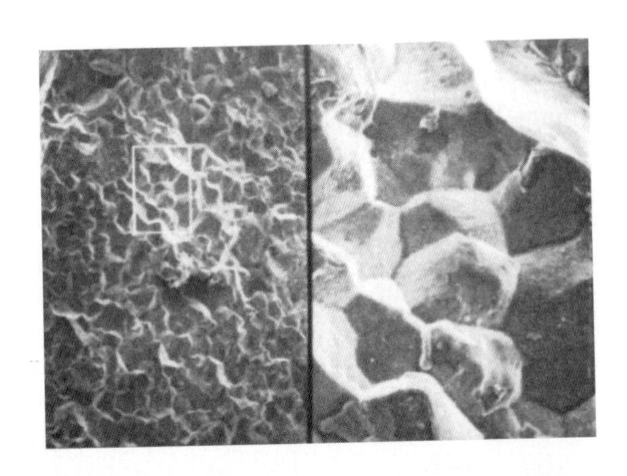

FIGURE 3.29 Intergranular fracture, at two different magnifications. Grains and grain boundaries are clearly visible in this micrograph. The fracture path is along the grain boundaries. Magnification: left, 100×; right, 500×. *Source*: Courtesy of Packer Engineering.

As Fig. 3.30 shows, cracks may be subjected to stresses in different directions. Mode I is tensile stress applied perpendicular to the crack. Modes II and III are shear stresses applied in two different directions. Tearing paper, cutting sheet metal with shears, and opening pop-top cans are examples of Mode III fracture.

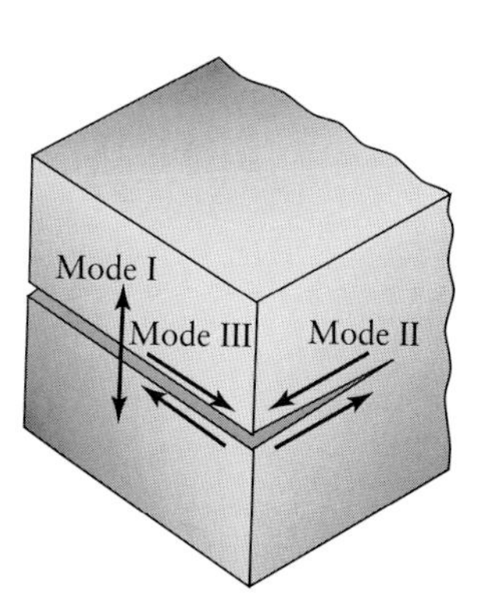

FIGURE 3.30 Three modes of fracture. Mode I has been studied extensively, because it is the most commonly observed in engineering structures and components. Mode II is rare. Mode III is the tearing process; examples include opening a pop-top can, tearing a piece of paper, and cutting materials with a pair of scissors.

Fatigue fracture. *Fatigue fracture* is basically of a brittle nature. Minute external or internal cracks develop at flaws or defects in the material. The cracks then propagate and eventually lead to total failure of the part. The fracture surface in fatigue is generally characterized by the term **beach marks**, because of its appearance (Fig. 3.31). With large magnification, such as higher than 1000×, a series of **striations** can be seen on fracture surfaces, where each beach mark consists of a number of striations.

Fatigue life is greatly influenced by the method of preparation of the surfaces of the part or specimen (Fig. 3.32). The fatigue strength of manufactured products can generally be improved by the following methods:

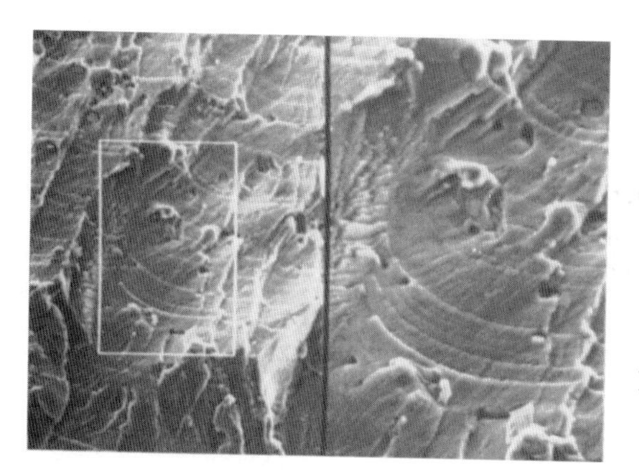

FIGURE 3.31 Typical fatigue fracture surface on metals, showing beach marks. Most components in machines and engines fail by fatigue and not by excessive static loading. Magnification: left, 500×; right, 1000×. *Source*: Courtesy of Packer Engineering.

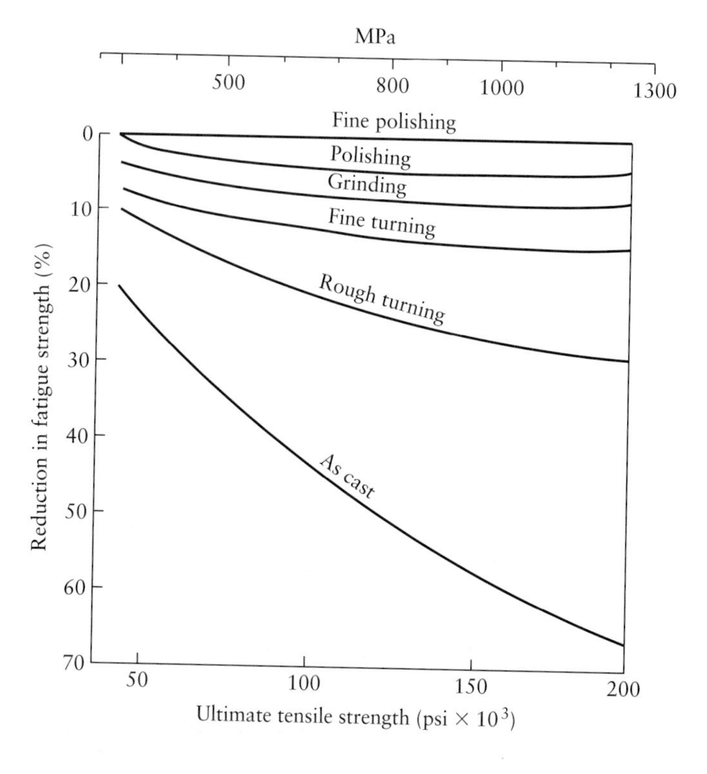

FIGURE 3.32 Reduction in fatigue strength of cast steels subjected to various surface-finishing operations. Note that the reduction is greater as the surface roughness and strength of the steel increase. *Source*: Shigley, J.E., and Mitchell, L.D., Mechanical Engineering Design, New York, McGraw-Hill, 1983.

1. Inducing compressive residual stresses on surfaces, such as by shot peening or roller burnishing (Section 4.5.1);
2. Surface (case) hardening by various means (Section 5.11.3);
3. Providing fine surface finish, thereby reducing the effects of notches and other surface imperfections;
4. Selecting appropriate materials and ensuring that they are free from significant amounts of inclusions, voids, and impurities.

Conversely, the following factors and processes can reduce fatigue strength: decarburization, surface pits due to corrosion that act as stress raisers, hydrogen embrittlement (see below), galvanizing, and electroplating. Two of these processes are described as follows:

Stress-corrosion cracking. An otherwise ductile metal can fail in a brittle manner by *stress-corrosion cracking* (also called *stress cracking* or *season cracking*). After forming, parts may, either over a period of time or soon after being made into a product, develop cracks. Crack propagation may be intergranular or transgranular.

The susceptibility of metals to stress-corrosion cracking depends mainly on the material, the presence and magnitude of tensile residual stresses, and the environment. Brass and austenitic stainless steels are among the metals that are highly susceptible to stress cracking. Environmental factors such as salt water or other chemicals could be corrosive to metals. The usual procedure to avoid stress-corrosion cracking is to stress relieve the part just after it is formed. Full annealing may also be done, but this treatment reduces the strength of cold-worked parts.

Hydrogen embrittlement. The presence of hydrogen can reduce ductility and cause severe embrittlement in many metals, alloys, and nonmetallic materials, leading to premature failure. This phenomenon is known as *hydrogen embrittlement* and is especially severe in high-strength steels. Possible sources of hydrogen are during melting of the metal, during pickling (removal of surface oxides by chemical or electrochemical reaction), through electrolysis in electroplating, from water vapor in the atmosphere, or from moist electrodes and fluxes used during welding. Oxygen can also cause embrittlement in metals, especially in copper alloys.

3.8.3 Size effect

The dependence of the properties of a material on its size is knows as *size effect*. Note from the foregoing discussions and from Eq. (3.8) that as the size decreases, defects, cracks, imperfections, and the like are less likely to be present; hence, the strength and ductility of a specimen increase with decreasing size. Strength is related to cross-sectional area, but the length of the specimen is also important: The greater the length, the greater the probability of defects. As an analogy, a long chain is more likely to be weaker than a shorter chain, because the probability of one of the links being weak increases with the number of links (i.e., the length of chain). Size effect is also demonstrated by **whiskers**, which, because of their small size, are either free from imperfections or do not contain the types of imperfections that affect their strength. The actual strength in such cases approaches the theoretical strength. (See also Sections 8.6.10 and 10.9.2 for some applications of whiskers.)

EXAMPLE 3.1 Brittle fracture of steel plates of the hull of the R.M.S. Titanic

A detailed analysis of the Titanic disaster in 1912 has indicated that the ship sank not so much because it hit an iceberg as because of structural weaknesses. The plates composing its hull were made of low-grade steel with a high sulfur content; they had low toughness (as determined by the Charpy test; see Section 2.9) when chilled, as was the case in the Atlantic Ocean, and subjected to an external impact loading. With such a material, a crack that starts in one part of a welded steel structure can propagate rapidly and completely around the hull and cause a large ship to split in two. Although Titanic was built with brittle plates, as we know from the physical and photographic observations of the sunken ship, not all ships of that time were built with such low-grade steel. Furthermore, better construction techniques could have been employed, among them better welding techniques (Chapter 12) to improve the structural strength of the hull.

3.9 Physical Properties

In addition to their mechanical properties, the *physical properties* of materials should also be considered in selecting and processing materials. Properties of particular interest are density, melting point, specific heat, thermal conductivity and expansion, electrical and magnetic properties, and resistance to oxidation and corrosion, as described next.

3.9.1 Density

The *density* of a material is its weight per unit volume; if expressed relative to the density of water, the density of a material is known as **specific gravity** and thus has no units. The density of materials depends on atomic weight, atomic radius, and the packing of the atoms. Alloying elements generally have a minor effect on the density of metals, where the strength of the effect depends on the density of the alloying elements. Table 3.2 lists the range of densities for a variety of materials at room temperature.

The most significant role that density plays is in the **specific strength** (strength-to-weight ratio) and **specific stiffness** (stiffness-to-weight ratio) of materials and structures. (See also Section 10.9.1.) Weight reduction is particularly important for aircraft and aerospace structures, automotive bodies and components, and other products for which energy consumption and power limitations are major concerns. Substitution of materials for weight saving and economy is a major factor in advanced equipment and machinery as well as in consumer products such as automobiles.

Density also is an important factor in the selection of materials for high-speed equipment, such as the use of magnesium in printing and textile machinery, many components of which usually operate at very high speeds. To obtain an exposure time of 1/4000 s in cameras without sacrificing accuracy, the shutters of some high-quality 35-mm cameras are made of titanium. The resulting light weight of the components in these high-speed operations reduces inertial forces that otherwise could lead to vibrations, inaccuracies, and even part failure over time. However, there are also applications for which higher levels of weight are desirable. Examples include counterweights for various mechanisms (using lead and steel), flywheels, and components for self-winding watches (using high-density materials such as tungsten).

3.9.2 Melting point

The *melting point* of a material depends on the energy required to separate its atoms. As shown in Table 3.2, the melting point of an alloy is a wide range of temperatures, unlike the case for pure metals, which have a definite melting point. The melting points of alloys depend on the alloys' particular composition. (See Section 5.2.) The melting point of a material has a number of indirect effects on manufacturing operations. The choice of a material for high-temperature applications is the most obvious consideration, such as in jet engines and furnaces. The choice of a material for applications where high frictional heat is generated, such as with high sliding speeds of machine components, is also affected by melting point. Sliding down a rope or rubbing your hands together fast is a simple demonstration of how high temperatures can become with increasing sliding speed. (See also Section 4.4.1.)

Since the recrystallization temperature of a metal is related to its melting point (Section 3.6), operations such as annealing, heat treating, and hot working require a

TABLE 3.2

Physical Properties of Various Materials at Room Temperature

	Density (kg/m^3)	Melting Point (°C)	Specific Heat (J/kg K)	Thermal Conductivity (W/m K)	Coefficient of Thermal Expansion (μm/m°C)
METAL					
Aluminum	2700	660	900	222	23.6
Aluminum alloys	2630–2820	476–654	880–920	121–239	23.0–23.6
Beryllium	1854	1278	1884	146	8.5
Columbium (niobium)	8580	2468	272	52	7.1
Copper	8970	1082	385	393	16.5
Copper alloys	7470–8940	885–1260	337–435	29–234	16.5–20
Gold	19300	1063	129	317	19.3
Iron	7860	1537	460	74	11.5
Steels	6920–9130	1371–1532	448–502	15–52	11.7–17.3
Lead	11350	327	130	35	29.4
Lead alloys	8850–11,350	182–326	126–188	24–46	27.1–31.1
Magnesium	1745	650	1025	154	26.0
Magnesium alloys	1770–1780	610–621	1046	75–138	26.0
Molybdenum alloys	10210	2610	276	142	5.1
Nickel	8910	1453	440	92	13.3
Nickel alloys	7750–8850	1110–1454	381–544	12–63	12.7–18.4
Silicon	2330	1423	712	148	7.63
Silver	10500	961	235	429	19.3
Tantalum alloys	16600	2996	142	54	6.5
Titanium	4510	1668	519	17	8.35
Titanium alloys	4430–4700	1549–1649	502–544	8–12	8.1–9.5
Tungsten	19290	3410	138	166	4.5
NONMETALLIC					
Ceramics	2300–5500	–	750–950	10–17	5.5–13.5
Glasses	2400–2700	580–1540	500–850	0.6–1.7	4.6–70
Graphite	1900–2200	–	840	5–10	7.86
Plastics	900–2000	110–330	1000–2000	0.1–0.4	72–200
Wood	400–700	–	2400–2800	0.1–0.4	2–60

knowledge of the melting points of the metals involved. These considerations, in turn, influence the selection of tool and die materials in manufacturing operations. Another major influence of the melting point is in the selection of the equipment and melting practice in casting operations; the higher the melting point of the material, the more difficult the operation becomes (Section 5.5). Also, the selection of die materials, as in die casting, depends on the melting point of the workpiece material. In the electrical-discharge machining process (Section 9.13), the melting points of metals are related to the rate of material removal and tool wear.

3.9.3 Specific heat

Specific heat is the energy required to raise the temperature of a unit mass of material by one degree. Alloying elements have a relatively minor effect on the specific heat of metals. The temperature rise in a workpiece, resulting from forming or machining operations, is a function of the work done and the specific heat of the workpiece material (see Section 2.12.1); thus, the lower the specific heat, the higher the temperature will rise in the material. If excessively high, temperature can have detrimental effects on product quality by adversely affecting surface finish and dimensional accuracy, can cause excessive tool and die wear, and can result in adverse metallurgical changes in the material.

3.9.4 Thermal conductivity

Whereas specific heat indicates the temperature rise in a material as a result of energy input, *thermal conductivity* indicates the rate with which heat flows within and through the material. Metallically bonded materials (metals) generally have high thermal conductivity, whereas ionically or covalently bonded materials (ceramics and plastics) have poor conductivity. Because of the large difference in their thermal conductivities, alloying elements can have a significant effect on the thermal conductivity of alloys, as can be seen in Table 3.2 by comparing the metals with their alloys.

When heat is generated by plastic deformation or friction, the heat should be conducted away at a high enough rate to prevent a severe rise in temperature. The main difficulty in machining titanium, for example, is caused by its very low thermal conductivity. Low thermal conductivity can also result in high thermal gradients and thus cause inhomogeneous deformation in metalworking processes.

3.9.5 Thermal expansion

Thermal expansion of materials can have several significant effects. Generally, the coefficient of thermal expansion is inversely proportional to the melting point of the material; alloying elements have a relatively minor effect on the thermal expansion of metals. Shrink fits use thermal expansion; other examples where relative expansion or contraction is important include electronic and computer components, glass-to-metal seals, metal-to-ceramic subassemblies (see also Section 11.8.2), struts on jet engines, and moving parts in machinery that require certain clearances for proper functioning.

Improper selection of materials and assembly can cause **thermal stresses**, *cracking, warping*, or loosening of components in the structure during their service life. Thermal conductivity, in conjunction with thermal expansion, plays the most significant role in causing thermal stresses, both in manufactured components and in tools and dies. This phenomenon is particularly important in a forging operation, for example, when hot workpieces are placed over relatively cool dies, making the die surfaces undergo thermal cycling. To reduce thermal stresses, a combination of high thermal conductivity and low thermal expansion is desirable.

Thermal stresses can lead to cracks in ceramic parts and in tools and dies made of relatively brittle materials. **Thermal fatigue** results from thermal cycling and causes a number of surface cracks; **thermal shock** is the term generally used to describe development of cracks after a single thermal cycle. Thermal stresses may be caused both by temperature gradients and by **anisotropy of thermal expansion**, which is generally observed in hexagonal close-packed metals and in ceramics.

The influence of temperature on dimensions and the modulus of elasticity can be a significant problem in precision instruments and equipment. A spring, for example, will have a lower stiffness as its temperature increases, because of reduced elastic modulus; similarly, a tuning fork or a pendulum will have different frequencies at different temperatures. Dimensional changes resulting from thermal expansion can be significant in measurements for quality control and in equipment such as measuring tapes for geodetic applications. To alleviate some of the problems of thermal expansion, a family of iron–nickel alloys that have very low thermal-expansion coefficients has been developed, called **low-expansion alloys.** Typical compositions are 64% Fe-36% Ni (Invar) and 54% Fe-28% Ni-18 Co (Kovar). Consequently, these alloys also have good thermal-fatigue resistance.

3.9.6 Electrical and magnetic properties

Electrical conductivity and dielectric properties of materials are of great importance, not only in electrical equipment and machinery, but also in manufacturing processes such as magnetic-pulse forming of sheet metals (Section 7.9.3) and electrical-discharge machining and electrochemical grinding of hard and brittle materials (Chapter 9).

The **electrical conductivity** of a material can be defined as a measure of the ease with which the material conducts electric current. Materials with high conductivity, such as metals, are generally referred to as **conductors**. The units of electrical conductivity are mho/m or mho/ft, where mho is the inverse of ohm (the unit for electrical resistance). **Electrical resistivity** is the inverse of conductivity. Materials with high resistivity are referred to as **dielectrics**, or **insulators**. The influence of the type of atomic bonding on the electrical conductivity of materials is the same as that for thermal conductivity. Alloying elements have a major effect on the electrical conductivity of metals: the higher the conductivity of the alloying element, the higher the conductivity of the alloy. **Dielectric strength** of materials is the resistivity to direct electric current and is defined as the voltage required per unit distance for electrical breakdown; the unit is V/m or V/ft.

Superconductivity is the phenomenon of almost zero electrical resistivity that occurs in some metals and alloys below a critical temperature. The highest temperature at which superconductivity has to date been exhibited, at about −123°C (−190°F), is with an alloy of lanthanum, strontium, copper, and oxygen, although other material compositions are also being studied. Developments in superconductivity indicate that the efficiency of electrical components such as large high-power magnets, high-voltage power lines, and various other electronic and computer components can be markedly improved.

The electrical properties of materials such as single-crystal silicon, germanium, and gallium arsenide make them extremely sensitive to the presence and type of minute impurities as well as to temperature. Thus, by controlling the concentration and type of impurities (**dopants**), such as phosphorus and boron in silicon, electrical conductivity can be controlled as well. This property is exploited in **semiconductor** (solid state) devices used extensively in miniaturized electronic circuitry as described in Chapter 13. Such devices are very compact, efficient, and relatively inexpensive; consume little power; and require no warm-up time for operation.

Ferromagnetism is the large and permanent magnetization resulting from an exchange interaction between neighboring atoms (such as iron, nickel, and cobalt) that aligns their magnetic moments in parallel. It has important applications in electric motors, generators, transformers, and microwave devices. **Ferrimagnetism** is the permanent and large magnetization exhibited by some ceramic materials, such as cubic ferrites.

The **piezoelectric** effect (*piezo* from Greek, meaning *to press*), in which there is a reversible interaction between an elastic strain and an electric field, is exhibited by some materials, such as certain ceramics and quartz crystals. This property is utilized in making *transducers*, which are devices that convert the strain from an external force to electrical energy. Typical applications of the piezoelectric effect include force or pressure transducers, strain gages, sonar detectors, and microphones.

Magnetostriction is the phenomenon of expansion or contraction of a material when subjected to a magnetic field. Some materials, such as pure nickel and some iron–nickel alloys, exhibit this behavior. Magnetostriction is the principle behind ultrasonic machining equipment. (See Section 9.7.)

3.9.7 Resistance to corrosion

Corrosion is the deterioration of metals and ceramics, while **degradation** is a similar phenomenon in plastics. Corrosion resistance is an important aspect of selection, especially for applications in the chemical, food, and petroleum industries. In addition to various possible chemical reactions from the elements and compounds present, environmental oxidation and corrosion of components and structures are a major concern, particularly at elevated temperatures and in automobiles and other transportation equipment.

Resistance to corrosion depends on the particular environment, as well as the composition of the material. Chemicals (acids, alkali, and salts), the environment (oxygen, pollution, and acid rain), and water (fresh or salt) may all act as corrosive media. Nonferrous metals, stainless steels, and nonmetallic materials generally have high corrosion resistance. Steels and cast irons generally have poor resistance and must be protected by various coatings and surface treatments.

Corrosion can occur over an entire surface, or it can be localized (such as in **pitting**). It can occur along grain boundaries of metals as **intergranular corrosion** and at the interface of bolted or riveted joints as **crevice corrosion**. Two dissimilar metals may form a galvanic cell (two electrodes in an electrolyte in a corrosive environment, including moisture) and cause **galvanic corrosion**. Two-phase alloys (Section 5.2.3) are more susceptible to galvanic corrosion, because of the two different metals involved, than are single-phase alloys or pure metals; thus, heat treatment can have an influence on corrosion resistance.

Corrosion may also act in indirect ways. **Stress-corrosion cracking** is an example of the effect of a corrosive environment on the integrity of a product that, as manufactured, had residual stresses in it. Likewise, metals that are cold worked are likely to contain residual stresses and hence are more susceptible to corrosion as compared with hot-worked or annealed metals.

Tool and die materials also can be susceptible to chemical attack by lubricants and coolants (Section 4.4.4). The chemical reaction alters their surface finish and adversely influences the metalworking operation. An example is carbide tools and dies that have cobalt as a binder (Section 8.6.4), where the cobalt is attacked by elements in the cutting fluid (**selective leaching**). Thus, compatibility of the tool, die, and workpiece materials and the metalworking fluid under actual operating conditions is an important consideration in the selection of materials. Note that chemical reactions should not be regarded as having adverse effects only. Nontraditional machining processes such as chemical machining and electrochemical machining are indeed based on controlled reactions (Chapter 9). These processes remove material by chemical reactions, similar to the etchings of metallurgical specimens.

The usefulness of some level of *oxidation* is exhibited by the corrosion resistance of aluminum, titanium, and stainless steel. Aluminum develops a thin (a few atomic layers), strong, and adherent hard oxide film (Al_2O_3) that protects the surface from further environmental corrosion. Titanium develops a film of titanium oxide (TiO_2). A similar phenomenon occurs in stainless steels, which, because of the chromium present in the alloy, develop a similar protective film on their surfaces. These processes are known as **passivation** (Section 4.2). When the protective film is scratched, thus exposing the metal underneath, a new oxide film forms in due time.

3.10 Properties and Applications of Ferrous Alloys

By virtue of their very wide range of mechanical, physical, and chemical properties, **ferrous alloys** are among the most useful of all metals. Ferrous metals and alloys contain iron as their base metal and are variously categorized as carbon and alloy steels, stainless steels, tool and die steels, cast irons, and cast steels. Ferrous alloys are produced as sheet steel for automobiles, appliances, and containers; as plates for ships, boilers, and bridges; as structural members (such as I-beams), bar products for leaf springs, gears, axles, crankshafts, and railroad rails; as stock for tools and dies; as music wire; and as fasteners such as bolts, rivets, and nuts.

A typical U.S. passenger car now contains about 816 kg (1800 lb) of steel, accounting for about 55% of its weight. As an example of their widespread use, ferrous materials comprise 70% to 85% by weight of virtually all structural members and mechanical components. Carbon steels are the least expensive of all metals, but stainless steels can be costly.

3.10.1 Carbon and alloy steels

Carbon and alloy steels are among the most commonly used metals and have a wide variety of applications. The composition and processing of these steels are controlled in a manner that makes them suitable for numerous applications. They are available in various basic product shapes: plate, sheet, strip, bar, wire, tube, castings, and forgings.

Several elements are added to steels to impart various properties such as hardenability, strength, hardness, toughness, wear resistance, workability, weldability, and machinability. Generally, the higher the percentages of these elements in steels, the higher the particular properties that they impart. Thus, for example, the higher the carbon content, the higher the hardenability of the steel and the higher its strength, hardness, and wear resistance. Conversely, ductility, weldability, and toughness are reduced with increasing carbon content.

Carbon steels. *Carbon steels* are generally classified as low, medium, and high:

1. **Low-carbon steel**, also called **mild steel**, has less than 0.30% carbon. It is generally used for common industrial products, such as bolts, nuts, sheet plates, tubes, and machine components that do not require high strength.

2. **Medium-carbon steel** has 0.30% to 0.60% carbon. It is generally used in applications requiring higher strength than those using low-carbon steels, such as machinery, automotive and agricultural equipment parts (e.g., gears, axles, connecting rods, crankshafts), railroad equipment, and parts for metalworking machinery.
3. **High-carbon steel** has more than 0.60% carbon. It is generally used for parts requiring strength, hardness, and wear resistance, such as cutting tools, cable, music wire, springs, cutlery, and rails. After being manufactured into shapes, the parts are usually heat treated and tempered. The higher the carbon content of the steel, the higher its hardness, strength, and wear resistance after heat treatment.
4. Carbon steels containing sulfur or phosphorus, known as **resulfurized** and **rephosphorized and resulfurized carbon steels**, are available. These steels have improved machinability, as described in Section 8.5.1.

Alloy steels. Steels containing significant amounts of alloying elements are called *alloy steels*, and they are usually made with more care than are carbon steels. **Structural-grade** alloy steels, as identified by ASTM specifications, are used mainly in the construction and transportation industries, because of their high strength. Other alloy steels are used in applications where strength, hardness, resistance to creep, fatigue, and toughness are required. These steels may also be heat treated to obtain the desired properties.

High-strength low-alloy steels. In order to improve the strength-to-weight ratio of steels, a number of *high-strength low-alloy* (HSLA) steels have been developed. These steels have a low carbon content (usually less than 0.30%), and are characterized by a microstructure consisting of fine-grain ferrite and a hard second phase of carbides, carbonitrides, or nitrides. First developed in the 1930s, HSLA steels are usually produced in sheet form by microalloying and controlled hot rolling. Plates, bars, and structural shapes are made from these steels. However, the ductility, formability, and weldability of HSLA steels are generally inferior to those of conventional low-alloy steels. To improve these properties, dual-phase steels have been developed.

Sheet products of HSLA steels typically are used in certain parts of automobile bodies to reduce weight (and hence fuel consumption) and in transportation equipment, mining equipments agricultural equipment, and various other industrial applications. Plates made of HSLA steel are used in ships, bridges, and building construction, and shapes such as I-beams, channels, and angles are used in buildings and various other structures.

Dual-phase steels. Designated by the letter D, these steels are processed specially and have a mixed ferrite and martensite structure. Developed in the late 1960s, dual-phase steels have high work-hardening characteristics (i.e., a high n value; see Section 2.2.3), which improve their ductility and formability.

3.10.2 Stainless steels

Stainless steels are characterized primarily by their corrosion resistance, high strength and ductility, and high chromium content. They are called *stainless* because, in the presence of oxygen (air), they develop a thin, hard adherent film of *chromium oxide* that protects the metal from corrosion (passivation). This protective film builds up again if the surface is scratched. For passivation to occur, the minimum chromium content of the steel should be 10% to 12% by weight.

In addition to chromium, other alloying elements in stainless steels typically are nickel, molybdenum, copper, titanium, silicon, manganese, columbium, aluminum, nitrogen, and sulfur. The higher the carbon content, the lower the corrosion resistance of stainless steels. The reason is that the carbon combines with chromium in the steel and forms chromium carbide, which lowers the passivity of the steel; the chromium carbides introduce a second phase in the metal, which promotes galvanic corrosion. The letter *L* is used to identify low-carbon stainless steels.

Developed in the early 1900s, stainless steels are made by techniques similar to those used in other types of steelmaking, using electric furnaces or the basic-oxygen process. The level of purity is controlled by various refining techniques. Stainless steels are available in a wide variety of shapes. Typical applications are for cutlery, kitchen equipment, health care and surgical equipment, and automotive trim and in the chemical, food-processing, and petroleum industries.

Stainless steels are generally divided into five types (Table 3.3): austenitic, ferritic, martensitic, precipitation-hardening, and duplex-structure steels. These types are described as follows:

1. The **austenitic steels** (200 and 300 series) are generally composed of chromium, nickel, and manganese in iron. They are nonmagnetic and have excellent corrosion resistance, but are susceptible to stress-corrosion cracking. Austenitic stainless steels are hardened by cold working; they are the most ductile of all stainless steels and hence can be formed easily. However, with increasing cold work, their formability is reduced. These steels are used in a wide variety of applications, such as kitchenware, fittings, welded construction, lightweight transportation equipment, furnace and heat-exchanger parts, and components for severe chemical environments.
2. The **ferritic steels** (400 series) have a high chromium content: up to 27%. They are magnetic and have good corrosion resistance, but have lower ductility than austenitic stainless steels. Ferritic stainless steels are hardened by cold working and are not heat treatable. They are generally used for nonstructural applications, such as kitchen equipment and automotive trim.

TABLE 3.3

Room-Temperature Mechanical Properties and Typical Applications of Annealed Stainless Steels

AISI (UNS)	Ultimate Tensile Strength (MPa)	Yield Strength (MPa)	Elongation (%)	Characteristics and Typical Applications
303 (S30300)	550–620	240–260	50–53	Screw-machine products, shafts, valves, bolts, bushings, and nuts; aircraft fittings; rivets; screws; studs.
304 (S30400)	565–620	240–290	55–60	Chemical and food-processing equipment, brewing equipment, cryogenic vessels, gutters, downspouts, and flashings.
316 (S31600)	550–590	210–290	55–60	High corrosion resistance and high creep strength. Chemical and pulp-handling equipment, photographic equipment, brandy vats, fertilizer parts, ketchup-cooking kettles, and yeast tubs.
410 (S41000)	480–520	240–310	25–35	Machine parts, pump shafts, bolts, bushings, coal chutes, cutlery, fishing tackle, hardware, jet engine parts, mining machinery, rifle barrels, screws, and valves.
416 (S41600)	480–520	275	20–30	Aircraft fittings, bolts, nuts, fire extinguisher inserts, rivets, and screws.

3. Most of the **martensitic steels** (400 and 500 series) do not contain nickel and are hardenable by heat treatment; their chromium content may be as high as 18%. These steels are magnetic and have high strength, hardness, and fatigue resistance, and good ductility, but moderate corrosion resistance. Martensitic stainless steels are used for cutlery, surgical tools, instruments, valves, and springs.
4. The **precipitation-hardening (PH) steels** contain chromium and nickel, along with copper, aluminum, titanium, or molybdenum. They have good corrosion resistance, good ductility, and high strength at elevated temperatures. Their main application is in aircraft and aerospace structural components.
5. The **duplex-structure steels** have a mixture of austenite and ferrite. They have good strength and higher resistance to corrosion in most environments and to stress-corrosion cracking than does the 300 series of austenitic steels. Typical applications of duplex-structure steels are in water-treatment plants and heat-exchanger components.

3.10.3 Tool and die steels

Tool and die steels are specially alloyed steels (Table 3.4) designed for high strength, impact toughness, and wear resistance at room and elevated temperatures. They are commonly used in forming and machining of metals. Table 3.5 lists various tool and die materials for a variety of manufacturing applications. The main categories of these materials are described as follows:

1. **High-speed steels** (HSS), first developed in the early 1900s, are the most highly alloyed tool and die steels and maintain their hardness and strength at elevated operating temperatures. There are two basic types of high-speed steels: the *molybdenum type* (M series) and the *tungsten type* (T series). The M-series steels contain up to about 10% molybdenum, with chromium, vanadium, tungsten, and cobalt as other alloying elements. The T-series steels contain 12% to 18% tungsten, with chromium, vanadium, and cobalt as other alloying elements. As compared with the T-series steels, the M-series steels generally have higher abrasion resistance, have less distortion in heat treatment, and are less expensive. The M-series steels constitute about 95% of all high-speed steels produced in the United States. High-speed steel tools can be coated with titanium nitride and titanium carbide for better resistance to wear.

TABLE 3.4

Basic Types of Tool and Die Steels

Type	AISI
High speed	M (molybdenum base) T (tungsten base)
Hot work	H1 to H19 (chromium base) H20 to H39 (tungsten base) H40 to H59 (molybdenum base)
Cold work	D (high carbon, high chromium) A (medium alloy, air hardening) O (oil hardening)
Shock resisting	S
Mold steels	P1 to P19 (low carbon) P20 to P39 (others)
Special purpose	L (low alloy) F (carbon–tungsten)
Water hardening	W

TABLE 3.5

Typical Tool and Die Materials for Various Processes

Process	Material
Die casting	H13, P20
Powder metallurgy	
Punches	A2, S7, D2, D3, M2
Dies	WC, D2, M2
Molds for plastic and rubber	S1, O1, A2, D2, 6F5, 6F6, P6, P20, P21, H13
Hot forging	6F2, 6G, H11, H12
Hot extrusion	H11, H12, H13
Cold heading	W1, W2, M1, M2, D2, WC
Cold extrusion	
Punches	A2, D2, M2, M4
Dies	O1, W1, A2, D2
Coining	52100, W1, O1, A2, D2, D3, D4, H11, H12, H13
Drawing	
Wire	WC, diamond
Shapes	WC, D2, M2
Bar and tubing	WC, W1, D2
Rolls	
Rolling	Cast iron, cast steel, forged steel, WC
Thread rolling	A2, D2, M2
Shear spinning	A2, D2, D3
Sheet metals	
Shearing	
Cold	D2, A2, A9, S2, S5, S7
Hot	H11, H12, H13
Pressworking	Zinc alloys, 4140 steel, cast iron, epoxy composites, A2, D2, O1
Deep drawing	W1, O1, cast iron, A2, D2
Machining	Carbides, high-speed steels, ceramics, diamond, cubic boron nitride

2. **Hot-work steels** (H series) are designed for use at elevated temperatures and have high toughness and high resistance to wear and cracking. The alloying elements are generally tungsten, molybdenum, chromium, and vanadium.
3. **Cold-work steels** (A, D, and O series) are used for cold-working operations. They generally have high resistance to wear and cracking. These steels are available as oil-hardening or air-hardening types.
4. **Shock-resisting steels** (S series) are designed for impact toughness and are used in applications such as header dies, punches, and chisels. Other properties of these steels depend on the particular composition.

3.11 Properties and Applications of Nonferrous Metals and Alloys

Nonferrous metals and alloys cover a wide range of materials, from the more common metals, such as aluminum, copper, and magnesium, to high-strength, high-temperature alloys, such as tungsten, tantalum, and molybdenum. Although more expensive than ferrous metals, nonferrous metals and alloys have important applications, because of their numerous mechanical, physical, and chemical properties.

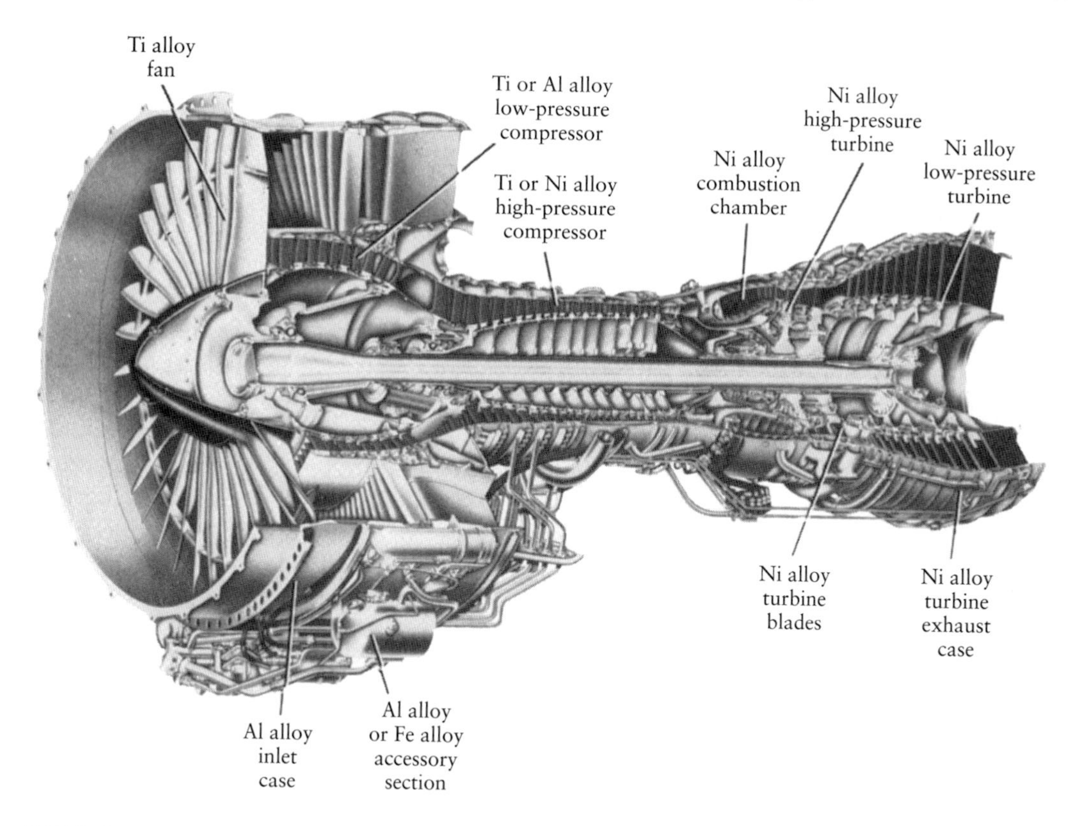

FIGURE 3.33 Cross-section of a jet engine (PW2037) showing various components and the alloys used in making them. *Source*: Courtesy of United Aircraft Pratt & Whitney.

A turbofan jet engine for the Boeing 757 aircraft typically contains the following nonferrous metals and alloys: 38% titanium, 37% nickel, 12% chromium, 6% cobalt, 5% aluminum, 1% niobium (columbium), and 0.02% tantalum. Without these materials, a jet engine (Fig. 3.33) could not be designed, manufactured, and operated at the required energy and efficiency levels.

Typical examples of the applications of nonferrous metals and alloys include aluminum for cooking utensils and aircraft bodies, copper wire for electricity and copper tubing for water in residences, titanium for jet-engine turbine blades and prosthetic devices (such as artificial joints), and tantalum for rocket engines. In this section, we discuss the general properties, production methods, and important engineering applications of nonferrous metals and alloys. We describe the manufacturing properties of these materials, such as formability, machinability, and weldability, in various chapters throughout this text.

3.11.1 Aluminum and aluminum alloys

The important factors in selecting *aluminum* (Al) and its alloys are their high strength-to-weight ratio, resistance to corrosion by many chemicals, high thermal and electrical conductivity, nontoxicity, reflectivity, appearance, and ease of formability and machinability; they are also nonmagnetic.

Principal uses of aluminum and its alloys, in decreasing order of consumption, include containers and packaging (aluminum cans and foil), buildings and other types of construction, transportation (aircraft and aerospace applications, buses, automobiles (see Fig. 1.5), railroad cars, and marine craft), electrical products (nonmagnetic

TABLE 3.6

Properties of Various Aluminum Alloys at Room Temperature

Alloy (UNS)	Temper	Ultimate Tensile Strength (MPa)	Yield Strength (MPa)	Elongation in 50 mm (%)
1100 (A91100)	O	90	35	35–45
1100	H14	125	120	9–20
2024 (A92024)	O	190	75	20–22
2024	T4	470	325	19–20
3003 (A93003)	O	110	40	30–40
3003	H14	150	145	8–16
5052 (A95052)	O	190	90	25–30
5052	H34	260	215	10–14
6061 (A96061)	O	125	55	25–30
6061	T6	310	275	12–17
7075 (A97075)	O	230	105	16–17
7075	T6	570	500	11

and economical electrical conductors), consumer durables (appliances, cooking utensils, and furniture), and portable tools (Tables 3.6 and 3.7). Nearly all high-voltage transmission wiring is made of aluminum. With respect to their structural (load-bearing) components, 82% of a Boeing 747 aircraft and 79% of a Boeing 757 aircraft is made of aluminum.

As in all cases of material selection, because each aluminum alloy has its particular properties, its selection can be critical with respect to manufacturing as well as economic considerations. For example, a typical aluminum beverage can consists of the following alloys, where all sheet is in the H19 condition, which is the highest cold-worked state: 3004 or 3104 for the can body, 5182 for the lid, and 5042 for the tab.

Aluminum alloys are available as mill products, that is, wrought product made into various shapes by rolling, extrusion, drawing, and forging. Aluminum ingots are available for casting, as are powder metals for powder metallurgy applications. There are two types of wrought alloys of aluminum: (1) alloys that can be hardened by cold

TABLE 3.7

Manufacturing Properties and Typical Applications of Wrought Aluminum Alloys

	Characteristics*			
Alloy	Corrosion Resistance	Machinability	Weldability	Typical Applications
1100	A	D–C	A	Sheet-metal work, spun hollow parts, tin-stock.
2014	C	C–B	C–B	Heavy-duty forgings, plate and extrusions for aircraft structural components, wheels.
3003	A	D–C	A	Cooking utensils, chemical equipment, pressure vessels, sheet-metal work, builders' hardware, storage tanks.
5054	A	D–C	A	Welded structures, pressure vessels, tube for marine uses.
6061	B	D–C	A	Trucks, canoes, furniture, structural applications.
7005	D	B–D	B	Extruded structural members, large heat exchangers, tennis racquets, softball bats.

* From A (excellent) to D (poor).

working (designated by the letter H) and are not heat treatable, and (2) alloys that are hardenable by heat treatment (designated by the letter T). The letter O indicates the annealed condition. Techniques have been developed whereby most aluminum alloys can be machined, formed, and welded with relative ease.

3.11.2 Magnesium and magnesium alloys

Magnesium (Mg) is the lightest engineering metal available and also has good vibration-damping characteristics. Its alloys are used in structural and nonstructural applications where weight is of primary importance. Magnesium is also an alloying element in various nonferrous metals.

Typical uses of magnesium alloys include aircraft and missile components, material-handling equipment, portable power tools (such as drills and sanders), ladders, luggage, bicycles, sporting goods, and general lightweight components. These alloys are available as either castings or wrought products, such as extruded bars and shapes, forgings, and rolled plate and sheet. Magnesium alloys are also used in printing and textile machinery to minimize inertial forces in high-speed components.

Because it is not sufficiently strong in its pure form, magnesium is alloyed with various elements (Table 3.8) to impart certain specific properties, particularly a high strength-to-weight ratio. A variety of magnesium alloys has good casting, forming, and machining characteristics. Because magnesium alloys oxidize rapidly (that is, they are *pyrophoric*) they are a potential fire hazard, and precautions must be taken when machining, grinding, or sand casting magnesium alloys. However, products made of magnesium and its alloys are not a fire hazard.

3.11.3 Copper and copper alloys

First produced in about 4000 BC, *copper* (Cu) and its alloys have properties somewhat similar to those of aluminum alloys. In addition, they are among the best conductors of electricity and heat and have good resistance to corrosion. Because of these properties, copper and its alloys are among the most important metals. They can be processed easily by various forming, machining, casting, and joining techniques.

Copper alloys often are attractive for applications where combined properties, such as electrical and mechanical properties, corrosion resistance, thermal conductivity, and wear resistance, are required. Such applications include electrical and electronic components, springs, cartridges for small arms, plumbing, heat exchangers, and marine hardware, as well as some consumer goods, such as cooking utensils, jewelry, and other decorative objects.

TABLE 3.8

Properties and Typical Forms of Various Wrought Magnesium Alloys

Alloy	Composition (%)				Condition	Ultimate Tensile Strength (MPa)	Yield Strength (MPa)	Elongation in 50 mm (%)	Typical Forms
	Al	Zn	Mn	Zr					
AZ31B	3.0	1.0	0.2		F	260	200	15	Extrusions
					H24	290	220	15	Sheet and plates
AZ80A	8.5	0.5	0.2		T5	380	380	7	Extrusions and forgings
HK31A*				0.7	H24	255	255	8	Sheet and plates
ZK60A		5.7		0.55	T5	365	365	11	Extrusions and forgings

* HK31A also contains 3%Th.

TABLE 3.9

Properties and Typical Applications of Various Wrought Copper and Brasses

Type and UNS Number	Nominal Composition (%)	Ultimate Tensile Strength (MPa)	Yield Strength (MPa)	Elongation in 50 mm (%)	Typical Applications
Oxygen-free electronic (C10100)	99.99 Cu	220–450	70–365	55–4	Bus bars, waveguides, hollow conductors, lead in wires, coaxial cables and tubes, microwave tubes, rectifiers.
Red brass, 85% (C23000)	85.0 Cu 15.0 Zn	270–72	70–435	55–3	Weather stripping, conduit, sockets, fasteners, fire extinguishers, condenser and heat-exchanger tubing.
Low Brass, 80% (C24000)	80.0 Cu 20.0 Zn	300–850	80–450	55–3	Battery caps, bellows, musical instruments, clock dials, flexible hose.
Free-cutting brass (C36000)	61.5 Cu, 3.0 Pb, 35.5 Zn	340–470	125–310	53–18	Gears, pinions, automatic high-speed screw-machine parts.
Naval brass (C46400 to C46700)	60.0 Cu, 39.25 Zn, 0.75 Sn	380–610	170–455	50–17	Aircraft turnbuckle barrels, balls, bolts, marine hardware, valve stems, condensor plates.

Copper alloys can acquire a wide variety of properties by the addition of alloying elements and by heat treatment to improve their manufacturing characteristics. The most common copper alloys are brasses and bronzes. **Brass**, which is an alloy of copper and zinc, was one of the earliest alloys developed and has numerous applications, including decorative objects (Table 3.9). **Bronze** is an alloy of copper and tin (Table 3.10). There are also other bronzes, such as aluminum bronze, an alloy of copper and aluminum, and tin bronzes. Beryllium–copper, or **beryllium bronze**, and **phosphor bronze** have good strength and hardness for applications such as springs and bearings. Other major copper alloys are copper nickels and nickel silvers.

3.11.4 Nickel and nickel alloys

Nickel (Ni), a silver-white metal discovered in 1751, is a major alloying element that imparts strength, toughness, and corrosion resistance to metals. It is used extensively in stainless steels and nickel-base alloys. These alloys are used for high-temperature applications, such as jet-engine components, rockets, and nuclear power plants, as well as in food-handling and chemical-processing equipment, coins, and marine applications. Because nickel is magnetic, nickel alloys are also used in electromagnetic applications, such as solenoids. The principal use of nickel as a metal is in electroplating for appearance and for improving resistance to corrosion and wear.

Nickel alloys have high strength and corrosion resistance at elevated temperatures. Alloying elements in nickel are chromium, cobalt, and molybdenum. The behavior of nickel alloys in machining, forming, casting, and welding can be modified by various other alloying elements. A variety of nickel alloys that have a range of strengths at different temperatures have been developed (Table 3.11). *Monel* is a nickel–copper alloy, and *Inconel* is a nickel–chromium alloy. A nickel–molybdenum–chromium alloy (*Hastelloy*) has good corrosion resistance and high strength at elevated temperatures. *Nichrome*, an alloy of nickel, chromium, and iron, has high oxidation and electrical resistance and is used for electrical-heating elements. *Invar*, an alloy of iron and nickel, has relatively low sensitivity to temperature. (See Section 3.9.5.)

TABLE 3.10

Properties and Typical Applications of Various Wrought Bronzes

Type and UNS Number	Nominal Composition (%)	Ultimate Tensile Strength (MPa)	Yield Strength (MPa)	Elongation in 50 mm (%)	Typical Applications
Architectural bronze (C38500)	57.0 Cu, 3.0 Pb, 40.0 Zn	415	140 (as extruded)	30	Architectural extrusions, storefronts, thresholds, trim, butts, hinges.
Phosphor bronze, 5% A (C51000)	95.0 Cu, 5.0 Sn, trace P	325–960	130–550	64–2	Bellows, clutch disks, cotter pins, diaphragms, fasteners, wire brushes, chemical hardware, textile machinery.
Free-cutting phosphor bronze (C54400)	88.0 Cu, 4.0 Pb, 4.0 Zn, 4.0 Sn	300–520	130–435	50–15	Bearings, bushings, gears, pinions, shafts, thrust washers, valve parts.
Low-silicon bronze, B (C65100)	98.5 Cu, 1.5 Si	275–655	100–475	55–11	Hydraulic pressure lines, bolts, marine hardware, electrical conduits, heat-exchanger tubing.
Nickel–silver, 65–18 (C74500)	65.0 Cu, 17.0 Zn, 18.0 Ni	390–710	170–620	45–3	Rivets, screws, zippers, camera parts, base for silver plate, nameplates, etching stock.

TABLE 3.11

Properties and Typical Applications of Various Nickel Alloys (All Alloy Names are Trade Names)

Alloy (Condition)	Principal Alloying Elements (%)	Ultimate Tensile Strength (MPa)	Yield Strength (MPa)	Elongation in 50 mm (%)	Typical Applications
Nickel 200 (annealed)	None	380–550	100–275	60–40	Chemical- and food-processing industry, aerospace equipment, electronic parts.
Duranickel 301 (age hardened)	4.4 Al, 0.6 Ti	1300	900	28	Springs, plastics-extrusion equipment, molds for glass, diaphragms.
Monel R-405 (hot rolled)	30 Cu	525	230	35	Screw-machine products, water-meter parts.
Monel K-500 (age hardened)	29 Cu, 3Al	1050	750	20	Pump shafts, valve stems, springs.
Inconel 600 (annealed)	15 Cr, 8 Fe	640	210	48	Gas-turbine parts, heat-treating equipment, electronic parts, nuclear reactors.
Hastelloy C-4 (solution treated and quenched)	16 Cr, 15 Mo	785	400	54	High-temperature stability, resistance to stress-corrosion cracking.

3.11.5 Superalloys

Superalloys are important in high-temperature applications; hence, they are also known as **heat-resistant**, or **high-temperature, alloys.** Major applications of superalloys are in jet engines and gas turbines, with other applications in reciprocating engines; rocket engines; tools and dies for hot working of metals; and in the nuclear, chemical, and petrochemical industries.

TABLE 3.12

Properties and Typical Applications of Various Nickel-Base Superalloys at 870°C (1600°F) (All Alloy Names Are Trade Names)

Alloy	Condition	Ultimate Tensile Strength (MPa)	Yield Strength (MPa)	Elongation in 50 mm (%)	Typical Applications
Astroloy	Wrought	770	690	25	Forgings for high-temperature applications.
Hastelloy X	Wrought	255	180	50	Jet-engine sheet parts.
IN-100	Cast	885	695	6	Jet-engine blades and wheels.
IN-102	Wrought	215	200	110	Superheater and jet-engine parts.
Inconel 625	Wrought	285	275	125	Aircraft engines and structures, chemical-processing equipment.
Inconel 718	Wrought	340	330	88	Jet-engine and rocket parts.
MAR-M 200	Cast	840	760	4	Jet-engine blades.
MAR-M 432	Cast	730	605	8	Integrally cast turbine wheels.
René 41	Wrought	620	550	19	Jet-engine parts.
Udimet 700	Wrought	690	635	27	Jet-engine parts.
Waspaloy	Wrought	525	515	35	Jet-engine parts.

These alloys are also referred to as **iron-base, cobalt-base,** or **nickel-base superalloys.** They contain nickel, chromium, cobalt, and molybdenum as major alloying elements. Other alloying elements are aluminum, tungsten, and titanium. Superalloys are generally identified by trade names or by special numbering systems and are available in a variety of shapes. Most superalloys have a maximum service temperature of about 1000°C (1800°F) for structural applications. The temperatures can be as high as 1200°C (2200°F) for non-load-bearing components. Superalloys generally have good resistance to corrosion, mechanical and thermal fatigue, mechanical and thermal shock, creep, and erosion at elevated temperatures.

Iron-base superalloys generally contain 32% to 67% iron, 15% to 22% chromium, and 9% to 38% nickel. Common alloys in this group are the *Incoloy* series. Cobalt-base superalloys generally contain 35% to 65% cobalt, 19% to 30% chromium, and up to 35% nickel. Cobalt (Co) is a white-colored metal that resembles nickel. These superalloys are not as strong as nickel-base superalloys, but they retain their strength at higher temperatures than do nickel-base superalloys. Nickel-base superalloys are the most common of the superalloys and are available in a wide variety of compositions (Table 3.12). The range of nickel is from 38% to 76%; they also contain up to 27% chromium and 20% cobalt. Common alloys in this group include the *Hastelloy, Inconel, Nimonic, René, Udimet, Astroloy,* and *Waspaloy* series.

3.11.6 Titanium and titanium alloys

Titanium (Ti) was discovered in 1791, but was not commercially produced until the 1950s. Although titanium is expensive, its high strength-to-weight ratio and its corrosion resistance at room and elevated temperatures make it attractive for applications such as aircraft, jet-engine, racing-car, chemical, petrochemical, and marine components; submarine hulls; and biomaterials such as orthopedic implants (Table 3.13). Titanium alloys have been developed for service at 550°C (1000°F) for long periods of time and at up to 750°C (1400°F) for shorter periods. Unalloyed titanium, known as commercially pure titanium, has excellent corrosion resistance for applications where strength considerations are secondary. Aluminum, vanadium, molybdenum, manganese, and other alloying elements are added to titanium alloys to impart properties such as improved workability, strength, and hardenability.

TABLE 3.13

Properties and Typical Applications of Wrought Titanium Alloys

			Room Temperature				Various Temperatures					
Nominal Composition (%)	UNS	Condition	Ultimate Tensile Strength (MPa)	Yield Strength (MPa)	Elongation (%)	Reduction of Area (%)	Temp (°C)	Ultimate Tensile Strength (MPa)	Yield Strength (MPa)	Elongation (%)	Reduction of Area (%)	Typical Applications
99.5 Ti	R50250	Annealed	330	240	30	55	300	150	95	32	80	Airframes; chemical, desalination, and marine parts; plate-type heat exchangers.
5 Al, 2.5 Sn	R54520	Annealed	860	810	16	40	300	565	450	18	45	Aircraft-engine compressor blades and ducting; steam-turbine blades.
6 Al, 4 V	R56400	Annealed	1000	925	14	30	300	725	650	14	35	Rocket motor cases; blades and disks for aircraft turbines and compressors; orthopedic implants; structural forgings; fasteners.
							425	670	570	18	40	
							550	530	430	35	50	
		Solution + age	1175	1100	10	20	300	980	900	10	28	
13 V, 11 Cr, 3 Al	R58010	Solution + age	1275	1210	8	–	425	1100	830	12	–	High-strength fasteners; aerospace components; honeycomb panels.

The properties and manufacturing characteristics of titanium alloys are extremely sensitive to small variations in both alloying and residual elements. Thus, control of composition and processing is important, including prevention of surface contamination by hydrogen, oxygen, or nitrogen during processing. These elements cause embrittlement of titanium, resulting in reduced toughness and ductility.

The body-centered cubic structure of titanium (beta-titanium, above 880°C[1600°F]) is ductile, whereas its hexagonal close-packed structure (alpha-titanium) is somewhat brittle and is very sensitive to stress corrosion. A variety of other structures (alpha, near alpha, alpha–beta, and beta) can be obtained by alloying and heat treating, such that the properties can be optimized for specific applications. **Titanium aluminide intermetallics** (TiAl and Ti_3Al) have higher stiffness and lower density and can withstand temperatures higher than conventional titanium alloys.

3.11.7 Refractory metals

Molybdenum, columbium, tungsten, and tantalum are *refractory metals*; they are called *refractory* because of their high melting point. Although refractory-metal elements were discovered about 200 years ago—and have been used as important alloying elements in steels and superalloys—their use as engineering metals and alloys did not begin until about the 1940s. More than most other metals and alloys, these metals maintain their strength at elevated temperatures; thus, they are of great importance and use in rocket engines, gas turbines, and various other aerospace applications; in the electronics, nuclear power, and chemical industries; and as tool and die materials. The temperature range for some of these applications is on the order of 1100°C to 2200°C (2000°F to 4000°F), where strength and oxidation are of major concern.

1. **Molybdenum.** *Molybdenum* (Mo), a silvery white metal, was discovered in the 18th century. It has a high melting point, a high modulus of elasticity, good resistance to thermal shock, and good electrical and thermal conductivity. Typical applications of molybdenum are in solid-propellent rockets, jet engines, honeycomb structures, electronic components, heating elements, and dies for die casting. Principal alloying elements for molybdenum are titanium and zirconium. Molybdenum is used in greater amounts than any other refractory metal. It is an important alloying element in cast and wrought alloys, such as steels and heat-resistant alloys, and imparts strength, toughness, and corrosion resistance. A major disadvantage of molybdenum alloys is their low resistance to oxidation at temperatures above 500°C (950°F), thus necessitating the use of protective coatings.
2. **Niobium.** *Niobium* (Nb), or *columbium* (after the mineral *columbite*), possesses good ductility and formability and has greater resistance to oxidation than do other refractory metals. With various alloying elements, niobium alloys can be produced with moderate strength and good fabrication characteristics. These alloys are used in rockets; missiles; and nuclear, chemical, and superconductor applications. Niobium, first identified in 1801, is also an alloying element in various alloys and superalloys.
3. **Tungsten.** *Tungsten* (W, from *wolframite*) was first identified in 1781 and is the most plentiful of all refractory metals. Tungsten has the highest melting point of any metal (3410°C [6170°F]), and thus it is characterized by high strength at elevated temperatures. On the other hand, it has high density, brittleness at low temperatures, and poor resistance to oxidation. Tungsten and its alloys are used for applications involving temperatures above 1650°C (3000°F), such as nozzle throat liners in missiles and in the hottest parts of jet and rocket

engines, circuit breakers, welding electrodes, and spark-plug electrodes. The filament wire in incandescent light bulbs is made of pure tungsten, using powder metallurgy and wire-drawing techniques. Because of its high density, tungsten is used in balancing weights and counterbalances in mechanical systems, including self-winding watches. Tungsten is an important element in tool and die steels, imparting strength and hardness at elevated temperatures. Tungsten carbide, with cobalt as a binder for the carbide particles, is one of the most important tool and die materials.

4. **Tantalum.** Identified in 1802, *tantalum* (Ta) is characterized by a high melting point (3000°C [5425°F]), good ductility, and good resistance to corrosion. However, it has high density and poor resistance to chemicals at temperatures above 150°C (300°F). Tantalum is also used as an alloying element. Tantalum is used extensively in electrolytic capacitors and various components in the electrical, electronic, and chemical industries, as well as for thermal applications, such as in furnaces and acid-resistant heat exchangers. A variety of tantalum-base alloys is available in many forms for use in missiles and aircraft.

3.11.8 Various other nonferrous metals

1. **Beryllium.** Steel gray in color, *beryllium* (Be) has a high strength-to-weight ratio. Unalloyed beryllium is used in nuclear and X-ray applications, because of its low neutron absorption characteristics, and in rocket nozzles, space and missile structures, aircraft disc brakes, and precision instruments and mirrors. Beryllium is also an alloying element, and its alloys of copper and nickel are used in applications such as springs (beryllium–copper), electrical contacts, and non-sparking tools for use in explosive environments, such as mines and in metal-powder production. Beryllium and its oxide are toxic and should not be inhaled.
2. **Zirconium.** *Zirconium* (Zr) is silvery in appearance, has good strength and ductility at elevated temperatures, and has good corrosion resistance because of an adherent oxide film. The element is used in electronic components and nuclear power reactor applications, because of its low neutron absorption characteristics.
3. **Low melting point metals.** The major metals in this category are lead, zinc, and tin and are described as follows:

   a. **Lead** *Lead* (Pb, after *plumbum*) has properties of high density, good resistance to corrosion (by virtue of the stable lead-oxide layer that forms and protects the surface), softness, low strength, high ductility, and good workability. Alloying with various elements, such as antimony and tin, enhances lead's properties, making it suitable for piping, collapsible tubing, bearing alloys, cable sheathing, roofing, and lead-acid storage batteries. Lead is also used for damping sound and vibrations, for radiation shielding against X rays, in printing (*type metals*), in weights, and in the chemical and paint industries. The oldest lead artifacts were made around 3000 BC. Lead pipes made by the Romans and installed in the Roman baths in Bath, England, two millennia ago are still in use. Lead is also an alloying element in solders, steels, and copper alloys and promotes corrosion resistance and machinability. Because of its toxicity, environmental contamination by lead is a significant concern.

   b. **Zinc** Industrially, *zinc* (Zn), which has a bluish–white color, is the fourth most utilized metal, after iron, aluminum, and copper. Although known about for many centuries, zinc was not studied and developed until the 18th century. It has two major uses: for galvanizing iron, steel sheet, and wire; and as an alloy base for casting. In **galvanizing**, zinc serves as the anode and protects the steel (cathode) from corrosive attack should the coating be scratched or punctured. Zinc is also used as an alloying element; brass, for example, is an alloy of copper and zinc.

Major alloying elements in zinc are aluminum, copper, and magnesium. They impart strength and provide dimensional control during casting of the metal. Zinc-base alloys are used extensively in die casting for making products such as fuel pumps and grills for automobiles, components for household appliances (such as vacuum cleaners, washing machines, and kitchen equipment), various other machine parts, and photoengraving plates. Another use for zinc is in superplastic alloys, which have good formability characteristics by virtue of their capacity to undergo large deformation without failure. Very fine-grained 78% Zn–22% Al sheet is a common example of a superplastic zinc alloy that can be formed by methods used for forming plastics or metals.

c. **Tin** Although used in small amounts, *tin* (Sn, after *stannum*) is an important metal. The most extensive use of tin, a silvery white, lustrous metal, is as a protective coating on steel sheet (**tin plate**), which is used in making containers (tin cans) for food and various other products. Inside a sealed can, the steel is cathodic and is protected by the tin (anode), so that the steel does not corrode. The low shear strength of the tin coatings on steel sheet also improves its performance in deep drawing and general pressworking operations.

Unalloyed tin is used in applications such as lining material for water distillation plants and as a molten layer of metal over which plate glass is made. Tin-base alloys (also called **white metals**) generally contain copper, antimony, and lead. The alloying elements impart hardness, strength, and corrosion resistance. Because of their low friction coefficients, which result from low shear strength and low adhesion, tin alloys are used as journal-bearing materials. These alloys are known as **babbitts** and contain tin, copper, and antimony. **Pewter** is an alloy of tin, copper, and antimony. It was developed in the 15th century and is used for tableware, hollowware, and decorative artifacts. Organ pipes are made of tin alloys. Tin is an alloying element for type metals; dental alloys; and bronze (copper–tin alloy), titanium, and zirconium alloys. Tin–lead alloys are common soldering materials, with a wide range of compositions and melting points. (See Section 12.12.3.)

4. **Precious metals.** Gold, silver, and platinum are the most important *precious* (costly) *metals* and are also called *noble metals*.

a. **Gold** (Au, after *aurum*) is soft and ductile and has good corrosion resistance at any temperature. Typical applications include jewelry, coinage, reflectors, gold leaf for decorative purposes, dental work, and electroplating, as well as such important applications as electric contact and terminals.

b. **Silver** (Ag, after *argentum*) is a ductile metal and has the highest electrical and thermal conductivity of any metal. It does, however, develop an oxide film that affects its surface properties and appearance. Typical applications for silver include tableware, jewelry, coinage, electroplating, and photographic film, as well as electrical contacts, solders, bearings, and food and chemical equipment. *Sterling silver* is an alloy of silver and 7.5% copper.

c. **Platinum** (Pt) is a soft, ductile, grayish–white metal that has good corrosion resistance, even at elevated temperatures. Platinum alloys are used as electrical contacts, spark-plug electrodes, catalysts for automobile pollution-control devices, filaments, nozzles, dies for extruding glass fibers, and thermocouples, and in the electrochemical industry; other applications include jewelry and dental work.

3.11.9 Special metals and alloys

1. **Shape-memory alloys.** Shape-memory alloys, after being plastically deformed at room temperature into various shapes, return to their original shapes upon heating. For example, a piece of straight wire made of this material can be wound into a helical spring; when heated with a match, the spring uncoils and returns to the original straight shape. A typical shape-memory alloy is 55% Ni and 45% Ti; other alloys under development are copper–aluminum–nickel, copper–zinc–aluminum, iron–manganese–silicon, and nickel–titanium alloys.

   The behavior of shape-memory alloys can be *reversible*, that is, the shape can switch back and forth repeatedly upon application and removal of heat. These alloys generally have good ductility and corrosion resistance and high electrical conductivity. Typical applications include temperature sensors, clamps, connectors, fasteners, and seals that are easy to install.

2. **Amorphous alloys.** *Amorphous alloys* are a class of metal alloys that, unlike ordinary metals, do not have a long-range crystalline structure. (See also Section 5.10.8.) These alloys have no grain boundaries, and the atoms are randomly and tightly packed. Because their structure resembles that of glasses (Section 11.10), these alloys are also called **metallic glasses**. Amorphous alloys typically contain iron, nickel, and chromium, alloyed with carbon, phosphorus, boron, aluminum, and silicon. They are available in the form of wire, ribbon, strip, and powder. These alloys exhibit excellent corrosion resistance, good ductility, high strength, and very low loss from magnetic hysteresis. The latter property is utilized in making magnetic steel cores for transformers, generators, motors, lamp ballasts, magnetic amplifiers, and linear accelerators, with greatly improved efficiency.

   The amorphous structure was first obtained in the late 1960s by extremely rapid cooling of the molten alloy. One such method is called *splat cooling* or *melt spinning* (see Fig. 5.35), in which the alloy is propelled at a very high speed against a rotating metal surface; since the rate of cooling is on the order of 10^6 K/s to 10^8 K/s the molten alloy does not have sufficient time to crystallize. If an amorphous alloy's temperature is raised and then cooled, the alloy develops a crystalline structure.

3. **Nanomaterials.** Important recent developments involve the production of materials with grains, fibers, films, and composites which have particles that are on the order of 1 nm to 100 nm in size. First investigated in the early 1980s and generally called *nanomaterials*, these materials have some properties that are often superior to those of traditional and commercially available materials. These characteristics include strength, hardness, ductility, wear resistance and corrosion resistance suitable for structural (load-bearing) and nonstructural applications, in combination with unique electrical, magnetic and optical properties. Current and potential applications for nanomaterials include cutting tools, metal powders, computer chips, flat-panel displays for laptops, sensors, and various electrical and magnetic components. (See Sections 8.6.10 and 11.8.1.)

   The composition of nanomaterials can be any combination of chemical elements; among the more important compositions are carbides, oxides, nitrides, metals and alloys, organic polymers, and various composites. Synthesis methods include inert-gas condensation, plasma synthesis, electrodeposition, sol-gel synthesis, and mechanical alloying or ball milling.

4. **Metal foams.** A recent development is the production of *metal foams*, usually of aluminum alloys, but also of titanium or tantalum, where the metal consists of only 5% to 20% of the structure's volume. Metal foams can be produced by blowing air into a molten metal and tapping the froth that forms at the surface, which solidifies into a foam. Alternative approaches to creating metal foams include chemical vapor deposition onto a polymer or carbon foam lattice, and doping molten or powder metals with titanium hydride (TiH_2), which then releases hydrogen gas at the elevated casting or sintering temperatures. Metal foams present unique combinations of strength- and stiffness-to-density ratios, but these ratios are not as high as that of the base metal itself. However, the foams are very lightweight, making them an attractive material for aerospace applications; other applications include filters, lightweight beams, and orthopedic implants.

SUMMARY

- Manufacturing properties of metals and alloys depend largely on their mechanical and physical properties. These properties, in turn, are governed mainly by their crystal structure, grain boundaries, grain size, texture, and various imperfections. (Sections 3.1, 3.2)
- Dislocations are responsible for the lower shear stress required to cause slip for plastic deformation. However, when dislocations become entangled with one another or are impeded by barriers such as grain boundaries, impurities, and inclusions, the shear stress required to cause slip increases; this phenomenon is called work hardening, or strain hardening. (Section 3.3)
- Grain size and the nature of grain boundaries significantly affect the properties of metals, including strength, ductility, and hardness, as well as increase the tendency for embrittlement with reduced toughness. Grain boundaries enhance strain hardening by interfering with the movement of dislocations. (Section 3.4)
- Plastic deformation at room temperature (cold working) of polycrystalline materials results in higher strength of the material as dislocations become entangled. The deformation also generally causes anisotropy, whereby the mechanical properties are different in different directions. (Section 3.5)
- The effects of cold working can be reversed by heating the metal within a specific temperature range and for a specific period of time, through the consecutive processes of recovery, recrystallization, and grain growth. (Section 3.6)
- Metals and alloys can be worked at room, warm, or high temperatures. Their overall behavior, force and energy requirements, and workability depend largely on whether the working temperature is below or above their recrystallization temperature. (Section 3.7)
- Failure and fracture of a workpiece when subjected to deformation in metalworking operations is an important consideration. Two types of fracture are ductile fracture and brittle fracture. Ductile fracture is characterized by plastic deformation preceding fracture and requires a considerable amount of energy. Because it is not preceded by plastic deformation, brittle fracture can be catastrophic and requires much less energy than does ductile fracture. Impurities and inclusions, as well as factors such as the environment, strain rate, and state of stress, can play a major role in the fracture behavior of metals and alloys. (Section 3.8)
- Physical and chemical properties of metals and alloys can significantly affect design considerations; service requirements; compatibility with other materials, including tools and dies; and the behavior of the metals and alloys during processing. (Section 3.9)

- A very wide variety of metals and alloys is now available, with a wide range of properties, such as strength, toughness, hardness, ductility, and resistance to high temperatures and oxidation. These materials can be classified generally as ferrous metals and alloys; nonferrous metals and alloys; superalloys; refractory metals and their alloys; and various other types, such as precious metals. More recently developed materials include amorphous alloys, shape-memory alloys, nanomaterials, and metal foams. (Sections 3.10, 3.11)

SUMMARY OF EQUATIONS

Theoretical shear strength of metals, $\tau_{max} = \frac{G}{2}\pi$

Theoretical tensile strength of metals, $\sigma_{max} = \sqrt{\frac{E\gamma}{a}} \simeq \frac{E}{10}$

Hall–Petch equation, $Y = Y_i + kd^{-1/2}$

ASTM grain-size number, $N = 2^{n-1}$

Tensile strength vs. crack length, $\sigma \propto \frac{1}{\sqrt{\text{crack length}}}$

BIBLIOGRAPHY

Aerospace Structural Metals Handbook, 5 vols., Metals and Ceramics Information Center, Battelle, 1987.

Ashby, M. F., and D. R. H. Jones, *Engineering Materials*, Vol. 1, "*An Introduction to Their Properties and Applications*," 2d. ed., Pergamon, 1996; Vol. 2, "*An Introduction to Microstructures, Processing and Design*," Pergamon, 1986; Vol. 3, "*Materials Failure Analysis: Case Studies and Design Implications*," Pergamon, 1993.

Ashby, M. F., *Materials Selection in Mechanical Design*, 2d. ed., Pergamon, 1999.

Askeland, D. R., *The Science and Engineering of Materials*, 5th ed., PWS-KENT, 1994.

ASM Handbook, various volumes, ASM International.

ASM Specialty Handbooks, various volumes. ASM International.

Boyer, H. E. (ed.), *Selection of Materials for Component Design: Source Book*, American Society for Metals, 1986.

Budinski, K. G., *Engineering Materials: Properties and Selection*, 5th ed. Prentice Hall, 1996.

Callister, W. D., Jr., *Materials Science and Engineering*, 5th ed., Wiley, 2000.

Dieter, G. E., *Engineering Design: A Materials and Processing Approach*, 3d. ed., McGraw-Hill, 1999.

———, *Mechanical Metallurgy*, 3d. ed., McGraw-Hill, 1986.

Farag, M. M., *Materials Selection for Engineering Design*. Prentice Hall, 1997.

Flinn, R. A., and P. K. Trojan, *Engineering Materials and Their Applications*, 4th ed., Houghton Mifflin, 1990.

Harper, C. (ed.), *Handbook of Materials for Product Design*, 3d. ed., McGraw-Hill, 2001.

Hertzberg, R. W., *Deformation and Fracture Mechanics of Engineering Materials*, 4th ed. Wiley, 1996.

Mangonon, P. C., *The Principles of Material Selection for Engineering Design*. Prentice Hall, 1999.

Material Selector, annual publication of *Materials Engineering Magazine*, Penton/IPC.

Meyers, M. A., and K. K. Chawla, *Mechanical Metallurgy: Principles and Applications*. Prentice-Hall, 1984.

Pollock, D. D., *Physical Properties of Materials for Engineers*, 2d. ed., CRC Press, 1993.

Roberge, P. R., and M. Tullmin, *Handbook of Corrosion Engineering*, McGraw-Hill, 1999.

Roberts, G. A., G. Krauss, R. Kennedy, and R. A. Cary, *Tool Steels*, 5th ed., ASM International, 1998.

Schaffer, J., A. Saxena, S. Antalovich, T. Sanders, and S. Warner, *The Science & Design of Engineering Materials*, 2d. ed., McGraw-Hill, 1999.

Shackelford, J. F., *Introduction to Materials Science for Engineers*, 5th ed., Macmillan, 2000.

Smith, W. F., *Principles of Materials Science and Engineering*, 3d. ed., McGraw-Hill, 1995.

Tool and Manufacturing Engineers Handbook, 4th ed., Vol. 3, "*Materials, Finishing and Coating*," Society of Manufacturing Engineers, 1985.

Woldman's Engineering Alloys, 9th ed., ASM International, 2000.

Wroblewski, A. J., and S. Vanka, *MaterialTool: A Selection Guide of Materials and Processes for Designers*. Prentice Hall, 1997.

Wulpi, D. J., *Understanding How Components Fail*, 2d. ed., ASM International, 1999.

QUESTIONS

3.1 What is the difference between a unit cell and a single crystal?

3.2 Explain why we should study the crystal structure of metals.

3.3 What effects does recrystallization have on the properties of metals?

3.4 What is the significance of a slip system?

3.5 Explain what is meant by structure-sensitive and structure-insensitive properties of metals.

3.6 What is the relationship between nucleation rate and the number of grains per unit volume of a metal?

3.7 Explain the difference between recovery and recrystallization.

3.8 (a) Is it possible for two pieces of the same metal to have different recrystallization temperatures? Explain. (b) Is it possible for recrystallization to take place in some regions of a workpiece before other regions in the same workpiece? Explain.

3.9 Make a list of various phenomena that result from the presence of grain boundaries in metals.

3.10 Describe why different crystal structures exhibit different strengths and ductilities.

3.11 Explain the difference between preferred orientation and mechanical fibering.

3.12 Name some analogies to mechanical fibering (e.g., layers of thin dough sprinkled with flour).

3.13 A cold-worked piece of metal has been recrystallized. When tested, it is found to be anisotropic. Explain the probable reason for its anisotropy.

3.14 Does recrystallization eliminate mechanical fibering in a workpiece? Explain.

3.15 Explain why we may have to be concerned with the orange-peel effect on metal surfaces.

3.16 Do you think it might be important to know whether a raw material for a manufacturing process has anisotropic properties? What about anisotropy in the finished product? Explain, using appropriate examples.

3.17 How can you tell the difference between two parts made of the same metal, one shaped by cold working and the other by hot working? Explain the differences you might observe. Note that there are several methods that can be used to determine the differences between the two parts.

3.18 Explain why the strength of a polycrystalline metal at room temperature decreases as its grain size increases.

3.19 Describe the technique by which you would reduce the orange-peel effect.

3.20 What is the significance of metals such as lead and tin having recrystallization temperatures at about room temperature?

3.21 You are given a deck of playing cards held together a rubber band. Which of the material-behavior phenomena described in this chapter could you demonstrate? What are the effects of increasing the number of rubber bands holding the cards together? Explain. (*Hint*: Inspect Figs. 3.4 and 3.6.)

3.22 Using the information given in Chapters 2 and 3, describe the conditions that induce brittle fracture in an otherwise ductile piece of metal.

3.23 A part fractures while being subjected to plastic deformation at an elevated temperature. List the factors that could have played a role in the fracture.

3.24 Make a list of metals that would be suitable for a (1) paper clip, (2) buckle for a safety belt, (3) nail, (4) battery cable, and (5) gas-turbine blade.

3.25 Explain the advantages and limitations of cold, warm, and hot working of metals, respectively.

3.26 Describe the significance of structures and machine components made of two materials with different coefficients of thermal expansion.

3.27 Explain why parts may crack when suddenly subjected to extremes of temperature.

3.28 From your own experience and observations, list three applications each for the following metals and their alloys: (1) steel, (2) aluminum, (3) copper, (4) lead, and (5) gold.

3.29 Name products that would not have been developed to their advanced stages as we find them today if alloys with high strength and corrosion and creep resistance at elevated temperatures had not been developed.

3.30 Inspect several metal products and components and make an educated guess as to what materials they are made from. Give reasons for your guess. If you list two or more possibilities, explain your reasoning.

3.31 List three engineering applications for which the following physical properties would be desirable: (1) high density, (2) low melting point, and (3) high thermal conductivity.

3.32 Two physical properties that have a major influence on the cracking of workpieces or dies during thermal cycling are thermal conductivity and thermal expansion. Explain why.

3.33 What might be the adverse effects of using a material with low specific heat in manufacturing?

3.34 Would a solid-metal shell with a metal-foam interior be useful in applications such as a piston in an internal combustion engine? Explain your answer.

3.35 Describe the advantages of nanomaterials over traditional materials.

3.36 Aluminum has been cited as a possible substitute material for steel in automobiles. What concerns, if any, would you have before purchasing an aluminum automobile?

3.37 Lead shot is popular among sportsmen for hunting, but birds commonly ingest the pellets (along with gravel) to help digest food. What substitute materials would you recommend for lead, and why?

PROBLEMS

3.38 The unit cells shown in Figs. 3.2a–c can be represented by hard-ball or hard-sphere models, similar to tennis balls arranged in various configurations in a box. In such an arrangement, the *atomic packing factor* (APF) is defined as the ratio of the volume of atoms to the volume of the unit cell. Show that the APF is 0.68 for the bcc structure and 0.74 for the fcc structure.

3.39 Show that the lattice constant a in Fig. 3.2b is related to the atomic radius by expression

$$a = 2\sqrt{2}R,$$

where R is the radius of the atom as depicted by the hard-ball model.

3.40 Show that for the fcc unit cell, the radius r of the largest hole is given by the expression

$$r = 0.414R.$$

Determine the size of the largest hole for the iron atom in the fcc structure.

3.41 Calculate the theoretical (1) shear strength and (2) tensile strength for aluminum, plain-carbon steel, and tungsten. Estimate the ratios of their theoretical strength to actual strength.

3.42 A technician determines that the grain size of a certain etched specimen is 4. Upon further checking, it is found that the magnification used was 150, instead of 100, the latter of which is required by ASTM standards. What is the correct grain size?

3.43 Estimate the number of grains in a regular paper clip if its ASTM grain size is 10.

3.44 Plot the following for the materials described in this chapter: (a) yield strength vs. density, and (b) elastic modulus vs. density. Explain your observations.

3.45 Determine the maximum crack velocity in aluminum, plain–carbon steel, and tungsten. (See Section 3.8.2 and Tables 2.1 and 3.3.)

3.46 Calculate the maximum theoretical shear stress τ_{max}, assuming that in Fig. 3.5 the shear stress τ varies linearly with x.

3.47 The natural frequency f of a cantilever beam is given by the expression

$$f = 0.56\sqrt{\frac{EIg}{wL^4}},$$

where E is the modulus of elasticity, I is the moment of inertia, g is the gravitational constant, w is the weight of the beam per unit length, and L is the length of the beam. How does the natural frequency of the beam change, if at all, as its temperature is increased?

3.48 A strip of metal is reduced in thickness by cold working from 25 mm to 15 mm. A similar strip is reduced from 25 mm to 10 mm. Which one of these strips will recrystallize at a lower temperature? Why?

3.49 Assume that you are asked to give a quiz to students on the contents of this chapter. Prepare three quantitative problems and three qualitative questions, and supply the answers.

3.50 A 1-m long, simply supported beam with a round cross-section is subjected to a load of 25 kg at its center. (a) If the shaft is made from AISI 303 steel and has a diameter of 20 mm, what is the deflection under the load? (b) For shafts made from 2024-T4 aluminum, architectural bronze, and 99.5% titanium, respectively, what must the diameter of the shaft be for the shaft to have the same deflection as in part (a)?

CHAPTER

4 Surfaces, Tribology, Dimensional Characteristics, Inspection, and Product Quality Assurance

4.1 Introduction

Because of the various mechanical, physical, thermal, and chemical effects induced by its processing history, the **surface** of a manufactured part generally has properties and behavior that are considerably different from those of its bulk. Although the **bulk material** generally determines a component's overall mechanical properties, the component's surface directly influences several important properties and characteristics of the manufactured part:

- Friction and wear properties of the part during subsequent processing when it comes in direct contact with tools and dies, or when it is placed in service;
- Effectiveness of lubricants during manufacturing processes, as well as throughout the part's service life;
- Appearance and geometric features of the part and their role in subsequent operations such as painting, coating, welding, soldering, and adhesive bonding, as well as corrosion resistance of the part;
- Initiation of cracks due to surface defects, such as roughness, scratches, seams, and heat-affected zones, which could lead to weakening and premature failure of the part by fatigue or other fracture mechanisms;
- Thermal and electrical conductivity of contacting bodies; for example, rough surfaces have higher thermal and electrical resistance than smooth surfaces.

Friction, wear, and lubrication (**tribology**) are surface phenomena. Friction influences forces, power requirements, and surface quality of parts. Wear alters the surface geometry of tools and dies, which, in turn, adversely affects the quality of manufactured products and the economics of production. Lubrication, with few exceptions, is an integral aspect of all manufacturing operations, as well as in the proper functioning of machinery and equipment. We describe those aspects of lubrication that are relevant to manufacturing operations, such as the types of wear encountered, how to reduce wear, and the proper selection and application of lubricants and coolants.

Another important aspect in surface technology is **surface treatments**, which modify the properties and characteristics of surfaces. Several mechanical, thermal, electrical, and chemical methods can be used to modify surfaces of parts to improve frictional behavior, effectiveness of lubricants, resistance to wear and corrosion, and surface finish and appearance.

Measurement of the relevant dimensions and features of parts is an integral aspect of interchangeable manufacture, the basic concept of standardization and mass production. In this chapter, we describe the principles involved and the various instruments used in measurement. Another important aspect is **testing and inspection** of manufactured products, using both destructive and nondestructive testing methods. One of the most critical aspects of manufacturing is **product quality**. This chapter emphasizes the technological and economic importance of *building quality into a product*, rather than inspecting the product after it is made, and it describes the statistical techniques and control charts in achieving these goals.

4.2 Surface Structure and Properties

Upon close examination of the surface of a piece of metal, we find that it generally consists of several layers (Fig. 4.1). The part of the metal starting at the interior and moving outward to the surface is called the *bulk metal*, or the **metal substrate**; its structure depends on the composition and processing history of the metal. Above this bulk metal is a layer that usually has been plastically deformed and work hardened during the manufacturing process.

The depth and properties of the work-hardened layer (the **surface structure**) depend on factors such as the processing method used and the extent of frictional sliding to which the surface was subjected. For example, sharp tools and selection of proper process parameters produce surfaces with little or no disturbance. If the surface is produced by machining with a dull tool or under poor cutting conditions, or is ground with a dull grinding wheel, this layer will be relatively thick. Also, nonuniform surface deformation or severe temperature gradients during manufacturing operations usually causes *residual stresses* in this work-hardened layer. A **Beilby, or amorphous, layer** may or may not be present on top of the work-hardened layer; it consists of a structure that is microcrystalline or amorphous. This layer develops in some machining and finishing operations where melting and surface flow, followed by rapid quenching, are encountered. Unless the metal is processed and kept in an inert (oxygen-free) environment, or it is a noble metal, such as gold or platinum, an **oxide layer** usually lies on top of the work-hardened or Beilby layer. Some examples of metals with an oxide layer include the following:

1. *Iron* has an oxide structure, with FeO adjacent to the bulk metal, followed by a layer of Fe_3O_4, and then a layer of Fe_2O_3, which is exposed to the environment.

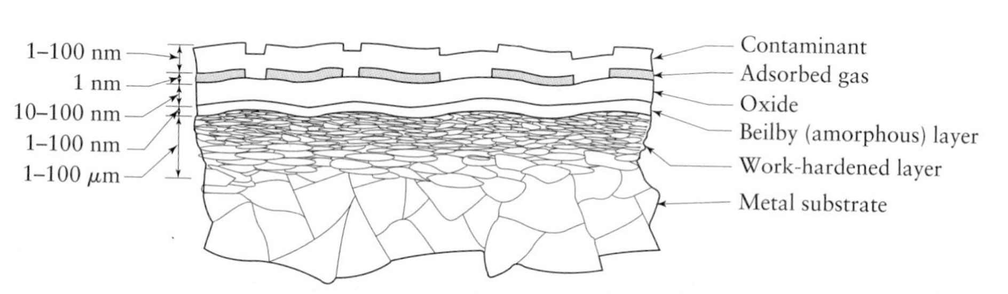

FIGURE 4.1 Schematic illustration of the cross-section of the surface structure of metals. The thickness of the individual layers depends on processing conditions and the environment. *Source*: After E. Rabinowicz and B. Bhushan.

2. *Aluminum* has a dense *amorphous* (i.e., without crystalline structure) layer of Al_2O_3, with a thick, porous hydrated aluminum-oxide layer over it.
3. *Copper* has a bright shiny surface when freshly scratched or machined. Soon after, however, it develops a Cu_2O layer, which is then covered with a layer of CuO; these layers give copper its somewhat dull color, such as seen in kitchen utensils.
4. *Stainless steels* are "stainless" because they develop a protective layer of chromium oxide, CrO (**passivation**).

Under normal environmental conditions, surface oxide layers are generally covered with *adsorbed* layers of gas and moisture. Finally, the outermost surface of the metal may be covered with contaminants, such as dirt, dust, grease, lubricant residues, cleaning-compound residues, and pollutants from the environment. Thus, surfaces generally have properties that are very different from those of the substrate. The oxide on a metal's surface, for example, is generally much harder than the base metal; hence, oxides tend to be brittle and abrasive. This surface characteristic has, in turn, several important effects on friction, wear, and lubrication in materials processing and subsequent coatings on products. The factors involved in the surface structure of metals are also relevant, to a great extent, to the surface structure of plastics and ceramics. The surface texture of these materials depends, as with metals, on the method of production; environmental conditions also influence the surface characteristics of these materials.

Surface integrity. *Surface integrity* describes not only the topological (geometric) aspects of surfaces, but also their mechanical and metallurgical properties and characteristics. Surface integrity is an important consideration in manufacturing operations, because it influences the properties of the product, such as its fatigue strength, resistance to corrosion, and service life.

Several **defects** caused by and produced during manufacturing can be responsible for lack of surface integrity. These defects are usually caused by a combination of factors, such as defects in the original material; the method by which the surface is produced; and lack of proper control of process parameters, which can result in excessive stresses and temperatures. The major surface defects found in practice usually consist of one or more of the following: *cracks*, *craters*, *folds*, *laps*, *seams*, *splatter*, *inclusions*, *intergranular attack*, *heat-affected zones*, *metallurgical transformations*, *plastic deformation*, and *residual stresses*.

4.3 Surface Texture

Regardless of the method of production, all surfaces have their own set of characteristics, which is referred to as **surface texture**. The description of surface texture as a geometrical property is complex; however, certain guidelines have been established for identifying surface texture in terms of well-defined and measurable quantities (Fig. 4.2):

1. **Flaws**, or **defects**, are random irregularities, such as scratches, cracks, holes, depressions, seams, tears, and inclusions.
2. **Lay**, or **directionality**, is the direction of the predominant surface pattern and is usually visible to the naked eye.
3. **Waviness** is a recurrent deviation from a flat surface, much like waves on the surface of water. It is measured and described in terms of (a) the space between

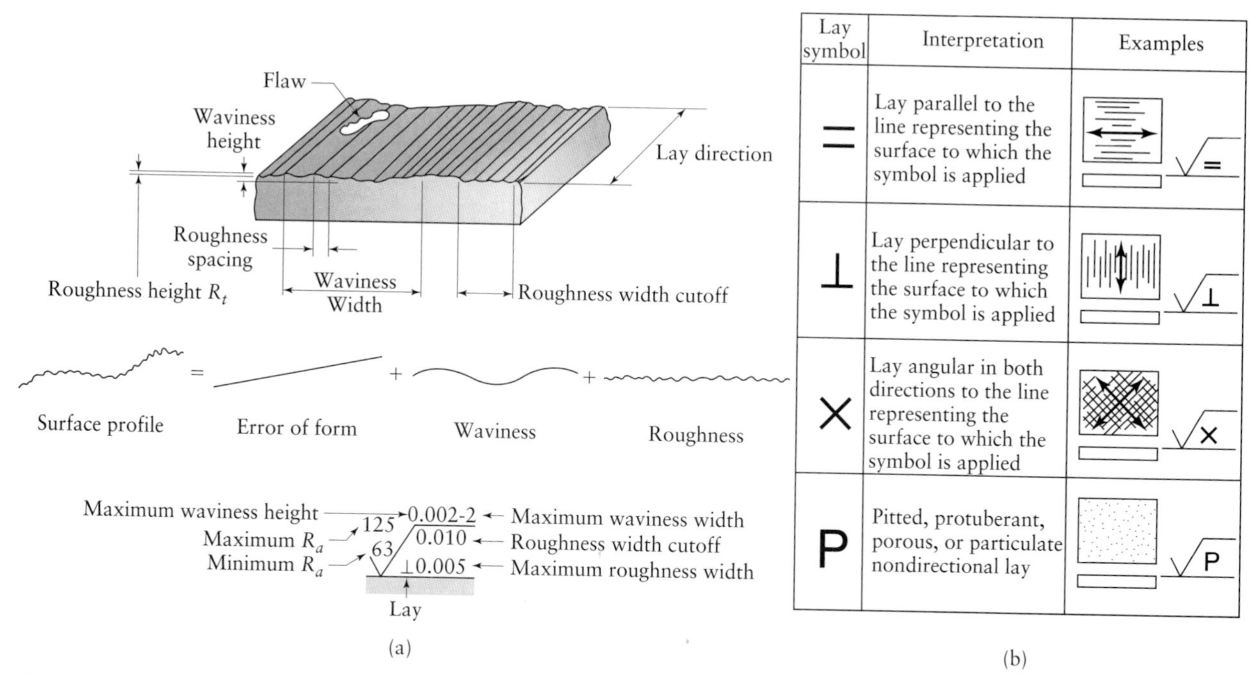

Lay symbol	Interpretation	Examples
=	Lay parallel to the line representing the surface to which the symbol is applied	
⊥	Lay perpendicular to the line representing the surface to which the symbol is applied	
X	Lay angular in both directions to the line representing the surface to which the symbol is applied	
P	Pitted, protuberant, porous, or particulate nondirectional lay	

FIGURE 4.2 (a) Standard terminology and symbols used to describe surface finish. The quantities are given in μ in. (b) Common surface-lay symbols.

adjacent crests of the waves (*waviness width*) and (b) the height between the crests and valleys of the waves (*waviness height*). Waviness may be caused by deflections of tools, dies, and the workpiece; warping from forces or temperature; uneven lubrication; and vibration or any periodic mechanical or thermal variations in the system during the manufacturing process.

4. **Roughness** consists of closely spaced, irregular deviations on a scale smaller than that for waviness; roughness may be superimposed on waviness. Roughness is expressed in terms of its height, its width, and the distance on the surface along which it is measured.

Surface roughness. Surface roughness is generally described by two methods: arithmetic mean value and root-mean-square average. The **arithmetic mean value**, R_a, formerly identified as AA (for *arithmetic average*) or CLA (for *center-line average*), is based on the schematic illustration of a rough surface, as shown in Fig. 4.3. The arithmetic mean value is defined as

$$R_a = \frac{y_a + y_b + y_c + \cdots + y_n}{n} = \frac{1}{n}\sum_{i=1}^{n} y_i = \frac{1}{\ell}\int_0^l |y|\, dx, \tag{4.1}$$

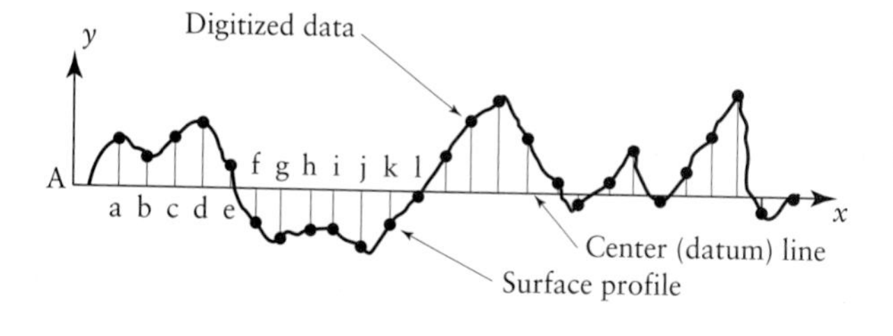

FIGURE 4.3 Coordinates used for measurement of surface roughness, using Eqs. (4.1) and (4.2).

where all ordinates $y_a, y_b, y_c, \ldots,$ are absolute values. The last term in Eq. (4.1) refers to the R_a of a continuous surface or wave, as is commonly encountered in signal processing. ℓ refers to the total profile length measured.

The **root-mean-square average,** R_q, formerly identified as RMS, is defined as

$$R_q = \sqrt{\frac{y_a^2 + y_b^2 + y_c^2 + \cdots + y_n^2}{n}} = \sqrt{\frac{1}{n}\sum_1^n y_i^2} = \left[\frac{1}{\ell}\int_0^l y^2\,dx\right]^{1/2}. \quad (4.2)$$

The center (*datum*) line in Fig. 4.3 is located so that the sum of the areas above the line is equal to the sum of the areas below the line. The units generally used for surface roughness are μm (micrometer, or micron) or μin. (microinch), where 1 μm = 40 μin. and 1 μin. = 0.025 μm. In addition, the **maximum roughness height,** R_t, may also be used as a measure of roughness. It is defined as the height from the deepest trough to the highest peak. It indicates the amount of material that has to be removed in order to obtain a smooth surface by polishing or other means.

Because of its simplicity, the arithmetic mean value was adopted internationally in the mid-1950s and is widely used in engineering practice. Equations (4.1) and (4.2) show that there is a relationship between R_q and R_a. For a surface roughness in the shape of a sine curve, R_q is larger than R_a by a factor of 1.11. This factor is 1.1 for most machining processes by cutting, 1.2 for grinding, and 1.4 for lapping and honing. R_q is more sensitive to the largest asperity peaks and deepest valleys, and since these peaks and valleys are important for friction and lubrication, R_q is often used even though it is slightly more complicated to calculate than R_a.

In general, a surface cannot be adequately described by its R_a or R_q value alone, since these values are averages. Two surfaces may have the same roughness value, but their actual topography may be quite different. A few deep troughs, for example, affect the roughness values insignificantly. However, such differences in the surface profile can be significant in terms of fatigue, friction, and the wear characteristics of a manufactured product.

Symbols for surface roughness. Acceptable limits for surface roughness are specified on technical drawings by the symbols shown around the check mark in the lower portion of Fig. 4.2, and their values are placed to the left of the check mark. Symbols used to describe a surface specify only the roughness, waviness, and lay; they do not include flaws. Whenever flaws are important, a special note is included in technical drawings to describe the method to be used to inspect for surface flaws.

Measuring surface roughness. Various commercially available instruments, called **surface profilometers,** are used to measure and record surface roughness. The most commonly used instruments feature a diamond stylus traveling along a straight line over the surface (Figs. 4.4a and b). The distance that the stylus travels, which can be varied, is called the **cutoff.** (See Fig. 4.2.) To highlight the roughness, profilometer traces are recorded on an exaggerated vertical scale (a few orders of magnitude greater than the horizontal scale; Figs. 4.4c–f), called **gain** on the recording instrument. Thus, the recorded profile is significantly distorted, and the surface appears to be much rougher than it actually is. The recording instrument compensates for any surface waviness and indicates only roughness. A record of the surface profile is made by mechanical and electronic means.

Surface roughness can be observed directly through interferometry, optical or scanning-electron microscopy, or atomic-force microscopy. Scanning-electron and atomic-force microscopy are especially useful for imaging very smooth surfaces for which features cannot be captured by the less sensitive instruments. Stereoscopic photographs are particularly useful for three-dimensional views of surfaces and can also be used to measure surface roughness.

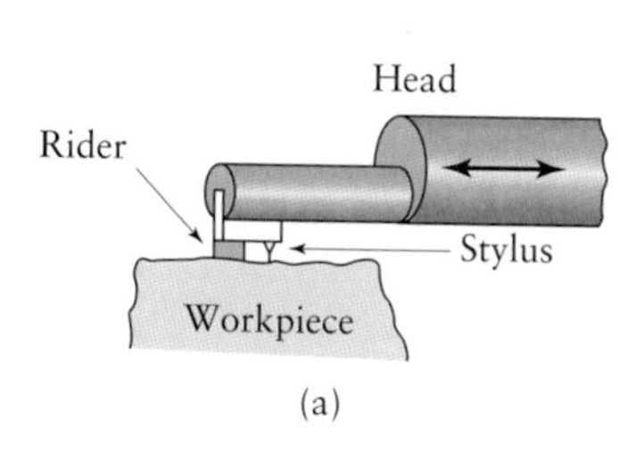

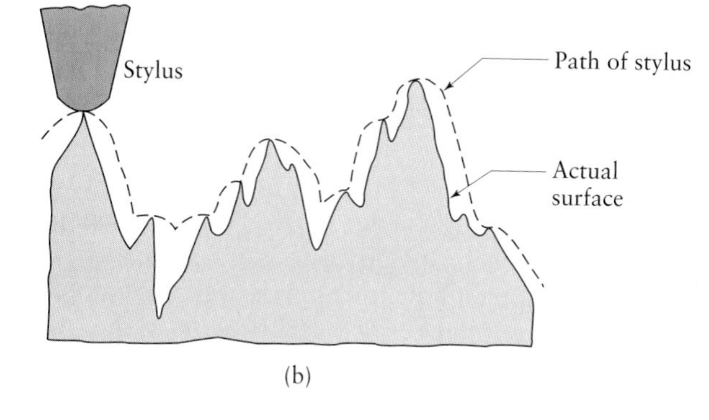

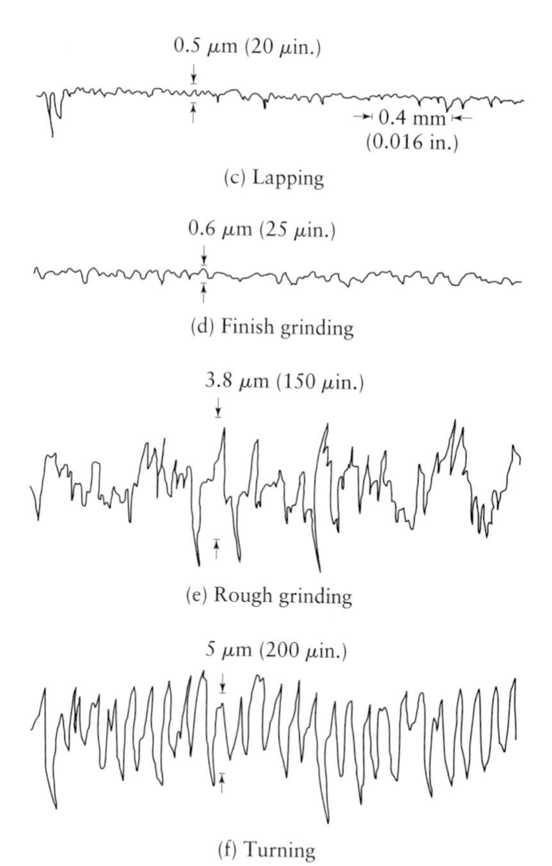

FIGURE 4.4 (a) Measuring surface roughness with a stylus. The rider supports the stylus and guards against damage. (b) Path of the stylus in measurements of surface roughness (broken line) compared with the actual roughness profile. Note that the profile of the stylus' path is smoother than the actual surface profile. *Source*: D. H. Buckley. Typical surface profiles produced by (c) lapping, (d) finish grinding, (e) rough grinding, and (f) turning processes. Note the difference between the vertical and horizontal scales. *Source*: from D. B. Dallas (ed.), *Tool and Manufacturing Engineers Handbook*, 3d ed.

Surface roughness in practice. Surface-roughness design requirements for typical engineering applications can vary by as much as two orders of magnitude for different parts. The reasons and considerations for this wide range include the following:

1. *Precision required on mating surfaces*, such as seals, fittings, gaskets, tools, and dies (for example, ball bearings and gages require very smooth surfaces, whereas surfaces for gaskets and brake drums can be quite rough);
2. *Tribological considerations*, that is, the effect of roughness on friction, wear, and lubrication;
3. *Fatigue and notch sensitivity*, because rougher surfaces often have shorter fatigue lives;
4. *Electrical and thermal contact resistance*, because the rougher the surface is, the higher the resistance will be;
5. *Corrosion resistance*, because the rougher the surface is, the greater will be the possibility that there are entrapped corrosive media;
6. *Subsequent processing*, such as painting and coating, in which a certain degree of roughness can result in better bonding;

7. *Appearance*, because, depending on the application, a rough or smooth surface may be preferred;
8. *Cost considerations*, because the finer the finish, the higher is the cost.

These factors should be carefully considered before a decision is made as to the surface-roughness recommendation for a given product. As in all manufacturing processes, the cost involved in the selection should also be a major consideration.

4.4 Tribology: Friction, Wear, and Lubrication

Tribology is the science and technology of interacting surfaces; thus, it involves friction, wear, and lubrication. All these factors can significantly influence the technology and economics of manufacturing processes and operations.

4.4.1 Friction

Friction is defined as the resistance to relative sliding between two bodies in contact under a normal load. Metalworking and manufacturing processes are significantly affected by friction, because of the relative motion and the forces present between tools, dies, and workpieces. Friction is an energy-dissipating process that results in the generation of heat. The subsequent rise in temperature can have a major effect on the overall operation. Furthermore, because it impedes movement at the tool, die, and workpiece interfaces, friction significantly affects the flow and deformation of materials in metalworking processes.

However, friction is not always undesirable; the presence of friction is necessary for the success or optimization of many operations, the best example of which is the rolling of metals to make sheet and plate. Without friction, it would be impossible to roll materials, just as it would be impossible to drive a car on the road. Various theories have been proposed to explain the phenomenon of friction. An early Coulomb model states that friction results from the mechanical interlocking of rough surface asperities, thereby requiring some force to slide the two bodies along each other.

For a particular theory of friction to be valid, it must explain the frictional behavior of two bodies under different conditions, including load, relative sliding speed, temperature, surface condition, environment, etc. Many models have been proposed, with varying degrees of success. The most commonly used theory of friction is based on adhesion, because of its reasonable agreement with experimental observations.

Adhesion theory of friction. The *adhesion theory of friction* is based on the observation that two clean, dry (unlubricated) metal surfaces, regardless of how smooth they are, contact each other at only a fraction of their apparent area of contact (Fig. 4.5). The static load at the interface is thus supported by the contacting asperities. The sum of the contacting areas is known as the **real area of contact, A_r**. For most engi-

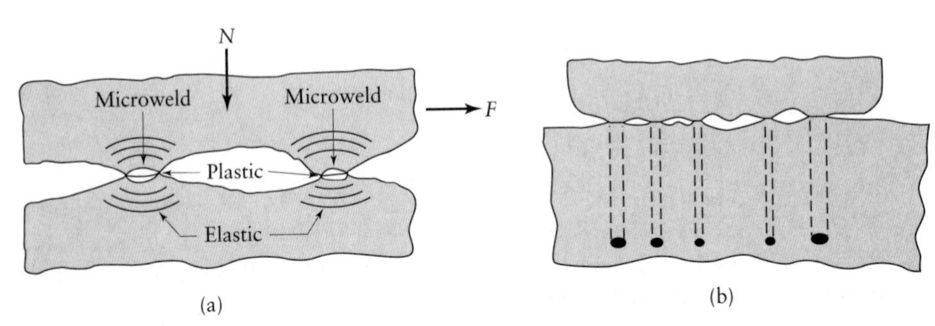

FIGURE 4.5 (a) Schematic illustration of the interface of two contacting surfaces, showing the real areas of contact. (b) Sketch illustrating the proportion of the apparent area to the real area of contact. The ratio of the areas can be as high as four to five orders of magnitude.

neering surfaces, the average angle that the asperity hill makes with a horizontal line is found to be between 5° and 15°. (Illustrations of surface roughness are highly exaggerated, because of the scales involved.)

Under light loads and with a large real area of contact, the normal stress at the asperity contacts is *elastic*. As the load increases, the stresses increase, and eventually, *plastic* deformation takes place at the **junctions**. Furthermore, with increasing load, the asperities in contact increase in area, and new junctions are formed with other asperities between the two surfaces in contact with each other. Because of the random asperity height distribution, some junctions are in elastic contact, while others are in plastic contact. The intimate contact of asperities creates an **adhesive bond.**

The nature of the adhesive bond is complex, involving atomic interactions, and may include mutual solubility and diffusion. The strength of the bond depends on the physical and mechanical properties of the metals in contact, the temperature, and the nature and thickness of any oxide film or other contaminants present on the surfaces. In many manufacturing processes (see Chapters 6 through 9), the load at the interface is generally high so that the normal stress on the asperity reaches the yield stress. Plastic deformation of the asperities then takes place, causing adhesion of the junctions. In other words, the asperities form **microwelds.** The cleaner the interface, the stronger are the adhesive bonds.

Coefficient of friction. Sliding between two bodies under a normal load, N, is possible only by the application of a tangential force F. According to the adhesion theory, F is the force required to *shear* the junctions (friction force). The *coefficient of friction*, μ, at the interface is defined as

$$\mu = \frac{F}{N} = \frac{\tau A_r}{\sigma A_r} = \frac{\tau}{\sigma}, \tag{4.3}$$

where τ is the shear strength of the junction and σ is the normal stress, which, for a plastically deformed asperity, is its yield stress. Thus, the numerator in Eq. (4.3) is a *surface* property, whereas the denominator is a *bulk* property. Because an asperity is surrounded by a large mass of material, the normal stress on an asperity is equivalent to the hardness of the material. (See Section 2.6.) The coefficient of friction can now be defined as

$$\mu = \frac{\tau}{\text{Hardness}}. \tag{4.4}$$

The nature and strength of the interface is the most significant factor in friction. A strong interface requires a high friction force for relative sliding, and a weak interface requires a low friction force. Equation (4.4) indicates that the coefficient of friction can be reduced not only by decreasing the numerator, but also by increasing the denominator. This observation suggests that placing thin films of low shear strength over a substrate with high hardness is the ideal method for reducing abrasive friction; in fact, this result is exactly what is achieved with a lubricant layer.

Two additional phenomena take place in asperity interactions under plastic stresses. For a strain-hardening material, the peak of the asperity is stronger than the bulk material, because of plastic deformation at the junction. Thus, under ideal conditions, the breaking of the bonds under a tensile load will likely follow a path below or above the geometric interface. Secondly, a tangential motion at the junction under load will cause *junction growth*; that is, the contact area increases. The reason is that, according to the yield criteria described in Section 2.11, the effective yield stress of the material decreases when the material is subjected to a shear stress. Hence, the junction has to grow in area to support the normal load. In other words, the two surfaces tend to approach each other under a tangential force.

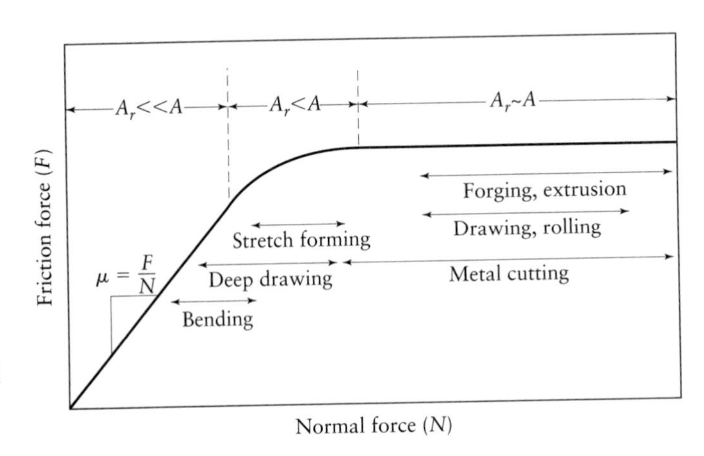

FIGURE 4.6 Schematic illustration of the relation between friction force F and normal force N. Note that as the real area of contact approaches the apparent area, the friction force reaches a maximum and stabilizes. Most machine components operate in the first region. The second and third regions are encountered in metalworking operations, because of the high contact pressures involved between sliding surfaces, i.e., die and workpiece.

If the normal load is increased further, the real area of contact, A_r, ideally continues to increase, especially with soft and ductile metals. In the absence of contaminants or fluids trapped at the interface, the real area of contact eventually reaches the apparent area of contact. This area is the maximum contact area that can be obtained. Because the shear strength at the interface is constant, the friction force (shearing force) reaches a maximum and levels off (Fig. 4.6).

This condition, known as **sticking**, creates difficulty in defining friction. Sticking in a sliding interface does not necessarily mean complete adhesion at the interface, as in welding. Rather, it means that the frictional stress at the surface has reached the shear yield stress, k, of the material. When two clean surfaces are pressed together with sufficiently high forces, *cold pressure welding* can take place, but usually the frictional stress is limited by contamination or by the presence of oxide layers. As the normal load, N, is increased further, the friction force, F, remains constant, and hence, by definition, the coefficient of friction decreases. This situation, which is an anomaly, indicates that there might be a different, and more realistic, means of expressing the frictional condition at an interface.

A more recent trend is to define a **friction factor**, or **shear factor**, m, as

$$m = \frac{\tau_i}{k}, \tag{4.5}$$

where τ_i is the shear strength of the interface and k is the *shear yield stress* of the softer material in a sliding pair. (The quantity k is equal to $Y/2$ according to Eq. (2.35) and to $Y/\sqrt{3}$ according to the distortion-energy criterion, where Y is the uniaxial yield stress of the material.) Equation (4.5) is often referred to as a **Tresca** friction model.

The value of τ_i is difficult to measure, because we have to know the real area of contact; average values will not be meaningful. However, note that when $m = 0$ in this definition, there is no friction, and when $m = 1$, complete sticking takes place at the interface. The magnitude of m is independent of the normal force or stress; the reason is that the shear yield stress of a thin layer of material is unaffected by the magnitude of the normal stress.

The sensitivity of friction to different factors can be illustrated by the following examples: A typical value for the coefficient of friction of steel sliding on lead is 1.0; for steel on copper, it is 0.9. However, for steel sliding on copper coated with a thin layer of lead, it is 0.2. The coefficient of friction of pure nickel sliding on nickel in a hydrogen or nitrogen atmosphere is 5; in air or oxygen, it is 3; and in the presence of water vapor, it is 1.6. Coefficients of friction in sliding contact, as measured experimentally, vary from as low as 0.02 to as high as 100, or even higher. This range is not surprising in view of the many variables involved in the friction process. In metalworking opera-

TABLE 4.1

Coefficient of Friction in Metalworking Processes

Process	Coefficient of Friction (μ) Cold	Hot
Rolling	0.05–0.1	0.2–0.7
Forging	0.05–0.1	0.1–0.2
Drawing	0.03–0.1	–
Sheet-metal forming	0.05–0.1	0.1–0.2
Machining	0.5–2	–

tions that use various lubricants, the range for μ is much narrower, as shown in Table 4.1. The effects of load, temperature, speed, and environment on the coefficient of friction are difficult to generalize, as each situation has to be studied individually.

Abrasion theory of friction. If the upper body in the model shown in Fig. 4.5(a) is harder than the lower one, or if its surface contains protruding hard particles, then as it slides over the softer body, it can scratch and produce grooves on the lower surface. This phenomenon is known as **plowing** and is an important aspect in abrasive frictional behavior; in fact, it can be a dominant mechanism for situations where adhesion is not strong.

Plowing may involve two different mechanisms. One is the generation of a *groove*, whereby the surface is deformed plastically. The other is the formation of a groove because the cutting action of the upper body generates a *chip* or *sliver*. Both of these processes involve work supplied by a force that manifests itself as friction force. The plowing force can contribute significantly to friction and to the measured coefficient of friction at the interface.

Measuring friction. The coefficient of friction is usually determined experimentally, either during actual manufacturing processes or in simulated tests using small-scale specimens of various shapes. The techniques used generally involve measurements of either forces or dimensional changes in the specimen. One test that has gained wide acceptance, particularly for bulk deformation processes such as forging, is the **ring compression test**. In this test, a flat ring is compressed plastically between two flat platens (Fig. 4.7). As its height is reduced, the ring expands radially outward.

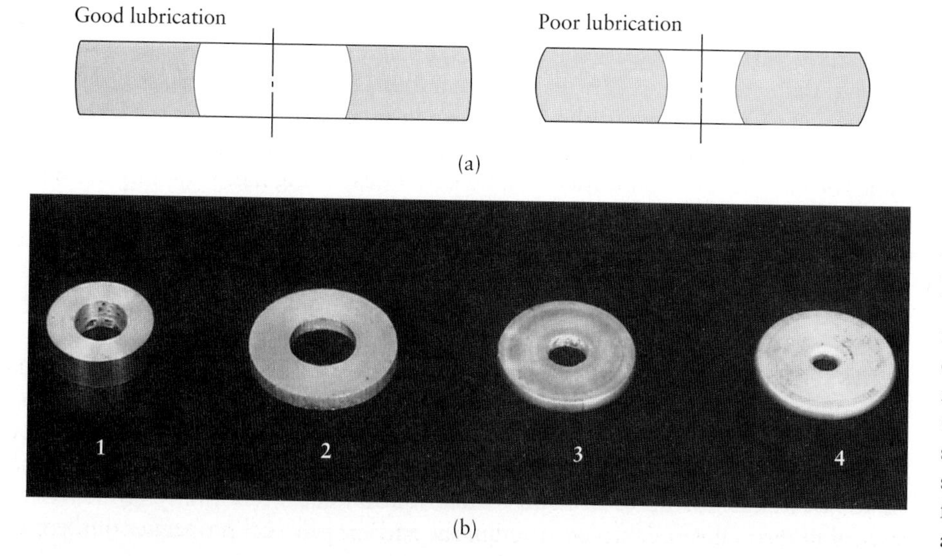

FIGURE 4.7 (a) The effects of lubrication on barreling in the ring compression test: (a) With good lubrication, both the inner and outer diameters increase as the specimen is compressed; and with poor or no lubrication, friction is high, and the inner diameter decreases. The direction of barreling depends on the relative motion of the cylindrical surfaces with respect to the flat dies. (b) Test results: (1) original specimen, and (2–4) the specimen under increasing friction. *Source*: A. T. Male and M. G. Cockcroft.

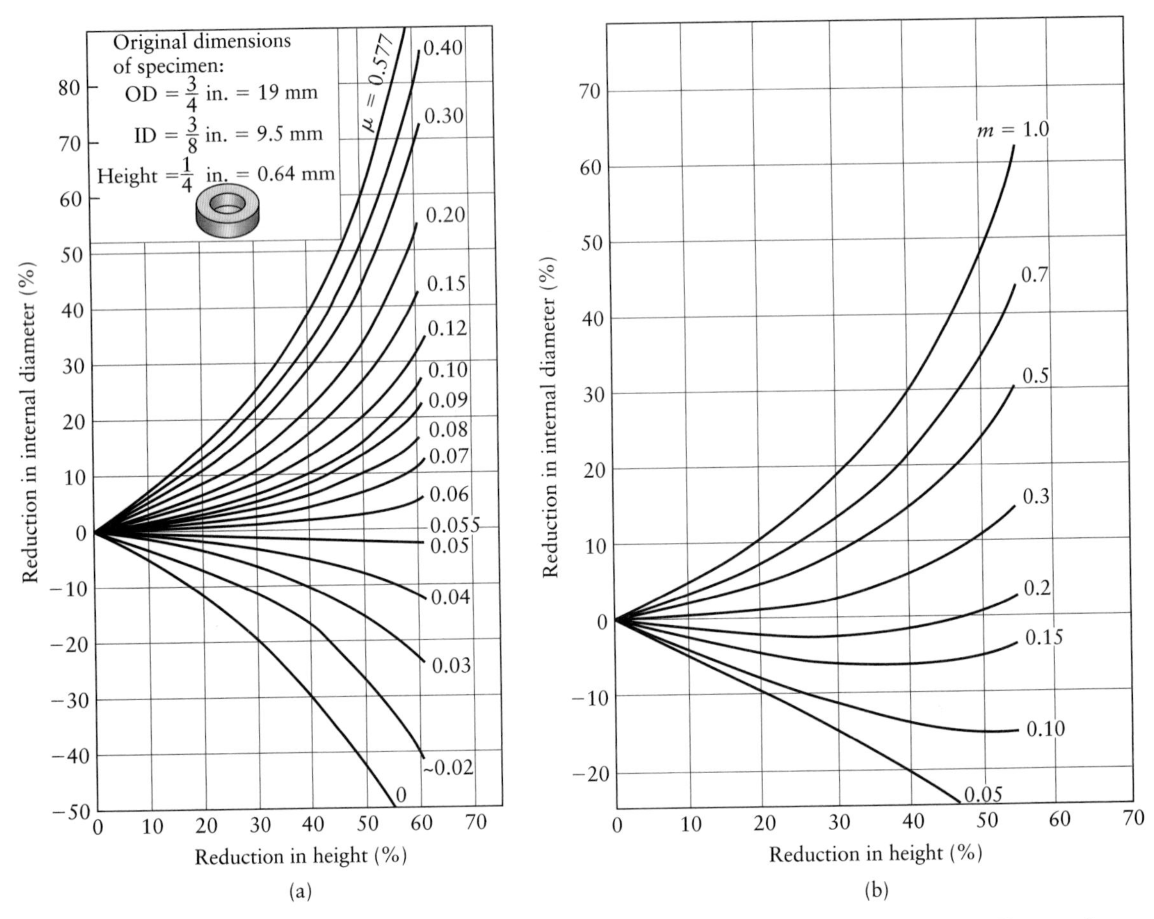

FIGURE 4.8 Charts to determine friction in ring compression tests: (a) coefficient of friction, μ; (b) friction factor, m. Friction is determined from these charts from the percent reduction in height and by measuring the percent change in the internal diameter of the specimen after compression.

If friction at the interfaces is zero, both the inner and outer diameters of the ring expand as if the ring were a solid disk; with increasing friction, the inner diameter of the ring becomes increasingly smaller.

For a particular reduction in height, there is a critical friction value at which the internal diameter increases (from the original) if μ is low and decreases if μ is high. By measuring the change in the specimen's internal diameter and using the curves shown in Fig. 4.8, which are obtained through theoretical analyses, we can determine the coefficient of friction. The geometry of each ring has its own specific set of curves; the most common geometry has specimen proportions of outer diameter to inner diameter to height of 6:3:2. The actual size of the specimen usually is not relevant in these tests; thus, once the percentage reductions in internal diameter and height are known, μ can be determined from the appropriate chart.

Temperature rise caused by friction. Almost all of the energy dissipated in overcoming friction is converted into heat; a small portion remains in the bodies as stored energy, and some of it goes into surface energy during generation of new surfaces and wear particles. Frictional heat raises the interface temperature. The magnitude of this temperature rise and its distribution depend not only on the friction force, but also on speed, surface roughness, and the physical properties of the materials, such as thermal conductivity and specific heat. The interface temperature increases with increasing friction and speed and with low thermal conductivity and specific heat of the materials. The interface temperature can be high enough to soften and melt the surface. (The temperature cannot, of course, exceed the melting point of the material.) A number of analytical expressions have been obtained to calculate the temperature rise in sliding; these expressions have been derived by methods including studies on the flash temperatures developed when a pair of asperities are sheared. The flash temperatures are highly localized and are much higher than the overall surface temperature during sliding.

Reducing friction. Friction can be reduced by selecting materials that exhibit low adhesion, such as relatively hard materials, like carbides and ceramics, and by using surface films and coatings. Lubricants, such as oils, or solid films, such as graphite, interpose an adherent film between tool, die, and workpiece. The film minimizes adhesion of one surface to the other and interactions between surfaces, thus reducing friction. Abrasive friction can be reduced by lowering the surface roughness of the harder of the two mating surfaces; a smooth surface inherently has less aggressive asperity peaks, and therefore, plowing will be less of a concern. However, it should be noted that adhesive friction is more of a concern with smooth surfaces. Friction can also be reduced significantly by subjecting the die–workpiece interface to **ultrasonic vibrations**, generally at 20 kHz. Depending on the amplitude, these vibrations momentarily separate the die and the workpiece, thus allowing the lubricant to flow more freely into the interface.

Friction in plastics and ceramics. Although their strength is low compared with that of metals, plastics (see Chapter 10) generally possess low frictional characteristics. This property makes polymers attractive for applications such as bearings, gears, seals, prosthetic joints, and general low-friction applications. In fact, polymers are sometimes called **self-lubricating** materials. The factors involved in metal friction are also generally applicable to polymers. In sliding, the plowing component of friction in thermoplastics and elastomers is a significant factor, because of their viscoelastic behavior (i.e., they exhibit both viscous and elastic behavior) and subsequent *hysteresis loss.*

An important factor in plastics applications is the effect of temperature rise at the sliding interfaces caused by friction. Thermoplastics lose their strength and become soft as temperature increases. Their low thermal conductivity and low melting points are thus significant in terms of heat generation by friction. If the temperature rise is not controlled, sliding surfaces can undergo plastic deformation and thermal degradation.

In view of the importance of ceramics (Section 11.9), their frictional behavior is now being studied intensively. Initial investigations indicate that the origin of friction in ceramics is similar to that in metals; thus, adhesion and plowing at interfaces contribute to the friction force in ceramics.

EXAMPLE 4.1 Determining the coefficient of friction

In a ring compression test, a specimen 10 mm in height with outside diameter (OD) = 30 mm and inside diameter (ID) = 15 mm is reduced in thickness by 50%. Determine the coefficient of friction, μ, and the friction factor, m, if the OD after deformation is 39 mm.

Solution. We first have to determine the new ID; this diameter is obtained from volume constancy, as follows:

$$\text{Volume} = \frac{\pi}{4}(30^2 - 15^2)10 = \frac{\pi}{4}(39^2 - \text{ID}^2)5.$$

From this equation, we find that the new ID = 13 mm. Thus,

$$\text{Change in ID} = \frac{15 - 13}{15} \times 100\% = 13\% \text{ (decrease)}.$$

For a 50% reduction in height and a 13% reduction in ID, the following values are interpolated from Fig. 4.8:

$$\mu = 0.09 \quad \text{and} \quad m = 0.4.$$

4.4.2 Wear

Wear is defined as the progressive loss or unwanted removal of material from a surface. Wear has important technological and economic significance, because it changes the shape of the workpiece, tool, and die interfaces; hence, it affects the process and the size and quality of the parts produced. The magnitude of wear is evident in the countless numbers of parts and components that continually have to be replaced or repaired. Examples of wear in manufacturing processes include dull drills that have to be reground, worn cutting tools that have to be indexed or resharpened, and forming tools and dies that have to be repaired or replaced. **Wear plates**, which are subjected to high loads, are important components in some metalworking machinery. These plates, also known as *wear parts*, because they are expected to wear, can easily be replaced.

Although wear generally alters the surface topography and can result in severe surface damage, it also has a beneficial effect: it can reduce surface roughness by removing the peaks from asperities (Fig. 4.9). Thus, under controlled conditions, wear may be regarded as a kind of smoothening or polishing process; the **running-in period** for various machines and engines produces this type of wear. Wear is usually classified into the following categories:

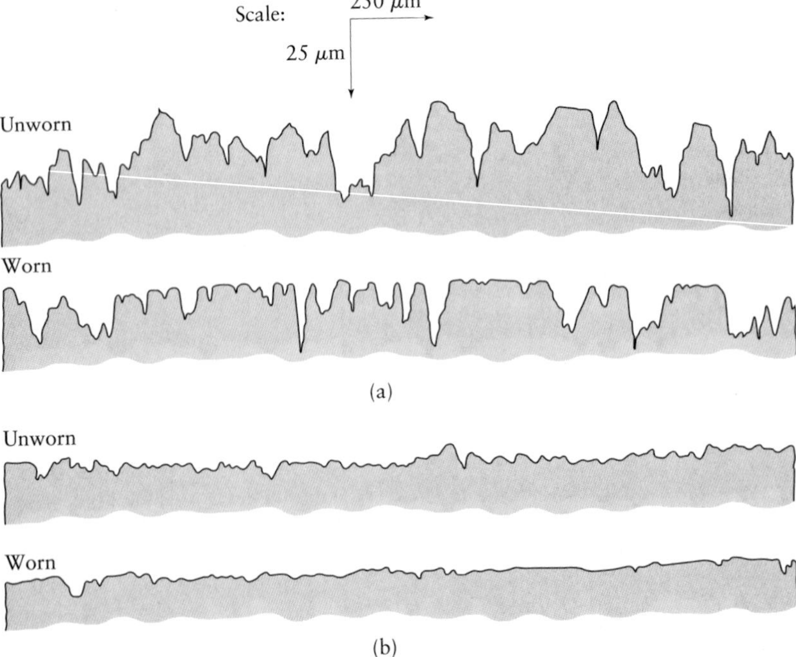

FIGURE 4.9 Changes in originally (a) wire-brushed and (b) ground-surface profiles after wear. *Source*: E. Wild and K. J. Mack.

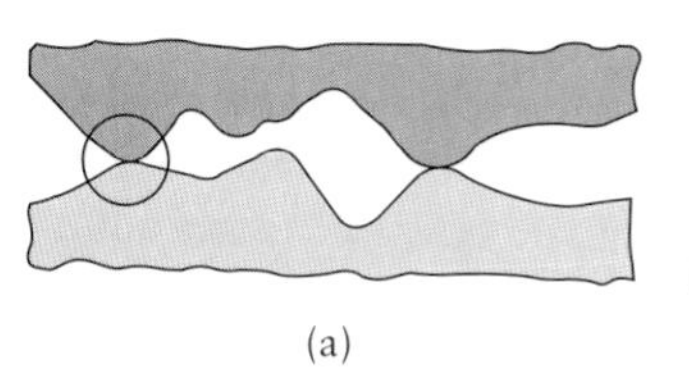

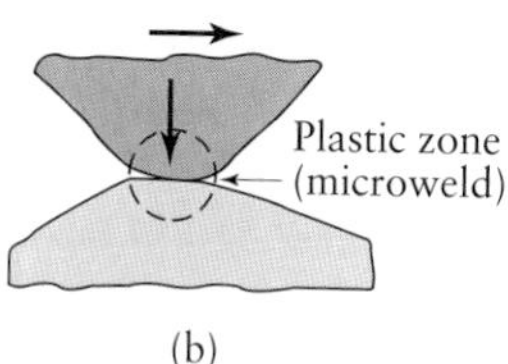

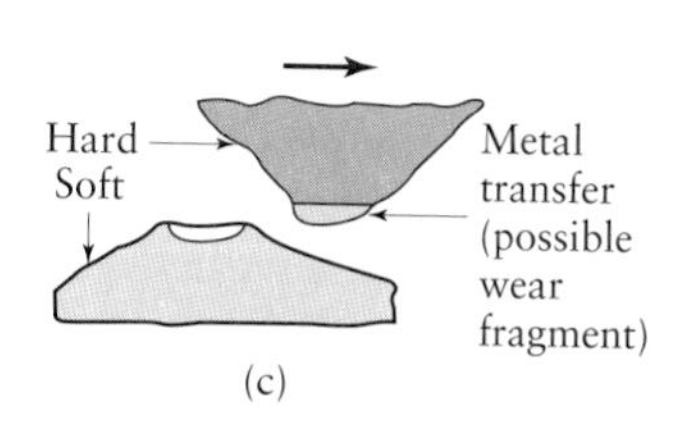

FIGURE 4.10 Schematic illustration of (a) two asperities contacting, (b) adhesion between two asperities, and (c) the formation of a wear particle.

Adhesive wear. If a tangential force is applied to the model shown in Fig. 4.5, shearing can take place either at the original interface or along a path below or above it (Fig. 4.10), causing *adhesive wear*. The fracture path depends on whether or not the strength of the adhesive bond of the asperities is greater than the cohesive strength of either of the two sliding bodies.

Because of factors such as strain hardening at the asperity contact, diffusion, and mutual solid solubility of the two bodies, the adhesive bonds are often stronger than the base metals. Thus, during sliding, fracture at the asperity usually follows a path in the weaker or softer component. A *wear fragment* is then generated. Although this fragment is attached to the harder component (the upper member in Fig. 4.10), it eventually becomes detached during further rubbing at the interface and develops into a loose wear particle. This process is known as *adhesive wear*, or *sliding wear*. In more severe cases, such as with high loads and strongly bonded asperities, adhesive wear is described as *scuffing, smearing, tearing, galling*, or *seizure*. Oxide layers on surfaces greatly influence adhesive wear; they can act as a protective film, resulting in what is known as **mild wear**, consisting of small wear particles.

Based on the probability that a junction between two sliding surfaces will lead to the formation of a wear particle, the **Archard wear law** provides an expression for adhesive wear,

$$V = k\frac{LW}{3p}, \tag{4.6}$$

where V is the volume of material removed by wear from the surface, k is the *wear coefficient* (dimensionless), L is the length of travel, W is the normal load, and p is the indentation hardness of the softer body. In some references, the factor 3 is deleted from the denominator and incorporated into the wear coefficient k.

Table 4.2 lists typical values of k for a combination of materials sliding in air. The wear coefficient for the same pair of materials can vary by a factor of 3, depending on whether wear is measured as a loose particle or as a transferred particle (to the other body). Loose particles have a lower k value. Likewise, when mutual solubility is a significant parameter in adhesion, similar metal pairs have higher k values than dissimilar pairs.

TABLE 4.2

Approximate Order of Magnitude for the Wear Coefficient *k* in Air

Unlubricated	k	Lubricated	k
Mild steel on mild steel	10^{-2} to 10^{-3}	52100 steel on 52100 steel	10^{-7} to 10^{-10}
60–40 brass on hardened tool steel	10^{-3}	Aluminum bronze on hardened steel	10^{-8}
Hardened tool steel on hardened tool steel	10^{-4}	Hardened steel on hardened steel	10^{-9}
Polytetrafluoroethylene (PTFE) on tool steel	10^{-5}		
Tungsten carbide on mild steel	10^{-6}		

Thus far, we have based the treatment of adhesive wear on the assumption that the surface layers of the two contacting bodies are clean and free from contaminants. Under these conditions, the adhesive-wear rate can be very high, in which case it is known as **severe wear**. As described in Section 4.2, however, metal surfaces are almost always covered with oxide layers, with thicknesses generally ranging between 10 nm and 100 nm. Although this thickness may at first be regarded as insignificant, the oxide layer has a profound effect on wear behavior.

Oxide layers are usually hard and brittle. When such a surface is subjected to rubbing, the layer can have the following effects: if the load is light and the oxide layer is strongly adhering to the bulk metal, the strength of the junctions between the asperities is weak, and wear is low. In this situation, the oxide layer acts as a protective film, and the type of wear is known as **mild wear**.

The oxide layer can be broken up under high normal loads if the layer is brittle and is not strongly adhering to the bulk metal, or if the asperities are rubbing against each other repeatedly, so that the oxide layer breaks up by fatigue. However, if the surfaces are smooth, the respective oxide layers are more difficult to break up. When one of the layers is eventually broken up, a wear particle is formed. An upper asperity can then form a strong junction with the lower asperity, which is now unprotected by the oxide layer; the wear rate will now be higher until a fresh oxide layer is formed.

In addition, a surface exposed to the environment is covered with adsorbed layers of gas and usually with other contaminants. (See Fig. 4.1.) Such films, even if they are very thin, generally weaken the interfacial bond strength of contacting asperities by separating the metal surfaces. The net effect of these films is to provide a protective layer by reducing adhesion and thus reducing wear. The magnitude of the effect of adsorbed gases and contaminants depends on many factors. For instance, it has been shown repeatedly that even small differences in humidity can have a profound influence on the wear rate.

EXAMPLE 4.2 Adhesive wear in sliding

The end of a rod made of 60–40 brass is sliding over the unlubricated surface of hardened tool steel with a load of 200 lb. The hardness of brass is 120 HB. What is the distance traveled to produce a wear volume of 0.001 in^3 by adhesive wear of the brass rod?

Solution. The parameters in Eq. (4.6) for adhesive wear are as follows:

$$V = 0.001 \text{ in}^3;$$
$$k = 10^{-3} \text{ (from Table 4.2)};$$
$$W = 200 \text{ lb};$$
$$p = 120 \text{ kg/mm}^2 = 170{,}700 \text{ lb/in}^2.$$

Therefore, the distance traveled is

$$L = \frac{3Vp}{kW} = \frac{(3)(0.001)(170{,}700)}{(10^{-3})(200)} = 2560 \text{ in.} = 213 \text{ ft.}$$

Abrasive wear. *Abrasive wear* is caused by a hard and rough surface, or a surface containing hard protruding particles, sliding along another surface. This type of wear removes particles by forming microchips or *slivers*, thereby producing grooves or

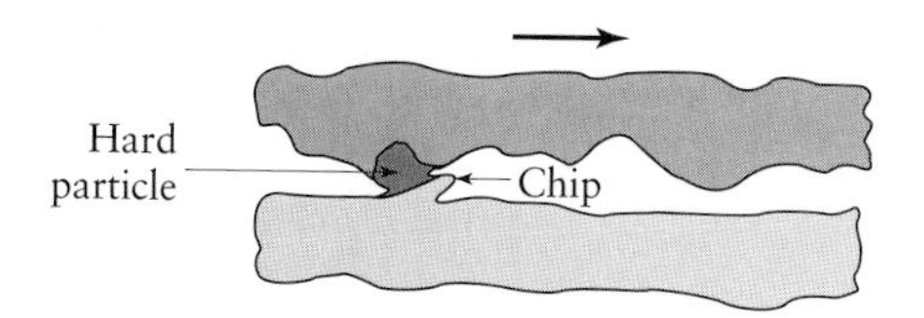

FIGURE 4.11 Schematic illustration of abrasive wear in sliding. Longitudinal scratches on a surface usually indicate abrasive wear.

scratches on the softer surface (Fig. 4.11). In fact, the abrasive processes described in Chapter 9, such as grinding, ultrasonic machining, and abrasive-jet machining, act in this manner. The difference is that in those operations, we control the process parameters to produce desired shapes and surfaces, whereas abrasive wear is unintended and unwanted.

The abrasive wear resistance of pure metals and ceramics is directly proportional to their hardness. Abrasive wear can thus be reduced by increasing the hardness of materials (such as by heat treating and microstructural changes) or by reducing the normal load. Other materials that resist abrasive wear are elastomers and rubbers, because they deform elastically and then recover when abrasive particles cross over their surfaces. The best example is automobile tires, which have long lives even though they are operated on abrasive road surfaces; even hardened steels would not last long under such conditions.

There are two basic types of abrasive wear. **Two-body wear** is the basis of **erosive wear**. The abrasive particles (such as sand) are usually carried in a jet of fluid or air, and they remove material from the surface by erosion. In **three-body wear**, the lubricant between the die and the workpiece may carry with it wear particles (generated over a period of time) and cause abrasive wear. Three-body wear is particularly important in metal-forming processes; thus, the necessity for proper inspection and filtering of the metalworking fluid is apparent. There is also the possibility that particles (from machining, grinding, the environment, etc.) may contaminate the lubricating system of the machine components and thus scratch and damage the surfaces.

Corrosive wear. *Corrosive wear*, also known as **oxidation**, or **chemical, wear**, is caused by chemical or electrochemical reactions between the surfaces and the environment. The fine corrosive products on the surface constitute the wear particles. When the corrosive layer is destroyed or removed, as by sliding or abrasion, another layer begins to form, and the process of removal and corrosive-layer formation is repeated. Among corrosive media are water (including seawater); oxygen, acids and chemicals, and atmospheric hydrogen sulfide and sulfur dioxide. Corrosive wear can be reduced by selecting materials that will resist environmental attack, controlling the environment, and reducing operating temperatures to lower the rate of chemical reaction.

Fatigue wear. *Fatigue wear*, also called **surface fatigue**, or **surface fracture, wear**, is caused when the surface of a material is subjected to cyclic loading, such as in rolling contact in bearings. The wear particles are usually formed by **spalling** or **pitting**. Another type of fatigue wear is **thermal fatigue**. In thermal fatigue, cracks on the surface are generated by thermal stresses from thermal cycling, such as a cool die repeatedly contacting hot workpieces (*heat checking*). These cracks then join, and the surface begins to spall, producing fatigue wear. This type of wear usually occurs in hot-working and die-casting dies. Fatigue wear can be reduced by lowering contact stresses; reducing thermal cycling; and improving the quality of materials by removing impurities, inclusions, and various other flaws that may act as local points for crack initiation.

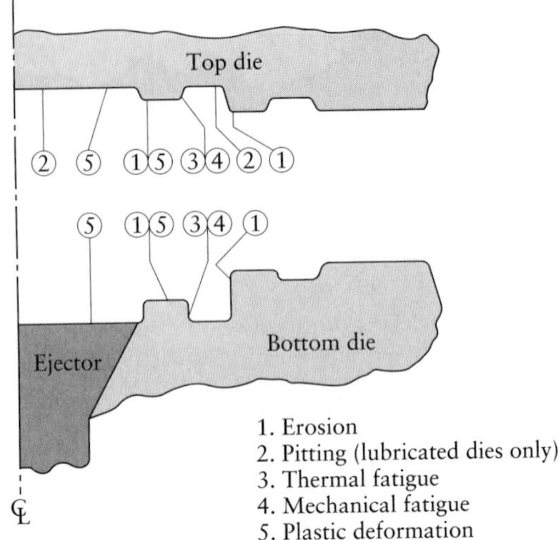

FIGURE 4.12 Types of wear observed in a single die used for hot forging. *Source*: T. A. Dean.

Other types of wear. Several other types of wear are important in manufacturing processes. **Erosion** is caused by loose abrasive particles abrading a surface. **Fretting** corrosion occurs at interfaces that are subjected to very small movements, such as in machinery. **Impact wear** is the removal of small amounts of materials from a surface by impacting particles. Deburring by vibratory finishing and tumbling (see Section 9.9) is an example of impact wear. In many cases, component wear is the result of a combination of different types of wear. Note in Fig. 4.12, for example, that even in the same forging die, various types of wear take place in different locations. A similar situation exists in cutting tools. (See Fig. 8.24.)

Wear of plastics and ceramics. The wear behavior of plastics is similar to that of metals; thus, wear may occur in ways similar to those described earlier. Abrasive-wear behavior depends partly on the ability of the polymer to deform and recover elastically, similar to the behavior of elastomers. There is evidence that the parameter describing this behavior may be the ratio of hardness to elastic modulus; thus, the polymer's resistance to abrasive wear increases as this ratio increases.

Typical polymers with good wear resistance include polyimides, nylons, polycarbonate, polypropylene, acetals, and high-density polyethylene (Chapter 10). These polymers are molded or machined to make gears, pulleys, sprockets, and similar mechanical components. Because plastics can be made with a wide variety of compositions, they can also be blended with internal lubricants, such as PTFE, silicon, graphite, and molybdenum disulfide, and with rubber particles that are interspersed within the polymer matrix.

Wear resistance of reinforced plastics (Section 10.9) depends on the type, amount, and direction of reinforcement in the polymer matrix. Carbon, glass, and aramid fibers all improve wear resistance. Wear takes place when fibers are pulled out of the matrix (*fiber pullout*). Wear is highest when the sliding direction is parallel to the fibers, because the fibers can be pulled out more easily in this direction. Long fibers increase the wear resistance of composites, because such fibers are more difficult to pull out, and cracks in the matrix cannot propagate to the surface as easily.

When ceramics slide against metals, wear is caused by small-scale plastic deformation and brittle surface fracture, surface chemical reactions, plowing, and possibly some fatigue. Metals can be transferred to the oxide-type ceramics surfaces, forming metal oxides; thus, sliding actually takes place between the metal and the metal-oxide surface. Conventional lubricants do not appear to influence ceramics wear significantly.

Measuring wear. Several methods can be used to observe and measure wear. The choice of a particular method depends on the accuracy desired and the physical constraints of the system, such as specimen or part size and difficulty of disassembly. Although not quantitative, the simplest method is visual and tactile (touching) inspection. Measuring dimensional changes, gaging the worn component, profilometry, and weighing are more accurate methods. However, for large workpieces or tools and dies, the weighing method is not accurate, because the amount of wear is usually very small compared with the overall weight of the components involved. Performance and noise level can also be monitored; worn machinery components emit more noise than new parts.

Radiography is a method in which wear particles from an irradiated surface are transferred to the mating surface, whose amount of radiation is then measured. An example is the transfer of wear particles from irradiated cutting tools to the back side of chips. In other situations, the lubricant can be analyzed for wear particles (spectroscopy); this method is precise and is used widely for applications such as checking jet-engine component wear.

General observations. It is apparent from the foregoing discussions that wear is a complex phenomenon and that each type of wear is influenced by several factors. Numerous theories have been advanced to explain and interpret the experimental data and observations made regarding wear. Because of the many factors involved, quantifying wear data is difficult; in fact, it is well known that wear data are difficult to reproduce.

The most significant aspects of wear are the nature and geometry of the contacting asperities. The strength of the bond at the asperity junction relative to the cohesive strength of the weaker body is an important factor. When a polymer such as PTFE (*Teflon*) is rubbed against steel, the surface of the polymer (which is much weaker than the steel) becomes embedded with steel wear particles. Also, when we rub gold with our fingers, wear particles of gold remain on our fingers; the mechanisms for this phenomenon are not clear.

The question also arises as to whether friction and wear for a pair of metals are directly related. Although, at first, one would expect such a relationship, experimental evidence for unlubricated metal pairs indicates that this is not always the case. These results are not surprising, however, in view of the many factors involved in friction and wear. For instance, ball bearings and races have very little friction, yet they undergo wear after a large number of cycles.

4.4.3 Lubrication

In manufacturing processes, the interface between tools, dies, and workpieces is subjected to a wide range of variables. Among the major variables are the following:

1. **Contact pressure**, ranging from low values of elastic stresses to multiples of the yield stress of the workpiece material;
2. **Speed**, ranging from very low (such as in some superplastic forming operations) to very high speeds (such as in explosive forming, thin-wire drawing, grinding, and high-speed metal cutting);
3. **Temperature**, ranging from ambient to almost melting, such as in hot extrusion and squeeze casting.

If two surfaces slide against each other under high pressure, speed, and/or temperature, and with no protective layers at the interface, friction and wear will be high. To reduce friction and wear, surfaces should be held as far apart as possible; this is generally done by using metalworking lubricants, which may be solid, semisolid, or liquid in nature. Metalworking lubricants are not only fluids or solids with certain desirable physical properties, such as viscosity, but also are chemicals that can react with the surfaces of the tools, dies, and workpieces and alter their properties.

Lubrication regimes. Four regimes of lubrication are relevant to metalworking processes (Fig. 4.13):

1. **Thick film**: The metal surfaces are completely separated by a fluid film that has a thickness of about one order of magnitude greater than that of the surface roughness; thus, there is no metal-to-metal contact. The normal load is light and is supported by the **hydrodynamic fluid film** formed by the wedge effect caused by the relative velocity of the two bodies and the viscosity of the fluid. The coefficient of friction is very low, usually ranging between 0.001 and 0.02, and wear is practically nonexistent.
2. **Thin film**: As the normal load increases or as the speed and viscosity of the fluid decrease (caused by, say, a rise in temperature), the thickness of the film is reduced to between 3 and 10 times that of the surface roughness. There may be some metal-to-metal contact at the higher asperities; this contact increases friction and leads to slightly increased wear.
3. **Mixed**: In this situation, a significant portion of the load is carried by the metal-to-metal contact of the asperities, and the rest of the load is carried by the pressurized fluid film present in hydrodynamic pockets, such as in the valleys of asperities. The average film thickness is less than three times that of the surface roughness. With proper selection of lubricants, a strongly adhering *boundary film* a few molecules thick can be formed on the surfaces; this film prevents direct metal-to-metal contact and thus reduces wear. Depending on the strength of the boundary film and other parameters, the friction coefficient in mixed lubrication may range up to about 0.4.
4. **Boundary lubrication**: The load here is supported by the contacting surfaces, which are covered with a *boundary layer*. A lubricant may or may not be present in the surface valleys, but it is not pressurized enough to support sig-

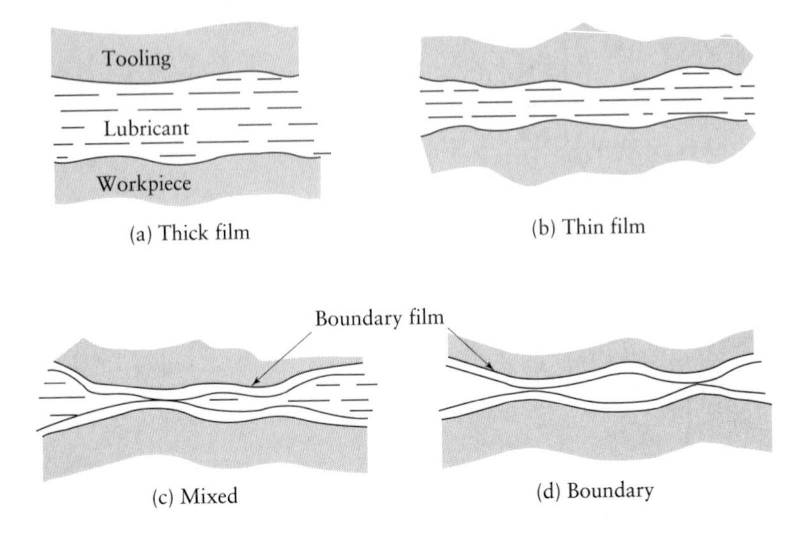

FIGURE 4.13 Types of lubrication generally occurring in metalworking operations. *Source*: After W. R. D. Wilson.

nificant load. Depending on the boundary film thickness and its strength, the friction coefficient ranges from about 0.1 to 0.4. Wear can be relatively high if the boundary layer is destroyed.

Typical *boundary lubricants* are natural oils, fats, fatty acids, and soaps. Boundary films form rapidly on metal surfaces. As the film thickness decreases and metal-to-metal contact takes place, the chemical aspects and roughness of surfaces become significant. (In thick-film lubrication, the viscosity of the lubricant is the important parameter in controlling friction and wear; chemical aspects are not particularly significant, except as they affect corrosion and staining of metal surfaces.) A boundary film can *break down*, or it can be removed by being disturbed or rubbed off during sliding, or because of *desorption* due to high temperatures at the interface. The metal surfaces may thus be deprived of this protective layer. The clean metal surfaces then contact each other, and, as a consequence, severe wear and scoring can occur; thus, adherence of boundary films is an important aspect in lubrication.

Surface roughness and geometric effects. Surface roughness, particularly in mixed lubrication, is important; roughness can serve to create local reservoirs or pockets for lubricants. The lubricant can be trapped in the valleys of the surface, and, because these fluids are incompressible, they can support a substantial portion of the normal load. The pockets also supply lubricant to regions where the boundary layer has been destroyed. There is an optimal roughness for lubricant entrapment. In metalworking operations, it is generally desirable for the workpiece, not the die, to have the rougher surface. Otherwise, the workpiece surface will be damaged by the rougher and harder die surface. The recommended surface roughness on most dies is about 0.40 μm (15 μin).

In addition to surface roughness, the overall *geometry* of the interacting bodies is an important consideration in lubrication. Figure 4.14 shows the typical metalworking processes of drawing, extrusion, and rolling. The movement of the workpiece into the deformation zone must allow a supply of the lubricant to be carried into the die–workpiece interfaces. Analysis of hydrodynamic lubrication shows that the *inlet angle* is an important parameter. As this angle decreases, more lubricant is entrained; thus, lubrication is improved, and friction is reduced. With proper selection of parameters, a relatively thick film of lubricant can be maintained at the die–workpiece interface.

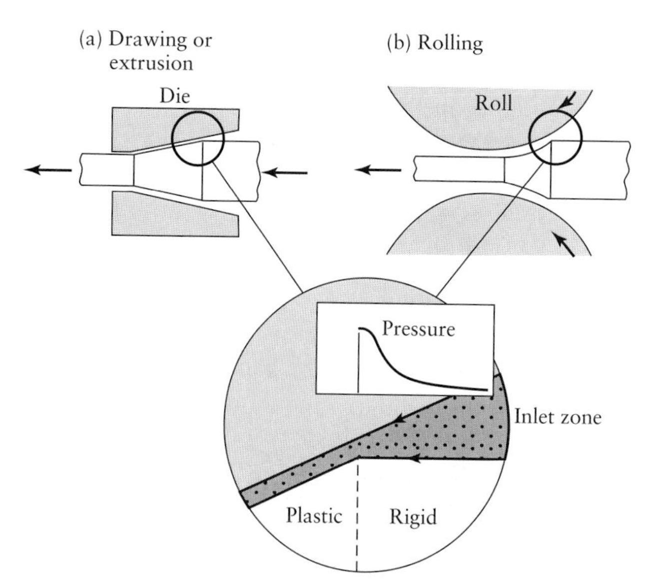

FIGURE 4.14 Inlet zones for entrainment of lubricants in drawing, extrusion, and rolling. Entrainment of lubricants is important in reducing friction and wear in metalworking operations. *Source*: After W. R. D. Wilson.

The reduction of friction and wear by *entrainment* or *entrapment of lubricants* may not always be desirable. The reason is that with a thick fluid film, the metal surface cannot come into full contact with the die surfaces and hence does not acquire the shiny appearance that is generally required on the product. (Note, for instance, the shiny appearance of a piece of copper wire.) A thick film of lubricant generates a grainy, dull surface on the workpiece, the degree of roughness depending on the grain size of the material. In operations such as coining and forging (Section 6.2), trapped lubricants are undesirable, because they interfere with the forming process and prevent precise shape generation.

4.4.4 Metalworking fluids

On the basis of the foregoing discussions, the functions of metalworking fluids can be summarized as follows:

- Reduce *friction*, thus reducing force and energy requirements and preventing increases in temperature;
- Reduce *wear, seizure*, and *galling*;
- Improve *material flow* in tools, dies, and molds;
- Act as a *thermal barrier* between the workpiece and tool and die surfaces, thus preventing workpiece cooling in hot-working processes;
- Act as a *release* or *parting agent* to help in the removal or ejection of parts from dies and molds.

Several types of metalworking fluids are now available to fulfill these requirements. Because of the lubricants' diverse chemistries, properties, and characteristics, the behavior and performance of lubricants can be complex. The general properties of the most commonly used lubricants are described as follows:

1. **Oils** have high film strength on the surface of a metal, as you well know if you have ever tried to clean an oily surface. Although they are very effective in reducing friction and wear, oils have low thermal conductivity and specific heat; thus, they are not effective in conducting away the heat generated by friction and plastic deformation in metalworking operations. Oils also are difficult to remove from component surfaces that subsequently are to be painted or welded, and they are difficult to dispose of. The sources of oils are **mineral** (petroleum), **animal, vegetable**, and **fish.** Oils may be **compounded** with a variety of additives or with other oils to impart special properties, such as lubricity, viscosity, detergents for cleaning ability, biocides, etc.
2. Metalworking fluids are usually blended with several **additives**, including oxidation inhibitors, rust preventatives, odor control agents, antiseptics, and foam inhibitors. Important additives in oils are sulfur, chlorine, and phosphorus. Known as **extreme-pressure (EP) additives** and used singly or in combination, they react chemically with metal surfaces and form adherent surface films of metallic sulfides and chlorides. These films have low shear strength and good antiweld properties and thus effectively reduce friction and wear. While EP additives are important in boundary lubrication, these lubricants may be undesirable in some cases. For example, they may preferentially attack the cobalt binder in tungsten carbide tools and dies (Section 8.6), causing undesirable changes in their surface roughness and integrity.
3. An **emulsion** is a mixture of two immiscible liquids, usually mixtures of oil and water in various proportions, along with additives. A typical emulsion will be

provided as a mixture of base oil and additives and then mixed with water by the user. Emulsions are of two types: direct and indirect. In a direct emulsion, oil is dispersed in water as very small droplets, with average diameters greater than 0.1 μm (4 μin.), but usually less than 30 μm (1200 μin.) in practice. In an indirect emulsion, water droplets are dispersed in the oil. Direct emulsions are important fluids, because the presence of water (usually 95% of the emulsion by volume) gives them high cooling capacity, yet they still provide good lubrication. They are particularly effective in high-speed metal-forming and cutting operations, where increases in temperature have detrimental effects on tool life, workpiece surface integrity, and dimensional accuracy.

4. **Synthetic solutions** are fluids that contain inorganic and other chemicals dissolved in water. Various chemical agents are added to impart different properties. **Semisynthetic solutions** are basically synthetic solutions to which small amounts of emulsifiable oils have been added.
5. **Soaps** are generally reaction products of sodium or potassium salts with fatty acids. Alkali soaps are soluble in water, but other metal soaps are generally insoluble. Soaps are effective boundary lubricants and can also form thick film layers at die–workpiece interfaces, particularly when applied on conversion coatings (see later) for cold metalworking applications.
6. **Greases** are solid or semisolid lubricants and generally consist of soaps, mineral oil, and various additives. They are highly viscous and adhere well to metal surfaces. Although used extensively in machinery, greases have limited use in manufacturing processes.
7. **Waxes** may be of animal or plant (*paraffin*) origin and have complex structures. Compared with greases, waxes are less "greasy" and are more brittle. They have limited use in metalworking operations, except for copper and, as chlorinated paraffin, for stainless steels and high-temperature alloys.

Solid lubricants. Because of their unique properties and characteristics, several solid materials are used as lubricants in manufacturing operations:

1. **Graphite.** The properties of *graphite* are described in Section 11.14. Graphite is weak in shear along its layers and has a low coefficient of friction in that direction; thus, it can be a good solid lubricant, particularly at elevated temperatures. However, its friction is low only in the presence of air or moisture. In a vacuum or an inert-gas atmosphere, its friction is very high; in fact, graphite can be quite abrasive in these environments. Graphite may be applied either by rubbing it on surfaces or as a *colloidal* (dispersion of small particles) suspension in liquid carriers, such as water, oil, or alcohols.

   A more recent development is the production of soccer-ball-shaped carbon molecules, called **buckyballs** or **fullerenes** (both after Buckminster Fuller). These chemically inert spherical molecules are produced from soot and act much like solid lubricant particles.
2. **Molybdenum disulfide.** *Molybdenum disulfide* (MoS_2), another widely used lamellar solid lubricant, is somewhat similar in appearance to graphite. However, unlike graphite, it has a high friction coefficient in an ambient environment. Oils are commonly used as carriers for molybdenum disulfide, which is used as a lubricant at room temperature. Molybdenum disulfide can also be rubbed onto the surfaces of a workpiece.
3. **Soft metals and polymer coatings.** Because of the low strength of *soft metals* and *polymer coatings*, thin layers of them are used as solid lubricants. Suitable metals are lead, indium, cadmium, tin, and silver, and suitable polymers

include PTFE, polyethylene, and methacrylates. However, these coatings have limited applications, because of their lack of strength under high stresses and at elevated temperatures. Soft metals are used to coat high-strength metals such as steels, stainless steels, and high-temperature alloys. Copper or tin, for example, is chemically deposited on the surface before the metal is processed. If the oxide of a particular metal has low friction and is sufficiently thin, the oxide layer can serve as a solid lubricant, particularly at elevated temperatures.

4. **Glass.** Although a solid material, *glass* becomes viscous at high temperatures and hence can serve as a liquid lubricant. Its viscosity is a function of temperature, but not of pressure, and depends on the type of glass. Poor thermal conductivity also makes glass attractive, since it acts as a thermal barrier between hot workpieces and relatively cool dies. Typical glass-lubrication applications include hot extrusion (Section 6.4) and forging (Section 6.2).

Lubricant selection. Selecting a lubricant for a particular manufacturing process and workpiece material involves consideration of several factors:

- The particular manufacturing process;
- Compatibility of the lubricant with the workpiece and tool and die materials;
- The surface preparation required;
- The method of the lubricant application;
- Removal of the lubricant after processing;
- Contamination of the lubricant by other lubricants, such as those used to lubricate the machinery;
- Treatment of waste lubricant;
- Storage and maintenance of lubricants;
- Biological and ecological considerations;
- Costs involved in all of the foregoing aspects.

In selecting an oil as a lubricant, the importance of its viscosity-temperature-pressure characteristics should be recognized. Low viscosity can have a significant detrimental effect on friction and wear. The different functions of a metalworking fluid, whether primarily a lubricant or a coolant, must also be taken into account. Water-base fluids are very effective coolants, but as lubricants are not as effective as oils.

Metalworking fluids should not leave any harmful residues that could interfere with machinery operations. The fluids should not stain or corrode workpieces or equipment. The fluids should be checked periodically for deterioration caused by bacterial growth, accumulation of oxides, chips, and wear debris and also for general degradation and breakdown due to temperature and time. A lubricant may carry wear particles within it and cause damage to the system, so proper inspection and filtering of metalworking fluids are important. These precautions are necessary for all types of machinery, such as they are for internal combustion and jet engines.

After completion of manufacturing operations, metal surfaces are usually covered with lubricant residues, which should be removed prior to further workpiece processing such as welding or painting. Various cleaning solutions and techniques can be used for this purpose (section 4.5.2). *Biological and ecological considerations*, with their accompanying health and legal ramifications, are also important. Potential health hazards may be involved in contacting or inhaling some metalworking fluids. Recycling and disposal of waste fluids are additional important factors to be considered.

4.5 Surface Treatments, Coatings, and Cleaning

After a component is manufactured, all or parts of its surfaces may have to be processed further in order to impart certain properties and characteristics. *Surface treatments* may be necessary to

- Improve resistance to wear, erosion, and indentation (e.g., for slideways in machine tools; wear surfaces of machinery; and shafts, rolls, cams, and gears);
- Control friction (sliding surfaces on tools, dies, bearings, and machine ways);
- Reduce adhesion (electrical contacts);
- Improve lubrication (surface modification to retain lubricants);
- Improve resistance to corrosion and oxidation (sheet metals for automotive or other outdoor uses, gas-turbine components, and medical devices);
- Improve fatigue resistance (bearings and shafts with fillets);
- Rebuild surfaces on components (worn tools, dies, and machine components);
- Improve surface roughness (appearance, dimensional accuracy, and frictional characteristics);
- Impart decorative features, color, or special surface texture.

4.5.1 Surface treatment processes

Numerous processes are used for surface treatments, based on mechanical, chemical, thermal, and physical methods. Their principles and characteristics are outlined as follows:

1. **Shot peening, water-jet peening, and laser peening.** In *shot peening*, the surface of the workpiece is hit repeatedly with a large number of cast-steel, glass, or ceramic shot (small balls), making overlapping indentations on the surface; this action causes plastic deformation of the surface to depths up to 1.25 mm (0.05 in.), using shot sizes ranging from 0.125 mm to 5 mm (0.005 in. to 0.2 in.) in diameter. Because plastic deformation is not uniform throughout a part's thickness, shot peening imparts compressive residual stresses on the surface, thus improving the fatigue life of the component. This process is used extensively on shafts, gears, springs, oil-well drilling equipment, and jet-engine parts (such as turbines and compressor blades).

   In *water-jet peening*, a water jet at pressures as high as 400 MPa (60 ksi) impinges on the surface of the workpiece, inducing compressive residual stresses to an extent similar to that of shot peening. This method has been used successfully on steels and aluminum alloys.

   In *laser peening*, the surface is subjected to laser shocks from high-powered lasers (up to 1 kW) and at energy levels of 100 J/pulse. This method has been used successfully on jet-engine fan blades and on materials such as titanium and nickel alloys, with compressive surface residual stresses deeper than 1 mm (0.04 in.).

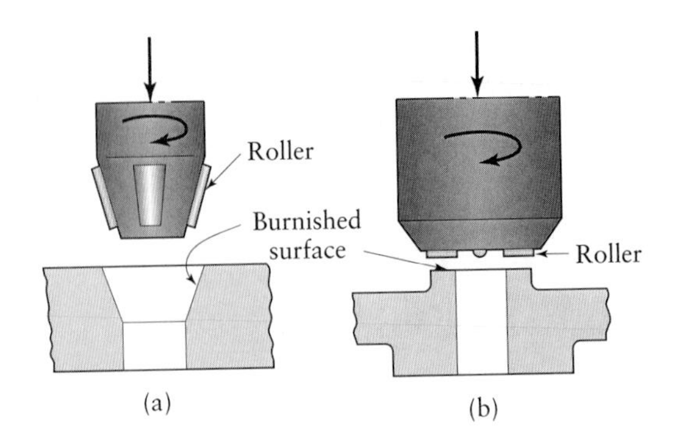

FIGURE 4.15 Examples of roller burnishing of (a) a conical surface and (b) a flat surface and the burnishing tools used. *Source*: Cogsdill Tool Products.

2. **Roller burnishing (surface rolling).** The surface of the component is cold worked by a hard and highly polished roller or series of rollers; this process is used on various flat, cylindrical, or conical surfaces (Fig. 4.15). Roller burnishing improves surface finish by removing scratches, tool marks, and pits. Consequently, corrosion resistance is also improved, since corrosive products and residues cannot be entrapped. Roller burnishing is used to improve the mechanical properties of surfaces, as well as the shape and surface finish of components. It can be used either singly or in combination with other finishing processes, such as grinding, honing, and lapping. Soft and ductile, as well as very hard, metals can be roller burnished. Typical applications of roller burnishing include hydraulic-system components, seals, valves, spindles, and fillets on shafts. Internal cylindrical surfaces are burnished by a similar process, called **ballizing**, or **ball burnishing**, in which a smooth ball slightly larger than the bore diameter is pushed through the length of the hole.
3. **Explosive hardening.** The surface is subjected to high transient pressures by placing a layer of explosive sheet directly on the workpiece surface and detonating it. Large increases in surface hardness can be obtained by this method, with very little change (less than 5%) in the shape of the component. Railroad rail surfaces can be hardened by this method.
4. **Cladding (clad bonding).** Metals are bonded with a thin layer of corrosion-resistant metal by applying pressure with rolls or other means. A typical application is cladding of aluminum (*Alclad*), in which a corrosion-resistant layer of aluminum alloy is clad over pure aluminum; other applications of cladding include steels clad with stainless steel or nickel alloys. The cladding material may also be applied through dies, as in cladding steel wire with copper, or by explosives.
5. **Mechanical plating (mechanical coating, impact plating, peen plating).** Fine particles of metal are compacted over the workpiece surfaces by impacting them with spherical glass, ceramic, or porcelain beads. The beads are propelled by rotary means. This process typically is used for hardened-steel parts for automobiles, with plating thickness usually less than 0.025 mm (0.001 in.).
6. **Case hardening (carburizing, carbonitriding, cyaniding, nitriding, flame hardening, induction hardening).** These processes are described in Section 5.11.3 and summarized in Tables 5.8. In addition to the common heat sources of gas and electricity, laser beams are also used as a heat source in surface hardening of both metals and ceramics. Case hardening, as well as some of the other surface-treatment processes, induces residual stresses on surfaces. The formation of martensite in case hardening of steels causes compressive residual stresses on surfaces. Such stresses are desirable, because they improve the fatigue life of components by delaying the initiation of fatigue cracks.

7. **Hard facing.** A relatively thick layer, edge, or point of wear-resistant hard metal is deposited on the surface by any of the welding techniques described in Chapter 12. A number of layers are usually deposited (*weld overlay*). Hard coatings of tungsten carbide, chromium, and molybdenum carbide can also be deposited using an electric arc (*spark hardening*). Hard-facing alloys are available as electrodes, rods, wire, and powder. Typical applications for hard facing include valve seats, oil-well drilling tools, and dies for hot metalworking. Worn parts are also hard faced for extended use.
8. **Thermal spraying.** In *thermal spraying*, also called **metallizing**, metal in the form of rod, wire, or powder is melted in a stream of oxyacetylene flame, electric arc, or plasma arc, and the droplets are sprayed on the preheated surface at speeds up to 100 m/s (20,000 ft/min) with a compressed-air spray gun. All of the surfaces to be sprayed should be cleaned and roughened to improve bond strength. Typical applications for thermal spraying include steel structures, storage tanks, and tank cars, sprayed with zinc or aluminum up to 0.25 mm (0.010 in.) in thickness.

   There are several types of thermal-spraying processes, listed in decreasing order of performance as follows: (a) **plasma**, either conventional, high energy, or vacuum. It produces temperatures on the order of 8300°C (15,000°F) and has very good bond strength with very low oxide content. (b) **Detonation gun**, in which a controlled explosion takes place using an oxyfuel gas mixture. It has a performance similar to that of plasma. (c) **High-velocity oxyfuel** (HVOF) **gas spraying**, which has a similarly high performance as those of the foregoing processes, but is less expensive. (d) **Wire arc**, in which an arc is formed between two consumable wire electrodes. It has good bond strength and is the least expensive process. (e) **Flame wire spraying**, in which the oxyfuel flame melts the wire and deposits it on the surface. Its bond is of medium strength, but the process is relatively inexpensive.
9. **Surface texturing.** Manufactured surfaces can be further modified by secondary operations for technical, functional, optical, or aesthetic reasons. These additional processes generally consist of (1) etching, using chemicals or sputtering techniques, (2) electric arcs, (3) laser pulses and (4) atomic oxygen, which reacts with surfaces and produces fine, conelike surface textures. The possible adverse effects of any of these processes on the properties and performance of materials should be considered.
10. **Ceramic coating.** Techniques have been developed for spraying ceramic coatings for high-temperature and electrical-resistance applications, such as to withstand repeated arcing. Powders of hard metals and ceramics are also used as spraying materials. Plasma-arc temperatures may reach 15,000°C (27,000°F), which is much higher than that obtained with flames. Typical applications are nozzles for rocket motors and wear-resistant parts.
11. **Vapor deposition.** In this process, the substrate is subjected to chemical reactions by gases that contain chemical compounds of the materials to be deposited; the thickness of the coating is usually a few μm. The deposited materials may consist of metals, alloys, carbides, nitrides, borides, ceramics, or various oxides. The substrate may be metal, plastic, glass, or paper. Typical applications include coatings for cutting tools, drills, reamers, milling cutters, punches, dies, and wear surfaces.

    There are two major deposition processes: physical vapor deposition and chemical vapor deposition. These techniques allow effective control of coating composition, thickness, and porosity. The three basic types of **physical vapor deposition** (PVD) processes are vacuum or arc evaporation (PV/ARC), sputtering, and ion plating. These processes are carried out in a high vacuum at temperatures in the range of 200°C–500°C (400°F–900°F). In physical vapor deposition, the particles to be deposited are carried physically to the workpiece, rather than by chemical reactions as in chemical vapor deposition.

In **vacuum evaporation**, the metal to be deposited is evaporated at high temperatures in a vacuum and deposited on the substrate, which is usually at room temperature or slightly higher. Uniform coatings can be obtained on complex shapes with this method. In PV/ARC, the coating material (cathode) is evaporated by a number of arc evaporators, using highly localized electric arcs. The arcs produce a highly reactive plasma consisting of ionized vapor of the coating material; the vapor condenses on the substrate (anode) and coats it. Applications for this process may be functional (e.g., oxidation-resistant coatings for high-temperature applications, electronics, and optics) or decorative (e.g., hardware, appliances, and jewelry).

In **sputtering**, an electric field ionizes an inert gas (usually argon). The positive ions bombard the coating material (cathode) and cause sputtering (ejecting) of its atoms; these atoms then condense on the workpiece, which is heated to improve bonding. In **reactive sputtering**, the inert gas is replaced by a reactive gas, such as oxygen, in which case the atoms are oxidized and the oxides are deposited. **Radiofrequency** (RF) **sputtering** is used for nonconductive materials such as electrical insulators and semiconductor devices. **Ion plating** is a generic term describing the combined processes of sputtering and vacuum evaporation. An electric field causes a glow discharge, generating a plasma; the vaporized atoms in this process are only partially ionized. **Dual ion-beam-assisted deposition** is a more recent development in which PVD is combined with simultaneous ion-beam bombardment. It results in good adhesion on metals, ceramics, and polymers; ceramic bearings and dental instruments are two examples of its applications.

Chemical vapor deposition (CVD) is a thermochemical process. In a typical application, such as for coating cutting tools with titanium nitride (TiN) (Fig. 4.16), the tools are placed on a graphite tray and heated to 950°C–1050°C (1740°F–1920°F) in an inert atmosphere. Titanium tetrachloride (a vapor), hydrogen, and nitrogen are then introduced into the chamber; the chemical reactions form titanium nitride on the tool's surfaces. For coating with titanium carbide, methane is substituted for the gases. Coatings obtained by chemical vapor deposition are usually thicker than those obtained via PVD. A more recent development is **medium-temperature CVD** (MTCVD), which produces coatings with higher resistance to crack propagation than CVD coatings.

In **ion implantation**, ions are introduced into the surface of the workpiece material. The ions are accelerated in a vacuum to such an extent that they penetrate the substrate to a depth of a few μm. Ion implantation (not to be confused

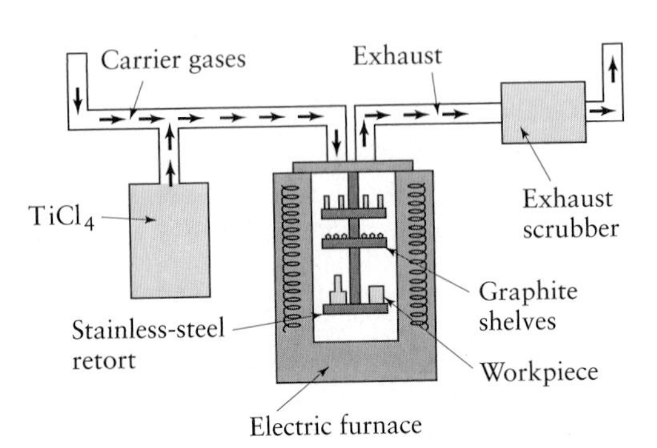

FIGURE 4.16 Schematic illustration of the chemical vapor deposition process.

with ion plating) modifies surface properties by increasing surface hardness and improving resistance to friction, wear, and corrosion. This process can be controlled accurately, and the surface can be masked to prevent ion implantation in unwanted places. When used in specific applications, such as semiconductors, this process is called **doping** (meaning alloying with small amounts of various elements).

12. **Diffusion coating.** In this process, an alloying element is diffused into the surface of the substrate, thus altering its properties. Such elements can be supplied in solid, liquid, or gaseous states. This process is given different names, depending on the diffused element, that describe diffusion processes, such as carburizing, nitriding, and boronizing. (See Table 5.8.)

13. **Electroplating.** The workpiece (cathode) is plated with a different metal (anode) while both are suspended in a bath containing a water-base electrolyte solution (Fig. 4.17). Although the plating process involves a number of reactions, the basic procedure is that the metal ions from the anode are discharged under the potential from the external source of electricity, combine with the ions in the solution, and are deposited on the cathode. All metals can be electroplated, with thicknesses ranging from a few atomic layers to a maximum of about 0.05 mm (0.002 in.). Complex shapes may have varying plating thicknesses. Common plating materials include chromium, nickel, cadmium, copper, zinc, and tin. **Chromium plating** is carried out by plating the metal first with copper, then with nickel, and finally with chromium. **Hard chromium plating** is done directly on the base metal and has a hardness of up to 70 HRC.

    Typical electroplating applications include copper plating aluminum wire and phenolic boards for printed circuits, chrome plating hardware, tin plating copper electrical terminals for ease of soldering, and plating various components for good appearance and resistance to wear and corrosion. Because they do not develop oxide films, noble metals (such as gold, silver, and platinum; see Section 3.11.8) are important electroplating materials for the electronics and jewelry industries. Plastics such as ABS, polypropylene, polysulfone, polycarbonate, polyester, and nylon also can have metal coatings applied through electroplating. Because they are not electrically conductive, plastics must be preplated by such processes as electroless nickel plating. (See below.) Parts to be coated may be simple or complex, and size is not a limitation.

14. **Electroless plating.** This process is carried out by chemical reactions and without the use of an external source of electricity. The most common application uses nickel, although copper is also used. In electroless nickel plating, nickel

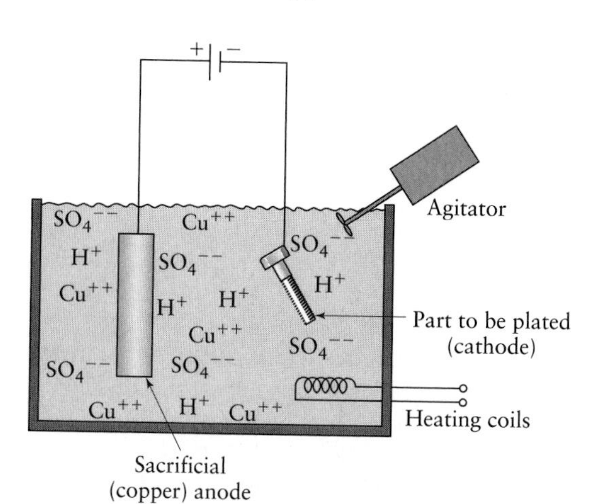

FIGURE 4.17 Schematic illustration of the electroplating process.

chloride (a metallic salt) is reduced (with sodium hypophosphite as the reducing agent) to nickel metal, which is then deposited on the workpiece. The hardness of nickel plating ranges between 425 HV and 575 HV, and the plating can be heat treated to 1000 HV. The coating has excellent wear and corrosion resistance.

15. **Electroforming.** A variation of electroplating, electroforming is actually a metal-fabricating process. Metal is electrodeposited on a *mandrel* (also called a *mold* or *matrix*), which is then removed; thus, the coating itself becomes the product. Simple and complex shapes can be produced by electroforming, with wall thicknesses as small as 0.025 mm (0.001 in.). Parts may weigh from a few grams to as much as 270 kg (600 lb). The electroforming process is particularly suited to low production quantities and intricate parts (such as molds, dies, waveguides, nozzles, and bellows) made of nickel, copper, gold, and silver; it is also suitable for aerospace, electronics, and electro-optics applications. Production rates can be increased with multiple mandrels.

16. **Anodizing.** Anodizing is an oxidation process (*anodic oxidation*) in which the surfaces of the workpiece are converted to a hard and porous oxide layer that provides corrosion resistance and a decorative finish. The workpiece is the anode in an electrolytic cell immersed in an acid bath, resulting in chemical adsorption of oxygen from the bath. Organic dyes of various colors (typically black, red, bronze, gold, and gray) can be used to produce stable, durable surface films. Typical applications for anodizing include aluminum furniture and utensils, architectural shapes, automobile trim, picture frames, keys, and sporting goods. Anodized surfaces also serve as a good base for painting, especially for aluminum, which otherwise is difficult to paint.

17. **Conversion coating.** In this process, also called **chemical-reaction priming,** a coating forms on metal surfaces as a result of chemical or electrochemical reactions. Various metals, particularly steel, aluminum, and zinc, can be conversion coated. Oxides that naturally form on their surfaces are a form of conversion coating; phosphates, chromates, and oxalates are used to produce conversion coatings. These coatings are used for purposes such as prepainting, decorative finishes, and protection against corrosion. The two common methods of coating are immersion and spraying. The equipment involved depends on the method of application, the type of product, and considerations of quality.

    An important application is in conversion coating of workpieces as a lubricant carrier in cold-forming operations (Section 4.4.4). Lubricants may not always adhere properly to workpiece surfaces, particularly when under high normal and shearing stresses; this condition is a problem particularly in forging, extrusion, and wire drawing of steels, stainless steels, and high-temperature alloys. For these applications, acids transform the workpiece surface by chemically reacting with it, leaving a somewhat rough and spongy surface that acts as a carrier for the lubricant. After treatment, borax or lime is used to remove any excess acid from the surfaces. A liquid lubricant, such as a soap, is then applied to the coated surface. The lubricant film adheres to the surface and cannot be scraped off easily. **Zinc phosphate** conversion coatings are often used on carbon and low-alloy steels. **Oxalate** coatings are used for stainless steels and high-temperature alloys.

18. **Coloring.** As the name implies, coloring involves processes that alter the color of metals, alloys, and ceramics. The change is caused by the conversion of surfaces (by chemical, electrochemical, or thermal processes) into chemical compounds, such as oxides, chromates, and phosphates. In **blackening,** iron and steels develop a lustrous black-oxide film, using solutions of hot caustic soda.

19. **Hot dipping.** In this process, the workpiece, usually steel or iron, is dipped into a bath of molten metal, such as zinc (for galvanized-steel sheet and plumbing supplies), tin (for tin plate and tin cans for food containers), aluminum (*aluminizing*), and *terne* (lead alloyed with 10% to 20% tin). Hot-dipped coatings on discrete parts or sheet metal provide galvanized pipe, plumbing supplies, and many other products with long-term resistance to corrosion. The coating thickness is usually given in terms of coating weight per unit surface area of the sheet, typically 150 g/m^2–900 g/m^2 (0.5 oz/ft^2–3 oz/ft^2). The service life depends on the thickness of the zinc coating and the environment to which it is exposed. Various **precoated sheet steels** are used extensively in automobile bodies. Proper draining to remove excess coating materials is an important consideration.

20. **Porcelain enameling.** Metals may be coated with a variety of glassy (vitreous) coatings to provide corrosion and electrical resistance and for service at elevated temperatures. These coatings are usually classified as porcelain enamels and generally include enamels and ceramics (Section 11.8). The word **enamel** is also used for glossy paints, indicating a smooth, hard coating. Porcelain enamels are glassy inorganic coatings consisting of various metal oxides. **Enameling** involves fusing the coating material on the substrate by heating them both to 425°C–1000°C (800°F–1800°F) to liquefy the oxides. Depending on their composition, enamels have varying resistances to alkali, acids, detergents, cleansers, and water and are available in different colors.

    Typical applications for porcelain enameling include household appliances, plumbing fixtures, chemical-processing equipment, signs, cookware, and jewelry; porcelain enamels are also used as protective coatings on jet-engine components. The coating may be applied by dipping, spraying, or electrodeposition, and thicknesses are usually 0.05 mm–0.6 mm (0.002 in.–0.025 in.). Metals that are coated with porcelain enamel are typically steels, cast iron, and aluminum. Glasses are used as lining for chemical resistance, and the thickness is much greater than in enameling. **Glazing** is the application of glassy coatings on ceramic and earthenwares to give them decorative finishes and to make them impervious to moisture.

21. **Organic coatings.** Metal surfaces may be coated or *precoated* with a variety of organic coatings, films, and laminates to improve appearance and corrosion resistance. Coatings are applied to the coil stock on continuous lines, with thicknesses generally of 0.0025 mm–0.2 mm (0.0001 in.–0.008 in.). Such coatings have a wide range of properties: flexibility, durability, hardness, resistance to abrasion and chemicals, color, texture, and gloss. Coated sheet metal is subsequently formed into various products, such as TV cabinets, appliance housings, paneling, shelving, siding for residential buildings, gutters, and metal furniture.

    More critical applications of organic coatings involve, for example, coatings for naval aircraft that are subjected to high humidity, rain, seawater, pollutants (such as from ship exhaust stacks), aviation fuel, deicing fluids, and battery acid and that are also impacted by particles such as dust, gravel, stones, and deicing salts. For aluminum structures, organic coatings have consisted typically of an epoxy primer and a polyurethane topcoat, with a lifetime of four to six years. Primer performance is very important for coating durability.

22. **Ceramic coatings.** Ceramics such as aluminum oxide and zirconium oxide are applied to a surface at room temperature, usually by thermal-spraying techniques. Such coatings serve as a thermal barrier in applications such as hot extrusion dies, diesel-engine components, and turbine blades.

23. **Painting.** Because of its decorative and functional properties, paint is widely used as a surface coating. Paints are basically classified as enamels, lacquers, and water-base paints, with a wide range of characteristics and applications. They are applied by brushing, dipping, spraying, or electrostatically.

24. **Diamond coating.** Important advances have been made in diamond coating of metals, glass, ceramics, and plastics, using various chemical and plasma-assisted vapor deposition processes and ion-beam enhanced deposition techniques. Techniques have also been developed to produce *free-standing diamond films* on the order of 1 mm (0.040 in.) thick and up to 125 mm (5 in.) in diameter, including smooth and optically clear diamond films. Development of these techniques, combined with important properties of diamonds, such as hardness, wear resistance, high thermal conductivity, and transparency to ultraviolet light and microwave frequencies, has enabled the production of various aerospace and electronic parts and components.

    Examples of diamond-coated products include scratchproof windows (such as for aircraft and missile sensors to protect against sandstorms), sunglasses, cutting tools (such as drills and end mills), calipers, surgical knives, razors, electronic and infrared heat seekers and sensors, light-emitting diodes, speakers for stereo systems, turbine blades, and fuel-injection nozzles. Studies are continuing on growing diamond films on crystalline copper substrate by implantation of carbon ions; an important application for such diamond films is in making computer chips. Diamond can be doped to form p- and n-type ends on semiconductors to make transistors (Chapter 13), and its high thermal conductivity allows closer packing of chips than with silicon or gallium–arsenide chips, thus significantly increasing the speed of computers.

25. **Diamond-like carbon** (DLC). By using a low-temperature, ion-beam-assisted deposition process, this relatively recently developed material is applied as a coating of a few nanometers in thickness. Less expensive than diamond films, but with similar properties as diamond, DLC has applications in such areas as tools and dies, gears, bearings, microelectromechanical systems, and microscale probes.

4.5.2 Cleaning of surfaces

We have stressed the importance of surfaces and the influence of deposited or adsorbed layers of various elements and contaminants on surfaces. A clean surface can have both beneficial and detrimental effects. Although an unclean surface may reduce the tendency for adhesion and galling, cleanliness generally is essential for more effective application of metalworking fluids, coating and painting, adhesive bonding, welding, brazing, soldering, reliable functioning of manufactured parts in machinery, manufacturing of food and beverage containers, effective storage, and in assembly operations.

Cleaning involves the removal of solid, semisolid, or liquid contaminants from a surface, and it is an important part of manufacturing operations and the economics of production. The word *clean*, or the degree of cleanliness of a surface, is somewhat difficult to define; how, for example, would you test the cleanliness of a dinner plate? Two simple and common tests are based on the following procedures:

1. Wiping the area with a clean cloth and observing any *residues* on the cloth; or

2. Observing whether water continuously coats the surface (called the *waterbreak test*). If water collects as individual droplets, the surface is not clean. Test this phenomenon yourself by wetting dinner plates that have been cleaned to varying degrees.

The type of cleaning process required depends on the type of contaminants to be removed. *Contaminants*, also called *soils*, may consist of rust, scale, chips and other metallic and nonmetallic debris, metalworking fluids, solid lubricants, pigments, polishing and lapping compounds, and general environmental elements.

Basically, there are two types of cleaning methods: mechanical and chemical. **Mechanical cleaning methods** consist of physically disturbing the contaminants, as with wire or fiber brushing, dry or wet abrasive blasting, tumbling, and steam jets. Many of these processes are particularly effective in removing rust, scale, and other solid contaminants. Ultrasonic cleaning may also be placed in this category.

Chemical cleaning methods usually involve the removal of oil and grease from surfaces. They consist of one or more of the following processes:

- *Solution:* The soil dissolves in the cleaning solution.
- *Saponification:* A chemical reaction that converts animal or vegetable oils into a soap that is soluble in water.
- *Emulsification:* The cleaning solution reacts with the soil or lubricant deposits and forms an emulsion; the soil and the emulsifier then become suspended in the emulsion.
- *Dispersion:* The concentration of soil on the surface is decreased by surface-active materials in the cleaning solution.
- *Aggregation:* Lubricants are removed from the surface by various agents in the cleaning fluid and collect as large dirt particles.

Some common **cleaning fluids** are used in conjunction with electrochemical processes for more effective cleaning. These fluids include *alkaline solutions, emulsions, solvents, hot vapors, acids, salts*, and mixtures of *organic compounds*.

In **vapor degreasing**, a heated, evaporated solvent collects on workpieces that are at room temperature. Dirt and lubricant on the workpiece are removed and dispersed, emulsified, or dissolved in the solvent. As the solvent drips off the workpiece into the solvent tank, the contaminants are removed. An advantage of vapor degreasing is that the solvent does not become dilute, since the vapor does not contain previously removed contaminant. A drawback is that the fumes can be irritating or toxic, so that environmental controls are needed.

Mechanical agitation of the surface can aid in removal of contaminants; examples of this method include wire brushing, abrasive blasting, abrasive jets, and ultrasonic vibration of a solvent bath (**ultrasonic cleaning**). In **electrolytic cleaning**, a charge is applied to the workpiece in an aqueous, often alkaline, cleaning solution. This charge results in bubbles of hydrogen or oxygen being released at the workpiece surface, depending on the polarity of the charge; these abrasive bubbles aid in the removal of contaminants.

Cleaning discrete parts that have complex shapes can be difficult. Design engineers should be aware of this difficulty; provide alternative designs, such as avoiding deep blind holes or making several smaller components instead of one large component that may be difficult to clean; and provide appropriate drain holes in the part to be cleaned.

4.6 Engineering Metrology and Instrumentation

Engineering metrology is defined as the measurement of dimensions such as length, thickness, diameter, taper, angle, flatness, and profiles. Traditionally, measurements have been made after the part has been produced, a procedure known as *postprocess inspection*. The practice in manufacturing now is to make measurements while the part is being produced on the machine, a procedure known as *in-process*, *on-line*, or *real-time inspection*. Here, the term *inspection* means to check the dimensions of what is produced or being produced, and to observe whether it complies with the specified dimensional accuracy. Much progress has been made in developing new and automated measuring instruments that are highly precise and sensitive.

4.6.1 Measuring instruments

1. **Line-graduated instruments.** Line-graduated instruments are used for measuring length (linear measurements) or angles (angular measurements). *Graduated* means marked to indicate a certain quantity. The simplest and most commonly used instrument for making linear measurements is a *steel rule* (*machinist's rule*), bar, or tape with fractional or decimal graduations. Lengths are measured directly, to an accuracy that is limited to the nearest division, usually 1 mm or $\frac{1}{64}$ in. Rules may be rigid or flexible and may be equipped with a hook at one end for ease of measuring from an edge. Rule-depth gages are similar to rules and slide along a special head.

   Vernier calipers have a graduated beam and a sliding jaw with a *vernier*; these instruments are also called **caliper gages**. The two jaws of the caliper contact the part being measured, and the dimension is read at the matching graduated lines. Vernier calipers can be used to measure inside or outside lengths. The vernier improves the sensitivity of a simple rule by indicating fractions of the smallest division on the graduated beam, usually to 25 μm (0.001 in.). Calipers are also equipped with *digital readouts*, which are easy to read and less subject to human error. Vernier height gages are vernier calipers with setups similar to that of a depth gage and have similar sensitivity as well.

   Micrometers, which have a graduated, threaded spindle, are commonly used for measuring the thickness and inside or outside diameters of parts. Circumferential vernier readings to a sensitivity of 2.5 μm (0.0001 in.) can be obtained. Micrometers are also available for measuring depths (*micrometer depth gage*) and internal diameters (*inside micrometer*) with the same sensitivity. Micrometers are also equipped with digital readouts to reduce errors in reading. The anvils on micrometers can be equipped with conical or ball contacts; they are used to measure inside recesses, threaded rod diameters, and wall thicknesses of tubes and curved sheets.

   Diffraction gratings consist of two flat optical glasses with closely spaced parallel lines scribed on their surfaces. The grating on the shorter glass is inclined slightly; as a result, interference fringes develop when it is viewed over the longer glass. The position of these fringes depends on the relative position of the two sets of glasses. With modern equipment, using electronic counters and photoelectric sensors, sensitivities of 2.5 μm (0.0001 in.) can be obtained with gratings that have 40 lines/mm (1000 lines/in.).

2. **Indirect-reading instruments.** Indirect-reading instruments typically are calipers and dividers without any graduated scales. They are used to transfer the size measured to a direct-reading instrument, such as a rule. After the legs of

the instrument have been adjusted to contact the part at the desired location, the instrument is held against a graduated rule, and the dimension is read. Because of both the experience required to use them and their dependence on graduated scales, the accuracy of this type of indirect measurement is limited. **Telescoping gages** are available for indirect measurement of holes or cavities.

Angles are measured in degrees, radians, or minutes and seconds of arc. Because of the geometry involved, angles are usually more difficult to measure than are linear dimensions. A **bevel protractor** is a direct-reading instrument similar to a common protractor, except that it has a movable member. The two blades of the protractor are placed in contact with the part being measured, and the angle is read directly on the vernier scale. The sensitivity of the instrument depends on the graduations of the vernier. Another type of bevel protractor is the **combination square,** which is a steel rule equipped with devices for measuring 45° and 90° angles.

Measuring with a **sine bar** involves placing the part on an inclined bar or plate and adjusting the angle by placing gage blocks on a surface plate. After the part is placed on the sine bar, a dial indicator (see later) is used to scan the top surface of the part. **Gage blocks** are added or removed as necessary until the top surface is parallel to the surface plate. The angle on the part is then calculated from geometric relationships. Angles can also be measured by using **angle gage blocks.** These blocks have different tapers that can be assembled in various combinations and used in a manner similar to that for sine bars. Angles on small parts can also be measured through microscopes, with graduated eyepieces, or with optical projectors.

3. **Comparative length-measuring instruments.** Unlike the instruments described thus far, instruments used for measuring *comparative lengths*, also called *deviation-type* instruments, amplify and measure variations or deviations in distance between two or more surfaces. These instruments compare dimensions, hence the word *comparative*. The common types of instruments used for making comparative measurements are described next.

   Dial indicators are simple mechanical devices that convert linear displacements of a pointer to rotation of an indicator on a circular dial. The indicator is set to zero at a certain reference surface, and the instrument or the surface to be measured (either external or internal) is brought into contact with the pointer. The movement of the indicator is read directly on the circular dial (as either plus or minus) to accuracies as high as 1 μm (40 μin.).

   Unlike mechanical systems, **electronic gages** sense the movement of the contacting pointer through changes in the electrical resistance of a strain gage or through inductance or capacitance. The electrical signals are then converted and displayed as linear dimensions. A commonly used electronic gage is the **linear variable differential transformer** (LVDT), used extensively for measuring small displacements. Although they are more expensive than other types of gages, electronic gages have advantages such as ease of operation, rapid response, digital readout, less possibility of human error, versatility, flexibility, and the capability to be integrated into automated systems through microprocessors and computers.

4. **Measuring straightness, flatness, roundness, and profile.** The geometric features of straightness, flatness, roundness, and profile are important aspects of engineering design and manufacturing. For example, piston rods, instrument components, and machine-tool slideways should all meet certain requirements with regard to these characteristics in order to function properly. Consequently, their accurate measurement is essential.

Straightness can be checked with straight edges or dial indicators. **Autocollimators**, resembling a telescope with a light beam that bounces back from the object, are used for accurately measuring small angular deviations on a flat surface. Optical means such as **transits** and **laser beams** are used for aligning individual machine elements in the assembly of machine components.

Flatness can be measured by mechanical means, using a surface plate and a dial indicator. This method can also be used for measuring perpendicularity, which can be measured with the use of precision steel squares as well. Another method for measuring flatness is by **interferometry**, using an **optical flat**. The device, a glass or fused quartz disk with parallel flat surfaces, is placed on the surface of the workpiece. When a monochromatic (i.e., of only one wavelength) light beam is aimed at the surface at an angle, the optical flat splits it into two beams, appearing as light and dark bands to the naked eye. The number of fringes that appear is related to the distance between the surface of the part and the bottom surface of the optical flat. Consequently, a truly flat workpiece surface (that is, when the angle between the two surfaces is zero) will not split the light beam, and fringes will not appear. When surfaces are not flat, fringes are curved. The interferometry method is also used for observing surface textures and scratches through microscopes for better visibility.

Roundness is usually described as deviations from true roundness (mathematically, a circle). The term **out of roundness** is actually more descriptive of the shape of the part. Roundness is essential to the proper functioning of rotating shafts, bearing races, pistons, cylinders, and steel balls in bearings. The various methods of measuring roundness fall into two categories. In the first method, the round part is placed on a V-block or between centers and is rotated, with the pointer of a dial indicator in contact with the surface. After a full rotation of the workpiece, the difference between the maximum and minimum readings on the dial is noted; this difference is called the **total indicator reading** (TIR), or **full indicator movement**. This method is also used for measuring the straightness (*squareness*) of shaft end faces. In the second method, called **circular tracing**, the part is placed on a platform, and its roundness is measured by rotating the platform. Conversely, the probe can be rotated around a stationary part to make the measurement.

Profile may be measured by several methods. In one method, a surface is compared with a template or profile gage to check shape conformity; radii or fillets can be measured by this method. Profile may also be measured with a number of dial indicators or similar instruments. Profile-tracing instruments are the latest development for the measurement of profile.

Threads and **gear teeth** have several features that require specific dimensions and tolerances. These dimensions must be produced accurately for the smooth operation of gears, reduction of wear and noise level, and part interchangeability. These features are measured by means of thread gages of various designs that compare the thread produced with a standard thread. Some of the gages used are threaded plug gages, screw-pitch gages (similar to radius gages), micrometers with cone-shaped points, and snap gages with anvils in the shape of threads. Gear teeth are measured with instruments that are similar to dial indicators, with calipers and with micrometers using pins or balls of various diameters.

Special profile-measuring equipment is also available, including optical projectors. **Optical projectors**, also called **optical comparators**, were first developed in the 1940s to check the geometry of cutting tools for machining screw threads, but are now used for checking all profiles. The part is mounted on a

FIGURE 4.18 A coordinate measuring machine, measuring dimensions on an engine block. *Source*: Courtesy of Sheffield Measurement Division, Giddings & Lewis.

table, or between centers, and the image is projected on a screen at magnifications up to 100× or higher. Linear and angular measurements are made directly on the screen, which is equipped with reference lines and circles. The screen can be rotated to allow angular measurements as small as 1 min, using verniers.

5. **Coordinate measuring machines and layout machines.** These machines consist of a platform on which the workpiece being measured is placed and moved linearly or rotated (Fig. 4.18); a stylus, attached to a head capable of lateral and vertical movements, records all measurements. *Coordinate measuring machines* (CMMs), also called **measuring machines**, are versatile in their capability to record measurements of complex profiles with high sensitivity (0.25 μm [10 μin.]).

   These machines are built rigidly and are very precise. They are equipped with digital readout or can be linked to computers for on-line inspection of parts. They can be placed close to machine tools for efficient inspection and rapid feedback for correction of processing parameters before the next part is made. They are also being made more rugged to resist environmental effects in manufacturing plants, such as temperature variations, vibration, and dirt. Dimensions of large parts are measured by **layout machines**, which are equipped with digital readout. These machines are equipped with scribing tools for marking dimensions on large parts, with an accuracy of 0.04 mm (0.0016 in.).

6. **Gages.** Thus far, we have used the word *gage* to describe some types of measuring instruments, such as a caliper gage, depth gage, telescoping gage, electronic gage, strain gage, and radius gage. The reason for this word choice is that the words *instrument* and *gage* (also spelled *gauge*) have traditionally been used interchangeably. However, *gage* also has a variety of other meanings, such as in pressure gage; gage length of a tension-test specimen; and gages for sheet metal, wire, railroad rail, and the bore of shotguns.

Gage blocks are individual square, rectangular, or round metal or ceramic blocks of various sizes, made very precisely from heat-treated and stress-relieved alloy steels or from carbides. Their surfaces are lapped and are flat and parallel within a range of 0.02 μm–0.12 μm (1 μin.–5 μin.). Gage blocks are available in sets of various sizes, some sets containing almost 100 gage blocks; the blocks can be assembled in many combinations to obtain desired lengths. Dimensional accuracy can be as high as 0.05 μm (2 μin.). Environmental-temperature control is important in using gages for high-precision measurements. Although their use requires some skill, gage-block assemblies are commonly utilized in industry as an accurate reference length. Angle blocks are made similarly and are available for angular gaging.

Fixed gages are replicas of the shapes of the parts to be measured. **Plug gages** are commonly used for holes. The *GO gage* is smaller than the *NOT GO* (or *NO GO*) *gage* and slides into any hole whose smallest dimension is less than the diameter of the gage. The NOT GO gage must not go into the hole. Two gages, one GO gage and one NOT GO gage, are required for such measurements, although both may be on the same device, either at opposite ends or in two steps at one end (*step-type gage*). Plug gages are also available for measuring internal tapers (in which deviations between the gage and the part are indicated by the looseness of the gage), splines, and threads (in which the GO gage must screw into the threaded hole).

Ring gages are used to measure shafts and similar round parts, and **ring thread gages** are used to measure external threads. The GO and NOT GO features on these gages are identified by the type of knurling on the outside diameter of the rings. **Snap gages** are commonly used to measure external dimensions. They are made with adjustable gaging surfaces for use with parts that have different dimensions. One of the gaging surfaces may be set at a different gap than that of the other, thus making a one-unit GO–NOT-GO gage. Although fixed gages are easy to use and inexpensive, they only indicate whether a part is too small or too large compared with an established standard; they do not measure actual dimensions.

There are several types of **pneumatic gages**, also called *air gages*. The gage head has holes through which pressurized air, supplied by a constant-pressure line, escapes. The smaller the gap between the gage and the hole, the more difficult it is for the air to escape, and hence the back pressure is higher. The back pressure, sensed and indicated by a pressure gage, is calibrated to read dimensional variations of holes.

7. **Microscopes.** Microscopes are optical instruments used to view and measure very fine details, shapes, and dimensions on small and medium-sized tools, dies, and workpieces. The most common and versatile microscope used in tool rooms is the **toolmaker's microscope**. It is equipped with a stage that is movable in two principal directions and can be read to 2.5 μm (0.0001 in.). Several models of microscopes are available with various features for specialized inspection, including models with digital readout. The **light-section microscope** is used to measure small surface details, such as scratches, and the thickness of deposited films and coatings. A thin light band is applied obliquely to the surface, and the reflection is viewed at 90°, showing surface roughness, contours, and other features. Unlike ordinary optical microscopes, the **scanning electron microscope** (SEM) has excellent depth of field. As a result, all regions of a complex part are in focus and can be viewed and photographed to show extremely fine detail. This type of microscope is particularly useful for studying surface textures and fracture patterns. Although expensive, such microscopes are capable of magnifications greater than 100,000×.

4.6.2 Automated measurement

With increasing automation in all aspects of manufacturing processes and operations, the need for *automated measurement* (also called **automated inspection**) has become much more apparent. Flexible manufacturing systems and manufacturing cells (see Chapter 15) have led to the adoption of advanced measurement techniques and systems. In fact, installation and use of these systems is now a necessary and not an optional manufacturing technology.

Traditionally, a batch of parts was manufactured and sent for measurement in a separate quality-control room, and if the parts passed measurement inspection, they were put into inventory. Automated inspection, however, is based on various on-line sensor systems that monitor the dimensions of parts being made and use the feedback from these measurements to correct the process, if necessary. To appreciate the importance of on-line monitoring of dimensions, let's find the answer to the following question: if a machine has been producing a certain part with acceptable dimensions, what factors contribute to subsequent deviation in the dimension of the same part produced by the same machine? The major factors are as follows:

1. Static and dynamic deflections of the machine because of vibrations and fluctuating forces, caused by variations such as in the properties and dimensions of the incoming material;
2. Deformation of the machine because of thermal effects, including changes in the temperature of the environment, of metalworking fluids, and of machine bearings and components;
3. Wear of tools and dies.

As a result of these factors, the dimensions of parts produced will vary, necessitating monitoring of dimensions during production. **In-process workpiece control** is accomplished by special gaging and is used in a variety of applications, such as high-quantity machining and grinding.

4.7 Dimensional Tolerances

Dimensional tolerance is defined as the permissible or acceptable variation in the dimensions (height, width, depth, diameter, angles) of a part. Tolerances are unavoidable, because it is virtually impossible (and unnecessary) to manufacture two parts that have precisely the same dimensions. Furthermore, because close dimensional tolerances substantially increase the product cost, a narrow tolerance range is undesirable economically. Tolerances become important only when a part is to be assembled or mated with another part. Surfaces that are free and not functional do not need close dimensional tolerance control; thus, for example, the accuracies of the dimensions of the holes and the distance between the holes for a connecting rod are far more critical than the accuracy of the rod's width and thickness at various locations along its length.

To illustrate the importance of dimensional tolerances, imagine that we want to assemble a simple shaft (axle) and a wheel with a hole, assuming that we want the axle's diameter to be 1 in. (Fig. 4.19a). We go to the hardware store and purchase a 1-in.-thick round rod and a wheel with a hole that has a diameter of 1 in. Will the rod fit into the hole without being forced, or will it be loose in the hole? We can't be sure. The 1-in. dimension is the *nominal size* of the shaft. If we purchase such a rod

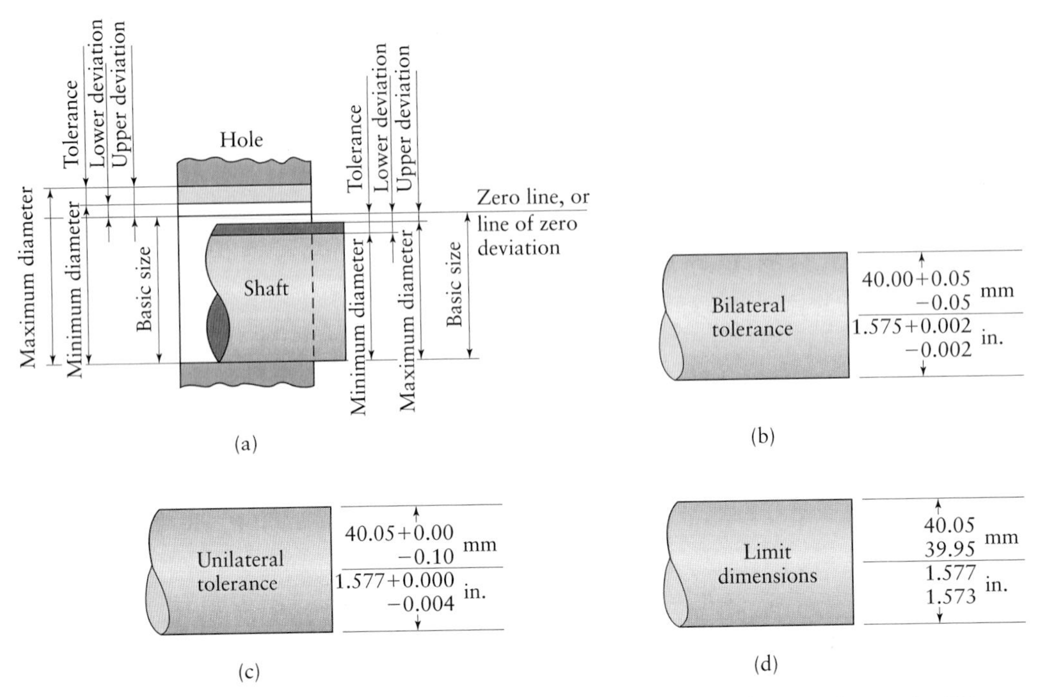

FIGURE 4.19 (a) Basic size, deviation, and tolerance on a shaft, according to the ISO system. (b)–(d) Various methods of assigning tolerances on a shaft. *Source*: L. E. Doyle.

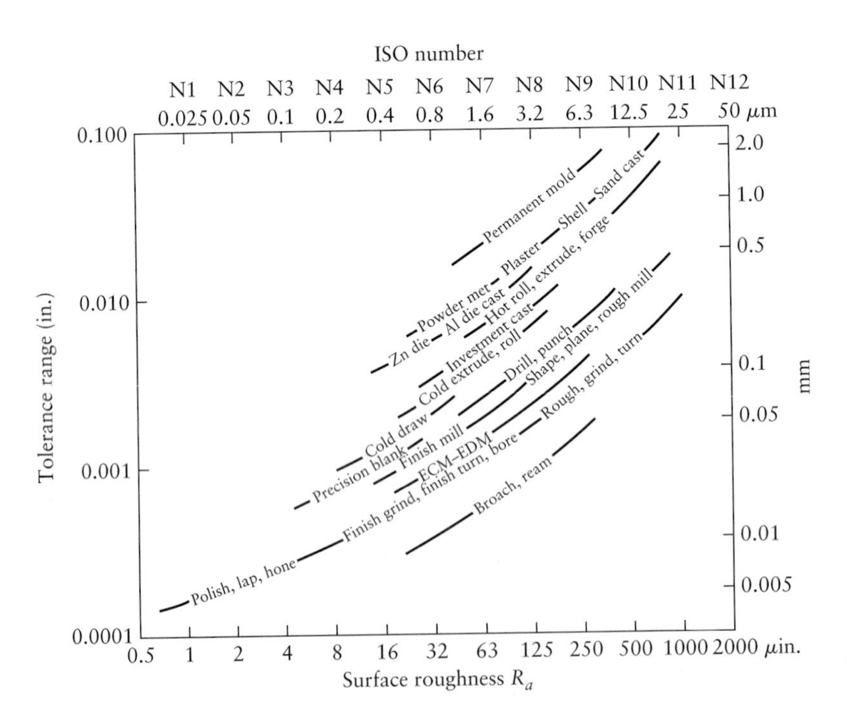

FIGURE 4.20 Tolerances and surface roughness obtained in various manufacturing processes. These tolerances apply to a 25-mm (1-in.) workpiece dimension. *Source*: Schey, J.A., *Introduction to Manufacturing Processes*, 3rd ed. New York, McGraw-Hill, 2000, p. 68.

from different stores or at different times, or select one randomly from a lot of, say, 100 shafts, the chances are that each rod will have a slightly different diameter. Machines may, with the same setup, produce rods of slightly different diameters, depending on a number of factors, such as speed of operation, temperature, lubrication, variations in the incoming material, and similar variables.

If we now specify a range of diameters for both the rod and the hole of the wheel, we can predict correctly the type of fit that we will have after assembly. Certain terminology has been established to clearly define these geometric quantities, such as the ISO system shown in Fig. 4.19a. Note that both the shaft and the hole have minimum and maximum diameters, respectively, the difference being the dimensional tolerance for each member. A proper engineering drawing should specify these parameters with numerical values, as shown in Fig. 4.19b.

The range of dimensional tolerances obtained in various manufacturing processes is given in Fig. 4.20. There is a general relationship between tolerances and surface finish of parts manufactured by various processes. Note the wide range of tolerances and surface finishes that can be obtained. Also, the larger the part, the greater the obtainable tolerance range becomes. Experience has shown that dimensional inaccuracies of manufactured parts are approximately proportional to the cube root of the size of the part; thus, doubling the size of a part increases the inaccuracies by $2^{1/3} = 1.26$ times, or 26%.

4.8 Testing and Inspection

Before they are marketed, manufactured parts and products are inspected for several characteristics. This inspection routine is particularly important for products or components whose failure or malfunction has potentially serious implications, such as bodily injury or fatality. Typical examples of such failures include cables breaking, switches malfunctioning, brakes failing, grinding wheels breaking, railroad wheels fracturing, turbine blades failing, pressure vessels bursting, and welded joints failing. In this section, we identify and describe the various methods that are commonly used to inspect manufactured products.

Product quality has always been one of the most important elements in manufacturing operations, and with increasing domestic and international competition, it has become even more important. Prevention against defects in products and on-line inspection are now major goals in all manufacturing activities. We again emphasize that quality must be built into a product and not merely checked after the product has been made. Close cooperation and communication between design and manufacturing engineers are thus essential.

4.8.1 Nondestructive testing

Nondestructive testing (*NDT*) is carried out in such a way that product integrity and surface texture remain unchanged. These techniques generally require considerable operator skill. Interpreting test results accurately may be difficult, because test results can be quite subjective. However, the use of computer graphics and other enhancement techniques has reduced the likelihood of human error in nondestructive testing. The basic principles of the more commonly used nondestructive testing techniques are described as follows:

1. In the **liquid-penetrants technique**, fluids are applied to the surfaces of the part and allowed to penetrate into surface openings, cracks, seams, and porosity; the penetrant can seep into cracks as small as 0.1 μm (4 μin.) in width. Two common types of liquids are (a) fluorescent penetrants with various sensitivities, which fluoresce under ultraviolet light; and (b) visible penetrants, using dyes usually red in color, which appear as bright outlines on the surface.

   The surface to be inspected is first thoroughly cleaned and dried. The liquid is brushed or sprayed on the surface to be inspected and allowed to remain long enough to seep into surface openings. Excess penetrant is then wiped off or washed away with water or solvent. A developing agent is then added to allow the penetrant to seep back to the surface and spread to the edges of openings, thus magnifying the size of defects. The surface is then inspected for defects, either visually in the case of dye penetrants or with fluorescent lighting. This method is capable of detecting a variety of surface defects and is used extensively. The equipment is simple and easy to use, can be portable, and is less costly to operate than other methods. However, this method can detect only defects that are open to the surface, not internal defects.

2. The **magnetic-particle inspection technique** consists of placing fine ferromagnetic particles on the surface of the part. The particles can be applied either dry or in a liquid carrier such as water or oil. When the part is magnetized with a magnetic field, a discontinuity (defect) on the surface causes the particles to gather visibly around it. The collected particles generally take the shape and size of the defect. Subsurface defects can also be detected by this method, provided that they are not deep. The ferromagnetic particles may be colored with pigments for better visibility on metal surfaces. Wet particles are used for detecting fine discontinuities, such as fatigue cracks.

3. In **ultrasonic inspection**, an ultrasonic beam travels through the part. An internal defect, such as a crack, interrupts the beam and reflects back a portion of the ultrasonic energy. The amplitude of the energy reflected and the time required for return indicate the presence and location of any flaws in the workpiece. The ultrasonic waves are generated by transducers, called *search units* or *probes*, of various types and shapes. They operate on the principle of piezoelectricity (Section 3.9.6), using materials such as quartz, lithium sulfate, and various ceramics. Most inspections are carried out at a frequency range of 1–25 MHz. Couplants are used to transmit the ultrasonic waves from the transducer to the test piece; typical couplants are water, oil, glycerin, and grease.

   The ultrasonic inspection method has high penetrating power and sensitivity. It can be used to inspect flaws in large volumes of material, such as railroad wheels, pressure vessels, and die blocks, from various directions. Its accuracy is higher than that of other nondestructive inspection methods. However, this method requires experienced personnel to carry out the inspection and interpret the results correctly.

4. The **acoustic-emission technique** detects signals (high-frequency stress waves) generated by the workpiece itself during plastic deformation, crack initiation and propagation, phase transformation, and sudden reorientation of grain boundaries. Bubble formation during boiling, and friction and wear of sliding interfaces, are other sources of acoustic signals. Acoustic-emission inspection is typically performed by stressing elastically the part or structure, such as bending a beam, applying torque to a shaft, and pressurizing a vessel; the technique is particularly effective for continuous surveillance of load-bearing structures. Acoustic emissions are detected by sensors consisting of piezoelectric ceramic elements.

5. The **acoustic-impact technique** consists of tapping the surface of an object and listening to and analyzing the signals in order to detect discontinuities and flaws. The principle is basically the same as tapping walls, desktops, or countertops in various locations with your fingers or a hammer and listening to the sound emitted. Vitrified grinding wheels are tested in a similar manner (using a ring test) to detect cracks in the wheel that may not be visible to the naked eye. The

acoustic-impact technique can be instrumented and automated and is easy to perform. However, the results depend on the geometry and mass of the part, thus requiring a reference standard to identify flaws.

6. **Radiography** involves X-ray inspection to detect internal flaws or variations in density and thickness in the part. The radiation source is typically an X-ray tube, and a visible permanent image is made on an X-ray film or radiographic paper.

7. The **eddy-current inspection method** is based on the principle of electromagnetic induction. The part is placed in or adjacent to an electric coil through which alternating current (exciting current) flows at frequencies ranging from 6 MHz to 60 MHz. This current induces eddy currents in the part. Defects in the part impede and change the direction of eddy currents, causing changes in the electromagnetic field. These changes affect the exciting coil (inspection coil), whose voltage is monitored to determine the presence of flaws.

8. **Thermal inspection** involves observing temperature changes by contact- or noncontact-type heat-sensing devices. Defects in the workpiece, such as cracks, debonded regions in laminated structures, and poor joints, cause a change in the temperature distribution. In **thermographic inspection**, materials such as heat-sensitive paints and papers, liquid crystals, and other coatings are applied to the surface of the part. Any changes in their color or appearance indicate defects.

9. The **holography technique** creates a three-dimensional image of the workpiece, using an optical system. This technique is generally used on simple shapes and highly polished surfaces, and the image is recorded on a photographic film. Use of this technique has been extended to inspection of parts (**holographic interferometry**) that have various shapes and surface conditions. Through double- and multiple-exposure techniques while the part is being subjected to external forces or time-dependent variations, changes in the images reveal defects in the part. In **acoustic holography**, information on internal defects is obtained directly from the image of the interior of the part. In **liquid–surface acoustical holography**, the workpiece and two ultrasonic transducers (one for the object beam and the other for the reference beam) are immersed in a water-filled tank. The holographic image is then obtained from the ripples in the tank. In **scanning acoustical holography**, only one transducer is used, and the hologram is produced by electronic-phase detection. This system is more sensitive, the equipment is usually portable, and very large workpieces can be accommodated by using a water column instead of a tank.

4.8.2 Destructive testing

As the name suggests, the part or product tested using *destructive testing methods* no longer maintains its integrity, original shape, or surface texture. *Mechanical testing methods* (see Chapter 2) are all destructive, in that a sample or specimen has to be removed from the product in order to test it. In addition to mechanical testing, other destructive tests include speed testing of grinding wheels to determine their bursting speed and high-pressure testing of pressure vessels to determine their bursting pressure. Hardness tests that leave large impressions may be regarded as destructive testing methods; however, microhardness tests may be regarded as nondestructive, because only very small permanent indentations are made. This distinction is based on the assumption that the material is not *notch sensitive* (Section 2.9). Most glasses, highly heat-treated metals, and ceramics are notch sensitive; that is, the small indentation produced by the indenter may lower the material's strength and toughness.

4.8.3 Automated inspection and sensors

Note that in all the preceding examples, the testing methods involve parts or products that have already been manufactured. Traditionally, individual parts and assemblies of parts have been manufactured in batches, sent to inspection in quality-control rooms, and, if approved, put in inventory. If products do not pass the quality inspection, they are either scrapped or kept on the basis of a certain acceptable deviation from the standard. Obviously, such a system lacks flexibility, requires maintenance of an inventory, and inevitably results in some defective parts going through the system. The traditional method actually tracks the defects after they occur (**postprocess inspection**), and in no way attempts to prevent defects.

In contrast, one of the most important practices in modern manufacturing is **automated inspection.** This method uses a variety of sensor systems that monitor the relevant parameters *during* the manufacturing process (**on-line inspection**). Then, using these measurements, the process automatically corrects itself to produce acceptable parts; thus, further inspection of the part at another location in the plant is unnecessary. Parts may also be inspected immediately after they are produced (**in-process inspection**).

The use of accurate sensors (see later) and computer-control systems (see Chapter 15) has integrated automated inspection into manufacturing operations. Such a system ensures that no part is moved from one manufacturing process to another (for example, a turning operation on a lathe followed by cylindrical grinding) unless the part is made correctly and meets the standards of the first operation. Automated inspection is flexible and responsive to product design changes. Furthermore, because the system uses automated equipment, less operator skill is required; productivity is increased; and parts have higher quality, reliability, and dimensional accuracy.

Sensors for automated inspection. Rapid advances in **sensor technology** are making on-line, or real-time, monitoring of manufacturing processes feasible. (See also Section 14.8.) Directly or indirectly, and with the use of various probes, sensors can detect dimensional accuracy, surface roughness, temperature, force, power, vibration, tool wear, and the presence of external or internal defects. Sensors operate on the same principles as those of strain gages, inductance, capacitance, ultrasonics, acoustics, pneumatics, infrared radiation, optics, lasers, and various electronic gages. Sensors may be tactile (touching) or nontactile.

Sensors, in turn, are linked to microprocessors and computers for graphical display of data. This capability allows rapid on-line adjustment of one or more processing parameters in order to produce parts that are consistently within specified standards of tolerance and quality. Such systems are already implemented as standard equipment on many metal-cutting machine tools and grinding machines.

4.9 Quality Assurance

Quality assurance is the total effort by a manufacturer to ensure that its products conform to a detailed set of specifications and standards. These standards cover several parameters, such as dimensions; surface finish; dimensional tolerances; composition; color; and mechanical, physical, and chemical properties. In addition, standards are usually written to ensure proper assembly using interchangeable, defect-free components and the fabrication of a product that performs as intended by its designers. Quality assurance is the responsibility of everyone involved with design and manufacturing. The often-repeated statement that quality must be built into a product reflects this important concept: Quality cannot be inspected into a finished product. Although product quality has always been important, increased domestic and international competition has caused quality assurance to become even more important. Every aspect of design and manufacturing operations, such as material selection, production, and assembly, is now analyzed in detail to ensure that quality is truly built into the final product.

If you were in charge of product quality for a manufacturing plant, how would you make sure that the final product is of acceptable quality? The best method is to control materials and processes in such a manner that the products are made correctly in the first place. However, 100% inspection is usually too costly to maintain. Therefore, several methods of inspecting smaller, statistically relevant sample lots have been devised. These methods all use **statistics** to determine the probability of defects occurring in the total production batch.

Inspection involves a series of steps:

1. Inspecting incoming materials to make sure that they meet certain property, dimension, and surface finish and integrity requirements;
2. Inspecting individual product components to make sure that they meet specifications;
3. Inspecting the product to make sure that individual parts have been assembled properly;
4. Testing the product to make sure that it functions as designed and intended.

Inspections must be continued during production, because there are always variations in the dimensions and properties of incoming materials; variations in the performance of tools, dies, and machines used in various stages of manufacturing; possibilities of human error; and errors made during assembly of the product. As a result, no two products are ever made exactly alike.

Another important aspect of quality control is the capability to analyze defects and promptly eliminate them or reduce them to acceptable levels. In an even broader sense, quality control involves evaluating the product design and customer satisfaction; the sum total of all these activities is referred to as **total quality management** (TQM). Thus, in order to control quality, we have to be able to (1) measure quantitatively the level of quality, and (2) identify all the material and process variables that can be controlled. The level of quality obtained during production can then be established by inspecting the product to determine whether it meets the specifications for dimensional tolerances, surface finish, defects, and other characteristics. The identification of material and process variables and their effect on product quality is now possible through the extensive knowledge gained from research and development activities in all aspects of manufacturing.

4.9.1 Statistical methods of quality control

Statistics deals with the collection, analysis, interpretation, and presentation of large amounts of numerical data. The use of statistical techniques in modern manufacturing operations is necessary because of the large number of material and process variables involved. Events that occur randomly, that is, without any particular trend or pattern, are called **chance variations**. Events that can be traced to specific causes are called **assignable variations**. The existence of **variability** in production operations has been recognized for centuries, but Eli Whitney (1765–1825) first grasped its full significance when he found that interchangeable parts were indispensable to the mass production of firearms. Modern statistical concepts relevant to manufacturing engineering were first developed in the early 1900s.

To understand **statistical quality control** (SQC), we first need to define some of the terms that are commonly used in this field.

a) **Sample size:** The number of parts to be inspected in a sample, whose properties are studied to gain information about the whole population.
b) **Random sampling:** Taking a sample from a population or lot in which each item has an equal chance of being included in the sample; thus, when taking samples from a large bin, an inspector does not take only those items that happen to be within reach.
c) **Population** (also called the **universe**): The totality of individual parts of the same design from which samples are taken.
d) **Lot size:** A subset of population. A lot or several lots can be considered subsets of the population and may be treated as representative of the population.

Samples are inspected for certain characteristics and features, such as dimensional tolerances, surface finish, and defects, with the instruments and techniques described earlier in this chapter. These characteristics fall into two categories: those that can be measured quantitatively (method of variables) and those that can be measured qualitatively (method of attributes).

The **method of variables** is the quantitative measurement of characteristics such as dimensions, tolerances, surface finish, and physical or mechanical properties. Such measurements are made for each of the units in the group under consideration, and the results are compared with the specifications for the part. The **method of attributes** involves observing the presence or absence of qualitative characteristics, such as external or internal defects in machined, formed, or welded parts, or dents in sheet-metal products, for each of the units in the group under consideration. Sample size for attributes-type data is generally larger than that for variables-type data.

During the inspection process, measurement results will vary. For example, assume that you are measuring the diameter of turned shafts as they are produced on a lathe, using a micrometer. You soon note that their diameters vary, even though ideally you want all the shafts to be exactly the same size. Let's now turn to consideration of statistical quality-control techniques, which allow us to evaluate these variations and set limits for the acceptance of parts.

If we list the measured diameters of the turned shafts in a given population, we note that one or more parts have the smallest diameter, and one or more have the largest diameter; the majority of the turned shafts have diameters that lie between these extremes. We can then group these diameters and plot them on a bar graph, where each bar represents the number of parts in each diameter group (Fig. 4.21a). The bars show a **distribution**, also called a **spread** or **dispersion**, of the shaft-diameter measurements. The bell-shaped curve in Fig. 4.21a is called **frequency distribution** and represents the frequency with which parts within each diameter group are being produced.

Data from manufacturing processes often fit curves represented by a mathematically derived **normal distribution curve** (Fig. 4.21b); these curves are also called *Gaussian curves* and are developed on the basis of probability. The bell-shaped normal distribution curve fitted to the data in Fig. 4.21b has two important features. First, it shows that most part diameters tend to cluster around an *average* value (**arithmetic mean**). This average is usually designated as $\bar{x}$ and is calculated from the expression

$$\bar{x} = \frac{x_1 + x_2 + x_3 + \cdots + x_n}{n}, \tag{4.7}$$

where the numerator is the sum of all measured values (diameters) and n is the number of measurements (the number of shafts in this case).

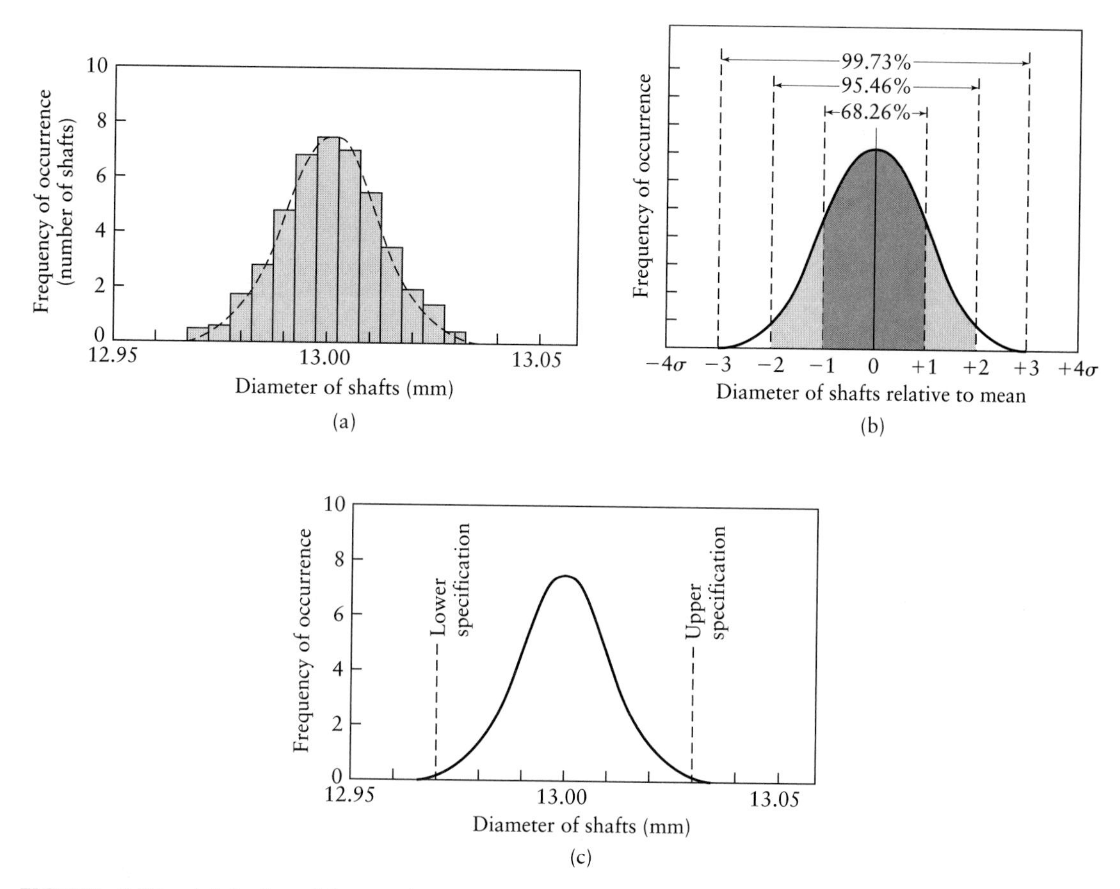

FIGURE 4.21 (a) A plot of the number of shafts measured and their respective diameters. This type of curve is called a *frequency distribution.* (b) A normal distribution curve indicating areas within each range of standard deviation. *Note*: The greater the range, the higher the percentage of parts that fall within it. (c) Frequency distribution curve, showing lower and upper specification limits.

The second feature of this curve is its width, indicating the *dispersion* of the diameters measured; the wider the curve, the greater is the dispersion. The difference between the largest value and smallest value is called the *range* R:

$$R = x_{\text{max}} - x_{\text{min}}. \tag{4.8}$$

Dispersion is estimated by the **standard deviation,** which is generally denoted as σ and is obtained from the expression

$$\sigma = \sqrt{\frac{(x_1 - \bar{x})^2 + (x_2 - \bar{x})^2 + (x_3 - \bar{x})^2 + \cdots + (x_n - \bar{x})^2}{n - 1}}, \tag{4.9}$$

where x is the measured value for each part. Note from the numerator in Eq. (4.9) that as the curve widens, the standard deviation becomes greater; also note that σ has units of linear dimension. In comparing Eqs. (4.8) and (4.9), we see that the range R is a simpler and more convenient measure of dispersion.

Since we know the number of parts that fall within each group, we can calculate the percentage of the total population represented by each group. Thus, Fig. 4.21b shows that the diameters of 99.73% of the turned shafts fall within the range of $\pm 3\sigma$, 95.46% within $\pm 2\sigma$, and 68.26% within $\pm 1\sigma$; only 0.2% fall outside the $\pm 3\sigma$ range.

Six sigma. An important trend in manufacturing operations, as well as in business and service industries, is the concept of *six sigma*. Note from the preceding discussion that maintaining a standard of three sigma in manufacturing would result in 0.27%, or 2700 defective parts per million parts. In modern manufacturing, this is an unacceptable rate; in fact, it has been estimated that at the three-sigma level, virtually no modern computer would function properly and reliably, and in the service industries, 270 million incorrect credit-card transactions would be recorded each year in the United States alone. It has further been estimated that companies operating at three- and four-sigma levels lose about 10% to 15% of their total revenue because of defects. Consequently, led by major companies such as Motorola and General Electric, extensive efforts are being made to eliminate defects in products and processes. The resulting savings are reported in billions of dollars.

Six sigma is a set of statistical tools, based on well-known total-quality-management principles, to continually measure the quality of products and services in selected projects. It includes considerations such as customer satisfaction, delivering defect-free products, and understanding process capabilities (see later). The approach consists of a clear focus on defining the problem; measuring relevant quantities; and analyzing, improving, and controlling processes and activities. Because of its major impact on business, six sigma is now well recognized as a management philosophy.

4.9.2 Statistical process control

If the number of parts that do not meet set standards (i.e., defective parts) increases during a production run, we must be able to determine the cause (e.g., incoming materials, machine controls, degradation of metalworking fluids, operator boredom) and take appropriate action. Although this statement at first appears to be self-evident, it was only in the early 1950s that a systematic statistical approach was developed to guide operators in manufacturing plants.

This approach advises the operator of certain measures to take and when to take them, in order to avoid producing further defective parts. Known as *statistical process control* (SPC), this technique consists of several elements: control charts and setting control limits, capabilities of the particular manufacturing process, and characteristics of the machinery involved.

Control charts. The frequency distribution curve in Fig. 4.21a shows a range of shaft diameters being produced that may fall beyond the predetermined design tolerance range. Figure 4.21c shows the same bell-shaped curve, which now includes the specified tolerances for the diameters of turned shafts. *Control charts* graphically represent the variations of a process over a period of time. They consist of data plotted during production, and there typically are two plots. The quantity $\bar{x}$ in Fig. 4.22a is the average for each subset of samples taken and inspected, say, a subset consisting of 5 parts. (A sample size of between 2 and 10 parts is sufficiently accurate, provided that sample size is held constant throughout the inspection.)

The frequency of sampling depends on the nature of the process. Some processes may require continuous sampling, whereas others may require only one sample per day. Quality-control analysts are best qualified to determine this frequency for a particular situation. Since the measurements shown in Fig. 4.22a are made consecutively, the abscissa of these control charts also represents time. The solid horizontal line in this figure is the **average of averages (grand average)**, denoted as $\bar{\bar{x}}$, and represents the *population mean*. The upper and lower horizontal broken lines in these control charts indicate the control limits for the process.

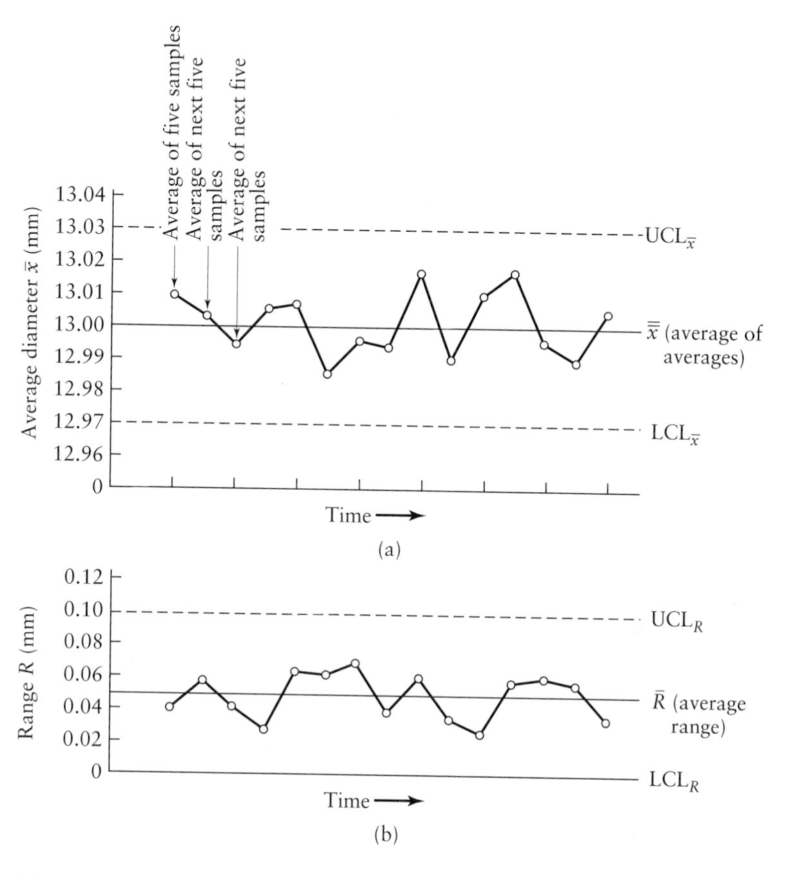

FIGURE 4.22 Control charts used in statistical quality control. The process shown is in good statistical control, because all points fall within the lower and upper control limits. In this illustration, the sample size is five, and the number of samples is 15.

The **control limits** are set on these charts according to statistical-control formulas designed to keep actual production within the usually acceptable $\pm 3\sigma$ range. Thus, for $\bar{x}$,

$$\text{Upper control limit } (\text{UCL}_{\bar{x}}) = \bar{x} + 3\sigma = \bar{\bar{x}} + A_2\bar{R} \tag{4.10}$$

and

$$\text{Lower control limit } (\text{LCL}_{\bar{x}}) = \bar{x} - 3\sigma = \bar{\bar{x}} - A_2\bar{R}, \tag{4.11}$$

where A_2 is read from Table 4.3 and $\bar{R}$ is the average of the R values.

TABLE 4.3

Constants for Control Charts

Sample Size	A_2	D_4	D_3	d_2
2	1.880	3.267	0	1.128
3	1.023	2.575	0	1.693
4	0.729	2.282	0	2.059
5	0.577	2.115	0	2.326
6	0.483	2.004	0	2.534
7	0.419	1.924	0.078	2.704
8	0.373	1.864	0.136	2.847
9	0.337	1.816	0.184	2.970
10	0.308	1.777	0.223	3.078
12	0.266	1.716	0.284	3.258
15	0.223	1.652	0.348	3.472
20	0.180	1.586	0.414	3.735

These limits are calculated based on the historical production capability of the equipment itself; they are generally not associated with design tolerance specifications and dimensions. They indicate the limits within which a certain percentage of measured values is normally expected to fall due to the inherent variations of the process itself, upon which the limits are based. The major goal of statistical process control is to improve the manufacturing process via the aid of control charts to eliminate assignable causes. The control chart continually indicates progress in this area.

The second control chart (Fig. 4.22b) shows the range R in each subset of samples. The solid horizontal line represents the average of R values in the lot, denoted as $\bar{R}$, and is a measure of the variability of the samples. The upper and lower control limits for R are obtained from the equations

$$\mathrm{UCL}_R = D_4\bar{R} \tag{4.12}$$

and

$$\mathrm{LCL}_R = D_3\bar{R}, \tag{4.13}$$

where the constants D_4 and D_3 are obtained from Table 4.3. This table also includes values for the constant d_2, which is used in estimating the standard deviation from the equation

$$\sigma = \frac{\bar{R}}{d_2}. \tag{4.14}$$

When the curve of a control chart is like that shown in Fig. 4.22b, the process is said to be **in good statistical control**. In other words, there is no clear discernible trend in the pattern of the curve, the points (measured values) are random with time, and they do not exceed the control limits. However, you can see that curves such as those shown in Figs. 4.23a–c indicate certain trends. For example, about halfway in Fig. 4.23a, the diameter of the shafts increases with time; the reason for this increase may be a change in one of the process variables, such as wear of the cutting tool. If, as in Fig. 4.23b, the trend is toward consistently larger diameters, hovering around the upper control limit, it could mean that the tool settings on the lathe may not be correct and, as a result, the parts being turned are consistently too large. Figure 4.23c shows two distinct trends that may be caused by factors such as a change in the properties of the incoming material or a change in the performance of the cutting fluid (for example, its degradation). These situations place the process **out of control.**

Analyzing control-chart patterns and trends requires considerable experience in order to identify the specific cause(s) of an out-of-control situation. **Overcontrol** of the manufacturing process, that is, setting upper and lower control limits too close to each other (and hence setting a smaller standard-deviation range), is a further reason for out-of-control situations. Avoidance of overcontrol is the reason that we calculate control limits on process capability rather than on a potentially inapplicable statistic.

It is evident that operator training is critical for successful implementation of SPC on the shop floor. Once process-control guidelines are established, operators also should, in the interest of efficiency of the operation, have some responsibility to make adjustments in processes that begin to become out of control. The capabilities of individual operators should be taken into account so that the operators are not overloaded with data input and interpretation. This task is, however, being made easier by a variety of software that is being continually developed. For example, digital readouts on electronic measuring devices are integrated directly into a computer system for real-time SPC. Figure 4.24 shows such a multifunctional computer system in which the output from a digital caliper or micrometer is analyzed in real time and displayed in several ways, such as frequency distribution curves and control charts.

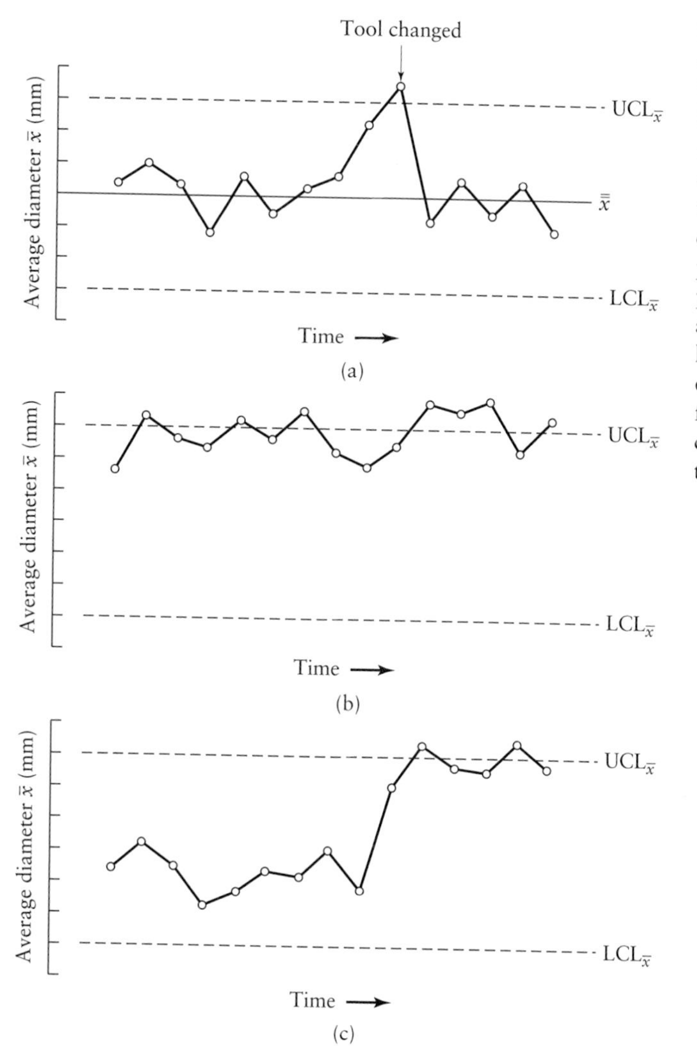

FIGURE 4.23 Control charts. (a) Process begins to become out of control, because of factors such as tool wear. The tool is changed, and the process is then in good statistical control. (b) Process parameters are not set properly; thus, all parts are around the upper control limit. (c) Process becomes out of control, because of factors such as a sudden change in the properties of the incoming material.

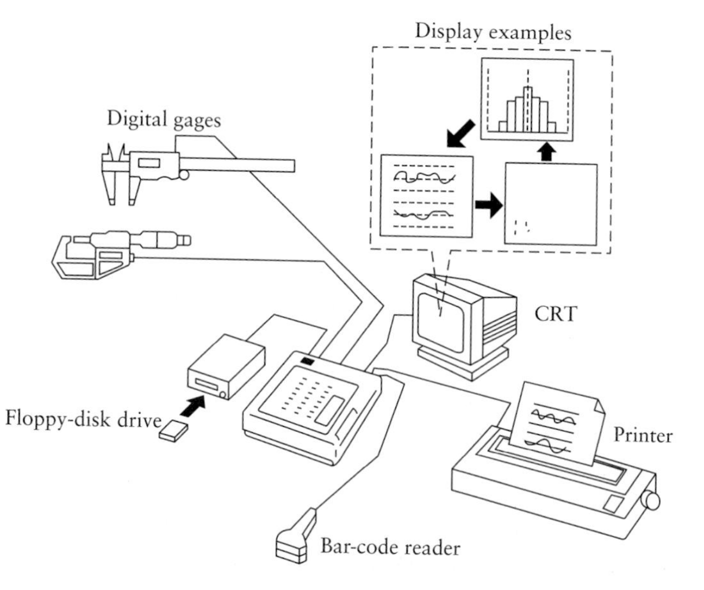

FIGURE 4.24 Schematic illustration showing integration of digital gages with a miniprocessor for real-time data acquisition and SPC/SQC capabilities. Note the examples on the CRT displays, such as frequency distribution and control charts. *Source*: Mitutoyo Corp.

Process capability. *Process capability* is defined as the limits within which individual measurement values resulting from a particular manufacturing process would normally be expected to fall when only random variation is present; it tells us that the process can produce parts within certain limits of precision. Since a manufacturing process involves materials, machinery, and operators, each of these aspects can be analyzed individually to identify a problem when process capabilities do not meet part specifications.

4.9.3 Total quality control

The *total quality control* (TQC) concept is a management system that emphasizes the fact that quality must be designed and built into a product; **defect prevention,** rather than defect detection, is the major goal. Total quality control is a systems approach in that both management and workers make an integrated effort to manufacture high-quality products consistently. All tasks concerning quality improvements and responsibilities in the organization are clearly identified. The TQC concept also requires 100% inspection of parts, usually by automated inspection systems; no defective parts are allowed to continue through the production line.

A related concept that has gained some acceptance is **the quality circle.** This activity consists of regular meetings by groups of workers who discuss how to improve and maintain product quality at all stages of the manufacturing process. Worker involvement and responsibility are emphasized. Comprehensive training is provided so that the worker can become capable of analyzing statistical data, identifying causes of poor quality, and taking immediate action to correct the situation. Putting this concept into practice recognizes the importance of quality assurance as a major companywide management policy, affecting all personnel and all aspects of production.

Quality engineering as a philosophy. Many of the quality-control concepts and methods described have been put into larger perspective by certain experts in quality control. Notable among these experts are W. E. Deming, G. Taguchi, and J. M. Juran, whose philosophies of quality and product cost have had a major impact on modern manufacturing.

During World War II, Deming and several others developed new methods of statistical process control in manufacturing plants for wartime industry. The need for statistical control arose from the recognition that there were variations in the performance of machines and people and the quality and dimensions of raw materials. Their efforts, however, involved not only statistical methods of analysis, but also a new way of looking at manufacturing operations to improve quality and lower costs. Recognizing the fact that manufacturing organizations are systems of management, workers, machines, and products, Deming emphasized communication, direct worker involvement, and education in statistics, as well as use of modern manufacturing technology. Fundamental to his philosophy is the dedication of management to quality and a shifting of emphasis from short-term profits to a longer term emphasis on improvement of products and processes.

Taguchi methods. In the *Taguchi methods*, high quality and low cost are achieved by combining engineering and statistical methods to optimize product design and manufacturing processes. *Loss of quality* is defined as the financial loss to society after the product is shipped, with the following results: (1) Poor quality leads to customer dissatisfaction; (2) costs are involved in servicing and repairing defective products, some in the field; (3) the manufacturer's credibility is diminished in the marketplace; and (4) the manufacturer eventually loses its share of the market.

The Taguchi methods of quality engineering emphasize the importance of the following functions:

1. Enhancing cross-functional team interaction. In this interaction, design engineers and process or manufacturing engineers communicate with each other in a common language. They quantify the relationships between design requirements and the manufacturing process.
2. Implementing experimental design, in which the factors involved in a process or operation and their interactions are studied simultaneously.
3. Minimizing variation between parts, so that a consistent, high quality level is achieved.

In **experimental design**, the effects of controllable and uncontrollable variables on the product are identified; this approach minimizes variations in product dimensions and properties, bringing the mean level of product quality to the desired level. The methods used for experimental design are complex, involving the use of fractional factorial design and orthogonal arrays, which reduce the number of experiments required. These methods are also capable of identifying the effect on the product of variables that cannot be controlled (called *noise*), such as changes in environmental conditions.

The use of these methods allows rapid identification of the controlling variables and determination of the best method of process control. These variables are then controlled, without the need for costly new equipment or major modifications to existing equipment. For example, variables affecting dimensional tolerances in machining a particular component can be readily identified, and the correct cutting speed, feed, cutting tool, and cutting fluids can be specified.

Taguchi loss function. Because traditional accounting practices had no real way of calculating losses on parts that met design specifications, the Taguchi loss function was introduced. In the traditional accounting approach, a part is defective and incurs a loss to the company when it exceeds its design tolerances; otherwise, there is no loss to the company. The Taguchi loss function calculates an increasing loss to the company the further the component is from the design objective.

4.9.4 The ISO 9000 and 14000 Standards

With increasing international trade, global manufacturing, and price-sensitive competition, a wide range of industrial and consumer products is now becoming available. Also, customers are increasingly demanding high-quality products and services at low prices and are looking for suppliers that can respond to this demand consistently and reliably. This strong trend, in turn, has created the need for international conformity and consensus regarding the establishment of quality-control methods and reliability and safety standards for products made in different countries and traded internationally.

The **ISO 9000** standard (**Quality Management and Quality Assurance Standards**) is a deliberately generic series of quality-system management standards. It has permanently influenced the way in which manufacturing companies conduct business in world trade and has become the world standard for quality. For certification, a company's plants are visited and audited by accredited and independent third-party teams to certify that the standard's 20 key elements are in place and functioning properly. Depending on the extent to which a company does not meet the requirements of the standard, registration may or may not be recommended at that time. The audit team does not advise or consult with the company on how to fix discrepancies, but merely describes the nature of the noncompliance. Periodic audits are required to maintain certification. The certification process can take from six months to a year or more and can cost tens of thousands of dollars, depending on the size, number of plants, and product line of the company.

The ISO 9000 standard is not a product certification, but a *quality-process certification*. Companies establish their own criteria and practices for quality. However, the documented quality system must be in compliance with the ISO 9000 standard; thus, a company cannot write into the system any criterion that opposes the intent of the standard. Registration means conformity to consistent practices, as specified by the company's own quality system (such as quality in design, development, production, installation, and servicing), including proper documentation of such practice. Thus, customers, including government agencies, are assured that specified practices are being followed by the supplier of the product or service (which may or may not be within the same country). In fact, manufacturing companies are themselves assured of such practice regarding their own suppliers who have ISO 9000 registration; they may even demand that their suppliers be registered.

ISO 14000 is a family of standards pertaining to international **Environmental Management Systems** (EMS) and concerns the way in which an organization's activities affect the environment throughout the life of its products. These activities may be internal or external to the organization; they range from production to ultimate disposal of the product and include effects such as depletion of natural resources, pollution, waste generation and disposal, noise, and efficiency of energy use.

QS 9000 was jointly developed by the U.S. automotive companies and is often described as an ISO 9000 standard with certain extras and additional requirements. The QS 9000 standards generally consist of production-part approval process, continuous improvement, manufacturing capabilities, and customer-specific requirements.

SUMMARY

- The surface of a workpiece can significantly affect its properties and characteristics, including friction and wear properties, the effectiveness of lubricants, appearance and geometric features, tendency for cracking, and thermal and electrical conductivity of contacting bodies. (Section 4.1)
- The surface structure of a metal has usually been plastically deformed and work hardened during prior processing. Surface roughness can be quantified by a variety of techniques. Surface integrity can be compromised by a variety of defects. Flaws, directionality, roughness, and waviness are measurable quantities that are used to describe surface texture. (Sections 4.2, 4.3)
- Tribology is the science and technology of interacting surfaces, and it encompasses friction, wear, and lubrication. Friction may be desirable or undesirable, depending on the specific manufacturing circumstances. The adhesion theory of friction is most commonly used to model and predict the frictional behavior between sliding metal surfaces, because it agrees reasonably well with experimental observations. (Section 4.4)
- Wear is defined as the progressive loss or removal of material from a surface. Wear alters the geometry of the workpiece surface and tool and die interfaces and thus affects the manufacturing process, dimensional accuracy, and the quality of parts produced. Friction and wear can be reduced by using various liquid or solid lubricants, as well as by applying ultrasonic vibrations. Four regimes of lubrication are relevant to metalworking processes. (Section 4.4)
- Several surface treatments are used to impart certain physical and mechanical properties to the workpiece, such as resistance to environmental attack, wear, and fatigue. The techniques employed typically include mechanical working, physical and chemical means, heat treatments, and coatings. Cleaning of a manufactured workpiece involves removal of solid, semisolid, and liquid contaminants from a surface by various means. (Section 4.5)

- Many parts are made with a high degree of dimensional accuracy, thus requiring measuring instruments with various features and characteristics. The selection of a particular instrument depends on factors such as the type of measurement, the environment, and the accuracy of measurement. Major advances have been made in automated measurement, linking measuring equipment to microprocessors and computers for accurate in-process control of manufacturing operations. (Section 4.6)
- Dimensional tolerances and their specification are important factors in manufacturing, as they not only affect subsequent assembly of the parts made and the accuracy and operation of all types of machinery and equipment, but also can significantly influence product cost. (Section 4.7)
- Several nondestructive and destructive testing techniques, each having its own applications, advantages, and limitations, are available for inspection of completed parts and products. Unlike the traditional approach, the practice now is towards 100% inspection of all parts being manufactured, using automated and reliable inspection techniques. (Section 4.8)
- Quality assurance is the total effort by a manufacturer to ensure that its products conform to a detailed set of specifications and standards. A variety of statistical quality-control and statistical process-control techniques are now employed in defect detection and prevention. The total-quality-control philosophy focuses on prevention, rather than detection, of defects. (Section 4.9)
- Management philosophies such as those developed by Deming summarize the characteristics of companies that manufacture high-quality products. Chief among these characteristics are dedication to long-term improvement in quality and performance, direct involvement of all employees associated with a product, and continuing education. (Section 4.9)
- The Taguchi methods describe a powerful approach for experimentally characterizing and optimizing a complex system such as a manufacturing process or product design. The Taguchi loss function is a tool that emphasizes minimization of product variation in order to optimize quality. Also, the ISO 9000 and 14000 standards are now universally employed. (Section 4.9)

SUMMARY OF EQUATIONS

Arithmetic mean value, $R_a = \dfrac{y_a + y_b + y_c + \cdots + y_n}{n} = \dfrac{1}{n}\sum_{i=1}^{n} y_i = \dfrac{1}{l}\int_0^l |y|dx$

Root-mean-square average, $R_q = \sqrt{\dfrac{y_a^2 + y_b^2 + y_c^2 + \cdots + y_n^2}{n}} = \sqrt{\dfrac{1}{n}\sum_{1}^{n} y_i^2} = \left[\dfrac{1}{l}\int_0^1 y^2\,dx\right]^{1/2}$

Coefficient of friction, $\mu = \dfrac{F}{N} = \dfrac{\tau}{\text{Hardness}}$

Friction (shear) factor, $m = \dfrac{\tau_i}{k}$

Adhesive wear, $V = \dfrac{kLW}{3p}$

Arithmetic mean, $\bar{x} = \dfrac{x_1 + x_2 + x_3 + \cdots + x_n}{n}$

Range, $R = x_{\max} - x_{\min}$

Standard deviation, $\sigma = \sqrt{\dfrac{(x_1 - \bar{x})^2 + (x_2 - \bar{x})^2 + (x_3 - \bar{x})^2 + \cdots + (x_n - \bar{x})^2}{n - 1}}$

Upper control limit, $\text{UCL}_{\bar{x}} = \bar{x} + 3\sigma = \bar{\bar{x}} + A_2\bar{R}$

Lower control limit, $\text{LCL}_{\bar{x}} = \bar{x} - 3\sigma = \bar{\bar{x}} - A_2\bar{R}$

BIBLIOGRAPHY

Aft, L. S., *Fundamentals of Industrial Quality Control*, 3d. ed., Addison-Wesley, 1998.

Arnell, R. D., P. B. Davies, J. Halling, and T. L. Whomes, *Tribology: Principles and Design Applications*, Springer, 1991.

ASM Handbook, Vol. 5: *Surface Engineering*, ASM International, 1994.

ASM Handbook, Vol. 17: *Nondestructive Evaluation and Quality Control*, ASM International, 1989.

ASM Handbook, Vol. 18: *Friction, Lubrication, and Wear Technology*, ASM International, 1992.

Bayer, R. G., *Mechanical Wear Prediction and Prevention*, Dekker, 1995.

Besterfield, D. H., *Quality Control*, Prentice Hall, 1997.

Bhushan, B., *Principles and Applications of Tribology*, Wiley, 1999.

Bhushan, B. (ed.), *Modern Tribology Handbook*, CRC Press, 2001.

Bhushan, B., and B. K. Gupta, *Handbook of Tribology: Materials, Coatings, and Surface Treatments*, McGraw-Hill, 1991.

Booser, E. R. (ed.), *Tribology Data Handbook*, CRC Press, 1998.

Bosch, J. A. (ed.), *Coordinate Measuring Machines and Systems*, Dekker, 1995.

Bothe, D. R., *Measuring Process Capability: Techniques and Calculations for Quality and Manufacturing Engineers*, McGraw-Hill, 1997.

Breyfogle, F., *Implementing Six Sigma: Smarter Solutions Using Statistical Methods*, Wiley, 1999.

Budinski, K. G., *Surface Engineering for Wear Resistance*, Prentice Hall, 1988.

Burakowski, T., and T. Wiershon, *Surface Engineering of Metals: Principles, Equipment, Technologies*, CRC Press, 1998.

Byers, J. P. (ed.), *Metalworking Fluids*, Dekker, 1994.

Cartz, L., *Nondestructive Testing*, ASM International, 1995.

Chattopadhyay, R., *Surface Wear: Analysis, Treatment, and Prevention*, ASM International, 2001.

Clausing, D., *Total Quality Development*, ASME Press, 1994.

Clements, R., *Handbook of Statistical Methods in Manufacturing*, Prentice Hall, 1991.

Deming, W. E., *Out of the Crisis*, MIT Press, 1982.

Dowson, D., *History of Tribology*, 2d. ed. ASME, 1999.

Drake, P. J., *Dimensioning and Tolerancing Handbook*, McGraw-Hill, 1999.

Edwards, J., *Coating and Surface Treatment Systems for Metals: A Comprehensive Guide to Selection*, ASM International, 1997.

Farrago, F. T., and M. A. Curtis, *Handbook of Dimensional Measurement*, 3d. ed., Industrial Press, 1994.

Feigenbaum, A. V., *Total Quality Control*, 3d. ed., McGraw-Hill, 1991.

Glasser, W. A., *Materials for Tribology*, Elsevier, 1992.

Grant, E. L., and R. S. Leavenworth, *Statistical Quality Control*, McGraw-Hill, 1997.

Henzhold, G., *Handbook of Geometric Tolerancing*, Wiley, 1995.

Holmberg, K., and A. Matthews, *Coating Tribology: Properties, Techniques, and Applications*, Elsevier, 1994.

Hutchings, I. M., *Tribology: Friction and Wear of Engineering Materials*, CRC Press, 1992.

Juran, J. M., and A. B. Godfrey (eds.), *Juran's Quality Handbook*, 5th ed., McGraw-Hill, 1999.

Juran, J. M., and F. M. Gryna, Jr., *Quality Planning and Analysis*, 4th ed., McGraw-Hill, 2000.

Kear, F. W., *Statistical Process Control in Manufacturing Practice*, Dekker, 1998.

Krulikowski, A., *Fundamentals of Geometric Dimensioning and Tolerancing*, Delmar, 1997.

Lansdown, A. R., and A. L. Price, *Materials to Resist Wear: A Guide to Their Selection and Use*, Pergamon Press, 1986.

Lindsay, J. H. (ed.), *Coatings and Coating Processes for Metals*, ASM International, 1998.

Ludema, K. C., *Friction, Wear, Lubrication: A Textbook in Tribology*, CRC Press, 1996.

Meadows, J. D., *Geometric Dimensioning and Tolerancing*, Dekker, 1995.

_____, *Measurement of Geometric Tolerances in Manufacturing*, Dekker, 1998.

Menon, H. G., *TQM in New Product Manufacturing*, McGraw-Hill, 1992.

Moore, D. F., *Principles and Applications of Tribology*, Pergamon, 1975.

Murphy, S. D., *In-Process Measurement and Control*, Dekker, 1990.

Nachtman, E. S., and S. Kalpakjian, *Lubricants and Lubrication in Metalworking Operations*, Dekker, 1985.

Neele, M. J. (ed.), *the Tribology Handbook*, 2d. ed., Butterworth-Heinemann, 1995.

Peterson, M. B., and W. O. Winer (eds.), *Wear Control Handbook*, American Society of Mechanical Engineers, 1980.

Puncochar, D. E., *Interpretation of Geometric Dimensioning and Tolerancing*, 2d. ed., Industrial Press, 1997.

Rabinowicz, E., *Friction and Wear of Materials*, 2d. ed., Wiley, 1995.

Robinson, S. L., and R. K. Miller, *Automated Inspection and Quality Assurance*, Dekker, 1989.

Roy, R., *A Primer on the Taguchi Method*, Society of Manufacturing Engineers, 1990.

Rudzki, G. J., *Surface Finishing Systems: Metal and Non-Metal Finishing Handbook-Guide*, Finishing Publications Ltd., 1983.

Schey, J. A., *Tribology in Metalworking: Friction, Lubricating and Wear*, ASM International, 1983.

Stachowiak, G. W., and A. W. Batchelor, *Engineering Tribology*, Butterworth-Heinemann, 2001.

Stern, K. H. (ed.), *Metallurgical and Ceramic Protective Coatings*, Chapman & Hall, 1996.

Sudarshan, T. S. (ed.), *Surface Modification Technologies*, ASM International, 1998.

Tool and Manufacturing Engineers Handbook, 4th ed., Vol. 3: *Materials, Finishing and Coating*, Society of Manufacturing Engineers, 1985.

Tool and Manufacturing Engineers Handbook, 4th ed., Vol. 4: *Assembly, Testing, and Quality Control*, Society of Manufacturing Engineers, 1986.

Wadsworth, H. M., *Handbook of Statistical Control Methods for Engineers and Scientists*, 2d. ed., McGraw-Hill, 1998.

Whitehouse, D. J., *Handbook of Surface Metrology*, Institute of Physics, 1994.

Winchell, W., *Inspection and Measurement in Manufacturing*, Society of Manufacturing Engineers, 1996.

QUESTIONS

4.1 Explain what is meant by surface integrity. Why should we be interested in it?

4.2 Why are surface-roughness design requirements in engineering so broad? Give appropriate examples.

4.3 We have seen that a surface has various layers. Describe the factors that influence the thickness of these layers.

4.4 What is the consequence of oxides of metals being generally much harder than the base metal? Explain.

4.5 What factors would you consider in specifying the lay of a surface?

4.6 Describe the effects of various surface defects (see Section 4.3) on the performance of engineering components in service. How would you go about determining whether or not each of these defects is important for a particular application?

4.7 Explain why the same surface roughness values do not necessarily represent the same type of surface.

4.8 In using a surface roughness measuring instrument, how would you go about determining the cutoff value? Give appropriate examples.

4.9 What is the significance of the fact that the stylus path and the actual surface profile are generally not the same?

4.10 Give two examples each in which waviness of a surface would be (1) desirable and (2) undesirable.

4.11 Explain why surface temperature increases when two bodies are rubbed against each other. What is the significance of a temperature rise due to friction?

4.12 To what factors would you attribute the fact that the coefficient of friction in hot working is higher than in cold working, as shown in Table 4.1?

4.13 In Section 4.4.1, we note that the values of the coefficient of friction can be much higher than unity. Explain why.

4.14 Describe the tribological differences between ordinary machine elements (such as meshing gears, cams in contact with followers, and ball bearings with inner and outer races) and elements of metalworking processes (such as forging, rolling, and extrusion, which involve workpieces in contact with tools and dies).

4.15 Give the reasons that an originally round specimen in a ring-compression test may become oval after deformation.

4.16 Can the temperature rise at a sliding interface exceed the melting point of the metals? Explain.

4.17 Explain why each of the terms in the Archard formula for adhesive wear, Eq. (4.6), should affect the wear volume.

4.18 How can adhesive wear be reduced? How can fatigue wear be reduced?

4.19 It has been stated that as the normal load decreases, abrasive wear is reduced. Explain why this is so.

4.20 Why is the abrasive wear resistance of a material a function of its hardness?

4.21 We have seen that wear can have detrimental effects on engineering components, tools, dies, etc. Can you visualize situations in which wear could be beneficial? Give some examples. (*Hint*: Note that writing with a pencil is a wear process.)

4.22 On the basis of the topics discussed in this chapter, do you think there is a direct correlation between friction and wear of materials? Explain.

4.23 You have undoubtedly replaced parts in various appliances and automobiles because they were worn. Describe the methodology you would follow in determining the type(s) of wear these components have undergone.

4.24 Why is the study of lubrication regimes important?

4.25 Explain why so many different types of metalworking fluids have been developed.

4.26 Differentiate between (1) coolants and lubricants, (2) liquid and solid lubricants, (3) direct and indirect emulsions, and (4) plain and compounded oils.

4.27 Explain the role of conversion coatings. Based on Fig. 4.13, what lubrication regime is most suitable for application of conversion coatings?

4.28 Explain why surface treatment of manufactured products may be necessary. Give several examples.

4.29 Which surface treatments are functional, and which are decorative? Give examples.

4.30 Give examples of several typical applications of mechanical surface treatment.

4.31 Explain the difference between case hardening and hard facing.

4.32 List several applications for coated sheet metal, including galvanized steel.

4.33 Explain how roller-burnishing processes induce residual stresses on the surface of workpieces.

4.34 List several products or components that could not be made properly, or function effectively in service, without implementation of the knowledge involved in Sections 4.2 through 4.5.

4.35 Explain the difference between direct- and indirect-reading linear measurements.

4.36 Why have coordinate measuring machines become important instruments in modern manufacturing? Give some examples of applications.

4.37 Give reasons that the control of dimensional tolerances is important.

4.38 Give examples where it may be preferrable to specify unilateral tolerances as opposed to bilateral tolerances in design.

4.39 Explain why a measuring instrument may not have sufficient precision.

4.40 Comment on the differences, if any, between (1) roundness and circularity, (2) roundness and eccentricity, and (3) roundness and cylindricity.

4.41 It has been stated that dimensional tolerances for nonmetallic stock, such as plastics, are usually wider than for metals. Explain why. Consider physical and mechanical properties.

4.42 Describe the basic features of nondestructive testing techniques that use electrical energy.

4.43 Identify the nondestructive techniques that are capable of detecting internal flaws and those that detect external flaws only.

4.44 Which of the nondestructive inspection techniques are suitable for nonmetallic materials? Why?

4.45 Why is automated inspection becoming an important part of manufacturing engineering?

4.46 Describe situations in which the use of destructive testing techniques is unavoidable.

4.47 Should products be designed and built for a certain expected life? Explain.

4.48 What are the consequences of setting lower and upper specifications closer to the peak of the curve in Fig. 4.23?

4.49 Identify factors that can cause a process to become out of control. Give several examples of such factors.

4.50 In reading this chapter, you will have noted that the specific term *dimensional tolerance* is often used, rather than just the word *tolerance*. Do you think this distinction is important? Explain.

4.51 Why is it important to minimize variation among manufactured parts?

4.52 What are the key elements of the Deming philosophy?

4.53 Does the presence of a lubricant affect abrasive wear? Explain.

4.54 Give an example of an assignable variation and a chance variation.

4.55 Explain why GO–NOT-GO gages are incompatible with the Taguchi-loss-function approach.

4.56 Explain how you would estimate the magnitude of the wear coefficient for a pencil writing on paper.

4.57 Describe a test method for determining the wear coefficient k in Eq. (4.9). What would be the difficulties in applying the results from this test to a manufacturing application, such as predicting the life of tools and dies?

PROBLEMS

4.58 Referring to the surface profile in Fig. 4.3, give some numerical values for the vertical distances from the center line. Calculate the R_a and R_q values. Then give another set of values for the same general profile, and calculate the same quantities. Comment on your results.

4.59 Using Fig. 4.8a, make a plot of the coefficient of friction versus the change in internal diameter for a reduction in height of (1) 20%, (2) 40%, and (3) 60%.

4.60 In Example 4.1, assume that the coefficient of friction is 0.30. If all other initial parameters remain the same, what is the new internal diameter of the ring specimen?

4.61 How would you go about estimating forces required for roller burnishing? (*Hint*: Consider hardness testing.)

4.62 Assume that a steel rule expands by 1% because of an increase in environmental temperature. What will be the indicated diameter of a shaft whose actual diameter is 75.00 mm?

4.63 Assume that in Example 4.3, the number of samples was 7 instead of 10. Using the top half of the data given in the table, recalculate the control limits and the standard deviation. Compare your results with the results obtained by using 10 samples.

4.64 Calculate the control limits for averages and ranges for the following: number of samples $= 5; \bar{\bar{x}} = 60$; $\bar{R} = 7$.

4.65 Calculate the control limits for the following: number of samples $= 5; \bar{\bar{x}} = 36.5$; $\text{UCL}_R = 4.75$.

4.66 In an inspection with a sample size of 7 and a sample number of 40, it was found that the average range was 10 and the average of averages was 75. Calculate the control limits for averages and ranges.

4.67 Determine the control limits for the data shown in the following table:

x_1	x_2	x_3	x_4
0.65	0.70	0.67	0.65
0.69	0.65	0.70	0.68
0.65	0.60	0.65	0.61
0.64	0.67	0.60	0.60
0.68	0.68	0.70	0.66
0.70	0.71	0.65	0.71

4.68 The average of averages of a number of samples of size 5 was determined to be 125. The average range was 17.82, and the standard deviation was 6. The following measurements were taken in a sample: 120, 132, 124, 130, 118, 132, 121, and 127. Is the process in control?

4.69 Assume that you are asked to give a quiz to students on the contents of this chapter. Prepare three quantitative problems and three qualitative questions, and supply the answers.

DESIGN

4.70 Design an experimental fixture that allows a conventional profilometer to be indexed across a surface and thereby measure the surface in three dimensions.

4.71 List several products or components that could not be designed properly, or function as effectively in service, without implementation of the information covered in this chapter.

4.72 Survey the available literature, and prepare a report on the environmental considerations regarding the application of the various processes and operations described in this chapter.

4.73 Comment on the design challenges that have to be faced in installing surface-measuring instruments directly on production machines. Consider the typical factory environment.

4.74 Are there product designs for which wear is actually desirable? Explain. (See also Question 4.21.)

4.75 Consider products that you are familiar with and for which wear is a serious problem. What design changes would you recommend to reduce wear in these products? Explain your reasoning.

4.76 An artificial implant has a porous surface so that bone will grow into the surface of the implant and produce a strong joint. Give recommendations for producing a porous surface.

4.77 Solar energy has been used as an energy source for posttreatment of coatings and deposited films. Design a suitable system to implement solar energy for this purpose.

4.78 As you know, vandalism of public places, such as graffiti, scratching, spray painting, etc., is a serious problem. Make a list of possible methods of minimizing such damage, and explain what design changes would be required to make these methods effective.

CHAPTER

5 Metal-Casting Processes and Equipment; Heat Treatment

5.1 Introduction

As described throughout this text, several methods can be used to shape materials into useful products. Making parts by **casting** molten metal into a mold and letting it solidify is an obvious choice. Indeed, casting is among the oldest methods of net-shape and near-net-shape manufacturing and was first used in about 4000 BC to make ornaments, copper arrowheads, and various other objects. Basically, casting processes involve the introduction of molten metal into a mold cavity; upon solidification, the metal takes the shape of the cavity. The casting process is thus capable of producing intricate shapes in a single piece, including shapes with internal cavities; very large, very small, and hollow parts can be produced economically by casting techniques. Typical cast products are engine blocks, cylinder heads, transmission housings, pistons, turbine disks, railroad and automotive wheels, and ornamental artifacts.

Almost all metals can be cast in (or nearly in) the final shape desired, often with only minor finishing required. With appropriate control of material and process parameters, parts can be cast with uniform properties throughout. As with all other manufacturing processes, a knowledge of certain fundamental relationships is essential to the production of good-quality and economical castings; these relationships help establish proper techniques for mold design and casting practice. The objective is to produce castings that are free from defects and that meet requirements for strength, dimensional accuracy, and surface finish.

Fundamentally, the important factors in casting operations are as follows:

a) **Solidification** of the metal from its molten state (which usually is accompanied by shrinkage);
b) **Flow** of the molten metal into the mold cavity;
c) **Heat transfer** during solidification and cooling of the metal in the mold;
d) **Mold material** and its influence on the casting process.

We first discuss these aspects of casting and then continue with a description of melting practice and the characteristics of casting alloys. The remainder of the chapter covers ingot and continuous casting; various expendable-mold and permanent-mold casting processes, including their advantages and limitations; processing of castings; heat treatment; casting design; and the economics of casting.

5.2 Solidification of Metals

This section presents an overview of the solidification of metals and alloys. We cover topics that are particularly relevant to casting and are essential to our understanding of the structures developed in casting and the structure–property relationships obtained in the casting processes described in this chapter.

Pure metals have clearly defined melting or freezing points, and solidification takes place at a constant temperature (Fig. 5.1a). When the temperature of a molten metal is reduced to the freezing point, the latent heat of fusion is given off while the temperature remains constant. At the end of this isothermal phase change, solidification is complete, and the solid metal cools to room temperature. Objects contract as they cool; with castings, this cooling-induced contraction is due to (a) contraction from a superheated state to the metal's solidification temperature, and (b) cooling as a solid from the solidification temperature to room temperature (Fig. 5.1b). In addition, a significant density change can occur as a result of phase change from liquid to solid.

Unlike pure metals, **alloys** solidify over a range of temperatures. Solidification begins when the temperature of the molten metal drops below the **liquidus**; it is completed when the temperature reaches the **solidus**. Within this temperature range, the alloy is in a mushy or pasty state; its composition and state are described by the particular alloy's phase diagram (Section 5.2.4).

5.2.1 Solid solutions

Two terms are essential in describing alloys: *solute* and *solvent*. **Solute** is the minor element (such as salt or sugar) that is added to the **solvent**, which is the major element (such as water). In terms of the elements involved in a metal's crystal structure, the solute is the element (*solute atoms*) added to the solvent (*host atoms*). When the particular crystal structure of the solvent is maintained during alloying, the alloy is called a *solid solution*.

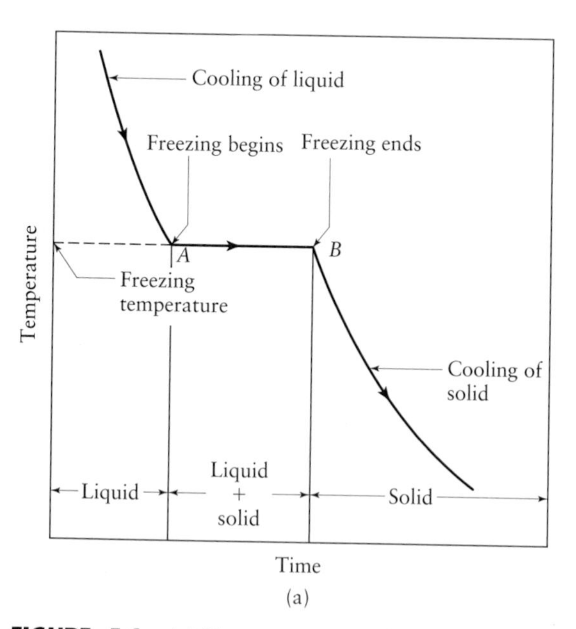

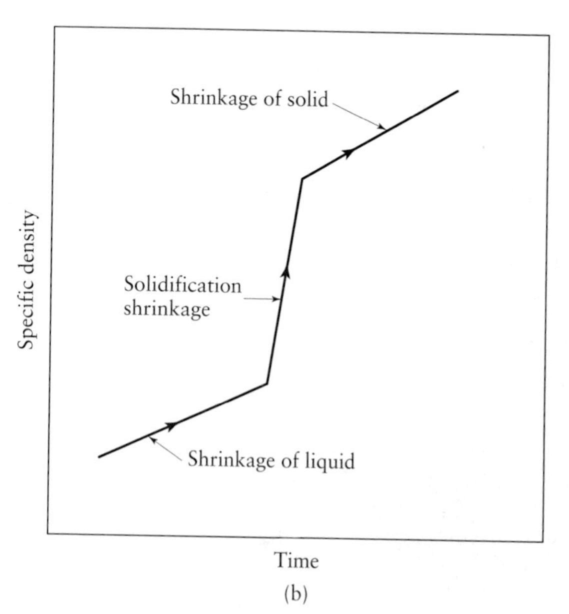

FIGURE 5.1 (a) Temperature as a function of time for the solidification of pure metals. Note that freezing takes place at a constant temperature. (b) Density as a function of time.

Substitutional solid solutions. If the size of the solute atom is similar to that of the solvent atom, solute atoms can replace solvent atoms and form a *substitutional solid solution*. (See Fig. 3.8.) An example of this phenomenon is brass, an alloy of zinc and copper, in which zinc (solute atoms) is introduced into the lattice of copper (solvent atoms). The properties of brasses can thus be altered over a certain range by controlling the amount of zinc in the copper.

Interstitial solid solutions. If the size of the solute atom is much smaller than that of the solvent atom, the solute atom occupies an interstitial position (See Fig. 3.8) and forms an *interstitial solid solution*. An important example of an interstitial solid solution is steel, an alloy of iron and carbon, in which carbon atoms are present in an interstitial position between iron atoms. As will be shown in Section 5.11, the properties of steel can be varied over a wide range by controlling the amount of carbon in iron. This is one reason that, in addition to being inexpensive, steel is such a versatile and useful material with a wide variety of properties and applications.

5.2.2 Intermetallic compounds

Intermetallic compounds are complex structures in which solute atoms are present among solvent atoms in certain specific proportions; thus, some intermetallic compounds have solid solubility. The type of atomic bonds may range from metallic to ionic. Intermetallic compounds are strong, hard, and brittle. An example is copper in aluminum, where an intermetallic compound of $CuAl_2$ can be made to precipitate from an aluminum–copper alloy; this process is an example of *precipitation hardening*. (See Section 5.11.2.)

5.2.3 Two-phase alloys

A solid solution is one in which two or more elements are soluble in a solid state, forming a single homogeneous solid phase in which the alloying elements are uniformly distributed throughout the solid mass. However, there is a limit to the concentration of solute atoms in a solvent-atom lattice, just as there is a solubility limit to sugar in water. Most alloys consist of two or more solid phases and may be regarded as mechanical mixtures. Such a system with two solid phases is called a **two-phase system**; each phase is a homogeneous part of the total mass and has its own characteristics and properties.

A typical example of a two-phase system in metals is lead added to copper in the molten state. After the mixture solidifies, the structure consists of two phases: one phase has a small amount of lead in solid solution in copper; in the other phase, lead particles, roughly spherical in shape, are *dispersed* throughout the matrix of the primary phase (Fig. 5.2a). This copper–lead alloy has properties that are different than those of either copper or lead alone. Alloying with finely dispersed particles (*second-phase particles*) is an important method of strengthening alloys and controlling their properties. Generally, in two-phase alloys, the second-phase particles present obstacles to dislocation movement, which increases the alloy's strength.

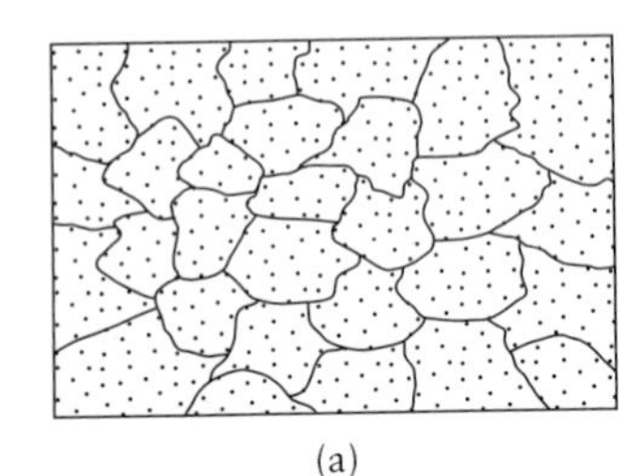
(a)

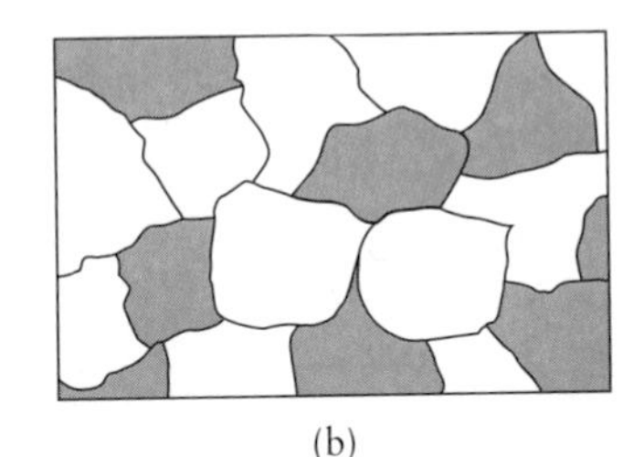
(b)

FIGURE 5.2 (a) Schematic illustration of grains, grain boundaries, and particles dispersed throughout the structure of a two-phase system, such as lead–copper alloy. The grains represent lead in a solid solution of copper, and the particles are lead as a second phase. (b) Schematic illustration of a two-phase system consisting of two sets of grains: dark and light. Dark and light grains have their own compositions and properties.

Another example of a two-phase alloy is the aggregate structure shown in Fig. 5.2b; this alloy system contains two sets of grains, each with its own composition and properties. The darker grains may, for example, have a different structure than the lighter grains and may be brittle, whereas the lighter grains may be ductile.

5.2.4 Phase diagrams

A *phase diagram*, also called an **equilibrium diagram** or **constitutional diagram**, graphically illustrates the relationships among temperature, composition, and the phases present in a particular alloy system. *Equilibrium* means that the state of a system remains constant over an indefinite period of time. *Constitutional* indicates the relationships among structure, composition, and physical makeup of the alloy.

An example of a phase diagram is shown in Fig. 5.3 for the nickel–copper alloy; it is called a **binary phase diagram** because of the two elements (nickel and copper) in the system. The left boundary of this phase diagram (100% Ni) indicates the melting point of nickel, and the right boundary (100% Cu) indicates the melting point of copper. (All percentages in this discussion are by weight.) Note that for a composition of, say, 50% Cu-50% Ni, the alloy begins to solidify at a temperature of

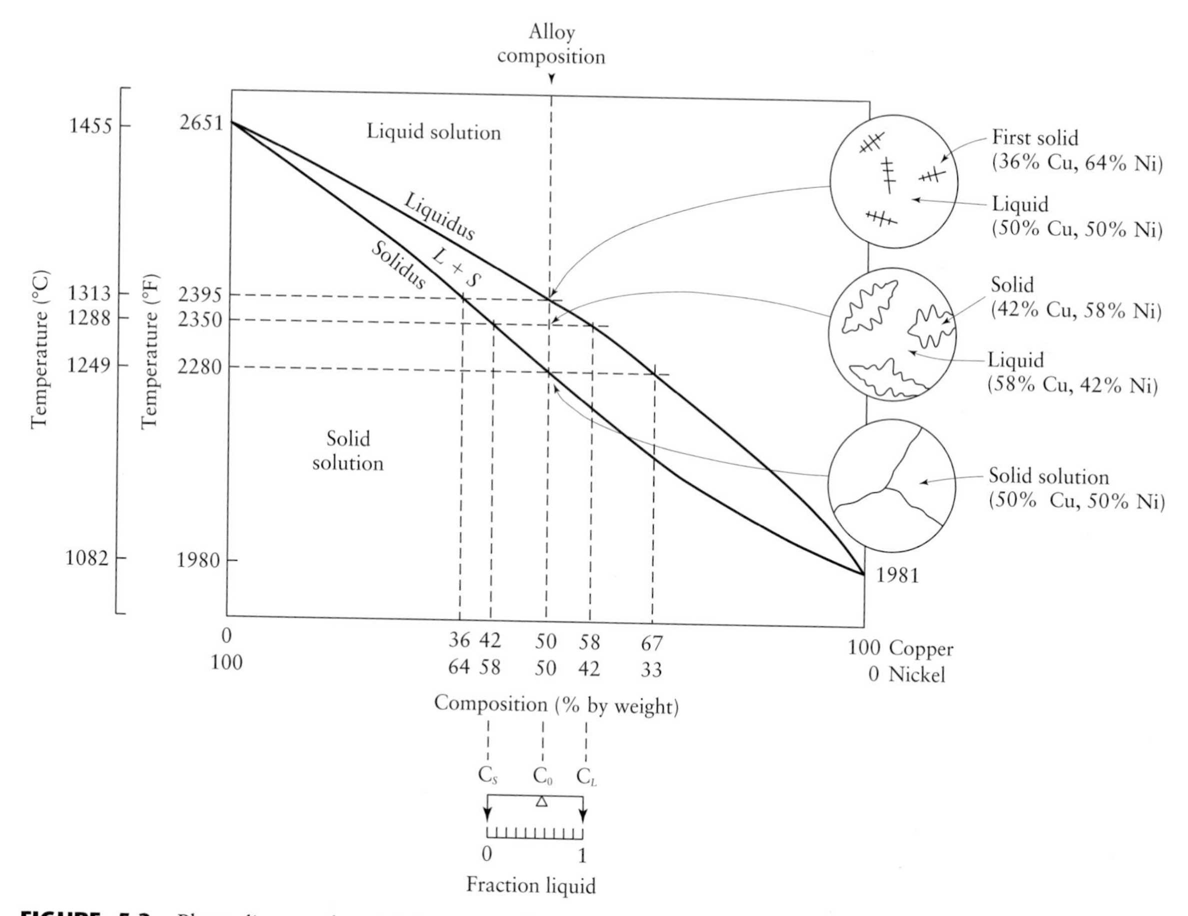

FIGURE 5.3 Phase diagram for nickel–copper alloy system obtained by a slow rate of solidification. Note that pure nickel and pure copper each have one freezing or melting temperature. The top circle on the right depicts the nucleation of crystals. The second circle shows the formation of dendrites. The bottom circle shows the solidified alloy with grain boundaries.

1313°C (2395°F), and solidification is complete at 1249°C (2280°F). Above 1313°C, a homogeneous liquid of 50% Cu-50% Ni exists. When cooled slowly to 1249°C, a homogeneous solid solution of 50% Cu-50% Ni results. However, between the liquidus and solidus curves, say, at a temperature of 1288°C (2350°F), is a two-phase region: a *solid phase* composed of 42% Cu-58% Ni, and a *liquid phase* of 58% Cu-42% Ni. To determine the composition of the solid phase, go left horizontally to the solidus curve and read down to obtain 42% Cu-58% Ni. The composition of the liquid phase can be determined similarly by going to the right to the liquidus curve (58% Cu-42% Ni).

The completely solidified alloy in the phase diagram shown in Fig. 5.3 is a **solid solution**, because the alloying element (Cu, the solute atom) is completely dissolved in the host metal (Ni, the solvent atom), and each grain has the same composition. The mechanical properties of solid solutions of Cu–Ni depend on composition. By increasing the nickel content, the properties of pure copper are improved. The improvements in properties result from pinning (blocking) of dislocations at solute atoms of nickel, which may also be regarded as impurity atoms. (See Fig. 3.8.) As a result, dislocations cannot move as freely, and consequently the strength of the alloy increases.

Lever rule. The composition of various phases in a phase diagram can be determined by a procedure called the *lever rule*. As shown in the lower portion of Fig. 5.3, we first construct a lever (*tie line*) between the solidus and liquidus lines; the lever is balanced (on the triangular support) at the nominal weight composition C_0 of the alloy. The left end of the lever represents the composition C_S of the solid phase, and the right end represents the composition C_L of the liquid phase. Note from the graduated scale in the figure that we also have indicated the liquid fraction along this tie line, ranging from zero at the left (fully solid) to unity at the right (fully liquid).

The lever rule states that the **weight fraction of solid** is proportional to the distance between C_0 and C_L:

$$\frac{S}{S+L} = \frac{C_0 - C_L}{C_S - C_L}. \tag{5.1}$$

Likewise, the **weight fraction of liquid** is proportional to the distance between C_S and C_0:

$$\frac{L}{S+L} = \frac{C_S - C_0}{C_S - C_L}. \tag{5.2}$$

Note that these quantities are fractions and must be multiplied by 100 to obtain percentages.

From inspection of the tie line in Fig. 5.3 (and for a nominal alloy composition of C_0 = 50% Cu-50% Ni), we note that, because C_0 is closer to C_L than it is to C_S, the solid phase contains less copper than does the liquid phase. Measuring on the phase diagram and using the lever-rule equations, we find that the composition of the solid phase is 42% Cu and the composition of the liquid phase is 58% Cu, as stated in the middle circle at the right in Fig. 5.3.

Note that these calculations refer to copper. If we now reverse the phase diagram in the figure, so that the left boundary is 0% nickel (whereby nickel becomes the alloying element in copper), these calculations give us the compositions of the solid and liquid phases in terms of nickel. The lever rule is also known as the *inverse lever rule*, because, as Eqs. (5.1) and (5.2) show, the amount of each phase is proportional to the length of the opposite end of the lever.

5.2.5 The iron–carbon system

Steel and cast iron are represented by the *iron–carbon binary system*. Commercially pure iron contains up to 0.008% C, steels up to 2.11% C, and cast irons up to 6.67% C, although most cast irons contain less than 4.5% C. In this section, we describe the iron–carbon system in order to evaluate and discuss how to modify the properties of these important materials for specific applications.

Figure 5.4 shows the **iron–iron-carbide phase diagram.** Although this diagram can be extended to the right (to 100% C, pure graphite), the range that is significant to engineering applications is up to 6.67% C, where cementite forms. Pure iron melts at a temperature of 1538°C (2800°F), as shown at the left in Fig. 5.4; as iron cools, it first forms δ-iron, then γ-iron, and finally α-iron, which are described as follows:

1. **Ferrite.** *Delta ferrite* is stable only at very high temperatures and has no significant or practical engineering applications. *Alpha ferrite*, or simply *ferrite*, is a solid solution of bcc iron and has a maximum solid solubility of 0.022% C at a temperature of 727°C (1341°F). Just as there is a solubility limit to salt in water (with any extra amount precipitating as solid salt at the bottom of the container), so there also is a solid solubility limit to carbon in iron.

   Ferrite is relatively soft and ductile and is magnetic from room temperature up to 768°C (1414°F). Although very little carbon can dissolve interstitially in bcc iron, the amount of carbon in the composition can significantly affect the mechanical properties of ferrite. Also, significant amounts of chromium, manganese, nickel, molybdenum, tungsten, and silicon can be contained in iron in solid solution, imparting certain desirable properties.

2. **Austenite.** Between 1394°C (2541°F) and 912°C (1674°F) iron undergoes an *allotropic transformation* from the bcc structure to the fcc structure, becoming what is known as *gamma iron*, or more commonly, *austenite*. This structure has a solid solubility of up to 2.11% C at 1148°C (2098°F); thus, the solid solubility of austenite is about two orders of magnitude higher than that of ferrite, with the carbon occupying interstitial positions in austenite (Fig. 5.5). Note that the atomic radius of Fe is 0.124 nm and of C is 0.071 nm. Austenite is an important phase in the heat treatment of steels (Section 5.11). It is denser than ferrite, and its single-phase fcc structure is ductile at elevated temperatures; thus, it possesses good formability. Large amounts of nickel and manganese can also be dissolved

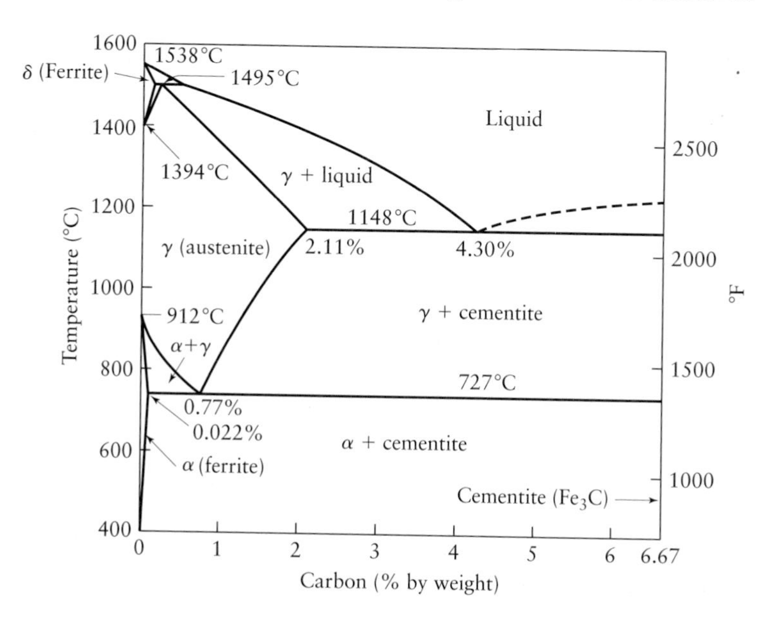

FIGURE 5.4 The iron–iron-carbide phase diagram. Because of the importance of steel as an engineering material, this diagram is one of the most important phase diagrams.

in fcc iron to impart various properties. Austenitic steel is nonmagnetic at high temperatures, and austenitic stainless steels are nonmagnetic at room temperature.

3. **Cementite.** The right boundary of Fig. 5.4 represents *cementite*, also called **carbide**, which is 100% iron carbide (Fe_3C), with a carbon content of 6.67%. (This carbide should not be confused with carbides used for tool and die materials, described in Section 8.6.4) Cementite is a very hard and brittle intermetallic compound and significantly influences the properties of steels. It can be alloyed with elements such as chromium, molybdenum, and manganese.

5.2.6 The iron–iron-carbide phase diagram

The region of the iron–iron-carbide phase diagram that is significant for steels is shown in Fig. 5.6, an enlargement of the lower left portion of Fig. 5.4. Various microstructures

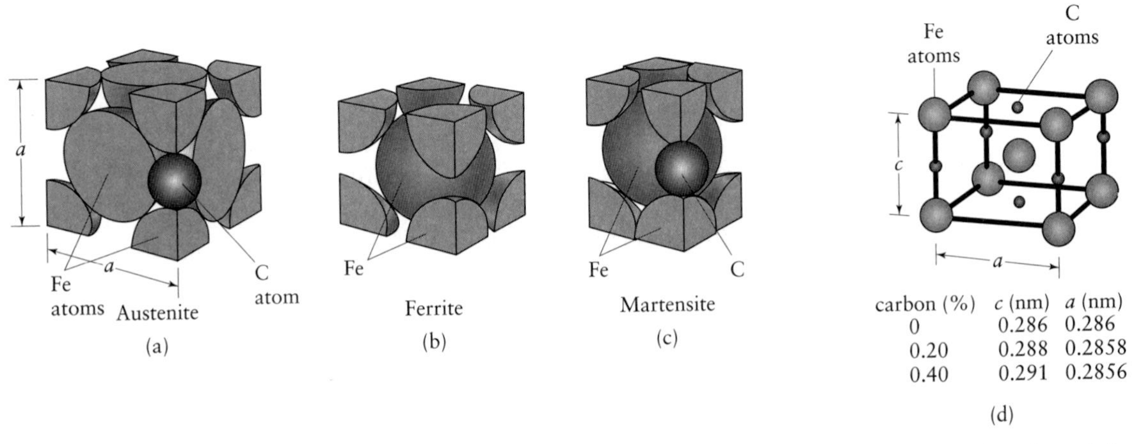

FIGURE 5.5 The unit cell for (a) austenite, (b) ferrite, and (c) martensite. The effect of the percentage of carbon (by weight) on the lattice dimensions for martensite is shown in (d). Note the interstitial position of the carbon atoms (see also Fig. 3.8) and the increase in dimension *c* with increasing carbon content. Thus, the unit cell of martensite is in the shape of a rectangular prism.

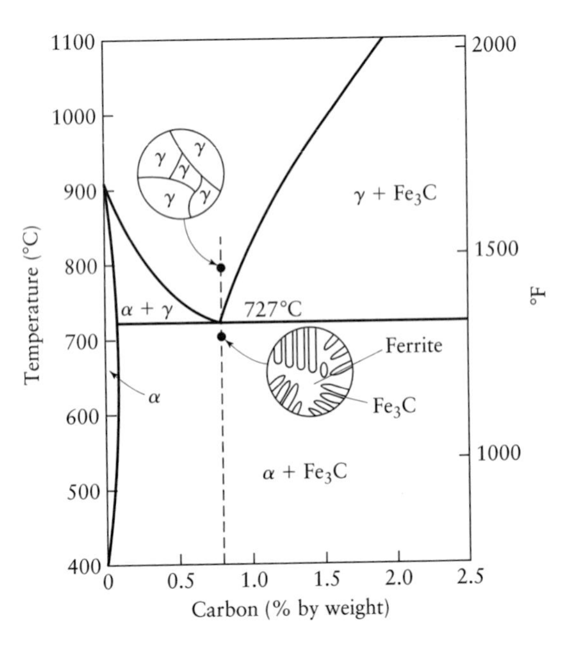

FIGURE 5.6 Schematic illustration of the microstructures for an iron–carbon alloy of eutectoid composition (0.77% carbon) above and below the eutectoid temperature of 727°C (1341°F).

can be developed in steels, depending on the carbon content and the method of heat treatment (Section 5.11). For example, let's consider iron with a 0.77% C content being cooled very slowly from a temperature of, say, 1100°C (2012°F) in the austenite phase. The reason for the slow cooling rate is to maintain equilibrium; higher rates of cooling are used in heat treating. At 727°C (1341°F), a reaction takes place in which austenite is transformed into alpha ferrite (bcc) and cementite. As the solid solubility of carbon in ferrite is only 0.022%, the extra carbon forms cementite.

This reaction is called a **eutectoid** (meaning *eutecticlike*) **reaction**, indicating that at a certain temperature, a single solid phase (austenite) is transformed into two other solid phases (ferrite and cementite). The structure of the eutectoid steel is called **pearlite**, because it resembles mother of pearl at low magnifications. The microstructure of pearlite consists of alternate layers (*lamellae*) of ferrite and cementite (Fig. 5.6). Consequently, the mechanical properties of pearlite are intermediate between ferrite (soft and ductile) and cementite (hard and brittle).

In iron with less than 0.77% C, the microstructure formed consists of a pearlite phase (ferrite and cementite) and a ferrite phase. The ferrite in the pearlite is called *eutectoid ferrite*. The ferrite phase is called *proeutectoid ferrite*, *pro* meaning before, because it forms at a temperature higher than the eutectoid temperature of 727°C (1341°F). If the carbon content is higher than 0.77%, the austenite transforms into pearlite and cementite. The cementite in the pearlite is called *eutectoid cementite*, and the cementite phase is called *proeutectoid cementite*, because it forms at a temperature higher than the eutectoid temperature.

Effects of alloying elements in iron. Although carbon is the basic element that transforms iron into steel, other elements are also added to impart various desirable properties. The effect of these alloying elements on the iron–iron-carbide phase diagram is to shift the eutectoid temperature and eutectoid composition (the percentage of carbon in steel at the eutectoid point). The eutectoid temperature may be raised or lowered from 727°C (1341°F), depending on the particular alloying element. However, alloying elements always lower the eutectoid composition; that is, the carbon content becomes less than 0.77%. Lowering the eutectoid temperature means increasing the austenite range; hence, an alloying element such as nickel is known as an **austenite former**. Because nickel has an fcc structure, it tends to favor the fcc structure of austenite. Conversely, chromium and molybdenum have a bcc structure, causing these elements to favor the bcc structure of ferrite; these elements are known as **ferrite formers**.

5.2.7 Cast irons

The term *cast iron* refers to a family of ferrous alloys composed of iron, carbon (ranging from 2.11% to about 4.5%), and silicon (up to about 3.5%). Cast irons (Section 5.6.1) are usually classified according to their solidification morphology:

1. **Gray cast iron,** or gray iron;
2. **Nodular cast iron,** ductile cast iron, or spheroidal graphite cast iron;
3. **White cast iron;**
4. **Malleable iron;**
5. **Compacted graphite iron.**

Cast irons are also classified by their structure: ferrite, pearlite, quenched and tempered, and austempered.

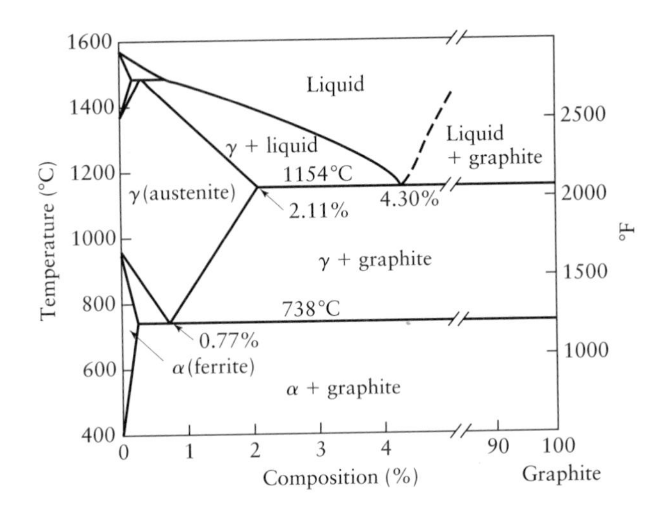

FIGURE 5.7 Phase diagram for the iron–carbon system with graphite, instead of cementite, as the stable phase. Note that this figure is an extended version of Fig. 5.4.

Figure 5.7 shows the equilibrium diagram relevant to cast irons, in which the right boundary is 100% C, that is, pure graphite. The horizontal liquid + graphite line is at 1154°C (2109°F); thus, cast irons are completely liquid at temperatures lower than those required for liquid steels. Consequently, cast irons have lower melting temperatures, which is why the casting process is so suitable for iron with high carbon content. Although cementite exists in steels almost indefinitely, it is not completely stable. That is, it is *metastable*, with an extremely low rate of decomposition. However, cementite can be made to decompose into alpha ferrite and graphite. The formation of graphite (*graphitization*) can be controlled, promoted, and accelerated by modifying the composition and the rate of cooling, and by the addition of silicon.

EXAMPLE 5.1 Determining the amount of phases in carbon steel

Determine the amount of gamma and alpha phases in a 10-kg, 1040 steel casting as it is being cooled slowly to the following temperatures: (a) 900°C, (b) 728°C, and (c) 726°C.

Solution. (a) Referring to Fig. 5.6, we draw a vertical line at 0.40% C at 900°C. We are in the single-phase austenite region, so the percent gamma phase is 100 (10 kg) and the percent alpha phase is zero. (b) At 728°C, the alloy is in the two-phase gamma–alpha field. When we draw the phase diagram in greater detail, we can find the weight percentages of each phase by the lever rule:

$$\text{Percent alpha} = \left(\frac{C_\gamma - C_0}{C_\gamma - C_\alpha}\right)100 = \left(\frac{0.77 - 0.40}{0.77 - 0.022}\right)100 = 50\%, \text{or } 5 \text{ kg};$$

$$\text{Percent gamma} = \left(\frac{C_0 - C_\alpha}{C_\gamma - C_\alpha}\right)100 = \left(\frac{0.40 - 0.022}{0.77 - 0.022}\right)100 = 50\%, \text{or } 5 \text{ kg}.$$

(c) At 726°C, the alloy is in the two-phase alpha–Fe_3C region. No gamma phase is present. Again, we use the lever rule to find the weight percentage of alpha phase present:

$$\text{Percent alpha} = \left(\frac{6.67 - 0.40}{6.67 - 0.022}\right)100 = 94\%, \text{or } 9.4 \text{ kg}.$$

5.3 Cast Structures

The type of *cast structure* developed during solidification of metals and alloys depends on the composition of the particular alloy, as well as on the rate of heat transfer and the flow of the liquid metal during the casting process. As described in other sections in this chapter, the structures developed, in turn, affect the properties of the finished castings.

5.3.1 Pure metals

Figure 5.8a shows the grain structure of a pure metal solidified in a square mold. At the walls of the mold, the metal cools rapidly, because the walls are at ambient temperature. As a result of cooling first at the mold walls, the casting develops a solidified **skin**, or *shell*, or fine **equiaxed grains**. The grains grow in the direction opposite to the direction of heat transfer from the mold. Grains that have a favorable orientation, called **columnar grains**, grow preferentially (Fig. 5.9); grains that have substantially different orientations are blocked from further growth.

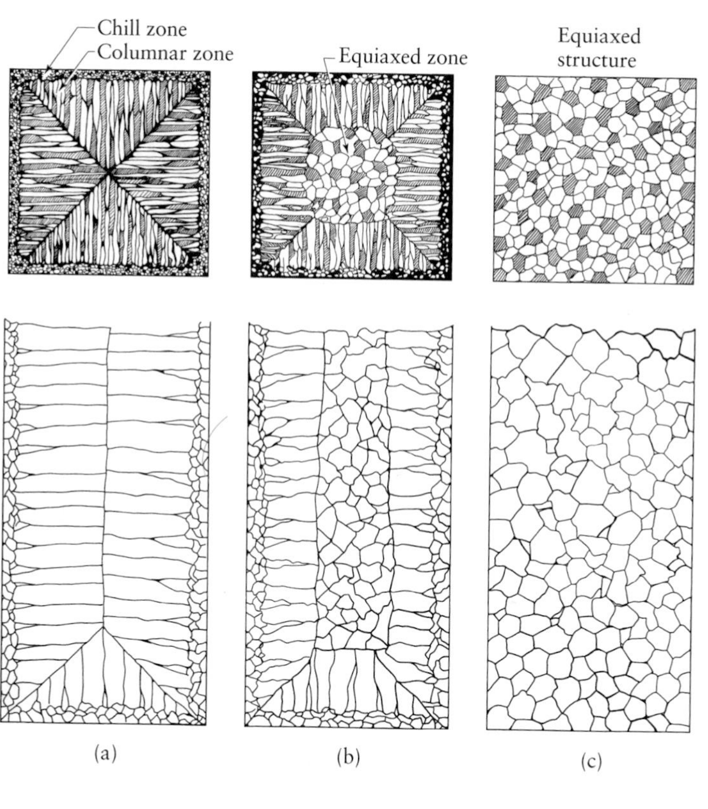

FIGURE 5.8 Schematic illustration of three cast structures of metals solidified in a square mold: (a) pure metals; (b) solid-solution alloys; and (c) the structure obtained by heterogenous nucleation of grains, using nucleating agents. *Source*: G. W. Form, J. F. Wallace, J. L. Walker, and A. Cibula.

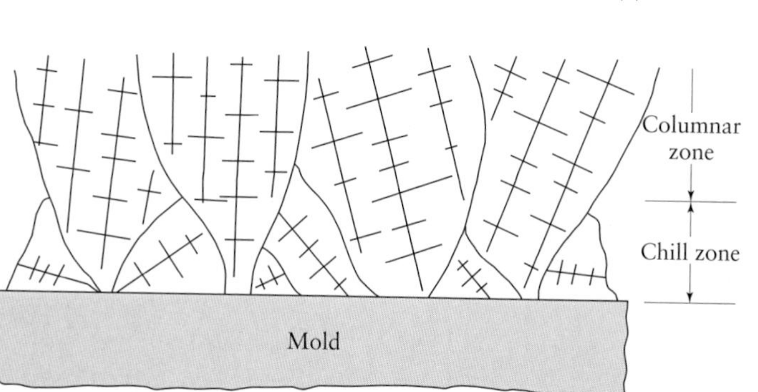

FIGURE 5.9 Development of a preferred texture at a cool mold wall. Note that only favorably oriented grains grow away from the surface of the mold.

5.3.2 Alloys

Pure metals have limited mechanical properties; however, their properties can be enhanced and modified by **alloying**. The vast majority of metals used in engineering applications are some form of an *alloy*, consisting of two or more chemical elements, at least one of which is a metal. Because many consumer and industrial products contain cast alloys, a study of their solidification is essential.

Solidification in alloys begins when the temperature drops below the liquidus, T_L, and is complete when it reaches the solidus, T_S (Fig. 5.10). Within this temperature range, the alloy is in a mushy or pasty state and has **columnar dendrites** (from the Greek words *dendron*, meaning *akin to*, and *drys*, meaning *tree*). Note the presence of liquid metal between the dendrites' arms. Dendrites have three-dimensional arms and branches (secondary arms), and the arms eventually interlock, as shown in Fig. 5.11. The width of the mushy zone, where both liquid and solid phases are pre-

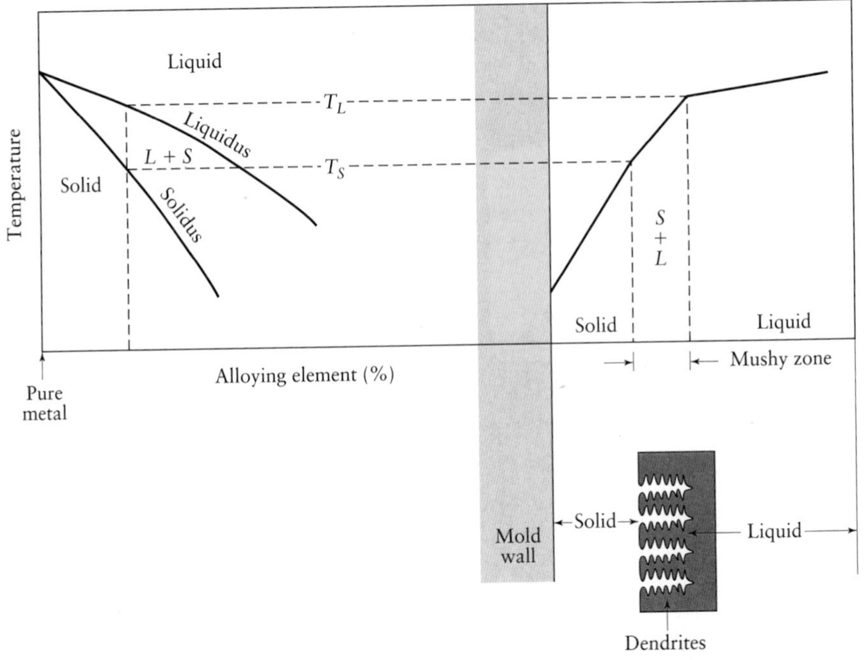

FIGURE 5.10 Schematic illustration of alloy solidification and temperature distribution in a solidifying metal. Note the formation of dendrites in the semisolid (mushy) zone.

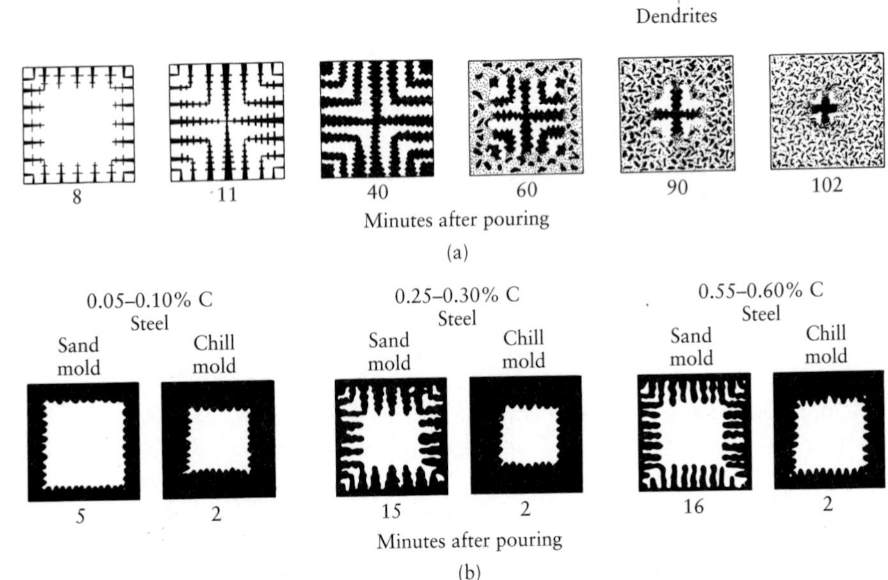

FIGURE 5.11 (a) Solidification patterns for gray cast iron in a 180-mm (7-in.) square casting. Note that after 11 minutes of cooling, dendrites reach each other, but the casting is still mushy throughout. It takes about two hours for this casting to solidify completely. (b) Solidification of carbon steels in sand and chill (metal) molds. Note the difference in solidification patterns as the carbon content increases. *Source*: H. F. Bishop and W. S. Pellini.

sent, is an important factor during solidification. This zone is described in terms of a temperature difference, known as the **freezing range**, as follows:

$$\text{Freezing range} = T_L - T_S. \tag{5.3}$$

Figure 5.10 shows that pure metals have a freezing range that approaches zero and that the *solidification front* moves as a plane front, without forming a mushy zone. Eutectics solidify in a similar manner, with an approximately planar front. The type of solidification structure developed depends on the composition of the eutectic. For alloys with a nearly symmetrical phase diagram, the structure is generally lamellar, with two or more solid phases present, depending on the alloy system. When the volume fraction of the minor phase of the alloy is less than about 25%, the structure generally becomes *fibrous*. These conditions are particularly important for cast irons.

For alloys, although it is not precise, a short freezing range generally involves a temperature difference of less than 50°C (90°F), and a long freezing range is generally at temperatures greater than 110°C (200°F). Ferrous castings typically have narrow semisolid (mushy) zones, whereas aluminum and magnesium alloys have wide mushy zones. Consequently, these alloys are in a semisolid state throughout most of the solidification process.

Effects of cooling rate. Slow cooling rates (on the order of 10^2 K/s) or long local solidification times result in coarse dendritic structures with large spacing between the dendrites' arms. For faster cooling rates (on the order of 10^4 K/s) or short local solidification times, the structure becomes finer, with smaller dendrite arm spacing. For still higher cooling rates (on the order of 10^6 to 10^8 K/s), the structures developed are *amorphous* (without any ordered crystalline structure), as described in Section 5.10.8. The structures developed and the resulting grain size, in turn, influence the properties of the casting. As grain size decreases, (a) the strength and ductility of the cast alloy increase (see the Hall-Petch equation, discussed in Section 3.4.1), (b) microporosity (interdendritic shrinkage voids) in the casting decreases, and (c) the tendency for the casting to crack (*hot tearing*) during solidification decreases. Lack of uniformity in grain size and distribution results in castings with anisotropic properties.

5.3.3 Structure–property relationships

Because all castings are expected to possess certain properties in order to meet design and service requirements, the relationships between these properties and the structures developed during solidification are important aspects of casting. In this section, we describe these relationships in terms of dendrite morphology and the concentration of alloying elements in various regions.

The compositions of dendrites and the liquid metal are given by the phase diagram of the particular alloy. When the alloy is cooled very slowly, each dendrite develops a uniform composition. Under normal cooling encountered in practice, however, **cored dendrites** are formed, which have a composition at their surface different from that at their centers (i.e., a *concentration gradient*). The surface has a higher concentration of alloying elements than does the core, because of solute rejection from the core toward the surface during solidification of the dendrite; this phenomenon is known as **microsegregation.** The darker shading in the interdendritic liquid near the dendrite roots shown in Fig. 5.12 indicates that these regions have a higher solute concentration; thus, microsegregation in these regions is much more pronounced than in others.

In contrast to microsegregation, **macrosegregation** involves differences in composition throughout the casting itself. In situations where the solidifying front moves

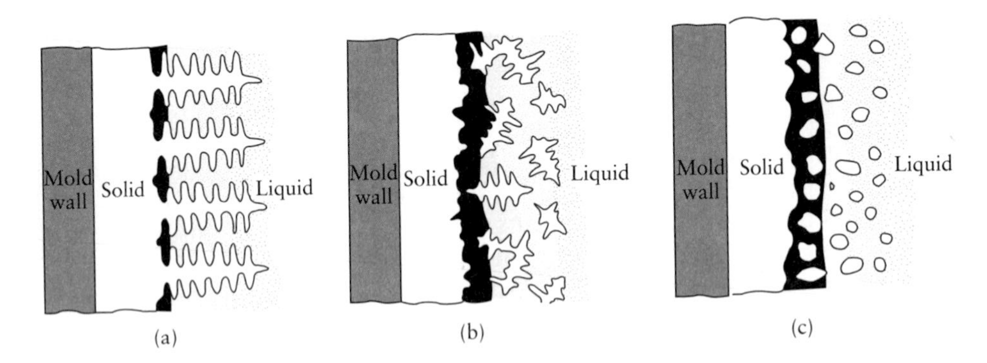

FIGURE 5.12 Schematic illustration of three basic types of cast structures: (a) columnar dendritic; (b) equiaxed dendritic; and (c) equiaxed nondendritic. *Source*: After D. Apelian.

away from the surface of a casting as a plane front (Fig. 5.13), constituents with a lower melting point in the solidifying alloy are driven toward the center (**normal segregation**). Consequently, such a casting has a higher concentration of alloying elements at its center than at its surfaces. In dendritic structures such as for solid-solution alloys (Fig. 5.8b), the opposite occurs: the center of the casting has a lower concentration of alloying elements (**inverse segregation**). The reason inverse segregation occurs is that liquid metal (which has a higher concentration of alloying elements) enters the cavities developed from solidification shrinkage in the dendrite arms, which have solidified sooner. Another form of segregation is due to gravity (**gravity segregation**), whereby higher density inclusions or compounds sink, and lighter elements (such as antimony in an antimony–lead alloy) float to the surface.

Figure 5.8b shows a typical cast structure of a solid-solution alloy with an inner zone of equiaxed grains. This inner zone can be extended throughout the casting, as shown in Fig. 5.8c, by adding an **inoculant** (nucleating agent) to the alloy. The inoculant induces nucleation of grains throughout the liquid metal (**heterogeneous nucleation**); an example is TiB_2 in AlSi alloys.

Because of the presence of thermal gradients in a solidifying mass of liquid metal, and because of gravity (and hence density differences), *convection* has a strong influence on the structures developed. Convection promotes the formation of an outer chill zone, refines grain size, and accelerates the transition from columnar to equiaxed grains. The structure shown in Fig. 5.12b can also be obtained by increasing convection within the liquid metal, whereby dendrite arms separate (**dendrite multiplication**). Conversely, reducing or eliminating convection results in coarser and longer columnar-dendritic grains. Convection can be enhanced by the use of mechanical or electromagnetic methods. The dendrite arms are not particularly strong, and they can be broken up by agitation or mechanical vibration in the early stages of solidification (**rheocasting**). This process results in finer grain size, with equiaxed-nondendritic grains distributed more uniformly throughout the casting (Fig. 5.12c). Experiments concerning the effects of gravity on the microstructure of castings are currently being conducted during space flights.

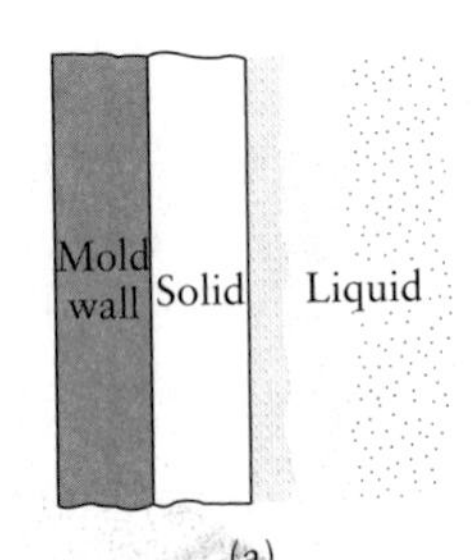

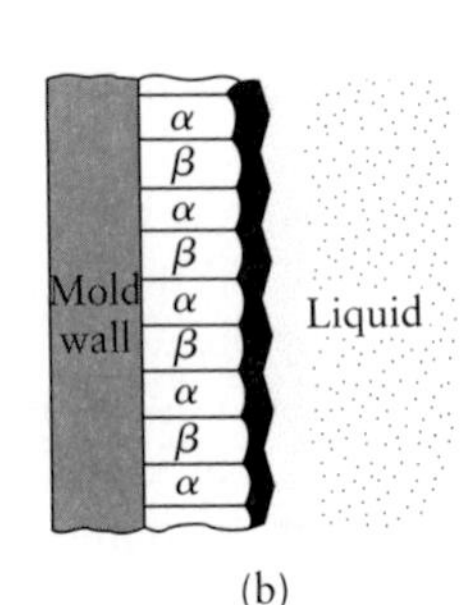

FIGURE 5.13 Schematic illustration of cast structures in (a) plane-front, single-phase alloys; and (b) plane-front, two-phase alloys. *Source:* After D. Apelian.

5.4 | Fluid Flow and Heat Transfer

5.4.1 Fluid flow

To emphasize the importance of fluid flow, let's briefly describe a basic gravity casting system, as shown in Fig. 5.14. The molten metal is poured through a **pouring basin**, or *cup*. Next, it flows through the **sprue** to the **well** and then into **runners** and into the mold cavity. **Risers**, also called *feeders*, serve as reservoirs of molten metal to supply the metal necessary to prevent porosity due to shrinkage, as described in Section 5.4.6. Although such a **gating system** appears to be relatively simple, successful casting requires careful design and control of the solidification process to ensure adequate fluid flow during solidification in the system. For example, one of the most important functions of the gating system is to trap contaminants (such as oxides and other inclusions) in the molten metal by having the contaminants adhere to the walls of the gating system, thereby preventing them from reaching the actual mold cavity. Furthermore, proper design of the gating system avoids or minimizes problems such as premature cooling, turbulence, and gas entrapment. Even before it reaches the mold cavity, the molten metal must be handled carefully to avoid the formation of oxides on molten metal surfaces from exposure to the environment or introduction of impurities into the molten metal.

Two basic principles of fluid flow are relevant to gating design: Bernoulli's theorem and the law of mass continuity.

Bernoulli's theorem. Bernoulli's theorem is based on the principle of conservation of energy and relates pressure, velocity, elevation of the fluid at any location in the system, and the frictional losses in a system that is full of liquid as

$$h + \frac{p}{\rho g} + \frac{v^2}{2g} = \text{constant}, \tag{5.4}$$

where h is the elevation above a certain reference plane, p is the pressure at that elevation, v is the velocity of the liquid at that elevation, ρ is the density of the fluid (assuming that it is incompressible), and g is the gravitational constant. Conservation of energy requires that, at any particular location in the system, the following relationship be satisfied:

$$h_1 + \frac{p_1}{\rho g} + \frac{v_1^2}{2g} = h_2 + \frac{p_2}{\rho g} + \frac{v_2^2}{2g} + f. \tag{5.5}$$

In this equation, the subscripts 1 and 2 represent two different elevations, respectively, and f represents the frictional loss in the liquid as it travels downward through the gating system. The frictional loss includes such factors as energy loss at the liquid–mold-wall interfaces and turbulence in the liquid.

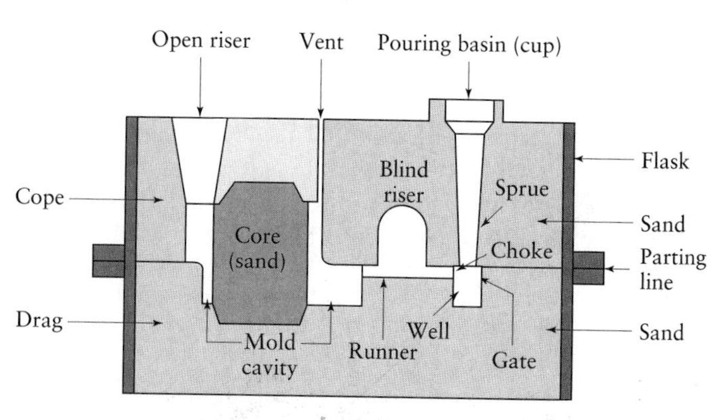

FIGURE 5.14 Schematic illustration of a sand mold, showing various features.

Mass continuity. The law of mass continuity states that, for incompressible liquids and in a system with impermeable walls, the rate of flow is constant and is expressed as

$$Q = A_1 v_1 = A_2 v_2, \tag{5.6}$$

where Q is the volumetric rate of flow, such as in m^3/s; A is the cross-sectional area of the liquid stream; v is the velocity of the liquid in that cross-sectional location; and the subscripts 1 and 2 pertain to two different locations in the system. Thus, the rate of flow must be maintained everywhere in the system. The permeability of the walls of the system is important, because otherwise some liquid will permeate through (for example, in sand molds) and the rate of flow will decrease as the liquid moves through the system. Coatings are often used to inhibit such behavior in sand molds.

Sprue design. An application of the two foregoing principles is the traditional tapered design of sprues (Fig. 5.14); we can determine the shape of the sprue by using Eqs. (5.5) and (5.6). Assuming that the pressure at the top of the sprue is equal to the pressure at the bottom and that there are no frictional losses, the relationship between height and cross-sectional area at any point in the sprue is given by the parabolic relationship

$$\frac{A_1}{A_2} = \sqrt{\frac{h_2}{h_1}}, \tag{5.7}$$

where, let's say, the subscript 1 denotes the top of the sprue and the subscript 2 the bottom. Moving downward from the top, the cross-sectional area of the sprue must decrease. Recall that in a free-falling liquid, such as water from a faucet, the cross-sectional area of the stream decreases as it gains velocity downward. If we design a sprue with a constant cross-sectional area and pour the molten metal, regions may develop where the liquid loses contact with the walls of the sprue. As a result, **aspiration** may take place, whereby air is sucked or entrapped in the liquid. A popular alternative to tapered sprues is the use of straight-sided sprues with a choking mechanism at the bottom, consisting of either a choke core or a runner choke (as shown in Fig. 5.14). The choke slows flow sufficiently to prevent aspiration.

Another application of the foregoing equations is in the *modeling of mold filling*. For example, consider the situation shown in Fig. 5.14, where molten metal is poured into a pouring basin; it flows through a sprue to a gate and runner and fills the mold cavity. If the pouring basin has a much larger cross-sectional area than that of the bottom of the sprue, then the velocity of the molten metal at the top of the pouring basin is very low. If frictional losses are due to viscous dissipation of energy, then f in Eq. (5.5) can be taken as a function of vertical distance, often approximated as a linear function, Therefore, the velocity of the molten metal leaving the gate is obtained from Eq. (5.5) as

$$v = c\sqrt{2gh}, \tag{5.8}$$

where h is the distance from the base of the sprue to the height of the liquid metal and c is a friction factor. For frictionless flow, c equals unity; it is always between 0 and 1. The magnitude of c varies with mold material, runner layout, and channel size and can include energy losses due to turbulence as well as viscous effects.

If the level of the liquid metal has reached a height x, then the gate velocity is

$$v = c\sqrt{(2g(h - x))}\,. \tag{5.9}$$

The rate of flow through the gate will be the product of this velocity and the gate area, according to Eq. (5.6). The shape of the casting will determine the height as a function of time. Integrating Eq. (5.9) gives the mean fill time and flow rate, and dividing the casting volume by this mean flow rate gives the fill time of the mold.

Simulation of mold filling assists designers in the specification of runner diameter, as well as the size and number of sprues and pouring basins. To ensure that the runners stay open, the fill time must be a small fraction of the solidification time (see Section 5.4.4), but the velocity should not be so high as to erode the mold material (*mold wash*) or to result in too high of a Reynolds number (see below), or else turbulence and associated air entrainment result. Fortunately, many computational tools are now available to evaluate gating designs and to assist in the sizing of components. (See Section 5.12.3.)

Flow characteristics. An important consideration in fluid flow in gating systems is the presence of **turbulence**, as opposed to **laminar flow** of fluids; the **Reynolds number** is used to characterize this aspect of fluid flow. The Reynolds number, Re, represents the ratio of the inertia to the viscous forces in fluid flow and is defined as

$$\text{Re} = \frac{vD\rho}{\eta}, \tag{5.10}$$

where v is the velocity of the liquid, D is the diameter of the channel, and ρ and η are the density and viscosity, respectively, of the liquid. The higher the Reynolds number, the greater is the tendency for turbulent flow. In ordinary gating systems, Re ranges from 2000 to 20,000. An Re value of up to 2000 represents laminar flow; between 2000 and 20,000, there is a mixture of laminar and turbulent flow, generally regarded as harmless in gating systems. However, Re values in excess of 20,000 represent severe turbulence, resulting in air entrainment and *dross* formation (the scum that forms on the surface of the molten metal) from the reaction of the liquid metal with air and other gases. Techniques for minimizing turbulence generally involve avoidance of sudden changes in flow direction and in the geometry of channel cross-sections in the design of the gating system.

The elimination of dross and *slag* (nonmetallic products from mutual dissolution of flux and nonmetallic impurities) is another important consideration in fluid flow. It can be achieved by skimming, using *dross traps*; properly designing pouring basins and gating systems; or using filters. Filters are usually made of ceramics, mica, or fiberglass, and their proper location and placement is important for effective filtering of dross and slag.

5.4.2 Fluidity of molten metal

Fluidity is a term commonly used to describe the capability of molten metal to fill mold cavities. This term consists of two basic factors: (1) characteristics of the molten metal and (2) casting parameters. The following characteristics of molten metal influence fluidity:

1. **Viscosity.** As viscosity and its sensitivity to temperature (*viscosity index*) increase, fluidity decreases.
2. **Surface tension.** A high surface tension of the liquid metal reduces fluidity. Oxide films developed on the surface of a molten metal thus have a significant adverse effect on fluidity; for example, the oxide film on the surface of pure molten aluminum triples the surface tension.

3. **Inclusions.** As insoluble particles, inclusions can have a significant adverse effect on fluidity. This effect can be verified by observing the viscosity of a liquid such as oil with and without sand particles in it; the latter has higher viscosity.
4. **Solidification pattern of the alloy.** The manner in which solidification occurs, as described in Section 5.3, can influence fluidity. Moreover, fluidity is inversely proportional to the freezing range [see Eq. (5.3)]; hence, the shorter the freezing range (as with pure metals and eutectics), the higher the fluidity becomes; conversely, alloys with long freezing ranges (such as solid-solution alloys) have lower fluidity.

The following casting parameters influence fluidity and also influence the fluid flow and thermal characteristics of the system:

1. **Mold design.** The design and dimensions of components such as the sprue, runners, and risers both influence fluidity.
2. **Mold material and its surface characteristics.** The higher the thermal conductivity of the mold and the rougher its surfaces, the lower the fluidity of the molten metal becomes. Heating the mold improves fluidity, even though it slows down solidification of the metal; also, the casting develops coarse grains, hence, it has less strength.
3. **Degree of superheat.** Defined as the increment of temperature above the melting point of an alloy, superheat improves fluidity by delaying solidification.
4. **Rate of pouring.** The slower the rate at which the molten metal is poured into the mold, the lower the fluidity becomes, because of the higher local rate of cooling.
5. **Heat transfer.** Heat transfer directly affects the viscosity of the liquid metal.

Although the interrelationships among these factors are complex, the term **castability** is generally used to describe the ease with which a metal can be cast to obtain a part with good quality. Obviously, this term includes not only fluidity, but casting practices as well.

Tests for fluidity. Although none is accepted universally, several tests have been developed to quantify fluidity. In one such test, the molten metal is made to flow along a channel at room temperature; the distance of metal flow before it solidifies and stops is a measure of the metal's fluidity.

5.4.3 Heat transfer

An important consideration in casting is the heat transfer during the complete cycle from pouring to solidification and cooling to room temperature. *Heat flow* at different locations in the system is a complex phenomenon; it depends on many factors relating to the casting material and the mold and process parameters. For instance, in casting thin sections, the metal flow rates must be high enough to avoid premature chilling and solidification. However, the flow rate must not be so fast as to cause excessive turbulence, with its detrimental effects on the properties of the casting.

Figure 5.15 shows a typical temperature distribution in the mold–liquid-metal interface. Heat from the liquid metal is given off through the mold wall and the surrounding air. The temperature drop at the air–mold and mold–metal interfaces is caused by the presence of boundary layers and imperfect contact at these interfaces. The shape of the curve depends on the thermal properties of the molten metal and the mold.

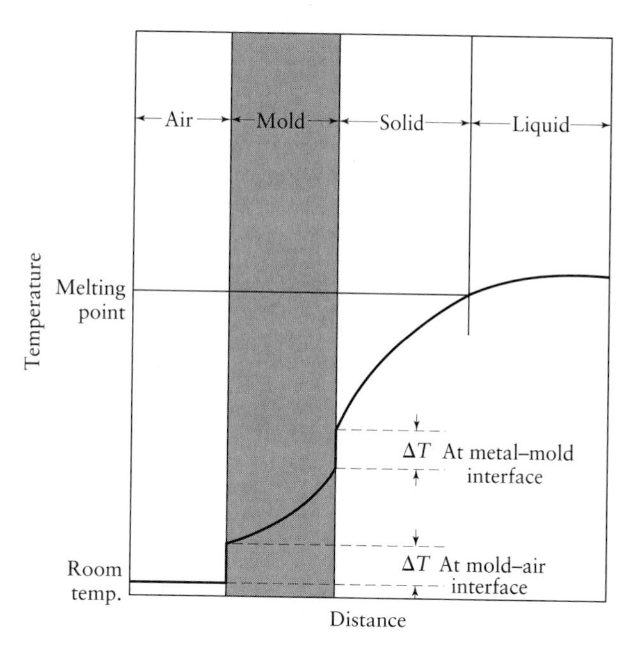

FIGURE 5.15 Temperature distribution at the mold wall and liquid–metal interface during solidification of metals in casting.

5.4.4 Solidification time

During the early stages of solidification, a thin solidified skin begins to form at the cool mold walls, and, as time passes, the skin thickens. With flat mold walls, this thickness is proportional to the square root of time; thus, doubling the amount of time will make the skin $\sqrt{2} = 1.41$ times, or 41%, thicker. The *solidification time* is a function of the volume of a casting and its surface area (**Chvorinov's rule**) and is expressed as

$$\text{Solidification time} = C\left(\frac{\text{volume}}{\text{surface area}}\right)^2, \tag{5.11}$$

where C is a constant that reflects the mold material, the properties of the metal (including latent heat), and temperature. Thus, a large sphere solidifies and cools to ambient temperature at a much slower rate than does a smaller sphere. The reason is that the volume of a sphere is proportional to the cube of its diameter, and the surface area is proportional to the square of its diameter. Similarly, it can be shown that molten metal in a cube-shaped mold will solidify faster than in a spherical mold of the same volume.

Figure 5.16 depicts the effects of mold geometry and elapsed time on skin thickness and shape. As illustrated, the unsolidified molten metal has been poured from the mold at different time intervals, ranging from five seconds to six minutes. Note that the skin thickness increases with elapsed time, but that the skin is thinner at internal angles (location A in the figure) than at external angles (location B). This latter condition is caused by slower cooling at internal angles than at external angles.

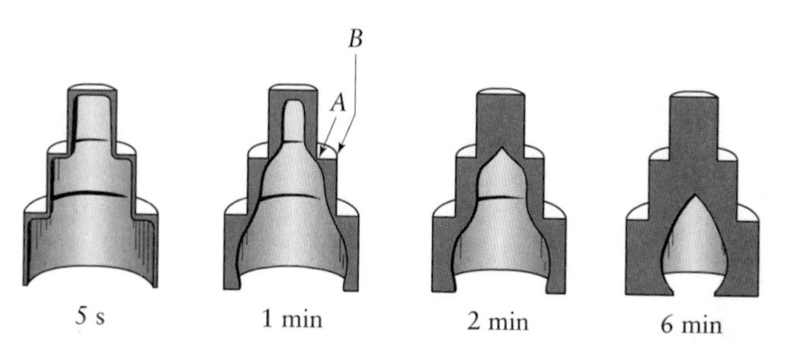

FIGURE 5.16 Solidified skin on a steel casting. The remaining molten metal is poured out at the times indicated in the figure. Hollow ornamental and decorative objects are made by a process called *slush casting*, which is based on this principle. *Source*: H. F. Taylor, J. Wulff, and M. C. Flemings.

EXAMPLE 5.2 Solidification times for various solid shapes

Three pieces of metal being cast have the same volume, but different shapes. One is a sphere, one a cube, and one a cylinder with a height equal to its diameter. Which piece will solidify the fastest, and which one will solidify the slowest?

Solution. Since all three pieces have the same volume, we can just set this volume at unity, so from Eq. (5.11), we have

$$\text{Solidification time} \propto \frac{1}{(\text{surface area})^2}.$$

The respective surface areas are calculated as follows:

Sphere: $V = \left(\frac{4}{3}\right)\pi r^3$, so $r = \left(\frac{3}{4\pi}\right)^{1/3}$; thus, $A = 4\pi r^2 = 4\pi\left(\frac{3}{4\pi}\right)^{2/3} = 4.84.$

Cube: $V = a^3$, so $a = 1$; thus, $A = 6a^2 = 6.$

Cylinder: $V = \pi r^2 h = 2\pi r^3$, so $r = \left(\frac{1}{2\pi}\right)^{1/3}$; thus

$$A = 2\pi r^2 + 2\pi rh = 6\pi r^2 = 6\pi\left(\frac{1}{2\pi}\right)^{2/3} = 5.54.$$

Therefore, the respective solidification times t are

$$t_{\text{sphere}} = 0.043C, \quad t_{\text{cube}} = 0.028C, \quad \text{and} \quad t_{\text{cylinder}} = 0.033C.$$

Therefore, the cube-shaped casting will solidify the fastest, and the sphere-shaped casting will solidify the slowest.

5.4.5 Shrinkage

Metals shrink (contract) during solidification and cooling. *Shrinkage*, which causes dimensional changes (and, sometimes, cracking) is the result of the following factors:

1. Contraction of the molten metal as it cools prior to solidification;
2. Contraction of the metal during phase change from liquid to solid (latent heat of fusion);
3. Contraction of the solidified metal (the casting) as its temperature drops to ambient temperature.

The largest amount of shrinkage occurs during cooling of the casting. Table 5.1 lists the percentages of contraction for various metals during solidification. Note, however, that some metals expand, including gray cast iron. The reason for the expansion of gray cast iron is that graphite has a relatively high specific volume, and when it precipitates as graphite flakes during solidification, a net expansion of the metal occurs.

5.4.6 Defects in castings

Depending on casting design and method, several defects can develop in castings. Because different names have been used to describe the same defect, the International Committee of Foundry Technical Associations has developed standardized nomenclature, consisting of seven basic categories of casting defects:

TABLE 5.1

Volumetric Solidification Contraction or Expansion Percentages for Various Cast Metals

Contraction (%)		Expansion (%)	
Aluminum	7.1	Bismuth	3.3
Zinc	6.5	Silicon	2.9
Al, 4.5% Cu	6.3	Gray iron	2.5
Gold	5.5		
White iron	4–5.5		
Copper	4.9		
Brass (70–30)	4.5		
Magnesium	4.2		
90% Cu, 10% Al	4		
Carbon steels	2.5–4		
Al, 12% Si	3.8		
Lead	3.2		

A) **Metallic projections,** consisting of fins, flash, or massive projections such as swells and rough surfaces.

B) **Cavities,** consisting of rounded or rough internal or exposed cavities, including blowholes, pinholes, and shrinkage cavities.

C) **Discontinuities,** such as cracks, cold or hot tearing, and cold shuts. If the solidifying metal is constrained from shrinking freely, cracking and tearing can occur. Although many factors are involved in tearing, coarse grain size and the presence of low-melting segregates along the grain boundaries (intergranular) increase the tendency for hot tearing. Incomplete castings result from the molten metal being at too low a temperature or from the metal being poured too slowly. *Cold shut* is an interface in a casting that lacks complete fusion because of the meeting of two streams of partially solidified metal.

D) **Defective surface,** such as surface folds, laps, scars, adhering sand layers, and oxide scale.

E) **Incomplete casting,** such as misruns (due to premature solidification), insufficient volume of metal poured, and runout (due to loss of metal from the mold after pouring).

F) **Incorrect dimensions or shape,** owing to factors such as improper shrinkage allowance, pattern-mounting error, irregular contraction, deformed pattern, or warped casting.

G) **Inclusions,** which form during melting, solidification, and molding. Generally nonmetallic, they are regarded as harmful because they act like stress raisers and reduce the strength of the casting. Hard inclusions (spots) also tend to chip or break tools in machining. They can be filtered out during processing of the molten metal. Inclusions may form during melting because of reaction of the molten metal with the environment (usually with oxygen) or the crucible material. Chemical reactions among components in the molten metal may produce inclusions; slags and other foreign material entrapped in the molten metal also become inclusions. Reactions between the metal and the mold material may produce inclusions as well. In addition, spalling of the mold and core surfaces produces inclusions, indicating the importance of maintaining melt quality and monitoring the conditions of the molds.

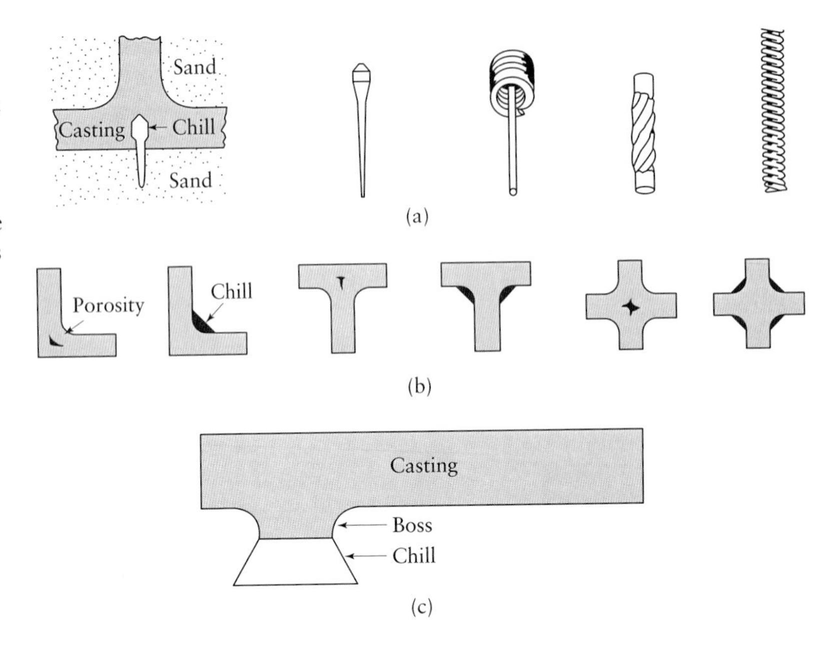

FIGURE 5.17 Various types of (a) internal and (b) external chills (dark areas at corners), used in castings to eliminate porosity caused by shrinkage. Chills are placed in regions where there is a large volume of metal, as shown in (c).

Porosity. *Porosity* in a casting may be caused by *shrinkage, trapped gases*, or both. Porosity is detrimental to the ductility of a casting and its surface finish, making it permeable and thus affecting pressure tightness of a cast pressure vessel. Porous regions can develop in castings because of shrinkage of the solidified metal. Thin sections in a casting solidify sooner than thicker regions. As a result, molten metal cannot be fed into the thicker regions that have not yet solidified. Because of contraction, as the surfaces of the thicker regions begin to solidify, porous regions develop at their centers. **Microporosity** can also develop when the liquid metal solidifies and shrinks between dendrites and between dendrite branches.

Porosity caused by shrinkage can be reduced or eliminated by various means. Basically, adequate liquid-metal feeding should be provided to all parts of the solidifying casting to avoid porosity. External or internal **chills**, used in sand casting (Fig. 5.17), also are an effective means of reducing shrinkage porosity. The function of chills is to increase the rate of solidification in critical regions. Internal chills are usually made of the same material as the castings. External chills may be made of the same material or may be made of iron, copper, or graphite. With alloys, porosity can be reduced or eliminated by making the temperature gradient steep, using, for example, mold materials that have high thermal conductivity. Subjecting the casting to **hot isostatic pressing** (see Section 11.3.3) is another method of reducing porosity; however, this process is costly and is used mainly for critical aircraft components.

Gases have much greater **solubility** in liquid metals than in solids (Fig. 5.18). When a metal begins to solidify, the dissolved gases are expelled from the solution. Gases may also result from reactions of the molten metal with the mold materials. Gases either accumulate in regions of existing porosity, such as in interdendritic areas, or they cause microporosity in the casting, particularly in cast iron, aluminum, and copper. Dissolved gases may be removed from the molten metal by flushing or purging with an inert gas or by melting and pouring the metal in a vacuum. If the dissolved gas is oxygen, the molten metal can be *deoxidized*. Steel is usually deoxidized with aluminum or silicon and copper-base alloys, where the copper alloy contains 15% phosphorus.

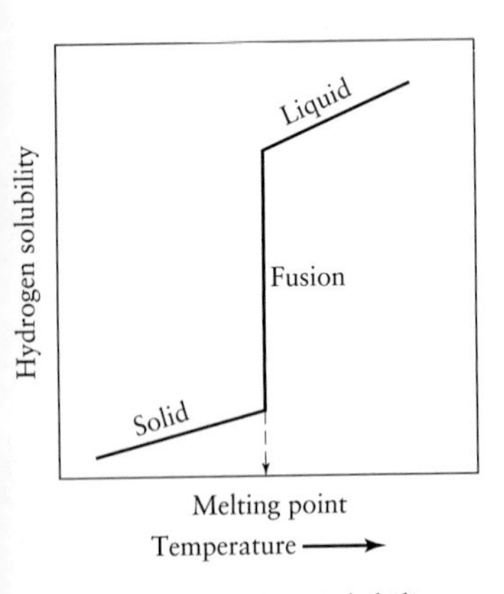

FIGURE 5.18 Solubility of hydrogen in aluminum. Note the sharp decrease in solubility as the molten metal begins to solidify.

Whether microporosity is a result of shrinkage or is caused by gases may be difficult to determine. If the porosity is spherical and has smooth walls, much like the shiny surfaces of holes in Swiss cheese, it is generally from trapped gases. If the walls are rough and angular, the porosity is likely to be from shrinkage between dendrites. Gross porosity, or *macroporosity*, occurs from shrinkage and is usually called **shrinkage cavities.**

5.5 Melting Practice and Furnaces

Melting practice is an important aspect of casting operations because it has a direct bearing on the quality of castings. Furnaces are charged with melting stock consisting of liquid and/or solid metal, alloying elements, and various other materials such as flux and slag-forming constituents. **Fluxes** are inorganic compounds that refine the molten metal by removing dissolved gases and various impurities. Fluxes have several functions, depending on the metal. For example, for aluminum alloys, there are cover fluxes (to form a barrier to oxidation), cleaning fluxes, drossing fluxes, refining fluxes, and wall-cleaning fluxes (because of the detrimental effect that some fluxes have on furnace linings, particularly in induction furnaces). Fluxes may be added manually or can be injected automatically into the molten metal.

To protect the surface of the molten metal against atmospheric reaction and contamination (and to refine the melt), the pour must be insulated against heat loss. Insulation is usually provided by covering the surface or mixing the melt with compounds that form a **slag.** In casting steels, the composition of the slag includes CaO, SiO_2, MnO, and FeO. A small quantity of liquid metal is usually tapped and its composition analyzed; necessary additions or inoculations are then made prior to pouring the metal into the molds.

The metal charge may be composed of commercially pure **primary metals,** which are remelted scrap; clean scrapped castings, gates, and risers may also be included in the charge. If the melting points of the alloying elements are sufficiently low, pure alloying elements are added to obtain the desired composition in the melt. If the melting points are too high, the alloying elements do not mix readily with the low-melting-point metals. In this case, **master alloys,** or **hardeners,** are often used. These alloys usually consist of lower-melting-point alloys with higher concentrations of one or two of the needed alloying elements. Differences in specific gravity of master alloys should not be great enough to cause segregation in the melt.

Melting furnaces. The melting furnaces commonly used in foundries are electric-arc furnaces, induction furnaces, crucible furnaces, and cupolas.

Electric-arc furnaces are used extensively in foundries and have advantages such as a high rate of melting (and hence high production rates), much less pollution output than that of other types of furnaces, and the ability to hold the molten metal for any length of time for alloying purposes.

Induction furnaces are especially useful in smaller foundries and produce composition-controlled smaller melts. There are two basic types of induction furnaces. The *coreless induction furnace* consists of a crucible completely surrounded with a water-cooled copper coil through which high-frequency current passes. Because there is a strong electromagnetic stirring action during induction heating, this type of furnace has excellent mixing characteristics for alloying and adding new charge of metal. The other type is called a *core furnace*, or *channel furnace*; it uses low frequency (as low as 60 Hz) and has a coil that surrounds only a small portion of the unit.

Crucible furnaces, which have been used extensively throughout history, are heated with various fuels, such as commercial gases, fuel oil, fossil fuel, and electricity. They may be stationary, tilting, or movable. Many ferrous and nonferrous metals are melted in these furnaces.

Cupolas are basically refractory-lined vertical steel vessels that are charged with alternating layers of metal, coke, and flux. Although they require major investments and are being replaced by induction furnaces, cupolas operate continuously, have high melting rates, and produce large amounts of molten metal.

Levitation melting involves magnetic suspension of the molten metal. An induction coil simultaneously heats a solid billet and stirs and confines the metal, thus eliminating the need for a crucible, which could be a source of contamination with oxide inclusions. The molten metal then flows downward into an investment-casting mold placed directly below the coil. Experiments indicate that investment castings made by this method are free of refractory inclusions and gas porosity and have uniform fine-grained structure.

Furnace selection requires careful consideration of several factors that can significantly influence the quality of castings, as well as the economics of casting operations. A variety of furnaces meets requirements for melting and casting metals and alloys in foundries.

Foundries and foundry automation. The casting operations described in this chapter are usually carried out in *foundries*. Although casting operations have traditionally involved much manual labor, modern foundries have automated and computer-integrated facilities for all aspects of their operations. They produce a wide variety and sizes of castings at high production rates, at low cost, and with excellent quality control.

Foundry operations initially involve two separate activities: (a) The first activity is pattern and mold making, which now regularly use computer-aided design and manufacturing and **rapid prototyping** techniques (see Section 10.12), minimizing the number of trial-and-error operations and improving efficiency. A variety of automated machinery is used to minimize labor costs, which can be significant in the production of castings. (b) The second activity is melting the metals while controlling their composition and impurities. The rest of the operations, such as pouring metals into molds carried along conveyors, shakeout, cleaning, heat treatment, and inspection, are also automated. Automation minimizes labor, reduces the possibility of human error, increases the production rate, and can be used to attain higher levels of quality. Industrial robots (Section 14.7) are being used extensively in various foundry operations such as removing die castings; cleaning; cutting risers; venting molds; spraying molds; pouring; sorting; inspecting; and automatically storing and retrieving systems for cores and patterns, using automated guided vehicles (Section 14.6).

The level of *automation* in foundries is an important economic consideration, particularly in light of the fact that many foundries are small businesses. The degree of automation depends on the type of products made. A die-casting facility or a foundry making parts for the automotive industry may involve production runs in the hundreds of thousands; thus, a high level of automation is desirable, and such facilities can afford automation. On the other hand, a jobbing foundry that has short production runs may not be as automated. Furthermore, foundries generally still tend to be somewhat dirty, hot, and labor-intensive operations. It is difficult to find qualified personnel to work in such an environment; consequently, automation has become increasingly necessary to compensate for the decline in worker availability. As in all manufacturing operations, safety in foundries is a major consideration.

5.6 Casting Alloys

The general properties of various casting alloys and processes are summarized in Fig. 5.19 and Tables 5.2 through 5.5.

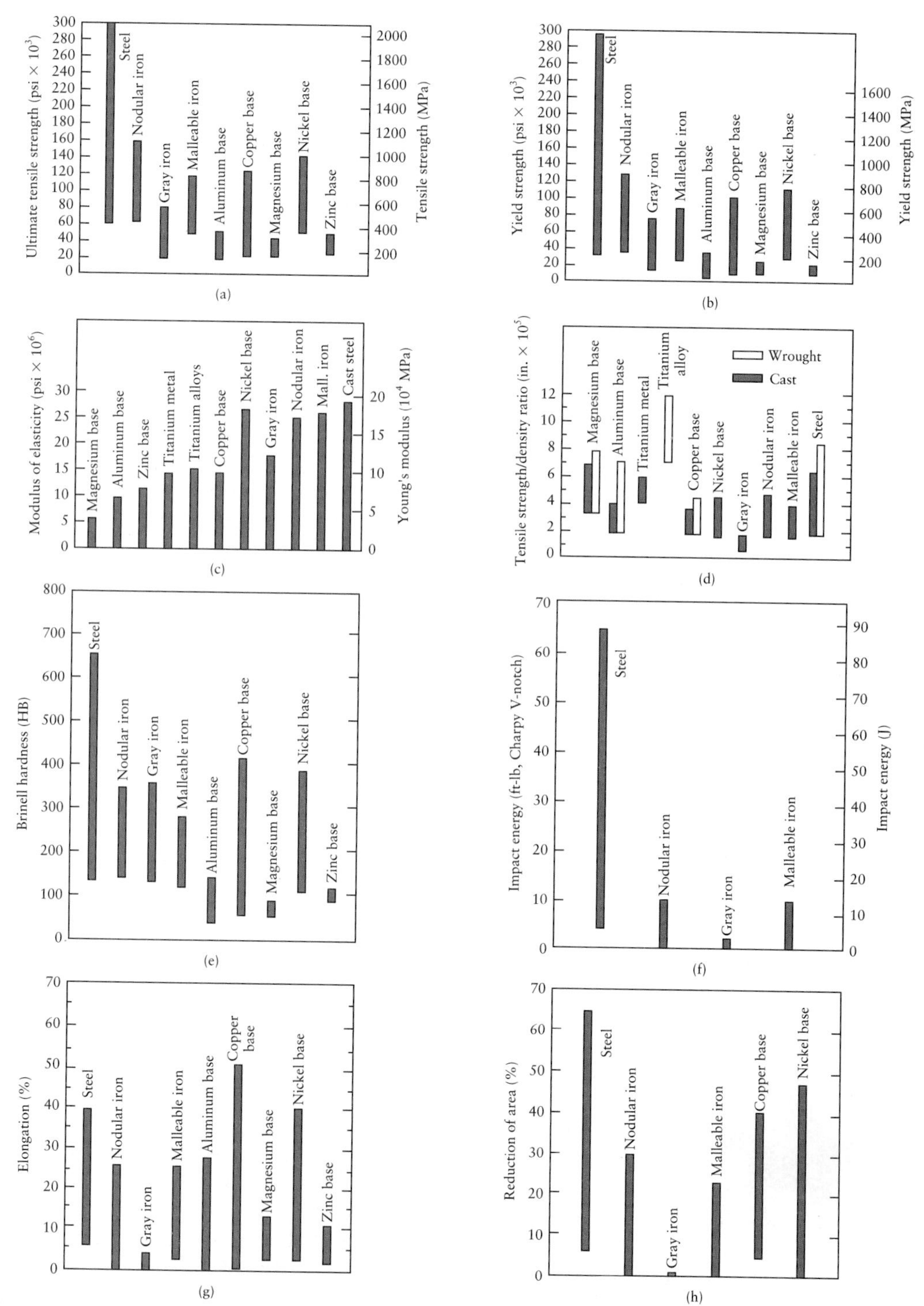

FIGURE 5.19 Mechanical properties of various groups of cast alloys. Compare this figure with the various tables of properties in Chapter 3. *Source*: Courtesy of Steel Founders' Society of America.

TABLE 5.2

General Characteristics of Casting Processes

	Sand	Shell	Evaporative pattern	Plaster	Investment	Permanent mold	Die	Centrifugal
Typical materials cast	All	All	All	Nonferrous (Al, Mg, Zn, Cu)	All	All	Nonferrous (Al, Mg, Zn, Cu)	All
Weight (kg):								
minimum	0.01	0.01	0.01	0.01	0.001	0.1	<0.01	0.01
maximum	No limit	100+	100+	50+	100+	300	50	5000+
Typ. surface finish (R_a, in μm)	5–25	1–3	5–25	1–2	0.3–2	2–6	1–2	2–10
Porosity[1]	3–5	4–5	3–5	4–5	5	2–3	1–3	1–2
Shape complexity[1]	1–2	2–3	1–2	1–2	1	2–3	3–4	3–4
Dimensional accuracy[1]	3	2	3	2	1	1	1	3
Section thickness (mm):								
minimum	3	2	2	1	1	2	0.5	2
maximum	No limit	—	—	—	75	50	12	100
Typ. dimensional tolerance (mm/mm)	1.6–4 mm (0.25 mm for small parts)	±0.003		±0.005–0.010	±0.005	±0.015	±0.001–0.005	0.015
Cost[1,2]								
Equipment	3–5	3	2–3	3–5	3–5	2	1	1
Pattern/die	3–5	2–3	2–3	3–5	2–3	2	1	1
Labor	1–3	3	3	1–2	1–2	3	5	5
Typical lead time[2,3]	Days	Weeks	Weeks	Days	Weeks	Weeks	Weeks–months	Months
Typical production rate[2,3] (parts/mold-hour)	1–20	5–50	1–20	1–10	1–1000	5–50	2–200	1–1000
Minimum quantity[2,3]	1	100	500	10	10	1000	10,000	10–10,000

Notes: 1. Relative rating, from 1 (best) to 5 (worst). For example, die casting has relatively low porosity, mid to low shape complexity, high dimensional accuracy, high equipment and die costs, and low labor costs. These ratings are only general; significant variations can occur, depending on the manufacturing methods used.
2. Approximate values without the use of rapid prototyping technologies.
Source: Data taken from J. A. Schey, *Introduction to Manufacturing Processes*, 3d. ed., 2000.

TABLE 5.3

Typical Applications and Characteristics for Castings

Type of Alloy	Application	Castability*	Weldability*	Machinability*
Aluminum	Pistons, clutch housings, intake manifolds, engine blocks, heads, cross members, valve bodies, oil pans, suspension components.	G–E	F	G–E
Copper	Pumps, valves, gear blanks, marine propellers.	F–G	F	G–E
Gray iron	Engine blocks, gears, brake disks and drums, machine bases.	E	D	G
Magnesium	Crankcases, transmission housings, portable computer housings, toys.	G–E	G	E
Malleable iron	Farm and construction machinery, heavy-duty bearings, railroad rolling stock.	G	D	G
Nickel	Gas-turbine blades, pump and valve components for chemical plants	F	F	F
Nodular iron	Crankshafts, heavy-duty gears.	G	D	G
Steel (carbon and low alloy)	Die blocks, heavy-duty gear blanks, aircraft undercarriage members, railroad wheels.	F	E	F–G
Steel (high alloy)	Gas-turbine housings, pump and valve components, rock-crusher jaws.	F	E	F
White iron (Fe_3C)	Mill liners, shot-blasting nozzles, railroad brake shoes, crushers, pulverizers.	G	VP	VP
Zinc	Door handles, radiator grills.	E	D	E

* E, excellent; G, good; F, fair; VP, very poor; D, difficult.

TABLE 5.4

Properties and Typical Applications of Cast Irons

Cast Iron	Type	Ultimate Tensile Strength (MPa)	Yield Strength (MPa)	Elongation in 50 mm (%)	Typical Applications
Gray	Ferritic	170	140	0.4	Pipe, sanitary ware.
	Pearlitic	275	240	0.4	Engine blocks, machine tools.
	Martensitic	550	550	0	Wearing surfaces
Ductile (Nodular)	Ferritic	415	275	18	Pipe, general service.
	Pearlitic	550	380	6	Crankshafts, highly stressed parts.
	Tempered martensite	825	620	2	High-strength machine parts, wear resistance.
Malleable	Ferritic	365	240	18	Hardware, pipe fittings.
	Pearlitic	450	310	10	Couplings.
	Tempered	700	550	2	Gears, connecting rods.
White	Pearlitic	275	275	0	Wear resistance, mill rolls.

TABLE 5.5

Typical Properties of Nonferrous Casting Alloys

Alloy	Condition	Casting Method*	UTS (MPa)	Yield Stress (MPa)	Elongation in 50 mm (%)	Hardness (HB)
Aluminum						
357	T6	S	345	296	2.0	90
380	F	D	331	165	3.0	80
390	F	D	279	241	1.0	120
Magnesium						
AZ63A	T4	S, P	275	95	12	—
AZ91A	F	D	230	150	3	—
QE22A	T6	S	275	205	4	—
Copper						
Brass C83600	—	S	255	177	30	60
Bronze C86500	—	S	490	193	30	98
Bronze C93700	—	P	240	124	20	60
Zinc						
No. 3	—	D	283	—	10	82
No. 5	—	D	331	—	7	91
ZA27	—	P	425	365	1	115

* S, sand; D, die; P, permanent mold.

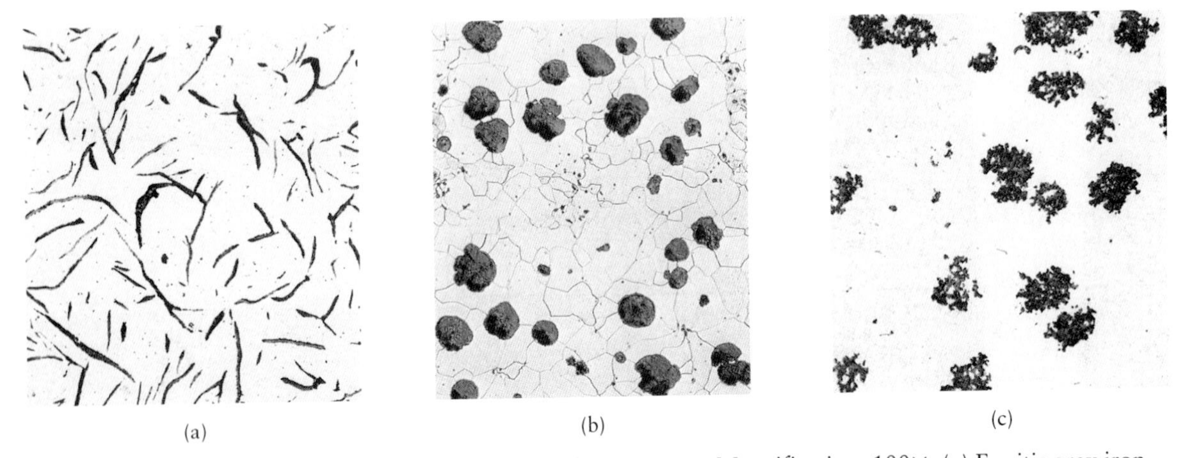

FIGURE 5.20 Microstructure for cast irons. Magnification: 100×. (a) Ferritic gray iron with graphite flakes. (b) Ferritic nodular iron (ductile iron), with graphite in nodular form. (c) Ferritic malleable iron. This cast iron solidified as white cast iron, with the carbon present as cementite (Fe_3C), and was heat treated to graphitize the carbon. *Source*: ASM International, Materials Park, OH.

5.6.1 Ferrous casting alloys

Cast irons represent the largest amount (by weight) of all metals cast. They generally possess several desirable properties, such as hardness, resistance to wear, and good machinability (Chapter 8). These alloys can easily be cast into complex shapes. The term **cast iron** refers to a family of alloys: gray cast iron (gray iron), nodular (ductile or spheroidal) iron, white cast iron, malleable iron, and compacted-graphite iron. These alloys are described as follows:

a) **Gray cast iron.** In this structure, graphite exists largely in the form of flakes (Fig. 5.20a). It is called *gray cast iron*, or *gray iron*, because when broken, the fracture path is along the graphite flakes, and hence it has a gray, sooty appearance. These flakes act as stress raisers; consequently, gray iron has negligible ductility and is weak in tension, although strong in compression, as are other brittle materials. The graphite flakes give this material the capacity to dampen vibrations by the internal friction (and hence energy dissipation) caused by the flakes. This capacity makes gray cast iron a suitable and commonly used material for constructing machine-tool bases and structures (Section 8.12).

The various types of gray cast iron are called *ferritic*, *pearlitic*, and *martensitic*. (See Section 5.11.) Because their structures are different, each has its own respective properties and applications. The structure of **ferritic gray iron**, also known as *fully gray iron*, consists of graphite flakes in an alpha-ferrite matrix. **Pearlitic gray iron** has a structure of graphite in a matrix of pearlite; although still brittle, it is stronger than gray iron. **Martensitic gray iron** is obtained by austenitizing a pearlitic gray iron, followed by rapid quenching to produce a structure of graphite in a martensite matrix; as a result, it is very hard.

Castings of gray iron have relatively few shrinkage cavities and little porosity. Typical uses of gray cast iron are for engine blocks, machine bases, electric-motor housings, pipes, and machine wear surfaces. Gray cast irons are specified by a two-digit ASTM designation. Class 20, for example, specifies that the material must have a minimum tensile strength of 140 MPa (20 ksi).

b) **Ductile iron (nodular iron).** In the structure of ductile iron, graphite is in *nodular* or *spheroid* form (Fig. 5.20b); this shape permits the material to be somewhat ductile and shock resistant. The shape of graphite flakes can be changed into nodules (spheres) by small additions of magnesium and/or cerium to the molten metal prior to pouring. Ductile iron can be made ferritic or pearlitic by heat treatment; it can also be heat treated to obtain a structure of tempered martensite. Typically used for machine parts, pipe, and crankshafts, ductile cast irons are specified by a set of two-digit numbers. Thus, class or grade 80-55-06, for example, indicates that the material has a minimum tensile strength of 550 MPa (80 ksi), a minimum yield strength of 380 MPa (55 ksi), and a 6% elongation in 50 mm (2 in.).

c) **White cast iron.** The structure of white cast iron is very hard, wear resistant, and brittle, because it contains large amounts of iron carbide instead of graphite. White cast iron is obtained either by rapid cooling of gray iron or by adjusting the material's composition by keeping the carbon and silicon content low. This type of cast iron is also called *white iron*, because the lack of graphite gives the fracture surface a white crystalline appearance. Because of its extreme hardness and wear resistance, white cast iron is used mainly for liners for machinery that processes abrasive materials, rolls for rolling mills, and railroad-car brake shoes.

d) **Malleable iron.** Malleable iron is obtained by annealing white cast iron in an atmosphere of carbon monoxide and carbon dioxide between 800°C and 900°C (1470°F and 1650°F) for up to several hours, depending on the size of the part. During this process, the cementite decomposes (*dissociates*) into iron and graphite. The graphite exists as *clusters* (Fig. 5.20c) in a ferrite or pearlite matrix and thus has a structure similar to that of nodular iron; this structure promotes ductility, strength, and shock resistance, hence the term *malleable*. Typical uses are for railroad equipment and various hardware. Malleable irons are specified by a five-digit designation; for example, 35018 indicates that the yield strength of the material is 35 ksi (240 MPa) and the elongation is 18% in 50 mm (2 in.).

e) **Compacted-graphite iron.** The compacted-graphite structure has short, thick, and interconnected flakes with undulating surfaces and rounded extremities. The mechanical and physical properties of this type of cast iron are intermediate between those of flake-graphite and nodular-graphite cast irons. Its machinability is better than that of nodular iron; it is easy to cast and has consistent properties throughout the casting. Typical applications include automotive engine blocks, crankcases, ingot molds, cylinder heads, and brake disks. It is particularly suitable for components at elevated temperatures and resists thermal fatigue.

f) **Cast steels.** Because of the high temperatures required to melt cast steels, up to about 1650°C (3000°F), their casting requires considerable knowledge and experience. The high temperatures involved present difficulties in the selection of mold materials (particularly in light of the high reactivity of steels with oxygen) in melting and pouring the metal. Steel castings possess properties that are more uniform (isotropic) than those made by mechanical working processes. Cast steels can be welded; however, welding alters the cast microstructure in the heat-affected zone (see Section 12.5), influencing the strength, ductility, and toughness of the base metal. Subsequent heat treatment must be performed to restore the mechanical properties of the casting. Cast weldments have gained importance where complex configurations or the size of the casting may prevent the part from being cast economically in one piece.

g) **Cast stainless steels.** Casting of stainless steels involves considerations similar to those for casting steels in general. Stainless steels generally have a long freezing range and high melting temperatures; they develop various structures, depending on their composition and the process parameters. Cast stainless steels are available in various compositions, and they can be heat treated and welded. These cast products have high resistance to heat and corrosion. Nickel-base casting alloys are used for severely corrosive environments and very high-temperature service.

5.6.2 Nonferrous casting alloys

1. **Aluminum-base alloys.** Alloys with an aluminum base have a wide range of mechanical properties, mainly because of various hardening mechanisms and heat treatments that can be used with them. Their fluidity depends on oxides and alloying elements in the metal. These alloys have high electrical and thermal conductivity and generally good corrosion resistance to most elements (except alkali). They are nontoxic and lightweight (and hence are called *light-metal castings*) and have good machinability; however, except for alloys with silicon, they generally have low resistance to wear and abrasion. Aluminum-base alloys have many applications, including architectural, decorative, aerospace, and electrical applications. Increasingly, engine blocks of automobiles are made of aluminum-alloy castings.
2. **Magnesium-base alloys.** Alloys with a magnesium base have good corrosion resistance and moderate strength. These characteristics depend on the particular heat treatment and the proper use of protective coating systems.
3. **Copper-base alloys.** Alloys with a copper base have the advantages of good electrical and thermal conductivity, corrosion resistance, nontoxicity (unless they contain lead), and wear properties suitable for bearing materials. Their mechanical properties and fluidity are influenced by the alloying elements. These alloys also possess good machinability characteristics.
4. **Zinc-base alloys.** Alloys with a zinc base have good fluidity and sufficient strength for structural applications. These alloys are commonly used in die casting.

5. **High-temperature alloys.** High-temperature alloys, which have a wide range of properties, typically require temperatures of up to 1650°C (3000°F) for casting titanium and superalloys; refractory alloys require higher temperatures. Special techniques are used to cast these alloys into parts for jet- and rocket-engine components. Some of these alloys are more suitable and economical for casting than for shaping by other manufacturing methods such as forging.

5.7 Ingot Casting and Continuous Casting

Traditionally, the first step in metal processing is the shaping of the molten metal into a solid form (an **ingot**) for further processing by rolling into shapes, casting into semi-finished forms, or forging. The shaping process is being rapidly replaced by *continuous casting*, which improves efficiency and eliminates the need for ingots. (See Section 5.7.2.) In ingot casting, the molten metal is poured (*teemed*) from the ladle into ingot molds in which the metal solidifies; this process is basically equivalent to sand casting or, more commonly, to permanent-mold casting.

Gas entrapment can be reduced by bottom pouring, using an insulated collar on top of the mold, or by using an exothermic compound that produces heat when it contacts the molten metal (hot top); these processes slow the cooling and result in a higher yield of high-quality metal. The cooled ingots are removed (*stripped*) from the molds and lowered into soaking pits, where they are reheated to a uniform temperature of about 1200°C (2200°F) for subsequent processing by rolling. Ingots may be square, rectangular, or round in cross-section, and their weight ranges from a few hundred pounds to 40 tons.

5.7.1 Ferrous-alloy ingots

Several reactions take place during solidification of an ingot; each reaction has an important influence on the quality of the steel produced. For example, significant amounts of oxygen and other gases can dissolve in the molten metal during steelmaking. However, much of these gases is rejected during solidification of the metal, because the solubility limit of gases in the metal decreases sharply as its temperature decreases. The rejected oxygen combines with carbon, forming carbon monoxide, which causes porosity in the solidified ingot. Depending on the amount of gas evolved during solidification, three types of steel ingots can be produced: killed, semikilled, and rimmed.

1. **Killed steel** is a fully deoxidized steel; that is, oxygen is removed, and thus porosity is eliminated. In the deoxidation process, the dissolved oxygen in the molten metal is made to react with elements such as aluminum (the most common such element) that are added to the melt. Other elements, such as vanadium, titanium, and zirconium, are also used. These elements have an affinity for oxygen and form metallic oxides. If aluminum is used, the product is called aluminum-killed steel. The term *killed* comes from the fact that the steel lies quietly after being produced into the mold.

   If sufficiently large, the oxide inclusions in the molten bath float out and adhere to, or are dissolved in, the slag. A fully killed steel is thus free from porosity caused by gases; it is also free from *blowholes* (large spherical holes near the surfaces of the ingot). Consequently, the chemical and mechanical properties of killed steels are relatively uniform throughout the ingot. However, because of metal shrinkage during solidification, an ingot of this type develops a *pipe* at its top. Also called a *shrinkage cavity*, the pipe has a funnellike appearance;

this pipe can constitute a substantial portion of the ingot and has to be cut off and scrapped.

2. **Semikilled steel** is a partially deoxidized steel. It contains some porosity, generally in the upper central section of the ingot, but has little or no pipe; thus, scrap is reduced. Piping in semikilled steel is less because it is compensated for by the presence of porosity in that region. Semikilled steels are economical to produce.
3. In a **rimmed steel**, which generally has a low carbon content (less than 0.15%), the evolved gases are only partially killed or controlled by the addition of elements such as aluminum. The gases form blowholes along the outer rim of the ingot, hence the term *rimmed*. Blowholes are generally not objectionable unless they break through the outer skin of the ingot. Rimmed steels have little or no piping and have a ductile skin with good surface finish.

5.7.2 Continuous casting

The traditional method of casting ingots is a batch process; that is, each ingot has to be stripped from its mold after solidification and processed individually. Piping and microstructural and chemical variations also are present throughout the ingot. These problems are alleviated by continuous casting processes, which produce higher quality steels for less cost. Conceived in the 1860s, *continuous casting*, or **strand casting**, was first developed for casting nonferrous metal strip. Now used for steel production, the process yields major improvements in efficiency and productivity and significant reductions in cost. Figure 5.21a shows schematically one system for continuous casting. The molten metal in the ladle is cleaned and equalized in temperature by blowing nitrogen gas through it for 5 to 10 min. The metal is then poured into a refractory-lined intermediate pouring vessel (**tundish**), where impurities are skimmed off. The tundish holds as much as three tons of metal. The molten metal travels through water-cooled copper molds and begins to solidify as it travels downward along a path supported by rollers (*pinch rolls*).

Before the casting process is started, a solid *starter*, or *dummy bar*, is inserted into the bottom of the mold. The molten metal is then poured and solidifies on the starter bar. (See the bottom of Fig. 5.21a.) The bar is withdrawn at the same rate the metal is poured. The cooling rate is such that the metal develops a solidified skin (shell) to support itself during its travel downward, at speeds typically of 25 mm/s (1 in./s). The shell thickness at the exit end of the mold is about 12–18 mm (0.5–0.75 in.). Additional cooling is provided by water sprays along the travel path of the solidifying metal. The molds generally are coated with graphite or similar solid lubricants to reduce friction and adhesion at the mold–metal interfaces. The molds are vibrated to further reduce friction and sticking.

The continuously cast metal may be cut into desired lengths by shearing or torch cutting, or it may be fed directly into a rolling mill for further reductions in thickness and for shape rolling of products such as channels and I-beams. In addition to costing less, continuously cast metals have more uniform compositions and properties than metals obtained by ingot casting. Although the thickness of steel strand is usually about 250 mm (10 in.), recent developments have reduced this thickness to about 15 mm (0.6 in.) or less. The thinner strand reduces the number of rolling operations required and improves the economy of the overall operation. After they are hot rolled, steel plates or shapes undergo one or more further processes, such as cleaning and pickling by chemicals to remove surface oxides, cold rolling to improve strength and surface finish, annealing, and coating (galvanizing or aluminizing) for resistance to corrosion.

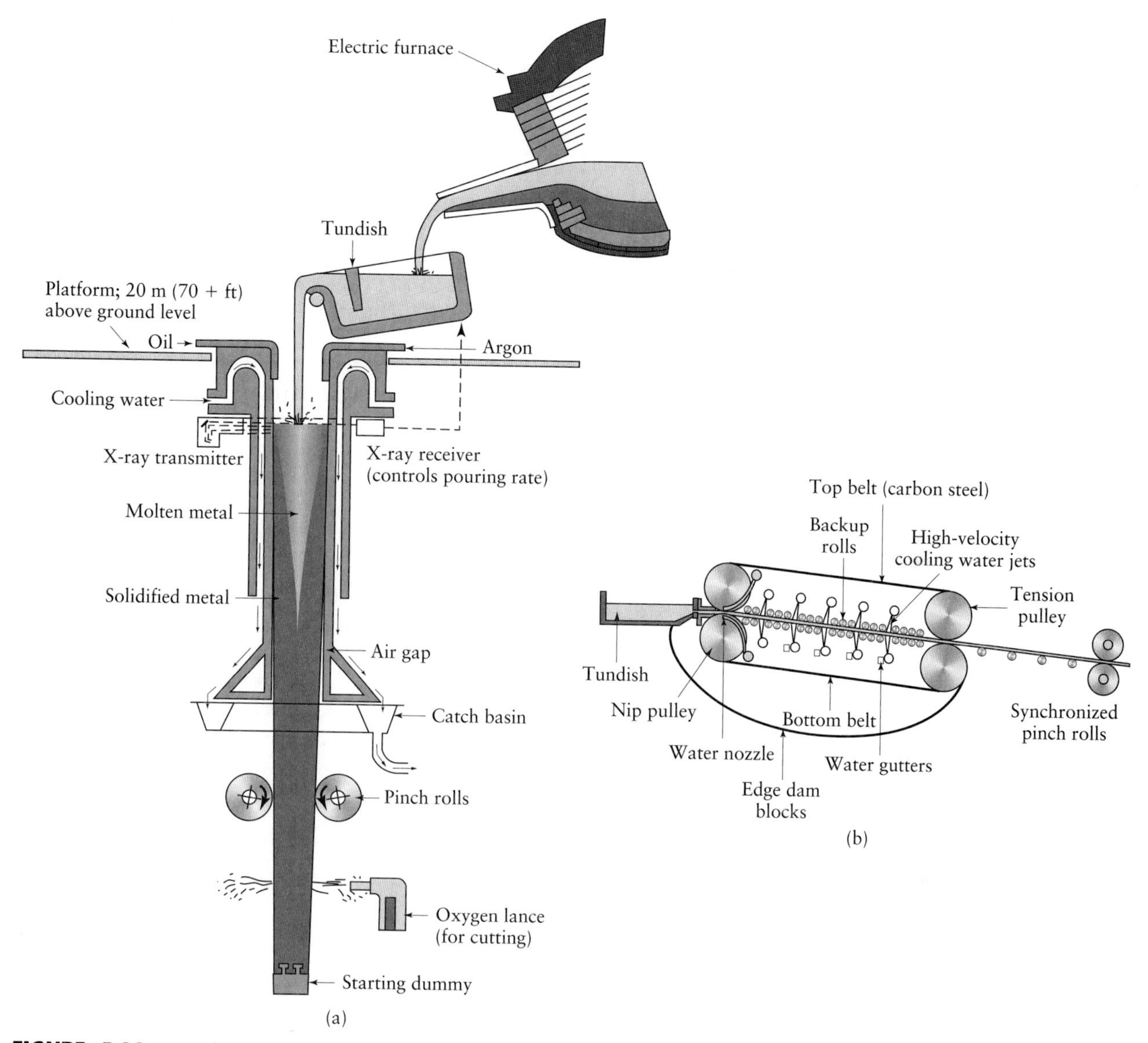

FIGURE 5.21 (a) The continuous-casting process for steel. Typically, the solidified metal decends at a speed of 25 mm/s (1 in/s). Note that the platform is about 20 m (65 ft) above ground level. *Source: Metalcaster's Reference and Guide*, American Foundryman's Society. (b) Continuous strip casting of nonferrous metal strip. *Source*: Courtesy of Hazelett Corporation.

5.7.3 Strip casting

In strip casting, thin slabs or strips are produced from molten metal; the metal solidifies in similar fashion to strand casting, but the hot solid is then rolled to form the final shape (Fig. 5.21b). The compressive stresses in rolling (see Section 6.3) serve to reduce porosity and provide better material properties. In effect, strip casting eliminates a hot-rolling operation in the production of metal strips or slabs. In modern facilities, final thicknesses on the order of 2–6 mm (0.08–0.25 in.) can be obtained for carbon, stainless and electrical steels, and other metals.

5.8 Casting Processes: Expendable Mold, Permanent Pattern

Casting processes are generally classified according to mold materials, molding processes, and methods of feeding the mold with the molten metal. The two major categories are *expendable-mold* and *permanent-mold* casting (Section 5.10). Expendable-mold processes are further categorized as permanent-pattern or expendable-pattern processes. Expendable molds are made of sand, plaster, ceramics, and similar materials, which generally are mixed with various **binders** or bonding agents. These materials are refractories; that is, they have the capability to withstand the high temperatures of molten metals. After the casting has solidified, the molds in these processes are broken up to remove the casting. Permanent molds and patterns are reused many times to make large production runs more economical.

5.8.1 Sand casting

The traditional method of expendable-mold casting is in sand molds. Simply stated, *sand casting* consists of placing a pattern (having the shape of the desired casting) in sand to make an imprint, incorporating a gating system, filling the resulting cavity with molten metal, allowing the metal to cool until it solidifies, breaking away the sand mold, and removing the casting. While the origins of sand casting date to ancient times, it is still the most prevalent form of casting. In the United States alone, about 15 million tons of metal are cast by this method each year. Typical parts made by sand casting include machine-tool bases, engine blocks, cylinder heads, and pump housings.

Sands. Most sand-casting operations use silica sands (SiO_2). Sand is the product of the disintegration of rocks over extremely long periods of time. It is inexpensive and is suitable as mold material because of its resistance to high temperatures. There are two general types of sand: *naturally bonded* (*bank sands*) and *synthetic* (*lake sands*). Because its composition can be controlled more accurately, synthetic sand is preferred by most foundries. Several factors are important in the selection of sand for molds. Sand that has fine, round grains can be closely packed and forms a smooth mold surface. Good permeability of molds and cores allows gases and steam evolved during casting to escape easily. The mold should have good **collapsibility** in order to avoid defects in the casting, such as hot tearing and cracking (because the casting shrinks while cooling). The selection of sand involves certain trade-offs with respect to properties. For example, fine-grained sand enhances mold strength, but the fine grains also lower mold permeability. Sand is typically conditioned before use. *Mulling* machines are used to uniformly mull (mix thoroughly) sand with additives; clay is used as a cohesive agent to bond sand particles, giving the sand better strength.

Types of sand molds. Sand molds are characterized by the types of sand and by the methods used to produce them; the major components of a sand mold are shown in Fig. 5.14. There are three basic types of sand molds: green-sand, cold-box, and no-bake molds. The most common mold material is *green molding sand*. The term *green* refers to the fact that the sand in the mold is moist or damp while the metal is being poured into it. Green molding sand is a mixture of sand, clay, and water. Green-sand molding is the least expensive method of making molds.

In the **skin-dried** method, the surfaces of the mold are dried, by either storing the mold in air or drying it with torches. Skin-dried molds are generally used for large castings, because of the higher strength. Sand molds are also oven dried (*baked*) prior to receiving the molten metal. They are stronger than green-sand molds and impart better dimensional accuracy and surface finish to the casting. However, distortion of the mold is greater; the castings are more susceptible to hot tearing, because of the lower collapsibility of the mold; and the production rate is slower, because of the drying time required.

In the **cold-box mold** process, various organic and inorganic binders are blended into the sand to bond the grains chemically for greater strength. These molds are dimensionally more accurate than green-sand molds, but are more expensive. In the **no-bake mold** process, a synthetic liquid resin is mixed with the sand, and the mixture hardens at room temperature. Because bonding of the mold in this and the cold-box process takes place without heat, they are called *cold-setting processes*.

Patterns. *Patterns* are used to mold the sand mixture into the shape of the casting. They may be made of wood, plastic, or metal. Patterns can also be made directly by **rapid prototyping** techniques. (See Section 10.12.) The selection of a pattern material depends on the size and shape of the casting, the dimensional accuracy and quantity of castings required, and the molding process to be used. Because patterns are used repeatedly to make molds, the strength and durability of the material selected for patterns must reflect the number of castings the mold will produce. Patterns may be made of a combination of materials in order to reduce wear in critical regions. Patterns are usually coated with a **parting agent** to facilitate their removal from the molds.

Patterns can be designed with a variety of features to fit application and economic requirements. **One-piece patterns** are generally used for simpler shapes and low-quantity production. They are generally made of wood and are inexpensive. **Split patterns** are two-piece patterns made so that each part forms a portion of the cavity for the casting. In this way, castings that have complicated shapes can be produced. **Match-plate patterns** are a popular type of mounted pattern in which two-piece patterns are constructed by securing each half of one or more split patterns to the opposite sides of a single plate. In such constructions, the gating system can be mounted on the drag side of the pattern. This type of pattern most often is used in conjunction with molding machines and large production runs.

Cores. Cores are used for castings with internal cavities or passages, such as in an automotive engine block or a valve body. Cores are placed in the mold cavity before casting to form the interior surfaces of the casting and are removed from the finished part during shakeout and further processing. Like molds, cores must possess strength, permeability, ability to withstand heat, and collapsibility. Therefore, cores are typically made of sand aggregates. The core is anchored by **core prints**, which are recesses that are added to the pattern to support the core and to provide vents for the escape of gases. A common problem with cores is that for certain casting requirements, as in the case for which a recess is required, they may lack sufficient structural support in the cavity. To keep the core from shifting, metal supports, known as **chaplets**, may be used to anchor the core in place.

Cores are generally made in a manner similar to that used for making molds, and most are made with shell, no-bake, or cold-box processes. Cores are formed in *core boxes*, which are used much as patterns are used to form sand molds. The sand can be packed into the boxes with sweeps or blown into the box by compressed air from *core blowers*; core boxes have the advantages of producing uniform cores and operating at very high production rates.

Sand-molding machines. The oldest known method of molding, which is still used for simple castings, is to compact the sand by hand hammering or ramming it around the pattern. For most operations, however, the sand mixture is compacted around the pattern by molding machines. These machines eliminate arduous labor, offer higher quality casting by improving the application and distribution of forces, manipulate the mold in a carefully controlled fashion, and increase the rate of production. Mechanization of the molding process can be further assisted by **jolting** the assembly. The flask, molding sand, and pattern are placed on a pattern plate mounted on an anvil and are jolted upward by air pressure at rapid intervals; the inertial forces compact the sand around the pattern.

In **vertical flaskless molding**, the halves of the pattern form a vertical chamber wall against which sand is blown and compacted. Then the mold halves are packed horizontally, with the parting line oriented vertically and moved along a pouring conveyor. This operation is simple and eliminates the need to handle flasks, making potential production rates very high, particularly when other aspects of the operation, such as coring and pouring, are automated.

Sandslingers fill the flask uniformly with sand under a stream of high pressure. Sandthrowers are used to fill large flasks and are typically operated by machine. An impeller in the machine throws sand from its blades or cups at such high speeds that the machine not only places the sand, but also rams it appropriately.

In **impact molding**, the sand is compacted by controlled explosion or instantaneous release of compressed gases. This method produces molds with uniform strength and good permeability. In **vacuum molding**, also known as the "V" *process*, the pattern is covered tightly by a thin sheet of plastic. A flask is placed over the coated pattern and filled with sand. A second sheet of plastic is placed on top of the sand, and a vacuum action hardens the sand so that the pattern can be withdrawn. Both halves of the mold are made this way and then assembled. During pouring, the mold remains under a vacuum, but the casting cavity does not. When the metal has solidified, the vacuum is turned off, and the sand falls away, releasing the casting. Vacuum molding produces castings that have very good detail and accuracy; however, it is quite expensive. It is especially well suited for large, relatively flat castings.

The sand-casting operation. After the mold has been shaped and the cores have been placed in position, the two halves (*cope* and *drag*) are closed, clamped, and weighted down. They are weighted to prevent the separation of the mold sections under the pressure exerted when the molten metal is poured into the mold cavity. The design of the gating system is important for proper delivery of the molten metal into the mold cavity. Turbulence must be minimized, air and gases must be allowed to escape by vents or other means, and proper temperature gradients must be established and maintained to eliminate shrinkage and porosity. The design of risers is also important for supplying the necessary molten metal during solidification of the casting; the pouring basin may also serve as a riser. After solidification, the casting is shaken out of its mold, and the sand and oxide layers adhering to the casting are removed by vibration (using a shaker) or by sand blasting. The risers and gates are cut off by oxyfuel-gas cutting, sawing, shearing, and abrasive wheels, or they are trimmed in dies.

Almost all commercially used alloys can be sand cast. The surface finish obtained is largely a function of the materials used in making the mold. The dimensional accuracy of sand casting is not as good as that of other casting processes. However, intricate shapes, such as cast-iron engine blocks and very large propellers for ocean liners and impellers, can be cast by this process. Sand casting can be economical for relatively small production runs as well as large production runs; equipment costs are generally low.

5.8.2 Shell-mold casting

The use of *shell-mold casting* has grown significantly, because it can produce many types of castings with close dimensional tolerances and good surface finishes at a low cost. In this process, a mounted pattern, made of a ferrous metal or aluminum, is heated to 175–370°C (350–700°F), coated with a parting agent such as silicone, and clamped to a box or chamber holding a fine sand containing a 2.5% to 4.0% thermosetting resin binder, such as phenol-formaldehyde, which coats the sand particles. The sand mixture is blown over the heated pattern, coating it evenly. The assembly is often, but not always, placed in an oven for a short period of time to complete the curing of the resin. The shell hardens around the pattern and is removed from the pattern, using built-in ejector pins. Two half-shells are made in this manner and are bonded or clamped together in preparation for pouring.

The shells are light and thin, usually 5–10 mm (0.2–0.4 in.) thick, and consequently their thermal characteristics are different than those for thicker molds. The thin shells allow gases to escape during solidification of the metal. The mold is generally used vertically and is supported by surrounding it with steel shot in a cart. The walls of the mold are relatively smooth, offering low resistance to flow of the molten metal and producing castings with sharper corners, thinner sections, and smaller projections than are possible in green-sand molds. With the use of multiple gating systems, several castings can be made in a single mold.

Shell-mold casting may be more economical than other casting processes, depending on various production factors, particularly energy cost. The relatively high cost of metal patterns becomes a smaller factor as the size of the production run increases. The high quality of the finished casting can significantly reduce cleaning, machining, and other finishing costs. Complex shapes can be produced with less labor, and the process can be automated fairly easily. Shell-mold casting applications include small mechanical parts that require high levels of precision, such as gear housings, cylinder heads, and connecting rods. Shell molding is also widely used in producing high-precision molding cores, such as engine-block water jackets.

Composite molds. *Composite molds* are made of two or more different materials and are used in shell molding and other casting processes. They are generally used for casting complex shapes, such as impellers for turbines. Figure 5.22 shows an example of composite molds. Commonly used molding materials include shells (made as previously described), plaster, sand with binder, metal, and graphite. These molds may also include cores and chills to control the rate of solidification in critical areas of castings. Composite molds have increased strength, improve the dimensional accuracy and surface finish of castings, and may help reduce overall costs and processing time.

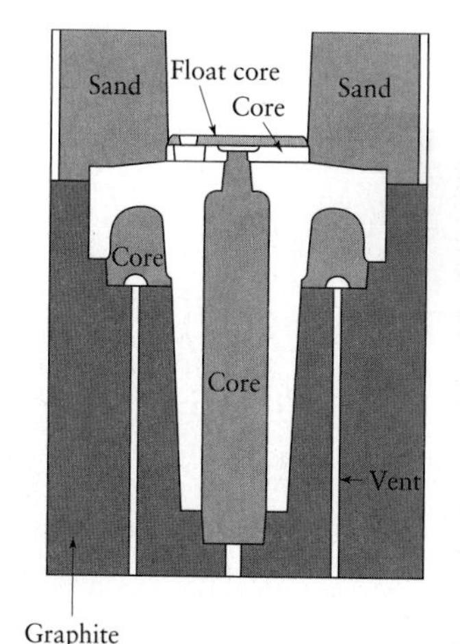

FIGURE 5.22 Schematic illustration of a semipermanent composite mold. *Source: Steel Castings Handbook*, 5th ed., Steel Founders' Society of America, 1980.

Sodium silicate process. The mold material in the *sodium silicate process* is a mixture of sand and 1.5% to 6% sodium silicate (*waterglass*) as the binder for sand. This mixture is then packed around the pattern and hardened by blowing carbon dioxide (CO_2) gas through it. This process, also known as *silicate-bonded sand* or the *carbon-dioxide process*, was first used in the 1950s and has been developed further, for example, by using various other chemicals for binders. Cores made by this process reduce the tendency to tear, because of their compliance at elevated temperatures.

Rammed-graphite molding. Rammed graphite, instead of sand, is used to make molds for casting reactive metals, such as titanium and zirconium. Sand cannot be used because these metals react vigorously with silica. In this specialty method, the molds are packed rather like sand molds and are then air dried, baked at 175°C (350°F), fired at 870°C (1600°F), and stored under controlled humidity and temperature. The casting procedures are similar to those for sand molds.

5.8.3 Plaster-mold casting

The *plaster-mold casting* process and the ceramic-mold and investment-casting processes are known as **precision casting**, because of the high dimensional accuracy and good surface finish obtained. Typical parts made by these methods include lock components, gears, valves, fittings, tooling, and ornaments, weighing as little as 1 g (0.035 oz). In the plaster-mold casting process, the mold is made of *plaster of paris* (gypsum, or calcium sulfate), with the addition of talc and silica flour to improve strength and to control the time required for the plaster to set. These components are mixed with water, and the resulting slurry is poured over the pattern. After the plaster sets, usually within 15 minutes, the pattern is removed, and the mold is dried to remove the moisture. The mold halves are then assembled to form the mold cavity and preheated to about 120°C (250°F) for 16 hours. Next, the molten metal is poured into the mold. Because plaster molds have very low permeability, gases evolved during solidification of the metal cannot escape. Consequently, the molten metal is poured either in a vacuum or under pressure. Plaster-mold permeability can be increased substantially by the *Antioch process*. In this process, the molds are dehydrated in an autoclave (pressurized oven) for 6 to 12 hours and are then rehydrated in air for 14 hours. Another method of increasing permeability is to use foamed plaster, which contains trapped air bubbles.

Patterns for plaster molding are generally made of aluminum alloys, thermosetting plastics, brass, or zinc alloys. Wood patterns are not suitable for making a large number of molds, because the patterns are repeatedly subjected to the water-based plaster slurry, which causes the wood to swell. Because there is a limit to the maximum temperature that the plaster mold can withstand, generally about 1200°C (2200°F), plaster-mold casting is used only for aluminum, magnesium, zinc, and some copper-base alloys. The castings have fine detail and good surface finish. Since plaster molds have lower thermal conductivity than other types of molds, the castings cool slowly, yielding more uniform grain structure with less warpage and better mechanical properties.

5.8.4 Ceramic-mold casting

The *ceramic-mold casting* is another precision-casting process and is similar to the plaster-mold process, with the exception that it uses refractory mold materials suitable for high-temperature applications; the process is also called **cope-and-drag investment casting**. The slurry is a mixture of fine-grained zircon ($ZrSiO_4$), aluminum oxide, and fused silica, which are mixed with bonding agents and poured over the pattern (Fig. 5.23), which has been placed in a flask. The pattern may be made of wood or metal. After setting, the molds (ceramic facings) are removed, dried, burned off to remove volatile matter, and baked. The molds are clamped firmly and used as all-ceramic molds. In the *Shaw process*, the ceramic facings are backed by fireclay (clay used in making firebricks that resist high temperatures) to give the molds strength. The facings are then assembled into a complete mold, ready to be poured.

The high-temperature resistance of the refractory molding materials allows these molds to be used in casting ferrous and other high-temperature alloys, stainless steels, and tool steels. The castings have good dimensional accuracy and surface finish over a wide range of sizes and intricate shapes, but the process is somewhat expensive. Typical parts made are impellers, cutters for machining, dies for metalworking, and molds for making plastic or rubber components, with some parts weighing as much as 700 kg (1500 lb).

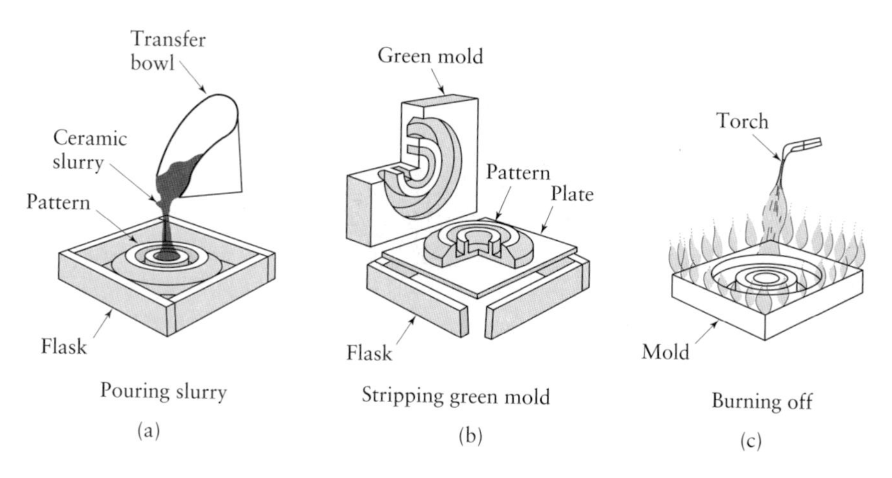

FIGURE 5.23 Sequence of operations in making a ceramic mold. *Source: Metals Handbook*, 8th ed., Vol. 5: *Forging and Casting*, ASM International, 1970.

5.8.5 Vacuum casting

Figure 5.24 shows a schematic illustration of the *vacuum-casting process*, or *counter-gravity low-pressure* (CL) *process* (not to be confused with the vacuum-molding process described in Section 5.8.1). A mixture of fine sand and urethane is molded over metal dies and cured with amine vapor. Then the mold is held with a robot arm and partially immersed into molten metal held in an induction furnace. The metal may be melted in air (*CLA process*) or in a vacuum (*CLV process*). The vacuum reduces the air pressure inside the mold to about two thirds of atmospheric pressure, drawing the molten metal into the mold cavities through a gate in the bottom of the mold. The molten metal in the furnace is usually at a temperature 55°C (100°F) above the liquidus temperature; consequently, the molten metal begins to solidify within a fraction of a second. After the mold is filled, it is withdrawn from the molten metal.

This process is an alternative to investment, shell-mold, and green-sand casting and is particularly suitable for thin-walled (0.75 mm [0.03 in.]) complex shapes with uniform properties. Carbon and low-and high-alloy steel and stainless-steel parts, weighing as much as 70 kg (155 lb), have been vacuum cast by this method. CLA parts are easily made at high volume and relatively low cost. CLV parts usually contain reactive metals, such as aluminum, titanium, zirconium, and hafnium. These parts, which are often in the form of superalloys for gas turbines, may have walls as thin as 0.5 mm (0.02 in.). The process can be automated, and production costs are similar to those for green-sand casting.

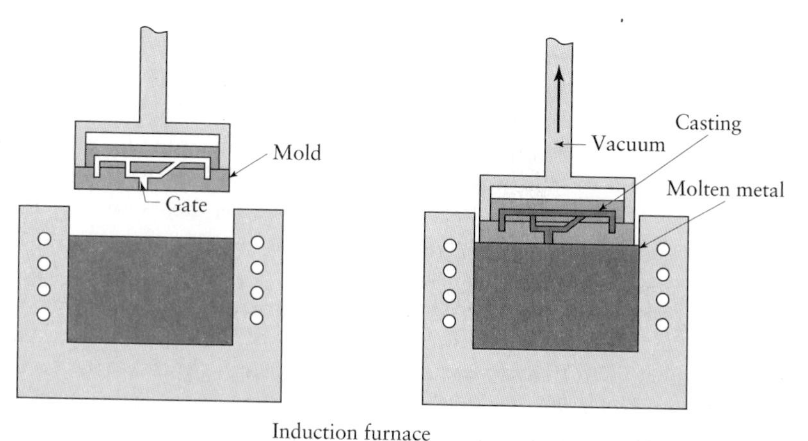

FIGURE 5.24 Schematic illustration of the vacuum-casting process. Note that the mold has a bottom gate. (a) Before and (b) after immersion of the mold into the molten metal. *Source*: After R. Blackburn.

5.9 Casting Processes: Expendable Mold, Expendable Pattern

5.9.1 Evaporative-pattern casting (lost foam)

The *evaporative-pattern casting process* uses a polystyrene pattern, which evaporates upon contact with molten metal to form a cavity for the casting. The process is also known as *lost-pattern casting* and under the trade name *Full-Mold process*. It was formerly known as the *expanded polystyrene process* and has become one of the more important casting processes for ferrous and nonferrous metals, particularly for the automotive industry. First, raw expandable polystyrene (EPS) beads, containing 5% to 8% pentane (a volatile hydrocarbon), are placed in a preheated die, usually made of aluminum. The polystyrene expands and takes the shape of the die cavity. Additional heat is applied to fuse and bond the beads together. The die is then cooled and opened, and the polystyrene pattern is removed. Complex patterns may also be made by bonding various individual sections of the pattern, using hot-melt adhesive. In such compound foam-mold assembly, precision of the bonding operation is critically important for maintaining overall dimensional accuracy.

The pattern is then coated with a water-base refractory slurry, dried, and placed in a flask. The flask is filled with loose fine sand, which surrounds and supports the pattern. The sand may be dried or mixed with bonding agents to give it additional strength. The sand is periodically compacted by various means. Then, without removing the polystyrene pattern, the molten metal is poured into the mold. This action immediately vaporizes the pattern (an ablation process) and fills the mold cavity, completely replacing the space previously occupied by the polystyrene pattern. The heat degrades (depolymerizes) the polystyrene, and the degradation products are vented into the surrounding sand.

The flow velocity in the mold depends on the rate of degradation of the polymer. Studies have shown that the flow of the molten metal is basically laminar, with Reynolds numbers in the range of 400 to 3000. (See Section 5.4.1.) The velocity of the molten metal at the metal–polymer pattern front is estimated to be in the range of 0.1–1.0 m/s (4–40 in./s). The velocity can be controlled by producing patterns with cavities or hollow sections; thus, the velocity will increase as the molten metal crosses these hollow regions, similar to pouring in an empty cavity, as is done in sand casting. Furthermore, because the polymer requires considerable energy to degrade, large thermal gradients are present at the metal–polymer interface; in other words, the molten metal cools faster than it would if it were poured into a cavity. Consequently, fluidity (see Section 5.4.2) is less than in sand casting. This characteristic has important effects on the microstructure throughout the casting and also leads to directional solidification of the metal.

Typical applications for this process include cylinder heads, crankshafts, brake components and manifolds for automobiles, and machine bases. The aluminum engine blocks and other components of the General Motors Saturn automobiles are made by this process. Recent developments include the use of polymethylmethacrylate and polyalkylene carbonate as pattern materials for ferrous castings. In a modification of the evaporative-pattern process, a polystyrene pattern is surrounded by a ceramic shell (*Replicast C-S process*). The pattern is burned out prior to pouring the molten metal into the mold. The principal advantage of this process over investment casting (with its wax patterns) is that carbon pickup into the metal is entirely avoided.

More recent developments in evaporative-pattern casting include production of metal-matrix composites (Section 11.14). During the process of molding the

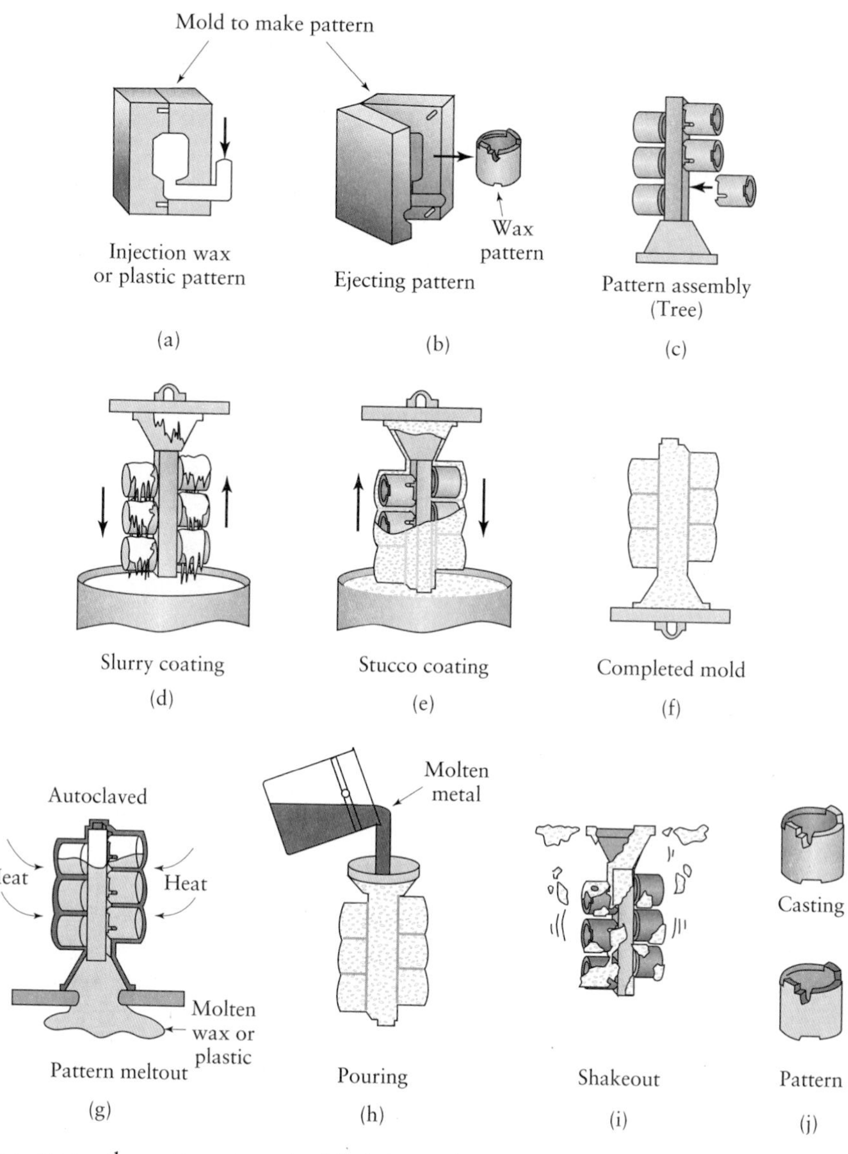

FIGURE 5.25 Schematic illustration of investment casting (lost-wax process). Castings by this method can be made with very fine detail and from a variety of metals. *Source*: Steel Founders' Society of America.

polymer pattern, the pattern is embedded throughout with fibers or particles, which then become an integral part of the casting. Further studies are being conducted on modification and grain refinement of the casting by the use of grain refiners and modifier master alloys (Section 5.5) within the pattern while it is being molded.

5.9.2 Investment casting (lost wax)

The *investment-casting process*, also called the *lost-wax process*, was first used during the period 4000–3000 BC. In this process, the pattern is made of wax or a plastic such as polystyrene. The sequences involved in investment casting are shown in Fig. 5.25.

Solid mold. The pattern is made by injecting semisolid or liquid wax or plastic into a metal die in the shape of the pattern. The pattern is then removed and dipped into a slurry of refractory material, such as very fine silica and binders, ethyl silicate, and acids. After this initial coating has dried, the pattern is coated repeatedly to increase its thickness. The term *investment* comes from investing the pattern with the refractory material. Wax patterns require careful handling, because they are not strong enough to withstand the forces involved during mold making. The one-piece mold is dried in air and heated to a temperature of 90–175°C (200–350°F) for about four hours, depending on the metal to be cast, to drive off the water of crystallization (chemically combined water). After the metal has been poured and has solidified, the mold is broken up and the casting is removed. A number of patterns can be joined to make one mold called a **tree**, thus increasing process productivity.

Although the labor and materials involved make the lost-wax process costly, it is suitable for casting high-melting-point alloys with a good surface finish and close dimensional tolerances; thus, little or no finishing is required, which otherwise would add significantly to the total cost of the casting. This process is capable of producing intricate shapes from a wide variety of ferrous and nonferrous metals and alloys, with parts generally weighing from 1 g to 100 kg (0.035 oz to 75 lb) and as much as 1140 kg (2500 lb). Typical parts made include components for office equipment; mechanical components such as gears, cams, valves, and ratchets; and jewelry.

Ceramic-shell investment casting. A variation of the investment-casting process is *ceramic-shell investment casting*. It uses the same type of wax or plastic pattern, which is dipped first in a slurry with colloidal silica or ethyl silicate binder and then into a fluidized bed of fine-grained fused silica or zircon flour. Next, the pattern is dipped into coarse-grain silica to build up additional coatings and thickness to withstand the thermal shock of pouring. This process is economical and is used extensively for precision casting of steels, aluminum, and high-temperature alloys.

If ceramic cores are used in the casting, they are removed by leaching with caustic solutions under high pressure and temperature. The molten metal may be poured in a vacuum to extract evolved gases and reduce oxidation, thus improving the quality of the casting. To further reduce microporosity, the castings made by this and other processes are subjected to hot isostatic pressing. Aluminum castings, for example, are subjected to a gas pressure of up to about 100 MPa (15 ksi) at 500°C (900°F).

EXAMPLE 5.3 Investment-cast superalloy components for gas turbines

Since the 1960s, investment-cast superalloys have been replacing wrought counterparts in high-performance gas turbines. Much development has been taking place in producing cleaner superalloys (nickel base and cobalt base). Improvements have been made in melting and casting techniques, such as by vacuum-induction melting, using microprocessor controls. Impurity and inclusion levels have continually been reduced, thus improving the strength, ductility, and overall reliability of these components. Such control is essential, because these parts operate at a temperature only about 50°C (90°F) below the solidus.

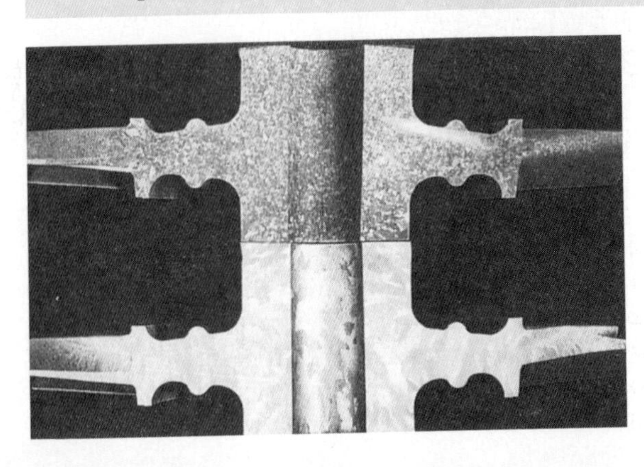

FIGURE 5.26 The top rotor was investment cast; the lower rotor was cast conventionally. *Source: Advanced Materials and Processes*, ASM International, October 1990, p. 25.

The microstructure of an integrally investment-cast, gas-turbine rotor is shown in the upper portion of Fig. 5.26. Note the fine, uniform, equiaxed grain size throughout the cross-section. Recent techniques to obtain this result include the use of a nucleant addition to the molten metal, as well as close control of its superheat, pouring techniques, and control of cooling rate of the casting. In contrast, the lower portion of Fig. 5.26 shows the same type of rotor cast conventionally; note the coarse grain structure. This rotor will have inferior properties compared with those of the fine-grained rotor. Due to developments in these processes, the proportion of cast parts to other parts in aircraft engines has increased from 20% to about 45% by weight.

5.10 | Casting Processes: Permanent Mold

Permanent molds, as the name implies, are used repeatedly and are designed so that the casting can be easily removed and the mold reused. These molds are made of metals that maintain their strength at high temperatures and therefore can be used repeatedly. Because metal molds are better heat conductors than expendable molds, the solidifying casting is subjected to a higher rate of cooling, which in turn affects the microstructure and grain size within the casting, as described in Section 5.3.

In permanent-mold casting, two halves of a mold are made from materials such as steel, bronze, refractory metal alloys, or graphite (**semipermanent mold**). The mold cavity and gating system are machined into the mold and thus become an integral part of it. To produce castings with internal cavities, cores made of metal or sand aggregate are placed in the mold prior to casting. Typical core materials are shell or no-bake cores, gray iron, low-carbon steel, and hot-work die steel. Gray iron is the most commonly used core material, particularly for large molds for aluminum and magnesium castings. Inserts are also used for various parts of the mold. In order to increase the life of permanent molds, the surfaces of the mold cavity are sometimes coated with a refractory slurry or sprayed with graphite every few castings. These coatings also serve as parting agents and thermal barriers, controlling the rate of cooling of the casting. Mechanical ejectors, such as pins located in various parts of the mold, may be needed for removal of complex castings. Ejectors usually leave small round impressions on castings and gating systems.

The molds are clamped together by mechanical means and heated to facilitate metal flow and reduce thermal damage to the dies. The molten metal is then poured through the gating system. After solidification, the molds are opened and the casting is removed. Special methods of cooling the mold include water or the use of air-cooled fins, similar to those found on motorcycle and lawnmower engines that cool the engine block and cylinder head. Although the permanent-mold casting operation can be performed manually, the process can be automated for large production runs. This process is used mostly for aluminum, magnesium, and copper alloys, because of their generally lower melting points; steels can also be cast in graphite or heat-resistant metal molds.

The permanent-mold process produces castings with good surface finish, close dimensional tolerances, and uniform and good mechanical properties at high production rates. Typical parts made include automobile pistons, cylinder heads, connecting rods, gear blanks for appliances, and kitchenware. Although equipment costs can be high because of die costs, the process can be mechanized, thus keeping labor costs low. Permanent-mold casting is not economical for small production runs. Furthermore, because of the difficulty in removing the casting from the mold, intricate shapes cannot be cast by this process. However, easily collapsed sand cores can be used and removed from castings to leave intricate internal cavities.

5.10.1 Slush casting

It was noted in Fig. 5.16 that a solidified skin develops first in a casting and that this skin becomes thicker with time. Using this principle, hollow castings with thin walls can be made by permanent-mold casting, a process called *slush casting*. The molten metal is poured into the metal mold, and after the desired thickness of solidified skin is obtained, the mold is inverted or slung, and the remaining liquid metal is poured out. The mold halves are then opened, and the casting is removed. The process is suitable for small production runs and is generally used for making ornamental and decorative objects and toys from low-melting-point metals, such as zinc, tin, and lead alloys.

5.10.2 Pressure casting

In the two permanent-mold processes just described, the molten metal flows into the mold cavity by gravity. In the *pressure-casting process*, also called *pressure pouring* or *low-pressure casting* (Fig. 5.27), the molten metal is forced upward by gas pressure into a graphite or metal mold. The pressure is maintained until the metal has completely solidified in the mold. The molten metal may also be forced upward by a vacuum, which removes dissolved gases and produces a casting with lower porosity as well.

5.10.3 Die casting

The *die-casting* process, developed in the early 1900s, is a further example of permanent-mold casting. The molten metal is forced into the die cavity at pressures ranging from 0.7 to 700 MPa (0.1 to 100 ksi). The European term *pressure die casting*, or simply *die casting*, described in this section, is not to be confused with the term *pressure casting* described in Section 5.10.2. Typical parts made by die casting include transmission housings, valve bodies, carburetors, motors, business machine and appliance components, hand tools, and toys. The weight of most castings ranges from less than 90 g (3 oz) to about 25 kg (55 lb).

a) **Hot-chamber process.** The *hot-chamber process* (Fig. 5.28) involves the use of a piston, which traps a certain volume of molten metal and forces it into the die cavity through a gooseneck and nozzle. The pressures range up to 35 MPa (5000 psi), with an average of about 15 MPa (2000 psi). The metal is held under

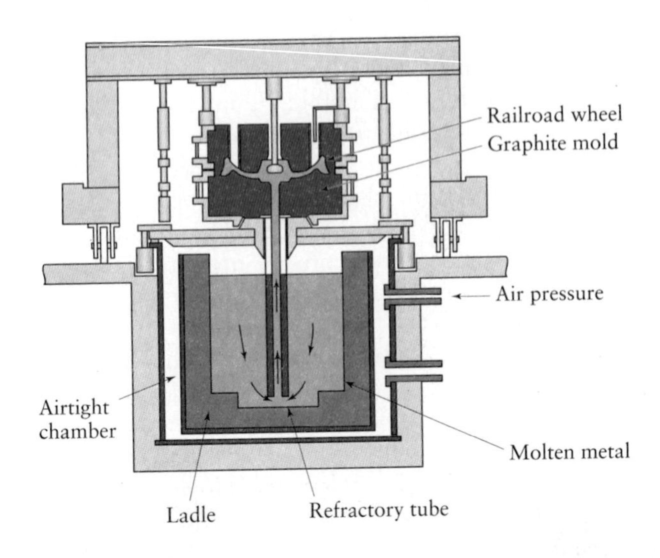

FIGURE 5.27 The pressure-casting process uses graphite molds for the production of steel railroad wheels. *Source*: Griffin Wheel Division of Amsted Industries Incorporated.

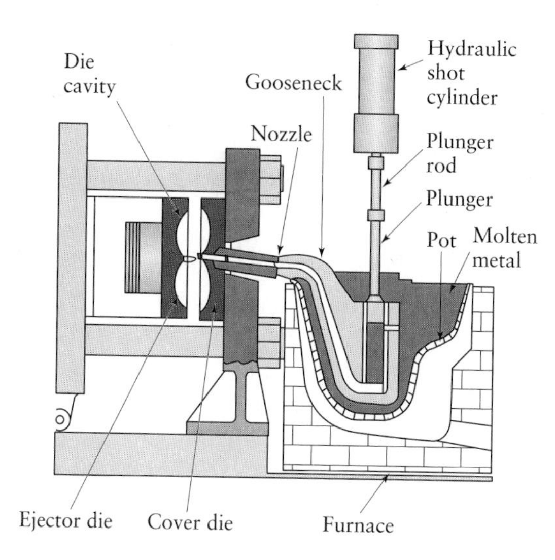

FIGURE 5.28 Sequence of steps in die casting of a part in the hot-chamber process. *Source*: Courtesy of Foundry Management and Technology.

pressure until it solidifies in the die. To improve die life and to aid in rapid metal cooling, thus reducing cycle time, dies are usually cooled by circulating water or oil through various passageways in the die block. Cycle times usually range up to 900 shots (individual injections) per hour for zinc, although very small components, such as zipper teeth, can be cast at 18,000 shots per hour. Low-melting-point alloys such as zinc, tin, and lead are commonly cast by this process.

b) **Cold-chamber process.** In the *cold-chamber* process (Fig. 5.29), molten metal is introduced into the injection cylinder (*shot chamber*). The shot chamber is not heated, hence the term *cold chamber*. The metal is forced into the die cavity at pressures usually ranging from 20 to 70 MPa (3 to 10 ksi), although they may be as high as 150 MPa (20 ksi). The machines may be horizontal or vertical; in the latter, the shot chamber is vertical, and the machine is similar to a vertical press.

High-melting-point alloys of aluminum, magnesium, and copper are normally cast by this method, although other metals (including ferrous metals) can also be cast in this manner. Temperatures of molten metals start at about 600°C (1150°F) for aluminum and magnesium alloys and increase considerably for copper-base and iron-base alloys.

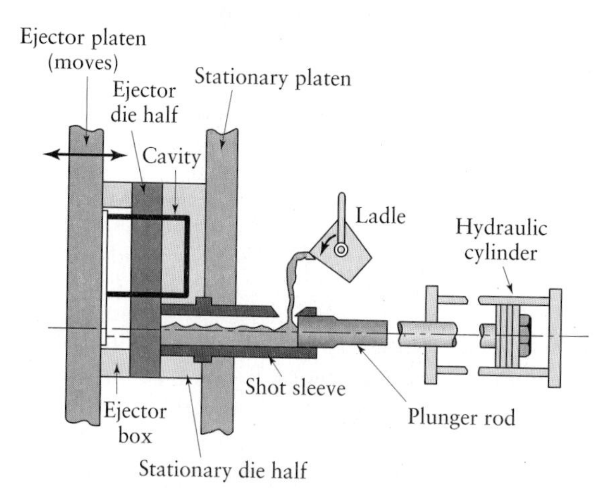

FIGURE 5.29 Sequence of operations in die casting of a part in the cold-chamber process. *Source*: Courtesy of Foundry Management and Technology.

Process capabilities and machine selection. Because of the high pressures involved, the dies have a tendency to part unless clamped together tightly. Die-casting machines are rated according to the clamping force that can be exerted to keep the dies closed. The capacities of commercially available machines range from about 25 to 3000 tons. Other factors involved in the selection of die-casting machines are die size, piston stroke, shot pressure, and cost.

Die-casting dies may be made as single-cavity, multiple-cavity (several identical cavities), combination-cavity (several different cavities), or unit dies (simple small dies that can be combined in two or more units in a master holding die). Dies are usually made of hot-work die steels or mold steels. Die wear increases with the temperature of the molten metal. **Heat-checking** of the dies (for surface cracking from repeated heating and cooling of the die) can be a problem. When die materials are selected and maintained properly, dies may last more than half a million shots before die wear becomes significant.

Die design includes taper (draft) to allow the removal of the casting. The sprues and runners may be removed either manually or by using trim dies in a press. The entire die-casting and finishing process can be highly automated. Lubricants (parting agents) are usually applied as thin coatings on die surfaces. Alloys, except magnesium alloys, generally require lubricants, which usually have a water base, with graphite or other compounds in suspension. Because of the high cooling capacity of water, these lubricants also are effective in keeping die temperatures low.

Die casting has the capability for high production rates of parts with good strength, high-quality parts with complex shapes, and good dimensional accuracy and surface detail, thus requiring little or no subsequent machining or finishing operations (*net-shape forming*). Components such as pins, shafts, and fasteners can be cast integrally (**insert molding**), a process similar to putting wooden sticks (the pin) in popsicles (the casting). Ejector marks remain, as do small amounts of *flash* (thin material squeezed out between the dies; see also Section 6.2.3) at the die parting line.

In the fabrication of certain parts, die casting can compete favorably with other manufacturing methods, such as sheet-metal stamping and forging, or other casting processes. In addition, because the molten metal chills rapidly at the die walls, the casting has a fine-grain, hard skin with higher strength than in the center. Consequently, the strength-to-weight ratio of die-cast parts increases with decreasing wall thickness. With good surface finish and dimensional accuracy, die casting can produce bearing surfaces that would normally be machined. Equipment costs, particularly the cost of dies, are somewhat high, but labor costs are generally low, because the process is semi-or fully automated. Die casting is economical for large production runs. Table 5.6 lists the properties and typical applications of common die-casting alloys.

5.10.4 Centrifugal casting

As its name implies, the *centrifugal casting process* uses the inertial forces caused by rotation to distribute the molten metal into the mold cavities. There are three types of centrifugal casting: true centrifugal casting, semicentrifugal casting, and centrifuging.

True centrifugal casting. In *true centrifugal casting*, hollow cylindrical parts, such as pipes, gun barrels, and lampposts, are produced by the technique shown in Fig. 5.30, in which molten metal is poured into a rotating mold. The axis of rotation is usually horizontal, but can be vertical for short workpieces. Molds are made of steel, iron, or graphite and may be coated with a refractory lining to increase mold life. Mold

TABLE 5.6

Properties and Typical Applications of Common Die-Casting Alloys

Alloy	Ultimate Tensile Strength (MPa)	Yield Strength (MPa)	Elongation in 50 mm (%)	Applications
Aluminum 380 (3.5 Cu, 8.5 Si)	320	160	2.5	Appliances, automotive components, electrical motor frames and housings, engine blocks.
13 (12Si)	300	150	2.5	Complex shapes with thin walls, parts requiring strength at elevated temperatures.
Brass 858 (60 Cu)	380	200	15	Plumbing fixtures, lock hardware, bushings, ornamental castings.
Magnesium AZ91B (9 Al, 0.7 Zn)	230	160	3	Power tools, automotive parts, sporting goods.
Zinc No. 3 (4 Al)	280	—	10	Automotive parts, office equipment, household utensils, building hardware, toys.
5 (4 Al, 1 Cu)	320	—	7	Appliances, automotive parts, building hardware, business equipment.

Source: Data taken from The North American Die Casting Association.

surfaces can be shaped so that pipes with various outer shapes, including square or polygonal, can be cast. The inner surface of the casting remains cylindrical, because the molten metal is uniformly distributed by centrifugal forces. However, because of differences in density, lighter elements such as dross, impurities, and pieces of the refractory lining tend to collect on the inner surface of the casting.

Cylindrical parts ranging from 13 mm (0.5 in.) to 3 m (10 ft) in diameter and 16 m (50 ft) long can be cast centrifugally, with wall thicknesses ranging from 6 to 125 mm (0.25 to 5 in.). The pressure generated by the centrifugal force is high, as much as 150 g's, and is necessary for casting thick-walled parts. Castings of good quality, dimensional accuracy, and external surface detail are obtained by this process. In addition to pipes, typical parts made include bushings, engine cylinder liners, and bearing rings with or without flanges.

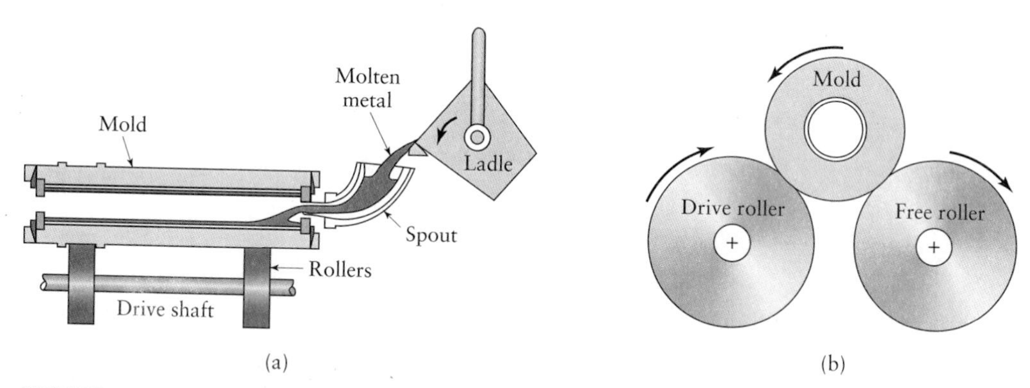

FIGURE 5.30 Schematic illustration of the centrifugal casting process. Pipes, cylinder liners, and similarly shaped parts can be cast by this process.

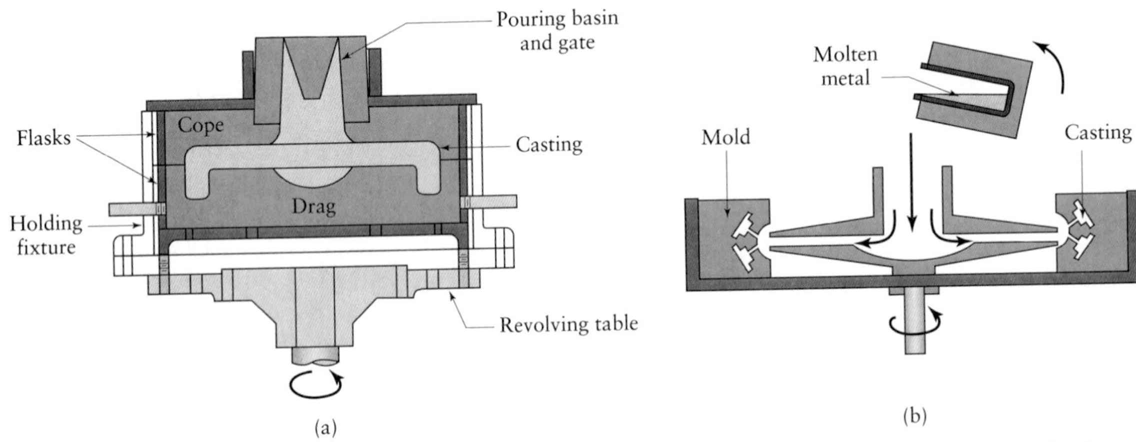

FIGURE 5.31 (a) Schematic illustration of the semicentrifugal casting process. Wheels with spokes can be cast by this process. (b) Schematic illustration of casting by centrifuging. The molds are placed at the periphery of the machine, and the molten metal is forced into the molds by centrifugal forces.

Semicentrifugal casting. Figure 5.31a shows an example of *semicentrifugal casting*. This method is used to cast parts with rotational symmetry, such as a wheel with spokes.

Centrifuging. In *centrifuging*, also called *centrifuge casting*, mold cavities of any shape are placed at a certain distance from the axis of rotation. The molten metal is poured at the center and is forced into the mold by centrifugal forces (Fig. 5.31b). The properties within the castings vary by the distance from the axis of rotation.

5.10.5 Squeeze casting

The *squeeze-casting process*, developed in the 1960s, involves solidification of the molten metal under high pressure; hence, the process is a combination of casting and forging (Fig. 5.32). The machinery includes a die, punch, and ejector pin. The pressure applied by the punch keeps the entrapped gases in solution (especially hydrogen in aluminum alloys), and the high-pressure contact at the die–metal interface promotes heat transfer, resulting in a fine microstructure with good mechanical properties and limited microporosity. Parts can be made to *near-net shape*, with complex shapes and fine surface detail, from both nonferrous and ferrous alloys. Typical products made include automotive wheels and mortar bodies (a short-barreled cannon). The pressures required in squeeze casting are typically higher than those used in pressure die casting, but lower than those for hot or cold forging.

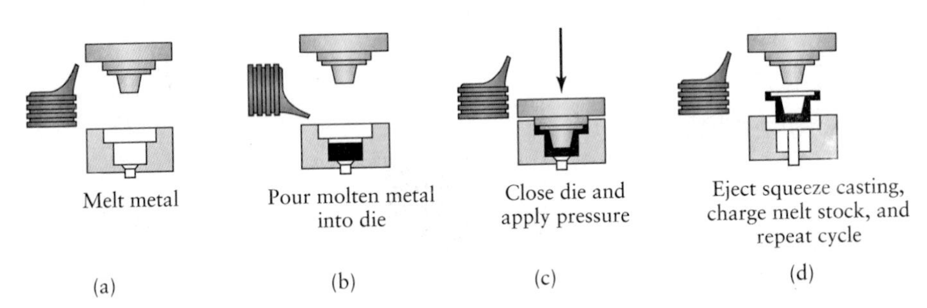

FIGURE 5.32 Sequence of operations in the squeeze-casting process. This process combines the advantages of casting and forging.

5.10.6 Semisolid metal forming

Also called *semisolid metalworking*, semisolid metal forming was developed in the 1970s. The metal or alloy has a nondentritic, roughly spherical, and fine-grained structure when it enters the die or mold. The alloy exhibits *thixotropic* behavior, that is, its viscosity decreases when agitated; hence, this process is also known as **thixoforming** or **thixocasting**. For example, at rest and above its solidus temperature, the alloy has the consistency of table butter, but when agitated vigorously, its consistency is more like that of machine oils. This behavior has been used in developing technologies that combine casting and forging of parts, with cast billets that are forged when 30% to 40% is liquid. Semisolid-metal-forming technology was in commercial production by 1981 and is also used in making cast metal-matrix composites. Another technique for forming in a semisolid state is **rheocasting**, in which a slurry is produced in a mixer and delivered to the mold or die. However, this process has not yet been commercially successful.

5.10.7 Casting techniques for single-crystal components

These techniques can be illustrated by describing the casting of gas-turbine blades, which are generally made of nickel-base superalloys. The procedures involved can also be used for casting other alloys and components.

Conventional casting of turbine blades. The conventional casting process involves investment casting using a ceramic mold. (See Fig. 5.23.) The molten metal is poured into the mold and begins to solidify at the ceramic walls. The grain structure developed is polycrystalline, and the presence of grain boundaries makes this structure susceptible to creep and cracking along those boundaries under the centrifugal forces at elevated temperatures.

Directionally solidified blades. In the *directional solidification process* (Fig. 5.33a), first developed in 1960, the ceramic mold is prepared basically by investment-casting techniques and is preheated by radiant heating. The mold is supported by a water-cooled chill plate. After the metal is poured into the mold, the assembly is lowered slowly. Crystals begin to grow at the chill-plate surface and upward. The blade is thus directionally solidified, with longitudinal, but no transverse, grain boundaries. Consequently, the blade is stronger in the direction of centrifugal forces developed in the gas turbine.

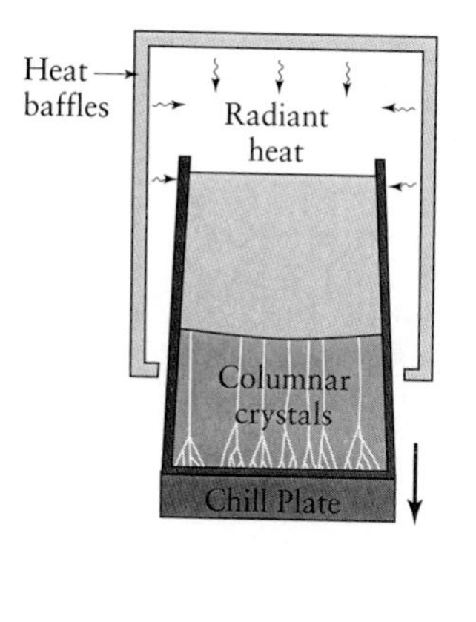

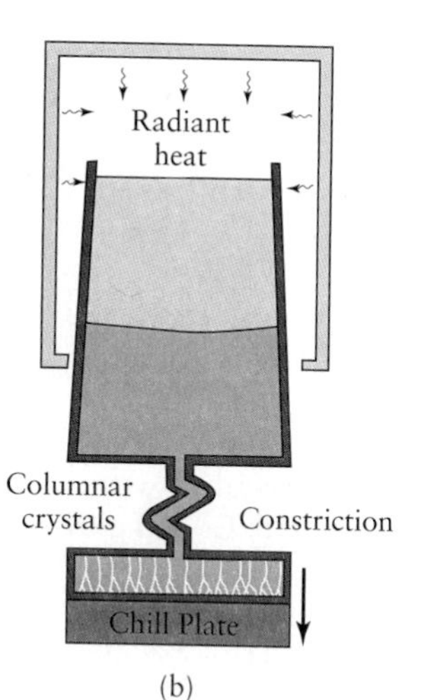

FIGURE 5.33 Methods of casting turbine blades: (a) directional solidification; (b) method to produce a single-crystal blade; and (c) a single-crystal blade with a constriction portion still attached. *Source*: (a) and (b) adapted from "Advanced Metal" by B. H. Kear, copyright © 1986 by Scientific American, Inc. All Rights Reserved. (c) *Advanced Materials and Processes*, ASM International, October 1990, p. 29.

Single-crystal blades. In the growing process for single-crystal blades, developed in 1967, the mold is prepared basically by investment-casting techniques and has a construction in the shape of a corkscrew (Figs. 5.33b and c), the cross-section of which allows only one crystal through. As the assembly is lowered slowly, a single crystal grows upward through the constriction and begins to grow in the mold. Strict control of the rate of movement is necessary. The solidified mass in the mold is a single-crystal blade. Although these blades are more expensive than other blades, the lack of grain boundaries makes these blades resistant to creep and thermal shock; thus, they have a longer and more reliable service life.

Single-crystal growing. With the advent of the semiconductor industry, *single-crystal growing* has become a major activity in the manufacture of microelectronic devices (Chapter 13). There are two methods of crystal growing. In the **crystal-pulling method**, known as the **Czochralski**, or **CZ**, **process**, (Fig. 5.34a), a seed crystal is dipped into the molten metal and then pulled out slowly, at a rate of about 10 μm/s (400 μin./s), while being rotated about 1 rev/s. The liquid metal begins to

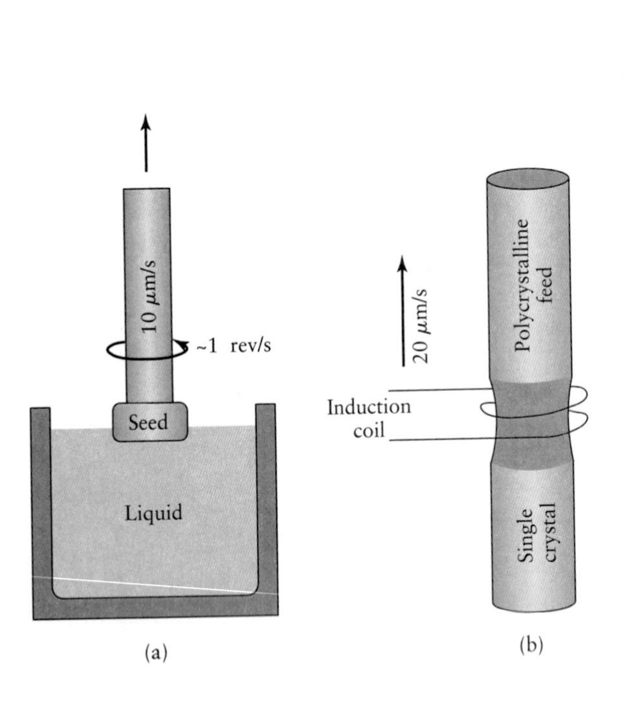

(c)

FIGURE 5.34 Two methods of crystal growing: (a) crystal-pulling method (Czochralski process) and (b) floating-zone method. Crystal growing is especially important in the semiconductor industry. *Source*: L. H. Van Vlack, *Materials for Engineering* (Fig. 18.5.1, p. 513), © 1982 Addison-Wesley Publishing Company, Inc. Reprinted by permission of Addison-Wesley Longman Publishing Co., Inc. (c) A single-crystal silicon ingot produced by the Czochralski process. *Source*: Intel Corp.

solidify on the seed, and the crystal structure of the seed is continued throughout. *Dopants* (alloying elements) may be added to the liquid metal to impart special electrical properties. Single crystals of silicon, germanium, and various other elements are grown by this process. Single-crystal ingots up to 225 mm (9 in.) in diameter and over 1 m (40 in.) in length have been produced by this technique, although 300-mm (12-in.) cylinders are under development (Fig. 5.34c).

The second technique for crystal growing is the **floating-zone method** (Fig. 5.34b). The process starts with a rod of polycrystalline silicon resting on a single crystal; an induction coil heats these two pieces while moving slowly upward. The single crystal grows upward while maintaining its orientation. Thin wafers are then cut from the rod, cleaned, and polished for use in microelectronic-device fabrication. Because of the limited diameters that can be produced through this process, it is being displaced by the Czochralski process.

5.10.8 Rapid solidification (amorphous alloys)

First developed in the 1960s, *rapid solidification* involves cooling of molten metal at rates as high as 10^6 K/s, whereby the metal does not have sufficient time to crystallize. These alloys are called **amorphous alloys** or **metallic glasses**, because they do not have a long-range crystalline structure. They typically contain iron, nickel, and chromium, which are alloyed with carbon, phosphorus, boron, aluminum, and silicon. Among other effects, rapid solidification results in a significant extension of solid solubility, grain refinement, and reduced microsegregation.

Amorphous alloys exhibit excellent corrosion resistance, good ductility, and high strength. Furthermore, they exhibit very little loss from magnetic hysteresis, high resistance to eddy currents, and high permeability. The latter properties are used in making magnetic steel cores for transformers, generators, motors, lamp ballasts, magnetic amplifiers, and linear accelerators, with greatly improved efficiency. Another major application is superalloys of rapidly solidified powders consolidated into near-net shapes for use in aerospace engines. These alloys are produced in the form of wire, ribbon, strip, and powder. In one method, called **melt spinning** (Fig. 5.35), the alloy is melted by induction in a ceramic crucible and propelled under high gas pressure at very high speed against a rotating copper disk (chill block) and chills rapidly (splat cooling).

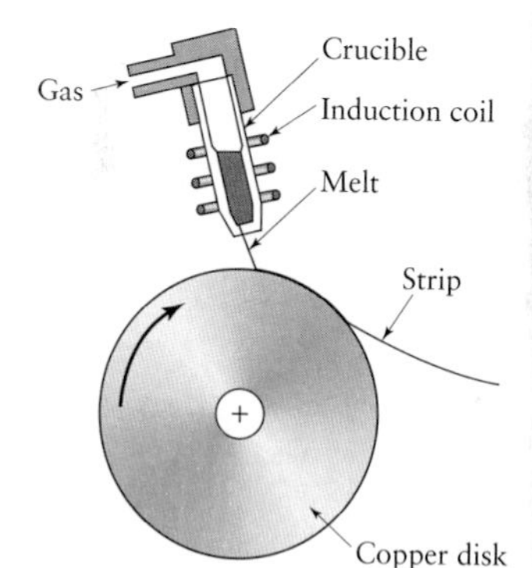

FIGURE 5.35 Schematic illustration of the melt-spinning process to produce thin strips of amorphous metal.

5.11 Heat Treatment

The various microstructures developed during metal processing can be modified by *heat treatment* techniques, that is, by controlled heating and cooling of the alloys at various rates (also known as **thermal treatment**). These treatments induce phase transformations that greatly influence mechanical properties such as strength, hardness, ductility, toughness, and wear resistance of the alloys. The effects of thermal treatment depend primarily on the alloy, its composition and microstructure, the degree of prior cold work, and the rates of heating and cooling during heat treatment. The processes of recovery, recrystallization, and grain growth (see Section 3.6) are examples of thermal treatment, involving changes in the grain structure of the alloy.

5.11.1 Heat treating ferrous alloys

The microstructural changes that occur in the iron–carbon system, discussed in Section 5.2.5, are described as follows:

Pearlite. If the ferrite and cementite lamellae in the pearlite structure (see Section 5.2.6) of the eutectoid steel are thin and closely packed, the microstructure is called **fine pearlite.** If the lamellae are thick and widely spaced, the microstructure is called **coarse pearlite.** The difference between the two depends on the rate of cooling through the eutectoid temperature, a reaction in which austenite is transformed into pearlite. If the rate of cooling is relatively high, as in air, fine pearlite is produced; if it is slow, as in a furnace, coarse pearlite is produced.

The transformation from austenite to pearlite (and for other structures) is best illustrated by Fig. 5.36b and c. These diagrams are called **isothermal transformation (IT) diagrams,** or **time-temperature-transformation (TTT) diagrams.** They are constructed from the data in Fig. 5.36a, which show the percentage of austenite transformed into pearlite as a function of temperature and time. The higher the temperature

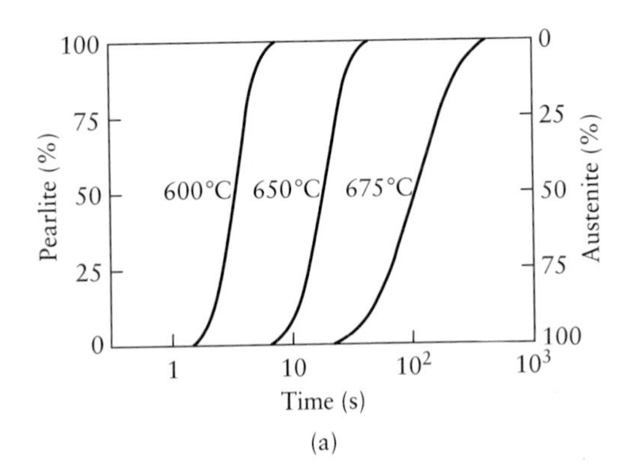

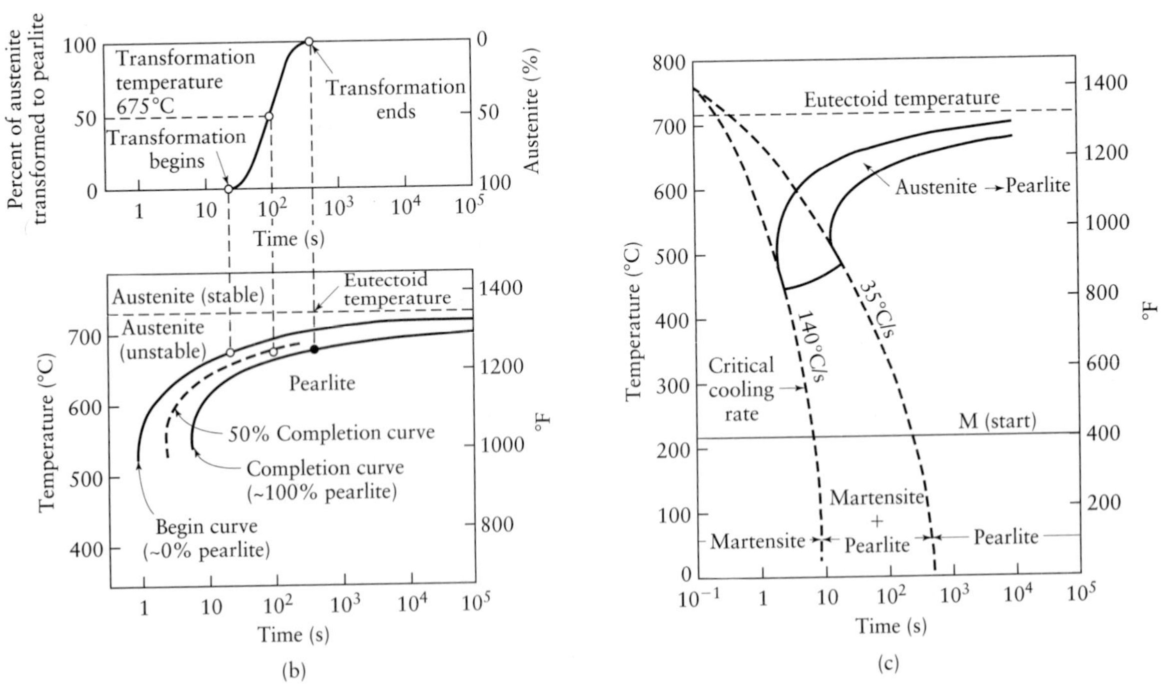

FIGURE 5.36 (a) Austenite-to-pearlite transformation of iron–carbon alloys as a function of time and temperature. (b) Isothermal transformation diagram obtained from (a) for a transformation temperature of 675°C (1247°F). (c) Microstructures obtained for a eutectoid iron–carbon alloy as a function of cooling rate. *Source*: ASM International, Materials Park, OH.

and/or the longer the time, the greater is the percentage of austenite transformed to pearlite. Note that for each temperature, a minimum amount of time is required for the transformation to begin and that sometime later, all the austenite is transformed to pearlite.

Spheroidite. When pearlite is heated to just below the eutectoid temperature and held at that temperature for a period of time, say, for a day at 700°C (1300°F), the cementite lamellae (see Fig. 5.6) transform to *spherical* shapes. Unlike the lamellar shape of cementite, which acts as stress raisers, *spheroidites* (spherical particles) are less conducive to stress concentration, because of their rounded shapes. Consequently, this structure has higher toughness and lower hardness than that of the pearlite structure. In this form, the structure can be cold worked, since the ductile ferrite has high toughness, and the spheroidal carbide particles prevent the propagation of cracks within the material.

Bainite. Visible only using electron microscopy, *bainite* has a very fine microstructure, consisting of ferrite and cementite. It can be produced in steels with alloying elements and at cooling rates that are higher than those required for transformation to pearlite. This structure, called *bainitic steel*, is generally stronger and more ductile than pearlitic steel at the same hardness level.

Martensite. When austenite is cooled rapidly, as by quenching in water, its fcc structure is transformed to a *body-centered tetragonal* (bct) structure. This structure, called *martensite*, can be described as a body-centered rectangular prism that is slightly elongated along one of its principal axes. Because it does not have as many slip systems as does a bcc structure, and the carbon is in interstitial positions, martensite is extremely hard and brittle, lacks toughness, and therefore has limited use. Martensite transformation takes place almost instantaneously (Fig. 5.36c), because it does not involve the diffusion process, a time-dependent phenomenon that is the mechanism in other transformations.

Transformations of martensite involve volume changes, because of the different densities of the various phases in the structure. For example, when austenite transforms to martensite, its volume increases (and hence its density decreases) by as much as 4%. A similar, but smaller, volume expansion also occurs when austenite transforms to pearlite. These expansions, and thus the thermal gradients present in a quenched part, can cause internal stresses within the body. These stresses may cause parts to crack during heat treatment, as in **quench cracking** of steels caused by rapid cooling during quenching.

Retained austenite. If the temperature to which the alloy is quenched is not sufficiently low, only a portion of the structure is transformed to martensite. The rest is *retained austenite*, which is visible as white areas in the structure along with dark needlelike martensite. Retained austenite can cause dimensional instability and cracking and lowers alloy hardness and strength.

Tempered martensite. *Tempering* is a heating process that reduces martensite's hardness and improves its toughness. The body-centered tetragonal martensite is heated to an intermediate temperature, where it transforms to a two-phase microstructure, consisting of body-centered cubic alpha ferrite and small particles of cementite. Longer tempering time and higher temperature decrease martensite's hardness. The reason is that the cementite particles coalesce and grow, and the distance between the particles in the soft ferrite matrix increases as the less stable, smaller carbide particles dissolve.

Hardenability of ferrous alloys. The capability of an alloy to be hardened by heat treatment is called its *hardenability*; it is a measure of the depth of hardness that can be obtained by heating and subsequent quenching. The term *hardenability* should not be confused with *hardness* (Section 2.6). Hardenability of ferrous alloys depends on the carbon content, the grain size of the austenite, and the alloying elements present in the material. A test (the **Jominy test**) has been developed in order to determine alloy hardenability.

Quenching media. Quenching may be carried out in water, brine (saltwater), oils, molten salts, or air. Caustic solutions, polymer solutions, and gases are also used. Because of the differences in the thermal conductivity, specific heat, and heat of vaporization of these media, the rate of cooling of the alloy (**severity of quench**) will also be different. In relative terms and in decreasing order from 5 to 0, the cooling capacity of several quenching media is as follows: agitated brine, 5; still water, 1; still oil, 0.3; cold gas, 0.1; and still air, 0.02. Agitation is also a significant factor in the rate of cooling. In tool steels, the quenching medium is specified by a letter (see Table 3.4), such as W for water hardening, O for oil hardening, and A for air hardening. The cooling rate also depends on the surface-area-to-volume ratio of the part. (See Eq. 5.11.) The higher this ratio, the higher is the cooling rate; thus, for example, a thick plate cools more slowly than a thin plate with the same surface area.

Water is a common medium for rapid cooling. However, the heated metal may form a **vapor blanket** along its surfaces from water-vapor bubbles that form when water boils at the metal–water interface. This blanket creates a barrier to heat conduction, because of the lower thermal conductivity of the vapor. Agitating the fluid or the part helps to reduce or eliminate the blanket. Also, water may be sprayed on the part under high pressure. Brine is an effective quenching medium, because salt helps to nucleate bubbles at the interfaces, thus improving agitation. However, brine can corrode the part. **Die quenching** is a term used to describe the process of clamping the part to be heat treated to a die, which chills selected regions of the part. In this way, cooling rates and warpage can be controlled.

5.11.2 Heat treating nonferrous alloys and stainless steels

Nonferrous alloys and some stainless steels generally cannot be heat treated by the techniques used with ferrous alloys. The reason is that nonferrous alloys do not undergo phase transformations as steels do. The hardening and strengthening mechanisms for these alloys are fundamentally different. Heat-treatable aluminum alloys, copper alloys, and martensitic and precipitation-hardening stainless steels are hardened and strengthened by a process called **precipitation hardening.** This is a technique in which small particles (of a different phase and called *precipitates*) are uniformly dispersed in the matrix of the original phase. (See Fig. 5.2a.) In this process, precipitate forms, because the solid solubility of one element (one component of the alloy) in the other is exceeded.

Three stages are involved in precipitation hardening; they can best be described by referring to Fig. 5.37, the phase diagram for the aluminum–copper system. For an alloy with the composition of 95.5% Al, 4.5% Cu, a single-phase (kappa) substitutional solid solution of copper (solute) in aluminum (solvent) exists between 500°C and 570°C (930°F and 1060°F). The kappa phase is aluminum rich, with an fcc structure, and is ductile. Below the lower temperature, that is, below the lower solubility curve, two phases are present: kappa and theta (a hard intermetallic compound of $CuAl_2$). This alloy can be heat treated and its properties modified by solution treatment or precipitation.

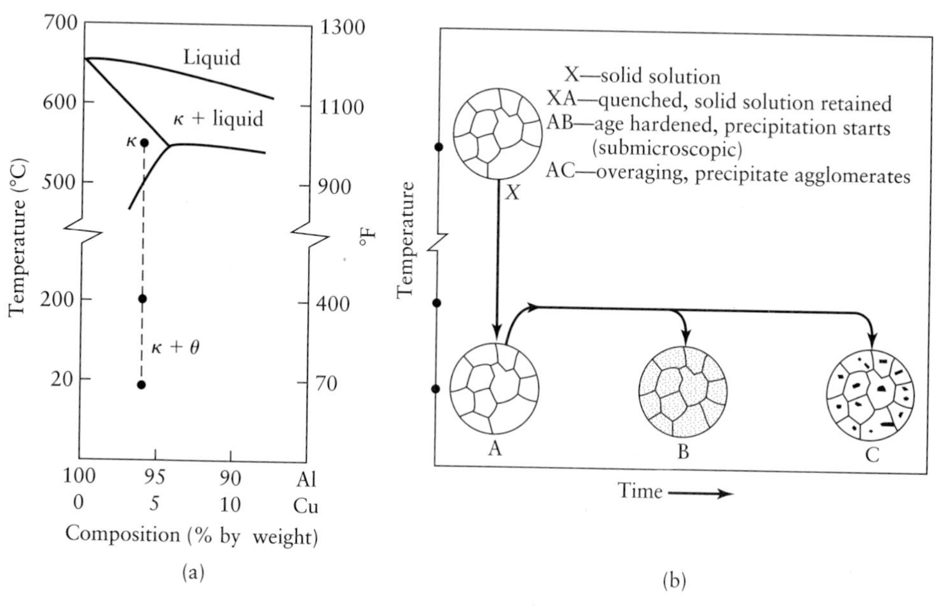

FIGURE 5.37 (a) Phase diagram for the aluminum–copper alloy system. (b) Various microstructures obtained during the age-hardening process. *Source*: L. H. Van Vlack, *Materials for Engineering* (Fig. 5.6.1, p. 123), © 1982 Addison-Wesley Publishing Company, Inc. Reprinted by permission of Addison-Wesley Longman Publishing Co., Inc.

Solution treatment. In *solution treatment*, the alloy is heated to within the solid-solution kappa phase, say 540°C (1000°F), and cooled rapidly, such as by quenching in water. The structure obtained soon after quenching (*A* in Fig. 5.37b), consists only of the single-phase kappa. This alloy has moderate strength and considerable ductility.

Precipitation hardening. The structure obtained in A in Fig. 5.37b can be strengthened by precipitation hardening. The alloy is reheated to an intermediate temperature and held there for a period of time, during which precipitation takes place. The copper atoms diffuse to nucleation sites and combine with aluminum atoms, producing the theta phase, which forms as submicroscopic precipitates, shown in B by the small dots within the grains of the kappa phase. This structure is stronger than that in A, although it is less ductile. The increase in strength is attributed to increased resistance to dislocation movement in the region of the precipitates.

Aging. Because the precipitation process is one of time and temperature, it is also called *aging*, and the property improvement is known as *age hardening*. If carried out above room temperature, the process is called **artificial aging**. However, several aluminum alloys harden and become stronger over a period of time at room temperature, by a process known as **natural aging**. Such alloys are first quenched and then, if desired, are formed at room temperature into various shapes and allowed to gain strength and hardness by natural aging. Natural aging can be slowed by refrigerating the quenched alloy.

In the precipitation process, if the reheated alloy is held at that elevated temperature for an extended period of time, the precipitates begin to coalesce and grow. They become larger, but fewer, as shown by the larger dots in C in Fig. 5.37b. This process is called **overaging**, which makes the alloy softer and weaker; thus, there is an optimal time–temperature relationship in the aging process for obtaining desired properties. Obviously, an aged alloy can be used only up to a certain maximum temperature in service; otherwise, it will overage and lose its strength and hardness. Although weaker, an overaged part has better dimensional stability.

Maraging. Maraging is a precipitation-hardening treatment for a special group of high-strength iron-base alloys. The word *maraging* is derived from the words martensite and aging. In this process, one or more intermetallic compounds are precipitated in a matrix of low-carbon martensite. A typical maraging steel may contain 18% nickel, in addition to other elements, and aging is done at 480°C (900°F). Hardening by maraging does not depend on the cooling rate; thus, full uniform hardness can be obtained throughout large parts with minimal distortion. Typical uses of maraging steels are for dies and tooling for casting, molding, forging, and extrusion.

5.11.3 Case hardening

The heat treatment processes described so far involve microstructural alterations and property changes in the *bulk* of the material or component by **through hardening.** In many situations, however, alteration of only the *surface* properties of a part (hence the term *case hardening*) is desirable. This method is particularly useful for improving resistance to surface indentation, fatigue, and wear. Typical applications for case hardening include gear teeth, cams, shafts, bearings, fasteners, pins, automotive clutch plates, and tools and dies. Through hardening of these parts would not be desirable, because a hard part lacks the necessary toughness for these applications; a small surface crack can propagate rapidly through the part and cause total failure.

Various surface-hardening processes are available (Table 5.7): **carburizing** (*gas*, *liquid*, and *pack carburizing*), **carbonitriding, cyaniding, nitriding, boronizing,** and **flame** and **induction hardening.** Basically, these processes are heat-treating operations in which the component is heated in an atmosphere containing elements (such as carbon, nitrogen, or boron) that alter the composition, microstructure, and properties of surfaces.

For steels with sufficiently high carbon content, surface hardening takes place without using any of these additional elements. Only the heat-treatment processes described in Section 5.10.1 are needed to alter the microstructures, usually by flame hardening or induction hardening. Laser beams and electron beams are also used effectively to harden both small and large surfaces and also for through hardening of relatively small parts.

Because case hardening is a localized heat treatment, case-hardened parts have a hardness gradient. Typically, the hardness is greatest at the surface and decreases below the surface, the rate of decrease depending on the composition of the metal and the process variables. Surface-hardening techniques can also be used for tempering, thus modifying the properties of surfaces that have been subjected to heat treatment. Various other processes and techniques for surface hardening, such as shot peening and surface rolling, improve wear resistance and various other characteristics. (See Section 4.5.1.)

Decarburization is the phenomenon in which alloys containing carbon lose carbon from their surfaces as a result of heat treatment or hot working in a medium, usually oxygen, that reacts with the carbon. Decarburization is undesirable, because it affects the hardenability of the surfaces of the part by lowering the carbon content. It also adversely affects the hardness, strength, and fatigue life of steels by significantly lowering their endurance limits. Decarburization is best avoided by processing the alloy in an inert atmosphere or a vacuum or by using neutral salt baths during heat treatment.

5.11.4 Annealing

Annealing is a general term used to describe the restoration of a cold-worked or heat-treated metal or alloy to its original properties, so as to increase ductility (and hence formability), reduce hardness and strength, or modify the microstructure. Annealing is also used to relieve residual stresses in a manufactured part for improved machinability

TABLE 5.7

Outline of Heat Treatment Processes for Surface Hardening

Process	Metals Hardened	Element Added To Surface	Procedure	General Characteristics	Typical Applications
Carburizing	Low-carbon steel (0.2% C), alloy steels (0.08–0.2% C)	C	Heat steel at 870–950°C (1600–1750°F) in an atmosphere of carbonaceous gasses (gas carburizing) or carbon-containing solids (pack carburizing). Then quench.	A hard, high-carbon surface is produced. Hardness of 55 to 65 HRC. Case depth <0.5–1.5 mm (<0.020 to 0.060 in.) Some distortion of the part during heat treatment.	Gears, cams, shafts, bearings, piston rings, sprockets, clutch plates.
Carbonitriding	Low-carbon steel	C and N	Heat steel at 700–800°C (1300–1600°F) in an atmosphere of carbonaceous gas and ammonia. Then quench in oil.	Surface hardness 55 to 62 HRC. Case depth 0.07–0.5 mm (0.003–0.020 in.) Less distortion than in carburizing.	Bolts, nuts, gears.
Cyaniding	Low-carbon steel (0.2% C), alloy steels (0.08–0.2% C)	C and N	Heat steel at 760–845°C (1400–1550°F) in a molten bath of solutions of cyanide (e.g., 30% sodium cyanide) and other salts.	Surface hardness up to 65 HRC. Case depth 0.025–0.25 mm (0.001–0.010 in.) Some distortion.	Bolts, nuts, screws, small gears.
Nitriding	Steels (1% Al, 1.5% Cr, 0.3% Mo), alloy steels (Cr, Mo), stainless steels, high-speed tool steels	N	Heat steel at 500–600°C (925–1100°F) in an atmosphere of ammonia gas or mixtures of molten cyanide salts. No further treatment.	Surface hardness up to 1100 HV. Case depth 0.1–0.6 mm (0.005–0.030 in.) and 0.02–0.07 mm (0.001–0.003 in.) for high-speed steel.	Gears, shafts, sprockets, valves, cutters, boring bars, fuel-injection pump parts.
Boronizing	Steels	B	Part is heated using boron-containing gas or solid in contact with the part.	Extremely hard and wear-resistant surface. Case depth 0.025–0.075 mm (0.001–0.003 in.).	Tool and die steels.
Flame hardening	Medium-carbon steels, cast irons	None	Surface is heated with an oxyacetylene torch and then quenched with water spray or other quenching methods.	Surface hardness 50 to 60 HRC. Case depth 0.7–6 mm (0.030–0.25 in.) Little distortion.	Gear and sprocket teeth, axles, crankshafts, piston rods, lathe beds and centers.
Induction hardening	Medium-carbon steels, cast irons	None	Metal part is placed in copper induction coils and is heated by high-frequency current and then quenched.	Surface hardness 50 to 60 HRC. Case depth 0.7–6 mm (0.030–0.25 in.) Little distortion.	Gear and sprocket teeth, axles, crankshafts, piston rods, lathe beds and centers.

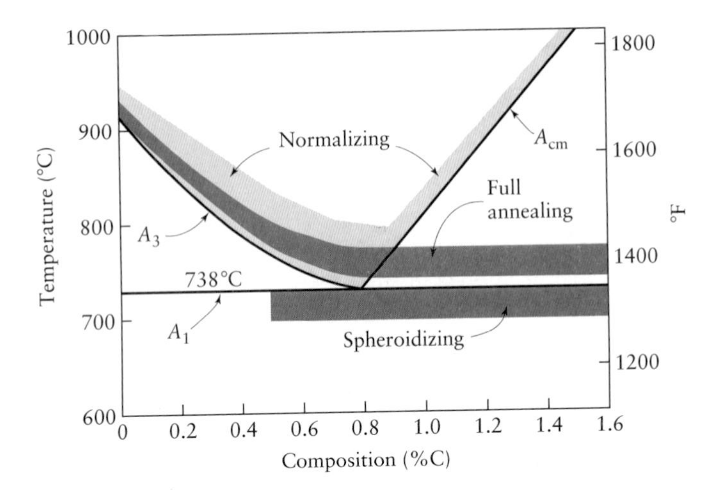

FIGURE 5.38 Heat-treating temperature ranges for plain-carbon steels, as indicated on the iron–iron-carbide phase diagram. *Source*: ASM International, Materials Park, OH.

and dimensional stability. The term *annealing* also applies to thermal treatment of glasses and similar products (Section 11.11.2) and weldments (Section 12.5.1).

The annealing process involves (1) heating the workpiece to within a specific range of temperatures, (2) holding it at that temperature for a period of time (soaking), and (3) cooling it slowly. The process may be carried out in an inert or controlled atmosphere or performed at low temperatures to prevent or minimize surface oxidation. Annealing temperatures may be higher than the recrystallization temperature, depending on the degree of cold work (and hence stored energy). For example, the recrystallization temperature for copper ranges between 200°C and 300°C (400°F and 600°F), whereas the annealing temperature needed to fully recover the original properties ranges from 260°C to 650°C (500°F to 1200°F), depending on the degree of prior cold work.

Full annealing is a term applied to annealing ferrous alloys, generally low- and medium-carbon steels. The steel is heated to above A_1 or A_3 in Fig. 5.38, and cooling takes place slowly, at a rate of, say, 10°C (20°F) per hour, in a furnace after it has been turned off. The structure obtained in full annealing is coarse pearlite, which is soft and ductile and has small, uniform grains. Excessive softness in the annealing of steels can be avoided if the entire cooling cycle is carried out in still air; this process is called **normalizing**, in which the part is heated to a temperature of A_3 or A_{cm} to transform the structure to austenite. It results in somewhat higher strength and hardness and lower ductility than in full annealing. The structure obtained is fine pearlite with small, uniform grains. Normalizing is generally done to refine the grain structure, obtain uniform structure (*homogenization*), decrease residual stresses, and improve machinability.

Process annealing. In *process annealing* (also called *intermediate annealing*, *subcritical annealing*, or *in-process annealing*), the workpiece is annealed to restore its ductility, part or all of which may have been exhausted by work hardening during cold working. In this way, the part can be worked further into the final desired shape. If the temperature is high and/or the time of annealing is long, grain growth may result, with adverse effects on the formability of annealed parts.

Stress-relief annealing. To reduce or eliminate residual stresses, a workpiece is generally subjected to *stress-relief annealing*, or simply **stress relieving**. The temperature and amount of time required for this process depend on the material and the magnitude of residual stresses present. The residual stresses may have been induced during forming, machining, or other shaping processes or caused by volume changes during phase transformations. For steels, the part is heated to below A_1, thus avoiding phase transformations. Slow cooling rates, such as in still air, are generally employed. Stress relieving promotes dimensional stability in situations where subsequent relaxing of residual stresses present may cause distortion of the part over a period of time when it is in service. It also reduces the tendency for stress-corrosion cracking.

5.11.5 Tempering

If steels are hardened by heat treatment, *tempering* is used in order to reduce brittleness, increase ductility and toughness, and reduce residual stresses. The term *tempering* is also used for glasses (Section 11.11.2). In tempering, the steel is heated to a specific temperature, depending on composition, and cooled at a prescribed rate. Alloy steels may undergo **temper embrittlement**, caused by the segregation of impurities along the grain boundaries at temperatures between 480°C and 590°C (900°F and 1100°F).

In **austempering**, the heated steel is quenched from the austenitizing temperature rapidly enough to avoid formation of ferrite or pearlite. It is then held at a certain temperature until isothermal transformation from austenite to bainite is complete. Next, it is cooled to room temperature, usually in still air, at a moderate rate to avoid thermal gradients within the part. The quenching medium most commonly used is molten salt, at temperatures ranging from 160°C to 750°C (320°F to 1380°F). Austempering is often substituted for conventional quenching and tempering, either to reduce the tendency for cracking and distortion during quenching or to improve ductility and toughness while maintaining hardness. Because of the relatively short cycle time, this process is economical for many applications. In *modified austempering*, a mixed structure of pearlite and bainite is obtained. The best example of this practice is **patenting**, which provides high ductility and moderately high strength, such as in the patented wire used in the wire industry (Section 6.5.3).

In **martempering (marquenching)**, the steel or cast iron is quenched from the austenitizing temperature into a hot fluid medium, such as hot oil or molten salt. It is held at that temperature until the temperature is uniform throughout the part and then cooled at a moderate rate, such as in air, to avoid temperature gradients within the part. The part is then tempered, because the structure thus far obtained is primarily untempered martensite and is not suitable for most applications. Martempered steels have less of a tendency to crack, distort, and develop residual stresses during heat treatment. In modified martempering, the quenching temperature is lower, and hence the cooling rate is higher. The process is suitable for steels with lower hardenability.

In **ausforming**, also called **thermomechanical processing**, the steel is formed into desired shapes within controlled ranges of temperature and amount of time to avoid formation of nonmartensitic transformation products. The part is then cooled at various rates to obtain the desired microstructures. Ausformed parts have superior mechanical properties.

5.11.6 Cryogenic treatment

In *cryogenic tempering*, steel is lowered from room temperature to −180°C (−300°F) at a slow rate of as low as 2°C per minute in order to avoid thermal shock. The material is then maintained at this temperature for 24 to 36 hours. At this temperature, the conversion of austenite to martensite occurs slowly, but almost completely, compared with typically only 50% to 90% in conventional quenching. Additional precipitates of carbon (with chromium, tungsten, and other elements) form, relieving residual stresses and leaving a refined grain structure. After the 24-to-36-hour "soak" time, the parts are tempered to stabilize the martensite.

Cryogenically treated steels have higher levels of hardness and wear resistance than those of untreated steels. For example, the wear resistance of D-2 tool steels (see Section 3.10.3) can increase by over 800% after cryogenic treatment to −180°C (−300°F), although most tool steels show a 100% to 200% increase in tool life. (See Section 8.3.) Similar property improvements have been observed in nonferrous alloys and polymers, but these improvements are attributable to refinement of grain structure, since no martensitic transformation occurs. Applications of cryogenic treatment include cutting tools and dies, dental instruments, aerospace materials, sporting goods such as golf club heads, and gun barrels.

5.11.7 Design for heat treating

In addition to the metallurgical factors described thus far, successful heat treating involves design considerations of avoiding problems such as cracking, warping, and nonuniform properties throughout the heat-treated part. The rate of cooling during quenching may not be uniform, particularly with complex shapes of varying cross-sections and thicknesses, thus producing severe temperature gradients. These gradients lead to variations in contraction, which induce thermal stress and may cause the part to crack. Furthermore, nonuniform cooling causes residual stresses in the part, which can lead to stress-corrosion cracking; hence, the method selected and care taken in quenching, as well as the proper choice of quenching media and temperatures, are important considerations.

As a general guideline for part design for heat treating, internal or external sharp corners should be avoided. Otherwise, stress concentrations at these corners raise stress levels high enough to cause cracking. Parts should have as nearly uniform thicknesses as possible, or the transition between regions of different thicknesses should be smooth. Parts with holes, grooves, keyways, splines, and unsymmetrical shapes may also be difficult to heat treat, because they may crack during quenching. Large surfaces with thin cross-sections are likely to warp. Hot forgings and hot steel-mill products may have a *decarburized skin* and thus may not successfully respond to heat treatment.

5.11.8 Cleaning, finishing, and inspection of castings

After solidification and removal from the mold or die, castings may be subjected to a variety of additional processes. In sand casting, the casting is shaken out of its mold, and the sand and oxide layers adhering to the casting are removed by vibration (using a shaker) or by sand blasting. In sand casting and other casting processes, and depending on size, the risers and gates are cut off by oxyfuel-gas cutting, sawing, shearing, and abrasive wheels, or they are trimmed in dies. Castings may be cleaned electrochemically or by pickling with chemicals to remove surface oxides. Maintaining a clean surface of castings is important in subsequent machining operations, because machinability can be adversely affected if the castings are not cleaned properly. If regions of the casting have not formed properly or have formed incompletely, the defects may be repaired by welding, thus filling them with weld metal. Sand castings generally have rough, grainy surfaces, although the extent depends on the quality of the mold and the materials used. *Finishing operations* for castings may include straightening or forging with dies and machining to obtain final dimensions.

Inspection is an important final step and is done to ensure that the casting meets all design and quality-control requirements. Several methods are available for inspection of castings to determine the quality of the castings and whether any defects are present. Castings can be inspected visually or optically for surface defects.

Subsurface and internal defects are investigated using various nondestructive techniques, described in Section 4.8. In destructive testing, test specimens are removed from various sections of a casting and tested for strength, ductility, and other mechanical properties and to determine the location of any defects.

Pressure tightness of cast components such as valves, pumps, and pipes is usually determined by sealing the openings in the casting and pressurizing it with water, oil, or air. The casting is then inspected for leaks while the pressure is maintained. Unacceptable or defective castings are melted down for reprocessing. Because of their major economic impact, the types of defects present in castings and their causes must be investigated. Control of all stages during casting, from mold preparation to removal of castings from molds or dies, is important in maintaining good quality.

5.12 Design Considerations

As in all engineering practice and manufacturing operations, certain guidelines and design principles pertaining to casting have been developed over many years. Although these principles were established primarily through practical experience, analytical methods and computer-aided design and manufacturing techniques are now coming into wider use, improving productivity and the quality of castings. Moreover, careful design can result in significant cost savings. Table 5.8 lists some of the advantages and limitations of casting processes that impact design.

TABLE 5.8

Casting Processes, and their Advantages and Limitations

Process	Advantages	Limitations
Sand	Almost any metal is cast; no limit to size, shape, or weight; low tooling cost.	Some finishing required; somewhat coarse finish; wide tolerances.
Shell mold	Good dimensional accuracy and surface finish; high production rate.	Part size limited; expensive patterns and equipment required.
Expandable pattern	Most metals cast with no limit to size; complex shapes may be cast.	Patterns have low strength and can be costly for low quantities.
Plaster mold	Intricate shapes may be cast; good dimensional accuracy and finish; low porosity.	Limited to nonferrous metals; limited size and volume of production; mold-making time relatively long.
Ceramic mold	Intricate shapes may be cast; close tolerance parts; good surface finish.	Limited size.
Investment	Intricate shapes may be cast; excellent surface finish and accuracy; almost any metal may be cast.	Part size limited; expensive patterns, molds, and labor.
Permanent mold	Good surface finish and dimensional accuracy; low porosity; high production rate.	High mold cost; limited shapes and intricacy may be cast; not suitable for high-melting-point metals.
Die	Excellent dimensional accuracy and surface finish; high production rate.	Die cost is high; part size limited; usually limited to nonferrous metals; long lead time.
Centrifugal	Large cylindrical parts with good quality; high production rate.	Equipment is expensive; part shape limited.

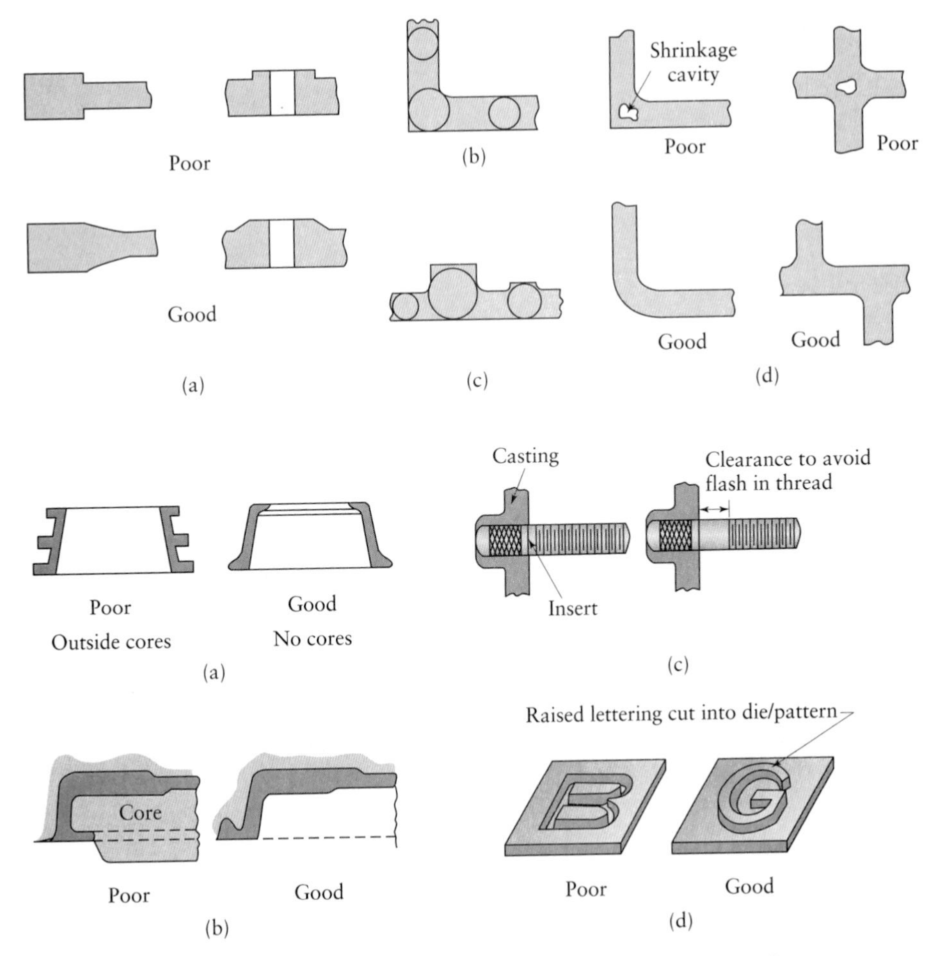

FIGURE 5.39 (a) Suggested design modifications to avoid defects in castings. Note that sharp corners are avoided to reduce stress concentrations. (b)–(d) Examples of designs that show the importance of maintaining uniform cross-sections in castings to avoid hot spots and shrinkage cavities.

FIGURE 5.40 Examples of casting design modifications. *Source*: Steel Castings Handbook, 5th ed., Steel Founders' Society of America, 1980. Used with permission.

5.12.1 Design principles for expendable-mold casting

The guidelines given in this section generally apply to all types of castings. In these guidelines, we identify and describe the most significant design considerations. Figures 5.39 and 5.40 show some typical design modifications for casting.

1. **Risers.** A major concern in the design of castings is the size and placement of risers. (See Fig. 5.14.) Traditionally, riser design has been based on experience and on considerations of fluid flow and heat transfer. More recently, commercial computer programs, based on finite-difference algorithms, have become available. Risers are designed according to six basic rules: (1) The riser must not solidify before the casting; (2) The riser volume must be large enough to provide sufficient liquid metal to compensate for shrinkage in the casting; (3) Junctions between the casting and feeder should not create a **hot spot** where shrinkage porosity can occur; (4) Risers must be placed so that liquid metal can be delivered to locations where it is most needed; (5) There must be sufficient pressure to drive liquid metal into locations in the mold where it is needed; (6) The pressure head from the riser should suppress cavity formation and encourage complete filling of cavities.
2. **Corners, angles, and section thickness.** Sharp corners, angles, and fillets should be avoided, because they may cause cracking and tearing during solidification of the metal. Fillet radii should be selected to reduce stress concentrations and to ensure proper liquid-metal flow during the pouring process. If the fillet radii are too large, the volume of the material in those regions will also be large, and, consequently, the rate of cooling will be lower. Section changes in

castings should smoothly blend into each other. The location of the largest circle that can be inscribed in a particular region is critical as far as shrinkage cavities are concerned (Fig. 5.39). Because the cooling rate in regions with larger circles is lower than in regions with smaller circles, regions with the larger circles are called *hot spots*. These regions could develop *shrinkage cavities* and *porosity*. Although they increase the cost of production, metal paddings or chills in the mold can eliminate or minimize hot spots.

3. **Flat areas.** Large flat areas (plain surfaces) should be avoided, as they may warp, because of temperature gradients during cooling, or develop poor surface finish, because of uneven flow of metal during pouring. Flat surfaces can be broken up with ribs and serrations.
4. **Parting lines.** In general, the parting line should be along a flat plane, rather than a contoured plane. Whenever possible, the parting line should be at the corners or edges of castings, rather than on flat surfaces in the middle of the casting. In this way, the **flash** at the parting line (i.e., the material running out between the two halves of the mold; see also Fig. 6.14) will not be as visible. The location of the parting line is important, because it influences mold design, ease of molding, number and shape of cores, method of support, and the gating system. Preparation of dry-sand cores requires additional time and cost, so these cores should be avoided or minimized; this consideration can usually be made by reviewing and simplifying the design of castings. Figure 5.40 shows two examples of casting design modifications.
5. **Draft.** A small draft (taper) is provided in sand-mold patterns to enable removal of the pattern without damaging the mold. Depending on the quality of the pattern, draft angles usually range from 0.5° to 2°. The angles on the inside surfaces of molds are typically twice this range; they have to be higher than those for outer surfaces, because the casting shrinks inward toward the core.
6. **Dimensional tolerances.** Dimensional tolerances depend on the particular casting process, size of the casting, and type of pattern used, with typical values given in Table 5.2. Tolerances are smallest within one part of the mold and, because they are cumulative, increase between different parts of the mold. Tolerances should be as wide as possible, within the limits of good part performance; otherwise, the cost of the casting increases. In commercial practice, tolerances usually are about ±0.8 mm (1/32 in.) for small castings and increase with size of castings, say, to 6 mm (1/4 in.) for large castings.
7. **Shrinkage.** Allowance for shrinkage during solidification should be provided for, so as to avoid cracking of the casting, particularly in permanent molds. Pattern dimensions should also provide for shrinkage of the metal during solidification and cooling. Allowances for shrinkage, also known as **patternmaker's shrinkage allowances**, usually range from about 10 to 20 mm/m (1/8 to 1/4 in./ft). The normal shrinkage allowance for metals commonly sand cast ranges approximately between 1% and 2%.
8. **Allowances.** A *shrinkage allowance* should be designed into castings so that the castings achieve the desired shape upon removal from the mold. Because most expendable-mold castings require some additional finishing operations, such as machining, allowances should be made in casting design for these operations. **Machining allowances** are included in pattern dimensions and depend on the type of casting. They increase with the size and section thickness of castings, usually ranging from about 2 to 5 mm (0.1 to 0.2 in.) for small castings to more than 25 mm (1 in.) for large castings. Since blowholes and other defects will occur in larger concentrations near the surface, larger machining allowances are needed for applications where surface finish is critical.

5.12.2 Design principles for permanent-mold casting

The design principles for permanent-mold casting are similar to those for expendable-mold casting. Typical design guidelines and examples for permanent-mold casting are shown schematically in Fig. 5.41 for die casting. Note that the cross-sections have been reduced in order to decrease solidification time and save material. Special considerations are involved in designing and tooling for die casting. Designs may be modified to eliminate the draft for better dimensional accuracy.

5.12.3 Computer modeling of casting processes

Because casting processes involve complex interactions among material and process variables, a quantitative study of these interactions is essential to the proper design of castings and the production of high-quality castings. In the past, such studies have presented major difficulties. However, rapid advances in computers and modeling techniques have led to important innovations in modeling various aspects of casting, including fluid flow, heat transfer, and microstructures developed during solidification under various casting-process conditions.

Modeling of *fluid flow* is based on Bernoulli's equation and the continuity equations (see Section 5.4). It predicts the behavior of the metal during pouring into the gating system and its travel into the mold cavity, as well as velocity and pressure distributions in the system. Progress has also been made in modeling of *heat transfer* in casting. Studies of this issue include investigation of the coupling of fluid flow and heat transfer and the effects of surface conditions, thermal properties of the materials involved, and natural and forced convection on cooling. Note that the surface conditions vary during solidification, as a layer of air develops between the casting and the mold wall as a result of shrinkage. Similar studies are being conducted on modeling the development of *microstructures* in casting. These studies encompass heat flow, temperature gradients, nucleation and growth of crystals, formation of dendritic and equiaxed structures, impingement of grains on each other, and movement of the liquid–solid interface during solidification.

Such models are now capable of predicting, for example, the width of the mushy zone (see Fig. 5.10) during solidification and grain size in castings. Similarly, the capability to calculate isotherms gives insight into possible hot spots and subsequent development of shrinkage cavities. With the availability of user-friendly computers and advances in computer-aided design and manufacturing (see Chapter 15), modeling techniques are becoming easier to implement. The benefits are increased productivity, improved quality, easier planning and cost estimation, and quicker response to design changes. Several commercial software programs are now available on modeling and casting processes, such as Magmasoft, ProCast, Solidia, and AFSsolid.

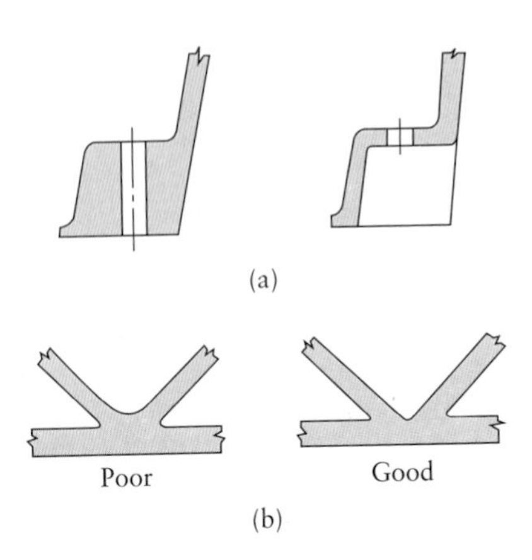

FIGURE 5.41 Examples of undesirable and desirable design practices for die-cast parts. Note that section-thickness uniformity is maintained throughout the part. *Source*: Courtesy of The North American Die Casting Association.

5.13 | Economics of Casting

When reviewing various casting processes, we noted that some require more labor than others, some require expensive dies and machinery, and some take a great deal of time to complete. Outlined in Table 5.2, each of these important factors affects to a greater or lesser degree the overall cost of a casting operation. As described in greater detail in Chapter 16, the total cost of a product involves the costs of materials, labor, tooling, and equipment. Preparations for casting a product include making molds and dies, which requires raw materials, time, and effort, all of which contribute to costs. Relatively little cost is involved in making molds for sand casting. At the other extreme, die-casting dies require expensive materials and a great deal of machining and preparation. In addition to molds and dies, facilities are required for melting and pouring the molten metal into the molds or dies. These facilities include furnaces and related machinery, their costs depending on the level of automation desired. Finally, costs are involved in cleaning and inspecting castings.

The amount of labor required for these operations can vary considerably, depending on the particular process and level of automation. Investment casting, for example, requires a great deal of labor, because of the large number of steps involved in the operation. Conversely, operations such as highly automated die casting can maintain high production rates with little labor required.

The cost of equipment per casting (*unit cost*), however, decreases as the number of parts cast increases (Fig. 5.42); thus, sustained high production rates can justify the high cost of dies and machinery. However, if demand is relatively small, the cost per casting increases rapidly. It then becomes more economical to manufacture the parts by sand casting or by other manufacturing processes. Note that the graphical comparison in Fig. 5.42 can include other casting processes suitable for making the same part. The sand and die-casting processes compared in the figure produce castings with significantly different dimensional and surface-finish characteristics; thus, not all manufacturing decisions are based purely on economic considerations. In fact, parts can usually be made by more than one or two processes. The final decision depends on both economic and technical considerations. The competitive aspects of manufacturing processes are discussed further in Chapter 16.

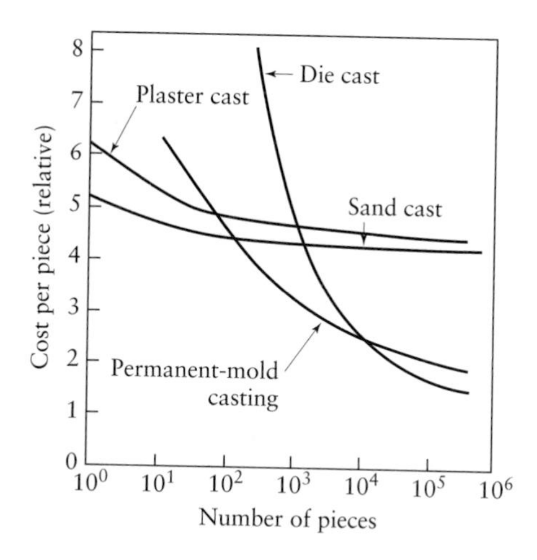

FIGURE 5.42 Economic comparison of making a part by different casting processes. Note that because of the high cost of equipment, die casting is economical for large production runs. *Source:* The North American Die Casting Association.

CASE STUDY | Permanent-Mold Casting of Aluminum Automotive Pistons

Figure 5.43 shows an aluminum piston used in automotive internal-combustion engines. These products must be manufactured at very high rates with very tight dimensional tolerances and strict material requirements in order to achieve proper operation. Economic concerns are obviously paramount, and it is essential that pistons be produced with a minimum of expensive finishing operations and with few rejected parts.

Aluminum pistons are manufactured by the permanent-mold casting process, because of its capability to produce near-net shaped parts at the required production rates. However, with poorly designed molds, underfills or excess porosity can cause parts to be rejected, adding to the cost of the process. These defects were traditionally controlled through the use of large machining allowances coupled with intuitive design of molds, based on experience.

Permanent-mold casting is a high-volume production, where low cycle times are achieved by using metallic molds with integral cooling water channels. Depending on the particular alloy cast and the size of the casting and mold, cycle times can vary from less than a minute to eight minutes. The molds are machined from H13 hot-work tool steel. (See Section 3.10.3.) Depending on the piston geometry, a permanent or expendable core may be required; expendable cores are constructed from sand or plaster, in which case the process is properly referred to as semipermanent-mold casting.

The pistons are produced from high-silicon alloys such as 413.0 aluminum alloy. This alloy has high fluidity and can create high-definition surfaces through permanent-mold casting; it also has high resistance to corrosion, good weldability, and low specific gravity. The universal acceptance of aluminum pistons for internal-combustion- engine applications is due mainly to the pistons' light weight and high thermal conductivity. Their low inertia allows for higher engine speeds and reduced counterweighting in the crankshaft, and their higher thermal conductivity allows more efficient heat transfer from the engine.

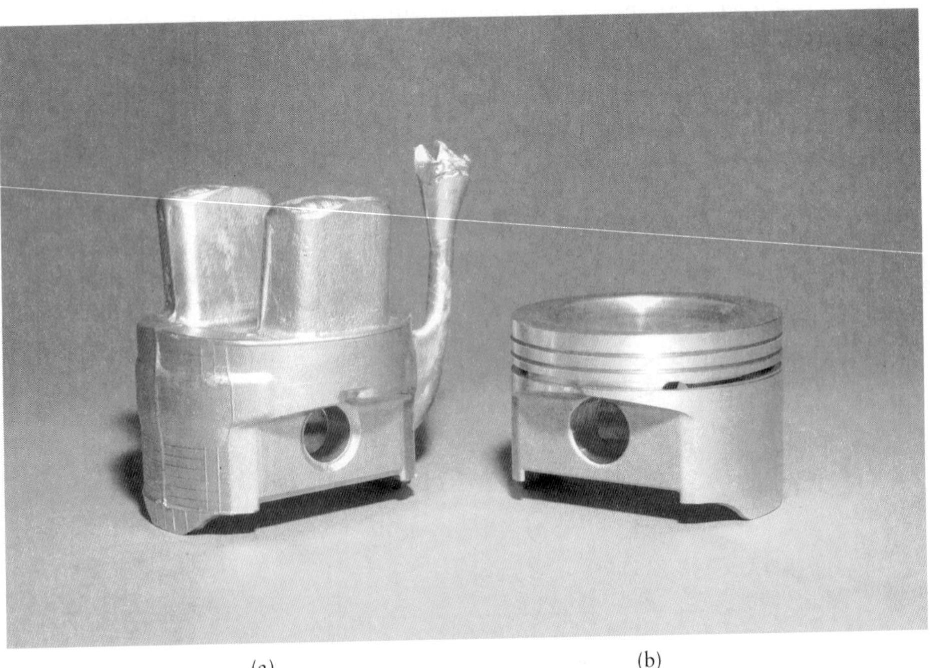

FIGURE 5.43 Aluminum piston for an internal combustion engine. (a) As cast; (b) after machining.

The mold is preheated to 200–450°C (400–850°F), depending on the cast alloy and part size. Initially, the preheat is achieved with a handheld torch, but after a few castings, the mold reaches a steady-state temperature profile. The molten aluminum is heated to 100–200°C (200–400°F) above its liquidus temperature, and then a shot is placed into the infeed section of the mold. Once the molten-metal shot is in place, a piston drives the metal into the mold. Because of the high thermal conductivity of the mold material, heat extraction from the molten metal is rapid, and the metal can solidify in small channels before filling the mold completely. Solidification usually starts at one end of the casting before the mold is fully filled.

As with most alloys, it is desired to begin solidification at one extreme end of the casting and have the solidification front proceed across the volume, resulting in a directionally solidified microstructure and elimination of gross porosity that arises when two solidification fronts meet inside a casting. Regardless, casting defects such as undercuts, hot spots, porosity, and cracking and entrapped air-zone defects such as blowholes and scabs can occur. With poor mold design, defects can make up 5% of the castings.

In order to improve the reliability and reduce the costs associated with permanent-mold casting, a mathematical modeling capability for the permanent-mold casting process was developed by researchers at the University of Notre Dame working with engineers at A. E. Goetze Corporation. The goal of the simulations is to identify conditions that lead to casting defects and allow evaluation of design alterations before the cost of mold production is incurred. Since molds are extremely expensive, numerical simulation is an essential step in evaluating the designs of new molds.

A typical problem is illustrated in Fig. 5.44, which shows a mold with cooling channels and the desired piston geometry. Initial modeling work was coupled with experiments that implemented thermocouples placed in the cavity and mold. The simulation of permanent-mold casting requires a coupling of the fluid mechanics equations of mold filling with the heat transfer and materials science equations of alloy solidification and is a very computationally intensive process. The goal of the model is to identify combinations of mold design, cooling-channel location and size, and cast-alloy superheat that reduce the numbers of defects.

Figure 5.45 illustrates some temperature profiles and solidification fronts that develop during casting in the part, with the mold removed for clarity. Note that the solidification starts during mold filling, and the developing mushy zone is very large.

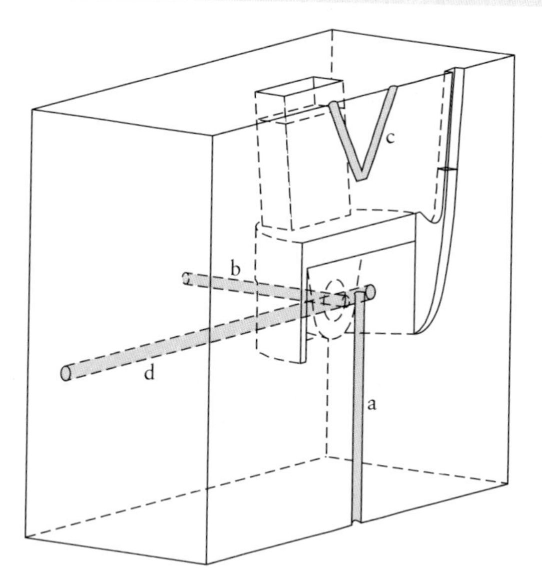

FIGURE 5.44 Schematic illustration of the permanent mold used to produce aluminum pistons, showing the position of four cooling channels.

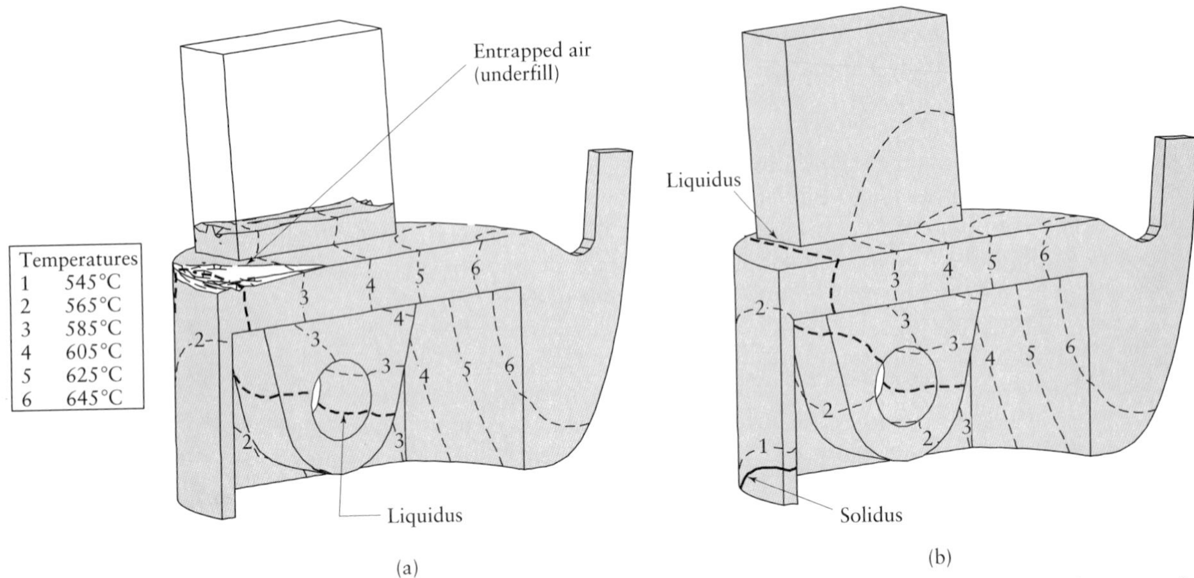

FIGURE 5.45 Simulation of mold filling and solidification. (a) 3.7 seconds after start of pour. Note that the mushy zone has been established before the mold is completely filled. (b) Using a vent in the mold for removal of entrapped air five seconds after pour.

Figure 5.45a shows a piston after 3.7 s from initial injection of metal into the mold, where a region of entrapped air has been identified that would normally lead to a systemic underfill in these castings. Figure 5.45b shows the piston with a venting system incorporated to correct the entrapped-air problem; it also shows the contours of constant temperature, or *isotherms*, that exist five seconds into the casting process. The temperature contours and their evolution can be used to predict locations of macrodefects, the size and shape of grains in the cast part, and the microporosity as a function of location in the part. Therefore, alternative designs can be quickly evaluated for their ability to address any of these issues. Numerical simulation has proven to be an invaluable tool in the evaluation of casting molds and is currently in widespread use.

Source: S. Shepel and S. Paolucci, University of Notre Dame.

SUMMARY

- Metal casting is among the oldest and most common manufacturing processes. Solidification of pure metals takes place at a clearly defined, constant temperature, whereas solidification of alloys occurs over a range of temperatures, depending on composition. Phase diagrams are useful tools for identifying the solidification point or points for metals and alloys. (Sections 5.1 and 5.2)
- The composition and cooling rate of the melt affect the size and shape of grains and dendrites in the solidifying metal and alloy and thus influence the properties of the solidified casting. (Section 5.3)
- Most metals shrink during solidification, although gray cast iron and a few others actually expand; either way, loss of dimensional control and, sometimes, cracking are difficulties that can arise during solidification and cooling. Seven basic categories of casting defects have been identified. (Section 5.3)
- In casting, the molten metal or alloy may flow through a variety of passages, including pouring basins, sprues, runners, risers, and gating systems, before reaching the final mold cavity. Bernoulli's theorem, the continuity law, and the Reynolds number are the analytical tools used in designing an appropriate system and

eliminating defects associated with fluid flow. Heat transfer affects fluid flow and solidification time in casting; solidification time is a function of the volume of a casting and its surface area (Chvorinov's rule). (Section 5.4)

- Melting practices have a direct effect on the quality of castings. Factors that affect melting include the following: (1) fluxes, which are inorganic compounds added to the molten metal to remove dissolved gases and various impurities; (2) the type of furnace used; and (3) foundry operations, that is, pattern and mold making, pouring of the melt, removal of cast items from molds, cleaning, heat treatment, and inspection. (Section 5.5)
- Several ferrous and nonferrous casting alloys are available, with a wide range of properties, casting characteristics, and applications. Because castings are often designed and produced to be assembled with other mechanical components and structures, various other considerations, such as weldability, machinability, and surface conditions, are also important. (Section 5.6)
- The traditional ingot-casting process has been replaced, for many applications, by continuous-casting methods for both ferrous and nonferrous metals. (Section 5.7)
- Casting processes are generally classified as either expendable-mold or permanent-mold casting. The most common expendable-mold methods are sand, shell-mold, plaster, ceramic-mold, and investment casting. Common permanent-mold methods include slush casting, pressure casting, and die casting. Compared with permanent-mold casting, expendable-mold casting usually involves lower mold and equipment costs, but produces parts with lower dimensional accuracy. (Sections 5.8–5.10)
- Castings, as well as wrought products made by other processes, may be subjected to further heat-treatment processes to enhance their properties and service life. As temperature is reduced at different rates, various transformations take place in microstructures that have widely varying characteristics and properties. Important mechanisms of hardening and strengthening involve thermal treatments, including quenching and precipitation hardening. (Section 5.11)
- Several general principles have been established to aid designers in producing castings that are free from defects and meet dimensional tolerance and service requirements; these principles typically concern the shape of the casting and various techniques to minimize hot spots in order to avoid shrinkage cavities. Because of the large number of variables involved, close control of all parameters is essential, particularly those related to the nature of liquid-metal flow into the molds and dies and the rate of cooling in different regions. (Section 5.12)
- Within the limits of good performance, the economic aspects of casting are just as important as the technical considerations. Factors affecting the overall cost of castings include the cost of materials, molds, dies, equipment, and labor, each of which varies with the particular casting process. An important parameter is the cost per casting, which, for large production runs, can justify major expenditures for automated machinery. (Section 5.13)

SUMMARY OF EQUATIONS

Bernoulli's theorem: $h + \dfrac{p}{\rho g} + \dfrac{v^2}{2g} = \text{constant}$

Sprue contour: $\dfrac{A_1}{A_2} = \sqrt{\dfrac{h_2}{h_1}}$

Reynolds number: $\text{Re} = \dfrac{vD\rho}{\eta}$

Chvorinov's rule: Solidification time $= C\left(\dfrac{\text{volume}}{\text{surface area}}\right)^2$

Continuity equation: $Q = A_1v_1 = A_2v_2$

Metal velocity after gate: $v = c\sqrt{2gh}$

BIBLIOGRAPHY

Abrasion-Resistant Cast Iron Handbook, American Foundry Society, 2000.

Alexiades, V., *Mathematical Modeling of Melting and Freezing Processes*, Hemisphere, 1993.

Allsop, D. F., and D. Kennedy, *Pressure Die Casting—Part II: The Technology of the Casting and the Die*, Pergamon, 1983.

An Introduction to Die Casting, American Die Casting Institute, 1981.

Analysis of Casting Defects, American Foundrymen's Society, 1974.

ASM Handbook, Vol. 3: *Alloy Phase Diagrams*, ASM International, 1992.

ASM Handbook, Vol. 4: *Heat Treating*, ASM International, 1991.

ASM Handbook, Vol. 15: *Casting*, ASM International, 1988.

ASM Specialty Handbook: Cast Irons, ASM International, 1996.

Bradley, E. F., *High-Performance Castings: A Technical Guide*, Edison Welding Institute, 1989.

Campbell, J., *Castings*, Butterworth-Heinemann, 1991.

Case Hardening of Steel, ASM International, 1987.

"Casting," in *Tool and Manufacturing Engineers Handbook*, Volume II: *Forming*, Society of Manufacturing Engineers, 1984.

Casting Defects Handbook, American Foundrymen's Society, 1972.

Clegg, A. J., *Precision Casting Processes*, Pergamon, 1991.

Davis, J. R. (ed.), *Cast Irons.*, ASM International, 1996.

Elliott, R., *Cast Iron Technology*, Butterworths, 1988.

Flemings, M. C., *Solidification Processing*, McGraw-Hill, 1974.

Heine, R. W., C. R. Loper, Jr., and C. Rosenthal, *Principles of Metal Casting*, 2d. ed., McGraw-Hill, 1967.

Investment Casting Handbook, Investment Casting Institute, 1997.

Johns, R., *Casting Design*, American Foundrymen's Society, 1987.

Karlsson, L. (ed.), *Modeling in Welding, Hot Powder Forming and Casting*, ASM International, 1997.

Kaye, A., and A. C. Street, *Die Casting Metallurgy*, Butterworth, 1982.

Krauss, G., *Steels: Heat Treatment and Processing Principles*, ASM International, 1990.

Kurz, W., and D. J. Fisher, *Fundamentals of Solidification*, Trans Tech Pub., 1994.

Liebermann, H. H. (ed.), *Rapidly Solidified Alloys*, Dekker, 1993.

Minkoff, I., *Solidification and Cast Structure*, Wiley, 1986.

Powell, G. W., S.-H. Cheng, and C. E. Mobley, Jr., *A Fractography Atlas of Casting Alloys*, Battelle Press, 1992.

Product Design for Die Casting, Diecasting Development Council, 1988.

Rowley, M. T. (ed.), *International Atlas of Casting Defects*, American Foundrymen's Society, 1974.

Society of Die Casting Engineers, *The Metallurgy of Die Castings*, 1986.

Street, A. C., *The Diecasting Book*, 2d ed., Portcullis Press, 1986.

Szekely, J., *Fluid Flow Phenomena in Metals Processing*, Academic Press, 1979.

Totten, G. E., and M. A. H. Howes, *Steel Heat Treatment*, Dekker, 1997.

Upton, B., *Pressure Die Casting-Part 1: Metals, Machines, Furnaces*, Pergamon, 1982.

Walton, C. F., and T. J. Opar (eds.), *Iron Castings Handbook*, 3d. ed., Iron Castings Society, 1981.

Wieser, P. P. (ed.), *Steel Castings Handbook*, 6th ed., ASM International, 1995.

Young, K. P., *Semi-solid Processing*, Chapman & Hall, 1997.

Yu, K.–O. (ed.), *Modeling for Casting and Solidification Processing*, Dekker, 2001.

QUESTIONS

5.1 Describe the characteristics of (1) an alloy, (2) pearlite, (3) austenite, (4) martensite, and (5) cementite.

5.2 What are the effects of mold materials on fluid flow and heat transfer?

5.3 How does the shape of graphite in cast iron affect its properties?

5.4 Explain the difference between short and long freezing ranges. How are they determined? Why are they important?

5.5 We know that pouring molten metal at a high rate into a mold has certain disadvantages. Are there any disadvantages to pouring it very slowly? Explain.

5.6 Why does porosity have detrimental effects on the mechanical properties of castings? Which physical properties are also affected adversely by porosity?

5.7 A spoked handwheel is to be cast in gray iron. In order to prevent hot tearing of the spokes, would you insulate the spokes or chill them? Explain.

5.8 Which of the following considerations is/are important for a riser to function properly? (1) Having a surface area larger than that of the part being cast; (2) being kept open to atmospheric pressure; (3) solidifying sooner. Why?

5.9 Explain why the constant C in Eq. (5.11) depends on mold material, metal properties, and temperature.

5.10 Explain why gray iron undergoes expansion, rather than contraction, during solidification.

5.11 How can you tell whether cavities in a casting are due to porosity or to shrinkage?

5.12 Explain the reasons for hot tearing in castings.

5.13 Would you be concerned about the fact that parts of internal chills are left within the casting? What materials do you think chills should be made of, and why?

5.14 Are external chills as effective as internal chills? Explain.

5.15 Do you think early formation of dendrites in a mold can impede the free flow of molten metal into the mold? Explain.

5.16 Is there any difference in the tendency for shrinkage void formation for metals with short freezing ranges and long freezing ranges, respectively? Explain.

5.17 It has long been observed by foundry-men that low pouring temperatures, i.e., low superheat, promote equiaxed grains over columnar grains. Also, equiaxed grains become finer as the pouring temperature decreases. Explain these phenomena.

5.18 What are the reasons for the large variety of casting processes that has been developed over the years?

5.19 Why can blind risers be smaller than open-top risers?

5.20 Would you recommend preheating molds in permanent-mold casting? Also, would you remove the casting soon after it has solidified? Explain.

5.21 In a sand-casting operation, what factors determine the time at which you would remove the casting from the mold?

5.22 Explain why the strength-to-weight ratio of die-cast parts increases with decreasing wall thickness.

5.23 We note that the ductility of some cast alloys is nil. (See Fig. 5.19.) Do you think that this phenomenon should be a significant concern in engineering applications of castings? Explain.

5.24 The modulus of elasticity, E, of gray iron varies significantly with its type, such as the ASTM class. Explain why.

5.25 List and explain the considerations for selecting pattern materials.

5.26 Why is the investment-casting process capable of producing fine surface detail on castings?

5.27 Explain why a casting may have a slightly different shape than the pattern used to make the mold.

5.28 Explain why squeeze casting produces parts with better mechanical properties, dimensional accuracy, and surface finish than that of parts produced by expendable-mold processes.

5.29 Why are steels more difficult to cast than cast irons?

5.30 What would you do to improve the surface finish in expendable-mold casting processes?

5.31 You have seen that even though die casting produces thin parts, there is a minimum thickness that can be produced. Why can't even thinner parts be made by this process?

5.32 What differences, if any, would you expect to find in the properties of castings made by permanent-mold versus sand-casting methods?

5.33 Which of the casting processes would be suitable for making small toys in large numbers? Why?

5.34 Why are allowances provided for in making patterns? What do the allowances depend on?

5.35 Explain the difference in the importance of drafts in green-sand casting versus permanent-mold casting.

5.36 Make a list of the mold and die materials used in the casting processes described in this chapter. Under each type of material, list the casting processes that are used, and explain why these processes are suitable for that particular mold or die material.

5.37 Explain why carbon is so effective in imparting strength to iron in the form of steel.

5.38 Describe the engineering significance of the existence of a eutectic point in phase diagrams.

5.39 Explain the difference between hardness and hardenability.

5.40 Explain why it may be desirable for castings to be subjected to various heat treatments.

5.41 Describe the differences between case hardening and through hardening insofar as engineering applications are concerned.

5.42 *Type metal* is a bismuth alloy used to cast type for printing. Explain why bismuth is ideal for this process.

5.43 Do you expect to see larger solidification shrinkage for a material with a bcc crystal structure or a material with an fcc crystal structure? Explain.

5.44 Describe the drawbacks to having a riser that is (a) too large and (b) too small.

5.45 If you were to incorporate lettering on a sand casting, would you make the letters protrude from the surface or recess into the surface? What if the part were to be made by investment casting?

5.46 List and briefly explain the three mechanisms by which metals shrink during casting.

5.47 Explain the significance of the "tree" in investment casting.

5.48 Sketch the microstructure you would expect for a slab cast through (a) continuous casting, (b) strip casting, and (c) melt spinning.

5.49 The general design recommendations for a well in sand casting are that (a) its diameter should be twice the exit diameter of the sprue, and (b) the depth should be approximately twice the depth of the runner. Explain the consequences of deviating from these rules.

PROBLEMS

5.50 Using Fig. 5.3, estimate the following quantities for a 20% Cu 80% Ni alloy: (1) liquidus temperature, (2) solidus temperature, (3) percentage of nickel in the liquid at 1400°C (2550°F), (4) the major phase at 1400°C, and (5) the ratio of solid to liquid at 1400°C.

5.51 Determine the amount of gamma and alpha phases (see Fig. 5.6) in a 5-kg, 1060 steel casting as it is being cooled to the following temperatures: (1) 750°C, (2) 728°C, and (3) 726°C.

5.52 A round casting is 0.2 m in diameter and 0.5 m in length. Another casting of the same metal is elliptical in cross-section, with a major-to-minor-axis ratio of 3, and has the same length and cross-sectional area as the round casting. Both pieces are cast under the same conditions. What is the difference in the solidification times of the two castings?

5.53 Derive Eq. (5.7).

5.54 Two halves of a mold (cope and drag) are weighted down to keep them from separating under the pressure exerted by the molten metal (buoyancy). Consider a solid, spherical steel casting with a diameter of 7 in. that is being produced by sand casting. Each flask (see Fig. 5.14) is 25 in. by 25 in. and 15 in. deep. The parting line is at the middle of the part. Estimate the clamping force required. Assume the metal's density is 500 lb/ft^3 and the sand density is 100 lb/ft^3.

5.55 Would changing the position of the parting line for the casting in Problem 5.54 change your answer to the problem? Explain.

5.56 Plot the clamping force in Problem 5.54 as a function of increasing diameter of the casting, from 10 in. to 20 in.

5.57 Sketch a graph of specific volume versus temperature for a metal that shrinks as it cools from the liquid state to room temperature. On the graph, mark the area where shrinkage is compensated for by risers.

5.58 A round casting has the same dimensions as that in Problem 5.52. Another casting of the same metal is rectangular in cross section, with a width-to-thickness ratio of 2, and has the same length and cross-sectional area as the round casting. Both pieces are cast under the same conditions. What is the difference in the solidification times of the two castings?

5.59 A 50-mm-thick square plate and a right circular cylinder with a radius of 100 mm and height of 50 mm each have the same volume. If each is to be cast using a cylindrical riser, will the two parts require the same-size riser to ensure proper feeding? Explain.

5.60 Assume that the top of a round sprue has a diameter of 4 in. and is at a height of 8 in. from the runner and 3 in. below the metal level in the pouring basin. Based on Eq. (5.7), plot the profile of the sprue's diameter as a function of the height of the sprue.

5.61 Estimate the clamping force for a die-casting machine in which the casting is rectangular, with projected dimensions of 100 mm × 150 mm. Would your answer depend on whether or not the process is a hot-chamber or cold-chamber process? Explain.

5.62 When designing patterns for casting, pattern makers use special rulers that automatically incorporate solid-shrinkage allowances into their designs. Therefore, a 12-in. pattern maker's ruler is longer than a foot. How long is a pattern maker's ruler designed for the making of patterns for (1) aluminum castings (2), phosphor bronze casting, and (3) high-manganese steel castings?

5.63 The blank for the spool shown in the figure accompanying this problem is to be sand cast out of A-319, an aluminum casting alloy. Make a sketch of the wooden pattern for this part. Include all necessary allowances for shrinkage and machining.

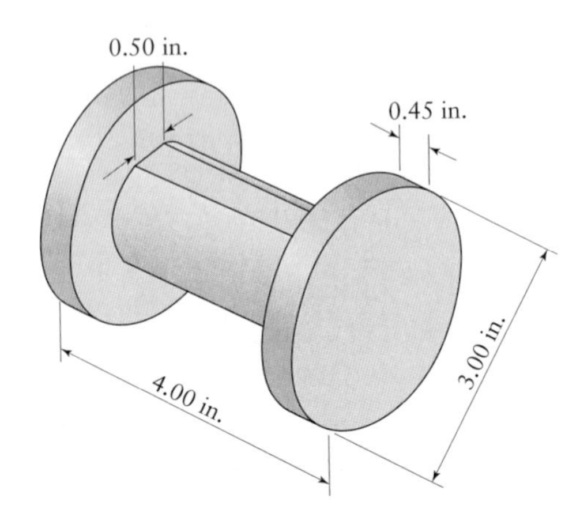

5.64 Repeat Problem 5.63, but assume that the aluminum spool is to be cast using expendable-pattern casting. Explain the important differences between the two patterns.

5.65 In sand casting, it is important that the cope mold half be held down with enough force to keep it from floating when the molten metal is poured in. For the casting shown in the accompanying figure, calculate the minimum amount of weight necessary to keep the cope from floating up as the molten metal is poured in. (*Hint*: The buoyancy force exerted by the molten metal on the cope is related to the effective height of the metal head above the cope.)

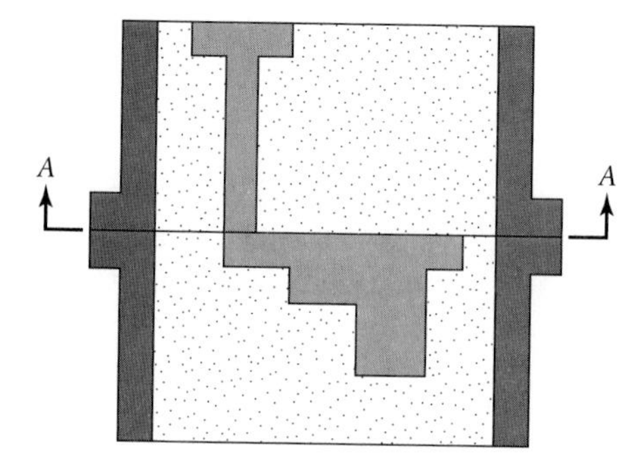

Section *A–A*

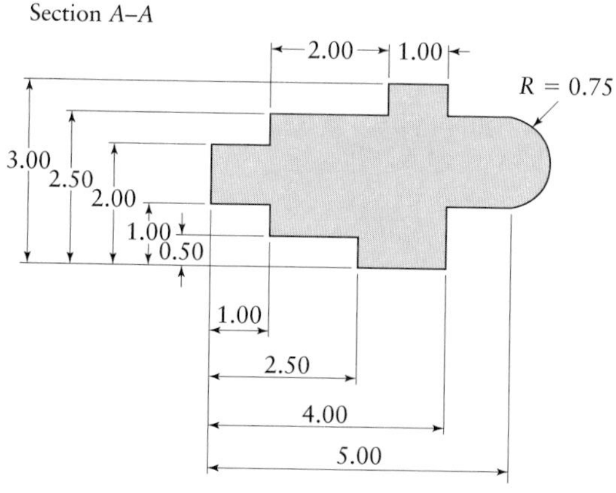

Material: Low-carbon steel
Density: 0.26 lb/in^3
All dimensions in inches

5.66 The optimum shape of a riser is spherical to ensure that the riser cools more slowly than the casting it feeds. Spherically shaped risers, however, are difficult to cast. (1) Sketch the shape of a blind riser that is easy to mold, but also has the smallest possible surface-area-to-volume ratio. (2) Compare the solidification time of the riser in part (a) to that of a riser shaped like a right circular cylinder. Assume that the volume of each riser is the same and that for each riser, the height is equal to the diameter. (See Example 5.2.)

5.67 Assume that you are asked to give a quiz to students on the contents of this chapter. Prepare three quantitative problems and three qualitative questions, and supply the answers.

5.68 The part shown in the accompanying figure is a semispherical shell used as an acetabular (mushroom shaped) cup in a total hip replacement. Select a casting process for this part, and provide a sketch of all patterns or tooling needed if the part is to be produced from cobalt–chrome alloy.

Dimensions in mm

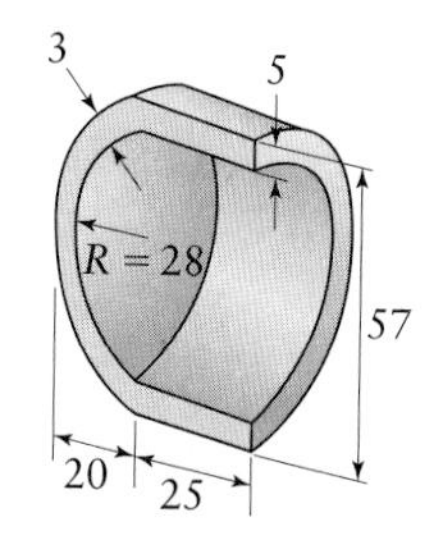

5.69 A cylinder with a height-to-diameter ratio of unity solidifies in three minutes in a sand-casting operation. What is the solidification time if the height of the cylinder is doubled? What is the solidification time if the diameter is doubled?

5.70 Steel piping is to be produced by centrifugal casting. The length of the pipe is to be 10 ft, the diameter 3 ft, and the thickness 0.5 in. Using basic equations from dynamics and statics, determine the rotational speed needed to have the centripetal force be 70 times the weight of the pipe.

5.71 A sprue is 10 in. long and has a diameter of 5 in. at the top, where the metal is poured. The molten metal level in the pouring basin is taken as 3 in. from the top of the sprue for design purposes. If a flow rate of 40 in^3/s is to be achieved, what should be the diameter of the bottom of the sprue? Will the sprue aspirate? Explain.

5.72 Small amounts of slag often persist after skimming and are introduced into the molten-metal flow in casting. Recognizing that the slag is much less dense than the metal, design mold features that will remove small amounts of slag before the metal reaches the mold cavity.

5.73 Pure lead is poured into a sand mold. The metal level in the pouring basin is 8 in. above the metal level in the mold, and the runner is circular, with a 0.5-in. diameter. What is the velocity and rate of the flow of the metal into the mold? Is the flow turbulent or laminar?

5.74 For the sprue described in Problem 5.73, what runner diameter is needed to ensure a Reynolds number of 2000? How long will a 25-in^3 casting take to fill with such a runner?

5.75 How long would it take for the sprue in Problem 5.73 to feed a casting with a square cross-section of 5 in. per side and a height of 4 in.? Assume that the sprue is frictionless.

DESIGN

5.76 Design test methods to determine the fluidity of metals in casting. (See Section 5.4.2.) Make appropriate sketches, and explain the important features of each design.

5.77 The accompanying figures indicate various defects and discontinuities in cast products. Review each one, and offer design solutions to avoid them.

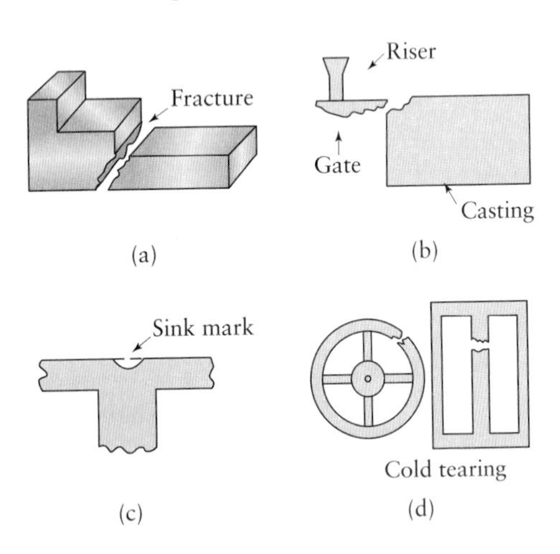

5.78 Using the equipment and materials available in a typical kitchen, design an experiment to reproduce results similar to those shown in Fig. 5.16.

5.79 Design a test method to measure the permeability of sand for sand casting.

5.80 Describe the procedures that would be involved in making a bronze statue. Which casting process(es) would be suitable? Why?

5.81 Porosity developed in the boss of a casting is illustrated in the accompanying figure. Show that by simply repositioning the parting line of this casting, this problem can be eliminated.

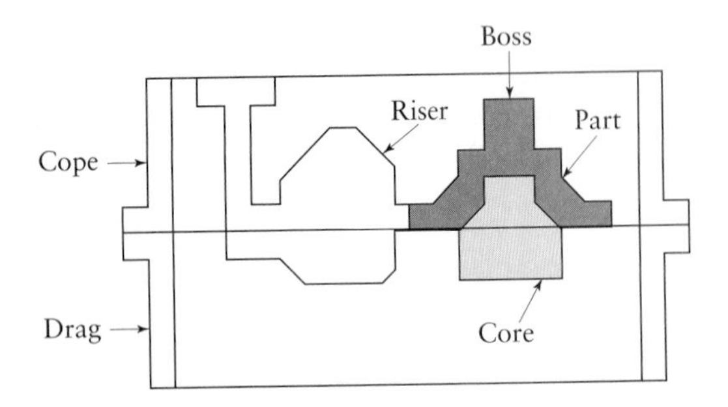

5.82 For the wheel illustrated in the accompanying figure, show how (a) riser placement, (b) core placement, (c) padding, and (d) chills may be used to help feed molten metal and eliminate porosity in the isolated hob boss.

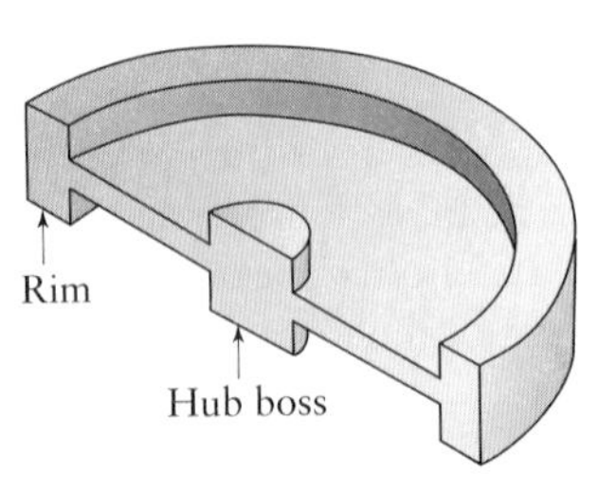

5.83 In the accompanying figure, the original casting design shown in (a) was changed to the design shown in (b). The casting is round, with a vertical axis of symmetry. What advantages do you think the new design, as a functional part, has over the old one?

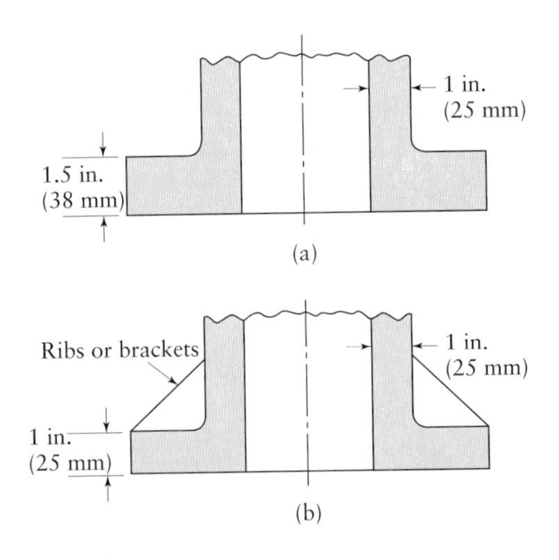

5.84 An incorrect and a correct design for casting are shown, respectively, in the accompanying figure. Review the changes made to the incorrect design, and comment on their advantages.

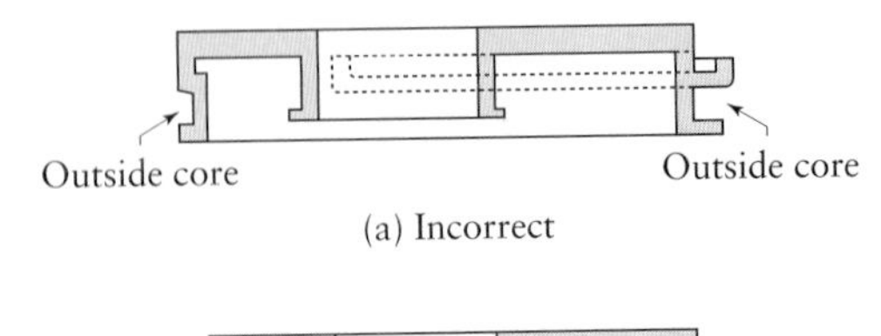

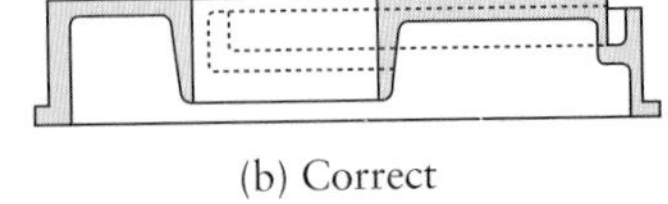

5.85 Three sets of designs for die casting are shown in the accompanying figure. Note the changes made to original die design (1), and comment on the reasons for these changes.

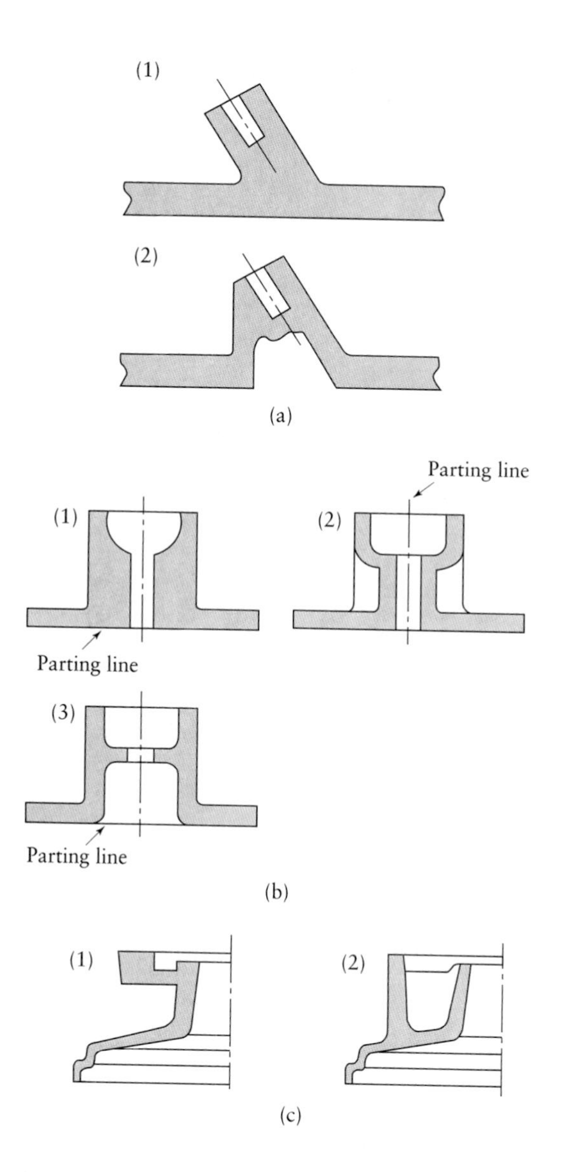

CHAPTER

6 Bulk Deformation Processes

6.1 Introduction

Deformation processes in manufacturing are operations that induce shape changes on the workpiece by plastic deformation under forces applied by various tools and dies. Deformation processes can be grouped by temperature, size, and shape of the workpiece and type of operation. For example, using temperature as a criterion, deformation processes may be divided into three basic categories of cold (room temperature), warm, and hot working. Also, deformation processes may be classified by type of operation as primary working or secondary working.

Primary-working operations are those that take a solid piece of metal (generally from a cast state, such as an ingot) and break it down successively into shapes such as slabs, plates, and billets. Traditionally, the primary-working processes are forging, rolling, and extrusion. **Secondary working** involves further processing of the products from primary working into final or semifinal products such as bolts, sheet-metal parts, and wire. These classifications are not rigid, because some operations belong to both primary and secondary working. For instance, some forgings result in a product as final as those that result from secondary working; likewise, in extrusion, some products are ready for use without any further processing.

A more recent trend has been to classify deformation processes according to the size and shape of the workpiece. **Bulk deformation** is the processing of workpieces that have a relatively small surface-area-to-volume (or surface-area-to-thickness) ratio, hence the term *bulk*. In all bulk-deformation processing, the thickness or cross-section of the workpiece changes. In **sheet forming**, the surface-area-to-thickness ratio is relatively large. In general, the material is subjected to shape changes by various dies. Changes in thickness are usually undesirable and, in fact, can lead to failure.

In this chapter, we describe the basic bulk-deformation processes for metal: forging, rolling, extrusion, and drawing of rod and wire (Table 6.1). Bulk-deformation processing of plastics and nonmetallic materials is covered in Chapters 10 and 11, respectively.

6.2 Forging

Forging denotes a family of processes by which plastic deformation of the workpiece is carried out by *compressive forces*. Forging is one of the oldest metalworking operations known, dating back to 5000 BC, and is used in making parts of widely varying sizes and shapes from a variety of metals. Typical parts made by forging today are crankshafts and connecting rods for engines, turbine disks, gears, wheels, bolt heads, hand tools, and many types of structural components for machinery and transportation equipment.

TABLE 6.1

General Characteristics of Bulk Deformation Processes

Process	General Characteristics
Forging	Production of discrete parts with a set of dies; some finishing operations usually necessary; similar parts can be made by casting and powder-metallurgy techniques; usually performed at elevated temperatures; dies and equipment costs are high; moderate to high labor costs; moderate to high operator skill.
Rolling	
Flat	Production of flat plate, sheet, and foil at high speeds and with good surface finish, especially in cold rolling; requires very high capital investment; low to moderate labor cost.
Shape	Production of various structural shapes, such as I-beams and rails, at high speeds; includes thread and ring rolling; requires shaped rolls and expensive equipment; low to moderate labor costs; moderate operator skill.
Extrusion	Production of long lengths of solid or hollow products with constant cross-sections, usually performed at elevated temperatures; product is then cut to desired lengths; can be competitive with roll forming; cold extrusion has similarities to forging and is used to make discrete products; moderate to high die and equipment cost; low to moderate labor cost; low to moderate operator skill.
Drawing	Production of long rod, wire, and tubing, with round or various cross-sections; smaller cross-sections than in extrusion; good surface finish; low to moderate die, equipment, and labor costs; low to moderate operator skill.
Swaging	Radial forging of discrete or long parts with various internal and external shapes; generally carried out at room temperature; low to moderate operator skill.

Forging can be carried out at room temperature (*cold working*), or at elevated temperatures, a process called *warm* or *hot forging*, depending on the temperature. The temperature range for these categories is given in Table 3.1 in terms of the homologous temperature, T/T_m, where T_m is the melting point of the workpiece material on the absolute scale. Note that the homologous recrystallization temperature for metals is about 0.5. (See also Fig. 3.16.) Simple forgings can be made with a heavy hammer and an anvil by techniques used by blacksmiths for centuries. Usually, though, a set of dies and a press are required. The three basic categories of forging are open die, impression die, and closed die.

6.2.1 Open-die forging

In its simplest form, open-die forging generally involves placing a solid cylindrical workpiece between two flat dies (platens) and reducing its height by compressing it (Fig. 6.1a). This operation is also known as **upsetting.** The die surfaces may be shaped, such as a conical or curved cavity, thereby forming the ends of the cylindrical workpiece during upsetting. Under ideal conditions, a solid cylinder deforms as shown in Fig. 6.1a; this is known as **homogeneous deformation.** Because the volume of the cylinder remains constant, any reduction in its height increases its diameter. For a specimen that has been reduced in height from h_0 to h_1,

$$\text{Reduction in height} = \frac{h_0 - h_1}{h_0} \times 100\%. \quad (6.1)$$

From Eqs. (2.1) and (2.9) and using absolute values (as is generally the case with bulk-deformation processes), we have

$$e_1 = \frac{h_0 - h_1}{h_0} \quad (6.2)$$

and

$$\varepsilon_1 = \ln\left(\frac{h_0}{h_1}\right). \quad (6.3)$$

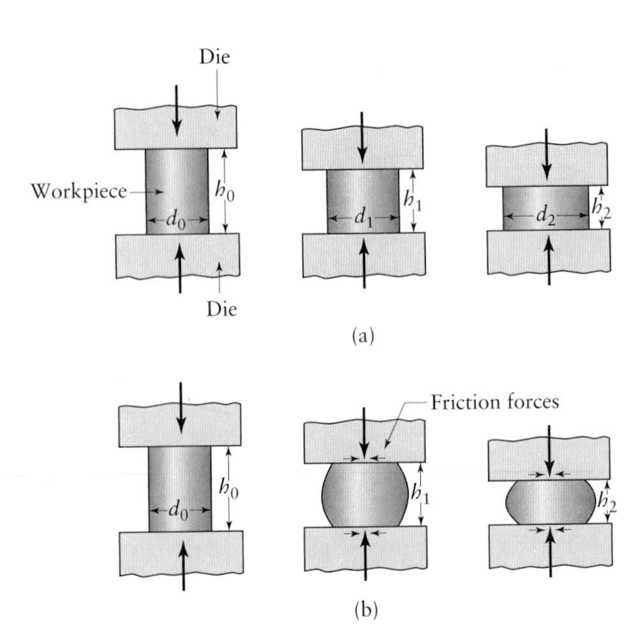

FIGURE 6.1 (a) Ideal deformation of a solid cylindrical specimen compressed between flat frictionless dies. This process is known as *upsetting*. (b) Deformation in upsetting with friction at the die–workpiece interfaces.

With a relative velocity v between the platens, the specimen is subjected to a strain rate, according to Eqs. (2.17) and (2.18), of

$$\dot{e}_1 = -\frac{v}{h_0} \tag{6.4}$$

and

$$\dot{\varepsilon}_1 = -\frac{v}{h_1}. \tag{6.5}$$

If the specimen is reduced in height from h_0 to h_2, the subscript 1 in the foregoing equations is replaced by subscript 2. Note that the true strain rate $\dot{\varepsilon}$ increases rapidly as the height of the specimen approaches zero. Positions 1 and 2 in Fig. 6.1a may be regarded as instantaneous positions during a continuous upsetting operation.

Actually, the specimen develops a **barrel** shape, as shown in Fig. 6.1b. Barreling is caused primarily by *frictional forces* at the die–workpiece interfaces that oppose the outward flow of the material at these interfaces. Barreling also occurs in upsetting hot workpieces between cool dies. The material at and near the interfaces cools rapidly, while the rest of the specimen is relatively hot. Because the strength of the material decreases with increasing temperature, the ends of the specimen show a greater resistance to deformation than does the center; thus, the central portion of the cylinder deforms to a greater extent than do its ends.

In barreling, the material flow within the specimen becomes *nonuniform*, or *inhomogeneous*, as can be seen from the polished and etched section of a barreled cylindrical specimen, as shown in Fig. 6.2. Note the roughly triangular stagnant (*dead*) zones at the top and bottom surfaces, which have a wedge effect on the rest of the material, thus causing barreling. This inhomogeneous deformation can also be observed in grid patterns, as shown in Fig. 6.1b, which are obtained experimentally using various techniques (Section 6.4.1).

In addition to the single barreling shown in Fig. 6.3, **double barreling** can also be observed. It occurs at large ratios of height to cross-sectional area and when friction at the die–workpiece interfaces is high. Under these conditions, the stagnant zone under the platen is sufficiently far from the midsection of the workpiece to allow the

FIGURE 6.2 Grain flow lines in upsetting a solid steel cylinder at elevated temperatures. Note the highly inhomogeneous deformation and barreling. The different shape of the bottom section of the specimen (as compared with the top) results from the hot specimen resting on the lower, cool die before deformation proceeded. The bottom surface was chilled; thus it exhibits greater strength and hence deforms less than the top surface. *Source*: J. A. Schey et al., IIT Research Institute.

midsection to deform uniformly while the top and bottom portions barrel out, hence the term *double barreling*. Barreling caused by friction can be minimized by applying an effective lubricant or **ultrasonic vibration** of the platens. The use of heated platens or a thermal barrier at interfaces will also reduce barreling in hot working.

Forces and work of deformation under ideal conditions. If friction at the interfaces is zero and the material is perfectly plastic with a yield stress of Y, then the normal compressive stress on the cylindrical specimen is uniform at a level Y. The force at any height h_1 is then

$$F = YA_1, \tag{6.6}$$

where A_1 is the cross-sectional area and is obtained from volume constancy:

$$A_1 = \frac{A_0 h_0}{h_1}.$$

The ideal work of deformation is the product of the volume of the specimen and the specific energy u [in Eq. (2.59)] and is expressed as

$$\text{Work} = \text{volume} \int_0^{\varepsilon_1} \sigma \, d\varepsilon, \tag{6.7}$$

where ε_1 is obtained from Eq. (6.3). If the material is strain hardening, with a true-stress–true-strain curve given by

$$\sigma = K\varepsilon^n,$$

then the expression for force at any stage during deformation becomes

$$F = Y_f A_1, \tag{6.8}$$

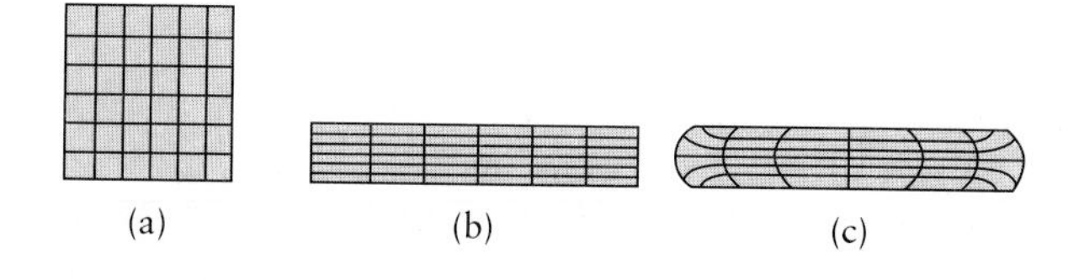

FIGURE 6.3 Schematic illustration of grid deformation in upsetting: (a) original grid pattern; (b) after deformation, without friction; (c) after deformation, with friction. Such deformation patterns can be used to calculate the strains within a deforming body.

where Y_f is the flow stress of the material, corresponding to the true strain given by Eq. (6.3) and Fig. 2.39.

The expression for the work done is

$$\text{Work} = (\text{volume})(\bar{Y})(\varepsilon_1), \tag{6.9}$$

where $\bar{Y}$ is the average flow stress and is given by

$$\bar{Y} = \frac{K\int_0^{\varepsilon_1} \varepsilon^n \, d\varepsilon}{\varepsilon_1} = \frac{K\varepsilon_1^n}{n+1}. \tag{6.10}$$

6.2.2 Methods of analysis

There are several methods of analysis to theoretically determine various quantities, such as stresses, strains, strain rates, forces, and local temperature rise, in deformation processing, outlined as follows:

Slab method. The *slab method* is one of the simpler methods of analyzing the stresses and loads in forging and other bulk-deformation processes. It requires the selection of an element in the workpiece and identification of all the normal and frictional stresses acting on that element.

a) **Forging of a rectangular workpiece in plane strain.** Let's take the case of simple compression with friction (Fig. 6.4), which is the basic deformation process in forging. As the flat dies compress the part, it is reduced in thickness, and, because the volume of the past remains constant, the part expands laterally. This relative movement at the die–workpiece interfaces causes frictional forces acting in opposition to the movement of the piece. These frictional forces are shown by the horizontal arrows in Fig. 6.4. For simplicity, let's also assume that the deformation is in plane strain; that is, the workpiece is not free to flow in the z-direction.

Let's now take an element and indicate all the stresses acting on it (Fig. 6.4b). Note the correct direction of the frictional stresses. Also note the difference between the horizontal stresses acting on the sides of the element; this difference is caused by the presence of frictional stresses on the element. We assume that the lateral stress distribution, σ_x, is uniform along the height, h.

The next step in this analysis is to balance the horizontal forces on this element, because it must be in static equilibrium. Assuming unit width, we have

$$(\sigma_x + d\sigma_x)h + 2\mu\sigma_y \, dx - \sigma_x h = 0,$$

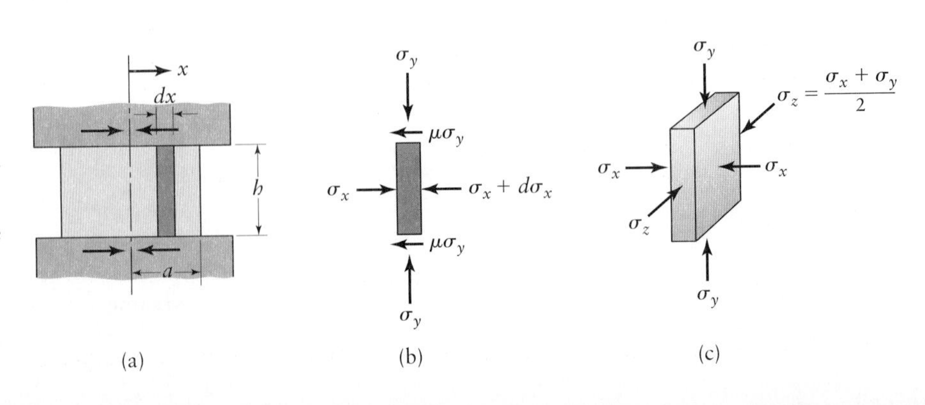

FIGURE 6.4 Stresses on an element in plane-strain compression (forging) between flat dies. The stress σ_x is assumed to be uniformly distributed along the height h of the element. Identifying the stresses on an element (slab) is the first step in the slab method of analysis for metalworking processes.

or

$$d\sigma_x + \frac{2\mu\sigma_y}{h}dx = 0.$$

Note that we now have one equation, but two unknowns: σ_x and σ_y. The necessary second equation is obtained from the yield criteria, as follows: As shown in Fig. 6.4c, this element is subjected to triaxial compression. Using the distortion-energy criterion for plane strain, we obtain

$$\sigma_y - \sigma_x = \frac{2}{\sqrt{3}}Y = Y'. \qquad (6.11)$$

Hence,

$$d\sigma_y = d\sigma_x.$$

Note that we assumed σ_y and σ_x to be *principal stresses*. In the strictest sense, σ_y cannot be a principal stress, because a shear stress is also acting on the same plane. However, this assumption is acceptable for low values of the coefficient of friction, μ, and is the standard practice in this method of analysis. Note also that σ_z in Fig. 6.4c is derived in a manner similar to that in Eq. (2.44). We now have two equations that can be solved by noting that

$$\frac{d\sigma_y}{\sigma_y} = -\frac{2\mu}{h}dx, \quad \text{or} \quad \sigma_y = Ce^{-2\mu x/h}. \qquad (6.12)$$

The boundary conditions are such that at $x = a$, $\sigma_x = 0$, and hence $\sigma_y = Y'$ at the edges of the specimen. (All stresses are compressive, so we may ignore negative signs for stresses, which is traditional practice in such analyses.) Hence, the value of C becomes

$$C = Y'e^{2\mu a/h},$$

and therefore,

$$p = \sigma_y = Y'e^{2\mu(a-x)/h}. \qquad (6.13)$$

Also,

$$\sigma_x = \sigma_y - Y' = Y'[e^{2\mu(a-x)/h} - 1]. \qquad (6.14)$$

Equation (6.13) is plotted qualitatively in Fig. 6.5 in dimensionless form. Note that the pressure increases exponentially toward the center of the part and also that it increases with the a/h ratio and increasing friction. For a strain-hardening material, Y' in Eqs. (6.13) and (6.14) is replaced by Y'_f. Because of its shape, the pressure-distribution curve in Fig. 6.5 is referred to as the *friction hill*. Note that the pressure with friction is higher than it is without friction. This result is expected, because the work required to overcome friction must be supplied by the upsetting force.

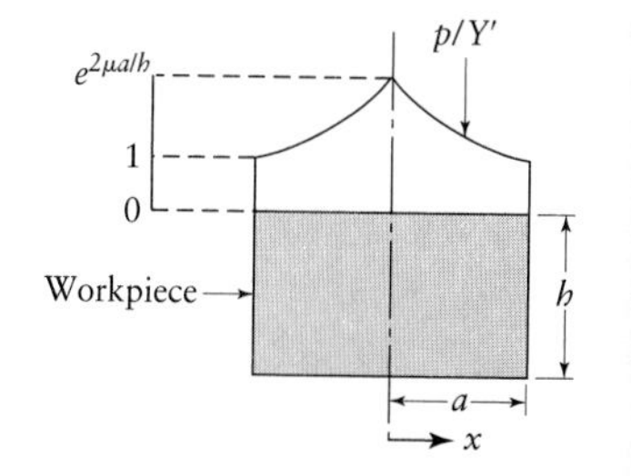

FIGURE 6.5 Distribution of die pressure, in terms of p/Y', in plane-strain compression with sliding friction. Note that the pressure at the left and right boundaries is equal to the yield stress in plane strain, Y'. Sliding friction means that the frictional stress is directly proportional to the normal stress.

The area under the pressure curve in Fig. 6.5 is the upsetting **force per unit width** of the specimen. This area can be obtained by integration, but an approximate expression for the *average pressure* p_{av} is

$$p_{av} \simeq Y'\left(1 + \frac{\mu a}{h}\right). \qquad (6.15)$$

Again note the significant influence of a/h and friction on the pressure required,

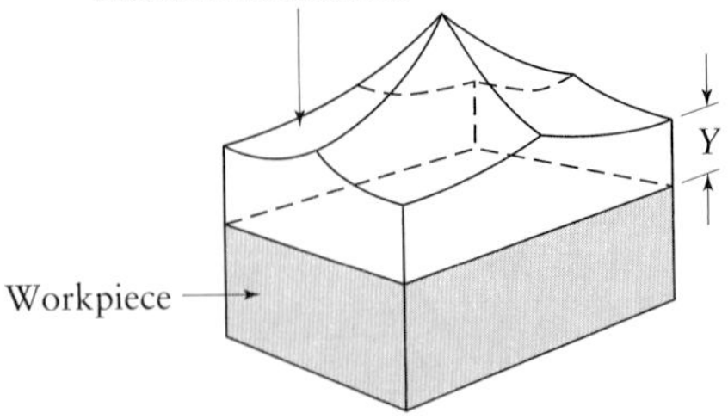

FIGURE 6.6 Normal stress (pressure) distribution in the compression of a rectangular workpiece with sliding friction under conditions of plane stress, using the *distortion-energy criterion*. Note that the stress at the corners is equal to the uniaxial yield stress, Y, of the material.

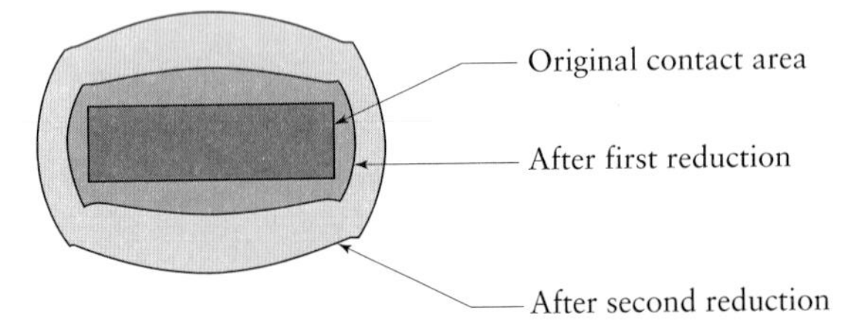

FIGURE 6.7 Increase in contact area of a rectangular specimen (viewed from the top) compressed between flat dies with friction. Note that the length of the specimen has increased proportionately less than its width. Likewise, a specimen in the shape of a cube acquires the shape of a pancake after deformation with friction.

especially at high a/h ratios. The *forging force*, F, is the product of the average pressure and the contact area; that is,

$$F = (p_{av})(2a)(\text{width}). \tag{6.16}$$

Note that the expressions for the pressure p are in terms of an instantaneous height h; thus, the force at any h during a continuous upsetting operation must be calculated individually.

A *rectangular specimen* can be upset without being constrained on its sides (*plane stress*). According to the distortion-energy criterion, the normal stress distribution can be given qualitatively by the plot in Fig. 6.6. The pressure is Y at the corners, because the elements at the corners are in uniaxial compression. A friction hill remains along the edges of the specimen, because of friction along the edges.

Figure 6.7 shows the lateral expansion (top view) of the edges of a rectangular specimen in actual plane-stress upsetting. Whereas the increase in the length of the specimen is 40%, the increase in width is 230%. The reason for the significantly larger increase in width is that, as expected, *the material flows in the direction of least resistance*. The width has less frictional resistance than the length. Likewise, after upsetting, a specimen in the shape of an equilateral cube acquires a pancake shape. The diagonal direction expands at a slower rate than the other directions.

b) **Forging of a solid cylindrical workpiece.** Using the slab method of analysis, we can also determine the pressure distribution in forging of a solid cylindrical specimen (Fig. 6.8). We first isolate a segment of angle $d\theta$ in a cylinder of radius r and height h, take a small element of radial length dx, and place on this element all the normal and frictional stresses acting on it. Following a similar approach as for the plane-strain case, we obtain an expression for the pressure, p, at any radius x as

$$p = Ye^{2\mu(r-x)/h}. \tag{6.17}$$

The *average pressure*, p_{av}, can be given approximately as

$$p_{av} \simeq Y\left(1 + \frac{2\mu r}{3h}\right). \tag{6.18}$$

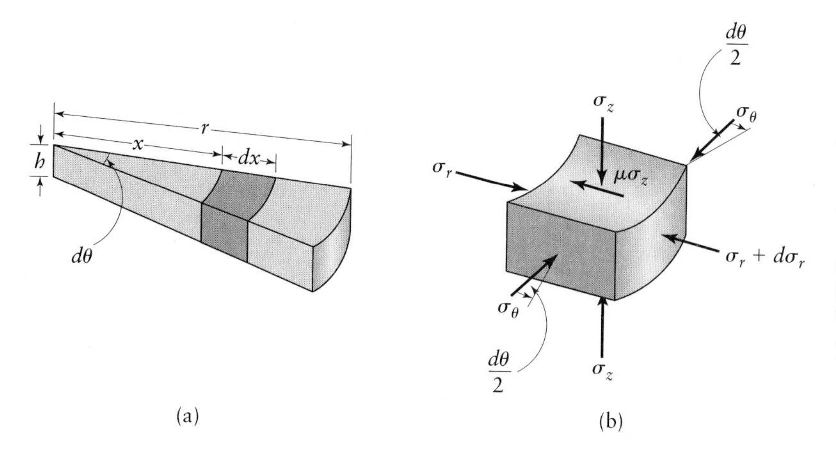

FIGURE 6.8 Stresses on an element in forging of a solid cylindrical workpiece between flat dies. Compare with Fig. 6.4. *Source:* (a) After J. F. W. Bishop, *J. Mech. Phys. Solids*, Vol. 6, 1958, pp. 132–144. (b) Adapted from W. Schroeder and D. A. Webster, *Trans. ASME*, Vol. 71, 1949, pp. 289–294.

The *forging force* is

$$F = (p_{av})(\pi r^2). \tag{6.19}$$

For strain-hardening materials, Y in Eqs. (6.17) and (6.18) is replaced by the flow stress, Y_f. In forging operations, the value of the coefficient of friction, μ, in Eqs. (6.17) and (6.18) can be estimated to be 0.05 to 0.1 for cold forging and 0.1 to 0.2 for hot forging. The exact value depends on the effectiveness of the lubricant and can be higher than stated, especially in regions where the workpiece surface is devoid of lubricant.

Figure 6.9 illustrates the effects of friction and the aspect ratio of the specimen (a/h or r/h) on the average pressure p_{av} in upsetting. The pressure is given in a dimensionless form and can be regarded as a pressure-multiplying factor. These curves are convenient to use and again show the significance of friction and the aspect ratio of the specimen on upsetting pressure.

c) **Forging under sticking condition.** The product of μ and p is the *frictional stress* (surface shear stress) at the interface at any location x from the center of the specimen. As p increases toward the center, μp also increases. However, the value of μp cannot be greater than the shear yield stress k of the material. When $\mu p = k$, **sticking** takes place. (In plane strain, the value of k is $Y'/2$; see Section 2.11.) Sticking does not necessarily mean adhesion at the interface; it reflects the fact that, relative to the platen surfaces, the material does not move.

For the sticking condition, the normal stress distribution in plane strain can be shown to be

$$p = Y'\left(1 + \frac{a - x}{h}\right). \tag{6.20a}$$

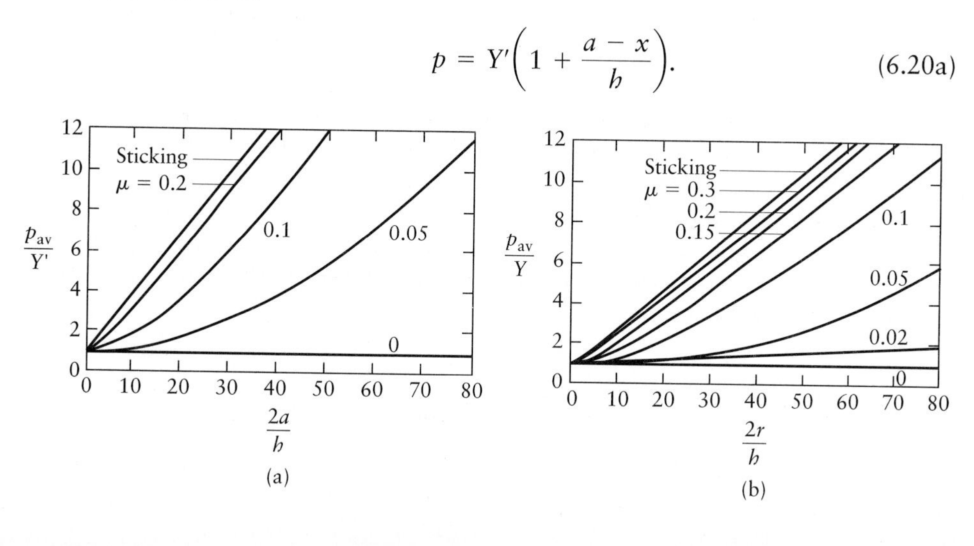

FIGURE 6.9 Ratio of average die pressure to yield stress as a function of friction and aspect ratio of the specimen: (a) plane-strain compression; and (b) compression of a solid cylindrical specimen. Note that the yield stress in (b) is Y, not Y', as in the plane-strain compression shown in (a). *Source*: Reprinted from J.F.W. Bishop, J. Mech. Phys. Solids, v. 6, 1958, pp. 132-144 with permission from Elsevier Science.

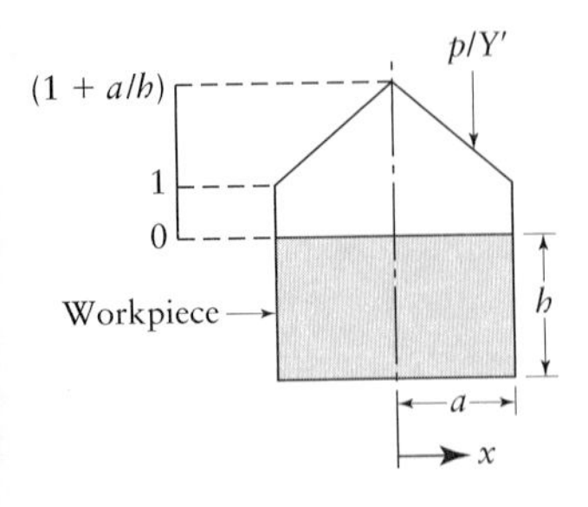

FIGURE 6.10 Distribution of die pressure, in terms of p/Y', in the compression of a rectangular specimen in plane strain and under sticking conditions. The pressure at the edges is the uniaxial yield stress of the material in plane strain, Y'.

The pressure varies linearly with x, as shown in Fig. 6.10. The normal stress distribution for a cylindrical specimen under sticking condition can be shown to be

$$p = Y\left(1 + \frac{r - x}{h}\right). \quad (6.20b)$$

The stress distribution again is linear, as shown in Fig. 6.10 for plane strain.

EXAMPLE 6.1 Upsetting force

A cylindrical specimen made of annealed 4135 steel has a diameter of 6 in. and is 4 in. high. It is upset by open-die forging with flat dies to a height of 2 in. at room temperature. Assuming that the coefficient of friction is 0.2, calculate the force required at the end of the stroke. Use the average-pressure formula.

Solution. The average-pressure formula, from Eq. (6.18), is

$$p_{av} \simeq Y\left(1 + \frac{2\mu r}{3h}\right),$$

where Y is replaced by Y_f, because the workpiece material is strain hardening. From Table 2.3, we have $K = 1015$ MPa (147,000 psi) and $n = 0.17$.

The absolute value of the true strain is

$$\varepsilon_1 = \ln\left(\frac{4}{2}\right) = 0.693.$$

Therefore,

$$Y_f = K\varepsilon_1^n = (147{,}000)(0.693)^{0.17} = 138{,}000 \text{ psi}.$$

The final height h_1 is 2 in. The radius r at the end of the stroke is found from volume constancy:

$$\frac{\pi 6^2}{4} \cdot 4 = \pi r_1^2 \cdot 2 \qquad \text{and} \qquad r_1 = 4.24 \text{ in.}$$

Therefore,

$$p_{av} \simeq 138{,}000\left[1 + \frac{(2)(0.2)(4.24)}{(3)(2)}\right] = 177{,}000 \text{ psi}.$$

The upsetting force is then

$$F = (177{,}000)\pi \cdot (4.24)^2 = 10^7 \text{ lb}.$$

[*Note*: Figure 6.9b can also be used to solve this problem.]

EXAMPLE 6.2 Transition from sliding to sticking friction

In plane-strain upsetting, the frictional stress cannot be higher than the shear yield stress, k, of the workpiece material; thus, there may be a distance x in Fig. 6.4 where a transition occurs from sliding to sticking friction. Derive an expression for x in terms of a, h, and μ only.

Solution. The shear stress at the interface due to friction can be expressed as

$$\tau = \mu p.$$

However, the shear stress cannot exceed the yield shear stress, k, of the material, which, for plane strain, is $Y'/2$. The pressure curve in Fig. 6.5 is given by Eq. (6.13); thus, in the limit, we have

$$\mu Y' e^{2\mu(a-x)/h} = Y'/2,$$

or

$$2\mu \frac{(a-x)}{h} = \ln\left(\frac{1}{2\mu}\right).$$

Hence,

$$x = a - \left(\frac{h}{2\mu}\right)\ln\left(\frac{1}{2\mu}\right).$$

Note that, as expected, the magnitude of x decreases as μ decreases. However, the pressures must be sufficiently high to cause sticking, i.e., the a/h ratio must be high. For example, let $a = 10$ mm and $h = 1$ mm. Then, for $\mu = 0.2$, $x = 7.71$ mm, and for $\mu = 0.4$, $x = 9.72$ mm.

Finite-element method. In the *finite-element method*, the deformation zone in an elastic–plastic body is divided into a number of elements interconnected at a finite number of nodal points. The actual velocity distribution is then approximated for each element. Next, a set of simultaneous equations is developed that represents unknown velocity vectors. From the solution of these equations, actual velocity distributions and the stresses are calculated. This technique can incorporate friction conditions at the die–workpiece interfaces and actual properties of the workpiece material. It has been applied to relatively complex geometries in bulk-deformation and sheet-forming problems. Its accuracy is influenced by the number and shape of the finite elements, the deformation increment, and the methods of calculation. To ensure accuracy, complex problems require extensive computations. The finite-element method gives a detailed picture of the actual stresses and strain distributions throughout the workpiece.

Figure 6.11 illustrates the results of applying the finite-element method of analysis on a solid cylindrical workpiece in impression-die forging. Only one quarter of the workpiece is shown, with the vertical and horizontal lines as the axes of symmetry. The grid pattern is known as a *mesh*, and its distortion is predicted from theory, aided by the use of computers, with the results displayed on the video output screen. Note how the workpiece deforms as it is being forged.

The finite-element method is also capable of predicting microstructural changes in the material during hot-working operations, temperature distribution throughout the workpiece, and the onset of defects without using experimental data. This knowledge is then used in modifying die design to avoid making defective parts. Application of this technique requires inputs such as the stress–strain characteristics of the material as a function of strain rate and temperature, and frictional and heat-transfer characteristics of the die and workpiece.

Other methods of analysis. A number of other methods of metal-forming analysis have been used, but are becoming less popular, because of the availability of powerful commercial software. The **slip-line analysis** method is generally applied to plane-strain conditions. The technique consists of the construction of a family of straight or curvilinear lines that intersect each other orthogonally. These lines, known

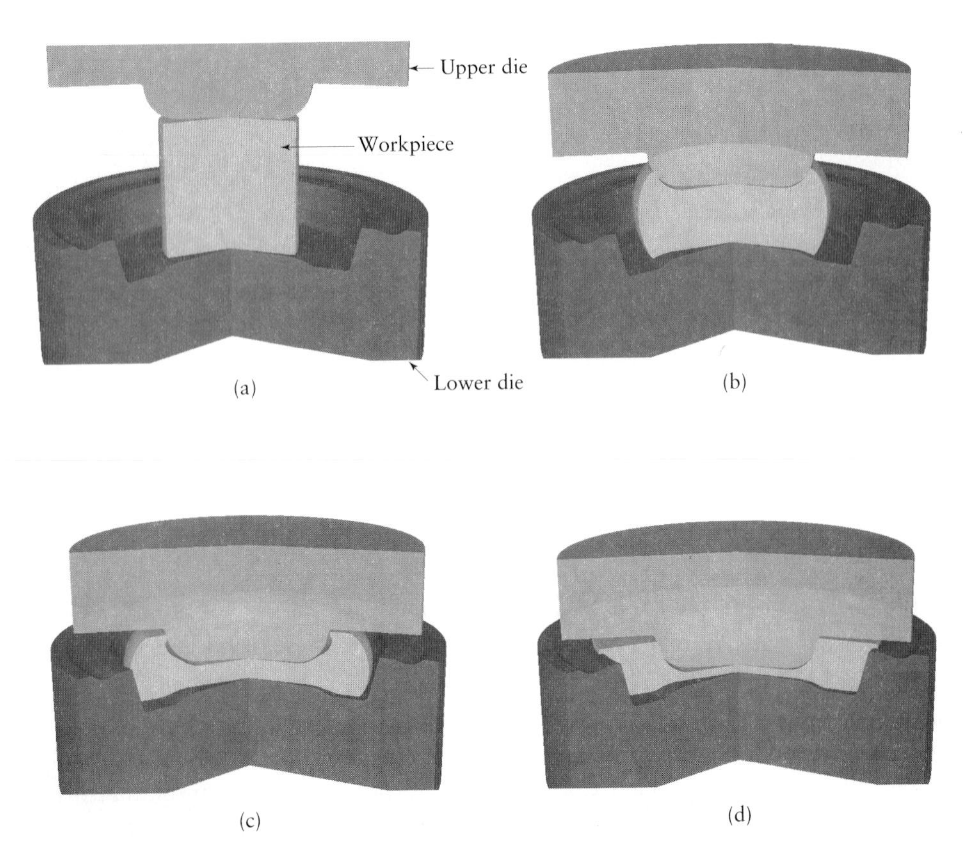

FIGURE 6.11 Plastic deformation in forging as predicted by DEFORM using the finite-element method of analysis. *Source*: Scientific Forming Technologies Corporation.

as a *slip-line field*, correspond to the directions of the maximum shear stress and allow calculation of stresses and strain rates in metal forming.

In the **upper-bound technique**, the overall deformation zone is approximated by a number of smaller zones where the velocity is constant or continuous. However, the velocity in adjacent zones may be different and discontinuous, as long as all discontinuities are tangential to the zone surface. In this approach, a velocity field that minimizes the total calculated power is found and taken as the actual velocity field and subsequently compared with experimental data.

Deformation-zone geometry. The observations made concerning Fig. 2.27 are important in estimating and calculating forces in forging and other bulk-deformation processes. As shown previously for hardness testing, the compressive stress required for indentation is, ideally, about three times the yield stress, Y, required for uniaxial compression. (See Section 2.6.8.) Moreover, (1) the deformation under the indenter is localized, making the overall deformation highly nonuniform; and (2) the deformation zone is relatively small compared with the size of the specimen.

However, in a simple frictionless compression test with flat dies, the top and bottom surfaces of the specimen are always in contact with the dies, and the specimen deforms uniformly. Various situations occur between these two extreme examples of specimen-deformation geometry. The *deformation zones* and the pressures required are shown in Fig. 6.12 for the frictionless condition as obtained from *slip-line analysis*. Note that the ratio h/L is the important parameter in determining the inhomogeneity of deformation. The frictionless nature of these examples is important, because friction significantly affects forces, particularly at small values of h/L. Deformation-zone geometry depends on the particular process and such parameters as die geometry and percent reduction of the material (Fig. 6.13).

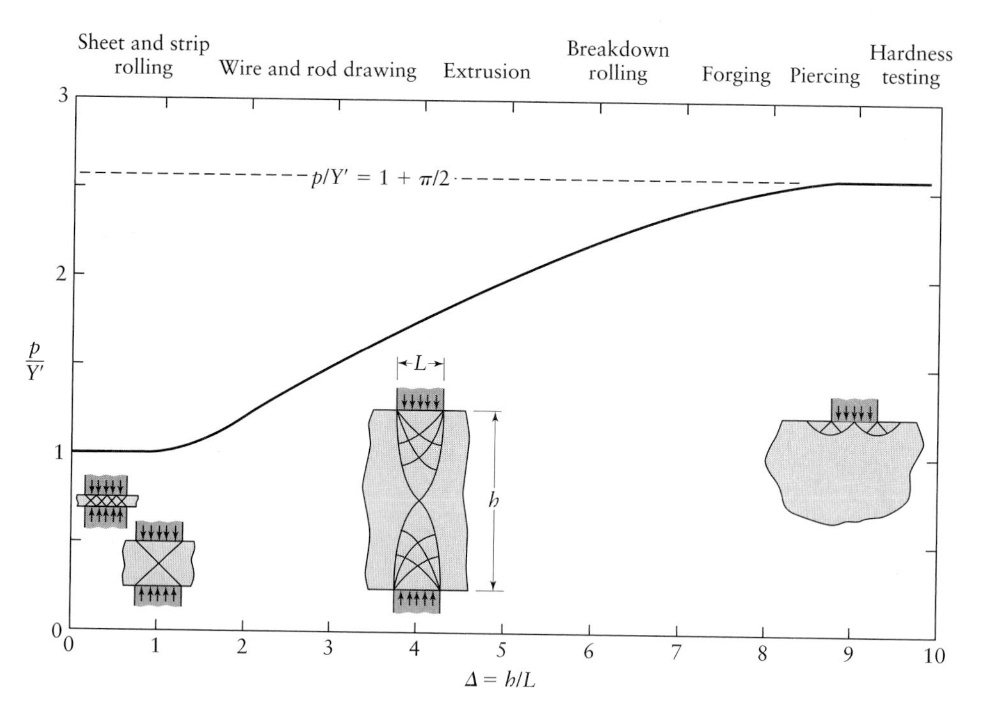

FIGURE 6.12 Die pressures required in *frictionless* plane-strain conditions for a variety of metalworking operations. The geometric relationship between contact area of the dies and workpiece dimensions is an important factor in predicting forces in plastic deformation of materials. *Source*: After W. A. Backofen.

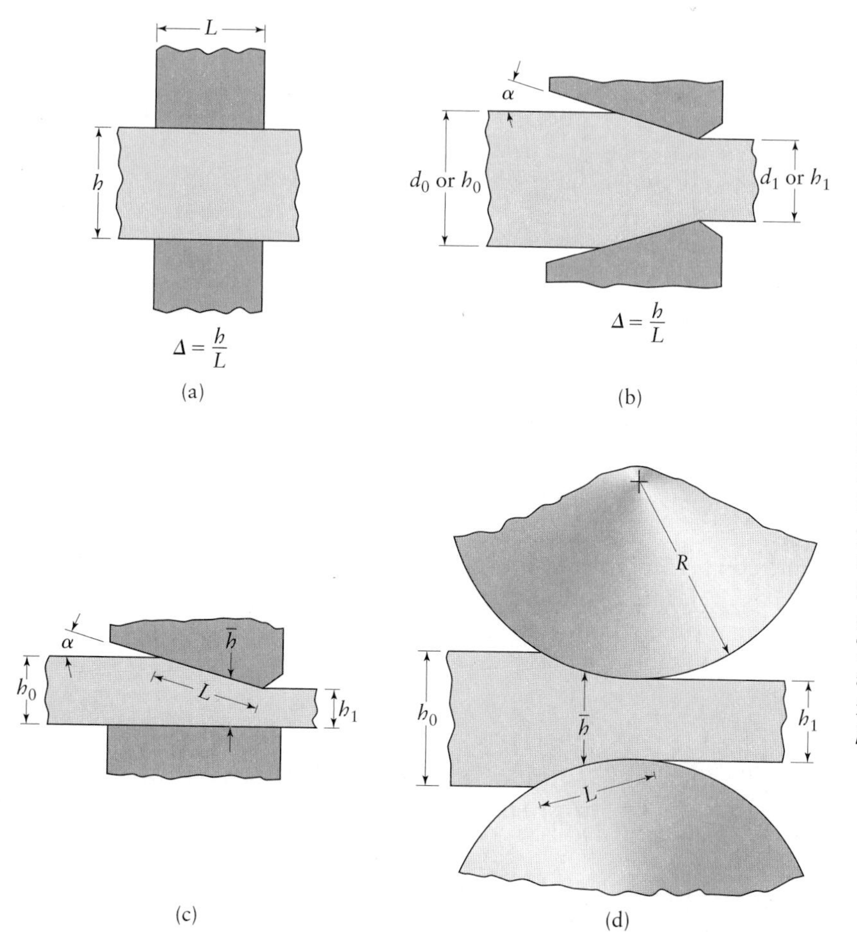

FIGURE 6.13 Examples of plastic deformation processes in plane strain, showing the h/L ratio. (a) Indenting with flat dies. This operation is similar to cogging, shown in Fig. 6.19. (b) Drawing or extrusion of strip with a wedge-shaped die, described in Sections 6.4 and 6.5. (c) Ironing; see also Fig. 7.54. (d) Rolling, described in Section 6.3. As shown in Fig. 6.12, the larger the h/L ratio, the higher the die pressure becomes. In actual processing, however, the smaller this ratio, the greater is the effect of friction at the die–workpiece interfaces. The reason is that contact area, and hence friction, increases with a decreasing h/L ratio.

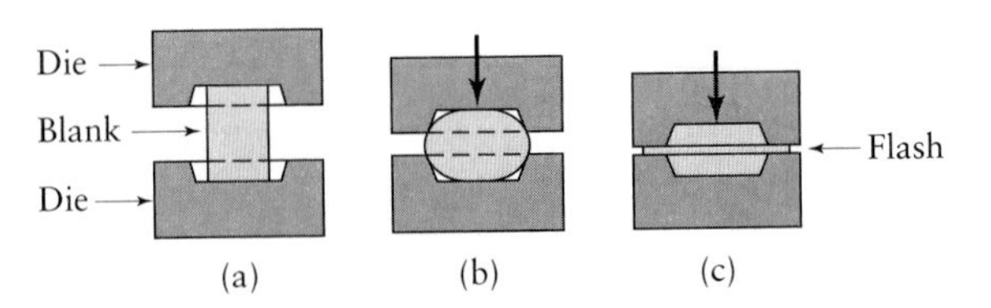

FIGURE 6.14 Schematic illustration of stages in impression-die forging. Note the formation of flash, or excess material that is subsequently trimmed off.

6.2.3 Impression-die forging

In *impression-die forging*, the workpiece acquires the shape of the die cavities (impressions) while it is being upset between the closing dies. A typical example is shown in Fig. 6.14. Some of the material flows radially outward and forms a **flash.** Because of its high length-to-thickness ratio (equivalent to a high a/h ratio), the flash is subjected to high pressure, which in turn signifies high frictional resistance to material flow in the radial direction in the flash gap. Since high friction encourages the filling of the die cavities, the flash has a significant role in the flow of material in impression-die forging. Furthermore, if the operation is carried out at elevated temperatures, the flash, because of its high surface-area-to-thickness ratio, cools faster than the bulk of the workpiece; consequently, the flash resists deformation more than the bulk does and helps fill the die cavities.

Because of the more complex shapes involved, accurate calculation of forces in impression-die forging is difficult. Depending on its position, each element within the workpiece is generally subject to different strains and strain rates. Consequently, the level of strength that the material exhibits at each location depends not only on the strain and strain rate, but also on the exponents n and m of the workpiece material in Eqs. (2.11) and (2.16). Because of the many difficulties involved in calculating forces in impression-die forging, certain pressure-multiplying factors K_p have been recommended, as shown in Table 6.2. These factors are to be used with the expression

$$F = (K_p)(Y_f)(A), \qquad (6.21)$$

where F is the forging load, A is the projected area of the forging (including the flash), and Y_f is the flow stress of the material at the strain and the strain rate to which the material is subjected.

A typical impression-die forging load as a function of the stroke of the die is shown in Fig. 6.15. For this axisymmetric workpiece, the force increases gradually as the cavity is filled (Fig. 6.15b). The force then increases rapidly as flash forms. In order for us to obtain the final dimensions and details on the forged part, the dies must close further, with an even steeper rise in the forging load. Note that the flash has a finite contact length with the die, called **land.** (See Fig. 6.27.) The land ensures that the flash generates sufficient resistance of the outward flow of the material to aid in die filling without contributing excessively to the forging load.

TABLE 6.2

Range of K_p Values in Eq. (6.21) for Impression-Die Forging

Simple shapes, without flash	3–5
Simple shapes, with flash	5–8
Complex shapes, with flash	8–12

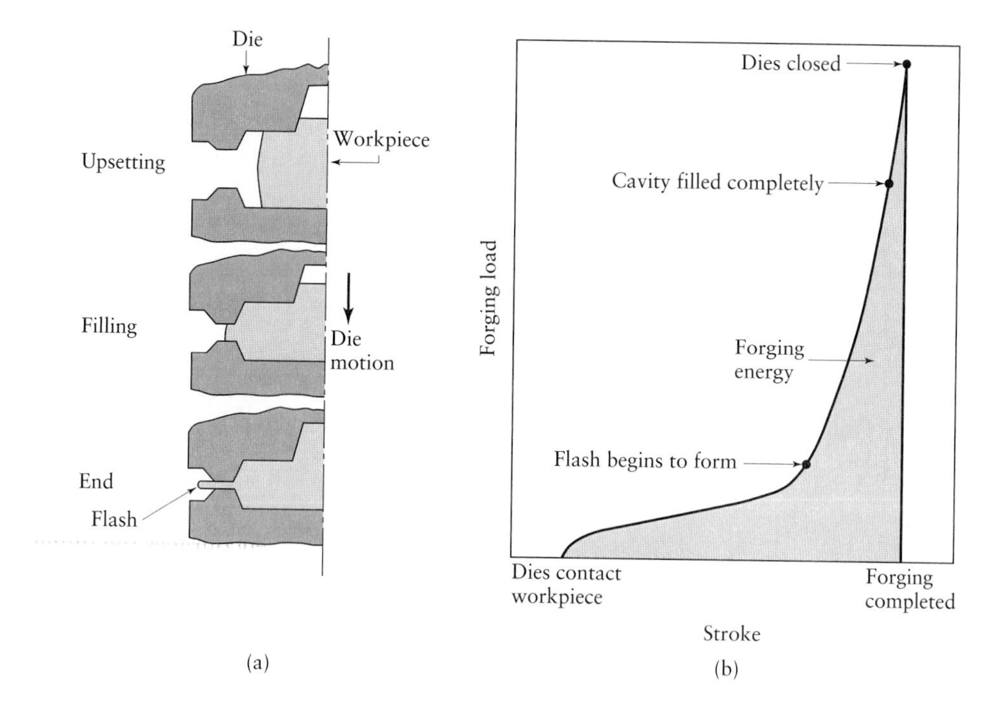

FIGURE 6.15 Typical load–stroke curve for impression-die forging. Note the sharp increase in load after the flash begins to form. In hot-forging operations, the flash requires high levels of stress, because it is thin—that is, it has a small h—and cooler than the bulk of the forging. *Source*: After T. Altan.

Closed-die forging. Although not quite correct, the example shown in Fig. 6.14 is also referred to as *closed-die forging*. In true closed-die forging, no flash is formed, and the workpiece is completely surrounded by the dies, while in impression-die forging, any excess metal in the die cavity is formed into a flash. Since no flash can be formed in closed-die forging, proper control of the volume of material is essential to obtain a forging of desired dimensions. Undersized blanks in closed-die forging prevent the complete filling of the die; oversized blanks may cause premature die failure or jamming of the dies.

Precision forging, or **flashless forging**, and similar operations where the part formed is close to the final dimensions of the desired component are also known as **near-net-shape production**. Any excess material is subsequently removed by various machining processes. In precision forging, special dies are made and machined to greater accuracy than in ordinary impression-die forging. Precision forging requires higher capacity forging equipment than do other forging processes. Aluminum and magnesium alloys are particularly suitable for precision forging, because of the low forging loads and temperatures required. Also, they result in little die wear and produce a good surface finish. Steels and other alloys are more difficult to precision forge. The choice between conventional forging and precision forging requires an economic analysis. Precision forging requires special dies. However, much less machining is involved, because the part is closer to the desired final shape.

In the **isothermal-forging** process, also known as **hot-die forging**, the dies are heated to the same temperature as the hot blank. In this way, cooling of the workpiece is eliminated, the low flow stress of the material is maintained, and material flow within the die cavities is improved. The dies are generally made of nickel alloys, and complex parts with good dimensional accuracy can be forged in one stroke in hydraulic presses. Isothermal forging generally is expensive; however, it can be economical for intricate forgings of expensive materials, provided that the quantity required is large enough to justify die costs.

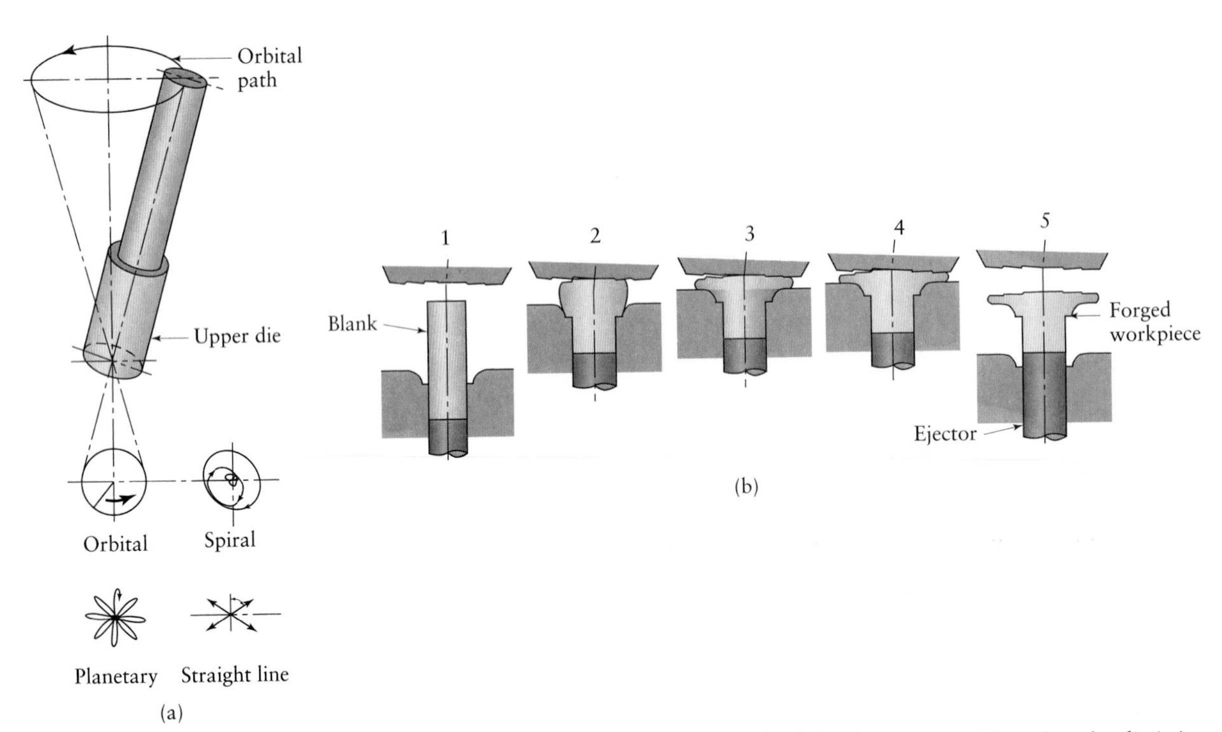

FIGURE 6.16 Schematic illustration of the orbital-forging process. Note that the die is in contact with only a portion of the workpiece surface. This process is also called *rotary forging, swing forging*, and *rocking-die forging* and can be used for forming bevel gears, wheels, and bearing rings.

Orbital forging is a relatively new process in which the die moves along an orbital path (Fig. 6.16) and forges or forms the part incrementally. Consequently the forces are lower, the operation is more quiet, and part flexibility can be achieved.

Incremental forging is a process in which a blank is forged into a shape with a tool or die that forms it in several small steps, a process somewhat similar to cogging (see Fig. 6.19). Consequently, this operation requires much lower forces as compared with conventional forging.

6.2.4 Miscellaneous forging operations

a) **Coining.** Another example of closed-die forging is the *minting of coins*, where the slug is shaped in a completely closed cavity. The pressures required can be as great as five to six times the flow stress of the material in order to produce the fine details of a coin or a medallion. Several coining operations may be necessary in order to obtain full detail on some parts. The coining process is also used with forgings and other products to improve surface finish and impart the desired dimensional accuracy (**sizing**) of the products. The pressures are large, and usually very little change in shape takes place. Lubricants cannot be tolerated in coining, because they can be trapped in die cavities and prevent reproduction of fine die-surface details.

b) **Heading.** Basically an upsetting operation, *heading* is performed at the end of a rod (usually round) to form a shape with a larger cross-section. Typical products formed by heading include the heads of bolts, screws, and nails (Fig. 6.17). An important aspect of the upsetting process during heading is the tendency for buckling if the length-to-diameter ratio is too high. Heading is done on

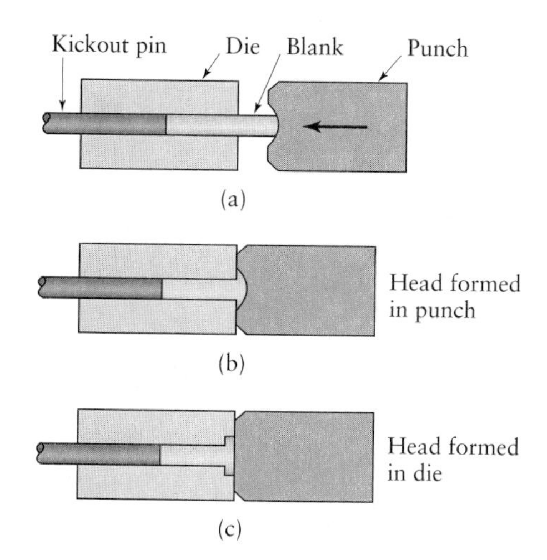

FIGURE 6.17 Forming heads on fasteners such as bolts and rivets. These processes are called *heading*.

machines called *headers*, which are usually highly automated horizontal machines with high production rates. The operation can be carried out cold, warm, or hot. Heading forces can be estimated reliably by using the slab method of analysis described in Section 6.2.2.

c) **Piercing.** In *piercing*, a punch indents the surface of a workpiece to produce a cavity or an impression (Fig. 6.18). The workpiece may be confined in a die cavity, or it may be unconstrained. The piercing force depends on the punch's cross-sectional area and tip geometry, the flow stress of the material, and the friction at the interfaces. Punch pressures may range from three to five times the flow stress of the material. The term *piercing* is also used to describe the process of cutting of holes with a punch and die (Section 7.3).

d) **Hubbing.** In *hubbing*, a hardened punch with a particular tip geometry is pressed into the surface of a block. This process is used to produce a die cavity, which is then used for subsequent forming operations. The cavity is usually shallow; for deeper cavities, some material may be removed from the surface by machining prior to hubbing. The pressure required to generate a cavity by hubbing is approximately three times the ultimate tensile strength (UTS) of the material of the block; thus, the force required is

$$\text{Hubbing force} = 3(\text{UTS})(A), \tag{6.22}$$

where A is the projected area of the impression. Note that the factor 3 is in agreement with the observations made with regard to hardness of materials in Section 2.6.8.

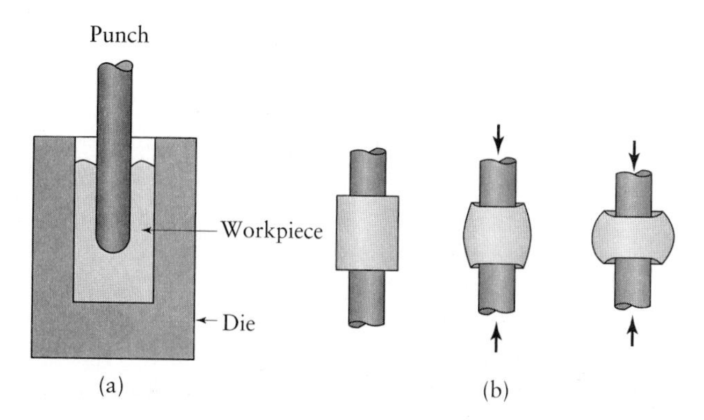

FIGURE 6.18 Examples of piercing operations.

e) **Cogging.** In *cogging*, also called *drawing out*, the thickness of a bar is reduced by successive steps at certain intervals (Fig. 6.19). A long section of a bar can be reduced in thickness without large forces, because the contact area is small.

f) **Fullering and edging.** These operations are performed, usually on bar stock, to distribute the material in certain regions of a forging. In *fullering*, the material is distributed away from an area; in *edging* (see Fig. 6.26b) it is gathered into a localized area. The parts are then formed into final shapes by other forging processes.

g) **Roll forging.** In *roll forging*, the cross-sectional area of a bar is reduced and altered in shape by passing it through a pair of rolls with grooves of various shapes (Fig. 6.20). This operation may also be used to produce a part that is basically the final product, such as tapered shafts, tapered leaf springs, table knives, and numerous tools. Roll forging is also used as a preliminary forming operation, followed by other forging processes. Typical products made by this method include crankshafts and other automotive components.

h) **Skew rolling.** A process similar to roll forging is *skew rolling* (Fig. 6.21), which is used for making ball bearings. Round wire or rod stock is fed into the roll gap, and spherical blanks are formed continuously by the rotating rolls. (Another method of forming blanks for ball bearings is by cutting pieces from a round bar and upsetting them, as shown in Fig. 6.22. The balls are then ground and polished in special machinery.)

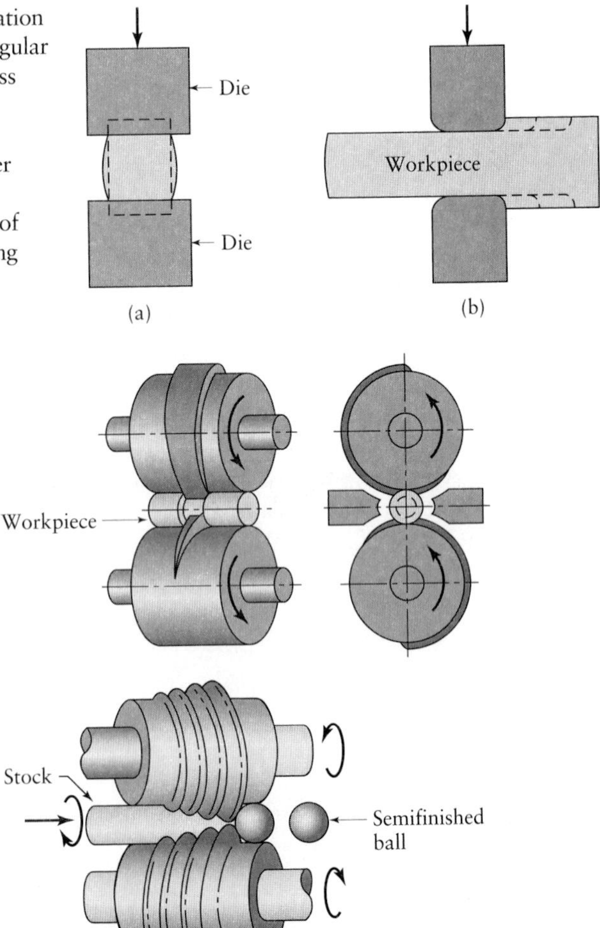

FIGURE 6.19 Schematic illustration of a cogging operation on a rectangular bar. With simple tools, the thickness and cross-section of a bar can be reduced by multiple cogging operations. Note the barreling after cogging. Blacksmiths use a similar procedure to reduce the thickness of parts in small increments by heating the workpiece and hammering it numerous times.

FIGURE 6.20 Schematic illustration of a roll forging (cross-rolling) operation. Tapered leaf springs and knives can be made by this process with specially designed rolls. *Source*: After J. Holub.

FIGURE 6.21 Production of steel balls for bearings by the skew-rolling process. Balls for bearings can also be made by the forging process shown in Fig. 6.22.

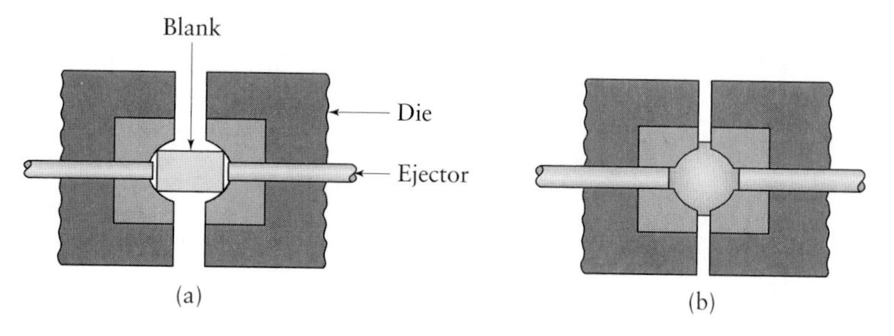

FIGURE 6.22 Production of steel balls by upsetting of a cylindrical blank. Note the formation of flash. The balls are subsequently ground and polished for use as ball bearings and in other mechanical components.

6.2.5 Defects in forging

In addition to *surface cracking*, other defects in forgings can be caused by the material flow patterns in the die cavity. As shown in Fig. 6.23, excess material in the web of a forging can buckle during forging and develop laps. If the web is thick, the excess material flows past the already forged portions and develops internal cracks (Fig. 6.24). These examples indicate the importance of properly distributing material and controlling the flow in the die cavity.

The various radii in the die cavity can significantly affect formation of defects. In Fig. 6.25, the material follows a large corner radius better than a small radius. With small radii, the material can fold over itself and produce a lap, called **cold shut**. During the service life of the forged component, such defects can lead to fatigue failure and other problems, such as excessive corrosion and wear. The importance of inspecting forgings before they are put into service, particularly for critical applications, is obvious. (For a discussion of inspection techniques, see Section 4.8.)

Although it may not be considered a flaw, another important aspect of quality in a forging is the **grain-flow pattern**. At times, the grain-flow lines reach a surface perpendicularly, exposing the grain boundaries directly to the environment. The exposed boundaries are known as **end grains**. In service, they can be attacked by the environment, develop a rough surface, and act as stress raisers. For critical components, end grains in forgings can be avoided by proper selection of the original workpiece orientation in the die cavity and by control of material flow.

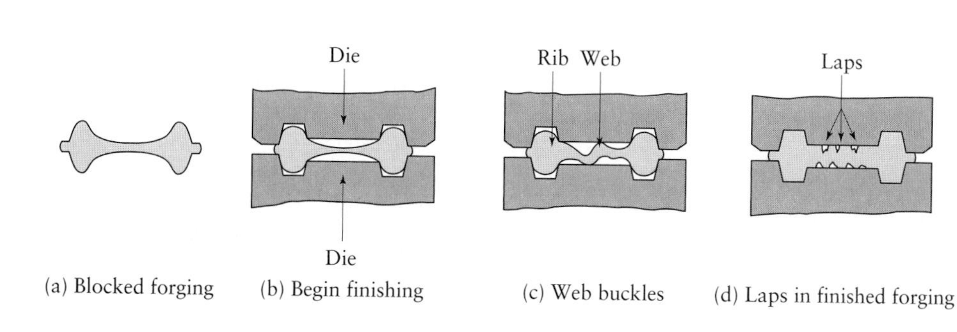

FIGURE 6.23 Laps formed by buckling of the web during forging.

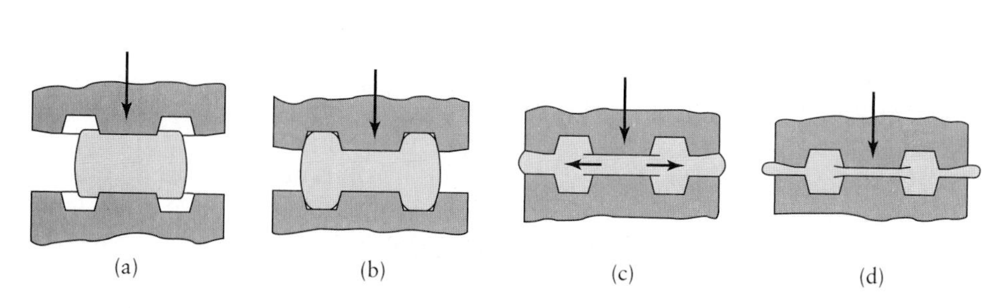

FIGURE 6.24 Internal defects produced in a forging because of an oversized billet. The die cavities are filled prematurely, and the material at the center of the part flows past the filled regions as deformation continues.

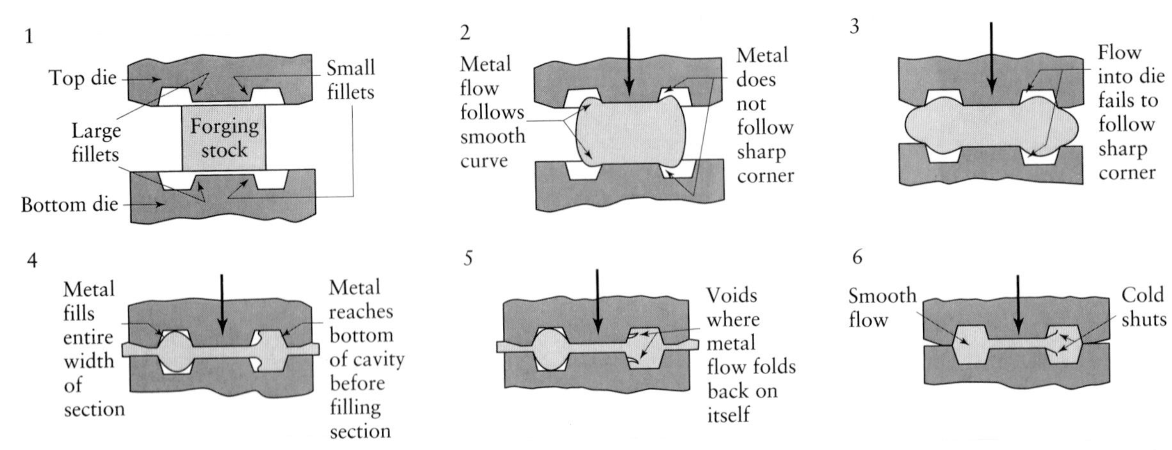

FIGURE 6.25 Effect of fillet radius on defect formation in forging. Small fillets (right side of drawings) cause the defects. *Source*: Aluminium Company of America.

Because the metal flows in various directions in a forging, and because of temperature variations, the properties of a forging are generally *anisotropic*. (See Section 3.5.) Strength and ductility vary significantly in test specimens taken from different locations and orientations in a forged part.

6.2.6 Forgeability

The forgeability of a metal can be defined as its capability to undergo deformation by forging without cracking. This definition can be expanded to include the flow strength of the metal; thus, a material with good forgeability is one that can be shaped with low forces without cracking. A number of tests have been developed to measure forgeability, although none is universally accepted. The more commonly used tests are described as follows:

Upsetting test. A standard test is to upset a solid cylindrical specimen and observe any cracking on the barreled surfaces (Fig. 3.20d). The greater the reduction in height prior to cracking, the greater is the forgeability of the metal. The cracks are caused by **secondary tensile stresses** on the barreled surfaces. They are referred to as *secondary* because no external tensile stress is applied to the material, and the stresses develop because of changes in the material's shape. Because barreling is a result of friction at the die–workpiece interfaces, friction has a marked effect on cracking in upsetting. As friction increases, the specimen cracks at a lower reduction in height. Also of interest is whether the crack is at a 45° angle or vertical; the angle depends on the sign of the axial stress σ_z on the barreled surface. If this stress is negative (compressive), the crack is at 45°; if the stress is positive (tensile), the crack is longitudinal. Depending on the notch sensitivity of the material, any surface defects will also affect the results by causing premature cracking. A typical surface defect is a **seam:** a longitudinal scratch, a string of inclusions, or a fold from prior working of the material.

Upsetting tests can be performed at various temperatures and strain rates. An optimal range for these parameters can then be specified for forging a particular material. Such tests can serve only as guidelines, since, in actual forging, the metal is subjected to a different state of stress as compared with a simple upsetting operation.

Hot-twist test. In the hot-twist test, a round specimen is twisted continuously until it fails. The test is performed at various temperatures, and the number of turns that each specimen undergoes before failure is observed. The optimal forging temperature is then determined. The hot-twist test is particularly useful in determining the forgeability of steels, although upsetting tests can also be used for that purpose. Impurities in the metal or small changes in the composition of the metal can have a significant effect on forgeability.

Effect of hydrostatic pressure on forgeability. As described in Section 2.2.8, hydrostatic pressure has a significant beneficial effect on the ductility of metals and nonmetallic materials. Experiments have indicated that the results of room-temperature forgeability tests, outlined previously, are improved (that is, cracking takes place at higher strain levels) if the tests are conducted in an environment of high *hydrostatic pressure*. Although such a test would be difficult to perform at elevated temperatures, techniques have been developed to forge metals in a high compressive environment in order to take advantage of this phenomenon. The pressure-transmitting medium is usually a low-strength ductile metal.

Forgeability of various metals. Based on the results of various tests and observations, the forgeability of several metals and alloys has been determined. It is based on such considerations as the ductility and strength of the metals, forging temperature required, frictional behavior, and quality of the forgings obtained. In general, aluminum, magnesium, copper and their alloys, carbon and low-alloy steels have good forgeability; high-temperature materials such as superalloys, tantalum, molybdenum, and tungsten and their alloys have poor forgeability.

6.2.7 Die design

Proper design of forging dies and selection of die materials require considerable experience and knowledge concerning the strength and ductility of the workpiece material, its sensitivity to strain rate and temperature, and its frictional characteristics. Die *distortion* under high forging loads can also be an important consideration in die design, particularly if close dimensional tolerances are required.

Complex forgings are produced within a number of die cavities, as shown in Fig. 6.26. Starting with a round bar stock, the bar is (a) first *preformed* (**intermediate shape**) by some of the techniques previously described, (b) formed into the final shape in two additional forging operations, and (c) trimmed. Note how proper distribution of the material is essential to fill the die cavities to produce the part successfully. The reason for the intermediate steps can be best understood from Eq. (4.6), with the understanding that it is desirable to maximize die life and, therefore, minimize wear. The approach used is to design the forging operation so that a cavity encounters

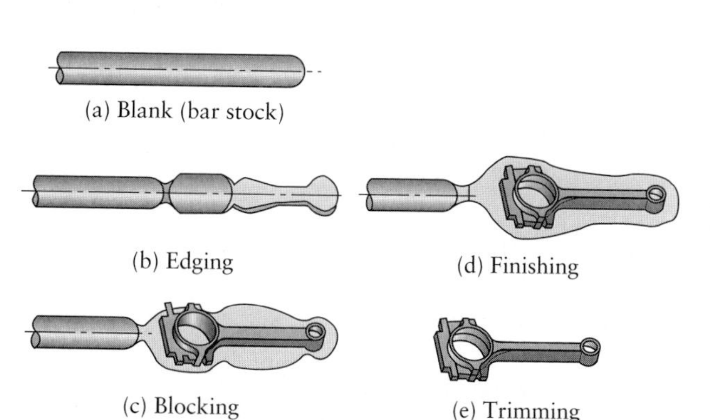

FIGURE 6.26 Stages in forging a connecting rod for an internal combustion engine. Note the amount of flash that is necessary to fill the die cavities properly.

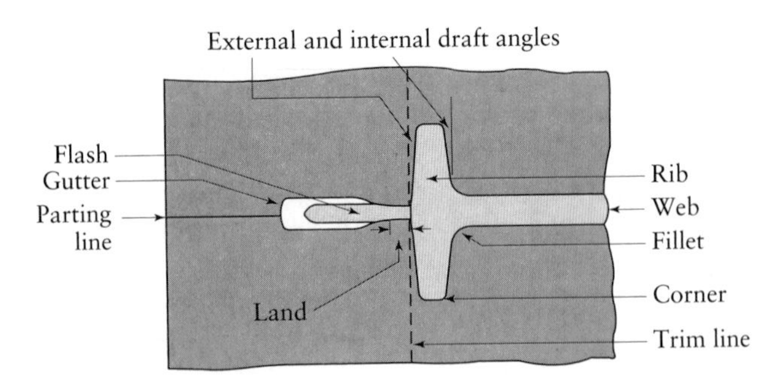

FIGURE 6.27 Standard terminology for various features of a typical forging die.

either high sliding speeds or high pressures, but not both. Thus, edging produces large deformations on a relatively thick workpiece, while finishing produces fine details and involves small strains, but requires large pressures to obtain the desired part definition.

Care is essential in calculating the volume of the material required to properly fill the die cavities and to enable efficient material flow through the cavities. Computer techniques are now being developed to expedite these calculations. Fig. 6.27 provides the terminology used in die design. The significance of various parameters are discussed next.

The **parting line** is where the two dies meet. For simple symmetrical shapes, the parting line is a straight line at the center of the forging; for more complex shapes, the line may be offset and may not be in a single plane. Selection of the proper location for the parting line is based on the shape of the part, flow of the metal, balance of forces, and the flash. The significance of **flash** is described in Section 6.2.3. After lateral flow has been sufficiently constrained (by the length of the land), the flash is allowed to flow into a **gutter**; thus, the extra flash does not unnecessarily increase the forging load. A general guideline for flash clearance (between the dies) is 3% of the maximum thickness of the forging. The length of the land is usually five times that of the flash clearance.

Draft angles are necessary in almost all forgings to facilitate removal of the part from the die. Draft angles usually range between 3° and 10°. Because the forging shrinks in its radial direction (as well as in other directions) as it cools, internal draft angles are made larger than external ones. Typically, internal angles are about 7° to 10° and external angles about 3° to 5°. Proper selection of the *radii* for corners and fillets ensures smooth flow of the metal in the die cavity and improves die life. Small radii are generally not desirable, because of their adverse effect on metal flow and their tendency to wear rapidly from stress concentration and thermal cycling. Small radii in fillets can cause fatigue cracking in dies.

Die materials. Because most forgings, particularly large ones, are performed at elevated temperatures, die materials generally must have strength and toughness at elevated temperatures, hardenability, resistance to mechanical and thermal shock, and resistance to wear (particularly to abrasive wear, because of the presence of scale on heated forgings). Selection of die materials depends on the size of the die, the composition and properties of the workpiece, the complexity of the workpiece shape, the forging temperature, the type of operation, the cost of die material, the number of forgings required, and the heat-transfer and distortion characteristics of the die material. Common die materials are tool and die steels, which contain chromium, nickel, molybdenum, and vanadium (see Section 3.10.3).

Forging temperatures and lubrication. Table 6.3 provides the range of temperatures for hot forging of various metals. *Lubrication* plays an important role in

TABLE 6.3

Hot-Forging Temperature Ranges for Various Metals

Metal	°C	°F	Metal	°C	°F
Aluminium alloys	400–450	750–850	Alloy steels	925–1260	1700–2300
Copper alloys	625–950	1150–1750	Titanium alloys	750–795	1400–1800
Nickel alloys	870–1230	1600–2250	Refractory alloys	975–1650	1800–3000

forging. It affects friction and wear and, consequently, the flow of metal into the die cavities. Lubricants can also serve as a *thermal barrier* between the hot forging and the relatively cool dies, thus slowing the rate of cooling of the workpiece. Another important role of a lubricant is to serve as a *parting agent*, that is, to prevent the forging from sticking to the dies. Lubricants can significantly affect the wear pattern in forging dies. A wide variety of metalworking fluids can be used in forging. For hot forging, graphite, molybdenum disulfide, and (sometimes) glass are commonly used as lubricants. For cold forging, mineral oils and soaps are common lubricants.

6.2.8 Equipment

Forging equipment of various designs, capacities, speeds, and speed-stroke characteristics is available (Fig. 6.28):

a) **Hydraulic presses.** Hydraulic presses have a constant low speed and are **load limited.** Large amounts of energy can be transmitted to the workpiece by a constant load available throughout the stroke. Ram speed can be varied during the stroke. These presses are used for both open-die and closed-die forging operations. The largest hydraulic press in existence has a capacity of 670 MN (75,000 tons).

b) **Mechanical presses.** Mechanical presses are **stroke limited.** They are of either the crank or the eccentric type, with speeds varying from a maximum at the center of the stroke to zero at the bottom. The force available depends on the stroke position and becomes extremely large at the bottom-dead-center position; thus, proper setup is essential to avoid breaking the dies or other equipment. The largest mechanical press has a capacity of 107 MN (12,000 tons).

c) **Screw presses.** Screw presses derive their energy from a flywheel. The forging load is transmitted through a vertical screw. These presses are **energy limited** and can be used for many forging operations. They are particularly suitable for producing small quantities, for parts requiring precision (such as turbine blades),

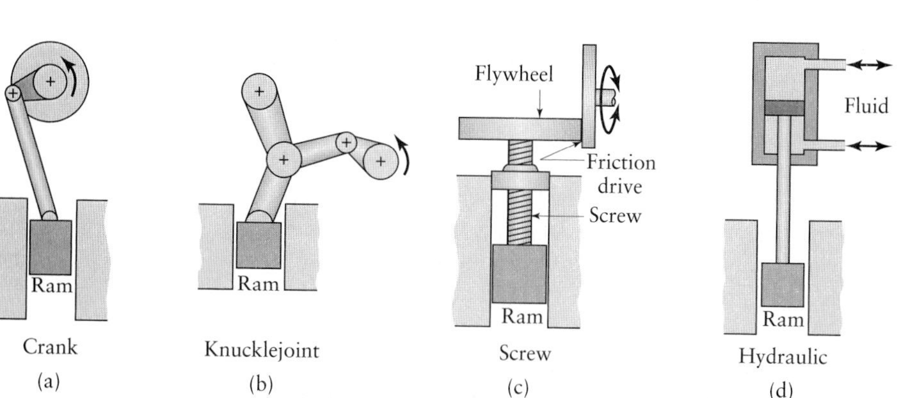

FIGURE 6.28 Schematic illustration of various types of presses used in metalworking. The choice of the press is an important factor in the overall operation.

and for control of ram speed. The largest screw press has a capacity of 280 MN (31,500 tons).

d) **Hammers.** Hammers derive their energy from the *potential energy* of the ram, which is then converted to kinetic energy; thus, hammers are energy limited. The speeds of hammers are high; therefore, the low forming times minimize cooling of the hot forging, allowing the forging of complex shapes, particularly with thin and deep recesses. To complete the forging, several blows may have to be made on the part. In power hammers, the ram is accelerated in the downstroke by steam or air, in addition to gravity. The highest energy available in power hammers is 1150 kJ (850,000 ft-lb).

e) **Counterblow hammers.** Counterblow hammers have two rams that simultaneously approach each other to forge the part. They are generally of the mechanical–pneumatic or mechanical–hydraulic type. These machines transmit less vibration to the foundation than other hammers. The largest counterblow hammer has a capacity of 1200 kJ (900,000 ft-lb).

Equipment selection. Selection of forging equipment depends on the size and complexity of the forging, the strength of the material and its sensitivity to strain rate, and the degree of deformation required. Generally, presses are preferred for aluminum, magnesium, beryllium, bronze, and brass; hammers are preferred for copper, steels, titanium, and refractory alloys. Production rate is also a consideration in equipment selection. The number of strokes per minute ranges from a few for hydraulic presses to as many as 300 for power hammers.

6.3 Rolling

Rolling is the process of reducing the thickness or changing the cross-section of a long workpiece by compressive forces applied through a set of rolls (Fig. 6.29), similar to rolling dough with a rolling pin to reduce its thickness. Rolling, which accounts for about 90% of all metals produced by metalworking processes, was first developed in the late 1500s. The basic operation is *flat rolling*, or simply *rolling*, where the rolled products are flat plate and sheet.

Plates, which are generally regarded as having a thickness greater than 6 mm ($\frac{1}{4}$ in.), are used for structural applications such as ship hulls, boilers, bridges, girders, machine structures, and nuclear vessels. Plates can be as much as 0.3 m (12 in.) thick for the supports for large boilers, 150 mm (6 in.) thick for reactor vessels, and 100–125 mm (4–5 in.) thick for battleships and tanks.

Sheets are generally less than 6 mm thick and are used for automobile bodies, appliances, containers for food and beverages, and kitchen and office equipment. Commercial-aircraft fuselages are usually made of about 1-mm-thick (0.040-in-thick) aluminum-alloy sheet, while beverage cans are made of 0.15-mm-thick (0.006-in-thick) aluminum sheet. Aluminum foil used to wrap candy and cigarettes has a thickness of 0.008 mm (0.0003 in.). Sheets are provided as flat pieces or as strip in coils to manufacturing facilities for further processing into products.

Traditionally, the initial form of material for rolling is an ingot. However, this practice is now being rapidly replaced by continuous casting and rolling, with their much higher efficiency and lower cost. Rolling is first carried out at elevated temperatures (hot rolling), wherein the coarse-grained, brittle, and porous structure of the ingot or continuously cast metal is broken down into a **wrought** structure, with finer grain size (Fig. 6.30).

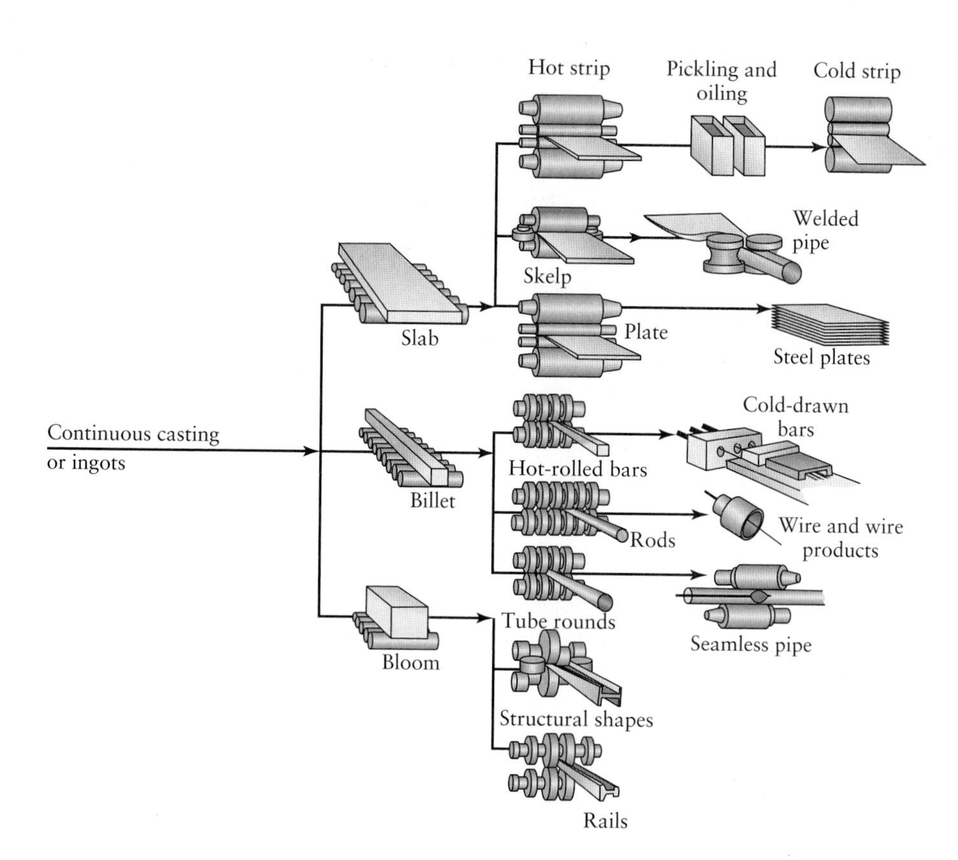

FIGURE 6.29 Schematic outline of various flat- and shape-rolling processes. *Source*: American Iron and Steel Institute.

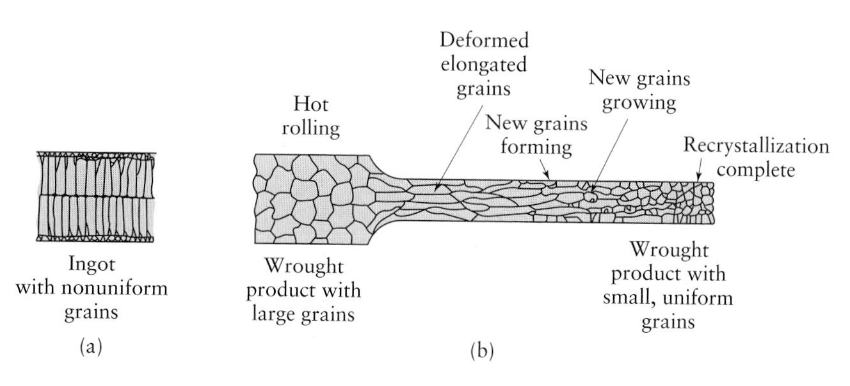

FIGURE 6.30 Changes in the grain structure of cast or large-grain wrought metals during hot rolling. Hot rolling is an effective way to reduce grain size in metals for improved strength and ductility. Cast structures of ingots or continuous castings are converted to a wrought structure by hot working.

6.3.1 Mechanics of flat rolling

The flat-rolling process is shown schematically in Fig. 6.31. A strip of thickness h_0 enters the roll gap and is reduced to a thickness of h_f by the rotating (powered) rolls. The surface speed of the roll is V_r. To keep the volume rate of metal flow constant, the velocity of the strip must increase as it moves through the roll gap (similar to fluid flow through a converging channel). At the exit of the roll gap, the velocity of the strip is V_f (Fig. 6.32). Because V_r is constant along the roll gap, sliding occurs between the roll and the strip.

At one point along the arc of contact, the two velocities are the same. This point is known as the **neutral point**, or **no-slip point**. To the left of this point, the roll moves

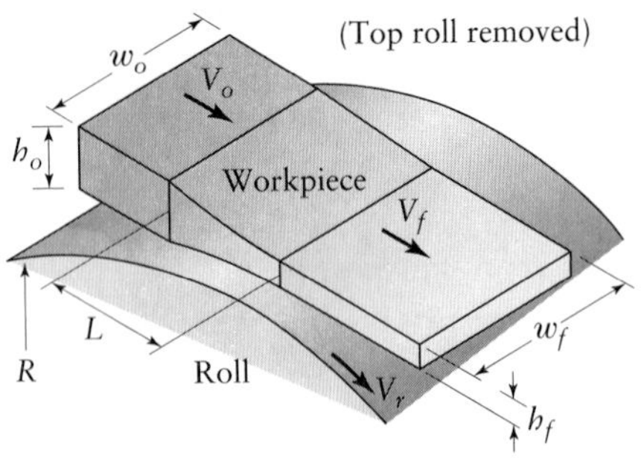

FIGURE 6.31 Schematic illustration of the flat-rolling process. A greater volume of metal is formed by rolling than by any other metalworking process.

(Top roll removed)

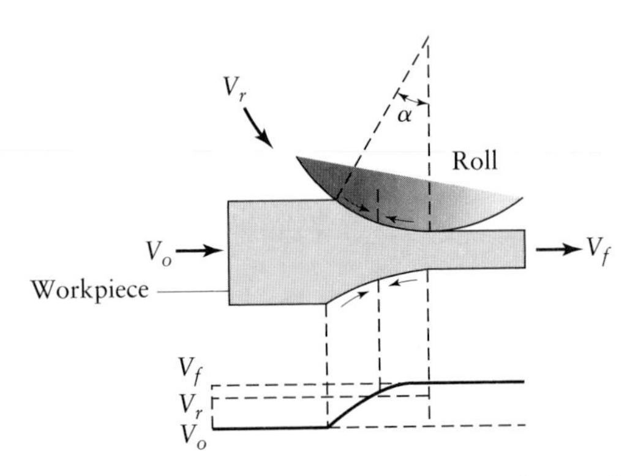

FIGURE 6.32 Relative velocity distribution between roll and strip surfaces. Note the difference in the direction of frictional forces. The arrows represent the frictional forces acting on the strip.

faster than the workpiece, and to the right the workpiece moves faster than the roll. Because of friction at the interfaces, the frictional forces (which oppose motion) act on the strip surfaces, as shown in Fig. 6.32.

Note the similarity between Figs. 6.32 and 6.1b. The frictional forces oppose each other at the neutral point. In upsetting, these forces are equal to each other, because of the symmetry of the operation. In rolling, the frictional force on the left of the neutral point must be greater than the frictional force on the right. This difference yields a net frictional force to the right, which makes the rolling operation possible by pulling the strip into the roll gap. Furthermore, the net frictional force and the surface velocity of the roll must be in the same direction in order to supply work to the system. Therefore, the neutral point should be located toward the exit in order to satisfy these requirements.

Forward slip in rolling is defined in terms of the exit velocity of the strip, V_f, and the surface speed of the roll, V_r, as

$$\text{Forward slip} = \frac{V_f - V_r}{V_r} \tag{6.23}$$

and is a measure of the relative velocities involved.

a) **Roll pressure distribution.** Although the deformation zone in the material is subjected to a state of stress similar to that in upsetting, the calculation of forces and stress distribution in flat rolling is more involved, because of the curved surface of contact. In addition, the material at the exit is strain hardened, so the flow stress at the exit is higher than that at the entry.

Figure 6.33 shows the stresses on an element in the entry and exit zones. Note that the only difference between the two elements is the direction of the friction force. Using the slab method of analysis for *plane strain* (described in Section 6.2.2), we may analyze the stresses in rolling as follows.

From the equilibrium of the horizontal forces on the element shown in Fig. 6.33, we have

$$(\sigma_x + d\sigma_x)(h + dh) - 2pR\, d\phi \sin\phi - \sigma_x h \pm 2\mu pR\, d\phi \cos\phi = 0.$$

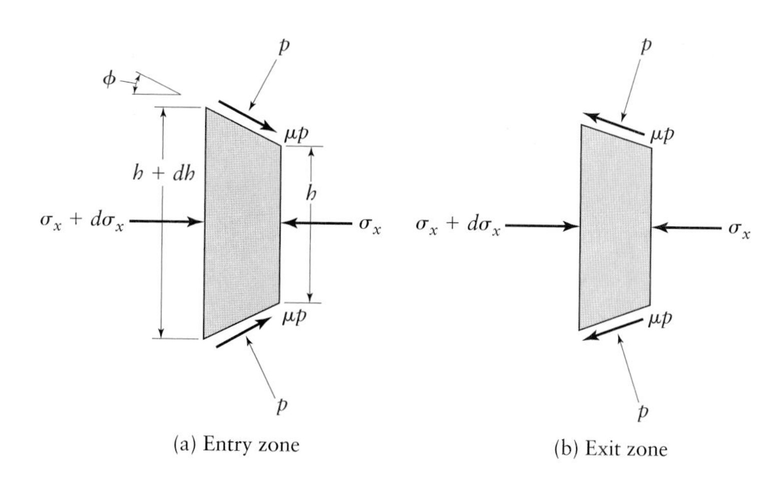

FIGURE 6.33 Stresses on an element in rolling: (a) entry zone and (b) exit zone.

When simplified, and if second-order terms are neglected, this expression reduces to

$$\frac{d(\sigma_x h)}{d\phi} = 2pR(\sin\phi \mp \mu\cos\phi).$$

In rolling, the angle α is only a few degrees; hence, we may assume that $\sin\phi = \phi$ and $\cos\phi = 1$. Thus,

$$\frac{d(\sigma_x h)}{d\phi} = 2pR(\phi \mp \mu). \tag{6.24}$$

Because the angles involved are small, we assume that p is a *principal stress*, σ_x being the other principal stress. The relationship between these two stresses and the flow stress, Y_f, of the material is given by Eq. (2.45) for plane strain, or

$$p - \sigma_x = \frac{2}{\sqrt{3}}Y_f = Y_f'. \tag{6.25}$$

For strain-hardening material, the flow stress, Y_f, in these expressions corresponds to the strain that the material has undergone at that particular location in the gap. Rewriting Eq. (6.24), we have

$$\frac{d[(p - Y_f')h]}{d\phi} = 2pR(\phi \mp \mu),$$

or

$$\frac{d}{d\phi}\left[Y_f'\left(\frac{p}{Y_f'} - 1\right)h\right] = 2pR(\phi \mp \mu),$$

which, upon differentiation, becomes

$$Y_f'h\frac{d}{d\phi}\left(\frac{p}{Y_f'}\right) + \left(\frac{p}{Y_f'} - 1\right)\frac{d}{d\phi}(Y_f'h) = 2pR(\phi \mp \mu).$$

The second term is very small, because as h decreases, Y_f' increases, due to cold working, thus making the product nearly a constant, and its derivative thus becomes zero. We now have

$$\frac{\frac{d}{d\phi}\left(\frac{p}{Y_f'}\right)}{\frac{p}{Y_f'}} = \frac{2R}{h}(\phi \mp \mu). \tag{6.26a}$$

Letting h_f be the final thickness, we have

$$h = h_f + 2R(1 - \cos\phi),$$

or, approximately,

$$h = h_f + R\phi^2. \tag{6.26b}$$

Substituting this expression for h in Eq. (6.26a) and integrating, we obtain

$$\ln\frac{p}{Y'_f} = \ln\frac{h}{R} \mp 2\mu\sqrt{\frac{R}{h_f}}\tan^{-1}\sqrt{\frac{R}{h_f}}\phi + \ln C,$$

or

$$p = CY'_f\frac{h}{R}e^{\mp\mu H},$$

where

$$H = 2\sqrt{\frac{R}{h_f}}\tan^{-1}\left(\sqrt{\frac{R}{h_f}}\phi\right). \tag{6.27}$$

At entry, $\phi = \alpha$; hence, $H = H_0$, with ϕ replaced by α. At exit, $\phi = 0$; hence, $H = H_f = 0$. Also, at entry and exit, $p = Y'_f$. Thus, in the **entry zone,**

$$C = \frac{R}{h_f}e^{\mu H_i}$$

and

$$p = Y'_f\frac{h}{h_0}e^{\mu(H_0 - H)}. \tag{6.28}$$

In the **exit zone,** we have

$$C = \frac{R}{h_f}$$

and

$$p = Y'_f\frac{h}{h_f}e^{\mu H}. \tag{6.29}$$

Note that the pressure p is a function of h and the angular position ϕ along the arc of contact. These expressions also indicate that the pressure increases with increasing strength of the material, increasing coefficient of friction, and increasing R/h_f ratio. The R/h_f ratio in rolling is equivalent to the a/h ratio in upsetting. (See Section 6.2.2.)

The dimensionless theoretical pressure distribution in the roll gap is shown in Fig. 6.34. Note the similarity of this curve to that in Fig. 6.5 (friction hill). Also note that the neutral point shifts toward the exit as friction decreases, which is to be expected. When friction approaches zero, the rolls begin to slip instead of pulling the strip in, so the neutral point must approach the exit. (This situation is similar to the spinning of automobile wheels when accelerating on wet pavement.)

Figure 6.35 shows the effect of reduction in thickness of the strip on the pressure distribution. As reduction increases, the length of contact in the roll gap increases, which in turn increases the peak pressure. The curves shown in Fig. 6.35 are theoretical; actual pressure distributions, as determined experimentally, have smoother curves, with their peaks rounded off.

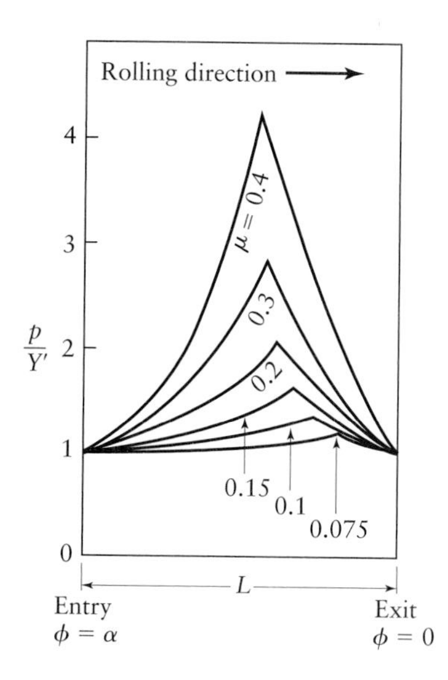

FIGURE 6.34 Pressure distribution in the roll gap as a function of the coefficient of friction. Note that as friction increases, the neutral point shifts toward the entry. Without friction, the rolls slip, and the neutral point shifts completely to the exit.

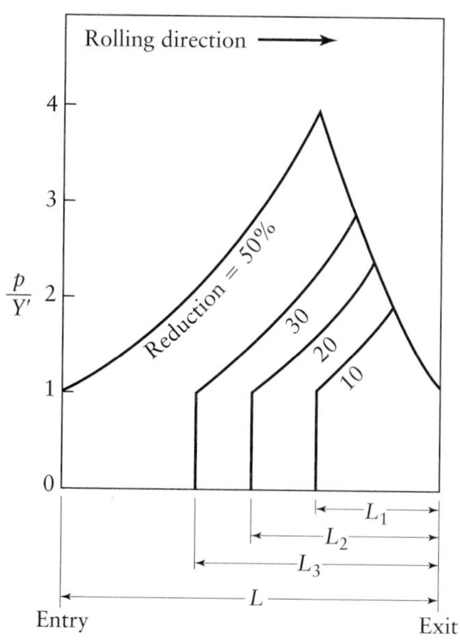

FIGURE 6.35 Pressure distribution in the roll gap as a function of reduction in thickness. Note the increase in the area under the curves with increasing reduction in thickness, thus increasing the roll force.

b) **Determination of the neutral point.** The neutral point can be determined simply by equating Eqs. (6.28) and (6.29); thus, at the neutral point,

$$\frac{h_0}{h_f} = \frac{e^{\mu H_0}}{e^{2\mu H_n}} = e^{\mu(H_0 - 2H_n)},$$

or

$$H_n = \frac{1}{2}\left(H_0 - \frac{1}{\mu}\ln\frac{h_0}{h_f}\right). \tag{6.30}$$

Substituting Eq. (6.30) into Eq. (6.27), we obtain

$$\phi_n = \sqrt{\frac{h_f}{R}}\tan\left(\sqrt{\frac{h_f}{R}}\cdot\frac{H_n}{2}\right). \tag{6.31}$$

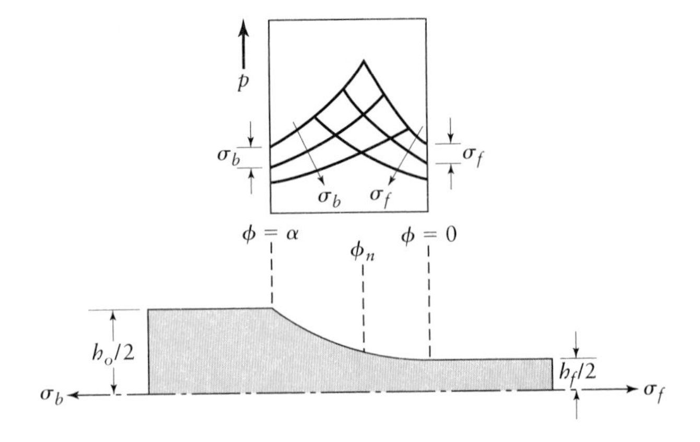

FIGURE 6.36 Pressure distribution as a function of front and back tension. Note the shifting of the neutral point and the reduction in the area under the curves with increasing tension.

c) **Front and back tension.** The roll force, F, can be reduced by various means, such as lowering friction, using rolls of smaller radii, taking smaller reductions, and raising workpiece temperature. A particularly effective method is to reduce the apparent compressive yield stress of the material by applying longitudinal *tension*. Recall from our discussion of yield criteria (Section 2.11) that if tension is applied to a strip (Fig. 6.36), the yield stress normal to the strip surface drops, and hence the roll pressure will decrease.

Tensions in rolling can be applied either at the entry (**back tension,** σ_b) or at the exit (**front tension,** σ_f) of the strip or at both. Equations (6.28) and (6.29) can then be modified to include the effect of tension for the entry and exit zones as follows, respectively:

$$\text{Entry zone: } p = (Y_f' - \sigma_b)\frac{h}{h_0}e^{\mu(H_0-H)} \tag{6.32}$$

and

$$\text{Exit zone: } p = (Y_f' - \sigma_f)\frac{h}{h_f}e^{\mu H}. \tag{6.33}$$

Depending on the relative magnitudes of the tensions applied, the neutral point may shift, as shown in Fig. 6.36. This shift affects the pressure distribution, torque, and power requirements in rolling. Front tension is controlled by the torque on the coiler (**delivery reel**) around which the rolled sheet is coiled. The back tension is controlled by a braking system in the uncoiler (**payoff reel**). Special instrumentation is available for these controls. Tensions are particularly important in rolling thin, high-strength materials, because such materials require high roll forces.

EXAMPLE 6.3 Back tension required to cause roll slip

It is to be expected that in rolling a strip, the rolls will begin to slip if the back tension, σ_b, is too high. Derive an expression for the magnitude of back tension required to make the rolls begin to slip.

Solution. Slipping of the rolls means that the neutral point has moved all the way to the exit in the roll gap; thus, the whole contact area becomes the entry zone, and Eq. (6.32) is applicable. Also, we know that when $\phi = 0, H = 0$; hence, the pressure at the exit is

$$p_{\phi=0} = (Y_f' - \sigma_b)\left(\frac{h_f}{h_0}\right)e^{\mu H_0}.$$

However, at the exit, the pressure is equal to Y_f'. Rearranging this equation, we obtain

$$\sigma_b = Y_f'\left[1 - \left(\frac{h_0}{h_f}\right)(e^{-\mu H_0})\right],$$

where H_0 is obtained from Eq. (6.27) with the condition $\phi = \alpha$. Since all the quantities are known, the magnitude of the back tension can be calculated.

d) **Roll forces.** The area under the pressure vs. contact–length curve multiplied by the strip width, w, is the *roll force, F,* (also called *roll–separating force*) on the strip. This area can be obtained graphically after the points are plotted. The force can also be calculated from the expression

$$F = \int_0^{\phi_n} wpR\,d\phi + \int_{\phi_n}^{\alpha} wpR\,d\phi. \tag{6.34}$$

A simpler method of calculating the roll force is to multiply the contact area by an average contact stress, as in

$$F = Lwp_{av}, \tag{6.35}$$

where L is the *arc of contact*. The dimension L can be approximated as the projected length by the equation

$$L = \sqrt{R\Delta h}, \tag{6.36}$$

where R is the roll radius and Δh is the difference between the original and final thicknesses of the strip (**draft**).

The estimation of p_{av} depends on the h/L ratio, where h is now the *average thickness* of the strip in the roll gap. (See Figs. 6.12 and 6.13.) For large h/L ratios (small reductions and/or large roll diameters), the rolls act as indenters in a hardness test. Friction is not significant, and p_{av} is obtained from Fig. 6.12, where $a = L/2$. Small h/L ratios (large reductions and/or large roll diameters) are equivalent to high a/h ratios; thus, friction is predominant, and p_{av} is obtained from Eq. (6.15). For strain-hardening materials, the appropriate flow stresses must be calculated.

As a rough approximation and for low frictional conditions, Eq. (6.35) can be simplified as

$$F = Lw\bar{Y}' \tag{6.37}$$

where $\bar{Y}'$ is the average flow stress in plane strain (see Fig. 2.39) of the material in the roll gap. For higher frictional conditions, we may write an expression similar to Eq. (6.15):

$$F = Lw\bar{Y}'\left(1 + \frac{\mu L}{2h_{av}}\right). \tag{6.38}$$

e) **Roll torque and power.** The *roll torque, T,* for each roll can be calculated analytically from the expression

$$T = \underbrace{\int_{\phi_n}^{\alpha} w\mu pR^2\,d\phi}_{\text{(entry zone)}} - \underbrace{\int_0^{\phi_n} w\mu pR^2\,d\phi}_{\text{(exit zone)}}. \tag{6.39}$$

Note that the minus sign indicates the change in direction of the friction force at the neutral point; thus, if the frictional forces are equal to each other, the torque is zero.

The torque in rolling can also be estimated by assuming that the roll force, F, acts in the *middle* of the arc of contact—that is, a moment arm of $0.5L$—and that F is perpendicular to the plane of the strip. (Whereas $0.5L$ is a good estimate for hot rolling, $0.4L$ is a better estimate for cold rolling.)

The **torque per roll** is then

$$T = \frac{FL}{2}.$$

The **power required per roll** is

$$\text{Power} = T\omega, \tag{6.40}$$

where $\omega = 2\pi N$, and N is the revolutions per minute of the roll. Consequently, the power per roll is

$$\text{Power} = \frac{\pi FLN}{60{,}000}\,\text{kW}, \tag{6.41}$$

where F is in newtons, L is in meters, and N is the rpm of the roll. The power per roll can also be expressed as

$$\text{Power} = \frac{\pi FLN}{33{,}000}\,\text{hp}, \tag{6.42}$$

where F is in lb and L is in ft.

EXAMPLE 6.4 Power required in rolling

A 9-in-wide 6061-O aluminum strip is rolled from a thickness of 1.00 in. to 0.80 in. If the roll radius is 12 in. and the roll rpm is 100, estimate the total horsepower required for this operation.

Solution. The power needed for a set of two rolls is given by Eq. (6.42) as

$$\text{Power} = \frac{2\pi FLN}{33{,}000}\,\text{hp},$$

where F is given by Eq. (6.37) and L by Eq. (6.36). Hence,

$$F = Lw\bar{Y}' \qquad \text{and} \qquad L = \sqrt{R\Delta h}.$$

Therefore,

$$L = \sqrt{(12)(1.0 - 0.8)} = 1.55\ \text{in.} = 0.13\ \text{ft} \qquad \text{and} \qquad w = 9\ \text{in.}$$

For 6061-O aluminum, $K = 30{,}000$ psi and $n = 0.2$ (from Table 2.3). The true strain in this operation is

$$\varepsilon_1 = \ln\left(\frac{1.0}{0.8}\right) = 0.223.$$

Thus, from Eq. (6.10), we have

$$\bar{Y} = \frac{(30{,}000)(0.223)^{0.2}}{1.2} = 18{,}500\ \text{psi}$$

and

$$\bar{Y}' = (1.15)(18{,}500) = 21{,}275 \text{ psi}.$$

Therefore,

$$F = (1.55)(9)(21{,}275) = 297{,}000 \text{ lb},$$

and for two rolls

$$\text{Power} = \frac{(2\pi)(297{,}000)(0.13)(100)}{33{,}000} = 735 \text{ hp}.$$

f) **Forces in hot rolling.** Because ingots and slabs are usually hot rolled, the calculation of forces and torque in hot rolling is important. However, there are two major difficulties. One is the proper estimation of the coefficient of friction, μ, at elevated temperatures. The other is the strain-rate sensitivity of metals at high temperatures.

The *average strain rate* in flat rolling can be obtained by dividing the strain by the time required for an element to undergo this strain in the roll gap. The time can be approximated as L/V_r, so

$$\dot{\varepsilon} = \frac{V_r}{L} \ln\left(\frac{h_0}{h_f}\right). \qquad (6.43)$$

The flow stress, Y_f, of the material corresponding to this strain rate must first be obtained and then substituted into the proper equations. These calculations are approximate; variations in μ and the temperature within the strip in hot rolling further contribute to the difficulties in calculating forces accurately.

g) **Friction.** In rolling, although the rolls cannot pull the strip into the roll gap without some friction, forces and power requirements rise with increasing friction. In cold rolling, μ usually ranges between 0.02 and 0.3, depending on the materials and lubricants used. The low ranges for the coefficient of friction are obtained with effective lubricants and regimes approaching *hydrodynamic lubrication*, such as in cold rolling of aluminum at high speeds. In hot rolling, μ may range from about 0.2, with effective lubrication, to as high as 0.7, indicating sticking, which usually occurs with steels, stainless steels, and high-temperature alloys.

The maximum possible draft, that is, $h_0 - h_f$, in flat rolling is a function of friction and radius

$$\Delta h_{\text{max}} = \mu^2 R. \qquad (6.44)$$

Therefore, the higher the friction and the larger the roll radius, the greater is the maximum draft. As expected, the draft is zero when there is no friction. The maximum value of the angle α (**angle of acceptance**) in Fig. 6.32 is geometrically related to Eq. (6.44). From the simple model of a block sliding down an inclined plane, it can be seen that

$$\alpha_{\text{max}} = \tan^{-1} \mu. \qquad (6.45)$$

If α_{max} is larger than this value, the rolls begin to slip, because the friction is not high enough to pull the material through the roll gap.

h) **Roll deflection and flattening.** Roll forces tend to bend the rolls, as shown in Fig. 6.37a, with the result that the strip is thicker at its center than at its edges (**crown**). The usual method of avoiding this problem is to grind the rolls

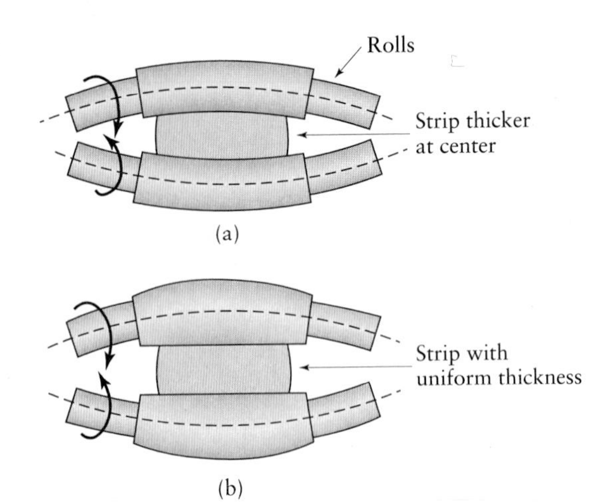

FIGURE 6.37 (a) Bending of straight cylindrical rolls (exaggerated) because of the roll force. (b) Bending of rolls, ground with camber, that produce a sheet of uniform thickness during rolling.

so that their diameter at the center is slightly larger than at the edges. This curvature is known as **camber**. In rolling sheet metals, the camber is generally less than 0.25 mm (0.01 in.) on the radius. Because of the heat generated during rolling, rolls can become slightly barrel-shaped, known as **thermal camber**. This effect can be controlled by varying the location of the coolant on the rolls.

When properly designed, such rolls produce flat strips, as shown in Fig. 6.37b. However, a particular camber is correct only for a certain load and width of strip. In hot rolling, uneven temperature distribution in the roll can also cause its diameter to vary along the length of the roll. In fact, camber can be controlled by varying the location of the coolant (lubricant) on the rolls in hot rolling.

Forces also tend to *flatten* the rolls elastically, much like the flattening of tires on automobiles. Flattening of a roll increases the roll's radius and hence yields a larger contact area for the same reduction in thickness; thus, the roll force, F, increases.

The new (distorted) roll radius R' is

$$R' = R\left(1 + \frac{CF'}{h_0 - h_f}\right). \quad (6.46)$$

In Eq. (6.46), C is 2.3×10^{-2} mm^2/kN (1.6×10^{-4} in^2/klb) for steel rolls and 4.57×10^{-2} (3.15×10^{-4}) for cast-iron rolls. F' is the **roll force per unit width** of strip, expressed in kN/mm (klb/in.). The higher the elastic modulus of the roll material, the less the roll distorts. Note from Eq. (6.46) that reducing the roll force, such as by using an effective lubricant or taking smaller reductions, also reduces roll flattening. We cannot determine R' directly from Eq. 6.46, because it is a function of the roll force, F, which is itself a function of the roll radius. The solution is obtained by trial and error. (Note that R in all previous equations should be replaced by R' when significant roll flattening occurs).

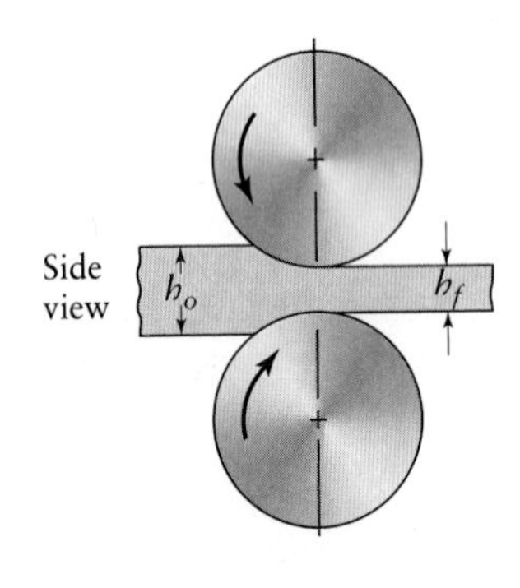

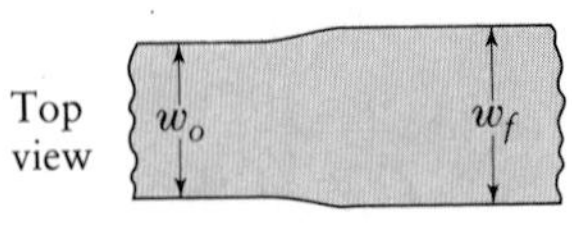

FIGURE 6.38 Increase in the width of a strip (spreading) in flat rolling. Spreading can be similarly observed when dough is rolled with a rolling pin.

i) **Spreading.** Although rolling plates and sheets with high width-to-thickness ratios is essentially a process of plane strain, with smaller ratios, such as a square cross-section, the width increases considerably during rolling. This increase in width is known as spreading (Fig. 6.38). Spreading decreases with increasing width-to-thickness ratios of the entering material, decreasing friction, and increasing ratios of roll radius to strip thickness. Spreading can be prevented by using vertical rolls that are in contact with the edges of the rolled product (**edger mill**).

6.3.2 Defects in rolling

Successful rolling practice requires balancing many factors, including material properties, process variables, and lubrication. There may be defects on the surfaces of the rolled plates and sheets, or there may be structural defects within the material. **Surface defects** may result from inclusions and impurities in the material, scale, rust, dirt, roll marks, and other causes related to the prior treatment and working of the material. In hot rolling blooms, billets, and slabs, the surface is usually preconditioned by various means, such as by torch (**scarfing**), to remove scale.

Structural defects are defects that distort or affect the integrity of the rolled product. Figure 6.39 shows some typical defects. **Wavy edges** are caused by bending of the rolls; the edges of the strip are thinner than the center. Because the edges elongate more than the center and are restrained from expanding freely, they buckle. The cracks shown in Fig. 6.39b and c are usually caused by low ductility and barreling. **Alligatoring** is a complex phenomenon that results from inhomogeneous deformation of the material during rolling or from defects in the original cast ingot, such as piping.

Residual stresses. Residual stresses can be generated in rolled sheets and plates because of inhomogeneous plastic deformation in the roll gap. Small-diameter rolls or small reductions tend to work the metal plastically at its surfaces (similarly to shot peening or roller burnishing). This working generates compressive residual stresses on the surfaces and tensile stresses in the bulk (Fig. 6.40a). Large-diameter rolls and high reductions, however, tend to deform the bulk to a greater extent than the surfaces, because of frictional constraint at the surfaces along the arc of contact. This situation generates residual stresses that are opposite to those of the previous case, as shown in Fig. 6.40b.

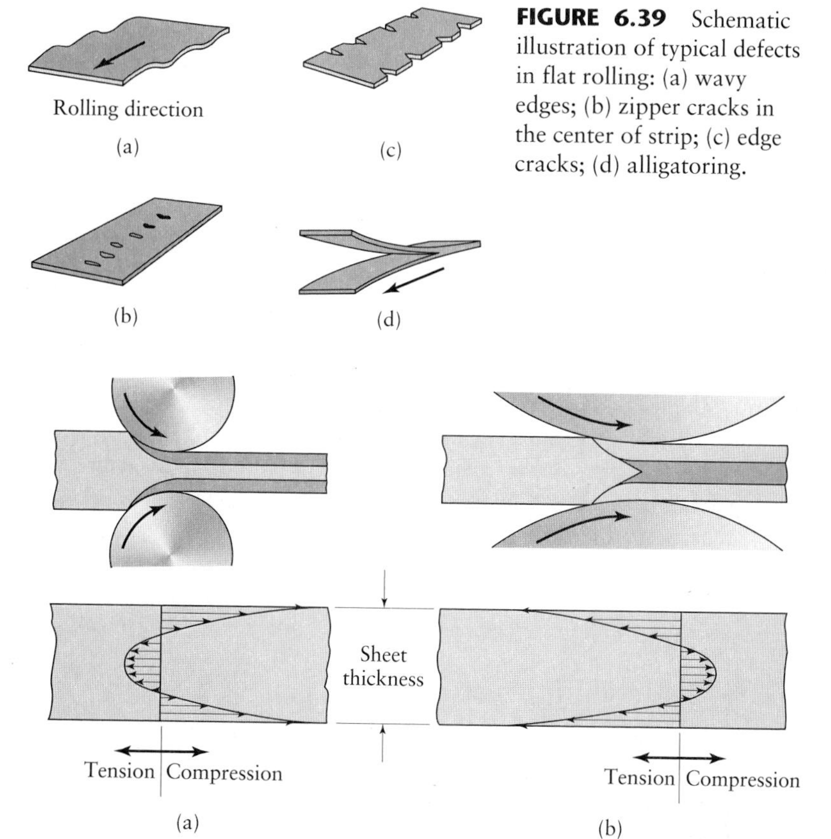

FIGURE 6.39 Schematic illustration of typical defects in flat rolling: (a) wavy edges; (b) zipper cracks in the center of strip; (c) edge cracks; (d) alligatoring.

FIGURE 6.40 The effect of roll radius on the type of residual stresses developed in flat rolling: (a) small rolls, or small reduction in thickness; and (b) large rolls, or large reduction in thickness.

6.3.3 Vibration and chatter in rolling

Vibration and *chatter* in metalworking processes can have significant effects on product quality and productivity of the operation. Chatter, generally defined as self-excited vibration, can occur in rolling as well as in extrusion, drawing, machining, and grinding operations. (See Sections 8.11 and 9.6.8.) In rolling, it leads to random variations in the thickness of the rolled sheet and its surface finish and, consequently, can lead to excessive scrap. Chatter in rolling has been found to occur predominantly in *tandem mills* (see Fig. 6.42e); it is a very complex phenomenon and results from interactions between the structural dynamics of the mill stand and the dynamics of the rolling operation. Rolling speed and lubrication are found to be the two most important parameters. Chatter is very important to productivity; it has been estimated, for example, that modern rolling mills could operate at speeds up to 50% higher were it not for chatter. Considering the very high cost of rolling mills, this issue is a major economic concern.

The vibration modes commonly encountered in rolling are classified as torsional, third-octave, and fifth-octave chatter (*octave* meaning a frequency on the eighth degree from a given frequency).

1. **Torsional chatter** is characterized by a low resonant frequency (approximately 5–15 Hz) and is usually a result of forced vibration (see Section 8.11), although it can also occur simultaneously with third-octave chatter. Torsional chatter leads to minor variations in gage thickness and surface finish and is usually not significant unless improper conditions exist in the machinery, such as malfunctioning speed controls, broken gear teeth, and misaligned shafts.
2. **Third-octave chatter** occurs in the frequency range of 125–240 Hz (the third musical octave is 128–256 Hz) and is *self-excited*; that is, energy from the rolling mill is transformed into vibratory energy, regardless of the presence of any external forcing (forced vibration). This mode of vibration is the most serious in rolling, leading to significant gage variations and fluctuations in strip tension between the stands and often resulting in strip breakage. Third-octave chatter is generally controlled by reducing the rolling speed, V. Although not always practical to implement, it has been suggested that this type of chatter can also be reduced by increasing the distance between the stands of the rolling mill; increasing the strip width, w; incorporating *dampers* (see Section 8.11.1) in the roll supports; decreasing the reduction per pass (draft); increasing the roll radius R; and increasing the coefficient of strip-roll friction, μ.
3. **Fifth-octave chatter** occurs in the frequency range of 550–650 Hz; it has been attributed to chatter marks that develop on work rolls and are caused by extended operation at certain critical speeds, surface defects on a backup roll or in the incoming strip, and/or improperly ground rolls. The chatter marks or striations are then imprinted onto the rolled sheet at any speed, adversely affecting its surface appearance. In addition, fifth-octave chatter is undesirable because of the objectionable noise generated. This type of chatter can be controlled by modulating the mill speed, using backup rolls of progressively larger diameters in the stands of a tandem mill, avoiding chatter in roll grinding, and eliminating the sources of other vibrations in the mill stand.

6.3.4 Flat-rolling practice

The initial breaking down of an ingot (or continuous casting) by hot rolling converts the coarse-grained, brittle, and porous cast structure to a wrought structure. (See Fig. 6.30.) This structure has finer grains and enhanced ductility, resulting from the breaking up of brittle grain boundaries and the closing up of internal defects, such as

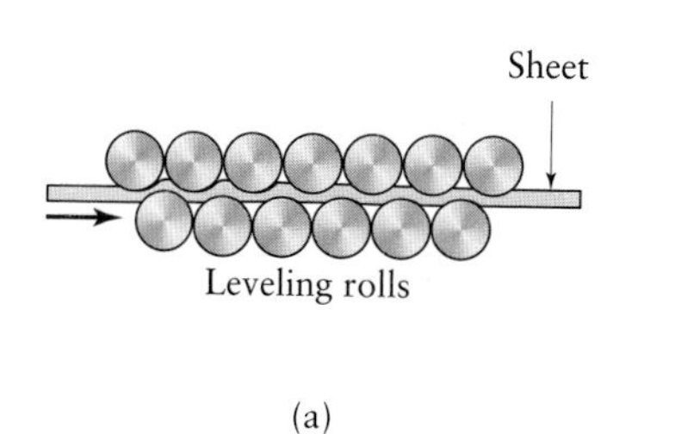

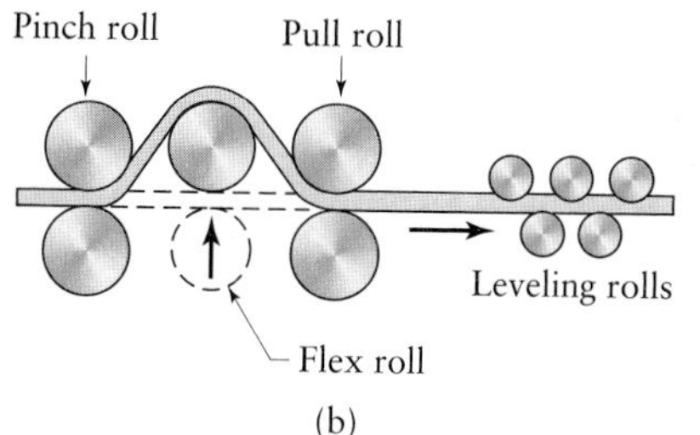

FIGURE 6.41 Schematic illustration of methods of roller leveling. These processes are used to flatten rolled sheets.

porosity. Traditional methods of rolling ingots are now being rapidly replaced by continuous casting. Temperature ranges for hot rolling are similar to those for forging. (See Table 6.3.)

The product of the first hot-rolling operation is called a **bloom** or **slab**. (See Fig. 6.29.) A bloom usually has a square cross-section, at least 150 mm (6 in.) on the side; a slab is usually rectangular in cross-section. Blooms are processed further by shape rolling into structural shapes, such as I-beams and railroad rails. Slabs are rolled into plates and sheet. **Billets** are usually square, with a cross-sectional area smaller than that of blooms, and are rolled into various shapes, such as round rods and bars, using shaped rolls. (Hot-rolled round rods are used as the starting material for rod and wire drawing and are called **wire rods**.)

In hot rolling blooms, billets, and slabs, the surface of the material is usually **conditioned** (i.e., prepared for a subsequent operation) prior to rolling. Conditioning is done by various means, such as with a torch (scarfing) to remove heavy scale, or by rough grinding to smoothen surfaces. Prior to cold rolling, the scale developed during hot rolling or other defects may be removed by pickling with acids; by mechanical means, such as blasting with water; or by grinding.

Pack rolling is a flat-rolling operation in which two or more layers of metal are rolled together, thus improving productivity. Aluminum foil, for example, is pack rolled in two layers. One side of aluminum foil is matte, and the other side is shiny. The foil-to-foil side has a matte and satiny finish, and the foil-to-roll side is shiny and bright, because it has been in contact with the polished roll.

Mild steel, when stretched during sheet-forming operations, undergoes **yield-point elongation**. This phenomenon causes surface irregularities called **stretcher strains** or **Lueder's bands**. (See also Section 7.2.) To avoid this situation, the sheet metal is subjected to a light pass of 0.5–1.5% reduction, known as **temper rolling** (*skin pass*).

A rolled sheet may not be sufficiently flat as it leaves the roll gap, because of variations in the material or in the processing parameters during rolling. To improve flatness, the strip is then passed through a series of **leveling rolls**. Each roll is usually driven separately with individual electric motors. The strip is flexed in opposite directions as it passes through the sets of rollers. Several different roller arrangements are used (Fig. 6.41).

Sheet thickness is identified by a **gage number**; the smaller the number, the thicker the sheet is. There are different numbering systems for different sheet metals. Rolled sheets of copper and brass are also identified by thickness changes during rolling, such as $\frac{1}{4}$ hard, $\frac{1}{2}$ hard, and so on.

Lubrication. Ferrous alloys are usually hot rolled without a lubricant, although graphite may be used. Aqueous solutions are used to cool the rolls and break up the scale on the workpiece. Nonferrous alloys are hot rolled with a variety of compounded oils, emulsions, and fatty acids. Cold rolling is done with low-viscosity lubricants, including mineral oils, emulsions, paraffin, and fatty oils.

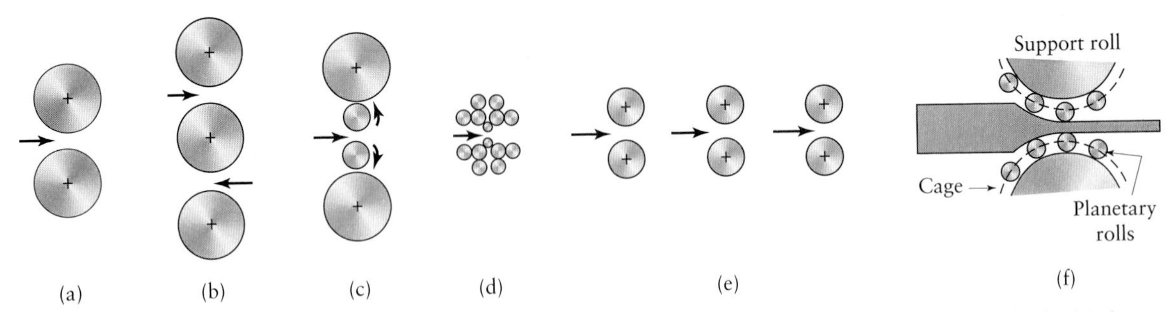

FIGURE 6.42 Schematic illustration of various roll arrangements: (a) two high; (b) three high; (c) four high; (d) cluster; (e) tandem rolling with three stands; (f) planetary.

Equipment. A wide variety of rolling equipment is available with a number of roll arrangements, the major types of which are shown in Fig. 6.42. Small-diameter rolls are preferable, because the smaller the roll radius, the lower the roll force is. However, small rolls deflect under roll forces and have to be supported by other rolls (Fig. 6.42c and 6.42d).

Two-high or **three-high** (developed in the mid-1800s) rolling mills typically are used for initial breakdown passes on cast ingots (primary roughing) with roll diameters ranging up to 1400 mm (55 in.). The **cluster mill (Sendzimir,** or **Z, mill)** is particularly suitable for cold rolling thin strips of high-strength metals. Figure 6.43 shows one arrangement for the rolls. The **work roll** (the smallest roll) can be as small as 6 mm ($\frac{1}{4}$ in.) in diameter and is usually made of tungsten carbide for rigidity, strength, and wear resistance. The rolled product obtained in a cluster mill can be as wide as 5000 mm (200 in.) and as thin as 0.0025 mm (0.0001 in.)

In **tandem rolling** (Fig. 6.42e) the strip is rolled continuously through a number of **stands**; a group of stands is called a **train**. Control of the gage and of the speed at which the strip travels through each roll gap is critical. Flat rolling can also be carried out with front tension only, using idling rolls (**Steckel rolling**); for this case, the torque on the roll is zero, assuming frictionless bearings.

Stiffness in rolling mills is important for controlling dimensions. Mills can be highly automated, with rolling speeds as high as 25 m/s (5000 ft/min). Requirements for roll materials are mainly strength and resistance to wear. Three common roll materials are cast iron, cast steel, and forged steel. For hot rolling, roll surfaces are generally rough and may even have notches or grooves in order to pull the metal through the roll gap at high reductions. Rolls for cold rolling are ground to a fine finish and, for special applications, are also polished.

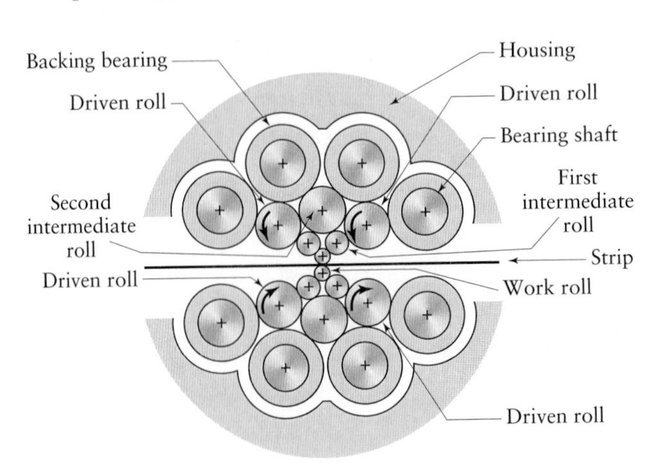

FIGURE 6.43 Schematic illustration of a cluster (*Sendzimir*) mill. These mills are very rigid and are used in rolling thin sheets of high-strength materials, with good control of dimensions.

Minimills. In minimills, scrap metal is melted in electric-arc furnaces, cast continuously, and rolled directly into specific lines of products. Each minimill produces essentially one kind of rolled product (rod, bar, or structural shapes) from one type of metal or alloy and is usually oriented to markets within the mill's particular geographic area. The scrap metal, which is obtained locally to reduce transportation costs, is usually old machinery, automobiles, and farm equipment.

Integrated mills. Integrated mills are large facilities that involve complete activities, from the production of hot metal in a blast furnace to casting and rolling of finished products that are ready to be shipped to the customer.

6.3.5 Miscellaneous rolling operations

a) **Shape rolling.** Straight structural shapes such as bars of various cross-sections, channel sections, I-beams, and railroad rails are rolled by passing the stock through a number of pairs of specially designed rollers (Fig. 6.44); these processes were first developed in the late 1700s. The original material is usually a bloom (Fig. 6.29). Designing a series of rolls (**roll-pass design**) requires experience in order to avoid defects and hold dimensional tolerances, although some defects may already exist in the material being rolled. The material elongates as it is reduced in cross-section. However, for a shape such as a channel, the reduction is different in various locations within the section; thus, elongation is not uniform, which can cause the product to warp or crack. Airfoil shapes can also be produced by shape-rolling techniques.

b) **Ring rolling.** In ring rolling, a small-diameter, thick ring is expanded into a larger-diameter, thinner ring. The ring is placed between two rolls, one of which is driven (Fig. 6.45). The thickness is reduced by bringing the rolls closer as they rotate. The reduction in thickness is compensated for by an increase in the diameter of the ring. A great variety of cross-sections can be ring rolled with shaped rolls. This process can be carried out at room or elevated temperatures, depending on the size and strength of the product. The advantages of ring rolling, compared with other processes for making the same part, are short production runs, material savings, close dimensional tolerances, and favorable

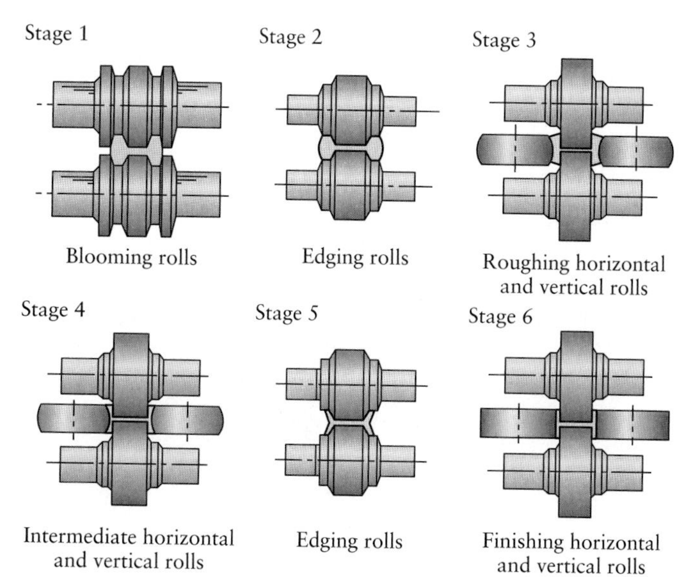

FIGURE 6.44 Stages in shape rolling of an H-section part. Various other structural sections, such as channels and I-beams, are also rolled by this process.

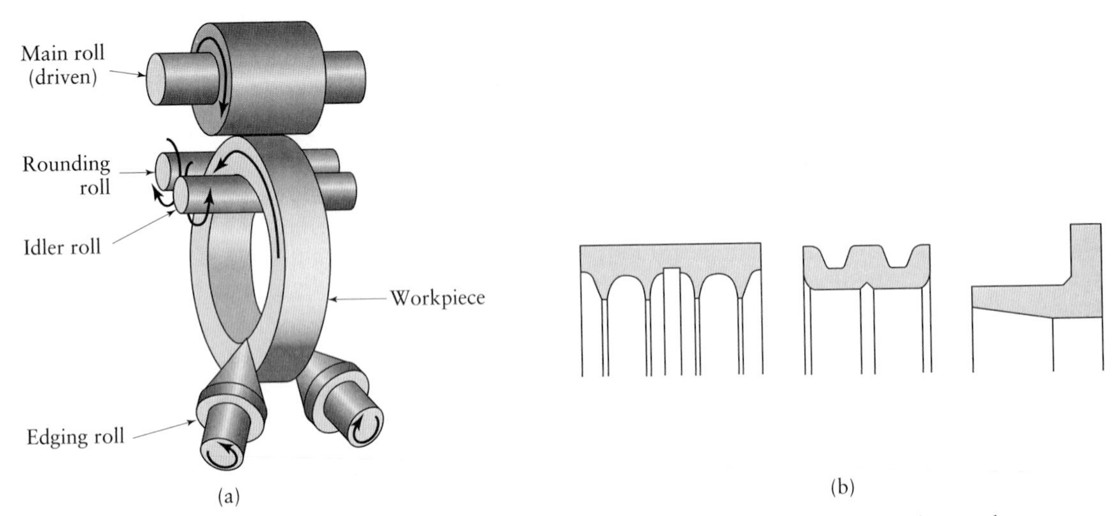

FIGURE 6.45 (a) Schematic illustration of a ring-rolling operation. Reducing the thickness results in an increase in the part's diameter. (b) Examples of cross-sections that can be formed by ring rolling.

grain-flow direction. Typical applications of ring rolling include large rings for rockets and turbines, gearwheel rims, ball- and roller-bearing races, flanges and reinforcing rings for pipes, and pressure vessels.

c) **Thread and gear rolling.** In the cold-forming thread-and-gear-rolling process, threads are formed on round rods or workpieces by passing them between reciprocating or rotating dies (Fig. 6.46a). Typical products made by this process include screws, bolts, and similar threaded parts. With flat dies, the threads are rolled on the rod or wire with each stroke of the reciprocating die. Production rates are very high, but depend on the diameter of the product. With products of small diameter, the rates can be as high as eight pieces per second, and with products of larger diameters (as much as 25 mm [1 in.]), rates are about one piece per second. Two- or three-roller thread-rolling machines are also available. In another design (Fig. 6.46b), threads are formed using a rotary die, with production rates as high as 80 pieces per second.

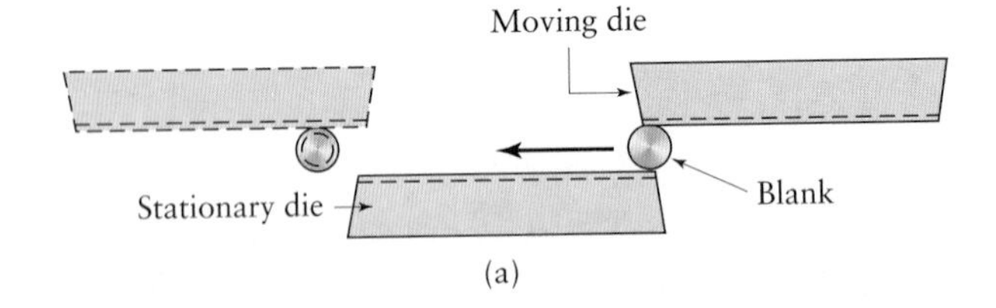

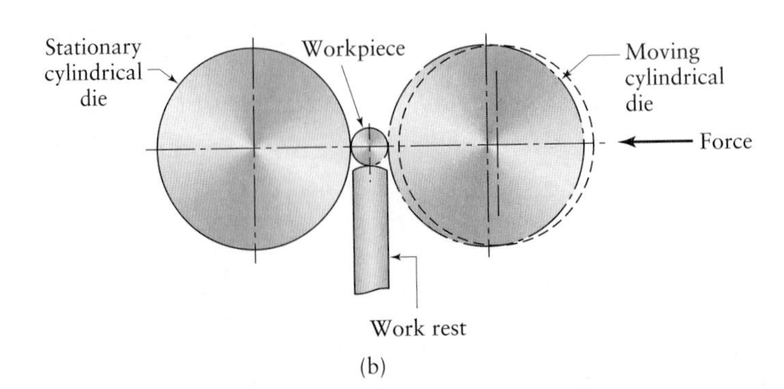

FIGURE 6.46 Thread-rolling processes: (a) flat dies and (b) two-roller dies. These processes are used extensively in making threaded fasteners at high rates of production.

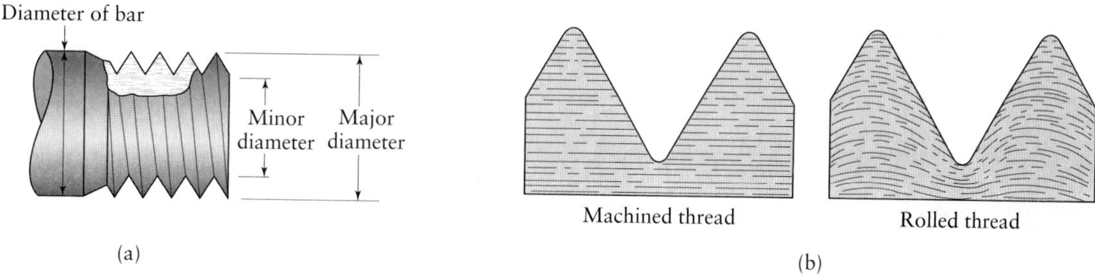

FIGURE 6.47 (a) Schematic illustration of threads. (b) Grain-flow lines in machined and rolled threads. Unlike machined threads, which are cut through the grains of the metal, rolled threads follow the grains and are stronger, because of the cold working involved.

The thread-rolling process generates threads without any loss of metal and with greater strength, because of the cold working involved. The surface finish is very smooth, and the process induces compressive residual stresses on the surfaces, which improve fatigue life. The product is superior to that made by thread cutting and is used in the production of almost all externally threaded fasteners.

Because of volume constancy in plastic deformation, a rolled thread requires a round stock of smaller diameter to produce the same major diameter as that of a machined thread (Fig. 6.47). Also, whereas machining removes material by cutting through the grain-flow lines of the material, rolled threads have a grain-flow pattern that improves the strength of the thread, because of the cold working involved. Thread rolling can also be carried out internally with a fluteless forming tap. The process is similar to external thread rolling and produces accurate threads with good strength. In all thread-rolling processes, it is essential that the material have sufficient ductility and that the rod or wire be of proper size. Lubrication is also important for good surface finish and to minimize defects.

Spur and helical gears are also produced by cold-rolling processes similar to thread rolling. The process may be carried out on solid cylindrical blanks or on precut gears. Helical gears can also be made by a direct extrusion process, using specially shaped dies. Cold rolling of gears has many applications in automatic transmissions and power tools.

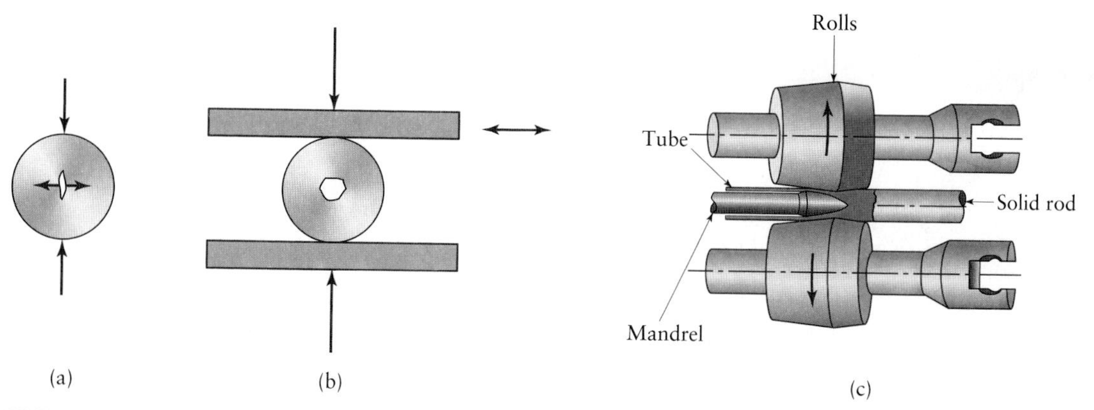

FIGURE 6.48 Cavity formation by secondary tensile stresses in a solid round bar and its use in the rotary-tube-piercing process. This is the principle of the Mannesmann mill for seamless tube making. The mandrel is held in place by the long rod, although techniques have been developed in which the mandrel remains in place without the rod.

d) **Rotary tube piercing.** A hot-working process, rotary tube piercing is used to make long, thick-walled seamless tubing, as shown in Fig. 6.48. The process is based on the principle that when a round bar is subjected to radial compression in the manner shown in Fig. 6.48a, tensile stresses develop at the center of the roll. When the rod is subjected to cycling compressive stresses, as shown in Fig. 6.48b, a cavity begins to form at the center of the rod.

The rotary-tube-piercing process (the **Mannesmann process**, developed in the 1880s) is carried out by an arrangement of rotating rolls, as shown in Fig. 6.48c. The axes of the rolls are skewed in order to pull the round bar through the rolls by the longitudinal-force component of their rotary action. A mandrel assists the operation by expanding the hole and sizing the inside diameter of the tube. Because of the severe deformation that the metal undergoes in this process, high-quality, defect-free bars must be used.

e) **Tube rolling.** The diameter and thickness of tubes and pipe can be reduced by *tube rolling* using shaped rolls, either with or without mandrels. In the **pilger mill**, the tube and an internal mandrel undergo a reciprocating motion, and the tube is advanced and rotated periodically.

6.4 Extrusion

In the basic extrusion process, developed in the late 1700s for lead pipe, a round billet is placed in a chamber and forced through a die opening by a ram. The die may be round or of various other shapes. There are four basic types of extrusion: direct, indirect, hydrostatic, and impact (Figs. 6.49 and 6.50).

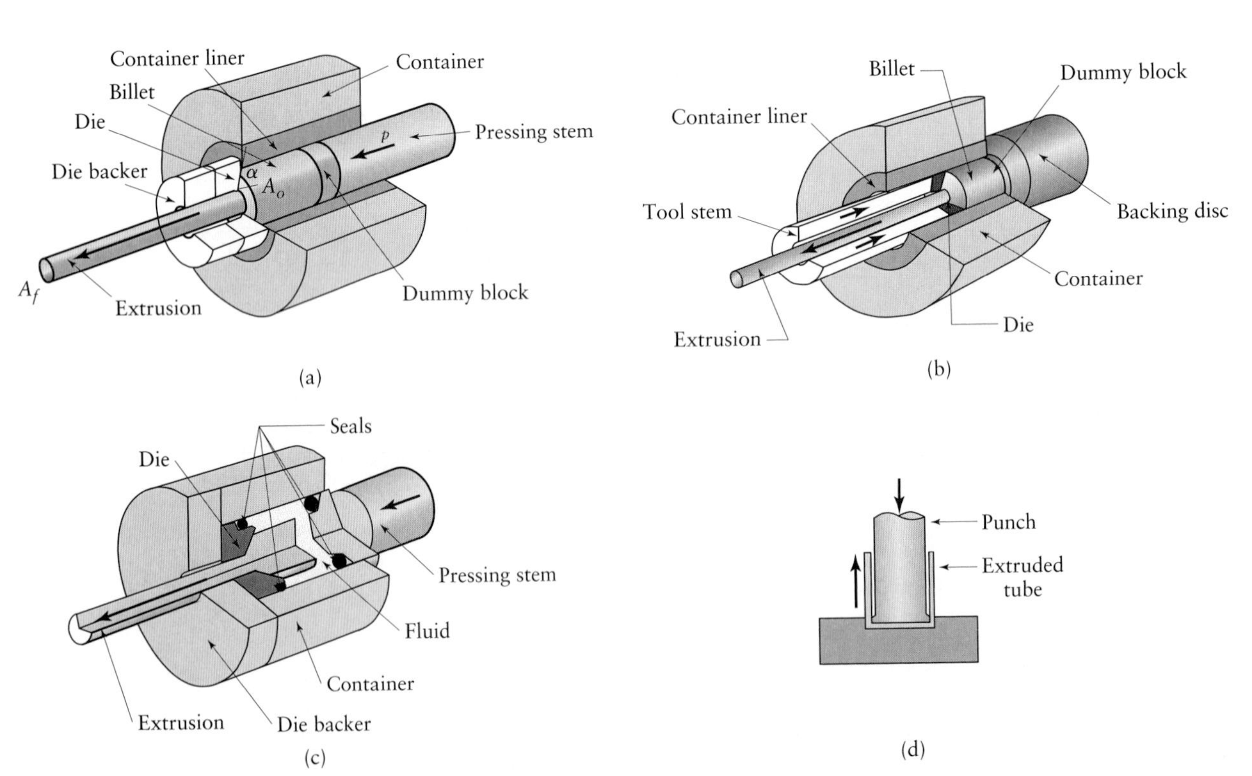

FIGURE 6.49 Types of extrusion: (a) direct; (b) indirect; (c) hydrostatic; (d) impact.

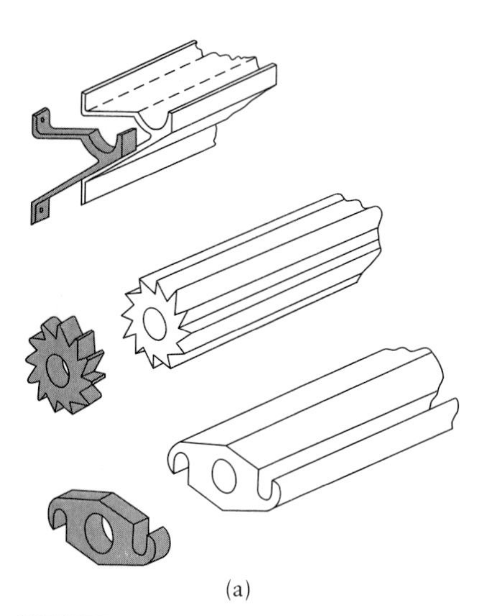

(a)

(b)

FIGURE 6.50 (a) Extrusions and examples of products made by sectioning off extrusions. *Source*: Kaiser Aluminum. (b) Examples of extruded cross sections. *Source*: © 2002. Photo courtesy of Plymouth Extruded Shapes, 1.270.886.6631.

Direct extrusion (*forward extrusion*) is similar to forcing the toothpaste through the opening of a toothpaste tube. The billet slides relative to the container wall; the wall friction increases the ram force considerably.

In **indirect extrusion** (*reverse*, *inverted*, or *backwards extrusion*), the die moves toward the billet; thus, except at the die, there is no relative motion at the billet–container interface. This process is especially advantageous for materials with high billet–container friction.

In **hydrostatic extrusion**, the chamber is filled with a fluid that transmits the pressure to the billet, which is then extruded through the die. There is no friction along the container walls.

Impact extrusion is a form of indirect extrusion and is particularly suitable for hollow shapes.

Extrusion processes can be carried out hot or cold. Because a chamber is involved, each billet is extruded individually, and hence extrusion is basically a *batch* process. **Cladding** by extrusion can also be carried out with **coaxial** billets (such as copper clad with silver), provided that the flow stresses of the two metals are similar.

The **circumscribing-circle diameter** (CCD), which is the diameter of the smallest circle into which the extruded cross-section will fit, is a parameter that describes the shape of the extruded product. Thus, the CCD for a square cross-section is the cross-section's diagonal dimension. The complexity of an extrusion is a function of the ratio of the perimeter to the cross-sectional area of the part, known as the **shape factor**; thus, a solid, round extrusion is the simplest shape.

6.4.1 Metal flow in extrusion

Because the billet is forced through a die, with a substantial reduction in its cross-section, the *metal flow pattern* in extrusion is an important factor in the overall process. A common technique for investigating the flow pattern is to halve the round billet lengthwise and mark one face with a grid pattern. The two halves are placed together in the container (they may also be fastened together, or *brazed* (Section 12.12), to keep the two halves intact) and extruded. They are then taken apart and inspected.

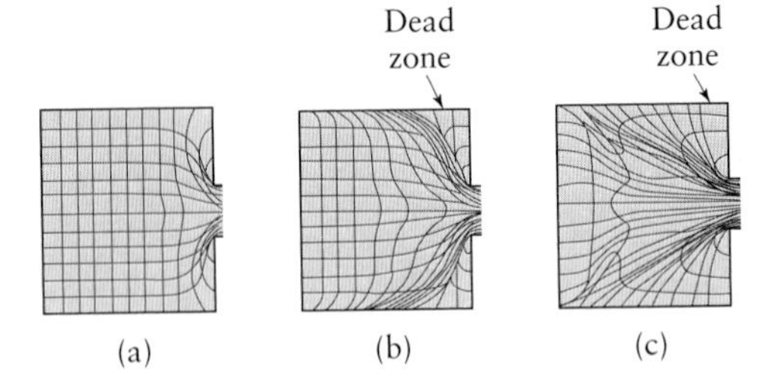

FIGURE 6.51 Schematic illustration of three different types of metal flow in direct extrusion.

Figure 6.51 shows three typical results of direct extrusion with **square dies.** The conditions under which these different flow patterns are obtained are described as follows:

1. The most homogeneous flow pattern is obtained when there is no friction at the billet–container–die interfaces (Fig. 6.51a). This type of flow occurs when the lubricant is very effective or with indirect extrusion.
2. When friction along all interfaces is high, a **dead-metal zone** develops (Fig. 6.51b). Note the high-shear area as the material flows into the die exit, somewhat like liquid flowing into a funnel. This configuration may indicate that the billet surfaces (with their oxide layer and lubricant) could enter this high-shear zone and be extruded, causing defects in the extruded product.
3. In the third configuration, the high-shear zone extends farther back (Fig. 6.51c). This extension can result from high container-wall friction, which retards the flow of the billet, or from materials in which the flow stress drops rapidly with increasing temperature. In hot working, the material near the container walls cools rapidly and hence increases in strength; thus, the material in the central regions flows toward the die more easily than that at the outer regions. As a result, a large dead-metal zone forms, and the flow is inhomogeneous. This flow pattern leads to a defect known as a *pipe*, or *extrusion, defect*.

Thus, the two factors that greatly influence metal flow in extrusion are the frictional conditions at billet–container–die interfaces and thermal gradients in the billet.

6.4.2 Mechanics of extrusion

The ram force in direct extrusion can be calculated in the following ways for different situations:

a) **Ideal deformation.** The extrusion ratio, R, is defined as

$$R = \frac{A_o}{A_f}, \tag{6.47}$$

where A_o is the cross-sectional area of the billet and A_f is the area of the extruded product (Fig. 6.49). The absolute value of the true strain is then

$$\varepsilon_1 = \ln\left(\frac{A_o}{A_f}\right) = \ln\left(\frac{L_f}{L_o}\right) = \ln R, \tag{6.48}$$

where L_o and L_f are the lengths of the billet and the extruded product, respectively. For a perfectly plastic material with a yield stress, Y, the energy dissipated in plastic deformation per unit volume, u, is

$$u = Y\varepsilon_1. \tag{6.49}$$

Hence, the work done on the billet is

$$\text{Work} = (A_o)(L_o)(u). \qquad (6.50)$$

This work is supplied by the ram force, F, which travels a distance L_o. Therefore,

$$\text{Work} = FL_o = pA_oL_o, \qquad (6.51)$$

where p is the extrusion pressure at the ram. Equating the work of plastic deformation to the external work done, we find that

$$p = u = Y \ln\left(\frac{A_o}{A_f}\right) = (Y)(\ln R). \qquad (6.52)$$

For strain-hardening materials, Y should be replaced by the *average flow stress*, $\bar{Y}$. Note that Eq. (6.52) is equal to the area under the true-stress–true-strain curve for the material.

b) **Ideal deformation and friction.** Equation (6.52) pertains to ideal deformation without any friction. Based on the slab method of analysis, when friction at the die–billet interface (but not the container-wall friction) is included and for small die angles, the pressure p is given by

$$p = Y\left(1 + \frac{\tan\alpha}{\mu}\right)[R^{\mu\cot\alpha} - 1]. \qquad (6.53)$$

Based on the assumption that the frictional stress is equal to the shear yield stress k, and that because of the dead zone formed, the material flows along a 45° "die angle," an estimate of p can be given as

$$p = Y\left(1.7R + \frac{2L}{D_o}\right). \qquad (6.54)$$

For strain-hardening materials, Y in these expressions should be replaced by $\bar{Y}$. Note that as the billet is extruded farther, L decreases, and thus the ram force decreases (Fig. 6.52), whereas in indirect extrusion, the ram force is not a function of billet length.

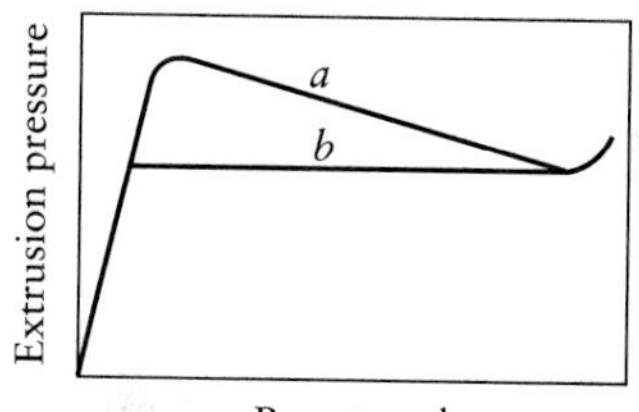

FIGURE 6.52 Schematic illustration of typical extrusion pressure as a function of ram travel: (a) direct extrusion and (b) indirect extrusion. The pressure in direct extrusion is higher because of frictional resistance in the chamber as the billet moves toward the die.

c) **Actual forces.** The derivation of analytical expressions, including those for friction, the die angle, and the redundant work due to inhomogeneous deformation of the material, can be difficult. Furthermore, there are difficulties in estimating the coefficient of friction, the flow stress of the material, and the redundant work in a particular operation. Consequently, a convenient *empirical* formula has been developed. The formula is

$$p = Y(a + b\ln R), \qquad (6.55)$$

where a and b are experimentally determined constants. An approximate value for a is 0.8, and b ranges approximately from 1.2 to 1.5. Again, note that for strain-hardening materials, Y is replaced by $\bar{Y}$.

d) **Optimum die angle.** The die angle has an important effect on forces in extrusion. Its relationship to work is as follows:

1. The **ideal work** of deformation is independent of the die angle (Fig. 6.53), because the work is a function only of the extrusion ratio.
2. The **frictional work** increases with decreasing die angle, because the length of contact at the billet–die interface increases, thus requiring more work.
3. The **redundant work** caused by inhomogeneous deformation increases with die angle.

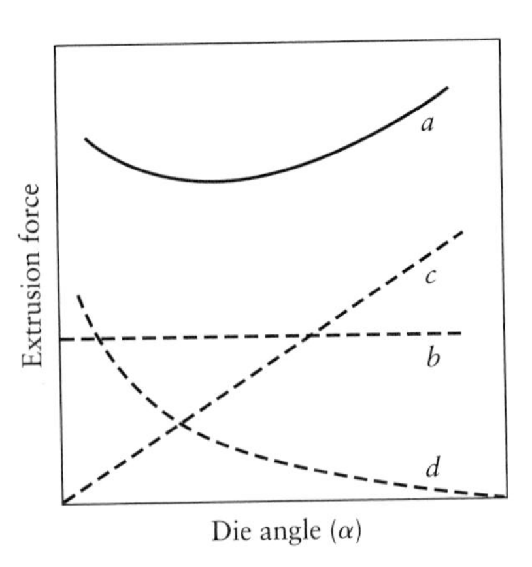

FIGURE 6.53 Schematic illustration of extrusion force as a function of die angle: (a) total force; (b) ideal force; (c) force required for redundant deformation; (d) force required to overcome friction. Note that there is an optimum die angle where the total extrusion force is a minimum.

Because the total ram force is the sum of these three components, there is an angle at which this force is a *minimum* (Fig. 6.53). Unless the behavior of each component as a function of the die angle is known, determination of this *optimum angle* is difficult.

EXAMPLE 6.5 Strain rate in extrusion

Determine the true strain rate in extruding a round billet of radius r_o as a function of distance x from the entry of the die.

Solution. From the geometry in the die gap, we have

$$\tan\alpha = \frac{(r_o - r)}{x}, \tag{6.56}$$

or

$$r = r_o - x\tan\alpha.$$

The incremental true strain can be defined as

$$d\varepsilon = \frac{dA}{A},$$

where $A = \pi r^2$. Therefore, $dA = 2\pi r\,dr$, and hence

$$d\varepsilon = \frac{2\,dr}{r},$$

where $dr = -\tan\alpha\,dx$. We also know that

$$\dot{\varepsilon} = \frac{d\varepsilon}{dt} = -\left(\frac{2\tan\alpha}{r}\right)\left(\frac{dx}{dt}\right).$$

However, $dx/dt = V$, which is the velocity of the material at any location x in the die. Hence, we have

$$\dot{\varepsilon} = -\frac{2V\tan\alpha}{r}$$

Because the flow rate is constant, we can write

$$V = \frac{V_o r_o^2}{r^2}$$

and hence we have

$$\dot{\varepsilon} = -\frac{2V_o r_o^2 \tan\alpha}{r^3}$$
$$= -\frac{2V_o r_o^2 \tan\alpha}{(r_o - x\tan\alpha)^3} \tag{6.57}$$

The negative sign is due to the fact that true strain is defined in terms of the cross-sectional area, which decreases as x increases.

e) **Forces in hot extrusion.** Because of the strain-rate sensitivity of metals at elevated temperatures, forces in hot extrusion are difficult to calculate. The average true-strain rate, $\bar{\dot{\varepsilon}}$ is

$$\bar{\dot{\varepsilon}} = \frac{6V_o D_o^2 \tan\alpha}{D_o^3 - D_f^3} \ln R, \tag{6.58}$$

where V_0 is the ram velocity. Note from this equation that for high extrusion ratios ($D_o \gg D_f$) and for $\alpha = 45°$, as may be the case with a square die (thus developing a dead zone) and poor lubrication, the strain rate reduces to

$$\bar{\dot{\varepsilon}} = \frac{6V_o}{D_o} \ln R. \tag{6.59}$$

Figure 6.54 shows the effect of ram speed and temperature on extrusion pressure. As expected, pressure increases rapidly with ram speed, especially at elevated temperatures. As extrusion speed increases, the rate of work done per unit time also increases. Because work is converted into heat, the heat generated at high speeds may not be dissipated fast enough. The subsequent rise in temperature can cause incipient melting of the workpiece material and cause defects. Circumferential surface cracks caused by **hot shortness** may also develop (see Section 3.4); in extrusion, this phenomenon is known as **speed cracking**. These problems can be eliminated by reducing the extrusion speed.

A convenient parameter that is used to estimate forces in extrusion is an experimentally determined extrusion constant K_e, which includes various factors and can be determined from

$$p = K_e \ln R. \tag{6.60}$$

Figure 6.55 gives some typical values of K_e for various materials.

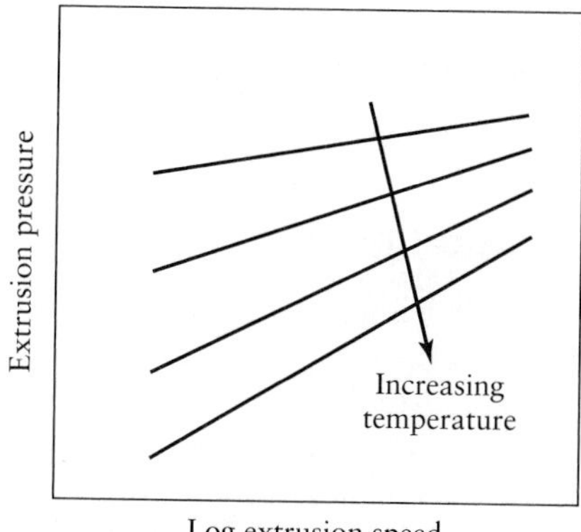

FIGURE 6.54 Schematic illustration of the effect of temperature and ram speed on extrusion pressure. Compare with Fig. 2.11.

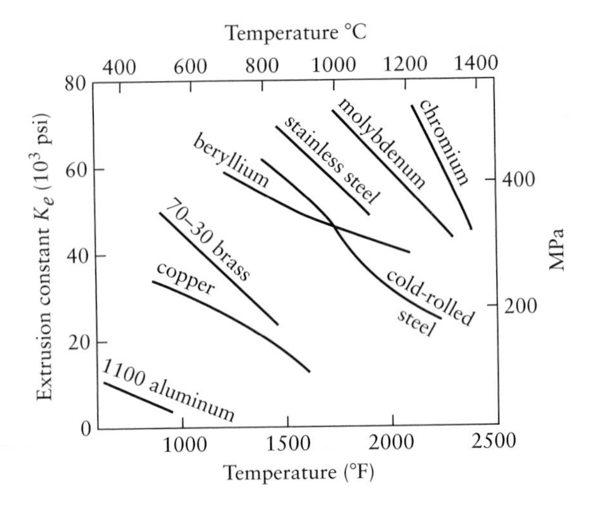

FIGURE 6.55 Extrusion constant, K_e, for various materials as a function of temperature. *Source*: After P. Loewenstein, ASTME Paper SP63-89.

EXAMPLE 6.6 Force in hot extrusion

A copper billet 5 in. in diameter and 10 in. long is extruded at 1500°F at a speed of 10 in./s. Using square dies and assuming poor lubrication, estimate the force required in this operation if the final diameter is 2 in.

Solution. The extrusion ratio is

$$R = \frac{5^2}{2^2} = 6.25.$$

The average true-strain rate, from Eq. (6.58), is

$$\dot{\varepsilon} = \frac{(6)(10)}{(5)} \ln 6.25 = 22 s^{-1}$$

From Table 2.5, let's assume that $C = 19{,}000$ psi (an average value) and $m = 0.06$. Then we have

$$\sigma = C\dot{\varepsilon}^m = (19{,}000)(22)^{0.06} = 22{,}870 \text{ psi}.$$

Assuming that $\bar{Y} = \sigma$, we have, from Eq. (6.54),

$$p = \bar{Y}\left(1.7 \ln R + \frac{2L}{D_o}\right) = (22{,}870)\left[(1.7)(1.83) + \frac{(2)(10)}{(5)}\right] = 162{,}630 \text{ psi}.$$

Hence,

$$F = (p)(A_o) = (162{,}630)\frac{(\pi)(5)^2}{4} = 3.2 \times 10^6 \text{ lb}.$$

6.4.3 Miscellaneous extrusion processes

a) **Cold extrusion.** *Cold extrusion* is a general term often used to denote a combination of processes, such as direct and indirect extrusion and forging (Fig. 6.56). Many materials can be extruded into various configurations, with the billet either at room temperature or at a temperature of a few hundred degrees. Cold extrusion has gained wide acceptance in industry, because of the following advantages over hot extrusion:

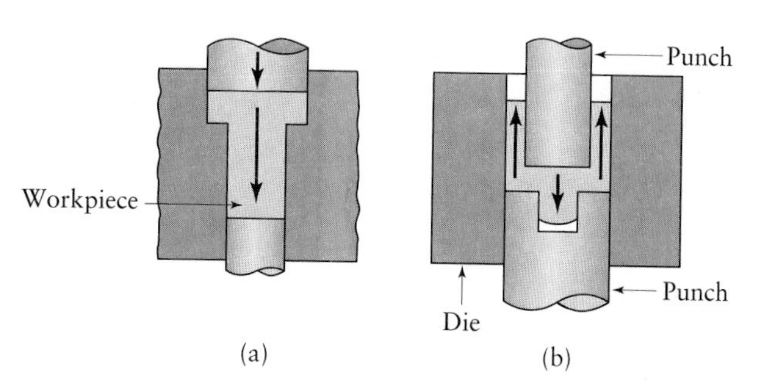

FIGURE 6.56 Examples of cold extrusion. Arrows indicate the direction of material flow. These parts may also be considered as forgings.

1. Improved mechanical properties resulting from strain hardening, provided that the heat generated by plastic deformation and friction does not recrystallize the extruded metal;
2. Good control of dimensional tolerances, thus requiring a minimum of machining operations;
3. Improved surface finish, provided that lubrication is effective;
4. Lack of oxide layers;
5. High production rates and relatively low costs.

However, the stresses on tooling in cold extrusion are very high, especially with steel workpieces. The levels on tooling are on the order of the hardness of the material, that is, at least three times its flow stress. The design of tooling and selection of appropriate tool materials are therefore crucial to success in cold extrusion. The hardness of tooling usually ranges between 60 and 65 HRC for the punch and 58 and 62 HRC for the die. Punches are a critical component; they must have sufficient strength, toughness and resistance to wear and fatigue.

Lubrication also is crucial, especially with steels, because of the generation of new surfaces and the possibility of seizure between the metal and the tooling, caused by the breakdown of lubrication. The most effective lubrication is provided by phosphate conversion coatings on the workpiece and soap (or wax in some cases) as the lubricant. Temperature rise in cold extrusion is an important factor, especially at high extrusion ratios. The temperature may be sufficiently high to initiate and complete the recrystallization process of the cold-worked metal, thus reducing the advantages of cold working.

b) **Impact extrusion.** A process often included in the category of cold extrusion, *impact extrusion* (Fig. 6.49d) is similar to indirect extrusion. In impact extrusion, the punch descends at a high speed and strikes the *blank* (*slug*), extruding it upward. The thickness of the extruded tubular section is a function of the clearance between the punch and the die cavity.

The impact-extrusion process usually produces tubular sections with wall thicknesses that are small in relation to their diameters. This ratio can be as small as 0.005. The concentricity between the punch and the blank is important for uniform wall thickness. A typical example of impact extrusion is the production of collapsible tubes, such as for toothpaste (Fig. 6.57a). The amount of distance traveled by the punch is determined by the setting of the press. A variety of nonferrous metals are impact extruded in this manner into various shapes (Fig. 6.57b), using vertical presses at production rates as high as two parts per second.

c) **Hydrostatic extrusion.** In *hydrostatic extrusion*, the pressure required for extrusion is supplied through a fluid medium surrounding the billet (Fig. 6.49c). Consequently, there is no container-wall friction. Pressures are usually about 1400 MPa (200 ksi). The high pressure in the chamber transmits some of the

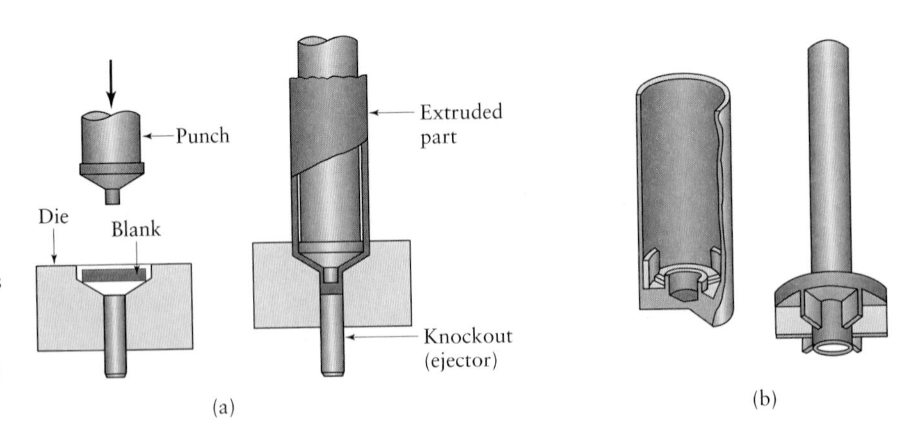

FIGURE 6.57 (a) Impact extrusion of a collapsible tube (Hooker process). (b) Two examples of products made by impact extrusion. These parts may also be made by casting, forging, and machining, depending on the dimensions and materials involved and the properties desired. Economic considerations are also important in final process selection.

fluid to the die surfaces, thus significantly reducing friction and forces (Fig. 6.58). Hydrostatic extrusion, which was developed in the early 1950s, has been improved by extruding the part into a second pressurized chamber, which is under lower pressure (**fluid-to-fluid extrusion**). This operation reduces the number of defects in the extruded product.

Because the hydrostatic pressure increases the ductility of the material, brittle materials can be extruded successfully by this method. However, the main reasons for this success appear to be low friction and use of low die angles and high extrusion ratios. Most commercial hydrostatic-extrusion operations use ductile materials. However, a variety of metals and polymers, solid shapes, tubes and other hollow shapes, and honeycomb and clad profiles have been extruded successfully.

Hydrostatic extrusion is usually carried out at room temperature, typically using vegetable oils as the fluid, particularly castor oil, because it is a good lubricant and its viscosity is not influenced significantly by pressure. For elevated-temperature extrusion, waxes, polymers, and glass are used as the fluid; these materials also serve as thermal insulators and help maintain the billet temperature during extrusion. In spite of the success obtained, hydrostatic extrusion has had limited industrial applications, largely because of the somewhat complex nature of tooling, the experience required in working with high pressures and design of specialized equipment, and the long cycle times required.

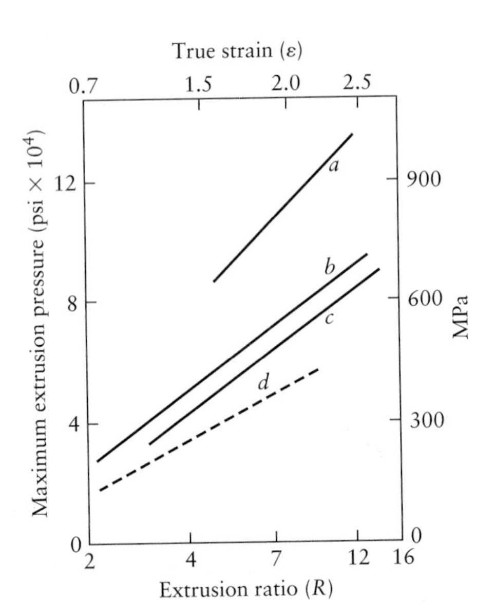

FIGURE 6.58 Extrusion pressure as a function of the extrusion ratio for an aluminum alloy. (a) Direct extrusion, $\alpha = 90°$. (b) Hydrostatic extrusion, $\alpha = 45°$. (c) Hydrostatic extrusion, $\alpha = 22.5°$. (d) Ideal homogeneous deformation, calculated. *Source*: After H. Li, D. Pugh, and K. Ashcroft.

6.4.4 Defects in extrusion

There are three principal defects in extrusion: surface cracking, extrusion defect, and internal cracking. These defects are described as follows:

a) **Surface cracking.** If the extrusion temperature, friction, or extrusion speed is too high, surface temperatures rise significantly, which can lead to surface cracking and tearing (*fir-tree cracking* or *speed cracking*). These cracks are intergranular and are usually the result of hot shortness; they occur especially with aluminum, magnesium, and zinc alloys, but are also observed with other metals, such as molybdenum alloys. This situation can be avoided by using lower temperatures and speeds.

   However, surface cracking may also occur at low temperatures and has been attributed to periodic sticking of the extruded product along the die land (**stick-slip**) during extrusion. When the product being extruded sticks to the die land, the extrusion pressure increases rapidly; shortly thereafter, the product moves forward again, and the pressure is released. The cycle is then repeated. Because of its appearance, this defect is known as a **bamboo defect.** This type of defect is especially common with hydrostatic extrusion where the pressure is sufficient to significantly increase the viscosity of the fluid. This increase in viscosity allows a thick lubricant film to develop; the billet then surges forward, which, in turn, relieves the pressure of the fluid and increases friction. Thus, the change in the physical properties of the metalworking fluid/lubricant is responsible for the stick-slip phenomenon. In such circumstances, it has been found that proper selection of a fluid is critical, but also that increasing the extrusion speed can eliminate stick-slip.

b) **Extrusion defect.** The type of metal flow observed in Fig. 6.51c tends to draw surface oxides and impurities toward the center of the billet, much like a funnel. This defect is known as *extrusion defect*, **pipe, tailpipe,** and **fishtailing.** A considerable portion of the material can be rendered useless as an extruded product because of it—as much as one third the length of the extrusion. This defect can be reduced by modifying the flow pattern to a less inhomogeneous one, such as by controlling friction and minimizing temperature gradients. Another method is to machine the surface of the billet prior to extrusion to eliminate scale and impurities. The extrusion defect can also be avoided by using a dummy block (Fig. 6.49a) that is smaller in diameter than the container, thus leaving a thin shell along the container wall as extrusion progresses.

c) **Internal cracking.** The center of an extruded product can develop cracks (variously known as **centerburst, center cracking, arrowhead fracture,** and **chevron cracking**), as shown in Fig. 6.59. These cracks are attributed to a state of *hydrostatic tensile stress* (also called *secondary tensile stress*) at the centerline of the deformation zone in the die. This situation is similar to the necked region in a uniaxial tensile-test specimen.

   The major variables affecting hydrostatic tension are the die angle, the extrusion ratio (reduction in cross-sectional area), and friction; we can understand them best by observing the extent of inhomogeneous deformation in extrusion. Experimental results indicate that, for the same reduction, as the die angle becomes larger, the deformation across the part becomes more inhomogeneous. In addition to the die angle, another factor in internal cracking is the die contact length. The smaller the die angle, the longer is the contact length. This situation is similar to a hardness test with a flat indenter. The size and depth of the deformation zone increases with increasing contact length. This condition is illustrated in Fig. 6.59a.

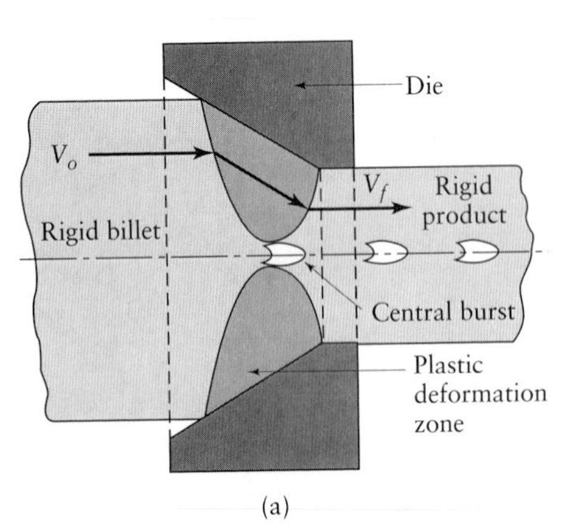

(a)

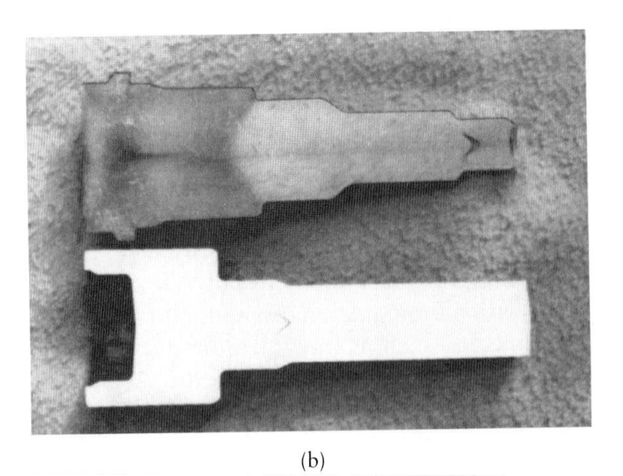

(b)

FIGURE 6.59 (a) Deformation zone in extrusion, showing rigid and plastic zones. Note that the plastic zones do not meet, leading to chevron cracking. The same observations are also made in drawing round bars through conical dies and drawing flat sheet or plate through wedge-shaped dies. *Source*: After B. Avitzur. (b) Chevron cracking in round steel bars during extrusion. Unless the part is inspected properly, such internal detects may remain undetected and possibly cause failure of the part in service.

As shown in Fig. 6.12, h/L ratio is an important parameter: The higher this ratio, the more inhomogeneous is the deformation. High ratios mean small reductions and large die angles. Inhomogeneous deformation indicates that the center of the billet is not in a fully plastic state. (It is rigid.) The reason is that the plastic deformation zones under the die contact lengths do not reach each other (Fig. 6.59a). Likewise, small reductions and high die angles retard the flow of the material at the surfaces, while the central portions are freer to move through the die.

The high h/L ratios described generate hydrostatic tensile stresses in the center of the billet, causing the type of defects shown in Fig. 6.59b. These defects form more readily in materials with impurities, inclusions, and voids, because they serve as nucleation sites for defect formation. As for the role of friction, high friction in extrusion apparently delays formation of these cracks. Such cracks have also been observed in tube extrusion and in spinning of tubes. The cracks appear on the inside surfaces for the reasons already given. In summary, the tendency for center cracking increases with increasing die angles and levels of impurities and decreases with increasing extrusion ratios. These observations are also valid for the drawing of rod and wire. (See Section 6.5.)

6.4.5 Extrusion practice

Numerous materials can be extruded to a wide variety of cross-sectional shapes and dimensions. Extrusion ratios can range from about 10 to over 100. Ram speeds may be up to 0.5 m/s (100 ft/min). Generally, slower speeds are preferable for aluminum, magnesium, and copper and higher speeds for steels, titanium, and refractory alloys. Presses for hot extrusion are generally hydraulic and horizontal; presses for cold extrusion are usually vertical.

Hot extrusion. In addition to the strain-rate sensitivity of the material at elevated temperatures, hot extrusion requires other special considerations. Cooling of the billet in the container can result in highly inhomogeneous deformation. Furthermore, because the billet is heated prior to extrusion, it is covered with an oxide layer (unless heated in an inert atmosphere). The resulting different frictional properties can affect the flow of the material and may produce an extrusion covered with an oxide layer.

In order to avoid this problem, the diameter of the dummy block ahead of the ram (Fig. 6.49a) is made a little smaller than that of the container. A thin cylindrical shell (skull), consisting mainly of the oxidized layer, is thus left in the container, and the extruded product is free of oxides. Temperature ranges for hot extrusion are similar to those for forging. (See Table 6.3.)

Lubrication. Lubrication is important in hot extrusion. For steels, stainless steels, and high-temperature materials, glass is an excellent lubricant (**Séjournet process**). Glass maintains its viscosity at elevated temperatures, has good wetting characteristics, and acts as a thermal barrier between the billet, the container, and the die, thus minimizing cooling. A circular glass pad is usually placed at the die entrance. This pad softens and melts away slowly as extrusion progresses and forms an optimal die geometry. The viscosity–temperature index of the glass is an important factor in this application. Solid lubricants such as graphite and molybdenum disulfide are also used in hot extrusion. Nonferrous metals are usually extruded without a lubricant, although graphite may be used.

For materials that have a tendency to stick to the container and the die, the billet can be enclosed in a jacket of a softer metal, such as copper or mild steel (**jacketing** or **canning**). Besides providing a low-friction interface, canning prevents contamination of the billet by the environment, or prevents the billet material from contaminating the environment if the material is toxic or radioactive. The canning technique is also used for processing metal powders.

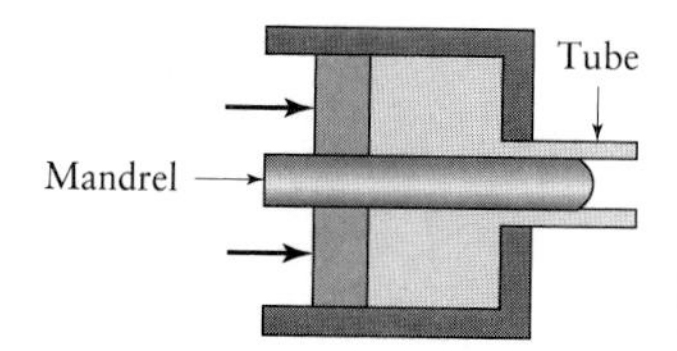

FIGURE 6.60 Extrusion of a seamless tube. The hole in the billet may be prepunched or pierced, or it may be generated during extrusion.

Die design. Designing dies requires considerable experience, because of the variety of products extruded. Dies with angles of 90° (**square dies**, or **shear dies**) can also be used in extrusion of nonferrous metals, especially aluminum. Tubing is also extruded with a ram fitted with a mandrel (Fig. 6.60). For billets with a pierced hole, the mandrel may be attached to the ram. If the billet is solid, it must first be pierced in the container by the mandrel.

Hollow shapes (Fig. 6.61a) can also be extruded by **welding-chamber methods**, using various special dies known as *spider, porthole*, and *bridge dies* (Fig. 6.61b).

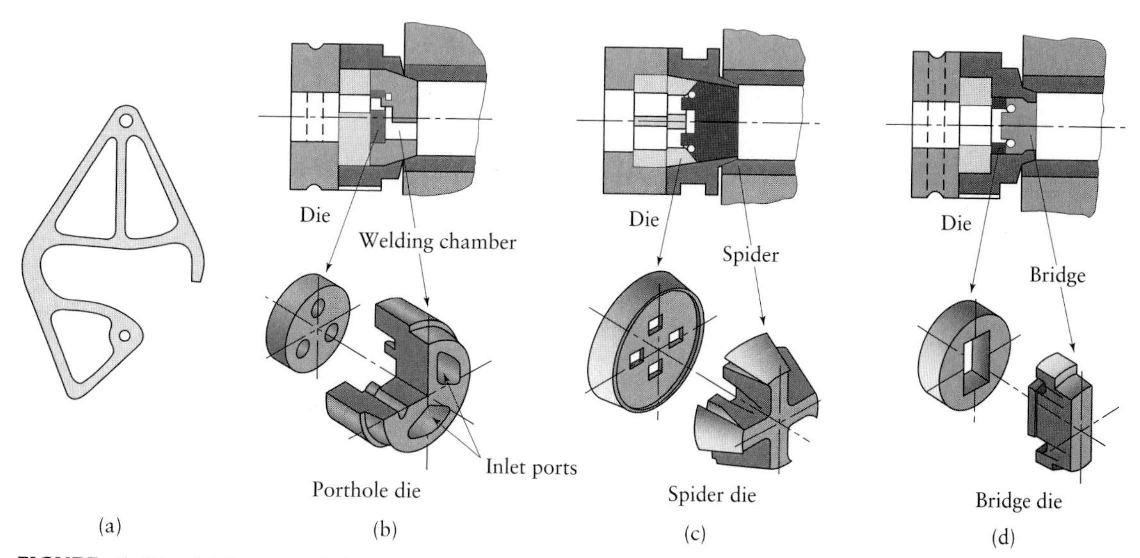

FIGURE 6.61 (a) An extruded 6063-T6 aluminum ladder lock for aluminum extension ladders. This part is 8 mm ($\frac{5}{16}$ in.) thick and is sawed from the extrusion. (See Fig. 6.50a.) (b) Components of various dies for extruding intricate hollow shapes. *Source* for (b)-(d): K. Laue and H. Stenger, *Extrusion-Processes, Machinery, Tooling*, ASM International, 1981. Used with permission.

The metal flows around the arms of the die into strands, which are then rewelded under the high die pressures at the exit. This process is suitable only for aluminum and some of its alloys, because of their capacity for pressure welding. Lubricants cannot be used, as they prevent rewelding during extrusion.

Die materials for hot extrusion are usually hot-work die steels. To extend die life, various coatings may be applied to the dies. (see Section 4.5.)

Equipment. The basic equipment for extrusion is a hydraulic press, usually horizontal; these presses are designed for a variety of extrusion operations. Crank-type mechanical presses are also used for cold extrusion and piercing and for mass production of steel tubing. The largest hydraulic press for extrusion has a capacity of 160 MN (16,000 tons).

6.5 Rod, Wire, and Tube Drawing

Drawing is an operation in which the cross-sectional area of a bar or tube is reduced by pulling it through a converging die (Fig. 6.62). The die opening may be any shape. Wire drawing involves materials of smaller diameter than those used for rod drawing, with sizes as small as 0.025 mm (0.001 in.). The drawing process, which was an established art by the 11th century AD, is somewhat similar to extrusion, except that in drawing, the bar is under tension, whereas in extrusion it is under compression.

Rod and wire drawing are usually finishing processes. The product is either used as produced or further processed into other shapes, usually by bending or machining. Rods are used for various applications, such as small pistons, tension-carrying structural members, shafts, and spindles, and as the raw material for fasteners such as bolts and screws. Wire and wire products have a wide range of applications, such as electrical and electronic equipment and wiring, cables, springs, musical instruments, fencing, bailing, wire baskets, and shopping carts.

6.5.1 Mechanics of rod and wire drawing

The major variables in the drawing process are reduction in cross-sectional area, die angle, and friction; these variables are illustrated in Fig. 6.62.

a) **Ideal deformation.** For a round rod or wire, the drawing stress σ_d for the simplest case of ideal deformation (no friction or redundant work) can be obtained by the same approach as that used for extrusion. Thus, we have

$$\sigma_d = Y \ln\left(\frac{A_o}{A_f}\right). \tag{6.61}$$

Note that this expression is the same as Eq. (6.52) and that it also represents the energy per unit volume, u. For strain-hardening materials, Y is replaced by an

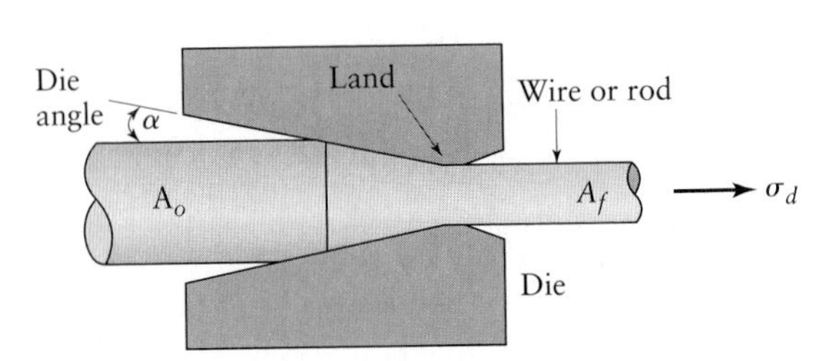

FIGURE 6.62 Variables in drawing round rod or wire.

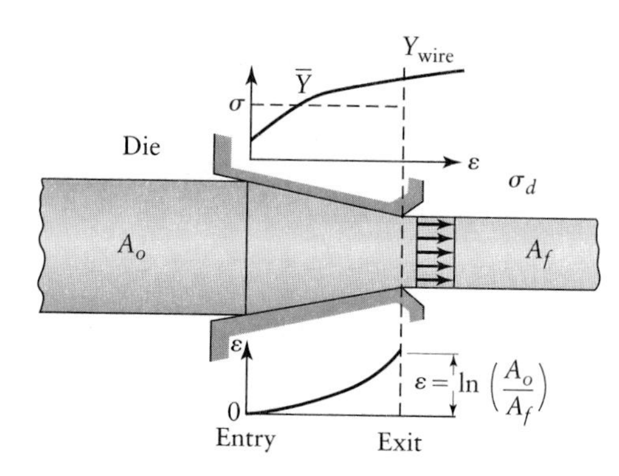

FIGURE 6.63 Variation in strain and flow stress in the deformation zone in drawing. Note that the strain increases rapidly toward the exit. The reason is that when the exit diameter is zero, the true strain reaches infinity. The point Y_{wire} represents the yield stress of the drawn wire.

average flow stress, $\bar{Y}$ in the deformation zone (Fig. 6.63). Therefore, for a material that exhibits the true-stress–true-strain behavior of

$$\sigma = K\varepsilon^n,$$

the average flow stress, $\bar{Y}$ is given as

$$\bar{Y} = \frac{K\varepsilon_1^n}{n + 1}. \tag{6.62}$$

The drawing force, F, is then given as

$$F = \bar{Y}A_f \ln\left(\frac{A_o}{A_f}\right). \tag{6.63}$$

Note that the greater the reduction in cross-sectional area and the stronger the material, the higher is the drawing force.

b) **Ideal deformation and friction.** In drawing, friction increases the drawing force, because work has to be supplied to overcome that friction. Using the slab method of analysis and on the basis of Fig. 6.64, we can obtain the following expression for the drawing stress, σ_d:

$$\sigma_d = Y\left(1 + \frac{\tan\alpha}{\mu}\right)\left[1 - \left(\frac{A_f}{A_o}\right)^{\mu\cot\alpha}\right]. \tag{6.64}$$

For a strain-hardening material, the yield stress, Y, is replaced by the average flow stress, $\bar{Y}$, as shown in Fig. 6.63. Investigations have shown that, even though Eq. (6.64) does not include the redundant work, it is in good agreement with experimental data for small die angles and for a wide range of reductions.

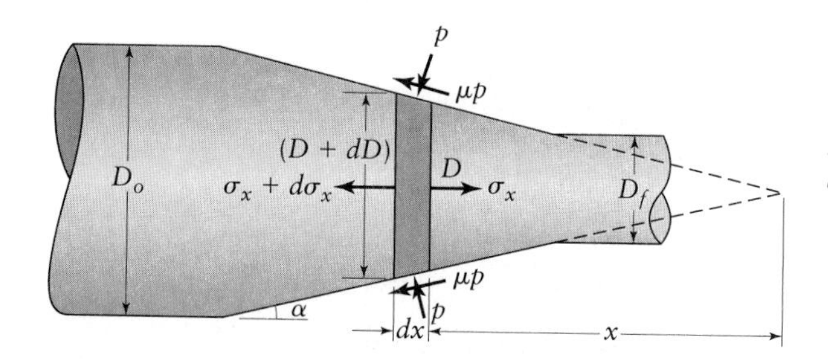

FIGURE 6.64 Stresses acting on an element in drawing of a solid cylindrical rod or wire through a conical converging die.

c) **Redundant work of deformation.** Depending on the die angle and reduction, the material in drawing undergoes *inhomogeneous deformation*, much as it does in extrusion; therefore, the redundant work of deformation also has to be included in the expression for the drawing stress. One such expression is

$$\sigma_d = \bar{Y}\left\{\left(1 + \frac{\tan\alpha}{\mu}\right)\left[1 - \left(\frac{A_f}{A_o}\right)^{\mu\cot\alpha}\right] + \frac{4}{3\sqrt{3}}\alpha^2\left(\frac{1-r}{r}\right)\right\}, \quad (6.65)$$

where r is the fractional reduction of area and α is the die angle in radians. The first term in Eq. (6.65) represents the ideal and frictional components of work. The second term represents the redundant-work component, which, as expected, is a function of the die angle. The larger the angle is, the greater is the inhomogeneous deformation and, hence, the greater is the redundant work.

Another expression for the drawing stress for small die angles (including all three components of work) is

$$\sigma_d = \bar{Y}\left[\left(1 + \frac{\mu}{\alpha}\right)\ln\left(\frac{A_o}{A_f}\right) + \frac{2}{3}\alpha\right]. \quad (6.66)$$

The last term in this expression is the redundant-work component; this component of work increases linearly with the die angle, as shown in Fig. 6.53. Because redundant deformation is a function of the h/L ratio (see Figs. 6.12 and 6.13), an **inhomogeneity** factor Φ can replace the last term in Eq. (6.66). For drawing of round sections this factor is, approximately,

$$\Phi = 1 + 0.12(h/L). \quad (6.67)$$

A simple expression for the drawing stress is then

$$\sigma_d = \Phi\bar{Y}\left(1 + \frac{\mu}{\alpha}\right)\ln\left(\frac{A_o}{A_f}\right). \quad (6.68)$$

Equations (6.64), (6.65), (6.66), and (6.68), while not always agreeing with experimental data, approximate reasonably well the stresses required for wire drawing. More important, they identify the effects of the various parameters involved. With good lubrication, the value of μ in these equations ranges from about 0.03 to 0.1.

d) **Die pressure.** The *die pressure, p,* at any diameter along the die contact length can be conveniently obtained from

$$p = Y_f - \sigma \quad (6.69)$$

where σ is the tensile stress in the deformation zone at any diameter. (Thus, σ is equal to σ_d at the exit and is zero at the entry). Y_f is the flow stress of the material at any diameter. Equation (6.69) is based on yield criteria for an element subjected to the stress system shown in Fig. 2.35b, where the compressive stresses in the two principal directions are both equal to p. Note that as the tensile stress increases toward the exit, the die pressure drops toward the exit. Tensile-stress and die-pressure distributions are shown qualitatively in Fig. 6.65.

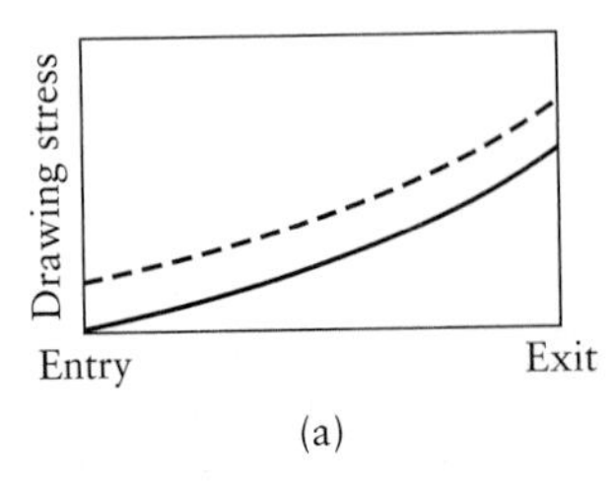

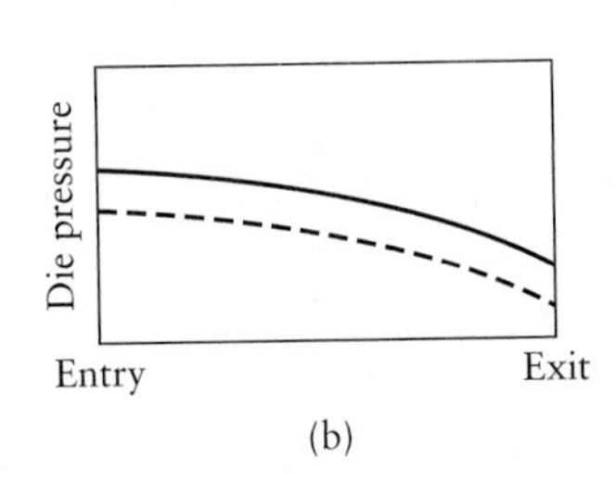

FIGURE 6.65 Variation in the drawing stress and die contact pressure along the deformation zone. Note that as the drawing stress increases, the die pressure decreases. This condition can be observed from the yield criteria, described in Section 2.11. Note the effect of back tension.

EXAMPLE 6.7 **Power and die pressure in rod drawing**

A round rod of annealed 302 stainless steel is being drawn from a diameter of 10 mm to a diameter of 8 mm at a speed of 0.5 m/s. Assume that the frictional and redundant work together constitute 40% of the ideal work of deformation. (1) Calculate the power required in this operation. (2) Calculate the die pressure at the exit of the die.

Solution. The true-strain in this operation is

$$\varepsilon_1 = \ln\left(\frac{10^2}{8^2}\right) = 0.446.$$

From Table 2.3, we know that $K = 1300$ MPa and $n = 0.30$. Hence,

$$\bar{Y} = \frac{K\varepsilon_1^n}{n+1} = \frac{(1300)(0.446)^{0.30}}{1.30} = 785 \text{ MPa}.$$

From Eq. (6.63), we have

$$F = \bar{Y}A_f \ln\left(\frac{A_o}{A_f}\right),$$

where

$$A_f = \frac{(\pi)(0.008)^2}{4} = 5 \times 10^{-5}\,\text{m}^2.$$

Hence,

$$F = (785)(5 \times 10^{-5})(0.446) = 0.0175 \text{ MN}.$$

So,

$$\begin{aligned}\text{Power} &= (F)(V_f) = (0.0175)(0.5)\\ &= 0.00875 \text{ MN}\cdot\text{m/s}\\ &= 0.00875 \text{ MW} = 8.75 \text{ kW},\end{aligned}$$

and

$$\text{Actual power} = (1.4)(8.75) = 12.25 \text{ kW}.$$

The die pressure can be obtained by noting from Eq. (6.69) that

$$p = Y_f - \sigma$$

where Y_f represents the flow stress of the material at the exit of the die. Thus,

$$Y_f = K\varepsilon_1^n = (1300)(0.446)^{0.30} = 1020 \text{ MPa}.$$

In this equation, σ is the drawing stress, σ_d. Hence, using the actual force, we have

$$\sigma_d = \frac{F}{A_f} = \frac{(1.4)(0.0175)}{0.00005} = 490 \text{ MPa}.$$

Therefore, the die pressure at the exit is

$$p = 1020 - 490 = 530 \text{ MPa}.$$

e) **Drawing at elevated temperatures.** At elevated temperatures, the flow stress of metals is a function of the strain rate. The *average true-strain rate* $\dot{\bar{\varepsilon}}$ in the deformation zone in drawing is

$$\dot{\bar{\varepsilon}} = \frac{6V_o}{D_o} \ln\left(\frac{A_o}{A_f}\right). \tag{6.70}$$

[Note that this expression is the same as Eq. (6.59).] For a particular drawing operation at an elevated temperature, we first calculate the average strain rate; from that quantity, we can determine the flow stress and the average flow stress, $\bar{Y}$, of the material.

f) **Optimum die angle.** Because of the various effects of the die angle on the three components of work (i.e., ideal, friction, and redundant work), there is an *optimum die angle* at which the extrusion force is a minimum. Drawing involves a similar type of deformation, so there is an optimum die angle in rod and wire drawing also, as shown in Fig. 6.53. Figure 6.66 shows a typical example from an experimental study. The optimum die angle for the minimum force increases with reduction. Note that the optimum angles in Fig. 6.66 are rather small.

g) **Maximum reduction per pass.** With greater reduction, the drawing stress increases, but, obviously, the magnitude of the drawing stress has a limit. If it reaches the yield stress of the material, the material will simply continue to yield further as it leaves the die. This result is unacceptable, because the product will undergo further deformation; thus, the maximum possible drawing stress is the yield stress of the exiting material.

In the ideal case of a perfectly plastic material with a yield stress Y, the limiting condition is

$$\sigma_d = Y \ln\left(\frac{A_o}{A_f}\right) = Y, \tag{6.71}$$

or

$$\ln\left(\frac{A_o}{A_f}\right) = 1.$$

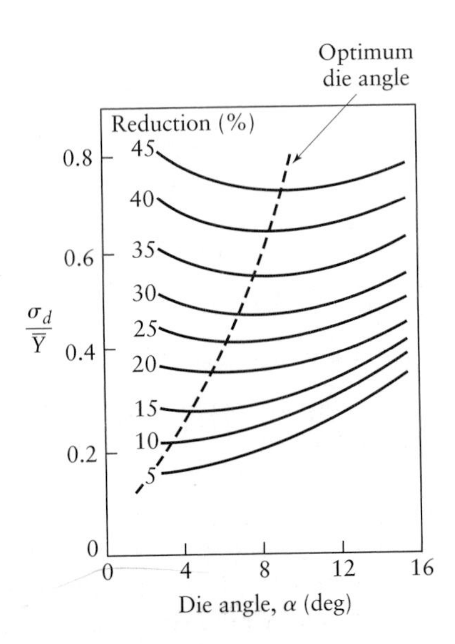

FIGURE 6.66 The effect of reduction in cross-sectional area on the optimum die angle. *Source*: After J. G. Wistreich.

Hence,

$$\frac{A_o}{A_f} = e,$$

and therefore,

$$\text{Maximum reduction per pass} = \frac{A_o - A_f}{A_o} = 1 - \frac{1}{e} = 0.63 = 63\%. \quad (6.72)$$

The effects of friction and die angle on maximum reduction per pass are similar to those shown in Fig. 6.53. Because both friction and redundant work increase the drawing stress, the maximum reduction per pass will be lower than the ideal. In other words, the exiting material must have a larger cross-sectional area to sustain the higher drawing forces. With strain hardening, the exiting material is stronger than the rest of the material, and the maximum reduction per pass increases.

EXAMPLE 6.8 Maximum reduction per pass for a strain-hardening material

Obtain an expression for the maximum reduction per pass for a material with a true-stress–true-strain curve of $\sigma = K\varepsilon^n$. Ignore friction and redundant work.

Solution. From Eq. (6.61), we have for this material

$$\sigma_d = \bar{Y} \ln\left(\frac{A_o}{A_f}\right) = \bar{Y}\varepsilon_1,$$

where

$$\bar{Y} = \frac{K\varepsilon_1^n}{n+1}$$

and σ_d, for this problem, can have a maximum value equal to the flow stress at ε_1, or

$$\sigma_d = Y_f = K\varepsilon_1^n.$$

Hence, we can write Eq. (6.71) as

$$K\varepsilon_1^n = \frac{K\varepsilon_1^n}{n+1}\varepsilon_1,$$

or

$$\varepsilon_1 = n + 1.$$

With $\varepsilon_1 = \ln(A_o/A_f)$ and a maximum reduction of $(A_o - A_f)/A_o$, these expressions reduce to

$$\text{Maximum reduction per pass} = 1 - e^{-(n+1)}. \quad (6.73)$$

Note that when $n = 0$ (perfectly plastic material), this expression reduces to Eq. (6.72). As n increases, the maximum reduction per pass increases.

h) **Drawing of flat strip.** Whereas drawing of round sections is an axisymmetric problem, drawing of flat strips with high width-to-thickness ratios can be regarded as a **plane-strain** problem. The dies are wedge shaped, and thus the process is somewhat similar to that of rolling wide strips. There is little or no

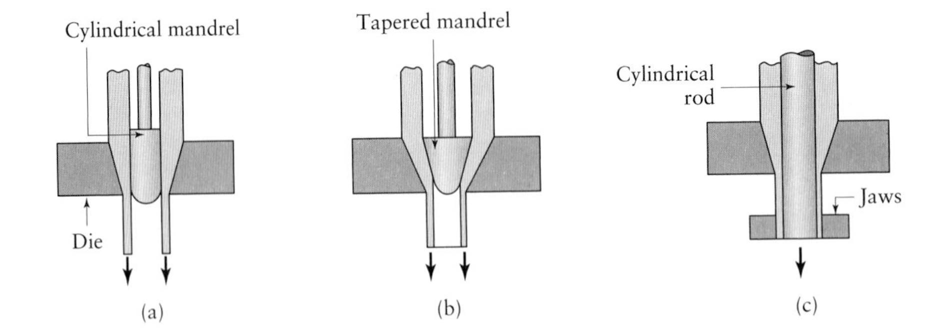

FIGURE 6.67 Various methods of tube drawing.

change in the width of the strip during drawing. This process, although not of any industrial significance in and of itself, is the fundamental deformation mechanism in **ironing**, as described in Section 7.12.

The treatment of this subject, as far as forces and maximum reductions are concerned, is similar to that for round sections. The drawing stress for the ideal condition is

$$\sigma_d = Y' \ln\left(\frac{t_o}{t_f}\right), \tag{6.74}$$

where Y' is the yield stress of the material in plane strain and t_o and t_f are the original and final thicknesses of the strip, respectively. The effects of friction and redundant deformation in strip drawing are similar to those for round sections.

The maximum reduction per pass can be obtained by equating the drawing stress [Eq. (6.74)] to the *uniaxial* yield stress of the material, because the drawn strip is subjected only to simple tension. Thus,

$$\sigma_d = Y' \ln\left(\frac{t_o}{t_f}\right) = Y, \qquad \ln\left(\frac{t_o}{t_f}\right) = \frac{Y}{Y'} = \frac{\sqrt{3}}{2}, \qquad \text{and} \qquad \frac{t_o}{t_f} = e^{\sqrt{3}/2}.$$

It can then be shown that

$$\text{Maximum reduction per pass} = 1 - \frac{1}{e^{\sqrt{3}/2}} = 0.58 = 58\%. \tag{6.75}$$

b) **Drawing of tubes.** Tubes produced by extrusion or other processes can be reduced in thickness or diameter (**tube sinking**) by the tube-drawing processes illustrated in Fig. 6.67. A variety of mandrels are used to produce different shapes. Shape changes can also be imparted by using dies and mandrels with various profiles. Drawing forces, die pressures, and the maximum reduction per pass in tube drawing can be calculated by methods similar to those described for round rods.

6.5.2 Defects in drawing

Defects in drawing are similar to those observed in extrusion, especially center cracking. The factors influencing these internal cracks are the same, namely, that the tendency for cracking increases with increasing die angle, with decreasing reduction per pass, with friction, and with the presence of inclusions in the material.

Another type of defect is the formation of **seams**, which are longitudinal scratches or folds in the material. Such defects can open up during subsequent forming operations by upsetting, heading, thread rolling, or bending of the rod or wire. Various surface defects can also result from improper selection of process parameters and lubrication.

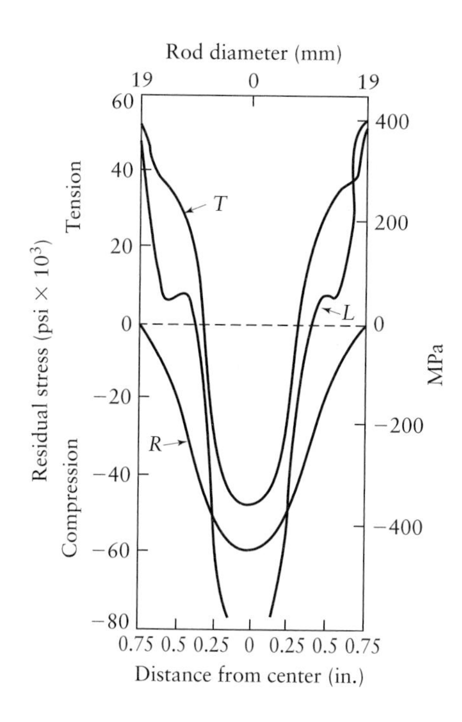

FIGURE 6.68 Residual stresses in cold-drawn AISI 1045 carbon steel round rod: T = transverse direction, L = longitudinal direction, and R = radial direction. *Source*: After E. S. Nachtman.

Because of inhomogeneous deformation, a cold-drawn rod, wire, or tube usually contains residual stresses. Typically, as shown in Fig. 6.68, a wide range of residual stresses is present within the rod in three principal directions. However, for very light reductions, the surface residual stresses are compressive. Light reductions are equivalent to shot peening or surface rolling, thus improving fatigue life. In addition to their effect on fatigue life, residual stresses can be significant in stress-corrosion cracking over a period of time and in warping of the component when a layer is subsequently removed, as by machining or grinding.

6.5.3 Drawing practice

Successful drawing operations require careful selection of process parameters and consideration of many factors. A typical die design for drawing, with its characteristic features, is shown in Fig. 6.69. Die angles usually range from 6° to 15°. The purpose of the land is to size and set the final diameter of the product. Also, when the die is reground after use, the land maintains the exit dimension of the die opening. Reductions in cross-sectional area per pass range from about 10% to 45%; usually, the smaller the cross-section, the smaller is the reduction per pass. Reductions per pass greater than 45% may result in breakdown of lubrication and deterioration of surface finish. Light reductions may also be made (**sizing pass**) on rods to improve surface finish and dimensional accuracy.

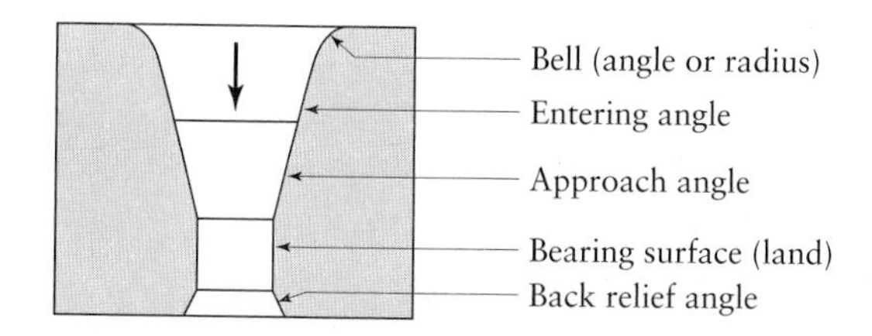

FIGURE 6.69 Terminology for a typical die for drawing round rod or wire.

A rod or wire is fed into the die by first **pointing** it by swaging (i.e., forming the tip of the rod into a conical shape, as discussed in Section 6.6). After the rod or wire is placed in the die, the tip is clamped into the jaws of the wire-drawing machine, and the rod or wire is drawn continuously through the die. In most wire-drawing operations, the wire passes through a series of dies (**tandem drawing**). In order to avoid excessive tension in the exiting wire, it is wound one or two turns around a capstan between each pair of dies. The speed of the capstan is adjusted so that it supplies not only tension, but also a small back tension to the wire entering the next die. Back tension reduces the die pressure and extends die life.

Drawing speeds depend on the material and cross-sectional area. They may be as low as 0.15 m/s (30 ft/min) for heavy sections and as high as 50 m/s (10,000 ft/min) for very fine wire. The temperature can rise substantially at high drawing speeds.

Rods and tubes that are not sufficiently straight (or are supplied in a coiled form) can be straightened by passing them through pairs of rolls placed at different axes. The rolls subject the material to a series of bending and unbending operations, similar to the method shown in Fig. 6.41 for rolling.

Large cross-sections can be drawn at elevated temperatures. In cold drawing, because of strain hardening, intermediate annealing between passes may be necessary to maintain sufficient ductility. Steel wires for springs and musical instruments are made by a heat-treatment process that precedes or follows the drawing operation (**patenting**). These wires have ultimate tensile strengths as high as 4800 MPa (700,000 psi), with tensile reduction of area of about 20%.

Bundle drawing. In this process, many wires (as much as several thousand) are drawn simultaneously as a bundle. To prevent sticking, the wires are separated from each other by a suitable material. The cross-section of the wires is somewhat polygonal.

Dies. Die materials are usually alloy tool steels, carbides, or diamond. A diamond die (used for drawing fine wires) may be a single crystal or a polycrystalline diamond in a metal matrix. Carbide and diamond dies are made as inserts or nibs, which are then supported in a steel casing. Figure 6.70 shows a typical wear pattern on a drawing die. Note that die wear is highest at the entry. Although the die pressure is highest in this region and may be partially responsible for wear, other factors are involved, including variation in the diameter of the entering wire and vibration (thus subjecting the entry contact zone to fluctuating stresses), and the presence of abrasive scale on the surface of the entering wire.

In addition to rigid dies, a set of idling rolls is also used in drawing of rods or bars of various shapes. This arrangement is known as **Turk's head** and is more versatile than using ordinary dies, since the rolls can be adjusted to various positions for different products.

Lubrication. Proper lubrication is essential in rod, tube, and wire drawing, regardless of whether the process is dry or wet drawing. In **dry drawing**, the surface of the wire is coated with various lubricants, depending on its strength and frictional characteristics.

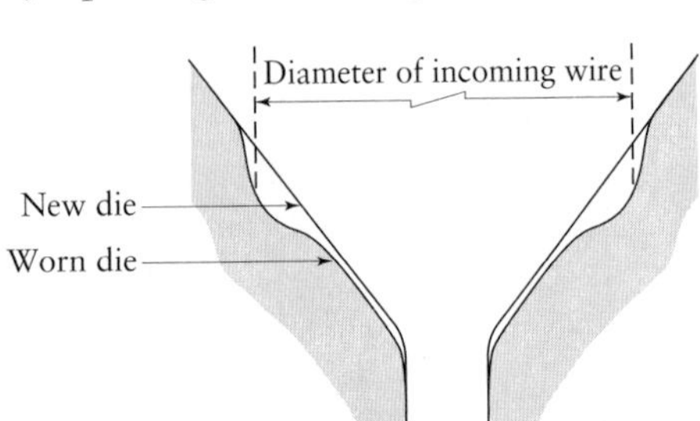

FIGURE 6.70 Schematic illustration of a typical wear pattern in a wire-drawing die.

A common lubricant used is soap. The rod to be drawn is first surface treated by **pickling**. This treatment removes the surface scale that could lead to surface defects and, being quite abrasive, could considerably reduce die life. The soap is picked up by the wire as it goes through a box filled with soap powder (*stuffing box*).

With high-strength materials, such as steels, stainless steels, and high-temperature alloys, the surface of the rod or wire may be coated either with a softer metal or with a **conversion coating**. Copper or tin can be chemically deposited on the surface of the metal. This thin layer of softer metal acts as a solid lubricant during drawing. Conversion coatings may consist of sulfate or oxalate coatings on the rod, which typically are then coated with soap, as a lubricant. Polymers are also used as solid lubricants, such as in drawing of titanium. In **wet drawing**, the dies and rod are completely immersed in a lubricant. Typical lubricants include oils and emulsions containing fatty or chlorinated additives, and various chemical compounds.

The technique of **ultrasonic vibration** of the dies and mandrels is also used successfully in drawing. When carried out properly, this technique reduces drawing forces, improves surface finish and die life, and allows higher reductions per pass.

Equipment. Two types of equipment are used in drawing. A **draw bench** is similar to a long horizontal tensile-testing machine. It is used for single draws of straight rods with large cross-sections and tubes with lengths up to 30 m (100 ft) with a hydraulic or chain-drive mechanism. Rod and wire of smaller cross-section are drawn by a **bull block (capstan)**, which is a rotating drum around which the wire is wrapped. The tension in this setup provides the force required to draw the wire.

6.6 Swaging

In *swaging*, also known as **rotary swaging** or **radial forging**, a solid rod or a tube is reduced in diameter by the reciprocating radial movement of two or four dies (Fig. 6.71). The die movements are generally obtained by means of a set of rollers in a cage. The

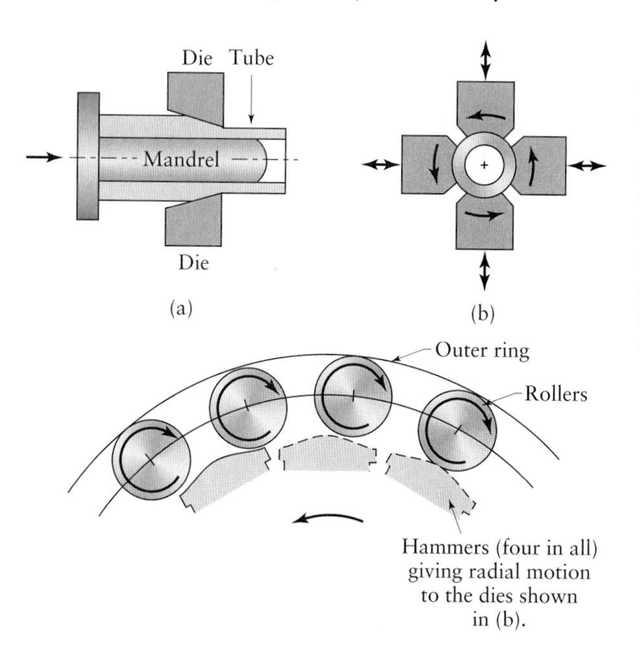

FIGURE 6.71 Schematic illustration of the swaging process: (a) side view and (b) front view. (c) Schematic illustration of roller arrangement, curvature on the four radial hammers (that give motion to the dies), and the radial movement of a hammer as it rotates over the rolls.

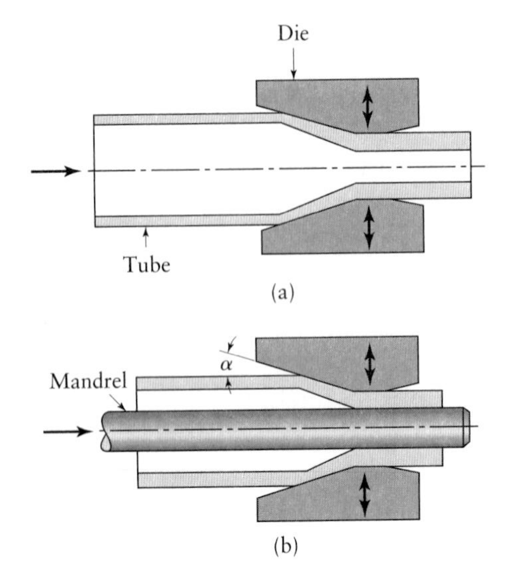

FIGURE 6.72 Reduction of outer and inner diameters of tubes by swaging. (a) Free sinking without a mandrel. The ends of solid bars and wire are tapered (pointing) by this process in order to feed the material into the conical die. (b) Sinking on a mandrel. Coaxial tubes of different materials can also be swaged in one operation.

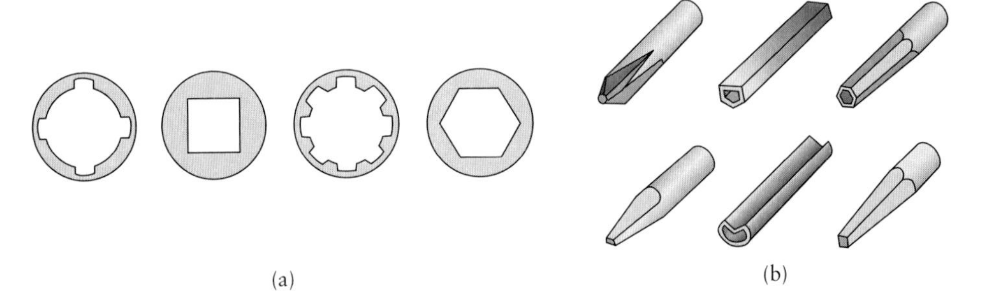

FIGURE 6.73 (a) Typical cross-sections produced by swaging tube blanks with a constant wall thickness on shaped mandrels. Rifling of small gun barrels can also be made by swaging, using a specially shaped mandrel. The formed tube is then removed by slipping it out of the mandrel. (b) These parts can also be made by swaging.

internal diameter and the thickness of the tube can be controlled with or without mandrels, as shown in Fig. 6.72. Mandrels can also be made with longitudinal grooves (similar in appearance to a splined shaft); thus, internally shaped tubes can be swaged (Fig. 6.73a). The rifling in gun barrels is made by swaging a tube over a mandrel with spiral grooves. Externally shaped parts can also be made by swaging (Fig. 6.73b).

The swaging process is usually limited to workpiece diameters of about 50 mm (2 in.), although special machinery has been built to swage gun barrels of larger diameter. The length of the product in this process is limited only by the length of the mandrel (if needed). Die angles are usually only a few degrees and may be compound, that is, the die may have more than one angle, for more favorable material flow during swaging. Lubricants are used for improved surface finish and longer die life. The process is generally carried out at room temperature. Parts produced by swaging have improved mechanical properties and good dimensional accuracy.

6.7 Die-Manufacturing Methods

Several manufacturing methods, either singly or in combination, are used in making dies for metal working operations, as well as for the molds used in the various processes described in Chapters 5, 7, 10, and 11. These processes include casting, forging, machining, grinding, and electrical and electrochemical methods of die sinking and finishing operations, such as honing, polishing, and coating. The selection of a die-manufacturing method depends primarily on the following parameters:

a) The particular process for which the die will be used;
b) The size and shape of the die;
c) The bulk and surface quality required;
d) Lead time to produce the die (e.g., some dies require months to prepare); and
e) Cost.

Economic considerations often dictate the process selected because tool and die costs can be significant in manufacturing operations. For example, the cost of a set of dies for pressworking automotive body panels (Figs. 1.5 and 7.68) may run up to $2 million; even small and simple dies can cost hundreds of dollars. On the other hand, because a large number of parts are typically produced by using the same die, the die cost per piece made is generally only a small fraction of a part's manufacturing cost (Chapter 16).

Dies may be classified as *male* and *female*; they may also be classified by their size. Small dies generally have a surface area of 10^3–10^4 mm^2 (2–15 in.^{2}), whereas large dies have a surface area of 1 m^2 (9 ft^2) or larger. Dies of various sizes and shapes can be made from steels, cast irons, and nonferrous alloys, as well as carbides, ceramics, and diamond (Chapters 8 and 11).

The processes used for producing die materials typically include *casting*, followed by various primary processing (forging, rolling, and extrusion) and secondary processing (machining, grinding, polishing and coating). Small dies and tools can also be made by the *powder metallurgy* techniques described in Chapter 11. (See also *rapid tooling*, Section 10.12). Sand casting is used for large dies weighing many tons and shell molding for small dies. Cast steels are generally preferred as die materials, because of their strength and toughness and the ease with which their composition, grain size, and properties can be controlled and modified. Depending on how they solidify, cast dies, unlike dies made of wrought metals, may not have directional properties; thus, they can exhibit isotropic properties on all working surfaces. However, because of shrinkage, the control of dimensional accuracy in cast-metal dies can be difficult compared with that in machined and finished dies, thus requiring further processing.

Most commonly, dies are *machined* from forged *die blocks* by processes such as milling, turning, grinding, electrical and electrochemical machining, and polishing (Chapters 8 and 9). Traditionally, dies have been machined by milling on an automatic copy machine that traces a master model (*pattern*), typically made of wood, epoxy resins, aluminum, or gypsum. Increasingly, however, dies are now machined with *computer-controlled machine tools* and *machining centers* using various software packages (Fig. 1.8 and Sections 8.10 and 14.3). For improved production, the cutter tool path is optimized.

Conventional machining can, however, be difficult and time consuming for high-strength, hard, tough, and wear-resistant die materials. Consequently, advanced machining processes such as *chemical* and *electrical-discharge machining* are used extensively (Sections 9.10 through Section 9.14), particularly for small or medium-sized dies; more recent developments include *hard machining* of dies (*hard turning*, Section 8.8.2). These processes are generally faster and more economical than traditional machining, and the dies may not require additional finishing. However, one must always consider any possible adverse effects that these processes may have on the properties of the die (including its fatigue life) because of possible surface damage and cracks.

Small dies with shallow cavities may also be produced by *hubbing* (Section 6.2.4). Diamond dies for drawing fine wire are manufactured by producing holes with a thin rotating needle, coated with diamond powder suspended in oil.

To improve their hardness, wear resistance, and strength, die steels (Section 3.10.3) are usually *heat treated* (Section 5.11). Improper heat treatment, however, is one of the most common causes of die failure, as described in Section 6.8. Particularly important are the condition and composition of the die surfaces; thus, the proper selection of temperatures and atmospheres for heat treatment, quenching media, quenching practice, tempering procedures, and handling is important. Dies may also be subjected to various surface treatments, including *coatings* (Sections 4.5.1 and 8.6.5) for improved frictional and wear characteristics, as described in Chapter 4.

After heat treatment, tools and dies are subjected to operations, such as grinding and polishing, to obtain the desired surface finish and dimensional accuracy. Note that heat treatment may distort dies because of microstructural changes or uneven thermal cycling. Also, if not controlled properly, the grinding process can cause surface damage by emitting excessive heat and inducing detrimental tensile residual stresses on the surface of the die, thus reducing its fatigue life. Scratches on a die's structure can act as stress raisers as well, and commonly used die-making and finishing processes, such as electrical-discharge machining (Section 9.13), can cause surface damage and cracks, unless the process parameters are carefully controlled. Finish grinding, deburring, and polishing may be carried out, either by hand or with programmable *industrial robots* (Section 14.7).

6.8 Die Failures

Failure of dies in metalworking operations generally results from one or more of the following causes:

a) Improper die design;
b) Defective die materials;
c) Improper heat treatment and finishing operations;
d) Improper installation, assembly, and alignment;
e) Overheating and heat checking (i.e., cracking caused by thermal cycling);
f) Excessive wear, and
g) Overloading, misuse, and improper handling.

The proper *design* of dies is as important as the appropriate selection of die materials. With rapid advances in *computer modeling and simulation* of processes and systems (Section 15.7), die design and optimization has now become an advanced technology. Basic design guidelines include the consideration that, in order to withstand the forces in manufacturing processes, a die must have appropriate cross sections and clearances. Sharp corners, radii, and fillets, as well as abrupt changes in cross section, act as stress raisers and can have detrimental effects on the life of the die. For improved strength, dies may be made in segments and may be prestressed during their assembly. Also, dies may be designed and constructed with inserts that can be replaced when worn or cracked. Overloading of dies can cause premature failure; for example, a common cause of failure of cold-extrusion dies (Section 6.4.3) is the operator's forgetting to remove an extruded part from the die before loading it with another blank. This type of failure can also occur in robots that automatically load and unload parts in dies.

In spite of their hardness and resistance to abrasion, die materials such as heat-treated steels, carbides, and diamond are susceptible to cracking and chipping from *impact forces* (as occur in mechanical presses and forging hammers) or from *thermal stresses* (caused by temperature gradients within the die in hot-working operations); thus, die surface preparation and finishing are important. Even metalworking fluids, often necessary, can adversely affect tool and die materials. For example, sulfur and chlorine additives in lubricants and coolants can attack the cobalt binder in tungsten carbide and lower its strength and toughness (*leaching*, Section 3.9.7).

During use, dies may also undergo *heat checking* from thermal cycling. To reduce heat checking (which has the appearance of parched land) and eventual breakage in hot-working operations, dies are usually preheated to temperatures of about 150 to 250°C (300 to 500°F). Cracked or worn dies may be repaired by welding and metal-deposition techniques. Even if they are manufactured properly, dies are subjected to stresses and temperature during their normal use, causing wear and, hence, changes in their shape and surface finish. When the shape of a die changes, the parts, in turn, have improper dimensions and surface finish; hence, the economics of the manufacturing operation are adversely affected.

Die failure and fracture in manufacturing plants can be hazardous to employees. It is not unusual for a set of dies resting on the plant floor or a shelf to disintegrate suddenly, because of their high internally stressed condition. The broken pieces are propelled at high speed and can cause serious injury or death. Highly stressed dies and tooling should always be surrounded by metal shielding, which should be properly designed and sufficiently strong to contain the fractured pieces in the event of sudden die failure.

CASE STUDY Manufacturing Solid Rocket-Motor Case Segments for the Space Shuttle

The U.S. Space Shuttle is a very complex spacecraft that has required extensive engineering to design and manufacture. The components on this space-launch vehicle are highly specialized and are subjected to extreme conditions during its flight. To lift the orbiter into space, the Space Shuttle employs two large, strap-on cylindrical solid rocket boosters, as seen in Fig. 6.74, which shows the launching of the Space Shuttle *Atlantis*. Each booster provides a 3.3-million-lb thrust force for the initial liftoff and assent. The rocket boosters consist of steel case segments (Fig. 6.75), which contain the solid rocket-fuel propellant and act as very large pressure vessels during launch.

The manufacture of the steel case segments is very challenging and requires a series of special processes and equipment (Fig. 6.76). The case segments are each approximately 3 m (120 in.) in diameter, 3 m (120 in.) long, and less than 13 mm (0.5 in.) thick; the dimensional tolerance on thickness is 0.25 mm (0.010 in.). A very thin and tightly controlled wall section is required for this pressure-vessel application. However, the case segments are so large that the weight of the final component is very close to the weight limit for the melting and casting of specialty-grade steel into ingots to be used as starting stock (blanks). Thus, it is imperative that very little material be wasted or discarded during the manufacturing process.

The mechanical-property requirements for these components dictate that ultraclean steel be used to eliminate melt-related material defects. (See Sections 5.3.3 and 5.5 and Fig. 3.24.) Ladish D6AC steel, with an ultimate tensile strength of 1500–1700 MPa (220–250 ksi), is first cast into large premium-quality vacuum-melted ingots. Because the weight of the steel ingots is so close to the weight of the finished components, material-efficient manufacturing steps have to be incorporated. To transform the cast ingot into final cylinder components, several different manufacturing steps are used, namely, (a) upsetting, (b) extrusion, (c) ring

FIGURE 6.74 The Space Shuttle *Atlantis* is launched by two strap-on solid rocket boosters. *Source*: NASA.

FIGURE 6.75 Assembly of steel case segments to form a solid rocket booster. Note that most of the rocket casing is below the platform level. *Source*: NASA.

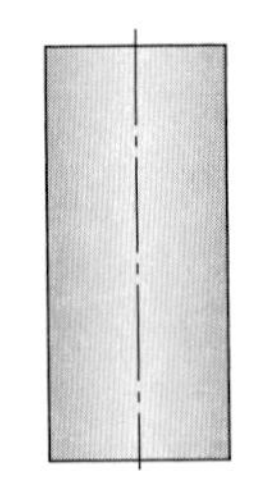

(a) As cast blank

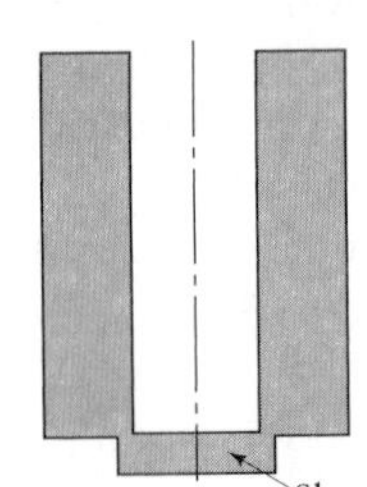

(b) Reverse extruded

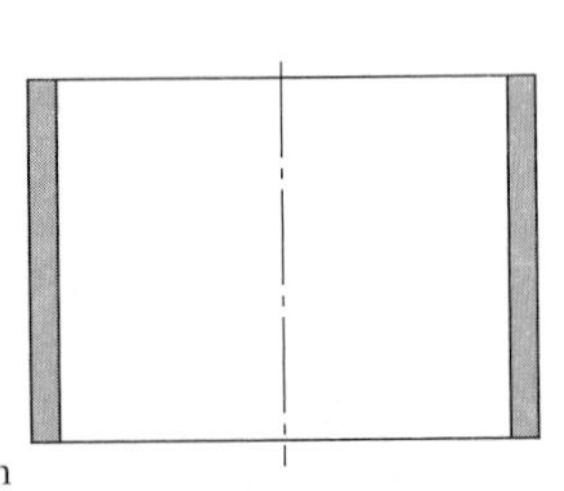

(c) Ring rolled

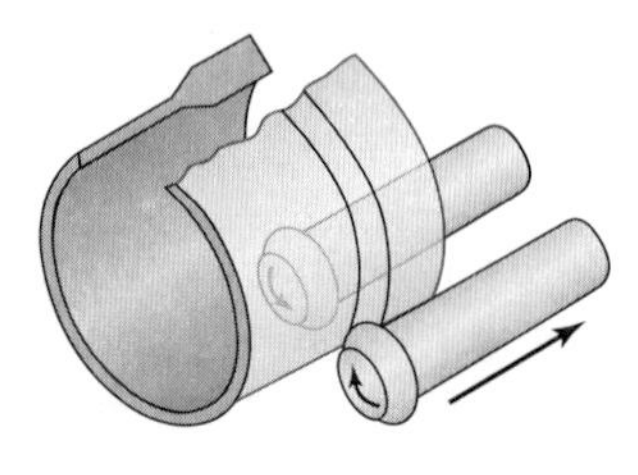

(d) Roll-formed casing

FIGURE 6.76 The forming processes involved in the manufacture of solid rocket casings for the U.S. Space Shuttle.

rolling, (d) roll forming, (e) machining, and (f) heat treating. Figure 6.76 shows a schematic of the forming steps used to make a case segment.

Indirect, or backward, extrusion (Section 6.4) is used first to produce a rough cylindrical shape for the case segments from the cast ingot. In this process, the ingot is upset (see Fig. 6.1), and then a large central punch is pressed into the upset steel blank. As the punch presses into the center of the heated metal, the metal is extruded backward between the moving punch and the container. The rough cylinder is removed from the container, and the thin slug of material that remains underneath the punch at the end of the operation is removed by hot shearing.

The hot cylinder is then placed on a large ring-rolling machine, where it is ring rolled (see Fig. 6.45a) to achieve the proper shape; this process reduces the thickness of the wall section of the cylinder while increasing its diameter. The length of the cylinder is maintained by containment or pinch rollers on the end faces. The rolled cylinder has a contour that contains the end flanges and the central pressure vessel. After this operation is completed, the cylinder is machined to remove all scale (from hot working) and to produce a smooth, uniform, bright surface finish. This step is designed to remove a minimum amount of material in order to produce a shape that can be ready for the final forming operation.

The machined cylinder is further processed by roll forming (also known as *flow forming* or *shear forming*, similar to spinning; see Section 7.8) to produce the final pressure-vessel walls. This process is performed at room temperature and by means of two opposing rollers, whereby the wall section is reduced, the diameter remains constant, and the length of the cylinder is increased. The cylinder wall is thus cold reduced to a net final thickness; it is then heat treated by normalizing, austenitizing, quenching, and tempering. (See Section 5.11.) Only the end flanges of the case segments require some final machining for proper assembly.

The aforementioned manufacturing steps have been optimized for the special requirements and constraints of this application, including (a) a clean, high-tensile-strength material; (b) a defect-free and inspectable component geometry; (c) a minimal loss of input material; and (d) a stable, robust, and reproducible process.

Source: D. Furrer, Ladish Co., Inc.

SUMMARY

- Bulk deformation processes, which generally consist of forging, rolling, extrusion, and drawing, involve major changes in the dimensions of the workpiece. Bulk properties as well as surface characteristics of the material are important. (Section 6.1)
- Forging denotes a family of processes by which deformation of the workpiece is carried out by compressive forces applied through dies. Forging is capable of producing a wide variety of parts with favorable characteristics of strength, roughness, dimensional accuracy, and reliability in service. Among important factors in forging are friction, material flow in dies, heat transfer, and lubrication. Various defects can develop, depending on die geometry, quality of the billet materials, and preform. Several forging machines are available, with various characteristics and capabilities. (Section 6.2)
- Currently used methods of analysis of the stresses, strains, strain rates, temperature distribution, and loads in forging and other bulk-deformation processes include the slab method and the finite-element method. (Section 6.2)
- Rolling is the process of reducing the thickness or changing the cross-section of a long workpiece by compressive forces applied through a set of rolls. It is a

continuous operation and involves several material and process variables, including the size of the roll relative to the thickness of the material, the amount of reduction per pass, speed, lubrication, and temperature. Rolled products include plate, sheet, foil, rod, pipe, and tube, as well as shape-rolled products such as I-beams, structural shapes, rails, and bars of various cross-sections. (Section 6.3)

- Extrusion is the process of forcing a material through the opening of a die, much like squeezing toothpaste from a tube. This process is capable of producing lengths of solid or hollow sections with constant cross-sectional area. Important factors include die design, extrusion ratio, lubrication, billet temperature, and extrusion speed. Cold extrusion, which is a combination of extrusion and forging operations, is capable of economically producing parts with good mechanical properties. (Section 6.4)
- Rod, wire, and tube drawing involve pulling the material through one or more dies. Proper die design and appropriate selection of materials and lubricants are essential to obtaining products of high quality and with good surface finish and product integrity. The major process variables in the mechanics of drawing are die angle, friction, and amount of reduction per pass. (Section 6.5)
- In swaging, a solid rod or a tube is reduced in diameter by the reciprocating radial movement of two or four dies. This process is suitable for producing short or long lengths of bar or tubing with various internal or external profiles. (Section 6.6)
- Because die failure has a major economic impact, die design, material selection, and the manufacturing methods employed are of major importance. A wide variety of die materials and manufacturing methods is available, including advanced material-removal processes and subsequent treatment, coating, and finishing operations. (Sections 6.7, 6.8)

SUMMARY OF EQUATIONS

Forging

Pressure in plane-strain compression, $p = Y' e^{2\mu(a-x)/h}$

Average pressure in plane-strain compression, $p_{av} \simeq Y'\left(1 + \dfrac{\mu a}{h}\right)$

Pressure in axisymmetric compression, $p = Y e^{2\mu(r-x)/h}$

Average pressure in axisymmetric compression, $p_{av} \simeq Y\left(1 + \dfrac{2\mu r}{3h}\right)$

Average pressure in plane-strain compression, sticking, $p_{av} = Y'\left(1 + \dfrac{a-x}{h}\right)$

Pressure in axisymmetric compression, sticking, $p = Y\left(1 + \dfrac{r-x}{h}\right)$

Rolling

Roll pressure in entry zone, $p = Y'_f \dfrac{h}{h_0} e^{\mu(H_0 - H)}$

with back tension, $p = (Y'_f - \sigma_b)\dfrac{h}{h_0} e^{\mu(H_0 - H)}$

Roll pressure in exit zone, $p = Y'_f \dfrac{h}{h_f} e^{\mu H}$

with front tension, $p = (Y'_f - \sigma_f)\dfrac{h}{h_f} e^{\mu H}$

Parameter $H = 2\sqrt{\frac{R}{h_f}}\tan^{-1}\left(\sqrt{\frac{R}{h_f}}\phi\right)$

Roll force, $F = \int_0^{\phi_n} wpRd\phi + \int_{\phi_n}^{\alpha} wpRd\phi$

Roll force, approximate, $F = Lw\bar{Y}'\left(1 + \frac{\mu L}{2h_{av}}\right)$

Roll torque, $T = \int_{\phi_n}^{\alpha} w\mu pR^2 d\phi - \int_0^{\phi_n} w\mu pR^2 d\phi$

Roll contact length, approximate, $L = \sqrt{R\Delta h}$

Power per roll $= \frac{\pi FLN}{60{,}000 \text{ kW}} = \frac{\pi FLN}{33{,}000 \text{ hp}}$

Maximum draft, $\Delta h_{max} = \mu^2 R$

Maximum angle of acceptance, $\alpha_{max} = \tan^{-1}\mu$.

Extrusion

Extrusion ratio, $R = A_o/A_f$

Extrusion pressure, ideal, $p = Y \ln R$

Extrusion pressure, with friction, $p = Y\left(1 + \frac{\tan\alpha}{\mu}\right)[R^{\mu\cot\alpha} - 1]$

Rod and wire drawing

Drawing stress, ideal, $\sigma_d = Y \ln\left(\frac{A_o}{A_f}\right)$

Drawing stress, with friction, $\sigma_d = Y\left(1 + \frac{\tan\alpha}{\mu}\right)\left[1 - \left(\frac{A_f}{A_o}\right)^{\mu\cot\alpha}\right]$

Die pressure, $p = Y_f - \sigma$

BIBLIOGRAPHY

Altan, T., S.-I. Oh, and H. Gegel, *Metal Forming—Fundamentals and Applications*, ASM International, 1983.

ASM Handbook, Vol. 14: *Forming and Forging*, ASM International, 1988.

Avitzur, B., *Handbook of Metalforming Processes*, Wiley-Interscience, 1983.

Blazynski, T. Z., *Plasticity and Modern Metal-Forming Technology*, Elsevier, 1989.

Boer, C. R., N. Rebelo, H. Rydstad, and G. Schroder, *Process Modelling of Metal Forming and Thermomechanical Treatment*, Springer, 1986.

Byrer, T. G. (ed.), *Forging Handbook*, ASM International, 1985.

Davis, J. R. (ed.), *Tool Materials*, ASM International, 1995.

Dieter, G. E., *Mechanical Metallurgy*, 3d. ed., McGraw-Hill, 1986.

— (ed.), *Workability Testing Techniques*, ASM International, 1984.

Forging Design Handbook, ASM International, 1972.

Frost, H. J., and M. F. Ashby, *Deformation-Mechanism Maps*, Pergamon, 1982.

Ginzburg, V. B., *High-Quality Steel Rolling: Theory and Practice*, Dekker,1993.

—, *Steel-Rolling Technology: Theory and Practice*, Dekker, 1989.

Hoffman, H. (ed.), *Metal Forming Handbook*, Springer Verlag, 1998.

Hosford, W. F., and R. M. Caddell, *Metal Forming, Mechanics and Metallurgy*, 2d. ed., Upper Saddle River, NJ: Prentice Hall, 1993.

Inoue, N., and M. Nishihara (eds.), *Hydrostatic Extrusion: Theory and Applications*, Elsevier, 1985.

Kobayashi, S., S.-I. Oh, and T. Altan, *Metal Forming and the Finite-Element Method*, Oxford, 1989.

Lange, K. (ed.), *Handbook of Metal Forming*, McGraw-Hill, 1985.

Laue, K., and H. Stenger, *Extrusion—Processes, Machinery, Tooling*, ASM International, 1981.

Lenard, J. G., M. Pietrzyk, and L. Cser, *Mathematical and Physical Simulation of the Properties of Hot Rolled Products*, Elsevier, 1999.

Mielnik, E. M., *Metalworking Science and Engineering*, McGraw-Hill, 1991.

Nachtman, E. S., and S. Kalpakjian, *Lubricants and Lubrication in Metalworking Operations*, Dekker, 1985.

Pietrzyk, M., and J. G. Lenoard, *Thermal-Mechanical Modelling of the Flat Rolling Process*, Springer, 1991.

Prasad, Y. V. R. K., and S. Sasidhara (eds.), *Hot Working Guide: A Compendium of Processing Maps*, ASM International, 1997.

Product Design Guide for Forging, Forging Industry Association, 1997.

Roberts, W. L., *Cold Rolling of Steel*, Dekker, 1978.

——, *Hot Rolling of Steel*, Dekker, 1983.

Sabroff, A. M., F. W. Boulger, and H. J. Henning, *Forging Materials and Practices*, Reinhold, 1968.

Saha, P. K., *Aluminum Extrusion Technology*, ASM International, 2000.

Sheppard, T., *Extrusion of Aluminum Alloys*, Chapman & Hall, 1998.

Tool and Manufacturing Engineers Handbook, Vol. II: *Forming*, Society of Manufacturing Engineers, 1984.

Wagoner, R. H., and Chenot, J. L., *Fundamentals of Metal Forming*, Wiley, 1996.

——, *Metal Forming Analysis*, Cambridge, 2001.

QUESTIONS

Forging

6.1 How can you tell whether a certain part is forged or cast? Describe the features that you would investigate to arrive at a conclusion.

6.2 Why is the control of volume of the blank important in closed-die forging?

6.3 What are the advantages and limitations of a cogging operation? Of die inserts in forging?

6.4 Explain why there are so many different kinds of forging machines.

6.5 Devise an experimental method whereby you can measure the force required for forging only the flash in impression-die forging. (See Fig. 6.15a.)

6.6 A manufacturer is successfully hot forging a certain part, using material supplied by Company *A*. A new supply of material is obtained from Company *B*, with the same nominal composition of the major alloying elements as that of the material from Company *A*. However, it is found that the new forgings are cracking even though the same procedure is followed as before. What is the probable reason?

6.7 Explain why there might be a change in the density of a forged product as compared to that of the blank.

6.8 Describe the role of surface oxide layers on a blank as it is being deformed by impression-die forging.

6.9 Since glass is a good lubricant for hot extrusion, would you use glass for impression-die forging as well? Explain.

6.10 Describe and explain the factors that influence spread in cogging operations on square billets.

6.11 In deriving the formula for forging, we had to use the yield criteria described in Chapter 2. Explain why.

6.12 Why are end grains generally undesirable in forged products? Give examples of such products.

6.13 Explain why one cannot obtain a finished forging in one press stroke, starting with a blank.

6.14 List the advantages and disadvantages of using a lubricant in forging.

Rolling

6.15 Explain how a cast structure is converted into a wrought structure by hot rolling.

6.16 As stated in Section 6.3.1, three factors that influence spreading in rolling are the width-to-thickness ratio of the strip, friction, and the ratio of the radius of the roll to the thickness of the strip. Explain how these factors affect spreading.

6.17 Explain how the residual stress patterns in Fig. 6.40 become reversed when the roll radius or reduction per pass is changed.

6.18 Explain how you would go about applying front and back tensions to sheet metals during rolling.

6.19 It was noted that rolls tend to flatten under roll forces. Which property(ies) of the roll material can be increased to reduce flattening?

6.20 Describe the methods by which roll flattening can be reduced.

6.21 Explain the technical and economic reasons for taking larger rather than smaller reductions per pass in flat rolling.

6.22 Surface roughness in hot-rolled products is higher than in cold-rolled products. Explain why.

6.23 List and explain the methods that can be used to reduce the roll force.

6.24 Explain the advantages and limitations of using small-diameter rolls in flat rolling.

6.25 Is it possible to adjust back and front tensions simultaneously so that the neutral point remains in the same position as in ordinary rolling without tensions? Explain.

6.26 A ring-rolling operation is successful for the production of bearing races. However, when the bearing race diameter is changed, the operation results in very poor surface finish. List the possible causes and the investigation you would perform to identify the parameters involved and correct the problem.

6.27 Describe the importance of controlling roll speed, roll gap, temperature, and other process variables in a tandem-rolling operation.

6.28 Is it possible to have a negative forward slip? Explain.

6.29 Rolling reduces the thickness of plates and sheets. However, it is also possible to reduce the thickness by simply stretching the material. Would this process be feasible? Explain.

6.30 In Fig. 6.34, explain why the neutral point moves towards the roll-gap entry as friction increases.

Extrusion

6.31 The extrusion ratio, die geometry, extrusion speed, and billet temperature all affect the extrusion pressure. Explain why.

6.32 How would you go about avoiding centerburst defects in extrusion? Explain why your methods would be effective.

6.33 Assume that you are reducing the diameter of two round rods, one by simple tension and the other by indirect extrusion. Which method will require more force? Why?

6.34 How would you go about making a stepped extrusion that has increasingly larger cross-sections along its length? Is it possible? Would your process be economical and suitable for high production runs? Explain.

6.35 It has been stated in this chapter that the temperature ranges for extruding various metals are similar to those for forging. (See Table 6.3.) Describe the consequences of extruding at a temperature (a) below and (b) above these ranges.

6.36 Note from Eq. (6.52) that, for low values of the extrusion ratio, such as $R = 2$, the ideal extrusion pressure p can be lower than the yield stress, Y, of the material. Explain whether or not this phenomenon is logical.

6.37 Explain why the three variables in Eq. (6.59) should influence the strain rate in extrusion.

6.38 In hydrostatic extrusion, complex seals are used between the ram and the container, but not between the extrusion and the die. Explain why.

Drawing

6.39 What changes would you expect in the strength, hardness, ductility, and anisotropy of annealed metals after they have been cold drawn through dies? Why?

6.40 We have seen that in rod and wire drawing, the maximum die pressure is at the die entry. Why?

6.41 Describe the conditions under which wet drawing and dry drawing, respectively, are desirable.

6.42 Name the important process variables in drawing, and explain how they affect the drawing process.

6.43 Assume that a rod-drawing operation can be carried out either in one pass or in two passes in tandem. If all die angles are the same and the total reduction is the same, will the drawing forces be different? Explain.

6.44 In Fig. 6.62, assume that the reduction in the cross-section is taking place by pushing the rod through the die instead of pulling it. Assuming that the material is perfectly plastic, sketch the die-pressure distribution, for the following situations: (a) frictionless; (b) with friction; (c) frictionless, but with front tension. Explain your answers.

6.45 In deriving Eq. (6.72), no mention is made of the ductility of the original material being drawn. Explain why.

6.46 Explain why the die pressure in drawing decreases toward the exit of the die.

6.47 What is the value of the die pressure at the exit when a drawing operation is being carried out at the maximum reduction per pass?

6.48 Explain why the maximum reduction per pass in drawing should increase as the strain-hardening exponent, n. increases.

6.49 Explain why the maximum reduction per pass is higher in drawing round bars (63%) as compared with plane-strain drawing of flat sheet or plate (58%).

6.50 If, in deriving Eq. (6.72), we include friction, will the maximum reduction per pass be the same (63%), higher, or lower? Explain.

6.51 Explain what effects back tension has on the die pressure in wire or rod drawing, and discuss why these effects occur.

6.52 Explain why the inhomogeneity factor ϕ in rod and wire drawing depends on the ratio h/L, as plotted in Fig. 6.12.

6.53 Describe the reasons for the development of the swaging process.

6.54 Occasionally, steel-wire drawing will take place within a sheath of a soft metal such as copper or lead. Why would this procedure be useful?

6.55 Recognizing that it is very difficult to manufacture a die with a submillimeter diameter, how would you produce a 10-μm-diameter wire?

General

6.56 With respect to the topics covered in this chapter, list and explain specifically two examples each where (a) friction is desirable and (b) friction is not desirable.

6.57 Take any three topics from Chapter 2 and, with a specific example for each, show their relevance to the topics covered in this chapter.

6.58 Take any three topics from Chapter 3 and, with a specific example for each, show their relevance to the topics covered in this chapter.

6.59 List and explain the reasons that there are so many different types of die materials used for the processes described in this chapter.

6.60 Explain the reasons that we should be interested in residual stresses developed in workpieces made by the forming processes described in this chapter.

6.61 Make a summary of the types of defects found in the processes described in this chapter. Indicate, for each type, methods of reducing or eliminating the defects.

PROBLEMS

Forging

6.62 In the free-body diagram in Fig. 6.4b, we have shown the incremental stress $d\sigma_x$ on the element, pointing to the left. Yet it would appear that, because of the direction of frictional stresses μp, the incremental stress should point to the right in order to balance the horizontal forces. Show that the same answer for the forging pressure is obtained regardless of the direction of this incremental stress.

6.63 Plot the force vs. reduction in height curve in open-die forging of a cylindrical, annealed copper specimen 2 in. high and 1 in. in diameter, up to a reduction of 70%, for the cases of (a) no friction between the flat dies and the specimen, (b) $\mu = 0.2$, and (c) $\mu = 0.4$. Ignore barreling. Use average-pressure formulas.

6.64 Use Fig. 6.9b to provide the answers to Problem 6.63.

6.65 Calculate the work done for each case in Problem 6.63.

6.66 Determine the temperature rise in the specimen for each case in Problem 6.63, assuming that the process is adiabatic and the temperature is uniform throughout the specimen.

6.67 To determine forgeability, a hot-twist test is performed on a round bar 30 mm in diameter and 300 mm long. It is found that bar underwent 200 turns before it fractured. Calculate the shear strain at the outer surface of the bar at fracture.

6.68 Derive an expression for the average pressure in plane-strain compression with sticking friction.

6.69 What is the value of μ when, for plane-strain compression, the forging load with sliding friction is equal to the load with sticking friction? Use average-pressure formulas.

6.70 Note that in cylindrical upsetting, the frictional stress cannot be greater than the shear yield stress k of the material. Thus, there may be a distance x in Fig. 6.8 where a transition occurs from sliding to sticking friction. Derive an expression for x in terms of r, h, and μ only.

6.71 Assume that a workpiece is being pushed to the right by a lateral force F while being compressed between flat dies. (See the accompanying figure.) (a) Make a sketch of the die-pressure distribution for the condition for which F is not large enough to slide the workpiece to the right. (b) Make a similar sketch, except with F now being large enough so that the workpiece slides to the right while being compressed.

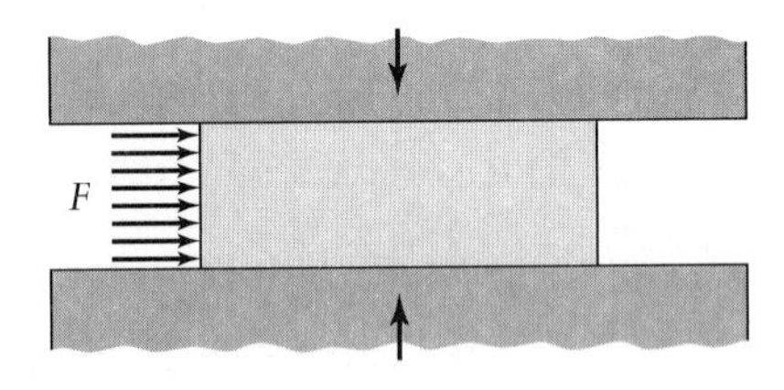

6.72 For the sticking example in Fig. 6.10, derive an expression for the lateral force F required to slide the workpiece to the right while the workpiece is being compressed between flat dies.

6.73 Two solid cylindrical specimens A and B, made of a perfectly plastic material, are being forged with friction and isothermally at room temperature to a reduction in height of 50%. Specimen A has a height of 2 in. and a cross-sectional area of 1 in^2, and specimen B has a height of is 1 in. and a cross-sectional area of 2 in^2. Will the work done be the same for the two specimens? Explain.

6.74 In Fig. 6.6, does the pressure distribution along the four edges of the workpiece depend on the particular yield criterion used? Explain.

6.75 Under what conditions would you have a normal pressure distribution in forging a workpiece as shown in the accompanying figure? Explain.

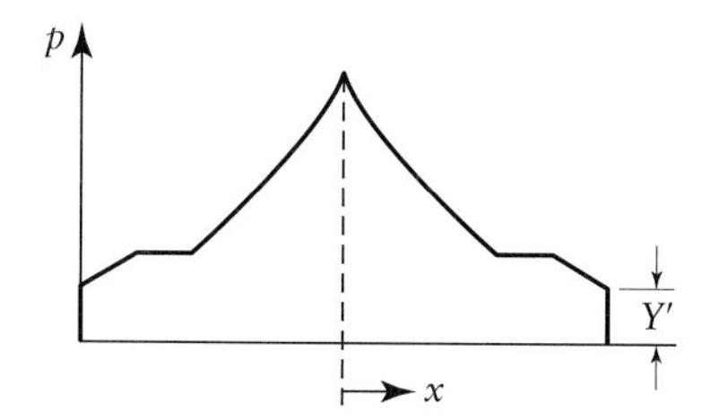

6.76 Derive the average die-pressure formula given by Eq. (6.15). (*Hint*: Obtain the volume under the friction hill over the surface by integration, and divide that quantity by the cross-sectional area of the workpiece.)

6.77 Take two solid cylindrical specimens of equal diameter, but different heights, and compress them (frictionless) to the same percent reduction in height. Show that the final diameters will be the same.

6.78 A rectangular workpiece has the following original dimensions: $2a = 120$ mm, $h = 30$ mm, and width $= 20$ mm. The metal has a strength coefficient of 400 MPa and a strain-hardening exponent of 0.3. It is being forged in plane strain with $\mu = 0.2$. Calculate the force required for the height to be reduced by 20%. Do not use average-pressure formulas.

6.79 Assume that in upsetting a solid cylindrical specimen between two flat dies with friction, the dies are rotated at opposite directions to each other. How, if at all, will the forging force change from that for nonrotating dies? (*Hint*: Note that the dies will now require torque, because of the change in the direction of frictional forces at the die–workpiece interfaces.)

6.80 A solid cylindrical specimen, made of a perfectly plastic material, is being upset between flat dies with no friction. The process is being carried out by a falling weight, as in a drop hammer. The downward velocity of the hammer is at a maximum when it first contacts the workpiece and becomes zero when the hammer stops at a certain height of the specimen. Establish quantitative relationships between workpiece height and velocity, and make a qualitative sketch of the velocity profile of the hammer. (*Hint*: The loss in the kinetic energy of the hammer is the plastic work of deformation; thus, there is a direct relationship between workpiece height and velocity.)

6.81 How would you go about estimating the swaging force acting on each die?

6.82 We note that in swaging a tube, the thickness of the tube increases. How would you go about calculating the thickness change without measuring it? You are allowed to make any other measurements.

6.83 A mechanical press is powered by a 30-hp motor and operates at 40 strokes per minute. It uses a flywheel, so that the rotational speed of the crankshaft does not vary appreciably during the stroke. If the stroke length is 6 in., what is the maximum contact force that can be exerted over the entire stroke length? To what height can a 5052-O aluminum cylinder with a diameter of 1 in. and a height of 2 in. be forged before the press stalls?

6.84 Estimate the force required to upset a 0.125-in-diameter C74500 brass rivet in order to form a 0.25-in-diameter head. Assume that the coefficient of friction between the brass and the tool-steel die is 0.25 and that the head is 0.125 in. in thickness.

6.85 Using the slab method of analysis, derive Eq. (6.17).

Rolling

6.86 In Example 6.4, what is the velocity of the strip leaving the rolls?

6.87 With appropriate sketches, explain the changes that occur in the roll-pressure distribution if one of the rolls is idling, i.e., power is shut off to that roll.

6.88 It can be shown that it is possible to determine μ in flat rolling without measuring torque or forces. By inspecting equations for rolling, describe an experimental procedure to do so. Note that you are allowed to measure any quantity other than torque or forces.

6.89 Derive a relationship between back tension σ_b and front tension σ_f in rolling such that when both tensions are increased, the neutral point remains in the same position.

6.90 Take an element at the center of the deformation zone in flat rolling. Assuming that all the stresses acting on this element are principal stresses, indicate the stresses qualitatively, and state whether they are tension or compression stresses. Explain the reasoning behind your devices. Is it possible for all three principal stresses to be equal to each other in magnitude? Explain.

6.91 It has been stated that in rolling a flat strip, the roll force is reduced about twice as effectively by back tension as it is by front tension. Explain this difference, using appropriate sketches. (*Hint*: Note the position of the neutral point.)

6.92 It can be seen that in rolling a strip, the rolls will begin to slip if the back tension, σ_b, is too high. Derive an analytical expression for the magnitude of the back tension in order to make the powered rolls begin to slip. Use the same terminology as applied in the text.

6.93 Prove Eq. (6.44).

6.94 In Steckel rolling, the rolls are idling, and thus there is no net torque, assuming frictionless bearings. Where, then, is the energy coming from to supply the work of deformation in rolling? Explain with appropriate sketches, and state the conditions that have to be satisfied.

6.95 Derive an expression for the tension required in Steckel rolling of a flat sheet, without friction, for a workpiece whose true-stress–true-strain curve is given by $\sigma = a + b\varepsilon$.

6.96 (a) Make a neat sketch of the roll-pressure distribution in ordinary rolling with powered rolls. (b) Assume now that the power to both rolls is shut off and that rolling is taking place by front tension only, i.e., Steckel rolling. Superimpose on your diagram the new roll-pressure distribution, explaining your reasoning clearly. (c) After completing part (b), further assume that the roll bearings are getting rusty and deprived of lubrication while rolling is still taking place by front tension only. Superimpose a third roll-pressure distribution diagram for this condition, explaining your reasoning.

6.97 Derive Eq. (6.26b), based on the equation preceding it. Comment on how different the h values are as the angle ϕ increases.

6.98 In Fig. 6.35, assume that $L = 2L_2$. Is the roll force, F, for L twice or more than twice the roll force for L_2? Explain.

6.99 A flat-rolling operation is being carried out where $h_o = 0.15$ in., $h_f = 0.100$ in., $w_o = 10$ in., $R = 10$ in., $\mu = 0.2$, and the average flow stress of the material is 40,000 psi. Estimate the roll force and the torque. Include the effects of roll flattening.

6.100 A rolling operation takes place under the conditions shown in the accompanying figure. What is the position x_n of the neutral point? Note that there are a front and back tension that have not been specified. Additional data are as follows: Material is 5052-O aluminum; $\mu = 0.1$; hardened steel rolls; surface roughness of the rolls = 0.02 μm; rolling temperature = 210°C.

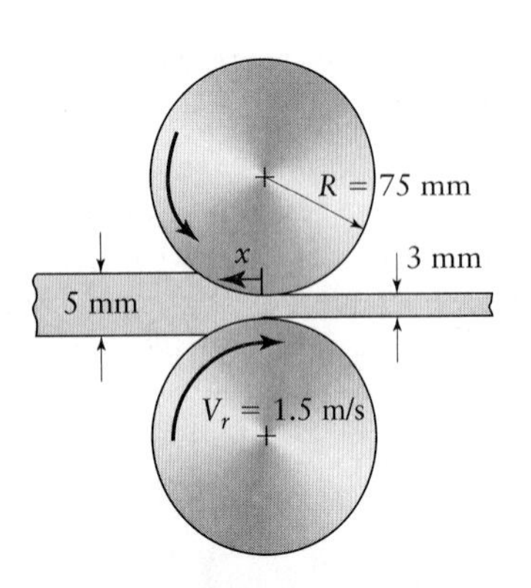

6.101 Estimate the roll force and power for annealed low carbon steel strip 400 mm wide and 10 mm thick, rolled to a thickness of 7 mm. The roll radius is 200 mm, and the roll rotates at 200 rpm. Use $\mu = 0.1$.

Extrusion

6.102 Derive an expression for the true-strain rate, $\dot{\varepsilon}$, in the deformation zone in extrusion of a square billet, in terms of the entering velocity V_o, the die angle α, an initial billet dimension t_o, and the distance x from entry.

6.103 Calculate the force required in direct extrusion of 1100-O aluminum from a diameter of 8 in. to a diameter of 2 in. Assume that the redundant work is 30% of the ideal work of deformation, and the friction work is 25% of the total work of deformation.

6.104 Prove Eq. (6.56).

6.105 Calculate the theoretical temperature rise in the extruded material in Example 6.6, assuming that there is no heat loss. (See Section 3.9 for information on the physical properties of the material.)

6.106 Using the same approach as that shown in Section 6.5 for wire drawing, show that the extrusion pressure is given by the expression

$$p = Y\left(1 + \frac{\tan\alpha}{\mu}\right)\left[1 - \left(\frac{A_o}{A_f}\right)^{\mu\cot\alpha}\right],$$

where A_o and A_f are the original and final workpiece diameters, respectively.

6.107 Derive Eq. (6.54).

6.108 A planned extrusion operation involves steel at 800°C, with an initial diameter of 100 mm and a final diameter of 20 mm. Two presses, one with a capacity of 20 MN and the other of 10 MN, are available for this operation. Obviously, the larger press requires greater care and more expensive tooling. Is the smaller press sufficient for the operation? If not, what recommendations would you make to allow use of the smaller press?

Drawing

6.109 Calculate the power required in Example 6.7 if the workpiece material is annealed 70–30 brass.

6.110 Using Eq. (6.61), make a plot similar to Fig. 6.66 for the following conditions: $K = 75$ MPa; $n = 0.3$; $\mu = 0.04$.

6.111 Using the same approach as that described in Section 6.5 for wire drawing, show that the drawing stress, σ_d, in plane-strain drawing of a flat sheet or plate is given by the expression

$$\sigma_d = Y'\left(1 + \frac{\tan\alpha}{\mu}\right)\left[1 - \left(\frac{h_f}{h_o}\right)^{\mu\cot\alpha}\right],$$

where h_o and h_f are the original and final thickness, respectively, of the workpiece.

6.112 Derive an analytical expression for the die pressure in wire drawing, without friction or redundant work, as a function of the instantaneous diameter in the deformation zone.

6.113 A linearly strain-hardening material with a true-stress–true-strain curve $\sigma = 5{,}000 + 25{,}000\varepsilon$ psi is being drawn into a wire. If the original diameter of the wire is 0.2 in., what is the minimum possible diameter at the exit of the die? Assume that there is no redundant work and that the frictional work is 25% of the ideal work of deformation. (*Hint*: The yield stress of the exiting wire is the point on the true-stress–true-strain curve that corresponds to the total strain that the material has undergone.)

6.114 In Fig. 6.68, assume that the longitudinal residual stress at the center is −60,000 psi. Using the distortion-energy criterion, calculate the minimum yield stress that this particular steel must have in order to sustain these levels of residual stresses.

6.115 What is the answer to Problem 6.114 for the maximum-shear-stress criterion?

6.116 Derive an expression for the die-separating force in frictionless wire drawing of a perfectly plastic material. Use the same terminology as in the text.

6.117 A material with a true-stress–true-strain curve $\sigma = 10{,}000\varepsilon^{0.3}$ is used in wire drawing. Assuming that the friction and redundant work compose a total of 60% of the ideal work of deformation, calculate the maximum reduction in cross-sectional area per pass that is possible.

6.118 Derive an expression for the maximum reduction per pass for a material of $\sigma = K\varepsilon^n$, assuming that the friction and redundant work contribute a total of 20% to the ideal work of deformation.

6.119 Prove that the true-strain rate $\dot{\varepsilon}$ in drawing or extrusion in plane strain with wedge-shaped dies is given by the expression

$$\dot{\varepsilon} = -\frac{2 \tan \alpha V_o t_o}{(t_o - 2x \tan \alpha)^2},$$

where α is the die angle, t_o is the original thickness, and x is the distance from die entry (*Hint*: Note that $d\varepsilon = dA/A$.)

6.120 In drawing a strain-hardening material with $n = 0.3$, what should be the percentage of friction plus redundant work, in terms of ideal work, so that the maximum reduction per pass is 63%?

6.121 Assume that you are asked to give a quiz to students on the contents of this chapter. Prepare three quantitative problems and three qualitative questions, and supply the answers.

DESIGN

6.122 Forging is one method of producing turbine blades for jet engines. Study the design of such blades and the relevant technical literature, and then prepare a step-by-step procedure for making the blades. Comment on the potential difficulties that may be encountered.

6.123 In comparing forged parts with cast parts, we have noted that the same part may be made by either process. Comment on the pros and cons of each process, considering factors such as part size, shape complexity, and design flexibility if a particular design has to be modified.

6.124 Review the technical literature, and make a detailed list of the manufacturing steps involved in the manufacture of hypodermic needles.

6.125 Figure 6.50a shows examples of products that can be obtained by slicing long extruded sections into small discrete pieces. Name several other products that can be made in a similar manner.

6.126 Make an extensive list of products that either are made of or have one or more components of (a) wire, (b) very thin wire, and (c) rods of various cross-sections.

6.127 Although extruded products typically are straight, it is possible to design dies whereby the extrudate has a constant radius of curvature. (a) What applications could you think of for such products? (b) Describe your ideas as to the shape that such a die should have.

6.128 Survey the technical literature, and describe the design features of the various roll arrangements shown in Fig. 6.42.

6.129 We have indicated the beneficial effects of using ultrasonic vibration to reduce friction in some of the processes described in this chapter. Survey the technical literature, and offer design concepts to apply such vibrations.

CHAPTER

7 Sheet-Metal Forming Processes

7.1 Introduction

It was noted in Chapter 6 that workpieces can be characterized by the ratio of their surface area to their volume or thickness. **Sheet forming**, unlike bulk-deformation processes, involves workpieces with a high ratio of surface area to thickness. A sheet thicker than 6 mm ($\frac{1}{4}$ in.) is generally called a **plate**. Although relatively thick plates, such as those used in boilers, bridges, ships, and nuclear power plants, may have smaller ratios of surface area to thickness, sheet forming usually involves relatively thin materials.

Typical examples of applications that use sheets include metal desks, appliance bodies, hubcaps, aircraft fuselage, beverage cans, car bodies, and kitchen utensils; thus, sheet forming, also called **pressworking** or **stamping**, is among the most important of metalworking processes. It dates back to as early as 5000 BC, when household utensils, jewelry, and other objects were made by hammering and stamping metals such as gold, silver, and copper. The products made by sheet-forming processes include a large variety of shapes and sizes, ranging from simple bends to double curvatures with shallow or deep recesses.

Sheet metal is produced by a rolling process, as described in Section 6.3. If the sheet is thin, it is generally coiled after rolling; if thick, it is available as flat sheets or plates, which may have been decoiled and flattened. Before a part is formed, a blank of suitable dimensions is first cut or removed from a large sheet. Removal is usually done by a shearing process; however, there are several other methods for cutting sheet and plates. Laser cutting has become an important process and is used with computer-controlled equipment to cut consistently a variety of shapes.

This chapter describes the principles and the technology of the wide variety of sheet-forming processes available to manufacturing industries. As shown in Table 7.1, each process has specific characteristics and uses different types of hard tools and dies, as well as soft tooling, such as rubber. The sources of energy typically involve mechanical means, but also include various other sources of energy, such as hydraulic, magnetic, and explosive.

7.2 Sheet-Metal Characteristics

Sheet metals are generally characterized by a high ratio of surface area to thickness. Forming of sheet metals generally is carried out by tensile forces in the plane of the sheet; otherwise, the application of compressive forces could lead to buckling, folding, and wrinkling of the sheet. In bulk-deformation processes such as forging, rolling,

TABLE 7.1

General Characteristics of Sheet-Metal Forming Processes

Process	Characteristics
Roll forming	Long parts with constant complex cross-sections; good surface finish; high production rates; high tooling costs.
Stretch forming	Large parts with shallow contours; suitable for low-quantity production; high labor costs; tooling and equipment costs depend on part size.
Drawing	Shallow or deep parts with relatively simple shapes; high production rates; high tooling and equipment costs.
Stamping	Includes a variety of operations, such as punching, blanking, embossing, bending, flanging, and coining; simple or complex shapes formed at high production rates; tooling and equipment costs can be high, but labor costs are low.
Rubber forming	Drawing and embossing of simple or complex shapes; sheet surface protected by rubber membranes; flexibility of operation; low tooling costs.
Spinning	Small or large axisymmetric parts; good surface finish; low tooling costs, but labor costs can be high unless operations are automated.
Superplastic forming	Complex shapes, fine detail, and close tolerances; forming times are long, and hence production rates are low; parts not suitable for high-temperature use.
Peen forming	Shallow contours on large sheets; flexibility of operation; equipment costs can be high; process is also used for straightening parts.
Explosive forming	Very large sheets with relatively complex shapes, although usually axisymmetric; low tooling costs, but high labor costs; suitable for low-quantity production; long cycle times.
Magnetic-pulse forming	Shallow forming, bulging, and embossing operations on relatively low-strength sheets; most suitable for tubular shapes; high production rates; requires special tooling.

extrusion, and wire drawing (described in Chapter 6), the thickness or the lateral dimensions of the workpiece are intentionally changed, whereas in most sheet-forming processes, any change in thickness is due to stretching of the sheet under tensile stresses (*Poisson's effect*). Thickness decreases in sheet forming generally should be avoided, as they can lead to necking and failure.

Because the mechanics of all sheet forming basically consists of the processes of stretching and bending, certain factors significantly influence the overall operation. As described next, the major factors are elongation, yield-point elongation, anisotropy, grain size, residual stresses, springback, and wrinkling.

7.2.1 Elongation

Although, as we shall see, sheet-forming operations rarely involve simple uniaxial stretching, the observations made in regard to simple tensile testing can be useful in understanding the behavior of sheet metals in forming operations. Recall from Fig. 2.2 that a specimen subjected to tension first undergoes **uniform elongation** (up to the UTS, after which necking begins); this elongation is then followed by additional nonuniform elongation (**postuniform elongation**) until fracture occurs. Because the material is being stretched during forming, high uniform elongation is, of course, desirable for good formability.

It was shown in Section 2.2.3 that, for a material that has a true-stress–true-strain curve that can be represented by the equation

$$\sigma = K\varepsilon^n, \tag{7.1}$$

the strain at which necking begins (**instability**) is given by

$$\varepsilon = n. \tag{7.2}$$

Thus, the true uniform strain in a simple stretching operation (uniaxial tension) is numerically equal to the strain-hardening exponent, n. A large value of n indicates large uniform elongation, and, thus, it is desirable for sheet forming.

Necking of a sheet specimen generally takes place at an angle ϕ to the direction of tension, as shown in Fig. 7.1a (**localized necking**). For an *isotropic* sheet specimen under simple tension, the Mohr's circle is constructed as shown in Fig. 7.1b. The strain ε_1 is the longitudinal strain, and ε_2 and ε_3 are the two lateral strains. Poisson's ratio in the plastic range is 0.5; hence, the lateral strains have the value $-\varepsilon_1/2$. The narrow neck band in Fig. 7.1a is in **plane strain**, because it is constrained by the material above and below the neck band.

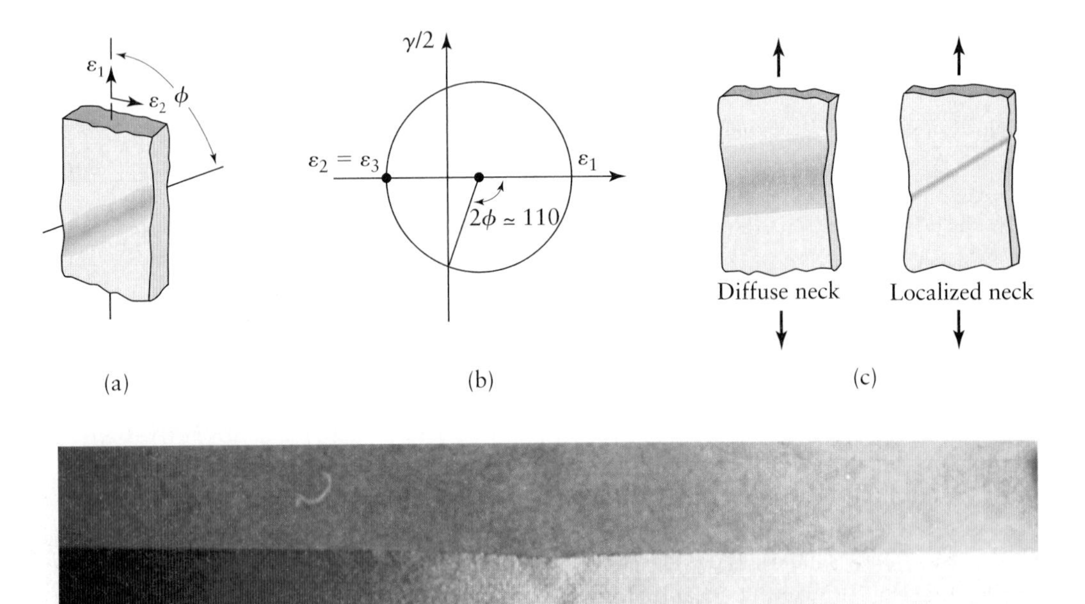

FIGURE 7.1 (a) Localized necking in a sheet specimen under tension. (b) Determination of the angle of neck from the Mohr's circle for strain. (c) Schematic illustrations for diffuse and localized necking. (d) Localized necking in an aluminum strip stretched in tension. Note the double neck.

The angle ϕ can be determined from the Mohr's circle by a rotation (either clockwise or counterclockwise) of 2ϕ from the ε_1 position (Fig. 7.1b). This angle is about 110°; thus, the angle ϕ is about 55°. Note that although the length of the neck band essentially remains constant during the test, its thickness decreases (because of volume constancy), and the specimen eventually fractures. The angle ϕ will be different for materials that are anisotropic in the plane of the sheet (planar anisotropy).

Whether necking is *localized* or **diffuse** (Fig. 7.1c) depends on the strain-rate sensitivity, m, of the material, as given by the equation

$$\sigma = C\dot{\varepsilon}^m. \tag{7.3}$$

As described in Section 2.2.7, the higher the value of m, the more diffuse the neck becomes. Figure 7.1d shows an example of localized necking on an aluminum strip in tension; note also the double localized neck. In other words, ϕ can be in either the clockwise or counterclockwise position in Fig. 7.1a.

In addition to uniform elongation, the total elongation of a tension-test specimen [at a gage length of 50 mm (2 in.)] is also a significant factor in formability of sheet metals. Note that total elongation is the sum of uniform elongation and postuniform elongation. As stated previously, uniform elongation is governed by the strain-hardening exponent, n, whereas postuniform elongation is governed by the strain-rate sensitivity index, m. The higher the m value, the more diffuse the neck is, and hence the greater the postuniform elongation becomes before fracture; thus, the total elongation of the material increases with increasing values of both n and m. The following are important factors that affect the sheet-metal forming process:

a) **Yield-point elongation.** Low-carbon steels exhibit a behavior called *yield-point elongation*, exhibiting *upper* and *lower* yield points, as shown in Fig. 7.2a; yield-point elongation is usually on the order of a few percent. This behavior indicates that, after the material yields, it stretches further in certain regions along the specimen, with no increase in the lower yield point, while other regions have not yet yielded. When the overall elongation reaches the yield-point elongation, the entire specimen has been deformed uniformly. The magnitude of the yield-point elongation depends on the strain rate (with higher rates, the elongation generally increases) and the grain size of the sheet metal. As grain size decreases, yield-point elongation increases.

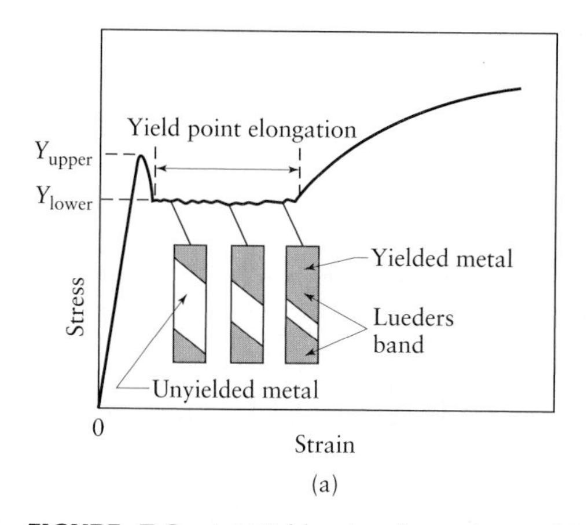

(a)

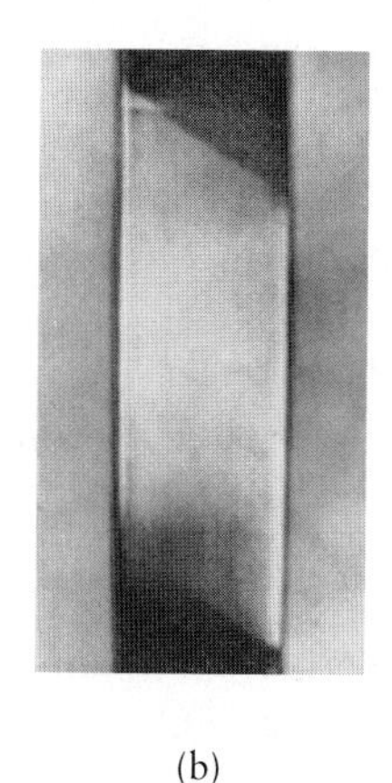
(b)

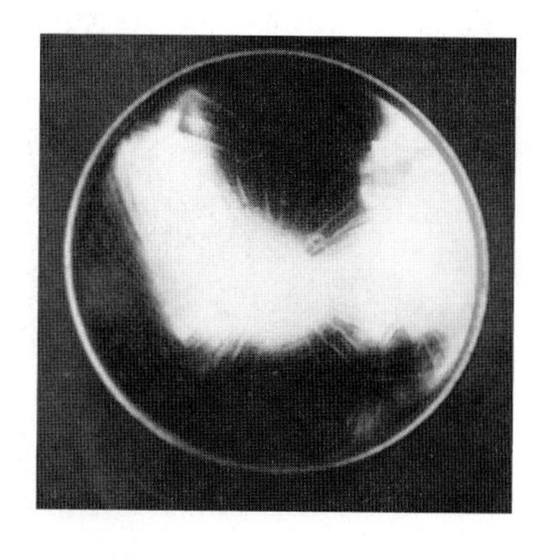
(c)

FIGURE 7.2 (a) Yield point elongation and Lueders bands in tension testing. (b) Lueder's bands in annealed low-carbon steel sheet. (c) Stretcher strains at the bottom of a steel can for household products. *Source*: (b) Reprinted Courtesy of Caterpillar Inc.

This behavior of low-carbon steels produces **Lueder's bands** (also called **stretcher strain marks** or *worms*) on the sheet, as shown in Fig. 7.2b. These bands are elongated depressions on the surface of the sheet and can be objectionable in the final product, because of its surface appearance. These bands can also cause difficulty in subsequent coating and painting operations. Stretcher strain marks can be seen on the curved bottom of steel cans for common household products, as shown in Fig. 7.2c; aluminum cans do not exhibit this behavior.

The usual method of avoiding this problem is to eliminate or to reduce yield-point elongation by reducing the thickness of the sheet 0.5% to 1.5% by cold rolling, known as **temper rolling** or **skin rolling**. However, because of strain aging (see Section 3.8.1), yield-point elongation reappears after even a few days at room temperature, or after a few hours at higher temperatures. Thus, the sheet metal should be formed within a certain period of time (from one to three weeks for rimmed steel; see Section 5.7.1) to avoid the reappearance of stretcher strains.

b) **Anisotropy.** Another important factor influencing sheet-metal forming is *anisotropy*, or **directionality**, of the sheet metal. Anisotropy is acquired during the thermomechanical-processing history of the sheet. Recall from Section 3.5 that there are two types of anisotropy: **crystallographic anisotropy** (from preferred grain orientation) and **mechanical fibering** (from alignment of impurities, inclusions, voids, and the like throughout the thickness of the sheet during processing). Anisotropy may be present not only in the plane of the sheet, but also in its thickness direction; the former is called **planar anisotropy**, and the latter is called **normal** or **plastic anisotropy**. These topics are particularly important in deep drawing of sheet metals, as described in Section 7.12.

c) **Grain size** *Grain size* of the sheet metal is important for two reasons: first, because of its effect on the mechanical properties of the material, and second, because of its effect on the surface appearance of the formed part. The coarser the grain, the rougher the surface appears (**orange peel**). An ASTM grain size of No. 7 or finer is preferred for general sheet-metal forming. (See Section 3.4.1.)

d) **Residual stresses.** *Residual stresses* can develop in sheet-metal parts because of the nonuniform deformation that the sheet undergoes during forming. When disturbed, such as by removing a portion of it, the part may distort. Tensile residual stresses on surfaces can also lead to **stress-corrosion cracking** of the part (Fig. 7.3) unless it is properly stress relieved. (See also Section 3.8.2.)

e) **Springback** Because they generally are thin and are subjected to relatively small strains during forming, sheet-metal parts are likely to experience considerable *springback*. This effect is particularly significant in bending and other forming operations where the bend radius-to-thickness ratio is high, such as in automotive body parts.

f) **Wrinkling.** Although, in sheet forming, the metal is generally subjected to tensile stresses, the method of forming may be such that compressive stresses are developed in the plane of the sheet; an example is the *wrinkling* of the flange in deep drawing (Section 7.12) because of circumferential *compression*. Other terms used to describe similar phenomena are **buckling, folding,** and **collapsing.** The tendency for wrinkling increases with (a) the unsupported or unconstrained length or surface area of the sheet, (b) decreasing thickness, and (c) nonuniformity of the thickness of the sheet. Lubricants that trapped or distributed unevenly between the surface of the sheet and the dies also can contribute to the initiation of wrinkling.

FIGURE 7.3 Stress-corrosion cracking in a deep-drawn brass part for a light fixture. The cracks developed over a period of time. Brass and austenitic (300 series) stainless steels are among metals that are susceptible to stress-corrosion cracking.

g) **Coated sheet.** Sheet metals, especially steel, can be **precoated** with a variety of organic coatings, films, and laminates; such coatings are used primarily for appearance and eye appeal, but they also offer corrosion resistance. Coatings are applied to the coil stock on continuous lines, at thicknesses generally ranging from 0.0025 to 0.2 mm (0.0001 to 0.008 in.). Coatings are available with a wide range of properties, such as flexibility, durability, hardness, resistance to abrasion and chemicals, color, texture, and gloss. Coated sheet metals are subsequently formed into products such as TV cabinets, appliance housings, paneling, shelving, residential siding, and metal furniture. Zinc is used extensively as a coating on sheet steel (**galvanized** steel; see Sections 3.9.7 and 4.5) to protect it from corrosion, particularly in the automotive industry. Galvanizing of steel can be done by hot dipping, electrogalvanizing, or galvannealing processes. (See also Section 4.5.1.)

7.3 Shearing

The *shearing* process involves cutting sheet metal by subjecting it to shear stresses, usually between a **punch** and a **die**, much like a paper punch (Fig. 7.4). The punch and die may be of any shape; for example, they may be circular or straight blades, similar to a pair of scissors. The major variables in the shearing process are the punch force, F, the speed of the punch, lubrication, the edge condition of the sheet, the punch and die materials, the corner radii of the punch and die, and the clearance between the punch and die.

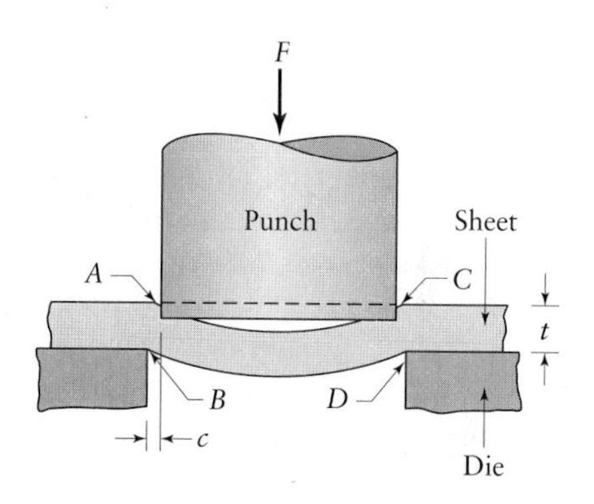

FIGURE 7.4 Schematic illustration of the shearing process with a punch and die. This process is a common method of producing various openings in sheet metals.

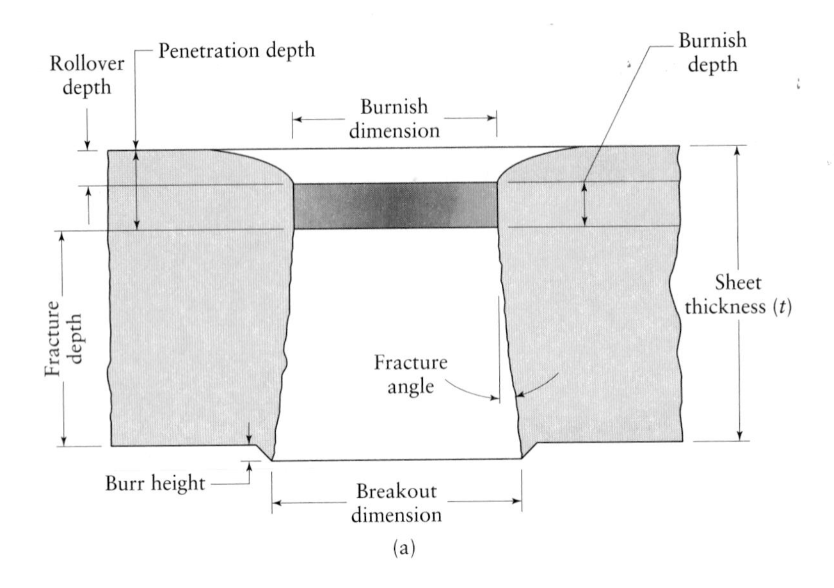

FIGURE 7.5 Characteristic features of (a) a punched hole and (b) the punched slug. Note that the slug has been scaled down as compared with the hole.

Figure 7.5 shows the overall features of a typical sheared edge for the two sheared surfaces (the sheet and the slug); note that the edges are neither smooth nor perpendicular to the plane of the sheet. The **clearance** *c* (Fig. 7.4) is the major factor that determines the shape and quality of the sheared edge. As shown in Fig. 7.6a, as clearance increases, the edges become rougher and the deformation zone becomes larger. Furthermore, the material is pulled into the clearance area, and the edges of the sheared zone become more and more rounded. In fact, if the clearance is too large, the sheet metal is bent and subjected to tensile stresses, instead of undergoing a shearing deformation. As Fig. 7.6b shows, sheared edges can undergo severe cold working, because of the high strains involved; this result can adversely affect the formability of the sheet during subsequent operations.

Observation of the shearing mechanism reveals that shearing usually starts with the formation of cracks on both the top and bottom edges of the workpiece (*A* and *B* in Fig. 7.4). These cracks eventually meet, and complete separation takes place. The **rough fracture surface** in the slug in Fig. 7.5 is caused by these cracks. The **smooth, shiny, and burnished surface** is from the contact and rubbing of the sheared edge against the walls of the die. In the slug shown in Fig. 7.5b, the burnished surface is in the lower region, because this region is the section that rubs against the die wall. On the other hand, inspection of the sheared surface on the sheet itself reveals that the burnished surface is on the upper region in the sheared edge and results from rubbing against the punch.

The ratio of the burnished to rough areas on the sheared edge increases with increasing ductility of the sheet metal; it decreases with increasing material thickness

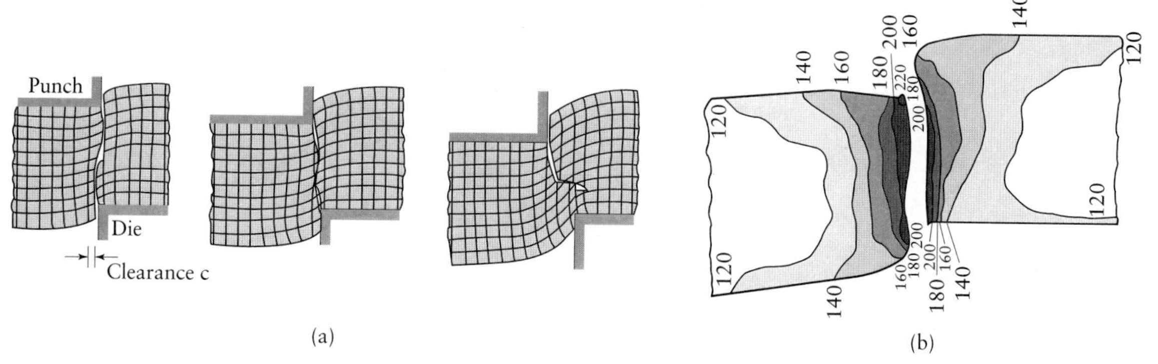

FIGURE 7.6 (a) Effect of clearance c between the punch and die on the deformation zone in shearing. As clearance increases, the material tends to be pulled into the die, rather than being sheared. In practice, clearances usually range between 2% and 10% of the thickness of the sheet. (b) Microhardness (HV) contours for a 6.4-mm (0.25-in.) thick AISI 1020 hot-rolled steel in the sheared region. *Source*: After H. P. Weaver and K. J. Weinmann.

and clearance. The amount of punch travel required to complete the shearing process depends on the maximum shear strain that the material can undergo before fracture. Hence, a brittle metal or a material that is highly cold worked requires little travel of the punch to complete shearing.

Note from Fig. 7.6 that the deformation zone is subjected to high shear strains. The width of this zone depends on the rate of shearing, that is, the punch speed. With increasing punch speed, the heat generated by plastic deformation is confined to a smaller zone (approaching a narrow adiabatic zone), and the sheared surface is smoother.

Note also the formation of a **burr** in Fig. 7.5. *Burr height* increases with increasing clearance and increasing ductility of the metal. Tools with dull edges are also a major factor in burr formation. (See also the upcoming discussion of *slitting*.) The height, shape, and size of the burr can significantly affect many subsequent forming operations, such as in flanging, where a burr can lead to cracks. Furthermore, burrs on parts may become dislodged under external forces during their movements in mechanisms and interfere with the operation or contaminate lubricants.

Punch force. The *punch force*, F, is basically the product of the shear strength of the sheet metal and the cross-sectional area being sheared. However, *friction* between the punch and the workpiece can increase this force substantially. Because the sheared zone is subjected to cracks, plastic deformation, and friction, the punch-force–stroke curves can have various shapes. Figure 7.7 shows one typical curve for a ductile material; the area under the curve is the *total work* done in shearing.

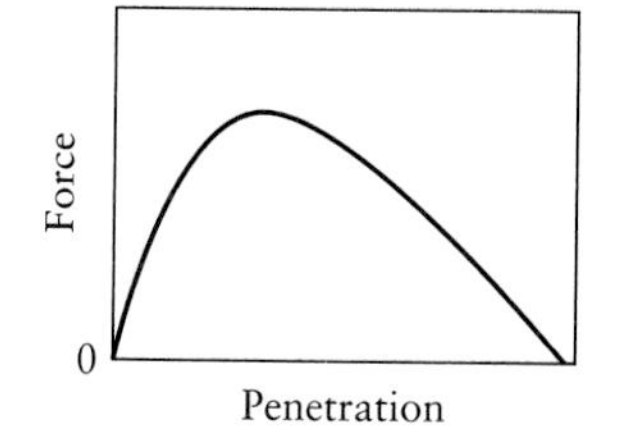

FIGURE 7.7 Typical punch–penetration curve in shearing. The area under the curve is the work done in shearing. The shape of the curve depends on process parameters and material properties.

An approximate empirical formula for estimating the **maximum punch force** F_{max} is

$$F_{max} = 0.7(\text{UTS})(t)(L), \tag{7.4}$$

where UTS is the ultimate tensile strength of the sheet metal, t is its thickness, and L is the total length of the sheared edge. Thus, for example, for a round hole of diameter D, $L = \pi D$. In addition to the basic punch force, a force is also required to strip the punch from the sheet during its return stroke. This force is difficult to calculate, however, because of the many factors involved, especially *friction* between the punch and the sheet.

EXAMPLE 7.1 Calculation of maximum punch force

Estimate the force required in punching a 1-in. (25 mm) diameter hole through a $\frac{1}{8}$-in. (3.2 mm) thick annealed titanium-alloy Ti-6Al-4V sheet at room temperature.

Solution. The force is estimated from Eq. (7.4), and the UTS for this alloy is found from Table 3.13 to be 1000 MPa, or 140,000 psi. Therefore,

$$F = 0.7\left(\frac{1}{8}\right)(\pi)(1)(140{,}000) = 38{,}500 \text{ lb}$$

$$= 19.25 \text{ tons} = 0.17 \text{ MN}.$$

7.3.1 Shearing operations

In this section, we describe various operations that are based on the shearing process. First, we must define two terms: in **punching**, the sheared slug is discarded (Fig. 7.8a); in **blanking**, the slug is the part itself, and the rest is scrap. The following processes are common shearing operations:

a) **Die cutting.** *Die cutting* typically consists of the following operations (Fig. 7.8b): (1) **perforating**, or punching a number of holes in a sheet; (2) **parting**, or shearing the sheet into two or more pieces, usually when the adjacent blanks do not have a matching contour; (3) **notching**, or removing pieces or various shapes from the edges; (4) **slitting** (see also the upcoming discussion), and (5) **lancing**, or leaving a tab on the sheet without removing any material. Parts produced by these processes have various uses, particularly in assembly with other components.

b) **Fine blanking.** Very smooth and square edges can be produced by *fine blanking* (Fig. 7.9a). One basic die design is shown in Fig. 7.9b, in which a V-shaped *stinger*, or *impingement*, locks the sheet metal tightly in place and prevents the type of distortion of the material shown in Fig. 7.6. This process involves clearances on the order of 1% of the sheet thickness, as compared with as much as 8% in ordinary shearing operations; the thickness of the sheet may range from 0.13 to 13 mm (0.005 to 0.5 in.). This operation is usually carried out on triple-action hydraulic presses where the movements of the punch, pressure pad, and die are controlled individually. Fine blanking usually involves parts that have

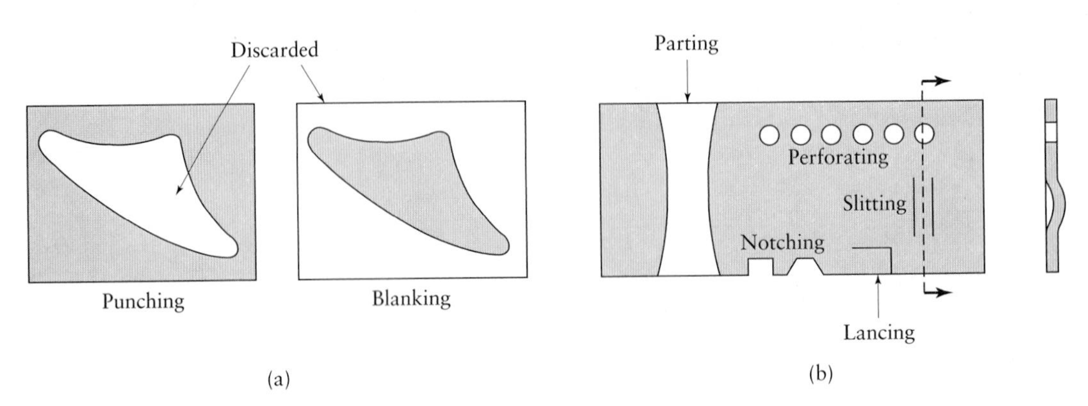

FIGURE 7.8 (a) Punching (piercing) and blanking. (b) Examples of various shearing operations on sheet metal.

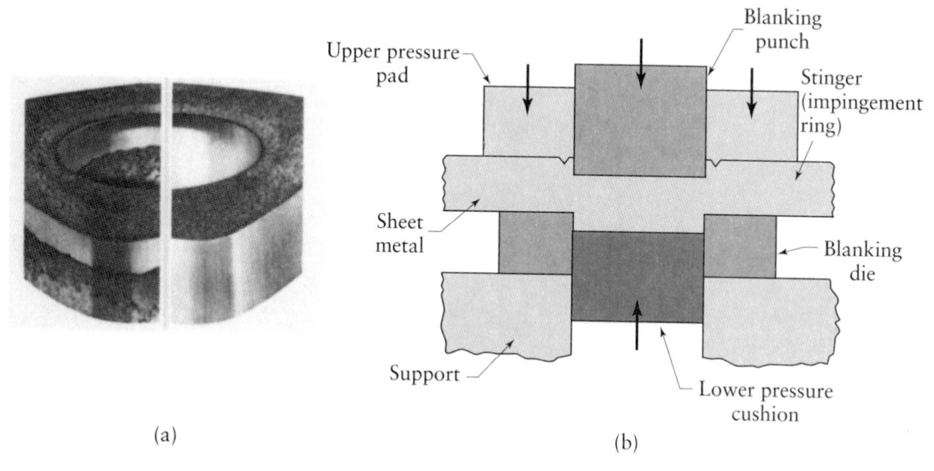

FIGURE 7.9 (a) Comparison of sheared edges by conventional (left) and fine-blanking (right) techniques. (b) Schematic illustration of the setup for fine blanking. *Source*: Feintool International Holding.

holes, which are punched simultaneously with blanking. Suitable sheet hardness is typically in the range of 50 to 90 HRB.

c) **Slitting.** Shearing operations can be carried out with a pair of circular blades (*slitting*), similar to the blade of a can opener (Fig. 7.10). The blades follow either a straight line or a circular or curved path. A slit edge normally has a burr, which may be removed by rolling it over the sheet's edge. There are two types of slitting equipment. In the *driven* type, the blades are powered; in the *pull-through* type, the strip is pulled through idling blades. Slitting operations, if not performed properly, may cause various planar distortions of the slit part or strip.

d) **Steel rules.** Soft metals, paper, leather, and rubber can be blanked with *steel-rule dies*. Such a die consists of a thin strip of hardened steel bent to the shape to be sheared (similar to a cookie cutter) and held on its edge on a flat wooden base. The die is pressed against the sheet, which rests on a flat surface, and cuts the sheet to the shape of the steel rule.

e) **Nibbling.** In *nibbling*, a machine called a *nibbler* moves a straight punch up and down rapidly into a die. The sheet metal is fed through the gap, and a number of overlapping holes are made. (Thus, the movements are similar to that of a sewing machine into which cloth is fed.) This operation is similar to making a large elongated hole by successively punching holes with a paper punch. Sheets can be cut along any desired path by manual control. The process is economical for small production runs, since no special dies are required; intricate slots and notches can be produced, using standard punches.

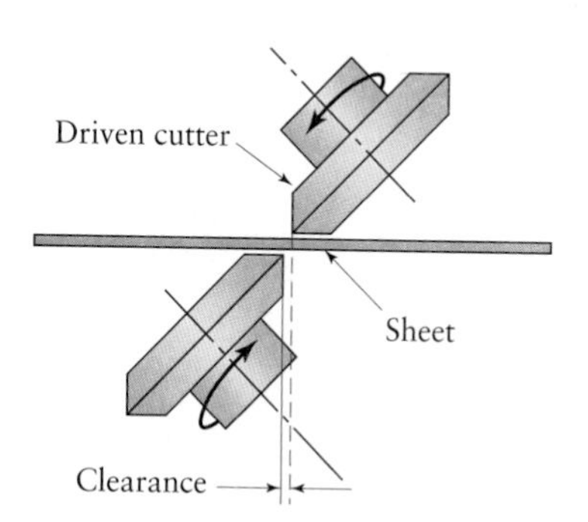

FIGURE 7.10 Slitting with rotary knives. This process is similar to opening cans.

Scrap in shearing. The amount of *scrap* produced, or **trim loss**, in shearing operations can be significant. On large stampings, it can be as high as 30% of the original sheet. A significant factor in manufacturing cost, scrap can be reduced significantly by proper arrangement of the shapes on the sheet to be cut, called **layout and nesting** (see Section 7.15). Computer-aided design techniques have been developed to minimize scrap in shearing operations.

7.3.2 Shearing dies

Because the formability of a sheared part can be influenced by the quality of its sheared edges, clearance control is important. In practice, clearances usually range between 2% and 8% of the sheet's thickness; generally, the thicker the sheet, the larger is the clearance (as much as 10%). However, the smaller the clearance, the better is the quality of the edge. In a process called **shaving** (Fig. 7.11), the extra material from a rough sheared edge is trimmed by cutting. Some common shearing dies are described as follows:

a) **Punch and die shapes.** Note in Fig. 7.4 that the surfaces of the punch and die are flat; thus, the punch force builds up rapidly during shearing, because the entire thickness of the sheet is sheared at the same time. However, the area being sheared at any moment can be controlled by beveling the punch and die surfaces, a shown in Fig. 7.12. The shape of a beveled punch is similar to that of a paper punch, which you can observe by looking closely at its tip. This geometry is particularly suitable for shearing thick blanks, because it reduces the total shearing force (since it is an incremental process); it also reduces the noise level during punching. Note also that because there will be a net lateral force in the punch shape shown in Fig. 7.12b, the punch and the press must have sufficient lateral rigidity.

b) **Compound dies.** Several operations on the same strip may be performed in one stroke with a *compound die* in one station. Although they have higher productivity than the simple punch and die processes, these operations are usually limited to relatively simple shearing, because they are somewhat slow, and the dies are more expensive than those for individual shearing operations.

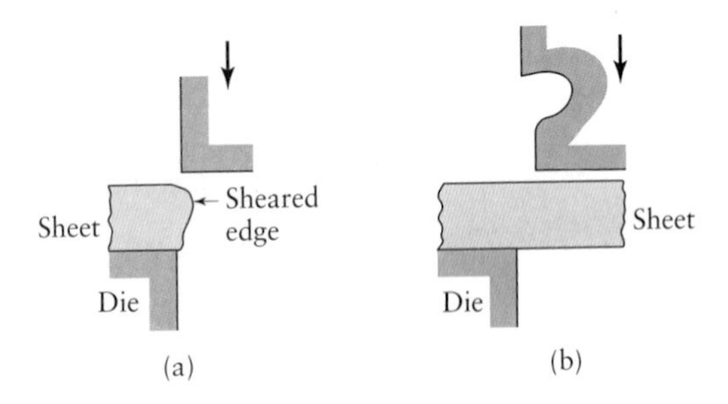

FIGURE 7.11 Schematic illustrations of shaving on a sheared edge. (a) Shaving a sheared edge. (b) Shearing and shaving, combined in one stroke.

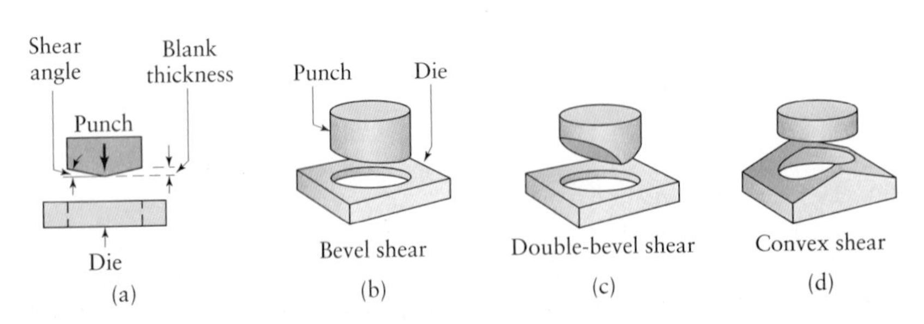

FIGURE 7.12 Examples of the use of shear angles on punches and dies.

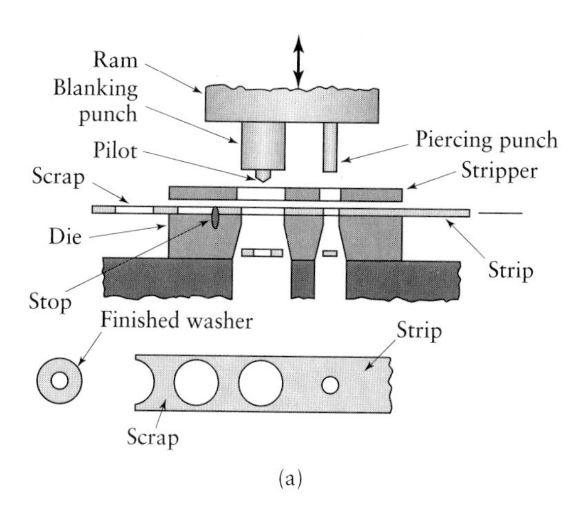

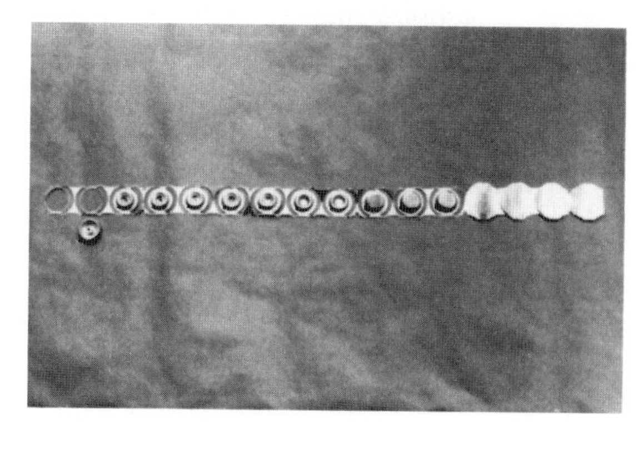

FIGURE 7.13 (a) Schematic illustration of the making of a washer in a progressive die. (b) Forming of the top piece of an aerosol spray can in a progressive die. Note that the part is attached to the strip until the last operation is completed.

c) **Progressive dies.** Parts requiring multiple operations, such as punching, bending, and blanking, are made at high production rates in *progressive dies*. The sheet metal is fed through a coil strip, and a different operation is performed at the same station with each stroke of a series of punches (Fig. 7.13a). Figure 7.13b shows an example of a part made in progressive dies.

d) **Transfer dies.** In a *transfer-die* setup, the sheet metal undergoes different operations at different stations, which are arranged along a straight line or a circular path. After each operation, the part is transferred (hence the name) to the next station for additional operations.

e) **Tool and die materials.** Tool and die materials for shearing are generally tool steels and, for high production rates, carbides. (See Table 3.5.) Lubrication is important for reducing tool and die wear and for improving edge quality.

7.3.3 Other methods of cutting sheet metal.

The following list describes several other methods of cutting sheets and, particularly, plates:

a) The sheet or plate may be cut with a **band saw**; see Section 8.9.5.

b) Particularly for thick plates, **oxyfuel-gas (flame) cutting** may be employed, as used widely in shipbuilding and heavy-construction industries. (See Section 9.14.2.)

c) **Laser-beam cutting** has become an important process; it is now used widely with computer-controlled equipment. (See Section 9.14.1.)

d) **Friction sawing** involves the use of a disk or blade that rubs against the sheet or plate at high surface speed. (See Section 8.9.5.)

e) **Water-jet** and **abrasive water-jet cutting** are effective on metals as well as on nonmetallic materials. (See Section 9.15.)

7.3.4 Tailor-welded blanks

In sheet-metal forming processes, the blank is typically made of one piece, is of a uniform thickness, and is usually cut from a large sheet. However, an important trend now involves laser butt welding (see Section 12.4) of pieces of sheet metal of different shapes and thicknesses (*tailor-welded blanks*, or *TWB*); the welded sheet is subsequently formed into a final shape. The most commonly used technique is laser-beam welding. (See Section 12.4.) Laser-welding techniques are now highly developed, and, as a consequence, weld joints are very strong. However, because of the small thicknesses involved, proper alignment of the sheets prior to welding is important. This method is becoming increasingly important, particularly in the automotive industry. Because each welded piece can have a different thickness (as guided by design considerations such as stiffness), grade of sheet metal, coating, or other characteristics, these blanks possess the needed characteristics in the desired locations of the formed part. As a result, productivity is increased, the need for subsequent spot welding of the product (e.g., a car body) is reduced or eliminated, and dimensional control is improved.

EXAMPLE 7.2 Use of tailor-welded blanks in the automotive industry

An example of the use of tailor-welded blanks is the production of an automobile outer side panel, shown in Fig. 7.14a. Note that five different pieces are first

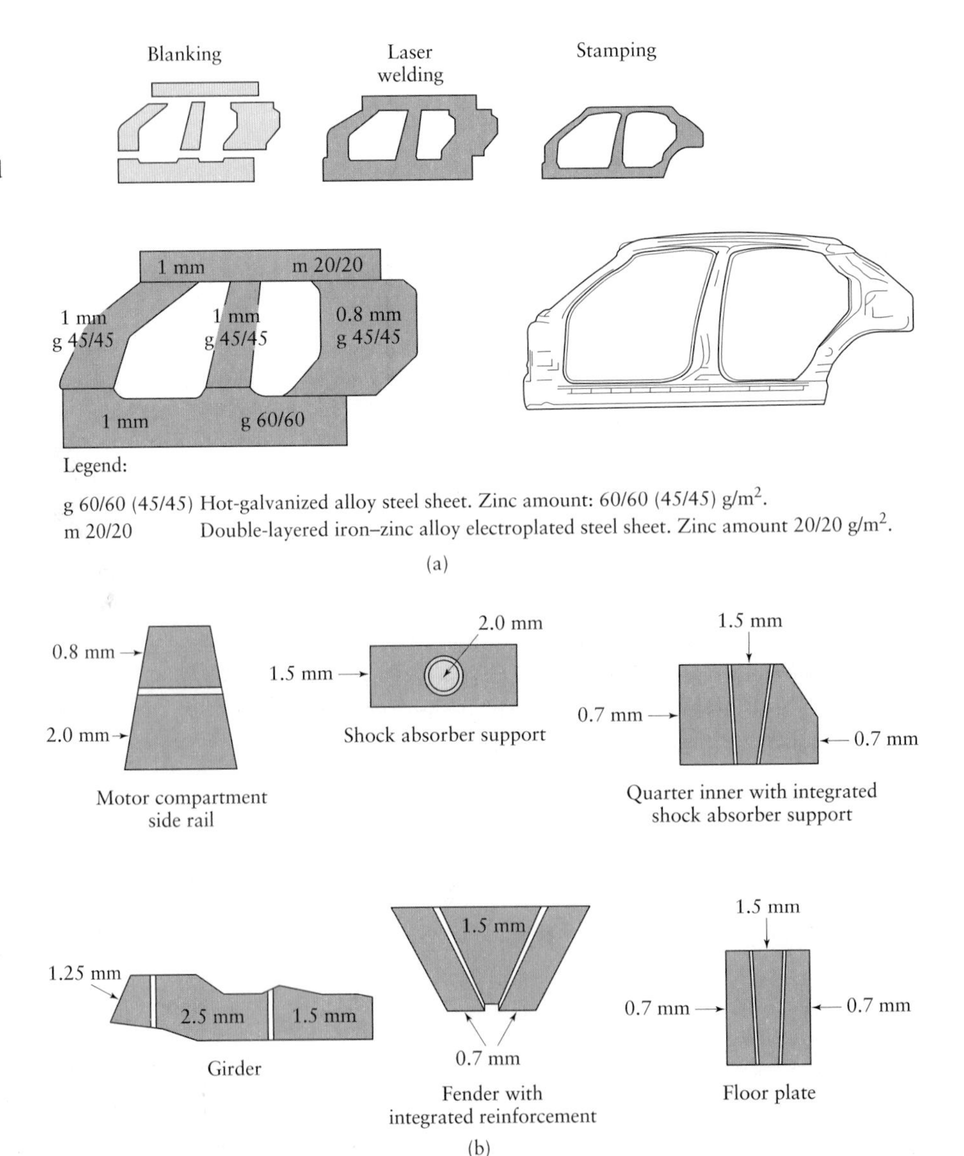

FIGURE 7.14 (a) Production of an outer side panel of a car body by laser welding and stamping. (b) Examples of laser-welded and stamped automotive body components. *Source*: After M. Geiger and T. Nakagawa.

blanked; four of the pieces are 1 mm (0.04 in.) thick, and one is 0.8 mm (0.03 in.) thick. The pieces are then laser butt welded and stamped into the final shape. Thus, the blanks can be tailored for a particular application, including not only sheets of different shape and thickness, but also sheets of different quality, and with or without coatings on one or both surfaces. This trend of welding and forming of sheet-metal pieces allows significant flexibility in product design, structural stiffness and crash behavior (*crashworthiness*), formability, and use of different materials in one component, as well as weight savings and cost reduction in materials, scrap, equipment, assembly, and labor.

There is a growing number of applications for this type of production in automotive companies. The various other components shown in Fig. 7.14b use the advantages outlined above. For example, note that the required strength and stiffness for supporting the shock absorber is achieved by welding a round piece on the surface of the large sheet. The sheet thickness in these components varies, depending on its location and contribution to characteristics such as stiffness and strength, while allowing significant weight and cost savings. *Source*: After M. Geiger and T. Nakagawa.

7.4 Bending of Flat Sheet and Plate

One of the most common metalworking operations is *bending*. This process is used not only to form parts such as flanges, curls, seams, and corrugations, but also to impart stiffness to a part by increasing its moment of inertia. Note, for instance, that a long strip of metal is much less rigid when it is flat than when it is formed into a *V* cross-section, as can be demonstrated with a piece of paper.

Figure 7.15a depicts the terminology used for bending. **Bend allowance** is the length of the *neutral axis* in the bend area and is used to determine the blank length for a bent part. However, the radial position of the neutral axis depends on the bend radius and bend angle, as described in texts on the mechanics of solids. An approximate formula for the bend allowance, L_b, is given by

$$L_b = \alpha(R + kt),$$

where α is the bend angle (in radians), R is the bend radius, k is a constant, and t is the sheet thickness. For the ideal case, the neutral axis remains at the center, and hence $k = 0.5$. In practice, k values usually range from 0.33 for $R < 2t$ to 0.5 for $R > 2t$.

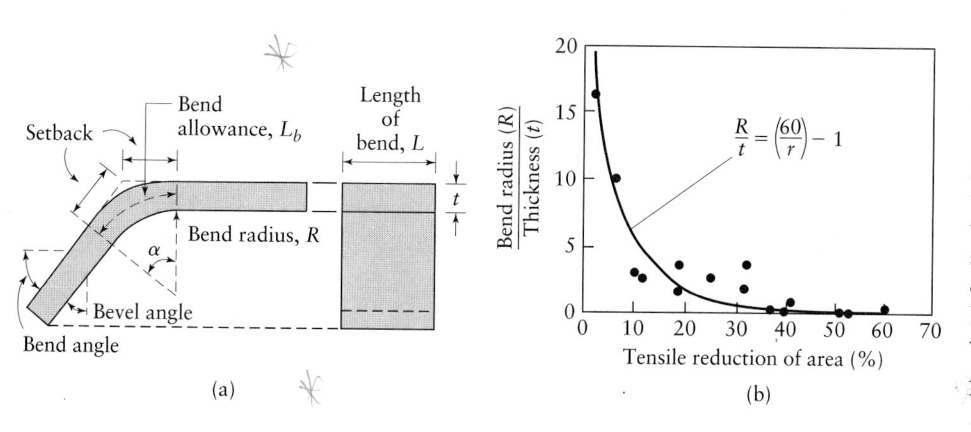

FIGURE 7.15 (a) Bending terminology. The bend radius is measured to the inner surface of the bend. Note that the length of the bend is the width of the sheet. Also note that the bend angle and the bend radius (sharpness of the bend) are two different variables. (b) Relationship between the ratio of bend radius to sheet thickness and tensile reduction of area for various materials. Note that sheet metal with a reduction of area of about 50% can be bent and flattened over itself without cracking. *Source*: After J. Datsko and C. T. Yang.

7.4.1 Minimum bend radius

The outer fibers of the part being bent are in tension, and the inner fibers are in compression. Theoretically, the strains at the outer and inner fibers are equal in magnitude and are given by the equation

$$e_o = e_i = \frac{1}{(2R/t) + 1}. \tag{7.5}$$

However, due to the *shifting of the neutral axis* toward the inner surface, the *length of bend* (the sketch on the right in Fig. 7.15a) is smaller in the outer region than in the inner region. (This phenomenon can easily be observed by bending a rectangular eraser.) Consequently, the outer and inner strains are different, and the difference increases with decreasing R/t ratio.

It can be seen from Eq. (7.5) that, as the R/t ratio decreases, the tensile strain at the outer fiber increases, and the material may crack after a certain strain is reached. The radius, R, at which a crack appears on the outer surface of the bend is called the *minimum bend radius*. The minimum bend radius to which a part can be bent safely is normally expressed in terms of its thickness, such as $2t$, $3t$, $4t$, and so on. Thus, for example, a bend radius of $3t$ indicates that the smallest radius to which the sheet can be bent without cracking is three times the thickness of the sheet. The minimum bend radii for various materials have been determined experimentally and are available in various handbooks; some typical results are given in Table 7.2.

Studies have also been conducted to establish a relationship between the minimum R/t ratio and a given mechanical property of the material. One such analysis is based on the following assumptions: (1) The true strain at cracking on the outer fiber in bending is equal to the true strain at fracture, ε_f, of the material in a simple tension test; (2) the material is homogeneous and isotropic; and (3) the sheet is bent in a state of plane stress—that is, its L/t ratio is small.

The true strain at fracture in tension is

$$\varepsilon_f = \ln\left(\frac{A_0}{A_f}\right) = \ln\left(\frac{100}{100 - r}\right),$$

TABLE 7.2

Minimum Bend Radii for Various Materials at Room Temperature

MATERIAL	MATERIAL CONDITION	
	SOFT	HARD
Aluminum alloys	0	$6t$
Beryllium copper	0	$4t$
Brass, low leaded	0	$2t$
Magnesium	$5t$	$13t$
Steels		
austenitic stainless	$0.5t$	$6t$
low carbon, low alloy, and HSLA	$0.5t$	$4t$
Titanium	$0.7t$	$3t$
Titanium alloys	$2.6t$	$4t$

where r is the percent reduction of area of the sheet in a tension test. From Section 2.2.2, we have for true strain

$$\varepsilon_o = \ln(1 + e_o) = \ln\left(1 + \frac{1}{(2R/t) + 1}\right) = \ln\left(\frac{R + t}{R + (t/2)}\right).$$

Equating the two expressions and simplifying, we obtain

$$\text{Minimum}\,\frac{R}{t} = \frac{50}{r} - 1. \tag{7.6}$$

The experimental data are shown in Fig. 7.15b; note that the curve that best fits the data is

$$\text{Minimum}\,\frac{R}{t} = \frac{60}{r} - 1. \tag{7.7}$$

The R/t ratio approaches zero (complete **bendability**—that is, the material can be folded over itself) at a tensile reduction of area of 50%. Interestingly, this percent is the same value obtained for spinnability of metals (described in Section 7.8); that is, a material with 50% reduction of area is found to be completely spinnable. A curve similar to that in Fig. 7.15b also is obtained when the data are plotted against the total elongation (such as percent in 50-mm gage length) of the material.

Factors affecting bendability. The bendability of a metal may be increased by increasing its tensile reduction of area, either by *heating* or by the application of *hydrostatic pressure.* Other techniques may also be employed to increase the compressive environment in bending, such as applying compressive forces in the plane of the sheet during bending to minimize tensile stresses in the outer fibers of the bend area.

As the length of the bend increases, the state of stress at the outer fibers changes from a uniaxial stress to a *biaxial stress*. The reason for this change is that the bend length L tends to become smaller due to stretching of the outer fibers, but it is constrained by the material around the bend area. Biaxial stretching tends to reduce ductility, that is, the amount of strain to fracture; thus, as L increases, the minimum bend radius increases (Fig. 7.16). However, at a bend length of about $10t$, the minimum bend radius increases no further, and a *plane-strain condition* is fully developed. As

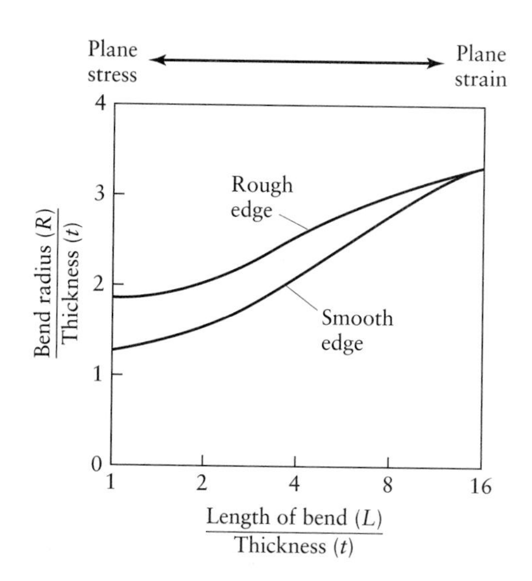

FIGURE 7.16 The effect of length of bend and edge condition on the ratio of bend radius to thickness for 7075-T aluminum. *Source:* After G. Sachs and G. Espey.

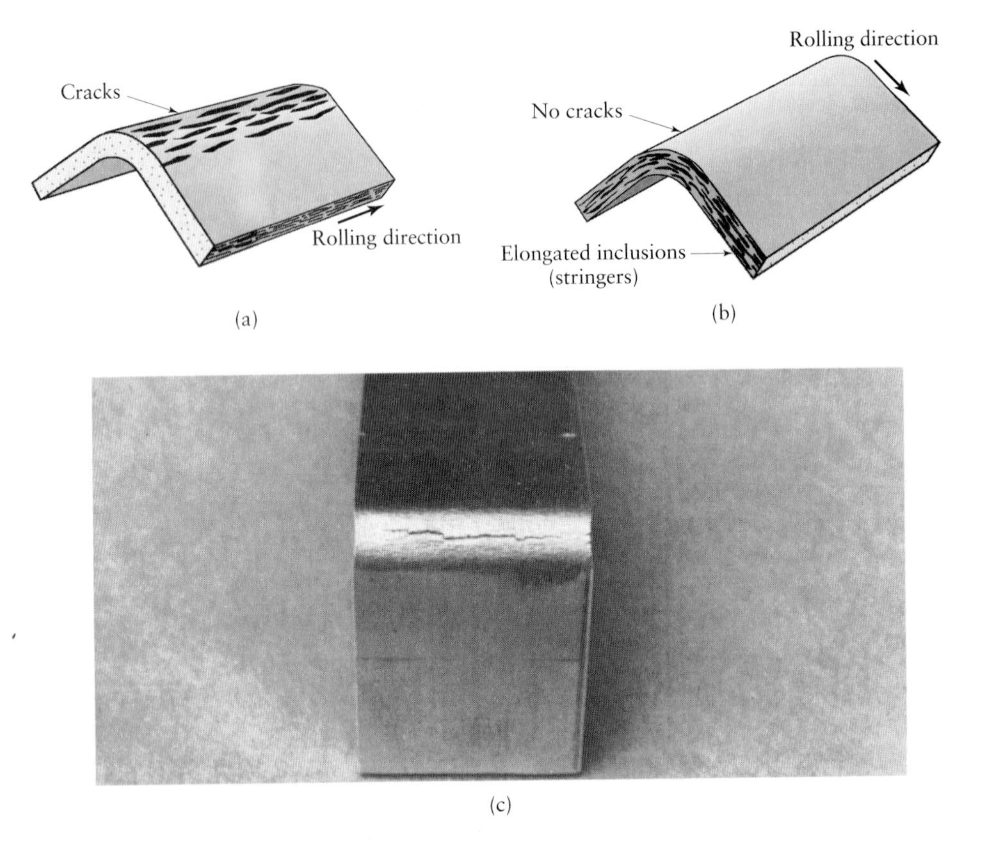

FIGURE 7.17 (a) and (b) The effect of elongated inclusions (stringers) on cracking as a function of the direction of bending with respect to the original rolling direction of the sheet. This example shows the importance of the direction of cutting from large sheets in workpieces that are subsequently bent to make a product. (c) Cracks on the outer radius of an aluminum strip bent to an angle of 90°.

the R/t ratio decreases, narrow sheets (smaller length of bend) crack at the edges, and wider sheets crack at the center, where the biaxial stress is the highest.

Bendability also depends on the **edge condition** of the sheet being bent; because rough edges are points of stress concentration, bendability decreases as edge roughness increases. Another important factor is the amount of **cold working** that the edges undergo during shearing, as can be observed from microhardness tests in the sheared region (Fig. 7.6b). Removal of the cold-worked regions, such as by shaving, machining, or heat treating, greatly improves the resistance to edge cracking during bending.

Another significant factor in edge cracking is the amount and shape of *inclusions* in the sheet metal. Inclusions in the form of **stringers** are more detrimental than globular-shaped inclusions; thus, anisotropy of the sheet is also important in bendability. As depicted in Fig. 7.17, cold rolling of sheets produces **anisotropy**, because of the alignment of impurities, inclusions, and voids (*mechanical fibering*), and thus transverse ductility is reduced, as shown in Fig. 7.17c; see also Fig. 3.15. In bending such a sheet, caution should be exercised in cutting or slitting the blank in the proper direction of the rolled sheet, although this may not always be possible in practice.

7.4.2 Springback

Because all materials have a finite modulus of elasticity, plastic deformation is followed by **elastic recovery** upon removal of the load; in bending, this recovery is known as *springback*. As shown in Fig. 7.18, the final bend angle after springback is smaller and the final bend radius is larger than before. This phenomenon can easily be observed by bending a piece of wire or a short strip metal. Springback occurs not only in flat sheets or plate, but also in bending bars, rod, and wire of any cross-section.

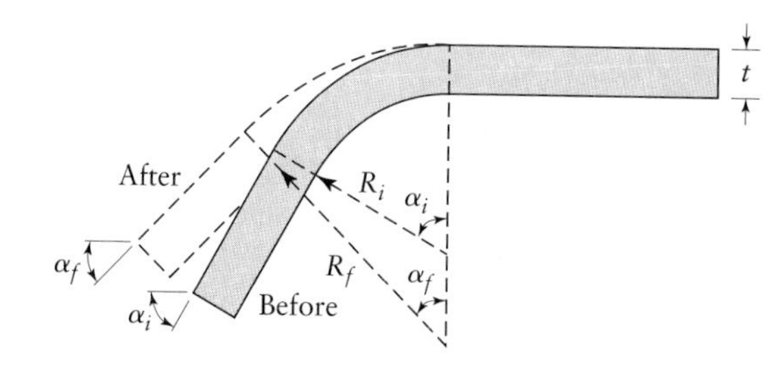

FIGURE 7.18 Terminology for springback in bending. Springback is caused by the elastic recovery of the material upon unloading. In this example, the material tends to recover toward its originally flat shape. However, there are situations where the material bends farther upon unloading (negative springback), as shown in Fig. 7.20.

A quantity characterizing springback is the **springback factor**, K_s, which is determined as follows. Because the bend allowance (see Fig. 7.15a) is the same before and after bending, the relationship obtained for pure bending is

$$\text{Bend allowance} = \left(R_i + \frac{t}{2}\right)\alpha_i = \left(R_f + \frac{t}{2}\right)\alpha_f. \quad (7.8)$$

From this relationship, K_s is defined as

$$K_s = \frac{\alpha_f}{\alpha_i} = \frac{(2R_i/t) + 1}{(2R_f/t) + 1}, \quad (7.9)$$

where R_i and R_f are the initial and final bend radii, respectively. Note from this expression that K_s depends only on the R/t ratio. A factor of $K_s = 1$ indicates that there is no springback, and $K_s = 0$ indicates complete elastic recovery (Fig. 7.19), as in a leaf spring. Thus, a true spring always has a springback factor of zero.

Recall from Fig. 2.3 that the amount of elastic recovery depends on the stress level and the modulus of elasticity, E, of the material; hence, elastic recovery increases with the stress level and with decreasing elastic modulus. Based on this observation, an approximate formula has been developed to estimate springback:

$$\frac{R_i}{R_f} = 4\left(\frac{R_iY}{Et}\right)^3 - 3\left(\frac{R_iY}{Et}\right) + 1. \quad (7.10)$$

In this equation, Y is the uniaxial yield stress of the material at 0.2% offset. (See Figure 2.2b.)

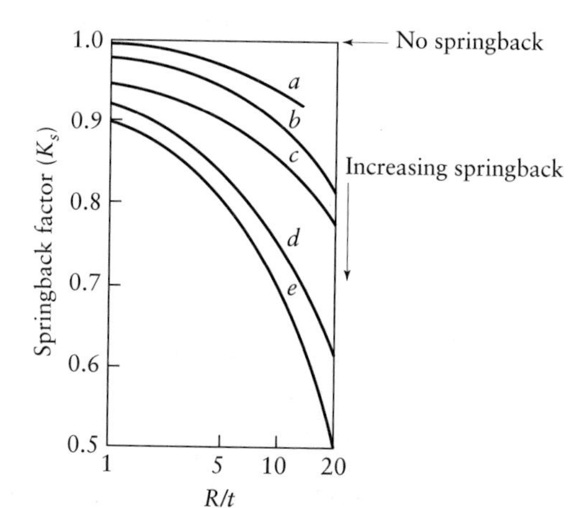

FIGURE 7.19 Springback factor K_s for various materials: (a) 2024-0 and 7075-0 aluminum; (b) austenitic stainless steels; (c) 2024-T aluminum; (d) $\frac{1}{4}$- hard austenitic stainless steels; (e) $\frac{1}{2}$-hard to full-hard austenitic stainless steels. *Source*: After G. Sachs.

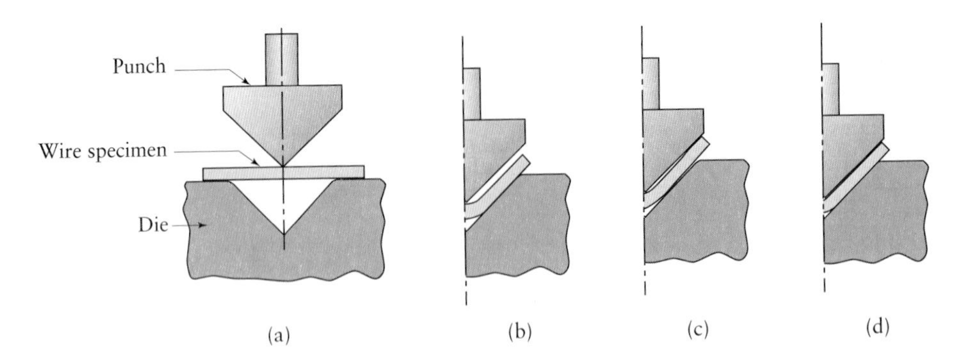

FIGURE 7.20 Schematic illustration of the stages in bending round wire in a V-die. This type of bending can lead to negative springback, which does not occur in air bending (shown in Fig. 7.24a). *Source*: After K. S. Turke and S. Kalpakjian.

Negative springback. The springback observed in Fig. 7.18 can be called *positive springback*. However, under certain conditions, *negative springback* is also possible. In other words, the bend angle in such cases becomes larger after the bend has been completed and the load is removed. This phenomenon is generally associated with V-die bending (Fig. 7.20).

The development of negative springback can be explained by observing the sequence of deformation in Fig. 7.20. If we remove the bent piece at stage (b), it will undergo positive springback. At stage (c), the ends of the piece are touching the male punch. Note that between stages (c) and (d), the part is actually being bent in the direction opposite to that between stages (a) and (b). Note also the lack of conformity of the punch radius and the inner radius of the part in both stage (b) and stage (c); in stage (d), however, the two radii are the same. Upon unloading, the part in stage (d) will spring back inwardly, because it is being *unbent* from stage (c), both at the tip of the punch and in the arms of the part. The amount of this inward (negative) springback can be greater than the amount of positive springback, because of the large strains that the material has undergone in the small bend area in stage (b). The net result is negative springback.

Compensation for springback. In practice, springback is usually compensated for by using various techniques:

1. **Overbending** the part in the die (Figs. 7.21a and b) can compensate for springback; overbending can also be achieved by the **rotary bending** technique shown in Fig. 7.21e. The upper die has a cylindrical rocker (with an angle of $< 90°$) and is free to rotate; as it travels downward, the sheet is clamped and bent by

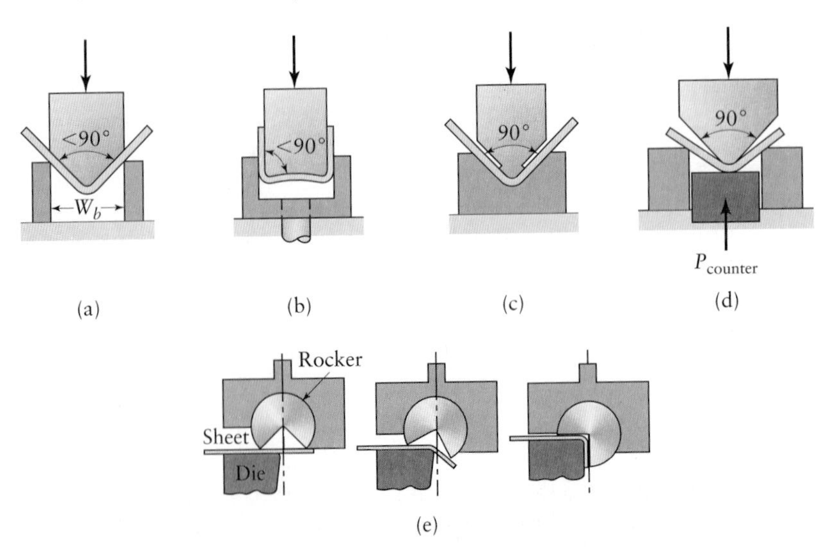

FIGURE 7.21 Methods of reducing or eliminating springback in bending operations. *Source*: V. Cupka, T. Nakagawa, and H. Tyamoto.

the rocker over the lower die (die anvil). A relief angle in the lower die allows overbending of the sheet at the end of the stroke, thus compensating for springback.

2. **Coining** the bend region by subjecting it to high localized compressive stresses between the tip of the punch and the die surface (Figs. 7.21c and 7.21d), known as **bottoming**.
3. **Stretch bending,** in which the part is subjected to tension while being bent, may be applied. The bending moment required to make the sheet deform plastically will be reduced as the combined tension (due to bending of the outer fibers and the applied tension) in the sheet increases. Therefore, springback, which is the result of nonuniform stresses due to bending, will also decrease. This technique is used to limit springback in stretch forming of shallow automotive bodies. (See Section 7.5.)
4. Because springback decreases as yield stress decreases [see Eq. (7.10)], all other parameters being the same, bending may also be carried out at *elevated temperatures* to reduce springback.

EXAMPLE 7.3 Estimating springback

A 20-gage steel sheet is bent to a radius of 0.5 in. Assuming that its yield stress is 40,000 psi, calculate the radius of the part after it is bent.

Solution. The appropriate formula is Eq. (7.10), where

$$R_i = 0.5 \text{ in.}, \quad Y = 40{,}000 \text{ psi}, \quad E = 29 \times 10^6 \text{ psi},$$

and $t = 0.0359$ in. for 20-gage steel, as found in handbooks. Thus, we have

$$\frac{R_i Y}{Et} = \frac{(0.5)(40{,}000)}{(29 \times 10^6)(0.0359)} = 0.0192,$$

and

$$\frac{R_i}{R_f} = 4(0.0192)^3 - 3(0.0192) + 1 = 0.942.$$

Hence,

$$R_f = \frac{0.5}{0.942} = 0.531 \text{ in.}$$

7.4.3 Forces

Bending forces can be estimated by assuming that the process is that of the simple bending of a rectangular beam. Thus, the bending force is a function of the strength of the material, the length and thickness of the part, and the width W of the die opening, as shown in Fig. 7.22. Excluding friction, the general expression for the **maximum bending force,** F_{max}, is

$$F_{max} = k\frac{(\text{UTS})Lt^2}{W}, \tag{7.11}$$

where the factor k ranges from about 1.2 to 1.33 for a V die. The k values for wiping and for U dies are 0.25 and 2 times, respectively, that of k for V dies. The effect

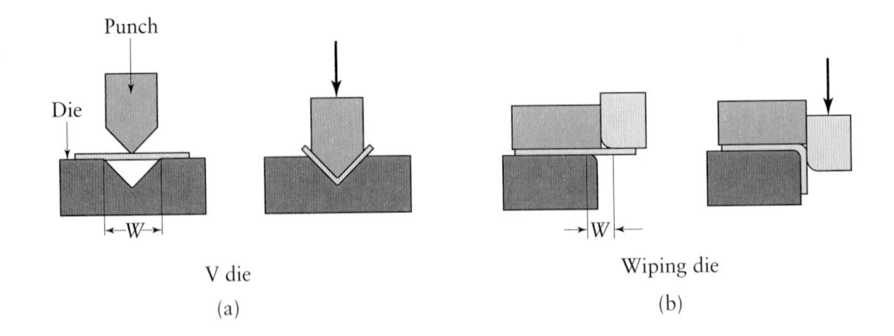

FIGURE 7.22 Common die-bending operations, showing the die-opening dimension W used in calculating bending forces. [See Eq. (7.11).]

of various factors on the bending force, such as friction, is included in the factor k. Equation (7.11) applies well to situations in which the punch radius and sheet thickness are small compared with the size of the die opening, W.

The bending force is also a function of punch travel; It increases from zero to a maximum and may decrease as the bend is completed. The force then increases sharply as the punch bottoms in the case of die bending. In *air bending*, or *free bending* (see Fig. 7.24a), the force does not increase again after it begins to decrease.

7.4.4 Common bending operations

In this section, we describe common bending operations. Some of these processes are performed on discrete sheet-metal parts; others are done continuously, as in roll forming of coiled sheet stock. The operations are described as follows:

a) **Press-brake forming.** Sheet metal or plate can be bent easily with simple fixtures, using a press. Parts that are long, i.e., 7 m (20 ft) or more, and relatively narrow are usually bent in a *press brake*. This machine uses long dies in a mechanical or hydraulic press and is suitable for small production runs. The tooling is simple and adaptable to a wide variety of shapes (Fig. 7.23); furthermore, the process can easily be automated. Die materials may range from hardwood (for low-strength materials and small production runs) to carbides; for most applications, carbon-steel or gray-iron dies are generally used.

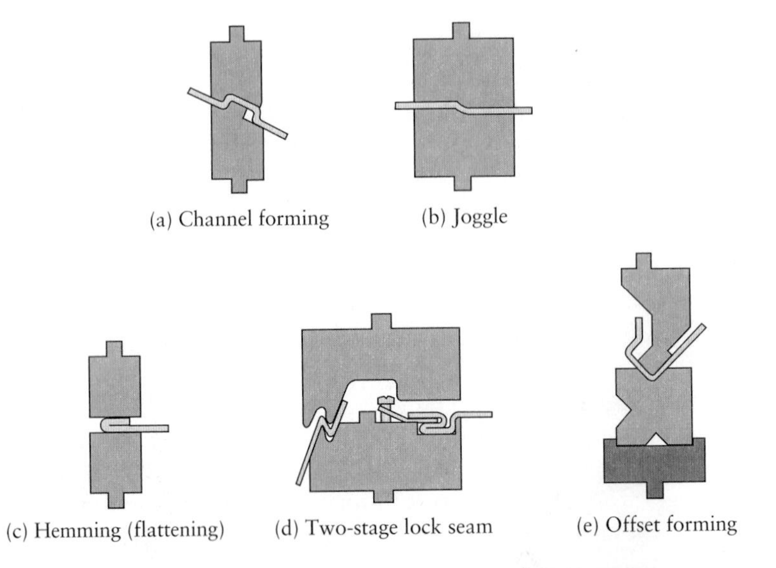

FIGURE 7.23 Schematic illustrations of various bending operations in a press brake.

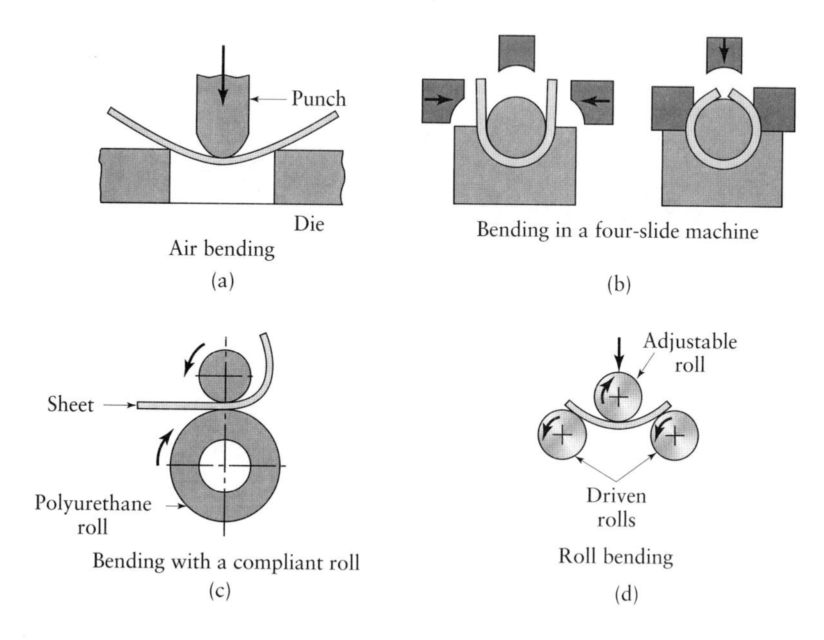

FIGURE 7.24 Examples of various bending operations.

b) **Other bending operations.** Sheet metal may be bent by a variety of processes (Fig. 7.24). In **air bending** (*free bending*), there is only one die. Bending of sheet metals can be carried out with a *pair of rolls*, the larger one of which is flexible and made of polyurethane. The upper roll pushes the sheet into the flexible lower roll and imparts a curvature to the sheet, the shape of which depends on the degree of indentation into the flexible roll. Hence, by controlling this depth, it is possible to form a sheet with a variety of curvatures on it. Plates are bent by the **roll-bending** process shown in Fig. 7.24d. Adjusting the distance between the three rolls produces various curvatures. Bending of relatively short pieces, such as a bushing, can also be done on highly automated **four-slide machines**, a process similar to that shown in Fig. 7.24b.

c) **Beading.** In *beading*, the edge of the sheet metal is bent into the cavity of a die (Fig. 7.25). The bead gives stiffness to the part by increasing the moment of inertia of the edges. Also, it improves the appearance of the part and eliminates exposed sharp edges, which may be a safety hazard.

d) **Flanging.** Flanging is a process of bending the edges of sheet metals, usually to 90°. In **shrink flanging** (Fig. 7.26a), the flange is subjected to compressive hoop stresses, which, if excessive, can cause the edges of the flange to wrinkle. The wrinkling tendency increases with decreasing radius of curvature of the flange. In **stretch flanging**, the flange edges are subjected to tensile stresses, which, if excessive, can lead to cracking at the edges.

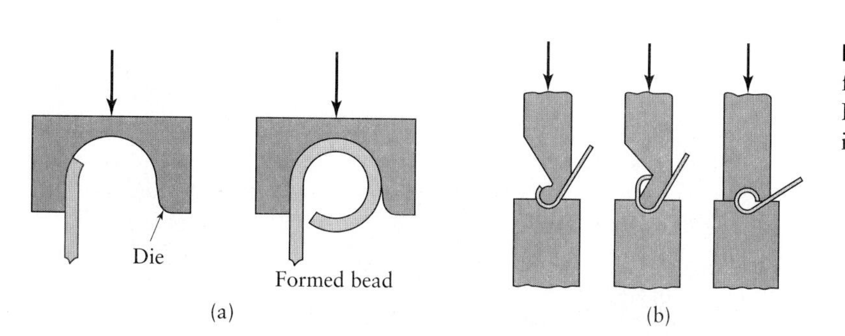

FIGURE 7.25 (a) Bead forming with a single die. (b) Bead forming with two dies in a press brake.

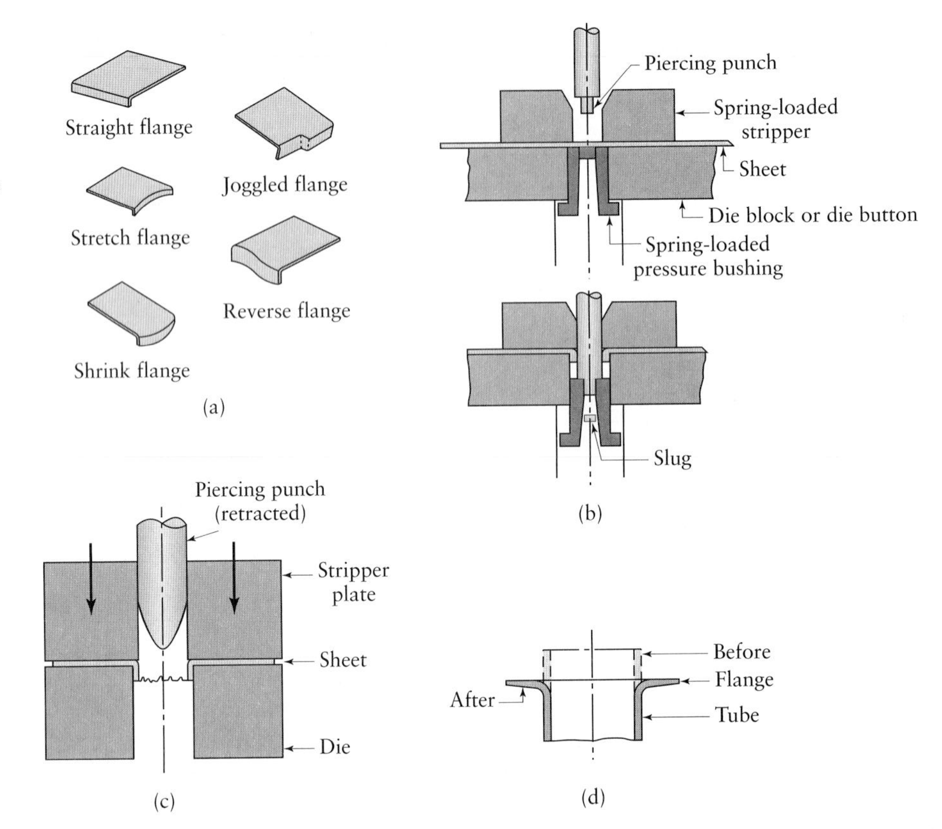

FIGURE 7.26 Various flanging operations. (a) Flanges on flat sheet. (b) Dimpling. (c) Piercing sheet metal to form a flange. In this operation, a hole does not have to be prepunched before the punch descends. Note, however, the rough edges along the circumference of the flange. (d) Flanging of a tube. Note the thinning of the edges of the flange.

e) **Dimpling.** In this operation (Fig. 7.26b), a hole is first punched and then expanded into a flange, or a shaped punch pierces the sheet metal and expands the hole. Flanges may be produced by **piercing** with a bullet-shaped punch (Fig. 7.26c). The ends of tubes are flanged by a similar process (Fig. 7.26d). When the angle of bend is less than 90°, as in fittings with conical ends, the process is called **flaring.** The condition of the edges is important in these operations; stretching the material causes high tensile stresses at the edges, which may lead to cracking and tearing of the flange. As the ratio of flange to hole diameter increases, the strains increase proportionately. The rougher the edge, the greater will be the tendency for cracking. Sheared or punched edges may be shaved with a sharp tool (see Fig. 7.11) to improve the surface finish of the edge and reduce the tendency for cracking.

f) **Hemming.** In the *hemming* process (also called **flattening**), the edge of the sheet is folded over itself. (See Fig. 7.23c.) Hemming increases the stiffness of the part, improves its appearance, and eliminates sharp edges. *Seaming* (Fig. 7.23d) involves joining two edges of sheet metal by hemming. Double seams are made by a similar process, using specially shaped rollers, for watertight and airtight joints, such as in food and beverage containers.

g) **Roll forming.** *Roll forming* is used for bending continuous lengths of sheet metal and for large production runs (**contour roll forming**, or *cold roll forming*). The metal strip is bent in stages by passing it through a series of rolls (Fig. 7.27). Typical products formed by this method include channels, gutters, siding, panels, frames (Fig. 7.28), and pipes and tubing with lock seams. The length of the part is limited only by the amount of material supplied from the coiled stock. The parts are usually sheared and stacked continuously. The thickness of the sheet used typically ranges from about 0.125 to 20 mm (0.005 to 0.75 in.). Forming speeds are generally below 1.5 m/s (300 ft/min), although they can be much higher for special applications.

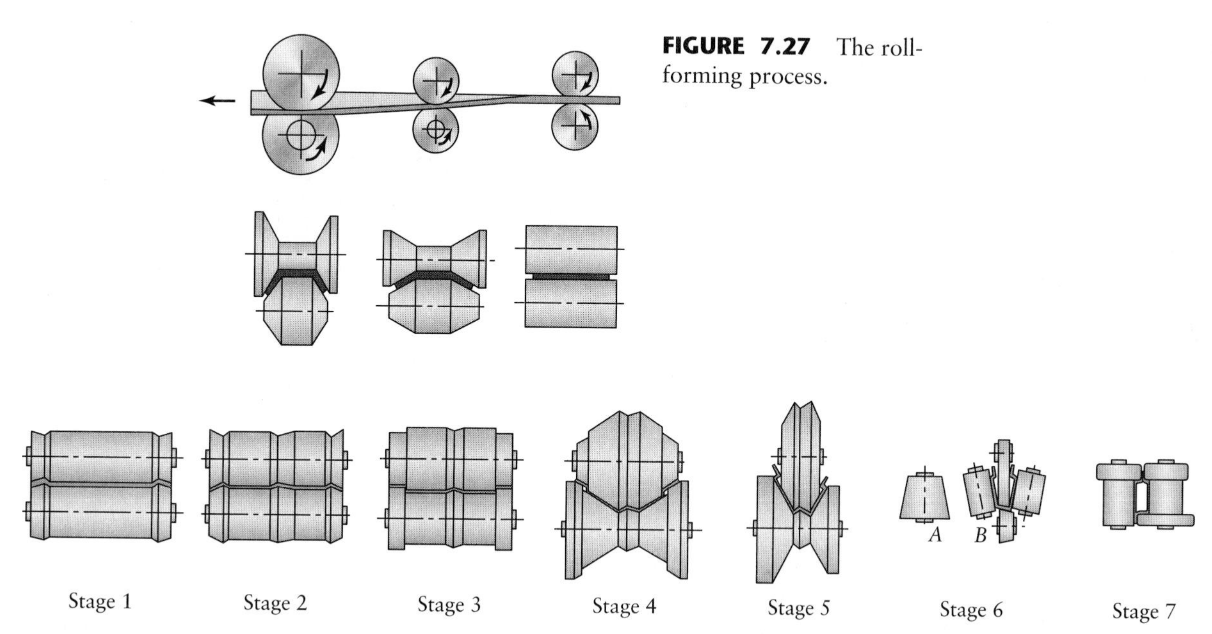

FIGURE 7.27 The roll-forming process.

FIGURE 7.28 Stages in roll forming of a sheet-metal door frame. In Stage 6, the rolls may be shaped as in *A* or *B*. *Source*: Oehler, G., Bending, Carl Hanser Verlag Munich, 1963.

Proper design and sequencing of the rolls, which usually are mechanically driven, requires considerable experience. Dimensional tolerances, springback, and tearing and buckling of the strip have to be considered. The rolls are generally made of carbon steel or gray iron and may be chromium plated, for better surface finish of the product and wear resistance of the rolls. Lubricants may be used to improve roll life and surface finish and to cool the rolls and the workpiece.

7.4.5 Tube bending

Bending and forming tubes and other hollow sections requires special tooling to avoid buckling and folding. The oldest and simplest method of bending a tube or pipe is to pack the inside with loose particles (such as sand) and bend the part in a suitable fixture. This technique does not allow the tube wall to buckle inward. After the tube is bent, the sand is shaken out. Tubes can also be *plugged* with various flexible internal mandrels, such as those shown in Fig. 7.29, which also illustrates various bending

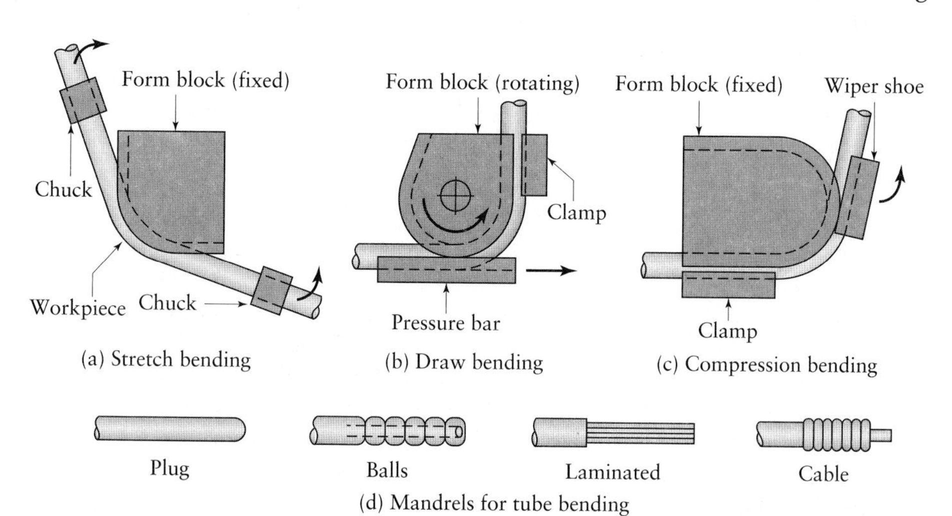

FIGURE 7.29 Methods of bending tubes. Using internal mandrels, or filling tubes with particulate materials such as sand, is often necessary to prevent collapsing of the tubes during bending. Solid rods and structural shapes are also bent by these techniques.

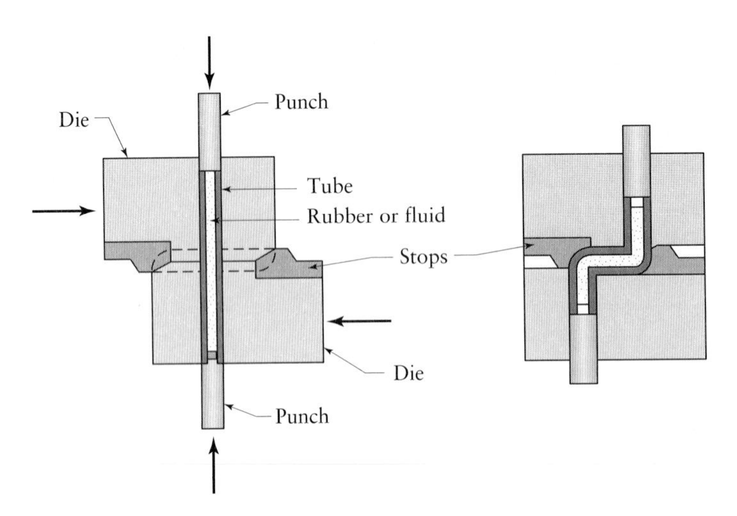

FIGURE 7.30 A method of forming a tube with sharp angles, using axial compressive forces. Compressive stresses are beneficial in forming operations because they delay fracture. Note that the tube is supported internally with rubber or fluid to avoid collapsing during forming. *Source*: After J. L. Remmerswaal and A. Verkaik.

methods and fixtures for tubes and sections. A relatively thick tube that has a large bend radius can be bent without filling it with particulates or using plugs, since it will have less of a tendency to buckle inward. The beneficial effect of forming metals under highly compressive stresses is demonstrated in Fig. 7.30 with respect to the bending of a tube with relatively sharp corners. Note that in this operation, the tube is subjected to longitudinal compressive stresses, which reduce the stresses in the outer fibers in the bend area, thus improving the bendability of the material.

7.5 Stretch Forming

In *stretch forming*, the sheet metal is clamped around its edges and stretched over a die or form block, which moves upward, downward, or sideways, depending on the particular machine (Fig. 7.31). Stretch forming is used primarily to make aircraft-wing skin panels, automobile door panels, and window frames. Aluminum skins for the Boeing 767 and 757 aircraft are made by stretch forming, with a tensile force of 9 MN (2 million lb). The rectangular sheets are 12 m × 2.5 m × 6.4 mm (40 ft × 8.3 ft × 0.25 in.). Although this process is generally used for low-volume production, it is versatile and economical.

In most operations, the blank is a rectangular sheet, clamped along its narrower edges and stretched lengthwise, thus allowing the material to shrink in its width; however, controlling the amount of stretching is important to avoid tearing. Stretch forming cannot produce parts with sharp contours or re-entrant corners (depressions of the surface of the die). Dies for stretch forming are generally made of zinc alloys, steel, hard plastics, or wood, and most applications require little or no lubrication. Var-

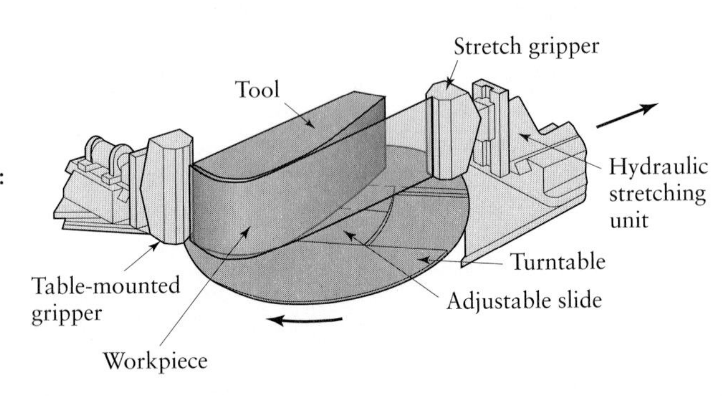

FIGURE 7.31 Schematic illustration of a stretch-forming process. Aluminum skins for aircraft can be made by this process. *Source*: Cyril Bath Co.

ious accessory equipment can be used in conjunction with stretch forming, including additional forming with both male and female dies while the part is in tension.

EXAMPLE 7.4 Work done in stretch forming

A 15-in.–long workpiece with a cross-sectional area of 0.5 in^2. (see the accompanying figure) is stretched with a force, F, until $\alpha = 20°$. The material has a true-stress–true-strain curve $\sigma = 100{,}000\ \varepsilon^{0.3}$. (a) Find the total work done, ignoring end effects or bending. (b) What is α_{max} before necking begins?

Solution.

a) This situation is equivalent to stretching a workpiece from 15 in. to a length of $a + b$. For $\alpha = 20°$, the final length is $L_f = 16.8$ in. The true strain is then

$$\varepsilon = \ln\left(\frac{L_f}{L_o}\right) = \ln\left(\frac{16.8}{15}\right) = 0.114.$$

The work done per unit volume is

$$u = \int_0^{0.114} \sigma\, d\varepsilon = 10^5 \int_0^{0.114} \varepsilon^{0.3} d\varepsilon = 10^5 \left[\frac{\varepsilon^{1.3}}{1.3}\right]_0^{0.114} = 4570 \text{ in} \cdot \text{lb/in}^3.$$

The volume of the workpiece is

$$V = (15)(0.5) = 7.5 \text{ in}^3.$$

Hence,

$$\text{Work} = (u)(V) = 34{,}275 \text{ in.-lb.}$$

b) The necking limit for uniaxial tension is given by Eq. (7.2), the problem being similar to a stretch-forming operation. Thus,

$$L_{max} = L_o\varepsilon^n = 15\varepsilon^{0.3} = 20.2 \text{ in.}$$

Therefore, $a + b = 20.2$ in., and from similarities in triangles, we obtain

$$a^2 - 10^2 = b^2 - 5^2,$$

or

$$a^2 = b^2 + 75.$$

Therefore, $a = 12$ in. and $b = 8.2$ in. Hence,

$$\cos\alpha = \tfrac{10}{12} = 0.833, \quad \text{or} \quad \alpha_{max} = 33.6°.$$

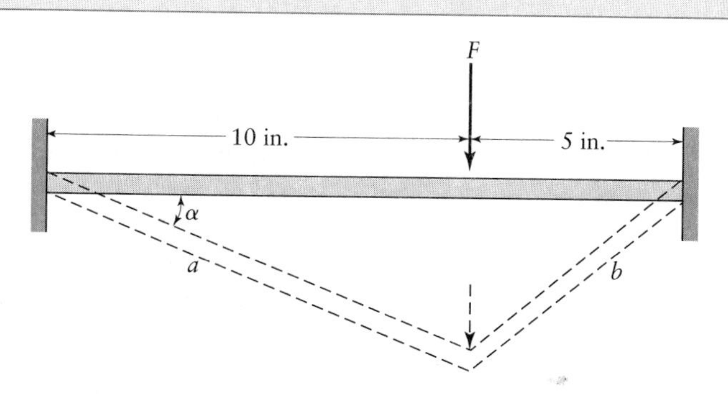

7.6 Bulging

The basic forming process of *bulging* involves placing a tubular, conical, or curvilinear hollow part in a split female die and expanding it with, say, a rubber or polyurethane plug (Fig. 7.32a). The punch is then retracted, the plug returns to its original shape, and the part is removed by opening the dies. Typical products made by this process include coffee and water pitchers, barrels, and beads on drums. For parts with complex shapes, the plug, instead of being cylindrical, may be shaped in order to apply greater pressure at critical points. The major advantage of using polyurethane plugs is that they are very resistant to abrasion, wear, and lubricants, and they do not damage the surface finish of the part being formed.

Formability in bulging processes can be enhanced by the application of longitudinal compressive stresses. *Hydraulic pressure* may also be used in bulging, although this technique requires sealing and needs hydraulic controls (Fig. 7.32b). *Bellows* are manufactured by a bulging process, as shown in Fig. 7.32c. After the tube is bulged at several equidistant locations, it is compressed axially to collapse the bulged regions, thus forming bellows. The tube material must, of course, be able to undergo the large strains involved during the collapsing process. **Segmented dies** that are expanded and retracted mechanically may also be used for bulging operations. These dies are relatively inexpensive and can be used for large production runs.

Embossing. This process involves forming a number of shallow shapes, such as numbers, letters, or designs, on a metal sheet. Parts may be embossed with male and female dies or by other means described throughout this chapter. The process is used principally for decorative purposes.

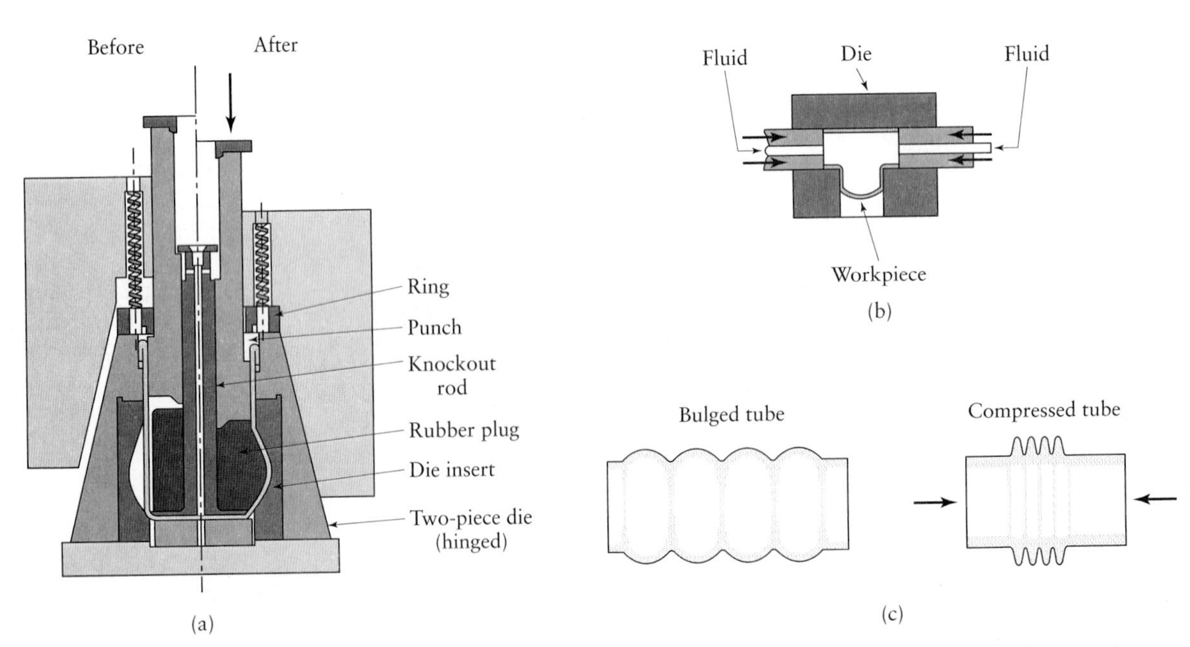

FIGURE 7.32 (a) Bulging of a tubular part with a flexible plug. Water pitchers can be made by this method. (b) Production of fittings for plumbing by expanding tubular blanks with internal pressure. The bottom of the piece is then punched out to produce a "T." *Source*: Schey J. A., *Introduction to Manufacturing Processes*, 2d ed., New York: McGraw-Hill, 1987. (c) Manufacturing of bellows.

7.7 Rubber Forming and Hydroforming

In describing the processes in the preceding sections, we noted that the dies are made of solid materials. However, in *rubber forming*, one of the dies in a set is made of flexible material, such as a rubber or polyurethane membrane. Polyurethanes are used widely because of their resistance to abrasion, long fatigue life, and resistance to damage by burrs or sharp edges of the sheet blank.

In bending and embossing sheet metal by the rubber-forming method, as shown in Fig. 7.33, the female die is replaced with a rubber pad. Note that the outer surface of the sheet is now protected from damage or scratches, because it is not in contact with a hard metal surface during forming. Parts can also be formed with laminated sheets of various nonmetallic materials or coatings. Pressures are typically on the order of 10 MPa (1500 psi).

In *hydroforming*, or **fluid-forming process** (Fig. 7.34), the pressure over the rubber membrane is controlled throughout the forming cycle, with maximum pressures reaching 100 MPa (15,000 psi). This procedure allows close control of the part during forming to prevent wrinkling or tearing. Deeper draws are obtained than in conventional deep drawing (see Section 7.12), because the pressure around the rubber membrane forces the cup against the punch. The friction at the punch–cup interface reduces the longitudinal tensile stresses in the cup being formed and thus delays fracture. The control of frictional conditions in rubber forming as well as in other sheet-forming operations can be a critical factor in making parts successfully. Selection of proper lubricants and application methods is important. In **tube hydroforming** (Fig. 7.35), steel or other metal tubing is formed in a die and pressurized by a fluid. This procedure can form simple tubes (Fig. 7.35a), or it can form intricate hollow tubes (Fig. 7.35b). Applications of tube-hydroformed parts include automotive exhaust and structural components.

When selected properly, rubber-forming and hydroforming processes have the advantages of (a) low tooling cost, (b) flexibility and ease of operation, (c) low die wear, (d) no damage to the surface of the sheet, and (e) capability to form complex shapes.

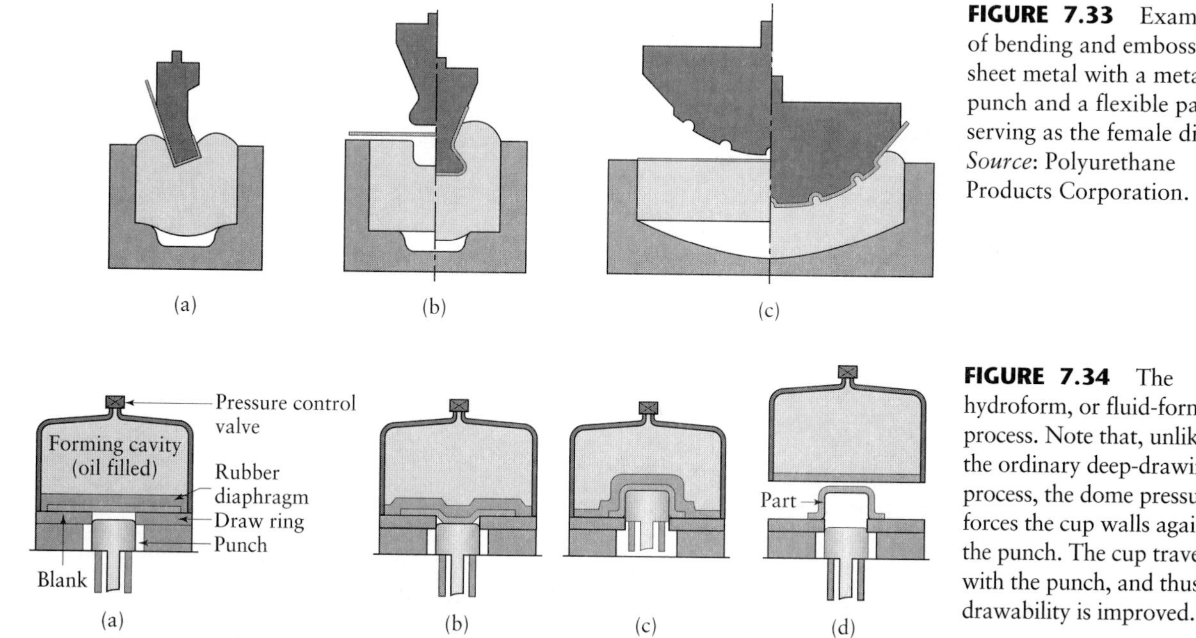

FIGURE 7.33 Examples of bending and embossing sheet metal with a metal punch and a flexible pad serving as the female die. *Source*: Polyurethane Products Corporation.

FIGURE 7.34 The hydroform, or fluid-forming, process. Note that, unlike in the ordinary deep-drawing process, the dome pressure forces the cup walls against the punch. The cup travels with the punch, and thus deep drawability is improved.

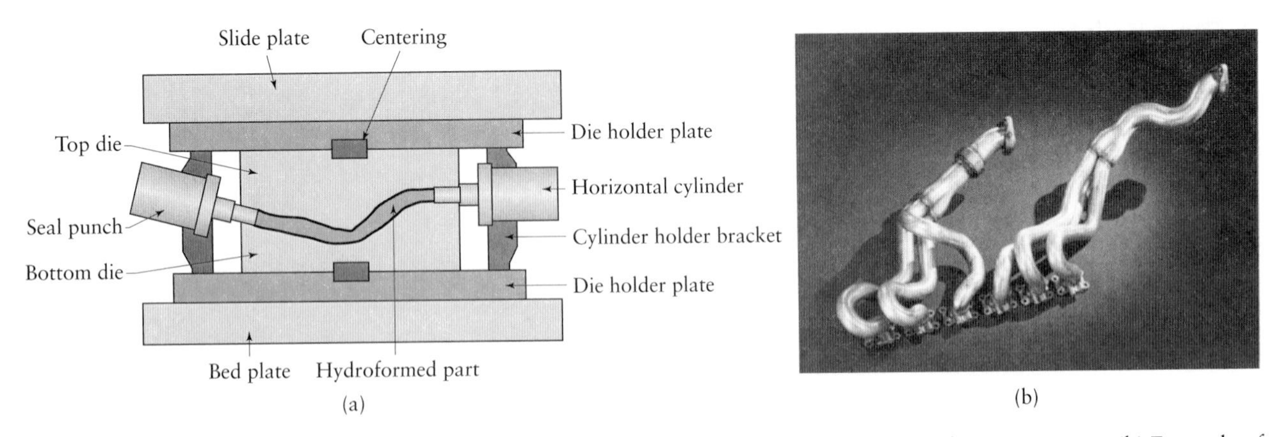

FIGURE 7.35 (a) Schematic illustration of the tube-hydroforming process. (b) Example of tube-hydroformed parts. Automotive exhaust and structural components, bicycle frames, and hydraulic and pneumatic fittings are produced through tube hydroforming. *Source*: Schuler GmBH.

7.8 Spinning

Spinning involves the forming of axisymmetric parts over a rotating mandrel, using rigid tools or rollers. There are three basic types of spinning processes: conventional (or manual), shear, and tube spinning. The equipment used for these processes is similar to a lathe (see Section 8.8.2) with various special features.

7.8.1 Conventional spinning

In the *conventional-spinning* process, a circular blank of flat or preformed sheet metal is held against a rotating mandrel while a rigid tool deforms and shapes the material over the mandrel (Fig. 7.36a). The tools may be actuated either manually or by a hydraulic mechanism. The operation involves a sequence of passes and requires considerable skill. Figure 7.37 shows some typical shapes made by conventional spinning; note that this process is particularly suitable for conical and curvilinear shapes, which would otherwise be difficult or uneconomical to form by other methods. Part diameters may range up to 6 m (20 ft). Although most spinning is performed at room temperature, thick parts or metals with low ductility or high strength require spinning at

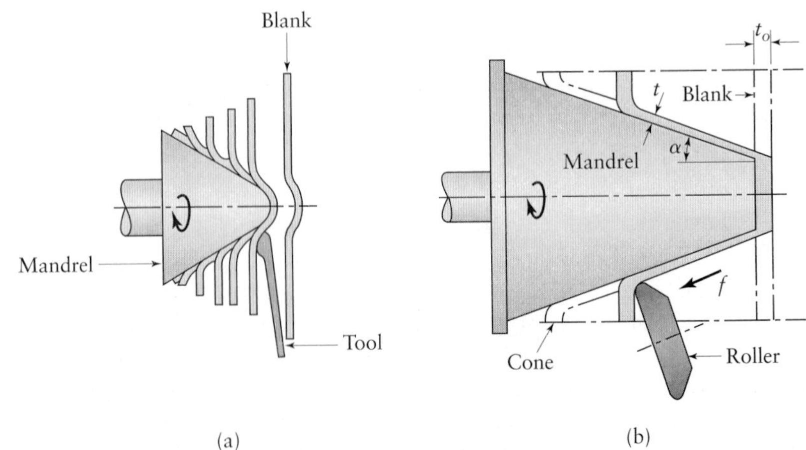

FIGURE 7.36 Schematic illustration of spinning processes: (a) conventional spinning and (b) shear spinning. Note that in shear spinning, the diameter of the spun part, unlike in conventional spinning, is the same as that of the blank. The quantity f is the feed (in mm/rev or in./rev).

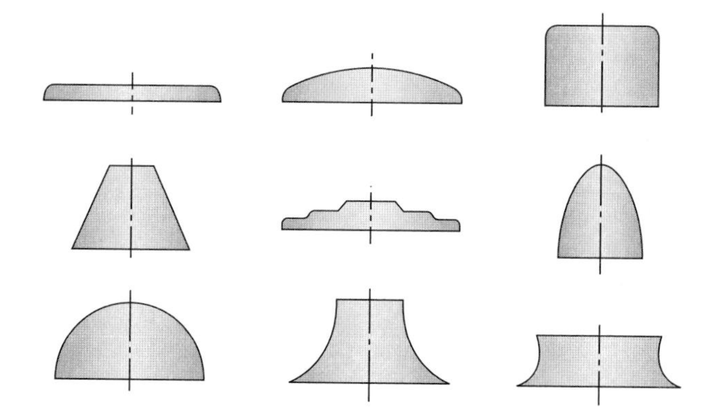

FIGURE 7.37 Typical shapes produced by the conventional-spinning process. Circular marks on the external surfaces of components usually indicate that the parts have been made by spinning. Examples include aluminum kitchen utensils and light reflectors.

elevated temperatures. Tooling costs in spinning are relatively low; however, because the operation requires multiple passes to form the final part, it is economical for relatively small production runs only. (See Section 7.16.)

7.8.2 Shear spinning

In the *shear-spinning* process (also known as **power spinning, flow turning, hydrospinning,** and **spin forging**), an axisymmetric conical or curvilinear shape is generated in a manner whereby the diameter of the part remains constant. (See Fig. 7.36b.) Parts typically made by this process include rocket-motor casings and missile nose cones. Although a single roller can be used, two rollers are desirable in order to balance the radial forces acting on the mandrel. Parts up to about 3 m (10 ft) in diameter can be spun to close dimensional tolerances. Little material is wasted, and the operation is completed in a relatively short time. If the operation is carried out at room temperature, the spun part has a higher yield strength than the original material, but lower ductility and toughness. A great variety of shapes can be spun with relatively simple tooling, generally made of tool steel. Because of the large plastic deformation involved, the process generates considerable heat; it is usually carried away by a coolant-type fluid applied during spinning.

In shear spinning over a conical mandrel, the thickness, t, of the spun part is simply

$$t = t_o \sin \alpha, \tag{7.12}$$

where t_o is the original sheet thickness. The force primarily responsible for supplying the energy required is the *tangential force*, F_t. For an *ideal* case in shear spinning of a cone, this force is

$$F_t = u\, t_o f \sin \alpha, \tag{7.13}$$

where u is the specific energy of deformation, from Eq. (2.59), and f is the feed. As described in Section 2.12, u is the area under the true-stress–true-strain curve that corresponds to a true strain related to the shear strain by

$$\varepsilon = \frac{\gamma}{\sqrt{3}} = \frac{\cot \alpha}{\sqrt{3}}. \tag{7.14}$$

(See also Section 2.11.7.) The actual force can be as much as 50% higher than that given by Eq. (7.13), because of factors such as redundant work and friction. (See Section 6.4.2.)

An important factor in shear spinning is the **spinnability** of the metal, that is, the maximum percent reduction to which a part can be spun without fracture. To determine spinnability, a simple test method has been developed (Fig. 7.38), in which a

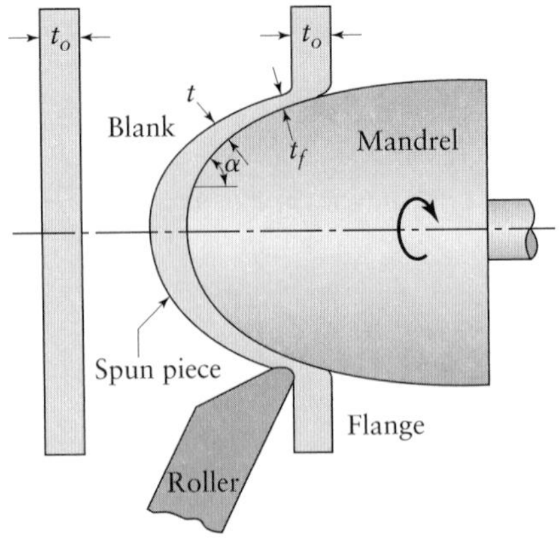

FIGURE 7.38 Schematic illustration of a shear-spinnability test. As the roller advances, the part thickness is reduced. The reduction in thickness at fracture is called the *maximum spinning reduction per pass*. *Source*: After R. L. Kegg.

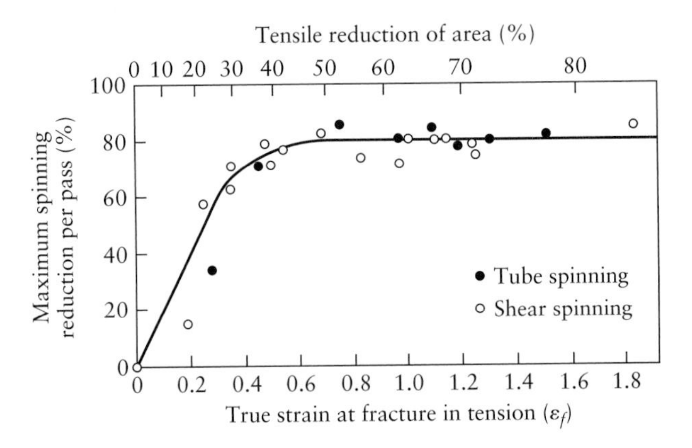

FIGURE 7.39 Experimental data showing the relationship between maximum spinning reduction per pass and the tensile reduction of area of the original material. Note that once a material has about 50% reduction of area in a tension test, any further increase in the ductility of the original material does not improve the material's spinnability. *Source*: S. Kalpakjian.

circular blank is spun over an ellipsoid mandrel, according to Eq. (7.12). As the thickness is eventually reduced to zero, all materials will fail at some thickness t_f. The maximum spinning reduction in thickness is defined as

$$\text{Maximum reduction} = \frac{t_o - t_f}{t_o} \times 100\% \tag{7.15}$$

and is plotted against the tensile reduction of area of the material tested (Fig. 7.39). Note that if a metal has a tensile reduction of area of about 50% or higher, it can be reduced in thickness by 80% by spinning in one pass. Note also that any further increase in the ductility of the original material does not improve its spinnability. Process variables such as feed and speed do not appear to have a significant effect on spinnability. In comparing Figs. 7.15b and 7.39, note that, interestingly, maximum bendability and spinnability both coincide with a tensile reduction of area of about 50%. For materials with low ductility, spinnability can usually be improved by forming at elevated temperatures, which improves the reduction of area.

7.8.3 Tube spinning

In the *tube-spinning* process, tubes are reduced in thickness by spinning them on a mandrel, using rollers. The operation may be carried out externally or internally (Fig. 7.40). Also, the part may be spun *forward* or *backward*, similar to a drawing or a backward extrusion process. (See Section 6.4.) In either case, the reduction in wall thickness results in a longer tube (because of volume constancy). Various internal

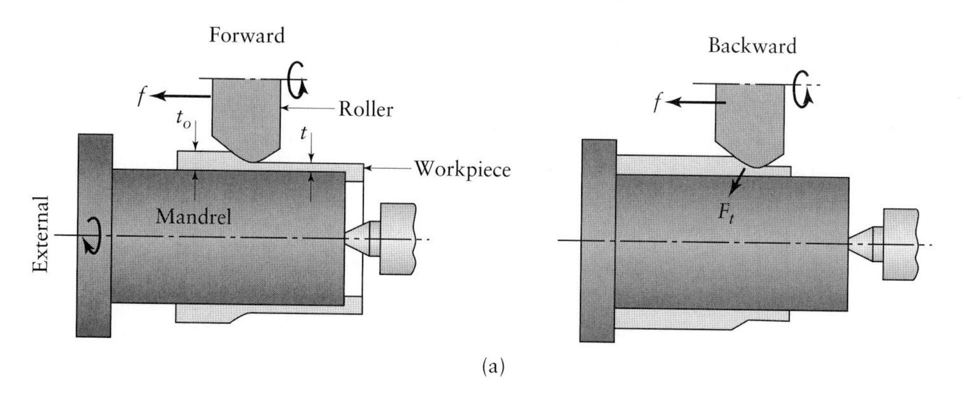

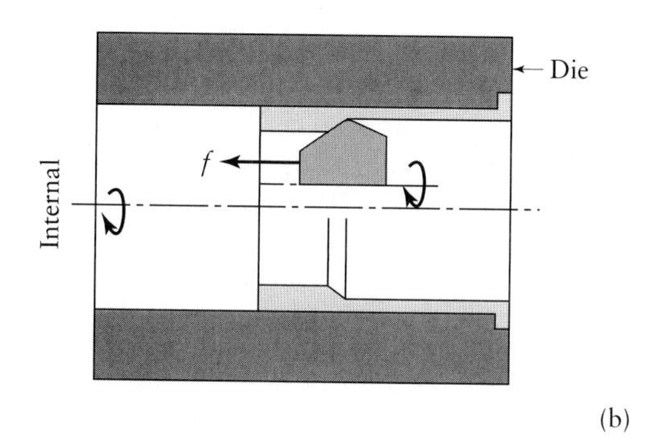

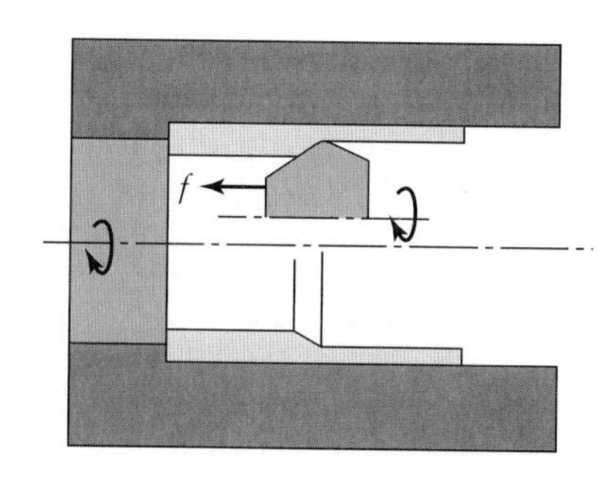

(b)

FIGURE 7.40 Examples of external and internal tube spinning and the process variables involved.

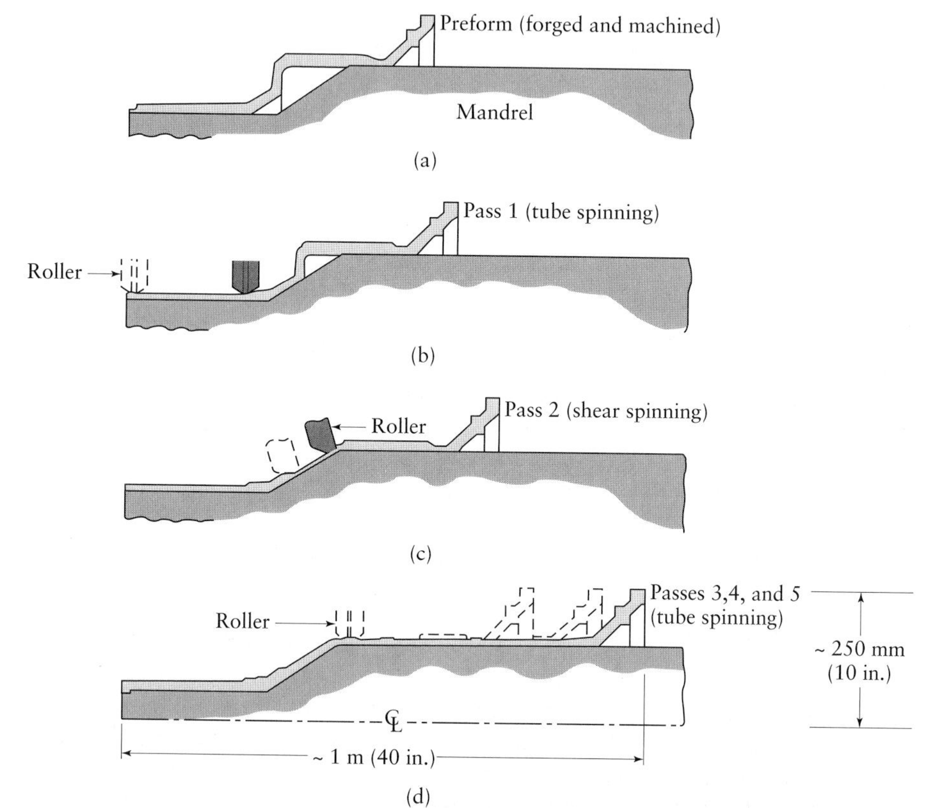

FIGURE 7.41 Stages in tube and shear spinning of a compressor shaft for the jet engine of the supersonic Concorde aircraft. Economic analysis indicated that the best method of manufacturing this part was to spin a preformed (forged and machined) tubular blank.

and external profiles can be produced by controlling the path of the roller during its travel along the mandrel. Tube spinning can be used for making pressure vessels, automotive components, and rocket and missile parts. It can also be combined with shear spinning, as is the case for the production of the compressor shaft for the Olympus engine (Rolls Royce) of the Concorde supersonic aircraft, as shown in Fig. 7.41.

Spinnability in this process is determined by a test method similar to that for shear spinning. Maximum reduction per pass in tube spinning is related to the tensile reduction of area of the material (see Fig. 7.39), with results very similar to those of shear spinning. As in shear spinning, ductile metals fail in tension after the reduction in thickness has taken place, whereas less ductile metals fail in the deformation zone *under* the roller.

The *ideal tangential force*, F_t, in *forward tube spinning* can be expressed as

$$F_t = \bar{Y}(t_o - t)f, \tag{7.16}$$

where $\bar{Y}$ is the *average flow stress* of the material. For *backward spinning*, the ideal tangential force is roughly twice that given by Eq. (7.16). Furthermore, friction and redundant work of deformation cause the actual forces to be about twice those calculated by this formula.

7.9 High-Energy-Rate Forming

This section describes sheet-metal forming processes that use chemical, electrical, or magnetic sources of energy. They are called *high-energy-rate processes*, because the energy is released in a very short period of time.

7.9.1 Explosive forming

Figure 7.42 depicts the most common *explosive-forming* process. In this process, the workpiece is clamped over a die, the air in the die cavity is evacuated, and the whole assembly is lowered into a tank filled with water. An explosive charge is then placed at a certain distance from the sheet surface and detonated. The rapid conversion of the explosive into gas generates a shock wave; the pressure of this wave is sufficiently high to force the metal into the die cavity.

The peak pressure generated in water is given by the expression

$$p = K\left(\frac{\sqrt[3]{W}}{R}\right)^a, \tag{7.17}$$

where p is the peak pressure in psi, K is a constant that depends on the type of explosive (e.g., 21,600 for TNT), W is the weight of the explosive in pounds, R is the

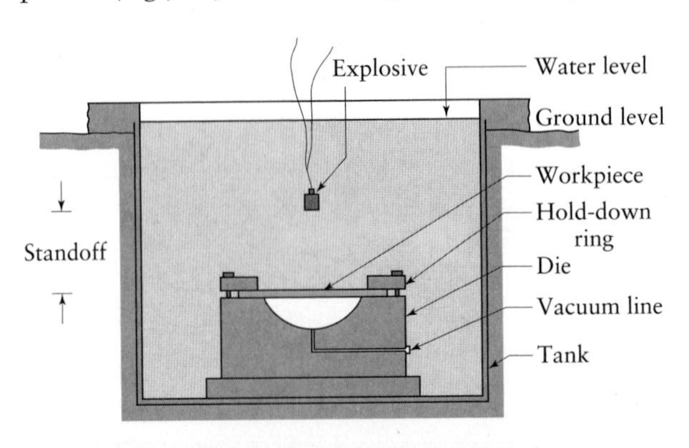

FIGURE 7.42 Schematic illustration of the explosive-forming process. Although explosives are generally used for destructive purposes, their energy can be controlled and employed in forming large parts that would otherwise be difficult or expensive to produce by other methods.

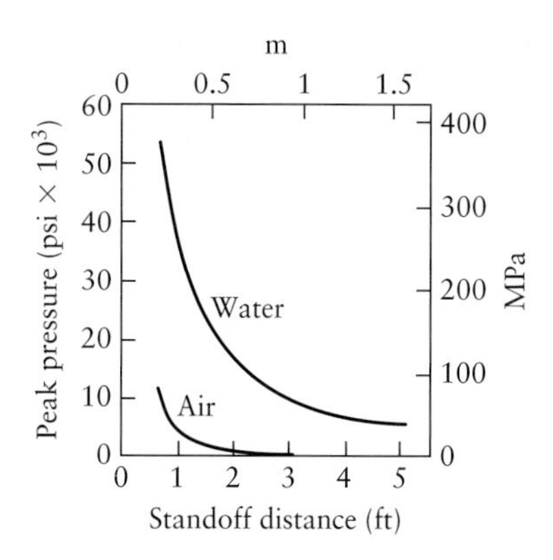

FIGURE 7.43 Influence of the standoff distance and type of energy-transmitting medium on the peak pressure obtained using 1.8 kg (4 lb) of TNT. To be effective, the pressure-transmitting medium should have high density and low compressibility. In practice, water is a commonly used medium.

distance of the explosive from the workpiece (*standoff*) in feet; and a is a constant that is generally taken as 1.15.

An important factor in determining peak pressure is the **compressibility** of the energy-transmitting medium (such as water) and the medium's **acoustic impedance** (the product of mass density and sound velocity in the medium). The lower the compressibility of the medium and the higher its density, the higher is the peak pressure (Fig. 7.43). Detonation speeds are typically 6700 m/s (22,000 ft/s), and the speed at which the metal is formed is estimated to be on the order of 30 to 200 m/s (100 to 600 ft/s).

A variety of shapes can be formed explosively, provided that the material is sufficiently ductile at high strain rates. Depending on the number of parts to be formed, dies used in this process may be made of aluminum alloys, steel, ductile iron, zinc alloys, reinforced concrete, wood, plastics, or composite materials. The final properties of the parts made by this process are basically the same as the properties of parts made by conventional methods. Safety is an important aspect in explosive forming.

Requiring only one die and being versatile, explosive forming is particularly suitable for small-quantity production runs of large parts. Steel plates 25 mm (1 in.) thick and 3.6 m (12 ft) in diameter have been formed by this method. Also, tubes with a wall thickness of 25 mm have been bulged by explosive-forming techniques. Another explosive-forming method, which uses a cartridge as the source of energy, is shown in Fig. 7.44; no other energy-transmitting medium is used. The process can be used for bulging and expanding of thin-walled short tubes; however, die wear and die failure can be a significant problem in this operation.

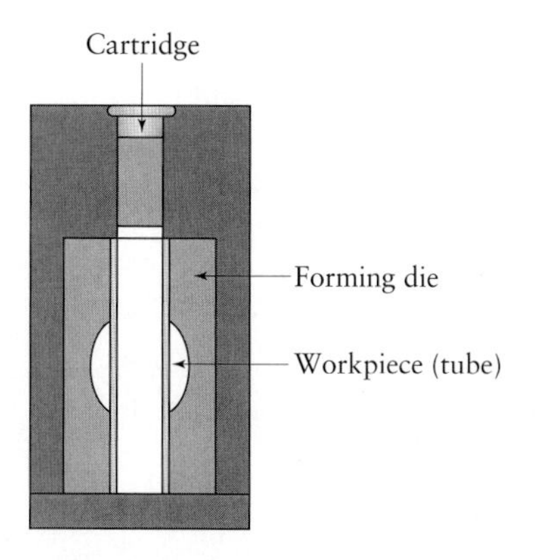

FIGURE 7.44 Schematic illustration of the confined method of explosive bulging of tubes. Thin-walled tubes of nonferrous metals can be formed to close tolerances by this process.

EXAMPLE 7.5 Peak pressure in explosive forming

(a) Calculate the peak pressure in water for 0.1 lb of TNT at a standoff of 1 ft. (b) Is this pressure sufficiently high for forming sheet metals?

Solution.

a) Using Eq. (7.17), we find that

$$p = (21{,}600)\left(\frac{\sqrt[3]{0.1}}{1}\right)^{1.15} = 9000 \text{ psi}.$$

b) This pressure is sufficiently high to form sheet metals. Note in Example 2.6, for instance, that the pressure required to expand a thin-walled spherical shell of a material similar to soft aluminum alloys is only 400 psi. Also note that a process such as hydroforming (Section 7.7) has a maximum hydraulic pressure in the dome of about 100 MPa (15,000 psi). Other rubber-forming processes use pressures from about 1500 psi to 7500 psi. Therefore, the pressure obtained in this problem is sufficient for most sheet-forming processes.

7.9.2 Electrohydraulic forming

In the *electrohydraulic-forming* process, also called **underwater-spark** or **electric-discharge forming**, the source of energy is a spark from two electrodes connected with a thin wire (Fig. 7.45). The energy is stored in a bank of condensers charged with direct current. The rapid discharge of this energy through the electrodes generates a shock wave, which is strong enough to form the part. This process is essentially similar to explosive forming, except that it utilizes a lower level of energy and is used with smaller workpieces; it is also a safer operation than explosive forming.

7.9.3 Magnetic-pulse forming

In the *magnetic-pulse-forming* process, the energy stored in a capacitor bank is discharged rapidly through a magnetic coil. In a typical example, a ring-shaped coil is placed over a tubular workpiece to be formed over another solid piece, thus making an integral part (Fig. 7.46). The magnetic field produced by the coil crosses the metal tube, generating an *eddy currents* in the tube; this current, in turn, produces its own magnetic field. The forces produced by the two magnetic fields oppose each other; thus, there is a repelling force between the coil and the tube, and the high forces generated collapse the tube over the inner piece.

The higher the electrical conductivity of the workpiece, the higher is the magnetic force. Note, however, that the metal does not need to have any special magnetic properties. Magnetic-pulse forming is used for a variety of operations, such as swaging of thin-walled tubes over rods, cables, and plugs (such as placing end fittings onto torque tubes for the Boeing 777 aircraft); bulging; and flaring. Flat coils are also used for forming sheet metal, as in embossing and shallow drawing operations.

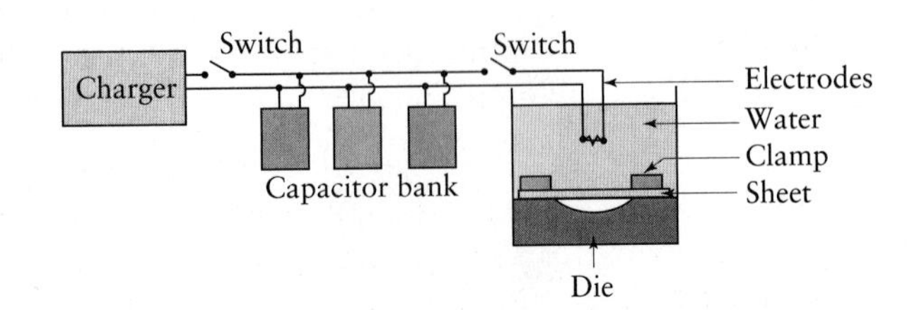

FIGURE 7.45 Schematic illustration of the electrohydraulic-forming process.

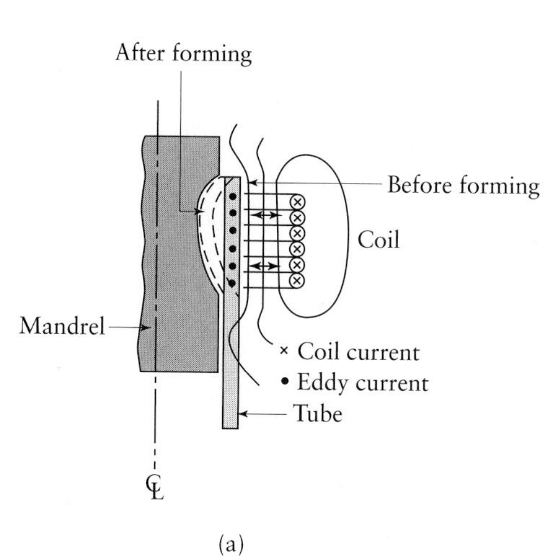

(a)

(b)

FIGURE 7.46
(a) Schematic illustration of the magnetic-pulse-forming process. The part is formed without physical contact with any object.
(b) Aluminum tube collapsed over a hexagonal plug by the magnetic-pulse-forming process.

7.10 | Superplastic Forming

Section 2.2.7 described the *superplastic behavior* of some very fine-grained alloys (normally less than 10 to 15 μm), where very large elongations (up to 2000%) are obtained at certain temperatures and low strain rates. These alloys, such as zinc–aluminum and titanium, can be formed into complex shapes by employing traditional metalworking or polymer-processing techniques. Commonly used die materials in superplastic forming include low-alloy steels, cast tool steels, ceramics, graphite, and plaster of paris; selection of the material depends on the forming temperature and strength of the superplastic alloy.

The high ductility and relatively low strength of superplastic alloys present the following advantages in superplastic forming:

1. Lower strength of tooling, because of the low strength of the material at forming temperatures, and, hence, lower tooling costs;
2. Ability to form complex shapes in one piece, with fine detail, close dimensional tolerances, and elimination of secondary operations;
3. Weight and material savings, because of the good formability of superplastic materials;
4. Little or no residual stresses in the formed parts.

The limitations are:

1. The material must not be superplastic at service temperatures;
2. Because of the extreme strain-rate sensitivity of the superplastic material, the material must be formed at sufficiently low rates (typically at strain rates of 10^{-4}/s to 10^{-2}/s). Forming times range anywhere from a few seconds to several hours, and thus cycle times are much longer than in conventional forming processes. Superplastic forming is therefore a batch-forming process.

Superplastic alloys (particularly Zn-22Al and Ti-6Al-4V) can be formed by bulk deformation processes, such as compression molding, closed-die forging, coining, hubbing, and extrusion. Sheet forming of these materials can also be done using operations such as thermoforming, vacuum forming, and blow molding.

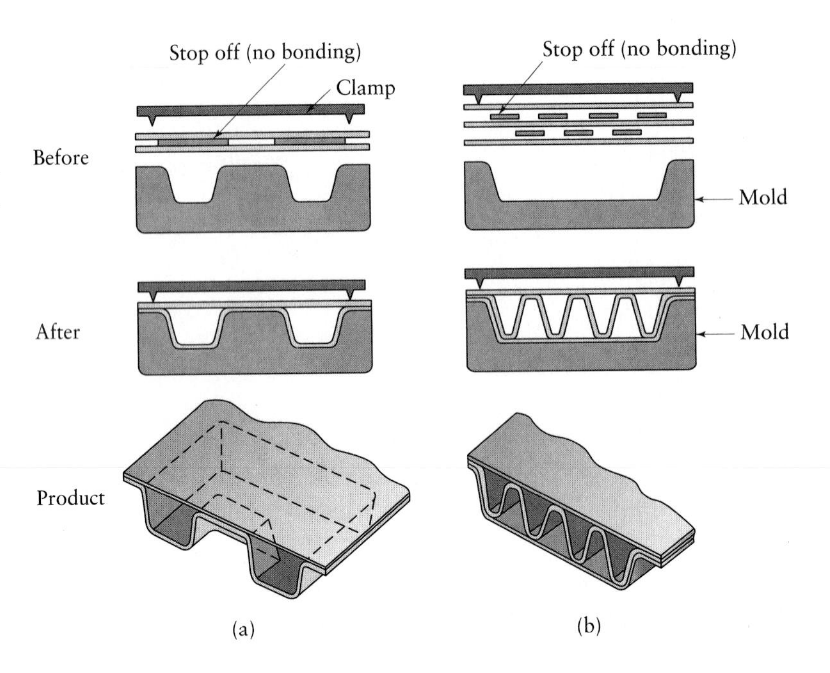

FIGURE 7.47 Two types of structures made by diffusion bonding and superplastic forming of sheet metal. Such structures have a high stiffness-to-weight ratio. *Source*: Rockwell Automation, Inc.

An important aspect is the ability to fabricate sheet-metal structures by combining **diffusion bonding** (see Section 12.11) *with* superplastic forming (SPF/DB). Typical structures in which flat sheets are diffusion bonded and formed are shown in Fig. 7.47. After diffusion bonding of selected locations of the sheets, the unbonded regions (*stop off*) are expanded into a mold by air pressure. These structures are thin and have high stiffness-to-weight ratios; hence, they are particularly important in aircraft and aerospace applications. This process improves productivity by eliminating mechanical fasteners and produces parts with good dimensional accuracy and low residual stresses. The technology is now well advanced for titanium structures (typically Ti-6Al-4V alloy) for aerospace applications. Structures made of 7475-T6 aluminum alloy continue to be developed using this technique; other metals suitable for superplastic forming include Inconel 100 and Incoloy 718 nickel alloys and iron-base, high-carbon alloys.

7.11 Various Forming Methods

a) **Peen forming** is used to produce curvatures on thin sheet metals by **shot peening** one surface of the sheet. (See Section 4.5.1.) Peening is done with cast-iron or steel shot, discharged either from a rotating wheel or with an air blast from a nozzle. Peen forming is used by the aircraft industry to generate complex and smooth curvatures on aircraft-wing skins (Fig. 7.48). Cast-steel shot about 2.5 mm (0.1 in.) in diameter at speeds of 60 m/s (200 ft/s) has been used to form wing panels 25 m (80 ft) long. For heavy sections, shot diameters as large as 6 mm $\left(\frac{1}{4}\text{ in.}\right)$ may be used.

In peen forming, the surface of the sheet is subjected to compressive stresses, which tend to expand the surface layer. Since the material below the peened surface remains rigid, the surface expansion causes the sheet to develop a curvature. The process also induces compressive surface residual stresses, which improve the fatigue strength of the sheet. The peen-forming process is also used for straightening twisted or bent parts; out-of-round rings, for example, can be straightened by this method.

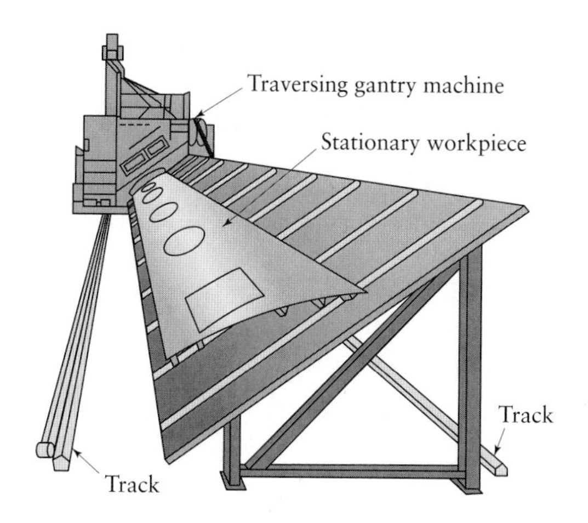

FIGURE 7.48 Peen-forming machine to form a large sheet-metal part, such as an aircraft-skin panel. The sheet is stationary, and the machine traverses it. *Source*: Metal Improvement Company.

b) **Gas mixtures** in a closed container have been used as an energy source in sheet-metal forming. When the gas is ignited, the pressures generated are sufficiently high to form parts. The principle is similar to the generation of pressure in an internal combustion engine.

c) **Liquefied gases**, such as liquid nitrogen, may be used to develop pressures high enough to form sheet metals. When allowed to reach room temperature in a closed container, liquified nitrogen becomes gaseous and expands, developing the necessary pressure to form the part.

d) **Laser forming** is a more recent technology that uses localized heating to induce thermal stress gradients through the thickness of the sheet. These stresses are sufficiently high to cause localized plastic deformation of the sheet without the use of external forces and result in, for example, a bent sheet.

e) **Laser-assisted forming** uses lasers as a local heat source in order to reduce the flow stress of the material at specific locations and to improve formability, thus increasing process flexibility. Examples of its uses include straightening, bending, embossing, and forming of complex flat or tubular components.

f) **Microforming** is a more recent term; it describes processes that are used to produce very small metallic parts and components. Examples of recently miniaturized products include a wristwatch with an integrated digital camera and a one-gigabyte hard-disk drive that fits into an eggshell. As discussed in greater detail in Chapter 13, several processes are involved in the fabrication of micromechanical devices. Through the use of specialized tooling, machines, and handling systems, microforming consists of miniaturization of traditional as well as more recent forming and machining processes, including the use of lasers. Other, similar terminology includes *microshaping*, *microcasting*, and *microsintering*. Typical components made by microforming include small shafts for micromotors; springs; screws; and a variety of cold-headed, extruded, bent, embossed, coined, punched, or deep-drawn parts, with dimensions typically in the submillimeter range and weights on the order of milligrams.

7.11.1 Manufacturing honeycomb structures

There are two principal methods of manufacturing honeycomb materials. In the **expansion** process (Fig. 7.49a), which is the most common method, sheets are cut from a coil, and an adhesive is applied at intervals (node lines). The sheets are stacked and cured in an oven, whereby strong bonds develop at the adhesive joints. The block is then cut into slices of the desired dimensions and stretched to produce a honeycomb

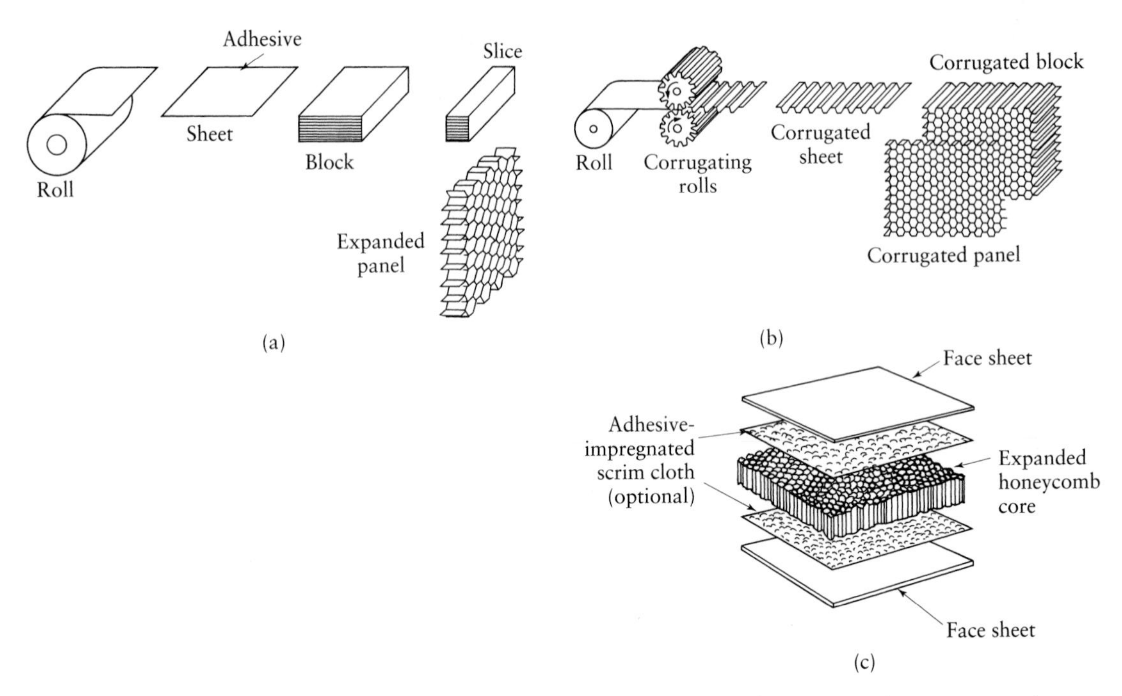

FIGURE 7.49 Methods of making honeycomb materials: (a) expansion process and (b) corrugation process. *Source: Materials Engineering*. Reprinted with permission. (c) Making a honeycomb sandwich.

structure. This procedure is similar to expanding folded paper structures into the shape of decorative objects such as lanterns.

In the **corrugation** process (Fig. 7.49b), sheets pass through a pair of specially designed rolls, which make them into corrugated sheets; the corrugated sheets are then cut into desired lengths. Again, adhesive is applied to the node lines, and the block is cured. Note that no expansion process is involved. The honeycomb material is then made into a sandwich structure (Fig. 7.49c). Face sheets are subsequently joined with adhesives to the top and bottom surfaces.

Honeycomb structures are most commonly made of 3000-series aluminum, but are also made of titanium, stainless steels, and nickel alloys. Because of their light weight and high resistance to bending forces, these structures are used for aircraft and aerospace components, as well as buildings and transportation equipment.

7.12 Deep Drawing

In *deep drawing*, a flat sheet-metal blank is formed into a cylindrical or box-shaped part by means of a punch that presses the blank into the die cavity (Fig. 7.50a). Although the process is generally called *deep drawing*, meaning *forming deep parts*, the basic operation also produces parts that are shallow or have moderate depth. Deep drawing, first developed in the 1700s, has been studied extensively and has become an important metalworking process. Typical parts produced by this method include beverage cans, pots and pans, containers of all shapes and sizes, kitchen sinks, and automobile panels.

Figure 7.50b shows the basic parameters in deep drawing a cylindrical cup. A circular sheet blank with a diameter D_o and thickness t_o is placed over a die opening with a corner radius R_d. The blank is held in place with a **blankholder**, or **hold-down ring**, with a certain force. A punch with a diameter D_p and a corner radius R_p moves downward and pushes the blank into the die cavity, thus forming a cup. The significant independent variables in deep drawing are

1. Properties of the sheet metal;
2. The ratio of the blank diameter to the punch diameter;
3. The thickness of the sheet;
4. The clearance between the punch and the die;
5. The corner radii of the punch and die;
6. The blankholder force;
7. Friction and lubrication at the punch, die, and workpiece interfaces;
8. The speed of the punch.

At an intermediate stage during the deep-drawing operation, the workpiece is subjected to the states of stress shown in Fig. 7.51. On element A in the blank, the radial tensile stress is due to the blank being pulled into the cavity, and the compressive stress normal to the element is due to the blankholder pressure. With a free-body

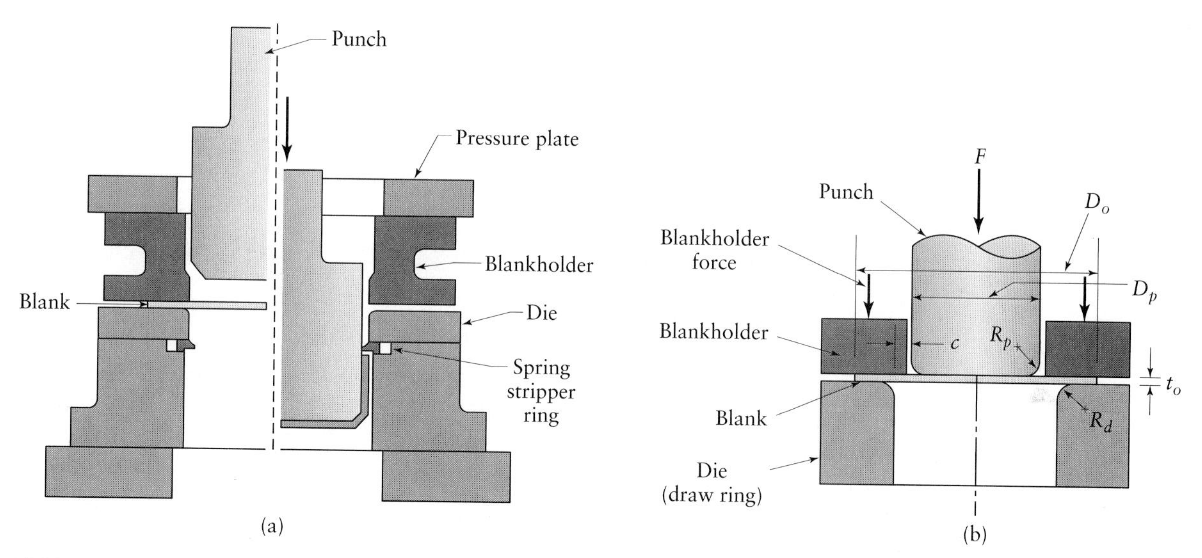

FIGURE 7.50 (a) Schematic illustration of the deep-drawing process. This procedure is the first step in the basic process by which aluminum beverage cans are produced today. The stripper ring facilitates the removal of the formed cup from the punch. (b) Variables in deep drawing of a cylindrical cup. Only the punch force in this illustration is a dependent variable; all others are independent variables, including the blankholder force.

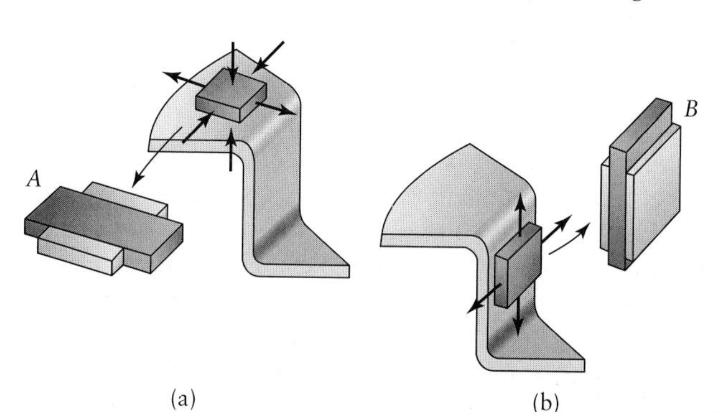

FIGURE 7.51 Deformation of elements in (a) the flange and (b) the cup wall in deep drawing of a cylindrical cup.

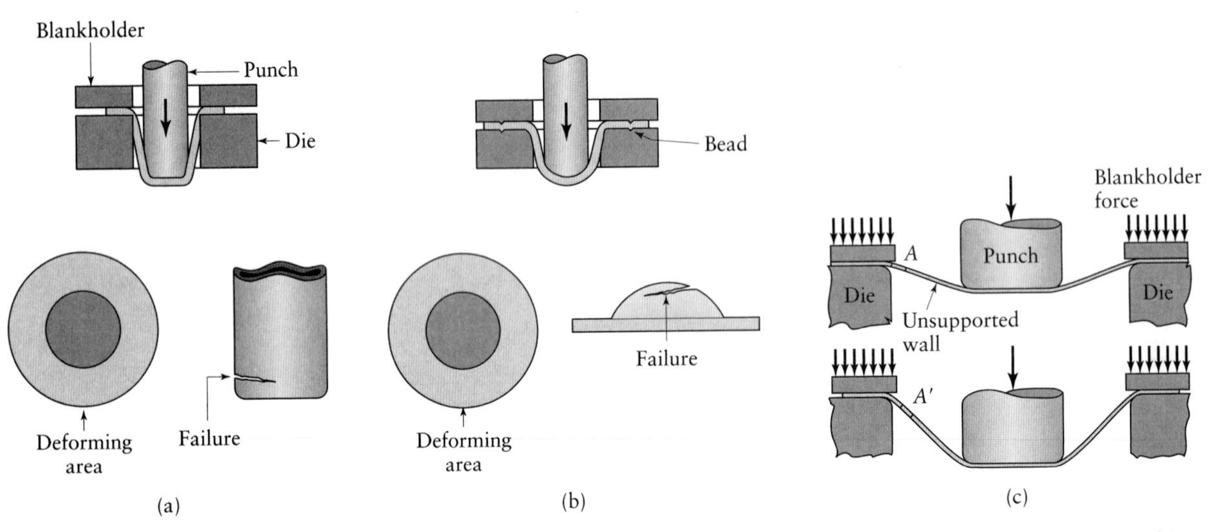

FIGURE 7.52 Examples of drawing operations: (a) pure drawing and (b) pure stretching. The bead prevents the sheet metal from flowing freely into the die cavity. (c) Possibility of wrinkling in the unsupported region of a sheet in drawing. *Source*: After W. F. Hosford and R. M. Caddell.

diagram of the blank along its diameter, it can be shown that the radial tensile stresses lead to compressive hoop stresses on element *A*. (It is these hoop stresses that tend to cause the flange to wrinkle during drawing, thus requiring a blankholder under a certain force.) Under this state of stress, element *A* contracts in the hoop direction and elongates in the radial direction.

The cup wall, which is already formed, is subjected principally to a longitudinal tensile stress, as shown in element *B* in Fig. 7.51. The punch transmits the drawing force, *F*, (see Fig. 7.50b) through the walls of the cup and to the flange that is being drawn into the die cavity. The tensile hoop stress on element *B* is caused by the cup being held tightly on the punch because of its contraction under tensile stresses in the cup wall. Note that a thin-walled tube, when subjected to longitudinal tension, becomes smaller in its diameter, as can be shown from the generalized *flow rule equations*; see Eq. (2.43). Thus, because it is constrained by the rigid punch, element *B* tends to elongate in the longitudinal (axial) direction, with no change in its width.

An important aspect of drawing is determining how much **stretching** and how much **pure drawing** is taking place (Fig. 7.52). Note that either with a high blankholder force or with the use of **draw beads** (Figs. 7.52b and 7.53), the blank can be prevented from flowing freely into the die cavity. The deformation of the sheet metal takes place mainly under the punch, and the sheet begins to *stretch*, eventually resulting in necking and tearing. Whether necking is localized or diffused depends on (a) the strain-rate sensitivity exponent, *m*, of the sheet metal (the higher the *m* value, the more diffuse the neck is), (b) geometry of the punch, and (c) lubrication.

Conversely, a low blankholder force will allow the blank to flow freely into the die cavity (*pure drawing*), whereby the blank diameter is reduced as drawing progresses. The deformation of the sheet is mainly in the flange, and the cup wall is subjected only to elastic stresses. However, these stresses increase with an increasing D_o/D_p ratio and can eventually lead to failure when the cup wall cannot support the load required to draw in the flange (Fig. 7.52a). Also note that, in pure drawing, element *A* in Fig. 7.51a will tend to increase in thickness as it moves toward the die cavity (because it is being reduced in diameter).

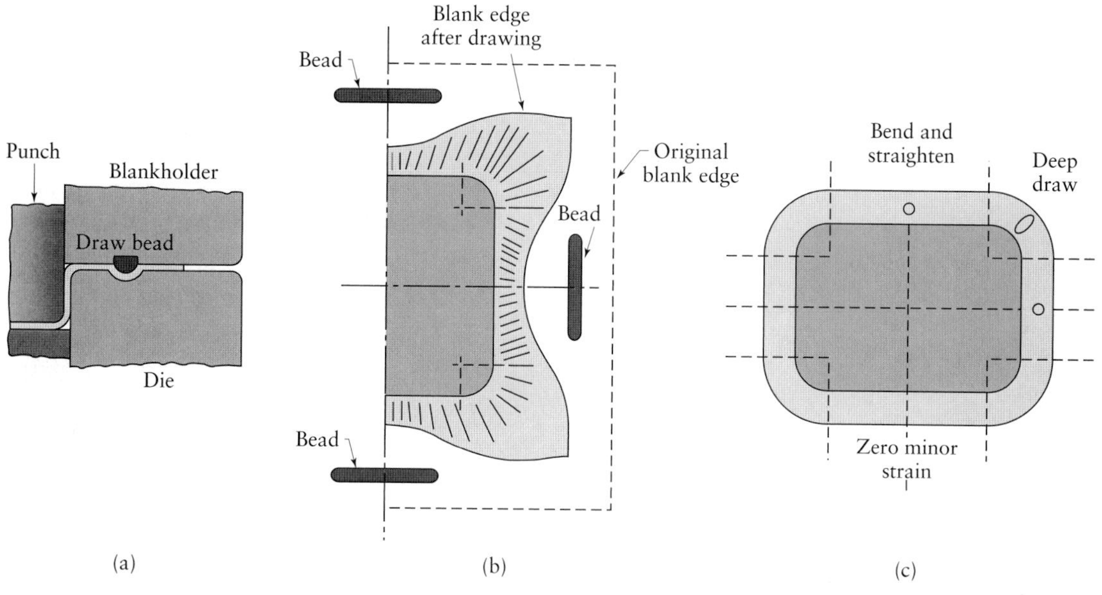

FIGURE 7.53 (a) Schematic illustration of a draw bead. (b) Metal flow during drawing of a box-shaped part, using beads to control the movement of the material. (c) Deformation of circular grids in drawing. (See Section 7.13.) *Source*: After S. Keeler.

In drawing operations, the length of the unsupported wall is significant in that it can lead to **wrinkling**. As shown in Fig. 7.52c, element A in the sheet is being pulled into the die cavity as the punch descends. Note, however, that the blank is becoming smaller in diameter and the circumference at the element is becoming smaller as the element moves to position A'. Thus, at this position, the element is being subjected to circumferential compressive strains and is unsupported by any tooling, unlike an element between the blankholder and the die surface. Because the sheet is thin and cannot support circumferential compressive strains to any significant extent, it tends to wrinkle in the unsupported region. This situation is particularly common in pure drawing, whereas it is less so as the process becomes more pure stretching and the blank diameter remains basically the same.

Ironing. If the thickness of the sheet as it enters the die cavity is more than the clearance between the punch and the die, the thickness will have to be reduced; this effect is known as *ironing*. Ironing produces a cup with constant wall thickness (Fig. 7.54); thus, the smaller the clearance, the greater is the amount of ironing. Obviously, because of volume constancy, an ironed cup will be longer than a cup produced with a large clearance. (See also the case study at the end of this chapter.)

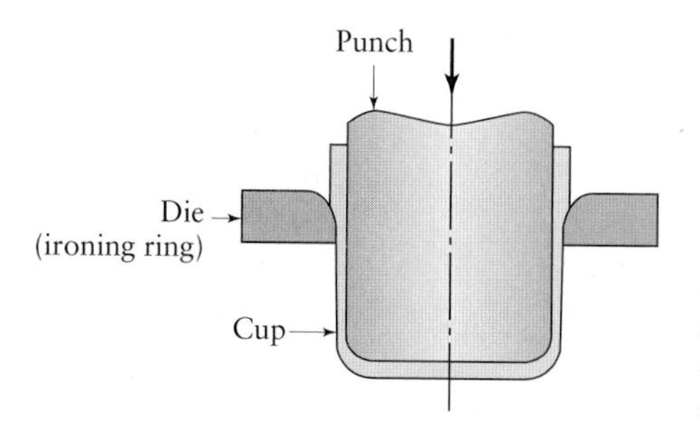

FIGURE 7.54 Schematic illustration of the ironing process. Note that the cup wall is thinner than its bottom. All beverage cans without seams (known as two-piece cans) are ironed, generally in three steps, after being deep drawn into a cup. (Cans with separate tops and bottoms are known as three-piece cans.)

7.12.1 Deep drawability (limiting drawing ratio)

The *limiting drawing ratio* (LDR) is defined as the maximum ratio of blank diameter to punch diameter that can be drawn without failure, or D_o/D_p. Many attempts have been made to correlate this ratio with various mechanical properties of the sheet metal. In an ordinary deep-drawing process, failure generally occurs by *thinning* in the cup wall under high longitudinal tensile stresses. By observing the movement of the material into the die cavity (see Fig. 7.51), we note that the material should be capable of undergoing a reduction in width (by being reduced in diameter), yet it should resist thinning under the longitudinal tensile stresses in the cup wall.

The ratio of width strain to thickness strain (Fig. 7.55) is defined as

$$R = \frac{\varepsilon_w}{\varepsilon_t} = \frac{\ln\left(\frac{w_o}{w_f}\right)}{\ln\left(\frac{t_o}{t_f}\right)}, \tag{7.18}$$

where R is known as the **normal anisotropy** of the sheet metal (also called **plastic anisotropy** or the **strain ratio**). The subscripts o and f refer to the original and final dimensions, respectively. An R value of unity indicates that the width and thickness strains are equal to each other; that is, the material is isotropic.

As sheet metals are generally thin compared with their surface area, errors in the measurement of small thicknesses is possible. Equation (7.18) can be modified, based on volume constancy, to

$$R = \frac{\ln\left(\frac{w_o}{w_f}\right)}{\ln\left(\frac{w_f \ell_f}{w_o \ell_o}\right)}, \tag{7.19}$$

where ℓ refers to the gage length of the sheet specimen. The final length and width in a test specimen are generally measured at an elongation of 15% to 20% or, for materials with lower ductility, below the elongation, where necking begins.

Rolled sheets generally have **planar anisotropy**; thus, the R value of a specimen cut from a rolled sheet (Fig. 7.55) will depend on the specimen's orientation with respect to the rolling direction of the sheet. (See also Fig. 3.15.) In this case, an *average* R value, $\bar{R}$, is calculated as

$$\bar{R} = \frac{R_0 + 2R_{45} + R_{90}}{4}, \tag{7.20}$$

where the subscripts 0, 45, and 90 refer to angular orientation (in degrees) of the test specimen with respect to the rolling direction of the sheet. Note that an isotropic

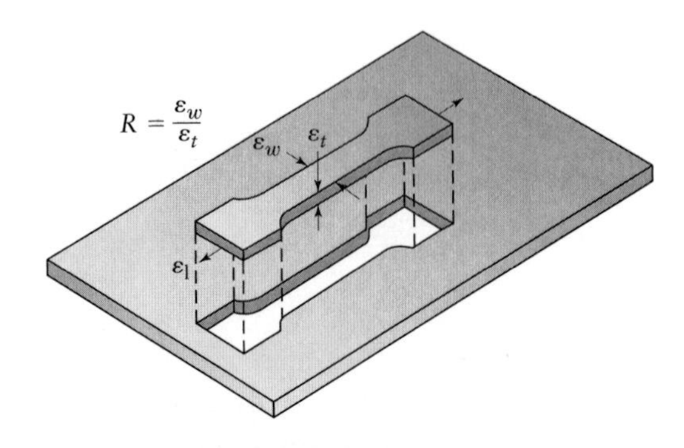

FIGURE 7.55 Definition of the normal anisotropy, R, in terms of width and thickness strains in a tensile-test specimen cut from a rolled sheet. Note that the specimen can be cut in different directions with respect to the length, or rolling direction, of the sheet.

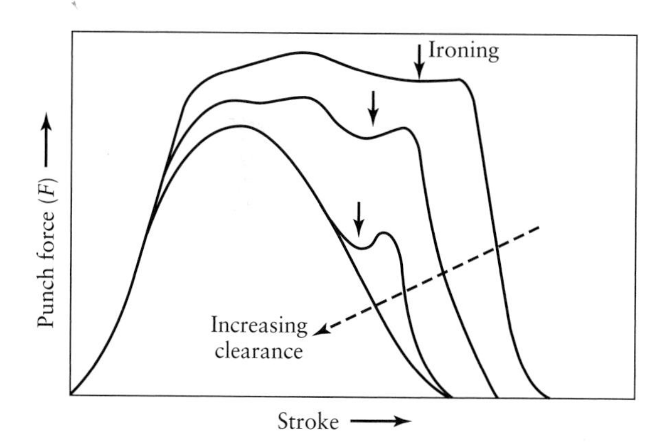

FIGURE 7.60 Schematic illustration of the variation of punch force with stroke in deep drawing. Note that ironing does not begin until after the punch has traveled a certain distance and the cup is formed partially. Arrows indicate the beginning of ironing.

Deep drawability is thus enhanced with a high $\bar{R}$ and a low ΔR. Generally, however, sheet metals with a high $\bar{R}$ also have a high ΔR. Attempts continue to be made to develop textures in sheet metals to improve their drawability. The controlling parameters in processing metals have been found to be alloying elements, additives, processing temperatures, annealing cycles after processing, thickness reduction in rolling, and cross (biaxial) rolling of plates when rolling them into sheets.

Maximum punch force. The *maximum punch force*, $F_{\max}$, supplies the work required in deep drawing. As in other deformation processes, the work consists of ideal work of deformation, redundant work, friction work, and, when present, the work required for ironing (Fig. 7.60). Because of the many variables involved in this operation, and because deep drawing is not a steady-state process, calculating the punch force is difficult. Various expressions have been developed; one simple and approximate formula for the punch force is

$$F_{\max} = \pi D_p t_o (\text{UTS})\left(\frac{D_o}{D_p} - 0.7\right). \quad (7.22)$$

Note that Eq. (7.22) does not include friction, the corner radii of the punch and die, or the blankholder force. However, this empirical equation makes approximate provisions for these factors.

The punch force is supported basically by the cup wall; if this force is excessive, **tearing** occurs, as shown in the lower part of Fig. 7.52a. Note that the cup was drawn to a considerable depth before failure occurred. This result can be expected by observing Fig. 7.60, where it can be seen that the punch force does not reach a maximum until after the punch has traveled a certain distance. The punch corner radius and die radius (if they are greater than 10 times the sheet thickness) do not significantly affect the maximum punch force.

EXAMPLE 7.6 Estimating the limiting drawing ratio

Estimate the limiting drawing ratio (LDR) that you would expect from a sheet metal that, when stretched by 23% in length, decreases in thickness by 10%.

Solution. From volume constancy of the test specimen, we have

$$w_o t_o l_o = w_f t_f l_f \quad \text{or} \quad \frac{w_f t_f l_f}{w_o t_o l_o} = 1.$$

From the information given, we obtain

$$\frac{\ell_f - \ell_o}{\ell_o} = 0.23 \qquad \text{or} \qquad \frac{\ell_f}{\ell_o} = 1.23$$

and

$$\frac{t_f - t_o}{t_o} = -0.10 \qquad \text{or} \qquad \frac{t_f}{t_o} = 0.90.$$

Hence,

$$\frac{w_f}{w_o} = 0.903.$$

From Eq. (7.18), we have

$$R = \frac{\ln\left(\dfrac{w_o}{w_f}\right)}{\ln\left(\dfrac{t_o}{t_f}\right)} = \frac{\ln 1.107}{\ln 1.111} = 0.965.$$

If the sheet has planar isotropy. then $R = \bar{R}$, and from Fig. 7.58 we estimate that

$$\text{LDR} = 2.4.$$

EXAMPLE 7.7 Theoretical limiting drawing ratio

Show that the theoretical limiting drawing ratio is 2.718.

Solution. In reviewing Fig. 7.50b and assuming that, ideally, there is no thickness change of the sheet during drawing, we note that the diametral change from a blank to a cup involves a true strain of

$$\varepsilon_{\text{max}} = \ln(\pi D_o/\pi D_p) = \ln(D_o/D_p).$$

The work required for plastic deformation is supplied by the radial stress on element A in Fig. 7.51a. Let's now refer to Fig. 6.62 for wire drawing and assume that the cross-section being drawn is thin and rectangular, with the thickness direction perpendicular to the page, and a thickness that ideally remains constant. From Eq. (6.61), we then note that the drawing stress (the radial stress in Fig. 7.51a), which we denote as σ_d, is given by

$$\sigma_d = Y\varepsilon_{\text{max}},$$

where Y is the yield stress of a perfectly plastic (ideal) material. Since, in the limit, the drawing stress is equal to the yield stress [see also the derivation of Eq. (6.72)], we have

$$Y = Y\varepsilon_{\text{max}},$$

or $\varepsilon_{\text{max}} = 1$. Consequently, $\ln(D_o/D_p) = 1$, and hence

$$D_o/D_p = 2.718.$$

EXAMPLE 7.8 Estimating cup diameter and earing

A steel sheet has R values of 0.9, 1.3, and 1.9 for the 0°, 45°, and 90° directions to rolling, respectively. If a round blank is 100 mm in diameter, estimate the smallest cup diameter to which it can be drawn. Will ears form?

Solution. Substituting the given values into Eq (7.20), we obtain

$$\bar{R} = \frac{0.9 + (2)(1.3) + 1.9}{4} = 1.35.$$

The limiting drawing ratio (LDR) is defined as the maximum ratio of the diameter of the blank to the diameter of the punch that can be drawn without failure, i.e., D_o/D_p. From Fig. 7.58, we estimate the LDR for this steel to be approximately 2.5.

To determine whether or not earing will occur, we substitute the R values into Eq (7.21). Thus, we obtain

$$\Delta R = \frac{0.9 - (2)(1.3) + 1.9}{2} = 0.1.$$

We know that ears will not form if $\Delta R = 0$, and since this is not the case here, ears will form in deep drawing this material.

7.12.2 Drawing practice

a) **Clearances and radii.** Generally, *clearances* are 7–14% greater than the original thickness of the sheet. The selection of clearance depends on the thickening of the cup wall (which is a function of the drawing ratio). As the clearance decreases, ironing increases. If the clearance is too small, the blank may simply be pierced and sheared by the punch. The *corner radii* of the punch and the die are important. If they are too small, they can cause fracture at the corners (Fig. 7.61). If they are too large, the unsupported area wrinkles. Wrinkling in this region (and from the flange area) causes a defect on the cup wall called **puckering**.

b) **Draw beads.** Draw beads (see Figs. 7.52b and 7.53) are useful in controlling the flow of the blank into the die cavity. They are essential in drawing box-shaped or nonsymmetric parts. Draw beads also help in reducing the blankholder forces required, because of the stiffness imparted to the flange by the bent regions along the beads. Proper design and location of draw beads requires considerable experience.

c) **Blankholder pressure.** Blankholder pressure is generally 0.7% to 1.0% of the sum of the yield and the ultimate tensile strength of the sheet metal. Too high a blankholder force increases the punch load (because of friction) and leads to tearing of the cup wall; on the other hand, if the force is too low, the flange wrinkles. Since the blankholder force controls the flow of the sheet metal within the die, new presses have been designed that are capable of applying a **variable blankholder force**; that is, the blankholder force can be programmed and varied with punch stroke. Additional features include the control of metal flow, which is achieved by using separate die cushions that allow the *local* blankholder force to be applied at the proper location on the sheet in order to improve drawability.

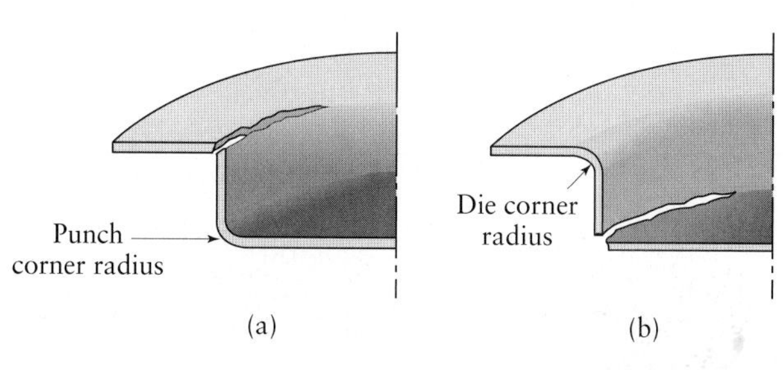

FIGURE 7.61 Effect of die and punch corner radii on fracture in deep drawing of a cylindrical cup. (a) Die corner radius too small. The die corner radius should generally be 5 to 10 times the sheet thickness. (b) Punch corner radius too small. Because friction between the cup and the punch aids in the drawing operation, excessive lubrication of the punch is detrimental to drawability.

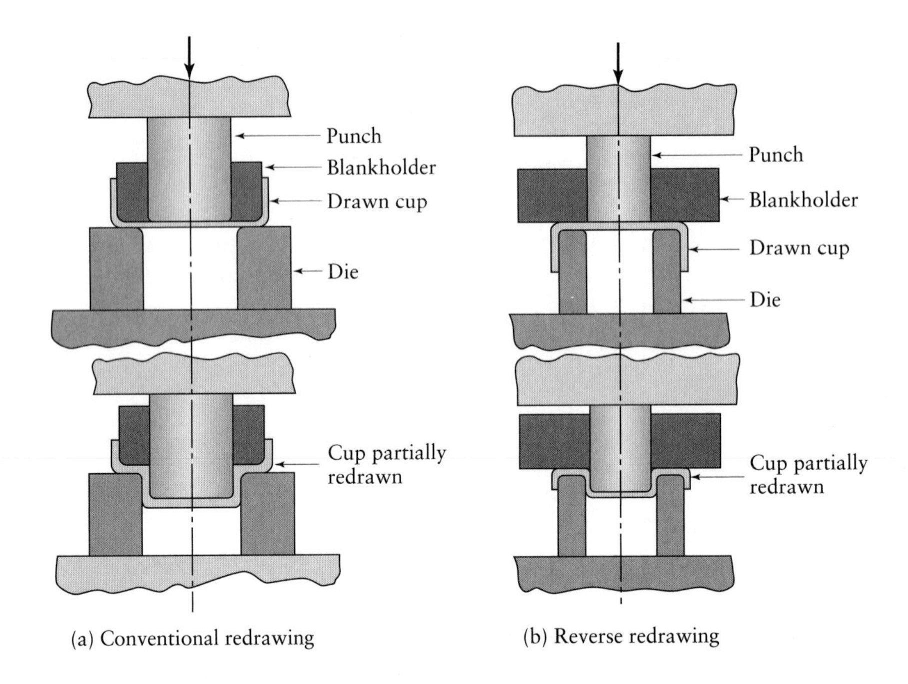

FIGURE 7.62 Reducing the diameter of drawn cups by redrawing operations: (a) conventional redrawing and (b) reverse redrawing. Small-diameter deep containers undergo many redrawing operations.

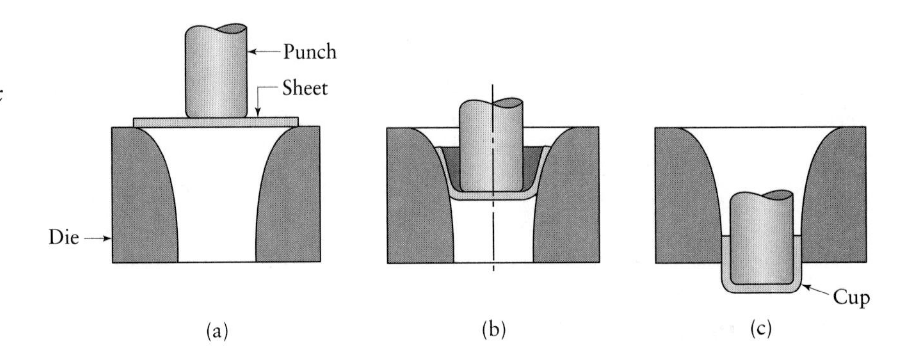

FIGURE 7.63 Deep drawing without a blankholder, using a *tractrix* die profile. The tractrix is a special curve, the construction for which can be found in texts on analytical geometry or in handbooks.

d) **Redrawing.** Containers or shells that are too difficult to draw in one operation are generally *redrawn*, as shown in Fig. 7.62a. In **reverse redrawing**, shown in Fig. 7.62b, the metal is subjected to bending in the direction opposite to its original bending configuration. This reversal in bending results in **strain softening** and is another example of the Bauschinger effect (see Section 2.3.2); this operation requires lower forces than direct redrawing, and the material behaves in a more ductile manner.

e) **Drawing without a blankholder.** Deep drawing may be carried out without a blankholder, provided that the sheet metal is sufficiently thick to prevent wrinkling. The dies are specially contoured for this operation; one example is shown in Fig. 7.63. An approximate limit for drawing without a blankholder is given by

$$D_o - D_p < 5t_o. \tag{7.23}$$

It is apparent from Eq (7.23) that the process can be used with thin materials for shallow draws. Although the punch stroke is longer than ordinary drawing, a major advantage of this process is the reduced cost of tooling and equipment.

f) **Tooling and equipment.** The most commonly used tool materials for deep drawing are tool steels and alloyed cast iron; other materials, including carbides and plastics, may also be used, depending on the particular application. (See Table 3.5.) A double-action mechanical press is generally used for deep drawing, although hydraulic presses are also used. The double-action press controls the punch and blankholder independently and forms the part at a constant speed. Punch speeds generally range between 0.1 and 0.3 m/s (20 and 60 ft/min). Speed is generally not important in drawability, although lower speeds are used for high-strength metals.

g) **Lubrication.** Lubrication in deep drawing is important in lowering forces, increasing drawability, reducing tooling wear, and reducing part defects. In general, lubrication of the punch should be minimized, as friction between the punch and the cup improves drawability. For general applications, commonly used lubricants include mineral oils, soap solutions, and heavy-duty emulsions. For more difficult applications, coatings, wax, and solid lubricants are used.

7.13 Formability of Sheet Metals

The formability of sheet metals has been of great and continued interest, because of its technological and economic significance. *Sheet-metal formability* is generally defined as the ability of a sheet to undergo the desired shape changes without failure such as necking, tearing, or splitting. Three factors have a major influence on formability: (a) properties of the sheet metal; (b) lubrication at various interfaces between the sheet, the dies, and the tooling; (c) characteristics of the equipment, tools, dies used in forming the sheet. Several techniques have been developed to test the formability of sheet metals. New developments include the ability to predict formability by modeling the particular forming process, using several inputs, including the crystallographic texture of the sheet metal.

7.13.1 Testing for formability

There are several tests used to determine the formability of sheet metal:

a) **Tension tests.** The *tension test* (see Section 2.2) is the most basic and common test used to evaluate formability. This test determines important properties of the sheet metal, such as total elongation of the sheet specimen at fracture, the strain hardening exponent, n, the planar anisotropy, ΔR, and the normal anisotropy, R.

b) **Cupping tests.** Because, as we have seen, sheet forming is basically a biaxial stretching process, the earliest tests developed to determine or to predict formability were *cupping tests*, such as the **Erichsen** and **Olsen** tests (*stretching*) and the **Swift** and **Fukui** tests (*drawing*). In the Erichsen test, a sheet-metal specimen is clamped over a circular flat die with a load of 1000 kg. A 20-mm-diameter steel ball is then hydraulically pushed into the sheet metal until a crack appears on the stretched specimen or until the punch force reaches a maximum. The distance D, in mm, is the Erichsen number. The greater the value of D, the greater is the formability of the sheet. Cupping tests measure the capability of the material to be stretched before fracturing and are relatively easy to perform; however, because the stretching under the ball is axisymmetric, they do not at all simulate the exact conditions of actual forming operations.

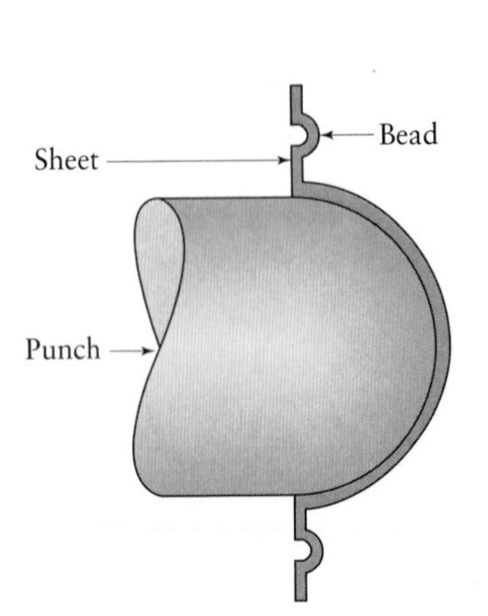

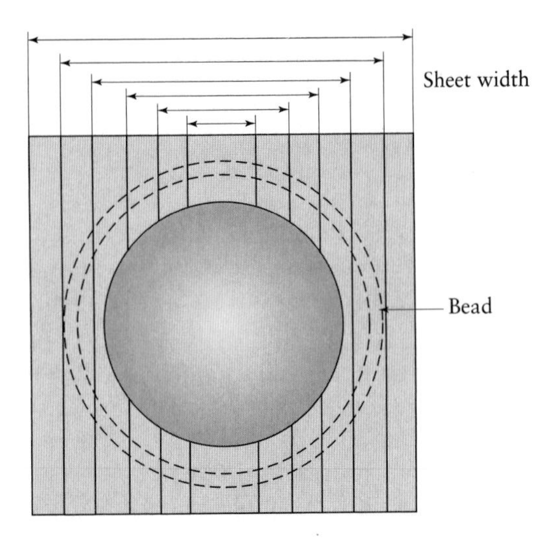

FIGURE 7.64 Schematic illustration of the punch–stretch test on sheet specimens with different widths, clamped at the edges. The narrower the specimen, the more uniaxial is the stretching. A large square specimen stretches biaxially under the hemispherical punch. (See also Fig. 7.65.)

c) **Bulge test.** Equal (balanced) biaxial stretching of sheet metals is also performed in the *bulge test*, which has been used extensively to simulate sheet-forming operations. In this test, a circular blank is clamped at its periphery and is bulged by *hydraulic* pressure, thus replacing the punch. The process is one of pure stretch forming, and no friction is involved, as would be in using a punch. Bulge tests can be used to obtain effective-stress–effective-strain curves for biaxial loading under frictionless conditions. The biaxial bulge limit (depth) is also a sensitive measure of sheet cleanliness (quality).

d) **Forming-limit diagrams.** An important development in testing formability of sheet metals is the construction of *forming-limit diagrams* (FLDs). In the punch–stretch test, the sheet blank is marked with a grid pattern of circles, typically 2.5 to 5 mm (0.1 to 0.2 in.) in diameter, or similar patterns, using chemical-etching or photoprinting techniques. The blank is then stretched over a punch (Fig. 7.64), and the deformation of the circles is observed and measured in regions where failure (necking and tearing) occurs. Note that the sheet is fixed by a draw bead, which prevents the sheet from being drawn into the die. For improved accuracy of measurement, the circles on the sheet are made as small as possible and the grid lines (if used) as narrow as possible. Because lubricants can significantly influence the test results, lubrication conditions should be stated in reporting the test results. In this manner, different lubricants can be evaluated for the same sheet metal, or different metals can be evaluated using the same lubricant. This test can also be run with an unlubricated punch.

In order to vary the deformation *strain path*, the specimens are prepared with varying widths, as shown in Fig. 7.64b. Thus, the center of the square specimen produces **equal (balanced) biaxial stretching** under the punch, whereas a specimen with a small width approaches a state of **uniaxial stretching** (simple tension). After a series of such tests is performed on a particular type of sheet metal (Fig. 7.65), the boundaries between *safe* and *failed* regions are plotted on the forming-limit diagram (Fig. 7.66a). Figure 7.67 shows another example of a failed part with round and square grid patterns. Note in Fig. 7.66a the various strain paths that are straight lines: (a) The 45° line on the right represents **equal biaxial strain**; (b) the vertical line at the center of the diagram represents **plane strain**, because the minor planar strain is zero (although the sheet does become thinner); (c) the **simple tension** line on the left has a 2 : 1 slope, because the Poisson's ratio in the plastic range is 0.5 (note that the minor strain is one half of the major strain); (d) the **pure-shear** line has a negative 45° slope, because of the nature of deformation, as can also be seen in Figure 2.21.

FIGURE 7.65 Bulge-test results on steel sheets of various widths. The first specimen (farthest left) stretched farther before cracking than the last specimen. From left to right, the state of stress changes from almost uniaxial to biaxial stretching. *Source*: Courtesy of Ispat Inland Inc.

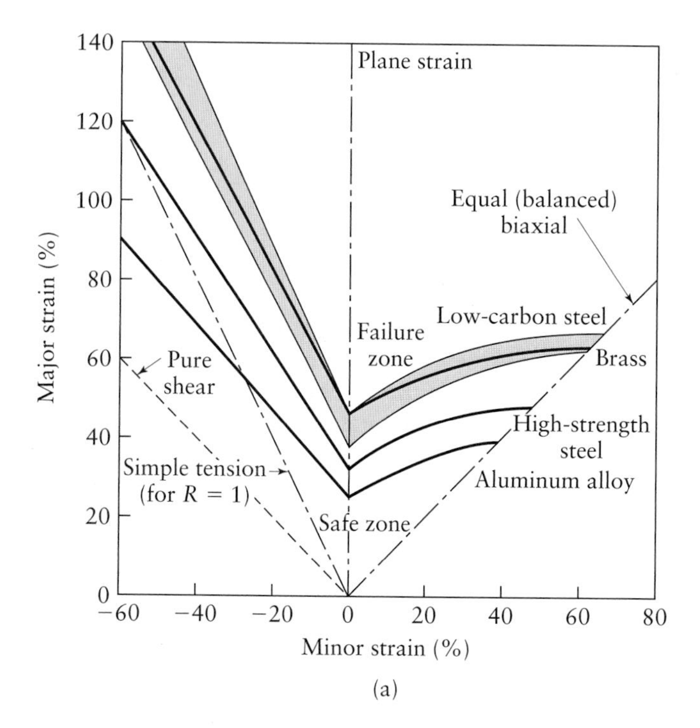

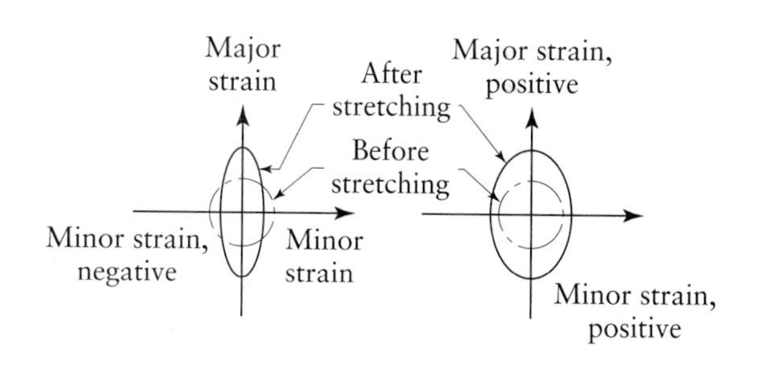

FIGURE 7.66 (a) Forming-limit diagram (FLD) for various sheet metals. The major strain is always positive. The region above the curves is the failure zone; hence, the state of strain in forming must be such that it falls below the curve for a particular material; R is the normal anisotropy. (b) Note the definition of positive and negative minor strains. If the area of the deformed circle is larger than the area of the original circle, the sheet is thinner than the original, because the volume remains constant during plastic deformation. *Source*: After S. S. Hecker and A. K. Ghosh.

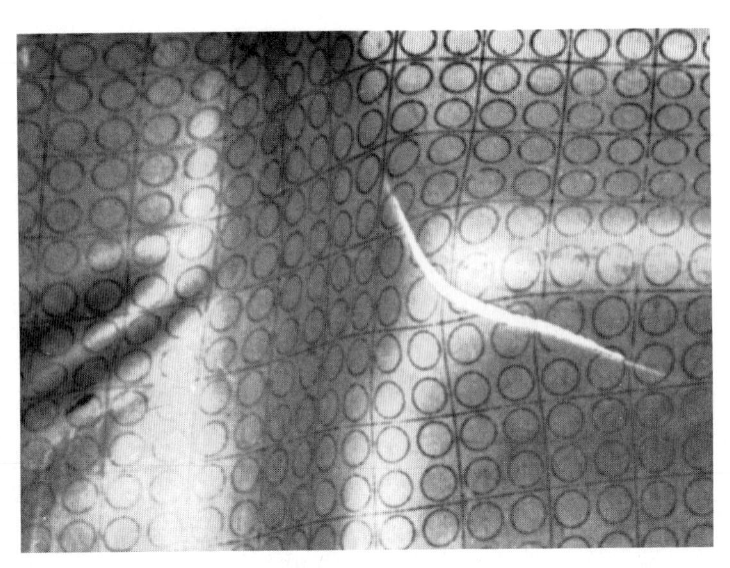

FIGURE 7.67 An example of the use of grid marks (circular and square) to determine the magnitude and direction of surface strains in sheet-metal forming. Note that the crack (tear) is generally perpendicular to the major (positive) strain.
Source: After S. P. Keeler.

In the forming-limit diagram, the **major strain** and the **minor strain** are obtained as follows: Note in Fig. 7.66b that after stretching, the original circle has deformed into an ellipse. The major axis of the ellipse represents the major direction and magnitude of stretching; the major strain plotted in Fig. 7.66a is the *engineering strain* (in percentage terms) in this direction. Likewise, the minor axis of the ellipse represents the magnitude of *stretching* (positive minor strain)or *shrinking* (negative minor strain) in the transverse direction. (The major strain is always positive, because forming sheet metal takes place by stretching in at least one direction.)

As an example, if we draw a circle in the center and on the surface of a sheet-metal tensile-test specimen and then stretch it plastically, the specimen becomes narrower; thus, the minor strain is negative. Because of volume constancy in the plastic range [see Eq. (2.48)], and assuming that the normal anisotropy is $R = 1$ (that is, the width and thickness strains are equal), we have $\varepsilon_w = -0.5\,\varepsilon_\ell$. As an experimental model, this phenomenon can easily be demonstrated by placing a circle on a wide rubber band (even though the deformation will only be in the elastic range) and stretching it. However, if we place a circle on a spherical rubber balloon and inflate the balloon, the minor strain is positive and equal in magnitude to the major strain; the circle simply becomes a larger circle. By observing the difference in surface area between the original circle and the ellipse, we can also determine whether the *thickness* of the sheet has changed. If the area of the ellipse is larger than that of the original circle, the sheet has become thinner (because volume remains constant in plastic deformation).

From Fig. 7.66a, it can be observed that, as expected, different materials have different forming-limit diagrams; note that the higher the curve, the better is the formability of the material. Also, it is important to note that for the same minor strain, say 20%, a compressive (negative) minor strain is associated with a higher major strain before failure than is a tensile (positive) minor strain. In other words, it is desirable for the minor strain to be *negative*, that is, to allow shrinking in the minor direction during sheet-forming operations. Special tooling has been designed for forming sheet metals and tubing that take advantage of the beneficial effect of negative minor strains on extending formability. (See, for example, Fig. 7.30.) The possible effect of the *rate of deformation* on forming-limit diagrams should also be assessed for each material. (See also Section 2.2.7.)

The effect of sheet-metal *thickness* on forming-limit diagrams is to *raise* the curves in Fig. 7.66a. Hence, the thicker the sheet, the higher its curve, and thus the more formable the sheet is. In actual forming operations, however, a thick blank may not bend easily around small radii and may develop cracks.

Friction at the punch–metal interface can also be significant in test results. With well-lubricated interfaces, the strains are more uniformly distributed over the punch. Also, depending on the notch sensitivity of the sheet metal, surface scratches, deep gouges, and blemishes can reduce formability and cause premature tearing and failure during testing and in actual forming operations. With effective lubrication, the coefficient of friction in sheet-metal forming generally ranges from about 0.05 to 0.1 for cold forming, and from 0.1 to 0.2 for elevated-temperature forming. (See also Table 4.1.)

e) **Limiting dome-height test.** Recall that in the punch–stretch test (illustrated in Fig. 7.64), the major and minor strains were measured and plotted on the forming-limit diagram. In the *limiting dome-height* (LDH) test, performed with similarly prepared specimens, the *height* of the dome at failure of the sheet, or when the punch force reaches a maximum, is measured. Because the specimens are clamped, the LDH test indicates the capability of the material to stretch without failure. It has been shown that high LDH values are related to sheet-metal properties such as high n and m values, as well as high total elongation (in percentage terms) of the sheet metal.

EXAMPLE 7.9 Estimating diameter of expansion

A thin-walled spherical shell made of the aluminum alloy whose forming-limit diagram is shown in Fig. 7.66a is being expanded by internal pressure. If the original diameter is 200 mm, what is the maximum diameter to which the shell can safely be expanded?

Solution. Because the material is being stretched in a state of equal (balanced) biaxial tension, we find from Fig. 7.66a that the maximum allowable engineering strain is about 40%. Thus,

$$e = \frac{\pi D_f - \pi D_o}{\pi D_o} = \frac{D_f - 200}{200} = 0.40$$

Hence,

$$D_f = 280 \text{ mm}.$$

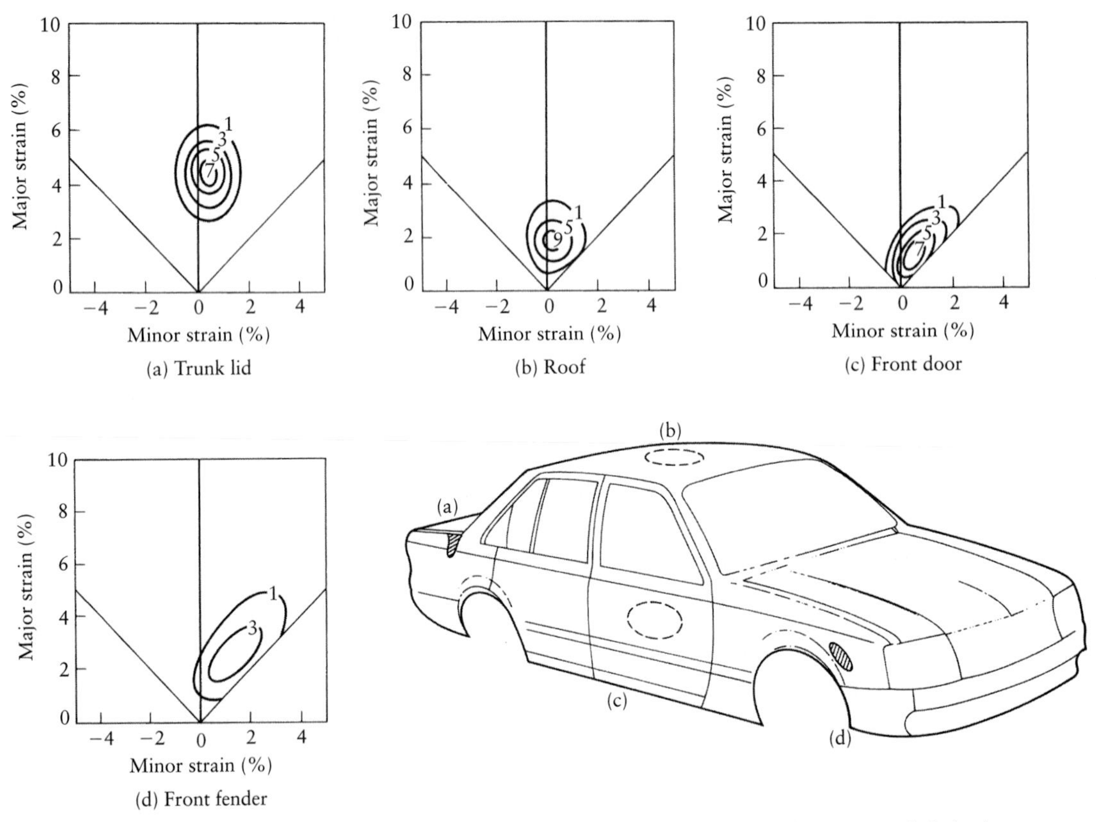

FIGURE 7.68 Major and minor strains in various regions of an automobile body.

EXAMPLE 7.10 Strains in the sheet-metal body of an automobile

Computer programs have been developed that compute the major and minor strains and their orientations from the measured distortions of grid patterns placed on the surface of sheet metal. Figure 7.68 shows one such application for the various panels of an automobile body. Note that the trunk lid and the roof are subjected mainly to plane strain (see the vertical line in Fig. 7.66a), whereas the front door and front fender are subjected to biaxial strains. The numbers in the strain paths indicate the frequency of occurrence. *Source*: After T. J. Nihill and W. R. Thorpe.

7.13.2 Dent resistance of sheet-metal parts

In certain applications involving sheet-metal parts, such as automotive body panels, appliances, and office furniture, an important consideration is the dent resistance of the sheet-metal panel. A *dent* is a small, but permanent, biaxial deformation. The factors significant in dent resistance are the yield stress, Y, the thickness, t, and the shape of the panel. *Dent resistance* is expressed by a combination of material and geometrical parameters as

$$\text{Dent resistance} \propto \frac{Y^2 t^4}{S}, \tag{7.24}$$

where S is the panel stiffness, which, in turn, is defined as

$$S = (E)(t^a)(\text{shape}), \tag{7.25}$$

where the value of a ranges from 1 to 2 for most panels. As for shape, the smaller the curvature (and hence, the flatter the panel), the greater the dent resistance is, because of the sheet's flexibility. Hence, dent resistance (a) increases with increasing strength and thickness of the sheet, (b) decreases with increasing elastic modulus and stiffness, and (c) decreases with decreasing curvature. Note that for $a = 2$, dent resistance is proportional to t^2. Dents are usually caused by *dynamic* forces, such as forces developed by falling objects or other objects that hit the surface of the panel from different directions. In typical automotive panels, for instance, impact velocities range up to 45 m/s (150 ft/s). Thus, the *dynamic yield stress* (that is, at high strain rates) rather than the static yield stress is the significant strength parameter. However, denting under quasistatic forces could also be important.

For materials in which yield stress increases with strain rate, denting requires higher energy levels than under static conditions. Furthermore, dynamic forces tend to cause more *localized* dents than do static forces, which tend to spread the dented area. Because a portion of the energy goes into elastic deformation, the modulus of resilience of the sheet metal [see Eq. (2.5)] is an additional factor to be considered.

7.13.3 Modeling of sheet-forming processes

In Section 6.2.2, we outlined the techniques for studying bulk-deformation processes and described *modeling* of impression-die forging, using the finite-element method (see Fig. 6.11). Such mathematical modeling is now also being applied to sheet-forming processes. The ultimate goal of such simulation techniques is rapid analysis of stresses, strains, flow patterns, wrinkling, and springback as functions of parameters such as material characteristics, friction, anisotropy, deformation speed, and temperature.

Through such interactive analyses, we can, for example, determine the optimum tool and die geometry to make a certain part, thus reducing or eliminating costly die tryouts. In addition, these simulation techniques can be used to determine the size and shape of blanks (including intermediate shapes), press characteristics, and process parameters to optimize the forming operation. The application of computer modeling, although requiring powerful computers and extensive software, has already proven to be a cost-effective technique in sheet-metal forming operations, particularly in the automotive industry.

7.14 Equipment for Sheet-Metal Forming

The basic equipment for most pressworking operations consists of mechanical, hydraulic, pneumatic, or pneumatic–hydraulic presses. (See also Section 6.2.8 and Fig. 6.28.) The traditional *C-frame* press structure, with an open front, has been widely used for ease of tool and workpiece accessibility. However, the *box-type* (O-type) pillar and double-column frame structures are stiffer, and advances in automation, industrial robots, and computer controls have made accessibility less important.

Press selection includes factors such as (1) the type of operation, (2) the size and shape of the workpiece, (3) the length of stroke of the slide(s), (4) the number of strokes per minute, (5) the press speed, (6) the *shut height* (i.e., the distance from the top of the press bed to the bottom of the slide, with the stroke down), (7) the number of slides (e.g., double action and triple action), (8) the press capacity and tonnage, (9) the type of controls, and (10) the safety features.

In addition, because changing dies in presses can involve a significant amount of effort and time, rapid die-changing systems have been developed. Called **single-minute exchange of dies** (SMED), such systems use automated hydraulic or pneumatic equipment to reduce die-changing times from hours down to as little as 10 minutes.

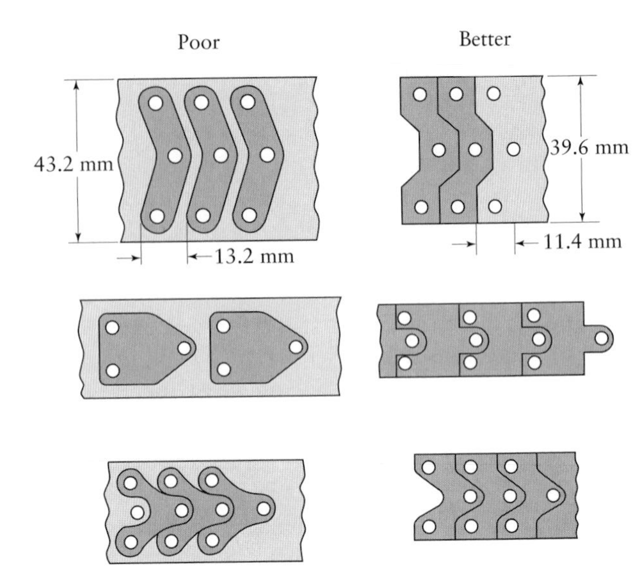

FIGURE 7.69 Efficient nesting of parts for optimum material utilization in blanking. *Source*: Society of Manufacturing Engineers.

7.15 Design Considerations

As with most other processes described throughout this text, certain design guidelines and practices have evolved with time. Careful design, using the best established design practices, computational tools, and manufacturing techniques, is the optimum approach for achieving high-quality designs and realizing cost savings. The following guidelines identify the most significant design issues of sheet-metal forming operations:

a) **Blank design.** Material scrap is the primary concern in blanking operations. Poorly designed parts will not nest, and there will be considerable amounts of scrap between successive blanking operations (Fig. 7.69). Some restrictions on the blank shape are made by the design application, but, when possible, blanks should be designed to reduce scrap to a minimum.

b) **Bending.** In bending operations, the main concerns are fracture of the material, wrinkling, and inability to form the bend. As shown in Fig. 7.70, bending a sheet-metal part with a flange that is to be bent will force the flange to undergo compression, which can cause buckling. This problem can be controlled by using a relief notch to limit the stresses from bending, as shown in Fig. 7.70. Right-angle bends have similar difficulties, and relief notches can be used to avoid tearing (Fig. 7.71).

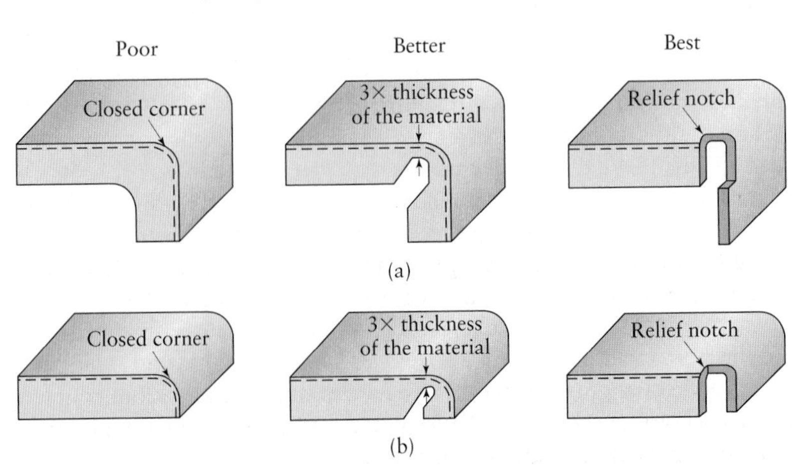

FIGURE 7.70 Control of tearing and buckling of a flange in a right-angle bend. *Source*: Society of Manufacturing Engineers.

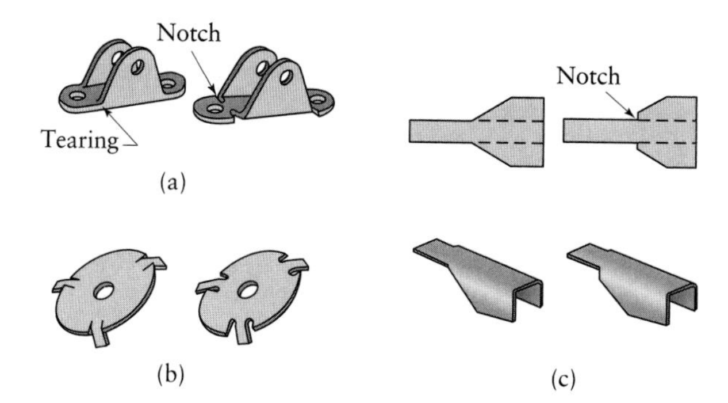

FIGURE 7.71 Application of notches to avoid tearing and wrinkling in right-angle bending operations. *Source*: Society of Manufacturing Engineers.

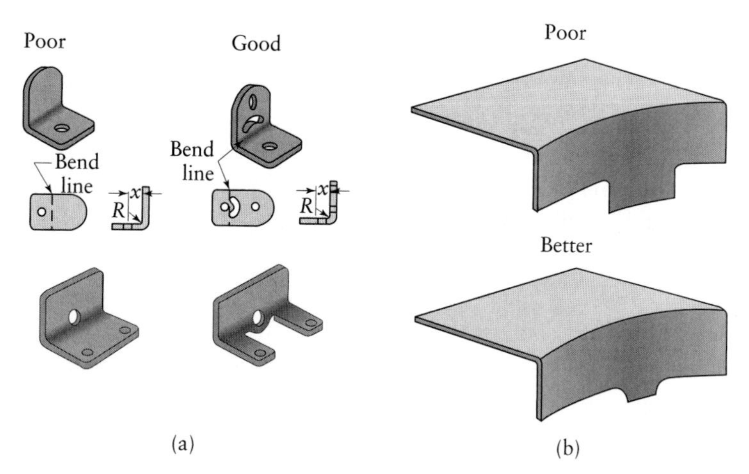

FIGURE 7.72 Stress concentrations near bends. (a) Use of a crescent or ear for a hole near a bend. (b) Reduction of the severity of a tab in a flange. *Source*: Society of Manufacturing Engineers.

The bend radius should be recognized as a highly stressed area, and all stress concentrations should be removed from the bend-radius location. An example is the placement of holes near bends; it is normally advantageous to remove the hole from the stress concentration, but when this is not possible, a crescent slot or ear can be used (Fig. 7.72a). Similarly, when bending flanges, tabs and notches should be avoided, since these stress concentrations will greatly reduce the formability. When tabs are necessary, large radii should be used to reduce the stress concentration (Fig. 7.72b).

When both bending and the production of notches are to be implemented, it is important to orient the notches properly with respect to the grain direction. As shown in Fig. 7.17, bends should ideally be perpendicular to the rolling direction (or oblique, if perpendicularity is not possible) in order to avoid cracking. Bending of sharp radii can be accomplished through scoring or embossing (Fig. 7.73), but it should be recognized that this procedure can result in fracture. Burrs should not be present in a bend allowance, since they are brittle (because of strain hardening) and can fracture, leading to a stress concentration that propagates the crack into the rest of the sheet.

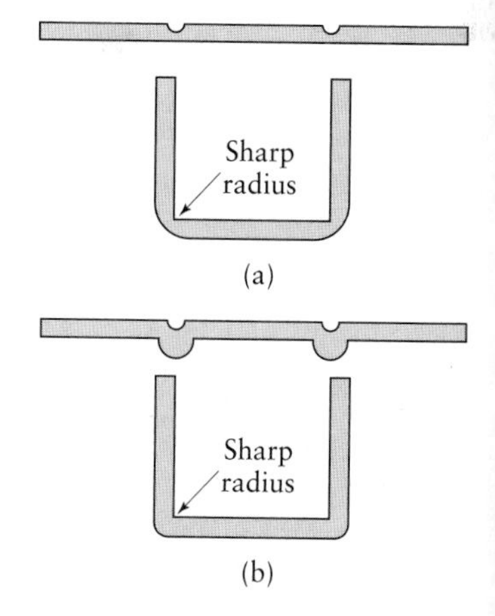

FIGURE 7.73 Application of (a) scoring or (b) embossing to obtain a sharp inner radius in bending. However, unless properly designed, these features can lead to fracture. *Source*: Society of Manufacturing Engineers.

c) **Stamping and progressive-die operations.** In progressive dies, the cost of the tooling and the number of stations are determined by the number of features and spacing of the features on a part. It is therefore advantageous to hold the number of features to a minimum in order to minimize tooling costs. Note that closely spaced features may provide insufficient clearance for punches and would thereby require two punches. Narrow cuts and protrusions are also problematic for forming with a single punch and die.

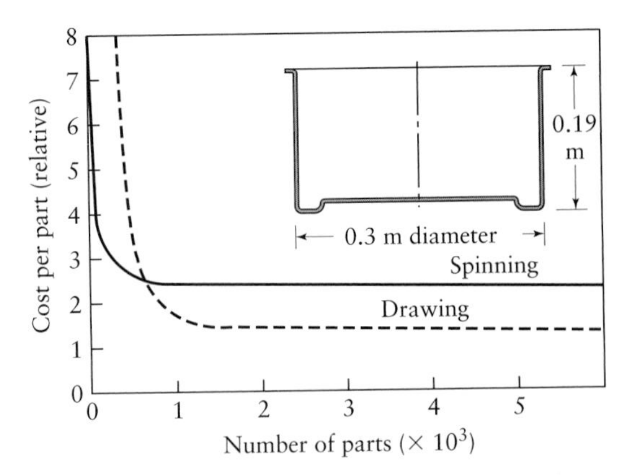

FIGURE 7.74 Cost comparison for manufacturing a round sheet-metal container by conventional spinning and deep drawing. Note that for small quantities, spinning is more economical.

d) **Deep drawing.** After a deep-drawing operation, a cup will invariably spring back towards its original shape. For this reason, designs that use a vertical wall in a deep-drawn cup are difficult to form. Relief angles of at least 3° on each wall are easier to produce. Cups with sharp internal radii are difficult to produce, and deep cups often require additional ironing operations.

7.16 Economics of Sheet-Metal Forming

Sheet-metal forming involves economic considerations similar to those for other metal working processes described in Chapter 6. Sheet-forming operations compete with each other, as well as with other processes, more than other processes compete with each other. As noted earlier, sheet-forming operations are versatile, and a number of different processes can be used to produce the same part. For example, a cup-shaped part can be formed by deep drawing, spinning, rubber forming, or explosive forming. Similarly, a cup can be formed by impact extrusion or casting, or by fabrication from different pieces.

As described in Section 16.5.5, almost all manufacturing processes produce some scrap, which can range up to 60% of the original material for machining operations and up to 25% for hot forging; in contrast, sheet-forming operations produce scrap on the order of 10% to 25% of the original material. The scrap should be recycled, although scrap with residual lubricant layers is less desirable than dry scrap.

The part shown in Fig. 7.74 can be made either by deep drawing or by conventional spinning; however, the die costs for the two processes are significantly different. Deep-drawing dies have many components and cost much more than the relatively simple mandrels and tools employed in spinning. Consequently, the die cost per part in drawing will be high if few parts are needed. On the other hand, this part can also be made by deep drawing in a much shorter time (seconds) than by spinning (minutes), even if the latter operation is highly automated. Furthermore, spinning requires more skilled labor. Considering all these factors, it is found that the *break-even point* for this particular part is at about 700 parts. Hence, deep drawing is more economical for quantities greater than that number.

CASE STUDY | Can Manufacturing

Can manufacturing is a major and competitive industry worldwide, with approximately 100 billion beverage cans and 30 billion food cans produced each year in the United States alone. These containers are strong and lightweight, typically weighing less than 0.5 oz.; they can reliably sustain 90 psi of internal pressure without leakage. In addition, there are stringent requirements on their surface finish, with brightly decorated and shiny cans preferred over dull, drab containers. Metal cans possessing these features are quite inexpensive: Can makers charge approximately \$40 per 1000 cans, or around 4 cents per can.

The most common food and beverage can designs are two-piece and three-piece cans. These cans can be produced in a number of ways. Two-piece cans consist of a body and a lid; because the body has been *drawn and ironed*, the industry refers to this style as D&I (after drawn and ironed; Section 7.12) cans. Three-piece cans are produced by attaching a lid and a bottom to a cylindrical sheet-metal body. This case study describes the production of D&I cans, although many of the manufacturing considerations apply equally to three-piece cans.

Drawn and ironed can bodies are produced from a number of alloys, but the two most common are 3004-H19 aluminum and electrolytic tin-plated (Section 4.5.1) ASTM A623 steel. The can lids are produced from 5182 aluminum, sometimes in the H19 condition and sometimes precoated in the H48 condition. Aluminum lids are used for both aluminum and steel cans. The reason can be seen by inspecting Fig. 7.75. The lids have an extremely demanding set of design requirements: Not only must the can top be easily scored, but an integral rivet is stretch formed and headed in the lid to hold the tab in place. The 5182 aluminum alloy has the unique characteristic of being sufficiently formable to make the integral rivet, as well as being able to be scored. Basically, the lids are stamped from 5182 aluminum sheet, then the pop top is scored (to provide a controlled fracture path when tearing it), and, finally, a plastic seal is placed around the periphery of the lid. The polymer seals the can's contents after the lid is seamed to the can body, as described below. Note that the pop top opens by means of a Mode III fracture (i.e., tearing), as shown in Fig. 3.30.

The traditional method of manufacturing the can bodies consists of several processes, as shown in the sequence of operations depicted in Fig. 7.76: (a) First, 5.5-in.-diameter *blanks* are produced from rolled sheet stock. (b) Then the blanks are *deep drawn* to a diameter of around 3.5 in. (c) Next, the cans are *redrawn* to the final diameter of around 2.6 in. (See also Fig. 7.62.) (d) After that, the redrawn cans are *ironed* through two or three ironing rings in one pass. (See also Fig. 7.54.) (e) Finally, the can bottom is *domed*. The first two operations are carried out in a short-stroke mechanical press (Fig. 6.28); the last three operations are performed in a special type of press that typically produces cans at speeds of over 400 strokes

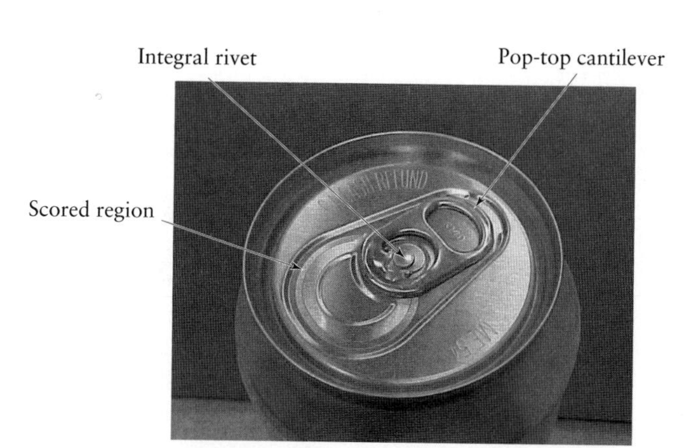

FIGURE 7.75 The lid of an aluminum beverage container.

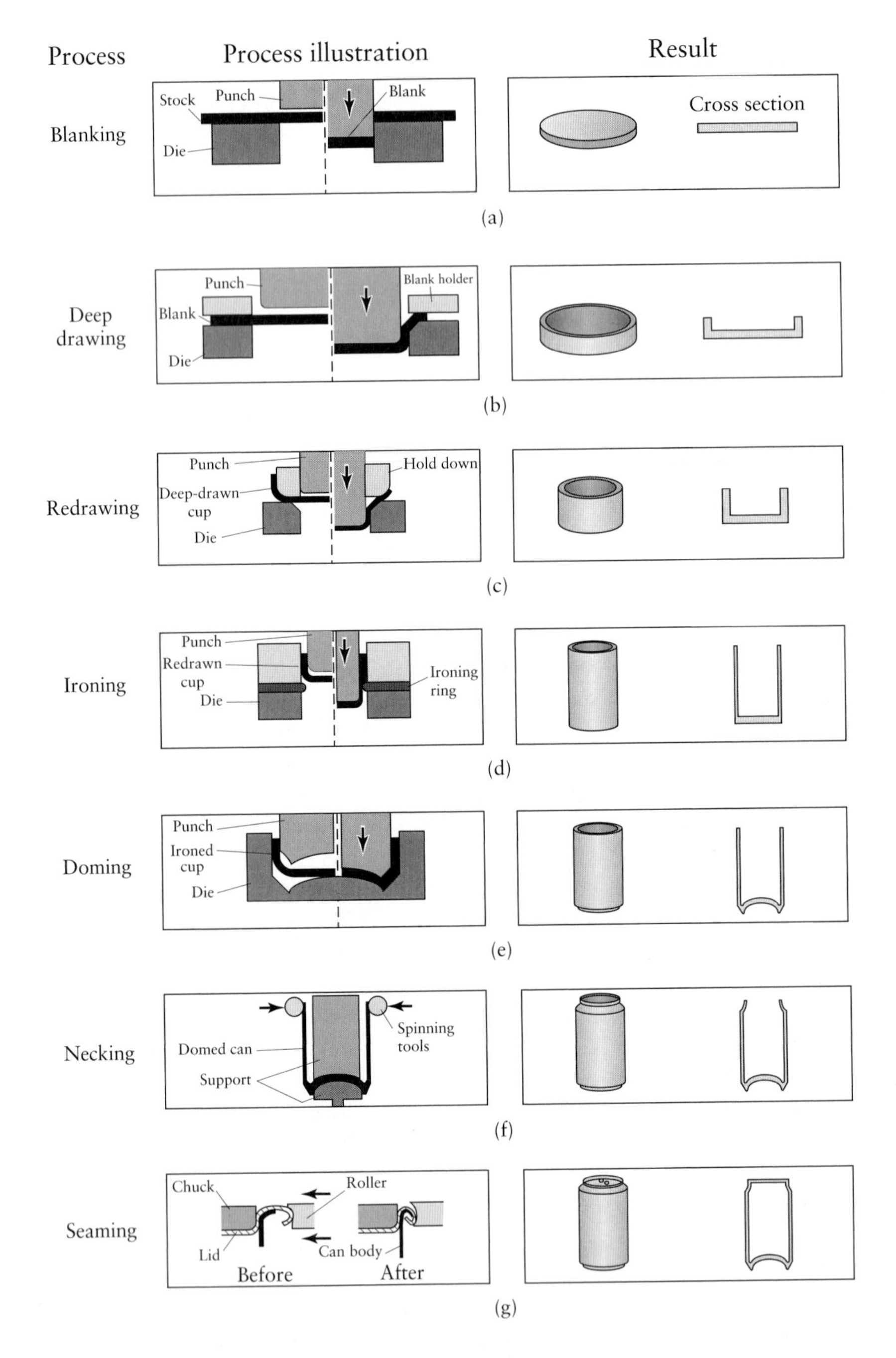

FIGURE 7.76 The metal-forming processes used to manufacture two-piece beverage cans.

per minute. Effective cooling and lubrication is necessary, and emulsion lubricants (Section 4.4.4) are widely used for this purpose.

Following this series of operations, a number of additional processes take place. *Necking* of the can body is performed either through spinning (Fig. 7.76f) or by die necking (an operation similar to that shown in Fig. 6.72a, in which the top of the can is pushed into a die). Then the body is spin-flanged. The reason for necking the can top is that the 5182 aluminum used for the lid is relatively expensive. By tapering the top of the can, a smaller volume of metal is required, thereby reducing the material cost. The cost of a can is often calculated to millionths of a dollar; hence, any design feature that reduces its cost will be exploit-

FIGURE 7.77 Aluminum two-piece beverage cans. Note the fine surface finish.

ed by this competitive industry. For example, reducing the blank diameter for the lid by 10% reduces the surface area, and hence the amount of metal, by $(1 - 0.9^2)/1 = 19\%$.

The interior surface of the can is not suitable for contact with food after forming, because the metal will rapidly adulterate its contents, either from metal-particulate contamination (resulting from the forming operations) or by metal–food interaction. To avoid contamination, a polymer solution is applied to the can's interior. However, lubricants used in the metal-forming operations must first be removed, because they can prevent adhesion of the coating to the metal. Cans are, therefore, subjected to a solvent wash operation (Section 4.5.2), after which the inner surfaces are coated with a thermoplastic or a thermosetting resin that is safe upon contact with food. Coating is performed by dissolving the polymer in a carrier fluid and applying it to the interior of the can.

After this application, an oven-curing operation allows the resin to bond to the interior surface, while the toxic carrier fluid boils off and is released as a volatile organic compound (VOC). In order to make this harmful by-product environmentally safe, an afterburner is used to reduce it to water vapor and carbon dioxide. Recent developments in coatings include the use of water-based carriers for the polymer coating and using prelaminated steels in the canmaking operation; these approaches eliminate the environmentally harmful VOCs without significantly affecting the cost of the can.

Decoration of the cans is a critical aspect of the overall manufacturing process. Aluminum cans are produced with an extremely smooth surface (Fig. 7.77) so that they appear shiny and vibrant. Lacquers and inks are printed directly onto the aluminum body, resulting in an aesthetically pleasing and attractive product. The final operation on a can is *seaming* the top over the body (Fig. 7.76g; see also Fig. 12.50). Note that, unlike aluminum cans, steel cans either have a base coating applied prior to the final decoration or have a paper or plastic label attached to the can.

Source: Courtesy of J. E. Wang, Texas A&M University.

SUMMARY

- Sheet-metal forming processes involve workpieces that have a high ratio of surface area to thickness. In these processes, tensile stresses are usually applied in the plane of the sheet through various punches and dies. The material is generally prevented from being reduced in thickness in order to avoid necking and subsequent tearing and fracture. Material properties and behavior important to these processes include elongation, yield-point elongation, anisotropy, and grain size. (Sections 7.1, 7.2)

- Blanks for subsequent forming operations are generally sheared from large sheets, using various processes. The quality of the sheared edge is important in subsequent processing, and important process parameters include the clearance between the dies, sharpness, and lubrication. (Section 7.3)
- Bending of sheet, plate, and tubing is a commonly performed operation, and important parameters include the minimum bend radius (without causing cracking in the bend area) and springback. Several techniques are available for bending and for minimizing springback. (Section 7.4)
- Discrete sheet-metal bodies can be made by stretch forming, bulging, rubber forming, and hydroforming, whereby the workpiece is shaped by stretching in different directions. In rubber forming and hydroforming, in contrast to the other two methods, one of the dies can be made of a flexible material such as rubber or polyurethane. (Sections 7.5, 7.6, 7.7)
- Spinning involves the forming of axisymmetric parts over a rotating mandrel, using rigid tools or rollers, and comprises conventional, shear, and tube-spinning processes. Various external and internal profiles can be produced. (Section 7.8)
- High-energy-rate forming processes use chemical, electrical, or magnetic means of energy. They include explosive forming, usually of large sheets; electrohydraulic forming; and magnetic-pulse forming of relatively small parts. (Section 7.9)
- Superplastic-forming methods utilize the superplastic behavior of materials and are capable of producing complex shapes with high strength-to-weight ratios. They are particularly effective when combined with diffusion bonding of two or more sheets prior to forming. (Section 7.10)
- Various other sheet-forming processes include peen forming, based on the principle of shot peening; use of gas mixtures or liquified gases; laser forming and laser-assisted forming; microforming; and manufacturing of honeycomb structures. (Section 7.11)
- Deep drawing, which uses a punch and a die, is one of the most important sheet-forming processes, because it can produce container-shaped parts with high depth-to-diameter ratios. Deep drawability has been found to depend on the normal anisotropy of the sheet metal. Earing also occurs in deep drawing, because of the planar anisotropy of the metal. Buckling and wrinkling in drawing can be significant problems; techniques have been developed to eliminate or minimize these problems. (Section 7.12)
- Tests have been developed to determine the formability of sheet metals in forming processes. The most comprehensive test is the punch–stretch (bulge) test, which enables the construction of a forming-limit diagram, indicating safe and failure regions as a function of the planar strains involved. Each type of metal, its condition and its thickness, has its own curve. Limiting dome-height tests have also been developed for determining formability. (Section 7.13)
- Several types of equipment are available for sheet-metal forming processes. Selection among these pieces of equipment depends on various factors concerning workpiece size and shape and equipment characteristics and their control features. Rapid die-changing techniques are important in the productivity and overall economics of sheet-metal forming. (Section 7.14)
- Design considerations in sheet-metal forming are based on factors such as shape of the blank, the type of operation to be performed, and the nature of the deformation. Several design rules and practices have been established to eliminate or minimize production problems. (Section 7.15)

- As in all manufacturing processes, economic considerations in sheet-forming operations often determine the specific processes and practices to be employed. Many processes can be competitive; consequently, a thorough knowledge of process characteristics and capabilities is essential for economical production. (Section 7.16)

SUMMARY OF EQUATIONS

Maximum punch force in shearing, $F_{max} = 0.7(\text{UTS})(t)(L)$

Strain in bending, $e_o = e_i = \dfrac{1}{(2R/t) + 1}$

Minimum $\dfrac{R}{t} = \dfrac{50}{r} - 1$

Springback, $\dfrac{R_i}{R_f} = 4\left(\dfrac{R_i Y}{Et}\right)^3 - 3\left(\dfrac{R_i Y}{Et}\right) + 1$

Maximum bending force, $F_{max} = k\dfrac{(\text{UTS})Lt^2}{W}$

Normal anisotropy, $R = \dfrac{\varepsilon_w}{\varepsilon_t}$

Average, $\bar{R} = \dfrac{R_0 + 2R_{45} + R_{90}}{4}$

Planar anisotropy, $\Delta R = \dfrac{R_0 - 2R_{45} + R_{90}}{2}$

Maximum punch force in drawing, $F_{max} = \pi D_p t_o(\text{UTS})\left(\dfrac{D_o}{D_p} - 0.7\right)$

Pressure in explosive forming, $p = K\left(\dfrac{\sqrt[3]{W}}{R}\right)^a$

BIBLIOGRAPHY

ASM Handbook, Vol. 14: *Forming and Forging*, ASM International, 1988.

Benson, S. D., *Press Brake Technology*, Society of Manufacturing Engineers, 1997.

Davis, J. R. (ed.), *Tool Materials*, ASM International, 1995.

Fundamentals of Tool Design, 4th ed., Society of Manufacturing Engineers, 1998.

Eary, D. F., and E. A. Read, *Techniques for Pressworking Sheet Metal*, 2d. ed. Prentice-Hall, 1974.

Hosford, W. F., and R. M. Caddell, *Metal Forming, Mechanics and Metallurgy*, 2d. ed., Prentice Hall, 1993.

Koistinen, D. P., and N. M. Wang (eds.), *Mechanics of Sheet Metal Forming*, Plenum, 1978.

Marciniak, Z., and J. L. Duncan, *The Mechanics of Sheet Metal Forming*, Edward Arnold, 1992.

Morgan, E., *Tinplate and Modern Canmaking Technology*, Pergamon, 1985.

Nachtman, E. S., and S. Kalpakjian, *Lubricants and Lubrication in Metalworking Operations*, Dekker, 1985.

Pacquin, J. R., and R. E. Crowley, *Die Design Fundamentals*, Industrial Press, 1987.

Pearce, R., *Sheet Metal Forming*, Adam Hilger, 1991.

Progressive Dies, Society of Manufacturing Engineers, 1994.

Sachs, G., *Principles and Methods of Sheet Metal Fabricating*, 2d. ed., Reinhold, 1966.

Smith, D. A. (ed.), *Die Design Handbook*, 3d. ed., Society of Manufacturing Engineers, 1990.

Suchy, I., *Handbook of Die Design*, McGraw-Hill, 1997.

Tool and Manufacturing Engineers Handbook, 4th ed., Vol. 2: *Forming*, Society of Manufacturing Engineers, 1984.

Wagoner, R. H., K. S. Chan, and S. P. Keeler (eds.), *Forming Limit Diagrams*, The Minerals, Metals and Materials Society, 1989.

QUESTIONS

7.1 Take any three topics from Chapter 2, and, with specific examples for each, show their relevance to the topics covered in this chapter.

7.2 Do the same as for Question 7.1, but for Chapter 3.

7.3 Describe (a) the similarities and (b) the differences between the bulk-deformation processes described in Chapter 6 and the sheet-metal forming processes described in this chapter.

7.4 Discuss the material and process variables that influence the shape of the curve of punch force vs. stroke for shearing, such as that shown in Fig. 7.7, including its height and width.

7.5 In preparing a large number of blanks for sheet-forming operations, the blanks are spaced (*layout and nesting*) such that material waste is minimized. Make some sketches of blanks, and describe the considerations involved in this procedure, with particular emphasis on subsequent forming operations that may be performed.

7.6 Describe your observations concerning Figs. 7.5 and 7.6.

7.7 Inspect a common paper punch, and comment on the shape of the tip of the punch as compared with those shown in Fig. 7.12.

7.8 Explain how you would go about estimating the temperature rise in the shear zone in a shearing operation.

7.9 As a practicing engineer in manufacturing, why would you be interested in the shape of the curve in Fig. 7.7? Explain.

7.10 Do you think the presence of burrs can be beneficial in certain applications? Give specific examples.

7.11 Explain the reasons that there are so many different types of die materials used for the processes described in this chapter.

7.12 Describe the differences between compound, progressive, and transfer dies.

7.13 It has been stated that the quality of the sheared edges can influence the formability of sheet metals. Explain why.

7.14 Explain why and how various factors influence springback in bending of sheet metals.

7.15 Can the hardness of a sheet metal have an effect on the metal's springback in bending? Explain.

7.16 We note in Fig. 7.16 that the state of stress shifts from plane stress to plane strain as the ratio of length of bend to sheet thickness increases. Explain why.

7.17 Describe the material properties that have an effect on the relative position of the curves shown in Fig. 7.19.

7.18 In Table 7.2, we note that hard materials have higher R/t ratios than soft ones. Explain why.

7.19 Why do tubes buckle when bent? Experiment with a soda straw, and describe your observations.

7.20 If you wish to bend a piece of copper tubing without collapsing it, you could go to a hardware store and buy a helical springlike gadget. Visit a hardware store, find out what this gadget is, and explain why you can use it successfully to bend a tube.

7.21 Based on Fig. 7.22, sketch the shape of a *U*-die used to produce channel-shaped bends.

7.22 Explain why negative springback does not occur in air bending of sheet metals.

7.23 Give examples of products in which the presence of beads is beneficial or even necessary.

7.24 Explain why cupping tests do not always predict the behavior of sheet metals in actual forming operations.

7.25 Assume that you are carrying out a sheet-forming operation and you find that the material is not sufficiently ductile. Make suggestions to improve its ductility.

7.26 Referring to Fig. 7.39, explain why a further increase in the ductility of the material (beyond about 50% tensile reduction of area) does not improve the maximum spinning reduction per pass.

7.27 Many missile components are made by spinning. What other methods would you use to make missile components if spinning processes were not available? Describe the relative advantages and limitations of each method.

7.28 In deep drawing of a cylindrical cup, is it always necessary for there to be tensile circumferential stresses on the element in the cup wall? (See Fig. 7.51b.) Explain.

7.29 When comparing the hydroforming process with the deep-drawing process, it has been stated that deeper draws are possible in the former method. With appropriate sketches, explain why.

7.30 We note in Fig. 7.51a that element *A* in the flange is subjected to compressive circumferential (hoop) stresses. Using a simple free-body diagram, explain why.

7.31 From the topics covered in this chapter, list and explain specifically several examples where (a) friction is desirable and (b) friction is not desirable.

7.32 Explain why increasing the normal anisotropy, R, improves the deep drawability of sheet metals.

7.33 What is the reason for the negative sign in the numerator of Eq. (7.21)?

7.34 If you had a choice whereby you could control the state of strain in a sheet-forming operation, would you rather work on the left or the right side of the forming-limit diagram? Explain.

7.35 Comment on the effect of lubrication of the punch on the limiting drawing ratio in deep drawing.

7.36 Comment on the size of the circles placed on the surfaces of sheet metals in determining the metals' formability. Are square grid patterns, as shown in Fig. 7.67, useful? Explain.

7.37 Make a list of the independent variables that influence the punch force in deep drawing of a cylindrical cup, and explain why and how the variables influence this force.

7.38 Explain why the simple tension line in the forming-limit diagram in Fig. 7.66a states that it is for $R = 1$, where R is the normal anisotropy of the sheet.

7.39 What are the reasons for developing forming-limit diagrams? Do you have any specific criticisms of such diagrams? Explain.

7.40 Explain the reasoning behind Eq. (7.20), for normal anisotropy, and Eq. (7.21), for planar anisotropy, respectively.

7.41 Describe why earing occurs. How would you avoid it?

7.42 What is the significance of the size of the grid patterns used in the study of sheet-metal formability?

7.43 We have stated in Section 7.13.1 that the thicker the sheet metal, the higher is the curve in the forming-limit diagram. Explain why.

7.44 Inspect the earing shown in Fig. 7.59, and estimate the direction in which the blank was cut.

7.45 Describe the factors that influence the size and length of beads in sheet-metal forming operations.

7.46 It is known that the strength of metals depends on the metals' grain size. Would you then expect strength to influence the R value of sheet metals? Explain.

7.47 Equation (7.23) gives a general rule for dimensional relationships for successful drawing without a blankholder. Explain what happens if this limit is exceeded.

7.48 Explain why the three broken lines (simple tension, plane strain, and equal biaxial stretching) in Fig. 7.66a have those particular slopes.

7.49 Identify specific parts on a typical automobile, and explain which of the processes described in Chapters 6 and 7 can be used to make that part. Then choose only one method for each part, and explain your reasoning.

7.50 We have seen that bendability and spinnability have a common aspect as far as properties of the workpiece material are concerned. Describe this common aspect.

7.51 Explain the reasons that such a wide variety of sheet-forming processes has been developed and used over the years.

7.52 Make a summary of the types of defects found in sheet-metal forming processes, and include brief comments on the reason(s) for each defects.

7.53 Which of the processes described in this chapter use only one die? What are the advantages of using only one die?

7.54 It has been suggested that deep drawability can be increased by (a) heating the flange and/or (b) chilling the punch by some suitable means. Comment on how these methods could improve drawability.

7.55 Offer designs whereby the suggestions given in Question 7.54 can be implemented. Would the required production rate affect your designs? Explain.

7.56 In the manufacture of automotive body panels from carbon-steel sheet, stretcher strains (Lueder's bands) are observed, which detrimentally affect surface finish. How can the stretcher strains be eliminated?

7.57 In order to improve its ductility, a coil of sheet metal is placed in a furnace and annealed. However, it is observed that the sheet has a lower limiting drawing ratio than it had before being annealed. Explain the reasons for this behavior.

7.58 What effects does friction have on a forming-limit diagram?

7.59 Why are lubricants generally used in sheet-metal forming? Explain, using examples.

7.60 Through changes in clamping, a sheet-metal forming operation can allow the material to undergo a negative minor strain. Explain how this effect can be advantageous.

7.61 How would you produce the part shown in Fig. 7.35 other than by tube hydroforming?

PROBLEMS

7.62 Referring to Eq. (7.5), it is stated that actual values of e_o are considerably higher than values of e_i, due to the shifting of the neutral axis during bending. With an appropriate sketch, explain this phenomenon.

7.63 Note in Eq. (7.11) that the bending force is a function of t^2. Why? (*Hint*: Consider bending-moment equations in mechanics of solids.)

7.64 Calculate the minimum tensile true fracture strain that a sheet metal should have in order to be bent to the following R/t ratios: (a) 2, (b) 3, and (c) 5. (See Table 7.2.)

7.65 Estimate the maximum bending force required for a $\frac{1}{8}$-in. thick and 10-in. wide Ti-5Al-2.5Sn titanium alloy in a *V*-die with a width of 8 in.

7.66 In Example 7.4, calculate the work done by the force–distance method, i.e., work is the integral product of the vertical force, *F*, and the distance it moves.

7.67 What would be the answer to Example 7.4 if the tip of the force, *F*, were fixed to the strip by some means, thus maintaining the lateral position of the force? (*Hint*: Note that the left portion of the strip will now be strained more than the right portion.)

7.68 Calculate the magnitude of the force *F* in Example 7.4 for $\alpha = 15°$.

7.69 How would the force in Example 7.4 vary if the workpiece were made of a perfectly plastic material?

7.70 Calculate the press force needed in punching 0.1-mm-thick 5052-0 aluminum foil in the shape of a square hole 30 mm on each side.

7.71 A straight bead is being formed on a 1-mm-thick aluminum sheet in a 20-mm-diameter die, as shown in the accompanying figure. (See also Fig. 7.25a.) Let Y = 120 MPa. Considering springback, calculate the outside diameter of the bead after it is formed and unloaded from the die.

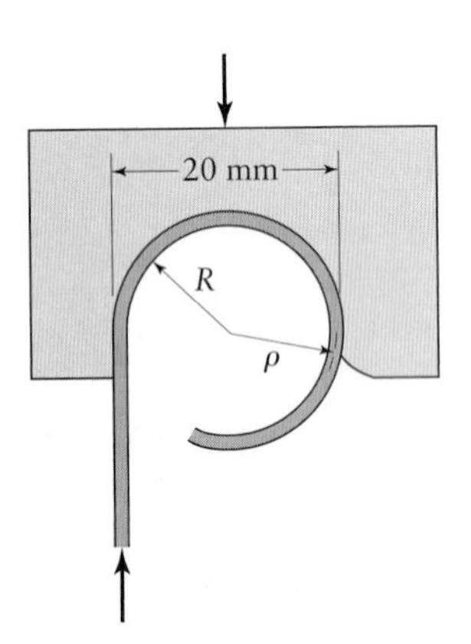

7.72 Inspect Eq. (7.10), and, substituting in some numerical values, show whether the first term can be neglected without significant error in calculating springback.

7.73 In Example 7.5, calculate the amount of TNT required to develop a pressure of 15,000 psi on the surface of the workpiece. Use a standoff of one foot.

7.74 Estimate the limiting drawing ratio (LDR) for the materials listed in Table 7.3.

7.75 For the same material and thickness as in Problem 7.65, estimate the force required for deep drawing with a blank of diameter 12 in. and a punch of diameter 9 in.

7.76 A cup is being drawn from a sheet metal that has a normal anisotropy of 3. Estimate the maximum ratio of cup height to cup diameter that can be drawn successfully in a single draw. Assume that the thickness of the sheet throughout the cup remains the same as the original blank thickness.

7.77 Obtain an expression for the curve shown in Fig. 7.58 in terms of the LDR and the average normal anisotropy, $\bar{R}$. (*Hint*: See Fig. 2.7b).

7.78 A steel sheet has *R* values of 1.0, 1.5, and 2.0 for the 0°, 45°, and 90° directions to rolling, respectively. If a round blank is 200 mm in diameter, estimate the smallest cup diameter to which it can be drawn.

7.79 In Problem 7.78, explain whether ears will form and, if so, why.

7.80 A 1-mm-thick isotropic sheet metal is inscribed with a circle 5 mm in diameter. The sheet is then stretched uniaxially by 25%. Calculate (a) the final dimensions of the circle and (b) the thickness of the sheet at this location.

7.81 Make a search of the literature, and obtain the equation for a tractrix curve, as used in Fig. 7.63.

7.82 In Example 7.4, assume that the stretching is done by two equal forces *F*, each at 5 in. from the ends of the workpiece. (a) Calculate the magnitude of this force for $\alpha = 10°$. (b) If we want stretching to be done up to $\alpha_{max} = 50°$ without necking, what should be the minimum value of n of the material?

7.83 Derive Eq. (7.5).

7.84 Estimate the maximum power in shear spinning a 0.5-in. thick annealed 304 stainless-steel plate that has a diameter of 15 in. on a conical mandrel of $\alpha = 45°$. The mandrel rotates at 100 rpm, and the feed is $f = 0.1$ in./rev.

7.85 Assume that you are asked to give a quiz to students on the contents of this chapter. Prepare five quantitative problems and five qualitative questions, and supply the answers.

7.86 Obtain an aluminum beverage can, and cut it in half lengthwise with a pair of tin snips. Using a micrometer, measure the thickness of the bottom of the can and of the wall of the can. Estimate the thickness reductions in ironing and the diameter of the original blank.

7.87 What is the force required to punch a square hole, 100 mm on each side, from a 1-mm-thick 5052-O aluminum sheet, using flat dies? What would be your answer if beveled dies were used instead?

DESIGN

7.88 Consider several shapes to be blanked from a large sheet (such as oval, triangle *L*-shape, and so forth) by laser-beam cutting, and sketch a nesting layout to minimize scrap generation.

7.89 Estimate the percent scrap in producing round blanks if the clearance between blanks is one tenth of the radius of the blank. Consider single and multiple-row blanking, as sketched in the accompanying figure.

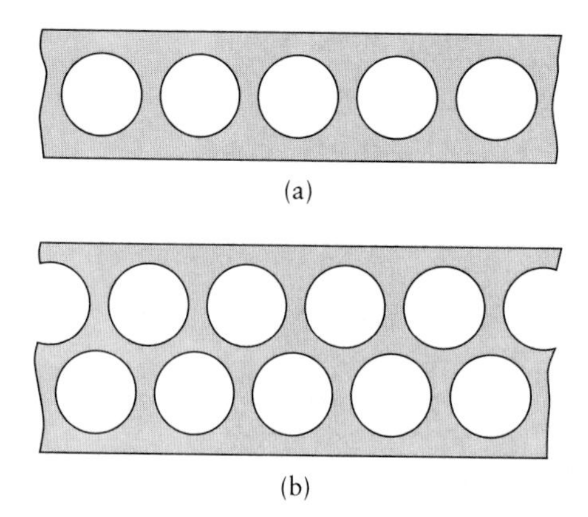

7.90 Give several structural applications in which diffusion bonding and superplastic forming are used jointly.

7.91 On the basis of experiments, it has been suggested that concrete, either plain or reinforced, can be a suitable material for dies in sheet-metal forming operations. Describe your thoughts regarding this suggestion, considering die geometry and any other factors that may be relevant.

7.92 Metal cans are of either the two-piece variety (in which the bottom and sides are integral) or the three-piece variety (in which the sides, the bottom, and the top are each separate pieces). For a three-piece can, should the seam be (a) in the rolling direction, (b) normal to the rolling direction, or (c) oblique to the rolling direction? Explain your answer, using equations from solid mechanics.

7.93 Find and inspect a brass drum cymbal, and describe the manufacturing processes used in its production.

7.94 Investigate methods for determining optimum shapes of blanks for deep-drawing operations. Sketch the optimally shaped blanks for drawing rectangular cups, and optimize their layout on a large sheet of metal.

7.95 The design shown in the accompanying figure is proposed for a metal tray, the main body of which is made from cold-rolled sheet steel. Noting its features and that the sheet is bent in two different directions, comment on various manufacturing considerations. Include factors such as anisotropy of the rolled sheet, the sheet's surface texture, the bend directions, the nature of the sheared edges, and the way the handle is snapped in for assembly.

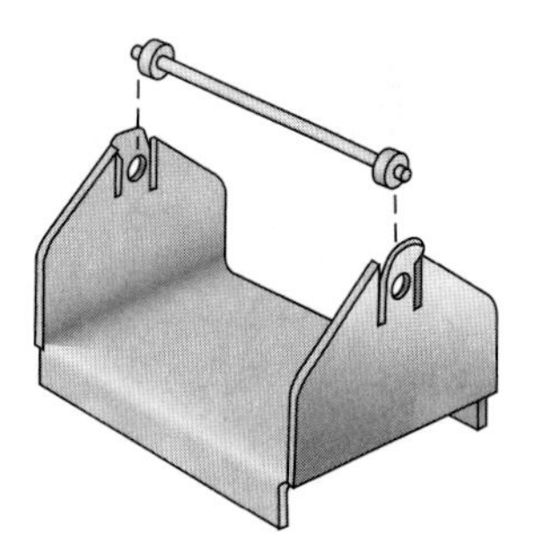

7.96 Design a box that will contain a 4-in. × 6-in. × 3-in. volume. The box should be produced from two pieces of sheet metal and require no tools or fasteners for assembly.

7.97 In opening a can using an electric can opener, you will note that the lid develops a scalloped periphery. (a) Explain why scalloping occurs. (b) What design changes for the can opener would you recommend in order to minimize or eliminate, if possible, this scalloping effect? (c) Since lids typically are discarded or recycled, do you think it is necessary or worthwhile to make such design changes? Explain.

CHAPTER

8 Material-Removal Processes: Cutting

8.1 Introduction

Parts manufactured by casting, forming, and various shaping processes often require further processing or finishing operations to impart specific characteristics, such as dimensional accuracy and surface finish, before the product is ready for use. This chapter and the following chapter describe in detail the various operations that are employed to obtain these characteristics; these processes are generally classified as **material-removal processes**. Because they are capable of producing shapes competitive with those produced by other methods, as described in the preceding chapters, critical choices have often to be made about the extent of shaping and forming versus the extent of machining to be done on a workpiece to produce an acceptable part.

Although **machining** is the general term used to describe *removal of material* from a workpiece, it covers several processes, which are usually divided into the following broad categories, the first of which is the subject of this chapter:

- **Cutting**, which generally involves single-point or multipoint cutting tools and processes, such as turning, boring, drilling, tapping, milling, sawing, and broaching;
- **Abrasive processes**, such as grinding, honing, lapping, and ultrasonic machining (Chapter 9);
- **Advanced machining processes**, which use electrical, chemical, thermal, hydrodynamic, and optical sources of energy to remove material from the workpiece surface (Chapter 9).

As described throughout this chapter, material-removal processes are desirable or even necessary in manufacturing operations for the following reasons:

1. Closer **dimensional accuracy** may be required than is available from casting, forming, or shaping processes alone. For example, in a forged crankshaft, the bearing surfaces and the holes cannot be produced with good dimensional accuracy and surface finish by forming and shaping processes alone.
2. Parts may possess external and internal **geometric features**, as well as sharp corners, flatness, etc., that cannot be produced by forming and shaping processes.
3. Some parts are heat treated for improved hardness and wear resistance. Since heat-treated parts may undergo distortion and surface discoloration, they generally require additional **finishing operations**, such as grinding, to obtain the desired characteristics.
4. Special **surface characteristics** or textures that cannot be produced by other means may be required on all or selected areas of the surfaces of the product. Copper mirrors with very high reflectivity, for example, are made by machining with a diamond cutting tool.

5. It may be more **economical** to machine the part than to manufacture it by other processes, particularly if the number of parts desired is relatively small.

Against these advantages, material-removal processes have certain limitations:

1. Removal processes inevitably **waste material** and generally require more energy, capital, and labor than forming and shaping operations; thus, they should be avoided or minimized whenever possible.
2. Removing a volume of material from a workpiece generally takes **longer time** than does shaping it by other processes.
3. Unless carried out properly, material-removal processes can have **adverse effects** on the surface integrity of the product.

In spite of these limitations, material-removal processes and machines, called **machine tools**, are indispensable to manufacturing technology. Ever since lathes were introduced in the 1700s, these processes have developed continuously; we now have available a wide variety of computer-controlled machines, as well as new techniques that use various sources of energy. It is important to view machining, as well as all manufacturing operations, as a *system* consisting of the workpiece, the tool, and the machine. As will be evident in reading this chapter, machining operations cannot be carried out efficiently and economically without a fundamental knowledge of the complex interactions among these elements. The importance of these considerations is even more evident from the estimate that, in the United States, labor and overhead costs alone for machining are about $300 billion per year.

8.2 Mechanics of Chip Formation

Cutting processes (Fig. 8.1) remove material from the surface of the workpiece by producing chips. The basic mechanics of chip formation is essentially the same for all cutting operations and is represented by the two-dimensional model shown in Fig. 8.2. In this model, a tool moves along the workpiece at a certain velocity, V, and a depth of cut, t_o. A **chip** is produced just ahead of the tool by *shearing* of the material continuously along the shear plane.

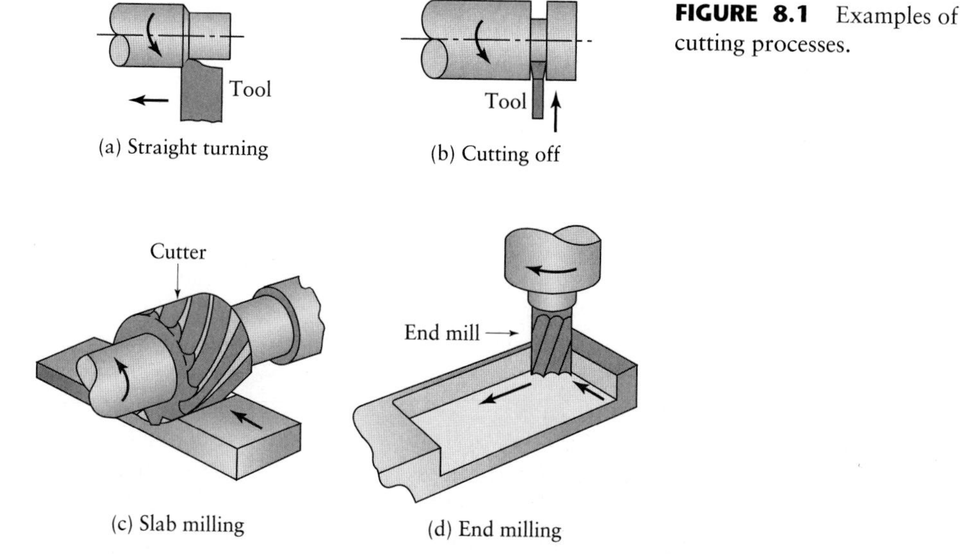

FIGURE 8.1 Examples of cutting processes.

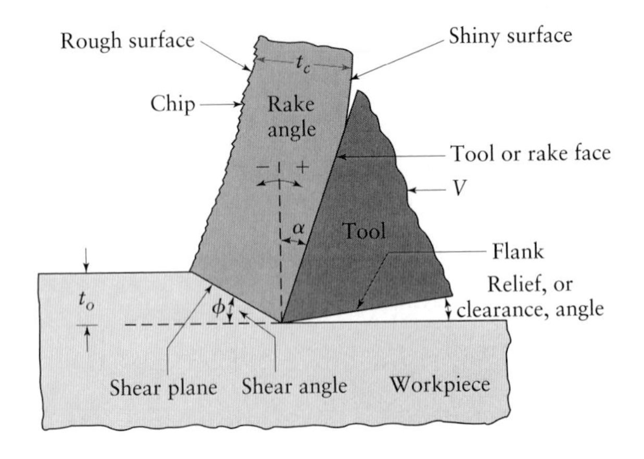

FIGURE 8.2 Schematic illustration of a two-dimensional cutting process, also called *orthogonal cutting*.

In the cutting process, the major *independent variables*, i.e., the variables that we can change and control directly, are

- Tool material and its condition;
- Tool shape, surface finish, and sharpness;
- Workpiece material and its condition and temperature;
- Cutting conditions, such as speed, feed, and depth of cut;
- Type of cutting fluids, if used;
- Characteristics of the machine tool, such as its structure, stiffness, and damping;
- Workholding and fixturing devices.

The *dependent variables*, which are the variables that are influenced by changes in the independent variables, are:

- Type of chip produced;
- Force and energy dissipated in the cutting process;
- Temperature rise in the workpiece, the chip, and the tool;
- Wear and failure of the tool;
- Surface finish and integrity of the workpiece after machining.

In order to appreciate the importance of studying the complex interrelationships among these variables, let's briefly ask the following questions: If, for example, the surface finish of the workpiece being cut is poor and unacceptable, which of the foregoing independent variables do we change first? The tool geometry? If so, which feature? If the workpiece becomes too hot, what do we do? If the tool wears and becomes dull rapidly, do we change the cutting speed, the depth of cut, or the tool material itself? If the cutting tool begins to vibrate and chatter, what should we do to eliminate this problem?

Although almost all cutting processes are three dimensional in nature, the two-dimensional model shown in Fig. 8.2 is very useful in studying the basic mechanics of cutting. In this model, known as **orthogonal cutting** (i.e., the tool edge is perpendicular to the movement of the tool), the tool has a **rake angle,** α, (positive, as shown in the figure) and a **relief,** or **clearance, angle.** Note that the sum of the rake, relief, and included angles of the tool is 90°.

Microscopic examinations reveal that chips are produced by the **shearing** process shown in Fig. 8.3a. Shearing takes place along a **shear plane,** which makes an angle ϕ, called the **shear angle,** with the surface of the workpiece. Note that this shearing

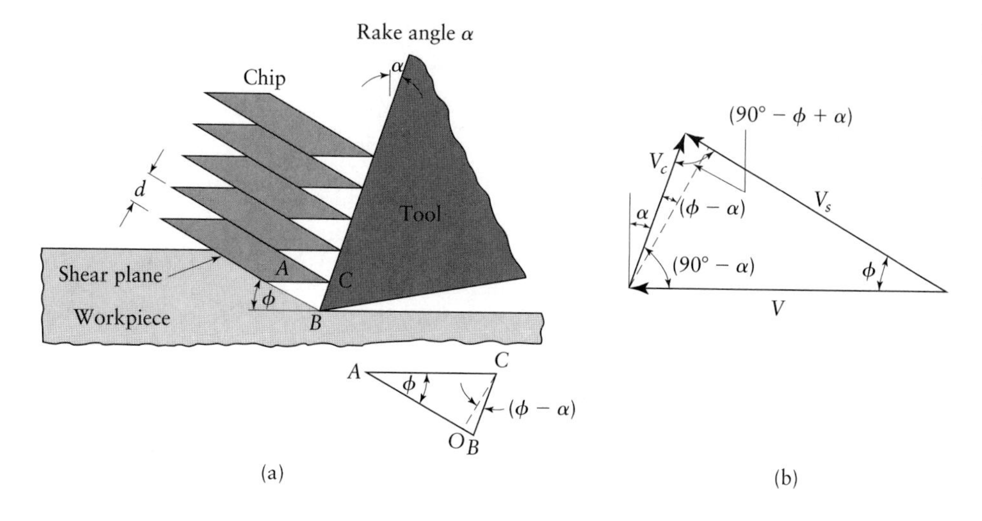

FIGURE 8.3 (a) Schematic illustration of the basic mechanism of chip formation in cutting. (b) Velocity diagram in the cutting zone.

process is like cards in a deck sliding against each other. Below the shear plane, the workpiece is undeformed (though it does undergo some elastic distortion), and above the shear plane, we have a chip already formed and moving up the face of the tool as cutting progresses. Because of the movement of the chip, there is friction between the chip and the rake face of the tool.

Note that we can determine the thickness of the chip, t_c, if we know t_o, α, and ϕ. The ratio of t_o to t_c is known as the **cutting ratio**, r, which can be expressed as

$$r = \frac{t_o}{t_c} = \frac{\sin \phi}{\cos(\phi - \alpha)}. \tag{8.1}$$

A review of this process indicates that the chip thickness is always greater than the depth of cut; hence, the value of r is always less than unity. The reciprocal of r is known as the **chip compression ratio** and is a measure of how thick the chip has become compared with the depth of cut; thus, the chip compression ratio is always greater than unity.

On the basis of Fig. 8.3a, we can write the **shear strain**, γ, that the material undergoes as

$$\gamma = \frac{AB}{OC} = \frac{AO}{OC} + \frac{OB}{OC}, \tag{8.2}$$

or

$$\gamma = \cot \phi + \tan(\phi - \alpha). \tag{8.3}$$

Note from this equation that large shear strains are associated with low shear angles and low or negative rake angles. Shear strains of 5 or higher have been observed in actual cutting operations; thus, compared with forming and shaping processes (see Chapter 6), the material undergoes greater deformation during cutting.

From Fig. 8.2, note that since chip thickness, t_c, is greater than the depth of cut, t_o, the *velocity of the chip*, V_c, has to be less than the cutting speed, V. Since mass continuity has to be maintained, we have

$$Vt_o = V_c t_c, \quad \text{or} \quad V_c = Vr. \tag{8.4}$$

Hence,

$$V_c = V\frac{\sin \phi}{\cos(\phi - \alpha)}. \tag{8.5}$$

We can construct a velocity diagram such as that shown in Fig. 8.3b, and from trigonometric relationships, we obtain the equations

$$\frac{V}{\cos(\phi - \alpha)} = \frac{V_s}{\cos \alpha} = \frac{V_c}{\sin \phi}, \tag{8.6}$$

where V_s is the velocity at which shearing takes place in the shear plane. The **shear-strain rate** is the ratio of V_s to the thickness d of the sheared element (shear zone), or

$$\dot{\gamma} = \frac{V_s}{d}. \tag{8.7}$$

Experimental evidence indicates that the magnitude of d is on the order of 10^{-2} mm to 10^{-3} mm (10^{-3} in. to 10^{-4} in.). This range means that, even at low cutting speeds, the shear-strain rate is very high, on the order of 10^3/s to 10^6/s. A knowledge of the shear-strain rate is essential, because of its effects on the strength and ductility of the material (see Section 2.2.7) and the type of chip.

8.2.1 Chip morphology

The type of chips produced (*chip morphology*) significantly influences surface finish and integrity, as well as the overall cutting operation. When actual chips are observed with a microscope and under different cutting conditions, there are some deviations from the ideal model shown in Figs. 8.2 and 8.3a. Figure 8.4 shows some types of metal chips commonly observed in practice; they are described in this subsection in the following order:

1. **Continuous;**
2. **Built-up edge;**
3. **Serrated, or segmented;**
4. **Discontinuous.**

Let's first note that a chip has two surfaces: One is in contact with the tool face (rake face), and the other is from the original surface of the workpiece. The tool side of the chip surface is shiny, or *burnished* (Fig. 8.5), caused by rubbing of the chip as it climbs up the tool face. The other surface of the chip does not come into contact with any solid body; this surface has a jagged, steplike appearance (see Fig. 8.4a), which is due to the shearing mechanism of chip formation. (See also Figs. 3.4 and 3.6 for an analogy.)

The basic types of chips produced in metal-cutting operations are described as follows:

1. **Continuous chips.** *Continuous chips* are usually formed at high cutting speeds and/or high rake angles (Fig. 8.4a); the deformation of the material takes place along a very narrow shear zone, called the **primary shear zone.** Continuous chips may develop a **secondary shear zone** at the tool–chip interface (Fig. 8.4b), caused by friction. As expected, the secondary zone becomes deeper as tool–chip friction increases. In continuous chips, deformation may also take place along a wide primary shear zone with *curved boundaries* (Fig. 8.4d). Note that the lower boundary is *below* the machined surface, which subjects the machined surface to distortion, as depicted by the distorted vertical lines to the right of the tool tip. This situation occurs particularly in machining soft metals at low speeds and low rake angles; it can produce poor surface finish and induce residual surface stresses, which may be detrimental to the properties of the machined part.

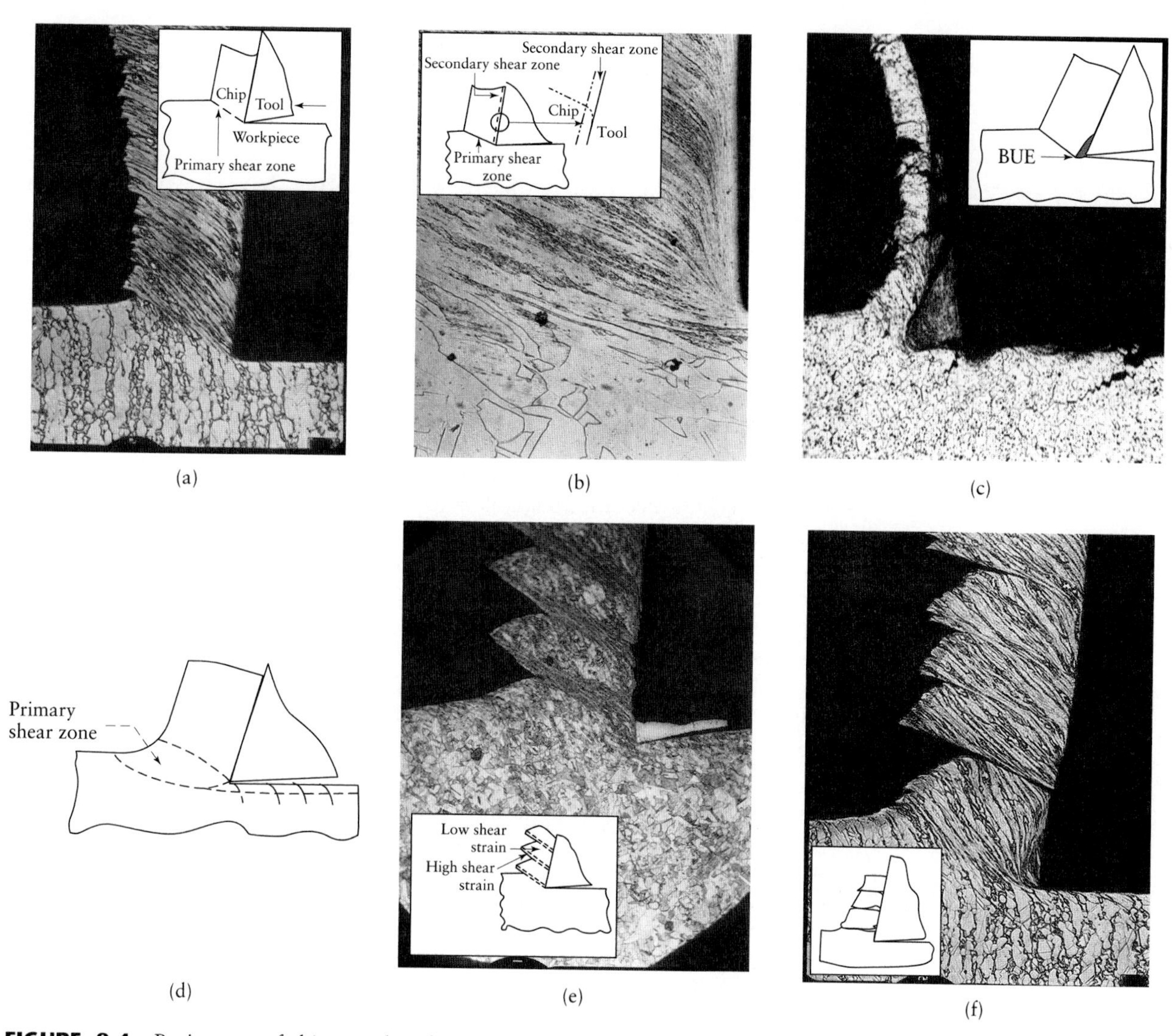

FIGURE 8.4 Basic types of chips produced in metal cutting and their micrographs: (a) continuous chip with narrow, straight primary shear zone; (b) secondary shear zone at the tool–chip interface; (c) continuous chip with built-up edge; (d) continuous chip with large primary shear zone; (e) segmented or nonhomogeneous chip; and (f) discontinuous chip. *Source*: After M. C. Shaw, P. K. Wright, and S. Kalpakjian.

FIGURE 8.5 Shiny (burnished) surface on the tool side of a continuous chip produced in turning.

Although they generally produce good surface finish, continuous chips are not always desirable, particularly in automated machine tools, because they tend to get tangled around the tool, and the operation has to be stopped to clear away the chips. This problem can be alleviated with **chip breakers** (discussed later).

As a result of strain hardening (caused by the shear strains to which the chip is subjected), the chip usually becomes harder, less ductile, and stronger than the original workpiece material. The increase in hardness and strength of the chip depends on the shear strain; as the rake angle decreases, the shear strain increases, and thus the chip becomes stronger and harder. With increasing strain, the chip tends to behave more like a rigid, perfectly plastic material. (See Fig. 2.8b.)

2. **Built-up-edge chips.** A *built-up edge* (BUE) may form at the tip of the tool during cutting (Fig. 8.4c); it consists of layers of material from the workpiece that are gradually deposited on the tool (hence the term *built-up*). As it becomes larger, the BUE becomes unstable and eventually breaks up; a portion of the BUE material is carried away on the tool side of the chip, and the rest is deposited randomly on the workpiece surface. The process of BUE formation and breakup is repeated continuously during the cutting operation. Although BUE is generally unwanted, a thin, stable BUE is usually regarded as desirable, because it protects the tool's surface.

   The built-up edge is commonly observed in practice and is one of the important factors that adversely affects surface finish and integrity in cutting, as Figs. 8.4 and 8.6 show. A built-up edge, in effect, changes the geometry of cutting. Note, for example, the large tip radius of the BUE and the rough surface finish produced. Because of work hardening and deposition of successive layers of material, BUE hardness increases significantly (Fig. 8.6a).

   The exact mechanism of formation of the BUE is not yet clearly understood; however, investigations have identified two distinct mechanisms that contribute to its formation. One is the adhesion of the workpiece material to the rake face of the tool; the strength of this bond depends on the affinity of the workpiece and tool materials. The second mechanism is the growth of the adhered metal layers. One of the important factors in forming the BUE is the tendency of the workpiece material for strain hardening; the higher the strain-hardening exponent, n, the greater is the probability for BUE formation. Experimental evidence indicates that as the cutting speed increases, the BUE decreases or is eliminated. In addition to making appropriate adjustments to the aforementioned factors, the tendency for BUE to form can be reduced by (a) decreasing the depth of cut, (b) increasing the rake angle, (c) using a tool with a small tip radius, and (d) using an effective cutting fluid.

3. **Serrated chips.** *Serrated chips*, also called **segmented** or *nonhomogeneous* chips, are semicontinuous chips, with zones of low and high shear strain (Fig. 8.4e). The chips have the appearance of saw teeth. Metals with low thermal conductivity and strength that decreases sharply with temperature, such as titanium, exhibit this behavior.

4. **Discontinuous chips.** *Discontinuous chips* consist of segments that may be either firmly or loosely attached to each other (Fig. 8.4f). These chips usually form under the following conditions: (a) when the workpiece materials are brittle, because they do not have the capacity to undergo the high shear strains in

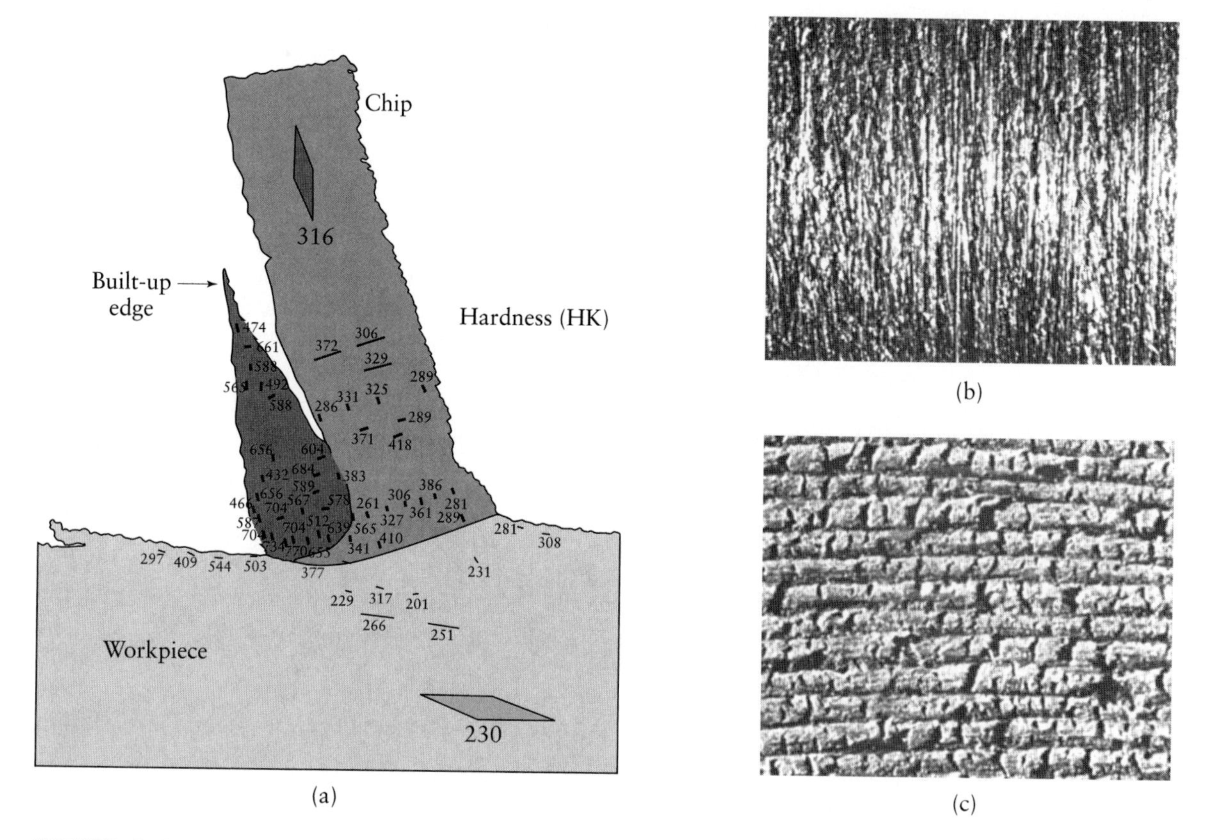

FIGURE 8.6 (a) Hardness distribution in the cutting zone for 3115 steel. Note that some regions in the built-up edge are as much as three times harder than the bulk workpiece. (b) Surface finish in turning 5130 steel with a built-up edge. (c) Surface finish on 1018 steel in face milling. Magnifications: 15×. *Source*: Courtesy of TechSolve, formerly the Institute of Advanced Manufacturing Sciences, Inc.

cutting; (b) when the workpiece materials contain hard inclusions and impurities (see Figs. 3.24 and 3.25) or have structures such as the graphite flakes in gray cast iron (Fig. 5.20a); (c) when the cutting speeds are very low or very high; (d) when the depths of cut are large or the rake angles are low; (e) when there is low stiffness and poor damping of the machine tool's structure; and (f) when there is a lack of an effective cutting fluid.

Impurities and hard particles act as sites for cracks, thereby producing discontinuous chips; as expected, a large depth of cut increases the probability of such defects being present in the cutting zone. Higher cutting speeds mean higher temperatures; hence, the material is more ductile and has less of a tendency to form discontinuous chips. Another factor in the formation of discontinuous chips is the magnitude of the compressive stresses on the shear plane; recall from Fig. 2.22 that the maximum shear strain at fracture increases with increasing compressive stress. Thus, if the normal stress is not sufficiently high, the material will be unable to undergo the large shear strain required to form a continuous chip.

Because of the nature of discontinuous-chip formation, forces continually vary during cutting. Consequently, the stiffness of the cutting-tool holder and the machine tool is important in cutting with discontinuous-chip as well as serrated-chip formation. If not sufficiently stiff, the machine tool may begin to vibrate and chatter. (See Section 8.11.) This effect, in turn, adversely affects the surface finish and dimensional accuracy of the machined component and may cause damage or excessive wear of the cutting tool.

Chip formation in cutting nonmetallic materials. Much of the discussion thus far for metals is also generally applicable to nonmetallic materials. A variety of chips is obtained in cutting thermoplastics, depending on the type of polymer and the process parameters, such as depth of cut, tool geometry, and cutting speed. Because they are brittle, thermosetting plastics and ceramics generally produce discontinuous chips. (See also the discussion on *ductile regime cutting* in Section 8.8.2.)

Chip curl. *Chip curl* (Figs. 8.5 and 8.7a) is common to all cutting operations with metals and nonmetallic materials such as plastics and wood. The reasons for chip curl are not clearly understood. Among possible factors contributing to it are the distribution of stresses in the primary and secondary shear zones, thermal gradients, the work-hardening characteristics of the workpiece material, and the geometry of the rake face of the tool. Process variables also affect chip curl; generally, the radius of curvature decreases (i.e., the chip becomes curlier) with decreasing depth of cut, increasing rake angle, and decreasing friction at the tool–chip interface. The use of cutting fluids and various additives in the workpiece material also influences chip curl.

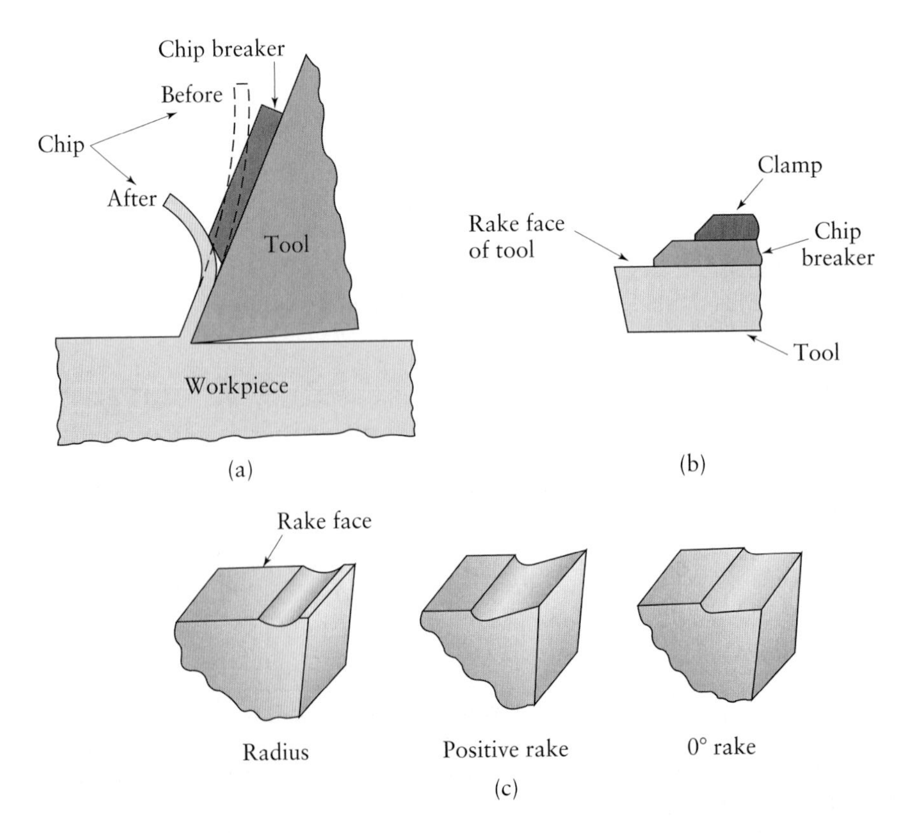

FIGURE 8.7 (a) Schematic illustration of the action of a chip breaker. Note that the chip breaker decreases the radius of curvature of the chip. (b) Chip breaker clamped on the rake face of a cutting tool. (c) Grooves in cutting tools, acting as chip breakers.

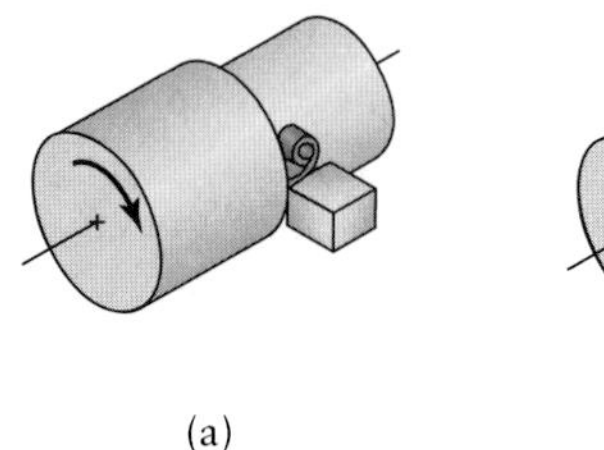

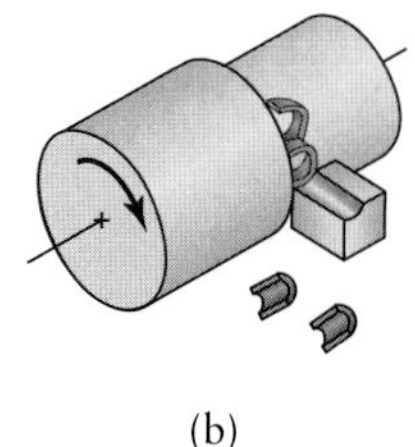

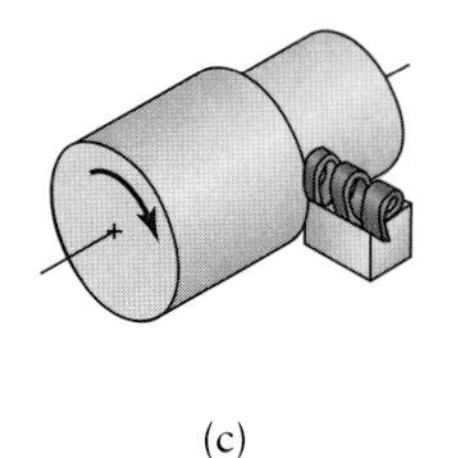

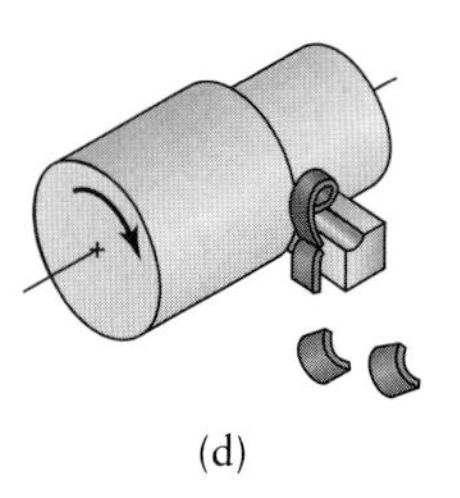

(a) (b) (c) (d)

FIGURE 8.8 Various chips produced in turning: (a) tightly curled chip; (b) chip hits workpiece and breaks; (c) continuous chip moving away from workpiece; and (d) chip hits tool shank and breaks off. *Source*: G. Boothroyd, *Fundamentals of Metal Machining and Machine Tools.*

Chip breakers. Long, continuous chips are undesirable, because they tend to become entangled and interfere with cutting operations and can be a safety hazard. This situation is especially troublesome in high-speed automated machinery. The usual procedure to avoid the formation of continuous chips is to break the chip intermittently with a *chip breaker*. Although the chip breaker has traditionally been a piece of metal clamped to the rake face of the tool (Fig. 8.7b), it is now an integral part of the tool itself (Fig. 8.7c). Chips can also be broken by changing the tool geometry, thus controlling chip flow, as in the turning operations shown in Fig. 8.8.

A wide variety of cutting tools and inserts with chip-breaker features are available. However, with soft workpiece materials, such as commercially pure aluminum and copper, chip breakers are generally not as effective; the remedy is usually to change the process parameters. In interrupted cutting operations, such as milling, chip breakers are generally not necessary, since the chips already have finite lengths, due to the intermittent nature of the operation. (See Fig. 8.54a.)

8.2.2 Mechanics of oblique cutting

Unlike the two-dimensional situations described thus far, the majority of cutting operations involve tool shapes that are three-dimensional (**oblique**). Figure 8.9a illustrates the basic difference between two-dimensional and oblique cutting. We have seen that in orthogonal cutting, the tool edge is perpendicular to the movement of the tool, and the chip slides directly up the rake face of the tool. In oblique cutting, however, the cutting edge is at an angle i, called the **inclination angle** (Fig. 8.9b). Note that the chip in Fig. 8.9a flows up the rake face of the tool at an angle α_c (**chip flow angle**), measured in the plane of the tool face. (Note that this situation is similar to an angled snowplow blade, which throws the snow sideways.) Angle α_n is known as the **normal rake angle**, which is a basic geometric property of the tool. The normal rake angle is the angle between the normal oz to the workpiece surface and the line oa on the tool face.

The workpiece material approaches the tool at a velocity V and leaves the workpiece surface (as a chip) with a velocity V_c. The effective rake angle, α_e, in the plane of these two velocities can be calculated as follows: Assuming that the chip flow angle, α_c, is equal to the inclination angle, i (an assumption that has experimentally been found to be approximately correct), the effective rake angle, α_e, is given by

$$\alpha_e = \sin^{-1}(\sin^2 i + \cos^2 i \sin \alpha_n). \tag{8.8}$$

Since i and α_n can both be measured directly, we can use this formula to calculate the value of the effective rake angle. Note that as i increases, the effective rake angle increases, and the chip becomes thinner and longer. Figure 8.9c illustrates the effect of the inclination angle on chip shape.

Figure 8.10 shows a typical single-point turning tool used on a lathe. Note the various angles, each of which has to be selected properly for efficient cutting. Various three-dimensional cutting tools are described in greater detail in Sections 8.8 and 8.9, including tools for drilling, tapping, milling, planing, shaping, broaching, sawing, and filing.

Shaving and skiving. Thin layers of material can be removed from straight or curved surfaces by a process similar to using a plane to shave wood. *Shaving* is particularly useful in improving the surface finish and dimensional accuracy of punched slugs or holes. (See Fig. 7.11.) Parts that are long or have a combination of shapes are shaved by *skiving* with a specially shaped cutting tool.

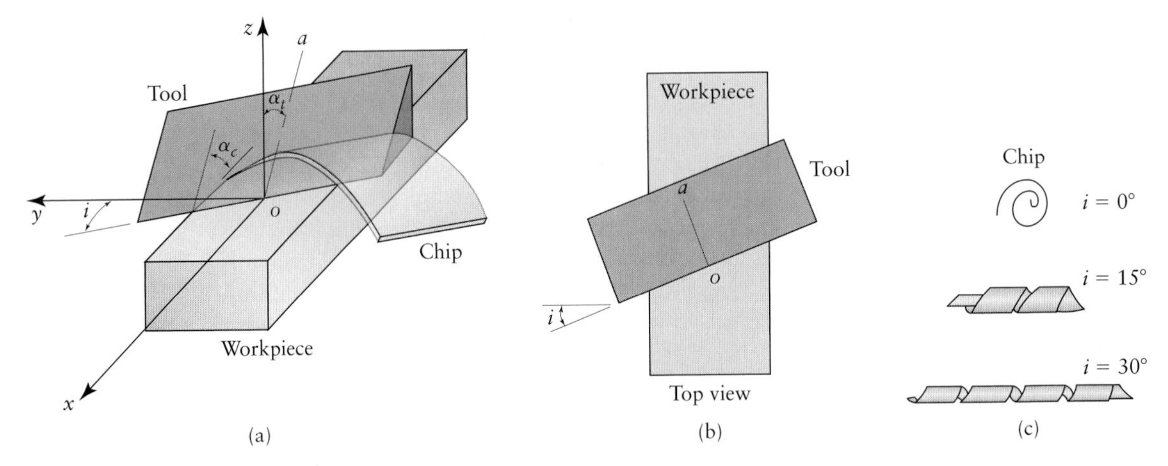

FIGURE 8.9 (a) Schematic illustration of cutting with an oblique tool. (b) Top view, showing the inclination angle, i. (c) Types of chips produced with different inclination angles.

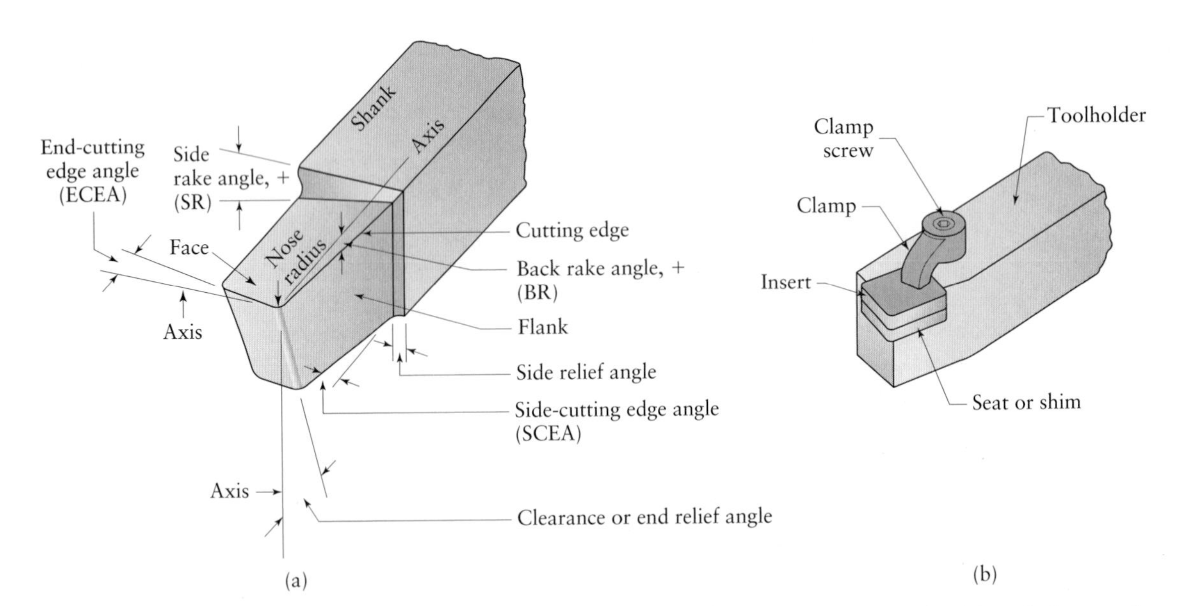

FIGURE 8.10 (a) Schematic illustration of a right-hand cutting tool. Although these tools have traditionally been produced from solid tool-steel bars, they have been largely replaced by carbide or other inserts of various shapes and sizes, as shown in (b).

8.2.3 Forces in orthogonal cutting

Knowledge of the forces and power involved in cutting operations is essential for the following reasons:

1. **Power requirements** must be determined so that a machine of suitable capacity can be selected;
2. Data on forces are necessary for the proper **design of machine tools** so that they are sufficiently stiff, thus maintaining desired dimensional accuracy for the machined part and avoiding excessive distortion of the machine components and vibration and chatter;
3. The workpiece must be able to withstand the cutting forces without excessive **distortion** in order to maintain dimensional tolerances.

The following list discusses the various factors that influence the forces and power involved in orthogonal cutting:

a) **Cutting forces.** Figure 8.11 depicts the forces acting on the tool in orthogonal cutting. **The cutting force,** F_c, acts in the direction of the cutting speed, V, and supplies the energy required for cutting. The **thrust force,** F_t, acts in the direction normal to the cutting velocity, that is, perpendicular to the workpiece. These two forces produce the **resultant force,** F_r. Note that the resultant force can be resolved into two components on the tool face: a **friction force,** F, along the tool–chip interface and a **normal force,** N, perpendicular to the interface. From Fig. 8.11, it can be shown the friction force is

$$F = R \sin \beta \tag{8.9}$$

and that the normal force is

$$N = R \cos \beta. \tag{8.10}$$

Note also that the resultant force is balanced by an equal and opposite force along the shear plane and is resolved into a **shear force,** F_s, and a **normal force,** F_n.

From Fig. 8.11, the cutting force can be shown to be

$$F_c = R \cos(\beta - \alpha) = \frac{wt_o \tau \cos(\beta - \alpha)}{\sin \phi \cos(\phi + \beta - \alpha)}, \tag{8.11}$$

where τ is the *average shear stress* along the shear plane.

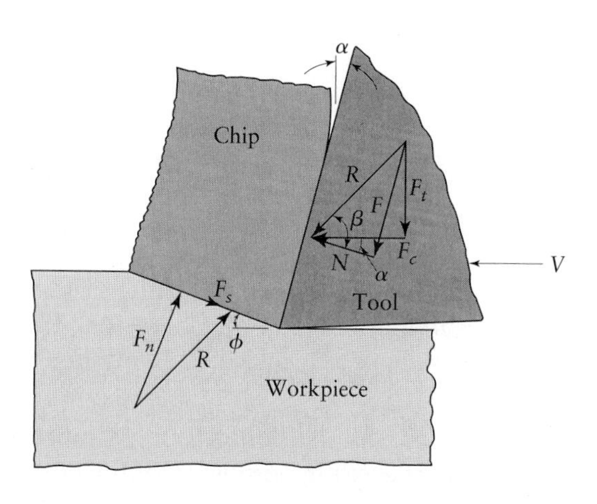

FIGURE 8.11 Forces acting on a cutting tool in two-dimensional cutting.

The ratio of F to N is the **coefficient of friction**, μ, at the tool–chip interface (see also Section 4.4.1), and the angle β is the **friction angle.** The coefficient of friction can be expressed as

$$\mu = \tan \beta = \frac{F_t + F_c \tan \alpha}{F_c - F_t \tan \alpha}. \tag{8.12}$$

In metal cutting, μ generally ranges from about 0.5 to 2, thus indicating that the chip encounters considerable frictional resistance while climbing up the rake face of the tool.

Although the magnitude of forces in actual cutting operations is generally on the order of a few hundred newtons, the *local* stresses in the cutting zone and the normal stresses on the rake face of the tool are very high, because the contact areas are very small. The tool–chip contact length (Fig. 8.2), for example, is typically on the order of 1 mm (0.04 in.); thus, the tool is subjected to very high stresses.

b) **Thrust force and its direction.** A knowledge of the magnitude of the thrust force in cutting is important, because the tool holder, the workholding and fixturing devices, and the machine tool must be sufficiently stiff to minimize deflections caused by this force. For example, if the thrust force (see Fig. 8.11) is too high, or if the machine tool is not sufficiently stiff, the tool will be pushed away from the workpiece surface. This movement will, in turn, reduce the actual depth of cut, leading to loss of dimensional accuracy of the machined part and possibly to vibration and chatter.

Note in Fig. 8.11, that the cutting force must always be in the direction of cutting in order to supply energy for cutting. Although the thrust force does not contribute to the total work done, it is important to know its magnitude. Also note from Fig. 8.11 that the thrust force is *downward.* However, it can be shown that this force can be upward (negative), by first observing that

$$F_t = R \sin(\beta - \alpha) \tag{8.13}$$

or

$$F_t = F_c \tan(\beta - \alpha). \tag{8.14}$$

Because F_c is always positive (as shown in Fig. 8.11), the sign of F_t can be either positive or negative. When $\beta > \alpha$ the sign of F_t is positive (downward), and when $\beta < \alpha$, it is negative (upward). Thus, it is possible to have an upward thrust force at low friction at the tool-chip interface and/or high rake angles.

This situation can be visualized by inspecting Fig. 8.11 and noting that when $\mu = 0$, $\beta = 0$, and consequently the resultant force, R, coincides with the normal force, N. In this case, then, F_r will have an upward thrust-force component; also note that for $\alpha = 0$ and $\beta = 0$, the thrust force is zero.

These observations have been verified experimentally, as shown in Fig. 8.12. The influence of the depth of cut is obvious, because as t_o increases, R must also increase so that F_c will increase as well. This action supplies the additional energy required to remove the extra material produced by the increased depth of cut. The change in direction and magnitude of the thrust force can play a significant role. Within a certain range of operating conditions, it can lead to *instability* problems in machining, particularly if the machine tool is not sufficiently stiff.

c) **Observations on cutting forces.** In addition to being a function of the strength of the workpiece material, forces in cutting are influenced by other

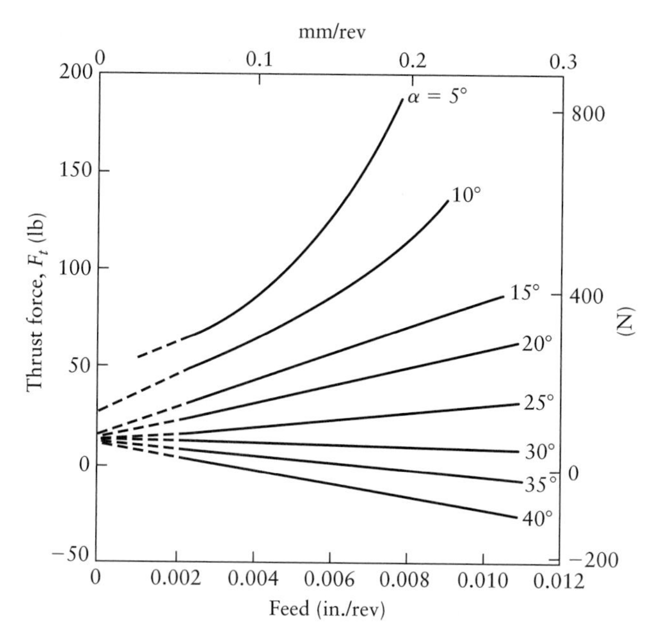

FIGURE 8.12 Thrust force as a function of rake angle and feed in orthogonal cutting of AISI 1112 cold-rolled steel. Note that at high rake angles, the thrust force is negative. A negative thrust force has important implications in the design of machine tools and in controlling the stability of the cutting process. *Source*: After S. Kobayashi and E. G. Thomsen.

variables. Extensive data are availables such as those shown in Tables 8.1 and 8.2, which show that the cutting force, F_c, increases with increasing depth of cut, decreasing rake angle, and decreasing speed. From analysis of the data in Table 8.2, we can attribute the effect of cutting speed to the fact that as speed decreases, the shear angle decreases and the coefficient of friction increases. Both effects increase the cutting force.

TABLE 8.1

Data on Orthogonal Cutting of 4130 Steel*

α	ϕ	γ	μ	β	F_c (lb)	F_t (lb)	u_t (in.-lb/in^3 × 10^3)	u_s	u_f	u_f/u_t (%)
25°	20.9°	2.55	1.46	56	380	224	320	209	111	35
35	31.6	1.56	1.53	57	254	102	214	112	102	48
40	35.7	1.32	1.54	57	232	71	195	94	101	52
45	41.9	1.06	1.83	62	232	68	195	75	120	62

* t_o = 0.0025 in.; w = 0.475 in.; V = 90 ft/min; tool: high-speed steel.
Source: After E. G. Thomsen.

TABLE 8.2

Data on Orthogonal Cutting of 9445 Steel*

α	V	ϕ	γ	μ	β	F_c	F_t	u_t	u_s	u_f	u_f/u_t (%)
+10	197	17	3.4	1.05	46	370	273	400	292	108	27
	400	19	3.1	1.11	48	360	283	390	266	124	32
	642	21.5	2.7	0.95	44	329	217	356	249	107	30
	1186	25	2.4	0.81	39	303	168	328	225	103	31
−10	400	16.5	3.9	0.64	33	416	385	450	342	108	24
	637	19	3.5	0.58	30	384	326	415	312	103	25
	1160	22	3.1	0.51	27	356	263	385	289	96	25

* t_o = 0.037 in.; w = 0.25 in.; tool: cemented carbide.
Source: After M. E. Merchant.

Another factor that can significantly influence the magnitude of the cutting force is the tip radius of the tool: The larger the radius (and hence the duller the tool), the higher is the force. Experimental evidence indicates that, for depths of cut on the order of five times the tip radius or higher, the effect of tool dullness on the cutting forces is negligible. (See also Section 8.4.)

d) **Shear and normal stresses in the cutting zone.** The stresses in the shear plane and at the tool–chip interface can be analyzed by assuming that they are uniformly distributed. The forces in the shear plane can be resolved into shear and normal forces and stresses. First, note that the area of the shear plane, A_s, is

$$A_s = \frac{wt_o}{\sin \phi}. \tag{8.15}$$

Therefore, the *average shear stress* in the shear plane is

$$\tau = \frac{F_s}{A_s} = \frac{F_s \sin \phi}{wt_o}, \tag{8.16}$$

and the *average normal stress* is

$$\sigma = \frac{F_n}{A_s} = \frac{F_n \sin \phi}{wt_o}. \tag{8.17}$$

Figure 8.13 provides some data pertaining to these average stresses. The rake angle is a parameter, and the shear plane area is increased by increasing the depth of cut. The following conclusions can be drawn from these curves:

1. The shear stress on the shear plane is independent of the rake angle.
2. The normal stress on the shear plane decreases with increasing rake angle.
3. Consequently, the normal stress in the shear plane has no effect on the magnitude of the shear stress; this phenomenon has also been verified by other mechanical tests. However, normal stress strongly influences the magnitude of the shear strain in the shear zone. The maximum shear strain to fracture increases with the normal compressive stress, as shown in Fig. 2.22.

Determining the stresses on the rake face of the tool presents considerable difficulties; one problem is accurately determining the length of contact at the tool–chip interface. This length increases with decreasing shear angle, indicating that the contact length is a function of rake angle, cutting speed, and

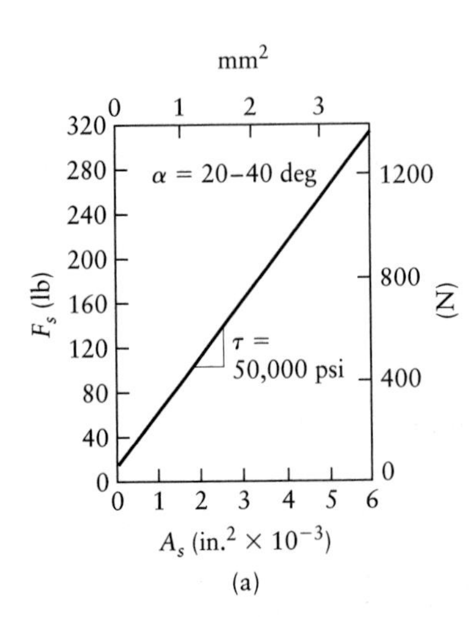

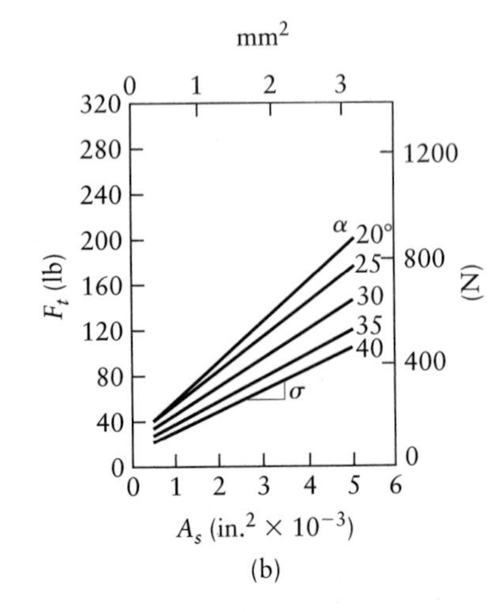

FIGURE 8.13 Shear force and normal force as a function of the area of the shear plane and the rake angle for 85–15 brass. Note that the shear stress in the shear plane is constant, regardless of the magnitude of the normal stress. Thus, normal stress has no effect on the shear flow stress of the material. *Source*: After S. Kobayashi and E. G. Thomsen, *J. Eng. Ind.*, 81: 251–262, 1959.

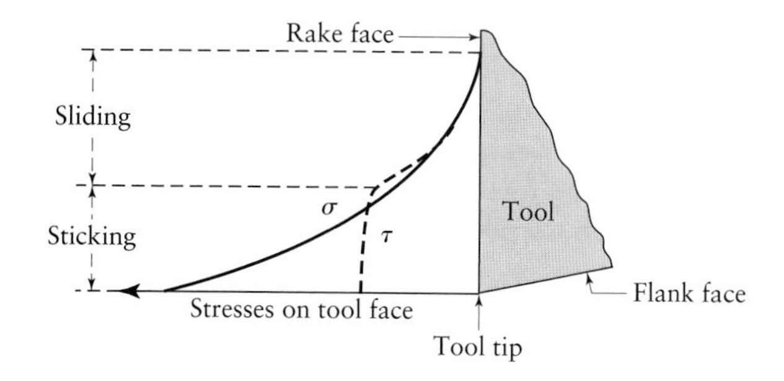

FIGURE 8.14 Schematic illustration of the distribution of normal and shear stresses at the tool–chip interface (rake face). Note that, whereas the normal stress increases continuously toward the tip of the tool, the shear stress reaches a maximum and remains at that value (a phenomenon known as *sticking*; see Section 4.4.1).

friction at the tool–chip interface. Another problem is that the stresses are not uniformly distributed on the rake face. *Photoelastic* studies have shown that the actual stress distribution is as shown in Fig. 8.14. Note that the stress normal to the tool face is a maximum at the tip and decreases rapidly toward the end of the contact length. The shear stress has a similar trend, except that it levels off after about the center of the contact length. This behavior indicates that *sticking* is taking place, whereby the shear stress has reached the shear yield strength of the workpiece material. (See Section 4.4.1.) Sticking regions have been experimentally observed on some chips. Sticking regions also can be observed in various other metal-forming processes; see, for example, Section 6.2.2.

e) **Measuring cutting forces.** Cutting forces can be measured by using suitable force **dynamometers** (with resistance-wire strain gages) or force **transducers** (such as piezoelectric crystals) mounted on the machine tool. Forces can also be calculated from the amount of **power consumption** during cutting, often measured by a power monitor, provided that the mechanical efficiency of the machine tool can be determined.

8.2.4 Shear-angle relationships

Because the shear angle and the shear zone have great significance in the mechanics of cutting, a great deal of effort has been expended to determine the relationship of the shear angle to material properties and process variables. One of the earliest analyses (by M. E. Merchant) is based on the assumption that the shear angle adjusts itself so that the cutting force is a minimum, or so that the maximum shear stress occurs in the shear plane. From the force diagram in Fig. 8.11, we can express the shear stress in the shear plane as

$$\tau = \frac{F_s}{A_s} = \frac{F_c \sec(\beta - \alpha)\cos(\phi + \beta - \alpha)\sin\phi}{wt_o}. \tag{8.18}$$

If we assume that β is independent of ϕ, we can find the shear angle corresponding to the maximum shear stress by differentiating Eq. (8.18) with respect to ϕ and equating to zero. Thus, we have

$$\frac{d\tau}{d\phi} = \cos(\phi + \beta - \alpha)\cos\phi - \sin(\phi + \beta - \alpha)\sin\phi = 0, \tag{8.19}$$

and hence,

$$\tan(\phi + \beta - \alpha) = \cot\phi = \tan(90° - \phi),$$

or

$$\phi = 45° + \frac{\alpha}{2} - \frac{\beta}{2}. \tag{8.20}$$

Note that Eq. (8.20) indicates that, as the rake angle decreases and/or as the friction at the tool–chip interface increases, the shear angle decreases, and the chip is thus thicker. This result is to be expected, because decreasing α and increasing β tend to lower the shear angle.

A second method of determining ϕ is based on slip-line analysis (by E. H. Lee and B. W. Shaffer) and gives an expression for the shear-plane angle as

$$\phi = 45^\circ + \alpha - \beta. \tag{8.21}$$

Note that this expression is similar to Eq. (8.20) and indicates the same trends, although numerically it gives different values.

In another study (by T. Sata and M. Mizuno), the following simple relationships have been proposed and serve as a simple guide for estimating the shear angle for practical purposes:

$$\phi = \alpha \quad \text{for} \quad \alpha > 15^\circ; \tag{8.22}$$

$$\phi = 15^\circ \quad \text{for} \quad \alpha < 15^\circ. \tag{8.23}$$

Several other expressions, based on various models and different assumptions, have been obtained for the shear angle; many of these expressions do not agree well with experimental data over a wide range of conditions (Fig. 8.15a), largely because shear rarely occurs in a thin plane. However, the shear angle always decreases with increasing $\beta - \alpha$, as shown in Fig. 8.15b. More recent and comprehensive studies appear to accurately predict the shear angle analytically, especially for continuous chips.

8.2.5 Specific energy

We note from Fig. 8.11 that the **total power** input in cutting is

$$\text{Power} = F_c V.$$

The **total energy per unit volume** of material removed (**specific energy**), u_t, is then

$$u_t = \frac{F_c V}{w t_o V} = \frac{F_c}{w t_o}, \tag{8.24}$$

where w is the width of the cut. In other words, u_t is simply the ratio of the cutting force to the projected area of the cut. Figures 8.3 and 8.11 also show that the power

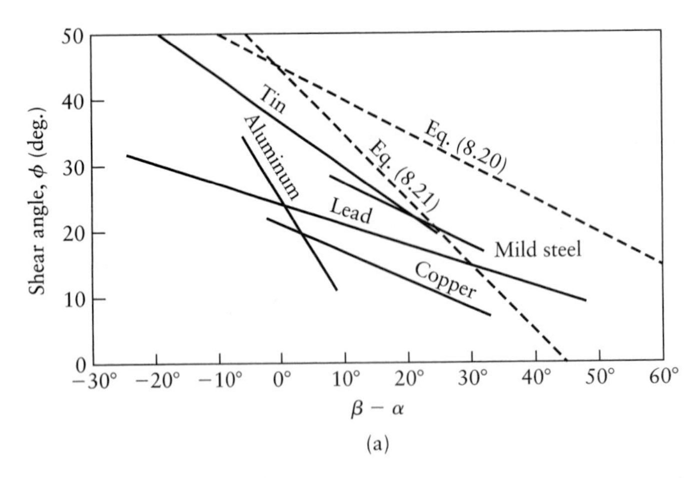

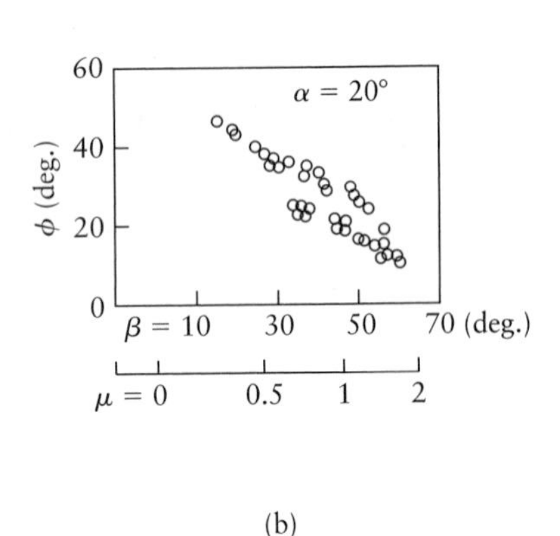

FIGURE 8.15 (a) Comparison of experimental and theoretical shear-angle relationships. More recent analytical studies have resulted in better agreement with experimental data. (b) Relation between the shear angle and the friction angle for various alloys and cutting speeds. *Source*: S. Kobayashi et al., *Trans. ASME, J. Eng. Ind.*, 82: 333–347, 1960.

required to overcome friction at the tool–chip interface is the product of F and V_c, or, in terms of **specific energy for friction,** u_f,

$$u_f = \frac{FV_c}{wt_oV} = \frac{Fr}{wt_o} = \frac{(F_c \sin\alpha + F_t \cos\alpha)r}{wt_o} \tag{8.25}$$

Likewise, the power required for shearing along the shear plane is the product of F_s and V_s. Hence, the **specific energy for shearing,** u_s, is

$$u_s = \frac{F_s V_s}{wt_o V}. \tag{8.26}$$

The **total specific energy,** u_t, is the sum of the two energies, or

$$u_t = u_f + u_s. \tag{8.27}$$

There are two additional sources of energy in cutting. One is the **surface energy** resulting from the formation of *two new surfaces* (the rake-face side of the chip and the newly machined surface) when a layer of material is removed by cutting. However, this energy represents a very small amount as compared with the shear and frictional energies involved. The other source is the energy associated with the **momentum change** as the volume of metal being removed suddenly crosses the shear plane. (This situation is similar to the forces involved in a turbine blade from momentum changes of the liquid or gas.) Although in ordinary metal-cutting operations, the momentum energy is negligible, it can be significant at very high cutting speeds, such as above 125 m/s (25,000 ft/min).

Tables 8.1 and 8.2 provide experimental data on specific energies. Note that as the rake angle increases, the frictional specific energy remains more or less constant, whereas the shear specific energy is rapidly reduced; thus, the ratio u_f/u_t increases significantly as α increases. This trend can also be predicted by obtaining an expression for the energy ratio as follows:

$$\frac{u_f}{u_t} = \frac{FV_c}{F_cV} = \frac{R\sin\beta}{R\cos(\beta-\alpha)} \cdot \frac{Vr}{V} = \frac{\sin\beta}{\cos(\beta-\alpha)} \cdot \frac{\sin\phi}{\cos(\phi-\alpha)}. \tag{8.28}$$

Experimental observations have indicated that as α increases, both β and ϕ increase, and inspection of Eq. (8.28) also indicates that the ratio u_f/u_t should increase with α. Obviously, u_f and u_s are related. Although u_f is not necessary for the cutting

TABLE 8.3

Approximate Specific-Energy Requirements in Cutting Operations

	SPECIFIC ENERGY*	
MATERIAL	W-s/mm^3	hp-min/in^3
Aluminum alloys	0.4–1.1	0.15–0.4
Cast irons	1.6–5.5	0.6–2.0
Copper alloys	1.4–3.3	0.5–1.2
High-temperature alloys	3.3–8.5	1.2–3.1
Magnesium alloys	0.4–0.6	0.15–0.2
Nickel alloys	4.9–6.8	1.8–2.5
Refractory alloys	3.8–9.6	1.1–3.5
Stainless steels	3.0–5.2	1.1–1.9
Steels	2.7–9.3	1.0–3.4
Titanium alloys	3.0–4.1	1.1–1.5

* At drive motor, corrected for 80% efficiency; multiply the energy by 1.25 for dull tools.

action to take place, it affects the magnitude of u_s. The reason is that as friction increases, the shear angle decreases; a decreasing shear angle, in turn, increases the magnitude of u_s.

The calculation of all the parameters involved in these specific energies in cutting presents considerable difficulties; good theoretical computations are available, but they are difficult to perform. Consequently, the reliable prediction of cutting forces and energies is still based largely on experimental data (as shown in Tables 8.1, 8.2, and 8.3). The wide range of values in Table 8.3 can be attributed to differences in strength within each material group and to the effects of various other variables, such as friction and operating conditions.

EXAMPLE 8.1 Relative energies in cutting

An orthogonal cutting operation is being carried out in which $t_o = 0.005$ in., $V = 400$ ft/min, $\alpha = 10°$, and the width of cut = 0.25 in. It is observed that $t_c = 0.009$ in., $F_c = 125$ lb, and $F_t = 50$ lb. Calculate the percentage of the total energy that is dissipated in friction at the tool–chip interface.

Solution. The percentage of energy can be expressed as

$$\frac{\text{Friction energy}}{\text{Total energy}} = \frac{FV_c}{F_cV} = \frac{Fr}{F_c},$$

where

$$r = \frac{t_o}{t_c} = \frac{5}{9} = 0.555,$$

$$F = R \sin \beta,$$

$$F_c = R \cos(\beta - \alpha)$$

and

$$R = \sqrt{F_t^2 + F_c^2} = \sqrt{50^2 + 125^2} = 135 \text{ lb}.$$

Thus,

$$125 = 135 \cos(\beta - 10°),$$

from which we find that

$$\beta = 32° \quad \text{and} \quad F = 135 \sin 32° = 71.5 \text{ lb}.$$

Therefore, the percentage of friction energy is calculated as

$$\text{Percentage} = \frac{(71.5)(0.555)}{125} = 0.32 = 32\%,$$

and similarly, the percentage of shear energy is calculated as 68%.

EXAMPLE 8.2 Comparison of forming and machining energies

You are given two pieces of annealed 304 stainless-steel rods, each with a diameter of 0.500 in. and a length of 6 in., and you are asked to reduce their diameters to 0.480 in., (a) for one piece by *pulling* it in tension and (b) for the other by *machining* it on a lathe (see Fig. 8.8) in one pass. Calculate the respective amounts of work involved, and explain the reasons for the difference in the energies dissipated.

Solution. (a) The work done in pulling the rod (see Section 2.12) is

$$W_{\text{tension}} = (u)(\text{volume}),$$

where

$$u = \int_0^{\varepsilon_1} \sigma \, d\varepsilon.$$

The true strain is found from the expression

$$\varepsilon_1 = \ln\left(\frac{0.500}{0.480}\right)^2 = 0.0816.$$

From Table 2.3, we obtain the following values for K and n for this material:

$$K = 1275 \text{ MPa} = 185{,}000 \text{ psi} \quad \text{and} \quad n = 045.$$

Thus,

$$u = \frac{K\varepsilon_1^{n+1}}{n+1} = \frac{(185{,}000)(0.0816)^{1.45}}{1.45} = 3370 \text{ in.} \cdot \text{lb/in}^3$$

and thus

$$W_{\text{tension}} = (3370)(\pi)(0.25)^2(6) = 3970 \text{ in.-lb.}$$

(b) From Table 8.3, let's estimate an average value for the specific energy in machining stainless steels as 1.5 hp-min/in^3. The volume of material machined is

$$\text{Volume} = \frac{\pi}{4}[(0.5)^2 - (0.480)^2](6) = 0.092 \text{ in}^3.$$

The specific energy, in appropriate units, is calculated as

$$\text{Specific energy} = (1.5)(33{,}000)(12) = 594{,}000 \text{ in.-lb/in}^3.$$

Hence, the work done in machining is

$$W_{\text{mach}} = (594{,}000)(0.092) = 54{,}650 \text{ in.-lb.}$$

Note that the work done in machining is about 14 times higher than that for tension. The reasons for the difference between the two energies are that tension involves very little strain and that there is no friction. Machining, however, involves friction, and the material removed (even though relatively small in volume) has undergone much higher strains than the bulk material undergoing tension. Assuming, from Tables 8.1 and 8.2, an average shear strain of 3, which is equivalent to an effective strain of 1.7 [see Eq. (2.58)], the material in machining is subjected to a strain of $1.7/0.0816 = 21$ times higher than that in tension.

These differences explain why machining consumes much more energy than reducing the diameter of this rod by stretching. It can be shown, however, that as the diameter of the rod decreases, the difference between the two energies becomes *smaller*, assuming that the same depth of material is to be removed. This result can be explained by noting the changes in the relative volumes involved in machining vs. tension as the diameter of the rod decreases.

8.2.6 Temperature

As in all metalworking operations, the energy dissipated in cutting operations is converted into heat, which, in turn, raises the temperature in the cutting zone. Knowledge of the temperature rise in cutting is important, because increases in temperature:

- Adversely affect the strength, hardness, and wear resistance of the cutting tool,
- Cause dimensional changes in the part being machined, making control of dimensional accuracy difficult, and
- Can induce thermal damage to the machined surface, adversely affecting its properties and service life.

In addition, the machine tool itself may be subjected to temperature gradients, causing distortion of the machine.

Because of the work done in shearing and in overcoming friction on the rake face of the tool, the main sources of heat generation are the primary shear zone and the tool–chip interface. In addition, if the tool is dull or worn, heat is also generated by the tool tip rubbing against the machined surface.

Variables affecting temperature. Various studies have been made of temperatures in cutting, based on heat transfer and dimensional analysis, using experimental data. Although Fig. 8.16 shows that there are severe temperature gradients in the cutting zone, a simple and approximate expression for the *mean temperature* for orthogonal cutting is

$$T = \frac{1.2Y_f}{\rho c}\sqrt[3]{\frac{Vt_o}{K}}, \tag{8.29}$$

where T is the mean temperature of the tool–chip interface (°F); Y_f is the flow stress of the material (psi); V is the cutting speed (in./s); t_o is the depth of cut (in.); ρc is the volumetric specific heat of the workpiece (in.-lb/in^3-°F); and K is the *thermal diffusivity* (ratio of thermal conductivity to volumetric specific heat) of the workpiece material (in^2/s). Note that because some of the parameters in Eq. (8.29) depend on temperature, it is important to use appropriate values that are compatible with the predicted temperature range.

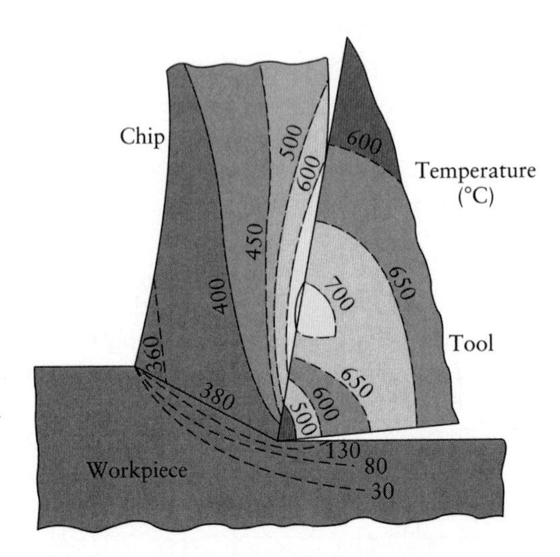

FIGURE 8.16 Typical temperature distribution in the cutting zone. Note that the maximum temperature is about halfway up the rake face of the tool and that there is a steep temperature gradient across the thickness of the chip. Some chips may become red hot, causing safety hazards to the operator and thus necessitating the use of safety guards. *Source*: After G. Vieregge.

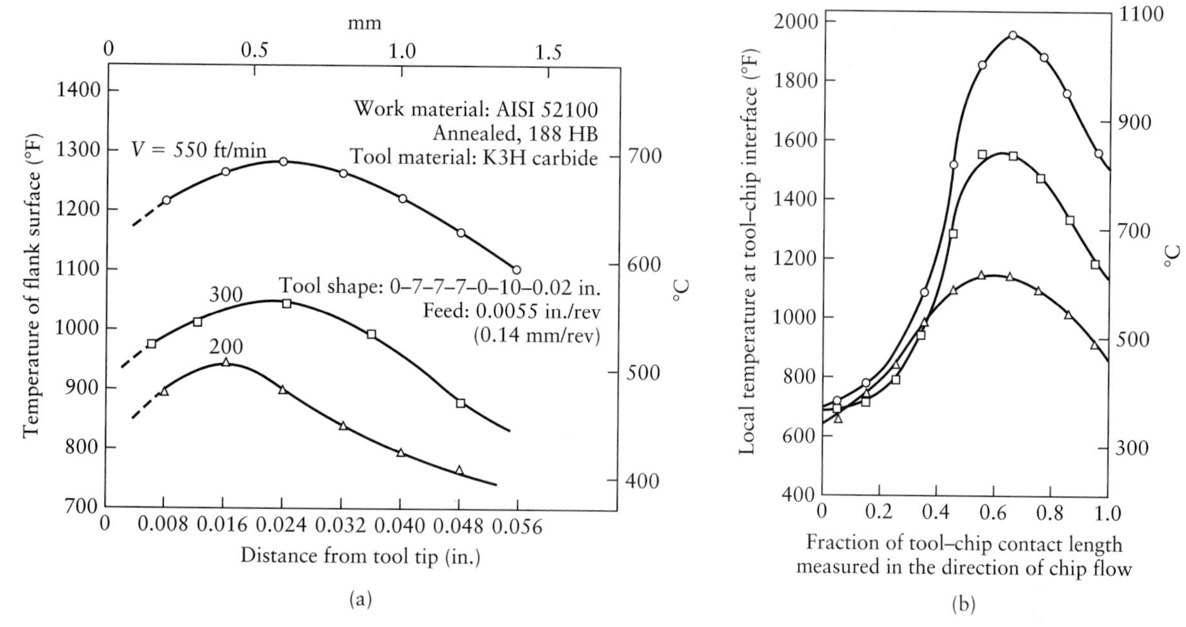

FIGURE 8.17 Temperature distribution in turning: (a) flank temperature for tool shape (see Fig. 8.41); (b) temperature of the tool–chip interface. Note that the rake face temperature is higher than that at the flank surface. *Source*: After B. T. Chao and K. J. Trigger.

The temperature generated in the shear plane is a function of the specific energy for shear, u_s, and the specific heat of the material. Hence, temperature rise is highest in cutting materials with high strength and low specific heat, as Eq. (2.65) indicates. The temperature rise at the tool–chip interface is, of course, also a function of the coefficient of friction. Flank wear (see Section 8.3) is another source of heat, caused by rubbing of the tool on the machined surface.

Figure 8.17 shows results from one experimental measurement of temperature (using thermocouples). Note that the maximum temperature is at a point away from the tip of the tool and that it increases with cutting speed. As the speed increases, there is little time for the heat to be dissipated, and hence temperature rises. As cutting speed increases, a larger proportion of the heat is carried away by the chip, as shown in Fig. 8.18. The chip is a good heat sink, in that it carries away most of the heat generated. (This action of the chip is somewhat similar to *ablation*, where layers of metal melt away from a surface, thus carrying away the heat.) It has been shown that thermal properties of the tool material (see Section 8.6) are relatively unimportant compared with those of the workpiece.

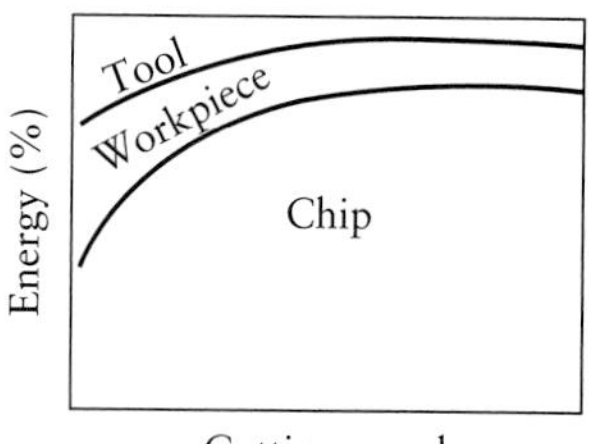

FIGURE 8.18 Typical plot of energy distribution as a function of cutting speed. Note that most of the cutting energy is carried away by the chip (in the form of heat), particularly as speed increases. For dimensional accuracy during cutting, it is important not to allow the workpiece temperature to rise significantly.

Note that Eq. (8.29) also indicates that the mean temperature increases with the strength of the workpiece material (because more energy will be required) and the depth of cut (because of the ratio of surface area to thickness of the chip; a thin chip cools faster than a thick chip). However, the depth of cut has negligible influence on the mean temperature for depths exceeding twice the tip radius of the tool. (See Fig. 8.29.)

Based on Eq. (8.29), another expression for the mean temperature in turning (on a lathe; see Fig. 8.42) is given as

$$T \propto V^a f^b, \tag{8.30}$$

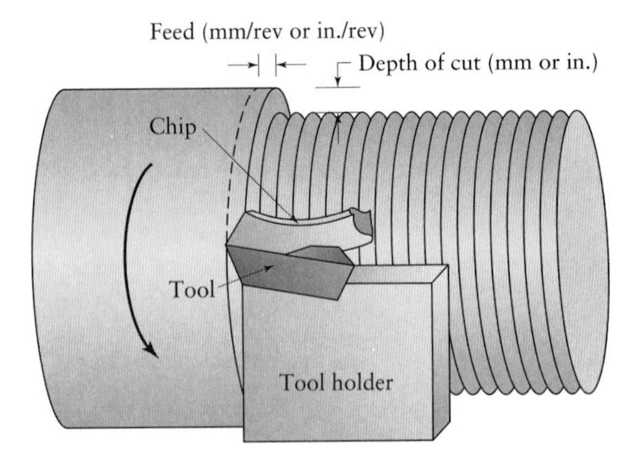

FIGURE 8.19 Terminology used in a turning operation on a lathe, where f is the feed (in in./rev or mm/rev) and d is the depth of cut. Note that feed in turning is equivalent to the depth of cut in orthogonal cutting (Fig. 8.2), and the depth of cut in turning is equivalent to the width of cut in orthogonal cutting. See also Fig. 8.42.

where a and b are constants, V is the cutting speed, and f is the feed of the tool (Fig. 8.19). Approximate values for a and b are given in the following table:

TOOL	a	b
Carbide	0.2	0.125
High-speed steel	0.5	0.375

Techniques for measuring temperature. Temperatures and their distribution in the cutting zone may be determined by using **thermocouples** embedded in the tool or the workpiece. This technique has been used successfully, although it involves considerable effort. A simpler technique for determining the average temperature is by the measuring **thermal emf** (electromotive force) at the tool–chip interface, which acts as a hot junction between two different (tool and chip) materials. **Infrared radiation** from the cutting zone may also be monitored with a *radiation pyrometer*; however, this technique indicates only surface temperatures, and the accuracy of the results depends on the emissivity of the surfaces, which is difficult to determine precisely.

8.3 Tool Wear and Failure

We have seen that cutting tools are subjected to forces, temperature, and sliding; all these conditions induce wear. (See Section 4.4.2.) Because of its effects on the quality of the machined surface and the economics of machining, *tool wear* is one of the most important and complex aspects of machining operations. Although cutting speed is an independent variable, the forces and temperatures generated in cutting are dependent variables and are functions of numerous parameters. Similarly, wear depends on tool and workpiece materials (i.e., their physical, mechanical, and chemical properties), tool geometry, properties of cutting fluids (if used), and various other operating parameters. The types of wear on a tool depend on the relative roles of the foregoing variables. Analytical studies of tool wear present considerable difficulties, and therefore our knowledge of wear is based largely on experimental observations.

The basic wear behavior of a cutting tool is shown in Fig. 8.20c for a two-dimensional cut and in Figs. 8.20a and 8.21b for a three-dimensional cut in which the position of the tool is shown in Fig. 8.19. The various regions of wear are identified as *flank wear*, *crater wear*, *nose wear*, and *chipping* of the cutting edge. The tool

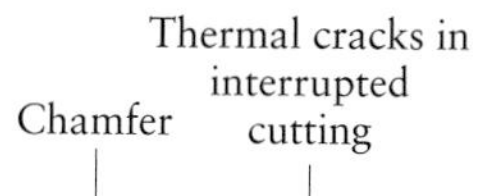

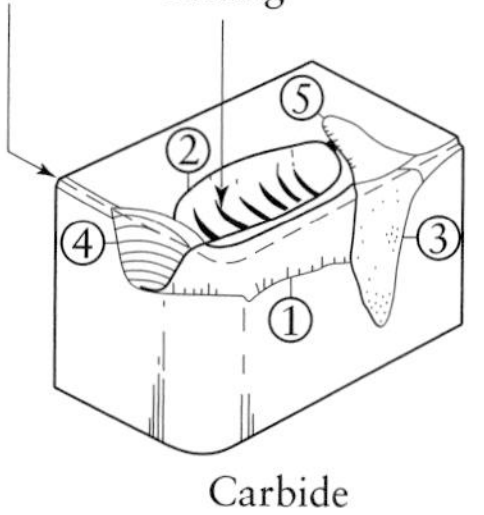

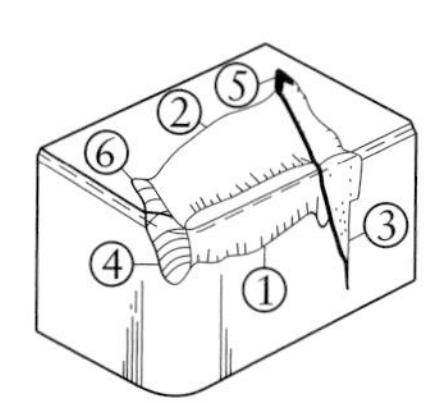

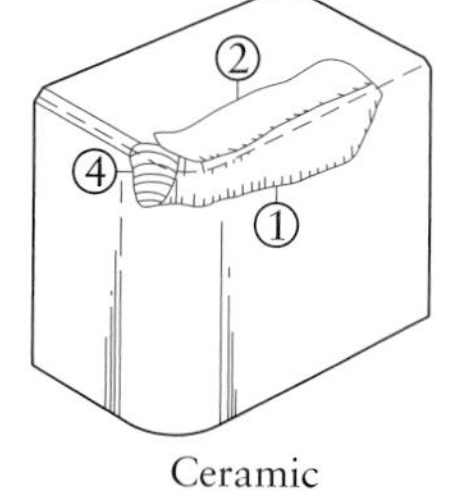

(a)

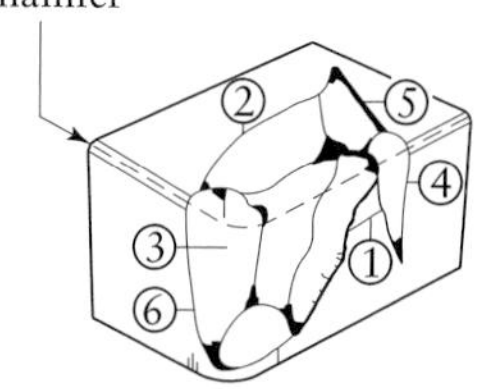

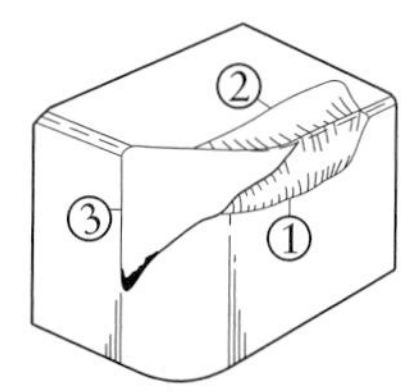

(b)

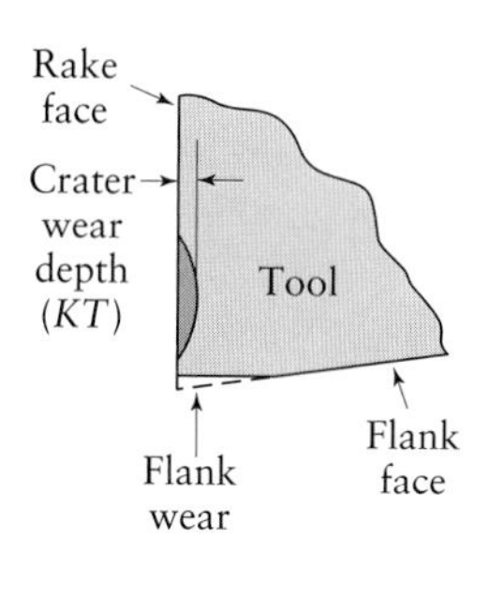

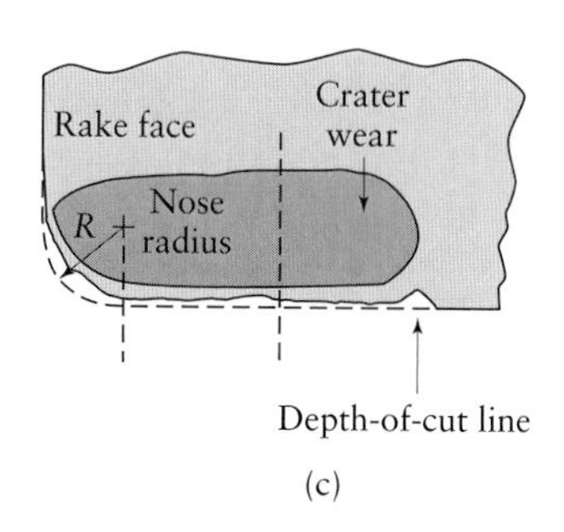

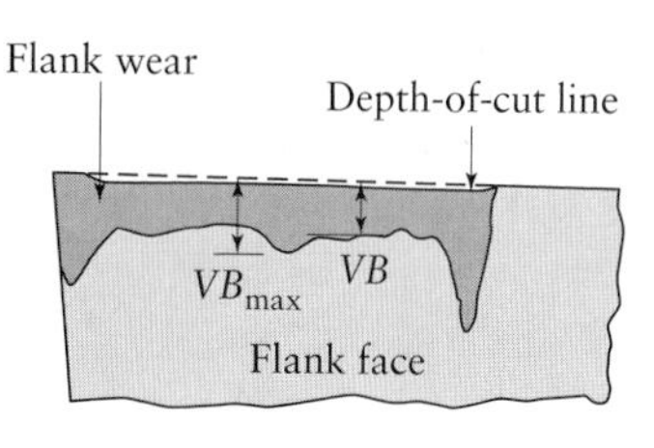

(c)

FIGURE 8.20 (a) Types of wear observed in cutting tools. The thermal cracks shown are usually observed in interrupted cutting operations, such as in milling. (b) Catastrophic failure of tools. (c) Features of tool wear in a turning operation. The *VB* indicates average flank wear. *Source*: (a) and (b) After V. C. Venkatesh. (c) Terms and definitions reproduced with the permission of the International Organization for Standardization, ISO. Copyright remains with ISO.

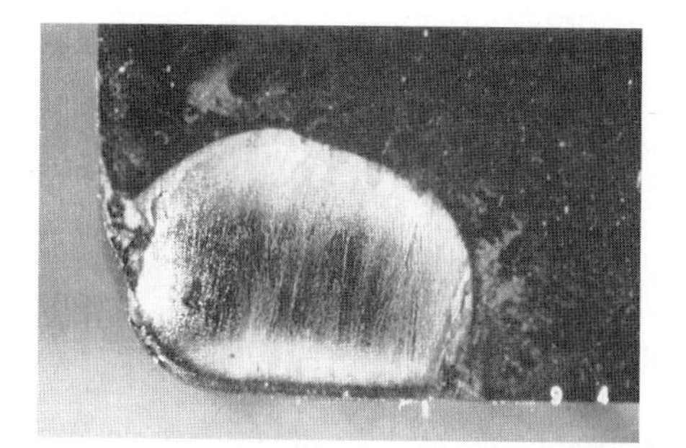

(a)

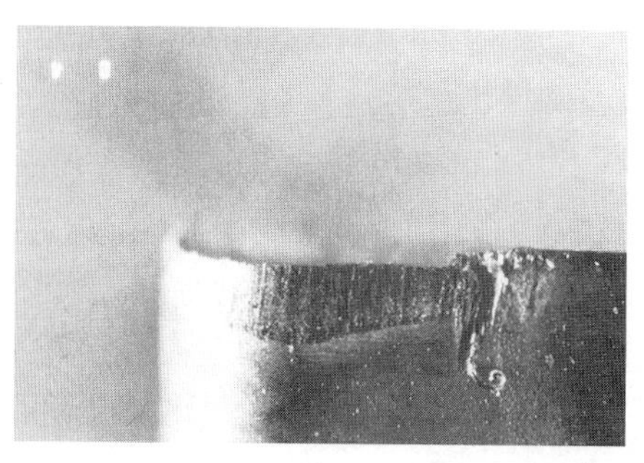

(b)

FIGURE 8.21 (a) Crater wear and (b) flank wear on a carbide tool. *Source*: J. C. Keefe, Lehigh University.

profile is of course, altered by these various wear and fracture processes, and this change influences the cutting operation. In addition to wear, *plastic deformation* of the tool can also take place to some extent, such as in softer tools at elevated temperatures. Gross chipping of the tool is called *catastrophic failure*, whereas wear is generally a gradual process. Because of the complex interactions among the numerous material and process variables involved, the proper interpretation of wear patterns and a careful study of the wear mechanisms are essential.

8.3.1 Flank wear

Flank wear has been studied extensively and is generally attributed to

1. Sliding of the tool along the machined surface, causing adhesive and/or abrasive wear, depending on the materials involved, and
2. Temperature rise, because of its influence on the properties of the tool material.

Following an extensive study by F. W. Taylor published in 1907, a relationship was established for cutting various steels:

$$VT^n = C. \tag{8.31}$$

In this equation, V is the cutting speed, T is the time (in minutes) that it takes to develop a flank wear land (VB in Fig. 8.20c), n is an exponent that depends on cutting conditions, and C is a constant. Equation (8.31) is a simple version of the many relationships developed by Taylor among the variables involved; each combination of workpiece and tool materials and each cutting condition has its own n value and a different constant C.

Tool-life curves. *Tool-life curves* are plots of experimental data obtained in cutting tests (Fig. 8.22). Note the rapid decrease in tool life as cutting speed increases and the

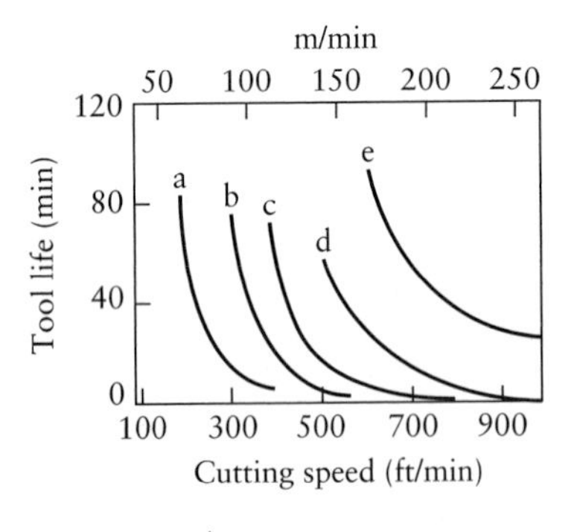

	Hardness (HB)	Ferrite	Pearlite
a. As cast	265	20%	80%
b. As cast	215	40	60
c. As cast	207	60	40
d. Annealed	183	97	3
e. Annealed	170	100	–

(a)

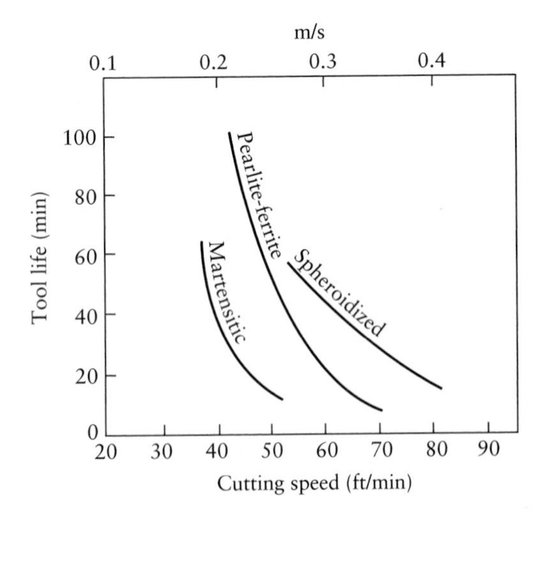

(b)

FIGURE 8.22 Effect of workpiece microstructure on tool life in turning. Tool life is given in terms of the time (in minutes) required to reach a flank wear land of a specified dimension. (a) Ductile cast iron. (b) Steels, with identical hardness. Note in both figures the rapid decrease in tool life as the cutting speed increases.

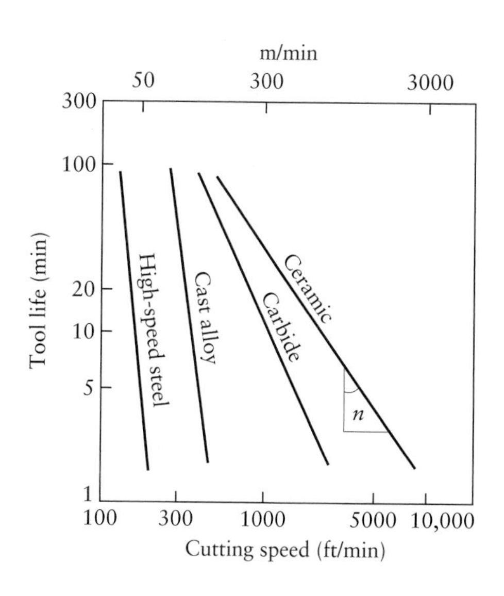

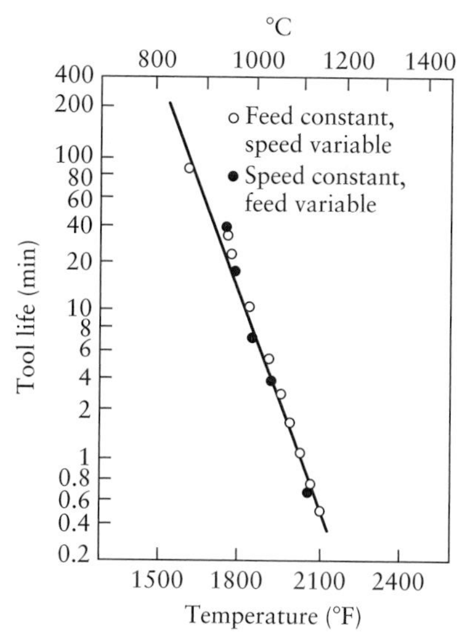

FIGURE 8.23 (a) Tool-life curves for a variety of cutting-tool materials. The negative inverse of the slope of these curves is the exponent n in tool-life equations. (b) Relationship between measured temperature during cutting and tool life (flank wear). Note that high cutting temperatures severely reduce tool life. See also Eq. (8.30). *Source*: After H. Takeyama and Y. Murata.

strong influence of the condition of the workpiece material on tool life; also note the large difference in tool life for different workpiece microstructures. Heat treatment is important largely because of increasing workpiece hardness; for example, ferrite has a hardness of about 100 HB, pearlite 200 HB, and martensite 300 HB to 500 HB. (See Section 5.11.) Impurities and hard constituents in the workpiece material are also important considerations, because they reduce tool life by their abrasive action.

Tool-life curves are usually plotted on log–log paper, from which we can easily determine the exponent n (Fig. 8.23a). Table 8.4 provides the range of n values that has been determined experimentally. Tool-life curves are usually linear over a certain range of cutting speeds, but are rarely so over a wide range. Moreover, the exponent n can indeed become negative at low cutting speeds; thus, tool-life curves may actually reach a maximum and then curve downward. Consequently, caution should be exercised when using tool-life equations beyond the range of cutting speeds for which they were developed.

TABLE 8.4

Range of *n* Values for Various Cutting Tools

High-speed steels	0.08–0.2
Cast alloys	0.1–0.15
Carbides	0.2–0.5
Ceramics	0.5–0.7

Because of the major influence of temperature on the physical and mechanical properties of materials, we would expect that wear is strongly influenced by temperature. Experimental investigations have shown that there is indeed a direct relationship between flank wear and the temperature generated during cutting (Fig. 8.23b). Although cutting speed has been found to be the most significant process variable in tool life, depth of cut and feed rate are also important; thus, Eq. (8.31) can be modified as

$$VT^n d^x f^y = C, \tag{8.32}$$

where d is the depth of cut and f is the feed rate (in mm/rev or in./rev) in turning. The exponents x and y must be determined experimentally for each cutting condition. For example, taking $n = 0.15$, $x = 0.15$, and $y = 0.6$ as typical values encountered in practice, we can see that cutting speed, feed rate, and depth of cut are of decreasing order of importance.

Equation (8.32) can be rewritten as

$$T = C^{1/n} V^{-1/n} d^{-x/n} f^{-y/n}, \tag{8.33}$$

or

$$T \simeq C^7 V^{-7} d^{-1} f^{-4}. \tag{8.34}$$

For a constant tool life, the following observations can be made from Eq. (8.34):

1. If the feed rate or the depth of cut is increased, the cutting speed must be decreased, and vice versa;
2. Depending on the exponents, a reduction in speed can then result in an increase in the volume of the material removed, because of the increased feed rate and/or depth of cut.

Allowable wear land. The *allowable wear land* (VB in Fig. 8.20c) for various conditions is given in Table 8.5; for improved dimensional accuracy and surface finish, the allowable wear land may be made smaller than the values given in the table. The recommended cutting speed for a high-speed-steel tool is generally the one that gives a tool life of 60–120 min (30–60 min for carbide tools).

Optimum cutting speed. We have stated previously that as cutting speed increases, tool life is rapidly reduced. On the other hand, if cutting speeds are low, tool life is long, but the rate at which material is removed is also low; thus, there is an *optimum* cutting speed.

TABLE 8.5

Allowable Average Wear Land for Cutting Tools for Various Operations

	Allowable Wear Land (mm)	
Operation	High-Speed Steels	Carbides
Turning	1.5	0.4
Face milling	1.5	0.4
End milling	0.3	0.3
Drilling	0.4	0.4
Reaming	0.15	0.15

The effect of cutting speed on the volume of metal removed between tool resharpenings or replacements can be appreciated by analyzing Fig. 8.22a. Let's assume that we are machining the material whose condition is represented by curve "a"—that is, as cast with a hardness of 265 HB. If the cutting speed is 1 m/s (200 ft/min), the tool life is about 40 min. Thus, the tool travels a distance of (1 m/s)(60 s/min)(40 min) = 2400 m before it is resharpened or replaced. If we now change the cutting speed to 2 m/s, tool life will be about 5 min, and the tool travels (2)(60)(5) = 600 m. Since the volume of material removed is directly proportional to the distance the tool has traveled, we see that by decreasing the cutting speed, we can remove more material between tool changes. Note, however, that the lower the cutting speed, the longer is the time required to machine a part; these variables have important economic impact. (See Section 8.14.)

EXAMPLE 8.3 Increasing tool life by reducing the cutting speed

Using the Taylor equation [Eq. (8.31)] for tool life and letting $n = 0.5$ and $C = 400$, calculate the percentage increase in tool life when the cutting speed is reduced by 50%.

Solution. Since $n = 0.5$, the Taylor equation can be rewritten as $V\sqrt{T} = 400$. Letting V_1 be the initial speed and V_2 the reduced speed, we note that, for this problem, $V_2 = 0.5V_1$. Because C is a constant, we have the relationship

$$0.5V_1\sqrt{T_2} = V_1\sqrt{T_1}.$$

Simplifying this expression, we find that

$$\frac{T_2}{T_1} = \frac{1}{0.25} = 4.0.$$

This relation indicates that the tool-life change is

$$\frac{T_2 - T_1}{T_1} = \left(\frac{T_2}{T_1}\right) - 1 = 4 - 1 = 3,$$

or that it is increased by 300%. Note that the reduction in cutting speed has resulted in a major increase in tool life and that, in this problem, the magnitude of C is not relevant.

8.3.2 Crater wear

The factors affecting flank wear also influence *crater wear*, but the most significant factors in crater wear are temperature and the degree of chemical affinity between the tool and the workpiece. As shown earlier, the rake face of the tool is subjected to high levels of stress and temperature, in addition to sliding at relatively high speeds. As shown in Fig. 8.17b, peak temperatures, for instance, can be on the order of 1100°C (2000°F). Interestingly, the location of *maximum depth* (KT) of crater wear generally coincides with the location of maximum temperature.

Experimental evidence indicates a direct relation between crater-wear rate and tool–chip interface temperature (Fig. 8.24); note the sharp increase in crater wear

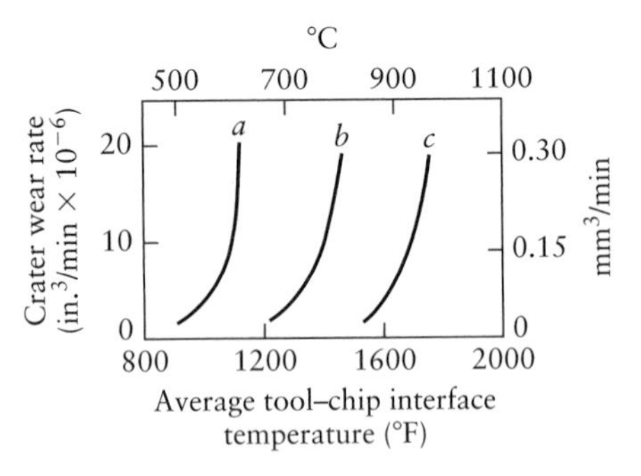

FIGURE 8.24 Relation between crater-wear rate and average tool–chip interface temperature in turning: (a) high-speed-steel tool; (b) C-1 carbide; (c) C-5 carbide. Note that crater wear increases rapidly within a narrow range of temperature. *Source*: After K. J. Trigger and B. T. Chao.

after a certain temperature range has been reached. Figure 8.25 shows the cross-section of the tool–chip interface in cutting steel at high speeds. Note the location of the crater-wear pattern and the discoloration of the tool (loss of temper) as a result of high temperatures; note also how well the discoloration profile agrees with the temperature profile shown in Fig. 8.16.

The effect of temperature on crater wear has been described in terms of a **diffusion** mechanism (the movement of atoms across the tool–chip interface). Diffusion depends on the tool–workpiece material combination and on temperature, pressure, and time; as these quantities increase, the diffusion rate increases. Unless these factors are favorable, crater wear will not take place by diffusion; wear will then be caused by other factors, such as those outlined for flank wear. High temperatures may also cause softening and *plastic deformation* of the tool, because of the decrease in the tool's yield strength and hardness with temperature. This type of deformation is generally observed in machining high-strength metals.

8.3.3 Chipping

The term *chipping* is used to describe the breaking away of a piece from the cutting edge of the tool. The chipped pieces may be very small (*microchipping* or *macrochipping*), or they may involve relatively large fragments (*gross chipping* or

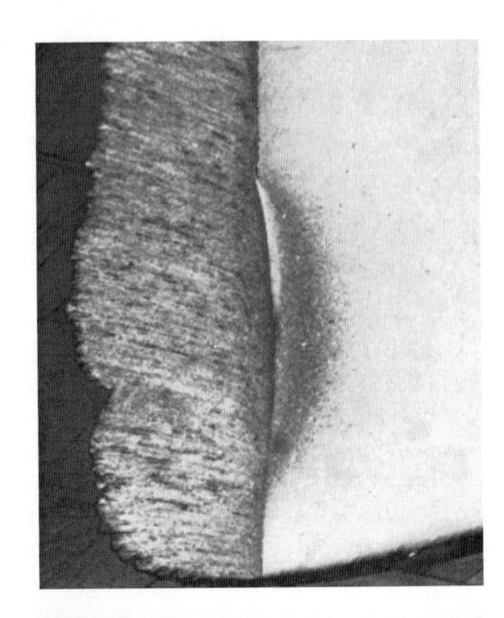

FIGURE 8.25 Interface of chip (left) and rake face of tool (right) and crater wear in cutting AISI 1004 steel at 3 m/s (585 ft/min). Discoloration of the tool indicates high temperature (loss of temper). Note how the crater-wear pattern coincides with the discoloration pattern. Compare this pattern with the temperature distribution shown in Fig. 8.16. *Source*: Courtesy of P. K. Wright.

fracture). Unlike wear, which is a more gradual process, chipping is a phenomenon that results in sudden loss of tool material. Two main causes of chipping are **mechanical shock** and **thermal fatigue** (both by interrupted cutting, as in milling operations).

Chipping by mechanical shock may occur in a region in the tool where a small crack or defect already exists. High positive rake angles also can contribute to chipping, because of the small included angle of the tool tip. Thermal cracks, which are generally perpendicular to the cutting edge (Fig. 8.20a), are caused by the thermal cycling of the tool in interrupted cutting. Thus, chipping can be reduced by selecting tool materials with high impact and thermal-shock resistance. Furthermore, crater wear may also progress toward the tool tip and weaken it, causing chipping.

8.3.4 General observations on tool wear

Because of the many factors involved, including the characteristics of the machine tool, the wear behavior of cutting tools varies significantly. In addition to the wear processes already described, other phenomena also occur in tool wear (Fig. 8.20a). The wear **groove** or **notch** on cutting tools (Fig. 8.20) has been attributed to the fact that this region is the boundary where the chip is no longer in contact with the tool. This boundary, or **depth-of-cut line** (DOC), oscillates, because of inherent variations in the cutting operation, and accelerates the wear process. Furthermore, this region is in contact with the machined surface from the previous cut. Since a machined surface may develop a thin work-hardened layer, this contact could contribute to the formation of the wear groove.

Because they are hard and abrasive, scale and oxide layers on the surface of a workpiece increase wear. In such cases, the depth of cut, d, (see Fig. 8.19) should be greater than the thickness of the oxide film or the work-hardened layer. In other words, light cuts should not be taken on rusted workpieces.

8.3.5 Tool-condition monitoring

With rapidly increasing use of computer-controlled machine tools and implementation of highly automated manufacturing, the reliable and repeatable performance of cutting tools has become an important consideration. Once programmed properly, most modern machine tools operate with little direct supervision by an operator; consequently, the failure of a cutting tool can have serious detrimental effects on the quality of the machined part as well as on the efficiency and economics of the overall machining operation. It is therefore essential to continuously and indirectly monitor the condition of the cutting tool, i.e., for wear, chipping, or gross failure. Most state-of-the-art machine controls are now equipped with tool-condition monitoring systems.

Techniques for condition monitoring typically fall into two general categories: direct and indirect. The **direct** method for observing the condition of a cutting tool involves *optical* measurement of wear, such as by periodically observing changes in the profile of the tool. This method is the most commonly employed and reliable technique and is usually done using a microscope (*toolmakers' microscope*). Continued research on computer vision systems (see Section 14.8.1) for this task is being conducted, mainly because conventional optical measurement techniques require that the cutting operation be stopped.

Indirect methods of measuring wear involve correlating the tool condition with process variables such as forces, power, temperature rise, surface finish, and vibrations. One such technique is **acoustic emission**, which uses a piezoelectric transducer attached to a tool holder. The transducer picks up acoustic-emission signals (typically above 100 kHz) resulting from the stress waves generated during cutting. By analyzing the signals, tool wear and chipping can be monitored (Fig. 8.26). This technique is

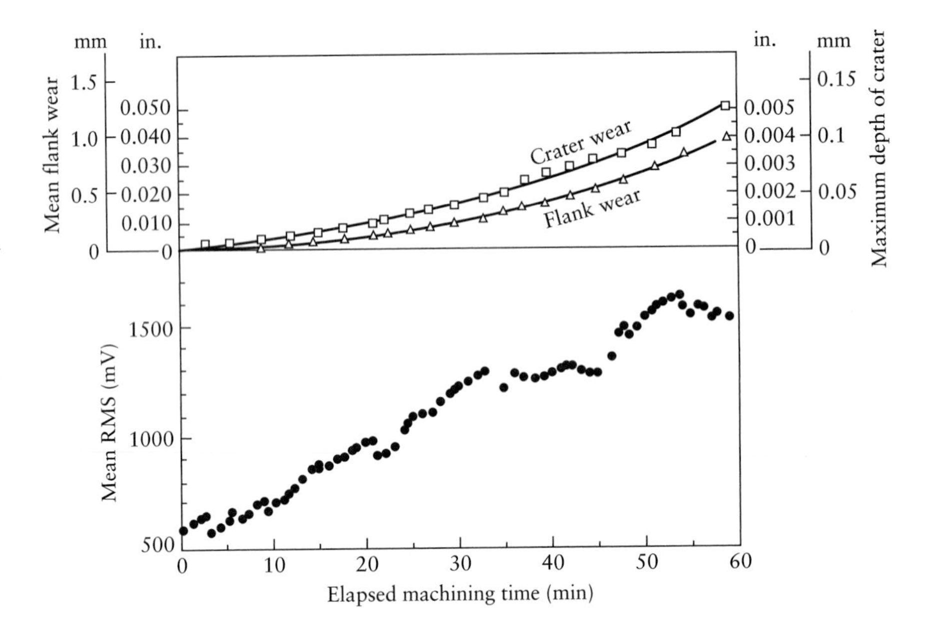

FIGURE 8.26 Relationship between mean flank wear, maximum crater wear, and acoustic emission (noise generated during cutting) as a function of machining time. This technique is being developed as a means for monitoring wear rate in various cutting processes without interrupting the operation. *Source*: After M. S. Lan and D. A. Dornfeld.

particularly effective in precision machining operations, where, because of the very small amounts of material removed, cutting forces are low.

A similar indirect tool-condition monitoring system consists of **transducers** that are installed in original machine tools or retrofitted on existing machines. They continually monitor spindle *torque* and tool *forces* in various directions in a variety of machining processes. The signals are preamplified, and a microprocessor analyzes the content of the signals and interprets it. This system is capable of differentiating the signals that come from tool breakage, tool wear, a missing tool, overloading of the machine, or collision of machine components such as spindles; it can also automatically compensate for tool wear.

The design of the transducers must be such that they are (a) nonintrusive to the machining operation, (b) accurate and repeatable in signal detection, (c) resistant to abuse and shop-floor environment, and (d) cost effective. Continued progress is being made in the development of such sensors (see Section 14.8), including the use of *infrared* and *fiber optic* techniques for temperature measurement in machining operations.

8.4 Surface Finish and Integrity

Surface finish influences not only the dimensional accuracy of machined parts, but also the properties of the parts, especially fatigue strength. Whereas **surface finish** describes the *geometric* features of surfaces, **surface integrity** pertains to *properties* such as fatigue life and corrosion resistance, which are strongly influenced by the type of surface produced. The factors that influence surface integrity are (a) temperatures generated during processing, (b) residual stresses, (c) metallurgical (phase) transformations, and (d) plastic deformation, tearing, and cracking of the surface. Figure 8.27 provides ranges of surface roughness for machining and other processes.

The built-up edge, with its significant effect on tool profile, has the greatest influence on surface roughness of all the factors. Figure 8.28 shows surfaces obtained

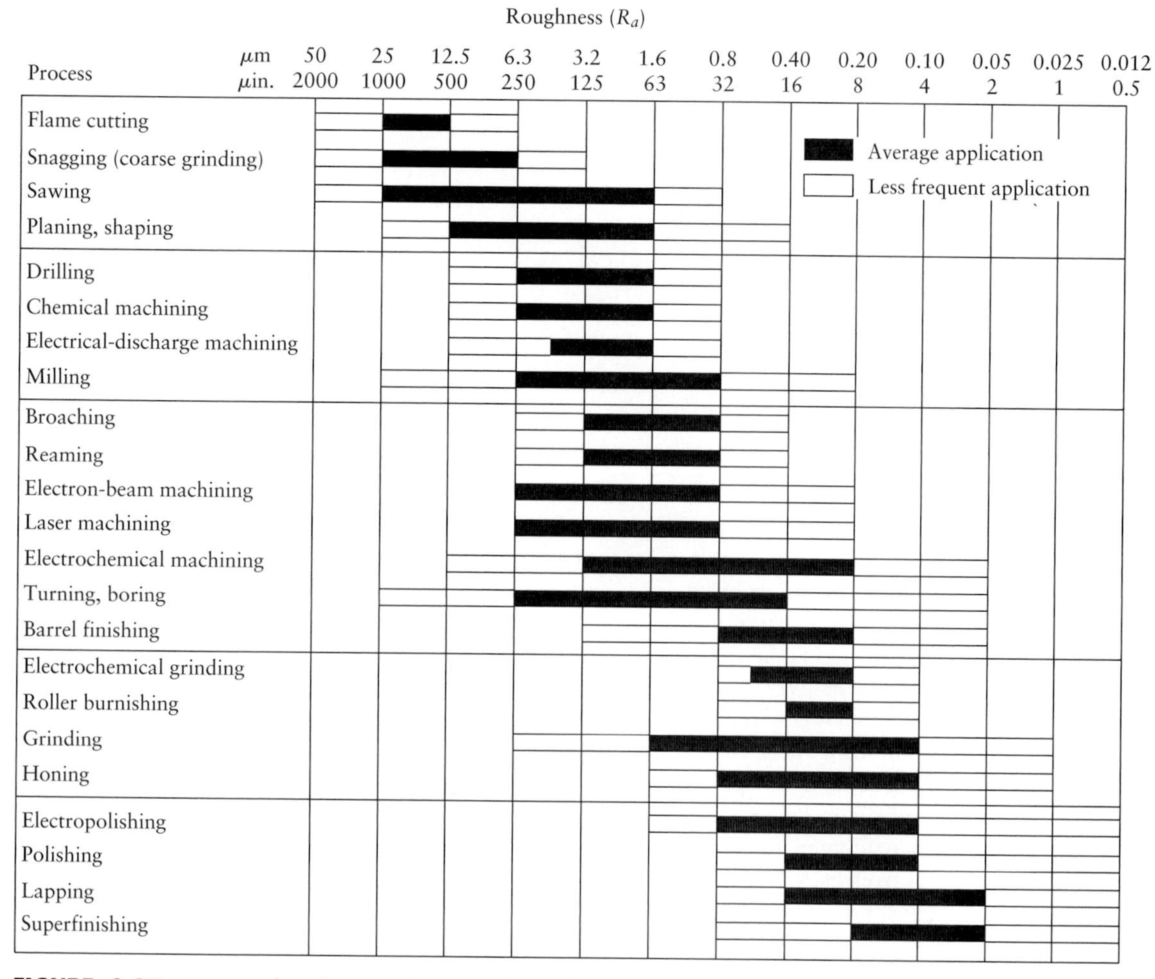

FIGURE 8.27 Range of surface roughnesses obtained in various machining processes. Note the wide range within each group. (See also Fig. 9.27).

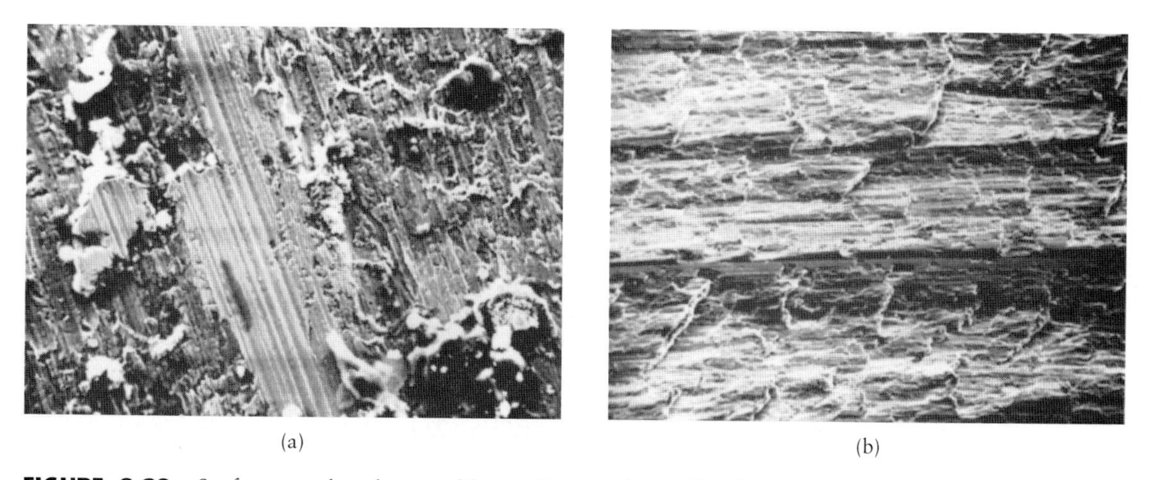

FIGURE 8.28 Surfaces produced on steel by cutting, as observed with a scanning electron microscope: (a) turned surface and (b) surface produced by shaping. *Source*: J. T. Black and S. Ramalingam.

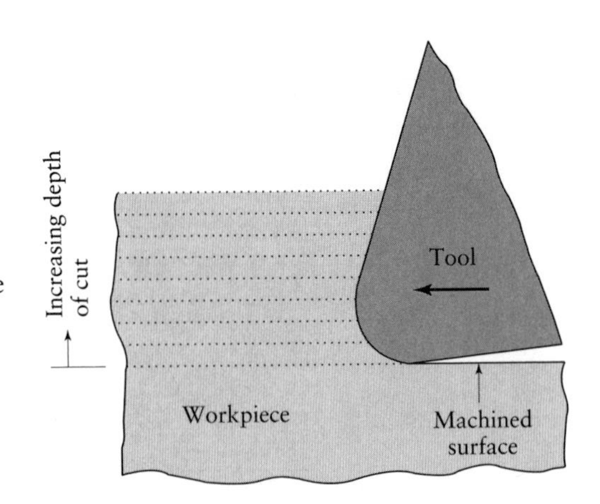

FIGURE 8.29 Schematic illustration of a dull tool in orthogonal cutting (exaggerated). Note that at small depths of cut, the rake angle can effectively become negative. In such cases, the tool may simply ride over the workpiece surface, burnishing it.

in two different cutting operations; note the considerable damage to the surfaces from BUE. Ceramic and diamond tools generally produce better surface finish than other tools, largely because of their much lower tendency to form BUE.

A tool that is not sharp has a large radius along its edges (see Fig. 8.20c), just as a dull pencil or knife does. Figure 8.29 illustrates the relationship between the radius of the cutting edge and depth of cut in orthogonal cutting. Note that at small depths of cut, the rake angle can effectively become negative, and thus the tool may simply ride over the workpiece surface and not remove any chips. You can simulate this behavior by trying to scrape the surface of a stick of butter along its length with a dull knife; note that you will not be able to remove a thin layer of butter, whereas if the knife is sharp, you will be able to do so. If this radius, not to be confused with radius R in Fig. 8.20c, is large in relation to the depth of cut, the tool will rub over the machined surface. Rubbing in cutting generates heat and induces surface residual stresses, which, in turn, may cause surface damage such as tearing and cracking. Thus, the depth of cut should generally be greater than the radius on the cutting edge.

Feed marks. In turning, as in other cutting operations, the tool leaves a spiral profile (*feed marks*) on the machined surface as it moves across the workpiece. (See Fig. 8.19.) The higher the feed, f, and the smaller the radius, R, the more prominent are these marks. Although not significant in rough machining operations, the marks are important in finish machining. In turning, the peak-to-valley roughness, R_t, can be expressed as

$$R_t = \frac{f^2}{8R}, \tag{8.35}$$

where f is the feed and R is the nose radius. If the nose radius is much smaller than the feed, the roughness is given by

$$R_t = \frac{f}{\tan \alpha_s + \cot \alpha_e}, \tag{8.36}$$

where α_s and α_e are the side and edge cutting angles, respectively (see Fig. 8.41). For face milling (see Fig. 8.57), the roughness is given by

$$R_t = \frac{f^2}{16(D \pm (2fn/\pi))}, \tag{8.37}$$

where D is the cutter diameter, f is the feed per tooth, and n is the number of inserts on the cutter. Equations (8.35) to (8.37) are derived considering geometry only and do not account for workpiece cracking (see Fig. 8.6c), thermal distortion, or chatter. Vibration and chatter are described in greater detail in Section 8.11. For now, we should recognize that if the tool vibrates or chatters during cutting, surface finish will be adversely affected, the reason being that a vibrating tool changes the dimensions of the cut periodically. Excessive chatter can also cause chipping and premature failure of the more brittle cutting-tool materials, such as ceramics and diamond.

8.5 Machinability

The *machinability* of a material is usually defined in terms of the following factors: (1) surface finish and integrity of the machined part, (2) tool life obtained, (3) force and power requirements, and (4) chip control. Thus, good machinability indicates *good surface finish and integrity, long tool life, low force and power* requirements, and easy collection of *chips* that does not interfere with the cutting operation. (As stated earlier, long, thin, and curled chips, if not broken up, can become entangled in the cutting zone.)

Because of the complex nature of machining operations, it is difficult to establish relationships to define quantitatively the machinability of a material. However, in manufacturing plants, tool life and surface roughness are generally considered to be the most important factors in machinability. Although not used much anymore, approximate **machinability ratings** are available. Machinability ratings are usually based on a tool life of $T = 60$ min. The standard material is AISI 1112 steel, which is given a rating of 100. Therefore, for a tool life of 60 min, this steel should be machined at a cutting speed of 100 ft/min (0.5 m/s); higher speeds will reduce tool life, and lower speeds will increase it. Note also that as the hardness of a material increases, its machinability rating generally decreases proportionately.

For example, (a) 3140 steel has a machinability rating of 55, which means that its tool life will be 60 min if it is machined at a cutting speed of 55 ft/min (0.275 m/s), and (b) nickel has a rating of 200, indicating that it should be machined at 200 ft/min (1 m/s) to obtain a tool life of 60 min. Machinability ratings for various other materials are as follows: free-cutting brass, 300; 2011 wrought aluminum, 200; pearlitic gray iron, 70; Inconel, 30; and precipitation-hardening 17–7 steel, 20.

8.5.1 Machinability of steels

Because steels are among the most important engineering materials, their machinability has been studied extensively. Their machinability has been improved mainly by adding lead and sulfur to obtain so-called **free-machining steels.**

Leaded steels. Lead is added to the molten steel and takes the form of dispersed fine lead particles. (See Fig. 5.2a.) Lead is insoluble in iron, copper, and aluminum and their alloys; thus, during cutting, the lead particles are sheared and smeared over the tool–chip interface. Because of their low shear strength, the lead particles act as a solid lubricant. This behavior has been verified by the presence of high concentrations of lead on the tool-side face of chips in machining leaded steels. In addition to this effect, it is believed that lead probably lowers the shear stress in the primary shear zone, thus reducing cutting forces and power consumption. Lead may be used in either nonsulfurized or resulfurized steels; however, because of environmental concerns, the trend now is towards eliminating the use of leaded steels in favor of bismuth and tin (*lead-free steels*). Leaded steels are identified by the letter L between the second and third numerals—for example, 10L45. However, similar use of the letter L in identifying stainless steels refers to a low carbon content, which improves the steels' resistance to corrosion.

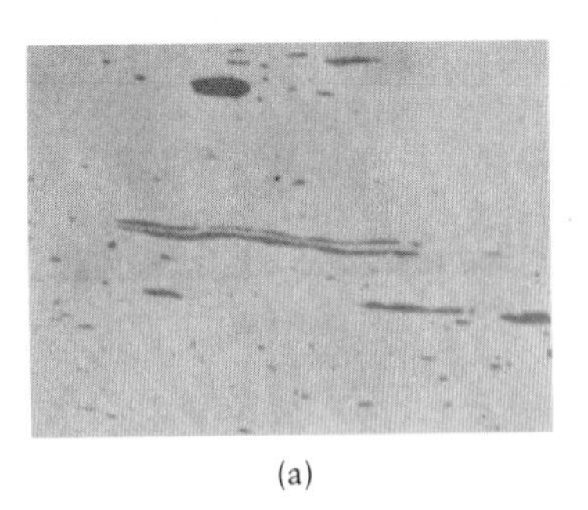
(a)

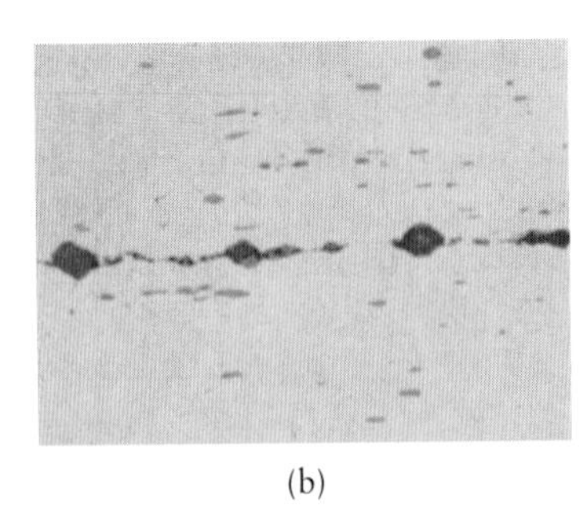
(b)

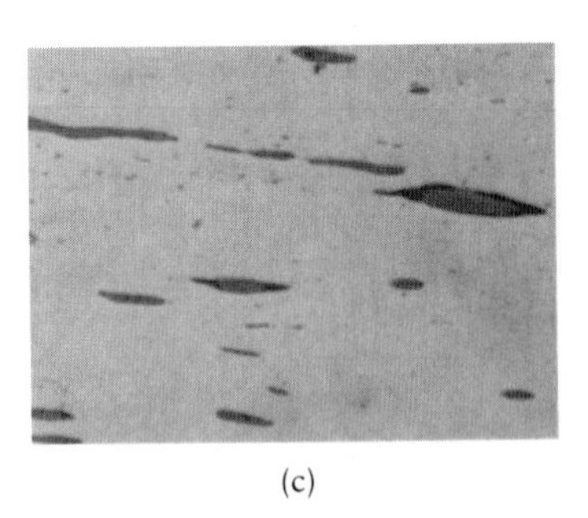
(c)

FIGURE 8.30 Photomicrographs showing various types of inclusions in low-carbon, resulfurized free-machining steels. (a) Manganese-sulfide inclusions in AISI 1215 steel. (b) Manganese-sulfide inclusions and glassy manganese-silicate-type oxide (dark) in AISI 1215 steel. (c) Manganese sulfide with lead particle as tails in AISI 12L14 steel. *Source*: Courtesy of Ispat Inland Inc.

Resulfurized and rephosphorized steels. Sulfur in steels forms manganese-sulfide inclusions (second-phase particles, see Fig. 8.30), which act as stress raisers in the primary shear zone. As a result, the chips produced are small and break up easily, thus improving machinability. The shape, orientation, distribution, and concentration of these inclusions significantly influence machinability. Elements such as tellurium and selenium (both chemically similar to sulfur) in resulfurized steels act as inclusion modifiers. Phosphorus in steels improves machinability by virtue of strengthening the ferrite, thereby increasing the hardness of the steels.

Calcium-deoxidized steels. In these steels, oxide flakes of calcium aluminosilicate (CaO, SiO_2, and Al_2O_3) are formed that, in turn, reduce the strength of the secondary shear zone, thus decreasing tool–chip interface friction and wear and, hence, temperature. Consequently, these steels cause less crater wear of the tool, especially at high cutting speeds.

Effects of other elements in steels on machinability. The presence of aluminum and silicon in steels is always harmful, because these elements combine with oxygen and form aluminum oxide and silicates. These compounds are hard and abrasive, thus increasing tool wear and reducing machinability.

Carbon and manganese have various effects on the machinability of steels, depending on their composition. As the carbon content increases, machinability decreases; however, plain low-carbon steels (less than 0.15% C) can produce poor surface finish by forming a built-up edge. Cast steels are more abrasive, although their machinability is similar to that of wrought steels. Tool and die steels are very difficult to machine and usually require annealing prior to machining. Machinability of most steels is generally improved by cold working, which reduces the tendency for built-up edge formation.

Other alloying elements, such as nickel, chromium, molybdenum, and vanadium, which improve the properties of steels, generally reduce machinability. The effect of boron is negligible. The role of gaseous elements such as oxygen, hydrogen, and nitrogen has not been clearly established; any effect that they may have would depend on the presence and amount of other alloying elements.

In selecting various elements to improve machinability, the possible detrimental effects of the elements on the properties and strength of the machined part in service also should be considered. At elevated temperatures, for example, lead causes embrittlement of steels (*hot shortness*; see Section 3.4.2), although at room temperature, it has no effect on mechanical properties. Sulfur can severely reduce hot workability of steels, because of the presence of iron sulfide, unless sufficient manganese is present to prevent the formation of iron sulfide. At room temperature, the mechanical properties of resulfurized steels depend on the orientation of the deformed manganese-sulfide inclusions (anisotropy). Rephosphorized steels are significantly less ductile than resulfurized steels and are produced solely to improve machinability.

Stainless steels. Austenitic (300 series) stainless steels are generally difficult to machine. Chatter could be a problem, which necessitates the use of machine tools with high stiffness and damping capacity; however, ferritic stainless steels (also 300 series) have good machinability. Martensitic (400 series) stainless steels are abrasive, tend to form built-up edge, and require tool materials with high hot hardness and resistance to crater wear. Precipitation-hardening stainless steels are strong and abrasive and thus require hard and abrasion-resistant tool materials.

8.5.2 Machinability of various other metals

Aluminum is generally easy to machine; however, the softer grades tend to form built-up edge. High cutting speeds, high rake angles, and high relief angles are recommended. Wrought alloys with high silicon content and cast-aluminum alloys may be abrasive and, hence, require harder tool materials. Dimensional control may be a problem in machining aluminum, because of its low elastic modulus and a relatively high thermal coefficient of expansion.

Beryllium is similar to cast irons, but is more abrasive and toxic; hence, it requires machining in a controlled environment.

Gray cast irons are generally machinable, but are abrasive. Free carbides in castings reduce the castings' machinability and cause tool chipping or fracture, thus requiring tools with high toughness. Nodular and malleable irons are machinable with hard tool materials.

Cobalt-base alloys are abrasive and highly work hardening; they require sharp and abrasion-resistant tool materials and low feeds and speeds.

Wrought copper can be difficult to machine, because of built-up edge formation, although cast-copper alloys are easy to machine. *Brasses* are easy to machine as well, especially those to which lead has been added (*leaded free-machining brass*). *Bronzes* are more difficult to machine than brass.

Magnesium is very easy to machine, has good surface finish, and prolongs tool life; however, care should be exercised when using it, because of its high rate of oxidation (*pyrophoric*) and the danger of fire.

Molybdenum is ductile and work hardening; hence, it can produce poor surface finish, thereby requiring the use of sharp tools.

Nickel-base alloys are work hardening, abrasive, and strong at high temperatures; their machinability is similar to that of stainless steels.

Tantalum is very work hardening, ductile, and soft; hence, it produces a poor surface finish, and tool wear is high.

Titanium and its alloys have poor thermal conductivity (the lowest of all metals), thus causing significant temperature rise and built-up edge; hence, it can be difficult to machine.

Tungsten is brittle, strong, and very abrasive; hence, its machinability is low. However, its machinability improves greatly at elevated temperatures.

Zirconium has good machinability; however, a coolant-type cutting fluid must be used, because of the danger of explosion and fire.

8.5.3 Machinability of various materials

Graphite is abrasive; thus, it requires hard, abrasion-resistant sharp tools.

Thermoplastics generally have low thermal conductivity, a low elastic modulus, and a low softening temperature. Consequently, machining them requires tools with a positive rake angle to reduce cutting forces; large relief angles; small depths of cut and feed; relatively high speeds; and proper support of the workpiece, because of lack of stiffness. Tools should be sharp, and external cooling of the cutting zone may be necessary to keep the chips from becoming "gummy" and sticking to the tools. Cooling can usually be done with a jet of air, vapor mist, or water-soluble oils. Residual stresses may develop during machining. To relieve residual stresses, machined parts can be annealed at temperatures ranging from 80°C to 160°C (175°F to 315°F) for a period of time and then cooled slowly and uniformly to room temperature.

Thermosetting plastics are brittle and sensitive to thermal gradients during cutting; their machinability is generally similar to that of thermoplastics.

Because of the fibers present, *reinforced plastics* are generally very abrasive and are difficult to machine. Fiber tearing and pulling is a significant problem. Furthermore, machining of these materials requires careful removal of machining debris to avoid human contact with and inhalation of the fibers.

Metal-matrix and *ceramic-matrix composites* can be difficult to machine, depending on the properties of the individual components in the material.

The machinability of *ceramics* has improved steadily, especially with the development of *nanoceramics* (see Section 11.8.1) and with the selection of appropriate machining parameters, such as in *ductile-regime cutting* (Section 8.8.2).

8.5.4 Thermally assisted machining

Materials that are difficult to machine at room temperature can be machined more easily at elevated temperatures, thus lowering cutting forces and increasing tool life. In *thermally assisted machining* (**hot machining**), the source of heat is a torch, high-energy beam (such as a laser or electron beam) or plasma arc, focused to an area just ahead of the cutting tool. Most applications for hot machining are in turning and milling. The process of heating to and maintaining a uniform temperature distribution within the workpiece may be difficult to control; also, the original microstructure of the workpiece may be adversely affected. Except in isolated cases, thermally assisted machining offers no significant advantage over machining at room temperature with the use of appropriate cutting tools and fluids. Some success has been obtained in laser-assisted machining of silicon-nitride ceramics and *Stellite*.

8.6 Cutting-Tool Materials

The proper selection of cutting-tool materials is among the most important factors in machining operations, as is the selection of mold and die materials for forming and shaping processes. We noted previously that in cutting, the tool is subjected to high temperatures, contact stresses, rubbing on the workpiece surface, and the effects of the chip climbing up the rake face of the tool. Consequently, a cutting tool must have the following characteristics in order to produce good-quality and economical parts:

- **Hardness**, particularly at elevated temperatures (**hot hardness**), so that the hardness and strength of the tool are maintained at the temperatures encountered in cutting operations (Fig. 8.31);

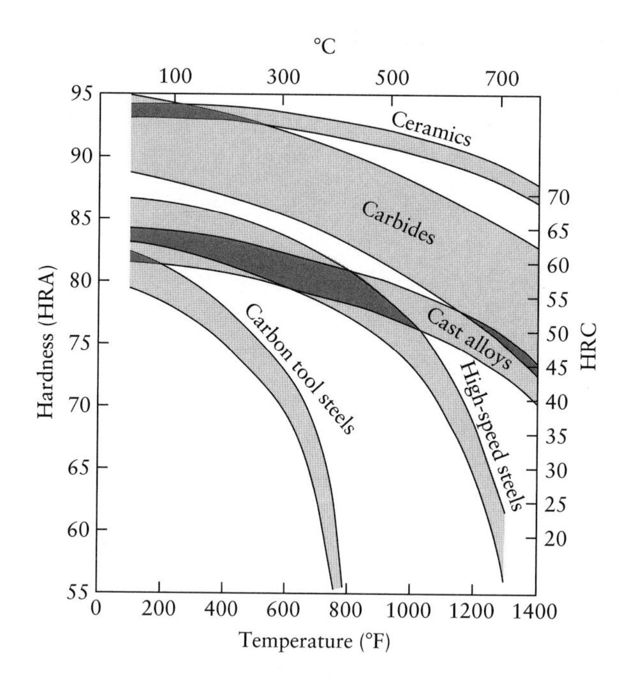

FIGURE 8.31 Hardness of various cutting-tool materials as a function of temperature (hot hardness). The wide range in each group of materials results from the variety of tool compositions and treatments available for that group.

- **Toughness**, so that impact forces on the tool in interrupted cutting operations, such as milling or turning a splined shaft, do not chip or fracture the tool;
- **Wear resistance**, so that an acceptable tool life is obtained before the tool is resharpened or replaced;
- **Chemical stability** or **inertness** with respect to the workpiece material, so that any adverse reactions contributing to tool wear are avoided or minimized.

Several cutting-tool materials having a wide range of these characteristics are now available. (See Table 8.6.) Tool materials are usually divided into the following general categories, which are listed in the approximate chronological order in which they were developed and implemented:

1. **Carbon and medium-alloy steels;**
2. **High-speed steels;**
3. **Cast-cobalt alloys;**
4. **Carbides;**
5. **Coated tools;**
6. **Alumina-base ceramics;**
7. **Cubic boron nitride;**
8. **Silicon-nitride-base ceramics;**
9. **Diamond;**
10. **Whisker-reinforced and nanocrystalline tool materials.**

Many of these materials are also used for dies and molds (Chapters 5, 6, 7, 10 and 11).

This section describes the characteristics, applications, and limitations of these tool materials; also discussed are characteristics such as hot hardness, toughness, impact strength, wear resistance, resistance to thermal shock, and costs, as well as the range of cutting speeds and depth of cut for optimum performance.

TABLE 8.6

Typical Range of Properties of Various Tool Materials

			CARBIDES				
PROPERTY	HIGH-SPEED STEEL	CAST ALLOYS	WC	TiC	CERAMICS	CUBIC BORON NITRIDE	SINGLE-CRYSTAL DIAMOND*
Hardness	83–86 HRA	82–84 HRA 46–62 HRC	90–95 HRA 1800–2400 HK	91–93 HRA 1800–3200 HK	91–95 HRA 2000–3000 HK	4000–5000 HK	7000–8000 HK
Compressive strength							
MPa	4100–4500	1500–2300	4100–5850	3100–3850	2750–4500	6900	6900
psi × 10^3	600–650	220–335	600–850	450–560	400–650	1000	1000
Transverse rupture strength							
MPa	2400–4800	1380–2050	1050–2600	1380–1900	345–950	700	1350
psi × 10^3	350–700	200–300	150–375	200–275	50–135	105	200
Impact strength							
J	1.35–8	0.34–1.25	0.34–1.35	0.79–1.24	<0.1	<0.5	<0.2
in.-lb	12–70	3–11	3–12	7–11	<1	<5	<2
Modulus of elasticity							
GPa	200	—	520–690	310–450	310–410	850	820–1050
psi × 10^6	30	—	75–100	45–65	45–60	125	120–150
Density							
kg/m^3	8600	8000–8700	10,000–15,000	5500–5800	4000–4500	3500	3500
lb/in^3	0.31	0.29–0.31	0.36–0.54	0.2–0.22	0.14–0.16	0.13	0.13
Volume of hard phase (%)	7–15	10–20	70–90	-	100	95	95
Melting or decomposition temperature							
°C	1300	—	1400	1400	2000	1300	700
°F	2370	—	2550	2550	3600	2400	1300
Thermal conductivity, W/mK	30–50	—	42–125	17	29	13	500–2000
Coefficient of thermal expansion, × 10^{-6}/°C	12	—	4–6.5	7.5–9	6–8.5	4.8	1.5–4.8

* The values for polycrystalline diamond are generally lower, except impact strength, which is higher.

8.6.1 Carbon and medium-alloy steels

Carbon steels are the oldest of tool materials and have been used widely for drills, taps, broaches, and reamers since the 1880s. Low-alloy and medium-alloy steels were developed later for similar applications, but with longer tool life. Although inexpensive and easily shaped and sharpened, these steels do not have sufficient hot hardness and wear resistance for cutting at high speeds, where, as we have seen, the temperature rises significantly. Note in Fig. 8.31, for example, how rapidly the hardness of carbon steels decreases as the temperature increases. Consequently, the use of these steels is limited to very low-speed cutting operations.

8.6.2 High-speed steels

High-speed-steel (HSS) tools are so named because they were developed to cut at speeds higher than previously possible. First produced in the early 1900s, high-speed steels are the most highly alloyed of the tool steels. They can be hardened to various depths, have good wear resistance, and are relatively inexpensive. Because of their high toughness and resistance to fracture, high-speed steels are especially suitable for high positive-rake-angle tools (small included angle), for interrupted cuts, and for use on machine tools with low stiffness that are subject to vibration and chatter. High-speed steels account for the largest tonnage of tool materials used today, followed by various die steels and carbides. They are used in a wide variety of cutting operations that require complex tool shapes such as drills, reamers, taps, and gear cutters. Their basic limitation is the relatively low cutting speeds when compared with those of carbide tools.

There are two basic types of high-speed steels: **molybdenum** (*M* series) and **tungsten** (*T* series). The *M* series contains up to about 10% molybdenum, with chromium, vanadium, tungsten, and cobalt as alloying elements. The *T* series contains 12% to 18% tungsten, with chromium, vanadium, and cobalt as alloying elements. The *M* series generally has higher abrasion resistance than the *T* series, undergoes less distortion during heat treating, and is less expensive. Consequently, 95% of all high-speed steel tools produced in the United States are made of *M*-series steels.

High-speed-steel tools are available in wrought, cast, and sintered (powder-metallurgy) forms. They can be **coated** for improved performance (see Section 8.6.5), and they may be subjected to surface treatments, such as case hardening for improved hardness and wear.

8.6.3 Cast-cobalt alloys

Introduced in 1915, *cast-cobalt alloys* have the following ranges of composition: 38% to 53% cobalt, 30% to 33% chromium, and 10% to 20% tungsten. Because of their high hardness, typically 58 HRC to 64 HRC, they have good wear resistance and maintain their hardness at elevated temperatures. Commonly known as *Stellite* tools, these alloys are cast and ground into relatively simple tool shapes. They are not as tough as high-speed steels and are sensitive to impact forces; consequently, they are less suitable than high-speed steels for interrupted cutting operations. Tools made of cast-cobalt alloys are now used only for special applications that involve deep, continuous roughing operations at relatively high feeds and speeds, as much as twice the rates possible with high-speed steels.

8.6.4 Carbides

The three groups of tool materials described previously have sufficient toughness, impact strength, and thermal shock resistance, but have important limitations on characteristics such as strength and hardness, particularly hot hardness. Consequently, they cannot be used as effectively where high cutting speeds, and hence high temperatures, are involved; thus, their tool life can be short. To meet the challenge of higher speeds for higher production rates, *carbides* (also known as **cemented** or **sintered carbides**) were introduced in the 1930s. Because of their high hardness over a wide range of temperatures (see Fig. 8.31), high elastic modulus and thermal conductivity, and low thermal expansion, carbides are among the most important, versatile, and cost-effective tool and die materials for a wide range of applications. However, stiffness of the machine tool is important, and light feeds, low cutting speeds, and chatter can be detrimental. The two basic groups of carbides used for machining operations are tungsten carbide and titanium carbide. In order to differentiate them from coated tools (see Section 8.6.5), plain carbide tools are usually referred to as **uncoated** carbides. (See also the discussion of **micrograin carbides** in Section 8.6.10.)

a) **Tungsten carbide.** *Tungsten carbide* (WC) is a composite material consisting of tungsten-carbide particles bonded together in a cobalt matrix; hence, it is also called *cemented carbide*. Tungsten-carbide tools are manufactured by powder-metallurgy techniques (Chapter 11) in which WC particles are combined together with cobalt in a mixer, resulting in a cobalt matrix surrounding the WC particles. These particles, which are 1–5 μm (40–200 μin.) in size, are then pressed and sintered (hence tungsten carbide is also called *sintered carbide*) into the desired insert shapes. (See Example 11.6.) WC is frequently compounded with carbides of titanium and niobium to impart special properties to the carbide.

    The amount of cobalt significantly affects the properties of carbide tools. As the cobalt content increases, the strength, hardness, and wear resistance decrease, while the toughness increases, because of the higher toughness of cobalt. Tungsten-carbide tools are generally used for cutting steels, cast irons, and abrasive nonferrous materials and have largely replaced HSS tools in many applications, due to the better performance of the tungsten-carbide tools.

b) **Titanium carbide.** *Titanium carbide* (TiC) has higher wear resistance than that of tungsten carbide, but is not as tough. With a nickel–molybdenum alloy as the matrix, TiC is suitable for machining hard materials, mainly steels and cast irons, and for cutting at speeds higher than those for which tungsten carbide may be used.

c) **Inserts.** Carbon-steel and high-speed-steel cutting tools can be shaped in one piece and ground to various shapes (see Fig. 8.33); other such tools include drills and milling cutters. However, after the cutting edge wears and becomes dull, the tool has to be removed from its holder and reground. The need for a more effective method has led to the development of *inserts*, which are individual cutting tools with a number of cutting edges (Fig. 8.32). Inserts are available in a variety of shapes, such as square, triangle, diamond, and round. A square insert, for example, has eight cutting edges, and a triangular insert has six.

    Inserts are usually clamped on the tool *shank* with various locking mechanisms (Fig. 8.32b and c). Less frequently, inserts are *brazed* to the tool shank (Fig. 8.32d); however, because of the difference in thermal expansion between the insert and the tool-shank materials, brazing must be done carefully in order to avoid cracking or warping. Clamping is the preferred method because each insert has a number of cutting edges, and after one edge is worn, it is *indexed* (i.e., rotated in its holder) so that another edge can be used. In addition to these examples, a wide variety of other toolholders is available for specific applications, including toolholders with quick insertion and removal features for efficient operation.

The strength of the cutting edge of an insert depends on the shape of the insert; the smaller the included angle (Fig. 8.33), the lower is the strength of the edge. In order to further improve edge strength and prevent chipping, insert

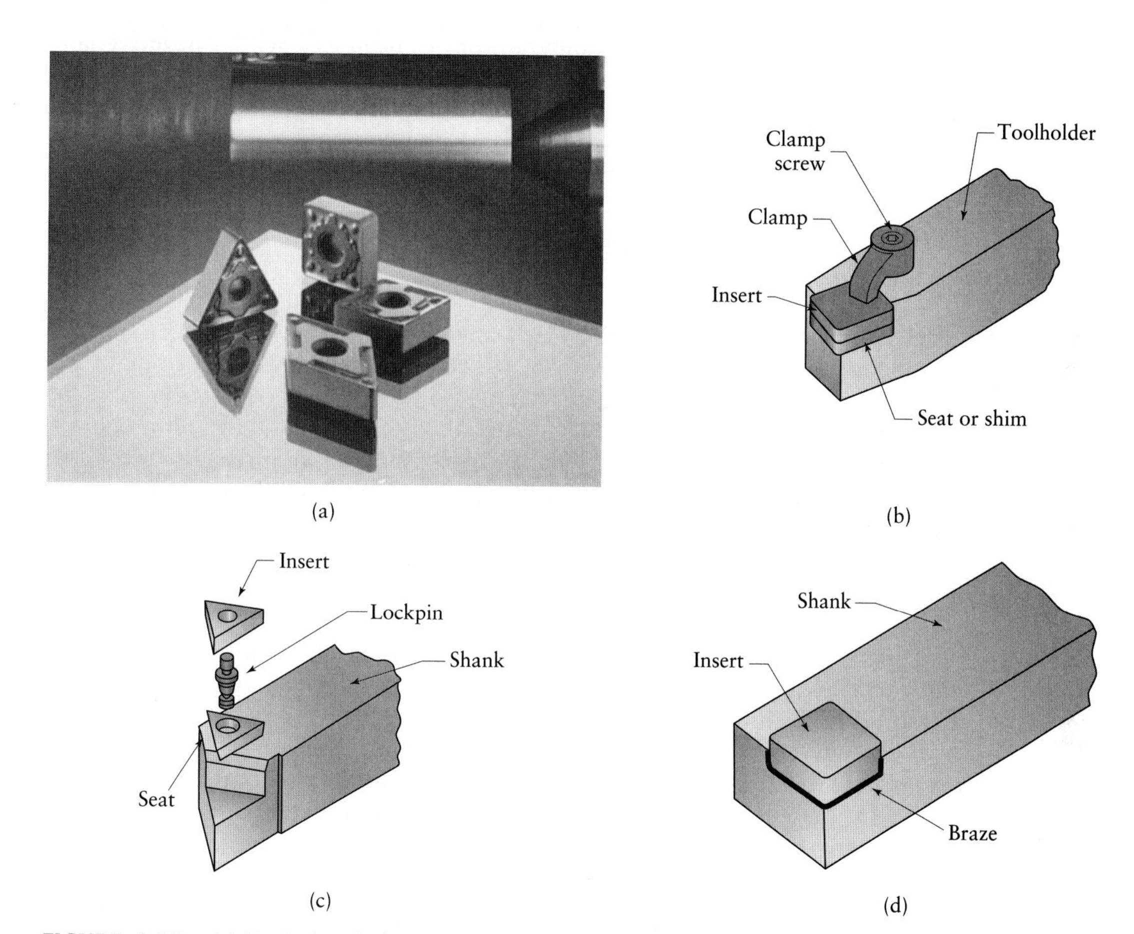

FIGURE 8.32 (a) Typical carbide inserts with various shapes and chip-breaker features. Round inserts are also available. The holes in the inserts are standardized for interchangeability. *Source*: Courtesy of Kyocera Engineered Ceramics, Inc., and *Manufacturing Engineering*, Society of Manufacturing Engineers. (b) Methods of attaching inserts to a tool shank by clamping, (c) with wing lockpins, and (d) with a brazed insert on a shank.

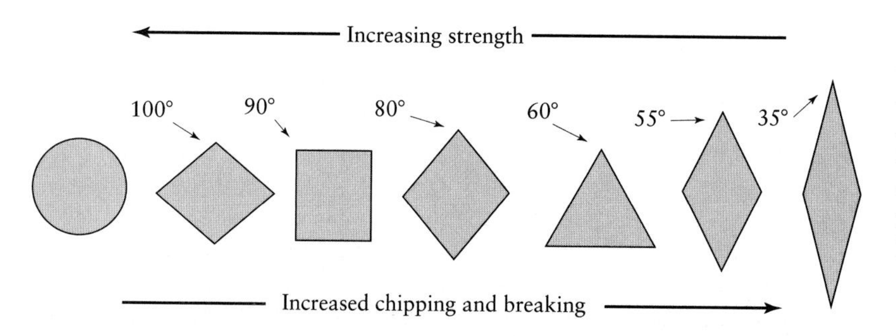

FIGURE 8.33 Relative edge strength and tendency for chipping and breaking of inserts with various shapes. Strength refers to that of the cutting edge shown by the included angles. *Source*: Kennametal, Inc.

FIGURE 8.34 Edge preparation of inserts to improve edge strength. *Source*: Kennametal, Inc.

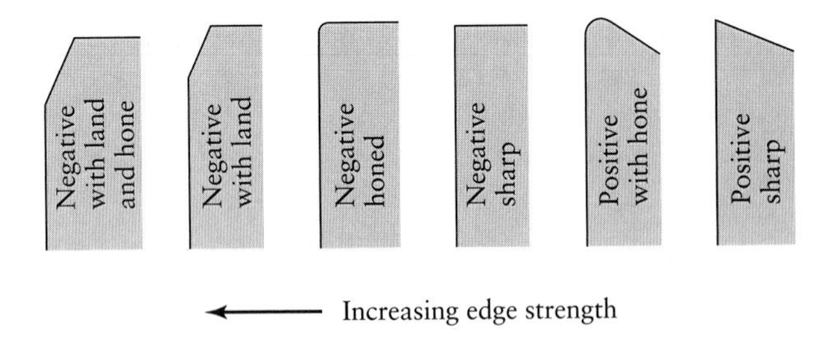

edges are usually honed, chamfered, or are produced with a negative land (Fig. 8.34). Most inserts are honed to a radius of about 0.025 mm (0.001 in.). Inserts are available with a wide variety of **chip-breaker** features for controlling chip flow and reducing vibration and the heat generated. Optimum chip-breaker geometries are now being developed using computer-aided design and finite-element techniques.

8.6.5 Coated tools

As described in Chapter 3, new alloys and engineered materials are being developed continuously, particularly since the 1960s. These materials have high strength and toughness, but are generally abrasive and highly chemically reactive with tool materials. The difficulty of machining these materials efficiently and the need to improve the performance in machining the more common engineering materials have led to important developments in coated tools. Because of their unique properties, coated tools can be used at high cutting speeds, thus reducing the time required for machining operations and, hence, costs. Note in Fig. 8.35 how the cutting time has been reduced by a factor of more than 100 since 1900. Coated tools can improve tool life by as much as 10 times that of uncoated tools.

Commonly used *coating materials* include titanium nitride, titanium carbide, titanium carbonitride, and aluminum oxide (Al_2O_3). Generally in the thickness range of 2–10 μm (80–400 μin.), these coatings are applied on tools and inserts by **chemical vapor deposition** (CVD) and **physical-vapor deposition** (PVD) techniques, described in Section 4.5. The CVD process is the most commonly used coating application method for carbide tools with multiphase and ceramic coatings. The PVD-coated carbides with TiN coatings, on the other hand, have higher cutting-edge strength, lower friction, lower tendency to form a built-up edge, and are smoother and more

FIGURE 8.35 Relative time required to machine with various cutting-tool materials, with indication of the year the tool materials were introduced. *Source*: Sandvik Coromant.

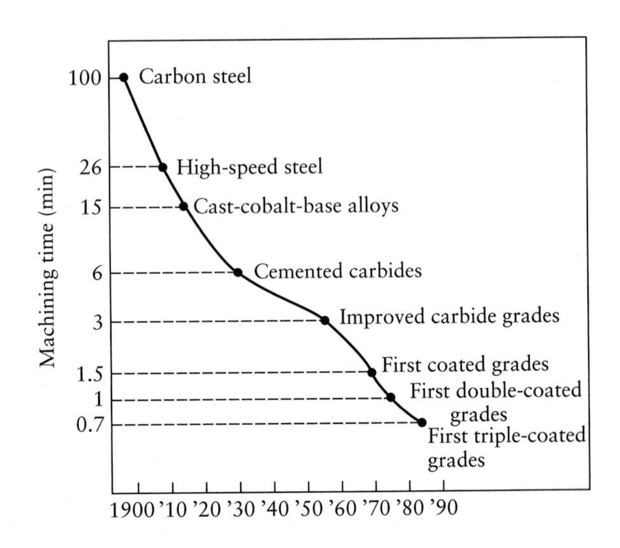

uniform in thickness, which is generally in the range of 2–4 μm (80–160 μin.). A more recent technology, particularly for multiphase coatings, is **medium-temperature chemical-vapor deposition** (MTCVD); it provides higher resistance to crack propagation than do CVD coatings.

Coatings should have the following general characteristics:

- High hardness at elevated temperatures,
- Chemical stability and inertness to the workpiece material,
- Low thermal conductivity,
- Good bonding to the substrate to prevent flaking or spalling, and
- Little or no porosity.

The effectiveness of coatings, in turn, is enhanced by hardness, toughness, and high thermal conductivity of the substrate, which may be carbide or high-speed steel. Honing (see Section 9.8) of the cutting edges is an important procedure to maintain the strength of the coating; otherwise, the coating may chip off at sharp edges and corners.

Some different types of coatings are discussed as follows:

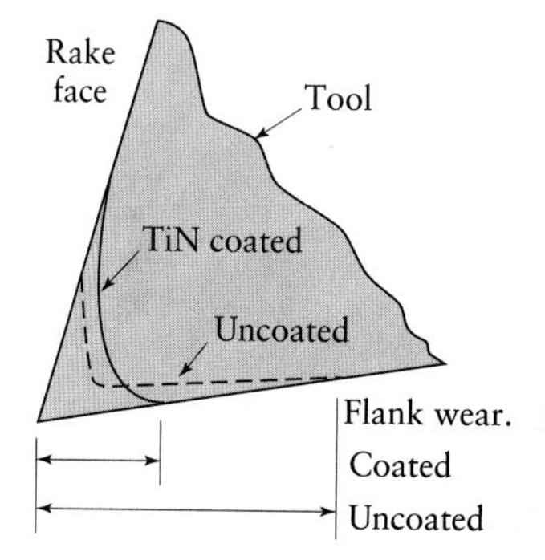

FIGURE 8.36 Wear patterns on high-speed-steel uncoated and titanium-nitride-coated tools. Note that flank wear is lower for the coated tool.

a) **Titanium nitride.** *Titanium nitride* (TiN) coatings have a low coefficient of friction, high hardness, high resistance to high temperature, and good adhesion to the substrate. Consequently, they greatly improve the life of high-speed-steel tools, as well as carbide tools, drills, and cutters. Titanium-nitride-coated tools (gold in color) perform well at higher cutting speeds and feeds. Flank wear is significantly lower than for uncoated tools (Fig. 8.36), and flank surfaces can be reground after use; regrinding does not remove the coating on the rake face of the tool. However, coated tools do not perform as well at low cutting speeds as uncoated tools, because the coating can be worn off by chip adhesion. Hence, the use of appropriate cutting fluids to discourage adhesion is important.

b) **Titanium carbide.** *Titanium carbide* (TiC) coatings on tungsten-carbide inserts have high resistance to flank wear in machining abrasive materials.

c) **Titanium carbonitride.** *Titanium carbonitride* (TiCN), which is deposited by physical-vapor deposition techniques, is harder and tougher than TiN. It can be used on carbides and high-speed-steel tools and is particularly effective in cutting stainless steels.

d) **Ceramic coatings.** Because of their resistance to high temperature, chemical inertness, low thermal conductivity, and resistance to flank and crater wear, ceramics are suitable coatings for tools. The most commonly used ceramic coating is aluminum oxide (Al_2O_3). However, because ceramic coatings are very stable (i.e., not chemically reactive), oxide coatings generally bond weakly with the substrate and thus may have a tendency to peel off.

e) **Multiphase coatings.** The desirable properties of the types of coating just described can be combined and optimized with the use of *multiphase coatings* (Fig. 8.37). Carbide tools are available with two or three layers of such coatings and are particularly effective in machining cast irons and steels.

The first layer over the substrate is TiC, followed by Al_2O_3, and then TiN. The first layer should bond well with the substrate; the outer layer should resist wear and have low thermal conductivity; the intermediate layer should bond well and be compatible with both layers. Typical applications of multiple-coated tools are listed as follows:

1. High-speed, continuous cutting: TiC/Al_2O_3.
2. Heavy-duty, continuous cutting: TiC/Al_2O_3/TiN.
3. Light, interrupted cutting: TiC/TiC + TiN/TiN.

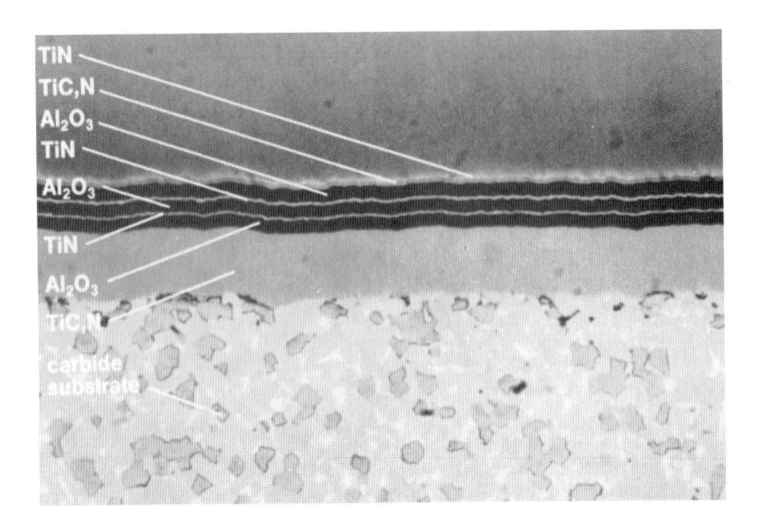

FIGURE 8.37 Multiphase coatings on a tungsten-carbide substrate. Three alternating layers of aluminum oxide are separated by very thin layers of titanium nitride. Inserts with as many as 13 layers of coatings have been made. Coating thicknesses are typically in the range of 2–10 μm (80–400 μin.). *Source*: Courtesy of Kennametal, Inc., and *Manufacturing Engineering*, Society of Manufacturing Engineers.

A more recent development in coatings is **alternating multiphase layers,** with layers that are thinner than in regular multiphase coatings (Fig. 8.37). The thickness of these layers is in the range of 2–10 μm (80–400 μin.). The reason for using thinner coatings is that coating hardness increases with decreasing grain size, a phenomenon that is similar to the increase in strength of metals with decreasing grain size (see Section 3.4.1); hence, thinner layers are harder than thicker layers.

f) **Diamond coatings.** A more recent development concerns the use of polycrystalline diamond as coatings for cutting tools, particularly tungsten-carbide and silicon-nitride inserts. However, difficulties exist regarding adherence of the diamond film to the substrate and the difference in thermal expansion between diamond and substrate materials (see Table 8.6).

   Thin-film diamond-coated inserts are available, as are thick-film diamond-brazed-tip cutting tools. Thin films are deposited on substrates by PVD and CVD techniques, whereas thick films are obtained by growing a large sheet of pure diamond, which is then laser cut to shape and brazed to a carbide shank. Diamond-coated tools are particularly effective in machining abrasive materials, such as aluminum alloys that contain silicon, fiber-reinforced and metal-matrix composite materials, and graphite. Improvements in tool life of as much as tenfold have been obtained over that of other coated tools.

g) **Other coating materials.** While titanium-nitride coatings made by chemical-vapor deposition processes are still common, advances are being made in developing and testing new coating materials. **Titanium aluminum nitride** (TiAlN) is effective in machining aerospace alloys. Chromium-based coatings such as **chromium carbide** (CrC) have been found to be effective in machining softer metals that tend to adhere to the cutting tool, such as aluminum, copper, and titanium. Other new coating materials include **zirconium nitride** (ZrN) and **hafnium nitride** (HfN). Much experimental data are still needed before these coatings and their behavior can be fully assessed. More recent developments include (a) **nanocoatings** with carbide, boride, nitride, oxide, or some combination thereof, and (b) **composite coatings**, using a variety of materials.

h) **Ion implantation.** In this process, ions are introduced into the surface of the cutting tool. The process does not change the dimensions of tools, but improves their surface properties. (See Section 4.5.1.) **Nitrogen-ion**-implanted carbide tools have been used successfully on alloy steels and stainless steels. **Xenon-ion** implantation of tools is also under development.

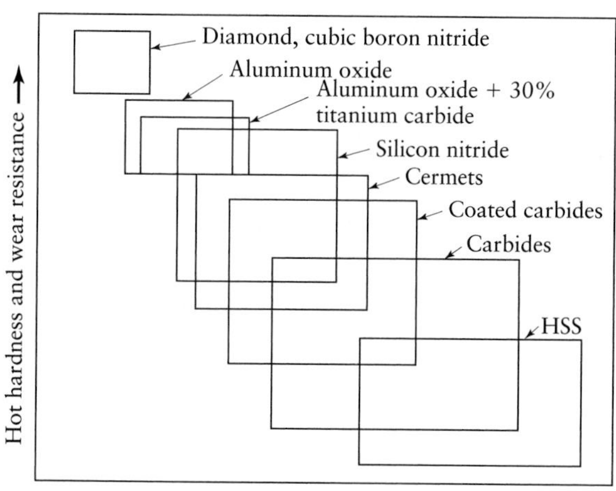

FIGURE 8.38 Ranges of properties for various groups of tool materials. (See also various tables in this chapter.)

8.6.6 Alumina-base ceramics

Ceramic tool materials, introduced in the early 1950s, consist primarily of fine-grained, high-purity **aluminum oxide.** They are pressed into insert shapes at room temperature and under high pressure, sintered at high temperature, and called **white,** or **cold-pressed, ceramics.** (See also Section 11.9.3.) Additions of titanium carbide and zirconium oxide improve properties such as toughness and resistance to thermal shock.

Alumina-base ceramic tools have very high abrasion resistance and hot hardness (Fig. 8.38). Chemically, they are more stable than high-speed steels and carbides; thus, they have less of a tendency to adhere to metals during cutting and, hence, less of a tendency to form a built-up edge. Consequently, good surface finish is obtained with ceramic tools in cutting cast irons and steels. However, ceramics lack toughness, which results in premature tool failure by chipping or catastrophic failure. (See Fig. 8.20.) The shape and setup of ceramic tools are important. Negative rake angles, and hence large included angles, are generally preferred in order to avoid chipping due to poor tensile strength. The occurrence of tool failure can be reduced by increasing the stiffness and damping capacity of machine tools, mountings, and workholding devices, thus reducing vibration and chatter (see Section 8.11).

Cermets. *Cermets* (from *cer*amic and *met*al), which are also called **black,** or **hot-pressed, ceramics** (carboxides), were introduced in the 1960s. They typically contain 70% aluminum oxide and 30% titanium carbide; other cermets may contain molybdenum carbide, niobium carbide, or tantalum carbide. The brittleness and high cost of cermets have been a problem when using them as cutting tools; however, further refinements have resulted in improved strength, toughness, and reliability. Their performance is between that of ceramics and carbides. Chip-breaker features are important when using cermets. Although cermets can be coated, the benefits of coatings are somewhat controversial, as the improvement in wear resistance appears to be marginal.

8.6.7 Cubic boron nitride

Next to diamond, *cubic boron nitride* (cBN) is the hardest material presently available. The cBN cutting tool was introduced in 1962; it is made by bonding a 0.5- to 1-mm (0.02- to 0.04-in.) layer of *polycrystalline cubic boron nitride* to a carbide substrate by sintering under pressure (Fig. 8.39). While the carbide provides shock

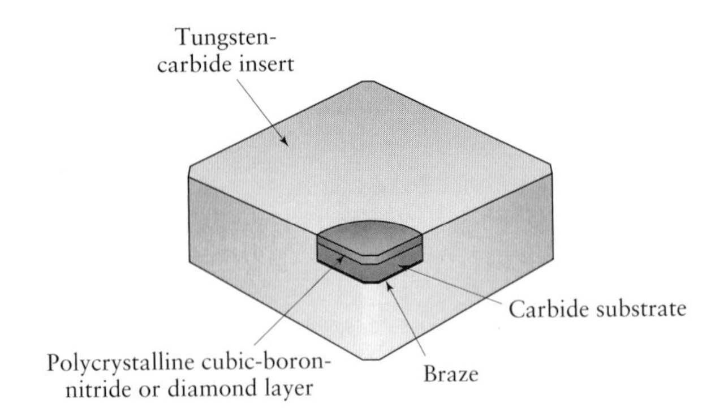

FIGURE 8.39 Construction of polycrystalline cubic-boron-nitride or diamond layer on a tungsten-carbide insert.

resistance, the cBN layer provides very high wear resistance and cutting-edge strength. Cubic-boron-nitride tools are also made in small sizes without a substrate. At elevated temperatures, cBN is chemically inert to iron and nickel, and its resistance to oxidation is high. It is therefore particularly suitable for cutting hardened ferrous and high-temperature alloys. (See the description of *hard turning* in Section 8.8.2.) Cubic boron nitride is also used as an abrasive. (See Chapter 9.) Because cBN tools are brittle, stiffness and damping capacity of the machine tool and fixturing is important to avoid vibration and chatter.

8.6.8 Silicon-nitride-base ceramics

Developed in the 1970s, *silicon-nitride* (SiN) *-base ceramic* tool materials consist of silicon nitride with various additions of aluminum oxide, yttrium oxide, and titanium carbide. These tools have high toughness, hot hardness, and good thermal-shock resistance. An example of an SiN-base material is **sialon**, so called after the elements *si*licon, *al*uminum, *o*xygen, and *n*itrogen in its composition. It has higher resistance to thermal shock than silicon nitride and is recommended for machining cast irons and nickel-base superalloys at intermediate cutting speeds. Because of their chemical affinity to steels, SiN-base tools are not suitable for machining steels.

8.6.9 Diamond

The hardest substance of all known materials is *diamond*, a crystalline form of carbon. (See also Section 11.13.) As a cutting tool, it has low tool–chip friction, high wear resistance, and the ability to maintain a sharp cutting edge. It is used when good surface finish and dimensional accuracy are required, particularly with soft nonferrous alloys and abrasive nonmetallic materials. **Single-crystal diamond** of various carats is used for special applications, such as machining copper-front high-precision optical mirrors. Because diamond is brittle, tool shape and sharpness are important; low rake angles (large included angles) are normally used to provide a strong cutting edge. Special attention should be given to proper mounting and crystal orientation in order to obtain optimum useage. Diamond wear may occur by microchipping (caused by thermal stresses and oxidation) and transformation to carbon (caused by the heat generated during cutting).

Single-crystal (single-point) diamond tools have been largely replaced by **polycrystalline diamond** tools (*compacts*), which are also used as wire-drawing dies for fine wire. These materials consist of very small synthetic crystals, fused by a high-pressure, high-temperature process to a thickness of about 0.5–1 mm (0.02–0.04 in.), and bonded to a carbide substrate, similar to cBN tools. (See Fig. 8.39.) The random orientation of the diamond crystals prevents the propagation of cracks through the structure, thus significantly improving its toughness.

Diamond tools can be used satisfactorily at almost any speed, but are suitable mostly for light, uninterrupted finishing cuts. In order to minimize tool fracture, the single-crystal diamond must be resharpened as soon as it becomes dull. Because of its strong chemical affinity, diamond is not recommended for machining plain-carbon steels and titanium, nickel, and cobalt-base alloys. Diamond is also used as an abrasive in grinding and polishing operations (see Chapter 9) and as coatings (see Section 4.5).

8.6.10 Whisker-reinforced and nanocrystalline tool materials

In order to further improve the performance and wear resistance of cutting tools, particularly in machining new workpiece materials and composites that are under continued development, new tool materials are being developed with enhanced properties such as (1) high fracture toughness, (2) resistance to thermal shock, (3) cutting-edge strength, and (4) hot hardness.

More recent developments consist of using *whiskers* (see Section 3.8.3) as reinforcing fibers in composite tool materials. Examples of **whisker-reinforced** materials include silicon-nitride-base tools reinforced with silicon-carbide (SiC) whiskers, and aluminum-oxide-base tools reinforced with silicon-carbide whiskers, sometimes with the addition of *zirconium oxide* (ZrO_2). However, the low inertness of silicon carbide to ferrous metals makes SiC-reinforced tools unsuitable for machining irons and steels. Progress in nanomaterials (Section 3.11.9) has led to the development of cutting tools made of very fine-grained (**micrograin**) carbides of tungsten, titanium, and tantalum. These materials are stronger, harder, and more wear resistant than are traditional carbides.

8.6.11 Cryogenic treatment of cutting tools

Studies are continuing on the beneficial effect of cryogenic treatment of tools and other metals on their performance in cutting operations. (See Section 5.11.6 for details.) In this procedure, the tool is cooled very slowly to temperatures of around −180°C (−300°F) and slowly returned to room temperature; it is then tempered. Claims have been made that, depending on the combinations of tool and workpiece material involved, tool-life increases of up to 300% can be achieved with this procedure.

8.7 Cutting Fluids

Also called *lubricants* and *coolants, cutting fluids* are used extensively in machining operations to

- Reduce friction and wear, hence improving tool life and surface finish;
- Reduce forces and energy consumption;
- Cool the cutting zone, thus reducing workpiece temperature and distortion and improving tool life;
- Wash away chips;
- Protect the newly machined surfaces from environmental corrosion.

A cutting fluid can interchangeably be a **coolant** and a **lubricant** (see Section 4.4.3); its effectiveness in cutting operations depends on a number of factors, such as the method of application, the temperature, the cutting speed, and the type of machining operation. As described in Section 8.2.6, temperature increases as cutting speed increases; therefore, *cooling* of the cutting zone is of major importance at high cutting speeds. Water is an excellent coolant; however, it causes rusting of workpieces and machine components and is a poor lubricant. On the other hand, if the cutting speed is low, such as in broaching or tapping, *lubrication*, not cooling, is the important factor. Lubrication reduces the tendency for the formation of built-up edge and thus improves surface finish.

There are situations, however, in which the use of cutting fluids can be detrimental. For example, in interrupted cutting operations, such as milling, the cooling action of the cutting fluid increases the extent of alternate heating and cooling (*thermal cycling*) to which the cutter teeth are subjected. This condition can lead to thermal cracks (*thermal fatigue* or *thermal shock*). Cutting fluids may also cause the chip to become more curled, thus concentrating the stresses near the tool tip. These stresses, in turn, concentrate the heat closer to the tool tip and reduce tool life.

Cutting fluids can be expensive and represent a **biological** and **environmental hazard** (see also Section 4.4.4) that requires proper recycling and disposal, thus adding to the cost of the machining operation. For these reasons, **dry cutting**, or **dry machining**, has become an increasingly important approach; in dry machining, no coolant or lubricant is used. Even though this approach would suggest that higher temperatures and more rapid tool wear would occur, some tool materials and coatings display a reasonable tool life at the higher temperatures. Dry cutting has sometimes been associated with high-speed machining, because higher cutting speeds transfer a greater amount of heat conveyed from cutting to the chip (see Fig. 8.18), which is a natural strategy for reducing the need for a coolant. (See also Section 3.9.7 on possible detrimental effects of cutting fluids on some cutting tools, called *selective leaching*, such as carbide tools with cobalt binders.)

8.7.1 Types of cutting fluids and methods of application

There are generally four types of cutting fluids commonly used in machining operations: **oils, emulsions, semisynthetics**, and **synthetics**. The general characteristics of these fluids are described in Section 4.4.4. Cutting-fluid recommendations for specific machining operations are given throughout the rest of this chapter. In selecting an appropriate cutting fluid, considerations should be given to its possibly detrimental effects on the workpiece material (e.g., corrosion, stress-corrosion cracking, staining), the components of the machine tool, biological and environmental effects, and recycling and disposal.

The most common method of applying cutting fluid is **flood cooling.** Flow rates range from 10 L/min (3 gal/min) for single-point tools to 225 L/min (60 gal/min) per cutter for multiple-tooth cutters, such as in milling. In operations such as gun drilling and end milling, fluid pressures of 700–14,000 kPa (100–2000 psi) are used to wash away the chips.

Mist cooling is another method of applying cutting fluids and is used particularly with water-base fluids. Although it requires venting (to prevent inhalation of fluid particles by the machine operator and others nearby) and has limited cooling capacity, mist cooling supplies fluid to otherwise inaccessible areas and provides better visibility of the workpiece being machined. It is particularly effective in grinding operations (see Chapter 9) and at air pressures of 70–600 kPa (10–80 psi).

With increasing speed and power of machine tools, heat generation in machining operations has become a significant factor. More recent developments include the use of **high-pressure refrigerated coolant** systems to improve the rate of heat removal from the cutting zone, as well as to reduce costs and avoid adverse environmental effects. High pressures as high as 35 MPa (5000 psi) are also used in delivering the cutting fluid to the cutting zone via specially designed nozzles that aim a powerful jet of fluid to the zone. This action breaks up the chips (thus acting as a chip breaker) in situations where the chips produced would otherwise be long and continuous and hence interfere with the cutting operation.

8.7.2 Cryogenic machining

Largely in the interest of reducing or eliminating the adverse environmental impact of using metalworking fluids, a recent technology is the use of *liquid nitrogen* as a coolant in machining, as well as in grinding. (See Section 9.6.9.) With appropriate small-diameter nozzles, liquid nitrogen at around −200°C (−320°F) is injected into the tool–workpiece interface, thus reducing its temperature. As a result, tool hardness and, hence, tool life is enhanced, thus allowing for higher cutting speeds; furthermore, the chips become more brittle and, thus, easier to remove from the cutting zone. Because no fluids are involved and the liquid nitrogen simply evaporates, the chips can be recycled more easily, thus also improving the economics of machining operations, without any adverse effects on the environment.

EXAMPLE 8.4 Effect of cutting fluids on machining

A machining operation is being carried out with a cutting fluid that is effective as a lubricant. Explain the changes in the mechanics of the cutting operation and total energy consumption if the fluid is shut off.

Solution. Since the cutting fluid is a good lubricant, the friction at the tool–chip interface will increase when the fluid is shut off. The following chain of events then takes place:

1. Fluid is shut off.
2. Friction at the tool–chip interface increases.
3. The shear angle decreases.
4. The shear strain increases.
5. The chip is thicker.
6. A built-up edge is likely to form.

As a consequence,

1. The shear energy in the primary zone increases.
2. The friction energy in the secondary zone increases.
3. Hence the total energy increases.
4. Surface finish is likely to deteriorate.
5. The temperature in the cutting zone increases, and hence tool wear increases and workpiece surface may lose its integrity.
6. Dimensional tolerances may be difficult to maintain, because of the increased temperature and expansion of the workpiece during machining.

8.8 Cutting Processes and Machine Tools for Producing Round Shapes

This section describes the processes that produce parts which are basically *round* in shape (Table 8.7); typical products made include parts as small as miniature screws for eyeglass-frame hinges and as large as shafts, pistons, cylinders, gun barrels, and turbines for hydroelectric power plants. The processes are usually performed by turning the workpiece on a lathe. *Turning* means that the part is rotating while it is being machined. The starting material is usually a workpiece that has been made by other processes, such as casting, shaping, forging, extrusion, and drawing. Turning processes are versatile and capable of producing a wide variety of shapes, as outlined in Fig. 8.40. The various types of turning processes are described as follows:

- **Turning** straight, conical, curved, or grooved workpieces, such as shafts, spindles, pins, handles, and various machine components;
- **Facing**, to produce a flat surface at the end of the part, such as for parts that are attached to other components, or to produce grooves for O-ring seats;
- Producing various shapes by **form tools**, such as for functional purposes or for appearance;

TABLE 8.7

General Characteristics of Machining Processes

Process	Characteristics	Commercial tolerances (±mm)
Turning	Turning and facing operations are performed on all types of materials; requires skilled labor; low production rate, but medium to high rates can be achieved with turret lathes and automatic machines, requiring less skilled labor.	Fine: 0.05–0.13 Rough: 0.13 Skiving: 0.025–0.05
Boring	Internal surfaces or profiles, with characteristics similar to those produced by turning; stiffness of boring bar is important to avoid chatter.	0.025
Drilling	Round holes of various sizes and depths; requires boring and reaming for improved accuracy; high production rate, labor skill required depends on hole location and accuracy specified.	0.075
Milling	Variety of shapes involving contours, flat surfaces, and slots; wide variety of tooling; versatile; low to medium production rate; requires skilled labor.	0.13–0.25
Planing	Flat surfaces and straight contour profiles on large surfaces; suitable for low-quantity production; labor skill required depends on part shape.	0.08–0.13
Shaping	Flat surfaces and straight contour profiles on relatively small workpieces; suitable for low-quantity production; labor skill required depends on part shape.	0.05–0.13
Broaching	External and internal flat surfaces, slots, and contours with good surface finish; costly tooling; high production rate; labor skill required depends on part shape.	0.025–0.15
Sawing	Straight and contour cuts on flats or structural shapes; not suitable for hard materials unless the saw has carbide teeth or is coated with diamond; low production rate; requires only low labor skill.	0.8

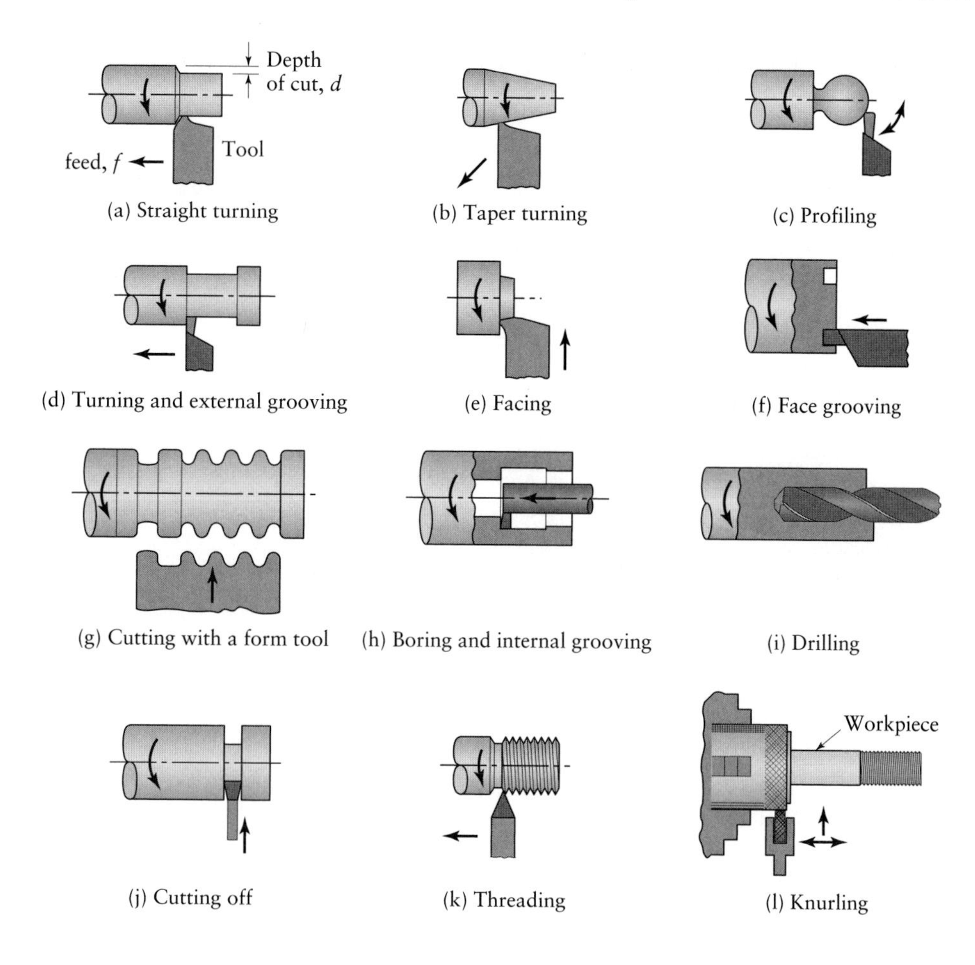

FIGURE 8.40 Various cutting operations that can be performed on a lathe.

- **Boring,** to enlarge a hole made by a previous process or in a tubular workpiece, or to produce internal grooves;
- **Drilling,** to produce a hole, which may be followed by tapping or boring to improve the accuracy of the hole and its surface finish;
- **Parting,** also called **cutting off,** to cut a piece from the end of a part, as in making slugs or blanks for additional processing into discrete parts;
- **Threading,** to produce external and internal threads in workpieces;
- **Knurling,** to produce a regularly shaped geometric roughness on cylindrical surfaces, as in making knobs.

These operations may be performed at various rotational speeds of the workpiece, depths of cut, d, and feed, f, (see Fig. 8.19), depending on the workpiece and tool materials, the surface finish and dimensional accuracy required, and the capacity of the machine tool. Cutting speeds are usually in the range of 0.15–4 m/s (30–800 ft/min). **Roughing cuts,** which are performed for large-scale material removal, usually involve depths of cut greater than 0.5 mm (0.02 in.) and feeds on the order of 0.2–2 mm/rev (0.008–0.08 *in./rev*). **Finishing cuts** usually involve lower depths of cut and feed.

8.8.1 Turning parameters

The majority of turning operations involve single-point cutting tools. Figure 8.41 shows the geometry of a typical right-hand simple cutting tool for turning; such tools

FIGURE 8.41 (a) Designations and symbols for a right-hand cutting tool; solid high-speed-steel tools have a similar designation. The designation "right hand" means that the tool travels from right to left, as shown in Fig. 8.19. (b) Square insert in a right-hand toolholder for a turning operation. A wide variety of toolholders is available for holding inserts at various angles. Thus, the angles shown in (a) can easily be achieved by selecting an appropriate insert and toolholder. *Source*: Kennametal, Inc.

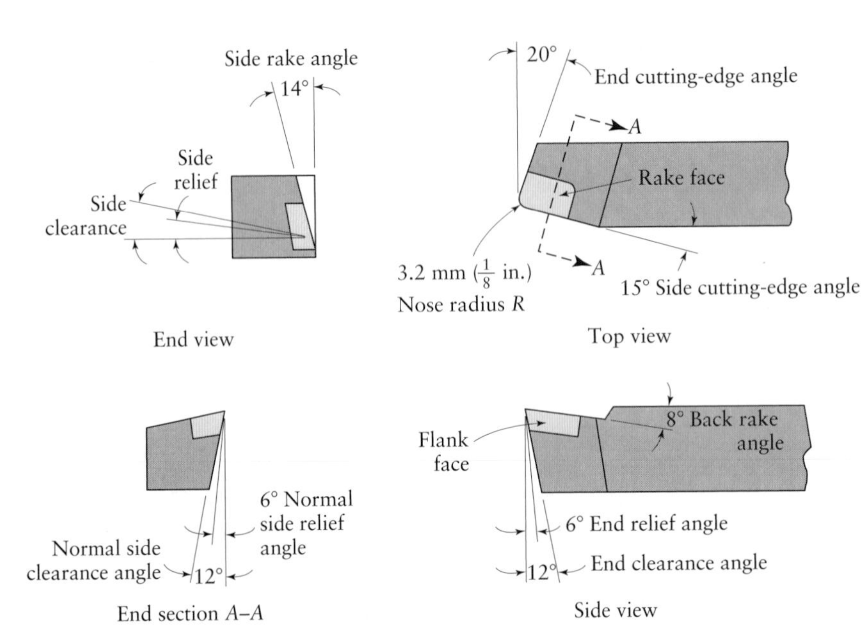

Tool Signature	Dimensions	Abbreviation
8	Back rake angle	BR
14	Side rake angle	SR
6	End relief angle	ER
12	End clearance angle	...
6	Side relief angle	SRF
12	Side clearance angle	...
20	End cutting-edge angle	ECEA
15	Side cutting-edge angle	SCEA
$\frac{1}{8}$	Nose radius	NR

(a)

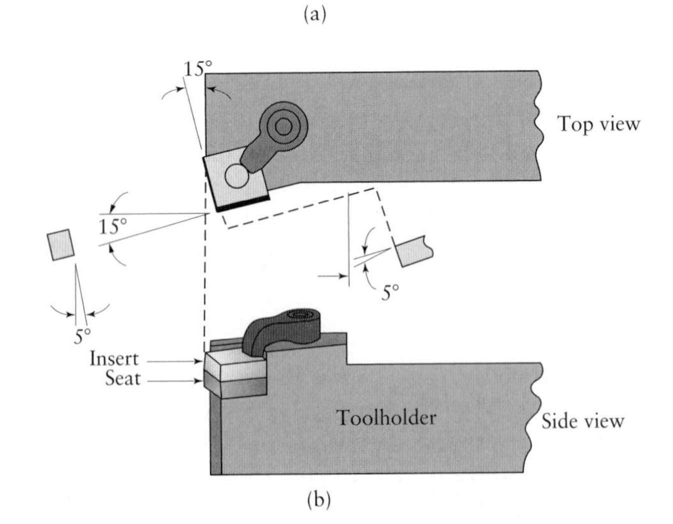

(b)

are identified by a standardized nomenclature. Each group of tool and workpiece materials has an optimum set of tool angles, which have been developed over many years, largely through experience. Data on tool geometry can be found in the references given in the bibliography at the end of this chapter.

a) **Tool geometry.** The various angles on a tool have important functions in cutting operations. **Rake angles** are important in controlling the direction of chip flow and the strength of the tool tip. Positive angles improve the cutting operation by reducing forces and temperatures; however, positive angles also produce a small included angle of the tool tip. (See Fig. 8.2.) Depending on the toughness of the tool material, this included angle may cause premature tool chipping and failure. **Side rake angle** is more important than **back rake angle,** although the latter usually controls the direction of chip flow. **Relief angles** control interference and rubbing at the tool–workpiece interface; if the relief angle is too large, the tool may chip off, and if it is too small, flank wear may be excessive. **Cutting-edge angles** affect chip formation, tool strength, and cutting forces to various degrees. **Nose radius** affects surface finish and tool-tip strength; also, the smaller the nose radius, the rougher will be the surface finish of the workpiece, and the lower will be the strength of the tool. On the other hand, large nose radii can lead to tool chatter.

b) **Material removal rate.** The *material removal rate* (MRR) is the volume of material removed per unit time, such as mm^3/min or in^3/min. Referring to Fig. 8.42a, note that for each revolution of the workpiece, we remove a ring-shaped layer of material whose cross-sectional area is the product of the distance the tool travels in one revolution (i.e., the feed f) and the depth of cut, d. The volume of this ring is the product of the cross-sectional area–that is, fd–and the average circumference of the ring–that is, πD_{avg}, where $D_{\text{avg}} = (D_o + D_f)/2$. For light cuts on large-diameter workpieces, the average diameter can be replaced by D_o. The rotational speed of the workpiece is N.

Thus, the material removal rate per revolution will be $\pi D_{\text{avg}} df$. Since we have N revolutions per minute, the removal rate is given by

$$\text{MRR} = \pi D_{\text{avg}} dfN. \tag{8.38}$$

The dimensional accuracy of this equation can be checked by substituting dimensions into the right-hand side; hence, (mm)(mm)(mm/rev)(rev/min) = mm^3/min, which indicates the volume rate of removal. Similarly, the cutting time T for a workpiece of length ℓ can be calculated by noting that the tool travels at a feed rate of fN = (mm/rev)(rev/min) = mm/min. Since the distance traveled is ℓ mm, the cutting time is

$$t = \frac{\ell}{fN}. \tag{8.39}$$

This time does not include the time required for *tool approach* and *retraction.* Because the time spent in *noncutting* cycles of a machining operation is

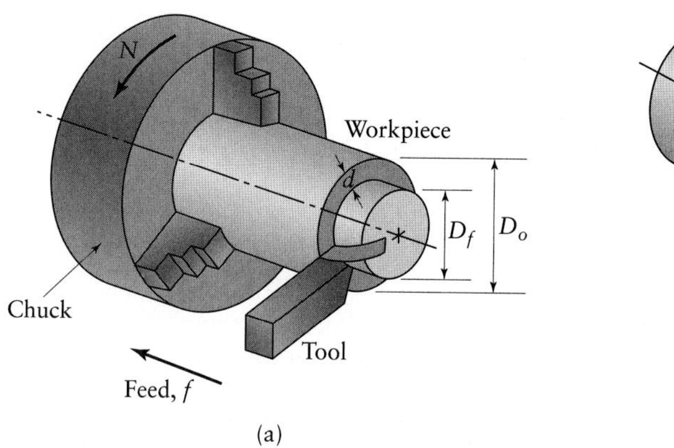

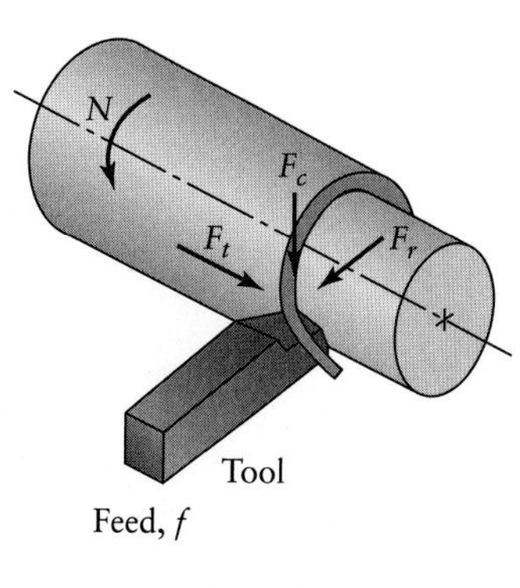

FIGURE 8.42 (a) Schematic illustration of a turning operation showing depth of cut, d, and feed, f. Cutting speed is the surface speed of the workpiece at the tool tip. (b) Forces acting on a cutting tool in turning. F_c is the cutting force; F_t is the thrust or feed force (in the direction of feed); and F_r is the radial force that tends to push the tool away from the workpiece being machined. Compare this figure with Fig. 8.11 for a two-dimensional cutting operation.

nonproductive and affects the overall economics of the process, the time involved in approaching and retracting tools to and from the workpiece is an important consideration. Machine tools are now designed and built to minimize this time, using computer controls; a typical method employed is first to rapidly traverse the tools and then to slow the movement as the tool engages the workpiece.

c) **Forces in turning.** Figure 8.42b illustrates the three forces acting on a cutting tool; these forces are important in the design of machine tools as well as in the deflection of tools for precision machining operations. The *cutting force*, F_c, acts downward on the tool tip and thus tends to deflect the tool downward. The cutting force is the force that supplies the energy required for the cutting operation. As can be seen in Example 8.5, the cutting force can be calculated from the energy per unit volume, described in Section 8.2.5, and by using the data in Table 8.3. The *thrust force*, F_t, acts in the longitudinal direction; this force is also called the *feed force*, because it is in the feed direction. The *radial force*, F_r, is in the radial direction and thus tends to push the tool away from the workpiece. The thrust and radial forces are difficult to calculate, because of the many factors involved in the cutting process, so they are determined experimentally.

d) **Tool materials, feeds, and cutting speeds.** Section 8.6 describes the general characteristics of cutting-tool materials. Figure 8.43 gives a broad range of cutting speeds and feeds applicable for these tool materials. Specific recommendations for cutting speeds for turning various workpiece materials and cutting tools are given in Table 8.8. In this table, **roughing** means cutting material without much consideration to dimensional tolerances and surface finish, and **finishing** means cutting to obtain final dimensions with acceptable dimensional tolerances and surface finish.

e) **Cutting fluids.** Although many metals and nonmetallic materials can be machined without a cutting fluid (dry machining), the application of a cutting fluid can improve the operation significantly in many cases. Specific recommendations for cutting fluids for various workpiece materials and operations can be found in the references in the bibliography at the end of this chapter.

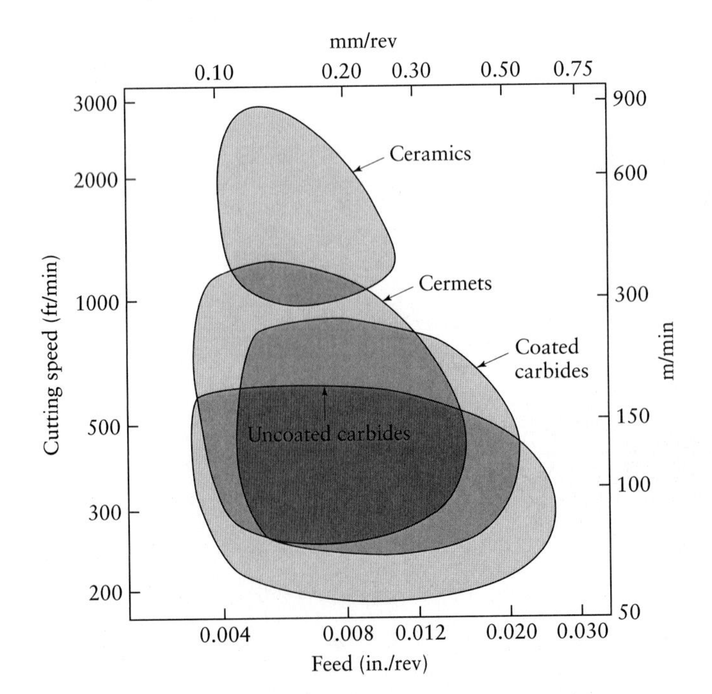

FIGURE 8.43 The range of applicable cutting speeds and feeds for a variety of tool materials. *Source*: Valenite, Inc.

TABLE 8.8

Approximate Ranges of Recommended Cutting Speeds for Turning Operations

WORKPIECE MATERIAL	CUTTING SPEED m/min	CUTTING SPEED ft/min
Aluminum alloys	200–1000	650–3300
Cast iron, gray	60–900	200–3000
Copper alloys	50–700	160–2300
High-temperature alloys	20–400	65–1300
Steels	50–500	160–1600
Stainless steels	50–300	160–1000
Thermoplastics and thermosets	90–240	300–800
Titanium alloys	10–100	30–330
Tungsten alloys	60–150	200–500

Note: (a) The speeds given in this table are for carbides and ceramic cutting tools. Speeds for high-speed-steel tools are lower than indicated. The higher ranges are for coated carbides and cermets. Speeds for diamond tools are significantly higher than any of the values indicated in the table.
(b) Depths of cut, d, are generally in the range of 0.5–12 mm (0.02–0.5 in.).
(c) Feeds, f, are generally in the range of 0.15–1 mm/rev (0.006–0.040 in./rev).

EXAMPLE 8.5 Material removal rate and cutting force in turning

A 6-in.-long, 0.5-in.-diameter 304 stainless-steel rod is being reduced in diameter to 0.480 in. by turning on a lathe. The spindle rotates at $N = 400$ rpm, and the tool is traveling at an axial speed of 8 in./min. Calculate the cutting speed, material removal rate, cutting time, power dissipated, and cutting force.

Solution. The cutting speed is the tangential speed of the workpiece. The maximum cutting speed is at the outer diameter, D_o, and is obtained from the expression

$$V = \pi D_o N.$$

Thus,

$$V = (\pi)(0.500)(400) = 628 \text{ in./min} = 52 \text{ ft/min}.$$

The cutting speed at the machined diameter is

$$V = (\pi)(0.480)(400) = 603 \text{ in./min} = 50 \text{ ft/min}.$$

From the information given, note that the depth of cut is

$$d = (0.500 - 0.480)/2 = 0.010 \text{ in.},$$

and the feed is

$$f = 8/400 = 0.02 \text{ in./rev}.$$

Thus, according to Eq. (8.38), the material removal rate is

$$\text{MRR} = (\pi)(0.490)(0.010)(0.02)(400) = 0.123 \text{ in}^3\text{/min}.$$

The actual time to cut, according to Eq. (8.39), is

$$t = 6/(0.02)(400) = 0.75 \text{ min}.$$

We can calculate the power required by referring to Table 8.3 and taking an average value for stainless steel as 4 W-s/mm^3 = 4/2.73 = 1.47 hp-min/in^3. Therefore, the power dissipated is

$$\text{Power} = (1.47)(0.123) = 0.181 \text{ hp},$$

and since 1 hp = 396,000 in.-lb/min, the power dissipated is 71,700 in.-lb/min. The cutting force, F_c, is the tangential force exerted by the tool. Since power is the product of torque, T, and rotational speed in radians per unit time, we have

$$T = \frac{(71{,}700)}{(400)(2\pi)} = 29 \text{ lb-in.}$$

Since $T = (F_c)(D_{avg}/2)$, we obtain

$$F_c = \frac{(29)(2)}{(0.490)} = 118 \text{ lb.}$$

8.8.2 Lathes and lathe operations

Lathes are generally considered to be the oldest machine tools. Although woodworking lathes were first developed during the period 1000–1 BC, metalworking lathes with lead screws were not built until the late 1700s. The most common type of lathe, shown in Fig. 8.44, was originally called an **engine lathe**, because it was powered with overhead pulleys and belts from nearby engines. Lathes are now equipped with various features, including computer controls, described as follows:

a) **Lathe components.** Lathes are equipped with a variety of components and accessories (Fig. 8.44). The *bed* supports all the other major components of the lathe. The *carriage*, or *carriage assembly*, which slides along the ways, consists

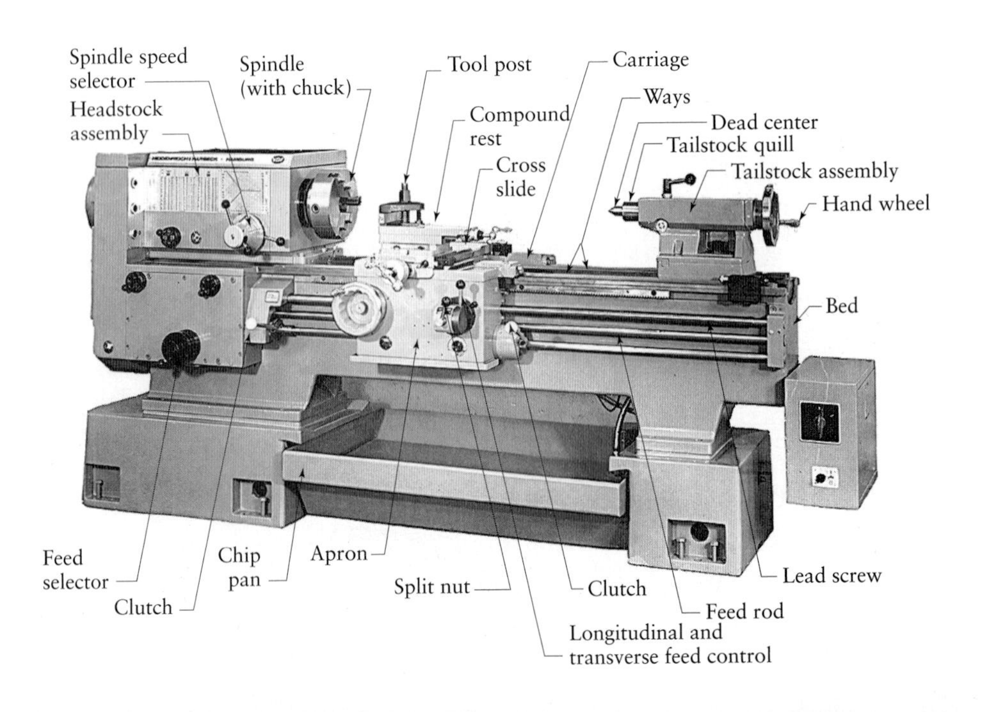

FIGURE 8.44 The components of a lathe. *Source*: Courtesy of MAKINO GmbH.

of an assembly of the *cross slide*, *tool post*, and *apron*. The cutting tool is mounted on the *tool post*, usually with a *compound rest* that swivels for tool positioning and adjustments. The *headstock*, which is fixed to the bed, is equipped with motors, pulleys, and V-belts that supply power to the *spindle* at various rotational speeds. Headstocks have a hollow spindle to which workholding devices, such as *chucks* and *collets*, are attached. The *tailstock*, which can slide along the ways and be clamped at any position, supports the other end of the workpiece. The *feed rod*, which is powered by a set of gears from the headstock, rotates during operation of the lathe and provides movement to the carriage and the cross slide. The *lead screw*, which is used for cutting threads accurately, is engaged with the carriage by closing a split nut around the lead screw.

A lathe is usually specified by its *swing*, that is, the maximum diameter of the workpiece that can be machined; by the maximum distance between the headstock and tailstock centers; and by the length of the bed. A variety of lathes is available, including *bench lathes*, *toolroom lathes*, *engine lathes*, *gap lathes*, and *special-purpose lathes*.

Tracer lathes, also called *duplicating lathes* or *contouring lathes*, are capable of turning parts with various contours; the cutting tool follows a path that duplicates the contour of a template through a hydraulic or electrical system. *Automatic lathes*, also called *chucking machines* or *chuckers*, are used for machining individual pieces of regular or irregular shapes. *Turret lathes* are capable of performing multiple cutting operations on the same workpiece, such as turning, boring, drilling, thread cutting, and facing. Several cutting tools are mounted on the hexagonal *main turret*. The lathe usually has a *square turret* on the cross slide, with as many as four cutting tools mounted on it.

b) **Computer-controlled lathes.** In the most advanced lathes, movement and control of the machine and its components are actuated by **computer numerical controls** (CNCs); the features of such a lathe are shown in Fig. 8.45. These lathes are usually equipped with one or more turrets; each turret is equipped with a variety of tools and performs several operations on different surfaces of the workpiece. These machines are highly automated, the operations are repetitive and maintain the desired dimensional accuracy, and less skilled labor is required (after the machine is set up) than with common lathes. They are suitable for low to medium volumes of production. The details of computer controls are presented in Chapters 14 and 15. (See also Section 8.10.)

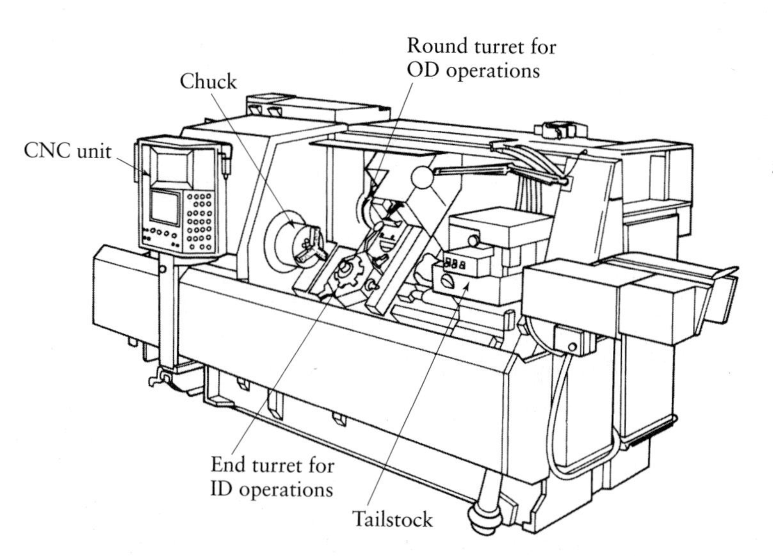

FIGURE 8.45 A computer-numerical-control lathe. Note the two turrets on this machine. *Source*: Jones & Lamson.

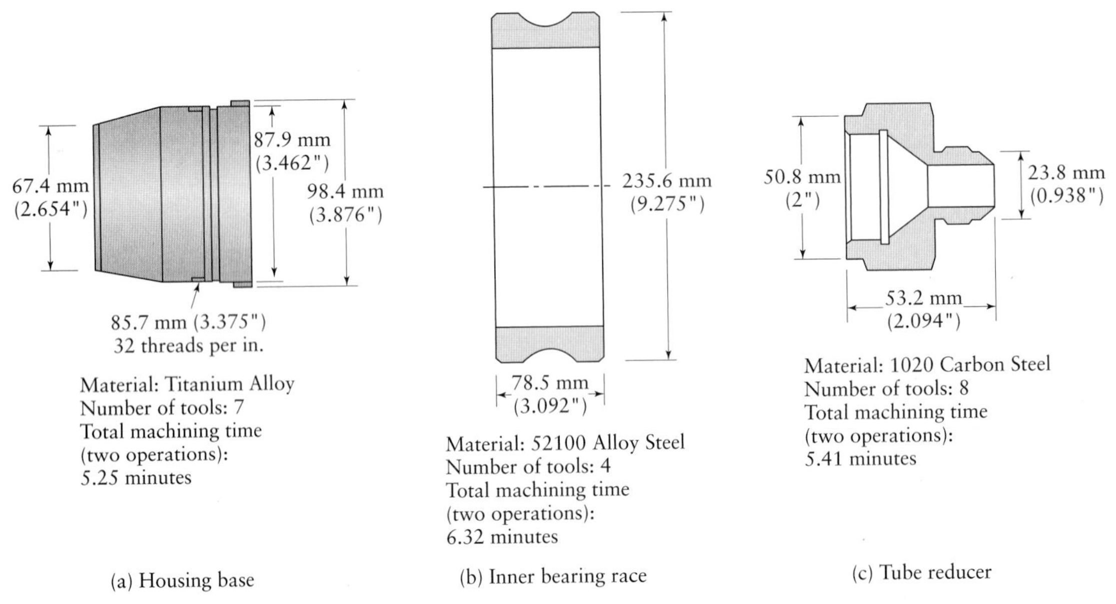

FIGURE 8.46 Typical parts made on computer-numerical-control machine tools.

EXAMPLE 8.6 Typical parts made on CNC turning machine tools

Figure 8.46 illustrates the capabilities of CNC turning machine tools. Workpiece materials, the number of cutting tools used, and machining times are indicated for each part. Although not as effectively or as consistently so, these parts can also be made on manual or turret lathes.

Source: Monarch Machine Tool Company.

c) **Turning process capabilities.** Table 8.9 provides relative *production rates* for turning, as well as for other cutting operations described in the rest of this chapter; these rates have an important bearing on productivity in machining operations. Note that there are major differences in the production rates of these processes. These differences are due not only to the inherent characteristics of the processes and machine tools, but also to various other factors, such as setup times and the types and sizes of the workpieces involved. The proper selection of a process and the machine tool for a particular product is important for minimizing production costs, as is also discussed in Section 8.14 and Chapter 16.

As stated previously, the ratings in Table 8.9 are relative, and there can be significant variations in specialized applications. For example, heat-treated, high-carbon cast-steel rolls (for rolling mills; see Section 6.3) can be machined on special lathes at material removal rates as high as 6000 cm^3/min (370 in^3/min), using multiple cermet tools. The important factor in this operation (also called **high-removal-rate machining**) is the very high stiffness of the machine tool (to avoid tool breakage due to chatter; see Section 8.11) and the machine tool's high power, which can be up to 450 kW (600 hp).

The surface finish and dimensional accuracy obtained in turning and related operations depend on such factors as the characteristics and condition of the machine tool, stiffness, vibration and chatter, process parameters, tool geometry and wear, cutting fluids, machinability of the workpiece material, and operator skill. As a result, a wide range of surface finishes can be obtained, as shown in Fig. 8.27. (See also Fig. 9.27.)

TABLE 8.9

Typical Production Rates for Various Cutting Operations

OPERATION	RATE
Turning	
Engine lathe	Very low to low
Tracer lathe	Low to medium
Turret lathe	Low to medium
Computer-control lathe	Low to medium
Single-spindle chuckers	Medium to high
Multiple-spindle chuckers	High to very high
Boring	Very low
Drilling	Low to medium
Milling	Low to medium
Planing	Very low
Gear cutting	Low to medium
Broaching	Medium to high
Sawing	Very low to low

Note: Production rates indicated are relative: *Very low* is about one or more parts per hour; *medium* is approximately 100 parts per hour; *very high* is 1000 or more parts per hour.

d) **High-speed machining.** With increasing demands for higher productivity and lower production costs, investigations have been carried out since the late 1950s to increase the material removal rate, particularly for applications in the aerospace and automotive industries. One obvious possibility is to increase the cutting speed. The term *high speed* is relative; as shown in Table 8.8, general recommendations for typical applications state that aluminum alloys, for example, should be machined at much higher speeds than titanium alloys or stainless steels. As a general guide, however, an approximate range of speeds may be defined as follows: **High speed**, 600–1800 m/min (2000–6000 ft/min); **very high speed**, 1800–18,000 m/min (6000–60,000 ft/min); **ultrahigh speed**: greater than 18,000 m/min (60,000 ft/min).

Much research is being carried out on *high-speed machining* (turning, milling, boring, and drilling), particularly on aluminum alloys, titanium alloys, steels, and superalloys. Much data have been collected regarding the effects of high speeds on the type of chips produced, the cutting forces, the temperatures generated, the tool wear, the surface finish, and the economics of the process. These studies have indicated that high-speed machining can be economical for certain applications, and consequently, it is now implemented for machining aircraft-turbine components and automotive engines, with 5 to 10 times the productivity of traditional machining. Important factors in these operations include (a) the selection of an appropriate cutting tool, (b) the power of the machine tools and their stiffness and damping capacity, (c) the stiffness of toolholders and the workholding devices, (d) special spindle designs for high power and high rotational speeds, (e) the inertia of the machine-tool components, (f) fast feed drives, and (g) the level of automation.

It is important to note, however, that high-speed machining should be considered mainly for situations in which *cutting time* is a significant portion of the floor-to-floor time of the operation. Other factors, such as *noncutting time* and labor costs, are important in the overall assessment of the benefits of high-speed machining for a particular application.

e) **Ultraprecision machining.** Beginning in the 1960s, increasing demands have been made for precision manufacturing of components for computer, electronics, nuclear energy, and defense applications. Examples include optical mirrors, computer memory disks, and drums for photocopying machines, with surface finish in the range of tens of nanometers [10^{-9} m, or 0.001μm (0.04 μin.)] and form accuracies in the μm and sub-μm range. The cutting tool for these *ultraprecision machining* applications is exclusively a single-crystal diamond (hence, the process is also called **diamond turning**), with a polished cutting-edge that has a radius as small as in the range of tens of nanometers. Wear of the diamond can be a significant problem; recent advances include **cryogenic diamond turning**, in which the tooling system is cooled by liquid nitrogen to a temperature of about −120°C (−184°F). (See also Section 8.7.2.)

The workpiece materials for ultraprecision machining to date include copper alloys, aluminum alloys, silver, gold, electroless nickel, infrared materials, and plastics (acrylics). The depths of cut involved are in the nanometer range. In this range, hard and brittle materials produce continuous chips (known as **ductile-regime cutting**; see also the discussion of *ductile-regime grinding* in Section 9.5.3); deeper cuts produce discontinuous chips.

The machine tools for ultraprecision machining are built with very high precision and high machine, spindle, and workholding-device stiffness. These machines, some parts of which are made of structural materials with low thermal expansion and good dimensional stability, are located in a dust-free environment (*clean rooms*) where the temperature is controlled within a fraction of one degree. Vibrations from external and internal sources are avoided as much as possible. Feed and position controls are made by laser metrology, and the machines are equipped with highly advanced computer control systems and with thermal and geometric error-compensating features.

f) **Hard turning.** As described in Chapter 9, there are several other mechanical (particularly grinding) processes and nonmechanical methods of removing material economically from hard or hardened metals. However, it is still possible to apply traditional cutting processes to hard metals and alloys by selecting an appropriate tool material and a machine tool with high stiffness. One common example is finish machining of heat-treated-steel (45 HRC to 65 HRC) machine and automotive components, using polycrystalline cubic-boron-nitride (PcBN) cutting tools. Called *hard turning*, this process produces machined parts with good dimensional accuracy, surface finish, and surface integrity. It can compete successfully with grinding of the same components, from both technical and economic aspects. A comparative example of hard turning vs. grinding is presented in Example 9.4.

g) **Cutting screw threads.** Threads are produced primarily by forming (*thread rolling*; see Fig. 6.46), which constitutes the process by which the largest quantity of threaded parts is produced, and cutting. *Casting* of threaded parts is also possible, although dimensional accuracy, surface finish, and production rate are not as good as obtained by other processes. Figure 8.40k shows that turning operations are capable of producing threads on round bar stock. When threads are produced externally or internally by cutting with a lathe-type tool, the process is called *thread cutting*, or *threading*. When the threads are cut internally with a special threaded tool (*tap*), the process is called *tapping*. External threads may also be cut with a die or by milling. Although it adds considerably to the cost of the operation, threads may be ground for improved accuracy and surface finish.

Automatic screw machines are designed for high-production-rate machining of screws and similar threaded parts. Because such machines are also capable of producing other components, they are now called *automatic bar machines*. All operations on these machines are performed automatically, with tools attached to a special turret. The bar stock is fed forward automatically through the headstock after each screw or part is machined to finished dimensions and cut off. The machines may be equipped with single or multiple spindles, and capacities range from 3- to 150-mm ($\frac{1}{8}$- to 6-in.)-diameter bar stock.

8.8.3 Boring and boring machines

The basic boring operation, as carried out on a lathe, is shown in Fig. 8.40h. *Boring* consists of producing circular internal profiles in hollow workpieces or on a hole made by drilling or another process and is carried out using cutting tools that are similar to those used in turning. However, because the boring bar has to reach the full length of the bore, tool deflection, and hence maintenance of dimensional accuracy, can be a significant problem. The boring bar must be sufficiently stiff—that is, made of a material with high elastic modulus, such as carbides—to minimize deflection and avoid vibration and chatter. Boring bars have been designed with capabilities for damping vibrations.

Although boring operations on relatively small workpieces can be carried out on a lathe, **boring mills** are used for large workpieces. These machines are either vertical or horizontal and are capable of performing operations such as turning, facing, grooving, and chamfering. A *vertical boring machine* (Fig. 8.47) is similar to a lathe, but with a vertical axis of workpiece rotation. In *horizontal boring machines*, the workpiece is mounted on a table that can move horizontally, both axially and radially. The cutting tool is mounted on a spindle that rotates in the headstock, which is capable of both vertical and longitudinal movements. Drills, reamers, taps, and milling cutters can also be mounted on the spindle.

8.8.4 Drilling, reaming, and tapping

One of the most common machining processes is *drilling*. **Drills** usually have a high length-to-diameter ratio (Fig. 8.48) and thus are capable of producing deep holes. However, they are somewhat flexible, depending on their diameter, and should be used with care in order to drill holes accurately and to prevent the drill from breaking. Moreover, chips are produced within the workpiece, and the chips have to move in a direction opposite to the axial movement of the drill. Consequently, chip disposal and the effectiveness of cutting fluids are important.

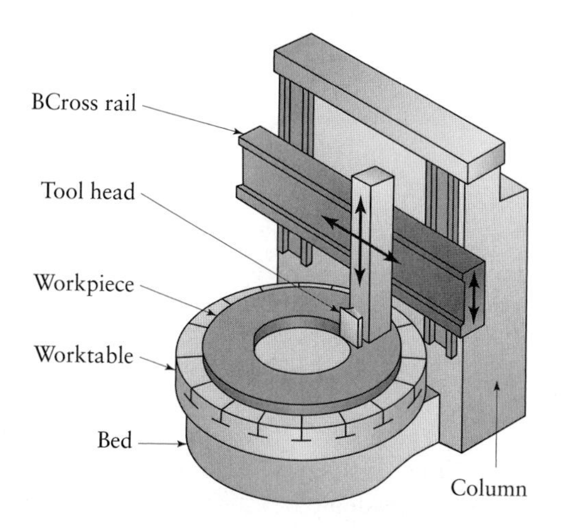

FIGURE 8.47 Schematic illustration of the components of a vertical boring mill.

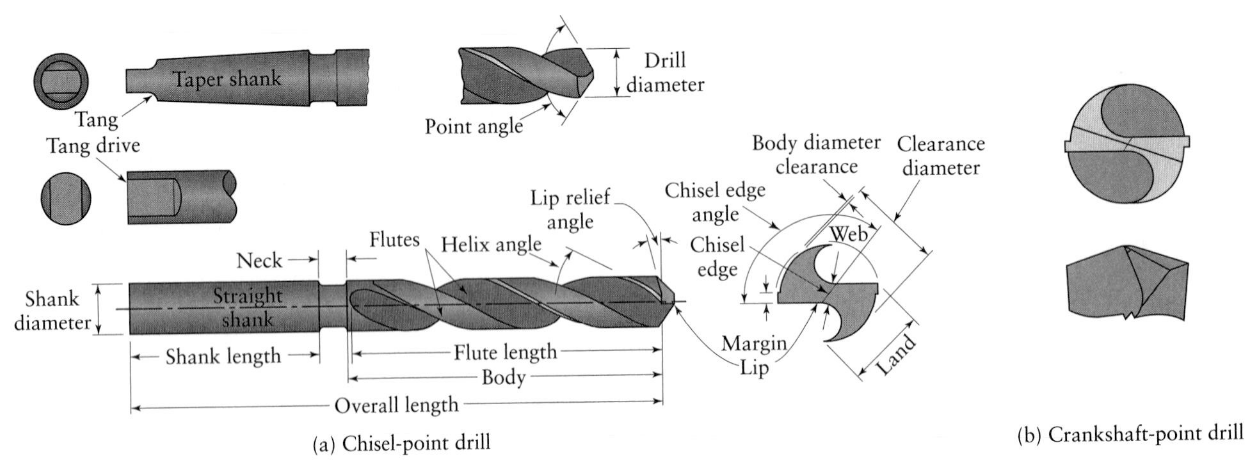

FIGURE 8.48 (a) Standard chisel-point drill, with various features indicated. (b) Crankshaft-point drill.

The most common drill is the standard-point **twist drill** (Fig. 8.48). The main features of the drill point are a *point angle*, a *lip-relief angle*, a *chisel-edge angle*, and a *helix angle*. The geometry of the drill tip is such that the normal rake angle and velocity of the cutting edge vary with the distance from the center of the drill. Other types of drills include the *step drill*, *core drill*, *counterboring* and *countersinking drills*, *center drill*, and *spade drill*, as illustrated in Fig. 8.49. *Crankshaft drills* have good centering ability, and because the chips they produce tend to break up easily, they are suitable for drilling deep holes. In *gun drilling*, a special drill is used for drilling deep holes; depth-to-diameter ratios of the holes produced can be 300 or higher. In the *trepanning* technique, a cutting tool produces a hole by removing a disk-shaped piece of material (*core*), usually from flat plates; thus, a hole is produced without reducing all the material to chips. The process can be used to make disks up to 150 mm (6 in.) in diameter from flat sheet or plate.

The *material removal rate* in drilling is the ratio of volume removed to time; thus, it can be expressed as

$$\text{MRR} = \frac{\pi D^2}{4} fN, \tag{8.40}$$

where D is the drill diameter, f is the feed, and N is the rpm of the drill. (See Table 8.10 for recommendations for speed and feed.)

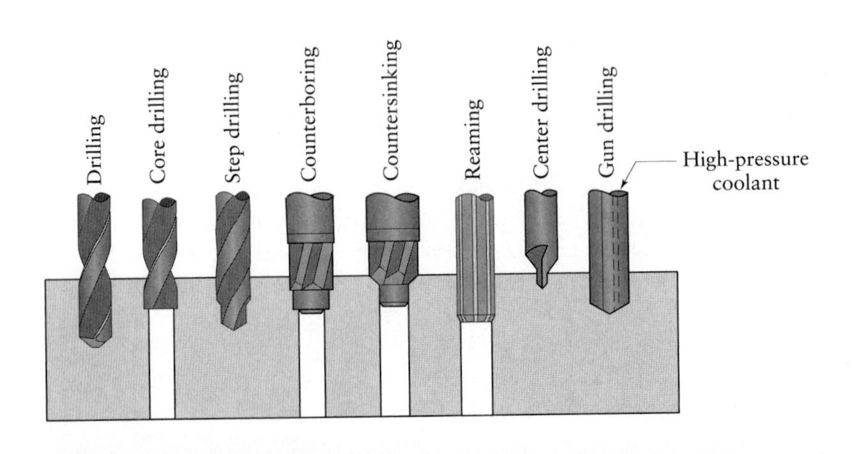

FIGURE 8.49 Various types of drills and drilling operations.

TABLE 8.10

General Recommendations for Speeds and Feeds in Drilling

	SURFACE SPEED		FEED, mm/rev (in./rev) DRILL DIAMETER		RPM	
WORKPIECE MATERIAL	m/min	ft/min	1.5 mm (0.060 in.)	12.5 mm (0.5 in.)	1.5 mm (0.060 in.)	12.5 mm (0.5 in.)
Aluminum alloys	30–120	100–400	0.025 (0.001)	0.30 (0.012)	6400–25,000	800–3000
Magnesium alloys	45–120	150–400	0.025 (0.001)	0.30 (0.012)	9600–25,000	1100–3000
Copper alloys	15–60	50–200	0.025 (0.001)	0.25 (0.010)	3200–12,000	400–1500
Steels	20–30	60–100	0.025 (0.001)	0.30 (0.012)	4300–6400	500–800
Stainless steels	10–20	40–60	0.025 (0.001)	0.18 (0.007)	2100–4300	250–500
Titanium alloys	6–20	20–60	0.010 (0.0004)	0.15 (0.006)	1300–4300	150–500
Cast irons	20–60	60–200	0.025 (0.001)	0.30 (0.012)	4300–12,000	500–1500
Thermoplastics	30–60	100–200	0.025 (0.001)	0.13 (0.005)	6400–12,000	800–1500
Thermosets	20–60	60–200	0.025 (0.001)	0.10 (0.004)	4300–12,000	500–1500

Note: As hole depth increases, speeds and feeds should be reduced. Selection of speeds and feeds also depends on the specific surface finish required.

The thrust force in drilling is the force that acts in the direction of the hole axis. If this force is excessive, it can cause the drill to break or bend. The thrust force depends on factors such as the strength of the workpiece material, feed, rotational speed, cutting fluids, drill diameter, and drill geometry. Although some attempts have been made, accurate calculation of the thrust force on the drill has proven to be difficult. Experimental data are available as an aid in the design and use of drills and drilling equipment. Thrust forces in drilling range from a few newtons for small drills to as high as 100 kN (22.5 klb) in drilling high-strength materials with large drills. Similarly, drill torque can range up to as high as 4000 N-m (3000 ft-lb).

The torque during drilling also is difficult to calculate; it can be obtained by using the data in Table 8.3 and by noting that the power dissipated during drilling is the product of torque and rotational speed. Therefore, by first calculating the MRR, we can calculate the torque on the drill. **Drill life**, as well as tap life, is usually measured by the number of holes drilled before the drill becomes dull and forces increase.

Drilling machines are used for drilling holes, tapping, reaming, and other general-purpose, small-diameter boring operations; they are generally vertical, the most common type being a **drill press**. The workpiece is placed on an adjustable table, either by clamping it directly into the slots and holes on the table or by using a vise that can be clamped to the table. The drill is then lowered manually by hand wheel or by power feed at preset rates. In order to maintain proper cutting speeds, the spindle speed on drilling machines has to be adjustable to accommodate different sizes of drills. Drill presses are usually designated by the largest workpiece diameter that can be accommodated on the table; sizes typically range from 150 to 1250 mm (6 to 50 in.).

Reaming and reamers. *Reaming* is an operation that makes an existing hole dimensionally more accurate than can be achieved by drilling alone; it also improves the hole's surface finish. The most accurate holes are produced by the following sequence of operations: centering, drilling, boring, and reaming. For even better accuracy and surface finish, holes may be internally ground and honed (Chapter 9). A *reamer* (Fig. 8.50) is a multiple-cutting-edge tool with straight or helically fluted edges that removes very little material. The shanks may be straight or tapered, as in drills. The basic types of reamers are *hand* and *machine* (*chucking*) reamers; other types include *rose* reamers with cutting edges that have wide margins and no relief and *fluted, shell, expansion*, and *adjustable* reamers.

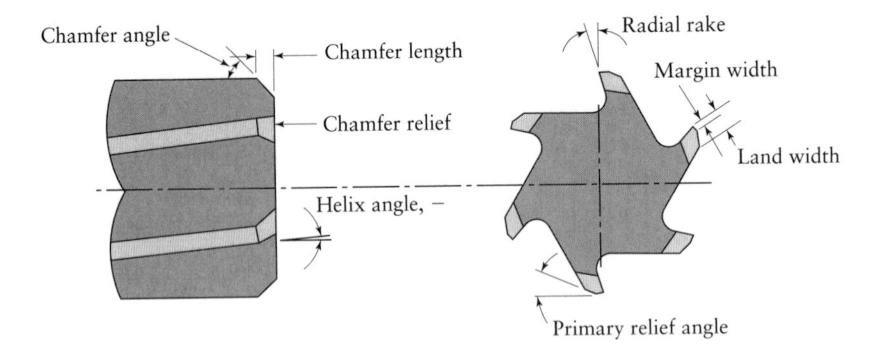

FIGURE 8.50 Terminology for a helical reamer.

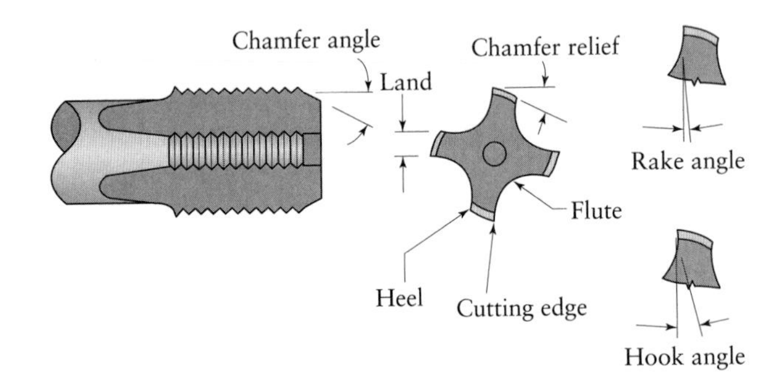

FIGURE 8.51 Terminology for a tap.

Tapping and taps. Internal threads in workpieces can be produced by tapping. A **tap** is basically a threading tool with multiple cutting teeth (Fig. 8.51). Taps are generally available with three or four flutes; three-fluted taps are stronger, because of the larger amount of material available in the flute. *Tapered* taps are designed to reduce the torque required for tapping through holes, and *bottoming* taps are designed for tapping blind holes to their full depth. *Collapsible* taps are used for large-diameter holes; after tapping is completed, the tap is mechanically collapsed and removed from the hole, without having to be rotated. Tap sizes range up to 100 mm (4 in.).

8.9 Cutting Processes and Machine Tools for Producing Various Shapes

Several cutting processes and machine tools are capable of producing complex shapes with the use of multitooth, as well as single-point, cutting tools (Fig. 8.52 and Table 8.7). We begin with one of the most versatile processes, called *milling*, in which a multitooth cutter rotates along various axes with respect to the workpiece. The planing, shaping, and broaching are then described, wherein flat and shaped surfaces are produced, and the tool or workpiece travels along a straight path. This section ends with discussion of sawing and filing processes.

8.9.1 Milling operations

Milling includes a number of versatile machining operations that use a **milling cutter**, a multitooth tool that produces a number of chips per revolution, to produce a variety of configurations. Parts such as the ones shown in Fig. 8.52 can be machined efficiently and repeatedly using various milling cutters.

The main types of milling operations are described as follows:

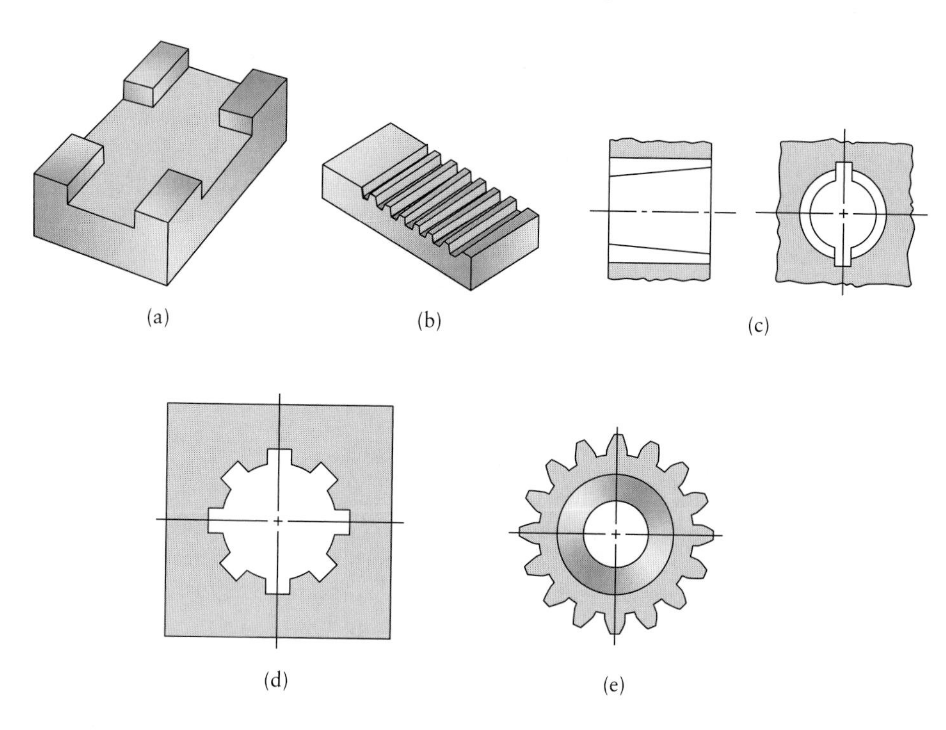

FIGURE 8.52 Typical parts and shapes produced by the cutting processes described in Section 8.9.

a) **Slab milling.** In *slab milling*, also called **peripheral milling**, the axis of cutter rotation is parallel to the surface of the workpiece to be machined (Fig. 8.53a). The cutter, generally made of high-speed steel, has a number of teeth along its circumference, each tooth acting like a single-point cutting tool called a *plain mill*. Cutters used in slab milling may have *straight* or *helical* teeth, producing orthogonal or oblique cutting action. Figure 8.1c shows the helical teeth on a cutter. In **conventional milling**, also called *up milling*, the maximum chip thickness is at the end of the cut (Fig. 8.53b). The advantages of conventional milling are that tooth engagement is not a function of workpiece surface characteristics, and contamination or scale on the surface does not affect tool life; this process is the dominant method of milling. The cutting process is smooth, provided that the cutter teeth are sharp. However, there is a tendency for the tool to chatter, and the workpiece has a tendency to be pulled upward; thus, proper clamping is important.

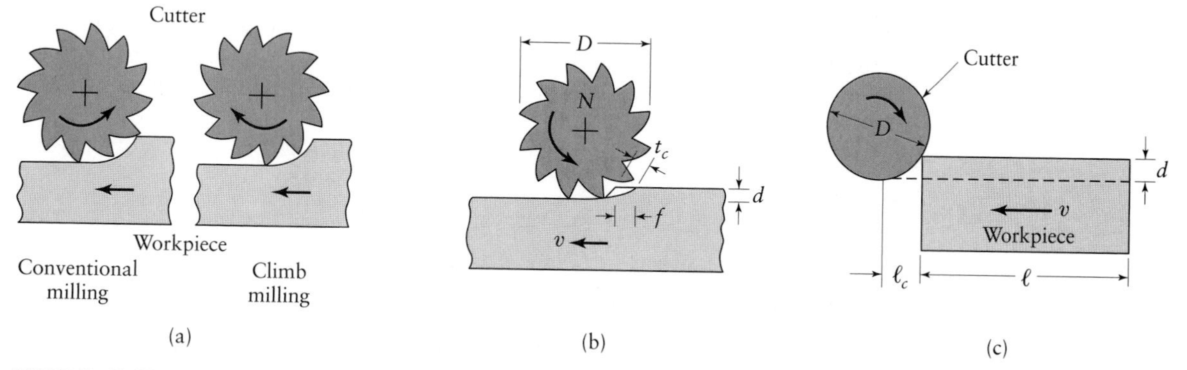

FIGURE 8.53 (a) Schematic illustration of conventional milling and climb milling. (b) Slab-milling operation, showing depth of cut, d; feed per tooth, f; chip depth of cut, t_c; and workpiece speed, v. (c) Schematic illustration of cutter travel distance ℓ_c to reach full depth of cut.

In **climb milling**, also called *down milling*, cutting starts with the thickest location of the chip. The advantage of climb milling is that the downward component of cutting forces holds the workpiece in place, particularly for slender parts. However, because of the resulting high-impact forces when the teeth engage the workpiece, this operation must have a rigid setup, and backlash must be eliminated in the table feed mechanism. Climb milling is not suitable for machining workpieces that have surface scale, such as hot-worked metals, forgings, and castings; the scale is hard and abrasive and causes excessive wear and damage to the cutter teeth, thus reducing tool life. Climb milling is recommended, in general, for maximum cutter life when CNC machine tools are used; a typical application is in finishing cuts on aluminum.

The cutting speed, V, in milling is the peripheral speed of the cutter, or

$$V = \pi DN, \tag{8.41}$$

where D is the cutter diameter and N is the rotational speed of the cutter (Fig. 8.53b). Note that the thickness of the chip in slab milling varies along its length, because of the relative longitudinal motion between cutter and workpiece. For a straight-tooth cutter, the approximate *undeformed chip thickness*, t_c (*chip depth of cut*) can be determined from the equation

$$t_c = 2f\sqrt{\frac{d}{D}}, \tag{8.42}$$

where f is the feed per tooth of the cutter, measured along the workpiece surface (that is, the distance the workpiece travels per tooth of the cutter, in mm/tooth or in./tooth), and d is the depth of cut. As t_c becomes greater, the force on the cutter tooth increases.

Feed per tooth is determined from the equation

$$f = \frac{v}{Nn}, \tag{8.43}$$

where v is the linear speed (feed rate) of the workpiece and n is the number of teeth on the cutter periphery. We can check the dimensional accuracy of this equation by substituting appropriate units for the individual terms; thus, (mm/tooth) = (m/min) (10^3 mm/m)/(rev/min)(number of teeth/rev), which checks out correctly. The cutting time, t, is given by the expression

$$t = \frac{(\ell + \ell_c)}{v}, \tag{8.44}$$

where ℓ is the length of the workpiece (Fig. 8.53c) and ℓ_c is the extent of the cutter's first contact with the workpiece. Based on the assumption that $\ell_c \ll \ell$ (although this relationship is not generally so), the *material removal rate* is

$$\text{MRR} = \frac{\ell wd}{t} = wdv, \tag{8.45}$$

where w is the width of the cut, which, for a workpiece narrower than the length of the cutter, is the same as the width of the workpiece. The distance that the cutter travels in the noncutting-cycle operation is an important economic consideration and should be minimized.

Although we can calculate the *power requirement* in slab milling, the *forces* acting on the cutter (tangential, radial, and axial; see also Fig. 8.42b) are difficult to calculate, because of the many variables involved, particularly tool geometry. These forces can be measured experimentally for a variety of conditions. The *torque* on the cutter spindle can be calculated from the power.

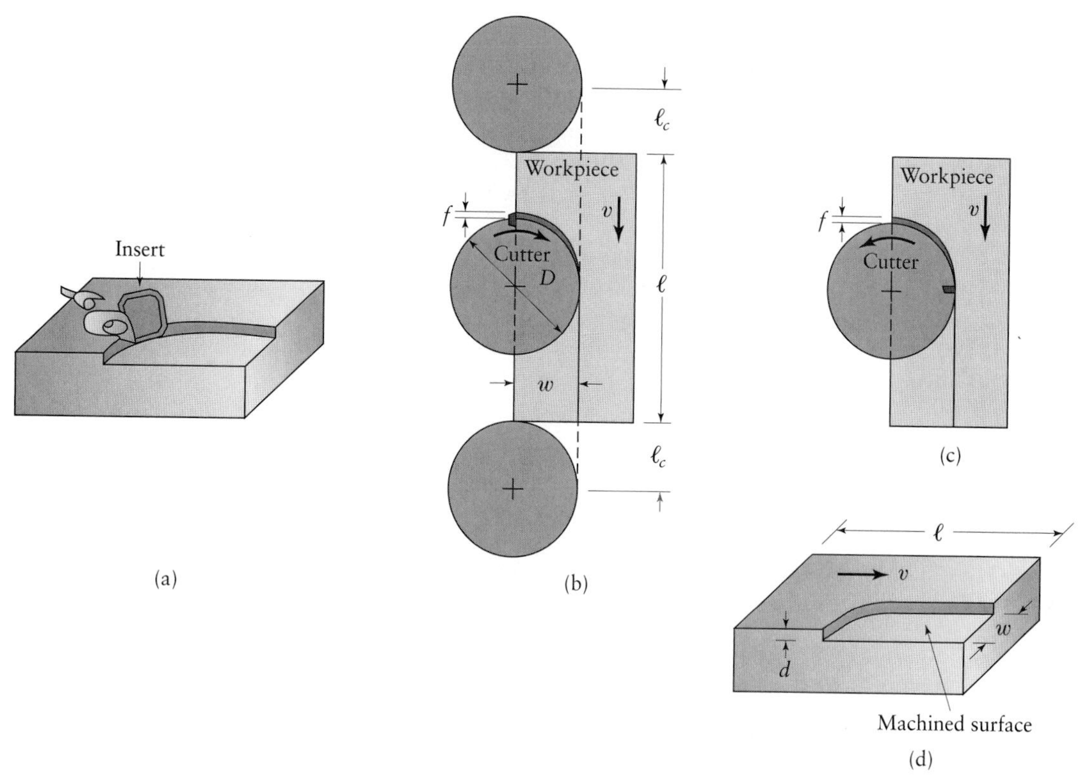

FIGURE 8.54 Face-milling operation showing (a) action of an insert in face milling; (b) climb milling; (c) conventional milling; (d) dimensions in face milling. The width of cut, w, is not necessarily the same as the cutter radius. *Source*: Courtesy of The Ingersoll Cutting Tool Company.

Although the torque is the product of the tangential force on the cutter and the cutter's radius, the tangential force per tooth depends on how many teeth are engaged during the cut.

b) **Face milling.** In *face milling*, the cutter is mounted on a spindle that has an axis of rotation perpendicular to the workpiece surface, and removes material in the manner shown in Fig. 8.54a. The cutter rotates at a rotational speed N, and the workpiece moves along a straight path at a linear speed, v. When the cutter rotates in the direction shown in Fig. 8.54b, the operation is *climb* milling; when it rotates in the opposite direction (Fig. 8.54c), it is *conventional* milling.

Because of the relative motion between the cutting teeth and the workpiece, a face-milling cutter leaves *feed marks* on the machined surface, much as in turning operations. Surface roughness depends on insert corner geometry and feed per tooth. [see also Eqs. (8.35) to (8.37).]

Figure 8.55 provides the terminology for a face-milling cutter and the various angles. The side view is shown in Fig. 8.56, where we note that, as in turning operations, the *lead angle* of the insert in face milling has a direct influence on the *undeformed chip thickness*. As the lead angle (positive as shown) increases, the undeformed chip thickness (thus, also the thickness of the actual chip) decreases, and the length of contact increases. The range of lead angles for most face-milling cutters is from 0° to 45°, thus influencing the axial and tangential forces. Note that the cross-sectional area of the undeformed chip remains constant. The lead angle also influences forces in milling. It can be seen

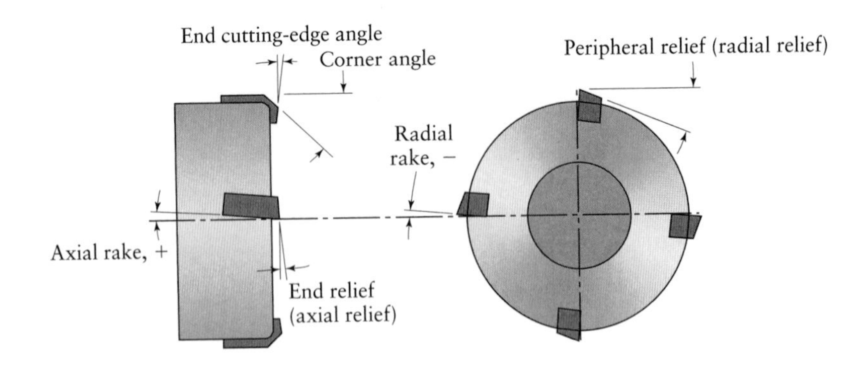

FIGURE 8.55 Terminology for a face-milling cutter.

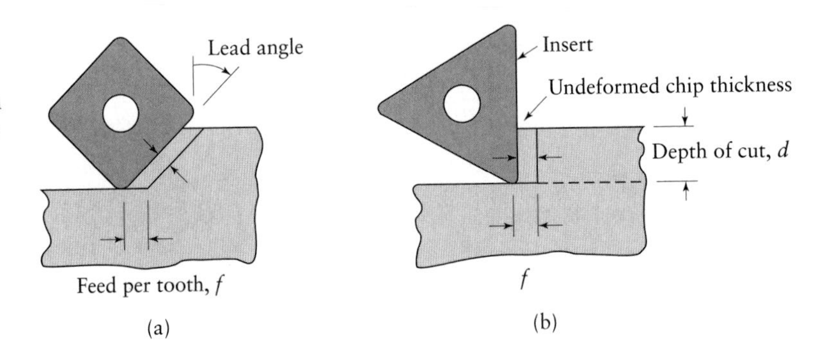

FIGURE 8.56 The effect of lead angle on the undeformed chip thickness in face milling. Note that as the lead angle increases, the undeformed chip thickness (and hence the thickness of the actual chip) decreases, but the length of contact (and hence the width of the chip) increases. The insert must be sufficiently large to accommodate the increase in contact length.

that as the lead angle decreases, there is less and less of a vertical force (axial force on the cutter spindle) component.

A wide variety of milling cutters is available. The cutter diameter should be chosen in such a manner that it will not interfere with fixtures and other components in the setup. In a typical face-milling operation, the ratio of the cutter diameter, D, to the width of cut, w, should be no less than 3:2. The cutting tools are usually carbide or high-speed-steel inserts and are mounted on the cutter body. (See Fig. 8.55.)

The relationship of cutter diameter and insert angles and their position relative to the surface to be milled is important, because they determine the angle at which an insert enters and exits the workpiece. Note in Fig. 8.54b for climb milling that if the insert has zero axial and radial rake angles (see Fig. 8.55), then the rake face of the insert engages the workpiece directly. However, as can be seen in Fig. 8.57a and b, the same insert engages the workpiece at different angles, depending on the relative positions of the cutter and the workpiece. In Fig. 8.57a, the tip of the insert makes the first contact, and hence there is potential for the cutting edge to chip off. In Fig. 8.57b, on the other hand, the contacts (i.e., at entry, reentry, and the two exits) are at an angle and away from the tip of the insert. There is, therefore, less of a tendency for the insert to fail, because the force on the insert increases and decreases gradually. Note from Fig. 8.55 that the radial and axial rake angles will also have an effect on this situation.

Figure 8.57c shows the exit angles for various cutter positions. Note that in the first two examples, the insert exits the workpiece at an angle, whereby the force on the insert diminishes to zero at a slower rate (desirable) than in the third example, where the insert exits suddenly (undesirable).

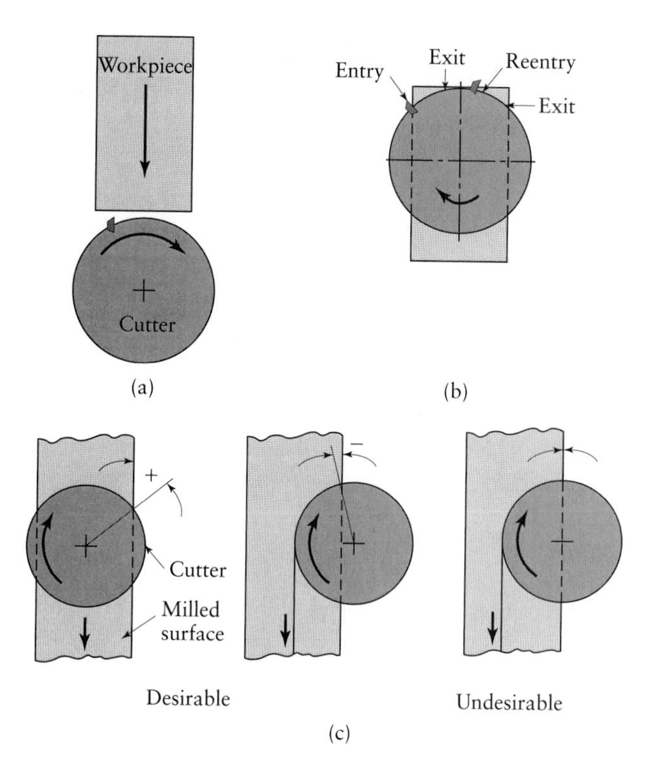

FIGURE 8.57 (a) Relative position of the cutter and insert as it first engages the workpiece in face milling, (b) insert positions toward the end of cut, and (c) examples of exit angles of insert, showing desirable (positive or negative angle) and undesirable (zero angle) positions. In all figures, the cutter spindle is perpendicular to the page.

EXAMPLE 8.7 Calculation of material removal rate, power required, and cutting time in face milling

Refer to Fig. 8.54, and assume that $D = 150$ mm, $w = 60$ mm, $\ell = 500$ mm, $d = 3$ mm, $v = 0.6$ m/min, and $N = 100$ rpm. The cutter has 10 inserts, and the workpiece material is a high-strength aluminum alloy. Calculate the material removal rate, cutting time, and feed per tooth, and estimate the power required.

Solution. The cross-section of the cut is $wd = (60)(3) = 180\text{ mm}^2$. Since the workpiece speed v is 0.6 m/min = 600 mm/min, we calculate the material removal rate as

$$\text{MRR} = (180)(600) = 108{,}000\text{ mm}^3/\text{min}.$$

The cutting time is given by

$$t = \frac{(\ell + 2\ell_c)}{v}.$$

Note from Fig. 8.54 that for this problem, $\ell_c = D/2$, or 75 mm. Thus, the cutting time is

$$t = \frac{(500 + 150)(60)}{600} = 65\text{ s} = 1.08\text{ min}.$$

We obtain the feed per tooth from Eq. (8.43). Noting that $N = 100$ rpm $= 1.67$ rev/s, we find

$$f = \frac{10}{(1.67)(10)} = 0.6\text{ mm/tooth}.$$

For this material, the unit power can be taken from Table 8.3 to be 1.1 W-s/mm^3; hence, the power can be estimated as

$$\text{Power} = (1.1)(1800) = 1980 \text{ W} = 1.98 \text{ kW}.$$

c) **End milling.** The cutter in *end milling* is shown in Fig. 8.1d; it has either straight or tapered shanks for smaller and larger cutter sizes, respectively. The cutter usually rotates on an axis perpendicular to the workpiece, although it can be tilted to machine tapered surfaces as well. Flat surfaces as well as various profiles can be produced by end milling. The end faces of some end mills have cutting teeth, and these teeth can be used as a drill to start a cavity; end mills are also available with hemispherical ends for producing curved surfaces, as in making dies. *Hollow end mills* have *internal* cutting teeth and are used for machining the cylindrical surface of solid round workpieces, as in preparing stock with accurate diameters for automatic machines. End mills are made of high-speed steels or have carbide inserts.

Section 8.8.2 has described high-speed machining and its applications. One of the more common applications is **high-speed milling**, using an end mill and with the same general provisions regarding the stiffness of machines, work-holding devices, etc., that have been described earlier. A typical application is milling aluminum-alloy aerospace components and honeycomb structures, with spindle speeds on the order of 20,000 rpm. Another application is in *die sinking*, i.e., producing cavities in die blocks. Chip collection and disposal can be a significant problem in these operations, because of the high rate of material removal.

d) **Other milling operations and cutters.** Several other types of milling operations and cutters are used to machine various surfaces. In *straddle milling*, two or more cutters are mounted on an arbor and are used to machine two *parallel* surfaces on the workpiece (Fig. 8.58a). *Form milling* is used to produce curved profiles, using cutters that have specially shaped teeth (Fig. 8.58b); such cutters are also used for cutting gear teeth (see Section 8.9.7).

Circular cutters are used for slotting and slitting. The teeth may be staggered slightly, as in a saw blade (see Section 8.9.5), to provide clearance for the cutter in making deep slots. *Slitting saws* are relatively thin, usually less than 5 mm $\left(\frac{3}{16}\text{ in.}\right)$. *T-slot cutters* are used to mill T-slots, such as those in machine-tool worktables for clamping of workpieces. The slot is first milled with an end mill; a T-slot cutter then cuts the complete profile of the slot in one pass. *Key-seat cutters* are used to make the semicylindrical (Woodruff) key sets for shafts. *Angle-milling cutters*, either with a single angle or double angles, are used to produce tapered surfaces with various angles.

Shell mills are hollow inside and are mounted on a shank, thus allowing the same shank to be used for different-sized cutters. The use of shell mills is

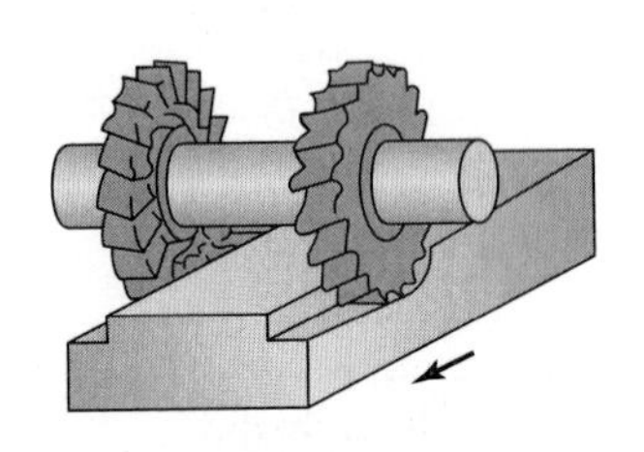

(a) Straddle milling

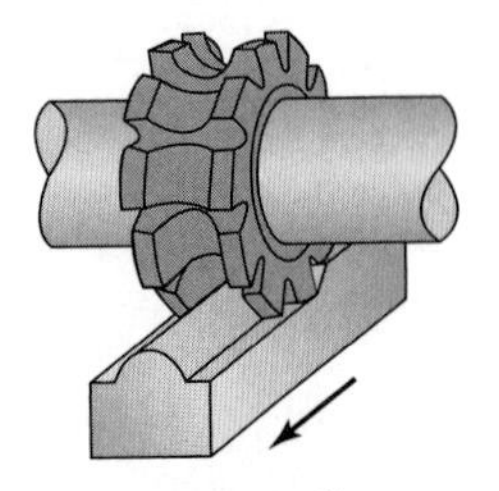

(b) Form milling

FIGURE 8.58 Cutters for (a) straddle milling and (b) form milling.

similar to that of end mills. Milling with a single cutting tooth mounted on a high-speed spindle is known as *fly cutting* and is generally used in simple face-milling and boring operations. The tool can be shaped as a single-point tool and can be placed in various radial positions on the spindle.

e) **Toolholders.** Milling cutters are classified into two basic types. *Arbor cutters* are mounted on an arbor, such as for slab, face, straddle, and form milling; in *shank-type cutters*, the cutter and the shank are one piece. The most common examples of shank cutters are end mills. Whereas small end mills have straight shanks, larger end mills have tapered shanks for better clamping to resist the higher forces and torque involved. Cutters with straight shanks are mounted in collet chucks or special end-mill holders; cutters with tapered shanks are mounted in tapered toolholders. In addition to mechanical means, hydraulic tool holders and arbors are also available. The stiffness of cutters and toolholders is important for surface quality and reducing vibration and chatter in milling operations. Conventional tapered toolholders have a tendency to wear and bellmouth under the radial forces in milling.

f) **Milling machines and process capabilities.** In addition to the various characteristics of milling processes described thus far, milling capabilities include parameters such as production rate, surface finish, dimensional tolerances, and cost considerations. Table 8.11 gives the typical conventional ranges of feeds and cutting speeds for milling. Cutting speeds vary over a wide range, from 30 to 3000 m/min (90 to 10,000 ft/min), depending on workpiece material, cutting-tool material, and process parameters. Because of their capability to perform a variety of operations, milling machines, which were first built in 1876, are among the most versatile and useful machine tools, and a wide selection of machines with numerous features is available. Used for general-purpose milling operations, **column-and-knee-type** machines are the most common milling machines. The spindle, to which the milling cutter is attached, may be *horizontal* (Fig. 8.59a) for slab milling or *vertical* for face and end milling, boring, and drilling operations (Fig. 8.59b). The various components are moved manually or by power.

In **bed-type** machines, the worktable is mounted directly on the bed, which replaces the knee, and can move only longitudinally. These milling

TABLE 8.11

Approximate Range of Recommended Cutting Speeds for Milling Operations

	CUTTING SPEED	
WORKPIECE MATERIAL	m/min	ft/min
Aluminum alloys	300–3000	1000–10,000
Cast iron, gray	90–1300	300–4200
Copper alloys	90–1000	300–3300
High-temperature alloys	30–550	100–1800
Steels	60–450	200–1500
Stainless steels	90–500	300–1600
Thermoplastics and thermosets	90–1400	300–4500
Titanium alloys	40–150	130–500

Note: (a) These speeds are for carbides, ceramic, cermets, and diamond cutting tools. Speeds for high-speed-steel tools are lower than those indicated in this table.
(b) Depths of cut, d, are generally in the range of 1–8 mm (0.04–0.3 in.).
(c) Feeds per tooth, f, are generally in the range of 0.08–0.46 mm/rev (0.003–0.018 in./rev).

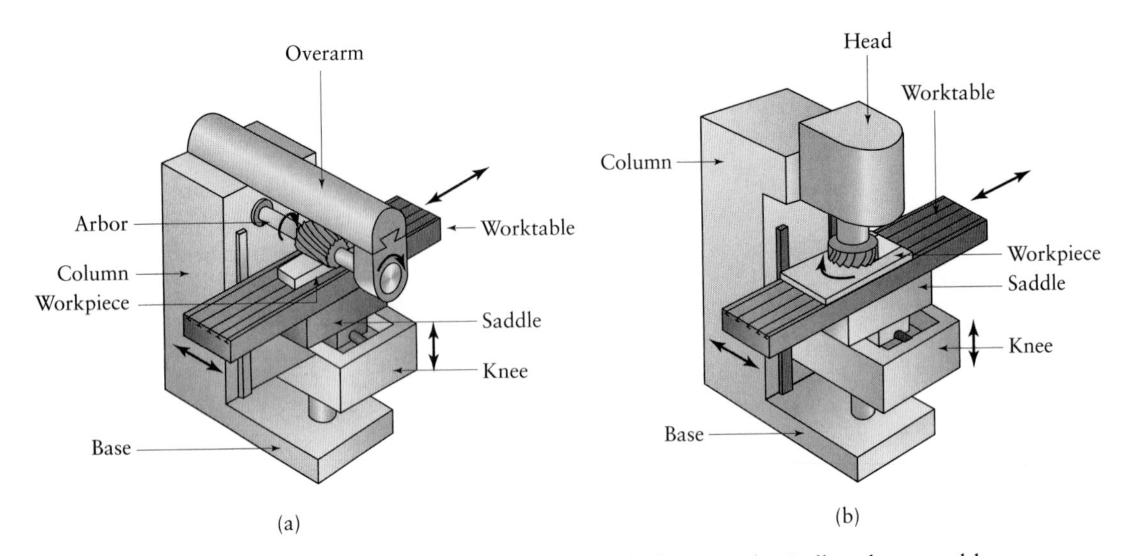

FIGURE 8.59 (a) Schematic illustration of a horizontal-spindle column-and-knee-type milling machine. (b) Schematic illustration of a vertical-spindle column-and-knee-type milling machine. *Source*: G. Boothroyd, *Fundamentals of Machining and Machine Tools*.

machines are not as versatile as others, but have high stiffness and are used for high-production-rate work. The spindles may be horizontal or vertical and of duplex or triplex types, that is, with two or three spindles for simultaneously milling two or three workpiece surfaces, respectively. **Planer-type** machines, which are similar to bed-type machines, are equipped with several heads and cutters to mill various surfaces. They are used for heavy workpieces and are more efficient than planers (see Section 8.9.2) when used for similar operations. *Rotary-table machines* are similar to vertical milling machines and are equipped with one or more heads for face milling. *Profile-milling machines* have five-axis movements. Using tracer fingers, *duplicating machines* (copy-milling machines) reproduce parts from a master model; they are used in the automotive and aerospace industries for machining complex parts and dies (die sinking), although they are being replaced rapidly with CNC machines. These machine tools are versatile and capable of milling, drilling, boring, and tapping with repetitive accuracy.

8.9.2 Planing and planers

Planing is a relatively simple cutting process by which flat surfaces, as well as various cross-sections with grooves and notches, are produced along the length of the workpiece. Planing is usually done on large workpieces, as large as 25 m × 15 m (75 ft × 45 ft). In a *planer*, the workpiece is mounted on a table that travels along a straight path. A horizontal cross rail, which can be moved vertically along the ways in the column, is equipped with one or more tool heads. The cutting tools are attached to the heads, and machining is done along a straight path. Because of the reciprocating motion of the workpiece, elapsed noncutting time during the return stroke is significant, both in planing and in shaping (see Section 8.9.3). Consequently, these operations are not efficient or economical, except for in low-quantity production. Efficiency of the operation can be improved by equipping planers with toolholders and tools that cut in both directions of table travel.

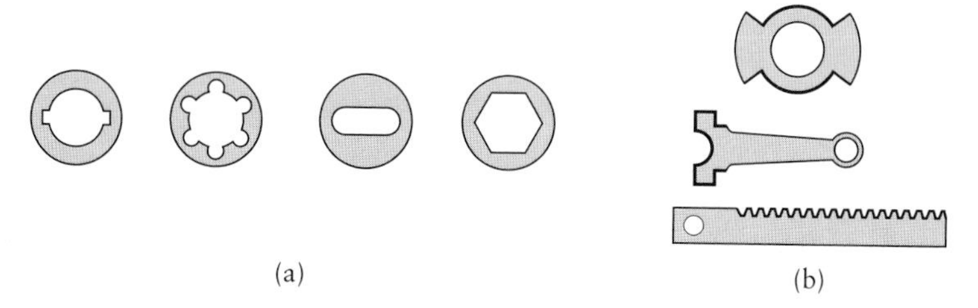

FIGURE 8.60 (a) Typical parts made by internal broaching. (b) Parts made by surface broaching. The heavy lines indicate broached surfaces. *Source*: General Broach and Engineering Company.

8.9.3 Shaping and shapers

Cutting by *shaping* is basically the same as in planing. In a *horizontal shaper*, the tool travels along a straight path, and the workpiece is stationary; the cutting tool is attached to the tool head, which is mounted on the ram. The ram has a reciprocating motion, and in most machines, cutting is done during the forward movement of the ram (*push cut*); in others, however, it is done during the return stroke of the ram (*draw cut*). *Vertical shapers* (*slotters*) are used for machining notches and keyways. Shapers are also capable of producing complex shapes, such as cutting a helical impeller, in which the workpiece is rotated during the cut through a master cam. Because of their low production rate, shapers are now generally used only in toolrooms, job shops, and for repair work.

8.9.4 Broaching and broaching machines

The *broaching* operation is similar to shaping and is used to machine internal and external surfaces (Fig. 8.60), such as holes of circular, square, or irregular section; keyways; teeth of internal gears; multiple spline holes; and flat surfaces. A *broach* (Fig. 8.61) is, in effect, a long multitooth cutting tool that makes successively deeper cuts; hence, the total depth of material removed in one stroke is the sum of the depths of cut of each tooth. A broach can remove material as deep as 6 mm (0.25 in.) in one stroke. Broaching can produce parts with good surface finish and dimensional accuracy; thus, it competes favorably with other processes to produce similar shapes. Although broaches can be expensive, the cost is justified, because of their use for high-quantity production runs.

Figure 8.61b provides the terminology for a broach. The rake (hook) angle depends on the material being cut, as in turning and other cutting operations, and usually ranges between 0° and 20°. The clearance angle is usually 1° to 4°; finishing teeth have smaller angles. Too small a clearance angle causes rubbing of the teeth

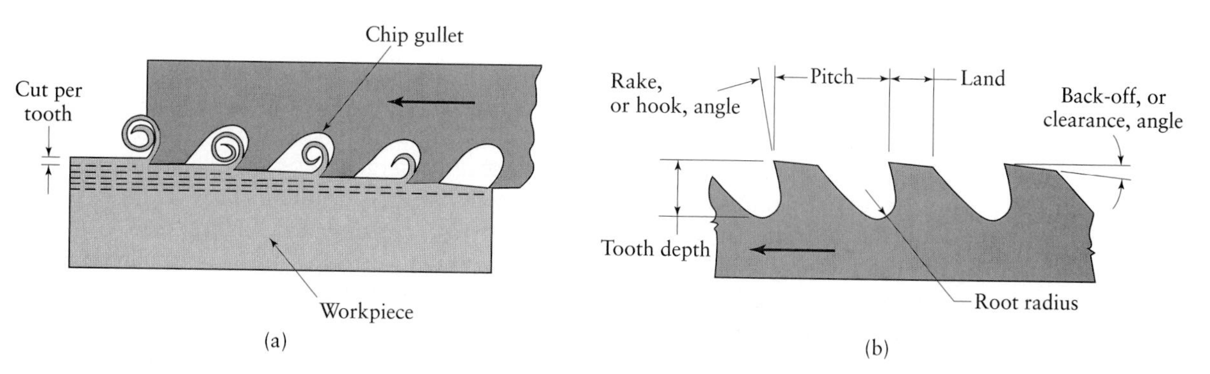

FIGURE 8.61 (a) Cutting action of a broach, showing various features. (b) Terminology for a broach.

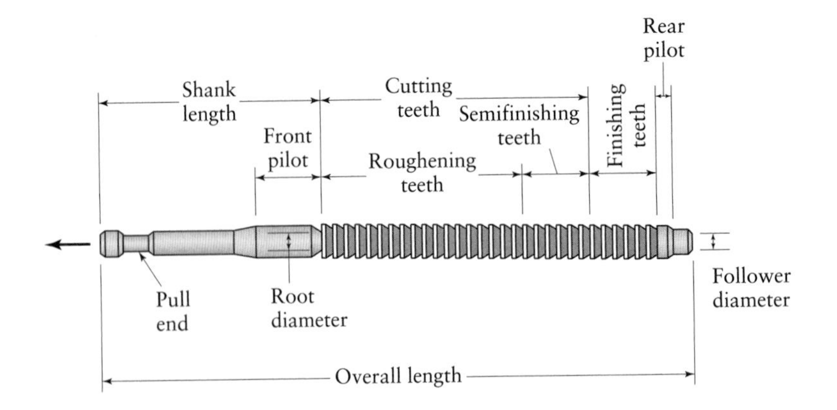

FIGURE 8.62 Terminology for a pull-type internal broach used for enlarging long holes.

against the broached surface. The pitch of the teeth depends on factors such as length of the workpiece (length of cut), tooth strength, and size and shape of chips. The tooth depth and pitch must be sufficiently large to accommodate the chips produced during broaching, particularly for long workpieces, but at least two teeth should be in contact with the workpiece at all times (similar to sawing).

Broaches are available with various tooth profiles, including some that have chip breakers. Broaches are also made round with circular cutting teeth and are used to enlarge holes (Fig. 8.62). Note that the cutting teeth on broaches have three regions: roughing, semifinishing, and finishing. Irregular internal shapes are usually broached by starting with a round hole in the workpiece, produced by drilling or boring.

In **turn broaching** of crankshafts, the workpiece rotates between centers, and the broach is equipped with multiple inserts and passes tangentially across the part; the process is thus a combination of broaching and skiving. There are also machines that broach a number of crankshafts simultaneously; main bearings for engines are broached in this way.

Broaching machines either pull or push the broaches and are made to be horizontal or vertical. *Push broaches* are usually shorter, generally in the range of 150–350 mm (6–14 in.) long. *Pull broaches* tend to straighten the hole, whereas pushing permits the broach to follow any irregularity of the leader hole. Horizontal machines are capable of longer strokes than are vertical machines. Many types of broaching machines are available, some with multiple heads, allowing a variety of shapes and parts to be produced, including helical splines and rifled gun barrels. Sizes range from machines for making needlelike parts to those used for broaching gun barrels. The capacities of broaching machines are as high as 0.9 MN (100 tons) of pulling force.

8.9.5 Sawing and saws

Sawing is a cutting operation in which the cutting tool is a blade that has a series of small teeth, with each tooth removing a small amount of material. Saws are used for all metallic and nonmetallic materials that are machinable by other cutting processes, and they are capable of producing various shapes. The width of cut (**kerf**) in sawing is usually narrow, and hence sawing wastes little material. Figure 8.63 illustrates typical saw-tooth and saw-blade configurations. Tooth spacing is usually in the range of 0.08 to 1.25 teeth per mm (2 to 32 teeth per in.).

Saw blades are made from carbon and high-speed steels; steel blades with tips of carbide steel or high-speed steel are used for sawing harder materials (Fig. 8.64). In order to prevent the saw from binding and rubbing during cutting, the teeth are alternately set in opposite directions so that the kerf is wider than the blade

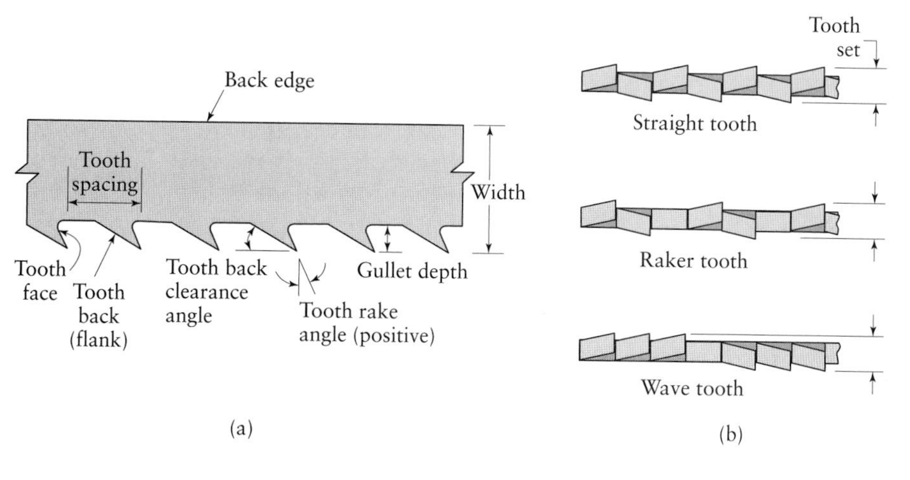

FIGURE 8.63 (a) Terminology for saw teeth. (b) Types of saw teeth, staggered to provide clearance for the saw blade to prevent binding during sawing.

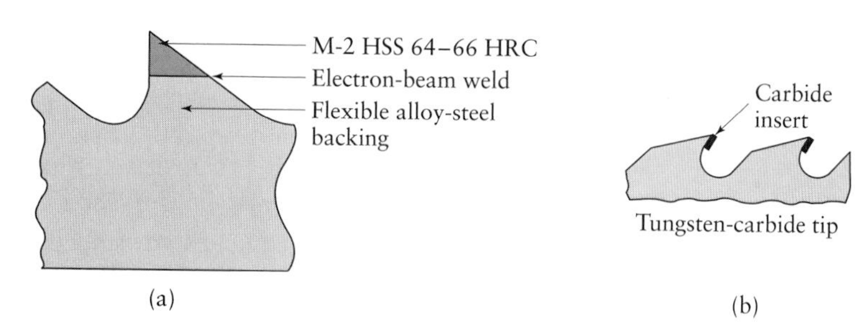

FIGURE 8.64 (a) High-speed-steel teeth welded on a steel blade. (b) Carbide inserts brazed to blade teeth.

(Fig. 8.63b). At least two or three teeth should always be engaged with the workpiece in order to prevent snagging (catching of the saw tooth on the workpiece); this requirement is the reason that sawing thin materials satisfactorily is difficult. Cutting speed in sawing usually ranges up to 1.5 m/s (300 ft/min), with lower speeds for high-strength metals; cutting fluids are generally used to improve the quality of cut and the life of the saw.

Hacksaws have straight blades and reciprocating motions; they may be manually or power operated. *Circular saws* (*cold saws*) are generally used for high-production-rate sawing of large cross-sections. *Band saws* have long, flexible, continuous blades and provide continuous cutting action. Blades and high-strength wire can be coated with diamond powder (*diamond-edged blades* and *diamond saws*). They are suitable for sawing hard metallic, nonmetallic, and composite materials.

Friction sawing. *Friction sawing* is a process in which a mild-steel blade or disk rubs against the work-piece at speeds of up to 125 m/s (25,000 ft/min); the frictional energy is converted into heat, which rapidly softens a narrow zone in the workpiece. The action of the blade or disk, which is sometimes provided with teeth or notches, pulls and ejects the softened metal from the cutting zone. The heat generated in the workpiece produces a *heat-affected zone* (see also Section 12.5) on the cut surfaces; thus, their properties can be adversely affected. Because only a small portion of the blade is engaged with the workpiece at any time, the blade cools rapidly as it passes through the air. Friction sawing is suitable for hard ferrous metals and fiber-reinforced plastics, but not for nonferrous metals, because they have a tendency to stick to the blade. Disks for friction sawing as large as 1.8 m (6 ft) in diameter are used to cut off large steel sections in rolling mills. Friction sawing can also be used to remove flash from castings.

8.9.6 Filing

Filing is small-scale removal of material from a surface, corner, edge, or hole. First developed in about 1000 BC, files are usually made of hardened steels and are available in a variety of cross-sections, including flat, round, half round, square, and triangular. Files have many tooth forms and grades of coarseness, such as smooth cut, second cut, and bastard cut. Although filing is usually done by hand, various machines with automatic features are available for high production rates, with files reciprocating at up to 500 *strokes/min*. *Band files* consist of file segments, each about 75 mm (3 in.) long, that are riveted to flexible steel bands and used in a manner similar to band saws. *Disk-type files* are also available. *Rotary files* and **burs** are used for special applications; they are usually conical, cylindrical, or spherical in shape and have various tooth profiles. The rotational speeds range from 1500 rpm, for cutting steel with large burs, to as high as 45,000 rpm, for cutting magnesium using small burs.

8.9.7 Gear manufacturing by cutting

Gears can be manufactured by casting, forging, extrusion, drawing, thread rolling, powder metallurgy, and blanking (for making thin gears such as those used in watches and small clocks); however, most gears are machined for better dimensional control and surface finish. Nonmetallic gears are usually made by injection molding and casting.

In **form cutting**, the cutting tool is similar to a form-milling cutter (see Fig. 8.58b) in the shape of the space between the gear teeth. The cutter travels axially along the length of the gear tooth at the appropriate depth to produce the gear tooth. After each tooth is cut, the cutter is withdrawn, the gear blank is rotated (indexed), and the cutter proceeds to cut another tooth. The process continues until all teeth are cut. *Broaching* can also be used to produce gear teeth and is particularly applicable to internal teeth. The process is rapid and produces fine surface finish with high dimensional accuracy. However, because broaches are expensive and a separate broach is required for each size of gear, this method is suitable mainly for high-quantity production.

In **gear generating,** the tool may be (a) a *pinion-shaped cutter*, (b) a *rack-shaped straight cutter*, or (c) a *hob*. The pinion-shaped cutter can be considered as one of the gears in a conjugate pair, where the other is the gear blank (Fig. 8.65a). This type of cutter is used in gear generating on machines called **gear shapers** (Fig. 8.65b). The cutter has an axis parallel to that of the gear blank and rotates slowly with the blank at the same pitch-circle velocity with an axial reciprocating motion. A train of gears provides the required relative motion between the cutter shaft and the gear-blank shaft. Cutting may take place at either the downstroke or the upstroke of the machine. Because the clearance required for cutter travel is small, gear shaping is suitable for gears that are located close to obstructing surfaces such as flanges (as in the gear blank in Fig. 8.65b). The process can be used for low-quantity as well as high-quantity production.

On a **rack shaper**, the generating tool is a segment of a rack (Fig. 8.65c), which reciprocates parallel to the axis of the gear blank. Because it is not practical to have more than 6 to 12 teeth on a rack cutter, the cutter must be disengaged at suitable intervals and returned to the starting point; the gear blank, meanwhile, remains fixed. A gear-cutting **hob** is basically a worm, or screw, that has been made into a gear-generating tool by machining a series of longitudinal slots, or gashes, into it to produce the cutting teeth (Fig. 8.65d). When hobbing a spur gear, the angle between the hob and gear-blank axes is 90° minus the lead angle at the hob threads. All motions in hobbing are rotary, and the hob and gear blank rotate continuously as in two gears meshing until all teeth are cut.

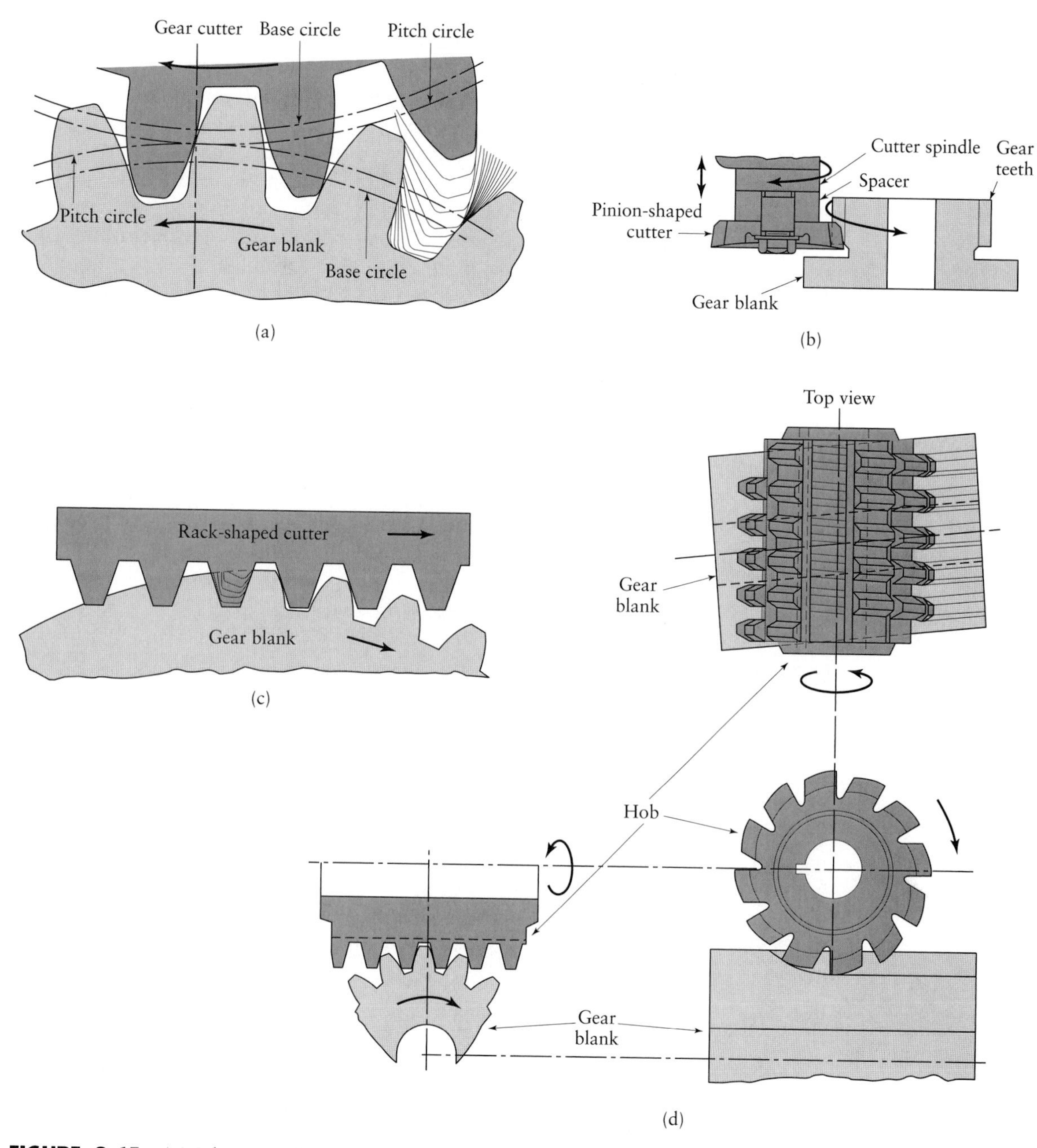

FIGURE 8.65 (a) Schematic illustration of gear generating with a pinion-shaped gear cutter. (b) Schematic illustration of gear generating in a gear shaper, using a pinion-shaped cutter. Note that the cutter reciprocates vertically. (c) Gear generating with a rack-shaped cutter. (d) Schematic illustration of three views of gear cutting with a hob. *Source*: After E. P. DeGarmo and Society of Manufacturing Engineers.

Through gear production by any of the processes previously described, the surface finish and dimensional accuracy of gear teeth may not be sufficiently accurate for certain specific applications; thus, several *finishing* operations are also performed, including shaving, burnishing, grinding, honing, and lapping. (See Chapter 9.) Modern gear-manufacturing machines are computer controlled, and multiaxis computer-controlled machines have capabilities of machining many types of gears, using indexable cutters.

8.10 Machining and Turning Centers

In the descriptions of individual machining processes and machine tools in the preceding chapters, we noted that each machine, regardless of how highly it is automated, is designed to perform basically one type of operation. Previous chapters have also shown that in manufacturing operations, most parts generally require a number of different machining operations on their various surfaces. Many parts have a variety of features and surfaces that require different types of machining operations, such as milling, facing, boring, drilling, reaming, and threading, to certain specified dimensional tolerances and surface finishes.

None of the processes and machine tools described thus far could individually produce these parts. Traditionally, these operations have been performed by moving the part from one machine tool to another until all machining is completed. This process is a viable manufacturing method that can be highly automated and is the principle behind **transfer lines**, consisting of numerous machine tools arranged in a sequence (described in Section 14.2.4). The workpiece, such as an engine block, moves from station to station, with a particular machining operation performed at each station; the part is then transferred automatically to the next machine for another operation, and so on. Transfer lines are commonly used in high-volume or mass production.

There are situations and products, however, for which such lines are not feasible or economical, particularly when the types of products to be machined change rapidly. An important concept, developed in the late 1950s, is machining centers. A **machining center** is a computer-controlled machine tool capable of performing a variety of cutting operations on different surfaces and in different orientations on a workpiece (Fig. 8.66). In general, the workpiece is stationary, and the cutting tools rotate, such as in milling and drilling. The development of machining centers is intimately related to advances in **computer control** of machine tools, details of which are described in Chapter 14. Recall that, as an example of advances in modern lathes, Fig. 8.45 illustrated a numerically controlled lathe (**turning center**), with two turrets that carry several cutting tools for operations such as turning, facing, boring, and threading. Because of the high productivity of machining centers, large amounts of chips are produced and must be collected and disposed of properly.

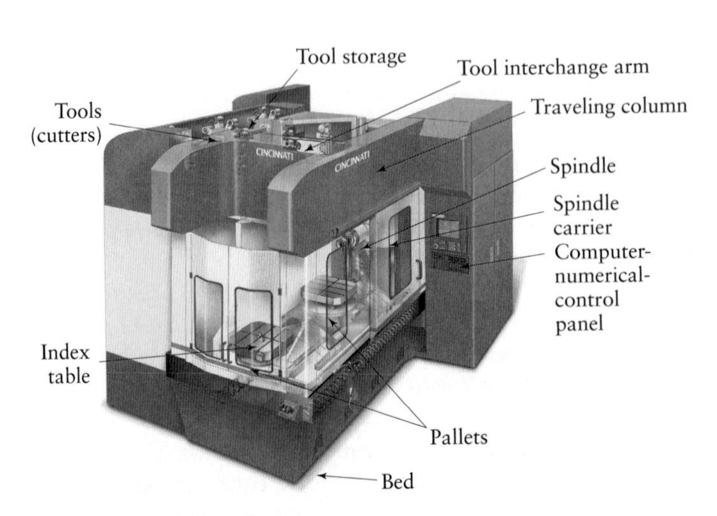

FIGURE 8.66 A horizontal-spindle machining center, equipped with an automatic tool changer. Tool magazines can store 200 cutting tools. *Source*: Courtesy of Cincinnati Machine, a UNOVA Company.

The workpiece in a machining center is placed on a **pallet** that can be oriented in three principal directions, as well as rotated around one or more axes on the pallet. Thus, after a particular cutting operation is completed, the workpiece does not have to be moved to another machine for subsequent operations, say, drilling, reaming, and tapping. In other words, the tools and the machine are brought to the workpiece. After all the cutting operations are completed, the pallet automatically moves away, carrying the finished workpiece, and another pallet containing another workpiece to be machined is brought into position by **automatic pallet changers.** All movements are guided by computer control, and pallet-changing cycle times are on the order of 10–30 s. Pallet stations are available with multiple pallets that serve the machining center. These machines can also be equipped with various automatic parts, such as loading and unloading devices.

Machining centers are equipped with a **programmable automatic tool changer.** Depending on the design, as many as 200 cutting tools can be stored in a magazine, drum, or chain (*tool storage*); auxiliary tool storage is available on some special machining centers for complex operations. The cutting tools are automatically selected, with random access for the shortest route to the machine spindle. A common design is a **tool-exchange arm** that swings around to pick up a particular tool (each tool has its own toolholder) and places it in the spindle. Tools are identified by coded tags, bar codes, or memory chips applied directly to the toolholders. Tool changing times are typically on the order of a few seconds.

Machining centers may be equipped with a **tool-checking** and/or **part-checking station** that feeds information to the computer numerical control to compensate for any variations in tool settings or tool wear. **Touch probes** are available that can be automatically loaded into a toolholder for determining reference surfaces of the workpiece, for selection of tool setting, and for on-line inspection of parts being machined. Several surfaces can be contacted, and their relative positions are determined and stored in the database of the computer software. The data are then used for programming tool paths and compensating for tool length and diameter, as well as compensating for tool wear in more advanced systems.

8.10.1 Types of machining and turning centers

Although there are several designs for machining centers, the two basic types are vertical spindle and horizontal spindle; many machines have the capability of using both axes. The maximum dimensions that the cutting tools can reach around a workpiece in a machining center are known as the *work envelope*; this term was first used in connection with industrial robots. (See Section 14.7.2.)

1. **Vertical-spindle machining centers,** or *vertical machining centers*, are suitable for performing various machining operations on flat surfaces with deep cavities, such as in mold and die making. Because the thrust forces in vertical machining are directed downward, these machines have high stiffness and produce parts with good dimensional accuracy. Also, they are generally less expensive than horizontal-spindle machines.
2. **Horizontal-spindle machining centers,** or *horizontal machining centers*, (see Fig. 8.66) are suitable for large or tall workpieces that require machining on a number of their surfaces. The pallet can be rotated on different axes and to various angular positions. Another category of horizontal-spindle machines is **turning centers,** which are computer-controlled *lathes* with several features. Figure 8.67 shows a three-turret computer-numerical-control turning center. This machine is designed with two horizontal spindles and three turrets, which are equipped with a variety of cutting tools to perform several operations on a rotating workpiece.

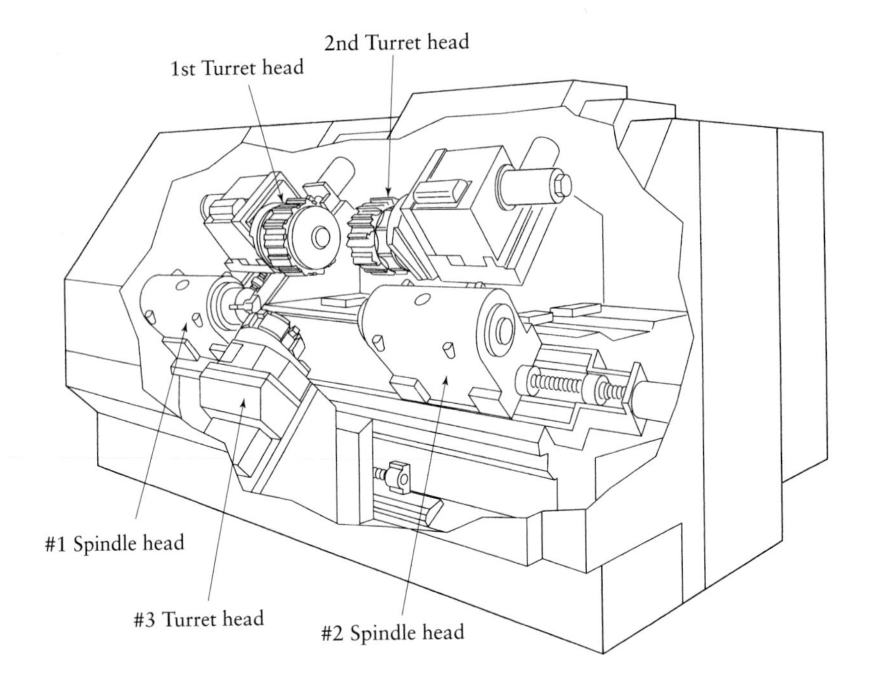

FIGURE 8.67 Schematic illustration of a three-turret, two-spindle computer-numerical-control turning center. *Source*: Hitachi Seiki USA, Inc.

3. **Universal machining centers** are equipped with both vertical and horizontal spindles. They have a variety of features and are capable of machining all surfaces of a workpiece (vertical, horizontal, and diagonal), hence the term *universal*.

Frequent product changes are a major consideration in computer-integrated manufacturing. Consequently, a more recent approach is the design of **reconfigurable machines**, which have flexibility. This flexibility is achieved by using **modular components** that can be arranged and rearranged in different ways. Those machines use advanced computer hardware and software and communications systems. (See also Chapters 14 and 15.)

8.10.2 Characteristics and capabilities of machining centers

The major characteristics of machining centers are described as follows:

- They are capable of handling a variety of part sizes and shapes efficiently, economically, and with repetitively high dimensional tolerances, on the order of ± 0.0025 mm (0.0001 in.).
- The machines are versatile, having as many as six axes of linear and angular movements and the capability of quick changeover from one type of product to another. Thus, the need for a variety of machine tools and a large amount of floor space is significantly reduced.
- The time required for loading and unloading workpieces, changing tools, gaging, and troubleshooting is reduced, thus improving productivity, reducing labor requirements (particularly for skilled labor), and minimizing overall costs.
- Machining centers are highly automated and relatively compact, so that one operator can attend to two or more machines at the same time.
- The machines are equipped with tool-condition monitoring devices for detecting tool breakage and wear, as well as probes for compensation for tool wear and for tool positioning.
- In-process and postprocess gaging and inspection of machined workpieces are now standard features of machining centers.

Machining centers are available in a wide variety of sizes and with a large selection of features, and their cost ranges from about $50,000 to $1 million and higher. Typical capacities range up to 75 kW (100 hp), and maximum spindle speeds are usually in the range of 4000–8000 rpm, but some are as high as 75,000 rpm for special applications and with small-diameter cutters. Some pallets are capable of supporting workpieces that weigh as much as 7,000 kg (15,000 lb), although higher capacities are available for special applications.

Machine-tool selection. Machining centers can require significant capital expenditures; therefore, to be cost effective, they have to be in operation for at least two shifts per day. Consequently, there must be sufficient and continued demand for products made on machining centers to justify this cost. However, because of their inherent versatility, machining centers can be used to produce a wide range of products, particularly in just-in-time manufacturing (see Section 15.11).

The selection of a particular type and size of machining center depends on several factors, among which are

- Type of products, their size, and their shape complexity;
- Type of machining operations to be performed and the type and number of tools required;
- Dimensional accuracy of the products;
- Production rate.

Although versatility is the key factor in selecting machining centers, these considerations must be weighed against the high capital investment required and compared with to those of manufacturing the same products by using a number of the more traditional machine tools.

EXAMPLE 8.8 Machining outer bearing races on a turning center

Outer bearing races (see Fig. 8.68) are machined on a turning center. The starting material is hot-rolled 52100 steel tube with an outside diameter of

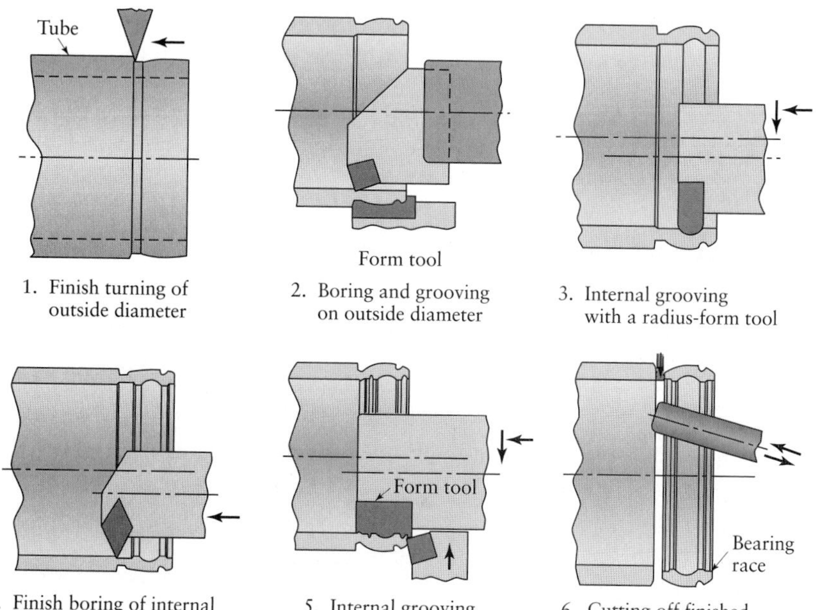

FIGURE 8.68 Machining outer bearing races on a turning center.

91 mm (3.592 in.) and an inside diameter of 75.5 mm (2.976 in.). The cutting speed is 95 m/min (313 ft/min) for all operations. All tools are carbide, including the cutoff tool (the last operation), which is 3.18 mm $\left(\frac{1}{8}\text{ in.}\right)$ wide instead of 4.76 mm $\left(\frac{3}{16}\text{ in.}\right)$, for the high-speed-steel cutoff tool that was formerly used. The material saved by this change is significant, because the width of the race is small. The turning center was able to machine these races at high speeds and with repeatable dimensional tolerances of ± 0.025 mm (0.001 in.).

8.11 Vibration and Chatter

In describing cutting processes and machine tools, we pointed out that machine stiffness is important in controlling dimensional accuracy and surface finish of parts. In this section, we describe the adverse effects of low stiffness on product quality and machining operations. Low stiffness affects the level of vibration and chatter in tools and machines. If uncontrolled, vibration and chatter can result in the following effects:

- Poor surface finish (right central region in Fig. 8.69);
- Loss of dimensional accuracy of the workpiece;
- Premature wear, chipping, and failure of the cutting tool, which is crucial with brittle tool materials, such as ceramics, some carbides, and diamond;
- Damage to machine-tool components from excessive vibrations;
- Objectionable noise generated, particularly if it is of high frequency, such as the squeal heard in turning brass on a lathe.

Extensive studies have shown that vibration and chatter in cutting are complex phenomena. Cutting operations cause two basic types of vibration: forced vibration and self-excited vibration. **Forced vibration** is generally caused by some periodic force present in the machine tool, such as from gear drives, imbalance of the machine-tool components, misalignment, or motors and pumps. In processes such as milling or turning a splined shaft or a shaft with a keyway, forced vibrations are caused by the periodic engagement of the cutting tool with the workpiece surface, including its exit from the surface.

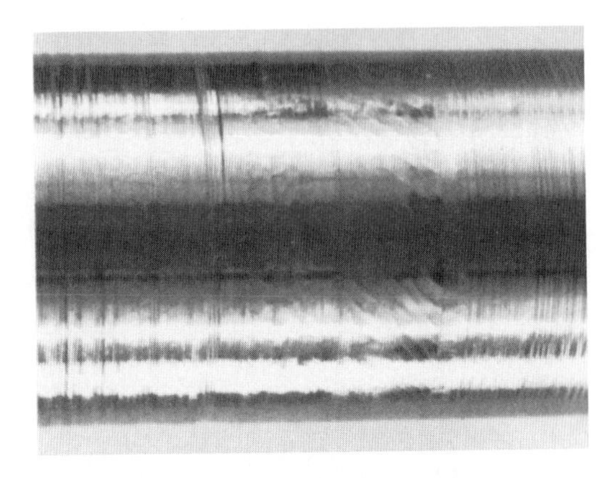

FIGURE 8.69 Chatter marks (right of center of photograph) on the surface of a turned part. *Source*: General Electric Company.

The basic solution to forced vibrations is to isolate or remove the forcing element. If the forcing frequency is at or near the natural frequency of a component of the machine-tool system, one of the frequencies may be raised or lowered. The amplitude of vibration can be reduced by increasing the stiffness or damping the system. Although changing the process parameters generally does not appear to greatly influence forced vibrations, changing the cutting speed and the tool geometry can be helpful.

Generally called **chatter, self-excited vibration** is caused by the interaction of the chip-removal process and the structure of the machine tool; these vibrations usually have very high amplitude. Chatter typically begins with a disturbance in the cutting zone. Such disturbances include lack of homogeneity in the workpiece material or its surface condition, changes in chip morphology, or a change in frictional conditions at the tool–chip interface, as also influenced by cutting fluids and their effectiveness. Self-excited vibrations can generally be controlled by increasing the dynamic stiffness of the system and by damping. **Dynamic stiffness** is the ratio of the amplitude of the force applied to the amplitude of vibration. Because a machine tool has different stiffnesses at different frequencies, changes in cutting parameters, such as cutting speed, can influence chatter.

The most important type of self-excited vibration in machining is **regenerative chatter**, which is caused when a tool cuts a surface that has a roughness or disturbances left from the previous cut. Because the depth of cut thus varies, the resulting variations in the cutting force subject the tool to vibrations; the process continues repeatedly, hence the term *regenerative*. You may observe this type of vibration while driving over a rough road (the so-called *washboard effect*).

8.11.1 Damping

The term *damping* is defined as the rate at which vibrations decay. Damping is an important factor in controlling machine-tool vibration and chatter.

Internal damping of structural materials. Damping results from the energy loss within materials during vibration. Thus, for example, steel has less damping than gray cast iron, and composite materials have more damping than gray iron (Fig. 8.70). You can observe this difference in the damping capacity of materials by striking them with a gavel and listening to the sound; try, for example, striking pieces of steel, concrete, and wood.

Joints in the machine-tool structure. Although less significant than internal damping, bolted joints in the structure of a machine tool are also a source of damping. Because friction dissipates energy, small relative movements along dry (unlubricated) joints dissipate energy and thus improve damping. In joints where oil is present, the internal friction of the oil layers also dissipates energy, thereby contributing to

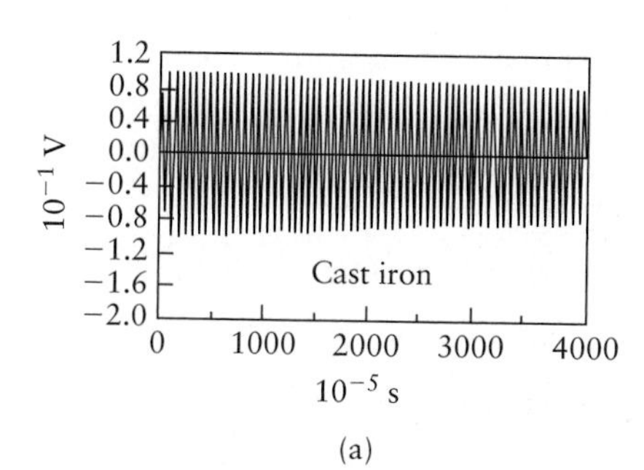

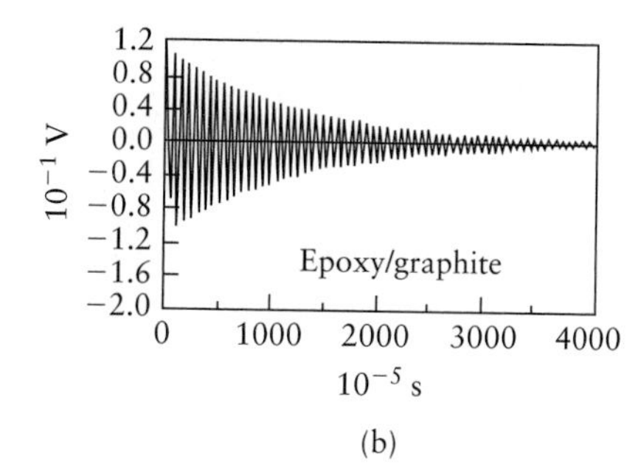

FIGURE 8.70 Relative damping capacity of gray cast iron and epoxy–granite composite material. The vertical scale is the amplitude of vibration, and the horizontal scale is time. *Source*: Courtesy of Cincinnati Machine, a UNOVA Company.

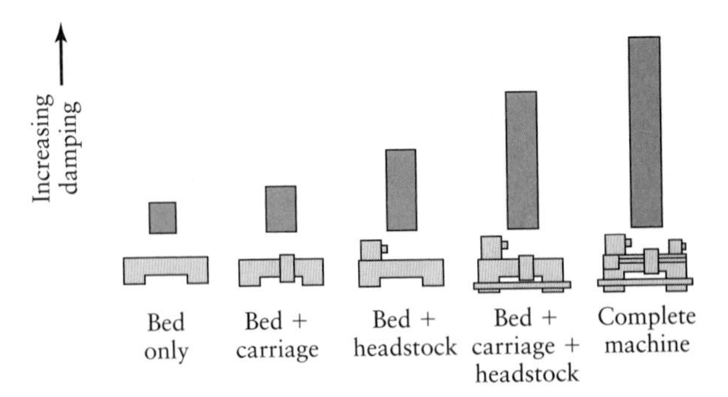

FIGURE 8.71 Damping of vibrations as a function of the number of components on a lathe. Joints dissipate energy; thus, the greater the number of joints, the higher the damping. *Source*: J. Peters, CIRP Annals, 1965.

damping. In describing the machine tools for various cutting operations, we noted that all machines consist of a number of large and small components, assembled into a structure. Consequently, this type of damping is cumulative, owing to the presence of a number of joints in a machine tool. Note in Fig. 8.71 how damping increases as the number of components on a lathe and their contact area increase; thus, the more joints, the greater is the amount of energy dissipated, and the higher is the damping.

External damping. External damping is accomplished with external dampers, which are similar to shock absorbers on automobiles. Special vibration absorbers have been developed and can be installed on machine tools for this purpose.

8.11.2 Factors influencing chatter

Studies have indicated that the tendency for a particular workpiece to chatter during cutting is proportional to the cutting forces and the depth and width of cut. Consequently, because cutting forces increase with strength (hardness), the tendency to chatter generally increases as the hardness of the workpiece material increases. Thus, for example, aluminum and magnesium alloys have less tendency to chatter than do martensitic and precipitation-hardening stainless steels, nickel alloys, and high-temperature and refractory alloys. An important factor in chatter is the type of chip produced during cutting operations. As we have described, continuous chips involve steady cutting forces. Consequently, they generally do not cause chatter. Discontinuous chips and serrated chips, on the other hand, may do so. These chips are produced periodically, and the resulting variations in force during cutting can cause chatter.

8.12 Machine-Tool Structures

The design and construction of machine tools are important aspects of manufacturing engineering. This section describes the *material* and *design* aspects of machine tools as structures with certain desired characteristics. The proper design of machine-tool structures requires a knowledge of the materials available for construction, their forms and properties, the dynamics of the particular machining process, and the forces involved. *Stiffness* and *damping* are important factors in machine-tool structures. Stiffness involves both the dimensions of the structural components and the elastic modulus of the materials used. Damping involves the type of materials used, as well as the number and nature of the joints in the structure.

Materials and design. Traditionally, the base and some of the major components of machine tools have been made of gray or nodular cast iron, which have the advantages of low cost and good damping capacity, but are heavy. Lightweight designs are desirable, because of ease of transportation, higher natural frequencies, and lower inertial forces of moving members. Lightweight designs and design flexibility require fabrication processes such as (a) mechanical fastening (bolts and nuts) of individual components and (b) welding. However, this approach to fabrication increases labor and material costs, because of the preparations involved.

Wrought steels are the likely choice for such lightweight structures, because of their low cost; availability in various section sizes and shapes (such as channels, angles, and tubes); desirable mechanical properties; and favorable characteristics, such as formability, machinability, and weldability. Tubes, for example, have high stiffness-to-weight ratios. On the other hand, the benefit of the higher damping capacity of castings and composites is not available with steels.

In addition to stiffness, another factor that contributes to lack of precision of the machine tool is the *thermal expansion* of its components, causing distortion. The source of heat may be internal, such as bearings, ways, motors, and heat generated from the cutting zone, or external, such as nearby furnaces, heaters, sunlight, and fluctuations in cutting fluid and ambient temperatures. Equally important in machine-tool precision are *foundations*, particularly their mass and how they are installed in a plant. For example, in installing a large grinding machine for grinding marine-propulsion gears 2.75 m (9 ft) in diameter with high precision, the concrete foundation is 6.7 m (22 ft deep). This large mass of concrete and the machine base reduce the amplitude of vibrations and their adverse effects.

Significant developments concerning the materials used for machine-tool bases and components have taken place. For example, **acrylic concrete** is a mixture of concrete and polymer (polymethylmethacrylate); it can easily be cast into desired shapes for machine bases and various components; several compositions are being developed. It can also be used for sandwich construction with cast iron, thus combining the advantages of each type of material. **Granite-epoxy composite** is a castable composite, with a typical composition of about 93% crushed granite and 7% epoxy binder. First used in precision grinders in the early 1980s, this composite material has several favorable properties: (1) good castability, which allows design versatility in machine tools; (2) high stiffness-to-weight ratio; (3) thermal stability; (4) resistance to environmental degradation; and (5) good damping capacity (see Fig. 8.70).

With the goal of improving the stiffness of machine tools and, therefore, attaining extremely fine dimensional tolerances on machined components, innovative experimental designs for machine-tool structures are continually being developed. Although still largely in a development stage, these designs are self-contained octahedral (eight-sided) machine frames, variously called **hexapods, parallel kinematic linked machines**, and **prism-shaped machines.** They are based on a mechanism called the **Stewart platform** (an invention used to position aircraft cockpit simulators), and the machines have high stiffness and high flexibility of operation. The workpiece is fixtured on a fixed table, and pairs of telescoping tubes (struts or legs), equipped with ballscrews and each with its own motor, are used to maneuver a rotating cutting tool. During machining, the machine controller shortens some tube while lengthening others, so that the cutting tool follows a specified contour around the workpiece. Design considerations include the coordinate system for machine movements and inertia effects resulting from rapid acceleration and deceleration of machine components.

8.13 Design Considerations

Major design considerations for the processes described in this chapter are outlined as follows:

a) **Turning.** Blanks to be machined should be as close to the final dimensions (such as by net or near-net-shape forming) as possible so as to reduce machining cycle time. The machinability of raw materials should be properly considered. Parts should be designed so that they can be fixtured and clamped in work-holding devices with relative ease. Thin and slender workpieces may be difficult to support properly. Dimensional accuracy and surface finish specified should be as broad as permissible for the part to function as desired. Sharp corners, tapers, and major dimensional variations along the part should be avoided. Design features should be such that commercially available standard cutting tools, inserts, and toolholders can be used.

b) **Screw thread cutting.** Designs should allow the termination of threads before the threads reach a shoulder. Internal threads in blind holes should have an unthreaded length at the bottom. Shallow blind-threaded holes should be avoided. Chamfers should be specified to avoid burr formation. Threaded sections should not be interrupted with holes, slots, or other discontinuities. Thread design should be such that standard threading tooling and inserts can be used. Parts should be designed so that all cutting operations can be completed in one setup.

c) **Drilling, reaming, and tapping.** Holes should be specified on flat surfaces and perpendicular to the direction of drill movement for ease of drilling. Hole bottoms should match standard drill-point angles, and holes with interrupted surfaces should be avoided. Through holes should be preferred to blind holes, and blind holes should be drilled deeper than subsequent reaming or tapping operations require. Part design and fixturing should be such that all operations can be performed in one setup.

d) **Boring.** Whenever possible, through holes rather than blind holes should be specified. The greater the length-to-diameter ratio of the hole, the more difficult is the boring operation. Interrupted internal surfaces should be avoided.

e) **Milling.** Designs should be such that standard milling cutters can be used. Chamfers should be preferred over radii, and internal cavities and pockets with sharp inner corners should be avoided. Parts should be sufficiently stiff, i.e., have a high section modulus, to minimize deflections from clamping as well as from cutting forces and to minimize vibration and chatter.

f) **Broaching.** Parts should be designed such that they can be fixtured securely and should have sufficient stiffness to withstand the forces involved. Chamfers should be preferred over radii. Holes, sharp corners, dovetail splines, and flat surfaces should be avoided.

g) **Gear machining.** Blank design is important for proper fixturing and to ease cutting operations. Wide gears are more difficult to machine than narrow gears. Designs should allow for sufficient clearance between gear teeth and flanges. Design of the part should allow the use of standard gear cutters. Dimensional accuracy and surface finish specified should be as broad as possible for economical production.

8.14 Economics of Machining

Section 8.1 outlined the advantages and limitations of machining, and the technical considerations were described in detail in various sections in this chapter. These considerations should be viewed in light of the competitive nature of manufacturing processes. Thus, for example, as compared with forming and shaping processes, machining involves more time and wastes more material; however, it can produce parts with dimensional control and surface finish better than that provided by other processes.

This section deals with the *economics* of machining. The two most important parameters are the *minimum cost per part* and the *maximum production rate*. It can be seen that the **total cost per piece** consists of four items, or

$$C_p = C_m + C_s + C_\ell + C_t, \tag{8.46}$$

where C_p is the cost (say, in dollars) per piece; C_m is the machining cost; C_s is the cost of setting up for machining, such as mounting the cutter and fixtures and preparing the machine for the particular operation; C_ℓ is the cost of loading, unloading, and machine handling; and C_t is the tooling cost, which includes tool changing, regrinding, and depreciation of the cutter. The **machining cost** is given by

$$C_m = T_m(L_m + B_m), \tag{8.47}$$

where T_m is the machining time per piece; L_m is the labor cost of production operator per hour; and B_m is the burden rate, or overhead charge, of the machine, including depreciation, maintenance, indirect labor, and the like. The **setup cost** is a fixed figure in dollars per piece. The **loading, unloading,** and **machine-handling cost** is given by

$$C_\ell = T_\ell(L_m + B_m), \tag{8.48}$$

where T_ℓ is the time involved in loading and unloading the part, changing speeds, changing feed rates, and so on. The **tooling cost** is expressed as

$$C_t = \frac{1}{N_p}[T_c(L_m + B_m) + T_g(L_g + B_g) + D_c], \tag{8.49}$$

where N_p is the number of parts machined per tool grind; T_c is the time required to change the tool; T_g is the time required to grind the tool; L_g is the labor cost of tool-grinder operator per hour; B_g is the burden rate of tool grinder per hour; and D_c is the depreciation of the tool in dollars per grind.

The **time** needed to produce one part is

$$T_p = T_\ell + T_m + \frac{T_c}{N_p}, \tag{8.50}$$

where T_m has to be calculated for each particular operation. For example, let's consider a turning operation; the machining time (see Section 8.8.1) is

$$T_m = \frac{L}{fN} = \frac{\pi LD}{fV}, \tag{8.51}$$

where L is the length of cut, f is the feed, N is the rpm of the workpiece, D is the workpiece diameter, and V is the cutting speed. Note that appropriate units must be used in all these equations. From the Taylor tool-life equation, we have

$$VT^n = C.$$

Hence,

$$T = \left(\frac{C}{V}\right)^{1/n}, \tag{8.52}$$

where T is the time, in minutes, required to reach a flank wear of certain dimension, after which the tool has to be reground or changed. The number of pieces per tool grind is thus simply

$$N_p = \frac{T}{T_m}. \tag{8.53}$$

The combination of Eqs. (8.51)–(8.53) gives

$$N_p = \frac{fC^{1/n}}{\pi LDV^{(1/n)-1}}. \tag{8.54}$$

The **cost per piece**, C_p, in Eq. (8.46) can now be defined in terms of several variables. To find the optimum cutting speed and the optimum tool life for **minimum cost**, we differentiate C_p with respect to V and set it to zero. Thus, we have

$$\frac{\partial C_p}{\partial V} = 0. \tag{8.55}$$

We then find that the optimum cutting speed, V_o, is

$$V_o = \frac{C(L_m + B_m)^n}{\{[(1/n) - 1][T_c(L_m + B_m) + T_g(L_g + B_g) + D_c]\}^n}, \tag{8.56}$$

and the optimum tool life, T_o, is

$$T_o = [(1/n) - 1]\frac{T_c(L_m + B_m) + T_g(L_g + B_g) + D_c}{L_m + B_m}. \tag{8.57}$$

To find the optimum cutting speed and the optimum tool life for **maximum production**, we differentiate T_p with respect to V and set the result to zero. Thus, we have

$$\frac{\partial T_p}{\partial V} = 0. \tag{8.58}$$

The **optimum cutting speed** now becomes

$$V_o = \frac{C}{\{[(1/n) - 1]T_c\}^n}, \tag{8.59}$$

and the **optimum tool life** is

$$T_o = [(1/n) - 1]T_c. \tag{8.60}$$

Qualitative plots of the minimum cost per piece and the minimum time per piece, that is, the maximum production rate, are given in Fig. 8.72. The cost of a machined surface also depends on the degree of finish required (see Section 9.17); the machining cost increases rapidly with finer surface finish.

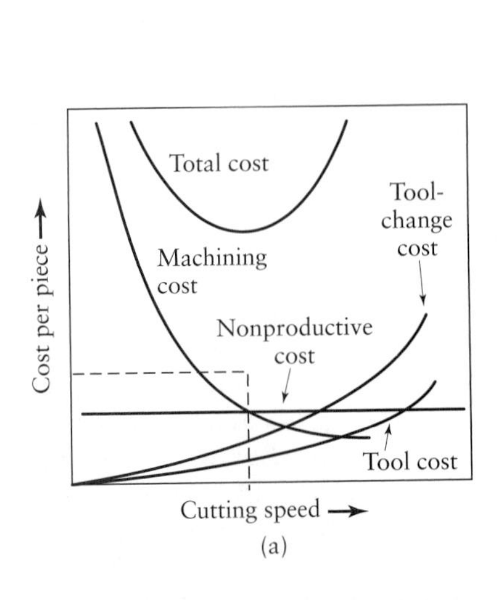

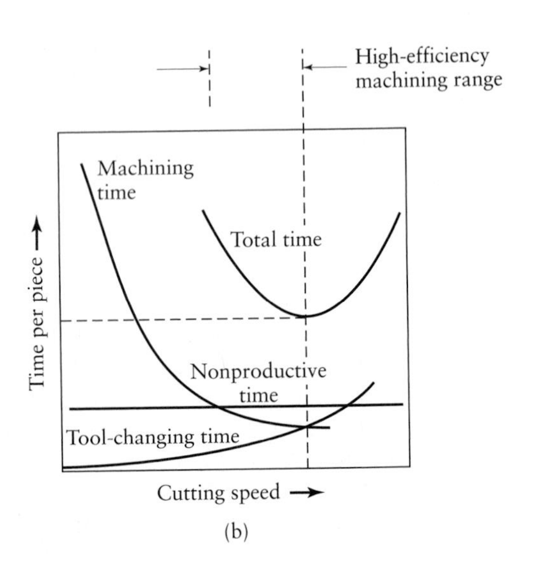

FIGURE 8.72 Graphs showing (a) cost per piece and (b) time per piece in machining. Note the optimum speeds for both cost and time. The range between the two optimum speeds is known as the *high-efficiency machining range*.

This analysis indicates the importance of identifying all relevant parameters in a machining operation, determining various cost factors, obtaining relevant tool-life curves for the particular operation, and properly measuring the various time intervals involved in the overall operation. The importance of obtaining accurate data is clearly shown in Fig. 8.72, as small changes in cutting speed can have a significant effect on the minimum cost or time per piece.

CASE STUDY | High-Speed Dry Machining of Cast-Iron Engine Blocks

High-speed machining has clear economic advantages as long as spindle damage and chatter are avoided. Similarly, dry machining is recognized as a promising approach to solving environmental problems associated with cutting fluids. However, it has often been perceived that the two approaches cannot be applied simultaneously, because the higher speeds without the use of cutting fluids would result in low tool life. However, high-speed dry machining is possible with aluminum and cast iron, and we examine it here for roughing, finishing, and hole-making operations on a cast-iron cylinder head for a 4.3-liter engine.

High-speed dry machining is a demanding operation. Heat dissipation without a liquid coolant requires high-performance coatings and heat-resistant tool materials. The cutting-tool materials must be either ceramic or cBN, because of the intense heat generated in the process. In addition, machine tools with high power are required, and spindles with high stiffness and power ratings are needed. Chatter can be a problem with high-speed machining; hence, CNC programs must be developed that maintain process parameters within a window of operation that is known to be chatter free.

The key to dry machining the cast-iron head is the use of two advanced techniques. The first strategy is *red-crescent machining*, where high temperatures are generated in the workpiece in front of the tool, creating a visible red arc (hence the name of the process). The second strategy is flush fine machining (see later in this case study), using in-spindle high-pressure air to remove chips that normally would be carried away by a cutting fluid. Red-crescent machining refers to milling operations that take place at high feed rates and high spindle speeds. This combination reduces the thrust force against the workpiece as compared with a standard milling operation. Red-crescent machining concentrates heat in the material in front of the tool and the workpiece can reach temperature ranges of 600–700°C (1100–1300°F) in some cases. This situation occurs because, at the high speeds, the heat generated is localized in the shear zone and does not have time to be conducted into the workpiece or the tool. This high-temperature zone generates a red glow. Red-crescent machining is beneficial with cast irons because the heat increases the ductility and reduces the strength of the cast iron, allowing for machining at much higher efficiency. Part distortion does not occur, because most of the heat is retained in the chip and removed before it can be conducted into the workpiece. Table 8.12 provides a comparison of red-crescent machining with conventional machining for this particular application.

Finish milling is accomplished by using low-thrust machining strategies, usually by using cutters with fewer inserts per cutter, at a lower feed rate per revolution and higher surface speeds than for conventional machining. Low-thrust-force machining has an additional advantage in that work-holder design is simplified, often allowing better access to complicated parts. cBN inserts are used for finish machining of the cast-iron engine blocks. Although cBN is more expensive than the alternatives of carbide or ceramic inserts, tool life is longer, and fewer

TABLE 8.12

Comparison of Conventional vs. Red-Crescent Machining Conditions

Parameter	Conventional	Red crescent
Surface speed (ft/min)	250	5000
Spindle speed (rpm)	160	3200
Feed (in./rev)	0.020	0.005
Number of inserts	10	5
Radial thrust force (lb)	4300	262
Metal removal rate (in^3/min)	32	80

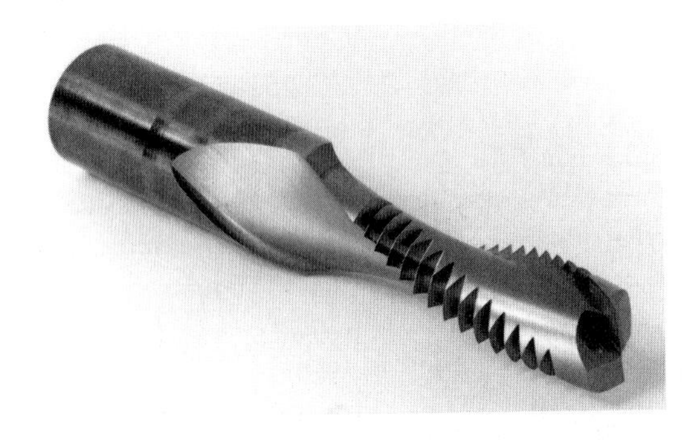

FIGURE 8.73 A high-speed tool for single-point milling, chamfering, counterboring, and threading of holes. *Source*: Courtesy of Makino, Inc.

setups are needed; thus the insert cost per machined part can be comparable with or lower than that for the conventional tool materials.

Flush fine machining is a process that uses high-pressure through-spindle air for chip disposal, removing heat from the cutting tool as well. (See also Section 8.7.1.) The air pressure required is in the range of 1100–1380 kPa (160–200 psi) in order to reduce the temperature of the workpiece by approximately 50%. Flush fine machining is suitable for most shallow holes (depth-to-diameter ratios of five or less), which accounts for most automotive applications. Deeper holes are produced by using more conventional drilling operations. For shallow holes, single-tool milling, chamfering, counterboring, and threading of holes can be accomplished by using tools such as shown in Fig. 8.73. The process uses high-speed, high-accuracy CNC control to move a small-diameter tool in a helical pattern to shape and thread the holes. In addition, this technique allows a single tool to machine holes of different diameters.

Source: Adapted from G. Hyatt, "High Speed, Dry Machining Can Cut Cycle Times and Cost," *Manufacturing Engineering*, 119(3), 1997.

SUMMARY

- Material removal processes are often necessary in order to impart the desired dimensional accuracy, geometric features, and surface finish characteristics to components, particularly those with complex shapes that cannot be produced economically or properly by other shaping techniques. On the other hand, these processes waste some material in the form of chips, generally cause the manufacturing process to take longer, and may have adverse effects on the surfaces produced. (Section 8.1)

- Important process variables include tool shape and material; cutting conditions such as speed, feed, and depth of cut; use of cutting fluids; and the characteristics of the machine tool and the workpiece material. Parameters influenced by these variables include forces and power consumption, tool wear, surface finish and integrity, temperature, and dimensional accuracy of the workpiece. The commonly observed chip types are continuous, built-up edge, discontinuous, and serrated. (Section 8.2)
- Temperature rise is an important consideration, as it can have adverse effects on tool life as well as on dimensional accuracy and surface integrity of the machined part. (Section 8.2)
- Tool wear depends primarily on workpiece and tool-material characteristics, cutting speed, and cutting fluids. Feeds, depth of cut, and machine-tool characteristics also have an effect. Two major types of wear are flank wear and crater wear. (Section 8.3)
- Surface finish of machined components is an important consideration, as it can adversely affect product integrity. Important variables that affect surface finish include the geometry and condition of the cutting tool, chip morphology, and process variables. (Section 8.4)
- Machinability is usually defined in terms of surface finish, tool life, force and power requirements, and chip type. Machinability of materials depends not only on their intrinsic properties and microstructure, but also on proper selection and control of process variables. (Section 8.5)
- A variety of cutting-tool materials has been developed, the most common ones being high-speed steels, carbides, ceramics, and cubic boron nitride. The materials have a broad range of mechanical and physical properties, such as hot hardness, toughness, chemical stability and inertness, and resistance to chipping and wear. Various tool coatings have also been developed, resulting in major improvements in tool life. (Section 8.6)
- Cutting fluids are important in machining operations. They reduce friction, forces, and power requirements and improve tool life. Generally, slower operations with high tool pressures require a fluid with good lubricating characteristics, whereas in high-speed operations, where temperature rise can be significant, fluids with cooling capacity are preferred. (Section 8.7)
- The cutting processes that produce external and internal circular profiles are turning, boring, drilling, tapping, and thread cutting. Because of the three-dimensional nature of these operations, chip movement and control are important considerations, since chips can interfere with the operation. Optimization of each process requires an understanding of the interrelationships among design and processing parameters. (Section 8.8)
- High-speed machining, ultraprecision machining, and hard turning are among more recent developments in cutting. They can help reduce machining costs, and they produce parts with exceptional surface finish and dimensional accuracy. (Section 8.8)
- Various complex shapes can be machined by slab, face, and end milling; broaching; and sawing. These processes use multitooth tools and cutters at various axes with respect to the workpiece. The machine tools are now mostly computer controlled, have various features and attachments, and possess considerable flexibility in operation. (Section 8.9)

- Because of their versatility and capability of performing a variety of cutting operations, machining and turning centers are among the most important developments in machine tools. Their selection depends on factors such as part complexity required, the number and type of cutting operations to be performed, the number of cutting tools needed, the dimensional accuracy required, and the production rate required. (Section 8.10)
- Vibration and chatter in machining are important considerations for workpiece dimensional accuracy and surface finish, as well as tool life. Stiffness and damping capacity of machine tools are important factors in controlling vibration and chatter. New materials are being developed and used for constructing machine-tool structures. (Sections 8.11 and 8.12)
- Several design guidlines have been developed for parts to be produced by machining. (Section 8.13)
- The economics of machining processes depend on various costs. Optimum cutting speeds can be determined for minimum machining time per piece and minimum cost per piece, respectively. (Section 8.14)

SUMMARY OF EQUATIONS

Cutting ratio, $r = \dfrac{t_o}{t_c} = \dfrac{\sin\phi}{\cos(\phi - \alpha)}$

Shear strain, $\gamma = \cot\phi + \tan(\phi - \alpha)$

Velocity relationships: $\dfrac{V}{\cos(\phi - \alpha)} = \dfrac{V_s}{\cos\alpha} = \dfrac{V_c}{\sin\phi}$

Friction force, $F = R\sin\beta$

Normal force, $N = R\cos\beta$

Coefficient of friction, $\mu = \tan\beta = \dfrac{F_t + F_c\tan\alpha}{F_c - F_t\tan\alpha}$

Thrust force, $F_t = R\sin(\beta - \alpha) = F_c\tan(\beta - \alpha)$

Shear-angle relationships, $\phi = 45^\circ + \dfrac{\alpha}{2} - \dfrac{\beta}{2}$

$$\phi = 45^\circ + \alpha - \beta$$

Total cutting power $= F_c V$

Specific energy, $u_t = \dfrac{F_c}{wt_o}$

Frictional specific energy, $u_f = \dfrac{Fr}{wt_o}$

Shear specific energy, $u_s = \dfrac{F_s V_s}{wt_o V}$

Mean temperature, $T = \frac{1.2Y_f}{\rho c}\sqrt[3]{\frac{Vt_o}{K}}$

$$T \propto V^a f^b$$

Tool life, $VT^n = C$

Material removal rate (MRR) in turning $= \pi D_{avg} d f N$

in drilling $= \pi\left(\frac{D^2}{4}\right)fN$

in milling $= wdv$

BIBLIOGRAPHY

Armarego, E. J. A., and R. H. Brown, *The Machining of Metals*, Prentice Hall, 1969.

Arnone, M., *High Performance Machining*, Hanser, 1998.

ASM Handbook, Vol. 1: *Properties and Selection: Irons, Steels, and High-Performance Alloys*, ASM International, 1990.

ASM Handbook, Vol. 16: *Machining*, ASM International, 1989.

ASM Specialty Handbook: Tool Materials, ASM International, 1995.

Astakhov, V. P., *Metal Cutting Mechanics*, CRC Press, 1998.

Boothroyd, G., and W. A. Knight, *Fundamentals of Machining and Machine Tools*, 2d. ed., Dekker, 1989.

Brown, J., *Advanced Machining Technology Handbook*, McGraw-Hill, 1998.

Byers, J. P. (ed.), *Metalworking Fluids*, Dekker, 1994.

DeVries, W. R., *Analysis of Material Removal Processes*, Springer, 1992.

Ewert, R. H., *Gears and Gear Manufacture: The Fundamentals*, Chapman & Hall, 1997.

Handbook of High-Speed Machining Technology, Chapman & Hall, 1985.

Hoffman, E. G., *Jig and Fixture Design*, 4th ed., Industrial Press, 1996.

Juneja, B. L., *Fundamentals of Metal Cutting and Machine Tools*, Wiley, 1987.

Komanduri, R., "Tool Materials," in *Kirk-Othmer Encyclopedia of Chemical Technology*, 4th ed., Vol. 24, Wiley, 1997.

Krar, S. F., and A. F. Check, *Technology of Machine Tools*, 5th ed., Glencoe Macmillan/McGraw-Hill, 1996.

Machinery's Handbook, Industrial Press, revised periodically.

Machining Data Handbook, 2 vols., 3d. ed., Machinability Data Center, 1980.

Metal Cutting Tool Handbook, 7th ed., Industrial Press, 1989.

Nachtman, E. S., and S. Kalpakjian, *Lubricants and Lubrication in Metalworking Operations*, Dekker, 1985.

Nakazawa, H., *Principles of Precision Engineering*, Oxford, 1994.

Oxley, P. L. B., *Mechanics of Machining: An Analytical Approach to Assessing Machinability*, Wiley, 1989.

Rechetov, D. N., and V. T. Portman, *Accuracy of Machine Tools*, ASME, 1989.

Rivin, E. I., *Stiffness and Damping in Mechanical Design*, Dekker, 1999.

Roberts, G. A., G. Krauss, and R. Kennedy, *Tool Steels*, 5th ed., ASM International, 1997.

Schey, J. A., *Tribology in Metalworking—Friction, Wear and Lubrication*, ASM International, 1983.

Schneider, A. F., *Mechanical Deburring and Surface Finishing Technology*, Dekker, 1990.

Shaw, M. C., *Metal Cutting Principles*, Oxford, 1984.

Sluhan, C. (ed.), *Cutting and Grinding Fluids: Selection and Application*, Society of Manufacturing Engineers, 1992.

Stephenson, D., and J. S. Agapiou, *Metal Cutting: Theory and Practice*, Dekker, 1996.

Stout, K. J, E. J. Davis, and P. J. Sullivan, *Atlas of Machined Surfaces*, Chapman & Hall, 1990.

Tool and Manufacturing Engineers Handbook, 4th ed., Vol. 1: *Machining*, Society of Manufacturing Engineers, 1983.

Townsend, D. P., *Dudley's Gear Handbook: The Design, Manufacturing, and Application of Gears*, 2d ed., McGraw-Hill, 1991.

Trent, E. M., and P. K. Wright, *Metal Cutting*, 4th ed., Butterworth Heinemann, 1999.

Venkatesh, V. C., and H. Chandrasekaran, *Experimental Techniques in Metal Cutting*, rev. ed. Prentice-Hall, 1987.

Walsh, R. A., *McGraw-Hill Machining and Metalworking Handbook*, McGraw-Hill, 1994.

Weck, M., *Handbook of Machine Tools*, 4 vols., Wiley, 1984.

QUESTIONS

8.1 In metal cutting, it has been observed that the shear strains obtained are higher than those calculated from the properties of the material. To what factors would you attribute this difference? Explain.

8.2 Describe the effects of material properties and process variables on serrated chip formation.

8.3 Explain why the cutting force, F_c, increases with increasing depth of cut and decreasing rake angle.

8.4 What are the effects of performing a cutting operation with a dull tool tip?

8.5 Describe the trends that you have observed in Tables 8.1 and 8.2.

8.6 To what factors would you attribute the large difference in the specific energies within each group of materials shown in Table 8.3?

8.7 Describe the effects of cutting fluids on chip formation. Explain why and how they influence the cutting operation.

8.8 Under what conditions would you discourage the use of cutting fluids?

8.9 Give reasons that pure aluminum and copper are generally classified as easy to machine.

8.10 Can you offer an explanation as to why the maximum temperature in cutting is located at about the middle of the tool–chip interface? (*Hint*: Note that there are two principal sources of heat: the shear plane and the tool–chip interface.)

8.11 State whether or not the following statements are true, explaining your reasons: (a) For the same shear angle, there are two rake angles that give the same cutting ratio. (b) For the same depth of cut and rake angle, the type of cutting fluid used has no influence on chip thickness. (c) If the cutting speed, shear angle, and rake angle are known, the chip velocity can be calculated. (d) The chip becomes thinner as the rake angle increases. (e) The function of a chip breaker is to decrease the curvature of the chip.

8.12 It is generally undesirable to allow temperatures to rise too high in cutting operations. Explain why.

8.13 Explain the reasons that the same tool life may be obtained at two different cutting speeds.

8.14 Inspect Table 8.6, and identify tool materials that would not be particularly suitable for interrupted cutting operations. Explain your choices.

8.15 Explain the possible disadvantages of a cutting operation in which the type of chip produced is discontinuous.

8.16 We have noted that tool life can be almost infinite at low cutting speeds. Would you then recommend that all machining be done at low speeds? Explain.

8.17 How would you explain the effect of cobalt content on the properties of carbides?

8.18 Explain why studying the types of chips produced is important in understanding machining operations.

8.19 How would you expect the cutting force to vary for the case of serrated-chip formation? Explain.

8.20 Wood is a highly anisotropic material; that is, it is orthotropic. Explain the effects of orthogonal cutting of wood at different angles to the grain direction on chip formation.

8.21 Describe the advantages of oblique cutting.

8.22 Explain why it is possible to remove more material between tool resharpenings by lowering the cutting speed.

8.23 Explain the significance of Eq. (8.8).

8.24 Why are there so many different types of cutting-tool materials? Explain, with appropriate examples.

8.25 How would you go about measuring the hot hardness of cutting tools?

8.26 Describe the reasons for making cutting tools with multiphase coatings of different materials.

8.27 Explain the advantages and limitations of inserts. Why were they developed?

8.28 Make a list of alloying elements used in high-speed-steel cutting tools. Explain why they are used.

8.29 What are the purposes of chamfers on cutting tools?

8.30 Why does temperature have such an important effect on cutting-tool performance?

8.31 Ceramic and cermet cutting tools have certain advantages over carbide tools. Why, then, are they not replacing carbide tools to a greater extent?

8.32 Why are chemical stability and inertness important in cutting tools?

8.33 What precautions would you take in machining with brittle tool materials?

8.34 Why do cutting fluids have different effects at different cutting speeds? Is the control of cutting-fluid temperature important? Explain.

8.35 Which of the two materials, diamond or cubic boron nitride, is more suitable for cutting steels? Why?

8.36 Describe the properties that the substrate for multiphase cutting tools should have.

8.37 List and explain the considerations involved in determining whether a cutting tool should be reconditioned, recycled, or discarded after use.

8.38 List the parameters that influence the temperature in metal cutting, and explain why and how they do so.

8.39 List and explain factors that contribute to poor surface finish in machining operations.

8.40 Explain the functions of different angles on a single-point lathe cutting tool. How does the chip thickness vary as the side cutting-edge angle is increased?

8.41 The helix angle for drills is different for different groups of workpiece materials. Why?

8.42 A turning operation is being carried out on a long, round bar at a constant depth of cut. Explain what changes, if any, may occur in the machined diameter from one end of the bar to the other. Give reasons for any changes that may occur.

8.43 Describe the relative characteristics of climb milling and up milling.

8.44 In Fig. 8.64a, high-speed-steel cutting teeth are welded to a steel blade. Would you recommend that the whole blade be made of high-speed steel? Explain your reasons.

8.45 Explain the technical requirements that led to the development of machining centers. What are the distinctive features of these centers? Why can their spindle speeds be varied over a wide range? Explain.

8.46 Describe the adverse effects of vibrations and chatter in machining.

8.47 Make a list of components of machine tools that could be made of ceramics, and explain why ceramics would be a suitable material for these components.

8.48 In Fig. 8.12, why do the thrust forces start at a finite value when the feed is zero?

8.49 Do you think temperature rise in cutting is related to the hardness of the workpiece material? Explain.

8.50 Describe the effects of tool wear on the workpiece and on the machining operation in general.

8.51 Explain whether or not it is desirable to have a high or low n value in the Taylor tool-life equation.

8.52 Are there cutting operations that cannot be performed on machining centers? Explain.

8.53 What is the importance of the cutting ratio?

8.54 Emulsion cutting fluids typically consist of 95% water and 5% soluble oil and chemical additives. Why is the ratio so unbalanced? Is the oil needed at all?

8.55 It is possible for the n value in the Taylor tool-life equation to be negative. Explain how.

8.56 Assume that you are asked to estimate the cutting force in slab milling with a straight-tooth cutter, but without running a test. Describe the procedure that you would follow.

8.57 Explain the possible reasons that a knife cuts better when it is moved back and forth. Consider factors such as the material being cut, interfacial friction, and the dimensions of the knife.

8.58 What are the effects of lowering the friction at the tool–chip interface, say with an effective cutting fluid, on the mechanics of cutting operations? Explain, giving several examples.

8.59 Why is it not always advisable to increase cutting speed in order to increase production rate? Explain.

8.60 It has been observed that the shear-strain rate in metal cutting is high even though the cutting speed may be relatively low. Why?

8.61 We note from the exponents in Eq. (8.30) that the cutting speed has a greater influence on temperature than does the feed. Why?

8.62 Describe the consequences of exceeding the allowable wear land (see Table 8.5) for cutting tools.

8.63 Explain whether it is desirable to have a high or low C value in the Taylor tool-life equation.

8.64 Comment on and explain your observations regarding Figs. 8.34, 8.38, and 8.43.

8.65 The tool-life curve for ceramic tools in Fig. 8.23a is to the right of those for other tools. Why?

8.66 In Fig. 8.18, we note that the percentage of the energy carried away by the chip increases with cutting speed. Why?

8.67 How would you go about measuring the effectiveness of cutting fluids? Explain.

8.68 Describe the conditions that are critical in using the capabilities of diamond and cubic-boron-nitride cutting tools.

8.69 In Table 8.6, the last two properties listed can be important to the life of the cutting tool. Why? Which of the properties listed are, in your opinion, the least important in machining? Explain.

8.70 In Fig. 8.31, we note that tool materials, especially carbides, have a wide range of hardnesses at a particular temperature. Why?

8.71 How would you go about recycling used cutting tools? Explain.

8.72 As you can see, there is a wide range of tool materials available and used successfully today, yet much research and development continues to be carried out on these materials. Why?

8.73 Drilling, boring, and reaming of large holes is generally more accurate than just drilling and reaming. Why?

8.74 A badly oxidized and uneven round bar is being turned on a lathe. Would you recommend a relatively small or large depth of cut? Explain your reasons.

8.75 Does the force or torque in drilling change as the hole depth increases? Explain.

8.76 Explain the advantages and limitations of producing threads by forming and cutting, respectively.

8.77 Describe your observations regarding the contents of Tables 8.8, 8.10, and 8.11.

8.78 The footnote to Table 8.10 states that as the depth of the hole increases, speeds and feeds should be reduced. Why?

8.79 List and explain the factors that contribute to poor surface finish in machining operations.

8.80 Make a list of the machining operations described in this chapter, according to the difficulty of the operation and the desired effectiveness of cutting fluids. (*Example*: Tapping of holes is a more difficult operation than turning straight shafts.)

8.81 Are the feed marks left on the workpiece by a face-milling cutter true segments of a true circle? Explain with appropriate sketches.

8.82 What determines the selection of the number of teeth on a milling cutter? (See, for example, Figs. 8.53 and 8.55.)

8.83 Explain the technical requirements that led to the development of machining and turning centers. Why do their spindle speeds vary over a wide range?

8.84 In addition to the number of components, as shown in Fig. 8.71, what other factors influence the rate at which damping increases in a machine tool? Explain.

8.85 Why is thermal expansion of machine-tool components important? Explain, with examples.

8.86 Would using the machining processes described in this chapter be difficult on nonmetallic or rubberlike materials? Explain your thoughts, commenting on the influence of various physical and mechanical properties of workpiece materials, the cutting forces involved, the parts geometries, and the fixturing required.

8.87 The accompanying illustration shows a part that is to be machined from a rectangular blank. Suggest the type of operations required and their sequence, and specify the machine tools that are needed.

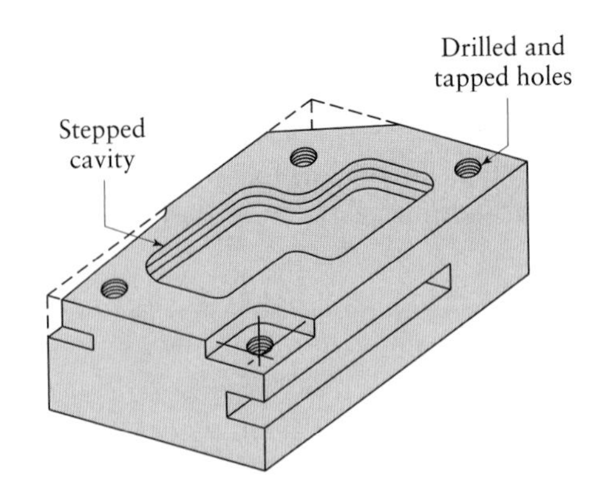

8.88 Select a specific cutting-tool material, and estimate the machining time for the parts shown in the accompanying figure: (a) pump shaft, stainless steel; (b) ductile (nodular) iron crankshaft; (c) 304 stainless-steel tube with internal rope thread.

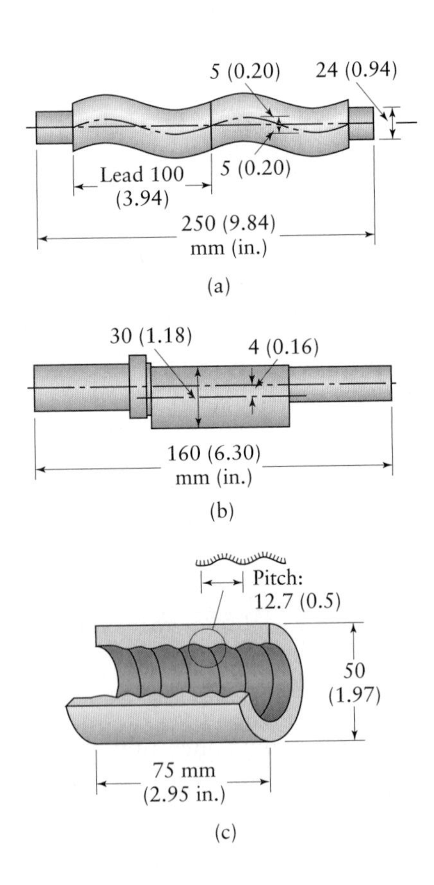

8.89 Why is the machinability of alloys difficult to assess?

8.90 What are the advantages and disadvantages of dry machining?

8.91 Can high-speed machining be performed without the use of cutting fluids? Explain.

8.92 If the rake angle is 0°, then the frictional force is perpendicular to the cutting direction and, therefore, does not consume machine power. Why, then, is there an increase in the power dissipated when machining with a rake angle of, say, 20°?

8.93 Would you recommend broaching a keyway on a gear blank before or after the teeth are machined? Explain.

PROBLEMS

8.94 Assume that you are an instructor covering the topics in this chapter, and you are giving a quiz on the quantitative aspects to test the understanding of the students. Prepare several numerical problems, and supply the answers to them.

8.95 Prove Eq. (8.1).

8.96 With a simple analytical expression to prove the validity of the last paragraph in Example 8.2.

8.97 Using Eq. (8.3), make a plot of the shear strain, γ, vs. the shear angle, ϕ, with the rake angle, α, as a parameter. Describe your observations.

8.98 Assume that in orthogonal cutting, the rake angle is 15° and the coefficient of friction is 0.25. Determine the percentage change in chip thickness when the friction is doubled.

8.99 Derive Eq. (8.11).

8.100 Determine the shear angle in Example 8.1. Is this calculation exact or an estimate? Explain.

8.101 Taking carbide as an example and using Eq. (8.30), determine how much the feed should be changed in order to keep the mean temperature constant when the cutting speed is tripled.

8.102 With appropriate diagrams, show how the use of a cutting fluid can affect the magnitude of the thrust force, F_t, in orthogonal cutting.

8.103 An 8-in-diameter stainless-steel bar is being turned on a lathe at 600 rpm and at a depth of cut, $d = 0.1$ in. If the power of the motor is 4 hp, what is the maximum feed that you can have before the motor stalls?

8.104 Using the Taylor equation for tool wear and letting $n = 0.4$, calculate the percentage increase in tool life if the cutting speed is reduced by (a) 20% and (b) 50%.

8.105 Determine the n and C values for the four tool materials shown in Fig. 8.23a.

8.106 Using Eq. (8.30) and referring to Fig. 8.18a, estimate the magnitude of the coefficient a.

8.107 (a) Estimate the machining time required in rough turning a 1.25-m-long, annealed aluminum-alloy round bar that is 75 mm in diameter, using a high-speed-steel tool. (b) Estimate the time for a carbide tool. Use a feed of 2 mm/rev.

8.108 A 200-mm-long, 75-mm-diameter titanium-alloy rod is being reduced in diameter to 65 mm by turning on a lathe. The spindle rotates at 400 rpm, and the tool is traveling at an axial velocity of 250 mm/min. Calculate the cutting speed, material removal rate, time of cut, power required, and cutting force.

8.109 Calculate the same quantities as in Example 8.5 for high-strength cast iron at $N = 300$ rpm.

8.110 A 0.75-in-diameter drill is being used on a drill press operating at 200 rpm. If the feed is 0.005 in./rev, what is the material removal rate? What is the MRR if the drill diameter is tripled?

8.111 A hole is being drilled in a block of magnesium alloy with a 10-mm drill at a feed of 0.1 mm/rev. The spindle is running at 800 rpm. Calculate the material removal rate, and estimate the torque on the drill.

8.112 Show that the distance ℓ_c in slab milling is approximately equal to $\sqrt{Dd}$ for situations where $D \gg d$.

8.113 Calculate the chip depth of cut in Example 8.7.

8.114 In Example 8.7, which of the quantities will be affected when the spindle speed is increased to 200 rpm?

8.115 A slab-milling operation is being carried out on a 30-in.-long, 6-in.-wide high-strength-steel block at a feed of 0.01 in./tooth and a depth of cut of 0.15 in. The cutter has a diameter of 3 in, has eight straight cutting teeth, and rotates at 150 rpm. Calculate the material removal rate and the cutting time, and estimate the power required.

8.116 Referring to Fig. 8.54, assume that $D = 250$ mm, $w = 50$ mm, $\ell = 600$ mm, $d = 2$ mm, $v = 1$ mm/s, and $N = 200$ rpm. The cutter has 10 inserts, and the workpiece material is 304 stainless steel. Calculate the material removal rate, cutting time, and feed per tooth, and estimate the power required.

8.117 Estimate the time required for face milling an 8-in.-long, 3-in.-wide brass block with an 8-in-diameter cutter that has 10 high-speed-steel teeth.

8.118 A 10-in-long, 2-in-thick plate is being cut on a band saw at 100 ft/min. The saw has 12 teeth per in. If the feed per tooth is 0.003 in., how long will it take to saw the plate along its length?

8.119 A single-thread hob is used to cut 40 teeth on a spur gear. The cutting speed is 150 ft/min, and the hob has a diameter of 4 in. Calculate the rotational speed of the spur gear.

8.120 In deriving Eq. (8.20), we assumed that the friction angle, β, was independent of the shear angle, ϕ. Is this assumption valid? Explain.

8.121 An orthogonal cutting operation is being carried out under the following conditions: depth of cut = 0.15 mm, width of cut = 5 mm, chip thickness = 0.2 mm, cutting speed = 2 m/s, rake angle = 15°, cutting force = 500 N, and thrust force = 200 N. Calculate the percentage of the total energy that is dissipated in the shear plane during cutting.

8.122 An orthogonal cutting operation is being carried out under the following conditions: depth of cut = 0.020 in., width of cut = 0.1 in., cutting ratio = 0.3, cutting speed = 300 ft/min, rake angle = 0°, cutting force = 200 lb, thrust force = 150 lb, workpiece density = 0.26 lb/in^3, and workpiece specific heat = 0.12 BTU/lb°F. Assume that (a) the sources of heat are the shear plane and the tool–chip interface; (b) the thermal conductivity of the tool is zero, and there is no heat loss to the environment; (c) the temperature of the chip is uniform throughout. If the temperature rise in the chip is 155°F, calculate the percentage of the energy dissipated in the shear plane that goes into the workpiece.

8.123 It can be shown that the angle ψ between the shear plane and the direction of maximum grain elongation (see Fig. 8.4a) is given by the expression

$$\psi = 0.5 \cot^{-1}\left(\frac{\gamma}{2}\right),$$

where γ is the shear strain, as given by Eq. (8.3). Assume that you are given a piece of the chip obtained from orthogonal cutting of an annealed metal. The rake angle and cutting speed are also given, but you have not seen the setup on which the chip was produced. Outline the procedure that you would follow to estimate the power required in producing this chip. Assume that you have access to a fully equipped laboratory and a technical library.

8.124 In Figs. 8.16 and 8.17b, we note that the maximum temperature is about halfway up the face of the tool. We have also described the adverse effects of temperature on various tool materials. Considering the mechanics of cutting operations, describe your thoughts on the technical and economic merits of embedding a small insert, made of materials such as ceramic or carbide, halfway up the rake face of a tool made of a material with lower resistance to temperature than ceramic or carbide.

8.125 A lathe is set up to machine a taper on a bar stock 100 mm in diameter; the taper is 1 mm per 10 mm. A cut is made with an initial depth of cut of 2 mm at a feed rate of 0.250 mm/rev and at a spindle speed of 150 rpm. Calculate the average metal removal rate.

8.126 Obtain an expression for optimum feed rate that minimizes the cost per piece if the tool life is as described by Eq. (8.34).

8.127 Assuming that the coefficient of friction in cutting is 0.3, calculate the maximum depth of cut for turning a hard aluminum alloy on a 20-hp lathe (mechanical efficiency of 80%) at a width of cut of 0.25 in, a rake angle of 10°, and a cutting speed of 300 ft/min. What is your estimate of the material's shear strength?

8.128 Assume that, using a carbide cutting tool, you measure the temperature in a cutting operation at a speed of 250 ft/min and feed of 0.0025 in./rev as 1200°F. What is the approximate temperature if the speed is doubled? What speed is required to lower the maximum temperature to 800°F?

8.129 A 4-in-diameter gray cast-iron cylindrical part is to be turned on a lathe at 600 rpm, with a depth of cut of 0.25 in. and a feed of 0.02 in./rev. What should be the minimum horsepower of the lathe?

8.130 (a) A 6-in-diameter aluminum cylinder with a length of 10 in. is to have its diameter reduced to 5 in. Estimate the machining time if an uncoated carbide tool is used. (b) What if a TiN-coated tool is used?

DESIGN

8.131 Tool life could be greatly increased if an effective means of cooling and lubrication were developed. Design methods of delivering a cutting fluid to the cutting zone, and discuss the advantages and shortcomings of your design.

8.132 Devise an experimental setup whereby you can perform an orthogonal cutting operation using a round tubular workpiece on a lathe.

8.133 Cutting tools are sometimes designed so that the chip–tool contact length is controlled by recessing it at the rake face some distance away from the tool tip. Explain the possible advantages of such a tool.

8.134 The accompanying illustration shows drawings for a cast-steel valve body before (left) and after (right) machining. Identify the surfaces that are to be machined. What type of machine tool would be suitable to machine this part? What type of machining operations are involved, and what should be the sequence of these operations? (Note that not all surfaces are to be machined.)

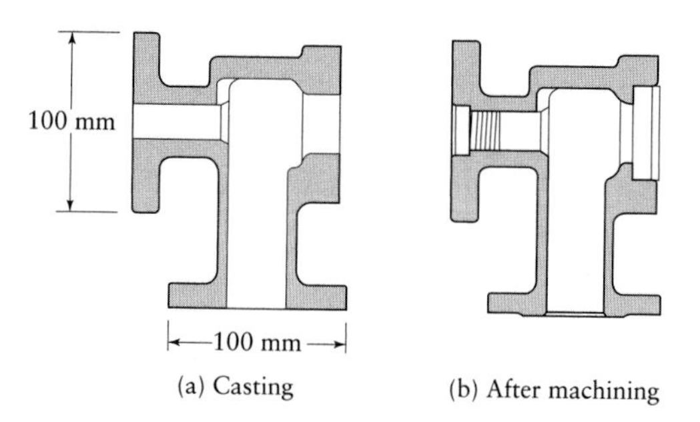

8.135 Make a comprehensive table of the process capabilities of the machining processes described in this chapter. Use several columns to describe the machines involved, the type of tools and tool materials used, the shapes of blanks and parts produced, the typical maximum and minimum sizes produced, the surface finish produced, the dimensional tolerances produced, and the production rates achieved.

8.136 A large bolt is to be produced from hexagonal bar stock by placing the hex stock into a chuck and machining the cylindrical shank of the bolt by turning on a lathe. List and explain the difficulties that may be presented by this operation.

8.137 The part shown in the accompanying figure is a power-transmitting shaft; it is to be produced on a lathe. List the operations that are needed to make this part, and estimate the machining time.

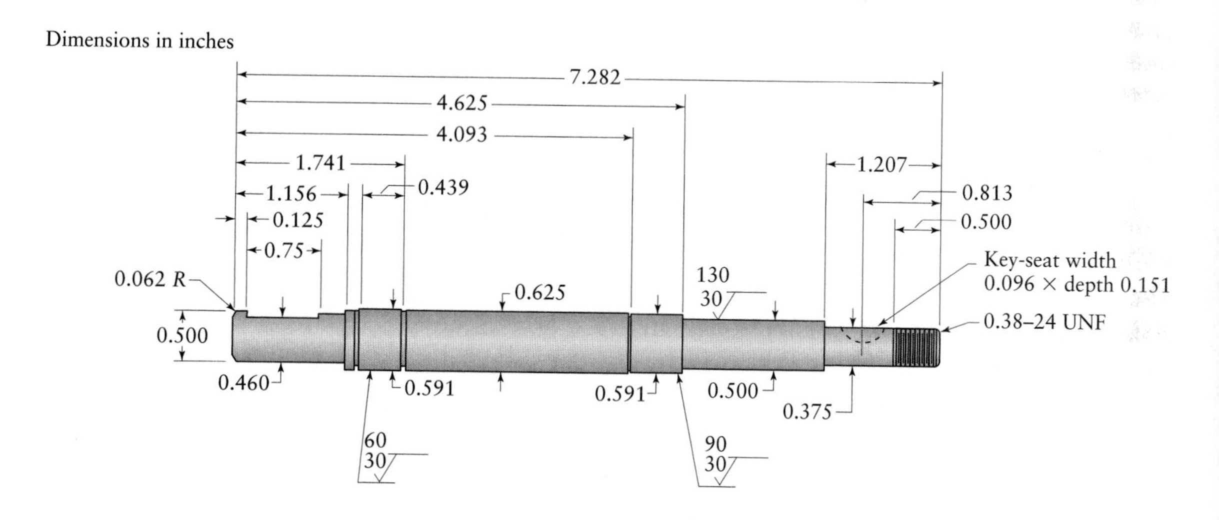

8.138 Design appropriate fixtures and describe the machining operations required to produce the aluminum piston shown in Fig. 5.43.

CHAPTER

9 Material-Removal Processes: Abrasive, Chemical, Electrical, and High-Energy Beams

9.1 Introduction

In all the cutting processes described in Chapter 8, the tool is made of a certain material and has a clearly defined geometry. Furthermore, the cutting process is carried out by chip removal, the mechanics of which are reasonably well understood. There are many situations in manufacturing, however, where the workpiece material is either too *hard* or too *brittle*, or its *shape* is difficult to produce with sufficient accuracy by any of the cutting methods described previously. One of the best methods for producing such parts is to use **abrasives.** An abrasive is a small, hard particle that has sharp edges and an irregular shape, unlike the cutting tools described earlier. Abrasives are capable of removing small amounts of material from a surface by a cutting process that produces tiny chips.

Abrasive machining processes are generally among the last operations performed on manufactured products. These processes, however, are not necessarily confined to fine or small-scale material removal. They are also used for large-scale removal operations and can indeed compete economically with some machining processes, such as milling and turning. Because they are hard, abrasives are also used in finishing very hard or heat-treated parts; shaping hard nonmetallic materials, such as ceramics and glasses; removing unwanted weld beads; cutting off lengths of bars, structural shapes, masonry, and concrete; and cleaning surfaces, for which jets of air or water that contain abrasive particles are used.

There are situations, however, where none of these processes is satisfactory or economical, for one or more of the following reasons:

1. The hardness and strength of the material is very high, typically above 400 HB.
2. The workpiece is too flexible or slender to support the cutting or grinding forces, or parts are difficult to clamp in workholding devices.
3. The shape of the part is complex—for example, it has internal and external profiles or small-diameter holes.

4. Surface finish and dimensional accuracy better than those obtainable by other processes are required.
5. Temperature rise or residual stresses in the workpiece is undesirable or unacceptable.

These requirements led to the development of chemical, electrical, and other means of material removal in the 1940s; these methods are now called **advanced machining**, but have also been called **nontraditional** or **unconventional machining**. When selected and applied properly, these processes offer significant economic and technical advantages over the traditional machining methods described in the preceding chapter.

9.2 Abrasives

The abrasives commonly used in manufacturing are as follows:

1. **Conventional abrasives:**
    - *Aluminum oxide*;
    - *Silicon carbide*.
2. **Superabrasives:**
    - *Cubic boron nitride*;
    - *Diamond*.

Abrasives are considerably harder than conventional cutting-tool materials (see Table 8.6 and Table 9.1); the last two of the four aforementioned abrasives are the two hardest materials known, hence the term *superabrasives*. In addition to hardness, an important characteristic of abrasives is **friability**, that is, the ability of abrasive grains to fracture (break down) into smaller pieces; this property gives abrasives *self-sharpening* characteristics, which are important in maintaining the sharpness of the abrasives in their use. High friability indicates low strength or low fracture resistance of the abrasive; thus, a highly friable abrasive grain fragments more rapidly under grinding forces than an abrasive grain with low friability. Aluminum oxide has lower friability than silicon carbide, hence it has less tendency to fragment.

The **shape** and **size** of the abrasive grain also affect its friability; blocky grains, for example, which may be analogous to negative-rake-angle cutters, are less friable than platelike grains. Furthermore, because the probability of defects in smaller grains is lower (due to the *size effect*; Section 3.8.3), they are stronger and less friable than larger grains. The importance of friability in abrasive processes is described in Section 9.5.

Types of abrasives. Abrasives found in nature include *emery*, *corundum* (*alumina*), *quartz*, *garnet*, and *diamond*. However, natural abrasives contain unknown amounts

TABLE 9.1

Knoop Hardness Range for Various Materials and Abrasives

Material	Hardness	Material	Hardness
Common glass	350–500	Titanium nitride	2000
Flint, quartz	800–1100	Titanium carbide	1800–3200
Zirconium oxide	1000	Silicon carbide	2100–3000
Hardened steels	700–1300	Boron carbide	2800
Tungsten carbide	1800–2400	Cubic boron nitride	4000–5000
Aluminum oxide	2000–3000	Diamond	7000–8000

of impurities and possess nonuniform properties; consequently, their performance is inconsistent and unreliable. As a result, aluminum oxides and silicon carbides used as abrasives are now made synthetically. These two abrasives and the two superabrasives are described as follows:

a) Synthetic **aluminum oxide** (Al_2O_3), first made in 1893, is obtained by fusing bauxite, iron filings, and coke. Aluminum oxides are divided into two groups: fused and unfused. **Fused** aluminum oxides are categorized as *white* (very friable), *dark* (less friable), and *monocrystalline*. **Unfused** alumina, also known as *ceramic aluminum oxides*, can be harder than fused alumina. The purest form of fused alumina is seeded gel. First introduced in 1987, **seeded gel** has a particle size on the order of 0.2 μm (8 μin.), which is much smaller than that of commonly used abrasive grains. These particles are sintered to form larger sizes. Because of their hardness and relatively high friability, seeded gels maintain their sharpness and are therefore used for difficult-to-grind materials.

b) **Silicon carbide** (SiC), first discovered in 1891, is made with silica sand, petroleum coke, and small amounts of sodium chloride. Silicon carbides are divided into *green* (more friable) and *black* (less friable) types and generally have higher friability than aluminum oxides; hence, they have a higher tendency to fracture and remain sharp.

c) **Cubic boron nitride** (cBN) was first discovered in the 1970s. Its properties and characteristics are described in Chapters 8 and 11.

d) **Diamond** was first used as an abrasive in 1955. It is also produced synthetically, whereby it is known as *synthetic* or *industrial diamond*. Its properties and characteristics are described in Chapters 8 and 11.

Grain size. As used in manufacturing processes, abrasives are generally very small compared with the size of the cutting tools and inserts described in Chapter 8. Also, abrasives have sharp edges, thus allowing the removal of very small quantities of material from the workpiece surface. Consequently, very fine surface finish and dimensional accuracy can be obtained. (See Figs. 8.27 and 9.27.)

The size of an abrasive grain is identified by a **grit number**, which is a function of sieve size. The smaller the sieve size, the larger is the grit number. For example, grit number 10 is rated as very coarse, 100 as fine, and 500 as very fine. Sandpaper and emery cloth are also identified in this manner, with the grit number printed on the back of the abrasive paper or cloth.

9.3 Bonded Abrasives

Because each abrasive grain usually removes only a very small amount of material at a time, high rates of material removal can be obtained only if a large number of these grains act together; this is accomplished by using *bonded abrasives* (as opposed to loose abrasives), typically in the form of a **grinding wheel**. A simple grinding wheel is shown schematically in Fig. 9.1. The abrasive grains are held together by a **bonding material** (various types of which are described in Section 9.3.1), which acts as supporting posts or braces between the grains. Some porosity is essential in bonded wheels to provide clearance for the minute chips being produced and to provide cooling; otherwise, the chips would interfere with the grinding process. (It is thus impossible to use a grinding wheel that is fully dense and solid with no porosity.) Porosity can easily be observed simply by looking at the surface of any grinding

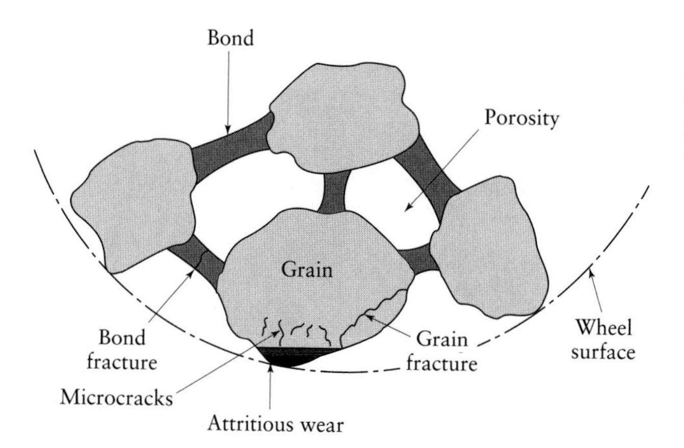

FIGURE 9.1 Schematic illustration of a physical model of a grinding wheel, showing its structure and wear and fracture patterns.

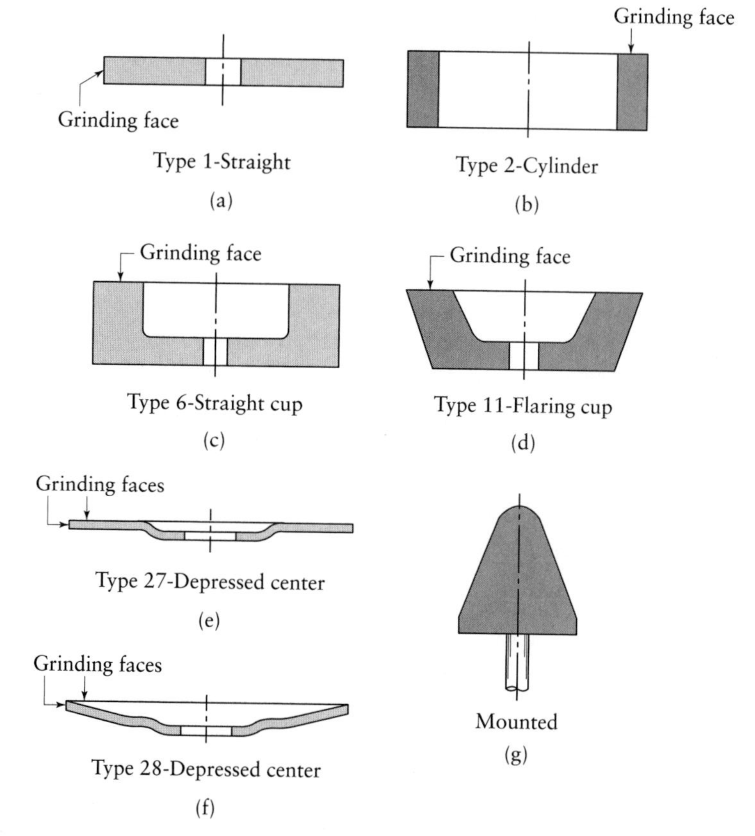

FIGURE 9.2 Some common types of grinding wheels made with conventional abrasives. Note that each wheel has a specific grinding face. Grinding on other surfaces is improper and unsafe.

wheel. Other features of the grinding wheel shown in Fig. 9.1 are described in Sections 9.4 and 9.5.

Some of the more commonly used types of grinding wheels are shown in Fig. 9.2 for conventional abrasives and in Fig. 9.3 for superabrasives. Note that due to the high cost of the latter, only a small percentage of the wheels used is of the superabrasives variety. Bonded abrasives are marked with a standardized system of letters and numbers, indicating the type of abrasive, grain size, grade, structure, and bond type.

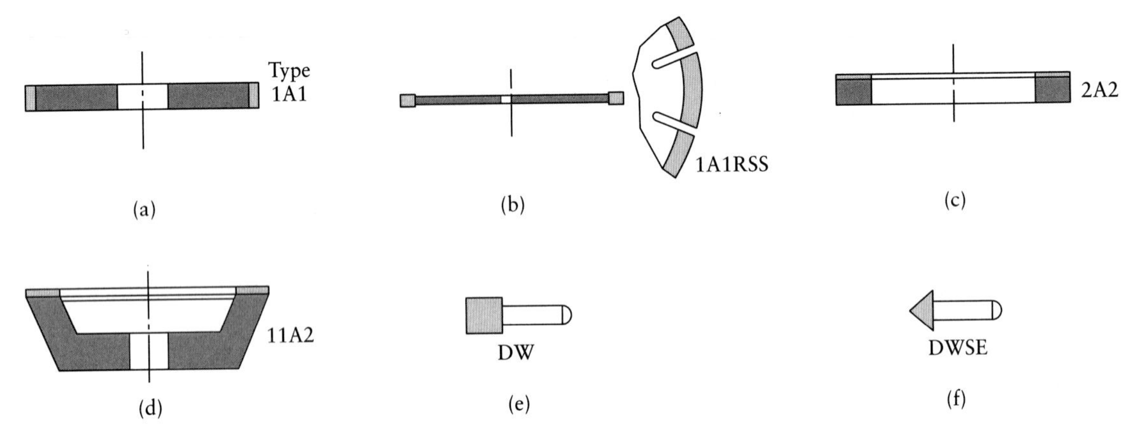

FIGURE 9.3 Examples of superabrasive-wheel configurations. The annular regions (rim) are superabrasive grinding surfaces, and the wheel itself (core) is generally made of metal or composites. Note that the basic numbering of wheel type is the same as shown in Fig. 9.2. The bonding materials for the superabrasives are (a), (d), and (e) resinoid, metal, or vitrified; (b) metal; (c) vitrified; and (f) resinoid.

Figure 9.4a shows the marking system for aluminum-oxide and silicon-carbide bonded abrasives, and Fig. 9.4b shows the marking system for diamond and cubic-boron-nitride bonded abrasives.

9.3.1 Bond types

The common bond types for bonded abrasives are vitrified, resinoid, rubber, and metal. These bonds are used for conventional abrasives as well as for superabrasives (except rubber) and are described as follows:

1. **Vitrified.** Essentially a glass, *vitrified bond* is also called a *ceramic bond*, particularly outside the United States; it is the most common and widely used bond. The raw materials consist of feldspar (a crystalline mineral) and clays. These materials are mixed with the abrasives, moistened, and molded under pressure into the shape of grinding wheels. These "green" products, which are similar to powder-metallurgy parts, are then fired slowly, up to a temperature of about 1250°C (2300°F), to fuse the glass and develop structural strength. The wheels are then cooled slowly to avoid thermal cracking, finished to size, inspected for quality and dimensional accuracy, and tested for defects. Vitrified bonds produce wheels that are strong, stiff, porous, and resistant to oils, acids, and water. Because the wheels are brittle, they lack resistance to mechanical and thermal shock. However, vitrified wheels are also available with steel backing plates or cups for better structural support during their use.
2. **Resinoid.** Resinoid bonding materials are *thermosetting resins* (see Section 10.4) and are available in a wide range of compositions and properties. Because the bond is an organic compound, wheels with *resinoid bonds* are also called **organic wheels**. The basic manufacturing technique consists of mixing the abrasive with liquid or powdered phenolic resins and additives, pressing the mixture into the shape of a grinding wheel, and curing it at temperatures of about 175°C (350°F). Because the elastic modulus of thermosetting resins is lower than that of glasses, resinoid wheels are more flexible than vitrified wheels. **Reinforced resinoid wheels** are widely used, in which one or more layers of fiberglass mats of various mesh sizes provide the reinforcement. The main purpose of the reinforcements is to retard the disintegration of the wheel should it break for

Prefix
Abrasive type
Abrasive grain size
Grade
Structure
Bond type
Manufacturer's record
51 - A - 36 - L - 5 - V - 23
Manufacturer's symbol, indicating exact kind of abrasive (optional)
A Aluminium oxide
C Silicon carbide
Coarse 8 10 12 14 16 20 24
Medium 30 36 46 54 60
Fine 70 80 90 100 120 150 180
Very fine 220 240 280 320 400 500 600
Dense
Open
1 2 3 4 5 6 7 8 9 10 11 12 13 14 15 16 etc. (optional)
Manufacturer's private marking to identify wheel (optional)
B Resinoid
BF Resinoid reinforced
E Shellac
O Oxychloride
R Rubber
RF Rubber reinforced
S Silicate
V Vitrified
Soft
Medium
Hard
A B C D E F G H I J K L M N O P Q R S T U V W X Y Z
Grade scale

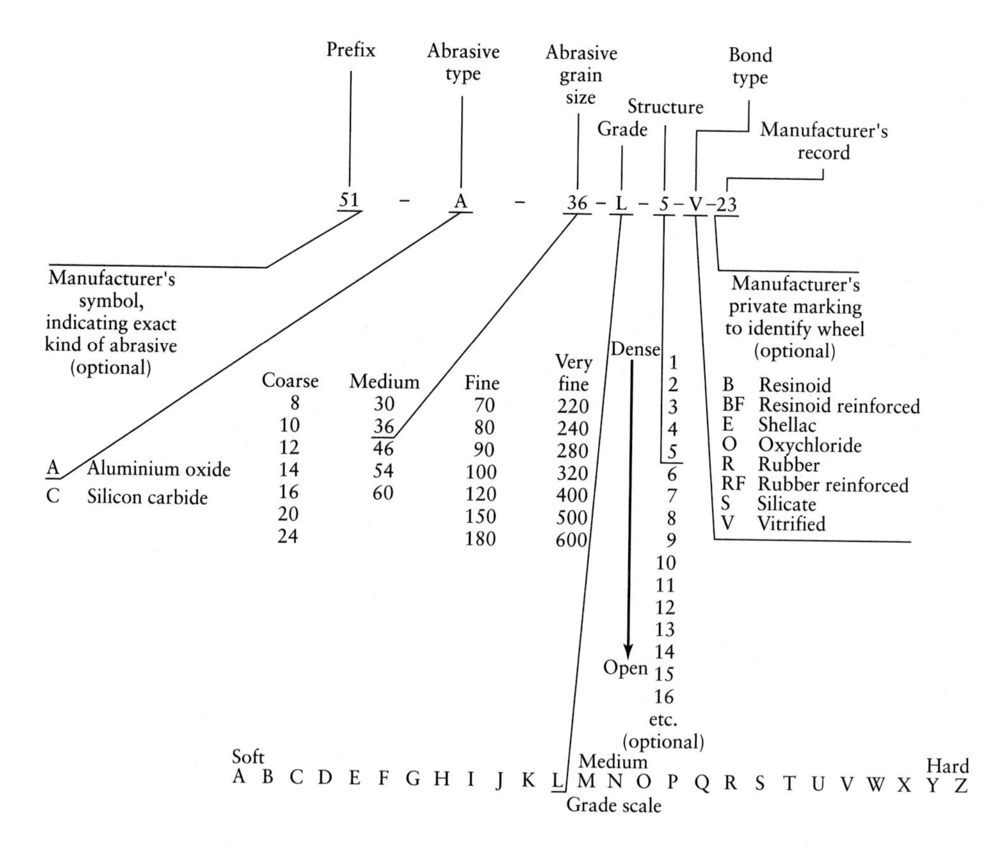

FIGURE 9.4a Standard marking system for aluminum-oxide and silicon-carbide bonded abrasives.

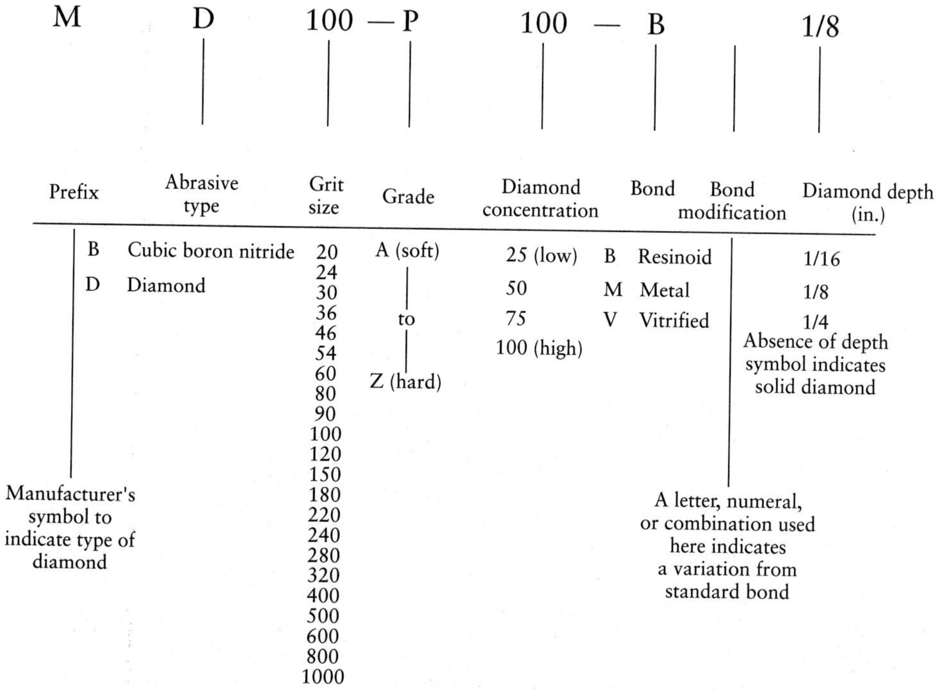

M D 100 — P 100 — B 1/8
Prefix
Abrasive type
Grit size
Grade
Diamond concentration
Bond
Bond modification
Diamond depth (in.)
B Cubic boron nitride
D Diamond
20 24 30 36 46 54 60 80 90 100 120 150 180 220 240 280 320 400 500 600 800 1000
A (soft)
to
Z (hard)
25 (low)
50
75
100 (high)
B Resinoid
M Metal
V Vitrified
1/16
1/8
1/4
Absence of depth symbol indicates solid diamond
Manufacturer's symbol to indicate type of diamond
A letter, numeral, or combination used here indicates a variation from standard bond

FIGURE 9.4b Standard marking system for diamond and cubic-boron-nitride bonded abrasives.

some reason, rather than to improve its strength. Large-diameter wheels can be supported additionally with one or more internal rings made of round steel bar, which are inserted during molding of the wheel.

3. **Rubber.** The most flexible bond used in abrasive wheels is *rubber*. The manufacturing process consists of mixing crude rubber, sulfur, and the abrasive grains together, rolling the mixture into sheets, cutting out circles, and heating the circles under pressure to vulcanize the rubber. Thin wheels can be made in this manner and are used like saws for cutting-off operations (cut-off blades).
4. **Metal bonds.** The abrasive grains, which are usually diamond or cubic boron nitride, are bonded, using powder-metallurgy techniques, to the periphery of a metal wheel, to depths of 6 mm (0.25 in.) or less. (See Fig. 9.3.) *Metal bonding* is carried out under high pressure and temperature. The wheel itself (core) may be made of aluminum, bronze, steel, ceramics, or composite materials, depending on requirements for the wheel, such as strength, stiffness, and dimensional stability.
5. **Other bonds.** In addition to those described previously, other types of bonds include *silicate*, *shellac*, and *oxychloride* bonds; however, these bonds have limited uses, and are not discussed further here. A new development is the use of *polyimide* (see Section 10.6) as a substitute for the phenolic in resinoid wheels. Polyimide is tough and has resistance to high temperatures. Superabrasive wheels may also be layered so that a single abrasive layer is plated or brazed to a metal wheel with a particular desired shape. Such wheels are lower in cost and are used for small production quantities.

9.3.2 Wheel grade and structure

The *grade* of a bonded abrasive is a measure of the bond's strength; it includes both the *type* and the *amount* of bond in the wheel. Because strength and hardness are directly related, the grade is also referred to as the *hardness* of a bonded abrasive. Thus, a hard wheel has a stronger bond and/or a larger amount of bonding material between the grains than a soft wheel. The *structure* is a measure of the *porosity* (the spacing between the grains in Fig. 9.1) of the bonded abrasive. Some porosity is essential to provide clearance for the grinding chips; otherwise, they would interfere with the grinding process. The structure of bonded abrasives ranges from dense to open.

9.4 Mechanics of Grinding

Grinding is basically a *chip removal* process in which the cutting tool is an individual abrasive grain. The following are major factors that differentiate the action of a single grain from that of a single-point tool:

a) The individual grain has an irregular geometry and is spaced randomly along the periphery of the wheel (Fig. 9.5).
b) The average rake angle of the grains is highly negative: $-60°$ or even lower. Consequently, the shear angles are very low.
c) The radial positions of the grains in a grinding wheel vary.
d) The cutting speeds of grinding wheels are very high (Table 9.2), typically 30 m/s (6000 ft/min).

Figure 9.6 shows an example of chip formation by an abrasive grain; note the negative rake angle, the low shear angle, and the small size of the chip. A variety of

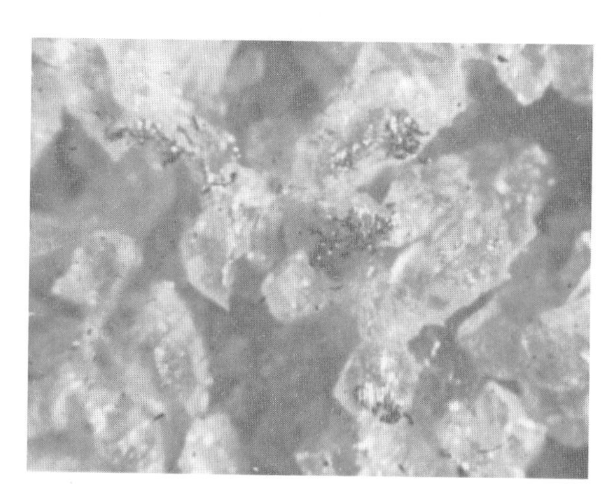

FIGURE 9.5 The grinding surface of an abrasive wheel (A46–J8V), showing grains, porosity, wear flats on grains (see also Fig. 9.9), and metal chips from the workpiece adhering to the grains. Note the random distribution and shape of the abrasive grains. Magnification: 50×.

TABLE 9.2

Typical Ranges of Speeds and Feeds for Abrasive Processes

Process Variable	Conventional Grinding	Creep-Feed Grinding	Buffing	Polishing
Wheel speed (m/min)	1500–3000	1500–3000	1800–3600	1500–2400
Work speed (m/min)	10–60	0.1–1	–	–
Feed (mm/pass)	0.01–0.05	1–6	–	–

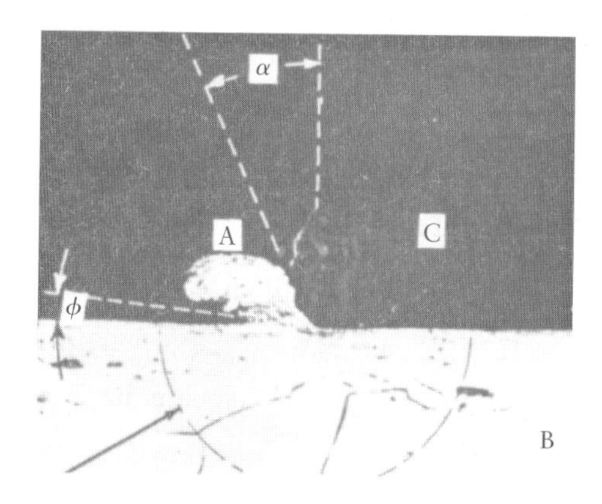

FIGURE 9.6 Grinding chip being produced by a single abrasive grain: (A) chip, (B) workpiece, (C) abrasive grain. Note the large negative rake angle of the grain. The inscribed circle is 0.065 mm (0.0025 in.) in diameter. *Source*: M. E. Merchant.

metal chips can be observed in grinding. (Chips are easily collected on a piece of adhesive tape held against the sparks of a grinding wheel.)

The mechanics of grinding and the variables involved can best be studied by analyzing the *surface-grinding* operation shown in Fig. 9.7. In this figure, a grinding wheel of diameter D is removing a layer of metal at a depth d, known as the **wheel depth of cut.** An individual grain on the periphery of the wheel is moving at a tangential velocity V (*up*, or *conventional*, *grinding*, as shown in Fig. 9.7), and the workpiece is moving at a velocity v. The grain is removing a chip whose *undeformed thickness*

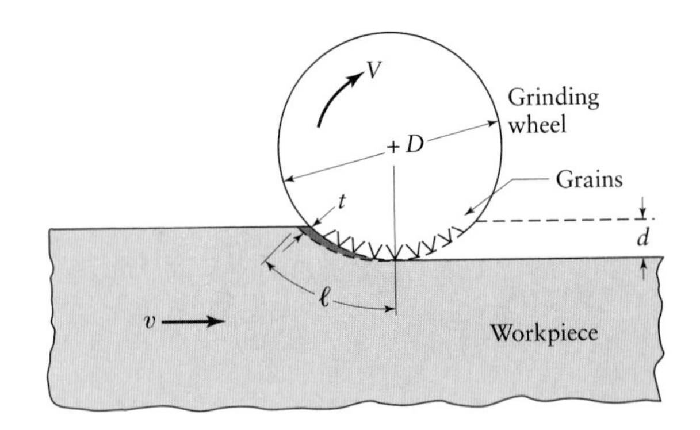

FIGURE 9.7 Variables in surface grinding. In actual grinding, the wheel depth of cut, d, and contact length, ℓ, are much smaller than the wheel diameter, D. The dimension t is called the *grain depth of cut.*

(**grain depth of cut**) is t and the undeformed length is ℓ. For $v \ll V$, the *undeformed-chip length*, ℓ is approximately

$$\ell \simeq \sqrt{Dd}. \tag{9.1}$$

For *external* (*cylindrical*) *grinding* (see Section 9.6),

$$\ell = \sqrt{\frac{Dd}{1 + (D/D_w)}}, \tag{9.2}$$

and for *internal grinding*,

$$\ell = \sqrt{\frac{Dd}{1 - (D/D_w)}}, \tag{9.3}$$

where D_w is the diameter of the workpiece.

The relationship between t and other process variables can be derived as follows: Let C be the number of cutting points per unit area of wheel surface; v and V are the surface speeds of the workpiece and the wheel, respectively (Fig. 9.7). If we let the width of the workpiece be unity, the number of chips produced per unit time is VC, and the volume of material removed per unit time is vd. Letting r be the ratio of the chip width w to the average chip thickness, we find that the volume of a chip with a rectangular cross-sectional area and constant width is

$$\text{Vol}_{\text{chip}} = \frac{wt\ell}{2} = \frac{rt^2\ell}{4}. \tag{9.4}$$

The volume of material removed per unit time, then, is the product of the number of chips produced per unit time and the volume of each chip, or

$$VC\frac{rt^2\ell}{4} = vd,$$

and because $\ell = \sqrt{Dd}$, the undeformed chip thickness in surface grinding is

$$t = \sqrt{\frac{4v}{VCr}\sqrt{\frac{d}{D}}}. \tag{9.5}$$

Experimental observations indicate the value of C to be roughly on the order of 0.1 per mm^2 to 10 per mm^2 (10^2 per in^2 to 10^3 per in^2); the finer the grain size of the wheel, the larger is this number. The magnitude of r is between 10 and 20 for most grinding operations. If we substitute typical values for a grinding operation into Eqs. (9.1) through (9.5), we obtain very small quantities for ℓ and t. For example, typical values for t are in the range of 0.3–0.4 μm (12–160 μin.).

EXAMPLE 9.1 Chip dimensions in grinding

Estimate the undeformed-chip length and the undeformed chip thickness for a typical surface-grinding operation.

Solution. The formulas for undeformed length and thickness, respectively, are

$$\ell = \sqrt{Dd} \quad \text{and} \quad t = \sqrt{\frac{4v}{VCr}\sqrt{\frac{d}{D}}}.$$

From Table 9.2, we select the following values:

$$v = 0.5 \text{ m/s} \quad \text{and} \quad V = 30\text{m/s}.$$

We also assume that

$$d = 0.05\text{mm} \quad \text{and} \quad D = 200 \text{ mm}$$

and let

$$C = 2 \text{ per mm}^2 \quad \text{and} \quad r = 15.$$

Therefore,

$$\ell = \sqrt{(200)(0.05)} = 3.2 \text{ mm} = 0.126 \text{ in.}$$

and

$$t = \sqrt{\frac{(4)(0.5)}{(30)(2)(15)}\sqrt{\frac{0.05}{200}}} = 0.006 \text{ mm} = 2.3 \times 10^{-4} \text{ in.}$$

Because of plastic deformation, the actual length of the chip is shorter and the thickness greater than these values. (See Fig. 9.6.)

9.4.1 Forces

A knowledge of forces is essential not only in the design of grinding machines, but also in determining the deflections that the workpiece and the machine will undergo, as well as in the design and use of workholding devices. Deflections, in turn, adversely influence dimensional accuracy and are thus critical in precision grinding. If we assume that the *force* on the grain (see the discussion of cutting force, F_c, in Section 8.2.3) is proportional to the cross-sectional area of the undeformed chip, we can show that the **relative grain force** is given by

$$\text{Relative grain force} \propto \frac{v}{VC}\sqrt{\frac{d}{D}}. \tag{9.6}$$

The actual force is the product of the relative grain force and the strength of the metal being ground.

The **specific energy** consumed in producing a grinding chip consists of three components:

$$u = u_{\text{chip}} + u_{\text{plowing}} + u_{\text{sliding}}. \tag{9.7}$$

In this equation, u_{chip} is the specific energy required for chip formation by plastic deformation, and u_{plowing} is the specific energy required for plowing, which is plastic

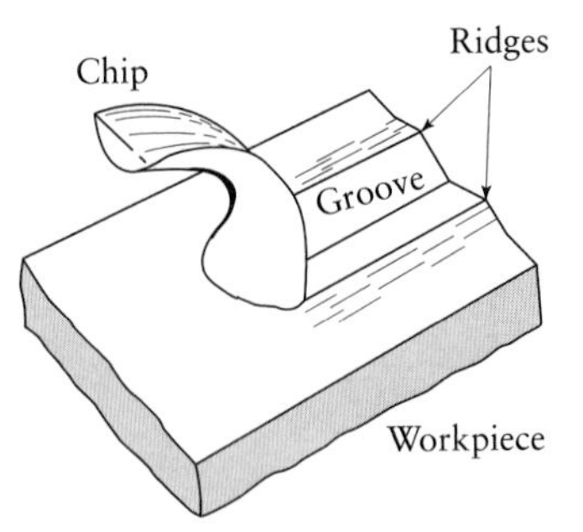

FIGURE 9.8 Chip formation and plowing of the workpiece surface by an abrasive grain.

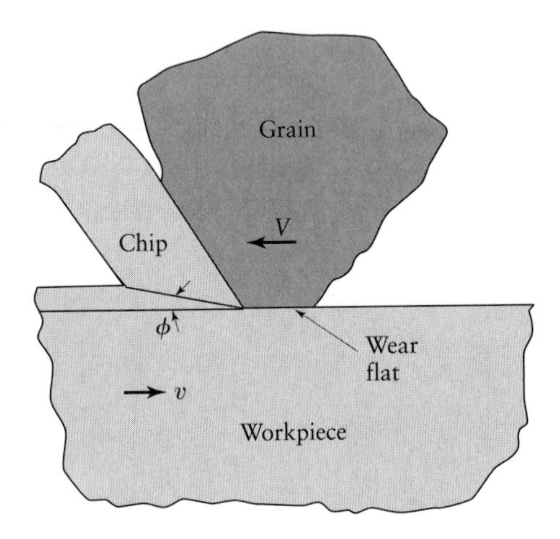

FIGURE 9.9 Schematic illustration of chip formation by an abrasive grain. Note the negative rake angle, the small shear angle, and the wear flat on the grain.

deformation without chip removal (Fig. 9.8). The last term, u_{sliding}, can best be understood by observing the grain in Fig. 9.9. The grain develops a **wear flat** as a result of the grinding operation (similar to flank wear in cutting tools). The wear flat slides along the surface being ground and, because of friction, requires energy for sliding; the larger the wear flat, the higher the grinding force is.

Table 9.3 provides typical specific-energy requirements in grinding. Note that these energy levels are much higher than those in cutting operations with single-point tools. (See Table 8.3.) This difference has been attributed to the following factors:

1. **Size effect**: As previously stated, the size of grinding chips is quite small compared with chips produced by other cutting operations, by about two orders of magnitude. As described in Section 3.8.3, the smaller the size of a piece of metal, the greater is its strength; consequently, grinding involves higher specific energy than cutting. Studies have indicated that extremely high dislocation densities (see Section 3.3.3) occur in the shear zone during chip formation, thus influencing the grinding energies involved.

TABLE 9.3

Approximate Specific-Energy Requirements for Surface Grinding

Workpiece Material	Hardness	Specific Energy W-s/mm^3	Specific Energy hp-min/in^3
Aluminum	150 HB	7–27	2.5–10
Cast iron (class 40)	215 HB	12–60	4.5–22
Low-carbon steel (1020)	110 HB	14–68	5–25
Titanium alloy	300 HB	16–55	6–20
Tool steel (T15)	67 HRC	18–82	6.5–30

2. **Wear flat**: Because a wear flat (see Fig. 9.9) requires frictional energy for sliding, this energy can contribute substantially to the total energy consumed. The size of the wear flat in grinding is much larger than the grinding chip, unlike in metal cutting by a single-point tool, where flank wear land is small compared with the size of the chip.
3. **Chip morphology**: Because the average rake angle of a grain is highly negative (see Fig. 9.6), the shear strains are very large. This indicates that the energy required for plastic deformation to produce a grinding chip is higher than in other cutting processes. Furthermore, plowing consumes energy without contributing to chip formation.

EXAMPLE 9.2 Forces in surface grinding

Assume that you are performing a surface-grinding operation on low-carbon steel with a wheel of diameter $D = 10$ in. that rotates at $N = 4000$ rpm, with a width of cut $w = 1$ in. The depth of cut is $d = 0.002$ in. and the feed rate v of the workpiece is 60 in./min Calculate the cutting force, F_c (the force tangential to the wheel), and the thrust force, F_n (the force normal to the workpiece).

Solution. We first determine the material removal rate as follows:

$$\text{MRR} = dwv = (0.002)(1)(60) = 0.12 \text{ in}^3/\text{min}.$$

The power consumed is given by

$$\text{Power} = (u)(\text{MRR}),$$

where u is the specific energy, as obtained from Table 9.3. For low-carbon steel, let's estimate u to be 15 hp-min/in^3. Hence, we have

$$\text{Power} = (15)(0.12) = 1.8 \text{ hp}.$$

By noting that 1 hp = 33,000 ft-lb/min = 396,000 in.-lb/min, we obtain

$$\text{Power} = (1.8)(396{,}000) = 712{,}800 \text{ in.-lb/min}.$$

Since power is defined as

$$\text{Power} = T\omega,$$

where T is the torque and is equal to $(F_c)(D/2)$ and ω is the rotational speed of the wheel in radians per minute, we also have $\omega = 2\pi N$. Thus,

$$712{,}800 = (F_c)\left(\frac{10}{2}\right)(2\pi)(4000),$$

and therefore, $F_c = 57$ lb. The thrust force, F_n, can be calculated by noting from experimental data in the technical literature that it is about 30% higher than the cutting force, F_c. Consequently,

$$F_n = (1.3)(57) = 74 \text{ lb}.$$

9.4.2 Temperature

Temperature rise in grinding is an important consideration, because it can adversely affect the surface properties and cause residual stresses on the workpiece. Furthermore, temperature gradients in the workpiece cause distortions by differential thermal expansion and contraction. When a portion of the heat generated is conducted into the workpiece, the heat expands the part being ground, thus making it difficult to control dimensional accuracy. The work expended in grinding is mainly converted into heat. The *surface temperature rise*, ΔT, has been found to be a function of the ratio of the total energy input to the surface area ground. Thus, in *surface grinding*, if w is the width and L is the length of the surface area ground, then

$$\Delta T \propto \frac{uwLd}{wL} \propto ud. \tag{9.8}$$

If we introduce size effect and assume that u varies inversely with the undeformed-chip thickness t, then the temperature rise is

$$\Delta T \propto \frac{d}{t} \propto d^{3/4}\sqrt{\frac{VC}{v}\sqrt{D}}. \tag{9.9}$$

The *peak temperatures* in chip generation during grinding may be as high as 1650°C (3000°F). However, the time involved in producing a chip is extremely short (on the order of microseconds); hence, melting may or may not occur. Because, as in metal cutting, the chips carry away much of the heat generated, only a fraction of the heat generated is conducted to the workpiece. Experiments indicate that in grinding, as much as one half the energy is conducted to the workpiece; this percentage is higher than that in metal cutting (see Section 8.2). The heat generated by sliding and plowing is conducted mostly into the workpiece.

Sparks. The sparks observed in metal grinding are actually glowing chips; the glowing occurs because of the exothermic reaction of the hot chips with oxygen in the atmosphere. Sparks have not been observed with any metal ground in an oxygen-free environment. The color, intensity, and shape of the sparks depend on the composition of the metal being ground. If the heat generated by the exothermic reaction is sufficiently high, the chip may melt and, because of surface tension, acquire a round shape and solidify as a shiny spherical particle. Observation of these particles under scanning electron microscopy has revealed that they are hollow and have a fine dendritic structure (see Fig. 5.8), indicating that they were once molten (by exothermic oxidation of hot chips in air) and resolidified rapidly. It has been suggested that some of the spherical particles may also be produced by plastic deformation and rolling of chips at the grit–workpiece interface during grinding.

9.4.3 Effects of temperature

The following list describes briefly the major effects of temperature in grinding:

a) **Tempering.** Excessive temperature rise caused by grinding can temper and soften the surfaces of steel components, which are often ground in the hardened state. Grinding process parameters must therefore be chosen carefully to avoid excessive temperature rise. The use of grinding fluids can effectively control temperatures.

b) **Burning.** If the temperature is excessive, the surface may burn; burning produces a bluish color on steels, which indicates oxidation at high temperatures. A burn may not be objectionable in itself; however, the surface layers may undergo metallurgical transformations, with martensite formation in high-carbon steels from reaustenization followed by rapid cooling (see Section 5.11). This effect is known as *metallurgical burn*, which also is a serious problem with nickel-base alloys. High temperatures in grinding may also lead to thermal cracking of the surface of the workpiece, known as **heat checking**. (See also

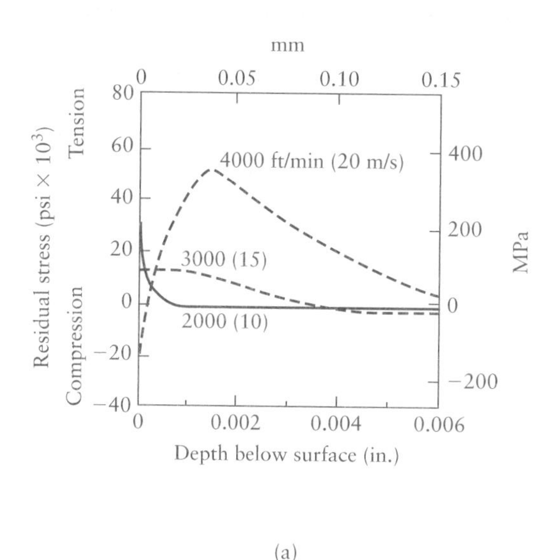

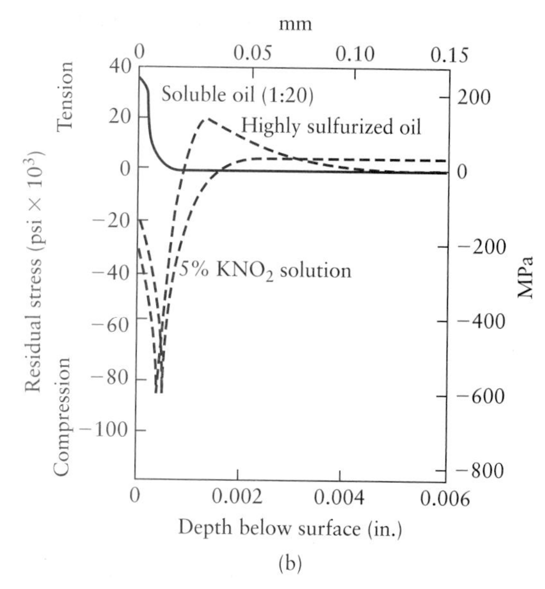

FIGURE 9.10 Residual stresses developed on the workpiece surface in grinding tungsten: (a) effect of wheel speed and (b) effect of grinding fluid. Tensile residual stresses on a surface are detrimental to the fatigue life of ground components. The variables in grinding can be controlled to minimize residual stresses, a process known as *low-stress grinding*. *Source*: After N. Zlatin et al., 1963.

Section 5.10.3.) Cracks are usually perpendicular to the grinding direction; however, under severe grinding conditions, parallel cracks may also develop.

c) **Residual stresses.** Temperature change and gradients within the workpiece are mainly responsible for residual stresses in grinding. Other contributing factors are the physical interactions of the abrasive grain in chip formation and the sliding of the wear flat along the workpiece surface, causing plastic deformation of the surface. Figure 9.10 shows two examples of residual stresses in grinding, demonstrating the effects of wheel speed and the type of grinding fluid used. The method and direction of the application of grinding fluid can also have a significant effect on residual stresses. Because of the deleterious effect of tensile residual stresses on fatigue strength, process parameters should be chosen carefully. Residual stresses can usually be lowered (**low-stress, or gentle, grinding**) by using softer grade wheels (*free-cutting wheels*), lower wheel speeds, and higher work speeds.

9.5 Grinding-Wheel Wear

Grinding-wheel wear is an important consideration, because it adversely affects the shape and accuracy of ground surfaces, as is the case with cutting tools. Grinding wheels wear by three different mechanisms: attritious grain wear, grain fracture, and bond fracture. These mechanisms are described as follows:

1. **Attritious wear.** In *attritious wear*, the cutting edges of a sharp grain become dull by attrition, developing a *wear flat* (see Fig. 9.9) that is similar to flank wear in cutting tools. Wear is caused by the interaction of the grain with the workpiece material, involving both physical and chemical reactions. These reactions are

complex and involve diffusion, chemical degradation or decomposition of the grain, fracture at a microscopic scale, plastic deformation, and melting.

Attritious wear is low when the two materials are chemically inert with respect to each other, much like with the use of cutting tools. The more inert the materials, the lower will be the tendency for reaction and adhesion to occur between the grain and the workpiece. For example, because aluminum oxide is relatively inert with respect to iron, its rate of attritious wear when it is used to grind steels is much lower than that for silicon carbide and diamond. On the other hand, silicon carbide can dissolve in iron, and hence it is not suitable for grinding steels. Cubic boron nitride has a higher inertness with respect to steels, and hence it is suitable for use as an abrasive. The selection of the type of abrasive for low attritious wear is, therefore, based on the reactivity of the grain and the workpiece and their relative mechanical properties, such as hardness and toughness. The environment and the type of grinding fluid used also have an influence on grit–workpiece-material interactions.

2. **Grain fracture.** Because abrasive grains are brittle, their fracture characteristics in grinding are important. If the wear flat caused by attritious wear is excessive, the grain becomes dull, and grinding becomes inefficient and produces undesirably high temperatures. Optimally, the grain should fracture or fragment at a moderate rate, so that new sharp cutting edges are produced continuously during grinding. This process is equivalent to breaking a piece of stone or dull chalk into two or more pieces in order to expose new sharp edges. Section 9.2 has described *friability* of abrasives, which gives them their self-sharpening characteristics, an important consideration in effective grinding.

   The selection of grain type and size for a particular application also depends on the attritious-wear rate. A grain–workpiece-material combination with high attritious wear and low friability causes dulling of grains and development of a large wear flat. Grinding then becomes inefficient, and surface damage such as burning is likely to occur.

   The following combinations are generally recommended:

   1. *Aluminum oxide*: for steels, ferrous alloys, and alloy steels.
   2. *Silicon carbide*: for cast iron, nonferrous metals, and hard and brittle materials, such as carbides, ceramics, marble, and glass.
   3. *Diamond*: for cemented carbide ceramics and some hardened steels.
   4. *Cubic boron nitride*: for steels and cast irons at 50 HRC or above and for high-temperature superalloys.

3. **Bond fracture.** The strength of the bond (*grade*) is a significant parameter in grinding. If the bond is too strong, dull grains cannot be dislodged so that other, sharp grains along the circumference of the grinding wheel can begin to contact the workpiece and remove chips; thus, the grinding process becomes inefficient. On the other hand, if the bond is too weak, the grains are dislodged easily, and the wear rate of the wheel increases; consequently, maintaining dimensional accuracy becomes difficult. In general, softer bonds are recommended for harder materials and for reducing residual stresses and thermal damage to the workpiece. Hard-grade wheels are used for softer materials and for removing large amounts of material at high rates.

9.5.1 Dressing, truing, and shaping of grinding wheels

Dressing is the process of conditioning worn grains on the surface of a grinding wheel in order to produce sharp new grains and for truing an out-of-round wheel. Dressing is necessary when excessive attritious wear dulls the wheel (a phenomenon called **glazing**, because of the shiny appearance of the wheel surface) or when the wheel becomes loaded. **Loading** occurs when the porosities on the grinding surfaces of the wheel (see Fig. 9.5) become filled or clogged with chips. Loading can occur in grinding soft materials or by improper selection of the grinding wheel, such as a wheel with low porosity, and improper selection of process parameters. A loaded wheel cuts very inefficiently, generating much frictional heat, causing surface damage and loss of dimensional accuracy.

Dressing is done by any of the following techniques:

1. A specially shaped diamond-point tool or diamond cluster is moved across the width of the grinding face of a rotating wheel, removing a small layer from the wheel surface with each pass. This method can be done either dry or wet, depending on whether the wheel is to be used dry or wet, respectively.
2. A set of star-shaped steel disks is manually pressed against the wheel. Material is removed from the wheel surface by crushing the grains. This method produces a coarse grinding surface on the wheel and is used only for rough grinding operations on bench or pedestal grinders.
3. Abrasive sticks may be held against the grinding surface of the wheel.
4. More recent developments in dressing techniques, for metal-bonded diamond wheels, involve the use of electrical-discharge and electrochemical material-removal techniques, which erode away very small layers of the metal bond, exposing new diamond cutting edges.

Dressing for *form grinding* involves **crush dressing**, or *crush forming*, and consists of pressing a metal roll on the surface of the grinding wheel, which is usually a vitrified wheel. The roll, which may be made of high-speed steel, tungsten carbide, or boron carbide, has a machined or ground profile and thus reproduces a replica of this profile on the surface of the grinding wheel being dressed. Dressing techniques and the frequency with which the wheel surface is dressed are significant, affecting grinding forces and surface finish. Modern computer-controlled grinders are equipped with automatic dressing features, which dress the wheel as grinding continues. Dressing can also be done to generate a certain shape or form on a grinding wheel for the purpose of grinding profiles on workpieces. (See Section 9.6.2.)

Truing is an operation by which a wheel is restored to its original shape; a round wheel is dressed to make its circumference a true circle (hence the word *truing*). Grinding wheels can also be *shaped* to the form to be ground on the workpiece. The grinding face on the Type 1 straight wheel shown in Fig. 9.2a is cylindrical and thus produces a flat surface; however, this surface can be shaped by dressing it to various forms. Although templates have been used for this purpose, new grinders are equipped with computer-controlled shaping features, whereby the diamond dressing tool traverses the wheel face automatically along a certain prescribed path (Fig. 9.11). Note that the axis of the diamond dressing tool remains normal to the wheel face at the point of contact.

9.5.2 Grinding ratio

Grinding-wheel wear is generally correlated with the amount of material ground by a parameter called the *grinding ratio*, G, which is defined as

$$G = \frac{\text{Volume of material removed}}{\text{Volume of wheel wear}}. \tag{9.10}$$

In practice, grinding ratios vary widely, ranging from 2 to 200 and higher, depending on the type of wheel, the workpiece material, the grinding fluid, and process

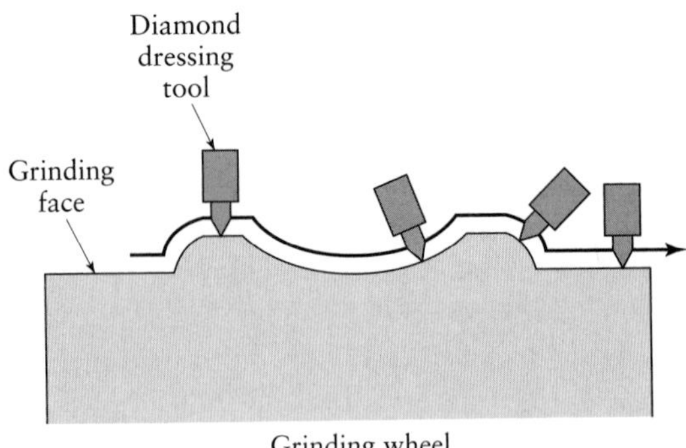

FIGURE 9.11 Shaping the grinding face of a wheel by dressing it with computer controlled shaping features. Note that the diamond dressing tool is normal to the surface at point of contact. *Source*: OKUMA America Corporation.

parameters such as depth of cut and speeds of the wheel and workpiece. Attempting to obtain a high grinding ratio in practice isn't always desirable, because high ratios may indicate grain dulling and possible surface damage. A lower ratio may be quite acceptable when an overall economic analysis justifies it.

Soft- or hard-acting wheels. During a grinding operation, a particular wheel may *act soft* (wear rate is high) or *hard* (wear rate is low), regardless of its grade. Note, for example, that an ordinary pencil acts soft when you write on rough paper and acts hard when you write on soft paper. This behavior is a function of the force on the grain. The greater the force, the greater is the tendency for the grains to fracture or be dislodged from the wheel surface, and hence the higher is the wheel wear and the lower is the grinding ratio. Equation (9.6) shows that the grain force increases with the strength of the workpiece material, the work speed, and the depth of cut and decreases with increasing wheel speed and wheel diameter. Thus, a wheel acts soft when v and d increase, or when V and D decrease.

EXAMPLE 9.3 Action of a grinding wheel

A surface-grinding operation is being carried out with the wheel running at a constant spindle speed. Will the wheel act soft or hard as the wheel wears down over a period of time?

Solution. Referring to Eq. (9.6), we note that the parameters that change with time in this operation are v and the wheel diameter D (assuming that the diameter remains constant and the wheel is dressed periodically). Hence, as D becomes smaller, the relative grain force increases, and therefore the wheel acts softer. Some grinding machines are equipped with variable-speed spindle motors to accommodate these changes and to make provisions for wheels of different diameter.

9.5.3 Wheel selection and grindability of materials

Selection of a grinding wheel for a certain application greatly influences the quality of surfaces produced, as well as the economics of the operation. The selection involves not only the shape of the wheel with respect to the shape of the part to be produced, but the characteristics of the workpiece material as well. The *grindability* of materials, like machinability (see Section 8.5), is difficult to define precisely. It is a general indication of how easy it is to grind a material and includes considerations such as the quality of the surface produced, surface finish, surface integrity, wheel wear, cycle time, and overall economics. It should also be pointed out that just as with machinability, grindability of a material can be enhanced greatly by proper selection of process parameters, the type of wheel, and grinding fluids, as well as machine characteristics and fixturing methods.

Grinding practices have been well established for a wide variety of metallic and nonmetallic materials, including newly developed materials for aerospace applications. Specific recommendations for selecting wheels and process parameters can be found in various handbooks. Some examples of such recommendations are C60-L6V for cast irons, A60-M6V for steels, C60-I9V or D150-R75B for carbides, and A60-K8V for titanium. Ceramics can be ground with relative ease by using diamond wheels, as well as by carefully selecting process parameters; a typical wheel selection for ceramics is D150-N50M.

It has been shown that with light passes and rigid machines with good damping capacity, it is possible to obtain continuous chips in grinding ceramics. (See Figs. 9.8 and 9.9.) This procedure is known as **ductile regime grinding**; it produces good workpiece surface integrity. However, ceramic chips are typically 1–10 μm (40–400 μin.) in size and hence are more difficult to remove from grinding fluids than metal chips, thus requiring fine filters and other effective methods.

9.6 Grinding Operations and Machines

Grinding operations are carried out with a variety of wheel–workpiece configurations. The selection of a grinding process for a particular application depends on part shape, part size, ease of fixturing, and the production rate required. The basic types of grinding operations are surface, cylindrical, internal, and centerless grinding. The relative movement of the wheel may be along the surface of the workpiece (*traverse grinding*, *through feed grinding*, or *cross-feeding*), or it may be radially into the workpiece (*plunge grinding*). Surface grinders constitute the largest percentage of grinders in use in industry, followed by bench grinders (usually with two wheels), cylindrical grinders, and tool and cutter grinders. The least used type of grinder is the internal grinder.

Grinding machines are available for various workpiece geometries and sizes. Modern grinding machines may be computer controlled, with features such as automatic workpiece loading and unloading, clamping, cycling, gaging, dressing, and wheel shaping. Grinders can also be equipped with probes and gages for determining the relative position of the wheel and workpiece surfaces, as well as with tactile sensing features whereby diamond-dressing-tool breakage, if any, can be detected during the dressing cycle. Because grinding wheels are brittle and are operated at high speeds, certain procedures must be carefully followed in their handling, storage, and use. Failure to follow these rules and the instructions and warnings printed on individual wheel labels may result in serious injury or death.

9.6.1 Surface grinding

Surface grinding involves grinding flat surfaces and is one of the most common grinding operations (Fig. 9.12). Typically, the workpiece is secured on a magnetic chuck attached to the worktable of a *surface grinder* (Fig. 9.13). Nonmagnetic materials generally are held by vises, special fixtures, vacuum chucks, or double-sided adhesive tapes. A straight wheel is mounted on the *horizontal spindle* of the grinder. Traverse grinding is done as the table reciprocates longitudinally and feeds laterally after each stroke. In *plunge grinding*, the wheel is moved radially into the workpiece, as in

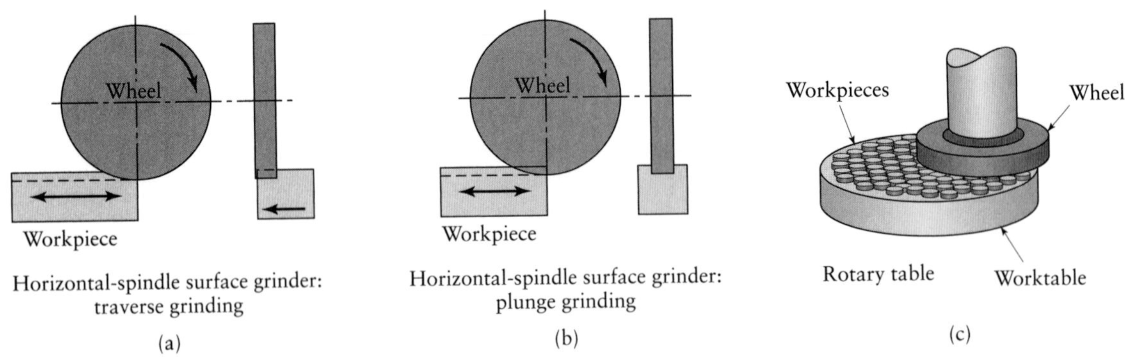

FIGURE 9.12 Schematic illustrations of surface-grinding operations. (a) Traverse grinding with a horizontal-spindle surface grinder. (b) Plunge grinding with a horizontal-spindle surface grinder, producing a groove in the workpiece. (c) Vertical-spindle rotary-table grinder (also known as the *Blanchard-type* grinder).

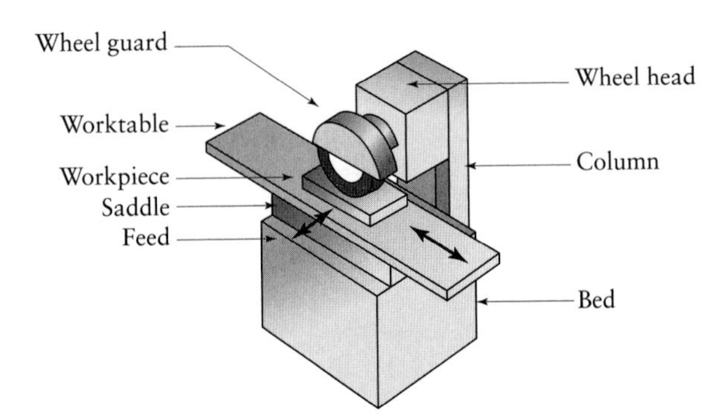

FIGURE 9.13 Schematic illustration of a horizontal-spindle surface grinder. The majority of grinding operations are done on such machines.

grinding a groove (Fig. 9.12b). The size of a surface grinder is determined by the surface dimensions that can be ground on the machine. Other types of setups for surface grinding have *vertical spindles* and *rotary tables* (Fig. 9.12c), also referred to as *Blanchard-type* grinders. These configurations allow a number of pieces to be ground in one setup.

9.6.2 Cylindrical grinding

In *cylindrical grinding*, also called *center-type grinding*, the workpiece's external cylindrical surfaces and shoulders are ground. Typical applications include crankshaft bearings, spindles, pins, bearing rings, and rolls for rolling mills. The rotating cylindrical workpiece reciprocates laterally along its axis, although in grinders used for large and long workpieces, the grinding wheel reciprocates. The latter design is called a *roll grinder* and is capable of grinding rolls as large as 1.8 m (72 in.) in diameter.

The workpiece in cylindrical grinding is held between centers or in a chuck, or it is mounted on a faceplate in the headstock of the grinder. For straight cylindrical surfaces, the axes of rotation of the wheel and workpiece are parallel. Separate motors drive the wheel and workpiece at different speeds. Long workpieces with two or more diameters are ground on cylindrical grinders. Cylindrical grinding can produce shapes in which the wheel is dressed to the form to be ground on the workpiece (*form grinding* and *plunge grinding*).

Cylindrical grinders are identified by the maximum diameter and length of the workpiece that can be ground, similar to engine lathes. In *universal grinders*, both the workpiece and the wheel axes can be moved and swiveled around a horizontal plane,

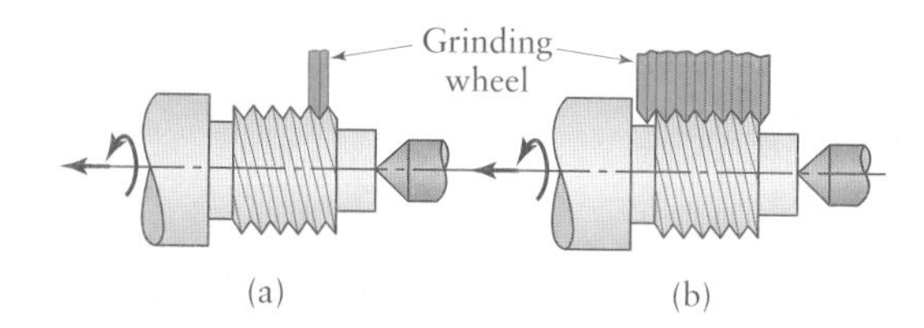

FIGURE 9.14 Thread grinding by (a) traverse and (b) plunge grinding.

thus permitting the grinding of tapers and other shapes. These machines are equipped with computer controls, thereby reducing labor and producing parts accurately and repetitively. Cylindrical grinders can also be equipped with computer-controlled features so that *noncylindrical* parts (such as cams) can be ground on rotating workpieces. The workpiece spindle speed is synchronized such that the distance between the workpiece and wheel axes is varied continuously to produce a particular shape.

Thread grinding is done on cylindrical grinders, as well as on centerless grinders (see Section 9.6.4), with specially dressed wheels that match the shape of the threads (Fig. 9.14). Although this process is costly, it produces threads more accurately than any other manufacturing process, and the threads have a very fine surface finish. The workpiece and wheel movements are synchronized to produce the pitch of the thread, usually in about six passes.

9.6.3 Internal grinding

In *internal grinding* (Fig. 9.15), a small wheel is used to grind the inside diameter of a part, such as bushings and bearing races. The workpiece is held in a rotating chuck, and the wheel rotates at 30,000 rpm or higher. Internal profiles can also be ground with profile-dressed wheels that move radially into the workpiece. The headstock of internal grinders can be swiveled on a horizontal plane to grind tapered holes.

9.6.4 Centerless grinding

Centerless grinding is a high-production process for continuously grinding cylindrical surfaces in which the workpiece is supported not by centers (hence the term *centerless*) or chucks, but by a blade (Figs. 9.16 and 9.17). Typical parts made by centerless grinding include roller bearings, piston pins, engine valves, camshafts, and similar components. This continuous-production process requires little operator skill. Parts with diameters as small as 0.1 mm (0.004 in.) can be ground using this process. Centerless grinders are now capable of wheel surface speeds on the order of 10,000 m/min (35,000 ft/min), using cubic-boron-nitride abrasive wheels.

In *through-feed grinding*, the workpiece is supported on a work-rest blade and is ground between two wheels. Grinding is done by the larger wheel, while the smaller wheel regulates the axial movement of the workpiece. The *regulating wheel*, which is rubber bonded, is tilted and runs at speeds of only about $\frac{1}{20}$ of those of the grinding wheel.

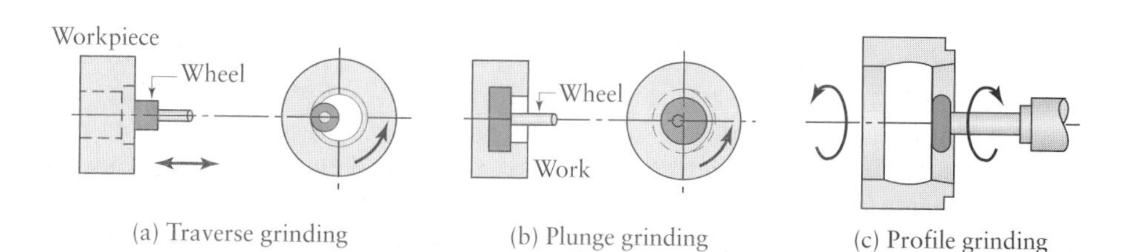

FIGURE 9.15 Schematic illustrations of internal-grinding operations.

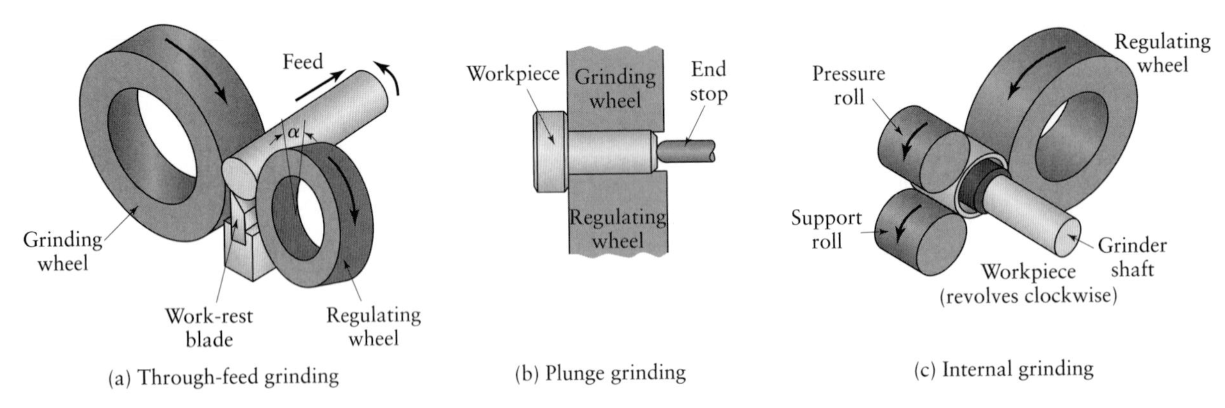

FIGURE 9.16 Schematic illustrations of centerless-grinding operations.

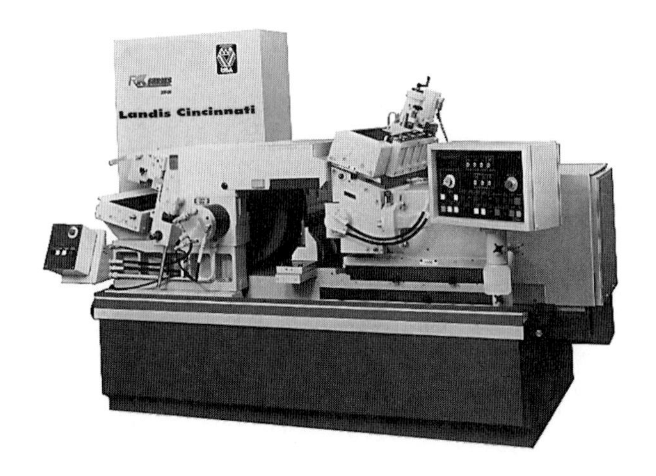

FIGURE 9.17 A computer-numerical-control centerless grinding machine. The movement of the workpiece is perpendicular to the page. *Source*: Courtesy of Cincinnati Milacron, Inc.

Parts with variable diameters, such as bolts, valve tappets, and distributor shafts, can be ground by centerless grinding; called *infeed*, or *plunge*, *grinding* ((Fig. 9.16b), the process is similar to plunge or form grinding with cylindrical grinders. Tapered pieces are centerless ground by *end-feed grinding*. High-production-rate thread grinding can be done with centerless grinders, using specially dressed wheels. In *internal centerless grinding*, the workpiece is supported between three rolls and is internally ground. Typical applications include sleeve-shaped parts and rings.

9.6.5 Other grinders

A variety of special-purpose grinders is available. *Bench grinders* are used for routine offhand grinding of tools and small parts. They are usually equipped with two wheels mounted on the two ends of the shaft of an electric motor. One wheel is usually coarse for rough grinding, and the other is fine for finish grinding. *Pedestal*, or *stand*, *grinders* are placed on the floor and used similarly to bench grinders.

Universal tool and cutter grinders are used for grinding single-point or multipoint tools and cutters. They are equipped with special workholding devices for accurate positioning of the tools to be ground. *Tool-post grinders* are self-contained units and are usually attached to the tool post of a lathe. The workpiece is mounted on the headstock and is ground by moving the tool post. These grinders are versatile, but the lathe should be protected from abrasive debris.

Swing-frame grinders are used in foundries for grinding large castings. Rough grinding of castings is called *snagging* and is usually done on floorstand grinders, using wheels as large as 0.9 m (36 in.) in diameter. *Portable grinders*, either air or electrically driven, or with a flexible shaft connected to an electric motor or gasoline engine, are available for operations such as grinding off weld beads and *cutting-off* operations using thin abrasive disks, usually on large workpieces.

9.6.6 Creep-feed grinding

Grinding has traditionally been associated with small rates of material removal and fine finishing operations; however, grinding can also be used for large-scale metal removal operations, similar to milling, broaching, and planing. In *creep-feed grinding*, developed in the late 1950s, the wheel depth of cut, d, is as much as 6 mm (0.25 in.), and the workpiece speed is low (Fig. 9.18). The wheels are mostly softer grade resin bonded with open structure to keep temperatures low and improve surface finish. Grinders with capabilities for continuously dressing the grinding wheel with a diamond roll are available. The machines used for creep-feed grinding have special features, such as high power of up to 225 kW (300 hp), high stiffness (because of the high forces due to the depth of material removed), high damping capacity, variable and well-controlled spindle and worktable speeds, and ample capacity for grinding fluids.

Its overall economics and competitive position with respect to other material removal processes indicate that creep-feed grinding can be economical for specific applications, such as in grinding shaped punches, key seats, twist-drill flutes, the roots of turbine blades (Fig. 9.18c), and various complex superalloy parts. The wheel is dressed to the shape of the workpiece to be produced. Consequently, the workpiece does not have to be previously milled, shaped, or broached; thus, near-net-shape castings and forgings are suitable parts for creep-feed grinding. Although a single pass generally is sufficient, a second pass may be necessary for improved surface finish.

9.6.7 Heavy stock removal by grinding

Grinding processes can be used for large-scale material removal by varying process parameters. However, this is a rough grinding operation, with possibly detrimental effects on the workpiece surface and its integrity. It can be economical in specific

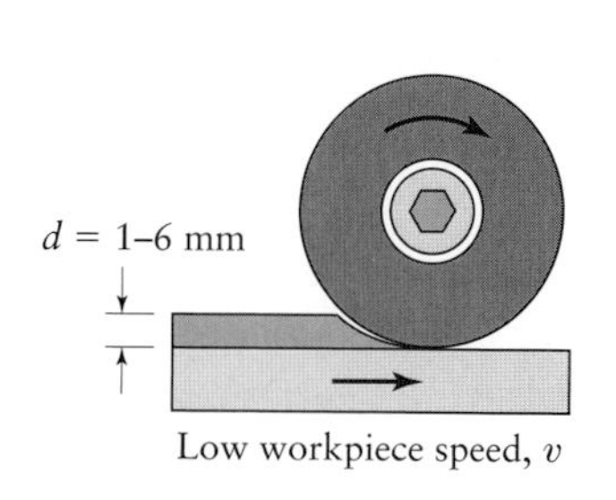

(a)

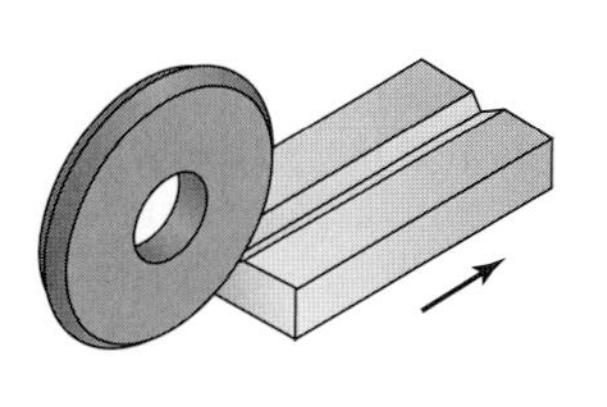

(b)

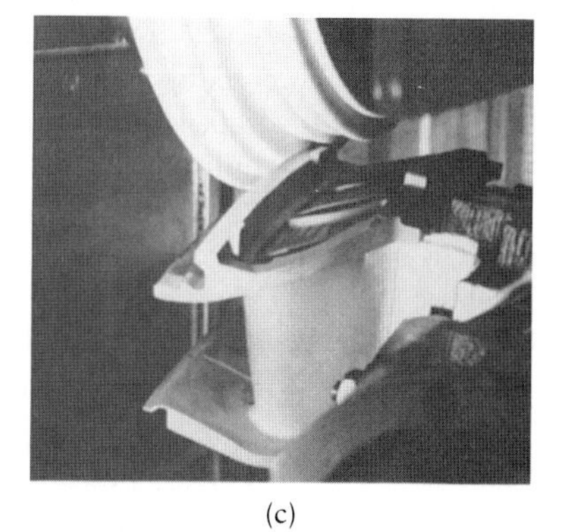

(c)

FIGURE 9.18 (a) Schematic illustration of the creep-feed grinding process. Note the large wheel depth of cut. (b) A shaped groove produced on a flat surface in one pass by creep-feed grinding. Groove depth can be on the order of a few mm. (c) An example of creep-feed grinding with a shaped wheel. *Source*: Courtesy of Blohm, Inc., and *Manufacturing Engineering*, Society of Manufacturing Engineers.

applications and compete favorably with machining processes, particularly milling, but also turning and broaching as well. In this operation, surface finish is of secondary importance, and the grinding wheel (or belt) is used to its maximum extent for minimum cost per piece. The geometric tolerances in this process are on the same order as those obtained by other machining processes. (See also Fig. 9.27.)

9.6.8 Grinding chatter

Chatter is particularly significant in grinding, because it adversely affects surface finish and wheel performance. Vibrations during grinding may be caused by bearings, spindles, and unbalanced wheels, as well as external sources, such as from nearby machinery; also, the grinding process can itself cause regenerative chatter. The analysis of chatter in grinding is similar to that for machining operations and involves *self-excited vibration* and *regenerative chatter* (see Section 8.11). Thus, the important variables are (a) stiffness of the machine tool and workholding devices and (b) damping. Additional factors that are unique to grinding chatter include nonuniformities in the grinding wheel itself, the dressing techniques used, and uneven wheel wear.

Because the foregoing variables produce characteristic **chatter marks** on ground surfaces, a study of these marks can often lead to the source of the problem. General guidelines have been established to reduce the tendency for chatter in grinding, such as (a) using soft-grade wheels, (b) dressing the wheel frequently, (c) changing dressing techniques, (d) reducing the material removal rate, and (e) supporting the workpiece rigidly. Advanced techniques include the use of sensors to monitor grinding chatter and control of it by chatter suppression systems.

9.6.9 Grinding fluids

The functions of grinding fluids are similar to those for cutting fluids (see also Section 8.7). Although grinding and other abrasive-removal processes can be performed dry, the use of a fluid is important and usually preferred, as the fluid prevents temperature rise in the workpiece and improves the part's surface finish and dimensional accuracy. Fluids also improve the efficiency of the operation by reducing wheel loading and wear and lowering power consumption.

Grinding fluids are typically *water-base emulsions* and *chemicals* and *synthetics; oils* may be used for thread grinding. The fluids may be applied as a stream (flood) or as mist, which is a mixture of fluid and air. Because of the high surface speeds involved, an air stream or air blanket around the periphery of the wheel usually prevents the fluid from reaching the cutting zone. Special *nozzles* that mate with the shape of the cutting surface of the grinding wheel have been designed for effective application of the grinding fluid under high pressure.

The temperature of water-base grinding fluids can rise significantly during their use as they remove heat from the grinding zone, in which case the workpiece can expand, making it difficult to control dimensional tolerances. The common method employed to maintain even temperature is to use refrigerating systems (chillers) through which the fluid is circulated. Also, as discussed with respect to cutting fluids in Section 8.7, the biological and ecological aspects, treatment, recycling, and disposal of grinding fluids are among the important considerations in their selection and use. The practices employed must comply with federal, state, and local laws and regulations.

Cryogenic grinding. As described in Section 8.7.2, a more recent technology is the use of liquid nitrogen as a coolant in grinding operations, largely in the interest of reducing or eliminating the adverse environmental impact of the use of metalworking fluids. With appropriate small-diameter nozzles, liquid nitrogen at around −200°C (−320°F) is injected into the wheel–workpiece grinding zone, thus reducing its temperature; as stated previously, high temperatures can have significant adverse effects on the part, such as poor surface finish and integrity. Experimental studies have indicated that, as compared with the use of traditional grinding fluids, use of cryogenic grinding is associated with a reduced level of surface (metallurgical) burn and oxidation, improved surface finish, lower tensile residual stress levels, and less loading of the grinding wheel (hence, less need for dressing). Cryogenic grinding appears to be particularly suitable for materials such as titanium, which has low thermal conductivity (see Table 3.2), low specific heat, and high reactivity. Other effects, such as on fatigue life of the ground pieces, as well as the economic advantages of cryogenic grinding, continue to be investigated.

EXAMPLE 9.4 Grinding vs. hard turning

Section 8.8.2 describes hard turning, an example of which is the machining of heat-treated steels (usually above 45 HRC), using a single-point polycrystalline cubic-boron-nitride cutting tool. In view of the discussions presented thus far in this chapter, it is evident that grinding and hard turning can be competitive in specific applications; consequently, there has been considerable debate regarding their respective merits.

Hard turning continues to be increasingly competitive with grinding, and dimensional tolerances and surface finish produced by hard turning are beginning to approach those for grinding. Turning requires much less energy than grinding (compare, for example, Tables 8.3 and 9.3), and with turning, thermal and other damage to the workpiece surface is less likely to occur, cutting fluids may not be necessary, and the machine tools are less expensive. In addition, finishing the part while it is still chucked in the lathe eliminates the need for material handling and setting the part in the grinder. On the other hand, workholding devices for large and slender workpieces for hard turning can present problems, since the cutting forces are higher than in grinding. Furthermore, tool wear and its control in hard turning can be a significant problem as compared with the automatic dressing of grinding wheels. It is thus evident that the competitive position of hard turning versus grinding must be evaluated individually for each application and in terms of product surface integrity, quality, and overall economics.

9.7 Ultrasonic Machining

In *ultrasonic machining* (UM), material is removed from a surface by microchipping or erosion with abrasive particles. The tip of the tool (Fig. 9.19a), called a *sonotrode*, vibrates at low amplitudes [0.05 to 0.125 mm (0.002 to 0.005 in.)] and at high frequency (20 kHz). This vibration, in turn, transmits a high velocity to fine abrasive grains between the tool and the surface of the workpiece. The grains are usually boron carbide, but aluminum oxide and silicon carbide are also used; grain size ranges from 100 (for roughing) to 1000 (for finishing). The grains are in a water slurry with concentrations ranging from 20% to 60% by volume; the slurry also carries away the debris from the cutting area.

Ultrasonic machining is best suited for hard and brittle materials, such as ceramics, carbides, glass, precious stones, and hardened steels. The tip of the tool is usually made of low-carbon steel and undergoes wear; it is attached to a transducer

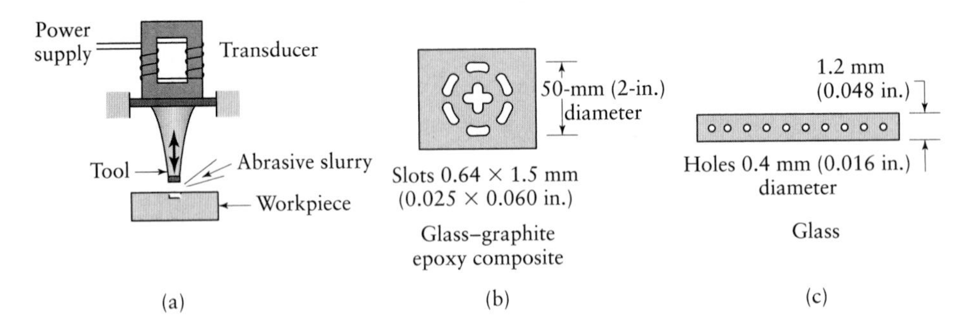

FIGURE 9.19 (a) Schematic illustration of the ultrasonic-machining process by which material is removed through microchipping and erosion. (b) and (c) Typical examples of holes produced by ultrasonic machining. Note the dimensions of cut and the types of workpiece materials.

through the toolholder. With fine abrasives, dimensional tolerances of 0.0125 mm (0.0005 in.) or better can be held in this process. Figures 9.19b and c show two applications of ultrasonic machining.

Microchipping in ultrasonic machining is possible because of the high stresses produced by particles striking a surface. The time of contact between the particle and the surface is very short (10 to 100μs), and the area of contact is very small. The time of contact, t_o, can be expressed as

$$t_o \simeq \frac{5r}{c_o}\left(\frac{c_o}{v}\right)^{1/5}, \tag{9.11}$$

where r is the radius of a spherical particle, c_o is the elastic wave velocity in the workpiece $\left(c_o = \sqrt{E/\rho}\right)$, and v is the velocity with which the particle strikes the surface. The force F of the particle on the surface is obtained from the rate of change of momentum. That is,

$$F = \frac{d(mv)}{dt}, \tag{9.12}$$

where m is the mass of the particle. The *average force*, F_{ave}, of a particle striking the surface and rebounding is

$$F_{ave} = \frac{2mv}{t_o}. \tag{9.13}$$

Substitution of numerical values into Eq. (9.13) indicates that even small particles can exert significant forces and, because of the very small contact area, produce very high stresses. In brittle materials, these stresses are sufficiently high to cause microchipping and surface erosion. (See also the discussion of *abrasive-jet machining* in Section 9.15.)

Rotary ultrasonic machining. In this process, the abrasive slurry is replaced by a tool with metal-bonded diamond abrasives that have been either impregnated or electroplated on the tool surface. The tool is rotated and ultrasonically vibrated, and the workpiece is pressed against it at a constant pressure; the process is similar to a face milling operation. (See Section 8.9.1.) This process is particularly effective in producing deep holes in ceramics and operating at high material removal rates.

9.8 Finishing Operations

In addition to the abrasive processes described thus far, several processes generally are used on workpieces as the final finishing operation. These processes use mainly abrasive grains. This section describes commonly used finishing operations, in the

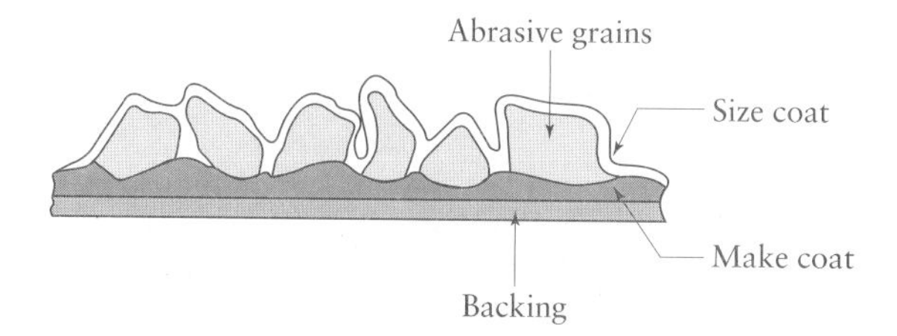

FIGURE 9.20 Schematic illustration of the structure of a coated abrasive. Sandpaper, developed in the 16th century, and emery cloth are common examples of coated abrasives.

order of improved surface finish produced. Finishing operations can contribute significantly to production time and product cost; they should therefore be specified with due consideration to their costs and benefits.

The commonly used finishing operations are described as follows:

1. **Coated abrasives.** Typical examples of *coated abrasives* include sandpaper and emery cloth. The grains used in coated abrasives are more pointed than those used for grinding wheels. The grains are electrostatically deposited on flexible backing materials, such as paper or cloth (Fig. 9.20), with their long axes perpendicular to the plane of the backing. The matrix (coating) is made of resins. Coated abrasives are available as sheets, belts, and disks and usually have a much more open structure than the abrasives used on grinding wheels. Coated abrasives are used extensively in finishing flat or curved surfaces of metallic and nonmetallic parts, in finishing metallographic specimens and in woodworking. The surface finish obtained depends primarily on grain size.

   Coated abrasives are also used as *belts* for high-rate material removal. **Belt grinding** has become an important production process and in some cases has replaced conventional grinding operations such as grinding of camshafts, with 8 to 16 lobes (of approximately elliptic shapes) per shaft. Belt speeds are usually in the range of 700–1800 m/min (2500–6000 ft/min) Machines for abrasive-belt operations require proper belt support and rigid construction to minimize vibrations.

   A more recent development is **microreplication**, in which aluminum-oxide abrasives in the shape of tiny pyramids are placed in a predetermined orderly arrangement on the belt surface. When they are used on stainless steels and superalloys, their performance is more consistent, and the temperature rise is lower then when using other coated abrasives. Typical applications include surgical implants, turbine blades, and medical and dental instruments.

2. **Wire brushing.** In the *wire-brushing* process, the workpiece is held against a circular wire brush that rotates at high speed. The tips of the wire produce longitudinal scratches on the workpiece surface. This process is used to produce a fine surface texture. Some efforts are taking place in developing wire brushing as a light material removal process.

3. **Honing.** *Honing* is an operation used primarily to give holes a fine surface finish. The honing tool (Fig. 9.21) consists of a set of aluminum-oxide or silicon-carbide bonded abrasives, called *stones*. The stones are mounted on a mandrel

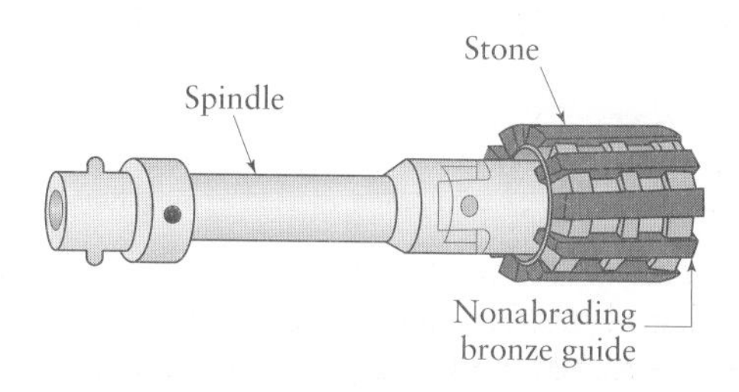

FIGURE 9.21 Schematic illustration of a honing tool to improve the surface finish of bored or ground holes.

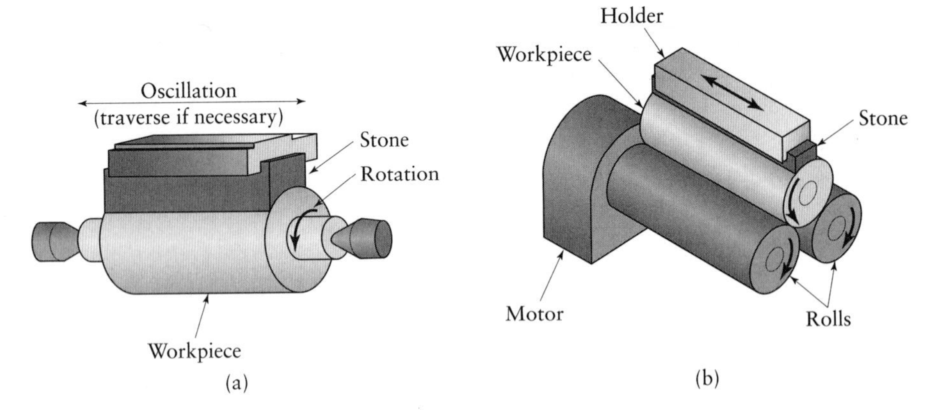

FIGURE 9.22 Schematic illustration of the superfinishing process for a cylindrical part: (a) cylindrical microhoning; (b) centerless microhoning.

that rotates in the hole, applying a radial force with a reciprocating axial motion, thus producing a crosshatched pattern. The stones can be adjusted radially for different hole sizes. The fineness of surface finish can be controlled by the type and size of abrasive used, the speed of rotation and the pressure applied. A fluid is used to remove chips and to keep temperatures low. If not implemented properly, honing can produce holes that are not straight and cylindrical, but rather that are bellmouthed, wavy, barrel shaped, or tapered. Honing is also done on external cylindrical or flat surfaces and to remove sharp edges on cutting tools and inserts. (See Fig. 8.34.)

In **superfinishing**, the pressure applied is very light, and the motion of the honing stone has a short stroke. The process is controlled so that the grains do not travel along the same path along the surface of the workpiece. Figure 9.22 shows examples of external superfinishing of a round part.

4. **Electrochemical honing.** This process combines the fine abrasive action of honing with electrochemical action. Although the equipment is costly, the process is as much as 5 times faster than conventional honing, and the tool lasts longer by as much as 10 times. Electrochemical honing is used primarily for finishing internal cylindrical surfaces.

5. **Lapping.** *Lapping* is a finishing operation used on flat or cylindrical surfaces. The lap (Fig. 9.23a) is usually made of cast iron, copper, leather, or cloth. The abrasive particles are embedded in the lap, or they may be carried through a slurry. Dimensional tolerances on the order of ± 0.0004 mm (± 0.000015 in.) can be obtained with the use of fine abrasives, up to size 900. Surface finish can be as smooth as $0.025–0.1\ \mu\text{m}(1–4\ \mu\text{in.})$. Production lapping on flat or cylindrical pieces is done on machines such as those shown in Figs. 9.23b and c. Lapping

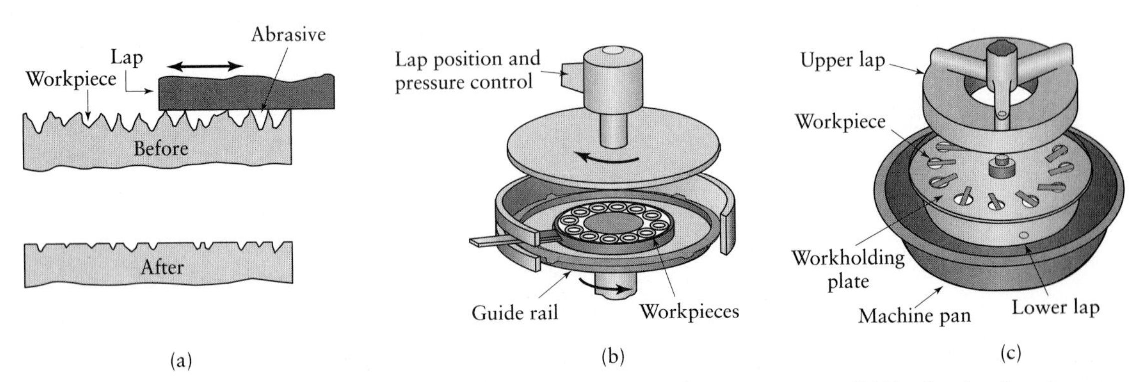

FIGURE 9.23 (a) Schematic illustration of the lapping process. (b) Production lapping on flat surfaces. (c) Production lapping on cylindrical surfaces.

is also done on curved surfaces, such as spherical objects and glass lenses, using specially shaped laps. Running-in of mating gears can be done by lapping. Depending on the hardness of the workpiece, lapping pressures range from 7–140 kPa (1–20 psi).

6. **Polishing.** *Polishing* is a process that produces a smooth, lustrous surface finish. Two basic mechanisms are involved in the polishing process: (1) fine-scale abrasive removal and (2) softening and smearing of surface layers by frictional heating. The shiny appearance of polished surfaces results from the smearing action. Polishing is done with disks or belts made of fabric, leather, or felt and coated with fine powders of aluminum oxide or diamond. Parts with irregular shapes, sharp corners, deep recesses, and sharp projections are difficult to polish.
7. **Buffing.** *Buffing* is similar to polishing, with the exception that very fine abrasives are used on soft disks made of cloth or hide. The abrasive is supplied externally from a stick of abrasive compound. Polished parts may be buffed to obtain an even finer surface finish.
8. **Electropolishing.** Mirrorlike finishes can be obtained on metal surfaces by *electropolishing*, a process that is the reverse of electroplating (see Section 4.5.1). Because there is no mechanical contact with the workpiece, this process is particularly suitable for polishing irregular shapes. The electrolyte attacks projections and peaks on the workpiece surface at a higher rate than for the rest of the surface, thus producing a smooth surface. Electropolishing is also used for deburring operations.
9. **Chemical mechanical polishing.** *Chemical mechanical polishing* is a process in which a chemically reactive surface is polished with a ceramic slurry in a sodium-hydroxide solution. Most of the material removal is due to chemical action, with the slurry serving mainly to ensure a good surface finish. A major application of this process is the polishing of silicon wafers. (See Section 13.9.)
10. **Polishing processes using magnetic fields.** A more recent development in polishing involves the use of magnetic fields to support abrasive slurries in finishing (polishing) ceramic balls and bearing rollers. **Magnetic float polishing** of ceramic balls is illustrated schematically in Fig. 9.24a. In this process, a

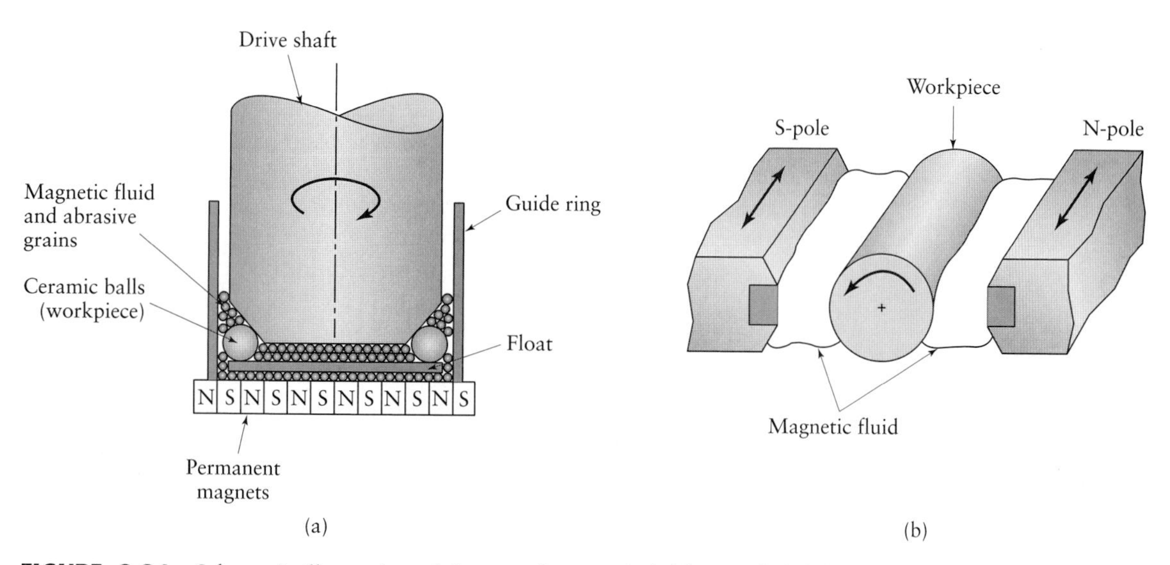

FIGURE 9.24 Schematic illustration of the use of magnetic fields to polish balls and rollers: (a) magnetic float polishing of ceramic balls; (b) magnetic-field-assisted polishing of rollers. *Source*: R. Komanduri, M. Doc, and M. Fox.

magnetic fluid, containing abrasive grains and extremely fine ferromagnetic particles in a carrier fluid such as water or kerosene, fills the chamber within a guide ring. The ceramic balls are located between a drive shaft and a float. The abrasive grains, ceramic balls, and the float (which is made of a nonmagnetic material) are all suspended by magnetic forces. The balls are pressed against the rotating drive shaft and are polished by the abrasive action. The forces applied by the abrasive particles on the balls are extremely small and controllable, and hence the polishing action is very fine. Polishing times are much lower than those for other polishing methods; thus, the process is very economical, and the surfaces produced have little or no defects.

Magnetic-field-assisted polishing of ceramic rollers is illustrated in Fig. 9.24b. In this process, a ceramic or steel roller (the workpiece) is clamped and rotated on a spindle. The magnetic poles are oscillated, introducing a vibratory motion to the magnetic–abrasive conglomerate. This action polishes the cylindrical roller surface. Bearing steels of 63 HRC have been mirror finished in 30 seconds by this process.

9.9 Deburring

Burrs are thin ridges, usually triangular in shape, that develop along the edges of a workpiece from shearing sheet materials (see Fig. 7.5), trimming forgings and castings, and machining. Burrs may interfere with the assembly of parts and can cause jamming of parts, misalignment, and short circuits in electrical components; furthermore, burrs may reduce the fatigue life of components. Because they are usually sharp, burrs can be a safety hazard to personnel. The need for deburring may be reduced by adding chamfers to sharp edges on parts. On the other hand, burrs on thin drilled or tapped components, such as tiny parts in watches, can provide extra thickness and hence improve the holding torque of screws.

Several *deburring processes* are available: (1) manually with files; (2) mechanically by cutting; (3) wire brushing; (4) flexible abrasive finishing, using rotary nylon brushes with the nylon filaments embedded with abrasive grits; (5) abrasive belts; (6) ultrasonics; (7) electropolishing; (8) electrochemical machining; (9) vibratory finishing; (10) shot blasting; (11) abrasive-flow machining; and (12) thermal energy. The last four processes are described as follows, and the other processes are covered in this and in other chapters:

1. **Vibratory** and **barrel-finishing** processes are used to improve the surface finish and remove burrs from large numbers of relatively small workpieces. In this batch-type operation, specially shaped *abrasive pellets* or media are placed in a container along with the parts to be deburred. The container is either vibrated or tumbled. The impact of individual abrasives and metal particles removes sharp edges and burrs from the parts. Depending on the application, this process is performed dry or wet, and liquid compounds may be added for requirements such as degreasing and providing resistance to corrosion.
2. **In shot blasting** (also called *grit blasting*), abrasive particles (usually sand) are propelled by a high-velocity jet of air or by a rotating wheel onto the surface of the workpiece. Shot blasting is particularly useful in deburring both metallic and nonmetallic materials and stripping, cleaning, and removing surface oxides. The surface produced has a matte finish. Small-scale polishing and etching can also be done by this process on bench-type units (*microabrasive blasting*).
3. **In abrasive-flow machining**, abrasive grains, such as silicon carbide or diamond, are mixed in a puttylike matrix, which is then forced back and forth through

the openings and passageways in the workpiece. The movement of the abrasive matrix under pressure erodes away burrs and sharp corners and polishes the part. The process is particularly suitable for workpieces with internal cavities that are inaccessible by other means. Pressures applied range from 0.7 MPa to 22 MPa (100 psi to 3200 psi). External surfaces can also be deburred using this process by containing the workpiece within a fixture that directs the abrasive media to the edges and areas to be deburred.

4. **The thermal-energy** method of deburring consists of placing the part in a chamber that is then injected with a mixture of natural gas and oxygen. This mixture is ignited, whereby a heat wave is produced with a temperature of 3300°C (6000°F). The burrs heat up instantly and are melted away, while the temperature of the part reaches only about 150°C (300°F). The process is effective in many applications on noncombustible parts. However, larger burrs or flashes tend to form beads after melting; the process can distort thin and slender parts; and the process does not polish or buff the workpiece surfaces as do many of the other deburring processes.

Robotic deburring. Deburring and flash removal from castings and forgings are now being performed increasingly by *programmable robots* (see Section 14.7), using a force-feedback system for control. This method eliminates tedious and costly manual labor and results in more consistent deburring.

9.10 | Chemical Machining

We know that certain chemicals attack metals and etch them, thereby removing small amounts of material from the surface. Thus, *chemical machining* (CM) was developed (Table 9.4), whereby material is removed from a surface by chemical dissolution, using chemical **reagents**, or **etchants**, such as acids and alkaline solutions. Chemical machining is the oldest of the nontraditional machining processes and has been used for many years for engraving metals and hard stones and, more recently, in the production of printed-circuit boards and microprocessor chips. Parts can also be deburred by chemical means.

9.10.1 Chemical milling

In *chemical milling*, shallow cavities are produced on sheets, plates, forgings, and extrusions for overall reduction of weight (Fig. 9.25). Chemical milling has been used on a wide variety of metals, with depths of removal to as much as 12 mm (0.5 in.). Selective attack by the chemical reagent on different areas of the workpiece surfaces is controlled by removable layers of *maskant* (Fig. 9.26) or by partial immersion in the reagent.

Chemical milling is used in the aerospace industry for removing shallow layers of material from large aircraft, missile skin panels, and extruded parts for airframes. Tank capacities for reagents are as large as 3.7 m × 15 m (12 ft × 50 ft). The process is also used to fabricate microelectronic devices (see Sections 13.8 and 13.13.) Figure 9.27 shows the range of surface finish and dimensional tolerances obtained by chemical machining and other machining processes. Some surface damage may result from chemical milling, because of preferential etching and intergranular attack (see Section 3.4), which adversely affect surface properties. Chemical milling of welded and brazed structures may produce uneven material removal. Chemical milling of castings may result in uneven surfaces caused by porosity and nonuniformity of the structure.

TABLE 9.4

General Characteristics of Advanced Machining Processes

Process	Characteristics	Process Parameters and Typical Material Removal Rate or Cutting Speed
Chemical machining (CM)	Shallow removal (up to 12 mm) on large flat or curved surfaces; blanking of thin sheets; low tooling and equipment cost; suitable for low production runs.	0.025–0.1 mm/min.
Electrochemical machining (ECM)	Complex shapes with deep cavities; highest rate of material removal; expensive tooling and equipment; high power consumption; medium to high production quantity.	V: 5–25 dc; A: 1.5–8 A/mm^2; 2.5–12 mm/min, depending on current density.
Electrochemical grinding (ECG)	Cutting off and sharpening hard materials, such as tungsten-carbide tools; also used as a honing process; higher material removal rate than grinding.	A: 1–3 A/mm^2; typically 1500 mm^3/min per 1000 A.
Electrical-discharge machining (EDM)	Shaping and cutting complex parts made of hard materials; some surface damage may result; also used for grinding and cutting; versatile; expensive tooling and equipment.	V: 50–380; A: 0.1–500; typically 300 mm^3/min.
Wire EDM	Contour cutting of flat or curved surfaces; expensive equipment.	Varies with workpiece material and its thickness.
Laser-beam machining (LBM)	Cutting and hole making on thin materials; heat-affected zone; does not require a vacuum; expensive equipment; consumes much energy; extreme caution required in use.	0.50–7.5 m/min.
Electron-beam machining (EBM)	Cutting and hole making on thin materials; very small holes and slots; heat-affected zone; requires a vacuum; expensive equipment.	1–2 mm^3/min.
Water-jet machining (WJM)	Cutting all types of nonmetallic materials to 25 mm (1 in.) and greater in thickness; suitable for contour cutting of flexible materials; no thermal damage; environmentally safe process.	Varies considerably with workpiece material.
Abrasive water-jet machining (AWJM)	Single or multilayer cutting of metallic and nonmetallic materials.	Up to 7.5 m/min.
Abrasive-jet machining (AJM)	Cutting, slotting, deburring, flash removal, etching, and cleaning of metallic and nonmetallic materials; tends to round off sharp edges; some hazard because of airborne particulates.	Varies considerably with workpiece material.

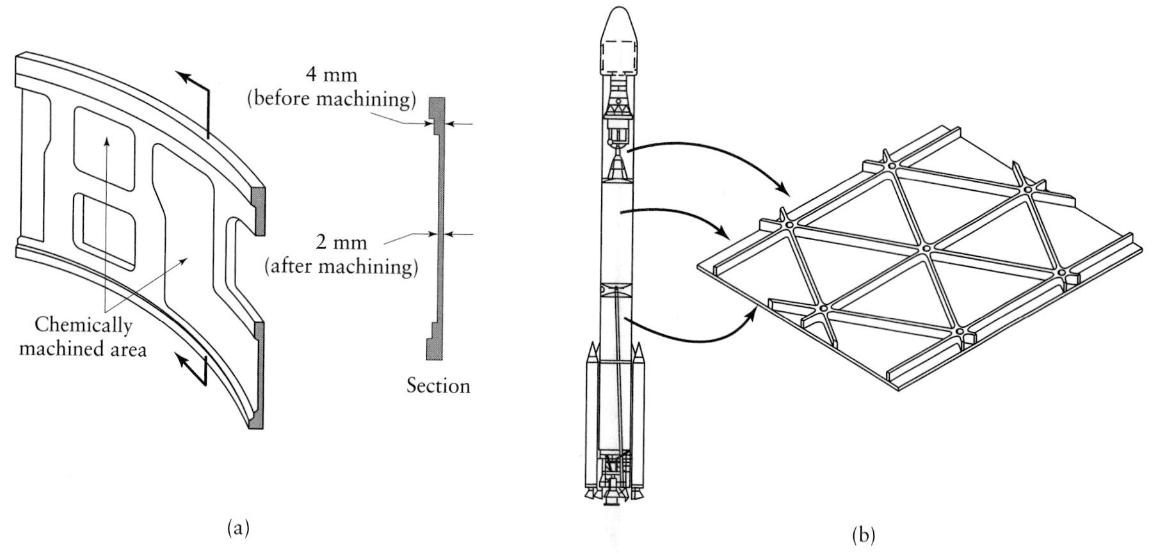

FIGURE 9.25 (a) Missile skin-panel section contoured by chemical milling to improve the stiffness-to-weight ratio of the part. (b) Weight reduction of space launch vehicles by chemical milling of aluminum-alloy plates. These panels are milled after the plates have first been formed into shape, such as by roll forming or stretch forming. The design of the chemically machined rib patterns can readily be modified at minimal cost. *Source: Advanced Materials and Processes*, ASM International, December 1990, p. 43.

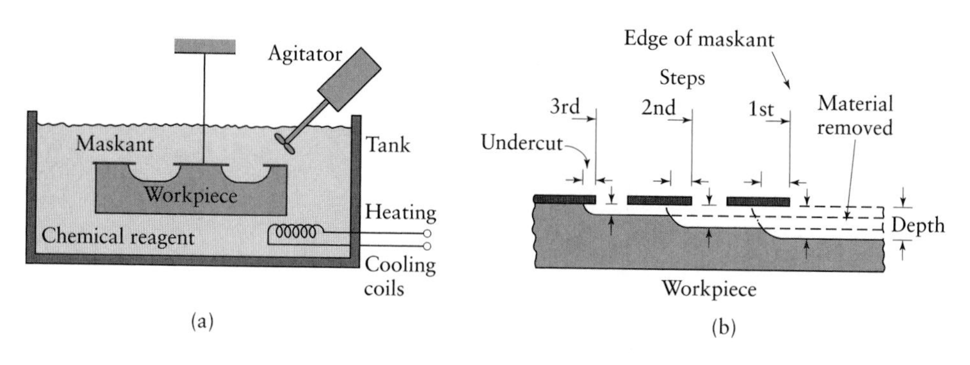

FIGURE 9.26 (a) Schematic illustration of the chemical-machining process. Note that no forces or machine tools are involved in this process. (b) Stages in producing a profiled cavity by chemical machining.

9.10.2 Chemical blanking

Chemical blanking is similar to blanking of sheet metal in that it is used to produce features that penetrate through the thickness of the material (see Fig. 7.8), with the exception that material is removed by chemical dissolution rather than by shearing. Typical applications for chemical blanking include burr-free etching of printed-circuit boards and the production of decorative panels, thin sheet-metal stampings, and complex or small shapes.

9.10.3 Photochemical blanking

Also called *photoetching*, *photochemical blanking* is a modification of chemical milling; material is removed, usually from flat thin sheet, by photographic techniques. Complex burr-free shapes can be blanked (Fig. 9.28) on metals as thin as 0.0025 mm (0.0001 in.); the process is also used for etching. Typical applications for photochemical blanking include fine screens, printed-circuit cards, electric-

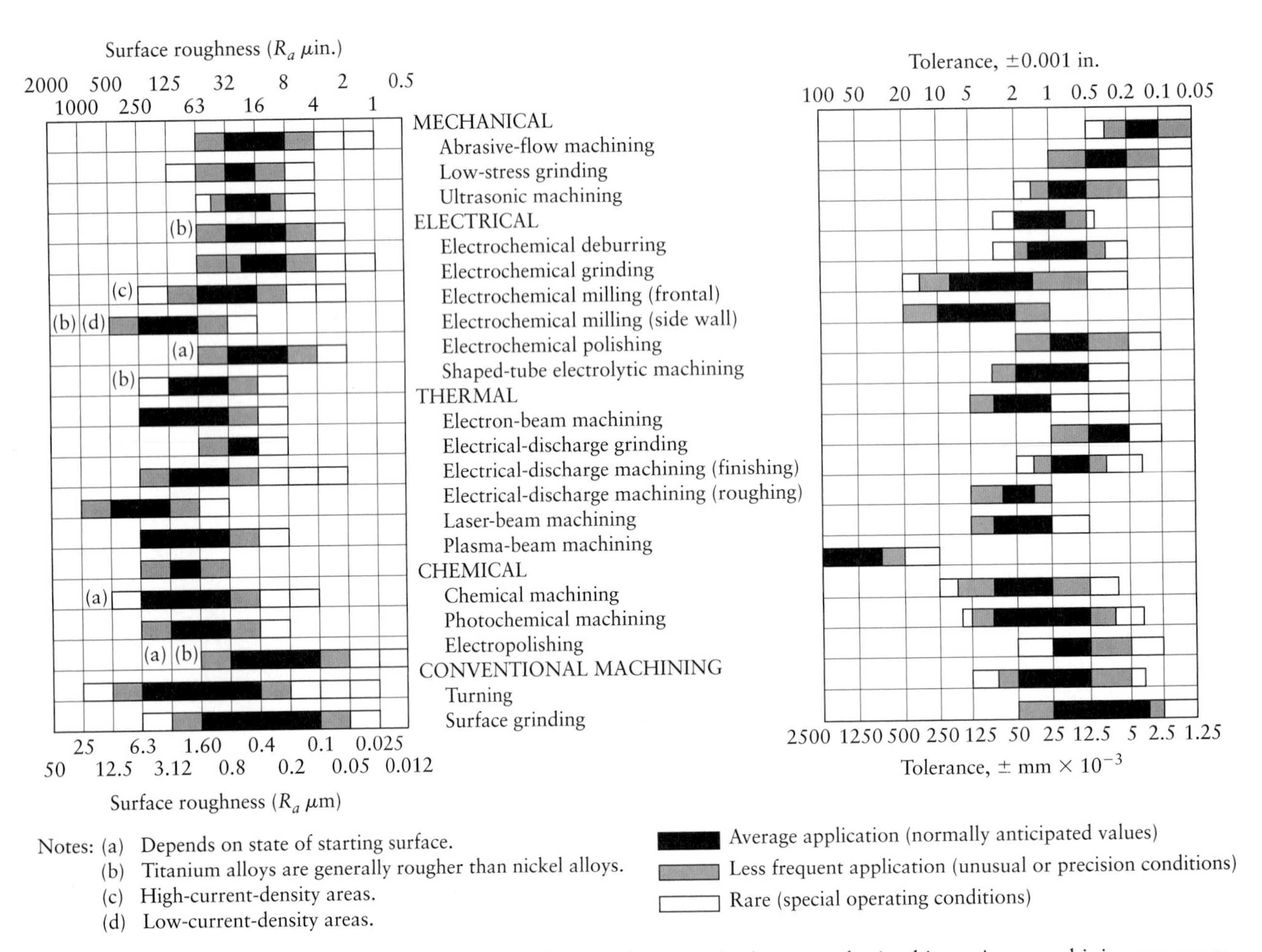

FIGURE 9.27 Surface roughness and tolerances obtained in various machining processes. Note the wide range within each process. (See also Fig. 8.27.) *Source*: Courtesy of TechSolve, formerly the Institute of Advanced Manufacturing Sciences'.

FIGURE 9.28 Various parts made by chemical blanking. Note the fine detail. *Source*: Courtesy of Buckabee-Mears St. Paul.

motor laminations, flat springs, and masks for color television. Although skilled labor is required for photochemical blanking, tooling costs are low, the process can be automated, and it is economical for medium to high production volume. Photochemical blanking is capable of forming very small parts for which traditional blanking dies (Section 7.3) are difficult to make. Also, the process is effective for blanking fragile workpieces and materials. Handling of chemical reagents requires precautions and special safety considerations to protect personnel against exposure to both liquid chemicals and volatile chemicals. Furthermore, disposal of chemical by-products from this process is a major consideration, although some by-products can be recycled.

9.11 Electrochemical Machining

Electrochemical machining (ECM) is basically the reverse of electroplating. An electrolyte (Fig. 9.29) acts as current carrier, and the high rate of electrolyte movement in the tool–workpiece gap washes metal ions away from the workpiece (*anode*) before they have a chance to plate onto the tool (*cathode*). Note that the cavity produced is the female mating image of the tool. Modifications of this process are used for turning, facing, slotting, trepanning, and profiling operations in which the electrode becomes the cutting tool.

The shaped tool is generally made of brass, copper, bronze, or stainless steel. The **electrolyte** is a highly conductive inorganic salt solution, such as sodium chloride mixed in water or sodium nitrate; it is pumped at a high rate through the passages in the tool. A dc power supply in the range of 5–25 V maintains current densities, which for most applications are 1.5–8 A/mm^2 (1000–5000 A/in^2) of active machined surface. Machines that have current capacities as high as 40,000 A and as small as 5 A are available. Because the metal removal rate is only a function of ion exchange rate, it is not affected by the strength, hardness, or toughness of the workpiece, which must be electrically conductive.

The material removal rate by electrochemical machining can be calculated from the equation

$$\text{MRR} = CI\eta, \tag{9.14}$$

where MRR = mm^3/min; I = current in amperes; and η = current efficiency, which typically ranges from 90% to 100%. C is a material constant in mm^3/A-min and, for pure metals, depends on valence; the higher the valence, the lower is its value. If a

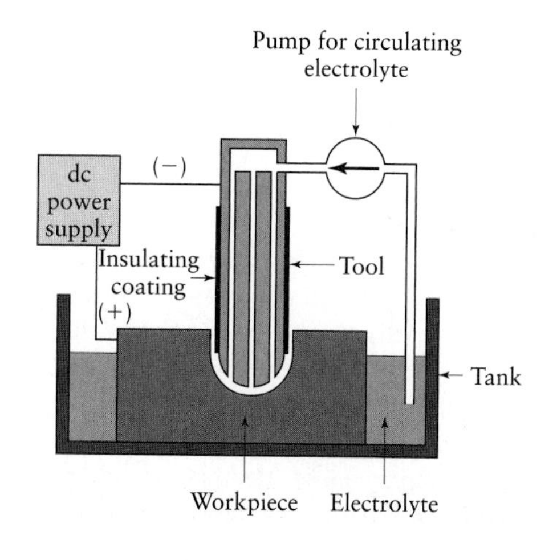

FIGURE 9.29 Schematic illustration of the electrochemical-machining process. This process is the reverse of electroplating, described in Section 4.5.1.

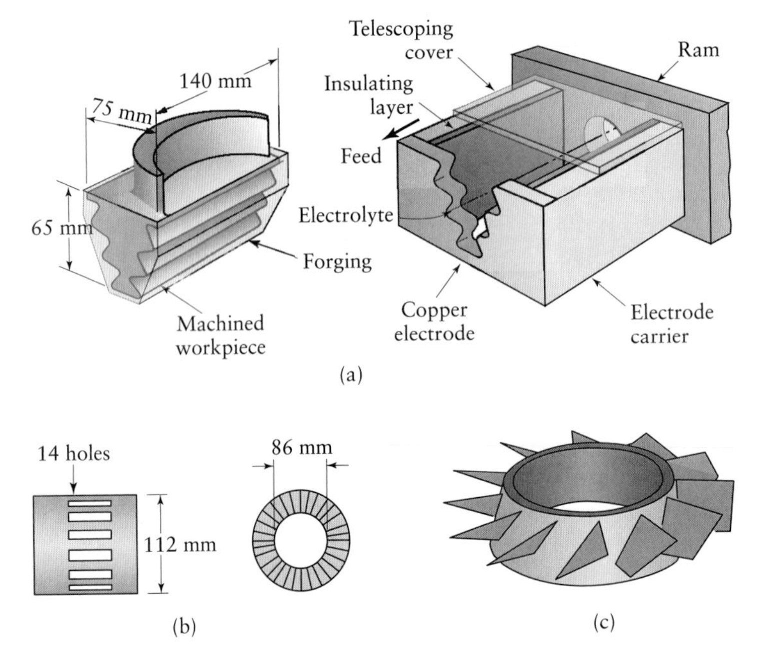

FIGURE 9.30 Typical parts made by electrochemical machining. (a) Turbine blade made of a nickel alloy, 360 HB. *Source: Metal Handbook*, 9th ed., Vol. 3, Materials Park, OH: ASM International, 1980, p. 849. (b) Thin slots on a 4340-steel roller-bearing cage. (c) Integral airfoils on a compressor disk.

cavity of uniform cross-sectional area A_o mm^2 is being electrochemically machined, the feed rate, f, in mm/min would be

$$f = \frac{\text{MRR}}{A_o}. \tag{9.15}$$

Note that the feed rate is the speed at which the electrode is penetrating the workpiece. For most metals, the value of C typically ranges between about 1 and 2.

Electrochemical machining is generally used for machining complex cavities in high-strength materials, particularly in the aerospace industry for mass production of turbine blades, jet-engine parts, and nozzles (Fig. 9.30). It is also used for machining forging-die cavities and producing small holes. The ECM process leaves a burr-free surface; in fact, it can also be used as a deburring process. It does not cause any thermal damage to the part, and the lack of tool forces prevents distortion of the part. Furthermore, there is no tool wear, and the process is capable of producing complex shapes as well as machining hard materials. However, the mechanical properties of components made by ECM should be compared carefully with those produced by other material removal methods. Electrochemical-machining systems are available as numerically controlled machining centers, with the capability for high production rates, high flexibility, and maintenance of close dimensional tolerances.

Pulsed electrochemical machining. *Pulsed electrochemical machining* (PECM) is a refinement of ECM; it uses very high current densities (on the order of 100 A/cm^2), but the current is pulsed rather than direct. The purpose of pulsing is to eliminate the need for high electrolyte flow rates, which limits the usefulness of ECM in die and mold making (die sinking). Investigations have shown that PECM improves fatigue life, and the process has been proposed as a possible method for eliminating the recast layer left on die and mold surfaces by electrical discharge machining. (See Section 9.13.)

9.12 Electrochemical Grinding

Electrochemical grinding (ECG) combines electrochemical machining with conventional grinding (Table 9.4). The equipment used in electrochemical grinding is similar to a conventional grinder, except that the wheel is a rotating cathode with abrasive particles (Fig. 9.31a). The wheel is metal bonded with diamond or aluminum-oxide abrasives and rotates at a surface speed of 1200–2000 m/min (4000–7000 ft/min.) The abrasives serve as insulators between the wheel and the workpiece and mechanically remove electrolytic products from the working area. A flow of electrolyte, usually sodium nitrate, is provided for the electrochemical-machining phase of the operation. The majority of metal removal in ECG is by electrolytic action, and typically less than 5% of metal is removed by the abrasive action of the wheel. Therefore, wheel wear is very low, and current densities range from 1–3 A/mm^2 (500–2000 A/in^2). Finishing cuts are usually made by the grinding action, but only to produce a surface with good finish and dimensional accuracy.

The material removal rate in electrochemical grinding can be calculated from the equation

$$\text{MRR} = \frac{GI}{\rho F} \tag{9.16}$$

where MRR is in mm^3/min, G = equivalent weight in grams, I = current in amperes, ρ = density in g/mm^3, and F = *Faraday's constant* in Coulombs. The speed of penetration, V_s, of the grinding wheel into the workpiece is given by the equation

$$V_s = \left(\frac{G}{\rho F}\right)\left(\frac{E}{gK_p}\right)K \tag{9.17}$$

where V_s is in mm^3/min; E = cell voltage in volts; g = wheel–workpiece gap in mm; K_p = coefficient of loss, which is in the range of 1.5 to 3; and K = electrolyte conductivity in Ω^{-1}mm^{-1}.

Electrochemical grinding is suitable for applications similar to those for milling, grinding, and sawing (Fig. 9.31b). It is not adaptable to cavity-sinking operations, such as die making; it has been successfully applied to carbides and high-strength alloys. The process offers a distinct advantage over traditional diamond-wheel grinding when processing very hard materials, for which wheel wear can be high. ECG machines are available with numerical controls, thus improving dimensional accuracy and providing repeatability and increased productivity.

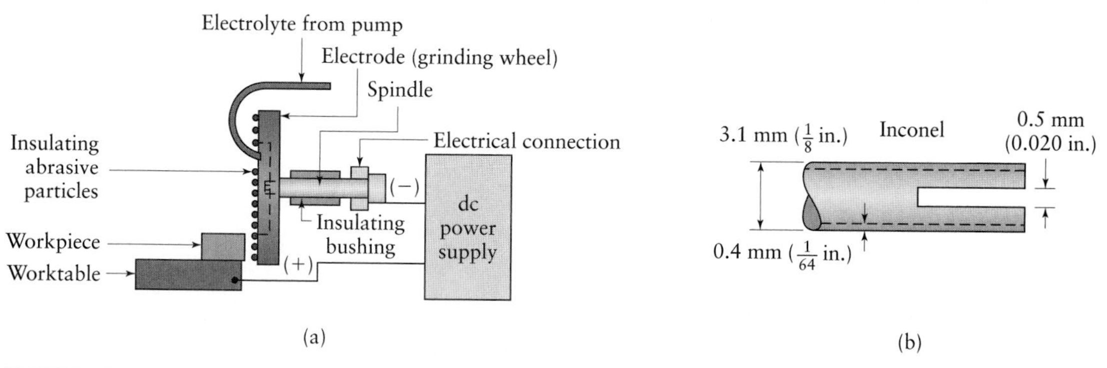

FIGURE 9.31 (a) Schematic illustration of the electrochemical-grinding process. (b) Thin slot produced on a round nickel-alloy tube by this process.

EXAMPLE 9.5 Machining time in electrochemical machining vs. drilling

A round hole 12.5 mm in diameter is being produced in a titanium-alloy block by electrochemical machining. Using a current density of 6 A/mm^2, estimate the time required for machining a 20-mm-deep hole. Assume that the efficiency is 90%. Compare this time with that required for ordinary drilling.

Solution. We note from Eqs. (9.14) and (9.15) that the feed rate can be expressed by the equation

$$f = \frac{CI\eta}{A_o}.$$

Letting $C = 1.6\ \text{mm}^3/\text{A}\cdot\text{min}$ and $I/A_o = 6\ \text{A/mm}^2$, we find that the feed rate is $f = (1.6)(6)(0.9) = 8.64$ mm/min. Since the hole is 20 mm deep,

$$\text{Machining time} = \frac{20}{8.64} = 2.3\,\text{min}.$$

To determine the drilling time, we refer to Table 8.10 and note the data for titanium alloys. Let's select the following values for a 12.5-mm drill: drill rpm = 300 and feed = 0.15 mm/rev. From these values, we calculate that the feed rate will be (300 rev/min)(0.15 mm/rev) = 45 mm/min. Since the hole is 20 mm deep,

$$\text{Drilling time} = \frac{20}{45} = 0.45\ \text{min},$$

which is about 20% of the time for ECM.

9.13 Electrical-Discharge Machining

The principle of *electrical-discharge machining* (EDM), also called *electrodischarge* or *sparkerosion machining*, is based on erosion of metals by spark discharges (Table 9.4). We know that when two current-conducting wires are allowed to touch each other, an arc is produced. If we look closely at the point of contact between the two wires, we note that a small portion of the metal has been eroded away, leaving a small crater. Although this phenomenon has been known since the discovery of electricity, it was not until the 1940s that a machining process based on this principle was developed. The EDM process has become one of the most important and widely accepted production technologies in manufacturing industries.

The EDM system (Fig. 9.32) consists of a shaped tool (**electrode**) and the workpiece, which are connected to a dc power supply and placed in a **dielectric** (electrically

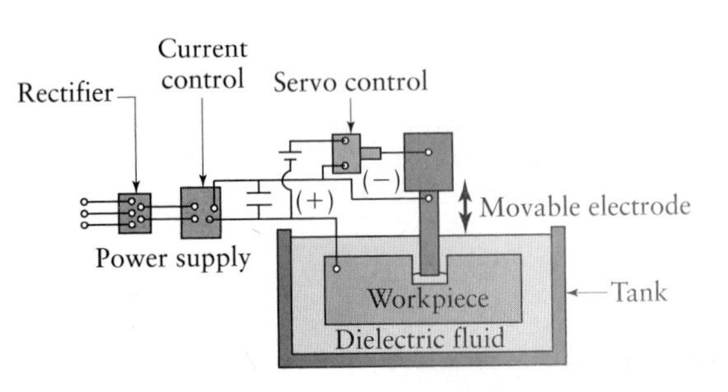

FIGURE 9.32 Schematic illustration of the electrical-discharge-machining process.

nonconducting) **fluid**. When a voltage is applied to the tool, a magnetic field causes suspended particles in the dielectric fluid to concentrate, eventually forming a bridge for current to flow to the workpiece. An intense electrical arc is then generated, causing sufficient heating to melt a portion of the workpiece and usually some of the tooling material as well. In addition, the dielectric fluid is heated rapidly, causing evaporation of the fluid in the arc gap; this evaporation, in turn, increases the resistance of the interface, until the arc can no longer be maintained. Once the arc is interrupted, heat is removed from the gas bubble by the surrounding dielectric fluid, and the bubble collapses (cavitates). The associated shock wave and flow of dielectric fluid flush debris from the surface and entrain any molten workpiece material into the dielectric fluid. The capacitor discharge is repeated at rates of between 50 kHz and 500 kHz, with voltages usually ranging between 50 V and 380 V and currents from 0.1 A to 500 A.

The dielectric fluid (1) acts as an insulator until the potential is sufficiently high, (2) acts as a flushing medium and carries away the debris in the gap, and (3) provides a cooling medium. The gap between the tool and the workpiece (called *overcut*) is critical; thus, the downward feed of the tool is controlled by a servomechanism, which automatically maintains a constant gap. The most common dielectric fluids are mineral oils, although kerosene and distilled and deionized water may be used in specialized applications. The workpiece is fixtured within the tank containing the dielectric fluid, and its movements are controlled by numerically controlled systems. The machines are equipped with a pump and filtering system for the dielectric fluid.

The EDM process can be used on any material that is an electrical conductor. The melting point and latent heat of melting are important physical properties that determine the volume of metal removed per discharge; as these values increase, the rate of material removal slows. The volume of material removed per discharge is typically in the range of 10^{-6} to 10^{-4} mm^3 (10^{-10} to 10^{-8} in^3). Since the process doesn't involve mechanical energy, the hardness, strength, and toughness of the workpiece material do not necessarily influence the removal rate. The frequency of discharge or the energy per discharge is usually varied to control the removal rate, as are the voltage and current. The rate and surface roughness increase with increasing current density and decreasing frequency of sparks.

Electrodes for EDM are usually made of graphite, although brass, copper, or copper–tungsten alloy may be used. The tools are shaped by forming, casting, powder metallurgy, or machining. Electrodes as small as 0.1 mm (0.005 in.) in diameter have been used. *Tool wear* is an important factor, since it affects dimensional accuracy and the shape produced. Tool wear can be minimized by reversing the polarity and using copper tools, a process called **no-wear EDM**.

Electrical-discharge machining has numerous applications, such as producing die cavities for large automotive-body components (die-sinking machining centers); narrow slots; turbine blades; various intricate shapes (Figs. 9.33a and b); and small-diameter deep holes (Fig. 9.33c), using tungsten wire as the electrode. Stepped cavities can be produced by controlling the relative movements of the workpiece in relation to the electrode (Fig. 9.34).

The material removal rate in electrical-discharge machining is basically a function of the current and the melting point of the workpiece material, although other process variables also have an effect. The following approximate empirical relationship can be used as a guide to estimate the metal removal rate in EDM:

$$\text{MRR} = 4 \times 10^4 I T_w^{-1.23}. \qquad (9.18)$$

In this equation, MRR is in mm^3/min, I is the current in amperes, and T_w is the melting point of the workpiece in °C.

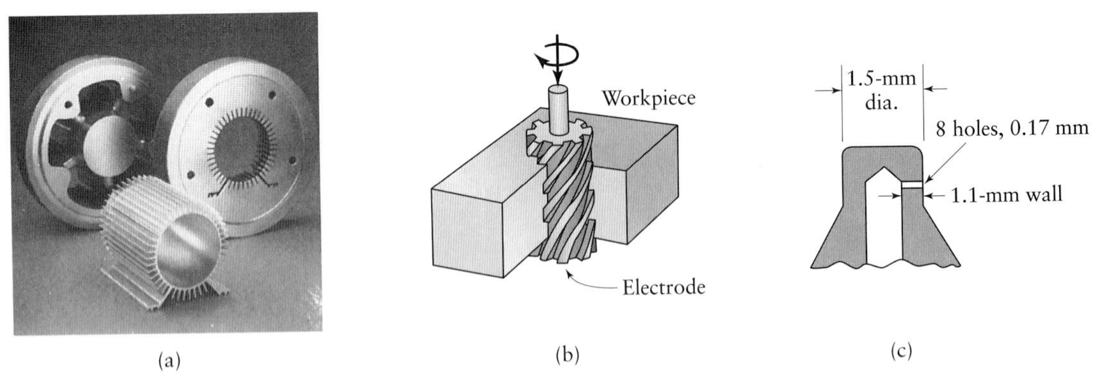

FIGURE 9.33 (a) Examples of cavities produced by the electrical-discharge-machining process, using shaped electrodes. The two round parts (rear) are the set of dies for extruding the aluminum piece shown in front. *Source*: Courtesy of AGIE USA Ltd. (b) A spiral cavity produced by a rotating electrode. *Source*: *American Machinist*. (c) Holes in a fuel-injection nozzle made by electrical-discharge machining. Material: Heat-treated steel.

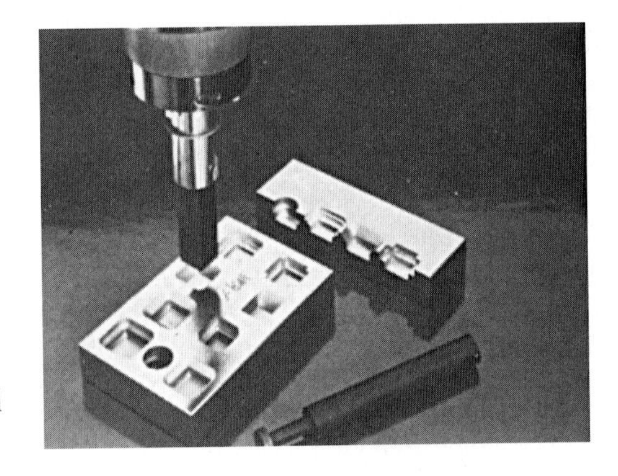

FIGURE 9.34 Stepped cavities produced with a square electrode by EDM. The workpiece moves in the two principal horizontal directions, and its motion is synchronized with the downward movement of the electrode to produce various cavities. Also shown is a round electrode capable of producing round or elliptical cavities. *Source*: Courtesy of AGIE USA Ltd.

The wear rate of the electrode, W_t, can be estimated from the empirical equation

$$W_t = 1.1 \times 10^{11} I T_t^{-2.38} \tag{9.19}$$

where W_t is mm^3/min and T_t is the melting point of the electrode material in °C. The wear ratio of the workpiece to electrode, R, can be estimated from the expression

$$R = 2.25\, T_r^{-2.38} \tag{9.20}$$

where T_r is the ratio of workpiece to electrode melting points. In practice, the value of R varies over a wide range, between 0.2 and 100. (Note that in abrasive processing, grinding ratios also are in the same range; see Section 9.5.2.)

Metal removal rates usually range from 2 to 400 mm^3/min. High rates produce a very rough finish, with a molten and resolidified (recast) structure with poor surface integrity and low fatigue properties; thus, finishing cuts are made at low metal removal rates, or the recast layer is removed later by finishing operations. A more recent technique is to use an oscillating electrode, which provides very fine surface finish, requiring significantly less benchwork to improve lustrous cavities.

EXAMPLE 9.6 Machining time in electrical-discharge machining vs. drilling

Calculate the machining time for producing the hole in Example 9.5 by EDM, and compare the time with that for drilling and for ECM. Assume that the titanium alloy has a melting point of 1600°C (see Table 3.2) and that the current is 100 A. Also calculate the wear rate of the electrode, assuming that the melting point of the electrode is 1100°C.

Solution. Using Eq. (9.18), we find that

$$\text{MRR} = (4 \times 10^4)(100)(1600^{-1.23}) = 458\ \text{mm}^3/\text{min}.$$

The volume of the hole is

$$V = \pi\left[\frac{(12.5)^2}{4}\right](20) = 2454\ \text{mm}^3.$$

Hence, the machining time for EDM is 2454/458 = 5.4 min; this time is 2.35 times that for ECM and 11.3 times that for drilling. Note, however, that if the current is increased to 300 A, then the machining time for EDM will only be 1.8 minutes, which is less than the time for ECM. The wear rate of the electrode is calculated using Eq. (9.19). Thus,

$$W_t = (11 \times 10^3)(100)(1100^{-2.38}) = 0.064\ \text{mm}^3/\text{min}.$$

9.13.1 Electrical-discharge grinding

The grinding wheel in *electrical-discharge grinding* (EDG) is made of graphite or brass and contains no abrasives. Material is removed from the surface of the workpiece by repetitive spark discharges between the rotating wheel and the workpiece. The material removal rate can be estimated from the equation

$$\text{MRR} = KI, \tag{9.21}$$

where MRR is in mm^3/min, I = current in amperes, and K = workpiece material factor in mm^3/A-min; for example, K = 16 for steel and 4 for tungsten carbide.

The EDG process can be combined with electrochemical grinding; the process is then called **electrochemical-discharge grinding** (ECDG). Material is removed by chemical action, with the electrical discharges from the graphite wheel breaking up the oxide film, which is washed away by the electrolyte flow. The process is used primarily for grinding carbide tools and dies, but can also be used for grinding fragile parts, such as surgical needles, thin-walled tubes, and honeycomb structures. The ECDG process is faster than EDG, but power consumption for ECDG is higher.

In **sawing** with EDM, a setup similar to a band or circular saw (but without any teeth) is used with the same electrical circuit as in EDM. Narrow cuts can be made at high rates of metal removal. Because the cutting forces are negligible, the process can be used on slender components.

9.13.2 Wire EDM

A variation of EDM is *wire EDM* (Fig. 9.35), or *electrical-discharge wire cutting*. In this process, which is similar to contour cutting with a band saw (Section 8.9.5), a slowly moving wire travels along a prescribed path, cutting the workpiece, with the discharge sparks acting like cutting teeth. This process is used to cut plates as thick

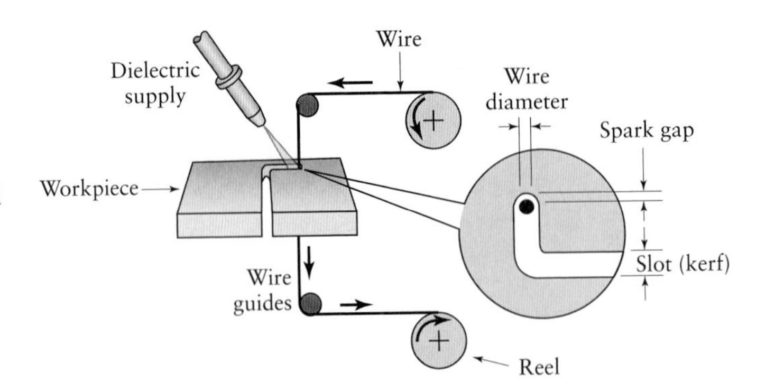

FIGURE 9.35 Schematic illustration of the wire EDM process. As much as 50 hours of machining can be performed with one reel of wire, which is then discarded or recycled.

as 300 mm (12 in.) and for making punches, tools, and dies from hard metals; it can also cut intricate components for the electronics industry (Table 9.4).

The wire is usually made of brass, copper, or tungsten and is typically about 0.25 mm (0.01 in.) in diameter, making narrow cuts possible; zinc- or brass-coated and multicoated wires are also used. The wire should have sufficient tensile strength and fracture toughness, as well as high electrical conductivity and capacity to flush away the debris produced during cutting. The wire is generally used only once, as it is relatively inexpensive. It travels at a constant velocity in the range of 0.15 to 9 m/min (6 to 360 in./min), and a constant gap (*kerf*) is maintained during the cut. The cutting speed is generally given in terms of the cross-sectional area cut per unit time; some typical examples are 18,000 mm^2/hr (28 in^2/hr) for 50-mm-thick (2-in-thick) D2 tool steel, and 45,000 mm^2/hr (70 in^2/hr) for 150-mm-thick (6-in-thick) aluminum. These removal rates indicate a linear cutting speed of 18,000/50 = 360 mm/hr = 6 mm/min, and 45,000/150 = 300 mm/hr = 5 mm/min, respectively.

The material removal rate for wire EDM can be obtained from the expression

$$\text{MRR} = V_f h b, \tag{9.22}$$

where MRR is in mm^3/min; V_f = feed rate of the wire into the workpiece in mm/min; h = workpiece thickness or height in mm; and the kerf (see Fig. 9.35) is denoted as $b = d_w + 2s$, where d_w = wire diameter in mm and s = gap, in mm, between wire and workpiece during machining.

Modern wire-EDM machines (**multiaxis EDM wire-cutting machining centers**) are equipped with (1) computer controls to control the cutting path of the wire, (2) automatic self-threading features in case of wire breakage, (3) multiheads for cutting two parts at the same time, (4) controls that prevent wire breakage, and (5) programmed machining strategies. Two-axis computer-controlled machines can produce cylindrical shapes in a manner similar to that of a turning operation or cylindrical grinding.

9.14 High-Energy-Beam Machining

9.14.1 Laser-beam machining

In *laser-beam machining* (LBM), the source of energy is a laser (an acronym for Light Amplification by Stimulated Emission of Radiation), which focuses optical energy on the surface of the workpiece (Fig. 9.36a). The highly focused, high-density energy melts and evaporates portions of the workpiece in a controlled manner. This process, which does not require a vacuum, is used to machine a variety of metallic and nonmetallic materials. There are several types of lasers used in manufacturing operations:

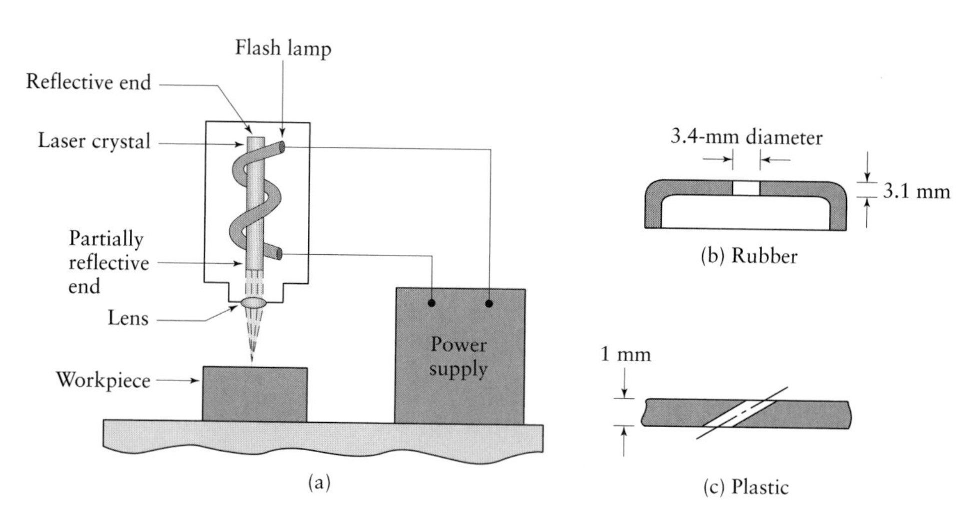

FIGURE 9.36 (a) Schematic illustration of the laser-beam-machining process. (b) and (c) Examples of holes produced in nonmetallic parts by LBM.

CO_2 (pulsed or continuous wave), Nd:YAG (neodymium: yttrium–aluminum–garnet), Nd:glass, ruby, and excimer (from the words *exci*ted and di*mer*, meaning two mers, or two molecules of the same chemical composition). The typical applications of laser-beam machining are outlined in Tables 9.4 and 9.5.

Important physical parameters in LBM are the *reflectivity* and *thermal conductivity* of the workpiece surface and its specific heat and latent heats of melting and evaporation; the lower these quantities, the more efficient is the process. The cutting depth t may be expressed as

$$t \propto \frac{P}{vd}, \tag{9.23}$$

where P is the power input, v is the cutting speed, and D is the laser-beam spot diameter. The surface produced by LBM is usually rough and has a heat-affected zone that, in critical applications, may have to be removed or heat treated. Kerf width is an important consideration, as it is in other cutting processes such as sawing, wire EDM, and electron-beam machining.

TABLE 9.5

General Applications of Lasers in Manufacturing

Application	Laser Type
Cutting	
Metals	PCO_2; $CWCO_2$; Nd:YAG; ruby
Plastics	$CWCO_2$
Ceramics	PCO_2
Drilling	
Metals	PCO_2; Nd:YAG; Nd:glass; ruby
Plastics	Excimer
Marking	
Metals	PCO_2; Nd:YAG
Plastics	Excimer
Ceramics	Excimer
Surface treatment (metals)	$CWCO_2$
Welding (metals)	PCO_2; $CWCO_2$; Nd:YAG; Nd:glass; ruby

Note: P = pulsed; CW = continuous wave.

Laser beams may be used in combination with a gas stream, such as oxygen, nitrogen, or argon (*laser-beam torch*), for cutting thin sheet materials. High-pressure inert-gas-assisted (e.g., nitrogen) laser cutting is used for stainless steel and aluminum; it leaves an oxide-free edge that can improve weldability. Gas streams also have the important function of blowing away molten and vaporized material from the workpiece surface.

Laser-beam machining is used widely in cutting and drilling metals, nonmetals, and composite materials (Figs. 9.36b and c). The abrasive nature of composite materials and the cleanliness of the operation have made laser-beam machining an attractive alternative to traditional machining methods. Holes as small as 0.005 mm (0.0002 in.), with hole depth-to-diameter ratios of 50 to 1, have been produced in various materials, although a more practical minimum is 0.025 mm (0.001 in.). Extreme caution should be exercised with lasers, as even low-power lasers can cause damage to the retina of the eye if proper precautions are not observed.

Laser-beam machining is being used increasingly in the electronics and automotive industries. Bleeder holes for fuel-pump covers and lubrication holes in transmission hubs, for example, are being drilled; the cooling holes in the first-stage vanes of the Boeing 747 jet engines are also produced by lasers. Significant cost savings have been achieved by laser-beam machining, such that it now competes with electrical-discharge machining in this respect.

Laser beams are also used for (1) *welding* (see Section 12.4), (2) small-scale *heat treating* of metals and ceramics to modify their surface mechanical and tribological properties (see Section 4.4), and (3) *marking* of parts with letters, numbers, codes, etc. Marking can be done by processes such as with ink; with mechanical devices such as punches, pins, styli, scroll rolls, or stamps; by etching; and with lasers. Although the equipment is more expensive than for other methods, marking and engraving with lasers has become increasingly common, due to the process' accuracy, reproducibility, flexibility, ease of automation, and on-line application in manufacturing.

The inherent flexibility of the laser-beam cutting process with fiber-optic-beam delivery, simple fixturing, low setup times, the availability of multi-kW machines and 2D and 3D computer-controlled systems are attractive features; thus, laser cutting can compete successfully with cutting of sheet metal using the traditional punching processes described in Section 7.3. There are now efforts to combine the two processes for improved overall efficiency.

9.14.2 Electron-beam machining and plasma-arc cutting

The source of energy in *electron-beam machining* (EBM) is high-velocity electrons, which strike the surface of the workpiece and generate heat (Fig. 9.37). The applications of this process are similar to those of laser-beam machining, except that EBM

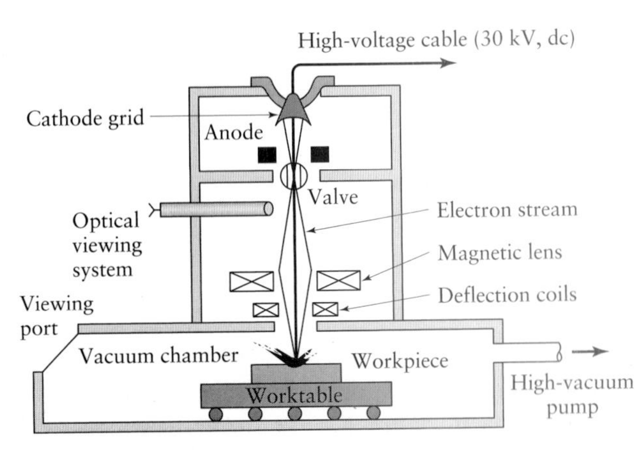

FIGURE 9.37 Schematic illustration of the electron-beam-machining process. Unlike LBM, this process requires a vacuum, and hence workpiece size is limited.

requires a vacuum. These machines use voltages in the range of 50–200 kV to accelerate the electrons to speeds of 50% to 80% of the speed of light. The interaction of the electron beam with the workpiece surface produces hazardous X rays; consequently, the equipment should be used only by highly trained personnel. Electron-beam machining is used for very accurate cutting of a wide variety of metals. Surface finish is better and kerf width is narrower than that for other thermal-cutting processes. (See also the discussion in Section 12.4 on electron-beam welding.)

In **plasma-arc cutting** (PAC), *plasma beams* (ionized gas) are used for rapid cutting of nonferrous and stainless-steel plates. The temperatures generated are very high (9400°C, or 17,000°F, in the torch for oxygen as a plasma gas); consequently, the process is fast, kerf width is small, and the surface finish is good. Material removal rates are much higher than in the EDM and LBM processes, and parts can be machined with good reproducibility. Plasma-arc cutting is highly automated today, through the use of programmable controls.

A more traditional method is **oxyfuel-gas cutting** (OFC), using a torch as in welding. It is particularly useful for cutting steels, cast irons, and cast steels. Cutting occurs mainly by oxidation and burning of the steel, with some melting taking place as well. Kerf widths are usually in the range of 1.5–10 mm (0.06–0.4 in.).

9.15 Water-Jet, Abrasive Water-Jet, and Abrasive-Jet Machining

We know that when we put our hand across a jet of water or air, we feel a considerable concentrated force acting on our hand. This force results from the momentum change of the stream and, in fact, is the principle on which the operation of water or gas turbines is based. This principle is also used in water-jet, abrasive water-jet, and abrasive-jet machining, described as follows:

1. **Water-jet machining** (WJM). In water-jet machining, also called *hydrodynamic machining* (Fig. 9.38a), the force from the jet is used in cutting and deburring operations. The water jet acts like a saw and cuts a narrow groove

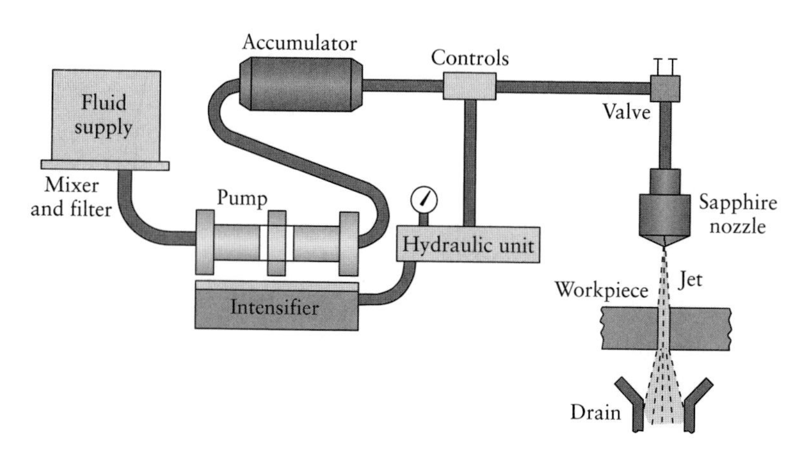

(a)

(b)

FIGURE 9.38 (a) Schematic illustration of water-jet machining. (b) Examples of various nonmetallic parts cut by a water-jet machine. *Source*: Jet Edge.

in the material. Although pressures as high as 1400 MPa (200 ksi) can be generated, a pressure level of about 400 MPa (60 ksi) is generally used for efficient operation. Jet-nozzle diameters usually range between 0.05 mm and 1 mm (0.002 in. and 0.040 in.).

A variety of materials can be cut with this technique, including plastics, fabrics, rubber, wood products, paper, leather, insulating materials, brick, and composite materials (Fig. 9.38b). Thicknesses range up to 25 mm (1 in.) and higher. Vinyl and foam coverings for some automobile dashboards, for example, are cut using multiple-axis, robot-guided water-jet machining equipment. Because it is an efficient and clean operation compared with other cutting processes, WJM is also used in the food-processing industry for cutting and slicing food products (Table 9.4).

The advantages of this process are that cuts can be started at any location without the need for predrilled holes; no heat is produced; no deflection of the rest of the workpiece takes place (hence the process is suitable for flexible materials); little wetting of the workpiece takes place; and the burr produced is minimal. It is also an environmentally safe manufacturing process.

2. **Abrasive water-jet machining.** In *abrasive water-jet machining* (AWJM), the water jet contains abrasive particles such as silicon carbide or aluminum oxide, thus increasing the material removal rate over that of water-jet machining. Metallic, nonmetallic, and advanced composite materials of various thicknesses can be cut in single or multiplelayers, particularly heat-sensitive materials that cannot be machined by processes in which heat is produced. Cutting speeds are as high as 7.5 m/min (2.5 ft/min) for reinforced plastics, but much lower for metals. With multiple-axis and robotic-control machines, complex three-dimensional parts can be machined to finish dimensions by this process. Nozzle life is improved by making the nozzle from rubies, sapphires, and carbide-base composite materials.

3. **Abrasive-Jet Machining.** In *abrasive-jet machining* (AJM), a high-velocity jet of dry air, nitrogen, or carbon dioxide containing abrasive particles is aimed at the workpiece surface under controlled conditions (Fig. 9.39). The impact of the particles develops sufficient concentrated force (see also Section 9.7) to perform operations such as (1) cutting small holes, slots, or intricate patterns in very hard or brittle metallic and nonmetallic materials; (2) deburring or removing small flash from parts; (3) trimming and beveling; (4) removing oxides and other surface films; and (5) general cleaning of components with irregular surfaces.

   The gas supply pressure is on the order of 850 kPa (125 psi), and the abrasive-jet velocity can be as high as 300 m/s (100 ft/s) and is controlled by a valve. The handheld nozzles are usually made of tungsten carbide or sapphire.

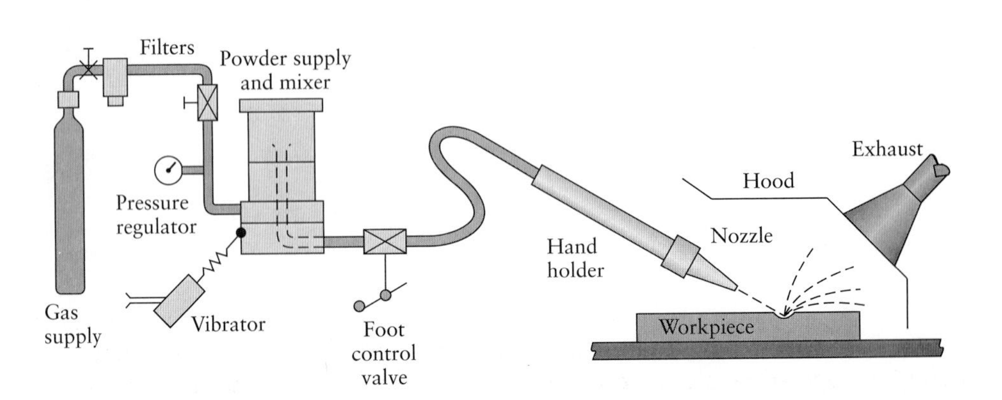

FIGURE 9.39 Schematic illustration of the abrasive-jet-machining process.

The abrasive size is in the range of 10–50 μm (400–2000 μin.). Because the flow of the free abrasives tends to round off corners, designs for abrasive-jet machining should avoid sharp corners; also, holes made in metal parts tend to be tapered. Because of airborne particulates, there is some hazard involved in using this process.

9.16 Design Considerations

The important design considerations for the processes and operations described in this chapter are outlined as follows:

1. **Abrasive processes.** Parts to be ground should be designed so that they can be held securely in workholding devices without causing part distortion. For high dimensional accuracy, interrupted surface features should be avoided, as they could cause vibration and chatter. In cylindrical grinding, parts should be balanced, and long, slender designs should be avoided to minimize deflections. Fillets and corner radii should be as large as possible, or relief should be provided for by prior machining. Designs requiring accurate form grinding should be kept as simple as possible to avoid frequent wheel dressing. Deep and small holes, and blind holes requiring internal grinding, should be avoided, or they should include a relief. In ultrasonic machining, sharp profiles should be avoided, and the bottom of brittle parts should be supported to avoid chipping; some taper should be expected for holes made by this process.
2. **Chemical, electrochemical, and electrical processes.** In chemical machining, sharp corners and deep and narrow cavities should be avoided; undercuts may occur under the edges of the maskant; the bulk of the workpiece should preferably be preshaped by other processes to reduce machining time; and size changes in artwork due to humidity and temperature can affect dimensional accuracy. In electrochemical machining, sharp corners or flat bottoms should be avoided; designs should make provision for a small taper for holes and cavities; and irregular cavities may not be produced to the desired dimensions. In electrical-discharge machining, the bulk of the workpiece should preferably be preshaped by other processes to reduce machining time; deep slots and narrow openings should be avoided; and, for economic production, the surface finish specified should not be too fine.
3. **High-energy-beam processes.** Any adverse effects of the processes on surface finish and integrity should be considered. In laser-beam machining, dull and unpolished workpiece surfaces are preferable, and designs should avoid deep cuts or very sharp profiles. In electron-beam machining, part size and shape are limited by the size of the vacuum chamber, and if a part requires machining on only a small portion, consideration should be given to making the part in a number of small components.

9.17 Process Economics

This chapter has shown that grinding may be used both as a finishing operation and for large-scale removal operations (as in creep-feed grinding). The use of grinding as a finishing operation is often necessary, because forming and machining processes

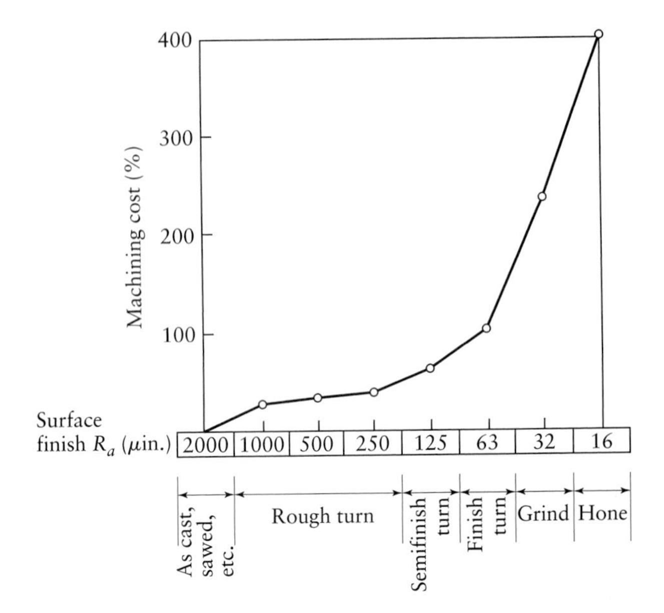

FIGURE 9.40 Increase in the cost of machining and finishing a part as a function of the surface finish required.

alone usually cannot produce parts with the desired dimensional accuracy and surface finish. However, because it is an additional operation, grinding contributes significantly to product cost. Creep-feed grinding, on the other hand, has proved to be an economical alternative to machining operations such as milling, even though wheel wear is high.

All finishing operations contribute to product cost. On the basis of the discussion thus far, it can be seen that as the surface finish improves, more operations are required, and hence the cost increases. Note in Fig. 9.40 how rapidly cost increases as surface finish is improved by processes such as grinding and honing.

Much progress has been made in automating the equipment involved in finishing operations, including computer controls. Consequently, labor costs and production times have been reduced, even though such machinery may require significant capital investment. If finishing is likely to be an important factor in manufacturing a particular product, the conceptual and design stages should involve an analysis of the degree of surface finish and dimensional accuracy required. Furthermore, all processes that precede finishing operations should be analyzed for their capability to produce a more acceptable surface finish and dimensional accuracy. This can be accomplished through proper selection of tools and process parameters and the characteristics of the machine tools involved.

The economic production run for a particular process depends on the cost of tooling and equipment, the material removal rate, operating costs, and the level of operator skill required, as well as the cost of secondary and finishing operations that may be necessary. In chemical machining, the costs of reagents, maskants, and disposal, together with the cost of cleaning the parts, are important factors; in electrical-discharge machining, the cost of electrodes and the need to replace them periodically are significant. The rate of material removal, and hence the production rate, can vary significantly in these processes. The cost of tooling and equipment also varies significantly, as does the operator skill required. If possible, the high capital investment for machines such as for electrical and high-energy-beam machining should be justified in terms of the production runs they are able to generate and the lack of feasibility of manufacturing the same part by other means.

CASE STUDY | Manufacture of Stents

Heart attacks, strokes, and other cardiovascular diseases claim one life every 33 seconds in the United States alone.[1] These illnesses are attributed mostly to coronary artery disease, which is the gradual buildup of fat (cholesterol) within the artery wall causing the coronary arteries to become narrowed or blocked. This condition reduces the blood flow to the heart muscle and eventually leads to a heart attack, stroke, or other cardiovascular diseases. One of the most popular methods today for keeping blocked arteries open is to implant a stent into the artery. The MULTI-LINK TETRA™, shown in Fig. 9.41, consists of a tiny mesh tube that is expanded using a balloon dilatation catheter and implanted into a blocked or partially blocked coronary artery. A stent serves as a scaffolding or mechanical brace to keep the artery open. A stent offers the patient a minimally invasive method for treating coronary heart disease for which the alternative is usually open-heart bypass surgery, a procedure with greater risk, pain, rehabilitation time, and cost to the patient.

Stent manufacturing is extremely demanding for which the accuracy and precision of its design are paramount for proper performance. The manufacture of the stent must allow satisfaction of all of the design constraints and provide an extremely reliable device; thus, strict quality control procedures must be in place throughout the manufacturing process. When designing a stent there are many material-selection factors to consider, such as radial strength, corrosion resistance, endurance limit, flexibility, and biocompatibility. Radial strength is important because the stent must withstand the pressure the artery exerts on the stent upon expansion. Because the stent is implantable in the body it must be corrosion resistant; it must also be able to withstand the stress undulations caused by heartbeats, and consequently, the material must also have high resistance to fatigue. In addition, the stent must have the appropriate wall thickness in order to be flexible enough to negotiate through the tortuous anatomy of the heart. Above all, the material must be biocompatible to the body, because it remains there for the rest

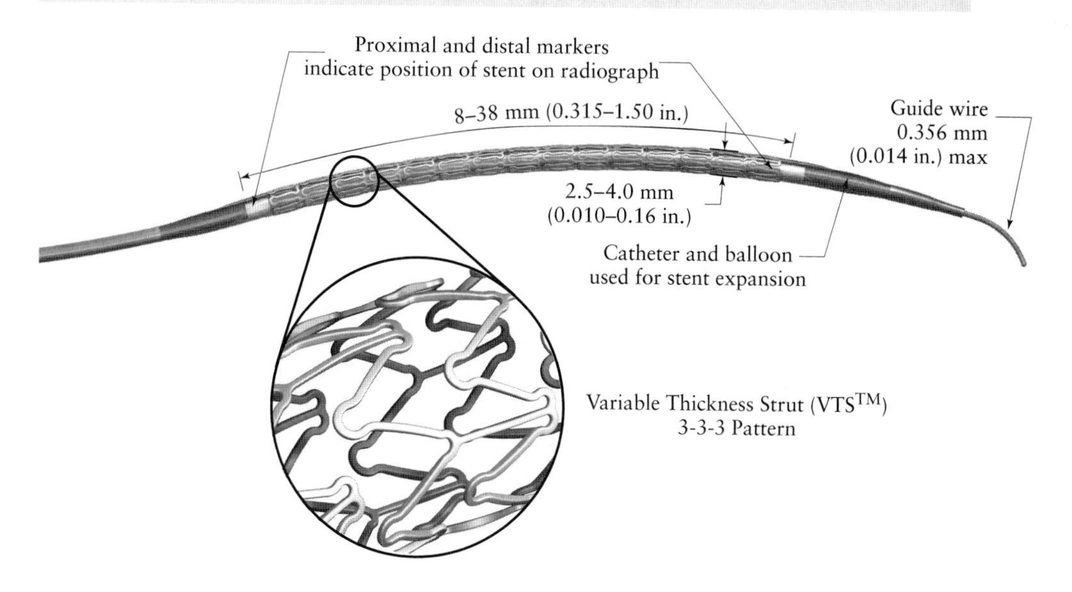

FIGURE 9.41 The Guidant MULTI-LINK TETRA™ coronary stent system.

[1] 2002 Heart and Stroke Statistical Update, American Heart Association, p. 4

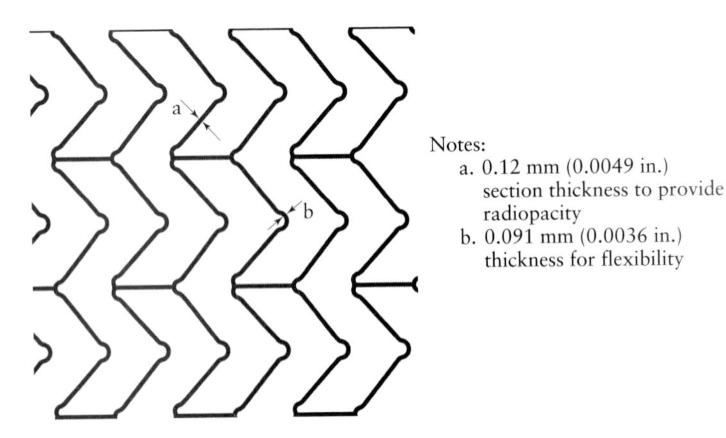

FIGURE 9.42 Detail of the 3-3-3 MULTI-LINK TETRA™ pattern.

of the patient's life. A standard material for stents, one that satisfies all of these requirements, is 316L stainless steel (Section 3.10.2).

The pattern of a stent also affects its performance. A finite-element analysis of the stent's design is first performed to determine how the stent will perform under the stresses induced from the beating heart. The strut pattern, tube thickness, and strut-leg width all affect the performance of the stent. Many different patterns are possible, such as the MULTI-LINK TETRA™ stent, which is shown in Figure 9.42 including some of the critical dimensions.

A stent starts out as a drawn stainless-steel tube with outer diameter that matches the final stent dimension, and a wall thickness selected to provide proper strength when expanded. The drawn tube is then laser machined (Section 9.14.1) to achieve the desired pattern (Fig. 9.43a). This method has proven to be very effective because of the small, intricate pattern of the stent and the tight dimensional tolerances that it must have. As the laser cuts the pattern of the stent, it leaves small metal slugs that need to be removed, therefore, it is very important that the laser cut all the way through the tube wall. A thick oxide layer or slag inevitably develops on the stainless steel due to the thermal and chemical attack from the air; in addition, weld splatter, burrs, and other surface defects result from laser machining. Finishing operations are therefore required to remove the splatter and oxide layer.

The first finishing operation after laser machining involves chemical etching to remove as much of the slag on the stent as possible; this operation is typically done in an acid solution, and the resulting surface is shown in Fig. 9.43b. Once the slag has been sufficiently removed, any residual burrs from the laser machining process need to be removed and the stent surface must be finished. It is very important that the surface finish of the stent be smooth to prevent possible forma-

(a)

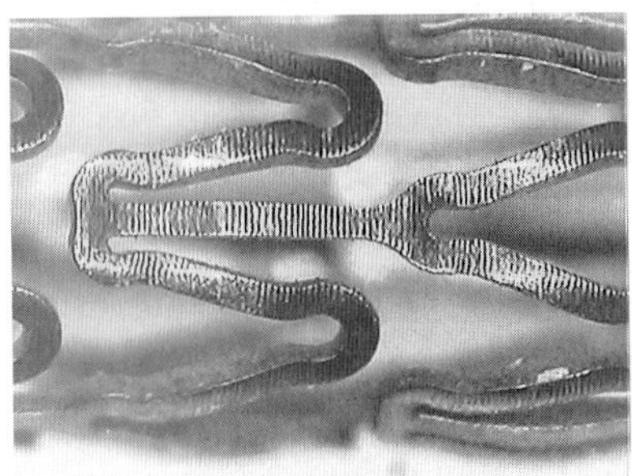
(b)

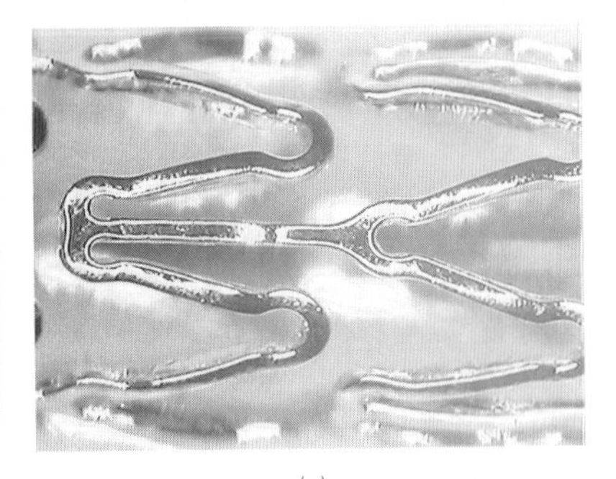
(c)

FIGURE 9.43 Evolution of the stent surface. (a) MULTI-LINK TETRA™ after lasing. Note that a metal slug is still attached. (b) After removal of slag. (c) After electropolishing.

tion of clots (thrombi) on the stent. To remove the sharp points, the stent is electropolished (Section 9.8) by passing an electrical current through an electrochemical solution, to ensure proper surface finish (Fig. 9.43c) that is both shiny and smooth. The stent is then placed onto a balloon catheter assembly, and sterilized and packaged for delivery to the surgeon.

Source: Courtesy of K. L. Graham, Guidant Corporation.

SUMMARY

- Abrasive machining and advanced machining processes are often necessary and economical when workpiece hardness is high, the materials are brittle or flexible, part shapes are complex, and surface finish and dimensional tolerance requirements are high. (Section 9.1)
- Conventional abrasives consist basically of aluminum oxide and silicon carbide, and superabrasives consist of cubic boron nitride and diamond. Friability of abrasive grains is an important factor, as are shape and size of the grains. (Section 9.2)
- Grinding wheels consist of a combination of abrasive grains and bonding agents. Important characteristics of wheels include abrasive-grain and bond types, grade, and hardness. Wheels may be reinforced to maintain wheel integrity. (Section 9.3)
- The mechanics of grinding processes are studied in order to establish quantitative relationships regarding chip dimensions, grinding forces and energy requirements, temperature rise, residual stresses, and any adverse effects on surface integrity of ground components. (Section 9.4)
- Wheel wear is an important consideration in ground surface quality and product integrity. Wear is usually monitored in terms of the grinding ratio, defined as the ratio of material removed to the volume of wheel wear. Dressing and truing of wheels are done by various techniques, with the aid of computer controls. (Section 9.5)
- A variety of abrasive machining processes and equipment is available for surface, external, and internal grinding, as well as for large-scale material removal processes. The proper selection of abrasives, process variables, and grinding fluids is important for obtaining the desired surface finish and dimensional accuracy and for avoiding damage such as burning, heat checking, harmful residual stresses, and chatter. (Section 9.6)
- Ultrasonic machining removes material by microchipping and is best suited for hard and brittle materials. (Section 9.7)
- Several finishing operations are available to improve surface finish. Because finishing processes can contribute significantly to product cost, proper selection and implementation of finishing operations are important. (Section 9.8)
- Deburring may be necessary for certain finished components. Commonly used deburring methods include vibratory and barrel finishing and shot blasting, although thermal-energy processes and other methods can also be used. (Section 9.9)
- Advanced machining processes involve chemical and electrical means and high-energy beams; they are particularly suitable for hard materials and complex shapes. The effects of these processes on surface integrity must be investigated, as they can damage surfaces considerably, thus reducing fatigue life. (Sections 9.10–9.14)
- Water-jet, abrasive water-jet, and abrasive-jet machining processes can be used effectively for cutting as well as deburring operations. Because they do not use hard tooling, they have inherent flexibility of operation. (Sections 9.15)

- As in all manufacturing processes, certain design guidelines should be followed for effective use of advanced machining processes. (Section 9.16)
- The processes described in this chapter have unique capabilities; however, because they involve different types of machinery, process variables, and cycle times, the competitive aspects of each process with respect to a particular product must be studied individually. (Section 9.17)

SUMMARY OF EQUATIONS

Undeformed chip length, ℓ:

Surface grinding, $\ell = \sqrt{Dd}$

External grinding, $\ell = \sqrt{\dfrac{Dd}{1 + D/D_w}}$

Internal grinding, $\ell = \sqrt{\dfrac{Dd}{1 - D/D_w}}$

Undeformed-chip thickness, surface grinding, $t = \sqrt{\dfrac{4v}{VCr}\sqrt{\dfrac{d}{D}}}$

Relative grain force $\propto \dfrac{v}{VC}\sqrt{\dfrac{d}{D}}$

Temperature rise in surface grinding, $\Delta T \propto \dfrac{d}{t} \propto d^{3/4}\sqrt{\dfrac{VC}{v}\sqrt{D}}$

Grinding ratio, $G = \dfrac{\text{volume of material removed}}{\text{volume of wheel wear}}$

Average force of particle in ultrasonic machining, where $F_{ave} = \dfrac{2mv}{t_o}$ where $t_o = \dfrac{5r}{c_o}\left(\dfrac{c_o}{v}\right)^{1/5}$

Material removal rate in

electrochemical machining, $\text{MRR} = CI\eta$

electrochemical grinding, $\text{MRR} = \dfrac{GI}{\rho F}$

electrical-discharge machining, $\text{MRR} = 4 \times 10^4 IT_w^{-1.23}$

electrical-discharge grinding, $\text{MRR} = KI$

wire EDM, $\text{MRR} = V_f hb$, where $b = d_w + 2s$

Penetration rate in electrochemical grinding, $V_s = \left[\dfrac{G}{dF}\right]\left[\dfrac{E}{gK_p}\right]K$

Wear rate of electrode in electrical-discharge machining, $W_t = 1.1 \times 10^{11} IT_t^{-2.38}$

Laser-beam cutting time, $t = \dfrac{P}{vd}$

BIBLIOGRAPHY

Andrew, C., T. D. Howes, and T. R. A. Pearce, *Creep-Feed Grinding*, Industrial Press, 1985.

ASM Handbook, Vol. 16: *Machining*, ASM International, 1989.

Borkowski, J., and A. Szymanski, *Uses of Abrasives and Abrasive Tools*, Ellis Horwood, 1992.

Brown, J., *Advanced Machining Technology Handbook*, McGraw-Hill, 1998.

Chryssolouris, G., and P. Sheng, *Laser Machining, Theory & Practice*, Springer, 1991.

Crafer, R. C., and P. J. Oakley, *Laser Processing in Manufacturing*, Chapman & Hall, 1993.

Farago, F. T., *Abrasive Methods Engineering*, Industrial Press, Vol. 1, 1976; Vol. 2, 1980.

Gillespie, L. K., *Deburring and Edge Finishing Handbook*, Society of Manufacturing Engineers/American Society of Mechanical Engineers, 2000.

Guitran, E. B., *The EDM Handbook*, Hanser-Gardner, 1997.

Hwa, L. S., *Chemical Mechanical Polishing in Silicon Processing*, Academic Press, 1999.

Jain, V. K., and P. C. Pandey, *Theory and Practice of Electrochemical Machining*, Wiley, 1993.

King, R. I., and R. S. Hahn, *Handbook of Modern Grinding Technology*, Chapman & Hall/Methuen, 1987.

Krar, S., and E. Ratterman, *Superabrasives: Grinding and Machining with CBN and Diamond*, McGraw-Hill, 1990.

Krar, S., *Grinding Technology*, 2d. ed., Delmar Publishers, 1995.

Machining Data Handbook, 3d. ed., 2 vols, Machinability Data Center, 1980.

Malkin, S., *Grinding Technology: Theory and Applications of Machining with Abrasives*, Wiley, 1989.

Marinescu, I. D. (ed.), *Handbook of Ceramics Grinding & Polishing*, Noyes, 1998.

Maroney, M. L., *A Guide to Metal and Plastic Finishing*, Industrial Press, 1991.

McGeough, J. A., *Advanced Methods of Machining*, Chapman & Hall, 1988.

McKee, R. L., *Machining with Abrasives*, Van Nostrand Reinhold, 1982.

Metzger, J. L., *Superabrasive Grinding*, Butterworths, 1986.

Momber, A. W., and R. Kovacevic, *Principles of Abrasive Water Jet Machining*, Springer, 1998.

Nachtman, E. S., and S. Kalpakjian, *Lubricants and Lubrication in Metalworking Operations*, Dekker, 1985.

Powell, J., *CO_2 Laser Cutting*, Springer, 1991.

Salmon, S. C., *Modern Grinding Process Technology*, McGraw-Hill, 1992.

Schneider, A. F., *Mechanical Deburring and Surface Finishing Technology*, Dekker, 1990.

Shaw, M. C., *Metal Cutting Principles*, Oxford, 1984.

———, *Principles of Abrasive Processing*, Oxford, 1996.

Sluhan, C. (ed.), *Cutting and Grinding Fluids: Selection and Application*, Society of Manufacturing Engineers, 1992.

Sommer, C., and S. Sommer, *Wire EDM Handbook*, Technical Advanced Publishing Co., 1997.

Steen, W. M., *Laser Material Processing*, Springer, 1991.

Szymanski, A., and J. Borkowski, *Technology of Abrasives and Abrasive Tools*, Ellis Horwood, 1992.

Taniguchi, N. (ed.), *Nanotechnology*, Oxford, 1996.

———, *Energy-Beam Processing of Materials*, Oxford, 1989.

Tool and Manufacturing Engineers Handbook, 4th ed., Vol. 1: *Machining*, Society of Manufacturing Engineers, 1983.

Webster, J. A., *Abrasive Processes: Theory, Technology, and Practice*, Dekker, 1996.

QUESTIONS

9.1 Why are grinding operations necessary for parts that have been machined by other processes?

9.2 Explain why there are so many different types and sizes of grinding wheels.

9.3 Explain the reasons for the large differences between the specific energies involved in grinding (Table 9.3) and in machining (Table 8.3).

9.4 Describe the advantages of superabrasives over conventional abrasives.

9.5 Give examples of applications for the grinding wheels shown in Fig. 9.2.

9.6 Explain why the same grinding wheel may act soft or hard.

9.7 Describe your understanding of the role of friability of abrasive grains on grinding-wheel performance.

9.8 Explain the factors involved in selecting the appropriate type of abrasive for a particular grinding operation.

9.9 What are the effects of wear flat on the grinding operation?

9.10 The grinding ratio, G, depends on the following factors: (1) type of grinding wheel, (2) workpiece hardness, (3) wheel depth of cut, (4) wheel and workpiece speeds, and (5) type of grinding fluid. Explain how.

9.11 List and explain the precautions you would take when grinding with high precision. Comment on the role of the machine, process parameters, the grinding wheel, and grinding fluids.

9.12 Describe the methods you would use to determine the number of active cutting points per unit surface area on the periphery of a straight (i.e., Type 1; see Fig. 9.2a) grinding wheel. What is the significance of this number?

9.13 Describe and explain the difficulties involved in grinding parts made of thermoplastics.

9.14 Explain why ultrasonic machining is not suitable for soft and ductile metals.

9.15 It is generally recommended that a soft-grade grinding wheel be used for hardened steels. Explain why.

9.16 Explain the reasons that the processes described in this chapter may adversely affect the fatigue strength of materials.

9.17 Describe the factors that may cause chatter in grinding operations, and give reasons that these factors cause chatter.

9.18 Outline the methods that are generally available for deburring parts. Discuss the advantages and limitations of each.

9.19 In which of the processes described in this chapter are physical properties of the workpiece material important? Why?

9.20 Give all possible technical and economic reasons that the material removal processes described in this chapter may be preferred, or even required, over those described in Chapter 8.

9.21 What processes would you recommend for die sinking in a die block? Explain.

9.22 The proper grinding surfaces for each type of wheel are shown with arrows in Fig. 9.2. Explain why grinding on other surfaces of the wheel is improper and unsafe.

9.23 Note that wheel (b) in Fig. 9.3 has serrations along its periphery, somewhat similar to circular metal saws. What does this design offer in grinding?

9.24 In Fig. 9.10, we note that wheel speed and grinding fluids can have a major effect on the type and magnitude of residual stresses developed in grinding. Explain the possible reasons for these phenomena.

9.25 Explain the consequences of allowing the workpiece temperature to rise too high in grinding operations.

9.26 Comment on any observations you have regarding the contents of Table 9.4.

9.27 Why has creep-feed grinding become an important process? Explain.

9.28 There has been a trend in manufacturing industries to increase the spindle speed of grinding wheels. Explain the possible advantages and limitations of such an increase.

9.29 Why is preshaping or premachining of parts generally desirable in the advanced machining processes described in this chapter?

9.30 Why are finishing operations sometimes necessary? How could they be minimized? Explain, with examples.

9.31 Why has the wire-EDM process become so widely accepted in industry?

9.32 Make a list of material removal processes described in this chapter that may be suitable for the following workpiece materials: (1) ceramics, (2) cast iron, (3) thermoplastics, (4) thermosets, (5) diamond, and (6) annealed copper.

9.33 Explain why producing sharp corners and profiles using some of the processes described in this chapter can be difficult.

9.34 How do you think specific energy, u, varies with respect to wheel depth of cut and hardness of the workpiece material? Explain.

9.35 In Example 9.2, it is stated that the thrust force in grinding is about 30% higher than the cutting force. Why is it higher?

9.36 Why should we be interested in the magnitude of the thrust force in grinding? Explain.

9.37 Why is the material removal rate in electrical-discharge machining a function of the melting point of the workpiece material?

9.38 Inspect Table 9.4, and, for each process, list and describe the role of various mechanical, physical, and chemical properties of the workpiece material that may affect performance.

9.39 Which of the processes listed in Table 9.4 would not be applicable to nonmetallic materials? Explain.

9.40 Why does the machining cost increase so rapidly as surface finish requirements become finer?

9.41 Which of the processes described in this chapter are particularly suitable for workpieces made of (a) ceramics, (b) thermoplastics, and (c) thermosets? Why?

9.42 Other than cost, is there a reason that a grinding wheel intended for a hard workpiece cannot be used for a softer workpiece?

9.43 How would you grind the facets on a diamond, as for an engagement ring, since diamond is the hardest material known?

9.44 Define dressing and truing, and describe the difference between them.

PROBLEMS

9.45 In a surface-grinding operation, calculate the chip dimensions for the following process variables: $D = 10$ in., $d = 0.001$ in., $v = 50$ ft/min, $V = 5000$ ft/min, $C = 500$ per in^2, and $r = 20$.

9.46 If the workpiece strength in grinding is doubled, what should be the percentage decreases in the wheel depth of cut, d, in order to maintain the same grain force, all other variables being the same?

9.47 Taking a thin Type 1 grinding wheel as an example and referring to texts on stresses in rotating bodies, plot the tangential stress, σ_t, and radial stress, σ_r, as a function of radial distance, i.e., from the hole to the periphery of the wheel. Note that because the wheel is thin, this problem can be regarded as a plane-stress problem. How would you determine the maximum combined stress and its location in the wheel?

9.48 Derive a formula for the material removal rate (MRR) in surface grinding in terms of process parameters. Use the same terminology as in the text.

9.49 Assume that a surface-grinding operation is being carried out under the following conditions: $D = 250$ mm, $d = 0.1$ mm, $v = 0.5$ m/s, and $V = 50$ m/s. These conditions are then changed to the following: $D = 150$ mm, $d = 0.1$ mm, $v = 0.3$ m/s, and $V = 25$ m/s. What is the difference in the temperature rise from the initial condition?

9.50 For a surface-grinding operation, derive an expression for the power dissipated in imparting kinetic energy to the chips. Comment on the magnitude of this energy. Use the same terminology as in the text.

9.51 The shaft of a Type 1 grinding wheel is attached to a flywheel only, which is rotating at a certain initial rpm. With this setup, a surface-grinding operation is being carried out on a long workpiece and at a constant workpiece speed, v. Obtain an expression for estimating the linear distance ground on the workpiece before the wheel comes to a stop. Ignore wheel wear.

9.52 Calculate the average impact force on a steel plate by a spherical aluminum-oxide abrasive particle with a 0.1-mm diameter, dropped from heights of (1) 1 m, (2) 2 m, and (3) 10 m.

9.53 A 50-mm-deep hole, 30 mm in diameter, is being produced by electrochemical machining. A high production rate is more important than the quality of the machined surface. Estimate the maximum current and the time required to perform this operation.

9.54 If the operation in Problem 9.53 were performed on an electrical-discharge machine, what would be the estimated machining time?

9.55 A cutting-off operation is being performed with a laser beam. The workpiece being cut is $\frac{3}{8}$ in. thick and 3 in. long. If the kerf width is $\frac{1}{6}$ in., estimate the time required to perform this operation.

9.56 Refering to Table 3.2, identify two metals or metal alloys that, when used as workpiece and electrode, respectively, in EDM would give the (1) lowest and (2) highest wear ratios, R. Calculate these quantities.

9.57 In Section 9.5.2, it was stated that, in practice, grinding ratios typically range from 2 to 200. Based on the information given in Section 9.13, estimate the range of wear ratios in electrical-discharge machining, and then compare them with grinding ratios.

9.58 Assume that you are an instructor covering the topics in this chapter, and you are giving a quiz on the quantitative aspects to test the understanding of the students. Prepare several numerical problems, and supply the answers.

9.59 It is known that heat checking occurs when grinding with a spindle speed of 4000 rpm, a wheel diameter of 10 in., and a depth of cut of 0.0015 in. for a feed rate of 50 ft/min. For this reason, the spindle speed should be kept at 3500 rpm. If a new, 8-in-diameter wheel is used, what spindle speed can be employed before heat checking occurs? What spindle speed should be used to keep the same grinding temperatures as those encountered with the existing operating conditions?

9.60 It is desired to grind a hard aerospace aluminum alloy. A depth of 0.003 in. is to be removed from a cylindrical section 10 in. long and with a 4-in. diameter. If each part is to be ground in not more than one minute, what is the approximate power requirement for the grinder? What if the material is changed to a hard titanium alloy?

9.61 A grinding operation is taking place with a 10-in. grinding wheel at a spindle rotational speed of 3000 rpm. The workpiece feed rate is 60 ft/min, and the depth of cut is 0.002 in. Contact thermometers record an approximate maximum temperature of 1800°F. If the workpiece is steel, what is the temperature if the spindle speed is increased to 4000 rpm? What if it is increased to 10,000 rpm?

9.62 The regulating wheel of a centerless grinder is rotating at a surface speed of 50 ft/min and is inclined at an angle of 5°. What is the feed rate of material past the grinding wheel?

9.63 Using some typical values, explain what changes, if any, take place in the magnitude of impact force of a particle in ultrasonic machining as the temperature of the workpiece is increased. Assume that the workpiece is made of hardened steel.

DESIGN

9.64 Would you consider designing a machine tool that combines, in one machine, two or more of the processes described in this chapter? Explain. For what types of parts would such a machine be useful? Make preliminary sketches for such machines.

9.65 With appropriate sketches, describe the principles of various fixturing methods and devices that can be used for each of the processes described in this chapter.

9.66 Surface finish can be an important consideration in the design of products. Describe as many parameters as you can that could affect the final surface finish in grinding, including the role of process parameters as well as the setup and the equipment used.

9.67 A somewhat controversial subject in grinding is size effect. (See Section 9.4.1.) Design a setup and a series of experiments whereby size effect can be studied.

9.68 Describe how the design and geometry of the workpiece affects the selection of an appropriate shape and type of a grinding wheel.

9.69 Prepare a comprehensive table of the capabilities of abrasive machining processes, including the shapes of parts ground, types of machines involved, typical maximum and minimum workpiece dimensions, and production rates.

9.70 How would you produce a thin, large-diameter round disk with a thickness that decreases linearly from the center outward?

9.71 Marking surfaces of manufactured parts with letters and numbers can be done not only with labels and stickers, but also by various mechanical and non-mechanical means. Make a list of some of these methods, explaining their advantages and limitations.

CHAPTER 10

Properties and Processing of Polymers and Reinforced Plastics; Rapid Prototyping and Rapid Tooling

10.1 Introduction

Although the word **plastics**, first used around 1909, is a commonly used synonym for **polymers**, plastics are one of numerous polymeric materials and have extremely large molecules (*macromolecules*). Consumer and industrial products made of polymers include food and beverage containers, packaging, signs, housewares, textiles, medical devices, foams, paints, safety shields, and toys. Compared with metals, polymers are generally characterized by low density, low strength and stiffness (Table 10.1), low electrical and thermal conductivity, good resistance to chemicals, and a high coefficient of thermal expansion. However, the useful temperature range for most polymers is generally low, up to about 350°C(660°F), and they are not as dimensionally stable in service, over a period of time, as metals.

The word *plastics* is from the Greek word *plastikos*, meaning "it can be molded and shaped." Plastics can be machined, cast, formed, and joined into many shapes with relative ease. Minimal or no additional surface-finishing operations are required, which is an important advantage over metals. Plastics are commercially available as sheet, plate, film, rods, and tubing of various cross-sections. The word *polymer* was first used in 1866. The earliest polymers were made of **natural organic materials** from animal and vegetable products, cellulose being the most common example. With various chemical reactions, cellulose is modified into cellulose acetate, used in making photographic films (celluloid), sheets for packaging, and textile fibers; cellulose is also modified into cellulose nitrate for plastics, explosives, rayon (a cellulose textile fiber), and varnishes. The earliest **synthetic polymer** was a phenol formaldehyde, a thermoset developed in 1906 and called *Bakelite* (a trade name, after L. H. Baekeland, 1863–1944).

The development of modern plastics technology began in the 1920s, when raw materials necessary for making polymers were extracted from coal and petroleum products. Ethylene was the first example of such a raw material, as it became the building block for polyethylene; it is the product of the reaction between acetylene and hydrogen, and acetylene is the product of the reaction between coke and methane.

TABLE 10.1

Approximate Range of Mechanical Properties for Various Engineering Plastics at Room Temperature

Material	UTS (MPa)	*E* (GPa)	Elongation in 50 mm (%)	Poisson's Ratio (ν)
ABS	28–55	1.4–2.8	75–5	–
ABS (reinforced)	100	7.5	–	0.35
Acetals	55–70	1.4–3.5	75–25	–
Acetals (reinforced)	135	10	–	0.35–0.40
Acrylics	40–75	1.4–3.5	50–5	–
Cellulosics	10–48	0.4–1.4	100–5	–
Epoxies	35–140	3.5–17	10–1	–
Epoxies (reinforced)	70–1400	21–52	4–2	–
Fluorocarbons	7–48	0.7–2	300–100	0.46–0.48
Nylon	55–83	1.4–2.8	200–60	0.32–0.40
Nylon (reinforced)	70–210	2–10	10–1	–
Phenolics	28–70	2.8–21	2–0	–
Polycarbonates	55–70	2.5–3	125–10	0.38
Polycarbonates (reinforced)	110	6	6–4	–
Polyesters	55	2	300–5	0.38
Polyesters (reinforced)	110–160	8.3–12	3–1	–
Polyethylenes	7–40	0.1–0.14	1000–15	0.46
Polypropylenes	20–35	0.7–1.2	500–10	–
Polypropylenes (reinforced)	40–100	3.6–6	4–2	–
Polystyrenes	14–83	1.4–4	60–1	0.35
Polyvinyl chloride	7–55	0.014–4	450–40	–

Likewise, commercial polymers including polypropylene, polyvinyl chloride, polymethylmethacrylate, polycarbonate, and others are all made in a similar manner; these materials are known as *synthetic organic polymers*. Figure 10.1 illustrates an outline of the basic process of making various synthetic polymers.

This chapter also covers the characteristics and processing of **polymer-matrix reinforced plastics** (also called *composite materials*), which exhibit a wide range of properties such as stiffness, strength, resistance to creep, and high strength-to-weight and stiffness-to-weight ratios. Their applications include numerous consumer and industrial products, as well as products in the automotive and aerospace industries. The properties and processing of metal-matrix and ceramic-matrix composites are covered in Chapter 11. Another important advance in manufacturing is **rapid prototyping and tooling**, also called *desktop manufacturing* or *free-form fabrication*. This method is a family of processes by which a solid physical model of a part is made directly from a three-dimensional CAD drawing, involving a time span that is much shorter than that for traditional prototype manufacturing. As described in detail in Section 10.12, rapid prototyping entails several different material-consolidation techniques, including resin curing, deposition, solidification, and sintering.

10.2 The Structure of Polymers

Many of a polymer's properties depend largely on (a) the structure of individual polymer molecules, (b) the shape and size of the molecules, and (c) how the molecules are arranged to form a polymer structure. Polymer molecules are characterized by their extraordinary size, a feature that distinguishes them from other organic chemical

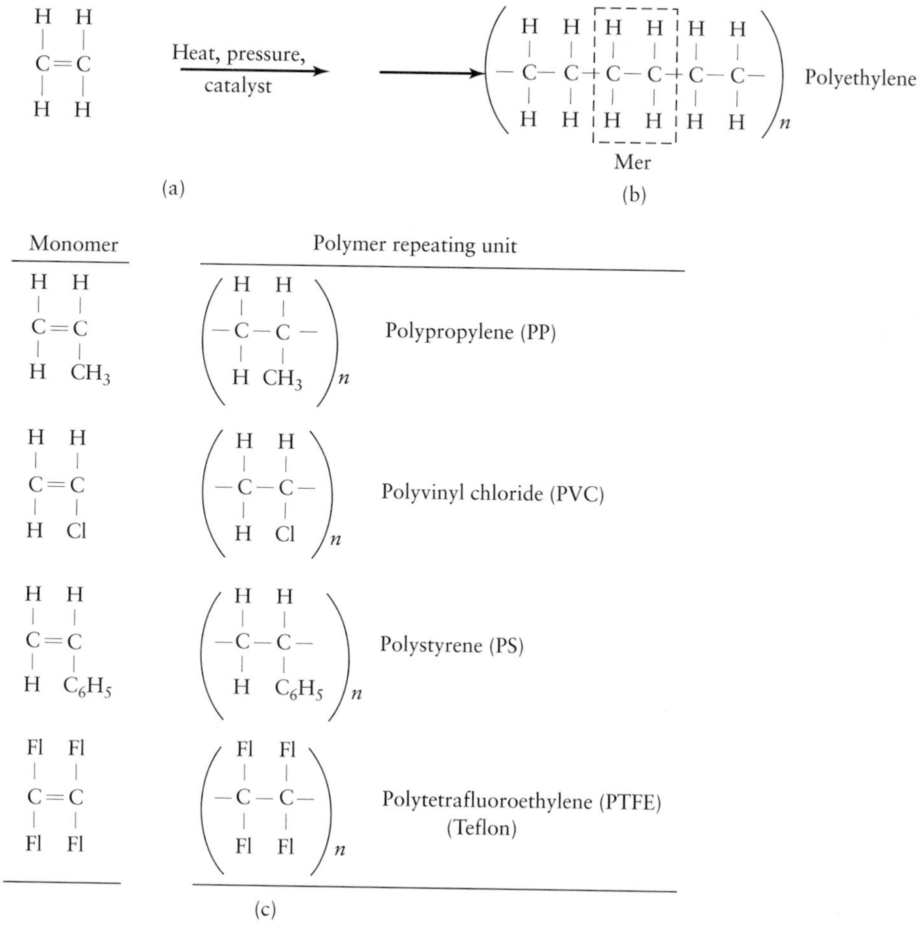

(a) Ethylene (C_2H_4) → Heat, pressure, catalyst → (b) Polyethylene $(-C_2H_4-)_n$, Mer

(c)

Monomer	Polymer repeating unit	
$CH_2=CH(CH_3)$	$(-CH_2-CH(CH_3)-)_n$	Polypropylene (PP)
$CH_2=CHCl$	$(-CH_2-CHCl-)_n$	Polyvinyl chloride (PVC)
$CH_2=CH(C_6H_5)$	$(-CH_2-CH(C_6H_5)-)_n$	Polystyrene (PS)
$CFl_2=CFl_2$	$(-CFl_2-CFl_2-)_n$	Polytetrafluoroethylene (PTFE) (Teflon)

FIGURE 10.1 Basic structure of polymer molecules: (a) ethylene molecule; (b) polyethylene, a linear chain of many ethylene molecules; (c) molecular structure of various polymers. These molecules are examples of the basic building blocks for plastics.

compositions. Polymers are *long-chain molecules*, also called **macromolecules** or **giant molecules**, that are formed by polymerization, that is, by linking and cross-linking different monomers. A **monomer** is the basic building block of polymers. The word **mer**, from the Greek *meros*, meaning "part," indicates the smallest repetitive unit, similar to the term *unit cell* used in connection with crystal structures of metals (see Section 3.2).

The term **polymer** means "many mers or units," generally repeated hundreds or thousands of times in a chainlike structure. Most monomers are organic materials in which carbon atoms are joined in *covalent* (electron-sharing) bonds with other atoms, such as hydrogen, oxygen, nitrogen, fluorine, chlorine, silicon, and sulfur. An ethylene molecule is a simple monomer consisting of carbon and hydrogen atoms (Fig. 10.1a).

10.2.1 Polymerization

Monomers can be linked in repeating units to make longer and larger molecules by a chemical reaction known as the *polymerization reaction*. Polymerization processes are complex and are described only briefly here. Although there are numerous variations, two basic polymerization processes are condensation polymerization and addition polymerization:

1. In **condensation polymerization**, polymers are produced by the formation of bonds between two types of reacting mers. One characteristic of this reaction is that reaction by-products such as water are condensed out, hence the term

condensation. This process is also known as *step-growth* or *step-reaction polymerization*, because the polymer molecule grows step by step until all of one reactant is consumed.

2. In **addition polymerization**, also known as *chain-growth* or *chain-reaction polymerization*, bonding takes place without reaction by-products; it is called *chain-reaction polymerization* because of the high rate at which long molecules form simultaneously, usually within a few seconds. This rate is much higher than that for condensation polymerization. In this reaction, an initiator is added to open the double bond between the carbon atoms and begins the linking process by adding many more monomers to a growing chain. For example, ethylene monomers (Fig. 10.1a) link to produce polyethylene (Fig. 10.1b); other examples of addition-formed polymers are shown in Fig. 10.1c.

Some basic characteristics and types of polymers are described below.

a) **Molecular weight.** The sum of the molecular weight of the mers in the polymer chain is the *molecular weight* of the polymer; the higher the molecular weight in a given polymer, the greater is the length of the chain. Because polymerization is a random event, the polymer chains produced are not all of equal length; however, the chain lengths produced fall into a traditional distribution curve (see Fig. 11.4). The average molecular weight of a polymer is determined and expressed on a statistical basis by averaging; the spread observed in the weight distribution is referred to as the **molecular weight distribution** (MWD). Molecular weight and MWD have a strong influence on the properties of the polymer. For example, tensile and impact strength, resistance to cracking, and viscosity in the molten state all increase with increasing molecular weight (Fig. 10.2). Most commercial polymers have a molecular weight between 10,000 and 10,000,000.

b) **Degree of polymerization.** In some cases, it is more convenient to express the size of a polymer chain in terms of the *degree of polymerization* (DP), defined as the ratio of the molecular weight of the polymer to the molecular weight of the repeat unit. In terms of polymer processing (see Section 10.10), the higher the DP, the higher is the polymer's viscosity, or resistance to flow (Fig. 10.2), which affects ease of shaping and overall cost of processing.

c) **Bonding.** During polymerization, the monomers are linked together in a *covalent bond*, forming a polymer chain. Because of their strength, covalent bonds are also called **primary bonds**. The polymer chains are, in turn, held

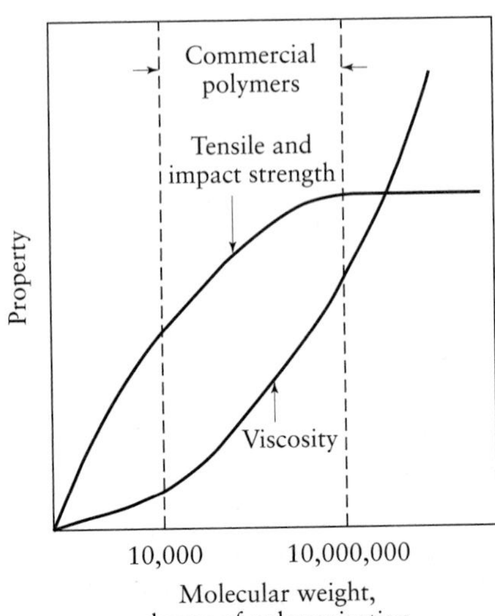

FIGURE 10.2 Effect of molecular weight and degree of polymerization on the strength and viscosity of polymers.

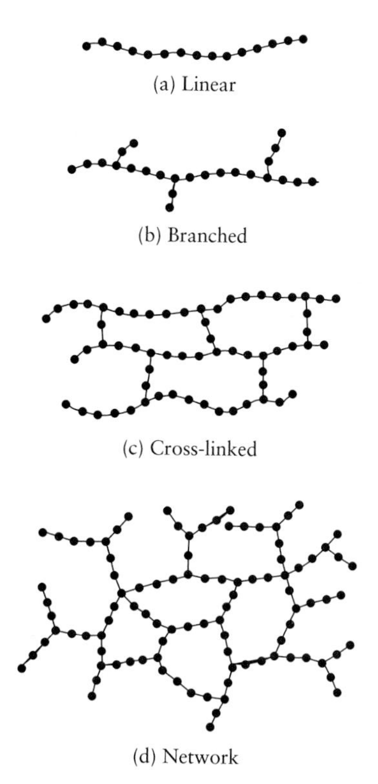

FIGURE 10.3 Schematic illustration of polymer chains. (a) Linear structure; thermoplastics such as acrylics, nylons, polyethylene, and polyvinyl chloride have linear structures. (b) Branched structure, such as in polyethylene. (c) Cross-linked structure; many rubbers and elastomers have this structure. Vulcanization of rubber produces this structure. (d) Network structure, which is basically highly cross-linked; examples include thermosetting plastics, such as epoxies and phenolics.

together by the secondary bonds, such as van der Waals bonds, hydrogen bonds, and ionic bonds. **Secondary bonds** are much weaker than primary bonds, by one to two orders of magnitude. In a given polymer, the increase in strength and viscosity with molecular weight comes in part from the fact that the longer the polymer chain, the greater is the energy needed to overcome the strength of the secondary bonds. For example, ethylene mers that have a DP of 1, 6, 35, 140, and 1350 are, respectively, in the form of gas, liquid, grease, wax, and hard plastic at room temperature.

d) **Linear polymers.** The chainlike polymers shown in Fig. 10.1 are called *linear polymers*, because of their linear structure (Fig. 10.3a). A linear molecule is not straight in shape. In addition to those shown in Fig. 10.3, other linear polymers include polyamides (nylon 6,6) and polyvinyl fluoride. Generally, a polymer consists of more than one type of structure; thus, a linear polymer may contain some branched and cross-linked chains. (See parts e and f of this list.) As a result of branching and cross-linking, the polymer's properties change.

e) **Branched polymers.** The properties of a polymer depend not only on the type of monomers in the polymer, but also on their arrangement in the molecular structure. In *branched polymers* (Fig. 10.3b), side-branch chains are attached to the main chain during the synthesis of the polymer. Branching interferes with the relative movement of the molecular chains; as a result, resistance to deformation increases and stress cracking resistance is affected. Also, the density of branched polymers is lower than that of linear-chain polymers, as branches interfere with the packing efficiency of polymer chains. The behavior of branched polymers can be compared with that of linear-chain polymers by making an analogy with a pile of tree branches (branched polymers) and a bundle of straight logs (linear-chain polymers). Note that it is more difficult to move a branch within the pile of branches than to move a log in its bundle. The three-dimensional entanglements of branches make movements more difficult, a phenomenon akin to increased strength.

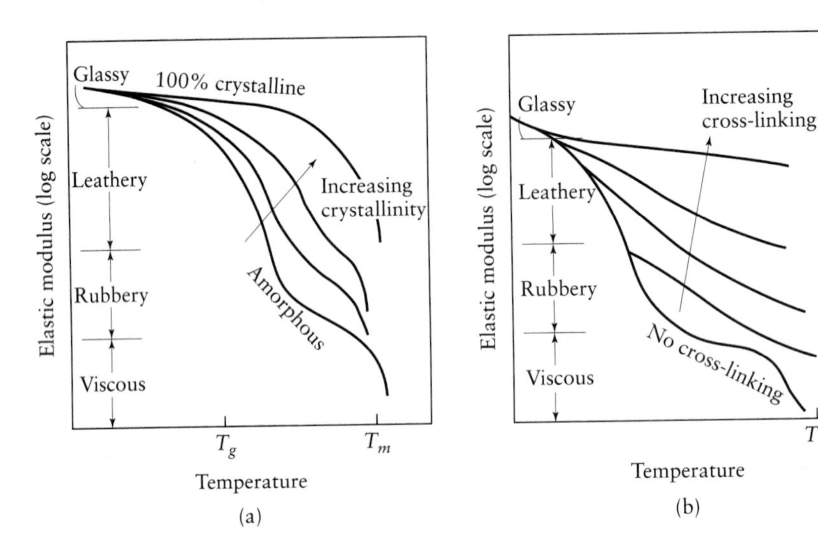

FIGURE 10.4 Behavior of polymers as a function of temperature and (a) degree of crystallinity and (b) cross-linking. The combined elastic and viscous behavior of polymers is known as *viscoelasticity*.

f) **Cross-linked polymers.** Generally three-dimensional in structure, *cross-linked polymers* have adjacent chains linked by covalent bonds (Fig. 10.3c). Polymers with a cross-linked chain structure are called **thermosets** or **thermosetting plastics**, such as epoxies, phenolics, and silicones. Cross-linking has a major influence on the properties of polymers (generally imparting hardness, strength, stiffness, brittleness, and better dimensional stability; see Fig. 10.4), as well as in the **vulcanization** of rubber. **Network polymers** consist of spatial (three-dimensional) networks of three active covalent bonds (Fig. 10.3d). A highly cross-linked polymer is also considered a network polymer. Thermoplastic polymers that have already been formed or shaped can be cross-linked to obtain greater strength by being subject to high-energy radiation, such as ultraviolet light, X rays, or electron beams. However, excessive radiation can cause *degradation* of the polymer.

g) **Copolymers and terpolymers.** If the repeating units in a polymer chain are all of the same type, the molecule is called a **homopolymer.** However, as with solid-solution metal alloys (see Section 5.2.1), two or three different types of monomers can be combined to impart certain special properties and characteristics to the polymer, such as improvement of both the strength and toughness as well as the formability of the polymer. *Copolymers* contain two types of polymers, such as styrene–butadiene, used widely for automobile tires. *Terpolymers* contain three types, such as ABS (acrylonitrile–butadiene–styrene), used for helmets, telephones, and refrigerator liners.

EXAMPLE 10.1 Degree of polymerization in polyvinyl chloride (PVC)

Determine the molecular weight of a polyvinyl-chloride mer. If a PVC polymer has an average molecular weight of 50,000, what is the degree of polymerization?

Solution. From Fig. 10.1c, note that each PVC mer has three hydrogen atoms, two carbon atoms, and one chlorine atom. Since the atomic number of these elements are 1, 12, and 35.5, respectively, the weight of a PVC mer is $(3)(1) + (2)(12) + (1)(35.5) = 62.5$; thus, the degree of polymerization is $50{,}000/62.5 = 800$.

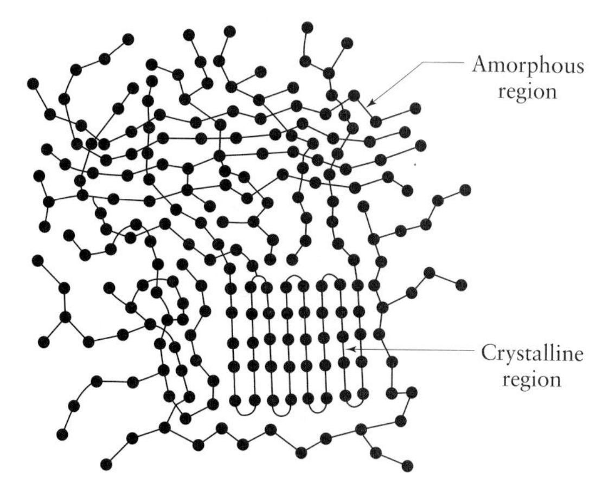

FIGURE 10.5 Amorphous and crystalline regions in a polymer. The crystalline region (crystallite) has an orderly arrangement of molecules. The higher the crystallinity, the harder, stiffer, and less ductile is the polymer.

10.2.2 Crystallinity

Polymers such as polymethylmethacrylate, polycarbonate, and polystyrene are generally *amorphous*; that is, the polymer chains exist without long-range order. (See also the discussion of *amorphous alloys* in Section 3.11.9.) The amorphous arrangement of polymer chains is often described as similar to a bowl of spaghetti, or worms in a bucket, all intertwined with each other. However, in some polymers, it is possible to impart some crystallinity and thereby modify their characteristics. This may be done either during the synthesis of the polymer or by deformation during its subsequent processing.

The crystalline regions in polymers are called **crystallites** (Fig. 10.5). The crystals are formed when the long molecules arrange themselves in an orderly manner, similar to folding a fire hose in a cabinet or facial tissue in a box. Thus, a partially crystalline (semicrystalline) polymer can be regarded as a two-phase material, one phase being crystalline and the other amorphous. By controlling the rate of solidification during cooling and chain structure, it is possible to impart different **degrees of crystallinity** to polymers, although a polymer can never be 100% crystalline. Crystallinity ranges from an almost complete crystal (up to about 95% by volume in the case of polyethylene) to slightly crystallized, but mostly amorphous polymers. The degree of crystallinity is also affected by branching. A linear polymer can become highly crystalline, but a highly branched polymer cannot. Although the latter may develop some low level of crystallinity, it will never achieve a high crystalline content, because the branches interfere with the alignment of the chains into a regular crystal array.

Effects of crystallinity. The mechanical and physical properties of polymers are greatly influenced by the degree of crystallinity; as crystallinity increases, polymers become stiffer, harder, less ductile, more dense, less rubbery, and more resistant to solvents and heat (Fig. 10.4). The increase in density with increasing crystallinity is caused by crystallization shrinkage and a more efficient packing of the molecules in the crystal lattice. For example, the highly crystalline form of polyethylene, known as high-density polyethylene (HDPE), has a specific gravity in the range of 0.941 to 0.970 (80% to 95% crystalline) and is stronger, stiffer, tougher, and less ductile than low-density polyethylene (LDPE), which is about 60% to 70% crystalline and has a specific gravity of about 0.910 to 0.925.

Optical properties also are affected by the degree of crystallinity. The reflection of light from the boundaries between the crystalline and amorphous regions in the polymer causes opaqueness. Furthermore, because the index of refraction is proportional to density, the greater the density difference between the amorphous and crystalline phases, the greater is the opaqueness of the polymer. Polymers that are completely amorphous can be transparent, such as polycarbonate and acrylics.

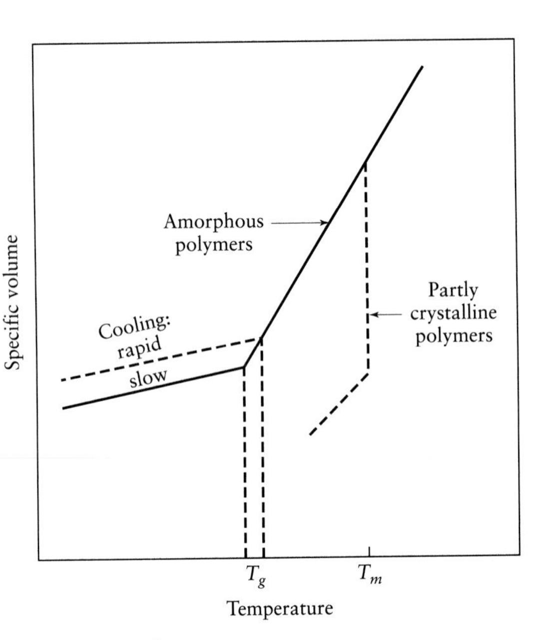

FIGURE 10.6 Specific volume of polymers as a function of temperature. Amorphous polymers, such as acrylic and polycarbonate, have a glass-transition temperature, T_g, but do not have a specific melting point, T_m. Partly crystalline polymers, such as polyethylene and nylons, contract sharply at their melting points during cooling.

10.2.3 Glass-transition temperature

Amorphous polymers do not have a specific melting point, but they undergo a distinct change in their mechanical behavior across a narrow range of temperature. At low temperatures, they are hard, rigid, brittle, and glassy, and at high temperatures, they are rubbery or leathery. The temperature at which this transition occurs is called the *glass-transition temperature*, T_g, or the *glass point* or *glass temperature*. The term *glass* is included in this definition because glasses, which are amorphous solids (see Section 3.11.9), behave in the same manner (as you can see by holding a glass rod over a flame and observing its behavior). Although most amorphous polymers exhibit this behavior, there are some exceptions, such as polycarbonate, which is not rigid or brittle below its glass-transition temperature. Polycarbonate is tough at ambient temperature and is therefore used for safety helmets and shields.

To determine T_g, the specific volume of the polymer is measured and plotted against temperature to find the sharp change in the slope of the curve (Fig. 10.6). However, in the case of highly cross-linked polymers, the slope of the curve changes gradually near T_g, making it difficult to determine T_g for these polymers. The glass-transition temperature varies for different polymers (Table 10.2); for example, room tem-

TABLE 10.2

Glass-Transition and Melting Temperatures of Some Polymers

Material	T_g(°C)	T_m(°C)
Nylon 6,6	57	265
Polycarbonate	150	265
Polyester	73	265
Polyethylene		
High density	−90	137
Low density	−110	115
Polymethylmethacrylate	105	–
Polypropylene	−14	176
Polystyrene	100	239
Polytetrafluoroethylene (Teflon)	−90	327
Polyvinyl chloride	87	212
Rubber	−73	–

perature is above T_g for some polymers and below it for others. Unlike amorphous polymers, partly crystalline polymers have a distinct melting point, T_m (Fig. 10.6; see also Table 10.2). Because of the structural changes (first-order changes) that occur the specific volume of the polymers drops rapidly as their temperature is reduced.

10.2.4 Polymer blends

To improve the brittle behavior of amorphous polymers below their glass-transition temperature, they can be blended, usually with small quantities of an *elastomer* (see Section 10.8). These tiny particles are dispersed throughout the amorphous polymer, enhancing its toughness and impact strength by improving its resistance to crack propagation; these blended polymers are known as **rubber-modified polymers.** Blending may involve several components (**polyblends**) that use the favorable properties of different polymers. Advances have been made in **miscible blends** (mixing without separation of two phases), a process similar to alloying of metals, enabling polymer blends to become more ductile. Polymer blends account for about 20% of all polymers produced.

10.2.5 Additives in polymers

In order to impart certain specific properties, polymers are usually compounded with *additives*. The additives modify and improve certain characteristics of the polymers, such as their stiffness, strength, color, weatherability, flammability, arc resistance for electrical applications, and ease of subsequent processing.

1. **Fillers** used are generally wood flour (fine sawdust), silica flour (fine silica powder), clay, powdered mica, and short fibers of cellulose, glass, and asbestos. Because of their low cost, fillers are important in reducing the overall cost of polymers. Depending on their type, fillers improve the strength, hardness, toughness, abrasion resistance, dimensional stability, and/or stiffness of plastics. These properties are greatest at various percentages of different types of polymer–filler combinations. As in reinforced plastics (Section 10.9), a filler's effectiveness depends on the nature and strength of the bond between the filler material and the polymer chains.
2. **Plasticizers** are added to some polymers to impart flexibility and softness by lowering their glass-transition temperature. Plasticizers are low-molecular-weight solvents with high boiling points (nonvolatile). They reduce the strength of the secondary bonds between the long-chain molecules, thus making the polymer soft and flexible. The most common use of plasticizers is in polyvinyl chloride (PVC), which remains flexible during its many uses. Other applications of plasticizers include thin sheet, film, tubing, shower curtains, and clothing materials.
3. Most polymers are adversely affected by **ultraviolet radiation** (sunlight) and oxygen, which weaken and break the primary bonds, resulting in the scission (splitting) of the long-chain molecules; the polymer then degrades and becomes brittle and stiff. On the other hand, degradation may be beneficial, as in the disposal of plastic objects by subjecting them to environmental attack. (See also Section 10.7.3.) A typical example of imparting protection against ultraviolet radiation is the compounding of certain plastics and rubber with *carbon black* (soot); the carbon black absorbs a high percentage of the ultraviolet radiation. Protection against degradation by oxidation, particularly at elevated temperatures, is achieved by adding antioxidants to the polymer. Applying various coatings is another means of protecting polymers against environmental attack.

4. The wide variety of colors available in plastics is obtained by adding **colorants;** these materials are either organic (dyes) or inorganic (pigments). The selection of a colorant depends on the polymer's service temperature and exposure to light. Pigments are dispersed particles and generally have greater resistance than dyes to temperature and light.
5. If the temperature is sufficiently high, most polymers will ignite. The **flammability** (ability to support combustion) of polymers varies considerably, depending on their composition (such as the chlorine and fluorine content). The flammability of polymers can be reduced either by making them from less flammable raw materials or by adding of **flame retardants**, such as compounds of chlorine, bromine, and phosphorus. Examples of polymers with different burning characteristics include the following: (a) Fluorocarbons (e.g., *Teflon*) do not burn; (b) carbonate, nylon, and vinyl chloride burn, but are self-extinguishing; and (c) acetal, acrylic, ABS, polyester, polypropylene, and styrene burn and are not self-extinguishing.
6. **Lubricants** may be added to polymers to reduce friction during their subsequent processing into useful products and to prevent parts from sticking to the molds. Lubrication is also important in preventing thin polymer films from sticking to each other.

10.3 Thermoplastics: Behavior and Properties

We noted earlier that the bonds between adjacent long-chain molecules (secondary bonds) are much weaker than the covalent bonds (primary bonds) within each molecule. Furthermore, it is the strength of the secondary bonds that determines the overall strength of the polymer; linear and branched polymers have weak secondary bonds. As the temperature is raised above the glass-transition temperature or the melting point certain polymers become easier to form or mold into desired shapes. The increased temperature weakens the secondary bonds (due to thermal vibration of the long molecules), and the adjacent chains can thus move more easily under the shaping forces. If the polymer is then cooled, it returns to its original hardness and strength; in other words the effects of the process are reversible. Polymers that exhibit this behavior are known as *thermoplastics*; typical examples include acrylics, cellulosics, nylons, polyethylenes, and polyvinyl chloride. Although the effects of the process are reversible, the repeated heating and cooling of thermoplastics can cause **degradation (thermal aging)**.

The behavior of thermoplastics, like their structure and composition, depends on numerous variables. Among the most important are temperature and rate of deformation. Below the glass-transition temperature, most polymers are glassy (described as rigid, brittle, or hard), and they behave like an elastic solid. If the load exceeds a certain critical value, the polymer fractures, just as a piece of glass does at room temperature. In the glassy region, the relationship between stress and strain is linear, or

$$\sigma = E\varepsilon. \tag{10.1}$$

If the polymer is tested in torsion, then

$$\tau = G\gamma. \tag{10.2}$$

The glassy behavior can be represented by a spring that has a stiffness equivalent to the modulus of elasticity of the polymer (Fig. 10.7a). Note that the strain is completely recovered when the load is removed at time t_1. When the applied stress is increased, the polymer eventually fractures, just as a piece of glass does at ambient

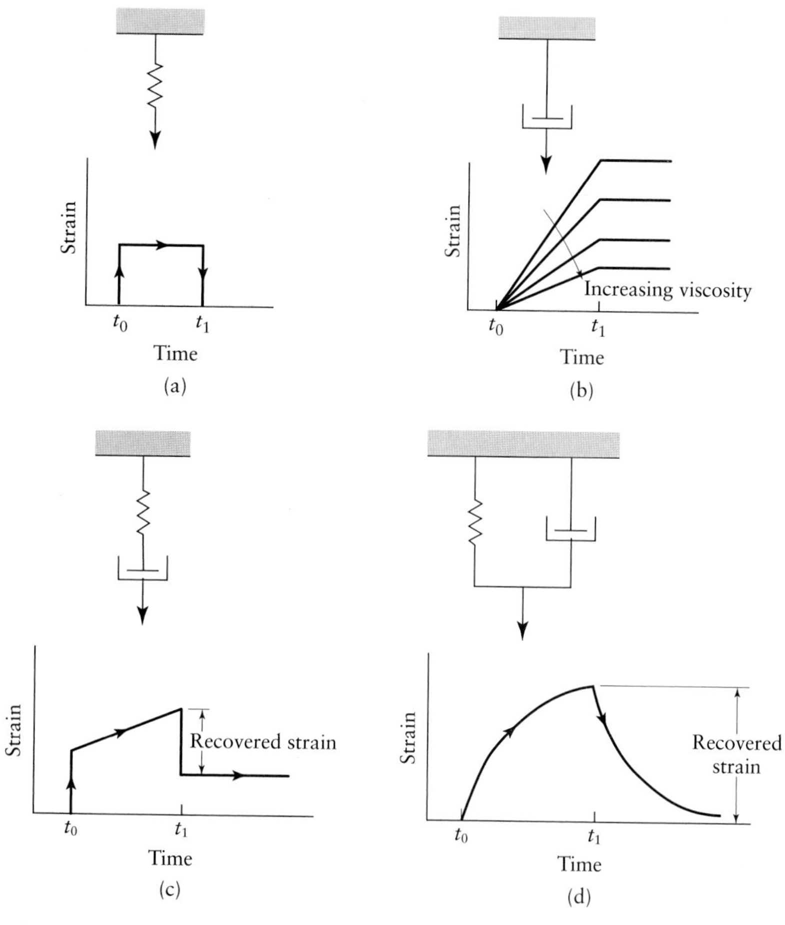

FIGURE 10.7 Various deformation modes for polymers: (a) elastic; (b) viscous; (c) viscoelastic (Maxwell model); and (d) viscoelastic (Voigt or Kelvin model).

temperature. The mechanical properties of several polymers listed in Table 10.1 indicate that thermoplastics are about two orders of magnitude less stiff than metals. Their ultimate tensile strength is about one order of magnitude lower than that of metals. (See Table 2.1.) Figure 10.8 shows typical stress–strain curves for some thermoplastics and thermosets at room temperature. Note that these plastics exhibit different behaviors, which can be described as rigid, soft, brittle, flexible, and so on. Plastics, like metals, undergo fatigue and creep phenomena as well.

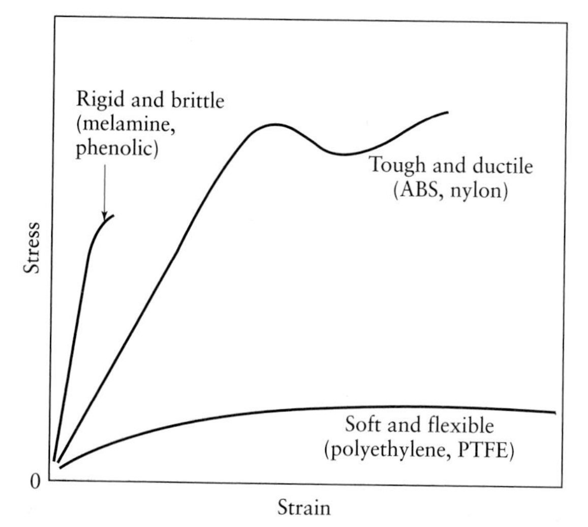

FIGURE 10.8 General terminology describing the behavior of three types of plastics. PTFE (polytetrafluoroethylene) is Teflon, a trade name. *Source*: R. L. E. Brown.

Some of the major characteristics of thermoplastics are described as follows.

a) **Effects of temperature and deformation rate.** The typical effects of temperature on the strength and elastic modulus of thermoplastics are similar to those for metals (see Sections 2.2.6 and 2.2.7); thus, with increasing temperature, the strength and modulus of elasticity decrease, and the toughness increases (Fig. 10.9). The effect of temperature on impact strength is shown in Fig. 10.10; note the large difference in the impact behavior of various polymers.

If we raise the temperature of a thermoplastic polymer above its T_g, it first becomes leathery and then rubbery with increasing temperature (Fig. 10.4). Finally, at higher temperatures, e.g., above T_m for crystalline thermoplastics, it becomes a viscous fluid, with viscosity decreasing with increasing temperature and strain rate. Since viscosity is not constant, polymers display **viscoelastic** behavior, as demonstrated by the spring and dashpot models in Figs. 10.7c and d, known as the *Maxwell* and *Kelvin* (or *Voigt*) models, respectively. When a constant load is applied, the polymer first stretches at a high strain rate and then continues to elongate over a period of time, because of its viscous behavior. Note in the models shown in Fig. 10.7 that the elastic portion of the elongation is reversible (elastic recovery), but the viscous portion is not.

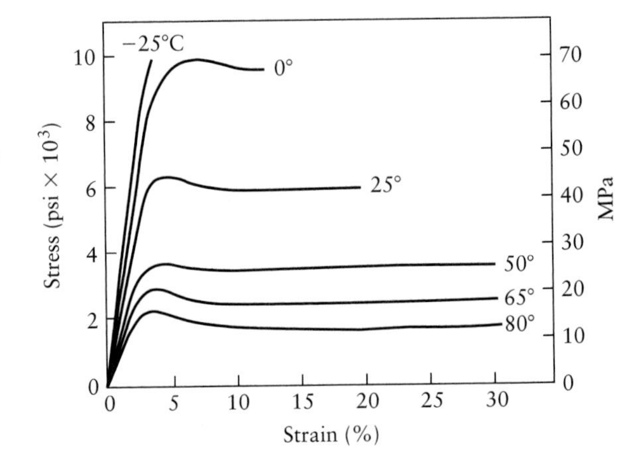

FIGURE 10.9 Effect of temperature on the stress–strain curve for cellulose acetate, a thermoplastic. Note the large drop in strength and increase in ductility with a relatively small increase in temperature. *Source*: After T. S. Carswell and H. K. Nason.

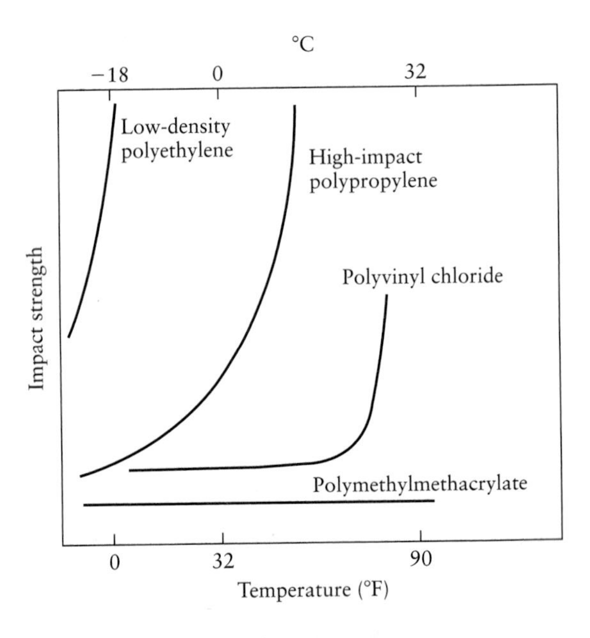

FIGURE 10.10 Effect of temperature on the impact strength of various plastics. Small changes in temperature can have a significant effect on impact strength. *Source*: P. C. Powell.

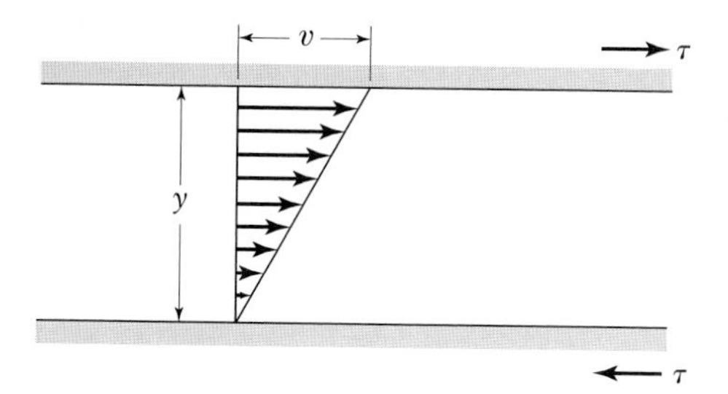

FIGURE 10.11 Parameters used to describe viscosity; see Eq. (10.3).

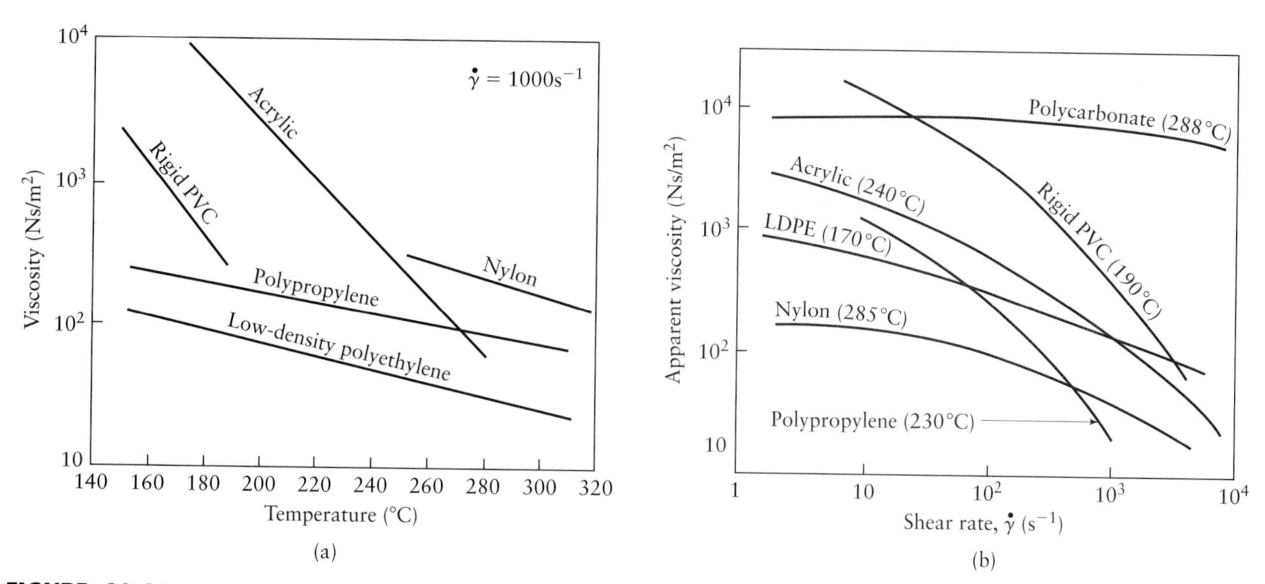

FIGURE 10.12 Viscosity of common thermoplastics as a function of (a) temperature and (b) shear rate. *Source*: From D. H. Morton-Jones, *Polymer Processing*, Chapman and Hall, 1989.

The viscous behavior is expressed by

$$\tau = \eta\left(\frac{dv}{dy}\right) = \eta\dot{\gamma}, \tag{10.3}$$

where η is the *viscosity* and dv/dy is the *shear-strain rate*, $\dot{\gamma}$, as shown in Fig. 10.11. Figure 10.12 provides the viscosities of some polymers. When the shear stress τ is directly proportional to the shear-strain rate, the behavior of the thermoplastic polymer is known as *Newtonian*. For many polymers, Newtonian behavior is not a good approximation, and Eq. (10.3) will give poor predictions of process performance. For example, polyvinyl chloride, polyethylene (low density and high density), and polypropylene have viscosities that decrease markedly with increasing strain rate (*pseudoplastic behavior*). Their viscosity as a function of strain rate can be expressed as

$$\eta = A\dot{\gamma}^{1-n}, \tag{10.4a}$$

where A is the *consistency index* and n is the *power-law index* for the polymer.

Note from the foregoing equations that the viscous behavior of thermoplastics is similar to the strain-rate sensitivity of metals (Section 2.2.7) and is given by the expression

$$\sigma = C\dot{\varepsilon}^{m}, \tag{10.4b}$$

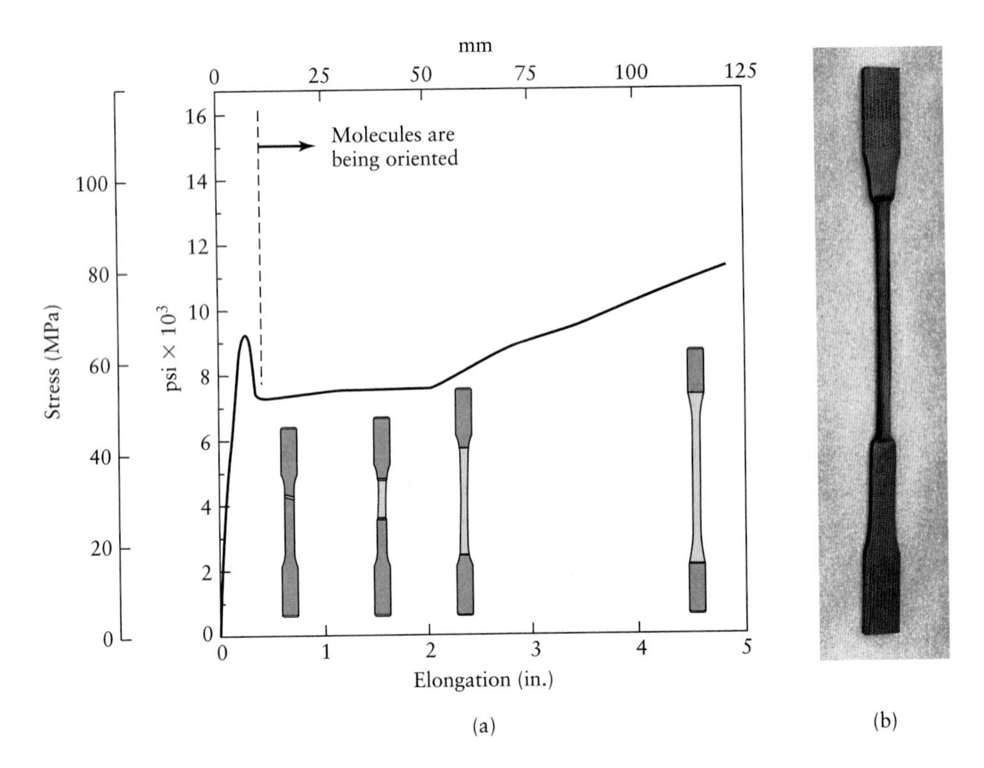

FIGURE 10.13 (a) Load–elongation curve for polycarbonate, a thermoplastic. *Source*: R. P. Kambour and R. E. Robertson. (b) High-density polyethylene tension-test specimen, showing uniform elongation (the long, narrow region in the specimen).

where, for Newtonian behavior, $m = 1$. Thermoplastics have high m values, indicating that they can undergo large uniform deformations in tension before fracture. Note how, unlike with ordinary metals, the necked region elongates considerably (Fig. 10.13). This behavior can easily be demonstrated by stretching a piece of the plastic (polypropylene) holder for six-pack beverage cans. This characteristic, which is the same as in superplastic metals (see Section 2.2.7), enables thermoplastics to be formed into complex shapes, such as bottles for soft drinks, cookie and meat trays, and lighted signs, as described in Section 10.10. Note also that the effect of increasing the rate of load application is that the strength of the polymer is increased.

Between T_g and T_m, thermoplastics exhibit leathery and rubbery behavior, depending on their structure and degree of crystallinity, as shown in Fig. 10.4. A term combining the strains caused by elastic behavior (e_e) and viscous flow (e_v) is called the *viscoelastic modulus*, E_r, and is expressed as

$$E_r = \frac{\sigma}{(e_e + e_v)}. \tag{10.5}$$

This modulus essentially represents a time-dependent elastic modulus.

The viscosity η of polymers (a measure of the resistance of their molecules in sliding along each other) depends on temperature and on the polymer's structure, molecular weight, and pressure. The effect of temperature can be represented by

$$\eta = \eta_0 e^{E/kT}, \tag{10.6}$$

where η_0 is a material constant, E is the activation energy (the energy required to initiate a reaction), k is Boltzmann's constant (the thermal-energy constant, or 13.8×10^{-24} J/K), and T is the temperature (in K). Thus, as the temperature of the polymer is increased, η decreases, because of the higher mobility of the molecules. Note that viscosity may be regarded as the inverse of fluidity of molten metals in casting. Increasing the molecular weight (and hence increasing the length of the chain) increases η, because of the greater number of secondary bonds present. As the molecular-weight distribution widens, there are more shorter chains, and η decreases. The effect of increasing pressure is that viscosity is increased, because of reduced *free volume*, or *free space*, defined as the volume in excess of the true volume of the crystal in the crystalline regions of the polymer.

Based on experimental observations that, at the glass-transition temperature, T_g, polymers have a viscosity η of about 10^{12} Pa-s, an empirical relationship between viscosity and temperature has been developed for linear thermoplastics:

$$\log \eta = 12 - \frac{17.5\Delta T}{52 + \Delta T}. \tag{10.7}$$

In this equation, $\Delta T = T - T_g$, in K or °C; thus, we can estimate the viscosity of the polymer at any temperature.

b) **Creep and stress relaxation.** The terms *creep* and *stress relaxation* have been defined in Section 2.8. Because of their viscoelastic behavior, thermoplastics are particularly susceptible to these phenomena. Referring back to the dashpot models in Figs. 10.7b, c, and d, note that under a constant load, the polymer undergoes further strain, and thus it creeps. The recovered strain depends on the stiffness of the spring and hence the modulus of elasticity. Stress relaxation in polymers occurs over a period of time, according to the relation

$$\tau = \tau_0 e^{-t/\lambda}, \tag{10.8}$$

where τ_0 is the shear stress at time zero, τ is the stress at time t, and λ is the *relaxation time*. This expression may also be given in terms of the normal stresses σ and σ_0. The relaxation time is defined as

$$\lambda = \frac{\eta}{G}, \tag{10.9}$$

where G is the shear modulus of the polymer and $\gamma = \tau/G$; thus, the relaxation time is a function of the material and its temperature. Because both η and G depend on temperature to varying degrees, the relaxation time also depends on temperature.

c) **Orientation.** When thermoplastics are permanently deformed, say by stretching, the long-chain molecules align in the general direction of elongation. This process is called *orientation*, and, just as with metals, the polymer becomes *anisotropic* (see Section 3.5). The specimen becomes stronger and stiffer in the elongated (stretched) direction than in the transverse direction. This process is an important technique to enhance the strength and toughness of polymers; however, orientation weakens the polymer in the transverse direction.

d) **Crazing.** Some thermoplastics, such as polystyrene and polymethylmethacrylate, develop localized, wedge-shaped narrow regions of highly deformed material when subjected to tensile stresses or to bending. This phenomenon is called *crazing*. Although they may appear to be like cracks, crazes are areas of spongy material, typically containing about 50% voids. With increasing tensile load on the specimen, the voids coalesce to form a crack and eventually lead to fracture of the polymer. Crazing has been observed in transparent glassy polymers as well as in other types of polymers. The environment and the presence of solvents, lubricants, and water vapor enhance the formation of crazes (**environmental stress cracking** and **solvent crazing**). Residual stresses in the material contribute to crazing and cracking of the polymer. Radiation and ultraviolet radiation also can adversely affect the crazing behavior in certain polymers.

A related phenomenon is **stress whitening.** When subjected to tensile stresses, such as by folding or bending, the plastic becomes lighter in color. This phenomenon is usually attributed to the formation of microvoids in the material. As a result, the material becomes less translucent (i.e., transmits less light), or more opaque; you can easily demonstrate this result by bending plastic components commonly found in household products and toys.

e) **Water absorption.** An important limitation of some polymers, such as nylons, is their ability to absorb water (*hygroscopy*). Water acts as a plasticizing agent—that is, it makes the polymer more plastic; thus, in a sense, it lubricates the chains in the amorphous region. Typically, with increasing moisture absorption, the glass-transition temperature, the yield stress, and the elastic modulus of the polymer are lowered severely. Dimensional changes also occur because of water absorption, such as in a humid environment.

f) **Thermal and electrical properties.** Compared with metals, plastics are generally characterized by low thermal and electrical conductivity, low specific gravity (ranging from 0.90 to 2.2), and a relatively high coefficient of thermal expansion—about one order of magnitude higher (see Table 3.2). Because polymers generally have low electrical conductivity, they can be used for electrical and electronic components. However, polymers can be made to be electrically conductive, as described in Section 10.7.2.

EXAMPLE 10.2 Lowering the viscosity of a polymer

In processing a batch of polycarbonate at 170°C to make a certain part, it is found that the polycarbonate's viscosity is twice that desired. Determine the temperature at which this polymer should be processed.

Solution. From Table 10.2, we know that the T_g for polycarbonate is 150°C. Its viscosity at 170°C is determined by Eq. (10.7):

$$\log \eta = 12 - \frac{17.5(20)}{52 + 20} = 7.14.$$

Hence, $\eta = 13.8$ MPa-s, and because this magnitude is twice what we want, the new viscosity should be 6.9 MPa-s. We substitute this new value to find the new temperature:

$$\log(6.9 \times 10^6) = 12 - \frac{17.5(\Delta T)}{52 + \Delta T}.$$

Thus, $\Delta T = 21.7$, or the new temperature is $150 + 21.7 = 171.7$°C. Note that the numbers for temperature are rounded and that viscosity is very sensitive to temperature.

EXAMPLE 10.3 **Stress relaxation in a thermoplastic member under tension**

A long piece of thermoplastic is stretched between two rigid supports at a stress level of 5 MPa. After 30 days, the stress level is found to have decayed to one half the original level. How long will it take for the stress level to reach one tenth of the original value?

Solution. Substituting these data into Eq. (10.8) as modified for normal stress, we have

$$2.5 = 5e^{-30/\lambda}.$$

Hence,

$$\ln\left(\frac{2.5}{5}\right) = \frac{-30}{\lambda}, \quad \text{or} \quad \lambda = 43.3 \text{ days}.$$

Therefore,

$$\ln\left(\frac{0.5}{5}\right) = \frac{-t}{43.3}, \quad \text{or} \quad t = 99.7 \text{ days}.$$

10.4 Thermosets: Behavior and Properties

When the long-chain molecules in a polymer are cross-linked in a three-dimensional arrangement, the structure in effect becomes one giant molecule with strong covalent bonds. As previously stated, such polymers are called **thermosetting polymers**, or *thermosets*, because during polymerization, the network is completed, and the shape of the part is permanently set. This **curing** (*cross-linking*) reaction, unlike that of thermoplastics, is irreversible. The response of a thermosetting plastic to temperature can be likened to baking a cake or boiling an egg; once the cake is baked and cooled, or the egg is boiled and cooled, reheating it will not change its shape.

Some thermosets, such as epoxy, polyester, and urethane, cure at room temperature. Although curing takes place at ambient temperature, the heat of the exothermic reaction cures the plastic. Thermosetting polymers do not have a sharply defined glass-transition temperature. The polymerization process for thermosets generally takes place in two stages: (1) The first stage occurs at the chemical plant, where the molecules are partially polymerized into linear chains, and (2) the second stage occurs at the parts-producing plant, where cross-linking is completed under heat and pressure during the molding and shaping of the part (Section 10.10).

Because of the nature of the bonds, strength and hardness of thermosets, unlike those of thermoplastics, are not affected by temperature or rate of deformation. A typical thermoset is phenolic, which is a product of the reaction between phenol and formaldehyde; common products made from this polymer are the handles and knobs on cooking pots and pans, and components of light switches and outlets. Thermosetting plastics (Section 10.6) generally possess better mechanical, thermal, and chemical properties, electrical resistance, and dimensional stability than do thermoplastics. However, if the temperature is increased sufficiently, the thermosetting polymer begins to burn up, degrade, and char.

10.5 Thermoplastics: General Characteristics and Applications

This section outlines the general characteristics and typical applications of major thermoplastics, particularly as they relate to manufacturing plastic products and their service life. Table 10.3 gives general recommendations for various plastics applications. The major thermoplastics are described as follows.

TABLE 10.3

General Recommendations for Plastic Products

Design Requirement	Typical Applications	Plastics
Mechanical strength	Gears, cams, rollers, valves, fan blades, impellers, pistons.	Acetals, nylon, phenolics, polycarbonates, polyesters, polypropylenes, epoxies, polyimides.
Wear resistance	Gears, wear strips and liners, bearings, bushings, roller-skate wheels.	Acetals, nylon, phenolics, polyimides, polyurethane, ultrahigh-molecular-weight polyethylene.
Frictional properties		
High	Tires, nonskid surfaces, footware, flooring.	Elastomers, rubbers.
Low	Sliding surfaces, artificial joints.	Fluorocarbons, polyesters, polyethylene, polyimides.
Electrical resistance	All types of electrical components and equipment, appliances, electrical fixtures.	Polymethylmethacrylate, ABS, fluorocarbons, nylon, polycarbonate, polyester, polypropylenes, ureas, phenolics, silicones, rubbers.
Chemical resistance	Containers for chemicals, laboratory equipment, components for chemical industry, food and beverage containers.	Acetals, ABS, epoxies, polymethylmethacrylate, fluorocarbons, nylon, polycarbonate, polyester, polypropylene, ureas, silicones.
Heat resistance	Appliances, cookware, electrical components.	Fluorocarbons, polyimides, silicones, acetals, polysulfones, phenolics, epoxies.
Functional and decorative features	Handles, knobs, camera and battery cases, trim moldings, pipe fittings.	ABS, acrylics, cellulosics, phenolics, polyethylenes, polpropylenes, polystyrenes, polyvinyl chloride.
Functional and transparent features	Lenses, goggles, safety glazing, signs, food-processing equipment, laboratory hardware.	Acrylics, polycarbonates, polystyrenes, polysulfones.
Housings and hollow shapes	Power tools, housings, sport helmets, telephone cases.	ABS, cellulosics, phenolics, polycarbonates, polyethylenes, polypropylene, polystyrenes.

1. **Acetals** (from acetic and alcohol) have good strength; stiffness; and resistance to creep, abrasion, moisture, heat, and chemicals. Typical applications include mechanical parts and components where high performance is required over a long period: bearings, cams, gears, bushings, rollers, impellers, wear surfaces, pipes, valves, showerheads, and housings. Common trade name: *Delrin*.
2. **Acrylics** (such as **polymethylmethacrylate**, or PMMA) possess moderate strength, good optical properties, and weather resistance. They are transparent, but can be made opaque and are generally resistant to chemicals and have good electrical resistance. Typical applications include lenses, lighted signs, displays, window glazing, skylights, bubble tops, automotive lenses, windshields, lighting fixtures, and furniture. Common trade names: *Orlon*, *Plexiglas*, and *Lucite*.
3. **Acrylonitrile–butadiene–styrene** (ABS) is dimensionally stable and rigid and has good resistance to impact, abrasion, chemicals, and electricity, strength and toughness, and low-temperature properties. Typical applications include pipes, fittings, chrome-plated plumbing supplies, helmets, tool handles, automotive components, boat hulls, telephones, luggage, housing, appliances, refrigerator liners, and decorative panels.
4. **Cellulosics** have a wide range of mechanical properties, depending on their composition. They can be made rigid, strong, and tough. However, they weather poorly and are affected by heat and chemicals. Typical applications include tool handles, pens, knobs, frames for eyeglasses, safety goggles, machine guards, helmets, tubing and pipes, lighting fixtures, rigid containers, steering wheels, packaging film, signs, billiard balls, toys, and decorative parts.
5. **Fluorocarbons** possess good resistance to temperature, chemicals, weather, and electricity; they also have unique nonadhesive properties and low friction. Typical applications include linings for chemical-processing equipment, nonstick coatings for cookware, electrical insulation for high-temperature wire and cable, gaskets, low-friction surfaces, bearings, and seals. Common trade name: *Teflon*.
6. **Polyamides** (from the words *poly*, *amine*, and *carboxyl acid*) are available in two main types—nylons and aramids:

   a) **Nylons** (a coined word) have good mechanical properties and abrasion resistance. They are self-lubricating and resistant to most chemicals. All nylons are hygroscopic (i.e., they absorb water). Moisture absorption reduces mechanical properties and increases part dimensions. Typical applications include gears, bearings, bushings, rollers, fasteners, zippers, electrical parts, combs, tubing, wear-resistant surfaces, guides, and surgical equipment.

   b) **Aramids** (aromatic polyamides) have very high tensile strength and stiffness. Typical applications include fibers for reinforced plastics (composite materials), bulletproof vests, cables, and radial tires. Common trade name: *Kevlar*.
7. **Polycarbonates** are versatile and have good mechanical and electrical properties; they also have high impact resistance and can be made resistant to chemicals. Typical applications include safety helmets, optical lenses, bullet-resistant window glazing, signs, bottles, food-processing equipment, windshields, load-bearing electrical components, electrical insulators, medical apparati, business-machine components, guards for machinery, and parts requiring dimensional stability. Common trade name: *Lexan*.
8. **Polyesters** (thermoplastics; see also Section 10.6) have good mechanical, electrical, and chemical properties; good abrasion resistance; and low friction. Typical applications include gears, cams, rollers, load-bearing members, pumps, and electromechanical components. Common trade names: *Dacron*, *Mylar*, and *Kodel*.

9. **Polyethylenes** possess good electrical and chemical properties. Their mechanical properties depend on their composition and structure. The three major classes of polyethylenes are low density (LDPE), high density (HDPE), and ultrahigh molecular weight (UHMWPE). Typical applications for LDPE and HDPE include housewares, bottles, garbage cans, ducts, bumpers, luggage, toys, tubing, bottles, and packaging material. UHMWPE is used in parts requiring high-impact toughness and abrasive wear resistance; examples include artificial knee and hip joints.
10. **Polyimides** have the structure of a thermoplastic, but the nonmelting characteristic of a thermoset; see Section 10.6. Common trade name: *Torlon*.
11. **Polypropylenes** have good mechanical, electrical, and chemical properties and good resistance to tearing. Typical applications include automotive trim and components, medical devices, appliance parts, wire insulation, TV cabinets, pipes, fittings, drinking cups, dairy-product and juice containers, luggage, ropes, and weather stripping.
12. **Polystyrenes** are inexpensive, have generally average properties, and are somewhat brittle. Typical applications include disposable containers; packaging; trays for meats, cookies, and candy; foam insulation; appliances; automotive, radio, and TV components; housewares; and toys and furniture parts (as a wood substitute).
13. **Polysulfones** have excellent resistance to heat, water, and steam and are highly resistant to some chemicals, but are attacked by organic solvents. Typical applications include steam irons, coffeemakers, hot-water containers, medical equipment that requires sterilization, power-tool and appliance housings, aircraft cabin interiors, and electrical insulators.
14. **Polyvinyl chloride** (PVC) has a wide range of properties, is inexpensive and water resistant, and can be made rigid or flexible. It is not suitable for applications that require strength and heat resistance. **Rigid PVC** is tough and hard and is used for signs and in the construction industry, such as for pipes and conduits. **Flexible PVC** is used in wire and cable coatings, low-pressure flexible tubing and hose, footwear, imitation leather, upholstery, records, gaskets, seals, trim, film, sheet, and coatings. Common trade names: *Saran* and *Tygon*.

10.6 Thermosets: General Characteristics and Applications

This section outlines the general characteristics and typical applications of major thermosetting plastics:

1. **Alkyds** (from *alkyl*, meaning alcohol, and *acid*) possess good electrical insulating properties, impact resistance, and dimensional stability and have low water absorption. Typical applications include electrical and electronic components.
2. **Aminos** (**urea** and **melamine**) have properties that depend on composition. Generally, aminos are hard and rigid and are resistant to abrasion, creep, and electrical arcing. Typical applications include small appliance housings, countertops, toilet seats, handles, and distributor caps. Urea is used for electrical and electronic components, and melamine is used for dinnerware.
3. **Epoxies** have excellent mechanical and electrical properties, dimensional stability, strong adhesive properties, and good resistance to heat and chemicals. Typical applications include electrical components that require mechanical strength and

high insulation, tools and dies, and adhesives. *Fiber-reinforced epoxies* have excellent mechanical properties and are used in pressure vessels, rocket motor casings, tanks, and similar structural components.

4. **Phenolics**, although brittle, are rigid; dimensionally stable; and have high resistance to heat, water, electricity, and chemicals. Typical applications include knobs, handles, laminated panels, telephones, bond material to hold abrasive grains together in grinding wheels, and electrical components, such as wiring devices, connectors, and insulators.
5. **Polyesters** (thermosetting plastics; see also Section 10.5) have good mechanical, chemical, and electrical properties. Polyesters are generally reinforced with glass or other fibers. Typical applications include boats, luggage, chairs, automotive bodies, swimming pools, and material for impregnating cloth and paper. Polyesters are also available as casting resins.
6. **Polyimides** possess good mechanical, physical, and electrical properties at elevated temperatures; they also have creep resistance and low friction and wear characteristics. Polyimides have the nonmelting characteristics of a thermoset, but the structure of a thermoplastic. Typical applications include pump components (bearings, seals, valve seats, retainer rings, and piston rings); electrical connectors for high-temperature use; aerospace parts; high-strength, impact-resistant structures; sports equipment; and safety vests.
7. **Silicones** have properties that depend on composition. Generally, they weather well, possess excellent electrical properties over a wide range of humidities and temperatures, and resist chemicals and heat. (See also Section 10.8.) Typical applications include electrical components that require strength at elevated temperatures, oven gaskets, heat seals, and waterproof materials.

10.7 High-Temperature Polymers, Electrically Conducting Polymers, and Biodegradable Plastics

This section describes three important trends in the development of polymers.

10.7.1 High-temperature polymers

During the past several years, there have been important advances in polymers and polymer blends for high-temperature applications, particularly in the aerospace industry. Polymers with properties such as high strength-to-weight ratio and fatigue strength are being further enhanced with high-temperature resistance. This resistance, as in metals, may be short term at relatively high temperatures or long term at lower temperatures. (See, for example, Section 3.11.6 for a similar consideration regarding titanium alloys.)

Short-term exposure to high temperatures generally requires *ablative* materials to dissipate the heat, such as phenolic–silicone copolymers, which have been used for rocket and missile components at temperatures of thousands of degrees. Long-term exposure for polymers is presently confined to temperatures on the order of 260°C (500°F). High-temperature thermoplastic polymers include fluorine-containing thermoplastics, polyketones, and polyimides. High-temperature thermosetting polymers include phenolic resins, epoxy resins, silicone-based thermosetting polymers, and phenolic–fiberglass systems. Several other polymers with complex structures are being developed for high-temperature applications.

10.7.2 Electrically conducting polymers

Most polymers are electrical insulators and hence are used as insulators for electrical equipment (e.g., light switches, outlets, and circuit breakers) and as packaging material for electronic components. In addition, the electrical conductivity of some polymers can be increased by **doping**, i.e., introducing certain impurities in the polymer, such as metal powder, salts, and iodides. The conductivity of polymers increases with moisture absorption, and the electronic properties of polymers can be altered by irradiation as well. Applications of electrically conducting polymers include microelectronic devices, rechargeable batteries, capacitors, fuel cells, catalysts, fuel-level sensors, de-icer panels, antistatic coatings, and as conducting adhesives for surface-mount technologies (see Section 12.13.4).

10.7.3 Biodegradable plastics

One third of plastics produced are in the disposable-products sector, such as bottles, packaging, and garbage bags. With the growing use of plastics and increasing concern over environmental issues regarding disposal of plastic products and limited landfills, major efforts are underway to develop biodegradable plastics. Three different **biodegradable plastics** are currently available: starch-based bioplastics, lactic-based bioplastics, and bioplastics made from the fermentation of sugar. These plastics degrade over various periods of time, from a few months to a few years, and have different degradability characteristics. They are designed to degrade completely when exposed to microorganisms in soil or water, without producing toxic by-products. It should also be noted that the costs of biodegradable plastics are substantially higher than those of synthetic polymers.

Three types of biodegradable plastics are described as follows:

a) The **starch-based system** is the farthest along of the three types of bioplastics in terms of production capacity. Starch may be extracted from potatoes, wheat, rice, and corn. In this system, starch granules are processed into a powder, which is heated and becomes a sticky liquid; various additives and binders are also blended in. The liquid is then cooled, formed into pellets, and processed in conventional plastic-processing equipment (Section 10.10). Various additives are blended with the starch to impart special characteristics to the bioplastic materials.

b) In the **lactic-based system**, fermenting corn or other feedstocks produce lactic acid, which is then polymerized to form a polyester resin.

c) In the third system, **organic acids** are added to a **sugar** feedstock. Through a specially developed process, the resulting reaction produces a highly crystalline and very stiff polymer that, after further processing, behaves in a manner similar to polymers developed from petroleum.

Recycling. Because the development of biodegradable plastics is relatively new, their long-range performance, both during their useful life cycle as products and in landfills, has not been fully assessed. There is concern that emphasis on biodegradability will divert attention from the issue of *recyclability* of plastics and the efforts for *conservation* of materials and energy. Recycled plastics are being used increasingly for a variety of products, including automotive-body components. Plastic products now carry the following numerals within a triangular mark, indicating the corresponding type of plastic: 1—PETE (polyethylene), 2—HDPE (high-density polyethylene), 3—V (vinyl), 4—LDPE (low-density polyethylene), 5—PP (polypropylene), 6—PS (polystyrene), 7—others.

10.8 Elastomers (Rubbers): General Characteristics and Applications

The terms *rubber* and *elastomer* are often used interchangeably. Generally, however, an **elastomer** is defined as being capable of recovering substantially in shape and size after a load has been removed; **rubber** is defined as being capable of recovering from large deformations quickly.

Elastomers comprise a large family of amorphous polymers that have a low glass-transition temperature. They have the characteristic ability to undergo large elastic deformations without rupture, and they are soft and have a low elastic modulus. The term *elastomer* is derived from the words *elastic* and *mer*. The structure of these polymers is highly kinked (tightly twisted or curled). They stretch, but then return to their original shape after the load is removed (Fig. 10.14). They can be cross-linked; the best example is the elevated-temperature **vulcanization** of rubber with sulfur, discovered by Charles Goodyear in 1839 and named for Vulcan, the Roman god of fire. Once the elastomer is cross-linked, it cannot be reshaped. For example, an automobile tire, which is one giant molecule, cannot be softened and reshaped.

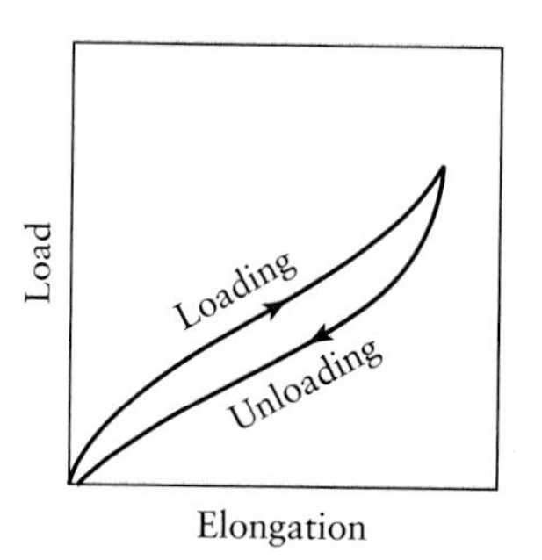

FIGURE 10.14 Typical load–elongation curve for rubbers. The clockwise loop, indicating loading and unloading paths, is the hysteresis loss. Hysteresis gives rubbers the capacity to dissipate energy, damp vibration, and absorb shock loading, as in automobile tires and vibration dampeners placed under machinery.

The hardness of elastomers, which is measured with a durometer (see Section 2.6.7), increases with increasing cross-linking of the molecular chains. A variety of additives can be blended with elastomers to impart specific properties, as with plastics. Elastomers have a wide range of applications, such as high-friction and nonskid surfaces, protection against corrosion and abrasion, electrical insulation, and shock and vibration insulation. Specific examples include tires, hoses, weather stripping, footwear, linings, gaskets, seals, printing rolls, and flooring. A characteristic of elastomers is their hysteresis loss in stretching or compression (Fig. 10.14). The clockwise loop in Fig. 10.14 indicates energy loss, whereby mechanical energy is converted into heat; this property is desirable for absorbing vibrational energy (damping) and sound deadening.

Some of the major types of elastomers are described below:

a) **Natural rubber.** The base for natural rubber is **latex**, a milklike sap obtained from the inner bark of a tropical tree. It has good resistance to abrasion and fatigue and high frictional properties, but low resistance to oil, heat, ozone, and sunlight. Typical applications include tires, seals, shoe heels, couplings, and engine mounts.

b) **Synthetic rubbers.** Further developed than natural rubbers, synthetic rubbers include synthetic natural rubber, butyl, styrene butadiene, polybutadiene, and ethylene propylene. Compared with natural rubbers, synthetic rubbers have improved resistance to heat, gasoline, and chemicals and a higher useful-temperature range. Examples of synthetic rubbers that are resistant to oil include neoprene, nitrile, urethane, and silicone. Typical applications of synthetic rubbers include tires, shock absorbers, seals, and belts.

c) **Silicones.** Silicones (see also Section 10.6) have the highest useful temperature range of all elastomers, up to 315°C (600°F), but their other properties (such as strength and resistance to wear and oils) are generally inferior to those of other elastomers. Typical applications include seals, gaskets, thermal insulation, high-temperature electrical switches, and electronic apparati.

d) **Polyurethane.** This elastomer has very good overall properties of high strength, stiffness, and hardness and exceptional resistance to abrasion, cutting, and tearing. Typical applications include seals, gaskets, cushioning, diaphragms for rubber forming of sheet metals (see Section 7.7), and auto-body parts.

10.9 Reinforced Plastics

Among the major developments in materials are *reinforced plastics* (**composite materials**); in fact, composites are now one of the most important classes of **engineered materials**, as they offer several outstanding properties as compared with conventional materials. These materials can be defined as a combination of two or more chemically distinct and insoluble phases whose properties and structural performance are superior to those of the constituents acting independently. It has been shown, for example, that plastics possess mechanical properties (particularly strength, stiffness, and creep resistance) that are generally inferior to those of metals and alloys.

These properties can be improved by embedding reinforcements of various types, such as glass or graphite fibers, to produce reinforced plastics. As shown in Table 10.1, reinforcements improve the strength, stiffness, and creep resistance of plastics, as well as their strength-to-weight and stiffness-to-weight ratios. Reinforced plastics have found increasingly wider applications in aircraft, space vehicles, offshore structures, piping, electronics, automobiles, boats, ladders, and sporting goods.

The oldest example of composites is the addition of straw to clay for making mud huts and bricks for structural use, dating back to 4000 BC. In that application, the straws are the reinforcing fibers, and the clay is the matrix. Another example of a composite material is the reinforcing of masonry and concrete with iron rods, begun in the 1800s. In fact, concrete itself is a composite material, consisting of cement, sand, and gravel. In reinforced concrete, steel rods impart the necessary tensile strength to the composite, since concrete is brittle and generally has little or no useful tensile strength.

10.9.1 Structure of polymer-matrix-reinforced plastics

Reinforced plastics consist of **fibers** (the *discontinuous* or *dispersed* phase) in a polymer **matrix** (the *continuous* phase). Commonly used fibers include glass, graphite, aramids, and boron (Table 10.4). These fibers are strong and stiff and have a high

TABLE 10.4

Typical Properties of Reinforcing Fibers

Type	Tensile Strength (MPa)	Elastic Modulus (GPa)	Density (kg/m^3)	Relative Cost
Boron	3500	380	2600	Highest
Carbon				
High strength	3000	275	1900	Low
High modulus	2000	415	1900	Low
Glass				
E type	3500	73	2480	Lowest
S type	4600	85	2540	Lowest
Kevlar				
29	2800	62	1440	High
49	2800	117	1440	High
129	3200	85	1440	High
Nextel				
312	1630	135	2700	High
610	2770	328	3960	High
Spectra				
900	2270	64	970	High
1000	2670	90	970	High

Note: These properties vary significantly, depending on the material and method of preparation. Strain to failure for these fibers is typically in the range of 1.5% to 5.5%.

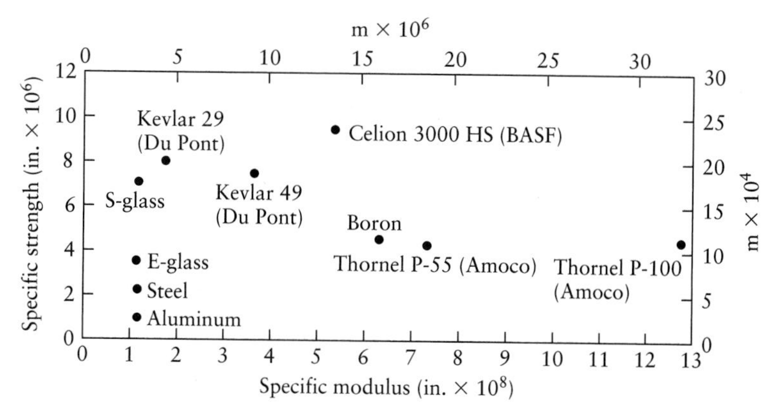

FIGURE 10.15 Specific tensile strength (ratio of tensile strength to density) and specific tensile modulus (ratio of modulus of elasticity to density) for various fibers used in reinforced plastics. Note the wide range of specific strengths and stiffnesses available.

specific strength (strength-to-weight ratio) and specific modulus (stiffness-to-weight ratio), as shown in Fig. 10.15. However, they are generally brittle and abrasive and lack toughness; thus, the fibers, by themselves, have little structural value. The plastic matrix is less strong and less stiff, but tougher than the fibers; reinforced plastics combine the advantages of each of the two constituents (Table 10.5). When more than one type of fiber is used in a reinforced plastic, the composite is called a **hybrid**, which generally has even better properties than a reinforced plastic with only one type of fiber.

In addition to high specific strength and specific modulus, reinforced plastic structures have improved fatigue resistance, greater toughness, and higher creep resistance than those of unreinforced plastics. These structures are relatively easy to design, fabricate, and repair. The percentage of fibers (by volume) in reinforced plastics usually ranges between 10% and 60%. Practically, the percentage of fiber in a matrix

TABLE 10.5

Types and General Characteristics of Reinforced Plastics and Metal-Matrix and Ceramic-Matrix Composites

Material	Characteristics
FIBER	
Glass	High strength, low stiffness, high density; E (calcium aluminoborosilicate) and S (magnesia–aluminosilicate) types are commonly used; lowest cost.
Graphite	Available typically as high modulus or high strength; less dense than glass; low cost.
Boron	High strength and stiffness; has tungsten filament at its center (coaxial); highest density; highest cost.
Aramids (Kevlar)	Highest strength-to-weight ratio of all fibers; high cost.
Other	Nylon, silicon carbide, silicon nitride, aluminum oxide, boron carbide, boron nitride, tantalum carbide, steel, tungsten, and molybdenum; see Chapters 3, 8, 9, and 10.
MATRIX	
Thermosets	Epoxy and polyester, with the former most commonly used; others are phenolics, fluorocarbons, polyethersulfone, silicon, and polyimides.
Thermoplastics	Polyetheretherketone; tougher than thermosets, but lower resistance to temperature.
Metals	Aluminum, aluminum–lithium alloy, magnesium, and titanium; fibers used are graphite, aluminum oxide, silicon carbide, and boron.
Ceramics	Silicon carbide, silicon nitride, aluminum oxide, and mullite; fibers used are various ceramics.

is limited by the average distance between adjacent fibers or particles. The highest practical fiber content is 65%; higher percentages generally result in diminished structural properties.

10.9.2 Reinforcing fibers: characteristics and manufacture

The major types of reinforcing fibers are described as follows:

1. **Polymer fibers.** Most polymer fibers are used in the textile industry and are mass produced. The most common reinforcing fibers are **aramids** (such as *Kevlar*); they are among the toughest fibers and have very high specific strength (Fig. 10.15 and Table 10.4). They can undergo some plastic deformation before fracture and thus have higher toughness than that of brittle fibers; however, aramids absorb moisture, which reduces their properties and complicates their application, as hygrothermal stresses must be considered. Other polymer reinforcements include rayon, nylon, and acrylics.

   A more recent high-performance polyethylene fiber is *Spectra* (a trade name); it has ultrahigh molecular weight and high molecular-chain orientation. The fiber has better abrasion resistance and flexural fatigue than the aramid fiber and at a similar cost. In addition, because of its lower density, it has higher specific strength and specific stiffness than the aramid fiber. However, a low melting point and poor adhesion characteristics as compared with those of other polymers are its major limitations. New polymer fibers are continually being introduced, including *Nextel*.

   Polymers are either melt spun or dry spun to make fibers. **Melt spinning** involves extruding a liquid polymer through small die holes, or *spinnerettes*. The fibers are then cooled before being gathered and wound onto bobbins. The fibers may be stretched to further orient the polymer in the fibers. In **dry spinning**, the polymer is dissolved in a liquid solution to form a partially oriented liquid-crystal form. As the polymer passes through the spinnerette, it is further oriented, and at this point, the fibers are washed, dried, and wound. Aramids are oriented in solution and are fully oriented when they pass through the spinnerette and therefore do not need to be further drawn.

2. **Glass fibers.** Glass fibers are the most widely used and least expensive of all fibers. (See also Section 11.10.2.) The composite material is called *glass-fiber reinforced plastic* (GFRP) and may contain between 30% and 60% glass fibers by volume. Glass fibers are made by drawing molten glass through small openings in a platinum die. The molten glass is then mechanically elongated, cooled, and wound on a roll. A protective coating or sizing may be applied to facilitate the passage of the glass fibers through machinery. The glass fibers are treated with **silane** (a silicon hydride) for improved wetting and bonding between the fiber and the matrix.

   The principal types of glass fibers are (1) the **E type**, a calcium aluminoborosilicate glass, which is used most often; (2) the **S type**, a magnesia–aluminosilicate glass, which has higher strength and stiffness and is more expensive than the other two types and (3) the **E-CR type**, which offers higher resistance to elevated temperature and acid corrosion than the other two types.

3. **Graphite fibers.** Graphite fibers (Fig. 10.16a), although more expensive than glass fibers, have a combination of low density, high strength, and high stiffness; the product is called *carbon-fiber reinforced plastic* (CFRP). Graphite fibers are made by pyrolysis of organic **precursors**, commonly polyacrylonitrile (PAN), because of its lower cost; rayon and pitch (the residue from catalytic crackers in petroleum refining) also can be used as precursors. **Pyrolysis** is the term used

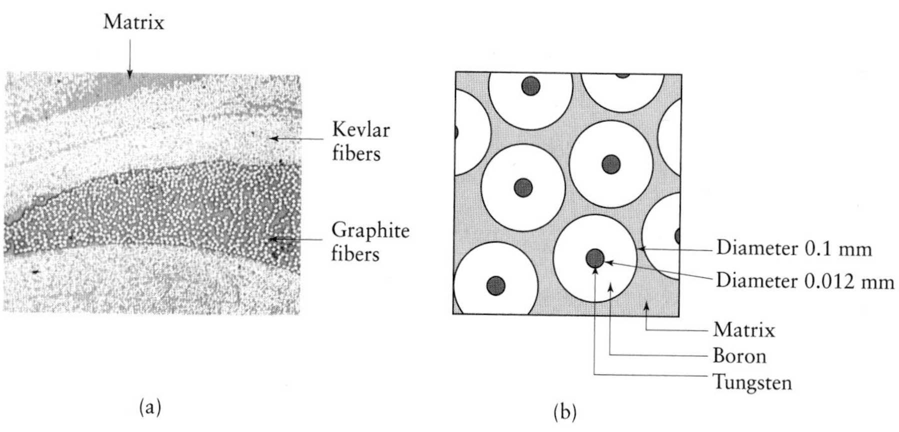

FIGURE 10.16 (a) Cross-section of a tennis racket, showing graphite and aramid (Kevlar) reinforcing fibers. *Source*: J. Dvorak, Mercury Marine Corporation, and F. Garrett, Wilson Sporting Goods Co. (b) Cross-section of boron-fiber-reinforced composite material.

to describe the process of inducing chemical changes by heat, such as burning a length of yarn, which becomes carbon and black in color. With PAN, the fibers are partially cross-linked at a moderate temperature in order to prevent melting during subsequent processing steps, and the fibers are simultaneously elongated. At this point, the fibers are carburized, that is, they are exposed to elevated temperature to expel the hydrogen (dehydrogenation) and nitrogen (denitrogenation) from the PAN. The temperatures range up to about 1500°C (2730°F) for carbonizing and to 3000°C (5400°F) for graphitizing.

The difference between carbon and graphite, although the terms are often used interchangeably, depends on the temperature of pyrolysis and the purity of the material. Carbon fibers are generally 80% to 95% carbon, and graphite fibers are usually more than 99% carbon; the rest is graphite and carbon, respectively. The fibers are classified by the magnitude of their elastic modulus, which typically ranges from 35 GPa to 800 GPa; tensile strengths typically range from 250 MPa to 2600 MPa.

Conductive graphite fibers are being produced to make it possible to enhance the electrical and thermal conductivity of reinforced plastics, such as in electromagnetic and radio-frequency shielding and lightning protection. These fibers are coated with a metal, usually nickel, typically to a thickness of 0.5 μm (20 μin.) on a graphite-fiber core 7 μm (280 μin.) in diameter, using a continuous electroplating process.

4. **Boron fibers.** Boron fibers consist of boron deposited (by chemical vapor-deposition techniques) on tungsten fibers (Fig. 10.16b), although boron can also be deposited on carbon fibers. These fibers have favorable properties, such as high strength and stiffness in tension and compression and resistance to high temperatures. However, because of the use of tungsten, they have high density and are expensive, thus increasing the cost and weight of the reinforced plastic component.

5. **Other fibers.** Other fibers that are used for composite materials include silicon carbide, silicon nitride, aluminum oxide, sapphire, steel, tungsten, molybdenum, boron carbide, boron nitride, and tantalum carbide. (See also Chapter 11.) Metallic fibers are drawn as described in Section 6.5, although at the smaller diameters, the wires are drawn in bundles. **Whiskers** (see Section 3.8.3) also are used as reinforcing fibers. Whiskers are tiny needlelike single crystals that grow to 1–10 μm (40–400 μin.) in diameter and have aspect ratios (defined as the ratio of fiber length to diameter) ranging from 100 to 15,000. Because of their small size, either the fibers are free of imperfections or the imperfections they contain do not significantly affect their strength, which approaches the theoretical strength of the material (Section 3.3.2).

10.9.3 Fiber size and length

The mean diameter of fibers used in reinforced plastics is usually less than 0.01 mm (0.0004 in.). The fibers are very strong and rigid in tension. The reason is that the molecules in the fibers are oriented in the longitudinal direction, and their cross-sections are so small that the probability is low that any defects exist in the fiber. Glass fibers, for example, can have tensile strengths as high as 4600 MPa (650 ksi), whereas the strength of glass in bulk form is much lower; thus, glass fibers are stronger than steel.

Fibers are classified as **short** or **long** fibers, also called **discontinuous fibers** or **continuous fibers**, respectively. Short fibers generally have an *aspect ratio* between 20 and 60, and long fibers have an aspect ratio between 200 and 500. The short- and long-fiber designations are, in general, based on the following observations: In a given fiber, if the mechanical properties improve as a result of increasing the fiber length, then the fiber is denoted as a short fiber. When no additional improvement in properties occurs, the fiber is denoted as a long fiber. In addition to the discrete fibers just described, reinforcements in composites may be in the form of continuous *roving* (slightly twisted strand of fibers), *woven* fabric (similar to cloth), *yarn* (twisted strand), and *mats* of various combinations. Reinforcing elements may also be in the form of particles and flakes.

10.9.4 Matrix materials

The matrix in reinforced plastics has three functions:

1. Support and transfer the stresses to the fibers, which carry most of the load;
2. Protect the fibers against physical damage and the environment;
3. Reduce propagation of cracks in the composite by virtue of the ductility and toughness of the plastic matrix.

Matrix materials are usually epoxy, polyester, phenolic, fluorocarbon, polyethersulfone, or silicon. The most commonly used matrix materials are epoxies (80% of all reinforced plastics) and polyesters, which are less expensive than epoxies. Polyimides, which resist exposure to temperatures in excess of 300°C (575°F), are being developed for use with graphite fibers. Some thermoplastics, such as polyetheretherketone (PEEK), also are used as matrix materials; they generally have higher toughness than thermosets, but their resistance to temperature is lower, being limited to the range of 100–200°C (200–400°F).

10.9.5 Properties of reinforced plastics

The properties of reinforced plastics depend on the type, shape, and orientation of the reinforcing material; the length of the fibers; and the volume fraction (percentage) of the reinforcing material. Short fibers are less effective than long fibers (Fig. 10.17), and their properties are strongly influenced by time and temperature. Long fibers transmit the load through the matrix better and thus are commonly used in critical applications, particularly at elevated temperatures. Fiber reinforcement also affects many other properties of composites.

A critical factor in reinforced plastics is the strength of the bond between the fiber and the polymer matrix, since the load is transmitted through the fiber–matrix interface. Weak bonding causes **fiber pullout** and **delamination** of the structure, particularly under adverse environmental conditions. Adhesion at the interface can be improved by special surface treatments, such as coatings and coupling agents. Glass fibers, for example, are treated with *silane* (see Section 10.9.2) for improved wetting

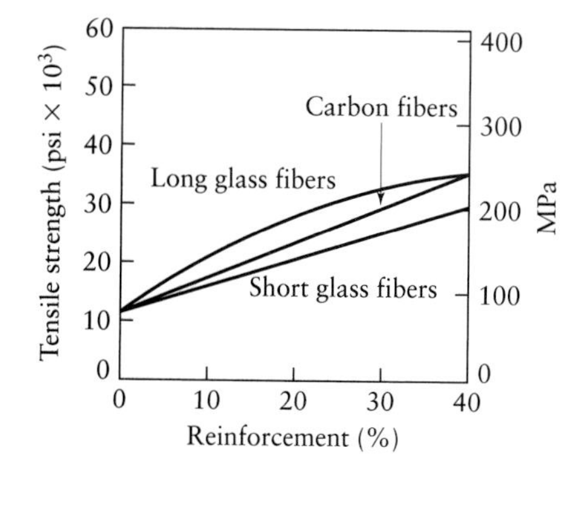

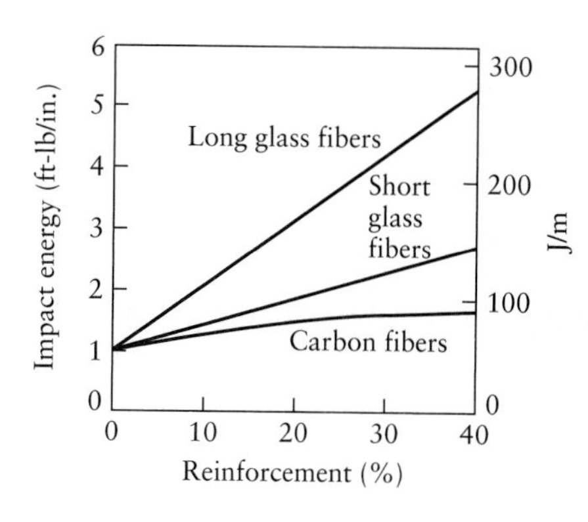

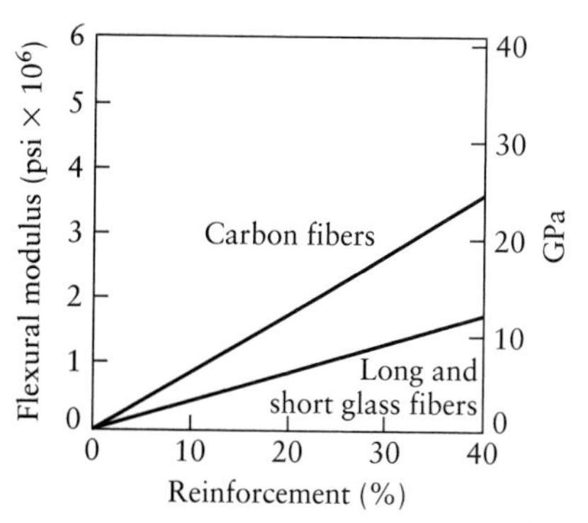

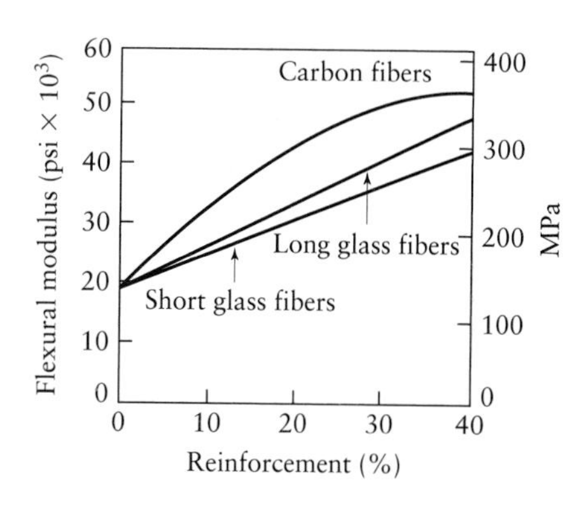

FIGURE 10.17 Effect of the percentage of reinforcing fibers and fiber length on the mechanical properties of reinforced nylon. Note the significant improvement with increasing percentage of fiber reinforcement. *Source*: Courtesy of Wilson Fiberfill International.

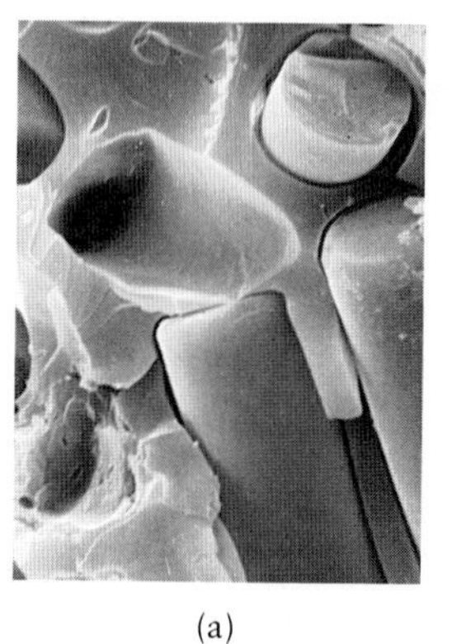
(a)

(b)

FIGURE 10.18 (a) Fracture surface of glass-fiber-reinforced epoxy composite. The fibers are $10 \mu m$ ($400 \mu in.$) in diameter and have random orientation. (b) Fracture surface of a graphite-fiber-reinforced epoxy composite. The fibers, 9–$11 \mu m$ in diameter, are in bundles and are all aligned in the same direction. *Source*: L. J. Broutman.

and bonding between the fiber and the matrix. The importance of proper bonding can be appreciated by inspecting Fig. 10.18, which shows the fracture surfaces of a reinforced plastic.

Generally, the highest stiffness and strength in reinforced plastics is obtained when the fibers are aligned in the direction of the tension force; this composite is, of course, highly anisotropic (Fig. 10.19). As a result, other properties of the composite, such as stiffness, creep resistance, thermal and electrical conductivity, and thermal expansion, are also anisotropic. The transverse properties of such a unidirectionally reinforced structure are much lower than the longitudinal properties. Note, for example, how easily you can split fiber-reinforced packaging tape, yet how strong it is when you pull on it (tension).

For a specific service condition, we can give a reinforced-plastic part an optimal configuration. For example, if the reinforced-plastic part is to be subjected to forces in different directions (such as thin-walled, pressurized vessels), the fibers are

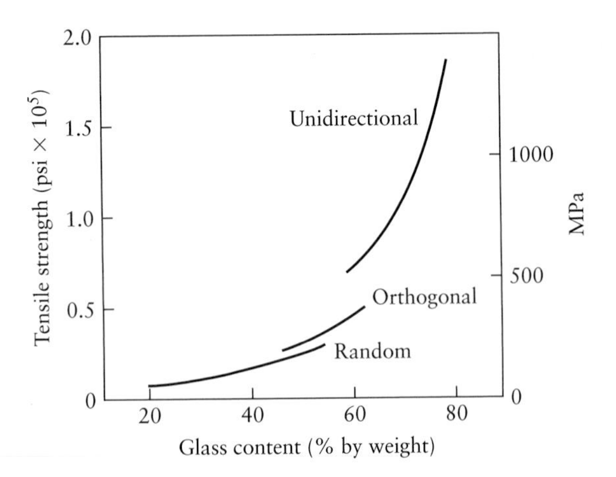

FIGURE 10.19 The tensile strength of glass-reinforced polyester as a function of fiber content and fiber direction in the matrix. *Source*: R. M. Ogorkiewicz, *The Engineering Properties of Plastics*, Oxford, U.K.: Oxford University Press, 1977.

crisscrossed in the matrix. (See the discussion of *filament winding* in Section 10.11.2.) Reinforced plastics may also be made with various other materials and shapes of the polymer matrix in order to impart specific properties (such as permeability and dimensional stability), to make processing easier, and to reduce costs.

Strength and elastic modulus of reinforced plastics. The strength of a reinforced plastic with longitudinal fibers can be determined in terms of the strength of the fibers and matrix, respectively, and the volume fraction of fibers in the composite. In the equations given next, c refers to the composite, f to the fiber, and m to the matrix. The total load F_c on the composite is shared by the fiber load F_f and the matrix load F_m; thus,

$$F_c = F_f + F_m, \tag{10.10}$$

which can be written as

$$\sigma_c A_c = \sigma_f A_f + \sigma_m A_m, \tag{10.11}$$

where A_c, A_f, and A_m are the cross-sectional areas of the composite, fiber, and matrix, respectively, and $A_c = A_f + A_m$. Let's now denote x as the area fraction of the fibers in the composite. (Note that x also represents the volume fraction, because the fibers are uniformly longitudinal in the matrix.) Equation (10.11) can now be rewritten as

$$\sigma_c = x\sigma_f + (1 - x)\sigma_m. \tag{10.12}$$

We can now calculate the fraction of the total load carried by the fibers. First, note that in the composite under a tension load, the strains sustained by the fibers and the matrix are the same (that is, $e_c = e_f = e_m$), and then recall from Section 2.2 that

$$e = \frac{\sigma}{E} = \frac{F}{AE}.$$

Consequently,

$$\frac{F_f}{F_m} = \frac{A_f E_f}{A_m E_m}. \tag{10.13}$$

Since we know the relevant quantities for a specific case, we can use Eq. (10.10) to determine the fraction F_f/F_c. Then, using the foregoing relationships, we can also calculate the elastic modulus E_c of the composite by replacing σ in Eq. (10.12) with E. Hence,

$$E_c = xE_f + (1 - x)E_m. \tag{10.14}$$

EXAMPLE 10.4 **Properties of a graphite–epoxy-reinforced plastic**

Let's assume that a graphite–epoxy-reinforced plastic with longitudinal fibers contains 20% graphite fibers with a strength of 2500 MPa and an elastic modulus of 300 GPa. The strength of the epoxy matrix is 120 MPa, and it has an elastic modulus of 100 GPa. Calculate the elastic modulus of the composite and the fraction of the load supported by the fibers.

Solution. The data given are $x = 0.2$, $E_f = 300$ GPa, $E_m = 100$ GPa, $\sigma_f = 2500$ MPa, and $\sigma_m = 120$ MPa. Using Eq. (10.14), we have

$$E_c = 0.2(300) + (1 - 0.2)100 = 60 + 80 = 140 \text{ GPa}.$$

The load fraction F_f/F_m is obtained from Eq. (10.13):

$$\frac{F_f}{F_m} = \frac{(0.2)(300)}{(0.8)(100)} = 0.75.$$

Since

$$F_c = F_f + F_m$$

and

$$F_m = \frac{F_f}{0.75},$$

we obtain

$$F_c = F_f + \frac{F_f}{0.75} = 2.33F_f,$$

or

$$F_f = 0.43F_c.$$

Hence, the fibers support 43% of the load, even though they occupy only 20% of the cross-sectional area (and hence volume) of the composite.

10.9.6 Applications

The first application of reinforced plastics (in 1907) was for an acid-resistant tank, made of a phenolic resin with asbestos fibers. Formica, commonly used for countertops, was developed in the 1920s. Epoxies were first used as a matrix in the 1930s. Beginning in the 1940s, boats were made with fiberglass, and reinforced plastics were used for aircraft, electrical equipment, and sporting goods. Major developments in composites began in the 1970s, and these materials are now called **advanced composites.** Glass- or carbon-fiber-reinforced hybrid plastics are being developed for high-temperature applications, with continuous use ranging up to about 300°C (550°F).

Reinforced plastics are typically used in military and commercial aircraft and rocket components, helicopter blades, automotive bodies, leaf springs, drive shafts, pipes, ladders, pressure vessels, sporting goods, sports and military helmets, boat hulls, and various other structures. Applications of reinforced plastics include components in the McDonnell Douglas DC-10 aircraft, the Lockheed Martin L-1011 aircraft, and the Boeing aircraft. The Boeing 777 is made of about 9% composites by total weight, which is triple the composite content of prior Boeing transport aircraft. The floor beams and panels and most of the vertical and horizontal tail are made of composite materials.

By virtue of the resulting weight savings, reinforced plastics have reduced aircraft fuel consumption by about 2%. The newly designed Airbus jumbo jet A380, with a capacity of 550 to 700 passengers and to be in service in 2006, will have horizontal stabilizers, ailerons, wing boxes and leading edges, secondary mounting brackets of the fuselage, and the deck structure made of composites with carbon fibers, thermosetting resins, and thermoplastics. The upper fuselage will be made of alternating layers of aluminum and glass-fiber-reinforced epoxy prepregs.

Substitution of aluminum in large commercial aircraft with graphite–epoxy-reinforced plastics could reduce both weight and production costs by 30%, with improved fatigue and corrosion resistance. The structure of the Lear Fan 2100 passenger aircraft is almost totally made of graphite–epoxy-reinforced plastic. Nearly 90% of the structure of the lightweight Voyager aircraft, which circled the Earth without refueling, was made of carbon-reinforced plastic. The contoured frame of the Stealth bomber is made of composites consisting of carbon and glass fibers, epoxy-resin matrices, high-temperature polyimides, and other materials. Boron-fiber-reinforced composites are used in military aircraft, golf-club shafts, tennis rackets, fishing rods, and sailboards.

The processing of polymer-matrix-reinforced plastics (Section 10.11) presents significant challenges; several innovative techniques have been developed to manufacture both large and small parts by molding, forming, cutting, and assembly. Careful inspection and testing of reinforced plastics is essential in critical applications in order to ensure that good bonding between the reinforcing fiber and the matrix has been obtained throughout. In some instances, the cost of inspection can be as high as one quarter of the total cost of the composite product.

10.10 Processing of Plastics

The processing of plastics involves operations similar to those used to form and shape metals, as described in the preceding chapters. Plastics can be molded, cast, and formed, as well as machined and joined, into many shapes with relative ease and with few or no additional operations required (Table 10.6). Plastics melt or cure at relatively low temperatures (Table 10.2) and thus, unlike metals, are easy to handle and require less energy to process. However, because the properties of plastic parts and components are influenced greatly by the method of manufacture and the processing conditions, the proper control of these conditions is important for part quality. Plastics are usually shipped to manufacturing plants as pellets or powders and are melted just before the shaping process. Plastics are also available as sheet, plate, rod, and tubing, which may be formed into a variety of products. Liquid plastics are especially used in making reinforced plastic parts.

10.10.1 Extrusion

In *extrusion*, raw materials in the form of thermoplastic pellets, granules, or powder are placed into a hopper and fed into the extruder barrel (Fig. 10.20). The barrel is equipped with a *screw* that blends and conveys the pellets down the barrel. The internal friction from the mechanical action of the screw, along with heaters around the extruder's barrel, heats the pellets and liquefies them. The screw action also builds up pressure in the barrel.

Screws have three distinct sections: (1) a *feed section*, which conveys the material from the hopper area into the central region of the barrel; (2) a *melt*, or *transition*, *section*, where the heat generated from shearing of the plastic causes melting to begin;

TABLE 10.6

Characteristics of Processing Plastics and Reinforced Plastics

Process	Characterisics
Extrusion	Long, uniform, solid or hollow, simple or complex cross-sections; wide range of dimensional tolerances; high production rates; low tooling cost.
Injection molding	Complex shapes of various sizes and with fine detail; good dimensional accuracy; high production rates; high tooling cost.
Structural foam molding	Large parts with high stiffness-to-weight ratio; low production rates; less expensive tooling than in injection molding.
Blow molding	Hollow thin-walled parts of various sizes; high production rates and low cost for making beverage and food containers.
Rotational molding	Large hollow shapes of relatively simple design; low production rates; low tooling cost.
Thermoforming	Shallow or deep cavities; medium production rates; low tooling costs.
Compression molding	Parts similar to impression-die forging; medium production rates; relatively inexpensive tooling.
Transfer molding	More complex parts than in compression molding, and higher production rates; some scrap loss; medium tooling cost.
Casting	Simple or intricate shapes, made with flexible molds; low production rates.
Processing of reinforced plastics	Long cycle times; dimensional tolerances and tooling costs depend on the specific process.

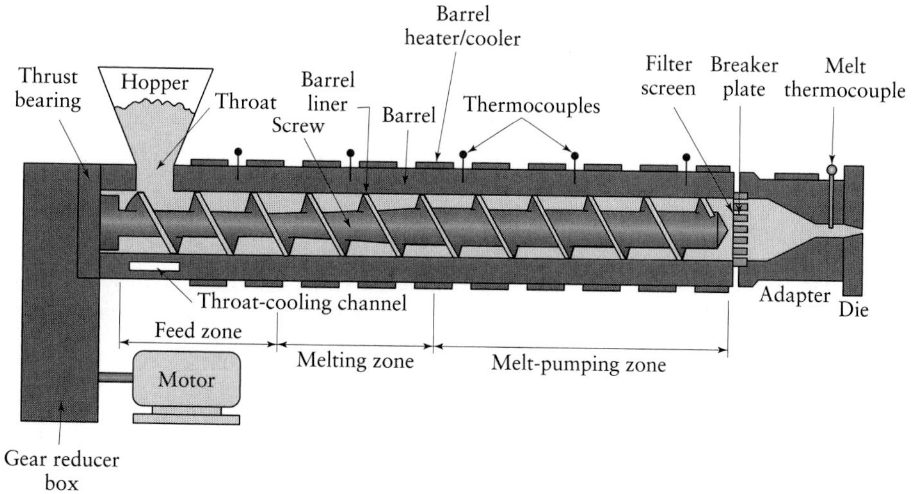

FIGURE 10.20 Schematic illustration of a typical extruder. *Source: Encyclopedia of Polymer Science and Engineering*, 2d ed. Copyright © 1985. This material is used by permission of John Wiley & Sons, Inc.

and (3) a *pumping section*, where additional shearing and melting occurs, with pressure buildup at the die. The lengths of these sections can be changed to accommodate the melting characteristics of different plastics.

a) **Mechanics of polymer extrusion.** The pumping section of the screw determines the rate of polymer flow through the extruder. Consider a uniform screw geometry with narrow *flights* and small clearance with the barrel (Fig. 10.21). At any point in time, the molten plastic is in the shape of a helical ribbon, and this ribbon is being conveyed towards the extruder outlet by the screw flights. If the pressure is constant along the pumping zone, then the volume flow rate of plastic out of the extruder, or *drag flow*, is given by

$$Q_d = \frac{vHW}{2}, \tag{10.15}$$

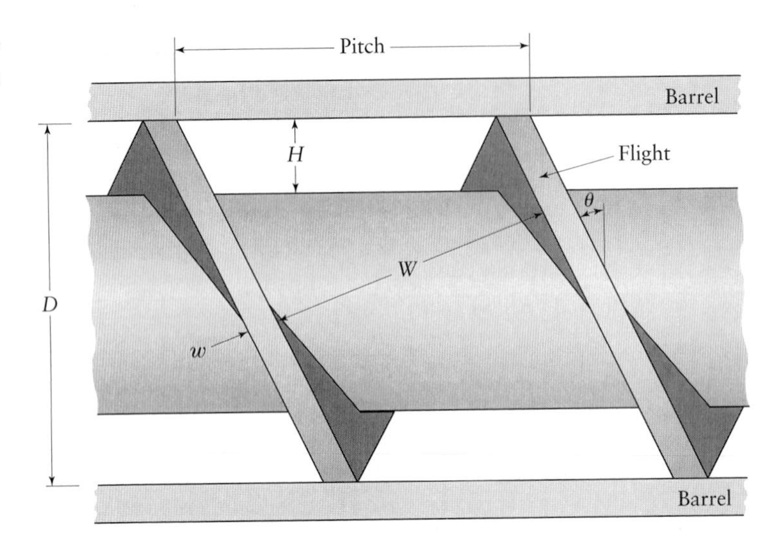

FIGURE 10.21 Geometry of the pumping section of an extruder screw.

where v is the velocity of the flight in the extrusion direction, H is the channel depth, and W is the width of the polymer ribbon. From the geometry defined in Fig. 10.21, we can write

$$V = \omega \cos\theta = \pi D N \cos\theta \tag{10.16}$$

and

$$W = \pi D \sin\theta - w, \tag{10.17}$$

where ω is the angular velocity of the screw, D is the screw diameter, N is the shaft speed (usually in rev/min), θ is the flight angle, and w is the flight width. If the flight width can be considered as negligibly small, then the drag flow can be simplified as

$$Q_d = \frac{\pi^2 H D^2 N \sin\theta\cos\theta}{2}. \tag{10.18}$$

The actual flow rate can be larger than this value if there is significant pressure buildup in the feed or melt sections of the screw, but it is usually smaller, because of the large pressure at the die end of the barrel. Therefore, the flow rate through the extruder can be taken as

$$Q = Q_d - Q_p, \tag{10.19}$$

where Q_p is a flow correction due to pressure. For Newtonian fluids (see Section 10.3), Q_p can be taken as

$$Q_p = \frac{WH^3 p}{12\eta(l/\sin\theta)} = \frac{p\pi D H^3 \sin^2\theta}{12\eta l}. \tag{10.20}$$

where l is the length of the pumping section. Hence, Eq. (10.19) becomes

$$Q = \frac{\pi^2 H D^2 N \sin\theta\cos\theta}{2} - \frac{p\pi D H^3 \sin^2\theta}{12\eta l}. \tag{10.21}$$

Equation (10.21) is known as the *extruder characteristic*. If the power-law index n from Eq. 10.4a is known, then the extruder characteristic is given by the following approximate relationship (after Rauwendaal, 1984):

$$Q = \left(\frac{4+n}{10}\right)(\pi^2 H D^2 N \sin\theta\cos\theta) - \frac{p\pi D H^3 \sin^2\theta}{(1+2n)4\eta}. \tag{10.22}$$

The die plays a major role in determining the output of the extruder. The *die characteristic* is the expression relating flow to the pressure drop across the die and in general form is written as

$$Q_{\text{die}} = Kp \tag{10.23}$$

where Q_{die} is the flow through the die, p is the pressure at the die inlet, and K is a function of the die's geometry. The determination of K is usually complicated and difficult to obtain analytically, although computer-based tools are becoming increasingly available for predictions. More commonly, K is determined experimentally, but one closed-form solution for extruding solid round cross-sections is given by

$$K = \frac{\pi D_d^4}{128\eta \ell_d}, \tag{10.24}$$

where D_d is the die opening diameter and ℓ_d is the die land. If both the die characteristic and the extruder characteristic are known, we have two simultaneous algebraic equations that can be solved for the pressure and flow rate during the operation.

b) **Process characteristics.** Once the extruded product exits the die, it is cooled, either by air or by passing through a water-filled channel. Controlling the rate and uniformity of cooling is important for minimizing product shrinkage and distortion. The extruded product can also be drawn (sized) by a puller after it has cooled; the extruded product is then coiled or cut into desired lengths. Complex shapes with constant cross-section can be extruded with relatively inexpensive tooling. This process is also used to extrude elastomers.

Because this operation is continuous, long products with cross-sections of solid rods, channels, pipe, window frames, and architectural components, as well as sheet, are extruded through dies of various cross-sectional shapes. Plastic-coated wire, cable, and strips for electrical applications are also extruded and coated by this process. The wire is fed into the die opening at a controlled rate with the extruded plastic.

Pellets, which are used for other plastic-processing methods described in this chapter, are also made by extrusion. In this case, the extruded product is a small-diameter rod and is chopped into short lengths, or pellets, as it is extruded. With some modifications, extruders can also be used as simple melters for other shaping processes, such as injection molding and blow molding.

Because the material being extruded is still soft as it leaves the die and the pressure is relieved, the cross-section of the extruded product is different than the shape of the die opening; this effect is known as **die swell**. (See Fig. 10.47b.) Thus, the diameter of a round extruded part, for example, is larger than the die diameter, the difference between the diameters depends on the type of polymer. Considerable experience is required to design proper dies for extruding complex cross-sections of desired shape and dimensions.

Extruders are generally rated by the diameter of the barrel and by the length-to-diameter (L/D) ratio of the barrel. Typical commercial units are 25 to 200 mm (1 to 8 in.) in diameter, with L/D ratios of 5 to 30. Production-size extrusion equipment costs \$30,000–\$80,000, with an additional \$30,000 cost for equipment for downstream cooling and winding of the extruded product; therefore, large production runs are generally required to justify such an expenditure.

c) **Sheet and film extrusion.** Polymer sheet and film can be produced using a flat extrusion die. The polymer is extruded by forcing it through a specially designed die, after which the extruded sheet is taken up first on water-cooled rolls and then by a pair of rubber-covered pull-off rolls.

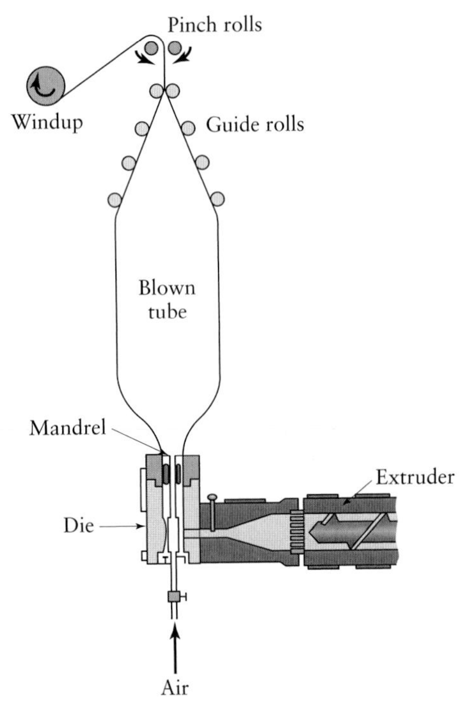

FIGURE 10.22 Schematic illustration of production of thin film and plastic bags from a tube produced by an extruder, and then blown by air. *Source*: D. C. Miles and J. H. Briston, *Polymer Technology*, Chemical Publishing Co., 1979. Reproduced by permission of Chemical Publishing Co., Inc.

Thin polymer films and common plastic bags are made from a tube produced by an extruder (Fig. 10.22). In this process, a thin-walled tube is extruded vertically and expanded into a balloon shape by blowing air through the center of the extrusion die until the desired film thickness is reached. The balloon is usually cooled by air from a cooling ring around it, which can also act as a barrier to further expansion of the balloon. **Blown film** is sold as wrapping film (after the cooled bubble has been slit) or as bags (where the bubble is pinched and cut off). Film is also produced by shaving solid round billets of plastics, especially polytetrafluoroethylene (PTFE), by *skiving* (shaving with specially designed knives).

EXAMPLE 10.5 Blown film

Assume that a typical plastic shopping bag, made by blown film, has a lateral (width) dimension of 400 mm. (a) What should be the extrusion die diameter? (b) These bags are relatively strong. How is this strength achieved?

Solution.

a) The perimeter of the bag is $(2)(400) = 800$ mm. Since the original cross-section of the film was round, the blown diameter should be $\pi D = 800$, or $D = 255$ mm. Recall that in this process a tube is expanded 1.5 to 2.5 times the extrusion die diameter. Taking the maximum value of 2.5, we calculate the die diameter as $255/2.5 = 100$ mm.

b) Note in Fig. 10.22 that, after being extruded, the bubble is being pulled upward by the pinch rolls. Thus, in addition to diametral stretching and the attendant molecular orientation, the film is stretched and oriented in the longitudinal direction as well. The biaxial orientation of the polymer molecules significantly improves the strength and toughness of the blown film.

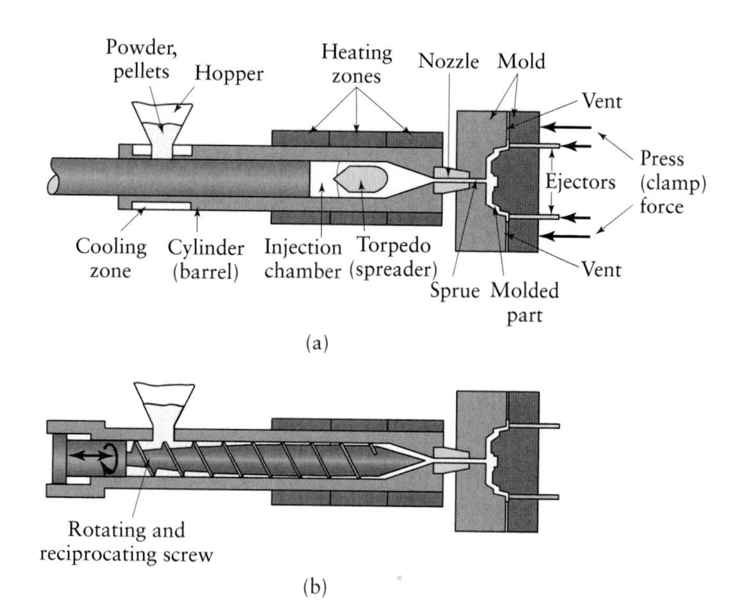

FIGURE 10.23 Injection molding with (a) a plunger and (b) a reciprocating rotating screw. Telephone receivers, plumbing fittings, tool handles, and housings are made by injection molding.

10.10.2 Injection molding

Injection molding is essentially the same process as hot-chamber die casting (see Fig. 5.28). The pellets or granules are fed into a heated cylinder, where they are melted. The melt is then forced into a split-die chamber (Fig. 10.23a), either by a hydraulic plunger or by the rotating screw system of an extruder. Newer equipment is of the *reciprocating-screw* type (Fig. 10.23b). As the pressure builds up at the mold entrance, the rotating screw begins to move backward under pressure to a predetermined distance, thus controlling the volume of material to be injected. The screw then stops rotating and is pushed forward hydraulically, forcing the molten plastic into the mold cavity. Injection-molding pressures usually range from 70 MPa to 200 MPa (10,000 psi to 30,000 psi).

Typical injection-molded products include cups, containers, housings, tool handles, knobs, electrical and communication components (such as cell phones), toys, and plumbing fittings. Although the molds are relatively cool for thermoplastics, thermosets are molded in heated molds, where *polymerization* and *cross-linking* take place. In either case, after the part is sufficiently cooled (for thermoplastics) or set or cured (for thermosets), the molds are opened, and the part is ejected. The molds are then closed, and the process is repeated automatically. Elastomers also are injection molded by these processes. Molds with moving and unscrewing mandrels are also used; they allow the molding of parts with multiple cavities and internal and external threads.

Because the material is molten when injected into the mold, complex shapes and good dimensional accuracy can be achieved. However, as in metal casting, the molded part shrinks during cooling. (Note also that plastics have higher thermal expansion than metals.) Shrinkage in plastics molding is compensated for by making the molds a little larger. Linear shrinkage for plastics typically ranges between 0.005 and 0.025 mm/mm or in./in., and the volumetric shrinkage is typically in the range of 1.5% to 7%. (See also Table 5.1 and Section 11.4.)

Various aspects of injection molding are discussed below.

a) **Molds.** Injection molds have several components, depending on part design, such as runners (as in metal-casting dies), cores, cavities, cooling channels, inserts, knockout pins, and ejectors. There are three basic types of molds: (1) *cold-*

runner two-plate mold, which is the basic and simplest mold design; (2) *cold-runner three-plate* mold, in which the runner system is separated from the part when the mold opens; and (3) *hot-runner* mold, also called *runnerless* mold, in which the molten plastic is kept hot in a heated runner plate. In cold-runner molds, the solidified plastic in the channels (that connect the mold cavity to the end of the barrel) must be removed, usually by trimming. This scrap can be chopped and recycled. In hot-runner molds, which are more expensive, there are no gates, runners, or sprues attached to the part. Cycle times are shorter, because only the injection-molded part must be cooled and ejected.

Metallic components, such as screws, pins, and strips, can also be placed in the mold cavity to become an integral part of the injection-molded product (**insert molding**; see Fig. 10.24); the most common examples are electrical components. *Multicomponent* injection molding (also called *coinjection* or *sandwich* molding) allows the forming of parts with a combination of colors and shapes. Examples include multicolor molding of rear-light covers for automobiles and ball joints made of different materials. Also, printed film can be placed in the mold cavity; thus, parts need not be decorated or labeled after molding.

Injection molding is a high-rate production process, with good dimensional control. Typical cycle times range from 5 to 60 seconds, but can be several minutes for thermosetting materials. The molds, generally made of tool steels or beryllium–copper, may have multiple cavities so that more than one part can be made in one cycle. Proper mold design and control of the flow of material in the die cavities are important factors in the quality of the product; other factors that affect quality include injection pressure, temperature, and condition of the resin. *Computer models* have been developed to study the flow of material in dies and thereby improve die design and establish appropriate process parameters. Modern machines used in injection molding are equipped with microprocessors and microcomputers in a control panel and monitor all aspects of the operation.

b) **Machines.** Injection-molding machines are usually horizontal, and the clamping force on the dies is supplied generally by hydraulic means, although electrical types are also available. Electrically driven models weigh less and are quieter than hydraulic machines. Vertical machines are used for making small, close-tolerance parts and for insert molding. Injection-molding machines are rated according to the capacity of the mold and the clamping force. Although in most machines, this force generally ranges from 0.9 MN to 2.2 MN (100 tons to 250 tons), the largest machine in operation has a capacity of 45 MN (5000 tons) and can

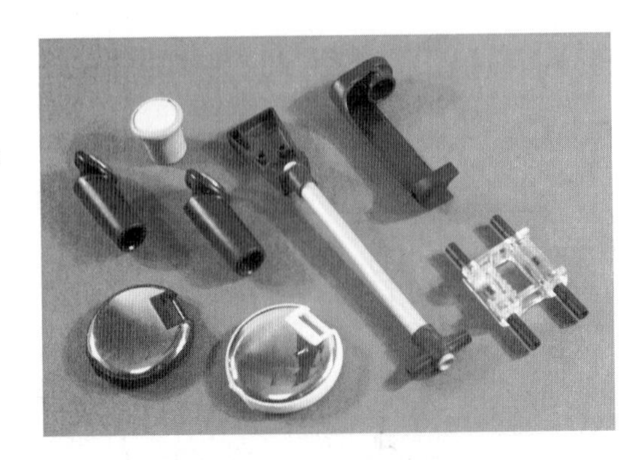

FIGURE 10.24 Products made by insert injection molding. Metallic components are embedded in these parts during molding. *Source*: Rayco Mold and Mfg. LLC.

produce parts weighing up to 25 kg (55 lb); however, parts typically weigh 100 g to 600 g (3 oz to 20 oz). Because of the high cost of dies, typically ranging from $20,000 to $200,000, high-volume production is required to justify such an expenditure.

c) **Overmolding and ice-cold molding.** *Overmolding* is a technique for producing hinge joints and ball-and-socket joints in one operation. Two different plastics are used to ensure that no bond will form between the two parts. *Ice-cold molding* uses the same kind of plastic to form both components, such as in a door hinge. It is done in one cycle and involves a two-cavity mold, using cooling inserts positioned such that no bond forms between the two pieces.

d) **Reaction-injection molding.** In the *reaction-injection-molding* (RIM) process, a mixture of two or more reactive fluids is forced into the mold cavity (Fig. 10.25). Chemical reactions take place rapidly in the mold, and the polymer solidifies, producing a thermoset part. Major applications include automotive bumpers and fenders, thermal insulation for refrigerators and freezers, and stiffeners for structural components. Various reinforcing fibers, such as glass or graphite, may also be used to improve the product's strength and stiffness. Mold costs are low for this process.

e) **Structural foam molding.** The *structural-foam-molding* process is used to make plastic products that have a solid skin and a cellular inner structure. Typical products include furniture components, TV cabinets, business-machine housings, and storage-battery cases. Although there are several foam-molding processes, they are basically similar to injection molding or extrusion. Both thermoplastics and thermosets can be used for foam molding, but thermosets are in the liquid-processing form, similar to polymers for reaction-injection molding.

f) **Injection foam molding.** In *injection foam molding*, also called *gas-assist molding*, thermoplastics are mixed with a **blowing agent** (usually an inert gas, such as nitrogen, or a chemical agent that produces gas during molding), which expands the material. The core of the part is cellular, and the skin is rigid. The thickness of the skin can be as much as 2 mm (0.08 in.), and part densities are

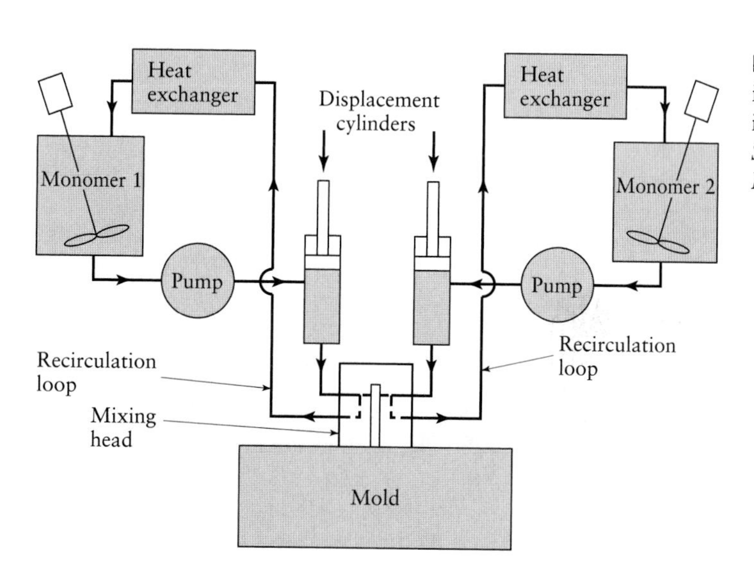

FIGURE 10.25 Schematic illustration of the reaction-injection-molding process. *Source*: *Modern Plastics Encyclopedia*.

as low as 40% of the density of the solid plastic; thus, parts have a high stiffness-to-weight ratio and can weigh as much as 55 kg (120 lb).

EXAMPLE 10.6 Injection molding of gears

A 250-ton injection-molding machine is to be used to make 4.5-in. diameter spur gears that are 0.5 in.-thick. The gears have a fine-tooth profile. How many gears can be injection molded in one set of molds? Does the thickness of the gears influence the answer?

Solution. Because of the fine detail involved, the pressures required in the mold cavity will probably be on the order of 100 MPa (15 ksi). The cross-sectional (projected) area of the gear is $\pi(4.5)^2/4 = 15.9 \text{ in}^2$. If we assume that the parting plane of the two halves of the mold is in the midplane of the gear, the force required is $(15.9)(15{,}000) = 238{,}500$ lb. The capacity of the machine is 250 tons, so we have $(250)(2000) = 500{,}000$ lb of clamping force available. Therefore, the mold can accommodate two cavities, thus producing two gears per cycle. Because it does not influence the cross-sectional area of the gear, the thickness of the gear does not directly influence the pressures involved, and hence the answer is the same for different gear thicknesses.

10.10.3 Blow molding

Blow molding is a modified extrusion and injection-molding process. In **extrusion blow molding,** a tube is extruded (usually vertically), clamped into a mold with a cavity much larger than the tube's diameter, and then blown outward to fill the mold cavity (Fig. 10.26a). Blowing is usually done with an air blast, at a pressure of 350 kPa to 700 kPa (50 psi to 100 psi). In some operations, the extrusion is continuous, and the molds move with the tubing. The molds close around the tubing, close off both ends (thereby breaking the tube into sections), and then move away as air is injected into the tubular piece. The part is then cooled and ejected. Corrugated pipe and tubing are made by continuous blow molding, whereby the pipe or tube is extruded horizontally and blown into moving molds. (See also Section 10.10.1).

In **injection blow molding,** a short tubular piece (**parison**) is first injection molded (Fig. 10.26b). The dies then open, and the parison is transferred to a blow-molding die. Hot air is injected into the parison, which then expands and fills the mold cavity. Typical products made by this process include plastic beverage bottles and hollow containers.

Multilayer blow molding involves the use of coextruded tubes or parisons, thus allowing the use of multilayer structures. Typical examples of multilayer structures include plastic packaging for food and beverages with characteristics such as odor and permeation barrier, taste and aroma protection, resistance to scuffing, printing capability, and ability to be filled with hot fluids. Other applications are in the cosmetics and pharmaceutical industries.

10.10.4 Rotational molding

Most thermoplastics and some thermosets can be formed into large hollow parts by *rotational molding*. The thin-walled metal mold is made of two pieces (*split female mold*) and is designed to be rotated about two perpendicular axes (Fig. 10.27). A premeasured quantity of powdered plastic material is placed inside a warm mold. The powder is obtained from a polymerization process that precipitates the powder from a liquid. The mold is then heated, usually in a large oven, while it is rotated about

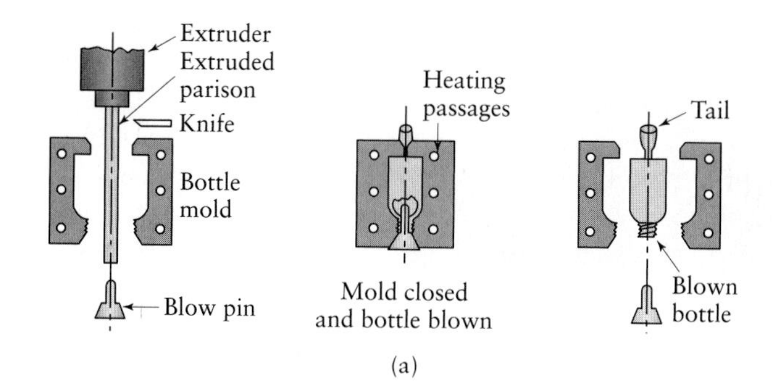

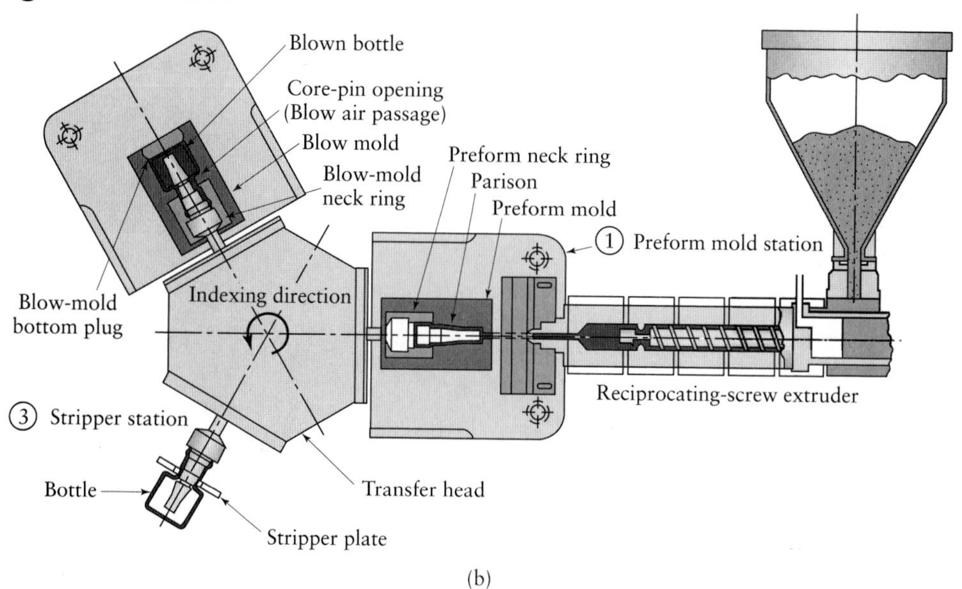

FIGURE 10.26 Schematic illustrations of (a) the blow-molding process for making plastic beverage bottles and (b) a three-station injection-blow-molding machine. *Source: Encyclopedia of Polymer Science and Engineering*, 2d ed. Wiley, Copyright © 1985. This material is used by permission of John Wiley & Sons, Inc.

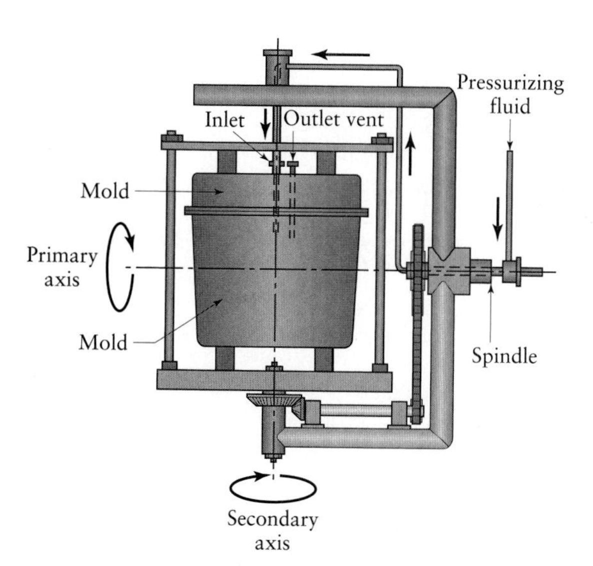

FIGURE 10.27 The rotational molding (rotomolding or rotocasting) process. Trash cans, buckets, and plastic footballs can be made by this process.

the two axes. This action tumbles the powder against the mold, where the heat fuses the powder without melting it. In some parts, a chemical cross-linking agent is added to the powder, and cross-linking occurs after the part is formed in the mold by continued heating. Typical parts made by rotational molding include tanks of various sizes, trash cans, boat hulls, buckets, housings, toys, carrying cases, and footballs. Various metallic or plastic *inserts* may also be molded into the parts made by this process.

Liquid polymers, called **plastisols** (vinyl plastisols being the most common), can also be used in a process called **slush molding.** The mold is simultaneously heated and rotated, and the particles of plastic material are forced against the inside walls of the heated mold by the tumbling action. Upon contact, the material melts and coats the walls of the mold. The part is cooled while still rotating and is then removed by opening the mold.

Rotational molding can produce parts with complex hollow shapes, with wall thicknesses as small as 0.4 mm (0.016 in.). Parts as large as 1.8 m × 1.8 m × 3.6 m (6 ft × 6 ft × 12 ft) have been formed using this method. The outer surface finish of the part is a replica of the surface finish of the mold walls. Cycle times are longer than in other processes, but equipment costs are low. Quality-control considerations usually involve proper weight of powder placed in the mold, proper rotation of the mold, and the temperature–time relationship during the oven cycle.

10.10.5 Thermoforming

Thermoforming is a family of processes for forming thermoplastic sheet or film over a mold with the application of heat and pressure differentials (Fig. 10.28). In this process, a sheet is heated in an oven to the *sag* (softening) *point*, but not to the melting point. The sheet is then removed from the oven, placed over a mold, and pulled against the mold through the application of a vacuum. Since the mold is usually at room temperature, the shape of the plastic is set upon contacting the mold. Because of the low strength of the materials formed, the pressure differential caused by the vacuum is usually sufficient for forming, although air pressure or mechanical means are also applied for some parts. The sheets used in thermoforming are made by sheet extrusion.

Typical parts made using this method include advertising signs, refrigerator liners, packaging, appliance housings, and panels for shower stalls; however, parts with openings or holes cannot be formed by this method, because the necessary pressure

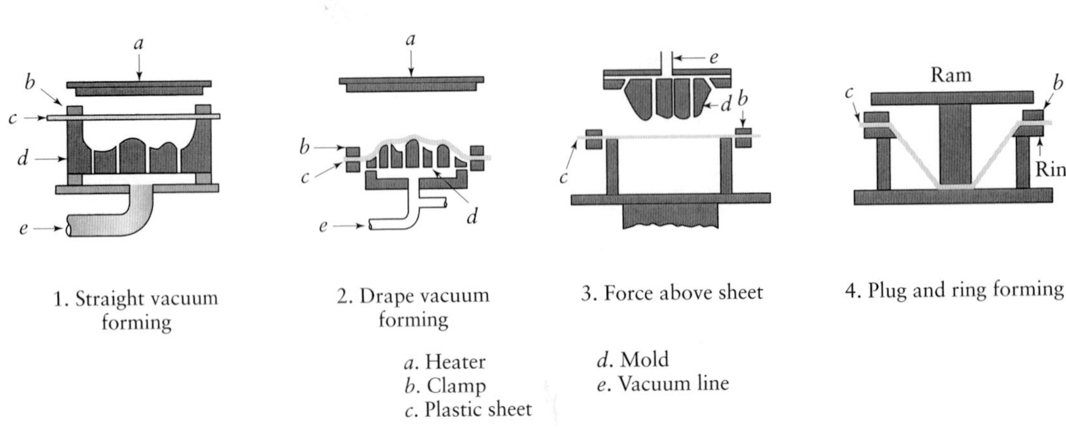

FIGURE 10.28 Various thermoforming processes for thermoplastic sheet. These processes are commonly used in making advertising signs, cookie and candy trays, panels for shower stalls, and packaging.

differential cannot be developed during forming. Because thermoforming is a combination of drawing and stretching operations, much like sheet-metal forming (see Chapter 7), the material should exhibit high uniform elongation; otherwise, it will neck and fail. Thermoplastics have a high capacity for uniform elongation by virtue of their high strain-rate sensitivity exponent, m. (See Section 2.2.7.)

Hollow parts can be produced by using twin sheets. In this operation, the two mold halves come together with the heated sheets between them. The sheets are drawn to the mold halves through a combination of vacuum from the mold and compressed air that is introduced between the sheets. The molds also fuse the sheet around the mold cavity. (See the discussion on *hot–plate welding* in Section 12.15.1.)

Molds for thermoforming are usually made of aluminum, since high strength is not a requirement. The holes in the molds are usually less than 0.5 mm (0.02 in.) in diameter in order not to leave any marks on the formed sheets. Tooling is inexpensive. Quality considerations include tears, nonuniform wall thickness, improperly filled molds, and poor part definition (surface details).

10.10.6 Compression molding

In *compression molding*, a preshaped charge of material, a premeasured volume of powder, or a viscous mixture of liquid resin and filler material is placed directly in a heated mold cavity. Forming is done under pressure with a plug or the upper half of the die (Fig. 10.29). Compression molding results in flash formation; the flash is removed by trimming or some other means. Typical parts made by this method include dishes, handles, container caps, fittings, electrical and electronic components, washing-machine agitators, and housings. Fiber-reinforced parts with long

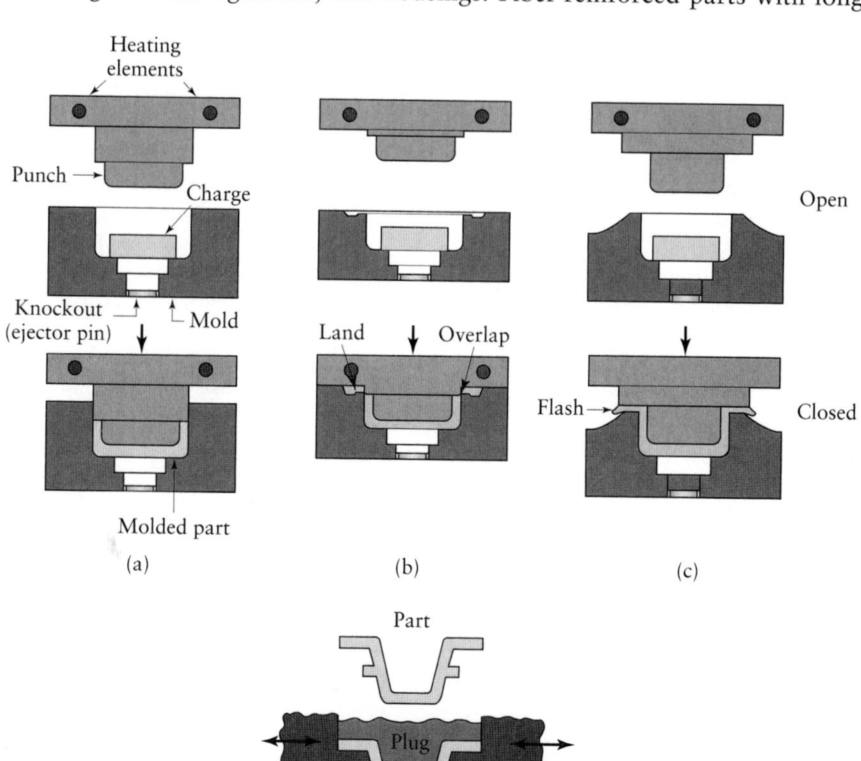

FIGURE 10.29 Types of compression molding, a process similar to forging: (a) positive, (b) semipositive, and (c) flash. The flash in part (c) has to be trimmed off. (d) Die design for making a compression-molded part with undercuts. Such designs are also used in other molding and shaping operations.

chopped fibers are formed exclusively by this process. Elastomers also are shaped by compression molding.

Compression molding is used mainly with thermosetting plastics, with the original material in a partially polymerized state. Cross-linking is completed in the heated die, with curing times typically ranging from 0.5 min to 5 min, depending on the material, part geometry, and thickness of the part. The thicker the material, the longer it will take to cure. Because of their relative simplicity, dies used in compression molding generally cost less than dies used in injection molding. Three types of compression molds are available: (1) **flash type** for shallow or flat parts, (2) **positive** for high-density parts, and (3) **semipositive** for high-quality production. Undercuts in parts are not recommended; however, dies can be designed to open sideways (Fig. 10.29d) to allow removal of the part. In general, the part complexity is less than with injection molding, and the dimensional control produced by compression molding is better.

10.10.7 Transfer molding

Transfer molding represents a further development of the compression-molding process. The uncured thermosetting material is placed in a heated transfer pot or chamber (Fig. 10.30). After the material is heated, it is injected into heated, closed molds. Depending on the type of machine used, a ram, plunger, or rotating screw feeder forces the material to flow through the narrow channels into the mold cavity. This flow generates considerable internal heat, which raises the temperature of the material and homogenizes it. Curing takes place by cross-linking. Because the resin is molten as it enters the molds, the complexity of the part and dimensional control approach those for injection molding.

Typical parts made by transfer molding include electrical and electronic components and rubber and silicone parts. The process is particularly suitable for intricate shapes that have varying wall thicknesses. The molds for this process tend to be more expensive than those for compression molding, and material is wasted in the channels of the mold during filling. (See also the discussion of *resin transfer molding* in Section 10.11.1.)

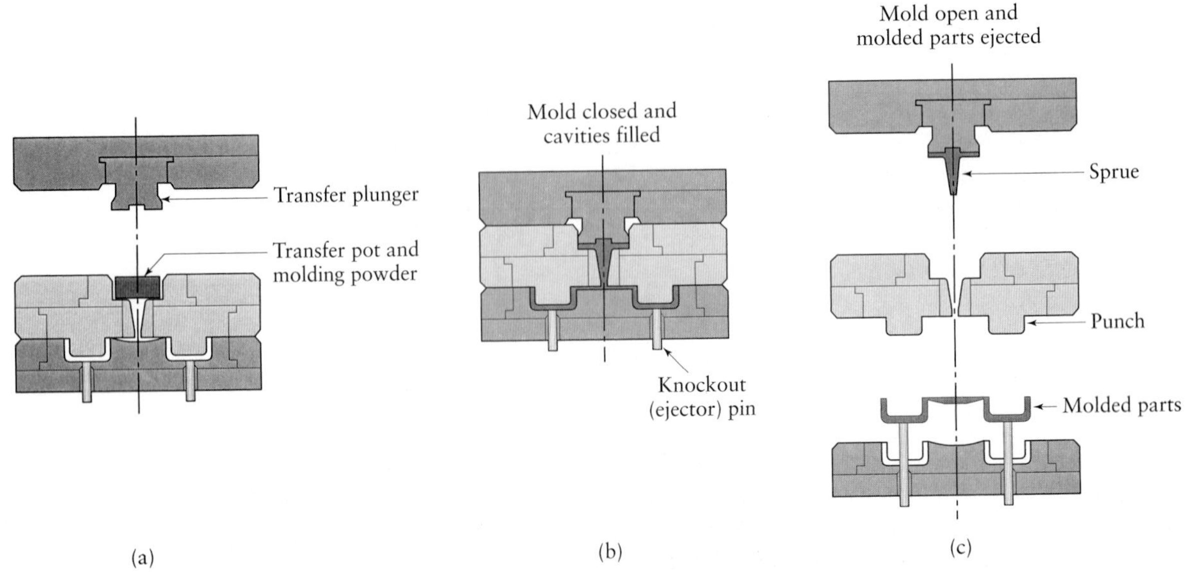

FIGURE 10.30 Sequence of operations in transfer molding for thermosetting plastics. This process is particularly suitable for intricate parts with varying wall thicknesses.

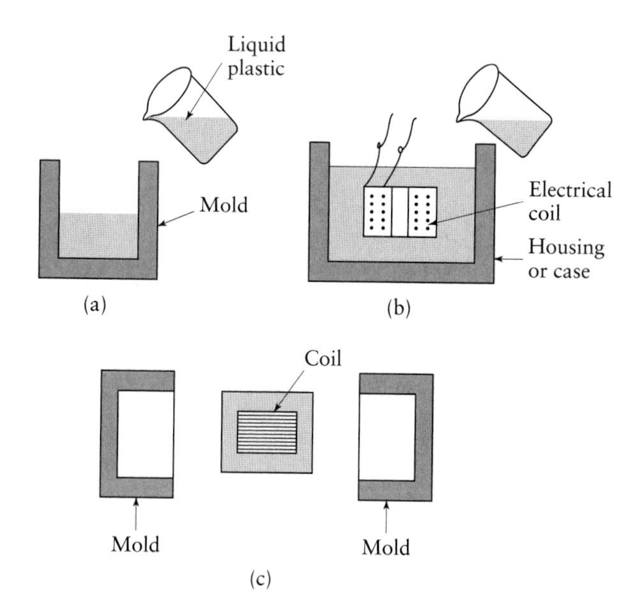

FIGURE 10.31 Schematic illustration of (a) casting, (b) potting, and (c) encapsulation of plastics.

10.10.8 Casting

Some thermoplastics, such as nylons and acrylics, and thermosetting plastics, such as epoxies, phenolics, polyurethanes, and polyester, can be *cast* in rigid or flexible molds into a variety of shapes (Fig. 10.31a). Typical parts cast include large gears, bearings, wheels, thick sheets, and components that require resistance to abrasive wear. In casting thermoplastics, a mixture of monomer, catalyst, and various additives is heated and poured into the mold. The part is formed after polymerization takes place at ambient pressure. Intricate shapes can be formed with flexible molds, which are then peeled off. *Centrifugal casting* (see Section 5.10.4) is also used with plastics, including reinforced plastics with short fibers. Thermosets are cast in a similar manner. Typical parts produced are similar to those made by thermoplastic castings. Degassing may be necessary for maintaining product integrity.

Potting and encapsulation. A variation of casting that is important to the electrical and electronics industry is potting and encapsulation. This process involves casting the plastic around an electrical component, thus embedding it in the plastic. *Potting* (Fig. 10.31b) is done in a housing or case, which is an integral part of the product. In *encapsulation* (Fig. 10.31c), the component is covered with a layer of the solidified plastic. In both applications, the plastic serves as a dielectric (nonconductor). Structural members, such as hooks and studs, may also be partly encapsulated.

Foam molding. Products such as *styrofoam* cups and food containers, insulating blocks, and shaped packaging materials (such as for cameras, appliances, and electronics) are made by *foam molding*. The material is made of expandable polystyrene, produced by placing **polystyrene beads** (obtained by polymerization of styrene monomer), containing a blowing agent, in a mold and exposing them to heat, usually by steam. As a result of their exposure to heat, the beads expand to as much as 50 times their original size and take the shape of the mold. The amount of expansion can be controlled through temperature and time. A common method of molding is using *preexpanded beads*, in which the beads are expanded by steam (or hot air, hot water, or an oven) in an open-top chamber. They are then placed in a storage bin and allowed to stabilize for a period of 3 to 12 hours. The beads can then be molded into shapes as described previously.

Polystyrene beads are available in three sizes: small for cups, medium for molded shapes such as containers, and large for molding of insulating blocks (which can then be cut to size); thus, the bead size chosen depends on the minimum wall thickness of the product. Beads can also be colored prior to expansion, or integrally colored beads can be used.

Polyurethane foam processing, for products such as cushions and insulating blocks, involves several processes that basically consist of mixing two or more chemical components. The reaction forms a cellular structure, which solidifies in a mold. Various low-pressure and high-pressure machines, with computer controls for proper mixing, are available.

10.10.9 Cold forming and solid-phase forming

Processes that are used in cold working of metals, such as rolling, deep drawing, extrusion, closed-die forging, coining, and rubber forming (Chapters 6 and 7), can also be used to form many thermoplastics at room temperature (*cold forming*). Typical materials formed include polypropylene, polycarbonate, ABS, and rigid PVC. The important considerations are that (a) the material be sufficiently ductile at room temperature (hence, polystyrenes, acrylics, and thermosets cannot be formed) and (b) the material's deformation must be nonrecoverable (to minimize springback and creep).

The advantages of cold forming of plastics over other methods of shaping are listed as follows:

1. Strength, toughness, and uniform elongation are increased.
2. Plastics with high molecular weight can be used to make parts with superior properties.
3. Forming speeds are not affected by the thickness of the part, since there is no heating or cooling involved.
4. Typical cycle times are shorter than those for molding processes.

Solid-phase forming is carried out at a temperature about 10°C to 20°C (20°F to 40°F) below the melting temperature of the plastic, if the plastic is a crystalline polymer, and it is formed while still in a solid state. The advantages of solid-phase forming over cold forming are that forming forces and springback effects are lower for the former. These processes are not as widely used as hot-processing methods and are generally restricted to special applications.

10.10.10 Processing elastomers

Elastomers can be formed by a variety of processes used for shaping thermoplastics. Thermoplastic elastomers are commonly shaped by extrusion and injection molding, extrusion being the most economical and the fastest process. In terms of its processing characteristics, a thermoplastic elastomer is a polymer; in terms of its function and performance, it is a rubber (Section 10.8). These polymers can also be formed by blow molding and thermoforming. Thermoplastic polyurethane can be shaped by all conventional methods; it can also be blended with thermoplastic rubbers, polyvinyl chloride compounds, ABS, and nylon. Dryness of the materials is important. For extrusion, the temperatures are in the range of 170°C to 230°C (340°F to 440°F), and for molding, they range up to 60°C (140°F). Typical extruded products include tubing, hoses, moldings, and inner tubes. Injection-molded products cover a broad range of applications, such as components for automobiles and appliances.

Rubber and some thermoplastic sheets are formed by the **calendering** process (Fig. 10.32), wherein a warm mass of the compound is fed through a series of rolls

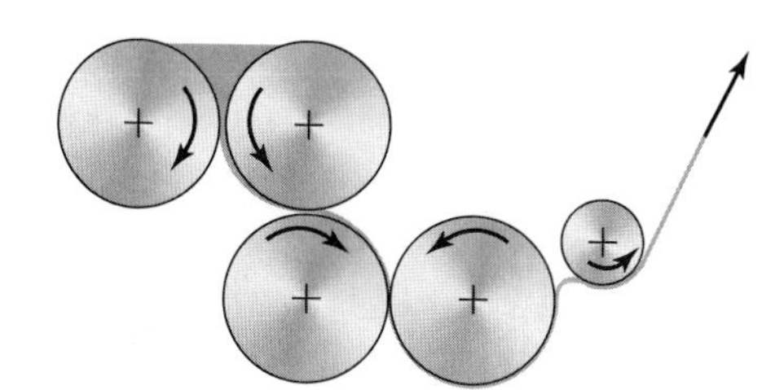

FIGURE 10.32 Schematic illustration of calendering. Sheets produced by this process are subsequently used in thermoforming.

(*masticated*) and is then stripped off in the form of a sheet. The rubber may also be formed over both surfaces of a fabric liner. Discrete rubber products, such as gloves, are made by dipping a form repeatedly into a liquid compound that adheres to the form; the material is then *vulcanized*, usually in steam, and stripped from the form.

10.11 Processing of Polymer-Matrix-Reinforced Plastics

As described in Section 10.9, reinforced plastics are among the most important materials and can be engineered to meet specific design requirements such as high strength-to-weight and stiffness-to-weight ratios and creep resistance. Because of their unique structure, reinforced plastics require special methods to shape them into useful products (Fig. 10.33).

The care required and the several steps involved in manufacturing reinforced plastics make processing costs substantial, so this process generally is not competitive with processing traditional materials and shapes at high production rates. This situation has necessitated the careful assessment and integration of the design and manufacturing processes (*concurrent engineering*) in order to minimize costs while maintaining product integrity and production rate. An important environmental concern with respect to reinforced plastics is the dust generated during processing, such as airborne carbon fibers, which are known to remain in the work area long after fabrication of parts has been completed.

Reinforced plastics can usually be fabricated by the methods described in this chapter, with some provision for the presence of more than one type of material in the composite. The reinforcement may be chopped fibers, woven fabric or mat, roving or yarn (slightly twisted fiber), or continuous lengths of fiber. In order to obtain good bonding between the reinforcing fibers and the polymer matrix, as well as to protect the fibers during subsequent processing, the fibers are surface treated by *impregnation* (*sizing*). Short fibers are commonly added to thermoplastics for injection

FIGURE 10.33 Reinforced-plastic components for a Honda motorcycle. The parts shown are front and rear forks, a rear swing arm, a wheel, and brake disks.

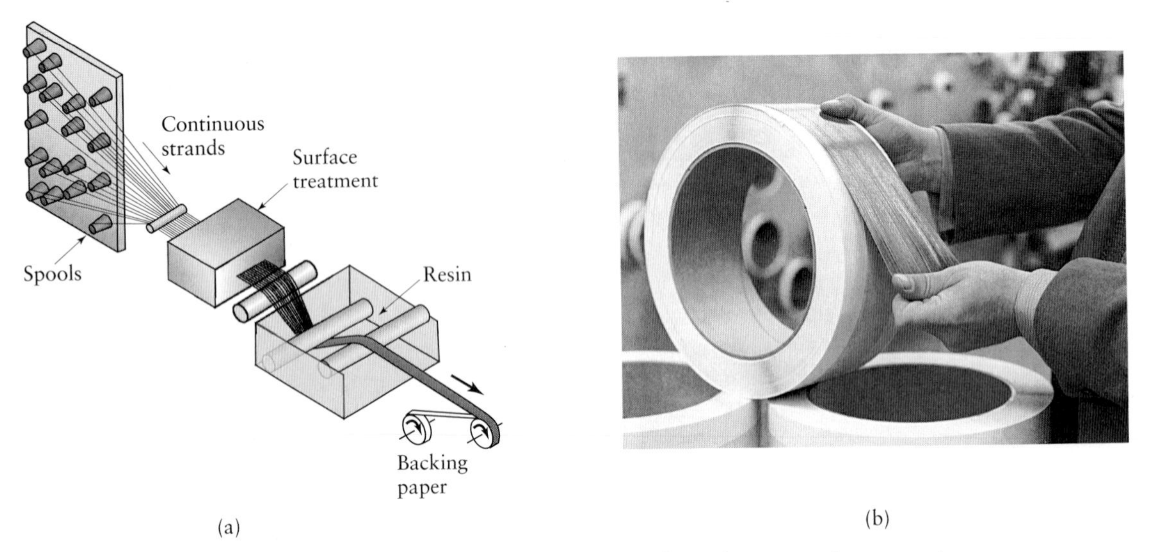

FIGURE 10.34 (a) Manufacturing process for polymer-matrix composite. *Source*: T.-W. Chou, R. L. McCullough, and R. B. Pipes. (b) Boron–epoxy prepreg tape. *Source*: Textron Systems.

molding; milled fibers can be used for reaction-injection molding; and longer chopped fibers are used primarily in compression molding of reinforced plastics.

When the impregnation is done as a separate step, the resulting partially cured sheets are referred to by various terms:

1. **Prepregs.** The continuous fibers are aligned (Fig. 10.34a) and subjected to surface treatment to enhance their adhesion to the polymer matrix. They are then coated by being dipped in a resin bath and made into a *sheet* or *tape* (Fig. 10.34b). Finally, individual pieces of the sheet are assembled into laminated structures, such as the horizontal stabilizer for the F-14 fighter aircraft. Special computer-controlled tape-laying machines have been developed for this purpose. Typical products include flat or corrugated architectural paneling, panels for construction and electric insulation, and structural components of aircraft that require good retention of properties and fatigue strength under hot or wet conditions.
2. **Sheet-molding compound** (SMC). Continuous strands of reinforcing fiber are chopped into short fibers (Fig. 10.35) and deposited over a layer of resin paste, usually a polyester mixture, carried on a polymer film such as polyethylene. A second layer of resin paste is deposited on top, and the sheet is pressed between rollers. The product is gathered into rolls, or placed into containers in layers, and stored until it undergoes a *maturation period*, reaching the desired

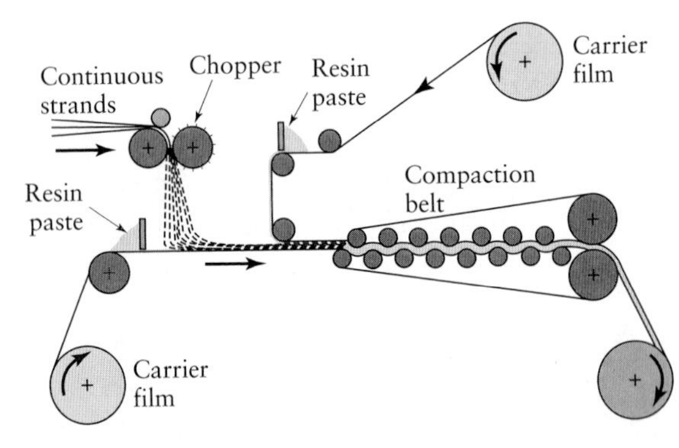

FIGURE 10.35 Manufacturing process for producing reinforced-plastic sheets. The sheet is still viscous at this stage and can later be shaped into various products. *Source*: T.-W. Chou, R. L. McCullough, and R. B. Pipes.

molding viscosity. The maturing process is under controlled temperature and humidity and usually takes one day. The matured SMC, which has a leatherlike feel and is tack free, has a shelf life of about 30 days and must be processed within this period. Alternatively, the resin and the fibers can be mixed together only at the time they are placed in the mold.

3. **Bulk-molding compound** (BMC). These compounds are in the shape of billets, generally up to 50 mm (2 in.) in diameter; they are made in the same manner as SMCs, that is, by the extrusion process. When processed into products, BMCs have flow characteristics similar to that of dough; hence, they are called *dough-molding compounds* (DMCs).
4. **Thick-molding compound** (TMC). This compound combines the characteristics of BMCs (lower cost) and SMCs (higher strength); it is usually injection molded, using chopped fibers of various lengths. One of its applications is for electrical components, because of the high dielectric strength of TMCs.

EXAMPLE 10.7 Tennis rackets made of composite materials

In order to impart certain desirable characteristics, such as light weight and stiffness, composite-material tennis rackets are being manufactured with graphite, fiberglass, boron, ceramic (silicon carbide), and Kevlar as the reinforcing fibers. Rackets have a foam core; some have unidirectional reinforcement, and others have braided reinforcement. Rackets with boron fibers have the highest stiffness, followed by rackets with graphite (carbon), glass, and Kevlar fibers. The racket with the lowest stiffness has 80%-fiberglass, whereas the stiffest racket has 95%-graphite and 5%-boron fibers; thus, the latter has the highest percentage of inexpensive reinforcing fiber and the smallest percentage of the most expensive fiber.

10.11.1 Molding

There are four basic methods for molding reinforced plastics, described as follows.

1. In **compression molding**, the material is placed between two molds, and pressure is applied. Depending on the material, the molds either are at room temperature or are heated to accelerate hardening. The material may be in bulk form (bulk-molding compound), which is a viscous, sticky mixture of polymers, fibers, and additives. It is generally shaped into a log, which is cut into the desired mass. Fiber lengths generally range from 3 mm to 50 mm (0.125 in. to 2 in.), although longer fibers [75 mm (3 in.)] may be used as well. Sheet-molding compounds can also be used in molding; these compounds are similar to BMCs, except that the resin–fiber mixture is laid between plastic sheets to make a sandwich that can be easily handled. The sheets are removed when the SMC is placed in the mold.
2. In **vacuum-bag molding** (Fig. 10.36), prepregs are laid in a mold to form the desired shape. The pressure required to form the shape and to develop good bonding is obtained by covering the layup with a plastic bag and creating a vacuum. If additional heat and pressure are desired, the entire assembly is put in an *autoclave*. Care should be exercised to maintain fiber orientation, if specific fiber orientations are desired. In materials with chopped fibers, no specific orientation is intended. In order to prevent the resin from sticking to the vacuum bag, and to facilitate removal of any excess, several sheets of various materials (*release cloth* or *bleeder cloth*) are placed on top of the prepreg sheets. The

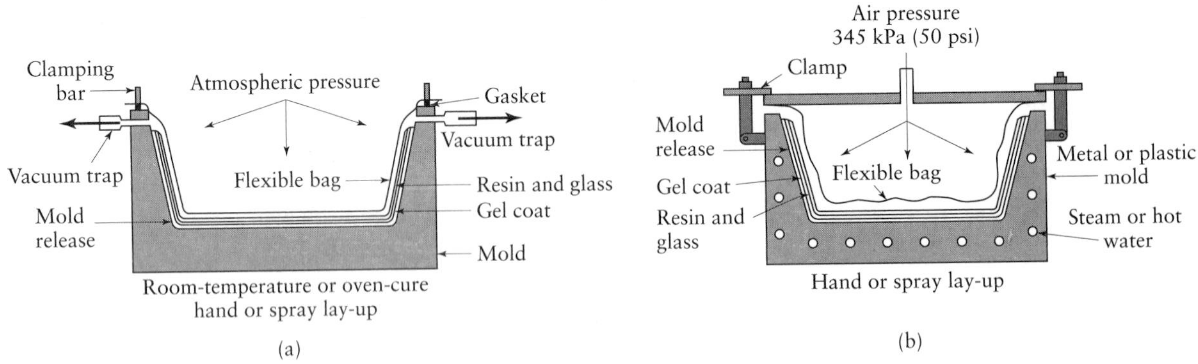

FIGURE 10.36 (a) Vacuum-bag forming. (b) Pressure-bag forming. *Source*: T. H. Meister.

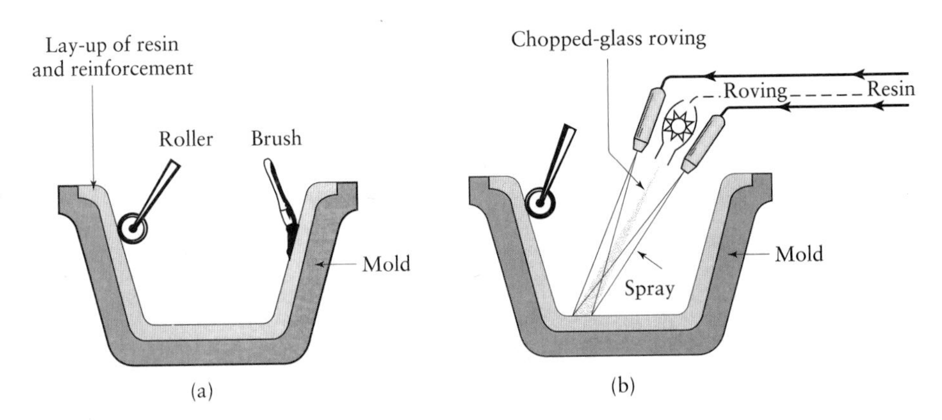

FIGURE 10.37 Manual methods of processing reinforced plastics: (a) hand lay-up and (b) spray-up. These methods are also called *open-mold processing*.

molds can be made of metal, usually aluminum, but more often are made from the same resin (with reinforcement) as the material to be cured. This condition eliminates any problem with differential thermal expansion between the mold and the part.

3. **Contact-molding** processes use a single male or female mold (Fig. 10.37) made of materials such as reinforced plastics, wood, or plaster. Contact molding is used in making products with high surface-area-to-thickness ratios, such as swimming pools, boats, tub and shower units, and housings. This process is a "wet" method, in which the reinforcement is impregnated with the resin at the time of molding. The simplest method is called **hand lay-up**. In this method, the materials are placed and formed in the mold by hand (Fig. 10.37a), and the squeezing action expels any trapped air and compacts the part.

   Molding may also be done by spraying (**spray-up**; see Fig. 10.37b). Although spraying can be automated, these processes are relatively slow, and labor costs are high. However, they are simple, and the tooling is inexpensive. Only the mold-side surface of the part is smooth, and the choice of materials is limited. Many types of boats are made by this process.

4. **Resin transfer molding** is based on transfer molding (Section 10.10.7) whereby a resin, mixed with a catalyst, is forced by a piston-type positive-displacement pump into a mold cavity filled with fiber reinforcement. The process is a viable alternative to hand lay-up, spray-up, and compression molding for low- or intermediate-volume production.

5. **Transfer/injection molding** is an automated process that combines compression molding, injection molding, and transfer molding. It combines the advantages of each process and produces parts with enhanced properties.

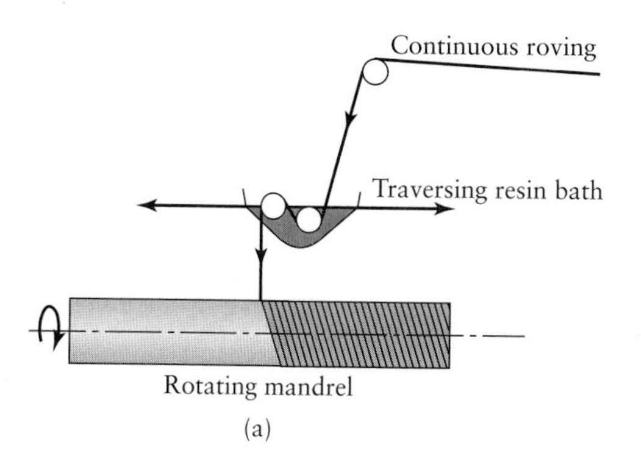

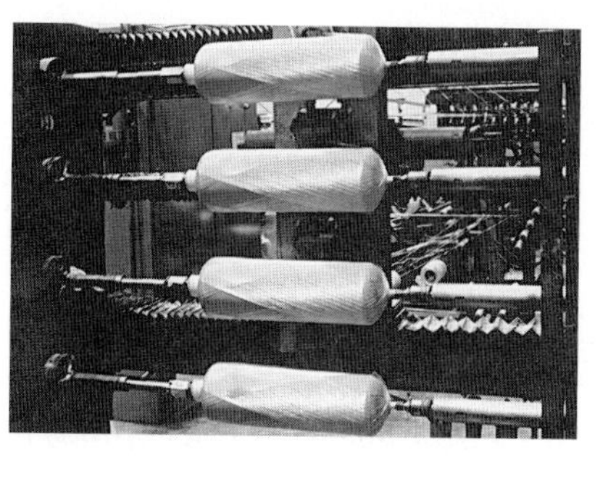

FIGURE 10.38 (a) Schematic illustration of the filament-winding process. (b) Fiberglass being wound over aluminum liners for slide-raft inflation vessels for the Boeing 767 aircraft. *Source*: Advanced Technical Products Group, Inc., Lincoln Composites.

10.11.2 Filament winding, pultrusion, and pulforming

Filament winding. *Filament winding* is a process whereby the resin and fibers are combined at the time of curing. Axisymmetric parts, such as pipes and storage tanks, as well as asymmetric parts are produced on a rotating mandrel. The reinforcing filament, tape, or roving is wrapped continuously around the form. The reinforcements are impregnated by passing through a polymer bath (Fig. 10.38a). The process can be modified by wrapping the mandrel with prepreg material.

The products made by filament winding are very strong, because of their highly reinforced structure. Filament winding has also been used for strengthening cylindrical or spherical pressure vessels (Fig. 10.38b) made of materials such as aluminum or titanium. The presence of a metal inner lining makes the part impermeable. Filament winding can be used directly over solid-rocket propellant forms. Seven-axis computer-controlled machines (see Chapter 14) have been developed for making asymmetric parts that automatically dispense several unidirectional prepregs. Typical asymmetric parts made include aircraft engine ducts, fuselages, propellers, blades, and struts.

Pultrusion. Long shapes with various constant profiles, such as rods, structural profiles, and tubing (similar to drawn metal products), are made by the *pultrusion* process. Typical products made by this process include golf clubs; drive shafts; and structural members such as ladders, walkways, and handrails. In this process, developed in the early 1950s, the continuous reinforcement (roving or fabric) is pulled through a thermosetting-polymer bath, and then through a long heated steel die (Fig. 10.39). The product is cured during its travel through the die and then cut into desired lengths. The most common material used in pultrusion is polyester with glass reinforcements.

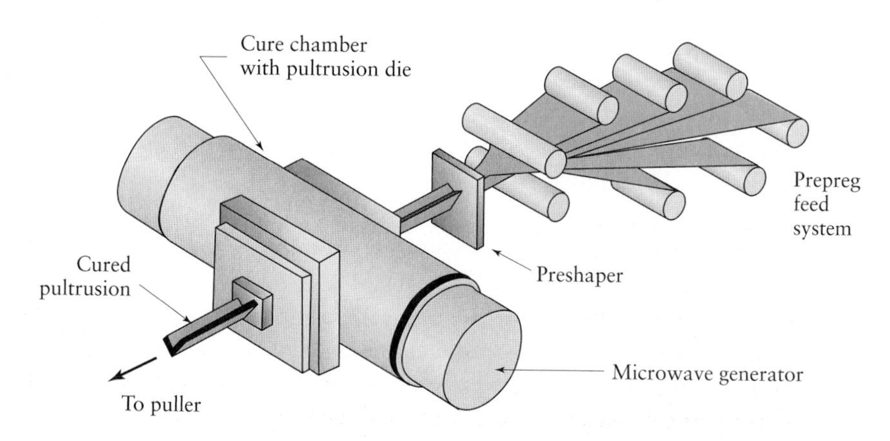

FIGURE 10.39 Schematic illustration of the pultrusion process. This figure shows a microwave heating arrangement, although curing can be performed in a heated die.

Pulforming. Continuously reinforced products other than profiles with a constant cross-section are made by *pulforming*. After being pulled through the polymer bath, the composite is clamped between the two halves of a die and cured into a finished product. The dies recirculate and shape the products successively. Common examples of products made by this method include glass-fiber-reinforced hammer handles and curved automotive leaf springs.

10.11.3 Product quality

The major quality considerations for the processes described thus far involve internal voids and gaps between successive layers of material. Volatile gases that develop during processing must be allowed to escape from the lay-up through the vacuum bag to avoid porosity due to trapped gases within the lay-up. Microcracks may develop due to improper curing or during transportation and handling. These defects can be detected using ultrasonic scanning and other techniques. (See Section 4.8.1.)

10.12 Rapid Prototyping and Rapid Tooling

Making a prototype, that is, a first full-scale model of a product (see Fig. 1.2), has traditionally involved actual manufacturing processes, using a variety of tooling and machines and usually taking weeks or months. An important advance in manufacturing is *rapid prototyping*, also called **desktop manufacturing** or **free-form fabrication**, a process by which a solid physical model of a part is made directly from a three-dimensional CAD drawing. Developed in the mid-1980s, rapid prototyping entails several different consolidation techniques: resin curing, deposition, solidification, and sintering.

The importance and economic impact of rapid prototyping can be appreciated by noting the following:

a) The conceptual design is viewed in its entirety and at different angles on the monitor through a three-dimensional CAD system, as described in Section 15.4;
b) A prototype from various nonmetallic and metallic materials is manufactured and studied thoroughly from functional, technical, and esthetic aspects; and
c) Prototyping is accomplished in a much shorter time and at lower cost than by traditional methods.

It should be noted, however, that modern CNC machine tools (Chapters 8 and 15) also provide the capability of producing complex shapes quickly and are a viable option for rapid prototyping.

This section emphasizes use of **additive manufacturing**, that is, processes that build parts in layers. This technique inherently involves integrated computer-driven hardware and software. As an example, Fig. 10.40 shows the computational steps involved in producing a part. Part production in rapid prototyping takes hours for most parts. Therefore, these processes are not viable for large production runs, especially since the workpiece material is often an expensive polymer or laminate. Table 10.7 summarizes characteristics of the rapid-prototyping process, and Table 10.8 lists typical material properties that are attainable via this method.

10.12.1 Stereolithography

The *stereolithography* (STL) process is based on the principle of curing (hardening) a *liquid photopolymer* into a specific shape. Consider what happens when a laser beam is focused on and translated across the surface of a liquid photopolymer. The laser serves to cure the polymer, providing the energy necessary for polymerization.

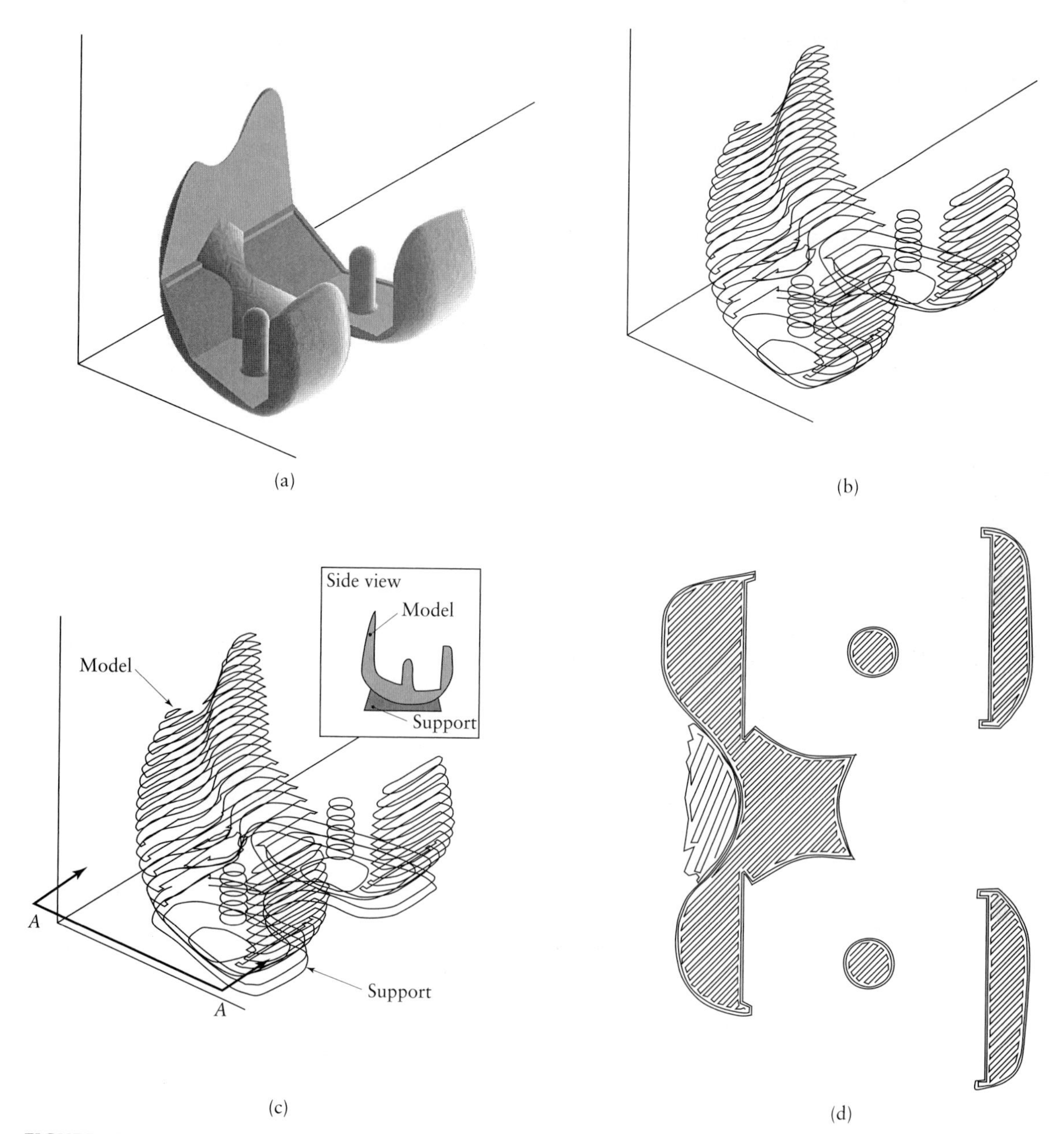

FIGURE 10.40 The computational steps in producing a stereolithography file. (a) Three-dimensional decription of the part. (b) The part is divided into slices. (Only 1 in 10 is shown.) (c) Support material is planned. (d) A set of tool directions is determined for manufacturing each slice. Shown is the extruder path at section *A–A* from (c), for a fused-deposition modeling operation.

Laser energy is absorbed by the polymer, but according to the Beer–Lambert law, the exposure decreases exponentially with depth according to the rule

$$E(z) = E_0^{-z/D_p}, \tag{10.25}$$

where E is the exposure in energy per area, E_0 is the exposure at the resin surface ($z = 0$), and D_p is the "penetration depth" at the laser wavelength and is a property of the resin. At the cure depth, the polymer is sufficiently exposed for it to gel, or

$$E_c = E_0^{-C_d/D_p}, \tag{10.26}$$

TABLE 10.7

Characteristics of Rapid-Prototyping Processes

Supply Phase	Process	Layer Creation Technique	Phase-Change Type	Materials
Liquid	Stereolithography	Liquid-layer curing	Photopolymerization	Photopolymers (acrylates, epoxies, colorable resins, and filled resins)
	Solid-base curing	Liquid-layer curing and milling	Photopolymerization	Photopolymers
	Fused-deposition modeling	Extrusion of melted plastic	Solidification by cooling	Thermoplastics (ABS, polycarbonate, and elastomer) and wax
	Ballistic-particle manufacturing	Droplet deposition	Solidification by cooling	Polymers and wax
Powder	Three-dimensional printing	Binder-droplet deposition onto powder layer	No phase change	Ceramic; polymer and metal powder with binder
	Selective laser sintering	Layer of powder	Laser-driven sintering or melting	Polymers, metals with binder, metals, ceramics, and sand with binder
Solid	Laminated-object manufacturing	Deposition of sheet material	No phase change	Paper and polymers

TABLE 10.8

Mechanical Properties of Selected Materials for Rapid Prototyping

Process	Material	Tensile Strength (MPa)	Elastic Modulus (GPa)	Elongation in 50 mm (%)
Stereo-lithography	SL 5180[a]	55–65	2.4–2.6	6–11
	SL 5195[a]	46.5	2.1	11
	SL 5510[b]	73	2.8	7.9
	SL 7940[b]	37–39	1.3	18–21
Fused-deposition modeling	Polycarbonate	62	–	100
	ABS	35	2.5	50
Selective laser sintering	Nylon	36	1.4	32
	Polycarbonate	23.4	1.2	5
	Polyamide	44	1.6	9
	SOMOS 201	17.3	14	130
	ST-100[c]	305	137	10

[a] After a 90-min UV cure.
[b] After a 90-min UV cure at 80°C.
[c] Sintered and bronze-infiltrated steel powder.

where E_c is the critical threshold exposure and C_d is the cure depth. Thus, the cure depth is given by

$$C_d = D_p \ln\left(\frac{E_0}{E_c}\right). \tag{10.27}$$

The cure depth represents the thickness in which the resin has polymerized into a gel, but this state does not have high strength. Recognizing this condition, the controlling

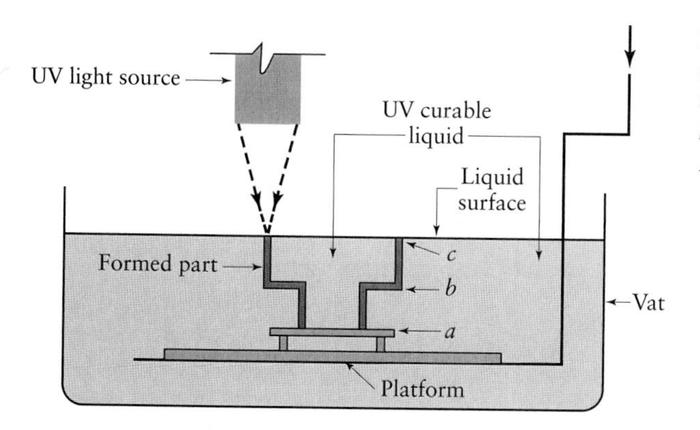

FIGURE 10.41 Schematic illustration of the stereolithography process. *Source*: Courtesy of A. S. Alpert, 3D Systems.

software slightly overlaps the cured volumes, but curing under fluorescent lamps is often necessary as a finishing operation.

The polymer at the periphery of the laser spot does not receive sufficient exposure to polymerize. It can be shown that the cured line width L_w at the surface is given by

$$L_w = B\sqrt{\frac{C_d}{2D_p}}, \tag{10.28}$$

where B is the diameter of the laser-beam spot.

Stereolithography (Fig. 10.41) involves a vat that contains a mechanism whereby a platform can be lowered and raised vertically and that is filled with a photocurable liquid acrylate polymer. The liquid is a mixture of acrylic monomers, oligomers (polymer intermediates), and a photoinitiator. When the platform is at its highest position, the layer of liquid above it is shallow. A *laser*, generating an ultraviolet beam, is then focused along a selected surface area of the photopolymer at surface a and moved in the x–y-direction.

The beam cures that portion of the photopolymer (say, a ring-shaped portion), producing a thin solid body. The platform is then lowered enough to cover the cured polymer with another layer of liquid polymer, and the sequence is repeated. In Fig. 10.41, the process is repeated until level b is reached. Hence, thus far we have a cylindrical part with a constant wall thickness. Note that the platform is now lowered by a vertical distance ab.

At level b, the x–y movements of the beam are wider, so that we now have a flange-shaped piece that is being produced over the previously formed part. After the proper thickness of the liquid has been cured, the process is repeated, producing another cylindrical section between levels b and c. Note that the surrounding liquid polymer is still fluid, because it has not been exposed to the ultraviolet beam, and that the part has been produced from the bottom up in individual "slices." The unused portion of the liquid polymer can be used again to make another part or another prototype. Note that the word *stereolithography*, as used to describe this process, comes from the fact that the movements are three dimensional and that the process is similar to lithography, in which the image to be printed on a flat surface is ink receptive and the blank areas are ink repellent. After completion, the part is removed from the platform, blotted, cleaned ultrasonically and with an alcohol bath, and subjected to a final curing cycle.

By controlling the movements of the beam and the platform through a servocontrol system, a variety of parts can be formed by this process. Total cycle times range from a few hours to a day. Progress is being made toward improvements in (1) accuracy and dimensional stability of the prototypes produced, (2) less expensive liquid modeling materials, (3) CAD interfaces to transfer geometrical data to model-making systems, and (4) strength so that the prototypes made by this process can be truly

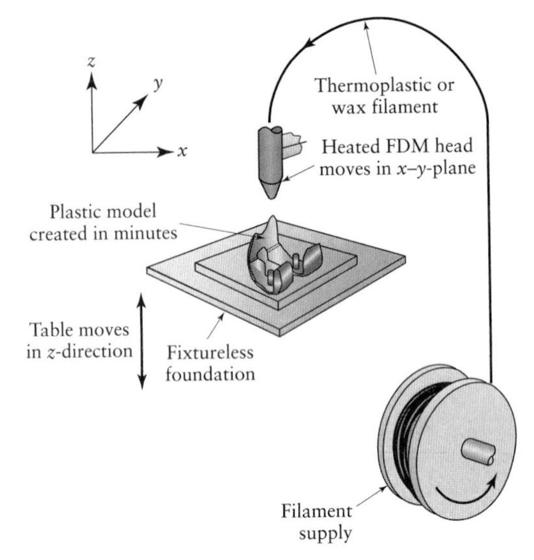

FIGURE 10.42 Schematic illustration of the fused-deposition modeling process. *Source*: Courtesy of Stratasys, Inc.

considered prototypes and models in the traditional sense. Sloping surfaces tend to be relatively rough, because of the manner in which layer-by-layer curing takes place.

Depending on capacity, the cost of machines for stereolithography ranges from \$100,000 to \$500,000, while the cost of the liquid polymer is on the order of \$300 per gallon. Maximum part size is 0.5 m × 0.5 m × 0.6 m (20 in. × 20 in. × 24 in.). The automotive, aerospace, electronics, and medical industries are among the industries that use stereolithography as a rapid and inexpensive method of producing prototypes and thus reducing product-development-cycle times. One major application of stereolithography is in the area of making molds and dies for casting and injection molding.

10.12.2 Fused-deposition modeling

In the *fused-deposition modeling* (FDM) process (Fig. 10.42), a gantry-robot-controlled extruder head moves in two principal directions over a table. The table can be raised and lowered as needed. A thermoplastic or wax filament is extruded through the small orifice of a heated die. The initial layer is placed on a foam foundation by extruding the filament at a constant rate while the extruder head follows a predetermined path (Fig. 10.40d). When the first layer is completed, the table is lowered so that subsequent layers can be superposed.

Occasionally, complicated parts are required, such as the part shown in Fig. 10.43a. This part is difficult to manufacture directly, because once the part has been constructed up to height *a*, the next slice would require the filament to be placed

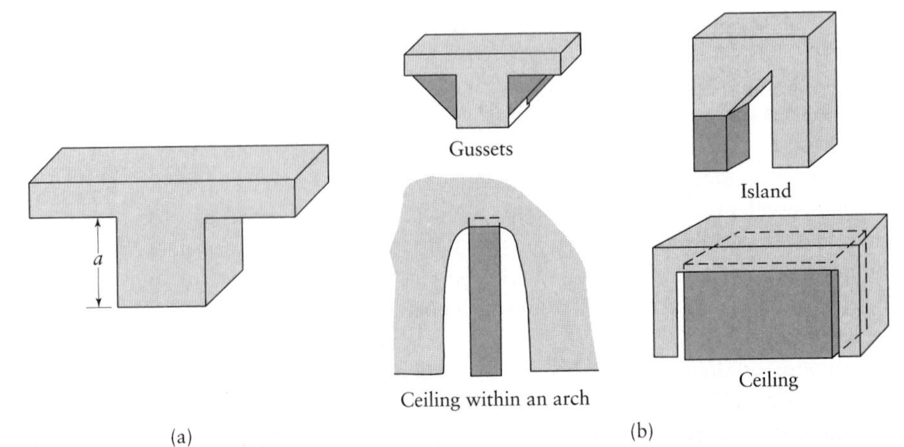

FIGURE 10.43 (a) A part with a protruding section that requires support material. (b) Common support structures used in rapid-prototyping machines. *Source*: Figure reprinted with permission of the Society of Manufacturing Engineers, from Rapid Prototyping & Manufacturing: Fundamentals of StereoLithography, copyright 1992.

in a location where no material exists to support it. The solution is to extrude a support material separately from the modeling material, so that a filament can be placed safely in the center of the part. The support material is extruded with a less dense spacing of filament on a layer, so that it is weaker than the model material and can be broken off or dissolved after the part is completed.

In the FDM process, the extruded layer's thickness is determined by the extruder-die diameter, the diameters typically range from 0.33 mm to 0.12 mm (0.013 in. to 0.005 in.). This thickness represents the best achievable dimensional tolerance in the vertical direction. In the x–y-plane, dimensional accuracy can be as fine as 0.025 mm (0.001 in.), as long as a filament can be extruded into the feature. Close examination of an FDM-produced part will indicate that a stepped surface exists on oblique exterior planes. If this surface roughness is unacceptable, chemical-vapor polishing or a heated tool can be used to smoothen the surface, or a coating can be applied, often in the form of a polishing wax. However, the overall dimensional tolerances are then compromised, unless care is taken in these finishing operations.

10.12.3 Selective laser sintering

Selective laser sintering (SLS) is a process based on the sintering of polymer (or, less commonly, metallic) powders selectively into an individual object. (See also Chapter 11.) The basic elements in this process are shown in Fig. 10.44. The bottom of the processing chamber is equipped with two cylinders: a part-build cylinder, which is lowered incrementally to where the sintered part is formed, and a powder-feed cylinder, which is raised incrementally to supply powder to the part-build cylinder through a roller mechanism. A thin layer of powder is first deposited in the part-build cylinder. A laser beam, guided by a process-control computer (using instructions generated by the 3D CAD program of the desired part), is then focused on that layer, tracing and melting (or, for metals, *sintering*; see Section 11.4) a particular cross-section, which then quickly resolidifies into a solid mass (after the laser beam moves to another section). The powder in other areas remains loose, yet supports the solid portion. Another layer of powder is then deposited; this cycle is repeated continuously until the entire three-dimensional part has been produced. The loose particles are then shaken off, and the part is recovered.

A variety of materials can be used in this process, including polymers (ABS, PVC, nylon, polyester, polystyrene, and epoxy), wax, and metals, as well as ceramics

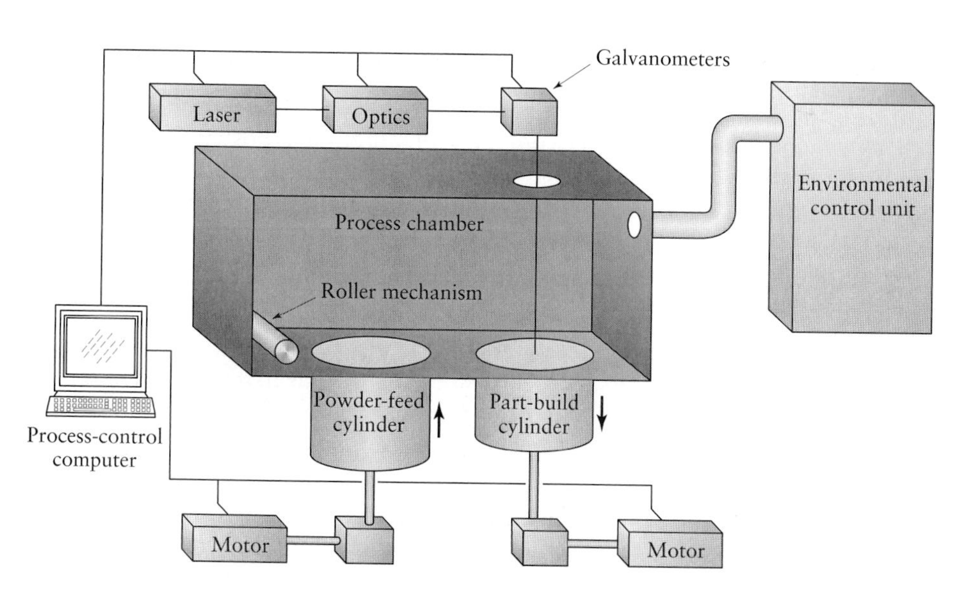

FIGURE 10.44 Schematic illustration of the selective-laser-sintering process. *Source*: C. Deckard and P. F. McClure.

with appropriate binders. It is most common to use polymers, because of the smaller, less expensive, and less complicated lasers required for sintering. With ceramics and metals, it is common to sinter only a polymer binder that has been blended with the ceramic or metal powders and then to complete the sintering in a furnace.

10.12.4 Laminated-object manufacturing

Lamination implies a laying down of layers that are adhesively bonded to one another. *Laminated-object manufacturing* (LOM) uses layers of paper or plastic sheets with a heat-activated glue on one side to produce parts. The desired shapes are burned into the sheet with a laser beam, and the parts are built layer by layer. Once the part is completed, the excess material must be removed manually. This process is simplified by programming the laser beam to burn perforations in crisscrossed patterns; the resulting grid lines make the part appear as if it had been constructed from gridded paper. LOM uses sheets as thin as 0.05 mm (0.002 in.), although a thickness of 0.5 mm (0.02 in.) is more commonly used, so that it can achieve dimensional tolerances similar to those available from stereolithography and fused-deposition modeling. The compressed paper has the appearance and strength of soft wood, and it is often mistaken for elaborate wood carvings. The paper parts are subsequently easy to finish or coat.

10.12.5 Ballistic-particle manufacturing

In the *ballistic-particle-manufacturing* process, a stream of a material such as a plastic, a ceramic, a metal, or wax, is ejected through a small orifice at a surface (target), by use of an ink-jet-style mechanism. The mechanism uses a piezoelectric pump, which operates when an electric charge is applied, generating a shock wave that propels 50-μm (0.002-in.) droplets at a rate of 10,000/s. The operation is repeated, in a manner similar to other processes, to form the part, with layers of wax deposited on top of each other. The ink-jet head is guided by a three-axis robot.

10.12.6 Solid-base curing

The *solid-base-curing* process, also called *solid-ground curing*, is unique in that entire slices of a part are manufactured at one time; as a result, a large throughput is achieved compared with that from other rapid-prototyping processes. This process is, however, among the most expensive, so its adoption has been less common than for other types of rapid prototyping.

The basic approach consists of the following steps:

1. Once a slice is created by the computer software, a mask of the slice is printed on a glass sheet, by use of an electrostatic printing process similar to that used in laser printers. A mask is required because the area of the slice where solid material is desired remains transparent.
2. While the mask is being prepared, a thin layer of photoreactive polymer is deposited on the work surface and spread evenly.
3. The mask is placed over the work surface, and an ultraviolet floodlight is projected through the mask. Wherever the mask is clear, the light shines through to cure the polymer and causes the desired slice to be hardened.
4. The unaffected resin, still liquid, is vacuumed off the surface.
5. Water-soluble liquid wax is spread across the work area, filling the cavities previously occupied by the unexposed liquid polymer. Since the workpiece is on a chilling plate and the workspace remains cool, the wax hardens quickly.

6. The layer is then milled (see Section 8.9.1) to achieve the correct thickness and flatness.
7. The process is repeated, layer by layer, until the part is completed.

Solid-base curing has the advantage of high production rate, both because entire slices are produced at once and because two glass screens are used concurrently. That is, while one mask is being used to expose the polymer, the next mask is already being prepared, and it is ready as soon as the milling operation is completed. The wax support is water soluble; it may be removed immediately, or it may remain in place as protection during shipping of the part. Use of the wax support means that any shape can be manufactured, including "ship-in-the-bottle" type of parts, for which removal of a solid support would be difficult. Usually, no finishing operation is required, but occasionally the polymer is cleaned in a slightly acidic solution to bleach the surface and to obtain a smoother finish.

10.12.7 Direct manufacturing and rapid tooling

The polymer parts that can be obtained from various rapid-prototyping operations are useful not only for design evaluation and troubleshooting, but also occasionally for manufacturing marketable products directly. Unfortunately, it is often desirable, for functional reasons, to use metallic parts, while the best developed and most available rapid-prototyping operations involve polymeric workpieces.

The solution is to use the components manufactured through rapid prototyping as aids in further processing. As an example, Fig. 10.45 shows an approach for investment casting. Here, the individual patterns are made in a rapid-prototyping operation (in this case, stereolithography) and then used as the blanks in assembling a tree for investment casting. It should be noted that this approach requires a polymer that will melt and burn from the ceramic mold completely; such polymers are available for all forms of polymer rapid-prototyping operations. Furthermore, the parts as drawn in CAD programs are usually software modified to account for shrinkage; it is the modified part that is produced in the rapid-prototyping machinery.

Rapid-prototyping processes are now being used to produce *engineering components* (*direct manufacturing*), but not economically for large quantities. The advantage of rapid prototyping is that it makes traditionally costly processes economical for very short production runs, often for merely one part. Several approaches have been devised for the rapid production of *tooling* by means of rapid-prototyping processes. For example, in the sand-casting operation, the pattern plates can be manufactured via rapid-prototyping approaches, which can then be applied to conventional sand casting (see Section 5.8.1). The advantage to this approach is the short time needed to produce the pattern plate compared with that in conventional manufacturing. The main shortcoming is the reduced pattern life, as compared with that obtained from machined high-strength metals.

Another common application is injection molding (see Section 10.10.2), where the mold (or, more typically, a *mold insert*) is manufactured via rapid prototyping and is used in injection-molding machines. Two approaches are used for the production of the tooling: Either a high-melting-point thermoplastic or a stable thermoset is used for low-temperature injection molding, or an investment-cast insert is produced.

Molds for slip casting of ceramics (see Section 11.9.1) can also be produced in this manner. To produce individual molds, rapid-prototyping processes are used directly, but the molds will be shaped with the desired permeability. For example, in fused-deposition modeling, this requirement mandates that the filaments be placed onto the individual slices with a small gap between adjacent filaments. These filaments are later positioned at right angles in adjacent layers. The advantage of rapid

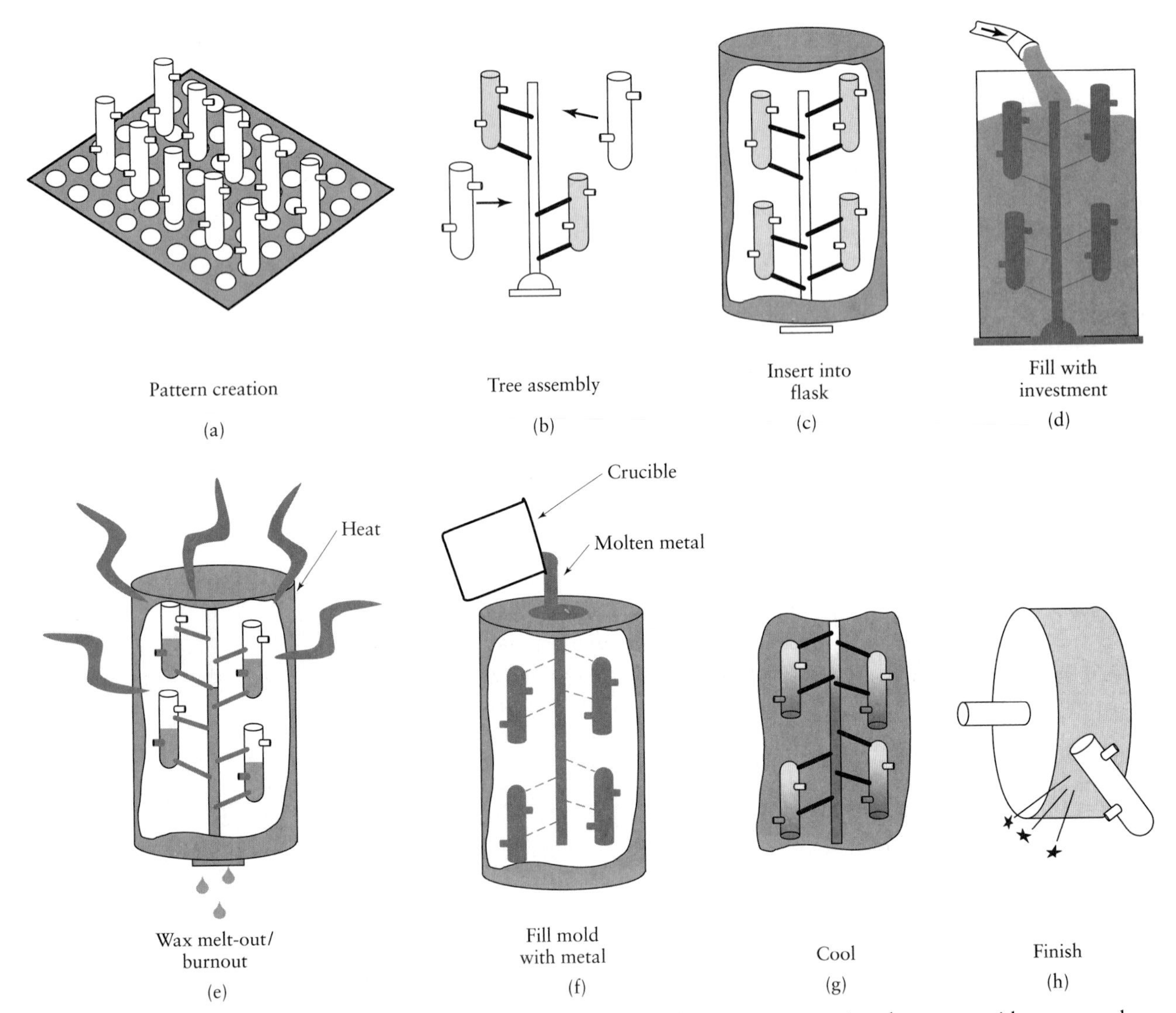

FIGURE 10.45 Manufacturing steps for investment casting that uses rapid-prototyped wax parts as blanks. This approach uses a flask for the investment, but a shell method can also be used. *Source:* 3D Systems, Inc.

tooling is the capability to produce a mold (or mold insert) that can be used for the manufacture of plastic components without the time lag (several months) traditionally required for the procurement of tooling. Furthermore, the design is simplified, because the designer need only analyze a CAD file of the desired part. Software then produces the tool geometry and automatically compensates for shrinkage.

EXAMPLE 10.8 Rapid prototyping of an injection manifold

The Rover Group, a subsidiary of British Aerospace, is well known for four-wheel-drive vehicles and performance automobiles. An innovative new injection-manifold design (Fig. 10.46), based on a plenum chamber for efficient movement of air into the cylinders, was designed for a new Rover power unit. However, the chamber is very intricate and has large trapped volumes, and consequently a prototype had to be constructed from one piece to provide accurate flow characteristics.

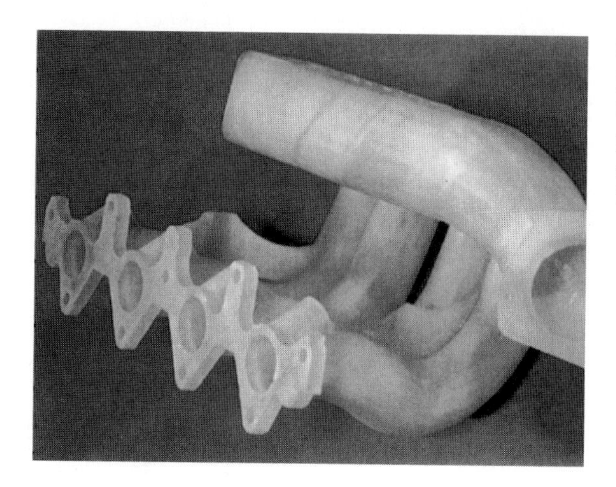

FIGURE 10.46 Rapid prototyped model of an injection-manifold design, produced through stereolithography. *Source*: 3D Systems.

Normally, Rover would have relied on traditional patternmaking followed by complex core-making and casting operations. However, this process is expensive and has a typical turnaround time of around 16 weeks. Using stereolithography, Rover produced a prototype multitrack injection manifold from a CAD file in only 39 hours, at less than 10% of the cost of the conventional prototype. More importantly, due to the model's strength and accuracy, it could be attached to an engine test bed, and its volumetric efficiency, i.e., its ability to allow the maximum air volume into the cylinder during one stroke under different conditions, could be directly measured. This procedure allowed optimization of the complex inlet tracts and plenum chamber and allowed Rover to incorporate design iterations into the product before committing to the purchase of expensive tooling.

Source: Courtesy of 3D Systems, Inc., and The Rover Group.

10.13 Design Considerations

The design considerations for forming and shaping plastics are to a large extent similar to those for processing metals. However, the mechanical and physical properties of plastics should carefully be considered during design, material, and process selection. Selection of an appropriate material from an extensive list requires consideration of mechanical and physical properties; service requirements; possible long-range effects of processing on properties and behavior, such as dimensional stability and degradation; and ultimate disposal at the end of the product's life cycle. Compared with metals, plastics have lower strength and stiffness, although the strength-to-weight and stiffness-to-weight ratios for reinforced plastics are higher than for many metals. Thus, section sizes should be selected accordingly, with a view to maintaining a sufficiently high section modulus (the ratio of moment of inertia to the distance from the neutral axis to the surface of the part) for desired stiffness. Improper part design or assembly can lead to warping and shrinking (Fig. 10.47a). Plastics can easily be molded around metallic parts and inserts; however, their compatibility with metals when so assembled should be assessed.

The overall part shape often determines the choice of the particular forming or molding process. Even after a particular process is selected, the design of the part

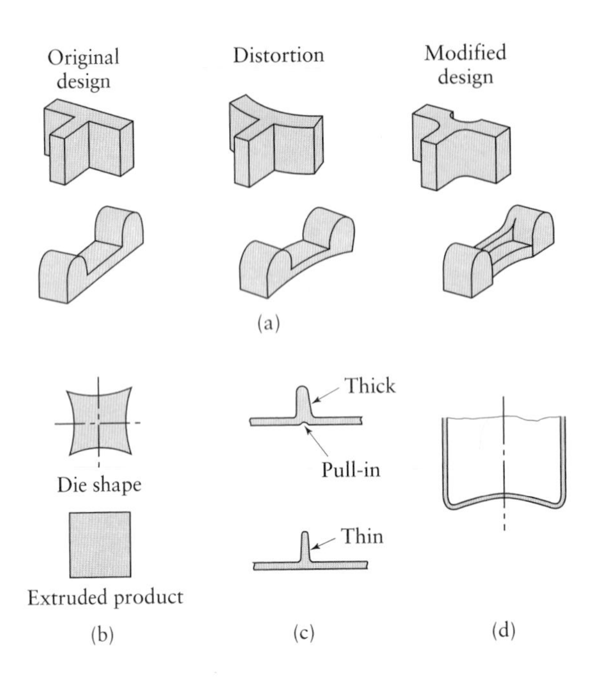

FIGURE 10.47 Examples of design modifications to eliminate or minimize distortion of plastic parts. (a) Suggested design changes to minimize distortion. *Source*: F. Strasser. (b) Die design (exaggerated) for extrusion of square sections. Without this design, product cross-sections would not have the desired shape, because of the recovery of the material, an effect known as *die swell*. (c) Design change in a rib to minimize pull-in caused by shrinkage during cooling. (d) Stiffening of the bottom of thin plastic containers by doming, which is similar to the process used to make the bottoms of

and die should be such that it will not cause problems concerning shape generation (Fig. 10.47b), dimensional control, and surface finish. As in casting metals and alloys (Chapter 5), material flow in the mold cavities should be properly controlled. The effects of molecular orientation during processing should also be considered, especially in extrusion, thermoforming, and blow molding.

Large variations in cross-section (Fig. 10.47c) and abrupt changes in geometry should be avoided in order to achieve better product quality and increased mold life. Furthermore, contraction in large cross-sections can cause porosity in plastic parts. Conversely, because of a lack of stiffness, removing thin sections from molds after shaping may be difficult. The low elastic modulus of plastics further requires that shapes be properly selected for improved stiffness of the component (Fig. 10.47d), particularly when it is important to save material. These considerations are similar to those in designing metal castings and forgings.

The properties of the final product depend on the original material and its processing history. Cold working of polymers improves their strength and toughness; on the other hand, because of the nonuniformity of deformation (even in simple rolling), residual stresses develop, as they do in metals. These stresses can also be generated by thermal cycling of the part. Also, the magnitude and direction of residual stresses, however the stresses are produced, can be important factors, as these stresses can relax over a period of time and cause distortion of the part during its service life, or during subsequent processing such as drilling; in addition, they can affect the part's resistance to adverse environmental conditions. (See Section 10.3.)

Reinforced plastics. The processing of reinforced plastics depends on the nature of the fibers, whether long, short, or chopped, as well as on the type of fibers and matrix materials. The presence of fibers makes processing of these composites more complex than that of thermoplastics and thermosets, including control of dimensional tolerances, and involves long cycle times. A major design advantage of reinforced plastics is the directional nature of the strength of the material. (See, for example, Fig. 10.19.) Forces applied to the material are transferred by the resin matrix to the fibers, which are much stronger and stiffer than the matrix. When fibers are all oriented in one direction, the resulting material is exceptionally strong in the fiber direction. This property is often used in designing reinforced plastic structures. For strength in two principal directions, the unidirectional materials are often laid at different angles to each other. If strength in the third (thickness) direction is desired, a different type of material is used to form a sandwich structure.

10.14 Economics of Processing Plastics

As in all processes, design and manufacturing decisions are ultimately based on performance and cost, including the costs of equipment, tooling, and production; the final selection of a process depends greatly on production volume. High equipment and tooling costs in plastics processing can be acceptable only if the production run is large, as is the case in casting and forging. Table 10.9 provides a general guide to economical processing of plastics and composite materials. Several types of equipment are used in plastics forming and shaping processes; the most expensive is injection-molding machines, with costs being directly proportional to the clamping force. A machine with a 2000-kN (225-ton) clamping force costs about $100,000, and one with a 20,000-kN (2250-ton) clamping force costs about $450,000. For composite materials, equipment and tooling costs for most molding operations are generally high, and production rates and economic production quantities vary widely.

The optimum number of cavities in the die for making the product in one cycle is an important consideration, as in die casting (Section 5.10.3). For small parts, a number of cavities can be made in a die, with runners to each cavity. If the part is large, then only one cavity may be accommodated. As the number of cavities increases, so does the cost of the die; larger dies may be considered for larger numbers of cavities, thus increasing die cost even further. On the other hand, more parts will be produced per machine cycle on a larger die, thereby increasing the production rate. Hence, a detailed analysis has to be made to determine the optimum number of cavities, die size, and machine capacity.

TABLE 10.9

Comparative Costs and Production Volumes for Processing of Plastics

	Equipment Capital Cost	Production Rate	Tooling Cost	Typical Production Volume, Number of Parts (10, 10^2, 10^3, 10^4, 10^5, 10^6, 10^7)
Machining	Med	Med	Low	10 – 10^2
Compression molding	High	Med	High	10^3 – 10^6
Transfer molding	High	Med	High	10^3 – 10^6
Injection molding	High	High	High	10^4 – 10^7
Extrusion	Med	High	Low	*
Rotational molding	Low	Low	Low	10^2 – 10^3 (approx.)
Blow molding	Med	Med	Med	10^3 – 10^7
Thermoforming	Low	Low	Low	10^2 – 10^3 (approx.)
Casting	Low	Very low	Low	10 – 10^2
Forging	High	Low	Med	10 – 10^2
Foam molding	High	Med	Med	10^4 – 10^7

*Continuous process.

Source: After R. L. E. Brown, *Design and Manufacture of Plastic Parts*. Copyright © 1980 by John Wiley & Sons, Inc. Reprinted by permission of John Wiley & Sons, Inc.

CASE STUDY | The EPOCH Hip

Each year, over one million osteoarthritic joints are replaced with artificial joints worldwide; these devices generally eliminate pain and greatly improve the quality of life of their recipients. A total hip prosthesis replaces both the ball and the socket of the natural joint, in most cases, with a femoral component inserted in the upper end of the femur and an acetabular component placed in the pelvis. These implants have been produced since the 1950s, and can last for years before a revision is needed. Most modern implants are constructed from (a) Ti-6Al-4V alloy, (b) ASTM F-75 cobalt-chrome alloy, or (c) 316 stainless steel alloy, with a separate head made of an appropriate alloy and attached via a tapered seat on the femoral component, and a Ti-6Al-4V shell with a polymer insert for the acetabular component.

Total hips may be classified in two basic types: cemented and press fit. Cemented implants use a polymethyl methacrylate-based grout for fixation, whereas press-fit implants have a rough, porous surface intended for bone to grow into. They are fixed directly to bone. This rough surface can be manufactured by plasma spray operations, by sintering metal beads to the surface, or by diffusion bonding wires to the implant. These kinds of implants are often used in patients that are younger and more active, since their bone is better able to grow than those of older, more sedentary patients.

One of the problems that has been encountered with total hip replacements is that of bone resorption (loss of bone). Healthy bone must be stressed to maintain its strength and vigor. However, metal implants that are very stiff tend to transfer loads to the femur differently than the patient's natural hip does, and much of the bone toward the top of the femur experiences very little stress. The body reacts to this low-stress state by resorbing the bone. This loss of bone mass can leave the implant in an unsupported loading environment. This unsupported condition may limit the service life of the implant, leading to revision sooner than would otherwise be necessary. The revision may also require the use of bone graft to replace some of the resorbed bone and support the revision of the implant.

The main cause of stress shielding is that the stiffness of the metals used for the implant is much greater than that of bone. For example, the elastic modulus of Ti-6Al-4V is around 105 GPa, whereas bone has a modulus nearer to 20 GPa. Many polymers have a stiffness close to that of bone, but their fatigue strengths are not high enough to be used directly as implants.

A new and innovative implant design is the EPOCH hip system shown in Fig. 10.48. It consists of three layers: (a) an outermost layer comprised of a

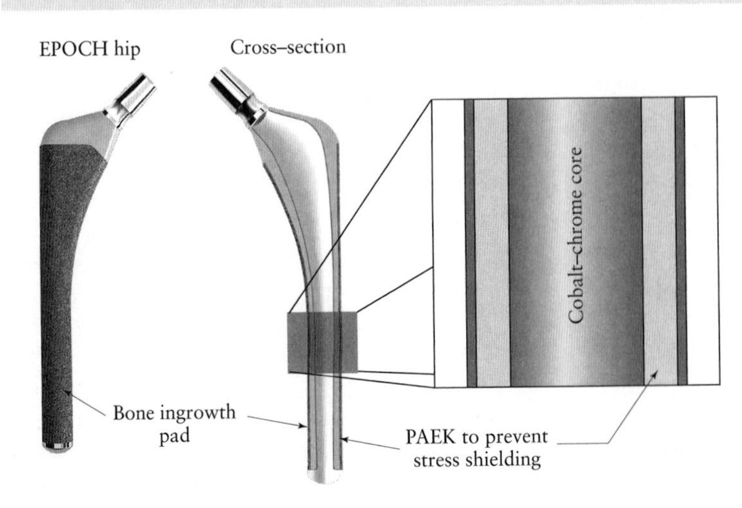

FIGURE 10.48 The EPOCH hip stem. This design uses a poly(aryletherketone) (PAEK) layer and bone ingrowth pad around a cobalt–chrome core in order to maximize bone ingrowth. The EPOCH Hip Stem is not available for sale in the United States. *Source*: Zimmer, Inc.

diffusion-bonded titanium-alloy mesh for bone ingrowth, (b) a thermoplastic polymer layer intended to provide a stiffness similar to bone, and (c) a cobalt-alloy core to provide static and fatigue strength. The polymer used in the Epoch hip is from the poly(aryletherketone) (PAEK) family, and the layers are proportioned to mimic bone stiffness while maximizing fatigue strength.

The Epoch hip is manufactured through an insert injection-molding operation (Fig. 10.49; see also Fig. 10.24). A vertical injection-molding machine is used (as opposed to a horizontal machine) in order to facilitate proper placement of the inserts. Two flexible pads are placed into seats on the mold halves. A machined cobalt-alloy insert is preheated and placed in the bottom half of the mold. The mold is then closed and the polymer is injected into the cavity defined by the mold halves, the flexible pads, and the insert. The polymer adheres to the cobalt-alloy insert and partially permeates the porous pad, integrally locking all components together. After the part is ejected from the mold, the flashing is trimmed and the hip stem is finished to obtain desired surface finish. A cross-section of the resulting structure is shown in Fig. 10.48.

In laboratory testing, the EPOCH hip has demonstrated flexibility near 20 GPa while meeting established strength requirements. This contribution of properties should assist in maintaining bone density without compromising performance expectations.

Source: M. Hawkins, Zimmer, Inc.

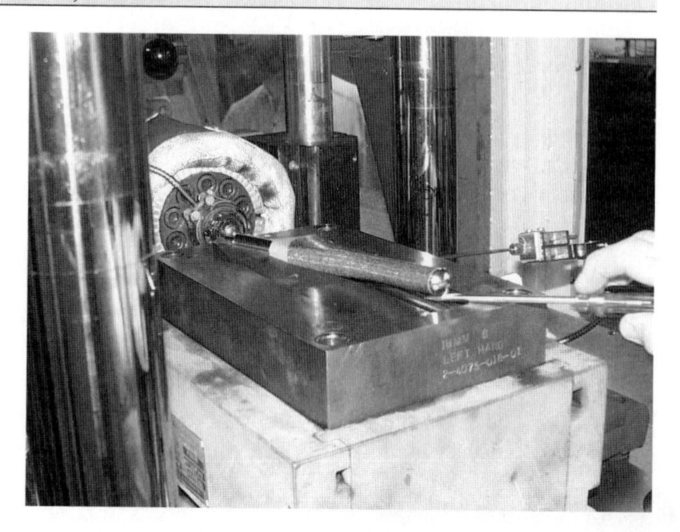

FIGURE 10.49 An EPOCH hip is removed from the mold after an insert-injection-molding operation. The EPOCH Hip Stem is not available for sale in the United States. *Source*: Zimmer, Inc.

SUMMARY

- Polymers are an important class of materials, because they possess a very wide range of mechanical, physical, chemical, and optical properties. Compared with metals, plastics are generally characterized by lower density, strength, elastic modulus, and thermal and electrical conductivity, and a higher coefficient of thermal expansion. (Section 10.1)
- Plastics are composed of polymer molecules and various additives. The smallest repetitive unit in a polymer chain is called a mer. Monomers are linked by polymerization processes to form larger molecules. The properties of the polymer depend on the molecular structure (linear, branched, cross-linked, or network), the degree of crystallinity, and additives (which perform various functions, such as improving strength, flame retardation, and lubrication, as well as imparting flexibility, color, and stability against ultraviolet radiation and oxygen). The glass-transition temperature separates the brittle and ductile regions of polymers. (Section 10.2)

- Thermoplastics become soft and are easy to form at elevated temperatures. Their mechanical behavior can be characterized by various spring and dashpot models and includes phenomena such as creep and stress relaxation, crazing, and water absorption. (Section 10.3)
- Thermosets, which are obtained by cross-linking polymer chains, do not become soft to any significant extent with increasing temperature. (Section 10.4)
- Thermoplastics and thermosets have a very wide range of consumer and industrial applications with a variety of characteristics. (Sections 10.5, 10.6)
- Among important developments in polymers is biodegradable plastics; several formulations have been developed. (Section 10.7)
- Elastomers have the characteristic ability to undergo large elastic deformations and return to their original shape when unloaded. Consequently, they have important applications as tires, seals, footwear, hoses, belts, and shock absorbers. (Section 10.8)
- Reinforced plastics are an important class of materials that have superior mechanical properties and are lightweight. The reinforcing fibers are usually glass, graphite, aramids, or boron; epoxies commonly serve as a matrix material. The properties of reinforced plastics depend on the composition of the plastic and the orientation of the fibers. (Section 10.9)
- Plastics can be formed and shaped by a variety of processes, such as extrusion, molding, casting, and thermoforming. Thermosets are generally molded or cast. (Section 10.10)
- Reinforced plastics are processed into structural components, using liquid plastics, prepregs, and bulk- and sheet-molding compounds. Fabricating techniques include various molding methods, filament winding, and pultrusion. The type and orientation of the fibers, and the strength of the bond between fibers and matrix and between layers of materials, are important factors. (Section 10.11)
- Rapid prototyping has become an increasingly important technology, because of its inherent flexibility and the much shorter times required for making prototypes. Typical techniques include stereolithography, fused-deposition modeling, selective laser sintering, laminated-object manufacturing, ballistic-particle manufacturing, and solid-base curing. Rapid production of tooling is a further development of these processes. (Section 10.12)
- The design of plastic parts includes considerations of their relatively low strength and stiffness, as well as high thermal expansion and generally low resistance to temperature. (Section 10.13)
- Because of the variety of low-cost materials and processing techniques available, the economics of processing plastics and reinforced plastics is an important consideration, particularly when compared with metal components, and includes die costs, cycle times, and production volume as important parameters. (Section 10.14)

SUMMARY OF EQUATIONS

Linearly elastic behavior

in tension, $\sigma = E\varepsilon$

in shear, $\tau = G\gamma$

Viscous behavior, $\tau = \eta\left(\dfrac{dv}{dy}\right) = \eta\dot{\gamma}$

Strain-rate sensitivity, $\sigma = C\dot{\varepsilon}^m$

Viscoelastic modulus, $E_r = \sigma/(e_e + e_v)$

Viscosity,

$$\eta = \eta_0 e^{E/kT}$$

Extruder Characteristic,

$$Q = \frac{\pi^2 H D^2 N \sin\theta\cos\theta}{2} - \frac{p\pi D H^3 \sin^2\theta}{12\eta l}.$$

Die Characteristic, $Q_{\text{die}} = Kp$

For round die, $K = \dfrac{\pi D_d^4}{128\eta \ell_d}$,

Cure depth in stereolithography, $C_d = D_p \ln\left(\dfrac{E_0}{E_c}\right)$.

$$\log \eta = 12 - \frac{(17.5\Delta T)}{(52 + \Delta T)}$$

Stress relaxation, $\tau = \tau_0 e^{-t/\lambda}$

Relaxation time, $\lambda = \dfrac{\eta}{G}$

Strength of composite, $\sigma_c = x\sigma_f + (1 - x)\sigma_m$

Elastic modulus of composite, $E_c = xE_f + (1 - x)E_m$

Line width in stereolithography, $L_w = B\sqrt{\dfrac{C_d}{2D_p}}$,

BIBLIOGRAPHY

Agarwal, B. D., and L. J. Broutman, *Analysis and Performance of Fiber Composites*, 2d. ed., Wiley, 1990.

Agassant, J. F., P. Avenas, J. Ph. Sergent, and P. J. Carreau, *Polymer Processing: Principles and Modeling*, Hanser, 1991.

Beaman, J. J., J. W. Barlow, D. L. Bourell, and R. Crawford, *Solid Freeform Fabrication*, Kluwer, 1997.

Belitskus, D. L., *Fiber and Whisker Reinforced Ceramics for Structural Applications*, Dekker, 1993.

Berins, M. L., *Plastics Engineering Handbook*, 5th ed., Chapman & Hall, 1995.

Bertholet, J.-M., *Composite Materials: Mechanical Behavior and Structural Analysis*, Springer, 1999.

Bhowmick, A. K., and H. L. Stephens, *Handbook of Elastomers*, Dekker, 1988.

______, *Rubber Products Manufacturing Technology*, Dekker, 1994.

Buckley, C. P., C. B. Bucknall, and N. G. McCrum, *Principles of Polymer Engineering*, 2d. ed., Oxford, 1997.

Burns, M., *Automated Fabrication: Improving Productivity in Manufacturing*, PTR Prentice Hall, 1993.

Campbell, P., *Plastic Component Design*, Industrial Press, 1996.

Chanda, M., and S. K. Roy, *Plastics Technology Handbook*, 3d. ed., Dekker, 1998.

Charrier, J.-M., *Polymeric Materials and Processing: I Plastics, Elastomers, and Composites*, Hanser, 1991.

Chawla, K. K., *Composite Materials: Science and Engineering*, 2d. ed., Springer, 1998.

Chua, C. K., and L. K. F. Leong, *Rapid Prototyping: Principles and Applications in Manufacturing*, World Scientific Co., 2000.

Cooper, K. G., *Rapid Prototyping Technology*, Dekker, 2001.

Daniel, I. M., and O. Ishai, *Engineering Mechanics of Composite Materials*, Oxford, 1994.

Dimov, S. S., and D. T. Pham, *Rapid Manufacturing: The Technologies and Applications of Rapid Prototyping and Rapid Tooling*, Springer, 2001.

Domininghaus, H., *Plastics for Engineers: Materials, Properties, Applications*, Hanser-Gardner, 1993.

Dumitriu, S., (ed.), *Polymeric Biomaterials*, Dekker, 1994.

Engineered Materials Handbook, Vol. 1: *Composites*, ASM International, 1987; Vol. 2: *Engineering Plastics*, ASM International, 1988.

Engineering Plastics and Composites, 2d. ed., ASM International, 1993.

Griskey, R. G., *Polymer Process Engineering*, Chapman & Hall, 1995.

Gutowski, T. G., *Advanced Composites Manufacturing*, Wiley, 1997.

Harper, C., *Handbook of Plastics, Elastomers, and Composites*, 3d. ed., McGraw-Hill, 1996.

Hollaway, L., *Handbook of Polymer Composites for Engineers*, Cambridge, 1994.

Hull, D., and T. W. Clyne, *An Introduction to Composite Materials*, Cambridge, 1996.

Jacobs, P., *Stereolithography and other RP&M Technologies: From Rapid Prototyping to Rapid Tooling*, American Society of Mechanical Engineers, 1995.

Jang, B. Z., *Advanced Polymer Composites: Principles and Applications*, ASM International, 1994.

Lee, N. C. (ed.), *Plastic Blow Molding Handbook*, Van Nostrand Reinhold, 1990.

Loewenstein, K. L., *The Manufacturing Technology of Continuous Glass Fibers*, 2d. ed., Elsevier, 1983.

Lu, L., J. Y. H. Fuh, and Y. S. Wong, *Laser-Induced Materials and Processes for Rapid Prototyping*, Kluwer, 2001.

MacDermott, C. P., and A. V. Shenoy, *Selecting Thermoplastics for Engineering Applications*, 2d. ed., Dekker, 1997.

Mallick, P. K. (ed.), *Composites Engineering Handbook*, Dekker, 1997.

______, *Fiber-Reinforced Composite Materials, Manufacturing, and Design*, 2d. ed., Dekker, 1993.

Malloy, R. A., *Plastic Part Design for Injection Molding: An Introduction*, Hanser-Gardner, 1994.

McCrum, N. G., C. P. Buckley, and C. B. Bucknall, *Principles of Polymer Engineering*, Oxford, 1989.

Mileiko, S. T., *Metal and Ceramic Based Composites*, Elsevier, 1992.

Miller, E., *Introduction to Plastics and Composites: Mechanical Properties and Engineering Applications*, Dekker, 1995.

Modern Plastics Encyclopedia, McGraw-Hill, published annually.

Modern Plastics Handbook, McGraw-Hill, 2000.

Muccio, E. A., *Plastic Part Technology*, ASM International, 1991.

_______, *Plastics Processing Technology*, ASM International, 1994.

Mustafa, N., *Plastics Waste Management: Disposal, Recycling, and Reuse*, Dekker, 1993.

Nielsen, L. E., and R. F. Landel, *Mechanical Properties of Polymers and Composites*, 2d. ed., Dekker, 1994.

Potter, K., *Introduction to Composite Products: Design, Development and Manufacture*, Chapman & Hall, 1997.

_______, *Resin Transfer Molding*, Chapman & Hall, 1997.

Progelhof, R. C., and J. L. Throne, *Polymer Engineering Principles: Properties, Tests for Design*, Hanser, 1993.

Rosato, D. V., *Plastics Processing Data Handbook*, 2d. ed., Chapman & Hall, 1997.

Rosato, D. V., and D. V. Rosato, *Injection Molding Handbook*, 3d. ed., Kluwer Academic Publishers, 2000.

Rosato, D. V., D. V. Rosato and M. G. Rosato, *Plastics Design Handbook*, Kluwer Academic Publishers, 2001.

Rudin, A., *Elements of Polymer Science and Engineering*, 2d. ed., Academic Press, 1999.

Schwartz, M., *Composite Materials*, Vol. 1: *Properties, Nondestructive Testing, and Repair*; Vol. 2: *Processing, Fabrication, and Applications*, McGraw-Hill, 1992.

Seymour, R. B., *Engineering Polymer Sourcebook*, McGraw-Hill, 1990.

_______, *Reinforced Plastics: Properties and Applications*, ASM International, 1991.

Shastri, R., *Plastics Product Design*, Dekker, 1996.

Skotheim, T. A., *Handbook of Conducting Polymers*, 2 vols., Dekker, 1986.

Strong, A. B., *Fundamentals of Composites Manufacturing: Materials, Methods, and Applications*, Society of Manufacturing Engineers, 1989.

_______, *Plastics: Materials and Processing*, Prentice Hall, 1996.

Tool and Manufacturing Engineers Handbook, 4th ed., Vol. 8: *Plastic Part Manufacturing*, Society of Manufacturing Engineers, 1996.

Tres, P. A., *Designing Plastic Parts for Assembly*, Hanser-Gardner, 1994.

Ulrich, H., *Introduction to Industrial Polymers*, Hanser, 1994.

Vollrath, L., and H. G. Haldenwanger, *Plastics in Automotive Engineering: Materials, Components, Systems*, Hanser-Gardner, 1994.

Zachariades, A. E., and R. S. Porter, *High-Modulus Polymers: Approaches to Design and Development*, Dekker, 1995.

QUESTIONS

10.1 Summarize the most important mechanical and physical properties of plastics.

10.2 What are the major differences between the properties of plastics and metals?

10.3 What properties are influenced by the degree of polymerization?

10.4 Give applications for which flammability of plastics would be a major concern.

10.5 What properties do elastomers have that thermoplastics, in general, do not have?

10.6 Is it possible for a material to have a hysteresis behavior that is the opposite of that shown in Fig. 10.14, whereby the arrows are counterclockwise? Explain.

10.7 Observe the behavior of the tension-test specimen shown in Fig. 10.13, and state whether the material has a high or low m value. (See Section 2.2.7.) Explain why.

10.8 Why would we want to synthesize a polymer with a high degree of crystallinity?

10.9 Add more to the applications column in Table 10.3.

10.10 Discuss the significance of the glass-transition temperature, T_g, in engineering applications.

10.11 Why does cross-linking improve the strength of polymers?

10.12 Describe the methods by which optical properties of polymers can be altered.

10.13 Explain the reasons that elastomers were developed. Are there any substitutes for elastomers? Explain.

10.14 Give several examples of plastic products or components for which creep and stress relaxation are important considerations.

10.15 Describe your opinions regarding recycling of plastics versus developing plastics that are biodegradable.

10.16 Explain how you would go about determining the hardness of the plastics described in this chapter.

10.17 Distinguish between composites and alloys.

10.18 Describe the functions of the matrix and the reinforcing fibers in reinforced plastics. What fundamental differences are there in the characteristics of the two materials?

10.19 What products have you personally seen that are made of reinforced plastics? How can you tell that they are reinforced?

10.20 Identify metals and alloys that have strengths comparable with those of reinforced plastics.

10.21 Compare the advantages and disadvantages of metal-matrix composites, reinforced plastics, and ceramic-matrix composites.

10.22 You have studied the many advantages of composite materials in this chapter. What limitations or disadvantages do these materials have? What suggestions would you make to overcome these limitations?

10.23 A hybrid composite is defined as a material containing two or more different types of reinforcing fibers. What advantages would such a composite have over other composites.

10.24 Why are fibers capable of supporting a major portion of the load in composite materials?

10.25 Assume that you are manufacturing a product in which all the gears are made of metal. A salesperson visits you and asks you to consider replacing some of the metal gears with plastic ones. Make a list of the questions that you would raise before making such a decision.

10.26 Review the three curves in Fig. 10.8, and name applications for each type of behavior. Explain your choices.

10.27 Repeat Question 10.26 for the curves in Fig. 10.10.

10.28 Do you think that honeycomb structures could be used in passenger cars? If so, where? Explain.

10.29 Other than those described in this chapter, what materials can you think of that can be regarded as composite materials?

10.30 What applications for composite materials can you think of in which high thermal conductivity would be desirable?

10.31 Make a survey of a variety of sports equipment, and identify the components that are made of composite materials. Explain the reasons for and advantages of using composites for these specific applications.

10.32 We have described several material combinations and structures in this chapter. In relative terms, identify those that would be suitable for applications involving one of the following: (1) very low temperatures; (2) very high temperatures; (3) vibrations; (4) high humidity.

10.33 Explain how you would go about determining the hardness of the reinforced plastics and composite materials described in this chapter. Are hardness measurements for these types of materials meaningful? Does the size of the indentation make a difference in your answer? Explain.

10.34 Describe the advantages of applying traditional metalworking techniques to the formation of plastics.

10.35 Describe the advantages of cold forming of plastics over other processing methods.

10.36 Explain the reasons that some forming processes are more suitable for certain plastics than for others.

10.37 Would you use thermosetting plastics for injection molding? Explain.

10.38 By inspecting plastic containers, such as for baby powder, you can see that the lettering on them is raised rather than sunk. Can you offer an explanation as to why they are molded in that way?

10.39 Give examples of several parts that are suitable for insert molding. How would you manufacture these parts if insert molding were not available?

10.40 What manufacturing considerations are involved in making a metal beverage container versus a plastic one?

10.41 Inspect several electrical components, such as light switches, outlets, and circuit breakers, and describe the process or processes used in making them.

10.42 Inspect several similar products that are made of metals and plastics, such as a metal bucket and a plastic bucket of similar shape and size. Comment on their respective thicknesses, and explain the reasons for their differences, if any.

10.43 Make a list of processing methods used for reinforced plastics. Identify which of the following fiber orientation and arrangement capabilities each has: (1) uniaxial, (2) cross-ply, (3) in-plane random, and (4) three-dimensional random.

10.44 Some plastic products have lids with integral hinges; that is, no other material or part is used at the junction of the two parts. Identify such products, and describe a method for making them.

10.45 Explain why operations such as blow molding and film-bag making are done vertically and why buildings that house equipment for these operations have ceilings 10 m to 15 m (35 ft to 50 ft) high.

10.46 Consider the case of a coffee mug being produced by rapid prototyping. How can the top of the handle be manufactured, since there is no material directly beneath the arch?

10.47 Make a list of the advantages and disadvantages of each of the rapid-prototyping operations.

10.48 Explain why finishing operations are needed for rapid-prototyping operations. If you are making a prototype of a toy automobile, list the post-rapid-prototyping finishing operations you would perform.

10.49 A current topic of research involves producing parts from rapid-prototyping operations and then using them in experimental stress analysis, in order to infer the strength of final parts produced by means of conventional manufacturing operations. List your concerns with this approach, and outline means of addressing these concerns.

10.50 Because of relief of residual stresses during curing, long unsupported overhangs in parts from stereolithography will tend to curl. Suggest methods of controlling or eliminating this curl.

10.51 One of the major advantages of stereolithography is that it can use semitransparent polymers, so that internal details of parts can be readily discerned. List parts for which this feature is valuable.

10.52 Based on the processes used to make fibers, explain how you would produce carbon foam. How would you make a metal foam?

10.53 Die swell in extrusion is radially uniform for circular cross-sections, but is not uniform for other cross-sections. Recognizing this fact, make a qualitative sketch of a die profile that will produce (a) square and (b) triangular cross-sections of extruded polymer.

10.54 What are the advantages of using whiskers as a reinforcing material?

10.55 By incorporating small amounts of blowing agent, it is possible to manufacture polymer fibers with gas cores. List some applications for such fibers.

10.56 With injection-molding operations, it is common practice to remove the part from its runner and then to place the runner into a shredder and recycle the resultant pellets. List the concerns you would have in using such recycled pellets as opposed to so-called virgin pellets.

PROBLEMS

10.57 Calculate the areas under the stress–strain curve (toughness) for the material in Fig. 10.9, plot them as a function of temperature, and describe your observations.

10.58 Note in Fig. 10.9 that, as expected, the elastic modulus of the polymer decreases as temperature increases. Using the stress–strain curves in the figure, make a plot of the modulus of elasticity versus temperature.

10.59 Calculate the percentage increase in mechanical properties of reinforced nylon from the data shown in Fig. 10.17.

10.60 A rectangular cantilever beam 100 mm high, 25 mm wide, and 1 m long is subjected to a concentrated force of 100 N at its end. Select two different unreinforced and reinforced materials from Table 10.1, and calculate the maximum deflection of the beam. Then select aluminum and steel, and for the same beam dimensions, calculate the maximum deflection. Compare the results.

10.61 In Sections 10.5 and 10.6, we listed several plastics and their applications. Rearrange this information, respectively, by making a table of products and the type of plastics that can be used to make the products.

10.62 Determine the dimensions of a tubular steel drive shaft for a typical automobile. If you now replace this shaft with shafts made of unreinforced and reinforced plastic, respectively, what should be the shaft's new dimensions to transmit the same torque for each case? Choose the materials from Table 10.1, and assume a Poisson's ratio of 0.4.

10.63 Calculate the average increase in the properties of the plastics listed in Table 10.1 as a result of their reinforcement, and describe your observations.

10.64 In Example 10.4, what would be the percentage of the load supported by the fibers if their strength is 1250 MPa and the matrix strength is 240 MPa? What if the strength is unaffected, but the elastic modulus of the fiber is 600 GPa while the matrix is 50 GPa?

10.65 Estimate the die clamping force required for injection molding ten identical 2-in.-diameter disks in one die. Include the runners of appropriate length and diameter.

10.66 A two-liter plastic beverage bottle is made from a parison with the same diameter as the threaded neck of the bottle and has a length of 5 in. Assuming uniform deformation during blow molding, estimate the wall thickness of the tubular section.

10.67 Estimate the consistency index and power-law index for the polymers in Fig. 10.12.

10.68 An extruder has a barrel diameter of 75 mm. The screw rotates at 100 rpm, has a channel depth of 6 mm, and a flight angle of 17.5°. What is the largest flow rate of polypropylene that can be achieved?

10.69 The extruder in Problem 10.68 has a pumping section that is 3 m long and is used to extrude round polyethylene rod. The die has a land of 1 mm and a diameter of 5 mm. If the polyethylene is at a mean temperature of 250°C, what is the flow rate through the die? What if the die diameter is 10 mm?

10.70 An extruder has a barrel diameter of 4 in., a channel depth of 0.2 in., a flight angle of 18°, and a pumping zone that is 6 ft long. It is used to pump a plastic with a viscosity of 100×10^{-4} lb-s/in^2. If the die characteristic is experimentally determined as $Q_x = (0.00210 \text{ in}^5/\text{lb-s})\, p$, what screw speed is needed to achieve a flow rate of 6 in^3/s from the extruder?

10.71 What flight angle should be used on a screw so that a flight translates a distance equal to the barrel diameter with every revolution?

10.72 For a laser providing 12.5 kJ of energy to a spot with diameter of 0.25 mm, determine the cure depth and the cured line width in stereolithography. Use $E_c = 6.36 \times 10^{10}$ J/m^2 and $D_p = 100\mu$m.

10.73 For the stereolithography system described in Problem 10.72, estimate the time required to cure a layer defined by a 50-mm (2 in.) circle if adjacent lines overlap each other by 10% and the power available is 10 MW.

10.74 The extruder head in a fused-deposition-modeling setup has a diameter of 1.25 mm (0.05 in.) and produces layers that are 0.25 mm (0.01 in.) thick. If the velocities of the extruder head and polymer extrudate are both 50 mm/s, estimate the production time for the generation of a 50-mm (2-in.) solid cube. Assume that there is a 15-s delay between layers as the extruder head is moved over a wire brush for cleaning.

DESIGN

10.75 Discuss the design considerations involved in replacing a metal beverage container with a container made of plastic.

10.76 Using specific examples, discuss the design issues involved in products made of plastics versus reinforced plastics.

10.77 Make a list of products, parts, or components that are not currently made of plastics, and offer some reasons that they are not.

10.78 In order to use a steel or aluminum container to hold an acidic material, such as tomato juice or sauce, the inside of the container is coated with a polymeric barrier. Describe the methods of producing such a container. (See also Chapter 7.)

10.79 Using the information given in this chapter, develop special designs and shapes for possible new applications of composite materials.

10.80 Would a composite material with a strong and stiff matrix and soft and flexible reinforcement have any practical uses? Explain.

10.81 Make a list of products for which the use of composite materials could be advantageous because of their anisotropic properties.

10.82 Name several product designs in which both specific strength and specific stiffness are important.

10.83 Describe designs and applications in which strength in the thickness direction of a composite is important.

10.84 Design and describe a test method to determine the mechanical properties of reinforced plastics in their thickness direction.

10.85 As described in this chapter, reinforced plastics can be adversely affected by environmental factors, such as moisture, chemicals, and temperature variations. Design and describe test methods to determine the mechanical properties of composite materials under these conditions.

10.86 As with other materials, the mechanical properties of composites are obtained by preparing appropriate specimens and testing them. Explain what problems you might encounter in preparing specimens for testing and in the actual testing process itself.

CHAPTER

11

Properties and Processing of Metal Powders, Ceramics, Glasses, Composites, and Superconductors

11.1 Introduction

In the manufacturing processes described in the preceding chapters, the raw materials used are either in a molten state or in solid form. This chapter describes how metal parts can be made by compacting metal powders in tool-steel or carbide dies and sintering them (heating without melting). This process is called **powder metallurgy (P/M)**. One of its first uses was in the early 1900s to make the tungsten filaments for incandescent light bulbs. The availability of a wide range of powder compositions, the capability to produce parts to net dimensions (**net-shape forming**), and the economics of the overall operation make this process attractive for many applications.

Typical products made by powder-metallurgy techniques include gears, cams, bushings, cutting tools, porous products such as filters and oil-impregnated bearings, and automotive components such as piston rings, valve guides, connecting rods, and hydraulic pistons (Fig. 11.1). Advances in technology now permit structural parts of aircraft, such as landing gears, engine-mount supports, engine disks, impellers, and engine nacelle frames, to be made by P/M.

Powder metallurgy has become competitive with processes such as casting, forging, and machining, particularly for relatively complex parts made of high-strength and hard alloys. Nearly 70% of P/M part production is for automotive applications. Parts made by this process have good dimensional accuracy, and their sizes range from tiny balls for ball-point pens to parts weighing about 50 kg (100 lb), although most parts made by P/M weigh less than 2.5 kg (5 lb). A typical family car now contains, on average, 13 kg (33 lb) of precision metal parts made by powder metallurgy, an amount that has been increasing by about 10% annually.

This chapter also describes the *structure, properties, and processing of ceramics, glasses, graphite, diamond, and metal-matrix and ceramic-matrix composites.* Because of the large number of possible combinations of elements, a great variety of these materials is now available for a wide range of consumer, industrial, and aerospace applications. There are several methods of producing these materials, as well as a number of techniques for subjecting them to additional processing for improved properties, dimensional tolerances, and surface finish.

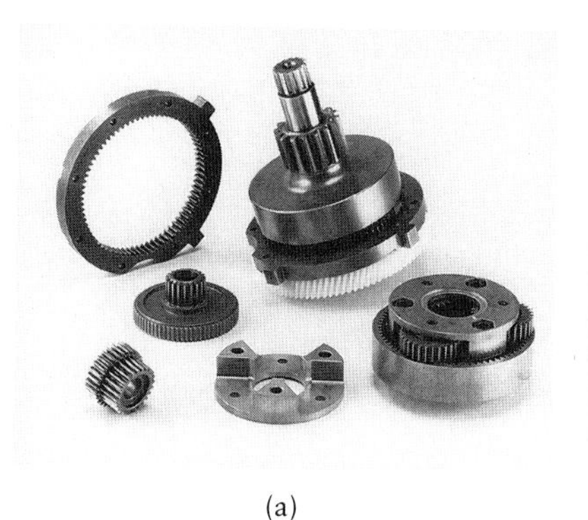

(a)

(b)

FIGURE 11.1 (a) Examples of typical parts made by powder-metallurgy processes. (b) Upper trip lever for a commercial irrigation sprnkler, made by P/M. This part is made of unleaded brass alloy; it replaces a die-cast part, with a 60% cost savings. *Source*: Reproduced with permission from *Success Stories on P/M Parts*, Princeton, NJ: Metal Powder Industries Federation, 1998.

11.2 Production of Metal Powders

The powder-metallurgy process basically consists of the following operations:

- **Powder production,**
- **Blending,**
- **Compaction,**
- **Sintering,** and
- **Finishing operations.**

For improved quality and dimensional accuracy, or for special applications, additional processing such as coining, sizing, machining, infiltration, and resintering may be carried out.

11.2.1 Methods of powder production

There are several methods of producing metal powder. Metal sources are generally bulk metals and alloys, ores, salts, and other compounds. Most metal powders can be produced by more than one method, the choice of which depends on the requirements of the end product. Particle sizes produced range from 0.1 μm to 1000 μm (0.04 μin. to 4 μin.). The shape, size distribution, porosity, chemical purity, and bulk and surface characteristics of the particles depend on the particular process used (Fig. 11.2). These characteristics are important, because they significantly affect flow during compaction and reactivity in subsequent sintering operations.

The methods of powder production are:

1. **Atomization.** *Atomization* produces a liquid-metal stream by injecting molten metal through a small orifice (Fig. 11.3a). The stream is broken up by jets of inert gas, air, or water. The size of the particles formed depends on the temperature of the metal, the rate of flow, nozzle size, and jet characteristics. Melt-atomization methods are widely used for production of powders for P/M. In one variation of this method, a consumable electrode is rotated rapidly in a helium-filled chamber (Fig. 11.3b). The centrifugal force breaks up the molten tip of the electrode, producing metal particles.
2. **Reduction.** *Reduction* of metal oxides (i.e., removal of oxygen) uses gases, such as hydrogen and carbon monoxide, as reducing agents; thus, very fine metallic oxides are reduced to the metallic state. The powders produced by this method are spongy and porous and have uniformly sized spherical or angular shapes.

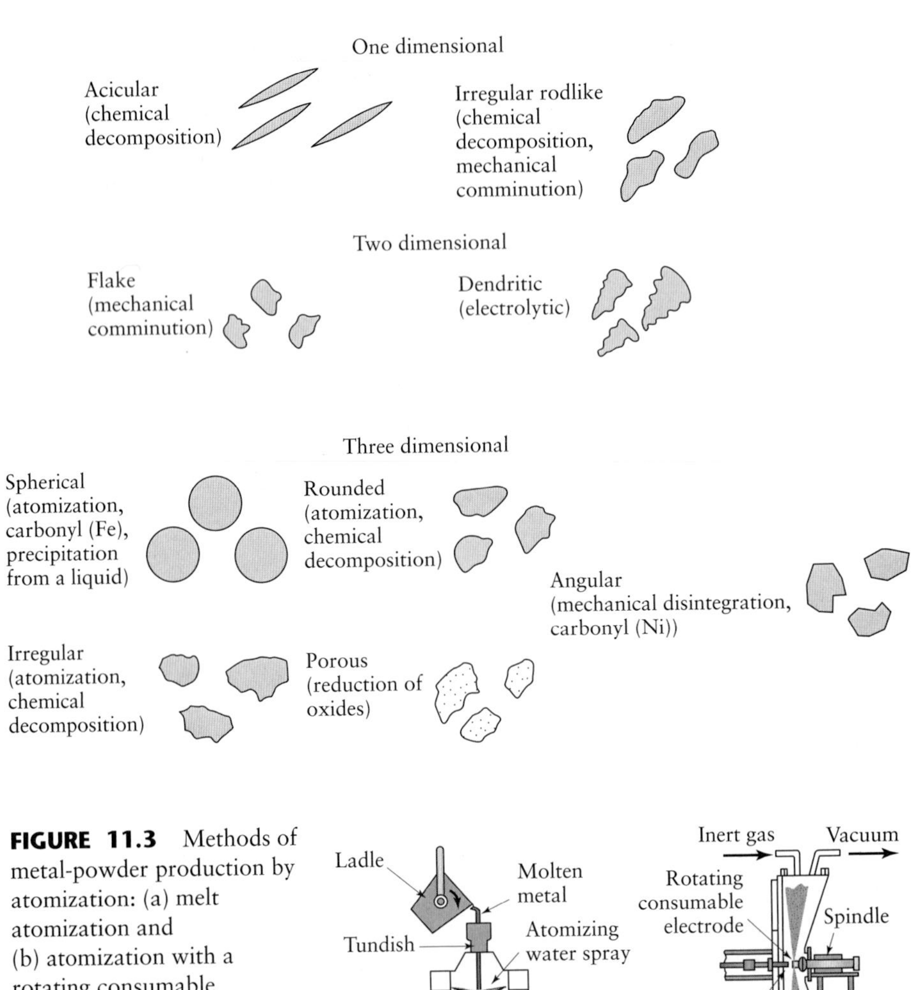

FIGURE 11.2 Particle shapes in metal powders and the processes by which they are produced. Iron powders are produced by many of these processes.

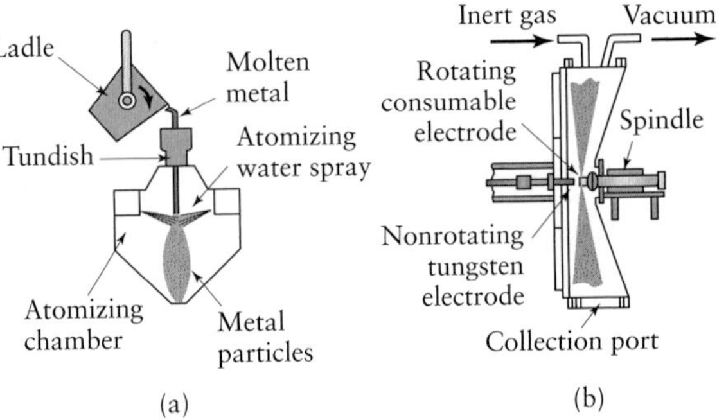

FIGURE 11.3 Methods of metal-powder production by atomization: (a) melt atomization and (b) atomization with a rotating consumable electrode.

3. **Electrolytic deposition.** *Electrolytic deposition* uses either aqueous solutions or fused salts. The powders produced are among the purest of all metal powders.
4. **Carbonyls.** Metal *carbonyls*, such as iron carbonyl [$Fe(CO)_5$] and nickel carbonyl [$Ni(CO)_4$], are formed by letting iron or nickel react with carbon monoxide. The reaction products are then decomposed to iron and nickel, producing small, dense, and uniform spherical particles of high purity.
5. **Comminution.** Mechanical *comminution* (*pulverization*) involves crushing, milling in a *ball mill*, or grinding brittle or less ductile metals into small particles. A ball mill is a machine with a rotating hollow cylinder that is partly filled with steel or white cast-iron balls. When made from brittle materials, the powder particles have angular shapes, whereas when they are made from ductile metals, they are flaky and are not particularly suitable for powder-metallurgy applications.

6. **Mechanical alloying.** In *mechanical alloying*, powders of two or more pure metals are mixed in a ball mill. Under the impact of the hard balls, the powders repeatedly fracture and weld together by diffusion, forming alloy powders.
7. **Other methods.** Other, less commonly used methods include (a) **precipitation** from a chemical solution, (b) production of fine metal chips by **machining,** and (c) **vapor condensation.** New developments include techniques based on high-temperature *extractive metallurgical processes.* Metal powders are being produced using high-temperature processing techniques based on (a) the reaction of volatile halides (a compound of halogen and an electropositive element) with liquid metals and (b) the controlled reduction and reduction/carburization of solid oxides.

Nanopowders. More recent developments include the production of *nanopowders* of various metals such as copper, aluminum, iron, and titanium. (See also the discussion of *nanomaterials* in Section 3.11.9.) When the metals are subjected to large plastic deformation at stress levels of 5500 MPa (800 ksi), their particle size is reduced, and the material becomes pore free and thus possesses enhanced properties.

Microencapsulated powders. *Microencapsulated powders* are completely coated with a binder and are compacted by warm pressing. (See also the discussion of *metal injection molding* in Section 11.3.4.) In electrical applications, the binder acts like an insulator, preventing electricity from flowing and thus reducing eddy-current losses.

11.2.2 Particle size, distribution, and shape

Particle *size* is measured usually by screening, that is, by passing the metal powder through screens (sieves) of various mesh sizes. Screening is achieved by using a vertical stack of screens with increasing mesh size as the powder flows downward through the screens. The larger the mesh size, the smaller is the opening in the screen. For example, a mesh size of 30 has openings of 600 μm, size 100 has openings of 150 μm, and size 400 has openings of 38 μm. (This method is similar to the technique used for numbering abrasive grains—the larger the number, the smaller is the size of the abrasive particle; see Section 9.2.)

In addition to screen analysis, several other methods are also used for particle-size analysis, particularly for powders finer than 45 μm (0.0018 in.):

1. **Sedimentation,** which involves measuring the rate at which particles settle in a fluid;
2. **Microscopic analysis,** including the use of transmission and scanning electron microscopy;
3. **Light scattering** from a laser that illuminates a sample consisting of particles suspended in a liquid medium—the particles cause the light to be scattered, which is then focused on a detector that digitizes the signals and computes the particle-size distribution;
4. **Optical means,** such as having particles blocking a beam of light, which is then sensed by a photocell;
5. **Suspension of particles** in a liquid and subsequent detection of particle size and distribution by *electrical sensors.*

The *size distribution* of particles is an important consideration, since it affects the processing characteristics of the powder. The distribution of particle size is given

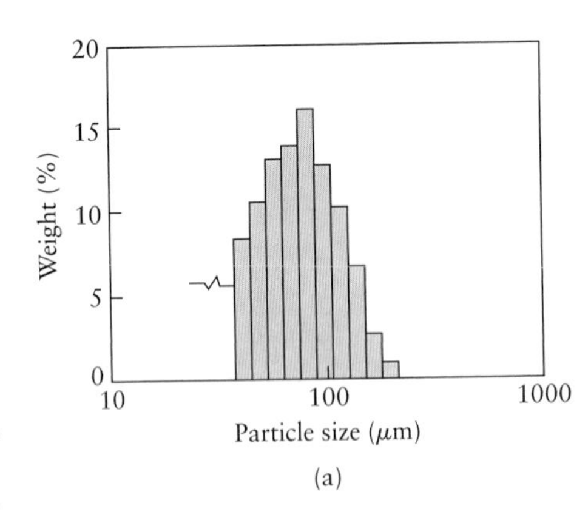

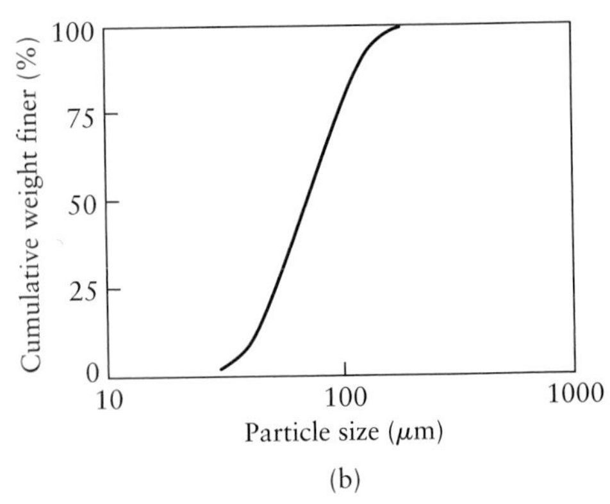

FIGURE 11.4 (a) A plot of the distribution of particle size, given as weight percentage. The most populous size is termed the *mode*. In this case, it is between 75 μm and 90 μm. (b) Cumulative particle-size distribution as a function of weight. *Source*: Reprinted with permission from Randall M. German, *Powder Metallurgy Science*, Princeton, NJ: Metal Powder Industries Federation, 1984.

in terms of a *frequency-distribution* plot, as shown in Fig. 11.4a. (See Section 4.9.1 and Fig. 4.21a for a detailed description of this type of plot.) In this plot, note that the highest percentage of the particles (by weight) has a size in the range of 75 μm to 90 μm. This maximum is called the *mode size*. The same particle-size data are also plotted in the form of a *cumulative distribution* (Fig. 11.4b); thus, for example, about 75% of the particles, by weight, have a size finer than 100 μm. As expected from the distribution plot in Fig. 11.4b, the cumulative weight reaches 100% at a particle size of about 200 μm.

The *shape* of particles has a major influence on the particles' processing characteristics. The shape is usually described in terms of aspect ratio or shape index. *Aspect ratio* is the ratio of the largest dimension to the smallest dimension of the particle. This ratio ranges from 1 for a spherical particle to about 10 for flakelike or needlelike particles. *Shape index*, or **shape factor (SF)**, is a measure of the surface area to the volume of the particle with reference to a spherical particle of equivalent diameter; thus, the shape factor for a flake is higher than that for a sphere.

EXAMPLE 11.1 Particle shape-factor determination

Determine the shape factor for (1) a spherical particle, (2) a cubic particle, and (3) a cylindrical particle with a length-to-diameter ratio of 2.

Solution.

(1) In general, the ratio of surface area A to volume V can be expressed as

$$\frac{A}{V} = \frac{k}{D_{eq}},$$

where k is the shape factor and D_{eq} is the diameter of a sphere that has the same volume as the particle or object being considered. In the case of a spherical particle, its diameter $D = D_{eq}$, its area $A = \pi D^2$, and its volume $V = \pi D^3/6$. Thus,

$$k = \left(\frac{A}{V}\right)D_{eq} = \left[\frac{(\pi D^2)}{(\pi D^3/6)}\right]D = 6.$$

(2) The surface area of a cube of length L is $6L^2$, and the cube's volume is L^3. Hence,

$$\frac{A}{V} = \frac{6L^2}{L^3} = \frac{6}{L}.$$

The particle's equivalent diameter is

$$D_{eq} = \left(\frac{6V}{\pi}\right)^{1/3} = \left(\frac{6L^3}{\pi}\right)^{1/3} = 1.24\,L.$$

Therefore, the shape factor is

$$k = \left(\frac{A}{V}\right)D_{eq} = \left(\frac{6}{L}\right)(1.24L) = 7.44.$$

(3) The surface area of a cylindrical particle with length-to-diameter ratio of 2 is

$$A = \left(\frac{2\pi D^2}{4}\right) + \pi DL = \left(\frac{\pi D^2}{2}\right) + 2\pi D^2 = 2.5\pi D^2.$$

The particle's volume is

$$V = \left(\frac{\pi D^2}{4}\right)(L) = \left(\frac{\pi D^2}{4}\right)(2D) = \frac{\pi D^3}{2},$$

and its equivalent diameter is

$$D_{eq} = \left(\frac{6V}{\pi}\right)^{1/3} = \left(\frac{6\pi D^3}{2\pi}\right)^{1/3} = 1.442D.$$

Hence, its shape factor is

$$k = \left(\frac{A}{V}\right)D_{eq} = \left[\frac{(2.5\pi D^2)}{(\pi D^3/2)}\right](1.442D) = 7.21.$$

11.2.3 Blending metal powders

Blending (mixing) powders is the second step in powder-metallurgy processing and is carried out for the following purposes:

a) Because the powders made by various processes may have different sizes and shapes, they must be mixed to obtain uniformity. The ideal mix is one in which all the particles of each material are distributed uniformly.

b) Powders of different metals and other materials may be mixed in order to impart special physical and mechanical properties and characteristics to the P/M product.

c) Lubricants may be mixed with the powders to improve the powders' flow characteristics. Such blends result in reduced friction between the metal particles, improved flow of the powder metals into the dies, and longer die life. The lubricants typically used are stearic acid and zinc stearate, in proportions of 0.25% to 5% by weight.

Powder mixing must be carried out under controlled conditions in order to avoid contamination and deterioration. Deterioration is caused by excessive mixing, which may alter the shape of the particles and work harden them, thus making the subsequent compacting operation more difficult. Powders can be mixed in air, in inert atmospheres (to avoid oxidation), or in liquids, which act as lubricants and make the mix more uniform; several types of blending equipment are available. These operations are being increasingly controlled by microprocessors to improve and maintain quality.

Because of their high surface-area-to-volume ratio, metal powders are explosive, particularly aluminum, magnesium, titanium, zirconium, and thorium powders. Great care must be exercised both during blending and during storage and handling. Precautions include grounding equipment, preventing the creation of sparks (by using nonsparking tools and avoiding the use of friction as a source of heat), and avoiding dust clouds, open flames, and chemical reactions.

11.3 Compaction of Metal Powders

Compaction is the step in which the blended powders are pressed into shapes in dies (Figs. 11.5a and b), using presses that are either hydraulically or mechanically actuated. The purposes of compaction are to obtain the required shape, density, and particle-to-particle contact and to make the part strong enough to be processed further. The as-pressed powder is known as a **green compact**. The powder must flow easily to feed properly into the die cavity. Pressing is generally carried out at room temperature, although it can be done at elevated temperatures as well.

In powder metallurgy, one refers to *density* at three different stages: (1) as loose powder, (2) as a green compact, and (3) after sintering. The particle shape, average size, and size distribution dictate the packing density of loose powder. Spherical powder with a wide size distribution gives a high packing density. But a compact made from such powder has poor green strength, and, hence, such powder is unsuitable for P/M parts to be made by the die-pressing method (which is the most commonly used method of fabricating P/M parts). For P/M parts production, powders with some irregularity of shape (for example, water-atomized ferrous powders) are preferred, even though the fill density of powder in the die is somewhat lower than that of spherical powder. For reproducibility of part dimensions, the fill density of the powder should

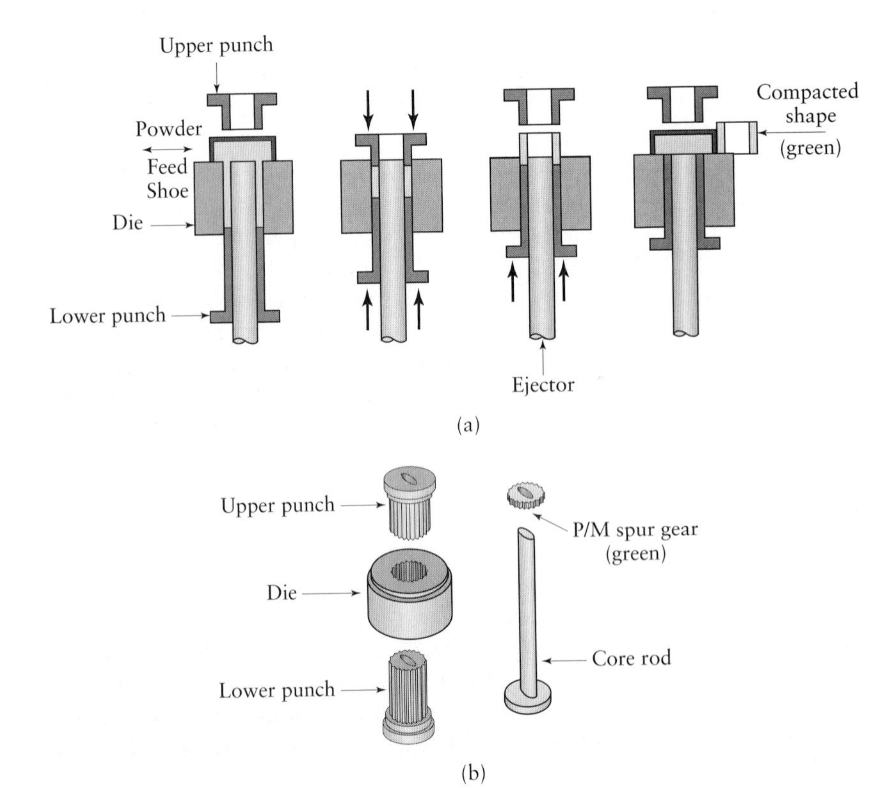

FIGURE 11.5 (a) Compaction of metal powder to form a bushing. The pressed-powder part is called *green compact*. (b) Typical tool and die set for compacting a spur gear. *Source*: Reprinted with the permission of the Metal Powder Industries Federation.

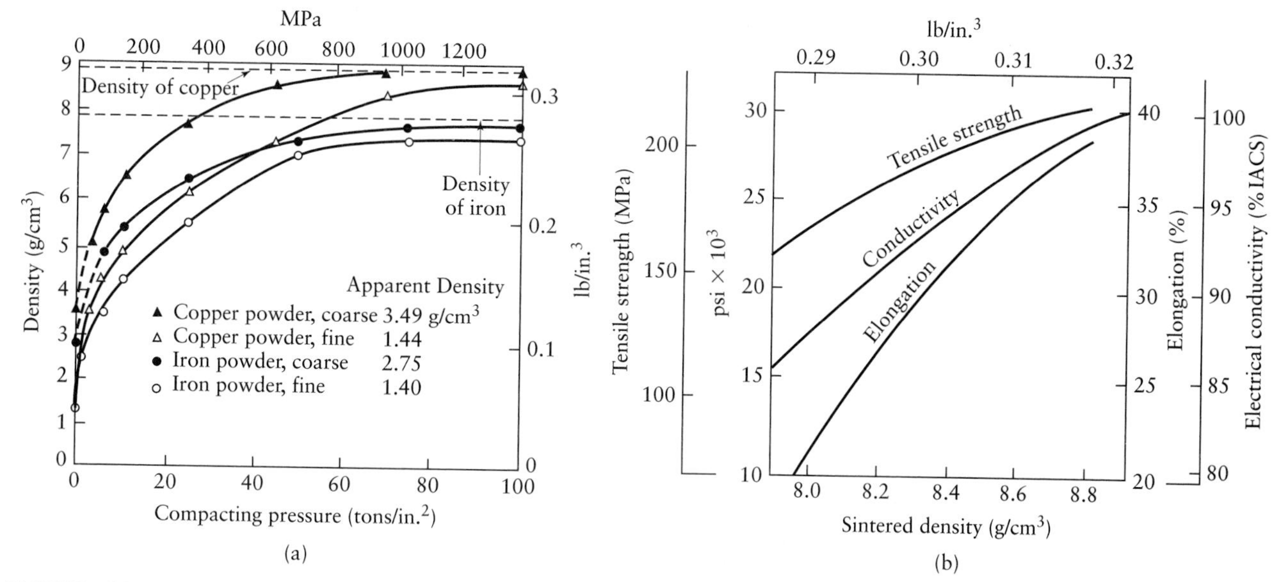

FIGURE 11.6 (a) Density of copper- and iron-powder compacts as a function of compacting pressure. Density greatly influences the mechanical and physical properties of P/M parts. *Source*: F. V. Lenel, *Powder Metallurgy: Principles and Applications*, Princeton, NJ: Metal Powder Industries Federation, 1980. Reprinted by permission of Metal Powder Industries Federation, Princeton, NJ. (b) Effect of density on tensile strength, elongation, and electrical conductivity of copper powder. IACS means International Annealed Copper Standard for electrical conductivity.

be consistent from one powder lot to the next. Spherical powder is preferred for hot isostatic pressing, a special technique that is described in Section 11.3.3.

The density after compaction (**green density**) depends primarily on the compaction pressure, the metal powder composition, and the hardness of the powder (Fig. 11.6a). Higher pressure and softer powder give a higher green density. Furthermore, pure-iron powder will compact to a higher density than powders composed of alloyed steels. The green density and its uniformity within a compact improve with the addition of a small quantity of admixed (blended-in) lubricant.

To understand the effect of particle shape on green density, consider two powder grades with the same chemical composition and hardness, one with a spherical particle shape and the other with an irregular particle shape. The spherical grade will have a higher apparent (or fill) density, but after compaction at a higher pressure, compacts from both grades will have similar green densities. When comparing two similar powders pressed under some standard conditions, the powder that gives a higher green density is said to have a higher *compressibility*. (See also the discussion of sintered density in Section 11.4.)

The higher the density, the higher will be the strength and elastic modulus of the part (Fig. 11.6b). The reason is that the higher the density, the higher will be the amount of solid metal in the same volume, hence the greater will be the part's resistance to external forces. Because of friction between the metal particles in the powder and between the punches and the die walls, the density can vary considerably within the part. This variation can be minimized by proper punch and die design and by friction control; for example, it may be necessary to use multiple punches, with separate movements in order to ensure that the density is more uniform throughout the part (Fig. 11.7).

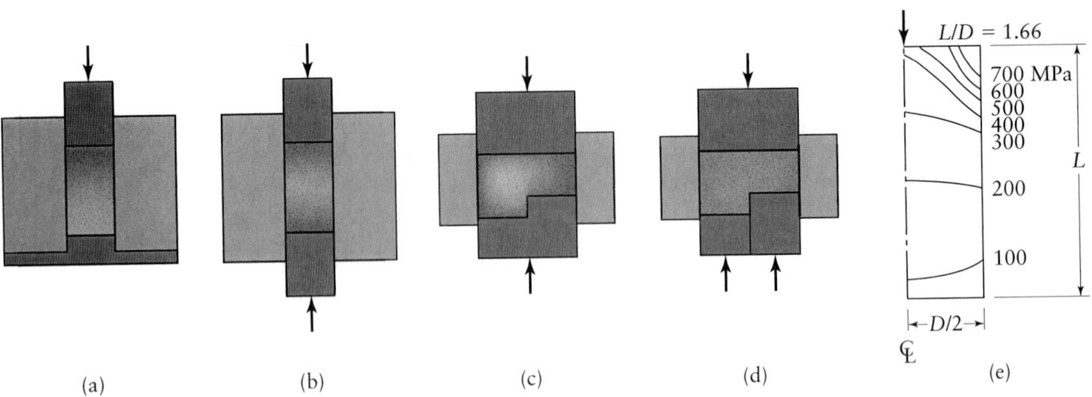

FIGURE 11.7 Density variation in compacting metal powders in different dies: (a) and (c) single-action press; (b) and (d) double-action press. Note in (d) the greater uniformity of density in pressing with two punches with separate movements as compared with (c). Generally, uniformity of density is preferred, although there are situations in which density variation, and hence variation of properties, within a part may be desirable. (e) Pressure contours in compacted copper powder in a single-action press. *Source*: P. Duwez and L. Zwell.

EXAMPLE 11.2 Density of metal-powder–lubricant mix

Zinc stearate is a lubricant that commonly is mixed with metal powders prior to compaction, in proportions up to 2% by weight. Calculate the theoretical and apparent densities of an iron-powder–zinc-stearate mix, assuming that (1) 1000 g of iron powder is mixed with 20 g of lubricant, (2) the density of the lubricant is 1.10 g/cm^3, (3) the theoretical density of the iron powder is 7.86 g/cm^3 (from Table 3.3), and (4) the apparent density of the iron powder is 2.75 g/cm^3 (from Fig. 11.6a).

Solution The volume of the mixture is

$$V = \left(\frac{1000}{7.86}\right) + \left(\frac{20}{1.10}\right) \simeq 145.41 \text{ cm}^3.$$

The combined weight of the mixture is 1020 g, so the mixture's theoretical density is 1020/145.41 = 7.01 g/cm^3. The apparent density of the iron powder is 2.75 g/cm^3, so it is (2.75/7.86)100 = 35% of the theoretical density. Assuming a similar percentage for the mixture, we can estimate the apparent density of the mix as (0.35)(7.01) = 2.45 g/cm^3. (*Note*: Although g and g/cm^3 are not SI units, they are commonly used in the powder-metallurgy industry.)

11.3.1 Pressure distribution in powder compaction

Figure 11.7e shows that, in single-action pressing, the pressure decays rapidly toward the bottom of the compact. The pressure distribution along the length of the compact can be determined by using the slab method of analysis of deformation processes, described in Section 6.2.2. As done in Fig. 6.4, we first describe the operation in terms of its coordinate system, as shown in Fig. 11.8: D is the diameter of the compact, L is the compact's length, and p_0 is the pressure applied by the punch. We take an element dx thick and place on it all the relevant stresses, namely, the compacting pressure, p_x, the die-wall pressure, σ_r, and the frictional stress, $\mu\sigma_r$. Note that the frictional stresses act upward on the element, because the punch movement is downward.

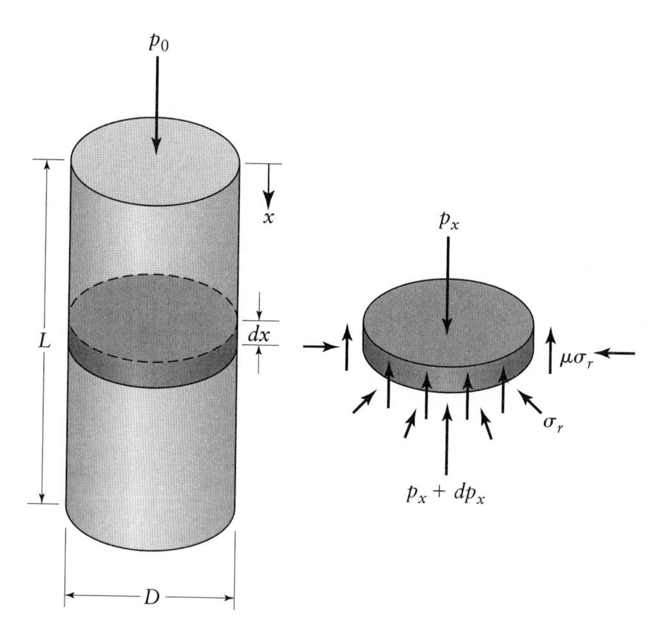

FIGURE 11.8 Coordinate system and stresses acting on an element in compaction of powders. The pressure is assumed to be uniform across the cross-section. (See also Fig. 6.4.)

Balancing the vertical forces acting on this element, we have

$$\left(\frac{\pi D^2}{4}\right)p_x - \left(\frac{\pi D^2}{4}\right)(p_x + dp_x) - (\pi D)(\mu\sigma_r)\,dx = 0, \qquad (11.1)$$

which we can simplify to

$$D\,dp_x + 4\mu\sigma_r\,dx = 0.$$

We have one equation, but two unknowns (p_x and σ_r). Let's now introduce the factor k, which is a measure of the interparticle friction during compaction:

$$\sigma_r = kp_x.$$

If there is no friction between the particles, $k = 1$, the powder behaves like a fluid, and we have $\sigma_r = p_x$, signifying a state of hydrostatic pressure. We now have the expression

$$dp_x + \frac{4\mu k p_x\,dx}{D} = 0,$$

or

$$\frac{dp_x}{p_x} = -\frac{4\mu k\,dx}{D}.$$

This expression is similar to that given in Section 6.2.2 for upsetting. In the same manner as in that section, we integrate this expression, noting that the boundary condition in this case is $p_x = p_0$ when $x = 0$:

$$p_x = p_0 e^{-4\mu kx/D}. \qquad (11.2)$$

Thus, the pressure within the compact decays as the coefficient of friction, the parameter k, and the length-to-diameter ratio increase.

EXAMPLE 11.3 Pressure decay in compaction

Assume that a powder mix has $k = 0.5$ and $\mu = 0.3$. At what depth will the pressure in a straight cylindrical compact 10 mm in diameter become (a) zero and (b) one half the pressure at the punch?

Solution. For case (a) from Eq. 11.2, $p_x = 0$. Consequently, we have the expression

$$0 = p_0 e^{-(4)(0.3)(0.5)x/10}, \quad \text{or} \quad e^{-0.06x} = 0.$$

The value of x must approach ∞ for the pressure to decay to 0. For case (b), we have $p_x/p_0 = 0.5$. Therefore,

$$e^{-0.06x} = 0.5, \quad \text{or} \quad x = 11.55 \text{ mm}.$$

In a practical case, a pressure drop of 50% is severe, as the compact density will then be unacceptably low. This example shows that, under the conditions assumed, uniaxial compaction of even a cylinder of length-to-diameter ratio of about 1.2 will be unsatisfactory.

11.3.2 Equipment

The pressure required for pressing metal powders ranges from 70 MPa (10 ksi) for aluminum to 800 MPa (120 ksi) for high-density iron parts (Table 11.1). The compacting pressure required depends on the characteristics and shape of the particles, the method of blending, and the lubrication.

Press capacities are on the order of 1.8–2.7 MN (200–300 tons), although presses with much higher capacities are used for special applications. Most applications require less than 100 tons. For small tonnage, crank or eccentric-type mechanical presses are used; for higher capacities, toggle or knucklejoint presses are employed. (See Fig. 6.28.) Hydraulic presses with capacities as high as 45 MN (5000 tons) can be used for large parts.

TABLE 11.1

Compacting Pressures for Various Metal Powders

	Pressure	
Metal	MPa	psi × 10^3
Aluminum	70–275	10–40
Brass	400–700	60–100
Bronze	200–275	30–40
Iron	350–800	50–120
Tantalum	70–140	10–20
Tungsten	70–140	10–20
Other Materials		
Auminum oxide	110–140	16–20
Carbon	140–165	20–24
Cemented carbides	140–400	20–60
Ferrites	110–165	16–24

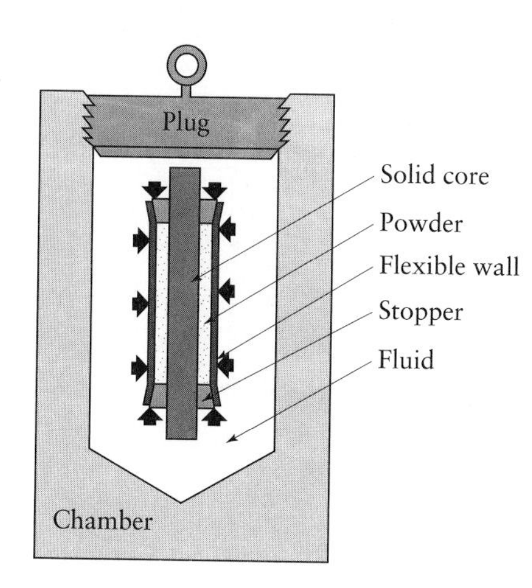

FIGURE 11.9 Schematic illustration of cold isostatic pressing as applied to formation of a tube. The powder is enclosed in a flexible container around a solid core rod. Pressure is applied isostatically to the assembly inside a high-pressure chamber. *Source:* Reprinted with permission from Randall M. German, *Powder Metallurgy Science*, Princeton, NJ: Metal Powder Industries Federation, 1984.

The selection of the press depends on part size and configuration, density requirements, and production rate. The higher the pressing speed, the greater is the tendency to trap air in the die cavity. Therefore, good die design, including provision of vents, is important so that trapped air does not hamper proper compaction.

11.3.3 Isostatic pressing

Compaction can also be carried out or improved by using a number of additional processes, such as isostatic pressing, rolling, and forging. Because the density of die-compacted powders can vary significantly, powders are subjected to *hydrostatic pressure* in order to achieve more uniform compaction. This process is similar to pressing your cupped hands together when making snowballs.

In **cold isostatic pressing** (CIP), the metal powder is placed in a flexible rubber mold made of neoprene rubber, urethane, polyvinyl chloride, or other elastomers (Fig. 11.9). The assembly is then pressurized hydrostatically in a chamber, usually with water. The most common pressure is 400 MPa (60 ksi), although pressures of up to 1000 MPa (150 ksi) have been used. The applications of CIP and other compacting methods, in terms of size and complexity of part, are shown in Fig. 11.10.

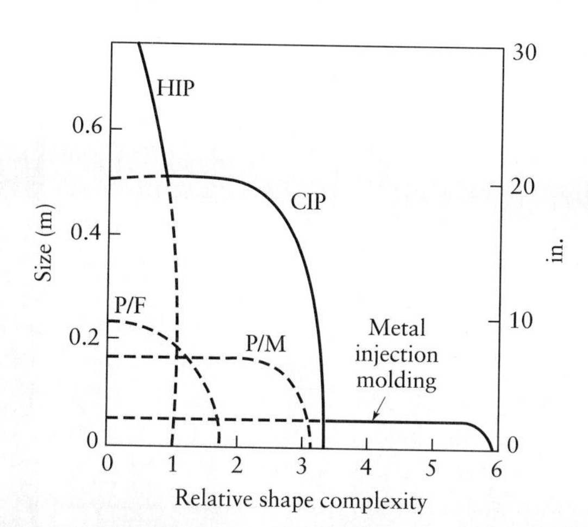

FIGURE 11.10 Capabilities of part size and shape complexity according to various P/M operations. *P/F* means powder forging. *Source:* Metal Powder Industries Federation.

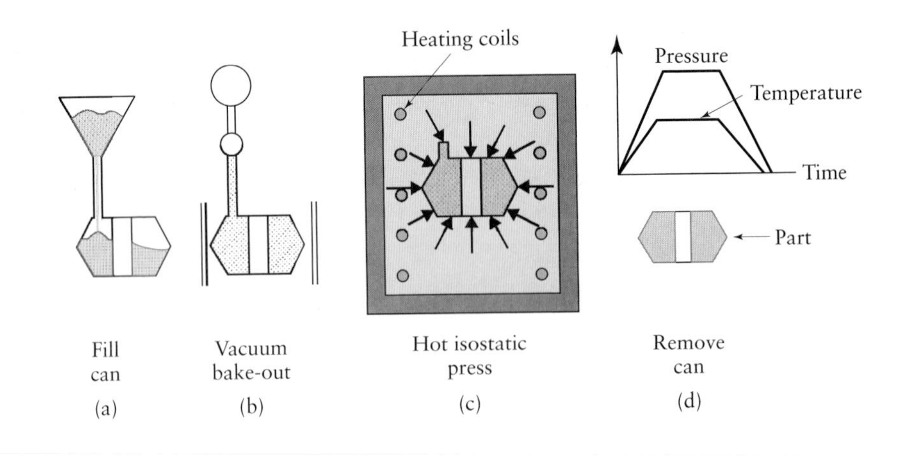

FIGURE 11.11 Schematic illustration of hot isostatic pressing. The pressure and temperature variation versus time are shown in the diagram. *Source*: Reprinted with permission from Randall M. German, *Powder Metallurgy Science*, Princeton, NJ: Metal Powder Industries Federation, 1984.

In **hot isostatic pressing** (HIP), the container is usually made of a high-melting-point sheet metal, and the pressurizing medium is inert gas or vitreous (glasslike) fluid (Fig. 11.11). A common condition for HIP is 100 MPa (15 ksi) at 1100°C (2000°F), although the trend is toward higher pressures and temperatures. The main advantage of HIP is its ability to produce compacts with essentially 100% density, good metallurgical bonding among the particles, and good mechanical properties. The HIP process is relatively expensive and is used mainly in making superalloy components for the aerospace industry. It is routinely used as a final densification step for tungsten-carbide cutting tools and P/M tool steels. The process is also used to close internal porosity and improve properties in superalloy and titanium-alloy castings for the aerospace industry.

The main advantage of isostatic pressing is that, because of the uniformity of pressure from all directions and the absence of die-wall friction, it produces compacts of practically uniform grain structure and density, irrespective of shape. Parts with high length-to-diameter ratios have been produced by this method, with very uniform density, high strength, high toughness, and good surface details. The limitations of this method include wider dimensional tolerances produced, higher cost and time required, and relatively small production quantities output.

11.3.4 Other compacting and shaping processes

Other compacting and shaping processes used in powder metallurgy include:

a) **Metal injection molding.** In *metal injection molding* (MIM), very fine metal powders (generally <45 μm, and often <10 μm) are blended with a polymer or a wax-based binder. The mixture then undergoes a process similar to injection molding of plastics. (See Section 10.10.2.) The molded green parts are placed in a low-temperature oven to burn off the plastic, or else the binder is partially removed by solvent extraction, and then the green parts are sintered in a furnace. The term *powder injection molding* (PIM) is used with a broader reference to molding of both metal and ceramic powders.

The major advantage of injection molding over conventional compaction is that complex shapes, with wall thicknesses as small as 0.5 mm (0.02 in.), can be molded and removed easily from the dies. The process is particularly well suited for small, complex parts required in large quantities; most commercially produced parts fabricated by this method weigh from a fraction of a gram to about 250 g (0.55 lb) each. Typical parts made by this method include components for guns, surgical instruments, automobiles, and watches. In spite of the high cost of fine metal powders, parts made by MIM are competitive with investment-cast parts (see Section 5.9.2), small forgings (Section 6.2), and complex machined parts (Chapter 8).

b) **Rolling.** In *powder rolling*, also called *roll compaction*, the powder is fed to the roll gap in a two-high rolling mill (see Fig. 6.42a) and is compacted into a continuous strip at speeds of up to 0.5 m/s (100 ft/min). The process can be carried out at room or elevated temperatures. Sheet metal for electrical and electronic components and for coins can be made by powder rolling.

c) **Extrusion.** Powders may be compacted by *hot extrusion*; in this process, the powder is encased in a metal container and extruded. (See Section 6.4.) Superalloy powders, for example, are hot extruded for improved properties. Preforms made from the extrusions may be reheated and forged in a closed die to their final shape.

d) **Pressureless compaction.** In *pressureless compaction*, the die is filled with metal powder by gravity, and the powder is sintered directly in the die. Because of the resulting low density, pressureless compaction is used principally for porous parts, such as filters.

e) **Ceramic molds.** Molds for shaping metal powders are made by the technique used in investment casting. After the ceramic mold is made, it is filled with metal powder and placed in a steel container; the space between the mold and the container is filled with particulate material. The container is then evacuated, sealed, and subjected to hot isostatic pressing. Titanium-alloy compressor rotors for missile engines have been made by this process.

f) **Spray deposition.** This method is a shape-generation process in which the basic components are an atomizer, a spray chamber with inert atmosphere, and a mold for producing preforms; although there are several variations, the best known is the *Osprey* process (Fig. 11.12). After the metal is atomized, it is deposited on a cooled preform mold, usually made of copper or ceramic, where it solidifies. The metal particles weld together, developing a density that is normally above 99% of the solid metal density. The mold may be composed of various shapes, such as billets, tubes, disks, and cylinders. Spray-deposited preforms may be subjected to additional shaping and consolidation processes, such as forging, rolling, and extrusion. The grain size of the part is fine, and the part's mechanical properties are comparable with those of wrought products of the same alloy.

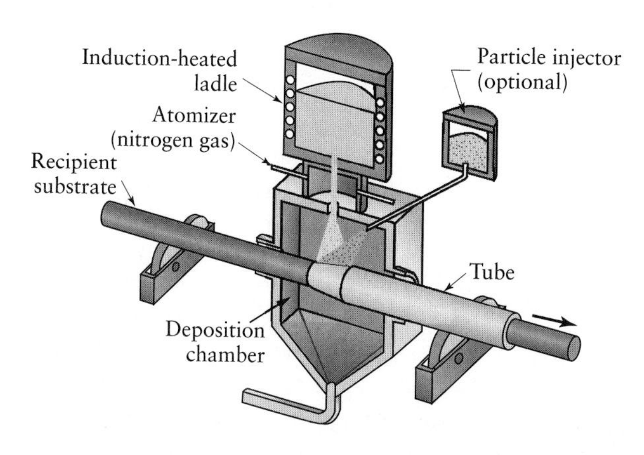

FIGURE 11.12 Spray casting (Osprey process) in which molten metal is sprayed over a rotating mandrel to produce seamless tubing and pipe. *Source*: After J. Szekeley, "Can Advanced Technology Save the U.S. Steel Industry," *Scientific American*, July 1987, © George V. Kelvin/Scientific American.

11.3.5 Punch and die materials

The selection of punch and die materials for P/M depends on the abrasiveness of the powder metal and the number of parts to be made. The most commonly used die materials are air- or oil-hardening tool steels, such as D2 or D3, with a hardness range of 60–64 HRC. Because of their greater hardness and wear resistance, tungsten-carbide dies are used for more severe applications. Punches are generally made of similar materials as those for dies. Close control of die and punch dimensions is essential for proper compaction and die life. Too large a clearance between the punch and the die will allow the metal powder to enter the gap, interfere with the operation, and result in eccentric parts. Diametral clearances are generally less than 25 μm (0.001 in.). Die and punch surfaces must be lapped or polished in the direction of tool movements for improved die life and overall performance.

11.4 Sintering

Sintering is the process whereby compressed metal powder is heated in a controlled-atmosphere furnace to a temperature below its melting point, but sufficiently high to allow bonding (fusion) of the individual particles. Prior to sintering, the compact is brittle, and its strength, known as **green strength**, is low. The nature and strength of the bond between the particles, and hence of the sintered compact, depend on the mechanisms of diffusion, plastic flow, evaporation of volatile materials in the compact, recrystallization, grain growth, and pore shrinkage.

The sintered density of a part depends mainly on the part's green density and on the sintering conditions, in terms of temperature, time, and furnace atmosphere. The sintered density increases with increasing values of temperature and time and usually with a more deoxidizing type of furnace atmosphere. For structural P/M parts, a higher sintered density is very desirable, as it leads to better mechanical properties. At the same time, in conventional P/M parts production, it is preferable to minimize the increase in density during sintering, for better dimensional accuracy. Better properties and accuracy can be achieved by using a powder with a high compressibility, i.e., a powder that will give a high green density and keep the sintering temperature moderate (not high). Such a powder also has another important benefit in that larger parts can be produced with a specific press tonnage. Therefore, a considerable amount of development work by powder producers is aimed at producing powder grades with higher compressibility.

Sintering temperatures (Table 11.2) are generally within 70% to 90% of the melting point of the metal or alloy. Sintering times range from a minimum of about 10 minutes for iron and copper alloys to as much as eight hours for tungsten and

TABLE 11.2

Sintering Temperature and Time for Various Metal Powders

Material	Temperature (°C)	Time (min)
Copper, brass, and bronze	760–900	10–45
Iron and iron graphite	1000–1150	8–45
Nickel	1000–1150	30–45
Stainless steels	1100–1290	30–60
Alnico alloys (for permanent magnets)	1200–1300	120–150
Ferrites	1200–1500	10–600
Tungsten carbide	1430–1500	20–30
Molybdenum	2050	120
Tungsten	2350	480
Tantalum	2400	480

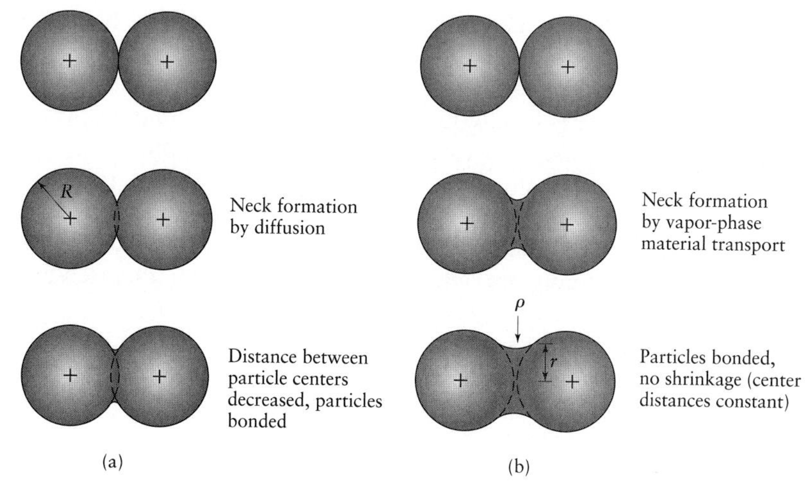

FIGURE 11.13 Schematic illustration of two mechanisms for sintering metal powders: (a) solid-state material transport; and (b) liquid-phase material transport. R = particle radius, r = neck radius, and ρ = neck profile radius.

tantalum. Continuous-sintering furnaces are used for most production today. These furnaces have three chambers: (1) a burn-off chamber to volatilize the lubricants in the green compact in order to improve bond strength and prevent cracking; (2) a high-temperature chamber for sintering; and (3) a cooling chamber.

Proper control of the furnace atmosphere is essential for successful sintering and to obtain optimum properties. An oxygen-free atmosphere is necessary in order to control the carburization and decarburization of iron and iron-base compacts and to prevent oxidation of powders. Oxide inclusions have a detrimental effect on mechanical properties. For the same volume of inclusions, the smaller inclusions have a larger effect because there are more of them per unit volume of the part. A vacuum is generally used for sintering refractory metal alloys and stainless steels. The gases most commonly used for sintering a variety of other metals are hydrogen, dissociated or burned ammonia, partially combusted hydrocarbon gases, and nitrogen.

Sintering mechanisms are complex and depend on the composition of the metal particles as well as the processing parameters (Fig. 11.13). As the temperature increases, two adjacent particles begin to form a bond by **diffusion** (*solid-state bonding*). As a result, the strength, density, ductility, and thermal and electrical conductivities of the compact increase (Fig. 11.14). At the same time, however, the compact shrinks; hence, allowances should be made for shrinkage, as in casting. (See Section 5.12.1.)

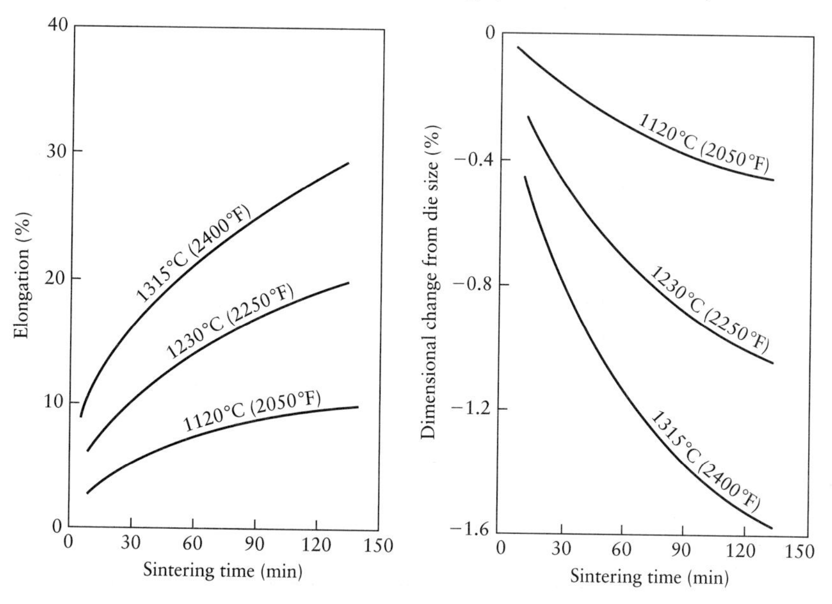

FIGURE 11.14 Effect of sintering temperature and time on elongation and dimensional change during sintering of type 316L stainless steel. *Source*: ASM International.

A second sintering mechanism is **vapor-phase transport**. Because the material is heated very close to its melting temperature, metal atoms will release to the vapor phase from the particles. At convergent geometries (the interface of two particles), the melting temperature is locally higher, and the vapor resolidifies; thus, the interface grows and strengthens, while each particle shrinks as a whole. If two adjacent particles are of different metals, *alloying* can take place at the interface of the two particles. One of the particles may have a lower melting point than the other. In that case, one particle may melt and, because of surface tension, surround the particle that has not melted (**liquid-phase sintering**). An example is cobalt in tungsten-carbide tools and dies. Stronger and denser parts can be obtained in this way.

Depending on temperature, time, and processing history, different structures and porosities can be obtained in a sintered compact. However, porosity cannot be completely eliminated, because voids remain after compaction, and gases evolve during sintering. Porosities can consist of either a network of interconnected pores or closed holes. Their presence is an important consideration in making P/M filters and bearings.

Another method, which is at an experimental stage, is **spark sintering.** In this process, loose metal powders are placed in a graphite mold, heated by an electric current, subjected to a high-energy discharge, and compacted, all in one step. The rapid discharge strips contaminants or any oxide coating, such as that found on aluminum, from the surfaces of the particles and thus encourages good bonding during compaction at elevated temperatures.

Table 11.3 gives typical mechanical properties for several sintered P/M alloys. Note the effect of heat treating on the properties of metals. To evaluate the differences between the properties of P/M, wrought, and cast metals and alloys, compare this table with those in other chapters. Table 11.4 shows the effects of various processes on the mechanical properties of a particular titanium alloy. Note that hot isostatically pressed (HIP) titanium has properties that are similar to those for cast and forged titanium. It should be remembered, however, that forged components are likely to require additional machining processes (unless the components are precision forged to net shape), while P/M components may not. Therefore, powder metallurgy can be a competitive alternative to casting and forging.

EXAMPLE 11.4 Shrinkage in sintering

In solid-state bonding during sintering of a powder-metal green compact, the linear shrinkage is 4%. If the desired sintered density is 95% of the theoretical density of the metal, what should be the density of the green compact? Ignore the small changes in mass that occur during sintering.

Solution. Linear shrinkage is defined as $\Delta L/L_o$, where L_o is the original part length. The volume shrinkage during sintering can then be expressed as

$$V_{\text{sint}} = V_{\text{green}}\left(1 - \frac{\Delta L}{L_o}\right)^3. \tag{11.3}$$

The volume of the green compact must be larger than that of the sintered part. However, the mass does not change during sintering, so we can rewrite this expression in terms of the density ρ as

$$\rho_{\text{green}} = \rho_{\text{sint}}\left(1 - \frac{\Delta L}{L_o}\right)^3. \tag{11.4}$$

Therefore,

$$\rho_{\text{green}} = 0.95(1 - 0.04)^3 = 0.84, \quad \text{or} \quad 84\%.$$

TABLE 11.3

Typical Mechanical Properties of Selected P/M Materials

Designation	MPIF type	Condition	Ultimate Tensile Strength (MPa)	Yield Stress (MPa)	Hardness	Elongation in 25 mm (%)	Elastic Modulus (GPa)
FERROUS							
FC-0208	N	AS	225	205	45 HRB	<0.5	70
		HT	295	—	95 HRB	<0.5	70
	R	AS	415	330	70 HRB	1	110
		HT	550	—	35 HRC	<0.5	110
	S	AS	550	395	80 HRB	1.5	130
		HT	690	655	40 HRC	<0.5	130
FN-0405	S	AS	425	240	72 HRB	4.5	145
		HT	1060	880	39 HRC	1	145
	T	AS	510	295	80 HRB	6	160
		HT	1240	1060	44 HRC	1.5	160
ALUMINUM							
601 AB,		AS	110	48	60 HRH	6	—
pressed bar		HT	252	241	75 HRH	2	—
BRASS							
CZP-0220	T	—	165	76	55 HRH	13	—
	U	—	193	89	68 HRH	19	—
	W	—	221	103	75 HRH	23	—
TITANIUM							
Ti-6Al-4V		HIP	917	827	—	13	—
SUPERALLOYS							
Stellite 19		—	1035	—	49 HRC	<1	—

Note: MPIF = Metal Powder Industries Federation; AS = as sintered; HT = heat treated; HIP = hot isostatically pressed.

TABLE 11.4

Mechanical Property Comparisons for Ti-6Al-4V Titanium Alloy

Process	Density (%)	Yield Stress (MPa)	Ultimate Tensile Strength (MPa)	Elongation (%)	Reduction of Area (%)
Cast	100	840	930	7	15
Cast and forged	100	875	965	14	40
Powder metallurgy					
Blended elemental (P + S)*	98	786	875	8	14
Blended elemental (HIP)*	>99	805	875	9	17
Realloyed (HIP)*	100	880	975	14	26

(*) P + S = pressed and sintered; HIP = hot isostatically pressed.
Source: R. M. German.

EXAMPLE 11.5 Production of automotive-engine main bearing caps by powder metallurgy

Main bearing caps (MBCs), as shown in Figure 11.15, are essential structural elements in internal combustion engines. Most MBCs are made from gray cast iron (see Section 5.6), which is cast and then followed by extensive machining operations. Grey cast iron is a very versatile material that exhibits excellent castability with good machinability (due to its low hardness and flake graphite in the microstructure), compressive strength, and sound-dampening properties. However, it has low fatigue strength and is very brittle, which limits its use for highly stressed applications.

Powder-metallurgy MBCs were used for the first time in 1993 in General Motors 3100 and 3800 V6 engines. P/M parts were chosen to replace the previous cast-iron parts because of the former's design flexibility and functional advantages, and, because of its net-shape capability, P/M eliminated the need for major investments in machining lines required for finishing the cast parts to final shape and dimensions. A high-strength, low-cost P/M alloy steel was developed for this application, involving liquid-phase sintering and using an optimized low-alloy combination in a carbon-steel matrix, designated as Zenith Material 833 (ZM833).

The P/M process molds individual MBCs into net shape, except for the new side bolt holes, which are drilled and tapped. Critical to the success of the P/M approach was the molding of the long bolt holes, which would be costly and difficult to machine. This molding step requires "upright" compaction, i.e., vertical bolt holes, which presents significant powder compaction, challenges due to the irregular shape conferred by the bearing arch. The advantages of upright compaction, however, include the ability to press in the bearing notch and the features of the top face, which provide identification and orientation for assembly. Upright compaction also permits use of a much smaller compacting press than for "flat" pressing, i.e., horizontal bolt holes, with associated cost savings. In addition, the 3800 engine's rear MBC has several features that would be impossible to compact in the flat orientation; thus, the use of upright compaction was inevitable. Principal features of the MBCs produced by this method include recessed bolt-hole bosses, a rectangular oil-drain hole, vertical side

FIGURE 11.15 Powder-metal main bearing caps for 3.8- and 3.1-liter GM engines. *Source*: Courtesy of Zenith Sintered Products, Inc., Milwaukee, WI.

grooves, a double arch to accept an oil seal, and a raised-arch feature for added strength.

Once the desired fatigue strength was achieved, the ZM833 was used to produce actual MBCs that were subjected to severe engine-endurance testing in both engines. It was found that the cast-iron engine block generally broke before the P/M cap cracked, and often several engine blocks had to be sacrificed to obtain a fatigue-test point, but eventually an *S–N* curve (see Fig. 2.28) was generated for the P/M material so that a safety margin could be determined. The P/M was proven to be a robust design by repeated successful fatigue testing, simulating 1.8 times the most severe running conditions predicted for the engine.

Material ductility is a functional benefit, since it is possible to induce cracks in a very brittle material during press-fit installation. The ZM833 P/M material had the following properties: (1) density of 6.6 g/cm^3; (2) tensile strength of 450 MPa (65,000 psi); (3) hardness of 70 HRB; (4) modulus of elasticity of 10^7 GPa (15.5×10^6 psi)—the P/M material developed for this application has a lower modulus of elasticity than that of cast iron, resulting in significantly lower initial tensile stresses in highly stressed areas of the MBC; (5) tensile elongation of 3% to 4%, which compares with that for gray iron, at 0.5%, and that for ductile iron, typically at 3%; and (6) it was found that the P/M alloy's fatigue endurance limit, determined by flexural bending fatigue tests, was intermediate between those of the gray and ductile irons, being double the gray iron's level.

Adapted from "Powder Metallurgy Main Bearing Caps for Automotive Engines," by T. M. Cadle, M. A. Jarrett, and W. L. Miller, *Int. Jr. of Powder Metallurgy*, 30(3): 275–282, 1994.

11.5 | Secondary and Finishing Operations

In order to further improve the properties of sintered P/M products or to give them special characteristics, several additional operations may be carried out after sintering:

1. **Re-pressing**, also called **coining** and **sizing**, is an additional compaction operation, performed under high pressure in presses. The purposes of this operation are to impart dimensional accuracy to the sintered part and to improve the part's strength and surface finish by additional densification.
2. **Forging** involves the use of unsintered or sintered alloy-powder preforms that are subsequently hot forged in heated, confined dies to the desired final shapes; the preforms may also be shaped by impact forging. The process is usually referred to as *powder-metallurgy forging* (P/F), and when the preform is sintered, the process is usually referred to as *sinter forging*. Ferrous and nonferrous powders can be processed in this manner, resulting in full-density products (on the order of 99.9% of the theoretical density of the material). The deformation of the preform may consist of the following two modes: *upsetting* and *re-pressing*. In upsetting, the material flows laterally outward, as shown in Fig. 6.1. The preform is subjected to compressive and shear stresses during deformation; any interparticle residual oxide films are broken up, as a result of which the part has better toughness, ductility, and fatigue strength. In *re-pressing*, the material flow is mainly in the direction of the movement of the punch, with little lateral flow; as a result, the mechanical properties of the part are not as high as those produced by upsetting.

P/M forged products have good surface finish, good dimensional tolerances, little or no flash formation, and uniform fine grain size. The superior properties obtained make this technology particularly suitable for applications such as highly stressed components for automotive (e.g., connecting rods), jet-engine, military, and off-road equipment.

3. The inherent porosity of powder-metallurgy components can be taken advantage of by **impregnating** the components with a fluid. A typical application is to impregnate the sintered part with oil, which is usually done by immersing the part in heated oil. Bearings and bushings that are internally lubricated, with up to 30% oil by volume, are made by this method. Internally lubricated components have a continuous supply of lubricant during their service lives. Universal joints are now being made with grease-impregnated P/M techniques; as a result, these parts no longer requiring grease fittings.
4. **Infiltration** is a process whereby a metal slug with a lower melting point than that of the part is placed against the sintered part, and the assembly is heated to a temperature sufficient to melt the slug. The molten metal infiltrates the pores by *capillary action*, resulting in a relatively pore-free part with good density and strength. The most common application is the infiltration of iron-base compacts with copper. The advantages of this technique are that hardness and tensile strength are improved and the pores of the part are filled, thus preventing moisture penetration, which could cause corrosion. Infiltration may also be done with lead, whereby, because of the low shear strength of lead, the infiltrated part has lower frictional characteristics than the uninfiltrated one; some bearing materials are formed in this manner.
5. Powder-metal parts may be subjected to other operations, including the following:
   a. **Heat treating** (quench and temper; steam treat), for improved strength, hardness, and resistance to wear:
   b. **Machining** (turning, milling, drilling, tapping, and grinding), for production of undercuts and slots, improved surface finish and dimensional accuracy, and production of threaded holes and other surface features;
   c. **Finishing** (deburring, burnishing, mechanical finishing, plating, and coating), for improved surface characteristics, corrosion and fatigue resistance, and appearance.

EXAMPLE 11.6 Production of tungsten carbide for tools and dies

Tungsten carbide is an important tool and die material, mainly because of its hardness, strength, and wear resistance to wear over a wide range of temperatures. Powder-metallurgy techniques are used in making these carbides. First, powders of tungsten and carbon are blended together in a ball mill or rotating mixer. The mixture (typically 94% tungsten and 6% carbon, by weight) is heated to approximately 1500°C (2800°F) in a vacuum-induction furnace. As a result of this process, the tungsten is carburized, forming tungsten carbide in fine-powder form. A binding agent (usually cobalt) is then added to the tungsten carbide (with an organic fluid such as hexane), and the mixture is ball milled to produce a uniform and homogeneous mix, a process that can take several hours, or even days.

The mixture is then dried and consolidated, usually by cold compaction, at pressures in the range of 200 MPa (30,000 psi). Finally, it is sintered in a hydrogen-atmosphere or vacuum furnace at a temperature of 1350°C to 1600°C (2500°F to 2900°F) depending on its composition. At this temperature, the cobalt is in a liquid phase and acts as a binder for the carbide particles. (Powders may also be hot pressed at the sintering temperature, using graphite dies.) During sintering, the tungsten carbide undergoes a linear shrinkage of about 16%, corresponding to a volume shrinkage of about 40%; thus, control of size and shape is important for producing tools with accurate dimensions. A combination of other carbides, such as titanium carbide and tantalum carbide, can also be produced, using mixtures made by the method described in this example.

11.6 Design Considerations for Powder Metallurgy

Because of the unique properties of metal powders, their flow characteristics in the die, and the brittleness of green compacts, there are certain design principles that should be followed when using them (Figs. 11.16–11.18):

- The shape of the compact must be kept as simple and uniform as possible. Sharp changes in contour, thin sections, variations in thickness, and high length-to-diameter ratios should be avoided.
- Provision must be made for ejection of the green compact from the die without damaging the compact; thus, holes or recesses should be parallel to the axis of punch travel. Chamfers should also be provided.
- As is the case for parts made by most other processes, P/M parts should be made with the widest dimensional tolerances, consistent with their intended applications, in order to increase tool and die life and reduce production costs.
- Walls should not be less than 1.5 mm (0.060 in.) thick, although walls as thin as 0.34 mm (0.0135 in.) have been pressed successfully on components 1 mm (0.04 in.) in length. Walls with length-to-thickness ratios greater than 8:1 are difficult to press, and density variations are virtually unavoidable.
- Simple steps can be produced if their size doesn't exceed 15% of the overall part length. Larger steps can be pressed, but require more complex, multiple-motion tooling.

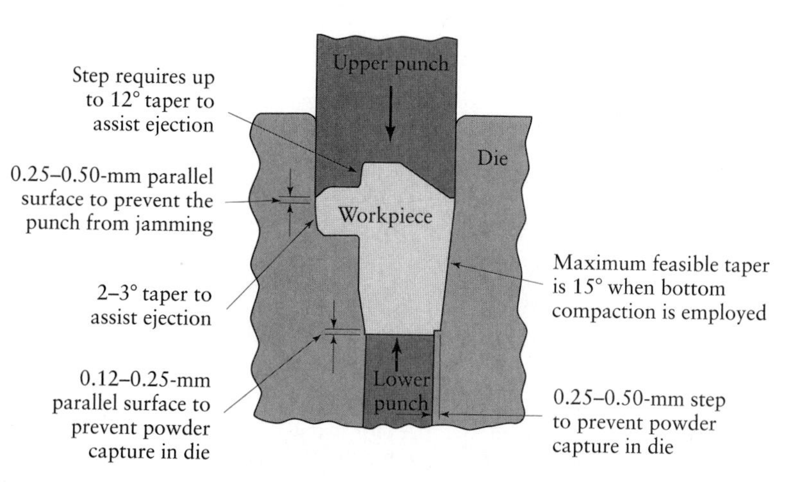

FIGURE 11.16 Die geometry and design features for powder-metal compaction. *Source*: Metal Powder Industries Federation.

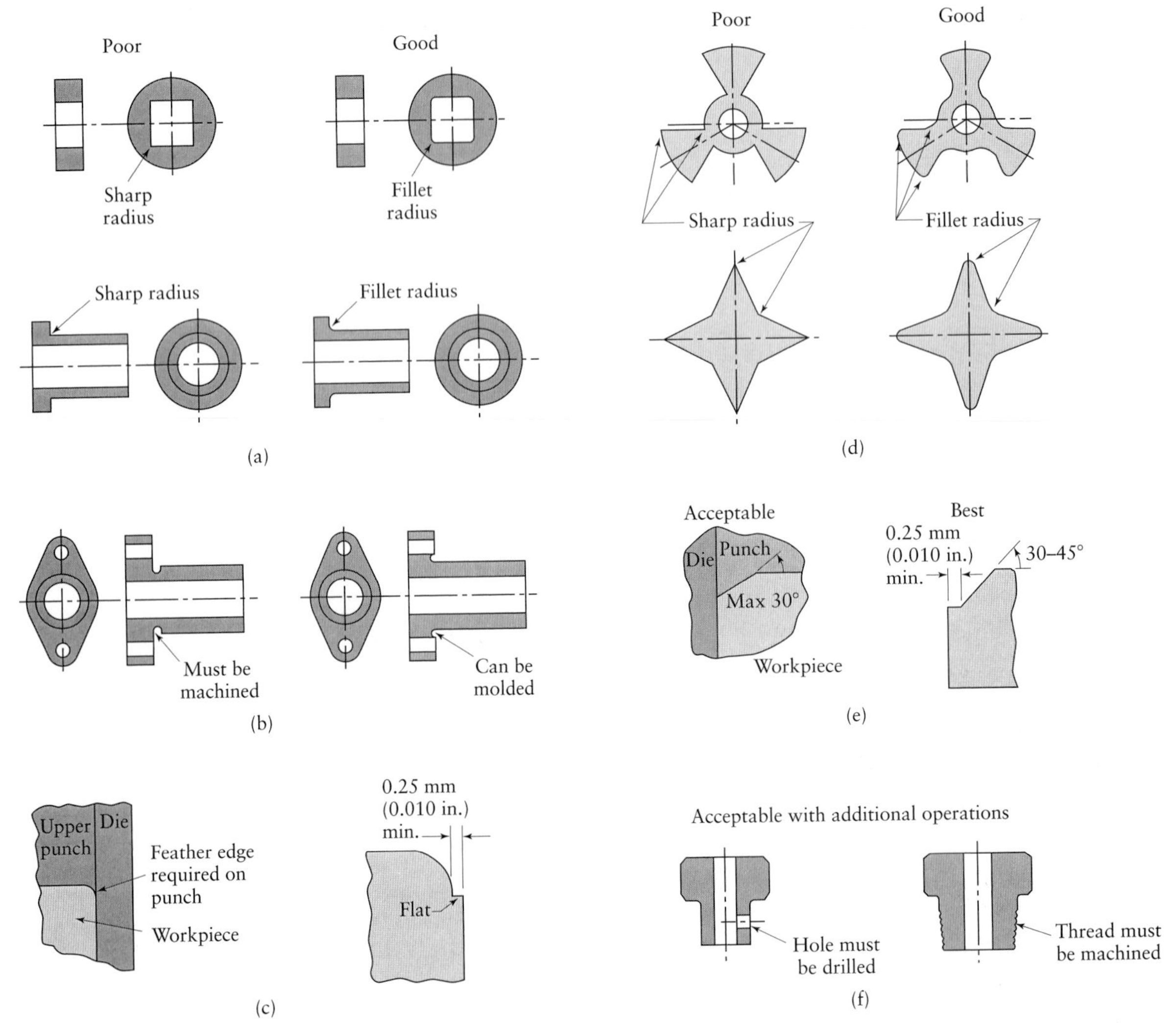

FIGURE 11.17 Examples of P/M parts, showing poor and good designs. Note that sharp radii and reentry corners should be avoided, and that threads and transverse holes have to be produced separately by additional machining operations. *Source*: Metal Powder Industries Federation.

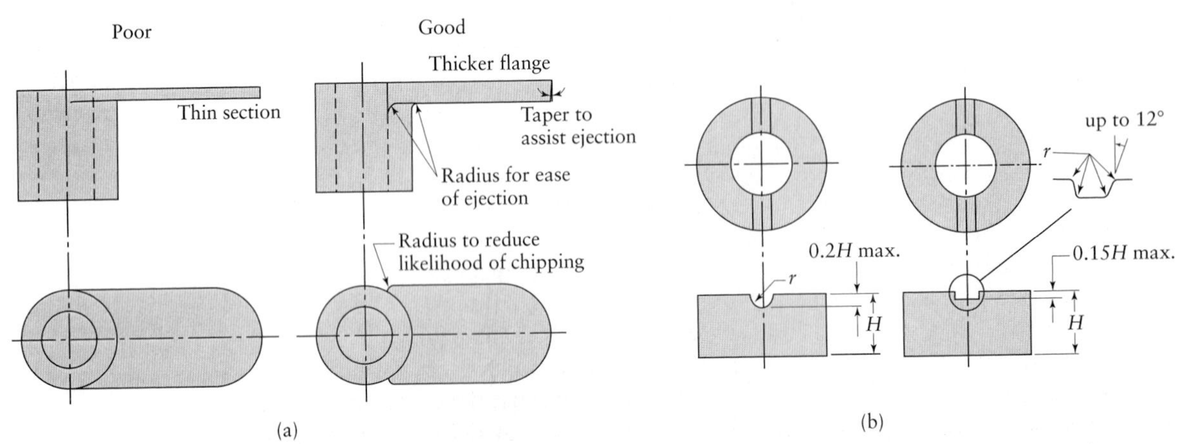

FIGURE 11.18 (a) Design features for use with unsupported flanges. (b) Design features for use with grooves. *Source*: Metal Powder Industries Federation.

- Letters can be pressed if they are oriented perpendicular to the direction of pressing; letters can also be raised or recessed. However, raised letters are more susceptible to damage in the green stage and prevent stacking during sintering.
- Flanges or overhangs can be produced by a step in the die. However, long flanges may be broken upon ejection and may require more elaborate tooling. A long flange should incorporate a draft around the flange, a radius at the bottom edge, and a radius at the juncture of the flange and/or component body to reduce stress concentrations and the likelihood of fracture.
- A true radius cannot be pressed into the edge of a part, because it would require the punch to be feathered to a zero thickness (Fig. 11.17c). Chamfers or flats are preferred for pressing; a common design approach is to use a 45° angle and 0.25-mm (0.010-in.) flat (Fig. 11.17e).
- Keys, keyways, and holes, used for transmitting torques on gears and pulleys, can be formed during compaction. Bosses can be produced if proper drafts are used and their length is small compared with the overall size of the component.
- Notches and grooves can be made if they are oriented perpendicular to the pressing direction (Fig. 11.17b). It is recommended that circular grooves not exceed a depth of 20% of the overall depth of the component. (Rectangular grooves shouldn't exceed a depth of 15% of the overall depth of the component - see Fig. 11.18b.)
- Parts produced through MIM have similar design constraints as injection molding. With MIM, wall thicknesses should be uniform in order to minimize distortion during sintering.

Dimensional tolerances of sintered P/M parts are usually on the order of ±0.05–0.1 mm (±0.002–0.004 in.). Tolerances improve significantly with additional operations such as sizing, machining, and grinding. (See Table 16.1.)

11.7 Economics of Powder Metallurgy

As we have seen, powder metallurgy can produce parts at or near net shape, thus eliminating many secondary manufacturing and assembly operations; consequently, P/M has become increasingly competitive with cast, forged, extruded, or machined parts (Table 11.5). Although some larger parts also are produced, the vast majority of conventional P/M parts weigh from a few grams to about a kilogram each. Many conventional die-pressed and sintered parts are used without any machining, although some may require minor finishing operations. Even in highly complex parts, finish machining of P/M parts involves relatively simple operations, such as drilling and tapping holes or grinding some surfaces; furthermore, a single P/M part may replace an assembly of several parts made by other manufacturing processes.

The net- or near-net-shape capability of P/M is an important factor because of its cost-effectiveness in numerous applications. Major cost elements in P/M parts production are the costs for metal powders, compacting dies, and equipment for compacting and sintering. The unit cost of metal powders is much higher than that for molten metal for casting or wrought bar stock for machining; the cost also depends on the method of powder production. The least expensive powder metal is iron, followed by aluminum, zinc, copper, chromium, stainless steel, molybdenum, tungsten, cobalt, niobium, zirconium, and tantalum. Due to the high initial cost of punches, dies,

TABLE 11.5

Competitive Features of P/M and Some Other Manufacturing Processes

Process	Advantages Over P/M	Limitations as Compared With P/M
Casting	Wide range of part shapes and sizes produced; generally low mold and setup cost.	Some waste of material in processing; some finishing required; may not be feasible for some high-temperature alloys.
Forging (hot)	High production rate of a wide range of part sizes and shapes; high mechanical properties through control of grain flow.	Some finishing required; some waste of material in processing; die wear; relatively poor surface finish and dimensional control.
Extrusion (hot)	High production rate of long parts; complex cross-sections may be produced.	Only a constant cross-sectional shape can be produced; die wear; poor dimensional control.
Machining	Wide range of part shapes and sizes; short lead time; flexibility; good dimensional control and surface finish; simple tooling.	Waste of material in the form of chips; relatively low productivity.

and various equipment for P/M processing, production volume must be sufficiently high to warrant this major expenditure, typically upwards of 50,000 parts per year. Today's equipment is highly automated, and P/M production is ideal for many automotive parts made in millions per year, for such large quantities, the labor cost per part is significantly low.

The cost-effectiveness of specialized P/M methods results from special considerations that are different from those for conventional die-press and sinter methods. For example, P/M forging is used mainly for critical applications in which full density and the accompanying superior fatigue resistance are essential. The technique is especially well suited for high-quantity production. For specific applications, P/M forging competes with processes such as conventional forging and casting, in terms of both product properties and production cost. Automotive connecting rods can be produced by both P/M forging and casting. A more recently developed process is metal injection molding (MIM), which uses finer, and therefore more expensive, powder; it also requires more production steps, such as feedstock preparation, molding, debinding, and sintering. Because of these higher cost elements, MIM is cost effective mainly for small (generally weighing less than 100 g), but highly complex, parts required in large quantities.

Relatively large aerospace parts can be produced in small quantities when there are critical property requirements or special metallurgical considerations. For example, some nickel-based superalloys are so highly alloyed that they are prone to segregation during casting and can be processed only by P/M methods; the powders may first be consolidated by hot extrusion and then hot forged in dies maintained at a high temperature. All beryllium processing is also based on powder metallurgy and involves either cold isostatic pressing and sintering or hot isostatic pressing. Many of these materials are difficult to process and may require much finish machining even when the parts are made by P/M methods; this is in sharp contrast to the numerous ferrous press-and-sinter P/M parts that require practically no finish machining. To a large extent, these special and rather expensive P/M methods are used for aerospace applications because no other production alternatives are technically feasible.

11.8 Ceramics: Structure, Properties, and Applications

Ceramics are compounds of metallic and nonmetallic elements. The term ceramics refers both to the material and to the ceramic product itself. In Greek, the word *keramos* means potter's clay, and *keramikos* means clay products. Because of the large number of possible combinations of elements, a great variety of ceramics is now available for widely different consumer and industrial applications.

The earliest use of ceramics was in pottery and bricks, dating back to before 4000 BC. Ceramics have been used for many years in automotive spark plugs as an electrical insulator and for high-temperature strength; they are becoming increasingly important in heat engines and various other applications (Table 11.6), as well as in tool

TABLE 11.6

Types and General Characteristics of Ceramics and Glasses

Type	General Characteristics
Oxide Ceramics	
Alumina	High hot hardness and abrasion resistance, moderate strength and toughness; most widely used ceramic; used for cutting tools, abrasives, and electrical and thermal insulation.
Zirconia	High strength and toughness; resistance to thermal shock, wear, and corrosion; partially-stabilized zirconia and transformation-toughened zirconia have better properties; suitable for heat-engine components.
Carbides	
Tungsten carbide	High hardness, strength, toughness, and wear resistance, depending on cobalt binder content; commonly used for dies and cutting tools.
Titanium carbide	Not as tough as tungsten carbide, but has a higher wear resistance; has nickel and molybdenum as the binder; used as cutting tools.
Silicon carbide	High-temperature strength and wear resistance, used for heat engines and as abrasives.
Nitrides	
Cubic boron nitride	Second hardest substance known, after diamond; high resistance to oxidation; used as abrasives and cutting tools.
Titanium nitride	Used as coatings on tools, because of its low frictional characteristics.
Silicon nitride	High resistance to creep and thermal shock; high toughness and hot hardness; used in heat engines.
Sialon	Consists of silicon nitrides and other oxides and carbides; used as cutting tools.
Cermets	Consist of oxides, carbides, and nitrides; high chemical resistance but is somewhat brittle and costly; used in high-temperature applications.
Nanophase ceramics	Stronger and easier to fabricate and machine than conventional ceramics; used in automotive and jet-engine applications.
Silica	High temperature resistance; quartz exhibits piezoelectric effects; silicates containing various oxides are used in high-temperature, nonstructural applications.
Glasses	Contain at least 50% silica; amorphous structure; several types available, with a wide range of mechanical, physical, and optical properties.
Glass ceramics	High crystalline component to their structure; stronger than glass; good thermal-shock resistance; used for cookware, heat exchangers, and electronics.
Graphite	Crystalline form of carbon; high electrical and thermal conductivity; good thermal-shock resistance; also available as fibers, foam, and buckyballs for solid lubrication; used for molds and high-temperature components.
Diamond	Hardest substance known; available as single-crystal or polycrystalline form; used as cutting tools and abrasives and as die insert for fine wire drawing; also used as coatings.

and die materials. More recent applications of ceramics are in automotive components, such as exhaust-port liners, coated pistons, and cylinder liners, with the desirable properties of strength and corrosion resistance at high operating temperatures.

Some properties of ceramics are significantly better than those of metals, particularly their hardness and thermal and electrical resistance. Ceramics are available as a single crystal or in polycrystalline form, consisting of many grains. Grain size has a major influence on the strength and properties of ceramics; the finer the grain size, the higher are the strength and toughness—hence, the term **fine ceramics.** Ceramics are general divided into the categories of **traditional ceramics** (whiteware, tiles, brick, pottery, and abrasive wheels) and **industrial ceramics,** also called **engineering** or **high-tech ceramics** (heat exchangers, cutting tool, semiconductors, and prosthetics).

11.8.1 Structure and types of ceramics

The structure of ceramic crystals is among the most complex of all materials, containing various elements of different sizes. The bonding between these atoms is generally covalent (electron sharing, hence strong bonds) and ionic (primary bonding between oppositely charged ions, hence strong bonds).

Among the oldest *raw materials* for ceramics is *clay,* a fine-grained sheetlike structure, the most common example being *kaolinite* (from Kaoling, a hill in China). It is a white clay, consisting of silicate of aluminum with alternating weakly bonded layers of silicon and aluminum ions (Fig. 11.19). When added to kaolinite, water attaches itself to the layers (adsorption), makes them slippery, and gives wet clay its well-known softness and plastic properties (*hydroplasticity*) that make it formable. Other major raw materials for ceramics that are found in nature are *flint* (rock of very fine-grained silica, SiO_2) and *feldspar* (a group of crystalline minerals consisting of aluminum silicates, potassium, calcium, or sodium). In their natural state, these raw materials generally contain impurities of various kinds, which have to be removed prior to further processing of the materials into useful products with reliable performance. Highly refined raw materials produce ceramics with improved properties.

The various types of ceramics are:

1. **Oxide ceramics**

Alumina. Also called *corundum* or *emery, alumina* (aluminum oxide, Al_2O_3) is the most widely used *oxide ceramic,* either in pure form or as a raw material to be mixed with other oxides. It has high hardness and moderate strength. Although alumina exists in nature, it contains varying amounts of impurities and possesses nonuniform properties in natural form. As a result, its behavior is unreliable. Aluminum oxide, as well as silicon carbide and many other ceramics, is now almost totally manufactured synthetically, so that its quality can be controlled.

First made in 1893, synthetic aluminum oxide is obtained by the fusion of molten bauxite (an aluminum-oxide ore that is the principal source of alu-

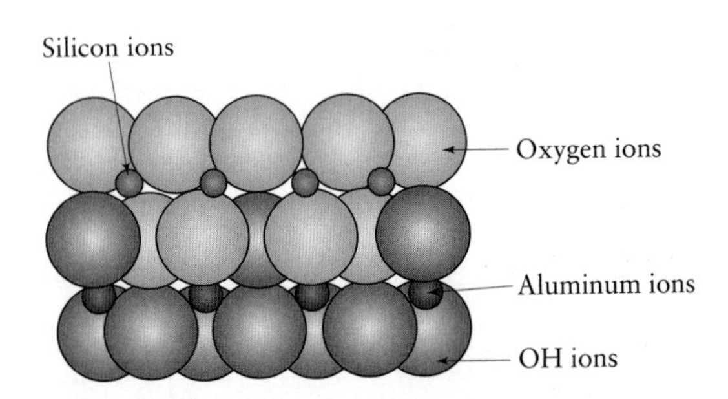

FIGURE 11.19 The crystal structure of kaolinite, commonly known as clay.

minum), iron filings, and coke in electric furnaces. The material is then crushed and graded by size by passing the particles through standard screens. Parts made of aluminum oxide are cold pressed and sintered (*white ceramics*). Their properties are improved by minor additions of other ceramics, such as titanium oxide and titanium carbide. Structures containing various alumina and other oxides are known as *mullite* and *spinel* and are used as refractory materials for high-temperature applications. The mechanical and physical properties of alumina are particularly suitable for applications such as electrical and thermal insulation, cutting tools, and abrasives.

Zirconia. *Zirconia* (zirconium oxide, ZrO_2), white in color, has good toughness; good resistance to thermal shock, wear, and corrosion; low thermal conductivity, and a low friction coefficient. A more recent development is **partially stabilized zirconia** (PSZ), which has high strength and toughness and performs more reliably than zirconia. It is obtained by doping the zirconia with oxides of calcium, yttrium, or magnesium. This process forms a material with fine particles of tetragonal zirconia in a cubic lattice. Typical applications include dies for hot extrusion of metals and zirconia beads for grinding and dispersion media for aerospace coatings, automotive primers and topcoats, and fine glossy print on flexible food packaging.

Another important characteristic of PSZ is the fact that its coefficient of thermal expansion is only about 20% lower than that of cast iron, and its thermal conductivity is about one third that of other ceramics. Consequently, it is very suitable for heat-engine components, such as cylinder liners and valve bushings, to keep the cast-iron engine assembly intact. New developments to further improve upon the properties of PSZ include **transformation-toughened zirconia** (TTZ), which has higher toughness than that of PSZ, because of dispersed tough phases in the ceramic matrix.

2. **Other ceramics**

Carbides. Typical examples of *carbides* are tungsten and titanium, used as cutting tools and die materials, and silicon carbide, used as abrasives (e.g., in grinding wheels):

a) **Tungsten carbide** (WC) consists of tungsten-carbide particles with cobalt as a binder. The amount of binder has a major influence on the material's properties: Toughness increases with the cobalt content, whereas hardness, strength, and resistance decrease.

b) **Titanium carbide** (TiC) has nickel and molybdenum as the binder and is not as tough as tungsten carbide.

c) **Silicon carbide** (SiC) has good wear, thermal shock, and corrosion resistance. It has a low friction coefficient and retains its strength at elevated temperatures. It is suitable for high-temperature components in heat engines and is also used as an abrasive. Synthetic silicon carbide is made from silica sand, coke, and small amounts of sodium chloride and sawdust. The process is similar to making synthetic aluminum oxide.

Nitrides. Another important class of ceramics is the *nitrides*:

a) **Cubic boron nitride** (cBN), the second hardest known substance, after diamond, has special applications, such as abrasives in grinding wheels and as cutting tools. It does not exist in nature and was first made synthetically in the 1970s, with techniques similar to those used for making synthetic diamond.

b) **Titanium nitride** (TiN) is used extensively as coatings on cutting tools; it improves tool life, because of its low frictional characteristics.

c) **Silicon nitride** (Si_3N_4) has high resistance to creep at elevated temperatures, low thermal expansion, and high thermal conductivity, and hence it resists thermal shock. It is suitable for high-temperature structural applications, such as in automotive engine and gas-turbine components, cam-follower rollers, bearings, sand-blast nozzles, and components for the paper industry.

Sialon. *Sialon* consists of silicon nitride, with various additions of aluminum oxide, yttrium oxide, and titanium carbide. (See Section 8.6.8.) The word *sialon* is derived from the words *silicon, aluminum, oxygen,* and *nitrogen.* Sialon has higher strength and thermal-shock resistance than silicon nitride does and thus far is used primarily as a cutting-tool material.

Cermets. *Cermets* are combinations of ceramics bonded with a metallic phase. Introduced in the 1960s and also called **black ceramics** or *hot-pressed ceramics*, they combine the high-temperature oxidation resistance of ceramics and the toughness, thermal-shock resistance, and ductility of metals. An application of cermets is cutting tools, a typical composition being 70% aluminum oxide and 30% titanium carbide; other cermets contain various oxides, carbides, and nitrides. They have been developed for high-temperature applications such as nozzles for jet engines and aircraft brakes. Cermets can be regarded as composite materials and can be used in various combinations of ceramics and metals bonded by powder-metallurgy techniques. (See also Section 11.14.)

3. **Silica**

Abundant in nature, *silica* is a polymorphic material; that is, it can have different crystal structures. The cubic structure is found in refractory bricks used for high-temperature furnace applications. Most glasses contain more than 50% silica. The most common form of silica is **quartz**, which is a hard, abrasive hexagonal crystal. It is used extensively as oscillating crystals of fixed frequency in communications applications, since it exhibits the piezoelectric effect (See Section 3.9.6.)

Silicates are products of the reaction of silica with oxides of aluminum, magnesium, calcium, potassium, sodium, and iron. Examples include clay, asbestos, mica, and silicate glasses. **Lithium aluminum silicate** has very low thermal expansion and thermal conductivity and good thermal shock resistance. However, it has very low strength and fatigue life, and thus it is suitable only for nonstructural applications, such as catalytic converters, regenerators, and heat-exchanger components.

4. **Nanophase ceramics**

A more recent development is *nanophase ceramics*, which consist of atomic clusters containing a few thousand atoms. Nanophase ceramics exhibit ductility at significantly lower temperatures than for conventional ceramics and are stronger and easier to fabricate, with fewer flaws; however, control of particle size and distribution and of contamination is important. Applications of nanophase ceramics in the automotive industry, for example, include valves, rocker arms, turbocharger rotors, and cylinder liners; they are also used in jet-engine components. Nanocrystalline second-phase particles, on the order of 100 nm or less, and fibers are also being used as reinforcement in composites (see Section 11.14); they have enhanced properties such as tensile strength and creep resistance (See also the discussion of *nanomaterials*, in Section 3.11.9.)

11.8.2 General properties and applications of ceramics

Compared with metals, ceramics have the following relative characteristics: brittleness, high strength and hardness at elevated temperatures, high elastic modulus, low toughness, low density, low thermal expansion, and low thermal and electrical conductivity. However, because of the wide variety of ceramic material composition and grain size, the mechanical and physical properties of ceramics vary significantly. For example, the electrical conductivity of ceramics can be modified from poor to good, which is the principle behind semiconductors. Because of their sensitivity to flaws, defects, and cracks (surface or internal); the presence of different types and levels of impurities; and different methods of manufacturing, ceramics can have a wide range of properties.

a) **Mechanical properties.** Table 11.7 presents the mechanical properties of several engineering ceramics. Note that the ceramics' strength in tension (transverse rupture strength) is approximately one order of magnitude lower than their compressive strength, because of their sensitivity to cracks, impurities, and porosity. Such defects lead to the initiation and propagation of cracks under tensile stresses, severely reducing tensile strength. Therefore, reproducibility and reliability (acceptable performance over a specified period of time) are important aspects in the service life of ceramic components. The tensile strength of polycrystalline ceramic parts increases with decreasing grain size.

Tensile strength is empirically related to porosity as

$$\text{UTS} \simeq \text{UTS}_0 e^{-nP}, \tag{11.5}$$

where P is the volume fraction of pores in the solid, UTS_0 is the tensile strength at zero porosity, and the exponent n ranges between 4 and 7.

The modulus of elasticity is likewise affected by porosity, as given by

$$E \simeq E_0(1 - 1.9P + 0.9P^2), \tag{11.6}$$

TABLE 11.7

Approximate Range of Properties of Various Ceramics at Room Temperature

Material	Symbol	Transverse Rupture Strength (MPa)	Compressive Strength (MPa)	Elastic Modulus (GPa)	Hardness (HK)	Poisson's Ratio (ν)	Density (kg/m^3)
Aluminum oxide	Al_2O_3	140–240	1000–2900	310–410	2000–3000	0.26	4000–4500
Cubic boron nitride	cBN	725	7000	850	4000–5000	–	3480
Diamond	–	1400	7000	830–1000	7000–8000	–	3500
Silica, fused	SiO_2	–	1300	70	550	0.25	–
Silicon carbide	SiC	100–750	700–3500	240–480	2100–3000	0.14	3100
Silicon nitride	Si_3N_4	480–600	–	300–310	2000–2500	0.24	3300
Titanium carbide	TiC	1400–1900	3100–3850	310–410	1800–3200	–	5500–5800
Tungsten carbide	WC	1030–2600	4100–5900	520–700	1800–2400	–	10,000–15,000
Partially stabilized zirconia	PSZ	620	–	200	1100	0.3	5800

Note: These properties vary widely, depending on the condition of the material.

where E_0 is the modulus of elasticity at zero porosity. Equation (11.5) is valid up to 50% porosity. Common earthenware has a porosity ranging between 10% and 15%, whereas the porosity of hard porcelain is about 3%.

Although there are exceptions, ceramics, unlike most metals and thermoplastics, generally lack impact toughness and thermal-shock resistance because of their inherent lack of ductility. Once initiated, a crack propagates rapidly. In addition to undergoing fatigue failure under cyclic loading, ceramics (particularly glasses) exhibit a phenomenon called **static fatigue:** When subjected to a static tensile load over a period of time, these materials may suddenly fail. This phenomenon occurs in environments where water vapor is present. Static fatigue, which does not occur in a vacuum or dry air, has been attributed to a mechanism similar to stress-corrosion cracking of metals.

Ceramic components that are to be subjected to tensile stresses may be **prestressed**, much like prestressed concrete. Prestressing shaped ceramic components subjects them to compressive stresses. Methods used include (1) heat treatment and chemical tempering; (2) laser treatment of surfaces; (3) coating with ceramics with different degrees of thermal expansion; and (4) surface-finishing operations, such as grinding, in which compressive residual stresses are induced on the surfaces. Significant advances are being made in improving the toughness and other properties of ceramics, including **machinable ceramics**. Among these advances are proper selection and processing of raw materials, control of purity and structure, and use of reinforcements, with particular emphasis during design on advanced methods of stress analysis in ceramic components.

b) **Physical properties.** Most ceramics have relatively low specific gravity, ranging from about 3 to 5.8 for oxide ceramics, compared with 7.86 for iron. They have very high melting or decomposition temperatures. The thermal conductivity of ceramics varies by as much as three orders of magnitude, depending on their composition, whereas the thermal conductivity of metals varies by one order. The thermal conductivity of ceramics, as well as of other materials, decreases with increasing temperature and porosity, because air is a poor thermal conductor.

The thermal conductivity k is related to porosity by

$$k = k_0(1 - P), \tag{11.7}$$

where k_0 is the thermal conductivity at zero porosity.

The thermal-expansion characteristics of ceramics are shown in Fig. 11.20. Thermal expansion and thermal conductivity induce thermal stresses that can lead to thermal shock or thermal fatigue. The tendency for *thermal cracking* (called *spalling* when a piece or a layer from the surface breaks off) is lower with low thermal expansion and high thermal conductivity. For example, fused silica has high thermal shock resistance because of its virtually zero thermal expansion.

A familiar example that illustrates the importance of low thermal expansion is heat-resistant ceramics for cookware and stove tops. These ceramics can sustain high thermal gradients, from hot to cold, and vice versa. Moreover, the relative thermal expansion of ceramics and metals is an important reason for the use of ceramic components in heat engines. The fact that the thermal conductivity of partially stabilized zirconia components is close to that of the cast iron in engine blocks is a further advantage in the use of PSZ in heat engines. An additional characteristic is the **anisotropy of thermal expansion** exhibited by oxide ceramics, whereby thermal expansion varies in different directions of the ceramic. This behavior causes thermal stresses that can lead to cracking of the ceramic component.

The *optical properties* of ceramics can be controlled by various formulations and control of structure, imparting different degrees of transparency and

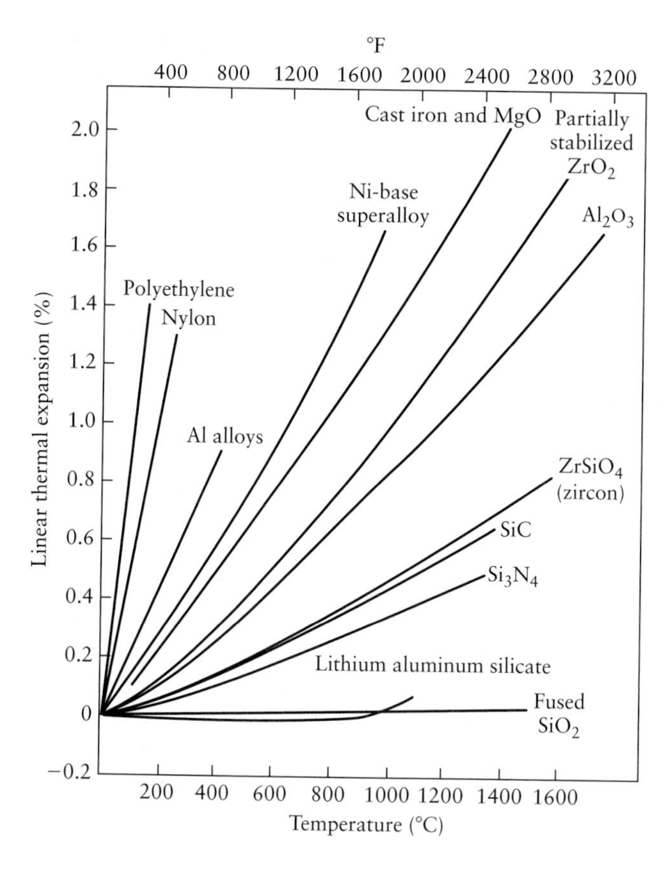

FIGURE 11.20 Effect of temperature on thermal expansion for several ceramics, metals, and plastics. Note that the expansions for cast iron and for partially stabilized zirconia (PSZ) are within about 20%.

colors. For example, single-crystal sapphire is completely transparent, zirconia is white, and fine-grained polycrystalline aluminum oxide is a translucent gray. Porosity influences the optical properties of ceramics, much like trapped air in ice cubes, which makes the ice less transparent and gives it a white appearance.

Although ceramics basically are resistors, it has been shown that they can be made to be electrically conductive by alloying them with certain elements, thus making the ceramic act like a semiconductor or even a superconductor.

c) **Applications.** As shown in Table 11.6, ceramics have numerous consumer and industrial applications. Several types of ceramics are used in the electrical and electronics industry, because of their high electrical resistivity, dielectric strength (the voltage required for electrical breakdown per unit thickness), and magnetic properties suitable for applications such as magnets for speakers. An example of such a ceramic is **porcelain**, which is a white ceramic composed of kaolin, quartz, and feldspar. Certain ceramics also have good piezoelectric properties.

The capability of ceramics to maintain their strength and stiffness at elevated temperatures (Figs. 11.21 and 11.22) makes them very attractive for high-temperature applications. Their high resistance to wear makes them suitable for applications such as cylinder liners, bushings, seals, and bearings. The higher operating temperatures made possible by the use of ceramic components enable more efficient fuel burning and reduced emissions. Currently, internal combustion engines are only about 30% efficient, but with the use of ceramic components, their operating performance can be improved by at least 30%. Ceramics that have been used successfully, especially in gasoline and diesel-engine components and as rotors, are silicon nitride, silicon carbide, and partially stabilized zirconia. Ceramics are also used to coat metal, which may be done to reduce wear, prevent corrosion, and provide a thermal barrier.

FIGURE 11.21 Effect of temperature on the strength of various engineering ceramics. Note that much of the strength is maintained at high temperatures.

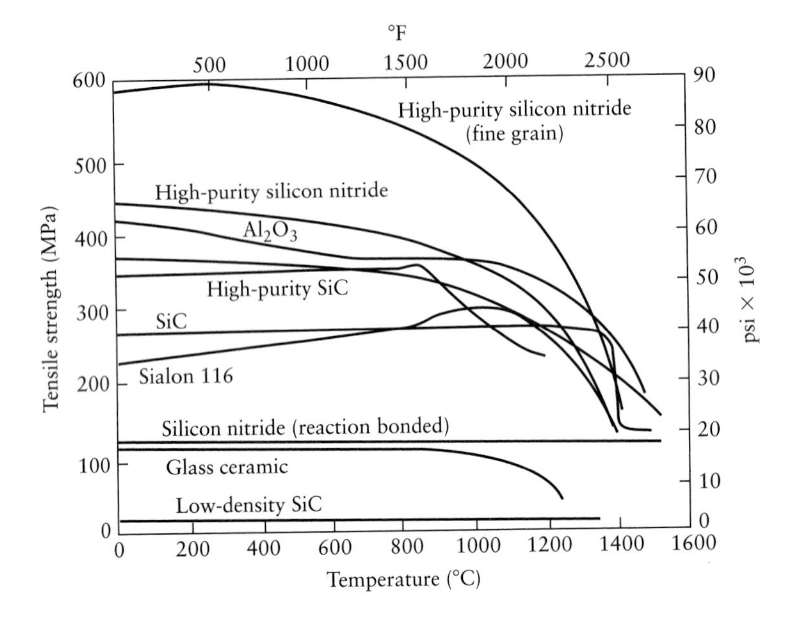

FIGURE 11.22 Effect of temperature on the modulus of elasticity for several ceramics. *Source*: D. W. Richerson, *Modern Ceramic Engineering*, New York: Marcel Dekker, Inc., 1982.

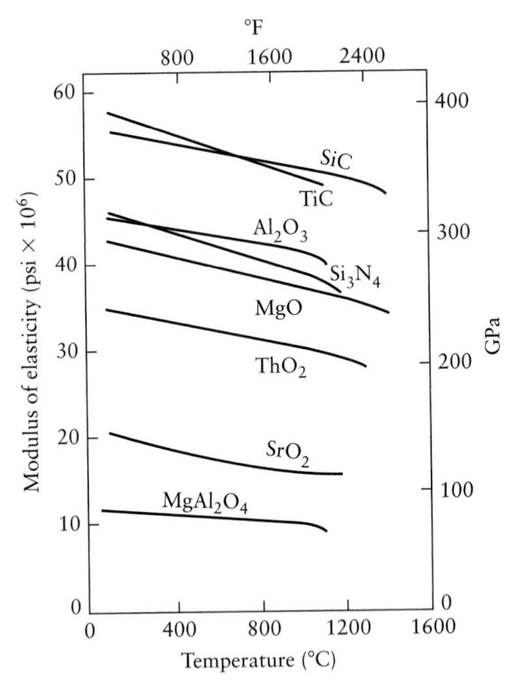

Other attractive properties of ceramics are their low density and high elastic modulus, which make it possible to reduce engine weight; also, in other applications that use ceramics, the inertial forces generated by moving parts are lower than with metals. High-speed components for machine tools, for example, are candidates for ceramics. Furthermore, the higher elastic modulus of ceramics makes them attractive for improving the stiffness, while reducing the weight, of machines. Silicon-nitride ceramics are also used as ball bearings and rollers. Because of their strength and inertness, ceramics are used as *biomaterials* to replace joints in the human body, as prosthetic devices, and for dental work. Commonly used bioceramics include aluminum oxide, silicon nitride, and various compounds of silica. Furthermore, ceramics can be made to be porous, thus allowing bone to grow into the porous surface and develop a strong mechanical bond.

EXAMPLE 11.7 Effect of porosity on properties

If a fully dense ceramic has the properties of $UTS_0 = 100$ MPa, $E_0 = 400$ GPa, and $k_0 = 0.5$ W/m · K, what are these properties at 10% porosity? Let $n = 5$ and $P = 0.1$.

Solution. Using Eqs. 11.5–11.7, we have

$$UTS = 100e^{-(5)(0.1)} = 61 \text{ MPa},$$

$$E = 400[1 - (1.9)(0.1) + (0.9)(0.1)^2] = 328 \text{ GPa},$$

and

$$k = 0.5(1 - 0.1) = 0.45 \text{ W/m} \cdot \text{K}.$$

11.9 Shaping Ceramics

Several techniques have been developed for shaping ceramics into useful products (Table 11.8). Generally, the procedure involves the following steps: crushing or grinding the raw materials into very fine particles; mixing the particles with additives to impart certain desirable characteristics; and shaping, drying, and firing the material. *Crushing* (also called *comminution* or *milling*) of the raw materials is generally done in a ball mill (Fig. 11.23b), either dry or wet. Wet crushing is more effective, because

TABLE 11.8

General Characteristics of Ceramics Processing

Process	Advantages	Limitations
Slip casting	Large parts; complex shapes; low equipment cost.	Low production rate; limited dimensional accuracy.
Extrusion	Hollow shapes and small diameters; high production rate.	Parts have constant cross-section; limited thickness.
Dry pressing	Close tolerances; high production rate with automation.	Density variation in parts with high length-to-diameter ratios; dies require high abrasive-wear resistance; equipment can be costly.
Wet pressing	Complex shapes; high production rate.	Limited part size and dimensional accuracy; tooling costs can be high.
Hot pressing	Strong, high-density parts.	Protective atmospheres required; die life can be short.
Isostatic pressing	Uniform density distribution.	Equipment can be costly.
Jiggering	High production rate with automation; low tooling cost.	Limited to axisymmetric parts; limited dimensional accuracy.
Injection molding	Complex shapes; high production rate.	Tooling costs can be high.

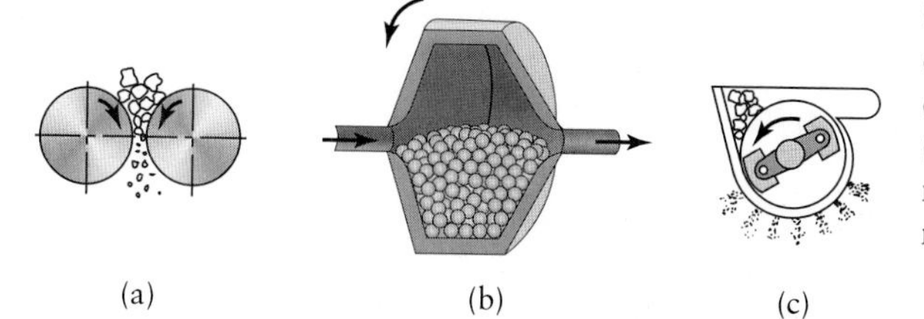

FIGURE 11.23 Methods of crushing ceramics to obtain very fine particles: (a) roll crushing, (b) ball milling, and (c) hammer milling.

it keeps the particles together and prevents the suspension of fine particles in air. The ground particles are then mixed with *additives*, the function of which is one or more of the following:

1. *Binder* for the ceramic particles;
2. *Lubricant* for mold release and to reduce internal friction between particles during molding;
3. *Wetting agent* to improve mixing;
4. *Plasticizer* to make the mix more plastic and formable;
5. *Deflocculent* to make the ceramic–water suspension uniform. The additive changes the electrical charges on the particles of clay so that they repel instead of attract each other. Water is added to make the mixture more pourable and less viscous. Typical deflocculants include Na_2CO_3 and Na_2SiO_3 in amounts of less than 1%.;
6. Various agents to *control foaming* and *sintering*.

The three basic shaping processes for ceramics are casting, plastic forming, and pressing, and are described in the following three subsections.

11.9.1 Casting

The most common casting process is **slip casting**, also called *drain casting* (Fig. 11.24). A **slip** is a suspension of ceramic particles in a liquid, generally water. In this process, the slip is poured into a porous mold made of plaster of paris. The slip must have sufficient fluidity and low viscosity to flow easily into the mold, much like the fluidity of molten metals. After the mold has absorbed some of the water from the outer layers of the suspension, it is inverted, and the remaining suspension is poured out (for making hollow objects, as in slush casting of metals - see Fig. 5.16). The top of the part is then trimmed, the mold is opened, and the part is removed.

Large and complex parts, such as plumbing ware, art objects, and dinnerware, can be made by slip casting. Although dimensional control is limited and the production rate is low, mold and equipment costs are also low. In some applications, components of the product (such as handles for cups and pitchers) are made separately and then joined, using the slip as an adhesive.

Thin sheets of ceramics, less than 1.5 mm (0.06 in.) thick, can be made by a casting technique called the **doctor-blade process.** In this operation, the slip is cast over a moving plastic belt, and its thickness is controlled by a blade. Other processes include *rolling* the slip between pairs of rolls and *casting* the slip over a paper tape, which is then burned off during firing. For solid ceramic parts, the slip is supplied continuously into the mold to replenish the absorbed water; the suspension is not drained from the mold. At this stage, the part is a soft solid or semirigid. The higher the concentration of solids in the slip, the less water has to be removed. The part, called *green*, as in powder metallurgy, is then fired.

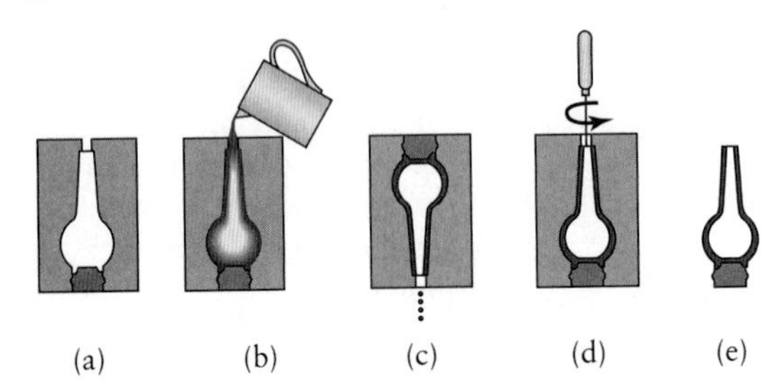

FIGURE 11.24 Sequence of operations in slip casting a ceramic part. After the slip has been poured, the part is dried and fired in an oven to give it strength and hardness. *Source*: F. H. Norton, *Elements of Ceramics.*

11.9.2 Plastic forming

Plastic forming (also called *soft*, *wet*, or *hydroplastic forming*) can be done by various methods, such as extrusion, injection molding, or molding and jiggering (as done on a potter's wheel). Plastic forming tends to orient the layered structure of clays along the direction of material flow. This orientation leads to anisotropic behavior of the material, both in subsequent processing and in the final properties of the ceramic product. In **extrusion**, the clay mixture, containing 20% to 30% water, is forced through a die opening by screw-type equipment. The cross-section of the extruded product is constant, but there are limitations on the wall thickness for hollow extrusions. Tooling costs are low, and production rates are high. The extruded products may be subjected to additional shaping operations.

11.9.3 Pressing

The various methods of pressing are:

a) **Dry pressing.** Similar to powder-metal compaction, *dry pressing* is used for relatively simple shapes. Typical parts made by this method include whiteware, refractories, and abrasive products. The process has the same high production rates and close control of dimensional tolerances as in P/M. The moisture content of the mixture is generally below 4%, but may be as high as 12%. Organic and inorganic binders, such as stearic acid, wax, starch, and polyvinyl alcohol, are usually added to the mixture; they also act as lubricants. The pressure for pressing is between 35 MPa and 200 MPa (5 ksi and 30 ksi). Modern presses used for dry pressing are highly automated. Dies, usually made of carbides or hardened steel, must have high wear resistance to withstand the abrasive ceramic particles and can be expensive.

Density can vary greatly in dry-pressed ceramics (Fig. 11.25), because of friction between particles and at the mold walls, as in P/M compaction. Density variations cause warping during firing. Warping is particularly severe for parts that have high length-to-diameter ratios; the recommended maximum ratio is 2 : 1. Several methods may be used to minimize density variations. Design of tooling is important in this respect. Vibratory pressing and impact forming are used, particularly for nuclear-reactor fuel elements. Isostatic pressing also reduces density variations.

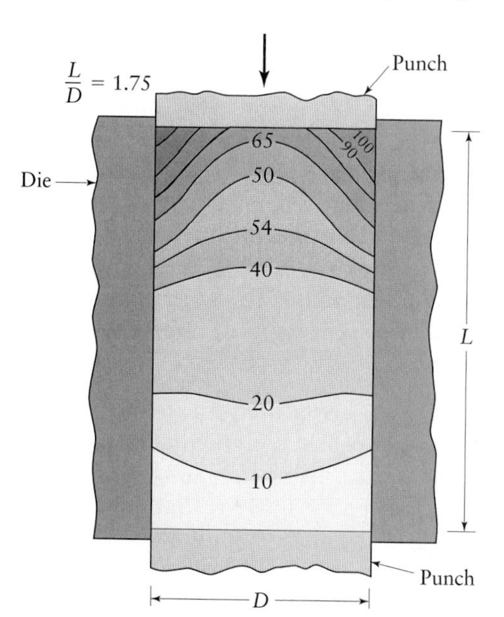

FIGURE 11.25 Density variation in pressed compacts in a single-action press. The variation increases with increasing *L/D* ratio. See also Fig. 11.7e. *Source*: Adapted from W. D. Kingery et al., *Ceramic Fabrication Processes*, MIT Press.

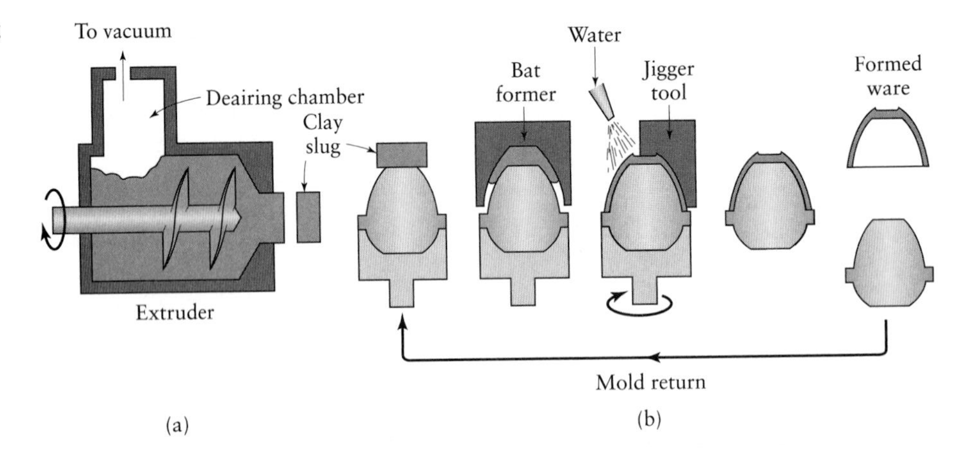

FIGURE 11.26 Extruding and jiggering operations. *Source*: R. F. Stoops.

b) **Wet pressing.** In *wet pressing*, the part is formed in a mold while under high pressure in a hydraulic or mechanical press. This process is generally used to make intricate shapes. The moisture content of the part usually ranges from 10% to 15%. Production rates are high, but part size is limited; dimensional control is difficult, because of shrinkage during drying; and tooling costs can be high.

c) **Isostatic pressing.** Used extensively in powder metallurgy, as you have seen, *isostatic pressing* is also used for ceramics in order to obtain uniform density distribution throughout the part. Automotive spark-plug insulators are made by this method. Silicon-nitride vanes for high-temperature use are also made by hot isostatic pressing.

d) **Jiggering.** A combination of processes is used to make ceramic plates. Clay slugs are first extruded, then formed into a *bat* over a plaster mold, and finally jiggered on a rotating mold (Fig. 11.26). *Jiggering* is an operation in which the clay bat is formed with templates or rollers. The part is then dried and fired. The process is limited to axisymmetric parts and has limited dimensional accuracy, but the operation can be automated.

e) **Injection molding.** The advantages of *injection molding* of plastics and powder metals were indicated earlier; this process is now being used extensively for precision forming of ceramics for high-technology applications, as in rocket-engine components. The raw material is mixed with a binder, such as a thermoplastic polymer (polypropylene, low-density polyethylene, ethylene vinyl acetate, or wax). The binder is usually removed by pyrolysis, and the part is sintered by firing. This process can produce thin sections, typically less than 10–15 mm (0.4–0.6 in.), using most engineering ceramics, such as alumina, zirconia, silicon nitride, and silicon carbide. Thicker sections require careful control of the materials used and of processing parameters in order to avoid internal voids and cracks, such as those due to shrinkage.

f) **Hot pressing.** In *hot pressing*, also called *pressure sintering*, pressure and temperature are applied simultaneously. This method reduces porosity, making the part denser and stronger. *Hot isostatic pressing* (see Section 11.3.3) may also be used in this operation, particularly to improve the quality of high-technology ceramics. Because of the presence of both pressure and temperature, die life in hot pressing can be short. Protective atmospheres are usually employed, and graphite is a commonly used punch and die material.

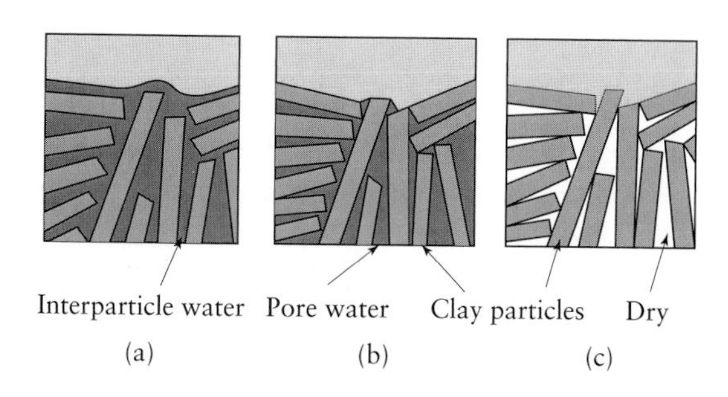

FIGURE 11.27 Shrinkage of wet clay caused by removal of water during drying. Shrinkage may be as much as 20% by volume. *Source*: After F. H. Norton.

11.9.4 Drying and firing

After the ceramic has been shaped by any of the methods described in this section, the next step is to dry and fire the part to give it the proper strength. *Drying* is a critical stage, because of the tendency for the part to warp or crack from variations in moisture content and thickness within the part and the complexity of its shape. Control of atmospheric humidity and temperature is important in order to reduce warping and cracking. Loss of moisture results in shrinkage of the part by as much as 15% to 20% of the original moist size (Fig. 11.27). In a humid environment, the evaporation rate is low, and consequently the moisture gradient across the thickness of the part is lower than that in a dry environment. The low moisture gradient, in turn, prevents a large, uneven gradient in shrinkage from the surface to the interior during drying. The dried part (called *green*, as in powder metallurgy) can be machined relatively easily to bring it closer to its final shape, although it must be handled carefully.

Firing, also called *sintering*, involves heating the part to an elevated temperature in a controlled environment, similar to sintering in powder metallurgy. Some shrinkage occurs during firing. Firing gives the ceramic part its strength and hardness. The improvement in properties results from (1) development of a strong bond between the complex oxide particles in the ceramic and (2) reduced porosity. *Nanophase ceramics* (see Section 11.8.1) can be sintered at lower temperatures than those for conventional ceramics. Nanophase ceramics are also easier to fabricate and can be compacted at room temperature to high densities, hot pressed to theoretical density, and formed into net-shape parts without binders or sintering aids.

11.9.5 Finishing operations

After firing, additional operations may be performed to give the part its final shape, remove surface flaws, and improve surface finish and dimensional tolerances. The processes that can be used are (a) grinding, (b) lapping, (c) ultrasonic machining, (d) electrical-discharge machining, (e) laser-beam machining, (f) abrasive water-jet machining, and (g) tumbling to remove sharp edges and grinding marks. The choice of the process is important in view of the brittle nature of most ceramics and the additional costs involved in these processes. The effect of the finishing operation on the properties of the product must also be considered; because of notch sensitivity, the finer the finish, the higher is the part's strength. To improve appearance and strength, and to make them impermeable, ceramic products are often coated with a **glaze** material, which forms a glassy coating after firing.

EXAMPLE 11.8 Dimensional changes during shaping of ceramic components

A solid cylindrical ceramic part is to be made whose final length must be $L = 20$ mm. It has been established that for this material, linear shrinkages during drying and firing are 7% and 6%, respectively, based on the dried dimension L_d. Calculate (1) the initial length L_o of the part and (2) the dried porosity P_d if the porosity of the fired part, P_f, is 3%.

Solution.

(1) On the basis of the information given, and remembering that firing is preceded by drying, we can write

$$\frac{(L_d - L)}{L_d} = 0.06,$$

or

$$L = (1 - 0.06)L_d;$$

hence,

$$L_d = \frac{20}{0.94} = 21.28 \text{ mm}$$

and

$$L_o = (1 + 0.07)L_d = (1.07)(21.28) = 22.77 \text{ mm}.$$

(2) Since the final porosity is 3%, the actual volume V_a of the ceramic material is

$$V_a = (1 - 0.03)V_f = 0.97V_f,$$

where V_f is the fired volume of the part. Since the linear shrinkage during firing is 6%, we can determine the dried volume V_d of the part as

$$V_d = \frac{V_f}{(1 - 0.06)^3} = 1.2V_f.$$

Hence,

$$\frac{V_a}{V_d} = \frac{0.97}{1.2} = 0.81 = 81\%.$$

Therefore, the porosity P_d of the dried part is 19%.

11.10 Glasses: Structure, Properties, and Applications

Glass is an amorphous solid with the structure of a liquid. In other words, it has been *supercooled*, that is, cooled at a rate too high for crystals to form. Generally, we define glass as an inorganic product of fusion that has cooled to a rigid condition without crystallizing. Glass has no distinct melting or freezing point; thus, its behavior is similar to that of amorphous polymers. At present, there are some 750 different types of commercially available glasses. The uses of glass range from windows, bottles, and cookware to glasses with special mechanical, electrical, high-temperature, chemical-resistant, corrosion-resistant, and optical characteristics. Special glasses are used in fiber optics for communication, with little loss in signal power, and in glass fibers with very high strength for reinforced plastics.

All glasses contain at least 50% silica, which is known as a *glass former*. The composition and properties of glasses, except strength, can be modified greatly by the addition of oxides of aluminum, sodium, calcium, barium, boron, magnesium, titanium, lithium, lead, and potassium. Depending on their function, these oxides are known as *intermediates* or *modifiers*. Glasses are generally resistant to chemical attack and are ranked by their resistance to acid, alkali, or water corrosion.

11.10.1 Types of glasses

Almost all commercial glasses are categorized by type (Table 11.9):

1. **Soda-lime glass** (the most common type);
2. **Lead-alkali glass;**
3. **Borosilicate glass;**
4. **Aluminosilicate glass;**
5. **96% silica glass;**
6. **Fused silica.**

Glasses are also classified as colored, opaque (white and translucent), multiform (variety of shapes), optical, photochromatic (darkens when exposed to light, as in some types of sunglasses), photosensitive (changes from clear to opal), fibrous (drawn into long fibers, as in fiberglass), and foam or cellular glass (containing bubbles, hence a good thermal insulator). Glasses are also referred to as **hard** or **soft**, usually in the sense of a thermal property rather than a mechanical property, as in hardness; thus, a soft glass softens at a lower temperature than does a hard glass. Soda-lime and lead-alkali glasses are considered soft, and the rest of the types of glass are considered hard.

TABLE 11.9

General Characteristics of Various Types of Glasses

	Soda-lime Glass	Lead Glass	Borosilicate Glass	96% Silica Glass	Fused Silica
Density	High	Highest	Medium	Low	Lowest
Strength	Low	Low	Moderate	High	Highest
Resistance to thermal shock	Low	Low	Good	Better	Best
Electrical resistivity	Moderate	Best	Good	Good	Good
Hot workability	Good	Best	Fair	Poor	Poorest
Heat treatability	Good	Good	Poor	None	None
Chemicals resistance	Poor	Fair	Good	Better	Best
Impact abrasion resistance	Fair	Poor	Good	Good	Best
Ultraviolet-light transmission	Poor	Poor	Fair	Good	Good
Relative cost	Lowest	Low	Medium	High	Highest

11.10.2 Mechanical properties

For all practical purposes, we regard the behavior of glass, as for most ceramics, as perfectly elastic and brittle. The range of modulus of elasticity for most commercial glasses is 55–90 GPa (8–13 million psi), and the Poisson's ratio of glasses ranges from 0.16–0.28. The hardness of glasses, as a measure of resistance to scratching, ranges from 5–7 on the Mohs scale, equivalent to a range of approximately 350–500 HK.

Glass in *bulk* form has a strength of less than 140 MPa (20 ksi). The relatively low strength of bulk glass is attributed to the presence of small flaws and microcracks on the surface of the glass, some or all of which may be introduced during normal handling of the glass by inadvertent abrading. These defects reduce the strength of glass by two to three orders of magnitude, compared with its ideal (defect-free) strength. Glasses can be strengthened by thermal or chemical treatments to obtain high strength and toughness.

The strength of glass can theoretically reach as high as 35 GPa (5 million psi). When molten glass is freshly drawn into fibers (*fiberglass*), its tensile strength ranges from 0.2 GPa to 7 GPa (30 ksi to 1000 ksi), with an average value of about 2 GPa (300 ksi). Thus, glass fibers are stronger than steel and are used to reinforce plastics in applications such as boats, automobile bodies, furniture, and sports equipment. The strength of glass is usually measured by bending it. The surface of the glass is first thoroughly abraded (roughened) to ensure that the test gives a reliable strength level in actual service under adverse conditions. The phenomenon of *static fatigue* observed in ceramics (see Section 11.8.2) is also exhibited by glasses. If a glass item must withstand a load for 1000 hours or longer, the maximum stress that can be applied to it is approximately one third the maximum stress that the same item can withstand during the first second of loading.

11.10.3 Physical properties

Glasses have low thermal conductivity and high electrical resistivity and dielectric strength. Their thermal expansion coefficient is lower than those for metals and plastics and may even approach zero. Titanium-silicate glass (a clear, synthetic high-silica glass), for example, has a near-zero coefficient of thermal expansion. Fused silica, a clear, synthetic amorphous silicon dioxide of very high purity, also has a near-zero coefficient of expansion. (See Fig. 11.20.) Optical properties of glasses, such as reflection, absorption, transmission, and refraction, can be modified by varying their composition and treatment.

11.10.4 Glass ceramics

Although glasses are amorphous, *glass ceramics* (such as Pyroceram, a trade name) have a high crystalline component to their microstructure. Glass ceramics contain large proportions of several oxides, and hence their properties are a combination of those for glass and ceramics. Most glass ceramics are stronger than glass. These products are first shaped and then heat treated, with **devitrification** (recrystallization) of the glass occurring. Unlike most glasses, which are clear, glass ceramics are generally white or gray in color.

The hardness of glass ceramics ranges approximately from 520 HK to 650 HK. They have a near-zero coefficient of thermal expansion; hence, they have good thermal shock resistance and high strength, because of the absence of the porosity usually found in conventional ceramics. The properties of glass ceramics can be improved by modifying their composition and by heat-treatment techniques. First developed in 1957, glass ceramics are suitable for cookware, heat exchangers for gas-turbine engines, radomes (housings for radar antenna), and electrical and electronics applications.

11.11 Forming and Shaping Glass

Glass products can generally be categorized as

- Flat sheet or plate, ranging in thickness from about 0.8 mm to 10 mm (0.03 in. to 0.4 in.), such as window glass, glass doors, and glass tabletops;
- Rods and tubing used for chemicals, neon lights, and decorative artifacts;
- Discrete products, such as bottles, vases, headlights, and television tubes;
- Glass fibers to reinforce composite materials and for fiber optics.

All glass forming and shaping processes begin with molten glass, which has the appearance of red-hot viscous syrup, supplied from a melting furnace or tank.

Traditionally, **flat-sheet glass** has been made by drawing or rolling from the molten state and, more recently, by the **float** method, all of which are continuous processes. In the **drawing** process, the molten glass passes through a pair of rolls, similar to an old-fashioned clothes wringer. The solidifying glass is squeezed between the rolls, forming a sheet, which is then moved forward over a set of smaller rolls. In the **rolling** process, the molten glass is squeezed between rollers, forming a sheet. The surfaces of the glass can be embossed with a pattern by shaping the surfaces of the rollers accordingly. Glass sheet produced by drawing and rolling has a rough surface appearance. In making plate glass, both surfaces have to be ground parallel and polished.

In the **float** method (Fig. 11.28), developed in the 1950s, molten glass from the furnace is fed into a bath of molten tin, under controlled atmosphere; the glass floats on the tin bath. The glass then moves over rollers into another chamber (called a *lehr*) and solidifies. *Float glass* has a smooth (*fire-polished*) surface and needs no further grinding or polishing.

Glass **tubing** is manufactured by the process shown in Fig. 11.29. In this process, molten glass is wrapped around a rotating hollow cylindrical or cone-shaped mandrel and is drawn out by a set of rolls. Air is blown through the mandrel to keep the

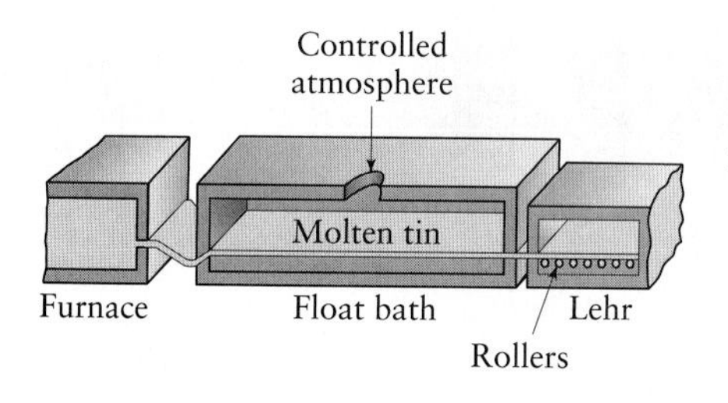

FIGURE 11.28 The float method of forming sheet glass. *Source*: Corning Glass Works.

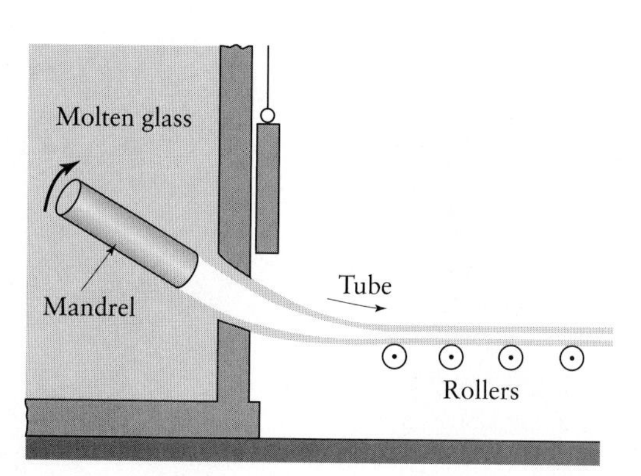

FIGURE 11.29 Manufacturing process for glass tubing. Air is blown through the mandrel to keep the tube from collapsing. *Source*: Corning Glass Works.

glass tube from collapsing. The machines used in this process may be horizontal, vertical, or slanted downward. Glass **rods** are made in a similar manner, but air is not blown through the mandrel; thus, the drawn product becomes a solid rod.

Continuous **fibers** are drawn through multiple (200 to 400) orifices in heated platinum plates, at speeds as high as 500 m/s (1700 ft/s). Fibers as small as 2 μm (80 μin.) in diameter can be produced by this method. In order to protect their surfaces, the fibers are subsequently coated with chemicals. Short glass fibers, used as thermal insulating material (**glass wool**) or for acoustic insulation, are made by a *centrifugal spraying process* in which molten glass is fed into a rotating head.

11.11.1 Making discrete glass products

Several processes are used for making discrete glass objects. These processes include blowing, pressing, centrifugal casting, and sagging.

Blowing. The *blowing* process is used to make hollow, thin-walled glass items, such as bottles and flasks, and is similar to blow molding of thermoplastics. Figure 11.30 shows the steps involved in the production of an ordinary glass bottle by the blowing

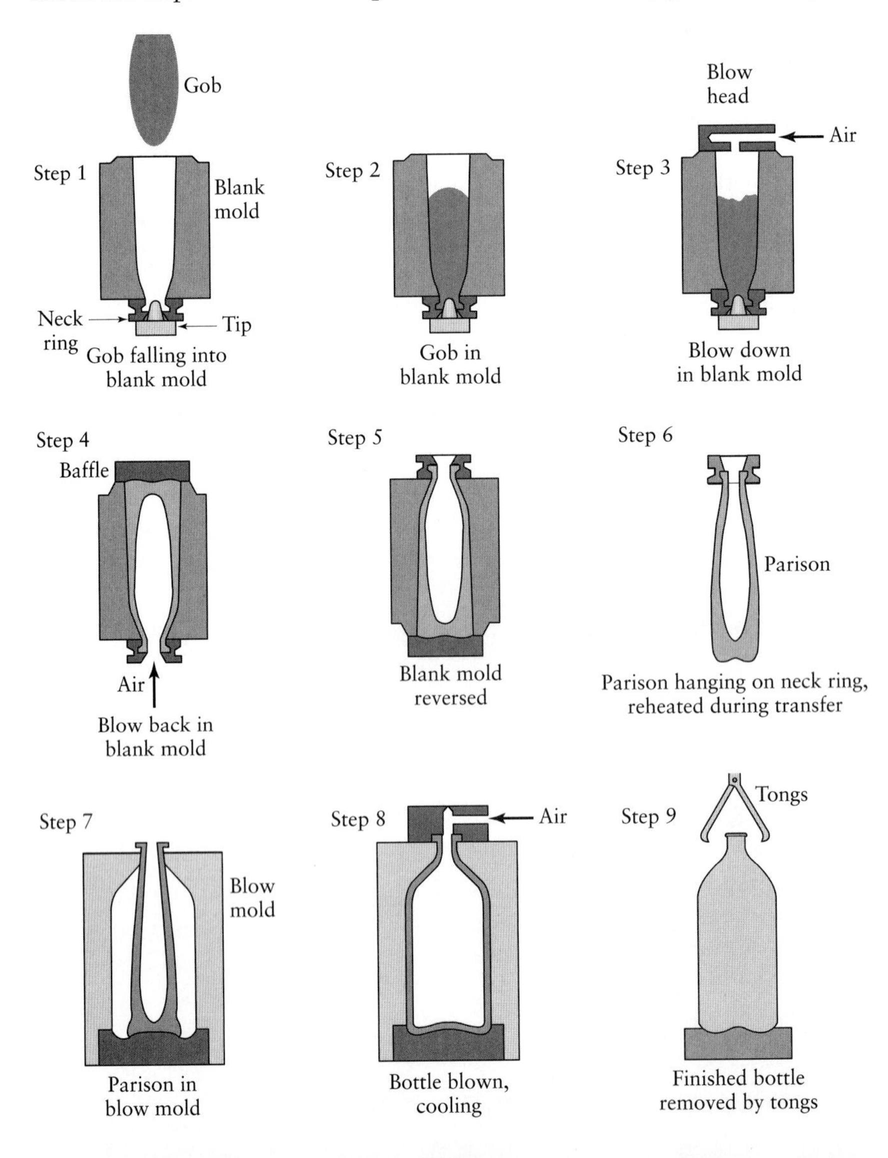

FIGURE 11.30 Stages in manufacturing a common glass bottle. *Source*: F. H. Norton.

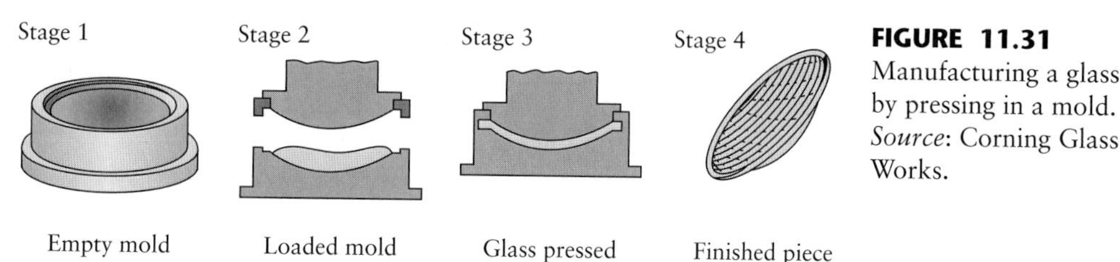

FIGURE 11.31 Manufacturing a glass item by pressing in a mold. *Source*: Corning Glass Works.

process. In this process, blown air expands a hollow gob of heated glass against the walls of a mold. The mold is usually coated with a parting agent, such as oil or emulsion, to prevent the part from sticking to it. The surface finish of products made by the blowing process is acceptable for most applications. Although it is difficult to control the wall thickness of the product, this process is used for high rates of production. Lightbulbs are made in automatic blowing machines, at a rate of over 1000 bulbs per minute.

Pressing. In *pressing*, a gob of molten glass is placed in a mold and is pressed into shape with the use of a plunger. The mold may be made in one piece (Fig. 11.31), or it may be a split mold (Fig. 11.32). After pressing, the solidifying glass acquires the shape of the cavity between the mold and the plunger. Because of the confined environment, the product has greater dimensional accuracy than can be obtained with blowing. However, pressing cannot be used on thin-walled items or for parts such as bottles from which the plunger cannot be retracted.

Centrifugal casting. Also known as *spinning* in the glass industry, the *centrifugal-casting* process is similar to that for metals. In this process, the centrifugal force pushes the molten glass against the mold wall, where it solidifies. Typical products include picture tubes for televisions and missile nose cones.

Sagging. Shallow dish-shaped or lightly embossed glass parts can be made by the *sagging* process. In this process, a sheet of glass is placed over the mold and is heated. The glass sags by its own weight and takes the shape of the mold. The process is similar to thermoforming with thermoplastics, but without pressure or a vacuum. Typical applications include dishes, lenses for sunglasses, mirrors for telescopes, and lighting panels.

11.11.2 Techniques for treating glass

As produced by the methods described in this chapter, glass can be strengthened by thermal tempering, chemical tempering, and laminating. Glass products may also be subjected to annealing and other finishing operations. These treatment techniques are described as follows.

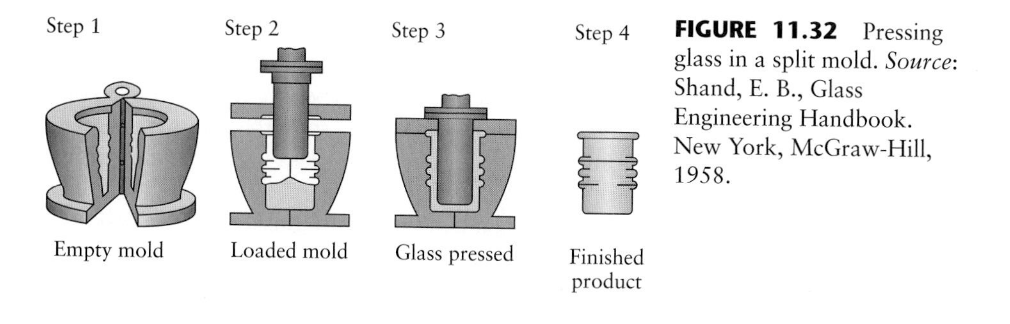

FIGURE 11.32 Pressing glass in a split mold. *Source*: Shand, E. B., Glass Engineering Handbook. New York, McGraw-Hill, 1958.

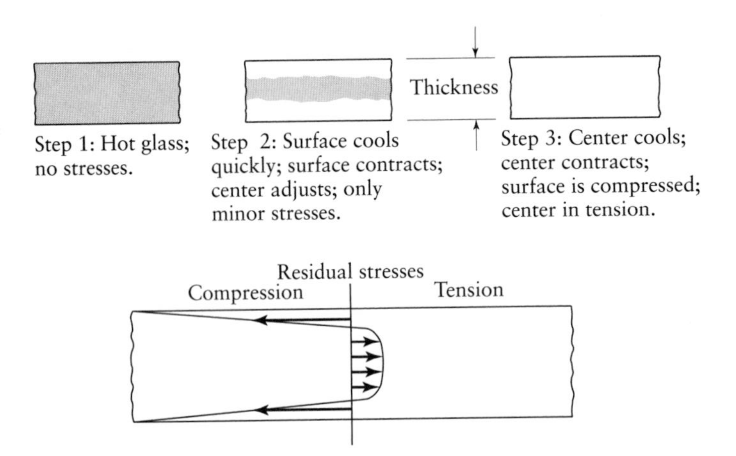

FIGURE 11.33 Residual stresses in tempered glass plate, and stages involved in inducing compressive surface residual stresses for improved strength.

a) In **thermal tempering** (also called *physical tempering* or *chill tempering*), the surfaces of the hot glass are cooled rapidly (Fig. 11.33). As a result, the surfaces shrink, and because the bulk is still hot, tensile stresses develop on the surfaces. As the bulk of the glass begins to cool, it contracts. The solidified surfaces are now forced to contract, thus developing residual compressive surface stresses and interior tensile stresses. Compressive surface stresses improve the strength of the glass, as they do in other materials. Note that the higher the coefficient of thermal expansion of the glass and the lower its thermal conductivity, the higher is the level of residual stresses developed and, hence, the stronger the glass becomes. Thermal tempering takes a relatively short time (measured in minutes) and can be applied to most glasses. Because of the large amount of energy stored from residual stresses, **tempered glass** shatters into a large number of pieces when broken.
b) **In chemical tempering**, the glass is heated in a bath of molten KNO_3, K_2SO_4, or $NaNO_3$, depending on the type of glass. Ion exchange takes place, with larger atoms replacing the smaller atoms on the surface of the glass. As a result, residual compressive stresses develop on the surface. This condition is similar to that created by forcing a wedge between two bricks in a wall. The time required for chemical tempering is longer (about one hour) than for thermal tempering. Chemical tempering may be performed at various temperatures. At low temperatures, distortion of the part is minimal, and complex shapes can be treated. At elevated temperatures, there may be some distortion of the part, but the product can be used at higher temperatures without loss of strength.
c) Another strengthening method is to make **laminated glass**, which is two pieces of flat glass with a thin sheet of tough plastic between them; the process is also called **laminate strengthening**. When laminated glass is broken, its pieces are held together by the plastic sheet. You probably have seen this phenomenon in a shattered automobile windshield.

Finishing operations. As in metal products, residual stresses can develop in glass products if the products are not cooled slowly enough. In order to ensure that the product is free from these stresses, it is *annealed* by a process similar to stress-relief annealing of metals. In this process, the glass is heated to a certain temperature and cooled gradually. Depending on the size, thickness, and type of glass, annealing times may range from a few minutes to as long as 10 months, as in the case of a 600-mm (24-in.) mirror for a telescope. In addition to annealing, glass products may be subjected to further operations, such as cutting, drilling, grinding, and polishing. Sharp edges and corners can be smoothened by grinding, as in glass tops for desks and shelves, or by holding a torch against the edges (**fire polishing**), which rounds them by localized softening and surface tension.

11.12 Design Considerations for Ceramics and Glasses

Ceramic and glass products require careful selection of composition, processing methods, finishing operations, and methods of assembly into other components. Limitations such as general lack of tensile strength, sensitivity to defects, and low impact toughness are important. These limitations have to be balanced against desirable characteristics, such as their hardness, resistance to scratching, compressive strength at room and elevated temperatures, and diverse physical properties. Control of processing parameters and the quality of and level of impurities in the raw materials used are important. As in all design decisions, there are priorities, limitations, and various factors that should be considered, including the number of parts needed and the costs of tooling, equipment, and labor.

The possibilities of dimensional changes, warping and cracking during processing are significant factors in selecting methods for shaping these materials. When a ceramic or glass component is part of a larger assembly, compatibility with other components is another important consideration. Particularly important in such cases are thermal expansion (as in seals) and the type of loading. The potential consequences of part failure are always a significant factor in designing ceramic products.

11.13 Graphite and Diamond

11.13.1 Graphite

Graphite is a crystalline form of carbon that has a *layered structure* of basal planes or sheets of close-packed carbon atoms. Consequently, graphite is weak when sheared along the layers. This characteristic, in turn, gives graphite its low frictional properties as a solid lubricant. However, its frictional properties are low only in an environment of air or moisture; graphite is abrasive and a poor lubricant in a vacuum.

Although brittle, graphite has high electrical and thermal conductivity and resistance to thermal shock and high temperature, although it begins to oxidize at 500°C, (930°F). It is therefore an important material for applications such as electrodes, heating elements, brushes for motors, high-temperature fixtures and furnace parts, mold materials such as crucibles for melting and casting of metals, and seals (because of its low friction and high wear resistance). Unlike other materials, the strength and stiffness of graphite increases with temperature.

An important use of graphite is as **fibers** in composite materials and reinforced plastics (Chapter 10). A characteristic of graphite is its resistance to chemicals; hence, it is used as filters for corrosive fluids. Furthermore, its low thermal-neutron-absorption cross-section and high scattering cross-section make graphite suitable for nuclear applications. Ordinary pencil "lead" is a mixture of graphite and clay.

Graphite is generally graded in terms of decreasing order of grain size: *industrial*, *fine grain*, and *micrograin*. As with ceramics, the mechanical properties of graphite improve with decreasing grain size. Micrograin graphite can be impregnated with copper and is used as electrodes for electrical-discharge machining and for furnace fixtures. Amorphous graphite is known as *lampblack* (black soot) and is used as a pigment. Graphite is usually processed by molding or forming, oven baking, and then machining to the final shape. It is available commercially in square, rectangular, or round shapes of various sizes.

A more recent development is the production of soccer-ball-shaped carbon molecules, called *buckyballs* (from *Buckminster*, after Buckminster Fuller, 1895–1983, inventor of the geodesic dome). Also called **fullerenes** (after *Fuller*), these chemically inert spherical molecules are produced from soot and act much like solid-lubricant particles. Fullerenes become superconductors when mixed with metals. Another development is **microcellular carbon foam**, which has uniform porosity and isotropic-strength characteristics. It is used, for example, as reinforcing components in aerospace structures that can be shaped directly.

11.13.2 Diamond

The second principal form of carbon is *diamond*, which has a covalently bonded structure. It is the hardest substance known (7000 HK to 8000 HK). This characteristic makes diamond an important cutting-tool material (Section 8.6.9), either as a single crystal or in polycrystalline form, as an abrasive in grinding wheels (Section 9.2); and for dressing of grinding wheels (sharpening of abrasive grains). Diamond is also used as a die material for drawing thin wire, with a diameter of less than 0.06 mm (0.0025 in.). Diamond is brittle and it begins to decompose in air at about 700°C (1300°F); in nonoxidizing environments it resists higher temperatures.

Synthetic diamond, or **industrial diamond**, was first made in 1955 and is used extensively for industrial applications. One method of manufacturing it is to subject graphite to a hydrostatic pressure of 14 GPa (2 million psi) and a temperature of 3000°C (5400°F). Synthetic diamond is identical to natural diamond for industrial applications and has superior properties because of its lack of impurities. It is available in various sizes and shapes, the most common abrasive grain size being 0.01 mm (0.004 in.) in diameter. *Gem-quality synthetic diamond* is now being made with electrical conductivity 50 times higher than that for natural diamond and 10 times more resistance to laser damage; potential applications include heat sinks for computers, products in the telecommunications and integrated-circuit industries, and windows for high-power lasers. Diamond particles can also be *coated* (with nickel, copper, or titanium) for improved performance in grinding operations.

A more recent development is **diamondlike carbon** (DLC). It is used as a coating, as described in Section 4.5.1.

11.14 Processing Metal-Matrix and Ceramic-Matrix Composites

New developments in composite materials are continually taking place, with a wide range and form of polymeric, metallic, and ceramic materials being used both as fibers and as matrix materials. Research and development activities in this area are concerned with improving strength, toughness, stiffness, resistance to high temperatures, and reliability in service.

11.14.1 Metal-matrix composites

The advantage of a metal matrix over a polymer matrix is the former's higher resistance to elevated temperatures and higher ductility and toughness. The limitations of the metal matrix are higher density and greater difficulty in processing components. The matrix materials in these composites are usually aluminum, aluminum–lithium, magnesium, and titanium, although other metals are also being investigated for this purpose. The fiber materials are graphite, aluminum oxide, silicon carbide, and boron, with beryllium and tungsten as other possibilities.

Because of their high specific stiffness, light weight, and high thermal conductivity, boron fibers in aluminum matrix have been used for structural tubular supports

TABLE 11.10

Metal-Matrix Composite Materials and Typical Applications

Fiber	Matrix	Typical Applications
Graphite	Aluminum	Satellite, missile, and helicopter structures
	Magnesium	Space and satellite structures
	Lead	Storage-battery plates
	Copper	Electrical contacts and bearings
Boron	Aluminum	Compressor blades and structural supports
	Magnesium	Antenna structures
	Titanium	Jet-engine fan blades
Alumina	Aluminum	Superconductor restraints in fusion power reactors
	Lead	Storage-battery plates
	Magnesium	Helicopter transmission structures
Silicon carbide	Aluminum, titanium	High-temperature structures
	Superalloy (cobalt base)	High-temperature engine components
Molybdenum, tungsten	Superalloy	High-temperature engine components

in the Space Shuttle Orbiter. Other applications include bicycle frames and sporting goods. Studies of techniques for optimum bonding of fibers to the metal matrix are in progress. Current applications of metal-matrix composites are in gas turbines, electrical components, and various structural components (Table 11.10).

There are three methods of manufacturing these composites into near-net-shape parts: liquid-phase processes, solid-phase processes, and two-phase (liquid–solid) processes. These methods are described as follows:

a) **Liquid-phase processing** consists basically of casting the liquid matrix and the solid reinforcement, using either conventional casting processes or pressure-infiltration casting. In the latter process, pressurized gas is used to force the liquid matrix metal into a perform (usually as sheet or wire) made of the reinforcing fibers.

b) **Solid-phase processing** consists basically of powder-metallurgy techniques, including cold and hot isostatic pressing. Proper mixing for homogeneous distribution of the fibers is important. A sample use of this technique, employed in tungsten-carbide tool and die manufacturing with cobalt as the matrix material, is discussed in Example 11.6. In making complex MMC parts with whisker or fiber reinforcement, die geometry and control of process variables are very important in ensuring proper distribution and orientation of the fibers within the part. MMC parts made by powder-metallurgy processes are generally heat treated for optimum properties.

c) The techniques used for **two-phase processing** consist of *rheocasting*, described in Section 5.10.6, and *spray atomization and deposition*. In the latter processes, the reinforcing fibers are mixed with a matrix that contains both liquid and solid phases.

EXAMPLE 11.9 Metal-matrix-composite brake rotors and cylinder liners

Brake rotors are currently being made of composites consisting of an aluminum matrix with 20% silicon-carbide (SiC) particles. The particles are stirred into molten aluminum alloys and cast into ingots. The ingots are then remelted and cast into shapes, such as brake rotors and drums, using processes such as green sand, bonded sand, investment, permanent mold, and squeeze casting. (See Chapter 5.) The

MMC brake rotors weigh about half as much as do gray-cast-iron rotors, conduct heat three times faster, exhibit the stiffness and wear-resistance of ceramics, and reduce noise and vibration.

To improve the wear- and heat-resistance of cast-iron cylinder liners in aluminum engine blocks, aluminum-matrix liners are being developed. The MMC layer consists of 12% aluminum-oxide (Al_2O_3) fiber and 9% graphite fiber, in a thickness that ranges from 1.5 mm to 2.5 mm (0.06 in. to 0.1 in.).

11.14.2 Ceramic-matrix composites

Ceramic-matrix composites are another important recent development in engineered materials. As described earlier, ceramics are strong and stiff and resist high temperatures, but generally lack toughness. Silicon carbide, silicon nitride, aluminum oxide, and mullite (a compound of aluminum, silicon, and oxygen) are new matrix materials that retain their strength to 1700°C (3100°F). Under development are carbon–carbon-matrix composites that retain much of their strength up to 2500°C (4500°F), although they lack oxidation resistance at high temperatures. Present applications for ceramic-matrix composites are in jet and automotive engines, equipment for deep-sea mining, pressure vessels, and various structural components.

There are several processes used to make these composites; three commons ones are briefly described next. In addition to these processes, new techniques such as melt infiltration, controlled oxidation, and hot-press sintering, which are still largely in the experimental stage, are being developed for improving the properties and performance of these composites. The three common processes for making ceramic-matrix composites are described as follows:

a) **Slurry infiltration**, the most common process, involves the preparation of a fiber preform that is hot pressed and then impregnated with a slurry that contains the matrix powder, a carrier liquid, and an organic binder. A further improvement is *reaction bonding* or *reaction sintering* of the slurry. High strength, toughness, and uniform structure are obtained by the slurry-infiltration process, but the product has limited high-temperature properties, due to the low melting temperature of the matrix materials used.

b) **Chemical-synthesis** processes involve the solgel and polymer-precursor techniques. In the *solgel* process, a sol (a colloidal fluid with the liquid as the continuous phase) containing fibers is converted to a gel, which is then subjected to heat treatment to produce a ceramic-matrix composite. The *polymer-precursor* method is analogous to the process used in making ceramic fibers.

c) In **chemical-vapor infiltration**, a porous fiber preform is infiltrated with the matrix phase, using the chemical vapor deposition technique (see Section 4.5.1). The product has very good high-temperature properties, but the process is costly and time consuming.

11.14.3 Other composites

Composites may also consist of coatings of various kinds on base metals or substrates. Examples include plating of aluminum and other metals over plastics for decorative purposes and enamels, dating to before 1000 BC, or applying similar vitreous (glasslike) coatings on metal surfaces for various functional or ornamental purposes.

Composites such as cemented carbides, usually tungsten carbide and titanium carbide, with cobalt and nickel, respectively, as a binder, are made into tools and dies. Other composites include grinding wheels made of aluminum-oxide, silicon-carbide, diamond, or cubic-boron-nitride abrasive particles, held together with various organic, inorganic, or metallic binders. Another category of composites is cermets.

A composite developed relatively recently is granite particles in an epoxy matrix. (See Section 8.12.) It has high strength, good vibration-damping capacity (better than that of gray cast iron), and good frictional characteristics. This composite is used as machine-tool beds for some precision grinders.

11.15 Processing Superconductors

Although superconductors (Section 3.9.6) have major energy-saving potential in the generation, storage, and distribution of electrical power, their processing into useful shapes and sizes for practical applications presents significant difficulties. Two basic types of superconductors are metals [**low-temperature superconductors** (LTSCs), including combinations of niobium, tin, and titanium] and ceramics [**high-temperature superconductors** (HTSCs), including various copper oxides]. Here, *high temperature* means closer to *ambient* temperature, hence the HTSCs are of more practical use.

Ceramic superconducting materials are available in powder form. The fundamental difficulty in manufacturing them is their inherent brittleness and anisotropy, which make it difficult to align the grains in the proper direction for high efficiency; the smaller the grain size, the more difficult it is to align the grains.

The basic manufacturing process for superconductors consists of the following steps:

a) Preparing the powder, mixing it, and grinding it in a ball mill to a grain size of 0.5 mm to 10 mm,
b) Forming it into shape, and
c) Heat treating to improve grain alignment.

The most common forming process is **oxide powder in tube** (OPIT). In this process, the powder is packed into silver tubes (because silver has the highest electrical conductivity of all metals) and sealed at both ends. The tubes are then mechanically worked, by such deformation processes as swaging, drawing, extrusion, isostatic pressing, and rolling, into final shapes, which may be wire, tape, coil, or bulk.

Other principal superconductor-shaping processes include (1) coating of silver wire with superconducting material, (2) deposition of superconductor films by laser ablation, (3) the doctor-blade process (see Section 11.9.1), (4) explosive cladding (see Section 12.10), and (5) chemical spraying.

CASE STUDY | Production of High-Temperature Superconducting Tapes

In recent years, significant progress has been made in understanding high-temperature superconducting materials and their potential use as conductors. Two bismuth-based oxides ($Bi_2Sr_2Ca_1Cu_2O_x$, also known as Bi-2212, and $Bi_2Sr_2Ca_2Cu_3O_x$, or Bi-2223) are superconducting, ceramic materials of choice for various military and commercial applications, such as electrical propulsion for ships and submarines, shallow-water and ground minesweeping systems, transmission cable generators, and superconducting magnetic-energy storage (SMES). However, wide variations in the electrical performance of test magnet coils using *monofilament* superconducting strips indicate that *multifilament* conductors are essential in reliably achieving the required critical current densities, so that property variations are averaged over a length. Tape stacking is a new approach to manufacturing multifilament tape. In this method, 7 to 10 monofilament tapes or strips are stacked, heated, and pressed together; the resulting laminate is then rolled to final thickness.

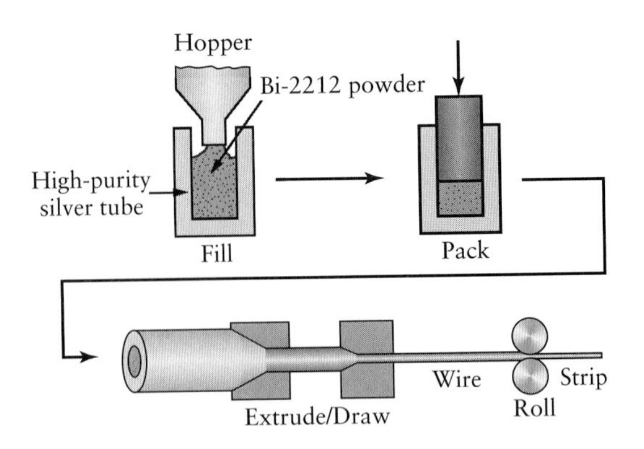

FIGURE 11.34 Schematic illustration of the powder-in-tube process. *Source*: Copyright © Concurrent Technologies Corporation (CTC) 1998 used with permission.

The quality and performance of multifilament tapes are directly related to that of the constituents. A variety of processes have been explored to produce wires and tapes, using Bi-2212 and Bi-2223. The powder-in-tube process (Fig. 11.34) has been used successfully to fabricate long lengths of Bi-based wires and tapes with desired properties. This case study discusses a prior implementation of this method for the production of high-temperature superconducting multifilament tapes. Although this case study concerns only Bi-2212, similar processing can be used for Bi-2223.

Production of the monofilament tapes involves the following steps:

a) Production of a composite billet, using a silver casing and Bi-2212 powder;
b) Extrusion and drawing of the billet to reduce the diameter and increase the powder density;
c) Rolling to final dimensions.

Composite-billet preparation. The composite billets were approximately 254 mm (10 in.) long, with an outside diameter of 6.35 mm (0.25 in.), and the wall thickness of the silver tube was 1 mm. The as-received high-purity silver tubes were cleaned and annealed at 400°C (750°F) for one hour to improve ductility. They were subsequently filled with fine-grained Bi-2212 powder (with a median particle size of 2.78 μm, or 110 μin.) that had been baked at 105°C (220°F) for an hour to remove moisture. The powder was poured into the silver container in an inert atmosphere and compacted to a 30% relative density, using a steel ram. In order to minimize density gradients (see Fig. 11.7), about one gram of powder was added to the billet for each stroke of the ram. Each billet was weighed and measured to verify the initial packing density. The ends of each billet were then sealed with a silver alloy to avoid contamination during subsequent deformation processing.

Wire drawing. A swaging machine was used for preparing the end of the composite wire for drawing. To measure elongation, two circumferential score markers, 25.4 mm apart, were placed on each billet and measured after each pass. Billets were drawn down to a final diameter of 1.63 mm on a drawing bench, using 12 passes of 20.7% reduction per pass. The drawing speed was approximately 1.4 m/min, and the dies had a semicone angle of 8°. A semisoluble oil and zinc-stearate spray were used as a lubricant.

Figure 11.35a shows the evolution of the silver-to-oxide (area) ratio with the drawing pass number while Fig. 11.35b shows the same for the powder density. Note that both quantities increase rapidly during the initial passes and tend to saturate towards the end of the process.

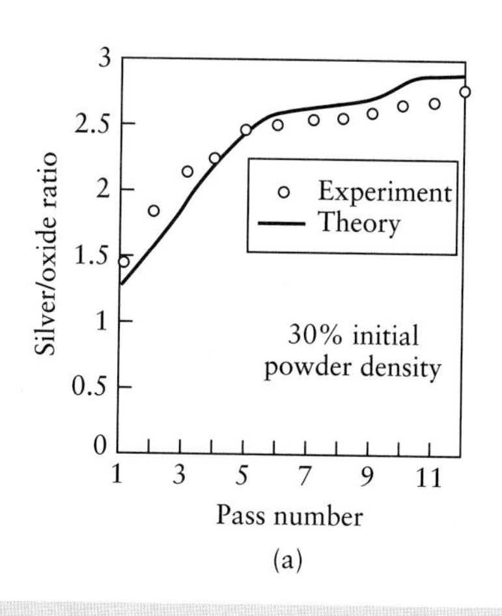

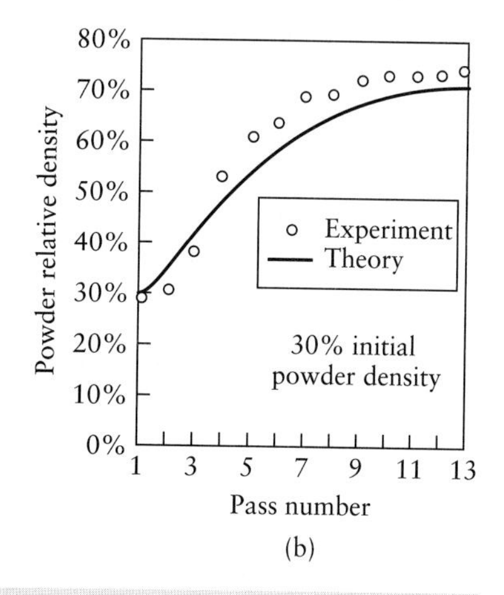

FIGURE 11.35 (a) Silver/oxide ratio as a function of drawing pass number. (b) Evolution of powder relative density with pass number. *Source:* Concurrent Technologies Corporation.

In addition to measuring these quantities directly from tests, a process model was developed using the finite-element method. This model enables prediction of stresses, strains and other quantities of interest resulting from the complex deformation the material undergoes. In the current case, it was used to model both the wire-drawing and tape-rolling (described later) processes. The evolution of the powder relative density and silver-to-oxide ratio obtained from the computer-based models is shown in Fig. 11.35. The models accurately predict these quantities, demonstrating their potential to conduct parametric studies and prescribe an optimum reduction schedule to achieve the desired makeup. The model can also be used to study the influence of die geometry, friction conditions, initial powder density, powder type, and initial silver tube thickness on dependent variables such as the drawing force, wire geometry, final core density, potential for sheath fracture, and mechanical properties of the superconducting wire. This information can reduce the number of experimental trial-and-error attempts needed for optimizing the process, thereby reducing the cost of the process.

Rolling. Following the drawing process, the wire was progressively transformed into tape, using a single-stand rolling mill in two-high and four-high configurations. (See Figs. 6.42a and c.) For the four-high case, the diameter of the backup rolls (which are the work rolls for the two-high configuration) was 213 mm, and the diameter of the work rolls was 63.5 mm. The mill was also equipped with payoff and take-up reel systems to maintain front and back tension, respectively, on the tape. Short samples were cut after each rolling pass for metallographic examination, including optical micrographs of polished, longitudinal, and transverse cross-sections, as shown in Fig. 11.36. The final tape was 100–200 μm thick and 2–3 mm wide, with a Bi-2212 core ranging from 40–80 μm in thickness and 1.0–1.5 mm in width.

The results depicted in Figs. 11.36 reflect a manufacturing sequence that was experimentally optimized with guidance from mathematical models. The performance of the high-temperature superconductor is strongly affected by the density distribution of the oxide core, the texture of the core, the integrity of the silver–oxide interface, the absence of microcracking, and variations in filament thickness (sausaging), all of which are strongly dependent on the work-roll diameters, rolling-reduction schedule, and frictional conditions.

Typically, the oxide's microstructure is dense, elongated, and well aligned only in the high-shear regions near the silver–oxide interface. The key to producing high-J_c tapes is to enhance the texture/grain structure and reduce the porosity in the oxide core during wire drawing and tape rolling. This case study has characterized four rolling schedules: two-high and four-high rolling configurations and

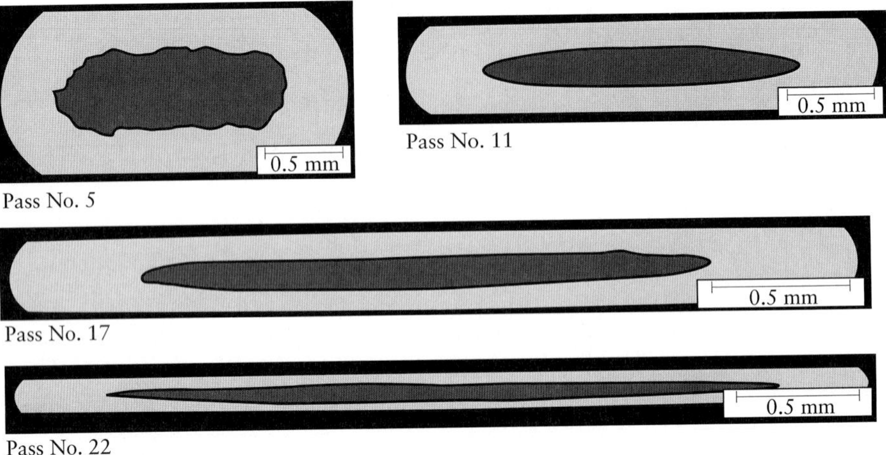

FIGURE 11.36 Selected cross-sections of rolled filament. *Source*: Copyright © Concurrent Technologies Corporation (CTC) 1998 used with permission.

10% and 25% reductions per pass. The longitudinal sections of tapes from the four rolling schedules show that the four-high + 10%-reduction schedule has the smoothest silver–oxide interface.

Source: Courtesy of S. Vaze and M. Pandheeradi, Concurrent Technologies Corporation.

SUMMARY

- The powder-metallurgy process is capable of economically producing relatively complex parts in net-shape form to close dimensional tolerances, from a wide variety of metal and alloy powders. (Section 11.1)
- The steps in powder-metallurgy processing are powder production, blending, compaction, sintering, and additional processing to improve dimensional accuracy, surface finish, mechanical or physical properties, or appearance. (Sections 11.2– 11.5)
- Design considerations in powder metallurgy include the shape of the compact, the ejection of the green compact from the die, and the acceptable dimensional tolerances of the application. (Section 11.6)
- From an economic standpoint, the powder-metallurgy process is suitable for medium- to high-volume production runs and for relatively small parts; it has certain competitive advantages over other processes. (Sections 11.7)
- Ceramics have typical characteristics such as brittleness, high strength and hardness at elevated temperatures, high modulus of elasticity, low toughness, low density, low thermal expansion, and low thermal and electrical conductivity. (Section 11.8)
- Three basic shaping processes for ceramics are casting, plastic forming, and pressing; the resulting product is dried and fired to give it the proper strength. Finishing operations, such as machining and grinding, may be performed to give the part its final shape, or the part may be subjected to surface treatments to improve specific properties. (Section 11.9)
- Almost all commercial glasses are categorized as one of six types that indicate their composition. Glass in bulk form has relatively low strength, but it can be strengthened by thermal or chemical treatments to obtain high strength and toughness. (Section 11.10)
- Continuous methods of glass processing are drawing, rolling, and floating. Discrete glass products can be manufactured by blowing, pressing, centrifugal casting, and

sagging. After initial processing, glasses can be strengthened by thermal or chemical tempering or by laminating. (Section 11.11)

- Design considerations for ceramics and glasses are guided by factors such as general lack of tensile strength and toughness, and sensitivity to external and internal defects. Warping and cracking are important considerations, as are the methods employed for production and assembly. (Section 11.12)
- Graphite, fullerenes, and diamond are forms of carbon that display unusual combinations of properties. These materials have unique and emerging applications. (Section 11.13)
- Metal-matrix and ceramic-matrix composites possess useful combinations of properties and have increasingly broad applications. Metal-matrix composites are processed via liquid-phase, solid-phase, and two-phase processes; ceramic-matrix composites can be processed via slurry infiltration, chemical synthesis, and chemical-vapor infiltration. (Sections 11.14)
- Manufacture of superconductors into useful products is a challenging area, because of the anisotropy and inherent brittleness of the materials involved. Although other processes are being developed, the basic and common practice involves packing the powder into a silver tube and deforming it plastically into desired shapes. (Section 11.15)

SUMMARY OF EQUATIONS

Shape factor of particles, $k = \left(\frac{A}{V}\right)D_{eq}$

Pressure distribution in compaction, $p_x = p_0 e^{-4\mu kx/D}$

Volume, $V_{sint} = V_{green}\left(1 - \frac{\Delta L}{L_0}\right)^3$

Density, $\rho_{green} = \rho_{sint}\left(1 - \frac{\Delta L}{L_0}\right)^3$

Tensile strength, $\text{UTS} \simeq \text{UTS}_0 e^{-nP}$

Elastic modulus, $E \simeq E_0(1 - 1.9P + 0.9P^2)$

Thermal conductivity, $k = k_0(1 - P)$

BIBLIOGRAPHY

ASM Engineered Materials Handbook, Desk Edition, ASM International, 1995.

ASM Handbook, Vol. 7: *Powder Metal Technologies and Applications*, ASM International, 1998.

ASM Handbook, Vol. 21: *Composites*, ASM International, 2001.

Barsoum, M. W., *Fundamentals of Ceramics*, McGraw-Hill, 1996.

Belitskus, D. L., *Fiber and Whisker Reinforced Ceramics for Structural Applications*, Dekker, 1993.

Brook, R. J., (ed.), *Concise Encyclopedia of Advanced Ceramic Materials*, Pergamon, 1991.

Buchanan, R. C., *Ceramic Materials for Electronics: Processing, Properties, and Applications*, 2d. ed., Dekker, 1991.

Burns, M., *Automated Fabrication: Improving Productivity in Manufacturing*, PTR Prentice Hall, 1993.

Chawla, K. K., *Composite Materials*, 2d. ed., Springer, 1998.

Cheremisinoff, N. P., *Handbook of Ceramics and Composites*, 3 vols., Dekker, 1991.

Chiang, Y.-M., *Physical Ceramics: Principles for Ceramics Science and Engineering*, Wiley, 1995.

Clyne, T. W., and P. J. Withers, *An Introduction to Metal Matrix Composites*, Cambridge University Press, 1993.

Concise Encyclopedia of Advanced Ceramics, The MIT Press, 1991.

Cranmer, D. C., and D. W. Richerson, *Mechanical Testing Methodology for Ceramic Design and Reliability*, Dekker, 1998.

Doresmus, R. H., *Glass Science*, 2d. ed., Wiley, 1994.

Dowson, G., *Powder Metallurgy: The Process and Its Products*, Adam Hilger, 1991.

Engineered Materials Handbook, Vol. 4: *Ceramics and Glasses*, ASM International, 1991.

German, R. M., *Powder Metallurgy Science*, 2d. ed., Metal Powder Industries Federation, 1994.

________, *Sintering Technology and Practice*, Wiley, 1996.

German, R. M., and A. Bose, *Injection Molding of Metals and Ceramics*, Metal Powder Industries Federation, 1997.

Gutowski, T. G. (ed.), *Advanced Composites Manufacturing*, Wiley, 1997.

Harper, C. A. (ed), *Handbook of Ceramics, Glasses, and Diamonds*, McGraw-Hill, 2001.

Hausner, H. H., and M. K. Mal, *Handbook of Powder Metallurgy*, 2d. ed., Chemical Publishing Co., 1982.

Hlavac, J., *The Technology of Glass and Ceramics*, Elsevier, 1983.

Hoa, S. V., *Computer-Aided Design for Composite Structures*, Dekker, 1996.

Holand, W., and G. H. Beall, *Design and Properties of Glass-Ceramics*, American Chemical Society, 2001.

Jahanmir, S., *Friction and Wear of Ceramics*, Dekker, 1994.

Jones, J. T., and M. F. Berard, *Ceramics—Industrial Processing and Testing*, 2d. ed., Iowa State University Press, 1993.

Karlsson, L. (ed.), *Modeling in Welding, Hot Powder Forming and Casting*, ASM International, 1997.

Kelly, A. (ed.), *Concise Encyclopedia of Composite Materials*, Pergamon, 1989.

King, A. G., *Ceramics Processing and Technology*, Noyes Pub., 2001.

Kingery, W. D., H. K. Bowen, and D. R. Uhlmann, *Introduction to Ceramics*, 2d. ed., Wiley, 1976.

Kubicki, B., *Sintered Machine Elements*, Prentice Hall, 1995.

Kuhn, H. A., and B. L. Ferguson, *Powder Forging*, Metal Powder Industries Federation, 1990.

Lawley, A., *Atomization: The Production of Metal Powders*, Metal Powder Industries Federation, 1992.

Lenel, F. V., *Powder Metallurgy: Principles and Applications*, Metal Powder Industries Federation, 1980.

Lewis, M. H. (ed.), *Glasses and Glass-Ceramics*, Chapman & Hall, 1989.

Loewenstein, K. L., *The Manufacturing Technology of Continuous Glass Fibers*, 2d. ed., Elsevier, 1983.

Lu, H. Y., *Introduction to Ceramic Science*, Dekker, 1996.

Mallick, P. K. (ed.), *Composites Engineering Handbook*, Dekker, 1997.

Nanoceramics, Institute of Materials, 1993.

Norton, F. H., *Elements of Ceramics*, 2d. ed., Addison-Wesley, 1974.

Ochiai, S., *Mechanical Properties of Metallic Composites*, Dekker, 1994.

Phillips, G. C., *A Concise Introduction to Ceramics*, Van Nostrand Reinhold, 1991.

Powder Metallurgy Design Guidebook, American Powder Metallurgy Institute, revised periodically.

Powder Metallurgy Design Manual, 3d. ed., Metal Powder Industries Federation, 1998.

Prelas, M. A., G. Popovici, and L. K. Bigelow (eds.), *Handbook of Industrial Diamonds and Diamond Films*, Dekker, 1997.

Rahaman, M. N., *Ceramic Processing Technology and Sintering*, Dekker, 1996.

Reed, J. S., *Principles of Ceramics Processing*, 2d. ed., Wiley, 1995.

Rhodes, M. J. (ed.), *Principles of Powder Technology*, Wiley, 1990.

Richerson, D. W., *Modern Ceramic Engineering: Properties, Processing, and Use in Design*, 2d. ed., Dekker, 1992.

Schwartz, M. M., *Handbook of Composite Ceramics*, McGraw-Hill, 1992.

________, *Handbook of Structural Ceramics*, McGraw-Hill, 1992.

Thummler, F., and K. Obercker, *An Introduction to Powder Metallurgy*, Institute of Materials, 1994.

Tooley, F. V. (ed.), *The Handbook of Glass Manufacture*, Ashlee Pub. Co., 1984.

Upadhyaya, G. S., *Sintering Metallic and Ceramic Materials: Preparation, Properties and Applications*, Wiley, 2000.

Vincenzini, P., *Fundamentals of Ceramic Engineering*, Elsevier, 1991.

Wilks, J., and E. Wilks, *Properties and Applications of Diamond*, Butterworth-Heinemann, 1991.

Yosomiya, R., *Adhesion and Bonding in Composites*, Dekker, 1990.

QUESTIONS

Powder metallurgy

11.1 Explain the advantages of blending metal powders.

11.2 Is green strength important in powder-metal processing? Explain.

11.3 Explain why density differences may be desirable within some P/M parts. Give specific examples.

11.4 Give the reasons that injection molding of metal powders has become an important process.

11.5 Describe the events that occur during sintering.

11.6 What is mechanical alloying, and what are its advantages over conventional alloying of metals?

11.7 It is possible to infiltrate P/M parts with various resins, as well as with metals. What possible benefits would result from infiltration?

11.8 What concerns would you have when electroplating P/M parts?

11.9 Describe the effects of different shapes and sizes of metal powders in P/M processing, commenting on the magnitude of the shape factor (SF) of the particles.

11.10 Comment on the shapes of the curves and their relative positions shown in Fig. 11.6.

11.11 Should green compacts be brought up to the sintering temperature slowly or rapidly? Explain the advantages and limitations of both methods.

11.12 Explain the effects of using fine powders and coarse powders, respectively, in making P/M parts.

11.13 Are the requirements for punch and die materials in powder metallurgy different than those for forging and extrusion, described in Chapter 6? Explain.

11.14 Describe the relative advantages and limitations of cold and hot isostatic pressing, respectively.

11.15 Why do mechanical and physical properties depend on the density of P/M parts? Explain with appropriate sketches.

11.16 What type of press is required to compact parts by the set of punches shown in Fig. 11.7d? (See Section 6.2.8.)

11.17 Explain the difference between impregnation and infiltration. Give some applications of each.

11.18 Tool steels are now being made by P/M techniques. Explain the advantages of this method over traditional methods such as casting and subsequent metal-working techniques.

11.19 Why do the compacting pressure and sintering temperature depend on the type of powder metal used?

Ceramics and other materials

11.20 Compare the major differences between ceramics, metals, thermoplastics, and thermosets.

11.21 Explain why ceramics are weaker in tension than in compression.

11.22 Explain why the mechanical and physical properties of ceramics decrease with increasing porosity.

11.23 What engineering applications could benefit from the fact that, unlike metals, ceramics generally maintain their modulus of elasticity at elevated temperatures?

11.24 Explain why the mechanical-property data in Table 11.7 have such a broad range. What is the significance of this range in engineering applications?

11.25 List the factors that you would consider when replacing a metal component with a ceramic component. Give examples of such substitutions.

11.26 How are ceramics made tougher? Explain.

11.27 Describe applications in which static fatigue can be important.

11.28 Explain the difficulties involved in making large ceramic components. What recommendations would you make to improve the process?

11.29 Explain why ceramics are effective cutting-tool materials. Would ceramics also be suitable as die materials for metal forming? Explain.

11.30 Describe applications in which the use of a ceramic material with a zero coefficient of thermal expansion would be desirable.

11.31 Give reasons for the development of ceramic-matrix components. Name some present and possible future applications of these components.

11.32 List the factors that are important in drying ceramic components, and explain why they are important.

11.33 It has been stated that the higher the coefficient of thermal expansion of the glass and the lower the glass' thermal conductivity, the higher is the level of residual stresses developed during processing. Explain why.

11.34 What types of finishing operations are performed on ceramics? Why are they done?

11.35 What should be the property requirements for the metal balls used in a ball mill? Explain why these conditions are required.

11.36 Which properties of glasses allow them to be expanded and shaped into bottles by blowing?

11.37 What properties should plastic sheet have when used in laminated glass? Why?

11.38 Consider some ceramic products that you are familiar with, and outline a sequence of processes performed to manufacture each of them.

11.39 Explain the difference between physical and chemical tempering of glass.

11.40 What do you think is the purpose of the operation shown in Fig. 11.24d?

11.41 As you have seen, injection molding is a process that is used for plastics and powder metals as well as for ceramics. Why is it used for all these materials?

11.42 Are there any similarities between the strengthening mechanisms for glass and those for other metallic and nonmetallic materials described throughout this text? Explain.

11.43 Describe and explain the differences in the manner in which each of the following flat surfaces would fracture when struck with a large piece of rock: (a) ordinary window glass, (b) tempered glass, and (c) laminated glass.

11.44 Describe the similarities and the differences between the processes described in this chapter and in Chapters 5 through 10.

PROBLEMS

11.45 Estimate the number of particles in a 500-g sample of iron powder, if the particle size is 100 μm.

11.46 Assume that the surface of a copper particle is covered with a 0.1-μm-thick oxide layer. What is the volume occupied by this layer if the copper particle itself is 50 μm in diameter? What would be the role of this oxide layer in subsequent processing of the powders?

11.47 Determine the shape factor for a flakelike particle with a ratio of surface cross-sectional area to thickness of $15 \times 15 \times 1$ and for an ellipsoid with an axial ratio of $5 \times 2 \times 1$.

11.48 We stated in Section 3.3 that the energy in brittle fracture is dissipated as surface energy. We also noted that the comminution process for powder preparation generally involves brittle fracture. What are the relative energies involved in making spherical powders of diameters 1, 10, and 100 μm, respectively?

11.49 Referring to Fig. 11.6a, what should be the volume of loose, fine iron powder in order to make a solid cylindrical compact 20 mm in diameter and 10 mm high?

11.50 In Fig. 11.7e, we note that the pressure is not uniform across the diameter of the compact. Explain the reasons for this lack of uniformity.

11.51 Plot the family of pressure-ratio p_x/p_0 curves as a function of x for the following ranges of process parameters: $\mu = 0$ to 1, $k = 0$ to 1, and $D = 5$ mm to 50 mm.

11.52 Derive an expression similar to Eq. (11.2) for compaction in a square die with dimensions a by a.

11.53 For the ceramic in Example 11.7, calculate (a) the porosity of the dried part if the porosity of the fired part is to be 9%, and (b) the initial length L_0 of the part if the linear shrinkages during drying and firing are 8% and 7%, respectively.

11.54 Plot the UTS, E, and k values for ceramics as a function of porosity P, and describe and explain the trends that you observe in their behavior.

11.55 Plot the total surface area of a 1-g sample of aluminum as a function of the natural log of particle size.

11.56 How large is the grain size of metal powders that can be produced in atomization chambers? Conduct a literature search to determine the answer.

11.57 A coarse copper powder is compacted in a mechanical press at a pressure of 30 tons/in^2. During sintering, the green part shrinks an additional 9%. What is the final density of the part?

11.58 A gear is to be manufactured from iron powder. It is desired that it have a final density that is 90% of that of cast iron, and it is known that the shrinkage in sintering will be approximately 7%. For a gear 2.5 in. in diameter and a 0.75-in. hub, what is the required press force?

11.59 What volume of powder is needed to make the gear in Problem 11.58?

11.60 The axisymmetric part shown in the accompanying figure is to be produced from fine copper powder and is to have a tensile strength of 175 GPa. Determine the compacting pressure and the initial volume of powder needed.

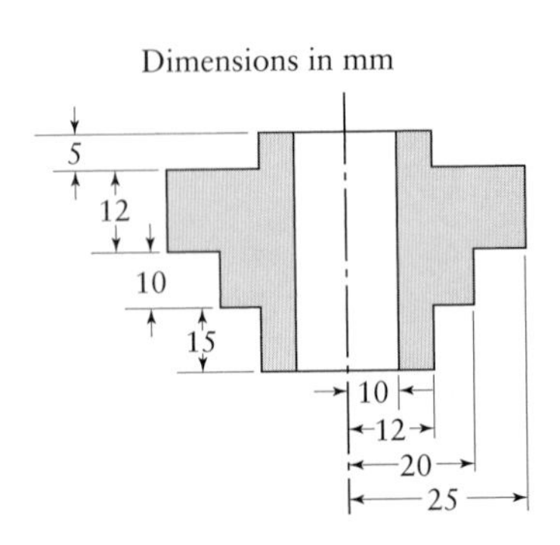

11.61 What techniques, other than the powder-in-tube process, could be used to produce superconducting monofilaments?

11.62 Describe other methods of manufacturing the parts shown in Fig. 11.1a. Comment on the advantages and limitations of these methods over P/M.

DESIGN

11.63 Describe the design features of several P/M products in which density variations would be desirable.

11.64 Compare the design considerations for P/M products with those for (a) products made by casting and (b) products made by forging. Describe your observations.

11.65 It is known that in the design of P/M gears, the outside diameter of the hub should be as far as possible from the root of the gear. Explain the reasons for this design consideration.

11.66 How are the design considerations for ceramics different, if at all, than those for the other materials described in this chapter?

11.67 Are there any shapes or design features that are not suitable for production by powder metallurgy? By ceramics processing? Explain.

11.68 What design changes would you recommend for the part shown in Problem 11.60.

11.69 The axisymmetric parts shown in the accompanying figure are to be produced through P/M. Describe the design changes that you would recommend.

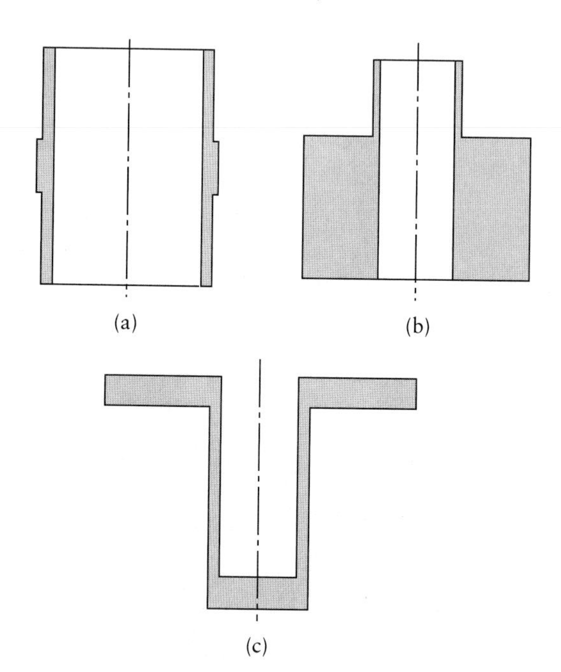

11.70 Assume that in a particular design, a metal beam is to be replaced with a beam made of ceramics. Discuss the differences in the behavior of the two beams, such as with respect to strength, stiffness, deflection, and resistance to temperature and to the environment.

11.71 Describe any special design considerations in products that use ceramics with a near-zero coefficient of thermal expansion.

11.72 Describe your thoughts regarding designs of internal combustion engines using ceramic pistons.

CHAPTER

12 Joining and Fastening Processes

12.1 Introduction

Joining is a general term covering numerous processes that are important aspects of manufacturing operations. The necessity of joining and assembly of components can be appreciated simply by inspecting various products around us such as automobile bodies, bicycle frames, machines, pipes and tubing, appliances, pots and pans, and printed circuit boards. Joining may be preferred or necessary for one or more of the following reasons:

- The product is impossible or uneconomical to manufacture as a single piece;
- The product is easier to manufacture in individual components, which are then assembled, than as a single piece;
- The product may have to be taken apart for repair or maintenance during its service life;
- Different properties may be desirable for functional purposes of the product—for example, surfaces subjected to friction and wear, or corrosion and environmental attack, typically require characteristics different than those of the component's bulk;
- Transportation of the product in individual components and subsequent assembly may be easier and more economical than transportation of the product as a single piece.

Although there are different ways of categorizing the wide variety of joining and fastening processes, this chapter will follow the sequence shown in the section headings listed to the left. Joining methods can be categorized in terms of their common principle of operation. **Fusion welding** involves melting together and coalescing materials by means of heat, usually supplied by electrical or high-energy means; filler metals may or may not be used. Fusion welding is a major category of welding and consists of consumable- and nonconsumable-electrode *arc welding*, and *high-energy-beam welding* processes. The high-energy processes are also used for cutting and machining applications, as described in Section 9.14. Because they generally are not used in a modern industrial setting, but are used mainly for maintenance and repair work, oxyfuel-gas welding and thermit welding processes are only briefly described in this text.

In **solid-state welding**, joining takes place without fusion; thus, there is no liquid (molten) phase in the joint. The basic categories of solid-state welding are *cold*, *ultrasonic*, *friction*, *resistance*, *explosion welding*, and *diffusion bonding*; the latter, combined with superplastic forming, has become an important manufacturing process for complex shapes. The joints produced by most of the aforementioned welding processes undergo metallurgical and physical changes; these changes, in turn, have major effects on the properties and performance of the welded components.

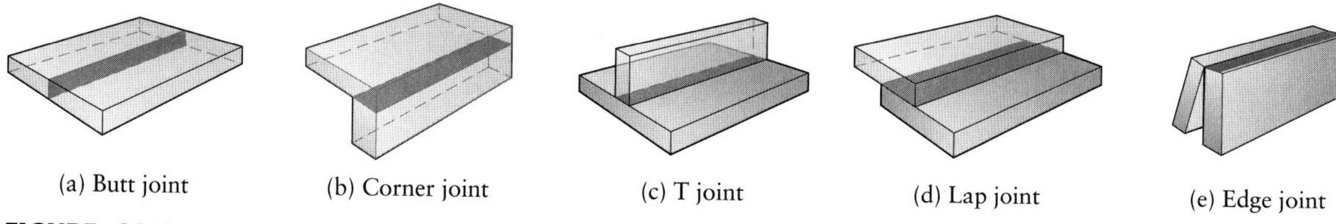

FIGURE 12.1 Examples of joints that can be made through the various joining processes described in Chapter 12.

Brazing and **soldering** use filler metals and involve lower temperatures than used in welding; the heat required is supplied externally. **Adhesive bonding** has been developed into an important technology, because of its various unique advantages for applications requiring strength, sealing, insulation, vibration damping, and resistance to corrosion between dissimilar or similar metals. Included in this category are electrically conducting adhesives for surface-mount technologies. **Mechanical-fastening** processes typically involve traditional methods, using a wide variety of fasteners, bolts, nuts, screws, and rivets. **Joining nonmetallic materials** can be accomplished by such means as adhesive bonding, fusion by various external or internal heat sources, diffusion, preplating with metal, and mechanical fastening.

An inspection of welded components in numerous household and industrial products will clearly indicate that a common feature of all the components is the type of joint. Although there are some variations, the basic and common joint designs are outlined in Fig. 12.1. The individual groups of joining processes briefly outlined previously have, as in all manufacturing processes, several important characteristics, such as joint design, size and shape of the parts to be joined, strength and reliability of the joint, cost and maintenance of equipment, and skill level of labor. These characteristics are outlined, in relative terms, in Tables 12.1 and 12.2 as a guide to process selection, as described in greater detail throughout this chapter.

TABLE 12.1

Comparison of Various Joining Methods

	Characteristics								
Method	Strength	Design Variability	Small Parts	Large Parts	Tolerances	Reliability	Ease of Maintenance	Visual Inspection	Cost
Arc welding	1	2	3	1	3	1	2	2	2
Resistance welding	1	2	1	1	3	3	3	3	1
Brazing	1	1	1	1	3	1	3	2	3
Bolts and nuts	1	2	3	1	2	1	1	1	3
Riveting	1	2	3	1	1	1	3	1	2
Fasteners	2	3	3	1	2	2	2	1	3
Seaming, crimping	2	2	1	3	3	1	3	1	1
Adhesive bonding	3	1	1	2	3	2	3	3	2

Note: 1, very good; 2, good; 3, poor.

TABLE 12.2

General Characteristics of Joining Processes

Joining Process	Operation	Advantage	Skill Level Required	Welding Position	Current Type	Distortion*	Cost of Equipment
Shielded metal arc	Manual	Portable and flexible	High	All	ac, dc	1 to 2	Low
Submerged arc	Automatic	High deposition	Low to medium	Flat and horizontal	ac, dc	1 to 2	Medium
Gas metal arc	Semiautomatic or automatic	Works with most metals	Low to high	All	dc	2 to 3	Medium to high
Gas tungsten arc	Manual or automatic	Works with most metals	Low to high	All	ac, dc	2 to 3	Medium
Flux-cored arc	Semiautomatic or automatic	High deposition	Low to high	All	dc	1 to 3	Medium
Oxyfuel	Manual	Portable and flexible	High	All	–	2 to 4	Low
Electron beam, laser beam	Semiautomatic or automatic	Works with most metals	Medium to high	All	–	3 to 5	High

* 1, highest; 5, lowest

12.2 Arc-Welding Processes: Consumable Electrode

In *arc welding*, developed in the mid-1800s, the heat required is obtained through electrical energy. Through use of either a *consumable* or *nonconsumable electrode* (rod or wire), an arc is produced between the tip of the electrode and the parts to be welded, using ac or dc power supplies. Arc welding includes several processes (Table 12.2), as described in the upcoming subsections.

12.2.1 Shielded metal arc-welding

Shielded metal-arc welding (SMAW) is one of the oldest, simplest, and most versatile joining processes; currently, about 50% of all industrial and maintenance welding is performed by this process. The electric arc (hence the term *electric-arc welding*) is generated by touching the tip of a coated electrode to the workpiece and then withdrawing it quickly to a distance sufficient to maintain the arc (Fig. 12.2a). The electrodes are in the shape of thin, long sticks (see Section 12.2.7); hence, this process is also known as **stick welding**. The heat generated melts a portion of the tip of the electrode, its coating, and the base metal in the immediate area of the arc. A weld forms after the molten metal (a mixture of the base metal, electrode metal, and substances from the coating on the electrode) solidifies in the weld area. The electrode coating deoxidizes and provides a shielding gas in the weld area to protect it from oxygen in the environment.

A bare section at the end of the electrode is clamped to one terminal of the power source, while the other terminal is connected to the part being welded (Fig. 12.2b). The current usually ranges between 50 A and 300 A, with power requirements generally less than 10 kW; too low a current causes incomplete fusion, and

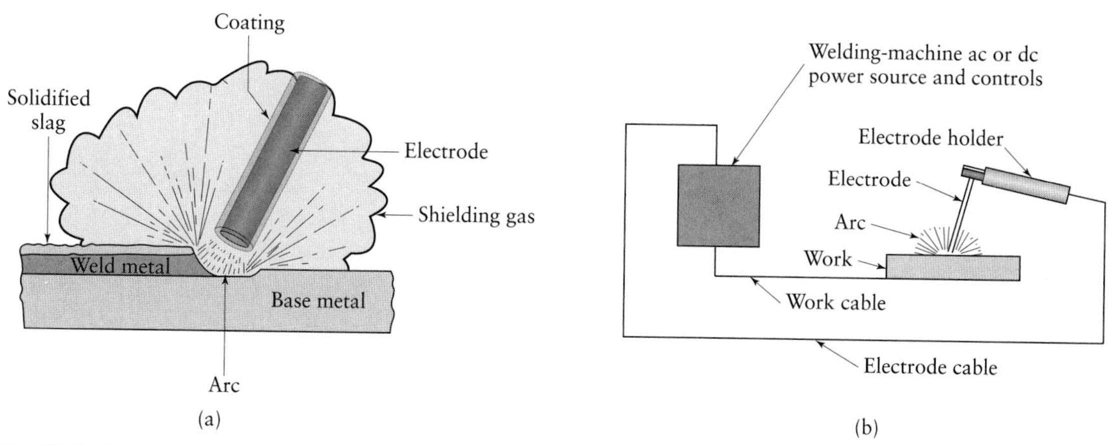

FIGURE 12.2 (a) Schematic illustration of the shielded metal-arc-welding process. About 50% of all large-scale industrial welding operations use this process. (b) Schematic illustration of the shielded metal-arc-welding operation, also known as *stick welding*, because the electrode is in the shape of a stick.

too high a current can damage the electrode coating and reduce its effectiveness. The current may be ac or dc; for sheet-metal welding, dc is preferred, because of the steady arc produced. The **polarity** (direction of current flow) of the dc current can be important, depending on factors such as the type of electrode, the metals to be welded, and the depth of the heated zone. In **straight polarity**, the workpiece is positive and the electrode negative; it is preferred for sheet metals, because of its shallow penetration, and for joints with very wide gaps. In **reverse polarity**, where the electrode is positive and the workpiece negative, deeper weld penetration is possible. When the current is ac, the arc pulsates rapidly; this condition is suitable for welding thick sections and using large-diameter electrodes at maximum currents.

The SMAW process requires a relatively small variety of electrodes, and the equipment consists of a power supply, power cables, and an electrode holder. This process is commonly used in general construction, shipbuilding, and for pipelines, as well as for maintenance work, since the equipment is portable and can be easily maintained. It is especially useful for work in remote areas, where portable fuel-powered generators can be used as the power supply. The SMAW process is best suited for workpiece thicknesses of 3–19 mm (0.12–0.75 in.), although this range can easily be extended by using multiple-pass techniques (Fig. 12.3). This process requires that *slag* (compounds of oxides, fluxes, and electrode coating materials) be cleaned after each weld bead; unless removed completely, the solidified slag can cause severe corrosion of the weld area and can lead to failure of the weld. Slag should also be completely removed, such as by wire brushing, before another weld is applied for multiple-pass welding; thus labor costs are high, as are the material costs.

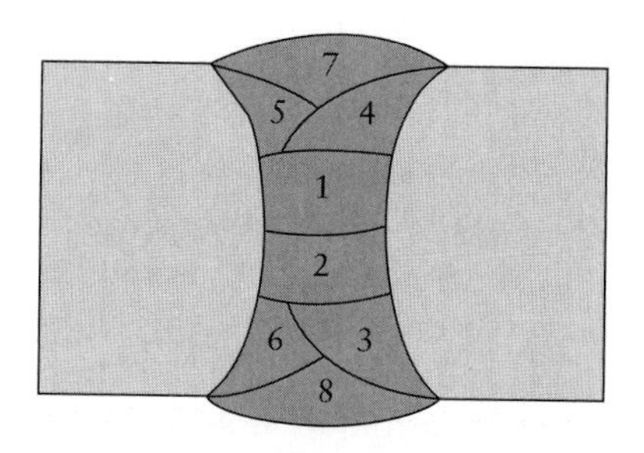

FIGURE 12.3 A weld zone showing the build-up sequence of individual weld beads in deep welds.

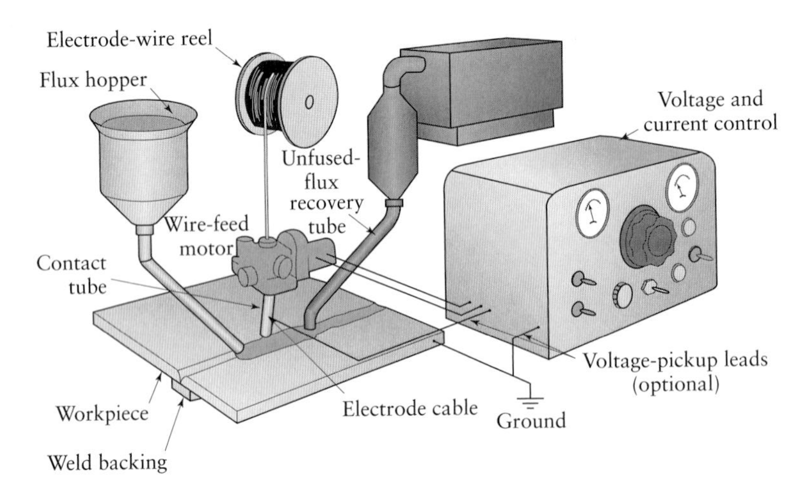

FIGURE 12.4 Schematic illustration of the submerged arc-welding process and equipment. Unfused flux is recovered and reused.

12.2.2 Submerged arc welding

In *submerged arc welding* (SAW), the weld arc is shielded by granular flux, consisting of lime, silica, manganese oxide, calcium fluoride, and other elements; the flux is fed into the weld zone by gravity flow through a nozzle (Fig. 12.4). The thick layer of flux completely covers the molten metal and prevents spatter and sparks and suppresses the intense ultraviolet radiation and fumes. The flux also acts as a thermal insulator, allowing deep penetration of heat into the workpiece; the unfused flux can be recovered (using a *recovery tube*), treated, and reused.

The consumable electrode is a coil of bare wire 1.5–10 mm $\left(\frac{1}{16}–\frac{3}{8}\text{ in.}\right)$ in diameter and is fed automatically through a tube (**welding gun**). Electric currents usually range between 300 A and 2000 A; the power supplies are usually connected to standard single- or three-phase power lines with a primary rating up to 440 V. Because the flux is fed by gravity, the SAW process is somewhat limited to welds in a flat or horizontal position, with a backup piece. Circular welds can be made on pipes, provided that the pipes are rotated during welding.

The SAW process can be automated for greater economy and is used to weld a variety of carbon- and alloy-steel and stainless-steel sheet or plate, at speeds as high as 5 m/min (16 ft/min). Typical applications include thick-plate welding for shipbuilding and fabrication of pressure vessels. The quality of the weld is very high, with good toughness, ductility, and uniformity of properties. The process provides very high welding productivity, depositing 4 to 10 times the amount of weld metal per hour as deposited by the SMAW process.

12.2.3 Gas metal-arc welding

In *gas metal-arc welding* (GMAW), the weld area is shielded by an external source, such as argon, helium, carbon dioxide, or various other gas mixtures (Fig. 12.5a). In addition, deoxidizers are usually present in the electrode metal itself, in order to prevent oxidation of the molten weld puddle. The consumable bare wire is fed automatically through a nozzle into the weld arc (Fig. 12.5b), and multiple weld layers can be deposited at the joint. In this process, formerly called *MIG welding* (for *metal inert gas*), the metal can be transferred in one of three ways:

a) In **spray transfer**, small droplets of molten metal from the electrode are transferred to the weld area, at rates of several hundred droplets per second; the transfer is spatter free and very stable. High dc current, high voltages, and large-diameter electrodes are used, with argon or argon-rich gas mixtures used as the shielding gas. The average current required in this process can be reduced by

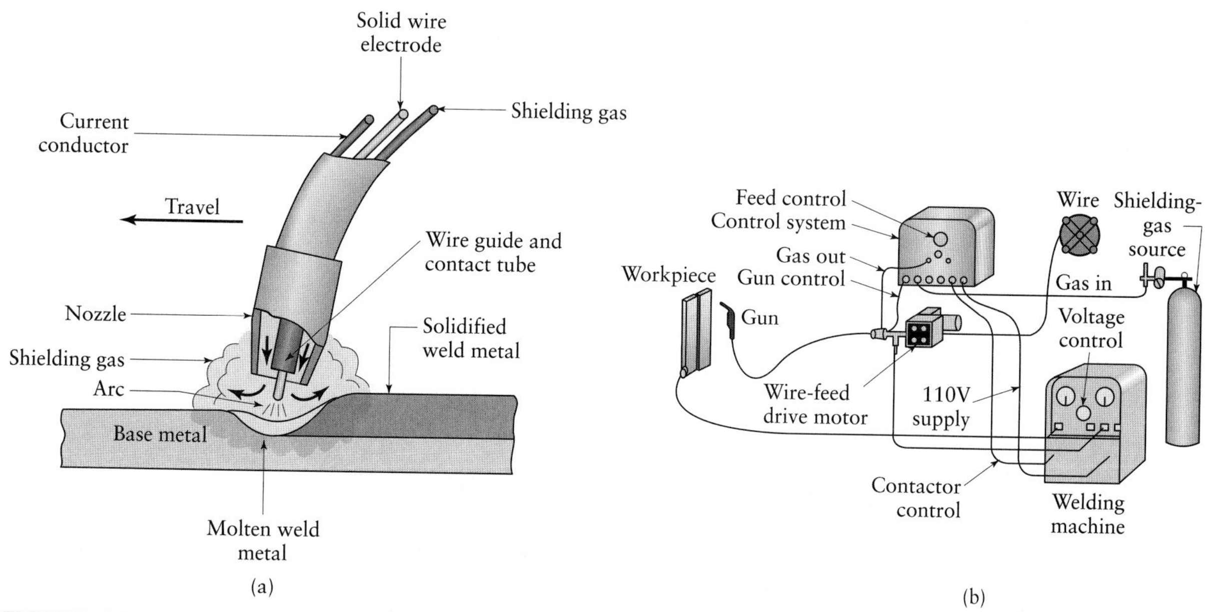

FIGURE 12.5 (a) Gas metal-arc-welding process, formerly known as MIG welding (for *metal inert gas*). (b) Basic equipment used in gas metal-arc-welding operations.

using *pulsed arcs*, which are high-amplitude pulses superimposed over a low, steady current. The process can be used for all welding positions.

b) In **globular transfer**, carbon-dioxide rich gases are used, and globules are propelled by the forces of the electric arc that transfer the metal, thus resulting in considerable spatter. High welding currents are used, with greater weld penetration and welding speed than in spray transfer. Heavier sections are commonly joined by this method.

c) In **short circuiting**, the metal is transferred in individual droplets, at rates of more than 50/s, as the electrode tip touches the molten weld pool and short circuits. Low currents and voltages are used, with carbon-dioxide rich gases and electrodes made of small-diameter wire. The power required is about 2 kW, and the temperatures involved are relatively low. Hence, this method is suitable only for thin sheets and sections [less than 6 mm (0.25 in.) thick]; with thicker materials, incomplete fusion may occur. This process is very easy to use and may be the most popular method for welding ferrous metals in thin sections; however, pulsed-arc systems are gaining wider usage for thin ferrous and nonferrous metals.

Developed in the 1950s, the GMAW process is suitable for a variety of ferrous and nonferrous metals and is used extensively in the metal-fabrication industry. The process is rapid, versatile, and economical; its welding productivity is double that of the SMAW process, and it can easily be automated and lends itself readily to robotics and flexible manufacturing systems (see Chapters 14 and 15).

12.2.4 Flux-cored arc welding

The *flux-cored arc welding* (FCAW) process (Fig. 12.6) is similar to gas metal-arc welding, with the exception that the electrode is tubular in shape and is filled with flux (hence the term *flux cored*). Cored electrodes produce a more stable arc, improve weld contour, and improve the mechanical properties of the weld metal. The flux in these electrodes is much more flexible than the brittle coating used on SMAW

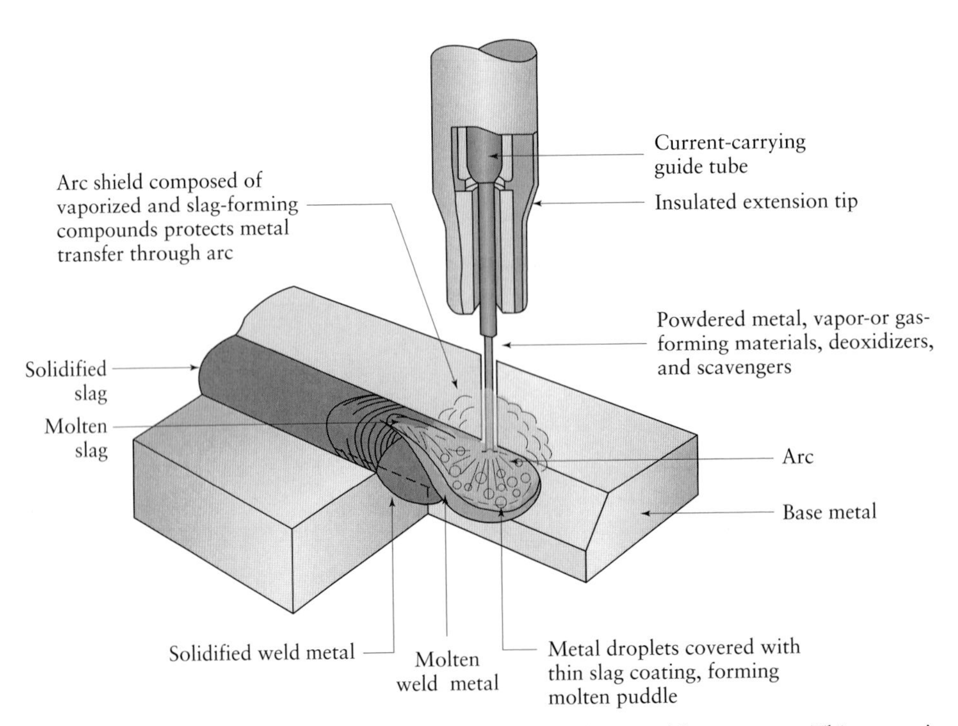

FIGURE 12.6 Schematic illustration of the flux-cored arc-welding process. This operation is similar to gas metal-arc welding, shown in Fig. 12.5.

electrodes; thus, the tubular electrode can be provided in long coiled lengths. The power required is about 20 kW.

The electrodes are usually 0.5–4 mm (0.020–0.15 in.) in diameter; small-diameter electrodes have made welding of thinner materials not only possible, but often desirable. Also, small-diameter electrodes make it relatively easy to weld parts out of position, and the flux chemistry enables welding of many base metals. Self-shielded cored electrodes are also available; these electrodes do not require external gas shielding, because they contain emissive fluxes that shield the weld area against the surrounding atmosphere. Advances in manufacturing of electrodes for FCAW, as well as in the chemistry of the flux, have made this process the fastest growing method in welding.

The flux-cored arc-welding process combines the versatility of SMAW with the continuous and automatic electrode-feeding feature of GMAW. It is economical and is used for welding a variety of joints, mainly with steels, stainless steels, and nickel alloys. The higher weld-metal deposition rate of FCAW over that of GMAW has led to the use of the former in joining sections of all thicknesses. The availability of tubular electrodes with very small diameters has extended the use of this process to smaller section sizes. A major advantage of FCAW is the ease with which specific weld-metal chemistries can be developed; by adding alloys to the flux core, virtually any alloy composition can be developed. This process is easy to automate and is readily adaptable to flexible manufacturing systems and robotics.

12.2.5 Electrogas welding

Electrogas welding (EGW) is used primarily for welding the edges of sections vertically in one pass with the pieces placed edge to edge (butt welding); it is classified as a machine-welding process because it requires special equipment (Fig. 12.7). The weld metal is deposited into a weld cavity between the two pieces to be joined; the space is enclosed by two water-cooled copper dams (*shoes*) to prevent the molten

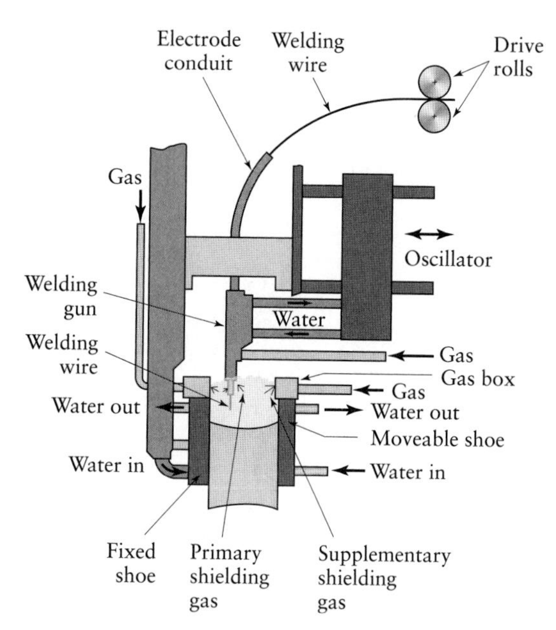

FIGURE 12.7 Schematic illustration of the electrogas-welding process. *Source*: Courtesy of the American Welding Society.

slag from running off. Mechanical drives move the shoes upward. Circumferential welds such as on pipes are also possible, with rotation of the workpiece.

Single or multiple electrodes are fed through a conduit, and a continuous arc is maintained, using flux-cored electrodes at currents up to 750 A, or solid electrodes at 400 A; power requirements are about 20 kW. Shielding is provided by an inert gas, such as carbon dioxide, argon, or helium, depending on the type of material being welded. The gas may be provided from an external source, or it may be produced from a flux-cored electrode, or both. Weld thickness ranges from 12 mm to 75 mm (0.5 in. to 3 in.) on steels, titanium, and aluminum alloys. Typical applications are in the construction of bridges, pressure vessels, thick-walled and large-diameter pipes, storage tanks, and ships.

12.2.6 Electroslag welding

Electroslag welding (ESW) was developed in the 1950s. Its applications are similar to those for electrogas welding; the main difference between the two processes is that in electroslag welding, the arc is started between the electrode tip and the bottom of the part to be welded (Fig. 12.8). Flux is added and melted by the heat of the arc. After the molten slag reaches the tip of the electrode, the arc is extinguished; energy is supplied continuously through the electrical resistance of the molten slag. However, because the arc is extinguished, ESW is not strictly an arc-welding process. Single or multiple solid as well as flux-cored electrodes may be used, and the guide may be nonconsumable (conventional method) or consumable.

Electroslag welding is capable of welding plates with thicknesses ranging from 50 mm to more than 900 mm (2 in. to more than 36 in.). Welding is done in one pass. The current required is about 600 A at 40–50 V, although higher currents are used for thick plates. The travel speed of the weld is 12–36 mm/min (0.5–1.5 in./min). The weld quality is good, and the process is used for heavy structural steel sections, such as heavy machinery and nuclear-reactor vessels.

12.2.7 Electrodes for arc welding

The *electrodes* for the consumable-electrode arc-welding processes described thus far are classified according to the strength of the deposited weld metal, the current (ac or dc), and the type of coating. Electrodes are identified by numbers and letters or by

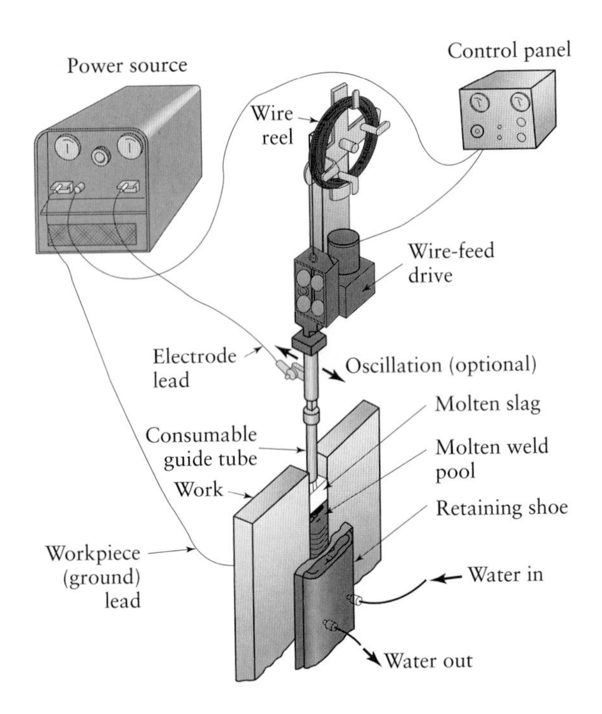

FIGURE 12.8 Equipment used for electroslag-welding operations. *Source*: Courtesy of the American Welding Society.

color code, particularly if they are too small to imprint with identification. Typical coated-electrode dimensions are a length of 150–460 mm (6–18 in.) and a diameter of 1.5–8 mm $\left(\frac{1}{16}-\frac{5}{16}\text{ in.}\right)$. The thinner the sections to be welded and the lower the current required, the smaller the diameter of the electrode should be.

The specifications for electrodes and filler metals, including tolerances, quality-control procedures, and processes, have been provided by the American Welding Society (AWS) and the American National Standards Institute (ANSI) and are stated in the Aerospace Materials Specifications (AMS) by the Society of Automotive Engineers (SAE). The specifications require, for examples that the diameter of the wire not vary more than 0.05 mm (0.002 in.) from nominal size and that the coatings be concentric with the wire. Electrodes are sold by weight and are available in a wide variety of sizes and specifications. Information on selection and recommendations for electrodes for a particular metal and its application can be found in supplier literature and various handbooks and references listed in the bibliography at the end of this chapter.

Electrode coatings. Electrodes are *coated* with claylike materials that include silicate binders and powdered materials such as oxides, carbonates, fluorides, metal alloys, and cellulose (cotton cellulose and wood flour). The coating, which is brittle and has complex interactions during welding, has the following basic functions:

a) Stabilize the arc;
b) Generate gases to act as a shield against the surrounding atmosphere (the gases produced are carbon dioxide and water vapor, and carbon monoxide and hydrogen in small amounts);
c) Control the rate at which the electrode melts;
d) Act as a flux to protect the weld against the formation of oxides, nitrides, and other inclusions and, with the slag produced, thus protect the molten weld pool;
e) Add alloying elements to the weld zone to enhance the properties of the weld, including deoxidizers to prevent the weld from becoming brittle.

The deposited electrode coating or slag must be removed after each pass in order to ensure a good weld. A manual or powered wire brush can be used for this purpose. Bare electrodes and wires made of stainless steels and aluminum alloys are also available; they are used as filler metals in various welding operations.

12.3 Arc-Welding Processes: Nonconsumable Electrode

Unlike the arc-welding processes that use consumable electrodes, described in the previous section, nonconsumable-electrode arc-welding processes typically use a **tungsten electrode**. As one pole of the arc, the electrode generates the heat required for welding; a shielding gas is supplied from an external source. The nonconsumable-electrode arc-welding process was first developed using carbon electrodes; hence, it was called *carbon-arc welding* (CAW). The three basic processes are described in the upcoming subsections.

12.3.1 Gas tungsten-arc welding

In *gas tungsten-arc welding* (GTAW), formerly known as *TIG welding* (for *tungsten inert gas*), a filler metal is typically supplied from a **filler wire** (Fig. 12.9a), although welding may be done without filler metals, as in welding close-fit joints. The composition of filler metals is similar to that of the metals to be welded, and flux is not used; the shielding gas is usually argon or helium, or a mixture of the two. Because the tungsten electrode is not consumed in this operation, a constant and stable arc gap is maintained at a constant level of current.

The power supply (Fig. 12.9b) ranges from 8 kW to 20 kW and is either dc at 200 A or ac at 500 A, depending on the metals to be welded. In general, ac is preferred for aluminum and magnesium, because the cleaning action of ac removes oxides and improves weld quality. Thorium or zirconium may be used in the tungsten electrodes to improve their electron-emission characteristics. Contamination of the tungsten electrode by the molten metal can be a significant problem, particularly in critical applications, as it can cause discontinuities in the weld; therefore, contact of the electrode with the molten metal pool should be avoided.

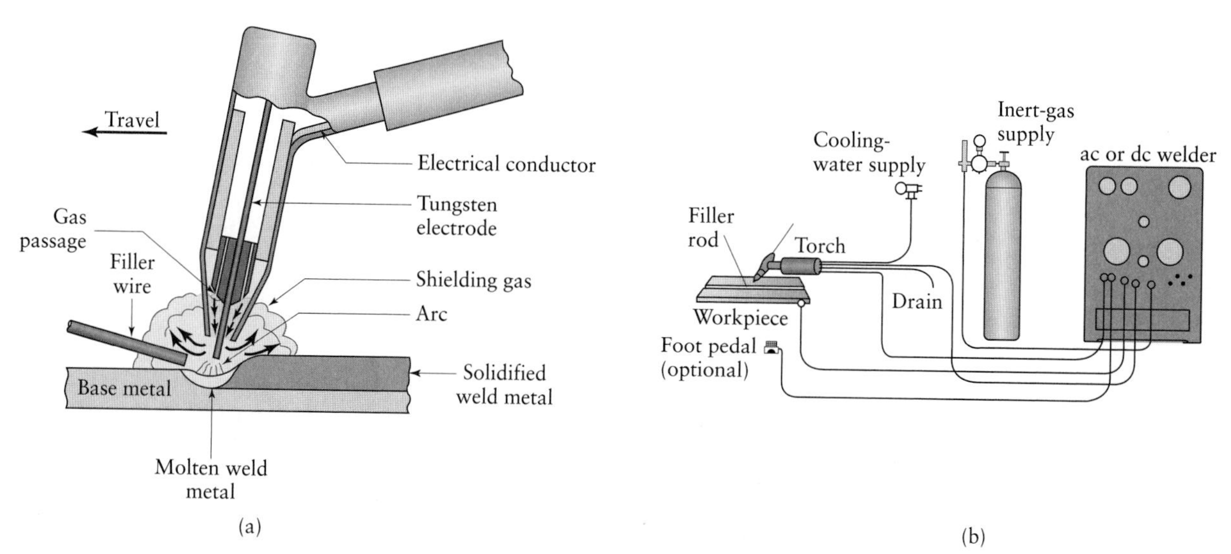

FIGURE 12.9 (a) Gas tungsten-arc-welding process, formerly known as TIG welding (for *tungsten inert gas*). (b) Equipment for gas tungsten-arc-welding operations.

The GTAW process is used for a wide variety of applications and metals, particularly aluminum, magnesium, titanium, and refractory metals, and it is especially suitable for thin metals. The cost of the inert gas makes this process more expensive than SMAW, but it provides welds with very high quality and surface finish.

12.3.2 Atomic-hydrogen welding

In *atomic-hydrogen welding* (AHW), an arc is generated between two tungsten electrodes in a shielding atmosphere of flowing hydrogen gas. The arc is maintained independently of the workpiece or parts being welded. The hydrogen gas is normally diatomic (H_2), but near the arc, where the temperatures are over 6000°C (11000° F), the hydrogen breaks down into its atomic form, simultaneously absorbing a large amount of heat from the arc. When the hydrogen strikes a relatively cold surface, i.e., the weld zone, it recombines into its diatomic form and releases the stored heat very rapidly. The energy in AHW can easily be varied by changing the distance between the arc stream and the workpiece surface. This process was previously much more commonly used than it is now; shielded arc-welding processes have become more popular, mainly because of the availability of inexpensive inert gases.

12.3.3 Plasma-arc welding

In *plasma-arc welding* (PAW), developed in the 1960s, a concentrated plasma arc is produced and aimed at the weld area; the arc is stable and reaches temperatures as high as 33,000°C (60,000° F). A *plasma* is ionized hot gas, composed of nearly equal numbers of electrons and ions. The plasma is initiated between the tungsten electrode and the orifice, using a low-current pilot arc. Unlike the arc in other processes, the plasma arc is concentrated, because it is forced through a relatively small orifice. Operating currents for this process are usually below 100 A, but they can be higher for special applications. When a filler metal is used, it is fed into the arc, as in GTAW. Arc and weld-zone shielding is supplied through an outer shielding ring by gases such as argon, helium, or mixtures of these gases.

There are two methods of plasma-arc welding. In the **transferred-arc** method (Fig. 12.10a), the workpiece being welded is part of the electrical circuit. The arc thus transfers from the electrode to the workpiece, hence the term *transferred*. In the **nontransferred** method (Fig. 12.10b), the arc is between the electrode and the nozzle, and the heat is carried to the workpiece by the plasma gas; this technique is also used for *thermal spraying*. (See Section 4.5.1.)

Compared with other arc-welding processes, plasma-arc welding has a greater energy concentration (hence, deeper and narrower welds can be made); better arc stability; less thermal distortion; and higher welding speeds, such as 120–1000 mm/min (5–40 in./min). A variety of metals can be welded, generally with part thicknesses less than 6 mm (0.25 in.). The high heat concentration can penetrate completely through

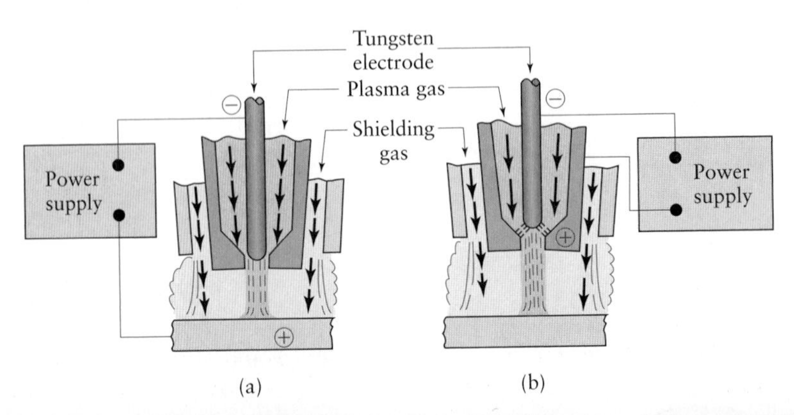

FIGURE 12.10 Two types of plasma-arc-welding processes: (a) transferred and (b) nontransferred. Deep and narrow welds are made by this process at high welding speeds.

the joint (**keyhole technique**), to thicknesses as much as 20 mm (0.75 in.) for some titanium and aluminum alloys. In the keyhole technique, the force of the plasma arc displaces the molten metal and produces a hole at the leading edge of the weld pool. Plasma-arc welding, rather than the GTAW process, is often used for butt and lap joints, because of the former's higher energy concentration, better arc stability, and higher welding speeds.

12.4 High-Energy-Beam Welding

Joining with high-energy beams, principally laser-beam and electron-beam welding, has increasingly important applications in modern manufacturing, because of its high quality and technical and economic advantages. The general characteristics of high-energy beams and their unique applications in machining operations are described in Section 9.14.

12.4.1 Electron-beam welding

In *electron-beam welding* (EBW), heat is generated by high-velocity, narrow-beam electrons; the kinetic energy of the electrons is converted into heat as the electrons strike the workpiece. This process requires special equipment to focus the beam on the workpiece and in a vacuum; the higher the vacuum, the more the beam penetrates and the greater is the depth-to-width ratio. The level of vacuum is specified as HV (high vacuum) or MV (medium vacuum), both of which are used commonly, or NV (no vacuum), which can be effective on some materials. Almost any metals, similar or dissimilar, can be butt or lap welded using this process, with thicknesses ranging from foil to plates as thick as 150 mm (6 in.). The intense energy is also capable of producing holes in the workpiece. Generally, no shielding gas, flux, or filler metal is required. Capacities of electron-beam guns range to 100 kW.

Developed in the 1960s, EBW has the capability to make high-quality welds that are almost parallel sided, are deep and narrow, and have small heat-affected zones. (See Section 12.5.) Depth-to-width ratios range between 10 and 30 (Fig. 12.11). By using servo controls, parameters can be precisely controlled, with welding speeds as high as 12 m/min (40 ft/min). Distortion and shrinkage in the weld area are minimal, and the weld quality is good, with very high purity. Typical applications of EBW include welding of aircraft, missile, nuclear, and electronic components, as well as gears and shafts in the automotive industry. Electron-beam welding equipment generates X rays, and hence proper monitoring and periodic maintenance are important.

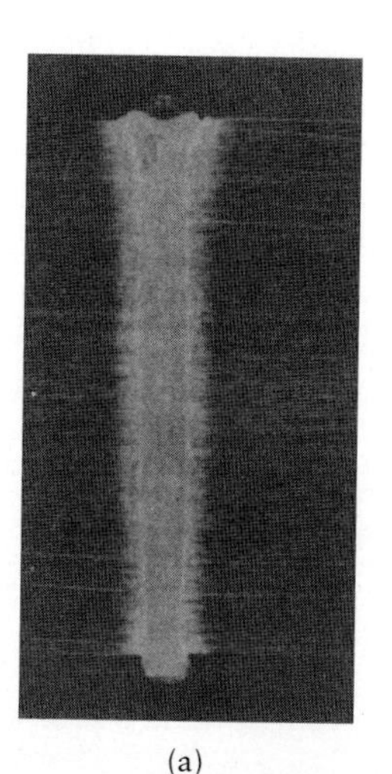

(a)

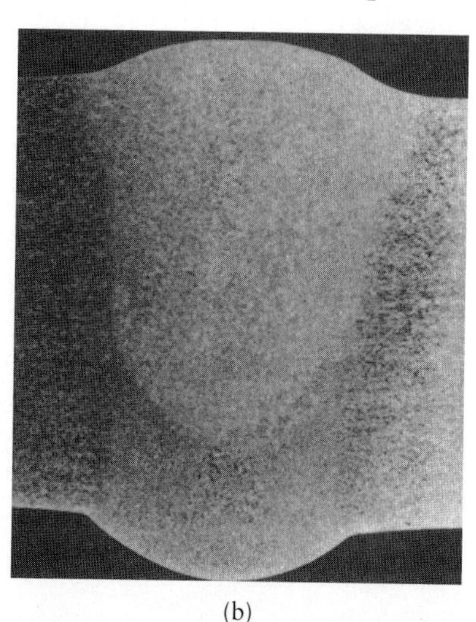

(b)

FIGURE 12.11 Comparison of the size of weld beads in (a) electron-beam or laser-beam welding with that in (b) conventional (tungsten-arc) welding. *Source*: American Welding Society, *Welding Handbook*, 8th ed., 1991.

12.4.2 Laser-beam welding

Laser-beam welding (LBW) uses a high-power laser beam as the source of heat (see Fig. 9.36 and Table 9.5) to produce a fusion weld; because the beam can be focused to a very small area, it has high-energy density and, therefore, has deep penetrating capability. The beam can be directed, shaped, and focused precisely on the workpiece; consequently, this process is particularly suitable for welding deep and narrow joints (Fig. 12.11a), with depth-to-width ratios typically ranging from 4 to 10. The laser beam may be **pulsed** (milliseconds) for applications such as spot welding of thin materials, with power levels up to 100 kW; **continuous** multi-kW laser systems are used for deep welds on thick sections. The efficiency of this process decreases with increasing reflectivity of the workpiece materials; to improve the process' performance, oxygen may be used for steels, and inert gases for nonferrous metals.

The LBW process can be automated and used successfully on a variety of materials with thicknesses of up to 25 mm (1 in.); it is particularly effective on thin workpieces. Typical metals and alloys welded by this process include aluminum, titanium, ferrous metals, copper, superalloys, and refractory metals. Welding speeds range from 2.5 m/min (8 ft/min) to as high as 80 m/min (250 ft/min) for thin metals. Because of the nature of the process, welding can be done in otherwise inaccessible locations. Safety is a particularly important consideration in laser-beam welding, because of the extreme hazards to the eye as well as the skin; solid-state (YAG) lasers are particularly harmful.

Laser-beam welding produces welds of good quality, with minimum shrinkage and distortion. Laser welds have good strength and are generally ductile and free of porosity. In the automotive industry, welding of transmission components is a widespread application of this process; among numerous other applications is the welding of thin parts for electronic components. As shown in Fig 7.14, another important application of laser welding is the forming of butt-welded sheet metal, particularly for automotive-body panels. (See the discussion of **tailor-welded blanks** in Section 7.3.4.)

The major advantages of LBW over electron-beam welding are:

- The laser beam can be transmitted through air, so a vacuum is not required, unlike in electron-beam welding.
- Because laser beams can be shaped, manipulated, and focused optically (using fiber optics), the process can easily be automated.
- The laser beams do not generate X rays, while the electron beams do.
- The quality of the weld is better with laser-beam welding, with less tendency for incomplete fusion, spatter, porosity, and distortion.

EXAMPLE 12.1 Laser-beam welding of razor blades

Figure 12.12 shows a close-up of the Gillette Sensor razor cartridge. Each of the two narrow, high-strength blades has 13 pinpoint welds, 11 of which can be seen, as darker spots about 0.5 mm in diameter, on each blade in the photograph. The welds are made with an Nd : YAG laser equipped with fiber-optic delivery, thus providing flexible beam manipulation to the exact locations along the length of the blade. With a set of these machines, production runs at a rate of 3 million welds per hour, with accurate and consistent weld quality.

Source: Courtesy of Lumonics Corporation, Industrial Products Division.

FIGURE 12.12 Gillette Sensor razor cartridge, with laser-beam welds.

12.5 The Fusion-Welded Joint

This section describes the following aspects of *fusion-welding* processes:

- The nature, properties, and quality of the welded joint;
- Weldability of metals;
- Testing welds.

The mechanical properties of a welded joint depend on many factors; for example, the rate of heat application and the thermal properties of metals are important, because they control the magnitude and distribution of temperature in a joint during welding. The microstructure and grain size of the welded joint depend on the magnitude of heat applied and of the temperature rise, the degree of prior cold work of the metals, and the rate of cooling after the weld is made. Weld quality depends on factors such as the geometry of the weld bead and the presence of cracks, residual stresses, inclusions, and oxide films; control of these factors is essential for producing reliable welds that have acceptable mechanical properties.

Figure 12.13 shows a typical fusion-welded joint, where three distinct zones can be identified:

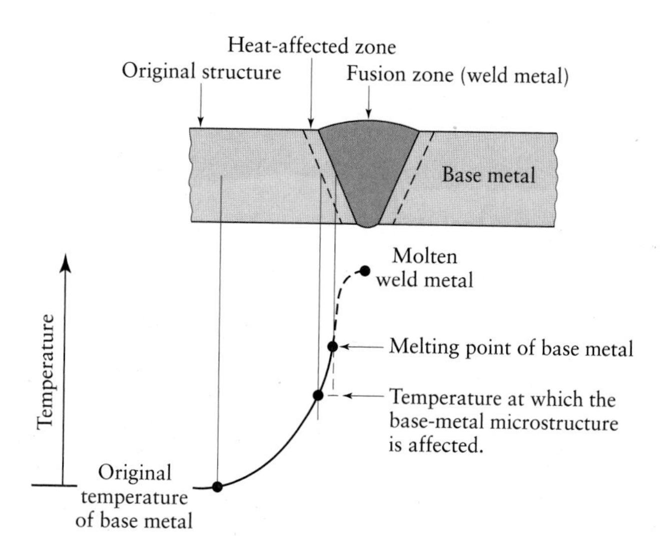

FIGURE 12.13 Characteristics of a typical fusion-weld zone in oxyfuel-gas and arc welding.

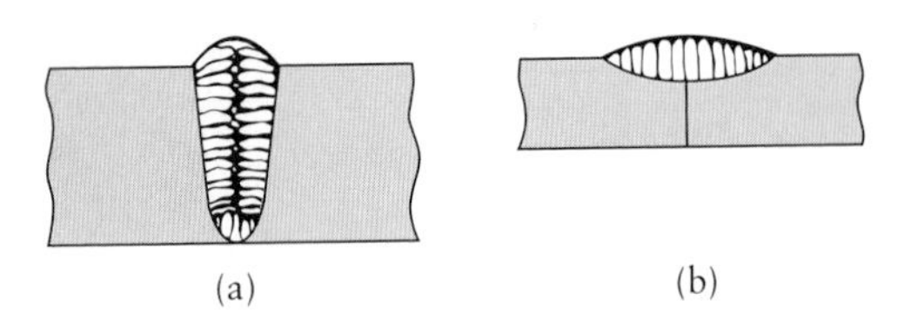

FIGURE 12.14 Grain structure in (a) a deep weld and (b) a shallow weld. Note that the grains in the solidified weld metal are perpendicular to their interface with the base metal. In a good weld, the solidification line at the center in the deep weld shown in (a) has grain migration, thus developing uniform strength in the weld bead.

1. The **base metal**, that is, the metal to be joined;
2. The **heat-affected zone** (HAZ);
3. The **fusion zone**, that is, the region that has melted during welding.

The metallurgy and properties of the second and third zones depend greatly on the metals joined (single phase, two phase, dissimilar), the welding process, the filler metals used (if any), and the process variables. A joint produced without a filler metal is called **autogenous**, and the weld zone is composed of the *resolidified* base metal. A joint made with a filler metal has a central zone called the **weld metal**, which is composed of a mixture of the base and filler metals.

Solidification of the weld metal. After heat has been applied and filler metal, if any, has been introduced into the weld area, the molten weld joint is allowed to cool to ambient temperature. The *solidification* process is similar to that in casting and begins with the formation of columnar (dendritic) grains; these grains are relatively long and form parallel to the heat flow. (See Fig. 5.9.) Because metals are much better heat conductors than the surrounding air, the grains lie parallel to the plane of the two plates or sheets being welded (Fig. 12.14a); the grains developed in a shallow weld are shown in Figs. 12.14b. Grain structure and size depend on the specific alloy, the welding process, and the filler metal used.

The weld metal is basically a *cast* structure, and, because it has cooled somewhat slowly, it typically has coarse grains; consequently, this structure has relatively low strength, hardness, toughness, and ductility (Fig. 12.15). However, with proper selection of filler-metal composition or with subsequent heat treatments, the joint's mechanical properties can be improved. The results depend on the particular alloy, its composition, and the thermal cycling to which the joint is subjected. Cooling rates may, for example, be controlled and reduced by *preheating* the weld area prior to welding; preheating is particularly important for metals with high thermal conductivity, such as aluminum and copper, as otherwise the heat during welding rapidly dissipates.

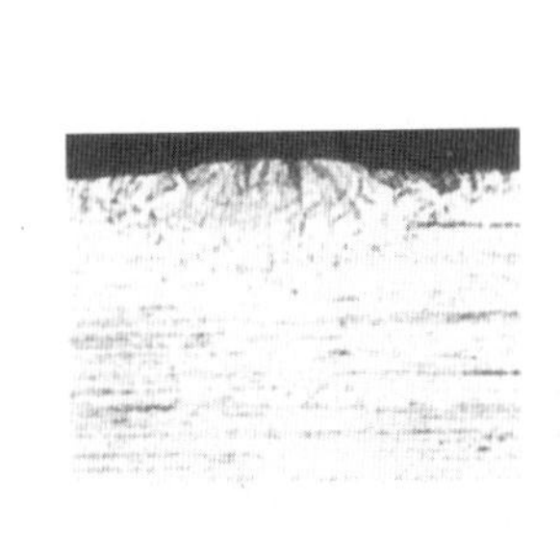

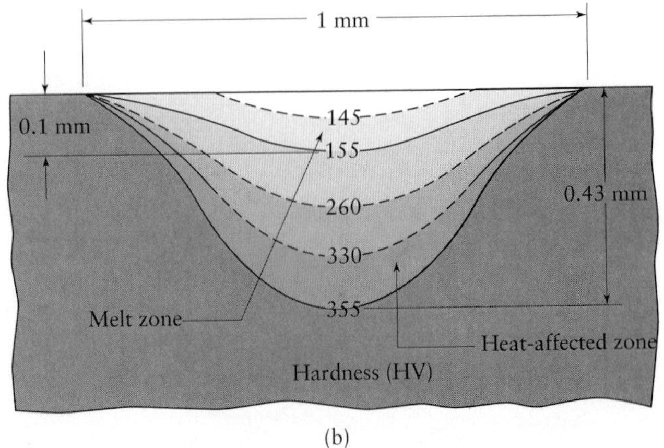

(a) (b)

FIGURE 12.15 (a) Weld bead on a cold-rolled nickel strip produced by a laser beam. (b) Microhardness profile across the weld bead. Note the softer condition of the weld bead compared with the base metal. *Source*: IIT Research Institute.

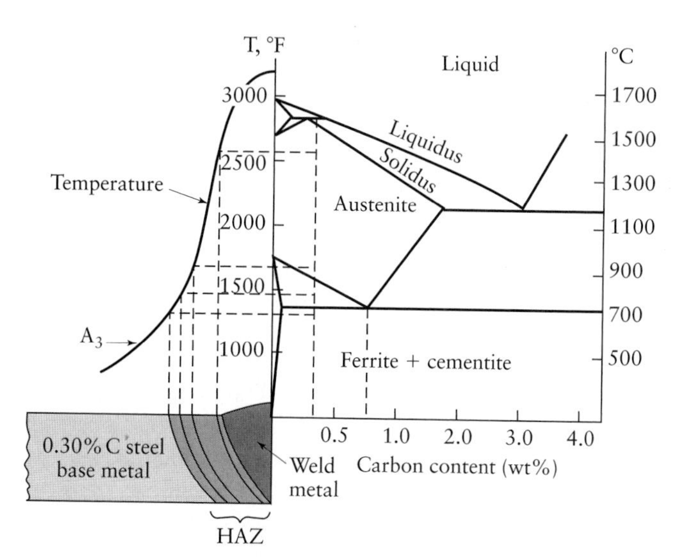

FIGURE 12.16 Schematic illustration of various regions in a fusion-weld zone and the corresponding phase diagram for 0.30% C steel. *Source*: Courtesy of the American Welding Society.

Heat-affected zone. The *heat-affected zone* is within the base metal itself. It has a microstructure different than that of the base metal before welding, because it has been subjected to elevated temperatures for a period of time during welding. The portions of the base metal that are far away from the heat source do not undergo any structural changes during welding. The properties and microstructure of the HAZ depend on (a) the rate of heat input and cooling and (b) the temperature to which this zone has been raised during welding. Figure 12.16 shows the heat-affected zone and the corresponding phase diagram for 0.30% C steel. In addition to metallurgical factors such as original grain size, grain orientation, and degree of prior cold work, the specific heat and thermal conductivity of the metals also influence the HAZ's size and characteristics.

The strength and hardness of the heat-affected zone depend partly on how the original strength and hardness of the particular alloy were developed prior to welding. As described in Chapters 3 and 5, they may have been developed by cold working, solid-solution strengthening, precipitation hardening, or various heat treatments. Of metals strengthened by these various methods, the simplest to analyze is base metal that has been cold worked, say, by cold rolling or forging.

The heat applied during welding *recrystallizes* the elongated grains (preferred orientation) of the cold-worked base metal. Grains that are away from the weld metal will recrystallize into fine equiaxed grains. However, grains close to the weld metal, having been subjected to elevated temperatures for a longer period of time, will grow. (See also Section 3.6.) This grain growth will result in a region that is softer and has less strength, and such a joint will be weakest in its heat-affected zone. The grain structure of such a weld exposed to corrosion by chemical reaction is shown in Fig. 12.17; the central vertical line is where the two workpieces originally met (butt welded). It is apparent that the effects of heat during welding on the HAZ for joints made with dissimilar metals, and for alloys strengthened by other mechanisms, are complex, and they are beyond the scope of this text.

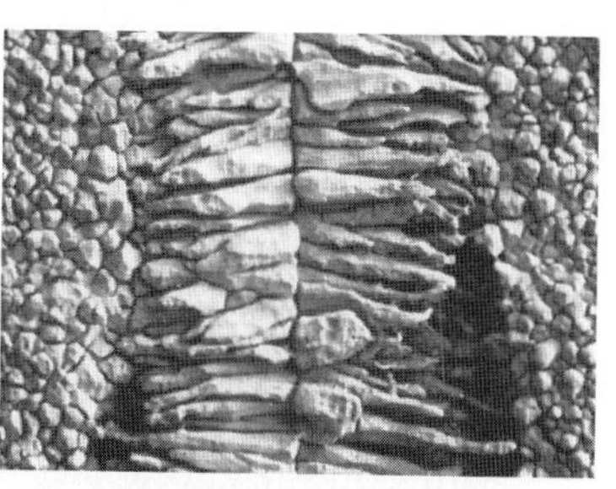

FIGURE 12.17 Intergranular corrosion of a weld in ferritic stainless-steel welded tube after exposure to a caustic solution. The weld line is at the center of the photograph. Scanning electron micrograph at 20×. *Source*: Courtesy of Allegheny Ludlum Corp.

12.5.1 Weld quality

Because of a history of thermal cycling and attendant microstructural changes, a welded joint may develop *discontinuities*. Welding discontinuities can also be caused by inadequate or careless application of established welding technologies or substandard operator training. The major discontinuities that affect weld quality are described as follows:

a) **Porosity.** *Porosity* in welds is caused by trapped *gases* that are released during melting of the weld zone, but trapped during solidification; by chemical reactions during welding; or by *contaminants*. Most welded joints contain some porosity, which is generally spherical in shape or in the form of elongated pockets. (See also Section 5.4.6.) The distribution of porosity in the weld zone may be random, or it may be concentrated in a certain region. Particularly important is the presence of hydrogen, which may be due to the use of damp fluxes or to environmental humidity; the result is porosity in aluminum-alloy welds and hydrogen embrittlement in steels. (See also Section 3.8.2.)

   Porosity in welds can be reduced by the following methods:

   1. Proper selection of electrodes and filler metals;
   2. Improvement of welding techniques, such as by preheating the weld area or increasing the rate of heat input;
   3. Proper cleaning of the weld zone and prevention of contaminants from entering the weld zone;
   4. Slowing the welding speed to allow time for gas to escape.

b) **Slag inclusions.** *Slag inclusions* are compounds, such as oxides, fluxes, and electrode coating materials, that are trapped in the weld zone. If the shielding gases used are not effective during welding, contamination from the environment may also contribute to such inclusions. Maintenance of proper welding conditions is also important, and when proper techniques are used, the molten slag will float to the surface of the molten weld metal and not be entrapped. Slag inclusions, or *slag entrapment*, may be prevented by the following methods:

   1. Cleaning the weld bead surface before the next layer is deposited (see Fig. 12.3), by using a hand or power wire brush;
   2. Providing adequate shielding gas;
   3. Redesigning the joint (see Section 12.16) to permit sufficient space for proper manipulation of the puddle of molten weld metal.

c) **Incomplete fusion and penetration.** *Incomplete fusion* (*lack of fusion*) produces poor weld beads, such as those shown in Fig. 12.18. A better weld can be obtained by the following methods:

   1. Raising the temperature of the base metal;
   2. Cleaning the weld area prior to welding;
   3. Changing the joint design and type of electrode;
   4. Providing adequate shielding gas.

   Incomplete penetration occurs when the depth of the welded joint is insufficient. Penetration can be improved by the following methods:

   1. Increasing the heat input;
   2. Lowering travel speed during welding;
   3. Modifying the design of the joint;
   4. Ensuring that the surfaces to be joined fit properly.

d) **Weld profile.** *Weld profile* is important not only because of its effects on the strength and appearance of the weld, but also because it can indicate incomplete

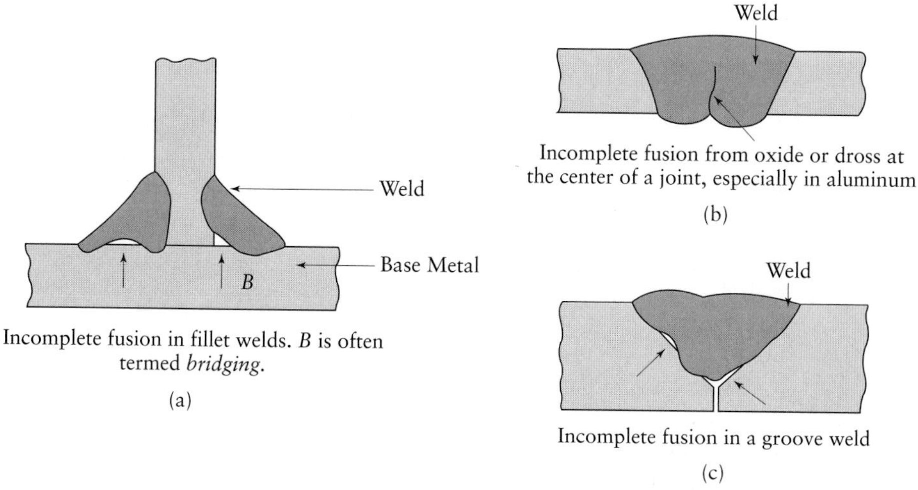

FIGURE 12.18 Examples of various discontinuities in fusion welds.

fusion or the presence of slag inclusions in multiple-layer welds. **Underfilling** results when the joint is not filled with the proper amount of weld metal (Fig. 12.19a); **undercutting** results from melting away of the base metal and subsequent generation of a groove in the shape of a sharp recess or notch. If deep or sharp, an undercut can act as a stress raiser and reduce the fatigue strength of the joint, which may lead to premature failure. **Overlap** (Fig. 12.19b) is a surface discontinuity generally caused by poor welding practice and selection of the wrong materials. A proper weld bead is shown in Fig. 12.19c.

e) **Cracks.** *Cracks* may occur in various locations and directions in the weld area. The typical types of cracks are longitudinal, transverse, crater, underbead, and toe cracks (Fig. 12.20). These cracks generally result from a combination of the following factors:

1. Temperature gradients that cause thermal stresses in the weld zone;
2. Variations in the composition of the weld zone that cause different contractions;
3. Embrittlement of grain boundaries by segregation of elements, such as sulfur, to the grain boundaries (see Section 3.4.2) as the solid–liquid boundary moves when the weld metal begins to solidify;
4. Hydrogen embrittlement (see Section 3.8.2);

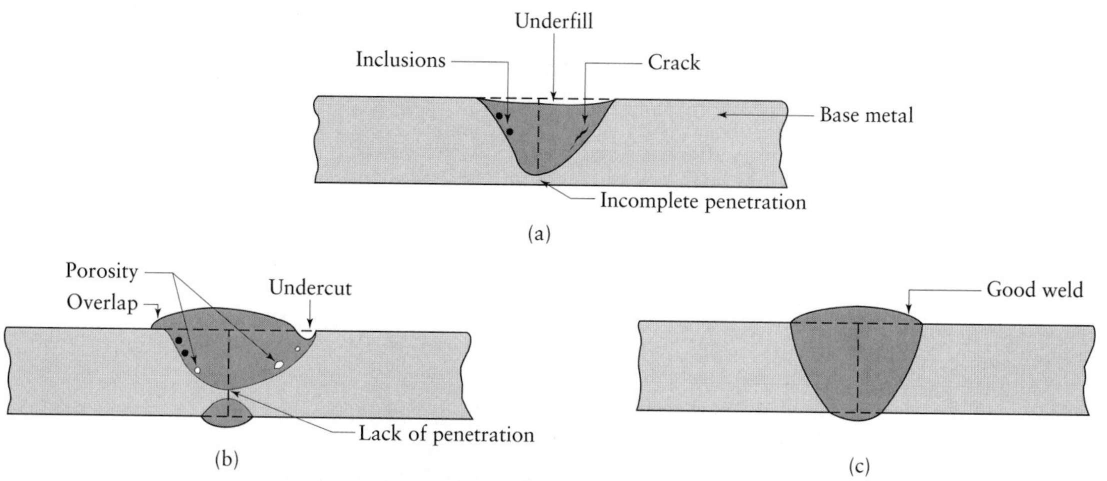

FIGURE 12.19 Examples of various defects in fusion welds.

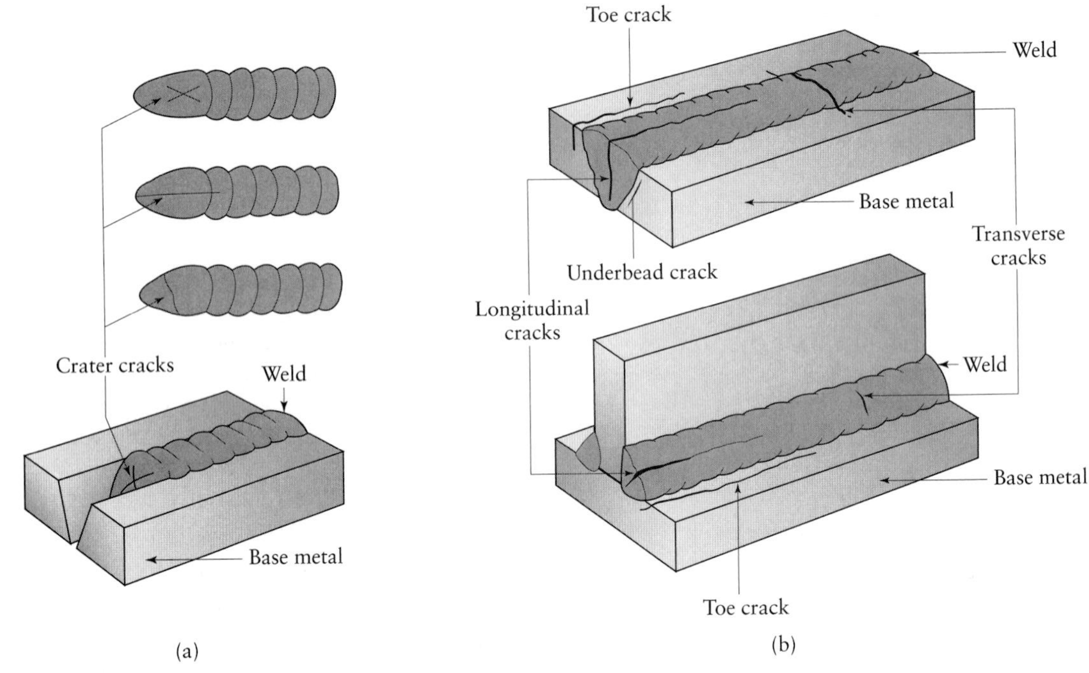

FIGURE 12.20 Types of cracks in welded joints. The cracks are caused by thermal stresses that develop during solidification and contraction of the weld bead and the welded structure: (a) crater cracks; (b) various types of cracks in butt and T joints.

5. Inability of the weld metal to contract during cooling (Fig. 12.21), a situation similar to the development of *hot tears* in castings, due to excessive restraint of the workpiece (see also Section 5.4.6).

Cracks are classified as follows: **Hot cracks** occur while the joint is still at elevated temperatures; **cold cracks** develop after the weld metal has solidified. Some crack-prevention measures are listed as follows:

1. Modify the design of the joint to minimize thermal stresses from shrinkage during cooling;
2. Change welding-process parameters, procedures, and sequence;
3. Preheat components being welded;
4. Avoid rapid cooling of the components after welding.

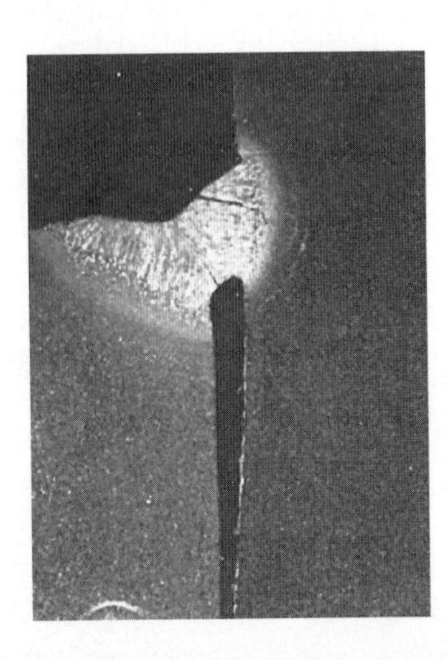

FIGURE 12.21 Crack in a weld bead. The two components were not allowed to contract after the weld was completed. *Source*: Courtesy of Packer Engineering.

f) **Lamellar tears.** In describing the anisotropy of plastically deformed metals (see Section 3.5), we stated that because of the alignment of nonmetallic impurities and inclusions (stringers), the workpiece is weaker when tested in its thickness direction than in other directions (*mechanical fibering*). This condition is particularly evident in rolled plates and structural shapes. In welding such components, *lamellar tears* may develop, because of shrinkage of the restrained members in the structure during cooling. Such tears can be avoided by providing for shrinkage of the members or by modifying the joint design to make the weld bead penetrate the weaker member more deeply.

g) **Surface damage.** During welding, some of the metal may spatter and be deposited as small droplets on adjacent surfaces; also, in arc welding, the electrode may inadvertently contact the parts being welded at places outside of the weld zone (**arc strikes**). Such surface discontinuities may be objectionable for reasons of appearance or subsequent use of the welded part. If severe, they may adversely affect the properties of the welded structure, particularly for notch-sensitive metals. Thus, following proper welding procedures is important in avoiding surface damage.

h) **Residual stresses.** Because of localized heating and cooling during welding, expansion and contraction of the weld area cause *residual stresses* in the workpiece. (See also Section 2.10.) Residual stresses can cause the following detrimental effects:

1. Distortion, warping, and buckling of the welded parts (Fig. 12.22);
2. Stress-corrosion cracking (see Section 3.8.2);
3. Further distortion if a portion of the welded structure is subsequently removed, say, by machining or sawing;
4. Reduced fatigue life.

The type and distribution of residual stresses in welds are best described by reference to Fig. 12.23a. When two plates are being welded, a long, narrow region is subjected to elevated temperatures, whereas the plates as a whole are essentially at ambient temperature. As the weld is completed and time elapses, the heat from the weld area dissipates laterally to the plates. The plates thus begin to expand longitudinally while the welded length begins to contract; these two opposing effects cause residual stresses that are typically distributed as shown in Fig. 12.23b. Note that the magnitude of compressive residual stresses in the plates diminishes to zero at a point away from the weld area. Because no external forces are acting on the welded plates, the tensile and compressive forces represented by these residual stresses must balance each other.

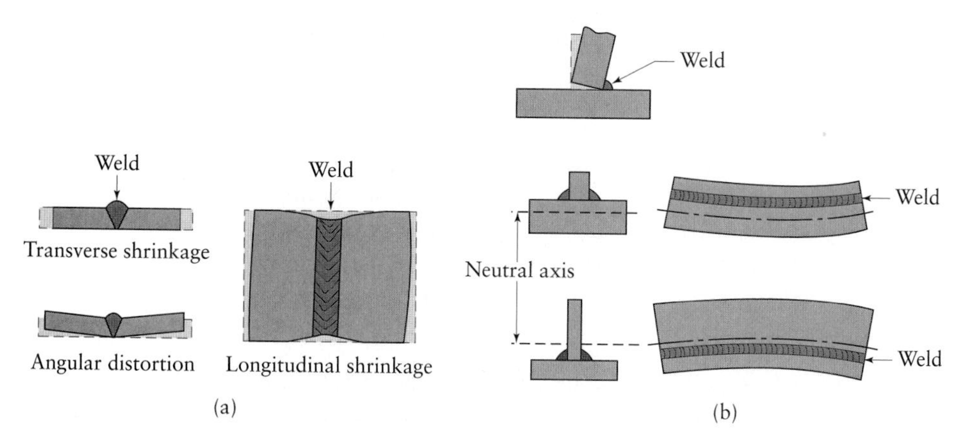

FIGURE 12.22 Distortion of parts after welding: (a) butt joints and (b) fillet welds. Distortion is caused by differential thermal expansion and contraction of different regions of the welded assembly. Warping can be reduced or eliminated by proper weld design and fixturing prior to welding.

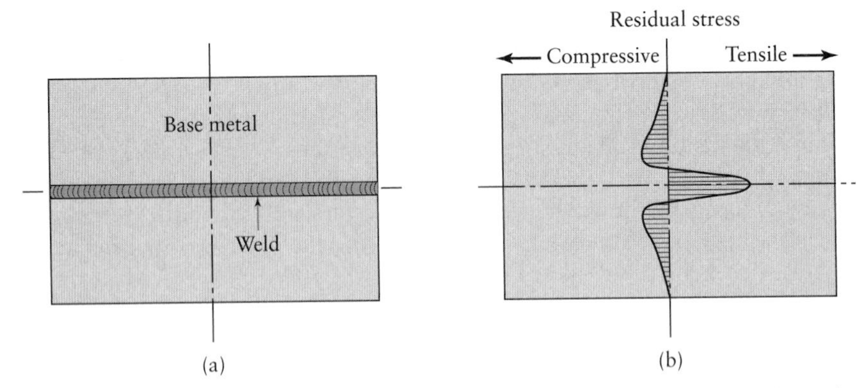

FIGURE 12.23 Residual stresses developed in a straight butt joint. *Source*: Courtesy of the American Welding Society.

In complex welded structures, residual-stress distributions are three dimensional and hence difficult to analyze. The preceding example involves only two plates that are not restrained from movement; in other words, the plates are not an integral part of a larger structure. If the plates are restrained, reaction stresses will be generated, because the plates are not free to expand or contract. This situation arises particularly in structures with high stiffness.

Stress relieving of welds. The problems caused by residual stresses, such as distortion, buckling, or cracking, can be reduced by *preheating* the base metal or the parts to be welded. Preheating reduces distortion by lowering the cooling rate and the magnitude of thermal stresses (by reducing the elastic modulus of the metals being welded); this technique also reduces shrinkage and possible cracking of the joint. Preheating is also effective in welding hardenable steels for which rapid cooling of the weld area would otherwise be harmful. The workpieces may be heated in a furnace or electrically or inductively; thin sections may be heated by radiant lamps or a hot-air blast. For optimum results, preheating temperatures and cooling rates must be controlled in order to attain acceptable strength and toughness in the welded structure.

Residual stresses can be reduced by *stress relieving* or *stress-relief annealing* the welded structure; the temperature and time required for stress relieving depend on the type of material and magnitude of the residual stresses developed. Other methods of stress relieving include peening, hammering, and surface rolling the weld bead area. (See Section 4.5.1.) These processes induce compressive residual stresses, thus reducing or eliminating tensile residual stresses in the weld. For multilayer welds, the first and last layers should not be peened, because peening may cause damage to those layers.

Residual stresses can also be relieved, or reduced, by plastically deforming the structure by a small amount. (See Fig. 2.34.) This technique can be used in some welded structures, such as pressure vessels, by pressurizing the vessels internally (*proof stressing*). In order to reduce the possibility of sudden fracture under high internal pressure, the weld must be free from notches or discontinuities, which could act as points of stress concentration.

Heat treatment of welds. In addition to stress relieving, welds may also be *heat treated* by various techniques in order to modify and improve properties such as strength and toughness. These techniques include (a) annealing, normalizing, or quenching and tempering of steels; and (b) solution treatment and aging of precipitation-hardenable alloys (see Section 5.11).

12.5.2 Weldability

Weldability of a metal is generally defined as the metal's capacity to be welded into a specific structure that has certain properties and characteristics and that will satisfactorily meet service requirements. As can be expected, weldability involves a large number of variables, making generalizations difficult. Material characteristics, such as alloying elements, impurities, inclusions, grain structure, and processing history of the base metal and filler metal, are important. Because of the melting, solidification, and microstructural changes involved, a thorough knowledge of the phase diagram and the response of the metal or alloy to elevated temperatures over a period of time also is essential. Other factors that influence weldability are the mechanical and physical properties of strength, toughness, ductility, notch sensitivity, elastic modulus, specific heat, melting point, thermal expansion, surface-tension characteristics of the molten metal, and corrosion.

Preparation of surfaces for welding is important, as are the nature and properties of surface oxide films and adsorbed gases. The welding process employed significantly affects the temperatures developed and their distribution in the weld zone. Other factors that influence weldability are shielding gases, fluxes, moisture content of the coatings on electrodes, welding speed, welding position, cooling rate, preheating, and postwelding techniques (such as stress relieving and heat treating).

The following list briefly summarizes, in alphabetic order, the general weldability of specific metals. The weldability of these materials can vary significantly, with some requiring special welding techniques and good control of processing parameters.

1. *Aluminum alloys*: weldable at a high rate of heat input; aluminum alloys containing zinc or copper generally are considered unweldable.
2. *Cast irons*: generally weldable.
3. *Copper alloys*: similar to that of aluminum alloys.
4. *Lead*: easy to weld.
5. *Magnesium alloys*: weldable with the use of protective shielding gas and fluxes.
6. *Molybdenum*: weldable under well-controlled conditions.
7. *Nickel alloys*: weldable by various processes.
8. *Niobium (columbium)*: weldable under well-controlled conditions.
9. *Stainless steels*: weldable by various processes.
10. *Steels, galvanized and prelubricated*: The presence of zinc coating and lubricant layer adversely affects weldability.
11. *Steels, high-alloy*: generally good weldability under well-controlled conditions.
12. *Steels, low-alloy*: fair to good weldability.
13. *Steels, plain-carbon*: excellent weldability for low-carbon steels; fair to good weldability for medium-carbon steels; poor weldability for high-carbon steels.
14. *Tantalum*: weldable under well-controlled conditions.
15. *Tin*: easy to weld.
16. *Titanium alloys*: weldable with the use of proper shielding gases.
17. *Tungsten*: weldable under well-controlled conditions.
18. *Zinc*: difficult to weld; soldering preferred.
19. *Zirconium*: weldable with the use of proper shielding gases.

12.5.3 Testing welded joints

As in all manufacturing processes, the *quality* of a welded joint is established by testing. Several standardized tests and test procedures have been established and are available from organizations such as the American Society for Testing and Materials (ASTM), the American Welding Society (AWS), the American Society of Mechanical

Engineers (ASME), the American Society of Civil Engineers (ASCE), and various federal agencies. Welded joints may be tested either destructively or nondestructively (see Section 4.8); each technique has its capabilities, sensitivity, limitations, reliability, and need for special equipment and operator skill.

a) **Destructive testing techniques.** Five methods of destructively testing welded joints are commonly used. A brief review of each method is given below.

Tension test. Longitudinal and transverse tension tests are performed on specimens removed from actual welded joints and from the weld-metal area; stress–strain curves are then obtained by the procedures described in Section 2.2. These curves indicate the yield strength (Y), ultimate tensile strength (UTS), and ductility of the welded joint in different locations and directions. Ductility is measured in terms of percentage elongation and percentage reduction of area. Tests of weld hardness may also be used to indicate weld strength and microstructural changes in the weld zone.

Tension-shear test. The specimens in the tension-shear test (Fig. 12.24a) are specially prepared to simulate actual welded joints and procedures. The specimens are subjected to tension, and the shear strength of the weld metal and the location of fracture are determined.

Bend test. Several bend tests have been developed to determine the ductility and strength of welded joints. In one test, the welded specimen is bent around a fixture (*wraparound bend test*; Fig. 12.24b). In another test, the specimens are tested in *three-point transverse bending* (Fig. 12.24c; see also Fig. 2.23). These tests help establish the relative ductility and strength of welded joints.

Fracture toughness test. Fracture toughness tests commonly use impact-testing techniques, described in Section 2.9. Charpy V-notch specimens are prepared and tested for toughness; other toughness tests include the drop-weight test, in which the energy is supplied by a falling weight.

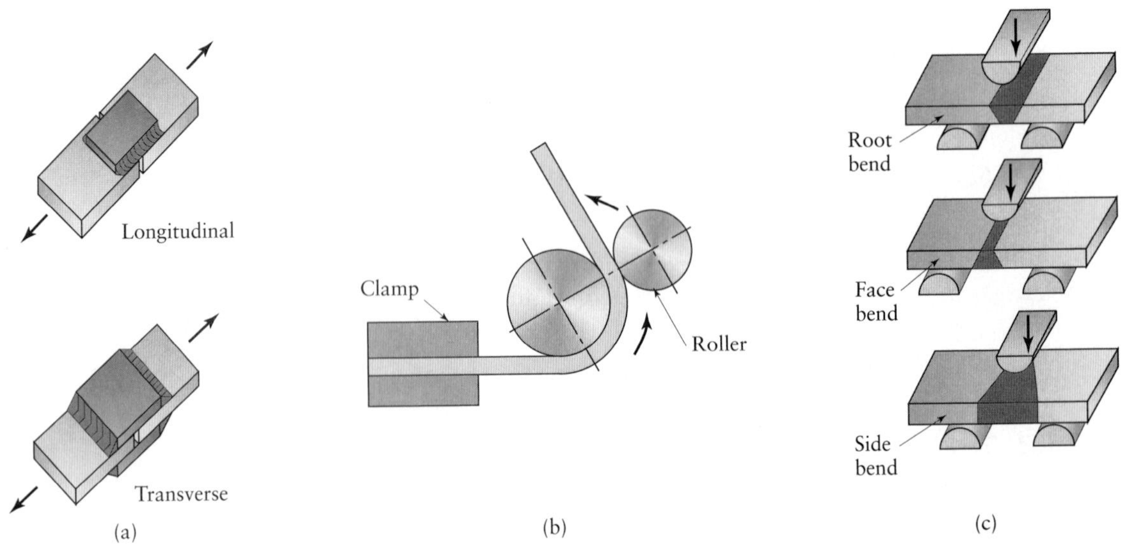

FIGURE 12.24 (a) Types of specimens for tension-shear testing of welds. (b) Wraparound bend test method. (c) Three-point bending of welded specimens. (See also Fig. 2.23.)

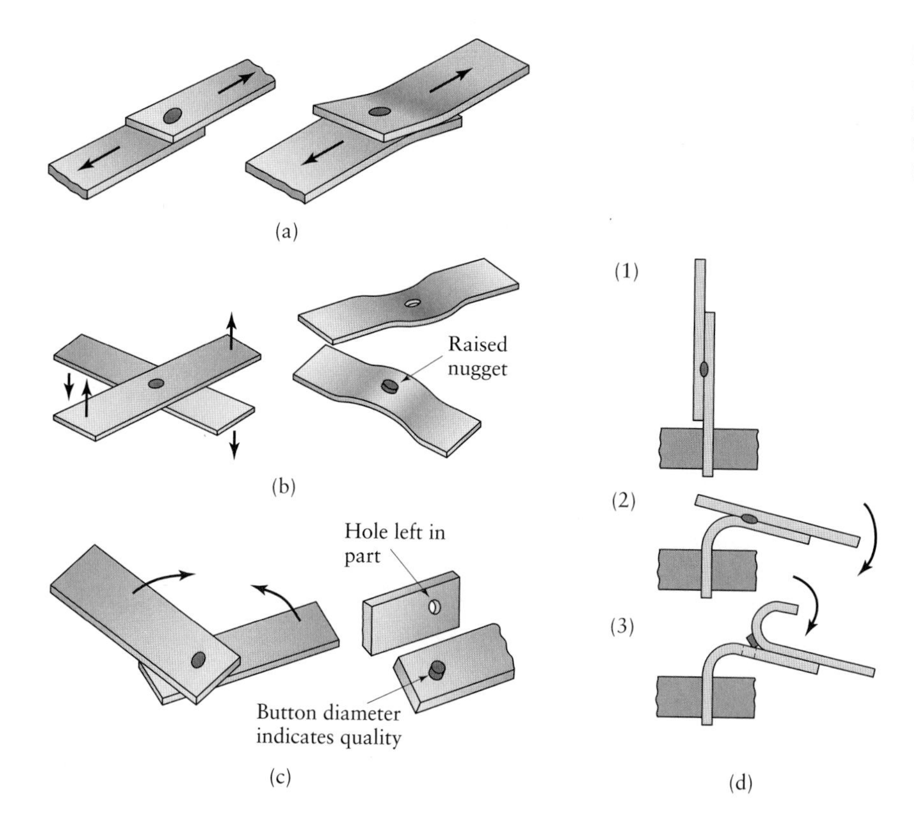

FIGURE 12.25 (a) Tension-shear test for spot welds. (b) Cross-tension test. (c) Twist test. (d) Peel test.

Corrosion and creep tests. In addition to mechanical tests, welded joints may be tested for corrosion and creep resistance. Because of the difference in the composition and microstructure of the materials in the weld zone, preferential corrosion may take place in this zone. (See Fig. 12.17.) Creep tests are essential in determining the behavior of welded joints at elevated temperatures.

Testing of spot welds. Spot-welded joints may be tested for weld-nugget strength (Fig. 12.25), using the (1) *tension-shear*, (2) *cross-tension*, (3) *twist*, and (4) *peel tests*. Because they are easy to perform and inexpensive, tension-shear tests are commonly used in fabricating facilities. The cross-tension and twist tests are capable of revealing flaws, cracks, and porosity in the weld area. The peel test is commonly used for thin sheets. After the joint has been bent and peeled, the shape and size of the torn-out weld nugget are evaluated.

b) **Nondestructive testing techniques.** Welded structures often have to be tested nondestructively, particularly for critical applications where weld failure can be catastrophic, such as in pressure vessels, pipelines, load-bearing structural members, and power plants. Nondestructive testing techniques for welded joints usually consist of visual, radiographic, magnetic-particle, liquid-penetrant, and ultrasonic testing methods (see Section 4.8.1).

12.5.4 Welding-process selection

In addition to the material characteristics described thus far, selection of an appropriate welding process involves the following considerations:

a) Configuration of the parts or structure to be welded and their thickness and size;

b) The methods used to manufacture component parts;

c) Service requirements, such as the type of loading and stresses generated;

d) Location, accessibility, and ease of welding;

e) Effects of distortion and discoloration;

f) Appearance;

g) Costs involved in edge preparation, welding, and postprocessing of the weld, including machining and finishing operations.

12.6 Cold Welding

In *cold welding* (CW), pressure is applied to the surfaces of the parts, through either dies or rolls. Because of the resulting plastic deformation, it is necessary that at least one, but preferably both, of the mating parts be sufficiently ductile. The interface is usually cleaned by wire brushing prior to welding. However, in joining two dissimilar metals that are mutually soluble, brittle *intermetallic compounds* may form (see Section 5.2.2), resulting in a weak and brittle joint. An example is the bonding of aluminum and steel, where a brittle intermetallic compound is formed at the interface. The best bond strength and ductility are obtained with two similar materials. Cold welding can be used to join small workpieces made of soft, ductile metals; applications include electrical connections and wire stock.

Roll bonding. In *roll bonding*, or *roll welding* (ROW), the pressure required for cold welding of long pieces or continuous strips can be applied through a pair of rolls (Fig. 12.26). This process, developed in the 1960s, is used for manufacturing some U.S. coins. (See Example 12.2.) Surface preparation is important for improved interfacial strength. Roll bonding can also be carried out at elevated temperatures (*hot roll bonding*) for improved interfacial strength; typical examples include *cladding* (Section 4.5.1) pure aluminum over precipitation-hardened aluminum-alloy sheet (Section 5.11.2) and cladding stainless steel over mild steel for corrosion resistance. Another common application of roll bonding is the production of bimetallic strips for thermostats and similar controls, using two layers of materials with different coefficients of thermal expansion. (See also Section 12.7.) Bonding in only selected regions in the interface can be achieved by depositing a parting agent such as graphite or ceramic (called *stop-off*). The use of this technique in *superplastic forming* of sheet-metal structures is shown in Fig. 7.47.

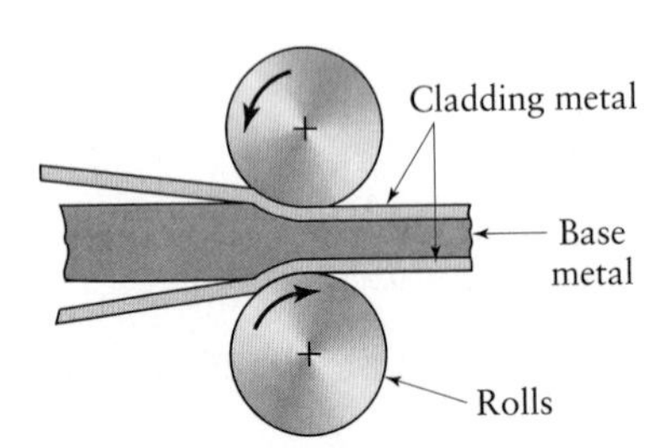

FIGURE 12.26 Schematic illustration of the roll-bonding, or cladding, process.

EXAMPLE 12.2 Roll bonding of the U.S. quarter

The technique used for manufacturing composite U.S. quarters is roll bonding of two outer layers of 75% copper-25% nickel (cupronickel), each 1.2 mm (0.048 in.) thick, with an inner layer of pure copper 5.1 mm (0.20 in.) thick. To obtain good bond strength, the mating surfaces (*faying surfaces*) are first chemically cleaned and wire brushed. The strips are then rolled to a thickness of 2.29 mm (0.090 in.); a second rolling operation reduces the final thickness to 1.36 mm (0.0535 in.). The strips thus undergo a total reduction in thickness of 82%. Because volume constancy is maintained in plastic deformation, there is a major increase in the surface area between the layers, thus generating clean interfacial surfaces and hence better bonding. This extension in surface area under the high pressure of the rolls, combined with the solid solubility of nickel in copper (see Section 5.2.1), produces a strong bond.

12.7 Ultrasonic Welding

In *ultrasonic welding* (USW), the faying surfaces of the two members are subjected to a static normal force and oscillating shearing (tangential) stresses. The shearing stresses are applied by the tip of a **transducer** (Fig. 12.27a) similar to that used for ultrasonic machining (see Fig. 9.19a); the frequency of oscillation generally ranges from 10 kHz to 75 kHz. The energy required increases with the thickness and hardness of the materials being joined. Proper coupling between the transducer and the tip (called *sonotrode*, from the word *sonic*) is important for effective operation. The welding tip can be replaced with rotating disks (Fig. 12.27b) for seam-welding structures (similar to those shown in Fig. 12.33 for resistance seam welding), one component of which can be a sheet or foil.

The shearing stresses cause small-scale plastic deformation at the workpiece interfaces, breaking up oxide films and contaminants, thus allowing good contact and producing a strong solid-state bond. Temperatures generated in the weld zone are usually in the range of one third to one half the melting point (absolute scale) of the

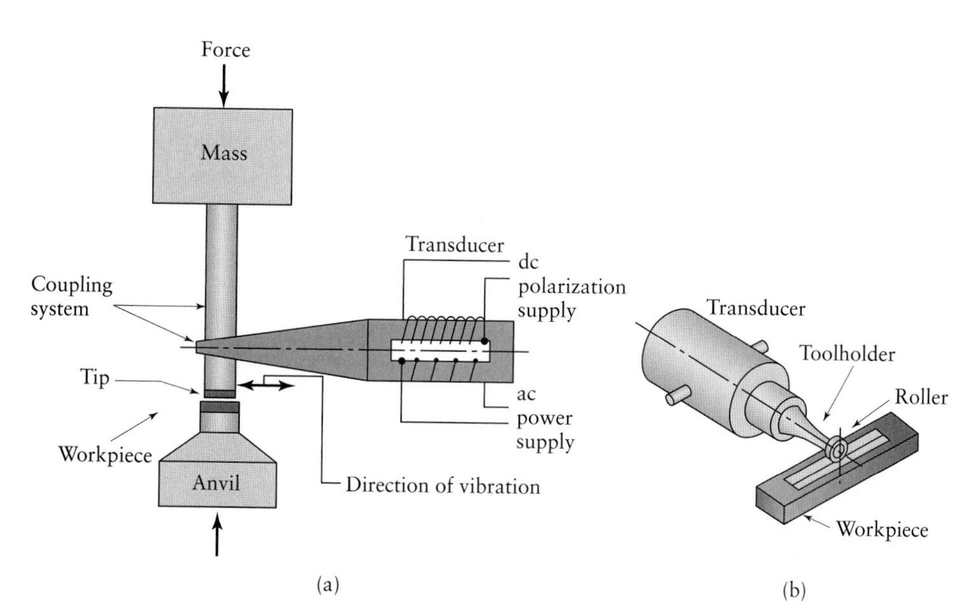

FIGURE 12.27 (a) Components of an ultrasonic-welding machine for lap welds. The lateral vibrations of the tool tip cause plastic deformation and bonding at the interface of the workpieces. (b) Ultrasonic seam welding using a roller.

metals joined; therefore, no melting and fusion take place. In certain situations, however, temperatures can be sufficiently high to cause metallurgical changes in the weld zone.

The ultrasonic-welding process is reliable and versatile; it can be used with a wide variety of metallic and nonmetallic materials, including dissimilar metals (*bimetallic strips*). It is used extensively in joining plastics and in the automotive and consumer-electronics industries for lap welding of sheet, foil, and thin wire and in packaging with foils. The mechanism of joining thermoplastics (see Section 12.15.1) by ultrasonic welding is different than that for metals, and interfacial melting takes place, due to the much lower melting temperature of plastics (see Table 10.2).

12.8 Friction Welding

In the joining processes described thus far, the energy required for welding is supplied externally. In *friction welding* (FRW), the heat required for welding is, as the name implies, generated through friction at the interface of the two members being joined; thus, the source of energy is mechanical. You can demonstrate the significant rise in temperature from friction simply by rubbing your hands together fast or sliding down a rope rapidly.

In the friction-welding process, one of the members remains stationary while the other is placed in a chuck or collet and rotated at a high constant speed; the two members to be joined are then brought into contact under an axial force (Fig. 12.28). The surface speed of rotation may be as high as 900 m/min (3000 ft/min). The rotating member must be clamped securely to resist both torque and axial forces without slip. After sufficient interfacial contact is established, the rotating member is brought to a sudden stop, so that the weld is not destroyed by shearing, while the axial force is increased. The pressure at the interface and the resulting friction produce sufficient heat for a strong joint to take place.

The weld zone is usually confined to a narrow region, depending on (a) the amount of heat generated, (b) the thermal conductivity of the materials, and (c) the mechanical properties of the materials at elevated temperature. The shape of the welded joint depends on the rotational speed and the axial force applied (Fig. 12.29); these factors must be controlled to obtain a uniformly strong joint. The radially outward movement of the hot metal at the interface pushes oxides and other contaminants out of the interface.

Developed in the 1940s, friction welding can be used to join a wide variety of materials, provided that one of the components has some rotational symmetry. Solid as well as tubular parts can be joined by this method, with good joint strength. Solid steel bars up to 100 mm (4 in.) in diameter and pipes up to 250 mm (10 in.) outside diameter have been welded successfully using this process. Because of the combined

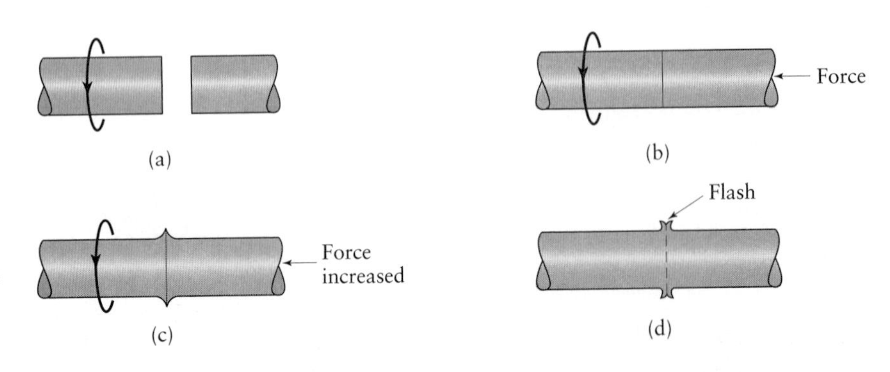

FIGURE 12.28 Sequence of operations in the friction-welding process. (a) The part on the left is rotated at high speed. (b) The part on the right is brought into contact under an axial force. (c) The axial force is increased; flash begins to form. (d) The part on the left stops rotating. The weld is completed. Flash can be removed by machining or grinding.

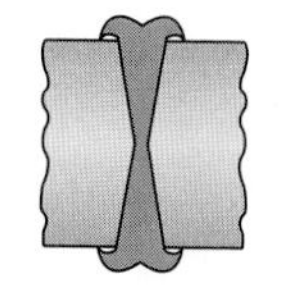

(1) High pressure or low speed

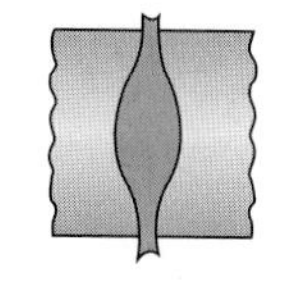

(2) Low pressure or high speed

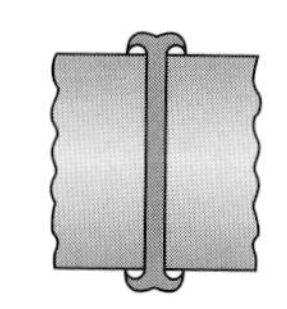

(3) Optimum

FIGURE 12.29 Shape of the fusion zone in friction welding as a function of the force applied and the rotational speed.

heat and pressure, the interface in FRW develops a flash by plastic deformation (upsetting) of the heated zone; if objectionable, this flash can easily be removed by subsequent machining or grinding.

Variations on the friction-welding process are described as follows:

a) **Inertia friction welding.** *Inertia friction welding* is a modification of FRW, although the term has been used interchangeably with *friction welding*. The energy required in this process is supplied through the kinetic energy of a flywheel. The flywheel is accelerated to the proper speed, the two members are brought into contact, and an axial force is applied; as friction at the interface begins to slow down the flywheel, the axial force is increased. The weld is completed when the flywheel comes to a stop. The timing of this sequence is important; if it is not properly controlled, the weld quality will be poor. The rotating mass of inertia friction-welding machines can be adjusted for applications requiring different levels of energy, which depend on workpiece size and its properties. In one application of inertia friction welding, a shaft 10 mm (0.4 in.) in diameter is welded to automotive turbocharger impellers at a rate of one every 15 s.

b) **Linear friction welding.** *Linear friction welding* is another development that involves subjecting the interface of the two parts to be joined to a linear reciprocating motion of at least one of the components to be joined; thus, the parts do not have to be circular or tubular in cross-section. In this process, one part is moved across the face of the other part, using a balanced reciprocating mechanism. Linear friction welding is capable of joining square or rectangular components, as well as round parts, made of metals or plastics. In one application, a rectangular titanium-alloy part was friction welded at a linear frequency of 25 Hz, an amplitude of ± 2 mm (0.08 in.), and a pressure of 100 MPa (15,000 psi) acting on a 240-mm^2 (0.38-in^2) interface. Various other metal parts have been welded successfully, with rectangular cross-sections as large as 50 mm $\times$ 20 mm (2 in. $\times$ 0.8 in.).

c) **Friction stir welding.** This technique is a more recent process, originally intended for welding of aerospace alloys, especially aluminum extrusions, although current research is being directed at extending this process towards polymers and composite materials. Whereas in conventional friction welding, heating of interfaces is achieved through friction by rubbing two contacting surfaces, in the *friction stir-welding* (FSW) process, a third body is rubbed against the two surfaces to be joined. The rotating tool is a small (5 mm to 6 mm in diameter, 5 mm in height) rotating member that is plunged into the joint (Fig. 12.30). The contact pressures cause frictional heating, raising the temperature to the range of 230°C to 260°C (450°F to 500°F). The probe at the tip of the rotating tool forces heating and mixing or stirring of the material in the joint.

The welding equipment can be a conventional, vertical-spindle milling machine (see Fig. 8.59b), and the process is relatively easy to implement. The thickness of the welded material can be as little as 1 mm (0.04 in.) and as much as 30 mm (1.2 in.) or so. The weld quality is high, with minimal pores and with

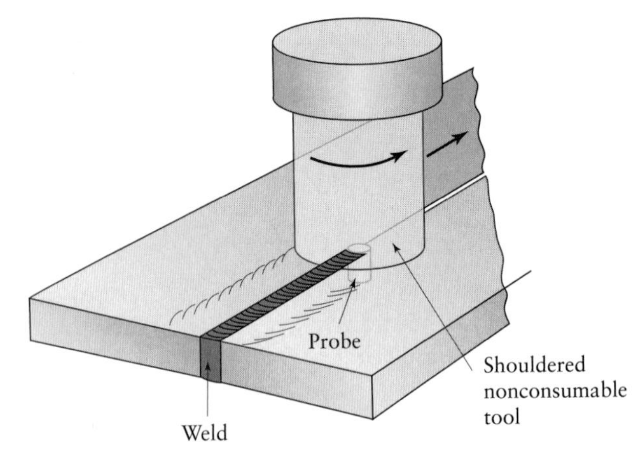

FIGURE 12.30 The principle of the friction stir-welding process. Aluminum-alloy plates up to 75 mm (3 in.) thick have been welded by this process. *Source*: TWI, Cambridge, United Kingdom.

uniform material structure. The welds are produced with low heat input and, therefore, low distortion and little microstructural changes; furthermore, there are no fumes or spatter produced, and the process is suitable for automation.

12.9 Resistance Welding

Resistance welding (RW) covers a number of processes in which the heat required for welding is produced by means of the *electrical resistance* between the two members to be joined; hence, these processes' major advantages include the fact that they don't require consumable electrodes, shielding gases, or flux. The heat generated in resistance welding is given by the general expression

$$H = I^2Rt, \tag{12.1}$$

where H = heat generated, in joules (watt-seconds); I = current, in amperes; R = resistance, in ohms; and t = flow time of the current, in seconds. This equation is often modified to represent the actual heat energy available in the weld by including a factor K that represents the energy losses through radiation and conduction. Thus, Eq. (12.1) becomes $H = I^2RtK$, where the value of K is less than unity.

The total resistance in these processes, such as in the resistance spot welding shown in Fig. 12.31, is the sum of the following:

1. Resistance of the electrodes;
2. Electrode–workpiece contact resistances;
3. Resistances of the individual parts to be welded;
4. Workpiece–workpiece contact resistances (faying surfaces).

The magnitude of the current in resistance-welding operations may be as high as 100,000 A, although the voltage is typically only 0.5–10 V. The actual temperature rise at the joint depends on the specific heat and thermal conductivity of the metals to be joined; consequently, because they have high thermal conductivity, metals such as aluminum and copper require high heat concentrations. Similar as well as dissimilar metals can be joined by resistance welding. Electrode materials should have high thermal conductivity and strength at elevated temperatures; they are typically made of copper alloys.

Developed in the early 1900s, resistance-welding processes require specialized machinery, and many are now operated with programmable computer control. The machinery is generally not portable, and the process is most suitable for use in

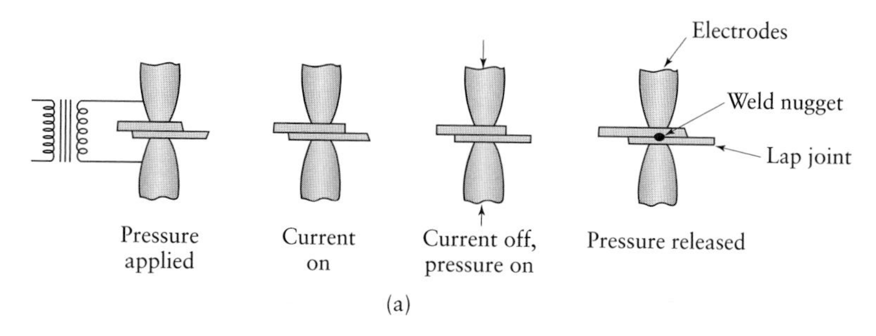

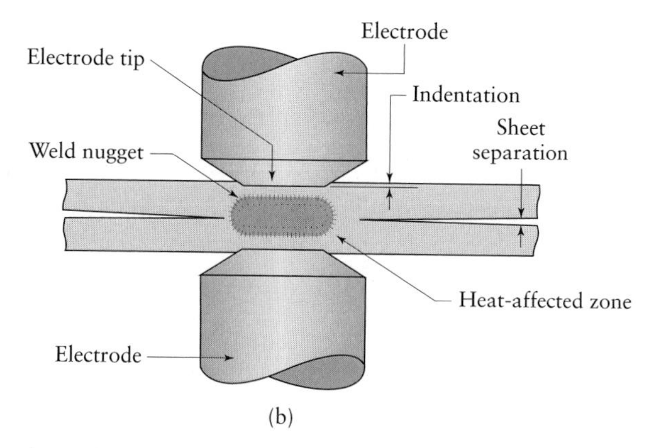

FIGURE 12.31 (a) Sequence in the resistance spot-welding process. (b) Cross-section of a spot weld, showing weld nugget and light indentation by the electrode on sheet surfaces. This method is one of the most common processes used in sheet-metal fabrication and automotive-body assembly.

manufacturing plants and machine shops. There are five basic methods of resistance welding: spot, seam, projection, flash, and upset welding. Lap joints are used in the first three processes and butt joints in the last two.

12.9.1 Resistance spot welding

In *resistance spot welding* (RSW), the tips of two opposing solid cylindrical electrodes contact the lap joint of two sheet metals, and resistance heating produces a spot weld (Fig. 12.31a). In order to obtain a good bond in the **weld nugget** (Fig. 12.31b), pressure is continually applied until the current is turned off; accurate control and timing of the electric current and pressure are essential. Currents usually range from 3000 A to 40,000 A, depending on the materials being welded and their thickness. Modern equipment for spot welding is computer controlled for optimum timing of current and pressure, and the spot-welding guns are manipulated by programmable robots (Section 14.7). Spot welding is widely used for fabricating sheet-metal products. Examples of its applications range from attaching handles to stainless-steel cookware to rapid spot welding of automobile bodies, using multiple electrodes; a typical automobile may have as many as 10,000 spot welds.

The strength of the weld depends on the surface roughness and cleanliness of the mating surfaces; thus oil, paint, and thick oxide layers should be removed before welding. However, the presence of uniform, thin oxide layers and other contaminants is not critical. The weld nugget is generally up to 10 mm (0.375 in.) in diameter. The surface of the weld spot has a slightly discolored indentation.

Spot welding is the simplest and most commonly used resistance-welding process. Welding may be performed by means of single or multiple electrodes, and the pressure required is supplied through mechanical or pneumatic means. The *rocker-arm type* of spot-welding machines is generally used for smaller parts, whereas the *press-type* machines is used for larger workpieces. The shape and surface condition of the electrode tip and accessibility of the area to be welded are important factors in spot welding; a variety of electrode shapes is used to spot weld areas that are difficult to reach (Fig. 12.32).

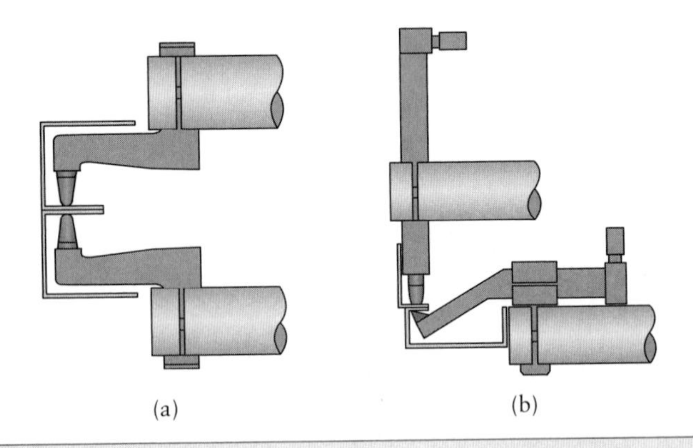

FIGURE 12.32 Types of special electrodes designed for easy access in spot-welding operations for complex shapes.

EXAMPLE 12.3 Heat generated in resistance spot welding

Assume that two 1-mm (0.04-in.)-thick steel sheets are being spot welded at a current of 5000 A and current flow time $t = 0.1$ s. Using electrodes 5 mm (0.2 in.) in diameter, estimate the amount of heat generated and its distribution in the weld zone.

Solution. Let's assume that the effective resistance in this operation is 200 $\mu\Omega$. Then, according to Eq. (12.1),

$$\text{Heat} = (5000)^2(0.0002)(0.1) = 500\ \text{J}.$$

From the information given, we estimate the weld-nugget volume to be 30 mm^3 (0.0018 in^3). If we assume a density for steel of 8000 kg/m^3 (0.008 g/mm^3), the weld nugget has a mass of 0.24 g. Since the heat required to melt 1 g of steel is about 1400 J, the heat required to melt the weld nugget is (1400)(0.24) = 336 J. Consequently, the remaining heat (164 J), or 33%, is dissipated into the metal surrounding the nugget.

12.9.2 Resistance seam welding

Resistance seam welding (RSEW) is a modification of resistance spot welding wherein the electrodes are replaced by rotating wheels or rollers (Fig. 12.33). With a continuous ac power supply, the electrically conducting rollers produce continuous spot welds whenever the current reaches a sufficiently high level in the ac cycle; these welds are actually overlapping spot welds and can produce a joint that is liquid tight and gas tight (Fig. 12.33b). The typical welding speed is 1.5 m/min (60 in./min) for thin sheet. With intermittent application of current to the rollers, a series of spot welds at

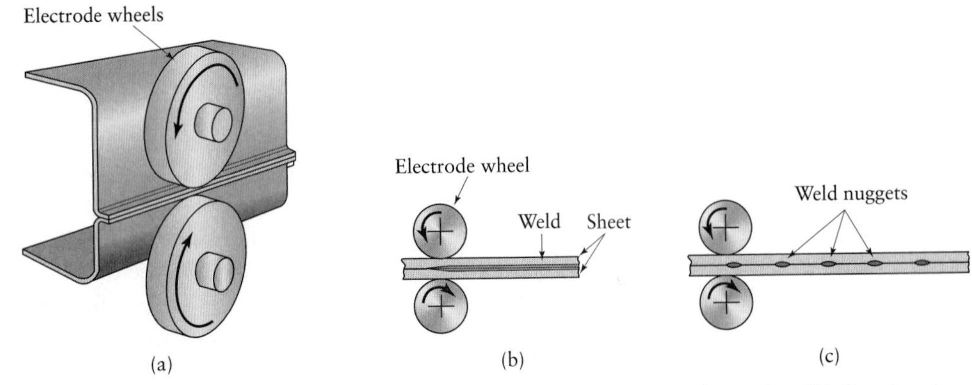

FIGURE 12.33 (a) Seam-welding process, with rolls acting as electrodes. (b) Overlapping spots in a seam weld. (c) Roll spot welds.

various intervals can be made along the length of the seam (Fig. 12.33c), a procedure called **roll spot welding.** The RSEW process is used to make the longitudinal (side) seam of cans for household products, mufflers, gasoline tanks, and other containers. In *mash-seam welding*, the overlapping welds are about one to two times the sheet thickness; tailor-welded blanks may be made by this process.

High-frequency resistance welding (HFRW) is similar to resistance seam welding, except that high-frequency (up to 450 kHz) current is employed. A typical application is making butt-welded tubing; structural sections such as I-beams can also be fabricated by HFRW, and spiral pipe and tubing, finned tubes for heat exchangers, and wheel rims may be made by these techniques as well. In **high-frequency induction welding** (HFIW), the tube or pipe to be welded is subjected to high-frequency induction heating, using an induction coil ahead of the rollers.

12.9.3 Resistance projection welding

In *resistance projection welding* (RPW), high electrical resistance at the joint is developed by embossing one or more projections (dimples) on one of the surfaces to be welded (Fig. 12.34). The projections may be round or oval for design or strength purposes. High localized temperatures are generated at the projections, which are in contact with the flat mating part. The electrodes, typically made of copper-base alloys and water cooled to keep their temperature low, are large and flat. Weld nuggets, similar to those produced in spot welding, are formed as the electrodes exert pressure to compress the projections.

Spot-welding equipment can be used for RPW by modifying the electrodes. Although embossing workpieces is an added expense, this process produces a number of welds in one stroke, extends electrode life, and is capable of welding metals with different thicknesses. Nuts and bolts are also welded to sheet and plate by this process, with projections that may be produced by machining or forging. The process used for joining a network of wires, such as metal baskets, grills, oven racks, and shopping carts, is considered as resistance projection welding as well, because of the small contact area between crossing wires (grids).

12.9.4 Flash welding

In *flash welding* (FW), also called **flash butt welding,** heat is generated from the arc as the ends of the two members begin to make contact, developing an electrical resistance at the joint (Fig. 12.35). Because of the arc's presence, this process is also classified as arc welding. After the proper temperature is reached and the interface begins to soften, an axial force is applied at a controlled rate, and a weld is formed by plastic deformation of the joint—a process of *hot upsetting*, hence the term **upset welding** (UW). Some metal is expelled from the joint as a shower of sparks during the flashing process. Because impurities and contaminants are squeezed out during this

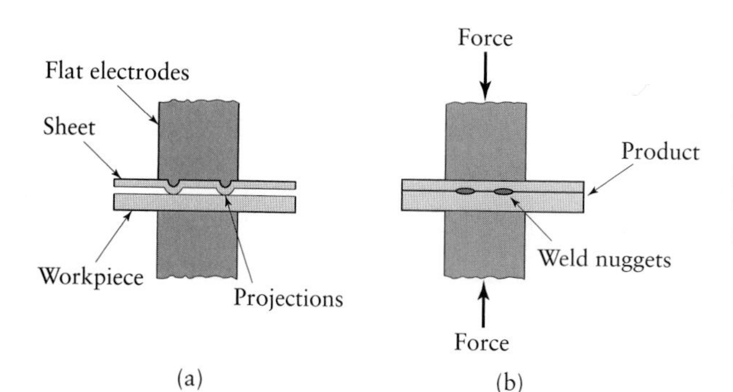

FIGURE 12.34 Schematic illustration of resistance projection welding: (a) before and (b) after. The projections are produced by embossing operations, as described in Section 7.6.

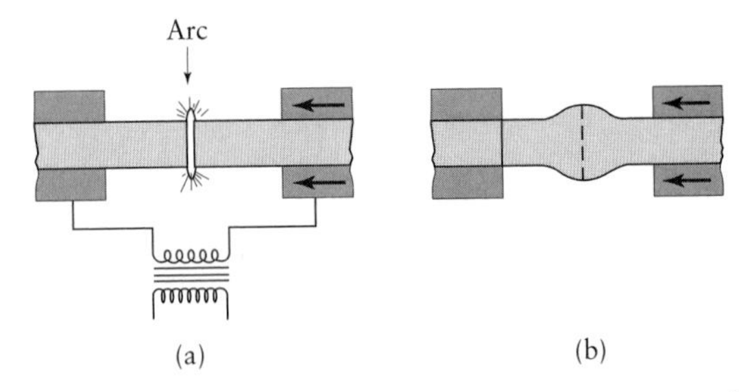

FIGURE 12.35 Flash-welding process for end-to-end welding of solid rods or tubular parts.

operation, the quality of the weld is good, although a significant amount of material may be burned off during the welding process. The joint may later be machined to improve its appearance. The machines for FW are usually automated and large, with a variety of power supplies ranging from 10 kVA to 1500 kVA.

The flash-welding process is suitable for end-to-end or edge-to-edge joining of similar or dissimilar metals 1–75 mm (0.05–3 in.) in diameter and sheet and bars 0.2–25 mm (0.01–1 in.) thick. Thinner sections have a tendency to buckle under the axial force applied during flash welding. Rings made by forming processes, such as those shown in Figs. 7.24b and c, can be flash butt welded as well. This process is also used to repair broken band-saw blades (Section 8.9.5), with fixtures that are attached to the band-saw frame. Flash welding can be automated for reproducible welding operations; typical applications include joining pipe and tubular shapes for metal furniture and windows. It is also used for welding high-speed steels to steel shanks and for welding the ends of coils of sheet or wire for continuous operation of rolling mills and wire-drawing equipment.

12.9.5 Stud arc welding

Stud arc welding (SW) is similar to flash welding. The stud, such as a small piece, bolt or threaded rod, hook, or hanger, serves as one of the electrodes while being joined to another member, which is typically a flat plate (Fig. 12.36). In order to properly concentrate the heat generated, prevent oxidation, and retain the molten metal in the weld zone, a disposable ceramic ring (*ferrule*) is placed around the joint. The equipment for stud welding can be automated, with various controls for arcing and applying pressure; portable equipment is also available. The process has numerous extensive applications in the automotive, construction, appliance, electrical, and shipbuilding industries. **Stud welding** is a more general term for joining a metal stud to a workpiece and can be done by various joining processes such as resistance welding, friction welding, or any other suitable means described in this chapter; shielding gases may or may not be used.

In **capacitor-discharge stud welding**, a dc arc is produced from a capacitor bank. No ferrule or flux is required, because the welding time is very short, on the order of one to six milliseconds. The process is capable of stud welding on thin metal sheets with coated or painted surfaces. The choice between this process and stud arc welding depends on factors such as the metals to be joined, part thickness, stud diameter, and shape of the joint.

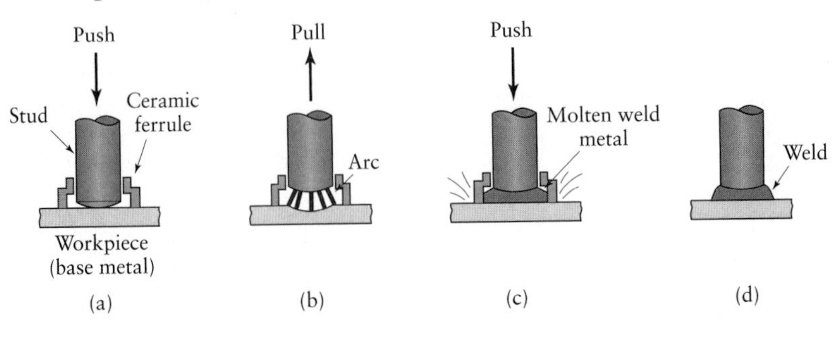

FIGURE 12.36 Sequence of operations in stud arc welding, which is used for welding bars, threaded rods, and various fasteners on metal plates.

12.9.6 Percussion welding

The resistance-welding processes described thus far require a transformer to supply power requirements; however, the electrical energy for welding may also be stored in a capacitor. *Percussion welding* (PEW) uses this technique, in which the power is discharged in a very short period of time (1–10 milliseconds), thus developing high localized heat at the joint. This process is useful where heating of the components adjacent to the joint is to be avoided, such as in electronic components.

12.10 | Explosion Welding

In *explosion welding* (EXW), pressure is applied by detonating a layer of explosive placed over one of the members being joined, called the *flyer plate* (Fig. 12.37). The contact pressures developed are extremely high, and the plate's kinetic energy striking the mating member produces a turbulent wavy interface (vortexes); this impact mechanically interlocks the two surfaces (Fig. 12.38). In addition, cold pressure welding by plastic deformation also takes place. The flyer plate is placed at an angle, and any oxide films present at the interface are broken up and propelled from the interface; as a result, bond strength in explosion welding is very high.

The explosive may be in the form of flexible plastic sheet, cord, granular solid, or liquid that is cast or pressed onto the flyer plate. Detonation speeds are usually 2400–3600 m/s (8000–12,000 ft/s), depending on the type of explosive, thickness of the explosive layer and its packing density. There is a minimum detonation speed for

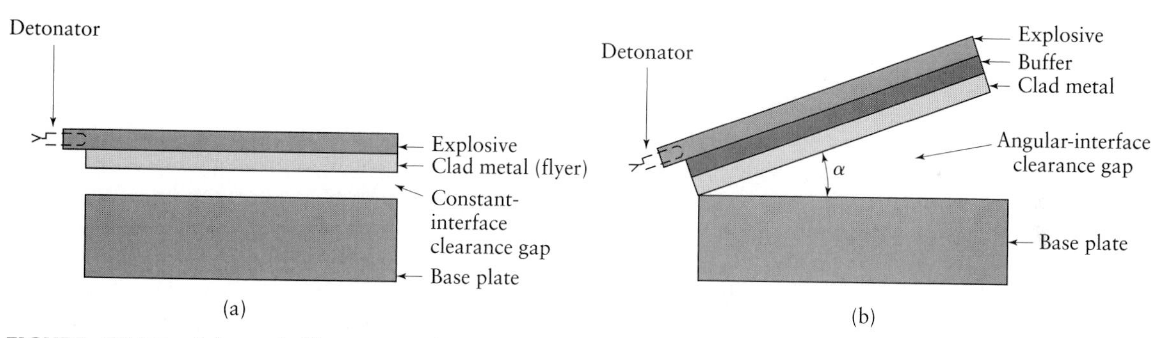

FIGURE 12.37 Schematic illustration of the explosion-welding process: (a) constant-interface clearance gap and (b) angular-interface clearance gap.

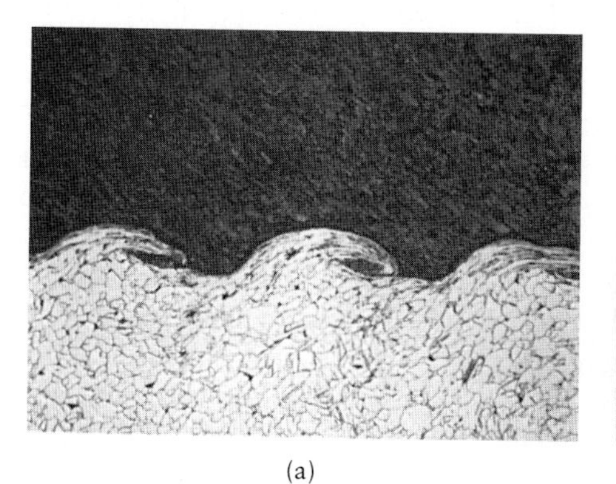

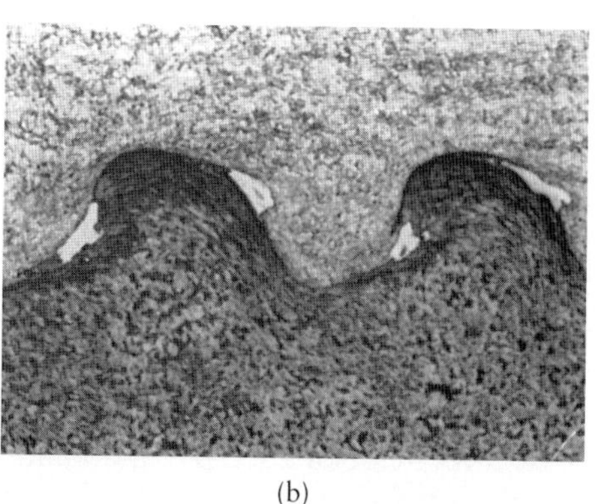

FIGURE 12.38 Cross-sections of explosion-welded joints: (a) titanium (top) on low-carbon steel (bottom) and (b) incoloy 800 (iron–nickel-base alloy) on low-carbon steel. The wavy interfaces shown improve the shear strength of the joint. Some combinations of metals, such as tantalum and vanadium, produce a much less wavy interface. If the two metals have little metallurgical compatibility, an interlayer may be added that has compatibility with both metals. *Source*: Courtesy of DuPont Company.

welding to occur in this process. Detonation is carried out using a standard commercial blasting cap. This process is particularly suitable for cladding plates and slab with dissimilar metals, such as for the chemical industry. Plates as large as 6 m × 2 m (20 ft × 7 ft) have been explosively clad; the resulting material may then be rolled into thinner sections. Tube and pipe are often joined to the holes in header plates of boilers and tubular heat exchangers by placing the explosive inside the tube; the explosion expands the tube and seals it against the plate.

12.11 Diffusion Bonding

Diffusion bonding, or **diffusion welding** (DFW), is a solid-state joining process in which the strength of the joint results primarily from diffusion (movement of atoms across the interface) and, to a lesser extent, some plastic deformation of the faying surfaces. This process requires temperatures of about $0.5T_m$ (where T_m is the melting point of the metal on the absolute scale) in order to have a sufficiently high diffusion rate between the parts to be joined. The bonded interface in DFW has essentially the same physical and mechanical properties as those of the base metal. Its strength depends on pressure, temperature, time of contact, and the cleanliness of the faying surfaces; these requirements can be lowered by using filler metal at the interfaces. The principle of diffusion bonding actually dates back centuries, when goldsmiths bonded gold over copper. To produce this material, called **filled gold**, a thin layer of gold foil is first made by hammering; the foil is placed over copper, and a weight is placed on top of it. The assembly is then placed in a furnace and left until a good bond is obtained; this process is also called *hot pressure welding* (HPW). The pressure required can also be achieved by using mechanisms with different coefficients of thermal expansion.

In diffusion bonding, the parts are usually heated in a furnace or by electrical resistance; pressure may be applied by dead weights, by a press, by using differential gas pressure, or from the relative thermal expansion of the parts to be joined. High-pressure autoclaves are also used for bonding complex parts. The process is generally most suitable for dissimilar metal pairs; however, it is also used for reactive metals, such as titanium, beryllium, zirconium, and refractory metal alloys. Diffusion bonding is also important in sintering in powder metallurgy and for processing composite materials (Chapter 11). Because diffusion involves migration of the atoms across the joint, the process is slower than other welding methods. Although DFW is used for fabricating complex parts in small quantities for the aerospace, nuclear, and electronics industries, it has been automated to make it suitable and economical for moderate-volume production.

Diffusion bonding/superplastic forming. An important development is the ability to fabricate complex sheet-metal structures by combining *diffusion bonding* with *superplastic forming* (DB/SPF; see also Section 7.10). After diffusion bonding of selected locations of the sheets, the unbonded regions are expanded into a mold by air pressure. Typical structures made by this process are shown in Fig. 12.39. The structures are thin and have high stiffness-to-weight ratios; hence, they are particularly important in aircraft and aerospace applications. This process improves productivity by eliminating mechanical fasteners, reducing the number of parts required, producing parts with good dimensional accuracy and low residual stresses, and reducing labor costs and lead times. First developed in the 1970s, this technology is well advanced for titanium structures (typically Ti-6Al-4V alloy) for aerospace applications; structures made of 7475-T6 aluminum alloy and various other alloys are also being developed.

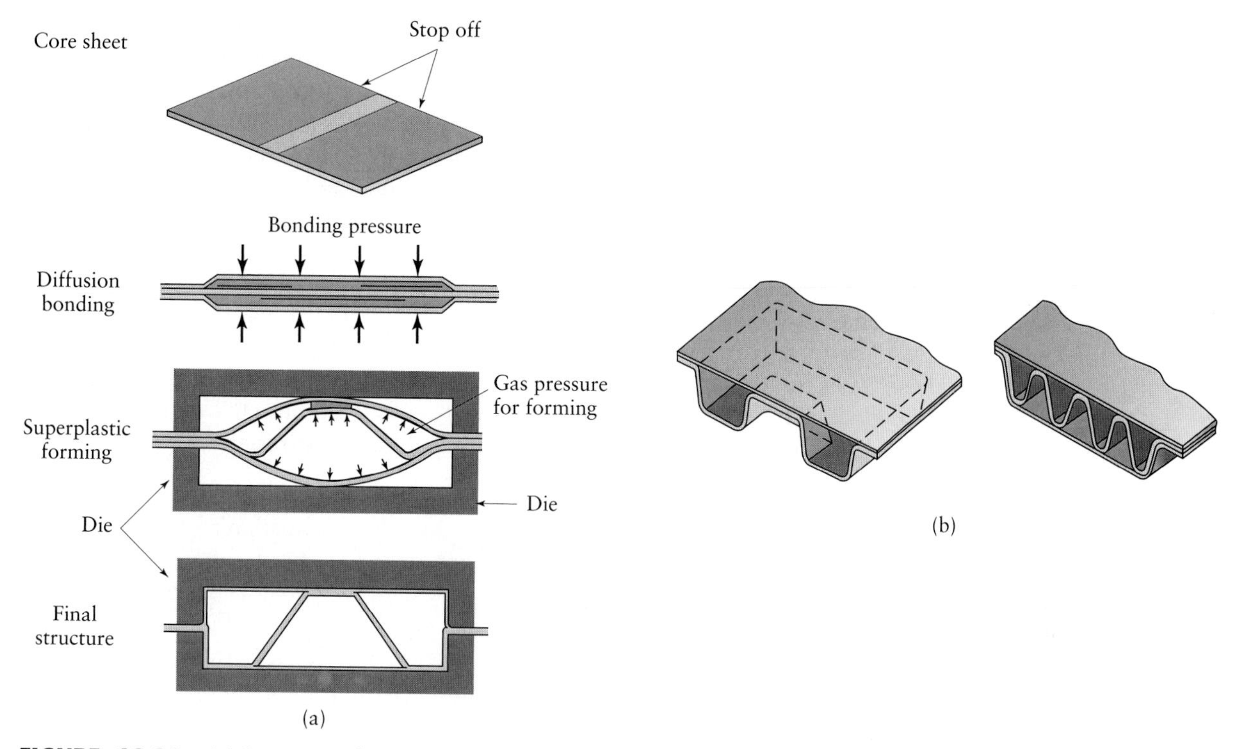

FIGURE 12.39 (a) Sequence of operations in diffusion bonding and superplastic forming of a structure with three flat sheets. *Source:* After D. Stephen and S. J. Swadling. (b) Typical structures fabricated. *Source:* Rockwell Automation, Inc.

12.12 Brazing and Soldering

There are two joining processes that involve lower temperatures than those required for welding: *brazing* and *soldering*. Brazing temperatures are higher than those for soldering, and the brazed joint has higher strength than thus the soldered joint; brazing and soldering are arbitrarily distinguished by temperature.

12.12.1 Brazing

In *brazing*, a process first used as far back as 3000 BC–2000 BC, a filler metal is placed at or between the faying surfaces to be joined, and the temperature is raised to melt the filler metal, but not the workpieces (Fig. 12.40a). The molten metal fills the closely fitting space by *capillary* action; upon cooling and solidification of the filler metal, a strong joint is developed. There are two types of brazing processes: (a) **brazing**, as we have already described, and (b) **braze welding**, in which the filler metal is deposited at the joint (Fig. 12.40b).

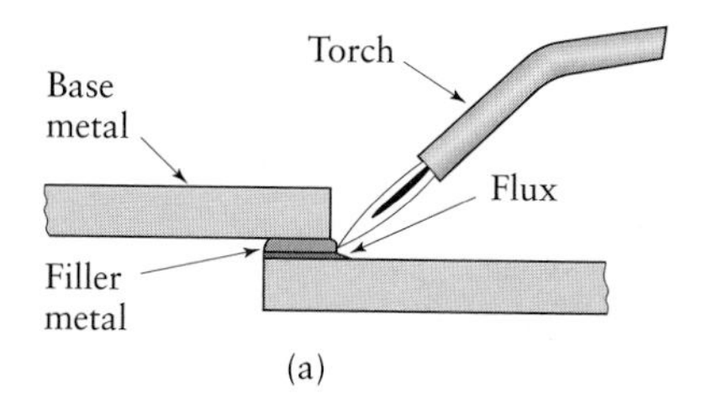

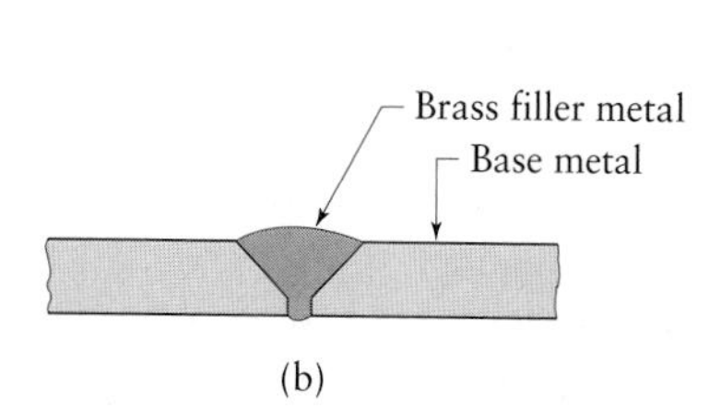

FIGURE 12.40 (a) Brazing and (b) braze-welding operations.

Filler metals for brazing generally melt above 450°C (840°F), but below the melting point (*solidus temperature*; see Fig. 5.3) of the metals to be joined; thus, this process is unlike liquid-state welding processes, in which the workpieces must melt in the weld area for fusion to occur. Any difficulties associated with heat-affected zones, warping, and residual stresses are, therefore, reduced in brazing. The strength of the brazed joint depends on joint design of the joint and adhesion at the interfaces of the workpiece and filler metal. Consequently, the surfaces to be brazed should be chemically or mechanically cleaned to ensure full capillary action; thus, the use of a flux is important.

The *clearance* between mating surfaces is an important parameter, as it directly affects the strength of the brazed joint. The smaller the gap, the higher is the *shear strength* of the joint; also, there is an optimum gap to achieve maximum *tensile strength*. The typical joint clearance ranges from 0.025 mm to 0.2 mm (0.001 in. to 0.008 in.), and because clearances are very small, surface roughness of the mating surfaces is also important.

Filler metals. Several *filler metals* (**braze metals**) are available, with a range of brazing temperatures (Table 12.3); they come in a variety of shapes such as wire, strip, rings, shims, preforms, and filings or powder. The choice of filler metal and its composition is important in order to avoid *embrittlement* of the joint (by grain-boundary penetration of liquid metal; see Section 3.4.2), formation of brittle intermetallic compounds at the joint, and galvanic corrosion in the joint. Note that filler metals for brazing, unlike those for other welding operations, generally have significantly different compositions than those of the metals to be joined.

Because of diffusion between the filler metal and the base metal, mechanical and metallurgical properties of joints can change during the service life of brazed components or in subsequent processing of the brazed parts. For example, when titanium is brazed with pure-tin filler metal, it is possible for the tin to completely diffuse into the titanium base metal by subsequent aging or heat treatment; when that happens, the joint no longer exists. The filler metal selected must also prevent galvanic corrosion of the brazed joint.

Fluxes. The use of a *flux* is essential in brazing in order to prevent oxidation and to remove oxide films from workpiece surfaces. Brazing fluxes are generally made of borax, boric acid, borates, fluorides, and chlorides and are available as paste, slurry, or powder. *Wetting agents* may also be added to improve both the wetting characteristics of the molten filler metal and capillary action. Surfaces to be brazed must be clean and free from rust, oil, lubricants, and other contaminants; clean surfaces are

TABLE 12.3

Typical Filler Metals for Brazing Various Metals and Alloys

Base Metal	Filler Metal	Brazing Temperature (°C)
Aluminum and its alloys	Aluminum–silicon	570–620
Magnesium alloys	Magnesium–aluminum	580–625
Copper and its alloys	Copper–phosphorus	700–925
Ferrous and nonferrous alloys (except aluminum and magnesium)	Silver and copper alloys, copper–phosphorus	620–1150
Iron-, nickel-, and cobalt-base alloys	Gold	900–1100
Stainless steels, nickel- and cobalt-base alloys	Nickel–silver	925–1200

essential to obtain the proper wetting and spreading characteristics of the molten filler metal in the joint, as well as maximum bond strength. Sand blasting may also be used to improve surface finish of the faying surfaces. Because they are corrosive, fluxes should be removed after brazing, especially in hidden crevices, usually by washing vigorously with hot water.

12.12.2 Brazing methods

The *heating methods* used in brazing also identify the various processes, as described in the upcoming list. A variety of special fixtures may be used to hold the parts together during brazing, some with provision for allowing thermal expansion and contraction.

a) **Torch brazing.** The heat source in *torch brazing* (TB) is oxyfuel gas with a carburizing flame. Brazing is performed by first heating the joint with the torch, and then depositing the brazing rod or wire in the joint. Suitable part thicknesses are usually in the range of 0.25–6 mm (0.01–0.25 in.). More than one torch may be used in this process. Although it can be automated as a production process, torch brazing is difficult to control and requires skilled labor. This process can also be used for repair work.

b) **Furnace brazing.** As the name suggests, *furnace brazing* (FB) is carried out in a furnace. The parts are precleaned and preloaded with brazing metal in appropriate configurations before being placed in the furnace (Fig. 12.41), and the whole assembly is heated uniformly in the furnace. Furnaces may be batch type for complex shapes or continuous type for high production runs, especially for small parts with simple joint designs. **Vacuum furnaces** or neutral atmospheres are used for metals that react with the environment.

c) **Induction brazing.** The source of heat in *induction brazing* (IB) is induction heating by high-frequency ac current. (See Section 5.5.) Parts are preloaded with filler metal and are placed near the induction coils for rapid heating. Unless a protective atmosphere is used, fluxes are generally needed. Part thicknesses are usually less than 3 mm (0.125 in.). Induction brazing is particularly suitable for brazing parts continuously.

d) **Resistance brazing.** In *resistance brazing* (RB), the source of heat is through electrical resistance of the components to be brazed. Electrodes are used for this purpose, as in resistance welding. Either parts are preloaded with filler metal or the filler metal is supplied externally during brazing. Parts that are commonly brazed by this process have a thickness of 0.1–12 mm (0.004–0.5 in.). As in induction brazing, the process is rapid, heating zones can be confined to very small areas, and the process can be automated to produce uniform quality.

e) **Dip brazing.** *Dip brazing* (DB) is carried out by dipping the assemblies to be brazed into either a molten filler-metal bath or a molten salt bath (at a temperature just above the melting point of the filler metal), which serves as the heat

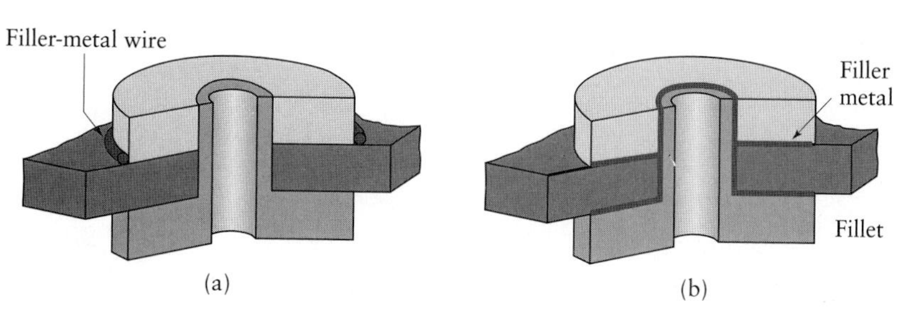

FIGURE 12.41 An example of furnace brazing: (a) before and (b) after. Note that the filler metal is a shaped wire.

source. All workpiece surfaces are thus coated with the filler metal. Dip brazing in metal baths is used only for small parts, such as sheet, wire, and fittings, usually of less than 5 mm (0.2 in.) in thickness or diameter. Molten salt baths, which also act as fluxes, are used for complex assemblies of various thicknesses. Depending on the size of the parts and the bath, as many as 1000 joints can be made at one time by dip brazing.

f) **Infrared brazing.** The heat source in *infrared brazing* (IRB) is a high-intensity quartz lamp. This process is particularly suitable for brazing very thin components, usually less than 1 mm (0.04 in.) thick, including honeycomb structures (see Fig. 7.49). The radiant energy is focused on the joint, and the process can be carried out in a vacuum. *Microwave* heating may also be used.

g) **Diffusion brazing.** *Diffusion brazing* (DFB) is carried out in a furnace, where, with proper control of temperature and time, the filler metal diffuses into the faying surfaces of the components to be joined. The brazing time required may range from 30 min to 24 hrs. Diffusion brazing is used for strong lap or butt joints and for difficult joining operations. Because the rate of diffusion at the interface does not depend on the thickness of the components, part thicknesses may range from foil to as much as 50 mm (2 in.).

h) **High-energy beams.** For specialized high-precision applications and with high-temperature metals and alloys, *electron-beam* and *laser-beam* heating may also be used.

i) **Braze welding.** The joint in *braze welding* is prepared as in fusion welding. Through use of an oxyacetylene torch with an oxidizing flame, filler metal is deposited at the joint, rather than by capillary action, as in brazing; thus, considerably more filler metal is used as compared with brazing. However, temperatures in braze welding are generally lower than in fusion welding, and part distortion is minimal. The use of a flux is essential in this process. The principal use of braze welding is to maintain and repair parts, such as ferrous castings and steel components; however, the process can be automated and used for mass production as well.

Figure 12.42 shows examples of typical brazed joints. In general, dissimilar metals can be assembled with good joint strength, including carbide drill bits (*masonry drill*) and carbide inserts on steel shanks (see Fig. 8.32d). The shear strength of brazed joints can reach 800 MPa (120 ksi), using brazing alloys containing silver (*silver solder*). Intricate, lightweight shapes can be joined rapidly and with little distortion by braze welding.

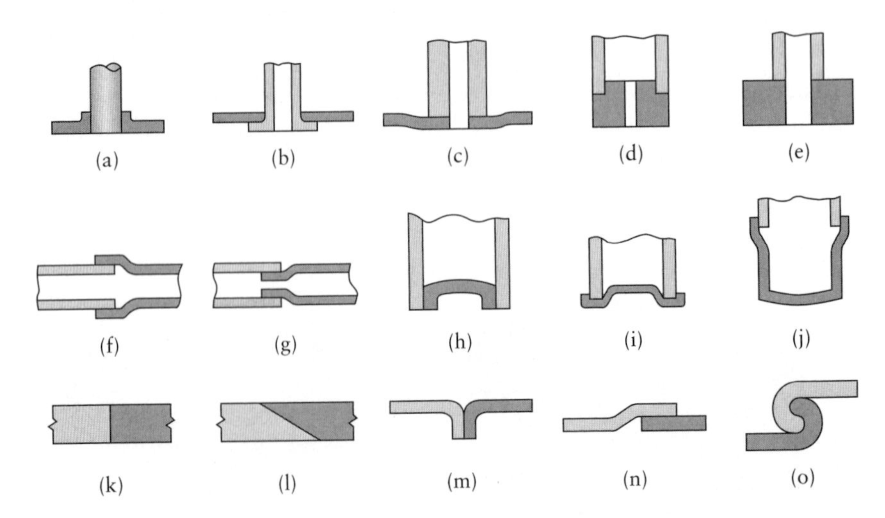

FIGURE 12.42 Joint designs commonly used in brazing operations.

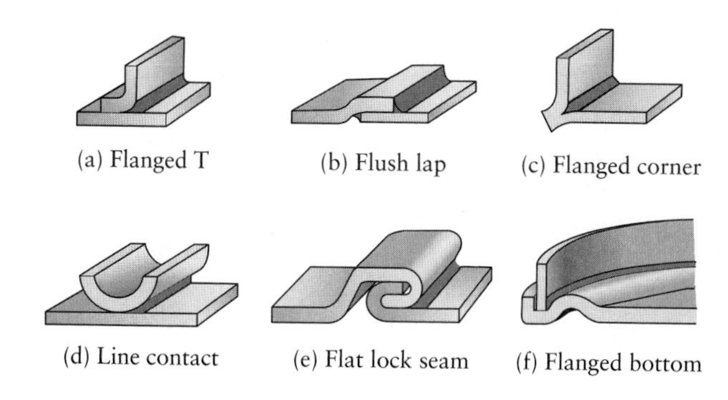

FIGURE 12.43 Joint designs commonly used for soldering.

12.12.3 Soldering

In soldering, the filler metal (**solder**) melts below 450°C (840°F); as in brazing, the solder fills the joint by capillary action between closely fitting or closely placed components (Fig. 12.43). Soldering with copper–gold and tin–lead alloys was first practiced as long ago as 4000 BC–3000 BC Heat sources for soldering are usually soldering irons, torches, or ovens. Soldering can be used to join various metals and part thicknesses. Although manual soldering operations require skill and are time consuming, soldering speeds can be high with automated equipment.

Types of solders and fluxes. Solders are usually tin–lead alloys in various proportions. For better joint strength and special applications, other solder compositions that can be used are tin–zinc, lead–silver, cadmium–silver, and zinc–aluminum alloys (Table 12.4). Because of the toxicity of lead and its adverse effects on the environment, **lead-free solders** are continually being developed and used; these solders are essentially *tin-based* solders, typical compositions being 96.5% Sn-3.5% Ag, and 42% Sn-58% Bi.

Fluxes are used in soldering as they are in welding and brazing and for the same purposes. Fluxes are generally of two types:

1. Inorganic acids or salts, such as zinc ammonium chloride solutions, which clean the surface rapidly. After soldering, the flux residues should be removed by washing thoroughly with water to avoid corrosion.
2. Noncorrosive resin-based fluxes, used in electrical applications.

Soldering is used extensively in the electronics industry. Unlike in brazing, temperatures in soldering are relatively low; consequently, a soldered joint has very limited use at elevated temperatures. Moreover, because solders do not generally have much strength, they are not used for load-bearing structural members. Because of the small faying surfaces, butt joints are rarely made with solders; in other situations, joint strength is improved by mechanical interlocking of the joint. (See also Section 12.16.)

TABLE 12.4

Types of Solders and their Applications

Tin–lead	General purpose
Tin–zinc	Aluminum
Lead–silver	Strength at higher than room temperature
Cadmium–silver	Strength at high temperatures
Zinc–aluminum	Aluminum; corrosion resistance
Tin–silver	Electronics
Tin–bismuth	Electronics

Solderability. *Solderability* may be defined in a manner similar to weldability, as described in Section 12.5.2. Copper and precious metals, such as silver and gold, are easy to solder; iron and nickel are more difficult to solder. Aluminum and stainless steels are difficult to solder, because of their strong, thin oxide film (see Section 4.2); however, these and other metals can be soldered by using special fluxes that modify surfaces. Other materials, such as cast irons, magnesium, and titanium, as well as nonmetallic materials such as graphite and ceramics, may be soldered by first plating the parts with metallic elements. (See also Section 12.15.3 for a discussion of the use of a similar technique for joining ceramics.) A common sheet metal is *tinplate*, which is steel coated with tin and used for food containers; the coating makes it very easy to solder.

Soldering methods. There are several soldering methods, which are similar to brazing methods:

a) **Torch soldering** (TS);
b) **Furnace soldering** (FS);
c) **Iron soldering** (INS), using a soldering iron;
d) **Induction soldering** (IS);
e) **Resistance soldering** (RS);
f) **Dip soldering** (DS);
g) **Infrared soldering** (IRS);
h) **Ultrasonic soldering,** in which a transducer subjects the molten solder to ultrasonic cavitation, which removes the oxide films from the surfaces to be joined—the need for a flux is thus eliminated;
i) **Wave soldering** (WS), used for automated soldering of printed circuit boards;
j) **Reflow (paste) soldering** (RS).

The last two techniques, which are significantly different from the other soldering methods, are described in more detail as follows:

a) **Reflow (paste) soldering.** Solder pastes are solder-metal particles bound together by flux and by binding and wetting agents; the pastes are semisolid in consistency. They have high viscosity, but are able to maintain a solid shape for relatively long periods of time—in this property, they are similar to greases and cake frostings. The paste is placed directly onto the joint, or on flat objects for finer detail, and it can be applied via a *screening* or *stenciling* process (Fig. 12.44). Stenciling is very commonly used during the mounting of electrical components to printed circuit boards. An additional benefit of this method is that the surface tension of the paste helps keep surface-mount packages aligned on their pads; this feature improves the reliability of these solder joints.

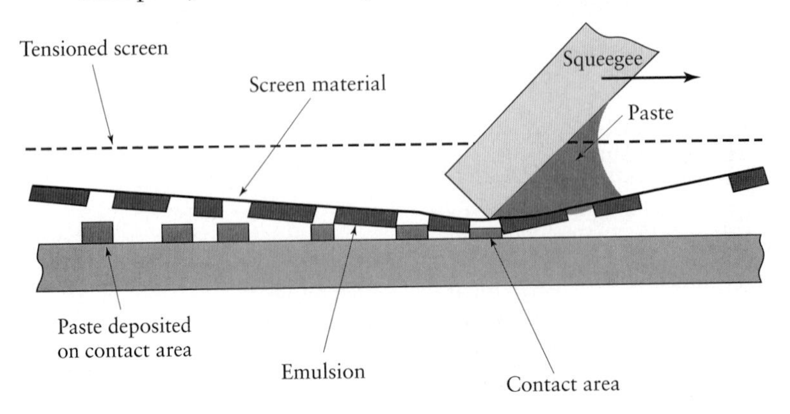

FIGURE 12.44 Screening solder paste onto a printed circuit board in reflow soldering. *Source:* V. Solberg, Design Guidelines for Surface Mount and Fine-Pitch Technology, 2d ed. New York, McGraw-Hill, 1996.

Once the paste has been placed and the joint assembled, the paste is heated in a furnace, and reflow soldering takes place. In *reflow soldering*, the product is heated in a controlled manner, so that the following sequence of events occurs:

1. Solvents present in the paste are evaporated.
2. The flux in the paste is activated, and fluxing action occurs.
3. The components are carefully preheated.
4. The solder particles are melted and wet the joint.
5. The assembly is cooled at a low rate to prevent thermal shock to and fracture of the solder joint.

While this process appears to be straightforward, there are several process variables for each stage, and good control over temperatures and exposures must be maintained at each stage to ensure proper joint strength.

b) **Wave soldering.** Wave soldering is a very common method for attaching circuit components to their boards (Section 13.12). To understand wave soldering, it is imperative to appreciate the fact that molten solder does not wet all surfaces; indeed, solder will not stick to most polymer surfaces and can easy be removed while molten. Also, as can be shown with a simple hand-soldering iron, the solder only wets metal surfaces and forms a good bond when the metal is preheated to a certain temperature. Therefore, wave soldering requires separate fluxing and preheating operations before it can be successfully completed.

Wave soldering is schematically illustrated in Fig. 12.45a. A standing laminar wave of molten solder is generated by a pump. Preheated and prefluxed circuit boards are conveyed over the wave; the solder wets the exposed metal surfaces, but it does not stay attached to the polymer package for the integrated circuits, and it does not stick to the polymer-coated circuit boards. An air knife (basically a high-velocity jet of hot air) blows excess solder from the joint, to prevent bridging between adjacent leads. When surface-mount packages are to be wave soldered, they must first be adhesively bonded to the circuit board before the soldering can commence. This bonding is usually accomplished by screening or stenciling epoxy onto the boards, placing the components in their proper locations, curing the epoxy, inverting the board, and performing wave soldering. Fig. 12.45b shows an SEM photograph of a typical surface-mount joint.

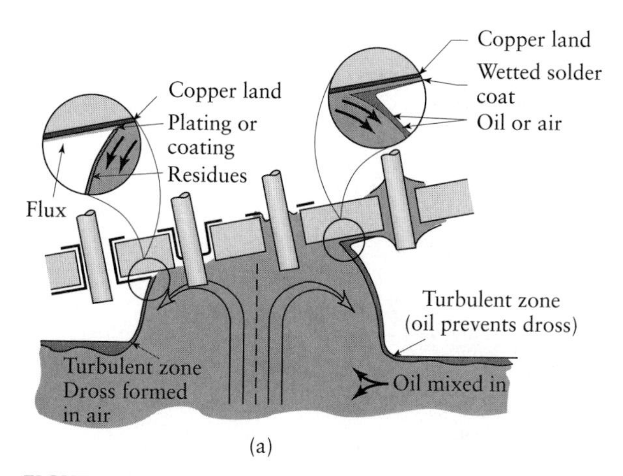

(a)

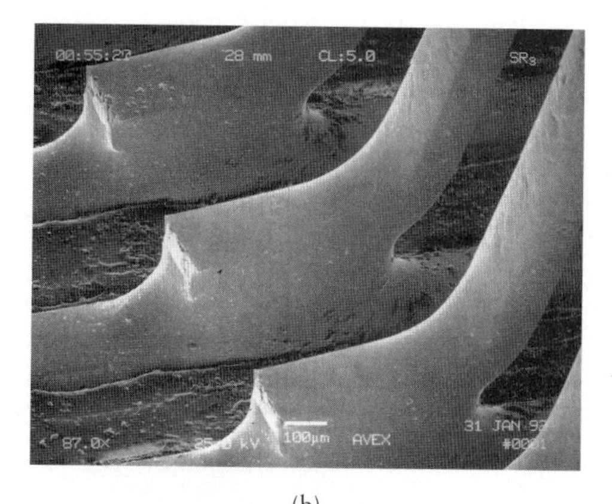

(b)

FIGURE 12.45 (a) Schematic illustration of the wave-soldering process. (b) SEM image of a wave-soldered joint on a surface-mount device.

12.13 Adhesive Bonding

Numerous components and products can be joined and assembled with an *adhesive*, rather than using any of the joining methods described thus far. **Adhesive bonding** has been a common method of joining and assembly for applications such as labeling, packaging, bookbinding, home furnishings, and footwear. Plywood, developed in 1905, is a typical example of adhesive bonding of several layers of wood with glue. Adhesive bonding has been gaining increased acceptance in manufacturing ever since its first use on a large scale in assembling load-bearing components in military aircraft manufactured during World War II (1939–1945). Figure 12.46 shows examples of adhesively bonded joints. Adhesives are available in various forms such as liquids, pastes, solutions, emulsions, powder, tape, and film; when applied, adhesives typically are on the order of 0.1 mm (0.004 in.) thick.

12.13.1 Adhesives

Numerous types of adhesives are available, and continue to be developed, that provide adequate joint strength, including fatigue strength. The three basic types of adhesives are listed as follows:

1. **Natural adhesives**, such as starch, dextrin (a gummy substance obtained from starch), soya flour, and animal products;
2. **Inorganic adhesives**, such as sodium silicate and magnesium oxychloride;
3. **Synthetic organic adhesives**, which may be thermoplastics (for nonstructural and some structural bonding) or thermosetting polymers (primarily for structural bonding).

Because of their cohesive strength, synthetic organic adhesives are the most important adhesives in manufacturing processes, particularly for load-bearing applications (*structural adhesives*). They are generally classified as follows:

1. **Chemically reactive**, such as polyurethanes, silicones, epoxies, cyanoacrylates, modified acrylics, phenolics, and polyimides—also included are anaerobics,

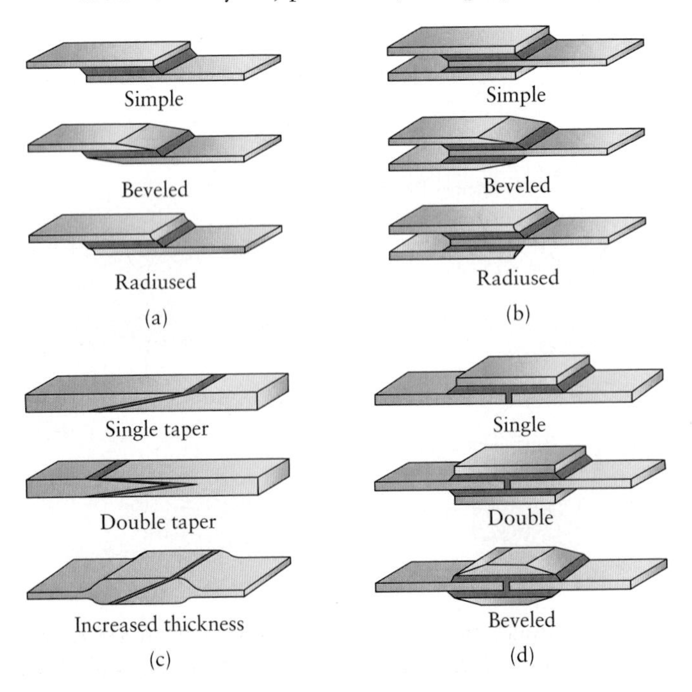

FIGURE 12.46 Various configurations for adhesively bonded joints: (a) single lap, (b) double lap, (c) scarf, and (d) strap.

which cure in the absence of oxygen (such as Loctite® for threaded fasteners; see the case study at end of this chapter);

2. **Pressure-sensitive**, including are natural rubber, styrene–butadiene rubber, butyl rubber, nitrile rubber, and polyacrylates;
3. **Hot melt**, which are thermoplastics such as ethylene–vinyl-acetate copolymers, polyolefins, polyamides, polyester, and thermoplastic elastomers;
4. **Reactive hot-melt**, which have a thermoset portion based on the urethane's chemistry, with improved properties;
5. **Evaporative or diffusion**, including vinyls, acrylics, phenolics, polyurethanes, synthetic rubbers, and natural rubbers;
6. **Film and tape**, such as nylon–epoxies, elastomer–epoxies, nitrile–phenolics, vinyl–phenolics, and polyimides;
7. **Delayed tack**, including styrene–butadiene copolymers, polyvinyl acetates, polystyrenes, and polyamides;
8. **Electrically** and **thermally conductive**, including epoxies, polyurethanes, silicones, and polyimides (see Section 12.13.4).

Adhesive systems may be classified based on their specific chemistries:

1. **Epoxy-based systems.** These systems have high strength and high-temperature properties, to as high as 200°C (400°F); typical applications include automotive brake linings and as bonding agents for sand molds for casting. (See Section 5.8.1.)
2. **Acrylics.** Suitable for applications with substrates that are not clean.
3. **Anaerobic systems.** Curing of these adhesives is by oxygen deprivation, and the bond is usually hard and brittle; curing times can be reduced by external heat or ultraviolet (UV) radiation.
4. **Cyanoacrylate.** The bond lines are thin, and the bond sets within 5 s to 40 s.
5. **Urethanes.** High toughness and flexibility at room temperature; widely used as sealants.
6. **Silicones.** Highly resistant to moisture and solvents, silicones have high impact and peel strength; however, curing times are typically in the range of one to five days.

Many of the adhesives listed previously can be combined to optimize their properties; examples include epoxy–silicon, nitrile–phenolic, and epoxy–phenolic. The least expensive adhesives are epoxies and phenolics, followed by polyurethanes, acrylics, silicones, and cyanoacrylates. Adhesives for high-temperature applications at up to about 260°C (500°F), such as polyimides and polybenzimidazoles, are generally the most expensive adhesives.

Depending on the particular application, an adhesive generally must have one or more of the following properties (Table 12.5):

- Strength (shear and peel);
- Toughness;
- Resistance to various fluids and chemicals;
- Resistance to environmental degradation, including heat and moisture;
- Capability to wet the surfaces to be bonded.

Adhesive joints are designed to withstand shear, compressive, and tensile forces, but they should not be subjected to peeling forces (Fig. 12.47). Note, for example, how

TABLE 12.5

Typical Properties and Characteristics of Chemically Reactive Structural Adhesives

	Epoxy	Polyurethane	Modified Acrylic	Cyanocrylate	Anaerobic
Impact resistance	Poor	Excellent	Good	Poor	Fair
Tension-shear strength, MPa (10^3 psi)	15.4 (2.2)	15.4 (2.2)	25.9 (3.7)	18.9 (2.7)	17.5 (2.5)
Peel strength, N/m (lb/in.)	< 525 (3)	14,000 (80)	5250 (30)	< 525 (3)	1750 (10)
Substrates bonded	Most	Most smooth, nonporous	Most smooth, nonporous	Most nonporous metals or plastics	Metals, glass, thermosets
Service temperature range, °C (°F)	−55 to 120 (−70 to 250)	−160 to 80 (−250 to 175)	−70 to 120 (−100 to 250)	−55 to 80 (−70 to 175)	−55 to 150 (−70 to 300)
Heat cure or mixing required	Yes	Yes	No	No	No
Solvent resistance	Excellent	Good	Good	Good	Excellent
Moisture resistance	Excellent	Fair	Good	Poor	Good
Gap limitation, mm (in.)	None	None	0.75 (0.03)	0.25 (0.01)	0.60 (0.025)
Odor	Mild	Mild	Strong	Moderate	Mild
Toxicity	Moderate	Moderate	Moderate	Low	Low
Flammability	Low	Low	High	Low	Low

Source: *Advanced Materials & Processes*, July 1990, ASM International.

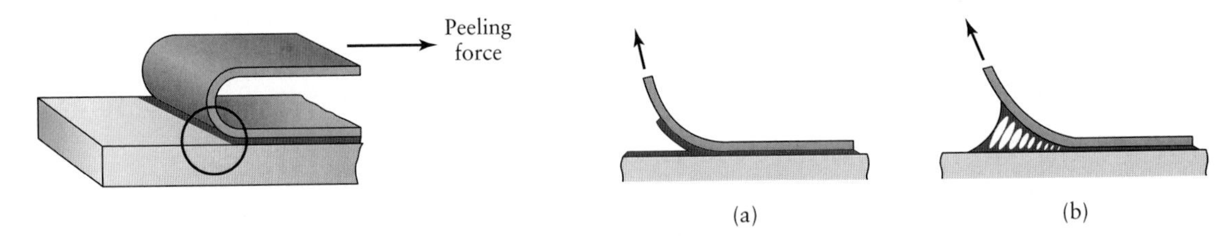

FIGURE 12.47 Characteristic behavior of (a) brittle and (b) tough adhesives in a peeling test. This test is similar to peeling adhesive tape from a solid surface. Adhesive joints should not be subjected to this type of loading in service.

easily an adhesive tape can be peeled from a surface. During peeling, the behavior of an adhesive may be brittle, or it may be ductile and tough, requiring large forces to peel it.

12.13.2 Surface preparation and application

Surface preparation is very important in adhesive bonding, as joint strength depends greatly on the absence of dirt, dust, oil, and various other contaminants; note, for example, that is virtually impossible to apply an adhesive tape over a dusty or oily surface. Contaminants also affect the wetting ability of the adhesive, and they can prevent spreading of the adhesive evenly over the interface; thick, weak, or loose oxide films on workpieces also are detrimental to adhesive bonding. On the other hand, a porous or thin and strong oxide film may be desirable, particularly one with some surface roughness to improve adhesion, by mechanical locking. However, the roughness must not be too high, or else air may be trapped and joint strength is reduced. Various compounds and primers are available to modify surfaces to increase the strength of adhesive bonds. Liquid adhesives may be applied by brushes, sprayers, and rollers.

12.13.3 Process capabilities

A wide variety of similar and dissimilar metallic and nonmetallic materials and components with different shapes, sizes, and thicknesses can be bonded to each other by adhesives. Adhesive bonding can be combined with mechanical fastening methods (see Section 12.14) to further improve the strength of the bond. An important consideration in the use of adhesives in production is curing time, which can range from a few seconds at high temperatures to many hours at room temperature, particularly for thermosetting adhesives; thus, production rates may be low. Furthermore, adhesive bonds for structural applications are rarely suitable for service above 250°C (500°F). Joint design and bonding methods require care and skill, and special equipment, such as fixtures, presses, tooling, and autoclaves and ovens for curing, is also usually needed.

Nondestructive inspection of the quality and strength of adhesively bonded components can be difficult; however, some of the techniques described in Section 4.8.1, such as acoustic impact (tapping), holography, infrared detection, and ultrasonic testing, are effective nondestructive testing methods. Major industries that use adhesive bonding extensively are aerospace, automotive, appliances, and building products. Applications include attaching rearview mirrors to windshields, automotive brake lining assemblies, laminated windshield glass (see Section 11.11.2), various appliances, helicopter blades, honeycomb structures, and aircraft bodies and control surfaces.

The advantages of adhesive bonding are listed as follows:

1. It provides a bond at the interface, either for structural strength or for nonstructural applications such as sealing, insulating, preventing electrochemical corrosion between dissimilar metals, and reducing vibration and noise through internal damping at the joints.
2. It distributes the load at an interface, thus eliminating localized stresses that typically result from joining the components with welds or mechanical fasteners, such as bolts and screws. Moreover, since no holes are required, structural integrity of the sections is maintained.
3. The external appearance of the joined components is unaffected.
4. Very thin and fragile components can be bonded without contributing significantly to weight.
5. Porous materials and materials with very different properties and sizes can be joined.
6. Because adhesive bonding is usually carried out between room temperature and about 200°C (400°F), there is no significant distortion of the components or change in their original properties; this factor is particularly important for materials that are heat sensitive.

The limitations of adhesive bonding are:

1. The service temperatures are relatively low.
2. The bonding time can be long.
3. Surface preparation is necessary.
4. It is difficult to test bonded joints nondestructively, particularly with large structures.

5. The reliability of adhesively bonded structures during their service life and under hostile environmental conditions (*degradation* by temperature, oxidation, radiation, stress corrosion, and dissolution) may be a significant concern.

12.13.4 Electrically conducting adhesives

Although the majority of adhesive bonding is for mechanical strength and structural integrity, a more recent advance is the development and application of *electrically conducting adhesives* to replace lead-base solder alloys, particularly in the electronics industry, as well as for economic considerations. These adhesives require curing or setting temperatures that are lower than that for soldering. Electrical conductivity in adhesives is obtained by the addition of *fillers*, such as silver (used most commonly, with up to 85% silver), copper, aluminum, nickel, gold, and graphite. (See also Section 10.7.2.) Recent developments in fillers include the use of polymeric particles, such as polystyrene, coated with thin films of silver or gold. Fillers are generally in the form of flakes or particles; fillers that improve electrical conductivity also improve thermal conductivity. The polymer is the matrix, and there is a minimum volume concentration of fillers in order to make the adhesive electrically conducting, typically in the range of 40%–70%. Matrix materials are generally epoxies, although various thermoplastics are also used.

The size, shape, and distribution of the conducting particles, as well as how heat and pressure are applied and the nature of contact among the individual particles, can be controlled so as to impart isotropic or anisotropic electrical conductivity to the adhesive. Applications of electrically conducting adhesives include calculators, remote controls, and control panels, as well as high-density use in electronic assemblies, liquid-crystal displays, pocket TVs, and electronic games.

12.14 Mechanical Fastening

Two or more components may have to be joined or fastened in such a way that they can be taken apart sometime during the product's service life or life cycle. Countless products, including mechanical pencils, watches, computers, bicycles, engines, and aircraft have components that are fastened mechanically. *Mechanical fastening* may be preferred over other methods because of its

1. Ease of manufacturing;
2. Ease of assembly and transportation;
3. Ease of parts replacement, maintenance, and repair;
4. Ease in creating designs that require movable joints, such as hinges, sliding mechanisms for drawers and doors, and adjustable components and fixtures;
5. Lower overall cost of manufacturing the product.

The most common method of mechanical fastening is by using bolts, nuts, screws, rivets, pins, and a wide variety of other **fasteners**; these techniques are also known as **mechanical assembly**. Most mechanical fastening typically requires that the components have holes through which fasteners are inserted. These assemblies may be structural (load-bearing) components and thus may be subjected to both shear and tensile stresses; consequently, they should be designed to resist these forces.

12.14.1 Hole preparation

Hole preparation is an important aspect of mechanical fastening. A hole in a solid body can be produced by punching, drilling, chemical and electrical means, and high-energy beams (Chapters 7, 8, and 9), depending on the type of material, its properties, and its thickness. Recall also from Chapters 5, 6, and 11 that holes may be produced as an *integral part* of the product during casting, forging, extrusion, and powder metallurgy. For improved accuracy and surface finish, many of these holemaking operations may be followed by finishing operations, such as shaving, deburring, reaming, and honing.

Because of their fundamental differences, each type of holemaking operation produces a hole with different surface finish and properties and dimensional characteristics. The most significant influence of a hole in a solid body is to act as a stress concentration, and thus the hole has a tendency to reduce the component's fatigue life. For holes, fatigue life can best be improved by inducing compressive residual stresses on the cylindrical surface of the hole. These stresses are usually developed by pushing a round rod (*drift pin*) through the hole and expanding it by a very small amount, a process that plastically deforms the surface layers of the hole in a manner similar to shot peening or roller burnishing (see Section 4.5.1).

12.14.2 Threaded fasteners

Bolts, screws, and nuts are among the most commonly used *threaded fasteners*. References on machine design describe in detail numerous standards and specifications, including thread dimensions, tolerances, pitch, strength, and the quality of materials used to make these fasteners. Bolts and screws may be secured with nuts or they may be *self-tapping*, whereby the screw either cuts or forms the thread into the part to be fastened; the latter method is particularly effective and economical in plastic products. If the joint is to be subjected to vibration, such as in aircraft and various types of engines and high-speed machinery, several specially designed nuts and lock washers are available. They increase the frictional resistance in the torsional direction, thus preventing vibrational loosening of the fasteners.

12.14.3 Rivets

The most common method of permanent or semipermanent mechanical joining is by *riveting* (Fig. 12.48). Hundreds of thousands of rivets may be used in the construction and assembly of large commercial aircraft. There are several types of rivets, and some may be solid or hollow; blind rivets are inserted from one side only. Installing a rivet basically consists of placing the rivet in the hole and deforming the end of its shank by upsetting (*heading*, similar to the process shown in Fig. 6.17). The operation may be performed by hand or by mechanized means, including the use of robots. Riveting may be done either at room temperature or hot, using special tools, or by using explosives in the cavity of the rivet.

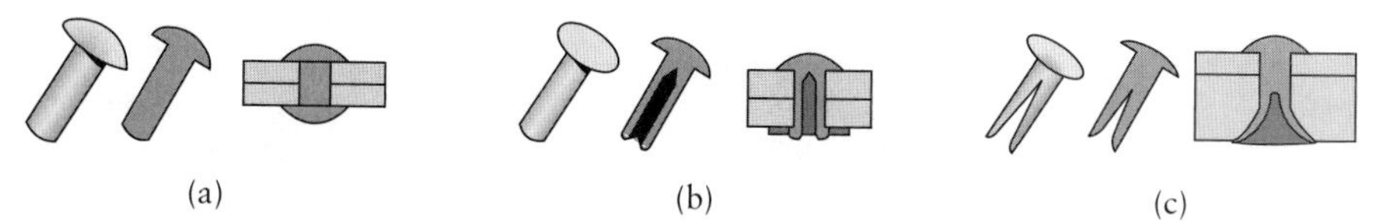

FIGURE 12.48 Examples of rivets: (a) solid, (b) tubular, (c) split, or bifurcated and (d) compression.

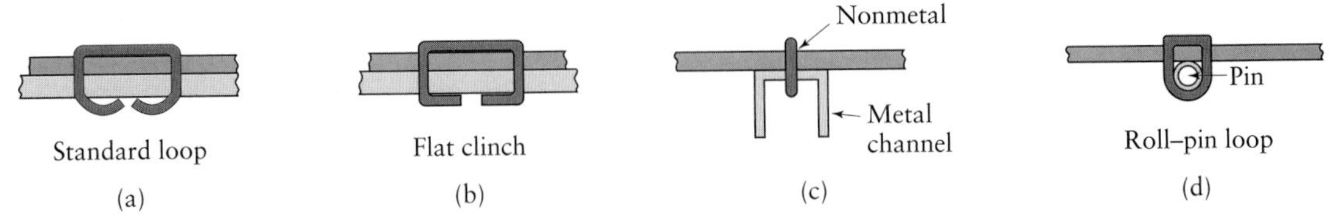

FIGURE 12.49 Examples of metal stitching.

12.14.4 Other methods of fastening

Several other fastening methods are also used in joining and assembly applications; the most common types are described as follows:

a) **Metal stitching or stapling.** The process of *metal stitching* or *stapling* (Fig. 12.49) is much like that of ordinary stapling of papers. This operation is fast and particularly suitable for joining thin metallic and nonmetallic materials, and it does not require holes to be made in the components; a common example is the stapling of cardboard containers for various products. In *clinching* (Fig. 12.49b), the fastening material should be sufficiently thin and ductile to withstand the large localized deformations during sharp bending.

b) **Seaming.** *Seaming* is based on the simple principle of folding two thin pieces of material together, much like folding two or more pieces of paper at their corners in the absence of a paper clip or a staple. The most common examples of seaming (*lock seams*) are the lids of beverage cans and containers for food and household products (Fig. 12.50). In seaming, the materials should be capable of undergoing bending and folding at very small radii (see Section 7.4.1); otherwise, they may crack, and the seams will not be airtight or watertight. The performance and airtightness of lock seams may be improved with adhesives, coatings, polymeric seals, or by soldering.

c) **Crimping.** The *crimping* process is a method of joining without using fasteners. It can be done with beads or dimples (Fig. 12.51), which can be produced by *shrinking* or *swaging* operations. Crimping can be used on both tubular and flat parts. Caps on glass bottles are attached by crimping, as are some connectors for electrical wiring.

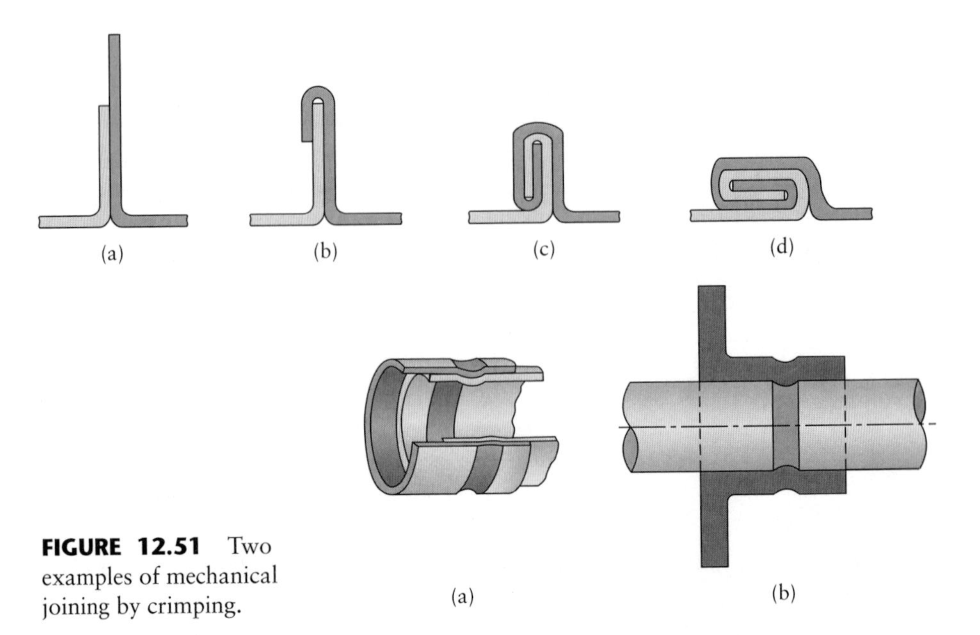

FIGURE 12.50 Stages in forming a double-lock seam.

FIGURE 12.51 Two examples of mechanical joining by crimping.

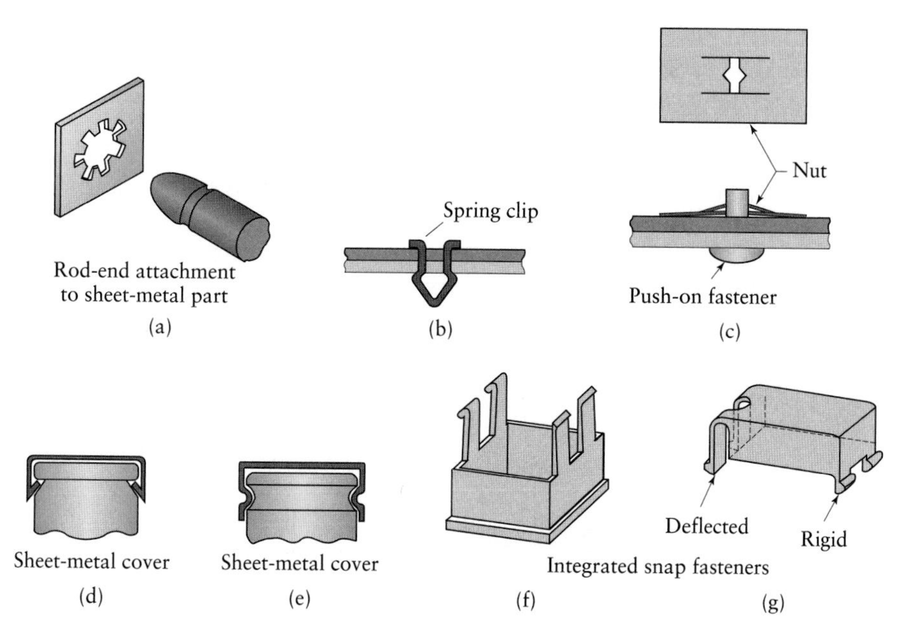

FIGURE 12.52 Examples of spring and snap-in fasteners to facilitate assembly.

d) **Snap-in fasteners.** Figure 12.52 shows various spring and *snap-in fasteners*. Such *snap-fit* joints are widely used in automotive bodies and household appliances; they are economical and permit easy and rapid component assembly, particularly for plastic products.

e) **Shrink and press fits.** Components may also be assembled by shrink fitting and press fitting. *Shrink fitting* is based on the principle of the differential thermal expansion and contraction of two components; typical applications include the assembly of die components and mounting gears and cams on shafts. In *press fitting*, one component is forced over another, resulting in high joint strength.

12.15 Joining Plastics, Ceramics, and Glasses

12.15.1 Joining thermoplastics

As described in Section 10.3, thermoplastics soften and melt as temperature increases; consequently, they can be joined by means whereby heat is generated at the interface, either through external means or internally. The heat softens the thermoplastic at the interface to a viscous or molten state and, with the application of pressure to ensure a good bond, allows *fusion* to take place. Filler materials of the same type of polymer may also be used.

Oxidation can be a problem in joining some polymers, such as polyethylene, causing degradation; typically, an inert shielding gas such as nitrogen is used to prevent oxidation. Because of the low thermal conductivity of thermoplastics, the heat source, if applied at too high a rate, may burn or char the surfaces of the components to be joined, possibly causing difficulties in obtaining sufficiently deep fusion for joint strength. *External heat sources* may consist of the following, depending on the compatibility of the polymers to be joined:

1. Hot air or gases, or infrared radiation using high-intensity focused quartz heat lamps;

2. Heated tools and dies, used in a process known as *hot-tool* or *hot-plate welding*, that contact and heat the surfaces to be joined by interdiffusion of molecular chains—this process is commonly used to butt-weld pipe and tubing;
3. Radio-frequency or dielectric heating, particularly useful for thin films;
4. Electrical-resistance wire or braids, or carbon-based tapes, sheet, and ropes, which are placed at the interface and pass electrical current, a process known as *resistive-implant welding*—these elements at the interface may be subjected to a radio-frequency field (*induction welding*) and must be compatible with the intended use of the joined product, since they remain in the weld zone;
5. Lasers, using defocused beams at low power to prevent degradation of the polymer.

Internal heat sources are developed through the following means:

1. Ultrasonic welding (see Section 12.7), which is the most commonly used process for thermoplastics, particularly amorphous polymers such as ABS and high-impact polystyrene;
2. Friction welding (see Section 12.8), also called *spin welding* for polymers—it includes linear friction welding, also called *vibration welding*, which is particularly useful for joining polymers with a high degree of crystallinity (see Section 10.2.2), such as acetal, polyethylene, nylon, and polypropylene;
3. *Orbital welding*, which is similar to friction welding, except that the rotary motion of one part is in an orbital path (see also Fig. 6.16a).

Other joining methods. *Adhesive bonding* of polymers is a versatile process, One of the best examples is the process for joining sections of polyvinyl chloride (PVC) pipe, used extensively in home plumbing systems, or ABS pipe, used in drain, waste, and vent systems. The liquid adhesive is applied to the connecting sleeve and pipe surfaces, sometimes using a primer to improve adhesion (much like using primers in painting). Adhesive bonding of polyethylene, polypropylene, and polytetrafluoroethylene (Teflon) can be difficult, because these polymers have low surface energy, and hence adhesives do not stick readily to their surfaces; the surfaces usually have to be treated chemically to improve bond strength. The use of adhesive primers or double-sided adhesive tapes can also be effective. Bonding may also be achieved by using solvents (*solvent bonding*).

Coextruded multilayer food wrappings consist of various types of films bonded by the *heat* generated during *extrusion*. Each film has a different function, such as to keep out moisture and oxygen or to facilitate heat sealing during packaging. Some wrappings have as many as seven layers, all bonded together during production of the film (*cocuring*).

Thermoplastics may also be joined by *mechanical* means, including the use of fasteners and self-tapping screws. The strength of the joint depends on the particular technique used and the inherent toughness and resilience of the plastic to resist tearing at the holes present for mechanical fastening. *Integrated snap fasteners* (Fig. 12.52f and g) are gaining wide acceptance for their simplified and cost-effective assembly operations.

Bonding may also be achieved by *magnetic* means, by embedding tiny [on the order of 1 μm (40 μin.)] particles in the polymer. A high-frequency field then causes induction heating of the polymer and melts it at the interfaces to be joined (*electromagnetic bonding*).

12.15.2 Joining thermosets

Because they do not soften or melt with increasing temperature, thermosetting plastics, such as epoxy and phenolics, are usually joined (1) by using threaded or other molded-in inserts (see, for example, Fig. 10.24), (2) by using mechanical fasteners, and (3) by solvent bonding. Bonding with solvents basically involves the following sequence: (a) roughening the surfaces of the thermoset parts with an abrasive cloth or paper, (b) wiping the surfaces with a solvent, and (c) pressing the surfaces together and maintaining the pressure until sufficient bond strength is developed.

12.15.3 Joining ceramics and glasses

A wide variety of types of ceramics and glasses is now available with unique and important properties; these ceramics and glasses are used either as products or components of products, or as tools, molds, and dies. As any other material used in products, these materials also have to be assembled into components, either with the same type of material or with different materials. These joints have to meet specific requirements, depending on the particular application. Nonmetallic parts can, of course, be joined using adhesive bonding; a typical example is assembling broken ceramic pieces, using two-component epoxy, mixed just prior to application. Other methods include mechanical means, such as fasteners, and shrink or press fitting. The relative thermal expansion of the two materials should be taken into consideration to avoid damage if the assembly is subjected to elevated temperature. (See Section 3.9.5.)

Ceramics. As described in Section 8.6.4, *tungsten carbide* has a matrix (binder) of cobalt, and *titanium carbide* has matrix of nickel–molybdenum alloy. Consequently, with both binders being metal, carbides can easily be brazed to other metals. Common applications include brazing carbide cutting tools to steel toolholders (see Fig. 8.32d), brazing carbide tips to masonry drills, and brazing cubic-boron-nitride or diamond tips to carbide inserts (see Fig. 8.39).

A commonly used technique that is effective in joining difficult-to-bond material combinations consists of first applying a coating of a material that bonds itself easily to one or both parts, thus acting as a bonding agent, just as an adhesive does between a piece of wood and a metal that otherwise would not bond to each other. Thus, the surface of *alumina ceramics*, for example, can be metallized (see Section 4.5.1); in this technique, known as the Mo–Mn process, the ceramic part is first coated with a slurry that, after being fired, forms a glassy layer. This layer is then plated with nickel, and, now that the part has a metallic surface, it can brazed to a metal surface. Depending on their particular structure, ceramics and metals can be joined by diffusion bonding (Section 12.11); it may be necessary to place a metallic layer at the joint to make the joint stronger.

It is also apparent that ceramic parts can be joined or assembled together during their shaping process (see Section 11.9); a common example is attaching the handle of a coffee mug. Thus, in a sense, the shaping of the product is done integrally, rather than as an additional operation after the part is already made.

Glasses. As evidenced by the availability of numerous glass objects, glasses can easily be bonded to each other, which is accomplished by first softening the surfaces to be joined and then pressing the two pieces together and cooling them. Bonding of glass to metals is possible because of the diffusion of metal ions into the amorphous surface structure of glass.

12.16 Design Considerations in Joining

Basic design guidelines for the variety of joining processes described throughout this chapter are outlined as follows:

12.16.1 Design for welding

As in all manufacturing processes, the optimum choice in welding is the one that satisfies all design and service requirements at minimum cost. Some examples of weld characteristics are shown in Fig. 12.53. General design guidelines for welding may be summarized as follows:

1. Product design should minimize the number of welds, as welding can be costly when it is not automated.
2. The weld location should be selected to avoid excessive stresses or stress concentrations in the welded structure, as well as for appearance.
3. The weld location should be selected so as not to interfere with subsequent processing of the part or with its intended use and appearance.
4. Parts should fit properly before welding; the method employed to produce the edges (sawing, machining, shearing, and flame cutting) can affect weld quality.
5. Modification of the design may avoid the need for edge preparation.
6. Weld-bead size should be kept to a minimum to conserve weld metal.
7. Mating surfaces for some processes may require uniform cross-sections at the joint, such as is shown for flash welding in Fig. 12.54.

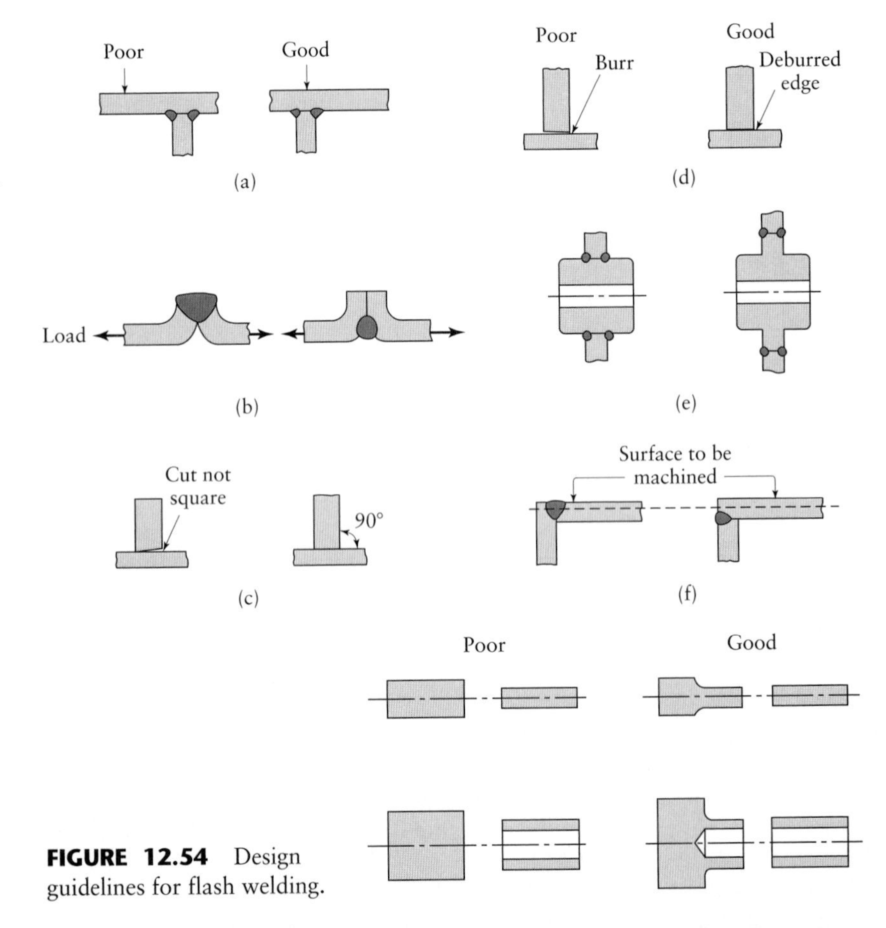

FIGURE 12.53 Design guidelines for welding. *Source*: Bralla, J. G. (ed.) *Handbook of Product Design for Manufacturing*, 2d ed. New York, McGraw-Hill, 1999.

FIGURE 12.54 Design guidelines for flash welding.

EXAMPLE 12.4 Weld design selection

Figure 12.55 shows three different types of weld design. In Fig. 12.55a, the two vertical joints can be welded either externally or internally. Obviously, full-length external welding takes a considerable amount of time and requires more weld material than the alternative design, which consists of intermittent internal welds. Moreover, in the alternative method, the appearance of the structure is improved, and distortion is reduced. Although both designs require the same amount of weld material and welding time (Fig. 12.55b), further analysis shows that the design on the right can carry three times the moment M than can the one on the left. In Fig. 12.55c, the weld on the left requires about twice the amount of weld material than that required by the design on the right. Note also that because more material must be machined, the design on the left will require more time for edge preparation, and more base metal will be wasted.

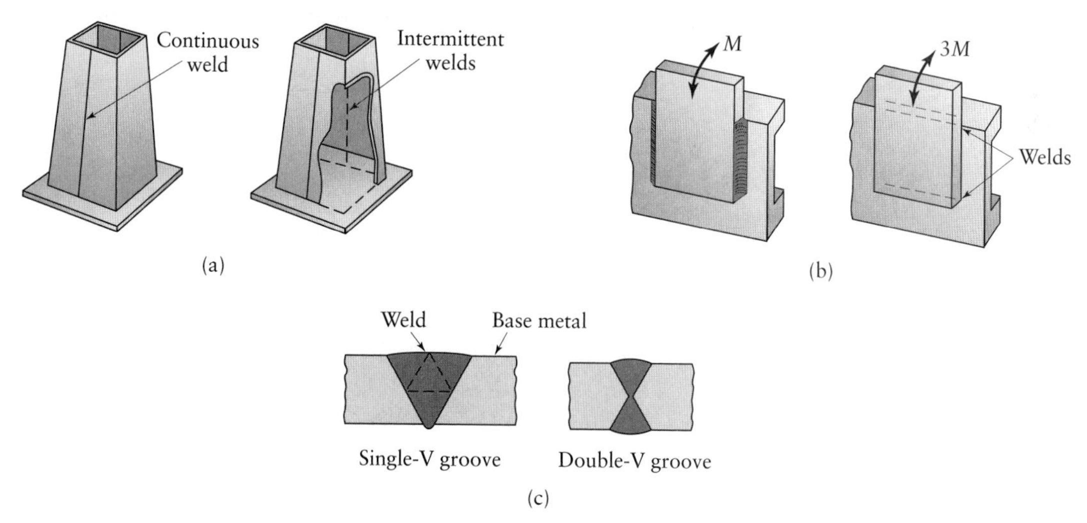

FIGURE 12.55 Weld designs for Example 12.4.

12.16.2 Design for brazing and joining

Some general design guidelines for *brazing* are given in Fig. 12.56; note that strong joints require a greater contact area for brazing than for welding. Design guidelines for *soldering* are similar to those for brazing. Figures 12.42 and 12.43 shows some examples of frequently used joint designs; note again the importance of large contact surfaces to the development of sufficient joint strength.

12.16.3 Design for adhesive bonding

Designs for *adhesive bonding* should ensure that joints are subjected to compressive, tensile, or shear forces, but not peeling or cleavage (see Fig. 12.47). Figure 12.57 shows several joint designs for adhesive bonding. They vary considerably in their strength; hence, selection of the appropriate design is important and should include considerations such as the type of loading and the environment to which the bonded structure will be subjected, particularly over a period of time.

Butt joints require large bonding surfaces, and lap joints tend to distort under tension, because of the force couple developed at the joint (see Fig. 12.25a). The coefficients of thermal expansion of the components to be bonded should preferably be

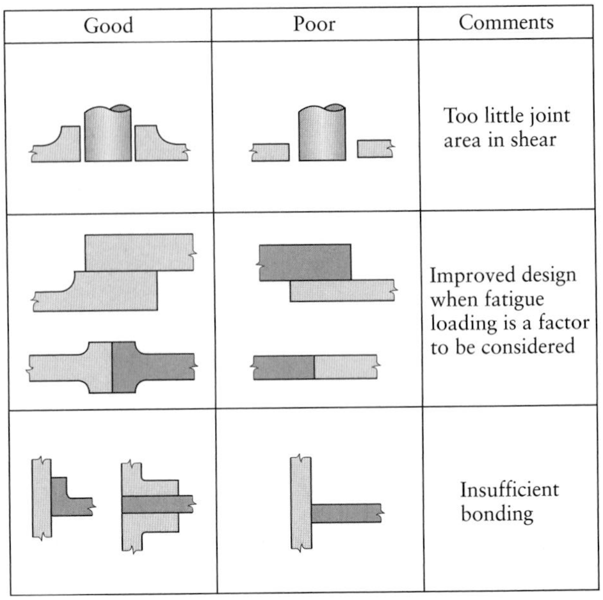

FIGURE 12.56 Examples of good and poor designs for brazing.

Good	Poor	Comments
		Too little joint area in shear
		Improved design when fatigue loading is a factor to be considered
		Insufficient bonding

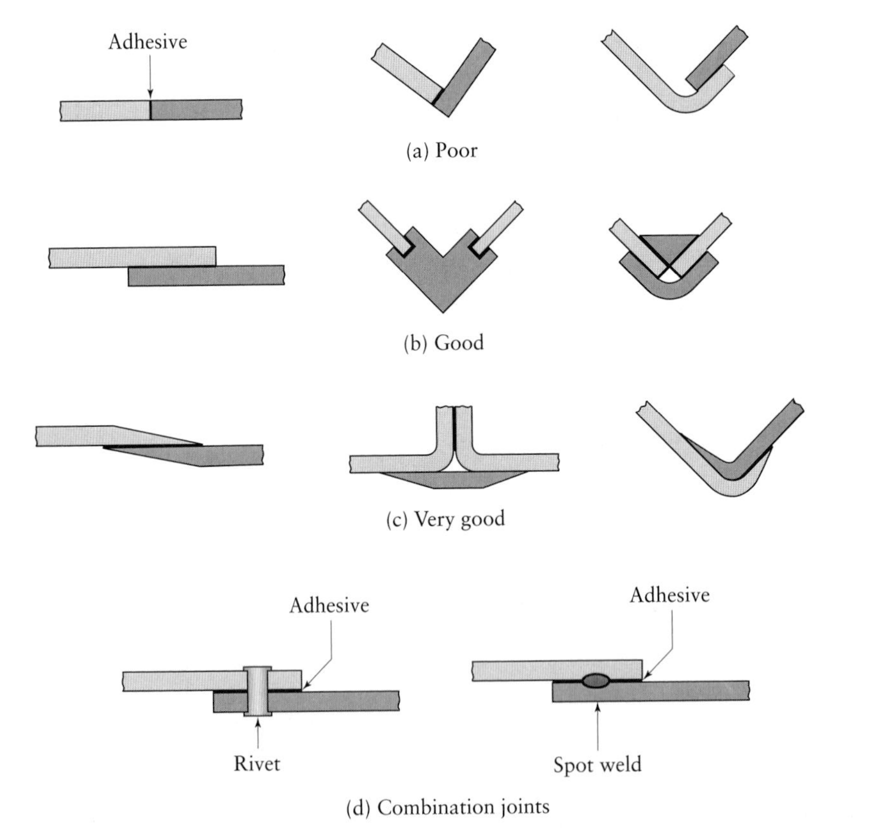

FIGURE 12.57 Various joint designs in adhesive bonding. Note that good design requires large contact areas between the members to be joined.

close, in order to avoid internal stress during adhesive bonding. Also, situations where thermal cycling can cause differential movement across the joint should be avoided.

12.16.4 Design for mechanical fastening

The design of *mechanical joints* requires consideration of (a) the type of loading, such as shear, tension, and moments, to which the structure will be subjected; and (b) the size and spacing of the holes. Also, the compatibility of the fastener material with the

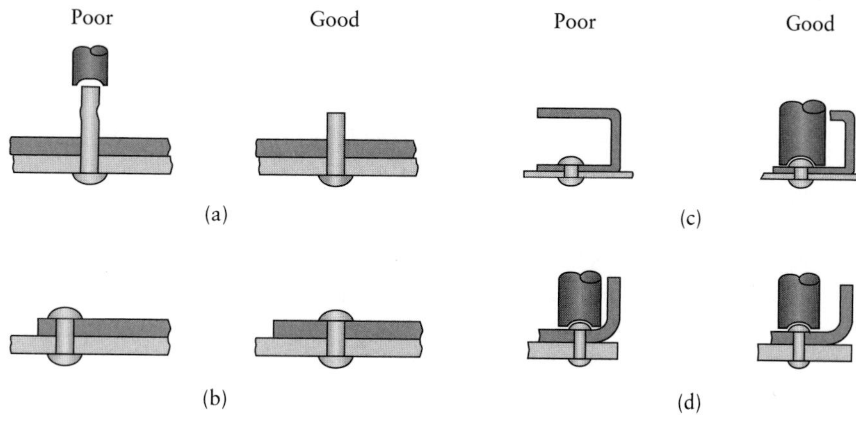

FIGURE 12.58 Design guidelines for riveting. *Source*: Bralla, J. G. (ed.) *Handbook of Product Design for Manufacturing*, 2d ed. New York, McGraw-Hill, 1999.

components to be joined is important; incompatibility may lead to galvanic corrosion, also known as *crevice corrosion*. In a system where, for example, a steel bolt or rivet is used to fasten copper sheets, the bolt is anodic and the copper plate cathodic, thus leading to rapid corrosion and subsequent loss of joint strength. Similar reactions take place when aluminum or zinc fasteners are used on copper products. Figure 12.58 illustrates some design guidelines for *riveting*.

Other general design guidelines for mechanical joining include the following (see also Section 14.10):

1. It is generally less costly to use fewer, but larger, fasteners than to use a large number of smaller fasteners.
2. Part assembly should be accomplished with a minimum number of fasteners.
3. The fit between parts to be joined should be as loose as permissible to reduce costs and facilitate the assembly process.
4. Fasteners of standard size should be used whenever possible.
5. Holes should not be too close to edges or corners in order to avoid the possibility that the material will tear when subjected to external forces.

12.17 Economic Considerations

We have included details on the general economic considerations for each particular joining process in our description of the process' capabilities; it can be seen that because of the wide variety of joining methods available, it is difficult to make generalizations regarding costs. The overall relative costs for these processes have been included in the last columns in Tables 12.1 and 12.2. Note in Table 12.1, for example, that because of the preparations involved, brazing and mechanical fastening can be the most costly methods. On the other hand, because of their highly automated nature, resistance welding and seaming and crimping processes are the least costly of the eight methods listed.

The cost of adhesive bonding depends on the particular operation. However, in many applications, the overall economics of the process makes adhesive bonding an attractive joining process; sometimes, it is the only process that is feasible or practical for a particular application.

Table 12.2 lists the cost of equipment involved for major categories of joining processes. Note that because of the specialized equipment required, laser-beam welding is the most expensive process in this respect, whereas the traditional oxyfuel-gas welding and shielded metal-arc welding are the least expensive processes. Typical ranges of equipment costs for some processes are listed as follows in alphabetical order, although the cost of some equipment exceeds these values significantly, depending on the size of the equipment and the level of automation and controls implemented:

Diffusion brazing: $50,000 to $300,000;
Electron-beam welding: $75,000 to over $1 million;
Electroslag welding: $15,000 to $25,000;
Flash welding: $5000 to $1 million;
Friction welding: $75,000 to $300,000;
Furnace brazing: $2000 to $300,000;
Gas metal-arc and flux-cored arc welding: $1000 to $3000;
Gas tungsten-arc welding: $1000 to $5000;
Laser-beam welding: $40,000 to $1 million;
Plasma-arc welding: $3000 to $6000;
Resistance welding: $20,000 to $50,000;
Shielded metal-arc welding: $1500;

An important aspect of the economics of joining is the major progress being made in automation of processes. (See Chapters 14 and 15.) These developments include computer control and optimization of processing parameters, as well as improvement in the manner in which various components in the operation are manipulated and controlled. Particularly important has been the extensive and effective use of industrial robots, programmed in such a manner that they can accurately track complex weld paths (such as in tailor-welded blanks), using machine vision systems and closed-loop controls. Consequently, the repeatability and accuracy of welding processes have improved greatly. The operations are optimized so that, unlike in manual operations, little time is wasted in manipulating various components in the system, just as such optimization has reduced the noncutting time in machining operations with advanced machine tools and machining centers. These developments have significantly reduced costs and have had a major impact on the overall economics of joining operations.

CASE STUDY | Light-Curing Acryclic Adhesives for Medical Products

Cobe Cardiovascular, Inc., is a leading manufacturer of blood-collection and blood-processing systems, as well as extracorporeal systems for cardiovascular surgery. In 1993, the company, like many other device manufacturers, used solvents for bonding several device components and subassemblies. However, several federal agencies began to encourage avoiding the use of solvents, and Cobe particularly wanted to eliminate its use of methylene chloride, for environmental and occupational safety reasons. To accomplish this goal, the company began to redesign most of its assemblies to accept light-curing (ultraviolet or visible) adhesives. Most of the company's devices were made of transparent plastics; consequently, its engineers determined that clear adhesive bonds were required for aesthetic purposes and because they have no tendency for stress cracking or crazing (see Section 10.3).

As an example of a typical product, Cobe's blood-salvage, or blood-collection, reservoir is an oval polycarbonate device that is approximately 300 mm (12 in.) tall, 200 mm (8 in.) in major diameter and 100 mm (4 in.) deep (Fig. 12.59). The reservoir is a one-time-use, or disposable, device; its purpose is to collect and hold blood during open-heart and chest surgery and arthroscopic and emergency-room procedures. Up to 3000 cc of blood may be stored in the reservoir while the blood awaits passage into a 250-cc centrifuge, which cleans the blood and returns it to the patient after the surgical procedure is completed. The collection reservoir consists of a clear, polycarbonate lid joined to a polycarbonate bucket. The joint is a tongue-and-groove

FIGURE 12.59 The Cobe Laboratories blood reservoir. The lid is bonded to the bowl with an airtight adhesive joint and tongue-in-groove joint. *Source*: Cobe Laboratories.

configuration; the goal was to create a strong, elastic joint that could withstand repeated stresses with no chance of leakage.

Light-cure acrylic adhesives offer a range of performance properties that make them well suited for this application (see the discussion of *anaerobic adhesives* in Section 12.13.1). First and foremost, they achieve high bond strength to the thermoplastics typically used to form medical-device housings. Loctite® 3211, for example, achieves shear strengths of 11 MPa (1600 psi) on polycarbonate. As important as the initial shear strength may be, it is even more important that the adhesive be able to maintain the high bond strength after sterilization; fortunately, disposable medical devices typically are subjected to very few sterilization cycles during manufacture. Also, these adhesives can endure a limited number of cycles of gamma irradiation, electron-beam irradiation, autoclaving, and ethylene-oxide or chemical immersion.

Another consideration that makes light-cure adhesives well suited for this application is their availability in formulations that allow them to withstand large strains prior to yielding; Loctite® 3211 yields at elongations in excess of 200%. This flexibility is critical, because the bonded joints are typically subjected to large amounts of bending and flexing when the devices are pressurized during qualification testing and use. If an adhesive is too rigid, it will fail this type of testing, even if it offers higher shear strength than that of a comparable and more flexible adhesive. Finally, light-cure acrylics are widely available in formulations that meet international quality standard certification (ISO; see Section 4.9.4), which means that when processed properly, they will not cause biocompatibility problems in the final assembly.

While these performance features are attractive, the adhesive must also meet certain processing characteristics during manufacturing. Light-cure acrylic adhesives have found wide use in medical-device assembly/joining operations because their processing characteristics are compatible with the high-speed automated manufacturing processes employed. These adhesives are available in a wide variety of viscosities and are easily dispensed through either pressure–time or positive-displacement dispensing systems. Once dispensed on the part, they can remain in contact with even highly-stressed plastic parts for several minutes or longer without causing stress cracking or degradation of the plastic. Liquid Loctite® 3211, for example, can remain in contact with polycarbonate that has been bent so as to induce stresses up to 17 MPa (2500 psi) for more than 15 minutes without stress cracking. Finally, the adhesive can be completely converted from liquid to solid state in seconds when exposed to light of the proper intensity and wavelength.

Since Loctite® 3211 absorbs light in the visible as well as the ultraviolet range, it can be used successfully on plastics that contain UV blockers, such as many grades of polycarbonate. The ability to have a long open time during which parts can be positioned, and yet cure the adhesive on demand, is a unique benefit of light-curing adhesives and dramatically reduces scrap costs. The equipment used to irradiate the part with high-intensity light typically requires a space of 10 ft^2 to 20 ft^2 on a production line, which is generally much less than that required for the ovens used by heat-cure adhesives and the racking shelves required for slower curing adhesives. Since floor space carries a cost premium in clean-room environments, this factor is a significant benefit.

It was also important that the joint be designed for maximum performance. If the enclosure were bonded with a joint consisting of two flat faces in intimate contact, peel stresses (see Fig. 12.47) would act on the bond when the vessel is pressurized. Peel stresses are the most difficult type of stress for an adhesive joint to withstand, due to the fact that the entire load is concentrated on the leading edge of the joint. The tongue-and-groove design used addressed this concern; in this design, the groove acts as a reservoir for holding the adhesive during the dispensing operation. When the parts are mated and the adhesive is cured, this design allows much of the load on the joint (when the device is pressurized) to be translated into shear and tensile forces, which the adhesive is much better suited to withstand. The gap between the tongue and the groove can vary widely, because most light-cure adhesives can quickly be cured to depths in excess of 5 mm (0.20 in.). This feature allows the manufacturer to have a robust joining process (meaning that wide dimensional tolerances can be accommodated).

With the new design and by using Loctite® 3211, Cobe thus eliminated the environmental concerns and the issues associated with solvent bonding while gaining the benefit of a safer, faster, and more consistent bond. The light-curing adhesive provided the aesthetic bond line the company wanted—one that was clear and barely perceptible; it also provided the structural strength needed and thus enabled Cobe to maintain a competitive edge in the marketplace.

Source: Courtesy of P. J. Courtney, Loctite Corporation.

SUMMARY

- Almost all products are an assemblage of many components; therefore, joining and fastening processes are an important and often necessary aspect of manufacturing, as well as of servicing and transportation. (Section 12.1)
- A major category of joining processes is fusion welding, in which the two pieces to be joined melt together and coalesce by means of heat; filler metals may or may not be used. Two common methods are consumable-electrode arc welding and nonconsumable-electrode are welding. Their selection of the particular method depends on factors such as workpiece material; shape complexity, size, and thickness; and type of joint. (Sections 12.2 and 12.3)
- Electron-beam and laser-beam welding are major categories of joining using high-energy beams. They produce small and high-quality weld zones and thus have important and unique applications in manufacturing. (Section 12.4)
- Because the joint undergoes important metallurgical and physical changes, the nature, properties, and quality of the weld joint are important factors to consider.

Weldability of metals, weld design, and process selection are factors that also should be considered. (Section 12.5)

- In solid-state welding, joining takes place without fusion, and pressure is applied either mechanically, using tools, electrodes, or rollers, or by explosives. Surface preparation and cleanliness can be important. Ultrasonic welding and resistance welding are two major examples of solid-state welding; these methods have a wide variety of applications, particularly for foil and sheet metals. (Sections 12.6–12.10)
- Diffusion bonding is another example of solid-state joining. Combined with superplastic forming, this method can be used to fabricate various complex sheet-metal structures with high strength-to-weight and stiffness-to-weight ratios. (Section 12.11)
- Brazing and soldering use filler metals at the interfaces to be joined. These processes require lower temperatures than those used in welding and are capable of joining dissimilar metals with intricate shapes and thicknesses. (Section 12.12)
- Adhesive bonding has gained wide acceptance in major industries such as aerospace, automotive, and building products. Adhesive-bonded joints have favorable characteristics such as strength, sealing, insulating, vibration damping, and resistance to corrosion between dissimilar metals. An important development is electrically conducting adhesives for surface-mount technologies. (Section 12.13)
- Mechanical fastening is among the oldest and most commonly used techniques. A wide variety of shapes and sizes of fasteners and fastening techniques have been developed for numerous permanent and semipermanent applications. (Section 12.14)
- With increasing use of thermoplastics and thermosetting plastics and various types of ceramics and glasses in a wide range of consumer and industrial applications, joining of these materials continue to undergo many advances. Several joining techniques are now available for these materials. (Section 12.15)
- General design guidelines have been established for the groups of processes described throughout this chapter. Some of these guidelines are applicable to a variety of processes, whereas other processes require special considerations, depending on their application. (Section 12.16)
- The economics of joining operations involves factors such as equipment cost, labor cost, and skill level required, as well as processing parameters such as time required, joint quality, and the need to meet specific requirements. The implementation of automation and computer controls can have a major impact on costs in these operations. (Section 12.17)

BIBLIOGRAPHY

Adams, R. D., J. Comyn, and W. C. Wake, *Structural Adhesive Joints in Engineering*, Chapman & Hall, 1997.

ASM Engineered Materials Handbook, Vol. 3: *Adhesives and Sealants*, ASM International, 1993.

ASM Handbook, Vol. 6: *Welding, Brazing and Soldering*, ASM International, 1993.

Baghdachi, J., *Adhesive Bonding Technology*, Dekker, 1996.

Bickford, J. H., and S. Nassar (eds.), *Handbook of Bolts and Bolted Joints*, Dekker, 1998.

Blake, A., *What Every Engineer Should Know about Threaded Fasteners*, Dekker, 1983.

Brazing Handbook, 4th ed., American Welding Society, 1991.

Cary, H. B., *Modern Welding Technology*, 4th ed., Prentice Hall, 1997.

Davies, A. C., *The Science and Practice of Welding*, 10th ed., 2 vols., Cambridge, 1993.

Dawes, C. T., *Laser Welding*, Woodhead, 1992.

Easterling, K. E., *Introduction to the Physical Metallurgy of Welding*, Butterworth, 1983.

Electronic Materials Handbook, Vol. 1: *Packaging*, ASM International, 1989.

Evans, G. M., and N. Bailey, *Metallurgy of Basic Weld Metal*, Wooodhead, 1997.

Frear, D. R., W. B. Jones, and K. R. Kinsman (eds.), *Solder Mechanics: A State of the Art Assessment*, The Minerals, Metals & Materials Society, 1991.

Galyen, J., G. Sear., and C. Tuttle, *Welding: Fundamentals and Procedures*, Wiley, 1985.

Granjon, H., *Fundamentals of Welding Metallurgy*, Woodhead Publishing, 1991.

Grong, O., *Metallurgical Modeling of Welding*, The Institute of Metals, 1994.

Handbook of Plastics Joining: A Practical Guide, William Andrew, Inc., 1996.

Haviland, G. S., *Machinery Adhesives for Locking, Retaining, and Sealing*, Dekker, 1986.

Hicks, J. G., *Welded Joint Design*, 2d. ed., Abington, 1997.

Houldcroft, P. T., *Welding and Cutting: A Guide to Fusion Welding and Associated Cutting Processes*, Industrial Press, 1988.

Humpston, G., and D. M. Jacobson, *Principles of Soldering and Brazing*, ASM International, 1993.

Hwang, J. S., *Modern Solder Technology for Competitive Electronics Manufacturing*, McGraw-Hill, 1996.

Introduction to the Nondestructive Testing of Welded Joints, 2d. ed., American Society of Mechanical Engineers, 1996.

Jeffus, L. F., *Welding: Principles and Applications*, 4th ed., Delmar, 1997.

Jellison, R., *Welding Fundamentals*, Prentice Hall, 1995.

Karlsson, I. (ed.), *Modeling in Welding, Hot Powder Forming and Casting*, ASM International, 1997.

Koellhoffer, L., *Welding Processes and Practices*, Prentice Hall, 1988.

Kou, S., *Welding Metallurgy*, Wiley, 1987.

Lambert, L., *Soldering for Electronic Assemblies*, Dekker, 1988.

Lancaster, J. F., *The Metallurgy of Welding*, Chapman & Hall, 1993.

Landrock, A. H., *Adhesives Technology Handbook*, Noyes, 1985.

Lee, L.-H., *Adhesive Bonding*, Plenum, 1991.

Lincoln, B., K. J. Gomes, and J. F. Braden, *Mechanical Fastening of Plastics*, Dekker, 1984.

Linnert, J. E., *Welding Metallurgy*, 4th ed., Vol. 1, American Welding Society, 1994.

Manko, H. H., *Soldering Handbook for Printed Circuits and Surface Mounting*, Van Nostrand Reinhold, 1995.

_______, *Solders and Soldering—Materials, Design, Production, and Analysis for Reliable Bonding*, McGraw-Hill, 1992.

Messler, R. W., Jr., *Joining of Advanced Materials*, Butterworth-Heinemann, 1993.

Mouser, J. D., *Welding Codes, Standards, and Specifications*, McGraw-Hill, 1997.

Nicholas, M. G., *Joining Processes: Introduction to Brazing and Diffusion Bonding*, Chapman & Hall, 1998.

North, T. H., *Advanced Joining Technologies*, Chapman & Hall, 1990.

Parmley, R. O. (ed.), *Standard Handbook of Fastening and Joining*, 3d. ed., McGraw-Hill, 1997.

Pecht, M. G., *Soldering Processes and Equipment*, Wiley, 1993.

Petrie, E. M., *Handbook of Adhesives and Sealants*, McGraw-Hill, 1999.

Powell, J., CO_2 *Laser Cutting*, Springer, 1992.

Rahn, A., *The Basics of Soldering*, Wiley Interscience, 1993.

Resistance Welding Manual, 4th ed., Resistance Welder Manufacturers' Association, 1989.

Satas, D., *Handbook of Pressure-Sensitive Adhesive Technology*, 3d. ed., Satas & Associates, 1999.

Schultz, H., *Electron Beam Welding*, Woodhead, 1994.

Schwartz, M. M., *Brazing for the Engineering Technologist*, Chapman & Hall, 1995.

_______, *Ceramic Joining*, ASM International, 1990.

_______, *Joining of Composite-Matrix Materials*, ASM International, 1994.

Shields, J., *Adhesives Handbook*, 3d. ed., Butterworth, 1984.

Skeist, I., *Handbook of Adhesives*, 3d. ed., Van Nostrand Reinhold, 1990.

Speck, J. A., *Mechanical Fastening, Joining, and Assembly*, Dekker, 1997.

Steen, W. M., *Laser Material Processing*, 2d. ed., Springer, 1998.

Stout, R. D., *Weldability of Steels*, Welding Research Council, 1987.

Swenson, L.-E, *Control of Microstructures and Properties in Steel Arc Welds*, CRC Press, 1994.

Tool and Manufacturing Engineers Handbook, Vol. 4: "Quality Control and Assembly," Society of Manufacturing Engineers, 1986.

Tres, P. A., *Designing Plastic Parts for Assembly*, 3d. ed., Hanser-Gardner, 1998.

Welding Handbook, 8th ed., 3 vols., American Welding Society, 1987.

Woodgate, R. W., *Handbook of Machine Soldering*, Wiley, 1996.

QUESTIONS

12.1 Explain the reasons that so many different welding processes have been developed.

12.2 List the advantages and disadvantages of mechanical fastening as compared with adhesive bonding.

12.3 What are the similarities and differences between consumable and nonconsumable electrodes?

12.4 What determines whether a certain welding process can be used for workpieces in horizontal, vertical, or upside-down positions, or for all types of positions? Explain, giving appropriate examples.

12.5 Comment on your observations regarding Fig. 12.3.

12.6 Discuss the need for and role of fixtures in holding workpieces in the welding operations described in this chapter.

12.7 Describe the factors that influence the size of the two weld beads in Fig. 12.11.

12.8 Why is the quality of welds produced by submerged arc welding very good?

12.9 Explain the factors involved in electrode selection in arc-welding processes.

12.10 Explain why the electroslag-welding process is suitable for thick plates and heavy structural sections.

12.11 What are the similarities and differences between consumable- and nonconsumable-electrode arc-welding processes?

12.12 In Table 12.2, there is a column on the distortion of welded components, ranging from lowest to highest. Explain why the degree of distortion varies among different welding processes.

12.13 Explain why the grains in Fig. 12.14 grow in the particular directions shown.

12.14 Prepare a table listing the processes described in this chapter and providing, for each process, the range of welding speeds as a function of workpiece material and thickness.

12.15 Explain what is meant by *solid-state welding*.

12.16 Describe your observations concerning Figs. 12.18, 12.19, and 12.20.

12.17 What advantages does friction welding have over the other joining methods described in this chapter?

12.18 Why is diffusion bonding, when combined with superplastic forming of sheet metals, an attractive fabrication process? Does it have any limitations?

12.19 Can roll bonding be applied to various part configurations? Explain.

12.20 Comment on your observations concerning Fig. 12.39.

12.21 If electrical components are to be attached to both sides of a circuit board, what soldering process(es) would you use? Explain.

12.22 Discuss the factors that influence the strength of (a) a diffusion-bonded component and (b) a cold-welded component.

12.23 Describe the difficulties you might encounter in applying explosion welding in a factory environment.

12.24 Inspect the edges of a U.S. quarter, and comment on your observations. Is the cross-section, i.e., the thickness of individual layers, symmetrical? Explain.

12.25 What advantages do resistance-welding processes have over others described in this chapter?

12.26 What does the strength of a weld nugget in resistance spot welding depend on?

12.27 Explain the significance of the magnitude of the pressure applied through the electrodes during resistance-welding operations.

12.28 Which materials can be friction stir welded, and which cannot? Explain your answer.

12.29 List the joining methods that would be suitable for a joint that will encounter high stresses and will need to be disassembled several times during the product life, and rank the methods.

12.30 Inspect Fig. 12.29, and explain why the particular fusion-zone shapes are developed as a function of pressure and speed. Comment on the influence of the properties of the material.

12.31 Which applications could be suitable for the roll spot-welding process shown in Fig. 12.33c? Give specific examples.

12.32 Give several examples concerning the bulleted items listed at the beginning of Section 12.1.

12.33 Could the projection-welded parts shown in Fig. 12.34 be made by any of the processes described in other parts of this text? Explain.

12.34 Describe the factors that influence flattening of the interface after resistance projection welding takes place.

12.35 What factors influence the shape of the upset joint in flash welding, as shown in Fig. 12.35b?

12.36 Explain how you would fabricate the structures shown in Fig. 12.39b with methods other than diffusion bonding and superplastic forming.

12.37 Make a survey of metal containers used for household products and foods and beverages. Identify those that have utilized any of the processes described in this chapter. Describe your observations.

12.38 Which process uses a solder paste? What are the advantages to this process?

12.39 Explain why some joints may have to be preheated prior to welding.

12.40 What are the similarities and differences between casting of metals (Chapter 5) and fusion welding?

12.41 Explain the role of the excessive restraint (stiffness) of various components to be welded on weld defects.

12.42 Discuss the weldability of several metals, and explain why some metals are easier to weld than others.

12.43 Must the filler metal be of the same composition as that of the base metal to be welded? Explain.

12.44 Describe the factors that contribute to the difference in properties across a welded joint.

12.45 How does the weldability of steel change as the steel's carbon content increases? Why?

12.46 Are there common factors among the weldability, solderability, castability, formability, and machinability of metals? Explain, with appropriate examples.

12.47 Assume that you are asked to inspect a weld for a critical application. Describe the procedure you would follow. If you find a flaw during your inspection, how would you go about determining whether or not this flaw is important for the particular application?

12.48 Do you think it is acceptable to differentiate brazing and soldering arbitrarily by temperature of application? Comment.

12.49 Loctite® is an adhesive (see the case study at the end of this chapter) used to keep metal bolts from vibrating loose; it basically glues the bolt to the nut once the nut is inserted in the bolt. Explain how this process is effective.

12.50 List the joining methods that would be suitable for a joint that will encounter high stresses and cyclic (fatigue) loading, and rank the methods in order of preference.

12.51 Why is surface preparation important in adhesive bonding?

12.52 Why have mechanical joining and fastening methods been developed? Give several specific examples of their applications.

12.53 Explain why hole preparation may be important in mechanical joining.

12.54 What precautions should be taken in mechanical joining of dissimilar metals?

12.55 What difficulties are involved in joining plastics? What about in joining ceramics? Why?

12.56 Comment on your observations concerning the numerous joints shown in the figures in Section 12.16.

12.57 How different is adhesive bonding from other joining methods? What limitations does it have?

12.58 Soldering is generally applied to thinner components. Why?

12.59 Explain why adhesively bonded joints tend to be weak in peeling.

12.60 Inspect various household products, and describe how they are joined and assembled. Explain why those particular processes were used.

12.61 Name several products that have been assembled by (a) seaming, (b) stitching, and (c) soldering.

12.62 Suggest methods of attaching a round bar made of thermosetting plastic perpendicularly to a flat metal plate.

12.63 Describe the tooling and equipment that are necessary to perform the double-lock seaming operation shown in Fig. 12.50, starting with flat sheet. (See also Fig. 7.23.)

12.64 What joining methods would be suitable to assemble a thermoplastic cover over a metal frame? Assume that the cover has to be removed periodically.

12.65 Repeat Question 12.64, but for a cover made of (a) a thermosetting plastic, (b) metal, and (c) ceramic. Describe the factors involved in your selection of methods.

12.66 Do you think the strength of an adhesively bonded structure is as high as that obtained by diffusion bonding? Explain.

12.67 Comment on workpiece size limitations, if any, for each of the processes described in this chapter.

12.68 Describe part shapes that cannot be joined by the processes described in this chapter. Gives specific examples.

12.69 Give several applications of electrically conducting adhesives.

12.70 Give several applications for fasteners in various household products, and explain why other joining methods have not been used instead.

12.71 Comment on workpiece shape limitations, if any, for each of the processes described in this chapter.

12.72 List and explain the rules that must be followed to avoid cracks in welded joints, such as hot tearing, hydrogen-induced cracking, lamellar tearing, etc.

12.73 If a built-up weld is to be constructed (see Fig. 12.3), should all of it be done at once, or should it be done a little at a time, with sufficient time allowed for cooling between beads?

12.74 Describe the reasons that fatigue failure generally occurs in the heat-affected zone of welds instead of through the weld bead itself.

12.75 If the parts to be welded are preheated, is the likelihood that porosity will form increased or decreased? Explain.

PROBLEMS

12.76 Two 1-mm-thick, flat copper sheets are being spot welded using a current of 5000 A and a current flow time of $t = 0.18$ s. The electrodes are 5 mm in diameter. Estimate the heat generated in the weld zone.

12.77 Calculate the temperature rise in Problem 12.76, assuming that the heat generated is confined to the volume of material directly between the two electrodes and that the temperature distribution is uniform.

12.78 In Fig. 12.23, assume that most of the top portion of the top piece is cut horizontally with a sharp saw. Thus, the residual stresses will be disturbed, and, as described in Section 2.10, the part will undergo shape change. For this case, how will the part distort? Explain.

12.79 The accompanying figure below shows a metal sheave that consists of two matching pieces of hot-rolled, low-carbon-steel sheets. These two pieces can be joined either by spot welding or by V-groove welding. Discuss the advantages and limitations of each process for this application.

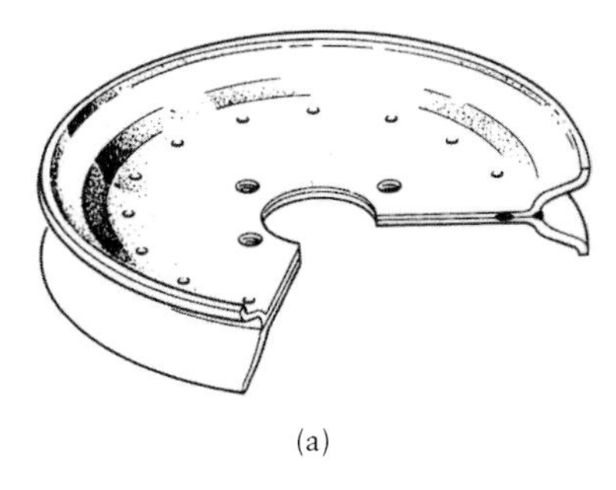

(a)

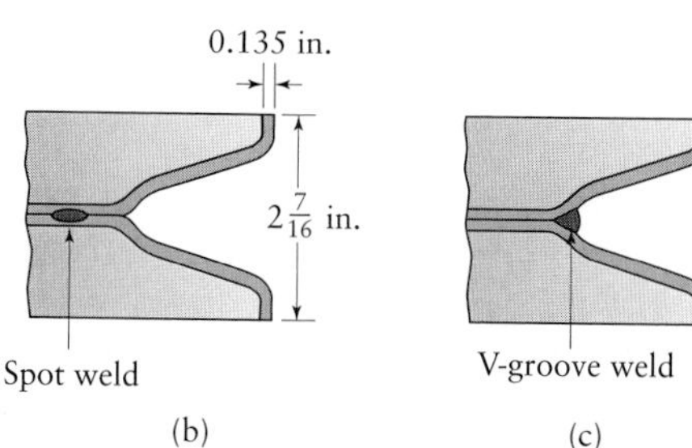

(b) (c)

12.80 A welding operation takes place on an aluminum-alloy plate. A pipe 2.5 in. in diameter, with a 0.20-in. wall thickness and a 2-in. length, is butt welded onto an angle iron 6 in. by 6 in. by 0.25 in thick. The angle iron is of an *L* cross-section and has a length of 1 ft. If the weld zone in a gas tungsten-arc welding process is approximately 0.5 in. wide, what would be the temperature increase of the entire structure due to the heat input from welding only? What if the process were an electron-beam welding operation, with a bead width of 0.08 in.? Assume that the electrode requires 1500 J and the aluminum alloy requires 1200 J to melt one gram.

12.81 An arc-welding operation is taking place on carbon steel. The desired welding speed is around 1 in./sec. If the power supply is 10 V, what current is needed if the weld width is to be 0.25 in.?.

12.82 The energy applied in friction welding is given by the formula $E = IS^2/C$, where I is the moment of inertia of the flywheel, S is the spindle speed in rpm, and C is a constant of proportionality. ($C = 5873$ when the moment of inertia is given in lb-ft^2.) For a spindle speed of 600 rpm and an operation in which a steel tube with a 3.5-in. outside diameter and a 0.25 in.-wall thickness is welded to a flat frame, what is the required moment of inertia of the flywheel if all of the energy is used to heat the weld zone, approximated as the material $\frac{1}{4}$ in. deep and directly below the tube? Assume that 1.4 ft-lbm is needed to melt the electrode.

DESIGN

12.83 Design a machine that can perform friction welding of two cylindrical pieces, as well as remove the flash from the welded joint. (See Fig. 12.28.)

12.84 How would you modify your design in Problem 12.83 if one of the pieces to be welded is noncircular?

12.85 Describe product designs that cannot be joined by friction-welding processes.

12.86 Make a comprehensive outline of joint designs relating to the processes described in this chapter. Give specific examples of engineering applications for each type of joint.

12.87 Review the two weld designs in Fig. 12.55a, and, based on the topics covered in courses on the strength of materials, show that the design on the right is capable of supporting a larger moment, as shown.

12.88 In the building of large ships, there is a need to weld large sections of steel together to form a hull. For this application, consider each of the welding operations described in this chapter, and list the benefits and drawbacks of that operation for this product. Which welding process would you select? Why?

12.89 Examine various household products, and describe how they are joined and assembled. Explain why those particular processes are used for these applications.

12.90 Repeat Problem 12.89 for a cover made of (a) a thermoset, (b) a metal, and (c) a ceramic. Describe the factors involved in your selection of methods.

12.91 A major cause of erratic behavior (hardware bugs) and failures of computer equipment is fatigue failure of the soldered joints, especially in surface-mount devices and devices with bond wires. (See Fig. 12.45.) Design a test fixture for cyclic loading of a surface-mount joint for fatigue testing.

12.92 Using two strips of steel 1 in. wide and 8 in. long, design and fabricate a joint that gives the highest strength in a tension test in the longitudinal direction.

CHAPTER 13

Fabrication of Microelectronic and Micromechanical Devices

13.1 Introduction

Although semiconducting materials have been used in electronics for many decades, it was the invention of the transistor in 1947 that set the stage for what would become one of the greatest technological advancements in all of history. Microelectronics has played an ever-increasing role since **integrated circuit** (IC) technology (Fig. 13.1) became the foundation for personal computers, cellular telephones, information systems, automotive control, and telecommunications.

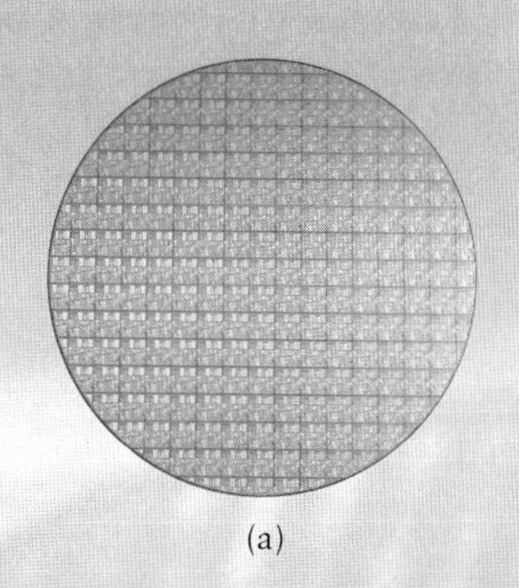

(a)

(b)

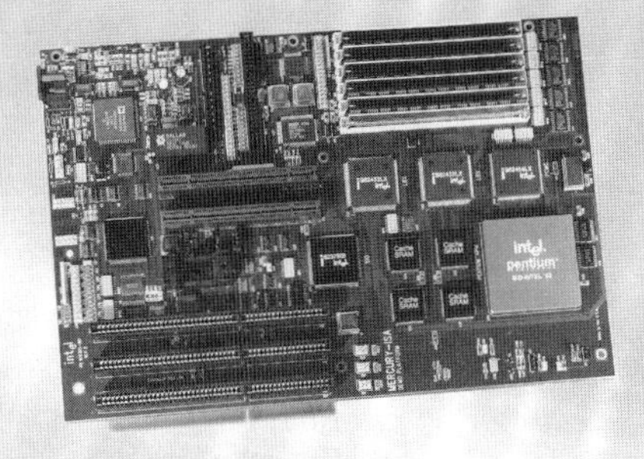

(c)

FIGURE 13.1 (a) The Pentium® 4 2.2GHz processor, as fabricated on a 300mm (11.8 in) wafer. (b) A Pentium® processor in a flip chip package with cover removed. (c) A Pentium® processor motherboard. *Source*: Courtesy Intel Corporation.

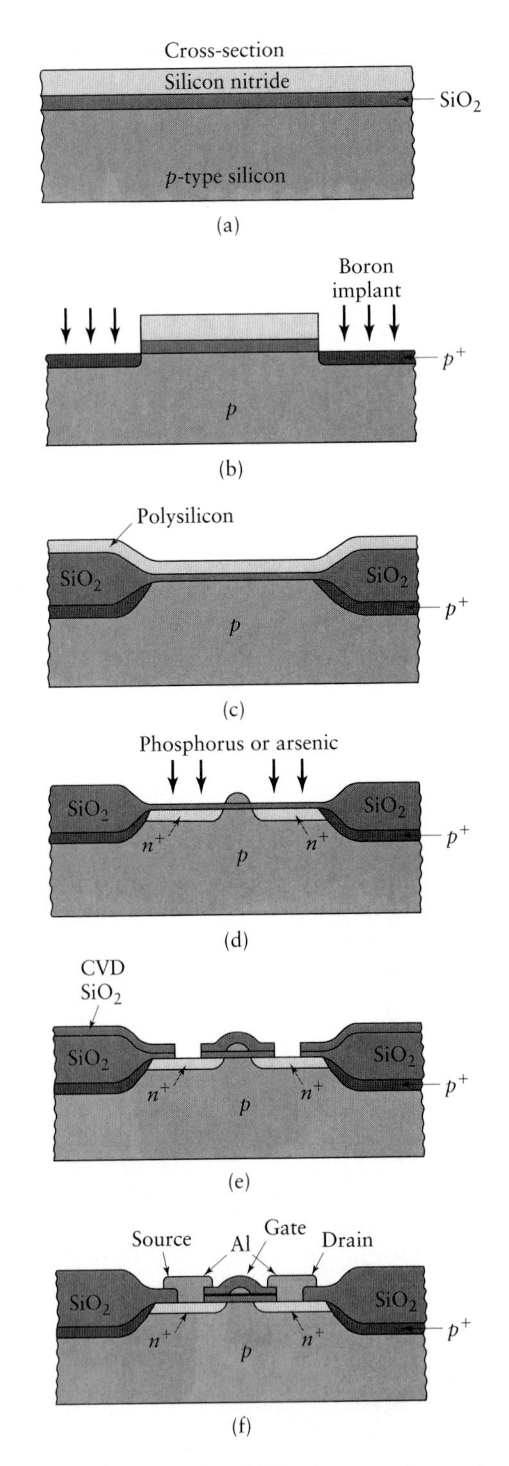

FIGURE 13.2 Cross-sectional views of the fabrication of a metal oxide semiconductor (MOS) transistor. *Source*: R. C. Jaeger.

The basic building block of a complex IC is the transistor. A transistor (Fig. 13.2) is a three-terminal device that acts as a simple on–off switch: If a positive voltage is applied to the "gate" terminal, a conducting channel is formed between the "source" and "drain" terminals, allowing current to flow between those two terminals (switched closed). If no voltage is applied to the gate, no channel is formed, and the source and drain are isolated from each other (switch opened). Figure 13.3 shows how the basic processing steps, described in this chapter, are combined to form a metal-oxide-semiconductor field effect transistor (MOSFET).

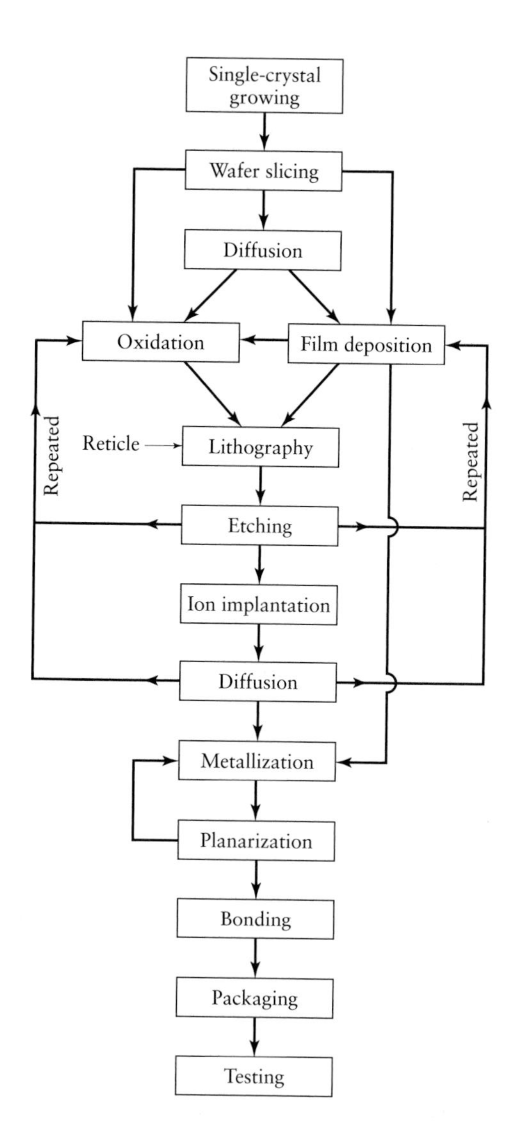

FIGURE 13.3 General fabrication sequence for integrated circuits.

In addition to the metal-oxide semiconductor structure, the **bipolar junction transistor** (BJT) is also used, but to a lesser extent. While the actual fabrication steps for these transistors are very similar to those for both the MOSFET and MOS technologies, their circuit applications are different. Memory circuits, such as RAMs (random access memory), and microprocessors consist primarily of MOS devices, whereas linear circuits, such as amplifiers and filters, are more likely to contain bipolar transistors. Other differences between these two types of devices include the faster operating speeds, higher breakdown voltage, and greater current drive of the BJT, and the lower current of the MOSFET.

The major advantage of today's ICs is their high degree of complexity, reduced size, and low cost. As fabrication technology becomes more advanced, the size of devices decreases, and, consequently, more components can be put onto a **chip** (a small slice of semiconducting material on which the circuit is fabricated). In addition, mass processing and process automation have greatly helped reduce the cost of each completed circuit. The components fabricated include transistors, diodes, resistors, and capacitors. Typical chip sizes produced today range from 0.5 mm × 0.5 mm (0.02 in. × 0.02 in.) to more than 50 mm × 50 mm (2 in. × 2 in.). In the past, no more than 100 devices could be fabricated on a single chip, but technology now allows densities in the tens of millions of devices per chip, commonly referred to as

very large-scale integration (VLSI). The Pentium 4 processor, for example, contains 42 million transistors. Some of the most advanced ICs may contain more than 100 million devices, termed **ultralarge-scale integration** (ULSI).

Because of the minute scale of microelectronic devices, all fabrication must take place in an extremely clean environment. *Clean rooms* are used extensively for this purpose and are rated by the maximum number of 0.5-μm (20-μin.) particles per cubic foot; most modern clean rooms are class-1 to class-10 facilities.

This chapter begins with a description of the properties of common semiconductors, such as silicon, gallium arsenide, and polysilicon. It then describes in detail the current processes employed (Fig. 13.3) in fabricating microelectronic devices and integrated circuits, including IC testing, packaging, and reliability. The chapter also describes a more recent and potentially more important development concerning the manufacture of **microelectromechanical systems** (MEMS) which are combinations of electrical and mechanical systems with characteristic lengths of less than 1 mm (0.040 in.). These devices utilize many of the batch-processing technologies used for the manufacture of electronic devices, although a number of other, unique processes have been developed as well.

MEMS devices have the potential of dramatically changing society; applications as varied as precise and rapid sensors, microrobots for nanofabrication, medical delivery systems, and artificial organs have been proposed. Given that common MEMS devices such as ink-jet printers, accelerometers, and computer hard-disk heads are already widespread, it is clear that MEMS are having a fundamental effect on society. Although "MEMS" is a term that came into use around 1987, it has been applied to a wide variety of applications. Currently, there are only a few microelectromechanical systems in use, such as accelerometers and some pressure sensors with on-chip electronics. Often, "MEMS" is a label applied to devices, such as pressure sensors, valves, and mirrors. There are many more device applications than integrated systems; even so, MEMS sales were in the range of $3 to $10 billion in 2001.

13.2 Semiconductors and Silicon

As the name suggests, **semiconductor materials** have electrical properties that lie between those of conductors and insulators; they exhibit resistivities between $10^{-3}\Omega$-cm and $10^{8}\Omega$-cm. Semiconductors have become the foundation for electronic devices, because their electrical properties can be altered by adding controlled amounts of selected impurity atoms into their crystal structures. These impurity atoms, also known as **dopants**, either have one more valence electron (n-type, or *negative dopant*) or one fewer valence electron (p-type, or *positive dopant*) than the atoms in the semiconductor lattice. For silicon, which is a Group IV element in the periodic table, typical n-type and p-type dopants include phosphorus and arsenic (Group V) and boron (Group III), respectively. The electrical operation of semiconductor devices is controlled by creating regions of different doping types and concentrations.

Although the earliest electronic devices were fabricated on *germanium*, **silicon** has become the industry standard. The abundance of silicon in its alternative forms in the earth's crust is second only to that of oxygen, thus making it economically attractive. Silicon's main advantage over germanium is its larger energy gap (1.1 eV) compared with that of germanium (0.66 eV); this larger energy gap allows silicon-based devices to operate at temperatures about 150°C (270°F) higher than the operating temperatures of devices fabricated on germanium [about 100°C(180°F).] Furthermore, the oxidized form of silicon, SiO_2, enables the production of metal-oxide-semiconductor (MOS) devices, which are the basis for MOS-Transistors. These materials are used in memory devices and processors and the like and account for by far the largest volume of semiconductor material produced worldwide.

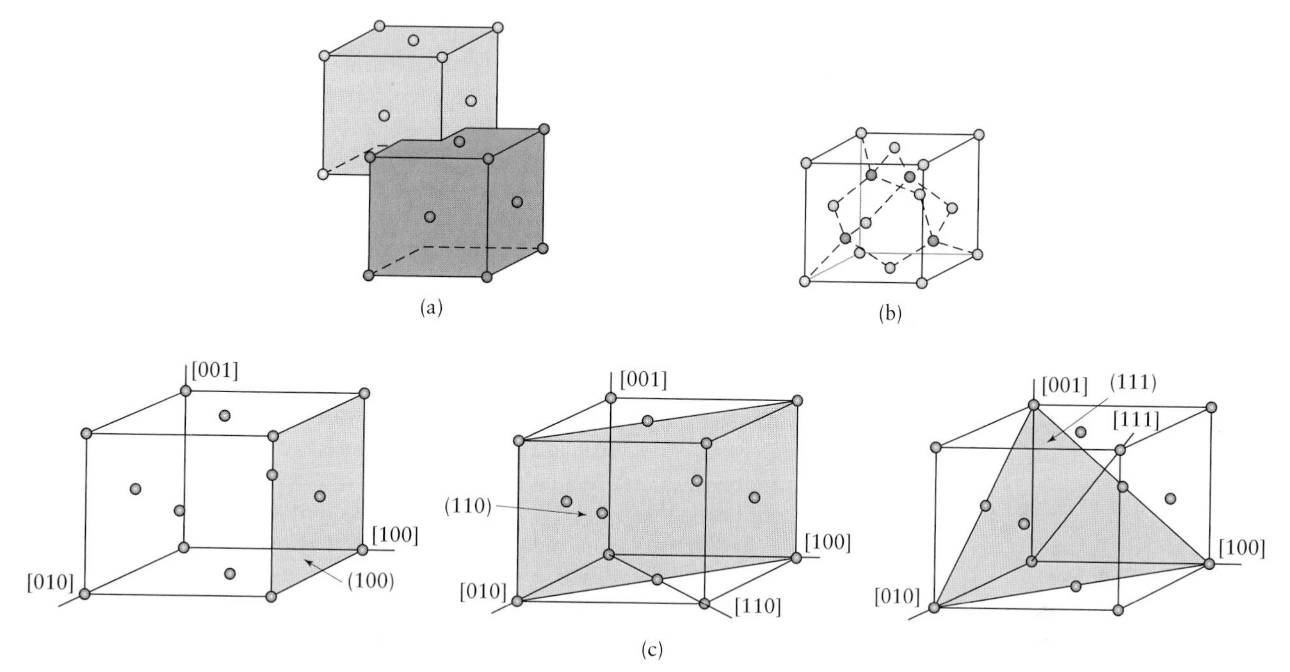

FIGURE 13.4 Crystallographic structure and Miller indices for silicon. (a) Construction of a diamond-type lattice from interpenetrating face-centered cubic cells (one of eight penetrating cells shown). (b) The diamond-type lattice of silicon. The interior atoms have been shaded darker than the surface atoms. (c) The Miller indices for a cubic lattice.

The crystallographic structure of silicon is a diamond-type fcc structure, as shown in Fig. 13.4, along with the Miller indices of an fcc material. Miller indices are a useful notation for identifying planes and directions within a unit cell. A crystallographic plane is defined by the reciprocal of its intercepts on the three axes. Since anisotropic etchants (see Section 13.8.1) preferentially remove material in certain crystallographic planes, the orientation of the silicon crystal in a wafer is important.

Silicon's important processing advantage is that its oxide (**silicon dioxide**) is an excellent insulator and is used for isolation and passivation purposes; conversely, germanium oxide is water soluble and unsuitable for electronic devices. However, silicon has some limitations, which has encouraged the development of compound semiconductors, specifically **gallium arsenide** (GaAs). Its major advantage over silicon is its capability for light emission [allowing fabrication of devices such as lasers and light-emitting diodes (LEDs)], in addition to its larger energy gap (1.43 eV) and, therefore, higher maximum operating temperature [to about 200°C (400°F).] Devices fabricated on gallium arsenide also have much higher operating speeds than those fabricated on silicon. Some disadvantages of gallium arsenide include its considerably higher cost, greater processing complications, and the difficulty of growing high-quality oxide layers, the need for which is emphasized throughout this chapter.

13.3 Crystal Growing and Wafer Preparation

Silicon occurs naturally in the forms of silicon dioxide and various silicates. It must undergo a series of purification steps in order to become a high-quality, defect-free, and single-crystal material needed for semiconductor-device fabrication. The purification process begins by heating silica and carbon together in an electric furnace, resulting in 95%- to 98%-pure polycrystalline silicon. This material is converted to an alternative form, commonly trichlorosilane, which in turn is purified and decomposed in a high-temperature hydrogen atmosphere. The result is an extremely high-quality **electronic-grade silicon** (EGS).

Single-crystal silicon is usually obtained by using the **Czochralski, or CZ, process** (see Fig. 5.34); this method uses a seed crystal that is dipped into a silicon melt and then slowly pulled out while being rotated. At this point, controlled amounts of impurities can be added to the system to obtain a uniformly doped crystal. Typical pull rates are on the order of 10 μm/s (400 μin./s). The result of this growing technique is a cylindrical single-crystal ingot, typically 100–300 mm (4–12 in.) in diameter and over 1 m (40 in.) long. However, this technique does not allow precise control of the ingot's diameter; therefore, ingots are commonly grown to a diameter a few millimeters larger than required and then ground to a precise diameter.

Next, the crystal is sliced into individual **wafers** by using an inner-diameter blade. This method uses a rotating blade with its cutting edge on the inner ring. While the substrate depth needed for most electronic devices is no more than several microns, wafers are typically cut to a thickness of about 0.5 mm (0.02 in.). This thickness provides the necessary mass to absorb temperature variations and provide mechanical support during subsequent fabrication steps. Finally, the wafers must be polished and cleaned to remove surface damage caused by the sawing process.

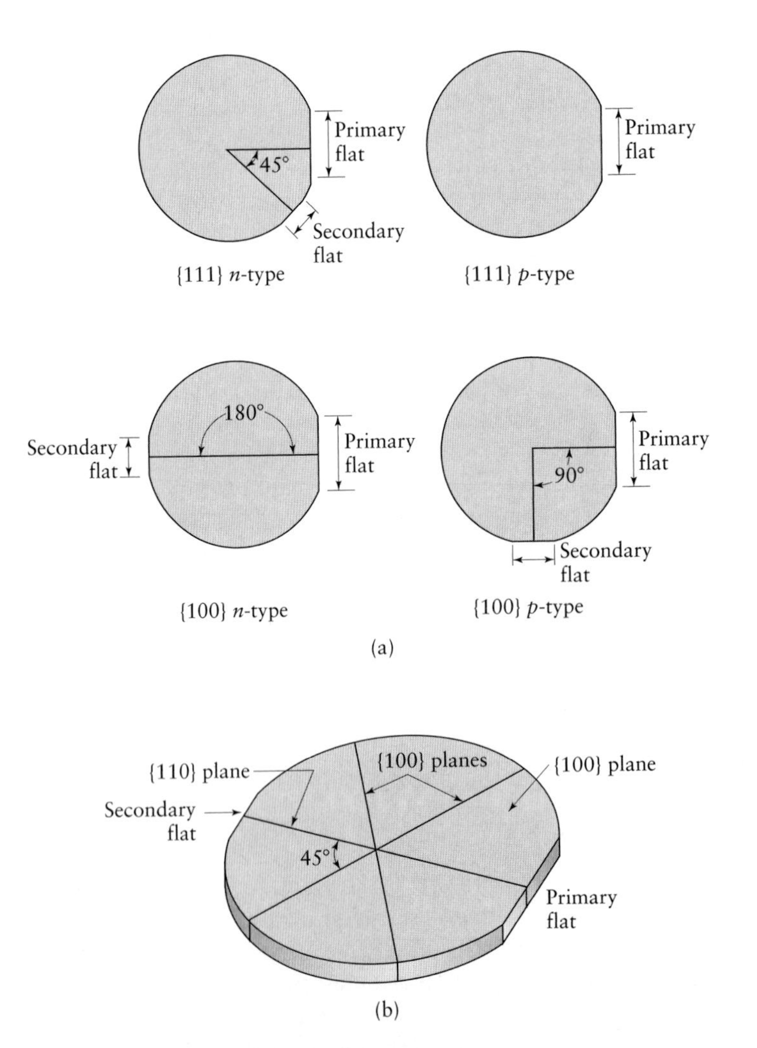

FIGURE 13.5 Identification of single-crystal wafers of silicon.

In order to properly control the manufacturing process, it is important to determine the orientation of the crystal in a wafer. The wafers, therefore, have notches and/or flats machined into them for identification, as shown in Fig. 13.5. Most commonly, the (100) or (111) plane of the crystal defines the wafer surface, although (110) surfaces can be used as well for micromachining applications. Wafers are also identified by a laser scribe marking produced by the manufacturer. Laser scribing of information can take place on the front or on the back side of the wafer. The front side of the wafer has an exclusion edge area from 3–10 mm in size reserved for the scribe information, including lot numbers, orientation, and a unique wafer identification code.

Device fabrication takes place over the entire wafer surface, and many identical circuits are generated at the same time. Wafers are typically kept in lots of 25 or 50 wafers with 6- to 8-in. diameters each or lots of 12–13 wafers with 12-in. diameters each, so that they can be easily handled and transferred during processing. Because of the small device size and large wafer diameter, thousands of individual circuits can be put on one wafer. Once processing is finished, the wafer is sliced into individual **chips**, each containing one complete integrated circuit.

13.4 Films and Film Deposition

Films of many different types, particularly insulating and conducting, are used extensively in microelectronic-device processing. Common deposition films include polysilicon, silicon nitride, silicon dioxide, tungsten, titanium, and aluminum. In some instances, single-crystal silicon wafers serve merely as a mechanical support on which custom *epitaxial layers* are grown (see Section 13.5). These silicon epitaxial films are of the same lattice structure as the substrate, so they are also single crystal materials. The advantages of processing on these deposited films, instead of on the actual wafer surface, include the inclusion of fewer impurities (notably carbon and oxygen), improved device performance, and the attainment of tailored material properties not obtainable on the wafers themselves.

Some of the major functions of deposited films include **masking** for diffusion or implants and protection of the semiconductor's surface. In masking applications, the film must effectively inhibit the passage of dopants while also being able to be etched into patterns of high resolution. Upon completion of device fabrication, films are applied to protect the underlying circuitry. Films used for masking and protection include silicon dioxide, phosphosilicate glass (PSG), and silicon nitride; each of these materials has distinct advantages, and they are often used in combination.

Other films contain dopant impurities and are used as doping sources for the underlying substrate. Conductive films are used primarily for device interconnection; these films must have a low resistivity, be capable of carrying large currents, and be suitable for connecting to terminal–packaging leads with wire bonds. Aluminum and copper are generally used for this purpose. To date, aluminum is the more popular material, because it can be dry etched more easily. However, aluminum is difficult to use for small structures and high current densities, and for this reason, copper, with its low resistivity, has been of significant research interest. Increasing circuit complexity has required up to six levels of conductive layers, which must all be separated by insulating films.

Films may be deposited by a number of techniques, which involve a variety of pressures, temperatures, and vacuum systems:

a) **Evaporation.** One of the simplest and oldest methods of film deposition is *evaporation*, used primarily for depositing metal films. In this process, the metal is heated in a vacuum until vaporization; upon evaporation, the metal forms a thin layer on the surface. The heat of evaporation is usually provided by a heating filament or electron beam.

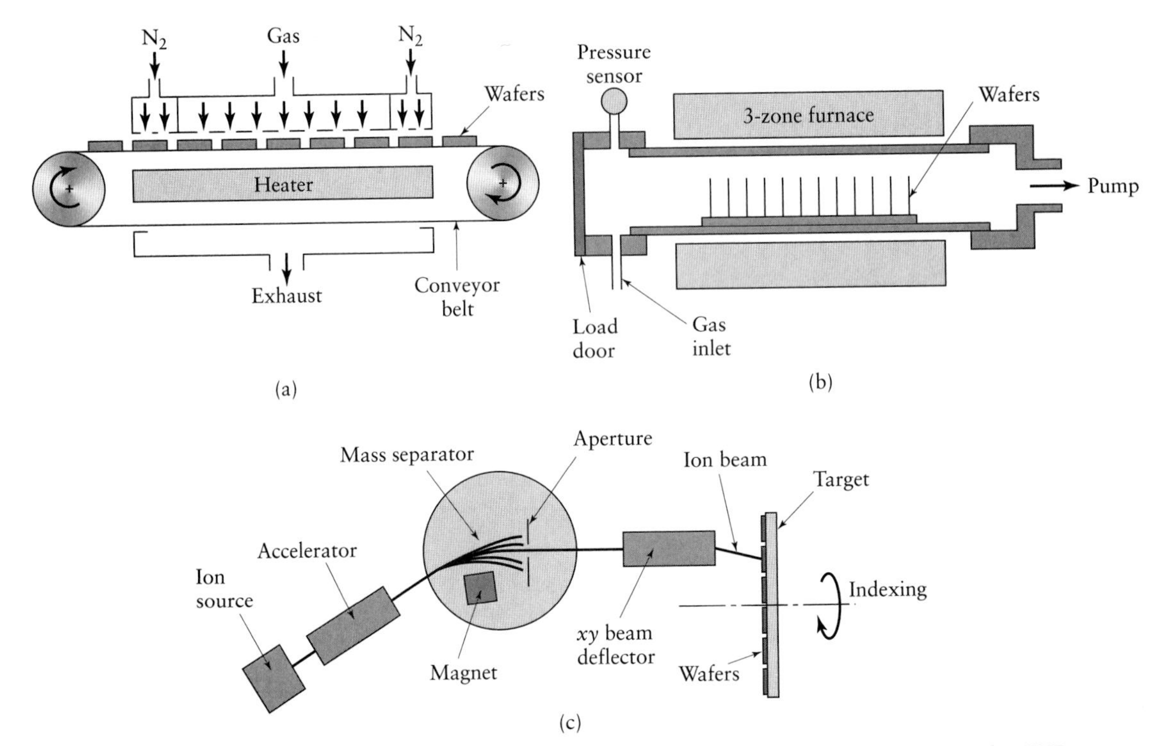

FIGURE 13.6 Schematic diagrams of (a) a continuous, atmospheric-pressure CVD reactor and (b) a low-pressure CVD reactor. *Source*: Sze, S. M., ed., VLSI Technology. New York, McGraw-Hill, 1983. (c) Apparatus for ion implantation. *Source*: J. A. Schey.

b) **Sputtering.** Another method of metal deposition is *sputtering*, which involves bombarding a target with high-energy ions, usually argon (Ar^+), in a vacuum. (See Section 4.5.1.) Sputtering systems usually include a dc power source to obtain the energized ions. As the ions impinge on the target, atoms are knocked off and subsequently deposited on wafers mounted within the system. Although some argon may be trapped within the film, this technique can provide very uniform coverage. Advanced sputtering techniques include use of a radio-frequency power source (**RF sputtering**) and introduction of magnetic fields (**magnetron sputtering**).

c) **Chemical vapor deposition.** In one of the most common techniques, *chemical vapor deposition* (CVD), film deposition is achieved by the reaction and/or decomposition of gaseous compounds. (See Section 4.5.1.) In this technique, silicon dioxide is routinely deposited by the oxidation of silane or a chlorosilane. Figure 13.6a shows a continuous CVD reactor that operates at atmospheric pressure.

d) **Low-pressure chemical vapor deposition.** Figure 13.6b shows the apparatus for a method similar to CVD that operates at lower pressures, referred to as *low-pressure chemical vapor deposition* (LPCVD). Capable of coating hundreds of wafers at a time, this method has a much higher production rate that than of atmospheric-pressure CVD and provides superior film uniformity with less consumption of carrier gases. This technique is commonly used for depositing polysilicon, silicon nitride, and silicon dioxide.

e) **Plasma-enhanced chemical vapor deposition.** The *plasma-enhanced chemical vapor deposition* (PECVD) process involves placing wafers in radio-frequency (RF) plasma containing the source gases and offers the advantage of maintaining low wafer temperature during deposition. However, the films deposited by this method generally incorporate hydrogen and are of lower quality than those deposited by other methods.

Silicon **epitaxy** layers, in which the crystalline layer is formed using the substrate as a seed crystal, can be grown via a variety of methods. If the silicon is deposited from the gaseous phase, the process is known as **vapor-phase epitaxy** (VPE). In another variation, called **liquid-phase epitaxy** (LPE), the heated substrate is brought into contact with a liquid solution containing the material to be deposited.

Another high-vacuum process uses evaporation to produce a thermal beam of molecules that deposit on the heated substrate. This process, called **molecular-beam epitaxy** (MBE), offers a very high degree of purity. In addition, since the films are grown one atomic layer at a time, excellent control of doping profiles is achieved, which is especially important in gallium-arsenide technology. However, MBE suffers from relatively low growth rates compared with those of other conventional film-deposition techniques.

13.5 Oxidation

Recall that the term *oxidation* refers to the growth of an oxide layer by the reaction of oxygen with the substrate material. Oxide films can also be formed by the deposition techniques described previously. Thermally grown oxides, described in this section, display a higher level of purity than deposited oxides, because the former are grown directly from the high-quality substrate. However, deposition methods must be used if the composition of the desired film is different than that of the substrate material.

Silicon dioxide is the most widely used oxide in integrated-circuit technology today, and its excellent characteristics are one of the major reasons for the widespread use of silicon. Aside from its functions of dopant masking and device isolation, silicon dioxide's most critical role is that of the *gate-oxide* material in MOSFETs. Silicon surfaces have an extremely high affinity for oxygen, and a freshly sawed silicon slice will quickly grow a native oxide of 30–40 Å (0.003–0.004 μm) in thickness. Modern IC technology requires oxide thicknesses from tens to thousands of angstroms.

a) **Dry oxidation.** Dry oxidation is a relatively simple process and is accomplished by elevating the substrate temperature, typically to 750°C–1100°C (1380°F–2020°F), in an oxygen-rich environment with variable pressure. Silicon dioxide is produced according to the chemical reaction

$$Si + O_2 \longrightarrow SiO_2. \tag{13.1}$$

Most oxidation is carried out in a batch process where up to 150 wafers are placed in a furnace. **Rapid thermal processing** (RTP) is a related process combining **rapid thermal oxidation** (RTO) with an anneal step [**rapid thermal annealing** (RTA)] and is used to produce thin oxides on a single wafer. As a layer of oxide forms, the oxidizing agents must be able to pass through the oxide layer and reach the silicon surface, where the actual reaction takes place. Thus, an oxide layer does not continue to grow on top of itself, but rather it grows from the silicon–silicon-dioxide interface outward. Some of the silicon substrate is consumed in the oxidation process (Fig. 13.7). The ratio of oxide thickness to the amount of silicon consumed is found to be 1 : 0.44. Thus, for example, to obtain an oxide layer 1000 Å (0.1 μm) thick, approximately 440 Å (0.044 μm) of silicon will be consumed; this condition does not present a problem, as substrates are always grown sufficiently thick.

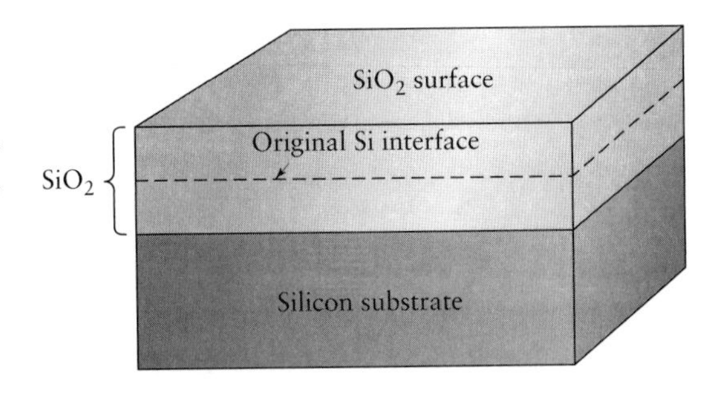

FIGURE 13.7 Growth of silicon dioxide, showing consumption of silicon. *Source*: Sze, S. M., Physics of Semiconductor Devices. New York: Wiley, 1981. This material is used by permission of John Wiley & Sons, Inc.

One important effect of this consumption of silicon is the rearrangement of dopants in the substrate near the interface. Some dopants deplete away from the oxide interface, while others pile up, and hence processing parameters have to be adjusted to compensate for this effect.

b) **Wet oxidation.** Another oxidizing technique uses a water-vapor atmosphere as the agent and is appropriately called *wet oxidation*. The chemical reaction involved in wet oxidation is

$$Si + 2H_2O \longrightarrow SiO_2 + 2H_2. \tag{13.2}$$

This method offers a considerably higher growth rate than that of dry oxidation, but it suffers from a lower oxide density and, therefore, a lower dielectric strength. The common practice in industry is to combine both dry and wet oxidation methods, growing an oxide in a three-part layer: dry, wet, and dry. This approach combines the advantages of the much higher growth rate in wet oxidation and the high quality obtained in dry oxidation.

c) **Selective oxidation.** The foregoing two oxidation methods are useful primarily for coating the entire silicon surface with oxide, but it is also necessary to oxidize only certain portions of the surface. The procedure used for this task, called *selective oxidation*, uses silicon nitride, which inhibits the passage of oxygen and water vapor. Thus, by masking certain areas with silicon nitride, the silicon under these areas remains unaffected while the uncovered areas are oxidized.

13.6 | Lithography

Table 13.1 provides a summary of lithographic techniques. *Lithography*, or *photolithography*, is the process by which the geometric patterns that define the devices are transferred from a **reticle** (also called a **photomask**, or **mask**) to the surface

TABLE 13.1

General Characteristics of Lithography Techniques

Method	Wavelength (nm)	Finest feature size (nm)
Ultraviolet (Photolithoraphy)	365	350
Deep UV	248	250
Extreme UV	10–20	30–100
X ray	0.01–1	20–100
Electron beam	–	80

Source: P. K. Wright, *21st Century Manufacturing*, Prentice Hall, Upper Saddle River, NJ, 2001.

of the chip. A reticle is a glass or quartz plate with a pattern of the chip deposited onto it usually with a chromium film, although iron oxide and emulsion are also used. The reticle image can be the same size as the desired structure on the chip, but it is often an enlarged image (usually 5× to 20× larger, although 10× magnification is most common). Enlarged images are then focused onto a wafer through a lens system in a process known as *reduction lithography*.

In current practice, the lithographic process is applied to microelectronic circuits several times, each time using a different reticle to define the different areas of the working devices. Typically designed at several thousand times their final size, reticle patterns undergo through a series of reductions before being applied permanently to a defect-free quartz plate. Computer-aided design (Section 15.4) has had a major impact on reticle design and generation. Cleanliness is especially important in lithography, and many manufacturers are turning to robotics and specialized wafer-handling apparati in order to minimize dust and dirt contamination.

Once the film deposition process is completed and the desired reticle patterns have been generated, the wafer is cleaned and coated with an organic **photoresist** (PR), which is sensitive to ultraviolet (UV) light. The wafers is then placed inside a resist spinner, and the photoresist is applied as a viscous liquid onto the wafer. Photoresist layers of 0.5–2.5 μm (20–100 μin.) thick are obtained by spinning at several thousand rpm for 30 to 60 s to give uniform coverage. The thickness of the resist is given by

$$t = \frac{kC^{\beta}\eta^{\gamma}}{\omega^{\alpha}}, \tag{13.3}$$

where t is the resist thickness, C is the polymer concentration in mass per volume, η is the viscosity of the polymer, ω is the angular velocity during spinning, and k, α, β, and γ are constants for the particular spinning system. Where masking levels are considered critical, a **barrier antireflective layer** (BARL) or **barrier antireflective coating** (BARC) is applied either beneath or on top of the photoresist to provide line width control, especially over aluminum.

The next step in lithography is **prebaking** the wafer to remove the solvent from the photoresist and harden it. This step is carried out in a convection oven at around 100°C (273°F) for a period of 10 to 30 min. The wafer is then aligned under the desired reticle in a *stepper*. In this crucial step, called **registration**, the reticle must be aligned correctly with the previous layer on the wafer. Once the reticle is aligned, it is stepped across the wafer and exposed to UV radiation. Upon development and removal of the exposed photoresist, a duplicate of the reticle pattern will appear in the photoresist layer. As can be seen in Fig. 13.8, the reticle can be either a negative or a positive image of the desired pattern. A positive reticle uses the UV radiation to break down the chains in the organic film, so that these chains are preferentially removed by the developer. Positive masking is more commonly used than negative masking, because with negative masking, the photoresist can swell and distort, making it unsuitable for small geometries. Newer negative photoresist materials do not have this problem, however.

Following the exposure and development sequence, **postbaking** the wafer is implemented to drive off solvent and toughen and improve the adhesion of the remaining resist. In addition, a deep UV treatment, which consists of baking the wafer to 150°C–200°C (300°F–400°F) in ultraviolet light, can also be used to further strengthen the resist against high-energy implants and dry etches. The underlying film not covered by the PR is then implanted or etched away (Sections 13.7 and 13.8). Finally, the photoresist is stripped by exposing it to wet stripper or an oxygen plasma, a technique also referred to as **ashing**. The lithography process may be repeated as many as 25 times in the fabrication of the most advanced ICs.

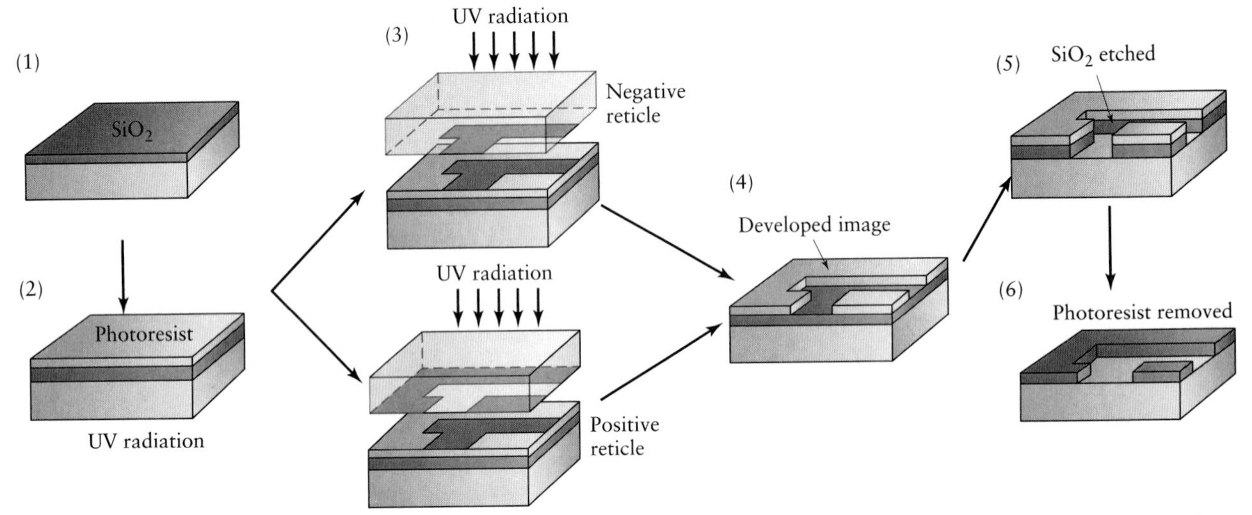

FIGURE 13.8 Pattern transfer by lithography. Note that the mask in Step 3 can be either a positive or a negative image of the pattern. *Source*: After W. C. Till and J. T. Luxon.

One of the major issues in lithography is **linewidth**, which refers to the width of the smallest feature obtainable on the silicon surface. As circuit densities have escalated over the years, device sizes and features have become smaller and smaller. Today, minimum commercially feasible linewidths are between 0.13 μm and 0.21 μm, with considerable research being conducted at smaller linewidths.

As pattern resolution, and therefore device miniaturization, is limited by the wavelength of the radiation source used, the need has arisen to move to wavelengths shorter than those in the ultraviolet range, such as "deep" UV wavelengths, "extreme" UV wavelengths, electron beams, and X rays (see Table 13.1). In these technologies, the photoresist is replaced by a similar resist that is sensitive to a specific range of shorter wavelengths.

Extreme-ultraviolet lithography. The pattern resolution in photolithography is limited by light diffraction. One method of reducing the effects of diffraction is to use ever shorter wavelengths. *Extreme-ultraviolet* (EUV) *lithography* uses light at a wavelength of 13 nm in order to obtain features around 30 to 100 nm in size. The waves are focused by highly reflective molybdenum/silicon mirrors (instead of glass lenses that absorb EUV light) through the mask to the wafer surface.

X ray lithography. Although photolithography is the most widely used lithography technique, it has fundamental resolution limitations associated with light diffraction. *X ray lithography* is superior to photolithography, because of the shorter wavelength of the radiation and the very large depth of focus involved in the former. This characteristic allows much finer patterns to be resolved, and X ray lithography is far less susceptible to dust. Furthermore, the aspect ratio (defined as the ratio of depth to lateral dimension) can be more than 100 with X ray lithography, but is limited to around 10 with photolithography. However, in order to achieve this benefit, synchrotron radiation is needed, which is expensive and available at only a few research laboratories. Given the large capital investment required for a manufacturing facility, industry has preferred to refine and improve optical lithography instead of investing new capital into X ray-based production. X ray lithography is not currently widespread; however, the LIGA process (see Section 13.14) fully exploits the benefits of X ray lithography.

Electron-beam and ion-beam lithography. Like X ray lithography, *electron-beam* (e-beam) and *ion-beam (i-beam)* lithography are superior to photolithography with respect to the resolutions attainable. These methods involve high current density in narrow electron or ion beams (*pencil sources*) that scan a pattern, one pixel at a time, onto a wafer. The masking is done by controlling the point-by-point transfer of the stored pattern and therefore is performed by software. These techniques have the advantages of accurate control of exposure over small areas of the wafer, large depth of focus, and low defect densities. Resolutions are limited to around 10 nm, because of electron scatter, although 2-nm resolutions have been reported for some materials. It should be noted that the scan time increases significantly as the resolution increases, because more highly focused beams are required. The main drawback of these techniques is that electron and ion beams need to be maintained in a vacuum, which significantly increases equipment complexity and production cost; furthermore, the scan time for a wafer for this technique is much slower than that for other lithographic methods.

13.7 Diffusion and Ion Implantation

As stated in Section 13.2, the operation of microelectronic devices often depends on regions of different doping types and concentrations. The electrical character of these regions can be altered by introducing dopants into the substrate, accomplished by *diffusion* and *ion implantation* processes. Since many different regions of microelectronic devices must be defined, this step in the fabrication sequence is repeated several times.

In the diffusion process, the movement of atoms results from thermal excitation. Dopants can be introduced to the substrate in the form of a deposited film, or the substrate can be exposed to a vapor containing the dopant source. This process takes place at elevated temperatures, usually 800°C–1200°C (1500°F–2200°F). Dopant movement within the substrate is a function of temperature, time, and the diffusion coefficient (or diffusivity) of the dopant species, as well as the type and quality of the substrate material. Because of the nature of diffusion, the dopant concentration is very high at the surface and drops off sharply away from the surface. To obtain a more uniform concentration within the substrate, the wafer is heated further to drive in the dopants, a process called **drive-in diffusion**. The fact that diffusion, desired or undesired, will always occur at high temperatures is invariably taken into account during subsequent processing steps. Although the diffusion process is relatively inexpensive, it is highly isotropic.

Ion implantation is a much more extensive process and requires specialized equipment (Fig. 13.6c). Implantation is accomplished by accelerating ions through a high-voltage beam of as much as one million volts and then choosing the desired dopant by means of a magnetic mass separator. In a manner similar to that used in cathode-ray tubes, the beam is swept across the wafer by sets of deflection plates, thus ensuring uniform coverage of the substrate. The complete implantation system must be operated in a vacuum.

The high-velocity impact of ions on the silicon surface damages the lattice structure, resulting in lower electron mobilities. This condition is undesirable, but the damage can be repaired by an annealing step, which involves heating the substrate to relatively low temperatures, usually 400°C–800°C (750°F–1500°F), for 15–30 min. This process provides the energy that the silicon lattice needs to rearrange and mend itself. Another important function of annealing is to allow the dopant to move from the interstitial to substitutional sites (see Fig. 3.8), where they are electrically active.

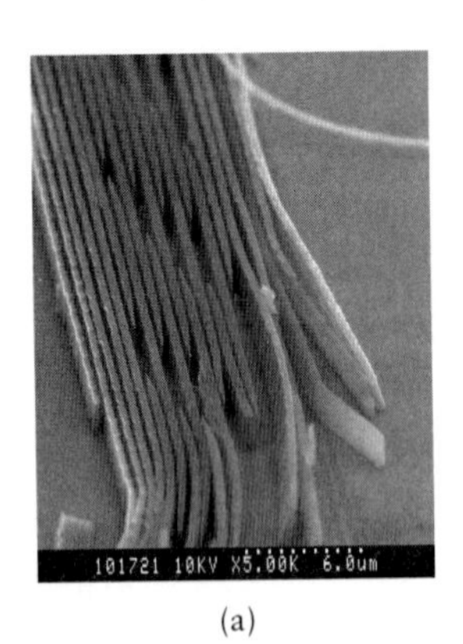

(a)

(b)

FIGURE 13.9 Effect of residual stress on patterns etched in a 2 μm thick carbon layer: (a) Patterns destroyed by high residual stress; (b) Patterns etched after thermal curing to relieve residual stresses. *Source*: P. Hudek, et al., *Journal of Vac. Sci. and Technology*, B Nov. 17 (6), 1999, pp. 3127–3131.

13.8 Etching

Etching is the process by which entire films, or particular sections of films or the substrate, are removed; it plays an important role in the fabrication sequence. One of the most important criteria in this process is **selectivity**, which refers to the ability to etch one material without etching another. Tables 13.2 and 13.3 provide a summary of etching processes. In silicon technology, an etching process must effectively etch the silicon-dioxide layer with minimal removal of the underlying silicon or the resist material. In addition, polysilicon and metals must be etched into high-resolution lines with vertical wall profiles and with minimal removal of the underlying insulating film. Typical etch rates range from tens to several thousands of nm/min, and selectivities (defined as the ratio of the etch rates of the two films) can range from 1 : 1 to 100 : 1.

It should be noted that etching is often performed in order to machine desired patterns into substrates or coatings. The resolution of the pattern is a critical concern, both for device performance and to satisfy the continuous goals of further device miniaturization. Residual stresses in the surface can be a major obstacle to pattern transfer. The effect of residual stresses can clearly be seen in Fig. 13.9a; as material is removed, residual stresses can cause slender features to warp or fracture (as also shown, for example, in Fig. 2.33). To reduce the likelihood of such defects, workpieces and coatings can be annealed or stress-relieved before etching, allowing good pattern transfer. When residual stresses are effectively controlled or eliminated, good pattern transfer can take place (Fig. 13.9b).

13.8.1 Wet etching

Wet etching involves immersing the wafers in a liquid, usually acidic, solution. The main drawback to most wet-etching operations is that they are *isotropic*, that is, they etch in all directions of the workpiece at the same rate. This condition results in undercuts beneath the mask material (see, for example, Fig. 13.10a) and limits the resolution of geometric features in the substrate.

Effective etching requires the following conditions:

1. Etchant transport to the surface;
2. A chemical reaction;
3. Transport of reaction products away from the surface;
4. Ability to stop the etching process rapidly in order to obtain superior pattern transfer (**etch stop**), usually using an underlying layer with high selectivity.

TABLE 13.2

Comparison of Etch Rates

Etchant	Target material	Etch rate (nm/min)[a]							
		Polysilicon n^+	Polysilicon, undoped	Silicon dioxide	Silicon nitride	Phosphosilicate glass, annealed	Aluminum	Titanium	Photoresist (OCG-820PR)
Wet etchants									
Concentrated HF (49%)	Silicon oxides	0	–	2300	14	3600	4.2	>1000	0
25:1 HF:H_2O	Silicon oxides	0	0	9.7	0.6	150	–	–	0
5:1 BHF[b]	Silicon oxides	9	2	100	0.9	440	140	>1000	0
Silicon etchant (126 HNO_3:60 H_2O:5 NH_4F)	Silicon	310	100	9	0.2	170	400	300	0
Aluminum etchant (16 H_3PO_4:1 HNO_3:1 HAc:2 H_2O)	Aluminum	<1	<1	0	0	<1	660	0	0
Titanium etchant (20 H_2O:1 H_2O_2:1 HF)	Titanium	1.2	–	12	0.8	210	>10	880	0
Piranha (50 H_2SO_4:1 H_2O_2)	Cleaning off metals and organics	0	0	0	0	0	180	240	>10
Acetone (CH_3COOH)	Photoresist	0	0	0	0	0	0	0	>4000
Dry etchants									
CF_4 + CHF_3 + He, 450W	Silicon oxides	190	210	470	180	620	–	>1000	220
SF_6 + He, 100W	Silicon nitrides	73	67	31	82	61	–	>1000	69
SF_6, 125W	Thin silicon nitrides	170	280	110	280	140	–	>1000	310
O_2, 400W	Ashing photoresist	0	0	0	0	0	0	0	340

Notes:

a. Results are for fresh solutions at room temperature unless otherwise noted. Actual etch rates will vary with temperature and prior use of solution, area of exposure of film, other materials present, and film impurities and microstructure.

b. Buffered hydrofluoric acid, 33% NH4F and 8.3% HF by weight.

Source: After K. Williams and R. Muller, *J. Microelectromechanical Systems*, Vol. 5, 1996, pp. 256–269.

TABLE 13.3

General Characteristics of Silicon Etching Operations

	Temperature (°C)	Etch rate (μm/min)	{111}/{100} selectivity	Nitride etch rate (nm/min)	SiO_2 etch rate (nm/min)	p++ etch stop
Wet etching						
$HF:HNO_3:CH_3COOH$	25	1–20	–	Low	10–30	No
KOH	70–90	0.5–2	100:1	<1	10	Yes
Ethylene-diamine pyrochatechol (EDP)	115	0.75	35:1	0.1	0.2	Yes
$N(CH_3)_4OH$ (TMAH)	90	0.5–1.5	50:1	<0.1	<0.1	Yes
Dry (plasma) etching						
SF_6	0–100	0.1–0.5	–	200	10	No
SF_6/C_4F_8 (DRIE)	20–80	1–3	–	200	10	No

Source: Adapted from N. Maluf, *An Introduction to Microelectromechanical Systems Engineering*, Artech House, 2000.

If the first or third steps limit the speed of the process, agitation or stirring of the solution can increase etching rates. If the second step limits the speed of the process, the etching rate will depend strongly on temperature, etching material, and solution composition. Therefore, reliable etching requires both good temperature control and repeatable stirring capability.

Isotropic etchants are widely used for the following procedures:

1. Removal of damaged surfaces;
2. Rounding of sharp etched corners to avoid stress concentrations;
3. Reduction of roughness after anisotropic etching;
4. Creation of structures in single-crystal slices; and
5. Evaluation of defects.

Microelectronic devices and MEMS (see Sections 13.13 through 13.15) require accurate machining of structures, and this task is done through masking. However, masking is a challenge with isotropic etchants. The strong acids etch aggressively (at a rate of 0.1–1 μm/min) and produce rounded cavities. Furthermore, the etch rate is very sensitive to agitation, and, therefore, lateral and vertical features are difficult to control.

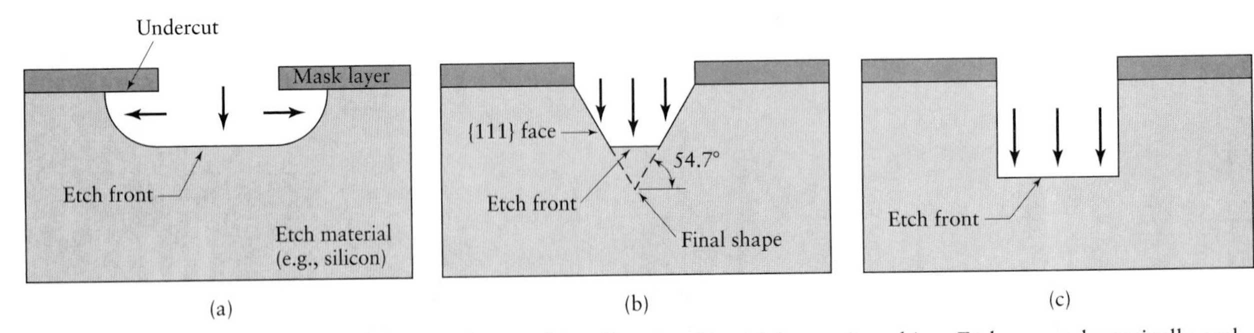

FIGURE 13.10 Etching directionality. (a) Isotropic etching: Etch proceeds vertically and horizontally at approximately the same rate, with significant mask undercut. (b) Orientation-dependant etching (ODE): Etch proceeds vertically, terminating on {111} crystal planes with little mask undercut. (c) Vertical etching: Etch proceeds vertically with little mask undercut. *Source*: Courtesy of K. R. Williams.

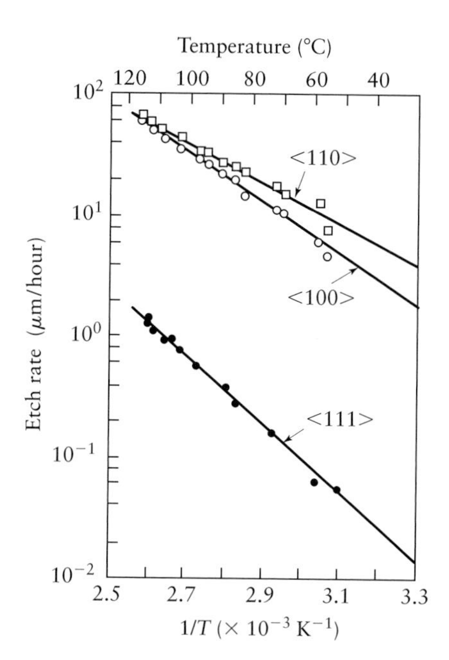

FIGURE 13.11 Etch rates of silicon in different crystallographic orientations, using ethylene-diamine/pyrocatechol-in-water as the solution. *Source*: After H. Seidel et al., *J. Electrochemical Society*, 1990, pp. 3612–3626.

The size of the features in an integrated circuit determines its performance, and for this reason, there is a strong desire to produce well-defined, extremely small structures. Such small features cannot be attained through isotropic etching, because of the poor definition that results from undercutting of masks.

Anisotropic etching takes place when etching is strongly dependent on compositional or structural variations in the material. There are two basic kinds of anisotropic etching—*orientation-dependent etching* (ODE) and *vertical etching*—although most vertical etching is done with dry plasmas and is discussed in Section 13.8.2. Orientation-dependent etching commonly occurs in a single crystal when etching takes place at different rates in different directions, as shown in Fig. 13.10b. When ODE is performed properly, the etchants produce geometric shapes with walls defined by the crystallographic planes that resist the etchants. For example, Fig. 13.11 shows the vertical etch rate for silicon as a function of temperature. As can be seen, etching is more than one order of magnitude slower in the [111] crystal direction than in other directions; therefore, well-defined walls can be obtained along the [111] crystal direction.

The **anisotropy ratio**, AR, for etching is defined by

$$AR = \frac{E_1}{E_2}, \tag{13.4}$$

where E is the etch rate and the subscripts refer to two crystallographic directions of interest. *Selectivity* is defined in a similar manner, but refers to the etch rates between materials of interest. The anisotropy ratio is unity for isotropic etchants and can be as high as 400/200/1 for (110)/(100)/(111) silicon. The $\{111\}$ planes always etch the slowest, but the $\{100\}$ and $\{110\}$ planes can be controlled through etchant chemistry.

Masking is also a concern for anisotropic etching, but silicon oxide is less valuable as a mask material for different reasons than in isotropic etching. Anisotropic etching is slower than isotropic etching (typically 3 μm/min; 120 μin./min), and thus anisotropic etching through a wafer may take several hours. Silicon oxide may etch too rapidly to use as a mask, and hence a high-density silicon-nitride mask may be needed.

Often, it is important to rapidly halt the etching process, a technique referred to as *etch stop*. This situation is typically the case when thin membranes are to be manufactured or when features with very precise thicknesses are needed. Conceptually, this task can be accomplished by removing the wafer from the etching solution. However, etching depends to a great extent on the ability to circulate fresh etchants to the desired locations. Since the circulation varies across a wafer surface, this strategy for halting the etching process would lead to large variations in etch depth.

The most common approach for uniform feature sizes across a wafer is to use a boron etch stop, where a boron layer is diffused or implanted into silicon. Examples of common etch stops include the placement of a boron-doped layer beneath silicon and the placement of silicon oxide (SiO_2) beneath silicon nitride (Si_3N_4). Since anisotropic etchants do not attack boron-doped silicon as aggressively as they do undoped silicon, surface features or membranes can be created by **back etching**. Figure 13.12 shows an example of the boron etch-stop approach.

A large number of etchant formulations have been developed. Some of the more common wet etchants are summarized as follows:

1. Silicon dioxide is commonly etched with hydrofluoric (HF) acid solutions. The driving chemical reaction for pure HF etching is

$$SiO_2 + 6HF \longrightarrow H_2SiF_6 + 2H_2O. \tag{13.5}$$

It is rare that silicon dioxide is etched purely through the reaction in Eq. (13.5). Hydrofluoric acid is a weak acid, and it does not completely dissociate into hydrogen and fluorine ions in water. HF_2^- is an additional ion that exists in hydrofluoric acid, and HF_2^- attacks oxide roughly 4.5 times faster than does HF alone. The reaction involving the HF_2^- ion is

$$SiO_2 + 3HF_2^- + H^+ \longrightarrow SiF_6^{2-} + 2H_2O. \tag{13.6}$$

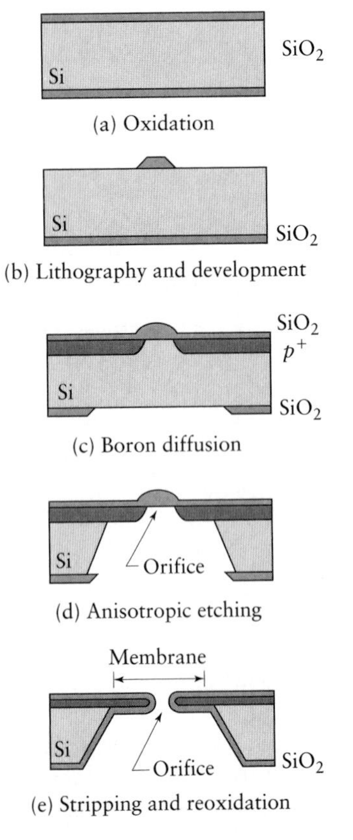

FIGURE 13.12 Application of a boron etch stop and back etching to form a membrane and orifice. *Source*: After Brodie, I., and Murray, J. J., *The Physics of Microfabrication*, Plenum Press, 1982.

The pH value of the etching solution is critical, because acidic solutions have sufficient hydrogen ions to dissociate the HF_2^- ions into HF ions. As HF and HF_2^- are consumed, the etch rate decreases. For that reason, a buffer of ammonium fluoride (NH_4F) is used to maintain the pH and thus keep the concentrations of HF and HF_2^- constant, stabilizing the etch rate. Such an etching solution is referred to as a *buffered hydrofluoric acid* (BHF) or *buffered oxide etch* (BOE) and has the reaction

$$SiO_2 + 4HF + 2NH_4F \longrightarrow (NH_4)_2SiF_6 + 2H_2O. \qquad (13.7)$$

2. Silicon nitride is etched with phosphoric acid (H_3PO_4), usually at an elevated temperature, typically 160°C (320°F.) The etch rate of phosphoric acid decreases with water content, so a *reflux system* is used to return condensed water vapor to the solution to maintain a constant etch rate.

3. Etching silicon often involves mixtures of nitric acid (HNO_3) and hydrofluoric acid (HF). Water can be used to dilute these acids; the preferred buffer is acetic acid because it preserves the oxidizing power of HNO_3; this system is referred to as an *HNA etching system*. A simplified description of this etching process is that the nitric acid oxidizes the silicon and then the hydrofluoric acid removes the silicon oxide. This two-step process is a common approach for chemical machining of metals. (See Section 9.10.) The overall reaction is

$$18HF + 4HNO_3 + 3Si \longrightarrow 3H_2SiF_6 + 4NO + 8H_2O. \quad (13.8)$$

The etch rate is limited by the silicon-oxide removal; therefore, a buffer of ammonium fluoride is usually used to maintain etch rates.

4. Anisotropic etching, or orientation-dependent etching, of single-crystal silicon can be done with solutions of potassium hydroxide, although other etchants have also be used. The reaction is

$$Si + 2OH^- + 2H_2O \longrightarrow SiO_2(OH)_2^{2-} + 2H_2. \qquad (13.9)$$

Note that this reaction does not require potassium to be the source of the OH ions; KOH attacks {111}-type planes much more slowly than other planes. Isopropyl alcohol is sometimes added to KOH solutions to reduce etch rates and increase uniformity of etching. KOH also is extremely valuable in that it stops etching when it contacts a very heavily doped *p*-type material (the boron etch stop described previously).

5. Aluminum is etched through a solution typically of 80% phosphoric acid (H_3PO_4), 5% nitric acid (HNO_3), 5% acetic acid (CH_3COOH), and 10% water. The nitric acid first oxidizes the aluminum, and the oxide is then removed by the phosphoric acid and water. This solution can be masked with a photoresist.

6. Wafer cleaning is accomplished through *Piranha solutions*, which have been in use for decades; these solutions are hot mixtures of sulfuric acid (H_2SO_4) and peroxide (H_2O_2) solutions. Piranha solutions strip photoresist and other organic coatings and remove metals on the surface, but do not affect silicon dioxide or silicon nitride, making it an ideal cleaning solution. Bare silicon forms a thin layer of hydrous silicon oxide, which is removed through a short dip in hydrofluoric acid.

7. Although photoresist can be removed through Piranha solutions, acetone is commonly used for this purpose instead. Acetone dissolves the photoresist, but it should be noted that if the photoresist is excessively heated during a process step, it will be significantly more difficult to remove with acetone; in such a case, the photoresist can be removed through an ashing plasma.

EXAMPLE 13.1 Processing of a *p*-type region in *n*-type silicon

Assume that we want to create a *p*-type region within a sample of *n*-type silicon. Draw cross-sections of the sample at each processing step in order to accomplish this task.

Solution See Fig. 13.13. This simple device is known as a *pn junction diode*, and the physics of its operation is the foundation for most semiconductor devices.

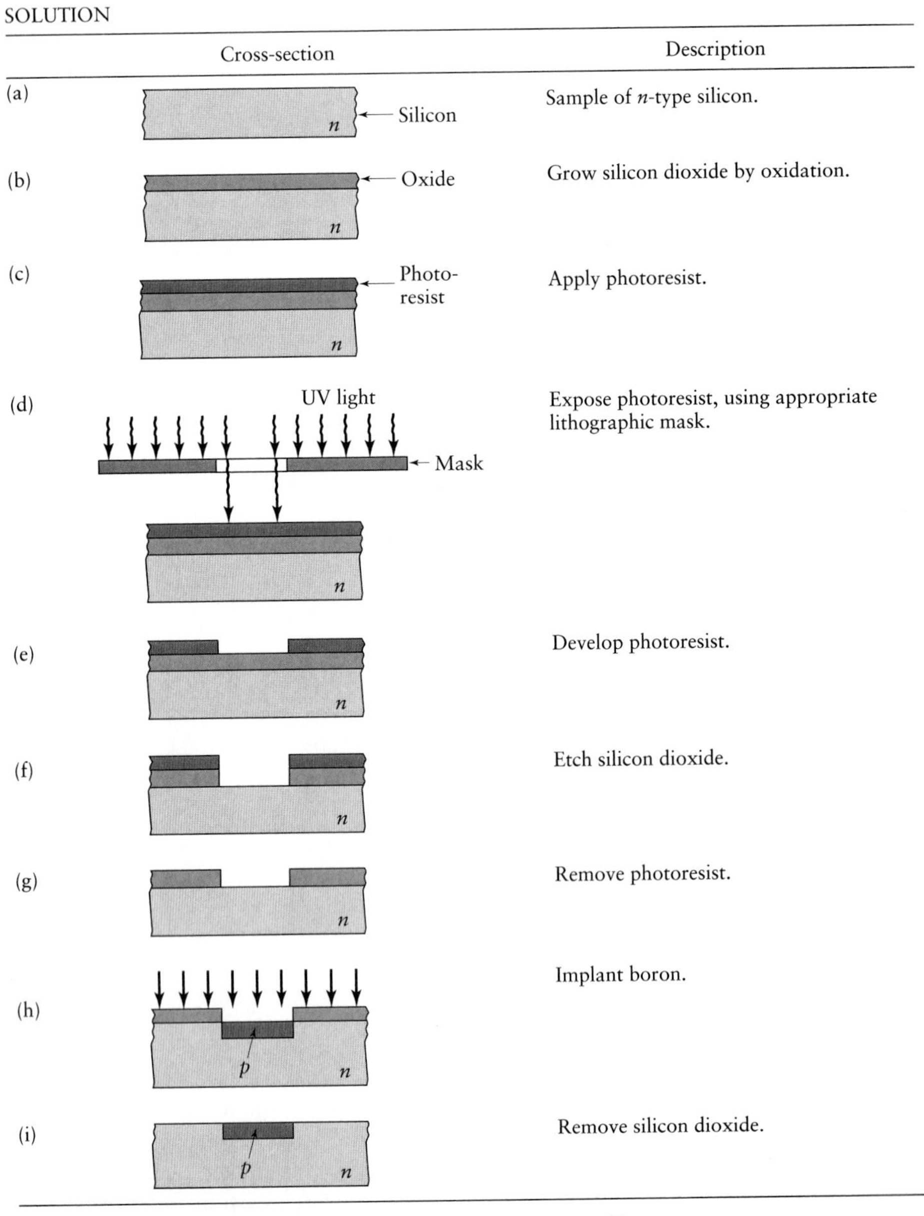

FIGURE 13.13 Processing of a *p*-type region in *n*-type silicon.

13.8.2 Dry etching

Modern integrated circuits are etched exclusively through *dry etching*, which involves the use of chemical reactants in a low-pressure system. In contrast to the wet process, dry etching can have a high degree of directionality, resulting in highly anisotropic etch profiles (Fig. 13.10c). Also, the dry process requires only small amounts of the reactant gases, whereas the solutions used in the wet process have to be refreshed periodically. Dry etching usually involves a plasma or discharge in areas of high electric and magnetic fields; any gases that are present are dissociated to form ions, electrons, or highly reactive molecules.

There are several specialized dry-etching techniques, described as follows:

a) **Sputter etching.** *Sputter etching* removes material by bombarding it with noble-gas ions, usually Ar^+. The gas is ionized in the presence of a cathode and anode (Fig. 13.14). If a silicon wafer is the target, the momentum transfer associated with bombardment of atoms causes bond breakage and material to be ejected, or sputtered. If the silicon chip is the substrate, then the material in the target is deposited onto the silicon after it has been sputtered by the ionized gas. Some of the concerns with sputter etching are listed as follows:

1. The ejected material can be redeposited onto the target, especially when the aspect ratios are large;
2. Sputter etching is not material selective (most materials sputter at about the same rate), and therefore masking is difficult;
3. Sputter etching is slow, with etch rates limited to tens of nm/min;
4. Sputtering can cause damage to or excessive erosion of the material;
5. The photoresist is difficult to remove.

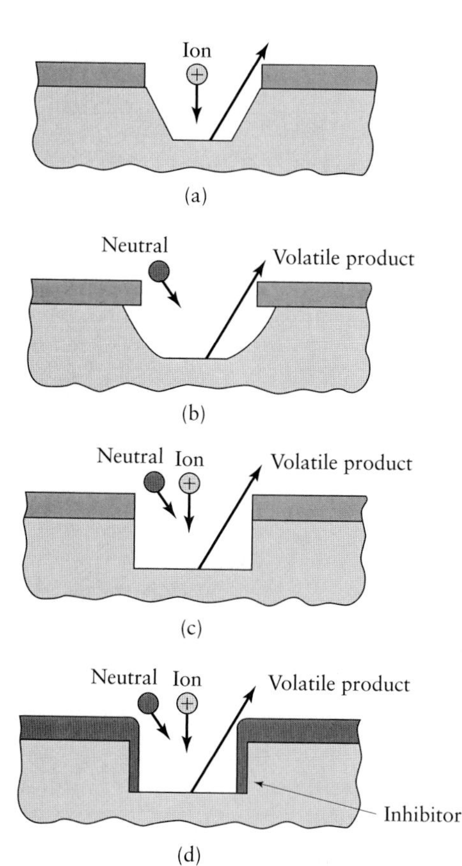

FIGURE 13.14 Machining profiles associated with different dry-etching techniques: (a) sputtering; (b) chemical; (c) ion-enhanced energetic; (d) ion-enhanced inhibitor. *Source*: After M. Madou.

b) **Reactive plasma etching.** Also referred to as *dry chemical etching, reactive plasma etching* involves chlorine or fluorine ions (generated by RF excitation) and other molecular species that diffuse to and chemically react with the substrate, forming a volatile compound that is removed by the vacuum system. The mechanism of reactive plasma etching is shown in Fig. 13.15. Here, a reactive species is produced, such as CF_4 dissociating upon impact with energetic electrons to produce fluorine atoms (step 1). The reactive species then diffuse to the surface (step 2), become adsorbed (step 3), and chemically react to form a volatile compound (step 4). The reactant then desorbs from the surface (step 5) and diffuses into the bulk gas, where it is removed by the vacuum system. Some reactants polymerize on the surface and require additional removal, either with oxygen in the plasma reactor or by an external ashing operation. The electrical charge of the reactive species is not great enough to cause damage through impact on the surface, so that no sputtering occurs. Thus, the etching is isotropic, and undercutting of the mask takes place (Fig. 13.10a). Table 13.2 lists some of the more common dry etchants, their target materials, and their typical etch rates.

c) **Physical–chemical etching.** Processes such as *reactive ion-beam etching* (RIBE) and *chemically assisted ion-beam etching* (CAIBE) combine the advantages of phys-

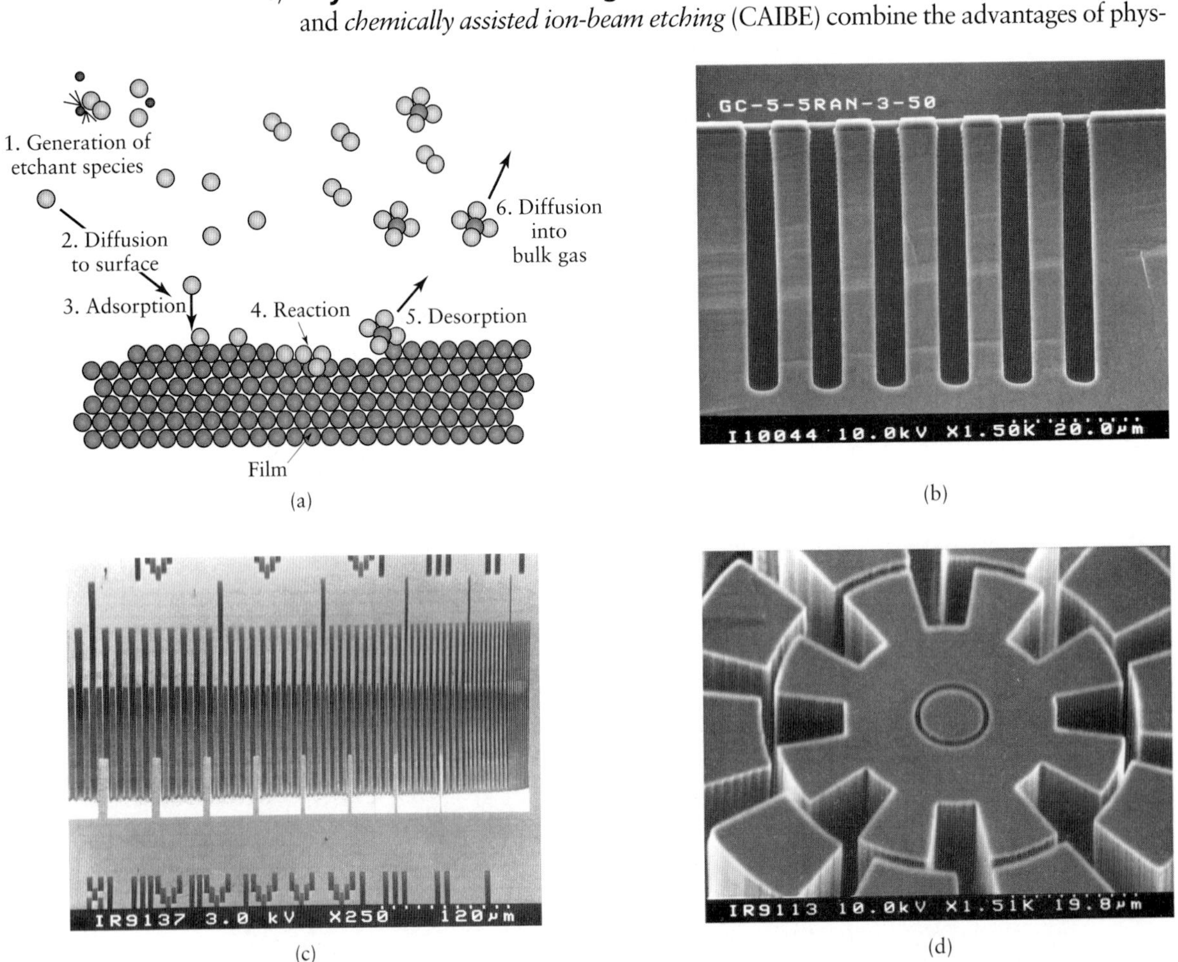

FIGURE 13.15 (a) Schematic illustration of reactive plasma etching. *Source*: (a) After M. Madou. (b) Example of deep reactive ion etched trench. Note the periodic undercuts, or scalloping. (c) Near vertical sidewalls produced through DRIE with an anisotropic etching process. (d) An example of cryogenic dry etching, showing a145 μm deep structure etched into Si using a 2.0 μm thick oxide masking layer. The substrate temperature was $-140°C$ during etching. *Source*: R. Kassing and I. W. Rangelow, University of Kassel, Germany.

ical and chemical etching. These processes use chemically reactive species to drive material removal, but this procedure is physically assisted by the impact of ions onto the surface. In RIBE, also known as *deep reactive ion etching* (DRIE), vertical trenches hundreds of micrometers deep can be produced by periodically interrupting the etch process and depositing a polymer layer. When done with an isotropic dry etching process, this results in scalloped sidewalls, as shown in Fig. 13.15b. Anisotropic DRIE can produce near vertical sidewalls, as shown in Fig. 13.15c.

In CAIBE, ion bombardment can assist dry chemical etching by

1. Making the surface more reactive;
2. Clearing the surface of reaction products and allowing the chemically reactive species access to the cleared areas;
3. Providing the energy to drive surface chemical reactions—however, the neutral species do most of the etching.

Physical–chemical etching is extremely useful, because the ion bombardment is directional, so that etching is anisotropic. Also, the ion bombardment energy is low and does not contribute much to mask removal. This factor allows generation of near vertical walls with very large aspect ratios. Since the ion bombardment does not directly remove material, masks can be used.

d) **Cryogenic dry etching.** This procedure is an approach used to obtain very deep features with vertical walls. The workpiece is lowered to cryogenic temperatures, and then chemically assisted ion-beam etching takes place. The low temperatures involved ensure that insufficient energy is available for a surface chemical reaction to take place, unless ion bombardment is normal to the surface. Oblique impacts, such as occur on side walls in deep crevices, cannot drive the chemical reactions; therefore, very smooth vertical walls can be produced, as shown in Fig. 13.15d.

Since dry etching is not selective, etch stops cannot be directly applied. Dry-etch reactions must be terminated when the target film is removed. Optical emission spectroscopy is often used to determine the "end point" of a reaction. Filters can be used to capture the wavelength of light emitted during a particular reaction. A noticeable change in light intensity at the point of etching will be detected.

EXAMPLE 13.2 Comparison of wet and dry etching

Consider a case where a ⟨100⟩ wafer has an oxide mask placed on it in order to produce square or rectangular holes. The sides of the square mask are precisely oriented with the ⟨110⟩ direction (see Fig. 13.5) of the wafer surface, as shown in Fig. 13.16.

Isotropic etching results in the cavity shown in Fig. 13.16a. Since etching occurs at constant rates in all directions, a rounded cavity is produced that undercuts the mask. An orientation-dependent etchant produces the cavity shown in Fig. 13.16b. Since etching is much faster in the ⟨100⟩ and ⟨110⟩ directions than in the ⟨111⟩ direction, sidewalls defined by the {111} plane are generated. For silicon, these sidewalls are at an angle of 54.74° to the surface.

The effect of a larger mask or shorter etch time is shown in Fig. 13.16c. The resultant pit is defined by ⟨111⟩ sidewalls and a bottom in the ⟨100⟩ direction parallel to the surface. A rectangular mask and resultant pit are shown in Fig. 13.16d. Deep reactive ion etching is depicted in Fig. 13.16e. Note that a polymer layer is periodically deposited onto the hole sidewalls to allow for deep pockets, but scalloping (greatly exaggerated in the figure) is unavoidable. A hole from chemically reactive ion etching is shown in Fig. 13.16f.

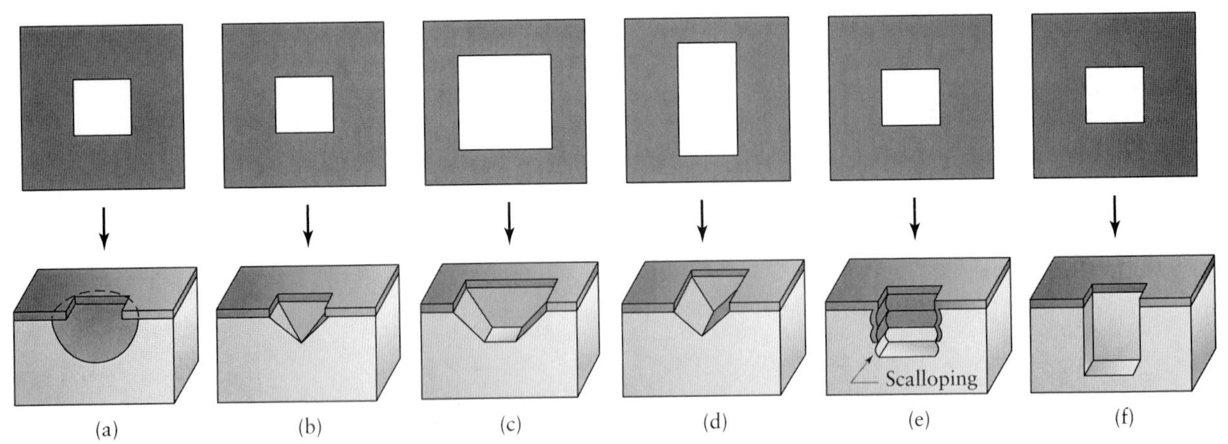

FIGURE 13.16 Different holes generated from a square mask in (a) isotropic (wet) etching; (b) ODE; (c) ODE with a larger hole; (d) ODE of a rectangular hole; (e) deep reactive ion etching; (f) vertical etching. *Source*: After M. Madou.

13.9 Metallization and Testing

The preceding sections have focused only on device fabrication; however, generation of a complete and functional integrated circuit requires that such devices be interconnected. **Interconnections** are made by metals that exhibit low electrical resistance and good adhesion to dielectric insulator surfaces. Aluminum and aluminum–copper alloys remain the most commonly used materials for this purpose in VLSI (very large scale integration) technology today. However, as device dimensions continue to shrink, electromigration has become more of a concern with aluminum interconnects.

Electromigration is the process by which metal atoms are physically moved by the impact of drifting electrons under high-current conditions. Low-melting-point metals such as aluminum are especially prone to electromigration. In extreme cases, this condition can lead to severed and/or shorted metal lines. Solutions to this problem include the addition of sandwiched metal layers such as tungsten and titanium, and more recently, the use of pure copper, which displays lower resistivity and significantly better antielectromigration performance than does aluminum.

Metals are deposited by standard deposition techniques, and interconnection patterns are generated by lithographic and etching processes. Modern ICs typically have one to four layers of metallization, in which case each layer of metal is insulated by a dielectric, either silicon oxide or borophosphosilicate glass. **Planarization** (producing a planar surface) of these interlayer dielectrics is critical to reducing metal shorts and linewidth variation of the interconnect. A common method for achieving a planar surface has been a uniform oxide etch process that smoothens out the peaks and valleys of the dielectric layer.

However, today's standard for planarizing high-density interconnects is **chemical mechanical polishing** (CMP; see Section 9.8), also called *chemical–mechanical planarization*. This process involves physically polishing the wafer surface similar to a disc or belt sander polishing the ridges on a piece of wood. A typical CMP process combines an abrasive medium with a polishing compound or slurry and can polish a wafer to within $0.03\ \mu m$ ($1.2\ \mu in.$) of being perfectly flat, with a R_q roughness on the order of 0.1 nm for a new, bare silicon wafer.

Different layers of metal are connected together by **vias**, and access to the devices on the substrate is achieved through **contacts** (Fig. 13.17). In recent years, as

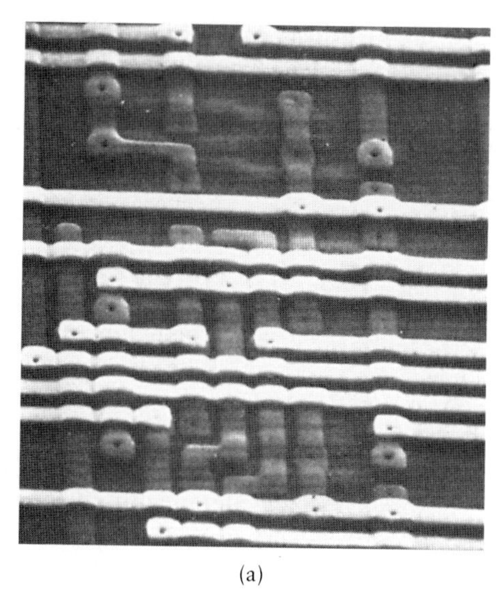

(a)

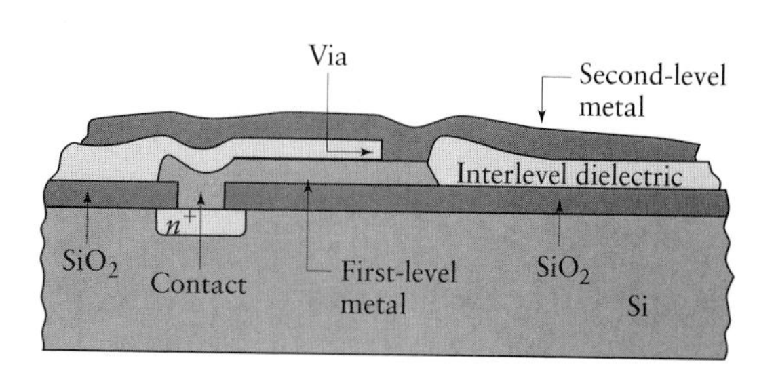

(b)

FIGURE 13.17 (a) Scanning electron microscope photograph of a two-level metal interconnect. Note the varying surface topography. (b) Schematic drawing of a two-level metal interconnect structure. *Source*: R. C. Jaeger.

devices have become smaller and faster, the size and speed of some chips have become dominated by the resistance of the metallization process itself and the capacitance of the dielectric and the transistor gate. Wafer processing is completed upon application of a *passivation layer*, usually silicon nitride (Si_3N_4), which acts as an ion barrier for sodium ions and also provides excellent scratch resistance.

The next step is to test each of the individual circuits on the wafer (Fig. 13.18). Each chip, also referred to as a **die**, is tested with a computer-controlled platform that contains needlelike probes to access the bonding pads on the die. The probes are of two forms:

FIGURE 13.18 A probe checks for defects, and an ink mark is placed on each defective die. *Source*: Intel Corp.

1. **Test patterns or structures.** The probe measures test structures, often outside of the active die, placed in the so-called *scribe line* (the empty space between dies). These structures consist of transistors and interconnect structures that measure various quantities such as resistivity, contact resistance, and electromigration.
2. **Direct probe.** This approach uses 100% testing on the bond pads of each die.

The platform steps across the wafer, testing whether each circuit functions properly, using computer-generated timing wave forms. If a defective chip is encountered, it is marked with a drop of ink. Up to one third of the cost of a micromachined part can be incurred during this testing.

After the wafer-level testing is complete, back grinding may be done to remove a large amount of the original substrate. Final die thickness is dependent on the packaging requirement, but anywhere from 25% to 75% of the wafer thickness can be removed. After back grinding, each die is separated from the wafer. Diamond sawing is a commonly used separation technique and results in very straight edges, with minimal chipping and cracking damage. The chips are then sorted, with the passing dice being sent on for packaging and the inked dice being discarded.

13.10 Wire Bonding and Packaging

The working dice must be attached to a more rugged foundation to ensure reliability. One simple method is to *bond* a die to its packaging material with an *epoxy cement* (see Section 12.13); another method uses a *eutectic bond*, made by heating metal-alloy systems. One widely used mixture is 96.4% gold and 3.6% silicon, which has a eutectic point at 370°C (700°F). Once the chip has been bonded to its substrate, it must be electrically connected to the package leads. This task is accomplished by *wire bonding* very thin [15 μm (600 μin.) in diameter] gold wires from the package leads to bonding pads located around the perimeter or down the center of the die (Fig. 13.19a). The bonding pads on the die are typically drawn at 50 μm (2000 μin.) or more per side, and the bond wires are attached using thermocompression, ultrasonic, or thermosonic techniques (Fig. 13.20).

The connected circuit is now ready for final *packaging*. The packaging process largely determines the overall cost of each completed IC, since the circuits are mass produced on the wafer, but are then packaged individually. Packages are available in

(a)

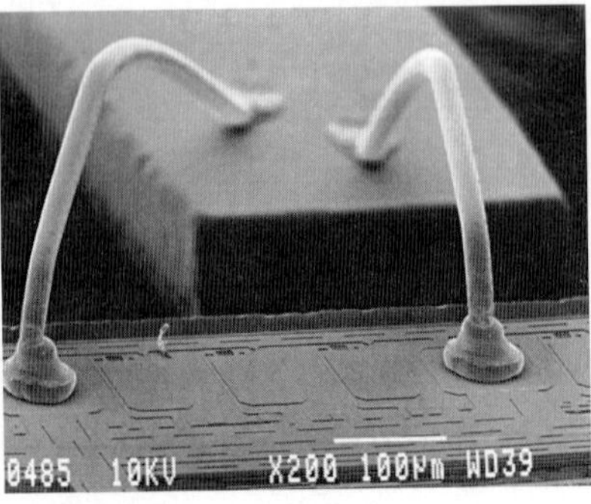

(b)

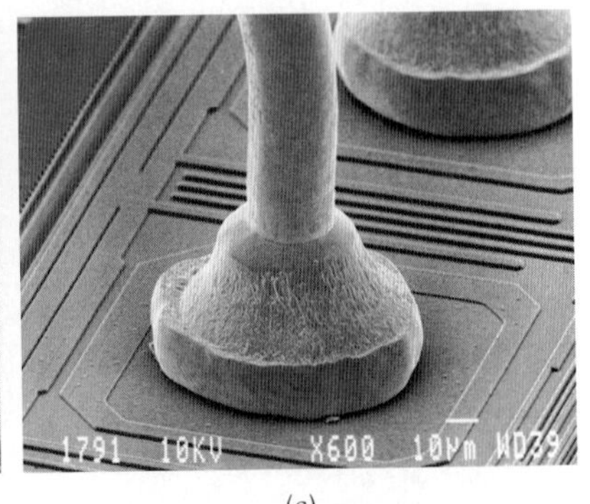

(c)

FIGURE 13.19 (a) SEM photograph of wire bonds connecting package leads (left-hand side) to die bonding pads. (b) and (c) Detailed views of (a). *Source*: Courtesy of Micron Technology, Inc.

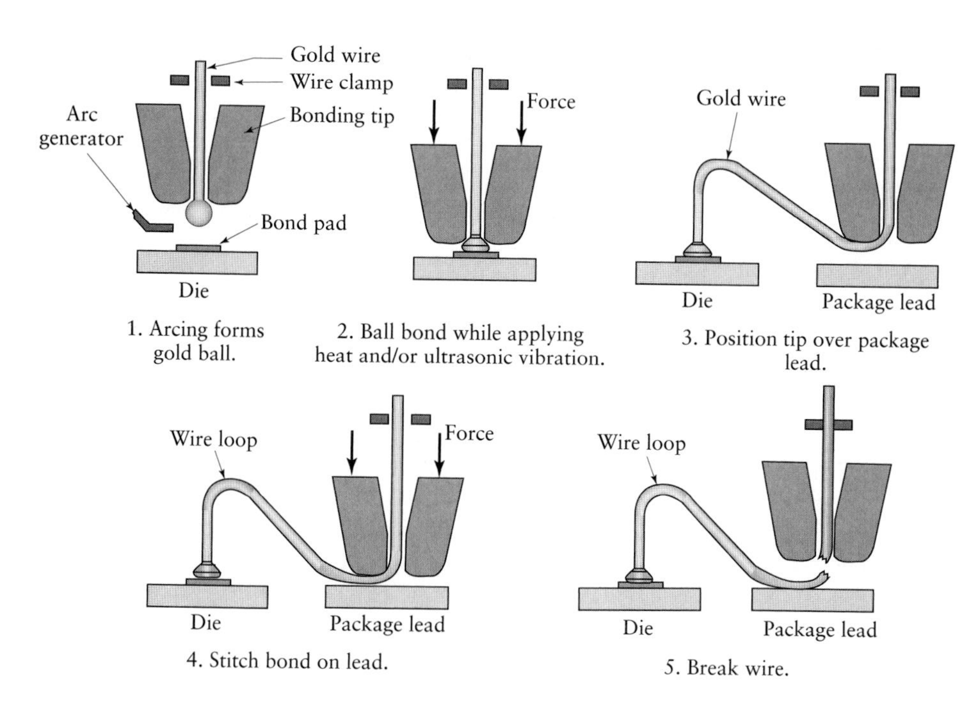

FIGURE 13.20 Schematic illustration of the thermosonic ball and stitch process. *Source*: After N. Maluf.

a variety of styles (Table 13.4), and selection of the appropriate one must take into account operating requirements. Consideration of a circuit's package includes consideration of chip size, the number of external leads, operating environment, heat dissipation, and power requirements. ICs that are used for military and industrial applications, for example, need packages of particularly high strength, toughness, hermicity and high-temperature resistance.

Packages are produced from polymers, metals, or ceramics. Metal containers are produced from alloys such as Kovar (an iron–cobalt–nickel alloy with a low coefficient of thermal expansion) and provide a hermetic seal and good thermal conductivity, but are limited in the number of leads that can be used. Ceramic packages are usually produced from Al_2O_3 and are hermetic and have good thermal conductivity, with higher lead counts than metal packages; however, they are also more expensive

TABLE 13.4

Summary of Molded-Plastic IC Packages

Package	Abbreviation	Pins min.	Pins max.	Description
Through-hole mount				
Dual in-line	DIP	8	64	Two in-line rows of leads.
Single in-line	SIP	11	40	One in-line row of leads.
Zigzag in-line	ZIP	16	40	Two rows with staggered leads.
Quad in-line package	QUIP	16	64	Four in-line rows of staggered leads.
Surface mount				
Small-outline IC	SOIC	8	28	Small package with leads on two sides.
Thin small-outline package	TSOP	26	70	Thin version of SOIC.
Small-outline J-lead	SOJ	24	32	Same as SOIC, with leads in a J-shape.
Plastic leaded chip carrier	PLCC	18	84	J-shaped leads on four sides.
Thin quad flat pack	TQFP	32	256	Wide, but thin, package with leads on four sides.

than metal packages. Plastic packages are inexpensive, with high lead counts, but they have high thermal resistance and are not hermetic.

An older style of packaging is the **dual in-line package** (DIP), shown schematically in Fig. 13.21a. Characterized by low cost (if plastic) and ease of handling, DIP packages are made of thermoplastic, epoxy, or ceramic, and they can have from 2 to 500 external leads. Ceramic packages are designed for use over a broader temperature range and in military applications and cost considerably more than plastic packages. A flat ceramic package is shown in Fig. 13.21b, in which the package and all the leads are in the same plane. This package style does not offer the ease of handling or the modular design of the DIP package; therefore, it is usually affixed permanently to a multiple-level circuit board in which the low profile of the flat pack is essential.

Surface-mount packages have become the standard for today's integrated circuits. Some common examples are shown in Fig. 13.21c, where it can be seen that the main difference among them is in the shape of the connectors. The DIP connection to the surface board is by way of prongs that are inserted into corresponding holes, whereas a surface mount is soldered onto specially fabricated pad or land designs. Package size and land layouts are selected from standard patterns and usually require adhesive bonding of the package to the board, followed by wave soldering of the connections (see Section 12.12.3).

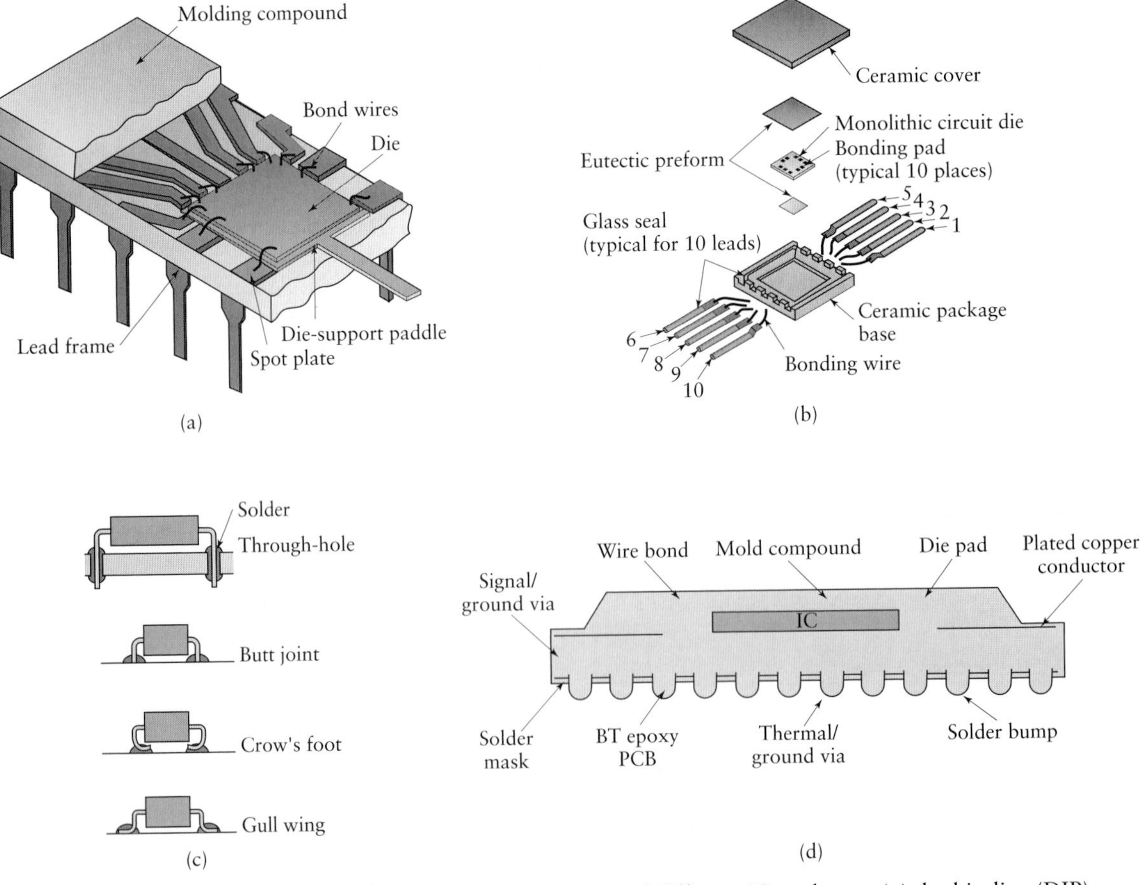

FIGURE 13.21 Schematic illustration of different IC packages: (a) dual in-line (DIP); (b) ceramic flat pack; (c) common surface-mount configurations; (d) ball-grid array (BGA). *Source*: After R. C. Jaeger, A. B. Glaser, and G. E. Subak-Sharpe. Coombs, C.F., Printed Circuits Handbook. New York, McGraw-Hill, 1996.

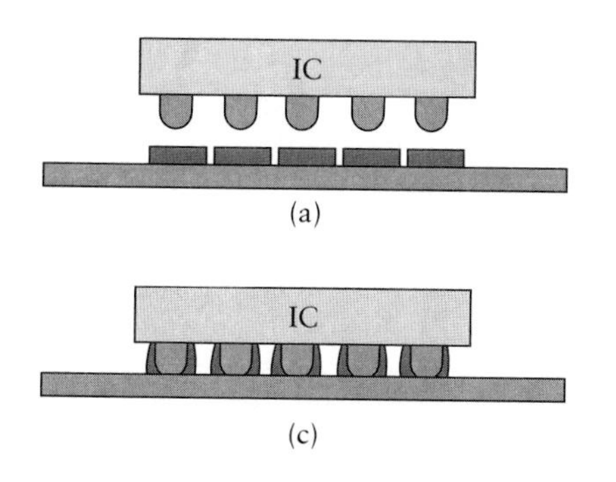

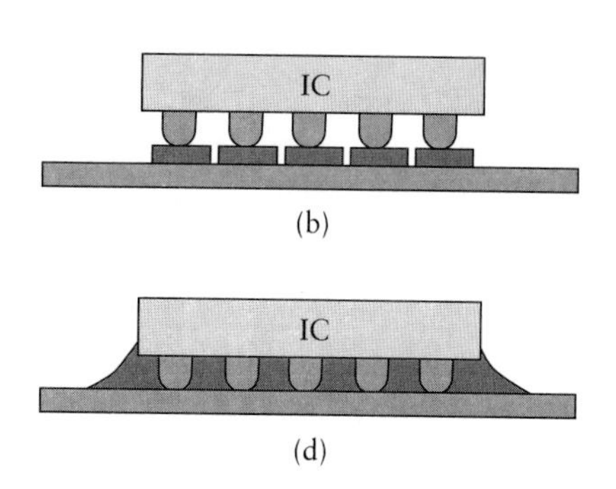

FIGURE 13.22 Illustration of flip-chip technology. (a) Flip-chip package with solder plated metal balls and pads on the printed circuit board; (b) flux application and placement; (c) reflow soldering; (d) encapsulation. *Source*: P. K. Wright, *21st Century Manufacturing*, Prentice Hall, Upper Saddle River, NJ, 2001.

Faster and more versatile chips require increasingly tightly spaced connections. **Pin-grid arrays** (PGAs) use tightly packed pins that connect by way of through-holes onto printed circuit boards. However, PGAs, and indeed other in-line and surface-mount packages, are extremely susceptible to plastic deformation of the wires and legs, especially with small-diameter, closely spaced wires. One way to achieve tight packing of connections and avoid the difficulties of slender connections is through **ball-grid arrays** (BGAs), shown in Fig. 13.21d. Such arrays have a plated solder coating on a number of closely spaced metal balls on the underside of the package. The spacing between balls can be as small as 50 μm (2000 μin.), but more commonly, the spacing is standardized as 1.0 mm (0.040 in.), 1.27 mm (0.050 in.), or 1.5 mm (0.060 in.).

BGAs can be designed with over 1000 connections, but such high numbers of connections are extremely rare; usually, 200–300 connections are sufficient for demanding applications. By using reflow soldering (Section 12.12.3), the solder serves to center the BGAs by surface tension, resulting in well-defined electrical connections for each ball. **Flip-chip technology** (FCT) refers to the attachment procedure of ball-grid arrays and is depicted in Fig. 13.22. The final encapsulation with an epoxy is necessary, not only to more securely attach the IC package to the printed circuit board, but also to evenly distribute thermal stresses during its operation.

After the chip has been sealed in the package, it undergoes final testing. Because one of the main purposes of packaging is isolation from the environment, testing at this stage usually encompasses heat, humidity, mechanical shock, corrosion, and vibration. Destructive tests are also performed to investigate the effectiveness of sealing.

13.11 Yield and Reliability of Chips

Yield is defined as the ratio of functional chips to the total number of chips produced. The overall yield of the total IC manufacturing process is the product of the wafer yield, bonding yield, packaging yield, and test yield. This value can range from only a few percent for new processes to above 90% for mature manufacturing lines. Most loss of yield occurs during wafer processing, due to its complex nature; in this stage, wafers are commonly separated into regions of good and bad chips. Failures at this stage can arise from point defects (such as oxide pinholes), film contamination, or metal particles, as well as from area defects, such as uneven film deposition or etch nonuniformity.

A major concern about completed ICs is their **reliability** and **failure rate**. Since no device has an infinite lifetime, statistical methods are used to characterize the expected lifetimes and failure rates of microelectronic devices. The unit for failure rate is the FIT (*failure in time*), defined as the number of failures per one billion device-hours. However, complete systems may have millions of devices, so the overall failure rate in entire systems is correspondingly higher. Failure rates greater than 100 FIT are generally unacceptable.

Equally important in failure analysis is determination of the failure mechanism, that is, the actual process that causes the device to fail. Common failures due to processing involve (1) diffusion regions (leading to nonuniform current flow and junction breakdown), (2) oxide layers (dielectric breakdown and accumulation of surface charge), (3) lithography (uneven definition of features and mask misalignment), and (4) metal layers (poor contact and electromigration resulting from high current densities). Other failures can originate from improper chip mounting, degradation of wire bonds, and loss of package hermeticity. Wire-bonding and metallization failures account for over one half of all integrated-circuit failures.

Because device lifetimes are very long (10 years or more; see also Section 16.3), it is impractical to study device failure under normal operating conditions. One method of studying failures efficiently is by **accelerated life testing**, which involves accelerating the conditions whose effects are known to cause device breakdown. Cyclic variations in temperature, humidity, voltage, and current are used to stress the components. Statistical data taken from these tests are then used to predict device failure modes and device life under normal operating conditions. Chip mounting and packaging are strained by cyclical temperature variations.

13.12 Printed Circuit Boards

Packaged integrated circuits are seldom used alone; rather, they are usually combined with other ICs to serve as building blocks of a yet larger system. A *printed circuit board* (PCB) is the substrate for the final interconnections among all completed chips and serves as the communication link between the outside world and the microelectronic circuitry within each packaged IC. In addition to the ICs, circuit boards also usually contain discrete circuit components, such as resistors and capacitors, that would take up too much "real estate" on the limited silicon surface, have special power dissipation requirements, or cannot be implemented on a chip. Other common discrete components include inductors, which cannot be integrated onto the silicon surface, and high-performance transistors, large capacitors, precision resistors, and crystals for frequency control.

A printed circuit board is basically a plastic resin material containing several layers of copper foil (Fig. 13.23). *Single-sided* PCBs have copper tracks on only one side of an insulating substrate, while *double-sided* boards have copper tracks on both sides. Multilayer boards also can be constructed from alternating copper and insulator layers. Single-sided boards are the simplest form of circuit board. Double-sided boards usually must have locations where electrical connectivity is established between the features on both sides of the board. This structure is accomplished with vias, as shown in Fig. 13.24. Multilayer boards can have partial, buried, or through-hole vias to allow for extremely flexible PCBs. Double and multilayer boards are beneficial in that IC packages can be bonded to both sides of the board, allowing more compact designs.

The insulating material is usually an epoxy resin 0.25 mm to 3 mm (0.01 in. to 0.12 in.) thick, reinforced with an epoxy/glass fiber, referred to as E-glass (see Section 10.9.2). They are produced by impregnating sheets of glass fiber with epoxy and then pressing the layers together between hot plates or rolls. The heat and pressure cure the board, resulting in a stiff and strong basis for printed circuit boards. Boards are

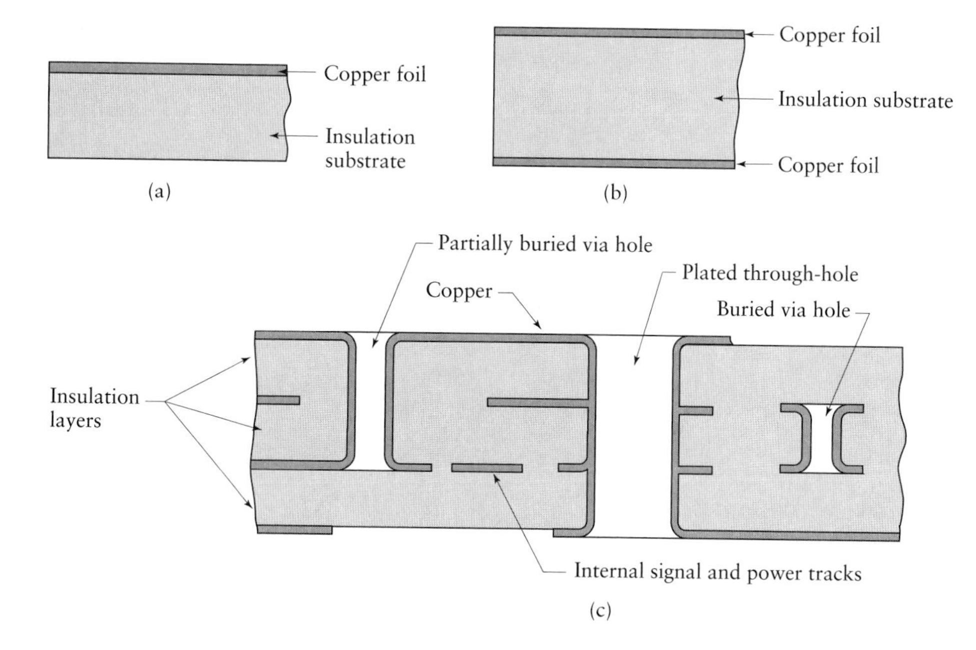

FIGURE 13.23 Types of circuit board structures: (a) single sided; (b) double sided; (c) multilayer, showing vias and pathways between layers. *Source*: M. Groover, *Fundamentals of Modern Manufacturing*, Prentice Hall, Upper Saddle River, NJ, 1996.

sheared to a desired size, and roughly 3-mm-diameter locating holes are then drilled or punched into the board corners to permit alignment and proper location of the board within chip-insertion machines. Holes for vias and connections are punched or produced through CNC drilling; stacks of boards can be drilled simultaneously to increase production rates.

The conductive patterns on circuit boards are defined by lithography, although originally they were produced through screen-printing technologies, hence the terms *printed circuit board* and *printed wiring board* (PWB). In the *subtractive method*, a copper foil is bonded to the circuit board. The desired pattern on the board is then defined by a positive mask developed through photolithography, and the remaining copper is removed through wet etching. In the *additive method*, a negative mask is placed directly onto an insulator substrate to define the desired shape. Electroless plating and electroplating of copper serve to define the connections, tracks, and lands on the circuit board.

The ICs and other discrete components are then fastened to the board by soldering. This procedure is the final step in making the integrated circuits and

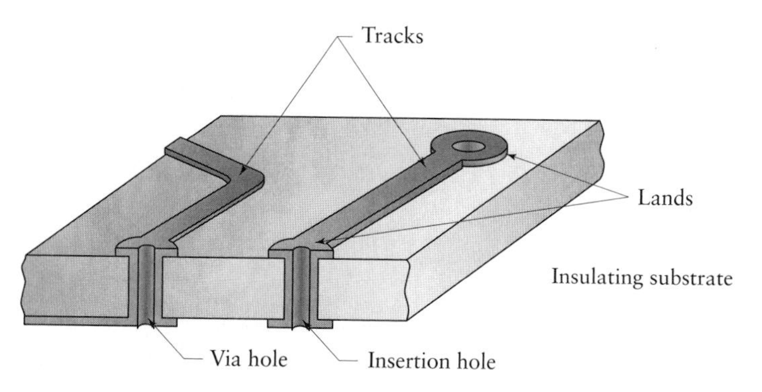

FIGURE 13.24 Design features in printed circuit boards. *Source*: M. Groover, *Fundamentals of Modern Manufacturing*, Prentice Hall, Upper Saddle River, NJ, 1996.

the microelectronic devices they contain to be accessible. *Wave soldering* and *reflow paste soldering* are the preferred methods of soldering ICs onto circuit boards.

Some of the design considerations in laying out PCBs are:

1. Wave soldering should be used only on one side of the board; thus, all through-hole mounted components should be inserted from the same side of the board. Surface-mount devices placed on the insertion side of the board must be reflow soldered in place; surface-mount devices on the leg side can be wave soldered.
2. To allow good solder flow in wave soldering, IC packages should carefully be laid out on the printed circuit board (Fig. 13.25). Inserting the packages in the same direction is advantageous for automated placing, whereas random orientations can cause problems in flow of solder across all of the connections.
3. The spacing of ICs is determined mainly by the need to remove heat during operation. Sufficient clearance between packages and adjacent boards is required to allow forced air flow and heat convection.
4. There should be sufficient space around each IC package to allow for rework and repair without disturbing adjacent devices.

FIGURE 13.25 Layout of IC packages and other components on circuit boards to facilitate wave soldering. *Source*: After G. Boothroyd, P. Dewhurst, and W. Knight, *Product Design for Manufacture and Assembly*, Marcel-Dekker, 1994.

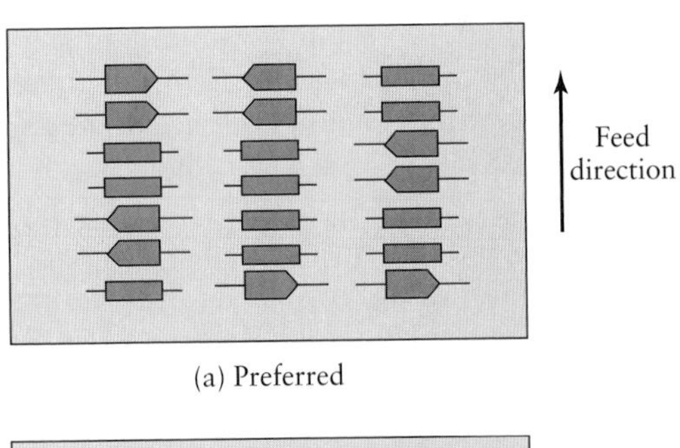

(a) Preferred

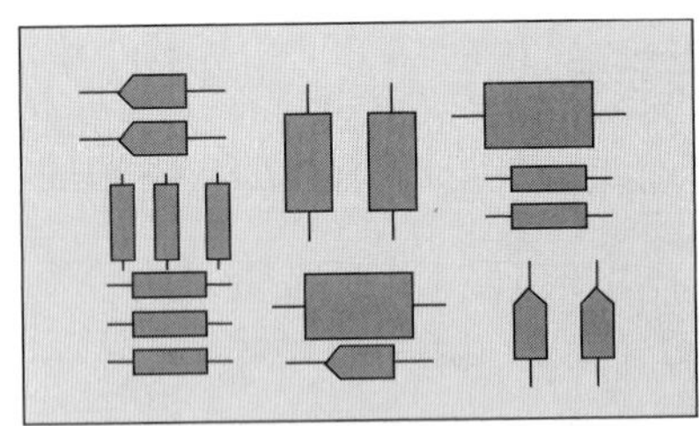

(b) Acceptable

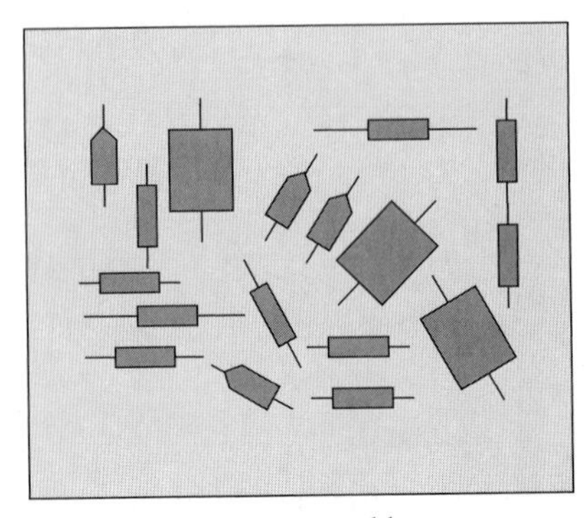

(c) Unacceptable

13.13 Micromachining of MEMS Devices

Although lithography and etching can be used in the manufacture of MEMS devices (see Fig. 1.7) to obtain 2D or $2\frac{1}{2}$-D features on wafer surfaces, three-dimensional features are often required. The production of features from micrometers to millimeters in size is called *micromachining*. MEMS devices have been constructed from **polycrystalline silicon (polysilicon)** and single-crystal silicon because the technologies for integrated-circuit manufacture, described earlier in this chapter, are well developed and exploited for these devices, and other, new processes have been developed that are compatible with the existing processing steps. The use of anisotropic etching techniques allows the fabrication of devices with well-defined walls and high aspect ratios, and for this reason, some single-crystal silicon MEMS devices have been fabricated.

One of the recognized difficulties associated with using silicon for MEMS devices is the high adhesion encountered at small length scales and the associated rapid wear. Most commercial devices are designed to avoid friction by, for example, using flexing springs instead of bushings; however, this approach complicates designs and makes some MEMS devices unfeasible. Therefore, significant research is being conducted to identify materials and lubricants that provide reasonable life and performance. Silicon carbide, diamond, and metals such as aluminum, tungsten, and nickel have been investigated as potential MEMS materials. Lubricants have also been investigated; it is known that surrounding the MEMS device in a silicone oil, for example, practically eliminates adhesive wear (see Section 4.4.2), but it also limits the performance of the device. Self-assembling layers of polymers also are being investigated, as well as novel and new materials with self-lubricating characteristics; however, the tribology of MEMS devices remains a main technological barrier to expansion of their already widespread use.

An example of a commercial MEMS-based product is the digital pixel technology (DPT™) device illustrated in Fig. 13.26; this device uses an array of *digital micromirror devices* (DMD™) to project a digital image, such as in computer-driven projection systems. The aluminum mirrors can be tilted so that light is directed into or away from the optics that focus light onto a screen, enabling each mirror to represent a pixel of an image's resolution. The mirror allows light or dark pixels to be projected, but levels of gray can also be accommodated. Since the switching time is about 15 microseconds, which is much faster than the human eye response time, the mirror will switch between the on and off states in order to reflect the proper light dose to the optics.

An array of such mirrors represents a gray-scale screen; using three mirrors (one each for red, green, and blue light) for each pixel results in a color image with millions of discrete colors. Digital pixel technology is widely applied for digital projection systems, high-definition television, and other optical applications. However, production of the device shown in Fig. 13.26 requires much more than $2\frac{1}{2}$-D features, and full three-dimensional, multipart assemblies have to be manufactured.

13.13.1 Bulk micromachining

Until the early 1980s, bulk micromachining was the most common form of machining at micrometer scales. This process uses orientation-dependent etches on single-crystal silicon. (See Fig. 13.10b.) The approach is based on etching down into a surface, stopping on certain crystal faces, doped regions, and etchable films to form the desired structure. As an example of the process, consider the fabrication of the silicon cantilever shown in Fig. 13.27. By using the masking techniques described previously, a rectangular patch of the *n*-type silicon substrate is changed to *p*-type silicon

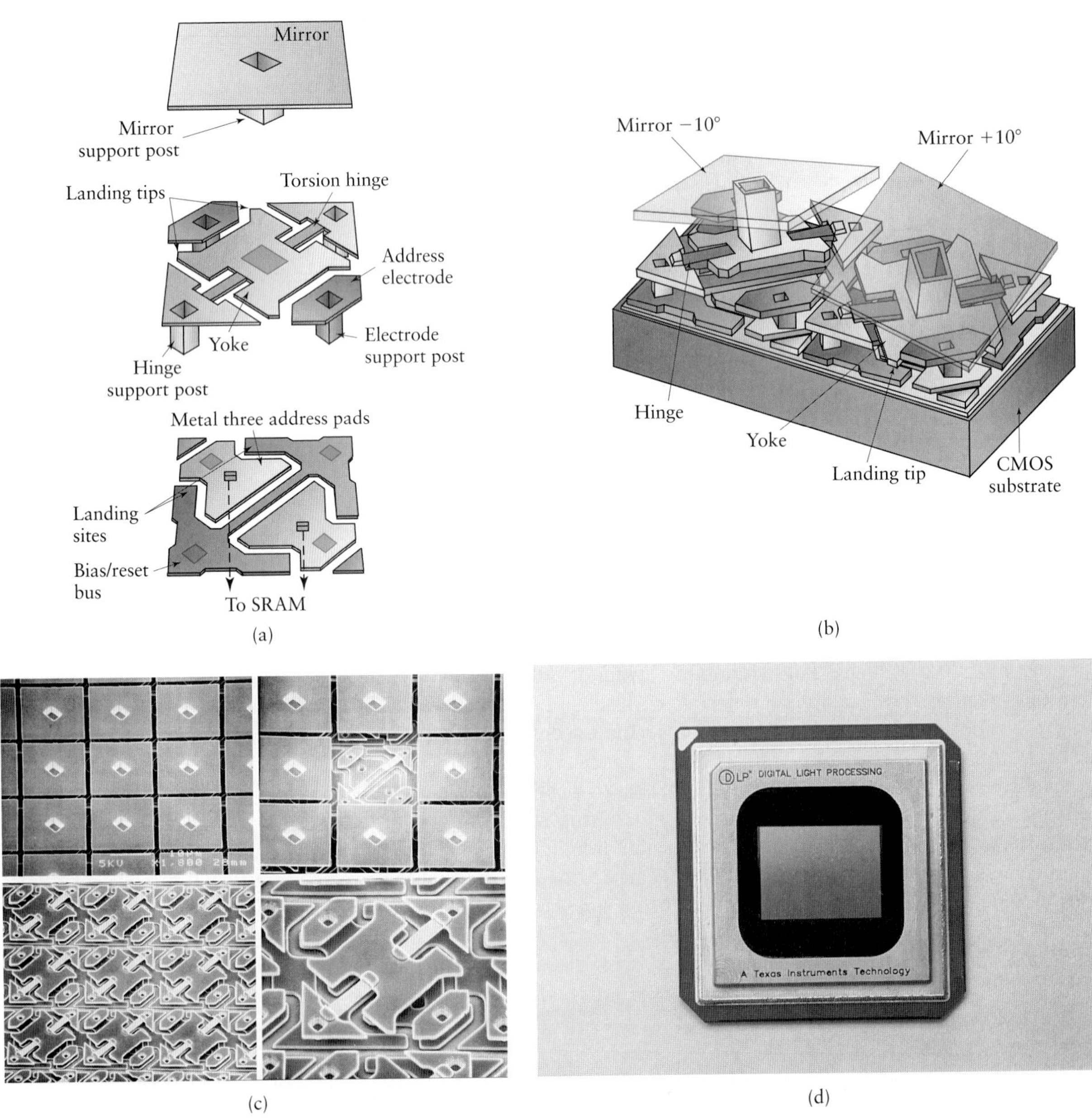

FIGURE 13.26 The Texas Instruments digital pixel technology (DPT™) device. (a) Exploded view of a single digital micromirror device (DMD™). (b) View of two adjacent DMD pixels (c) Images of DMD arrays, with some mirrors removed for clarity. Each mirror measures approximately 17 μm (680 μin.) on a side. (d) A typical DPT device, used for digital projection systems, high-definition televisions, and other image-display systems. The device shown contains 1,310,720 micromirrors and measures less than two inches per side. *Source*: Texas Instruments Corp.

through boron doping. Recall that ODE etchants, such as potassium hydroxide, will not be able to etch heavily boron-doped silicon; hence, this patch will not be etched.

A mask is then produced, such as with silicon nitride on silicon. When etched with potassium hydroxide, the undoped silicon will be removed rapidly, while the mask and the doped patch will essentially be unaffected. Etching progresses until the (111) planes are exposed in the n-type silicon substrate, and they undercut the patch, leaving a suspended cantilever, as shown in the figure.

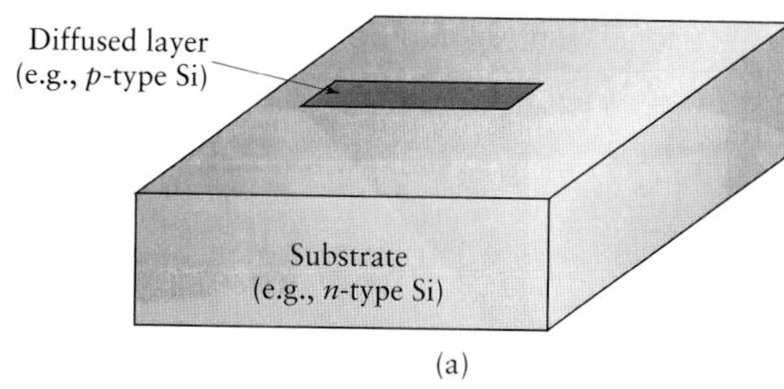

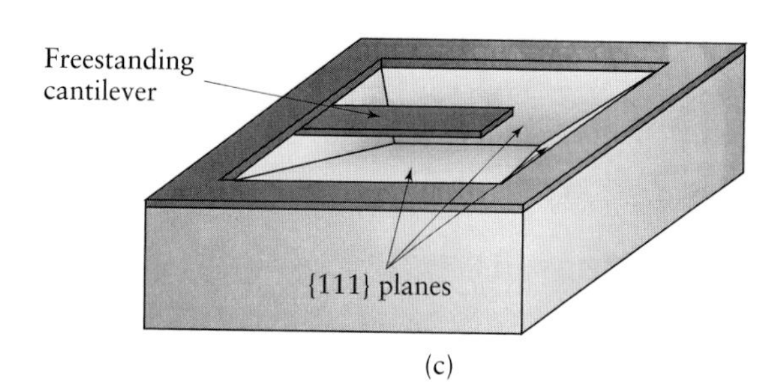

FIGURE 13.27 Schematic illustration of bulk micromachining. (a) Diffuse dopant in desired pattern. (b) Deposit and pattern masking film. (c) Orientation-dependent etch, leaving behind a freestanding structure. *Source*: K. R. Williams.

13.13.2 Surface micromachining

The basic steps in surface micromachining are illustrated in Fig. 13.28 for silicon devices. A spacer, or sacrificial, layer is deposited onto a silicon substrate coated with a thin dielectric layer (isolation, or buffer, layer). Phosphosilicate glass deposited by chemical-vapor deposition is the most common material for a spacer layer, because it etches very rapidly in hydrofluoric acid. Figure 13.28b shows the spacer layer after application of masking and etching. At this point, a structural thin film is deposited onto the spacer layer; the film can be polysilicon, metal, metal alloy, or dielectric (Fig. 13.28c). The structural film is then patterned, usually through dry etching in order to maintain vertical walls and tight dimensional tolerances. Finally, wet etching of the sacrificial layer leaves a free-standing, three-dimensional structure (Fig. 13.28e). It should be noted that the wafer is usually annealed to remove the residual stresses in the deposited metal before it is patterned. If this procedure is not done, the structural film may severely warp once the spacer layer is removed.

Figure 13.29 shows a microlamp that emits a white light when current is passed through it. It has been produced through a combination of surface and bulk micromachining. The top patterned layer is a 2.2-μm-thick layer of plasma-etched tungsten forming a meandering filament and bond pad. The rectangular overhang is dry-etched silicon nitride. The steeply sloped layer is wet-HF-etched phosphosilicate glass. The substrate is silicon, which is ODE etched.

The etchant used to remove the spacer layer must be carefully chosen; it must preferentially etch the spacer layer while leaving the dielectric, silicon, and structural film as intact as possible. With large features and narrow spacer layers, this task

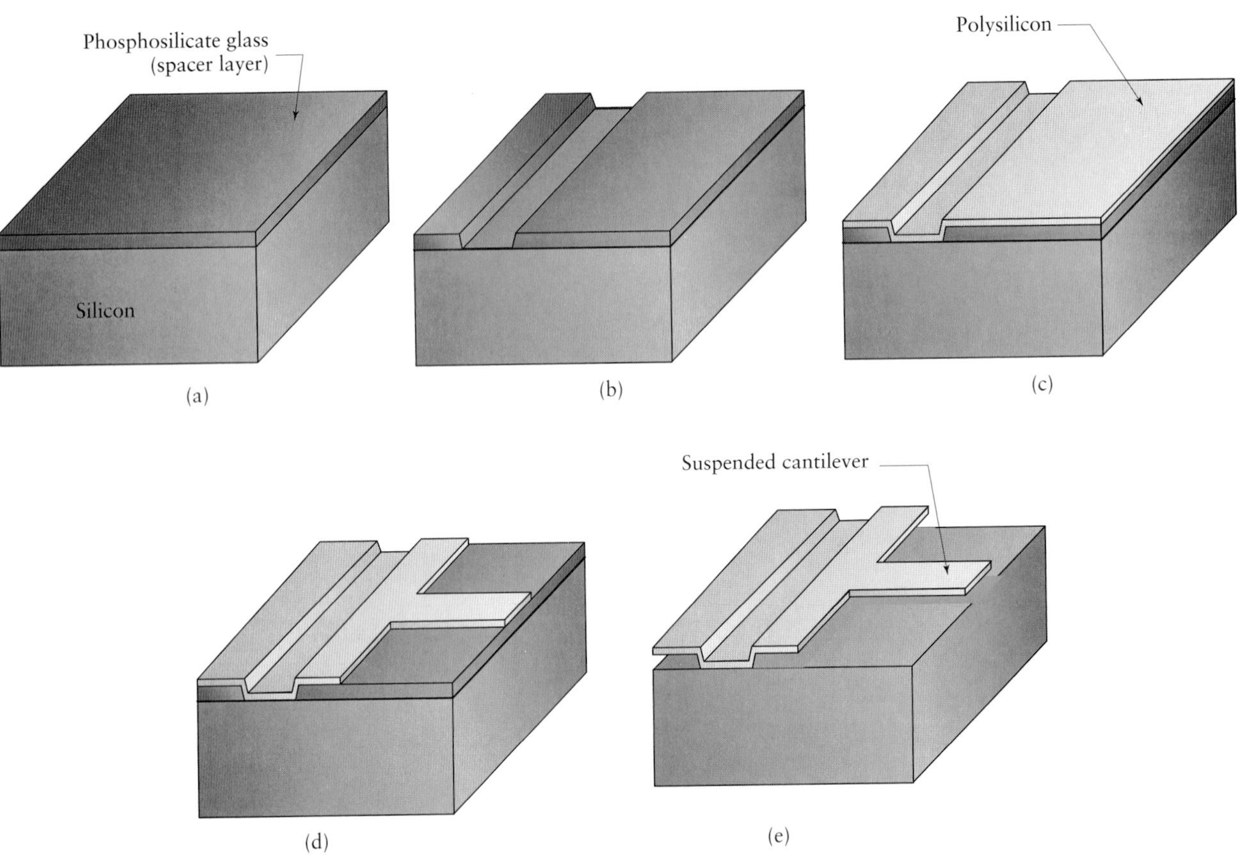

FIGURE 13.28 Schematic illustration of the steps in surface micromachining: (a) deposition of a phosphosilicate glass (PSG) spacer layer; (b) etching of the spacer layer; (c) deposition of polysilicon; (d) etching of polysilicon; (e) selective wet etching of PSG, leaving the silicon substrate and deposited polysilicon unaffected.

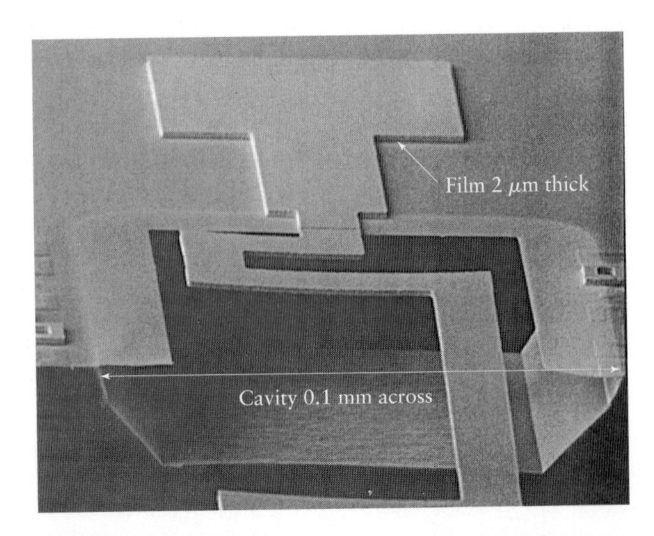

FIGURE 13.29 A microlamp produced from a combination of bulk and surface micromachining. *Source*: Courtesy of K. R. Williams.

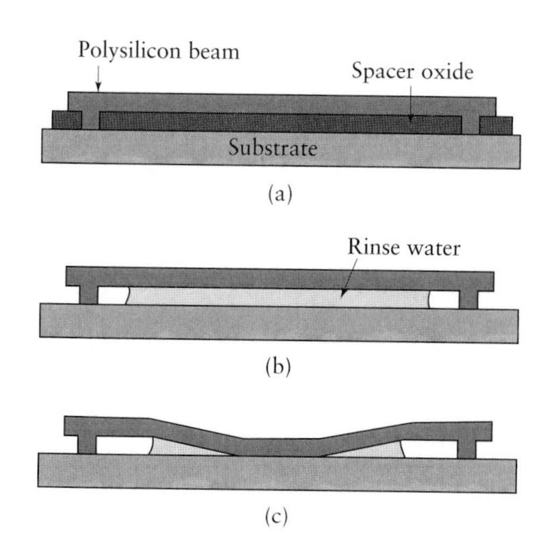

FIGURE 13.30 Stiction after wet etching: (a) unreleased beam; (b) released beam before drying; (c) released beam pulled to the surface by capillary forces during drying. Once contact is made, adhesive forces prevent the beam from returning to its original shape. *Source*: After B. Bhushan.

becomes very difficult to do, and etching can take hours. To reduce the etch time, additional etch holes can be designed into the microstructures to increase access to the spacer layer.

Another difficulty that must be overcome is **stiction** after wet etching. Consider the situation illustrated in Fig. 13.30: After the spacer layer has been removed, the liquid etchant is rinsed from the wafer surface. The rinse water meniscus formed between layers then results in capillary forces that can deform the film and cause it to contact the substrate as the liquid evaporates. Since adhesion forces are more significant at small length scales, it is possible that the film may permanently stick to the surface, and the desired three-dimensional features will not be produced.

Surface micromachining is a very widespread approach for the production of MEMS. Its applications include accelerometers, pressure sensors, micropumps, micromotors, actuators, and microscopic locking mechanisms. Often, these devices require very large vertical walls that cannot be directly manufactured, because the high vertical structure is difficult to deposit. This problem is overcome by machining large flat structures horizontally and then rotating or folding them into an upright position, as shown in Fig. 13.31.

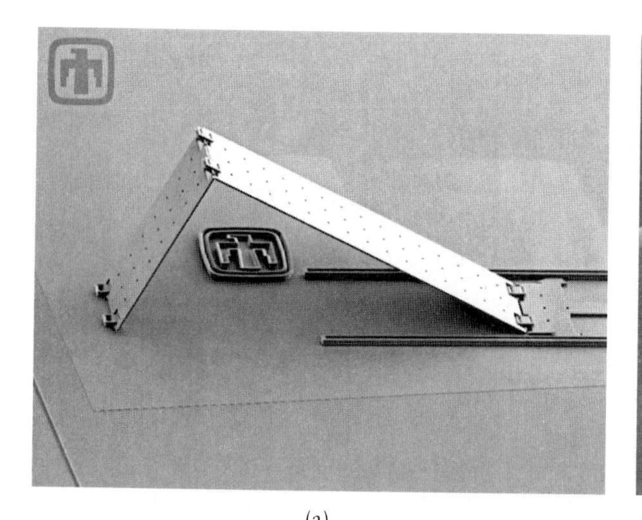

(a)

(b)

FIGURE 13.31 (a) SEM image of a deployed micromirror. (b) Detail of the micromirror hinge. *Source*: Sandia National Laboratories.

EXAMPLE 13.3 Surface micromachining of a hinge for a mirror actuation system

Figure 13.31a shows a micromirror that has been inclined with respect to the surface on which it was manufactured. Such systems can be used for reflecting light (that is oblique to a surface) onto detectors or towards other sensors. It is apparent that a device that has such depth and has the aspect ratio of the deployed mirror is very difficult to machine directly. Instead, it is easier to surface micromachine the mirror along with a linear actuator and then fold the mirror into a deployed position. In order to do so, special hinges, as shown in Fig. 13.31b, are integrated into the design.

Figure 13.32 shows the cross-section of a hinge during manufacture. The following steps are required to produce the hinge:

1. A 2-μm-thick layer of phosphosilicate glass (PSG) is deposited onto the substrate material.
2. A 2-μm-thick layer of polysilicon (Poly1 in Fig. 13.32a) is deposited onto the PSG, patterned by photolithography, and dry etched to form the desired structural elements, including hinge pins.
3. A second layer of sacrificial PSG, with a thickness of 0.5 μm, is deposited (Fig. 13.32b).
4. The connection locations are etched through both layers of PSG (Fig. 13.32c).
5. A second layer of polysilicon (Poly2 in Fig. 13.32d) is deposited, patterned, and etched.
6. The sacrificial layers of PSG are removed through wet etching.

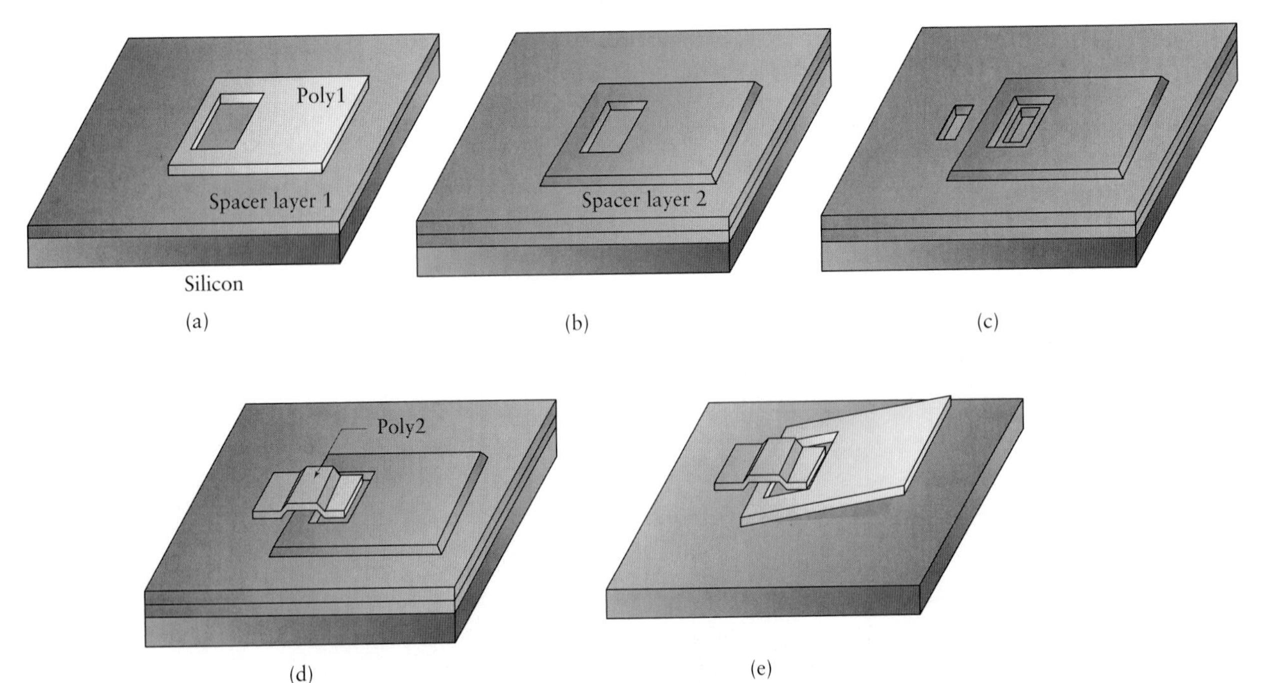

FIGURE 13.32 Schematic illustration of the steps required to manufacture a hinge. (a) Deposition of a phosphosilicate glass (PSG) spacer layer and polysilicon layer. (See Fig. 13.28.) (b) Deposition of a second spacer layer. (c) Selective etching of the PSG. (d) Depostion of polysilicon to form a staple for the hinge. (e) After selective wet etching of the PSG, the hinge can rotate.

Hinges such as those discussed in this example have very high friction. If mirrors such as that shown in Fig. 13.31 are manually and carefully manipulated with probe needles, they will remain in position. Often, such mirrors will be combined with linear actuators to enable precise control of their deployment.

SCREAM. Another approach for making very deep MEMS structures is the *single-crystal silicon reactive etching and metallization* (SCREAM) process, depicted in Fig. 13.33. In this technique, standard lithography and etching processes produce trenches 10–50 μm (400–2000 μin.) deep, which are then protected by a layer of chemically vapor-deposited silicon oxide. An anisotropic etch step removes the oxide only at the bottom of the trench, and the trench is then extended through dry etching. An isotropic etch, using sulfur hexafluoride (SF_6), laterally etches the exposed sidewalls at the bottom of the trench; this undercut, when it overlaps adjacent undercuts, releases the machined structures.

SIMPLE. An alternative to SCREAM is the *silicon micromachining by single-step plasma etching* (SIMPLE) technique, as depicted in Fig. 13.34. This technique uses a chlorine-gas-based plasma etch process that machines *p*-doped, or lightly doped, silicon anisotropically, but heavily *n*-doped silicon isotropically. A suspended MEMS device can thus be produced in one plasma etching step, as shown in the figure.

Some of the concerns with the SIMPLE process are as follows:

1. The oxide mask is machined, although at a slower rate, by the chlorine-gas plasma; therefore, relatively thick oxide masks are needed.
2. The isotropic etch rate is low, typically 50 nm/min; consequently, this process is very slow.

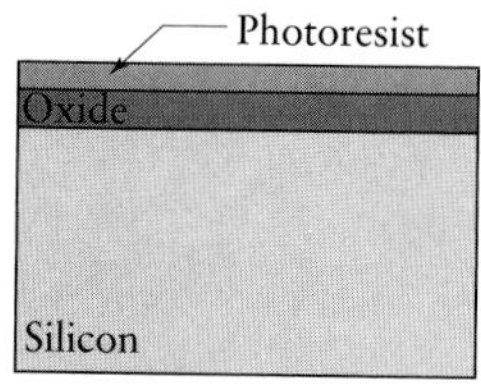

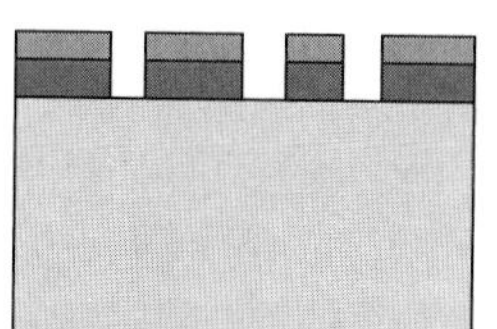

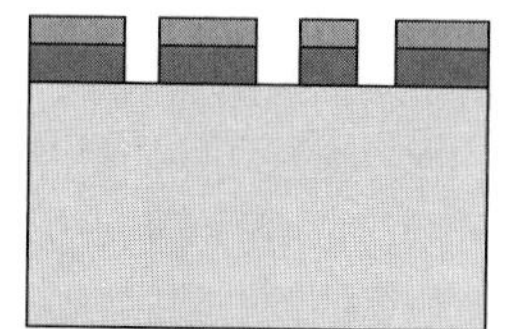

3. Silicon etch.

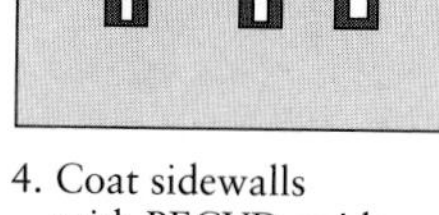

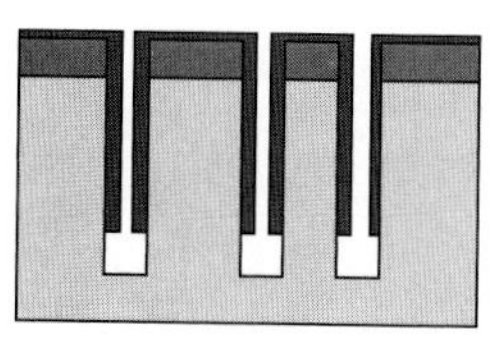

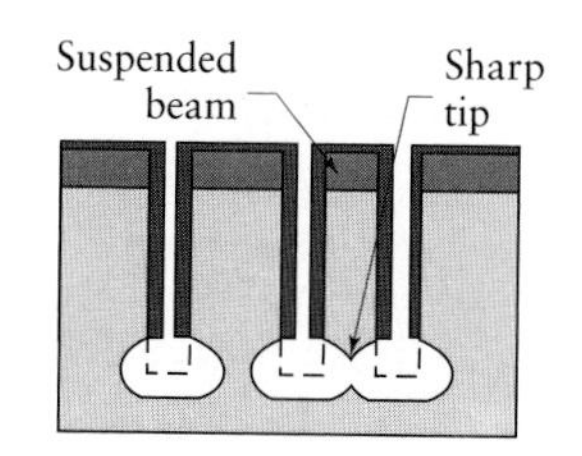

FIGURE 13.33 Steps involved in the SCREAM process. *Source*: After N. Maluf.

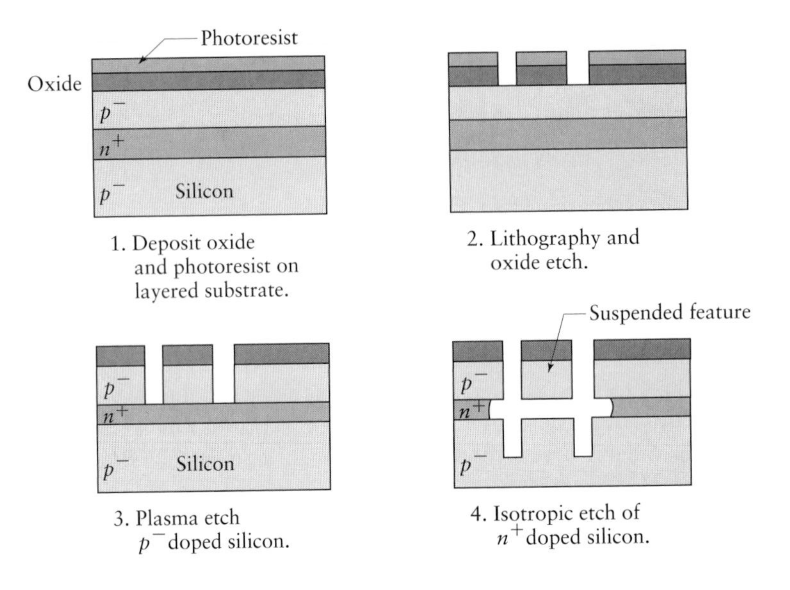

FIGURE 13.34 Schematic illustration of silicon micromachining by the single-step plasma etching (SIMPLE) process. Reprinted with permission of Cambridge University Press.

3. The layer beneath the structures will contain deep trenches, which may affect the motion of free-hanging structures.

Etching combined with fusion bonding. Very tall structures can be produced in crystalline silicon through a combination of *silicon fusion bonding* and *deep reactive ion etching* (SFB–DRIE), as illustrated in Fig. 13.35. First, a silicon wafer is prepared with an insulating oxide layer, with the deep trench areas defined by a standard lithography procedure. This step is followed by conventional wet or dry etching to form a large cavity. A second layer of silicon is fusion bonded to this layer, which can then be ground and polished to the desired thickness, if necessary. At this stage, integrated circuitry is manufactured through the steps outlined in Fig. 13.3. A protective resist is applied and exposed, and the desired trenches are etched by deep reactive ion etching through to the cavity in the first layer of silicon.

EXAMPLE 13.4 Operation and fabrication sequence for a thermal ink-jet printer

Thermal ink-jet printers are perhaps the most successful application of MEMS to date. These printers operate by ejecting nano- or picoliters (10^{-12} liters) of ink from a nozzle towards paper. Ink-jet printers use a variety of designs, but silicon machining technology is most applicable to high-resolution printers. It should be realized that a resolution of 1200 dots per inch (dpi) requires a nozzle pitch of approximately 20 μm (800 μin.).

Figure 13.36 depicts the mode of operation of an ink-jet printer. When an ink droplet is to be generated and expelled, a tantalum resistor below a nozzle is heated. This resistor heats a thin film of ink so that a bubble forms within five microseconds. The bubble expands rapidly, with internal pressures reaching 1.4 MPa (200 psi), and as a result, fluid is forced rapidly out of the nozzle. Within 24 microseconds, the tail of the ink droplet separates, because of surface tension. The heat source is then removed (turned off), and the bubble collapses inside the nozzle. Within 50 microseconds, sufficient ink has been drawn into the nozzle from a reservoir to form the desired meniscus for the next droplet.

Traditional ink-jet printer heads were produced with electroformed nickel nozzles, fabricated separately from the integrated circuitry, and required a bonding

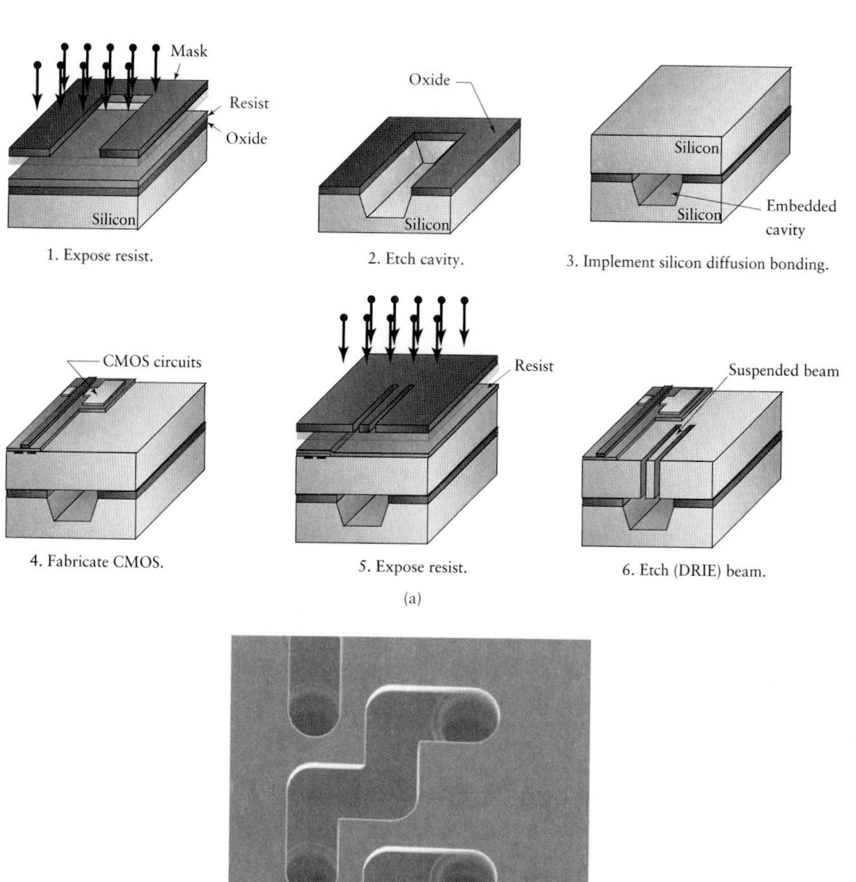

FIGURE 13.35 (a) Silicon fusion bonding combined with deep reactive ion etching to produce large suspended cantilevers. *Source*: After N. Maluf. (b) A micro–fluid-flow device manufactured by applying the DRIE process to two separate wafers and then aligning and silicon-fusion bonding them together. Afterward, a Pyrex layer (not shown) is anodically bonded over the top to provide a window for observing fluid flow. *Source*: K. R. Williams.

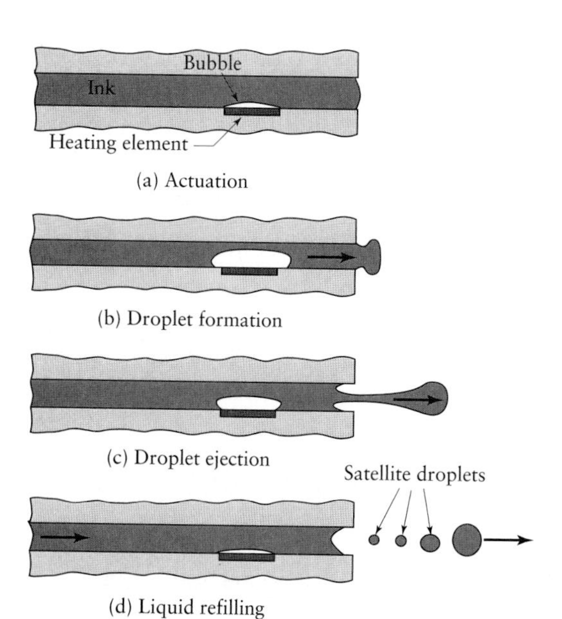

FIGURE 13.36 Sequence of operation of a thermal ink-jet printer. (a) Resistive heating element is turned on, rapidly vaporizing ink and forming a bubble. (b) Within five microseconds, the bubble has expanded and displaced liquid ink from the nozzle. (c) Surface tension breaks the ink stream into a bubble, which is discharged at high velocity. The heating element is turned off at this time, so that the bubble collapses as heat is transferred to the surrounding ink. (d) Within 24 microseconds, an ink droplet (and undesirable satellite droplets) are ejected, and surface tension of the ink draws more liquid from the reservoir. *Source*: From F. G. Tseng, "Microdroplet Generators," in M. Gad-el-hak (ed.), *The MEMS Handbook*, CRC Press, 2002.

FIGURE 13.37 The manufacturing sequence for producing thermal ink-jet printer heads. *Source*: From F. G. Tseng, "Microdroplet Generators," in M. Gad-el-hak (ed.), *The MEMS Handbook*, CRC Press, 2002.

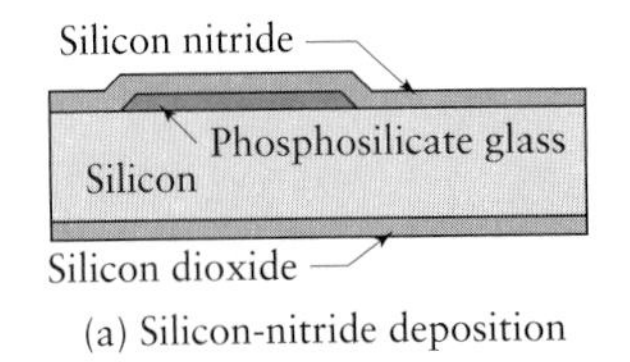

(a) Silicon-nitride deposition

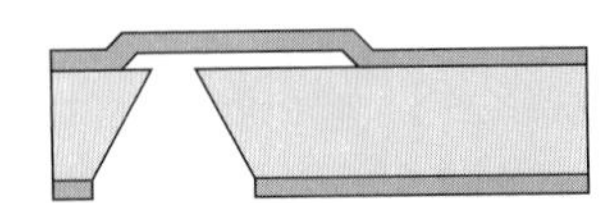

(b) Wet etch manifold; remove PSG

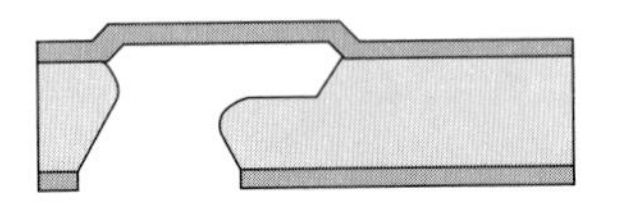

(c) Wet etch, enlarge chamber

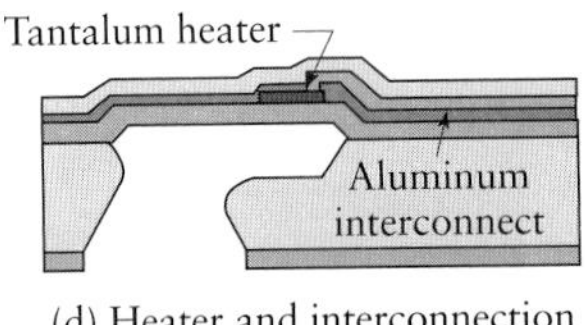

(d) Heater and interconnection formulation

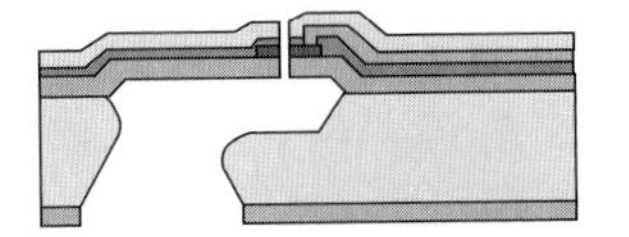

(e) Laser ablade nozzle

operation to attach these two components. With increasing printer resolution, however, it is more difficult to bond the components with a tolerance under a few micrometers. For this reason, single-component, or monolithic, fabrication is of interest.

Figure 13.37 shows the fabrication sequence for a monolithic ink-jet printer head. A silicon wafer is prepared and coated with a phosphosilicate-glass (PSG) pattern and low-stress silicon-nitride coating. The ink reservoir is obtained by isotropically etching the back side of the wafer, followed by PSG removal and then enlargement of the reservoir. The required CMOS controlling circuitry is then produced (this step is not shown in Fig. 13.37), and a tantalum heater pad is deposited. The aluminum interconnection between the tantalum pad and the CMOS circuit is formed, and the nozzle is produced through laser ablation. An array of such nozzles can be placed inside an ink-jet printing head, and resolutions of 2400 dpi or higher can be achieved.

13.14 The LIGA Microfabrication Process

LIGA is a German acronym for the combined processes of X ray lithography, electrodeposition, and molding (**X ray Lithographie, Galvanoformung und Abformung**); a schematic illustration of this process is given in Fig. 13.38.

The LIGA process involves the following steps:

1. A very thick (up to hundreds of microns) resist layer of polymethylmethacrylate (PMMA) is deposited onto a primary substrate.
2. The PMMA is exposed to collimated X rays and developed.
3. Metal is electrodeposited onto the primary substrate.
4. The PMMA is removed or stripped, resulting in a freestanding metal structure.
5. Plastic injection molding is done in the metal structure, which acts as a mold.

Depending on the application, the final product from a LIGA process may be:

a) A freestanding metal structure, resulting from the electrodeposition process;
b) A plastic injection-molded structure;
c) An investment-cast metal part, where the injection molded structure was used as a blank;
d) A slip-cast ceramic part, produced with the injection-molded parts as the molds.

The substrate used in LIGA is a conductor or a conductor-coated insulator. Examples of primary substrate materials include austenitic steel plate; silicon wafers with a titanium layer; and copper plated with gold, titanium, or nickel. Metal-plated ceramic and glass have also been used. The surface may be roughened by grit blasting to encourage good adhesion of the resist material.

Resist materials must have high X ray sensitivity, dry- and wet-etching resistance when unexposed, and thermal stability. The most common resist material is polymethylmethacrylate, which has a very high molecular weight (more than 10^6

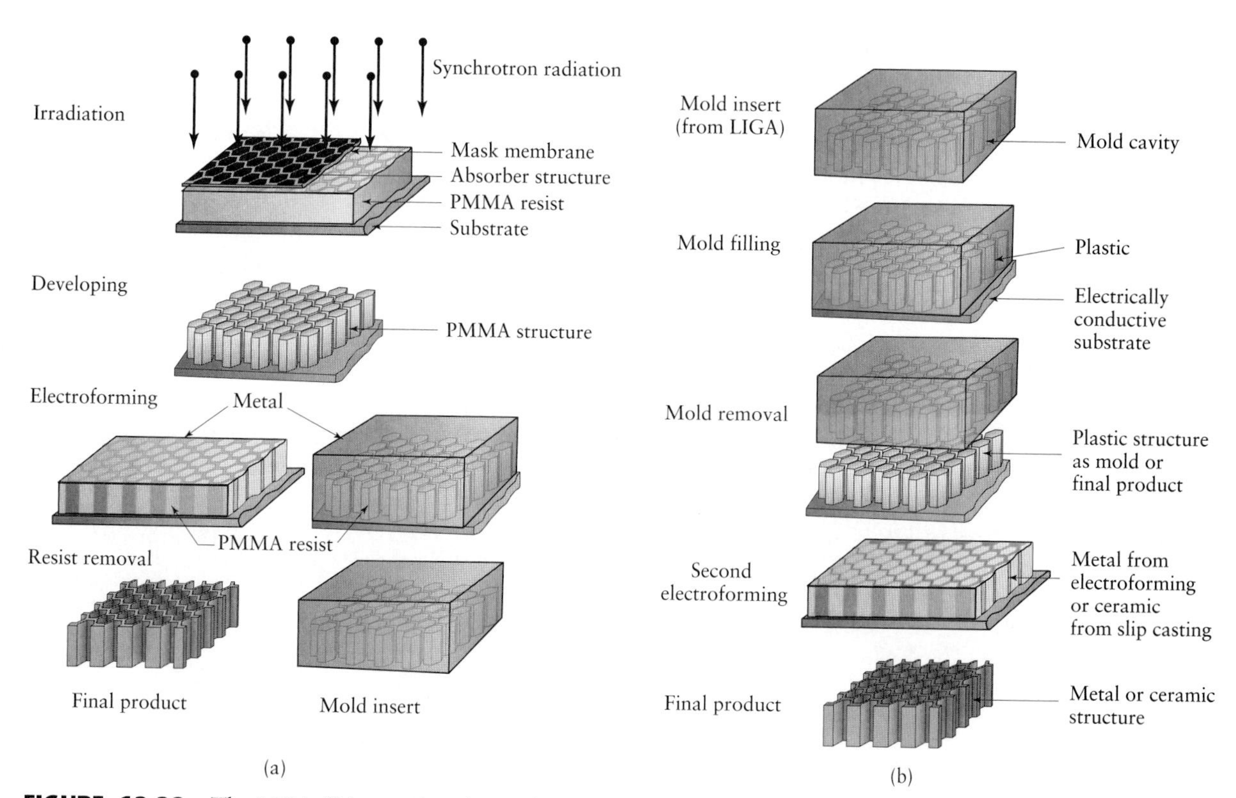

FIGURE 13.38 The LIGA (lithography, electrodeposition, and molding) technique. (a) Primary production of a metal final product or mold insert. (b) Use of the primary part for secondary operations, or *replication*. *Source*: Courtesy of IMM Institute für Mikrotechnik.

grams per mole; see Section 10.2.1). The X rays break the chemical bonds, leading to the production of free radicals and significantly reduced molecular weight in the exposed region. Organic solvents then preferentially dissolve the exposed PMMA in a wet-etching process. After development, the remaining three-dimensional structure is rinsed and dried, or it is spun and blasted with dry nitrogen.

Electrodeposition of metal usually involves electroplating of nickel (see Section 4.5.1 and Fig. 4.17). The nickel is deposited onto exposed areas of the substrate; it fills the PMMA structure and can even coat the resist (Fig. 13.38a). Nickel is the material of choice because of the relative ease in electroplating with well-controlled deposition rates and residual stress control. Electroless plating of nickel is also possible, and the nickel can be deposited directly onto electrically insulating substrates. However, because nickel displays high wear rates in MEMS, significant research has been directed towards the use of other materials or coatings.

After the metal structure has been deposited, precision grinding either removes the substrate material or a layer of the deposited nickel, a process referred to as *planarization* (see also Section 13.9). The need for planarization is clear when it is recognized that three-dimensional MEMS devices require micrometer tolerances on layers many hundreds of micrometers thick. Planarization is difficult to achieve; conventional lapping (Section 9.8) leads to preferential removal of the soft PMMA and smearing of the metal. Planarization is usually accomplished with a diamond lapping procedure referred to as *nanogrinding*. Here, a diamond slurry-loaded soft metal plate is used to remove material in order to maintain flatness within 1 μm (40 μin.) over a 75-mm (3-in.) diameter substrate.

If cross-linked (Section 10.2.1), the PMMA resist is then exposed to synchrotron X ray radiation and removed by exposure to an oxygen plasma or through solvent extraction; the result is a metal structure, which may be used for further processing. Examples of freestanding metal structures produced through electrodeposition of nickel are shown in Fig. 13.39.

The processing steps used to make freestanding metal structures are extremely time consuming and expensive. The main advantage of LIGA is that these structures serve as molds for the rapid replication of submicron features through molding operations. Table 13.5 lists and compares the processes that can be used for producing micromolds; it can be seen that LIGA provides some clear advantages. Reaction injection molding, injection molding, and compression molding (Section 10.10) have also been used to make these micromolds.

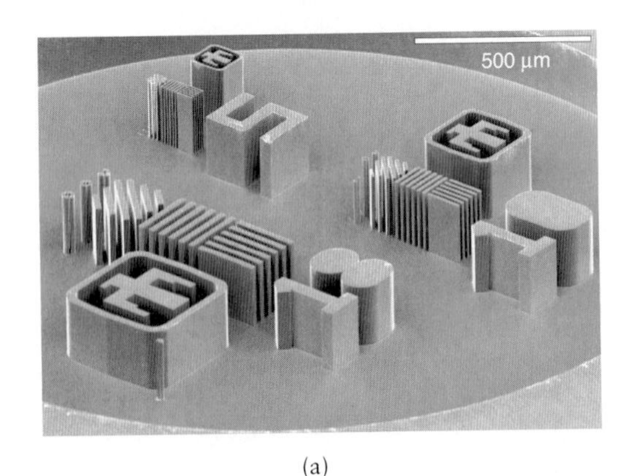

(a)

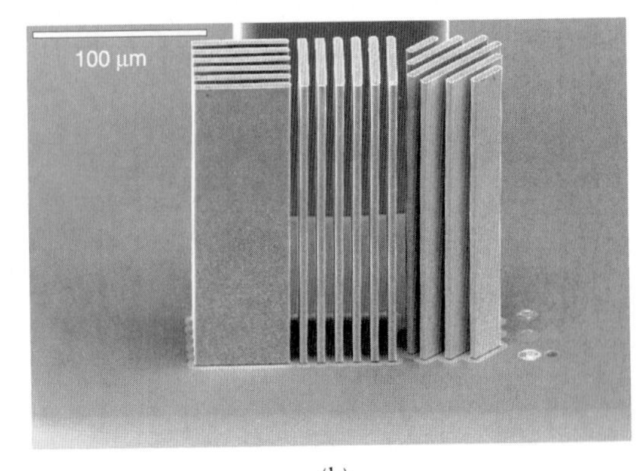

(b)

FIGURE 13.39 (a) Electroformed 200-μm-tall nickel structures. (b) Detail of 5-μm-wide nickel lines and spaces. *Source*: After T. Christenson, *The MEMS Handbook*, CRC Press, 2002.

TABLE 13.5

Comparison of Micromold Manufacturing Techniques

	Production technique		
	LIGA	Laser machining	EDM
Aspect ratio	10–50	10	up to 100
Surface roughness	<50 nm	100 nm	0.3–1 μm
Accuracy	<1 μm	1–3 μm	1–5 μm
Mask Required?	Yes	No	No
Maximum height	1–500 μm	200–500 μm	μm to mm

Source: L. Weber, W. Ehrfeld, H. Freimuth, M. Lacher, M. Lehr, and P. Pech, *SPIE Micromachining and Microfabrication Process Technology II*, Austin, TX, 1996.

EXAMPLE 13.5 Production of rare-earth magnets

A number of scaling issues in electromagnetic devices indicate that there is an advantage in using rare-earth magnets from the samarium cobalt (SmCo) and neodymium iron boron (NdFeB) families. These materials are available in powder form and are of interest because they can produce magnets that are an order of magnitude more powerful than conventional magnets (Table 13.6). Therefore, these materials can be used when effective miniature electromagnetic transducers are to be produced.

Figure 13.40 shows the processing steps used to manufacture these magnets. The polymethylmethacrylate mold is produced by exposure to X ray radiation and solvent extraction. The rare-earth powders are mixed with a binder of epoxy and then applied to the PMMA mold through a combination of calendering (see Fig. 10.32) and pressing. After curing in a press at a pressure around 70 MPa (10 ksi), the substrate is planarized. The substrate is then subjected to a magnetizing field of at least 35 kilooersteds in the desired orientation. Once the material has been magnetized, the PMMA substrate is dissolved, leaving behind the rare-earth magnets, as shown in Fig. 13.41.

Source: T. Christenson, Sandia National Laboratories.

TABLE 13.6

Comparison of Properties of Permanent-Magnet Materials

Material	Energy product (Gauss–Oersted $\times 10^{-6}$)
Carbon steel	0.20
36% cobalt steel	0.65
Alnico I	1.4
Vicalloy I	1.0
Platinum–cobalt	6.5
$Nd_2Fe_{14}B$, fully dense	40
$Nd_2Fe_{14}B$, bonded	9

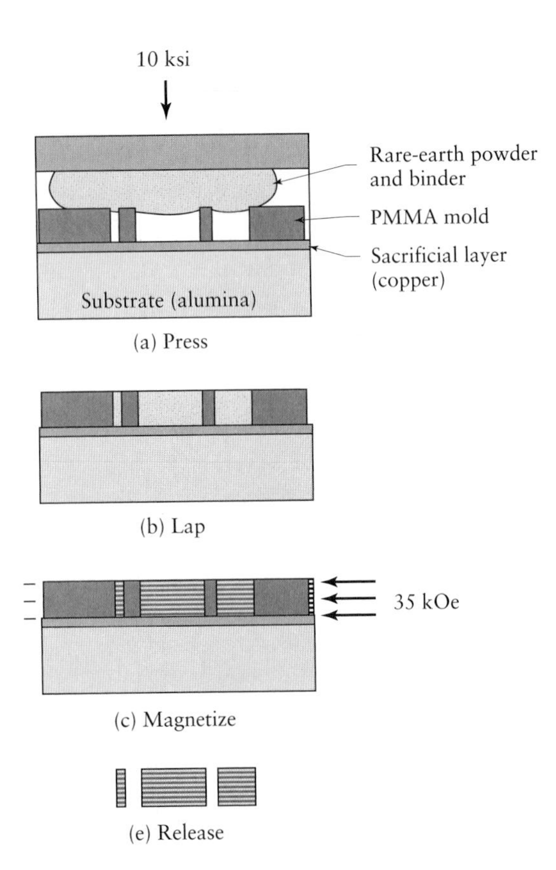

FIGURE 13.40 Fabrication process used to produce rare-earth magnets for microsensors. *Source*: T. Christenson, Sandia National Laboratories.

FIGURE 13.41 SEM images of $Nd_2Fe_{14}B$ permanent magnets. Powder particle size ranges from 1 to 5 μm (40 to 200 μin.) and the binder is a methylene-chloride resistant epoxy. Mild distortion is present in the image, due to magnetic perturbation of the imaging electrons. Maximum-energy products of 9 MGOe have been obtained with this process. *Source*: T. Christenson, in Gad-el-Hak, ed., The MEMS Handbook, CRC Press, 2002.

Multilayer X ray lithography. The LIGA technique is very suitable for producing MEMS devices with large aspect ratios and reproducible shapes. Often, a multilayer stepped structure that cannot be made directly through LIGA is required. For nonoverhanging geometries, direct plating can be applied. In this technique, a layer of electrodeposited metal with surrounding PMMA is produced as described previously. A second layer of PMMA resist is then bonded to this structure and X ray exposed with an aligned X ray mask.

It is often useful to have overhanging geometries within complex MEMS devices. A batch diffusion bonding and release procedure has been developed for this purpose and is schematically illustrated in Fig. 13.42a. This process involves the preparation of two PMMA patterned and electroformed layers, with the PMMA being subsequently removed. The wafers are then aligned face to face with guide pins that press-fit into complimentary structures on the opposite surface. Next, the substrates

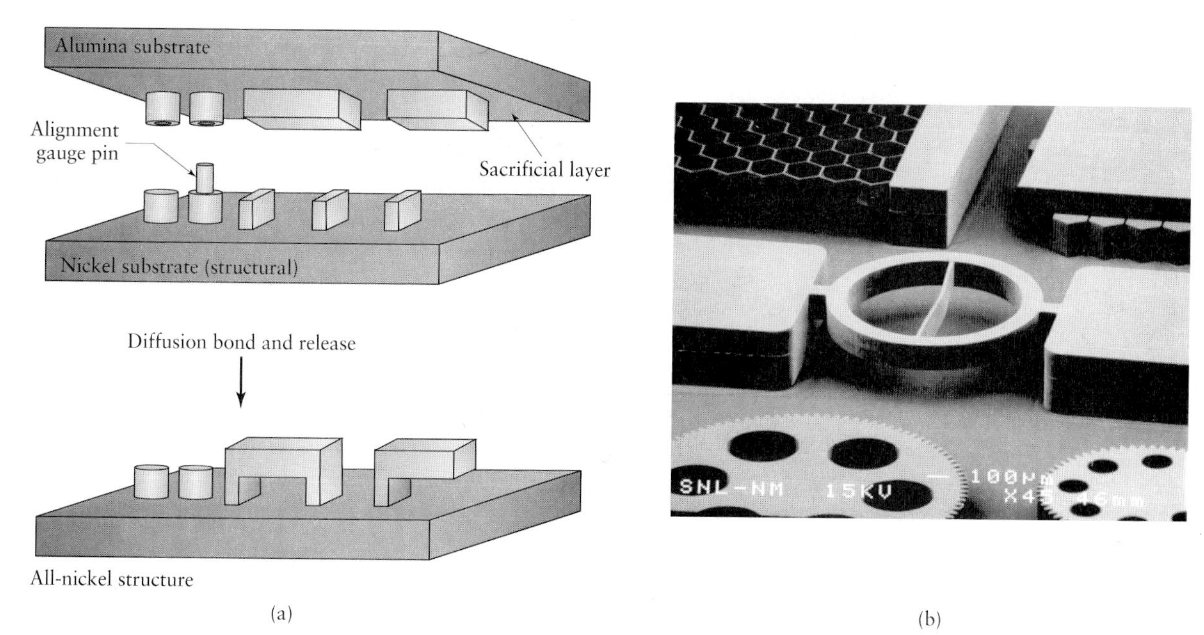

FIGURE 13.42 (a) Multilevel MEMS fabrication through wafer-scale diffusion bonding. (b) A suspended ring structure for measurement of tensile strain, formed by two-layer wafer-scale diffusion bonding. *Source*: After T. Christenson, Sandia National Laboratories.

are joined in a hot press, and a sacrificial layer on one substrate is etched away, leaving one layer bonded to the other. Fig. 13.42b shows an example of such a structure.

13.15 Solid Freeform Fabrication of Devices

Solid freeform fabrication is another term for rapid prototyping, described in Section 10.12. This method is unique in that complex three-dimensional structures are produced through additive manufacturing, as opposed to material removal. Many of the advances in rapid prototyping also are applicable to MEMS manufacture for processes with sufficiently high resolution. Recall that *stereolithography* involves curing of a liquid thermosetting polymer, using a photoinitiator and a highly focused light source. Conventional stereolithography uses layers between 75 μm and 500 μm (0.003 in. and 0.02 in.) in thickness, with a laser dot focused to a 0.25-mm (0.01 in.) diameter.

In *microstereolithography*, the process uses the same approach, but the laser is more highly focused, to a diameter as small as 1 μm (40 μin.), and layer thicknesses are around 10 μm (400 μin.). This technique has a number of cost advantages, but the MEMS devices are difficult to integrate with the controlling circuitry, because stereolithography produces nonconducting polymer structures.

Instant masking is a technique for producing MEMS devices (Fig. 13.43). The solid freeform fabrication of MEMS devices by instant masking is known as **electrochemical fabrication** (EFAB). A mask of elastomeric material is first produced through the conventional photolithography techniques described in Section 13.6. The mask is pressed against the substrate in an electrodeposition bath, so that the elastomer conforms to the substrate and excludes plating solution in contact areas. Electrodeposition takes place in areas that are not masked, eventually producing a mirror image of the mask. By using a sacrificial filler, made of a second material, complex three-dimensional shapes can be produced, complete with overhangs, arches, and other features, using instant masking technology.

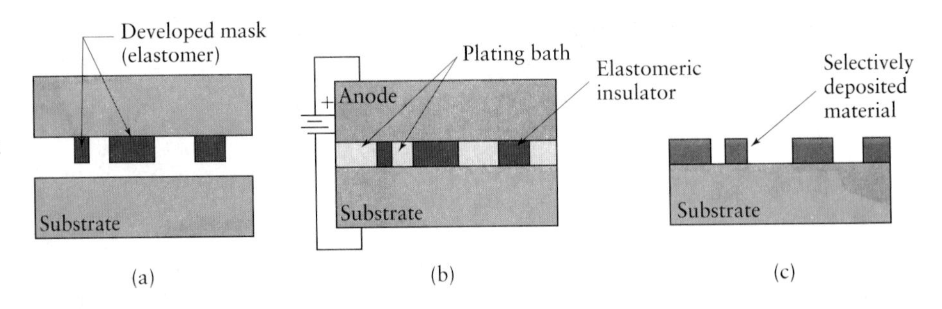

FIGURE 13.43 The instant masking process: (a) bare substrate; (b) during deposition, with the substrate and instant mask in contact; (c) the resulting pattern deposit. *Source*: After A. Cohen, MEMGen Corporation.

CASE STUDY | Accelerometer for Automotive Air Bags

Accelerometers based on lateral resonators represent the largest commercial application of surface micromachining today and are widely used as sensors for automotive air-bag deployment systems. The sensor portion of such an accelerometer is shown in Fig. 13.44. A central mass is suspended over the substrate, but anchored through four slender beams that act as springs to center the mass under static-equilibrium conditions. An acceleration causes the mass to deflect, reducing or increasing the clearance between the fins on the mass and the stationary fingers on the substrate. By measuring the electrical capacitance between the mass and fins, the deflection of the mass, and therefore the acceleration or deceleration of the system, can be directly measured. Figure 13.44 shows an arrangement for the measurement of acceleration in one direction, but commercial sensors employ several masses so that accelerations can be measured in multiple directions simultaneously.

Figure 13.45 shows the 50-g surface-micromachined accelerometer (ADXL-50) with onboard signal conditioning and self-diagnostic electronics. The polysilicon sensing element (visible in the center of the die) occupies only 5% of the total die area, and the whole chip measures 500 μm by 625 μm (20 μin. by 25 μin.) The mass is approximately 0.3 μg, and the sensor has a measurement accuracy of 5% over the $\pm$50-g range.

Fabrication of the accelerometer proved to be a challenge, since it required a complimentary metal-oxide semiconductor (CMOS) fabrication sequence to be closely integrated with a surface-micromachining approach. Analog Devices, Inc., was able to modify a CMOS production technique to directly incorporate surface micromachining. In the sensor design, $n+$ underpasses connect the sensor area to the electronic circuitry, replacing the usual heat-sensitive aluminum connect lines.

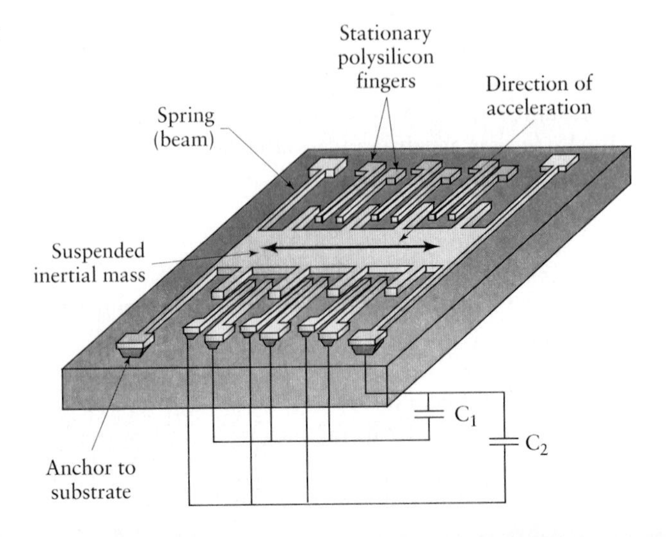

FIGURE 13.44 Schematic illustration of a micro–acceleration sensor. *Source*: After N. Maluf.

FIGURE 13.45 Photograph of Analog Devices's ADXL-50 accelerometer with a surface-micromachined capacitive sensor (center), on-chip excitation, self-test, and signal-conditioning circuitry. The entire chip measures 0.500 by 0.625 mm. *Source*: From R. A. Core et al., *Solid State Technol.*, October 1993, pp. 39–47.

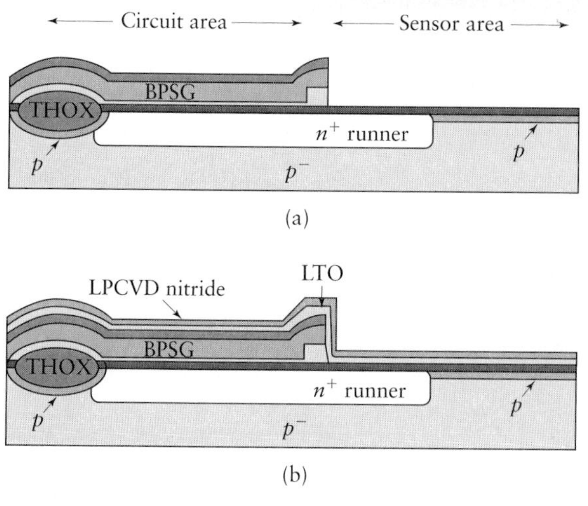

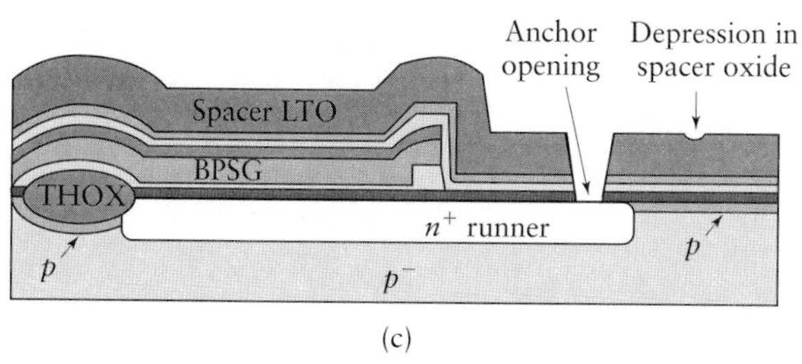

FIGURE 13.46 Preparation of IC chip for polysilicon. (a) Sensor area post-BPSG planarization and moat mask. (b) Blanket deposition of thin oxide and thin nitride layer. (c) Bumps and anchors made in LTO spacer layer. *Source*: From R. A. Core et al., *Solid State Technol.*, October 1993, pp. 39–47.

Most of the sensor processing is inserted into the fabrication process right after borosilicate-glass planarization. After planarization, a designated sensor region, or *moat*, is cleared in the center of the die (Fig. 13.46a). A thin oxide is then deposited to passivate the $n+$ underpass connects, followed by a thin, low-pressure chemical-vapor-deposited (LPCVD) nitride to act as an etch stop for the final polysilicon release etch (Fig. 13.46b). The spacer, or sacrificial oxide, used is a 1.6-μm (64-μin.)

densified low-temperature oxide (LTO) deposited over the whole die (Fig. 13.46c).

In a first etch, small depressions that will form bumps, or dimples, on the underside of the polysilicon sensor are created in the LTO layer. These bumps will limit adhesive forces and sticking in case the sensor comes in contact with the substrate. A subsequent etch cuts anchors into the spacer layer to provide regions of electrical and mechanical contact (Fig. 13.46c). The 2-μm-thick (80-μin.-thick) sensor polysilicon layer is then deposited, implanted, annealed, and patterned (Fig. 13.47a).

Metallization follows, which starts with the removal of the sacrificial spacer oxide from the circuit area, along with the LPCVD nitride and LTO layer. A low-temperature oxide is deposited on the polysilicon sensor part, and contact openings appear in the integrated circuit part of the die where platinum is deposited to form platinum silicide (Fig. 13.47b). The trimmable thin film material, TiW barrier metal, and Al/Cu interconnect metal are sputtered on and patterned in the integrated-circuit area.

The circuit area is then passivated in two separate deposition steps. First, plasma oxide is deposited and patterned (Fig. 13.47c), followed by a plasma nitride (Fig. 13.48a) to form a seal with the previously deposited LPCVD nitride. The nitride acts as a hydrofluoric-acid barrier in the subsequent etch release in surface micromachining. The plasma oxide left on the sensor acts as an etch stop for the removal of the plasma nitride (Fig. 13.48a). The sensor area is then prepared for the final release etch. The dielectrics are removed from the sensor, and the final protective resist mask is applied. The photoresist protects the circuit area from the long-term buffered oxide etch (Fig. 13.48b). The final device cross-section is shown in Fig. 13.48c.

Source: Adapted from M. Madou, *Fundamentals of Microfabrication*, 2d. ed., CRC Press, 2002.

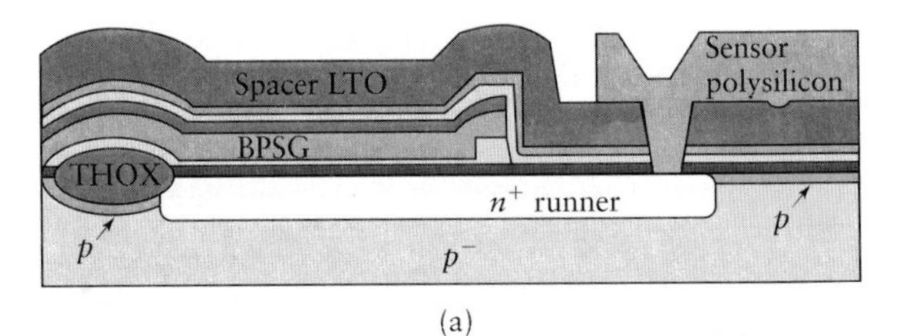

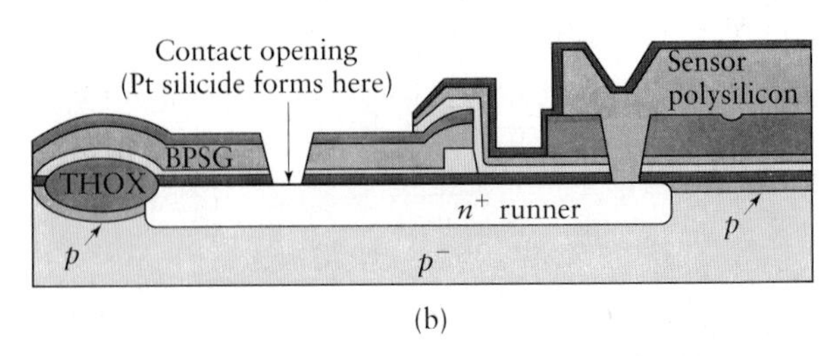

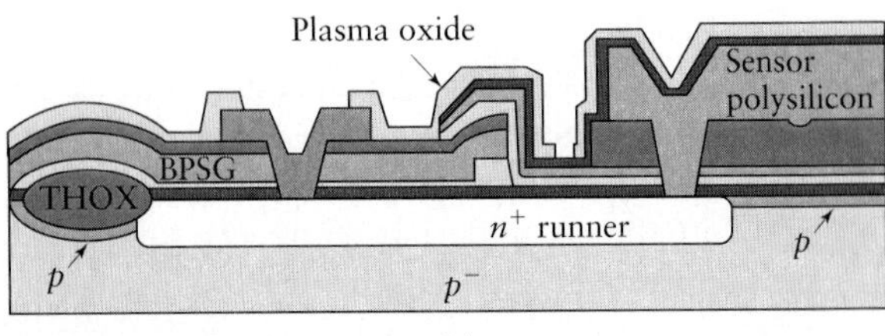

FIGURE 13.47 Polysilicon deposition and IC metallization. (a) Cross-sectional view after polysilicon deposition, implant, anneal, and patterning. (b) Sensor area after removal of dielectrics from circuit area, contact mask, and platinum silicide. (c) Metallization scheme and plasma-oxide passivation and patterning. *Source*: From R. A. Core et al., *Solid State Technol.*, October 1993, pp. 39–47.

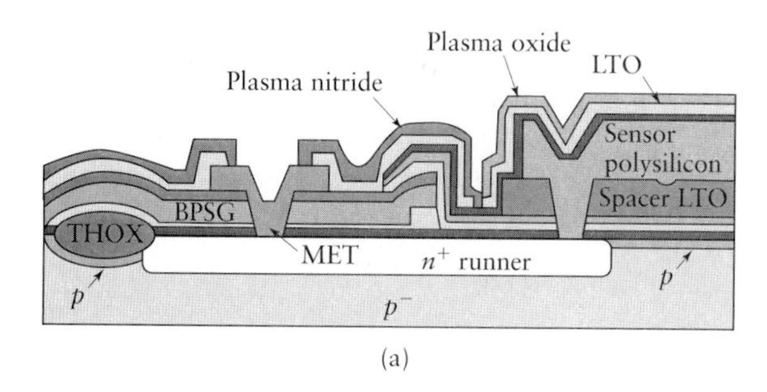

(a)

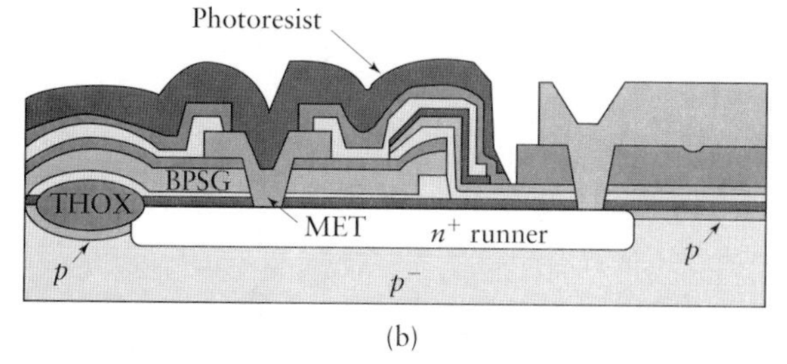

(b)

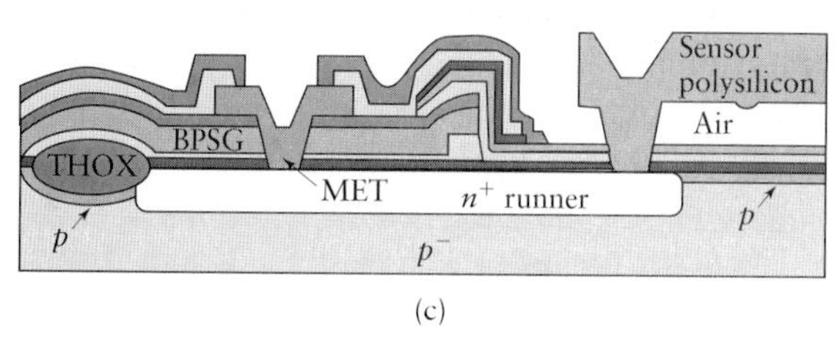

(c)

FIGURE 13.48 Prerelease preparation and release. (a) Post-plasma-nitride passivation and patterning. (b) Photoresist protection of the IC. (c) Freestanding, released, polysilicon beam. *Source*: From R. A. Core et al., *Solid State Technol.*, October 1993, pp. 39–47.

SUMMARY

- The microelectronics industry is well developed, but still rapidly changing. The possibilities for new device concepts and circuit designs appear to be endless. MEMS devices have become important technologies. (Section 13.1)
- Semiconductor materials, such as silicon, gallium arsenide, and polysilicon, have unique properties, and their electrical properties lie between those of conductors and insulators. (Section 13.2)
- Techniques have been developed for single crystal growing and wafer preparation; after fabrication, the wafer is sliced into individual chips, each containing one complete integrated circuit. (Section 13.3)
- The fabrication of microelectronic devices and integrated circuits involves many different types of processes. After bare wafers have been prepared, they undergo repeated oxidation, film deposition, lithographic, and etching steps to open windows in the oxide layer in order to provide access to the silicon substrate. (Sections 13.4–13.8)
- Dopants are introduced into various regions of the silicon structure to alter their electrical characteristics; this task is done by diffusion and ion implantation. (Section 13.9)
- Microelectronic devices are interconnected by multiple metal layers, and the completed circuit is packaged and made accessible through electrical connections. A variety of packages is available. (Section 13.10)
- Yield in devices is important for economic considerations, and reliability has become increasingly important, because of the long expected life of these devices. (Section 13.11)

- Requirements for more flexible and faster integrated circuits demand packaging that places many closely spaced contacts on a printed circuit board. (Section 13.12)
- MEMS devices are manufactured through techniques and with materials that, for the most part, have been pioneered in the microelectronics industry. Bulk and surface micromachining, LIGA, SCREAM, and fusion bonding of multiple layers are the most prevalent methods. (Sections 13.13–13.15.)

BIBLIOGRAPHY

Bakoglu, H. B., *Circuits, Interconnections, and Packaging for VLSI*, Addison-Wesley, 1990.

Berger, L. I., *Semiconductor Materials*, CRC Press, 1997.

Brar, A. S., and P. B. Narayan, *Materials and Processing Failures in the Electronics and Computer Industries: Analysis and Prevention*, ASM International, 1993.

Campbell, S. A., *The Science and Engineering of Microelectronic Fabrication*, Oxford, 1996.

Chandrakasan, A., and R. Brodersen (eds.), *Low Power CMOS Design*, IEEE, 1998.

Chandrakasan, A., and R. Brodersen, *Low Power Digital CMOS Design*, Kluwer, 1995.

Chang, C.-Y., and S. M. Sze (eds.), *ULSI Devices*, Wiley-Interscience, 2000.

Colclaser, R. A., *Microelectronics: Processing and Device Design*, Wiley, 1980.

Electronic Materials Handbook, Vol. 1: *Packaging*, ASM International, 1989.

Elwenspoek, M., and H. Jansen, *Silicon Micromachining*, Cambridge University Press, 1998.

Elwenspoek, M., and R. Wiegerink, *Mechanical Microsensors*, Springer, 2001.

Gad-el-Hak, M. (ed.), *The MEMS Handbook*, CRC Press, 2002.

Ghandhi, S. K., *VLSI Fabrication Principles*, 2d. ed., Wiley-Interscience, 1993.

Griffin, P. B., J. D. Plummer, and M. D. Deal, *Silicon VLSI Technology: Fundamentals, Practice and Modeling*, Prentice Hall, 2000.

Harper, C. A. (ed.), *Electronic Packaging and Interconnection Handbook*, 3d. ed., McGraw-Hill, 2000.

———, *High-Performance Printed Circuit Boards*, McGraw-Hill, 2000.

Hwang, J. S., *Modern Solder Technology for Competitive Electronics Manufacturing*, McGraw-Hill, 1996.

Javits, M. W. (ed.), *Printed Circuit Board Materials Handbook*, McGraw-Hill, 1997.

Judd, M., and K. Brindley, *Soldering in Electronics Assembly*, 2d. ed., Newnes, 1999.

Kovacs, G. T. A., *Micromachined Transducers Sourcebook*, McGraw-Hill, 1998.

Madou, M. J., *Fundamentals of Microfabrication*, CRC Press, 1997.

Mahajan, S., and K. S. S. Harsha, *Principles of Growth and Processing of Semiconductors*, McGraw-Hill, 1998.

Maluf, N., *An Introduction to Microelectromechanical Systems Engineering*, Artech House, 2000.

Manko, H. H., *Soldering Handbook for Printed Circuits and Surface Mounting*, Van Nostrand Reinhold, 1995.

Matisoff, B. S., *Handbook of Electronics Manufacturing*, 3d. ed., Chapman & Hall, 1996.

Nishi, Y., and R. Doering (eds.), *Handbook of Semiconductor Manufacturing Technologies*, Dekker, 2000.

Pierret, R. F., and G. W. Neudeck, *Semiconductor Device Fundamentals*, Addison-Wesley, 1996.

Quirk, M., and J. Serda, *Semiconductor Manufacturing Technology*, Prentice Hall, 2000.

Sze, S. M. (ed.), *Modern Semiconductor Devices Physics*, Wiley, 1997.

———, *Semiconductor Devices: Physics and Technology*, Wiley, 2001.

Taur, Y., and T. H. Ning, *Fundamentals of Modern VLSI Devices*, Cambridge, 1998.

Wise, K. D. (ed.), *Proceedings of the IEEE: Special Issue on Integrated Sensors, Microactuators and Microsystems (MEMS)*, 1998.

Wolf, S., and R. N. Tauber, *Silicon Processing for the VSLI Era: Process Technology*, 2d. ed., Lattice Press, 1999.

van Zant, P., *Microchip Fabrication: A Practical Guide to Semiconductor Processing*, 4th ed., McGraw-Hill, 2000.

Varadan, V. K., X. Jiang, and V. V. Varadan, *Microstereolithography and Other Fabrication Techniques for 3D MEMS*, Wiley, 2001.

Yeap, G., *Practical Low Power Digital VLSI Design*, Kluwer, 1997.

QUESTIONS

13.1 Define the terms *wafer, chip, device,* and *integrated circuit.*

13.2 Why is silicon the most commonly used semiconductor in IC technology?

13.3 What do VLSI, IC, CVD, CMP, and DIP stand for?

13.4 How do *n*-type and *p*-type dopants differ?

13.5 How is epitaxy different than other forms of film deposition?

13.6 Compare wet and dry etching.

13.7 How is silicon nitride used in oxidation?

13.8 What are the purposes of prebaking and postbaking in lithography?

13.9 Define selectivity and isotropy and their importance in relation to etching.

13.10 What do the terms *linewidth* and *registration* refer to?

13.11 Compare diffusion and ion implantation.

13.12 What is the difference between evaporation and sputtering?

13.13 What is the definition of yield? How important is yield?

13.14 What is accelerated life testing? Why is it practiced?

13.15 What do BJT and MOSFET stand for?

13.16 Explain the basic process of surface micromachining.

13.17 What is LIGA? What are its advantages?

13.18 What is the difference between isotropic and anisotropic etching?

13.19 What is a mask? Of what materials is it composed?

13.20 What is the difference between chemically reactive ion etching and dry plasma etching?

13.21 Which process(es) in this chapter allow(s) fabrication of products from polymers?

13.22 What is a PCB?

13.23 Explain the process of thermosonic stitching.

13.24 What is the difference between a die, a chip, and a wafer?

13.25 Why are flats machined onto silicon wafers?

13.26 What is a via? What is its function?

13.27 What is a flip chip? What are its advantages over a surface-mount device?

13.28 Explain how IC packages are attached to a printed circuit board if both sides will contain ICs.

13.29 In a horizontal epitaxial reactor (see the accompanying figure), the wafers are placed on a stage (susceptor) that is tilted by a small amount, usually 1°–3°. Why is this procedure done?

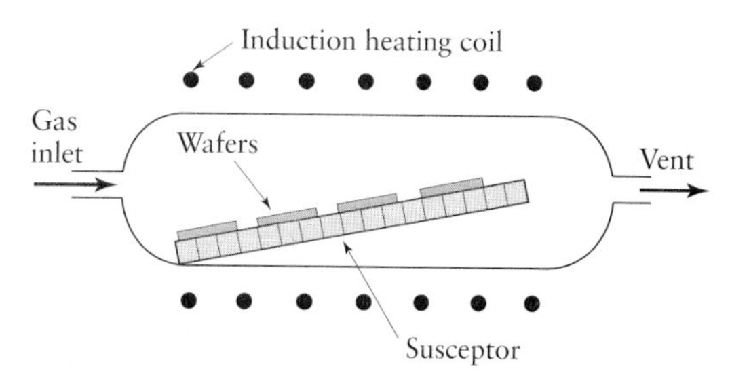

13.30 The accompanying table describes three changes in the manufacture of a wafer: increase of the wafer diameter, reduction of the chip size, and increase of the process complexity. Complete the table by filling in the words "increase," "decrease," or "no change" to indicate the effect that each change would have on the wafer yield and on the overall number of functional chips.

Effects of manufacturing changes

Change	Wafer yield	Number of functional chips
Increase wafer diameter		
Reduce chip size		
Increase process complexity		

13.31 The speed of a transistor is directly proportional to the width of its polysilicon gate, with a narrower gate resulting in a faster transistor, and a wider gate resulting in a slower transistor. With the knowledge that the manufacturing process has a certain variation for the gate width, say $\pm 0.1\ \mu m$, how might a designer alter the gate size of a critical circuit in order to minimize its speed variation? Are there any penalties for making this change? Explain.

13.32 A common problem in ion implantation is channeling, in which the high-velocity ions travel deep into the material through channels along the crystallographic planes before finally being stopped. What is one simple way to stop this effect?

13.33 The MEMS devices discussed in this chapter apply macroscale machine elements, such as spur gears, hinges, and beams. Which of the following machine elements can and cannot be applied to MEMS, and why?

a. ball bearing;
b. bevel gears;
c. worm gears;
d. cams.
e. helical springs;
f. rivets;
g. bolts;

13.34 Figure 13.5b shows the Miller indices on a wafer of (100) silicon. Referring to Fig. 13.4, identify the important planes for the other wafer types illustrated in Fig. 13.5a.

13.35 Referring to Fig. 13.16, sketch the holes generated from a circular mask.

13.36 Explain how you would produce a spur gear if its thickness were one tenth its diameter and its diameter were (a) 10 μm, (b) 100 μm, (c) 1 mm, (d) 10 mm, and (e) 100 mm.

PROBLEMS

13.37 A certain wafer manufacturer produces two equal-sized wafers, one containing 500 chips and the other containing 300 chips. After testing, it is observed that 50 chips on each wafer are defective. What are the yields of the two wafers? Can any relationship be drawn between chip size and yield?

13.38 A chlorine-based polysilicon etch process displays a polysilicon:resist selectivity of 4:1 and a polysilicon:oxide selectivity of 50:1 How much resist and exposed oxide will be consumed in etching 3500 Å of polysilicon? What would the polysilicon:oxide selectivity have to be in order to lose only 40 Å of exposed oxide?

13.39 During a processing sequence, four silicon-dioxide layers are grown by oxidation: 4000 Å, 1500 Å, 400 Å, and 150 Å. How much of the silicon substrate is consumed?

13.40 A certain design rule calls for metal lines to be no less than 2 μm wide. If a 1-μm-thick metal layer is to be wet etched, what is the minimum photoresist width allowed (assuming that the wet etch is perfectly isotropic)? What would be the minimum photoresist width if a perfectly anisotropic dry-etch process were used?

13.41 Obtain mathematical expressions for the etch rate as a function of temperature, using Fig. 13.11.

13.42 If a square mask of side length 100 μm is placed on a {100} plane and oriented with a side in the ⟨110⟩ direction, how long will it take to etch a hole 4 μm deep at 80°C using thylene-diamine/pyrocatechol? Sketch the resulting profile.

13.43 Obtain an expression for the width of the trench bottom as a function of time for the mask shown in Fig. 13.10b.

13.44 Estimate the time of contact and average force when a fluorine atom strikes a silicon surface with a velocity of 1 mm/s. *Hint*: See Eqs. 9.11 and 9.13.

13.45 Calculate the undercut in etching a 10-μm-deep trench if the anisotropy ratio is (a) 200, (b) 2, and (c) 0.5. What is the sidewall slope for these cases?

13.46 Calculate the undercut in etching a 10-μm-deep trench for the wet etchants listed in Table 13.3. What would the undercut be if the mask were silicon oxide?

13.47 Estimate the time required to etch a spur-gear blank from a 75-mm-thick slug of silicon.

13.48 A resist is applied in a resist spinner spun operating at 2000 rpm, using a polymer resist with viscosity of 0.05 $N \cdot s/m^2$. The measured resist thickness is 1.5 μm. What is the expected resist thickness at 6000 rpm? Assume that $\alpha = 1.0$ in Eq. (13.3).

DESIGN

13.49 The accompanying figure shows the cross-section of a simple *npn* bipolar transistor. Develop a process flow chart to fabricate this device.

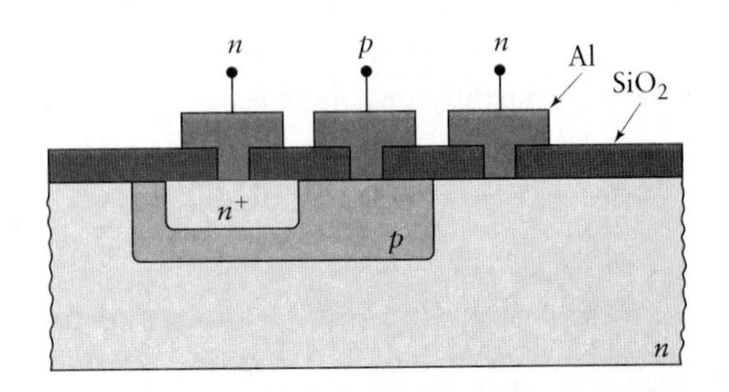

13.50 Referring to the MOS transistor cross-section in the accompanying figure and the given table of design rules, what is the smallest obtainable transistor size *W*? Which design rules, if any, have no impact on *W*?

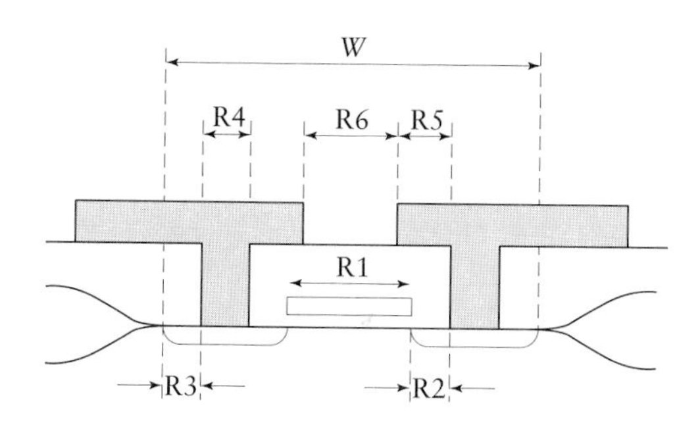

Table of Design Rules

Rule #	Rule name	Value (μm)
R1	Minimum polysilicon width	0.50
R2	Minimum poly-to-contact spacing	0.15
R3	Minimum enclosure of contact by diffusion	0.10
R4	Minimum contact width	0.60
R5	Minimum enclosure of contact by metal	0.10
R6	Minimum metal-to-metal spacing	0.80

13.51 The accompanying figure shows a mirror that is suspended on a torsional beam and that can be inclined through electrostatic attraction by applying a voltage on either side of the mirror at the bottom of the trench. Make a flow chart of the manufacturing operations required to produce this device.

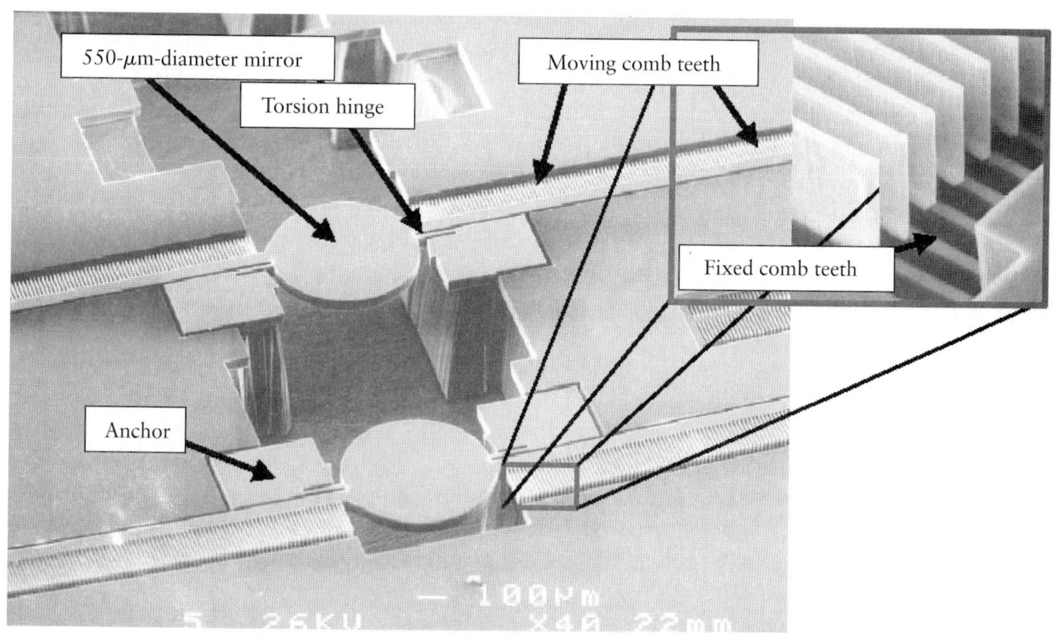

Source: R. Muller, University of California at Berkeley.

13.52 Referring to Fig. 13.30, design an experiment to find the critical dimensions of an overhanging cantilever that will not stick to the substrate.

13.53 Explain how you would manufacture the device shown in Fig. 13.27.

CHAPTER

14 Automation of Manufacturing Processes and Systems

14.1 Introduction

Until the early 1950s, most manufacturing operations were carried out on traditional machinery, such as lathes, milling machines, and presses, which required skilled labor and lacked flexibility. Each time a different product was manufactured, the machinery had to be retooled, and the movement of materials had to be rearranged. Furthermore, the development of new products and parts with complex shapes required numerous trial-and-error attempts by the operator in order to set the proper processing parameters on the machine. Also, because of the human involvement, making parts that were exactly alike was difficult and time consuming. These circumstances meant that processing methods were generally inefficient and that labor costs were a significant portion of the overall production costs. The necessity for reducing the labor share of product cost gradually became apparent, as did the need to improve the efficiency and flexibility of manufacturing operations.

Productivity also became a major concern. Defined as the optimum use of all resources, such as materials, energy, capital, labor, and technology, or as output per employee per hour, productivity basically measures operating efficiency. **Mechanization** refers to the use of various mechanical, hydraulic, pneumatic, or electrical devices to run a process or operation. In mechanized systems, the operator still directly controls the particular process and must check each step of the machine's performance. For example, if a cutting tool breaks during machining, parts are overheated during heat treatment, surface finish begins to deteriorate during grinding, or dimensional tolerances become too large in sheet-metal forming, the operator must intervene and change one or more of the relevant process parameters.

Mechanization of machinery and operations had, by and large, reached its peak by the 1940s, although with rapid advances in the science and technology of manufacturing, the efficiency of manufacturing operations began to improve, and the percentage of total cost represented by labor costs began to decline. The next step in improving productivity was **automation**, from the Greek word *automatos*, meaning self-acting; this word was coined in the mid-1940s by the U.S. automobile industry to indicate automatic handling and processing of parts in production machines. During the past six decades, major advances and breakthroughs in the types and in the extent of automation have occurred, made possible largely through rapid advances in the capacity and sophistication of **computers and control systems**.

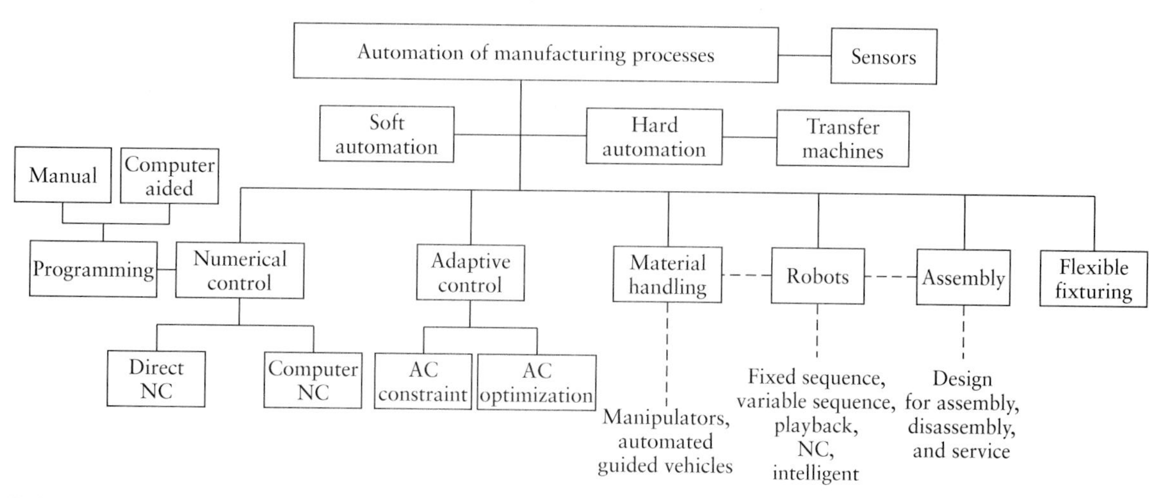

FIGURE 14.1 Outline of topics described in this chapter.

This chapter follows the outline shown in Fig. 14.1. First, it reviews the history and principles of automation and how they have helped us integrate various operations and activities in manufacturing plants to improve productivity. Then, it introduces the important concept of control of machines and systems through **numerical control** and **adaptive-control** techniques. Next, an essential aspect of manufacturing is discussed: material handling—that is, the movement of raw materials and parts in various stages of completion throughout the plant. Material handling has been developed into various systems, particularly those that include the use of **industrial robots.** The subject of **sensor technology** is then described; this technology is an essential element in the control and optimization of machinery, processes, and systems. Other developments discussed include **flexible fixturing** and **assembly operations.** These methods enable us to take full advantage of advanced manufacturing technologies, particularly **flexible manufacturing systems** (described in Section 15.10), and of major developments in **computer-integrated manufacturing systems** and their impact on all aspects of manufacturing operations.

14.2 Automation

Automation is generally defined as the process of having machines follow a predetermined sequence of operations with little or no human labor, using specialized equipment and devices that perform and control manufacturing processes. As described in Section 14.8 and in Chapter 15, automation is achieved through the use of a variety of devices, sensors, actuators, and specialized equipment that are capable of (a) observing all aspects of the manufacturing operation, (b) making decisions concerning the changes that should be made, and (c) controlling all aspects of the operation.

Automation is an *evolutionary* rather than a revolutionary concept. In manufacturing plants, it has been implemented successfully in the following basic areas of activity:

- **Manufacturing processes.** Machining, grinding, forging, cold extrusion, casting, and plastics molding are typical examples of processes that have been extensively automated.
- **Material handling.** Materials and parts in various stages of completion are moved throughout a plant by computer-controlled equipment, with little or no human guidance.

- **Inspection.** Parts are automatically inspected for quality, dimensional accuracy, and surface finish, especially as they are being made.
- **Assembly.** Individually manufactured parts are automatically assembled into subassemblies and, finally, into products.
- **Packaging.** Products are packaged automatically.

14.2.1 Evolution of automation

Some metalworking processes were developed as early as 4000 BC (see Table 1.1); it was not, however, until the beginning of the Industrial Revolution, in the 1750s, that automation began to be introduced in the production of goods. Machine tools, such as turret lathes, automatic screw machines, and automatic bottle-making equipment, were in development starting in the late 1890s. *Mass-production* techniques and transfer machines were developed in the 1920s; these systems had *fixed* automatic mechanisms and were designed to produce *specific products*. These developments were best represented in the automobile industry, which produced passenger cars at high production rates (mass production) and at low cost.

The major breakthrough in automation began with numerical control (NC) of machine tools, in the early 1950s. Since this historic development, rapid progress has been made in automating most aspects of manufacturing (Table 14.1), including the

TABLE 14.1

Developments in the History of Automation and Control of Manufacturing Processes. (See also Table 1.1)

Date	Development
1500–1600	Water power for metalworking; rolling mills for coinage strips.
1600–1700	Hand lathe for wood; mechanical calculator.
1700–1800	Boring, turning, and screw cutting lathe, drill press.
1800–1900	Copying lathe, turret lathe, universal milling machine; advanced mechanical calculators.
1808	Sheet-metal cards with punched holes for automatic control of weaving patterns in looms.
1863	Automatic piano player (Pianola).
1900–1920	Geared lathe; automatic screw machine; automatic bottle-making machine.
1920	First use of the word *robot*.
1920–1940	Transfer machines; mass production.
1940	First electronic computing machine.
1943	First digital electronic computer.
1945	First use of the word *automation*.
1947	Invention of the transistor.
1952	First prototype numerical control machine tool.
1954	Development of the symbolic language APT (Automatically Programmed Tool); adaptive control.
1957	Commercially available NC machine tools.
1959	Integrated circuits; first use of the term *group technology*.
1960	Industrial robots.
1965	Large-scale integrated circuits.
1968	Programmable logic controllers.
1970s	First integrated manufacturing system; spot welding of automobile bodies with robots; microprocessors; minicomputer-controlled robot; flexible manufacturing system; group technology.
1980s	Artificial intelligence; intelligent robots; smart sensors; untended manufacturing cells.
1990–2000s	Integrated manufacturing systems; intelligent and sensor-based machines; telecommunications and global manufacturing networks; fuzzy-logic devices; artificial neural networks; Internet tools; virtual environments; high-speed information systems.

introduction of computers and their use in automation; computerized numerical control (CNC); adaptive control (AC); industrial robots; and computer-integrated manufacturing (CIM) systems, including computer-aided design, engineering, and manufacturing (CAD/CAE/CAM).

14.2.2 Goals of automation

Automation has the following primary goals:

a) **Integrate** various aspects of manufacturing operations so as to improve product quality and uniformity, minimize cycle times and effort involved, and, thus, reduce labor costs.
b) **Improve productivity** by reducing manufacturing costs through better control of production: Raw materials and parts are loaded, fed, and unloaded on machines faster and more efficiently; machines are used more effectively; and production is organized more effectively.
c) **Improve quality** by improving the repeatability of manufacturing processes.
d) **Reduce human involvement**, boredom, and thus the possibility of human error.
e) **Reduce workpiece damage** caused by manual handling of parts.
f) **Economize on floor space** in the manufacturing plant by arranging machines, material movement, and auxiliary equipment more efficiently.
g) **Raise the level of safety** for personnel, especially under hazardous working conditions.

Automation and production quantity. Production quantity is crucial in determining the type of machinery and the level of automation required to produce parts economically. Let's first define some basic production terms. **Total production quantity** is defined as the total number of parts to be made; this quantity can be produced in individual batches of various **lot sizes**. Lot size greatly influences the economics of production, as described in Chapter 16. **Production rate** is defined as the number of parts produced per unit time—for example, per day, per month, or per year. The approximate and generally accepted ranges of production volumes are shown in Table 14.2 for some typical applications. As expected, **experimental** and **prototype** products represent the lowest volume. (See also Section 10.12.)

Small quantities per year can be manufactured in **job shops** (Fig. 14.2), using various standard general-purpose machine tools (**stand-alone machines**) or **machining centers** (see Section 8.10). These operations have high part variety, meaning that different parts can be produced in a short time without extensive changes in tooling and in operations. On the other hand, machinery in job shops generally requires skilled labor, and the production quantity and rate are low; as a result, the cost per part can

TABLE 14.2

Approximate Annual Volume of Production

Type of production	Number produced	Typical products
Experimental or prototype	1–10	All types
Piece or small batch	<5000	Aircraft, machine tools, dies
Batch or high volume	5000–100,000	Trucks, agricultural machinery, jet engines, diesel engines, orthopedic devices
Mass production	100,000+	Automobiles, appliances, fasteners, bottles, food and beverage containers

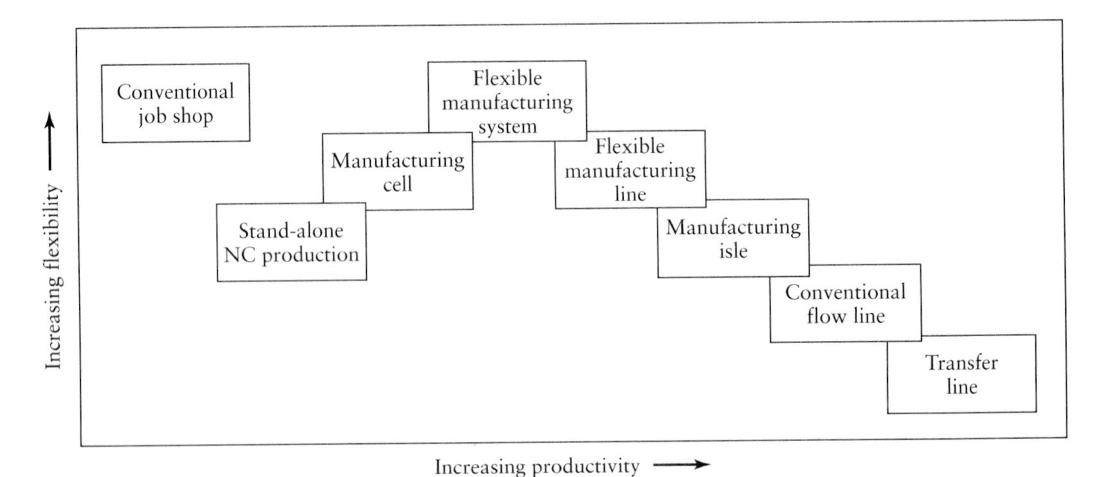

FIGURE 14.2 Flexibility and productivity of various manufacturing systems. Note the overlap between the systems, which is due to the various levels of automation and computer control that are possible in each group. See also Chapter 15 for more details. *Source*: U. Rembold et al., *Computer Integrated Manufacturing and Engineering*, Addison-Wesley, 1993.

be high (Fig. 14.3). When products involve a large labor component, their production is referred to as **labor intensive**. (See also Section 14.12.)

Piece-part production usually involves very small quantities and is suitable for job shops. The majority of piece-part production is in lot sizes of 50 or less. Quantities for **small-batch production** typically range from 10 to 100, using general-purpose machines and machining centers with various computer controls. **Batch production** usually involves lot sizes between 100 and 5000; it utilizes machinery similar to that used for small-batch production, but with specially designed fixtures for higher production rates.

Mass production often involves quantities over 100,000; it requires special-purpose machinery (**dedicated machines**) and automated equipment for transferring materials and parts. Although the machinery, equipment, and specialized tooling are expensive, both the labor skills required and the labor costs are relatively low, because of the high level of automation. However, these production systems are organized for a specific type of product and, hence, lack flexibility. Most manufacturing facilities operate with a variety of machines in combination and with various levels of automation and computer controls.

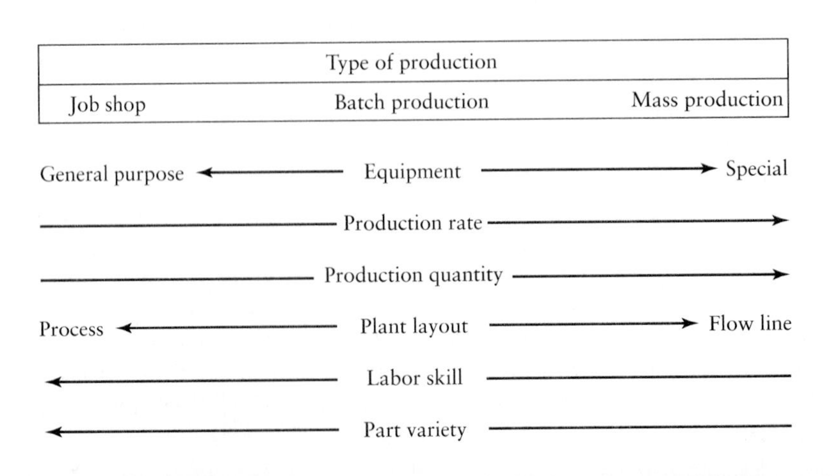

FIGURE 14.3 General characteristics of three types of production methods: job shop, batch production, and mass production.

14.2.3 Applications of automation

Automation can be applied to the manufacture of all types of goods, from raw materials to finished products, and in all types of production, from job shops to large manufacturing facilities. The decision to automate a new or existing facility requires that the following additional considerations be taken into account:

- Type of product manufactured;
- Quantity and the rate of production required;
- Particular phase of the manufacturing operation to be automated;
- Level of skill in the available workforce;
- Reliability and maintenance problems that may be associated with automated systems;
- Economics.

Because automation generally involves a high initial cost of equipment and requires a knowledge of the principles of operation and maintenance, a decision about the implementation of even low levels of automation must involve a study of the true needs of an organization. It is not unusual for a company to begin the implementation of automation with great enthusiasm and with high across-the-board goals, only to discover that the economic benefits of automation largely were illusory rather than real and that, in the final assessment, automation was not cost effective. In many situations, **selective automation**, rather than *total* automation, of a facility is desirable. Generally, the higher the level of skill available in the workforce, the lower is the need for automation, provided that labor costs are justified and that there is a sufficient number of workers available. Conversely, if a manufacturing facility is already automated, the skill level required is lower.

14.2.4 Hard automation

In *hard automation*, also called **fixed-position automation**, the production machines are designed to produce a standardized product, such as an engine block, a valve, a gear, or a spindle. Although product size and processing parameters (such as speed, feed, and depth of cut in machining) can be changed, these machines are specialized and thus lack flexibility. They cannot be modified to any significant extent to accommodate a variety of products with different shapes and dimensions. (See also the discussion of *group technology* in Section 15.8.) Because these machines are expensive to design and construct, their economical use requires the production of parts in very large quantities.

The machines that are used in hard-automation applications are usually fabricated on the **building-block (modular) principle**; they are generally called **transfer machines** and consist of two major components: power-head production units and transfer mechanisms. **Power-head production units** consist of a frame or bed, electric drive motors, gearboxes, and tool spindles and are self-contained; their individual components are commercially available in various standard sizes and capacities. Because of their intrinsic modularity, these units can easily be regrouped to produce a different part and thus have some adaptability and flexibility. Transfer machines consisting of two or more power-head units can be arranged on the shop floor in *linear, circular*, or U *patterns*. The weight and shape of the workpieces influence the arrangement selected. The arrangement is also important for continuity of operation in the event of tool failure or machine breakdown in one or more of the units. **Buffer storage** features are often incorporated in these systems to permit continued operation in the case of such an event.

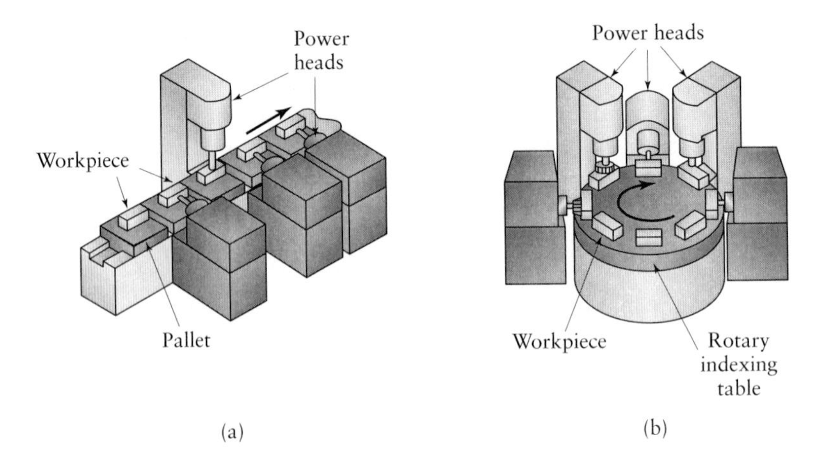

FIGURE 14.4 Two types of transfer mechanisms: (a) straight, and (b) circular patterns.

Transfer mechanisms and transfer lines are used to move the workpiece from one station to another in the machine, or from one machine to another, in order to enable various operations to be performed on the part. Workpieces are transferred by several methods: (1) *rails* along which the parts, usually placed on pallets, are pushed or pulled by various mechanisms (Fig. 14.4a); (2) *rotary indexing tables* (Fig. 14.4b); and (3) *overhead conveyors*. Transfer of parts from station to station is usually controlled by sensors and other devices. Tools on transfer machines can be easily changed in toolholders that have quick-change features. These machines can also be equipped with various automatic gaging and inspection systems; these systems are used between consecutive operations to ensure that a part produced in one station is within acceptable dimensional tolerances before it is transferred to the next station. (As described in Section 14.10, transfer machines are also used extensively in automatic assembly.)

Figure 14.5 shows the **transfer lines**, or **flow lines**, in a very large system for producing cylinder heads for engine blocks, consisting of a number of transfer machines. This system is capable of producing 100 cylinder heads per hour. Note the various machining operations performed: milling, drilling, reaming, boring, tapping, and honing, as well as washing, and gaging.

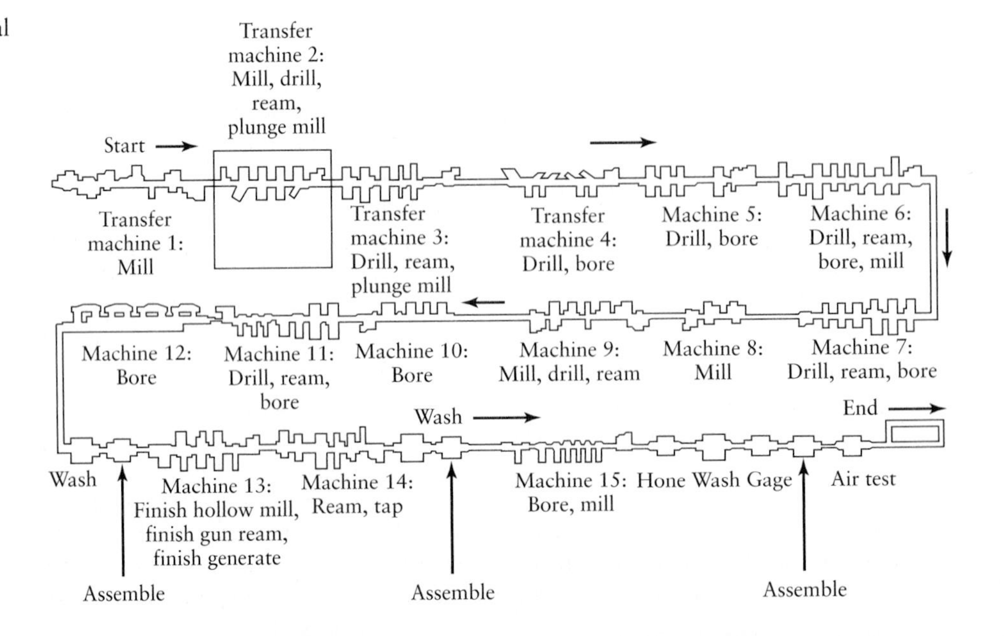

FIGURE 14.5 A traditional transfer line for producing engine blocks and cylinder heads. *Source*: Ford Motor Company.

14.2.5 Soft automation

In contrast to hard automation, *soft automation*, also called *flexible automation* and *programmable automation*, has greater flexibility through the use of computer control of the machine and its functions by various programs; details are given in Sections 14.3 and 14.4. The machine can be easily and readily reprogrammed to produce a part that has a shape or dimensions different than those of the one produced just before it; because of this characteristic, soft automation can produce parts that have complex shapes. Advances in flexible automation have led to the development of **flexible manufacturing systems** (Section 15.10), which have high levels of efficiency and productivity.

14.2.6 Programmable controllers

The control of a manufacturing process in the proper sequence, especially a process that involves groups of machines and material-handling equipment, has traditionally been performed by timers, switches, relays, counters, and similar hard-wired devices based on mechanical, electromechanical, and pneumatic principles. Beginning in 1968, **programmable logic controllers** (PLCs) were introduced to replace these devices. Because PLCs eliminate the need for relay control panels and because they can be reprogrammed and take up less space than relay control panels, they have been widely adopted in manufacturing systems and operations. Their basic functions are (a) on–off motion, (b) sequential operations, and (c) feedback control. PLCs are also used in system control, with high-speed digital-processing and communication capabilities; they perform reliably in industrial environments and improve the overall efficiency of the operation.

Because of advances in numerical control machines, PLCs have become less popular in new installations, but they still represent a very large existing installation base. There is a growing trend toward using microcomputers instead of PLCs, because the former are less expensive, easier to program, and easy to network. This advance has been made possible by the new breeds of "real-time" operating systems such as Windows NT and SCADA (Supervisory Control and Data Acquisition Software).

14.2.7 Total productive maintenance

The management and maintenance of a wide variety of machines, equipment, and systems are among the significant aspects that affect the productivity of a manufacturing organization. Consequently, the concepts of *total productive maintenance* (TPM) and *total productive equipment management* (TPEM) are now being advanced. These concepts include continued analysis of such factors as (a) equipment breakdown and equipment problems; (b) monitoring and improving of equipment productivity; (c) implementation of preventive and predictive maintenance; (d) reduction of setup time, idle time, and cycle time; (d) full utilization of machinery and equipment and the improvement of their effectiveness; and (e) reduction of product defects. Teamwork—for example, as implemented by continuous improvement action teams—is an important component of this activity and involves the full cooperation of machine operators, maintenance personnel, engineers, and management of the organization.

14.3 Numerical Control

Numerical control (NC) is a method of controlling the movements of machine components by directly inserting coded instructions, in the form of numbers and letters, into the system. The system automatically interprets these data and converts them to output signals. These signals, in turn, control various machine components, such as

(a) turning spindles on and off, (b) changing tools, (c) moving the workpiece or the tools along specific paths, and (d) turning cutting fluids on and off.

In order to appreciate the importance of numerical control of machines, let's briefly review how a process such as machining has traditionally been carried out. After studying the working drawings of a part, the operator sets up the appropriate process parameters, determines the sequence of the machining operations to be performed, clamps the workpiece in a workholding device (such as a chuck or collet), and proceeds with the fabrication of the part. Depending on the part's shape and on the dimensional accuracy specified, this approach usually requires skilled operators. The machining procedure followed may depend on the particular operator; because of the possibilities of human error, even parts produced by the same operator may not all be identical. Part quality may, therefore, depend on the particular operator—or, even with the same operator, on the hour of the day or on the day of the week. Because of increased concern with improving product quality and reducing manufacturing costs, such variability and its adverse effects on product quality are no longer acceptable; this situation can be eliminated by numerical control of the machining operation.

The importance of numerical control can be further illustrated by the following example: Assume that several holes are to be drilled on a part in the positions shown in Fig. 14.6. In the traditional manual method of machining this part, the operator positions the drill bit with respect to the workpiece, using reference points given by any of the three methods shown in the figure. The operator then proceeds to drill the holes. Let's first assume that 100 parts, all having exactly the same shape and dimensional accuracy, are to be drilled. Obviously, this operation is going to be tedious, because the operator has to go through the same motions repeatedly. Moreover, the probability is high that, for a variety of reasons, some of the parts machined will be different than others. Let's now assume that during this production run, the order for these parts is changed, and 10 of the parts now require holes in different positions. The machinist now has to reposition the worktable; this operation will thus be time consuming and is also subject to error.

Such operations can be performed easily by numerical control machines that are capable of producing parts repeatedly and accurately and of handling different parts by simply loading different part programs, as described in Section 14.4. Data concerning all aspects of the machining operation can now be stored on hard disks; specific information is then relayed to the machine tool's control panel. On the basis of input information, relays and other devices can be actuated to obtain a desired machine setup. Complex operations, such as turning a part that has various contours or die sinking in a milling machine, can be carried out easily.

14.3.1 Computer numerical control

In the next step in the development of numerical control, the control hardware (mounted on the NC machine) was converted to *local* computer control by software. Two types of computerized systems have been developed: direct numerical control and computer numerical control.

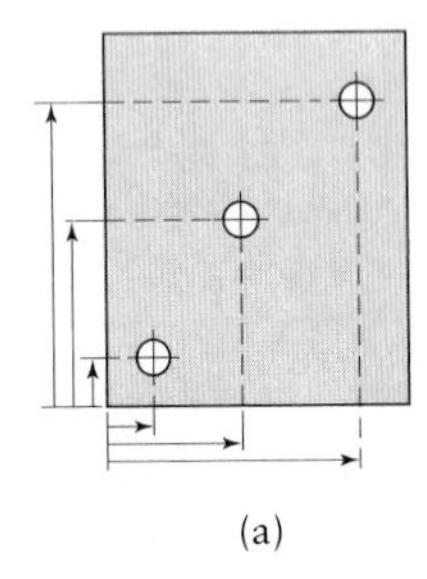
(a)

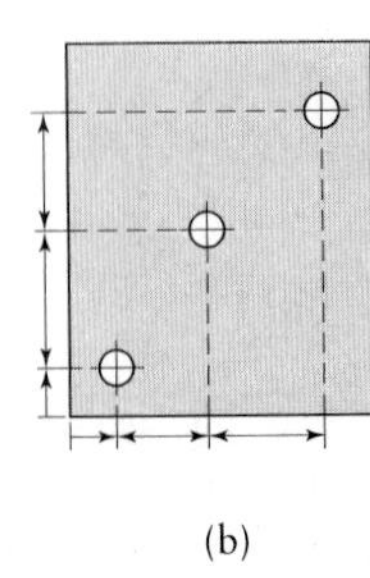
(b)

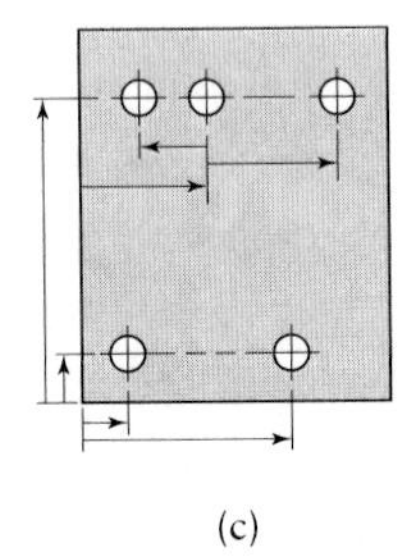
(c)

FIGURE 14.6 Positions of drilled holes in a workpiece. Three methods of measurements are shown: (a) absolute dimensioning, referenced from one point at the lower left of the part; (b) incremental dimensioning, made sequentially from one hole to another; and (c) mixed dimensioning, a combination of both methods.

In **direct numerical control** (DNC), several machines are directly controlled, step by step, by a central mainframe computer. In this system, the operator has access to the central computer through a remote terminal, and the status of all machines in a manufacturing facility can be monitored and assessed from the central computer. However, DNC has a crucial disadvantage: If the computer shuts down, all the machines become inoperative. **Distributed numerical control** covers the use of a central computer serving as the control system over a number of individual computer numerical control machines that have onboard microcomputers. This system provides large memory and computational capabilities and offers flexibility while overcoming the disadvantage of direct numerical control.

Computer numerical control is a system in which a control microcomputer is an integral part of a machine or piece of equipment (onboard computer). The part program may be prepared at a remote site by the programmer, and it may incorporate information obtained from drafting software packages and from machining simulations, in order to ensure that the part program is bug free. The machine operator can, however, easily and manually program onboard computers; the operator can modify the programs directly, prepare programs for different parts, and store the programs. Because of the availability of small computers that have a large memory, microprocessor(s), and program-editing capabilities, as well as the increased flexibility, accuracy, and versatility of such computers, CNC systems are widely used today.

14.3.2 Principles of NC machines

Fig. 14.7 depicts the basic elements and operation of a typical NC machine. The functional elements involved are:

a) **Data input:** The numerical information is read and stored in computer memory.

b) **Data processing:** The programs are read into the machine control unit for processing.

c) **Data output:** This information is translated into commands (typically pulsed commands) to the servomotor (Fig. 14.8). The servomotor then moves the table (onto which the workpiece is mounted) to specific positions (through linear or rotary movements) by means of stepping motors, lead screws, and other devices.

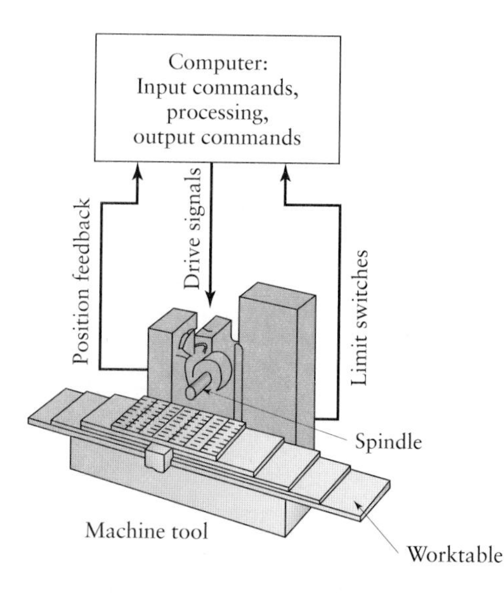

FIGURE 14.7 Schematic illustration of the major components of a numerical control machine tool.

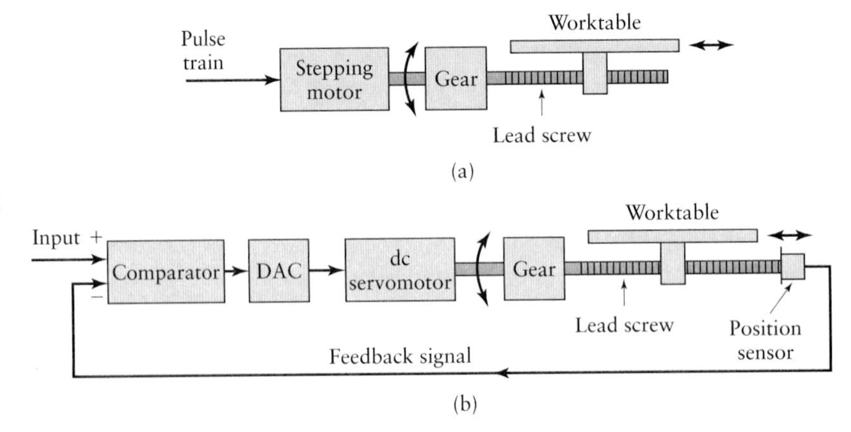

FIGURE 14.8 Schematic illustration of the components of (a) an open-loop, and (b) a closed-loop control system for a numerical control machine. DAC means digital-to-analog converter.

Types of control circuits. In the **open-loop** system (Fig. 14.8a), the signals are sent to the servomotor by the controller, but the movements and final positions of the worktable are not checked for accuracy. In contrast, the **closed-loop** system (Fig. 14.8b) is equipped with various transducers, sensors, and counters that accurately measure the position of the worktable. Through **feedback control**, the position of the worktable is compared against the signal; table movements terminate when the proper coordinates are reached. The closed-loop system is more complicated and more expensive than the open-loop system.

Position measurement in NC machines can be accomplished through direct or indirect methods. In *direct measuring systems*, a sensing device reads a graduated scale on the machine table or slide for linear movement (Fig. 14.9a). This system is more accurate than those used for indirect methods, because the scale is built into the machine, and backlash (the play between two adjacent mating gear teeth) in the mechanisms is not significant. In *indirect measuring systems*, **rotary encoders** or **resolvers** (Figs. 14.9b and c) convert rotary movement to translation movement. In this system, backlash can significantly affect measurement accuracy. Position feedback mechanisms use various sensors that are based mainly on magnetic and photoelectric principles.

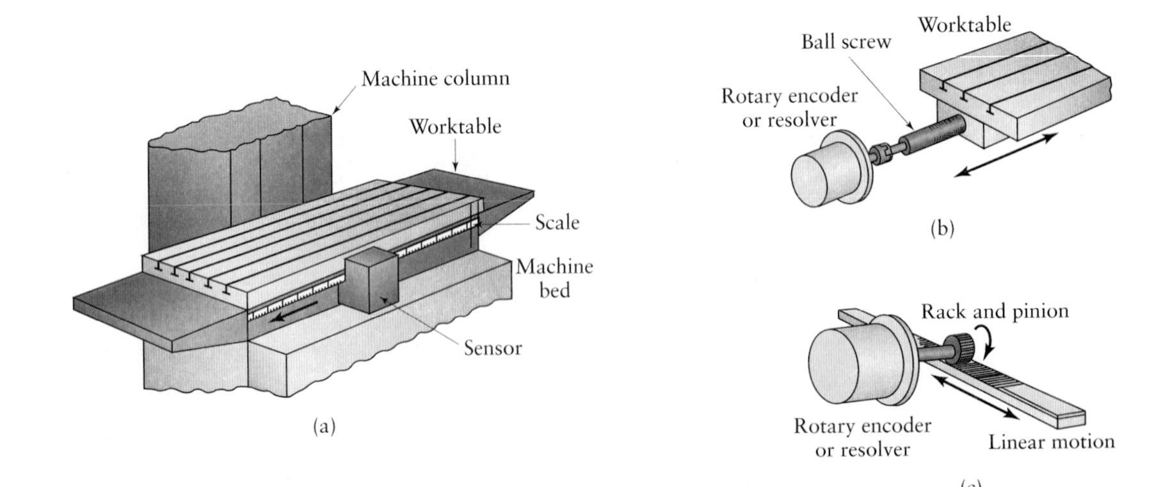

FIGURE 14.9 (a) Direct measurement of the linear displacement of a machine-tool worktable. (b) and (c) Indirect measurement methods.

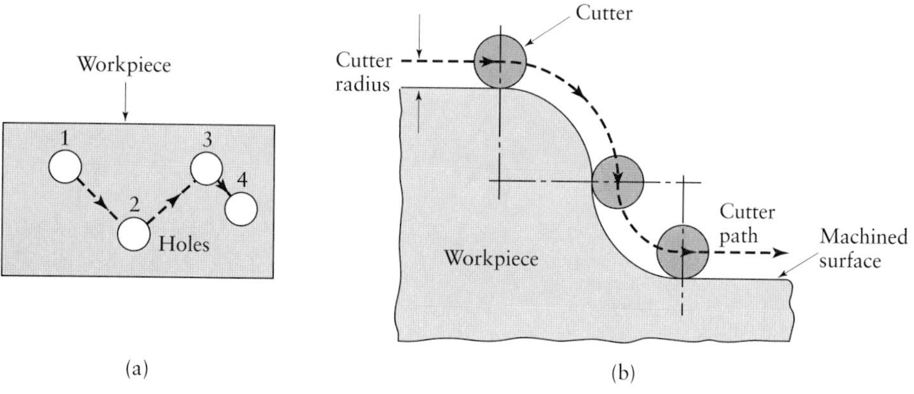

FIGURE 14.10 Movement of tools in numerical control machining. (a) Point-to-point system: The drill bit drills a hole at position 1, is then retracted and moved to position 2, and so on. (b) Continuous path by a milling cutter. Note that the cutter path is compensated for by the cutter radius. This path can also compensate for cutter wear.

14.3.3 Types of control systems

There are two basic types of control systems in numerical control: point-to-point and contouring. These types of control systems are described as follows:

a) In the **point-to-point system**, also called the **positioning system**, each axis of the machine is driven separately by lead screws and, depending on the type of operation, at different velocities. The machine moves initially at maximum velocity in order to reduce nonproductive time, but decelerates as the tool approaches its numerically defined position. Thus, in operations such as drilling or punching a hole, the positioning and cutting take place *sequentially* (Fig. 14.10a). After the hole is drilled or punched, the tool retracts upward and moves rapidly to another position, and the operation is repeated.

b) In the **contouring system**, also known as the **continuous-path system**, the positioning and the operations are *both* performed along controlled paths, but at different velocities. Because the tool performs as it travels along a prescribed path (Fig. 14.10b), accurate control and synchronization of velocities and movements are important. The contouring system is typically used on lathes, milling machines, grinders, welding machinery, and machining centers.

Interpolation. Movement along a path (*interpolation*) occurs *incrementally* by one of several basic methods (Fig. 14.11). Figure 14.12 illustrates examples of actual paths in drilling, boring, and milling operations. In all interpolations, the path controlled is that of the *center of rotation* of the tool. Compensation for different types of tools, different diameters of tools, or for tool wear during machining can be made in the NC program.

1. In **linear interpolation**, the tool moves in a straight line from start to end (Fig. 14.11a), on two or three axes. Theoretically, all types of profiles can be produced by this method, by making the increments between the points small (Fig. 14.11b); however, a large amount of data has to be processed in order to do so.

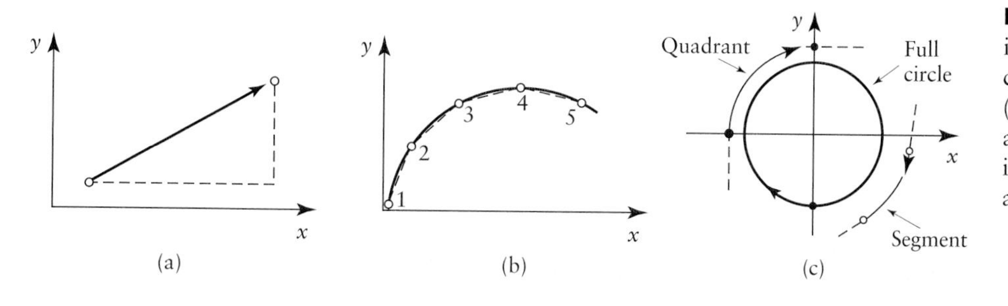

FIGURE 14.11 Types of interpolation in numerical control: (a) linear; (b) continuous path approximated by incremental straight lines; and (c) circular.

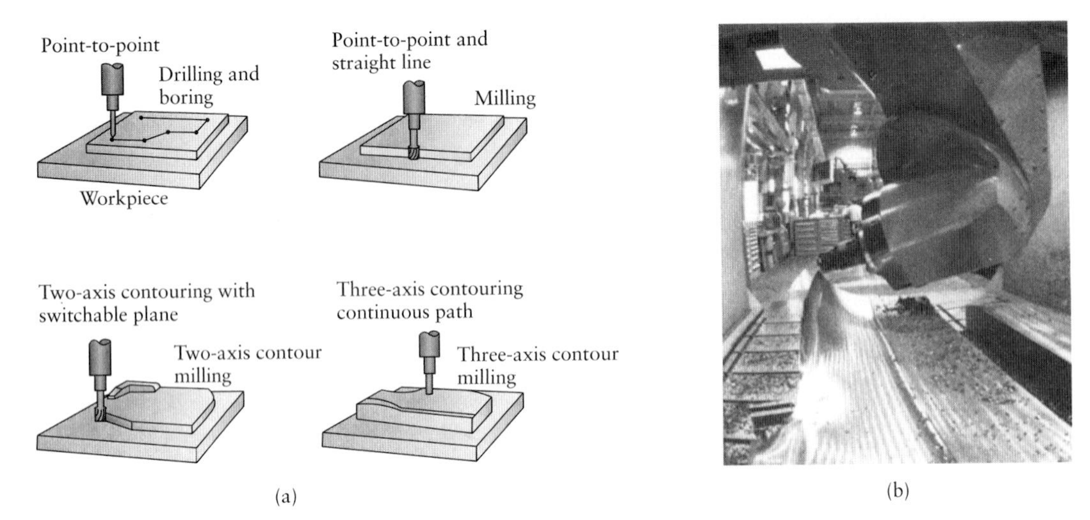

FIGURE 14.12 (a) Schematic illustration of drilling, boring, and milling operations with various cutter paths. (b) Machining a sculptured surface on a five-axis numerical control machine. *Source*: The Ingersoll Milling Machine Co.

2. In **circular interpolation** (Fig. 14.11c), the inputs required for the tool path are the coordinates of the end points, the coordinates of the center of the circle and its radius, and the direction of the tool along the arc.
3. In **parabolic interpolation** and **cubic interpolation**, the tool path is approximated by curves, using higher order mathematical equations. This method is effective in five-axis machines and is particularly useful in die-sinking operations for sheet forming of automotive bodies. These interpolations are also used for the movements of industrial robots. (See Section 14.7.)

14.3.4 Positioning accuracy of NC machines

Positioning accuracy of numerical control machines is defined by how accurately the machine can be positioned to a certain coordinate system; an NC machine usually has a positioning accuracy of at least $\pm$ 3 μm (0.0001 in.). **Repeatability** is defined as the closeness of agreement of repeated position movements under the same operating conditions of the machine; it is usually about $\pm$ 8 μm (0.0003 in.). **Resolution** is defined as the smallest increment of motion of the machine components; it is usually about 2.5 μm (0.0001 in.).

The *stiffness* of the machine tool (see Section 8.11) and elimination of the *backlash* in its gear drives and lead screws are essential for dimensional accuracy. Although in older machines, backlash has been eliminated with special backlash take-up circuits (whereby the tool always approaches a particular position on the workpiece from the same direction), backlash in modern machines is eliminated by means such as using preloaded ball screws. Also, rapid response to command signals requires that friction and inertia be minimized, such as, for example, by reducing the mass of moving components of the machine.

14.3.5 Advantages and limitations of numerical control

Numerical control has the following advantages over conventional methods of machine control:

1. Flexibility of operation is improved, as well as the ability to produce complex shapes with good dimensional accuracy, good repeatability, reduced scrap loss, high production rates, high productivity, and high product quality.

2. Tooling costs are reduced, because templates and other fixtures are not required.
3. Machine adjustments are easy to make with microcomputers and digital readouts.
4. More operations can be performed with each setup, and the lead time for setup and machining required is less as compared with that for conventional methods; furthermore, design changes are facilitated, and inventory is reduced.
5. Programs can be prepared rapidly, and they can be recalled at any time, by using microprocessors; less paperwork is involved.
6. Faster prototype production is possible.
7. Operator skill required is less than that for a qualified machinist, and the operator has more time to attend to other tasks in the work area.

The major limitations of NC are (a) the relatively high initial cost of the equipment, (b) the need for and cost of programming and computer time, and (c) the special maintenance that requires trained personnel. Because NC machines are complex systems, breakdowns can be costly, so preventive maintenance is essential. These limitations are often easily outweighed by the overall economic advantages of NC.

14.4 Programming for Numerical Control

A program for numerical control consists of a sequence of directions that causes an NC machine to carry out a certain operation, machining being the most common process. *Programming for NC* may be performed by an internal programming department, be done on the shop floor, or be purchased from an outside source. The program contains instructions and commands. *Geometric instructions* pertain to relative movements between the tool and the workpiece. *Processing instructions* concern spindle speeds, feeds, cutting tools, cutting fluids, and so on. *Travel instructions* pertain to the type of interpolation and to the speed of movement of the tool or the worktable. *Switching instructions* relate to the on–off position for coolant supplies, direction or lack of spindle rotation, tool changes, workpiece feeding, clamping, and so on.

Manual part programming consists first of calculating the dimensional relationships of the tool, workpiece, and worktable, on the basis of the engineering drawings of the part (including CAD), the manufacturing operations to be performed, and the sequence of the operations. A program sheet is then prepared, detailing the necessary information to carry out the particular operation. The part program is then prepared on the basis of this information. Manual programming can be done by someone who is knowledgeable about the particular manufacturing process and is able to understand, read, and change part programs. Because they are familiar with machine tools and process capabilities, skilled machinists can, with some training in programming, also do manual programming. However, the work involved is tedious, time consuming, and uneconomical; consequently, manual programming is used mostly in simple point-to-point applications.

Computer-aided part programming involves special symbolic **programming languages** that determine the coordinate points of corners, edges, and surfaces of the part. A programming language is a means of communicating with the computer; it involves the use of symbolic characters. The programmer describes the component to be processed in this language, and the computer converts that description to commands for the NC machine. Several programming languages, with various features and applications, are commercially available. The first language that used English-like statements (called **APT**, for **Automatically Programmed Tools**) was developed in the late 1950s; this language, in its various expanded forms, is still the most widely used language for both point-to-point and continuous-path programming.

Complex parts are machined using graphics-based, computer-aided machining programs. A tool path is created in a largely graphic environment that is similar to a CAD program. The machine code (**G-Code**) is created automatically by the program. Before production begins, the programs are verified, either by viewing a simulation of the process on a monitor or by making the part from an inexpensive material, such as aluminum, wood, wax, or plastic, rather than from the actual material specified for the finished part. The use of programming languages (**compilers**) not only results in higher part quality, but also allows for more rapid development of machining instructions. In addition, simulations can be run on remote computer terminals to ensure that the program functions as intended; this method prevents unnecessary use of time on expensive machinery for debugging procedures.

14.5 Adaptive Control

In *adaptive control* (AC), the operating parameters automatically adapt themselves to conform to new circumstances, such as changes in the dynamics of the particular process and any disturbances that may arise. You will readily appreciate that this approach is basically a feedback system. You may recognize that human reactions to occurrences in everyday life already contain dynamic feedback control. For example, driving your car on a smooth road is relatively easy, and you need to make few, if any, adjustments. However, on a rough road, you may have to steer to avoid potholes by visually and continuously observing the condition of the road. Also, your body feels the car's rough movements and vibrations; you then react by changing the direction and the speed of the car to minimize the effects of the rough road and to increase the comfort of your ride. An **adaptive controller** would check these conditions, adapt an appropriate desired braking profile (for example, antilock brake system and traction control), and then use feedback to implement it.

One may contrast dynamic feedback control with adaptive control as follows: Dynamic feedback control has a fixed controller mechanism that adapts or adjusts controller signals in response to measured changes in system behavior. A constant-gain control is a special case of dynamic feedback control, the term **gain** being defined as the ratio of output to input in an amplifier. Adaptive control, on the other hand, adjusts not only the controller signals, but also the controller mechanism. Research in adaptive control was first concerned with the design of autopilots for high-performance aircraft, which operate over a wide range of altitudes and speeds. During tests, it was observed that constant-gain feedback control systems would work well under some operating conditions, but not under others. **Gain scheduling** is perhaps the simplest form of adaptive control. In gain scheduling, a different gain for the feedback is selected, depending on the measured operating conditions. A different gain is assigned to each region of the system's operating space. With advanced adaptive controllers, the gain may vary continuously with changes in operating conditions.

Several adaptive-control systems are commercially available for applications such as ship steering, chemical-reactor control, rolling mills, and medical technology. In manufacturing engineering, the purposes of adaptive control are the following: (a) optimize *production rate*, (b) optimize *product quality*, and (c) minimize *cost*. Although adaptive control has, for some time, been widely used in continuous processing in the chemical industry and in oil refineries, its successful application to manufacturing processes is relatively recent. Application of AC in manufacturing is particularly important in situations where workpiece dimensions and quality are not uniform, such as a poor casting or an improperly heat-treated part.

Adaptive control is a logical extension of computer numerical control systems. As described in Section 14.4, the part programmer sets the processing parameters, based on the existing knowledge of the workpiece material and various data on the

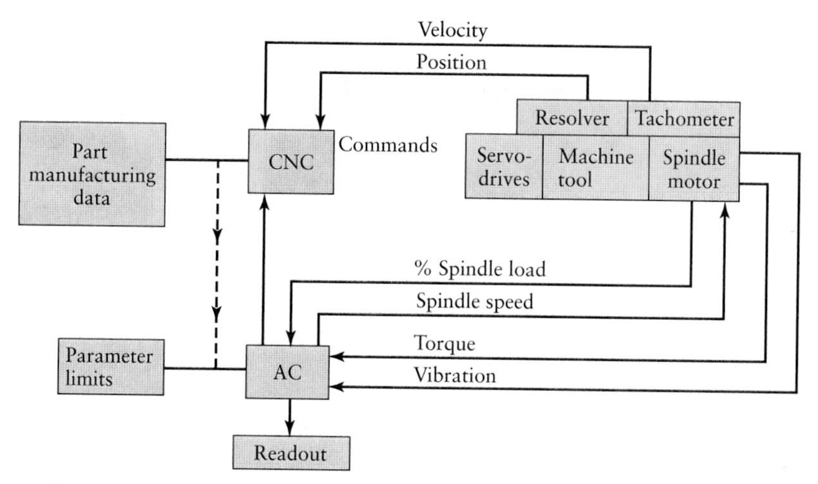

FIGURE 14.13 Schematic illustration of the application of adaptive control (AC) for a turning operation. The system monitors such parameters as cutting force, torque, and vibrations; if they are excessive, it modifies process variables such as feed and depth of cut to bring them back to acceptable levels.

particular manufacturing process. In CNC machines, these parameters are held constant during a particular process cycle. In AC, on the other hand, the system is capable of automatic adjustments during processing, through closed-loop feedback control (Fig. 14.13).

Principles and applications of adaptive control. The following are the basic functions common to adaptive-control systems:

1. Determine the operating conditions of the process, including measures of performance. This information is typically obtained by using sensors that measure process parameters, such as force, torque, vibration, and temperature.
2. Configure the process control in response to the operating conditions. Large changes in the operating conditions may provoke a decision to make a major switch in control strategy. More modest alterations may include the modification of process parameters, such as the speed of operation or of the feed in machining.
3. Continue to monitor the process, making further changes in the controller as needed.

In an operation such as turning on a lathe (Section 8.8.2), the adaptive-control system senses real-time cutting forces, torque, temperature, tool-wear rate, tool chipping or tool fracture, and surface finish of the workpiece. The system then converts this information into commands that modify the process parameters on the machine tool to hold the parameters constant (or within certain limits) or to optimize the cutting operation.

Those systems that place a constraint on a process variable (such as forces, torque, or temperature) are called **adaptive-control constraint** (ACC) systems. Thus, if the thrust force and the cutting force (and hence the torque) increase excessively (because, for example, of the presence of a hard region in a casting), the adaptive-control system changes the speed or the feed to lower the cutting force to an acceptable level (Fig. 14.14). Without adaptive control or without the direct intervention of the operator (as in traditional machining operations), high cutting forces may cause the tools to chip or break or cause the workpiece to deflect or distort excessively. As a result, the dimensional accuracy and surface finish deteriorate. Those systems that optimize an operation are called **adaptive-control optimization** (ACO) systems. Optimization may involve maximizing material-removal rate between tool changes (or resharpening) or improving surface finish. Currently, most systems are based on ACC, because the development and the proper implementation of ACO is complex.

Response time must be short for AC to be effective, particularly in high-speed operations (see Section 8.8.2). Assume, for example, that a turning operation is being

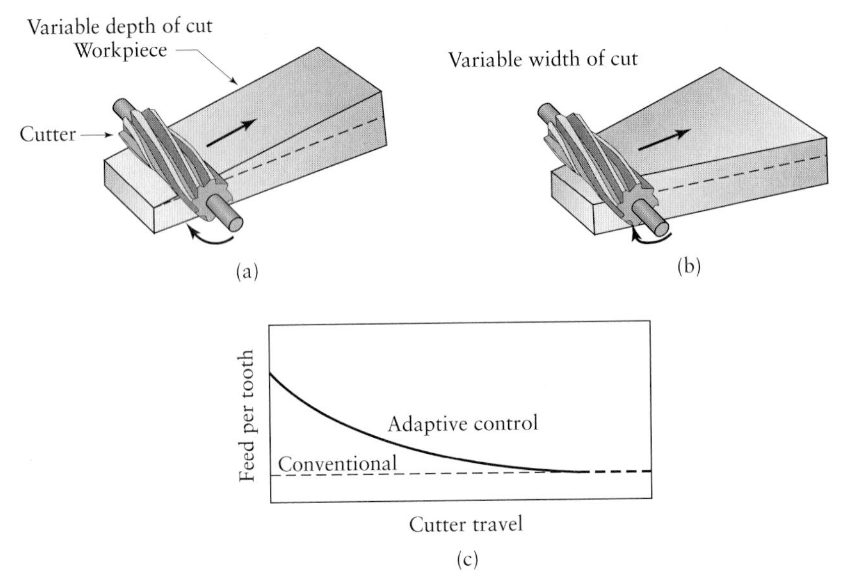

FIGURE 14.14 An example of adaptive control in milling. As the depth of cut or the width of cut increases, the cutting forces and the torque increase. The system senses this increase and automatically reduces the feed to avoid excessive forces or tool breakage, in order to maintain cutting efficiency. *Source*: Y. Koren.

performed on a lathe at a spindle speed of 1000 rpm, and the tool suddenly breaks, adversely affecting the surface finish and dimensional accuracy of the part. In order for the AC system to be effective, the sensing system must respond within a very short time, as otherwise the damage to the workpiece will be extensive.

For adaptive control to be effective in manufacturing operations, *quantitative relationships* must be established and stored in the computer software as mathematical models. For example, if the tool-wear rate in a machining operation is excessive, the computer must be able to decide how much of a change in speed or feed is necessary (and whether to increase it or decrease it) in order to reduce the wear rate to an acceptable level. The system should also be able to compensate for dimensional changes in the workpiece due to such causes as tool wear and temperature rise (Fig. 14.15). If the operation is, for example, grinding (Chapter 9), the computer software must reflect the desired quantitative relationships among process variables (wheel and work speeds, feed, type of wheel) and such parameters as wheel wear, dulling of abrasive grains, grinding forces, temperature, surface finish, and part deflections. Similarly, for the bending of a sheet in a V-die (Section 7.4), data on the dependence of springback on punch travel and on other material and process variables must be stored in the computer memory.

It is apparent that, because of the many factors involved, mathematical equations for such quantitative relationships in manufacturing processes are difficult to establish. Compared to the various other parameters involved, forces and torque in machining have been found to be the easiest to monitor by AC. Various solid-state power controls are commercially available, in which power is displayed or interfaced with data-acquisition systems.

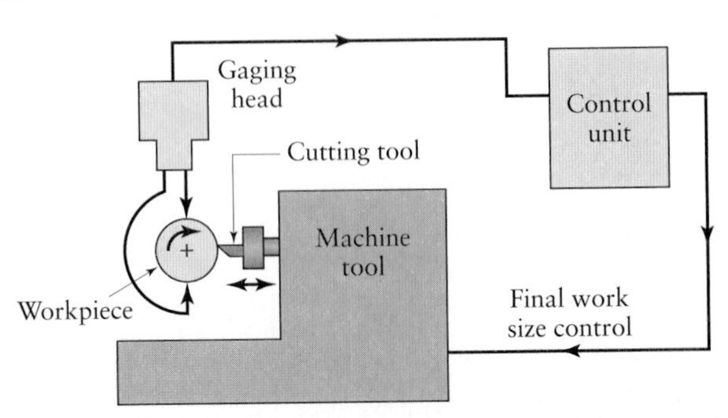

FIGURE 14.15 In-process inspection of workpiece diameter in a turning operation. The system automatically adjusts the radial position of the cutting tool in order to produce the correct diameter.

14.6 | Material Handling and Movement

During a typical manufacturing operation, raw materials and parts are moved from storage to machines, from machine to machine, from inspection to assembly and to inventory, and finally to shipment. For example, a forging is mounted on a milling-machine bed for further processing; the machined forging may subsequently be ground for better surface finish and dimensional accuracy, and it is inspected prior to being assembled into a finished product. Similarly, cutting tools are mounted on lathes, either manually or automatically; dies are placed in presses or hammers; grinding wheels are mounted on spindles; and parts are mounted on special fixtures for dimensional measurement and inspection. *Material handling* is defined as the functions and systems associated with the transportation, storage, and control of materials and parts in the total manufacturing cycle of a product.

Material handling should be repeatable and reliable. Consider, for example, what happens if a part or workpiece is loaded improperly into a die, mold, or the chuck or collet on a lathe. The consequences of such action may well be broken dies and tools or parts that are improperly made or are out of dimensional tolerance. Moreover, this action can also present safety hazards and possibly cause injury to the operator and nearby personnel.

The total amount of time required for manufacturing a part depends on the part's size, complexity of shape, surface finish and dimensional tolerances specified, and the resulting set and sequence of operations required. Depending on the particular operations performed, idle time and the time required for transporting materials can constitute the majority of the production time consumed. **Plant layout** is an important aspect of the orderly flow of materials and components throughout the manufacturing cycle. Obviously, the time and distances required for moving raw materials and parts should be minimized, and storage areas and service centers should be organized accordingly. For parts requiring multiple operations, equipment should be grouped around the operator or the industrial robot. (See the discussion of *cellular manufacturing* in Section 15.9.) Material handling should, therefore, be an integral part of the planning, implementing, and controlling of manufacturing operations.

Important aspects of material handling are discussed as follows:

a) **Methods of material handling.** Several factors have to be considered in selecting a suitable material-handling method for a particular manufacturing operation:

1. Shape, weight, and characteristics of the parts;
2. Types and distances of movements, and the position and orientation of the parts during movement and at their final destination;
3. Conditions of the path along which the parts are to be transported;
4. Level of automation and control desired, and integration with other systems and equipment;
5. Operator skill required;
6. Economic considerations.

For small batch operations, raw materials and parts can be handled and transported by hand, but this method can be time consuming and hence costly. Furthermore, because it involves human beings, this practice can be unpredictable and unreliable; it can even be unsafe to the operator, because of the weight and shape of the parts to be moved and because of environmental factors, such as heat and smoke in older foundries and forging plants.

b) **Equipment.** Several types of equipment can be used to move materials, such as conveyors; rollers; self-powered monorails; carts; forklift trucks; and various mechanical, electrical, magnetic, pneumatic, and hydraulic devices and manipulators. **Manipulators** can be designed to be controlled directly by the operator, or they can be automated for repeated operations (such as the loading and unloading of parts from machine tools, presses, dies, and furnaces). Manipulators are capable of gripping and moving heavy parts and of orienting them as may be required in manufacturing and assembly operations. Machinery combinations that have the capability of conveying parts without the use of additional material-handling equipment are called **integral transfer devices.**

In order to use warehouse space efficiently and to reduce labor costs, **automated storage/retrieval systems** (AS/RS) have been developed. There are several such systems available, with varying degrees of automation and complexity. These systems are currently considered undesirable, because of the current focus on minimal or zero inventory and on just-in-time production methods (see Section 15.11).

c) **Automated guided vehicles.** Flexible material handling and movement, with real-time control, has become an integral part of modern manufacturing. *Automated guided vehicles* (AGVs) are used extensively in flexible manufacturing (Fig. 14.16); this transport system has high flexibility and is capable of random delivery to different workstations. The movements of AGVs can be planned so that they interface with automated storage/retrieval systems.

First developed in the 1950s, AGVs were put into extensive use in the 1970s to reduce direct-labor costs and improve productivity by automation. Although similar in their basic design and operation, there are several types of AGVs commonly in use, with a variety of designs, load-carrying capacities, and features for specific applications. Among these types of AGVs are (a) unit-load vehicles with pallets that can be handled from both sides (see Fig. 14.16); (b) light-load vehicles, equipped with bins or trays for light manufacturing; (c) tow vehicles; (d) forklift-style vehicles; and (e) assembly-line vehicles to carry

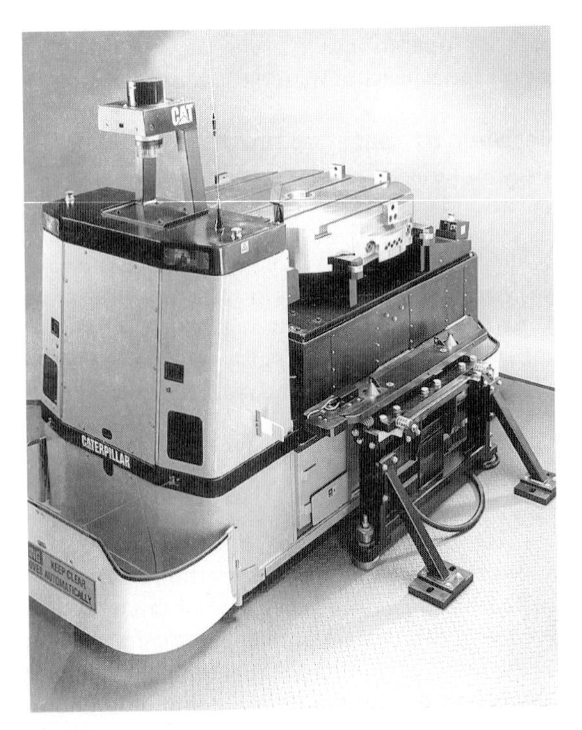

FIGURE 14.16 A self-guided vehicle (Caterpillar Model SGC-M) carrying a machining pallet. The vehicle is aligned next to a stand on the floor. Instead of following a wire or stripe path on the factory floor, this vehicle calculates its own path and automatically corrects for any deviations. *Source*: Courtesy of Caterpillar Industrial, Inc.

subassemblies to final assembly. Pallet sizes vary, and loading and unloading are accomplished either manually or automatically, by using various transfer mechanisms. The load capacity of AGVs can range up to thousands of pounds.

AGVs are guided automatically along pathways with in-floor wiring (for magnetic guidance) or tapes or fluorescent painted strips (for optical guidance, called *chemical guide path*). Some systems may require additional operator guidance. More recent developments include *autonomous guidance* (without any wiring or tapes) and use of various optical, ultrasonic, inertial, and dead-reckoning techniques with onboard controllers. *Routing* of the AGV can be controlled such that the system optimizes the movement of materials and parts in case of congestion around workstations, machine breakdown, or the failure of one section of the manufacturing system. Proper traffic management on the factory floor is thus important; sensors and various controls on the vehicles are designed to ensure that collisions with other AGVs or machines do not occur.

d) **Coding systems.** Various *coding systems* have been developed to locate and identify parts and subassemblies throughout the manufacturing system and to correctly transfer them to their appropriate stations:

1. **Bar coding** is the most widely used and least costly system. The codes are printed on labels, which are attached to the parts themselves and read by fixed or portable bar code readers, using light pens.
2. **Magnetic strips**, such as those on the back of a credit card, constitute the second most common coding system.
3. The third system uses **RF** (radio frequency) **tags.** Although expensive, the tags do not require the clear line of sight needed by the two systems discussed previously; in addition, they have a long range and are rewritable.

Other identification systems are based on acoustic waves, optical character recognition, and machine vision. (See Section 14.8.1.)

14.7 Industrial Robots

The word *robot* was coined in 1920 by the Czech author K. Čapek in his play *R.U.R.* (Rossum's Universal Robots); it is derived from the Czech word *robota*, meaning worker. An *industrial robot* has been defined as a reprogrammable multifunctional manipulator designed to move materials, parts, tools, or other devices by means of variable programmed motions and to perform a variety of other tasks. In a broader context, the term *robot* also includes manipulators that are activated directly by an operator. Introduced in the early 1960s, the first industrial robots were used in hazardous operations, such as the handling of toxic and radioactive materials and the loading and unloading of hot workpieces from furnaces and in foundries. Some simple rule-of-thumb applications for robots are the three D's (*dull, dirty, and dangerous, including demeaning, but necessary, tasks*) and the three H's (*hot, heavy, and hazardous.*) Industrial robots have become important components in manufacturing operations and systems and have greatly helped improve productivity and product quality and reduce labor costs.

14.7.1 Robot components

To appreciate the functions of robot components and their capabilities, we might simultaneously observe the flexibility and capability of the diverse movements of our arm, wrist, hand, and fingers in reaching for and grabbing an object from a shelf, in using a hand tool, or in operating a car or a machine. The basic components of an industrial robot are described as follows (Fig. 14.17a):

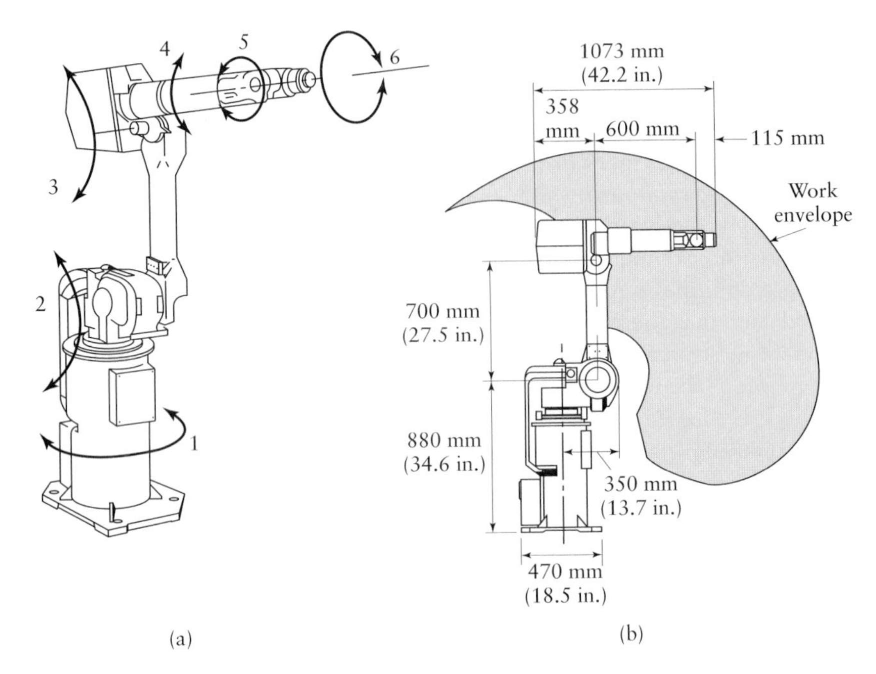

FIGURE 14.17 (a) Schematic of a six-axis S-10 GMF robot. The payload at the wrist is 10 kg (22 lb.) and repeatability is ±0.2 mm (±0.008 in.) The robot has mechanical brakes on all its axes, which are coupled directly. (b) The work envelope of a robot, as viewed from the side. *Source*: GMFanuc Robotics Corporation.

1. **Manipulator.** Also called **arm and wrist,** the *manipulator* is a mechanical unit that provides motions (trajectories) similar to those of a human arm and hand; manipulation is carried out using devices such as linkages, gears, and various joints. The end of the wrist can reach a point in space with a specific set of coordinates and in a specific orientation. Most robots have six rotational joints. (See Fig. 14.17a.) There are also four-degrees-of-freedom (d.o.f.) and five-d.o.f. robots, but these kinds are not fully dexterous, because full dexterity, by definition, requires six d.o.f. Seven-d.o.f. (or *redundant*) robots for special applications are also available.
2. **End Effector.** The end of the wrist in a robot is equipped with an *end effector*, also called **end-of-arm tooling.** End effectors are generally custom made to meet special handling requirements. Mechanical grippers are the most commonly used end effectors and are equipped with two or more fingers. The selection of an appropriate end effector for a specific application depends on such factors as the *payload* (weight of the object to be lifted and moved; load-carrying capacity), environment, reliability, and cost. Depending on the type of operation, conventional end effectors may be equipped with any of the following:

   a) Grippers, hooks, scoops, electromagnets, vacuum cups, and adhesive fingers, for material handling (Fig. 14.18a);
   b) Spray guns, for painting;
   c) Various attachments, such as for spot and arc welding and for arc cutting;
   d) Power tools, such as drills, nut drivers, and burrs;
   e) Measuring instruments, such as dial indicators.

   Compliant end effectors are used to handle fragile materials or to facilitate assembly. They can use elastic mechanisms to limit the force that can be applied to a workpiece or part, and they can be designed with a specific required stiffness. For example, the end effector shown in Fig. 14.18b is stiff in the axial direction, but it is very compliant in lateral directions; this arrangement prevents damage to parts in those assembly operations in which slight misalignments can occur.

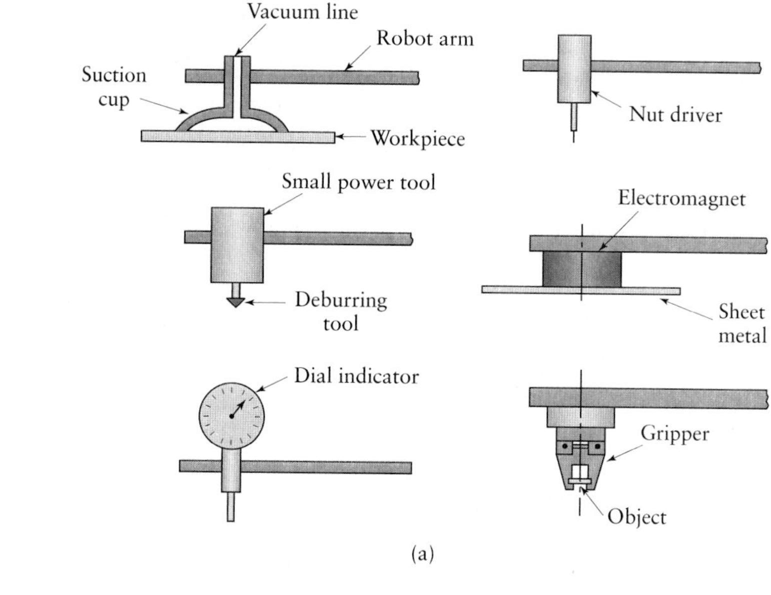

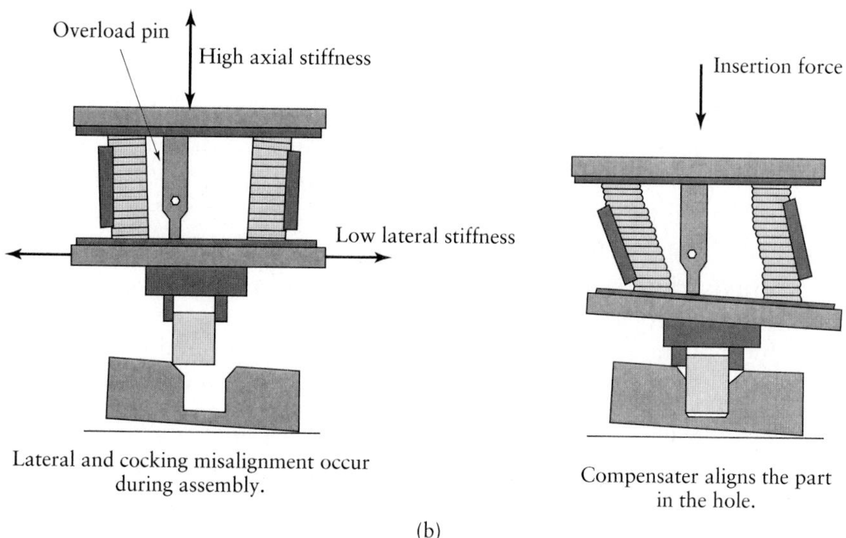

FIGURE 14.18 (a) Various devices and tools attached to end effectors to perform a variety of operations. (b) A system that compensates for misalignment during automated assembly. *Source*: ATI Industrial Automation.

3. **Power supply.** Each motion of the manipulator (in linear and rotational axes) is controlled and regulated by independent actuators that use an electrical, pneumatic, or hydraulic power supply. Each source of energy and each type of motor has its own characteristics, advantages, and limitations.
4. **Control system.** Whereas manipulators and end effectors are the robot's arms and hands, the control system is the *brain* of a robot. Also known as the **controller**, the control system is the communications and information-processing system that gives commands for the movements of the robot; it stores data to initiate and terminate movements of the manipulator. It also acts as the *nerves* of the robot; it interfaces with computers and other equipment such as manufacturing cells or assembly systems. **Feedback devices**, such as transducers, are an important part of the control system. Robots with a *fixed* set of motions have **open-loop control**. In this system, commands are given, and the robot arm goes through its motions; unlike in **closed-loop systems**, where feedback is given, accuracy of the movements is not monitored; consequently, this system does not have a self-correcting capability.

As in numerical control machines, the types of control in industrial robots are *point to point* and *continuous path*. (See also Section 14.3.3.) Depending on the particular task, the *positioning repeatability* required may be as small as 0.050 mm (0.002 in.), as in assembly operations for electronic printed circuitry. Specialized robots can reach such accuracy, although most robots are unable to do so. Accuracy and repeatability vary greatly with payload and with position within the *work envelope* (see Section 14.7.2) and, as such, are very difficult to quantify for most robots.

14.7.2 Classification of robots

Robots may be classified by basic type, as follows (Fig. 14.19):

a) **Cartesian**, or **rectilinear**;
b) **Cylindrical**;
c) **Spherical**, or **polar**;
d) **Articulated**, or **revolute**, **jointed**, or **anthropomorphic**.

Also, robots may be attached permanently to the floor of a manufacturing plant, move along overhead rails (**gantry robots**), or be equipped with wheels to move along the factory floor (**mobile robots**).

1. **Fixed- and variable-sequence robots.** The *fixed-sequence robot* (also called a **pick-and-place robot**) is programmed for a specific sequence of operations. Its movements are from point to point, and the cycle is repeated continuously; these robots are simple and relatively inexpensive. The *variable-sequence robot* can be programmed for a specific sequence of operations, but can be reprogrammed to perform another sequence of operations.
2. **Playback robot.** An operator leads or walks the *playback robot* and its end effector through the desired path; in other words, the operator *teaches* the robot by showing it what to do. The robot memorizes and records the path and sequence of motions and can then repeat them continually without any further action or guidance by the operator. The **teach pendant** uses handheld button boxes that are connected to the control panel; these button boxes are used to control and guide the robot and its tooling through the work to be performed. These movements are then registered in the memory of the controller and are automatically reenacted by the robot whenever needed.
3. **Numerically controlled robot.** The *numerically controlled robot* is programmed and operated much like a numerically controlled machine; the robot is servocontrolled by digital data, and its sequence of movements can be changed with relative ease. As in NC machines, there are two basic types of controls: point to point and continuous path. Point-to-point robots are easy to program and have a higher payload and a larger **work envelope**, also called the **working**

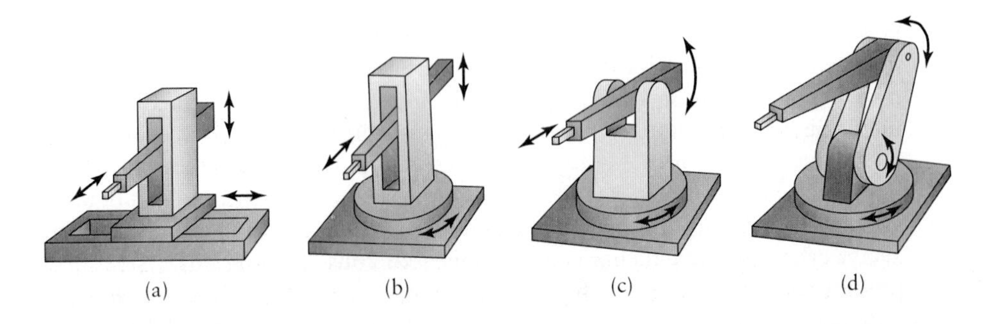

FIGURE 14.19 Four types of industrial robots: (a) Cartesian (rectilinear); (b) cylindrical; (c) spherical (polar); and (d) articulated (revolute, jointed, or anthropomorphic).

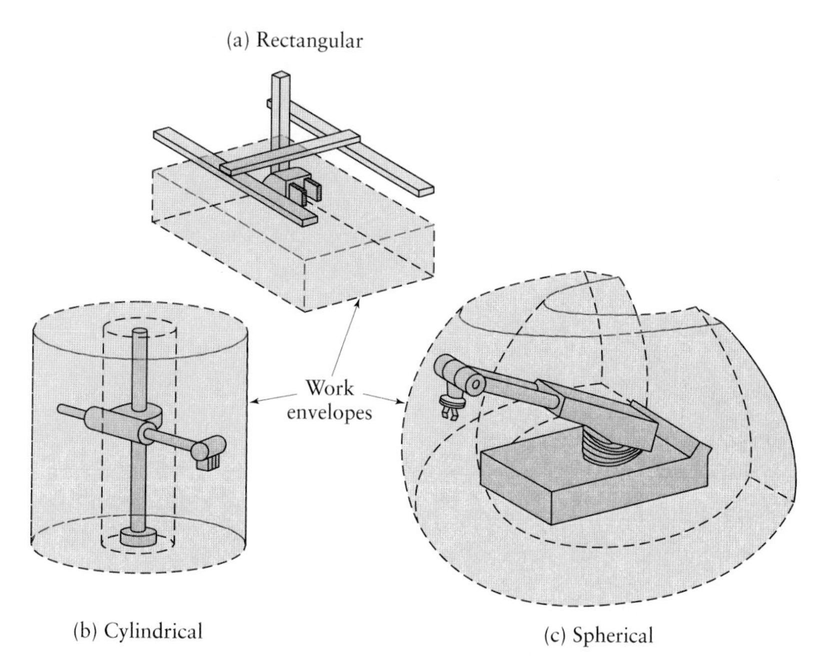

FIGURE 14.20 Work envelopes for three types of robots. The choice depends on the particular application (See also Fig. 14.17b.)

envelope (the maximum extent or reach of the robot hand or working tool in all directions; see Figs. 14.17b and 14.20). Continuous-path robots have a higher accuracy than point-to-point robots, but they have a lower payload. More advanced robots have a complex system of path control, enabling high-speed movements with high accuracy.

4. **Intelligent (sensory) robot.** The *intelligent robot* is capable of performing some of the functions and tasks carried out by human beings; it is equipped with a variety of sensors with *visual* (*computer vision*) and *tactile* (touching) capabilities. (See Section 14.8.) Much like humans, the robot observes and evaluates the immediate environment and its proximity to other objects, especially machinery in its path, by *perception* and *pattern recognition*; it then makes appropriate decisions for the next movement and proceeds accordingly. Because its operation is very complex, powerful computers are required to control this type of robot. Developments in intelligent robots include capabilities so that they will:

   a) Behave more and more like humans, performing tasks such as moving among a variety of machines and equipment on the shop floor and avoiding collisions;

   b) Recognize, select, and properly grip the correct raw material or workpiece;

   c) Transport a part to a machine for further processing or inspection;

   d) Assemble components into subassemblies or the final product.

14.7.3 Applications and selection of robots

Major applications of industrial robots include the following:

1. *Material handling* consists of the loading, unloading, and transferring of workpieces in manufacturing facilities. These operations can be performed reliably and repeatedly with robots, thereby improving product quality and reducing scrap losses. Some examples of material-handling applications are (a) casting and molding operations, in which molten metal, raw materials, lubricants, and parts in various stages of completion are handled without operator interference;

FIGURE 14.21 Spot welding automobile bodies with industrial robots. *Source*: Courtesy of Cincinnati Milacron, Inc.

(b) heat treating, in which parts are loaded and unloaded from furnaces and quench baths; and (c) forming operations, in which parts are loaded and unloaded from presses and various other types of machinery.

2. *Spot welding* by robots unitizes automobile and truck bodies, producing welds of good and reliable quality (Fig. 14.21); robots also perform other, similar operations, such as arc welding, arc cutting, and riveting (see Chapter 12).
3. *Finishing operations* such as deburring, grinding, and polishing (Chapter 9) can be done by using appropriate tools attached to end effectors.
4. *Application of adhesives and sealants* by industrial robots has widespread usefulness—for example, in the automobile frame shown in Fig. 14.22.
5. *Spray painting*, particularly of parts with complex shapes, and cleaning operations are frequent applications of industrial robots, because the motions for one piece repeat very accurately for the next.
6. *Automated assembly*, a very repetitive operation, is performed by industrial robots (Fig. 14.23); see also Section 14.10.
7. *Inspection and gaging* in various stages of manufacture as performed by industrial robots allows speeds that are much higher than those achieved by humans.

Factors that influence the *selection of robots* in manufacturing are as follows: (a) payload, (b) speed of movement, (c) reliability, (d) repeatability, (e) arm configuration, (f) number of degrees of freedom, (g) control system, (h) program memory, and (i) work envelope.

FIGURE 14.22 Sealing joints of an automobile body with an industrial robot. *Source*: Courtesy of Cincinnati Milacron, Inc.

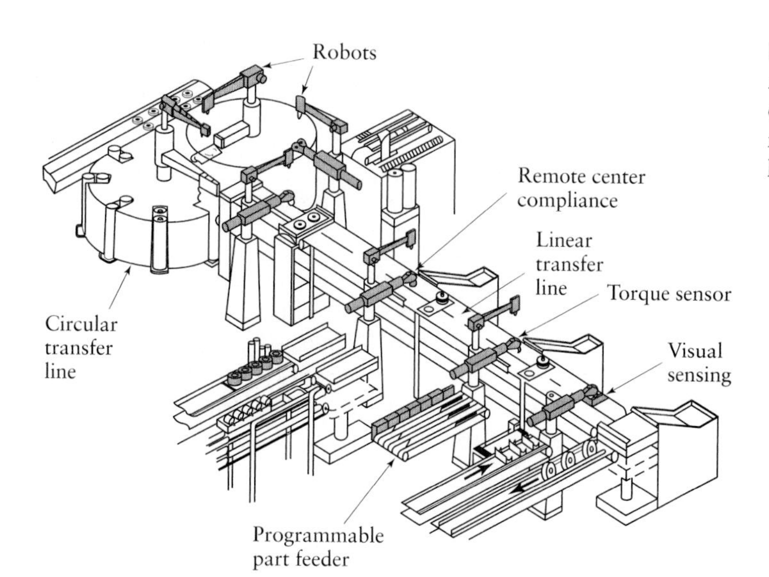

FIGURE 14.23 Automated assembly operations using industrial robots and circular and linear transfer lines.

14.8 Sensor Technology

A *sensor* is a device that produces a signal in response to its detection or measurement of a certain quantity or property, such as position, force, torque, pressure, temperature, humidity, speed, acceleration, or vibration. Traditionally, sensors, actuators, and switches have been used to set limits on the performance of machines, such as stops on machine-tool slideways to restrict worktable movements, pressure and temperature gages with automatic shutoff features, and governors on engines to prevent excessive speed of operation. Sensor technology is an important aspect of manufacturing processes and systems, and sensors are essential for data acquisition, monitoring, communication, and computer control of machines and systems (Fig. 14.24). Because they convert one quantity to another, sensors are also often referred to as **transducers. Analog sensors** produce a signal, such as voltage, that is proportional to the measured quantity; **digital sensors** have numeric or digital outputs that can directly be transferred to computers. **Analog-to-digital converters** (ADCs) are used for interfacing analog sensors with computers.

14.8.1 Sensor classification

Sensors that are of interest in manufacturing may be classified generally as follows:

a) Mechanical sensors, which measure such quantities as position, shape, velocity, force, torque, pressure, vibration, strain, and mass;

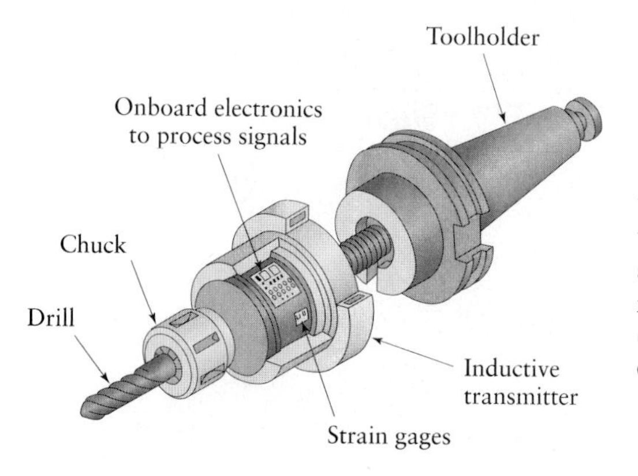

FIGURE 14.24 A tool holder equipped with thrust-force and torque sensors (*smart tool holder*), capable of continuously monitoring the cutting operation. Such tool holders are necessary for adaptive control of manufacturing operations. (See Section 14.5). *Source:* Cincinnati Milacron, Inc.

b) Electrical sensors, which measure voltage, current, charge, and conductivity;

c) Magnetic sensors, which measure magnetic field, flux, and permeability;

d) Thermal sensors, which measure temperature, flux, conductivity, and specific heat;

e) Other types of sensors, such as acoustic, ultrasonic, chemical, optical, radiation, laser, and fiber-optic sensors.

Depending on its application, a sensor may be made of metallic, nonmetallic, organic, or inorganic materials and fluids, gases, plasmas, or semiconductors. Using the special characteristics of these materials, sensors convert the quantity or property measured to analog or digital output. The operation of a common mercury thermometer, for example, is based on the difference between the thermal expansion of mercury and that of glass. Similarly, a machine part or a physical obstruction or barrier in a space can be detected by a break a beam of light, sensed by a photoelectric cell. A proximity sensor, which senses and measures the distance between it and an object or a moving member of a machine, can be based on acoustics, magnetism, capacitance, or optics. Other actuators physically touch the object and take appropriate action, usually by electromechanical means. Sensors are essential to the control of intelligent robots; they continue to be developed with capabilities that resemble those of the senses of human beings (*smart sensors*; see later).

Sensors are also classified as follows:

1. **Tactile sensing** is the continuous sensing of varying contact forces, commonly by an *array* of sensors; such a system is capable of performing within an arbitrary three-dimensional space. Fragile parts, such as glass bottles and electronic devices, can be handled by robots with *compliant* (*smart*) *end effectors*. These effectors can sense the force applied to the object being handled, using piezoelectric devices, strain gages, magnetic induction, ultrasonics, and optical systems of fiber optics and light-emitting diodes. Tactile sensors that are capable of measuring and controlling gripping forces and moments in three axes have been designed (Fig. 14.25).

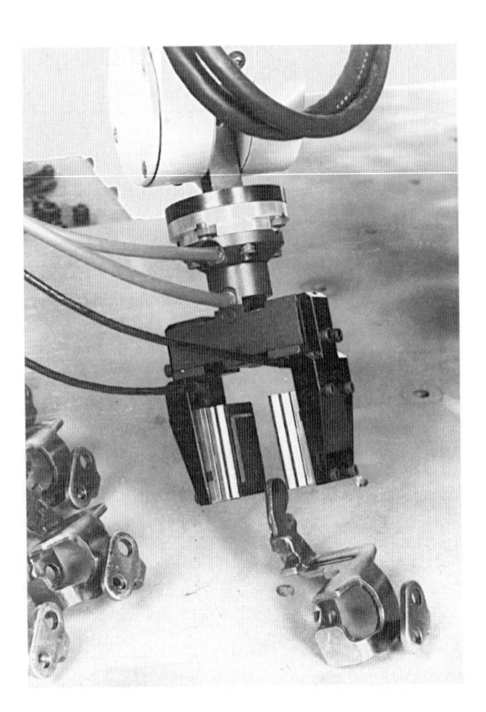

FIGURE 14.25 A robot gripper with tactile sensors. In spite of their capabilities, tactile sensors are now being used less frequently, because of their high cost and their low durability in industrial applications. *Source*: Courtesy of Lord Corporation.

The gripping force of an end effector is sensed, monitored, and controlled through closed-loop feedback devices. Compliant grippers that have force-feedback capabilities and sensory perception can, however, be complicated and require powerful computers; thus, they can be costly. *Anthropomorphic end effectors* continue to be designed to simulate the human hand and fingers and to have the capability of sensing touch, force, movement, and pattern. The ideal tactile sensor must also sense slip (such as a heavy glass slipping between your fingers), a motion that human fingers sense so readily.

2. In **visual sensing (machine vision; computer vision)**, cameras optically sense the presence (and hence also the absence) and shape of the object (Fig. 14.26). A microprocessor then processes the image, usually in less than one second; the image is then measured, and the measurements are digitized (**image recognition**). There are two basic systems of machine vision: **Linear arrays** sense only one dimension, such as the presence of an object or some feature on its surface. **Matrix arrays** sense two, or even three, dimensions and are capable of detecting, for example, a properly inserted component in a printed circuit or a properly made solder joint, known as **assembly verification.** When used in automated inspection systems (Section 4.8.3), these sensors can also detect cracks and flaws.

   Machine vision is particularly suitable for parts that are otherwise inaccessible, in hostile manufacturing environments, for measuring a large number of small features, and in situations where physical contact with the part may cause damage. Examples of applications of machine vision include (a) on-line,

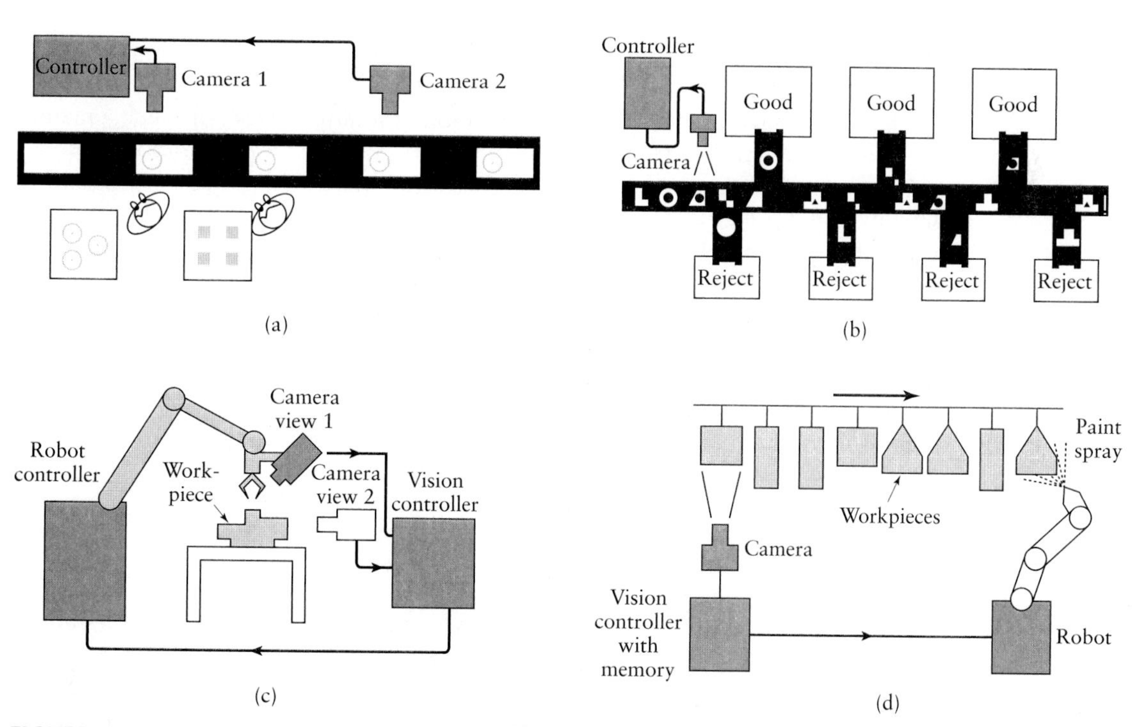

FIGURE 14.26 Examples of machine vision applications. (a) In-line inspection of parts. (b) Identification of parts with various shapes, and inspection and rejection of defective parts. (c) Use of cameras to provide positional input to a robot relative to the workpiece. (d) Painting of parts that have different shapes by means of input from a camera. The system's memory allows the robot to identify the particular shape to be painted and to proceed with the correct movements of a paint spray attached to the end effector.

real-time inspection in sheet-metal stamping lines; and (b) sensors for machine tools that can sense tool offset and tool breakage, verify part placement and fixturing, and monitor surface finish; see also Fig. 14.26. Machine vision is capable of in-line identification and inspection of parts and of rejection of defective parts. With visual sensing capabilities, end effectors can pick up parts and grip them in the proper orientation and location.

3. **Smart sensors** have the capability to perform a logic function, to conduct two-way communication, and to make decisions and take appropriate actions. The necessary input, and the knowledge required to make decisions, can be built into a smart sensor. For example, a computer chip with sensors can be programmed to turn a machine tool off in the event that a cutting tool fails. Likewise, a smart sensor can stop a mobile robot or a robot arm from accidentally coming in contact with an object or people by sensing quantities such as distance, heat, and noise.

The *selection* of a sensor for a particular application depends on such factors as (a) the particular quantity to be measured or sensed, (b) the sensor's interaction with other components in the system, (c) its expected service life, (d) the required level of its performance, (e) the difficulties associated with use of the sensor, and (f) its cost. Another important consideration is the type of environment in which the sensors are to be used. Rugged sensors have been developed to withstand extremes of temperature, shock and vibration, humidity, corrosion, dust and various contaminants, fluids, electromagnetic radiation, and other interferences. Industrial environments, in particular, require robust sensors. (See also the discussion of *robust design* in Section 16.2.3.)

14.8.2 Sensor fusion

Sensor fusion basically involves the integration of multiple sensors in such a manner that the individual data from each of the sensors (such as force, vibration, temperature, and dimensions) are combined to provide a higher level of information and reliability. It has been suggested that an example of sensor fusion occurs when someone drinks from a cup of hot tea or coffee; although we take such a common practice for granted, it can readily be seen that this process involves data input from the person's eyes, hands, lips, and tongue. Through our various senses (sight, hearing, smell, taste, and touch), there is real-time monitoring of temperature, movements, and positions. Thus, for example, if the fluid is too hot, the hand movement of the cup toward the lip is controlled accordingly.

The earliest applications of sensor fusion were in robot-movement control and in missile-flight tracking and similar military applications, primarily because these activities involve movements that mimic human behavior. In manufacturing, an example of sensor fusion is a machining operation in which a set of different, but integrated, sensors continuously monitors (a) the dimensions and surface finish of the workpiece; (b) cutting-tool forces, vibrations, wear, and fracture; (c) the temperatures in various regions of the tool–workpiece system; and (d) the spindle power.

An essential aspect in sensor fusion is **sensor validation**, in which the failure of one particular sensor is detected so that the control system retains high reliability. For this application, the receipt of redundant data from different sensors is essential. It can be seen that the receipt, integration, and processing of all data from various sensors can be a complex problem. With advances in sensor size, quality, and technology, and with new developments in computer-control systems, artificial intelligence, expert systems, and artificial neural networks (see Chapter 15), sensor fusion is becoming practical and available at relatively low cost.

14.9 | Flexible Fixturing

In describing workholding devices for the manufacturing operations covered in Chapters 6 through 9, the words *fixture*, *clamp*, and *jig* are often used interchangeably and sometimes in pairs, such as in *jigs and fixtures*. Common workholding devices include chucks, collets, and mandrels, many of which are usually operated manually. Other workholding devices are designed and operated at various levels of mechanization and automation, such as *power chucks*, which are driven by mechanical, hydraulic, or electrical means. **Fixtures** are generally designed for specific purposes; **clamps** are simple multifunctional devices; **jigs** have various reference surfaces and points for accurate alignment of parts and tools and are widely used in mass production (see also the discussion of *pallets* in Section 8.10). These devices may be used for actual manufacturing operations (in which case the forces exerted on the part must maintain the part's position in the machine without slipping or distortion), or they may be used to hold workpieces for purposes of measurement and inspection, where the part is not subjected to any forces.

Workholding devices have certain ranges of capacity; for example, (a) a particular collet can accommodate bars only within a certain range of diameters; (b) four-jaw chucks can accommodate square or prismatic workpieces that have certain dimensions; and (c) various other devices and fixtures are designed and made for specific workpiece shapes and dimensions and for specific tasks—such devices are called **dedicated fixtures.** It is very simple to reliably fixture a workpiece in the shape of, say, a brick, such as by clamping it between the parallel jaws of a vise. If the part has curved surfaces, it is possible to shape the contacting surfaces of the jaws themselves by machining them (*machinable jaws*) to conform to the workpiece surfaces.

The emergence of flexible manufacturing systems has necessitated the design and use of workholding devices and fixtures that have built-in **flexibility**. There are several methods of *flexible fixturing*, based on different principles (also called **intelligent fixturing systems**), and the term itself has been defined somewhat loosely. Basically, however, these devices are capable of quickly accommodating a range of part shapes and dimensions without the necessity of making extensive changes and adjustments or requiring operator intervention, both of which would adversely affect productivity.

A challenging example of the need for proper fixturing of a workpiece for further processing, such as grinding or deburring, is a series of turbine blades of various shapes and sizes. (See Fig. 3.1.) In addition to the use of some of the methods outlined previously, including the use of robots and computer controls for manipulating part orientation for proper clamping, other techniques for fixturing are described as follows:

1. **Modular fixturing.** *Modular fixturing* is often used for small or moderate lot sizes (Fig. 14.27), especially when the cost of dedicated fixtures and the time required to make them are difficult to justify. Complex workpieces can be located within machines through fixtures produced quickly from standard components and can be disassembled when a production run is completed. These modular fixtures are usually based on tooling plates or blocks configured with grid holes or T-slots upon which a fixture is constructed. A number of other standard components, such as locating pins, adjustable stops, workpiece supports, V-blocks, clamps, and springs, can be mounted onto the base plate or block to quickly produce a fixture. By computer-aided fixture planning for specific situations, such fixtures can be assembled and modified using robots. As compared with dedicated fixturing, modular fixturing has been shown to be

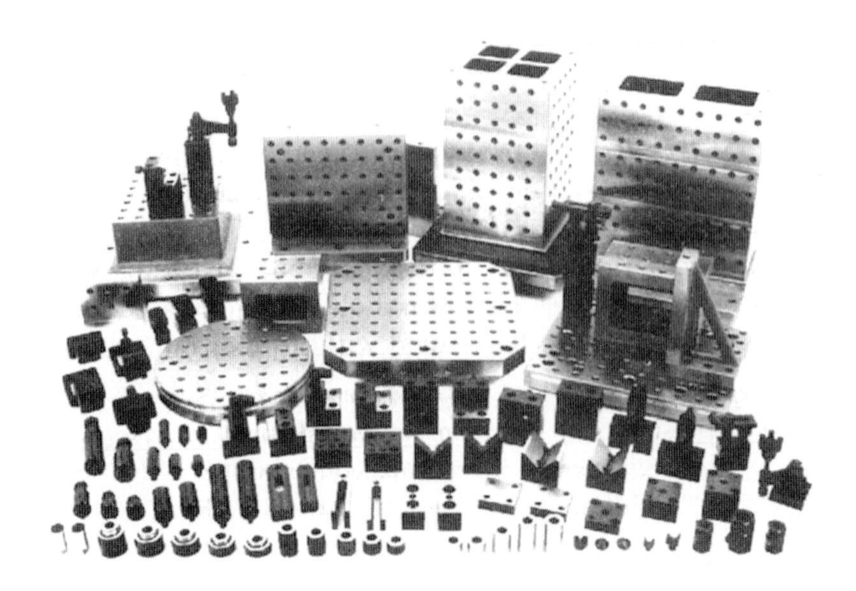

FIGURE 14.27 Typical components of a modular workholding system. *Source*: Carr Lane Manufacturing Co.

low in cost, have a shorter lead time, provide greater ease of repair of damaged components, and offer a more intrinsic flexibility of application.

2. **Bed-of-nails device.** This fixture consists of a series of air-actuated pins that conform to the shape of the external surfaces of the part. Each pin moves as necessary to conform to the shape at its point of contact with the part; the pins are then mechanically locked against the part. The fixture is compact and has high stiffness, and it is reconfigurable.
3. **Adjustable-force clamping.** Figure 14.28 shows a schematic illustration of another flexible fixturing system. In this system, referred to as an *adjustable-force clamping* system, the strain gage attached to the clamp senses the magnitude of the clamping force; the system then adjusts this force to keep the workpiece securely clamped to the workpiece.
4. **Phase-change materials.** There are two methods that hold irregularly shaped or curved workpieces in a medium, other than hard tooling:

   a) In the first, and older, method, a *low-melting-point metal* is used as the clamping medium. For example, an irregularly shaped workpiece is dipped

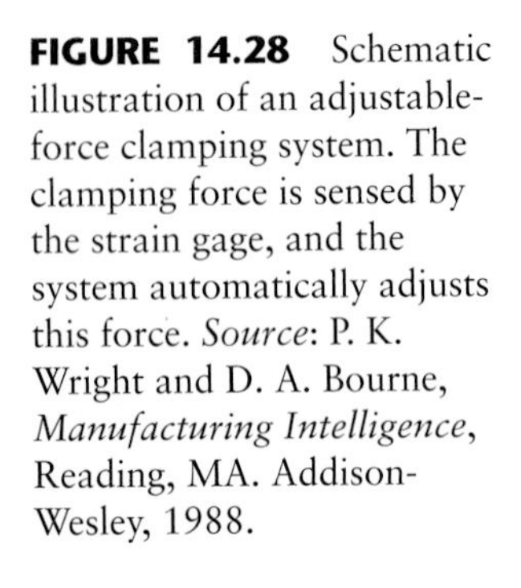

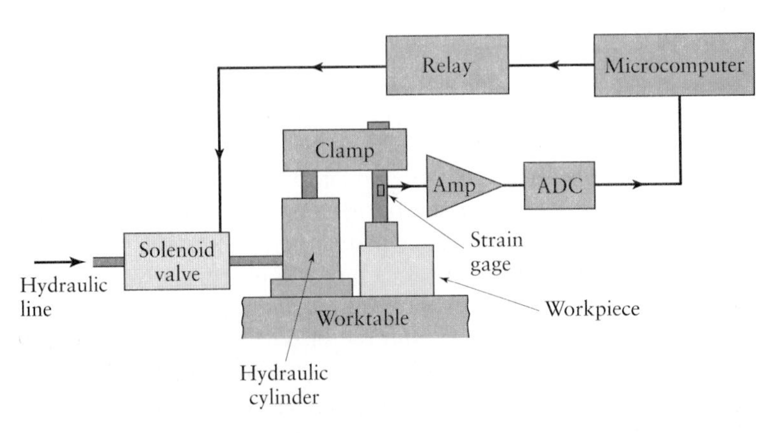

FIGURE 14.28 Schematic illustration of an adjustable-force clamping system. The clamping force is sensed by the strain gage, and the system automatically adjusts this force. *Source*: P. K. Wright and D. A. Bourne, *Manufacturing Intelligence*, Reading, MA. Addison-Wesley, 1988.

into molten lead and allowed to set (like the wooden stick in a Popsicle, a process similar to *insert molding*; see Section 10.10.2 and Fig. 10.24). After setting, the body of the solid metal itself is clamped in a simple fixture. This method is particularly common in the aerospace industry. Note, however, the possible adverse effects of such materials as lead (see the discussion of *liquid-metal embrittlement* in Section 3.4.2) on human health and the environment.

There is a similar application of this method in machining the honeycomb structures shown in Fig. 7.49. Note that because the walls of the hollow structure are very thin, the forces exerted by the cutting tool would easily distort it or damage it. One method of stiffening this structure is to fill the cavities with water and freeze it. Thus, the hexagonal cavities are now filled with ice, whose strength is sufficient to resist the cutting forces; after machining, the ice is allowed to melt away.

b) In the second method, which is still in experimental stages, the supporting medium is a *magnetorheological* (MR) or *electrorheological* (ER) *fluid*. In the MR application, the particles, which have sizes measurable in micrometers or are nanoparticles (see Sections 3.11.9, 11.2.1, and 11.8.1), are ferromagnetic or paramagnetic in a nonmagnetic fluid; surfactants are added to maintain dispersal of powders. After the workpiece is immersed in the fluid, an external magnetic field is applied, whereby the particles are polarized and the behavior of the fluid changes from a liquid to a solid. Afterwards, the workpiece is retrieved by removing the external magnetic field. The process is particularly suitable for nonferrous metal parts. In the ER application, the fluid is a suspension of fine dielectric particles in a liquid with a low dielectric constant. After application of an electrical field, the liquid becomes a solid.

Studies continue on (a) improving the shear strength (currently around 80 kPa for MR) and yield stress (currently up to 60 kPa for ER) of the solidified fluid, (b) ensuring that the fluids do not react with the workpieces, (c) improving the speed of the operation, (d) maintaining the level of positioning accuracy that can be achieved in industrial applications, and (e) integration into CIM systems.

14.10 Assembly, Disassembly, and Service

Some products are simple and have only two or three components to *assemble*; such an operation can be done with relative ease. Examples include the manufacture of an ordinary pencil with an eraser, a frying pan with a wooden handle, and an aluminum beverage can. Most products, however, consist of many parts, and their assembly requires considerable care and planning.

Traditionally, assembly has generally involved much *manual* work and, thus, has contributed significantly to product cost. The total assembly operation is usually broken into individual assembly operations (**subassemblies**), with an operator assigned to carry out each step. Assembly costs are now typically 25% to 50% of the total cost of manufacturing, with the percentage of workers involved in assembly operations ranging from 20% to 60%. In the electronics industries, for example, some 40% to 60% of total wages are paid to assembly workers. As costs increase, the necessity for **automated assembly** become obvious. Beginning with the hand assembly of muskets in the late 1700s (and the introduction of *interchangeable parts* in the early 1800s), assembly methods have been vastly improved over the years. The first application of large-scale modern assembly was for the flywheel magnetos for the Model T Ford; this activity eventually led to mass production of the automobile itself.

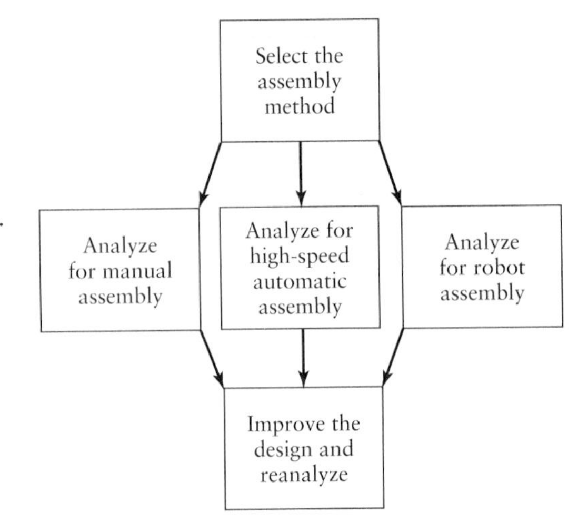

FIGURE 14.29 Stages in the design-for-assembly analysis. *Source*: *Product Design for Assembly*, 1989 edition, by G. Boothroyd and P. Dewhurst. Reproduced with permission.

The choice of an assembly method and system depends on the required production rate, the total quantity to be produced, the product's market life, labor availability, and cost.

As we have seen throughout this text, parts are manufactured within certain dimensional tolerance ranges. (See also Section 4.7.) Taking ball bearings as an example, we know that although they all have the same nominal dimensions, some balls in a particular lot will be smaller than others by a very small amount; likewise, some inner and outer races produced will be smaller than others in the lot. There are two methods of assembly for such high-volume products:

1. In *random assembly*, parts are put together by selecting them randomly from the lots produced.
2. In *selective assembly*, the balls and the races are segregated by groups of sizes, from smallest to largest. The parts are then selected to mate properly. Thus, the smallest diameter balls are mated with inner races that have the largest outside diameters and with outer races that have the smallest inside diameters.

14.10.1 Assembly systems

There are three basic methods of assembly: *manual*, *high-speed automatic*, and *robotic*. These methods can be used individually or, as is the case for most applications in practice, in combination. Before assembly operations commence, an analysis of the product design (Fig. 14.29) should be made to determine the most appropriate and economical method(s) of assembly.

a) **Manual assembly** uses simple tools and is generally economical for relatively small lots. Because of the dexterity of the human hand and fingers and their capability for feedback through various senses, workers can assemble even complex parts without much difficulty. Note, for example, that the alignment and placement of a simple square peg into a square hole, involving small clearances, can be difficult in automated assembly—yet the human hand is capable of doing this simple operation with relative ease. There can, however, be potential problems of *cumulative trauma disorders* (*carpal tunnel syndrome*) associated with this method.
b) **High-speed automated assembly** uses **transfer mechanisms** designed specially for assembly. Two examples of such assembly are shown in Fig. 14.30, in which individual assembly is carried out on products that are *indexed* for proper positioning during assembly. In **robotic assembly**, one or two general-purpose

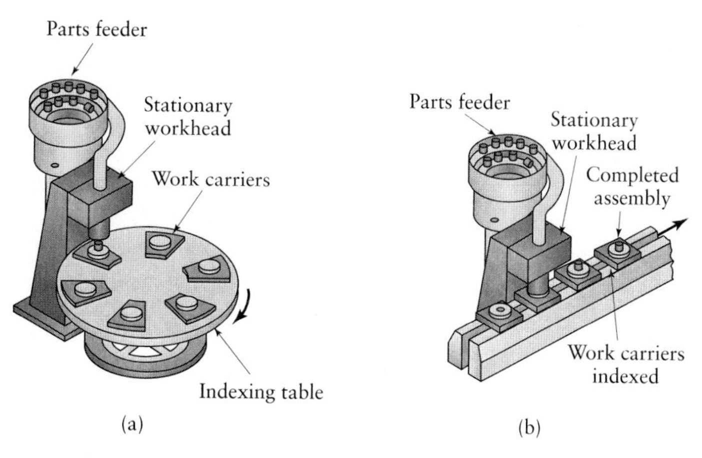

FIGURE 14.30 Transfer systems for automated assembly: (a) rotary indexing machine; (b) in-line indexing machine. *Source*: G. Boothroyd.

robots operate at a single workstation, or the robots operate at a multistation assembly system.

There are three basic types of assembly systems: synchronous, nonsynchronous, and continuous.

1. **In synchronous systems**, also called **indexing systems**, individual parts and components are supplied and assembled at a constant rate at fixed individual stations. The rate of movement is based on the station that takes the longest time to complete its portion of the assembly. This system is used primarily for high-volume, high-speed assembly of small products.

   Transfer systems move the partial assemblies from workstation to workstation by various mechanical means; two typical transfer systems are **rotary indexing** and **in-line indexing systems** (Fig. 14.30). These systems can operate in either a fully automatic mode or a semiautomatic mode. However, a breakdown of one station will shut down the whole assembly operation. The part feeders supply the individual parts to be assembled and place them on other components, which are secured on work carriers or fixtures. The feeders move the individual parts (by vibratory or other means) through delivery chutes and ensure their proper orientation by various ingenious means (Fig. 14.31). Proper orientation of parts and avoidance of jamming are essential in all automated assembly operations.

2. **In nonsynchronous systems**, each station operates independently, and any imbalance is accommodated in storage (**buffer**) between stations. The station continues operating until the next buffer is full or the previous buffer is empty. Furthermore, if one station becomes inoperative for some reason, the assembly line continues to operate until all the parts in the buffer have been used up. Nonsynchronous systems are suitable for large assemblies with many parts to be assembled. For types of assembly in which the times required for individual assembly operations vary widely, the rate of output will be constrained by the slowest station.

3. **In continuous systems**, the product is assembled while moving at a constant speed on pallets or similar workpiece carriers. The parts to be assembled are brought to the product by various workheads, and their movements are synchronized with the continuous movement of the product. Typical applications of this system are in bottling and packaging plants, although this method has also been used on mass-production lines for automobiles and appliances.

Assembly systems are generally set up for a certain product line; however, they can be modified for increased flexibility in order to be used in product lines that have a variety of models. Such *flexible assembly systems* (FAS) use computer controls,

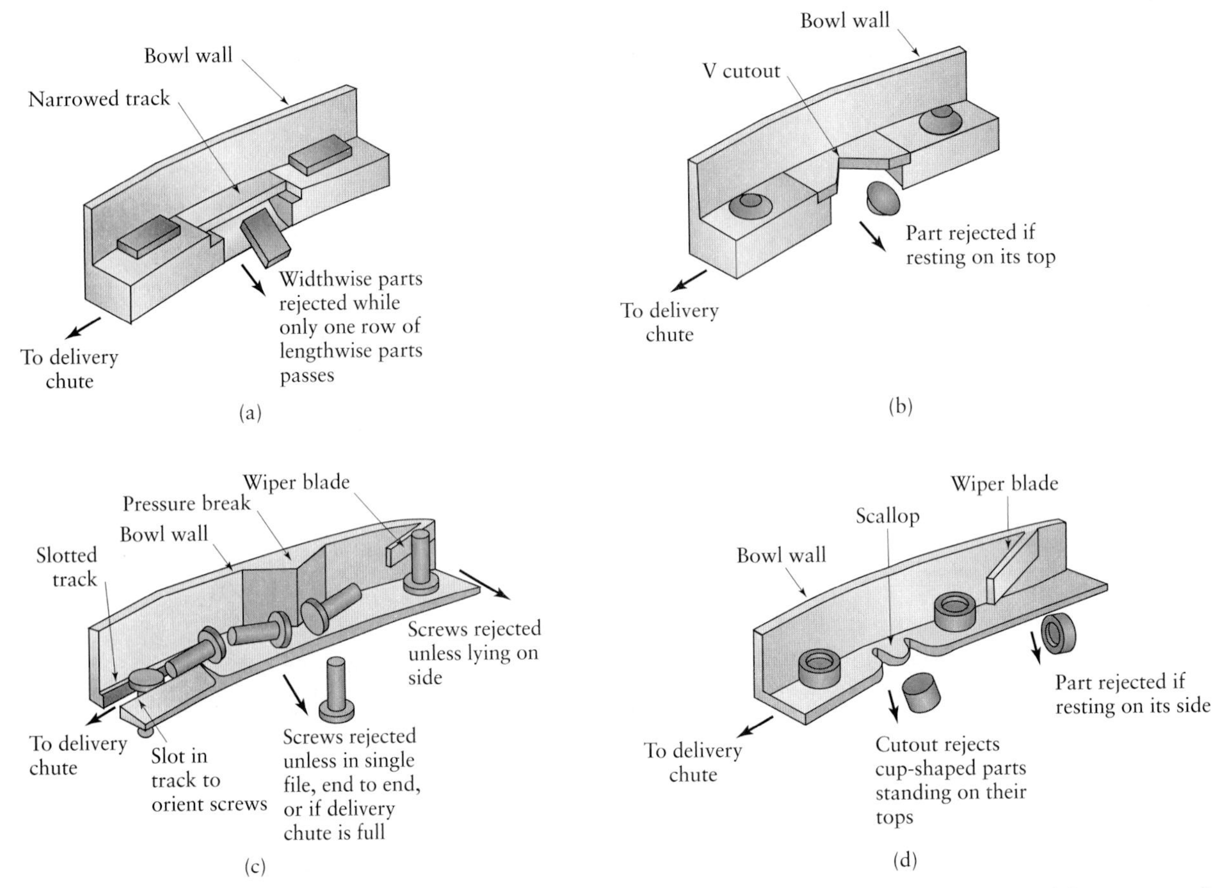

FIGURE 14.31 Various guides that ensure that parts are properly oriented for automated assembly. *Source*: G. Boothroyd.

interchangeable and programmable workheads and feeding devices, coded pallets, and automated guiding devices. The General Motors plant for the Saturn subcompact automobile, for example, is designed with a flexible assembly system. The system is capable of assembling up to a dozen different transmission and engine combinations and power-steering and air-conditioning units.

14.11 Design Considerations

As in many aspects of manufacturing processes and systems, design is an integral part of the topics described in this chapter. Two of these topics require special design considerations, as described in this section.

14.11.1 Design for fixturing

The proper design, construction, and operation of flexible workholding devices and fixtures are essential to the operation and efficiency of advanced manufacturing systems. The following is a list of the major design issues involved:

1. Workholding devices must position the workpiece automatically and accurately; they must maintain its location precisely and with sufficient clamping force to withstand the particular manufacturing operation. Fixtures should also be able to accommodate parts placed repeatedly in the same position.

2. The fixtures must have sufficient stiffness to resist, without excessive distortion, the normal and shear stresses developed at the workpiece–fixture interfaces.
3. The presence of loose chips and other debris between the locating surfaces of the workpiece and the fixture can be a serious problem. Chips are most likely to be present where cutting fluids are used, because small chips tend to stick to the wet surfaces, due to surface-tension forces.
4. A flexible fixture should accommodate parts to be made by various processes and cases for which dimensions and surface features vary from part to part. These considerations are even more important when the workpiece (a) is fragile or made of a brittle material; (b) is made of a relatively soft and flexible material, such as plastic or rubber part; or (c) has a relatively soft coating on its contacting surfaces.
5. Fixtures and clamps should have low profiles, so as to avoid collision with cutting tools. Collision avoidance is also an important factor in programming tool paths in machining operations; see Sections 14.3 and 14.4.
6. Flexible fixturing must also meet special requirements in order to function properly in manufacturing cells and flexible manufacturing systems. The time required to load and unload parts on modern machinery, for example, should be minimal in order to reduce cycle times.
7. Workpieces should be designed to allow locating and clamping within a fixture. Flanges, flats, or other locating surfaces should be incorporated into the design in order to simplify fixture design and to aid in part transfer into machinery.

14.11.2 Design for assembly, disassembly, and service

Design for assembly. Although the functions of a product and its design for manufacture have been matters of major interest for some time, *design for assembly* (DFA) has attracted special attention (particularly design for automated assembly), because of the need to reduce assembly costs. In **manual assembly**, a major advantage is that humans can easily pick the correct similar or different parts from bulk, such as from a nearby bin, and the human senses guide the hands for proper assembly. In **high-speed automated assembly**, however, automatic handling generally requires that parts be separated from the bulk, conveyed by hoppers or vibratory feeders (see Fig. 14.30), and assembled in the proper locations and orientations.

Based on analyses of assembly operations as well as on experience, general guidelines for DFA have been developed through the years and are summarized as follows:

1. Reduce the number and variety of parts in a product; incorporate multiple functions into a single part. Also, consider subassemblies that could serve as modules.
2. Parts should have a high degree of symmetry (such as round or square) or a high degree of asymmetry (such as oval or rectangular) so that they cannot be installed incorrectly and so that they do not require location, alignment, or adjustment. Design parts for easy insertion.
3. Designs should allow parts to be assembled without obstructions or the lack of a direct line of sight. Assemblies should not be turned over for insertion of parts.
4. Consider methods such as snap fits, and avoid the need for fasteners such as bolts, nuts, and screws. If such fasteners must be used, minimize the variety used, and space and locate them such that tools can be used without obstruction.
5. Part designs should consider such factors as size, shape, weight, flexibility, abrasiveness, and tangling with other parts.

6. Parts should be inserted from a single direction, preferably vertically (that is, from above) to take advantage of gravity; assembly from two or more directions can be difficult.
7. Products should be designed, or existing products redesigned, so that there are no physical obstructions to the free movement of the parts during assembly (see Fig. 1.3); thus, for example, sharp external and internal corners should be replaced with chamfers, tapers, or radii.
8. Color-code parts that may appear to be similar, but are different.

Robotic-assembly design guidelines have rules similar to those for manual and high-speed automated assembly. The development of *compliant end effectors* and *dexterous manipulators* has somewhat relaxed the inherent inflexibility of robots. Some distinguishing additional guidelines are as follows:

1. Parts should be designed so that they can be gripped and manipulated by the same gripper (end effector) of the robot (see Fig. 14.18); such a design avoids the need for different grippers. Also, parts should be made available to the gripper in the proper orientation.
2. Assembly that involves threaded fasteners, such as bolts, nuts, and screws, may be difficult for robots to perform; one exception is the use of self-threading screws for sheet metal, plastics, and wooden parts. Also, robots easily can handle snap fits, rivets, welds, and adhesives.

Design for disassembly. The manner and ease with which a product may be taken apart for maintenance or replacement of its parts is another important consideration in product design. Consider, for example, the difficulties one has in removing certain components from under the hood of some automobiles; similar difficulties exist in the disassembly of several other products. Although there is no established set of guidelines, the general approach to design for disassembly requires the consideration of factors that are similar to those outlined previously for design for assembly. Analysis of computer or physical models of products and their components with regard to disassembly generally indicates any potential problems, such as obstructions, narrow and long passageways, lack of line of sight, and the difficulty of firmly grasping and guiding objects.

An important aspect of design for disassembly is how, after its life cycle, a product is to be taken apart for *recycling*, especially with respect to the more valuable components. Note, for example, that depending on (a) their design and location, (b) the type of tools used for disassembly and (c) whether the tools are manual or power tools, rivets may take longer to remove than screws or snap fits, and that a bonded layer of valuable material on a component would be very difficult, if not impossible, to remove for recycling or reuse. Obviously, the longer it takes to disassemble components, the higher is the cost of doing so; it is then possible that this cost becomes prohibitive. Consequently, the time required for disassembly has been studied and measured; although it depends on the manner in which disassembly takes place, some examples are as follows: cutting wire, 0.25s; disconnecting wire, 1.5 s; removing snap fits and clips, 1–3 s; and removing screws and bolts, 0.15–0.6 s per revolution.

Design for service. Design for assembly and disassembly includes taking into account the ease with which a product can be serviced and, if necessary, repaired. Design for service is essentially based on the concept that the elements that are most likely to need servicing should be at the outer layers of the product. In this way, individual parts are easier to reach and to service, without the need to remove various other parts in order to do so.

14.12 Economic Considerations

As described in greater detail in Chapter 16, and as we have seen throughout the preceding chapters, there are numerous considerations involved in determining the overall economics of production operations. Because all production systems are essentially a combination of machines and people, important factors influencing the final decisions include the types and cost of machinery, cost of its operation, the skill level and amount of labor required, and production quantity. We have seen that lot size and production rate greatly influence the economics of production. Small quantities per year can be manufactured in job shops; however, the type of machinery in job shops generally requires skilled labor to operate. Furthermore, production quantity and rate in job shops are low, and as a result, cost per part can be high. (See Fig. 14.3.)

At the other extreme, there is the production of very large quantities of parts, using conventional flow lines and transfer lines and involving special-purpose machinery and equipment, specialized tooling, and computer control systems. Although all these components constitute major investments, the level of labor skill required and the labor costs are both relatively low, because of the high level of automation implemented. However, these production systems are organized for a specific type of product and hence lack flexibility.

Because most manufacturing operations are between these two extremes, an appropriate decision has to be made regarding the desirable level of automation to be implemented. In many situations, *selective automation* rather than *total* automation of a facility has been found to be cost effective. Generally, the higher the level of skill available in the workforce, the lower is the need for automation, provided that higher labor costs are justified and that there is a sufficient number of qualified workers available. Conversely, if a manufacturing facility has already been automated, the skill level required is lower.

In addition, the manufacture of some products must have a large labor component, and thus their production is *labor intensive*; this is especially the case with products that require assembly. Examples of labor-intensive products include aircraft, software, bicycles, pianos and other musical instruments, furniture, toys, shoes, textiles, and garments. This high labor requirement is a major reason that so many household as well as high-tech products are now made or assembled in countries where labor costs are low, such as Mexico, China, the Philippines, and other Pacific-rim countries. (See Section 16.9).

In manual assembly, relatively simple tools are used, and the process is economical for relatively small lots. Because of the dexterity of the human hand and fingers and their capability for feedback through various senses, workers can manually assemble even complex parts without much difficulty. As described in Section 14.7, labor cost and benefit considerations are also significant aspects of robot selection and use. The increasing availability and reliability and the reduced cost of intelligent robots are having a major economic impact on manufacturing operations, causing human labor to be gradually replaced.

CASE STUDY | Development of a Modular Fixture

Figure 14.32 shows a round cast-iron housing that requires a fixture for milling the housing flat, counterboring the center hole, drilling the four corner holes, and machining the circular mounting area. These operations are to be performed in moderate lot sizes on a CNC milling machine; thus, a fixture that accurately and

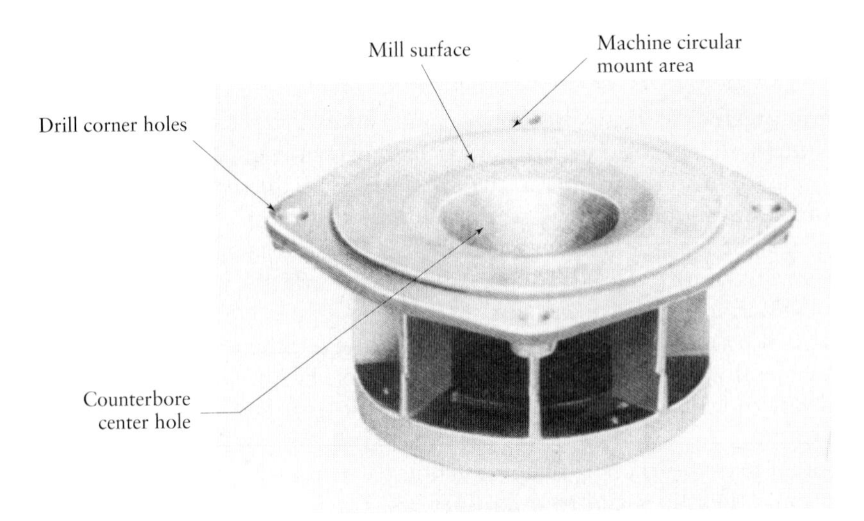

FIGURE 14.32 Cast-iron housing and the machining operations required.

consistently locates the workpiece is required. The lot size is not large enough to justify the design and fabrication of dedicated fixtures, and hence a modular fixture is constructed from the components illustrated in Fig. 14.33.

The first step in designing such a fixture is to select a tooling plate. Tooling plates with T-slots or gridded holes are the most common alternatives for modular fixtures. For this case, a rectangular plate with gridded holes is selected that has a sufficiently large surface area to accommodate both the workpiece and the fixturing elements.

The workpiece must not be clamped on the surface that is to be machined; however, there is a lower flange that is suitable for clamping. Clamping is generally accomplished by exerting a force with a clamp against a locating button. For this round workpiece, it is desirable to locate the lower flange at three points, spaced ap-

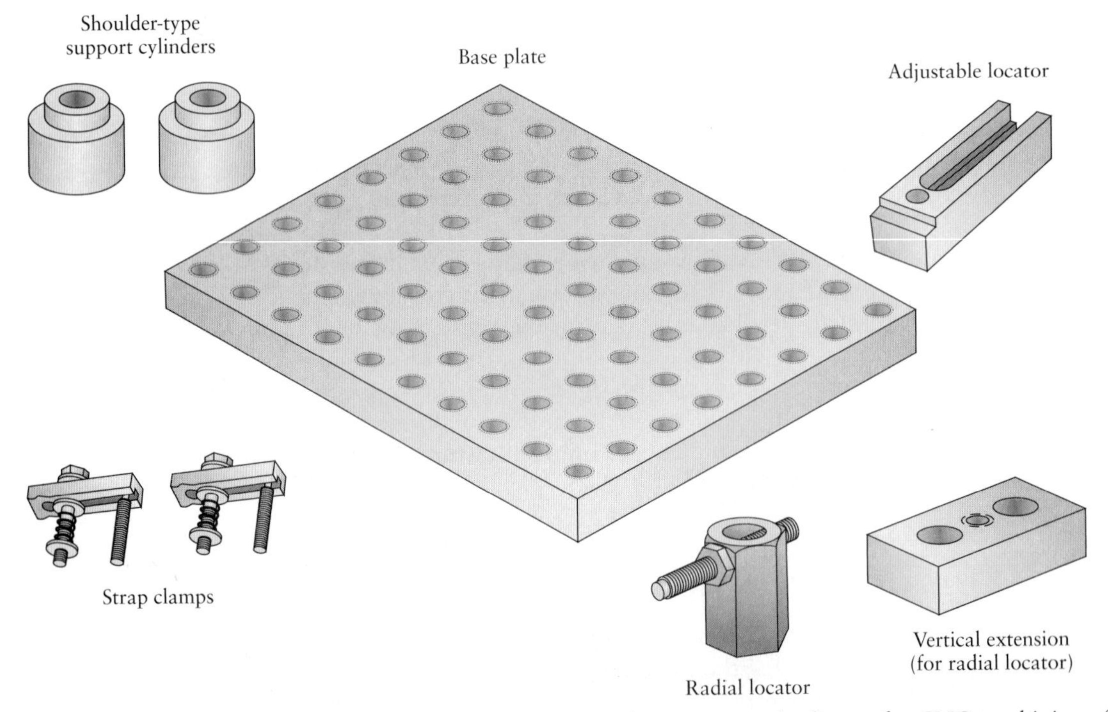

FIGURE 14.33 Modular components used to construct the fixture for CNC machining of the cast-iron housing depicted in Fig. 14.32.

FIGURE 14.34 Completed modular fixture with cast-iron housing in place, as would be assembled for use in a machining center or CNC milling machine.

proximately 120° apart, for reasons of stability. The first locating elements selected are shoulder-type support cylinders. The flange rests on the shoulders, and the cylinders support the workpiece on its diameter and also raise the workpiece above the tooling plate. This method has the advantage of eliminating the adverse effects of chips (from machining) on the tooling plate, which could interfere with the orientation of the workpiece. The support cylinders are mounted onto the tooling plate, using locating screws. An adjustable extension support is then mounted to the tooling plate in a position where the correct diameter is set between the three locators. The extension support is positioned to contact only the workpiece's bottom edge, and sufficient space is allowed so that the locator will not bind the housing on the locating diameter.

The workpiece must be consistently oriented in order to maintain the required tolerances on the corner holes. An adjustable stop is utilized to locate the workpiece; this stop uses a threaded locating button that is adjusted to orient the workpiece. An extension support is used so that the stop is vertically located directly below the surface to be machined.

Strap clamps are then used to hold the workpiece in place. The springs and washers on the clamping stud permit the strap clamp to rise automatically when loosened, so that the clamp will not fall when the workpiece is removed. Figure 14.34 shows the modular fixture with the workpiece in place. Note that planar, concentric, and radial locations have all been accurately defined by the fixture.

Source: Courtesy of Carr Lane Manufacturing Company.

SUMMARY

- There are several levels of automation, from simple automation to untended manufacturing cells. Automation has been implemented successfully in manufacturing processes, material handling, inspection, assembly, and packaging. Production quantity and production rate are major factors in selecting the most economic level of automation for a process. (Sections 14.1, 14.2)
- True automation began with the numerical control of machines, achieving flexibility of operation, lower cost, and ease of making different parts with lower operator skill. (Sections 14.3, 14.4)
- The quality and cost of manufacturing operations are further optimized by adaptive-control techniques, which continuously monitor the operation and automatically make appropriate adjustments in the processing parameters. (Section 14.5)

- Significant advances have been made in material handling, particularly with the implementation of industrial robots and automated guided vehicles. (Sections 14.6 and 14.7)
- Sensors are essential in the implementation of modern manufacturing technologies and computer-integrated manufacturing, and a wide variety of sensors based on various principles has been developed and successfully installed. (Section 14.8)
- Other advances include flexible fixturing and automated assembly techniques that reduce the need for worker intervention and lower manufacturing costs. The effective and economic implementation of these techniques requires that design for assembly, disassembly, and servicing be recognized as an important factor in the total design and manufacturing process. (Sections 14.9, 14.10)
- As in all manufacturing processes, there are certain design considerations regarding the implementation of the various topics described in this chapter. (Section 14.11)
- Economic considerations in automation include decisions regarding the level of automation to be implemented; such decisions, in turn, involve parameters such as the quantity and rate of production. (Section 14.12)

BIBLIOGRAPHY

Amic, P. J., *Computer Numerical Control Programming*, Prentice Hall, 1996.

Astrom, K. J., and B. Wittenmark, *Adaptive Control*, 2d. ed., Addison-Wesley, 1994.

Bolhouse, V., *Fundamentals of Machine Vision*, Robotic Industries Association, 1997.

Boothroyd, G., *Assembly Automation and Product Design*, Dekker, 1991.

Boothroyd, G., P. Dewhurst, and W. Knight, *Product Design for Manufacture and Assembly*, 2d. ed., Dekker, 2001.

Brooks, R. R., and S. Iyengar, *Multi-Sensor Fusion: Fundamentals and Applications with Software*, Prentice Hall, 1997.

Burke, M., *Handbook of Machine Vision Engineering*, Chapman & Hall, 1999.

Busch-Vishniac, I., *Electromechanical Sensors and Actuators*, Springer, 1999.

Chow, W., *Assembly Line Design: Methodology and Applications*, Dekker, 1990.

Fraden, J., *Handbook of Modern Sensors: Physics, Designs, and Applications*, 2d. ed., Springer, 1996.

Galbiati, L. J., *Machine Vision and Digital Image Processing Fundamentals*, Prentice Hall, 1997.

Gibbs, D., and T. Crandell, *CNC: An Introduction to Machining and Part Programming*, Industrial Press, 1991.

Ioannu, P. A., *Robust Adaptive Control*, Prentice Hall, 1995.

Jain, R. (ed.), *Machine Vision*, Prentice Hall, 1995.

Lim, S. C. J., *Computer Numerical Control*, Delmar, 1994.

Lynch, M., *Computer Numerical Control for Machining*, McGraw-Hill, 1992.

Molloy, O., E. A. Warman, and S. Tilley, *Design for Manufacturing and Assembly: Concepts, Architectures and Implementation*, Kluwer, 1998.

Myler, H. R., *Fundamentals of Machine Vision*, Society of Photo-optical Instrumentation Engineers, 1998.

Nof, S. Y., W. E. Wilhelm, and H.-J. Warnecke, *Industrial Assembly*, Chapman & Hall, 1998.

Rampersad, H. K., *Integral and Simultaneous Design for Robotic Assembly*, Wiley, 1995.

Rehg, J. A., *Introduction to Robotics in CIM Systems*, 5th ed., Prentice Hall, 2002.

Sandler, B.-Z., *Robotics: Designing the Mechanisms for Automated Machinery*, 2d. ed., Prentice Hall, 1999.

Sava, M., and J. Pusztai, *Computer Numerical Control Programming*, Prentice Hall, 1997.

Seames, W., *Computer Numerical Control: Concepts and Programming*, 3d. ed., Delmar, 1995.

Smid, P., *CNC Programming Handbook*, Industrial Press, 2000.

Soloman, S., *Sensors Handbook*, McGraw-Hill, 1997.

Stenerson, J., and K. S. Curran, *Computer Numerical Control: Operation and Programming*, 2d. ed., Prentice Hall, 2000.

Tool and Manufacturing Engineers Handbook, 4th ed., Vol. 4: *Assembly, Testing, and Quality Control*, Society of Manufacturing Engineers, 1986.

_____, Vol. 9: *Material and Part Handling in Manufacturing*, Society of Manufacturing Engineers, 1998.

Valentino, J. V., and J. Goldenberg, *Introduction to Computer Numerical Control*, 2d. ed., Prentice Hall, 1999.

Williams, D. J., *Manufacturing Systems*, 2d. ed., Chapman & Hall, 1994.

Zuech, N., *Understanding and Applying Machine Vision*, 2d. ed., Dekker, 1999.

QUESTIONS

14.1 Describe the differences between mechanization and automation. Give several specific examples for each.

14.2 Why is automation generally regarded as evolutionary rather than revolutionary?

14.3 Are there activities in manufacturing operations that cannot be automated? Explain, and give specific examples.

14.4 Explain the difference between hard and soft automation. Why are they named as such?

14.5 Describe the principle of numerical control of machines. What factors led to the need for and development of numerical control? Name some typical applications of NC.

14.6 Explain the differences between direct numerical control and computer numerical control. What are their relative advantages?

14.7 Describe open-loop and closed-loop control circuits.

14.8 What are the advantages of computer-aided NC programming?

14.9 Describe the principle and purposes of adaptive control. Give some examples of present applications in manufacturing and other areas that you think can be implemented.

14.10 What factors have led to the development of automated guided vehicles? Do automated guided vehicles have any limitations? Explain your answers.

14.11 List and discuss the factors that should be considered in choosing a suitable material-handling system for a particular manufacturing facility.

14.12 Make a list of the features of an industrial robot. Why are these features necessary?

14.13 Discuss the principles of various types of sensors, and give two applications for each type.

14.14 Describe the concept of design for assembly. Why has it become an important factor in manufacturing?

14.15 Is it possible to have partial automation in assembly operations? Explain.

14.16 Describe your thoughts on adaptive control in manufacturing operations.

14.17 What are the two kinds of robot joints? Give applications for each.

14.18 What are the advantages of flexible fixturing over other methods of fixturing? Are there any limitations to flexible fixturing? Explain.

14.19 How are robots programmed to follow a certain path?

14.20 Giving specific examples, discuss your observations concerning Fig. 14.2.

14.21 What are the relative advantages and limitations of the two arrangements for power heads shown in Fig. 14.4?

14.22 Discuss methods of on-line gaging of workpiece diameters in turning operations other than that shown in Fig. 14.15. Explain the relative advantages and limitations of the methods.

14.23 Is drilling and punching the only application for the point-to-point system shown in Fig. 14.10a? Are there others? Explain.

14.24 Describe possible applications for industrial robots not discussed in this chapter.

14.25 What determines the number of robots in an automated assembly line such as that shown in Fig. 14.23?

14.26 Describe situations in which the shape and size of the work envelope of a robot (see Fig. 14.20) can be critical.

14.27 Explain the functions of each of the components of the robot shown in Fig. 14.17a. Comment on their degrees of freedom.

14.28 Explain the difference between an automated guided vehicle and a self-guided vehicle.

14.29 It has been commonly acknowledged that, at the early stages of development and implementation of industrial robots, the usefulness and cost effectiveness of the robots were overestimated. What reasons can you think of to explain this situation?

14.30 Describe the type of manufacturing operations (see Fig. 14.2) that are likely to make the best use of a machining center (see Section 8.10). Comment on the influence of product quantity and part variety.

14.31 Give a specific example of a situation in which an open-loop control system would be desirable, and give a specific example of a situation in which a closed-loop system would be desirable. Explain.

14.32 Why should the level of automation in a manufacturing facility depend on production quantity and production rate?

14.33 Explain why sensors have become so essential in the development of automated manufacturing systems.

14.34 Why is there a need for flexible fixturing for holding workpieces? Are there any disadvantages to such flexible fixturing? Explain.

14.35 Describe situations in manufacturing for which you would not want to apply numerical control. Explain your reasons.

14.36 Table 14.2 shows a few examples of typical products for each category of production by volume. Add several other examples to this list.

14.37 Describe situations for which each of the three positioning methods shown in Fig. 14.6 would be desirable.

14.38 Describe applications of machine vision for specific parts, similar to the examples shown in Fig. 14.26.

14.39 Add examples of guides other than those shown in Fig. 14.31.

14.40 Sketch the work envelope of each of the robots in Fig. 14.19. Describe its implications in manufacturing operations.

14.41 Give several applications for the types of robots shown in Fig. 14.19.

14.42 Name some applications for which you would not use a vibratory feeder. Explain why vibratory feeding is not appropriate for these applications.

14.43 Give an example of a metal-forming operation (from Chapters 6 and 7) that is suitable for adaptive control similar to that shown in Fig. 14.15.

14.44 Give some applications for the systems shown in Fig. 14.26a and c.

14.45 Comment on your observations regarding the system shown in Fig. 14.18b.

14.46 Give examples for which tactile sensors would not be suitable. Explain why tactile sensors are unsuitable for these applications.

14.47 Give examples for which machine vision cannot be applied properly and reliably. Explain why machine vision is inappropriate for these applications.

14.48 Comment of the effect of cutter wear on the profiles produced, such as those shown in Figs. 14.10b and 14.12.

14.49 Although future trends are always difficult to predict with certainty, describe your thoughts as to what new developments in the topics covered in this chapter could possibly take place as we move through the early 2000s.

14.50 Describe the circumstances under which a dedicated fixture, a modular fixture, and a flexible fixture would be preferable.

DESIGN

14.51 Design two different systems of mechanical grippers for widely different applications.

14.52 For a system similar to that shown in Fig. 14.27, design a flexible fixturing setup for a lathe chuck. (See Figs. 8.42 and 8.44.)

14.53 Add other examples to those shown in Fig. 1.3.

14.54 Give examples of products that are suitable for the type of production shown in Fig. 14.3.

14.55 Choose one machine each from Chapters 6 through 12, and design a system for the machine in which sensor fusion can be used effectively. How would you convince a prospective customer of the merits of such a system? Would the system be cost effective?

14.56 Does the type of material (metallic or nonmetallic) used for the parts shown in Fig. 14.31 have any influence on the effectiveness of the guides? Explain.

14.57 Think of a product, and design a transfer line for it similar to that shown in Fig. 14.5. Specify the types of and the number of machines required.

14.58 Section 14.9 has described the basic principles of flexible fixturing. Considering the wide variety of parts made, prepare design guidelines for flexible fixturing. Make simple sketches illustrating each guideline for each type of fixturing, describing its ranges of applications and limitations.

14.59 Describe your thoughts on the usefulness and applications of modular fixturing, consisting of various individual clamps, pins, supports, and attachments mounted on a base plate.

14.60 Inspect several household products, and describe the manner in which they have been assembled. Comment on any design changes you would make so that assembly, disassembly, and servicing are simpler and faster.

14.61 Review the last design shown in Fig. 14.18a, and design grippers that would be suitable for gripping the following products: (a) an egg; (b) an object made of soft rubber; (c) a metal ball with a very smooth and shiny surface; (d) a newspaper; and (e) tableware, such as forks, knives, and spoons.

14.62 Obtain an old toaster, and disassemble it. Make recommendations on its redesign, using the guidelines given in Section 14.11.2.

14.63 Comment on the design and materials used for the gripper shown in Fig. 14.25. Why do such grippers have low durability on the shop floor?

14.64 Comment on your observations regarding Fig. 14.27, and offer designs for similar applications in manufacturing. Also comment on the usefulness of the designs in actual production on the shop floor.

14.65 Review the various toolholders used in the machining operations described in Chapter 8, and design sensor systems similar for them to that shown in Fig. 14.24. Comment on the features of the sensor systems and discuss any problems that may be associated with their use on the factory floor.

14.66 Design a guide that operates in the same manner as those shown in Fig. 14.31, but to align U-shaped parts so that they are inserted with the open end down.

14.67 Design a flexible fixture that uses powered workholding devices for a family of parts, as in the case study, but that allows for a range of diameters and thicknesses.

14.68 Design a modular fixturing system for the part shown in the figure accompanying Problem 8.134.

CHAPTER

15 Computer-Integrated Manufacturing Systems

15.1 Introduction

The implementation, benefits, and limitations of mechanization, automation, and computer control of various stages of manufacturing operations have been described in the preceding chapter; this chapter focuses its attention on the *integration of manufacturing activities*. This concept means that processes, equipment, operations, and their management are treated as a **manufacturing system**, one that makes possible total control of the manufacturing facility, thereby increasing productivity, product quality, and product reliability and reducing manufacturing costs. As described in this chapter, in **computer-integrated manufacturing** (CIM), the traditionally separate functions of product design, research and development, production, assembly, inspection, and quality control are all linked. Consequently, integration requires that quantitative relationships among product design, materials, manufacturing processes, process and equipment capabilities, and related activities be well understood. In this way, changes in, for example, material requirements, product types, or market demand can be properly accommodated.

The importance of product quality was emphasized in Section 4.9, along with the necessity of commitment of a company's management to total quality control. Recall the statements that (a) quality must be built into the product, (b) higher quality does not necessarily mean higher cost, and (c) marketing poor-quality products can indeed be very costly to the manufacturer. It has been shown that high quality is far more attainable and less expensive via the proper integration of design and manufacturing than if the two were separate activities. Integration is now successfully and effectively accomplished through **computer-aided design, engineering, manufacturing, process planning,** and **simulation of processes and systems.**

This chapter also emphasizes the importance of **flexibility** in machines, equipment, tooling, and production operations to the ability to respond to market and product changes and to ensure **on-time delivery** of high-quality products to the customer's satisfaction. As described in detail, important developments during the past 30 years or so have had a major impact on modern manufacturing, especially in an intensely competitive global marketplace. Among these advances are **group technology, cellular manufacturing, flexible manufacturing systems,** and **just-in-time production** (also called **zero inventory, stockless production,** or **demand scheduling**) with no excess production such as was previously stored in inventory for future sales. Furthermore, because of the extensive use of computer controls and hardware and software in integrated manufacturing, the planning and effective implementation of **communication networks** soon became an essential component of these activities.

The chapter concludes with a review of **artificial intelligence**, consisting of expert systems, natural-language processing, machine vision, artificial neural networks, and fuzzy logic, and how these developments impact manufacturing activities, as well as a discussion of the concept of **the factory of the future**, including its possibilities and existing concerns about it.

15.2 Manufacturing Systems

As we have seen throughout this text, manufacturing consists of a large number of interdependent activities with distinct entities, such as materials, tools, machines, and controls; it should therefore be regarded as a *system*. Manufacturing is indeed a complex system, because it consists of numerous diverse physical and human elements. Some of these factors can be difficult to predict and to control, such as the supply and cost of raw materials, market changes, and global trends, as well as human behavior and performance.

Ideally, we should be able to represent a system by **mathematical and physical models**, which can identify the nature and extent of the interdependence of the variables involved. In a manufacturing system, a change or disturbance anywhere in the system requires that it adjust itself systemwide in order to continue functioning efficiently. For example, if the supply of a particular raw material is reduced (for example, by geopolitical maneuvers, wars, or strikes) and, consequently, its cost increases, alternative materials must be selected. This selection should be made only after careful consideration of the effect this change may have on processing conditions, production rate, product quality, and, especially, manufacturing costs.

Similarly, the demand for a product may fluctuate randomly and rapidly, or even unexpectedly, on account of its style, size, or capacity. Note, for example, the downsizing of automobiles during the 1980s in response to fuel shortages, and note as well the current popularity of sport-utility vehicles with comparatively low gas mileage. The manufacturing system must be capable of producing the modified product on a short **lead time** and, preferably, with relatively small major capital investment in machinery and tooling. Lead time is defined as the length of time between the creation of the product as a concept (or receipt of an order for the product) and the time that the product first becomes available in the marketplace (or is delivered to the customer).

Computer simulation and modeling of such a complex system can be difficult, because of a lack of comprehensive and reliable data on some of the numerous variables involved; furthermore, it is not always easy to predict correctly and to control some of these variables. The following are some examples of the problems that are encountered:

a) Machine tools' characteristics, performance, and response to random external disturbances cannot always be modeled precisely.
b) Raw-material costs over a period of time may be difficult to predict accurately.
c) Market demands and human behavior and performance are difficult to model reliably.

In spite of these challenges, however, much progress continue to be made in modeling and simulation of manufacturing systems. (See also Section 15.7.)

15.3 Computer-Integrated Manufacturing

The various levels of automation in manufacturing operations, as described in Chapter 14, have been extended further by including *information-processing* functions, using an extensive network of interactive computers. The result is *computer-integrated manufacturing* (CIM), which is a broad term used to describe the computerized integration of product design, planning, production, distribution, and management. Computer-integrated manufacturing is a *methodology* and a *goal*, rather than merely an assemblage of equipment and computers.

The effectiveness of CIM greatly depends on the use of a large-scale **integrated communications system**, involving computers, machines, and their controls. The problems that may arise in such systems are described in Section 15.12. Furthermore, because CIM ideally should involve the total operation of a company, it requires an extensive database concerning the technical as well as business aspects of the operation. Consequently, if planned all at once, CIM can be prohibitively expensive, particularly for small and medium-size companies.

Implementation of CIM in existing plants may begin with the use of modules in selected phases of the operation; for new manufacturing plants, on the other hand, comprehensive and long-range strategic planning covering all phases of the operation is essential in order to benefit fully from CIM. Such plans must take into account considerations such as (a) the availability of resources; (b) the mission, goals, and culture of the organization; (c) existing as well as emerging technologies; and (d) the level of integration desired.

Computer-integrated manufacturing systems consist of *subsystems* that are integrated into a whole (Fig. 15.1); these subsystems consist of the following:

a) Business planning and support;
b) Product design;
c) Manufacturing process planning;
d) Process automation and control;
e) Factory-floor monitoring systems.

The subsystems are designed, developed, and implemented in such a manner that the output of one subsystem serves as the input of another subsystem, as shown by the various arrows in Fig. 15.1. Organizationally, the subsystems are generally divided into two functions: (a) **business-planning functions**, which include activities such as forecasting, scheduling, material-requirements planning, invoicing, and accounting; and (b) **business-execution functions**, which include production and process control, material handling, testing, and inspection.

The benefits of CIM include the following:

a) Responsiveness to shorter product life cycles (see Section 16.4), changing market demands, and global competition;
b) Emphasis on product quality and uniformity, implemented through better process control;
c) Better use of materials, machinery, and personnel and the reduction of *work-in-progress* (WIP) inventory, all of which improve productivity and lower product cost;
d) Better control of the production, scheduling, and management of the total manufacturing operation, resulting in lower product cost.

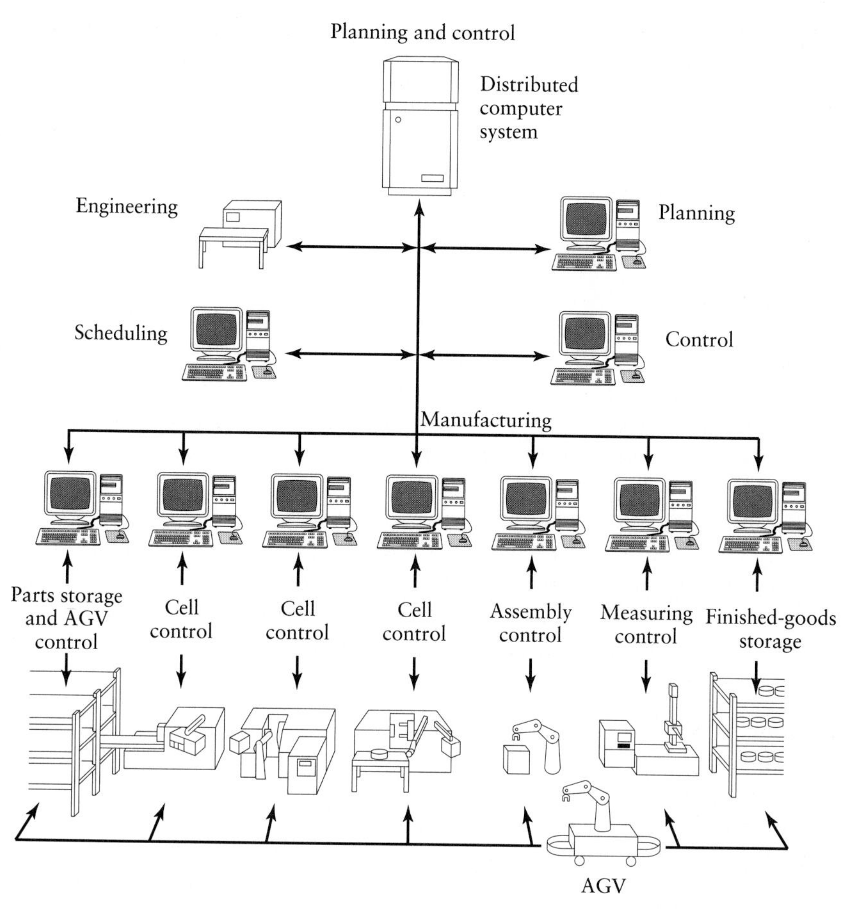

FIGURE 15.1 A schematic illustration of a computer-integrated manufacturing system. *Source*: U. Rembold et al., *Computer-Integrated Manufacturing and Engineering*, Addison-Wesley, 1993.

15.3.1 Databases

An effective computer-integrated manufacturing system requires a single, large database that is shared by members of the entire organization. *Databases* consist of up-to-date, detailed, and accurate data on products, designs, machines, processes, materials, production, finances, purchasing, sales, marketing, and inventory. This vast array of data is stored in computer memory and randomly recalled or modified as necessary, either by individuals in the organization or by the CIM system itself while it is controlling various aspects of design and production.

A database generally consists of the following information:

a) **Product data**, such as part shape, dimensions, tolerances, and specifications;
b) **Data management attributes**, such as owner, revision level, and part number;
c) **Production data**, such as the manufacturing processes used in making parts and products;
d) **Operational data**, such as scheduling, lot sizes, and assembly requirements;
e) **Resources data**, such as capital, machines, equipment, tooling, and personnel, and the capabilities of these resources.

Databases are built by individuals and through the use of various sensors in the machinery and equipment employed in production. Data from the latter are collected automatically by a **data acquisition system** (DAS) that can report, for example, the number of parts being produced per unit time, dimensional accuracy, surface finish, and weight, at specified rates of sampling. The components of DASs include microprocessors, transducers, and analog-to-digital converters (ADCs). Data acquisition systems are also capable of analyzing the data and transferring them to other computers for such purposes as statistical analysis, data presentation, and forecasting of product demand.

Several factors are important in the use and implementation of databases:

a) Databases should be up to date, accurate, user friendly, and easily accessible and shared.

b) In the event that there is a malfunction in data retrieval, the original, correct data must be recovered and restored in the database.

c) Because it is used for many purposes and by many people, the database must be flexible and responsive to the needs of users with different backgrounds and interests.

d) Databases should be accessed by designers, manufacturing engineers, process planners, financial officers, and the management of the company by using appropriate access codes. Obviously, companies must protect data against tampering or unauthorized use.

15.4 Computer-Aided Design and Engineering

Computer-aided design (CAD) involves the use of computers to create design drawings and geometric models of products and components (see also Fig. 1.2) and is usually associated with **interactive computer graphics**, known as a **CAD system**. *Computer-aided engineering* (CAE) simplifies the creation of the database by allowing several applications to share the information in the database. It basically consists of a comprehensive computer-based analysis and evaluation of engineering design. Its diverse applications include, for example, (a) finite-element analysis of stresses, strains, deflections, and temperature distribution in machine elements, load-bearing members, and structures; (b) generation, storage, and retrieval of numerical-control data for machining operations; and (c) design of integrated circuits and other microelectronic and micromechanical devices (see Chapter 13).

In CAD, the traditional drawing board is replaced by **workstations**, where the user can generate drawings or sections of a drawing on the computer monitor. The design is continuously displayed on the monitor and in different colors for its various components; the final drawing is printed or plotted on appropriate hardware connected to the computer. When using a CAD system, the designer can conceptualize the object to be designed on the graphics screen and can consider alternative designs or quickly modify a particular design to meet specific design requirements.

Through the use of powerful software, such as CATIA (after Computer-Aided Three-Dimensional Interactive Applications), the design can be subjected to *engineering analysis*, which can identify potential problems, such as an excessive load or deflection, and interference at mating surfaces during assembly. In addition to the design's geometric and dimensional features, other information, such as a list of materials, specifications, and manufacturing instructions, is stored in the CAD database. Using such information, the designer can also analyze the economics of alternative designs.

15.4.1 Exchange specifications

Because of the availability of a wide variety of CAD systems with different characteristics, supplied by different vendors, proper communication and exchange of data between these systems is essential. (See also Section 15.12.) **Drawing exchange format**

(DFX) was developed for use with Autodesk and has become a de facto standard, because of the long-term success of this software package. DFX is limited to transferring geometry information only. Similarly, **STL** (STereo Lithography) formats are used to export 3D geometries, initially to rapid prototyping systems; recently, however, STL has become a format for data exchange between CAD systems.

The need for a single, neutral format for better compatibility and for the transfer of more information than geometry alone is currently filled mainly by the **Initial Graphics Exchange Specification** (IGES). Vendors need only to provide translators for their own systems, to preprocess outgoing data into the neutral format, and to postprocess incoming data from the neutral format into their system. IGES is used for translation in two directions (into and out of a system) and is also used widely for translation of 3D line and surface data. Because IGES is evolving, there are many variations of IGES in existence.

Another specification is a solid-model-based standard called **Product Data Exchange Specification** (PDES), which is based on the Standard for the Exchange of Product Model Data (STEP) developed by the International Standards Organization. PDES allows information on shape, design, manufacturing, quality assurance, testing, maintenance, etc., to be transferred between CAD systems.

15.4.2 Elements of CAD systems

The design process in a CAD system consists of four stages, described as follows:

a) **Geometric modeling.** In *geometric modeling*, a physical object or any of its parts is described mathematically or analytically. The designer first constructs a geometric model by giving commands that create or modify lines, surfaces, solids, dimensions, and text that, together, compose an accurate and complete two- or three-dimensional representation of the object. The results of these commands are then displayed; the images can be moved around on the screen, and any section can be magnified in order to view its details. The models are stored in the database contained in computer memory.

The models can be presented in different ways, described as follows:

1. In **line representation** (also called **wire-frame representation**; see Fig. 15.2), all edges of the model are visible as solid lines. However, this type of image can be ambiguous, particularly for complex shapes; hence, various colors are generally used for different parts of the object. The three types of wire-frame representations are 2D, $2\frac{1}{2}$-D, and 3D. A 2D image shows the profile of the object, and a $2\frac{1}{2}$-D image can be obtained by *translational sweep*, that is, by moving the 2D object along the z-axis. For round objects, a $2\frac{1}{2}$-D model can be generated by simply *rotating* a 2D model around its axis.

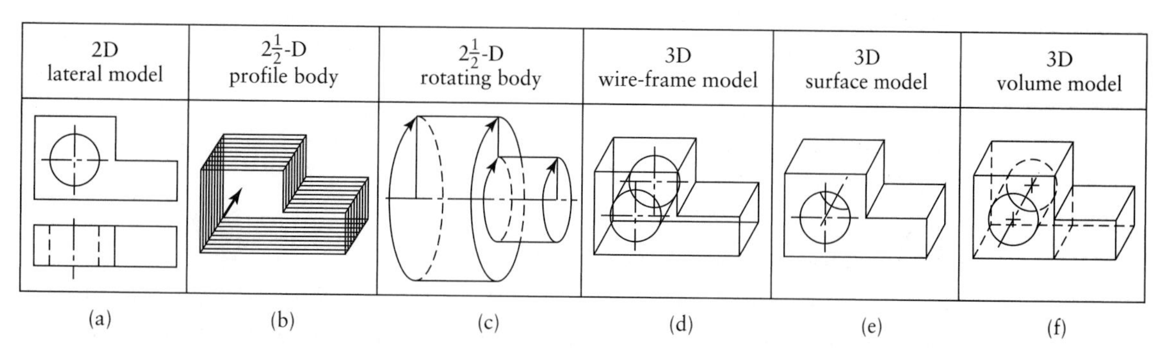

FIGURE 15.2 Various types of modeling for CAD.

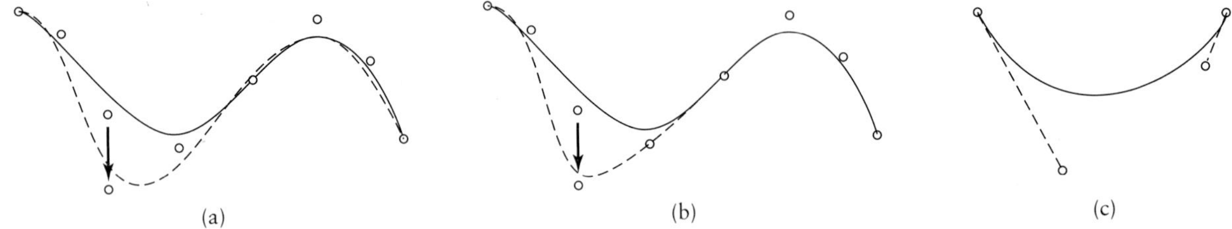

FIGURE 15.3 Types of splines. (a) A Bezier curve passes through the first and last control point and generates a curve from the other points. Changing a control point modifies the entire curve. (b) A B-spline is constructed piecewise, so that changing a vertex affects the curve only in the vicinity of the changed control point. (c) A third-order piecewise Bezier curve constructed through two adjacent control points, with two other control points defining the curve slope at the end points. A third-order piecewise Bezier curve is continuous, but its slope may be discontinuous.

2. In the **surface model**, all visible surfaces are shown in the model. Surface models define surface features and edges of objects. Modern CAD programs use Bezier curves, B-splines, or nonuniform rational B-splines (NURBS) for surface modeling. Each of these methods uses control points to define a polynomial curve or surface. A Bezier curve passes through the first and last vertex and uses the other control points to generate a blended curve. The drawback to Bezier curves is that modification of one control point will affect the entire curve. B-splines are a blended piecewise polynomial curve, where modification of a control point affects the curve only in the area of the modification. Figure 15.3 shows examples of two-dimensional Bezier curves and B-splines. A NURBS is a special kind of B-spline where each control point has a weight associated with it.
3. In the **solid model**, all surfaces are shown, but the data also describe the interior volume. Solid models can be constructed from *swept volumes* (Figs. 15.2b and c) or by the techniques shown in Fig. 15.4. In **boundary representation** (BREP), surfaces are combined to develop a solid model (Fig. 15.4a); in **constructive solid geometry** (CSG), simple shapes (called **primitives**

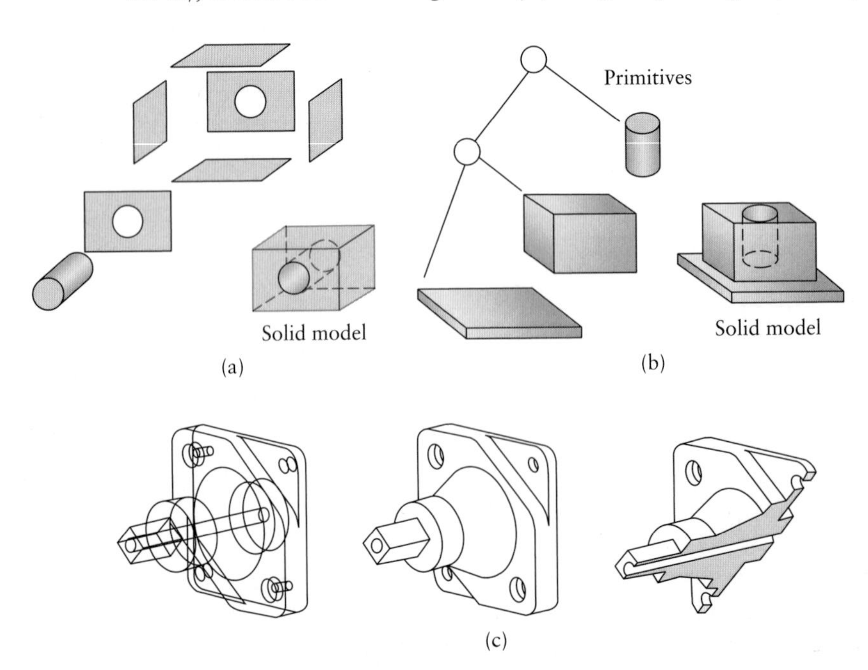

FIGURE 15.4 (a) Boundary representation of solids, showing the enclosing surfaces and the generated solid model. (b) A solid model represented as compositions of solid primitives. (c) Three representations of the same part by CAD. *Source*: P. Ranky.

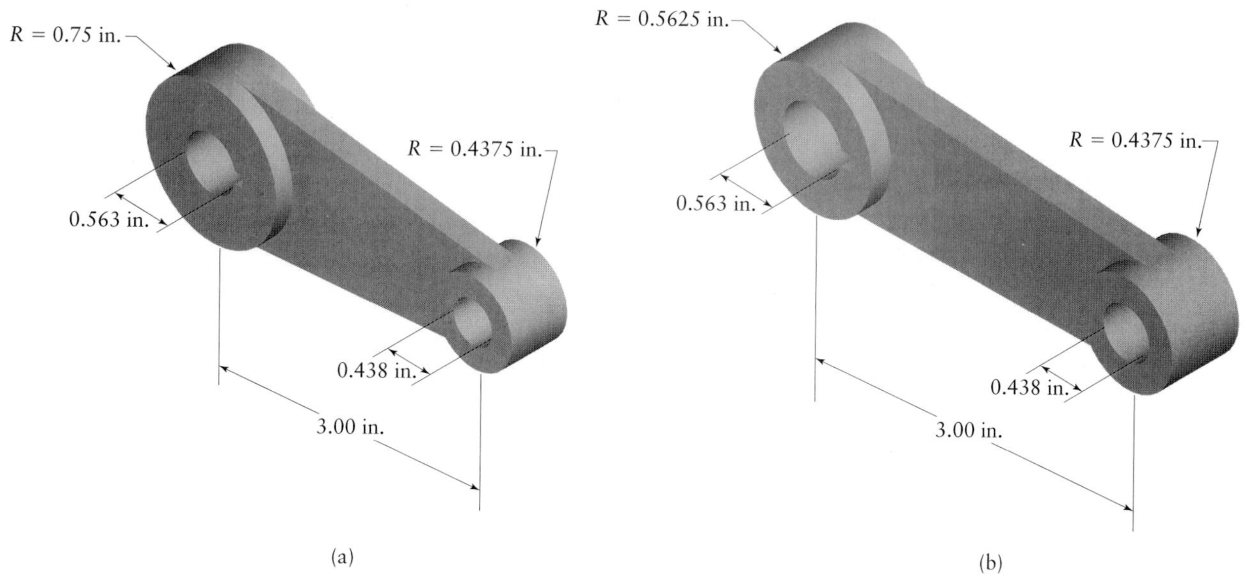

FIGURE 15.5 An example of parametric design. Dimensions of part features can be easily modified to quickly obtain an updated solid model.

of solids) such as spheres, cubes, blocks, cylinders, and cones are combined to develop a solid model (Fig. 15.4b). The user selects any combination of primitives and their sizes and combines them into the desired solid model. Although solid models have certain advantages, such as ease of design analysis and ease of preparation for manufacture of the part, they require more computer memory and processing time than the wire-frame and surface models.

A special kind of solid model is a **parametric model**, where a part not only is stored in terms of a BREP or CSG definition, but also is derived from the dimensions and constraints that define the features (Fig. 15.5). Whenever a change is made, the part is re-created from these definitions. This feature allows simple and straightforward updates and changes to be made to models.

4. Figure 15.6 shows the **octree representation** of a solid object; this type of model is basically a three-dimensional analog to pixels on a television screen or monitor. Just as any area can be broken down into quadrants, any volume

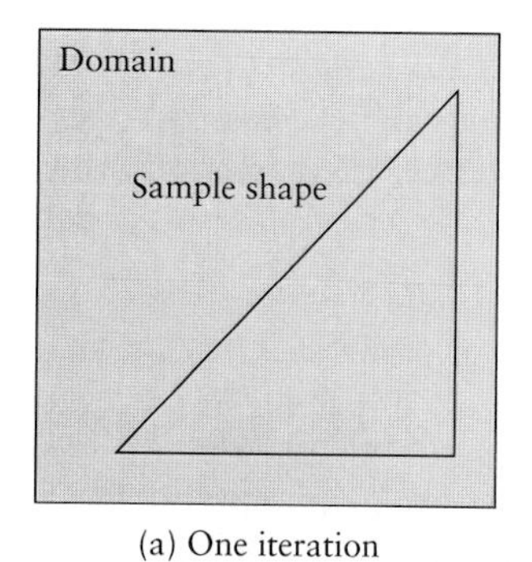

(a) One iteration

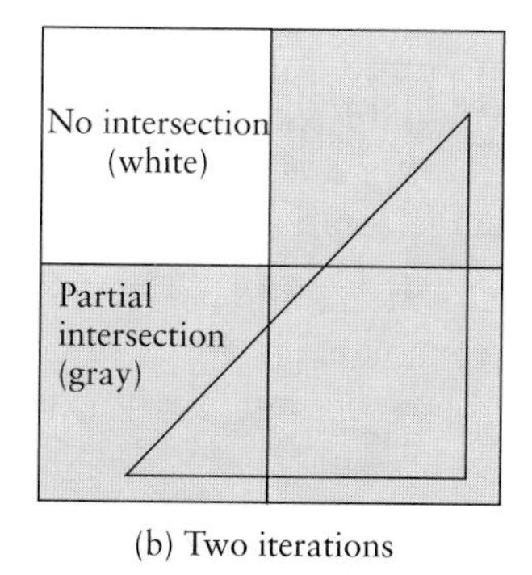

(b) Two iterations

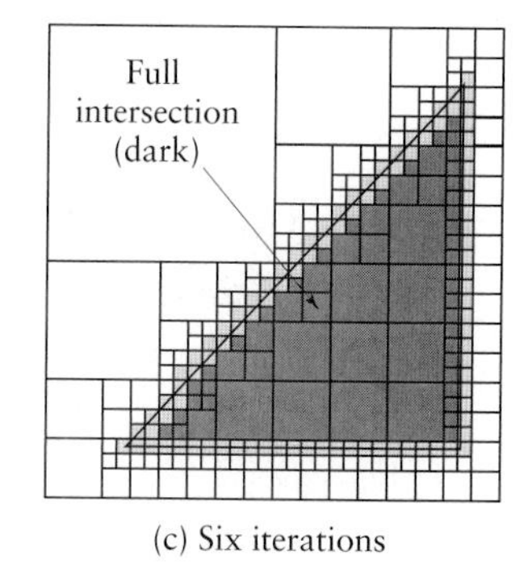

(c) Six iterations

FIGURE 15.6 The octree representation of a solid object. Any volume can be broken down into octants, which are then identified as solid, void, or partially filled. Shown is two-dimensional, or quadtree, version, for representation of shapes in a plane.

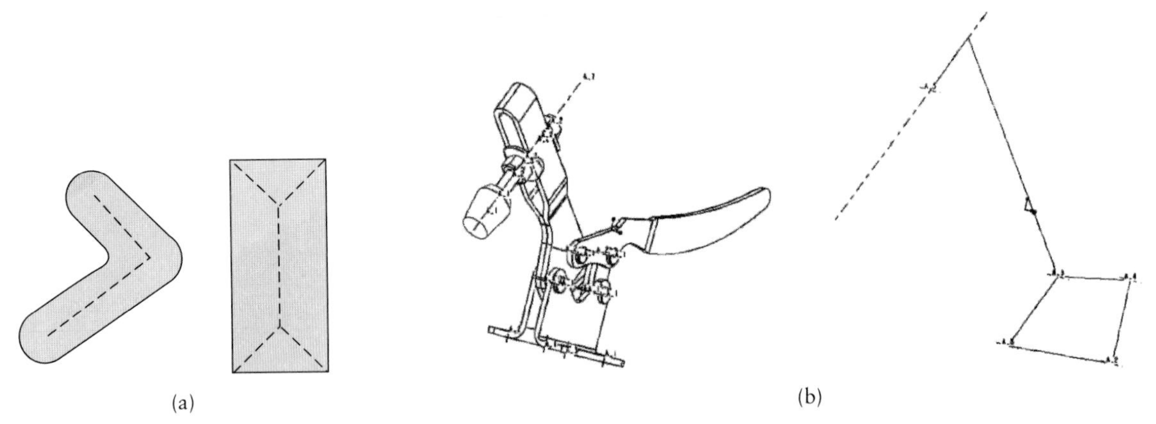

FIGURE 15.7 (a) Illustration of the skeleton data structure for solid objects. The skeleton is the dashed line in the object's interior. (b) A skeleton model used for kinematic analysis of a clamp. *Source*: S. D. Lockhart and C. M. Johnson, *Engineering Design Communication*, Prentice Hall, 2000.

can be broken down into *octants*, which are then identified as solid, void, or partially filled. Partially filled *voxels* (from *v*olume pi*xels*) are broken into smaller octants and reclassified. With increasing resolution, exceptional part detail can be achieved. This process may appear to be somewhat cumbersome, but it allows for accurate description of complex surfaces; it is used particularly in biomedical applications, such as modeling bone geometries.

5. A **skeleton** (Fig. 15.7) is commonly used for kinematic analysis of parts or assemblies. A skeleton is the family of lines, planes, and curves that describe a part without the detail of surface models. Conceptually, a skeleton can be constructed by fitting the largest circles (or spheres for three-dimensional objects) within the geometry; the skeleton is the set of points that connect the centers of the circles (or spheres). A current area of research involves using skeleton models instead of conventional surface or solid models.

b) **Design analysis and optimization.** After its geometric features have been determined, the design is subjected to an *engineering analysis*. This phase may consist of analyzing, for example, stresses, strains, deflections, vibrations, heat transfer, temperature distribution, or dimensional tolerances. Several software packages are available that have the capabilities to compute these quantities accurately and rapidly. Because of the relative ease with which such analyses can be done, designers can study a design more thoroughly before it moves on to production. Experiments and measurements in the field may nonetheless be necessary to determine and verify the actual effects of loads, temperature, and other variables on the designed components.

c) **Design review and evaluation.** An important design stage is *review and evaluation*, to check for any interference among various components. This step is necessary in order to avoid difficulties during assembly or use of the part and to determine whether moving members (such as linkages, gears, and cams) will operate and function as intended. Software with animation capabilities is available to identify potential problems with moving members and other dynamic situations. During the review and evaluation stage, the part is also precisely dimensioned and set within the full range of tolerance required for its production (Section 4.7).

d) **Documentation.** After the preceding stages have been completed, the design is reproduced by automated drafting machines, for documentation and reference. At this stage, detail and working drawings are also developed and printed. The CAD system is also capable of developing and drafting sectional views of the part, scaling the drawings, and performing transformations in order to present various views of the part.

e) **Database.** Many components are either standard components that are mass produced according to a given design specification (such as bolts or gears) or are identical to parts used in previous designs. Modern CAD systems therefore have a built-in database management system that allows designers to identify, view, and adopt parts from a library of stock parts. These parts can be parametrically modeled to allow cost-effective updating of the geometry. Some databases with extensive parts libraries are commercially available; many vendors make their part libraries available on the World Wide Web.

15.5 | Computer-Aided Manufacturing

Computer-aided manufacturing (CAM) involves the use of computers and computer technology to assist in all phases of manufacturing a product, including process and production planning, scheduling, manufacture, quality control, and management. Because of the increased benefits, computer-aided design and computer-aided manufacturing are often combined into **CAD/CAM systems**. This combination allows the transfer of information from the design stage to the planning stage for the manufacture of a product, without the need to manually reenter the data on part geometry. The database developed during CAD is stored; then it is processed further by CAM into the necessary data and instructions for operating and controlling production machinery and material-handling equipment and for performing automated testing and inspection for product quality (Section 4.8.3).

The emergence of CAD/CAM has had a major impact on manufacturing by standardizing product development and by reducing design effort, evaluation, and prototype work; it has made possible significant cost reductions and improved productivity. The two-engine Boeing 777 passenger airplane, for example, was designed completely by computer (**paperless design**), with 2000 workstations linked to eight computers. The plane is constructed directly from the CAD/CAM software developed (an enhanced CATIA system), and no prototypes or mock-ups were built, such as were required for previous models. The cost for this development was on the order of $6 billion.

An example of an important feature of CAD/CAM in machining is its capability to describe the *cutting-tool path* for various operations such as NC turning, milling, and drilling (see Section 14.4). The instructions (programs) are computer generated, and they can be modified by the programmer to optimize the tool path. The engineer or technician can then display and visually check the tool path for possible tool collisions with clamps or fixtures or for other interferences. The tool path can be modified at any time to accommodate other part shapes to be machined. CAD/CAM systems are also capable of *coding and classifying parts* into groups that have similar shapes, using alphanumeric coding. (See the discussion of *group technology* in Section 15.8.)

15.6 | Computer-Aided Process Planning

For a manufacturing operation to be efficient, all of its diverse activities must be planned and coordinated; this task has traditionally been done by process planners. *Process planning* involves selecting methods of production, tooling, fixtures, machinery,

FIGURE 15.8 An example of a simple routing sheet. These operation sheets may include additional information on materials, tooling, estimated time for each operation, processing parameters (such as cutting speeds and feeds), and other details. The routing sheet travels with the part from operation to operation. The current trend is to store all relevant data in computers and to affix to the part a bar code that serves as a key into the database of parts information.

ROUTING SHEET

CUSTOMER'S NAME: Midwest Valve Co. PART NAME: Valve body

QUANTITY: 15 PART NO.: 302

Operation no.	Description of operation	Machine
10	Inspect forging, check hardness	Rockwell tester
20	Rough machine flanges	Lathe No. 5
30	Finish machine flanges	Lathe No. 5
40	Bore and counterbore hole	Boring mill No. 1
50	Turn internal grooves	Boring mill No. 1
60	Drill and tap holes	Drill press No. 2
70	Grind flange end faces	Grinder No. 2
80	Grind bore	Internal grinder No. 1
90	Clean	Vapor degreaser
100	Inspect	Ultrasonic tester

the sequence of operations, the standard processing time for each operation, and methods of assembly; these choices are all documented on a **routing sheet** (Fig. 15.8). When done manually, this task is highly labor intensive and time consuming; also, it relies heavily on the experience of the process planner.

Computer-aided process planning (CAPP) accomplishes this complex task by viewing the total operation as an integrated system, so that the individual operations and steps involved in making each part are coordinated with each other and are performed efficiently and reliably. CAPP is thus an essential adjunct to CAD and CAM. Although it requires extensive software and good coordination with CAD/CAM as well as with other aspects of integrated manufacturing systems (described throughout the rest of this chapter), CAPP is a powerful tool for efficiently planning and scheduling manufacturing operations. It is particularly effective in small-volume, high-variety parts production requiring machining, forming, and assembly operations.

15.6.1 Elements of CAPP systems

There are two types of computer-aided process-planning systems: variant and generative process-planning systems. These types of systems are described as follows:

1. In the **variant system**, also called the **derivative system**, the computer files contain *a standard process plan* for a particular part to be manufactured. The search for a standard plan is made in the database, using a code number for the part; the plan is based on the part's shape and its manufacturing characteristics. (See Section 15.8.) The standard plan is retrieved, displayed for review, and printed as a routing sheet. The process plan includes information such as the types of tools and machines to be used, the sequence of operations to be performed, the speeds, the feeds, the time required for each sequence, and so on. Minor modifications to an existing process plan, which generally are necessary, can also be made. If the standard plan for a particular part is not in the computer files, a plan that is similar to it and that has a similar code number and an existing routing sheet is retrieved. If a routing sheet does not exist, one is prepared for the new part and stored in computer memory.

2. In the **generative system**, a process plan is automatically generated on the basis of the same logical procedures that would be followed by a traditional process planner in making that particular part. However, the generative system is complex, because it must contain comprehensive and detailed knowledge of the part's shape and dimensions, process capabilities, the selection of manufacturing methods, machinery, tools, and the sequence of operations to be performed. (Computers with such capabilities, known as **expert systems**, are described in Section 15.13.) The generative system is capable of creating a new plan instead of having to use or modify an existing plan, as the variant system must do. Although it currently is used less commonly than is the variant system, the generative system has such advantages as (a) flexibility and consistency for process planning for new parts and (b) higher overall planning quality, because of the capability of the decision logic in the system to optimize the planning and to use up-to-date manufacturing technology.

The process-planning capabilities of computers can be integrated into the planning and control of production systems. Process-planning activities are a subsystem of computer-integrated manufacturing, as described in Section 15.3. Several functions can be performed using these activities, such as **capacity planning** for plants to meet production schedules, control of *inventory*, *purchasing*, and *production scheduling*.

Advantages of CAPP systems. The advantages of CAPP systems over traditional process-planning methods include the following:

a) The standardization of process plans improves the productivity of process planners, reduces lead times and costs of planning, and improves the consistency of product quality and reliability.

b) Process plans can be prepared for parts that have similar shapes and features, and the plans can be easily retrieved to produce new parts.

c) Process plans can be modified to suit specific needs.

d) Routing sheets can be prepared more quickly, and compared with the traditional handwritten routing sheets, computer printouts are neater and much more legible.

e) Other functions, such as cost estimation and work standards, can be incorporated into CAPP.

15.6.2 Material-requirements planning systems and manufacturing resource planning systems

Computer-based systems for managing inventories and delivery schedules of raw materials and tools are called *material-requirements planning* (MRP) systems. This activity, sometimes regarded as a method of inventory control, involves keeping complete records of inventories of materials, supplies, parts in various stages of production (work in progress), orders, purchasing, and scheduling. Several files of data are usually involved in a master production schedule. These files pertain to the raw materials required (**bill of materials**), product structure levels (i.e., individual items that compose a product, such as components, subassemblies, and assemblies), and scheduling.

A further development is *manufacturing resource planning* (MRP-II) systems, which, through feedback, control all aspects of manufacturing planning. Although the system is complex, MRP-II is capable of final production scheduling, monitoring actual results in terms of performance and output, and comparing those results against the master production schedule.

15.6.3 Enterprise resource planning

Beginning the 1990s, *enterprise resource planning* (ERP) became an important trend; it is basically an extension of MRP-II. Although there are variations, it generally has been defined as a method for effective planning and control of all the resources needed in a business enterprise (i.e., companies) to take orders for products, produce them, ship them to the customer, and service them. ERP thus attempts to coordinate, optimize, and dynamically integrate all information sources and the widely diverse technical and financial activities in a manufacturing organization, as outlined in Chapters 14, 15, and 16. Among its major goals are to improve productivity, reduce manufacturing cycle times, and optimize processes, benefiting not only the organization, but also the customer, in an increasingly competitive global marketplace.

However, truly effective implementation of ERP is a difficult and challenging task. The main reasons are the following:

a) Difficulties are encountered in attempting timely, effective, and reliable communication among all parties involved, especially in a global business enterprise—thus teamwork is essential;
b) There is a need for changing and evolving business practices in an age where information systems and e-commerce have become highly relevant to the successful future of business organizations;
c) Extensive and specific hardware and software requirements must be met for ERP.

15.7 Computer Simulation of Manufacturing Processes and Systems

With increasing power and sophistication of computer hardware and software, one area that has grown rapidly is *computer simulation* of manufacturing processes and systems. Process simulation takes two basic forms:

a) It is a model of a specific operation, intended to determine the viability of a process or to optimize or improve its performance, and
b) It models multiple processes and their interactions, and it helps process planners and plant designers in the layout of machinery and facilities.

Individual processes have been modeled using various mathematical schemes. Finite-element analysis has been increasingly applied in software packages (**process simulation**) that are commercially available and inexpensive. Typical problems addressed by such models include **process viability** (such as assessing the formability and behavior of a certain sheet metal in a certain die) and **process optimization** (such as analyzing the metal-flow pattern in forging in a given die to identify potential defects, or improving mold design in casting to reduce or eliminate hot spots and minimize defects by promoting uniform cooling).

Simulation of an entire manufacturing system, involving multiple processes and equipment, helps plant engineers to organize machinery and to identify critical machinery elements. In addition, such models can assist manufacturing engineers with scheduling and routing by **discrete-event simulation**. Commercially available software packages are often used for these simulations, but the use of dedicated software programs written for a particular company and line of products is not unusual.

EXAMPLE 15.1 Simulation of plant-scale manufacturing

Several examples and case studies presented in this text have focused upon simulation of individual manufacturing processes. However, the proliferation of low-cost, high-performance computer systems and the development of advanced software have allowed the simulation of entire manufacturing systems and have led to the optimization of manufacturing and assembly operations.

As an example, Envision software, produced by the Delmia Corporation, allows the simulation of manufacturing processes in three dimensions, including the use of human manikins, to identify safety hazards, manufacturing problems, or bottlenecks; improve machining accuracy; or optimize tooling organization. Since simulation can be performed prior to building an assembly line, it can significantly reduce development times and cost. For example, Fig. 15.9 shows a simulation of a robotic welding line in an automotive plant, where a collision between a robot and a vehicle has been detected. The program for the robot can thereby be modified to prevent such collisions before the line is put into operation. While this example is a powerful demonstration of the utility of system simulation, the more common application is to optimize the sequence of operations and organization of machinery to reduce manufacturing costs.

Software also has the capability of conducting ergonomic analysis of various operations and machinery setups and therefore of identifying bottlenecks in the movement of parts, equipment, or personnel. The bottlenecks can then be relieved by the process planner by adjusting the automated or manual procedures at these locations. Using such techniques, a Daimler-Chrysler facility in Rastatt, Germany, was able to balance its production lines so that each worker is busy an average of 85% to 95% of the time.

Another application of systems simulation is planning of manufacturing operations to optimize production and to prepare for just-in-time production (Section 15.11). For example, if a car manufacturer needs to produce 1000 vehicles in a given period of time, production can be optimized by using certain strategies, such as by distributing the number of vehicles with sunroofs throughout the day

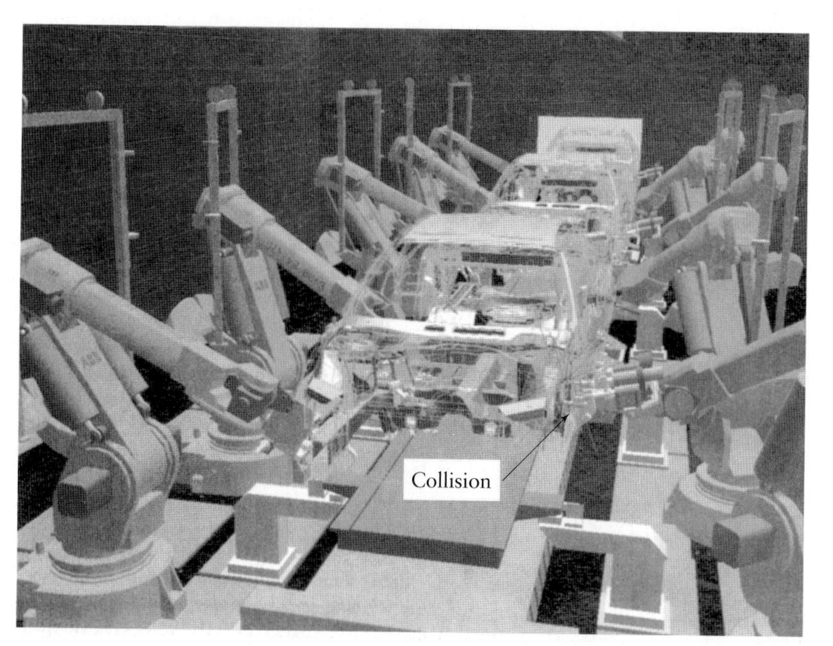

FIGURE 15.9 Simulation of a robotic welding station. A collision has been detected that production engineers can rectify before building the assembly line, thereby reducing development time and cost. *Source*: Mechanical Engineering, March 2001.

or by grouping vehicles by color, so that the number of paint changes in paint booths are minimized. With respect to just-in-time production, software such as that produced by ILOG Corporation can plan and schedule plant operations far enough in advance to order material as it is needed, thus eliminating stockpiled inventory.

The Nissan Corporation used the ILOG software to expand production at its Sunderland, England, facility. The plant was already producing 7,000 vehicles per week when Nissan decided to increase production by 3,000 vehicles per week. The conventional solution would have been to construct an additional assembly line, at great capital investment. However, it was noted that the existing scheduling software, based on the input of plant managers, did not take into account variables such as part availability and tool changeover, so that the plant adhered to its schedule only 3% of the time. The ILOG software, on the other hand, allows optimization of the manufacturing sequence and schedule, enabling cars to cross over from assembly lines to paint lines and tooling lines as needed to increase throughput. The result was that the Sunderland plant met the vehicle scheduling targets 85% of the time, eliminating much of the time-consuming task of rescheduling vehicles while they wait in storage buffers between the main sections of the plant. This approach led to increased production, and the cost of the software was recovered in three days of production.

Source: Adapted from *Mechanical Engineering*, March 2001.

15.8 Group Technology

It can be seen throughout this text that many parts produced have certain similarities in their shape and in their method of manufacture. Traditionally, each part has been viewed as a separate entity, and each has been produced in individual batches. *Group technology* (GT) is a concept that seeks to take advantage of the **design and processing similarities** among the parts to be produced. This concept, first developed in Europe in the early 1900s, starts by categorizing the parts and recording them; the designs are then retrieved as needed. The term *group technology* was first used in 1959, but not until the use of interactive computers became widespread in the 1970s did this technology develop significantly.

The similarity in the characteristics of a group of parts (Fig. 15.10) suggests that benefits can be obtained by *classifying* and *coding* these parts into *families*. Surveys in manufacturing plants have repeatedly indicated the commonness of similarity in parts. Such surveys consist of breaking each product into its components and then identifying the similar parts; one survey, for instance, found that 90% of the 3000 parts made by a particular company fell into only five major families of parts. As an example, a pump can be broken into such basic components as the motor, the housing, the shaft, the seals, and the flanges. In spite of the variety of pumps manufactured, each of these components is basically the same in terms of design and manufacturing methods. Consequently, all shafts, for example, can be categorized into one family of shafts. Also, questions have been raised, for example, about why a particular product has so many different sizes of fasteners, such as bolts and nuts.

The group-technology approach becomes even more attractive in view of consumer demand for an ever-greater variety of products, each in smaller quantity, thus involving *batch production* (see Section 14.2.2). Maintaining high efficiency in batch operations can be difficult, and overall manufacturing efficiency can be adversely affected because nearly 75% of manufacturing today is batch production. The traditional

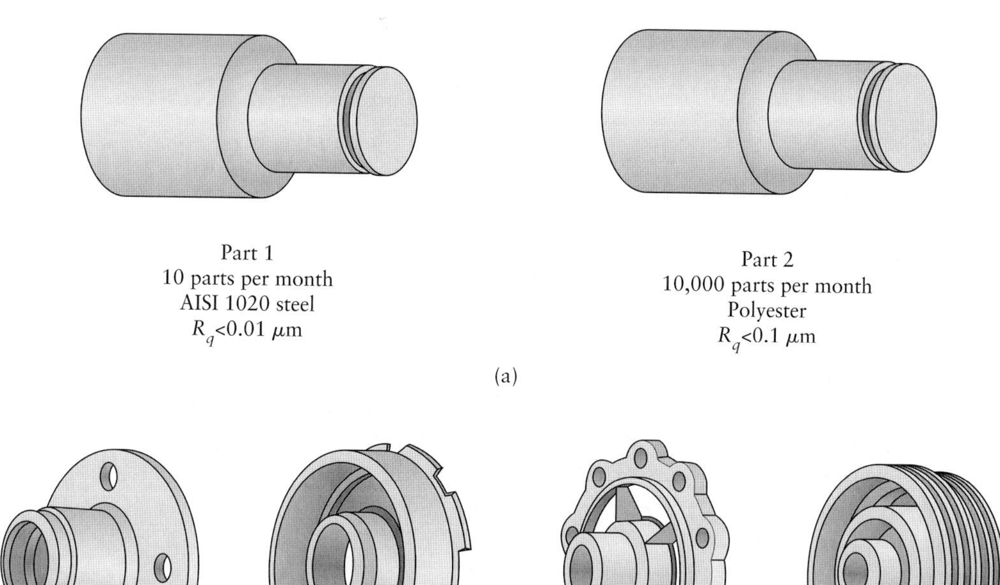

(b)

FIGURE 15.10 (a) Two parts with identical geometries but different manufacturing attributes. (b) Parts with similar manufacturing attributes but different geometries.

product flow in a batch manufacturing operation is shown in Fig. 15.11a. Note that machines of the same type are arranged in groups, that is, groups of lathes, of milling machines, of drill presses, and of grinders. In such a layout (**functional layout**), there is usually considerable random movement on the production floor, as shown by the arrows, which indicate movement of materials and parts. Such an arrangement is not efficient, because it wastes time and effort. The machines in *cellular manufacturing* (Section 15.9) are arranged in a more efficient product flow line (**group layout**; Fig. 15.11b).

15.8.1 Advantages of group technology

The major advantages of group technology are the following:

a) Group technology makes possible standardization of part designs and minimization of design duplication. New part designs can be developed using similar, but previously specified, designs, and in this way, a significant amount of time and effort can be saved. The product designer can quickly determine whether data on a similar part already exist in the computer files.

b) Data that reflect the experience of the designer and the process planner are stored in the database. A new and less experienced engineer can, therefore, quickly benefit from that experience by retrieving any of the previous designs and process plans.

c) Manufacturing costs can be estimated more easily, and the relevant statistics on materials, processes, number of parts produced, and other factors can be more easily obtained.

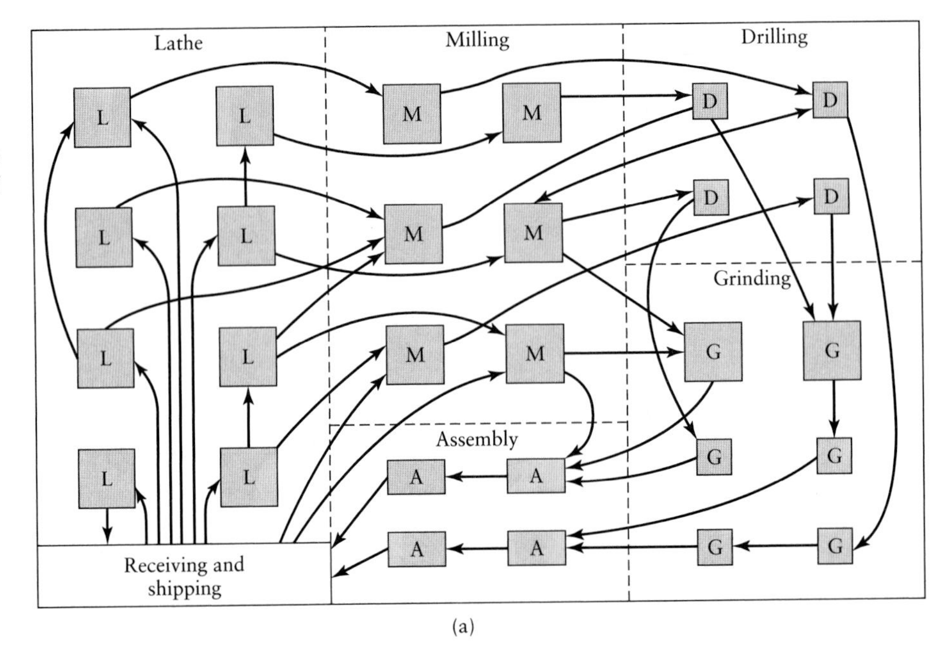

(a)

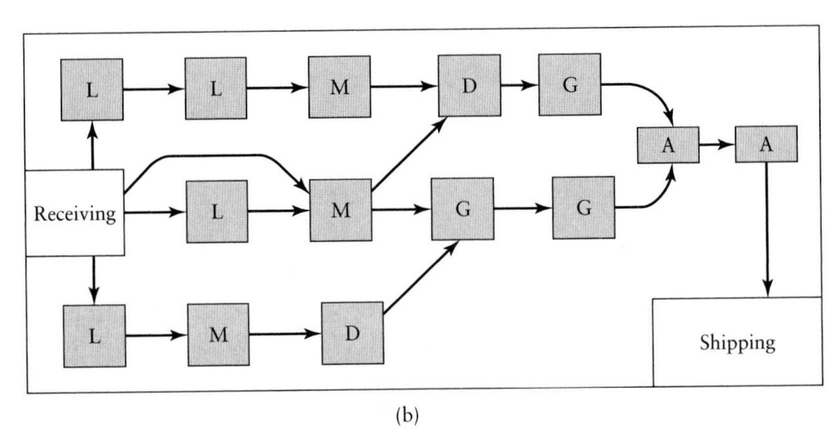

(b)

FIGURE 15.11 (a) Functional layout of machine tools in a traditional plant. Arrows indicate the flow of materials and parts in various stages of completion. (b) Group-technology (cellular) layout. Legend: L = lathe; M = milling machine; D = drilling machine; G = grinding machine; A = assembly. *Source*: M. P. Groover.

d) Process plans are standardized and scheduled more efficiently, orders are grouped for more efficient production, and machine utilization is improved. Furthermore, setup times are reduced, and parts are produced more efficiently and with better and more consistent product quality. Also, similar tools, fixtures, and machinery are shared in the production of a family of parts, and programming for NC is more automated.

e) With the implementation of CAD/CAM, cellular manufacturing, and CIM, group technology is capable of greatly improving productivity and reducing costs in small-batch production (almost approaching those of mass production). Depending on the level of implementation, potential savings in each of the various design and manufacturing phases have been estimated to range from 5% to as much as 75%.

15.8.2 Classification and coding of parts

In group technology, parts are identified and grouped into families by **classification and coding (C/C) systems.** This process is a critical and complex first step in GT, and it is done according to the part's design and manufacturing attributes (see Fig. 15.10):

1. **Design attributes** pertain to similarities in geometric features and consist of the following:
   a) External and internal shapes and dimensions;
   b) Aspect ratio (length-to-width ratio or length-to-diameter ratio);
   c) Dimensional tolerance;
   d) Surface finish;
   e) Part function.
2. **Manufacturing attributes** pertain to similarities in the methods and the sequence of the operations performed on the part. As described throughout this text, selection of a manufacturing process or processes depends on numerous factors, including the workpiece; consequently, manufacturing and design attributes are interrelated. The manufacturing attributes of a particular part or component consist of the following:
   a) Primary processes;
   b) Secondary and finishing processes;
   c) Dimensional tolerances and surface finish;
   d) Sequence of operations performed;
   e) Tools, dies, fixtures, and machinery;
   f) Production quantity and production rate.

In reviewing these lists, it can be appreciated that the coding can be time consuming and that it requires considerable experience in the design and manufacture of products. In its simplest form, coding can be done by viewing the shapes of the parts in a generic way and then classifying the parts accordingly, such as in the categories of parts having rotational symmetry, parts having rectilinear shape, and parts having large surface-to-thickness ratios. Parts may also be classified by studying their production flow throughout the total manufacturing cycle, called **production-flow analysis** (PFA). Recall from Section 15.6 that routing sheets clearly show process plans and the operations to be performed. One drawback of PFA is that a particular routing sheet does not necessarily indicate that the total operation is optimized. In fact, depending on the experience of the particular process planner, routing sheets for manufacturing the same part can be quite different. The benefits of computer-aided process planning as regards avoiding such problems are obvious.

15.8.3 Coding

Coding of parts may be based on a particular company's own coding system, or it may be based on one of several commercially available classification and coding systems. Because of widely varying product lines and organizational needs, none of the C/C systems has been universally adopted. Whether it was developed in-house or it was purchased, the system must be compatible with the company's other systems, such as NC machinery and CAPP systems. The *code structure* for part families typically consists of numbers, of letters, or of a combination of the two. Each specific component of a product is assigned a code; this code may pertain to design attributes only (generally less than 12 digits) or to manufacturing attributes only, although most advanced systems include both, using as many as 30 digits.

The three basic **levels of coding** outlined as follows vary in their degree of complexity:

1. **Hierarchical coding.** In this type of coding, also called **monocode**, the interpretation of each succeeding digit depends on the value of the preceding

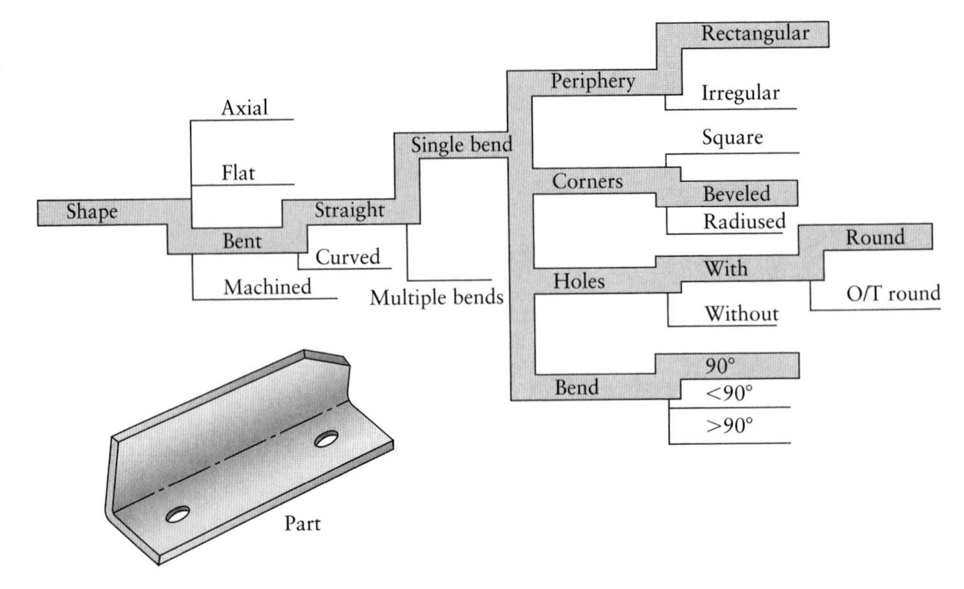

FIGURE 15.12 Decision-tree classification for a sheet-metal bracket. *Source*: G. W. Millar.

digit. Each symbol amplifies the information contained in the preceding digit, so a single digit in the code cannot be interpreted alone. The advantage of this system is that a short code can contain a large amount of information. This method is, however, difficult to apply in a computerized system.

2. **Polycodes.** Each digit in this code, also known as **chain-type code**, has its own interpretation, which does not depend on the preceding digit. This structure tends to be relatively long, but it allows the identification of specific part attributes and is well suited to computer implementation.
3. **Decision-tree coding.** This system, also called **hybrid codes**, is the most advanced, and it combines both design and manufacturing attributes (Fig. 15.12).

15.8.4 Coding systems

Three major industrial coding systems are described as follows:

a) The **Opitz system** was developed in the 1960s in Germany by H. Opitz (1905–1977), and it was the first comprehensive coding system ever presented. The basic code consists of nine digits (in the format 12345 6789) that represent design and manufacturing data (Fig. 15.13); four additional codes (in the format ABCD) may be used to identify the type and sequence of production operations. This system has two drawbacks: (a) It is possible to have different codes for parts that have similar manufacturing attributes, and (b) several parts with different shapes can have the same code.

b) The **MultiClass system** was originally developed under the name MICLASS, for Metal Institute Classification System (Fig. 15.14), for the purpose of helping automate and standardize several design, production, and management functions in an organization. This system involves up to 30 digits; it is used interactively with a computer that asks the user a number of questions. On the basis of the answers given, the computer automatically assigns a code number to the part.

c) The **KK-3 system** is a general-purpose classification and coding system for parts that are to be machined or ground. It uses a 21-digit decimal system. This code is much greater in length than the two coding systems described previously, but it classifies dimensions and dimensional ratios such as the length-to-diameter ratio of the part. The structure of a KK-3 system for rotational components is shown in Fig. 15.15.

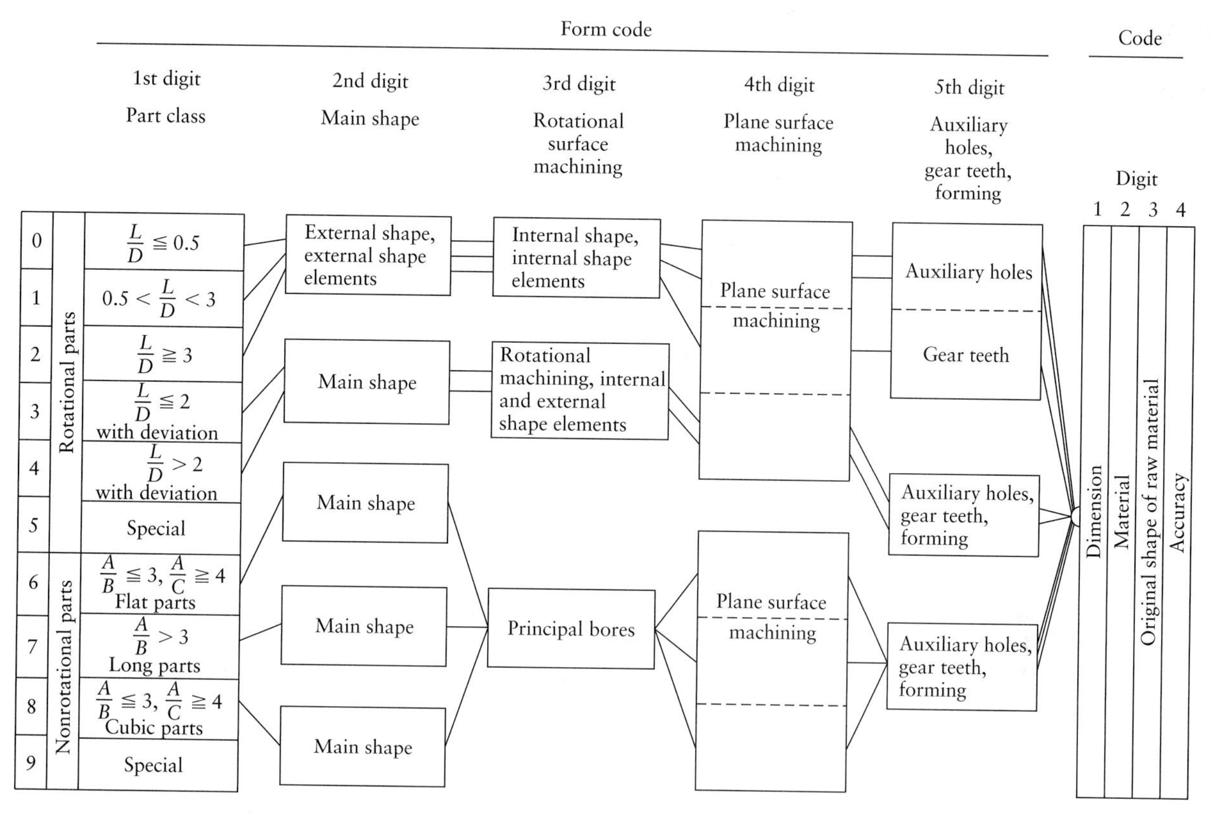

FIGURE 15.13 A classification and coding scheme using the Opitz system, consisting of five digits and a supplementary code of four digits.

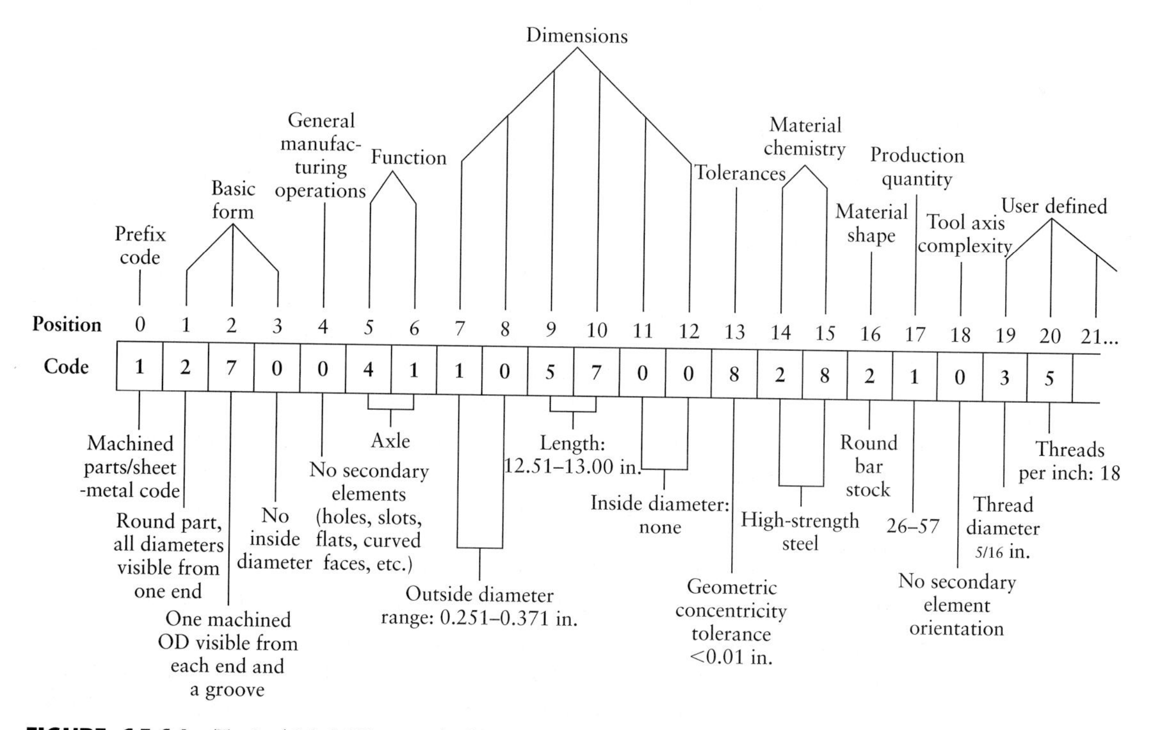

FIGURE 15.14 Typical MultiClass code for a machined part. *Source*: Organization for Industrial Research.

FIGURE 15.15 The structure of a KK-3 system for rotational components. *Source*: Japan Society for the Promotion of Machine Industry.

Digit	Items			(Rotational component)
1	Part name			General classification
2				Detail classification
3	Materials			General classification
4				Detail classification
5	Major dimensions			Length
6				Diameter
7	Primary shapes and ratio of major dimensions			
8	Shape details and kinds of processes	External surface		External surface and outer primary shape
9				Concentric screw-threaded parts
10				Functional cut-off parts
11				Extraordinary shaped parts
12				Forming
13				Cylindrical surface
14		Internal surface		Internal primary shape
15				Internal curved surface
16				Internal flat surface and cylindrical surface
17		End surface		
18		Nonconcentric holes		Regularly located holes
19				Special holes
20		Noncutting process		
21	Accuracy			

15.9 Cellular Manufacturing

The concept of group technology can be effectively implemented in *cellular manufacturing*, which consists of one or more manufacturing cells. A **manufacturing cell** is a small unit, composed of one to several *workstations*, within a manufacturing system. A workstation usually contains either one machine (**single-machine cell**) or several machines (**group-machine cell**) that each perform a different operation on the part. The machines can be modified, retooled, and regrouped for different product lines within the same family of parts. Manufacturing cells are particularly effective in producing families of parts for which there is relatively constant demand.

Cellular manufacturing has thus far been used primarily in material removal (Chapters 8 and 9) and in sheet-metal forming operations (Chapter 7). The machine tools commonly used in manufacturing cells are lathes, milling machines, drills, grinders, electrical-discharge machines, and machining centers; for sheet-metal forming, the equipment consists of shearing, punching, bending, and other forming machines. Cellular manufacturing has some degree of automatic control for the following typical operations:

1. Loading and unloading of raw materials and workpieces at workstations;
2. Changing of tools and dies at workstations, thus reducing labor and time;
3. Transfer of workpieces and tools and dies between workstations;
4. Scheduling and controlling of the total operation in the manufacturing cell.

Central to these activities is a *material-handling system* for transferring materials and parts among workstations. In attended (manned) machining cells, materials can be moved and transferred manually by the operator or by an industrial robot (Section 14.7) located centrally in the cell. Automated inspection and testing equipment may also be components of such a cell.

The significant benefits of cellular manufacturing are the economics of reduced work in progress, improved productivity, and the fact that product quality problems are detected right away. Furthermore, because of the variety of the machines and the processes involved, the operator becomes multifunctional and is not subjected to the tedium experienced with always working on the same machine.

Manufacturing cell design. Because of the unique features of manufacturing cells, their design and implementation in traditional plants inevitably requires the reorganization of the plant and the rearrangement of existing product flow lines. The machines may be arranged along a line, a *U* shape, an *L* shape, or in a loop. For a group-machine cell, where the materials are handled by the operator, the *U*-shaped arrangement is convenient and efficient, because the operator can easily reach various machines. With mechanized or automated material handling, the linear arrangement and the loop layout are more efficient. Selecting the best arrangement of machines and material-handling equipment also involves taking into account such factors as the production rate, the type of product, and its shape, size, and weight.

Flexible manufacturing cells. In the introduction to this chapter and in Chapter 14, it was stressed that in view of rapid changes in market demand and the need for more product variety in smaller quantities, flexibility of manufacturing operations is highly desirable. Manufacturing cells can be made flexible by using CNC machines and machining centers (see Section 8.10) and by means of industrial robots or other mechanized systems for handling materials. Figure 15.16 shows an example of a *flexible manufacturing cell* (FMC) for machining operations.

Flexible manufacturing cells are usually untended, so their design and operation are more exacting than those for other cells. The selection of machines and robots, including the types and capacities of end effectors and their control systems, is important to the proper functioning of the FMC. The likelihood of significant change in demand for part families should be considered during design of the cell, to ensure that the equipment involved has the proper flexibility and capacity. As is the case for other flexible manufacturing systems (described in Section 15.10), the cost of flexible manufacturing cells is high. Cellular manufacturing generally requires more machine tools and thus increases manufacturing cost; however, this disadvantage is

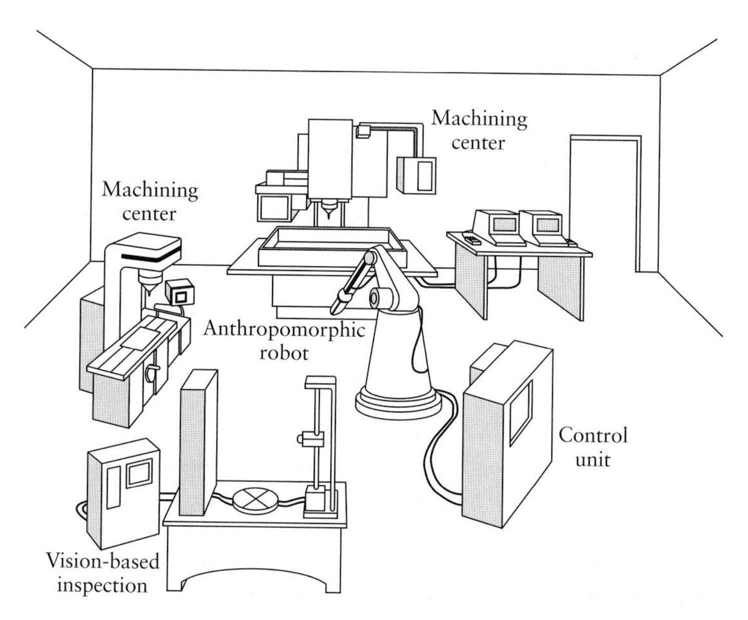

FIGURE 15.16 Schematic view of a flexible manufacturing cell, showing two machine tools, an automated part inspection system, and a central robot serving these machines. *Source*: P. K. Wright.

outweighed by increased speed of manufacture, flexibility, and controllability. Proper maintenance of the tools and the machinery is essential, as is the implementation of two- or three-shift operation of the cells.

EXAMPLE 15.2 Manufacturing cells in small shops

Beginning in the mid-1990s, large companies have increasingly outsourced machining operations to their first-tier suppliers. This trend has created pressure for small job shops to develop the capability to perform sophisticated projects, and it also has created a demand for larger firms to take on lower volume, or turnkey, operations. These two basic market factors have driven a number of trends in the design and implementation of flexible manufacturing cells:

- Manufacturing cells are being developed with a smaller number of machines, but greater versatility. Instead of four or five machines, many new manufacturing cells consist of only one or two machines.
- Advanced CAD systems are being used by smaller firms. While high-end CAD systems running on computer workstations are available at larger corporations, increasing computational power and software accessibility has allowed smaller firms to use less expensive, but still very powerful, PC-based packages.
- PC controllers are becoming more popular and are driving a push to open architecture. Historically, machine manufacturers maintained full control over their control systems, and PC-based control was limited. Open architecture enables machine control to take place from a PC and therefore allows much greater flexibility in defining environments for machine operators.
- PC-based software is now available to translate solid-model part descriptions into surfaces for machining.

15.10 Flexible Manufacturing Systems

A *flexible manufacturing system* (FMS) integrates all major elements of manufacturing into a highly automated system. First used in the late 1960s, FMS consists of a number of manufacturing cells, each containing an industrial robot (serving several CNC machines) and an automated material-handling system, all interfaced with a central computer. Different computer instructions for the manufacturing process can be downloaded for each successive part that passes through the workstation.

This system is highly automated and is capable of optimizing each step of the total manufacturing operation. These steps may involve one or more processes and operations (such as machining, grinding, cutting, forming, powder metallurgy, heat treating, and finishing), as well as handling of raw materials, inspection, and assembly. The most common applications of FMS to date have been in machining and assembly operations. A variety of FMS technology is available from machine-tool manufacturers. Flexible manufacturing systems represent the highest level of efficiency, sophistication, and productivity that has been achieved in manufacturing plants (Fig. 15.17). The flexibility of FMS is such that it can handle a variety of part configurations and produce them in any order.

FMS can be regarded as a system that combines the benefits of two other systems: (a) the highly productive, but inflexible, transfer lines (Section 14.2.4) and (b)

FIGURE 15.17 A general view of a flexible manufacturing system, showing several machine tools and an automated guided vehicle. *Source*: Courtesy of Cincinnati Machine, a UNOVA Company.

job-shop production, which can fabricate a large variety of products on stand-alone machines, but is inefficient. Table 15.1 compares some of the characteristics of transfer lines and FMS. Note that in FMS, the time required for changeover to a different part is very short. The quick response to product and market-demand variations is a major benefit of FMS.

Elements of FMS. The basic elements of a flexible manufacturing system are (a) workstations, (b) automated handling and transport of materials and parts, and (c) control systems. The workstations are arranged to yield the greatest efficiency in production, with an orderly flow of materials, parts, and products through the system.

The types of machines in workstations depend on the type of production. For machining operations (Chapter 8), they usually consist of a variety of three- to five-

TABLE 15.1

Comparison of the Characteristics of Transfer Lines and Flexible Manufacturing Systems

Characteristic	Transfer line	FMS
Types of parts made	Generally few	Infinite
Lot size	> 100	1–50
Part changing time	$\frac{1}{2}$ to 8 hr	1 min
Tool change	Manual	Automatic
Adaptive control	Difficult	Available
Inventory	High	Low
Production during breakdown	None	Partial
Efficiency	60–70%	85%
Justification for capital expenditure	Simple	Difficult

axis machining centers, CNC lathes, milling machines, drill presses, and grinders. Also included is various other equipment, such as that for automated inspection (including coordinate measuring machines), assembly, and cleaning. Other types of operations suitable for FMS include sheet-metal forming, punching and shearing, and forging systems for these operations incorporate furnaces, forging machines, trimming presses, heat-treating facilities, and cleaning equipment.

Because of the flexibility of FMS, the proper operation of material-handling, storage, and retrieval systems is very important. Material handling is controlled by a central computer and performed by automated guided vehicles, conveyors, and various transfer mechanisms. The system is capable of transporting raw materials, blanks, and parts in various stages of completion to any machine, in any order, at any time. Prismatic parts are usually moved on specially designed **pallets**. Parts having rotational symmetry (such as parts for turning operations) are usually moved by mechanical devices and robots.

The computer control system of FMS is its brains and includes various software and hardware. This subsystem controls the machinery and equipment in workstations and the transportation of raw materials, blanks, and parts in various stages of completion from machine to machine. It also stores data and provides communication terminals that display the data visually.

Scheduling. Because FMS involves a major capital investment, efficient machine utilization is essential: Machines must not stand idle. Consequently, proper scheduling and process planning are crucial. Scheduling for FMS is *dynamic*, unlike that in job shops, where a relatively rigid schedule is followed to perform a set of operations. The scheduling system for FMS specifies the types of operations to be performed on each part, and it identifies the machines or manufacturing cells to be used. Dynamic scheduling is capable of responding to quick changes in product type and thus is responsive to real-time decisions.

Because of the flexibility provided by FMS, no setup time is wasted in switching between manufacturing operations; the system is capable of performing different operations in different orders and on different machines. However, the characteristics, performance, and reliability of each unit in the system must be checked to ensure that parts moving from workstation to workstation are of acceptable quality and dimensional accuracy.

15.11 Just-in-Time Production

The *just-in-time production* (JIT) concept was first implemented in Japan to eliminate waste of materials, machines, capital, manpower, and inventory throughout the manufacturing system. The JIT concept has the following goals:

1. Receive supplies just in time to be used;
2. Produce parts just in time to be made into subassemblies;
3. Produce subassemblies just in time to be assembled into finished products;
4. Produce and deliver finished products just in time to be sold.

In traditional manufacturing, the parts are made in batches, placed in inventory, and used whenever necessary. This approach is known as a **push system**, meaning that parts are made according to a schedule and are in inventory to be used if and when they are needed. In contrast, just-in-time manufacturing is a **pull system**, meaning that parts are produced to order, and the production is matched with demand for the final assembly of products. There are no stockpiles, and thus the extra motions and expenses involved in stockpiling parts and then retrieving them from storage are eliminated. Hence, the ideal production quantity is one product; therefore, this system is also called **zero inventory, stockless production**, and **demand scheduling**. Moreover, parts are inspected by the worker as they are being manufactured and are used within a short period of time. In this way, the worker maintains continuous production control, immediately identifying defective parts and reducing process variation in order to produce quality products. The worker takes pride in good product quality.

Implementation of the JIT concept requires that all aspects of manufacturing operations be continuously reviewed and monitored, so that all operations and resources that do not add value are eliminated. This approach emphasizes pride and dedication in producing high-quality products; elimination of idle resources; and the use of teamwork among workers, engineers, and management to quickly solve any problems that arise during production or assembly. The ability to detect production problems has been likened to the level of water (representing the inventory levels) in a lake covering a bed of boulders (representing production problems). When the water level is high (the high inventories associated with push production), the boulders are not exposed. By contrast, when the level is low (the low inventories associated with pull production), the boulders are exposed; consequently, the problems can be identified and removed. This analogy indicates that high inventory levels can mask quality and production problems involving parts that are already stockpiled.

An important aspect of the JIT concept is the delivery of supplies and parts from outside sources and from other divisions of the company, so as to significantly reduce in-plant inventory. As a result, major reductions in storage facilities can take place, and storage space can be reclaimed for productive purposes; in fact, the concept of building large warehouses for parts has now become obsolete. Suppliers are expected to deliver, often on a daily basis, preinspected goods as they are needed for production. Consequently, this approach requires reliable suppliers, close cooperation and trust between the company and its vendors, and a reliable system of transportation. Also important for smoother operation is reduction of the number of suppliers. In one example, an Apple Computer plant reduced the number of suppliers from 300 to 70.

Kanban. Although the basic concept of JIT originated in the United States decades ago, it was first demonstrated on a large scale in 1953 at the Toyota Motor Company, under the name *kanban*, meaning "visible record." These records usually consist of two types of cards (kanbans): (a) the **production card**, which authorizes the *production* of one container or cart of identical, specified parts at a workstation, and (b) the **conveyance card**, or **move card**, which authorizes the *transfer* of one container or cart of parts from that particular workstation to the workstation where the parts will be used next. The cards, which now consist of bar-coded plastic tags or other devices, contain information about the type of the part, the place of issue, the part number, and the number of items in the container. The number of containers in circulation at any time is completely controlled and can be scheduled as desired for maximum production efficiency.

Advantages of JIT. The advantages of just-in-time production may be summarized as follows:

- Low inventory-carrying costs;
- Fast detection of defects in the production or the delivery of supplies and, hence, low scrap loss;
- Reduced need for inspection and reworking of parts;
- Production of high-quality parts at low cost.

Although there are significant variations, implementation of just-in-time production has resulted in reductions estimated at 20% to 40% in product cost; 60% to 80% in inventory; up to 90% in rejection rates; 90% in lead times; and 50% in scrap, rework, and warranty costs. Furthermore, increases of 30% to 50% in direct labor productivity and of 60% in indirect labor productivity have also been achieved.

15.12 Communications Networks in Manufacturing

In order to maintain a high level of coordination and efficiency of operation in integrated manufacturing, an extensive, high-speed, and interactive **communications network** is required. A major advance in communications technology is the **local area network** (LAN). In this hardware and software system, logically related groups of machines and equipment, such as those in a manufacturing cell, communicate with each other; the network links these groups to each other, bringing different phases of manufacturing into a unified operation. A local area network can be very large and complex, linking hundreds, or even thousands, of machines and devices in several buildings; various network layouts (Fig. 15.18) of fiber-optic or copper cables are typically used, over distances ranging from a few meters to as much as 32 km (20 mi). For longer distances, **wide area networks** (WANs) are used. Different types of networks can be linked or integrated through "gateways" and "bridges."

Access control to the network is important; otherwise, collisions can occur when several workstations are transmitting simultaneously. Thus, continuous scanning of the transmitting medium is essential. In the 1970s, a *carrier sense multiple access with collision detection* (CSMA/CD) system was developed and implemented in **Ethernet.** Other access-control methods are **token ring** and **token bus,** in which a "token" (special message) is passed from device to device; the device that has the token is allowed to transmit, while all other devices only receive.

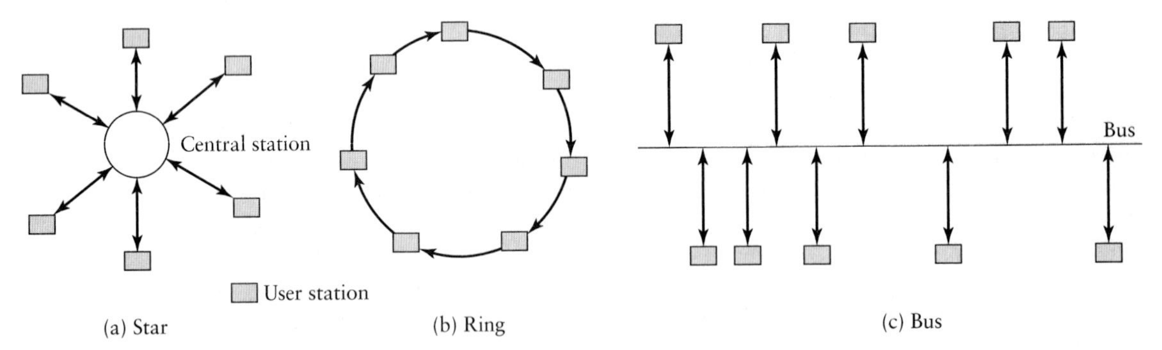

FIGURE 15.18 Three basic types of topology for a local area network (LAN). (a) The star topology is suitable for situations that are not subject to frequent configuration changes. All messages pass through a central station. Telephone systems in office buildings usually have this type of topology. (b) In the ring topology, all individual user stations are connected in a continuous ring. The message is forwarded from one station to the next until it reaches its assigned destination. Although the wiring is relatively simple, the failure of one station shuts down the entire network. (c) In the bus topology, all stations have independent access to the bus. This system is reliable and is easier to service than the other two. Because its arrangement is similar to the layout of the machines in a factory, its installation is relatively easy, and it can be rearranged when the machines are rearranged.

A new trend is the use of **wireless local area networks** (WLAN). Conventional LANs require routing of wires, often through masonry walls or other permanent structures, and require computers or machinery to remain stationary. WLANs allow equipment such as mobile test stands or data collection devices, such as bar code readers, to easily maintain a network connection. A communication standard (IEEE 802.11) currently defines frequencies and specifications of signals, and defines two radio frequency and one infrared methods. It should be noted that wireless networks are slower than those that are hard wired, but flexibility makes them desirable, especially for situations where slow tasks such as machine monitoring are the main application.

Personal area networks (PAN) are under development. PANs are based on communications standards such as Bluetooth, IrDA, and HomeRF, and are designed to allow data and voice communication over short distances. For example, a short-range Bluetooth device will allow communication over a 10 m (32 ft) distance. PANs are undergoing major changes, and communications standards are continually being refined.

15.12.1 Communications standards

Often, one manufacturing cell is built with machines and equipment purchased from one vendor, another cell with machines purchased from another vendor, a third cell with machines purchased from yet a third vendor, and so on. As a result, a variety of programmable devices is involved, driven by several computers and microprocessors purchased at various times from different vendors and having various capacities and levels of sophistication. Each cell's computers have their own specifications and proprietary standards, and they cannot communicate beyond the cell with others unless equipped with custom-built interfaces. This situation creates *islands of automation*, and in some cases, up to 50% of the cost of automation has been related to overcoming difficulties in the communications between individual manufacturing cells and other parts of the organization.

The existence of automated cells that could function only independently of each other, without a common base for information transfer, led to the need for a communications standard to improve communications and efficiency in computer-integrated manufacturing. After considerable effort and on the basis of existing national and international standards, a set of communications standards known as **manufacturing automation protocol** (MAP) was developed. The capabilities and effectiveness of MAP were first demonstrated in 1984 with the successful interconnection of devices from a number of vendors.

The International Organization for Standardization/Open System Interconnect (ISO/OSI) reference model is accepted worldwide. This model has a hierarchical structure in which communication between two users is divided into seven layers (Fig. 15.19). Each layer has a special task: (a) mechanical and electronic means of data transmission, (b) error detection and correction, (c) correct transmission of the message, (d) control of the dialog between users, (e) translation of the message into a common syntax, and (f) verification that the data transferred have been understood.

The operation of this system is complex. Basically, each standard-sized chunk of message or data to be transmitted from user *A* to user *B* moves sequentially through the successive layers at user *A*'s end, from layer 7 to layer 1. More information is added to the original message as it travels through each layer. The complete packet is transmitted through the physical communications medium to user *B* and then moves through the layers from layer 1 to layer 7 at user *B*'s end. The transmission takes place through fiber-optic cable, coaxial cable, microwaves, and similar devices.

Communication protocols have been extended to *office automation* as well with the development of **technical and office protocol** (TOP), which is based on the ISO/OSI reference model. In this way, total communication (MAP/TOP) can be established

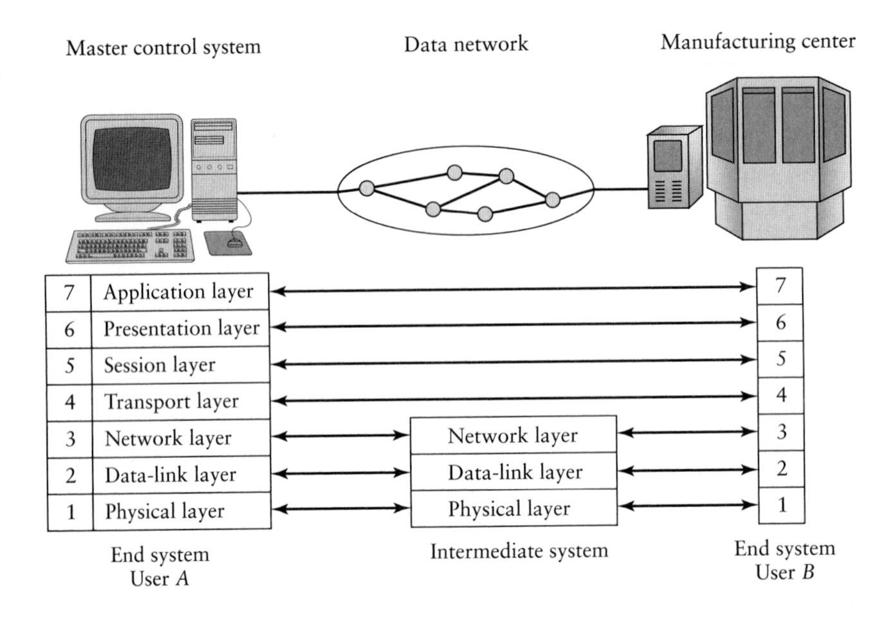

FIGURE 15.19 The ISO/OSI reference model for open communication. *Source*: U. Rembold et al., *Computer Integrated Manufacturing and Engineering*, Addison-Wesley, 1993.

across the factory floor and among offices at all levels of an organization. **Internet tools** (hardware, software, and protocols) within a company now link all departments and functions into a self-contained and fully compatible **Intranet**. Several tools for implementing this linkage are available commercially; they are inexpensive and easy to install, integrate, and use.

15.13 Artificial Intelligence

Artificial intelligence (AI) is an area of computer science concerned with systems that exhibit some characteristics which are usually associated with intelligence in human behavior, such as learning, reasoning, problem solving, and the understanding of language. The goal of AI is to simulate such human endeavors on the computer. The art of bringing relevant principles and tools of AI to bear on difficult application problems is known as **knowledge engineering**. Artificial intelligence is having a major impact on the design, the automation, and the overall economics of manufacturing operations, in large part because of advances in computer-memory expansion (VLSI chip design) and decreasing costs. Artificial-intelligence packages costing on the order of a few thousand dollars have been developed, many of which can now be run on personal computers; thus, AI has become accessible to office desks and shop floors.

Artificial-intelligence applications in manufacturing generally encompass the following:

a) **Expert systems;**
b) **Natural language;**
c) **Machine (computer) vision;**
d) **Artificial neural networks;**
e) **Fuzzy logic.**

15.13.1 Expert systems

Also called a **knowledge-based system**, an *expert system* (ES) is generally defined as an intelligent computer program that has the capability to solve difficult real-life problems, using **knowledge-base** and **inference** procedures (Fig. 15.20). The goal of an expert system is to develop the capability to conduct an intellectually demanding task in the way that a human expert would. The field of knowledge required to perform this task is called the **domain** of the expert system. Expert systems use a knowledge base containing facts, data, definitions, and assumptions. They also have the capacity to follow a **heuristic** approach, that is, to make good judgments on the basis of *discovery* and *revelation* and to make high-probability *guesses*, just as a human expert would. The knowledge base is expressed in computer codes, usually in the form of **if–then rules**, and can generate a series of questions; the mechanism for using these rules to solve problems is called an **inference engine**. Expert systems can also communicate with other computer software packages.

To construct expert systems for solving the complex design and manufacturing problems encountered, one needs (a) a great deal of knowledge and (b) a mechanism for manipulating this knowledge to create solutions. Because of the difficulty involved in accurately modeling the many years of experience of an expert or a team of experts and the complex inductive reasoning and decision-making capabilities of humans, including the capacity to learn from mistakes, the development of knowledge-based systems requires much time and effort.

Expert systems operate on a *real-time* basis, and their short reaction times provide rapid responses to problems. The programming languages most commonly used for this application are C++, LISP, and PROLOG; other languages can also be used. A significant development is expert-system software **shells**, or **environments**, also called **framework systems**. These software packages are essentially expert-system outlines that allow a person to write specific applications to suit special needs. Writing these programs requires considerable experience and time.

Several expert systems have been developed and used for such specialized applications as

1. Problem diagnosis in machines and equipment and determination of corrective actions;
2. Modeling and simulation of production facilities;
3. Computer-aided design, process planning, and production scheduling;
4. Management of a company's manufacturing strategy.

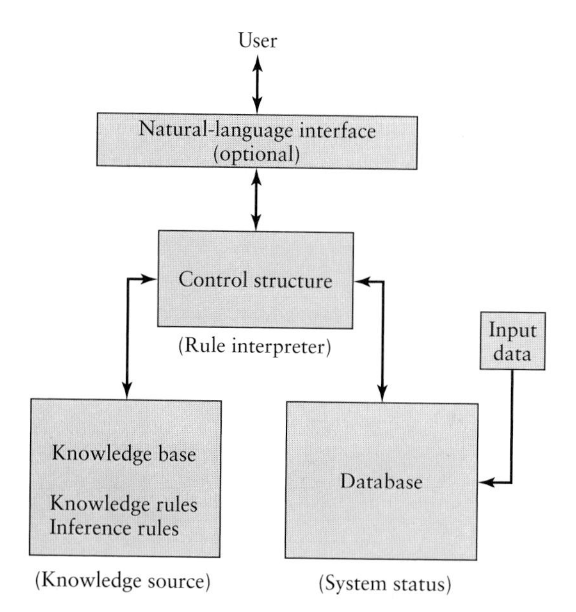

FIGURE 15.20 The basic structure of an expert system. The knowledge base consists of knowledge rules (general information about the problem) and inference rules (the way conclusions are reached). The results may be communicated to the user through the natural-language interface. *Source*: K. W. Goff, *Mechanical Engineering*, October 1985.

15.13.2 Natural-language processing

Traditionally, obtaining information from a database in computer memory has required the use of computer programmers to translate questions expressed in natural language into "queries" in some machine language. Natural-language interfaces with database systems are in various stages of development; these systems allow a user to obtain information by entering English-language commands in the form of simple, typed questions. Natural-language software shells are used in such applications as scheduling of material flow in manufacturing and analyzing information in databases. Major progress continue to be made in the development of computer software that will have speech-synthesis and recognition (**voice recognition**) capabilities, thus eliminating the need to type commands on keyboards.

15.13.3 Machine vision

The basic features of *machine vision* are described in Section 14.8. In systems that incorporate machine vision, computers and software implementing artificial intelligence are combined with cameras and other optical sensors. These machines then perform such operations as inspecting, identifying, and sorting parts and guiding robots (*intelligent robots*; Fig. 15.21); in other words, operations that would otherwise require human involvement and intervention.

15.13.4 Artificial neural networks

Although computers are much faster than the human brain at *sequential* tasks, humans are much better at pattern-based tasks that can be attacked with *parallel processing*, such as recognizing features (e.g., in faces and voices, even under noisy conditions), assessing situations quickly, and adjusting to new and dynamic conditions. These advantages are also due partly to the ability of humans to use, in real time, their **senses** (i.e., sight, hearing, smell, taste, and touch) simultaneously, a process called *data fusion*. The branch of artificial intelligence called *artificial neural networks* (ANN) attempts to gain some of these human capabilities through computer imitation of the way that data are processed by the human brain.

The human brain has about 100 billion linked **neurons** (cells that are the fundamental functional units of nervous tissue) and more than a thousand times that many connections. Each neuron performs only one simple task: It receives input signals from a fixed set of neurons, and when those input signals are related in a certain way (specific to that particular neuron), it generates an electrochemical output signal that goes to a fixed set of neurons. It is now believed that human learning is accomplished by changes in the strengths of these signal connections between neurons.

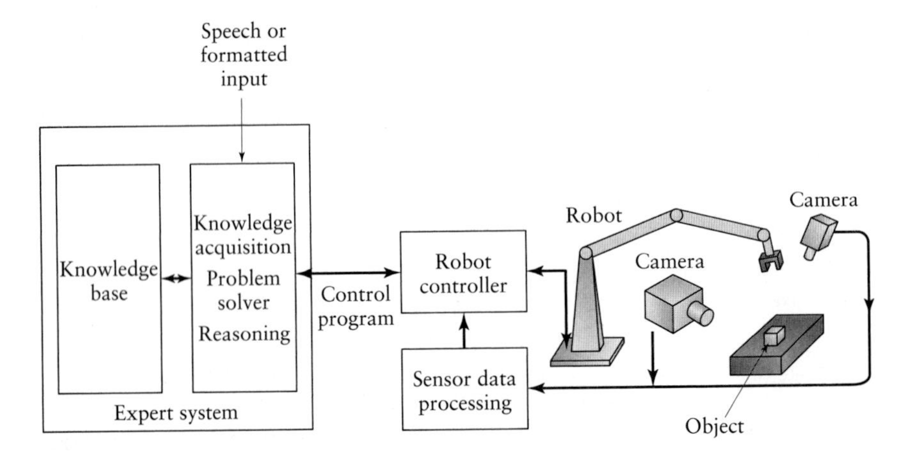

FIGURE 15.21 Expert system as applied to an industrial robot guided by machine vision.

A fully developed, feed-forward network is the most common type of ANN, and it is built according to this principle from several layers of processing elements (simulating neurons). The elements in the first (input) layer are fed with input data—for example, forces, velocities, or voltages. Each element sums up all its inputs—one per element in the input layer, and many per element in succeeding layers. Each element in a layer then transfers the data (according to a transfer function) to all the elements in the next layer. Each element in that next layer, however, receives a different signal, because of the different connection weights between the elements.

The last layer is the output layer, within which each element is compared with the desired output—that of the process being simulated. The difference between the desired output and the calculated output (*error*) is fed back to the network by changing the weights of the connections in a way that reduces this error. After this procedure has been repeated several times, the network has been "trained," and it can thus be used on input data that have not previously been presented to the system.

Other kinds of artificial neural networks are (a) *associative memories*, (b) *self-organizing ANN*, and (c) *adaptive-resonance ANN*. The feature common to these neural networks is that they must be trained with concrete exemplars. It is, therefore, very difficult to formulate input–output relations mathematically and to predict an ANN's behavior with untrained inputs.

Artificial neural networks are used in such applications as noise reduction (in telephones), speech recognition, and process control in manufacturing. For example, they can be used to predict the surface finish of a workpiece obtained by end milling (see Fig. 8.1d) on the basis of input parameters such as cutting force, torque, acoustic emission, and spindle acceleration. Although this field still controversial, the opinion of many is that true artificial intelligence will evolve only through advances being made in ANN.

15.13.5 Fuzzy logic

An element of artificial intelligence that has important applications in control systems and pattern recognition is *fuzzy logic* (**fuzzy models**). Introduced in 1965, and based on the observation that people can make good decisions on the basis of imprecise and nonnumerical information, fuzzy models are mathematical means of representing vagueness and imprecise information (hence the term *fuzzy*). These models have the capability to recognize, represent, manipulate, interpret, and use data and information that are vague or lack precision. Fuzzy-logic methods deal with reasoning and decision making at a level higher than do neural networks. Typical linguistic examples of concepts used in fuzzy logic are the following words and terms: *few*, *very*, *more or less*, *small*, *medium*, *extremely*, and *almost all*.

Fuzzy-logic technologies and devices have been developed and successfully applied in areas such as robotics and motion control, image processing and machine vision, machine learning, and the design of intelligent systems. Some applications of fuzzy logic include the following: the automatic transmission of Lexus automobiles; a washing machine that automatically adjusts the washing cycle for load size, fabric type, and amount of dirt; and a helicopter that obeys vocal commands to go forward, up, left, and right; to hover; and to land.

15.14 | The Factory of the Future

On the basis of the advances made in all aspects of manufacturing technology and computer integration, we may envision the *factory of the future* as a fully automated facility in which human beings will not be directly involved with production on the shop floor (hence the term **untended factories**). All manufacturing, material handling, assembly, and inspection will be done by computer-controlled machinery and equipment. Similarly, activities such as the processing of incoming orders, production planning and scheduling, cost accounting, and various decision-making processes (usually performed by management) will also be done automatically by computers. The role of human beings will be confined to activities such as overall supervision, maintenance (especially preventive maintenance), and the upgrading of machines and equipment; the shipping and receiving of supplies and finished products; the provision of security for the plant facilities; the programming, upgrading, and monitoring of computer programs; and the monitoring, maintenance, and upgrading of computer hardware.

Some businesses in the food, petroleum, and chemical industries already operate automatically, with little human intervention. These industries involve continuous processes that are easier to automate fully than is piece-part manufacturing. Even so, the direct involvement of fewer people in the manufacturing of products is already apparent: Surveys of modern manufacturing facilities show that only 10% to 15% of the workforce is directly involved in production; the majority is involved in the gathering and processing of information. Virtually untended manufacturing cells already make products such as engine blocks, axles, and housings for clutches and air compressors. For large-scale flexible manufacturing systems, however, highly trained and skilled personnel will always be needed in order to plan, maintain, and oversee operations.

The reliability of machines, equipment, control systems, power supplies, and communications networks is crucial to full factory automation. Without rapid human intervention, a local or general breakdown in even one of these components can cripple production. The computer-integrated factory of the future must be capable of automatically rerouting materials and production flows to other machines and to the control of other computers in the case of such emergencies.

In the implementation of advanced concepts of manufacturing, with their attendant impact on the operation of the factory of the future, certain developments and improvements are bound to continue at a rapid pace:

a) Common communications networks, software, and standards for every aspect of manufacturing technology, from product design to manufacturing;
b) Reduction in the size and increase in the efficiency of knowledge-rich controllers, and their integration into machines instead of functioning as separate units;
c) Increasing availability of off-the-shelf control hardware;
d) Increase in the reliability of voice-recognition capabilities, so that machines can understand and implement spoken messages;
e) Tighter sensor fusion, employing more rugged sensors, and improvement of feedback methods for monitoring and diagnosing all aspects of manufacturing operations (both developments are augmented by advances in artificial-intelligence technology);
f) Cost-effective system integration at all levels of an organization.

Concerns and opportunities. An important consideration in fully automating factory is the nature and extent of its impact on employment, including all the social and political ramifications. Although economic forecasts indicate that there will be a decline in the number of machine-tool operators and tool-and-die workers, there will be major increases in the number of people working in service occupations (such as computer service technicians and maintenance electricians). Thus, the generally low-skilled, manual-effort labor force traditionally required in manufacturing will evolve into a knowledge-effort labor force that possesses the specialized training or retraining required in such activities as computer programming, information processing, implementation of CAD/CAM, application of information science and technology, and other high-technology tasks. The development of more user-friendly computer software is making retraining of the workforce much easier.

There are many widely divergent opinions about the impact of untended factories. Consequently, predicting the nature of future manufacturing strategies with any certainty is difficult. Although economic considerations and trade-offs are crucial, companies that do not install computer-integrated operations where they will be cost-effective are in obvious jeopardy. It is now widely recognized that, in a highly competitive global marketplace, rapid adaptability is crucial to the survival of a manufacturing organization.

The continuous advances in all aspects of the science, the engineering, and the technology of manufacturing are constantly being analyzed by corporate management, with a view to the near-term and long-term economic impact on their operations. Over the past few years, much has been written, discussed, and debated concerning the relative manufacturing technologies and industrial strengths among many industrialized nations. There are various complex issues stemming from the following considerations:

a) The characteristics of each industrialized nation—in particular, its history, its social and demographic structure, and its standard of living;
b) The nature, educational level, and loyalty of its workforce;
c) The relationships among management, labor, and government;
d) Interactions among companies, research universities, and social organizations;
e) Productivity and the effective utilization of capital equipment;
f) The missions and the operational philosophies of the management of industrial and business organizations;
g) Stockholder attitudes and preferences.

The term **world-class** is frequently used to describe a certain quality and level of manufacturing activities, signifying the fact that products must meet international standards and be acceptable worldwide. It should also be recognized that the designation of world-class, like quality, is not a fixed target for a manufacturing company or a country to reach, but rather a moving target, rising to higher and higher levels as time passes. Manufacturing organizations must be aware of this moving target and plan and execute their programs accordingly.

15.15 Economic Considerations

The economic considerations in implementing the various computer-integrated activities described in this chapter are crucial in view of the complexities and the high costs involved. Installations of flexible manufacturing systems are very capital intensive; consequently, a thorough cost–benefit analysis must be conducted before a final decision is made. This analysis should include such factors as (a) the cost of capital, energy, materials, and labor; (b) the expected markets for which the products are to be produced; (c) anticipated fluctuations in market demand and product type; and (d) the time and effort required for installing and debugging the system.

Typically, an FMS system can take two to five years to install and at least six months to debug. Although FMS requires few, if any, machine operators, the personnel in charge of the total operation must be trained and highly skilled. These personnel include manufacturing engineers, computer programmers, and maintenance engineers. The most effective FMS applications have been in medium-volume batch production. (See Fig. 14.2.) When a variety of parts is to be produced, FMS is suitable for production volumes of 15,000–35,000 aggregate parts per year. For individual parts of the same configuration, production may reach 100,000 units per year. In contrast, high-volume, low-variety parts production is best obtained from transfer machines (dedicated equipment).

As indicated in Chapter 14, low-volume, high-variety parts production can best be done on conventional standard machinery (with or without NC) or by machining centers. Compared with conventional manufacturing systems, FMS has the following benefits:

1. Parts can be produced in any order, in batch sizes as small as one item, and at lower unit cost.
2. Direct labor and inventories are reduced, yielding major savings over conventional systems.
3. Lead times required for product changes are shorter.
4. Production is more reliable, because the system is self-correcting and thus product quality is uniform.
5. Work-in-progress inventories are reduced.

CASE STUDY | CAD Model Development for Automotive Components

CAD models are used extensively for a great variety of tasks, as described in Section 15.4. In the automotive industry, for example, it is especially important to have a detailed CAD model of a particular component in the product database in order to ensure that all those who will be working on it have all the data they need to perform their tasks. Special care is taken to build very precise CAD models of the automobile components that passengers will see and interact with on a regular basis; examples of such components are outer-body panels, handles, seats, and the instrument panel (Fig. 15.22). The quality of the *visible* (Class I) surfaces has a major impact on overall vehicle quality and customer perception of the look and feel of the automobile.

The route from product concept to full working CAD models of visible components involves several steps, as well as the efforts of a large number of highly skilled designers, engineers, and technicians. The personnel constructing the CAD models are in constant interaction with the workers who will be using them, from concept to final design, to support the activities that rely on CAD data throughout the life of the vehicle.

FIGURE 15.22 CAD description of an automobile. Every vehicle component, from body panels to knobs on the instrument panel, has a solid model associated with it. *Source*: Ford Motor Company.

2D concept sketches. Stylists with a background and experience in industrial design and/or art first develop 2D concepts through a series of sketches. These sketches are most frequently drawn by hand, although software may be used, especially if the stylist starts with a photograph or a scanned drawing that needs to be modified. Concept sketches provide an overall feel for the aesthetics of the object and are frequently very detailed and show texture, color, and the areas at which individual surfaces on a vehicle should meet. Most often, stylists are given a set of packaging constraints such as (a) how the component should be assembled with other components, (b) what the size of the component should be, and (c) what the size and shape of any structure lying behind the visible surfaces should be. The time involved in producing a series of such concept sketches for an individual component or a set of components typically ranges from a few days to several weeks.

3D concept model. It is difficult to have an accurate sense of how the vehicle or a component will look and feel from a 2D sketch alone; thus, the next step is to make a 3D model. Both physical and computer models are built. First, further sketches of the component (such as cross-sections, perspective views, and orthographic views) are drawn. In many cases, a rough digital surface model is built to better represent the 3D object. A skilled clay sculptor then constructs a full-scale mock-up of the component from clay (Fig. 15.23) that is spread over a foam core, either by hand, using the sketches as a guide, or by a combination of CNC milling driven by a computer model (see, for example, Fig. 1.8); this activity is then followed by hand finishing. Several iterations may be necessary to achieve the shape that the stylist intended; hence, constant interaction between the sculptor and stylist is essential.

Alternatively, the stylist or clay sculptor can use 3D sculpting software, such as FreeForm,[1] to construct a digital clay model of the component. This digital model is then projected onto a large screen so that designers can study the component in full scale, or the digital model is used to generate tool paths for CNC machining of a full-scale clay mock-up.

At this point, all information from the sketches, rough computer drawings, and the concept clay model begins to be fed to downstream processes, such as engineering analysis and manufacturing. Although the final surfaces have not yet been decided upon, the basic shape and size of the component can now be used to begin tooling design and to analyze component feasibility.

FIGURE 15.23 A clay sculptor working on an instrument-panel concept. *Source*: Ford Motor Company.

[1] FreeForm is a product of SensAble Technologies, Inc.

3D surface model. As a concept is being reviewed and refined, several very accurate surface models of the component are constructed. To start the surface model, a computer-controlled optical scanner scans the concept clay model, producing a *cloud of points* organized along scan lines. Depending on the size of the component, a point cloud may consist of hundreds of thousands to millions of points. The point cloud is read into point-processing software such as Paraform[2] or ICEM Surf[3] to further organize the points and filter out noise. Scanning can take anywhere from several hours to a day to be completed. If, however, a digital 3D clay model of the component is already available, it is converted to a point cloud organized into scan lines, without the need for physical scanning and point-cloud postprocessing.

Next, the scan lines from the point cloud are used to construct mathematical surfaces, using software such as ICEM Surf and Alias/Wavefront StudioTools.[4] First, to construct the surfaces, freeform NURBS (*nonuniform rational B-spline,* see Section 15.4.2) curves that interpolate or approximate the scan lines are constructed; then a NURBS surface patch is fit through the curves. An individual surface patch models a small region of a single component's face; several patches are constructed and joined smoothly at common edges to form the entire face. Faces join each other at common edges to model an entire component. A great deal of experience is required to determine how to divide a face into a collection of patches that can be fit with the simplest low-order surfaces possible and still meet smoothly at common boundaries. A surfacing specialist performs this task in conjunction with the stylist to ensure that the surfaces are of high quality and that they capture the stylist's intent. A single component may take as long as a week to model.

Surface models are passed along to the various downstream departments to be used for tooling design, feasibility checks, analysis, and for the design of *nonvisible* (Class II) surfaces. As the design evolves, the dimensional tolerances on the surfaces are gradually tightened. When the component surfaces are within 3 to 4 mm (0.12 to 0.16 in.) of their final shape, a great deal of engineering that relies on them has already been accomplished.

When designing outer-body panels, a major milestone is finalizing what is called *first flange and fillet,* in which the edges of the body panels are turned under, or hemmed (see Fig. 7.23c), to provide a connecting flange for the inner panel (Fig. 15.24). The shape of the flange and fillet affects the overall aesthetics and shape of the body panel, and hence their careful design is important. After the shape of the fillet and flange has been decided, inner-panel designs can be completed.

Surface verification. Once the final surface model is completed, it must be verified and evaluated for surface quality and aesthetics. NC tool paths are generated automatically from the surfaces, either by the surfacing software or through specialized machining software; the tool paths are then used to CNC-machine the sur-

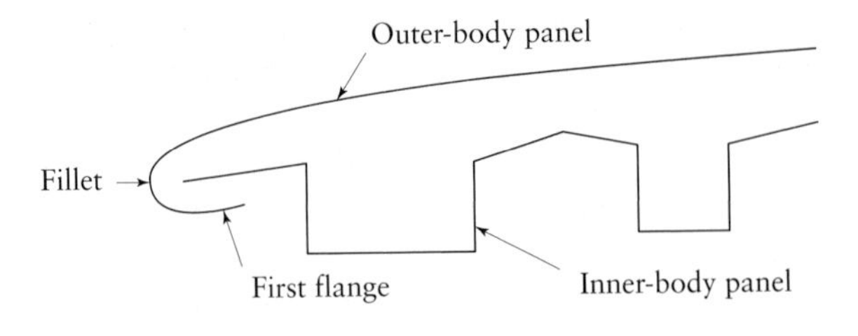

FIGURE 15.24 Schematic illustration of first flange and fillet, where the edges of the body panel are hemmed to provide a flange for connection. The shape of this flange and fillet is important for overall vehicle aesthetics.

[2] Paraform is a product of Paraform, Inc.
[3] ICEM Surf is a product of ICEM Technologies, Inc.
[4] StudioTools is a product of Alias/Wavefront.

faces in clay. If the component is small, then instead of machining a clay model, an STL file can be generated and the mock-up can be built on a rapid prototyping machine (see Section 10.12). It takes anywhere from several hours to several days to machine a component and perform any hand-finishing that may be required. Clay models can be coated with a thin layer of latex material and painted to make them look more realistic and to help evaluate surface quality. The coating may be modified to improve surface smoothness between patches or change the way that light reflects off the surface of the model. All changes made to the clay must be translated back to the digital surface model, either through rescanning and refitting the surface patches or by tweaking the shape of the existing patches. At this stage of surface verification, only very minor changes, to within ± 0.5 mm (0.04 in.), can be made.

Solid-model construction. After the surface model is finalized, it is used to develop a solid model. For sheet-metal components, such as body panels, body specialists offset the surfaces to form a solid. For other components, such as instrument panels, door handles, and wheels, manufacturing features are added to the surface model, such as flanges, bosses, and ribs. Also, it is confirmed that all surfaces form a closed, watertight solid object. Suppliers are contacted and consulted to determine what features must be added and where they should be placed to ensure that the component can be fabricated from the desired materials and at the target cost.

In the process of making a solid model, it may be discovered that the surface model must be modified because (a) of changes in packaging constraints, (b) the component fails to meet minimal manufacturing requirements, or (c) surfaces do not match up properly at common edges with the appropriate smoothness. These changes are communicated back to the surface modeler and the stylist so that the surfaces can be modified and reverified. Finally, the solid model is entered into a product database, where it is now available to suppliers and engineers for further analysis and manufacturing.

Source: Courtesy of A. Marsan and P. Stewart, Ford Motor Company.

SUMMARY

- Computer-integrated manufacturing systems have become the most important means of improving productivity, responding to changing market demands, and better controlling manufacturing operations and management functions. (Sections 15.1–15.3)
- With rapid developments in advanced and sophisticated software, together with the capability for computer simulation and analysis, computer-aided product design and engineering have become much more efficient, detailed, and comprehensive. (Section 15.4)
- Computer-aided manufacturing is often combined with computer-aided design to transfer information from the design stage to the planning stage and to production. (Section 15.5)
- Continuing advances in manufacturing operations such as computer-aided process planning, computer simulation of manufacturing processes and systems, group technology, cellular manufacturing, flexible manufacturing systems, and just-in-time manufacturing contribute significantly to improved productivity. (Sections 15.6–15.11)
- Communications networks and their global standardization are essential to CIM strategies. (Section 15.12)
- Artificial intelligence is likely to create new opportunities in all aspects of manufacturing science, engineering, and technology. (Section 15.13)

- A fully automated factory of the future appears to be theoretically possible. However, there are important issues to be considered regarding its impact on employment. (Section 15.14)
- Economic considerations in the design and implementation of computer-integrated manufacturing systems, especially flexible manufacturing systems, are particularly crucial, because of the major capital expenditures involved. (Section 15.15)

BIBLIOGRAPHY

Amirouche, F. M. L., *Computer-Aided Design and Manufacturing*, Prentice Hall, 1993.

Askin, R. G., and C. R. Standridge, *Modeling and Analysis of Manufacturing Systems*, Wiley, 1993.

Badiru, A. B., *Expert Systems Applications in Engineering and Manufacturing*, Prentice Hall, 1992.

Bedworth, D. D., M. R. Henderson, and P. M. Wolf, *Computer-Integrated Design and Manufacturing*, McGraw-Hill, 1991.

Burbidge, J. L., *Production Flow Analysis for Planning Group Technology*, Oxford, 1997.

Chang, T.-C., R. A. Wysk, and H. P. Wang, *Computer-Aided Manufacturing*, 2d. ed., Prentice Hall, 1997.

Cheng, T. C. E., and S. Podolsky, *Just-in-Time Manufacturing: An Introduction*, 2d. ed., Chapman & Hall, 1996.

Chorafas, D., *Expert Systems in Manufacturing*, Van Nostrand Reinhold, 1992.

Corbett, J., M. Dooner, J. Meleka, and C. Pym, *Design for Manufacture: Strategies, Principles and Techniques*, Addison-Wesley, 1991.

Driankov, D., H. Hellendoorn, and M. Reinfrank, *Introduction to Fuzzy Control*, 2d. ed., Springer, 1996.

Famili, A., D. S. Nau, and S. H. Kim, *Artificial Intelligence Applications in Manufacturing*, MIT Press, 1992.

Fausett, L. V., *Fundamentals of Neural Networks*, Prentice Hall, 1994.

Foston, A. L., C. L. Smith, and T. Au, *Fundamentals of Computer-Integrated Manufacturing*, Prentice Hall, 1991.

Gu, P., and D. H. Norrie, *Intelligent Manufacturing Planning*, Chapman & Hall, 1995.

Hannam, R., *CIM: From Concept to Realisation*, Addison-Wesley, 1998.

Haykin, S. S., *Neural Networks: A Comprehensive Foundation*, 2d. ed., Prentice Hall, 1998.

Higgins, P., L. R. Roy, and L. Tierney, *Manufacturing Planning and Control: Beyond MRP II*, Chapman & Hall, 1996.

Hitomi, K., *Manufacturing Systems Engineering*, 2d. ed., Taylor & Francis, 1996.

Irani, S. (ed.), *Handbook of Cellular Manufacturing Systems*, Wiley, 1999.

Jamshidi, M., *Design and Implementation of Intelligent Manufacturing Systems: From Expert Systems, Neural Networks, to Fuzzy Logic*, Prentice Hall, 1995.

Kasabov, N. K., *Foundations of Neural Networks, Fuzzy Systems, and Knowledge Engineering*, MIT Press, 1996.

Koenig, D. T., *Manufacturing Engineering: Principles for Optimization*, 2d. ed., Taylor and Francis, 1994.

Krishnamoorty, C. S., and S. Rajeev, *Artificial Intelligence and Expert Systems for Engineers*, CRC Press, 1996.

Kusiak, A., *Concurrent Engineering: Automation, Tools, and Techniques*, Wiley, 1993.

Lee, K., *Principles of CAD/CAM Systems*, Addison-Wesley, 1999.

Leondes, C. T. (ed.), *Fuzzy Logic and Expert Systems Applications*, Academic Press, 1998.

Liebowitz, J. (ed.), *The Handbook of Applied Expert Systems*, CRC Press, 1997.

Louis, R. S., *Integrating Kanban with MRP II: Automating a Pull System for Enhanced JIT Inventory Management*, Productivity Press, 1997.

Luggen, W. W., *Flexible Manufacturing Cells and Systems*, Prentice Hall, 1991.

Maus, R., and J. Keyes (eds.), *Handbook of Expert Systems in Manufacturing*, McGraw-Hill, 1991.

McMahon, C., and J. Browne, *CADCAM—Principles, Practice and Manufacturing Management*, Addison-Wesley, 1999.

Mitchell, F. H., Jr., *CIM Systems: An Introduction to Computer-Integrated Manufacturing*, Prentice Hall, 1991.

Monden, Y., *Toyota Production System: An Integrated Approach to Just-in-Time*, 3d. ed., Institute of Industrial Engineers, 1998.

Nee, A. Y. C., K. Whybrew, and A. S. Kumar, *Advanced Fixture Design for FMS*, Springer, 1994.

Popovic, D., and V. Bhatkar, *Methods and Tools for Applied Artificial Intelligence*, Dekker, 1994.

Rehg, J. A., *Introduction to Robotics in CIM Systems*, 4th ed., Prentice Hall, 2000.

Rehg, J. A., and H. W. Kraebber, *Computer-Integrated Manufacturing*, 2d. ed., Prentice Hall, 2001.

Rembold, U., B. O. Nnaji, and A. Storr, *Computer Integrated Manufacturing and Engineering*, Addison-Wesley, 1993.

Sandras, W. W., *Just-in-Time: Making It Happen*, Wiley, 1997.

Singh, N., *Systems Approach to Computer-Integrated Design and Manufacturing*, Wiley, 1995.

Singh, N., and D. Rajamani, *Cellular Manufacturing Systems: Design, Planning and Control*, Chapman & Hall, 1996.

Tempelmeier, H., and H. Kuhn, *Flexible Manufacturing Systems*, Wiley, 1993.

Vajpayee, S. K., *Principles of Computer-Integrated Manufacturing*, Prentice Hall, 1995.

Vollmann, T. E., W. L. Berry, and D. C. Whybark, *Manufacturing Planning and Control Systems*, 4th ed., Irwin, 1997.

Williams, D. J., *Manufacturing Systems*, 2d. ed., Chapman & Hall, 1994.

Wu, J.-K., *Neural Networks and Simulation Methods*, Dekker, 1994.

QUESTIONS

15.1 In what ways have computers had an impact on manufacturing?

15.2 What advantages are there in viewing manufacturing as a system? What are the components of a manufacturing system?

15.3 Discuss the benefits of computer-integrated manufacturing operations.

15.4 What is a database? Why is it necessary in manufacturing? Why should the management of a manufacturing company have access to databases?

15.5 Explain how a CAD system operates.

15.6 What are the advantages of CAD systems over traditional methods of design? Do CAD systems have any limitations?

15.7 Describe the purposes of process planning. How are computers used in such planning?

15.8 Explain the features of two types of CAPP systems.

15.9 Describe the features of a routing sheet. Why is a routing sheet necessary in manufacturing?

15.10 What is group technology? Why was it developed? Explain its advantages.

15.11 What is a manufacturing cell? Why was it developed?

15.12 Describe the principle of flexible manufacturing systems. Why do they require major capital investment?

15.13 Why is a flexible manufacturing system capable of producing a wide range of lot sizes?

15.14 What are the benefits of just-in-time production? Why is it called a *pull system*?

15.15 Explain the function of a local area network.

15.16 What are the advantages of a communications standard?

15.17 What is meant by the term *factory of the future*?

15.18 What are the differences between ring and star networks? What is the significance of these differences?

15.19 What is kanban? Why was it developed?

15.20 What is an FMC, and what is an FMS? What are the differences between them?

15.21 Describe the elements of artificial intelligence. Why is machine vision a part of it?

15.22 Explain why humans will still be needed in the factory of the future.

15.23 How would you describe the principle of computer-aided manufacturing to an older worker in a manufacturing facility who is not familiar with computers?

15.24 Give examples of primitives of solids other than those shown in Fig. 15.4b.

15.25 Explain the logic behind the arrangements shown in Fig. 15.11b.

15.26 Describe your observations regarding Fig. 15.17.

15.27 What should be the characteristics of an effective guidance system for an automated guided vehicle?

15.28 Give examples in manufacturing for which artificial intelligence could be effective.

15.29 Describe your opinions concerning the voice-recognition capabilities of future machines and controls.

15.30 Would machining centers be suitable for just-in-time production? Explain.

15.31 Give an example of a push system and of a pull system, to clarify the fundamental difference between the two methods.

15.32 Give a specific example in which the variant system of CAPP is desirable and an example in which the generative system is desirable.

15.33 Artificial neural networks are particularly useful where the problems are ill defined and the data are fuzzy. Give examples in manufacturing where this is the case.

15.34 Is there a minimum to the number of machines in a manufacturing cell? Explain.

15.35 List as many three-letter acronyms used in manufacturing (such as CNC) as you can, and give a brief definition of each, for your future reference.

15.36 What are the disadvantages of zero inventory?

15.37 Why are robots a major component of an FMC?

15.38 Is it possible to exercise JIT in global companies?

15.39 A term sometimes used to describe factories of the future is *untended factories*. Can a factory ever be completely untended? Explain your answer.

15.40 What are the advantages of hierarchical coding?

15.41 Assume that you are asked to rewrite Section 15.14, on the factory of the future. Briefly outline your thoughts regarding this topic.

15.42 Assume that you own a manufacturing company and that you are aware that you have not taken full advantage of the technological advances in manufacturing. Now, however, you would like to do so, and you have the necessary capital. Describe how you would go about analyzing your company's needs and how you would plan to implement these technologies. Consider technical as well as human aspects.

15.43 With specific examples, describe your thoughts concerning the state of manufacturing in the United States as compared with its state in other industrialized countries.

15.44 It has been suggested by some that artificial intelligence systems will ultimately be able to replace the human brain. Do you agree? Explain your response.

DESIGN

15.45 Review various manufactured parts described in this text, and group them in a manner similar to that shown in Fig. 15.10.

15.46 Think of a specific product, and make a decision-tree chart similar to that shown in Fig. 15.12.

15.47 Describe the trends in product designs and features that have had a major impact on manufacturing.

15.48 Think of a specific product, and design a manufacturing cell for making it (see Fig. 15.16), describing the features of the machines and equipment involved. Explain how the cell arrangement would change, if at all, if design changes are made to the product.

15.49 Surveys have indicated that 95% of all the different parts made in the United States are produced in lots of 50 or less. Comment on this observation, and describe your thoughts regarding the implementation of the technologies outlined in Chapters 14 and 15.

15.50 Think of a simple product, and make a routing sheet for its production, similar to that shown in Fig. 15.8. If the same part is given to another person, what is the likelihood that the routing sheet developed will be the same? Explain.

15.51 Review Fig. 15.8, and then suggest a routing sheet for the manufacture of each of the following: (a) an automotive connecting rod, (b) a compressor blade, (c) a glass bottle, (d) an injection-molding die, and (e) a bevel gear.

15.52 What types of production machines would not be suitable for a manufacturing cell? What design or production features make them unsuitable? Explain.

CHAPTER 16

Product Design and Competitive Manufacturing

16.1 Introduction

In an increasingly competitive global marketplace, manufacturing high-quality products at the lowest possible cost requires an understanding of the often complex relationships among many factors. We have seen that product design, selection of materials, and selection of manufacturing processes are all interrelated. Furthermore, designs are periodically modified to improve product performance, to take advantage of available new materials or use cheaper materials, to make the products easier and faster to manufacture and assemble, and to strive for zero-based rejection and waste.

Because of the very wide variety of materials and manufacturing processes now available, the task of producing a high-quality product by selecting the best materials and the best processes, while minimizing costs, has become a major challenge, as well as an opportunity. This chapter begins with various considerations in **product design**, and the concept of **robust design**. Product design involves numerous aspects of not only the design itself, but also the ease of manufacturing that product and the time required to produce it. Again, there are many opportunities to simplify designs, reduce the number of components, and reduce the size and certain dimensions of the components to save money on especially costly materials.

Product quality and **life expectancy** are then discussed, outlining relevant parameters involved, including the concept of return on quality. Increasingly important are **life-cycle assessment** and **life-cycle engineering** of products, services, and systems, particularly as regards their potential adverse impact on the environment (air, water, and ground). The major emphasis in sustainable manufacturing is to reduce or eliminate any adverse effects of manufacturing on the environment and society in general, while allowing a company to be profitable.

Although the *selection of materials* for products traditionally requires much experience, there are now several databases and expert systems available that facilitate the selection process in order to meet several specific requirements. Also, in reviewing the materials used in existing products, from simple hand tools to automobiles and aircraft, there are numerous opportunities for *substitution of materials* for better performance and, especially, for cost savings.

In the production phase, it is imperative that we properly assess the *capabilities of manufacturing processes* as a basis of and an essential guide to the ultimate selection of an appropriate process or series of processes. As described throughout this text, depending on the product design and the materials specified, there is usually more than one method of manufacturing a product, with its usual number of components and subassemblies. An improper selection can have a major effect not only on product quality, but also on product cost.

Although the *economics* of particular manufacturing processes have been described at the end of individual chapters, this chapter takes a broader view and summarizes the important overall manufacturing cost factors in a competitive marketplace. The **cost** of a product often, but not always, determines its marketability and its customer acceptance. Meeting this challenge requires not only a thorough and up-to-date knowledge of the characteristics of materials, state-of-the-art processes and manufacturing systems, and economics, but also innovative and creative approaches to product design and manufacturing. Finally, we describe the method of **value analysis**, which is a powerful tool that can be used to evaluate the cost of each step in production relative to its contribution to the value of the product.

16.2 Product Design and Robust Design

Although it is beyond the scope of this book to describe the principles and methodology of product design and various optimization techniques, we have highlighted, throughout various chapters, those aspects that are relevant to **design for manufacture and assembly** (DFMA), as well as to competitive manufacturing. The references listed in Table 16.1 give design guidelines for manufacturing processes.

Major advances are being made in design for manufacture and assembly, for which software packages are now available. Although their use requires considerable training and knowledge, they help designers develop products that require fewer components, reduced assembly time, and reduced time for manufacture and thus that have reduced total cost.

16.2.1 Product design considerations

In addition to the design guidelines outlined throughout this text, there are general product design considerations. Designers often must ask themselves if they've addressed considerations such as the following:

a) Can the product design be simplified and the number of its components reduced without adversely affecting its intended functions and performance?

b) Have all alternative designs been investigated?

c) Can unnecessary features of the product, or some of its components, be eliminated or combined with other features?

d) Have modular design and building-block concepts been considered for a family of similar products and for servicing and repair, upgrading, and installation options?

e) Can the design be made smaller and lighter?

f) Are the specified dimensional tolerances and surface finish excessively stringent? Can they be relaxed without any adverse effects?

g) Will the product be difficult or excessively time consuming to assemble and disassemble for maintenance, servicing, or recycling?

h) Have subassemblies been considered?

i) Is the use of fasteners, and their quantity and variety, minimized?

j) Are some of the components commercially available?

TABLE 16.1

References to Various Topics in this Text

PROCESSES: DESIGN CONSIDERATIONS	
Metal casting	Section 5.12
Bulk deformation	Various sections in Ch. 6
Sheet-metal forming	Section 7.15
Machining	Section 8.13
Abrasives	Section 9.16
Polymers	Section 10.13
Powder metallurgy and ceramics	Sections 11.6 and 11.12
Joining	Section 12.16

MATERIAL PROPERTIES
Tables 2.1, 2.3, and 2.5; Figs. 2.4, 2.6, 2.12, 2.13, and 2.29
Tables 3.2, 3.3, 3.6, 3.8, and 3.9 through 3.13
Tables 5.4, 5.5, and 5.6; Fig. 5.19
Table 7.3
Table 8.6; Fig. 8.31
Table 9.1
Tables 10.1, 10.4, 10.5, and 10.8; Fig. 10.15
Tables 11.3, 11.4, 11.7 and 11.9; Figs. 11.20, 11.21, and 11.22
Table 12.5

MANUFACTURING CHARACTERISTICS OF MATERIALS
Table 3.7
Tables 5.2 and 5.3
Table 7.2
Section 8.5
Tables 11.5, 11.8, and 11.9
Section 12.5
Table 16.6

CAPABILITIES OF MANUFACTURING PROCESSES
Fig. 4.20
Tables 5.2, 5.7, and 5.8
Table 6.1
Table 7.1
Table 8.7; Fig. 8.27
Table 9.4; Fig. 9.27
Tables 10.6, 10.7, and 10.9
Tables 11.5 and 11.8
Tables 12.1, 12.2, and 12.5
Table 13.5; Fig. 14.3; Table 15.1; Fig. 16.3

DIMENSIONAL TOLERANCES AND SURFACE FINISH
Fig. 4.20
Table 5.2
Tables 8.7 and 8.27
Fig. 9.27
Fig. 16.4

GENERAL COSTS
Table 5.2; Fig. 5.42
Fig. 7.74
Fig. 9.40
Table 10.9
Table 11.9
Tables 12.1 and 12.2
Table 16.7; Fig. 16.5

MATERIAL COSTS
Table 10.4
Table 11.9
Table 16.3

As some examples of the foregoing considerations, note how (1) the size of products such as radios, electronic equipment, cameras, computers, calculators, and cell phones has been reduced; (2) repair of products is now often done by replacing subassemblies and modules; and (3) there are fewer traditional fasteners and more snap-on type of assemblies used in products.

16.2.2 Product design and quantity of materials

With the high production rates and reduced labor permitted by automation, the cost of materials becomes a significant portion of a product's cost. Although the material cost cannot be reduced below the market level, reductions can be made in the *quantity* of materials used in the components to be produced. The overall shape of the product is usually optimized during the design and prototype stages, through techniques such as finite-element analysis, minimum-weight design, design optimization, and computer-aided design and manufacturing. These methods have greatly facilitated design analysis, material selection, material usage, and optimization.

Typically, reductions in the amount of material used can be achieved by decreasing the component's volume. This approach may require the selection of materials that have high strength-to-weight or stiffness-to-weight ratios. (See Section 3.9.1.) Obviously, higher ratios can also be obtained by improving the product's design and by selecting different cross-sections, such as ones that have a high moment of inertia (as in I-beams), or by using tubular or hollow components instead of solid bars.

Implementing such design changes and minimizing the amount of materials used can, however, lead to thin cross-sections and present significant problems in manufacturing. Consider the following examples:

a) Casting or molding of thin sections can present difficulties in mold filling and in maintaining the specified dimensional accuracy and surface finish.
b) Forging of thin sections requires high forces, due to effects such as friction and chilling of thin sections.
c) Impact extrusion of thin-walled parts can become difficult, especially when high dimensional accuracy is required.
d) Formability in sheet-metal working may be reduced as sheet thickness decreases; furthermore, reducing the thickness of the sheet can lead to wrinkling, due to compressive stresses developed in the plane of the sheet during forming.
e) Machining and grinding of thin workpieces can present difficulties such as part distortion, poor dimensional accuracy, and chatter; consequently, advanced material-removal processes have to be considered.
f) Welding of thin sheets or structures can cause distortion, due to thermal gradients developed during the welding cycle.

Conversely, making parts that have thick cross-sections can also have adverse effects. Some examples are as follows:

a) In processes such as casting and injection molding, the production rate can be slower, because of the increased cycle time required to allow for part cooling in the mold.
b) Unless controlled, porosity can develop in thicker sections of castings.
c) The bendability of sheet metals decreases as their thickness increases.
d) In powder metallurgy, there can be significant variations in density and properties throughout thick parts.
e) Welding of thick sections can present problems.
f) In die-cast parts, thinner sections will have higher strength per unit thickness than will thicker sections, because of the smaller grain size of the former.
g) Processing of polymer parts requires increased cycle times as their thickness or volume increases, because of the longer time required for the parts to cool sufficiently to be removed from the molds.

16.2.3 Robustness and robust design

An increasingly important trend in product design is *robustness*. Originally introduced by G. Taguchi (see also Section 4.9.3), robustness can be defined as a design, a process, or a system that continues to function within acceptable parameters despite variabilities in its environment. These variabilities are called *noise*, defined as those factors that are difficult or impossible to control. Examples of noise in manufacturing include (a) ambient temperature and humidity in a production facility, (b) random and unanticipated vibrations of the shop floor, (c) batch-to-batch variations in the dimensions and properties of the incoming raw materials, and (d) the performance of different operators and machines at different times during the day or on different days. Thus, a robust process does not change with changing noise, and in a *robust design*, the part will function sufficiently well even if unanticipated events occur.

As a simple illustration, consider a sheet-metal mounting bracket to be attached to a wall with two bolts (Fig. 16.1a). The positioning of the two mounting holes will, of course, include some error, such as caused by the type of manufacturing process and machine involved (e.g., an error due to punching, drilling, or rimming). This error will, in turn, prevent the top edge of the bracket from being horizontal. In a more robust design (Fig. 16.1b), the mounting holes are twice as far apart from each other. Even though the same manufacturing method is used and the production cost remains the same, the robust design has a variability that is one-half of that of the original design. In an even more robust design, the type of bolts used would be such that they will not loosen in case of vibration or during their use over a period of time.

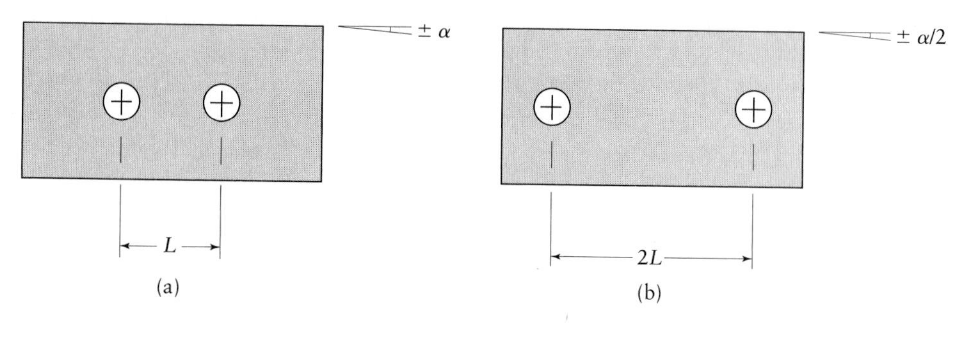

FIGURE 16.1 A simple example of robust design. (a) Location of two mounting holes on a sheet-metal bracket, where the deviation of the top surface of the bracket from being perfectly horizontal is $\pm\ \alpha$. (b) New location of holes, whereby the deviation of the top surface of the bracket from being perfectly horizontal is reduced to $\pm\ \alpha/2$.

EXAMPLE 16.1 An example of DFMA application

Numerous examples can be given regarding the benefits of applying DFMA principles in the early stages of product concept and development. These principles can also be applied to the modification of existing designs and the selection of appropriate production methods. In this example, we consider the redesign of the pilot's instrument panel for a military helicopter designed and built by McDonnell Douglas.

The components of the panel consisted of sheet metal, extrusions, and rivets. Through the use of DFMA software and analysis of the panel, it was estimated that the redesign would lead to the following changes: The number of parts would be reduced from 74 to 9; the weight of the panel would be reduced from 3.00 kg to 2.74 kg; fabrication time would be reduced from 305 hrs to 20 hrs; assembly time would be reduced from 149 hrs to 8 hours; and total time to produce the part would be reduced from 697 hrs to 181 hrs. It was estimated that, as a result of redesign, the cost savings would be 74%. Based on these results, other components of the instrument panel also were subjected to similar analysis.

16.3 Product Quality and Life Expectancy

Product quality and the techniques involved in quality assurance and control have been described in Section 4.9. The word *quality* is difficult to define precisely, partly because it includes not only well-defined technical considerations, but also human, and hence subjective, opinion. However, a high-quality product generally is considered to have the following characteristics:

1. It satisfies the needs and expectations of the customer.
2. It is compatible with the customer's working environment.
3. It functions reliably over its intended life.
4. Its aesthetics are pleasing.
5. It provides a high level of safety.
6. Installation, maintenance, and future improvements are easy to perform.
7. Maintenance and upgrade costs are low.

Product quality has always been a major consideration in manufacturing. In view of the global economy and competition, a major priority is the concept of *continuous improvement in quality*, as exemplified by the Japanese term **kaizen**, meaning never-ending improvement. Still, the level of quality that a manufacturer chooses for its products depends on the market for which the products are intended. Thus, for example, low-cost, low-quality products have their own market niche (as is easily proved by a visit to a large department store or a home-improvement center), just as there is a market, especially in prosperous times, for high-quality, expensive products, such as a Rolls-Royce automobile, audio equipment, and sporting equipment like tennis rackets, golf clubs, and skis.

We have seen that, through *concurrent engineering* (see Section 1.2), design and manufacturing engineers have the responsibility and latitude to select and specify materials for the particular products to be made. Quality considerations are always a part of such tasks. Let's take a simple case of selecting materials for a screwdriver stem. Based on the functions of a screwdriver, one can specify stem materials that have high yield and tensile strength, torsional stiffness, and resistance to wear and corrosion. A screwdriver made from such materials will perform better and last longer than one made of materials with inferior properties. On the other hand, materials with better properties may generally be more expensive and may even be more difficult to produce than others. Consequently, the manufacturing operations and systems involved must thoroughly be reviewed to keep final product costs low, which can be done based on the numerous technical and economic considerations discussed throughout this text. Unfortunately, design engineers are often faced with this dilemma, which has been succinctly stated as "Good, fast, or cheap; pick any two."

Because of the significant costs that can be incurred in manufacturing activities, the concept of **return on quality** (ROQ) has been advanced. The basic components are: (a) because of its major influence on customer satisfaction, quality should be viewed as an investment; (b) there has to be a certain limit on how much money should be spent on quality improvements; (c) the specific area or issue for which the expenditure should be made toward quality improvement must be properly assessed; and (d) the incremental improvement in quality must be carefully reviewed with regard to the additional costs involved.

As has been discussed previously, the concept of quality is fairly subjective. Clearly, parts or products that do not serve their intended purpose detract from quality, and there is a clearly defined cost that can be calculated for such defects. For example, if a critical part tolerance exceeds a design specification, the part is usually discarded or reworked, and the inspection, material, machining, labor, handling, and other costs can be accounted for and a ROQ quantitatively expressed.

Contrary to a common perception, high-quality products do not necessarily cost more. In some industries, the ROQ is minimized at a value of zero defects, while in others, the cost of eliminating the final few defects is very high. However, indirect factors still may encourage elimination of these final few defects. For example, customer satisfaction is a qualitative factor that is difficult to include in calculations, but satisfaction is increased and customers are more likely to be retained when there are no defects in the product they purchase. Although there can be significant variations, it has been estimated that the relative costs involved in identifying and repairing defects in products grow by orders of magnitude, a concept called the *rule of ten* (see also Section 16.9.2), as follows:

Stage	Relative cost of repair
Fabrication of the part	1
Subassembly	10
Final assembly	100
At product distributor	1000
At the customer	10,000

Life expectancy of products. Numerous surveys have indicated that the *average life expectancies* of certain products, in years, are as follows (in alphabetical order):

Car battery	4	Machinery	30
Central air-conditioning unit	15	Motor vehicles	10
Dishwasher	10	Refrigerator	17
Disposal	10	Vacuum cleaner	10
Dryer, gas	13	Washer	13
Furnace, gas	18	Water heater, electric	14
Hair dryer	5	Water heater, gas	12

As described in various sections in this chapter, numerous examples can be given of situations where a choice has to be made between different processes and materials to manufacture a product. Consider, for example, (a) sheet-metal vs. cast frying pans, (b) carbon-steel vs. stainless-steel exhaust systems for automotive vehicles, (c) wood vs. metal handles for hammers, (d) plastic vs. metal outdoor chairs, and (e) wood vs. reinforced-plastic ladders. Depending on the materials and processes employed, and hence on the quality of the product, the life expectancy of products can vary significantly; life expectancy is also determined in part by the frequency of maintenance.

16.4 Life-Cycle Assessment, Life-Cycle Engineering, and Sustainable Manufacturing

Although there are a number of similar definitions, according to the ISO 14000 standard (Section 4.9.4), *life-cycle assessment* (LCA) may be defined as a systematic set of procedures for compiling and examining the inputs and outputs of materials and energy, and the associated environmental impacts or burdens directly attributable to the functioning of a product, process, or service system throughout its entire life cycle. Like LCA, *life-cycle engineering* (LCE) is concerned with environmental factors, but the latter also deals in greater detail with design, optimization, and various technical considerations regarding each component of a product or process life cycle. A major aim of LCE is to give thought to the reuse and recycling of products from the earliest stage of the design process (*green design*, or *green engineering*).

Life cycle, also called *cradle-to-grave*, can be defined as consecutive and interlinked stages of a product or service system, including the (a) extraction of natural resources; (b) processing of raw materials; (c) manufacture of products; (d) transportation and distribution of the product to the customer; (e) use, reuse, and maintenance of the product; and (f) disposal of the product or recovery and recycling of its components.

Note that all these factors are equally applicable to each of the various products listed in foregoing table. Each product has its own share of metallic and nonmetallic materials that have been processed into individual parts and assembled, and each product has its own life cycle. Note also that some products are disposable, while others are reusable. In addition to materials, other important considerations include the use of fluids (such as lubricants, coolants, toxic solutions, and various fluids used in heat treating and plating) that can have serious and adverse environmental effects. Government regulations in many countries are now moving towards the direction of making manufacturers life-cycle accountable.

Although life-cycle analysis and engineering is a comprehensive, powerful, and necessary tool, its implementation can be expensive, challenging, and time consuming. This is largely because of uncertainties in the input data regarding materials, processes, long-term effects, costs, etc., and the time required to gather reliable data to properly assess the often-complex interrelationships among various components of the whole system. A number of software packages are being developed to expedite such analysis in certain industries, particularly the chemical and process industries, because of the higher potential for environmental damage in their operations.

Sustainable manufacturing. In recent years, we all have become increasingly aware that, ultimately, our natural resources are limited, which clearly necessitates the need for conservation of materials and energy. The term *sustainable manufacturing* is now being used to emphasize the need for conserving these resources, particularly through maintenance and reuse. This conservation is to be undertaken in order to (a) increase the life cycle of products; (b) eliminate damage to the environment; and (c) increase our collective social well-being, especially of future generations, while (d) maintaining a company's profitability.

EXAMPLE 16.2 Sustainable manufacturing in the production of Nike athletic shoes

Among various examples, the production of Nike athletic shoes has indicated the benefits of sustainable manufacturing. The athletic shoes are assembled using adhesives (Section 12.13). Up to around 1990, the adhesives used contained petroleum-based solvents; these adhesives pose health hazards to humans and contribute to petrochemical smog. In order to improve this situation, the company worked with chemical suppliers to successfully develop water-based adhesive technology, which is now used for the majority of the assembly operations. As a result, solvent use in all manufacturing processes in Nike's subcontracted facilities in Asia was reduced by 67% since 1995, and in 1997, 834,000 gallons of hazardous solvents were replaced with 1290 tons of water-based adhesives.

As another example, the rubber outsoles of the shoe were made by a process that resulted in significant amount of extra rubber around the periphery of the sole, called *flashing*, similar to the flash shown in Figs. 6.15a and 10.29c. With about 40 factories using thousands of molds and producing over a million outsoles a day, this flashing constitutes the largest source of waste in the manufacturing process for the shoes. In order to reduce this waste, the company developed a technology that grinds the flashing into 500-μm rubber powder, which is then added to the rubber mixture used to make subsequent outsoles. As a result of this process, waste was reduced by 40%, and furthermore, it was found that the mixed rubber had better abrasion resistance, durability, and overall performance than the highest premium rubber.

16.5 Selection of Materials for Products

A clear understanding of the functional requirements for each of the individual components of a product is essential. Although the general criteria for selecting materials have been described in Section 1.5, this chapter will discuss them in further detail.

16.5.1 General properties of materials

As described in Chapter 2, mechanical properties of materials include strength; toughness; ductility; stiffness; hardness; and resistance to fatigue, creep, and impact. Physical properties (see Section 3.9) include density, melting point, specific heat, thermal and electrical conductivity, thermal expansion, and magnetic properties. The chemical properties (see Section 3.9.7) that are of primary concern in manufacturing are susceptibility to oxidation and susceptibility to corrosion. The relevance of these properties to product design and manufacturing has been described in various chapters, and several tables relating to properties of metallic and nonmetallic materials are included therein. Table 16.1 lists the sections in this text that are relevant to material properties.

Selection of materials is now easier and faster, because of the availability of computerized and extensive *databases*, which provide much greater accessibility to information than previously available. However, to facilitate the selection of materials and the determination of the parameters described later in this section, expert-system software (*smart databases*) has been developed. With proper input of product design and functional requirements, these systems are capable of identifying appropriate materials for a particular application, just as an expert or a team of experts would.

The following considerations are important in materials selection for products:

a) Do the materials selected have properties that unnecessarily exceed minimum requirements and specifications?

b) Can some materials be replaced by others that are less expensive?

c) Do the materials selected have the appropriate manufacturing characteristics?

d) Are the raw materials (stock) to be ordered available in standard shapes, dimensions, tolerances, and surface finishes?

e) Is the material supply reliable?

f) Are there likely to be significant price increases or fluctuations for the materials?

g) Can the materials be obtained in the required quantities in the desired time frame?

16.5.2 Shapes of commercially available materials

For obvious economic reasons, the shapes and sizes in which raw materials are commercially available are an important consideration (Table 16.2). Materials are generally available in various forms: castings, extrusions, forgings, bar, plate, sheet, foil, rod, wire, and metal powders. Purchasing materials in shapes that require the least additional processing is an additional consideration. However, such characteristics as surface quality, dimensional tolerances, and straightness of the raw materials must also be taken into account. Obviously, the better and the more consistent these characteristics are, the less additional processing effort and time are required.

For example, if we want to produce simple shafts that have good dimensional accuracy, roundness, straightness, and surface finish, we could purchase round bars that are already turned and centerless ground to the dimensions specified. Unless our facilities are capable of producing round bars economically, it is cheaper to purchase them. On the other hand, if we need to make a stepped shaft (a shaft that has different diameters along its length), we could purchase a round bar that has a diameter at least equal to the largest diameter of the final stepped shaft and turn it on a lathe. We may also process it by some other means to reduce the diameter. If the stock has broad dimensional tolerances or is warped or out of round, we must order a larger size to ensure proper dimensions of the final shaft.

As we have seen throughout various chapters, each manufacturing operation produces parts that have specific shapes, dimensional tolerances, and surface finishes. Consider the following examples:

a) Hot-rolled and hot-drawn products have a rougher surface finish and greater dimensional tolerances than cold-rolled and cold-drawn products (Chapter 6).

b) Round bars turned on a lathe have a rougher surface finish than bars that are ground on cylindrical or centerless grinding machines (Chapters 8 and 9).

c) The wall thickness of seamless tubing made by the tube-rolling process is less uniform than that of welded tubing (Chapters 6 and 12).

d) Extrusions have smaller cross-sectional dimensional tolerances than parts made by roll forming (Chapters 6 and 7).

e) Castings generally have less dimensional accuracy and poorer surface finish than parts made by cold extrusion or powder metallurgy (Chapters 5, 6, and 11).

TABLE 16.2

Commercially Available Forms of Materials

Material	Available as
Aluminum	B, F, I, P, S, T, W,
Ceramics	B, p, s, T
Copper and brass	B, f, I, P, s, T, W
Elastomers	b, P, T
Glass	B, P, s, T, W
Graphite	B, P, s, T, W
Magnesium	B, I, P, S, T, w
Plastics	B, f, P, T, w
Precious metals	B, F, I, P, t, W
Steels and stainless steels	B, I, P, S, T, W
Zinc	F, I, P, W

Note: B = bar and rod; F = foil; I = ingots; P = plate and sheet; S = structural shapes; T = tubing; W = wire. Lowercase letters indicate limited availability. Most of the metals are also available in powder form, including prealloyed powders.

16.5.3 Manufacturing characteristics of materials

Manufacturing characteristics of materials typically include castability, workability, formability, machinability, grindability, weldability, and hardenability by heat treatment. Because raw materials have to be formed, shaped, machined, ground, fabricated, or heat treated into individual components having specific shapes and dimensions, these properties are crucial to the proper selection of materials. Table 16.1 lists references to general manufacturing characteristics of materials.

Recall also that the quality of a raw material can greatly influence its manufacturing characteristics. The following are some examples:

a) A rod or bar with a longitudinal seam (lap) will develop cracks during simple upsetting and heading operations.

b) Bars with internal defects and inclusions will crack during seamless-tube production.

c) Porous castings will produce poor surface finish when machined.

d) Blanks that are heat treated nonuniformly and bars that are not stress relieved will distort during subsequent operations, such as machining or drilling a hole.

e) Incoming stock that has variations in composition and microstructure cannot be heat treated or machined consistently.

f) Sheet-metal stock that has variations in its cold-worked condition or thickness will exhibit springback during bending and other forming operations, because of differences in yield stress and strains in bending.

g) If prelubricated sheet-metal blanks have nonuniform lubricant distribution and thickness, their formability, surface finish, and overall quality will be adversely affected.

16.5.4 Reliability of material supply

It is well known that geopolitical factors can significantly affect the supply of strategic materials. Other factors, such as strikes, shortages, and the reluctance of suppliers to produce materials in a particular shape, quality, or quantity, also affect reliability of supply. Even though the availability of materials may not be a problem throughout a country as a whole, it can cause difficulties for a certain business, because of the location of a particular manufacturing plant.

16.5.5 Cost of materials and processing

Because of a raw material's processing history, the unit cost of the raw material depends not only on the material itself, but also on its shape, size, and condition. For example, because more operations are involved in the production of thin wire than in that of round rod, the unit cost of the wire is much higher. Similarly, powder metals generally are more expensive than bulk metals. Furthermore, the cost of materials typically decreases as the quantity purchased increases. As a result, certain segments of industry, such as automotive companies, purchase materials in very large quantities; the larger the quantity, the lower is the cost per unit weight (bulk discount).

The cost of materials may be determined in terms of cost per unit weight or cost per unit volume. Table 16.3 shows the cost per unit volume for wrought metals and plastics relative to that for carbon steel. The benefit of using cost per unit volume

TABLE 16.3

Approximate Cost Per Unit Volume for Wrought Metals and Plastics Relative to the Cost of Carbon Steel

Gold	60,000	Magnesium alloys	2–4
Silver	600	Aluminum alloys	2–3
Molybdenum alloys	200–250	High-strength low-alloy steels	1.4
Nickel	35	Gray cast iron	1.2
Titanium alloys	20–40	Carbon steel	1
Copper alloys	5–6	Nylons, acetals, and silicon	1.1–2
Stainless steels	2–9	Rubber*	0.2–1
		Other plastics and elastomers*	0.2–2

* As molding compounds.
Note: Costs vary significantly with the quantity of puchase, supply and demand, size and shape, and various other factors.

can be seen by the following simple example: In the design of a steel cantilevered rectangular beam that is to support a certain load at its end, a maximum deflection is specified. Using equations developed in the mechanics of solids and assuming that the weight of the beam can be neglected, we can determine an appropriate cross-section of the beam. Since all dimensions are now known, the volume of the beam can be calculated, and if we know the cost of the material per unit volume, we can easily calculate the cost of the beam. If the cost is measured per unit weight, however, we first have to calculate the beam's weight and then determine the cost.

The cost of a particular material is subject to fluctuations caused by factors as simple as supply and demand or as complex as geopolitics. If a product is no longer cost competitive, alternative and less costly materials can be selected. For example, the copper shortage in the 1940s led the U.S. government to mint pennies from zinc-plated steel. Similarly, when the price of copper increased substantially during the 1960s, the electrical wiring being installed in homes was, for a time, made of aluminum. It should be noted, however, that this substitution led to the redesign of switches and outlets to avoid excessive heating at the junctions.

When scrap is produced during manufacturing, as in sheet-metal fabrication, forging, and machining, the value of the scrap is deducted from the material's cost, in order to obtain net material cost. Table 16.4 lists the percentages of scrap produced in selected manufacturing processes. Note that in machining, scrap can be very high, whereas rolling, ring rolling, and powder metallurgy (all of which are net- or near-net-shape processes) produce the least scrap. As expected, the value of the scrap depends on the type of metal and on the demand for the scrap; typically, it is between 10% to 40% of the original cost of the material.

TABLE 16.4

Typical Scrap Produced in Various Manufacturing Processes

Process	Scrap (%)	Process	Scrap (%)
Machining	10–60	Hot extrusion	15
Hot closed-die forging	20–25	Permanent-mold casting	10
Sheet-metal forming	10–25	Powder metallurgy	<5
		Rolling and ring rolling	<1

16.6 Substitution of Materials in Products

Although new products continually appear on the market, the majority of the design and manufacturing effort is concerned with improving existing products. Major product improvements can result from (a) substitution of materials, (b) implementation of new or improved processing techniques, (c) better control of the processing parameters, and (d) increased plant automation.

There is hardly a product on the market today for which substitution of materials has not played a major role in helping companies maintain their competitive positions. Automobile and aircraft manufacturing are particular examples of major industries in which substitution of materials is an important and ongoing activity. A similar trend is evident in the manufacture of sporting goods and in various other consumer products.

There are several reasons for substituting materials in existing products:

1. Reduce the costs of materials and processing;
2. Improve manufacturing and assembly, installation, and conversion to automated assembly;
3. Improve the performance of products—for example, by reducing weight and by improving resistance to wear, fatigue, and corrosion;
4. Increase the stiffness-to-weight and strength-to-weight ratios of structures;
5. Reduce the need for maintenance and repair;
6. Reduce vulnerability to the unreliability of domestic and overseas supply of certain materials;
7. Improve compliance with legislation and regulations prohibiting the use of materials that have adverse environmental impact;
8. Improve the perceived aesthetics of a product;
9. Reduce performance variations or environmental sensitivity in the product, i.e., to improve *robustness* (see Section 16.2.3).

From a brief review of this list, it is evident that there are numerous important technological and economic factors involved in material substitution, and access to much reliable and extensive data is essential before a decision can be made to substitute materials. Note, for instance, that an important factor in substitution is the compatibility of the materials used; a typical example of incompatibility is galvanic corrosion in metal pairs, as described in Section 3.9.7. Another major factor is the influence that the materials selected for substitution may have on the manufacturing processes used. Thus, for example, in spite of some similarities that exist, the production of ceramic or plastic parts and the production of metal parts typically involve different processes, machinery, assembly, production rates, and costs.

a) **Substitution of materials in the automobile industry.** The automobile industry provides a number of good examples of the effective substitution of materials in order to achieve one or more of the foregoing objectives:
   1. Several parts of the metal body of automobiles have been replaced with plastic or reinforced-plastic parts.
   2. Metal bumpers, gears, pumps, fuel tanks, housings, covers, clamps, and various other components have been replaced with plastic substitutes.
   3. Engine components have been replaced with ceramic and reinforced-plastic parts.
   4. All-metal drive shafts have been replaced with composite-material drive shafts.
   5. Cast-iron engine blocks have been changed to cast-aluminum blocks; forged crankshafts to cast crankshafts; and forged connecting rods to cast, powder-metallurgy, or composite-material connecting rods.

6. Steel structural elements have been replaced with extruded aluminum sections, and rolled sheet-steel body panels have been changed to rolled aluminum sheet (see Fig. 1.5).

Because the automobile industry is a major consumer of both metallic and nonmetallic materials, there is constant and rigorous competition among suppliers, particularly for steel, aluminum, and plastics. Industry engineers and management are continually investigating the relative advantages and limitations of these principal materials in their applications, recycling and other environmental considerations, and, in particular, the materials' relative costs and benefits.

b) **Substitution of materials in the aircraft industry.** In the aircraft and aerospace industries, conventional aluminum alloys (2000 and 7000 series) are being replaced with aluminum–lithium alloys and titanium alloys, because of the latter two alloys' higher strength-to-weight ratios. Forged parts are being replaced with powder-metallurgy parts that are manufactured with better control of impurities and microstructure; the powder-metallurgy parts also require less machining and produce less scrap of expensive materials. Furthermore, advanced composite materials and honeycomb structures are replacing traditional aluminum airframe components (Fig. 16.2), and metal-matrix composites are replacing some of the aluminum and titanium parts used previously in structural components.

EXAMPLE 16.3 Changes made in materials between C-5A and C-5B military cargo aircraft

Table 16.5 shows the changes made in materials for various components of the C-5A and C-5B military cargo aircraft and the reasons for the changes.

Source: H. B. Allison.

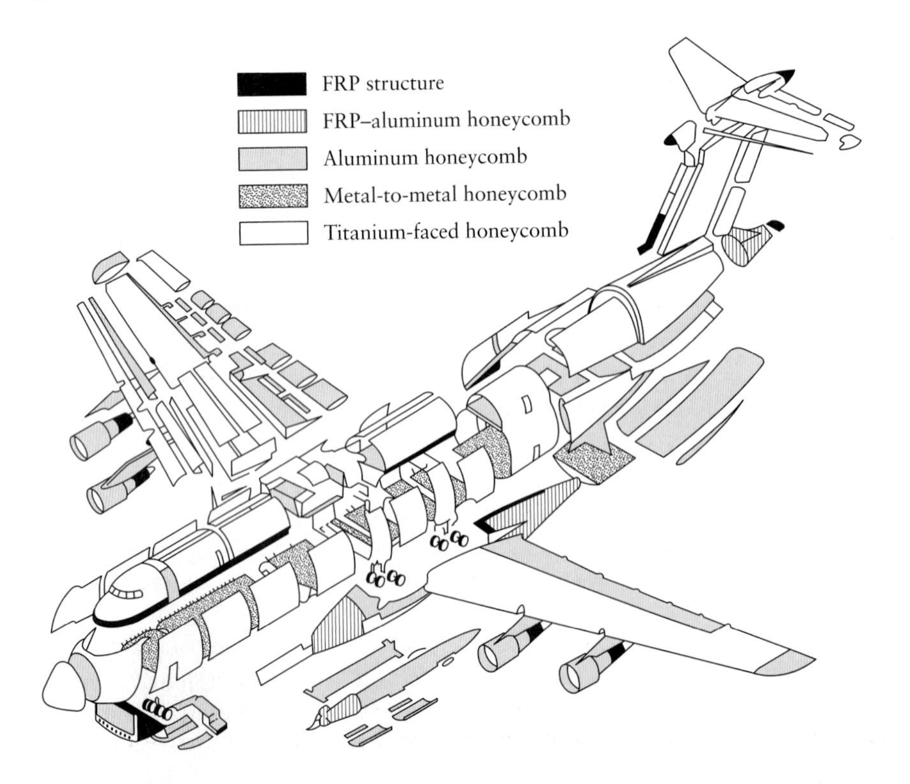

FIGURE 16.2 Advanced materials used on the Lockheed C-5A transport aircraft. (FRP stands for *fiber-reinforced plastic.*)

TABLE 16.5

Changes in the Materials From C-5A to C-5B Military Cargo Aircraft

Item	C-5A material	C-5B material	Reason for change
Wing panels	7075-T6511	7175-T73511	Durability
Main frame			
Forgings	7075-F	7049-01	Stress-corrosion resistance
Machined frames	7075-T6	7049-T73	
Frame straps	7075-T6 plate	7050-T7651 plate	
Fuselage skin	7079-T6	7475-T61	Material availability
Fuselage under-floor end fittings	7075-T6 forging	7049-T73 forging	Stress-corrosion resistance
Wing/pylon attach fitting	4340 alloy steel	PH13-8Mo	Corrosion prevention
Aft ramp's lock hooks	D6-AC	PH13-8Mo	Corrosion prevention
Hydraulic lines	AM350 stainless steel	21-6-9 stainless steel	Improved field repair
Fuselage fail-safe straps	Ti-6Al-4V	7475-T61 aluminum	Titanium strap debonding

16.7 Capabilities of Manufacturing Processes

We have seen that each manufacturing process has its particular advantages and limitations. Casting and injection molding, for example, can generally produce more complex shapes than can forging and powder metallurgy (because the molten metal or plastic is capable of filling complex mold and die cavities). On the other hand, forgings can generally be made into complex shapes by subsequent machining and finishing operations, and they have a toughness that is generally superior to that of castings and powder-metallurgy products. References to the general characteristics and capabilities of manufacturing processes are listed in Table 16.1.

Recall that the shape of a product may be such that it can best be fabricated from several parts, by joining them with fasteners or with such techniques as brazing, welding, and adhesive bonding. The reverse may be true for another product: Manufacturing it in one piece may be more economical, because of the significant assembly costs otherwise involved. Other factors that must be considered in process selection are the minimum section size and dimensions that can be satisfactorily produced (Fig. 16.3). For example, very thin sections can be produced by cold rolling, but processes such as sand casting and forging prohibit the fabrication of thin sections.

a) **Dimensional tolerances and surface finish.** The *dimensional tolerances* and *surface finish* produced are particularly important in subsequent assembly operations and in the proper operation of machines and instruments. The dimensional tolerances and surface finishes obtained by various manufacturing processes are qualitatively illustrated in Fig. 16.4, and references to the capabilities of manufacturing processes with respect to these features are provided in Table 16.1.

In order to obtain closer dimensional tolerances and better surface finish, additional finishing operations, better control of processing parameters, and the use of higher quality equipment and controls may be required. However, the closer the tolerance and the finer the surface finish specified, the higher is the

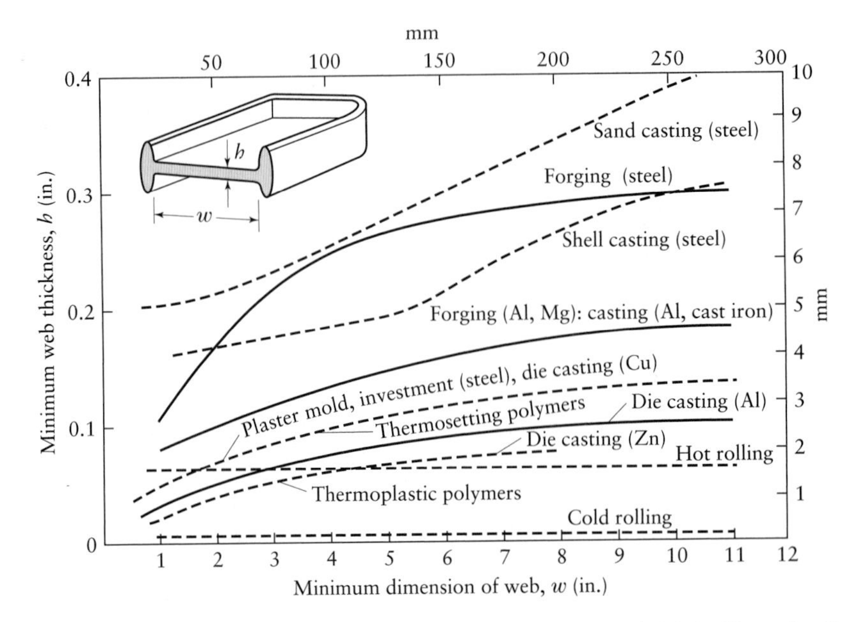

FIGURE 16.3 Minimum part dimensions obtainable by various manufacturing processes. *Source*: J. A. Schey, *Introduction to Manufacturing Processes*, 2d. ed., McGraw-Hill, 1987.

cost of manufacturing, as shown in Figs. 9.40 and 16.5. Also, the finer the surface finish required, the longer is the manufacturing time (Fig. 16.6), the higher is the number of processes involved, and, hence, the higher is the product cost (Fig. 16.5). In the machining of aircraft structural members made of titanium alloys, for example, as much as 60% of the cost of machining the part may be expended in the final machining pass, in order to maintain proper tolerances and surface finish. Unless it is specifically required otherwise by proper technological and economic justification, parts should be made with as rough a surface finish and as wide a tolerance as will be functionally and aesthetically acceptable. In this regard, the importance of continual interaction and communication between the product designer and the manufacturing engineer becomes obvious.

b) **Production volume.** Depending on the type of product, the *production volume*, or *quantity* (*lot size*), can vary widely. For example, paper clips, bolts, washers, spark plugs, bearings, and ballpoint pens are produced in very large quantities. On the other hand, jet engines for large commercial aircraft, diesel engines for locomotives, and propellers for cruise ships are manufactured in limited quantities. Production quantity plays a significant role in process and equipment selection. In fact, an entire manufacturing discipline is devoted to determining mathematically the optimum production quantity, called the *economic order quantity*.

c) **Production rate.** A significant factor in manufacturing process selection is the *production rate*, defined as the number of pieces to be produced per unit of time (such as per hour, per month, or per year). Processes such as powder metallurgy, die casting, deep drawing, and roll forming are high-production-rate operations. By contrast, sand casting, conventional and electrochemical machining, spinning, superplastic forming, adhesive and diffusion bonding, and the processing of reinforced plastics are relatively slow operations. These production rates can, of course, be increased by automation or by using multiple machines. Note, however, that a lower production rate does not necessarily mean that the manufacturing process is inherently uneconomical.

d) **Lead time.** *Lead time* is defined as the length of time between receipt of an order and the delivery of the product to the customer. The selection of a manufacturing process is greatly influenced by the time required to start production.

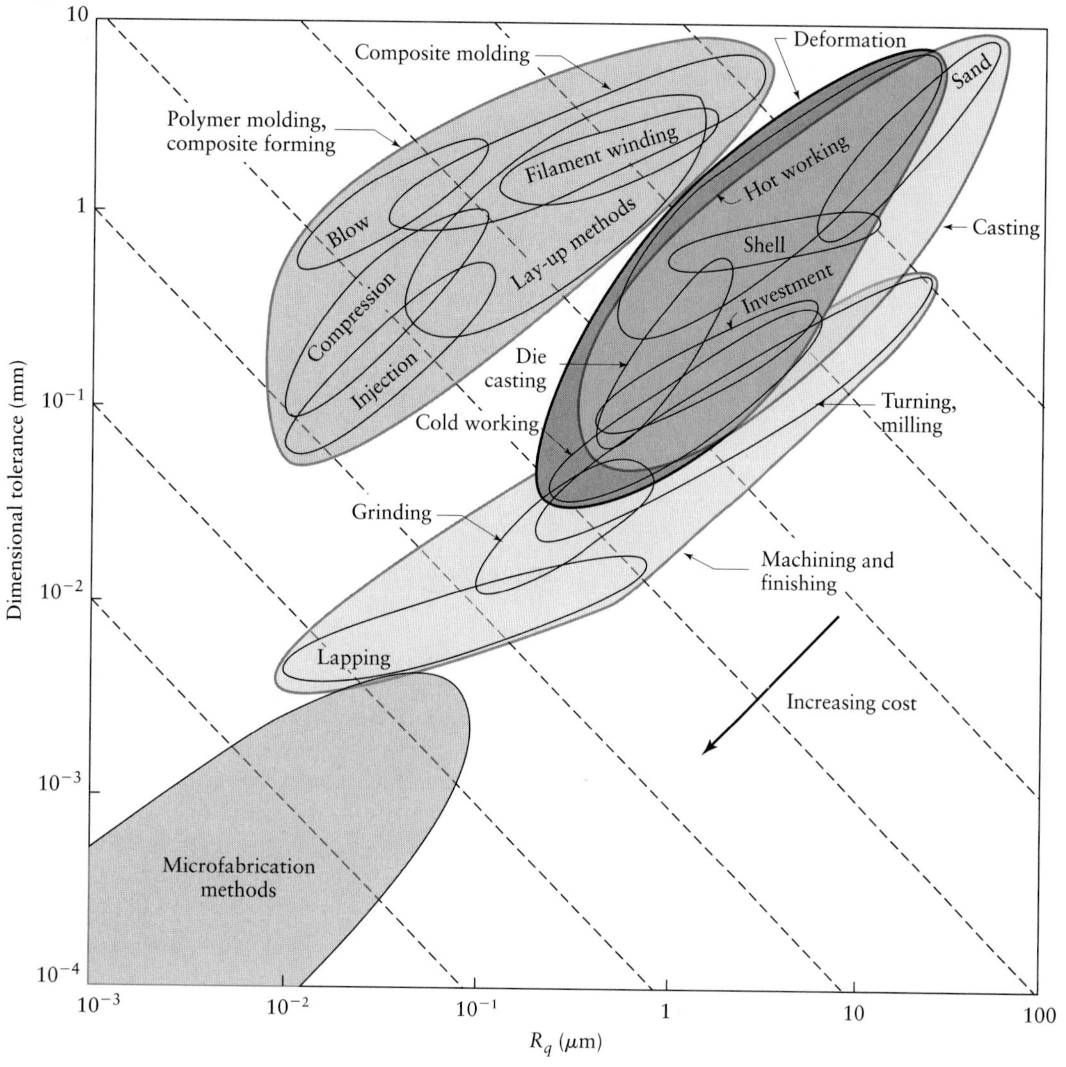

FIGURE 16.4 A plot of achievable dimensional tolerance versus surface roughness for assorted manufacturing operations. The dashed lines indicate cost factors; an increase in precision corresponding to the separation of two neighboring lines gives an increase in cost, for a given process, of a factor of two. *Source*: M. F. Ashby, *Materials Selection in Design*, Butterworth-Heineman, 1999.

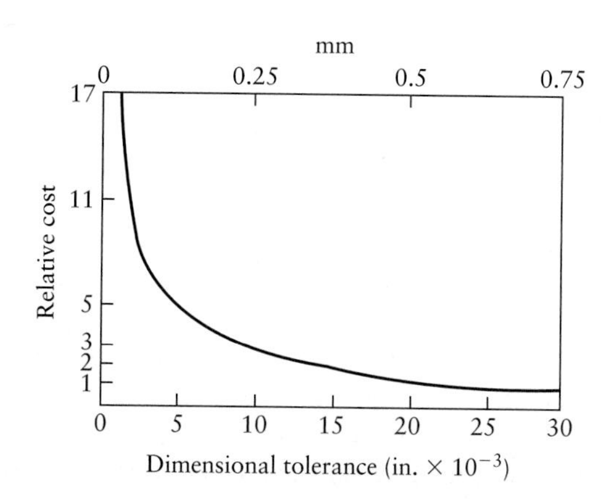

FIGURE 16.5 Relationship between relative manufacturing cost and dimensional tolerance.

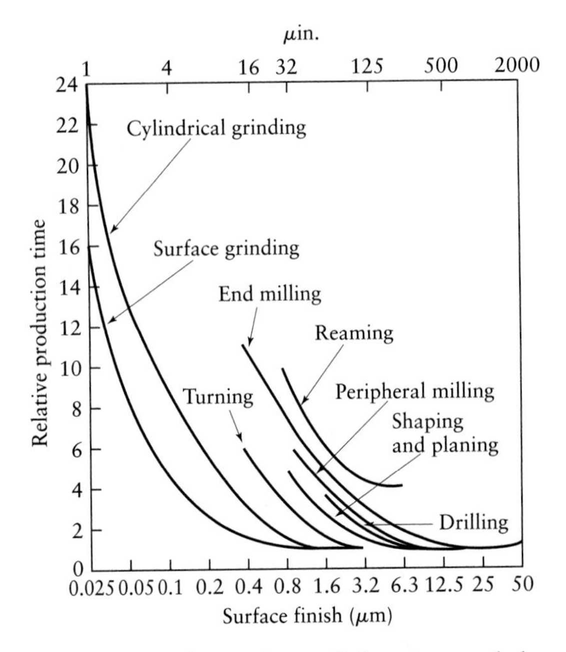

FIGURE 16.6 Relative production time as a function of surface finish produced by various manufacturing processes. *Source: American Machinist*. See also Fig. 9.40.

Such processes as forging, extrusion, die casting, roll forming, and sheet-metal forming typically require dies and tooling that can take a considerable amount of time to produce. Depending on the complexity of the shape of the die, its size, and the material from which the die is to be made (see Table 3.5), lead times can range from weeks to months. In contrast, material-removal processes such as machining and grinding have significant built-in flexibility. As described in Chapters 8 and 9, these processes use tooling and abrasive wheels that can be adapted to most requirements in a very short time. Note also the capabilities of machining centers, flexible manufacturing cells, and flexible manufacturing systems, which are able to respond quickly and effectively to product changes.

16.7.1 Robustness in manufacturing processes and machinery

Robustness has been described in Section 16.2.3 in terms of a design, a process, or a system. In order to appreciate its importance in manufacturing processes, let's consider a simple injection-molded plastic gear, and let's assume that there are significant variations in quality as the products are being made. As described in Section 10.10.2, there are several well-understood variables in injection molding of plastics, including the quality of the raw material (pellets), temperature, and time. These are independent variables and hence they can be controlled. However, as noted in Section 16.2.3, there are certain other variables (*noise*) that are largely beyond our control, such as variations in the ambient temperature and humidity in the plant, dust in the air entering the plant from an open door (and thus contaminating the pellets being fed into the hoppers of the injection-molding machine), and the variability in performance of different operators during different shifts. Obviously, these variables are difficult or impossible to control.

In order to obtain sustained good quality, it is first necessary to understand the effects, if any, of each element of this noise on product quality. Why and how does the ambient temperature affect the quality of molded gears? Why and how does the dust coating on a pellet affect its performance in the machine chamber? How different are the performances of different operators during different shifts? After we answer questions such as the foregoing, it will be possible to establish new operating parameters so that variations in, say, ambient temperature do not adversely affect gear quality.

16.8 | Selection of Manufacturing Processes

This section describes the importance of proper selection of manufacturing processes and machinery, and how the selection process relates to the characteristics of materials, the dimensional tolerances and surface finish obtained, and manufacturing cost. As described in various chapters, especially Chapters 14 and 15, most manufacturing processes have been automated and are increasingly computer controlled, in order to optimize all aspects of operations. Computerization is also effectively increasing product reliability and product quality and reducing labor costs.

The choice of a manufacturing process is dictated by various considerations (Table 16.6):

1. Characteristics and properties of the workpiece material;
2. Shape, size, and thickness of the part;
3. Dimensional-tolerance and surface-finish requirements;
4. Functional requirements of the part;
5. Production volume (quantity);
6. Level of automation required to meet production volume and production rate;
7. Costs involved in individual and combined aspects of the manufacturing operation.

Some materials can be processed at room temperature, whereas others require elevated temperatures and thus need furnaces and appropriate tooling. Some materials are easy to work, because they are soft and ductile; others, being hard, brittle, and abrasive, require special processing techniques and tool and die materials.

Different materials have different manufacturing characteristics, such as castability, forgeability, workability, machinability, and weldability, and few materials have favorable characteristics in all the relevant categories. For example, a material that is castable or forgeable may later present difficulties in the machining, grinding, or finishing operations that may be required in order to produce a product with acceptable surface finish, dimensional accuracy, and quality.

Process selection considerations. A summary of the factors involved in process selection is given as follows, in the form of questions to be asked during this stage:

a) Have all alternative manufacturing processes been investigated?
b) Are the methods chosen economical for the type of material, the shape to be produced, and the required production rate?
c) Can the requirements for dimensional tolerances, surface finish, and product quality be met consistently?
d) Can the part be formed and shaped to final dimensions without the use of additional processes?
e) Are secondary processes such as machining, grinding, and finishing necessary?
f) Is the tooling required available in the plant? Can it be purchased as a standard item?
g) Is scrap produced? If so, is it minimized? What is the value of the scrap?
h) Are all processing parameters optimized?
i) Have all the automation and computer-control possibilities been explored for all phases of the manufacturing cycle?
j) Can group technology be implemented for parts with similar geometric and manufacturing attributes?

TABLE 16.6

General Applications of Manufacturing Processes for Various Metals and Alloys

	Carbon Steels	Alloy steels	Stainless steels	Tool and die steels	Aluminum alloys	Magnesium alloys	Copper alloys	Nickel alloys	Titanium alloys	Refractory alloys
Casting										
Sand	A	A	A	B	A	A	A	A	B	A
Plaster	-	-	-	-	A	A	A	-	-	-
Ceramic	A	A	A	A	B	B	A	A	B	A
Investment	A	A	A	-	A	B	A	A	A	A
Permanent	B	B	-	-	A	A	A	-	-	-
Die	-	-	-	-	A	A	A	-	-	-
Forging, hot	A	A	A	A	A	A	A	A	A	A
Extrusion										
Hot	A	A	A	B	A	A	A	A	A	A
Cold	A	B	A	-	A	-	A	B	-	-
Impact	-	-	-	-	A	A	A	-	-	-
Rolling	A	A	A	-	A	A	A	A	A	B
Powder metals	A	A	A	A	A	A	A	A	A	A
Sheet-metal forming	A	A	A	-	A	A	A	A	A	B
Machining	A	A	A	A	-	A	A	A	B	A
Chemical	A	B	A	B	A	A	A	B	B	B
ECM	-	A	B	A	-	-	B	A	A	A
EDM	-	B	B	A	B	-	B	B	B	B
Grinding	A	A	A	A	A	A	A	A	A	A
Welding	A	A	A	-	A	A	A	A	A	A

Note: (A) Generally processed by this method; (B) can be processed by this method, but may present some difficulties; (-) usually not processed by this method. Product quality and productivity depend greatly on the techniques and equipment used, operator skill, and proper control of processing variables.

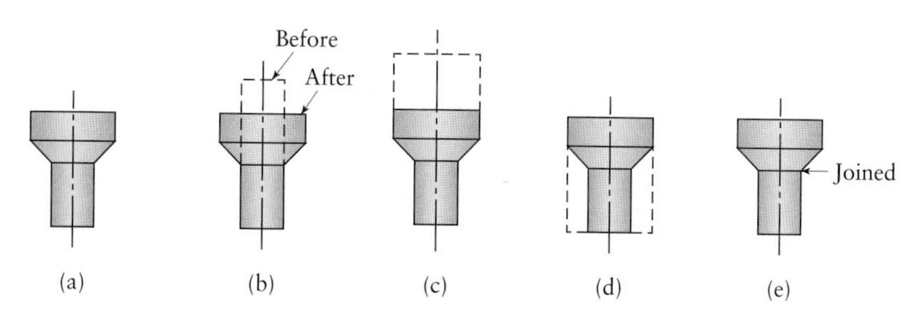

FIGURE 16.7 Various methods of making a simple part: (a) casting or powder metallurgy, (b) forging or upsetting, (c) extrusion, (d) machining, and (e) joining two pieces.

k) Are inspection techniques and quality control being implemented properly?

l) Does each component of the product have to be manufactured in the plant? Are some of its parts commercially available as standard items from external sources?

An example of process selection in making a simple part. Assume that you are asked to make the part shown in Fig. 16.7. You should first determine the part's function, the types of load and environment to which it is to be subjected, the dimensional tolerances and surface finish required, and so on. For the sake of discussion, let's assume that the part is round, that it is 125 mm (5 in.) long, and that the large and small diameters are 38 mm and 25 mm (1.5 in. and 1.0 in.), respectively. Let's further assume that, because of functional requirements (such as stiffness, hardness, and resistance to elevated temperatures), this part should be made of metal.

Which manufacturing process would you choose, and how would you organize the production facilities to manufacture a cost-competitive, high-quality product? Recall that, as much as possible, parts should be produced at or near their final shape (net- or near-net-shape manufacturing), an approach that largely eliminates much secondary processing (such as machining, grinding, and other finishing operations) and thus reduces the total manufacturing time and manufacturing cost.

This part is relatively simple and could be suitably manufactured by different methods: (a) casting or powder metallurgy, (b) upsetting, (c) extrusion, (d) machining, and (e) joining two separate pieces together.

For net-shape processing, the two logical methods are casting and powder metallurgy; each process has its own characteristics, need for specific tooling and labor skill, and costs. This part can also be made by cold, warm, or hot forming. One method, for example, is the upsetting of a round bar 25 mm (1 in.) in diameter in a suitable die to form the larger end (heading). Another possibility is the partial extrusion of a bar 38 mm (1.5 in.) in diameter in order to reduce its diameter to 25 mm. Note that each of these processes shapes the material while yielding little or no material waste.

This part can also be made by machining a 38-mm-diameter bar stock to obtain the 25-mm-diameter section. Machining, however, will take a much longer time than forming, and some material will inevitably be wasted as metal chips. On the other hand, machining does not require special tooling (unlike net-shape processes, which generally require special dies), and this operation can easily be carried out on a lathe. Note, finally, that this part could be made in two separate pieces that are later joined by welding, brazing, or adhesive bonding.

Because of the different operations required in producing the raw materials, costs depend not only on the type of material (e.g., ingot, powder, drawn rod, extrusion), but also on its size and shape. Thus, per unit weight, (a) square bars are more expensive than round bars, (b) cold-rolled plate is more expensive than hot-rolled plate or sheet, and (c) hot-rolled bars are much less expensive than powders of the same metal.

In addition to technical requirements, process selection depends on factors such as the required production quantity and production rate, as described in Section 16.6. In sum, it appears that if only a few parts are needed, machining of this part is the most economical method. As the production quantity increases, however, producing this part by a heading operation or by cold extrusion would be the proper choice. Joining would be the appropriate choice if it were required that the top and bottom pieces of this part be made of different metals.

EXAMPLE 16.4 Manufacturing a sheet-metal part by various methods

As shown throughout this text, there is often more than one method of manufacturing a part. Consider, for instance, forming of a simple dish-shaped part from sheet metal (Fig. 16.8). Such a part can be formed by placing a flat piece of sheet metal between a pair of male and female dies and closing the dies by applying a vertical force in a press. Parts can be formed at high production rates by this method, generally known as *stamping* or *pressworking*.

Assume now that the size of the part is very large, say, 2 m (80 in.) in diameter, and that only 50 parts are required. We now have to reconsider the total operation and ask a number of questions: Is it economical to manufacture a set of dies 2 m in diameter (which are very costly) when the production run (quantity) is so low? Are machines available with sufficient capacity to accommodate such large dies? Does the part have to be made in one piece? What are the alternative methods of manufacturing it?

This part can also be made by welding together smaller pieces of metal formed by other methods. Note, for example, that ships and municipal water tanks are made by this method. Would a part manufactured by welding be acceptable for its intended purpose? Will it have the required properties and the correct shape after welding, or will it require additional processing?

This part can also be made by explosive forming, as shown in Fig. 16.8b. It should be noted, however, that deformation of the material in explosive forming takes place at a very high rate. Consequently, several questions have to be asked:

a) Is the material capable of undergoing deformation at high rates without fracture?

b) Does the high rate of deformation have any detrimental effects on the final properties of the formed part?

c) Can the dimensional tolerances be held within acceptable limits?

d) Is the life of the die sufficiently long under the high transient pressures generated in this process?

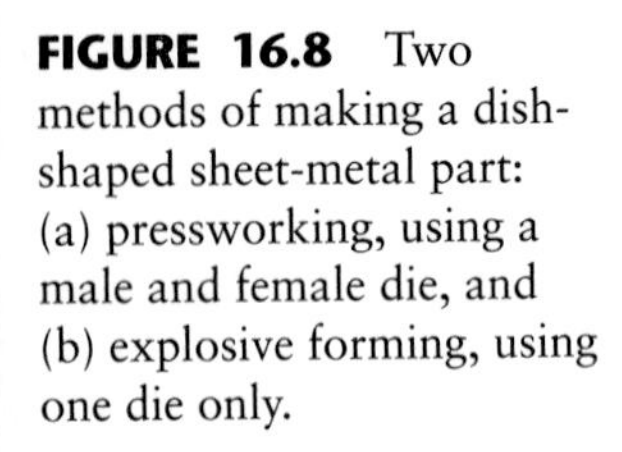

FIGURE 16.8 Two methods of making a dish-shaped sheet-metal part: (a) pressworking, using a male and female die, and (b) explosive forming, using one die only.

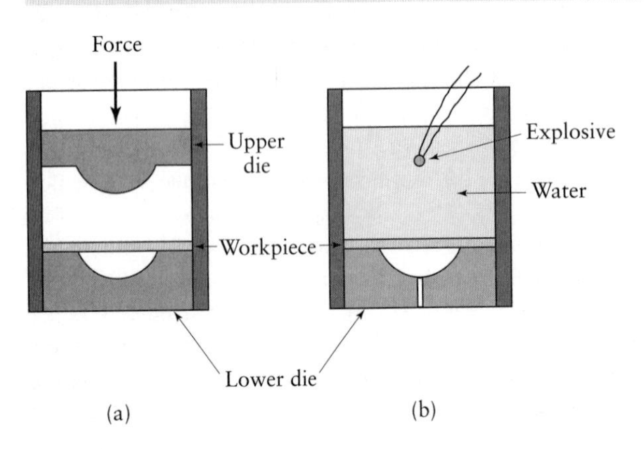

e) Can this operation be performed in a manufacturing plant within a city, or should it be carried out in open land?

f) Although explosive forming requires only one die (an obvious advantage), is the overall operation economical?

From this general discussion, it can be appreciated that for each part or component, a similar approach is required to arrive at a conclusion as to which process is the most suitable and economical.

16.9 Manufacturing Costs and Cost Reduction

In order for a product to be successful, its cost must be competitive with that of similar products, particularly in the global marketplace. The total cost of a product is spread over several categories, such as material costs, tooling costs, fixed costs, variable costs, direct labor costs, and indirect labor costs. References to some cost data are listed in Table 16.1.

Manufacturing organizations use several methods of cost accounting. The methodologies employed by these accounting procedures can be complex (even controversial), and their selection depends on the particular company and its type of operations. Furthermore, because of the technical and operational factors involved, calculating the individual cost factors is difficult, time consuming, and not always accurate or reliable.

Trends in costing systems (*cost justification*) include the following considerations: (a) the intangible benefits of quality improvements and inventory reduction, (b) life-cycle costs; (c) machine usage; (d) the cost of purchasing, compared with that of leasing, machinery; (e) the financial risks involved in implementing automation and the new technologies available. The costs to a manufacturer that are directly attributable to *product liability* (Section 1.9) and defense claims have been a matter of concern and discussion among all parties involved. Every modern product has a built-in added cost to cover possible liability claims. For example, it is estimated that liability suits against car manufacturers in the United States add about $500 to the indirect cost of an automobile and that 20% of the price of a ladder is attributed to potential product liability costs.

a) **Material costs.** *Material costs* are described throughout earlier sections; see also Table 16.1.

b) **Tooling costs.** *Tooling costs* are the costs involved in making the tools, dies, molds, patterns, and special jigs and fixtures necessary for manufacturing a product. Tooling costs are greatly influenced by the production process selected. For example, the tooling cost of die casting is higher than that of sand casting. Similarly, the tooling cost of machining or grinding is much lower than that of powder metallurgy, forging, or extrusion. Still, extrusion dies can sometimes be easily produced by electrical discharge machining, and, depending on the material and part size, it can be far more economical to extrude a complicated cross section than invest in expensive tooling required for roll forming.

In machining operations, carbide tools are more expensive than are high-speed steel tools, but the life of carbide tools is longer. If a part is to be manufactured by spinning, the tooling cost for conventional spinning is much lower than that of shear spinning. Tooling for rubber-forming processes is less expensive than that of the die sets (male and female) used for the deep drawing and stamping of sheet metals. High tooling costs might nonetheless be justified by the high-volume production of a single item. The expected life of tools and dies and their obsolescence because of product changes are also important considerations.

c) **Fixed costs.** *Fixed costs* include the costs of electric power, fuel, taxes on real estate, rent, insurance, and capital (including depreciation and interest). The company must meet these costs, regardless of whether it actually made a particular product. Consequently, fixed costs are not sensitive to production volume.

d) **Capital costs.** *Capital costs* represent investments in buildings, land, machinery, tooling, and equipment; these are major expenses for most manufacturing facilities (Table 16.7). Note the wide range of processes in each category and the fact that some machines cost a million dollars or more. (Coincidentally, the cost per pound of traditional machine tools has historically been approximately equal to that of beefsteak; however, this cost can easily double for machinery that has computer controls and sophisticated auxiliary equipment.)

As a simple example of capital costs, let's assume that a company has decided to manufacture a variety of valves. A new plant has to be built or an old plant has to be remodeled, and all the necessary machinery, support equipment, and facilities must be purchased. In order to cast and machine the valve bodies, the following items must be purchased: melting furnaces, casting equipment, machine tools, quality control equipment, and related equipment and machinery. If valve designs are to be modified often or if product lines are likely to vary greatly, the production machinery and equipment must have sufficient flexibility to accommodate these requirements. Machining centers (Section 8.10) and flexible manufacturing cells and systems (Chapter 15) are especially suitable for this purpose.

TABLE 16.7

Approximate Ranges of Machinery Base Prices

Type of machinery	Price range ($000)	Type of machinery	Price range ($000)
Broaching	10–300	Machining center	50–1000
Drilling	10–100	Mechanical press	20–250
Electrical discharge	30–150	Milling	10–250
Electromagnetic	50–150	Ring rolling	>500
Extruder	30–80	Robot	20–200
Fused-deposition modeling	60–120	Roll forming	5–100
Gear shaping	100–200	Rubber forming	50–500
Grinding		Stereolithography	100–500
Cylindrical	40–150	Stretch forming	>1000
Surface	20–100	Transfer line	>1000
Headers	100–150	Welding	
Injection molding	20–200	Electron beam	75–1000
Jig boring	50–150	Gas tungsten arc	1–5
Horizontal boring mill	100–400	Laser beam	40–1000
Flexible manufacturing system	>1000	Resistance, spot	20–50
Lathe	10–100	Ultrasonic	50–200
Automatic	30–250		
Vertical turret	100–400		

Note: Prices vary significantly, depending on size, capacity, options, and level of automation and computer controls.

The equipment and machinery listed are capital cost items, and they require major investment. In view of the generally high equipment costs (see Table 16.7), particularly those involving transfer lines and flexible manufacturing cells and systems, high production rates and quantities are necessary to justify such large expenditures and to maintain product costs at or below the all-important competitive level. Lower unit costs can be achieved by continuous production involving around-the-clock operation, but only as long as demand warrants it. Proper equipment maintenance is essential to ensure high productivity (Section 14.2.7). Any breakdown of machinery leading to downtime can be very expensive, costing typically from a few hundred dollars per hour to thousands of dollars per hour.

e) **Labor costs.** *Labor costs* are generally divided into direct and indirect costs. *Direct labor costs* are for the labor directly involved in manufacturing a part (productive labor). These costs include the cost of all labor, from the time raw materials are first handled to the time the product is finished. This period is generally referred to as *floor-to-floor time*. For example, it is the time during which a machine operator picks up a round bar from a bin, machines it into the shape of a threaded rod, and places it into another bin.

Direct labor costs are calculated by multiplying the labor rate (hourly wage, including benefits) by the time the worker spends producing the part. The time required for producing a particular part depends not only on its size, shape, and dimensional accuracy, but also on the workpiece material. For example, as can be seen in Table 8.8, the highest recommended cutting speeds for high-temperature alloys are lower than those for aluminum, cast iron, or copper alloys. Consequently, the cost of machining aerospace materials is higher than that for machining the more common metals and alloys.

Labor costs in manufacturing and assembly vary greatly from country to country. For the year 2000, the relative hourly compensation for production workers in manufacturing, based on a scale of 100 for the United States, are as follows:

Germany	116	Italy	74
Japan and Norway	111	Korea	41
Switzerland	107	Taiwan	30
United States	100	Mexico	12
Canada	81	China	5

It is not surprising that many of the products one purchases today, from clothing to high-tech toys, are either made or assembled in countries where labor costs are low. In contrast, firms located in countries with high labor rates tend to emphasize high value-added manufacturing tasks or high automation levels, so that the labor component of the cost is significantly reduced.

Indirect labor costs are costs that are generated in the servicing of the total manufacturing operation. The total is composed of such activities as supervision, repair, maintenance quality control, engineering, research, and sales; it also includes the cost of office staff. Because these costs do not contribute directly to the production of finished parts or are not chargeable to a specific product, they are referred to as overhead (burden rate) and charged proportionally to all products. The personnel involved in these activities are categorized as nonproductive labor.

16.9.1 Manufacturing costs and production volume

One of the significant factors in manufacturing costs is *production volume*. Obviously, large production volume requires high production rates, which in turn, requires the use of mass-production techniques that involve special machinery (dedicated machinery) and employ proportionally less direct labor. High production rates also require the use of plants that operate on two or three shifts. At the other extreme, small production volume usually means a larger direct labor involvement.

As described in Section 14.2, small batch production is usually done on general-purpose machines, such as lathes, milling machines, and hydraulic presses. The equipment is versatile, and parts with different shapes and sizes can be produced by appropriate changes in the tooling. Direct labor costs are, however, high, because these machines are usually operated by skilled labor.

For larger quantities (medium-batch production), these same general-purpose machines can be equipped with various jigs and fixtures or else be computer controlled. To reduce labor costs further, machining centers and flexible manufacturing systems are important alternatives. Generally, for quantities of 100,000 or more, the machines are designed for specific purposes and they perform a variety of specific operations with very little direct labor.

16.9.2 Cost reduction

Cost reduction first requires an assessment of how the costs just described are incurred and interrelated, with relative costs depending on numerous factors. Consequently, the unit cost of the product will vary widely, depending on design and manufacturing choices. For example, some parts may be made from expensive materials, but ones that require very little processing, such as minted gold coins. Here, the cost of materials relative to that of direct labor is high. By contrast, some products may require several complex, expensive production steps to process relatively inexpensive materials, such as carbon steels. For instance, an electric motor is made of relatively inexpensive materials, yet many different manufacturing processes are involved in the production of the housing, rotor, bearings, brushes, wire windings, and various other components. Note also that, in such cases, assembly operations can become a significant portion of the overall cost.

An approximate, but typical, breakdown of costs in manufacturing today is as follows:

Design	5%
Material	50%
Direct labor	15%
Overhead	30%

In the 1960s, labor accounted for as much as 40% of the production cost; today, it can be as low as 5%, depending on the type of product and the level of automation. Note, in the preceding breakdown, the very small contribution of the design phase; yet, in design for manufacture and assembly, including concurrent engineering, the design phase generally has the largest influence on the cost and success of a product in the marketplace. The engineering changes that often are made in the development of products can have a significant influence on costs. In addition to the nature and extent of the changes made, the stage at which they are made is very significant. The cost of engineering changes made (from design stage to the final production) increases by orders of magnitude (*rule of ten*) when changes are made at later stages. (See also Section 16.3.)

Overall cost reductions can be achieved by a thorough analysis of all the costs incurred in each step in manufacturing a product. The methods employed are described in detail in some of the references in the bibliography at the end of the chapter. The various opportunities for cost reduction have been stated throughout this text; among them are the following:

1. Simplifying part design and reducing the number of subassemblies required;
2. Specifying broader dimensional tolerances and allowing rougher surface finish;
3. Using less expensive materials;
4. Investigating alternative methods of manufacturing;
5. Using more efficient machines, equipment, and controls.

The introduction of more automation and of up-to-date technology in a manufacturing facility is an obvious means of reducing some costs. However, this approach must be undertaken with due care and only after a thorough cost–benefit analysis, with reliable data input and a consideration of the technical as well as the human factors involved. Implementing advanced technology, which can be very expensive, should occur only after a complete analysis of the more obvious cost factors, known as **return on investment** (ROI). (See also *return on quality*, Section 16.3.)

It is commonly observed that, over a period of time, the prices of some products (such as calculators, computers, and digital watches) have decreased, while the prices of other products (such as automobiles, aircraft, houses, and books) have gone up. Such differences generally are due to changes in the various costs over time, including changes due to labor, machinery, domestic and international competition, and worldwide economic trends (such as demand, exchange rates, and tariffs), and to the inevitable impact of computers on all aspects of product design, manufacturing, and marketing.

16.10 Value Analysis

There are several areas of activity in manufacturing in which cost reduction is possible. Manufacturing adds *value* to materials as they become discrete products and are then marketed. Because this value is added in individual stages during the creation of the product, the utilization of value analysis (also called **value engineering** or **value management**) is important.

First developed at the General Electric Company in the 1940s, *value analysis* (VA) is a methodology that evaluates in detail each step in design, materials, processes, and operations so as to manufacture a product that performs all its intended functions and does so at the lowest possible cost. In this analysis, a monetary value is established for each of two product attributes:

a) *Use value*, reflecting the functions of the product, and
b) *Esteem*, or *prestige, value*, reflecting the attractiveness of the product that makes its ownership desirable.

The *value* of a product is then defined as the following ratio:

$$\text{Value} = \frac{\text{Product function and performance}}{\text{Product cost}} \tag{16.1}$$

Thus, the goal of value analysis is to obtain maximum performance per unit cost. Value analysis generally consists of the following phases:

1. *Information phase*, to determine the costs of materials and labor and to gather data on specifications, quality, and other relevant topics;
2. *Analysis phase*, to define functions and identify problem areas and opportunities and to prioritize opportunities;
3. *Creativity phase*, to seek new ideas and ways to respond to problems and opportunities, without judging the value of each of these ideas;
4. *Evaluation phase*, to select and refine the best ideas to be developed and evaluated and to identify the costs involved;
5. *Implementation phase*, to present facts, costs, and values to the company management; to develop a plan; and to motivate positive action, all in order to obtain management's commitment to provide the resources necessary to accomplish the task;
6. *Review* the overall value analysis process and identify any adjustments that need to be made.

Value analysis is an important and all-encompassing interdisciplinary activity. It is usually coordinated by a value engineer and is conducted jointly by designers and engineers, by quality control, purchasing, and marketing personnel, and by managers. In order for value analysis to be effective, it must have the full support of a company's top management.

The implementation of value analysis in manufacturing can result in one or more of the following benefits:

a) Significant cost reduction;
b) Reduced lead times;
c) Reduction in product weight and size;
d) Reduction in manufacturing times;
e) Better product quality and product performance.

CASE STUDY | Concurrent Engineering for Intravenous Solution Containers

Baxter Healthcare Corporation manufactures over 1 million intravenous (IV) solution containers every day in the United States, providing critical therapies within the health care industry. Recognizing that patients' lives depend upon the safe delivery of medical solutions, the introduction of new or improved products is highly regulated by internal company standards, and external government agencies. A well-defined product development process provides the framework to meet consistently the regulated quality, reliability, and manufacturing design requirements on a consistent basis. More importantly, a concurrent engineering environment catalyzes the development process to minimize development cost and time to market.

In the 1990s, Baxter focused development efforts on a new set of materials for flexible IV containers. The container system being developed was expected to be more environmentally friendly, be compatible with a larger variety of new critical-care drugs, exceed all quality requirements, and remain cost effective. Some of the key design issues shown in the table below would allow the product to be safe for patients and maintain economic viability.

Container product requirements	Container processing requirements
• Provide physical and chemical shelf-life of up to several years without compromising the solution. • Allow for addition and mixing of various drug solutions in the hospital or alternate site pharmacy. • Provide a surface for printing of fade-, flake-, and smear-proof labeling. • Provide a surface for adherence of various pressure-sensitive adhesive labels at room, refrigerated, and elevated temperatures. • Withstand pressurization when an infusion device (cuff) is placed on the container for controlled solution delivery to the patient. • Maintain compatibility with available	administration set devices. • Implement the new product with existing equipment and plant personnel. • Allow product manufacturing at rates up to one million per day without additional production cells. • Maintain fabrication throughput at over 60 containers per minute per machine. • Maintain printing and filling speeds over 120 per minute. • Withstand steam sterilization temperatures, pressures, and times. • Allow packing in cartons that provide fullest density of pallet loading. • Withstand air and ground shipping and handling through areas as diverse as Arizona and Alaska.

The matrixed team. Given the company's extensive requirements, Baxter formed a multifunctional team of over 25 individuals. As the requirements and goals indicated, marketing, manufacturing, and development team members worked side by side with specialists in materials science, regulatory affairs, clinical affairs, toxicology, chemical stability, and sterility assurance. The product team members each recognized their responsibility for the success or failure of the product design effort.

The active team. All team members contributed to the master requirements definition during the product conceptualization phase. All team members led and communicated the test and development activities within their respective fields to the team at large during the development phase. All disciplines offered and accepted peer-review criticism of product or process designs at major development steps. All team members made sure that the product designs, quality assessment methods, and fabrication techniques were transferred efficiently to the designated plants during the implementation phase. The product team avoided costly design iterations and revisions and minimized duplicative efforts by maintaining a matrixed, active team throughout the process.

The virtual team. Many team members were located at manufacturing plants or development centers throughout northern Illinois and the greater United States. By necessity, team meetings leveraged teleconference and videoconference capabilities, which also minimized travel expense. Electronic mail access provided rapid and broad communication of impending issues and resolutions. A companywide intranet site was developed to share documents and design prints, which assured that all locations were utilizing the most recent project advancements. The team customized emerging communication technologies to enhance the concurrent development of new materials, processing, and fabrication technologies.

Finished goods. The engineering effort put forth by the matrixed, active, and virtual teams resulted in a product within three years. New materials and components were developed that met all product and process requirements and ultimately satisfied patient needs. The new product was transparent, and patient-care professionals continued to use their existing training and well-practiced techniques. Disposal of the product was streamlined; the new materials can be safely disposed of or recycled.

Reliability and user satisfaction remained high. The new materials were compatible with an extended array of solution format drugs packaged in IV containers. Indeed, team success will be measured by the number of additional products introduced in subsequent years.

Source: K. Anderson, Baxter Healthcare Corporation, and Shelly Petronis, formerly Baxter Healthcare Corporation, currently Zimmer Holdings, Inc.

SUMMARY

- Competitive aspects of production and costs are among the most significant considerations in manufacturing. Regardless of how well a product meets design specifications and quality standards, it must also meet economic criteria in order to be competitive in the domestic and global marketplace. (Section 16.1)
- Various guidelines for designing products for economical production have been established, including a reduction in material volume. Robust design is an important concept. (Section 16.2)
- Product quality and life expectancy are significant concerns because of their impact on customer satisfaction and the marketability of the product. (Section 16.3)
- Life cycle assessment and life cycle engineering are among increasingly important considerations in manufacturing, particularly as regards reducing any adverse impact the product might have on the environment. Sustainable manufacturing is another concept aimed at reducing waste of natural resources such as materials and energy. (Section 16.4)
- Selecting an optimal material from among numerous candidates is one competitive aspect of manufacturing and depends on factors such as properties, commercially available shapes, reliability of supply, and costs of materials and processing. (Section 16.5)
- Substitution of materials, modification of product design, and relaxing of dimensional tolerance and surface finish requirements are important methods of reducing costs. (Section 16.6)
- The capabilities of manufacturing processes vary widely; consequently, their proper selection for a particular product to meet specific design and functional requirements is critical. (Sections 16.7 and 16.8)
- The total cost of a product includes several elements. Material costs can be reduced through careful selection, so that the least expensive material can be identified and used, without compromising design, service requirements, and quality. Although labor costs are continually becoming an increasingly smaller percentage of production costs, they can be reduced further through the use of automated and computer-controlled machinery. (Sections 16.9)
- Value analysis is a methodology that evaluates each step in design, materials and process selection, and production operations to manufacture a product that performs its intended functions at the lowest possible cost. (Section 16.10)

BIBLIOGRAPHY

Anderson, D. M., *Design for Manufacturability, Optimizing Cost, Quality, and Time-to-Market*, 2d. ed., CIM Press, 2001.

Ashby, M. F., *Materials Selection in Mechanical Design*, 2d. ed., Pergamon, 1999.

ASM Handbook, Vol. 20: *Materials Selection and Design*, ASM International, 1997.

Baxter, M., *Product Design: A Practical Guide to Systematic Methods of New Product Development*, Chapman & Hall, 1995.

Billatos, S., and N. Basaly, *Green Technology and Design for the Environment*, Taylor & Francis, 1997.

Boothroyd, G., P. Dewhurst, and W. Knight, *Product Design for Manufacture and Assembly*, 2d. ed., Dekker, 2001.

Bralla, J. G., *Design for Excellence, DFX*, McGraw-Hill, 1996.

———, *Design for Manufacturability Handbook*, 2d. ed., McGraw-Hill, 1999.

Brown, J., *Value Engineering: A Blueprint*, Industrial Press, 1992.

Cattanach, R. E. (ed.), *The Handbook of Environmentally Conscious Manufacturing*, McGraw-Hill, 1994.

Demaid, A., and J. H. W. DeWit (eds.), *Case Studies in Manufacturing with Advanced Materials*, North-Holland, 1995.

Dettmer, W. H., *Breaking the Constraints to World-Class Performance*, ASQ Quality Press, 1998.

Dieter, G. E., Jr., *Engineering Design: A Materials and Processing Approach*, 3d. ed., McGraw-Hill, 1999.

Dorrf, R. C., and A. Kusiak (eds.), *Handbook of Design, Manufacturing and Automation*, Wiley, 1995.

Farag, M. M., *Materials Selection for Engineering Design*, Prentice Hall, 1997.

Fleischer, M., and J. K. Liker, *Concurrent Engineering Effectiveness: Integrating Product Development Across Organizations*, Hanser Gardner, 1997.

Graedel, T. E., and B. R. Allenby, *Design for the Environment*, Prentice Hall, 1997.

Halevi, G., *Restructuring the Manufacturing Process: Applying the Matrix Method*, St. Lucie Press, 1998.

Harper, C. A. (ed.), *Handbook of Materials for Product Design*, McGraw-Hill, 2001.

Hartley, J. R., and S. Okamoto, *Concurrent Engineering: Shortening Lead Times, Raising Quality, and Lowering Costs*, Productivity Press, 1998.

Helander, M., and M. Nagamachi, *Design for Manufacturability*, Taylor and Francis, 1992.

Kusiak, A. (ed.), *Concurrent Engineering: Automation, Tools and Techniques*, Wiley, 1993.

Lesco, J., *Industrial Design: Guide to Materials and Manufacturing*, Van Nostrand Reinhold, 1998.

Lindbeck, J. R., *Product Design and Manufacturing*, Prentice Hall, 1995.

Magrab, E. B., *Integrated Product and Process Design and Development: The Product Realization Process*, CRC Press, 1997.

Mahoney, R. M., *High-Mix Low-Volume Manufacturing*, Prentice Hall, 1997.

Mangonon, P. C., *The Principles of Materials Selection for Design*, Prentice Hall, 1999.

Mather, H., *Competitive Manufacturing*, 2d. ed., CRC Press, 1998.

Meyers, F. E., *Motion and Time Study for Lean Manufacturing*, Prentice Hall, 1998.

Otto, K. N., and K. L. Wood, *Product Design: Techniques in Reverse Engineering and New Product Development*, Prentice Hall, 2001.

Paashuis, V., *The Organization of Integrated Product Development*, Springer, 1997.

Pahl, G., and W. Beitz, *Engineering Design: A Systematic Approach.*, 2d. ed., Springer, 1996.

Park, R. J., *Value Engineering: A Plan for Invention*, St. Lucie Press, 1999.

Park, S., *Robust Design and Analysis for Quality Engineering*, Chapman & Hall, 1997.

Poli, C., *Design for Manufacturing: A Structured Approach*, Butterworth-Heinemann, 2001.

Prasad, B., *Concurrent Engineering Fundamentals*, 2 vols, Prentice Hall, 1995.

Pugh, S., D. Clausing, and R. Andrade (eds.), *Creating Innovative Products Using Total Design*, Addison-Wesley, 1996.

Pugh, S., *Total Design: Integrated Methods for Successful Product Engineering*, Addison-Wesley, 1991.

Rhyder, R. F., *Manufacturing Process Design and Optimization*, Dekker, 1997.

Roosenburg, N. F. M., and J. Eekels, *Product Design: Fundamentals and Methods*, Wiley, 1995.

Salomone, T. A., *What Every Engineer Should Know about Concurrent Engineering*, Dekker, 1995.

Shina, S. G. (ed.), *Successful Implementation of Concurrent Engineering Products and Processes*, Wiley, 1997.

Sims, E. R., Jr., *Precision Manufacturing Costing*, Dekker, 1995.

Stoll, H. W., *Product Design Methods and Practices*, Dekker, 1999.

Swift, K. G., and J. D. Booker, *Process Selection: From Design to Manufacture*, Wiley, 1997.

Tool and Manufacturing Engineers Handbook, 4th ed., Vol. 5: *Manufacturing Engineering Management*, Society of Manufacturing Engineers, 1988.

———, Vol. 6: *Design for Manufacturability*, 1992.

_____, Vol. 7: *Continuous Improvement*, 1993.

Ullman, D. G., *The Mechanical Design Process*, 2d. ed., McGraw-Hill, 1997.

Ulrich, K. T., and S. D. Eppinger, *Product Design and Development*, 2d. ed., McGraw-Hill, 1999.

Urban, G., and J. Hauser, *Design and Marketing of New Products*, 2d. ed., Prentice Hall, 1993.

Walker, J. M. (ed.), *Handbook of Manufacturing Engineering*, Dekker, 1996.

Wang, B., *Integrated Product, Process, and Enterprise Design*, Chapman & Hall, 1997.

Wenzel, H., M. Hauschild, and L. Alting, *Environmental Assessment of Products*, Vol. 1, Chapman & Hall, 1997.

Wenzel, H., and M. Hauschild, *Environmental Assessment of Products*, Vol. 2, Chapman & Hall, 1997.

Wroblewski, A. J., and S. Vanka, *MaterialTool: A Selection Guide of Materials and Processes for Designers*, Prentice Hall, 1997.

QUESTIONS

16.1 List and describe the major considerations involved in selecting materials for products.

16.2 Why is a knowledge of available shapes of materials important? Give five different examples.

16.3 Describe what is meant by the *manufacturing characteristics* of materials. Give three examples demonstrating the importance of this information.

16.4 Why is material substitution an important aspect of manufacturing engineering? Give five examples from your own experience or observations.

16.5 Why has material substitution been particularly critical in the automotive and aerospace industries?

16.6 What factors are involved in the selection of manufacturing processes? Explain why these factors are important.

16.7 What is meant by *process capabilities*? Select four different, specific manufacturing processes, and describe their capabilities.

16.8 Is production volume significant in process selection? Explain your answer.

16.9 Discuss the advantages of long lead times, if any, in production.

16.10 What is meant by an *economic order quantity*?

16.11 Describe the costs involved in manufacturing. Explain how you could reduce each of these costs.

16.12 What is value analysis? What are its benefits?

16.13 What is meant by *trade-off*? Why is it important in manufacturing?

16.14 Explain the difference between direct labor cost and indirect labor cost.

16.15 Explain why the larger the quantity per package of food products, the lower is the cost per unit weight.

16.16 Explain why the value of the scrap produced in a manufacturing process depends on the type of material.

16.17 Comment on the magnitude and range of scrap shown in Table 16.4.

16.18 Describe your observations concerning the information given in Table 16.6.

16.19 Other than the size of the machine, what factors are involved in the range of prices in each machine category shown in Table 16.7?

16.20 Explain how the high cost of some of the machinery listed in Table 16.7 can be justified.

16.21 Explain the reasons for the relative positions of the curves shown in Fig. 16.3.

16.22 What factors are involved in the shape of the curve shown in Fig. 16.5?

16.23 Make suggestions as to how to reduce the dependence of production time on surface finish (shown in Fig. 16.6).

16.24 Is it always desirable to purchase stock that is close to the final dimensions of a part to be manufactured? Explain your answer, and give some examples.

16.25 What course of action would you take if the supply of a raw material selected for a product line became unreliable?

16.26 Describe the potential problems involved in reducing the quantity of materials in products.

16.27 Present your thoughts concerning the replacement of aluminum beverage cans with steel ones.

16.28 There is a period, between the time that an employee is hired and the time that the employee finishes with training, during which the employee is paid and receives benefits, but produces nothing. Where should such costs be placed among the categories given in this chapter?

16.29 Why is there a strong desire in industry to practice near-net-shape manufacturing? Give several examples.

16.30 Estimate the position of the following processes in Fig. 16.4: (a) centerless grinding, (b) electrochemical machining, (c) chemical milling, and (d) extrusion.

16.31 From your own experience and observations, comment on the size, shape, and weight of specific products as they have changed over the years.

16.32 In Section 16.9.2, a breakdown of costs in today's manufacturing environment suggests that design costs contribute only 5% to total costs. Explain why this suggestion is reasonable.

16.33 Make a list of several (a) disposable and (b) reusable products. Discuss your observations, and explain how you would go about making more products that are reusable.

16.34 Describe your own concerns regarding life-cycle assessment of products.

DESIGN

16.35 As you can see, Table 16.6 on manufacturing processes includes only metals and their alloys. On the basis of the information given in this book and other sources, prepare a similar table for nonmetallic materials, including ceramics, plastics, reinforced plastics, and metal–matrix and ceramic–matrix composite materials.

16.36 Review Fig. 1.2, and present your thoughts concerning the two flowcharts. Would you want to make any modifications to them and if so, what would the modifications be and why?

16.37 Over the years, numerous consumer products have become obsolete, or nearly so, such as rotary-dial telephones, analog radio tuners, turntables, and vacuum tubes. By contrast, many new products have entered the market. Make a comprehensive list of obsolete products and one of new products. Comment on the possible reasons for the changes you have observed. Discuss how different manufacturing methods and systems have evolved in order to make the new products.

16.38 Select three different household products, and make a survey of the changes in their prices over the past 10 years. Discuss the reasons for the changes.

16.39 Figure 2.2a shows the shape of a typical tension-test specimen having a round cross section. Assuming that the starting material (stock) is a round rod and that only one specimen is needed, discuss the processes and the machinery by which the specimen can be made, including their relative advantages and limitations. Describe how the process you selected can be changed for economical production as the number of specimens required increases.

16.40 Table 16.2 lists several materials and their commercially available shapes. By contacting suppliers of materials, extend the list to include (a) titanium, (b) superalloys, (c) lead, (d) tungsten, and (e) amorphous metals.

16.41 Select three different products commonly found in homes. State your opinions about (a) what materials were used in each product, and why, and (b) how the products were made, and why those particular manufacturing processes were used.

16.42 Inspect the components under the hood of your automobile. Identify several parts that have been produced to net-shape or near-net-shape condition. Comment on the design and production aspects of these parts and on how the manufacturer achieved the near-net-shape condition for the parts.

16.43 Comment on the differences, if any, between the designs, the materials, and the processing and assembly methods used to make such products as hand tools, ladders for professional use, and ladders for consumer use.

16.44 Other than powder metallurgy, which processes could be used (singly or in combination) in the making of the parts shown in Fig. 11.1? Would they be economical?

16.45 Discuss production and assembly methods that can be employed to build the presses used in sheet-metal forming operations described in Chapter 7.

16.46 The shape capabilities of some machining processes are shown in Fig. 8.40. Inspect the various shapes produced, and suggest alternative processes for producing them. Comment on the properties of workpiece materials that would influence your suggestions.

16.47 The following figure shows a steel bracket that can be made either by casting or by the stamping of sheet metal:

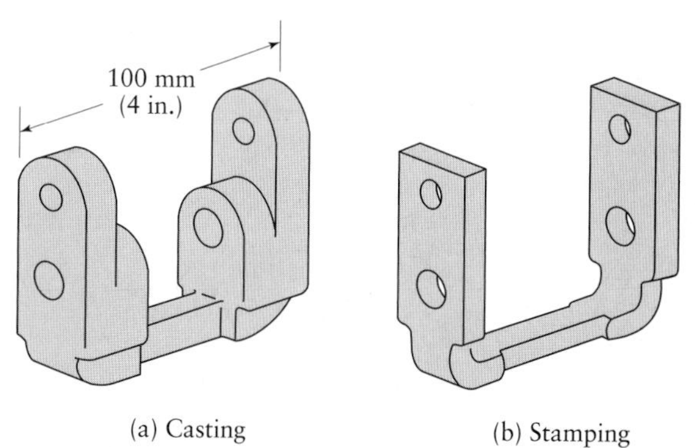

(a) Casting (b) Stamping

Describe the manufacturing processes and the sequences involved in each method. Assume that the approximate dimensions of the bracket are 100 mm by 100 mm by 25 mm wide (4 in. by 4 in. by 1 in.).

16.48 If the dimensions of the part in Question 16.47 were (a) 10 times larger and (b) 20 times larger, how different would your answer be? Explain.

16.49 A new model of an internal combustion engine is shown in Fig. 1.1. On the basis of the topics covered in this text, select any three individual components of such an engine, and describe the materials and processes that you would use in making those components. Remember that the parts must be manufactured at very large numbers and at minimum cost, yet maintain their quality, integrity, and reliability during service.

16.50 Discuss the trade-offs involved in selecting between the two materials for each of the applications listed:

a) Sheet metal vs. reinforced plastic chairs;
b) Forged vs. cast crankshafts;
c) Forged vs. powder-metallurgy connecting rods;
d) Plastic vs. sheet-metal light-switch plates;
e) Glass vs. metal water pitchers;
f) Sheet-metal vs. cast hubcaps;
g) Steel vs. copper nails;
h) Wood vs. metal handles for hammers.

Also, discuss the typical conditions to which these products are subjected in their normal use.

16.51 Discuss the manufacturing process or processes suitable for making the products listed in Question 16.50. Explain whether the products would require additional operations (such as coating, plating, heat treating, and finishing). If so, make recommendations and give the reasons for them.

16.52 Discuss the factors that influence the choice between the following pairs of processes to make the products indicated:

a) Sand casting vs. die casting of a fractional electric-motor housing;
b) Machining vs. forming of a large-diameter bevel gear;
c) Forging vs. powder-metallurgy production of a cam;
d) Casting vs. stamping a sheet-metal frying pan;
e) Making outdoor summer furniture from aluminum tubing vs. cast iron;
f) Welding vs. casting of machine-tool structures;
g) Thread rolling vs. machining of a bolt for high-strength application;
h) Thermoforming a plastic vs. molding a thermoset to make the blade for an inexpensive household fan.

16.53 The following figure shows a sheet-metal part made of steel:

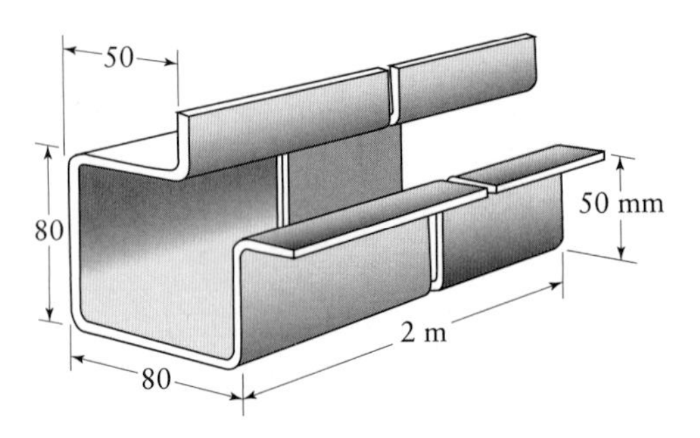

Discuss how this part could be made and how your selection of a manufacturing process may change (a) as the number of parts required increases from 10 to thousands and (b) as the length of the part increases from 2 m to 20 m.

16.54 The part shown in the following figure is a carbon-steel segment (partial) gear:

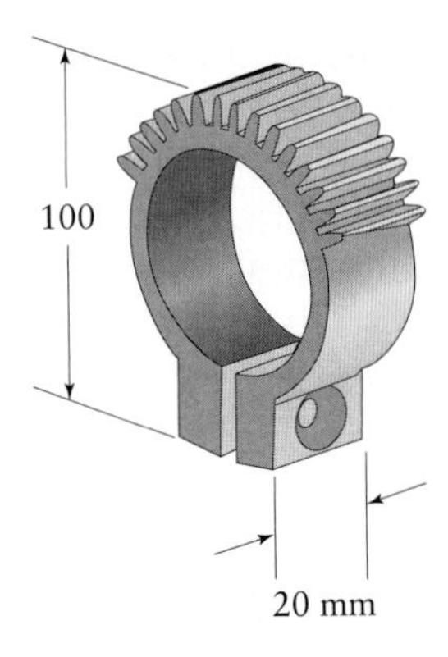

The smaller hole at the bottom is for clamping the part onto a round shaft, using a screw and a nut. Suggest a sequence of manufacturing processes to make this part. Consider such factors as the influence of the number of parts required, dimensional tolerances, and surface finish. Discuss such processes as machining from a bar stock, extrusion, forging, and powder metallurgy.

16.55 Several methods can be used to make the part shown in the following figure:

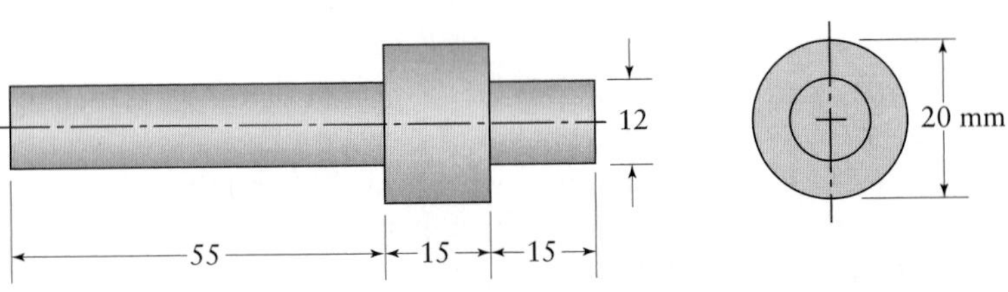

List these methods, and for each one, explain such factors as the machinery and equipment required, the amount of scrap produced, the strength of the part, etc. What would be the influence of the sharpness of the corners at the two diameters on the choice of processes or the need for secondary processing? Also, comment on how different your answers would be if the material were (a) ferrous, (b) nonferrous, (c) thermoplastic, (d) thermoset, and (e) ceramic.

16.56 The following illustration shows various components of a modular artificial knee, which allows surgeons to choose the particular components that best fit an individual patient's needs:

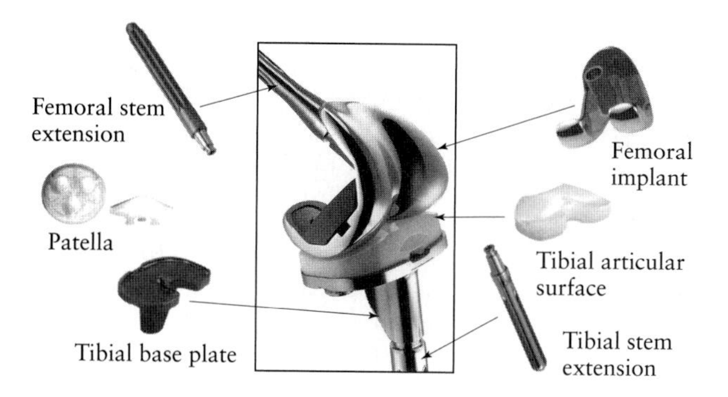

Choose any three components, and describe suitable materials and manufacturing processes for producing them. (*Source for illustration*: Zimmer, Inc.)

16.57 Inspect several small and large products that you are familiar with. Present a plan on how you would disassemble them and recycle their individual components. Discuss any difficulties you may encounter in the process of doing so. Are there any components that you don't think could be recycled, and if so, why?

Index

The New International Dictionary of the Christian Church

The New International Dictionary

Revised Edition

of the Christian Church

J. D. Douglas
GENERAL EDITOR

Earle E. Cairns
CONSULTING EDITOR

James E. Ruark
COPY EDITOR

THE NEW INTERNATIONAL DICTIONARY
OF THE CHRISTIAN CHURCH

ISBN 0-310-23830-7

THE NEW INTERNATIONAL DICTIONARY OF THE CHRISTIAN CHURCH
Copyright © 1974, 1978 by The Zondervan Corporation
Grand Rapids, Michigan

Requests for information should be addressed to:
Zondervan Publishing House
Academic and Professional Books
Grand Rapids, Michigan 49530

Library of Congress Cataloging in Publication Data

Douglas, James Dixon.
The new international dictionary of the Christian church.
Includes bibliographical references.

1. Theology—dictionaries. I. Title.
BR95.D68 203 74-8999
ISBN 0-310-23830-7

Printed in the United States of America

93 94 / AF / 14 13 12 11 10 9

Preface
to the Second Edition

We are gratified that the first edition of this work should have been so quickly taken up as to make this second edition necessary. Publisher and editor are both deeply appreciative of the friendly letters that have come from all over the worldwide Christian fellowship. Many, in response to the editorial invitation, have made comments critical and constructive, as indeed have numerous newspapers and journals. All have been carefully noted, not a few of the points have been incorporated in this new improved edition, and others have been kept for future reference. Through the time and trouble taken by these thoughtful correspondents, this project has been made an even more notable piece of international and ecumenical teamwork.

Because of the short time that has elapsed since the initial publication, and because we wanted to ensure that the book remained the same convenient size, major additions have been precluded at this time. Many new entries and features have nonetheless been introduced, others updated. All this has involved several additions to our list of contributors, all of whom helpfully completed assignments at short notice. We apologize to the Reverend Reginald Kissack whose name was inadvertently omitted from the earlier list of contributors; we remember with affection and thankfulness to God our colleagues Dr. Harold R. Cook, Dr. Arthur Fawcett, and Principal Robin Nixon who have died since the book enriched by their pens was published.

While it is manifestly impossible to include everything and everyone with a claim to appear in such a volume, we have tried to be sensitive even at the expense of departing from strict historical method. Thus, for example, we have included some less well-known missionaries as in some sense representative of all the noble army of men and women who down the ages for Christ's sake have taken the Gospel throughout the world.

This new edition has meant urgent editorial demands on busy people. We are grateful for timely responses and tasks readily undertaken. Finally, of the warm cooperation given by the publisher, renewed acknowledgment is made by the editor across the water.

J.D. Douglas

Preface

Church history cannot be discussed in isolation; it is not something that is happening in a vacuum while the rest of the world goes by untouched and untouching. To make a distinction between spiritual and natural phenomena is on one view to fall into that dualism which Christianity itself has roundly condemned down the centuries. Opinions will nonetheless differ on what then falls within the scope of a dictionary of the Christian Church, especially one audacious enough to restrict treatment to a single volume.

The editor who is given a million words and carte blanche will quickly learn the futility of approaching his task in terms of what he is *not* producing. He is not aiming at a theological word-book, but what record of Christian history could exclude reference to such subjects as Justification and the Atonement? His brief ostensibly excludes the major non-Christian religions, but how can he ignore, say, Islam, which has profoundly affected the course of Christianity and been a perennial challenge to Christian missions? Similar difficulties attend any attempt to reject other areas, and the editor finds himself reluctantly fathering something which is variously gazetteer, archaeological record, compendium of saintly lore, and liturgical primer.

An even thornier problem concerns how much he should do the work of a Bible dictionary. He cannot exclude an entry on the Old Testament and features of Judaism. Various aspects of the New Testament must be covered, and major New Testament characters call for separate entries. Despite the extraordinary tendency of kindred publications, moreover, he might feel that a volume on the Christian Church should have entries on God and Jesus Christ. The fact that he cannot treat a subject comprehensively is no valid reason for ignoring it altogether.

In handling the articles, clarity has been preferred to consistency in cases where to combine both was impracticable, and conciseness has precluded polish. To a large degree writers have been given a free hand to treat the space allotted to them in the way which they as experts have felt best. In biographical entries this may result in only the briefest allusion to the subject's life history so that more consideration can be given his work. In the case of subjects still alive some information inevitably will be out of date by the time of publication; under this category there are, however, very few entries, and such as there are usually concern only those who have retired from their main life's work. The rules governing matters of style and layout were modified as the project developed—hopefully always in the interests of greater lucidity.

This volume sets out to give information not easily available elsewhere in such convenient form, and thus to encourage the reader to marvel at the richness, diversity, and wholeness of the Christian tradition. Church history lends itself to very different interpretations, but a concerted attempt has been made here to be factual rather than apologetic, and to avoid a misguided manipulation of history that would result in a partisan manifesto where it has no right to be. At the same time, few historians in reality do much worshiping at the high altar of impartiality: even the attempt to present the facts may, through selectivity and omission, reflect the interests and prejudices of the author or editor.

The aim has been to steer a middle path between academic textbook and popular introduction. In projects such as this, allocation of space confronts the organizer with countless pitfalls and trip-wires. There is a colossal arrogance about sitting down and cold-bloodedly assessing at 150 words the lifetime labors of some bygone saint who knew incredible hardship in taking the Gospel to desolate places—and even more about excluding one of his colleagues altogether. No two editors will have the same views on how space should be apportioned, and any enterprising reviewer will find what to his mind are glaring anomalies. Why should C.S. Lewis have more lines than Lollards? Why write at all about Witchcraft or American Indians? Those who get the scent of too many peripheral hares rushing down the byways of history might at least be led on, as it were, to marvel at how colorful and exciting and many-faceted is our Christian record, and how musicians and missionaries, seers and soldiers, kings and councils, poets and persecutors, humanists and heretics have all, for better or worse, made a contribution that ought to be delineated.

In certain subjects where hard facts have been lost in the mists of history and only legend or speculation remains, only the merest mention is given of them, or they are omitted altogether. In other areas where history has thrown up a significant question mark, recourse has been made to the policy of Bishop Gilbert Burnet: "Where things appear doubtful ... deliver them with the same uncertainty to the world." If an inordinately large amount of space seems to have been given an entry, it may mean that there are few if any complementary articles on that subject in the dictionary.

Of cross-references there could be no end; because no hard-and-fast policy was practicable here, we have tried to be sensible about them and to indicate them as economically as possible. Thus, for example, a cross-reference asterisk in the text against "Monophysite" will be understood as alluding to the entry headed "Monophysitism." The asterisk is generally found at the end of a complete name—e.g., "William of Malmesbury*"—even where it is the first part that determines alphabetical order, but exceptions have here and there been made to avoid more serious ambiguities.

In dealing with several thousand entries covering twenty centuries of history, no gift short of infallibility would prevent occasional editorial slips. We would not only ask the forgiveness of readers for such, but invite their cooperation in pointing them out, so that a future edition of the volume may benefit.

The editor is keenly and humbly aware that this dictionary has been a piece of teamwork, an ecumenical enterprise which has joined Christian writers from many lands and denominations. Their faithful and knowledgeable labors have immeasurably enriched the project, and their understanding letters and friendship have lightened the editorial load. Four of our colleagues have died since completing their contributions; we thank God for these tangible reminders which are a small part of the total work and witness of Dr. S. Richey Kamm, Dr. Carl S. Meyer, Dr. Matthew Spinka, and Dr. D.P. Thomson.

The project is indebted particularly to scholars, notably Peter Toon and Jim Norman, who came to the editor's rescue with eleventh-hour works of supererogation after many articles were disconcertingly orphaned. To his American associate, Dr. Earle E. Cairns, the editor owes a great debt for counsel in the initial stages and for giving time amid a busy life to read so many manuscripts, and to help and encourage in so many areas. Dr. Robert K. DeVries and Mr. James E. Ruark, fellow-laborers at The Zondervan Corporation, have shown considerable expertise both in preparing for press a most demanding piece of work and in their cheerful and patient coping with an editor 4,000 miles away. The task of proofreading, finally, has been considerably aided by the technical ability and the initiative with which the printers have carried out their task.

It is our prayer that this volume will give readers a renewed sense of history; an identification and feeling of fellowship with those who have carried the torch before them—many of them, in David Livingstone's words, "the watchmen of the night ... who worked when all was gloom"; and most of all an appreciation of the priceless heritage which is ours in Christ.

J.D. DOUGLAS

Contributors

MARVIN W. ANDERSON
Ph.D., Professor of Ecclesiastical History, Bethel Theological Seminary, St. Paul, Minnesota.

JOHN S. ANDREWS
Ph.D., Sub-Librarian (Reader Services), University of Lancaster, England.

G.T.D. ANGEL
M.A., Dean of Studies, Trinity College, Bristol, England.

BRIAN G. ARMSTRONG
Ph.D., Associate Professor of History, Georgia State University, Atlanta, Georgia.

STEVEN BARABAS
Th.D., Professor of Theology Emeritus, Wheaton College, Illinois.

PAUL M. BECHTEL
Ph.D., Professor of English Emeritus, Wheaton College, Illinois.

DARREL E. BIGHAM
Ph.D., Assistant Professor of History, Indiana State University, Evansville, Indiana.

J.N. BIRDSALL
Ph.D., Reader in New Testament Studies and Textual Criticism, University of Birmingham, England.

HUGH J. BLAIR
Ph.D., Minister of Ballymoney Reformed Presbyterian Church and Professor of Old Testament Language and Literature, Reformed Presbyterian Theological Hall, Belfast, Northern Ireland.

IAN BREWARD
Ph.D., Professor of Church History and History of Doctrine, Knox College, and Dean of the Faculty of Theology, Otago University, New Zealand.

COLIN BROWN
Ph.D., Professor of Systematic Theology, Fuller Theological Seminary, Pasadena, California.

ROBERT I. BROWN
Minister of Highgate Road Chapel, London, England.

F.F. BRUCE
D.D., F.B.A., Professor of Biblical Criticism and Exegesis Emeritus, University of Manchester, England.

COLIN O. BUCHANAN
M.A., Principal of St. John's College, Nottingham, England.

PHILIP H. BUSS
M.A., Vicar of Christ Church, Fulham, London, England.

EARLE E. CAIRNS
Ph.D., Professor of History Emeritus and formerly Chairman of the Division of Social Sciences, Wheaton College, Illinois.

R.H. CAMPBELL
M.A., Professor of Economic History, University of Stirling, Scotland.

G.L. CAREY
Ph.D., Principal of Trinity College, Bristol, England.

H.M. CARSON
B.A., B.D., Minister of Hamilton Road Baptist Church, Bangor, Northern Ireland.

GORDON A. CATHERALL
Ph.D., Minister of Hamlet Baptist Church, Liverpool, England.

J.W. CHARLEY
M.A., Member of The Beacon Group Ministry, Liverpool, England.

ROBERT E. D. CLARK
Ph.D., Editor of *Faith and Thought,* Cambridge, England.

ARTHUR CLARKE
B.A., B.D., Minister of First Holywood Presbyterian Church, County Down, Northern Ireland.

ERNEST F. CLIPSHAM
M.A., B.D., Minister of Cottingham Road Baptist Church, Hull, England.

ROBERT G. CLOUSE
Ph.D., Professor of History, Indiana State University, Terre Haute, Indiana.

The late HAROLD R. COOK
M.A., formerly Professor of Missions, Moody Bible Institute, Chicago, Illinois.

GEOFFREY S.R. COX
M.A., Vicar of Hucclecote, Gloucester, England.

D.G.L. CRAGG
D. Phil., Tutor in John Wesley College, Alice, Cape Province, Republic of South Africa.

JAMES DAANE
Th.D., Professor of Pastoral Theology, Fuller Theological Seminary, Pasadena, California.

MARTIN B. DAINTON
M.A., Missionary with the Overseas Missionary Fellowship in Indonesia.

G.C.B. DAVIES
D.D., Canon and Director of Pre-Ordination Studies, Diocese of Worcester, England.

KENNETH R. DAVIS
Ph.D., Vice-Principal, Trinity Western College, Langley, British Columbia, Canada.

PETER S. DAWES
B.A., Archdeacon of West Ham, Essex, England.

DONALD W. DAYTON
B.D., M.S., Director, Mellander Library, North Park Theological Seminary, Chicago, Illinois.

JAMES A. DE JONG
Th.D., Assistant Professor of Theology, Trinity Christian College, Palos Heights, Illinois.

A. MORGAN DERHAM
Editorial Secretary, The Leprosy Mission, London, England.

WAYNE DETZLER
Ph.D., Minister of Kensington Baptist Church, Bristol, England.

JOHN P. DEVER
Th.D., Assistant Professor of Sociology and Religion, Averett College, Danville, Virginia.

J.D. DOUGLAS
Ph.D., St. Andrews, Fife, Scotland, Editor-at-Large, *Christianity Today*.

RICHARD DOWSETT
M.A., Missionary with the Overseas Missionary Fellowship in the Philippines.

G.E. DUFFIELD
M.A., Editor and Publisher, The Sutton Courtenay Press, Abingdon, England.

PAUL ELLINGWORTH
M.A., European Translation Coordinator, United Bible Societies, London, England.

H.L. ELLISON
B.D., B.A., Dawlish, Devon, England, Lecturer and Writer on the Old Testament.

ROBERT H. ELMORE
Mus.B., LL.D., L.H.D., Organist-Director of Music, Tenth Presbyterian Church, Philadelphia, Pennsylvania.

H. CROSBY ENGLIZIAN
B.D., Th.D., Professor of Historical Theology and Director of Postgraduate Studies, Western Conservative Baptist Seminary, Portland, Oregon.

ROBERT P. EVANS
Ph.D., European Director, Greater Europe Mission.

BARBARA L. FAULKNER
Ph.D., Associate Professor of History, Eastern Nazarene College, Wollaston, Massachusetts.

The late **ARTHUR FAWCETT**
Ph.D., formerly Minister of Johnstone High Church, Renfrewshire, Scotland.

LAWRENCE FEEHAN
M.A., Lecturer in Edge Hill College of Education, Ormskirk, England.

STIG-OLOF FERNSTRÖM
Missionary with the Finnish Lutheran Mission in Senegal.

ALBERT H. FREUNDT, JR.
B.A., B.D., Professor of Church History and Polity, Reformed Theological Seminary, Jackson, Mississippi.

EDWARD J. FURCHA
Ph.D., Professor of Church History, Serampore College, Serampore, W.B., India.

FRANK E. GAEBELEIN
Litt.D., D.D., LL.D., Arlington, Virginia, General Editor, *The Expositor's Bible Commentary*, and Headmaster-Emeritus, The Stony Brook School, New York.

R.F.R. GARDNER
F.R.C.O.G., Consultant Obstetrician and Gynaecologist, Sunderland Group of Hospitals, England.

W. WARD GASQUE
Ph.D., President and Professor of New Testament, New College, Berkeley, California.

GEORGE GIACUMAKIS, JR.
Ph.D., Associate Professor of History, California State University, Fullerton, California.

ROBERT P. GORDON
Ph.D., Lecturer in Hebrew, Cambridge University, England.

RICHARD L. GREAVES
Ph.D., Associate Professor of History, Florida State University, Tallahassee, Florida.

G.W. GROGAN
M.Th., Principal, Bible Training Institute, Glasgow, Scotland.

JOHN E. GROH
Ph.D., Assistant Professor of Historical Theology, Christ Seminary, St. Louis, Missouri.

DONALD GUTHRIE
Ph.D., Vice-Principal, London Bible College, Northwood, England.

KEITH J. HARDMAN
Ph.D., Assistant Professor of Philosophy and Religion, Ursinus College, Collegeville, Pennsylvania.

RUDOLPH W. HEINZE
Ph.D., Associate Professor of History, Concordia Teachers College, River Forest, Illinois.

PAUL HELM
B.A., Lecturer in Philosophy, University of Liverpool, England.

COLIN J. HEMER
Ph.D., Lecturer in Biblical Studies, University of Sheffield, England.

CARL F.H. HENRY
Th.D., Ph.D., Arlington, Virginia, President of the Directors, Institute for Advanced Christian Studies.

ANTHONY A. HOEKEMA
Th.D., Professor of Systematic Theology Emeritus, Calvin Theological Seminary, Grand Rapids, Michigan.

EDWIN A. HOLLATZ
Ph.D., Chairman of the Department of Speech Communication, Wheaton College, Illinois.

JOYCE M. HORN
M.A., B.Litt., Publications Secretary and Assistant Editor, *Bulletin* of the University of London Institute of Historical Research, England.

JAMES M. HOUSTON
D.Phil., Chancellor of Regent College, Vancouver, British Columbia, Canada.

The late G.C.D. HOWLEY
Formerly Editor of *The Witness*, Purley, England.

DAVID ALLAN HUBBARD
Ph.D., President, Fuller Theological Seminary, Pasadena, California.

PHILIP EDGCUMBE HUGHES
D.Litt., Th.D., Visiting Professor of New Testament, Westminster Theological Seminary, Philadelphia, Pennsylvania.

ROBERT B. IVES
Ph.D., College Pastor, Messiah College, Grantham, Pennsylvania.

DIRK JELLEMA
Ph.D., Professor of History, Calvin College, Grand Rapids, Michigan.

PAUL KING JEWETT
Ph.D., Professor of Systematic Theology, Fuller Theological Seminary, Pasadena, California.

GEOFFREY JOHNSTON
M.A., B.D., Lecturer in the United Theological College, Kingston, Jamaica.

R. TUDUR JONES
D.Phil, D.D., Principal of Coleg Bala-Bangor, Bangor, Wales.

The late S. RICHEY KAMM
Ph.D., formerly Professor of History, Wheaton College, Illinois.

GILBERT W. KIRBY
M.A., formerly Principal of London Bible College, Northwood, England.

R. KISSACK
M.A., B.D., Chairman of the Liverpool District, Methodist Church, England.

DAVID KUCHARSKY
M.A., Editor, *Christian Herald*, Chappaqua, New York.

MICHAEL KYRIAKAKIS
B.D., Minister of the First Evangelical Church, Athens, Greece.

DONALD M. LAKE
Ph.D., Associate Professor of Theology, Wheaton College, Illinois.

DAVID LAZELL
Writer and Researcher, Bristol, England.

ROBERT D. LINDER
Ph.D., Professor of History, Kansas State University, Manhattan, Kansas.

HAROLD LINDSELL
Ph.D., formerly Editor, *Christianity Today*, Wheaton, Illinois.

MARCUS L. LOANE
K.B.E., D.D., Archbishop of Sydney and Anglican Primate of Australia, retired.

ADAM LOUGHRIDGE
D.D., Professor in the Reformed Presbyterian Theological Hall, Belfast, Northern Ireland.

LESLIE T. LYALL
M.A., London, England, formerly Editorial Secretary, Overseas Missionary Fellowship.

H.D. McDONALD
Ph.D., D.D., formerly Vice-Principal of London Bible College, Northwood, England.

OONAGH McDONALD
Ph.D., Member of Parliament, Lecturer in Philosophy of Religion, University of Bristol, England.

C.T. McINTIRE
Ph.D., Assistant Professor of History, Institute for Christian Studies, Toronto, Ontario, Canada.

R.J. McKELVEY
D.Phil., formerly Principal of the Federal Theological Seminary of South Africa, Alice, Cape Province, Republic of South Africa.

ROBERT J. McMAHON
B.D., Minister of Crossford and Kirkfieldbank, Lanarkshire, Scotland.

J. BUCHANAN MacMILLAN
Ph.D., Professor of History and Theory of Music, Nyack College, New York.

GEORGE MARSDEN
Ph.D., Associate Professor of History, Calvin College, Grand Rapids, Michigan.

I. HOWARD MARSHALL
Ph.D., Professor of New Testament Exegesis, University of Aberdeen, Scotland.

RALPH P. MARTIN
Ph.D., Professor of New Testament, Fuller Theological Seminary, Pasadena, California.

D.C. MASTERS
D.Phil., Professor of History, University of Guelph, Ontario, Canada.

J.W. MEIKLEJOHN
M.B.E., M.A., formerly Schools' Secretary, The Scripture Union, Scotland.

The late CARL S. MEYER
Ph.D., formerly Professor of Church History, Concordia Seminary, St. Louis, Missouri.

DAVID MICHELL
B.D., Missionary with the Overseas Missionary Fellowship in Japan.

PAUL E. MICHELSON
Ph.D., Romania.

SAMUEL J. MIKOLASKI
D.Phil., Professor of Historical Theology and Christian Heritage, North American Baptist Seminary, Sioux Falls, South Dakota.

WATSON E. MILLS
Th.D., Associate Professor of Philosophy and Religion, Averett College, Danville, Virginia.

SAMUEL HUGH MOFFETT
Ph.D., Dean of the Graduate School and Professor of Historical Theology, Presbyterian Seminary, Seoul, Korea.

LEON MORRIS
Ph.D., formerly Principal of Ridley College, Melbourne, Australia.

R.N. MUGFORD
B.A., S.T.B., Professor of Pastoral Theology, Vancouver School of Theology, Vancouver, British Columbia, Canada.

GORDON C. NEAL
M.A., M.Litt., Lecturer in Greek and Latin, University of Manchester, England.

ROBERT C. NEWMAN
M.A., B.D., Pastor of Faith Baptist Church, Winfield, Illinois.

ALAN NICHOLS
Th. Schol., Director of Information, Anglican Diocese of Sydney, Australia.

The late R.E. NIXON
M.A., formerly Principal of St. John's College, Nottingham, England.

The late J.G.G. NORMAN
M.Th., formerly Minister of Rosyth Baptist Church, Fife, Scotland.

GOTTFRIED OSEI-MENSAH
B.Sc., Executive Secretary, Lausanne Continuation Committee for World Evangelism.

JOAN OSTLING
M.A., M.A., Writer, Teaneck, New Jersey.

D.F. PAYNE
M.A., Head of the Department of Semitic Studies, The Queen's University, Belfast, Northern Ireland.

E.K. VICTOR PEARCE
M.A., Vicar of Audley, Stoke-on-Trent, England.

ROYAL L. PECK
B.A., M. Div., General Director, Istituto Biblico Evangelico, Rome, Italy.

P.W. PETTY
B.A., B.D., formerly Minister of Prestwick North Church, Ayrshire, Scotland.

RICHARD V. PIERARD
Ph.D., Professor of History, Indiana State University, Terre Haute, Indiana.

DONALD E. PITZER
Ph.D., Professor and Chairman of the Department of History, Indiana State University, Evansville, Indiana.

ARTHUR POLLARD
B.A., B.Litt., Professor of English, University of Hull, England.

NOEL S. POLLARD
M.A., B.D., Lecturer in St. John's College, Nottingham, England.

JOHN C. POLLOCK
M.A., South Moulton, Devon, England.

N.O. RASMUSSEN
Lecturer in Biblical Studies and Church History, Danish Lutheran Mission's Bible College, Hillerod, Denmark.

W. STANFORD REID
Ph.D., Professor of History, University of Guelph, Ontario, Canada.

IAN S. RENNIE
Ph.D., Associate Professor of History, Regent College, Vancouver, British Columbia, Canada.

MARY E. ROGERS
M.A., Assistant Professor of History, University of Guelph, Ontario, Canada.

DAISY D. RONCO
Dott.Lett., Senior Lecturer in Italian, University College of North Wales, Bangor, Wales.

HAROLD H. ROWDON
Ph.D., Lecturer in Church History and Christian Ethics, London Bible College, Northwood, England.

ERWIN RUDOLPH
Ph.D., Professor of English, Wheaton College, Illinois.

HOWARD SAINSBURY
M.A., Lecturer in Religion, Edge Hill College of Education, Ormskirk, England.

MICHAEL SAWARD
B.A., Vicar of Ealing, London, England.

DANIEL C. SCAVONE
Ph.D., Associate Professor of Ancient History, Indiana State University, Evansville, Indiana.

MILLARD SCHERICH
Ph.D., Professor of Education, Wheaton College, Illinois.

ROBERT V. SCHNUCKER
Ph.D., Professor of History and Religion, Northeast Missouri State University, Kirksville, Missouri.

MARTIN H. SCHRAG
Ph.D., Professor of the History of Christianity, Messiah College, Grantham, Pennsylvania.

CALVIN G. SEERVELD
Ph.D., Professor of Aesthetics, Institute for Christian Studies, Toronto, Ontario, Canada.

HENRY R. SEFTON
Ph.D., Senior Lecturer in Church History, University of Aberdeen, Scotland.

IAN SELLERS
Ph.D., Senior Lecturer in Padgate College, Warrington, England.

ERIC J. SHARPE
Teol.D., Senior Lecturer in Religious Studies, University of Lancaster, England.

BRUCE L. SHELLEY
Ph.D., Professor of Church History, Conservative Baptist Theological Seminary, Denver, Colorado.

E. MORRIS SIDER
Ph.D., Professor of History and English Literature, Messiah College, Grantham, Pennsylvania.

JOHN A. SIMPSON
M.A., Vicar of Ridge, Hertfordshire, England.

C. GREGG SINGER
Ph.D., Professor of Church History, Catawba College, Salisbury, North Carolina.

HARRY SKILTON, JR.
M.A., M. Div., Minister of First Presbyterian Church, Coalport, Pennsylvania.

STEPHEN S. SMALLEY
M.A., B.D., Canon-residentiary and Precentor of Coventry Cathedral, England.

CLYDE CURRY SMITH
Ph.D., Associate Professor of History, Wisconsin State University, River Falls, Wisconsin.

The late MATTHEW SPINKA
D.D., Th.D., formerly Waldo Professor of Church History, Hartford Theological Seminary, Connecticut.

ALVA STEFFLER
B.D., M.A.T., Associate Professor of Art, Wheaton College, Illinois.

DAVID C. STEINMETZ
Ph.D., Divinity School, Duke University, Durham, North Carolina.

ROY A. STEWART
B.D., M.Litt., formerly Minister of Muirkirk, Ayrshire, Scotland.

TIMOTHY C.F. STUNT
M.A., Head of the Department of History, Aiglon College, Chesières, Switzerland.

JAMES TAYLOR
M.A., Minister of Stirling Baptist Church, Scotland.

JOHN B. TAYLOR
M.A., Bishop of St. Albans, England.

JOHN A. THOMPSON
Ph.D., Senior Lecturer, Department of Middle Eastern Studies, University of Melbourne, Australia.

The late D.P. THOMSON
D.D., formerly Evangelist of the Church of Scotland.

C.G. THORNE, JR.
B.Phil., M.Litt., Ephrata, Pennsylvania.

JOHN TILLER
M.Litt., Vicar of Christ Church, Bedford, England.

DONALD G. TINDER
Ph.D., Professor of Church History, New College, Berkeley, California.

A.S. TOON
M.A., London, England.

PETER TOON
Th.D., Ph.D., Tutor in Doctrine, Oak Hill Theological College, London, England.

T.L. UNDERWOOD
Ph.D., Associate Professor of History, University of Minnesota, Morris, Minnesota.

HOWARD F. VOS
Th.D., Ph.D., Professor of History, The King's College, Briarcliff Manor, New York.

C. PETER WAGNER
Ph.D., Associate Professor of Church Growth, School of World Missions, Fuller Theological Seminary, Pasadena, California.

RONALD S. WALLACE
Ph.D., formerly Professor of Systematic Theology, Columbia Theological Seminary, Decatur, Georgia.

ANDREW F. WALLS
M.A., B.Litt., Professor of Religious Studies, University of Aberdeen, Scotland.

ROBERT C. WALTON
Ph.D., Associate Professor of History, Wayne State University, Detroit, Michigan.

DAVID F. WELLS
Ph.D., Professor of Historical and Systematic Theology, Gordon-Conwell Theological Seminary, South Hamilton, Massachusetts.

J.C. WENGER
Th.D., Professor of Historical Theology, Associated Mennonite Bible Seminaries, Elkhart, Indiana.

HOWARD A. WHALEY
A.M., Chairman of the Division of Missions, Moody Bible Institute, Chicago, Illinois.

JOHN WILKINSON
B.D., M.D., Presbyterian Church of East Africa.

C. PETER WILLIAMS
M.A., Lecturer in Trinity College, Bristol, England.

DAVID J. WILLIAMS
Ph.D., Tutor in Ridley College, Melbourne, Australia.

HADDON WILLMER
Ph.D., Lecturer in Theology, University of Leeds, England.

ROBERT S. WILSON
Ph.D., Lecturer in History and Dean of Arts, Atlantic Baptist College, Moncton, New Brunswick, Canada.

CARL FR. WISLØFF
Teol.D., Professor of Church History, Free Faculty of Theology, Oslo, Norway.

CARLTON O. WITTLINGER
Ph.D., Professor of History and Chairman of the Division of Social Sciences, Messiah College, Grantham, Pennsylvania.

A. SKEVINGTON WOOD
Ph.D., Principal, Cliff College, Calver, England.

JOHN D. WOODBRIDGE
Ph.D., Associate Professor and Chairman of the Division of Church History, Trinity Evangelical Divinity School, Deerfield, Illinois.

PAUL WOOLLEY
D.D., Professor of Church History Emeritus, Westminster Theological Seminary, Philadelphia, Pennsylvania.

DAVID F. WRIGHT
M.A., Senior Lecturer in Ecclesiastical History and Associate Dean of the Faculty of Divinity, University of Edinburgh, Scotland.

J. STAFFORD WRIGHT
M.A., Canon of Bristol Cathedral, formerly Principal of Tyndale Hall, Bristol.

EDWIN M. YAMAUCHI
Ph.D., Associate Professor of History, Miami University, Oxford, Ohio.

WILLIAM G. YOUNG
M.A., B.D., Minister of Resolis and Urquhart, Scotland, and formerly Bishop of Sialkot, Pakistan.

Notes and Abbreviations

Quotations from the New Testament are from *The New International Version, The New Testament,* copyright © 1973 by the New York Bible Society International, unless otherwise designated. Quotations from the Old Testament are from the *King James Version* unless otherwise designated.

Contributors' names have generally been deleted from entries of less than fifteen lines of type.

An asterisk (*) designates a subject for which there is an entry under an identical or closely similar heading elsewhere in the dictionary.

Abbreviations are used sparingly in this volume. Books of the Bible are abbreviated when employed parenthetically. Besides the common literary abbreviations, the following appear:

E	east, eastern (used in a geographical, nonpolitical noneccleslastical sense)
ed.	editor, edited by, edition
ET	English translation
fl.	flourished
Ger.	German
Gr.	Greek
KJV, AV	King James (Authorized) Version
Lat.	Latin
LXX	Septuagint
MS	manuscript
NE	northeast, northeastern (geographical)
NEB	New English Bible
NT	New Testament
OT	Old Testament
Port.	Portuguese
pub.	published
rep.	reprinted
RSV	Revised Standard Version
RV	(English) Revised Version
tr.	translator, translated by

Bibliographical Abbreviations

ACW	Ancient Christian Writers. The Works of the Fathers in Translation (1946ff.).
BA	*Biblical Archaeologist.*
BJRL	*Bulletin of the John Rylands Library.*
CAH	*The Cambridge Ancient History* (12 vols., 1923–39).
CHB	*Cambridge History of the Bible* (1970–).
CSEL	Corpus Scriptorum Ecclesiasticorum Latinorum (1866ff.).
DACL	*Dictionnaire d'Archéologie Chrétienne et de Liturgie* (15 vols., 1970–53).
DCA	*Dictionary of Christian Antiquities* (2 vols., 1875–80).
DHGE	*Dictionnaire d'Histoire et de Géographie Ecclésiastiques* (1912ff.).
FC	*Fathers of the Church* (1947ff.).
HERE	*Hastings' Encyclopaedia of Religion and Ethics* (12 vols., Index, 1908–26).
HTR	*Harvard Theological Review.*
HTS	*Harvard Theological Studies.*
JBL	*Journal of Biblical Literature.*
JEH	*Journal of Ecclesiastical History* (1950ff.).
JTS	*Journal of Theological Studies* (1900ff.).
LCC	*Library of Christian Classics* (26 vols., 1953–70).
NBD	*New Bible Dictionary.*
NPNF	Nicene and Post-Nicene Fathers (1887–1900).
NTS	*New Testament Studies.*
PG	*Patrologia Graeca* (162 vols., 1857–66).
PL	*Patrologia Latina* (221 vols., 1844–64).
RHPR	*Revue d'Histoire et de Philosophie religieuses* (1921ff.).
RU	*Religiongeschichtliche Untersuchungen.*
SJT	*Scottish Journal of Theology.*
TDNT	Kittel, *Theological Dictionary of the New Testament.*
TU	*Texte und Untersuchungen zur Geschichte der altchristlichen Literatur* (1882ff.).
ZPEB	*Zondervan Pictorial Encyclopedia of the Bible.*
ZTK	*Zeitschrift für Theologie und Kirche* (1891ff.).

AACHEN, SYNODS OF. Church assemblies held between 789 and 1023. Aachen (Aix-la-Chapelle) had political importance in the empire of Charlemagne and his successors and thus was a natural venue. There were synods, meetings, or councils of clergy (with state officials) there in 789, 797, 799, 801-2, 809, 816-17, 819, 825, 1000, and 1023. Their pronouncements related mainly to the ecclesiastical discipline of the parish clergy, monks, and nuns. Also discussed were the faithful performance of duty by servants of the state, and doctrinal issues. Charlemagne himself saw that Adoptianism, which had its roots in Spain and which had been condemned at Regensburg (792) and Frankfurt (794), was again condemned at Aachen in 799. In addition, he disagreed with the pope on the doctrine of Double Procession,* and had this discussed in 809. Alcuin of York was present in both 802 and 816 and contributed to discussion about implementing the Benedictine Rule. The last synod of Aachen was called to decide which diocese, Cologne or Liège, should control the monastery at Burtscheid.

PETER TOON

ABA I, THE GREAT (d.552). Patriarch of the East from 540. Persian-born sometime before 500, he was converted about 520 through the courtesy and testimony of a catechist Yusuf and took the name Aba. He studied at Nisibis, went to the Byzantine Empire possibly for safety, and visited Constantinople. Around 533 he returned, a distinguished scholar, and taught at Nisibis, revising the Syrian Bible. Appointed patriarch, he toured the church with two metropolitans and seven bishops, taking vigorous and practical steps to end schism and restore Christian morals and church discipline. Accused in 543 of enforcing Christian standards in church courts and of converting Zoroastrians, he was arrested, tried, and finally exiled to Azarbaijan. Under house arrest he continued to administer the church, and he held a synod in 544. After attempted assassination by an apostate Christian in 549, he fled to Ctesiphon and threw himself on the shah's mercy. Arrested and heavily chained by Magians in 550, he was formally pardoned and released by the shah in 551. Worn out by long suffering, he died the following year.

WILLIAM G. YOUNG

ABAILARD, PETER, see ABELARD

ABBÉ. The title given to a class of unbeneficed secular clerks in France and Italy. Originally French for "abbot", an extension of meaning took place in the sixteenth century, when the Concordat of 1516 authorized Francis I to nominate secular priests *in commendam.* Laymen were often appointed and, not being bound to residence, many *abbés commendataires* never saw the monasteries of which they were titular rulers. The term is now applied to secular clerics in general as a title of courtesy.

ABBESS (Lat. *abbatissa*). The female head of a community of women known as a nunnery or convent. She is elected to her position of authority, which is similar to that of an abbot* over the monks, by the secret votes of the sisters or nuns of the community. It is customary for a bishop to install her in her office with an abbatial cross, staff, and ring. She usually holds her office for life. In the Middle Ages she was often of noble or royal birth and so played an important part in the life of the church—e.g., St. Hilda in England who ruled over a double monastery* and took a leading part at the Synod of Whitby. The Council of Trent tried to regularize the position of the abbess and to bring her under the control of the diocesan bishop. In the German Lutheran Church the title remained in use for the head of collegiate foundations of unmarried women known as *stiftsdamen.*

NOEL S. POLLARD

ABBEY. A building occupied by, or the group name for, a particular house of a religious order of monks or nuns. The name was originally confined to one of the orders of the Benedictine family (e.g., Cistercian, Carthusian, Trappist). Under the rule of St. Benedict, each abbey is to be regarded as a family unit under the nearly autonomous authority of its abbot. It is now also used for a building once used as an abbey.

ABBO (945?-1004). French Benedictine abbot. Born near Orléans, he became a Benedictine monk at the great abbey of Fleury. He studied at Paris, Rheims, and Orléans and subsequently became widely recognized as an authority on astronomy, mathematics, and philosophy. In 986 he took charge of an English monastery at Ramsey, but two years later returned as abbot to Fleury, where he installed the Cluniac observance. Under him a flourishing culture arose, marking perhaps the beginning of a renaissance movement which came to fruition in the eleventh century. Abbo may have been responsible for a treatise on Aristotle's *Categories,* and certainly the earliest manuscript containing Aristotle's *Analytics* dates from Fleury during this period. Much in demand as an arbitrator in monastic disputes, he was killed while attempting to separate two groups of quarreling monks in Gascony. J.G.G. NORMAN

ABBOT. The title given to the head of a community of monks of the Benedictine Order or some of the regular canons. The name comes from the Hebrew word for "father", and was commonly used in the Eastern churches for all the older monks. In the West, where it derives from the Latin *abbas,* it was applied to the head of the community alone. At first the abbot was a layman and was under the control of the local bishop. During the Middle Ages abbots became responsible to the pope and assumed authority sometimes greater than that of the bishop. At first the abbot was appointed by the bishop, but in time the monks elected the head of their house. The bishop now confirmed and blessed the new abbot, giving him a mitre, crosier, and ring. These symbols recognized his semi-episcopal power, which was wielded in both church and state. At the end of the Middle Ages his authority over his own house had become such that he often lived in great state. In the sixteenth century the long struggle to gain exemption from episcopal control ended, when all orders of monks gained immunity. The abbot is now directly responsible to the pope. He is elected to his office for a period of years, or more usually for life.

NOEL S. POLLARD

ABBOT, GEORGE (1562-1633). Archbishop of Canterbury from 1611. Born in Guildford, son of a cloth worker, he was educated at Oxford and became successively master of University College and vice-chancellor (three times), dean of Winchester (1600), and bishop of London (1609) before becoming primate. His rise to power followed his defense of the hereditary monarchy (1606) and his efforts to join the English and Scottish churches (1608). For many years he was the recognized leader of the English Calvinists and showed pronounced Puritan sympathies. He took a leading part in the translation of the Authorized Version and is regarded as one of the first to establish Anglicanism as a militant force based on the concept of a godly king. He was less tolerant toward Roman Catholics, insisting that the designation was nonsensical, Rome being a local place and "Catholic" meaning "universal". Though he was often in favor with James I and Charles I, he firmly stood his ground against them when they demanded compromise with conscience, especially on matters of divorce. Throughout much of his career he was bitterly opposed by the Oxford High Churchmen, especially by Laud,* who was to succeed him at Canterbury. Abbot was temporarily under a cloud when in 1622 he accidentally shot a gamekeeper while hunting, but the king was responsible for his exoneration from blame.

See P.A. Welsby, *George Abbot the Unwanted Archbishop* (1962). R.E.D. CLARK

ABBOT, GEORGE (1603-1648). English Puritan. He was born in Yorkshire, but little is known of his early life. He later excelled in Hebrew studies and patristics. Although he remained a layman, he is chiefly known for his theological writings, which were clear and succinct for his age. *A Paraphrase of the Whole Book of Job* appeared in 1640, and an important contribution on the Sabbatarian controversy in *Vindiciae Sabbath* in 1641. A volume published posthumously in 1651 contained his brief notes upon the whole Book of Psalms.

ABBOT OF UNREASON/MISRULE. A person selected in medieval times to preside over the Feast of Fools* or other revels and games celebrated at the Christmas/New Year season. He received a staff of office, retaining his authority throughout the feast. In a secularized form the institution survived into the seventeenth century.

ABBOTT, EDWIN ABBOTT (1838-1926). Educationist and religious writer. Distinguished student in classics and mathematics at Cambridge, he became fellow of St. John's and was ordained in 1862, was appointed headmaster of City of London School in 1865, and resigned in 1889 to devote himself to study and writing. Though chiefly remembered for his educational work, Abbott (a Broad Churchman) also wrote biographies of Francis Bacon, Cardinal Newman, and Thomas à Becket, books on textual criticism, and several religious romances. Reflecting learning, piety, and originality, his works still repay study. *Philomythus* (1891) is a superb psychological treatment of theological rationalizing, with Newman as the butt. His science-fiction novel *Flatland* (by A. Square, 1884) describes a two-dimension world into which the world of three dimensions impinges seemingly miraculously. Abbott suggests that miracles in our world may be incursions from a fourth dimension. R.E.D. CLARK

ABBOTT, LYMAN (1835-1922). Congregational minister. Born in Massachusetts, he was educated at New York University and practiced law before deciding to enter the ministry of the Congregational Church. He became a pastor in Terre Haute, Indiana, in 1860 and after the Civil War served as an executive of the American Union Commission which promoted reconstruction in the South. He wrote for *Harper's Magazine,* then became editor of the *Illustrated Christian Weekly* in 1870. In 1876 he joined H.W. Beecher* as an editor of the *Christian Union.* In 1888 he was called to succeed Beecher as pastor of Plymouth Church in Brooklyn. Abbott was one of the most influential American religious thinkers of his time. Until the 1880s he remained fairly orthodox, then gradually accepted radical biblical criticism and became a theological liberal. He accepted Darwinism and applied the evolutionary principle to religious questions: even God could be conceived as an immanent evolutionary power; history is the record of divinity out of humanity; "what Jesus was, humanity is becoming". Abbott's books include *The Theology of an Evolutionist* (1897) and *Reminiscences* (1915).

See EVOLUTION. HARRY SKILTON

ABBREVIATORS. Papal chancery officials who drafted the pope's written statements. The name came from the highly developed system of abbreviations used in papal documents. Pius II (1458-64) fixed their number at seventy; Pius VII (1800-1823) reduced it to seventeen. Their duties were transferred in 1908 to the *Collegium Pronotariorum Apostolicorum.*

ABECEDARIANS. A name given to Anabaptists who scorned normal methods of education and affirmed that God could provide enlightenment by more direct methods such as visions and ecstasies. Academic study was repudiated as idolatrous, and learned preaching was regarded as falsifying God's Word. Some went to such lengths of obscurantism as to assert that it was necessary to be ignorant even of the letters of the alphabet—hence A-B-C-darians. Among the group were Nicholas Storch and the Zwickau* Prophets, and even Carlstadt* was influenced to the extent of renouncing his title of doctor of divinity.

ABELARD (Abailard), PETER (1079-1142). Scholastic philosopher and theologian. Born in Pallet, Brittany, he studied successively under the Nominalist Roscellinus, the extreme Realist William of Champeaux (whom he made appear inconsistent on the issue of universals), and Anselm of Laon. A brilliant debater and lecturer, Abelard attracted large numbers of enthusiastic students, first in dialectics and later in theology. His arrogance, however, and his celebrated love affair with the beautiful and talented Héloïse almost ruined his professorial career. About 1115, Abelard was in Paris where he lived in the home of Fulbert, canon of Notre Dame, whose teenage niece Héloïse he had agreed to tutor; but the relationship became too personal, and they had a son whom they named Astrolabe. To pacify Fulbert, Abelard secretly married Héloïse after the son's birth. When calumnious rumors began to circulate, Héloïse agreed to retire to the convent of Argenteuil rather than further damage Abelard's teaching career. Fulbert in anger hired a band of men who broke into Abelard's quarters one night and castrated him.

After this humiliation Abelard entered the monastery of St.-Denis, at the age of forty. In 1121 he was condemned unheard by the Council of Soissons for his view of the Trinity, and his book on the subject was burned. Pursued from place to place by both the authorities and large numbers of students, Abelard finally became abbot of the secluded monastery of St. Gildas in Brittany in 1125. Conditions at St. Gildas were unbearable, and he soon went back to Paris where he once again became a popular lecturer.

At about the same time he incurred the animosity of Bernard of Clairvaux* because of alleged heretical statements about the Trinity in his writings. In 1141, several propositions selected from his works were condemned at the Council of Sens. On his way to Rome to appeal his case, Abelard stopped at Cluny where he was convinced by Peter the Venerable of the hopelessness of any further attempts to defend himself. He died at a Cluniac priory.

He left a considerable body of works on logic and theology, including his famous *Sic et Non* (1122), in which he arranged contradictory statements from the Scriptures and the Church Fathers to force students to reconcile them; his autobiography, *Story of My Misfortunes,* personal letters, and a number of poems, sermons, and letters. His influence lived on through his students, among whom were a number of future popes and cardinals, John of Salisbury, and Otto of Freising.

A century ago historians generally hailed Abelard as the precursor of modern free thought, but recent scholarship has challenged this view, emphasizing rather that he was an intellectual who approached his faith with new methods and who sought to understand faith by the use of reason. Although he is best remembered for his affair with Héloïse, his greatest contribution to medieval Christian history was to help initiate the task of reconciling faith and reason. He held to the existence of individual things, but added that man had a mental idea of common elements in things as well as the existence of ultimate universals in the mind of God.

See ATONEMENT concerning his "moral theory" of that doctrine.

BIBLIOGRAPHY: J.G. Sikes, *Peter Abailard* (1932); J.R. McCallum, *Abelard's Christian Theology* (1949); E. Gilson, *Heloise and Abelard* (tr. L.K. Shook, 1951); R. Pernoud, *Héloïse et Abelard* (1970).

ROBERT D. LINDER

ABELONIANS. An obscure sect in Roman North Africa known of only from Augustine (*Heresies* 87), formerly active in the country around Hippo, but defunct when he wrote in 428 through the recent conversion of its last adherents to the Catholic Church. Both marriage and total sexual abstinence were obligatory for its members. Each couple was required to adopt a boy and a girl, who after the death of both adoptive parents formed a new pair and themselves adopted children. Augustine believed the sect's name derived from Punic, but knew that others connected it with Abel (whence they were called also Abelians or Abeloites), presumably because in Jewish, Christian, and Gnostic legends Abel died not only childless but also in unsullied chastity though (in some versions) married (cf. L. Ginzberg, *The Legends of the Jews,* 5, 1925). The sect was perhaps related to the Gnostic-Manichaean tradition in Africa.

D.F. WRIGHT

ABERCIUS, INSCRIPTION OF (c.182). In 1883 Sir William Ramsay discovered an incomplete epitaph of Avircius of Hieropolis in Phrygia. Without mentioning Christ or the Church, Avircius speaks of the all-seeing Shepherd who taught faithful Scriptures, and of Paul, of faith, of the fish from the spring, of the Virgin, of the wine and loaf, which guided him through the plains of Syria, across the Euphrates to Nisibis and to Rome. At each place he met "brethren." He invites prayers for himself, warning against using his tomb for others, on pain of paying gold into the treasuries of Rome and Hieropolis. Interpretations vary, but the symbolism suggests a Christian context. The mutilated word *basil?* may refer to the Roman emperor or to the sovereign church at Rome, the latter suggesting growing imperialism.

If Avircius Marcellus, addressed favorably in an anti-Marcionite tract in Eusebius, wrote the inscription, his warning was possibly against Montanist activity. The legendary fourth-century *Vita Abercii* identifies Avircius as bishop of Hieropolis. G.T.D. ANGEL

ABERDEEN DOCTORS. A group of early seventeenth-century Scottish divines, so called because they had all taken the degree of doctor of divinity by thesis at King's College, Aberdeen. (The degree had been in abeyance since the Reformation, but was revived by James VI.) They came into prominence because of their opposition to the National Covenant* of 1638. The Covenant had revived the anti-Romanist Negative Confession of 1581 and had given it an anti-episcopacy and anti-Prayer Book interpretation. The Doctors objected to this and to the undermining of the royal authority implicit in the Covenant, to take which, they asserted, was to separate the Church of Scotland from the other Reformed churches and from the early church. They had hoped to dispel the notion that the Fathers were on the side of Rome and to show the essential catholicity of the Reformed Church, but despite them episcopacy was abolished, Presbyterianism restored. Most famous of the Doctors was John Forbes.* The effectiveness of their advocacy of episcopacy is seen in the comparative strength of the Episcopal Church in NE Scotland even today. HENRY R. SEFTON

ABERHART, WILLIAM (1878-1943). Baptist lay preacher and politician. Born in Ontario, he graduated from Queen's University in 1906 and as a young schoolteacher migrated to Calgary in 1910. From 1915 to 1935 he was principal of Crescent Heights High School. He was a lay preacher and in 1918 began a Bible class in Westbourne Baptist Church, soon outgrowing the building. After a period in downtown theaters, the Bible class developed into the Prophetic Bible Institute. This was the period of the modernist-fundamentalist controversy, and Aberhart's preaching of the Gospel, with a strong premillennial emphasis, drew a large following. Realizing the potential of radio, he began broadcasting in 1925 and soon was on the air several hours each Sunday. By the mid-thirties, radio evangelism had resulted in a movement of near-revival in parts of Alberta and western Saskatchewan, bringing an unparalleled proliferation of Bible institutes and an army of volunteers for overseas missions. As the worldwide economic depression of the 1930s settled upon the Canadian prairies, with the concomitant of a ten-year drought, Aberhart came across the "social credit" theory. He began to weave this theme into his Sunday broadcasts, broadening his appeal to include the Mormons and many who had not been sympathetic with his theology. In 1935 his Social Credit Party won a landslide victory in the Alberta provincial election. Aberhart became premier, and Social Credit became an orthodox and conservative administration, with strong religious and evangelical overtones, all of which was carried on after Aberhart's death in 1943 by his outstanding pupil, E.C. Manning.* IAN S. RENNIE

ABERNETHY, JOHN (1680-1740). Irish Presbyterian minister. Trained at Glasgow and Edinburgh, he became minister of the Presbyterian congregation at Antrim in 1703. In 1718 he refused the general synod's appointment to a Dublin congregation—an unprecedented exercise of independent judgment which led to a division between "Subscribers" and "Non-Subscribers." The latter, led by Abernethy, were excommunicated in 1726. From 1730 he ministered in Wood Street, Dublin. Abernethy asserted that terms of Communion, fixed in the NT, do not include subscription to humanly devised confessions of faith, and he opposed the exclusion of men from the service of the state on religious grounds. His published works include *Discourses concerning the Being and Natural Perfections of God.* HAROLD H. ROWDON

ABGAR, LEGEND OF. Eusebius quotes two letters and a story from the record office at Edessa, in which King Abgar V Ukkama invites Jesus to visit and heal him. Jesus commends Abgar's faith, explains that He must accomplish His work where He is, but promises to send a disciple after the Ascension. The story relates the mission of the Apostle Thaddaeus (alias Addai*) to Edessa. An expanded Syriac version, *Doctrina Addai*, discovered in 1876, mentions a portrait of Jesus painted by Abgar's messenger and the reply of Jesus promising safety to Edessa. This letter of Jesus became a charm to avert evil and appeared on doorposts, sepulchers, and city gates throughout the empire, even as far afield as twelfth-century Britain. G.T.D. ANGEL

ABJURATION. In ecclesiastical usage the term denotes the renunciation on oath of heresy, made when the penitent is reconciled with the church. Gregory the Great (d.604) had occasion to set out the practice of the early church when it dealt with heretics such as Montanists, Eunomians, and others. There was added to the abjuration a solemn profession of faith, especially after the rise of Nestorianism* and Eutychianism.* The procedure was further elaborated during the time of the Inquisition,* according to the degree of heresy. In modern times abjuration in its formal sense is normally restricted to receptions into, or reconciliation with, Greek and Roman Catholic churches. J.D. DOUGLAS

ABJURATION, OATH OF. Originating under William III in 1690 and made compulsory in 1701, this required holders of public office in England, such as members of Parliament, lawyers, and clergy, to renounce the claims of the recently overthrown Stuart dynasty. It was reimposed under Hanoverian kings, when in addition Roman Catholics had to reject papal claims to jurisdiction in England. The oath was later modified in order to give relief to Roman Catholics, Jews, and others.

ABLUTIONS. These are of two types: first and most common, the washing of the fingers and chalice by the celebrant after the Communion in the Mass. The custom became part of the Eucharist by the eleventh century, but was regulated by the Missal of Pius V, which prescribed a double ablution: the chalice with wine, then the chalice and fingers with wine and water. Most Eastern rites have a similar procedure, but in the Greek rite ablutions are made privately by the celebrating priest after the Mass. The need for such ablutions is necessarily connected with the belief that the bread and wine are truly the body and blood of Christ. The second type of ablution is the rinsing of the mouth with wine after the reception of the sacrament by newly ordained priests and the rinsing of the mouth with water after the communion of the sick. Both these customs probably originated in the medieval custom of giving communicants unconsecrated wine after the actual Communion; but they may go back to the time when the Lord's Supper was part of a larger Christian meal (1 Cor. 11). PETER TOON

ABOLITIONISM. The movement which opposed slavery and the slave trade in North America prior to the American Civil War (1861-65). Although the practice of slavery was criticized by a few Christians, notably Quakers and other radical sectarians, the importation of Negro slaves from Africa into the American colonies was supported by the vast majority of the churches and churchmen up to the end of the first decade of the nineteenth century, on the basis of Scripture, tradition, and economic necessity. The beginning of the nineteenth century saw the growth of public opinion against slavery, especially in the Northern States, though it was an uphill struggle; at the same time the position in the Southern States was hardened in the defense of slavery, adding to the traditional arguments the theory of the racial inferiority of the Negro. This difference of white opinion on both sides eventually led to the war between the states. Notable centers of abolitionism were the religion-oriented colleges of the North and Midwest, such as Knox, Oberlin, Western Reserve, and Wheaton. The movement was spread by a host of newspaper editors (some of whom were killed), lecturers, clergymen, and authors, including William Lloyd Garrison, C.G. Finney,* T.D. Weld,* Horace Greeley, and Wendell Phillips. Garrison's *Liberator* (founded in 1831) early gave teeth to the movement so effectively advanced two decades later by Harriet Beecher Stowe's* *Uncle Tom's Cabin.*

See also AMERICAN ANTI-SLAVERY SOCIETY.

W. WARD GASQUE

ABORTION. Criminal abortion of illegitimate and unwanted legitimate pregnancy is widespread. Its adverse effect on the health, future fertility, and even survival of the woman is a main cause in the demand for the more ready availability of legal abortion.

Therapeutic abortion, often without legal sanction, has long been practiced by reputable gynecologists in situations where the woman's life or health is endangered by pregnancy. The problem of defining "health," especially "mental health," is great. In practice it is impossible to separate physical, mental, and socioeconomic factors. Despite this, there is a real place for termination of pregnancy, especially in the worn-out, defeated mother, although adverse medical sequelae and remorse occur in probably five to twenty percent of women.

Recent liberalizing legislation in Scandinavia (from 1935), in Britain (1967), and in many of the United States (from 1967) has resulted in a flood of abortions far beyond the aim of the legislatures —the vast majority of abortions being requested for merely emotional disturbance due to situational conflicts. Safeguards for hospital staff who have conscientious objection to participating have proved inadequate. Abortion on demand or by negotiation between mother and doctor free of all legal restraint had been allowed, or at least practiced, in parts of the Soviet bloc and in Japan. Following decisions of the Supreme Court in 1973 this is so throughout the United States. In some countries of Eastern Europe four pregnancies out of every five have been aborted; Romania has now severely restricted its earlier permissive laws.

The commandment "Thou shalt not murder" is not appropriate to therapeutic abortion, the intention of which is towards the health of the woman, not against the fetus. Exodus 21:22f. refers to unintentional miscarriage and is generally held to indicate that the fetus has a different status from the person. The "sanctity of life," based on such Scriptures as Genesis 9:5f., is an important concept, but is relevant not only for the fetus but also for the life of the mother which may be overwhelmed by a further pregnancy. Scripture places a greater emphasis on the quality of life than on mere existence. Whether the fetus has a soul, and when that is implanted, has been debated since pre-Christian times. With the success of *in vitro* impregnation of ova by sperm in the laboratory it becomes an acute problem. Similarly, the realization that from fifteen to fifty percent of all pregnancies are lost by spontaneous miscarriage, often unrecognized even by the mother, increases the difficulties of visualizing the afterlife if all such are ensouled. Many would rather consider that the person *is* a soul and therefore esteem the fetus as a potential person, its value and importance increasing with gestational age.

The moral dilemma of weighing the value of a fetus with all its human and spiritual potentialities against the cost to the life of the mother and existing family is a real one, felt by all who have to give permission or have to operate. This dilemma will soon be felt by all women if the abortion "pill" in process of development comes into use as a standard contraceptive and self-abortive technique.

BIBLIOGRAPHY: K. Barth, *Church Dogmatics*, Part III, vol. 4; H. Thielicke, *The Ethics of Sex* (1964); *Abortion: An Ethical Discussion* (Church Assembly Board for Social Responsibility, 1965); W.O. Spitzer and C.L. Saylor (eds.) *Birth Control and the Christian* (1969); D. Callahan, *Abortion: Law, Choice and Morality* (1970); J. Noonan (ed.), *The Morality of Abortion* (1971); R.F.R. Gardner,

Abortion: The Personal Dilemma (1972); O. O'Donovan, *The Christian and the Unborn Child* (1973); Report of the committee on the Working of the Abortion Act (Lane Report) (1974).

R.F.R. GARDNER

ABRAHAM, APOCALYPSE OF. This Jewish apocryphal work survives only in a Slavonic version, translated from a lost Greek version dependent possibly on a Hebrew or Aramaic original. The extant form was edited by a Christian and has Christian editions. The original was composed in the period A.D. 70-130. Chapters 1-8 expound the conversion of Abraham from idolatry. The remaining chapters (9-32) contain the apocalypse proper, where God tells Abraham of the fall of man and of the idolatry of his seed which leads to the destruction of the Temple (A.D. 70). When, however, the present age, lasting twelve "hours," is over, the End will come when the heathen will be destroyed, the apocalyptic trumpet will sound, and the people of God will be gathered together.

G.T.D. ANGEL

ABRAHAM, TESTAMENT OF. An apocryphal writing originating among first-century Jews (Köhler) or second-century Jewish-Christians (M.R. James). It describes how the patriarch is shown the universe and told that it will survive seven thousand years. Then the archangel Michael at the first gate of heaven shows him the paths to hell and paradise, and three different judgments taking place. Praying for the forgiveness of sinners, Abraham is returned to earth and on to paradise. The major interest is in the fate of individual souls and not with the impending crisis characteristic of conventional Jewish apocalyptic. The Greek text, which includes Christian passages (interpolations?), survives in a longer and a shorter recension, with translations in Coptic, Arabic, Ethiopic, Slavonic, and Romanian.

G.T.D. ANGEL

ABRAHAM ECCHELLENSIS (1600-1664). Maronite* scholar. Surnamed from his birthplace Hekel (Ecchel) on the Syrian slopes of Mt. Hermon, he was a brilliant student at the Maronite college in Rome, where he obtained philosophical and theological doctorates. His valuable contributions to Le Jay's Polyglot Bible* (Arabic, Latin of Ruth, Arabic of III Maccabees) were cut short by an unfortunate jealous quarrel with his collaborator Gabriel Sionita. His immense labors included compiling, translation, philosophy, church history, and Christian biography (see L. Petit in *Dictionnaire d'histoire et de géographie ecclésiastiques*, vol. I). Though a man of prodigious scholarship, he is expertly judged (like many students of minutiae) to have failed in critical faculty or total synthetic grasp.

ROY A. STEWART

ABSALON (1128-1201). Archbishop of Lund. Born in Denmark, he studied and taught in Paris. In 1158 he became bishop of Roskilde, served as papal legate to Scandinavia, and after 1178 was appointed to Lund. He introduced into Denmark Western religious customs like clerical celibacy and monasticism. In fortifying his diocese against an invasion of Slavs (Wends) he built at Havn a castle-keep which became the nucleus of the city of Copenhagen. He served as counselor to Waldemar I and Canute IV, doing much to arrange wise legislation. He patronized the arts and encouraged Saxo Grammaticus to write the history of Denmark. He founded a monastery at Soro, where he was buried.

J.G.G. NORMAN

ABSOLUTION. There is no discussion of this subject as such in the Bible, and the word never occurs there. But the sinner's need of forgiveness, of which absolution speaks, is entirely scriptural. So is the truth that forgiveness of sin is the gracious work of God in Jesus Christ, who died for our sins and was raised for our justification (Rom. 4:25; cf. 1 John 1:9). He alone has the authority to absolve (Luke 7:47f.; cf. Col. 1:13f.).

In the early church, when postbaptismal sin became a problem, the power to announce forgiveness to the penitent (stemming from the word of Jesus in Matt. 16:19; 18:18; cf. John 20:23, "if you forgive anyone his sins, they are forgiven") was associated with the clergy. At first this declaration was public; but later, and with the development of sacramental theory during the scholastic period, absolution was mostly given after auricular confession in private, by means of the formula "I absolve you," or (especially in the East) in the course of a prayer. The former tended to obscure the real source of absolution.

Since the Reformation and the decline of private confession, absolution in the technical sense has tended, in the Protestant but not in the Roman Catholic Church, to be confined to public worship. The Anglican Book of Common Prayer, for example, returns to a biblical emphasis by following the general confession with an announcement of God's forgiveness in Christ.

STEPHEN S. SMALLEY

ABSTINENCE. The practice within the Christian Church of abstaining from the consumption of certain kinds of food. The principle behind it has often been traced to the custom of fasting* in the OT and its continuation in the NT. The latter considers the question in a new light. While it is encouraged as a means of defeating sin or helping the weak in conscience, it is condemned when it is set up as a new law (Col. 2:20-23; 1 Tim. 4:1-3). Abstinence was made a prominent ideal in the early church, especially in the growing monastic communities. The ideal was translated for lay people into a rule of abstinence from meat on Fridays, which continued in the Roman Catholic and other churches until modern times. The Eastern Orthodox churches have made even stricter rules. The sixteenth-century Reformers tried to restore the balanced view of the NT and attacked medieval views of abstinence. The evangelical revival in the eighteenth and nineteenth centuries fostered the growth of a specialized use of the word. Symptomatic of a return to a more legalistic view in Protestantism, societies in the USA and Britain formed to advocate total abstinence from alcohol.

NOEL S. POLLARD

ABSTINENTS. The title given to several sects that flourished in SW Europe in the third and fourth centuries. Partly a revolt against the worldliness of the established churches, the movement shared the Gnostic concept of matter as intrinsically evil. Strict asceticism was demanded, including the proscription of marriage. There was total abstinence from animal food, and constant prayers, fasts, and vigils were further evidence of the adherents' austere devotion. Priscillian,* their most prominent member, was executed at Trier in 385 for alleged magic and heresy.

ABUNA ("our father"). The name used of the patriarch of the Ethiopian Church.*

ABYSSINIAN CHURCH, see ETHIOPIA

ACACIAN SCHISM (484-519). A schism between the Eastern Church and Rome during the Monophysite* controversy, arising from the highhanded excommunication of Acacius of Constantinople* by Felix III of Rome. Attempts by succeeding patriarchs and emperors to heal the breach failed because of Rome's extravagant demands. On the accession of Justin I, however, Patriarch John yielded to Pope Hormisdas, and removed the names of Zeno, Acacius, and his five successors from the "diptychs," ending the schism.

ACACIUS (d.366). Semi-Arian theologian and bishop of Caesarea in Palestine. A friend of Eudoxios of Antioch and George of Alexandria, he opposed the dogmatic (*homoousios*) theology of the Council of Nicea (325). He enjoyed the patronage of the emperor and in 340, after the death of Eusebius, became bishop of Caesarea and inherited the great library of his predecessor, of which he made much use. He became leader of the Court Party or Homoeans*—those who asserted that Jesus Christ was like (*homoios*) the Father but not necessarily of the same essence. He played a major role at the Council of Seleucia (359), but was deposed by those who were defenders of the Nicene orthodoxy. Constantius nevertheless stood by him. When Jovian was emperor, Acacius found it expedient to accept the Creed of Nicea, but he later returned to Arianism* when Valens began to rule. He wrote much in his lifetime, but only fragments now remain.
PETER TOON

ACACIUS OF CONSTANTINOPLE (d.489). Patriarch from 471. He opposed the anti-Chalcedon encyclical of the usurping Emperor Basiliscus (475) and, on Zeno's restoration (476), collaborated in the deposition of Monophysite* bishops. In 479 he consecrated Calanaion, a Chalcedonian, as bishop of Antioch, and was criticized by Pope Simplicius for interfering in another see. At the instigation of the Monophysite Peter Mongus he encouraged Zeno to promulgate the *Henoticon** in the interests of theological peace (482), for which Felix of Rome excommunicated him. Rejecting this with contempt, Acacius erased Felix's name from the "diptychs." Thus arose the Acacian Schism.*
J.G.G. NORMAN

ACADEMIES, DISSENTING, see DISSENTING ACADEMIES

ACCIDENT. Ancient and medieval philosophy distinguished between accident ("nonessential property") and substance, as both Plato and Aristotle considered the former as secondary at most and Aquinas emphasized its relative and dependent aspects. Philosophical understanding prepared the way for its use in theology, in the doctrine of the Eucharist, to elucidate the nature of the divine presence in the bread and wine. As early as about 1200 Alain of Lille spoke of transubstantiation *(Theologicae regulae),* and Aquinas taught that after the consecration the physical accidents existed with the divine substance but without inhering. This was accepted until the widespread rejection by the Reformers, who thought not at all of accidents but much more symbolically rather than literally about the substance itself. The Council of Trent continued the medieval teaching, though without mention of accidents, and this continues to be the Roman position. Reformed theology has never ceased to disagree fundamentally, though it has its differences in interpreting the nature of the actual presence.
C.G. THORNE, JR.

ACCIDIE (Gr. *akēdia,* "negligence"). Used in both the Septuagint (Isa. 61:3) and in Cicero (*Att.* xii:45) in its basic meaning, it later also became a technical term for the mental prostration, induced by fasting and bodily discipline, experienced by hermits, monks, and recluses. As such it is described by Cassian (360-435) in his *Institutes* and by medieval writers such as Aquinas. The word has, however, commonly been used to describe the fourth cardinal sin and is thus a synonym for sloth and sluggishness. As a description of a spiritual condition it was given a new lease of life in the English-speaking world by F. Francis, bishop of Oxford, because of his treatment of it in *The Spirit of Discipline* (1891).
PETER TOON

ACCOMMODATION. As a technical term this refers to the adaptation by God of the manner (including the language) in which revelation has taken place to suit the finitude, needs, and culture of men. In Calvin's phrase, God "lisps" His word to man. The logical and epistemological problems such a view raises are similar to those raised by any attempt to speak of divine reality in human language. God may also be said to adapt the content of His revelation to suit the historical progressiveness of His purposes. Regarding the limits of this process of accommodation, some argue that the Old Testament incorporates pagan ideas and mythical legends. It is also held by some that Christ in "emptying himself" of His glory (Phil. 2:17) took on the erroneous beliefs and thought patterns of His age on such matters as demon possession and the authority of the Old Testament. But it is hard to credit that, while fiercely condemning the Pharisees on so many matters, He should have left other errors untouched. If it is said He was unaware of the extent of His ignorance, this involves not accommodation but outright error, and is of course incompatible with His true deity. H.D. MC DONALD

ACOEMETAE (Gr. = "the sleepless ones"). A monastic group, founded by the abbot Alexander in Constantinople, which flourished in the Eastern Church in the middle of the fifth century. Their name was derived from the fact that in their monasteries the members were divided into choirs which engaged alternately in psalm-singing without intermission, day and night, the whole year round. Alexander met with opposition and was forced to flee from Constantinople, but later the Studite monastery, founded in Constantinople by Studius, a Roman consul, became an influential center of the Acoemetae. Possibly through the influence of Studius, the group had some imitators in the Western Church, and in the sixth century they were established in the abbey of St. Maurice of Agaune in Valois, by King Sigismund of Burgundy. Theologically their main contribution was their strong defense of the orthodox faith against the Monophysites, but this laid them open to a charge of Nestorianism,* for which they were excommunicated by Pope John II in 534, after which their influence became negligible.
HUGH J. BLAIR

ACOLYTES. First attested about 250 at Rome, these were listed after subdeacons among the clergy, and in Cyprian's *Epistles* in Africa as couriers for letters and gifts. Their Greek title ("attendants") indicates an origin in the Greek-speaking Roman Church, probably earlier in the century by devolution from the diaconate. For some time observed only in major churches, they later acquired eucharistic functions, especially as candlebearers. They became increasingly prominent in Rome from the seventh century as the chief of the four "minor orders"* of Latin clergy (officially under Innocent III in 1207), with special papal duties. Today their liturgical role is often performed by laity. In the East they appear only briefly as a distinct order (cf. Justinian, *Novels*, 59), except in the Armenian Church.
D.F. WRIGHT

ACONCIO, GIACOMO (1492-c.1566). Advocate of toleration. Known also as Jacobus Acontius and Jacopo Aconzio, he was an evangelical Catholic (following the teaching of Juan de Valdés) who later supported Protestantism. On the accession of the strict Pope Paul IV in 1557 he had fled to Basle and joined the circle of Castellio and Curione. In 1559 he went to England, was naturalized, and remained there until his death. For services as an ordinance officer he received a state pension. He attached himself to the "Strangers' Church" in London, but was excommunicated for defending Adrian Haemstede, pastor of the Dutch congregation, who had befriended Anabaptists. Aconcio's famous *Satanae Stratagemata* (1565) was a powerful plea for tolerance, stressing that religious strife was devil-inspired and that peaceful discussion was the best way to defeat Satan. He supported this by six articles of faith, imprecise but orthodox. J.G.G. NORMAN

ACRELIUS, ISRAEL (1714-1800). Lutheran clergyman and author. Born in Öster-Aker, Sweden, son of the local pastor, Acrelius studied at the University of Uppsala and was ordained in 1743. He served Swedish pastorates for six years, then sailed for missionary service among the Swedes on the Delaware River south of Philadelphia. Finding church discipline at Christina indecently lax, he put his church in order. He gave active help to the German Lutherans in Pennsylvania and during leisure hours gathered material for his *History of New Sweden* which, published in 1759, is the chief literary monument of the Swedes on the Delaware. Poor health forced his return to Sweden in 1756. BRUCE L. SHELLEY

ACTA SANCTORUM. The monumental collection of biographies and legends of the saints, organized according to the ecclesiastical calendar of saints' days, published by the Bollandists,* a group of Belgian Jesuit scholars named for John Bolland, under whom the first volume appeared in 1643. Work continued until interrupted in 1773 by the dissolution of the Society of Jesus and halted in 1796 by the French Revolution, not to be renewed until 1837. The critical attitudes of the Bollandists brought them under the censures of the Spanish Inquisition* in 1695 and caused some of their publications to be placed on the Index. The project is now complete through 10 November. Earlier sections have been supplemented in *Analecta Bollandiana*, a review devoted to hagiography published since 1882.
MARY E. ROGERS

ACTION FRANÇAISE. The name of both a political league and its journal, it was founded in 1898 in the aftermath of the Dreyfus Affair with the intention of restoring French national unity under monarchical rule. Hostile to parliamentary government, never a mass movement, it attracted many Roman Catholic students and intelligentsia. Its leaders, notably Charles Maurras, were atheistic and on a naturalist basis held that the national interest had absolute primacy in moral matters. They valued Catholicism for its social function, regardless of its religious truth. Roman Catholic authorities were soon worried by the movement. A prohibition of Catholic membership was prepared in 1914 but not published, for political reasons. Public condemnation in 1926 spelled the eventual end of Action Française's appeal to Catholics, though only at the cost of much conflict in the 1930s. HADDON WILLMER

ACTION SERMON. A term formerly used in Scottish Presbyterian churches to denote the sermon immediately preceding the sacrament of the Lord's Supper. It has been suggested that the term derived from *actio gratiarum*, the offering of thanks.

ACTS OF THE APOSTLES. The first history of the Christian Church, covering certain important phases of the first thirty years of its existence, was originally the second and concluding part of a record of Christian origins, the first part of which is the gospel according to Luke. The traditional titles borne by the two parts were given them in the church of the second century; originally the complete work may have been called "Luke to Theophilus." Luke,* to whom the twofold work is ascribed without dispute from the second century onwards, is probably identical with Paul's friend of that name, his "beloved physician" of Colossians 4:14. The sections of Acts which are narrated in the first person plural, the "we" passages (Acts 16:10-17; 20:5-21:18; 27:1-28:16), are best regarded as extracts from his travel diary on occasions when he was present at the events related.

Contents. The book may be divided into six parts, of which *Part 1* (1:1–5:42) deals with the beginnings of the church in Jerusalem. After the last appearance of the risen Christ, the disciples in that city waited for the fulfillment of His promise that the Holy Spirit would come upon them to empower them for their world-mission. This fulfillment was experienced on the Day of Pentecost, when, led by Peter, they bore such effective witness to the act of God in the resurrection of Christ that 3,000 believed and were baptized in Christ's name, thus forming the nucleus of the church. The life of this primitive community is then illustrated by a series of incidents showing its public triumphs and private problems.

Part 2 (6:1–9:31) tells how the peace of the community was shattered by the campaign of repression launched by the Jewish establishment against the Hellenistic members of the church in particular, after the capital conviction of Stephen, one of their leaders, before the Sanhedrin on a charge of blasphemy against the Temple. In its earliest days the Jerusalem church included Hellenists as well as Hebrews, and the Hellenists appear to have taken up a radical attitude towards the temple order and religious tradition in general, whereas the Hebrews, led by the apostles, were more conservative in these matters. It is significant that in the persecution which followed the stoning of Stephen the apostles were immune from molestation. The Hellenists in the church were forced to leave Jerusalem and Judea, and in their dispersion they spread the Gospel, not only among their fellow Jews, but among the Samaritans (8:4-25). In this work Philip, another leader of the Hellenists, played an important part. The prime agent in the campaign of repression was Saul of Tarsus, an alumnus of the school of Gamaliel; with his sudden conversion to faith in the risen Lord, who confronted him when he was in midcareer as a persecutor and called him into his service, the campaign quickly collapsed.

Part 3 (9:32–12:24) records the beginnings of Gentile evangelization. The first step in this was taken—hesitantly—by Peter, who accepted an invitation to visit the house of the Roman centurion Cornelius in Caesarea and explain the way of salvation. Peter, as he spoke, was presented with a divine *fait accompli* when the Holy Spirit came on his hearers; Cornelius and his family were baptized as those who, despite their being formerly outsiders, had now been manifestly admitted to the believing community. This fraternizing with Gentiles may have lost the apostles much of the general good will which they had formerly enjoyed in Jerusalem, which explains how Herod Agrippa I could attack them with popular approval (12:1ff.) and how James replaces them in the leadership of the mother church (12:17; 15:13; 21:18). Not long afterward the Gospel was presented to Gentiles on a much larger scale in Syrian Antioch,* by Hellenists who made their way there from the persecution in Jerusalem. Barnabas,* sent by the Jerusalem leaders to supervise this forward movement, secured the help of Saul of Tarsus in his work, and under their guidance the church of Antioch flourished. It gave evidence of its positive Christianity by sending a gift to the Jerusalem church in time of famine.

Part 4 (12:25–15:35) continues the record of Gentile evangelization. Antioch became a base for missionary outreach: Barnabas and Saul (to whom Luke henceforth refers mostly by his Roman cognomen Paul*) were released by the church there to carry the Gospel to Cyprus and Asia Minor. Churches were planted in the Phrygian and Lycaonian cities of Pisidian Antioch, Iconium, Lystra, and Derbe. The rapid increase of Gentile members in the church caused misgivings in Judea, but the Council of Jerusalem (15:6-29), at which Barnabas and Paul were present with a delegation from Syrian Antioch, turned down a proposal that Gentile converts should be circumcised and taught to keep the Jewish law, contenting itself with some simple provisions which made it easier for Jewish and Gentile Christians to live together as fellow members of one fellowship.

Part 5 (15:36–19:4) records a new advance in the Gentile mission—the evangelization of the provinces west and east of the Aegean—Macedonia, Achaia, and Asia. West of the Aegean, Paul's principal base was Corinth, where he spent eighteen months founding and consolidating one of his most important churches (18:1-18); east of the Aegean, his base was Ephesus where, during nearly three years, he and his colleagues planted Christianity not only in that city but throughout the province of Asia. In Corinth the proconsul Gallio's refusal to take up the charge that Paul was propagating an illegal religion must for several years thereafter have provided other Roman magistrates with a more powerful precedent than appears on the surface of Luke's narrative (18:12-17). In Ephesus the demonstration in the theater illustrates the sensitivity of property interests when threatened by the Gospel (19:23-41).

In 19:21 Luke records Paul's plan, conceived towards the end of his Ephesian ministry, to make for Rome after visiting Jerusalem. *Part 6* (20:1–28:31) tells how this plan was realized by means unforeseen by Paul—his rescue by Roman soldiers from a hostile mob in the temple court at Jerusalem, his two years' custody at Caesarea, his appearances before Felix, Festus, and Agrippa II, his appeal to Caesar, and his consequent voyage to Rome (in the course of which he and his company were shipwrecked off Malta). The book ends with his spending two years under house arrest in Rome while he waited for his appeal to be heard, preaching the Gospel unhindered to all who visited him.

Purpose and Perspective. In Paul's last words to the Roman Jews, "God's salvation has been sent to the Gentiles and they will listen!" (28:28), we have one dominant theme of Luke's narrative. He is concerned to trace the advance of the Gospel throughout the world as a further stage in the continuous history of salvation. He concentrates on its advance along the road from Jerusalem to Rome, ignoring its advance in other directions. The greater part of his account is bound up with Paul's missionary career, but reference to Paul's letters shows that there are several phases of Paul's career which are not recorded in Acts. Even so, our indebtedness to Luke may be measured by our difficulty in constructing a history of Christian advance in the subsequent generations, for which no such record as Acts is available. After Luke we have no church historian until Eusebius* in the fourth century.

Luke lays distinctive stress on the presence, power, and guidance of the Holy Spirit in the Church from His descent at Pentecost onward. The current age is the age of the Spirit. In this, as in his emphasis on salvation, with special reference to the Gentile mission, Luke shows himself a disciple of Paul. The Parousia, resurrection, and judgment are fixed by divine appointment, but they are not so imminent as to foreclose the irresistible progress of the Gospel in the world. Yet there is no hint that their "delay" was felt to be a problem.

Luke anticipates the later Apologists,* not only by presenting Christianity as the true knowledge of God, as in Paul's *Areopagitica* (Acts 17:22-31), but more particularly by arguing that Christianity is no menace to imperial law and order. It has come to stay, as a factor to be increasingly reckoned with in public life. Several responsible officials throughout the provinces give their witness to this effect, and the note on which Acts ends, with Paul discharging his missionary task at the heart of the empire without interference, clinches the argument. The rioting which follows the arrival of the Gospel in city after city is mostly instigated by local Jews who refuse it themselves and resent its being offered to the Gentiles. Like Paul, Luke holds that blindness has befallen Israel, but he does not add, like Paul, that this condition is partial and temporary.

The date of Acts cannot be determined with certainty. A date about the outbreak of the Jewish War of A.D. 66 has its attractions, and is probably not ruled out by the consideration that this is very early for Luke's historical perspective to have taken shape.

BIBLIOGRAPHY: F.F. Bruce, *The Acts of the Apostles* (1952); H.J. Cadbury, *The Book of Acts in History* (1955); *idem, The Making of Luke-Acts* (1958); M. Dibelius, *Studies in the Acts of the Apostles* (ET 1956); H. Conzelmann, *The Theology of Luke* (ET 1960); J. Dupont, *The Sources of Acts* (ET 1964); L.E. Keck and J.L. Martyn (eds.), *Studies in Luke-Acts* (1966); A. Ehrhardt, *The Acts of the Apostles* (1970); W.W. Gasque and R.P. Martin (eds.), *Apostolic History and the Gospel* (1970); I.H. Marshall, *Luke: Historian and Theologian* (1970); E. Haenchen, *The Acts of the Apostles* (ET 1971); W. Gasque, *A History of the Criticism of the Acts of the Apostles* (1975).

F.F. BRUCE

ACTS OF THE MARTYRS. Accounts of early Christian martyrdoms, divisible into several categories:

(1) Official transcripts of court proceedings (*acta* proper; *gesta*), deposited in archives: e.g., Justin, the Scillitans, Cyprian.

(2) Broader, more literary narratives compiled by Christians from the personal testimony of participants (e.g., Perpetua) or spectators (Gr. *martyria;* Lat. *passiones*): e.g., Polycarp, martyrs of Lyons and Vienne, Perpetua and Felicitas.

Both these genres were liable to interpolation (e.g., the Scillitans' *acta*), normally with embellishment of the miraculous, a feature not always absent from the original versions.

(3) Largely legendary stories constructed around a slender historical core, sometimes solely the martyr's name: e.g., Ignatius, Vincent, Lawrence, George, Catherine of Alexandria. As the golden age of martyrdom faded into the past, this literature multiplied and became the Christians' novels and romances. Their reflection of popular Christianity gives them a secondary historical value.

These various accounts served for apologia ("the blood of the martyrs is seed") and edification (the martyr's *imitatio Christi*) as well as commemoration (in the West, especially Africa, the acts were read liturgically on the martyr's "birthday," *natalitia*). There existed models both Jewish (e.g., *2 Maccabees*) and pagan (H.A. Musurillo, *The Acts of the Pagan Martyrs,* 1954).

BIBLIOGRAPHY: H. Delehaye, *Les Passions des Martyrs et les genres littéraires* (1921); *idem, Les Origines du culte des Martyrs* (2nd ed., 1933); H. von Campenhausen, *Die Idee des Martyriums in der Alten Kirche* (1936); H. Leclercq in *DACL* 1, pp. 373-446; J. Quasten, *Patrology* 1, pp. 176-85.

Texts: R. Knopf, G. Krüger, G. Ruhbach, *Ausgewählte Martyrerakten* (4th ed. 1965). With translation: H. Musurillo, *The Acts of the Christian Martyrs* (1972).

D.F. WRIGHT

ADALBERT OF BREMEN (c.1000-1072). Archbishop of Bremen from 1045. Of noble birth, able and ambitious, Adalbert was influential during the reign of Henry III. He consecrated bishops for the Orkneys and Iceland, sent a missionary to the Lapps, and divided the Slavic diocese of Oldenburg into three. Made papal vicar and legate of the Nordic nations in 1053, he attempted to create a patriarchate for the north, in a vain attempt to counter the desire of the church in Denmark for independence from Hamburg. After the death of Henry III in 1056, Adalbert had to face strong opposition from the Saxon dukes, and although he held a dominating position during the minority of Henry IV he was banished from court (1066-69) and never regained his former influence. Pagan Wends ravaged the Christian north during Adalbert's closing years, and their destruction of Hamburg prevented his being buried there.

HAROLD H. ROWDON

ADALBERT OF PRAGUE (956-997). Missionary and martyr. Born in Bohemia, he was educated in Germany by Adalbert of Magdeburg, taking his name at confirmation. His native name was Woytiech. He became the second bishop of Prague about 982, but incurred much opposition through his attempts at moral reformation. In 990, under political pressure, he went to Rome and became a Benedictine. Twice he returned to Prague, but finally left in 996 when nobles seized a prisoner in the cathedral and thus violated the law of sanctuary. He shared in the conversion of Hungary and is reputed to have baptized its great king, Stephen. At the request of Duke Boleslaus of Poland, he led missions to Pomerania and Danzig and possibly Russia. He was martyred by pagans along the Nogat River. He was buried at Gnesen in Poland and his shrine attracted many pilgrims, but after an expedition by Duke Bratislav of Bohemia in 1039 his bones were solemnly transferred to Prague. J.G.G. NORMAN

ADALBERT OF UTRECHT (fl. 700). Northumbrian-born missionary. Possibly the grandson of Oswald, king of Deira, he went to Ireland with Egbert, then joined Willibrord,* the famous Northumbrian missionary, and ten others on a mission to Friesland. After the conversion of the Frisians, he served as the first archdeacon of Utrecht. His name occurs among a list of preachers sent into western Germany and Kennemaria in 702 by order of the Council of Utrecht, with Egmont specially mentioned as the scene of his labors. The abbey of Egmont founded by the counts of Holland in his honor was completely destroyed by the Spaniards at the siege of Alkmaar in 1573. J.G.G. NORMAN

ADAMANTIUS (probably late third century). The unknown writer of a dialogue, *De Recta In Deum Fide* ("On the Correct Faith in God"). In the Greek version the author's name is given as "Adamantius who is also Origen." The Cappadocian Fathers mistakenly but understandably identified him with the great Alexandrian teacher, but this is impossible as Adamantius draws freely on Methodius who was one of Origen's chief opponents. The dialogue, an anti-Gnostic work in five books, deals with the heresies of Bardesanes* and Marcion.* The Latin translation is the work of Rufinus. Precise dating of the dialogue is impossible, but it can be no later than the end of the Great Persecution in 313. G.L. CAREY

ADAMITES. A term given to small groups on the fringes of Christianity at various periods of church history, who were said to have practiced nudity or indulged in sexual promiscuity. Epiphanius gives an account on hearsay of a sect of *Adamiani* who were said to worship naked and call their meetings paradise. Among later groups were a sect of Bohemian Taborites,* said to have committed wild excesses in nocturnal dances, who were suppressed in 1421, and an Anabaptist sect in the Netherlands about 1580, who were alleged to require candidates for admission to appear naked before the congregation. The followers of an Anabaptist called Adam Pastor were also given the name Adamites.

HAROLD H. ROWDON

ADAMNAN (624-704). Abbot of Iona. Born in County Donegal of the same royal blood as Columba,* he was educated at the monastic school at Clonard and converted during a visit to Northumbria about 676. Three years later he became the ninth abbot of Iona, which position he held to his death, proving himself to be the most distinguished incumbent since Columba. Adamnan wrote two books. His *De Locis Sanctis* is an account of a visit to the Holy Land and Constantinople about the year 690 by Arculf, a French bishop, and is of interest for the light it throws on similarities between monastic settlements in the Celtic* and Syriac Churches. His *Life of Saint Columba* is not a regular biography, but a compilation of the prophecies, miracles, and visions of the saint, interspersed with anecdotes that give us a fair picture of the man. Deeply concerned about the detachment of the Celtic Church from Rome, he persuaded the monks of the Irish Church to accept the Roman method of determining Easter, but could not convince his own monks to do so. A schism on this issue that occurred at his death was settled in 716 when Iona acknowledged the authority of Rome.

ADAM LOUGHRIDGE

ADAM OF MARSH (Adam de Marisco) (d.1258). English Franciscan monk. Born in Somerset, he studied under Robert Grossteste at Oxford, was ordained and received the living of Wearmouth, County Durham, and about 1232 became a Franciscan monk at Worcester. He accompanied Grosseteste, now bishop of Lincoln, to the Council of Lyons (1244-46), and thereafter taught at Oxford for some three years. Known in his lifetime as "Doctor Illustris," he had such learning and sagacity that his advice was sought by Henry III, Boniface of Savoy (archbishop of Canterbury), and the pope. In addition to some 250 letters, he is said to have been the author of commentaries on Canticles and Hebrews. J.D. DOUGLAS

ADAM OF ST.-VICTOR (c.1110-c.1180). Liturgical poet. Born probably in Britain or Brittany, he was educated in Paris and about 1130 entered the monastery of St.-Victor, where he stayed until his death. His theological ideas are Augustinian. He is chiefly renowned for his composition of about forty-five "sequences," rhythmic pieces to be used in the liturgy of the Mass preceding the Gospel. These were approved by the Fourth Lateran Council in 1215. He is credited with having brought to perfection the sequence poetry initiated by Notker Balbulus (d.912) and nurtured at St.-Victor even before Adam's time. He used ideas borrowed from contemporary theology, biblical allusions, and legendary material, and presented a highly developed system of allegory, typology, and symbolism. The melodies for his sequences were composed by a fellow monk, H. Spanke. The sequences have been translated into English, and some have been used as hymns (e.g., four in *Hymns Ancient and Modern*). He was thought to be the author of scholastic and biblical works, but this has been disputed.

See D.S. Wrangham, *The Liturgical Poetry of Adam of St.-Victor* (1881). J.G.G. NORMAN

ADAMS, THOMAS (d.1653). "The prose Shakespeare of puritan theologians." Little is known about his early life, but he appears as a preacher in London and the Home Counties from about 1612. He shows the typical Puritan pastoral interests, but no interest in sectarian or separatist Puritan groups. He wrote a commentary on 2 Peter and published sixty-five sermons and some meditations on the creed. His style is somewhat allegorical. Adams was probably a Royalist in the Civil War, which may explain his sequestration under the Commonwealth.

ADAMSON, PATRICK (1537-1592). Archbishop of St. Andrews. Born in Perth and graduate of St. Andrews, he was briefly minister of Ceres before going to the Continent for eight years. He studied theology in Geneva under Beza.* He was evidently in hiding for seven months after the massacre of St. Bartholomew's Day.* On returning to Scotland he became chaplain in the household of the Regent Morton, and minister of Paisley. The regent, his patron, made him archbishop of St. Andrews in 1576, though according to Robert Keith (*Scottish Bishops,* 1824, p.40) he was not consecrated. He served for a time also as James VI's ambassador to England. An accomplished scholar, he debated much with the Presbyterians, against whom he instituted strong legislation. He was accused of heresy and other offenses, but two attempts to excommunicate him failed. It is said that he recanted just before his death and condemned episcopacy as unlawful, but no evidence of this exists. It was Adamson who, before gaining high office, made the famous distinction between "my lord bishop," "my lord's bishop," and "the Lord's bishop." His collected works, including Latin verse translations of Job and Revelation, were printed in 1614. J.D. DOUGLAS

ADDAI. Founder of the church at Edessa. He was allegedly one of the seventy disciples of Jesus sent to King Abgar V Ukkama by the Apostle Thomas after the Ascension. Eusebius records the legend and calls Addai "Thaddaeus." The Syriac *Doctrina Addaei* was published in 1876. The conversion of King Abgar IX (179-216) points more reliably to evangelism at Edessa, possibly in the late second century, by a Jewish Christian Addai (Baus, *Handbook of Church History,* vol. 1, p.207).

See ABGAR, LEGEND OF.

ADDAI AND MARI, LITURGY OF. The extant manuscripts contain the ancient rite of the Syrian Church of Edessa, traditionally founded by the apostle Addai* and his disciple Mari. Nestorians claim that Theodore of Mopsuestia redacted the liturgy. Its Greek interpolations are not universally recognized by liturgical scholars. The Nestorian schism about 431 is a commonly held date of origin of the liturgy, but Jewish traits and its kinship with Hippolytus* suggest a date about 200. These traits are: the eucharistic prayer is addressed to the Son, not to the Father; there is no account of the institution of the Lord's Supper nor even of the eucharistic words; the invocation of the Spirit is not a prayer of "consecration" but a petition for blessings such as forgiveness and new life, as in the rite of Hippolytus; peculiarly Jewish is the stress on the Name and on divine activity in the doxology. G.T.D. ANGEL

ADDAMS, JANE (1860-1935). Social reformer and intellectual. Born in Cedarville, Illinois, she was greatly influenced by the moral earnestness of her Quaker father. She attended Women's Medical College in Philadelphia, but poor health forced her to discontinue medical studies. She traveled and studied in Europe, and after a long period of dissatisfaction and uncertainty returned in 1889 with her friend Ellen Gates Starr to establish Hull House, a settlement house in a poor immigrant neighborhood in Chicago, patterned after Toynbee Hall in London. She worked tirelessly to serve the needs of people, identifying and attempting solutions to numerous problems encountered by poor city dwellers. She contributed to efforts to secure adequate welfare legislation, mothers' pensions, juvenile courts, tenement house codes, workmen's compensation, and improved sanitation. Although she was not an orthodox Christian, she inspired people of all kinds to become concerned about the lot of the poor. She participated in the women's suffrage movement which was committed to pacifism as a way of life. She received the Nobel Peace Prize in 1931. Her books include *Twenty Years at Hull House* (1910) and *Democracy and Social Ethics* (1902).
HARRY SKILTON

ADDISON, JOSEPH (1672-1719). English essayist. Graduate of Oxford, he had a distinguished political career (being a member of Parliament for the last eleven years of his life), but is usually remembered as one of the founders of the *Spectator,* which set out to restore good sense and high standards to a society that had not yet recovered from the imbalance of the Restoration. Although no friend of Puritanism, he rated Milton above Homer and Virgil. To the Christian Church he bequeathed the magnificent hymns "When all thy mercies, O my God" and "The spacious firmament on high."

ADELARD OF BATH (twelfth century). English philosophical writer. Little is known of his career, but he evidently studied at French universities, traveled extensively in Europe, Africa, and the Near East, and wrote a book on Arabic science. One of the most versatile men of his age, he was distinguished in many fields, including philosophy, theology, natural sciences, and mathematics. Of his published works the best known are *De Eodem et Diverso (Identity and Difference),* and *Quaestiones naturales* in which he used Aristotle's argument from motion to establish the existence of God.

ADELOPHAGI (fourth century). A sect that adhered to the doctrine attributed to Macedonius of Constantinople: that the Holy Spirit is a created being, ranking with the angels. They were said by Philastrius, the antiheretical writer, to be unsocial at meals, hence their name which indicates that a Christian should eat in secret.

ADHEMAR LE PUY (de Monteil) (d.1098). French bishop. A nobleman who became bishop of Le Puy about 1080, he seems to have made a pilgrimage to the Holy Land about 1086-87. Thus, when Urban II proclaimed the First Crusade* at Clermont in 1095, Adhemar was named as the pope's legate and deputy with the crusading armies. His functions included giving military advice, mediating between the rival Christian princes in the crusading force, and establishing harmonious relations between the Crusaders and the Eastern Christians in the territories conquered by the Crusaders. He also tried to integrate Eastern and Western clergy into a single church in the Holy Land. ROBERT G. CLOUSE

ADHERENTS. Used in the Church of Scotland, the term covers parishioners who wish to be permanently connected with a particular congregation, and in the case of whom there would be no reason for refusing to admit them to communicant membership if they should apply for such. Adherents, who must be over twenty-one and not in membership of any other church or congregation, may on formal application be added to the electoral register and thus be permitted to participate and to vote in congregational meetings. In the Highlands and the Western Isles many adherents, including those of smaller Presbyterian bodies, maintain this status because of a feeling of unworthiness to come to the Communion table.
J.D. DOUGLAS

ADIAPHORISTS (Gr. *adiaphora*, "things indifferent"). Those who supported Philip Melanchthon* when, on the ground of necessity, he maintained that concessions made by Protestants in the Leipzig Interim (1548-52) were in any case *adiaphora.* The terms included the necessity of good works, and the restoration of the Mass with most of its ceremonies. Many Protestants, such as Illyricus Flacius* and John Calvin,* believed that Melanchthon had sacrificed too much. The *adiaphora* argument has recurred often in Christian thought. It can concern actions that are indifferent (neither bad nor good, being neither commanded nor forbidden by God), ceremonies (neither forbidden nor commanded so they may be used or discarded), and doctrines (although taught in the Word of God, they are of such minor importance that they may be disbelieved without injury to the faith). The Interim was never in fact implemented, but it caused bitter controversy. The Peace of Augsburg* (1555) removed the occasion of the argument. ROBERT G. CLOUSE

ADMONITION TO PARLIAMENT (1572). An anonymous English tract, secretly printed, which probably represents an extreme reaction to Archbishop Parker's attempts to secure conformity and Queen Elizabeth I's suppression of parliamentary attempts to reform the Prayer Book. The first part was written by Thomas Wilcox in clear, concise style, with many biblical references and without abuse. It asserted that the church had not yet been fully reformed since the biblical model had not been followed in such matters as the appointment of elders and maintenance of discipline. The second part, written in cutting phrases by John Field, objected to papal associations in the Prayer Book and vestments. The *Admonition* quickly ran to three editions, but its authors were apprehended and imprisoned for a year. Later in 1572 a second *Admonition* appeared, possibly written by Thomas Cartwright. This outlined a Presbyterian structure for the church.
HAROLD H. ROWDON

ADO (799?-876?). Archbishop of Vienne. Successively a monk at Ferrières, head of the monastic school at Prum, and priest of a parish church at Lyons, he became archbishop of Vienne in 859. Charles the Bald sent him to the pope to argue (successfully) the cause of Theutberga, after a compliant Synod of Metz had allowed her husband, Louis II of Lorraine, to marry his mistress. So that Vienne could boast apostolic foundation, Ado claimed spuriously that Vienne's first bishop, Crescens (died first century), was the same Crescens Paul had sent—but not, he argued, to Galatia (2 Tim. 4:10), but to Gaul. Clerical reformer, author of saints' lives and a chronicle of world history, Ado is best known for his martyrology *Passionum Codices Undecumque Collecti* (c.858), a model for later martyrologies, especially that by his contemporary Usuard.* The Roman martyrology also has certain links with Ado's, which was mostly copies from Florus of Lyons and partly from an ancient martyrology he claimed to have found at Ravenna *(Martyrologium Romanum Parvum),* though modern authorities believe he wrote it himself. L. FEEHAN

ADOPTIANISM. The view that Jesus was a man of blameless life who became the adoptive Son of God. In the early centuries some maintained that the divine Spirit descended upon Jesus—a man of perfect virtue, sometimes granted to have been born of a virgin—at His baptism, and that He was deified after His resurrection. A form of Adoptianist theology was expounded by Dynamic Monarchians, e.g., Theodotus and Paul of Samosata. The Antiochene* School, particularly Theodore of Mopsuestia and Nestorius, expressed themselves in ways which appear Adoptianist, though their language is insufficiently precise to make this certain. Adoptianism was canvassed in Spain during the eighth century, possibly due to the influence of Latin translations of works of Theodore of Mopsuestia. Alcuin attributed it to the influence of Nestorian writings, and Leo III suggested that it was derived from contact with the Muslims. It may have originated out of the Arianism* from which Spain had been converted at the Third Council of Toledo* (589). The Spanish liturgy still spoke of "the man who was assumed." The immediate cause seems to have been reaction against the teaching of Migetius,* who held that Jesus was one of the divine persons of the Trinity. Elipandus* of Toledo reacted by drawing a sharp distinction between the second person of the Trinity and the human nature of Christ. The Logos, eternal Son of God, had adopted the humanity—not the person—with the result that Jesus in his human nature became the adoptive Son of God. Such views met with opposition in Spain, particularly from the Asturian monks Beatus and Etherius, whose appeal to Rome led to the doctrine's condemnation by Adrian I.

Support for Elipandus came from other Spanish bishops, particularly Felix of Urgel, who defended Adoptianism at the Council of Regensburg (792) but signed an orthodox confession at Rome (which he subsequently repudiated). The Council of Frankfurt (794) accepted two memoranda drawn up by Italian and Frankish bishops, and the English scholar Alcuin* wrote several treatises against him. In 798 Felix agreed to meet his opponents, including Alcuin, in the presence of Charlemagne at Aix-la-Chapelle. He eventually acknowledged defeat and was received back into the church, but not to his bishopric. Whether he fully repudiated Adoptianism is doubtful. A Roman council under Leo III reiterated the orthodox view and anathematized Adoptianism. The latter died out in Spain, but the Scholastic theology (see SCHOLASTICISM) of men like Abelard and Peter Lombard caused them to draw a distinction between the two natures in order to safeguard the immutability of God. This involved them in a position rather like Adoptianism.

HAROLD H. ROWDON

ADRIAN I (d.795). Pope from 772. A Roman of noble birth, he was an able diplomat, administrator, and builder. Following Stephen III's policy, he aligned himself with Charlemagne.* By persuading Charlemagne to conquer the Lombards and depose their king, Desiderius, he freed the papacy from a formidable threat. He also secured Charlemagne's help against Adoptianism. Adrian supplied the emperor with the Dionysio-Hadriana collection and a copy of the Gregorian sacramentary which were to provide the basis for Western church law and worship. Charlemagne failed to implement his promise to grant to the pope all the territory named in the Donation of Pepin, occasionally intervened in purely ecclesiastical matters, and at Frankfurt (794), apparently due to a misunderstanding, condemned a decision of the Council of Nicea (787). It was probably Adrian who first used the years of the pontificate and the pope's name and image on papal documents and Roman coins.

HAROLD H. ROWDON

ADRIAN IV (Nicholas Breakspear) (c.1100-1159). Pope from 1154. The only English pope, he was born on the estates of St. Alban's, studied at Paris and Arles, and entered the house of secular canons of St. Rufus near Avignon where he became abbot in 1137. A visit to Rome in connection with a dispute with his canons brought him to the notice of Eugenius III. He was made cardinal bishop of Albana and was sent as papal legate to Norway and Sweden (1152). He reformed the rudimentary canon law of the Scandinavian churches, made Trondheim a metropolitan bishopric for Norway, created a new bishopric, Hamar, and introduced the payment of Peter's Pence.* As pope he was confronted by the reforming zeal of Arnold of Brescia and the relentless enmity of Frederick Barbarossa. The former combined moral indignation against clerical abuses with advocacy of a Roman republic independent of the pope. Adrian banished Arnold from Rome, and subsequently secured his repudiation by Frederick and execution (1155). He withstood Frederick, exacting full homage before consenting to crown him as emperor, and insisting that his crown was a *beneficium* held from the pope. In similar vein the Benevento Treaty (1156), which recognized the territorial rights of the Sicilian kingdom, was granted on condition that William of Sicily did homage to the pope. The claim that Adrian granted overlordship of Ireland to the English king, Henry II, is based on the bull *Laudabiliter,* which may be a forgery.

See E.M. Almedingen, *The English Pope* (1925).

HAROLD H. ROWDON

ADRIAN VI (1459-1523). Pope from 1522. Born in Utrecht, son of a ship's carpenter, he was educated by the Brethren of the Common Life* and at Louvain, where he became a teacher (1488) and a doctor of theology (1492). He published several theological works. In 1507 he became tutor to the future Charles V. Sent to Spain to prepare the way for Charles's succession, he became bishop of Tortosa (1516), Inquisitor of Spain and cardinal (1517), and regent of Spain (1520). As pope he attempted to reform the Roman Curia, unite the powers of Europe in defense of Christendom against the Turks, and combine the crushing of Luther with the reform of the church. His attempts at reform were blocked by inertia and vested interest. Despite strenuous efforts, Rhodes (which he regarded as strategically vital) fell to the Turks in 1522. Although he at first instructed his legate at the Diet of Nuremberg to adopt a conciliatory stance, and expressed willingness to modify the theory of indulgences, his resolute opposition to the teachings of Luther—in 1521 he had demanded the suppression of Lutheran literature in Spain—made compromise impossible.

HAROLD H. ROWDON

ADVENT. The liturgical season of preparation for Christmas (i.e., advent of Christ into the world). It thus marks the start of the Christian Year,* and serves also to complete the cycle by drawing attention to Christ's "second coming to judge the world." Historically this season followed the development of the parallel one of preparation for Easter, known as Lent. Accordingly Advent grew in the East, and in parts of western Europe, to a full six weeks in length. In the late fifth century in Gaul a fast began on St. Martin's Day (11 November), but there is no evidence for Advent at Rome until the time of Gregory the Great,* a century later, when it contained only four weeks; nor was the period regarded as a time of fasting there. In fact, the origin of the fast in the East seems to have been (as in Lent) a preparation for baptism, in this case at Epiphany* (6 January). There is strictly no liturgical season of Advent in the East (i.e., allusion to the Advent theme in prayers and readings). The more limited Roman season gradually prevailed in the West, where Advent Sunday is now always the one nearest to St. Andrew's Day (30 November).

JOHN TILLER

ADVENTISTS. Christian groups holding as a distinctive tenet the expectation of an imminent and literal Second Advent of Christ. Adventism has been found in all periods of church history; early noteworthy proponents include Polycarp, Ignatius, Papias, Hermas, Justin Martyr, and Montanists. Reaction against Montanism and Donatism led to a lull until about 1000, when interest revived. Exponents in later centuries include Joachim of Floris, the Hussite Taborites, and the Anabaptists. In Britain, John Napier, Joseph Mede, Isaac Newton, and in Germany, the Pietists, Campegius Vitringa, and J.A. Bengel held adventist views, as did eighteenth- and nineteenth-century sects such as the Ronsdorf sect, the Shakers, Irvingites, and Mormons. A well-defined interchurch movement in the USA came into being with William Miller,* who taught that Christ would return in 1843 and 1844. When the predictions failed, Miller frankly admitted his error and gave up the movement. Dissensions arose, leading to the formation of the Seventh-day Adventists,* and a later splinter group, the "Church of God (Abrahamic Faith)." Another Adventist group, the Advent Christian Church, was organized about 1855. Smaller bodies include the Life and Advent Union, the Primitive Advent Christian Church, and the United Seventh-day Brethren. Adventists in the strict sense of the definition are found in most denominations, and there are also interchurch organizations such as the Prophetic Witness Movement International.

J.G.G. NORMAN

ADVOCATUS DIABOLI, see DEVIL'S ADVOCATE

AELFRIC (c.955-c.1020). Abbot of Eynsham and creator of a new level of grammatical anglo-Saxon prose. A product of the monastic revival in England, he was first a disciple at Winchester of the famous Ethelwold, whose life he later wrote. In 987 he transferred to the new foundation of Cerne Abbas and there published his most influential prose works in English, including two sets of homilies (c.991) and his *Lives of the Saints* (before 998). In 1005 he became abbot of Eynsham in Oxfordshire. Apart from his importance for English literature, he made a notable contribution to the education of the rural clergy of his own time; and he also received fresh attention for his theological views in the sixteenth century, when the Reformers, in their search for an ancient English church free from the errors of Rome, noticed that his teaching on the Eucharist, probably influenced by Ratramnus, excluded transubstantiation. Archbishop Parker published Aelfric's "Paschal Homily" in 1567.

JOHN TILLER

AELRED, see AILRED

AERIUS (fourth century). Companion of Eustathius* in pioneering monasticism in Pontus, he was appointed presbyter to supervise a hostel for the poor when Eustathius became bishop of Sebaste in Armenia Minor about 355, but, thus frustrated in his own episcopal ambitions (according to Epiphanius, *Heresies* 75), quickly assumed hostility towards Eustathius. He abandoned the hostel about 360, and his propaganda campaign issued in ostracism and heretical condemnation for himself and numerous followers. He was still active when Epiphanius wrote about 375. Viewed in later controversy as anticipating Reformation protests, he advocated the parity of bishops and presbyters, and the rejection of Easter, of prescribed fasts as relics of Jewish bondage, and of prayers and almsgiving for the dead as futile and detrimental to sanctification during life. Like Eustathius, he was Arian in inclination, but was not condemned as such. Philaster (*Heresies* 72) wrongly confuses the Aerians with the Encratites.*

D.F. WRIGHT

AETHERIA, see PILGRIMAGE OF ETHERIA

AETIUS (d. c.367). Intellectual leader of Anomoean* heretics. A Christian tutor to Julian, later the "open-minded" emperor, Aetius was exiled in 358 by Constantius II for having unwelcome contacts at court and for opposition to the *Homoiousion* doctrine. In 360 he was banished for denying that God and the Son are in any way alike. Recalled in 362 by Julian, he helped the Anomoeans form a distinct group. In his *Syntagmation* Aetius teaches that the uncreated God and his offspring can be neither identical nor alike. Logic forbids someone transcendent submitting to change or being located in time. Nothing can be two things in essence. God is entirely uncaused, absolute, and indivisible; the Son is entirely caused and relative. They must be unlike one another *(anomoios).*

G.T.D. ANGEL

AFFIRMATION. The right to affirm is provided for those giving evidence in English civil courts who, with or without religious faith, object on conscientious grounds to taking an oath. Affirmation is distinguishable from oath-taking in that no penalty can be invoked for false witness. The practice probably arose originally because of the scruples of witnesses in court who were afraid to invoke the name of God (as in oath-taking) in case false evidence should be given unconsciously.

Historically, affirmation is associated particularly with the Quakers,* who were persecuted in the seventeenth century for refusing to take oaths. By an Act of 1696 they were eventually allowed to affirm instead, except in criminal proceedings; but even this restriction was in due course removed. The Common Law Procedure Act of 1854 extended the right of affirmation to any who for conscientious reasons objected to being sworn; but this option did not explicitly include atheists until the Oaths Act of 1888.

Affirmation in this sense is to be distinguished from the (Auburn) Affirmation in America (1924), the Presbyterian document pleading for a liberal statement of Christian truth.

STEPHEN S. SMALLEY

AFFUSION, see BAPTISM

AFRICA, see SOUTHERN/WEST/NORTH/EAST AFRICA; also ZAIRE; ETHIOPIA

AFRICA, ROMAN. Obscurity shrouds Christian beginnings in Africa, i.e., the Roman provinces of Proconsular Africa, Numidia, and Mauretania (with later subdivisions), extending today from Tripolitania in Libya through Tunisia and Algeria into Morocco. Carthage, second only to Rome in the West and seat of a strong Jewish population, must quickly have attracted missionaries. They may have been Jewish Christians, whether from Palestine via the Jewish colonies of Cyrenaica, Asia Minor, or Rome (alternatively, an African group may early have influenced Roman Christianity; Pope Victor was an African); but Christianity did not prosper there until the later second century. The story begins in martyrdom, with the Scillitans in 180 at Carthage. Most if not all of the NT was already in Latin. Christianity also took root in Greek-speaking communities (perhaps from the first), but there is scant evidence that its coming rejuvenated native Punic or Berber cultures.

Around the turn of the second/third centuries, the *Passion of Perpetua* and the works of Tertullian* and Minucius Felix* (perhaps later third century) portray a thriving church reaching well to the south and west of Carthage and the more Romanized towns and littoral. The Latin Bible was now complete, close ties bound Africa to Rome, and some seventy bishops attended the Carthage council about 220 which under Agrippinus (the first known bishop of Carthage, unless the Optatus in the *Passion of Perpetua* was a predecessor) decreed the rebaptism of heretics. African Christianity owed its distinctive features largely to Tertullian, who created an African Latin Christian culture out of the raw materials of the Greek tradition. It became a religion of the sacred Book, interpreted literally, and eschewed profane literature. Apocalyptic and "enthusiastic" in tone and uncompromising in its defiance of the rulers of this world, it prized the martyr and confessor as the true Christians, possessed of and by the Spirit. Its church was the elect remnant; its holiness, in Judaistic fashion, legal; and its discipline severe. Tertullian attests the presence of Gnostic groups and was himself the channel of a significant Montanist influence.

In Cyprian* this rigorist Christianity was wedded to the episcopate, which in his hands through frequent councils exercised a collegial authority as yet unparalleled elsewhere. Provincial organization was well developed; in Proconsular Africa the primate was the bishop of Carthage, in other provinces the senior bishop. The church suffered severely in the persecutions of Decius* and Valerian,* in whose reign the martyrs of the Massa Candida at Utica probably died. Cyprian's handling of the ensuing controversies—over the lapsed, schism, and rebaptism—demonstrated a typically African combination of deference and ultimate independence towards the Roman see.

Late in the century Manichaeism reached Africa where Augustine,* its most famous adherent, was to crown its successes, despite proscription from 297 onwards. Diocletian's Great Persecution was foreshadowed by military martyrdoms (295-c.300) in North Africa, where alone in the West it hit really hard, the victims including all forty-seven Christians of Abitina, SW of Carthage. Two important writers were produced by Africa in these years, Arnobius the Elder* and Lactantius.* The disruption of persecution gave birth to the Donatist schism which for over a century divided African Christianity in its characteristic preoccupation with ecclesiology and baptism. The Donatist appeal to earlier African tradition in support of a rigorist purity was countered by the inclusivist vision of the church of the empire developed by Optatus of Milevis* and Augustine, but the eclipse of Donatism through the leadership of Aurelius and Augustine early in the fifth century after its dominance in the fourth was achieved only with imperial assistance. Against the Pelagians the African bishops secured Roman support on their own terms, but soon rejected anew Rome's claims of superior jurisdiction in the case of Apiarius.* Augustine's contributions to African Christianity were manifold, e.g., the propagation of monasticism and the supply of bishops from his own community of the "servants of God" in Hippo.

The century of Vandal rule in Africa (429-533) brought renewed persecution under three kings, Geiseric (429-77), Huneric (477-84), and Thrasamund (496-523). To the Arian Vandals at war with the empire the Catholics appeared like a Roman fifth column. Organized church life was overthrown, monasteries dissolved, bishops exiled (among them Quodvultdeus of Carthage [d. c.453], a writer and preacher of some note), new appointments barred, and Arian rebaptism imposed. In 484, at Huneric's conference of Arians and Catholics at Carthage, 466 Catholic bishops were present. But there were interludes of peace, especially under Gunthamund (484-96) and Hilderic (523-30). In 525 an all-African council could meet in Carthage, and Catholics and Donatists, probably treated alike by the Vandals, learned to live together, with Moorish pressure from the south an added incentive. Writers of this era mostly dealt with Arianism.* They included Victor of Vita, Vigilius of Thapsus, Fulgentius, and Christian Africa's only poet of merit, Dracontius of Carthage (fl. under Gunthamund).

The Byzantine reconquest in 533 enabled some recovery of earlier vigor but hardly of peace. The Moors were repeatedly disruptive, and the bishops and monks who defended the Three Chapters* against the condemnations of Justinian, the Council of Constantinople, and Pope Vigilius, whom they excommunicated, experienced signal imperial repression for thus maintaining Africa's independence vis-à-vis the imperial and papal authorities. Writers favoring the Three Chapters included Ferrandus (d.546/7)—a Carthaginian deacon, biographer of Fulgentius, and canonist—and Facundus of Hermiane, while Primasius of Hadrumetum opposed them. Gregory the Great found frequent occasion to rebuke African bishops, especially for tolerating a resurgent Donatism,* now perhaps scarcely distinguishable from Catholicism, but they naturally thwarted attempts at more direct papal control.

In the seventh century, Monothelitism* was rejected by African churchmen (except briefly Fortunatus, bishop of Carthage about 639-46), and again for a time the imperial will was resisted.

The Muslim Saracens' conquest of Africa, begun in 642/3 and completed by 709 with Carthage falling in 698, meant flight, slavery, or apostasy for many Christians and the reduction of Africa's bishoprics to three by the mid-eleventh century and to none by the thirteenth. The Turkish conquest of the late sixteenth century effaced the last vestiges of a Christian presence in North Africa, except, that is, for the rich archaeological remains which have so illuminated its church history.

BIBLIOGRAPHY: L.R. Holme, *The Extinction of the Christian Churches in North Africa* (1898); P. Monceaux, *Histoire Littéraire de l'Afrique Chrétienne* (7 vols., 1901-1923); A. Harnack, *The Expansion of Christianity,* vol. 2 (1904), pp. 411-35; H. Leclercq, *L'Afrique Chrétienne* (2 vols., 1904), and in *DACL* 1, pp. 575-775; E. Buonaiuti, *Il Cristianesimo nell' Africa Romana* (1928); J.J. Gavignan, *De Vita Monastica in Africa Septentrionali* (1962); A. Audollent in *DHGE* 1, pp. 705-861.

D.F. WRIGHT

AFRICANUS, SEXTUS, JULIUS, see JULIUS AFRICANUS, SEXTUS

AGAPE (Gr. = "love"; technically "love feast" in Jude 12). The communal religious meal or "love feast" of the early church, closely associated with the Lord's Supper. The fullest account occurs in 1 Corinthians 11:17-34. Normally held in the afternoon, rich and poor met together for the occasion. Ignatius and the *Didache** join Agape and Eucharist together, but during the second century they were separated, as Pliny apparently implies (*Epp.* 10:96). Tertullian linked the Agape with monetary contributions for poor relief (*Apology* 39) and speaks of the Eucharist as celebrated before daylight (*De Corona Militis* 3). Clement of Alexandria still associated the two, held in the evening, publicly at church and privately at home (*Paedagogus* 2). To Augustine it was just a charity supper, and it fell into disuse. The Trullan Council (692) excommunicated those holding love feasts in churches. It has persisted in sections of the Orthodox Church, in the Mar Thoma Church in India, in the Unitas Fratrum and the Moravians (whence John Wesley introduced it to Methodism), and among small groups like the "Peculiar People."

J.G.G. NORMAN

AGAPEMONISM. A religious movement founded by Henry James Prince (1811-99), an evangelical perfectionist. Ordained in 1840, Prince became a curate first in the Bath and Wells diocese and later in the diocese of Ely. Both bishops inhibited him. It was probably in 1843 that he began to make extravagant statements which gave the impression that he was claiming to be in some sense an incarnation of God. A community was formed at Spaxton where a magnificent residence was acquired and called Agapemone (Abode of Love). Prince declared that community of goods was binding upon believers, and numerous devotees handed over their property to him. The legal case *Nottidge* v. *Prince* revealed grave disorders, and the movement was generally discredited, though Prince and a number of followers continued to live in the Agapemone. In the 1890s the movement enjoyed a revival under J.H. Smyth-Pigott, formerly a curate of St. Jude's, Mildmay Park. Calling themselves "Children of the Resurrection," his followers built a meeting place known as the "Ark of the Resurrection." In 1902 Smyth-Pigott proclaimed himself to be Jesus Christ, and the movement lost its vogue. Some of Prince's writings breathe a spirit of devotion to Christ, but they are marred by an erotic element. Regarding himself and Samuel Starky, his former Somerset rector, as the two witnesses of Revelation 11, Prince proclaimed the doom of Christendom, for example in *The Council of God in Judgment.*

HAROLD H. ROWDON

AGAPETAE (Gr. = "beloved") or **Syneisaktoi** ("brought in together"; Lat. *subintroductae*), women who lived with men (or men with women) under vows of continence in "spiritual marriages" (which otherwise resulted from the "conversion" of ordinary marriages). First condemned, mostly for its actual or potential abuse, about 250-60 in Syria and Africa, and regularly in canons dealing with clergy from the early fourth century, the practice remained widespread and was repeatedly censured, notably by Chrysostom. Its original inspiration was union in asceticism,* but the setting determined the form: the anchorite and his female servant; a wealthy lady keeping a steward-cum-chaplain; the cleric and his housekeeper (especially as clerical celibacy became normative). It was common in monastic circles, even on a communal basis (with a precedent among the Jewish Therapeutae described by Philo). Its prevalence prior to the third century is variously estimated. Many discern it in 1 Corinthians 7:36-38 (cf. NEB) and in Hermas's symbolism. The first unambiguous reference is Irenaeus's rebuke of the Valentinians (*Against Heresies* 1:6:3). The Montanist Tertullian favored it.

D.F. WRIGHT

AGAPETUS (d.536). Bishop of Rome from 535. A Roman by birth, he succeeded John II as pope in June 535. One of his first acts was to repeal the anathemas of Boniface II against Dioscorus* (530). The following year he visited Constantinople, arriving there in February 536. His mission was to avert the war threatened by Justianian against Theodahad, and although he failed in this he was successful in another respect. In 535 Anthimus, a Monophysite, had been appointed patriarch of Constantinople through the influence of Theodora. Agapetus refused to acknowledge him, demanding proof of the emperor's orthodoxy and denouncing Anthimus's heresy. Anthimus was deposed and Mennas became patriarch. Agapetus followed his success against Anthimus by denouncing the other heretics who had come to Constantinople through Theodora's influence. He died in Constantinople on 21 April and was buried at St. Peter's, Rome, on 17 September.

DAVID JOHN WILLIAMS

AGATHANGELUS. Author of the first two books of the *History of Armenia.* He claims to have been the royal secretary of King Trdat (or Tiridates) II (289-317), but the legendary nature of much of his work suggests a date about the late fifth century. The *History* begins with the official conversion of Armenia to Christianity through the missionary work of Gregory the Illuminator.* Trdat, his persecutor, was converted and empowered Gregory to organize Christianity as the national religion of Armenia. Pagan temples and priests were replaced by Christian churches and clergy.

AGATHO (c.576-681). Pope from 678. At the invitation of Constantine IV, who hoped to effect a reconciliation between Western and Eastern churches, Agatho sent legates to the sixth ecumenical council* at Constantinople in 680. The legates read Agatho's letter in which he declared the inerrancy of the Roman Church and held that the Roman bishops had always strengthened their brethren in terms of Christ's injunction to Peter (Luke 22:32). He then set forth the orthodox doctrine of the "two wills." The letter was accepted with acclamation by all present, including the emperor, but unfortunately for Agatho's interpretation of Luke 22:32 the council condemned as a heretic one of his predecessors, Honorius (d.638).

DAVID JOHN WILLIAMS

AGDE, COUNCIL OF (506). A synod held at Agde, S France, under the presidency of Caesarius of Arles. Thirty-five bishops were present. Of the seventy-one canons published, only forty-seven were regarded as genuine by Sirmond. They deal, *inter alia,* with clerical celibacy, the canonical age of ordination (twenty-five for deacons, thirty for priests or bishops; nuns must be forty before receiving the veil), relations of bishops with their diocesan synods, public peace, and the religious obligations of the faithful. No bishop shall alienate buildings, slaves, or furniture belonging to the church; if it is absolutely necessary, the consent of two or three neighboring bishops is required. If a bishop grants liberty and property to slaves, his successors must respect the act. A clergyman becoming intoxicated is excommunicated for thirty days or corporally chastened. J.G.G. NORMAN

AGGREY, JAMES E.K. (1875-1927). African orator and educator. Born at Anamabu, Ghana, he attended as a young Christian convert a Methodist mission school at Cape Coast and himself taught school until he went to the USA in 1898 to study at Livingstone College, Salisbury, North Carolina, from which he graduated with honors in 1902. He taught there for several years and served two small rural churches. He twice toured Africa with the Phelps-Stokes African Education Commission. In 1924 he became vice-principal of the new Achimota College in Ghana (then the Gold Coast). He died while in the USA to write his doctoral dissertation. Aggrey was a gifted Christian educator, an advocate of cooperation between the races, and a mediator between African and Western cultures.

ALBERT H. FREUNDT, JR.

AGLIPAY CRUZ Y LABAYAN, GREGORIO (1860-1940). First bishop of the Philippine Independent Church. A Catholic priest in Manila, he accepted in 1898 a post as military chaplain of the revolutionary army of General Aguinaldo. He was the only priest-member of the revolutionary congress. The latter appointed him vicar-general, for which he was later excommunicated. He convened the Paniqui Assembly of clergy which set up a provisional government for the church until such time as the pope would name Filipino bishops. When Rome refused, the Philippine Independent Church was formed and Aglipay was consecrated supreme bishop. A militant nationalist, he ran unsuccessfully for the presidency of the Commonwealth of the Philippines in 1935. He remained Supreme Bishop of the "Aglipayan" Church until his death. RICHARD DOWSETT

AGNELLUS OF PISA (1194-1236). Franciscan provincial minister. Recruited personally by Francis of Assisi in 1211, he became guardian at the Paris house till 1224. Then, as first English provincial with eight brethren, he introduced the order into England. After a brief stay at Canterbury, they rented a house in Cornhill, London. Ordained priest in 1229, Agnellus exhibited tremendous humility and personal charm. He adhered strictly to the Rule's ideals of poverty and its rigid observance, but he was one of the group of officials whose policies quickly transformed the order from the original absolute simplicity of Francis's ideals through their enthusiasm for learning and papal privileges. He established a school at Oxford with Robert Grosseteste as lecturer about 1229 (the precursor of the university) and successfully defended his order's rights against the bishops in 1231. L. FEEHAN

AGNES. Roman Christian martyr. She was renowned for the heroic defense of her chastity, probably in the persecution by Maximian in the West (304-5). Constantina, daughter of the Emperor Constantine, built a basilica about 350 over her reputed grave near the Via Nomentana. She is depicted on several cups with eyes uplifted and arms raised to God in prayer. Fiction is mixed with fact in her *Passio,* falsely attributed to Ambrose of Milan.

AGNOETAE (Gr. *agnoeō,* "to be ignorant"). Monophysite sect. The Monophysite* issue in the Eastern Church blossomed forth following the Council of Chalcedon* (451). Even though the Monophysites were agreed that Christ had only one nature and not two, they themselves divided internally into factional parties. The two most important were the Severians—from Severus, patriarch of Antioch—and the Julianists—from Julian of Halicarnassus. These split further into subdivisions. The Severians divided into the Agnoetae, sometimes called the Themistians, and the Theodosians. The Themistians—from Themistius, deacon of Alexandria—were known as Agnoetae because of their particular emphasis about Christ's nature: Jesus Christ as a man was not totally omniscient, but shared man's ignorance of many things. The strict Monophysites would naturally reject this view.

GEORGE GIACUMAKIS, JR.

AGNOSTICISM. The term (derived from Acts 17:23, "to the unknown god") was introduced by T.H. Huxley in 1869 to denote the doctrine that man does not and cannot know whether God exists. He based his case on Locke's maxim that man ought not to accept "propositions with more certainty than the evidence warrants," and this dogma in opposition to faith characterizes agnosticism. The term is now used in several senses in addition to the above: (1) as the view that we should suspend judgment on all ultimate issues, such as God, free will, immortality; (2) to describe a secular attitude to life, i.e., that God is irrelevant to modern man; (3) for an emotionally tinged anti-Christian and anticlerical attitude; (4) as a synonym for atheism.

The modern agnostic often argues on the basis of the open-mindedness required in science, but if we cannot in science know the answer until we have collected evidence, how can we expect to know the answer beforehand in religion? The basic principle of agnosticism is belied by what Koestler has called "the act of creation": a study of creativity and invention in the artistic, literary, and scientific fields proves that confidence in the truth of propositions beyond what is intellectually justified by available facts is usually essential to success.

BIBLIOGRAPHY: T.H. Huxley, *Science and the Christian Tradition* (1900); A.O.J. Cockshut, *The Unbelievers* (1946); A.W. Brown, *The Metaphysical Society* (1947); B. Ghiselin (ed.), *The Creative Process* (1952); A. Koestler, *The Act of Creation* (1966).

R.E.D. CLARK

AGNUS DEI, see LAMB OF GOD

AGOBARD (c.779-840). Archbishop of Lyons. A refugee from Moorish Spain, he became a priest in France and in 816 was appointed to the Lyons post. He had an unrealistic vision of a Carolingian Empire encompassing all Christendom whose subjects, despite their racial origins, would be Christian, under one law and all called Franks. Participation in Carolingian political struggles led to his temporary deposition as archbishop at the Council of Thionville, from 835 to 837, by order of the Emperor Louis the Pious. His replacement, Amalric of Metz, was condemned by the Synod of Quiercy (838) for his interpretation of the ceremonies of the Mass in his book *De Ecclesiasticiis Officiis* (820) which Agobard subsequently attacked. He also wrote against the Jews (*De Insolentia Judaeorum*) and against superstitious beliefs and practices *(Liber de Grandine et Tonitruis)* but did not write the *Liber de Imaginibus* once attributed to him. His theological writings were mainly against the Adoptianists,* especially Felix, bishop of Urgel. Agobard was a severe critic of the German proprietory church system which was to cause great troubles in the future.

L. FEEHAN

AGONIZANTS (Lat. = "to be at the point of death"). A religious order founded by Camillus of Lellis* in 1584. His followers (also known as Camillians) vowed to devote themselves to the plague-stricken. In 1591 Gregory XIV gave them the privileges of a Mendicant Order.

AGRAPHA (Gr. = "unwritten sayings"). Generally understood to mean sayings of Jesus not found in our four canonical gospels. Sources are (1) the New Testament (i.e., 1 Thess. 4:16ff.; Acts 20:35; Codex D after Luke 6:4); (2) Christian writers, Papias and after; (3) papyri, especially Oxyrhynchus papyri;* (4) Arabic and Islamic agrapha.

See E. Hennecke, *New Testament Apocrypha* (ET 1963), I, p.85 for full bibliography.

AGRICOLA, JOHANN (1494-1566). German Reformer. Born in Eisleben in 1494, he studied at Wittenberg under Luther. In 1519 he served as Luther's recording secretary at the Leipzig Disputation. In 1525 he went for a short time to Frankfurt and was later appointed director of the school in Eisleben and preacher in the church of St. Nicolai. In 1527 he had a dispute with Melanchthon* on the relation between repentance and faith. There may have been personal reasons behind their theological differences, as in 1526 Melanchthon was preferred to him for appointment to a chair in the University of Wittenberg. Melanchthon's view, shared by the other Reformers, was that the moral law was needed to bring the sinner to repentance, leading on to faith in Christ. Agricola held that the law has no place in Christian experience. Luther, who gave Agricola and his followers the title of Antinomians,* refuted his arguments and elicited some form of recantation, but bitterness remained. In 1540 Agricola went to Berlin, where he was appointed court preacher and general superintendent. In addition to theological works he compiled a collection of German proverbs.

HUGH J. BLAIR

AGRICOLA, MIKAEL (c.1510-1557). Finnish Reformer. After schooling in Viborg (Viipuri) he came to Abo (Turku), where he became acquainted with the Reformation through Peter Särkilax.* Working as secretary to the bishop, Agricola had the opportunity to preach the evangelical truth in Abo as well as in the countryside. He studied in Wittenberg (1536-39), and returning to Finland brought with him a letter of recommendation from Luther. For nine years he was principal of the school in Abo, the main institution for theological training. He strove to renew the inner life of the church in the spirit of the Reformation. In 1554 he was appointed bishop of Abo. Agricola did outstanding and significant work as author and publisher of books in Finnish. He published an ABC-book (1542) and a prayer book (1544) of prayers taken from the Bible and from the writings of Luther and other Reformers. His translation of the NT was finished in 1543 and published in 1548. The translation is from the original Greek, but the influence of Luther is evident. Agricola also translated parts of the OT into Finnish, among others the Psalms. In 1549 he published a church manual and a missal which greatly determined the development of the Finnish church services. These books are in the main based on the Swedish manual and missal by Olavus Petri. Agricola, who has been called "the Father of Finnish Literature," laid the foundation of evangelical Christianity, theological training, preaching, education, and liturgy in Finland.

STIG-OLOF FERNSTROM

AGRICOLA, RUDOLPHUS FRISIUS (Rudolf Husmann) (1444-1485). Dutch humanist. Born at Baflo, Groningen, he studied at St. Martin's School there under the influence of the Brethren of the Common Life,* and at Erfurt, Louvain, and Cologne universities. He experienced Italian Renaissance culture in Pavia and Ferrara (1469-79) and formed an excellent Latin style and mastered Greek. He spent his last three years lecturing informally on rhetoric and the classics at Heidelberg. His major work, *De Inventione Dialectica* (1479), was concerned with the relation of logic to rhetoric and harmonized new and traditional views. In *De Formando Studio* (1484) he put forward enlightened views of education. His ideas greatly influenced Erasmus* and the sixteenth-century German humanists. He was also a famous lutist and composer of part-songs.

J.G.G. NORMAN

AGRIPPA VON NETTESHEIM, HEINRICH CORNELIUS (1486-c.1535). Soldier and wandering scholar. Born in Cologne, he wandered all his life from one place to another in Europe, including England, seeking patronage. His polemical attitude, combined with tactlessness, created many enemies. At the age of twenty-three, disillusioned with the learning of the day, Agrippa sought to make a synthesis of Christianity, Neoplatonism, Pythagoreanism, and cabalism. The result *(De Occulto)* was a standard work still in use by magicians two centuries after his death. Later he published *De Vanitate,* a ruthless satire of Scholasticism, monkishness, belief in witchcraft, indulgences, and image worship in the Roman Church. Though he had much in common with the Reformers, he did not seek the destruction of the church; and his attitude to the Bible, which he believed to contain mistakes, was different. His books, often reprinted, found acceptance in many quarters. *De Vanitate* became a favorite of atheists and skeptics on account of its anticlericalism and misgivings about human reason; his work on magic lined him up with occultism, and his exposure of Roman Catholic abuses with the Reformation. Though his writings sometimes show him to be inconsistent, empirical, and opportunist, he was also prepared to fight for the right and suffer as a result.

BIBLIOGRAPHY: His two books mentioned above have appeared in a number of English editions under the titles *Occult Philosophy* and *The Vanity of Arts and Sciences;* see also H. Morley, *The Life of H.C. Agrippa* (2 vols., 1856); L. Spence in *Three Famous Alchemists* (ed. Rider, 1939); C.G. Nauert, *Agrippa and the Crisis of Renaissance Thought* (1965, with bibliography).

R.E.D. CLARK

AIDAN (d.651). First bishop of Lindisfarne, England. After his victory at Heavenfield near Hexham in 633, Oswald,* king of Northumbria, requested the monks of Iona* to send someone to teach his people the Christian faith. The first missionary made no impression and on his return to Iona complained of the barbarous nature of the people. In his place was appointed another monk, Aidan, who was consecrated bishop. He chose for his headquarters the small island of Lindisfarne not far from Oswald's capital at Bamburgh. He founded a monastic community and adapted himself fully to the new situation, having close links with Oswald and his successor Oswin, and identifying himself with the common people. He made many missionary journeys on the mainland, largely on foot, sometimes accompanied by Oswald as interpreter. He founded a school for twelve boys, including Eata, Wilfrid, Cedd, and Chad, to continue his work. He displayed notably the Celtic virtues of simplicity, humility, and gentleness, and such was his missionary success that J.B. Lightfoot could say: "Not Augustine, but Aidan, is the true apostle of England."

R.E. NIXON

AILLY, PIERRE D', see D'AILLY, PIERRE

AILRED (Aelred) (1109-1167). Abbot of Rievaulx in England. He was born at Hexham, son of a priest at the abbey there. He spent his youth at the court of King David of Scotland who wished to make him a bishop. Ailred preferred the monastic life and entered the Cistercian house at Rievaulx in Yorkshire in 1131. He became master of the novices and in 1143 was appointed abbot of Revesby in Lincolnshire. In 1147 he was recalled to Rievaulx to become abbot. He was present at the canonization of Edward the Confessor* in Westminster Abbey and later wrote on his life and miracles. Ailred made various evangelistic journeys, including one among the Picts of Galloway in 1164. He wrote a prose eulogy of Cuthbert,* and a number of treatises on the Christian faith and life from a mystical viewpoint. He came to be known as "the English St. Bernard."

R.E. NIXON

AINSWORTH, HENRY (1571-1622). Puritan divine. Educated at Caius College, Cambridge, he became a separatist* and fled to Holland, where he became first a porter to a bookseller and then teacher to a London congregation which had reassembled in Amsterdam under the pastorship of Francis Johnson. He drew up a confession of faith, and with Johnson wrote in 1604 a *Defence* of the Brownists.* Ainsworth was a good scholar, learned particularly in Hebrew and rabbinism. Though a controversialist, he was gentle by the standards of the time in his replies to John Ainsworth who had joined the Roman Church, to Richard Bernard, and to John Smyth. Later he parted from Johnson over principles of church government (Ainsworth leaned to congregationalist principles). He established a considerable scholarly reputation for his *Annotations* on various OT books, the whole Pentateuch, the Psalms, and the Song of Solomon, which first appeared separately and then in collected form in 1627, and was reprinted later. G.E. DUFFIELD

AIRAY, HENRY (1560-1616). Puritan divine. Born in Westmoreland, he was educated at the expense of Bernard Gilpin,* "the Apostle of the North," both at a local school founded by him and at Queen's College, Oxford. He became provost of his college in 1598 and vice-chancellor of the university in 1606. William Laud* was accused of Romish tendencies before him in 1606 and narrowly escaped public censure. "A frequent and zealous preacher," Airay delivered a series of lectures on Philippians in which he denounced popery and expounded evangelical Calvinism at its best.

AISLE, see ARCHITECTURE

AKHMÎM FRAGMENT. A portion of the *Gospel of Peter** found at Akhmîm in Upper Egypt in 1886-87 in the grave of a Christian monk. It is clear from the ornamentation of the manuscript that the scribe possessed no more of it than has now been discovered. Some scholars unite this incomplete fragment with the *Apocalypse of Peter* because a Greek text of part of the latter was discovered in the same grave. The earliest witness to the *Gospel of Peter* is Serapion, bishop at Antioch at the end of the second century, who agreed to the reading of this book in the church at Rhossus until he discovered its Docetic* teaching. From this it may be deduced that the *Gospel of Peter* probably came from Gnostic circles in Syria, but the Akhmîm fragment is evidence that it enjoyed circulation over a much wider area. Origen refers to such a gospel, but may not have been acquainted with its contents.

DONALD GUTHRIE

ALACOQUE, MARGUERITE MARIE (1647-1690). Founder of the devotion to the Sacred Heart of Jesus. She entered the Visitation convent at Paray-le-Moniae, central France, in 1671, and after a regime of severe austerities declared that Christ had revealed to her His heart burning with love for man and had commanded her to establish the Holy Hour, Communion on the first Friday of each month, and the feast of the Sacred Heart to be observed on the Friday after the octave of Corpus Christi. The skepticism with which her visions were at first regarded was gradually dispelled, and devotion to the Sacred Heart quickly spread throughout the Christian world. She was pronounced venerable in 1824, beatified in 1864, canonized in 1920. J.D. DOUGLAS

ALAIN OF LILLE (1125-c.1203). Theologian and eclectic philosopher; sometimes called "Doctor Universalis" because of his encyclopedic learning. He may have been an auditor of Gilbert de la Porrée. He taught in Paris (c.1157-70) and Montpellier (c.1171-85), took part in the Third Lateran Council (1179), and eventually entered the monastery at Cîteaux. He held a rational-mystical view of the relation of philosophy and religion, maintaining that the truths of religion are discoverable by unaided reason. Belonging to no particular school, his outlook was largely Neoplatonic, influenced especially by Boethius and Proclus, and his method was an accentuated dialectic. His works include sermons, poems, proverbs, besides philosophical and theological writings. He wrote *Contra Haereticos* against the Waldenses, Albigenses, Jews, and Muslims. His epic *Anticlaudianos* (c.1184) inspired Dante and Chaucer. J.G.G. NORMAN

ALANE, ALEXANDER, see ALESIUS

ALARIC (d.410). Visigothic king. German nomads settled in Romania in the second and third centuries, where Arians later evangelized them. About 375, Huns from the northeast forced these Visigoths into Bulgaria, where they became federated with the Roman Empire, but Alaric, their elected king, led their search for land and wealth around Greece. Though given land in Albania and the prestige of the imperial title *magister militum,* Alaric demanded Hungary and then Austria, and on refusal marched on Rome, which he ransacked in August 410. The sack disillusioned both pagans and Christians who had believed that *piety* guaranteed the political freedom and social security of Rome. Augustine* tackled this dilemma in *The City of God.* Within a month of the sack Alaric died in southern Italy.

G.T.D. ANGEL

ALB. A white linen garment reaching from the neck to the ankles, with tight-fitting sleeves, usually gathered in at the waist by a girdle, worn by the ministers at the Mass or Eucharist in the Roman and some Anglican churches. Derived from the classical undertunic it was not at first regarded as a specifically eucharistic vestment.

ALBAN (latter third century?). First known British martyr. According to Bede's *Ecclesiastical History* he was a pagan of Verulamium (later named after him St. Albans) who sheltered a fugitive priest and as a result was converted to Christianity. When soldiers came in search of the priest, Alban disguised himself as his guest, and suffered martyrdom in his place. Bede places the episode in the Great Persecution around 303, but since there is little evidence for persecution in Britain at this time, scholars have suggested that it may have taken place earlier, either under Decius or Valerian (250-60) or even at the beginning of the third century. Evidence for the authenticity of the event is strong, since its undoubtedly early record in the *Acta Martyrum* shows detailed knowledge of the topography of Verulamium. The cult of St. Alban can be traced back to 429. HAROLD H. ROWDON

ALBERT OF BRANDENBURG (1490-1545). Elector of Mainz and cardinal. Younger son of the elector of Brandenburg, he is a notorious example of the multiplication of ecclesiastical benefices in one person. Before becoming archbishop of Mainz (1514) and later cardinal, he had held two bishoprics and a number of rich abbeys. To meet his debts, Pope Leo X permitted Albert to sell indulgences in his diocese, the proceeds to be divided between him and the pope, and this work, entrusted to John Tetzel,* led to Luther's historic protest. Yet there were times when Albert seemed well-disposed to the work of reformation. He had many friends among the humanists, notably Ulrich von Hutten, and up to the time of the Peasants' War it seemed possible that he might be won over to the Reformed faith. As late as 1532 he accepted and rewarded the dedication of Melanchthon's commentary on Romans, but from 1525 he had ranged himself definitely on the side of the papacy. He was one of the princes who met at Dessau in 1525 to unite in the defense of Romanism, and he was a member of the League of Nuremberg which was formed in 1538 to counteract the Smalcald League. Realizing the threat posed by the Reformation, he helped to muster the forces against it, and particularly in his later years lent his support to the new Jesuit order in its work of counter-Reformation.

HUGH J. BLAIR

ALBERT OF PRUSSIA (1490-1568). Grand Master of the Teutonic Order. He is noted for his success in bringing the Prussian state under the influence of the Reformation. Initially his aims were largely political—he endeavored to secure Prussia's independence from Poland. He was defeated and in 1521 consented to a truce for four years. Anticipating a renewal of the war, he sought allies and visited Nuremberg in 1522. There he came under the influence of the Reformer Andreas Osiander,* who won him over to the Reformed faith. On the advice of Luther he dissolved the Teutonic Order and determined to make Prussia a hereditary duchy for himself. The spread of Lutheran teachings in Prussia furthered this aim, and in 1525 he became duke of Prussia, though still under the suzerainty of the Polish king. He did much to make Prussia a Protestant state, encouraged education, founded the University of Königsberg in 1544 (he appointed Osiander professor in 1549). Theological differences between Osiander and Melanchthon* led to political disputes, but a strict form of Lutheranism was adopted and declared binding on all teachers and preachers in Albert's dominions.

HUGH J. BLAIR

ALBERTUS MAGNUS (1193-1280). Dominican theologian. Born in Swabia of noble parents, he entered the Dominican Order (1223) and lectured in Dominican schools in Germany (1228-45). He then taught at Paris (1245-48) and Cologne (1248-55), after which he was successively provincial governor of the order in Germany and bishop of Regensberg until 1262, when he retired to the Dominican convent at Cologne. Albert was the first medieval Christian scholar to master the whole corpus of Aristotle.* During his active and varied career he wrote twenty-one massive volumes consisting mainly of commentaries on Aristotle's works and theological books based on Aristotelian philosophy. His main interest lay in the natural science of Aristotle and the problem of reconciling philosophy and Christianity. In natural science he did not blindly follow the teaching of Aristotle. He stated clearly that he did not believe that Aristotle was a god but that he was a man who was liable to error as are others. Albert was such a careful student of nature that he was accused of neglecting sacred studies, and a host of incredible legends were circulated about his miraculous power. Although his theological work was not so successful as that of Aquinas,* Albert did defend the distinction between the realm of revelation and that of human reason. He maintained that no truth could contradict revelation. At the same time, he taught the superiority of revelation and the right of scholars to use all of human knowledge in the investigation of divine mysteries.

ROBERT G. CLOUSE

ALBIGENSIANS (Albigenses). Adherents of a religion derived from the teaching of Mani who lived in Persia in the third century. In a modified form his ideas spread into Asia Minor in the late Roman period, and from there into the Balkans in the early Middle Ages. By following the trade routes they appeared in northern Italy and southern France by the eleventh century. Although given various names such as Cathari* and Bogomiles,* in the West they were generally called Albigensians because the center of their greatest strength was the town of Albi in Languedoc. This religion was dualistic, with a god of light (Truth, the god of the NT), and a god of darkness (Error, the god of the OT). Life on earth was a struggle between these gods and their principal forces, spirit, and matter. The good life for man was a gradual purification from matter. Hence the Albigensians condemned marriage, procreation, eating food, war and the use of anything material in worship. Because they refused to take oaths they were subversive to a society that rested on the oath of a vassal to his feudal lord. They also believed that human government was wicked and evil. All of these positions represent the extreme of Albigensian doctrine and in most cases the teaching was compromised. For example, a good Albigensian did not have to stop eating, but he was required to be a vegetarian.

The followers of this religion were divided into the few (*perfecti*, their clergy) and the many (*credentes*, the believers). The *perfecti* lived up to the rigid asceticism of their faith, and the *credentes* tried to become *perfecti.* This was accomplished by receiving the only sacrament allowed, the *consolamentum.* If the newly consecrated *perfectus* showed signs of not being able to live up to the ascetic discipline of his calling, his friends could insure the salvation of his soul through a ceremonial death by self-starvation (the *endura*). The Albigensians, however, did not believe in hell or purgatory. The only "hell" they taught was imprisonment of the soul within the body. This led some to licentious living, so many of their leaders began to teach that the souls of those who were not saved transmigrated into the bodies of lower animals.

The Albigensian homeland, S France, at the end of the twelfth century was a pleasant, tolerant, prosperous land, the center of a flourishing Provençal civilization. One of the rulers of this area, Count Raymond VI of Toulouse, was also the leading supporter of the Albigensian cause. Raymond's agents murdered a papal legate, thus provoking a crusade which crushed the religion and the Provençal civilization (1209-1229).

See S. Runciman, *The Medieval Manichee* (1947). ROBERT G. CLOUSE

ALBRIGHT, JACOB (1759-1808). Founder of the Evangelical Church.* Born near Pottstown, Pennsylvania, son of German immigrants, he received little formal education. He grew up in the Lutheran Church, served as a soldier in the Continental Army, and became a prosperous brickmaker. The unexpected death of several of his children led to his conversion in 1790. He joined a Methodist class and was later licensed as a lay preacher. By 1800 several classes had been formed, and three years later Albright was ordained to the ministry by his congregation. In 1807 the Albright People were organized as "The Newly Formed Methodist Conference" with Albright as the first bishop. Later they became known as "the Evangelical Association" and then as the Evangelical Church (1922). The church adopted Arminian* doctrine and Methodist* polity. HARRY SKILTON

ALBRIGHT BRETHREN, see EVANGELICAL CHURCH

ALCAREZ, PEDRO RUIZ DE, see ILLUMINATI

ALCUIN OF YORK (d.804). Medieval Christian scholar. Educated at the cathedral school of York where his teacher had been a pupil of Bede, Alcuin was the most important and influential of the scholars invited to Charlemagne's* court (782) where his teaching and writing provided the basis on which later Carolingian writers were to build. A prolific author, he wrote commentaries on the Bible, dogmatic treatises, and controversial manuscripts. After his work at the palace school he became abbot of the monastery of St. Martin of Tours, where in addition to his writing he added many volumes to the library and supervised the monastic school. At Tours, Alcuin labored to raise the intellectual level of the monks. One of the results of his work was the development of the style of handwriting called Caroline minuscule. This system, which used both small and capital letters, was easier to read than the earlier Merovingian cursive or the Italian hand and it has influenced the printing of books today through roman type. Alcuin also led a group of scholars in the revision of the Vulgate* text of the Bible. These men collated the oldest manuscripts, correcting many errors and discrepancies in the text. Although a uniform text was not attained, it halted further scribal corruptions.

ROBERT G. CLOUSE

ALDRED (Ealdred) (d.1069). Archbishop of York. Previously monk and abbot, he became bishop of Worcester in 1044, was Edward the Confessor's ambassador to Emperor Henry III (1054), and went to York in 1060. He was degraded from the episcopate for various offenses, but submitting to William the Conqueror after Hastings (1066), he was chosen to crown the new king that year and Queen Matilda in 1068. He was buried in his own cathedral.

ALEANDER, GIROLAMO (1480-1542). Roman Catholic scholar. He headed the opposition to Luther at the Diet of Worms.* Born in Venice, he studied medicine and theology, and was said to have been one of the most learned men of his day. He taught Latin, Hebrew, and Greek at the University of Paris from 1508 and was for a time rector of the university. In 1519 Leo X appointed him librarian of the Vatican, and thereafter he was employed on various papal missions. Evidence of the virulence of his attacks on Luther is provided by copies of his letters which are in the Vatican Library. At the Diet of Worms he made a lengthy speech advocating the sternest possible measures against Luther and his teaching, and it was he who drew up the edict, accepted by Charles V and the Diet, by which Luther was condemned. Later, in the Netherlands, it was at Aleander's instigation that two monks were burned at the stake in Brussels in 1523, the first martyrs of the Reformation. HUGH J. BLAIR

ALESIUS, ALEXANDER (1500-1565). Scottish Reformer. Born Alexander Alane in Edinburgh, he graduated at St. Andrews and became canon of the priory there. Applauded for public refutation of Luther's arguments (he admitted later to have borrowed his major points from Bishop Fisher of Rochester), he was selected in 1528 to reclaim Patrick Hamilton* from Lutheranism. This produced only a change of heart in himself and his own imprisonment and subsequent flight to Germany (1532), where he met Luther and Melanchthon. About 1535 he went to England, was warmly welcomed by Cranmer and Latimer, and by Henry VIII himself who secured for Alesius a teaching post at Cambridge where he was said to have been the first to deliver lectures on the Hebrew Scriptures. Finding his life endangered there, however, he returned to London and practised medicine. When the king changed direction once more and issued decrees upholding transubstantiation and clerical celibacy, Alesius went back to Germany (1540). He taught briefly at Frankfurt, then spent the rest of his life at Leipzig University, where he was rector at least twice. It was Melanchthon* (with whom he was a great favorite) who gave him the name Alesius, because of his earlier wanderings. Alesius revisited England and translated into Latin Edward VI's first liturgy, and published also many exegetical, dogmatic, and controversial works. In Germany he was continually active in the Protestant cause, and arranged many disputations. He was the first to plead in his writings for free circulation of the Scriptures in Scotland—a land which inexplicably has failed to realize the full significance of the contribution made by this great native son.

BIBLIOGRAPHY: A.F. Mitchell, *The Scottish Reformation* (1900), pp. 239-83; J.H. Baxter, "Alesius and Other Reformed Refugees in Germany," *Records of the Scottish Church History Society*, V, 2 (1934), pp. 93-102; F.S. Pearson,

"Alexander Alesius and the English Reformation," ibid., X, 2 (1949), pp. 57-87.

J.D. DOUGLAS

ALEXANDER (d.328). Bishop of Alexandria (c.313-328). He held that the Son is *eternally* the Son of the Father. Both in a local clerical debate and at a council of about 100 bishops from Egypt and Libya, he accused (perhaps mistakenly) a presbyter in his see named Arius* of following Paul of Samosata and opposed his views that "the Son had a beginning" and that the Son "was from nothing." Rejecting the pleas of Eusebius of Nicomedia and others on behalf of Arius, he in council anathematized Arius and his adherents about 321, prior to the embargo on synods by Licinius. Although regarded by Constantine in late 324 as overscrupulous, Alexander was upheld by the synods of Antioch and Nicea in 325, and consistently opposed Arians and Melitians until his death. He believed that the Father and the Son are exactly alike except that the Father is unbegotten. This exception made Arius stress "being begotten" as a temporal moment and as a sign of distinct identity between Father and Son.

G.T.D. ANGEL

ALEXANDER II (d.1073). Pope from 1061. The former Anselm, bishop of Lucca, was one of the leading popes of the Hildebrandine or Gregorian reform. He was elected without the participation of the German king and against the will of many of the Roman nobles. Hence these groups elected Cadalus of Parma (antipope Honorius II). Later they withdrew allegiance to the antipope. Under Alexander's pontificate the reform of the Church continued through correspondence and the activity of legates. The latinization of Greek sees proceeded with the Norman conquest of S Italy and the pope encouraged moves against the Muslims in Spain and the Saxons in England. He also condemned the maltreatment of the Jews in S France and Spain. During the later part of his reign a dispute over the appointment to the bishopric of Milan increased tension with the Roman emperors. This argument continued into the reign of Gregory VII and led to the Investiture Controversy.*

ROBERT G. CLOUSE

ALEXANDER III (c.1105-1181). Pope from 1159. Orlando (Roland) Bandinelli, born in Siena, was a professor at Bologna celebrated as theologian and canonist. Becoming cardinal in 1153, he advised Adrian IV, whom he succeeded. A strong antagonist of Emperor Frederick I (Barbarossa), he was opposed by three antipopes* set up by Frederick (Victor IV, Paschal III, Callistus III). The schism lasted seventeen years, ending only when Frederick was defeated by the Lombard league at Legnano in 1176, followed by the Peace of Venice (1177). During the schism Alexander lived mainly in France. He supported Thomas à Becket* and imposed penance on Henry II of England for Becket's murder. First of the great lawyer-popes, he convoked the Third Lateran Council (1179), which vested the exclusive right of electing a pope in a two-thirds majority of the cardinals.

J.G.G. NORMAN

ALEXANDER IV (d.1261). Pope from 1254. A nephew of Gregory XI, he was cardinal bishop of Ostia from 1231. As pope he continued Innocent IV's policy of implacable hostility to the House of Hohenstaufen. In 1255 he attempted to secure English help in return for granting the kingdom of Sicily to Edmund, second son of Henry III of England. He vainly attempted to unite the powers of Europe in crusade. He was a notable friend of the Franciscans. Early in his reign he canonized Clare.* He took the friars of Paris under his protection in the violent controversies aroused by William of St. Amour.

ALEXANDER V (c.1339-1410). Pope from 1409. Peter of Crete, Greek by birth, became a Franciscan and studied at Oxford and Paris where he became master of theology (1381) and lectured on the *Sentences* of Peter Lombard. His *Commentary*, which has survived, shows Nominalistic leanings and suggests that he played a role of some importance in the development of medieval thought. He became bishop of Piacenza (1386), Vicenza (1387), Novara (1389), and archbishop of Milan (1402). In an effort to end the Great Schism he was chosen by the Council of Pisa (1409) in place of two rival popes. He died less than a year later, and there is some doubt as to the validity of his reign. The council had no authority in strict canon law, as it was summoned not by papal order but by certain cardinals who had deserted the two rival popes.

HAROLD H. ROWDON

ALEXANDER VI (1431-1503). Pope from 1492. While he was a young man in the service of the church, the scandalous private life of Rodrigo Borgia called forth a rebuke from Pius II. Ruling in an age of trouble for both Italy and the church, he pursued a papal, Italian, and family policy until 1498 when he became interested in greater cohesion for the Papal States* and supported the political activities of his son Cesare. Alexander can be credited with several achievements during his pontificate. He supported the work of Pinturiccio and subsidized the *Pietà* by Michelangelo. He also encouraged evangelization in the New World and ensured peace between Portugal and Castile by arbitrating a line of demarcation. Despite these achievements, he is remembered as the father of Lucrezia Borgia and as the pontiff with whom Girolamo Savonarola* struggled. He was a type of pope that gave the early Protestants reason to condemn the Roman Church.

ROBERT G. CLOUSE

ALEXANDER VII (1599-1667). Pope from 1655. Fabio Chigi was born at Siena, where he studied philosophy, law, and theology, receiving his doctorate in theology in 1626. He entered papal service at Rome. After being Inquisitor of Malta, he became nuncio at Cologne (1639-51). During negotiations leading to the Peace of Westphalia* (1648), he urged Catholic princes not to sacrifice the rights of the church. Secretary of state to Innocent X (1651) and cardinal (1652), after his election as pope he at first strongly opposed nepotism, but in 1656 he gave way to pressure and called his brother and nephews to Rome.

His pontificate was marked by disputes with Catholic powers, notably with France. When he resisted French claims, Louis XIV seized Avignon and Venaissin, and threatened to invade the states of the church, and Alexander had to accept the humiliating Peace of Pisa (1664). Theologically strongly anti-Jansenist, in 1665 he condemned the Jansenists'* views, but they refused to submit. Following the controversy on Probabilism,* he condemned in 1665-66 forty-five Laxist propositions, though not the Probabilist system. He befriended the Jesuits and procured the readmission of the order to the republic of Venice. He encouraged foreign missions and did much to modernize and embellish Rome. J.G.G. NORMAN

ALEXANDER VIII (1610-1691). Pope from 1689. Pietro Vito Ottoboni was born at Venice, where his father was chancellor. At seventeen he won a doctorate in civil and canon law at Padua. He went to Rome in 1630 and was made governor of Terni, Rieti, and Spoleto and auditor of the Rota. Innocent X made him cardinal in 1652 and bishop of Brescia two years later. Under Innocent XI he became Grand Inquisitor of Rome and secretary of the Holy Office. As pope he was conspicuous for his nepotism. He sought to diminish tensions with France, and persuaded Louis XIV to restore Avignon and Venaissin and to renounce the privilege of diplomatic residence. As a result of this reconciliation with Louis, however, his relations with Emperor Leopold I worsened. He was interested in a possible Stuart restoration in England and established a group to study English affairs. He supported Venice in the Turkish wars. He condemned thirty-one Jansenist propositions and the Four Gallican Articles* of 1682 in 1690. J.G.G. NORMAN

ALEXANDER, ARCHIBALD (1772-1851). Presbyterian theologian and educator. Born near Lexington, Virginia, he studied at Liberty Hall Academy (now Washington and Lee University). Shortly after confessing faith in Christ, he began theological studies under William Graham, who encouraged him to preach. Two years after ordination in 1794 he assumed the presidency of Hampden-Sydney College, a position he held for almost a decade. Early in 1807 he became minister of Pine Street Church, Philadelphia, one of the largest congregations in the nation. In 1812 the general assembly (whose moderator he had been) established a theological seminary at Princeton and selected Alexander as its first professor. BRUCE L. SHELLEY

ALEXANDER, CECIL FRANCES (1823?-1895). Irish hymnwriter. Born in Dublin, she married William Alexander, rector of a country parish in Tyrone and later archbishop of Armagh. On a complaint from her godsons that they found the catechism difficult, she was prompted to write verses illustrating the creed. Well-known hymns resulting from this include "All things bright and beautiful," "Once in royal David's city," and "There is a green hill." She became a prolific writer, and nearly 400 hymns and poems came from her pen, many of them written for her Sunday school class who heard them before they were published.

ALEXANDER, CHARLES McCALLON (1867-1920). Evangelistic songleader. Born on a Tennessee farm of devout Presbyterian parents, he was educated in the state and at Moody Bible Institute, where he developed unusual ability in leading large groups of people in the singing of gospel songs. He conducted evangelistic meetings from 1902 with two of the most famous American revivalists, R.A. Torrey* and J.W. Chapman,* not only in America but also in Great Britain, Australia, and Asia. Both men were dignified and serious in their preaching, but "Charlie" Alexander preceded their evangelistic appeals with a period of "warming up" the audience with jovial humor and lively singing. His style was copied by many others and continues to influence evangelical churches in America. HARRY SKILTON

ALEXANDER, JOSEPH ADDISON (1809-1860). Linguist, educator, and author. Born in Philadelphia, he received most of his early education from his minister father, Archibald, and quickly revealed remarkable linguistic gifts. At fifteen he entered the junior class of Princeton College, graduating with highest honors in 1826. After two years' private study of ancient languages, he became adjunct professor of ancient languages and literature at the college in 1830. In 1834 he moved to Princeton Seminary, instructing first in oriental and biblical literature and after 1851 in biblical and ecclesiastical history. During these years he was the author of several biblical commentaries, which established his reputation in Europe as well as America. BRUCE L. SHELLEY

ALEXANDER, MICHAEL SOLOMON (1799-1845). Anglican bishop in Jerusalem. Born in Germany of an Orthodox Jewish family, he was educated there and became a rabbi. In 1820 he went to England, but five years later lost his status as rabbi when he was converted to Christianity. He taught Hebrew in Dublin, was ordained in the Anglican Church in 1827, and for three years served as missionary among the Jews in Danzig. He then continued his work in London, adding to it in 1832 the chair of Hebrew at King's College. He participated in the revision of the NT in Hebrew. In 1841 he became the first Anglican bishop in Jerusalem, an appointment that did not go undisputed. Four years later he died in Egypt on his way back to England. J.D. DOUGLAS

ALEXANDER, SAMUEL (1859-1938). Jewish philosopher. Born in Sydney, Australia, he became a fellow of Lincoln College, Oxford, and (in 1893) professor of philosophy at Manchester. Alexander is best known for his work *Space, Time and Deity* (1920), in which he sought to reconcile philosophy with the ideas of his day, dominated by materialism, evolution, belief in progress, and (later) relativity. Starting with Spinoza's* static concept of the god-world, he made it dynamic by "taking time seriously." Matter, he taught, "emerges" from motion in space-time; mind from brain structure; deity from cosmic structure (in fact, several deities may emerge). God exists only in the sense that the universe tends, in evolution, to produce a deity-quality. R.E.D. CLARK

ALEXANDER OF ABONUTICHUS IN PAPHLAGONIA (c.150-170). Popular cult-founder hostile to Christians in Pontus. He is described as a charlatan by the apostate satirist Lucian of Samosata. His oracles relied on glossolalia, extrasensory perception and trickery; staged "theophanies" established his revival of Aesculapius in the form of a new "god" Glycon; elaborate ritual acted out the mythical claims of the prophet, whose clientèle came from Bithynia, Galatia, and Thrace.

ALEXANDER OF HALES (c.1170-1245). Theologian and philosopher. Born at Hales in Gloucestershire, he probably studied first at Oxford and then at Paris where he took his master's degree in arts and theology. In 1220 he became a teacher in Paris and soon was one of the most widely celebrated theologians of the university. He created a sensation in the scholastic world when about 1236 he joined the Franciscans.* This in itself marked a highly important landmark in the involvement of the Minors with the university and hence with academic theology. He at once made a mark. He helped to instigate opposition to the very unsatisfactory Franciscan minister-general Elias. He became regent master of the Franciscan convent in Paris and soon had a famous school with students of the calibre of John de la Rochelle, William of Melitonia, and Bonaventura. Alexander and his school are important because they attempted to understand the rediscovered Aristotelian* philosophy and its implications for theology. Alexander himself wrote a commentary on the *Sentences* and his name is associated with the *Summa Theologica*—but this was a composite work finished by his followers. It is therefore difficult to identify with certainty his own work, but the evidence is that he did not assimilate Aristotle and remained faithful to his Augustinian theological roots. At the same time he indicated the problems posed by the new learning and also paved the way for the more satisfactory solutions of his followers.

C. PETER WILLIAMS

ALEXANDER OF LYCOPOLIS (Nile Valley). Third-century writer of an anti-Manichaean tract entitled "of Alexander of Lycopolis who turned from pagans against the opinions of Mani." One interpretation is that he turned from paganism to Christ. The tract favors Christian orthodoxy against Manichaeism* for its practical concern to make mankind virtuous and for its plausible interpretation of the Crucifixion. According to Photius, Alexandria gained archieratic (episcopal?) rights. On another interpretation (e.g., Brinkmann), Alexander was pagan. After sketching the Manichaean system he criticizes it from the standpoint of Greek philosophy; his admiration of Christianity simply highlights the defects of Manichaeism. Alexander received information from acquaintances of Mani, who died in 272. A date for the tract about 275 seems appropriate.

G.T.D. ANGEL

ALEXANDRIAN THEOLOGY. Alexandria even in pre-Christian times was a center of learning. Philo* flourished there at the turn of the first century, and subsequently several streams of thought flowed together to give vogue in Alexandria to the Neoplatonism of Photinus and the Gnosticism of Basilides and Valentinus. The coming of Christianity to Alexandria is generally attributed to the preaching of Mark the Evangelist; the organization of the church seems to have been simple and to have accommodated itself somewhat to the prevailing climate of opinion. It was not until the beginning of the third century that Alexandria became important as a seat of Christian theology. Pantaenus* is generally regarded as the first head of the school there, which seems to have continued the ancient catechetical school. The latter combined aspects of the Hellenistic "Museum" and the Jewish schools. The general program of the Alexandrian School found expression in Clement's trilogy: *Protrepticus* (Exhortation, addressed to the heathen), *Paedogogus* (Instruction in Christian morals), and the *Stromata* (Miscellanies, training in Divine Wisdom, the true gnosis for the mature believer).

The Alexandrian School reached its highest peak of influence under Origen*; later leaders include Pierius, Theognostus, Serapion, Peter the Martyr, and Didymus the Blind. Arius held high office in the church of Alexandria, and this may have been one reason for the decline of the school, since he could use one aspect of Origen's thought as high authority for his own theory of Christ as a created Son of God. During the fourth century the school passed more and more into obscurity, though Alexandria had as bishops Alexander and Athanasius,* who led the attack on Arianism* and were foremost in the establishment of Christian orthodoxy.

During its heyday the school greatly influenced the leaders of the Palestinian Church, notably Julius Africanus* and Alexander of Jerusalem. Early in the fifth century a new school of Neoplatonism arose in Alexandria under the leadership of the learned woman Hypatia. The main features of the Alexandrian School are three: (1) *the use of the weapon of philosophy.* In contrast with the North African, Tertullian, the Alexandrians were Christian philosophers par excellence, using its method and terms in the interest of the Faith, seeking to beat opponents with their own weapons; (2) *the supremacy of Logology.* The Alexandrians stressed the Logos*-doctrine in an effort to bridge the gap between God and the world, and as the bond of union between the religion of the gospel and Gentile science; (3) *the radical application of the allegorical method of biblical exegesis* (see ALLEGORY). The main opponent of the Alexandrian School was that of Antioch,* in contrast to which the Alexandrians tended towards a Logos-flesh Christology and towards Monophysitism.

BIBLIOGRAPHY: J. Simon, *Histoire critique de l'école d'Alexandrie* (1845); C. Kingsley, *Alexandria and Her Schools* (1854); C. Biggs, *The*

Christian Platonists of Alexandria (1886); E. Molland, *The Conception of the Gospel in Alexandrian Theology* (1938); G. Bardy, "Pour l'histoire de l'école d'Alexandrie," *Vivre et Penser,* 2 (1942), pp. 80-109; E.A. Parsons, *The Alexandrian Library* (1952); J.E.L. Oulton and H. Chadwick, *Alexandrian Christianity* (1954).

H.D. MC DONALD

ALFORD, HENRY (1810-1871). Dean of Canterbury. Born in London, he early showed precociousness by writing Latin odes and a history of Jews before he was ten. In 1829 he entered Trinity College, Cambridge, and in 1834 was elected fellow. In 1835 he became vicar of Wymeswold in Leicestershire. From 1853-57 he ministered to a large congregation at the Quebec Chapel, Marylebone, until he became dean of Canterbury. Alford edited the works of John Donne (1839) and was the first editor of the *Contemporary Review.* He wrote hymns including "Come ye thankful people come" and "Ten thousand times ten thousand." He is chiefly known, however, for his monumental edition of the Greek New Testament which appeared over the years 1849-61.

See his wife's *Life, Journal and Letters* (2 vols., 1873).

P.H. BUSS

ALFRED THE GREAT (849-899). King of Wessex from 871. Although he was concerned to restore the condition of the English Church after the devastation of the Danish raids, the continuing financial burden of military defense severely limited what he could achieve. A general revival of monasticism was still out of the question, although communities were precariously established at Shaftesbury and Athelney, for women and men respectively. Plans for diocesan reform likewise had to be delayed beyond Alfred's death. The king concentrated on restoring the most important treasure that had been lost: education. Even here, Latin scholarship could not flourish without the abbeys and minsters. But something of greater significance at the time was possible. With the help of an international band of scholars, Alfred translated into English some of the fundamental works of theology, philosophy, history, and spiritual direction, by authors such as Augustine, Boethius, Orosius, and Gregory the Great. As one who had traveled on the Continent and visited Rome, Alfred understood the importance of maintaining these links with wider Christendom in an age when disruption of the monasteries could quickly mean the end of civilization.

BIBLIOGRAPHY: Asser, *Life of King Alfred* (ed. W.H. Stevenson, 1904); R.H. Hodgkin, *A History of the Anglo-Saxons* (3rd ed., 2 vols., 1952); M. Deanesly, *The Pre-Conquest Church in England* (1961); J. Godfrey, *The Church in Anglo-Saxon England* (1962).

JOHN TILLER

ALISON, FRANCIS (1705-1779). Presbyterian minister and educator. Born in Ireland, he studied at Glasgow University and in 1735 moved to America, settling first in Maryland, then in New London, Pennsylvania. In New London he was licensed to preach and shortly afterwards opened a school (1743). In 1752 he became rector of a new academy in Philadelphia. At his suggestion the trustees of the school approved the granting of degrees, and in 1755 Alison was chosen vice-provost of the college, with which his name was linked for over twenty-five years. He founded the Presbyterian Society for the Relief of Ministers and their Widows.

ALLAH. The unpluralizable, monotheistic Islamic proper name for God, like Hebrew Yahweh. It designates omnipotent creator, merciful, unbegotten, unbegetting, fiercely rejecting the incarnation of Christ. The name is used also by modern Arab Christians who say concerning future contingencies: *"in sha' Allah"* as other Christians might say "D.V." The Koran* and traditions use ninety-nine names for God, yet a cardinal point of Islam proclaimed from every minaret is *"la ilah illa' Allah"*—"there is no God but Allah."

ALLEGIANCE, OATH OF. The oath to the sovereign taken by the clergy of the Church of England at ordination and on admission to a benefice. Such an oath existed before the Reformation; but at that time a supplementary Oath of Supremacy was introduced, recognizing the Crown as supreme in spiritual as well as temporal matters, and renouncing allegiance to any foreign jurisdiction (i.e., the pope). The equivalent declaration enjoined in the Elizabethan Supremacy Act was later incorporated in the 1604 Canons as part of clerical subscription to the Anglican establishment, alongside acceptance of the Book of Common Prayer and the Thirty-Nine Articles.* In 1865 the form of clerical subscription laid down in this canon was altered, and all reference to royal supremacy thereby omitted. Consequently, the form of the oath of allegiance was altered by Parliament in 1868. The occasions when the oath is required remain unaltered, but exceptions are made in the new Canons of 1969 to cover overseas clergy serving in England.

JOHN TILLER

ALLEGORY. The use of language to convey a deeper and a different meaning from that which appears on the surface. The methodology was elaborated in the rhetorical schools of Greece, originally to relieve Homer of any charge of impiety or ignorance. The Jews of the Diaspora, influenced by Hellenistic culture, adopted the allegorical canon of exegesis in the interpretation of Scripture. The Jew Aristobulus (first half of the second century B.C.) appears to have been the first to apply the Stoic method to the Old Testament, but the Alexandrian Philo* is the Jewish allegorist par excellence. Any passage of Scripture where the literal sense would impugn the transcendent and holy character of God, or which suggests a contradiction, must be interpreted allegorically. In Palestinian usage the allegorical principle was less marked, less radical, and sought to keep close to the literal meaning of the text.

In biblical usage a distinction must be drawn between allegory as a medium of revelation and allegory as a method of interpretation. There are undoubtedly allegorical passages in Scripture; Paul explicitly declares his use of the method in

Galatians 4:21-31 (cf. 1 Cor. 10:1-4), but evidently this was a departure from his usual practice. In the early church, allegory found expression, e.g., in the works of Clement of Rome,* Irenaeus,* and Tertullian*; it was carried to excess in the Alexandrian School.* Jerome, Hilary, Ambrose, and Augustine gave more or less prominence to the allegorical hermeneutic. Bernard of Clairvaux* was the supreme allegorist of the Middle Ages. Aquinas took up the earlier fourfold system of interpretation and made it normative for Catholicism. At all periods there were those who felt uneasy about or were openly opposed to allegory—e.g., the Antiochene School.* Theodore of Mopsuestia* wrote five books, *Against the Allegorists.* It was not, however, until the time of the Reformation that the allegorical method was seriously challenged; Reformed theologians generally rejected it, subscribing instead to the principle "Do not carry a meaning into but draw it out of (the Scriptures)."

BIBLIOGRAPHY: F.W. Farrar, *History of Interpretation* (1886), pp. 127ff.; J. Tate, "The Beginning of Greek Allegory," *Classical Quarterly,* LXI (1927), pp. 214f.; P.K. Jewett, "Concerning Allegorical Interpretation," *Westminster Theological Journal* (1954), pp. 1f.; R.P.C. Hanson, *Allegory and Event* (1959); E.C. Blackman, "Allegory—Plato to Augustine," *Biblical Interpretation* (1957).

H.D. MC DONALD

ALLEINE, JOSEPH (1634-1668). Nonconformist* divine. Born at Devizes in Wiltshire, he was converted during a spiritual crisis provoked by the death of his eldest brother. He began studies at Lincoln College, Oxford, in 1649, then received a scholarship to Corpus Christi College, where he became tutor and chaplain. Oxford was ruled at the time by John Owen and other Puritans. Alleine preached in the villages around Oxford and at the prison. In 1655 he received Presbyterian ordination and became assistant to George Newton, minister of St. Mary Magdalene, Taunton, where he visited and catechized assiduously. In 1662 he was ejected and subsequently suffered under the Clarendon Code.* In 1663 he was imprisoned at Ilchester for singing psalms in his own house and preaching to his family. He evangelized with John Wesley, grandfather of John and Charles. Alleine was one of the best-known Nonconformist preachers. A lost work, *Theologica Philosophica,* was esteemed, according to Richard Baxter, for its harmonizing of revelation and natural theology. But Alleine is best remembered for his *Alarm to the Unconverted* which was published in 1672 after his death and sold 20,000 copies. Republished in 1675 as the *Sure Guide to Heaven,* it sold a further 50,000. Later debtors to this work included George Whitefield and Charles Spurgeon. He wrote other books, including an explanation of the Westminster Catechism.

P.H. BUSS

ALLEINE, RICHARD (1611-1681). Nonconformist* divine. Born at Ditcheat in Somerset where his father was rector, he was educated at St. Alban's Hall and New Inn Hall, Oxford. After ordination he first assisted his father, then in 1641 became rector of Batcombe, Somerset. He signed the Presbyterian manifesto "Testimony of the Ministers of Somerset," as well as the Solemn League and Covenant.* During the Protectorate of Cromwell he served as a ministerial assistant to the commissioners for ejecting scandalous ministers. In 1662 he refused to conform and was ejected. Because of the Five Mile Act* he moved to nearby Frome Selwood, where he preached in private homes. His writings are all in the tradition of Puritan practical divinity. The *Vindiciae Pietatis* appeared in four parts between 1663 and 1668, and his *Instructions about Heart-Work* in 1681.

P.H. BUSS

ALLELUIA, see HALLELUJAH

ALLEN, ETHAN (1737/8-1789). American soldier and exponent of Deism.* Born in Connecticut, he became military leader of the Green Mountain Boys of Vermont in their struggle during the 1770s to maintain their land grants against the efforts of New York to take possession of the disputed land. During the American Revolution Allen captured Fort Ticonderoga. He then identified himself with the ideals of the French Enlightenment and became Deistic in outlook. In 1784 he wrote the first book to be published in the USA openly attacking the Christian religion, *Reason the Only Oracle of Man,* in which he rejected the Christian claim that the Bible is the special revelation of God to man. He asserted that priestcraft and superstition in the churches had imposed a tyranny upon mankind, but that their time was coming to an end. Knowledge of nature and science would exalt reason and bring men "back to the religion of nature and truth," free from clergy-imposed ignorance.

HARRY SKILTON

ALLEN, RICHARD (1760-1831). Founder of the African Methodist Episcopal Church. Born a slave, he was sold to a farmer near Dover, Delaware. Converted under Methodist influence, he was permitted to hold services in his home which resulted in the conversion of his master—and freedom for Allen and his family. He educated himself and preached while working at woodcutting and hauling. He was accepted as a Methodist preacher at Baltimore in 1784 and made preaching journeys with Richard Watcoat and Bishop Asbury.* He preached occasionally at St. George Methodist Church, Philadelphia, where his forceful approach attracted many Negroes, resulting in white protests. The former withdrew and formed the "Free African Society" (1787). From this body Allen influenced the majority to form the African Methodist Episcopal Church, which Bishop Asbury dedicated in 1794. Fifteen other Negro churches joined them. Allen was ordained in 1799 and became first bishop in 1816. Before his death he won national standing for the denomination.

J.G.G. NORMAN

ALLEN, WILLIAM (1532-1594). Scholar and cardinal. Born in Rossall, Lancashire, he went to Oriel College, Oxford, and in 1556 was chosen principal of St. Mary's Hall. When Elizabeth acceded, Allen was deprived and left England in 1561 for refusing the Oath of Supremacy. He was an ardent Romanist and soon gathered around him, at Louvain, other English refugees to study theology and reestablish the Roman faith. Allen himself returned secretly to England to encourage the recusants. He moved from place to place to avoid arrest and finally left England in 1565 for Flanders. After ordination as priest he lectured at the Benedictine College in Malines. He established a college at Douai in 1568, in response to the call of the Council of Trent,* and became a professor there. In 1575 he founded the English College in Rome and in 1589 a further college at Valladolid. He was the overseer of the Douai version* of the Bible project. All the while Allen continued his theological study and writing, and he regarded Protestantism in England as a temporary and passing phase to be hurried on as quickly as possible. His meeting with the Jesuit Robert Parsons brought a new element to his activities—that of political intrigue. Parsons dominated Allen; they both went to Rome in 1585, where Allen remained for the rest of his life. The Jesuit mission to England begun in 1580 was now under Parsons' control. Allen was made cardinal in 1587 and was intended to be the first Roman Catholic archbishop of Canterbury if English Protestantism were undone. He took part in numerous intrigues, writing a defense of the traitorous surrender of Deventer to the Spaniards by Sir William Stanley. He urged Roman Catholics to rebel against Elizabeth, but lost his influence after the defeat of the Armada. Later he came to regret his intrigue and Parsons' methods. He ended his days as Vatican librarian and in revising the Vulgate. While his scholarship and integrity are not in dispute, it cannot be doubted that his activities, with those of Parsons, gave the English real grounds for suspecting Jesuit institutions of sedition and treason.

G.E. DUFFIELD

ALLINE, HENRY (1748-1784). Leader of the Great Awakening* in Nova Scotia. Born in Rhode Island, he received little formal education. In 1775 he underwent an unusually powerful conversion experience which gave him a new sense of wholeness. At once he sensed a call to become a preacher of the Gospel. Initially hindered by his lack of theological education, a second moving experience gave him the conviction that he "needed nothing to qualify me but Christ." He began preaching as he wandered through the countryside and started the revival which became known as the "New Light" movement. He was considered a fanatic and destroyer by many in the established denominations, but he indirectly fostered the growth of the Baptist churches in Nova Scotia. He wrote hymns, a journal of the revival, and *Two Mites Cast Into the Offering of God for the Benefit of Mankind* (1804).

HARRY SKILTON

ALLIX, PIERRE (1641-1717). French Reformed pastor. He ministered at St. Agobile, then at Charenton, but revocation of the Edict of Nantes* in 1685 compelled him to flee to England, where he began a church for French exiles in London. He was a prolific writer, replying to Bossuet and seeking to show that the Albigensians* were true Christians and not heretics. Bishop Gilbert Burnet appointed him canon of Salisbury cathedral, and he received doctorates from both Oxford and Cambridge.

ALL SAINTS' DAY. A feast kept on 1 November in the West, and on the first Sunday after Pentecost in the East, to celebrate the fellowship of all Christians, in the Church Triumphant as well as on earth. Its origins are uncertain. A hymn by Ephraem (in 359) refers to a commemoration of all martyrs on 13 May, while a sermon of John Chrysostom (d.407) shows that Antioch remembered its martyrs on the Sunday after Pentecost. Such feasts soon included other saints besides martyrs. On 13 May, 609 or 610, Boniface IV received the Roman Pantheon from the Emperor Phocas (d.610) and dedicated it under the title *S. Maria ad Martyres.* The anniversary of this event was later observed as a major festival and may have been the origin of All Saints' Day. How there came to be a feast of all the saints on 1 November is unknown; but it possibly stems from the dedication on that day in St. Peter's basilica by Gregory III (731-41) of an oratory to "all the saints."

STEPHEN S. SMALLEY

ALMONER. One who is an official distributor of the alms of a person or institution. He may be acting for a religious house, a bishop, a prince, or a person of rank. In England the royal almoner distributes the royal almony on Maundy Thursday.* The term is sometimes applied to the chaplain of a hospital, infirmary, or orphanage.

ALMS. Almsgiving constituted standard OT righteousness (Deut. 15:7-11; Prov. 25:21f.; 28:27; Isa. 58:7-11), emphasized dominical precept (Matt. 5:42; Luke 12:33), and was a familiar apostolic virtue (2 Cor. 9:5-7; Heb. 13:16). The Hebrew term for "righteousness" had acquired by Mishnaic times (A.D. 200) the secondary meaning "almsgiving." The Talmud constantly advocates charity: impoverished, even aid-receiving, persons thus acquire virtue. *Didache* XV, 4, the earliest postcanonical Christian document emphasizing almsgiving, claims dominical authority. Generosity is never the root of justification, but merely the fruit of redeemed life (Rom. 5:1; cf. James 2:14-17). Good pagans practiced almsgiving, as is demonstrated by Egyptian tomb inscriptions about 2400 B.C., and by Confucianism and Buddhism nearly 2,000 years later. The good Buddhist seeking the "Noble Eightfold Path" of holy living also cultivates *Dana* (generosity, renunciation). Islam (A.D. 650 onward) requires legal and recommends voluntary almsgiving.

True poverty pleads for kindly benevolence; human hearts are naturally warm. Dangers exist, however: ostentatious pride in the benefactor (Matt. 6:1-4), habitual parasitism on the part of the recipient. The Mishnah pungently states that he who takes yet does not need will come to real hardship; he who needs yet does not take will live to endow others.

ROY A. STEWART

ALOGI. An obscure Christian group in Asia Minor about 175 which, in reaction to Montanism,* questioned the authority of those sacred books on which they based their claims. Hence they rejected *en bloc* the gospel of John and the Apocalypse which, according to them, were written by Cerinthus. They also objected to the Logos* theology of the Apologists. The nickname Alogi (Gr. *alogoi*) was scornfully applied to them by Epiphanius, who used it in a double sense to denote that they were "irrational" people who were without the "Logos."

ALOPEN (seventh century). First known Christian missionary to China. Native of Syria, he arrived in China in 635, according to the inscription on the "Nestorian Tablet," first erected in 781 and uncovered in 1625 by workmen near Sian. Alopen was received with honor by the Emperor T'ai-tsung, a monastery was built, sacred books were translated, and a measure of success gained. But Nestorian Christianity failed to survive persecution under subsequent emperors and was finally suppressed in 845, leaving no permanent influence on Chinese life and thought.

ALPHONSUS LIGUORI, see LIGUORI, ALPHONSUS

ALSTED, JOHANN HEINRICH (1588-1638). German Calvinist. Trained at the Reformed Academy of Herborn, he then studied at Marburg, Frankfurt, Heidelberg, Strasbourg, and Basle. Returning to his alma mater, he taught in the preparatory school and later joined the faculty of the philosophy department at Herborn. When dissension broke out between Reformed and Arminian, and the Synod of Dort* (1618) was called to settle the dispute, Alsted was chosen to represent his area. After the synod he became professor of theology. The Thirty Years' War brought devastation to the Rhineland, causing Alsted to leave Herborn and take a position as a teacher at Stuhl-Weissenburg in Transylvania, where he remained until his death. A prolific writer, he tried to unify all knowledge through an approach that combined Aristotelianism, Lullianism, and Ramism. The finest illustration of this work is his *Encyclopedia Septem Tomis Distincta* (1630). He was also a premillenarian and his *Diatribe de mille annis Apocalypticis* (1627; ET *Beloved City,* 1643) was a major influence in seventeenth-century English apocalyptic speculation.

See R.G. Clouse, "Johann Heinrich Alsted and English Millennialism," *HTR* 62 (1969), pp. 189-207; and F.W.E. Roth, "Johann Heinrich Alsted," *Monatshefte der Comenius-Gesellschaft* (1895), IV, pp. 29ff.

ROBERT G. CLOUSE

ALTAR (Lat. *altus,* "high"). A place where sacrifice is offered. There is frequent mention in the early period of the OT of altars at which various animals were ritually killed. In later years sacrifice was centered on the altar in the Temple at Jerusalem. The death of Christ, being "for all time one sacrifice for sins" (Heb. 10:12 NEB), put an end to sacrificial worship and hence any need for altars. The reference to an altar in Hebrews 13:10 is a reference to the death of Christ. Although both Greek and Latin writers in the early church used the term "table" for the place where the Eucharist was celebrated, the word gradually became replaced by "altar." This was a consequence of the Eucharist itself being regarded as sacrifice, which partly arose from patristic exegesis of Malachi 1:11, partly from the habit of signs being called by the names of that which they represented.

Altars were originally made of wood, but while the material varied in the East, in the West stone became increasingly the rule. As the material changed, so did the shape from a table to that of a tomb, possibly because the Eucharist was often celebrated at the tomb of a martyr. This latter practice is reflected in the way in which later altars were consecrated by having relics of martyrs placed within them, a practice supported by an interpretation of Revelation 6:9. As altars became fixed they acquired greater architectural importance; churches were built so that the altar was the focal point. In the beginning there was only one altar in each church, but gradually, perhaps under the influence of private masses, altars were multiplied. In Eastern Orthodox Churches there is only one altar, and only one Eucharist may be celebrated at it during any one day.

At the Reformation, change from the sacrifice of the Mass to the Lord's Supper meant that the altar was replaced both in name and fact by tables usually of wood. The rubrics of the Book of Common Prayer refer only to tables, and these have to be movable. The word "altar," however, is often popularly used to refer to the Communion table. Where reference is made to a high altar this means the principal altar of the church.

PETER S. DAWES

ALTAR FELLOWSHIP. Some denominations allow only those to receive Holy Communion at their altars who are from their own denomination or from denominations which are in doctrinal agreement with them. Churches which practice "close communion" or "fencing the table" have such restrictive policies. The Roman Catholic Church does not permit altar fellowship with other churches. In Lutheranism in America, the Akron (1872) and Galesburg (1875) rules were adopted which said, "Lutheran altars for Lutherans only."

ALTAR LIGHTS, see CANDLES

ALTHAUS, PAUL (1888-1966). Lutheran scholar. Born in Obershagen, he taught in Göttingen (1914), Rostock (1920), and Erlangen (1925). He was editor of *Das Neue Testament Deutsch: Göttinger Bibelwerk* and contributed extensively to New Testament studies by his expositions particularly of the Pauline Epistles. He became involved in the Synoptic problem for which he wrote *Fact and Faith in the Kerygma Today,* a translation of his *Die sogenannte Kerygma und der historische Jesus.* His chief contribution to systematic theology was *Die christliche Wahrheit: Lehrbuch der Dogmatik,* which reached its eighth edition in 1969. The doctrines of justification by faith in the theology of Martin Luther, law and gospel, and the problem of the relationship between church and state according to Luther are some of Althaus's major concerns. *Die Ethik Martin Luthers* combined his interests in Luther and ethics, the latter evidenced also in his *Grundriss der Ethik.* He investigated also Luther's doctrine of the Lord's Supper *(Die lutherische Abendsmahlslehre).* Although a Lutheran, Althaus frequently differed with Luther, not least in the doctrine of the Lord's Supper. One of his major emphases was on eschatology. His work on this subject, *Die letzten Dinge,* was in its ninth edition in 1964. His sermons, some of which were published, and his devotional writings added to his fame. CARL S. MEYER

ALUMBRADOS (or *Illuminati,* "enlightened"). A mystical Spanish sect of the sixteenth and seventeenth centuries. First appearing among the Franciscan friars about 1512, the movement emphasized passive surrender to, and personal sinless unity with, God as the object of the spiritual life. Sacraments and good works were consequently undermined. The criticisms it brought against the organized church gave it some common ground with Erasmianism and Lutheranism. In the mind of the Inquisition* it became identified in some measure with the latter, and its beliefs were condemned in 1525. From then the Alumbrados were persecuted, and indeed the sexual excesses of some of its leaders, such as Francisca Hermández, made them an easy target. Francisca exercised a fatal fascination for many, and the involvement of some of the Erasmian leaders with her helped to secure their downfall also. Many of the Illuminati, however, lived in a morally orthodox fashion. Ignatius Loyola* was temporarily imprisoned for suspected sympathy with them in 1527. C. PETER WILLIAMS

ALYPIUS (latter fourth/early fifth century). Bishop of Thagaste. Friend of Jerome and Augustine, he is mentioned as collaborating with Augustine* in the conversion of an Arian physician, Maximus, of the town of Thenae in Byzacena. With Augustine and five others he was a spokesman for the Catholic bishops at a conference between Catholics and Donatists at Carthage in 411. Again with Augustine he represented Numidia at a council at Carthage in 418, at which the Catholic view of original sin and of grace was set forth in nine canons. Alypius is mentioned also among the African bishops summoned to a council at Spoleto in 419 to settle the question of the rival claims of Eulalius and Boniface I to the papacy. He was also a friend and adviser of Pinianus and Melania, a notable Roman couple who had fled before the threat of Alaric and had finally settled at Thagaste. DAVID JOHN WILLIAMS

AMALAR OF METZ (also Amalaric or Amalarius) (c.780-c.850). Liturgical writer. His work as a disciple of Alcuin* furthered that fusion of Roman and Gallican practice which produced the medieval Mass. He attempted in his chief work *De ecclesiasticis officiis* and elsewhere a thoroughgoing allegorical interpretation of the Mass which gave every prayer, chant, and ceremony a symbolic reference to the life and work of Christ. Allusions to the OT and other teaching were also introduced. Much of the symbolism was highly artificial. When Amalar was appointed in 835 to administer the see of Lyons in the absence of Archbishop Agobard,* he found considerable opposition to his views, some of which were condemned by the Synod of Quiercy (838). Yet he set the trend for the future and actually influenced developments in ceremonial which were modified to fit in better with his overall pattern of symbolism. Apart from this he remains a fundamental source for the history of the liturgy, being the first to give evidence, for example, of the practices of incensing the altar and of reading the Gospel from a higher place than the Epistle.
JOHN TILLER

AMALRIC (d. c.1207). French mystic and philosopher. Born at Bena, near Chartres, he lectured at Paris in theology and philosophy, and enjoyed the favor of Louis VIII. His teaching, influenced by his study of J. Scotus Erigena,* contained pantheistic elements, and he was condemned in his own diocese. He was also summoned to Rome to appear before Innocent III to give an account of his beliefs. Returning to Paris, he recanted and died soon afterward. He evidently held that "God is all things" and that Christians are to accept that they are in the body of Christ and to walk in love in order to be forgiven. His followers, the Amalricians, extended his teaching, and seven of them were burned at the stake as heretics soon after his death. His teaching and that of his disciples was condemned at a synod in Paris in 1210 and five years later at the Fourth Lateran Council. PETER TOON

AMANA CHURCH SOCIETY. A Pietistic sect in Iowa, called also "The Community of True Inspiration," it includes about 730 members in seven congregations. The Amana Society arose in 1714 when a company of German Pietists was awakened by the message of Johann Rock and Ludwig Grüber, who claimed that the days of true and direct inspiration from God had not ended. After Rock's death in 1749 the movement waned until 1817, when three new "instruments of true inspiration," Michael Krausert, Christian Metz, and Barbara Heinemann, led in a renewal of the fellowship. To escape government persecution, over 800 migrated from Germany in 1842 and settled in a village called Ebenezer near Buffalo, New York.

In 1855 the society moved to Iowa, where Amana and six other small communities were established and incorporated in 1859 under the Amana name. These constituted for a time an outstanding experiment in communal living. Their worship was simple, consisting of hymns, testimonies, prayers, reading of Scripture (or the writings of the "Inspired"), and occasional exhortations by the elders. In 1932 the society was reorganized—many of the communistic practices were abandoned—and reconstituted as a corporation for profit. Today members might be called cooperative rather than communistic. They are stockholders in a multimillion dollar corporation, conducting fifty different businesses. Though still a dominant influence, the church is now separated from the business affairs of the community. Many of the society's traditional rules, such as those regarding pacifism and worldly amusements, have been surrendered. BRUCE L. SHELLEY

AMANDUS (584?-679). A founder of Belgian monasticism. Apparently at twenty he became a monk at Ye, and later he lived many years as a hermit. In 629, after consecration as a missionary bishop, he evangelized in Flanders and Carinthia. There is confusion about the next phase of his life. Temporary exile for criticism of King Dagobert I's moral behavior (628-39) might explain subsequent missionary work which could have been in the Danube or in the Basque country. Ultimately Dagobert forgave him and asked him to baptize his son Sigebert. Amandus then proselytized in the Scheldt and Scarpe river regions, becoming bishop of Tongeres-Maasricht about 649. Founder of eight abbeys, he retired to one at Elnon near Tournai as abbot in 675 and there wrote his *Testamentum,* the only reliable evidence about his life. L. FEEHAN

AMBO. The raised platform in a basilica from which the Scriptures were read. Subsequently there were two ambos, for the Epistle and Gospel respectively, on the south and north sides. The pulpit replaced them after the fourteenth century.

AMBROSE (c.339-397). Bishop of Milan. Born at Trier in Gaul into the Christian family of Aurelius Ambrosius, the praetorian prefect of Gaul, he trained in law, followed his father into an administrative career, and about 370 was appointed governor of the province of Aemilia-Liguria, the leading town of which was Milan. When Auxentius, bishop of Milan, died in 374, Ambrose (an unbaptized catechumen) was baptized, ordained, and consecrated bishop. His first act as bishop was to distribute his great wealth among the poor. He was outstanding as preacher and teacher; his *De Fide, De Spiritu Sancto,* and *De Mysteriis* are testimony to his diligence in teaching the faith and refuting heresy. Among the many influenced by him was Augustine,* whose fame and ability were one day to eclipse even his own. Ambrose was also a fearless church leader. Events brought him into contact with the rulers of the West. When Theodosius had put down a seditious movement in Thessalonica with exceptional severity by killing thousands of people, Ambrose wrote to him refusing the sacrament of Holy Communion until he had openly made penance. Ambrose's attitude was to affect profoundly relationships between church and state for generations to come. "The emperor," he declared, "was within the church and not over it." Ambrose was influential also in encouraging monasticism in Italy and molding psalmody and hymnody in the direction of congregational participation. His main work, *De Officiis Ministrorum,* was a book on Christian ethics for the clergy.

See F. Homes Dudden, *The Life and Times of St. Ambrose* (2 vols., 1935); and W.G. King, *The Emperor Theodosius and the Establishment of Christianity* (1961). G.L. CAREY

AMBROSE, ISAAC (1604-1664). Nonconformist* minister. Born in Ormskirk, Lancashire, he was educated at Brasenose College, Oxford, and ordained to the parish of Castleton, Derbyshire. Appointed in 1634 as one of the king's four itinerant preachers in Lancashire, he became vicar of Preston, where he developed Presbyterian leanings. In the Civil War he was twice captured by the royalists, but served on the committee for the ejection of scandalous and ignorant clergy and schoolmasters. In 1654 he left Preston for Garstang, but was turned out in the 1662 ejection. He was a vivid writer and was much given to prayer and meditation. After recovering from an illness he determined to write his experiences, and in 1658 his *Looking unto Jesus* was published and achieved considerable popularity.

G.E. DUFFIELD

AMBROSIANS. Anabaptist* sect. One of the biblical doctrines rediscovered at the Reformation was the priesthood of all believers: every believer could have direct access to God without the intervention of a human priest, and every believer was called to Christian witness and service. Inevitably, reaction against the clericalism of the medieval church became in some instances overreaction. Many of the Anabaptist groups denounced by Luther overstressed the direct operation of the Holy Spirit in the individual soul. Among these were the Ambrosians, named after their leader Ambrosius, who based their theology on their interpretation of John 1:9. If there were in fact direct illumination from God in every soul, then there was no need of a formal order of priests or ministers to interpret the Bible. The Bible itself was not the only authoritative medium of divine revelation—Ambrose held that the spiritual and direct revelations given to him had a higher authority than that of the Scriptures. To some extent the Quakers were the lineal descendants of groups such as the Ambrosians.

The name of Ambrosians had been used earlier by an order founded under the patronage of Ambrose of Milan. Given the Rule of Augustine in 1375, the order was dissolved in 1650.

HUGH J. BLAIR

AMBROSIASTER. A pseudonym applied since Erasmus to a commentary on the Pauline epistles which appeared in Rome about 375, falsely attributed to Ambrose* of Milan. The pseudo-Augustinian *Quaestiones Veteris et Novi Testamenti* are also attributed (e.g., by J.N.D. Kelly) to this author. His identity is not yet established. Augustine attributes part of the commentary on Romans 5:12 to "sanctus Hilarius," possibly the Christian *praefectus urbi* at Rome in 383. Ambrosiaster relates Pauline teaching to contemporary legal institutions (Heggelbacher). Like Ambrose and Jerome, he believed the Cross to have broken the hold which the devil gained on men at the Fall, and the eucharistic elements he saw as "types" of

Christ's body and blood. His eschatology included a Millennium. Original guilt he did not hold, but his understanding of universal sin in Adam *quasi in massa,* based on the Old Latin mistranslation of Romans 5:12, did lead on to the Augustinian doctrine. G.T.D. ANGEL

AMERICAN ANTI-SLAVERY SOCIETY. Established in 1833 in Philadelphia by members of state and local abolition* societies. About 1830 the abolitionist movement began to organize and spread quickly as a religious and humanitarian crusade. William Lloyd Garrison, an outspoken radical leader of the abolitionists in New England, in 1832 organized the New England Anti-Slavery Society. Abolition of slavery in the British Empire by the British government in 1833 led American abolitionists to unite quickly to form a national organization. Western leaders of the revivalist reform movement which had arisen from the work of C.G. Finney* joined the generally more radical New Englanders to form the American Anti-Slavery Society with the announced goal of immediate abolition, but with the expectation by many members that the process of emancipation would be gradually and moderately accomplished. Influential evangelical supporters of the society were the philanthropists Arthur and Lewis Tappan, and T.D. Weld,* an eloquent spokesman, organizer, and traveling agent of the national movement. Weld at Lane Seminary was successful in training effective antislavery agitators who converted entire communities to an awareness of the sinfulness of slavery and the need to abolish it. But the national society lacked unity and was considered too radical by many advocates of abolition. Differences among the factions within it led the anti-Garrison abolitionists to leave it in 1840, thus bringing its effectiveness to an end. HARRY SKILTON

AMERICAN BAPTIST CHURCHES. Fourth largest of a dozen major and scores of minor Baptist denominations in the USA, with over 6,000 congregations. From 1950 to 1972 the group was known as the American Baptist Convention, and from 1907 to 1950 as the Northern Baptist Convention. Before 1907 the congregations were joined only in local and state associations and were served on the national level by specialized agencies, the most important of which were the American Baptist Foreign Mission Society (founded 1814), the publication society (1824), and the home mission society (1832). By the close of the nineteenth century, the American Baptist membership and ministry was reluctantly but effectively confined chiefly to the northern and western states. For the ABC, the twentieth century has brought three major developments: considerable theological diversification; increased organizational coordination; and resulting loss of the overwhelming predominance it had long held among Baptists outside the South.

Northern Baptists in the twenties shared in the modernist–fundamentalist controversies, and tension has continued on various fronts. Conservative seminaries were formed near the older, changing schools. Northern Baptist Seminary began in 1913; Eastern started in 1925 near Crozer in the Philadelphia area; and a seminary began in Covina, California, in 1944 even though there was already one in the state, at Berkeley. Extremists on both sides have left the convention over the years, but the ABC still includes probably a greater diversity of theologies with strength than any other American denomination. This diversity is on fundamental issues such as the deity and the second coming of Christ.

Financial pressures and desires for efficiency led many to seek the formation of the convention in 1907 and even greater coordination of the various districts and specialized agencies. Conservatives tended to distrust such moves because theological liberals and moderates effectively controlled most of the denominational machinery and were able to repel the major attempts to wrest it from them in the annual conventions of 1922 and 1946.

Unlike the Baptist groups in the South, the ABC has always been comparatively small. Only in Maine and West Virginia have Baptists been as much as ten percent of the churchgoing population. In the Midwest and West the approximately 2,500 American Baptist congregations are now greatly outnumbered by Southern Baptist* congregations (3,400 outside the South), and former ABC-linked groups including the General Association of Regular Baptist Churches (1,400), Conservative Baptist Association (1,100), Baptist General Conference (600), and North American Baptist General Conference (300).

Although other Baptist denominations, including some that withdrew from the SBC, have shown greater growth rates, the American Baptists did unite most Arminian-rooted northern congregations beginning in 1910, in recent years have increased black participation in their activities, and have gained some congregations from the South. The ABC and many of its members have made important contributions to the ecumenical movement, though not participating in the Consultation on Church Union. Burma, China, Haiti, India, Japan, the Philippines, Puerto Rico, Thailand, and Zaire are some of the countries where ABC missionaries and funds have played a significant role in developing Protestantism.

BIBLIOGRAPHY: P.M. Harrison, *Authority and Power in the Free Church Tradition: A Social Case Study of the American Baptist Convention* (1959); R.G. Torbett, *A History of the Baptists* (rev. ed., 1963); D.C. Woolley, *Baptist Advance: The Achievement of the Baptists of North America for a Century and a Half* (1964).

DONALD TINDER

AMERICAN BIBLE SOCIETIES, see BIBLE SOCIETIES

AMERICAN BOARD OF COMMISSIONERS FOR FOREIGN MISSIONS. The first American foreign missionary society. Organized in 1810 in Massachusetts in response to a request for guidance by a group of missionary-minded students, it sent out the first contingent of missionaries to India in 1812, resulting in missions at Bombay and in Ceylon. In 1818 it appointed the first two Protestant missionaries to the Near East, and in 1819 sent out the first missionary group to Hawaii. Within fifty years it had also established missions in other parts of Asia, including China and Japan, in Africa and Micronesia. It engaged also in home missions on the frontiers, among Indians and Negroes. Though started by Congregationalists, for many years the board served also Presbyterians, Dutch Reformed, and German Reformed, until these churches set up their own mission societies. When the Congregational-Christian churches and the Evangelical and Reformed Church merged in 1961, the ABCFM became part of the United Church Board for World Ministries.

HAROLD R. COOK

AMERICAN COLONIZATION SOCIETY. Formed in 1816 to return freed slaves to Africa, the society established the country of Liberia. Started by minister Robert Finley with government help, its members hoped that Christian freedmen might help evangelize Africa. Even some slaveholders supported the scheme. It never, however, gained a broad base of support. After initial difficulties, a tract of land was secured for a colony. Disease almost wiped out the first group of 114 settlers in 1820-21. In 1822 a group of fifty-three under minister Jehudi Ashmun made a permanent settlement near Monrovia. The Maryland branch of the society in 1833 established a colony at Cape Palmas. When Britain refused to recognize the authority of the society to govern, an independent government was set up in 1847, patterned after and sponsored by the United States. By 1867 about 10,000 freedmen had been transported to the colony. HAROLD R. COOK

AMERICAN COUNCIL OF CHRISTIAN CHURCHES. An association of militant fundamentalists organized under the leadership of Carl McIntire in 1941 to promote and defend the historic orthodox Protestant faith, and to counter the activities of the Federal (since 1950 the National) Council of Churches. Its membership includes conservative separatist denominations such as the Bible Presbyterian Church and the Bible Protestant Church, as well as individuals from churches which belong to the National Council. The ACCC seeks to bring about a reformation of the doctrine and practice of the American churches on the basis of acceptance of the entire Bible as the verbally inspired, inerrant, authoritative Word of God. The NCC is considered apostate because it is tolerant of denials of orthodox doctrine and promotes programs of a communist and pacifist nature. True believers are urged to leave their corrupt denominations and join or form pure churches. In recent years the movement has been disrupted by serious internal squabbles.

HARRY SKILTON

AMERICAN INDIANS, CHRISTIANITY AND. The motivation to discover and develop the New World derived from several sources, political and economic as well as religious. When in June 1523 Charles V instructed Lucas Vasquez de Ayllon regarding his mission to the New World, he stated that "the chief motive you are to bear and hold in this affair" was the conversion of the Floridian Indians. Although Vasquez de Ayllon was not successful, by 1634 Florida had forty-four missions conducted by thirty-five Franciscans, and the converts from the Indian population numbered more than 25,000. In the Far West, much the same development occurred. In New Mexico in 1630, the number of Christian Indians numbered more than 35,000. In 1609, the governor and councillors of Virginia, an Anglican colony, issued a proclamation that "the principal and Maine Endes . . . were first to preach and baptize into Christian Religion, and by propagation of the Gospell, to recover . . . a number of poore and miserable soules."

The Puritans likewise sought to extend their theocratic kingdom to the Indians. The charter of the Massachusetts Bay Colony included the Macedonian call of Acts 16:9 as a description of the spiritual plight of the Indians. But after direct contact with the Indians and especially after the Pequot War of 1637, the Puritan attitude was one of mixed pity and hatred. The Puritans were fascinated by the natives and speculated as to their origins. Some held they were a cursed race and were therefore prime subjects for slaves. Many were sold into bondage. Thomas, John, and Experience Mayhew did carry on successful work among the Indians of Martha's Vineyard; Experience translated the Psalms and the gospel of John into the Indian language and published a book, *Indian Converts* (1727). John Eliot* translated the Bible and published a *Catechism* (1653), the first book to be printed in the Indian language. Through the influence of Eliot and Thomas Shepard, the Long Parliament established the "Society for the Propagation of the Gospell in New England" in 1649. Periodic uprisings constantly set back the progress of these missions, and few Indians were able to gain the fame of Pocahontas.

But preaching and churches were not the only attempts to improve the state of the natives. In 1618 in Virginia, separate schools, even a college, had been established to educate the Indians. Rev. Eleazer Wheelock in the 1750s organized Moor's Indian Charity School (now Dartmouth College) in Lebanon, Connecticut, to train Indians to minister to their own people. Throughout the colonial period the Indians were caught in the cross-fire of either their own conflicts with warring tribes or the struggles for power in the New World among the Spanish, French, and British. With the westward expansion the Indians were gradually moved to government-established reservations, and during the late nineteenth and most of the twentieth century, the major American denominations as well as many independent missionary agencies have attempted not only to convert the Indians to the Christian faith but to help them adjust to the modern world. One peculiar fact about the Indians always stimulated discussion: whence did they come? Joseph Smith in his *Book of Mormon* identified them with the "Ten Lost Tribes of Israel."

See W. Howitt, *Colonization and Christianity: A Popular History of the Treatment of the Natives* (n.d.); A.T. Vaughn, *New England Frontier: Indians and Puritans, 1620-1675* (1965).

DONALD M. LAKE

AMERICANISM. A condemnatory term for the adaptation of church doctrine and practice to American culture which provoked controversy (and condemnation by Pope Leo XIII) within the Roman Catholic Church in the late nineteenth century. The trouble originated in conflict between progressives and traditionalists in the American Church over the value of parochial schools, and whether it was wise to try to preserve the native language and culture of immigrants in order to protect their faith, or rather to help them adopt American customs. When the biography of an American progressive priest, I.T. Hecker,* by W. Elliott was translated into French, conservative French priests denounced the ideas contained in it as "Americanism." American progressive bishops such as James Gibbons* were accused of subverting the Faith. In his apostolic letter of 1899 addressed to Gibbons, Leo condemned such errors as the rejection of religious vows, the assertion that external religious authority is unnecessary in a time of liberty, and the view that natural and active virtues are more valuable in the modern world than are supernatural and passive. Gibbons denied that such views were held by American Catholics.

HARRY SKILTON

AMERICAN LUTHERAN CHURCHES, see LUTHERAN CHURCH BODIES IN THE USA

AMERICAN METHODISTS, see METHODIST CHURCHES, AMERICAN

AMERICAN NEGRO CHURCHES. Removed from their African environment, slaves in colonial America were introduced to the Christian religion which helped them adjust to the social structures of a new civilization. At first there was little organized effort to evangelize them. Slaves ordinarily attended the church of their master or were provided minimal religious instruction by master, pastor, or missionary. After the Great Awakening* in the eighteenth century, aggressive evangelists, particularly Baptists and Methodists, reached them with a simple, personal, and emotional Gospel that injected new meaning and hope into their lives.

Except among free Negroes, the Negro church rarely emerged as an independent institution before the Civil War. Separate congregations existed, some under white and some under black leadership, but most Negro church members belonged to congregations where membership was shared with white members who were frequently in the minority. After the Revolutionary War there were numerous secessions by free Negroes from white churches.

The first known Negro church in America was a Baptist church founded at Silver Bluff, South Carolina, in 1775. Soon Baptist congregations were formed in Savannah (1788), Boston (1805), New York (1807), Philadelphia (1809), and subsequently in many other places. Several Methodist congregations were established at the end of the eighteenth century, but they were soon organized into Negro denominations. The African Methodist Episcopal Church was founded in Philadelphia (1816) and the African Methodist Episcopal Zion Church in New York City (1821).

After the Civil War, Negro church organizations grew rapidly as ex-slaves, withdrawing from white churches, were for the most part absorbed by the institutions begun by free Negroes before the war. In 1870 the Christian Methodist Episcopal Church, originally part of the Methodist Episcopal Church, South, was constituted a separate denomination. In 1886 the majority of Baptists were brought together in what was to become the National Baptist Convention, U.S.A., Inc., from which the National Baptist Convention of America separated in 1916; these are the two largest Negro denominations in America. The Negro church—the one institution in which Negroes could find self-expression, develop leadership, and provide social services—became and still remains for many the most important agency for the achievement of a sense of community and status.

More than two-thirds of Negro church members are concentrated in five denominations: National Baptist Convention, U.S.A., Inc. with 5.5 million (1958); National Baptist Convention of America with 2.669 million (1956); African Methodist Episcopal Church with 1.66 million (1951); African Methodist Episcopal Zion Church with 940,000 (1970); and Christian Methodist Episcopal Church with 467,000 (1965). Negroes are also found in smaller denominations and in predominantly white denominations, though usually in black congregations. Of white Protestant denominations, the United Methodist Church has the largest Negro membership. Since World War II, the Roman Catholic Church has become one of the leading religious bodies among Negroes, and its Negro membership is believed now to equal that in predominantly white Protestant denominations.

The migration to the cities after World War I contributed to the rise of "storefront" churches and numerous organized cults for Negroes who did not feel at home in more conventional churches. Typical of many groups that were hostile to traditional religious expressions are the Black Muslims, Black Jews, and Father Divine's Peace Mission.

Negro churches follow American religious patterns, reflecting in large measure the other churches of corresponding or parent denominations and of the same educational and economic level. Their theology is typically fundamentalist, pietist, and evangelical. The "otherworldly" emphasis of the "spirituals" is still prominent; the Church remains a refuge and a source of hope in a hostile world. As Negro education and economy improve, their theology and worship become more intellectual, sophisticated, and emotionally disciplined.

The Negro community is "overchurched," with a higher ratio of ministers and churches than the general population. The minister has the historic role of leadership. Unfortunately, too few possess adequate formal education, and the supply does not appear to be increasing.

The Negro church is predominantly urban, with more and more of its members sharing the middle-class ideals and common secular attitudes. Increasing emphasis upon this world has induced many church leaders to head movements for civil rights, economic justice, and better educational opportunities. The expectation among Negroes is substantial and growing that churches should be involved in community improvement through cooperation with other organizations and through social and political action. The struggle for racial and social justice has created a crisis of role and identity for the church and its leaders.

The Negro church exists for the purposes of a yet deprived and troubled people. As long as Negroes are excluded from or are uncomfortable in white churches, the Negro church will remain an instrument for the expression of racial identity and a medium for shaping and expressing their aspirations.

BIBLIOGRAPHY: W.E.B. Dubois, *The Negro Church* (1903); C.G. Woodson, *History of the Negro Church* (1921); B.E. Mays and J.W. Nicholson, *The Negro's Church* (1933); A.H. Fauset, *Black Gods of the Metropolis* (1944); H.V. Richardson, *Dark Glory: A Picture of the Church among Negroes in the Rural South* (1947); F.S. Loescher, *The Protestant Church and the Negro* (1948); R.F. Johnston, *The Development of Negro Religion* (1954); E.F. Frazier, *The Negro Church in America* (1964); J.R. Washington, Jr., *Black Religion: The Negro and Christianity in the United States* (1964); H.V. Richardson, "The Negro in American Religious Life," in *The American Reference Book* (ed. J.P. Davis, 1966), pp. 396-413.

ALBERT H. FREUNDT, JR.

AMERICAN REVISED VERSION, see BIBLE (ENGLISH VERSIONS)

AMERICAN STANDARD VERSION, see BIBLE (ENGLISH VERSIONS)

AMES, WILLIAM (1576-1633). Puritan theologian. Educated at Christ's College, Cambridge, where he was tutored and greatly influenced by W. Perkins,* he was suspended for a sermon attacking card-playing and for refusing to wear the surplice, and prevented from seeking a pastorate at Colchester by the bishop of London. He became chaplain to Sir H. Vere, English governor of Brill in Holland. After attending the Synod of Dort*—where his theological acumen became apparent—as an English observer, he became professor of theology at Franeker in 1622 and rector in 1626. Ill health led to his retirement, and he died within a year. His great reputation as a theologian and marked ability as a teacher attracted students from all over Europe, but in contemporary opinion his genius was better adapted to the professor's chair than to the pulpit. A considerable controversialist, against Anglicanism *(Fresh Suit against Roman Ceremonies)*, Arminianism *(Medulla Theologiae)*, and Remonstrants *(Animadversiones in Synodalia)*, he was also a careful casuist. His *De Conscientia, eius Iure et Casibus* was one of the first systematic Protestant attempts to clarify general principles.

G.S.R. COX

AMILLENNIALISM (Amillenarianism). A particular interpretation of Revelation 20. The premillennialist maintains this chapter teaches a thousand-year reign of Christ after His second advent, the postmillennialist before the second advent, while the amillennialist denies such a thousand-year reign: he stresses that the Apocalypse normally treats numbers symbolically. The binding of Satan for a thousand years simply means that he is completely bound; this has been effected through the victory of Calvary. Some amillennialists hold the expression to refer to the Church's rest from spiritual conflict beyond death. Most apply it, however, to her present victory over Satan in Christ crucified and exalted. Many Reformed and Lutheran theologians hold this view, and elements of it can be traced in Augustine.

G.W. GROGAN

AMISH, see MENNONITES

AMMONIAN SECTIONS. Most Greek and Latin manuscripts of the four gospels contain divisions into longer or shorter sections, which can be collocated in parallel columns to reveal synoptic correspondences. These are called "Ammonian Sections," since Eusebius* of Caesarea attributes to Ammonius of Alexandria an edition of Matthew containing parallel passages from the remaining three gospels in the margin. Eusebius himself numbered the sections and arranged them into his ten "canons." This Ammonius (c.220?) was neither the Neoplatonist philosopher nor the commentator on Daniel, John, and Acts. It is uncertain how far Eusebius himself was responsible for the divisions.

G.T.D. ANGEL

AMMONIUS SACCAS (c.174-c.242). Philosopher and teacher. Thought originally to have been a porter, he taught rhetoric and an eclectic and esoteric form of Platonism at Alexandria. His pupils included Origen, Plotinus the famous Neoplatonist, and a pagan Origen. The pagan biographer Porphyry claimed that learning made Ammonius reject his Christian upbringing for traditional paganism and that he left no writings. Eusebius is regarded as mistaken in his counterclaim that Ammonius was consistently a Christian philosopher and left such writings as *On the Agreement of Moses and Jesus.* His theories on Providence, the soul, and the cosmos echo the Christian Origen.

G.T.D. ANGEL

AMOUN (d. c.348). Ascetic. A native Egyptian Christian, he began semi-eremitic asceticism about 325 at Nitria in the Wadi Natron, where his followers traded niter. Out of sight or sound of one another during the week, his hermits met for worship both on the Sabbath and on the Lord's Day, according to Egyptian custom, at a central church governed by a disciplinary "Sanhedrin." His example inspired Macarius of Scete and was admired by Antony.

AMPHILOCHIUS (d. after 394). Bishop of Iconium. A lawyer at Constantinople, he became bishop in 373 at the instigation of his friend Basil (the Great)* of Caesarea. Associated also with Gregory of Nyssa and Gregory of Nazianzus, he affirmed the Cappadocian Trinitarian model of one substance and three modes of existence or relation. His Christology held fast, without change or confusion in either, both a divinity consubstantial with that of the Father and a humanity preserving free will. The surviving remains of his large literary output reveal wide interests. They are (1) thirty-three *Laudi ad Seleucum* on devout living and successful study, a list of biblical books putting the Apocalypse outside the canon; (2) eight sermons on church feasts and texts of Scripture; (3) a Coptic treatise against the Apotactites and Gemellites, hyperascetic sects which, in addition to the Arians and Messalians, Amphilochius opposed vigorously. G.T.D. ANGEL

AMSDORF, NICHOLAS VON (1483-1565). German Reformer. Born at Torgau near Leipzig, he began his education at Leipzig (1500), then was one of the first students at Wittenberg (1502). There he came under the influence of Luther and became his close friend and ardent defender. Lecturer in theology, philosophy, and canon at Wittenberg from 1508 and professor from 1511, he accompanied Luther to the Disputation at Leipzig in 1519 and to Worms in 1521. He was ordained in 1524 and became pastor and superintendent at Magdeburg, where he introduced the Reformation along the lines established under Luther in Wittenberg. He helped to reform Goslar (1531) and Einbeck (1534). In 1539 he disagreed with Luther's advice on the bigamous marriage of Philip Landgrave of Hesse. He attended the Regensburg Conference in 1541 and allegedly was partly responsible for its failure, his position being described as "fearless as it was narrow." John Frederick I, elector of Saxony, over the objections of the chapter but with the support of Luther, placed Amsdorf in the position of Lutheran bishop in Naumberg-Zeitz in 1542. After the Protestant defeat at Muhlberg in 1547, Amsdorf went to Magdeburg where he was a counselor to the dukes of Eisenbach.

Much of Amsdorf's life was spent in acrimonious theological disputation. Among others, he wrote against Melanchthon, Bucer, Melchior Hoffman, George Major, the Zwinglians, and anyone he considered outside the pale of pure Lutheran doctrine. This was apparently part of his motivation for founding Jena University, the calling of Matthias Flacius* to teach and assist him there, and the issuance of the Jena edition of Luther's works. He was sure the Wittenberg edition was full of error. During the Synergist controversy (see SYNERGISM) his contentiousness carried him to the extreme of arguing that good works were not only useless but harmful. This was criticized in the Formula of Concord.* Many of the letters and works of Amsdorf survive. When Flacius and his followers were forced out of Jena, Amsdorf was allowed to stay due to his advanced age and former association with Luther.

BIBLIOGRAPHY: Biography by E.J. Meier in M. Meurer, *Das Leben der Altväter der lutherischen Kirche,* III (1863); Amsdorf's *Ausgewahlte Schriften* (ed. O. Lerche, 1938); studies of Amsdorf by O.H. Nebe (1935), H. Stille (1937), and P. Brunner (1961). ROBERT SCHNUCKER

AMSTERDAM ASSEMBLY. Following the Oxford Conference* of 1937 it was decided to set up a committee to plan to inaugurate a World Council of Churches which would carry on the interests of both the Life and Work and the Faith and Order conferences. World War II prevented any early realization of this plan, although member churches had approved the formation of such a council, and a provisional committee existed. Fears that the ecumenical movement might be hindered by the war guilt question were dispelled by the German delegation's acceptance of this with the Stuttgart declaration.

In 1948 the long-postponed assembly met at Amsterdam, and on 23 August the WCC came into being. The basis was: "The World Council of Churches is a Fellowship of churches which accept our Lord Jesus Christ as God and Saviour." As before, it was stressed that the WCC was not a superchurch and that its pronouncements would carry no external authority but only "the weight it carries with the churches by its wisdom." The conference was attended by 351 delegates from 147 churches. The Roman Catholic Church was invited, but permission was not given by that communion for any to attend. The Orthodox were only partly represented. Although representation from the younger churches would by modern standards be considered poor, it was considerably more than at any previous conference. The conference was still, like its predecessors, overwhelmingly Western in outlook. Amsterdam was a landmark in that the churches accepted responsibility as churches for the ecumenical movement, and vice versa the ecumenical movement became more fully rooted in the participating churches.

See the Official Reports, ed. W.A. Visser 't Hooft; and R. Rouse and S.C. Neill (eds.), *A History of the Ecumenical Movement: 1517-1948* (rev., 1967). PETER S. DAWES

AMYRALDISM. The doctrine that God wills all men to be saved, on condition that they believe. Expressed by Moses Amyrald (1596-1664), a French Protestant pastor, the doctrine was designed to be Calvinistic rather than Arminian and to provide the basis for conciliation between Reformed and Lutheran theology. Agreeing with Calvin on the absolute sovereignty of God, expressed in history through the realization of God's sovereign purpose, Amyrald affirmed that individual redemption and the establishment of the kingdom of God were wholly divine prerogatives and could in no case be either accidental or contingent. Nevertheless, he found scriptural warrant for a universalism in the divine decree to salvation, averring that God wills all men to be saved.

But this universalism in the system of Amyrald does not issue in the actual salvation of all men—

by virtue of the corresponding universalism of man's sinfulness. The commonness of man's sin has destroyed the true purpose and end of life provided by God through His providence, in accordance with His love. Hence the universalism is seen to be purely ideal or hypothetical. Yet, though man's sin is universally corruptive, God's goodness remains infinite, expressing itself to all persons and throughout all history, since His *desire* to save remains forever unabated. This means that the salvation of individual persons is quite unaffected by the ideal universalism explicit in the system. The result is a combination of ideal universalism and of actual particularism.

Though it had a wide following during the latter half of the seventeenth century, Amyraldism was much opposed, especially by French and Swiss scholars. The opponents of the scholastic Calvinists were in no way satisfied by the Amyrald interpretation, arguing that the universalism, being only hypothetical, to no degree mitigated the absolute predestinarianism of Calvinism. The purer Calvinists tended to look upon Amyraldism as an inconsequential, even misleading, addendum to the *Institutes*, creating, rather than resolving, problems relating to the basis of salvation: God wills all to be saved, but man's sin prevents any from being saved. The view, however, may be more in keeping with Calvin's own theology than with that of his scholastic interpreters.

MILLARD SCHERICH

ANABAPTISTS. These groups, variously called the radicals or left wing of the Reformation, agreed in denouncing the baptism of infants. They held that only those who were old enough to understand the meaning of faith and repentance should be baptized. The majority of Christians regarded the baptism of infants as a most important Christian ordinance and as initiation into a state church. Generally the Anabaptists showed a deep moral earnestness, insisting on the primacy of Scripture and the separation of church and state. Some were millennialists, and others were pacifists and distrusted the state. They believed in a pure believers' church and strict church discipline.

The most biblical Anabaptism appeared and flourished in Switzerland, where it developed in Zurich in the time of Zwingli under the leadership of Conrad Grebel* and Felix Manz*; south Germany, where Balthasar Hubmaier and Hans Denck* were the leaders; Moravia, where the Hutterites* were located; and the Netherlands and N Germany, where the Mennonite* movement grew. The movement began in 1523 in Zurich, where the Reformation caused the questioning of traditional values including the rite of baptism. Grebel and Manz preached and baptized adults in the Zurich area. Their success brought both converts and official persecution in 1526. Manz was drowned, and many of his followers and fellow preachers were exiled. Those who remained in Switzerland went underground, where the movement continued until the seventeenth century. Anabaptists who left this area spread the movement into S Germany and Moravia.

Strasbourg became the center for Anabaptism in Germany from 1527 until the establishment of the Magisterial Church. In 1533 both Capito and Bucer, leading reformers of that city, upset by the separatist tendencies of the Anabaptists, agreed in opposing them. Melchior Hoffman was another Anabaptist preacher who lived for a time in S Germany. His preaching bore little fruit in Strasbourg, but it led in Münster to the Anabaptists' gaining control. A man named John Matthys became leader, claiming that he was Enoch who should prepare the way for Christ by establishing the community of goods and doing away with all law codes. Many hundreds in the city were baptized, and "the ungodly" who would not submit to rebaptism had to flee or be slaughtered. Despite a siege and the death of Matthys, the Münsterites held out for more than a year before the defense collapsed; there followed the slaughter and torture of the defenders. This episode discredited the Anabaptist movement, and a wave of persecution swept the Low Countries. Tens of thousands of Dutch Anabaptists died during the sixteenth century, but from this persecution emerged the Mennonites.

The Anabaptists developed along with the Magisterial Reformation, but rejecting some of the important features of this reform, such as infant baptism and the state church, they became prophetic of free church life in our own time, and ancestors of the Baptists, Mennonites, and Schwenkfelders.* The Left Wing Reformation upheld the need for toleration and sealed this testimony with blood.

BIBLIOGRAPHY: F.H. Littell, *The Origins of Sectarian Protestantism* (1964); G.R. Elton (ed.), *The New Cambridge Modern History,* II, pp. 119ff.; G.H. Williams, *The Radical Reformation* (1964).

ROBERT G. CLOUSE

ANACLETUS (Gr. *anegklētos*, "blameless"). According to tradition the third bishop of Rome, following Linus (64-76) and preceding Clement (88-96). Anacletus or Anencletus is to be identified with Cletus, although the Liberian Catalogue assumes that Anencletus and Cletus were different persons. The ascription of twelve years to his office and that of Linus may indicate that tradition was anxious to provide a link between Peter (d. 64) and Clement of Rome. It is unfortunate for this theory that no trace of monepiscopacy can be found in Rome until the middle of the second century.

ANAPHORA (Gr. = "offering"). The central prayer in the Eucharist, containing the Consecration, the Anamnesis, and the Communion. The traditional order is Sursum Corda, Sanctus, Memorial of the Incarnation, Words of Institution, Epiclesis, and sometimes an Intercession.

ANASTASIA. (1) Daughter of Constantius Chlorus (293-306), father of the sole emperor, Constantine. Her name, based on the Greek for "resurrection," is paralleled elsewhere in the late third century only among Jews and Christians, and therefore she is regarded generally as evidence of Christian influence within the home of Constantius.

(2) The name given to an oratory by Gregory of Nazianzus after his translation to the see of Constantinople in 379, to mark the resurrection of the Nicene faith after the supremacy of Arianism from 360 onwards.

(3) A fourth-century saint apparently martyred at Sirmium in Pannonia. G.T.D. ANGEL

ANASTASIUS BIBLIOTHECARIUS (c.810-c.880). Antipope. Family connections and a knowledge of Greek learned from Greek monks resident in Italy helped his early promotion to cardinal-priest by about 847. Unexplained events led to his excommunication (850), anathematization, and deposition (853). In 855 Benedict III was elected pope, but a rival group chose Anastasius, who captured the Lateran Palace and Benedict himself. The imperial legate mediated in favor of Benedict, who forgave Anastasius, and henceforward the latter remained a loyal and influential official. Abbot of Santa Maria, Trastevere (858-67), he then became papal librarian (hence "Bibliothecarius"). In 868 a cousin murdered Pope Adrian II's daughter and her mother; this was a temporary setback to Anastasius's fortunes, but by 869 he was back in papal favor and acting unsuccessfully as Louis II's negotiator in Constantinople for a marriage between his daughter Ermengard and the Eastern emperor Basil I's son. In Constantinople Anastasius successfully championed the papal claims of supremacy at the Eighth Ecumenical Council (Fourth Council of Constantinople, 869-70) which upheld Nicholas I's deposition of the Byzantine Patriarch Photius* and condemned his heretical teachings. Later Anastasius translated into Latin the Acts of this council and of the Seventh Council at Nicea (787). He also wrote a number of saints' lives, the *Chronographia Tripartita* (derived from Byzantine chronicles), and translated many Greek religious works. L. FEEHAN

ANATHEMA. The Hellenistic Greek word means literally "something set up" or "placed" for a divinity. Both this and the rather stronger classical Greek form were originally used of a votive offering (cf. Luke 21:5). In the LXX the word *anathema* corresponds to the Hebrew term "consecrated (to God") or "accursed." Becoming anathema in the OT period could involve extermination (Deut. 7:1f., etc.). The NT use of the word implies exclusion, being banned, rather than complete extinction (Rom. 9:3; 1 Cor. 16:22, *lect. vid.;* Gal. 1:8f.; cf. 1 Cor. 12:3; Acts 23:14). The early church extended the biblical meaning (see A.F. Walls, *NBD*, p.35) to make it synonymous with excommunication. The earliest example of anathematizing beyond the NT occurs in the legislation of the Council of Elvira.* Normally the conciliar anathema was invoked against heresy (cf. the twelve anti-Nestorian anathemas of Cyril of Alexandria*). From the sixth century onward, anathematizing (as complete banning from the church) is distinguished from excommunication (as exclusion from worship and the sacraments).

See also EXCOMMUNICATION.

STEPHEN S. SMALLEY

ANATOLIUS (d.458). Bishop of Constantinople. In November 449, through the Monophysite* sympathies of Dioscorus of Alexandria and ambition for his see's primacy in the East, Flavian was deposed from the see of Constantinople. His successor, Anatolius, an Alexandrian *apocrisiarius* at Constantinople, was consecrated probably early in 450. Pressure from Pope Leo and the emperor Marcian in May 451 made Anatolius accept the *Tome.* At Chalcedon (451) he denied that Dioscorus had been deposed for heresy, and he headed a committee which drew up a pro-Cyrilline *Definitio.* This Leo rejected, also refusing later to recognize canons secured by Anatolius reaffirming the primacy of Constantinople in the East and securing rights of jurisdiction and consecration over neighboring metropolitans. In February 457 Anatolius was the first Christian bishop to crown a Christian monarch, Emperor Leo I. Shortly before his death, Anatolius raised clerical support at Constantinople for the Chalcedonian party against the violent Monophysite Timothy Aelurus who had usurped the see of Alexandria.

G.T.D. ANGEL

ANCHORITE; ANCHORESS. A person who becomes a hermit in order to triumph over the flesh by prayer, contemplation, and mortification. Such a way of life became respected with the great escape to solitude of the fourth and fifth centuries. Gradually the solitary asceticism of Antony developed into the organized monasticism of Pachomius and Basil of Caesarea. Though technically the term "anchorite" could be applied to the monk who had withdrawn from society, it became more precisely used of those who lived as hermits —usually after a period of probation in a monastery. Some, especially in Syria, and most notoriously Simeon the Stylite,* engaged in amazing acts of asceticism in their anchorite existence. As a way of life it was generally considered superior to monasticism. The word is not now quite synonymous with "hermit" and is used of those who live in very confined quarters. C. PETER WILLIAMS

ANCYRA (modern Ankara). Capital of the Roman province of Galatia, prominent ecclesiastically from the second century, and scene of important councils:

(1) In 314, after the Great Persecution, a synod of between twelve and eighteen bishops widely representative of Syria and Asia Minor promulgated canons apportioning penalties to the different categories of the lapsed, and regulating sundry ecclesiastical cases, mainly clerical, and standard penitential discipline. Canon 13 seems to allow presbyters to ordain, an interpretation perhaps unknown to antiquity. These canons have special significance in the elaboration of canonical penance, and as the earliest canons of a provincial synod included in the universal code of canons. The council which met at Nicea in 325 was originally planned for Ancyra. Constantine changed its venue.

(2) In 358 Basil of Ancyra presided over a synod of twelve bishops acknowledged as voicing conservative Eastern opinion, misleadingly called semi-Arians, more accurately Homoiousions,* and even deserving the description semi-Nicenes, since their deliberations marked an advance in theological reconstruction in reaction against the extreme Arianism of the "Blasphemy" of Sirmium and the Anomoeans like Aetius and Eudoxius. Their synodal letter and nineteen anathemas still rejected Nicea's *homoousios* ("of one substance") but declared the Son to be "like in essence" (equivalent to *homoiousios*) to the Father. Hilary of Poitiers and Athanasius welcomed this development and sought to overcome the unhappiness with *homoousios.*

(3) An Arian synod in 375 deposed several bishops, including Gregory of Nyssa, who had earlier convened synods at Ancyra to support the endeavors of his brother Basil. D.F. WRIGHT

ANDERSON, SIR ROBERT (1841-1918). Lay theologian and Bible teacher. He was a barrister in Dublin and London, adviser to the British Home Office in matters relating to political crime (1868-88), and then assistant police commissioner for the metropolis and head of the criminal investigation department, Scotland Yard (1888-1901). A Presbyterian layman, he was active as a speaker and writer; his work centered principally on apologetics and Bible prophecy (i.e., eschatology). A leading popularizer of the dispensational* interpretation of Scripture, he taught that there was a radical distinction between the Pauline gospel (the gospel for the church) and the gospel of the kingdom (the gospel given to the apostles as a message for Jews), between the authority for the church of the Pauline letters and the rest of the Bible. His books, which were very popular, include *The Gospel and Its Ministries* (1876); *The Coming Prince* (1882); *The Silence of God* (1897); *The Bible and Modern Criticism* (1902); and *Christianized Rationalism and the Higher Criticism.* W. WARD GASQUE

ANDOVER CONTROVERSY. The debate over the doctrine of "future probation" involving the faculty of Andover Theological Seminary from about 1886 until 1893. The seminary had been established by New England Congregationalists in 1808 to counter the Unitarian tendencies of Harvard. Attempting to preserve Andover's orthodoxy, the founders required faculty subscription to the Andover Creed, summarizing Edwardsean theology restated by Samuel Hopkins.* After the Civil War, however, faculty members joined other New England progressives in restating their faith along the lines of the liberal emphasis upon the immanence of God, emerging biblical criticism, and the doctrine of progress. Future probation developed when the Andover men applied the "new theology" to missions. In a series of articles in the *Andover Review,* E.C. Smyth and colleagues argued that heathen who die without knowledge of the Gospel will have an opportunity in the future life either to accept or to reject the Gospel before facing final judgment. In 1887 Smyth was deprived of his chair, but in 1891 his dismissal was voided by the Supreme Court of Massachusetts. BRUCE L. SHELLEY

ANDREAE, JAMES (1528-1590). Lutheran scholar. Born in Weiblingen, Württemberg, he studied in Stuttgart and Tübingen, and in 1546 became pastor in Stuttgart but was deposed for his refusal to subscribe to the Interim (1548). He then went to Tübingen as pastor. In 1553 he became successively pastor and superintendent at Göttingen. He was appointed professor in 1561 and later chancellor of the University of Tübingen. Between 1567 and 1580 he was adviser to the elector August of Saxony. He participated in many of the religious colloquies of the time, among them the conference with Farel and Beza in 1557. He sided with Joachim Westphal against John Calvin in the controversy regarding the Lord's Supper. Andreae probably prepared the Strasbourg *Formula of Concord* (1563). Besides his reformatory activities, his work was focused on reconciling the Lutheran factions. With men like John Brenz, Martin Chemnitz, and David Chrytaeus, he belonged to the center party, between the Gnesio-Lutherans and the Philippists. His *Six Sermons* (1572) treated the points in controversy: original sin, justification and good works, free will, the Lord's Supper, adiaphora, predestination, and the person of Christ. He drafted the *Swabian Concordia.* He favored the *Maulbronn Formula* and was responsible in part for the *Torgau Book* and the *Bergic Book,* together with Chemnitz and Selneccer. He is the editor of the official German edition of the *Book of Concord,* published on June 25, 1580. His works run to over 200 titles. CARL S. MEYER

ANDREAE, JOHANN VALENTIN (1586-1654). Lutheran theologian, grandson of James Andreae.* Born at Herrenberg, he studied at Tübingen University, but was dismissed in 1607. During his subsequent journeyings he gained some sympathy for the Calvinistic churches. In 1614 he was ordained deacon in Veihingen and six years later was appointed superintendent minister in Calw. Here he showed courage in the ravages of the Thirty Years' War. In 1639 he went to Stuttgart as a court chaplain and was also a member of the consistorial court. By 1650 he was general superintendent and abbot of Bebenhausen. In 1654 he was appointed abbot of Adelberg; here he sought to reorganize and help the churches in Württemberg. When the storms of theological controversy raged around him, he took an irenic stance and hopefully wrote of a form of utopian state based on the example of Geneva—see his *Rei publicae Christianopolitanae descriptio* (1619). He died at Stuttgart. PETER TOON

ANDREW. Apostle. The brother of Simon Peter, he was a native of Bethsaida, but carried on business as a fisherman at Capernaum. He became a disciple of John the Baptist, who directed his attention to Jesus as the Lamb of God (John 1:29). When later Jesus called several disciples to follow Him, Andrew was among the first to do so. It was he who introduced his brother Simon to Jesus and was later responsible for directing some Greek inquirers (John 12:21,22). He was of a practical turn of mind, as is seen from John 6:8,9 on the occasion of the feeding of the five thousand. His name appears in all the lists of the twelve apostles. Nothing is known of his contribution to the developing church, but the many apocryphal works which were later ascribed to him testify to the respected place he held in popular tradition even of an unorthodox kind. DONALD GUTHRIE

ANDREW, ACTS OF, see APOCRYPHAL NEW TESTAMENT

ANDREW OF CRETE (c.660-740). Theologian and hymnwriter. Born at Damascus, he became a monk at Jerusalem. Attending the Council of Constantinople in 680, he was ordained deacon of Hagia Sophia and became warden of the orphanage. Appointed archbishop of Gortyna, Crete, in 692, he participated in the Monothelite Synod of Constantinople in 712, but in 713 recanted his Monothelitism.* He was later involved in the Iconoclastic Controversy.* He helped to introduce in Constantinople the liturgy of Jerusalem, based upon the Rule of St. Sabas and favoring hymn-singing. He composed the "Great Canon," containing 250 stanzas, and many other canons and hymns. J.G.G. NORMAN

ANDREW OF LONGJUMEAU (d.1270). French Dominican missionary. Born at Longjumeau (Lonjumel), he spent much of his life in the East as a missionary. In 1238 Louis IX of France sent him to Constantinople to collect a relic, the Crown of Thorns, which the Latin emperor Baldwin II had given to Louis in return for help and money. Before Andrew's arrival, debt forced Baldwin II to sell the crown to the Venetians. Andrew was thus obliged to take the relic first to Venice, where a high price was wanted for it, before he could take it to France. In 1245 Innocent IV sent him to the Holy Land to heal a reported schism there. Between 1248 and 1252 he accompanied Louis IX's ill-fated Crusade to Egypt, and he also visited the Great Khan, who rumor wrongly alleged had become a Christian. L. FEEHAN

ANDREWES, LANCELOT (1555-1626). Bishop of Winchester. Born in London and educated at Pembroke Hall, Cambridge, he was in 1576 elected fellow of his college and on his ordination in 1580 was appointed catechist. At the beginning of his academic career he held views similar to those of the predominant Puritanism of the Cambridge of his day. During the 1580s, while the struggle with the Puritans was at its height, Andrewes sided with the episcopal leaders of the Church of England. He became chaplain to both Queen Elizabeth and Archbishop Whitgift about 1587, and this led to his appointment as master of Pembroke Hall and other appointments in London in 1589. His opposition to Whitgift's Lambeth Articles* of 1595 showed that he was starting to oppose Calvinism in the Church of England.

His considerable reputation as a preacher in a new and ornate style brought him episcopal office under James I. He became bishop of Chichester in 1605, of Ely in 1609, and of Winchester in 1619. He was responsible for the Pentateuch and the historical books of the Old Testament in the Authorized Version of 1611. He was taken by James to Scotland to help enforce his policy of episcopal church government in that country in 1617. While William Laud* regarded Andrewes as his master, he should be linked with his friends Richard Hooker and George Herbert. Andrewes held a strong view of episcopacy, but was unwilling to unchurch continental churches which had lost the episcopate. Though he favored a more ordered ceremonial in his private chapel, he did not try to enforce these ideas in his dioceses. He is famous for his book of private devotions, the *Preces Privatae,* which was published posthumously in 1648. His reputation for saintliness has somewhat overshadowed his activities as a court ecclesiastic. NOEL S. POLLARD

ANDREWS, CHARLES FREER (1871-1940). Missionary to India and friend of the oppressed. Brought up in the Catholic Apostolic Church* (Irvingites) of which his father was a minister, he went to Cambridge after an intense conversion experience and subsequently joined the Church of England. In 1904 he went to the Cambridge Mission in Delhi. The Indian people he found very congenial, and his many friends included Sadhu Sundar Singh,* whose biography he wrote. Over the years, Andrews's doubts about the Athanasian Creed made him unhappy about his Anglican orders. He was helped greatly by Albert Schweitzer's* writings. In 1913 he went to South Africa to help Indian laborers who were being penalized by the notorious indenture system, and there he met Gandhi, who became a firm friend. Returning to India, Andrews left the Anglican Mission and joined Rabindranath Tagore at his ashram at Santiniketan. The Bengali poet's words might well be applied to Andrews, whose life was given to the service of "the poorest, the lowliest, and the lost." J.G.G. NORMAN

ANGEL. By derivation the term means "messenger," from God manward. Judaism and Christianity have full angelologies, like their corrupt syncretistic derivative, Islam. Biblical angels are created, higher than unfallen man, spiritual, incorporeal, rational, moral, immortal, all originally good and perfectible. Satan with his hosts revolted through vainglorious pride, desiring equality with God (Isa. 14:12-15; Ezek. 28:12-17; 2 Pet. 2:4; Jude 6), and became the implacable foe of God and man, seducer and accuser of the brethren, yet restricted by the permissive will of God. Familiar throughout the OT, angels gain firmer contour in exilic and later prophets, and in the Apocrypha, Pseudepigrapha, and rabbinical sources.

The NT references speak of angels as primarily created for the ceaseless praise of Father and Son in heavenly surroundings (Heb. 1:6; Rev. 5:11ff.), but the subsequent creation and sin of man gave them commissions on earth. Guardian angels are popularly attributed both to individuals (Matt. 18:10; Acts 12:15) and to nations (cf. Dan. 10:13), though this should perhaps be interpreted broadly and not in exclusive or individualistic terms.

The six-winged seraphim are probably of the highest angelic order, the four-winged cherubim perhaps of slightly lesser rank. Other titles used in Scripture include: watchers, holy ones, princes, thrones, dominions, principalities, powers, morning stars; also the "living creatures" of Revelation 4:6-8. Scriptural authority for hierarchical detail is minimal.

The Roman Catholic doctrine of angels is dubiously indebted to the Greek writings of Dionysius the Pseudo-Areopagite, through the Latin rendering of John Erigena. This totally unscriptural elaboration has prompted widespread abuse. The Council of Trent* taught that angels intercede for men and that "it is good and profitable to invoke them suppliantly . . . for the purpose of obtaining benefits from God through His Son Jesus Christ."

Orthodox Protestant views are copiously quoted in Heppe, *Reformed Dogmatics*, pp. 201-219: these follow familiar biblical lines. While angels are incorporeal and immortal, they were originally created *ex nihilo* and therefore lack true eternity, the prerogative of deity, which has to be retrospective as well as anticipatory. Though finite and therefore limited spirits, they possess powers far outstripping those of man, with knowledge based on "nature, use and revelation."

ROY A. STEWART

ANGELA MERICI (1474-1540). Foundress of the Order of Ursulines.* A tertiary* of St. Francis, she began religious schools for girls in her native Desenzano and at Brescia. In 1524 she was a pilgrim to the Holy Land, becoming blind at Crete but cured there on her return. In 1525 Clement VII unsuccessfully invited her to establish her work in Rome. In 1535 she founded the Order of Ursulines at Brescia, having had a vision of such in 1506, and she was superior until her death. The order existed to combat immorality and to train wives and mothers in the faith. With no formal vows, the sisters lived outside to exercise their apostolate widely. Angela wrote a Testament, Counsels, and the order's Rule. She was canonized in 1807.

C.G. THORNE, JR.

ANGELA OF FOLIGNO (c.1248-1309). Umbrian mystic. Born into a wealthy family at Foligno, she spent most of her life there. She married and, when nearly forty, underwent sudden conversion and became a Franciscan tertiary.* After the death of her husband and children she lived an austere, cloistered life surrounded by disciples. It is recorded that she received frequent visions, which she dictated later to her confessor Arnold and circulated eventually as *Liber Visionum et Instructionum.* She analyzed the "twenty steps of penitence" by which she was initiated into the mystical life, culminating in a vision of herself in God. She was beatified by Innocent XII in 1693.

J.G.G. NORMAN

ANGELICO, FRA (1387-1455). Florentine painter. Known also as Giovanni da Fiesola or Guido di Pietro, he entered when twenty the Dominican monastery at Fiesola. Between 1409 and 1418 the Great Schism forced him to go first to Foligno and then to Cortona, where some of his greatest paintings are to be found. He decorated the convent of San Marco in Florence and did frescoes in two chapels in the Vatican. It is not known where or when Angelico began his training as a painter, but his early work (1418-30) shows a close relationship to international Gothic style. Examples of this type of painting include his *Coronation of the Virgin* and *The Last Judgment.* His later work is characterized by Renaissance realism with the use of more natural backgrounds rather than an abstract gold setting. His works of this period, including *Deposition from the Cross* and the Crucifixion and other scenes from Christ's life, decorate the convent of San Marco.

ROBERT G. CLOUSE

ANGELUS. The practice in the Roman Catholic Church of reciting, at morning, noon, and evening, three "Hail Marys" together with versicles and responses, and a collect of the Annunciation. It is intended as a continual reminder of the Incarnation. Warning of the observance is given at the appropriate hours by the ringing of the Angelus bell. It is rung three times for each "Ave" and nine times for the collect. The name of the rite is derived from the first word of the opening versicle: "Angelus Domini nuntiavet Mariae." Different words are used, however, at Eastertide. The practice of the Angelus was begun in Italy during the thirteenth century and did not spread into general use until after the Reformation.

JOHN TILLER

ANGELUS SILESIUS, see SCHEFFLER, JOHANN

ANGLICAN COMMUNION. It is not possible to give an exact date at which the Anglican Communion came into being. The Anglican churches of England, Scotland, Ireland, and Wales certainly formed the original provinces, to which was added in 1789 the Protestant Episcopal Church of the USA. Thereafter, chiefly through the pioneering work of the Church of England's missionary societies and their later counterparts in North America and Australasia, the structure developed until in 1971 there were some 365 dioceses. These are mainly to be found in countries which were, and in many cases still are, parts of the British Commonwealth, though China, Japan, and Brazil are notable exceptions.

It is also difficult to define precisely what constitutes a sufficient qualification for membership in the Anglican Communion. The 1930 Lambeth Conference adopted a descriptive resolution which set out its understanding of the situation thus: "The Anglican Communion is a fellowship, within the one Holy Catholic and Apostolic Church, of those duly constituted dioceses, provinces or regional Churches in communion with the See of Canterbury, which have the following characteristics in common: (a) They uphold and propagate the catholic and apostolic faith and order as they are generally set forth in the Book of Common Prayer as authorised in their several Churches; (b) They are particular or national Churches, and, as such, promote within each of their territories a national expression of Christian faith, life and worship; (c) They are bound together not by a central legislative and executive authority, but by mutual loyalty sustained through the common council of the bishops in conference."

This became the classical statement, but is misleading not least with regard to the reference to the Book of Common Prayer, since this is no longer the touchstone to which modern Anglican liturgies are brought when revisions are taking place. Nevertheless, the above definition was sufficiently flexible to allow for the very considerable range of opinion, theology, and liturgical practice which has marked the Anglican churches during the nineteenth and twentieth centuries. In large measure the "color" of the originating missionary organizations determined the practice of the dioceses which emerged. Some provinces (for example, South Africa) were almost monochrome in their Anglo-Catholicism; others (such as Kenya) were marked by their evangelicalism. In West Africa, where the four former British Crown Colonies of Nigeria, the Gold Coast, Sierra Leone, and the Gambia were associated with different missionary societies, a province ultimately came into being with traditionally Catholic churches in Ghana and the Gambia, and traditionally evangelical churches in Nigeria and Sierra Leone. Between 1945 and 1970 it became evident that the various traditions were slowly modifying, at least at episcopal level, due to the increasing contacts which were developing on a worldwide scale.

The focal point of the Anglican Communion since 1867 has been the succession of Lambeth Conferences* which in a consultative capacity guided and interpreted Anglican thinking. Lacking any formal constitution, their resolutions carry no mandatory power within the various provinces, but their influence cannot be denied; at times it has virtually determined Anglican policy (for example, the Lambeth Quadrilateral of 1888, in which Scripture, the two Creeds, the dominical Sacraments, and the Historic Episcopate were laid down as the basis of reunion for churches within the Anglican family). Some indication of growth is to be seen in the increase in the number of bishops attending Lambeth Conferences—from 70 in 1867 to 310 in 1958. By tradition the conference is chaired by the archbishop of Canterbury, but this office is one of honor and carries no legal rights *vis à vis* provinces outside Canterbury.

During the twentieth century three Pan-Anglican Congresses were held (London, 1908; Minneapolis, 1954; Toronto, 1963), at which both clerical and lay representatives were present, but the 1968 Lambeth Conference recommended that these should be discontinued. As a result of the 1958 Lambeth Conference, the post of Anglican Executive Officer was created. His task was to be that of informing and coordinating the various provinces in order to avoid duplication and waste. Three bishops (Stephen Bayne, Jr., Ralph Dean, and John Howe) successively held the office between 1959 and 1971, in which latter year the title was changed to that of Secretary General to the Anglican Consultative Council, a body set up at the request of the 1968 Lambeth Conference to supersede the Lambeth Consultative Body and Advisory Council on Missionary Strategy, which were composed entirely of primates and other archbishops and metropolitans.

The council held its first meeting at Limuru, Kenya, in 1971 and meets on alternate years. It has a membership of fifty-five, drawn from all the Anglican provinces. Each province sends two representatives—one bishop and one clergyman or layman; the larger churches (Australia, Canada, England, and the United States of America) have a representation of three—one bishop, one clergyman, one layman. The council has an advisory function.

The developments which have taken place in ecumenical relationships since 1947, and which have led in some places to united churches and in others to reunion schemes yet unfulfilled, have posed obvious questions about the future of the Anglican Communion in general and its relationship with united churches in particular. This has led to a continuous reconsideration by the Anglican churches of their standing in relation to such united and uniting bodies.

In 1966 an Anglican Centre was set up in Rome to facilitate better understanding between the Anglican churches, the Roman Catholic Church, and others; this action was endorsed by the 1968 Lambeth Conference which also called for the Anglican presence in Geneva to be strengthened.

The Anglican Communion grew out of the work of mission. Since the Toronto Congress of 1963 it has made a special effort to integrate the worldwide mission of the Anglican churches through the implementation of a document entitled "Mutual Responsibility and Interdependence in the Body of Christ." Certain aspects of this scheme have received widespread criticism, and its operation has not been universally successful.

BIBLIOGRAPHY: H.A. Wilson (ed.), *The Anglican Communion* (1929); J.W.C. Wand (ed.), *The Anglican Communion* (1948); G.F.S. Gray, *The Anglican Communion* (1958); J.S. Higgins, *One Faith One Fellowship* (1958); H.G.G. Herklots, *Frontiers of the Church* (1961); W.E. Leidt (ed.), *Anglican Mosaic* (1963); S.F. Bayne, Jr., *Mutual Responsibility and Interdependence in the Body of Christ* (1963); S. Neill, *Anglicanism* (1965); A.T. Hanson, *Beyond Anglicanism* (1965); T. Wilson (ed.), *All One Body* (1969); *Lambeth Conference Reports 1867-1968; The Time is Now—Limuru 1971.*

MICHAEL SAWARD

ANGLICANISM, see ENGLAND, CHURCH OF

ANGLO-CATHOLICISM, see ENGLAND, CHURCH OF

ANGLO-SAXON CHURCH. Christianity had existed in Britain as early as A.D. 156, but the fourth- and fifth-century pagan Anglo-Saxon invasions drove the Britons, and with them Christianity, into ever-diminishing enclaves. Through hatred of the Anglo-Saxons, the Britons refused to bring them the consolation of Christianity. Their conversion finally came through two main channels, of which the more famous one was launched by

Gregory the Great who sent Augustine of Canterbury* with forty monks to England. Augustine landed in Kent in 597 and its king, Ethelbert, gave him land and a disused church at Canterbury. This is the historical origin of Canterbury's subsequent claims to primacy in the English Church. Much of N England was converted from Ireland and Scotland by missionaries of the Celtic Church.* The conversion of Anglo-Saxon England was a slow process; paganism was never very far below the surface. Famine or natural disaster might see a kingdom relapse into paganism; kings relapsed as a result of personal quarrels with the church, and the ninth- and tenth-century pagan Scandinavian invasions ensured that paganism remained a problem up to and beyond 1066.

By the eleventh century a parish system had developed, and there were sixteen often large and unwieldy dioceses, some of which were coterminous with the boundaries of ancient Anglo-Saxon folk-groups and kingdoms. Kings and nobles played a great part in the affairs of the Anglo-Saxon Church. Missionaries always tried to convert and gain the support of the king first of all. A converted king would order his subjects to accept baptism—however tenuous their subsequent level of belief and understanding. Royal protection was very necessary when evangelizing in pagan areas. Grateful kings and nobles gave rich gifts of land and buildings to the church. In return, the kings expected and usually got the support of the church, which preached loyalty to the king and placed religious sanctions on those who disobeyed or plotted against the kings and generally buttressed law and order. The church assumed responsibility for the Ordeal.* Kings naturally assumed that they could appoint their nominees as bishops and abbots. Similarly, nobles who founded local parish churches or monasteries claimed the right to appoint. The church had a civilizing effect upon the Anglo-Saxons, and it gradually replaced the anarchic private wars of the blood feud by a *wergild,* or money compensation, and it brought literate government. The Celtic and Roman churches brought a fusion of two outstanding cultural traditions which continued right up to the Norman conquest, despite the turbulence of the Scandinavian invasions.

Monasticism was very popular, though apparently monasteries and nunneries admitted only nobility. The Scandinavian invasions supposedly caused a deterioration in monasticism, but informed critics claim that wealth and idleness were the main culprits in the eighth and ninth centuries. As it became institutionalized, the church had become also very wealthy, and some like Bede looked back nostalgically to the pioneer days of the early conversion as a "golden age." There was a monastic revival in the tenth century, Dunstan and Aethelwold being the key figures, and it lasted into the eleventh century when books produced in English monasteries were in demand throughout the Continent. Subsequent Norman allegations of the eleventh-century church's corruption were largely propaganda and generally unfair.

See C.J. Godfrey, *The Church in Anglo-Saxon England* (1962).

L. FEEHAN

ANNA COMNENA (1083-after 1148). Byzantine historian. Eldest daughter of Emperor Alexius Comnenus, she sought to commemorate "with truth dear and sacred" the excellences of her father's reign in the *Alexiad.* He had saved the empire against hostile tribes, participated in the First Crusade (he later became antagonistic toward it as a threat to his rule), and was zealous in persecuting the Bogomiles.* Aided by her mother, Empress Irene, Anna sought in vain to persuade her father during his last illness to appoint her husband, Nicephorus Bryennius, his successor in place of her brother John. Failing in a conspiracy to overthrow her brother, she was retired in 1118 to a convent where she used her considerable literary skill in eulogizing the exploits of her father's reign. The *Alexiad* is an *ex parte* and uncritical statement of the Byzantine conception of government and the religious and intellectual outlook of the period.

ARTHUR FAWCETT

ANNATES (Lat. *annatae* from *annus,* "year"). The first year's revenue of an ecclesiastical benefice paid to the pope. Originating in a bishop's right to the first year's profits of the living from a newly inducted incumbent, first mentioned in the thirteenth century, popes under financial stress later claimed the privilege for themselves, temporarily at first. Thus Clement V in 1305 claimed them from all vacant benefices in England, and John XXII in 1319 from all Christendom. Protests became frequent, e.g., from England at the Council of Lyons in 1245. Henry VIII transferred the English annates to the Crown in 1534, and these were in 1704 converted into "Queen Anne's Bounty." With the gradual transformation of the system of benefices, annates as such fell into disuse.

J.G.G. NORMAN

ANNE (1665-1714). Queen of Great Britain and Ireland from 1702. Second daughter of James II, she was brought up as a Protestant member of the Church of England and was married in 1683 to the Lutheran Prince George of Denmark. She differed from her family, who adopted Roman Catholicism. Under the influence of Sarah Churchill (later the duchess of Marlborough) she did not follow her family into exile in 1689. As a result she succeeded William III to the throne in 1702. Anne supported the High Church party in the Church of England. Her appointments to the episcopal bench replaced Latitudinarians with High Church bishops. By 1710 her ecclesiastical policy led to the overthrow of the Whig party in the state, of which the fall of Marlborough was a sign. The creation of "Queen Anne's Bounty" was typical of her great loyalty to the Church of England. The reverse side of this loyalty appeared in the growing intolerance shown to the Dissenters during the short-lived supremacy of the Tories in the last years of her reign. The brevity of her support of the High Church party and her failure to produce an heir meant that her policy died with her.

NOEL S. POLLARD

ANNIHILATIONISM, see CONDITIONAL IMMORTALITY

ANNO DOMINI. Meaning "in the year of the Lord," this anchors the modern Gregorian calendar to the calculation of the birth of Christ made in 527 by Dionysius Exiguus. NT exegesis cannot confirm the calculation; Matt. 2:19 sets the Nativity before Herod the Great died (4 B.C.). Some interpreters of Luke 2:3 associate it with the Augustan census of Quirinius in A.D. 6-7.

ANNUNCIATION. The account given in Luke 1:26-38 of the visit of the angel Gabriel to Mary to tell her that she was to be the mother of the Messiah. The message is largely cast in terms of one who is to bring to true fulfillment the promises made to David. In reply to the hesitation of Mary, the angel makes clear that the conception will take place through the Holy Spirit without the agency of a human father and that the child will be the Son of God. Mary's attitude of wondering and willing submission is in contrast to that of Zechariah when the coming birth of John was announced to him (Luke 1:8-23). In Matthew's gospel there is no record of the annunciation to Mary, but Joseph is told by an angel in a dream what has happened.

The Feast of the Annunciation has been observed from early times. It is dated March 25 (sometimes known as Lady Day), as that comes nine months before the date chosen for the observance of Christmas. At one time it was observed on April 6 (nine months before the Epiphany, which celebrated the birth of Christ in the East).

R.E. NIXON

ANOINTING. The application of oil (or similar substance) as a religious ceremony. In the OT, objects thus anointed included battleshields, sacred rocks, and tabernacle appurtenances, using unguents sternly forbidden elsewhere. Kings, priests, and prophets were solemnly anointed in God's name and presumed thereby to receive the Spirit. The enormity of slaying, even by command, the Lord's anointed remained. Symbolically, anointing is every believer's portion. Messiah is the Anointed One *par excellence.* In both Testaments, anointing symbolizes outpoured Spirit; experience remains potential, outward symbols are changed. Paganism too had its sacral anointings. Chrism, episcopally consecrated oil, is still used by Roman and Eastern churches for baptism, confirmation, ordination, and extreme unction, and by Anglicans for coronations and sporadically for the requesting sick. Roman Catholic extreme unction* is invalidly based on James 5:14f. which intended counsel to qualified apostles.

ROY A. STEWART

ANOMOEANS. Radical Arians. Led by Aetius* and Eunomius (see EUNOMIANISM) in the period 357-61, they held that the Son is unlike (Gr. *anomoios*) the Father. Aetius maintained "unlikeness" consistently and later influenced the Pneumatomachi, who excluded the Son and the Spirit from the Godhead. Other Anomoeans, however, such as Eudoxius of Antioch and Acacius, supported the Creeds of Sirmium (357) and of Constantinople (360) which excluded "substance" (*ousia*) from its formulae but affirmed that the Son is like the Father. Aetius was excommunicated by this group in 361, but according to the historian Socrates some of these reverted to Aetian views, interpreting "God from God" in the sense that all things are from God. Anomoeans were anathematized at the Council of Constantinople in 381.

G.T.D. ANGEL

ANSELM OF CANTERBURY (c.1033-1109). Archbishop of Canterbury from 1093. Born in Aosta, Italy, he quarreled with his father as a youth and left home. After years of wandering, at the age of twenty-six he settled in Normandy at Bec, becoming a monk under the influence of Lanfranc. When he was sixty years old he left the abbey and was made archbishop of Canterbury, a post he held until his death. Anselm took part in the intellectual development of eleventh- and twelfth-century Europe when, due to the increase in wealth and the challenge of new ideas, the Scholastic* tradition was formed. At the beginning of this process, the monastic communities took the lead. They had various advantages over the secular schools, the greatest of these being the close and continuing contact between the teacher and the student. Given a bright teacher, a tradition of learning, and the unhurried pace of a monastic community, the results could be very impressive. This was the situation at Bec while Anselm was prior and then abbot. Most of his works take the form of dialogues with students as he attempted to answer the worried questions of the young men in his care.

While archbishop of Canterbury, Anselm worked to apply the Hildebrandine reforms to the English Church. This led to conflict with William II (Rufus). Anselm refused to cooperate with lay investiture and so was forced to leave England. At the death of Rufus in 1100 Anselm was asked by Henry I to return to England, but he argued with the English monarch and so went into exile once more (1103). Finally, by 1107, a compromise was reached between the pope and the English king, and Anselm returned to his see. The remaining years of his life were spent in enforcing clerical celibacy and other Gregorian reforms on the church in England.

His writings are divided into (1) systematic works, (2) prayers and meditations, and (3) letters. In the first category one should note the *Monologion,* the *Proslogion, Cur Deus Homo* (Why God Became Man), *On the Virginal Conception and Original Sin, On the Procession of the Holy Spirit,* and *De Concordia.* Although he did not work out a complete system of theology as the later medieval scholars were to do, his treatises cover much of Christian thought. Anselm believed that faith was a necessary foundation and support for philosophic speculation. As he wrote, "I do not seek to understand that I may believe, but I believe that I may understand: for this I also believe, that unless I believe I will not understand." He proceeded to formulate the "ontological proof" for God's existence. He held also that the Atonement was necessary to satisfy the majesty of God rather than the older view held since Origen's time that Christ died to pay a ransom to the devil. Anselm's dependence upon Platonic thought made him a leader among the medieval realists.

The most recent edition of Anselm's works is F.S. Schmitt, *S. Anselmi Opera Omnia* (5 vols. and index, 1942). For excellent biographical detail see R.W. Southern, *Saint Anselm and His Biographer* (1963). ROBERT G. CLOUSE

ANSELM OF LAON (d.1117). A famous Schoolman known as the "Doctor Scholasticus." He was educated at Bec under Anselm of Canterbury. From 1076 he taught in Paris where his pupils included William of Champeaux and then, towards the end of the century, with his brother Ralph established a renowned school at Laon. Its celebrity was sufficient to attract Abelard as a pupil and, though Abelard had a low estimate of his teacher, it is now clear that Anselm did much to develop the *quaestio*—the setting up of opposing authorities and attempting to reconcile them—which became the standard method of later thinkers, including Abelard's *Sic et Non.* Also very important was Anselm's contribution to biblical exegesis in terms of commentary glosses. He was responsible for part of the exegetical work *Glossa Interlinearis.* C. PETER WILLIAMS

ANSELM OF LUCCA (1036-1086). Bishop of Lucca. Of a noble Milanese family (his uncle of the same name became Alexander II), he firmly supported Pope Gregory VII and became bishop in 1073. Soon he resigned to retire to the Cluniac monastery at Polirone. At Gregory's behest he returned, but the austerities he imposed brought expulsion by Henry IV and the antipope Clement III, and he spent the rest of his life as spiritual director to Countess Matilda of Tuscany and as papal legate in Lombardy. His *Collectio canonum,* compiled about 1083, aided the Gregorian reform by discussing ecclesiastical privileges, episcopal elections, lapsed clergy, sacraments, and excommunication. Based on several sources, it was later incorporated in Gratian's *Decretum.* C.G. THORNE, JR.

ANSKAR (Ansgar) (801-865). "Apostle of the North." Frankish-born, at the age of five he was placed in the Benedictine monastery at Corbie where he remained until 822, when he was transferred as master of the daughter school at Corvey in Saxony. About four years later he and a colleague went to Denmark, where a few important Christian converts seem to have been made, and established a small wooden church at Hedeby. By 829, however, he had returned to the Frankish court when there arrived from Björn of Sweden a request for help. With a fellow monk, Witmar, Anskar again went north, to the trading center of Birka, where a bishopric was established under Simeon-Gauzbert. In 831 he became archbishop of Hamburg, which post he held until his death despite the disruptive plunder of the city by Horik in 845. Anskar maintained his interest in Birka, and when Simeon-Gauzbert was driven out by his own Frankish people, Anskar spent two years in the diocese from 852 under the protection of Horik. On his way he again visited Hedeby. Schleswig along with Hamburg and Bremen (he had assumed pastoral oversight also of the latter see in 847) remained the area of his movements until he died—not by the martyrdom he sought. He was buried in St. Peter's Church, Bremen.

Anskar is sometimes considered to have failed, since Christianity was not firmly established in Scandinavia at his death, yet a major factor in his relative accomplishments is that no stable dynasty had been formed in any of the nation-states there. But foundations had been laid, and Christianity was officially to come within a century with the conversion of Harald ("Bluetooth") Gormsen (c.940-86).

See St. Rimbert, *Anskar, the Apostle of the North, 801-865* (ET by C.H. Robinson, 1921). CLYDE CURRY SMITH

ANTELAPSARIANISM, see SUPRALAPSARIANISM

ANTHEM, see MUSIC, CHRISTIAN

ANTHONY, see ANTONY

ANTHROPOMORPHISM. Literally "in the form of man," it is usually applied to God either (1) in the strict sense that God or the gods (e.g., the Olympian gods of Greek mythology) has or have a body like man's, or (2) that God's mental or spiritual qualities can only be understood in terms of those of man. The term may also be applied to the natural world, both inorganic (e.g., force, conceived as an extension of feeling in our muscles) and organic (e.g., the mental life of animals). The charge that theism is anthropomorphic in sense (1) above is often leveled against Christianity. It is urged (e.g., by Winwoode Reade in the *Martyrdom of Man,* 1872) that God is infinitely greater than man and cannot therefore be conceived in human terms. But this does not follow. If, whether directly or indirectly, God created man, then God must at least be familiar with the qualities inherent in man. Man can only think with his mind, the tool given him for thought, and this limits his conception of God to an anthropomorphic one. Man can and does know God intuitively, but if he *reasons* about God he must do so in anthropomorphic terms such as those used in the Bible. R.E.D. CLARK

ANTHROPOSOPHY. A system of Christian mystical philosophy developed by the Austrian, Rudolf Steiner (1861-1925). Son of a stationmaster, he studied science at Vienna University and thereafter worked till 1897 on the Weimar edition of Goethe's works. He joined the Theosophists,* but finding them overinfluenced by oriental religious ideas, he left them to develop a Christianized version of theosophy which he called "anthroposophy." In 1913 he established an institute at Dornach near Basle, to incorporate the newly formed Anthroposophical Society and to provide a publishing house. Similar institutes were later set up in other countries as were also the Rudolf Steiner schools which specialize in the education of maladjusted children.

Steiner's thought encompassed the whole range of speculative philosophy. Basically his aim was "to raise the faculties of the soul to develop organs of spiritual insight." He held that the story of evolution represented the various stages by which man as a created spiritual being became clothed in flesh. Though spiritual man was made in God's image, this image was soon distorted and Christ often intervened (e.g., by giving man an upright posture) to aid his restoration. The last and most important spiritual intervention was at Calvary by which means man is finally offered salvation from earthly entanglement. Steiner held that the Parousia, heralding the completion of man's redemption, started in our own century with the appearance of Christ in the etheric sphere. The anthroposophical movement owed much to Steiner's magnetic personality: he attracted many, some "cranky" and unbalanced, but many too of acknowledged ability who were unconventional but religiously and mystically inclined. Perhaps his most enduring work was in the field of education.

BIBLIOGRAPHY: Steiner's autobiography (ET 1928) is the main source of information about the man; his views are contained in his numerous books, notably *Spiritual Science and Medicine* (1948 ed.); see also *Rudolph Steiner: Recollections by some of his Pupils* (1958); works by G. Kaufmann (1922) and A.P. Shepherd (1954); and particularly the centenary volume, *The Faithful Thinker* (ed. A.C. Harward, 1961).

R.E.D. CLARK

ANTIBURGHERS, see BURGHERS

ANTICHRIST. A biblical term found only in 1 John 2:18,22; 4:3 and 2 John 7. Its meaning has often been enriched by concepts from Daniel 7:8; 8:8-14; 2 Thessalonians 2:3; Revelation 13:4-18, etc. Antichrist has generally been understood as a person (sometimes an institution) opposed to Christ and even a deliberate counterfeit. Early Christian writers (e.g., Chrysostom) were agreed that he would appear immediately before the second advent of Christ and be a person under the direct inspiration of Satan. With Tertullian the idea was put forward that he would not appear so long as the Roman Empire remained intact, but that he would arise to reunite the ten kingdoms into which it would have disintegrated. Sometimes it was thought that he would be Nero resuscitated (Lactantius, Jerome, Augustine) or that he would be a Jew of the tribe of Dan (Irenaeus). He would simulate the powers and functions of Christ (Hippolytus) and be an incarnation of the devil (Theodoret). The Tiburtine Sibylline introduced the idea of a great emperor who would arise prior to the appearance of antichrist. Pseudo-Methodius, written under the shadow of Islam, envisaged such an emperor overcoming Islam. Such writings became influential in the West. The Crusades intensified apocalyptic speculation which found a focus in Joachim of Floris. Antichrist, or his forerunner, was seen in numerous ecclesiastical, political, national, or social opponents. The Spiritual Franciscans viewed the pope of Rome as antichrist, or at least his forerunner. Similar ideas were held by men like Wycliffe and Huss. Luther held that every pope was antichrist, since antichrist is collective, the institution of the papacy. Modern interpreters of the idealist school view antichrist as the timeless personification of evil. Futurists believe in a personal antichrist who will initiate a period of tribulation prior to the return of Christ.

See W. Bousset, *Antichrist* (ET 1896); C. Hill, *Antichrist in 17th Century England* (1971).

HAROLD H. ROWDON

ANTICLERICALISM. A basic attitude of hostility directed against clerical power and civic privilege—especially that of the hierarchy—which in modern times has found expression at two levels, the religious and the political, and has been of two types, active and passive. Anticlericalism dates from the early days of institutionalized Christianity. Medieval anticlericalism was sporadic and most unorganized, but the Reformation stimulated its growth as Protestants assailed clerical abuses and proclaimed the priesthood of all believers. The impetus for modern political anticlericalism, however, came principally from the Enlightenment,* with its opposition to organized religion and its teaching that the clergy were a bulwark of political reaction. Thus nineteenth-century Europe saw numerous fierce battles between anticlericals and defenders of the traditional role of the established churches. This conflict intensified with the growth of liberalism and nationalism, especially in largely Roman Catholic countries like France, Italy, and Spain.

As a result of the Revolution of 1789 in France, the homeland of modern anticlericalism, Roman Catholicism was abolished as the state religion. Restored in 1815, the Roman Church became the focal point of a bitterly divisive political struggle which culminated in the permanent separation of church and state in France in 1905. In Italy, political anticlericalism climaxed with the liquidation of the Papal States in 1870, while in Germany the Kulturkampf* in the 1870s resulted in a large body of anticlerical legislation. Similar attempts to limit clerical power and privileges made at about the same time in Spain and Spanish America met with varying degrees of success.

In the USA, anticlerical feeling among radical political thinkers and religious nonconformists expressed itself in the constitutional separation of church and state after 1787. A similar alliance led to the reforms of the 1830s in Britain which virtually disestablished the Church of England. In the USA, however, anticlericalism subsided in the nineteenth century with the growth of civic religion; but since World War II there has been a resurgence of anticlerical feeling among American Christians with the "revolt of the laity" in many denominations and the advent of the Jesus Movement and Catholic Pentecostalism. Much recent anticlericalism is passive rather than active, religious rather than political. Nonpracticing Catholics reflect this trend, as do many active Protestants who exhibit considerable indifference toward the clergy. Such passive anticlericalism usually stems from the rejection by the laity of the traditional authoritarian role of pastor and priest. ROBERT D. LINDER

ANTILEGOMENA. This term relates to those NT books over which there was some dispute within the Christian Church during the first four centuries. It was used by Eusebius* in his classification of Christian books to distinguish certain books from those admitted by all, which he called the *homologoumena.* Among the Antilegomena he placed James, Jude, 2 Peter, 2 and 3 John of the canonical books, together with others which were revered but not included in the NT canon. Among these latter he included the *Acts of Paul,* the *Shepherd of Hermas,* the *Apocalypse of Peter,* the *Epistle of Barnabas,* and the *Didache,* all of which attained only a local significance, mainly in Egypt. The Antilegomena were sharply distinguished by Eusebius from spurious works which were emphatically rejected.

DONALD GUTHRIE

ANTI-MARCIONITE PROLOGUES. Many manuscripts of the Vulgate contain prologues to individual books of Scripture. But these early Latin translations of prologues to Mark, Luke (also extant in Greek), and John were dated by De Bruyne within A.D. 160-80 and interpreted as anti-Marcionite, as possible imitations of Marcionite prologues to some Pauline epistles. W.G. Kummell (*Introduction to the New Testament,* ET 1966) disputes the conclusion by De Bruyne that the four prologues (Matthew is lost) reveal the united expression of an orthodox NT canon in the late second century. Their origin is unknown, their date uncertain, and the opposition to the heretical Marcion* is most explicit in the Johannine prologue. They claim to describe biographical details of the evangelists, the place and order of writing the gospels, and their relationship to their sources. G.T.D. ANGEL

ANTINOMIANISM (Gr. *anti,* "against"; *nomos,* "law"). While Luther was apparently the first to use the term "Antinomian" in his controversy with Johann Agricola* to describe the rejection of the moral law as a relevant part of Christian experience, Antinomianism clearly goes back to the time of the NT. Paul refutes the suggestion that the doctrine of justification by faith alone leaves room for persistence in sin, and frequently in the NT Epistles the view that the Gospel condones licentiousness is forthrightly condemned. Such counterattacks make it evident that antinomian views were current in the apostolic age. It is probably wise not to apply the title "antinomian" to the Gnostic heresies whose libertinism was based not on any supposed implications of the doctrine of justification by faith alone, but on a philosophical view of matter as intriniscally evil. Omitting these, there are two main forms of antinomian rejection of the law. Some Antinomians, like Agricola, maintain that the moral law is not needed to bring the sinner to repentance. This runs counter to Paul's experience and teaching (Rom. 7:7; Gal. 3:24). Others who accept the pedagogic use of the law, to convince the sinner of his sin and lead him to Christ, insist that the moral law has no place in the life of the believer, who is not under the law but under grace, and so not bound by the law as the rule for Christian living. Some of the English Puritans, notably Tobias Crisp and John Saltmarsh, held this view. The Brethren movement, consistent with the teaching of J.N. Darby* on the sharply contrasted dispensations of law and grace, is antinomian in the same sense. This form of Antinomianism seems to arise from a misunderstanding of the teaching of Paul who, while he utterly rejected the law as means of salvation, nevertheless affirmed the continuing validity of the law for the Christian (Rom. 3:31; 8:4).

For the protracted antinomian controversy in England between 1690-1700, see P. Toon, *Emergence of Hyper-Calvinism* (1967), chap. 3.

HUGH J. BLAIR

ANTINOMISTIC CONTROVERSY. This arose from a dispute between Johann Agricola* and Philip Melanchthon* about the relationship between repentance and faith, and the place of the moral law in the experience of the believer. Melanchthon held that the moral law was required to produce conviction of sin and repentance, as the prelude to faith. Agricola's view, to which Luther gave the name "Antinomian,"* maintained that repentance is the fruit not of the law but of the Gospel, and that the law has no relevance for the Christian. The dispute was temporarily settled at Torgau in 1527 in a conference between Luther, Melanchthon, and Agricola, but was revived later and, becoming acute in 1537, was never really resolved. HUGH J. BLAIR

ANTIOCH, COUNCILS OF. They include:

(1) in 268: Paul of Samosata, bishop of Antioch, was condemned for his Adoptianist* theology at a series of councils culminating in his deposition in 268. It is noteworthy that for the first time bishops in council passed judgment by asking the accused to sign a doctrinal formulation.

(2) In 325: a council met to elect a new bishop of Antioch but took the opportunity to severely condemn Arianism* and to announce that certain bishops, including Eusebius of Caesarea, had not signed their credal statements.

(3) About 327-330: Eustace, bishop of Antioch, was deposed by the supporters of Eusebius of Nicomedia because of his anti-Origenist theology.

(4) In 341: the Dedication Council of Antioch was a meeting of ninety-seven Eastern bishops at the time of the dedication of a new cathedral and in the presence of the Eastern emperor Constantius. They put forward four creeds. In the first they denied Athanasius's accusation that they were Arians ("how should we who are bishops follow a presbyter?"). Further, though they avoided the use of *homoousios,* they did insist that the Son was begotten before all ages and coexisted with the Father. They clearly condemned Marcellus, bishop of Ancyra (who taught that one day Christ's lordship would be handed back to the Father), by asserting that Jesus "continues King and God for ever." In the "Second Antiochene Creed" they stated that there were three separate

hypostases united by a common will ("in agreement one"). Here again they were attacking Marcellus. If the council showed that the Western picture of the East, as being totally Arian, was untrue, it also reflected the Eastern conviction that the West was wholly Marcellan. (See ANCYRA.)

(5) In 375: 153 Eastern bishops met under Melitius of Antioch and agreed to reconciliation with the Western Church, thus paving the way for the ecumenical Council of Constantinople.

(6) Throughout the fifth and sixth centuries there were frequent synods concerning the Nestorian* and Monophysite* controversies.

(7) Under Latin rule there were synods in 1139 and 1204. The former deposed an arrogant patriarch, Radulf, while the latter decided that the count of Tripolis had claims to the principality of Antioch. C. PETER WILLIAMS

ANTIOCH (Syrian). A Hellenistic city in NW Syria (modern Antakya), it was situated on the Orontes River some twenty miles from the sea. As the third of the greatest cities in the Greco-Roman world, it had been founded about 300 B.C. by Seleucus Nicator in honor of his father—one of some sixteen cities of that name. Its main outlet to the sea was Seleucia Pieria, one of the finest harbors on the coast. Antioch also lay on the most important land route between Asia Minor, Syria, and Palestine. As a royal city it had a splendid palace, with spacious boulevards, parks, and gardens. It was the only city of its time that had street lighting at night. The population of early Antioch is not known, but by the end of the fourth century A.D. it was estimated as high as 800,000. It had a mixed ethnic character, with large numbers of Jews. Its interest in mystery cults also gave it an eclectic intellectual spirit and an interest in religious inquiry. After the persecution of Stephen some followers of Jesus fled as far as Antioch (Acts 11:19), and many of its inhabitants were converted (Acts 11:21); it was here that the followers of Christ were first called "Christians" (Acts 11:26) following the year's ministry of Paul and Barnabas. The conversion of these Gentiles raised the question of the application of the Jewish law (Acts 15:1ff.) and some understanding was reached (Acts 15:19-35; Gal. 2:1-10). Following the visit of Judas Barsabbas and Silas as emissaries of James, who sought to win over the Gentile converts to the view that the law must be enforced, Peter and Barnabas disputed and broke away from Paul (Acts 15:22-29; Gal. 2:11-13). In time, the Jewish Christian community in Antioch disappeared. In subsequent debates over Roman primacy, it has been said that Peter was "the founder" and "the first bishop" of the church at Antioch. But the problem of his episcopacy is obscure, as is the rabbinical character of the Antiochene School of Theology (see ANTIOCHENE THEOLOGY).

See Downey and Glanville, *Ancient Antioch* (1963). JAMES M. HOUSTON

ANTIOCHENE THEOLOGY. Malchion, a converted Sophist of the second half of the third century A.D., is sometimes regarded as the founder of the Antiochene School of theology. He gained prominence as an opponent of Paul of Samosata, whose heretical views were condemned at Antioch in 268. But the originator of the distinctive Antiochene emphasis was Diodore, later bishop of Tarsus (d. c.390), the instructor of John Chrysostom and Theodore of Mopsuestia. Earlier Lucian, one of the ablest biblical scholars of his time, went to Antioch (c.260-65) and became the teacher of Arius and Eusebius of Nicomedia (he cannot with certainty be held responsible for their heretical views).

In Scripture and Christology the Antiochene theology's methodology was more rational, historical, and literal than that of Alexandria, opposing the latter's mystical and allegorical treatment of the biblical text. The Antiochene approach to Scripture was critical insofar as some parts of it were regarded as having more doctrinal and spiritual value than others. Philosophically Antioch favored Aristotle as more empirical and down-to-earth, whereas Alexandria allied itself to Plato's more mystical views. A less ontological view of the Trinity distinguished the Antiochene from the Alexandrian theology.* The tendency was toward Sabellianism,* due to its recoil from the tritheistic drift of the Alexandrian Trinitarianism (but the charge of modalism was strongly repudiated).

In Christology the divergence between Antioch and Alexandria is sharpest. The Antiochene teachers generally insisted upon Christ's true humanity and approached an understanding of his person from the human end. But their more radical advocates, Theodore and Nestorius especially, tended to destroy the concrete unity of Christ's person and to see him, not as the God-man, but a man indwelt by God. Lacking any clear doctrine of a substantive Logos,* interest centered on the historical Jesus. The general movement was thus towards an "adoptive" understanding Christ's person. On the other hand, His full humanity was denuded of a human soul. At this point the two schools overlap. Some Antiochenes developed their Christology in a *Logos-sarx* ("Word-flesh") framework, while some Alexandrians presented theirs in *Logos-anthropos* ("Word-man") terms.

The emphasis on moral achievement in the Antiochene Christology, in its attempt to solve the problem of the relation of human and divine in Christ, found prominence in its soteriology which admitted a significant place to human merit. This fact may explain Nestorius's sympathy for Pelagius. The Antiochene School continued to exercise a powerful influence until its decline in the eighth century, but meanwhile its christological doctrine was carried by its zealous missionaries to the utmost bounds of Asia. H.D. MC DONALD

ANTIPOPE. A pope elected in opposition to one held to have been canonically chosen. Although on general reckoning there have been some twenty-five antipopes in the history of the Church, the term is especially applied to the Great Schism* in the Western Church (1378-1417), during which period there was always a pope and an antipope (latterly two antipopes).

ANTI-SEMITISM. A term coined in the late nineteenth century which signifies hostility toward Jews (not all Semites). In the ancient world, Jews were ridiculed and often persecuted, particularly in Alexandria and later throughout the Roman Empire, on two main grounds: religious exclusiveness seen in their rejection of idolatry, and social exclusiveness arising from their stress on food laws and ritual purity. Christian antagonism to Jews—they were assailed for their unbelief and even regarded as deicides—ensured that the conversion of the Roman Empire brought them little relief. In the Middle Ages, the wealth of many Jews, gained through money-lending as well as trade, provoked hostility. Wild accusations were made: that Jews murdered Christian children at Eastertime (this developed into the "blood accusation," current from the thirteenth century, that they used Christian blood for ritual purposes); that they desecrated the Host; that they poisoned wells, etc.

The French Revolution worked in the Jews' favor. The National Assembly repealed all repressive measures against them (1791), and the Constitution of Year iii gave them equal rights (1795). Some repressive legislation subsequently enacted was set aside by the July Revolution of 1830. Absolute religious equality was granted throughout the N. German Federation (1869), and this was subsequently extended to the German Empire. Nevertheless, modern anti-Semitism arose in Germany in the 1870s. This seems to have resulted from growing Jewish prosperity, influence, sensitivity, and belligerence. The situation was exacerbated by signs of unified Jewish influence in the Alliance Israëlite Universelle and by the migration of Jews from E Europe, especially Russia and Romania, to Germany (and then to England and America).

Anti-Semitism flourished in Germany (where it came to a head in the Nazi atrocities), Austria-Hungary, France (e.g., the Dreyfus case), Romania (where a resolution of the Berlin Congress of 1878 that all Romanian citizens should enjoy equal civil rights was circumvented by the declaration that Jews were foreigners), and Russia (where massacres took place in the early twentieth century). Jews were treated as scapegoats in both Germany and Russia, where millions were liquidated. Today, anti-Semitism often takes the form of anti-Zionism. In the Middle East, the influence of European anti-Semitism and endemic intolerance of ethnic and religious minorities has strengthened anti-Zionism in Arab countries.

Anti-Semitism has causes deeper than Jewish particularism and allegations of ritual killings, secret literature, low Jewish standards of behavior, etc. Its deepest roots lie in Christian, especially fundamentalist, soil. It may well be true, as James Daane has argued, that the belief that the Jews were responsible for the death of Christ and that the Jewish nation lies under the total and final curse of God are the causes of its strength. Jules Isaac sees a third factor in belief that the spiritual life of the Jews was degenerate at the time of Christ.

See J. Daane, *The Anatomy of Anti-Semitism* (1965).

HAROLD H. ROWDON

ANTI-TRINITARIANISM, see UNITARIANISM

ANTONELLI, GIACOMO (1806-1876). Cardinal and Vatican secretary of state to Pius IX, 1848-76. He was the principal political executive of the States of the Church, known as the Temporal Power of the pope. His work and policy consisted chiefly in resisting, unsuccessfully, the final and revolutionary overthrow of the pope's political rule and its incorporation into the secular kingdom of Italy (1859-61, 1870). Although not a priest but only a deacon, he formulated an active ultramontane policy in internal and foreign affairs, relying upon, and resenting, the occupation by French and Austrian troops from 1850 to maintain basic order and support the papacy. His education was to the rank of doctor in law and philosophy. His character remains much debated, but his competence was never questioned.

C.T. MC INTIRE

ANTONIANS. The name used by several communities claiming descent from the Egyptian hermit Antony:

(1) The original disciples of Antony organized by him around 305 into the first hermit community to possess a rule.

(2) The Hospital Brothers of St. Antony, founded in 1095 by Gaston de Dauphine. It survived in France, Italy, and Spain until the time of the French Revolution.

(3) An order of the Armenian Church founded in the seventeenth century to maintain the connection with the Roman Catholic Church.

(4) A community founded in Flanders in 1615.

The Antonians are to be found also among the Chaldean and Maronite Uniats of Eastern Christendom. The Orthodox monastery of Mt. Sinai, dedicated to Catherine of Alexandria, claims to follow the Rule of Antony. "Antonians" was also the name used by a Swiss sect founded by Anton Unterhäher (1759-1824), who claimed to be the ruler of the world.

JAMES TAYLOR

ANTONINUS (1389-1459). Archbishop of Florence. Born in Florence, he joined the Dominican Order at the age of sixteen. Successively he became prior at Cortona (1418), at Friesole (1421), at Naples (1428), and at Rome, the Convent of St. Maria Sopra Minerva (1430), and was made auditor-general of the Rota under Eugenius IV in 1431 and vicar-general of the Dominican Order (1432-45). He participated in the Council of Florence (1439), dealing with the reunion with the Eastern Church. In 1446 he was appointed archbishop of Florence, and he was canonized by Adrian VI in 1523. He is the author of a *Summa confessionalis,* which included writings for those who were making confession, and a manual for priests who heard confessions and made absolutions. He also wrote a *Summa Theologica.* His *Chronicon* printed in Venice (1474-79) is a compilation of world history up to the year 1457.

A noted preacher, living as a simple friar also as archbishop, he was much concerned about the poor, was a ready counselor, and was deeply disturbed about corruption in the church. The repair of church buildings, the regularity of church services, pastoral ministration, and the reform of religious communities were major concerns. In his theology he was influenced by Thomas Aquinas. The humanism of the Italian Renaissance affected him little. His *Opera a ben vivere* was printed in 1858. His statue is the only statue of a priest in the Uffizi Palace in Florence. CARL S. MEYER

ANTONINUS PIUS (86-161). Roman emperor. Succeeding Hadrian and entitled "Pius" by the Roman Senate, he was a mildly progressive ruler, eager to relieve taxation and to foster public building in Italy. He sought to centralize the government with senatorial cooperation. During his principate but not at his instigation Bishop Polycarp of Smyrna was executed.

ANTONY (c.251-356). Pioneer of anchoritic* monasticism. Born at Coma in middle Egypt, his well-to-do Christian parents not long dead, at about the age of twenty Antony heard read in church, "If you want to be perfect, go, sell, . . ." followed by "Do not worry about tomorrow" (Matt. 19:21; 6:34). Bequeathing his property to the poor and his sister to a convent, he became an ascetic devotee, directed by an older solitary—first near his house, later outside Coma, and then in a tomb further afield. Seeking isolation about 285, he crossed the Nile eastward to his Outer Mountain, where for twenty years he occupied a disused fort at Pispir. Finally, after 312 he retreated to his remote Inner Mountain, Mt. Colzim, near the Red Sea. Although opting out of civilization to escape from one's troubles was not uncommon in third-century Egypt, Antony was the first to attract influential publicity in "taking to the bush" *(anachōrein),* for Christian reasons. His utterances enjoy first place in the *Sayings of the Fathers.* He was beset by visitors, people seeking help and imitators, whose attachment to "Father" (Abba, Apa) Antony created colonies of hermit cells *(monastēria)* around Pispir.

Knowledge of Antony depends largely on the Life written soon after his death by Athanasius and translated at least twice into Latin by 379. Inspired partly by classical or Hellenistic lives of heroes and sages, it was influential in disseminating monasticism in both East and West and became a model for later Christian hagiography. Antony is depicted as the pattern of anchoritic life, one of severe austerity, incessant prayer, supernatural healings and perceptions, and above all perpetual warfare with the demons peopling the deserts. This individualistic quest for perfection, i.e., the recovery of the soul's created nature, bypassed the Church, although Antony remained a champion of episcopal orthodoxy, hostile only to schismatics (Melitians) and heretics. In 338 he visited Alexandria to disavow any sympathy towards Arianism. In 311 he was there during the persecution of Maximin Daia, sustaining the confessors but being denied the martyrdom he desired. A Copt who knew no Greek, untutored (as a boy too shy for school) and perhaps barely literate, he left eight extant letters.

BIBLIOGRAPHY: J. Quasten, *Patrology* 3, pp. 39-45, 148-53; J. David in *DHGE* 3, pp. 726-34; H. Queffebe, *Saint Anthony of the Desert* (1954); B. Steidle (ed.), *Antonius Magnus Eremita 356-1956 (Stud. Anselm.* 38) (1956); D.J. Chitty, *The Desert a City* (1966), chaps. 1-2; translation of Life by R.T. Meyer (*ACW* 10, 1950).

D.F. WRIGHT

ANTONY OF PADUA (1195-1231). Patron saint of the poor, of Portugal, and of Padua, Italy. Born of noble parents at Lisbon, he joined the Augustinian Canons* in 1210. In 1212 he entered the Augustinian study house at Coimbra, where he took his doctorate, gaining a reputation for preaching skill and biblical and theological learning. Seeking to emulate the first Franciscan martyrs in Morocco, he joined that order, exchanging his baptismal name of Ferdinand for Antony, and left for Africa, but illness forced his return. His ship was carried in a storm to Sicily, from which he traveled to the general chapter at Assisi in 1221. He then retreated for further study and contemplation with the celebrated mystic, Thomas Gallo, translator and commentator on the so-called Dionysius the Areopagite. Summoned from this life of seclusion at the behest of St. Francis,* who at the urging of Elias of Cortona reluctantly commissioned him the first teacher of the order, Antony subsequently taught at Bologna (1222), Montpellier (1224), Toulouse, and Padua, and is credited with introducing Augustinian theology among the Franciscans. He served his order at Puy (1224), Limoges (1226), and in the Romagna. His zeal against the Cathari,* Patarines,* and most probably the Albigensians,* earned him the title *Maleus hereticorum* ("Hammer of heretics"), while his persuasiveness led him to establish a brotherhood of penitents among his converts at Padua. From 1230 he devoted himself to preaching, with repentance and contempt for the world providing frequent topics. His theology was mystical, and his interpretation of Scripture allegorical rather than literal. His gifts as a preacher were extraordinary including, in addition to a clear voice and compelling manner, prophetic powers and miracles, accounts of which have been embellished since his death. His Lenten series in Padua in 1231 reached the proportions of a revival, with 30,000 reported auditors at one time in an open field. The response was massive reconciliations and restitutions, such that the clergy were insufficient for the needs of the people. Antony was canonized by Gregory IX in the year following his death at thirty-six.

BIBLIOGRAPHY: J. de La Haye (ed.), *Sancti Francisci Assisiatis . . . nec non Sancti Antonii Paduani Opera Omnia* (1641); Antony's *Moral Concordances* (ET ed. J.M. Neale, 1856); biographical studies by J. Rigauld (in French, 1899; A. Lepitre (tr. E. Guest) (1913); E. Gilliat-Smith (1926); R.M. Huber (1945); S. Clasen (tr. I. Brady, 1961).

MARY E. ROGERS

ANTWERP, SYNOD OF (1566). A clandestine conventicle held by the Reformed Church. Its chief importance was the adoption of a Latin recension of the Belgic Confession* as the Netherlandish statement of faith. This confession, originally composed in 1561 by Guido de Bres, pastor at Tournay, was revised for the synod by an Antwerp pastor, Francis Junius of Bourges, a pupil of Calvin, who later became professor at Leyden. This adoption marked the final acceptance of Calvinistic principles in the Netherlands.

ANUSIM (Maranos). Spanish or Portuguese crypto-Jews outwardly professing Christianity through expediency, compulsion, or fear. The subterfuge began with a frightful anti-Semitic pogrom in Seville in 1391, when 4,000 Jews were massacred. Others simulated faith and accepted baptism to save their lives. Some were cynical apostates who lacked Jewish faith, but the majority conformed secretly. The persecution continued when, with imaginary atrocities charged against them, the Maranos were hounded by the Spanish Inquisition.* Many outstanding men were among the refugees who from that time were scattered all over the globe. The excommunicated Jew, Baruch Spinoza,* was of Marano stock. The bloodcurdling record stresses the brutality of bigoted persecution, the impervious disregard of human liberty, and the final impossibility of proselytizing by force.

ROY A. STEWART

APHRAATES (Aphrahat). The twenty-three Syriac Tractates of Aphrahat, written 337-45, are the earliest extant evidence of Syrian church life and thought. A Bible handed on from Jewish Christian sources and gospel citations from the *Diatessaron* of Tatian combine with a background of virile apologetic against rabbinical Judaism to produce a doctrinal outlook almost unaffected by Greek speculation and Nicene theology. An ascetic from Mosul in Persia, Aphrahat taught God in Christ, the Spirit and Resurrection, baptism and asceticism. God the Creator gave the Law to Moses; Christ is the Son of God, whose Holy Spirit enters a man at baptism and helps him attain resurrection. Baptism commits the believer to moral virtues and ascetic practice. In principle Aphrahat associates postbaptismal life with renunciation of the world, direct warfare with Satan, and celibate asceticism. He advises priests to discriminate against baptismal candidates who have little ascetic potential. G.T.D. ANGEL

APIARIUS. A priest from Sicca Veneria in Proconsular Africa, he is a test case for relations between the African and Roman churches in the fifth century. Excommunicated and deposed by the pro-Augustinian diocesan Urbanus, Apiarius appealed to Zosimus of Rome (417-19), who demanded his reinstatement by the African bishop (Council of Carthage, 418) on the basis of canons of the Western Council of Sardica (343) which, he claimed, were Nicene. The Africans after inquiry to the East failed to find them among Nicene canons, Apiarius confessed, and the case was closed. Excommunicated later for offenses at Tabraca, Apiarius appealed to Celestine of Rome (422-32), who ordered him restored. At another Council of Carthage (424), Apiarius confessed and the African bishops asserted their right to judge their affairs exclusively, while conceding to Rome primacy of honor. G.T.D. ANGEL

APOCALYPSE, see REVELATION, BOOK OF

APOCALYPTIC LITERATURE. The Greek word *apokalypsis,* from which "apocalyptic" is derived, means "unveiling" (Lat. *revelatio*) and indicates the unfolding of things hitherto hidden or secret. In the literature thus described, the subject-matter concerns the future rather than the present, the spiritual rather than the material world, the purposes of God rather than the actions of men. It transports the reader out of his immediate existence and allows him to share in the mysteries of what God will finally do with His universe. It is no accident that the period when such writings flourished (c.200 B.C.–A.D 150) was an age of persecution for both Jews and Christians, spanning the Maccabean Wars, the fall of Jerusalem, and the Bar-Kochba uprising, and including the persecutions of Nero and Domitian.

Although a few complete books like Daniel, 2 *Esdras* (in the Apocrypha*), and Revelation* can be firmly classified under this term, apocalyptic describes a style of writing, elements of which can be discerned in quite different literary categories, such as Isaiah (24-27), Ezekiel (38, 39), Joel (3:9ff.), Zechariah (9-14), and Mark (13). Most apocalyptic writings are noncanonical, however, and are to be found among the Pseudepigrapha* and the sectarian literature of the Qumran* community. Of these the best-known examples are the two *Books of Enoch,* the *Book of Jubilees,* the *Assumption of Moses,* the *Apocalypse of Baruch,* and (from Qumran) the *War of the Sons of Light against the Sons of Darkness,* though many other writings with marked apocalyptic features could be added to the list, notably the Christian *Shepherd of Hermas.*

Among the characteristics of apocalyptic are these four elements:

(1) Its esoteric framework: the secrets of the universe and of the last days are revealed to the author or chief character of the book in a series of visions, often through angelic mediation.

(2) Its developed imagery: great use is made of symbols and symmetrical patterns. Numbers play an important part in the unfolding of world history, especially seven (as in Daniel's "week of years" [Dan. 9:24]), forty-nine (as in the *Book of Jubilees*) and a thousand (cf. the many millennial references). Wild animals appear as representing nations, as also do angels (cf. Dan. 10:13). Apocalyptic literature thus develops a language of its own, which must be interpreted in the light of earlier usage. This is especially important for the correct interpretation of Revelation.

(3) Its eschatological setting: in many visions the whole of history is surveyed from the Creation onward, but the prime interest is always the end of the age. This is frequently preceded by the most dreadful persecution of the faithful and a period of conflict between the forces of good and evil. Satan and his demonic satellites feature strongly, as do messianic and angelic figures. There is great stress on divine judgment, the day of the Lord and the messianic age, and doctrines of resurrection and the afterlife.

(4) Its pseudonymity: because of the ascendance of the Mosaic law in the Judaism of this pe-

riod, these apocalyptic revelations were often attributed to great men of the past, who were either exponents of the law like Moses and Ezra, or precursors of it like Enoch, Abraham, and the patriarchs. It does not follow from this, however, that the two canonical books, Daniel and Revelation, are also pseudonymous. Daniel is a composite work, combining narrative episodes with visions, and has unique problems regarding authorship which have to be studied separately. In Revelation, the exiled disciple is using conventional apocalyptic imagery to reveal truths that could be published in no other form without fear of reprisal.

Whereas most apocalyptic writers leaned heavily on the ideas and expressions of their predecessors in the genre, it is less easy to say what actually gave rise to this style of writing. To regard Daniel as the prototype of all subsequent apocalyptic writing ignores both the uniqueness of that book and also the many apocalyptic patterns within prophetic literature. There is a continuity between prophecy and apocalyptic which B.W. Anderson catches well in his description of the latter as "prophecy in a new idiom." On the other hand, Von Rad traces its origins, not to prophecy, but to wisdom, on the grounds of its totally different theological standpoint and the primacy it gives to inner knowledge and illumination. Certainly there are significant differences between the theologies of prophecy and apocalyptic: the first saw God's activity as being primarily in this world, the latter despaired of the present age and looked to His judgment beyond the limits of time and human history; the first endeavored to apply the ethical demands of God's righteousness to the immediate situations of Israel and her neighbors, the latter applied the principle of God's determinism to the wider setting of national destinies in the last days. Apocalyptic shows signs of a cosmic dualism between good and evil, light and darkness, God and Satan, which the prophets would never have allowed and which may be attributable to Persian influences. It is noteworthy, however, that in the canonical writings at any rate this dualism never degenerates into a power struggle between two equally balanced forces. In both prophecy and biblical apocalyptic God is at all times supreme and the Lord of His creation.

BIBLIOGRAPHY: H.H. Rowley, *The Relevance of Apocalyptic* (1947); D.S. Russell, *The Method and Message of Jewish Apocalyptic* (1964).

JOHN B. TAYLOR

APOCRYPHA. Derived from a Greek neuter plural adjective meaning "hidden things," the word "apocrypha" has a different meaning for different church traditions. The Protestant use refers to the books which are sometimes printed in Bibles as a separate block of literature between the OT and the NT. Apart from *2 Esdras,* which originates from the Vulgate, these books together with the OT books constituted the Septuagint. In Roman Catholic usage these writings are called "deuterocanonical," and the term "apocrypha" is reserved for those books wholly outside the canon which Protestants call the pseudepigrapha."* This article will restrict itself to the books of the Protestant tradition. The books are:

(1) *1 Esdras* (= *3 Esdras* in Vulgate, which uses *1* and *2 Esdras* for Ezra and Nehemiah; = *Esdras B* in Septuagint). This consists of material, paralleled in 2 Chronicles, Ezra, and Nehemiah, but rearranged by the editor/author, dealing with the story of the Jerusalem temple from Josiah to Zerubbabel and Ezra's restoration. An interesting fictional addition is the debate of the three youths, which Zerubbabel, a guardsman of Darius, wins by showing that truth is the strongest power on earth.

(2) *2 Esdras* (= *4 Esdras* in Vulgate) is a Christian expansion of a Jewish apocalyptic work consisting of seven visions about the age to come. In its present form it dates from about A.D. 90.

(3) *Tobit* is a delightful short story about the adventures of Tobit's son, Tobias, who journeys to Media in company with the angel Raphael disguised as a mortal. With his help two healing miracles take place, one on Sarah, Tobias's betrothed, who has been tormented by a demon, and the other on the aged Tobit, whose sight is restored.

(4) *Judith* tells the equally fictitious story of how a young widow of Bethulia (= Bethel), which was besieged by Holofernes's army, delivered the city by enticing the enemy general to his death.

(5) *The rest of Esther.* These are some additions to the canonical book, designed to deepen its religious content and to strengthen its claim to canonicity.

(6) *The Wisdom of Solomon* is in the finest tradition of Hebrew wisdom, though it betrays considerable Hellenistic influence and is one of the few parts of the Apocrypha written originally in Greek and not translated from a Hebrew or Aramaic original.

(7) *Ecclesiasticus,* or the Wisdom of Jesus, Ben-Sira, written about 180 B.C., is a guidebook to the good life, summed up in the word "wisdom," and is probably the most significant of the apocryphal books.

(8) *Baruch,* a composite work including the letter of Jeremiah, is a first-century B.C. compilation attributed to Jeremiah's scribe and companion.

(9) The additions to Daniel, viz., *The Song of the Three Holy Children* in the fiery furnace (known as the Benedicite) and the legendary tales of *Susanna* and *Bel and the Dragon.* (Alternative accountings list these as separate books, resulting in enumeration of up to fourteen apocryphal books as compared with this listing of twelve.)

(10) *The Prayer of Manasseh* is a brief but moving expression of penitence for sin.

(11) *1 Maccabees* is a historical survey of the events from 175 to 134 B.C. in which the Maccabean house is exalted as the means of bringing salvation to Israel. It is a reliable account of the period and was written in Hebrew about 100 B.C., after the death of John Hyrcanus.

(12) *2 Maccabees* covers roughly the same period but is much less reliable as history, containing a number of highly colored episodes from the story of the uprising.

None of these books was accepted into the Hebrew canon by the Jewish synod of Jamnia, which met at a time (c. A.D. 100) when the authentic

Jewish heritage was thought to be in danger of erosion from the syncretistic tendencies of apocalyptic writing and from the increasing influence of Christianity. The Septuagint tradition reflects an earlier, pre-Christian stage of development where the need for rigid norms had not yet arisen. The early Christian Church never resolved its attitude to these divergent approaches: Clement, Cyprian, and Augustine were among those who followed the Septuagint canon, while Origen, Cyril, and Jerome held to the Hebrew books. Although Jerome's Bible, the Vulgate (which became the official Roman Catholic text of Holy Scripture), incorporated the Apocrypha, Jerome himself wrote that these *libri ecclesiastici* (as distinct from the *libri canonici* of Hebrew tradition) could be read for edification, but not for confirming the authority of church dogmas.

In the Reformation Luther incorporated the Apocrypha into his translation of the Bible (1534), adding that the books were not equal to Scripture but nevertheless were "profitable and good to read." The Reformed churches went further and excised them altogether from the canon of Scripture,* Article 3 of the Westminster Confession* (1647) explicitly rejecting their inspiration, authority, and spiritual usefulness.

The Roman Catholic Church at the Council of Trent* (1546) anathematized those who did not regard as sacred and canonical all the books contained in the Vulgate, and this view was substantially upheld by the Vatican Council of 1870. The Greek Church, after a period of uncertainty, eventually settled at the Synod of Jerusalem (1672) for the Jamnia canon, with the addition of *Tobit, Judith, Wisdom of Solomon,* and *Ecclesiasticus.*

Questions of canonicity apart, the Apocrypha has considerable value for biblical scholarship for the light it sheds on the intertestamental period—the development of apocalyptic thought, of wisdom and nomistic theology, and the increasing impact of Hellenistic ideas on Judaism.

BIBLIOGRAPHY: R.H. Charles, *The Apocrypha and Pseudepigrapha of the Old Testament* (2 vols., 1913); W.O.E. Oesterley, *An Introduction to the Books of the Apocrypha* (1935); R.H. Pfeiffer, *History of the New Testament Times, with an Introduction to the Apocrypha* (1949); B.M. Metzger, *An Introduction to the Apocrypha* (1957); L.H. Brockington, *A Critical Introduction to the Apocrypha* (1961). JOHN B. TAYLOR

APOCRYPHAL NEW TESTAMENT. This is a general description of those books circulating during the first centuries of the Christian era which purported to relate details about Christ and the apostles, but which were never considered to be canonical. The title probably developed on the analogy of the OT Apocrypha,* which was, however, a more specific collection of books. In the NT apocrypha there was never a collection of books which offered an alternative to the NT canon, but rather a motley variety of literature whose only unity was its common noncanonical status. The popularity of these books is demonstrated by the number which are extant in whole or in part and the wide geographical distribution of their use. Many of them have been preserved only in versions, although a number of originals are known.

One of the most striking features about the Christian apocrypha is the fact that the majority are attempts to produce literary forms parallel to those of the NT books. It is possible therefore to classify them under gospels, acts, epistles, and apocalypses, plus a small group of miscellaneous works. Of these literary forms the most prolific was the acts and the least prolific was the epistolary form. This reflects the differing degrees of difficulty surrounding their production. To produce a narrative of events in which various apostles figured as heroes was clearly a simpler procedure than to produce an epistle which possessed some air of validity.

It will be possible here only to mention the major examples of these apocryphal works to illustrate both their variety and their characteristics. Among the gospels there were three main types. The first shows some influence from the Synoptic gospels in its literary form. There is a small fragment known as *Papyrus Oxyrhynchus** 840 which consists of only a few verses but belongs to the early second century. Of about the same date is the *Egerton Papyrus** 2 which combines Synoptic-type material with Johannine. But these may be instances of the combining of written material. The *Gospel of Peter* draws some material from the Synoptics, but mixes it with Gnostic overtones. Another work which may belong to this type is the *Gospel of the Egyptians,* but the remains of this are too fragmentary to provide an accurate picture of its original form.

The second type of gospel is essentially Gnostic, imparting Gnostic doctrine in the form of teaching attributed to the exalted Lord. Examples of this kind are the *Apocryphon of John,* the *Gospel of Thomas,* and the *Sophia Jesu Christi.* Two other Gnostic works which have the word "gospel" in their title, but which bear no relationship to the canonical gospels, are the *Gospel of Truth* and the *Gospel of Philip.*

Because of the fact that the canonical gospels confine themselves almost wholly to the ministry of Jesus, it is not surprising that there was an urge in the third type of apocryphal gospels to fill in some of the gaps by recourse to imagination. This is particularly true of the Infancy gospels, which offered considerable scope for descriptions of the early years of Jesus about which the canonical books are silent. A work like the *Gospel of Nicodemus* is an example of an attempt to fill out the Pilate story. The proportion of legendary material in works of this type is high.

The apocryphal acts offer a more varied form of literature, bound together by what was loosely considered as possibly "apostolic." Sometimes the emphasis was on polemic, sometimes on apologetic. Some of these apocryphal acts made attempts to edify, some only to entertain. There is no easy way of classifying these works. All of them are attributed to apostles. The earliest books of this character which are extant are those which circulated under the names of John, Peter, Paul, Andrew, and Thomas. There is abundant evidence to show the high esteem in which apostolic names were held in the earliest period of church history, and this literature corroborates this evidence. It is significant that although these books purport to have the same form as the canonical Acts, the attribution to individual apostles at once sets them apart. Moreover the frequency with which heterodox doctrine occurs further reveals the gulf between the canonical and apocryphal books.

In addition to these books, epistles and apocalypses circulated. Examples of the former are few because of the difficulty of producing this type of literature with any appearance of authenticity. The most notable is that known as *3 Corinthians* which appeared as part of the *Acts of Paul* but circulated separately at least in the Syriac-speaking church where at one time it seems to have been accorded canonical status. The only other epistle which deserves special mention is the *Epistle to the Laodiceans* which had a wide circulation during the Middle Ages, although there is nothing distinctive about its contents. In fact it is almost wholly a plagiarization from the canonical Philippians. Such works as the pseudocorrespondence between Paul and Seneca and the pseudo-*Epistle of Titus* make no attempt to imitate the epistolary form.

Among the apocalypses, the most notable are those ascribed to *Peter,* which in certain quarters achieved some sort of semicanonical status (it is mentioned in the Muratorian Canon, although doubts regarding it are reported) and the *Ascension of Isaiah,* which shares the form of Jewish apocalypses but has clear Christian allusions. The *Apocalypse of Paul* and another ascribed to *Thomas* are both later productions.

It is important to examine the motives which prompted these apocryphal books. It is worth observing the comment of Tertullian that truth precedes forgery, for it is clear that the widespread acknowledgment of the canonical books was a necessary prelude to the production of imitations, at least in title. By means of pseudonymous literature the producers hoped to gain acceptance for their ideas. The importance of apostolicity in relation to Christian tradition largely dictated the pseudonyms which were chosen.

Some of the literature was simply the result of the desire to satisfy imagination. It is not difficult to see that such a reference as Colossians 4:16 could have proved sufficient impulse for someone to produce an epistle to the Laodiceans. The author of this epistle may well have thought that nothing written by Paul should have been lost and therefore an epistle to suit this reference was desirable. Many of the narrative forms in these apocrypha are fanciful and clearly fictional. In an age which was mainly uncritical, particularly among members of the general public, there was an ever-ready market for romances about the earliest Christian leaders.

Another motive which was particularly dominant was the desire to add details which are missing from the canonical books. The most prevalent source of such a motive was Gnosticism,* which by this means introduced its own particular tenets into much of the pseudepigraphical literature. A book like the *Gospel of Thomas,* for instance, contains a medley of sayings, some closely parallel to the synoptic gospels and some couched in the language of Gnosticism. Other books, like the *Gospel of Peter,* which is in the main orthodox but has Docetic implications in its account of the crucifixion, are less pronouncedly Gnostic (see DOCETISM). In many cases the introduction of heterodox doctrine is subtly done.

Apocryphal literature offered a suitable medium for those who wished to claim a secret source for their doctrines as many of the authors of this kind of literature did. There was no reasonable alternative when pseudonymous works were published a considerable time after the putative author had died. Moreover, a favorite device was to concentrate on the post-resurrection appearances of Jesus, which offered most opportunity for the creation of speeches containing deviating doctrine. Indeed, Gnostics generally showed little interest in the humanity of Jesus, and thus the resurrection experiences came into greater prominence.

The history of the canon shows that the orthodox Christians had a discerning approach to the mass of apocryphal literature. None of the books came to be generally received, although some enjoyed extensive popularity. There is a wide gap between the canonical books and their apocryphal imitations. In spite of their use of apostolic names they completely lack the apostolic content. They are nevertheless a witness to the unrestrained character of much that passed for popular Christianity. The vigilance of the leaders of early Christian thought deprived them of any authority.

BIBLIOGRAPHY: M.R. James, *The Apocryphal New Testament* (1924); R.M. Wilson (ed.), *New Testament Apocrypha,* I (1963), II (1964)—based on E. Henneeke-W. Schneemelcher's *Neutestamentliche Apokryphen.* DONALD GUTHRIE

APOLLINARIS, SIDONIUS, see SIDONIUS APOLLINARIS

APOLLINARIUS; APOLLINARIANISM. Born about 310 at Laodicea in Syria, Apollinarius became a reader under the Arian bishop Theodotus and shared with his priestly father a delight in pagan literature. When Julian deprived Christians of pagan classics, they restyled parts of the Bible in poetic meters or as philosophical dialogues. He had welcomed Athanasius back from exile in 346, supported the *homoousion* (see ANCYRA) and became bishop of the Nicene church at Laodicea about 361. His views were opposed when the Council of Alexandria, chaired by Athanasius in 362, attributed a human soul to Christ, and about 375 he seceded from the orthodox church. By 377 the Western Council of Rome under Bishop Damasus condemned him, followed by the Eastern councils of Alexandria (378), Antioch (379), and Constantinople (381). Theodosian decrees (383-88) forbade Apollinarian worship and outlawed his adherents.

Apollinarius wrote extensively, but few writings remain. Some are attributed to orthodox writers: to Gregory Thaumaturgus, a detailed creed; to Athanasius, a sermon *Quod unus sit Christus, De Incarnatione Dei Verbi,* and a creed addressed to the emperor Jovian; to Julius I of Rome (337-52) *De Unione Corporis et Divinitatis in Christo, De Fide et Incarnatione,* and a letter to Dionysius. Two works can be reconstructed from his opponents: a *Demonstratio de Divina Incarnatione* from Gregory of Nyssa's *Antirrheticus,* and a brief *Recapitulatio* from a dialogue attributed to Athanasius. Only fragments of his other works can be gleaned from patristic writers and catenae. These included commentaries on both OT and NT, apologetic works against Porphyry and Emperor Julian, and dogmatic polemical works against Origen, Dionysius of Alexandria, Eunomius of Cyzicus, Marcellus of Ancyra, Diodore of Tarsus, and Flavian of Antioch. Of the metrical version of the Bible, only the Psalms are extant, and their authenticity is suspect.

Apollinarius belongs to the tradition of Alexandrian* Christology seen earlier in Athanasius and later in Cyril of Alexandria. Like the former, he held that Christ has one active principle, the divine Logos,* and that the essential attribute of His humanity (flesh) is its capacity for experience, not for initiative. His error was to exclude even the potential for initiative from the humanity of Christ.

Christ had one active principle alone, because according to the biblical evidence Christ is one and never experienced volitional conflict. The Logos was that principle, since only God (not man) can redeem, resurrect, and avoid error; only God could have performed miracles, displayed authority, and created. To see Christ as an inspired man, as Paul of Samosata did, or to attribute both divine and human active principles, is to rob Him of worship and to risk His being fallible. Christ became one in the union of the Logos and the flesh of Mary. The Spirit of God "sanctified" (cf. John 10:33-36) her flesh and formed Christ. Independently the Logos and flesh were incomplete, but together in the union they became someone living, "a mixture of God and man." This vital union alone distinguishes the flesh of Christ from human flesh in general, since in the latter case the "soul" of man (variously described as *pneuma* or *psyche + nous*) unites with flesh.

The Logos alone motivated Christ. His flesh, like Solomon's temple, had no independent life, mind, or will, but it "experienced" passively. Their dynamic unity was so close that Christians worship the flesh of Christ, becoming divine as they assimilate it in the Eucharist. Christ had no human source of initiative, no human soul, for the Logos alone saves, the flesh passively experiencing human conditions. Critics like Gregory of Nyssa pointed to the biblical evidence of Christ's human experience, to the principle of Hebrews 2:17, and to the presupposition that full salvation requires identification with full humanity, soul and flesh. Was Christ *bound* to sin if he had a human will?

BIBLIOGRAPHY: H. Lietzmann, *Apollinaris von Laodicea und seine Schule* (1904); B. Altaner, *Patrology* (ET 1960), pp. 363-65 for bibliography; P.A. Norris, *Manhood and Christ* (1963), pp. 81-122; J.N.D. Kelly, *Early Christian Doctrines* (3rd ed., 1965), pp. 289-95; A. Grillmeier, *Christ in Christian Tradition* (1965), pp. 220-33; M.F. Wiles, "The Nature of the Early Debate about Christ's Human Soul," *Journal of Ecclesiastical History.* XVI (1965), pp. 139-51. G.T.D. ANGEL

APOLLINARIUS, CLAUDIUS. Bishop of Hierapolis, c.175. An apologist for the faith and critic of heresy, he wrote four apologies (to Emperor Marcus Aurelius, against the Greeks, on Truth, and against the Jews), an encyclical against the Montanists* who became active during his episcopate, and a treatise on Easter. All the writings are lost, but his loyalty to the Roman state and his fierce opposition to the Montanists are clear from the use made of his writings by Serapion of Antioch and the historian Eusebius.

APOLLONIUS OF TYANA (d. c.98). A Neopythagorean philosopher whose virtuous life and ascetic practices attracted widespread attention after his death. Philostratus, sophist at the emperor's court, was commanded by Julia Domna, wife of Severus, to write a "Life of Apollonius," possibly as a pagan counterpart to the life of Christ. In Philostratus's account, leading points of the gospel story are followed: Apollonius was the son of Jupiter and conscious of his filial sonship, and he went around doing good and performing miracles. Like Christ he was brought to trial, but unlike Christ he was miraculously delivered. Subsequently he ascended to heaven. This "Life" is the story of the gospels revised and made more palatable by non-Christians who were impressed by Christ. It is what the enlightened pagans around Julia Domna thought Christ should have been. It is significant that they omit the Crucifixion. G.L. CAREY

APOLOGETICS. The use of theology in order to justify Christianity before men, in the claims it makes to be ultimate truth, in the demands it makes on its followers, and in its universal mission. Jesus Himself was often ready to answer objections and insinuations made against Him and His teaching (cf. John 8:41-58; 18:19-24), which latter He developed and justified against His opponents (cf. Mark 2:6-12; 10:2-9; Luke 4:22-28, etc.). Paul also tried to speak about the wisdom and power of the Cross in the light of deep-seated objections (cf. 1 Cor. 1:18-31). "Always be prepared," wrote Peter (1 Pet. 3:15), "to give an answer to everyone who asks you to give the reason for the hope that you have."

Defensive statements of faith, or "apologies," appear as early as the second century when a group known as the "Apologists"* (Justin, Tatian, Athenagoras, Tertullian, etc.) took up the task of answering current slanders against Christianity—e.g., that it encouraged cannibalism and impiety, that it discouraged loyalty to state religion—and was therefore atheistic—and that its central doctrines were ridiculous and offensive. The Apologists stressed the antiquity of the Gospel, the genuineness of its miracles, and its striking fulfillment of prophecy. They had to show Christianity, not only as a superior religion, but as the ultimate truth. They tried to bridge the gap between their opponents and themselves by laying hold of what they believed were similarities as well as differences between the Gospel and pagan philosophy.

After the establishment of the Church under Constantine, apologetics became an aspect of the work of great constructive theologians such as Augustine and, later, Aquinas. Even Calvin's *Institutes* was presented with a noble introductory letter to the French king Francis I as a defensive statement of the faith he was mistakenly persecuting. Wherever theology has pursued its main task, the apologetic aim has never been lacking: demonstrating the validity of the claims it makes for Christ, and showing that the faith is not unreasonable but has its own inner logic and consistency. In the late eighteenth and nineteenth centuries, however, apologetics developed as a special branch of theology dealing with the defense and proof of Christianity.

Among other matters, apologetics has sought to meet questions about the historicity of the main events on which Christianity is based, and of the Bible. It has discussed miracles, the existence of God, the knowledge of God, the harmonizing of the biblical account of Creation with that of science (i.e., "Christian Evidences").

There are dangers in a too specific and conscious apologetic approach in our statement of the Gospel: overmuch attention to specific objections can lead to an unhealthy one-sided emphasis. A defensive program, moreover, tends to produce the mentality that prefers seclusion to open Christian warfare. In defending the Gospel, theology must never change what is essential either in its content or form or message, or remove the offense of the Cross. But dialogue need not involve compromise. Growth in understanding can come by struggling with the questions and even the unjust accusations of opponents (cf. 2 Sam. 16:9-12). An alien world needs to be shown that the Gospel has also the only teaching and power that can enable man to recover and express his true humanity.

BIBLIOGRAPHY: A.B. Bruce, *Apologetics* (1892); J. Baillie, *Our Knowledge of God* (1939); A. Richardson, *Christian Apologetics* (1947); B. Ramm, *Problems of Christian Apologetics* (1949); C. Van Til, *The Defense of the Faith* (1955).

RONALD S. WALLACE

APOLOGISTS. The term given to a number of early Christian writers (c.120-220) who belonged to a period in history when the growing Christian Church was meeting with ever-increasing hostility in every department of public life. They include Quadratus, Aristides, Justin, Tatian, Theophilus, Minucius Felix, and Tertullian. They worked on the frontier of the church, seeking to defend the Faith from misrepresentation and attack, commending it to the inquirer and demonstrating the falsity of both Judaism and polytheism. These writers did not need to create a literary form for their purpose, because it already existed in the legal speech for the defense *(apologia)* which was delivered before the judicial authorities and subsequently published. There was also the literary form of the dialogue which was usually based upon fictitious circumstances. As the person of Christ was the central difficulty to pagan thinkers, the Apologists found the Logos concept common to both Platonism and Christianity a welcome means of making this doctrine acceptable to Hellenistic philosophy.

G.L. CAREY

APOPHTHEGMATA PATRUM. An anonymous collection of ascetic sayings and anecdotes which illustrate early monastic thought, especially in Egypt. A Greek edition, produced probably in the sixth century, ranks the material in alphabetical order according to the names of speakers. The basic core stems from Coptic* monks in the Wadi Natron, and their words and deeds have been added to from other areas. The sayings or examples arose when inquirers induced ascetics to break their accustomed silence with the request, "Grant me a word." The reply was recorded orally and later in writing with the veneration given to divine revelation. Corresponding disregard for biblical teaching is illustrated by Amoun: "We prefer to use the Sayings of the Fathers and not passages of the Bible; it is very dangerous to quote the Bible."

G.T.D. ANGEL

APOSTASY. The abandonment or renunciation of Christianity, either voluntarily or by compulsion. The use of the term for religious apostasy in the Hebrew-Christian tradition derives probably from Septuagint usage. Both voluntary (Josh. 22:22; 2 Chron. 22:9) and involuntary aspects occur (1 *Macc.* 2:15). Mattathias's refusal to apostatize to pagan rites was the occasion for the Maccabean revolt; it denoted deserting from, rebellion against, or abandonment of the Mosaic teaching. While the term does not occur in the KJV, it does in the Greek (Acts 21:21; 2 Thess. 2:3).

There are frequent biblical allusions to the evils and the dangers of apostasy. It is described as departure from the faith (1 Tim. 4:1-3), being carried away by the error of lawless men (2 Pet. 3:17), and falling away from the living God (Heb. 3:12). The great apostasy, "The Rebellion" of 2 Thess. 2:3, is associated with the return of Christ. The serious consequences of apostasy are stressed in Hebrew 6:4-6; 10:26 (cf. 2 Pet. 2:20). It occurs through the subverting activities of false teachers (Matt. 24:11; Jude, etc.), but it may also occur because of persecution and stress (Matt. 24:9, 10; Luke 8:13). Thus the NT warns against both voluntary and involuntary apostasy so identified.

Church history reflects the activities of apostates and alleged apostates, and as well the problems of persecution, involuntary recanting, and what to do with the lapsed. The use of the civil power by both Catholics and Protestants to punish those charged with apostasy resulted in great cruelties during the Middle Ages and later. The Anabaptist concept of a religiously composite society prevailed in the New World and later in the Old also. Compositism does not diminish the seriousness of doctrinal error, but it does tolerate divergent views within society under law in the belief that persuasion not coercion reflects the Christian ideal. This in no way abrogates the responsibility of the Church to maintain and defend its doctrinal purity in relation to the norms of biblical teaching.

SAMUEL J. MIKOLASKI

APOSTLE. The Greek word *apostolos* means "one sent out." In the New Testament it derives part of its meaning from the Hebrew *shaliah,* who acted as a representative for others. It is conferred on Jesus in Hebrew 3:1, but is normally reserved for those appointed for a special function in the church. From the large number of disciples who followed Him in His ministry, Jesus chose twelve whom He called apostles (Luke 6: 13). They were to act in His name (Mark 9:38-41). After Jesus' resurrection there had to be found a replacement for the traitor Judas so that the number could be made up to twelve, and Matthias was chosen (Acts 1:15-26). There is no record of a replacement for James bar Zebedee (Acts 12:2). The qualification was to have been with Jesus from John's baptism to the Ascension and to have been a witness of the Resurrection. This meant an experience of the saving events and therefore the ability to preach the *kerygma* firsthand.

To Paul, who did not have the same contact with Jesus during the ministry, there was given a special resurrection appearance and a special commission to go to the Gentiles (Acts 26:16-18; 1 Cor. 9:1; 15:8). There was a division of spheres of responsibility arranged between him and the Jerusalem apostles (Gal. 2:1-10). The apostles were seen as a gift of the Spirit to the church (1 Cor. 12:28; Eph. 4:11), and the work of the true apostle was accompanied by signs and wonders and mighty works (2 Cor. 12:12; cf. Acts 8:14-19). The apostles were seen as part of the foundation of the church (Eph. 2:20), and it was through them as witnesses to the saving events and as interpreters of them, as well as chief ministers in the church and propagators of the Gospel, that the mission of Christ was completed. The term is also used of others such as James the brother of Jesus apparently (Gal. 1:19; 2:9; cf. 1 Cor. 15:7), Barnabas (Acts 14:4, 14), Silvanus and Timothy (1 Thess. 2:6), and Andronicus and Junias (Rom. 16:-7).

BIBLIOGRAPHY: J.B. Lightfoot, *Epistle to the Galatians* (1902), pp. 92-101; K.H. Rengstorf in *TDNT* I (1964), pp. 398-447; C.K. Barrett, *The Signs of an Apostle* (1970). R.E. NIXON

APOSTLES' CREED. A statement of faith used by both Roman Catholic and Protestant churches in the West. Originally treated with suspicion by the Eastern churches, it is now accepted as orthodox, but not used in public services. The origin of the Apostles' Creed is to be found in the form learned by the catechumen in the course of his preparation for baptism in the early church. This kind of confession was called a *symbolum* and was not intended to be a complete summary of Christian doctrine, but rather a brief statement about the Trinity and the person and work of Christ. Since an important part of the catechesis consisted of memorizing the *symbolum* and repeating it to the bishop after scrutiny, it is not surprising that few written specimens of the early baptismal creed have survived. The old Roman form is known, however, from a commentary by Rufinus (c.404). Since virtually the same creed exists in a Greek version by Marcellus of Ancyra (c.340), it is supposed to date from the time when the liturgical language at Rome was still Greek (i.e., before c.250). The baptism service described in the *Apostolic Tradition* of Hippolytus (c.215) puts a similar creed in the form of a question to the candidate in three parts, demanding a threefold response of faith.

The structure of the Old Roman Creed was Trinitarian, with a considerable expansion of the second article about Christ to include a list of His saving acts which were proclaimed in the primitive *kerygma.* Three German scholars (K. Holl, A. von Harnack, and H. Lietzmann) have argued strongly that in fact the creed was produced by welding together a Trinitarian formula (perhaps derived from Matt. 28:19) with an originally independent christological summary. Other scholars (e.g., J.H. Crehan) believe the christological part was the primitive baptismal confession.

The Old Roman Creed became the standard pattern for the church throughout the West. Nicetas of Remesiana was using a similar text at the end of the fourth century. In the sixth century Caesarius of Arles gives evidence of a process of elaboration which took place in Gaul and eventually gained acceptance at Rome itself. These additional phrases are seven in number: "maker of heaven and earth"; "conceived"; "dead"; "He descended into hell"; "almighty"; "catholic"; and, "the communion of saints." The earliest example of the Latin text in exactly its modern form dates from the eighth century. While there is no foundation for the story of Rufinus that each apostle contributed an article to this creed, it has commended itself as a useful and succinct statement of faith which is today used frequently in worship. Although no longer confined to baptism, it should serve as a constant reminder of the Christian's baptismal confession.

BIBLIOGRAPHY: P. Schaff, *The Creeds of Christendom* (3 vols., 1877); J. de Ghellinck, *Patristique et Moyen âge—I: Les recherches sur les origines du symbole des Apôtres* (2nd ed., 1949); O. Cullmann, *The Earliest Christian Confessions* (ET 1949); J.N.D. Kelly, *Early Christian Creeds* (1950); J.H. Crehan, *Early Christian Baptism and the Creed* (1950). JOHN TILLER

APOSTOLIC CANONS. Eighty-five canons attributed to the apostles are contained in book 8 of the *Apostolic Constitutions** (c.381). They deal both with the election, ordination, official responsibilities, and moral conduct of clergy and with Christian life in general. Deposition and excommunication of offenders are given as sanctions. They are first referred to as a set at the Council of Constantinople in 394, but some are paralleled by canons of the Synod of Antioch (341). Canon 85 provides a list of sacred books; the Apocalypse is omitted, but included are three books of the Maccabees, three letters of Clement, and the *Apostolic Constitutions* themselves. Dionysius Exiguus* translated canons 1-50 into Latin and included them among his larger collections of canons. This formed the basis of Western canon law. The East accepted all the canons while rejecting the *Apostolic Constitutions* as a whole at the Trullan Council (692). G.T.D. ANGEL

APOSTOLIC CONSTITUTIONS. Eight books on church pastoral and liturgical practice, they are attributed to Clement of Rome, but were compiled by an Eastern Arian in the late fourth century. Rejected because of heretical influence by the Trullan Council in 692, they had little regulative influence on the Greek Church, although excerpts are found in Eastern collections of canon law. Material from earlier works such as the *Didascalia** (early third century) and the *Apostolic Tradition* of Hippolytus (c.200-220) is taken over and brought up to date. For example, the epiclesis in the *Apostolic Constitutions* is more concrete than the invocation of the Spirit in the *Apostolic Tradition,* and a forty-day fast now precedes Easter, whereas the *Didascalia* appoints a fast for Holy Week alone. Again, minor orders* have increased to subdeacon, janitor, and psaltes. Significant inclusions are the *Gloria* and the so-called Clementine liturgy of the Mass. Book 8 contains the *Apostolic Canons.* G.T.D. ANGEL

APOSTOLIC DELEGATE. An official permanent papal representative from the Holy See to the Roman Catholic Church of a given area, usually where no Nuncio* is appointed. His office is ecclesiastical, not diplomatic, with the duties of observing and reporting to the Holy See on the life of the church and of communicating instructions of all kinds from Rome to the church. A delegate was appointed to the American Catholic Church in 1892-93, the British in 1938, the Canadian in 1899.

APOSTOLIC FATHERS. A group of early Christian writers believed at one time to have had direct contact with apostles. J.B. Cotelier's edition (1672) of the *Epistle of Barnabas, 1* and *2 Clement,* the *Shepherd of Hermas,* and the *Epistles* of Ignatius and Polycarp spoke of "the Fathers who flourished in Apostolic Times," while L.T. Ittig published Clement, Ignatius, and Polycarp as "Apostolic Fathers" in 1699. Severus of Antioch had used the phrase similarly in the sixth century. Other works have featured among later collections: the fragments of Papias and Quadratus, the *Epistle to Diognetus,* the *Didache,** and the *Martyrdoms* of Clement, Ignatius, and Polycarp. Recent editors have generally omitted Quadratus, the *Martyrdoms* except for Polycarp's, and often the *Epistle to Diognetus.* The designation "apostolic" is problematic in every case, but is most appropriately applied, if at all, to Clement, Ignatius, and Polycarp. As used today of the earliest noncanonical writings of the late first and early second centuries, it is more conventional than descriptive.

In emphasis they are broadly pastoral and practical rather than theological or speculative, concerned with the internal life of the Christian communities moving toward "early Catholicism." Their alleged decline from apostolic Christianity (e.g., T.F. Torrance, *The Doctrine of Grace in the Apostolic Fathers,* 1948) appears less flagrant when their limited aims and changed circumstances are taken into account, but remains inescapable. They frequently recall NT books, especially the Pauline epistles, but not always as Scripture on a par with the OT. Their access to written gospels rather than other forms of the gospel tradition is often difficult to demonstrate, but has been too readily denied by H. Koester. Some of these writings enjoyed for varying periods localized recognition on the fringe of the NT canon.

See individual entries for each writer.

BIBLIOGRAPHY: Texts: J.B. Lightfoot (5 vols., 2nd ed., 1889-90—Clement, Ignatius, Polycarp); J.B. Lightfoot and J.R. Harmer (1891); K. Lake (2 vols., 1912-13); K. Bihlmeyer and W. Schneemelcher, vol. I (3rd ed., 1970)—all except Hermas.

Translations: Lightfoot; Lightfoot-Harmer; Lake; J.A. Kleist (2 vols., 1946-48); C.C. Richardson (1953); R.M. Grant et al. (6 vols.), vol. I: Introduction (1964-68).

Studies: *The New Testament in the Apostolic Fathers* (ed. Oxford Society of Historical Theology, 1905); H. Koester, *Synoptische Überlieferungen bei den Apostolischen Vätern* (1957); J. Lawson, *A Theological and Historical Introduction to the Apostolic Fathers* (1961); H. Kraft, *Clavis Patrum Apostolicorum* (1963)—vocabulary; L.W. Barnard, *Studies in the Apostolic Fathers and Their Background* (1966). D.F. WRIGHT

APOSTOLICI. Several bodies and sects have used this name. Epiphanius (c.315-403), bishop of Salamis, used the name for several Gnostic communities of his day. The title was also adopted by ascetic groups which flourished in the twelfth century near Cologne and in France. They claimed to imitate the apostles in renouncing the world, believing all matter to be absolutely corrupt. Several groups rejected marriage. They attacked the hierarchy of the church, believing it was so corrupt as to have vitiated all the sacraments except baptism. The name was used also for a sect founded in Parma in 1260 by Gerard Segarelli who, after his sect had twice been condemned by Rome, was burnt to death in 1300. His successor, Fra Dolcino, who expounded apocalyptic doctrines, was also put to death. The name has further been used by some Anabaptist sects.
JAMES TAYLOR

APOSTOLIC SUCCESSION. The theory of a continuing line of descent from the apostles to the present-day church transmitted through episcopal consecration. The death of the apostles left a problem of continuity for future generations because they had been the representatives of the ascended Christ as witnesses and interpreters of the saving events. At first there were claims to a succession of doctrine. Where the Gnostics* claimed a secret tradition traceable to the apostles, catholic Christians asserted that the succession of bishop to bishop in a see would mean that the teaching originally given by the apostles was faithfully preserved. The idea which gained greater currency seems to be first found in the West in the third century among Christians with legal minds, such as Tertullian and Cyprian. This was that the apostles had by consecration appointed bishops as their sucessors and that they in turn had consecrated other bishops. In this way the apostolate was kept alive in the episcopate, and this became a guarantee of truth and grace. In modern times this view has had particular attraction for "Catholic" Christians who did not acknowledge the see of Rome, because it seemed to ensure their "catholicity." It was held particularly strongly among a number of Anglicans from the time of Newman and has strongly influenced the practice of the Anglican Communion in relations with other churches. Its last major defense was in *The Apostolic Ministry* (ed. K.E. Kirk, 1946). Since then it has been demonstrated to the satisfaction of most scholars that the argument derived from the concept of *shaliah* is invalid because the *shaliah* could not pass on his commission, and that the NT evidence is strongly against there being monepiscopal succession throughout the Church.

BIBLIOGRAPHY: K.M. Carey (ed.), *The Historic Episcopate* (1954); E.M.B. Green, *Called to Serve* (1964); A.T. Hanson, *The Pioneer Ministry* (1961); T.W. Manson, *The Church's Ministry* (1948).

R.E. NIXON

APPIAN WAY (Lat. *Via Appia*). The road from Rome to southern Italy. It is named after the censor Appius Claudius Caecus who built the section from Rome to Capua in 312 B.C. By 244 B.C. it had been extended to Brundisium (Brindisi), a total of 234 miles. Acts 28 tells how Paul, on his journey to Rome, landed at Puteoli and presumably from there went to join the Appian Way at Capua. He met groups of Christians on the road at the Forum of Appius and the Three Taverns. The section of the road from Rome to Beneventum is well preserved, but beyond there it came to be neglected because of the building of the Via Traiana. By the roadside there are many tombs—some of them of famous Romans—and some ancient pavement, bridges, and milestones. Christian monuments along this part of the Appian Way include the catacombs of St. Callistus, the burying place of most of the third-century bishops of Rome

R.E. NIXON

AQUARIANS. A description of certain people and sects in the early church (e.g., Tatian the Syrian and the Encratites*) who used water (*aqua*) instead of wine in the celebration of Holy Communion. Such people were also called *Hydroparastatae* (Gr. = "those who advocate water"). This practice was attacked by such men as Cyprian, Augustine, and Philaster.

AQUAVIVA, CLAUDIUS (1543-1615). Fifth general of the Jesuit order. Born into the aristocratic family of the duke of Atri, in Abruzzi, he became a student of civil and canon law at Perugia. He was a Jesuit novitiate in Rome in 1567, served as a provincial at Naples and Rome, and was elected superior general in 1581. A gifted leader and statesman, he consolidated the work of the order and promoted its wider influence. His leadership was tested in the early days of his generalship: persecution in England, trouble with the Huguenots in France, unpopularity of the order resulting in expulsion from Venice, and division within the society in Spain, organized by C. Vasque, backed by Philip II and the pope. Aquaviva placated the pope and withstood a demand for an examination into the order's constitution by the Inquisition (a possible indication of the rivalry between Dominicans and Jesuits). An attempt by Sixtus V to alter the order's organization was averted by the pope's death. During Aquaviva's leadership the society developed its missionary work, e.g., substantial work in Japan and educational work in the Philippines. The order was also involved in theological controversy, in particular with the Dominicans on the matter of grace. Aquaviva stressed the spiritual life and discipline with the establishment of a system of rules that have been the basis of the Jesuit system of education, *Ratio Studiorum*, issued finally in 1599, as well as a comprehensive interpretation of Ignatius's *Spiritual Exercises*.

GORDON A. CATHERALL

AQUINAS, THOMAS (1224-1274). The greatest philosopher and theologian of the medieval church. Born in Italy, he studied at the University of Naples and became a Dominican in 1244. Later he studied under Albertus Magnus at Paris, and also at Cologne. Most of the remainder of his life was spent as a teacher in Paris. In 1273 he had to discontinue his *Summa Theologica* due to ill health. He died the following year.

His thought is expressed in an enormous literary output, not only the *Summa contra Gentiles* (1261-64), intended as a manual of apologetics and doctrine for missionaries, and the *Summa Theologica* (1265-73), on which his reputation as a theologian and philosopher chiefly rests, but also in commentaries on Scripture, and on Aristotle,* and a variety of miscellaneous discussions. The *Summa Theologica* must be seen as a marvelous systematizing of the data of Christian revelation (as understood by Aquinas) along Aristotelian lines, impressive in the thoroughness and the success with which the program was executed. The Augustinian contrast between the certainty of the intelligible order (known through intellectual illumination) and the uncertainty (and hence/unreliability) of sense impressions was replaced by the Aristotelian contrast between "form" and "matter." All human knowledge is regarded as being sensory in origin, and the human understanding, through abstraction, is able to build up knowledge of the forms of things. But if knowledge is sensory in origin, how may God be known? Much of Thomas's work may be considered an answer to that question.

He made a sharp distinction between "sacred doctrine" and philosophy. Sacred doctrine proceeds from the data of revelation, philosophy from data accessible to (and acceptable to) all men. It is a mistake to think of Aquinas as a Christian philosopher if by this is meant someone who elaborates answers to philosophical questions on the basis of Christian revelation. The different places assigned to philosophy and theology may be vividly illustrated by Thomas's view of creation. Philosophically the universe might be eternal. But the Christian believes from revelation that creation is an act of God.

Aquinas claimed that God's existence can be established *philosophically*. His famous "Five

Ways" are five *a posteriori* arguments (some say, five variants of one basic argument) based on God's effects in the world, data open to all men. The Five Ways may be seen as an effort to fill a gap in explanation, to show that if certain contingent states of affairs exist, there must be some necessary ground for their existence. How much Thomas's arguments depend on outmoded science, whether the arguments are sound, and whether, if at least one of them is sound, the God whose existence is established is the god of Christian revelation—these are all debatable and much-debated questions.

The Five Ways provide a part of one answer to the question, How may God be known? He is knowable only by His effects. Yet not only by His effects in nature, but also by His revelation in Scripture, to the acceptance of which the Five Ways are a natural, rational preamble. Not only was Aquinas a natural theologian, he was an eminent philosopher of revelation, employing philosophical concepts in elucidating the propositions and notions of revealed truth.

Despite the elaborateness of his theological discussions, Thomas's stress was on how little God is known. God is categorically distinct from His creatures, unique, transcendent. How then can He be spoken about? Only imperfectly. Such speech is by means of analogy (God's wisdom is in some respects the same as Solomon's wisdom, in some respects different) and negation (God is *not* embodied, He does *not* exist in time) from speech about finite things. Hence the doctrine of analogy (in particular) plays a crucial role in Thomas's account of the knowledge of God.

He distinguished between faith, opinion, and knowledge. Faith, an act of the mind, is stronger than opinion. It involves a firm assent to its object. But because it lacks full comprehension (vision) it is less than knowledge. Religious faith is a species of this genus. (Aquinas knows nothing of faith as a "leap in the dark" or as "personal but not propositional.") The disposition to have faith in divinely revealed matters is a product of God's grace.

God's sovereignty is expressed by saying that He is the "first cause" of all that is, evil (regarded by Aquinas, following Augustine, as a privation of goodness) excepted. Does this mean that God is the efficient cause of all that is? If so, what about human freedom? Aquinas is firmly Augustinian as well in his insistence on God's providential ordering of human actions, and of his foreknowledge of them. Foreknowledge does not causally necessitate actions, it makes them certain, as from God's point of view all actions take place in an "eternal present."

In ethics Thomas's thinking stresses the purposive, end-directed character of human action. He distinguishes between moral theology (action viewed in accordance with divinely revealed law) and general ethical principles which, through an appreciation of natural law, are accessible to all. In his account of human action Aquinas stresses voluntariness (not freedom in the sense of indeterminacy) as a necessary condition of fully human (i.e., responsible) action.

The influence of Aquinas on the Christian Church has been enormous, although the acclaim his work received was not instantaneous (several of his teachings were condemned as errors after his death, a decision later reversed). The strong modern revival of his influence and of "Thomism" dates from the publication of the encyclical *Aeterni Patris* by Leo XIII in 1879, praising and endorsing Thomism and giving it an "official" (though not exclusive) place in the thinking of the Roman Catholic Church.

The influence of Aquinas on Protestantism must not be minimized. Though what were regarded as his (and others') speculative excesses and unbiblical errors were repudiated at the Reformation, the Augustinian character of much of his theology was gratefully recognized. Post-Tridentine Protestant systematic theologians such as F. Turretin were obviously indebted to him both methodologically and for detailed arguments on points held in common. Similarly, Thomas's view of apologetics—the sharp distinction between "nature" (accessible to all) and "grace" (derived from revelation, but perfecting, not repudiating, the conclusions of reason)—has been a recurring theme in Protestantism, which the strong post-Kantian tradition, culminating in the work of Karl Barth, ought not to be allowed to hide. The influence of Thomas will always be felt where philosophical theology is pursued vigorously. But Thomism, pursued uncritically, can have an intellectually cramping effect. And where religion is understood primarily in terms of *values* and not of *truths*, his influence can be expected to be much less.

BIBLIOGRAPHY: *Opera omnia* (Parma ed., 25 vols., 1852-72); A. Pegis (ed.), *Basic Writings* (1945); G.K. Chesterton, *Thomas Aquinas* (1947); F.C. Copleston, *Aquinas* (1955); T. Gilby (ed.), *Philosophical Texts* (1951), *Theological Texts* (1955), and *Summa Theologica* (Lat. text with ET, vol. I, 1964); E. Gilson, *The Christian Philosophy of St. Thomas Aquinas* (1956); J. Maritain, *St. Thomas Aquinas* (ET, rev. ed., 1958); K. Foster, *The Life of St. Thomas Aquinas: Biographical Documents* (1959).

PAUL HELM

ARABIC VERSIONS OF THE NEW TESTAMENT. Christianity was established in Arabia well before the rise of Islam, but evidence is lacking of any attempt to translate the Bible into Arabic in the pre-Islamic era. The first complete Arabic Bible is attributed to Hunayn b. Ishâq (ninth century), but it is not extant. Parts of the New Testament were translated before then; the Mt. Sinai monastery has provided a manuscript of the gospels dating from perhaps as early as the eighth century (as well as a tenth-century MS of some of Paul's epistles). The great majority of Arabic New Testament MSS are late, however, few of them antedating the sixteenth century. The lateness of the MSS limits their usefulness for NT textual criticism; moreover they are of very mixed ancestry, some of the material being translated, not from Greek, but from Syriac, Latin, and Coptic sources. The two surviving Arabic recensions of Tatian's *Diatessaron* are of greater importance. The first printed Arabic NT appears in the Paris Polyglot (1629-45). The Society for the Propagation of Christian Knowledge first published an Arabic NT in 1727, the British and Foreign Bible Society in 1816. More recently, translations have been produced under both Roman Catholic and Protestant auspices. Some parts of the NT are also available in certain of the Arabic colloquials.

See POLYGLOT BIBLES.

BIBLIOGRAPHY: G. Graf, *Geschichte der arabischen Literatur,* i (1944), pp. 88-101, 139-85; J. Henninger, "Arabische Bibelübersetzungen," *Neue Zeitschrift für Missionswissenschaft* 17 (1961), pp. 201-223; P.P. Saydon, "Arabic versions," *New Catholic Encyclopedia,* ii, pp. 461f.

D.F. PAYNE

ARAMAIC. A Semitic language closely related to Hebrew and the language of certain parts of the OT (Ezra 4:8–6:18; 7:12-26; Jer. 10:11; Dan. 2:4–7:28). Numerous references to the Aramaeans occur in Assyrian inscriptions of the second millennium B.C., but the earliest Aramaic inscriptions come from the tenth or ninth centuries B.C. Aramaic was a diplomatic language in the Assyrian Empire (cf. 2 Kings 18:26) and consolidated its position in the Babylonian and Persian empires, becoming the official language of the latter (cf. the fifth-century Aramaic papyri from the Jewish colony on Elephantine). In the postexilic period Aramaic took over from Hebrew as the language of the common people in Palestine (cf. Acts 22:2). Included among the Dead Sea finds are some fragments of early Aramaic Targums and the so-called Genesis Apocryphon dating perhaps from the first century B.C. Aramaic, in the Galilean dialect, was spoken by our Lord and His disciples (cf. Matt. 26:73; Mark 5:41; 7:34, etc.); the amount of Aramaic spoken in first-century Palestine is currently a matter of debate. Aramaic idiom can frequently be detected beneath the Greek form of the gospels (cf. Mark 4:12), but theories of Aramaic originals are generally regarded as untenable. By the beginning of the Christian era Aramaic had divided into two branches: West Aramaic (Nabatean, Palmyrene, Jewish Palestinian) and East Aramaic (Syriac, Babylonian Aramaic, Mandaean). The Aramaic Palestinian Targums to the Pentateuch, and the residual Palestinian material in the Babylonian versions of the Pentateuch and Prophets, preserve much valuable material roughly contemporary with the NT writings.

BIBLIOGRAPHY: H.H. Rowley, *The Aramaic of the Old Testament* (1929); F. Rosenthal, *Die Aramaistische Forschung* (1939); W.B. Stevenson, *Grammar of Palestinian Jewish Aramaic* (2nd ed., 1962); R. Le Déaut, *Introduction à la Littérature Targumique* (1966); M. Black, *An Aramaic Approach to the Gospels and Acts* (3rd ed., 1967); H. Ott, "Um die Muttersprache Jesu," *Novum Testamentum* IX (1967), pp. 1-25; J. Barr, "Which language did Jesus speak?" *BJRL* LIII (1970), pp. 9-29,

ROBERT P. GORDON

ARBROATH, DECLARATION OF (1320). A letter drawn up by a large number of Scottish nobles and barons and "the whole community of the realm," and sent to the pope who had declared against Robert the Bruce. The document stated that Providence, the laws, the customs of the country, and the choice of the people, had made Bruce their king, and that if he betrayed his country they would elect another. They cared not for glory, continued the declaration, but for that liberty which no man renounces until death.

ARBUTHNOTT MISSAL. Transcribed between 1471 and 1491 for use in Arbuthnott Church, Kincardineshire, it is the only missal of the pre-Reformation Scottish use now extant. The Sarum use is followed, but there are also masses for St. Columba, St. Ternan (patron of Arbuthnott), and St. Ninian. Prepared at the expense of Sir Robert Arbuthnott by James Sibbald, vicar of Arbuthnott, it includes an obituary of the Arbuthnott family from 1314 to 1551 and is notable for its ornaments of flowers, leaves, scrolls and fruit, showing the beginnings of the Renaissance style. It is preserved along with the Arbuthnott Prayer Book and the Arbuthnott Psalter in Paisley Museum.

HENRY R. SEFTON

ARCHAEOLOGY AND THE EARLY CHRISTIAN CHURCH. Some definition of the period under study in this article is necessary because of the vast field. The Christian Church was born at Pentecost in the days of the Roman procurator Pontius Pilate (A.D. 26-36). The upper limit of the term "early" is presumably at the start of the medieval period. Discussion here is limited, therefore, to the period between the beginning of the second century and the first half of the seventh century. The archaeological evidence from the first century is normally covered by the archaeology of the NT. In Palestine the year 638 marks the time that Jerusalem fell to the Muslims. The Muslim conquest of the Middle East marked the end of an age, although in the West no such transition is recognized. In the interests of brevity this article shall examine the period c.100-650—i.e., to the later Roman and earlier Byzantine periods.

In the early decades after the death and resurrection of Jesus Christ, the Christian Church spread rapidly and by the early second century had taken root in Egypt, Palestine, Syria, Mesopotamia, Asia Minor, Greece, Rome, and even further to the west. During the centuries that followed, significant structures of many varieties were erected throughout these areas—churches, chapels, monasteries. There were also tombs, catacombs, and various memorial structures. The churches contained a vast range of items of furniture used in their services: floor mosaics, painted frescoes, jeweled crosses, manuscripts, etc.

An important aid in the interpretation of the archaeological finds is the writings that have come down to us from these early centuries—church histories like that of Eusebius* (c.265-c.339), accounts of pilgrims like that of the Bordeaux Pilgrim* (c.333), important geographical mosaics like the one from Madaba in Transjordan from the second half of the sixth century, and a wide variety of inscriptions.

We commence with a brief review of the churches of *Palestine* up to the seventh century. Approximately two hundred of these remain for study although some are in very fragmentary form. The earliest churches were in houses and have disappeared. During the second and third centuries in both East and West, the church adopted the basilica* type of structure for worship. This was an oblong building with interior colonnades. The advent of Constantine the Great as emperor of Rome (312-37) brought a tremendous impetus to church building everywhere in the Roman world. In Palestine there is little evidence of pre-Constantine structures. Constantine remains have now been identified in the Church of the Holy Sepulchre in Jerusalem and in the Church of the Nativity in Bethlehem.

Sacred edifices of the post-Constantine period are known both from excavation and from literary sources. Excavation and archaeological survey show that there were three main types of church —the basilica with one, three, or five naves; the circular or octagonal church; and the mixed type which combined these two into a cruciform church.

In Palestine the majority of the larger churches which were built from the fourth to the seventh centuries were basilicas with a nave and two aisles. At the end of the nave stood an apse, and at the end of the aisles a small chamber or an apse. The original Church of the Nativity at Bethlehem had three aisles and a single apse. The later church built by Justinian (527-65) had five aisles and a central apse in the east wall and an apse in each of the north and south walls. Other early Palestinian churches are St. Lazarus at Bethany, the Church of the Probatic Pool in Jerusalem, the Church of the Finding of the Head of John the Baptist at Sebaste (Samaria), and the Church of the Multiplying of the Loaves and Fishes at Tabgha near the Sea of Galilee—all fifth-century churches.

Some basilicas had one protruding external apse. This was common in the West but not in Palestine. In some cases the apse was polygonal. Side apses took the place of side chambers in many churches. In a few churches the apses were arranged in a trefoil as at Bethlehem and in the Church of St. Theodosius near Jerusalem.

A few churches with a central plan, circular, octagonal, or trefoil, are known. The oldest is from the end of the fourth century—the Church of the Ascension on the Mount of Olives. The old cathedral of Beisan and the church on Mt. Gerizim (fifth century) are examples. The cruciform plan is rare. Probably there was one such church at Shechem and another at the Tomb of the Virgin in Jerusalem.

Both the contemporary literature and modern excavations provide evidence of a flourishing church in *Transjordan* in the early Christian centuries. The most important center is Jerash. In the days of Justinian no fewer than seven churches were built, but thirteen are known in the area, mostly closely dated. The cathedral (c.350-75) is pre-Justinian and, probably the earliest; it was a three-aisled basilica with an enclosed apse. The Church of St. Theodore built between 474 and 476 is nearby, and just to the west are three churches side by side. The central one of St. John the Baptist is circular; the two others, that of St. George and that of SS. Cosmas and Damianus, are basilicas, all built between 529 and 533. All these churches are rich in mosaic floors. Several other fine churches existed in Jerash.

Further south at Ras Siyagh on Mt. Nebo stood the fine fifth-century church and adjacent monastery. Other churches including traces of a fourth-century structure have been found in the same area. At Madaba stood the Church of Theotokos going back to the sixth century. The floor of another sixth-century church yielded the famous Madaba map which, though partly destroyed, has preserved valuable geographical information about Palestine and an excellent map of Jerusalem. At Petra there were Christian churches. One tomb has an inscription referring to its conversion into a church in the fifth century. At Umm al-Jimal, northeast of Amman, numerous Christian churches go back to the Byzantine period. The cathedral dates to 557. There are at least nine other churches in the city.

Syria too had its quota of ancient churches. According to Eusebius, a magnificent church was built in Antioch. This has disappeared, but two other fine churches are known in the same area. In the village of Kaoussie the Church of St. Babylas was built in 387, according to information preserved in the floor mosaics. Then at Seleucia, the port of Antioch, may be seen the ruins of the Martyrion, a quatrefoil-shaped church with an ambulatory adorned with rich mosaics and a chancel projecting east. It was built originally in the late fifth century.

In the interior of Syria many early churches were lost and rediscovered a century ago. The greatest was Qalat Siman, the Church of St. Simeon Stylites, northeast of Antioch, built at the end of the fifth century. In southern Syria at Bosra (Bostra), once the seat of an archbishop, there stood an impressive cathedral, circular in shape, enclosed in a square with circular apses at the corners and a chancel and apse. There are other old churches in the region. The house church of Dura-Europos on the Euphrates is of particular interest. A room in a private house was transformed into a church in the early third century. Later a wall was removed to include a second room. A baptistery stood in the northwest corner of the house. Wall paintings and graffiti identify the building as a church. Further east at Edessa and Nisibis and also in Persia there are the remains of numerous churches in use long before the Muslim conquest.

Anatolia was opened to the Christian gospel following Paul's work. By the time of the Council of Nicea in 325 there was a network of bishoprics all over the area. Significant archaeological work has been done this century.

*Constantinople,** the former Byzantium, was named by Constantine as his new capital in 330. He adorned the city with many structures, among them houses of prayer and memorials to martyrs. At least two churches were founded by him, the Church of the Apostles, which has now disappeared, and the Church of St. Eirene, formerly a Christian sanctuary but considerably enlarged. The original church was damaged by fire in 532, restored, and again damaged by earthquake in 740. Some elements of these older churches remain. Another famous church, Hagia Sophia, has likewise suffered through the ages. It became a mosque but is today a museum of Byzantine art. Ancient mosaics have been excavated.

The city of *Rome* is of very great importance in the field of Christian archaeology. Here Constantine built the most famous basilicas of the fourth century, chief among them being the Church of St. Peter. Due to many changes over the centuries little remains of the original basilica, but there are traces, and its outlines are clear. The whole area of St. Peter's has been subject to intense archaeological investigation. The question of Peter's tomb has been raised many times. The site was formerly a pagan cemetery but later used by Christians also. Constantine built churches at places associated with strong Christian traditions. A tradition about Peter's burial here must have existed when Constantine built the basilica.

The *Liber Pontificalis,* a series of biographies of the popes from the seventh century, refers to many churches in Rome, among them the Church of St. Paul Outside the Walls, originally built by Constantine. Little of Constantine's church remains, yet there is a vast task for the archaeologist to unravel the many reconstructions in this church. Numerous other churches in Rome have a long history of destruction, extension, and reconstruction. All of them are of interest to the archaeologist.

Space does not permit discussion of other churches in Greece, Italy, France, Britain, Spain, Egypt, North Africa—all of which have many Christian remains from the early Christian centuries.

Brief mention must now be made of several other aspects of the archaeology of the early Christian Church—mosaics, church furniture and utensils, tombs, catacombs, inscriptions, papyri, Christian symbols, and art.

Following secular practice, the Christian Church in every land made great use of mosaics in the floors of its churches, courtyards, and other structures, Palestinian towns and villages during these centuries paved their churches with mosaics, generally with geometrical designs but sometimes with beautiful compositions of plants, animals, and human figures. Important centers for this craft lay in the regions of Nebo, Jerash, Bethlehem, Et Tabgha, Beit Jibrin, 'Amwas. Sometimes inscriptions woven into the mosaic work preserve valuable geographical and historical information.

The recovery of church furnishings including altars, reliquaries, ciboria, basins, crosses, chalices, pattens, candlesticks, thuribles, and lamps is a feature of early Christian archaeology.

The tombs and catacombs of the early Christian Church have contributed greatly to our understanding of the burial practices and the art of the early church. The best known of the Christian catacombs are in Rome. The four oldest are those of Lucina, Callistus, Domitilla, and Priscilla. In 1867 it was estimated that Christian catacombs covered a surface area of 615 acres. The total length of the corridors was 500 miles. Today thirty-five or more Christian catacombs are known around Rome alone. One section of the catacomb of Praetextatus goes back to the second century. Other sections belong to the third and fourth centuries. In the catacomb of Sebastian, among numerous graffiti scratched on the walls in Greek and Latin are more than a hundred short prayers addressed to Peter and Paul. Apart from the catacombs, hundreds of Christian sarcophagi, ossuaries, and burial chambers have been excavated. Many bear inscriptions or graffiti and display a distinctive Christian art.

Not the least exciting of the discoveries from the early Christian centuries are important biblical manuscripts. A fragment of John's gospel found in Egypt and dated to the first half of the second century attests the presence of Christians there at an early date. The Chester Beatty Papyri* dating from the second to the fourth century, also from Egypt, contain parts of nine OT and fifteen NT books. The Bodmer Papyri from about the second century include the gospels of Luke and John and some of the NT epistles, parts of Genesis, and some apocryphal works.

Finally, in the area of art and Christian symbols there is a wealth of detail to be obtained from church architecture and ornamentation, floor mosaics, frescoes, catacomb and church paintings, church statuary, sarcophagi, and ossuaries.

Clearly the contribution of archaeology to our understanding of the early Christian Church is enormous and demands the attention of specialists for every land and for each significant area of the subject.

See also articles on ARCHITECTURE and ART.

BIBLIOGRAPHY: A.L. Frothingham, *The Monuments of Christian Rome* (1908); A. Van Millingen, *Byzantine Churches in Constantinople* (1912); H.C. Butler (ed.), *Publications of the Princeton Archaeological Expeditions to Syria in 1904-5 and 1909:* Architecture, A. Southern Syria (1919), B. Northern Syria (1920); A. Obadiah, *Corpus of the Byzantine Churches in the Holy Land* (1920); H.C. Butler in *Early Churches in Syria, Fourth to Seventh Centuries* (ed. E. Baldwin Smith, 1929); C. Hopkins, "The Christian Church," in *The Excavations at Dura-Europos* (ed. M.I. Rostovtzeff—preliminary report of 1931-32 season of work; W. Harvey, *Church of the Holy Sepulchre, Jerusalem, Structural Survey, Final Report* (1935); W.A. Campbell, "The Martyrion," in *Antioch-on-the-Orontes, III: The Excavations 1937-1939,* pp. 35-54; A.M. Schneider, *The Church of the Multiplying of the Loaves and Fishes* (ET 1937); J.W. Crowfoot, *Churches at Bosra and Samaria—Sebaste* (1937); *idem,* "The Christian Churches," in *Gerasa, City of the Decapolis* (ed. C.H. Kraeling, 1938); J. Lassus, "L'Église Cruciforme," in *Antioch-on-the-Orontes, II: The Excavations 1933-1936* (ed. R. Stilwell, 1938), pp. 5-44; J.W. Crowfoot, *Early Churches in Palestine* (1941); C.R. Morey, *Early Christian Art* (1942); J. Lassus, *Sanctuaires Chrétiens de Syrie* (1947); R.T. O'Callaghan, "Recent Excavations underneath the Vatican Crypts," *The Biblical Archaeologist,* XII, No. 1 (Feb. 1949), pp. 1-23; J.G. Davies, *The Origin and Development of Early Christian Church Architecture* (1952); R.T. O'Callaghan, "Vatican Excavations and the Tomb of St. Peter," *The Biblical Archae-*

ologist, XVI, No. 4 (Dec. 1953), pp. 70-87; M. Avi-Yonah, *The Madaba Mosaic Map with Introduction and Commentary* (1954); J. Finegan, *Light from the Ancient Past* (2nd ed., 1959); G.L. Harding, *The Antiquities of Jordan* (1959); F.V. Filson, "The Bodmer Papyrus," *The Biblical Archaeologist,* XXII, No. 2 (May 1959), pp. 48-51; *idem,* "More Bodmer Papyri," *The Biblical Archaeologist,* XXV, No. 2 (May 1962), pp. 50-57; E. Kitzinger, *Israeli Mosaics in the Byzantine Period* (1965); E.R. Goodenough, *Jewish Symbols in the Greco-Roman Period,* XXX (1968)—see index for references to Christian symbolism; J. Finegan, *The Archaeology of the New Testament, The Life of Jesus and the Beginning of the Early Church* (1969); B. Bagatti, *The Church from the Gentiles in Palestine, History and Archaeology* (1971).

JOHN A. THOMPSON

ARCHBISHOP. The title given to a bishop who has jurisdiction over the other bishops of a province besides exercising episcopal authority in his own diocese. The title seems originally to have been given in the fourth and fifth centuries to prelates of outstanding sees such as Rome or Alexandria, or when provincial synods were held in the chief town of the province under the presidency of the bishop.

ARCHDEACON. An Anglican cleric who has administrative authority of part or all of a diocese delegated to him by his bishop. He is particularly concerned with the proper maintenance of church property and also with a general disciplinary role toward the parish clergy. He must be in priest's orders. In the Eastern and Roman churches the office now has virtually no significance. From the third/fourth century and archdeacon was the chief deacon at the bishop's church with responsibilities for preaching, supervision of the deacons' pastoral and administrative work, and the distribution of alms. With the expansion of the church from the fourth century and the increase in revenue, the archdeacon's role became increasingly important. In the Middle Ages in the West he was second only to the bishop until rivalry led to a radical curtailment of his power.

HOWARD SAINSBURY

ARCHIMANDRITE (Gr. *archimandritēs,* "head of the fold"). In patristic Greek the word *mandra* served as a designation for the temple or the church, but in the late fourth century it was applied to emerging monasticism, which was legally regulated by Justinian I (527-65) following the Basilian pattern, placing each monastery or group of monasteries under the control of the archimandrite. Though he was usually a presbyter, he could be a deacon or even a lay monk elected by majority vote of the community, though receiving the staff of his office from the diocesan bishop.

ARCHITECTURE, CHRISTIAN. As the science of building for distinctively Christian activities, Christian architecture has as its principal task to provide accommodation for the different forms of worship. The style of the resulting structures has varied according to the needs and wishes of the worshiping community of believers.

The earliest buildings for Christian worship were the homes of individual believers. When the need was felt for a building set apart as a "church," a house such as that excavated at Dura-Europos (232) in Syria was adapted for the purpose. A single entrance door in the north wall opens into a vestibule and thence into a court on the east side of which is a portico. In the northwest corner a room containing a cistern has been made into a baptistery* decorated with frescoes. The cistern has become a font with a canopy. Two rooms on the south side have been made into one and a small platform provided for an altar. A room leading off this one may have been used for preparations for the Lord's Supper. The room on the west side of the court may have been used for the instruction of catechumens. This was the period of persecution which varied in intensity and duration from place to place. The Christians in Rome found it safer to worship in suitably appointed places near subterranean burial chambers or catacombs. Thus was established the association of Christian worship with the remains of the faithful departed.

With the publication of the Edict of Milan* (313), the Christian Church led by the emperor himself became an established community worshiping openly. The natural result was the building of magnificent places of worship. The model which was followed was the basilica.* The elevated bishop's throne placed in the apse and surrounded by the seats of the presbyters took the place of the judge's seat in the civil basilica. In front of these stood the altar, below which were often housed the remains of a saint. Beyond this was the main part of the building where the rest of the worshipers assembled.

Besides the basilica two other types of Christian building developed, the baptistery and the martyrium (chapel), housing the tomb or relics of a martyr. Both were constructed with their focal point in the center and not, as in the basilica, at one end. The martyria were often round in shape; in the center was the tomb over which was a dome. These buildings had a considerable effect on the development of church building in the East. The basic plan of a Byzantine church was frequently a combination of a domed superstructure with a squarish basilican plan.

In both East and West the altar became more and more distant from the ordinary worshiper. In the East it became customary to separate the altar from the rest of the church by a massive solid screen (iconostasis), adorned with icons (pictorial representations) of the saints. In the West the basic basilican plan was retained, but as in the East the church was divided by a screen, though not usually solid. The chancel, or room for the clergy, contained the high altar. The nave, or room for the lay people, contained the pulpit and sometimes a second altar. The baptismal font was placed at the back of the nave at the entrance to the church. These arrangements prevailed throughout the Middle Ages and still influence the design of churches today. The links of worship with the faithful departed were strengthened with the establishment of side altars, often containing relics, for the saying of masses for the souls of the founder and his friends.

In the thirteenth and fourteenth centuries, immediately preceding the dawn of the Renaissance, the focus of Christian architecture took on the lofty, overpowering exaggerations of the Gothic style: cavernous vaulted arches, flying buttresses, soaring spires. It effected feelings which further distorted church life and worship as a relevant, comforting source of life and thought. At the time of the Reformation the buildings inherited by the Reformers were considerably adapted. The screen was either removed so that the church became one room again, or else it was made into a proper wall and either the chancel or the nave used by the worshiping community. Sometimes each was used by a different congregation when the population increased. In this single-chamber place of worship, the pulpit was placed on one of the long sides and the people gathered around it. The altar was removed and long tables set up when the Lord's Supper was observed. A bracket attached to the pulpit held the basin for the administration of baptism. Not content with adaptation, the Reformers soon began to build new churches and often adopted the circular plan. The congregation faced the center, and the seats were tiered in circular rows. The pulpit was placed near the center, where there was space reserved for baptisms and for the communion table which was set up when the sacrament was observed.

Up to this point Christian architecture was strictly functional. The design and arrangement of a church building reflected the theology of worship. The Eastern iconostasis separated the mystery of the Eucharist from the eyes of the layman. The distant altar, only dimly discerned, made the same point in the churches of the West. The division of the church also reflected the strict separation of clergy from laity. And so the removal of the screen at the Reformation emphasized the unity of the body of Christ, the Church. Likewise, the gathering of the people around the pulpit and the table emphasized the corporate nature of Reformed worship, as opposed to the individualism of the many side altars. The Anglican middle road was seen most clearly in the churches built after the Great Fire of London in 1666. Although the congregation could not gather around the pulpit and table, both were related to each other and to the font in what was clearly one room.

A radical change came over Christian architecture with the Romantic revival of the late eighteenth and early nineteenth centuries. The appearance rather than the function of a church became determinative. The revived appreciation of the great medieval cathedrals* led to a desire for a "cathedral" in every community. Up to now churches had been built in a contemporary style, but henceforth Gothic became the "right" style for churches. The distant altar was restored to the end of a long chancel because of the vista, and it was further separated from the ordinary worshiper by stalls for the clergy and a robed choir. In Reformed churches a "cathedral" exterior often bore no relation to the interior, where an organ might occupy the apse. A high proportion of existing church buildings were erected during this period.

The twentieth century has been slow to discard this Gothic inheritance, preferring to modernize it. Where contemporary styles of architecture have been adopted, the medieval arrangements of the interior have often been retained. There have recently been signs of radical rethinking of the ways in which the functions of a church building can best be expressed and provided for.

Christian architecture has never been concerned exclusively with church buildings.The house-church at Dura-Europos had a room which may have been used for the instruction of catechumens, and church schools have existed in various forms throughout the Church's history. The early Christians lived together, and this ideal has never been lost. Monasteries and nunneries, retreat houses, lay academies, and church conference centers have been a concern of Christian architecture. The growth of the church hall and the parish house has challenged the position of the church as the principal Christian building in many parishes.

It has been pointed out that the Church can baptize in a river, preach in the open air, and celebrate the Lord's Supper on any table, and many young people are critical of the money spent on Church buildings. It is not without significance that in many lively parishes much of the Church's worship is offered in the houses of the members.

See ARCHAEOLOGY AND THE EARLY CHRISTIAN CHURCH and ART, CHRISTIAN (especially in relation to Gothic architecture).

BIBLIOGRAPHY: E. Short, *A History of Religious Architecture* (4th ed., 1955); P. Hammond (ed.), *Towards a Church Architecture* (1962); A. Bieler, *Architecture in Worship* (1965); J. Rykwerk, *Church Building* (1966); S.S. Smalley, *Building for Worship* (1967); K. Lindley, *Chapels and Meeting Houses* (1969); W. Swaan, *The Gothic Cathedral* (1969); G. Frere-Cook (ed.), *Art and Architecture of Christianity* (1972); R. Krautheimer, *Early Christian and Byzantine Architecture* (1975).

HENRY R. SEFTON

ARCHPRIEST. From Herodotus on, Greek designates by *archiereus* any "high priest" of whatever religion—classical, Jewish, or the Roman *pontifex maximus.* The NT follows Septuagint usage for various members of particular priestly families (plural) and contemporary practice for the president of the Sanhedrin (singular), while extending figurative meaning to Christ (Heb. 2:17, etc.). Patristic citations retain all these values while broadening their range to include archangels, the Byzantine emperor, Christian ministers (especially the bishop), and metaphorically Christians in general. In traditions where the title "dean" has not come into usage, archpriest identifies the one who services his bishop's cathedral. As a special title from 1598 to 1623 an archpriest headed the Roman Catholic seminarians who were sent to England.

CLYDE CURRY SMITH

AREOPAGUS, see ATHENS

ARESSON, JON (1484-1550). Icelandic bishop and poet. The son of poor parents, he rose quickly to eminence in the church and was consecrated bishop of Holar, the northern diocese, in 1524. He administered his diocese prosperously until Christian III of Denmark began to impose Lutheranism. With Bishop Ögmundr Palsson of Skalholt he protested vigorously, and he continued his resistance after his colleague was deported in 1541. He captured the Lutheran Bishop Marteinn and seized his see (1549-50), but soon after he was captured and beheaded at Skalholt with two of his sons. He was the author of several religious and satirical poems, notably one entitled "Lamentations on the Passion." He invited a Swedish printer to set up the first printing press in Iceland about 1530; he published the *Breviarium Holense* in 1534. J.G.G. NORMAN

ARGENTINE, see LATIN AMERICA

ARGUE, ANDREW HARVEY (1868-1959). Pioneer of Pentecostalism and evangelist in Canada. Born in Ontario of a line of Methodist lay preachers, beginning with George Argue who went to Canada from Ireland in 1821, he was a successful young businessman in Winnipeg when he came under the influence of such holiness teachers as A.B. Simpson.* Thus when news of a "Pentecostal" revival in California (1906) and Chicago (1907) reached him, he went to Chicago to investigate. There he received an ecstatic religious experience including "tongues-speaking" which he interpreted as "the baptism of the Holy Spirit." He became an apostle of this Pentecostal experience: first opening a mission in Winnipeg, which grew to be one of the largest Pentecostal churches in Canada, then touring Canada and the USA to conduct large revivals and to establish churches. In 1908 he published *The Apostolic Messenger,* known later as *The Revival Broadcast.* KENNETH R. DAVIS

ARGYLL, FIRST MARQUIS OF, see CAMPBELL, ARCHIBALD

ARIANISM. A heresy that denied the eternality of Jesus Christ the Son of God as the Logos.* It was condemned at the Council of Nicea* in 325. Very little of the written work of Arius, presbyter of Alexandria (d.336), remains, but the Arian controversy (c.318-81) was strategic to the crystallization and development of Christian doctrine. Along with Eusebius of Nicomedia, Arius studied under Lucian of Antioch, whose views foreshadowed Arius's Christology. Arius's genius was to push the christological question back to the origin of the pre-incarnate Logos. The controversy seems to have arisen in a dispute between Arius and his bishop, Alexander of Alexandria,* though after Nicea it was the young Athanasius,* deacon to Alexander, who carried the argument against Arius and whose defense of biblical Christology* eventually triumphed over the Arians in the fourth century.

Affirming a univocal sense of "begetting" with reference to our Lord's being the "only begotten Son," Arius said (to quote Socrates Scholasticus): "If the Father begat the Son, he that was begotten had a beginning of existence: and from this it is evident, that there was (a time) when the Son was not. It therefore necessarily follows, that he had his subsistence from nothing."

On the basis of a certain logic of terms, Arius's subordinationist Christology is consistent, but it is also patently heretical judged by the apostolic witness. If God is indivisible and not subject to change, then, on one reading of "begotten," whatever is begotten of God must derive from a creative act, not from the being of God. Hence it has a beginning of existence. Therefore the Son is not coeternal with the Father.

Fastening upon the term "begotten," Arius said that because Christ is begotten He must have had a beginning. Athanasius countered that because Christ is begotten of the Father, He could not have had a beginning. To say that a father begets a child is one thing, but to say that the Father begat the Son is another. The one is temporal, the other eternal; the one is of the will, the other from the being of the Father. Thus the Nicene Creed insisted that Christ is of the substance of the Father, thereby sacrificing neither the impassibility of God nor the deity of the Son. To say that the Son is begotten from the Father from eternity is not to divide the indivisible God but to accept the testimony of the apostles.

Crucial to the question are the doctrines of Creation and the Trinity. At Nicea, Christians adopted the teaching that the one Lord Jesus Christ from eternity is of one substance with the Father (note John's prologue, 1:1-18). This marked the end of the period in which Christ could be thought of as God's intermediary in His work of creation and redemption. Thus was vindicated the OT doctrine of the direct creation of the world by God, rather than the Greek concept of an intermediary or intermediaries who linked the world to God but not God to the world. The concept of intermediaries (as in Gnosticism) was formulated to overcome the antinomy of how God could be ingenerate and impassible yet act to create the world. Against Arius, Athanasius insisted there is no room in Christian thought for any being of intermediate status between Creator and creature, and because redemption is a divine prerogative, only God in Christ, not some intermediate being, could redeem.

The Arian controversy was protracted and involved many complicated documents circulated in the fourth century. The Arians achieved great popularity after the Council of Nicea, especially following the death of Constantine in 337, because his son and successor Constantius was fond of Arius (see separate articles on ANOMOEANS, and HOMOEANS). Eventually the force of Arian teaching was dissipated, though only through fierce struggle involving Athanasius. The Nicene Symbol was confirmed at the Council of Constantinople in 381.

The most noteworthy Arian-like Christology in modern times is the teaching of the Jehovah's Witnesses, who deny the eternality of the Son of

God, the doctrine of the Trinity, and who, like Arius, posit the Logos as an intermediate being between the Creator and creation.

BIBLIOGRAPHY: Athanasius, *Orations Against the Arians* (1873) and *On the Incarnation of the Word of God* (1944); G.L. Prestige, *Fathers and Heretics* (1940), chap. 6; H. Bettenson, *Documents of the Christian Church* (1946) and *Early Christian Fathers* (1956); E.R. Hardy, *Christology of the Later Fathers* (1954); J.N.D. Kelly, *Early Christian Doctrines* (1958) and *Early Christian Creeds* (1960); B. Altaner, *Patrology* (1960).

SAMUEL J. MIKOLASKI

ARIMINUM (Rimini), COUNCIL OF. Constantius II in 359 summoned two parallel councils for the West and East, at Ariminum and Seleucia.* The Homoiousion group led by Basil of Ancyra wished to gain universal episcopal support at a time when Constantius had expelled the extreme Anomoeans Aetius and Eunomius. Less radical Anomoeans* (also called Homoeans*) such as Valens of Mursa, having persuaded Hosius of Cordova and Constantius to recognize their position (357), suggested with success that Constantius should summon two councils, facilitating their attempt to outmaneuver Basil. The Homoeans had constructed a creed "dated on the 22 May," indicating, according to Athanasius, its novelty. Bishops at the Council of Arminum rejected it at first in favor of the creed of Nicea. When their delegates announced the decision to Constantius, however, he induced them to accept the "Dated Creed," which was formally ratified at Constantinople in 360.

G.T.D. ANGEL

ARISTIDES (second century). Christian Apologist* and philosopher of Athens. Until the last century Aristides was only a name in the writings of Eusebius and Jerome. Then in 1878 the Armenian Fathers of the Lazarist Monastery at Venice published an Armenian version of his "Apology" and in 1889 Rendel Harris discovered the Syriac version in a monastery on Mt. Sinai.

Shortly afterward, J.A. Robinson made the astonishing discovery that the Greek version of Aristides has been taken over almost wholly into a popular Oriental Christian romance, "Barlaam and Josaphat." The Apologist opens his "Apology" with an outline demonstration of God's existence based upon Aristotle's argument from motion. He states that mankind is divided into four races—Barbarians, Greeks, Jews, and Christians—and that Christians have the most complete understanding of the nature of God and a correspondingly satisfactory moral code. The treatise was addressed, not as Eusebius said to Emperor Hadrian, but to his successor Antoninus Pius (138-61).

G.L. CAREY

ARISTION. Papias reports that Aristion along with John the Elder was a primary witness for the early tradition about Christ. Nothing further is known for certain of him, but in an Armenian MS of the gospels dated 986, discovered by F.C. Conybeare in 1891, Mark 16:9-20 is attributed to the "Elder Ariston." This may be an authentic tradition, but further proof is lacking. Aristion must be distinguished from Aristo of Pella.

ARISTO OF PELLA (second century). According to Maximus the Confessor in the seventh century, Aristo wrote a "Dialogue between Jason and Papiscus concerning Christ," in which proofs from the OT (version of Aquila) adduced by the Jewish Christian Jason led to a request for baptism from the Jew Papiscus. Celsus, Clement of Alexandria, Jerome, and apparently Tertullian knew the work which, dated about 140, might make Aristo the earliest known Apologist against Judaism. Eusebius found in Aristo information about the defeat of Bar-Kochba and of the exclusion of Jews from Jerusalem by Hadrian.

ARISTOTLE (384-322 B.C.). Greek philosopher. Born at Stagira, he went to Plato's school at Athens (367-347). After the death of Plato he lived in the Troad and on Lesbos, eventually becoming tutor to the son (the future Alexander the Great) of Philip II of Macedon. In 335 he returned to Athens to open a new school called the Lyceum. When Alexander died in 323, the school was in danger from the anti-Macedonian forces, so Aristotle took refuge on Euboea.

Although a student of Plato, Aristotle came to differ with his teacher. Werner Jaeger (*Aristotle, Fundamentals of the History of His Development*, ET 1934) has distinguished three periods in his life. In the first (to 347 B.C.) he was a defender of Platonism, presenting his material in dialogues holding the Platonic view of the soul and adhering to the doctrine of form. In the second period (347-335) he became increasingly critical of Platonism, especially of the idea of forms. Finally, in the period after 335, he became an exponent of empirical science, and by the end of his life had come to reject all the essential features of Platonic otherworldly metaphysics.

The material on which Aristotle's fame rests is not in dialogue form, but apparently in lecture notes preserved perhaps by his students. His work is encyclopedic, which may help to explain why it was so popular in the Middle Ages when knowledge sources were limited. Among his major works are *Ethics; Physics; Metaphysics;* the works on logic known as the *Organon;* a variety of writings on natural science such as *On the Heavens, On the Soul, On the Parts of Animals; Politics; Rhetoric;* and *Poetics.*

His teachings were not very influential among Christians until the high Middle Ages (A.D. 1050-1100). During these years knowledge of his work was gradually gained from Arabic translation made by Jews and Muslims. These were then translated into Latin during the twelfth century. The logical works were recovered first, and then the entire Aristotelian metaphysical system became available. The intellectual shock this work caused may be likened to that occasioned by Copernican cosmology in the sixteenth and seventeenth centuries, or to Darwinian biology in the nineteenth and twentieth centuries. Aristotle presented a complete explanation of reality without any reference to the Christian God. His "unmoved mover" or "first cause" was a principle of existence, not a personal being. The universe he described was eternal, without beginning or end, and man had no individual immortality in the Aristotelian system. Some medieval scholars wished to ban his works, while others opted for a "double truth" theory in which Aristotle should guide logic and Christianity should be supreme in revelation, but the two fields could not be reconciled. The future of Western thought, however, lay in the work of scholars such as Albertus Magnus* and Thomas Aquinas* who attempted to harmonize Aristotelian philosophy with Christianity and then developed the Scholastic movement.

See also RAMUS, PETER.

BIBLIOGRAPHY: W.D. Ross (ed.), *The Works of Aristotle Translated into English* (12 vols., 1908-52); *idem, Aristotle* (1953); J.H. Randall, Jr., *Aristotle* (1960). ROBERT G. CLOUSE

ARIUS, see ARIANISM

ARLES, SYNODS OF. The most important were:

(1) In 314. The Donatists* refused to accept the decision of the Synod of Rome against them and, as a result, Constantine who regarded Christian unity as very important ordered a new synod at Arles to give judgment. No previous gathering had had such a representative section of Western clergy—it included three British bishops. The synod decided that Caecilian* had been rightfully elected, and the Donatists were condemned. In addition, twenty-two canons were issued dealing with various aspects of church life, including the fixing of one day to observe Easter and the repudiation of the Cyprianic practice of rebaptizing heretics.

(2) In 353. One of the Arian synods in the West which condemned Athanasius.

(3) In 813. Charlemagne called five councils, one of them at Arles, to ensure adequate ecclesiastical education for the clergy and to emphasize the preaching and teaching of the Catholic faith.

(4) In 1234. Canons were issued against the Albigensian* heresy, emphasizing observation of the decrees of the Lateran (1215) and Toulouse (1229) councils and urging bishops to counter heresy by closer surveillance of their dioceses.

(5) About 1260. An attempt to order details of church life and to condemn the doctrines of Joachim of Fiore. C. PETER WILLIAMS

ARMAGH. Founded by St. Patrick* in the fifth century, it is the ecclesiastical capital of Ireland in the sense that it is the seat of the primates of the Roman Catholic Church and the Church of Ireland. The Church of Ireland cathedral dates probably from the thirteenth century, though it was completely rebuilt in the eighteenth century; the Roman Catholic cathedral was consecrated in 1875. Traditions of Patrick's links with Armagh gave it a prominent place in Irish ecclesiastical history from an early date: it was a seat of learning from the fifth century, and its college achieved considerable international fame. Patrick's successors in Armagh were called the heirs of Patrick and claimed some measure of jurisdiction over other churches in Ireland. Up to the early years of the twelfth century Armagh maintained the independence of the Celtic Church from the Roman and English churches. In 1152 at the Synod of Kells it was made the primatial see of Ireland.

HUGH J. BLAIR

ARMENIAN CHURCH. Ancient Armenia, south of the Caucasus chain, nurtured a highly intelligent Indo-Germanic race, later dispersed and dispossessed like the Jews. Her native, autonomous dynasty ruled 300 years before Persian/Byzantine partition (387-190), thereafter surviving in the Persian sector until A.D. 428. Apostolic Christianization is claimed from A.D. 34—legends enshrining factual substratum abound in the considerable early indigenous literature. The secular background was Roman/Iranian, the religious affiliation Syriac/Greek. Gregory the Illuminator* inaugurated hereditary religious rule. The Adoptianist* heresy was widespread and may have affected Gregory. The native Armenian alphabet and vernacular Bible proved stabilizing influences.

Armenians consistently repudiated both Nestorianism,* upholding the Councils of Ephesus (431, 449), and Cyril of Alexandria who encouraged Monophysitism.* Abhorring the overhuman Nestorian Jesus, they posited an altogether suprahuman Christ. Struggling against Persia for religious independence, they missed the Council of Chalcedon (451), were out of touch and misunderstood, and the result was a hopeless split with the West that weakened the entire church and facilitated Islam's conquest.

The independent national Gregorian Church, repudiating Chalcedon, began about 506 and suffered two later splits: (1) the Romanizing Armenian Uniat Society, founded in 1335, with its renowned monastery of Mechitarists at St. Lazaar's Island, Venice; and (2) the lively evangelical Protestant Church, fruit of the American Mission after 1831, and speedily anathematized by the parent body.

Like Palestine the battleground of greater powers, and the victim of Muslim expansionism, Armenia suffered centuries of devastation by Persians, Turks, Russians, and others. The 1895 Turkish massacres shocked the world; Gregorian nationals, together with the hated Protestants, were the marked victims, while the Uniats enjoyed protection under Rome (see UNIAT CHURCHES). Dispersed Armenians still prosper in the Mediterranean basin and beyond.

BIBLIOGRAPHY: H.A. Chakmakjian, *Armenian Christology and Evangelization of Islam* (1965); K. Sarkissian, *The Council of Chalcedon and the Armenian Church* (1965). ROY A. STEWART

ARMENIAN VERSION. Armenia (or Hayastan) was converted in the late third or early fourth century under King Tiridates by Gregory the Illuminator.* The beginnings of biblical translation are attributed to Mashtotz (otherwise Mesrob*) who created an alphabet for this purpose. As some traditions say and as scholars deduce, the basis of his work was the Syriac versions. No MS of this earliest stage of the version survives, and the scholar must work from the data of quotations in the earliest writers. In the case of the gospels, it is debated whether the earliest form was a harmony of the gospels related to the *Diatessaron** of Tatian. The canon of Scripture continued to bear the mark of its Syriac origin for a long time in its retention of the apocryphal *3 Corinthians*, and in the absence of the Book of Revelation, which although translated as early as the fifth century did not figure in canonical lists until late. From about the sixth century there were increasing movements of revision to a Greek model, which in the gospels was akin to the textual use of Caesarean scholars such as Origen and Eusebius. The complexion of the OT Armenian text is basically Lucianic with some hexaplaric readings. This former may derive from a Syriac base akin to the Peshitta,* the latter from the stage of revision to Greek norms. The data in the two Armenian forms of *Ecclesiasticus* throw this double line of descent into high relief. The version in its revised form is notable for its careful technique and accuracy.

BIBLIOGRAPHY: A. Vöobus, *Early Versions of the New Testament* (1954); L. Leloir, *(Versions) Orientales de la Bible*, II; "Versions armeniennes," in *Dictionnaire de la Bible*, Supplement VI (1960).

J.N. BIRDSALL

ARMINIANISM. A theological system named after Jacobus Arminius (Jakob Hermandszoon), a Dutch theologian (1560-1609) who was educated at Leyden, Basle, and Geneva. After studying under Beza he went to Amsterdam to serve as minister of the Reformed congregation (1588). Holland had become a center of Calvinism during the sixteenth century, but during his fifteen years as pastor Arminius came to question some of the teachings of Calvinism. Disputes arose, and he left the pastorate and became professor of theology at the University of Leyden. Here he gave a series of lectures on the doctrine of predestination* which led to a violent controversy with his colleague, Francis Gomar.* This conflict continued until it divided the student body as well as the ministers of the Reformed Church. The Gomarists or Strict Calvinists wished to have the matter settled by a national synod, but Jan van Oldenbarneveldt, a liberal Dutch politician then in control of the government, did not wish such a meeting. The protagonists even debated their ideas before the States General of the Dutch Provinces, but still no agreement could be reached.

After the death of Arminius, his followers issued the Remonstrance of 1610 which outlines the system known as Arminianism. The major points of departure from strict Calvinism are that (1) the decree of salvation applies to all who believe on Christ and who persevere in obedience and faith; (2) Christ died for all men; (3) the Holy Spirit must help men to do things that are truly good (such as having faith in Christ for salvation); (4) God's saving grace is not irresistible; (5) it is possible for those who are Christians to fall from grace.

In an attempt to stop this teaching, the Calvinist party made an alliance with Maurice of Nassau, son of William the Silent. Their desires for a synod coincided with Maurice's policy of centralizing the United Provinces and transforming them into a monarchy. For eight years after the Remonstrance, the political forces of Oldenbarneveldt and Maurice struggled for supremacy. Finally Maurice won and his opponent was accused of treason and beheaded (1619). This cleared the way for Maurice to try to use religious ideology to centralize the state. Consequently the Synod of Dort,* one of the most famous meetings in the history of the Reformed Church, met (1618-19). The synod passed a point-by-point refutation of the Remonstrance. The Belgic Confession* and the Heidelberg Catechism* were confirmed as standards of orthodoxy, and the Arminians were condemned.

Following the synod, many of the disciples of Arminius, among them such able men as Hugo Grotius,* were imprisoned or banished. By 1625 there was a reaction against this severity, and a limited toleration was extended to the Arminian Remonstrants. Although the Arminians were not numerous in Holland, their teaching has exercised considerable influence in other lands. In seventeenth-century England, the Laudian anti-Calvinist movement was influenced by the Arminians. John Wesley also followed this belief, and so it has left its mark on the Methodist Church. Among groups with a Calvinist heritage the debate over the points that Arminius raised still continues.

BIBLIOGRAPHY: A.W. Harrison, *The Beginnings of Arminianism to the Synod of Dort* (1926); *idem, Arminianism* (1937); C. Bangs, *Arminius, A Study in the Dutch Reformation* (1971).

ROBERT G. CLOUSE

ARNAUD, HENRI (1641-1721). Savoyard pastor and Waldensian leader. Born in France of Piedmontese parents, he became the organizer of the Waldensians* (also called Vaudois) on their historic return to their ancestral home in Savoy. Going back to Piedmont, he became a Waldensian pastor there in 1685. He and many others were forced to flee a short time later during the persecution inaugurated by Victor Amadeus II of Savoy. Arnaud went to Switzerland where he helped plan the so-called Glorious Return of the Waldensians in 1689. Financed and encouraged by numerous English and Dutch Protestants, Arnaud led the Waldensian band of about 1,000 as it fought its way back home. Shifting international political alliances brought relief from persecution in 1690 and its revival in 1696. Once again several thousand Waldensians went into exile, many of them to Württemberg in Germany, Arnaud founded a Waldensian settlement there and served as its pastor until his death.

ROBERT D. LINDER

ARNAULD, ANTOINE (1612-1694). French theologian and philosopher. He is generally considered the foremost member of a distinguished French family, a number of whom embraced Jansenism* and exercised a formative influence over the movement. Born in Paris, he first studied law and later theology at the Sorbonne. In 1638, influenced by his mother, he placed himself under the spiritual tutelage of the Abbé de Saint-Cyran* (d.1643) who was a close friend of Cornelis Jansen (d.1638) and at that time chief apostle in Paris of the movement Jansen founded. Saint-Cyran encouraged Arnauld's ordination as priest in 1641 and inspired him to write his first major work entitled *De la fréquente communion* in 1643. The book provoked a storm of protest from the Jesuits and established Arnauld as the head of Jansenism in France after Saint-Cyran's death. Arnauld was caught up in the struggle between Jansenists and Jesuits for control of the French Church. After suffering numerous indignities but seeming to enjoy controversy, he finally left France and settled in Brussels where he spent the remainder of his life after 1682, writing against the Jesuits and in defense of Jansenism. His 320 works now collected in 43 volumes stand as an eloquent testimony of his intellectual vigor and reforming zeal. ROBERT D. LINDER

ARNAULD, JACQUELINE-MARIE-ANGELIQUE (1591-1661). Abbess of Port-Royal, France, and a leading figure in Jansenism.* Born in Paris, she was the daughter of the distinguished lawyer and fervent Gallican, Antoine Arnauld the elder. She entered the lax Cistercian abbey of Port-Royal-des-Champs at the age of eight, became abbess within three years, and began her thoroughgoing reform work after a conversion experience in 1608. In 1625 she moved Port-Royal to Paris and, though removed as abbess in 1630, she helped to bring Saint-Cyran* to Port-Royal as spiritual leader. Under his influence the convent became a hotbed of Jansenism. During her second period as abbess (1642-55) she helped spread Jansenist ideas widely, especially since the Fronde filled the convent with refugees. She died shortly after signing Louis XIV's formulary against Jansenism. BRIAN G. ARMSTRONG

ARNDT, ERNST MORITZ (1769-1860). German hymnwriter and historian. Having studied theology at Greifswald and Jena, he lectured in history at Greifswald, but fled to Sweden because of his anti-Napoleonic *Vom Geist der Zeit* (1806-1818). He wrote patriotic pamphlets and poems, including *Des Deutschen Vaterland,* before becoming in 1818 professor of history at the University of Bonn. He was suspended two years later because of alleged republicanism, but was reinstated in 1840. After doubts caused by contemporary philosophy, he adopted from 1817 (under Schleiermacher's influence) a more Christian position. His attack, *Von dem Wort und dem Kirchenliede* (1819), on the rationalistic bowdlerization of evangelical hymns helped to restore original versions. Of his own eighty-three hymns, notably *Ich weiss, woran ich glaube,* fourteen have been translated into English; no translation is in common use now. When hopes of a Protestant-Catholic union had foundered he became a staunch Protestant (see his *Über den gegenwärtigen Stand des Protestantismus,* 1844). JOHN S. ANDREWS

ARNDT, JOHANN (1555-1621). German Lutheran mystic. After studying theology at Helmstedt, Wittenberg, Strasbourg, and Basle, he became in 1583 minister at Padeborn, but an argument with the lay rulers over the place of pictures and ceremonies in the church caused him to leave. Later he served churches at Quedlinburg, Brunswick, and Eisleben. In 1611 he became general superintendent of the church at Celle, where he remained until his death. Here he exercised an important influence on the development of the Lüneburg church system. His fame rests upon his writings, especially *Four Books Concerning True Christianity* (1606-9). Arndt emphasized mysticism in his interpretations of the Christian life by asserting that orthodox belief is not enough to attain true Christianity, but moral purification made possible by righteous living and communion with God is also necessary. Though remaining within the Lutheran Church he nevertheless helped prepare the way for the Enlightenment* and for Pietism.* ROBERT G. CLOUSE

ARNDT, WILLIAM FREDERICK (1880-1957). American Lutheran scholar. Born at Mayville, Wisconsin, and educated at various Lutheran seminaries, he was ordained in 1903 and pastored churches in Bluff City, St. Joseph, and Brooklyn. He taught theology at St. Paul's College, Missouri (1912-21), and at Concordia Seminary, St. Louis (1921-51). A keen advocate of Lutheran unity and missions, he wrote *Does the Bible Contradict Itself?* (1926) and a number of popular apologetical works, a commentary on Luke, a life of Paul, and with F.W. Gingrich edited a *Greek-English Lexicon of the New Testament* (1957). He edited also the St. Louis *Theological Monthly* (1926-30) and its successor, the *Concordia Theological Magazine* (1938-50). IAN SELLERS

ARNOBIUS JUNIOR (fifth century). Attested only by his writings from Rome in the fifth century, his biblical works were allegorical commentaries on the Psalms (criticizing predestination at Psalm 108) and scholia on gospel texts. About 440 he wrote *Praedestinatus,* in which he catalogues ninety heresies largely described in Augustine's *De Haeresibus,* including Pelagians, Nestorians, and the *Praedestinati.** The views of the last are reviewed in a sermon circulating falsely under the name of Augustine; in the end the author refutes them. After 454 Arnobius wrote *The Conflict of Arnobius the Catholic with the Egyptian Serapion,* in which he refutes Sabellian, Arian, and Pelagian views and defends the Leonine doctrine of the Two Natures. He gives the writings of Augustine* the respect due to the Scriptures. About 440 he probably represented anti-Augustinian moderates supporting Julian of Eclanum. G.T.D. ANGEL

ARNOBIUS THE ELDER (fl. c.304-310). A teacher of rhetoric from Sicca in Numidia, he was converted from paganism to Christ. His defense of Christianity, entitled *Ad Nationes,* concentrated on exposing the errors of pagan worship and mythology. The divinity of Christ he based primarily on the miracles, and he advocated hope in Christ as the only real basis for immortality. Despite his respect for pagan philosophers like Plato, he argued that the soul is not immortal by nature. As a recent convert he lapsed occasionally into unorthodoxy; Christ, for instance, is not coequal with the Father, he said.

ARNOLD, GOTTFRIED (1666-1714). Lutheran theologian and devotional writer. He was educated at Wittenberg and afterward, through the influence of Philip Spener, became a teacher at Quedlinburg. He became identified with exponents of mystic and separatist tendencies and in 1696 published *Die erste Liebe,* a eulogy on the simplicity and poverty of the primitive church and a condemnation of what he considered the later addition of dogma and ecclesiasticism. The next year he was invited to Giessen as professor of church history, but finding himself out of sympathy with the school resigned and returned to his former position. There he wrote the monumental work, *Unparteiische Kirchen-und Ketzer-Historie* (1699-1700) in which he showed more impartiality to heresy than to the church. In this study of heretical movements, Arnold refused to accept as evidence the statements of hostile contemporaries and based his work on the writings of the sectarians themselves. This, with his presuppositions about the weaknesses of the orthodox position, led him to favor the separatists of various ages and caused controversies which forced him deeper into a mystical position. In 1704 he became pastor and inspector at Werben in Prussia and was reconciled with establishment Christianity. In 1707 he became inspector at Perleberg. In addition to his church history he wrote over fifty works and composed many beautiful religious songs, some of which are still used.

ROBERT G. CLOUSE

ARNOLD, MATTHEW (1822-1888). English poet. Eldest son of Thomas Arnold,* he was born at Laleham and educated at Winchester, Rugby, and Balliol College, Oxford. He was government inspector of schools from 1851 to 1886, and from 1857 to 1867 he was also professor of poetry at Oxford. Though popular as a poet in his lifetime, his poetic ability has been severely criticized in this century. He attacked many of the religious attitudes of his time, particularly tendencies to rely on unprovable assumptions and neglect of reason. He saw "culture" as man's greatest need, deplored bibliolatry as alien to the scientific spirit, stressed the personal and moral sides of Christianity, and denied miracles. His important religious works were *Culture and Anarchy* (1869); *St. Paul and Protestantism* (1870); *Literature and Dogma* (1873); *God and the Bible* (1875); and *Last Essays on the Church and Religion* (1877).

JOHN A. SIMPSON

ARNOLD, THOMAS (1795-1842). Anglican teacher and Broad Churchman. Educated at Warminster, Winchester, and Corpus Christi College, Oxford, he was ordained and in 1828 was appointed headmaster of Rugby School. There he laid the foundations of the modern public school system in England, with its emphasis on religious training, moral character, and public service. He became in 1841 regius professor of modern history at Oxford. He opposed the Oxford Movement,* and his reaction to the church crisis was to propose a comprehensive English Church, excluding only Roman Catholics, Unitarians, and Quakers. The essence of Christianity was to him practical goodness, which both church and state exist to realize, each needing the other. He stressed the universal priesthood of the laity, and considered matters of doctrine and ritual inessential. He wrote *Principles of Church Reform* (1834) and *Fragment on the Church* (1844).

JOHN A. SIMPSON

ARNOLD OF BRESCIA (1100-1155). Radical church reformer. After studying under Peter Abelard,* he joined the Augustinian order. He returned to Italy where he advocated the necessity of apostolic poverty by the church. His proposals were condemned by the Second Lateran Council and he was banished. Arnold then went to France, where he helped his teacher Abelard defend himself at Sens (1141). They were unsuccessful and were condemned to confinement in separate monasteries. Arnold resumed his teaching in Paris, but was banished from France and lived for a time in Zurich, afterward in Bohemia. In 1145 he was reconciled with the pope, but this was short-lived. After condemning the papal power once more, he allied himself with a rebel political party that wished to abolish the temporal power of the pope and establish a commune at Rome. The community had created a senate and appointed a patrician in place of the city prefect, who was dependent on the pope. In this situation Arnold advocated the idea of reclaiming for Rome her ancient powerful position in the world. He encouraged his followers to appeal to Frederick I with a statement that included a condemnation of papal approval of the emperor, the allegation that the Donation of Constantine* was a fable, and a claim that the empire belonged to the city of Rome. The more extreme schemes in this statement repelled many, and in the election of 1152 Arnold's group was defeated. Later he was excommunicated and expelled. Frederick I captured him and returned him to Rome, where he was condemned by the prefect to be hanged and his body burned (1155). A movement grew in memory of Arnold called the Arnoldists. They stressed apostolic poverty and repudiated the power of the hierarchy, holding as invalid sacraments administered by clerics who had worldly goods.

ROBERT G. CLOUSE

ARNOT, FREDERICK STANLEY (1858-1914). Pioneer Brethren missionary. Born in Glasgow, Scotland, and acquainted with the family of David Livingstone,* he wished to follow the latter's example and in 1881 set out for Africa, making his way alone from Durban to the Zambezi. In Barotseland (1882-84) he preached to Chief Lewanika and his people, but suffered severely from malaria and dysentery. He started work in Benguela, Angola (1884), then moved to Katanga where he gained the confidence of Chief Msidi, warning him against signing papers submitted by Europeans (1886). After a furlough in Britain (1888), where he married, he opened new work in Northern Rhodesia (now Zambia), and spent his remaining years establishing mission stations from Benguela to Katanga. He was the foremost architect of Brethren missionary work in central Africa.

See SOUTHERN AFRICA. J.G.G. NORMAN

ARNOT, WILLIAM (1808-1875). Scottish preacher and author. Born at Scone, Perthshire, a farmer's son, he was apprenticed in early life to a gardener. He studied for the ministry in Glasgow, gaining distinction especially in Greek. After an assistantship at Dunipace, he became minister of St. Peter's Church, Glasgow, continuing the ministry in connection with the Free Church after 1843. He was ejected from the church by decision of the Court of Session (1849), thereafter opening a new church in Main Street, Glasgow (1850). He was called to the Free High Church, Edinburgh, in 1863 and ministered there until his death. He visited the USA three times. His many books include *Illustrations of the Book of Proverbs* (1857-58). J.G.G. NORMAN

ART, CHRISTIAN. Art done by nominal or professing Christians is not a sufficient condition for calling the product "Christian art," since a Christian's right hand of faith sometimes does not know what his artistic left hand is doing. Christian art must be bona fide art, and the art product itself must by the spirit at work in its colors or sculptured form bear witness to the Lord God revealed in Jesus Christ, if it would legitimately be called "Christian art." Secular treatment of a biblical topic, like the Crucifixion, does not make the painting Christian art. Church use of a wooden sculpture—an icon of a woman as the Madonna, for example—does not qualify it as intrinsically Christian art. It is also not helpful to constrict the notion of "Christian art" to a given historic style, as if "Gothic" or "baroque" or "Pre-Raphaelite" contours be the definitive model. It depends rather upon whether a spirit of compassion for creation plagued by sin and hope for reconciliation of life to God through Jesus Christ is embodied in the painterly elements.

Within this distinctive guideline one's sense of Christian art should be catholic enough to include: (1) a Roman (Catholic), worldly wise Christian appeal in the line, color, and design of an art product to a higher realm of grace beyond what is naturally visible; (2) an ascetic, (Greek) Orthodox Christian temper to the composition, texture, and color of a painting appealing somewhat mystically to a heavenly spirituality beyond the earthly; (3) the full-orbed, Reformational Christian spirit showing up within the layout, shape, and color of art products calling for a renewed cosmic life under Jesus Christ's rule; and (4) an internalized, evangelical Christian cast to pieces of art that breathe devotional piety and withdrawal from worldliness. The committed Christian spirit that distinguishes painting and sculpture as Christian art from pagan and secular art is not always readily discernible in a given artifact (no more than it is always easily discernible in the daily walk of a Christ-believer), but it is a matter of historical fact that Christian art, as herewith defined, has been extant, in varyingly corrupted forms, especially in Western civilization.

Historical development. Early Christians decorated the walls of their catacomb burial and worship places with little figures representing the Good Shepherd, Jonah with the whale, and the like, in a manner derived from pagan Roman imagery. When the Christian Church came above ground in the fourth century with the conversion of Constantine, its architectural art was pressed to be commensurate with the status of a grand, official state religion. From its beginnings, then, "Christian art" was more a concessive modification of current pagan fashion than a radically new start in plastic art internally demanded by faith in the new Gospel of Jesus Christ. Original features showed up, nonetheless: delicate, miniature forms carved into the marble of sarcophagi replaced the larger-than-life, Greco-Roman sculptural tendency; and instead of floor mosaics of marble common to Rome, now shiny glass tesserae, especially glittering gold pieces, were used to cover whole vaulted ceilings and walls of churches, producing an unreal, luminous, ethereal effect.

Byzantine art continued the use of sumptuous, jewellike colors so typical of an Eastern feeling for enthroned majesty and, more importantly, it discarded the illusionistic perspective of pictorial art rooted in pagan Hellenism. The glorious works at Ravenna (the Pantocrator Christ apse of Sant' Apollinaire in Classe, for example) show something new in the West. A Plotinian, Neoplatonizing aesthetic helped Christian craftsmen in the sixth century replace the old, natural, plastic realism with dematerialized figures, flattened space, and luxuriant, green colors that testified, without spiritualizing, to a new earth reality. The mimetic ideal which had chained art to what is three-dimensionally visible got broken; now there was an opening for disproportionately large eyes, abstractly schematized trees, and animals with ornamental colors to celebrate ceremonially, as it were, a life that had conquered and gone beyond our normal creaturely existence weighted down by burdens.

Imaging art was banned by imperial edict in 726. Western Christendom, especially with the rise of Charlemagne, by mid-ninth century did not honor such iconoclasm; but the attendant severity did seem to introduce depiction of Christ's suffering passion into Western art. Independent Celtic monasticism, meanwhile, contented itself with the enormously intricate, decorative embellishment of capital letters copied in biblical manuscripts, spilling over sometimes into margins and whole pages with colorful plants, flowers, and strange, allegorical beasts (related, perhaps, to the later famous gargoyles of Gothic church architecture). Only with the rise of "lay" pilgrimages and beginning of the Crusades in 1095 did sculptured human figures, with flexibly slenderized, softened classical contours, appear in stone again on church exteriors. "Christian art" during this Romanesque period was channeled mainly into cathedral architecture north of the Alps; craftsmen worked with the base plan of a cross to fashion places of height and light proper for worship. (Dogmatic concern on the matter of representation and spirituality was simply unimportant, academic, to them.)

Architecture, sculpture, and (stained glass) painting of the thirteenth and fourteenth centuries canonized the gradual restriction of Christian art to churchly art in a particular way: whether it was the overwhelming grandeur of Notre Dame in Paris, of Chartres, Amiens, the Reims cathedral, or the piteous *Andachtsbilder*, such "Gothic" art conspired to drown the human believer in an ocean of deeply moving, contemplative stupor. Both the explicit Babel of ribs and flying buttresses and countless pinnacles outside the church and the endlessly soaring lines of ascent inside, as well as the cavernous *Pietà* sculptures with relic overtones, worked at losing a man in quasi-mystical ecstasy, touched always by a feeling of unworthiness and an uncertain closeness to death. (The Black Death killed half the urban population in some areas of Europe, 1347-51.) So Gothic art is only questionably "Christian," arousing pity rather than recognition of mercy, exalting an unattainable sublimity rather than testifying of grace freely given, quickening apprehensive awe instead of a firm hope. Gothic art is the culturally powerful embodiment of a devotionalistic disintegration of Scholastic theology and the imperialistic church, and the pious, nondescript beginning of a frank humanism. A less introverted and very chaste humanization of Christian sentiment during the same period is found in Giotto's* murals, where the presence of grace on earth shows up in Christ's solemnity as well as in the golden halos worn by the saints.

The tremendous secularizing spirit of Renaissance culture which arose in Italy and northern Europe during the fifteenth century tended to rub out further the fits and starts toward Christian art begun in the so-called Middle Ages. Masaccio's (1400-1428) deep allegiance to the truths of the Christian faith still showed through in his Bible story frescoes bursting with the rediscovery of actually naked bodies and a foreshortening perspective that spelled a sturdy, this-worldly reality coming into focus. Fra Angelico's* (1387-1455) limpid and breathlessly still Madonnas, Annunciation, and Adorations hold on more tightly to the passing order, the devout severity of monastic vow and quiet retirement away from the bustle of commerce and worldwide exploration. Flemish Jan van Eyck (c.1390-1441), however, typifies the new fascination with perceptible things, with people as personages, with painting in oil the very atmospheric tangibleness of one's natural surroundings (see WEYDEN, VAN DER). Of course, the "supernatural double truths" were iconographically dubbed in, but this left-handed acknowledgment to the realm of grace became an increasingly specious mannerism as the fabulous quattrocento century progressed. Renaissance masterpieces by Donatello, Botticelli,* Leonardo da Vinci,* Michelangelo,* Raphael,* Titian,* and others may have had biblical titles, Christian motifs, and been under the patronage of popes; but their bold, exciting lines, voluptuous colors, and confident show of three-dimensional perspective—greatly extending the expressive reach of painting, sculpture, and architecture—was by and large driven by an utterly worldly spirit, a this-worldly, man-honoring commitment. Art was no longer a means sanctified by the end of instructing the illiterate people of God in otherworldly matters; now art was an autonomous glory itself, wielding its own lordly (secular) authority among the rich and powerful.

Grünewald's (c.1485-1528) Isenheim altarpiece is unique. Technically pre-Reformation, it is unspoiled by a worldly Renaissance grandeur, too rough-hewn and dynamic to be caught in Gothic introspective mysticism; Grünewald's spread-eagled Christ hanged on the cross against a stark blue blackness of sky portrays unforgettably the cursed dead end of sin, while his panel on the Resurrection uses strange, unearthly colors and shifting masses of rock to suggest with startling power the God-man's triumph over death. Other painters in the Reformation period, like Hans Holbein the Younger (1497-1543), also treat the reality of death and distortion, not with the delight of earlier Hieronymous Bosch and his grotesqueries, but with a biblical matter-of-factness that catches the awful finality and punishment character of death without any hope. Holbein, like his German contemporaries (see DURER; CRANACH), served the taste of his royal patrons too (e.g., the magnificent portrait of Henry VIII). Like Luther's makeshift alliances with German princes, this development freed artists from the dominance of ecclesiastical commissions, but it opened wide the door for compromises with secular humanism.

The Council of Trent (1545-63) tried to pull art back into liturgical bondage by giving its sanction to "sacred art," that is, art which by its reverent realism or clear allegorical touches could be used to instruct people in tenets of the Roman Catholic Church. The great El Greco (1541-1614), situated in Spain, inflamed by a visionary, Byzantine ecstasy, epitomizes at its best what the Counter-Reformation wanted (cf. *The Burial of Count Orgaz*): admission of secular opulence at a lower level, topped by a nervous, metallically colored, austere affirmation of floating, celestial glory. The Jesuits baptized use of baroque art similarly, accepting its sensuous, rhetorical luxuriousness (see RUBENS) favored by the aristocracy and so imposing to the masses, so long as it ended by pointing to the sins-absolving office of the church. Art oriented more toward the Reformation, however, as it developed in the Netherlands, took a quite different, Christian direction. The late portraits done by Rembrandt,* the landscapes and interiors of Vermeer (1632-75), Pieter de Hooch, and other Dutch masters, presented in loving detail, pellucid color, and with a quiet, panoramic dignity, both the outdoor world and intimate daily life as a creation able to be filled with *shalom*. Their canvases never monumentalized scenes, but gave both the cosmos and ordinary, homely activities the deepened dimension of God's ordering presence.

Current problems and options. A hardening secularism of the eighteenth-century Enlightenment* simultaneously robbed "Christian" of definite, biblical meaning, and divorced "fine art" (cf. rococo style) from the realm of a normal workaday task. With the nineteenth-century advent of a dominant positivism's facticity, art seemed to get lost in esoteric Romantic genius, *l'art pour l'art* escapism, and the brilliant, sensational irrelevance of Impressionism; "Christian" became increasingly a dirty weasel-word for past holy wars and simple bigotry. Christian art, therefore, was at best considered an anachronism. The Pre-Raphaelite movement (see HUNT, W.H.) seemed to confirm that judgment in its abortive attempt to reconstruct Bible-pure art by undoing the Renaissance. An intense sculptor like Ernest Barlach (1870-1933) and the important painter Georges Rouault* (1871-1958) demonstrate, however, that bona fide Christian art is possible in our post-Christian age, even though it is desperately difficult for an artist to escape being formed by the total, secularistic, cultural matrix.

Significant art that is harnessed for ecclesiastical service in our day (like some work of Matisse, Leger, Chagall, and many others) is really a forced surrogate for what was done more naturally in the medieval period, and may be less fruitful for Christian art than it appears, for today the hybrid is less integral—the artist seeks a rooted context without bowing to all the dogma of Mother Church, and the institutional church seeks the engagement sometimes merely to show its tolerant modernity. Devotionalistic art consciously conceived for outright evangelistic propaganda, however well intentioned and executed, shall necessarily fall short of Christian art in our differentiated culture, because it denatures the norming symbolic quality of art into a matter of pop illustration. The most fruitful alternatives for Christian art to pursue may well be: (a) a visionary reach, in sorrow, for the real heavenly certainties beyond our technocratized, broken world, pulling on the mask of a Byzantine tradition, El Greco, William Blake, and some of Rouault. Salvador Dali exemplifies a slick, effete secular engulfment of this option. And (b) the Christ-transforming-culture perspective initiated by seventeenth-century Dutch art—too clean then, perhaps, of historical struggle—tuned with grit to the meanness of our hollowed-out lives and the pied beauty visible to the faithful. The earthy insight of certain contemporary Jewish artists, like Abraham Rattner and sculptor Chaim Gross, may give body to the largess of this option.

Christian art is a calling for professionally competent craftsmen, not a *fait accompli* if one applies certain formulae. That means Christian art has been begun, but will not be wholly perfected until the Lord returns to rule completely. To support the communal, generation-building development of Christian art, for the well-being of society at large, is a mark of the obedient body of Christ.

BIBLIOGRAPHY: S.R. Hopper (ed.), *Spiritual Problems in Contemporary Literature* (1952); J. Maritain, *Creative Intuition in Art and Poetry* (1953); "Christianity and the Arts," special issue of *The Christian Scholar* 40 (December 1957), no. 4; E. Gilson, *Painting and Reality* (1957); E. Panofsky, *Gothic Architecture and Scholasticism* (1957); H.R. Rookmaaker, *Gauguin and Nineteenth Century Art Theory* (1959); F. Glendenning, *The Church and the Arts* (1960); P. Courthion, *Rouault* (1962); C. Seerveld, *A Christian Critique of Art and Literature* (1964); D.J. Bruggink and C.H. Droppers, *Christ and Architecture* (1965); D. Whittle, *Christianity and the Arts* (1966); R. Huyghe (ed.), Introduction and chap. 1, "The First Centuries of the Christian Era," in *Larousse Encylopedia of Byzantine and Medieval Art* (rev. ed., 1968); H.R. Jauss, Jacob Taubes et al., "Die klassische und die christliche Rechtfertigung des Hässlichen in Mittelalterlicher Literatur," "Die Rechtfertigung des Hässlichen in urchristlicher Tradition," in *Die nicht mehr schoenen Kuenste. Grenzphaenomene des Aesthetischen* (ed. H.R. Jauss, 1968), pp. 143-85, 583-609; H.R. Rookmaaker, *Art and the Public Today* (1968); F.A. Schaeffer, *Escape from Reason* (1968); P.D. van der Walt, *Die Calvinis en die Kuns* (c.1968); C.D. Carls, *Barlach* (1969); H.R. Rookmaaker, *Modern Art and the Death of a Culture* (1970); W.A. Dyrness, *Rouault: A Vision of Suffering and Salvation* (1971); C. Seerveld, "The relation of the arts to the presentation of the truth," in *Truth and Reality* (1971).

CALVIN SEERVELD

ARTEMAS (third century). A Monarchian heretic, he taught that "the Saviour was a mere man." He maintained that his view had been orthodox at Rome until the time of Bishop Zephyrinus (198-217). Associated with Theodotus at Rome about 195, he lived on to influence Paul of Samosata about 260. *The Little Labyrinth* attributed to Hippolytus of Rome was written to refute him.

ARTICLES, THIRTY-NINE, see THIRTY-NINE ARTICLES

ARTICLES OF RELIGION. The doctrinal standard of the United Methodist Church of America. In 1784 John Wesley prepared a revised and shortened version of the Thirty-Nine Articles* for use in American Methodism.* Fifteen were eliminated altogether, three were rewritten, and the remainder subjected to minor verbal alterations and omissions. Apart from abbreviation, Wesley's objective was to remove from the Anglican formulary whatever inclined toward either ritualism or Calvinism. It is significant that no additions were made to cover distinctive Methodist emphases, but Wesley offered the Twenty-Four Articles to complement his sermons and *Notes on the New Testament* which were already accepted in England as indicating the theological standpoint of Methodism. The American Methodists added an article of their own affirming their loyalty to the American government, following the war of independence. The Twenty-Five Articles were adopted by the Baltimore Conference of 1784.

A. SKEVINGTON WOOD

ARTICLES OF WAR. Before being "sworn in" publicly under the Salvation Army* flag, a soldier must be converted, serve a period as a recruit, be accepted by the local corps board, and sign the *Articles of War.* This document blends Methodist doctrines of holiness with a measure of military discipline. The recruit declares that he or she "now, and forever, renounces the world with all its sinful pleasures, companionships, treasures and objects ... no matter what I suffer, do or lose...." Other articles include abstinence from alcohol, baneful drugs, "low and profane language," "impurity," and "the reading of any obscene paper or book"; renunciation of "deceit or dishonesty" and of "oppressive, cruel or cowardly" behavior; and pledge to give everything to the "salvation war," and to obey the lawful orders of officers and carry out "all the orders and regulations of the Army." Within this framework of total obedience the Army keeps strict discipline with the aim of setting its members free to concentrate on their Christian service. In recent years there has been some evidence of restiveness, particularly on the part of the younger generation.

A. MORGAN DERHAM

ARUNDEL, THOMAS (1353-1414). Archbishop of Canterbury from 1396. Third son of the earl of Arundel, he was consecrated bishop of Ely in 1374 and was thereafter lord chancellor (1386-89) to Richard II, during which time he became also archbishop of York. A determined opponent of the Lollards* and of reform movements within the church, he was impeached in Parliament in 1397 and banished. He returned to power with Henry IV, both as chancellor and in his see, which he retained till his death. Arundel opposed the "Lacklearning" Parliament of 1404 and its successor of 1410 over attempts to disendow the church; in 1408 he helped a provincial council at Oxford against the Lollards, being instrumental for the trial and burning of Lord Cobham for that heresy in 1413.

G.S.R. COX

ASAPH (d.600?). Welsh bishop. Little is known about him. He was possibly descended from an important N Welsh family, and was related to Saints Deiniol and Tysilio. He apparently became bishop and abbot of Llanalwy (Denbighshire) and later established a monastery at Llanasa (Flintshire) which was renamed St. Asaph by the Normans when they erected a diocese in the area.

ASBURY, FRANCIS (1745-1816). American Methodist bishop. Born and reared near Birmingham, England, in a deeply religious home, he experienced religious awakening at the age of thirteen or fourteen and soon joined the Methodists. Of limited schooling, he was apprenticed for about six years in a now unknown trade. For five years (1766-71) he had various appointments as an itinerant minister before volunteering to serve in America in response to an appeal by John Wesley. During the Revolutionary War he alone of Wesley's appointees stayed in America and after some internal struggle identified with the emerging nation. In 1784 Wesley appointed Asbury and Thomas Coke* joint superintendents, though Asbury insisted that his appointment be ratified by the Conference of Methodist preachers. Against Wesley's wishes he assumed the title of bishop, and in the frequent absence of Coke was until his death the major force shaping American Methodism. Never well and constantly beset by numerous maladies, he nonetheless traveled nearly 300,000 miles, mostly on horseback, and endured the rigors of the American frontier to nurture the emerging denomination. Viewed by some as autocratic, he emphasized discipline and the values of the itinerant ministry. Under his leadership Methodist membership grew from a few hundred to over 200,000.

See E.T. Clark et al. (eds.), *The Journal and Letters of Francis Asbury* (3 vols., 1958); L.C. Rudolph, *Francis Asbury* (1966).

DONALD W. DAYTON

ASCENSION. The only narrative of the Ascension which can with confidence be ascribed to the original text of the NT is found in Acts 1:4-11. The reference in Mark 16:19 comes in a passage which is a late addition to the gospel, and the reading in Luke 24:51 is uncertain. The story does not suggest that Jesus was taken on a long journey upward into the sky. It simply states that "he was lifted" (a symbolic movement but not necessarily involving any great distance) and that "a cloud took him out of their sight" (again a symbolic way of veiling His divine presence as it was removed in that mode from earth). The ascension brings to a close the post-resurrection appearances of Jesus (apart from the special one to Saul in Acts 9) and so marks the end of the once-for-all revelation of God in Christ and opens the way for His universal presence through the Holy Spirit.

Theologically the Ascension is closely associated with the Resurrection as demonstrating the vindication and exaltation of Jesus (Acts 2:32f., 5:30f.; Rom. 8:34; Eph. 1:20). He is present in heaven in His glorified humanity as a pledge of the completion of His act of redemption and of the final salvation of His people (John 14:2; Heb. 1:3; 6:20). There He continues His priestly work in interceding for His people (Rom. 8:34; Heb. 7:25). The present sovereignty of Christ over all will be demonstrated clearly at the *parousia* (1 Cor. 15:24-26).

Ascension Day has been celebrated since at least the fourth century; following, perhaps overliterally, the "forty days" of Acts 1:3, it has been kept on the fifth Thursday after Easter.

R.E. NIXON

ASCETICISM. The Greek *askēsis*, "training," was used of both athletic exercises and, especially among Stoics and Cynics, moral training through education, mastery of passions, and beneficence. Greek antecedents for ascetic renunciation and privation, however, are few and probably insignificant. The OT, particularly the Wisdom tradition, also emphasizes self-discipline, but asceticism is only marginally evidenced in Judaism. Fasting was thus a mark not of asceticism but of piety.

In the NT, detachment rather than abandonment (e.g., of property) is the basis for asceticism, though apocalyptic-eschatological contexts import a puritan radicalism, e.g., in warnings to the wealthy. The NT attacks dualist-Gnostic contempt for the body and foods (cf. Col. 2:21-23; 1 Tim. 4:3-4). Jesus, an example for ascetic imitation in several regards, e.g., homelessness (cf. ascetic ideals of wandering, exile, and removal to alien territory), taught that self-denial might involve celibacy, but as a charisma, not self-willed (Matt. 19:11,12). More influential for Christian asceticism was the Lukan form of His statement that "those who are considered worthy of taking part in . . . the resurrection from the dead neither marry nor be given in marriage, and they can no longer die; for they are like the angels" (20:35,36), which restricted marriage to this (fallen) world and was readily applied to *present* experience of the Resurrection. Paul's letters, as well as instilling spiritual discipline (e.g., 1 Cor. 9:25-27), had the effect of boosting virginity and depreciating marriage as belonging to the old aeon, but they condemned an anticipated resurrection-celibacy at Corinth (cf. 1 Cor. 7:1, NEB mg., and *passim*).

By the second century, virginity had become the basis of asceticism. The superiority of the unmarried is prominent in apocryphal writings, especially acts of apostles. In Syria celibacy was required for baptism. The underlying theology, discernible in the Coptic *Gospel of Thomas*, ascribes to Adam's fall the origins of sexual differentiation and marriage, which Christ therefore came to abolish (cf. the agraphon in the *Gospel of the Egyptians*, "I am come to undo the works of woman"). The paradisal state to which the baptized is restored approximates to angelic existence (cf. Luke 20:36), which includes bodilessness, asexuality, and absence of need of food. Marcion's prohibition of marriage reflects such beliefs, but his case illustrates the difficulty of assessing the alternative influence on asceticism of a Hellenistic or Gnostic flesh-spirit dualism.

The eschatological orientation, whether of imminence (Paul and the Montanists) or realization (Syria), easily fuses with a dualist hostility to the body that was virtually universal in Gnostic and Manichaean groups and tended to develop in Hellenistic philosophy. Such enmity towards the flesh may be an introjection of disgust for the material world as a whole. Mortification of the flesh, which among the hermits of Egypt, Syria, and Celtic Ireland assumed bizarre forms, even self-destruction (cf. too the Circumcellions*), releases the imprisoned soul and prepares for discarnate angelic life. The anchorite* asserts, "I am killing the body because it is killing me." The ascetic—especially in the desert, the demons' home territory—fights the same battle as the martyr. He prepares for death by despising the body and may see himself hastening the kingdom by conquering the flesh.

Encratite* (Gr. *enkrateia*, "self-restraint") was the title given to heretical groups which insisted on continence and practices like abstinence from animal flesh held to mark life before the Fall, but encratism was more widespread than condemned sects. Behind Syrian encratism probably lay a Jewish-Christian outlook elaborating traditions concerning Adam. The Qumran* covenanters were perhaps crucial generators of Jewish eschatological asceticism.

It has been argued that the Greek *monachos* in pre-monastic contexts meant "celibate," solitary as lacking a spouse, and the Syriac *īhīdāya*, literally "beloved," and Greek *monogenēs*, "only (begotten)," and *agapētos*, "beloved," were similarly used to designate the ascetic's true following of Christ, the Only Beloved Son who was unmarried. The *Gospel of Thomas* uses *monachos* also in the sense of "unified," i.e., restored to asexuality.

Clement of Alexandria* and Origen,* drawing on Philo (who described the Therapeutae and depicted Jacob as the model ascetic), developed a more permanent basis for asceticism, while rejecting the dualistic foundation for Gnostic antagonism to the world. Under Stoic influence, Clement stressed *apatheia*, the "passionlessness" of inward detachment and purification from passions as a condition for the soul's ascent to God. Origen's approach was more radical (cf. his self-castration and zeal for martyrdom), and his works were widely read by monks. The Alexandrian understanding of salvation as divinization (cf. 2 Pet. 1:4) provided a ground similar to the angelic view for purification from everything corruptible in anticipation of divine life. The Alexandrians perpetuated a double standard by regarding ascetics as the *pneumatikoi*, the "spiritual" elite. In the long tradition of Byzantine mystical theology, Origen and Neoplatonism were major sources of ascetic themes, and the latter also influenced Augustine* and others in the West. The holy ascetic's conquest of the material order gave him an exalted role in Eastern society, while from the mid-fourth century most church leaders in East and West were committed to ascetic ideals.

Others of note among ancient writers on asceticism are Methodius of Olympus,* Evagrius of Pontus,* Ps-Macarius and Gregory of Nyssa,* John Cassian,* Nilus the Ascetic,* and Dorotheus* of Gaza, with Aphraates* and Ephraem* among the Syrians and the historians Palladius* and Theodoret of Cyrrhus.*

Later centuries displayed forms of asceticism more refined or simple (cf. the friars), more Christ-centered (cf. the *imitatio Christi* motif), more "secular" (cf. the Calvinist-Puritan tradition), but the early centuries provided the patterns basic to all subsequent manifestations.

BIBLIOGRAPHY: J. de Guibert et al. in *Dictionnaire de Spiritualité*, I (1937), pp. 936-1010; texts in M.J. Rouet de Journel and J. Dutilleul, *Enchiridion Asceticum* (2nd ed., 1947); and H. Koch, *Quellen zur Geschichte der Askese und des Mönchtums in der alten Kirche* (1933); A. Vöobus, *History of Asceticism in the Syrian Orient* (2 vols., 1958, 1960); G. Kretschmar, "Ein Beitrag zur Frage nach dem Ursprung frühchristlicher Askese," in *ZTK* 61 (1964), pp. 27-67; P. Nagel, *Die Motivierung der Askese in der alten Kirche und der Ursprung des Mönchtums* (*TU* 95; 1966); H. von Campenhausen, *Tradition and Life in the Church* (1968), chs. 4,11; R. Murray, "Features of the Earliest Christian Asceticism," in *Christian Spirituality* (ed. P. Brooks, 1975).

D.F. WRIGHT

ASH WEDNESDAY. The first day of Lent.* It is so called from the custom in the ancient church and continued in the Roman Catholic Church of marking the foreheads of worshipers with ashes which have been previously blessed. The appointment of Ash Wednesday as the first day of

Lent occurred sometime in the seventh century and the custom of imposing the ashes upon the congregation probably from sometime in the eighth century. The significance of this rite is based on the OT where we often find imposition of ashes as a sign of penitence and mourning. In the early church, Christians who had fallen into grave sin were admitted to the "order of penitents" to do their penances so that they could be reconciled to the church in Holy Week ready for their Easter Communion. The widespread use of this form of public penance decreased in the early Middle Ages, but the ceremonies associated with it—notably the ashes—were extended to the whole congregation. In the Roman Church the ashes are obtained by burning the palms from the previous Palm Sunday. They are then placed on the heads of the worshipers with the words "Remember, man, that thou art dust and to dust shalt thou return." The ceremony was abolished by the Reformers, and in the Anglican Book of Common Prayer there is provided a service of Scripture readings and prayers for Ash Wednesday known as the Commination Service. The theme of the service is indicated in its subtitle, a "denouncing of God's anger and judgements against sinners."

See H. Thurston, *Lent and Holy Week* (1904).

PETER S. DAWES

ASKE, ROBERT (c.1501-1537). Leader of the "Pilgrimage of Grace."* Lawyer and fellow of Gray's Inn, he came from an old Yorkshire family. The suppression of the monasteries and the rumors that further changes were imminent, together with dislike of the emerging Reformation doctrines and also dislike of new taxes, were the cause of discontent first in Lincolnshire in 1536. This was soon quelled by threats and promises from Henry VIII. The Lincolnshire rising, however, proved to be the spark to a much more serious uprising in Yorkshire, and Aske came south with 30,000 men. Henry broke up this formidable threat by promising to listen to their cause, and the pilgrimage disbanded. Aske came to London and stayed at court, receiving promises from the king that most grievances would be put right. Relying on Henry's word, Aske kept the north quiet, even assisting at putting down another riot in Yorkshire. The king, however, now felt strong enough to repudiate his promises and to have Aske hanged.

PETER S. DAWES

ASKEW, ANNE (c.1521-1546). Protestant martyr. Born at Stallingborough, Lincolnshire, she read the Bible avidly and came as a result to dispute transubstantiation, causing a stir in her home area. She moved to London and there made friends with Joan Bocher. In March 1545 Anne was examined for heresy and committed to prison. She was then examined by Bishop Edmund Bonner,* though the frequent Roman assertion that she recanted is untrue. In June she was charged as a sacramentarian under the Six Articles, but the jury acquitted her. Shortly after, she was again brought before Bonner and others, condemned on her own confession, and sent to the Tower to be racked into recantation. But her courage overcame her torturers, and in July 1546 she was burned at Smithfield, having refused consistently to recant. The facts about Anne Askew are to be found in Bale and Foxe.

G.E. DUFFIELD

ASPERGES. This ceremony of sprinkling with holy water is observed, chiefly by the Roman Catholic Church, when consecrating churches or purifying houses, as an aid to unction and immediately before the principal Mass. Sprinkling the congregation at Mass was ordered by Pope Leo IV in 847, but the practice is attested in the eighth century. The words of Psalm 51:7 usually accompany the ceremony, although the original objective of the psalmist was inward purification by the merciful Lord.

ASPERSION, see BAPTISM

ASSEMBLIES OF GOD. The largest denomination to stem from the Pentecostal movement of the early twentieth century. Pentecostals were at first reluctant to form separate churches, hoping to transform existing churches. As early as 1914, however, American Pentecostal leaders agreed to form a simple fellowship of churches with the name "Assemblies of God" as a scriptural designation. A Presbyterian type of structure was adopted in 1918, with headquarters at Springfield, Missouri. In Canada, assemblies were incorporated in 1917 as "The Pentecostal Assemblies of Canada" (see PENTECOSTAL CHURCHES).

In Britain, the sole Pentecostal organization till 1924 was the Pentecostal Missionary Union (1909). The opposition aroused by the movement, its growing self-consciousness, its lack of large assemblies to act as points of concentration (as in Scandinavia), and concern over the development of the Apostolic Church combined to produce growing desire for some form of organized fellowship. An application by Welsh Pentecostal assemblies for acceptance as a Welsh District Council of AOG, USA, precipitated action. On the initiative of Thomas Myerscough and J. Nelson Parr, a preliminary conference in Birmingham (1924) recognized the need for unity and fellowship, and agreed to the name "Assemblies of God." Three months later a conference in London adopted a constitution which created an executive presbytery of seven with Parr as chairman and secretary and as editor of a paper, *Redemption Tidings*. The first annual conference followed in London. In 1925 the Pentecostal Missionary Union merged with AOG which also undertook responsibility for the annual Pentecostal Whitsuntide convention, then held in Kingsway Hall, London. The 1925 conference agreed to compile a hymnbook, *Redemption Tidings*. Two independent Pentecostal Bible colleges, at Hampstead and Bristol, were united in 1951 and handed over to conference. Situated on a new site at Kenley, this institution became the official AOG Bible school in Britain.

AOG support a large missionary force which has established assemblies in most parts of the world. They are particularly strong in France and Italy, Congo and Nigeria, and especially in Brazil.

AOG describe themselves as "Pentecostal in experience, evangelical in outlook, and fundamental in their approach to the Bible." They include baptism in the Holy Spirit with the initial evidence of speaking in tongues among the fundamental truths of Christianity. In organization they use a Presbyterian structure, accepting the autonomy of the local church, but also utilizing district councils and a general council which acts as "the controlling body." This latter operates through a general conference which meets annually.

See D. Gee, *Wind and Flame* (1967).

HAROLD H. ROWDON

ASSEMBLY, GENERAL, see GENERAL ASSEMBLY

ASSOCIATIONS, LAW OF (1901). This defined the legal status in France of all voluntary societies including Catholic religious orders. It formed part of a new wave of secular anti-Catholic legislation during the Third Republic, culminating in the rupture of relations between France and the Vatican (1904) and between church and state (1905). Promoted by the ministry of Pierre Waldeck-Rousseau, the law was believed by its supporters to be necessary to check the influence of a rival religious force within the state threatening to break the "moral unity" of France which, they asserted, rested on "revolutionary and republican principles." The law provided, *inter alia,* that no religious congregation or any of its dependent institutions (e.g., schools, hospitals) may be formed, or continue to exist, without state authorization, that any congregation may be closed by simple decree, that no member of an unauthorized order could teach, that the state will determine the liquidation of property of dissolved orders. Some 615 congregations complied while 215 did not, including the Benedictines and Jesuits. The successor ministry of Emile Combes used the law severely and extended its provisions to prohibit teaching by any religious order. By 1904, Combes boasted, he had closed 13,904 schools.

C.T. MC INTIRE

ASSUMPTION. The belief that "Mary, immaculately conceived by God and ever virgin, when the course of her earthly life had been finished had been taken up in body and soul to heavenly glory," which was defined as an article of faith in the Roman Catholic Church by Pius XII (*Munificentissimus Deus,* 1950). Previously the belief had the status of a pious and probable opinion which Benedict XIV refused to make into an article of faith in 1740. There had been pressure since 1870 for some such definition to be made. Nothing is said in the NT about the death of Mary, and the belief is first found in some apocryphal documents with a Gnostic flavor in the late fourth century. They include such titles as *The Passing of Mary, The Obsequies of Mary,* and *The Book of the Passing of the Blessed Virgin.* These writings vary in their accounts of when the assumption took place and indeed when Mary's death took place. One work was condemned in the *Decretum Gelasianum* and they were not accepted in orthodox circles in the West until the time of Gregory of Tours in the late sixth century. There are texts in Greek, Latin, Syriac, Coptic, Arabic, and Ethiopic. The Coptic texts may be the most important, as the legend was probably first elaborated in Egypt. The standard Greek writing is one attributed to John the Evangelist and the standard Latin writing one attributed to Melito of Sardis. In the East, Andrew of Crete in the eighth century believed, on the evidence of the pseudonymous work *Concerning the Divine Names,* that Dionysius the Areopagite had witnessed the assumption; John of Damascus presented it as an ancient catholic doctrine. The idea has been less precisely defined in the East and is known as the *Koimēsis* ("falling asleep").

R.E. NIXON

ASSUMPTIONISTS (Augustinians of the Assumption). A congregation of priests with simple vows, founded at Nîmes about 1843-45 by the diocesan vicar-general, Emmanuel Daudé d'Alzon. The purpose was "to restore higher education according to the mind of St. Augustine and St. Thomas, to fight the Church's enemies in secret societies under the revolutionary flag; to fight for the unity of the Church. . . ." They were expelled from France in 1900 by an anticlerical government, but spread to many parts of the world. Their work includes care of asylums and schools, the dissemination of literature, and missionary endeavor. Their "Institute for Byzantine Studies," established in 1897, has especially fostered the study and theology of the Eastern Church, notably through the publication *Revue des Études Byzantine.*

J.G.G. NORMAN

ASSURANCE. The English word is used to translate two Greek words: *pistis,* "faith" (Acts 17:31), and *plērophoria,* "full confidence" (1 Thess. 1:5). That assurance was necessary for the full enjoyment of salvation was dimly recognized in the OT (cf. Isa. 32:17). In its NT context, the word has both objective and subjective references. As objective it denotes the ground of the believer's confidence and certainty (Acts 17:31; cf. 1 Tim. 1:15). This external ground of assurance is the "so great salvation" wrought by Christ, His heavenly session at the right hand of God, and the Scriptures of truth which make wise unto salvation. As subjective, assurance has reference to the actual experience of the believer. Saving faith brings an experience by the Holy Spirit of the Gospel with "full conviction" (1 Thess. 1:5 RSV) and a firm certainty (Col. 2:2). Living in "full assurance of hope" to the end (Heb. 6:11 KJV), the trusting soul can come to God with a true heart in "full assurance of faith" (Heb. 10:22; cf. Eph. 3:12). But inward assurance must be checked by moral and spiritual tests (cf. e.g., 1 Cor. 6:9; Eph. 4:17; 1 John 2:3-5, etc) by which we know that we are of the truth and that our hearts are assured before God (1 John 3.19).

In contrast with the teachings of the Roman Catholic Church and the decrees of the Council of Trent,* and with some Arminian declarations, Reformed theology generally has stressed the possibility of, and blessing attendant upon, the assurance of salvation. Recent Lutheranism, however, has weakened, if not denied altogether, full assurance, while the influence of Schleiermacher and Ritschl on modern theology has been in the direction of a revived Pelagianism.* The subjectivism of many present-day charismatic movements has, in some evangelical circles, tended to move the ground of assurance from the objective Word to the inward state of the believer. The result has led, in several instances, to an unhealthy introspection and a clouding of the full biblical perspective.

BIBLIOGRAPHY: J. Arminius, *Works* (1825), II, pp. 725-26; W. Cunningham, "The Reformers and the Doctrine of Assurance," in *The Reformers and the Theology of the Reformation* (1862), pp. 111-48; J. Calvin, *Institutes*, III, 2; A.S. Yates, *The Doctrine of Assurance* (1952); G.C. Berkouwer, *Faith and Perseverance* (1958); I.H. Marshall, *Kept by the Power of God* (1969).

H.D. MC DONALD

ASSYRIAN CHURCH. Born of the controversy over the Incarnation, christologically unorthodox but fervent in mission and evangel, the Assyrian (Syriac-speaking, Nestorian) Church carried its particular version of the Gospel from its first center in Mesopotamian Edessa Callirrhoe, and its second in Nisibis, to earth's remotest known ends. Repudiated and anathematized for Nestorianism* by Western Christendom (Council of Ephesus, 431), and harried out of the Roman Empire, these sectarians advanced ultimately to China and India, making tremendous impact. They suffered innumerable persecutions, culminating in the Kurdistan mountains in World War I. Settled in Edessa long before 431, they expanded to Nisibis in 435, but lost the Edessa foothold in 489—which merely sparked astounding missionary expansion. Celibacy was originally essential, even for laymen seeking baptism. By a decree in 499, clergy, including bishops, received sanction to marry, and some did this, much to the scandal of the Roman Church. The mendicant praying monks (Euchites) were probably lazy rather than immoral, as enemies unkindly insinuated. Lofty in ideals, they have similarities with later Quietists and perfectionists.

ROY A. STEWART

ASTERIUS THE SOPHIST (d. after 341). Arian* theologian. Fragments remain in the works of Athanasius and of Marcellus of Ancyra from his *Syntagmation*, which was published possibly before the Council of Nicea (325). A disciple of Lucian, the teacher at Antioch, Asterius was too moderate in his Arianism for the historian Philostorgius. Recently M. Richard has edited his homilies on the Psalms, which provide liturgical evidence for the turn of the third century. He confirms that baptism was a regular part of the Easter vigil leading up to the Eucharist on Easter Day. An intriguing hymn to Easter night survives in his homilies.

ASTROLOGY. The prediction of events on the basis of the positions of heavenly bodies started in ancient Babylonia, but did not reach Greece until about 410 B.C. After Alexander's campaigns it spread rapidly, and from the second century B.C. it influenced the entire intellectual world until the time of Newton. Its greatest protagonist was Ptolemy (second century A.D.), whose book *Tetrabiblos* became the standard astrological text of the Middle Ages. Astrology still exerts enormous influence throughout the world; it is particularly prevalent in India and Ceylon. In the West it is usually taken half-seriously, but not always; in World War II Hitler and Himmler maintained regular contact with astrologers.

Astrology takes three forms: (1) meteorological or astronomical phenomena are regarded as omens relating to rulers and their peoples; (2) planets are supposed to exert influences at birth as in the germination of seeds or the birth of children (such influences are commonly depicted in horoscopes); (3) planets are regarded as ruling over particular geographical regions. Horoscopes and geographical astrology date only from the Greek period.

According to the medieval view, man the microcosm is influenced by the external universe, the macrocosm. The planets in particular are linked with both the parts of man's anatomy and his dispositions. Though now regarded as a superstition, horoscope astrology was once regarded as the zenith of intellectual achievement; universities maintained chairs of astrology while even professors of astronomy (as in early seventeenth-century Oxford) were compelled to lecture on astrology. Even Kepler* cast horoscopes.

The orthodox Christian view was that though man is influenced by his stars, his actions are not determined by them. Paracelsus* speaks of man controlling his stars, meaning that despite the influences they exert, man is not deprived of freedom. Paul (Rom. 8:39) insists that neither zenith ("height") nor nadir ("depth")—believed to be astrological terms—can separate us from the love of God.

Astrology steadily declined as the heliocentric system gained credence. Proper evidence for the older view was not forthcoming, and the difficulty of understanding how planets could act in the manner supposed proved insuperable. Nevertheless, the principle underlying astrological beliefs is not entirely wrong; astrology rightly insists on the interrelatedness of the natural order. In addition there is scientific evidence favoring the view that the positions of planets may affect the earth physically (e.g., conjunctions of the major planets may trigger earthquakes).

Astrological predictions are particularly dangerous because they can be self-fulfilling, e.g., if astrologers predict a financial crisis many businessmen may sell their shares and precipitate the crisis predicted. Similarly, an astrological prediction of death on a particular day may suggest suicide.

BIBLIOGRAPHY: F. Cumont, *Astrology and Religion Among the Greeks and Romans* (1912); D.C. Allen, *The Star-Crossed Renaissance* (1941); R. Eisler, *The Royal Art of Astrology* (1946); F.H. Cramer, *Astrology in Roman Law and Politics* (1954); P.H. Kocher, *Science and Religion in Elizabethan England* (1954), chap. 10; D.R. Dicks, *The Geographical Fragments of Hipparcus*

(1960); W. Knappich, *Geschichte der Astrologie* (1967); J. Lindsay, *Origins of Astrology* (1971).

R.E.D. CLARK

ASTRUC, JEAN (1684-1766). Physician; founder of modern Pentateuchal criticism. Possibly of Jewish extraction, he was born at Sauve, Languedoc. His father, a Huguenot* pastor, became Roman Catholic when the Edict of Nantes was revoked (1685), so Jean was brought up a Catholic. He studied at Montpellier, taught medicine there (1707), at Toulouse (1716), and at Paris (1731) and was court physician to Augustus II of Poland (1729) and to Louis XV of France (1730). In 1753 he published anonymously his *Conjectures sur les mémoires originaux dont il paraît que Moise s'est servi pour composer le livre de la Genèse.* He concluded that Moses made use of earlier documents; the two primary sources used *YHWH* and *Elohim* respectively as divine names. Though some ridiculed his hypothesis, J.G. Eichhorn* later established its importance.

J.G.G. NORMAN

ASYLUM, see SANCTUARY

ATHANASIAN CREED. Two creeds need to be distinguished: (1) the Nicene Creed*; (2) the Athanasian Creed or the *Quicunque Vult,* known also as the *Fides Catholica.*

How the latter became known as the Athanasian Creed (beyond the fact that it expresses Nicene sentiments) is unknown, but it was apparently written originally in Latin, then translated into Greek, and is later than Athanasius.* It has been widely used in the West among Anglicans (in the Book of Common Prayer), Catholics, and Protestants. The late medieval controversy between the East and the West on the double procession* of the Holy Spirit (that the Holy Spirit is from the Father and the Son) intensified its use in the liturgy of Western churches. But its use is now diminishing.

In the preface and conclusion, belief in the truths it declares is said to be necessary to salvation and it anathematizes divergent faith. It is made up of forty rhythmical sentences and is thus more a sermon or instructional hymn than a creed. It expounds the doctrine of the Trinity and the divine relationships, the Incarnation, and the two natures of Christ, and includes statements about our Lord's work as Savior and Judge. It is a valuable compendium of orthodox faith and contains one of the best Christian confessions on the Trinity, "we worship one God in Trinity, and Trinity in unity; neither confounding the persons, nor dividing the substance."

As to its origin, Babcock suggests it be dated either in the latter half of the fourth or the fifth century, but not later than the sixth century. It seems to reflect Augustinian views or was known to him. Recently some attribute it to Ambrose, while others attribute it to writers from Gaul such as Hilary. The errors it opposes are primarily Arian,* Apollinarian, and Sabellian, rather than Nestorian and Eutychian. Parallels between the *Quicunque Vult* and the letters sent from the Council of Constantinople in 382 seem to confirm the period 381-428 as the time of its writing. It appears in the handbooks of certain Eastern Orthodox churches, including the Greek Horologium and Russian service books from the seventeenth century, but as translated from the Greek version omitting the *Filioque* clause.

BIBLIOGRAPHY: F.J. Babcock, *History of the Creeds* (1930); J.F. Bethune-Baker, *Early History of Christian Doctrine* (1954); J.N.D. Kelly, *The Athanasian Creed* (1964).

SAMUEL J. MIKOLASKI

ATHANASIUS (c.296-373). Champion of orthodoxy against Arianism.* Born to wealthy parents, he was Egyptian by birth but Greek by education. In the excellent catechetical school of Alexandria he was deeply moved by the martyrdoms of Christians during the last persecutions and was profoundly influenced by Alexander,* bishop of Alexandria, by whom he was ordained deacon. Of small stature but keen mind, Athanasius took no official part in the proceedings of the Council of Nicea (325), but as secretary to Alexander his notes, circulars, and encyclicals written on behalf of his bishop had an important effect on the outcome. He was a clear-minded and skilled theologian, a prolific writer with a journalist's instinct for the power of the pen, and a devout Christian—which endeared him to the large Christian public of Alexandria and the the vast majority of the clergy and monks of Egypt.

Athanasius contested Arius and the Arians during most of the fourth century. Arius taught that Christ the Logos* was not the eternal Son of God, but a subordinate being, which view attacked the doctrines of the Trinity, the Creation, and redemption. Athanasius said the Scriptures teach the eternal Sonship of the Logos, the direct creation of the world by God, and the redemption of the world and men by God in Christ. *On the Incarnation of the Word of God,* written while Athanasius was in his twenties, expounds these truths.

Alexander died in 328, and by public demand Athanasius was enthroned as bishop when he was only thirty-three. The victory at Nicea remained in political jeopardy for two generations, and Athanasius was the focal point of Arian attack. Arianism had a wide following in the empire and also the sympathies of Constantius, Constantine's successor in 337. The history of the Church in the fourth century parallels the events of Athanasius's life and his public ministry. He was hounded through five exiles embracing seventeen years of flight and hiding, not only among the monks of the desert, but often in Alexandria where he was shielded by the people. During one exile, at Rome in 339, he established firm links with the Western Church which supported his cause. His later years were spent peacefully at Alexandria. G.L. Prestige declares that almost single-handedly Athanasius saved the Church from pagan intellectualism, that "by his tenacity and vision in preaching one God and Saviour, he had preserved from dissolution the unity and integrity of the Christian faith."

The volume and scope of his writings is impressive. *Contra Gentes,* a refutation of paganism, and *de Incarnatione,* the exposition of the incarnation and work of Christ, were both written early (c.318) and are really two parts of one work. *De Decretis* and *Expositio Fidei* are also important doctrinal writings. The polemical and historical essays include *Apologia Contra Arianos, ad Episcopos Aegypti,* and *de Synodis.* He wrote many commentaries on biblical books. There are numerous other writings, including letters, many of which are readily accessible (*The Nicene and Post-Nicene Fathers,* Series 2, IV). Key doctrines which he discusses include Creation, the Incarnation, the Holy Spirit and the Trinity, the work of Christ, and baptism and the Eucharist. Athanasius greatly influenced the monastic movement, especially in Egypt.

See also ATHANASIAN CREED and CHRISTOLOGY.

BIBLIOGRAPHY: G.L. Prestige, *Fathers and Heretics* (1940); E.R. Hardy, *Christology of the Later Fathers* (1954); H.E.W. Turner, *The Pattern of Christian Truth* (1954); J.N.D. Kelly, *Early Christian Doctrines* (1958); B. Altaner, *Patrology* (1960); J. Quasten, *Patrology,* III (1966).

SAMUEL J. MIKOLASKI

ATHANASIUS (the Athonite) (c.920-1003). Byzantine founder of cenobite* monasticism (in 961) on Mt. Athos. After studies at Constantinople, he first went to Mt. Athos as a hermit to avoid court honors. He established there the first of the famous monasteries (the Lavra) through the support of the emperor Nicephorus Phocas. He introduced a rule for cenobites based on the common-life ideals of Basil the Great and Theodore the Studite. Opposition, on the grounds that his experiment was a result of imperial influence, followed the death of the emperor in 969 and drove him into exile. The new emperor, John I Tzimisces, supported his return. Athanasius became abbot-general of all the communities on Mt. Athos before his death in a building accident.

JAMES TAYLOR

ATHEISM. The denial that God exists. Atheism in the strict sense is rare; self-styled atheists are often agnostics or secularists. Tillich (1948) defines atheism as the view that "life has no depth, that it is shallow." For Toulmin (1957) atheism is the view that cosmic powers are indifferent or "positively callous" to man. Atheists have been classified as "tough" and "tender," or "crude" and "sensitive." The tender atheist accepts higher values (empathy, truth, beauty, etc.) and so has a good deal in common with the Christian, even though his intellectual position differs radically.

Arguments for atheism are based on: (1) logical positivism and linguistics: God's existence cannot be empirically verified, and the word "God" is asserted to be meaningless; (2) alleged fallacies in the traditional proofs of God's existence; (3) the claim that theism does actual harm, e.g., by encouraging persecution; (4) the claim that science offers an adequate explanation of the world without need for the supernatural; (5) the claim that accepted psychological laws or principles explain belief in God.

To these arguments, taken in order, it may be replied: (1) we cannot determine truth by the way we talk which, in any case, is often equivocal; a criterion of meaning which makes God meaningless cannot be sustained—no one doubts the reality of the past even though it cannot be perceived or empirically verified; (2) the classical "proofs" can be restated in modern terms, and one or two of them are remarkably cogent; (3) persecutions by atheists and pagans (Communists, Nazis) have been more cruel in our time than those perpetrated in the past by theists; cruelty is not the result of theism—genuine belief in a forgiving God predisposes to kindness; (4) this is not true. Science gives no account of the very small, of the very large, or of origins. Nor can it be shown that scientific explanations exclude explanations of other kinds (artistic, teleological as applied to machines, as well as theological); (5) psychological explanations are double-edged weapons. Atheism can be explained psychologically by disappointment in the father; alternatively a masochistic desire *not* to be happy may result in a repudiation for oneself of the happiness associated with belief.

BIBLIOGRAPHY: E.T. Whittaker, *The Beginning and End of the World* (1942); H. de Lubac, *The Drama of Atheist Humanism* (1949); R.E.D. Clark, *The Universe, Plan or Accident?* (3rd ed., 1961); A.O.J. Cockshut, *The Unbelievers* (1964); R. Robinson, *An Atheist's Values* (1964); F.H. Cleobury, *A Return to Natural Theology* (1967); P. Edwards, "Atheism," in *Encyclopaedia of Philosophy* (1967); H.D. Lewis, *The Elusive Mind* (1969).

R.E.D. CLARK

ATHENAGORAS (second century). Christian Apologist of Athens who presented Christian doctrine within the framework provided by a Middle Platonic epitome of Plato's philosophy. His *Apology* petitions Emperor Marcus Aurelius* and his son Commodus on behalf of the Christians and refutes the calumnies leveled against them, namely, atheism, eating human flesh, and practicing incest. He draws attention to the peaceful and blameless living of Christians and claims equal rights for them with other citizens. His treatise *On the Resurrection of the Body* defends a doctrine which the cultured pagans of his time found most difficult to accept. Athenagoras's lucid discussion is addressed to the philosophers, and the argument is kept entirely on their ground. It is therefore deficient in that the incarnation and resurrection of Christ are ignored. He also stressed the divinity of the Logos* and the triadic nature of God. The description of him as the "Christian Philosopher of Athens" is most appropriate.

G.L. CAREY

ATHENS. Capital of Attica, metropolis of ancient Greek culture, and capital of modern Greece. Its name seems to have derived from that of its patron goddess, Athene. The city has a striking situation with the rock of the Acropolis rising more than 500 feet above the plain. A monarchy in its early days and then an oligarchy, it became a pioneer in democracy from about 500 B.C. It played a leading part in the Persian wars and reached the peak of its political and cultural power in the fifth century B.C. After its conquest by the Macedonians in 338 B.C., there began the Hellenistic period in which Athens was still revered for her culture and learning though it had passed its peak. Athens was taken by the Romans in 86 B.C., but was made a free city under Augustus and was adorned with new buildings, particularly by Hadrian.

Acts 17 records the visit of Paul and his preaching the Gospel in the Areopagus. He showed himself able to present it in terms which were related to current philosophy, but he made only a few converts, among them Dionysius the Areopagite and a woman named Damaris. Apart from an allusion in passing in 1 Thess. 3:1, there is no further mention of Athens in the New Testament. The first reference to the church in Athens comes from Melito of Sardis who states (according to Eusebius) that the emperor Antoninus Pius tried to stop the harassment of Christians which was going on there in the middle of the second century.

Because of the importance of its schools of philosophy Athens was the home of a number of Christian Apologists* such as Quadratus, Aristides, and Athenagoras. Among those who studied at the philosophical schools were Julian, Basil, and Gregory Nazianzus, but in 529 Justinian forbade the study of philosophy. During the Byzantine period it had the status of a provincial town, and the ancient temples were used as Christian churches. In 869 it was given its own archbishop. From 1204 it came under Latin rule until its conquest by the Turks in 1456. The Greek Church was allowed to continue during the period of Turkish domination, which ended in 1833 with the setting up of the modern kingdom of Greece, of which Athens became the capital. At the same time the Orthodox Church of Greece became independent under the archbishop of Athens.

R.E. NIXON

ATHOS, MOUNT. The "holy mountain" on which developed a famous monastic community. During the tenth century there were two dominant monastic centers in the Byzantine Church. One was the monastery at Studium, the other was the monastic community on Mt. Athos. The Athos community went on to become the more important center of the two. The mountain is 6,350 feet above sea level, located on a rocky peninsula protruding into the Aegean Sea in northern Greece. There are over 900 places of worship (churches, chapels, etc.) on this mountain. As early as the fourth century there appears to have been hermit monks carrying out their asceticism at Athos. The style of monastic life went from the secluded hermit existence, to the monastic community, and then to very regulated monasteries.

Today the mountain contains about twenty semi-independent monasteries, plus smaller houses and hermit cells. Seventeen of the twenty identify themselves as Greek. Also included are a Russian house, a Serbian house, and a Bulgarian one. In Byzantine times there were Georgian and Latin houses also. It is said that at one time the population of Athos was about 40,000 monks, but by 1965 the numbers had dwindled to a little less than 1,500.

When the Ottoman Turks conquered Greece, Mt. Athos submitted quickly to their conquest. The monks were well treated, and the monastic community was allowed its independence subject to a tribute.

GEORGE GIACUMAKIS, JR.

ATONEMENT. This is one of the few theological terms of Anglo-Saxon origin. It means "at-one-ment" and signifies the process of making God and man one after the tragedy of man's sin had separated them (Isa. 59:2) and made them enemies (Col. 1:21). The NT has much to say about the way Christ's death brings them together, and in the literal sense the Atonement is the *crucial* doctrine of Christianity.

The Christian Church has never accepted any one way of viewing the Atonement as the orthodox way. There is no doctrine of the Atonement equivalent to the two-natures doctrine in Christology, for example. The result is that there are many ways in which Christians have answered the question, "How does the death of Christ long ago and so far away save me here and now?" We can detect three broad trends in the multiplicity of theories of atonement emerging during nineteen centuries of church history.

The first trend is seen in what Gustav Aulén has called the "classic" or "dramatic" view. It leans heavily on those biblical passages which speak of the Atonement as a ransom. It sees sinners as justly belonging to Satan because of their sin. But in the death of His Son God paid the price of their redemption. Satan accepted Jesus in place of sinners but he could not hold Him. On Easter Day Jesus rose triumphant, leaving Satan without either his original captives or their ransom. Aulén maintains that the essential point is not the grotesque imagery in which the Fathers expressed this theory but the authentic note of victory. He sees the essence of the Atonement as a process of victory over all the forces of death and evil. Most agree that victory is important, but they do not see this as the whole story.

The second group of theories may be said to have originated with Anselm of Canterbury, who saw sin as dishonor to the majesty of God. On the cross the God-man rendered satisfaction for this dishonor. Along similar lines the Reformers thought that Christ paid the penalty sinners incurred when they broke God's law. The strong points of this theory are its agreement with biblical teaching (e.g., on justification) and its insistence that the moral law cannot be disregarded in the process of forgiveness.

The third group of theories (especially linked with the name of Abelard) sees the Atonement in the effect on man of what Christ did. When we contemplate the love of God shown in the death of his Son we are moved to repent and to love Him in return. We are thus transformed. All is subjective.

All three theories have something to say to us. Each is inadequate by itself (especially the third, for it sees Christ as doing nothing except setting an example; the real salvation is worked out by sinners themselves). But taken together they help us to see a little of Christ's great work for men.

BIBLIOGRAPHY: J. Denney, *The Death of Christ* (1905); *idem, The Christian Doctrine of Reconciliation* (1918); R.S. Franks, *The Work of Christ* (1962); L. Morris, *The Apostolic Preaching of the Cross* (1965); *idem, The Cross in the New Testament* (1965). LEON MORRIS

ATTICUS. Patriarch of Constantinople, 406-425. From Sebaste in Armenia, he was educated among Eustathian monks (Pneumatomachi*) but joined the Catholic Church. According to the historian Socrates, Atticus was kind, courtly, scholarly, and able; he did not persecute. This is confirmed by his protection of Novatianists but denied by his attitude towards Pelagians, Messalians, and followers of John Chrysostom. He expelled Celestius about 413, and Julian of Eclanum about 418. He urged the bishops of Pamphylia and Amphilochus of Side to suppress the Messalians. In 406 he secured a rescript expelling and dispossessing bishops who refused communion with himself and Theophilus of Alexandria and Porphyry of Antioch. Such bishops, including Innocent of Rome, supported John Chrysostom* and pressed that his name should appear on the diptychs of Constantinople, where several congregations were refusing to acknowledge Atticus. Eventually popular pressure on Theodotus of Antioch (420-29) and the generous influence of Acacius of Beroea persuaded Atticus to restore the name of John; his defense of this failed to convince Cyril of Alexandria. Atticus's kindness was shown in shelter given to Persian refugees about 420 and in financial aid sent to famine-struck Nicea. At court he had advised the regent of the young Theodosius II, who in 421 transferred the ecclesiastical jurisdiction of Illyricum to Constantinople. G.T.D. ANGEL

ATTILA (d.453). King of the Huns. From victories in the East his hordes swept westward, were checked in Gaul (451), but in 452 invaded Italy and were halted only when Pope Leo I somehow convinced Attila of the wisdom of remaining beyond the Danube. There was no persecution of Christianity as such—the invaders were not without religion, and Attila himself is said to have besought the prayers of Lupus, bishop of Troyes—but the invasions erased or eroded the Christian presence in a large area of central Europe. In Latin Christendom, tales of Attila's ferocity grew, prompting the description "the Scourge of God."

ATTO II (c.885-961). Bishop of Vercelli from 924. Of a distinguished Lombard family, son of Viscount Aldegarius, he was also grand chancellor to Hugh of Provence, king of Italy. *De pressuris ecclesiasticis* (about 940) is his earliest major work, on refutation of charges against clergy, filling of clerical posts especially bishoprics, and lay seizure of church property after a bishop's death. He also wrote *Commentary on the Epistles of Paul, Letters and Sermons, Canones statutaque Vercellensis ecclesiae,* and *Polipticum* on moral philosophy. He should not be confused with Atto of Milan (d.1085/6), who wrote *Breviarium,* which was the basis for Gregorian reforms.

C.G. THORNE, JR.

ATTWOOD, THOMAS (1765-1838). English composer. Studies at Naples and with Mozart at Vienna brought him into contact with the finest trends in composition on the Continent. While he devoted his energies chiefly to operatic works, his small number of anthems shows a freshness of melody and a refinement of taste in advance of his English contemporaries. Several pieces, such as "Teach me, Lord, the way of Thy statutes," remain in the repertory of many choirs today. Attwood served as organist of the Chapel Royal and of St. Paul's Cathedral, London. He was a friend of Mendelssohn and was one of those influential in introducing the music of J.S. Bach into England. J.B. MAC MILLAN

AUBIGNÉ, J.H.M. D', see D'AUBIGNÉ

AUBURN AFFIRMATION (1924). Issued by a group of Presbyterian ministers meeting in Auburn, New York, this was "designed to safeguard the unity and liberty" of the Presbyterian Church. It asserted that the General Assembly had acted unconstitutionally when in 1923 it declared that all candidates for the ministry must affirm five "essential and necessary" doctrines prior to ordination. These were: (1) the inspiration and inerrancy of the Scriptures; (2) the virgin birth of Christ; (3) "that Christ offered up himself a sacrifice to satisfy divine justice"; (4) that Jesus arose from the dead "with the same body in which he suffered"; (5) that Jesus worked "mighty miracles" which made "changes in the order of nature." The Affirmation held that this was an attempt to commit the church to "certain theories concerning the inspiration of the Bible, the Incarnation, the Atonement, the Resurrection, and the Continuing Life and Supernatural Power of our Lord Jesus Christ." The signatories of the Affirmation claimed to "hold most earnestly to these great facts and doctrines," yet believed that "these are not the only theories allowed by the Scriptures and our standards as explanations of these facts and doctrines." The Auburn Affirmation was essentially a plea for toleration of theological diversity. Many conservative Presbyterians viewed the Affirmation as a shocking revelation of the growth of liberal influence within the denomination. HARRY SKILTON

AUBURN DECLARATION (1837). Declaration of belief issued by a strategy convention of the New School within the Presbyterian Church in the USA. Earlier that year the General Assembly, which was controlled by an Old School majority, exscinded the predominantly New School synods of Genesee, Geneva, Utica, and Western Reserve. The Old School believed that the New School party had departed from the Calvinistic theology of the Westminster Confession and were overly tolerant of the New Haven Theology* of Nathaniel Taylor* which placed a greater stress on human initiative in the process of salvation than did orthodox Calvinism. The Auburn Declaration rejected these accusations, reaffirmed the main, distinctive points of Calvinism, and asserted that "God permitted the introduction of sin, not because he was unable to prevent it . . . but for wise and benevolent reasons which he has not revealed." Because of the sin of Adam all mankind became morally corrupt and liable to eternal death. Saving faith is "an effect of the special operations of the Holy Spirit." The reason why some embrace the Gospel while others reject it is that "God has made them to differ." The Auburn Declaration established the New School's case for its Calvinist orthodoxy and eventually helped bring about reunion with the Old School in 1869.

HARRY SKILTON

AUDIANI. A group which took its name from Audi, an ascetic deacon near Edessa about 325, who criticized worldliness in the church and clergy. Maltreated by his opponents, he left the church to become bishop of monastic communities in suburbs and deserts, from Antioch to Arabia and Mesopotamia. The exile of Audi and his followers to Scythia by Constantine won Gothic adherents.

AUDOIN, see OUEN

AUFKLÄRUNG, see ENLIGHTENMENT

AUGSBURG, INTERIM OF (1548). The attempt by the emperor Charles V* to establish religious unity in Germany. His sympathies were with the Roman Catholic Church, but the strength of the Protestant cause, openly manifested in the League of Smalcald,* could not be ignored. After the defeat of the Protestant princes in the Smalcald War, Charles felt that he could impose some measure of religious uniformity and drew up the Interim, which was to be a provisional arrangement until the Council of Trent* had completed its work of investigating possible reforms. In the Interim Charles sought to reimpose the Roman Catholic hierarchy on the German Church and to reestablish the old fasts, feasts, and ceremonies. To allay the discontent of the Protestants, he introduced certain external reforms, permitting the marriage of the clergy and the giving of the cup to the laity in the Lord's Supper. Inevitably such a compromise satisfied no one; force of arms in the person of Spanish troops was needed to compel the Protestants in particular to accept it. The reaction to this compulsion led to the defeat of Charles and to the Peace of Augsburg* in 1555, when each state was given liberty to choose the creed which it would adopt—Lutheran or Roman Catholic.

HUGH J. BLAIR

AUGSBURG, PEACE OF (1555). An agreement reached after the defeat of the emperor Charles V* by Protestant princes in Germany (1552). A preliminary settlement negotiated with his brother Ferdinand at Passau in 1552 recognized all secularizations of church lands and approved the principle of a religious peace. Failing to capture Metz in 1553, Charles left Germany and commissioned Ferdinand to settle affairs at the Diet of Augsburg.

The terms of the peace were: (1) Lutheran princes, imperial knights, and free cities were guaranteed security equal to that of the Catholic estates. Cities could permit both faiths if they were already established; (2) the peace applied only to Catholics and those Protestants who adhered to the Augsburg Confession.* "Sacramentarians" (Calvinists) and "Sectarians" (Anabaptists) were excluded; (3) each estate or prince determined the religion of his domain and all subjects must conform (*Cuius regio, eius religio* was the term later used by jurists). Dissenters could sell their property and emigrate with their families; (4) all church lands secularized prior to 1552 would remain in Protestant hands; (5) an "ecclesiastical reservation" provided that archbishops, bishops, and abbots who turned Protestant lost their dignity and rights, and the chapters would select an orthodox successor.

The peace is significant because it meant that both the political unity of Germany and the medieval unity of Christendom were permanently shattered. The power granted secular rulers to control religious matters in their domains weakened genuine Christianity in Germany. The exclusion of Calvinism and vagueness of the ecclesiastical reservation made the settlement fragile, but its principles held until the Peace of Westphalia* in 1648.

See M. Simon, *Der Augsburger Religionsfriede* (1955), and H. Holborn, *A History of Modern Germany*, vol. I, *The Reformation* (1959).

RICHARD V. PIERARD

AUGSBURG CONFESSION (1530). A summary of the evangelical faith presented to Emperor Charles V* for the Diet of Augsburg. Commissioned by John, elector of Saxony, it was written by Luther, Justus Jonas, Bugenhagen, and Melanchthon. The group, which met at Torgau, had before it the Schwabach Articles written in 1529 and the articles presented during the Marburg Colloquy. The call for the diet indicated the emperor's hope that some conciliation between Catholics and Protestants might be achieved. This, with the recent failure at Marburg with the Zwinglians, might explain the conciliatory and irenic wording of the document.

The Torgau Articles were reworked by Melanchthon. One of his objectives in writing the Confession was to refute the charges of Johann Eck in his book, *404 Articles*, that Lutheranism was reviving certain ancient heresies. The draft was sent to Luther for his perusal and some alterations were made after consultation with Jonas, the Saxon Chancellor Brück, Bishop Stadion, and Alfonso Valdez, the imperial secretary. On 23 June 1530, the Confession was approved by John, elector of Saxony; Philip Landgrave of Hesse; George, margrave of Brandenburg; Dukes Francis and Ernest of Lüneburg; representatives from the cities of Nuremberg and Reutlingen; and other counselors and theologians. Ultimately four more cities accepted it during the meeting of the diet. On the insistence of the Protestant princes and against the objection of Charles V, the Confession was read publicly in German during the diet on 25 April 1530. It took two hours.

The German and Latin texts were then given to a group of twenty Catholic theologians chosen by Campeggius for examination and refutation. The reply, called the Papalist Confutation, approved without qualification nine of the articles; six were approved with qualifications or in part; thirteen were condemned. A revised form of the reply was later adopted by Charles V as his own confession. The emperor demanded that the Protestants conform to the Confutation, but they sought the opportunity to reply to it. This was done by Melanchthon,* and the reply or apology was then affixed to the Confession. The German translation by Jonas in 1532 helped to make this the principal confession of the Lutheran Church. Due to the unauthorized publication of the Confession in 1530, Melanchthon issued the *Editio Princeps* in 1531, presenting the authorized text. The first twenty-one articles dealt with similarities and dissimilarities between Lutherans and Catholics, the last seven articles with abuses in the church, such as failure to give both bread and cup during the Lord's Supper; celibacy; paying for masses; compulsory confession; equating grace with fasts and festivals; lack of monastic discipline; abuse of ecclesiastical power. Melanchthon issued another text in 1540 called the *Variata.* There is no sure evidence Luther rejected it. The *Variata* was used by the Crypto-Calvinists. When the Book of Concord* was adopted, the Latin text of 1531 was chosen over the 1540 *Variata* text. The Confession was the earliest of the formal creedal statements and became the authoritative confessional standard for the Lutheran Church, having influence upon other confessions.

BIBLIOGRAPHY: J.M. Reu, *The Augsburg Confession* (1930); W.D. Allbeck, *Studies in the Lutheran Confessions* (1952); T.G. Tappert (ed.), *The Book of Concord* (1959); M. Lackmann, *The Augsburg Confession and Catholic Unity* (1963); G.W. Forell, *The Augsburg Confession* (1968); A. Kimme, *Theology of the Augsburg Confession* (1968).

ROBERT SCHNUCKER

AUGUSTA, JOHN (Jan) (1500-1575). Leader of the Bohemian Brethren.* The latter, following the suppression of the Taborites in 1453, possessed many of the evangelical and social views of the early Hussites, It was Augusta's hope to encourage his group to initiate the formation of a large evangelical party in Bohemia lacking doctrinal differences. This ambition led several Bohemian nobles to the Brethren. Negotiations, which were protracted and unsuccessful, were begun in 1533 with Luther, Bucer, and Calvin when Augusta assumed leadership of the Brethren. During the persecution of the Brethren, Augusta was held in prison from 1547 to 1564 by Ferdinand of Hapsburg. In 1564 Augusta tried to unite the Brethren with the Utraquists* in order to form a national evangelical church in Bohemia, but the attempt failed.

JAMES TAYLOR

AUGUSTANA SYNOD, see LUTHERAN CHURCH BODIES IN THE USA

AUGUSTINE OF CANTERBURY (d.604?). First archbishop of Canterbury. Previously the prior of a monastery at Rome, he was sent by Gregory the Great* in 596 on a mission to convert the pagan English. He was somewhat reluctant and asked permission to turn back before finally crossing the Channel in 597 and landing in Thanet. Gregory knew, however, that the time was ripe. Ethelbert, king of Kent, whose territory lay closest to the Continent, had married a Christian princess of the Franks named Bertha, and she had brought to England with her Bishop Liuthard as her chaplain. Moreover, Ethelbert at this time was the dominant ruler among the Anglo-Saxon tribes south of the Humber. Within four years, and perhaps much sooner, Ethelbert received baptism. Augustine was then made archbishop, and Bede says the consecration took place at Arles, although the authority for this statement is unknown.

It had been Gregory's intention to make the old Roman centers of London and York the metropolitan sees of the English Church. London belonged, however, to the East Saxons, and so Augustine fixed his seat at Canterbury. Later he sent his companions, Justus (to preach west of the Medway as bishop of Rochester) and Mellitus (to convert the East Saxons as bishop of London).

Gregory's sustained interest in the details of the English mission are apparent in the very specific instructions he sent to Augustine. The latter failed (c.603) in his attempt to carry out Gregory's order to reach agreement with the leaders of the ancient Celtic Church in the west of Britain. Beyond Kent, the successful conversion of the English was carried out by other missions unrelated to that of Augustine, the most important of which was the Celtic mission from Iona.* Augustine also established at Canterbury with Ethelbert's help the monastery of St. Peter and St. Paul, where the first ten archbishops and several kings were buried.

BIBLIOGRAPHY: Bede, *Opera Historica* (ed. C. Plummer, 1896); *idem, The Ecclesiastical History of the English People* (ed. B. Colgrave and R.A.B. Mynors, 1967); M. Deanesley, *The Pre-Conquest Church in England* (1961); J. Godfrey, *The Church in Anglo-Saxon England* (1962).

JOHN TILLER

AUGUSTINE OF HIPPO (354-430). Aurelius Augustinus (Austin), bishop of Hippo Regius in Numidia in Roman North Africa, and greatest of the Latin Fathers. Born at Tagaste of middle-class parents, Patricius (converted only shortly before his death in 372) and the devout, domineering Monica, he was educated locally, then at Madaura and Carthage (371-75). He excelled in the rhetoric-centered training of late antiquity, but failed to master Greek. A catechumen since infancy, he nevertheless indulged a passion for the theater and disciplined his sexuality only by an unofficial marriage (372) which lasted until 385 and produced a son, Adeodatus ("gift of God," died c.390).

Cicero's (lost) *Hortensius* converted Augustine to philosophy, the pursuit of divine wisdom (373), which precipitated him, disillusioned with the Bible's style and substance, into Manichaeism.* Attracted by its claim to rational demonstration of wisdom, its rejection of the OT, and the lofty spirituality of its "elect," he remained a Manichaean *auditor* ("hearer") for a decade; at first an enthusiastic proselytizer but progressively disabused of its intellectual pretensions, he was yet unable to abandon its philosophical materialism and dualistic resolution of the nagging problem of evil. At Tagaste (375-76) and Carthage, Augustine became a teacher of rhetoric and wrote his first (lost) work, *The Beautiful and the Fitting* (c.380). Ambition took him briefly to Rome (383) and then —through the patronage of Symmachus, the city prefect and doyen of the pagan aristocracy—to Milan, the imperial seat, as professor of rhetoric (384), soon pursued by Monica to mother his professional advance and Catholic recovery. In anticipation of a suitable marriage, his unnamed concubine was dismissed.

Though tempted by academic skepticism, Augustine succumbed to the learned eloquence and allegorism of Ambrose,* bishop of Milan. Prominent Milanese Christian intellectuals introduced him to writings by Neoplatonists Plotinus* and Porphyry* (early 386), which consummated his emancipation from Manichaeism and fanned his devotion to spiritual philosophy. Their vision of transcendent, immaterial being and evil as privation of goodness dissolved his difficulties and exposed him to the impact of repeated accounts of Christian "conversion from the world." His struggles to follow these *exempla* climaxed in the famous garden with the reading of Romans 13:13,14 (late summer 386). Abandoning a public career, he retreated with family, friends, and pupils to an estate at Cassiciacum for the untrammeled pursuit of wisdom, which the classical dialogues and *Soliloquies* written there delineate as the quest of a confident Catholic Neoplatonist. Baptism by Ambrose in Milan followed at Easter 387.

On his return to Tagaste, following Monica's death at Ostia after their ecstatic (Neoplatonic?) vision, Augustine formed a fellowship of "the servants of God" committed to contemplative philosophy (cf. *True Religion*, 389-91). He had already started writing against Manichaeism. While visiting Hippo for ascetic purposes, he was pressganged into the priesthood (391), first begging time to improve his biblical knowledge. He founded a monastery there, and on succeeding Bishop Valerius in 396 (having been consecrated in advance in 395), turned the episcopal house into a clerical monastic community, a nursery of future African bishops. As Augustine reconciled himself to being Catholic bishop in the teeth of Donatist* ascendancy, Manichaean persistence, and "catholicized" paganism, his outlook changed decisively. A more biblically radical diagnosis of man and history (even church history) progressively displaced an optimistic, Neoplatonic "humanism." *The Confessions* (c.397-401) interpret his past up to Monica's death in this severer light. Though in emphasis they diverge from his pre-Hippo writings, their basic historicity is unquestioned.

He became deeply involved in a bishop's usual duties—liturgical, pastoral, disciplinary, administrative, judicial. He preached assiduously, with long series on the Psalms and John's gospel, in a style as appealing and profound as any of the Fathers. He traveled often, around his diocese, to Carthage and elsewhere, especially for synods, consultations, and disputations, but never again left Africa. He propagated the monastic life and in league with Aurelius of Carthage reinvigorated African Catholicism.

His theology ripened in controversy. Against the Manichaeans he defended the goodness of creation *qua* being, defining evil as absence of good and ascribing sin to abuse of free will, and he developed a rationale of faith as evoked by the impressive authority of the universal church and leading to understanding.

The Donatists evaded the personal confrontations in which Augustine worsted Manichaean notables, but his tireless historical and theological refutation, popularized in rhyme, slogan, and poster, advanced African views of the church and sacraments yet left them still underdeveloped. Building on Tyconius's ecclesiology, he stressed that the church's purity was eschatological, incapable of present realization, and its universality as certain as prophecy. Sacraments outside the church are real, because their minister is Christ, but profitless until their recipients rejoin the only body wherein the Spirit, the bond of love and unity, bestows life. Pragmatic considerations led Augustine to abandon his disapproval of the coercion of heretics and schismatics, but he justified it as a corrective, not a punitive, function of the Christian magistrate, and merely one facet of divine discipline for man's good ("Love, and do what you will").

His typically African preoccupation with the church and baptism persisted against Pelagius* and his followers. He refined the African conception of original sin, as the inherited guilt and corruption of Adam, and taught the necessity of inward grace to enable man to obey God, apparently denying his anti-Manichaean voluntarism which Pelagius cited. The impossibility of sinlessness on earth, the indispensability of baptismal forgiveness for infants, and the inclusive "hospitality" of the church were stressed against Pelagius's perfectionist élitism. The inscrutable predestination and perseverance of the elect (but not the reprobate) came to the fore against Julian of Eclanum, and the semi-Pelagian (better, semi-Augustinian) monks of S Gaul.

The City of God (c.413-27) began as an apologia against allegations that Christianity was ultimately answerable for the sack of Rome in 410. It became a disorderly review of Roman and Christian history, interpreted theologically and thus eschatologically, through the entangled earthly fortunes of two "cities" created by conflicting loves.

Augustine died as Roman Africa succumbed to the Vandals besieging Hippo. His friend Possidius, bishop of Calama, compiled a Life and a catalogue of his works. One of the last was the *Revisions* (*Retractationes*, 426-27), in which Augustine listed his writings, correcting and defending himself at points. *The Trinity* (399-419) sums up the patristic Trinitarian achievement and advances psychological analogies, while *Christian Instruction* (396-426) became an influential manual of Christian culture and hermeneutics. In addition to numerous contributions to controversies, there survive extensive exegetical works, and hundreds of letters (including the monastic Rule) and sermons.

These voluminous writings massively influenced almost every sphere of Western thought in later centuries. In many conflicts, including the Reformation, both sides could claim his patronage, appealing to selected facets of his ever-shifting mind. One of the four Doctors of the Latin Church, he became *post apostolos omnium ecclesiarum magister* (Gottschalk).

BIBLIOGRAPHY: Editions: *PL* 32-47 (by Maurist Benedictines, 1679-1700) and *PL Suppl.* 2; collected editions in process in *CSEL* and *Corpus Christianorum* and in Latin-French *Bibliothèque Augustinienne;* convenient editions of *Confessions* by J. Gibb and W. Montgomery (2nd ed., 1927), and of *City of God* by J.E.C. Welldon (2 vols., 1924).

English translations of many works in *NPNF;* more recently, of a few in *LCC,* of increasing numbers in *FC* and *ACW;* of *Confessions* by F.J. Sheed (1943), of *City of God* by H. Bettenson (1972); of Possidius's *Life* in F.R. Hoare, *The Western Fathers* (1954).

Selected studies in English: J. Burnaby, *Amor Dei: A Study of the Religion of St. Augustine* (1938); J.H.S. Burleigh, *The City of God* (1949); R.H. Barrow, *Introduction to St. Augustine's "The City of God"* (1950); G.G. Willis, *St. Augustine and the Donatist Controversy* (1950); J.J. O'Meara, *The Young Augustine* (1954) and *The Charter of Christendom: The Significance of the "City of God"* (1961); H.I. Marrou, *Saint Augustine and His Influence Through the Ages* (1957); S.J. Grabowski, *The Church: An Introduction to the Theology of St. Augustine* (1957); A.D.R. Polman, *The Word of God According to Saint Augustine* (1961); E. Gilson, *The Christian Philosophy of St. Augustine* (1961); F. Van der Meer, *Augustine the Bishop* (1961); H.A. Deane, *The Political and Social Ideas of St. Augustine* (1963); G. Bonner, *St. Augustine of Hippo: Life and Controversies* (1963) and *Augustine and Pelagianism in the Light of Modern Research* (1973); A.H. Armstrong, *St. Augustine and Christian Platonism* (1967); R.A. Markus in *The Cambridge History of Later Greek and Early Medieval Philosophy* (ed. A.H. Armstrong, 1967); P. Brown, *Augustine of Hippo: A Biography* (1967, with chronological tables, listing works, editions, and translations) and *Religion and Society in the Age of Saint Augustine* (1972); R.A. Markus, *Saeculum: History and Society in the Theology of St. Augustine* (1970); E. TeSelle, *Augustine the Theologian* (1970). See also B. Altaner, *Patrology* (1960), pp. 487-534; and *Revue des Études Augustiniennes* for current literature.

D.F. WRIGHT

AUGUSTINIAN CANONS. Also known as "Black," "Regular," and "Austin" Canons. Though Augustine of Hippo did try to establish houses wherein there was a complete monastic rule, the origins of these canons who bear his name were connected with the reform movement of Pope Gregory VII's time (c.1021-85). The Lateran Synods of 1059 and 1063, taking note of the revival of the common life in such areas as N Italy and S France, discussed and recommended monastic poverty without making it compulsory—i.e., monks were still allowed to own some property. With the discovery and implementation of the Rule of Augustine by the mid-eleventh century, the title "regular canons" came to be virtually synonymous with Augustinian canons as the monks adopted the Augustinian Rule. The canons did not, however, belong to a single order but were organized into various houses, which in turn subdivided into congregations. Well-known examples of the latter were at Prémontré, St. Ruf, and Windesheim. Thomas à Kempis* and Gerhard Groote belonged to Windesheim. Erasmus* was an Augustinian canon. From the fifth century many of the houses disappeared, but the Canons Regular of the Lateran and the Premonstratensian Canons still survive, as do several convents of canonesses. Two famous London teaching hospitals, St. Bartholomew's and St. Thomas's, owe their origins to the Augustinian canons.

See J.C. Dickinson, *The Origins of the Austin Canons and Their Introduction into England* (1950).

PETER TOON

AUGUSTINIAN HERMITS (Friars). A mendicant order formed from several Italian congregations of hermits by Pope Alexander IV in 1256. While the constitution was based on the Dominicans, the rule was that of Augustine of Hippo. Increasing rapidly, they lost their eremitical character and became an important mendicant order. The title "Hermit" was preserved in order to distinguish them from the canons. The order comprised clerical and lay members (who later included women). Its head is the prior general, who is assisted by a council. Each province has a provincial, and each monastery a prior. They wear a black habit, long pointed cowl, and black leather cincture. Two important theological schools—those connected with Giles Colonna (d.1316) and Cardinal Noris (d.1704)—came from their ranks. Martin Luther was a member of the reformed German congregation. Though they experienced many setbacks in the sixteenth century, they remain today. There are two discalced (barefoot) congregations whose origins go back to the Counter-Reformation of the late sixteenth century.

PETER TOON

AULD LICHTS. A name given to the group within the Original Secession Church* in Scotland which held that the civil magistrate had a duty to impose the true faith on the people. Those who opposed them, who claimed to have seen a "New Light," rejected such compulsion.

AULÉN, GUSTAV (1879-1978). Swedish theologian. Born in S Sweden, he became doctor of theology at Uppsala in 1915 and was professor of systematic theology at Lund (1913-33) and bishop of Strängnäs (1933-52), thereafter living in retirement at Lund. A great influence on Swedish theology, Aulén's books included *Den kristna forsoningstanken* (abridged ET *Christus Victor,* 1931), in which he expounded the "dramatic" or "classic" theory of the Atonement.* During

World War II he worked against Nazism, writing *Kyrkan och Nationalsocialismen.* A determined ecumenical churchman, he was vice-president of the Edinburgh Faith and Order Conference (1937), and later books reflect this interest, e.g., *Reformation and Catholicity* (1959).

J.G.G. NORMAN

AURELIUS. Catholic bishop of Carthage, and thus primate of Africa, 391/2-c.430. A deacon when Augustine, returning from Italy, met him in 388, and thereafter closely associated with Augustine, especially in measures against Donatism and the Pelagians, he presided over a series of regular African councils, mostly at Carthage but first in Hippo in 393. His contributions to their debates displayed neither erudition nor eloquence. Though a faithful if unsophisticated pillar of the church, he has been eclipsed by Augustine, who regarded him highly and for whose ideas and initiatives Aurelius's authority and organizing ability provided an executive arm.

AURELIUS, MARCUS, see MARCUS AURELIUS

AURICULAR CONFESSION. Private confession of sins before the priest alone, who is authorized to pronounce absolution on the basis of Matthew 16:19 and 18:18. Resulting from the abuse and dangers of public confessions in the early church, private confession became more frequent until the Fourth Lateran Council* (1215) made annual confession compulsory.

AUSTIN, see AUGUSTINE

AUSTRALIA. The white population of Australia, from the first settlement in 1788 to the end of World War II in 1945, was almost entirely of British origin. The Church of England, which is in process of changing its name to the Anglican Church of Australia, was the predominant religious force throughout this period. It was, until the 1850s, a distant branch of the "established" church in England, but from the 1860s adopted a synodical form of government. The first synod of the diocese of Sydney was held in 1866. It continued, however, to import its bishops from England; the first Australian-born archbishop, Marcus Lawrence Loane, was elected to Sydney in 1966.

The origins of Christian ministry in the colony of New South Wales go back to the First Fleet, mainly of convicts, whose chaplain was Richard Johnson.* Samuel Marsden, the second chaplain, became involved in the police administration of the colony as a magistrate and in the establishment of Parramatta, west of Sydney. History does not substantiate his reputation as "the hanging parson."

Until the 1830s all education in the colony, of both convicts and free settlers, was in the hands of the Church of England, which was paid in land grants known as "glebes." By the 1970s, such lands as the church retained were valued in millions of dollars, providing an endowment inconceivable when the grants were made. In 1836 W.G. Broughton* became first bishop of the diocese of Australia, thus severing a connection with the diocese of Calcutta.

In the early days of the colony, ministrations by Roman Catholic priests to the convict settlers were forbidden, though there were many Irish political prisoners. This situation was changed by law in 1820, but until 1844 in some places all prisoners were still forced to attend Anglican services. J.B. Polding, OSB, was appointed bishop of Hiero-Caesarea (titular) with jurisdiction over Australia in 1834, and an independent see was set up in 1841 and soon subdivided. Many Irish Roman Catholics entered Australia in the gold rush of the 1850s.

Immigration after World War II brought great changes to the Australian situation. Many immigrants were of S European origin, and with their higher birth rate this vastly increased the proportion of Roman Catholics. By the 1972 census the proportion of Anglicans to Roman Catholics had leveled. Whereas in 1851 fifty-two percent were Anglicans, twenty-six percent Catholics, five percent Methodist, and ten percent Presbyterian, in 1901 the figure was thirty-nine for Anglicans, twenty-three for Catholics; in 1966 it was just over thirty-three for Anglicans, twenty-six for Catholics, as a result of immigration.

A more realistic assessment than census figures, however, needs to be applied to appreciate the relative strength of the churches in Australia. Only one definitive sociological survey has been taken of Australian attitudes toward religion, and it revealed that of those who said at the 1966 census that they belonged to a particular church, actual attendance figures were: twenty-one percent of Anglicans went to church usually (more than once a month), sixty-nine of Catholics, forty-one of Methodists, thirty-four of Presbyterians. While this may lead to the conclusion that in fact Australia has become a Catholic country, the figures need to be amended to allow for Sunday school attendance.

Today the Church of England has organized itself into twenty-seven dioceses, of which the largest are the metropolitan dioceses of Sydney, Melbourne, Brisbane, and Perth. A national general synod brings together representatives of the dioceses (which are autonomous). The church is self-supporting, receiving government grants only for independent schools and for some kinds of missionary work. The church maintains missions in the north of Australia and scattered inland towns among aborigines, but has come under strong criticism for allegedly neglecting aboriginal customs and civil rights. In fact, anthropologists suggest that the full-blooded aborigines would have died out in the 1930s if the missions had not sustained them with food and medicine at a time when the majority of Australians were expressing very little concern. Control of the mission towns has now passed to the federal government, and the church maintains a pastoral ministry and some medical work. Australia as a nation now feels much greater responsibilty for its original inhabitants, and in 1967 a referendum by a huge majority granted full citizenship rights and equality to the aborigines.

The church supports also missionary dioceses in the South Pacific and in Papua New Guinea. Similar support is extended through the other denominations, Protestant and Catholic, throughout the Pacific and Southeast Asian region. The Australian churches are now feeling more a part of Asia and the Third World than of Europe and America.

One of the unusual problems of church extension in Australia is the extraordinary distance and isolation of many people in the outback. The Bush Brotherhoods, founded in 1903, serve the outback by providing low-cost and sacrificial ministry and evangelism in those areas. The Bush Church Aid Society provides clergy and medical staff for the new mining towns of Western Australia and Queensland as well as some outback towns.

Theological training is now indigenous, and relatively few Australians seek overseas degrees. All denominations conduct their own training courses, and most maintain their own colleges in each state. Theological scholarship tends to be conservative, partly because of the distance from overseas radical thought, partly because of the conservative nature of the Evangelical tradition within the Anglican Church and the Irish conservatism of the Roman Catholic hierarchy.

BIBLIOGRAPHY: R. Hamilton, *A Jubilee History of the Presbyterian Church of Victoria* (1888); P.F. Moran, *History of the Catholic Church in Australasia* (1894); E. Symonds, *The Story of the Australian Church* (1898); J. Colwell, *The Illustrated History of Methodism* (1904); J. Cameron, *Centenary History of the Presbyterian Church in New South Wales* (1905); A.E. David, *Australia* (1908); H.N. Birt, *Benedictine Pioneers in Australia* (2 vols., 1911): F.W. Cox, *Three Quarters of a Century*—South Australia Congregationalism (1912); E.M. O'Brien, *The Dawn of Catholicism in Australia* (2 vols., 1928); R.A. Giles, *The Constitutional History of the Australian Church* (1929); H.E. Hughes, *Our First Hundred Years*—Baptists in S Australia (1937); F.J. Wilkin, *Baptists in Victoria* (1939); J.G. Murtagh, *Australia: The Catholic Chapter* (1946); J.C. Robinson, *The Free Presbyterian Church of Australia* (1947); R.S.C. Dingle (ed.), *Annals of Achievement*—Queensland Methodism (1947); R. Bardon, *The Centenary History of the Presbyterian Church of Queensland* (1949); M. Loane, *History of Moore College* (1955); H. Mol, *Religion in Australia* (1971); P. Hollingworth, *The Powerless Poor* (1972).

ALAN NICHOLS

AUTHORITY. In biblical teaching the source of all authority is God Himself (Rom. 13:1; cf. Dan. 4:34; John 19:11). We must distinguish between authority and power, and between religious or ecclesiastical authority and civil authority and power. Christianity claims to be based upon divine revelation, to which reason and conscience must be subject. This does not jettison reason in apprehending the revelation or discovering truth. Reason itself, however, is not autonomous, for one cannot begin thinking—even to examining his own perceptions and thoughts—without making the act of faith that the things he is thinking about make sense. The distinction between natural and special revelation is not absolute. The concept of the revelation of God as Creator and the revelation of God as Redeemer is more comprehensive because all truth is from God and all truth must be grasped by men who have the gift of reason from God. Christians believe that men cannot discover truths behind God's back or without God's assistance and that there is no use in God's giving revelations to creatures incapable of receiving them. In contrast to claims of totally subjective revelatory authority, the Christian claim to historical revelation involves historical events and narratives as the actual form the eternal realities take.

The biblical revelation comprises the utterances of prophets and apostles and the record of the life and teaching of our Lord, which have authority because they are inspired by the Spirit of God (2 Tim. 3:16). For Christians, the biblical writings transcend all other claims to religious authority. Some claims to the authority of church tradition and the episcopacy (including the papacy) have been made, especially in the Orthodox and Roman Catholic traditions, but these have been played down recently in favor of discovering biblical and early church roots of authority for faith.

At the Reformation, the Bible as the Word of God interpreted to faith through the inner witness of the Spirit was reestablished as the norm of faith and practice in Protestant and evangelical churches. The magisterial Word of God was moved to center as the judge of the faith and life of the church, not the church as the judge of Scripture. The canonical Scriptures without supplement from church tradition were seen to be self-interpreting and complete.

The Reformed and Lutheran theologies of the Word were complemented by the Anabaptist personal religion of the Spirit, in which English and American evangelicalism and independency have their roots. Their view that the church is essentially nondynastic, nonterritorial, and a spiritual democracy of believing people has profoundly influenced Western Christianity including rejection of the enforcement of church sanctions by civil powers.

The evangelical principle entails Word and Spirit in which the authoritative Word of God is the chief agency of the Holy Spirit and the chief function of the church. It is the Holy Spirit who makes the Word to be revelation, and it is the Word that makes revelation historic and concrete. Theology is not the mold but the image of the church's spiritual life. Political democracy recognizes no authority but what it creates, but the church as a spiritual democracy recognizes no authoritative principle but that which creates it as Christ's body, namely the Word, the Gospel, and the Spirit under Jesus Christ's lordship.

BIBLIOGRAPHY: P.T. Forsyth, *Faith, Freedom and the Future* (1912); *idem, The Principle of Authority* (1952); J. Oman, *Vision and Authority* (1928); H.E.W. Turner, *The Pattern of Christian Truth* (1954); L. Hodgson, *For Faith and Freedom* (1956).

SAMUEL J. MIKOLASKI

AUTHORIZED VERSION, see BIBLE (ENGLISH VERSIONS)

AUTOCEPHALOUS (Gr. = "himself the head"). In current usage this term has two meanings. First, it describes any national church which is a part of the Eastern Orthodox Church and in communion with Constantinople but which is governed by its own national synod. Second, it describes an independent monastery as, for example, that on Mt. Sinai. In earlier usage it also had two meanings. First, it described bishops in the early church who were independent of the jurisdiction of either a patriarch or a metropolitan. Such were the bishops of Cyprus. Second, it also described certain Eastern bishops who were directly responsible to the patriarch without any reference to the metropolitan. PETER TOON

AUTO-DA-FÉ (Port. = "act of faith"). A ceremony of the Spanish Inquisition at which, after a procession, Mass, and sermon, sentences were read and the execution process inaugurated. Heretics were dressed in the ceremonial *san-benito,* a yellow penitential garment with a red cross front and back, grotesquely embroidered for the unrepentant, and they wore a yellow miter. Those sentenced to death were handed over to the secular authority for execution within five days, usually by burning. Most of the great *auto-da-fés* occurred when Tomás de Torquemada* was head of the Inquisition. The last in Spain was at Seville in 1781, though one was celebrated in Mexico as late as 1815. J.G.G. NORMAN

AUTOSACRAMENTAL. A peculiarly Spanish one-act play produced mostly between the sixteenth and eighteenth centuries on the Feast of Corpus Christi* to elucidate the meaning of the Eucharist. It evolved from clergy bearing the host in a street procession, through floats depicting biblical scenes and choreography and drama, to written sacramental plays. From this emerged the secular and religious theater of the Spanish "golden age." Competitions were held, with poets, actors, and prizes, leading to intercity rivalries and extravagance while the citizenry was entertained. As the cities' contribution to the feast, these plays tried to help the unknown be discovered through the known. Lope de Vega and José de Valdivielso brought this genre to maturity, while Calderón (1600-1681) raised the play to a near sacrament itself: "sermons set in verse, problems of Sacred Theology set in representable ideas. . . ." Rationalism brought the demise of autosacramentals; they were banned under Charles III (1765), but abroad have survived in some places. C.G. THORNE, JR.

AUXENTIUS (d.374). Bishop of Milan. A Cappadocian ordained by Gregory of Alexandria about 343, he succeeded the exiled Dionysius in the see of Milan in 355. Supporting Valens at the Council of Rimini (359), he subscribed to the "Dated Creed." Homoousion* censures on him there and at Paris (360) did not affect him. Toleration of religion under Julian, Jovian, and Valentinian I encouraged Hilary (of Poitiers)* to oppose Arian bishops in Gaul; he overthrew the bishops of Arles and Perigueux, and aided by Eusebius of Vercilli, he sought to dislodge Auxentius while Valentinian was residing at Milan (364-65). Auxentius secured a favorable verdict both by counseling popular quiet in the face of agitation by Hilary and by glossing his affirmation of "one divinity and substance" with a subtle formula capable of either an Arian or a Nicene interpretation. Influenced by Athanasius, Pope Damasus in 372 condemned the "Dated Creed" and its supporters, but Valentinian would not depose Auxentius. G.T.D. ANGEL

AUXILIARY BISHOP. Functionally the bishop was *archiereus* or *archipresbuteros* of the incipient Christian congregation, and the remaining priest-presbyters were his assistants. The multiplication of congregations, however, necessitated new measures, and the special meaning of coadjutor* and suffragan* reflect these. Again in modern times with the growth of urban areas under the bishop's jurisdiction, special assistance is often required; those elevated to meet these needs, but without right of succession, can be designated "auxiliary." While in the United States this is Roman Catholic usage, the Protestant Episcopal Church employs "suffragan" in this sense.
CLYDE CURRY SMITH

AUXILIARY SAINTS. A group of fourteen saints traditionally venerated together, especially in Germany. They comprise three bishops (Denis of Paris, Erasmus or Elmo, Blaise), three virgins (Barbara, Margaret, Catherine of Alexandria), three knightly patrons (George, Achatius, Eustace), the physician Pantaleon, the deacon Cyriacus, the martyr Vitus, the monk Giles, the travelers' patron Christopher. There are local variations, e.g., Leonard of Noblat, Dorothy, Nicholas ("Santa Claus"), Pancras, the English king Oswald. The cult, first advanced by the Dominicans, reached its climax in mid-fifteenth century.

AVERROISM. The doctrine that man's soul is mortal or, more specifically, that the souls of all men are part of a single soul-substance out of which individuals arise at birth and into which they return at death. The name comes from (Ibn Rushd) Averroes (1126-98), a learned jurist of Cordova who became friend and physician to the ruling caliph when Islam was ascendant. For three centuries Islamic Ash'arite philosophers had denied causality in nature on the ground that it implies the presence of principles other than God in the universe, so making God less than supreme. Averroes saw clearly that the denial of subsidiary causes endangers all knowledge and even reason itself. In support of this view he appealed to Aristotle, who had accepted causality wholeheartedly.

Averroes wrote commentaries on Aristotle* and came to be known to posterity as "The Commentator." To reconcile his views with Islam he interpreted the Koran allegorically and perhaps not overseriously; in late life he was accused of heresy. Apart from his views on immortality and the soul, Averroes identified God with Aristotle's remote and impersonal Prime Mover, denied free will, and taught that both the world and mankind are eternal. In 1253 Aristotle was prescribed for study by the University of Paris, and the Commentaries of Averroes were accepted as the standard text. There was consternation in Christendom when many students accepted their anti-Christian teachings, and the study of Aristotle was prohibited by Urban IV in 1263. Albert the Great in 1256 and Thomas Aquinas in 1257 and again in 1270 wrote works directed against the Averroist heresies, but despite prohibitions the latter survived in places, notably in Padua to the time of the Renaissance.

See M. Fakhry, *Islamic Occasionalism* (1958).

R.E.D. CLARK

AVICEBRON (Ibn Gabirol) (c.1021-c.1058). Spanish Jewish philosopher and poet. His biographical details are uncertain, but he was a threshold figure of the brilliant demi-millennium of Jewish learning when it shifted to western Europe. In ethics he was an original; virtues and vices he held to be linked in pairs to five senses, with other physiological details. He probably compiled *Choice of Pearls,* sixty-four chapters of aphorisms, some originally Arabic. His poems, used in the synagogue, were pure biblical Hebrew and theology, with multiple rhymings. In metaphysics his Arabic dialogue treatise, *Fons Vitae,* methodologically ignores Bible and Talmud and so is seriously underestimated by Judaism. A monist thinker, his views (Lat. version, 1150) nevertheless profoundly affected scholastic Christendom; the Franciscans considered him to be Muslim or Christian. S. Munk demonstrated (1846) his identity with Ibn Gabirol.

ROY A. STEWART

AVICENNA (Abu Sina) (980-1037). Persian Muslim philosopher. Born in Bokhara, he was a voluminous commentator on Aristotle. A versatile man, his *Canon* was the standard medical textbook for five centuries in Europe and beyond. Like other Arabic Aristotelians, he helped restore to Judaism and Christendom that Greek learning which had been lost in the Dark Ages. He gave Aristotle a Neoplatonic twist, with his nine descending mediating Intelligences between deity and man, the ninth occupying the lunar sphere. He postulated an eternal universe, with an inbuilt necessity system and a remote deterministic God debarred from direct creative action. Science, metaphysics, and theology are deeply indebted to Avicenna; Dante mentions him respectfully (*Inferno* iv, 143). He was an eclectic thinker in whose work experts detect unresolved internal inconsistencies. He has been charged with originating a disreputable compromise doctrine of twofold truth to harmonize reason with revelation, which introduces a double standard before which all intellectual integrity would finally crumble.

ROY A. STEWART

AVIGNON. In 1309 Clement V moved to Avignon in SE France and so began the "Babylonian Captivity"* of the papacy. Considerations of security led the popes to prefer Avignon to Rome until 1377. Avignon was in a local political pressure area removed from the tensions of Italian politics and was geographically much more convenient than Rome. Its position made it suitable as a center of judicial affairs, and it became the base of a great bureaucratic organization efficient at raising funds for papal purposes. When the papal court returned to Rome, Avignon continued to be the seat, until 1408, of two antipopes, Clement VII and Benedict XIII. By the end of the eleventh century the nearby Abbey of St. Ruf had become famous as a pioneering community living under the rule of Augustine.

JAMES TAYLOR

AVITUS (d. c.519). Bishop of Vienne. Born Alcumus Edicius Avitus in Auvergne of a Roman senatorial family, he succeeded to the see of Vienne in Gaul about 490, on the death of his father Isychius. He became very influential in the ecclesiastical life of Burgundy and won King Sigismund from Arianism to orthodoxy. He was a strong contender against Arianism,* ardently defended the primacy of Rome, and strongly advocated closer ecclesiastical union between Gaul and Rome. His reputation for learning much impressed his generation, including the then pagan King Clovis. Among his surviving writings are five poems inspired by the Book of Genesis *(De Mosaicae Historiae Gestis),* homilies, a poem in praise of virginity *(De Virginitate),* and about a hundred epistles.

J.G.G. NORMAN

AVVAKUM (1620-1682). Leader and martyr of the Raskolniki or Old Believers.* Son of a Russian village priest, he was himself ordained and in his early pastoral activity and his family life showed himself an eager exponent of ascetic piety. He became an intimate of Czar Alexis and at length archpriest of Our Lady of Kazan in Moscow. He opposed the liturgical reforms of Patriarch Nikon* and was deported to Siberia in 1653. After the fall of Nikon he returned in 1664, but was soon exiled again when the czar elected to continue the reforms. The council of 1666 excommunicated Avvakum and established the new liturgical practices. In 1670 his companions were punished by mutilation, and Avvakum condemned to imprisonment underground. He continued to direct the dissidents, and finally Alexis's successor Theodore condemned him and his companions to death at the stake. He is regarded as the greatest saint and martyr of the Old Believers. His autobiography, a masterpiece of Russian literature, is still highly regarded.

See P. Pascal, *Avvakum et les débuts du Rasko* (1938).

J.N. BIRDSALL

AWAKENING, THE GREAT, see GREAT AWAKENING, THE

AYLMER, JOHN (1521-1594). Bishop of London. Educated at Cambridge through Henry Grey, duke of Suffolk, to whose daughter (Lady Jane Grey) he then became chaplain and tutor, he was appointed archdeacon of Stow in 1553, but was forced to flee to the Continent for his opposition to transubstantiation. There he assisted Foxe* in translating his *Acts and Monuments* into

Latin. On his return to England he was chosen as one of the eight disputants against the Roman Catholics. He became archdeacon of Lincoln in 1562 and bishop of London in 1576. His arbitrary and unconciliatory disposition became apparent, and he was bitterly attacked in the Marprelate Tracts* for his exceptional severity in fining and imprisoning all who disagreed doctrinally, whether Puritan or Catholic. Similar in temperament to W. Laud,* he is to be commended for his learning and his discerning patronage of scholars. His only notable work was a reply to John Knox's *Monstrous Regiment of Women.* G.S.R. COX

AYLWARD, GLADYS (1902-1970). Missionary to China. London parlormaid turned down on educational grounds by missionary societies, she saved most of her very modest wages for several years till she could pay her fare to China, leaving England in 1932. After an incredible journey through Siberia and complications raised by the Russo-Chinese war, she finally reached her destination through Japan and joined Jeannie Lawson in remote Yangcheng. They opened an inn where they told Bible stories. After her colleague's death Miss Aylward continued and extended her work, finding an unexpected friend in the local mandarin. When the Japanese invaded in 1940, she led 100 children to safety on an epic journey, the dramatic appeal of which was realized and exploited by Hollywood moviemakers. After a serious illness she returned to England in 1947. She stayed for eight years, and finally opened an orphanage in Formosa, where she worked till her death. J.G.G. NORMAN

AZARIAH, VEDNAYAKAM SAMUEL (1874-1945). First Indian bishop in the Anglican Church. He was born at Vellalanvillai in Tinnevelly district of Madras state; his father, a convert from Hinduism, was an Anglican minister. After education at Madras—he could not take his degree because of illness—Azariah became in 1895 a YMCA secretary, working among students. His concern for evangelism showed in his helping to form the indigenous Indian Missionary Society of Tinnevelly (1903) and the interdenominational National Missionary Society (1905). He was first secretary of the Tinnevelly society, which sent a missionary to a hitherto neglected district called Dornakal in the Telugu-language part of the Nizam of Hyderabad's territory. In 1909 he himself was ordained and went to Dornakal, having resigned from the YMCA. The Dornakal church expanded and in 1912 Azariah was consecrated missionary bishop. Later Dornakal was separated as a diocese from Madras. Azariah welcomed the mass movements which augmented the Telugu church, including those from depressed classes.

He was a leader at the meeting of Indian ministers in Tranquebar in 1919 which marked the beginning of the church union movement in India. To him, "unhappy divisions" were in fact a "sin and a scandal" in the Indian setting. He did not live to see the two major church unions which resulted from the decades of negotiation. As chairman of the National Christian Council (1929-45) and host at the World Missionary Conference at Tambaram (Madras) in 1938, Azariah had a special place in Indian Christian leadership during the period when the "younger churches" assumed importance and replaced foreign missions. He died at Dornakal.

See J.Z. Hodge, *Bishop Azariah of Dornakal* (1946). ROBERT J. MC MAHON

AZYMITES (Lat. *infermentarii,* "unleavenders"). An Eastern (non-Armenian) Church term of reproach against the Western Church for the eucharistic use of unleavened bread. It furnished a pretext for the 1054 schism of East from West under Michael Cerularius.* The Council of Florence (1437) authorized (transubstantiationally) wheaten bread, leavened or unleavened. Modern Anglicans use either indifferently; other Protestants prefer leavened.

B

BABYLAS (d.250). Bishop of Antioch. Noted for his courage, he would not admit Philip the Arabian* to public worship without penance. He was imprisoned and martyred in 250 during the Decian persecution; his martyr cult grew at Antioch. Caesar Gallus removed his corpse to a chapel near the Temple of Apollo in the suburb of Daphne. In 362 the Emperor Julian, frustrated by abortive attempts to acquire oracles from a spring at the temple, ordered the coffin of Babylas to be removed. Local Christians, antagonized by the order, buried the body again in a Christian cemetery six miles away. The temple was destroyed by fire within the year.

BABYLONIAN CAPTIVITY. A pejorative term originating in the inferences of Italian patriots such as Dante and Petrarch that the popes at Avignon* were the captives of the French kings (cf. 2 Kings 24:14-16; 25:11). The popes in fact remained there because Italy was unsafe. The papacy only legally owned Avignon from 1348. Benedict XII (1334-42) and Clement VI (1342-52) built the papal palace to house their expanding administrations—to which development John XXII's pontificate (1316-34) had greatly contributed. Clement VI was responsible for the biggest and finest rooms. The Avignon period was not sterile; apart from administrative developments it removed the popes from incessantly distracting Italian strife. Geographically, moreover, Avignon was more central. The popes never relinquished the desire to return to Italy, devoting large budgets, diplomacy, and force to this end. Though individual popes returned temporarily, it was Cardinal Albornoz's pacification of the Papal States* which decided for Gregory XI that Rome was safe enough for permanent residence in 1377.

L. FEEHAN

BACH, JOHANN SEBASTIAN (1685-1750). German composer. Born in Eisenach, where he went to a school once attended by Luther, this greatest member of Germany's greatest musical family was grounded in the strict Lutheran orthodoxy to which he clung throughout his life. Orphaned at ten, he was taken to live with his brother, the organist at Ohrdruf. Here he was subjected to the Comenian principles of education and probably to Pietistic influences. At fifteen he fended for himself as choir boy and violinist and soon as church organist. At twenty-three he was court organist to the pious duke of Weimar. Here he met and absorbed the Italian concerto and operatic style, which he fused with his N German heritage of churchly choral and organ music.

For six years as *Kapellmeister* of the princely court at Köthen, his concern was secular chamber music that differed in function but not in essence from that of the church. Many a movement from works written here reappeared in later masterpieces refurbished with sacred words. From 1723 till his death he was *cantor* of the historic *Thomasschule* in Leipzig.

Bach was known to his own generation for his transcendent skill at the organ. The worth of his compositions, especially his choral ones, filled with intricate contrapuntal craftsmanship and baroque musical rhetoric, went unrecognized in an age of changing musical taste. They were, however, cherished and perpetuated by a small circle of pupils and connoisseurs. *The Welltempered Clavier* in particular profoundly influenced the great classical masters, Haydn,* Mozart,* and Beethoven.*

Bach was unquestionably the greatest composer of all time for the organ. His toccatas, preludes, and fugues, and over 100 pieces based on Lutheran chorales were conceived for various functions in the church. He composed five cycles of cantatas for the Sundays and feasts of the ecclesiastical year—about 300 in all, of which almost 200 survive. These were functional service music, related to the Gospel of the day. The *Christmas Oratorio* is a series of six such cantatas. Today they are largely relegated to the concert hall. The *Mass in B minor* and the *St. Matthew Passion* are his most monumental works. The revival of the latter in 1829 by Mendelssohn marked the beginning of a deep and continuing appreciation of Bach's significance as a sacred composer. Every major composer of sacred works since has been in varying degree his debtor.

See H.T. David and A. Mendel, *The Bach Reader* (1945); and K. Geiringer, *Johann Sebastian Bach, the Culmination of an Era* (1966).

J.B. MAC MILLAN

BACKUS, ISAAC (1724-1806). Baptist minister, historian, and champion of religious liberty. Born in Connecticut, he came under conviction of sin during the Great Awakening* in 1741, and finally experienced evangelical conversion. He joined a "New Light" or Separatist Congregationalist church, but remained rather passive for several years. Aroused by the preaching of G. Whitefield,* he felt a definite call to become a preacher in 1746, and promptly started out on the first of many preaching tours. He was ordained in 1748, and in 1751 adopted Baptist principles and was immersed with his wife before his Middleborough congregation, of which he was pastor until his death. He contributed much to the growth of the Baptist movement in New England and was an organizer of the Warren Association of Baptists. He became the most persistent and effective advocate for the cause of religious freedom and separation of church and state, and traveled widely and wrote extensively to improve the status of Baptists. His three-volume work, *A History of New England with Particular Reference to the ... Baptists (1777-96),* contains valuable source material for historians. HARRY SKILTON

BACON, FRANCIS (1561-1626). English statesman and philosopher of science. A versatile genius, he was distinguished in law, literature, philosophy, and science. His father, Sir Nicholas Bacon, died when Francis was eighteen, leaving him virtually penniless since he was the youngest son. He then turned to a career in law and at the age of twenty-three gained a seat in the House of Commons. After holding a succession of political appointments he became lord chancellor under James I. His rise to power alienated men such as Sir Edward Coke, and this led to Bacon's indictment for accepting a bribe. He was found guilty and removed from his offices (1621). Bacon spent his remaining years writing books and devising schemes for the advancement of science. Among his works are *Advancement of Learning, Essays, Novum Organum,* and *New Atlantis.* Bacon emphasized the empirical approach to science and because of this he has been called "the Prophet of Modern Science." The fact that these books appeared during the course of his busy life demonstrates that his interest in science was never entirely separate from his activities as a lawyer and statesman. He believed that science was necessary to improve the lot of mankind and that the state should finance this work. He was never able, however, to interest James I in this goal.

See F.H. Anderson, *The Philosophy of Francis Bacon* (1948), and I. Levine, *Francis Bacon, Viscount of St. Albans* (1925).

ROBERT G. CLOUSE

BACON, LEONARD (1802-1881). Congregational pastor, educator, and editor. Educated at Yale and Andover Theological Seminary, he was ordained as an evangelist to the western frontier in 1824. He accepted a call to the First Church of New Haven the following year, and served there with distinction (1825-66). Although sympathetic to the New Haven School* of theology, he tried to serve as peacemaker in the major theological debates within Congregationalism. He was active in the slavery issue, including among his writings editorials in the *Independent* that often led to bitter debates within his own congregation. He also wrote a number of hymns. In semi-retirement from 1866 he lectured in Yale Divinity School on revealed theology, church polity, and American church history. DONALD M. LAKE

BACON, ROGER (c.1214-1292). English Franciscan philosopher and scientist. He studied at Oxford and taught at Paris, where he was among the first to lecture on the books of Aristotle. Returning to Oxford, he became familiar with the work of Robert Grosseteste.* Bacon joined the Franciscans about 1257 so that he might secure the experimental equipment he needed. He also became involved with the followers of Joachim of Fiore* and consequently was sent to Paris and forbidden to circulate his writings. Pope Clement IV, learning of Bacon's attempt to construct a universal science, sent for his encyclopedia (1266). Bacon did not have the work ready, but in the short space of eighteen months he composed a preliminary draft, his *Opus Maius*. The pope died before anything could come of this contact, but Bacon was allowed to return to Oxford where he continued his scholarship until his death.

Many have believed that his fame rested on his emphasizing the need for experimental science. This was probably overstated because, in addition to observation, he believed that a study of the Bible in the original languages would help one to understand nature better. Bacon surpassed his contemporaries in his knowledge of lenses and mirrors. He even foresaw the practical application of science in various ways.

BIBLIOGRAPHY: A.C. Little (ed.), *Roger Bacon: Essays ... on the Occasion of ... the Seventh Centenary of his Birth* (1914); L. Thorndike, *A History of Magic and Experimental Science,* vol. 2 (1929), pp. 616-91; S.C. Easton, *Roger Bacon and His Search for a Universal Science* (1952).

ROBERT G. CLOUSE

BADBY, JOHN (d.1410). Lollard* martyr. A tailor of Evesham, Badby denied transubstantiation, asserting that the consecrated host remains material bread (although a sacramental sign of the living God). He declared that one John Rackier of Bristol had as much power as any priest to make God. Declared an incorrigible heretic by the bishop of Worcester, he spent more than a year in prison. Tried thereafter before Archbishop Arundel and found to be obdurate, he was committed to the secular arm. Henry, Prince of Wales, was present at his execution in Smithfield. Hearing cries which he took to be pleas for mercy, he ordered the fire to be quenched and offered Badby pardon if he would recant. On his refusal he was burnt to death. HAROLD H. ROWDON

BADER, JESSE MOREN (1886-1963). Disciples of Christ minister and ecumenist. Born in Illinois, he graduated from Drake University, and after various pastorates and YMCA service he was an executive in the evangelism department first of his own denomination (1920-31), then of the Federal Council of Churches (1932-50). He was very active in the ecumenical movement and the World Council of Churches,* attending the founding sessions at Amsterdam in 1948, the Faith and Order conferences at Oxford and Edinburgh in 1937, and the Third Assembly at New Delhi in 1961. He served also as an observer for the Disciples at Vatican II.* He was author of *Evangelism in a Changing America* (1957) and numerous articles in Christian periodicals.

DONALD M. LAKE

BAEDEKER, FREDERICK WILLIAM (1823-1906). Evangelist. Son of a Westphalian ornithologist and cousin of the traveler to whose guidebooks he contributed, he settled in 1862 as a schoolmaster in Weston-super-Mare. Together with his English wife, he was converted in 1866 through the ministry of Lord Radstock, who soon encouraged him to evangelize on the Continent. In 1877 he settled for three years with his family in St. Petersburg where in association with Count Bobrinsky, Princess Lieven, and others he played a prominent part in the current "drawing-room" awakening. His evangelistic journeys continued thereafter, at first in the universities and prisons of Scandinavia where he worked with the Finnish Baroness von Wrede, and later in Russia from the Caucasus to Siberia. In 1889, in spite of Pobiedonostzeff's official disfavor, he obtained a unique permit (renewed every two years until his death) enabling him to preach and distribute Bibles in any Russian prison. His associations were mainly with Stundists on the Continent and the Brethren in England.

See R.S. Latimer, *Dr Baedeker and his Apostolic Work in Russia* (1907).

TIMOTHY C.F. STUNT

BAHA'I. A religious movement founded in Iran by Bahā'u'llāh. Its origins are to be found in Shi'a Islam, and in the group of disciples of Sayyid Alī Muhammed, or "the Bāb" (1819-50), who is now regarded by Bahā'īs as the forerunner of the Prophet. Bahā'u'llāh (1817-92) was a disciple of the Bāb, though he had never met him. After the latter's death, the former had a mystical experience in prison (1852), which he interpreted as a divine call. He was exiled to Baghdad, but gathered followers and eventually proclaimed himself as God's manifestation to the present age, not only to his disciples, but to many world rulers. Until 1957 leadership of the movement remained within the Prophet's family, though in the meantime his teachings had spread to many parts of the world.

The main teaching of the Bahā'ī faith is that it is the crown and culmination of all the religions of the world, which it does not seek to overthrow, but to fulfill. Bahā'ī temples contain no accepted religious symbols; they are domed, circular, and have nine doors, representing the existing traditions (there are only four such temples, in Evanston, Illinois; Frankfurt, Germany; Sydney, Australia; and Kampala, Uganda). The movement's principles may be summarized thus: (1) the unity of mankind; (2) the duty of each individual to seek for truth independently; (3) equality between the sexes; (4) the essential unity of all religions; (5) science and religion are not contradictory; (6) religion must be the source of love and unity; (7) all prejudice should be abolished; (8) universal education; (9) the solving of economic problems; (10) the encouragement of international language; (11) peace; and (12) the setting up of an international court of justice.

There are followers of Bahā'ī in most parts of the world, though their numbers are relatively small.

See *Dictionary of Comparative Religion* (1970), s.v. Bābis, Bahā'īs.

ERIC J. SHARPE

BAILLIE, DONALD MacPHERSON (1887-1954). Scottish theologian and ecumenist. He served various Scottish parishes for sixteen years and was then professor of systematic theology in the University of St. Andrews for a further twenty. An outstanding teacher, he attracted to St. Andrews students from many parts of the world, especially from the USA. His published works are few, as teaching for him took precedence over writing, but his *God Was in Christ* (1948) went through many impressions and has been described as the most significant book of its time in Christology. *The Theology of the Sacraments* and two volumes of sermons were published posthumously. Joint editor of a volume on *Intercommunion* (1952), he was convener of the Church of Scotland committee on inter-church relations and was engaged in the Anglican-Presbyterian conversations at the time of this death.

HENRY R. SEFTON

BAILLIE, JOHN (1886-1960). Scottish theologian. He held various chairs in theology in the USA and Canada (1919-34) and was professor of divinity in the University of Edinburgh (1934-56). Convener during World War II of a commission appointed by the Church of Scotland "for the interpretation of God's will in the present crisis," and moderator of the general assembly (1943), he was a keen ecumenist like his brother, Donald MacPherson Baillie*; was a president of the World Council of Churches,* and a signatory of the "Bishops Report" which in 1957 suggested the introduction of bishops-in-presbytery into the Church of Scotland. Deeply concerned about the doubts which people felt about the Christian faith, he excelled as an apologist (cf. his *Invitation to Pilgrimage,* 1942). His *And the Life Everlasting* (1933) and *A Diary of Private Prayer* (1936) reached a wide public. His Gifford Lectures, *The Sense of the Presence of God,* were published posthumously.

HENRY R. SEFTON

BAILLIE, ROBERT (1599-1662). Scottish divine. Born and educated in Glasgow, he was episcopally ordained in 1622 and after a brief period of teaching was inducted to the Ayrshire parish of Kilwinning. He protested against the service-book which Archbishop Laud* was trying to impose on Scotland, and was a member of the famous Glasgow Assembly of 1638 which reestablished Presbyterianism in Scotland. Baillie became divinity professor in Glasgow University (1642) and a year later attended the Westminister Assembly* in London, where he was to spend much of the ensuing three years. In 1649 he was one of those sent to Holland to invite Charles II to sign the National Covenant,* and after the Restoration he became principal of Glasgow University. A learned, moderate, and modest man, Baillie produced *Letters and Journals* which are among the most important materials of the century. He died disillusioned after episcopacy was once more forced upon the country and he saw how his temperate policies had been mistaken.

J.D. DOUGLAS

BAINBRIDGE, CHRISTOPHER (1464?-1514). Archbishop of York. Born in Westmoreland, he was educated at Queen's College, Oxford, of which he became provost in 1495. Two years later he was treasurer of St. Paul's, then successively dean of York (1503), dean of Windsor (1505), bishop of Durham (1508), and archbishop of York (1509). He was sent to Rome as ambassador by Henry VIII* in the same year, and in 1511 rewarded for his services to Pope Julius II by a cardinal's hat and the command of a papal army against Ferrara. There are two conflicting accounts of the manner of his death—by stabbing or poisoning—but only one agent, his chaplain, allegedly at the instigation of the bishop of Worcester (permanent English ambassador in Rome) who was finally acquitted. G.S.R. COX

BAIUS, MICHEL (Michel de Bay) (1513-1589). Flemish Catholic theologian. Augustinian in outlook, forerunner of and an influence on Cornelius Jansen and Jansenism,* he had been educated at Louvain and taught philosophy (1544) and later theology there. In his thirties he turned to serious study of Augustine, and developed a radical Augustinian position which denied many of the doctrinal positions developed by medieval Scholasticism.* Attacking the idea of merit in good works, the immaculate conception, papal infallibility, the limited effect of the fall, any conditions on predestination, and similar views, his teachings soon aroused violent controversy. To his opponents, especially the Jesuits, they seemed close to Calvinism. By 1560 the theologians at the Sorbonne denounced his version of Augustinianism, and theological faculties at the Spanish universities followed suit. A trip by Baius and his colleague John Hessels to the Council of Trent* to defend their views proved abortive, and in 1567 Pius V issued a condemnation of several propositions Baius was alleged to have taught. Denying that he had in fact done so, he continued to present his Augustinianism as before. Controversy continued, and in 1579 Gregory XIII issued a further and more specific condemnation, to which Baius submitted, though still claiming he was misunderstood. His teachings were surely contrary to the dominant understanding of the post-Trent church. The controversy did not end with Baius's death, but broke out again and more importantly with the Jansenist movement a short time later.

BIBLIOGRAPHY: N.J. Abercrombie, *The Origins of Jansenism* (1936); H.J.D. Denzinger, *Enchiridon symbolorum* (tr. R.J. Deferrari as *The Sources of Catholic Dogma,* 1957); M.J.P. Van Doozen, *Michael Baius, zijn leer over de mens* (1958).

DIRK JELLEMA

BAKER, GEORGE, see DIVINE, MAJOR J.

BAKER, SIR HENRY WILLIAMS (1821-1877). English baronet and hymnwriter. For most of his ordained life (i.e., from 1851) vicar of Monkland near Leominster, he had distinctly High Church sympathies, and this is reflected not only in his hymns but also in his published tracts and devotional books. The original *Hymns Ancient and Modern* (1861) owed its existence to him, and to the volume he contributed twenty-five hymns, some original and some translated. His hymns have a freshness, simplicity, and rhythmic ease, the best known being "The King of love my Shepherd is" and "Lord, Thy word abideth."

BALDACHINO. The canopy over a throne, couch, pulpit, altar, or other sacred object. It is used to describe also a square umbrella, supported on four poles, borne over a priest who carries the Host. In vaulted stone, it often forms a protective and decorative cover over recumbent figures on medieval tombs, especially in England, originating in the twelfth century.

BALDWIN (d.1190). Archbishop of Canterbury from 1184. He was already archdeacon of Totnes when he decided to enter the Cistercian house at Ford in Dorset. He soon became abbot and then from 1180 was bishop of Worcester. His primacy was concerned with two major issues. First, his dispute with the monks of Christ Church, Canterbury, concerning the grant of revenues to the monks and their privileges in electing the archbishop. Baldwin, who wished to found a college of secular priests, was supported by the king, while the monks were upheld by the pope. Baldwin could count also on a general mood of hostility to the Black Monks, whom not only Cistercians regarded as relaxed. Then, too, Baldwin preached the Crusade following Saladin's capture of Jerusalem in 1187, and he himself went to the Holy Land, where he died. He had been saddened by the conduct of the Christian armies.

JOHN TILLER

BALE, JOHN (1495-1563). Protestant controversialist. Educated in a Carmelite house at Norwich and at Jesus College, Cambridge, he renounced his vows in the early thirties and served the Protestant cause by writing miracle plays and prose works. He enjoyed the patronage of Thomas Cromwell,* but was a refugee in Germany, 1540-47. He subsequently became rector of Bishopstoke, and in 1552 bishop of Ossory. His Protestant zeal created hostility, and during the reign of Mary he was in Holland. After Elizabeth's accession he was canon and prebendary of Canterbury. Of his twenty-one plays, five have survived. *King John,* the most important, never rises above doggerel and is unashamedly polemical. Prose works include *Brief Chronicle concerning Sir John Oldcastle. Illustrium Majoris Britanniae Scriptorum Summarium* and *Catalogus* constitute important catalogues of British writers. His works are marred by the bitter and somewhat coarse invective which gained him the name "Bilious Bale."

HAROLD H. ROWDON

BALFOUR, ARTHUR JAMES (1848-1930). British statesman. His religious views were influential because of his political position. In *A Defence of Philosophic Doubt* (1879), he attempted to justify faith by arguing that all men's basic convictions rest on the nonrational ground of religious faith. He developed this view in *Foundations of Belief* (1895), and in two sets of Gifford Lectures—*The-*

ism and Humanism (1915) and *Theism and Thought* (1923). During his premiership (1902-5) the Education Act (1902) aroused hostility from Nonconformists and secularists, by placing "non-provided" schools on the rates. This relieved such schools of financial pressure, but involved Free Churchmen in supporting Anglican schools. In 1917 he produced the "Balfour Declaration" which committed Great Britain to securing "a national home for the Jewish people" in Palestine. Balfour was a communicant in both the Anglican and Scottish Presbyterian churches. At one time he took an interest in spiritualism, but later abandoned it. JOHN A. SIMPSON

BALL, JOHN (d.1381). A leader of the 1381 Peasants' Revolt. An unbeneficed priest in York and Colchester, Ball evidently began to attack the structures of society. He was forbidden to preach in 1366, but continued to denounce clerical prelates and to proclaim the right not to pay tithes to unworthy clergy and also the equality of bondsmen and gentry. In 1381 he was in prison, but was released by the Kentish rebels and quickly became a widely known leader of the revolt. Basing his notorious Blackheath sermon on the rhyme "When Adam dalf, and Eve span, Who was then the gentilman?" he encouraged the killing of everyone who was harmful to the community. When the rebellion collapsed he was captured and sentenced to be hanged, drawn, and quartered. He cannot be regarded as a disciple of Wycliffe; his views were produced before Wycliffe was well known, they did not include doctrinal grievances—and Wycliffe in any case strongly condemned the rebellion. C. PETER WILLIAMS

BALL, JOHN (1585-1640). Presbyterian divine. Educated at Brasenose College, Oxford, he later became a private tutor. After conversion he was ordained in 1610 but without the usual subscription, and he continued to stand loose to all "Romish" elements within the Anglican Church. Inevitably he came to the attention of the authorities and was "deprived" and imprisoned for periods. As a considerable scholar and divine he enjoyed sanction among Puritan sympathizers when free. Most learned in the field of Roman controversy, especially as represented by Bellarmine,* his writings were particularly admired in New England. His *Trial of the New Church Way in New England and Old* was, however, an attack on the independent church system developed there, and also accused those who emigrated of being deserters afraid of the difficulties of living in England. This roused some resentment and produced several replies. C. PETER WILLIAMS

BALLARD, GUY (1878-1939). Founder of the I AM Movement.* Born in Kansas, he worked as a mining engineer before claiming in 1930 that the great Ascended Master of the spirit world St. Germain appeared to him on Mount Shasta in California and revealed to him the secret mysteries of the universe, and showed him himself, his wife Edna, and their son Donald in their previous existence. The parents had long been involved in occult activities, and in 1934 they formally launched the I AM Movement. Ballard published *Unveiled Mysteries* and *The Magic Presence*, books which told of the revelations from Germain and other Masters. Meetings were at first open to the public, but the Ballards drew criticism and adopted more secretive tactics. Mrs. Ballard gradually became the dominant figure of I AM, though it was believed that Guy would never die but would ascend directly to heaven. When he did die, numerous followers left the movement, but Mrs. Ballard claimed he had ascended and continued to lead the faithful. HARRY SKILTON

BALLOU, HOSEA (1771-1852). American Universalist. Born in New Hampshire, son of a Baptist minister, he at first accepted Calvinist doctrine, but under the influence of Universalist preaching and deistic writings he became a leader of the Universalist Church and was ordained in 1794. On the basis of his own independent study of Scripture he decided that it was impossible to defend the doctrines of the Trinity, the deity of Christ, human depravity, vicarious atonement, or eternal punishment on the basis of either Scripture or human reason. When Ballou arrived on the scene, most Universalists were orthodox in theology except for their belief that all men would be saved. But his extensive preaching, writing, and training of ministerial students influenced them toward Unitarianism. He edited Universalist journals and wrote hymns. Among his books are *Treatise on the Atonement* (1805) and *Examination of the Doctrine of a Future Retribution* (1834).
HARRY SKILTON

BAMPTON LECTURES. Canon John Bampton of Salisbury died in 1751 and endowed in his will an annual lectureship at St. Mary's Church, Oxford. The lectures were first delivered in 1780, and from 1895 they have been given on alternate years. The will specified that the lectures shall cover the exposition and defense of the Christian faith as set out in the creeds, and on the authority of the Scriptures and the Fathers.

BANCROFT, RICHARD (1544-1610). Archbishop of Canterbury from 1604. Born at Farnworth in Lancashire, he was educated at Cambridge and after ordination became chaplain to Bishop Richard Cox* of Ely and rector of Teversham. He was successively rector of St. Andrew's, Holborn, treasurer of St. Paul's Cathedral, and canon of Westminster. At St. Paul's Cross in 1589 he launched a stern attack on Presbyterian Puritanism, strongly reasserting episcopacy to such an extent that one of the queen's councillors feared he was threatening the royal supremacy. In 1590 he was made a prebendary of St. Paul's, then chaplain to Archbishop Whitgift. Seven years later he became bishop of London, exercising great influence due to Whitgift's age and infirmity. He had to deal with the Marprelate Tracts.* In 1604 he succeeded Whitgift as primate. He attended the Hampton Court Conference* with the Presbyterians and all but wrecked it by his belligerence and intransigence. He was largely responsible for the 1604 Canons which received the royal assent, but which Parliament was soon to set aside. He sought unsuccessfully to make the ecclesiastical courts independent of the law. In his closing months he was involved in the scheme to reestablish episcopacy in Scotland. He was also responsible for overseeing the Authorized Version of the Bible, though he died before its completion.

Bancroft's iron discipline, authoritarianism, and intransigence must be seen in their historical context. They have sometimes been erroneously used to bolster Tractarian innovations on episcopacy as if they were part of earlier Anglicanism, but in historical reality he was dealing with the truculent wing of extreme Puritans epitomized in the Marprelate Tracts. This extreme wing should not be confused with mainstream Puritanism which in England remained for the most part during Bancroft's life loyal to the national church, although pressing for reform and improvement.

See S.B. Babbage, *Puritanism and Richard Bancroft* (1962); A. Peel (ed.), *Tracts ascribed to Richard Bancroft* (1953). G.E. DUFFIELD

BAÑEZ, DOMINGO (1528-1604). Spanish Dominican theologian. Staunch defender of Thomistic doctrine which he sought to follow even in its minutest detail, he held various professorships for ten years, then in 1580 was elected to the chief chair in the University of Salamanca where he remained for twenty years. He was prominently involved in the controversy with the Jesuits concerning the merit of Christ's death, predestination, and justification. He was interested in logic, well-versed in metaphysics, and never hesitated to treat a subject in depth if it was important or useful. He was director and confessor of St. Teresa.

BANGORIAN CONTROVERSY (1717). A dispute in England which brought to a head the conflict between the High Church and Jacobite lower clergy and the Latitudinarian and Erastian bishops, leading to the effective suppression of Convocation for 135 years. Bishop Hoadly of Bangor preached a sermon before George I, supposedly at the king's suggestion, on "The Nature of the Kingdom of Christ," from John 18:36. A committee of the Lower House of Convocation accused Hoadly of denying the visible nature of the true church and of ignoring the working of the Holy Spirit, upon which the whole House requested the Upper House (of bishops) to make a definite censure. A considerable war of pamphlets followed the publication of the sermon, involving such as William Sherlock and William Law,* who attacked Hoadly strongly. To save him, the king prorogued Convocation, which was to transact no more business until 1852. G.S.R. COX

BAPTISM. The Christian rite of initiation whereby Christians confess their faith in Christ and are admitted into membership in the Christian Church. The origin of the rite probably is in the universal practice of sacred lustrations. The rite is variously interpreted as a sacrament essential to salvation; as merely a sign of one's Christian commitment; as a symbol picturing the believer's identification with Christ in his death, burial, and resurrection; or, as in the case of some spiritualist groups, as being entirely spiritual without material form.

Apostolic age. Christian baptism has its specific background in the OT acts of ritual purification as well as the Jewish practice of proselyte baptism and the ritual lustrations of the Qumran community. Although Jewish proselyte baptism has been questioned, the references in Epictetus, the Sibylline Oracles, and the Mishnah confirm a first-century practice. The practice of Qumran differs from the NT practice in that it was frequently observed, and although the mode and specific directions for its administration are lacking, its association with the Holy Spirit, purity of life, messianism, and the judgment of Yahweh has led scholars to find parallels with both the baptism of John the Baptist and early Christian baptism.

The following facts may be noted regarding the baptism of John the Baptist and early Christian baptism: (1) John's baptism was primarily an ethical act whereby one prepared for the coming kingdom of God and associated oneself with the herald of that kingdom (cf. Matt. 3:1-17; 11:2-15; Acts 19:1-7). (2) Jesus began his public ministry by association with John's baptism, probably not as a sign of repentance but as the King of the kingdom (cf. Matt. 3:13-17; cf. John 1:19-34), and according to John 4:2 Jesus Himself probably did not baptize. (3) The apostolic history of the Book of Acts reveals a developing practice and theology for the rite of baptism. In the early chapters beginning with the first Christian Pentecost, baptism is closely associated with repentance as a qualification for salvation and membership in the Christian community (cf. Acts 2:37-41), whereas the later passages place greater emphasis upon repentance and faith (cf. 3:16; 16:31) with baptism following (cf. 9:1-19, 16:31-34). The references to baptism "in the name of Jesus" are probably attempts to distinguish Christian baptism from Jewish proselyte baptism rather than a specific baptismal formula (cf. Acts 2:38; 8:16). Also, the ritual act was closely associated with the Holy Spirit and the confirmatory sign of the *glōssai* (cf. Acts 2:1-13; 10:44-48; 19:1-7). (4) Paul's baptismal theology is stated in several key passages: Romans 6:1-4; 1 Corinthians 12:12-13; Galatians 3:26-29; Colossians 2:9-13. Baptism is primarily an act of identification with the dead, buried, and resurrected Christ, but it is also a sign of the covenant and is therefore corresponded to the OT rite of circumcision. That Paul did not conceive of baptism as an essential saving sacrament is clearly indicated by 1 Corinthians 1:10-18. For Paul there seems to be one essential baptism, the baptism of the Holy Spirit by which we are incorporated into the body of Christ (cf. 1 Cor. 12:12-13; Eph. 4:4). (5) The rest of the NT corpus does not present a unified picture regarding the meaning and administration of baptism. Hebrews 6:2 speaks of baptisms (RSV "ablutions"); 1 Peter 3:21 distinguishes between a bodily and a psychological cleansing associated with baptism; and Paul speaks of a mysterious "baptism for the dead" in 1 Corinthians 15:29 (cf. Matt. 28:18-20).

Patristic and medieval period. Probably our earliest references to baptism outside the NT are to be found in the Didache.* Here the mode is clearly a tri-immersion in the name of the Father, Son, and Holy Spirit, but the Didache permits baptism by affusion if insufficient water is available. It seems clear that up until about the end of the fifth century, adult believer's baptism was the normal practice of the church. This fact is demonstrated by the emphasis in the Early Fathers on careful preparation for baptism, and the necessity to live a sinless life after baptism. As a result of this latter teaching, many postponed baptism until their deathbed, which came to be known as "Clinical Baptism" (cf. Gr. *klinē*, "bed"). Doctrinally, baptism very early came to be understood as a means of grace or a sacrament, in the sense of an instrumental means of regeneration. Justin Martyr bases his doctrine of baptismal regeneration on John 3:3,5 and Isaiah 1:16-20. Although Irenaeus says almost nothing about the mode or practice of baptism, he is a strong defender of the Trinitarian formula and baptismal regeneration.

Schism within and heresy without forced the church to raise the question of the validity of the baptismal rite. In the second century, Tertullian denied the validity of baptism administered by heretics. Normally it was done by the bishop; however, on special occasions deacons and presbyters might be permitted to administer the rite. Even women were forbidden to perform it, according to the decrees of the Fourth Council of Carthage (c.255). The issue was raised in a somewhat different way with the Novatian* and Donatist* controversies. From these controversies the idea, if not the term, of *ex opere operato* developed.

The practice of infant baptism completes the early patristic developments. Infant baptism was practiced in the second century, but only with the aid of an adult sponsor. A full defense of this custom came to expression in the theology of Augustine in the late fourth and early fifth centuries. Only in the fifth century did the Syrian Church make infant baptism obligatory; prior to this time it was the exception rather than the rule. But with the developing conceptions of original sin,* a theological foundation for infant baptism was found. And it was Augustine more than anyone else who lent his theological genius to this issue. Basing his conception of an original act of sin by Adam in which all humanity participated, Augustine drew the conclusion that each child is not only born with an inherited tendency toward sin, concupiscence, but also shares in the guilt of Adam's sin. It is noteworthy that Augustine based this position largely upon the faulty text of Rom. 5:12—a Latin text which translated the Greek phrase "because that" as *"in quo"* ("in whom"). Prior to the fifth century, the normal practice was to administer baptism at Easter and Pentecost, but from the fifth century the importance of infant baptism placed great stress upon the rite's being administered prior to the eighth day after birth.

During the Middle Ages, the theology and practice of baptism were largely refinements of the earlier developments. The major concern during the period from the fifth to the eleventh centuries was the issue of heretical baptismal practices. At the Fourth Lateran Council,* baptism and the six other sacraments received their basic dogmatic definition. Hugh of St.-Victor modified the sign emphasis of Augustine, and Aquinas* provided the theological framework not only for baptism but for all seven official sacraments. Aquinas is the first to trace all seven back to Christ, and he taught that baptism, confirmation, and ordination stamp an indelible mark upon the soul; consequently, these three sacraments cannot be repeated. Greater stress was also given to the concept of *ex opere operato,* and the earlier teaching—which appeared prior to Augustine but which Augustine gave greater recognition, regarding a baptism of blood for martyrs and the baptism of desire for those who had not been officially baptized but who gave evidence of intending to do so—was further refined by Aquinas and the Scholastics. In the debate between Nominalism and Realism, both baptism and the Eucharist were involved in the manner in which grace was bestowed. The creedal formulations of the Fourth Lateran Council and the theological foundations of Thomistic theology were restated at the Council of Trent* (1545-63) and have remained to this day the essential position of the Roman Catholic Church.

Reformation and contemporary developments. During the sixteenth century, baptism along with the Eucharist became one of the major divisive issues not only separating reforming groups from Catholicism, but also dividing the rival sects. Nearest the Roman Catholic position was Luther's view, which reduced the number of sacraments finally to two—baptism and the Lord's Supper—but which stressed infant baptism on Augustinian grounds. In his *Babylonian Captivity of the Church,* Luther recognized the meaning of *baptizō* as immersion, but qualified this by regarding the mode as a matter of indifference to the sacramental power of the rite; and he further admitted that the normal NT pattern was adult believer's baptism. The absolutely essential element of faith was to be supplied by the sponsoring adult; Luther sometimes stated that God gave the infant faith in the baptismal act. Calvin treats the subject of baptism in his *Institutes* (IV, 15). For him, baptism is a sacrament, but its efficacy is limited to the elect. Later Calvinism made greater use of covenant theology, the effect of which was to minimize the sacramental nature of baptism and replace sacramentalism with a covenantal sign significance. Luther and Calvin's contemporary Zwingli had already reduced baptism to a mere sign, making it unnecessary to salvation. Zwingli's ideas were also represented by the Anabaptists,* whose major emphasis was upon believer's baptism rather than the mode. Only with the English Baptists about 1633 did the issue of immersion arise among the Particular Baptists. Prior to this, even the Baptists practiced affusion or sprinkling, since the issue was believer's baptism as opposed to paedo-baptism. Among the spiritualists, especially the seventeenth-century Quakers, baptism and the Lord's Supper were rejected as irrelevant to the age of the Spirit. On other grounds, eighteenth-century rationalism also set aside not only sacramentalism but also ecclesiastical institutionalism which included baptism. The Tractarian Movement* and High Church Anglican theologians (cf. E.B. Pusey) sought to restore baptism to its earlier sacramentalism. American theology has tended more and more in the direction of Zwinglianism, largely because of the influence of revivalistic emphases. In the twentieth century there has been a revival of interest in baptism mainly as a result of the influence of Karl Barth* and his rejection of infant baptism, although he stood in a paedobaptist tradition. The ecumenical movement has also stimulated interest in baptismal and sacramental theology as the various branches of Christendom have attempted to work out a rapprochement. More often than not it has been the sacraments that have provided one of the greatest hindrances to achieving ecumenicity in spirit and form.

BIBLIOGRAPHY: J. Corblet, *Historie dogmatique, liturgique et archéologique du sacrament de baptême* (2 vols., 1881-82); K. Barth, *The Teaching of the Church Regarding Baptism* (1948); G.W.H. Lampe, *The Seal of the Spirit* (1951); J. Murray, *Christian Baptism* (1952); P. Ch. Marcel, *The Biblical Doctrine of Infant Baptism* (ET 1953); J. Warns, *Baptism: its History and Significance* (ET 1958); T.F. Torrance, *Conflict and Agreement in the Church*, vol. 2 (1959), pp. 93-132; R.E.O. White, *The Biblical Doctrine of Initiation* (1960); W. Carr, *Baptism: Conscience and Clue for the Church* (1964); D. Moody, *Baptism: Foundation for Christian Unity* (1967).

DONALD M. LAKE

BAPTISM OF CHRIST. The baptism of Jesus by John the Baptist is recorded in the synoptic gospels (Matt. 3:13-17; Mark 1:9-11; Luke 3:21f.) and is implied in John (1:32-34). John's role was to be a link between the old and new covenants. Where the prophets of the OT had spoken in general terms of the Messiah, it was John who was able to identify him. The baptism of Jesus by John was the beginning point of the ministry, and therefore also of the apostolic witness and the *kerygma* (Acts 1:22; 10:37; 13:24f.).

Because John's baptism was connected with repentance for the forgiveness of sins, Matthew records John's hesitation to baptize Jesus. John is persuaded to do so when Jesus says, "It is proper for us to do this to fulfil all righteousness." This seems to mean that He accepted the divine plan and was willing to identify Himself with the faithful remnant in Israel. The opening of the heavens shows the direct contact between the Father and Jesus in the descent of the Spirit and is both in line with the coming of the Spirit upon the prophets and a continuation of what He had done as the agent of the Father in the conception of Jesus. The words which refer to Jesus as God's "beloved Son" are often taken to be a combination of Isaiah 42:1 and Psalm 2:7, implying that his Sonship is to be worked out in the role of the Servant of the Lord. The fact that the baptism immediately precedes the temptation in the wilderness suggests that there is typology involved as Jesus recapitulates the experience of Israel in the Red Sea and the desert. The answer Jesus gives to the chief priests about the source of His authority (Mark 11:27-33) suggests that on the historical level the authority for His mission is to be found in His baptism.

R.E. NIXON

BAPTIST CHURCHES, AMERICAN, see AMERICAN BAPTIST CHURCHES

BAPTISTERY. A building or part of a building in which baptism is administered. One of the rooms in the third-century house-church at Dura-Europos was furnished as a baptistery with font and canopy. Separate baptisteries appear in the fourth century, sometimes containing a suite of rooms; candidates assembled in one, disrobed in another, were baptized in the next, dressed in white robes in a fourth—and thence entered the church. During the Middle Ages, baptisteries ceased to be built, a font at the back of the church being regarded as sufficient provision for baptism. There have been some modern attempts to give greater prominence to the place of baptism, in architectural terms.

HENRY R. SEFTON

BAPTISTS. So named from their practice of baptizing only those who have made a personal profession of faith in Jesus Christ, Baptists constitute one of the largest Free Church communions, with a world membership in 1971 of over thirty-one million, and a total community strength considerably higher. Twenty-seven million are to be found in the USA, and there are also substantial groups in India (633,000), Russia (550,000), Congo (450,000), Brazil (342,000), the British Isles (269,000), Burma (249,000), Canada (175,000), and Romania (120,000). They are evangelical in outlook, with a strong emphasis on the necessity of personal commitment to Christ and a personal experience of His grace, and an accompanying understanding of the Christian life in terms of personal faith and discipleship. Believer's baptism, they maintain, expresses more clearly than any alternative practice the NT teaching concerning the nature of both the Gospel and the church. It is administered in the name of the Trinity and is normally by immersion.

The soil out of which the modern Baptist movement arose was that of seventeenth-century English Separatism. Whether there were any direct links with the continental Anabaptists* of the sixteenth century is a difficult and much disputed question; some indirect Anabaptist influence, however, was probable. In 1609 John Smyth's* English Separatist congregation in exile in Amsterdam was led by a study of the NT to disband and reorganize itself, with believer's baptism as the basis of church fellowship. Smyth and most of his congregation applied to join the Mennonites, and were accepted by them in 1615, three years after Smyth's death. Meanwhile, in 1612, a small group under Thomas Helwys* returned to England, forming the first Baptist church on English soil, at Spitalfields. They were General (or Arminian) Baptists. The first Particular (or Calvinist) Baptist church came into being between 1633 and 1638, as a secession from the Independent Jacob-Lathrop-Jessey church, so named from its succession of pastors. Both streams made considerable progress, especially during the Commonwealth period, and by 1660 there were between

200 and 300 Baptist churches in England and Wales, most of them in London, the Midlands, and the South.

At first the mode of baptism practiced was affusion, but from the 1640s immersion became general. From earliest times there were Baptists who believed in "mixed" or open communion. John Bunyan* was pastor of a church consisting of Baptists and paedobaptists. From the seventeenth century a few Baptists have observed the seventh day as their sabbath. Seventh Day Baptists are still to be found in America, but have virtually disappeared from England. America's earliest Baptist churches were formed in the seventeenth century, the first probably being Providence, Rhode Island (1639), in the establishment of which Roger Williams* played a leading part.

Concern at the widespread influence of Unitarian views among General Baptists during the eighteenth century led Dan Taylor (1738-1816) to organize those churches remaining orthodox and evangelical in a New Connexion, in 1770. Particular Baptists too were experiencing theological and spiritual renewal at this time. The writings of Andrew Fuller,* particularly *The Gospel Worthy of All Acceptation* (1785), helped to break down the rigid and extreme Calvinism then common among them, and in 1792, spurred on by William Carey,* the ministers of the Northamptonshire Association formed the Baptist Missionary Society, Carey himself going to India as a missionary the following year. Despite strong resistance from those who came to be known as Strict Baptists,* their evangelical Calvinism gained a wide acceptance on both sides of the Atlantic.

A "General Union" of Particular Baptist ministers and churches (known since 1873 as the Baptist Union of Great Britain and Ireland) was formed in 1813. Gradually Baptists within the Union and those of the New Connexion moved closer together, and in 1891 the two organizations merged. John Clifford (1836-1923), who played a leading part in the merger, was noted for his emphasis on evangelism and the social implications of the Gospel. His contemporary, C.H. Spurgeon,* was widely influential as a preacher, especially in London and the South. The reshaping of the Union to meet the demands of the twentieth century was largely the work of J.H. Shakespeare, its secretary from 1898 till his death in 1928. It was mainly due to his vision that the Federal Council of Evangelical Free Churches was formed in 1919, and in *The Churches at the Cross Roads* (1918) he pleaded for wider Christian reunion. Under his successors, M.E. Aubrey (secretary, 1928-51), E.A. Payne (1951-67), and D.S. Russell (appointed 1967), the Union has taken a full share in the British and World Councils of Churches. In 1970 there were 2,192 churches with a membership of 106,767 affiliated to the Union. There are separate unions in Scotland and Wales, with which the British Union maintains close fraternal relationships. Ireland too has its own union.

For a century after Roger Williams, progress in America was slow. From 1740, however, under the influence of the Great Awakening,* the Baptist cause made considerable headway. The nineteenth century was a period of outreach in rural and frontier areas and among Indians, Negroes, and various immigrant groups, the number of Baptists increasing from 700,000 to well over four million between 1850 and 1900. By the mid-1960s there were some thirty separate groups of Baptists in the USA, the largest being the eleven-million strong Southern Baptist Convention,* two Negro conventions (see AMERICAN NEGRO CHURCHES) with a combined numerical strength of nearly ten million, and the American Baptist Churches* with approximately 1.5 million members. The Baptist community in the USA has produced such outstanding figures as Walter Rauschenbusch,* theologian of the Social Gospel; Martin Luther King,* Negro advocate of nonviolence and Nobel Prize winner; and Billy Graham,* world-famous evangelist.

The first Baptist church in mainland Europe was established in Hamburg in 1834 by J.G. Oncken (1800-84), whose influence was felt far beyond Germany, especially in Scandinavia and eastern Europe. Work in Australia and New Zealand also began in the nineteenth century, while the twentieth saw striking advance in Africa, Asia, and Latin America.

Baptists acknowledge Christ as the sole and absolute authority in all matters of faith and practice, viewing the Scriptures as the principal means by which He speaks to the church. Not surprisingly, they stress the prophetic rather than the priestly aspects of religion, and are conscious of the dangers of uniformity and invariable "official" forms in matters relating to worship, church government, and theological definition. From their earliest days they have been keen advocates of religious freedom.

Worship is largely nonliturgical, with emphasis on the reading and preaching of the Word. In addition to its minister, a Baptist church is served by deacons, elected from the membership, who share with the minister in leadership and administration, and assist him in the celebration of the Lord's Supper.

Probably few Baptists would be prepared to recognize infant baptism as having theological validity, though in practice most (with the notable exception of those in the Strict Baptist tradition) welcome to the Lord's Table Christians not baptized as believers, and "open membership" churches also admit them to membership. A small number of Union churches, affiliated to more than one denomination, and practicing both forms of baptism, are to be found in several countries, including England. Baptists are divided in their attitude to the ecumenical movement. Some, like British Baptists and those of the American Baptist Churches, are prepared for ecumenical involvement; others, principally those of the Southern Convention, take a more critical attitude. During the present century Baptists in China and Japan have participated in schemes of union or federation, and in 1970 Baptists joined with non-Baptists to form the Church of North India.

Baptists regard the church as a society of believers, the local church having an especially important part in their thinking. Though not committed theologically to one particular type of polity, they generally favor a congregationalist form of church government. Belief in the autonomy of the local church, however, is balanced by a belief in the necessity of fellowship and interdependence. All except the most rigidly independent churches are linked together both in regional groupings known as associations, which have been common since the seventeenth century, and in national unions or conventions. A Baptist World Alliance* was formed in 1905.

BIBLIOGRAPHY: T. Crosby, *The History of the English Baptists* (4 vols., 1738-40); J. Ivimey, *A History of the English Baptists* (4 vols., 1811-30); A. Taylor, *The History of the English General Baptists* (2 parts, 1818); T. Armitage, *A History of the Baptists* (1888); E.R. Fitch, *The Baptists of Canada* (1911); J.H. Rushbrooke, *The Baptist Movement in the Continent of Europe* (rev. ed. 1923); *idem, Some Chapters in European Baptist History* (1929); G. Yulle (ed.), *History of the Baptists in Scotland from Pre-Reformation Times* (1926); W.T. Whitley, *A History of British Baptists* (2nd ed. 1932); F.T. Lord, *Achievement: A Short History of the Baptist Missionary Society, 1792-1942* (1942); *idem, Baptist World Fellowship: A Short History of the Baptist World Alliance* (1955); H.W. Robinson, *Baptist Principles* (1945); *idem, The Life and Faith of the Baptists* (rev. ed. 1946); A.C. Underwood, *A History of the English Baptists* (1947); E.A. Payne, *The Fellowship of Believers* (enlarged, 1952); *idem, The Baptist Union: A Short History* (1959); A. Gilmore (ed.), *Christian Baptism* (1959); W.S. Hudson (ed.), *Baptist Concepts of the Church* (1959); W.L. Lumpkin, *Baptist Confessions of Faith* (1959); C.C. Goen, *Revivalism and Separatism in New England, 1740-1800* (1962); N.H. Maring and W.S. Hudson, *A Manual of Baptist Polity and Practice* (1963); R.G. Torbet, *A History of the Baptists* (rev. ed. 1966).

ERNEST F. CLIPSHAM

BAPTIST WORLD ALLIANCE. A voluntary and fraternal world association of Baptists founded in 1905 to show, according to its constitution, "the essential oneness of Baptist people in the Lord Jesus Christ, to impart inspiration to the brotherhood, and to promote the spirit of fellowship, service and co-operation among its members." The alliance serves as an agent of communication between Baptists (in 1970 it represented just over twenty-three million in twenty-five countries), a forum for study and discussion, a channel for relief and mutual help, a vigilant force for safeguarding religious liberty, and a sponsor of gatherings for the spread of the Gospel. A Baptist World Congress and a World Conference for Baptist Youth are held every five years. The women's department organizes an annual Baptist Women's Day of Prayer. Between meetings the work of the Congress is carried on by an executive meeting annually, with six continents represented in its membership, and by a series of study commissions on such subjects as Christian teaching and training, evangelism and missions, Baptist doctrine, religious liberty, and human rights and cooperative Christianity.

JAMES TAYLOR

BARADAEUS, JACOB, see JACOB BARADAEUS

BARBAROSSA, see FREDERICK I

BARCLAY, JOHN (1734-1798). Founder of the Bereans.* He trained for the Presbyterian ministry at St. Andrews, where he came under the influence of Dr. Archibald Campbell. A powerful preacher, Barclay served as assistant minister at Errol (1759-63) and Fettercairn (1763-72). He came under censure for the views expressed in his book *Rejoice Evermore, or Christ All in All* (1766). An appeal to the General Assembly (which was dismissed) enabled him to air his views in Edinburgh, where he gained a number of adherents who resolved to secede and invited Barclay to become their minister. He went to England for ordination at Newcastle in 1773. His church in Edinburgh became known as the Berean Assembly. From 1776 to 1778 he was in England, where churches of his persuasion were formed in London and Bristol. Returning to Edinburgh, he resumed his ministry there and in other churches which had been formed in Scotland.

Barclay diverted from current Calvinism in a number of particulars. He amplified the assertion of Doctor Campbell that man is unable to reach belief in God by using his rational powers, and he questioned the value of theistic arguments, even that from design. The revealed truth of the Bible can be received only by the illumination of the Holy Spirit. Faith is neither a subjective emotion nor personal appropriation of Christ, but the intellectual acceptance of biblical revelation. Barclay therefore rejected both the moralism of the Moderates* and the "soul-struggles" of the Evangelicals, and asserted that assurance of salvation is the hallmark of Christianity. Unbelief is the sin against the Holy Ghost. Unbelievers cannot even pray for their own conversion. The Lord's Supper requires no special soul preparation.

See J. Thomson and D. McMillan (eds.), *The Works of John Barclay* (1852).

HAROLD H. ROWDON

BARCLAY, ROBERT (1648-1690). Scottish Quaker theologian and apologist. Son of a professional soldier who had fought with Gustavus Adolphus as well as in the Scottish army, Barclay was sent to France at the age of ten. There he spent four years studying with his uncle, who was rector of the Scots College (Roman Catholic) at Paris. Returning to Scotland, he followed his father in accepting Quakerism (1667). In 1670 he married a convert, Christina Mollison, settled at his estate at Ury, and began to write the apologetic treatises which were to bring him fame. Barclay's persecution began in 1672 after he walked through the streets of Aberdeen clothed with sackcloth and with ashes on his head. He was imprisoned the same year and several times thereafter, including his longest sentence during the winter of 1676-77. He also traveled to the Continent in the service of his faith. His first trip in 1676 brought him into contact with his distant relative Princess Elizabeth of the Rhine. She had taken an interest in Quaker ideals, and when Barclay left he carried a letter from her to Prince Rupert, asking him to use his influence on behalf of Quaker prisoners. In 1677 William Penn,* George Fox,* and Barclay went to Germany, and Robert once more interviewed Elizabeth. After his return to Scotland he became a favorite of the duke of York (later James II). This friendship led to the granting of the colony of East New Jersey to a group of Friends,* including Barclay who was made governor of the territory in 1683. Although he never went to the New World, his brother settled there. The province was meant to be a refuge for the persecuted and to provide a practical application of Quaker ideals of toleration. Among his many works the better known are *A Catechism and Confession of Faith* (1673), *The Anarchy of the Ranters* (1676), and *An Apology for the True Christian Divinity, Being an Explanation and Vindication of the People Called Quakers* (1678). Insisting upon divine inward revelation as necessary for true faith, Barclay also formulated the humanitarian and pacifist ideals still followed by the Society of Friends.

BIBLIOGRAPHY: *Truth Triumphant* (Barclay's works) (1692); W. Armistead, *Life of Robert Barclay* (1850); M.C. Cadbury, *Robert Barclay* (1912); D.E. Trueblood, *Robert Barclay* (1968).

ROBERT G. CLOUSE

BARDESANES (Bar-Daisan) (154-222). Edessene Christian. A man of remarkable versatility, he was the most outstanding representative of the early Christian community in Edessa, Syria. A friend of court, and particularly of King Abgar IX, he was a philosopher able to write in Greek and Syriac, and a poet of considerable powers. Prior to his conversion he had been interested in astrology, and this remained with him, contributing to his reputation for unorthodoxy and even heresy. He is famous for *The Book of the Laws of the Countries,* written by a disciple, Philip. In this he argued, against the astrologers, that there is a free will in the universe. Although he refuted Marcion* and Valentinus,* and was not a Gnostic, he did not eradicate dualism from his system. He was sufficiently faithful to Christianity to be on the point of martyrdom. As a poet he can be considered one of the founders of Syriac literature.

C. PETER WILLIAMS

BAR HEBRAEUS (Grigor Abu-l-Farag Bar-Hebraya) (1225-1286). Bishop of the Syrian Monophysite (Jacobite) Church, and Mafrian (primate) from 1264. Writer of prodigious output, his works include a Bible commentary in scholia ("Storehouse of Secrets"), systematic theology ("Candelabrum of the Sanctuary") and in summary ("Book of Lightning"), canon law or nomocanon ("Book of Guidance"), ethics ("Book of Ethics"), ascetic spirituality ("Book of the Dove"), exposition of Aristotelian philosophy ("The Cream of Wisdom"), a sketch of logic, physics, and metaphysics ("Book of the Conversation of Wisdom"), grammatical exposition ("Book of Splendors"), astronomy ("The Ascent of Reason"), chronography (both secular and church history), prose and poetry, and a number of translations from Arabic. His works are important for their preservation of textual, lexical, and historical information, excerpted with encyclopedic enthusiasm by this widely learned scholar.

J.N. BIRDSALL

BARING-GOULD, SABINE (1834-1924). Hymnwriter and medieval scholar. Born in Exeter and educated at Cambridge, he took holy orders in 1864 and was for three years curate at Horbury Bridge, Yorkshire. In 1865 he wrote his famous hymn "Onward, Christian soldiers" for a Whit Monday Sunday school procession. After incumbencies in Dalton, Yorkshire, and East Mersea, Essex, he appointed himself in 1881 rector of Lew Trenchard, Devon, which had been the family seat for 300 years. He was a versatile author, and his chief work was *The Lives of the Saints* in fifteen volumes, which had the distinction of being placed in the Roman *Index Expurgatorius.* Other hymns include "Now the day is over" and the translation "Through the night of doubt and sorrow."

J.G.G. NORMAN

BARKER, FREDERIC (1808-1882). Bishop of Sydney. An English clergyman educated at Cambridge (he was influenced by Charles Simeon*), his evangelicalism was colored by his contacts with Ireland and his incumbency in a Liverpool parish. His service under Archbishop Sumner led to Barker's appointment to the bishopric of Sydney in 1854. He reestablished the evangelical tradition first impressed on the Australian diocese by the early chaplains. His influence spread through the founding of Moore College, the oldest theological college in Australia, in 1856. Barker promoted church extension by means of a church society, which paved the way for the setting up of synodical government by laity and clergy in 1866. With his support a general synod was formed in 1872 for the whole Church of England in Australia. During his twenty-eight-year episcopate he saw the division of his vast diocese and a considerable strengthening of the parish system.

NOEL S. POLLARD

BARLOW, WILLIAM (d.1568). Anglican bishop. A member of the reforming party under Henry VIII, his writings which showed Protestant inclinations were condemned in 1529, but later he was again accepted, and in 1536 became bishop of St. Davids, in 1548 bishop of Bath and Wells. When Mary came to the throne, he fled abroad and probably spent most of her reign in Poland. He returned on her death and became bishop of Chichester. Barlow has been the subject of much ecclesiastical controversy between Roman Catholics and High Anglicans since he was a consecrator of Archbishop Parker, without there being any record of his own earlier consecration. Barlow translated part of the Apocrypha for the Bishops' Bible,* contributed to the *Institution of a Christian Man,* and was a member of the commission for reforming church law.

G.E. DUFFIELD

BARMEN DECLARATION (1934). A declaration made by the free Synod of Barmen attended by representatives of the Protestant churches in Germany as a response to the German-Christians.* It was the theological rallying point of the Confessing Church.* Largely written by Karl

Barth,* its Reformed exclusion of natural theology, though necessary in the circumstances, limited its long-term appeal to Lutherans. The declaration quotes the Nazi-approved church constitution of 1933 which stated that the German Evangelical Church rested on the Gospel of Jesus Christ revealed in Holy Scripture and brought to light again in the Reformation confessions; and that the church was a federal union of equal territorial churches. So it opposed both German-Christian theology and church government, by which, it believed, the church ceased to be the church. In each of six main paragraphs NT texts are given, then positively expounded; and contrasting errors are repudiated. Its thrust is that since Jesus Christ is the one Word of God, the Church is not to recognize other events, powers, or images alongside Him as divine revelation.

HADDON WILLMER

BARNABAS. A man of Cyprus who became a Christian in the earliest period of the Christian Church, Barnabas first came to notice for his liberality in contributing the proceeds of the sale of his property for the common support of the Jerusalem Christians (Acts 4:36). He had relatives in Jerusalem, since the home of his cousin Mark was situated there. He was willing to befriend the recently converted Saul of Tarsus when the other Jerusalem Christians were afraid to do so, having recognized more readily than others the genuineness of Saul's conversion. He was sent by the Jerusalem church to investigate reports of the remarkable growth of the Antioch church. His reaction was encouraging (Acts 11:24ff.). On recognizing the need for a teaching ministry there, he called Saul from Tarsus to join him at Antioch. After a year's ministry, both were comissioned by that church to undertake a missionary journey. In the Acts narrative describing the course of this mission they are both described as apostles (Acts 14:14).

When the problem of Gentile circumcision arose, Barnabas with Paul opposed it and the Jerusalem church vindicated their policy. But when a second missionary journey was proposed, Barnabas clashed with Paul over the position of Mark (Acts 15:34ff.). On a later occasion Paul expressed his regret that even Barnabas was carried away by the hypocrisy of the Judaizers (Gal. 2:13). In his Corinthian correspondence Paul mentions that, like himself, Barnabas supported himself rather than depend on the churches (1 Cor. 9.6).

DONALD GUTHRIE

BARNABAS, GOSPEL OF. A pseudogospel containing portions from the Koran and the four canonical gospels. Its date is in doubt, though internal evidence points to the first half of the fourteenth century. Written in Italian by one who had renounced Christianity for Islam, its essential feature in teaching religious toleration is a belief that God's message of salvation is for all.

BARNABITES. The congregation of "Clerks Regular of St. Paul" (or "Paulines"), founded in Milan in 1533 by Antonio Maria Zaccaria, a doctor of Cremona, with Bartolommeo Ferrari, a lawyer, and Giacomo Antonio Morigia, a mathematician, to preach missions and conduct educational work in the city. They were given the church of St. Barnabas, hence their name. A staunch supporter was Charles Borromeo,* whose confessor, Alessandro Sauli, was general of the order (1565). During the seventeenth century they spread into France and central Europe, and today there are about thirty-five houses.

BARNARDO, THOMAS JOHN (1845-1905). Converted in 1862, he became a member of the Plymouth Brethren and went to London from his Dublin home in 1866, determined to become a missionary doctor. Visiting the Stepney slums, the plight of a homeless waif, Jim Jarvis, inspired him to establish his first home for destitute boys in 1870. Equipped with amazing, if rather autocratic, nervous energy, great organizational flair, and journalistic skill in appealing to the public, he quickly built up a vast system. In 1873 a former public house provided the base for a church and a coffee palace; in 1876 he built a village at Ilford to provide homes for girls in small, less institutional, units; in 1882 he started to send children to Canada because of the better employment prospects, and in 1886 he began to arrange boarding-out for children. The organization so expanded that by the time of his death he had admitted 59,384 children to his homes, helped 20,000 to emigrate, and materially assisted a further 250,-000.

C. PETER WILLIAMS

BARNBY, SIR JOSEPH (1838-1896). One of the most influential choral conductors of nineteenth-century England. He did much to introduce the works of Bach and Dvorák into England, wrote over 200 hymn tunes, edited five hymnals, and was for years musical adviser to Novello and Company. He is best known today for his tune *Laudes Domini* for the hymn "When morning gilds the skies." His numerous anthems and part-songs are in a sweet, romantic idiom that is out of favor today.

J.B. MAC MILLAN

BARNES, ALBERT (1798-1870). Presbyterian minister and commentator. Born in Rome, New York, of Methodist background, he was educated at Hamilton College and Princeton Seminary and became a Presbyterian pastor first at Morristown, New Jersey, then at First Church in Philadelphia. He was a dynamic evangelical preacher who sought to challenge the human will to respond to God's free offer of salvation. He became a leader of the New School revivalist party and was considered a dangerous radical by the strictly Calvinistic Old School party. In 1830 he was charged with doctrinal error by his presbytery, acquitted by the general assembly, but admonished about the objectionable passages in a sermon, "The Way of Salvation." He argued in his book *An Inquiry into the Scriptural Views of Slavery* (1846) that the Bible and common sense alike condemned slavery. His *Notes* on the NT, Psalms, and Isaiah were read widely.

HARRY SKILTON

BARNES, ERNEST WILLIAM (1874-1953). Bishop of Birmingham. Educated as a scientist at

Trinity College, Cambridge, he was ordained in 1902 and lectured at Trinity in mathematics. He became master of the Temple in 1915, canon of Westminster in 1918, bishop of Birmingham in 1924. He wrote various papers on mathematics, but is chiefly remembered as a prominent modernist with views that included serious questioning of the historic Christian faith on matters such as the virgin birth and the Resurrection, on one occasion provoking criticism from Archbishop Fisher in public. A consistent pacifist with a horror of atomic weapons, Barnes was also a writer; his two main works were *Should such a Faith offend?* (1928) and *Scientific Theory and Religion* (1933). G.E. DUFFIELD

BARNES, ROBERT (1495-1540). Reformer and martyr. He graduated D.D. (1523) from Cambridge, where he met with others at the White Horse Tavern for Bible study. After preaching a sermon considered to be unorthodox, he was examined in 1526 by five bishops, Thomas Wolsey among them, and condemned to be burnt or abjure. He abjured and was imprisoned, but escaped to Antwerp. Later he met Luther at Wittenberg and also Stephen Vaughan, who wrote kindly of him to Thomas Cromwell* in England. Returning there, he became an intermediary between Henry VIII's government and the German Protestants. In 1540 the Protestant and Roman Catholic factions fought bitterly in Henry's council. The king had earlier refused to embrace the Protestant faith, and in the end the Catholic party gained dominance. Cromwell was beheaded, and Barnes with two others (William Jerome and Thomas Gerrard) was burnt for heresy.
R.E.D. CLARK

BARNETT, SAMUEL AUGUSTUS (1844-1913). Anglican social reformer. After reading law and modern history at Oxford, he was ordained in 1867. In 1869 he founded the Charity Organization Society, and from 1872 to 1893 was vicar of St. Jude's, Whitechapel, where his unusual methods aroused criticism—evening schools, entertainments, Oxford dons to lecture parishioners, etc. He aimed at improving the East End cultural level, encouraging Christians to study social problems, and promoting Christian social action. In 1884 he helped form the Education Reform League; in 1885 he promoted the Artisans' Dwellings Act; and from 1884 to 1896 he was the first warden of Toynbee Hall. Barnett House at Oxford was founded in his memory, for the study of social problems. JOHN A. SIMPSON

BARNHOUSE, DONALD GREY (1895-1960). Presbyterian minister and radio preacher. Born in Watsonville, California, he studied at various institutions including the universities of Chicago and Princeton, culminating in a Th.D. degree from Aix-en-Provence, France. In 1919 he joined the Belgian Gospel Mission. From 1919 to 1921 he was director of a Bible school in Brussels, and later he worked in several small French Reformed churches. Returning to the USA (1927), he began a thirty-three-year pastorate at Tenth Presbyterian Church, Philadelphia. He became a noted radio preacher and just prior to his death had completed eleven years of weekly broadcasts devoted to the epistle to the Romans. He was editor of *Revelation* magazine (1931-49), and of *Eternity* magazine (1950-60). He was the author of numerous books. BRUCE L. SHELLEY

BARO, PETER (1534-1599). French pastor and theologian. Born at Etampes, he studied law at Bourges, where he came under Reformed influences that led him to the study of theology. In Geneva he studied under Calvin, by whom he was ordained in 1560. He returned to France in 1572, but later had to flee from persecution to England and was appointed Lady Margaret professor of divinity at Cambridge in 1574. He became critical of the Reformed doctrine of predestination and entered into controversy with its advocates. He retired in 1596 and lived in London for the rest of his life.

BARONIUS, CESARE (1538-1607). Cardinal and church historian. In 1568, in reply to the Centuries of Magdeburg,* he began his *Annales Ecclesiastici*, a history of the church, each chapter corresponding to a year. His purpose was to demonstrate that the practices and beliefs claimed by Flacius* to be corruptions of the early faith were in fact found in the church from the beginning. He attempted to reassure those who feared lest historical criticism had shaken the evidence for early legends. Unfortunately the documents used by Baronius, who became librarian of the Vatican in 1597, were not yet recognized as spurious, and his work was marred by inaccuracies. JAMES TAYLOR

BARRIER ACT (1697). Passed by the general assembly of the Church of Scotland, its aim was the prevention of hasty and ill-considered legislation which, although appearing at the time to be of value, might prove in due course to be prejudicial to the church's best interests. Thus it was decreed that before any acts were passed which are "to be binding rules and constitutions to the Church," these after approval of the assembly should be remitted to presbyteries for consideration, and for report to the following year's assembly, "if the more general opinion of the Church thus had agreed thereto." The Barrier Act is relevant only in matters affecting the doctrine, government, worship, and discipline of the church.
J.D. DOUGLAS

BARROW, HENRY (d.1593). Church reformer. Graduate of Cambridge (1570) and member of Gray's Inn (1576), he was converted from a riotous and dissolute life through the chance hearing of a "loud-voiced preacher . . . and made the leap from a vain and libertine youth to preciseness in the highest degree" (Bacon). Giving himself to the study of the Bible, he became friends with John Greenwood,* with whom he shared a great respect for the works of Robert Browne.* He was detained in 1586 on orders of Archbishop Whitgift* while visiting the imprisoned Greenwood, tried in 1590 for circulating seditious books, and three years later was sentenced and hanged. His works, including *A True Description of the Visible Congregation of the Saints* (1589) and *A Brief Discovery of the False Church* (1590) were printed, as were most of his other works, including an account of his trials and investigations smuggled from prison, by his friends in Holland. He is sometimes claimed as the father of modern Congregationalism, but authorities differ as to his precise beliefs, and he himself vigorously disavowed the title of a mere "sectary." G.S.R. COX

BARROW, ISAAC (1630-1677). Mathematician and Anglican divine. A royalist in politics, he traveled widely in Europe and the East (1655-59). Successively professor of Greek at Cambridge (1660), professor of geometry at Gresham College, and first Lucasian professor of mathematics at Cambridge (1663), he resigned in favor of his pupil, Isaac Newton,* and gave himself to theology. Charles II made him royal chaplain, D.D. by royal mandate, and master of Trinity (1672). He became vice-chancellor of Cambridge University in 1675. His theological works (ed. A. Napier, 1859) include *A Treatise of the Pope's Supremacy,* a skillful piece of controversial writing. His mathematical works (ed. W. Whewell, 1860) were highly esteemed in his day.

HAROLD H. ROWDON

BARSUMAS (d.458). Monophysite* leader. An archimandrite, he promoted in Syria the "one nature" heresy of Eutyches,* whose cause he supported at the Council of Constantinople (448) and, with 1,000 of his militant Monophysite monks, at the Robber Synod of Ephesus* (449). There, at the invitation of Emperor Theodosius II, he represented the Eastern monks and voted with the bishops. His promotion of Eutychian views and his subscription to the condemnation of Flavian and Eusebius of Dorylaeum at Ephesus earned for him at Chalcedon (451) the reputation "assassin" who "has upset all Syria." Still loyal to Eutyches, he was exiled.

BARSUMAS (d.493). Bishop of Nisibis. The Monophysite* purge of Nestorians* at the Council of Ephesus (431) caused Barsumas (or Bar Sauma) to flee to Nisibis in Persia, where he became bishop. Under his influence Monophysite priests were massacred, and the Synod of Beth Lapat in 484 passed measures designed to dissociate the Nestorian cause from churches in the Roman Empire. All doctrines other than Nestorian and maintained by these churches were condemned; marriage of priests was legalized. Barsumas himself married a nun. He was assassinated by Persian monks in 493. He founded the Nestorian theological school at Nisibis. Six of his letters survive. G.T.D. ANGEL

BARTH, KARL (1886-1968). Swiss theologian. Born in Basel, he studied in Switzerland and Germany under some of the great liberal scholars of the day. His commentary on the epistle to the Romans, however, written in 1919 while he was a pastor of the Swiss town of Safenwil amid the tumult of World War I, broke with liberalism and established him as the leader of the new Dialectical Theology.* The liberal gospel of the fatherhood of God and the brotherhood of man was too shallow. Barth wanted to get beyond treating Christianity merely as an institution or a phenomenon in the history of religions, and to recover the reality witnessed to by the prophets and apostles. His understanding of Scripture was, however, colored by his reading of Dostoevsky* and Kierkegaard.* He stressed the transcendent hiddenness of God who reveals Himself in Christ. In so doing, God reveals both His grace and His judgment. The revelation of grace illuminates the sin and guilt of man even, or rather especially, in his religion.

Barth subsequently taught at the universities of Göttingen (1921), Münster (1925), and Bonn (1930) until his ejection from Germany when the Nazis came to power. He returned to his native town of Basel, where he taught until his retirement in 1962. Barth wrote over 500 books, papers, and articles, many of which have been translated into English, French, and other languages. They include important studies of Anselm, nineteenth-century theology and several NT commentaries. But his most enduring work is likely to be the monumental, but incomplete, *Church Dogmatics,* written largely at Basel and embodying his teaching there. The thirteen books are divided into four "volumes" dealing with: I. Prolegomena, the doctrine of the Word of God; II. the doctrine of God; III. Creation; and IV. Reconciliation. A fifth volume dealing with Redemption (eschatology) remained unwritten.

Barth's teaching changed and evolved over the years. But the theme of God's sovereignty in revelation through His Word can be traced from the commentary on Romans to his last works. In the *Church Dogmatics* it forms the basis for all theology. The knowledge of God occurs in the revelation of the Father, through the Son by the Holy Spirit. The basis of theology is thus the living Trinity Itself. The Word of God is not a thing or an object, but God Himself speaking. The Word of God has a threefold form: the Son as the Word of the Father, Scripture as the commissioned witness to that Word, and Christian proclamation. The three forms are in practice inseparable. The Word of God is known only through Scripture, and consequently all Christian proclamation (whether it be sermons, books, or any other form of testimony) must be tested by Scripture. Since God has chosen to reveal Himself in this way, Barth rejects as pointless, uninteresting, and sinful all forms of natural theology which attempt to find God by other means.

In the later volumes of the *Church Dogmatics* the thought of God's sovereign grace in His revealing Word is amplified by Barth's understanding of the incarnate Word. Barth came to regard the incarnation as the establishment of a covenant union of God and mankind in view of the union of God and man in Jesus Christ. This idea became the determining factor of Barth's later teaching. It became the key to his understanding of God, who is above all the kind of God who takes man into partnership with Himself. It was basic to his doctrine of Creation. God created the world with this union of God and man in view. Man is not to be understood in the abstract. The deepest truth about him can only be grasped by seeing man as he is in Christ.

Barth thought of sin, not as the transgression of an abstract law, but as man's attempt to break free from the grace in which he already stands. It is the attempt to live as if he were not God's covenant partner in Christ. Barth's teaching on redemption was a form of double predestination in which Christ is both the reprobate and the elect for all. On the cross He suffered rejection for all, so that all might be redeemed in Him. This brought Barth to the brink of universalism, though he refused to identify himself with that position.

Although he enjoyed a worldwide fame, Barth found himself increasingly isolated in later life. Conservative scholars complained that he had not done justice to the Bible's teaching about itself as the Word of God. Others complained that he was too biblicist and narrow in his conception of revelation. To many charges of his critics Barth had adequate answers. He had, moreover, insights which many contemporary theologians lacked. But he cannot be entirely exonerated of the charge of being more christocentric than the Bible, and (despite his own polemic against natural theology) of having erected a theology which was in many ways speculative on the basis of a biblical core. Barth's main contribution to theology was not the system that he constructed but the many profound insights and incentives to further thought that are to be found in his writings.

BIBLIOGRAPHY: C. Van Til, *The New Modernism* (1946); G.C. Berkouwer, *The Triumph of Grace in the Theology of Karl Barth* (1956); H. Bouillard, *Karl Barth* (3 vols., 1957); F.H. Klooster, *The Significance of Barth's Theology* (1961); K. Runia, *Karl Barth's Doctrine of Holy Scripture* (1962); T.F. Torrance, *Karl Barth: An Introduction to His Early Theology, 1910-1931* (1962); C. Van Til, *Christianity and Barthianism* (1962); G.H. Clark, *Karl Barth's Theological Method* (1963); H. Meynell, *Grace Versus Nature* (1965); H. Küng, *Justification: The Doctrine of Karl Barth and a Catholic Reflection* (1966); C. Brown, *Karl Barth and the Christian Message* (1967); C. O'-Grady, *The Church in the Theology of Karl Barth* and *The Church in Catholic Theology* (1968-69); J. Bowden, *Karl Barth* (1971); T.H.L. Parker, *Karl Barth* (1970).

For the most complete list of Barth's writings see *Antwort. Karl Barth zum siebzigsten Geburtstag* (1956), pp. 945-960; and *Parrhesia. Karl Barth zum achtzigsten Geburtstag* (1966), pp. 709-723.

COLIN BROWN

BARTHOLOMEW. Apostle. The name is a patronymic meaning "son of Tolmai" and may not have been his only name. He is mentioned only in the lists of the twelve apostles and not in connection with any incident in or after the ministry of Jesus. Attempts have often been made to identify him with Nathanael, who is mentioned in John's gospel but not in the synoptics. There are a number of difficulties connected with this view. Eusebius reports that he took the Gospel to India.

BARTHOLOMEW, GOSPEL OF, see APOCRYPHAL NEW TESTAMENT

BARTHOLOMEW OF THE MARTYRS (1514-1590). Archbishop of Braga, in Portugal. He became a Dominican friar in 1528 and later taught philosophy and theology for twenty years before, against his wishes, he was elevated to the archiepiscopate (1558). In the Trent sessions of 1562-63 he introduced clerical reforms, enforcing the conciliar decrees strictly in his provincial council of 1566. He began a seminary in his palace, instituted chairs of moral theology in Braga and Viana do Castelo, composed a catechism, preached vigorously, and visited methodically his nearly 1,300 parishes. Resigning his see in 1582, he retired to the Dominican priory of Viana do Castelo. He was the author of more than thirty works.

C.G. THORNE, JR.

BARTHOLOMEW'S DAY, MASSACRE OF ST. (1572). The massacre of Huguenots* in Paris and other French cities took place on the night of 23-24 August, the eve of St. Bartholomew's Day. It was instigated chiefly by Catherine de Medici, queen mother and for thirty years the real ruler of France. Despite papal demands that heretics should be killed, Catherine, who was a member of the Italian nobility and a strong admirer of Machiavelli, at first granted privileges to Protestants and sought to harmonize Protestant and Roman Catholic interests. When, however, Admiral G. de Coligny,* the acknowledged leader of the Huguenots, sought to use the French army in the Protestant cause by declaring war on Spain and later aided William of Orange in his revolt against Spain, Catherine arranged for his murder. The king's killer Maurevert fired at Coligny, but only wounded him. Catherine's complicity in the plot was suspected, but the king, expressing sympathy for Coligny, ordered an enquiry. Catherine, fearing reprisals, ordered the army to kill Coligny and with him all the Huguenots. Estimates of those murdered vary from 5,000 to 100,000, but the lower figures are now favored. The horror of the event was deepened by the circumstances that Protestant leaders had been invited to Paris to celebrate the wedding of Henry of Navarre, a Protestant, to Charles IX's sister Margaret, a Catholic.

R.E.D. CLARK

BARTHOLOMITES. Two groups have borne this name:

(1) A group of Armenian monks who fled to Genoa in 1306 and had a church dedicated to St. Bartholomew built for them. As their numbers were increased by further Armenian refugees, they spread throughout Italy. Initially Clement V authorized them to celebrate divine service according to their own rites, but later Innocent VI (1356) approved their adoption of the Roman liturgy and the Rule of St. Augustine. Boniface IX granted them the privileges of the Dominicans. After a period of decline they were suppressed by Innocent X in 1650.

(2) A German congregation of secular priests founded in 1640 by Bartholomew Holzhauser (1613-58) for the purpose of preaching, teaching,

and reviving the morals and discipline of the clergy and laity. Also known as the "United Brethren" and "Communists," they lived in communities under a superior but without vows. They never worked alone, but always two together. Following papal approval in 1680, they spread to many European countries, including England, Poland, and Italy. They became extinct in 1803, though several attempts have been made to revive them. JAMES TAYLOR

BARTOLOMMEO, FRA (1475-1517). Italian painter. A student of Cosimo Rosselli, he was later converted under Savonarola,* took orders in 1501, and joined the Dominican monastery of San Marco. For a time he abandoned the practice of painting, but resumed in 1504. By 1509 he assumed the leading role in the contemporary school of painting in Florence. He became the master of a shop that was the principal source of supply in the city of major church altar pieces. His mature paintings are grave and noble, an effect achieved more from impressiveness of structure than from any exposition of human content. His works include *Assumption of the Virgin, St. Catherine of Siena and the Magdalene,* and the *Marriage of St. Catherine.* ROBERT G. CLOUSE

BARTON, ELIZABETH (c.1506-1534). "Maid of Kent." Born at Aldington, Kent, England; she appears to have been neurotic, but her trances attracted attention and local admiration. She claimed to be in touch with the Virgin Mary, but her utterances had a political character. She condemned Henry VIII when he divorced Catherine of Aragon. Her later declaration that he was no longer king in the sight of God was regarded as likely to provoke rebellion. She was tried, confessed imposture (some say under duress), and was condemned and executed.

BASHFORD, JAMES WHITFORD (1849-1919). First resident bishop of the American Methodist Episcopal Church in China. Born in Fayette, Wisconsin, from college days he had aspired to go to China as a missionary. From 1876 to 1889, however, he served pastorates in Massachusetts, Maine, and New York. Then from 1889 to 1904 he was president of Ohio Wesleyan University. When in 1904 he was elected bishop, he asked to be assigned to China. For fourteen years he served in that land, speaking largely through interpreters but making a strong impact on the Christian movement. In 1907 he organized the China Centennial Thank Offering, which raised $600,000 for missionary work. He also assisted in organizing relief measures for famine sufferers. HAROLD R. COOK

BASILICA. The form of building used for Christian worship when Christianity was recognized as the official religion of the Roman Empire. It was a large, rectangular room divided into three sections by two rows of columns parallel with the longer side. The entrance was in one of the shorter sides.

See ARCHAEOLOGY AND THE EARLY CHRISTIAN CHURCH.

BASILIDES. Gnostic* thinker who taught in Alexandria during the reign of Hadrian (117-38). Different accounts of his teaching are given by Irenaeus and Hippolytus, but most scholars agree that Hippolytus more accurately represents Basilides, and Irenaeus the popularized system of his school. Basilides's system, though philosophically expressed, is a characteristic Gnostic myth which he claimed was descended from Peter. A nonexistent God generated out of nonexistence a Triple Sonship from which emerged Archons with authority over the universe and the world. These Archon-gods were ignorant of the nonexistent God, but the Gospel of Light descended to them and thence to Jesus. Through Jesus all "men of the Sonship" return above, but those left behind have no hope of salvation. The school of Basilides adopted a churchlike form and was characterized by practices of magical ritual. It survived in Alexandria at least until the end of the second century. C. PETER WILLIAMS

BASIL OF ANCYRA (fourth century). Arian* bishop of Ancyra. Succeeding Marcellus in the see in 336, Basil was deposed by the Western bishops at Sardica in 342. Reinstated by Constantius about 348, he came to share with George of Laodicea* leadership of the Homoiousion group of bishops. In 358 this group composed at Ancyra a memorandum describing the Son as neither "like" nor "identical with" the Father, but "like according to substance." Basil hoped to unite the episcopate behind the creed and encouraged Constantius to convene a united synod of bishops. At the twin synods of Arminum (Rimini) and Seleucia, the "Dated Creed" retaining the term "like" but excluding "substance" was favored by the Emperor Constantius. Its advocates, led by Acacius,* secured Basil's deposition in 360, and he was exiled to Illyricum, where he died. Coauthor with George of *A Memoir on the Doctrine of the Trinity,* Basil is credited with a treatise *De Virginitate.* G.T.D. ANGEL

BASIL OF SELEUCIA. Bishop of Seleucia in Asia Minor (c.448-458). He probably contributed to the Definition of Chalcedon the phrase "known in two natures." Although a correspondent of the pro-Monophysite Dioscorus, he condemned Eutyches* at the Home Synod (Constantinople, 448), but at the Ephesian Robber Synod* (449), where Dioscorus was prominent, he supported Eutyches and withdrew his earlier claim to worship "our one Lord Jesus Christ made known in two natures." Almost deposed at Chalcedon (451) for this, he based his vacillation on intimidation and recalled his orthodoxy in 448, adding that he had already signed the Tome of Leo. His extant writings include forty-one sermons, seven homilies, and a legendary account of St. Thecla. G.T.D. ANGEL

BASIL THE GREAT (c.329-379). Cappadocian Father.* Eldest son of Christian parents, and brother of Gregory of Nyssa* and Macrina, he was educated at home in Caesarea (Cappadocia) and Constantinople before going in 351 to the university of Athens. There friendships were formed with the young prince Julian and Gregory, another student from Cappadocia who was later to become famous as Gregory of Nazianzus.* Following his study at Athens, Basil returned to Caesarea about 356 and taught rhetoric with conspicuous success. He resisted attractive offers to undertake educational work in the city because he had already determined to devote himself to an ascetic and devotional life. About 357 he was baptized and ordained reader. This was followed by visits to monastic settlements in Palestine, Syria, and Egypt, which at a later stage helped him decide the nature of the community he wanted to establish. On his return to Pontus he retired to a small hermitage by the river Iris not far from his home. He left his seclusion in 364 at the request of his bishop, Eusebius, who was facing much opposition from extreme Arians,* was ordained presbyter, and proceeded to write books against Eunomius. After Eusebius's death in 370, Basil was made bishop of Caesarea, a role which was to bring him into controversy not only with the Arians but also with the Pneumatomachi* and the emperor. When Emperor Valens visited the province eager to impose Arianism upon a defiant Catholic Church, he was outclassed by the eloquent, forceful arguments of a dignified Basil.

Basil's contributions to the church and theology are threefold: (1) As an *ascetic,* he devoted much time to introducing and establishing the monastic system into Pontus, and extensive institutions sprang up under his fostering care. A novel feature of this was the Coenobium (Gr. *koinobios,* "living in community"), for hitherto ascetics had either lived in solitude or in groups of two or three. (2) As a *bishop,* he showed a genuine gift of leadership which is seen not only in his able directorship of the ecclesiastical affairs of Cappadocia but also in the application of the Gospel to the social needs of his people. On the outskirts of the city he built an elaborate and complex unit of hostels for the poor, a hospital, bishop's house, and clergy dwellings with an imaginative system of oversight. This was so comprehensive at the time that it was called the Newtown and was afterwards known as the Basilead. (3) As a *theologian* and *teacher,* he showed determination to uphold Nicene doctrine. *De Spiritu Sancto* and *Adversus Eunomium* attack Arian doctrines, but his chief contribution in this field was in fact his towering personality and popularity which made him an ideal mediator between East and West. Through his conciliatory influence, together with that of Gregory of Nazianzus and Gregory of Nyssa, the confusion over terminology was eventually resolved. Basil's many letters reveal him as a warm pastor who was concerned for the spiritual and physical well-being of his people. He died at fifty, a prematurely old man, worn out by his self-inflicted privations.

BIBLIOGRAPHY: W.K. Clarke, *The Ascetic Works of St. Basil* (1925); E.R. Hardie (ed.), *The Christology of the Later Fathers* (1954); J.W.C. Wand, *Doctrines and Councils* (1962).

G.L. CAREY

BASLE, CONFESSION OF (1534). Published by the city council with a preface by the burgomaster, Adelberg Meyer, and signed by Heinrich Ryhiner, the clerk, it was authored by Oswald Myconius,* but he probably used a draft composed by his predecessor, John Oecolampadius (d.1532). The confession, having not more than 2,500 words, is divided into twelve articles: Concerning God, Man, God's Care for us, Christ, the Church, our Lord's Supper, the use of Excommunication, Government, Faith and Works, the Day of Judgment, Things commanded and not commanded, Against the error of the Anabaptists. The citizens of Basle subscribed to it under oath, and their practice was followed later by the city of Mühlhausen—hence the title *Confessio Muhlhusana.* From 1534 to 1826 the confession was read each year in Holy Week from the pulpits of the churches of Basle, and until 1872 all ministers had to subscribe to it.

For the text see P. Schaff, *Creeds of Christendom,* vol. 3; and A. Cochrane, *Reformed Confessions* (1966).

PETER TOON

BASLE, COUNCIL OF (1431-49). Its claim to being the seventeenth ecumenical council is contested by Roman Catholic theologians on the grounds that papal recognition was not given to its decrees and because they consider it did not truly represent the universal church. The council was called by Martin V on the eve of his death and took place during the papacy of Eugenius IV. Presided over by Julian Cesarini,* it was widely representative, including members of the lower and higher clergy. It was well supported by the universities and the European princes. The council inherited the tasks of the previous council at Constance,* i.e., the extirpation of heresy, reform of the church, and the peace of Christendom. There was conflict from the beginning between council and pope. Eugenius IV, apprehensive because of reports on the council, issued a bull (18 December 1431) dissolving the assembly. In reply, the council, supported by Cesarini and influential men including Nicholas of Cusa,* confirmed the decrees of the Council of Constance on the superiority of the general council over the pope. Eugenius, aware that considerable opinion in the church and in Europe was against him, yielded and recognized the legality of the council (15 December 1433). The council then attempted to carry out reforms including the imposition of restrictions on papal legates, decrees on the nominations of cardinals, and other matters affecting the Curia.

Angered by these interferences in what he regarded as his domain, Eugenius denounced the council in 1436 in a memorandum to all Catholic princes. Dissension grew within the council, caused by a revolutionary spirit of defiance and criticism against the papacy. Eugenius was able to utilize this spirit in his efforts to unite the Latin and Greek wings of the church. He called the council to Ferrara to discuss the proposals of Emperor John VII Palaeologus, and the patriarch of Constantinople. The minority of the council, who supported the pope, came to Ferrara, and on 5 July 1439 the union was proclaimed. Those who had remained at Basle deposed Eugenius IV and appointed Amadeus VIII of Savoy as Felix V (the last of the antipopes). The council dragged on for ten more years, moving to Lausanne in 1448; it dissolved on 25 April 1449. Felix V abdicated, and the council recognized Nicholas V. The church had kept its monarchical form. As a result of the council, the individual states became increasingly independent in ecclesiastical affairs. Princes had exploited the quarrels between popes and council and had restricted the intervention of Rome within their states. The church's authority over the temporal domain had been decreased. In dealing with heresy the council had a measure of success. By negotiation and compromise it reached agreement with the moderate wing of the Hussites (Utraquists*). In 1433 the Compactata of Prague was ratified, conceding Communion in both kinds to the laity.

JAMES TAYLOR

BASNAGE, JACQUES (1653-1723). Huguenot scholar. Educated at the Huguenot seminary at Saumur, and at Geneva (under F. Turretin) and Sedan (under P. Jurieu), he was ordained in 1676 but nine years later was forced to flee to the Netherlands when Louis XIV revoked the Edict of Nantes.* He settled in Rotterdam as pastor of the Walloon (French) church there. The Dutch statesman Heinsius (Heyns), impressed with his talents, asked him to become preacher in The Hague, where he could also assist in contacts with the French government; he aided Heinsius in preparing the Peace of Utrecht, and later the Triple Alliance (1717) with France and England. His unceasing attempts to gain toleration for the Huguenots were unsuccessful; despite this, he opposed the last-ditch revolt of the Huguenot Camisards (1702) and counseled obedience to the French monarchy. As a scholar, Basnage wrote several works in church history, especially *Histoire de la religion des églises réformées* (1690), a skillful rejoinder to Bossuet's derogatory history of the Reformation (1688).

DIRK JELLEMA

BASSENDYNE BIBLE. The first edition of the Bible in English published in Scotland. Thomas Bassendyne (d.1577), the king's printer in Edinburgh, was granted a license by the privy council to print it. The NT appeared in 1576, the whole Bible in 1579. The version used was that of the Geneva Bible.*

BATIFFOL, PIERRE (1861-1929). Roman Catholic scholar. Born at Toulouse, he studied at the *Seminaire de St. Sulpice* under the Abbé Duchesne.* For two years he worked in Rome, collaborating in the publication of *Melanges d'Archaeologie et d'Histoire,* dealing with papal history. In 1888 he was appointed to teach at the *Collège Ste Barbe* in Paris, where he worked for the next decade. Afterward he became *recteur* of *L'Institut Catholique de Toulouse.* During the time since his residence in Rome he was constantly doing research and writing on Roman Catholic history. He took a strong stand against modernism, but his book on the Eucharist was placed on the Index,* and although he made required corrections he found it necessary to leave Toulouse for Paris. He later took part with Cardinal Mercier at the conversations at Malines,* and in 1928 represented the pope at the International Historical Congress. He died in Paris.

W.S. REID

BAUER, BRUNO (1809-1882). German radical scholar. He reacted from his early right-wing Hegelian Christian position to become one of the most negative NT and theological critics of his day. Deprived of his license as a university teacher in 1842 (he had taught at Berlin and Bonn), he retired from the academic theological world, and through engaging in politics became known as the "Hermit of Rixdorf." His bitterness against "the theologians" influenced his criticism. Against D.F. Strauss's view that the gospels came out of the mythopoeic imagination of the community, Bauer argued that they were creations of individual artists, Mark's gospel being the source. He came to believe that there never was a historical Jesus, and that Christianity originated from Greco-Roman civilization in the second century A.D., when all the NT writings were composed. He thought too that Christianity was "the misfortune of the world" standing in the way of free full humanity. Neglected by theologians of his day, Bauer has received more attention recently; Schweitzer* believed the value of his questions for gospel criticism outweighed the inadequacy of his own answers.

HADDON WILLMER

BAUER, WALTER (1877-1960). German evangelical theologian and lexicographer. Professor at Göttingen (1916-45), he was the author of *Rechtglaubigkeit und Ketzerei im ältesten Christentum* (1934; ET 1972), which is still provoking fruitful discussion, though its thesis is not universally accepted. In contrast with the traditional view that heresy was a departure from an already existing orthodoxy, Bauer suggested that what later became known as heresies were at first widely held forms of Christianity, and that orthodoxy was the outcome of the centralizing orderly influence of Rome overcoming various less coordinated but not necessarily less legitimate types of Christianity. Bauer was responsible also for the *Griechischdeutsches Wörterbuch zu den Schriften des NT und der übrigen urchristlichen Literatur,* a definitive NT lexicon which was the basis of the English *Greek-English lexicon* ... edited by W.F. Arndt and F.W. Gingrich (1957).

HADDON WILLMER

BAUR, FERDINAND CHRISTIAN (1762-1860). Leader of the "Tübingen School"* of German radical biblical criticism. Almost his entire career was spent at the University of Tübingen. His vast writings centered on NT criticism, church history, and historical theology. Among them was his *Paul the Apostle of Jesus Christ* (2 vols., 1845; ET 1875), in which he argued that of the epistles attributed to Paul, only Romans, Galatians, and 1 and 2 Corinthians were genuine. He reached this conclusion on the basis of a hypothesis which posited a fundamental conflict in the early church between the Jewish party led by Peter and the Hellenist party led by Paul. The four "authentic" letters are regarded as genuine because of their anti-Judaizing tendencies (their rejection of the law, circumcision, etc.) and their wider conception of God. The remaining letters are regarded as late and inauthentic because they do not reflect these tendencies, but exhibit others which Baur placed in a later period. The gospels were all late, John being placed in the second half of the second century.

It has often been claimed that Baur's biblical criticism was the result of the application of the Hegelian dialectic of thesis, antithesis, and synthesis in history (the thesis being primitive Jewish Christianity, the antithesis Hellenistic Christianity, and the synthesis the catholicism of the early church). Examination of his writings does not bear this out. This dialectic was not even characteristic of Baur's philosophy. The decisive factor in Baur's approach was not a preconceived philosophy, but his tendency criticism which anticipated a good deal of contemporary redaction criticism. It may be faulted, however, for seeing tendencies where there were none, and mistaking others. Baur pioneered an anti-theistic, nonsupernatural approach to history and Christian origins. In common with Hegel he insisted that history could not be understood without philosophical speculation and that the proper way to understand it was to see it as the forward movement of the Spirit in temporal and particular forms. But his interpretation of NT theology had much in common with Schleiermacher* also. It saw Christianity as an expression of human self-consciousness and Jesus as the mediator of a higher religious and moral awareness.

BIBLIOGRAPHY: W. Geiger, *Spekulation und Kritik: Die Geschichtstheologie Ferdinand Christian Baur* (1964); P.C. Hodgson, *The Formation of Historical Theology: A Study of Ferdinand Christian Baur* (1966): full bibliography; H. Liebing, "Historical Critical Theology. In Commemoration of the One Hundredth Anniversary of the Death of Ferdinand Christian Baur," *Journal for Theology and Church* (1967), pp. 55-69; P.C. Hodgson (ed.), *Ferdinand Christian Baur on the Writing of Church History* (1968). A five-volume selection of Baur's writings in German is being edited by K. Scholder, *Ausgewahlte Werke in Einzelausgaben* (1963-); H. Harris, *The Tübingen School* (1975).

COLIN BROWN

BAVINCK, J.H. (1895-1964). Dutch Calvinist writer. Educated at the Free University and at Erlangen, he was early attracted by the field of religious psychology. In 1921 he went to the Dutch East Indies and, apart from a pastorate in the Netherlands (1926-29), spent the next two decades there, increasingly involved in mission work and writing extensively. In 1939 he became professor of missions at Kampen and at the Free University; worked in the anti-Nazi underground during World War II; and thereafter resumed his teaching and writing career. He founded the Calvinist *(Gereformeerde Kerk)* center for missions at Baarn. His best-known work is *Inleiding in de Zendingswetenschap* (1954), translated as *An Introduction to the Science of Missions* (1960). Other writings deal with religious consciousness, mysticism, and related topics. DIRK JELLEMA

BAXTER, RICHARD (1615-1691). Puritan divine. Born in Rowton, Shropshire, he attended the free school at Wroxeter, but attained his education largely through self-instruction, private study, and introspection. In his late teens he was in London under the tutelage of Sir Henry Herbert, Master of Revels. Baxter left London, displeased with the quality of life. He then came under the influence of two Nonconformists, Joseph Symonde and Walter Craddock. By 1638 he was ordained by the bishop of Worcester, and the following year was nominated to the mastership of the Free Grammar School at Bridgnorth. The Et Cetera Oath (1640) brought about his rejection of episcopal polity. From 1641 to 1660 he served a parish in Kidderminster where his Latitudinarian* views became more evident as he tried to put them in practice in working with his parish and the other clergy in the area.

During the Civil War his sympathies were basically with the Parliamentarians, but he came to oppose the Solemn League and Covenant* and some of Cromwell's aims. He sought also to curb the views of sectarians and republicans. By 1647 he left the army and returned to Rouse Lench, where he wrote *The Saints' Everlasting Rest* (1650). He welcomed the Restoration and was made chaplain to Charles II and offered the bishopric of Bedford, which he rejected. At the Savoy Conference* (1661) he presented a revision of the Book of Common Prayer for Nonconformists and served as their leader. In 1662 the Act of Uniformity deprived him of an ecclesiastical living, but the wealth, position, and love of the bride he took that year sustained him until her death in 1681. Although excluded from the Church of England, Baxter continued to preach and was imprisoned as a result in 1685 and 1686. He took part in the overthrow of James II and welcomed the Toleration Act of William and Mary. His many other works include *The Reformed Pastor* (1656) and *Reliquiae Baxterianae* (1696).

BIBLIOGRAPHY: *Works* (23 vols., ed. W. Orme, 1830); F.J. Powicke, *A Life of the Reverend Richard Baxter* (1924); G.F. Nuttall, *Richard Baxter and Philip Doddridge* (1951); H. Martin, *Puritanism and Richard Baxter* (1954); R. Schlatter (ed.), *Richard Baxter and Puritan Politics* (1957); G.F. Nuttall, *Richard Baxter* (1962).

ROBERT SCHNUCKER

BAYLE, PIERRE (1647-1706). French writer. Son of a Reformed minister, he became an educator, teaching first at the Huguenot Academy of Sedan and then, when that institution was closed (1681), at Rotterdam. In 1682 he published a work on comets that indicated religion and morality were not related. In later books he advocated universal toleration and a conciliatory attitude toward the French government (even after the revocation of the Edict of Nantes*). These ideas antagonized important French Protestant leaders such as Pierre Jurieu,* who supported the school where Bayle taught, and led to his dismissal (1693). Thereafter he devoted himself to his *Dictionnaire historique et critique* (1695-97; rev. ed., 1702). This encyclopedic work contains many notes informed by Bayle's skeptical philosophy which made the work popular during the eighteenth-century Age of Enlightenment.*

ROBERT G. CLOUSE

BAYLY, LEWIS (1565-1631). Bishop of Bangor. Native of Wales and educated at Oxford, he graduated D.D. in 1613. His series of sermons while vicar of Evesham became the basis of his famous book, *The Practice of Piety,* a devotional handbook used in many Puritan households. The date of the first edition is unknown, but the third appeared in 1613, and the fifty-ninth in 1735. It was translated into many languages, including an edition for some American Indians. Despite his outspoken Puritan views, Bayly became a chaplain to Prince Henry. A period of disfavor after the latter's death ended in 1616 when he was appointed first chaplain to the king, then bishop of Bangor. He was imprisoned briefly in 1621 for opposition to royal policy. He resided in his see and spoke Welsh, both of which were unusual for Welsh bishops at the time. But his own conduct was not above suspicion, and many charges were brought against him later in life. JOHN TILLER

BAY PSALM BOOK. The earliest book known to have been printed in English in the British North American colonies. Compiled by John Eliot,* Richard Mather, and Thomas Weld, and printed by Stephen Day at Cambridge, Massachusetts in 1640, it was intended to give a more literal translation of the Hebrew Psalms than that of the King James Version, and yet be suitable for congregational singing. *The Whole Book of Psalmes Faithfully Translated into English Metre* became the approved hymnal of the Massachusetts Bay Colony and was popular in other colonies and in Britain.

BEACH, HARLAN PAGE (1854-1933). Congregational missionary and professor of missions. Born in New Jersey, he graduated from Yale (1878) and Andover Theological Seminary (1883). Sent by the American Board of Foreign Missions to Tung Chou, China in 1883, he returned to the USA in 1890 because of his wife's poor health. During his term in China he developed a system of shorthand for the Mandarin language. He served successively as home representative for the American Board; pastor of Lowrie Hill Congregational Church, Minneapolis; superintendent for the School for Christian Workers, Springfield, Massachusetts; educational secretary of the Student Volunteer Movement*; and finally professor of the theory and practice of missions at Yale (1906-21) and Drew Seminary (1921-28). His most important works concerned missions, such as the *World Missionary Atlas* (1925).

DONALD M. LAKE

BEADLE. Sometimes called the church officer, the beadle has been for four centuries one of the most distinctive personalities in the life and lore of the Scottish churches. "Minister's man," custodian of the church buildings, guide, philosopher and friend to old and young alike, he has figured in more anecdotes and been etched by more pens than anyone except the minister himself.

BEARDS. These have been generally characteristic of the Caucasian races. In the Orient in particular it was a sign of male dignity. In the OT the cutting of the corners of the beard was forbidden (Lev. 19:27), and the removal of the beard was a disgrace (2 Sam. 10:4,5). This led in later Judaism to religious prescription for the beard. Since Roman times there have been many arbitrary switches of beard fashions in society in general, from no beards to beards in a multiplicity of styles. For example, rulers such as Louis XIII, Francis I, and Henry VIII were notable trendsetters. These variations in fashion without any other significance were largely reflected in church custom. The painter James Ward (1769-1859), however, gave eighteen scriptural reasons "why man was bound to grow a beard unless he was indifferent as to offending the Creator and good taste," in his *Defence of the Beard.* In 1860, "Theologos" produced a book entitled *Shaving a breach of the Sabbath and a hindrance to the spread of the Gospel.* HOWARD SAINSBURY

BEATIFICATION. The papal decree permitting the public ecclesiastical veneration of a faithful Catholic after his death. This veneration is limited to a particular locality and is not binding on the whole church. It is unlawful to pay to the person known as the Blessed (Beatified) public reverence outside of the place for which permission is granted. Up to the seventeenth century the public service of beatification was performed by the bishops in their local churches. Unlike decrees of canonization, those of beatification are not regarded as infallible: they only grant permission and do not give commands. Also, the cultus permitted by beatification is restricted to a determined province, city, or religious order.

S. TOON

BEATIFIC VISION. In heaven the just see God by direct intuition, clearly and distinctly. Knowledge so gained is called a vision because it is immediate and direct in contrast with the knowledge of God, which is attained in this life indirectly in creation. The primary object of the Beatific Vision is God as He is. In beholding God face to face the intelligence finds perfect happiness. To enable it to see God, the intellect of the blessed is supernaturally perfected by the light of glory. According to Roman Catholic theology, the Beatific Vision is the ultimate destiny of the redeemed. It is believed by some theologians that this vision is bestowed in exceptional circumstances for brief periods in this life. For example, Thomas Aquinas held that it was granted to Moses (Exod. 34:28-35) and to Paul (2 Cor. 12:2-4).

S. TOON

BEATON, DAVID (1494-1546). Archbishop of St. Andrews and later cardinal. After attending St. Andrews and Glasgow universities, he probably graduated from the latter before pursuing his studies for some years at Paris and Orléans. By 1525, through the influence of his uncle James Beaton, archbishop of St. Andrews, he sat in Parliament as abbot of Arbroath. Sent as emissary to Francis, king of France, in 1533, he made such an impression that in 1537 Francis had him consecrated bishop of Mirepoix in Languedoc. One month later Beaton was made cardinal. In 1539 he succeeded his uncle at St. Andrews, and became also lord high chancellor in 1540. His French connections contributed to his dislike of the English seen in his political policies. Beaton had the Scottish Reformer George Wishart* burnt at the stake in 1545, but was himself assassinated in St. Andrews the following year.

J.D. DOUGLAS

BEATUS (d.798). Spanish presbyter, abbot of the Benedictine monastery of St. Martin's at Liébana, near Santander, diocese of León, and counselor of Queen Adosinda of León. Expecting the sixth age of the world to end in 800, he compiled, largely from earlier writers from Irenaeus to Isidore, a *Commentary on the Apocalypse,* in three recensions (776, 784, 786). This catena gathers up the Western tradition of Apocalyptic commentary, and in particular preserves much of Tyconius's* lost work and possibly unknown sections of Apringius's (mid-sixth-century bishop of Beja in S Portugal). The richly illustrated MSS of Beatus's *Commentary,* ranging from the ninth to the thirteenth centuries, are of major significance for medieval Spanish art. Beatus opposed the Adoptianist* Christology of Elipandus, and with his disciple and fellowmonk Etherius, bishop of Osma, wrote in 785 *Against Elipandus.* He was also a hymnwriter.

D.F. WRIGHT

BEAUFORT, HENRY (c.1375-1447). Bishop of Winchester and cardinal. Born at Beaufort, Anjou, France, the illegitimate son of John of Gaunt and Catherine Swynford, he received rapid ecclesiastical preferment, and became bishop of Lincoln in 1398, and of Winchester in 1404. Under Henry V he became chancellor, and loans from his large fortune financed Henry's military expeditions. At the Council of Constance* in 1417 he was largely responsible for the election of Martin V as pope, but was prevented by the king from accepting the pope's offer of a cardinal's hat. As a guardian of the infant Henry VI and as chancellor, Beaufort virtually ruled the realm, but over the next twenty years he was involved in continual conflict with the duke of Gloucester. He became cardinal in 1426.

JOYCE HORN

BEAUFORT, MARGARET (1443-1509). English patroness of learning. Countess of Richmond and Derby, she was born at Bletsoe, daughter of the duke of Somerset, and was married four times. She helped Henry VII gain the throne and arranged his marriage. Guided by her confessor, John Fisher, she founded the Lady Margaret professorships of divinity in Oxford (1502) and Cambridge (1503), a preachership at Cambridge, and Christ's and St. John's Colleges, Cambridge.

BECK, JOHANN TOBIAS (1804-1878). German theologian. Ordained in 1827, he taught systematic theology at Basle (1836-43), then in Tübingen where he gradually won considerable influence in an alien Hegelian atmosphere. He was a great Christian personality, developing the biblicism of J.A. Bengel.* Free from a mechanical view of biblical inspiration, he saw the Bible as the history of the work of the Spirit progressing toward the salvation of man, and the revelation of the kingdom of God as a supernatural reality in history, shared wherever men gave themselves to practical discipleship. He incurred the hostility of confessional Lutherans for his view that in justification man was made as well as reckoned righteous. In many ways he stood in the Pietist tradition, though freely critical of some of its organizations.

HADDON WILLMER

BECKET, THOMAS (c.1118-1170). Archbishop of Canterbury from 1162. Born of a well-to-do Norman family in London and educated in England and France, he received in 1141 a position in the court of Archbishop Theobald of Canterbury and was sent to university for legal training. After serving for a time as archdeacon of Canterbury, he became chancellor to Henry II. The king and his minister became good friends and spent their time, not only in government work, but also in drinking and carousing. When Theobald died in 1162, Henry appointed Thomas archbishop of Canterbury. Since at the time Becket was still in minor orders, it was necessary to ordain him priest and consecrate him bishop on the same day. As primate of England, Becket was transformed from a stalwart supporter of royal policy to an ardent champion of the church. He resigned his position with the king's government because he considered it a conflict of interest and adopted a pious lifestyle. Henry was disgusted and angry with this turn of events and so in addition to the conflict between the ecclesiastical and royal interests there was a showdown between two forceful personalities.

Henry wished to recover the royal authority over the church which had been lost during the reign of Stephen, his predecessor. The immediate occasion for conflict was the king's efforts to prosecute in royal courts clergymen already tried and convicted in church courts, in order to sentence them to harsher penalties. Thomas refused to allow this on the grounds that it was double jeopardy. In response, Henry issued the Constitutions of Clarendon* which declared that the king, not the pope, was to have authority over the English Church and that he recognized the pope's authority only in a nominal way. Thomas agreed at first to these ideas, but then recanted and fled to the Continent. There he persuaded the pope to condemn many of the Constitutions of Clarendon as uncanonical and to excommunicate churchmen who followed them.

Becket returned to England, and in 1170 was murdered by four overzealous knights who carried out the king's wish, indiscreetly murmured, that someone would rid him of the archbishop. All Europe was outraged at Becket's martyrdom and Henry was forced to do penance at Avranches, France (1172) for the act. The king also found it necessary to allow the exercise of church powers that Thomas had insisted upon. Although canonized by the Roman Catholic Church in 1173, historians have debated as to whether Becket should be characterized as a saint, traitor, fanatic, or politician.

See W.H. Hutton, *Thomas Becket, Archbishop of Canterbury* (1926); and M.D. Knowles, "Archbishop Thomas Becket: A Character Study," in *Proceedings of the British Academy*, vol. XXXV (1949), pp. 177-205. ROBERT G. CLOUSE

BECON, THOMAS (c.1512-1567). English Reformer. Born in Norfolk and educated at Cambridge, he became vicar of Brenzet in Kent and of St. Stephen's, Walbrook, and subsequently chaplain to Cranmer* and Protector Somerset. On the death of Edward VI (1553) he was committed to the Tower of London, but was released soon afterward due, according to Foxe, to mistaken identity. He fled with Bishop Ponet to Strasbourg, and later moved on to Frankfurt and Marburg. On the accession of Elizabeth he became a canon of Canterbury and held various benefices. His writings extended to three volumes in the Parker Society series (ed. J. Ayre, 1843-44), and he may have contributed "The Homily against Adultery" to the *Book of Homilies.* Becon, who had studied under Latimer,* was not perhaps the most original thinker of his time, but he was a vigorous writer, popular, and knew how to express his views. He influenced Cranmer over the Black Rubric.

See D.S. Bailey, *Thomas Becon and the Reformation of the Church in England* (1952).

G.E. DUFFIELD

BEDE (Baeda) (c.673-735). Monk of Jarrow and "the Father of English history." Born at Monkton on Tyne, County Durham, he was taken at the age of seven to the newly founded monastery of Wearmouth a few miles away, moving almost at once to become one of the first members of the community at Jarrow, near his birthplace. He spent the whole of the rest of his life there, never traveling outside Northumbria so far as is known, and yet he became one of the most learned men in Europe. The scholarship and culture of Italy had been brought to England by Theodore of Tarsus,* who was archbishop of Canterbury in the early years of Bede's life, and it was introduced into Wearmouth and Jarrow by Benedict Biscop.* Here it coalesced with the simpler traditions of devotion and evangelism which came from the Celtic Church.* This caused Northumbria to be a beacon of Christian learning while darkness was gathering on the Continent, and Bede was the foremost example and promoter of that learning.

He grew up at Jarrow under Ceolfrith, from whom he learned the love of scholarship and personal devotion and discipline. When an epidemic swept the monastery, only Ceolfrith and Bede were left, and he records how they managed to maintain the regular divine worship. He was made deacon at the age of nineteen and priest at thirty by John of Beverley, bishop of Hexham. He learned Latin, Greek, and Hebrew. He had a good knowledge of classical authors, which often had to be acquired from books of extracts or quotations in other people's writings. He was familiar with the works of Ambrose, Jerome, Augustine, and Gregory the Great, and he knew something of Anglo-Saxon poetry.

Bede's writings cover a wide range, including natural history, chronology, biblical translation, and exposition. Most important was his *Historia Ecclesiastica Gentis Anglorum* ("Church History of the English People"). He is described as "the Father of English history" partly because he was the first to try to write any kind of history of England at all, for he sets the story of the church in the general history of the nation. But this title is also due to him because of his methodology. His thorough scholarship is known, for example, by his asking friends to search the archives of the Roman Church and bring him copies of documents which he needed to see. He also had copies made of epitaphs. Where there was nothing in writing he tried to consult the best available oral tradition. He was not always critical, but his work is nonetheless invaluable and his stories are told with great charm. The account of his finishing his translation of John's gospel before his death is deservedly famous. His fame continued after his death, when he began to be known as "the Venerable Bede." His bones were removed to Durham to the coffin of Cuthbert, and in 1370 placed in a special tomb in the cathedral.

See his *History*, ed. B. Colgrave and R.A.B. Mynors (1969); and P.H. Blair, *The World of Bede* (1970). R.E. NIXON

BEECHER, HENRY WARD (1813-1887). American Congregational minister. Son of Lyman Beecher,* he was born in Litchfield, Connecticut, and graduated from Amherst College and Lane Theological Seminary, Cincinnati, of which his father was president. He was refused ordination by the Old School Miami (Ohio) Presbytery and was ordained by the New School Presbytery of Cincinnati in 1838. After two Indiana Presbyterian frontier charges he was pastor of the Plymouth Church of Brooklyn (Congregational) from 1847 to 1887. He became world famous as a dramatic and witty preacher, and he was also a political activist and reformer. Theologically he departed from Calvinism and radically reinterpreted the Bible in moralistic terms, laying great stress on God's love. His disbelief in a literal hell and his acceptance of the evolutionary hypothesis brought criticism, but none so great as an unproven adultery charge which overshadowed his later years. His works include *The Plymouth Pulpit* (10 vols.) and *Yale Lectures on Preaching* (1872-74). He was founder and for over ten years editor of the *Christian Union*, later *The Outlook.*

ROBERT C. NEWMAN

BEECHER, LYMAN (1775-1863). American minister and educator. Born in New Haven, he graduated from Yale in 1797, was ordained two years later, and was pastor first of the East Hampton Presbyterian Church, New York (1799-1810), then of Congregational churches at Litchfield, Connecticut (1810-26), and Hanover Street, Boston (1826-32). He then became president and professor of theology at Lane Theological Seminary, which post he combined until 1842 with the pastorate of Cincinnati's Second Presbyterian Church. He retired from Lane in 1852. Liberal in theology, he rejected Unitarianism, rigid Calvinism, and Roman Catholicism, and was an active foe of intemperance, slavery, and dueling. While his moderate Calvinism had made the going hard in New England, it was not strong enough for Ohio, where his allegedly heretical views led to arraignment but acquittal before presbytery and synod. A founder of the American Bible Society, he was an eloquent preacher with revivalistic emphases that brought conversions. He was said to have been "the father of more brains than any other man in America" (among his thirteen children were Henry Ward Beecher* and Harriet Beecher Stowe*). J.D. DOUGLAS

BEETHOVEN, LUDWIG VAN (1770-1827). German composer. His extraordinary gifts were early recognized by Neefe, the cathedral organist in Bonn, who permitted the boy to deputize at the organ, guided his early efforts at composition, and introduced him to some of Bach's* music. In early manhood Beethoven moved to Vienna, where he soon attained fame as a pianist and composer. Before he was thirty he started losing his hearing, but he persevered to become one of the greatest instrumental composers of the nineteenth century. While he cannot be classed as a church composer, he wrote some very outstanding sacred compositions: six sacred songs to texts by the Christian poet Gellert; an oratorio, *Christ on the Mount of Olives,* from which the "Hallelujah" chorus is often performed; the *Mass in C,* a fine example of the classical symphonic Mass; and the *Missa solemnis,* one of the most monumental choral works of all time. J.B. MAC MILLAN

BÉGUINES; BÉGHARDS. Béguines were members of sisterhoods founded in the Netherlands in the twelfth century. Without common rule or hierarchy, free to hold private property, and with leave to marry, they lived austerely without vows, save chastity, and they emphasized manual work. Béghards were their male counterparts—usually weavers, dyers, or fullers—who held no private property and had a common fund. Their names might be derived from Lambert le Bègue ("the Stammerer"; d.1177), a revivalist preacher at Liége. Mainly Netherlandish, there were communities in Germany and France (Louis IX founded a béguinage in Paris in 1264). Called extra-regulars as they were neither lay nor monastic, they served the sick and indigent together with their contemplation. Both groups were long suspected of heresy, chiefly because of their association with the Spiritual Franciscans, and were condemned by the Council of Vienne* (1311). Many Béghards reformed and were permitted to carry on (by John XXII) and survived until the French Revolution. The Béguines were long persecuted, being little more than charitable institutions from the fifteenth century. They became rehabilitated in Belgium, even to establishing closer relations with approved orders and adopting the Austin rule. Some still exist in the Low Countries, notably in Bruges and Ghent. C.G. THORNE, JR.

BEHMENISTS, BEHMENITES, see BOEHME, JAKOB

BEISSEL, JOHN KONRAD (1690-1768). Mystic and founder of the Ephrata Society.* Born in Eberbach (Palatinate), Germany, he was educated at the University of Halle, early adopted the views of the Dunkards, and because of religious persecution emigrated to America in 1720. He settled in Pennsylvania where he came under the influence of the Seventh-Day Baptists. Highly respected for his piety, he soon gathered a following which he rebaptized, constituting the Ephrata Society. Members changed their names, and Beissel was given the name Gottrecht Friedsam ("godly and peaceable"). Several of his hymns were published by Benjamin Franklin. After Beissel's death the Ephrata Society gradually declined. DONALD M. LAKE

BELGIAN CONGO, see ZAIRE

BELGIC CONFESSION *(Confessio Belgica).* A major Calvinist creed, one of the three standards used in the Dutch Reformed Church and its offshoots (the two others are the Heidelberg Catechism and the Canons of Dort). It was written primarily by Guido de Bres* (Guy de Bray) in 1561; at the time, de Bres was a hunted man in his late thirties, preaching illegally as the Lowlands neared revolt against Spanish rule. Modifications in detail were made by Adrian de Saravia, H. Modetus, G. Wingen, and F. Junius. The confession was written by de Bres in French, but was immediately translated into Dutch, and soon into German (1566). Received enthusiastically by the Calvinist churches of the Lowlands, it was adopted by a synod at Antwerp (1566), by synods at Dort (1574) and Middelburg (1581), and finally by the national synod of the Netherlands (northern Lowlands) after independence from Spain had been gained, in 1619. It draws heavily on the 1559 Gallic Confession,* written for the Huguenot churches by Calvin and by Antoine de la Roche Chandieu. DIRK JELLEMA

BELGIUM, see LOW COUNTRIES

BELL, GEORGE KENNEDY ALLEN (1883-1958). Prominent ecumenist and Anglican bishop. Ordained in 1907, he became an Oxford don in 1910 and chaplain to Archbishop R.T. Davidson* in 1914. He was at Lambeth during World War I and the reconstruction afterward. In particular he was involved in the early stages of the ecumenical movement and in the 1920 Lambeth Conference.* He was later Davidson's biographer. In 1924 he became dean of Canterbury, and in 1929 bishop of Chichester, where he remained until his retirement only a few months before his death. He was secretary of the 1930 Lambeth Conference and a member of the 1948 and 1958 conferences.

His ecumenical concerns date from his time at Lambeth. He was present at the Oud Wassenaar Conference in Holland in 1919, at which Christians from European and other nations involved in the war consulted about peace and reconciliation. This led by several steps to the Life and Work movement which was inaugurated by the Stockholm Conference* of 1925. Bell was one of those who drafted the conference message. In 1932 he became a president of the movement and chairman of its council, and despite his own concern for peace and disarmament thus became involved in the movement's confrontation with Nazi anti-Semitism and with the "German-Christians."* In the course of this he became deeply sympathetic with Martin Niemöller* and Dietrich Bonhoeffer.* After 1939 his understanding of the anti-Hitlerite elements in the German nation made him advocate a more open approach to concluding the war than was possible with the policy of "unconditional surrender," and also to deplore the saturation bombing of German cities in his speeches in the House of Lords. In these he stood alone. He even met Bonhoeffer in Sweden in 1942 to discuss the possibilities of ending the war from within Germany, and he incurred the strong hostility of Churchill. Rumor has credited the latter with blocking Bell's appointment to Canterbury when Temple died in 1944, but it was by no means probable that he would have been appointed in any case.

Bell's continuing ecumenical work helped to lead Life and Work on from the Oxford Conference of 1937 to the World Council of Churches, which was constituted in Amsterdam in 1948. His interest remained in the church and international order. At Evanston in 1954 he was elected an honorary president of the WCC.

His main publications are the four series of *Documents on Christian Unity* (1924, 1930, 1948, 1958); *Randall Davidson* (1935); *Christianity and World Order* (1940); *Christian Unity: The Anglican Position* (1948); and *The Kingship of Christ* (1954). His own life is the subject of the definitive biography *George Bell, Bishop of Chichester*, by R.C.D. Jasper (1967).

COLIN BUCHANAN

BELLAMY, JOSEPH (1719-1790). Congregational clergyman, theologian, and educator. Born in Connecticut, he graduated from Yale in 1735, but took his theological training personally under Jonathan Edwards.* Ordained as pastor of Bethlehem, Connecticut, church in 1740, he traveled extensively throughout Connecticut, Massachusetts, and New York, and preached almost daily during a two-year span. A Calvinist of the Edwardsean school, he taught a universal atonement based on divine moral government like that set forth by the Arminian Hugo Grotius,* and this despite the fact that his two most frequent opponents were Arminians and Antinomians. He is author of more than a dozen major works, the most important of which is his *True Religion Delineated* (1750).

DONALD M. LAKE

BELLARMINE, ROBERT (1542-1621). Roman Catholic cardinal and theologian. He entered the Jesuit order in 1560 and studied theology at Padua and Louvain. Ordained in 1570, he taught at Louvain for seven years, then for eleven more he lectured on theological controversies at the New Roman College (later the Gregorian University). Here he produced his famous *Disputationes de Controversiis Christianae Fidei*, which is generally recognized as one of the best statements of Roman Catholic theology as it was defined by the Council of Trent.* In 1589 the pope sent him on an important diplomatic mission to the French court, and in 1592 he took part in the Sistine revision of the Vulgate. In 1597 he became personal theologian to the pope, and in 1608 was involved in a controversy with James I of England over the authority of the Roman Church. Not until 1930 was he canonized and made a Doctor of the Church.

C. GREGG SINGER

BELLS. A metallic structure assuming the campaniform shape; to be distinguished in musical history from the gong, cymbal, struck tube, or bar. In the East, the bell appears as early as 2500 B.C. Biblical reference may not always meet the campaniform structure (cf. Exod. 28:33-35; Zech. 14:20). Josephus writes that King Solomon used large bells on the roofs of his dwellings to keep the birds away. Ecclesiastically, the first Christian writer to speak of bells in any significant way is Gregory of Tours about 585. Paulinus, bishop of Nola in Campania about 420, has traditionally been given credit for introducing the bell; however, the historicity of these accounts is debatable and may be an attempt to justify the two Latin words used to denote bells—*campana* and *nola*. The famous bell of St. Patrick in Dublin dating from the sixth century still survives, and since the seventh century there is evidence of widespread use of bells in England and the Continent. About the eighth century, the custom and rite of blessing bells with holy water and chrism came to be known as the "baptism of the bells." There seems to be greater justification for attributing the use of bells to functional rather than symbolic reasons. It could summon the congregation for worship or announce some special occasions such as the death of a church member, a special religious holy day, or set times for prayer. Since late medieval times it has been a custom to inscribe bells with a dedicatory statement of significance. Perhaps as a result of the OT association of bells with the high priest (Exod. 28:33-35), gradually certain small bells began to be associated with the eucharistic celebrations in the liturgy of the church. The bells of the carillon are a seventeenth-century French attempt to make musical instruments from the bell. Most of the largest bells in the world are church bells: Notre Dame (1680), 17 tons; "Great Paul" of St. Paul's Cathedral (1716), 16.25 tons; Cathedral (*Duomo*) of Milan, 15 tons.

See S.N. Coleman, *Bells, Their History, Legends, Making and Uses* (1928).

DONALD M. LAKE

BENDER, CARL JACOB (1869-1935). Pioneer missionary in Africa. Born in Germany, he emigrated to America at age twelve. After studying at the German Baptist Seminary in Rochester, New York, he was appointed to the Cameroons in 1899 under the Berlin-based German Baptist Missionary Society and supported by American churches. He served first in Douala, and in 1909 took charge of the Soppo station on Mount Cameroon. Besides establishing a school and missionary rest center, he engaged in evangelistic work in the area, and founded more than two dozen outstations during the next ten years. Since he held American citizenship, the Benders were the only German missionaries permitted to remain in the Cameroons during World War I. When they left in 1919 the work was left entirely in indigenous hands. After pastoring churches in the United States, he returned in 1929 to supervise reconstruction at Soppo, where he died. RICHARD V. PIERARD

BENEDICT IX (d. c.1055). Pope, 1032-45. Perhaps the most discreditable representative of the papacy before it was reformed in the eleventh century. From the powerful Tusculan faction, he reputedly became pope at the age of twelve, and certainly under him the moral tone of the papal court was very low. The Romans, disliking his spiritual deficiencies and even more his political affiliations, expelled him in 1044 and appointed Sylvester III. Benedict returned in the following year, but then abdicated after a money payment. His successor was the much more creditable Gregory VI. In 1046 the reforming emperor, Henry III, deposed Gregory for simony and also (lest they enter their claims again) both Benedict and the antipope Sylvester. Nine months later the new pope, Clement II, was poisoned, perhaps through the agency of Benedict who again became pope, briefly, in 1047 before being finally deposed by Henry. C. PETER WILLIAMS

BENEDICT XI (1240-1304). Pope from October 1303 until July 1304. Born Nicholas Boccasini, he was a Dominican friar, notable theologian, and former master general of his order. Native of the province of Venetia, he had been papal legate at the court of Emperor Albert of Hapsburg. Successor to Boniface VIII, he inherited the results of a long period of avaricious secular policies. The papacy had lost prestige and power after the struggle between Boniface and Philip the Fair of France, and Benedict tried to rectify matters. He attempted to reconcile factions in wealthy Florence which, with its European revenue, made a profitable ally for the papacy. He absolved the Colonna ex-cardinals and tried to make peace between them and the Gaetani. He acquitted Philip the Fair, but was determined to see justice done for the outrage at Anagni. William de Nogaret (who was Philip's minister), Sciarra Colonna, and others were therefore accused of criminal attack on Boniface VIII, and summoned to appear for judgment and sentence, but Benedict died at Perugia before judgment was made. He was beatified in 1736. JAMES TAYLOR

BENEDICT XII (1285-1342). Pope from 1334. Born Jacques Fournier, he was a distinguished theologian who became master of the University of Paris, and as a young man entered the Cistercian Abbey of Boulbonne. He was eventually created cardinal and, until his election as the third Avignon pope, was cardinal-priest of South Priscia. At this time he played a leading part in the controversy caused by John XXII's teaching on the fate of departed souls. He showed great zeal in the pursuit of heretics. Austere in his public and private life, he continued to wear his habit when he became pope.

A conscientious reformer within the church, Benedict required all diocesan bishops and all clergy who had benefices with the care of souls to be sent back from Avignon to their duties. He abolished the granting of abbeys to nonresident abbots; reorganized the office of penitentiary and the administration of papal correspondence; and attempted to reform the religious orders by a series of constitutions (1330-39) aimed at renewing their fervor and strictness. The constitution which he imposed on the Franciscans was not welcomed and was abolished by his successor. Benedict's efforts were openly resisted by the master general of the Dominicans. The radical reforms introduced into the Benedictine Order included the restoration of common life in monasteries and courses of monastic study. Benedict also had little success in his efforts to improve the political scene. Relations between papacy and emperor were not helped by his tendency to ally himself with the policies of Philip VI of France. The electors eventually declared at Rense that the emperor's rights needed no papal confirmation. Benedict's main theological work was the *Benedictus Deus* (1336) in which he pronounced that the souls of the just who have no sins to expiate will, on dying, experience immediately the Beatific Vision.* He began the building of the Palace of the Popes at Avignon.

JAMES TAYLOR

BENEDICT XIII (c.1328-1423). Antipope. Born Pedro de Luna, he taught canon law in France and was created cardinal in 1375. Having sided with antipope Clement VII during the Avignonese papacy, he was the first legate to his native Aragon, Castile, Navarre, and Portugal, then in 1393 to France, Flanders, and Scotland. Succeeding Clement in 1394 because of his canonical knowledge and successful Spanish legations, and having promised to return to Rome, he proved unworthy. As a result Charles VI of France denied him his right of provision to major offices and advowson to minor benefices, and other provinces refused him obedience as did cardinals. Imprisoned in the papal palace at Avignon, then seeking refuge in Provence in 1403, he had his obediences restored but his plans for compromise with Boniface IX, Innocent VII, and Gregory XII at Rome failed. Weary of him, France's support waned, and in 1409 the Council of Pisa* declared Gregory and Benedict to have forfeited their pontifical rank. The Council of Constance* confirmed this deposition in 1417. Backed by Aragon, Benedict contested this and, still confident of his proper claim, named four cardinals in 1422. He died at Peniscola, near Valencia, the following year. C.G. THORNE, JR.

BENEDICT XIII (1649-1730). Pope from 1724. Born Pierfrancesco Orsini into a wealthy ducal family, he renounced title and wealth to join a religious order. His promotion in the church was rapid: he became a cardinal at twenty-three, an archbishop at twenty-six. He accepted the pontificate with great reluctance and tried unsuccessfully to reform clerical morals. At the Vatican he refused to live in pomp, choosing modest rooms and living in simplicity. Despite Jansenist* opposition he allowed Dominicans to preach Augustine's doctrine of grace. Though personally saintly, scholarly, and well-intentioned, he showed no aptitude for practical affairs. Administration of the Papal States* was entrusted to the unscrupulous Nicolo Cardinal Coscia, who did great harm to the papacy. R.E.D. CLARK

BENEDICT XIV (1675-1758). Pope from 1740. From an old Bolognese noble family, Prospero Lorenzo Lambertini was educated at the Collegium Clementinum, Rome. Beginning as an assistant lawyer in Rome, he became Consistoral Advocate (1701), Promoter Fidei (1708), assessor of the Congregation of Rites (1712), and secretary of the Congregation of the Council (1718). He was appointed bishop of Ancona in 1727, cardinal in 1728, and archbishop of Bologna in 1731. He was elected pope following a struggle between the moderates and those resisting the takeover of ecclesiastical privileges by secular authorities.

A man of wide outlook, common sense and integrity, he promoted good understanding between the papacy and European rulers. He conceded to the Spanish and Portuguese crowns the right to nominate most benefices, made concordates with Naples and Sardinia, and recognized Frederick the Great as sovereign of Prussia in return for Frederick's acknowledgment of the bishop of Breslau's jurisdiction over Prussian Catholics. He settled controversies concerning Indian and Chinese rites. He issued an encyclical easing the position of Jansenists* in France (1756). Ruling the Papal States* in an enlightened manner, he established free trade, developed agriculture, and reduced taxation. He patronized the arts and learning, founded four academies for historical study, purchased MSS and books for the Vatican library, inaugurated the catalogue of Vatican MSS, and improved the University of Rome. He promulgated a bill restraining hasty and unnecessary prohibitions of books on the Index,* protecting in particular L.A. Muratori when attacked by the Jesuits (see MURATORIAN CANON). He himself was a brilliant canonist, writer, and controversialist. His writings included an important book on canonization, *De servorum Dei beatificatione et beatorum canonizatione* (1734-38), and a treatise on the Sacrifice of the Mass (1748). His conciliatory policy was interpreted as weakness by some anticlericals, but he was the outstanding pope in a period when papal authority was in decline. J.G.G. NORMAN

BENEDICT XV (1854-1922). Pope from 1914. Born Giacomo della Chiesa, he became archbishop of Bologna in 1907 and succeeded Pius X shortly after the outbreak of World War I. He consistently urged peace on both the Allies and Central Powers, especially in the encyclical *Ad beatissimi* (1 November 1914) and in a seven-point peace note to the governments involved (1 August 1917). His policy held the papacy to be above the conflict, the voice of moral authority, and thus frequently condemned what he considered to be violations of morality and right, and he established avenues of Christian charity to locate missing persons, care for the sick and wounded. The Vatican and some agencies were accused, probably falsely, by anticlericals in France and Italy of espionage activities for the Central Powers, while Germans and Austrians called him "the French Pope." Generally his perspective was neo-Thomist, following Leo XIII. He condemned modernism, led a codification of canon law (1917), greatly promoted Catholic missions, worked to establish better relations with the Eastern Orthodox Church. He sought to augment the papal position through establishing diplomatic relations with additional states, increasing the number so represented at the Vatican from fourteen to twenty-six, including Britain. Continuing to resist the Italian-imposed solution of the Roman Question, he maintained the need for clear papal temporal sovereignty, and established secret negotiations between his secretary of state and Benito Mussolini.

BIBLIOGRAPHY: F. Hayward, *Un Pape méconnu: Benoît XV* (1955); W.H. Peters, *The Life of Benedict XV* (1959); *idem,* "Benedict XV," *New Catholic Encyclopedia,* II, pp. 278-280; J. Schmidlin, *Papstgeschichte der neusten Zeit* (4 vols., 1933-39); G. Dalla Torre, "Benedetto XV," *Enciclopedia Cattolica,* II (1950), pp. 1285-94.

C.T. MC INTIRE

BENEDICT BISCOP (c.628-689). Founder of the monasteries of Wearmouth and Jarrow. Of a noble Northumbrian family, he entered the service of King Oswy. He made a number of visits to Rome, the first in 653 with Wilfrid.* In 666 he made his profession as a monk at Lerins. In 669 he was sent from Rome with Theodore of Tarsus* to Canterbury and was for two years abbot of the monastery of St. Peter and St. Paul. On his return from his fourth visit to Rome in 674, he founded the monastery at Wearmouth (Sunderland in County Durham). In 682 he founded the sister monastery at Jarrow. Bede was a novice in his charge at both monasteries. He was responsible for introducing glaziers and other craftsmen to work in the monastic buildings and for bringing the precentor of St. Peter's, Rome, to teach Gregorian chants. The churches in both places still stand with substantial Saxon parts.

R.E. NIXON

BENEDICTINES. Monks living under the Rule of St. Benedict* (of Nursia). The original foundation was at Subiaco, but shortly afterwards Benedict founded twelve monasteries with twelve monks each. It is believed that the third abbot of Monte Cassino began to spread knowledge of the Rule beyond the circle of Benedict's own foundations. The Rule developed by Benedict became the "constitution" for the order. Each monastery was to be economically self-supporting. There is no general or common superior over the whole order other than the pope himself, and the order consists of "congregations," each of which is autonomous, united only by the spiritual bond of allegiance to the same Rule, which may be modified according to the circumstances of each particular congregation. Each monastery was to have an abbot elected for life and other officers elected for limited terms. The vow taken by each monk emphasized perpetuity, poverty, chastity, and especially obedience to the abbot. Those who were admitted were novitiates for one year. Monks were expelled for serious offenses and though a penitent monk might be restored twice, he was permanently expelled for a third offense.

Pope Gregory the Great,* himself a Benedictine, encouraged the movement which gradually spread throughout Western Christendom. Augustine and his forty monks came from the Benedictine monastery of St. Andrew in Rome and established the first English Benedictine monastery at Canterbury soon after their arrival in 597. Various reform movements arose because of the abuses which had crept into the order. The first attempt to confederate the monastic houses of a single kingdom was made in the ninth century by Benedict of Aniane* under the auspices of Charlemagne and Louis the Pious. The most noteworthy reform movements were those of Cluny (910), which by the twelfth century had become a center and head of an order embracing some 314 monasteries in all parts of Europe. The Benedictine Order became noted for its literary achievements, e.g., the works of Bede,* and helped to preserve learning through the Middle Ages. S. TOON

BENEDICTION, see BLESSING

BENEDICT OF ANIANE (c.750-821). Monastic reformer. Born Witiza (Euticius), son of a Visigoth count in Languedoc, he distinguished himself in Charlemagne's wars. A narrow escape from drowning led him to enter the St.-Seine Monastery in Burgundy. He found the Benedictine monasteries in a parlous state. In 779 he established a monastery at Aniane, Languedoc, where he enforced the Benedictine Rule strictly, emphasizing manual labor rather than study, contrary to Charlemagne's wishes. He vigorously opposed the Adoptianism* of Felix of Urgel* (c.795). When Louis the Pious became emperor (814), he made Benedict his ecclesiastical adviser, calling him to found a monastery near Aix-la-Chapelle, the seat of the court. The Council of Aix-la-Chapelle (817), under Benedict's influence, enjoined a rigid uniformity on all monasteries. The reform was, however, short-lived. J.G.G. NORMAN

BENEDICT OF NURSIA (c.480-c.547). Founder of monasteries. Born at Norcia (Nursia) in Umbria, he was early sent to Rome to study. Seemingly, the degenerate life of the city caused him to flee to the country and live in a cave as a hermit at Subiaco. After a brief period in a monastery he returned to Subiaco where he set up twelve small monastic communities. About 529 he was pressured to leave these groups, and so with a small nucleus of men he moved to what is now Monte San Germano (halfway between Rome and Naples) to establish the monastery of Monte Cassino, where he remained until his death. The basis of the monastery was two chapels dedicated to John the Baptist and Martin of Umbria. Making use of previous rules (e.g., those of John Cassian* and Basil of Caesarea*), he composed his own rule. The wide usage of this has ensured his fame and gained for him the title "Patriarch of Western Monasticism." He was buried in the grave of his sister Scholastica in the Chapel of John the Baptist. Our main source of information about him is contained in the *Dialogues* of Gregory the Great. There are Lives in English by J. McCann (1938) and T.F. Lindsay (1949). PETER TOON

BENEDICTUS. In Luke 1:68-79, Zacharias sings a hymn of praise and thanksgiving at the birth of his son, John the Baptist. In form it is a typical OT Psalm in two parts: the first praises God for the fulfilment of promises which are recounted, the second is addressed to the child in whom these promises will be fulfilled. This psalm is used liturgically in the Roman Catholic Church at Lauds, and in the Church of England in the 1662 Book of Common Prayer in Morning Prayer, where it is known as a Canticle (together with the *Te Deum* and *Benedicite,* and the *Magnificat* and *Nunc Dimittis* at Evening Prayer).

BENEFICE. The *beneficium* was the legal term for a grant of land for life as a reward for services rendered. This developed under canon law into an ecclesiastical office under the twofold heading of "spiritualities" (the duties involved) and "temporalities" (the emoluments provided). In the Church of England there are three kinds of parochial benefice: rectories, vicarages, and perpetual curacies. The difference is largely one of who is the recipient of the various tithes between the first two, and of the date of establishment of the benefice between these and the last. In return for his freehold, the holder or incumbent was, before the Pastoral Measure, removable only for grave misconduct or dereliction of duty. G.S.R. COX

BENEFIT OF CLERGY. One sequel to Becket's* politically disastrous murder was Henry II's concession to all English tonsured clergy and nuns of trial for all criminal offenses, except forest offenses, in a church court instead of in a secular court with its heavier punishments (1176). Edward III (1327-77) included certain lay first offenders who could read—usually the "Neck verse" (Ps. 51:1). Many in later medieval England considered benefit of clergy was open to abuse, and the sixteenth-century English Reformation saw its systematic erosion. Major crimes were excepted: guilty clerics could be imprisoned (and under George I [1714-27] transported). In 1827 the privilege was abolished. L. FEEHAN

BENGEL, JOHANN ALBRECHT (1687-1752). Lutheran minister and theologian. Trained at Tübingen, he taught at a seminary in Denkendorf (1713-41), then became superintendent of Herbrechtingen (1741) and Alpirsbach (1749). His chief work was a critical edition of the NT (1734) which became the starting point for modern textual criticism. This work was followed by the *Gnomon novi testamenti* (1742), a commentary in which he gives a word-by-word explanation of the Greek text in capable fashion. John Wesley* translated most of its notes and incorporated them into *Notes Upon the New Testament* (1755).

Bengel wrote also several apocalyptic works, among them *Erklärte Offenbarung Johannis* (1740), in which he tried to fix the number of the beast and set the date of the beginning of the Millennium at 1836. Among evangelical scholars, Bengel's *Gnomon* is still in use.

ROBERT G. CLOUSE

BENNO (1010-1106). Bishop of Meissen in E Germany. Of noble Saxon birth, he was canon in the imperial collegiate church at Goslar before going to Meissen in 1066. Imprisoned by Henry IV (1075-76) for nonsupport during the Saxon nobles' revolt, and involved in the election of Rudolph of Swabia to replace Henry, he was temporarily removed from his see (1085-88) by imperial prelates. He was called "Apostle of the Wends" for preaching to Slavonic peoples in his diocese. His cult was established in 1285 when his relics were honored in Meissen cathedral. Chronicles record many miracles at his tomb. His canonization (1523) and relics evoked much protest, including Luther's. He is regarded as patron of fishermen and drapers. C.G. THORNE, JR.

BENSON, CLARENCE HERBERT (1879-1954). Presbyterian minister and author. Educated at the University of Minnesota, Macalester College, and Princeton Seminary (1908), he served pastorates in New York, Pennsylvania, and at Union Church, Kobe, Japan (1919-22). In 1922 he became director of the Christian education department of Moody Bible Institute. Between 1925 and 1933 he supervised the development of the All Bible Graded Series of Sunday school lessons, and with Victor Cory founded Scripture Press to publish them (1934). He helped to found what became the Evangelical Teacher Training Association (1930), founded *The Church School Promoter*, forerunner of *Christian Life*, and was associate editor of *Moody Monthly* (1926-41). His most influential books were *A Popular History of Christian Education, The Church at Work*, and *The Sunday School in Action*. Benson helped also to found the National Sunday School Association and its Uniform Bible Lesson Series.

HOWARD A. WHALEY

BENSON, EDWARD WHITE (1829-1896). Archbishop of Canterbury from 1882. Born at Birmingham, he studied at King Edward's School and at Trinity, Cambridge. An assistant master at Rugby (1852), he was ordained later and became first headmaster of Wellington College (1859-72). He was consecrated first bishop of Truro (1877), and was appointed to Canterbury partly because of his friendship with W.E. Gladstone.* A zealous churchman, he defended the principle of establishment, vigorously opposing the disestablishment of the Welsh Church (1893). In 1890 he gave an important judgment at the trial of Edward King, bishop of Lincoln, charged with ritual offenses, in which he countenanced High Church usages, e.g., the "eastward position," and use of candles at the Eucharist. J.G.G. NORMAN

BERDYAEV, NIKOLAI (1874-1948). Religious philosopher. Born in Kiev, he was early attracted to Marxism, and although he never repudiated his commitment, he was also a member of the Russian Orthodox Church. He was brought to trial by the church in 1914 for his nonconformist position in religious matters, and was saved from sentencing only by the onset of the Russian Revolution. Like other intellectuals during the early purges, he was expelled from his post as a professor of philosophy at Moscow University, and from the USSR in 1922, He went first to Berlin and then to Paris (1924) where he founded an academy for the study of philosophical problems.

Although not a systematic thinker or philosopher, he was a prolific writer. In more than twenty books and many articles, Berdyaev emphasizes freedom, creativity, and the reality of the transcendent. He is often referred to as a "Christian existentialist" (see EXISTENTIALISM). In his thinking, the truth is a gleam of light which penetrates the objective world from the transcendent realm of the spirit. Man's glory is in his ability to appropriate this order of the spirit and to become creative; man's deterioration comes about through the loss of these capabilities. He was indebted for some of his ideas to Jakob Boehme,* Kant,* Nietzsche,* and Dostoevsky.* Among his books are *Freedom and the Spirit* (1935), *The Destiny of Man* (1937), and *The Beginning and the End* (1952). Though he was not sympathetic to Marxist materialism and denounced Soviet terrorism, he hoped that the true Russian spirit would ultimately emerge in the new state. PAUL M. BECHTEL

BEREANS. Originating in Edinburgh in 1773, they were followers of John Barclay,* who took their name from Acts 17:10,11. When Barclay was inhibited from preaching in Fettercairn, seceders built a chapel to which James Macrea was called as minister. Barclay himself led the Berean assembly in Edinburgh, and another was formed at Crieff, near Barclay's birthplace. During a prolonged visit to England (1776-78) he formed congregations in London and Bristol. He ordained and sent to lead the Bereans in Edinburgh William Nelson, surgeon and Calvinistic Methodist,* who had been trained for the Anglican ministry. On Barclay's return, Nelson was sent to strengthen new Berean churches at Glasgow, Kirkcaldy, Dundee, Arbroath, Montrose, and Brechin. Barclay himself visited his churches extensively and wrote numerous works expounding the distinctive elements in his otherwise Calvinistic theology. After his death the Berean church in Edinburgh flourished for twenty-five years under James Donaldson, originally pastor of the church in London, then in Dundee. When he died, the Berean church in Edinburgh split and, like the other Berean churches, eventually merged with the Congregationalists.

See J. Campbell, *The Berean Church—Especially in Edinburgh* (1937), offprint from the Proceedings of the Scottish Church History Society.

HAROLD H. ROWDON

BERENGAR OF TOURS (c.1000-1088). French theologian. Student of Fulbert of Chartres, he was canon and director of the cathedral school at

Tours (1031) and later archdeacon of Angers (1041). A man of learning and piety, he came (1040-45) to question the eucharistic interpretation of Paschasius* Radbertus (a ninth-century monk who taught transubstantiation). A series of controversies resulted from Berengar's teachings, which prompted the development of the Roman Catholic teaching about the Eucharist. In the course of these arguments Berengar was forced to sign several statements, one even asserting that when a believer partakes of the element he actually masticates the body of Christ. He maintained that one cannot literally eat and drink Christ's body and blood, but that nevertheless by faith the Christian can have real spiritual communion with the flesh, that is, the glorified humanity of Christ in heaven. In his teaching, the elements remain in substance as well as appearance, after the consecration. They are, however, endowed with new value, for whatever is consecrated is lifted to a higher sphere and transformed. Perhaps if Berengar had been willing to die for his convictions he would have won more adherents. As it was, his repeated recantations served to prompt men such as Lanfranc* to articulate in a more detailed manner the teaching of transubstantiation.

See A.J. MacDonald, *Berengar and the Reform of Sacramental Doctrine* (1930).

ROBERT G. CLOUSE

BERGGRAV, EIVIND (1884-1959). Norwegian Lutheran bishop. Son of a clergyman (who later became a bishop), he studied theology and was ordained in 1908. He was a teacher in various types of schools from 1909, became pastor in a rural parish in 1918, penitentiary chaplain in 1924, bishop of Tromso in 1928, bishop of Oslo 1937-50. Originally influenced by liberal theology, he nevertheless took no part in debates on the subject, and as bishop won the confidence of the majority of active church people. Berggrav was a prolific writer who published about thirty books on various subjects: psychology of religion, philosophy of education, the relation between church and state, devotional works, etc. He wanted to establish a fruitful relationship between the church and contemporary cultural life, and for this purpose he edited from 1909 the magazine *Kirke og Kultur.* During the German occupation (1940-45) he was the leader of the church in its controversy with the Nazi government. He took a prominent part in the writing of the pastoral letters and declarations which became vital factors in this struggle. From Easter 1942 until the close of the war he was placed in heavily guarded police internment. In 1950 because of poor health he resigned his position as bishop, but remained active. He took a leading part in the ecumenical efforts of this period. From 1950 to 1954 he was one of the presidents of the World Council of Churches.*

CARL FR. WISLOFF

BERKELEY, GEORGE (1685-1753). Irish philosopher. The most brilliant theistic philosopher of his age, he was dean of Derry and later bishop of Cloyne. In an era when the sufficiency of reason, at the expense of revelation, was confidently asserted, he demonstrated that the church could outthink its critics. In his major work, *Alciphron* (1732), he showed that the problems of the Age of Reason could be solved only through a new understanding of the role of reason. He believed that ideas are the things which really exist. Material things exist only in so far as they are perceived. It is God who is responsible for the existence of ideas; they are not the products of our own minds. Berkeley was also a missionary enthusiast and unsuccessfully proposed a college in Bermuda.

JAMES TAYLOR

BERKHOF, LOUIS (1873-1957). American Calvinist theologian. Born in the Netherlands, he emigrated as a child to the USA, where his family joined the Christian Reformed Church, at that time still small and using the Dutch language. Attracted to the ministry, he attended and graduated from the church's fledgling Calvin Seminary in Grand Rapids, Michigan. He went on to postgraduate work at Princeton and after a brief pastorate was called to teach at Calvin Seminary (1906). There he spent three decades as teacher, and from him almost every Christian Reformed preacher learned systematic theology. During the 1920s he played a leading role in the controversies of the day, including the debates over Herman Hoeksema's "hyper-Calvinism," which led to Hoeksema's formation of the Protestant Reformed Church. Berkhof's main interest was in systematics, and he followed closely the Dutch Calvinist theologians A. Kuyper* and H. Bavinck,* though also influenced by orthodox Calvinist theologians in America such as C. Hodge* and B.B. Warfield.* Berkhof wrote extensively; his major work is his *Reformed Dogmatics* (3 vols., 1932), which in popularized form (*Manual of Reformed Doctrine,* 1933) was used extensively in the schools of the Christian Reformed Church. His last book, written at eighty, was *The Second Coming of Christ* (1953).

DIRK JELLEMA

BERNADETTE (1844-1879). Roman Catholic visionary. Eldest child of a poor miller, surnamed Subirous, of Lourdes, France, she had a series of visions between 11 February and 16 July 1858, in a cave by the river Gave. She believed that the young, beautiful lady who spoke to her in some of the eighteen apparitions was the Virgin Mary, who had the title of "the Immaculate Conception." At first the church authorities disbelieved her, and their distrust was heightened by an epidemic of false visionaries. In 1866 she was admitted to the convent of the Sisters of Charity at Nevers. She suffered from chronic asthma, but her life was seen as holy, and she died at the age of thirty-five. She was beatified by Pius XI in 1925 and canonized in 1933. Lourdes has become one of the greatest centers of pilgrimage in Western Christendom.

PETER TOON

BERNARD, JOHN HENRY (1860-1927). Archbishop of Dublin. Born in India, eldest son of a civil engineer, he was educated at Trinity College, Dublin, proved himself a brilliant scholar, and thereafter lectured at Trinity, where he was provost from 1919 until 1927. An outstanding administrator, he was dean of St. Patrick's Cathedral, Dublin (1902-11), bishop of Ossory, Ferns and Leighlin (1911-15), and archbishop of Dublin from 1915. His scholastic gifts are seen in some twenty books, including a commentary on John's gospel (1928), his popularity as a preacher in his university sermons, and his devotion to the arts in his presidency of the Royal Irish Academy from 1916 to 1921. ADAM LOUGHRIDGE

BERNARD DE MONTFAUCON (1655-1741). French scholar. Born of a noble family in Soulage, he joined the Maurist Benedictines in Toulouse (1676) after a brief period of military service. Having studied Greek, Hebrew, and Syriac at Saint-Germain-des-Prés, Paris, he researched in Italian libraries (1698-1701), and this led to his *L'Antiquité expliquée et representée en figures* (10 vols., 1719) and *Les Monuments de la monarchie francaise* (5 vols., 1729-33). He produced excellent editions of Athanasius (3 vols., 1698), Origen's *Hexapla* (2 vols., 1713) and John Chrysostom (13 vols., 1718-38). His *Palaeographia graeca* (1708) virtually created the science of paleography. *Bibliotheca Bibliothecarum,* a manuscripts survey, is perhaps best known. He defended the Benedictine edition of Augustine against Jesuit attacks (1699). The Bernardines, a younger generation of Benedictine students, he influenced heavily. C.G. THORNE, JR.

BERNARDINES, see FEUILLANTS

BERNARDINO OF SIENA (1380-1444). Franciscan friar and reformer. Born of noble parents at Massa di Carrera, where his father was governor, he became a friar in 1402. Thereafter he preached for many years both inside and outside churches throughout Italy. He made use of the monogram IHS* and stressed devotion to the holy name of Jesus. He constantly attacked usury and the party strife of the Italian cities. His attitude toward Jews, however, did not rise above the prejudice of his day. His general example in life and preaching had a beneficial effect on the number of friars and the rigor of their discipline. In 1438 he was elected provincial of the Friars of the Strict Observance. He took part in the Council of Florence* when union with the Greek Church was debated. He died at Aquila in Abruzzi while on a preaching tour; he was canonized by Pope Nicholas in 1450. His works have been printed several times, latterly in 1950, and are composed of sermons and tracts on morals, asceticism, and mysticism.

See Life by I. Origo (1963). PETER TOON

BERNARD OF CLAIRVAUX (1090-1153). Monastic reformer, mystic, and theologian. Born to a noble family in Fontaines, near Dijon, France, he joined the Cistercian* monastery at Cîteaux about 1111, where he soon was asked to found a new house. In 1115 the young abbot established a monastic community at Clairvaux which shortly became a principal center of the order. A firm believer in strict observance, in 1119 he attacked Cluny for its alleged disciplinary decadence. In 1128 he secured recognition for the order of Knights Templar,* whose rules he drafted himself. In the 1130 papal election controversy he sided with Innocent II; the new pope responded by bestowing privileges upon the Cistercians, and Bernard's influence was further enhanced with the election of Eugenius III, who had been his disciple at Clairvaux, as pope in 1145. He engaged in controversies with Abelard* in 1140 and Gilbert de la Porrée in 1148. He was officially charged with preaching the Second Crusade in 1146-47, and its outcome bitterly disappointed him. He obtained the condemnation of Arnold of Brescia's* reformist doctrines and attacked the heretical teaching of Henry of Lausanne.* He was canonized in 1174 and proclaimed a Doctor of the Church in 1830.

Because of his personality rather than any force of intellect, Bernard was the dominant figure in twelfth-century Latin Christendom, yet he was just as controversial in his day as now. He was both rigidly orthodox and aggressively self-righteous, and deeply pious and ascetic. He was simultaneously a contemplative mystic and an activist man of affairs in the world. In hundreds of sermons and letters and several treatises on theology and liturgy he expressed hostility to rationalism and set forth the value of contemplation and mystical experiences. In his theology he shifted the emphasis from God's judgment to His infinite love and mercy and the hope of redemption for even the worst sinner. As a mystic he stressed a christocentric union, the Word as the spouse of the soul. Because of his deeply felt devotion to the Mother of God, he gave impetus to the heretofore insignificant cult of the Virgin in the West.

A number of hymns are attributed to him, some translated as "Jesus, the very thought of Thee," "O sacred Head now wounded," and "Jesus, Thou joy of loving hearts."

BIBLIOGRAPHY: *Works* (tr. S.J. Eales, 5 vols., 1889-96); E. Vacandard, *Vie de Saint Bernard* (2 vols., 1895); E. Gilson, *The Mystical Theology of St. Bernard* (1940); W.W. Williams, *Studies in St. Bernard of Clairvaux* (1952); B. Scott-James, *Saint Bernard of Clairvaux* (1957).

RICHARD V. PIERARD

BERNARD OF CLUNY (fl. 1140). Monk and poet. Little is known of the author of *De Contemptu Mundi* except that he lived in the famous monastery of Cluny and was possibly of British extraction. His poem of about 3,000 lines, written in dactylic hexameters and beginning *Hora novissima,* satirizes contemporary monastic corruption and contrasts the transient pleasures of this life with the peace and glory of heaven. He attributed his mastery of the difficult meter to divine inspiration. First found in a thirteenth-century manuscript (now in the Bodleian), extracts from it were translated in ballad metre by Neale* in 1851 and 1858. Stanzas from these form the basis of the hymns "Brief life is here our portion"; "For thee, O dear, dear country"; "Jerusalem the golden"; and "The world is very evil."

JOHN S. ANDREWS

BERNE, THESES OF (1528). In November 1527 the Bernese city council resolved to hold a disputation upholding the sole authority of the Word of God in religious affairs. Bishops from Constance, Basle, Lausanne, and Wallis with delegates from all the Swiss cantons were invited. The Zurich council agreed on 7 December, with Zwingli* announcing to Oecolampadius* that all was ready. Delegates from Schaffhausen, St. Gall, and Constance assembled in Zurich on 1 January 1528, setting out with 300 armored men to the borders of Berne. The disputation began on 5 January and lasted until the twenty-sixth. Berthold Haller and Francis Kolb of Berne with Zwingli's aid prepared ten theses for debate, and these were begun with Kolb reading the first thesis: "The Holy Christian Church whose only Head is Christ, is born and nourished out of God's Word and hears not the voice of a stranger."

The daily sessions ended with the Bernese city council abolishing the Mass and church images. Zwingli preached two sermons attacking them; Bullinger recounts that a priest robed for the Mass heard Zwingli and refused to hold it any longer. The Acts were published by Christopher Froschouer on 23 April 1528. The heart of the matter is found in thesis III: "Christ is the one wisdom, justice, redemption and satisfaction for the sins of the entire world: therefore to confess any other ground of salvation or merit for sin is to deny Christ." In spite of Luther's letter to Zwingli of 7 March that nothing had been done, these theses became the vehicle of Bernese entry into Swiss Protestantism. The Reformation came to that Swiss city, as to so many others, in a formal debate. MARVIN W. ANDERSON

BERSIER, EUGENE (1831-1889). Swiss pastor. He was a leader of the schismatic Free Reformed Church between 1855 and 1877, when he persuaded his congregation to rejoin the larger Reformed Church. His theological studies at Geneva, Göttingen, and Halle were followed by a ministry in the Paris churches of Faubourg St. Antoine (1855-61), Taitbout (1861-74), and the Étoile (1874-77). Bersier's published sermons were popular in France and abroad. *The Gospel in Paris: Sermons* (1884) and a liturgy (1874) used in the Reformed Church both had a wide vogue. He also wrote the authoritative *Histoire du synode de 1872* (2 vols., 1872), *Coligny, the Earlier Life of the Great Huguenot* (1884), and other historical works. ROBERT P. EVANS

BERTHOLD OF REGENSBURG (c.1210-1272). Franciscan preacher. Educated probably at Magdeburg *Studienanstalt,* he became lector then preacher (1240) in the Franciscan monastery in Regensburg. Thereafter he preached throughout Bavaria, the rest of Germany, Switzerland, Czechoslovakia, and in Hungary (1262-63) where he was later, according to legend, reckoned a saint. In 1263 Urban IV ordered him to preach against heresy. He went through Germany and Switzerland, reaching Paris and meeting Louis IX. Knowing the writings of Augustine, Gregory the Great, and Bernard of Clairvaux, as well as having an interest in natural science, he preached didactic, moral sermons to huge crowds, mostly outdoors. He sought repentance and moral betterment by emphasizing the negative: seven capital sins, violations of the Commandments. Some of his sermons were taken down in Latin by clerics and religious present, and later edited by himself as *Sermones de Dominicis, de Sanctis, de Communi, ad Religiosos et quosdam alios, speciales et extravagantes,* though not all of these are his. C.G. THORNE, JR.

BÉRULLE, PIERRE DE (1575-1629). French spiritual director and diplomat. Born at Sérilly in the Champagne, educated by Jesuits and at the Sorbonne, he was ordained priest in 1599, and was for a time honorary almoner to Henry IV, becoming well known as a spiritual director. At Paris he founded the French congregation of the Oratorians* (1611). As confidant and counselor of Queen Marie de Médicis and friend of Louis XIII he had profound influence at court. Hoping for the conversion of England, he negotiated with Rome the dispensation required for the marriage of Henrietta Maria with the Anglican Charles I (1625). He was made cardinal in 1627. His spirituality was characterized by devotion to Christ's human personality, so that Urban VIII called him *Apostolis Verbi Incarnati.* He wrote extensively, his best-known work being *Discours de l'état et de la grandeur de Jésus* (1623), which was much used by J.B. Bossuet.* J.G.G. NORMAN

BERYLLUS OF BOSTRA (third century). Reputed heretic. Present knowledge of him comes from Eusebius, Jerome, and Origen. Beryllus lived in Bostra of Arabia Petraea (present Sinai, S Jordan and part of NW Arabia). He was considered a heretic because he did not accept the preexistence and independent divinity of Christ; however, he felt that the divinity of the Father was in Christ during His earthly life. Origen, who at the time was also considered heretical by some, was said to have convinced Beryllus that he should change his mind about Christ and recognize His divinity. This took place at a synod meeting in Arabia in 244. GEORGE GIACUMAKIS, JR.

BESANT, ANNIE (1847-1933). British theosophist and educator. Born in London, she was educated by private tutor and at London University. She was married to the vicar of Sibsey, the Rev. Frank Besant, but it ended in divorce after six years. During her spiritual pilgrimage she moved successively from Anglican to atheist to spiritualist and finally to theosophist. Most of her life after 1889 was spent in India, where she established a number of educational institutions, such as the Central Hindu College of Benares (1898) and the University of India (1907), and even rose to the presidency of the National Congress. She proclaimed her adopted son and spiritual mystic Jidder Krishnamurti the new Messiah, a claim which he later repudiated. A prolific writer of major works on Eastern religion, Mrs. Besant became president of the Theosophical Society upon the death of its founder, Mme. Helena P. Blavatsky. Her major works include *The Religious Problems of India, The Wisdom of the Upanishads, The Basis of Morality,* and *India: Bond or Free?*

See THEOSOPHY. DONALD M. LAKE

BESSARION, JOHN (Basilius) (1403-1472). Greek theologian. Born in Trebizond and adopted by Metropolitan Dositheus, he studied rhetoric, philosophy, and asceticism at Constantinople. He became a Basilian monk, taking the name Bessarion in 1423, was ordained deacon, and after study with the Neoplatonist George Gemistos Plethos was made abbot of St. Basil's monastery. In preparation for the Council of Ferrara/Florence, he was created archbishop of Nicea (1437) by Emperor John VII Palaeologus, whom he accompanied to the council. He labored for the union of Greek and Latin churches, which gained him unpopularity at Constantinople. Pope Eugenius IV made him cardinal in 1439, and he eventually resided in Italy. He fulfilled important ecclesiastical missions, and in 1463 received the title of Latin patriarch of Constantinople. He presented his library to the senate of Venice, which formed the nucleus of St. Mark's Library. An enthusiastic scholar and patron of learning, he played a crucial part in the development of the Italian Renaissance. J.G.G. NORMAN

BETHEL INSTITUTIONS. From a farmhouse near Bielefeld, Germany, with five epileptic boys, opened in 1867 as the fruit of a revival in Ravensberg, Westphalia, have grown institutions housing over 10,000 people. Besides homes for epileptics, mental patients, tramps, refugees, and youths in need of guidance, there are institutes for deacons and deaconesses, a mission in East Africa, a theological college, and secondary schools. The driving force was Pastor ("Father") Friedrich von Bodelschwingh,* who became director in 1872 and named the institutions "Bethel." The work is part of the Innere Mission.*

BETHLEHEMITES. Obscure hospital and military orders, sometimes under Austin rule. Matthew of Paris attests to such in England in 1257, at Cambridge. There was also the hospital of St. Mary of Bethlehem in London for mental patients (established in 1247), which in 1547 became a royal establishment for the care of lunatics: hence the word "bedlam." Pius II founded a military order, dedicated to the BVM of Bethlehem, for protecting the Aegean after 1453. There were other institutions in Scotland, Italy, and France. Brothers and sisters wore a habit with a red star, much confused with the Bohemian hospital order established in Prague in the thirteenth century. C.G. THORNE, JR.

BETHUNE, DAVID, see BEATON, DAVID

BETTING, see GAMBLING

BEVAN, (EMMA) FRANCES (1827-1909). Hymnwriter. Daughter of P.N. Shuttleworth, the anti-Tractarian bishop of Chichester, Frances was a High Churchwoman until her marriage in 1856 to the Evangelical Anglican, R.C.L. Bevan. He was a banker and came of distinguished Quaker and Low Church stock. Soon afterwards she became associated with Open Brethren. From 1858 onward she published several collections of hymns, mainly paraphrases from the German. Although she ranged widely, she preferred the medieval mystics, the Pietists,* and Tersteegen* (see her *Hymns of Ter Steegen, Suso and Others,* 2 vols., 1894-97, rep. 1920). Her best-known, but not most characteristic, translation was "Sinners Jesus will receive," from E. Neumeister. She wrote anonymously. JOHN S. ANDREWS

BEYSCHLAG, WILLIBALD (1823-1900). German Evangelical Church leader and publicist. He fought for a broadminded biblical Christianity, much concerned with national questions and living piety. He asked why the return to the traditional confessions had done so little to renew the spiritual life in the church. As a mediating theologian, he rejected both the Chalcedonian Christology and the rationalism of Strauss's* and Renan's* presentations of Jesus. Active as a religious journalist, he also founded the *Evangelischen Vereinigung* in 1876, an organization for the middle party in the church. He was always concerned about the problem of Roman Catholicism, having met it in aggressive form when in Trier (1850-56); he welcomed the Old Catholics, supported the *Kulturkampf,** and in 1886 founded the *Evangelischen Bundes zur Wahrung deutsch-protestanten Interessen* to counter catholicizing influences within the Evangelical Church. Beyschlag was a strong supporter of the rights of the laity and of the autonomy of the church. In his day he had great power. HADDON WILLMER

BEZA, THEODORE (1519-1605). Calvin's successor in Geneva as the head of Reformed Protestantism. Born at Vezelay, Burgundy, of a lesser noble family, his formal education was made possible by his uncle Nicholas, counselor to the Parlement of Paris. In 1534 he went to Orleans to study law, which licentiateship he received in 1539. He then went to Paris to practice law and there pursued his affinity for the classics. In 1548 he published a book of poems, *Poemata Juvenilia,* which reflected his interest in classicism and humanism. Later in life he edited these poems, expurgated some, and reissued the volume. While in Paris he was under some pressure from his family to be ordained, but his involvement with Claude Desnoz (whom he had privately married) complicated his situation.

After a severe illness in 1548—a physical as well as a spiritual crisis—he renounced Catholicism, became a Protestant, went to Geneva, and there publicly married Claude. In Lausanne he visited Pierre Viret,* who got him the position of professor of Greek in the academy there. Beza taught and wrote extensively for the next decade. He sided with Calvin against Bolsec* on the doctrine of predestination, and came to Calvin's defense after the death of Servetus* in the pamphlet *Haeriticiss a civili Magistratu Puniendus (Concerning Heretics Who Should Be Punished by the Magistrate,* 1554). In 1556 he published an annotated Latin translation of the Greek NT. He was to continue this interest in biblical textual problems throughout his life. During 1557 he visited, with Farel,* the Waldensians* and other Protestant groups, hoping to help them gain some security through intercessions of the German princes with the king of France.

Upon the invitation of Calvin, Beza went to Geneva in 1558 as a professor of Greek. In 1559 he was named rector and eventually taught theology in the Genevan Academy.* Earlier Calvin had suggested to Beza that he might complete Marot's translation of the Psalms into French, and in 1561 after translating about a hundred Psalms, they were published. In that same year he represented the French Protestants at the Colloquy of Poissy,* and later supported and advised the Huguenots during the wars of religion in France. He returned to Geneva in 1563, and on the death of Calvin (1564) the full weight of Calvin's responsibility came upon Beza. Beza was the head of the academy, a teacher there, moderator of the Company of Pastors, a powerful influence with the magistrates of Geneva, and the spokesman and defender of the Reformed Protestant position.

Throughout his life he maintained wide interests. In 1565 he published a Greek text of the NT, to which he added the Vulgate and his own translation. This biblical textual interest is further seen in his use of Codex Bezae and Codex Claromantus. He continued to defend the Reformed position as evidenced by his vigorous polemics with Ochino, Castellio, Morel, Ramus, the Zwinglians, Arminius, and others. He continued his activities in the Huguenot movement by serving as an adviser, and in 1571 by presiding over the National Synod of La Rochelle. After the St. Bartholomew's Day massacre* in 1572, he published *De Jure Magistratu* which argued for the right of the inferior magistrates to revolt against the government. In 1580 he published a history of the Reformed movement in France, and in 1582 he again published a work dealing with textual criticism, his second edition of the Greek NT. His biblical criticism influenced the King James Version of 1611. His works appeared in French, Latin, and English and had a wide and deep impact upon the Reformation movement during the last half of the sixteenth century. His strong defense of double predestination, biblical literalism, church discipline, and other Calvinistic ideas did much to harden the movement, and to begin the period of Reformed Scholasticism.

BIBLIOGRAPHY: F. Aubert et al. (eds.), *Correspondence de Theodore de Beze* (1960-); H.M. Baird, *Theodore Beza, the Counsellor of the French Reformation* (1899); F.L. Gardy and A. Dufour, *Bibliographie des oeuvres theologiques* (1960); P. Geisendorf, *Theodore de Beze* (1967); R.M. Kingdon, *Geneva and the Consolidation of the French Protestant Movement 1564-1572* (1967).

ROBERT SCHNUCKER

BIARD, PIERRE (1567?-1622). Jesuit missionary. Born in Grenoble, France, he entered the Jesuit order in 1583. His studies qualified him for a teaching post in theology at Lyons. In 1610 he was commissioned to join the French mission in Acadia, together with Fr. Enemond Masse. Because of difficulties with Huguenot merchants who rigged the ships, Biard did not arrive at the mission post until 1611. His name is intimately connected with Port Royal (Annapolis Royal), Nova Scotia, and with the founding of St. Saveur (Bar Harbour). He was later captured by the English and made a dramatic escape back to France. The famed *Jesuit Relations* contains his accounts for the year 1616. His writing is of value to students of Indian ethnology. A defense of his somewhat unconventional missionary endeavors, allegedly written around 1620, was apparently never published. He died in Avignon.

EDWARD J. FURCHA

BIBLE, see OLD TESTAMENT; SYNOPTIC GOSPELS; JOHN, GOSPEL OF; ACTS OF THE APOSTLES; EPISTLES, GENERAL and PAULINE; HEBREWS; REVELATION

BIBLE (ENGLISH VERSIONS). *Old English.* The oldest surviving Bible versions in English are metrical paraphrases of the history of salvation. It is uncertain if any of these can be ascribed to Caedmon,* the seventh-century poet of Whitby. A ninth-century poem, Cynewulf's *Crist,* relates the Gospel story; from the following century we have a metrical version of *Judith.* Aldhelm, first bishop of Sherborne (c.700), is said to have translated the Psalter into English. Bede* of Jarrow was engaged on an English version of the gospels at his death on Ascension Day, 735. Alfred the Great* (d.901) included an English version of the Decalogue and of some other Pentateuchal laws in his national law-code, and is credited with the translation of part of the Psalter. From the tenth century come the English interlinear glosses in the Lindisfarne and Rushworth Gospels; the Wessex Gospels, an independent version, belong to the same period. A little later Aelfric,* abbot of Eynsham, translated parts of the Heptateuch.

Middle English. Versions and paraphrases of the Psalter and most of the NT have come down to us from the thirteenth and fourteenth centuries, but the first complete versions of the English Bible are the two associated with John Wycliffe* and his school. The earlier Wycliffite version—the work of Nicholas of Hereford from Genesis 1:1 to *Baruch* 3:20 and then of others (possibly including Wycliffe himself)—appeared about 1384. It was a literal rendering of the Latin Bible and may have been conceived as a codification of God's law, replacing traditional canon law. The later Wycliffite version was a thorough revision of the earlier one, couched in idiomatic English. It was produced several years after Wycliffe's death by his secretary John Purvey,* who followed up his revision with a tract commonly called the *General Prologue* (1395-6), in which he set forth the aims and principles of his work. The *Prologue* defends the right of the common people to have access to God's law in a form which they can understand, and shows an impressive grasp of the methodology of translation, beginning with the establishment of a sound text (in this case a Latin text) to serve as a basis.

Because of the association of Purvey's version with the proscribed Lollard* movement, it incurred official disapproval. In 1408 the Constitutions of Oxford forbade the production or use of vernacular Scriptures without the permission of the diocesan bishop or a provincial council. But Purvey's work retained its popularity until the beginning of the sixteenth century. About that time a Scottish adaptation of it was produced by Murdoch Nisbet.

Tyndale and his successors. The Bible translation of William Tyndale* (1494/5-1536) was preceded by three crucial events or movements in western Europe—the invention of printing (c.1450), the spread of Greek learning, and the beginning of the Protestant Reformation (1517ff.), one incident in which was the appearance of Luther's German NT (1522), followed by the complete German Bible (1534). Tyndale, who had studied Greek at Cambridge, was ambitious, following Luther's precedent, to give his fellow countrymen the Scriptures in the vernacular, but found it impossible to do so in England. Accordingly, he settled on the Continent, and after an abortive start at Cologne (1525) succeeded in publishing the first printed edition of the English NT at Worms (the "Worms octavo") in February 1526. Not only was it the first printed edition; it was also the first English NT translated directly from Greek. The Greek text used was Erasmus's* third edition (1522). In the following years, while engaging vigorously in theological polemic, Tyndale produced English versions of the Pentateuch, translated from Hebrew (1530; Genesis revised, 1534), Jonah (1531), a second and much improved edition of the NT with an appendix containing translations of the OT "Epistles" (1534), and a third edition (1535). He also appears to have left in manuscript a translation of the OT historical books from Joshua to 2 Chronicles.

Tyndale spent the last seventeen months of his life in captivity, and probably never heard that in October 1535, a year before his death, a complete printed Bible in English, dedicated to Henry VIII, was circulating freely in his native land. This was the edition of Miles Coverdale,* as committed a Lutheran as Tyndale—but the political climate in England had changed radically. This edition was basically Tyndale's, as far as Tyndale's translation had been published; for the rest, Coverdale (who made no claim to scholarship) depended on Latin and German versions. Coverdale's Bible introduced chapter headings and printed the Apocrypha as an appendix to the OT. Two further editions appeared in 1537; the second of these bore on its title page the words "Set forth with the king's most gracious licence," as also in the same year did the version of Thomas Matthew (a pen name concealing the identity of the editor, John Rogers), which included, in addition to Tyndale's published work, his translation of the books from Joshua to 2 Chronicles. "Matthew's Bible"* also contained the first English version of the Prayer of Manasseh.

These versions were superseded by the "Great Bible" of 1539, published by royal authority to be made available in every parish church in England. It was basically Coverdale's revision of Matthew's Bible, and appeared in several successive editions in 1540 and 1541.

Elizabethan versions. The "Geneva Bible"* (1560), produced by the community of Protestant exiles in Geneva during the reign of Mary I (1553-58), was the first English Bible translated throughout from the original text. Preliminary editions of the NT and Psalter were published in 1557, associated with the name of William Whittingham,* one of the team of Geneva translators. In these editions, as in the complete Geneva Bible, verses were printed as separate paragraphs and numbered, and roman type was used, except that words having no direct equivalent in the original text but necessary to give the sense in English were set in italics. The Geneva Bible was dedicated to Elizabeth I, whose accession naturally brought great relief to its sponsors. But the notes and comments with which it was equipped represented a more radical Reformation ideal than the Elizabethan Settlement and were uncongenial to the leaders in church and state. The OT was a thoroughgoing revision of the Great Bible, the language being brought carefully into line with the Hebrew text and even with Hebrew idiom. The NT was a revision of Tyndale's latest edition. The work reflected the best scholarship of the day; not until 1881 did an English version pay such attention in its margins to variant readings.

The Geneva version became the household Bible of English-speaking Protestants. From its inception it was the Bible appointed to be read in churches in Scotland, where the Reformation owed much to the pattern of Geneva. It was the Bible of Shakespeare and of the Pilgrim Fathers; it was the Bible from which excerpts—mainly from the OT!—were made for Cromwell's "Soldier's Pocket Bible" (1643). It was frequently reprinted until the last edition appeared in 1644.

The church leaders in England who found the Geneva Bible unacceptable were compelled to provide an alternative, and in 1568 produced the "Bishops' Bible"—a much better version than the Great Bible which it was designed to replace, but not so good as the Geneva Bible. It did not establish itself in popular esteem and was given no formal recognition by Elizabeth.

In her reign the Bible was translated (from the Latin Vulgate*) into English for the benefit of Roman Catholics. The translator was Gregory Martin, professor in the expatriate English College in Flanders. Although he translated the OT before the New, the NT was published first (at Reims, in 1582); the OT was not published until many years later (at Douai, in 1609-10). This "Douai-Reims"* version in its original form was a highly latinate rendering.

The King James Version. Both the Geneva and Bishops' Bibles were superseded by the "Authorized Version" of 1611, a work which proved so acceptable that it remained for three centuries, without a serious rival, the Bible of English-speaking Protestants. It was the one tangible result of the Hampton Court Conference,* convened by James I of England in 1604 "for the hearing, and for the determining, things pretended to be amiss in the Church." A suggestion by John Rainolds, president of Corpus Christi College, Oxford, that a new translation of the Bible be undertaken, was eagerly seized on by the king, and the work was entrusted to three panels of scholars, some forty-seven translators in all. Formally, it was a revision of the Bishops' Bible (1602 edition); actually the translators drew on the work of all their predecessors and on versions in other languages, with constant reference to the original text. Annotations relating to theological or ecclesiastical controversies were excluded; this greatly facilitated the widespread acceptance of the version. The translators' feeling for prose rhythm made their work admirably suited for reading aloud. When published, it was probably authorized by order in council. Probably—because the Privy Council registers from 1600 to 1613 were destroyed by a fire in January 1618/19, so that no record of the authorization survives.

The Authorized Version set a standard of "Bible English" (going back in essence to Tyndale) which has exercised a profound literary influence. Richard Challoner's* successive revisions of the Douai-Reims Bible (1749-72) brought its style largely into line with the Authorized Version, so that Roman Catholics shared substantially the same "Bible English" as Protestants.

Authorized to Revised. The eighteenth and early nineteenth centuries witnessed a succession of private ventures in Bible translation, especially of the NT. But more important for the history of the English Bible during this period was the progressive study of the NT text, to which substantial contributions were made by Brian Walton* (1654-57), John Mill (1707), Richard Bentley (1717), J.A. Bengel* (1734), J.J. Griesbach* (1774-1806) and J.M.A. Scholz (1830-36). The unsatisfactory character of the text of the early printed editions of the Greek NT (substantially the same as that of the later Byzantine manuscripts), on which the translations of Tyndale and his successors were based, was increasingly appreciated. Karl Lachmann* (1842-50) disregarded the bulk of later manuscripts and concentrated on those of earlier date. The discovery of Codex Aleph (Sinaiticus) in 1844 and the greater accessibility of Codex B (Vaticanus) further stimulated the hope of establishing the primitive NT text (see MANUSCRIPTS OF THE BIBLE). The effects of this stimulus were seen in such private translations as Henry Alford's* NT for English Readers (1869) and J.N. Darby's* (1871), and preeminently in the NT in the "Revised Version" (1881).

The Revised Version was undertaken in consequence of a resolution adopted by the Upper House of the Convocation of the Province of Canterbury in 1870. Two companies of translators, including scholars of various churches, were set up—one for the OT and one for the NT. Despite the cautious rules of procedure, which were weighted in favor of conservatism, the basic text adopted by the NT revisers was largely in line with that of B.F. Westcott* and F.J.A. Hort,* which was published five days before the Revised NT. The Revised OT, on the other hand, was based on the Massoretic text, just as its predecessors had been. The complete Revised Bible appeared in 1885; the Revised Apocrypha followed in 1895. Parallel companies of translators in the USA published their counterpart to the Revised Version, the "American Standard Version," in 1901. (The ASV did not include the Apocrypha.) An updated edition of the ASV, the "New American Standard Bible," was completed in 1971.

While the textual basis of the Revised NT represented a great advance on the Authorized Version, the revision failed to gain wide acceptance. The qualities which made it such an admirable version for the study—its almost pedantic precision, sometimes straining English idiom to the limit in the endeavor to reproduce the minutiae of the original—made it less suitable for liturgical or devotional use.

Twentieth century. From the beginning of the twentieth century a succession of private translations has appeared, many of them aiming at modern, and in some instances colloquial, English. Such were "The Twentieth Century New Testament" (1902); R.F. Weymouth's "New Testament in Modern Speech" (1903); Ferrar Fenton's "Holy Bible in Modern English" (1882-1903), individual to the point of eccentricity; James Moffatt's* "New Translation of the Bible" (1913-28), for many years the most popular of modern versions in Britain; "The Complete Bible: An American Translation," by E.J. Goodspeed* and others (1923-38). Simplified vocabularies were used for "The Bible in Basic English," by S.H. and B.E. Hooke (1940-49), and "The New Testament: A New Translation in Plain English," by C.K. Williams (1952). New versions for Roman Catholics were produced in Britain by R.A. Knox (1945-49), based on the Latin Vulgate, and in America by scholars working under the sponsorship of the Episcopal Confraternity of Christian Doctrine (1941-69). The Jewish Publication Society of America, which produced in "Bible English" in 1917 "The Holy Scriptures according to the Masoretic Text," published in 1963 the first volume ("The Torah") of "A New Translation of the Holy Scriptures according to the Masoretic Text," in contemporary idiom. "The Authentic New Testament" (1955) is the work of an English Jewish scholar, H.J. Schonfield.

The "Revised Standard Version" (NT, 1946; complete Bible, 1952; Apocrypha, 1957) is the latest, and perhaps the last, version of the English Bible in the Tyndale tradition. It is formally a revision of the AV and ASV, carried through by committees of American and Canadian scholars. While it was sponsored by the National Council of Churches in the USA, its intrinsic qualities have won it wide acceptance in other English-speaking countries where it enjoys no such sponsorship, and it would probably be true to say that for the English-speaking world as a whole, the RSV comes nearer than any other version to making the all-purpose provision which the AV made for so many years.

A Catholic edition of the RSV was published in 1965 (NT) and 1966 (complete Bible), but was soon outstripped in popularity by the English edition of the "Jerusalem Bible" (1966). In the USA a thorough revision of the "Confraternity Version" appeared in 1970, with the new title "The New American Bible."

The same year (1970) saw the publication of the complete "New English Bible," the fruit of a project mooted in the general assembly of the Church of Scotland in 1948 and taken up by most of the other non-Roman churches in Great Britain and Ireland. This broke away from the Tyndale tradition and aimed at rendering the original texts into a "timeless" English, "avoiding equally both archaisms and transient modernisms," to quote the general director of the work, C.H. Dodd.* Like the RSV, the NEB is based on an eclectic text in both Testaments, readings being assessed on their merits as they occur.

Other translations which have appeared recently are "The Berkeley Version in Modern English" (1945-59, revised 1969), the work of American evangelicals; J.B. Phillips' "New Testament in Modern English" (1947-58, revised 1972), the best kind of paraphrase; William Barclay's "The New Testament" (1968-69); "Today's English Version" ("The Good News Bible"), translated for the American Bible Society (1966-76); and "The New International Version" (completed 1978), translated by American evangelicals for the New York Bible Society. The "Anchor Bible," an interconfessional translation with commentary, began to appear in 1964 and is to be completed in thirty-eight volumes.

BIBLIOGRAPHY: W.F. Moulton, *The History of the English Bible* (5th ed., 1911); M. Deanesly, *The Lollard Bible* (1920) and *The Significance of the Lollard Bible* (1951); B.F. Westcott, *A General View of the History of the English Bible* (3rd ed., 1927); J.F. Mozley, *William Tyndale* (1937) and *Coverdale and His Bibles* (1953); H.W. Robinson (ed.), *The Bible in Its Ancient and English Versions* (2nd ed., 1954); G. MacGregor, *The Bible in the Making* (1961); *Cambridge History of the Bible,* vols. 2 (ed. G.W.H. Lampe, 1969) and 3 (ed. S.L. Greenslade, 1963); A.S. Herbert (ed.), *Historical Catalogue of Printed Editions of the English Bible, 1525-1961* (1968); W. Allen, *Translating for King James* (1969); F.F. Bruce, *The English Bible* (2nd ed., 1970); G. Hunt, *About the New English Bible* (1970).

F.F. BRUCE

BIBLE CHRISTIANS. A Methodist body stemming from the unauthorized itinerant preaching of William O'Bryan (1778-1868), fervent Methodist preacher in Devon, England. Expelled for his refusal to confine himself to preaching within the circuit, O'Bryan formed a society in Shebbear, N Devon (1815). Aided by men of the caliber of James Thorne (1795-1872) and numerous women preachers, and concentrating on areas devoid of Gospel preaching, the Bible Christians spread not only in the west of England but also in the Isle of Wight, Channel Islands, Kent, and Northumberland. The first conference was held near Launceston (1819), with O'Bryan president and Thorne secretary. O'Bryan's autocracy aroused opposition which led to his withdrawal from the movement (1829-35). The deed enrolled by conference in 1831 made conference, composed of superintendents of districts together with ministerial and lay representatives, the supreme organ of government. Every fifth conference was to be composed of equal numbers of ministers and laymen. District meetings were to be attended by itinerant preachers with one steward, and every fifth year both stewards, from each circuit. Members were to be received by ministers with the approbation of church members.

Bible Christians were sometimes called Quaker Methodists. The influence of Quakerism* is to be seen in the simplicity of style, reliance upon inner illumination, and the importance of the role played by women (in 1823 there were about 100 women preachers in the movement). A missionary society was formed in 1821, but expansion overseas did not begin till 1831 and never reached large proportions. Notable preachers include William Read (1800-1858), the somewhat eccentric William Bailey (1795-1873), and the immortal Billy Bray.* F.W. Bourne (1830-1905), upon whom fell the mantle of James Thorne, became an influential member of the National Free Church Council. He guided the Bible Christians into the United Methodist Church,* formed in 1907 in Great Britain. The Bible Christians then numbered 206 ministers, 1,515 local preachers and 32,202 members, with much smaller numbers overseas.

HAROLD H. ROWDON

BIBLE MANUSCRIPTS, see MANUSCRIPTS OF THE BIBLE

BIBLE SCHOOLS (AMERICAN). These appeared in the United States after 1880 to restore biblical authority and to fulfill the Great Commission. The Grossner Mission, started by Johannes Grossner in 1842 to train missionary candidates, probably preceded all other institutions. H. Grattan Guinness organized the East London Institute for Home and Foreign Missions in 1872. This work inspired A.B. Simpson to found in New York City the first American Bible school in 1882. The school was moved to Nyack in 1897 as the Missionary Training Institute (now Nyack College). Bible studies and practical training for Christian service were coupled with a disciplined life. Moody Bible Institute began as the Chicago Evangelization Society in 1886 to provide what D.L. Moody* called "gap men," trained to fill the gap between the laity and the ministers. This institute flourished after R.A. Torrey* became superintendent in 1889, and the first building was erected. The school now has about 1,000 day students with night and correspondence schools, radio, science films, and other ministries. Other schools followed these in rapid succession, with the greatest growth coming between 1941 and 1960. There are now over 250 such schools in Canada and the USA. Two-thirds of these are denominational. These schools have a Bible-centered curriculum. They seek to cultivate the spiritual life by developing prayer, faith, and self-denial. All demand some practical work, such as teaching Sunday school, street meetings, or personal witnessing. They, unlike seminaries, accept high school graduates, and, unlike liberal arts colleges, train for lay church vocations and Christian ministries rather than for the professions. Many graduates later attend college, seminary, or graduate school. Large numbers become missionaries. This Bible school movement helped to swell what had been called the "Third Force," those evangelicals loyal to the Scriptures as God's fully inspired Word, and to Christ as Savior and Lord.

EARLE E. CAIRNS

BIBLE SOCIETIES. The movement can be traced back to the Pietist movement in Germany with the founding of the Van Canstein Bible Society in 1710. The modern movement began in 1804 with the founding of the British and Foreign Bible Society in London by a group of evangelicals, mostly Anglicans. Its stated aim was "to encourage the wider circulation of the Holy Scriptures, without note or comment." The movement spread quickly to Europe and the British colonies; 1816 saw the origin of the American Bible Society, and by 1819 a Russian Bible Society had produced the NT in a new translation.

A major division occurred in 1825-26 over the question of the Deuterocanonical books (the Apocrypha) which a number of the European societies wished to publish, but which the BFBS felt it should not handle. This led to duplication of organization in a number of European countries. In more recent times this division has been overcome.

By the end of the nineteenth century the BFBS was the center of a worldwide network of agencies and offices. The rapid growth of the American Bible Society resulted in the establishment of rival agencies in many countries, in addition to agencies of the National Bible Society of Scotland and the Netherlands Bible Society.

Then followed the era of constructive reorganization, with the setting up of "joint agencies" primarily responsible to one or other of the major societies. Europe still remained outside the mainstream of development. Following conversations in the late thirties, interrupted by World War II, plans for a closer partnership were discussed; in 1946 at Haywards Heath a conference of the major societies led to the establishment of the United Bible Societies organization. In 1947 Olivier Beguin was appointed general secretary. National societies or offices either are autonomous or are being encouraged to move toward that state; each center aims to draw the widest possible representation of the total Christian community into the Bible cause.

The UBS structure consists of four regional centers (Nairobi, Africa; Mexico City, the Americas; Singapore, Asia; Bassersdorf in Switzerland, Europe). Each of these has a secretariat, including regional consultants on such matters as Bible translation, production, distribution, and office management. Translations consultants serve within each region under a world UBS translations coordinator. Currently over fifty are involved globally in some 800 projects. Similarly, there is a group to coordinate worldwide production of Scriptures. The UBS produced the generally accepted agreed text of the Greek NT and is currently working on OT textual variants.

The UBS Council meets every six years; its general committee meets every three years and takes overall responsibility through an executive committee which meets at least once a year. Financial resources are pooled and shared through a World Service Budget; in 1971 this amounted to almost $7.5 million (of which just over half came from the USA, and close to $1 million from the BFBS). Each society prepares a budget early in the year; it is then considered by the regional secretaries, and finally passed by the executive committee.

Following the launching of the "God's Word for a New Age" campaign in 1963, and the further development, "The Book for New Readers," annual distribution of Scriptures by the Bible societies increased from 54.1 million in 1963 to about 173.4 million (comprised of Bibles, New Testaments, portions—i.e., complete books of the Bible —and selections—short extracts on particular themes). Scriptures are now available in 1,526 languages. Bible society work is conducted on a noncommercial basis, and many Scriptures are heavily subsidized. Normally the selling price of a Bible or NT meets only the bare production cost of a book, leaving nothing for translation, revision, storage, transport, free distribution, and so on.

The UBS publishes *The Bible Translator* and *The Bulletin* quarterly, together with a French equivalent to the Bulletin, and a general information document for member societies, with occasional press releases. There is also an annual UBS prayer booklet.

A. MORGAN DERHAM

BIBLIANDER, THEODOR (c.1504-1564). Protestant scholar. Born at Bischofzell, he studied at Zurich and then at Basle under C. Pellican, J. Oecolampadius,* and W. Capito.* He taught in Leugnitz (1527-29), then returned to Zurich where he succeeded Zwingli* as professor. An accomplished linguist who called himself *homo grammaticus*, he was reputed to be master of thirty languages and was one of the most important Swiss biblical exegetes of his day. Many of his works remained in manuscript, but he published a Hebrew grammar (1535), commentaries, and a notorious edition of the Koran (1543) which the magistrates of Basle attempted to ban, until a number of scholars like Luther intervened on Bibliander's behalf. A pioneer advocate of mission to the heathen, he was also a critic of Calvin's doctrine of predestination. His reputation rests on his biblical exegesis and his ministry in Zurich.

IAN BREWARD

BIBLICAL INTERPRETATION. The Bible comes to us from different ages and cultures from our own and in different languages from our own. It is necessary, therefore, that it should be interpreted so each generation of Christians in their own setting may understand its meaning. By the first century A.D. there were a number of different approaches to the interpretation of the OT. Philo* of Alexandria used the method of allegory to bring out of the text of the OT, including what appear to be literal historical narratives, the philosophies of Platonism, Stoicism, and neo-Pythagoreanism.

The Qumran community used the *pesher* method to suggest that the real meaning of general statements in the OT was to be found in the particular personalities and circumstances of their own sect. The rabbinic school of Shammai was generally literal in its own understanding and harsh in its application, while that of Hillel was more subtle and had a number of rules designed to apply the Law to the contemporary situation. In the postapostolic period similar methods were used in interpreting both Testaments. Origen* in particular emphasized the allegorical approach, believing that all Scripture had a threefold sense, the "body" for the simple, the "soul" for beginners, and the "spirit" for the mature. The school of Antioch was more literal and grammatical in its approach than the Alexandrians*; Jerome and, to some extent, Augustine in the West tended to follow the Antiochene* way.

In the Middle Ages it was fairly generally accepted that there were four senses: literal, allegorical, moral, and anagogical. The approach of Nicholas of Lyra* who laid more stress on a literal and christological interpretation prepared the way for Luther and the Reformers. They denied the supreme authority of the church and believed that by proper grammatical study and comparing Scripture with Scripture the true meaning would be given through the internal testimony of the Holy Spirit. The enormous advance in knowledge of the text, background, and languages of the Bible over the last century or so has helped greatly toward more accurate interpretation, but this has often been nullified by the tendency to read into the Bible modern philosophies which have no place there.

The correct way to understand the Bible involves seeing it as both a collection of human documents and also a unified, divine book. To the first end must be applied all available knowledge about the background, circumstances, intention, and language of the individual writings. It is often of special importance to know what type of literature a biblical book is supposed to be. The meaning of the smaller parts can be discovered by noting the usage of words and phrases. Where there is apparent symbolism, we must ask whether that is in addition to or instead of a literal meaning. It must then be remembered that the biblical canon has been defined because its constituent parts provide a unified witness to the saving purposes of God, particularly as they are centered in his actions in Christ. This means that not only do many OT passages provide a background to the NT, but also the principles which they are setting forth find their fullest meaning in what God was to do later through Christ. This is the way in which the OT is interpreted in the NT, and it should be the pattern for us. There is no one, cast-iron hermeneutical system into which the meaning of the Bible can be forced for all time. Each generation and each culture must seek, using the wisdom of Christians throughout the ages and depending upon the guidance of the Holy Spirit, to find as fully as possible for itself the meaning of the faith once for all delivered to the saints and enshrined in the Holy Scriptures.

BIBLIOGRAPHY: R.V.G. Tasker, *The Old Testament in the New Testament* (1954); B. Ramm, *Protestant Biblical Interpretation* (1956); J.D. Wood, *The Interpretation of the Bible* (1958); D.E. Nineham (ed.), *The Church's Use of the Bible* (1963); R.N. Longenecker, *Biblical Exegesis in the Apostolic Period* (1975). R.E. NIXON

BICKERSTETH, EDWARD (1786-1850). English Evangelical divine. Born in Westmoreland, he worked for the post office before moving into the legal field, finally becoming a solicitor in Norwich. In 1805 he had a deep spiritual experience, and was later to write *Help to Studying the Scriptures* which went through twenty-one editions. Ordained in 1815, he went briefly to Sierra Leone to report on Church Missionary Society work there, and afterward became one of the society's secretaries, during which time he wrote several books. In 1830 he accepted the living of Watton, Hertfordshire, but maintained his CMS connection. Shortly afterwards he compiled his *Christian Psalmody,* comprising over 700 hymns, to which he subsequently added another 200; this work went through fifty-nine editions in seven years. In 1814 he took part in the formation of the Parker Society (for republishing the work of the English Reformers). He was active also in the formation of both the Evangelical Alliance and Irish Church Missions. His son, Bishop Edward Henry Bickersteth (1825-1906), was prominent as a clergyman and hymnwriter. J.D. DOUGLAS

BIDDLE, JOHN (1615-1662). Unitarian. Son of a Gloucestershire tailor, he showed precocious talent as a youth. He went to Magdalen Hall, Oxford in 1634, where he became a tutor before returning to teach in Gloucester. There his Trinitarian orthodoxy was suspected by the Presbyterian party, and in 1645, his views having been betrayed by a friend, he was imprisoned. In the following year he appeared before a House of Commons committee and in 1647 his *Twelve Arguments,* clearly refuting accepted teaching about the divinity of the Holy Spirit, was published. In the next five years he was in and out of prison, but when released in 1652 under the act of oblivion, he began to organize a congregation, write catechisms, and publish Socinian books. He came before the authorities again and was sent to the Scilly Isles in 1655, but pleading by powerful friends obtained his release in 1658. In 1662 he was sent once more to prison, where he died.

C. PETER WILLIAMS

BIEDERWOLF, WILLIAM EDWARD (1867-1939). American Presbyterian evangelist and educator. Born in Indiana, he was probably the best educated evangelist of the period, having studied at Wabash (Indiana) College, Princeton University and Seminary, and several schools in Europe. Ordained in 1897, he served for three years in the pastorate at Logansport, Indiana, and one year as a chaplain in the Spanish-American War. The rest of his life was spent in evangelistic work and direction of the famous Winona Lake Bible School of Theology. His leadership saved the Bible Conference from bankruptcy. In 1909 he organized the Family Altar League, which he directed. From 1929 to 1939 he was a pastor in Palm Beach, Florida. His more important works are *The Millennium Bible* and *The New Paganism.* DONALD M. LAKE

BIEL, GABRIEL (1420-1495). German philosopher. Born at Speyer and educated at Heidelberg and Erfurt, he became a noted preacher first at Mainz and then at Urach. He was responsible with Count Eberhard of Württemberg for the founding of the University of Tübingen, where he held the chair of theology from 1484. In old age he joined the Brethren of the Common Life.* A follower of William of Ockham and one of the last great Scholastic thinkers, he held to a very high, if mechanical, sacramentalism, betrayed certain semi-Pelagian tendencies, and apologized for developing capitalist ethics of his age. His best known works are his *Epitome* of Ockham's writings (1495), his *Lecture* on and *Exposition* of the Canon of the Mass (1488, 1499), and his *Sermons* (1499). IAN SELLERS

BIGG, CHARLES (1840-1908). Classical scholar and theologian. He had a varied career as a teacher in Oxford University and public schools, as a minister, and as regius professor of ecclesiastical history at Oxford from 1901. Besides a commentary on Peter and Jude (1901) and editions of several classics of Christian spirituality, he wrote works on early Christian history which, though dated, all still valuable. His Bampton Lectures, *The Christian Platonists of Alexandria* (1886), reveal much of his own humane piety and critical freedom, as well as his erudition. *The Origins of Christianity* (1909) is also noteworthy.

BILDERDIJK, WILLEM (1756-1831). Dutch poet. Born into a strong Calvinist and monarchist family, he early developed habits of study as a result of incapacitation following a foot injury. Later he studied law at Leiden and practiced as an advocate at The Hague till his exile in 1795, in which year the French established a republic to which he refused the required oath of allegiance. He moved to Hamburg and then to London, not returning to the Netherlands till 1806. There he prospered for a time, but suffered privation after the accession of William of Orange in 1813. A man of deeply Christian conviction, his quiet testimony led to the conversion to Christianity of Isaak da Costa,* who later edited his works in sixteen volumes (1856-59). R.E.D. CLARK

BILLING, EINAR MAGNUS (1871-1939). Swedish theologian. Born in Lund, he was professor in dogmatic and moral theology in Uppsala (1909-20), then bishop of Vasteras. He was a leading figure in the revival of Luther studies, the so-called "Luther Renaissance," which is generally considered to date from the publication of Billing's book on Luther's teaching on the state (1900). He laid down the lines for what was to become the distinctive Scandinavian method of Luther study—a historical, systematic approach crystallizing in the characteristic Scandinavian "motif research." Other books included one on the Swedish national church.

BILLINGS, WILLIAM (1746-1800). American composer. A musical amateur and a tanner by trade, he gained by his unbounded enthusiasm and innate vein of originality a unique place in the development of American church music. He published six collections during his life, the last being his *Continental Harmony* (1794), which was very influential in New England. "When Jesus wept" is a piece of both beauty and poignancy, and his tune "Chester" to patriotic words of the revolutionary period has attracted wide attention in recent years.

BILNEY, THOMAS (c.1495-1531). Protestant martyr. Born near Norwich, he went to Trinity Hall, Cambridge to study law. In 1519 he was ordained to St. Bartholomew's Priory, Smithfield, and the next year he became a fellow of Trinity Hall. After a long period of searching for spiritual peace he read Erasmus's* Latin NT and was converted through reading 1 Timothy 1:15. He became a central figure in the group of theologians who started to meet at the White Horse Inn. Most famous of his converts was Hugh Latimer,* to whom he went to make his confession in 1524 and who was brought to a sense of forgiveness through it. He was arrested in 1527 for heresy and was released only after promising to stop preaching Reformed doctrine. He could not keep himself in check for long and was arrested again that year, and after a trial before Bishop Tunstall* and others he was persuaded by some of his friends to recant. Again Bilney could not hold himself to what he had done under pressure, and early in 1531 he set off on a preaching tour of Norfolk. Later that year he was arrested, tried, and sentenced to die at the stake. A shy and retiring man, he was the first of the early Cambridge Protestants to be martyred. R.E. NIXON

BINGHAM, HIRAM, JR. (1831-1908). Congregational missionary to Micronesia. Born of missionary parents in Honolulu, he received most of his education in the USA, in keeping with missionary policy. After serving briefly as principal of Northampton (Massachusetts) High School, the American Board in 1856 sent him out as a missionary. He opened work in the Gilbert Islands, reduced the language to writing, and in 1860 published the first Bible portion in Gilbertese. By 1890 he had finished translating the whole Bible. For two years he commanded the missionary brig *Morning Star* that maintained contact between the widely scattered islands. For three years he was corresponding secretary of the Hawaiian Evangelical Association, which sponsored Hawaiian missionary activity. From 1880 to 1882 he was Hawaiian government protector of South Sea immigrants. HAROLD R. COOK

BINGHAM, HIRAM, SR. (1789-1869). Pioneer missionary to Hawaii. Born in Bennington, Vermont, he reached Hawaii (then known as the Sandwich Islands) under the American Board in 1820. He helped create a written language for the people and set up schools for their instruction. With other missionaries he completed a translation of the Bible in 1839. He built the first church in Honolulu in 1821. As trusted adviser to the chiefs, he came into conflict with godless Americans and others who flocked to the islands for personal satisfaction and gain. After twenty-one years, his wife's health forced return to the States. Bingham has been much maligned by recent writers unsympathetic to the strict New England morality he sought to inculcate.

HAROLD R. COOK

BINGHAM, JOSEPH (1668-1723). Church historian. Born at Wakefield and educated at University College, Oxford, where he became fellow (1689) and tutor (1691), he was expelled for heresy in 1695 after preaching a sermon on the terms "Person" and "Substance" concerning the Godhead. He became rector of Headbournworthy, near Winchester, where he remained until 1712 when he removed to Havant, near Portsmouth. He lost most of his meager savings in the South Sea Bubble (a speculation hoax) in 1720. He spent much of his life writing *Antiquities of the Christian Church* (10 vols., 1708-22), a work of lasting value. Later he advocated reunion of the Christian churches on the basis of apostolic episcopacy. His collected works were published in 1855.
R.E.D. CLARK

BINNEY, THOMAS (1798-1874). Minister and writer. Son of a Presbyterian, he was apprenticed to a bookseller before studying for the Congregational ministry. He held pastorates at Bedford and on the Isle of Wight before going to Weigh House Chapel, London (1829-69). His impressive preaching appealed to youth, for whom he wrote the most popular of his more than fifty books: *Is it possible to make the best of both worlds?* (1853). He attacked the established church, pioneered Nonconformist liturgical services (see his edition of and appendix to C.W. Baird's *Chapter on Liturgies: Historical Sketches,* 1856), wanted better music in such services (see his *Service of Song in the House of the Lord,* 1849), and was one of the first Nonconformists to introduce anthems and chants. Of his many poems, only the hymn "Eternal Light!" has survived. JOHN S. ANDREWS

BIRD, WILLIAM, see BYRD, WILLIAM

BIRETTA. A hard square cap worn by Roman and some Anglican clergy. In the Middle Ages it was worn only by higher graduates of universities, but in the sixteenth century its use was permitted to all clergy. It is colored black for priests, purple for bishops, and red for cardinals.

BIRGITTA, see BRIDGET OF SWEDEN

BIRTHRIGHT MEMBERSHIP. This was adopted for Quakers by the London yearly meeting of 1737. If the father was a member of a meeting, the wife and all their children were to be enrolled as members of the meeting and eligible for financial aid if needed. No profession of faith was required of such persons, and membership continued after the death of the husband and father. This plan enrolled many members who had no change of heart, and it diminished evangelism as a mode of winning members. Birthright membership was replaced after 1900 by associate membership.

BISCHOP, SIMON, see EPISCOPIUS

BISHOP. From the vulgar Latin *biscopus,* the word is often given as a translation of *episkopos* in the NT. An alternative translation is "overseer." Within the NT it seems to have denoted a function of the ministry, and to be an alternative for presbyter (cf. Acts 20:17; Phil. 1:1; 1 Tim. 3; Titus 1:7ff.). Christ himself was regarded as *the* Bishop (1 Pet. 2:25). The origins of the monarchical bishop and the threefold ministry of bishop, presbyters, and deacons are wrapped in some mystery. Among the Apostolic Fathers* only Ignatius speaks of monarchical episcopacy, and with him the emphasis is on unity around the bishop in perilous times, not on the divine institution of the office. Gradually, with the disappearance of the charismatic ministry, the opposition from Gnosticism, and the imperial recognition of the church in the fourth century, the single bishop in charge of a diocese or group of churches emerged. Normally he was the head of a city or town church. Furthermore, with the adoption by the church of the divisions within the empire, there also evolved bishops among bishops—that is, pope, patriarch, metropolitan, and archbishop. The division of Eastern and Western Christendom, the close association of church and state, and the rise to power in the West of the see of Rome, had an important effect upon the development of episcopacy. During and after medieval times bishops were both spiritual and temporal lords. This tradition is still reflected in England, where a number of bishops have seats in the House of Lords.

At the time of the Reformation, Protestants wished to reform or to abolish the office of bishop, since its medieval accretions alarmed them. The Calvinist churches equated the office of bishop with that of pastor or parish minister. Lutherans saw the continuance of the office of bishop (if understood as a superintendent minister) as among the *adiaphora* (see ADIAPHORISTS). This resulted in the retention of bishops in Scandinavia and their abolition in Germany. In recent times the office has been revived in Germany, but no apostolic succession is claimed. The Church of England retained a succession of bishops in the transition from Catholicism to Protestantism, and this has continued to the present. In some more recent denominations the title of bishop has been given to superintendent ministers—e.g., in American Methodism.

Within the Orthodox Churches bishops are chosen from the celibate priests in the monasteries by election at a synod on the advice of the patriarch. In the Roman Catholic Church the pope has the last word and he actually does the appointing—and they are responsible to him. Within the Church of England the chapters of cathedrals elect a bishop on the advice of the monarch. (Anglican churches elsewhere have a more democratic system.) In churches which claim the apostolic succession, consecration is normally performed by one archbishop and two bishops, a rule which was first agreed upon at the Council of Arles in 314. In other churches not claiming apostolic succession, choice is usually by a synod and installation into office by representatives of the synod.

Traditionally since early times, the bishop's ministry is seen to involve ruling, sacramental, and pastoral aspects. He rules both clergy and people in his diocese; he alone can confirm and ordain; and he is the chief pastor of the flock. Often a bishop is assisted by an assistant, suffragan, auxiliary, or coadjutor bishop. The insignia of a bishop include miter, pastoral staff, pectoral cross, ring, and *caligae.* Within ecumenical dialogue in recent times there has been much discussion as to whether episcopacy is of the *esse,* the *bene esse,* or the *plene esse* of the church.

BIBLIOGRAPHY: W. Telfer, *The Office of Bishop* (1962); A.G. Hebert, *Apostle and Bishop* (1963); R.B. Kuiper, *The Glorious Body of Christ* (1966); L. Berkhof, *Systematic Theology* (1966, or earlier editions); see also documents of Vatican II on Church and Ministry. PETER TOON

BISHOPS' BIBLE, see BIBLE, ENGLISH VERSIONS

BISHOPS' BOOK. The *Institution of a Christen Man* or the "Bishops' Book" was a compromise doctrinal statement of the English bishops and divines in 1537. More conservative than the Ten Articles,* it reestablished the seven sacraments though not all on the same level. Luther's influence was still apparent—transubstantiation was not mentioned, the definition of justification was moderately Protestant, images were attacked, and the supremacy of Rome was denied, while the freedom and equality of national churches were upheld. Henry VIII did not give it official authority, and indeed in 1538 began to correct it extensively and conservatively, and was courageously opposed in private by Cranmer.*

C. PETER WILLIAMS

BISHOPS' WARS, THE (1639-1640). The two confrontations between Charles I* and the Scottish Covenanters* caused by the king's determination to force a full-fledged episcopacy upon their country. The first phase, which produced little fighting, ended with the disbandment of the army the Scots had raised, on Charles's undertaking to call a general assembly (with the bishops excluded) in August 1639, followed by a parliament that would ratify acts passed by the assembly. Charles was, however, playing for time in the hope that the English Parliament would finance a second attempt to subdue the Scots, and he continually prorogued the Scottish parliament. The exasperated Scots met in defiance of him, abolished episcopacy (June 1640), defeated the king's army at Newburn, and occupied Newcastle and Durham. The English "Long Parliament," called by Charles in desperation (November 1640), was friendly to the Scots and took advantage of the king's dilemma to limit powers he had claimed and exercised in England. This was the beginning of the king's downfall, climaxed by his execution in 1649. J.D. DOUGLAS

BLACK DEATH, THE. The name given to the bubonic and pneumonic infections that swept unchecked across Europe from 1347 to 1351. Its origins were evidently in central Asia, where headstones dating from 1338/39 in Nestorian graveyards in Kirgiz commemorate plague victims. From there the outbreak evidently spread to India, China, and Europe. Reaching Italy in late 1347, it went through the peninsula and into Switzerland, Germany, and parts of eastern Europe, before going on to France, Spain, and England. London was reached by the spring of 1349. During the following year plague had got to Scotland, Scandinavia, and the Baltic countries. In parts of western Europe, fatalities numbered thirty to forty percent of the urban population. Medical knowledge was hopelessly inadequate in the face of this greatest disaster in European history, which had widespread and ruinous effects on economic, political, and social life. Prices increased, incomes shrank, the peasants demanded lower rents, and with lords impoverished and entire manors abandoned, the breakdown of the manorial system was hastened. Popular religion responded with renewed piety and preoccupation with death. In some places Jews were blamed for poisoning the wells, and many of them were murdered. The population of Siena was so reduced that the enlargement of the city's cathedral was abandoned. Many of the best people who had not fled from their posts to uninfected areas (local officials, physicians, priests, scholars) died in the public interest, so that the following generation had a surfeit of the incompetent. The Dominican Order suffered such casualties that it was forced to admit semiliterate postulants who probably made no small contribution to the superstitious and heretical accretions that grew up in communities previously noted for the quality of their scholarship. C.G. THORNE, JR.

BLACK FRIARS, see DOMINICANS

BLACK MASS. A blasphemous, obscene parody of the Roman Catholic Mass based on the belief that the worship of Satan is a complete reversal of the worship of God. It is addressed to hell and the devil. The cross is suspended upside down, and the Lord's Prayer recited backward. It is often celebrated by a renegade priest, using a woman's naked back as an altar and using his left instead of his right hand.

BLACK RUBRIC, see KNEELING

BLACKSTONE, WILLIAM EUGENE (1841-1935). Friend of the Jews and writer on Christ's second coming. Born in Adams, New York, he was converted in boyhood. He served in the Civil War with the U.S. Christian Commission, then started a flourishing business in Oak Park, Illinois. Impressed by the lack of literature on Christ's return, he wrote *Jesus Is Coming,* a book still in print. In 1887 he helped start the Chicago Hebrew Mission, and in 1890 headed the first conference between Jews and Christians in Chicago. The next year he presented President Harrison with a memorial signed by 413 Christian and Jewish leaders urging an international conference on the Jewish situation. Israel celebrated the seventy-fifth anniversary of this memorial and dedicated a forest in his name. HAROLD R. COOK

BLACKWELL, ANTOINETTE LOUISA BROWN (1825-1921). Reformer and one of the first ordained American women. Born in Henrietta, New York, she graduated from Oberlin in 1847, then completed a theological course (1850). Refused a preaching license due to her sex, she finally became pastor of the Congregational Church of South Butler, New York. She resigned after four years because of theological problems, became a Unitarian, and gained fame as a temperance speaker. She also espoused abolitionism and women's rights.

BLACKWELL, GEORGE (c.1545-1613). Roman Catholic archpriest. Educated at Trinity College, Oxford, where he was made a fellow in 1566, he was attracted to the Roman Catholic Church and entered the English College at Douai in 1574. Ordained in 1575, he was sent back to England in the following year as one of the first missionary priests. After the death of William Cardinal Allen in 1594, he was appointed archpriest in 1598 to provide leadership for the English Roman Catholics. As a friend of the Jesuits in England he lost the support of some of the secular clergy. For a number of years there were serious divisions about the attitude of Roman Catholics to the English crown. In 1606 James I imposed an oath of loyalty, and Blackwell after a period of imprisonment took the oath despite papal opposition. In 1608 he was deprived of his office as archpriest. He died a supporter of Catholic loyalty to the Crown. NOEL S. POLLARD

BLAIR, JAMES (1655-1743). Episcopalian minister. Scottish-born, he graduated from Edinburgh in 1673 and was subsequently ordained. He ministered in Scotland and later worked in a London library until, at the request of the bishop of London, he went to Virginia. Here he served the Henrico parish from 1685 to 1694. After 1689, as the bishop's commissary, he secured better-trained ministers for the colony. From 1694 to 1710 he was pastor at Jamestown, and then until his death he served the Bruton parish church in Williamsburg. He obtained the charter for, and was the first president of, William and Mary College in 1693. He was also a member of the Virginia Council until his death.

EARLE E. CAIRNS

BLAKE, EUGENE CARSON (1906-). Ecumenical leader. Born in St. Louis and educated at Princeton and Edinburgh, he taught for a time in India, then in 1932 began a successful career as a pastor until 1951, when he was appointed stated clerk of the general assembly of the Presbyterian Church, USA. In 1966 he became general secretary of the World Council of Churches,* which post he held until his retirement in 1972, after which he engaged in lecturing and preaching. The holder of several outstanding awards, Blake became famous for a sermon delivered in 1960 at Grace Episcopal Cathedral, San Francisco, at the invitation of Bishop James Pike,* in which he suggested a union of the Methodist, United Church of Christ, Protestant Episcopal, and Presbyterian churches. Other denominations have since expressed interest in the venture, and annual meetings are held, but the movement has to date produced no genuine organic union. Blake's writings include *The Church in the Next Decade* (1966).

ROBERT C. NEWMAN

BLAKE, WILLIAM (1757-1827). English poet. Best known for his *Songs of Innocence* (1789) *and Experience* (1794), he published also many other poems elaborating his very private mythology and system of belief, culminating in the final symbolic works *Milton* (1804-19) and *Jerusalem* (1804-20). All his works are illustrated by his own engravings. He himself declared the idiosyncrasy of his position:

> The Vision of Christ that thou dost see
> Is my Vision's Greatest Enemy.
> ("The Everlasting Gospel")

In *Jerusalem* he wrote: "I must create a system or be enslav'd by another man's." This extreme Romantic individualism was in part the product of Blake's own rebellious nature, but it was also his reaction against the "single vision," scientific, materialistic, and rationalistic, of Bacon, Newton, and Locke. Against this Blake wished to assert imagination, the "divine vision," by which "God is Man and exists in us and we in him" ("Annotations to Berkeley"). At its best this led him into that absorption with God when "Self is lost in the contemplation of faith / And wonder at the Divine Mercy," but, despite his grasp of paradox, he never attained the fullness of the mystical experience, for he was never able to hold in tension immanence and transcendence, justice and mercy, righteousness and peace.

He saw the narrow moralistic emphasis of much contemporary orthodox religion, and typically he rejected it in extravagant terms. He rightly saw that "If Morality was Christianity, Socrates was the Savior" ("Annotations to Thornton's Lord's Prayer"), and he nobly emphasized the uniqueness of Christianity as lying in its doctrine of the forgiveness of sins; but his rejection of Christ's alleged virtues in "The Everlasting Gospel" is outrageous. Characteristically, he associates forgiveness of sins with an act of love, but rejects any idea of atonement in the death of Christ.

See *Poems of Blake* (ed. W.H. Stevenson, 1971) and J.G. Davies, *The Theology of William Blake* (1948). ARTHUR POLLARD

BLANCHARD, CHARLES ALBERT (1848-1925). President of Wheaton College, Illinois, for forty-three years. Born in Galesburg, Illinois, he graduated from the college during the presidency of his father, Jonathan. When called to succeed him, Charles had already been associated with the school for ten years. In a real sense Wheaton College was his life; he it was who maintained its conservative evangelical character and gave it stature as an educational institution. From 1891 to 1893 he was also pastor of the Chicago Avenue Church in Chicago (now Moody Memorial Church).

BLANDINA (d.177). Christian martyr. Eusebius records a letter to Asian and Phrygian churches from Christians at Lyons and Vienne, describing their martyrs in a local anti-Christian outburst in 177. Emphasis is laid on the tortures, courage, and

Christlikeness of a servant girl, Blandina, who inspired a fellow-sufferer and refused allegiance to pagan gods.

BLASPHEMY. For the ancient Greek, to blaspheme was to use "abusive words" by which to destroy another's reputation. In Judaism the object of such blasphemy was finally always God, against whom it was so serious a sin that the penalty was death (Lev. 24:11f.). Used more broadly in the NT, the concept is controlled throughout by the thought of reviling God's name, a discrediting of His Word, or an abuse of His majesty (e.g., 1 Tim. 6:1; Titus 2:5; Rev. 16:11, 21). Jesus was accused of blasphemy by the Jews, and it was on this charge that they called for His death. Stephen and Paul also were so accused. The most heinous sin of all, according to Mark 3:29, is blasphemy against the Holy Spirit.*

In the Middle Ages, to vilify the church, the Virgin Mary, the saints, or the sacraments was blasphemy. The penalty of death for such offenders against God was sanctioned by the Council of Aachen in 818, but this was seldom in fact carried out. In the post-Reformation period enactments against blasphemy continued in force in Protestant countries (e.g., in Britain from the seventeenth to the nineteenth centuries). With the growing secularization of society following the Enlightenment,* blasphemy came to be regarded rather as a crime against the good order of the state. It remains on the statute books in many countries as punishable by law to deny God, ridicule Christ, and profane the Bible, but in practice the law is not easily invoked where God, Christ, and the Bible are no longer highly revered.

H.D. MC DONALD

BLASS, FRIEDRICH WILHELM (1843-1907). German philologist and grammarian. He taught at the universities of Kiel and Halle and wrote extensively in the areas of classical studies and NT criticism. His *Grammatik des neutestamentlichen Griechisch* (1896; ET 1898) was foundational and is still in use today, though in a radically revised form. Blass argued that the so-called Western and non-Western texts of Acts originated as different editions by Luke himself, a view which has not commended itself to many scholars.

BLAURER, AMBROSE (1492-1564). Zwinglian pastor. Son of a member of the town council in Constance, Germany, where he attended Latin school, he went to Tübingen University (where he met Melanchthon*) and in 1509/10 he took orders at the Benedictine monastery in Alpirsbach, where he remained until, convinced by Luther's writings in 1522, he returned to Constance. In 1525 he began to preach in St. Stephan Church. Back in Württemberg (1534-38), his Zwinglian supper view and original tolerance for Anabaptists brought him into conflict with the duke. He preached in Augsburg in 1539, but his moralizing and adamant Zwinglianism brought disfavor. He returned to Constance, only to flee before the emperor's troops in 1548. The next years, embittered and pessimistic, he spent in Biel and Winterthur.

ROBERT B. IVES

BLAUROCK, GEORG (c.1492-1529). Early Anabaptist* evangelist. Called "the Blue Coat" and "Sturdy George," Blaurock (his real name was Jörg Cajakob) was a priest from Chur, Switzerland, who responded to Zwingli's evangelical preaching sometime before 1523. He eventually embraced Anabaptist views after a period of Bible study. Blaurock apparently initiated the practice of believer's baptism in Zurich, and in January 1525 he founded the first Anabaptist congregation at nearby Zollikon where he won more than 150 converts with his powerful preaching. The local authorities, however, soon intervened and arrested him and other Anabaptist leaders. Exiled from Zurich in 1527, Blaurock became an itinerant evangelist, winning several thousand to Christ and planting the Anabaptist faith over much of central Europe. He became an important link with later Anabaptist work by establishing many congregations in Tyrol which afterward supplied thousands of members for Anabaptist colonies in Moravia. Hapsburg officials finally caught Blaurock and burned him for heresy.

ROBERT D. LINDER

BLAVATSKY, H.P., see THEOSOPHY

BLEEK, FRIEDRICH (1793-1859). German biblical scholar. Born at Ahrensböck, Holstein, he studied in Berlin under W.M.L. de Wette,* Johann Neander,* and F.D.E. Schleiermacher* (1814-17), and was "Repetent" in theology under de Wette (1818-19). He was professor of theology at Bonn (1829-59), and was raised to the office of consistorial councillor (1843) and rector of the university. He worked wholly in biblical criticism and exegesis, and while he held advanced views on the OT he took a conservative position in NT studies, opposing the Tübingen School* and defending the traditional authorship of the fourth gospel. His magnum opus was a three-part commentary on Hebrews (1828-40).

J.G.G. NORMAN

BLEMMYDES, NICEPHORUS (c. 1197-1272). Greek theologian. The Fourth Crusade ended with the capture and sacking of Constantinople in 1204. The Byzantine Empire disintegrated, and three independent Greek centers were formed while the Latins controlled Constantinople. One of the three was Nicea, which became an important Greek political and cultural center. Blemmydes became an outstanding figure in the cultural life there, founding a monastery, establishing a school, and becoming a teacher of philosophy. He wrote in various areas of church and state, and a number of theological works in the fields of dogmatics, polemics, poetry, sermons, etc. He was known also for his secular works, such as his political writing *The Imperial Statue,* and for his philosophical and geographical writings. He died in his monastery.

GEORGE GIACUMAKIS, JR.

BLESSING (Benediction). Blessing is God's imparting of divine favor; when a blessing is given by a man it is a human act invoking divine favor. It can mean also to give thanks to God. In the OT,

God blesses people directly—see Genesis 1:22; 2 Samuel 6:11—and there are many examples of men blessing others—Genesis 27:27ff. Such blessing is usually of the greater to the less (cf. Heb. 7:7). Things are also blessed in the OT, God blesses the Sabbath Day (Gen. 2:3), and Moses blesses a man's property (Deut. 33:11).

In the Beatitudes, our Lord gives His well-known teaching. There is also the example of our Lord's blessing children who were brought to Him. The blessing of the loaves and fishes, and of the bread and wine at the Last Supper, is almost certainly indicative of thanksgiving.

It has become the custom to end most church services with a blessing, although this is by no means a primitive practice and, for instance, in the new Communion services in the Church of England, is optional. In the Roman Catholic and Greek Orthodox churches there is a wide use of the blessing of objects. In general the Reformed churches have been reluctant to admit the blessing of things, though the blessing of water in the baptismal service has been made plain in the Church of England Series II services. Curiously enough, side by side with this reluctance, there has been no similar hesitation about dedicating various objects.

Benediction is also the name for a service in the Roman Catholic Church where, at the climax, the congregation is "blessed" by the consecrated host. This developed in the Middle Ages with popular devotion to the host as a sign of Christ's presence, but is now falling into disuse through the impact of the Liturgical Movement and the practice of celebrating the Eucharist in the evening. PETER S. DAWES

BLISS, PHILIPP (1838-1876). Baptist hymnwriter. Born in Pennsylvania where he was converted at twelve, he joined the Elk Run Baptist Church. With little financial resources, he was forced to work at farming and woodcutting until 1860, when he entered the Normal Academy of Music in New York. He soon gained a wide reputation as a bass of great range and beauty. In 1865 he was hired by the Root & Cady Company of Chicago to conduct musical conventions in the northwestern states. During this same period he made the acquaintance of famed evangelist D.L. Moody.* Bliss assisted in the musical ministry of Moody's campaigns as well as those of Major D.W. Whittle. Bliss and his wife met an untimely death in a train accident. He produced more than twenty familiar hymns and songs, such as "Almost persuaded," "Brightly beams our Father's mercy," and "Hallelujah! what a Saviour." His tombstone bears the title of another of his hymns, "Hold the fort." DONALD M. LAKE

BLONDEL, DAVID (1590-1655). French Protestant church historian. Born at Châlons-sur-Marne, he was educated at Sedan and the Genevan Academy. He was a country pastor at Roucy for most of his life, refusing a chair of theology at Saumur in 1631, though he was created "honorary professor" there in 1645. In 1650 he left Roucy for a professorship at Amsterdam. He wrote *Pseudo-Isidorus et Turrianus Vapulans,* which finally discredited the historicity of the False Decretals,* and this was probably his most important work. He also wrote *De la primauté en l'Église* (1641), a defense of Reformed ecclesiastical polity, and *Apologia pro Sententia Hieronymi de Presbyteris et Episcopus* (1646), an attack on episcopacy. All his works are on the Roman Catholic Index* of Forbidden Books.

J.G.G. NORMAN

BLONDEL, MAURICE (1861-1949). French Roman Catholic philosopher. He taught at Aix-en-Provence (1896-1927), and then though blind wrote much in retirement. His resistance to traditional Scholastic theology made him suspect of Modernism, but though associating with some Modernists, he rejected Loisy's* views on history and dogma, and was never condemned. He reached orthodox conclusions by non-Scholastic routes, and gained considerable influence among Catholics. Often obscure, the essence of his position was stated in his doctoral thesis, *L'Action. Essai d'une critique de la vie et d'une science de la pratique* (1893). Truth cannot be found by intellect alone, but only by the whole of being, including willing and feeling, totally involved in the movement of life. Faith therefore is not accepting dogmas, but is coming to realize the supernatural within human experience, which Blondel tried to show required a goal beyond the natural order (i.e., God). So he offered a "method of immanence" which led compellingly to transcendence —a position quite different from the immanentism for which Modernism was condemned.

HADDON WILLMER

BLUMHARDT, CHRISTOPH FREDERICK (1842-1919). German evangelical leader. He became assistant to his father, J.C. Blumhardt,* at Bad Boll in 1869 and succeeded him as head of that establishment in 1880. His theology included a strong emphasis on the righteousness of God and His judgment against "the flesh." Because of his sympathy with the workers of Württemberg, Blumhardt was elected to the diet of that province during the years 1900-1906. This step was misunderstood by many of his contemporaries, but he justified it as being in the spirit of Jesus who associated with "publicans and sinners."

BLUMHARDT, JOHANN CHRISTOPH (1805-1880). German evangelical leader. After theological studies at Tübingen he became a tutor in 1830 at the missionary training institution in Basle of which his uncle, Christian Gottlieb Blumhardt, was founder. In 1838 he was installed as pastor at Möttlingen, where a revival occurred which was accompanied by healings of bodily and mental diseases. The most notable exorcism was that of Gottliebin Dittus. So renowned did the Möttlingen revival become that on Good Friday 1845, no fewer than 176 communities were represented in the services. Blumhardt resigned from his pastorate in 1852 to purchase and operate a center for sufferers from all kinds of illnesses and ranks of society at Bad Boll. Gottliebin Dittus assisted in the work, as did Blumhardt's two sons, who joined

him in 1869 and 1872 respectively.
WAYNE DETZLER

BOCSKAY, STEPHEN (1557-1606). Hungarian Protestant leader. He led the Protestant opposition to the attempt of Rudolph II of Austria to destroy civil and religious liberty in Hungary. Roused to action by the outrages inflicted on the people of Transylvania by the Austrian general Basta (1602-4), he sought the help of the Turks and drove Basta from Transylvania. He was elected prince in 1605, and in 1606 concluded the Peace of Vienna with the Archduke Matthias, who had succeeded Rudolph. By this all the religious and constitutional rights of the Hungarian people were guaranteed. Bocskay did not live to enjoy his victory for long: he is said to have been poisoned by his chancellor, Mihaly Katay, who was hacked to pieces by Bocskay's supporters.
HUGH J. BLAIR

BODELSCHWINGH, FRIEDRICH (1831-1910). Lutheran pastor. A Westphalian, he was led by Pietist influence into the ministry. In teaching children, the Word of God in the Bible came alive for him and the experience of the death of four of his own children within a few weeks in 1869 deeply moved him. From 1872 he took charge at Bielefeld of an institution for epileptics, later known as Bethel.* Its work, connected with the Innere Mission,* grew to include the training of deaconesses, a workers' colony, etc. Bodelschwingh held a strong view of the church's social responsibility and entered the Prussian Landtag in 1903. But he stood closer to C.F. Blumhardt,* whom he knew, than to Adolf Stöcker, believing that all welfare activities were limited by eschatology: they do not build God's kingdom but prepare for the returning Lord. The epileptics, on the fringes of society and having their future only in God, were thus seen to be teaching concretely the truth about the kingdom. At the same time, hope for all in the Gospel meant that even the most disabled were found useful work, and theological students were expected to learn from paroled convicts.
HADDON WILLMER

BODELSCHWINGH, FRIEDRICH (1877-1946). Reichsbishop of the German Evangelical *Landeskirchen,* and son of the above. Having succeeded his father at Bethel,* he expanded its work in education and research (the relation of medicine and theology). He took care that its growth did not detract from its central concern with compassion for the unfortunate. His high standing in the German Evangelical Church was shown by his election as reichsbishop in 1933, though he was displaced by Hitler's nominee, L. Müller. Quietly active rather than prominent in the church conflict, he refused to surrender the sick of Bethel to the Nazi euthanasia program and thus was, along with other protesters, influential in ending it.
HADDON WILLMER

BODMER PAPYRI, see ARCHAEOLOGY

BOEHLER, PETER (1712-1775). Moravian* missionary. Born at Frankfurt-am-Main, son of an innkeeper, he studied at Frankfurt Gymnasium (1722) and Jena University (1731), where he was influenced by Spangenberg* and Zinzendorf.* Through the latter he became a missionary of the English Society for the Propagation of the Gospel among the slaves of South Carolina, and pastor of some Moravians at Savannah, leading the Moravians in migration to Bethlehem, Pennsylvania. While in England he met and profoundly influenced the Wesleys. After a second stay in America, he spent six years in England as superintendent of the Moravian Church, being consecrated bishop in 1748. He returned to Bethlehem (1753) for eleven years. Later he was a member of the Unity Elders' Conference.
J.G.G. NORMAN

BOEHM, MARTIN (1725-1812). Co-founder of the Church of the United Brethren in Christ.* Born in Lancaster County, Pennsylvania, of Swiss Mennonite ancestry, he was chosen by lot at the age of thirty-one to become pastor of the Mennonite congregation to which he belonged. He prayed for God's help and was granted assurance of salvation and the desire to share his faith with others. He began his itinerant ministry at once among the German-speaking settlers of Pennsylvania and Virginia. His revivalistic activities and his willingness to conduct his work in English when necessary led the Mennonites to denounce and expel him. He continued his work in cooperation with other likeminded preachers, and met Philip Otterbein* at a great preaching meeting near Lancaster by 1768. Upon hearing Boehm preach, Otterbein embraced him saying, "We are brethren." They became co-workers in the growing pietistic movement which was organized in 1800 as the Church of the United Brethren in Christ, with Boehm and Otterbein as bishops.
HARRY SKILTON

BOEHME, JAKOB (1575-1624). German Lutheran mystic and theosophist. He was born at Altseidenberg near Goerlitz, where he lived nearly all his life working as a shoemaker. Among his mystical experiences, the most important occurred in 1600, when he looked at a dish reflecting the sunlight and in an ecstatic state saw "the Being of Beings, the byss and the abyss, the eternal generation of the trinity, the origin and descent of this world, and of all creatures through the divine wisdom." In 1612 he published some of his insights in a work, *The Beginning of Dawn,* followed by a devotional treatise, *The Way to Christ* (1623). His other writings were published posthumously.

Although not formally educated, Boehme read widely in the books of Paracelsus* and Valentin Weigel* and shows the influence of their mystical, alchemical, and astrological ideas in his use of obscure and difficult terminology. Mostly, however, he seemed to rely on his own mystical experiences. He believed that God Himself contains both good and evil. The "abyss" is God considered as the *Ungrund,* from which erupt the "fiery will of love" and the "sinister will of wrath." Despite such statements, at times he wrote as if evil were not necessary. In general he shifts his position, and no single theory fits all his work.

Boehme taught also that there are qualities in nature which he coordinated with his ideas of God in differing ways. The seven qualities divide into two triads, a higher and a lower, between which there is creative energy called "the flash." The lower group consists of contraction (or individualisation), diffusion (or attraction), and rotation (the struggle between the two foregoing). The higher triad is in effect the lower transformed including love, expression, and the kingdom of God which achieves a harmony between the spiritual and material world. Man must make a choice between the world of sensation represented by the oscillation of nature or the "dying" to self and living on a higher plane. This makes the true Christian life a mystical imitation of Christ's suffering and triumph. Boehme was critical of the Protestantism of his day because of its bibliolatry, doctrine of election, and notions of heaven. His influence was very great, not only in Germany where the Pietist, Romantic, and Idealist movements all owed something to his teachings, but also in England where the Cambridge Platonists, William Law,* and the Behmenists accepted his ideas.

BIBLIOGRAPHY: *Works* (4 vols., tr. J. Ellistone and J. Sparrow, 1644-62; reedited 1764-81; rep. 1909-24); H.L. Martinsen, *Jacob Boehme: His Life and Teaching* (tr. T. Rievans, 1885); R.M. Jones, *Spiritual Reformers in the Sixteenth and Seventeenth Centuries* (1914), chaps. 9-11; J.J. Stoudt, *Jacob Boehme, His Life and Thought* (1968).

ROBERT G. CLOUSE

BOETHIUS, ANICIUS MANLIUS TORQUATUS SEVERINUS (c.480-c.524). Philosopher and statesman. Educated at Athens and Alexandria, he was accused of treason and imprisoned in Italy, where his most famous work *De Consolatione Philosophiae,* in five books, was written in prison. It concerned philosophy's leading the soul to God and was the ground for debating whether he was a Christian. This was answered affirmatively on the basis of his *De Trinitate,* studying Augustine and adopting "new and unaccustomed words," and addressing to John the Deacon a shorter tract on the Trinity and a treatise against Eutyches* and Nestorius. It is believed that he wrote *De fide catholica,* which summarizes central doctrines and rejects tenets of Arius,* Sabellians,* and Manichaeans,* teaching that all corruptible things will perish and men will rise for future judgment. His philosophical writings and translations are numerous: translations of Aristotle's *De Interpretatione* and *Categories* with commentary, and his own *An Introduction to Categorical Syllogisms* as well as two books each on categorical and hypothetical syllogisms, and many others. He planned to translate into Latin all of Plato and Aristotle, and reconcile their thinking, as Cassiodorus had employed Epiphanius to make the Greek fathers available to Latin readers. He wrote on the trivium and quadrivium *(On Arithmetic, On Music)*, and translated and commented on Porphyry. About 520 came the theological *Tractates,* known as *Opuscula sacra,* which established his theological authority, even to getting a commentary by Aquinas. His work was taken seriously throughout the Middle Ages, as many commentaries witness, and he was canonized.

See H.M. Barrett, *Boethius: Some Aspects of His Time and Works* (1940); and H.R. Patch, *The Tradition of Boethius* (1935). C.G. THORNE, JR.

BOGOMILES. A group that arose toward the latter part of the eleventh century, especially in Bulgaria, and was considered heretical by the Eastern Orthodox Church. Much that is known about them comes from hostile writers sympathetic to the established church of the day. Their views of the sacraments were in some ways similar to several Protestant groups in later centuries both in western Europe and in America. The Bogomiles opposed the sacramental materialism of the church. They rejected water baptism as well as the material elements used in the sacrament of the Lord's Supper. Instead, they advocated a spiritual baptism which was conferred by the laying of a gospel of John on the head of the recipient and the chanting of the Lord's Prayer. Likewise, the Lord's Supper was to be practiced spiritually, for the bread and the wine could not be transubstantiated into the body and blood of Christ.

The nonacceptance of almost all of the OT by the Bogomiles bears close resemblance to the views of the Paulicians,* another heretical group in Eastern Christianity. The patriarchs in the Pentateuch, the Bogomiles stated, were in reality inspired by Satan. Satan originally was at the right hand of God, but was expelled because of his revolutionary planning. The creation of the earth was given to Satan, so that all of creation becomes basically evil, including the human body. Only things of the Spirit would be considered good. Birth is the imprisonment of the good spirit in evil flesh as punishment for sins in a preexistent state. As a part of this logic, sex would also be considered wrong. The natural result of such an incorrect interpretation of Scripture led to a definite dualism of two world principles, good and evil, in contention throughout creation.

GEORGE GIACUMAKIS, JR.

BOHARIC, see EGYPTIAN VERSIONS

BOHEMIAN BRETHREN. Later known as Unitas Fratrum and Moravian Brethren.* In Prague during 1453-54, the preaching of Archbishop Rokycana (who gave Holy Communion in both kinds) led to the foundation of a community in the city guided by his nephew Gregory. This congregation then associated itself with the followers of Peter Chelcicky* (d. 1460). The members sought to fulfill the law of Christ as given in the gospels, and also rejected military service and many aspects of town life. Yet they still believed in the celibacy of priests, the seven sacraments, and other Catholic doctrine; they nonetheless required that their priests be men of integrity, that they give Communion in both kinds, and that they emphasize the place of faith in God through the sacraments.

In 1457 they settled in the village of Kunwald. In the community were three groups: the beginners or penitents, the advanced, and the perfected (the priests). The latter preached and heard confessions in addition to administering the sacraments. Eventually the priesthood was wholly separated from that of the Roman Catholic and Utraquist* churches. The supreme power in the community was vested legally in the synod, comprising all the clergy, but the "Close Council," made up of ten members of the synod, exercised the real power. As the movement grew there were area synods and dioceses. The latter had a bishop with priests and deacons at the parish level. Discipline and respect for others was instilled at all levels. Schools were regarded as important. A significant holder of the post of presiding bishop in the formative years of the Brethren was Lukas, who dominated the activity from 1496 to 1528, though he was bishop only from 1517. He sought to unify the community and systematically express in writing its doctrines.

Despite persecution, the Brethren increased in numbers and influence in Bohemia. After the death of Lukas the leadership was in the hands of men who were pro-Lutheran; the nobility also assumed greater prominence within the movement—which gave King Ferdinand the excuse he wanted to crush the Brethren in 1547. The seat of government of the Brethren was transferred to Moravia. Many Brethren escaped to Poland, where they helped the cause of reform and eventually joined the new Calvinist church. The leader in Bohemia, John Augusta,* was tortured and kept in prison for sixteen years. By 1609 the Brethren who had returned to their homeland managed to gain state recognition for their religion, but only through a confederation with the Utraquists, who were now Lutheran in theology. The Brethren were thus able to retain their own organization and regulations, and even their own Creed (1564), while the Bohemian Lutherans held to the Augsburg Confession.* Both groups claimed their confession to be in harmony with the Bohemian Confession of 1575.

Definitive form was given to the polity and discipline of the Brethren at the Synod of Zeravic in 1616, but it was never fully implemented: the Battle of the White Mountain (1620) virtually destroyed Protestantism in Bohemia and Moravia for over 150 years. Scattered groups of Brethren managed to survive, and these accepted the invitation of Count Zinzendorf* to join the Herrnhuter in 1721. One famous seventeenth-century bishop was J.A. Comenius.*

BIBLOGRAPHY: J.T. Müller, *Geschichte der Böhmischen Brüder* (3 vols., 1922-31); E. Langton, *A History of the Moravian Church* (1956); M. Spinka, *John Hus and the Czech Reform* (1941).

PETER TOON

BÖHMER, HEINRICH (1869-1927). German church historian. Born at Zwickau, he was successively professor at Bonn (1906), Marburg (1912), and Leipzig (1915) and established himself as a leading authority on Luther and the Reformation period. His meticulous scholarship is shown in his studies of Loyola* and the Jesuits (1914, 1907), while two of his works on Luther, written in 1906 and 1925 respectively, were translated into English as *Luther in the Light of Modern Research* (1930) and *Road to Reformation* (1946). His third major Luther study was *Luthers erste Vorlesung* (1924). He wrote also in German a number of studies in Anglo-Norman church history of the eleventh and twelfth centuries.

IAN SELLERS

BOLLANDISTS. An association of ecclesiastical scholars engaged in editing the *Acta Sanctorum* and named after the editor of the first volume, John van Bolland (1596-1665). The idea, however, did not originate with him but was first conceived by Heribert Rosweyde (1569-1629) who was professor of philosophy in the Jesuit College at Douai during the last years of the sixteenth century, yet devoted his leisure time to exploring the libraries and numerous monasteries scattered throughout Hainault and French Flanders. Bolland went further and made appeal to collaborators, either Jesuits or others, residing in all the different countries of Europe. Their aim was to produce a critical edition of the lives of the saints, based on authentic sources. Their researches led them to combat the Carmelite* tradition that the origin of their order went back to the prophet Elijah, who was regarded as its founder. The undertaking soon attracted others, among whom was Godfrey Henschen (1601-81), an invaluable contributor. The 1773 suppression of the Jesuits severely affected their work, which was not resumed fully until 1837.

S. TOON

BOLOGNA, CONCORDAT OF (1516). This pact between Francis I of France and Leo X was the result of a sweeping military victory won by the French king at Marignano over the combined military forces of the Italian states and the papacy. Through it Francis made himself master of the French Church. The concordat allowed Francis to nominate candidates for bishoprics, abbeys, and priories in France, with only a very few exceptions. But the right of canonical investiture remained with the papacy. The concordat thus gave a degree of independence to the French Church and reflected the growing spirit of nationalism in France.

BOLSEC, JEROME HERMES (d.1585). Controversialist. Member of the order of Carmelites* in Paris, he had been driven from the city as a result of a sermon which he had preached favoring Protestant doctrine. He took up medicine, was converted to Protestantism, and went to Geneva only to find himself in dispute with Calvin on the doctrine of predestination.* He had begun to question the doctrine in private at first, but later ventured to express his opposition in the presence of the congregation. He was prosecuted and banished from Geneva in 1551. He returned to Paris, but was compelled to leave again on his refusal to accept a ruling of the Council of Orleans (1563) ordering him to recant. He eventually returned to the Roman Catholic Church, and died at Lyons. He took revenge in the publication of slanderous biographies on Calvin (1577) and Beza (1582).

HUGH J. BLAIR

BOLTON, ROBERT (1572-1631). Puritan preacher. He went to Lincoln College, Oxford, in 1592 and became a fellow of Brasenose in 1602. He took part in a disputation before James I in 1605. Before entering the ministry, Bolton was brought to an experience of conversion. He became rector of Broughton, Northamptonshire, in 1610 and remained there for the rest of his life. He was a typical Jacobean Puritan. Baxter* later named him with others of his contemporaries such as Preston and Sibbes as "an excellent sort of conforming ministers." They were men who accepted the basis of the national church in episcopacy and a liturgy, but preached Puritan theology which they applied in their pastoral work, and desired the reform of some of the ceremonies. Robert's son, Dr. Samuel Bolton, conformed at the Restoration and became a prebend of Westminster and chaplain-in-ordinary to Charles II.

JOHN TILLER

BOMPAS, WILLIAM CARPENTER (1834-1906). Pioneer Anglican bishop in the Canadian North. Educated privately in London, he was ordained deacon in 1859, and in 1865 volunteered for service in the missionary circuit of the Yukon. As an itinerant clergyman he quickly earned the admiration and respect of the Indian and Eskimo and in 1874 became the first bishop of Athabasca. In 1884 he was translated to the see of Mackenzie River, and in 1891 to that of Selkirk (now Yukon). In 1905 he resigned his bishopric, but continued to live in the Yukon until his death. A doughty Evangelical, he translated the NT into a multitude of Indian dialects and founded several schools and hostels for native children. His many travels and his endurance of cold and famine earned him the title "Apostle of the North" (the name of the biography by H.A. Cody, 1908), and left a permanent mark on missionary work in that area.

RICHARD N. MUGFORD

BONAR, ANDREW ALEXANDER (1810-1892). Scottish preacher and author. Seventh son of a solicitor of excise in Edinburgh, he with his brothers John and Horatius* made a famous trio of ministers. His great friend was R.M. McCheyne,* whose memoir he wrote, regarded widely as a Christian classic. They both belonged to the Non-Intrusion movement which led to the Disruption* of 1843, and also to a revival movement which culminated in the Kilsyth Revival of 1838-39. Greatly concerned with the evangelization of the Jews, he and McCheyne were members of a "Mission of Inquiry" to Palestine in 1839. After short ministries in Jedburgh and Edinburgh, he was ordained at Collace, Perthshire, and remained there after the Disruption, preaching in a tent until a Free Church was built. In 1856 he started a new Free Church at Finnieston, Glasgow, exercising a fine ministry till his death. Among his writings, his edition of Rutherford's *Letters* and his own *Diary* have become devotional classics.

J.G.G. NORMAN

BONAR, HORATIUS (1808-1889). "The prince of Scottish hymn-writers" (James Moffatt). In Edinburgh University one of his tutors was Thomas Chalmers.* While assistant minister at St. James's, Leith, he wrote hymns to familiar tunes to interest the children in worship. In 1837 he became minister of the North Parish Church, Kelso, after the Disruption* of 1843 remaining in Kelso as minister of the Free Church. In 1866 he was translated to the Chalmers Memorial Church in Edinburgh. He was moderator of the Free Church Assembly in 1883. He was the author of many books, but is now chiefly remembered as a hymnwriter. He wrote over 600, of which nearly 100 are in common use. Probably the best known is "I heard the voice of Jesus say." Ira D. Sankey called him "my ideal hymn-writer." He issued eight collections of hymns.

J.G.G. NORMAN

BONAVENTURA (1221-1274). "Prince of Mystics" (according to Leo XIII). Born near Viterbo, Italy, and baptized John ("of Fidanza"), he believed that St. Francis's (of Assisi*) intercession rescued him from a dangerous childhood illness. In 1238 or 1243 he entered the Friars Minor, then read arts and theology at Paris, lecturing on Holy Scripture there (1248-55). After some difficulty with the secular doctors and finally by order of Alexander IV, he was awarded the doctorate there for his commentary on the *Sentences* of Peter Lombard* and the treatise *De Paupertate Christi.* Though not yet thirty-six, he was elected minister-general of the Friars Minor and preserved the order from division between the "Spirituals" and Observants. Having declined the archbishopric of York (1265), he was compelled to accept the see of Albano (1273) and a cardinalship; and he was responsible for Gregory X's election in 1271. He attended the Council of Lyons* (1274) and contributed to the short-lived reunion with the Greek schismatics. Alexander of Hales,* his mentor at Paris, noted "Adam did not seem to have sinned in Bonaventure." His works are christocentric, saturated with Scripture, and learned in the Fathers. *Itinerarium mentis in Deum* emphasizes the folly of human wisdom when compared with the illumination God waits to give the Christian. Bonaventura's mysticism is founded on dogmatic and moral theology, believing contemplative prayer to be no extraordinary grace. Union with God turns on wanting to pay the price. Holding simple sanctity high, he also emphasizes the gift of knowledge, and rejects the doctrine of the immaculate conception.

See E. Gilson, *La Philosophie de Saint Bonaventure* (in Études de Philosophie Médiévale, iv) (1924; ET 1938).

C.G. THORNE, JR.

BONHOEFFER, DIETRICH (1906-1945). German pastor and ecumenist. Son of a famous neurologist, he studied philosophy and theology at Tübingen and Berlin, coming under the influence of such men as Deissmann,* Harnack,* Lietzmann,* Seeberg, and Karl Barth.* Ordained as a Lutheran pastor, he ministered to German congregations in Barcelona and London and became acquainted with G.K.A. Bell,* bishop of Chichester, with whom he shared his concern about the Nazification of the German Church. He took a leading part in drafting the Barmen Declaration* and thus became a leader of the Confessing Church* which refused the notorious Aryan Clauses (1933) imposed by the Nazi ideology. The seminary which he founded for training pastors for the Confessing Church was short-lived; his license to teach was revoked in 1936; Himmler closed the seminary in 1937. Bonhoeffer traveled much to inspire concern for the plight of the German Church. His opposition to Hitlerism involved him in the Resistance movement and led to his arrest by the Gestapo in April 1943. He was executed, on a charge of treason, at Flossenbürg on 9 April 1945; a simple tablet in the village church is inscribed: "Dietrich Bonhoeffer a witness of Jesus Christ among his brethren."

His later works which have been translated into English include *The Cost of Discipleship* (1948); *Letters and Papers from Prison* (1953); *Life Together* (1954); *Creation and Fall* (1959); *No Rusty Swords* (1965); *Christology* (1966: in USA *Christ the Center*); and *Way to Freedom* (1966). So varied and opposing are the theories deriving from his writings that a meaningful sketch of his ideas presents difficulty. Among his most fruitful insights were his total rejection of natural theology, and of a "religious apriori" in man; the reality of God's absolute self-disclosure in Christ; the historical and present Christ as God revealed incognito; Christ interpreted in terms of "the-man-for-others"; and particularly, his much discussed and misunderstood concepts of "religionless" and "worldly Christianity," and "man come of age."

BIBLIOGRAPHY: J.D. Godsey, *The Theology of Dietrich Bonhoeffer* (1960); M.E. Marty (ed.), *The Place of Bonhoeffer: Problems and Possibilities in His Thought* (1962); E.H. Robertson, *Dietrich Bonhoeffer* (1966); W.-D. Zimmerman and R.G. Smith (eds.), *I Knew Dietrich Bonhoeffer* (1966); J. Moltmann and J. Weissbach, *Two Studies in the Theology of Bonhoeffer* (1967); M. Bosanquet, *The Life and Death of Dietrich Bonhoeffer* (1968); E. Bethge, *Dietrich Bonhoeffer: Man of Vision, Man of Courage* (1970).

H.D. MC DONALD

BONIFACE (680-754). Missionary bishop and martyr. Born near Crediton, Devon, he was trained in abbeys at Exeter and Nursling in Hampshire, and later himself refused an English abbacy. He went to Frisia to serve under the English missionary Willibrord,* and in 719 received Gregory II's authority for his work. After laboring successfully in Thuringia and Bavaria, and seeing thousands of baptisms among the Hessians, he again went to Rome to be consecrated bishop (722). It is unlikely that his name was changed from Wynfrith to Boniface then, but much earlier. He returned to Hesse as a missionary statesman, having also the authority of Charles Martel,* and from there he went on to Thuringia, being made archbishop in 732 by Gregory III, and papal legate in 739—the first one to be sent beyond the Alps. Now he divided Bavaria, Hesse, and Thuringia into dioceses, and began founding Benedictine monasteries, of which Fulda (c.743) was the most famous. The connection with England remained, even to English monastics joining him, whereupon the English reverence for the papacy was transmitted to the new German Church. With Charles Martel's death (741), he was called to Francia by the new mayors, Carloman and Pepin,* to reform the church; this began under the new pope, Zacharias, in 742, and was accomplished through a series of councils. About 747 he became archbishop of Mainz, but resigned after a few years to return to Frisia, where his career ended in martyrdom. His felling of the pagan Oak of Thor at Geismar to make a chapel of its timber comments profoundly on his ministry; perhaps his devotion exceeded that of the popes he served. He renewed their authority beyond the Alps and extended the boundaries of Latin Christendom which had already begun to shrink in Spain owing to Muslim conquest. With Boniface, unity in the Western Church and the empire took form.

See more recent Lives by G.F. Brown (1910); J.J. Laux (1922); and W. Lampen (1949).

C.G. THORNE, JR.

BONIFACE I (d.422). Pope from 418. Unlike his predecessor Zosimus,* he sought to establish and defend respect for Roman authority in the West. He withdrew the unpopular papal vicar from S Gaul. A Council of Carthage in 419, reacting against rash claims by Zosimus, urged upon Boniface that the bishop of Rome should accept appeals from bishops alone, not from clerics. This Boniface did when he restored the pro-Augustinian Anthony to his see in the Donatist* stronghold of Fussala, but he accepted the appeal with caution, on the condition "... if he has truthfully told us the facts." Although Illyricum had been made part of the Eastern Empire, Boniface defended the traditional Roman ecclesiastical control through the bishop of Thessalonica. His nine extant letters, concerned with his rights in the consecration of the bishop of Corinth, introduce papal claims later institutionalized by Leo I.*

G.T.D. ANGEL

BONIFACE V (d.625). Pope from 619. A Neapolitan, he was noted for his organizing ability and for his concern to spread Christianity in England, especially in Northumbria. In Rome he endeavored to conform ecclesiastical usage to civil law in the matter of bequests, established the principle of asylum, and issued laws concerning the liturgical functions of various orders of clerics. He wrote letters to Mellitus and Justin, archbishops of Canterbury, sending Justin the pallium as a symbol of honor and jurisdiction. He also wrote to Edwin, king of Northumbria, and to his Christian queen, Aethelberga, thus supporting the work of Paulinus of York* for the conversion of Northumbria. During the time of Boniface and his successors, the three ancient patriarchates of Jerusalem, Antioch, and Alexandria capitulated to the rule of Islam.

J.G.G. NORMAN

BONIFACE VIII (c.1234-1303). Pope from 1294. During his pontificate relations between the papacy and the Western monarchs reached a crisis. Philip IV* and Edward I claimed the right to tax without papal consent. Unfortunately Boniface had come to the papal throne under circumstances that cast doubt on the legality of his title. His predecessor, Celestine V, was a holy hermit who could not adjust to his papal role and resigned after five months; many argued that abdication from the papacy was impossible and that therefore Boniface's election was invalid. Also, Celestine's experience seemed to prove that no holy man could be pope. Boniface seemed to confirm that view. Against the claim of Philip and Edward he issued the bull *Clericis Laicos* (1296), stating that the clergy were prohibited from paying taxes without papal approval. Any ruler, moreover, who levied such taxes would be automatically excommunicated. In retaliation, Philip forbade the export of money from his realm, thus halting papal revenue from France, and Edward withdrew royal protection from the clergy, in effect outlawing them. These actions forced Boniface to declare that the bull did not apply to emergencies and that the king could decide when an emergency existed.

Later, Philip and Boniface argued over the trial of a bishop accused of treason. In 1301 the pope issued the bull *Ausculta fili* which, in addition to asserting the power of the pope over all kings, contained a list of specific charges against Philip's government. By changing some of the bull's statements, Philip made it appear that it was an affront to all Frenchmen. Then he called the Estates of France and secured the support of that assembly in his protest against Boniface's claim. The pope replied with the bull *Unam sanctam,* which closes with the statement: "Furthermore, we declare, state, define, and proclaim that it is altogether necessary to salvation for every human creature to be subject to the Roman pontiff." Philip's advisers answered this by drawing up an indictment of Boniface, accusing him of murder, heresy, simony, adultery, schism, and keeping a demon as a pet. A Frenchman, William Nogaret, led an armed band into Italy to arrest Boniface and return him to France for trial. The expedition proved to be a failure, but Boniface died soon after and Philip used the charges against Boniface to pressure the papacy into doing his will. Thus the medieval popes were defeated by the national monarchs.

See T.S.R. Boase, *Boniface VIII* (1933).

ROBERT G. CLOUSE

BONIFACE IX (c.1355-1404). Pope from 1389. Pietro Tomacelli was born in Naples of a poor but ancient family. Created cardinal-deacon of St. George while still a young man, he was made cardinal-priest of St. Anastasia (1385) by Urban VI. His election to the papacy came in the middle of the Great Schism.* More amiable than his predecessor, he was nevertheless convinced of his papal rights and excommunicated the Avignon pope, Clement VII; declared sinful the proposal to end the Schism through a general council (1391); and successfully resisted Anglo-French and German pressure to abdicate. He regained control of the Papal States* (lost by Urban) and reestablished papal authority in Rome. His authority was tenuous, however, and weakened by the neutrality of much of Europe and the loss of Sicilian and Genoese support. Because they had sided with him against Avignon, Boniface was forced to support Ladislaus as king of Naples, and Rupert of Bavaria as German emperor. To raise funds for these political activities he had to resort to indulgences and simony, and in 1399 transformed the annates* into a permanent tax. In this he was assisted by Baldassare Cossa, later antipope John XXIII, whom he made cardinal. His pontificate was called "the crooked days of Boniface IX."

J.G.G. NORMAN

BONIFACE OF QUERFURT (970?-1009). Hagiographer, missionary, and martyr. Born in Saxony of influential parents, Boniface (otherwise known as Bruno) became canon of Magdeburg at an early age. He was a close friend of Emperor Otto III, whom he accompanied to Rome in 996. There he entered a Benedictine abbey and took for himself the religious name of Boniface. There also, after the martyrdom in 997 of Adalbert,* bishop of Prague, at the hands of the heathen Russians, he planned further missionary activity in Russia. Following much frustration due to war, he visited Kiev and then worked among the pagan tribesmen in the area north of the Black Sea, whom he Christianized at least nominally. Later he visited Poland and attempted unsuccessfully to act as peacemaker in the war between the Poles and Germans. Determined to continue his missionary work he then, with eighteen other missionaries, crossed the border into Russia (Lithuania) where the whole party was killed by the heathens. Boniface wrote several lives of Christian missionaries in eastern Europe, notably that of Adalbert, whose pioneering work he continued.

R.E.D. CLARK

BONIFACE OF SAVOY (d.1270). Archbishop of Canterbury. Son of a count of Savoy, he was nearly related by marriage to Henry III of England. While still a boy he showed ascetic inclinations by entering the Carthusian Order.* He became bishop of Belley in Burgundy in 1234; and in 1241 was elected archbishop by the monks of Canterbury, but papal confirmation was delayed. He was eventually consecrated by Pope Innocent IV at the Council of Lyons* (1245). His enthronement was further delayed until his return to England in 1249. A visitation of his province was strongly resisted by the clergy, and he went to Rome for a time. Boniface spent relatively little time in England, nevertheless his provincial constitutions, presented between 1257 and 1261, were a notable assertion of clerical privileges. He died in Savoy while waiting to join Prince Edward's crusade.

JOHN TILLER

BONN CONFERENCES. Two international conferences on church reunion were held in 1874-75 under the presidency of Ignaz Von Döllinger,* precipitated by the *Munich Manifesto* (1871) of the first Congress of the Old Catholics.* The *Manifesto* expressed the hope of reunion with the Greek-Oriental and Russian Churches. In March 1872 Von Döllinger delivered a series of lectures on *The Reunion of the Churches,* in which the Scriptures and the ecumenical creeds of the early church were proposed as a basis for church unity. The first Bonn Conference was an informal meeting of theologians representing

Germany, the Eastern Churches, the Anglican and the Dutch Churches. When the *Filioque* clause concerning the Holy Spirit was debated, a serious division appeared. One year later, the second and larger Bonn Conference took place. Orthodox representation was considerably larger and included the Ecumenical Patriarch, representatives of the Church of Romania and the Church of Greece, and the metropolitan of Belgrade. The crucial question again emerged as the gap between Eastern and Western doctrines of the Holy Spirit. After a prolonged and rather strained discussion, the view of John of Damascus* was agreed upon as having been the prevailing one at the time of the Ecumenical Councils* of the early church. Furthermore, Orthodox representatives refused to commit themselves on the subject of the validity of Anglican orders.

WAYNE DETZLER

BONNER, EDMUND (c.1500-1569). Bishop of London. Born in Worcestershire and educated at Oxford, he was ordained priest in 1519 and with his training as a canon lawyer entered the service of Cardinal Wolsey.* As a supporter of Henry VIII's "divorce," he obtained the king's favor after the fall of Wolsey. Bonner's service to Henry on the Continent and his acceptance of the break with Rome led to his appointment to the bishopric of London in 1540. Under Edward VI he opposed the Protestant regime and was deprived of his bishopric in 1549. Under Mary he was restored and became a chief agent of the royal persecution. As London was closest to the court, the persecution was fiercest there, and Bonner was hated intensely by the people of London. When Elizabeth came to the throne, he was again deprived and spent the rest of his life in the Marshalsea prison. NOEL S. POLLARD

BONOSUS (d. c.400). Bishop of Naissus in Yugoslavia. He was deposed after the Council of Capua (391) instructed the bishop of Thessalonica and the Illyrian bishops to examine him, and they found him guilty of teaching that Mary had children by Joseph after the birth of Jesus; he might also have taught Photinian Adoptianism.* The Bonosiani followed him into schism and survived, especially among Goths, in Spain and Gaul down to the seventh century. Controversy arose in Illyricum concerning the validity of his ordinations.

BOOK OF COMMON ORDER, see COMMON ORDER

BOOK OF COMMON PRAYER, see COMMON PRAYER

BOOK OF DEER, THE. Ninth-century Gospelbook, so-called because of the Gaelic notes added to it describing the foundation and grants of land made to the monastery of Deer in Aberdeenshire, Scotland. It is notable for its illustrations, its evidence about social status and land-tenure, and for a Celtic liturgical fragment.

BOOK OF KELLS, THE. A copy of the four gospels in Jerome's* Latin Version with prefaces, summaries, and a partial glossary of Hebrew names and their interpretation. The book has 340 leaves, each 13 by 9 1/2 inches of thick glazed parchment. The copy is made with brownish black ink. Many letters are ornamented with an amazing blend of colors, and the first two words, *Liber generationis,* fill a well-illuminated page. The work was possibly begun at Iona* and finished at Kells at the beginning of the ninth century. Remarkably preserved from loss and damage for seven centuries, it was deposited in the library of Trinity College, Dublin, in 1660. A facsimile was made in 1950.

ADAM LOUGHRIDGE

BOOK OF SPORTS. The name given to a declaration of James I in 1617/8 in which he authorized, but did not command, the continuance of the old English Sunday—e.g., morris dancing, archery, vaulting, etc., following morning worship and midday meal. It was aimed at the "Puritan Sabbath" to which the declaration ascribed two evils—the hindering of the conversion of Roman Catholics because Protestantism seemed to be too austere, and the fact that it did not keep men healthy and ready to fight in war, but rather caused them to become drunken. The declaration was reissued in 1633 by Charles I when a determined Archbishop Laud* forced it upon many unwilling clergy: they had to read it in their churches or else face expulsion. It was burned in 1643 by the Long Parliament. PETER TOON

BOOKS OF DISCIPLINE, see DISCIPLINE, BOOKS OF

BOOS, MARTIN (1762-1825). German preacher. After studying at Dillingen under J.M. Sailer,* he came under the influence of Michael Feneberg at Seeg Allgäu. Failing to find forgiveness through extreme asceticism, Boos adopted a doctrine of justification by faith which came very near to Lutheranism. Revival crowned the preaching at Wiggensbach in 1776/7. Put on trial for heterodoxy in 1797, he was acquitted. Accused again in 1798, he was compelled to flee to the diocese of Linz Austria. There he remained until he was forced to leave in 1816, when he became a religion teacher at a school in Düsseldorf. His final years were spent as parish priest at Sayn/Rhineland. The motto of Boos's life was "Christ for us and in us." WAYNE DETZLER

BOOTH, BALLINGTON (1857-1940). Founder of the Volunteers of America.* Born in Brighouse, England, second son of William Booth,* he early began his Salvation Army* work. An outstanding orator, musician, and leader, at twenty-three he was in charge of the first training home for officers. In 1883 he was sent to Australia as joint commander. Four years later he and his bride took over the recently established work in the United States. Strongly evangelistic, he proved a popular and effective leader. As a naturalized American he disagreed with his father about the British and authoritarian character of the Army. When ordered to relinquish his position, he complied, but then withdrew from the Army. In March 1896 he began the Volunteers, with similar objectives but more democratic organization.

HAROLD R. COOK

BOOTH, CATHERINE (1829-1890). "Mother of the Salvation Army"; wife of William Booth.* Born Catherine Mumford in Derbyshire, she was the daughter of a Wesleyan preacher. She was educated at home by her deeply religious mother. Later she lived in London, where she joined the Wesleyan Church at Brixton. She was expelled for her religious zeal, as was also William Booth, who preached there. They married in 1855, and had eight children. Husband and wife traveled widely preaching the Gospel. In 1864 they returned to London, and in 1865 started the Christian Revival Association (variously named) in Whitechapel; this is generally regarded as the start of the Salvation Army.* Catherine commenced the women's work, a prominent feature of the movement. For many years she continued her labors, though never out of pain. In 1890 she died of cancer; 36,000 people attended her funeral at Olympia.

See F. de L. Booth-Tucker, *The Life of Catherine Booth* (2 vols., 1892). R.E.D. CLARK

BOOTH, EVANGELINE CORY (1865-1950). Salvation Army* leader. Seventh child of William Booth,* the founder, she was only fifteen when she became a sergeant. Conspicuous for her dedicated activity, she came to be a trouble-shooter in difficult situations, as when her brother Ballington* quit the Army in the USA. She was field commissioner of operations in London for five years and principal of the international training colleges. She was made commander in Canada in 1896, and opened work in the Klondike. In 1904 she became commander-in-chief in the USA; in thirty years she led it to record achievements in social services, including World War I activities. She was elected general of the worldwide organization in 1934 and retired to the USA in 1938.

HAROLD R. COOK

BOOTH, WILLIAM (1829-1912). Founder and first general of the Salvation Army.* Born in Nottingham, he was converted in 1844 and became first a minister, then an evangelist, in the Methodist New Connexion Church. In 1861, however, he resigned because its leaders wanted to restrict him to a limited circuit. He became a free-lance evangelist and in 1865 began meetings in London's East End, where extreme poverty and hardship were the rule for most people. Gradually the work grew, was named "The Christian Mission," and spread to other centers. He was aided by his wife Catherine,* herself a gifted preacher. Both tackled social evils alongside his direct evangelism; by 1872 he was running five "Food-for-the-Million" shops, selling cheap meals. He and his colleagues were often attacked physically when preaching, but there were remarkable instances of lives transformed by the Gospel.

Military terminology was then the vogue ("Onward, Christian soldiers" was written in 1865), and one of Booth's leading helpers, Elijah Cadman, in 1877 advertised meetings in Whitby of "The Hallelujah Army Fighting for God." They labeled Booth "General," which was fitting because control of the Mission was centralized in Booth's hands. Then in 1878 the process was taken a final stage: the "Salvation Army" was born. In 1879 it had 81 mission stations manned by 127 full-time evangelists; and it had the first Salvation Army band, formed in Salisbury. Very soon the policy of setting sacred words to secular tunes was adopted; as Booth put it, "Why should the Devil have all the best tunes?"

In 1880 came the uniform, preceded in 1878 by the first volume of "Orders and Regulations for the Salvation Army." The military system was rigidly enforced, and Booth was undoubted commander-in-chief. As the Army grew, Booth and his tireless wife (helped by their children) set up training homes for cadets, and initiated overseas advances, to the USA, the Continent, and India. A new headquarters was opened in London in 1881, and by 1884 the Army had 900 corps, more than 266 of them outside Britain.

Booth was totally absorbed in the growth of the Army, which was fiercely and at times brutally opposed. Money was scarce, debts grew, scandals threatened to destroy the cause. In 1886 he toured the USA, pulling together an organization that had fragmented into three parts. In 1887 the sight of homeless men on London Bridge prompted Booth to start the Army's social work; careful inquiries showed desperate need. Cheap-food centers, night shelters, an unemployment exchange—all these he set up as his wife lay dying of cancer. The need was focused in his book *In Darkest England—and the Way Out,* finished in 1890 just before Catherine died. It was a best seller, a storm center of controversy. Booth planned Farm Colonies, a Missing Persons Bureau, a Poor Man's Bank, Legal Aid for the poor; he even set up a match factory to help expose the evils of a private enterprise firm in that same line of business.

Now virtually alone, Booth traveled the world, a figure of international renown. In 1904, aged seventy-five, he did a twenty-nine-day "automobile evangelistic" tour of Britain—1,224 miles and 164 meetings long. In 1907 he toured the USA. In 1908, nearly blind, he was in Scandinavia. In 1910 he visited Holland, Denmark, Germany, Switzerland, Italy. At last, in August 1912, he died; 150,000 people filed past his coffin; 40,000 attended his funeral. He had traveled five million miles, preached nearly 60,000 sermons, and drawn some 16,000 officers to serve in his Army.

BIBLIOGRAPHY: Biographies by H. Begbie (1920), W.H. Nelson (1929), S.J. Ervine (2 vols., 1934), R. Collier (1965). See also bibliography under SALVATION ARMY. A. MORGAN DERHAM

BORBORIANS. A heretical sect with a reputation for licentious living and disbelief in future judgment, they are associated by Epiphanius about 370 with Ophite* Gnostics. *Borboriani* or *Borboritae* (i.e., "Mud Men") is commonly thought to be a nickname for *Barbelitae,* Gnostic worshipers of the female aeon Barbelo. In 428 they were forbidden to assemble or pray; in the sixth-century laws of Justinian against heretics they are described as a subsection of the Manichaeans.*

BORDEAUX PILGRIM. In A.D. 333/4 a man traveled from Bordeaux in Gaul to Jerusalem and back to Milan. He recorded his route and described the memorial sites which were shown to him in the Holy Land. He also described his stopping-places in Northern Italy, in the Balkan peninsula, in Constantinople, in Asia Minor, and in Syria. Most of his itinerary is concerned with the stages on the way, and his descriptions give little information, although he does tell of the buildings which Constantine and his family gave to the holy places. This firsthand source is the earliest record of a Western pilgrimage to the Holy Land.

BORGIA, FRANCIS (1510-1572). Jesuit.* Great-grandson of Pope Alexander VI and son of the duke of Gandia, he married Eleanor de Castro and had eight children. Charles V appointed him, after he had succeeded to his father's title, as viceroy of Catalonia, where he sought to reform the administration from corruption. His wife died in 1546 and a year later he was received privately into the Society of Jesus, being ordained priest in 1551. He was sent as commissary by Ignatius Loyola* to Spain and Portugal in 1554; he also had oversight of Jesuit work in the overseas empires. In 1561 he was summoned to Rome, where four years later he became the third general of the society. He had a deep interest in education and used his wealth to help found and build colleges or to improve what is now the Gregorian University of Rome. He was canonized by Clement X in 1671.

See M. Yeo, *The Greatest of the Borgias* (1936).

PETER TOON

BORIS (d.907). Khan (king) of Bulgaria, 852-889. Christian missionary activity had taken place in Bulgaria as early as the seventh century, and perhaps even earlier. This early work in the Balkans was primarily under the direction of the Byzantine Church. The missionaries were effective, for by the eighth century there were Christians within the society and in the palaces of the princes. Boris became a convert to Orthodox Christianity by receiving baptism by the patriarch of Constantinople about 864. Because Byzantine theology linked church and state in a union, Boris's baptism carried overtones of entrance into the state as well as the church. Boris desired the church in Bulgaria to be independent, but Byzantium rejected this. The king of Bulgaria then encourged Western missionary activity, primarily German, to take place in his land. Pope Nicholas I sent Latin priests, and the Greek clergy were driven out. The Khan's alliance with Rome was short-lived, for Rome also did not approve his desire for Bulgarian church independence. The Byzantines in turn were forced to agree to a Bulgarian archbishopric which was semiautonomous.

GEORGE GIACUMAKIS, JR.

BORNHOLMERS; BORNHOLMIANS. A popular designation of the *Luthersk Missionsforening i Danmark* (the Danish Lutheran Mission), which is an evangelical laymen's home mission movement within the Danish national church and, at the same time a society of foreign missions. It originates from a revival movement on the island of Bornholm in the 1860s and was officially formed in 1868. The influence from the nineteenth-century Swedish spiritual leader C.O. Rosenius has been of decisive importance for the history of the Lutheran Mission. Like that of Rosenius, the preaching of the latter has been characterized by a strong emphasis upon the total depravity of man, reconciliation solely through the vicarious atonement of Christ, justification by pure and free grace and through faith alone, and sanctification, not as a condition, but as a consequence of salvation. In addition, the Lutheran Mission is characterized by a special accentuation of the spiritual gifts and the universal priesthood of all believers, and a consequent craving for full freedom for the development of autonomous and independent lay activities within the framework of the national church.

N.O. RASMUSSEN

BORROMEO, CHARLES (1538-1584). Archbishop of Milan. Of noble birth, he received his first benefice at twelve. He studied civil and canon law under Alciati, but when his uncle became Pius IV, Borromeo was called to Rome and appointed archbishop of Milan, cardinal secretary of state, protector of the Low Countries, Portugal, the Swiss Catholic cantons, and several religious orders. His humanist ideals were decisively channeled in the direction of reform by the death of his elder brother (1562) and his ordination to the priesthood (1563), which combined to inspire him to a more austere life. Besides playing an important role in the final session at Trent, he helped reform the College of Cardinals and was a reviser of the Missal and Breviary. His notable achievement was diocesan reform, and he has continued to be regarded as a model bishop, some of whose works were reprinted during Vatican II. Borromeo called his first provincial council in 1565, but was not permitted by the pope to take up residence in his see until 1566.

Skillful use of provincial and diocesan synods created a constitutional framework for reform and discipline. Careful attention was given to education of the clergy, and Borromeo founded six seminaries. Monasteries were reformed and clergy encouraged to join the Oblates of St. Ambrose (1578) in order to raise standards of pastoral care. Corrupt clergy and religious were disciplined, and feeling ran so high that an unsuccessful assassination attempt was made in 1569. Borromeo also had to contend with opposition from the Spanish governors of the city. Nevertheless he pressed on with the work of reform and education, made extensive use of the Jesuits, and founded a Confraternity of Christian Doctrine to help with the instruction of the young. Orphanages, refuges for deserted wives, and *montes pietatis* were founded through his pastoral zeal; his own work during the plague of 1576, when the civic officials had all fled, showed that he was himself a selfless pastor of the highest order. He continued to play an important role in Rome, advised neighboring dioceses, pursued heretics, consolidated the Catholic reformation in Switzerland, and personally visited neglected and remote Alpine valleys. Canonized in 1610, he was one of the greatest Catholic reformers.

BIBLIOGRAPHY: G.A. Sassi, *Opere* (5 vols., 1747); A. Ratti (ed.), *Acta ecclesiae Mediolanensis* (1890-92); S. Caroli, Borromaei, *Orationes XII* (1963); A. Sala, *Biografia di S. Carlo Borromeo* (3 vols., 1857-61); T. Schwegler, *Geschichte der katholische Kirche in der Schweiz* (1935); M. Yeo, *A prince of pastors* (1938); A. Deroo, *Saint Charles Borromée* (1963); P. Brodi, "Charles Borromée archevêque de Milan et la Papaute," *Rev. Hist. Eccl.* 62:2 pp. 379-411 (1967).

IAN BREWARD

BORROW, GEORGE HENRY (1803-1881). Author, linguist, traveler, and friend of gypsies. From 1818 he was articled to a firm of Norwich solicitors. He learned languages in his spare time, and in 1824 went to London where he carried out translation work for a publisher, for which he was grossly underpaid. Later, in great poverty, he left London as a tramp. From 1827 to 1840, and again in 1844 and 1854, Borrow wandered on foot in Europe and the East, working at various times for a newspaper and, both in Spain and in Russia, for the British and Foreign Bible Society.* In Spain he was arrested several times. He anticipated the dangers; he knew, he said, that "very possibly the fate of St. Stephen might overtake me." In 1840 he married and settled in England to write. Of his colorful books on travel, transcribed by his wife Mary from random jottings, *The Bible in Spain* (1841) brought him immediate fame. His later studies were much concerned with gypsies: they include a complete Romany dictionary. Over a dozen biographies of him have been written, the most recent those by M.D. Armstrong (*George Borrow,* 1950) and E. Bigland (*In the Steps of George Borrow,* 1951). R.E.D. CLARK

BORTHWICK, JANE LAURIE (1813-1897). She and her sister Sarah (Mrs. E.J. Findlater, 1823-1907), staunch supporters of the Free Church of Scotland, published anonymously *Hymns from the Land of Luther* (4 vols., 1854-62). The title of these translations supplied the initials "H.L.L." over which many of the hymns first appeared. A number were published as *Thoughts for Thoughtful Hours* (1857). A complete edition of *H.L.L.* (113 translations) was published in 1862; an 1884 edition added thirty-three "Alpine Lyrics" from the German-Swiss Meta Heusser-Schweizer (issued separately in 1875). The sisters' translations represented relatively more hymns for the Christian life and fewer for the Christian year than those of C. Winkworth.* Miss Borthwick's best-known translations are "Be still, my soul!" (K.A.D. von Schlegel); "How blessed, from the bonds of sin" (Spitta); "Jesus, still lead on" (Zinzendorf); and "Jesus, Sun of Righteousness" (Knorr von Rosenroth). Mrs. Findlater's are "O happy home" (Spitta), which reflects the type of home enjoyed by both author and translator, and "God calling yet!" (Tersteegen). They wrote also some original hymns. JOHN S. ANDREWS

BORTHWICK, SARAH, see previous entry

BOSCH, HIERONYMOUS (Jeroen den Bosch) (c.1450-c.1516). Dutch painter. Born in Brabant at Den Bosch, he did most of his work in that area. He received commissions from the Hapsburg ruler of much of the Low Countries, Philip, duke of Burgundy, grandfather of Philip II of Spain, who also liked Bosch's paintings and collected many of them at Madrid. Bosch is best known for his mature works (*The Haywagon,* the *Garden of Delights,* the *Ship of Fools,* etc.), which contain dreamlike vistas filled with fantastic details—weird hybrid figures, mixtures of human, mechanical, animal, vegetable—done with striking technique and vibrant color. His grotesquerie was imitated by Jan Mandijn and others; his landscape and color techniques were an important influence on Pieter Breughel as well as on lesser figures. Despite suggestions that he was connected with heretical sectarianism, his sometimes strange paintings can more easily be explained in terms of the more general late-Gothic use of fantasy in sculpture, mystery plays, and the like.

See L. Baldass, *Hieronymus Bosch* (1960).

DIRK JELLEMA

BOSCO, JOHN (Giovanni Melchior Bosco) (1815-1888). Roman Catholic educationist and founder of the Salesian Order. He was born in near poverty at Becchi, Italy, was ordained priest in 1841, and worked in anticlerical Turin where the plight of poor boys in the city led him to pledge his life to them. He especially hoped to prepare some for the priesthood. His persistence overcame many difficulties of finance and accommodation, and his early efforts led to the establishment of night schools and eventually technical schools, workshops, and a church in Turin. In 1859 he founded the Society of St. Francis of Sales (Salesians*), which has now spread all over the world. Reason, kindness, and Christian faith were the basis of his educational philosophy, known as the "preventive" system. "As far as possible avoid punishing," he said. "Try to gain love before inspiring fear." He was canonized in 1934.

HOWARD SAINSBURY

BOSSUET, JACQUES BENIGNE (1627-1704). French Roman Catholic bishop and writer. Born at Dijon, he was educated in Jesuit schools there until sent to the Parisian Collège de Navarre at the age of fifteen. He became a doctor of the Sorbonne in 1652 and was ordained the same year. For seven years he served as archdeacon of the cathedral chapter at Metz, then moved to Paris in 1659 where within two years he had become preacher in the royal chapel. In 1670 Louis XIV appointed him tutor to the dauphin, a task he performed for twelve years with great diligence. In 1681 he became bishop of Meaux, a post he held until his death.

Bossuet had a many-faceted career, but was particularly renowned for his great oratorical skill, his controversialist ability, and his energetic but nonschismatic espousal of Gallicanism.* He also provided a classical statement of the Divine Right of Kings* theory in his *Politique tirée de l'Ecriture*

Sainte (1679). He apparently, however, regarded his *Discours sur l'histoire universelle* (1681) as his most important work, and most modern scholars would agree. It is a classic statement of the philosophy of history which sees Providence as the key to historical causation. His latter years saw him more and more entrenched as the guardian of orthodoxy in the face of biblical criticism, rationalism, skepticism, and various sectarian groups.

BIBLIOGRAPHY: *Oeuvres complètes* (10 vols., 1877); *Correspondance* (15 vols., 1909-25); A. Rébelliau, *Bossuet, historien du Protestantisme* (1892); W.J. Simpson, *A Study of Bossuet* (1937); P. Hazard, *The European Mind* (1939); A.G. Martimort, *Le Gallicanisme de Bossuet* (1953).

BRIAN G. ARMSTRONG

BOSTON, THOMAS (1676-1732). Scottish Presbyterian minister and scholar. Born in Duns, Berwickshire, he read arts and divinity at Edinburgh and was a recognized Hebraist. After ordination he held pastorates in his native county and (most notably) at Ettrick, Selkirkshire, where he was installed on the day of the union with England in 1707. An English Puritan work, *The Marrow of Modern Divinity,* greatly influenced him, and despite its ban because of its Arminianism, his own writings popularized its doctrines. Implicated in the "Marrow case," he emerged without stain, demonstrating his profound theological thought. Leading a life of deep prayer, he performed many exemplary parochial tasks. Illness never prevented his preaching, and even on his last two Sundays, too feeble for the pulpit, he preached from the manse window on self-examination. His books are *Human Nature in its Fourfold Estate* (1720), *Notes to the Marrow of Modern Divinity* (1726), *A View of the Covenant of Grace* (1734), *An Illustration of the Doctrines of the Christian Religion* (1773), and many other treatises and volumes of sermons. C.G. THORNE, JR.

BOTTICELLI, SANDRO (1445-1510). Florentine painter. He epitomizes in his masterpieces of painting the changes from the period of scholastic Christendom during the early Renaissance: grace no longer means a realm of God's favor superadded on beyond nature; now grace is simply a human posture of beautiful, leisurely dalliance, a concatenated quality of feminine loveliness, flowers, love apples and embroidered clothes. The whole Medici world of Lorenzo the Magnificent and Marsilio Ficino's ideal of blending pagan culture and biblical truth into one glorious, new feast of human purity and effortless radiance is captured by Botticelli's work. He painted portraits of the Medici family as wise men into his *Magi Adoring the Christ Child,* immortalizing his patrons as it were with a sanctified splendor. In both the famous *Spring* painting (c.1478) and *Birth of Venus* (c.1485) the figures semi-float off the ground in a weightless, nymphlike dance that has a flutter of decorative delicacy to it. Botticelli's Venus may duplicate the ancient, voluptuous Aphrodite sculptures, but his decolorized treatment and the lilting, well-outlined, suspended movement of the bodies chastens to innocence any sensuousness, with an allegorical, ritual character. That is Botticelli's significant achievement: canonizing the pagan *kalokagathon* ethic in the Renaissance mentality without offending Christian sensibility of the time, spiritualizing the gods and goddesses of pre-Christian mythology to cultural sweetness and light. Late in life, however, while Columbus was discovering the Americas for Spain, Botticelli was converted by Savonarola's* evangelistic crusade in Florence; his latest paintings (cf., e.g., *Nativity* in National Gallery of London) began to forsake the aura of wispy, nostalgic finery and introduce a more rough-hewn, zigzagged, earthy reality. CALVIN SEERVELD

BOUNDS, EDWARD McKENDREE (1835-1913). American Methodist minister and devotional writer. Born in Missouri, he studied law and at twenty-one was admitted to the bar. After practicing three years, he was called to preach in the Methodist Episcopal Church South. During the Civil War he served as captain in the Confederate army and was captured. The war over, he again served pastorates in Tennessee, Alabama, and Missouri. For nine years he was editor of the *St. Louis Christian Advocate,* the official organ of the church. His "Spiritual Life Books" come from the last decades of his life, when he was free from the day-to-day responsibilities of the pastorate. They have had a wide ministry in many denominations. HAROLD R. COOK

BOURCHIER, THOMAS (1404?-1486). Archbishop of Canterbury from 1455. Educated at Oxford, where he lived in Nevill's Inn, he was ordained and eventually, not of full canonical age, received the see of Worcester in 1434. That same year he became chancellor of Oxford, a post he held three years. In 1444 he moved to the bishopric of Ely, and eleven years later, after the death of John Kempe, he was promoted to Canterbury. For a short time he acted as lord chancellor, with his brother Henry as lord treasurer. In 1457 he took part in the deprivation of Reginald Pecock,* bishop of Chichester, for supposed heresy. Inevitably he was involved in the struggles between the Houses of Lancaster and York, and he became such a decided Yorkist that he crowned Edward IV and his queen. Pope Paul made him a cardinal in 1467, but his red hat did not arrive until 1473. Not long before his death he officiated at the marriage of Henry VII and Elizabeth of York, thus joining the red with the white rose.

PETER TOON

BOURDALOUE, LOUIS (1632-1704). "The king of orators and the orator of kings." Born in Bourges, he became a Jesuit in 1648. After teaching in provincial houses, he began preaching in Amiens in 1666 and speedily made a reputation. In 1669, he went to Paris and made a deep impression on the court with his splendid oratory, skillful exposition of Catholic orthodoxy, and passion to win his hearers' consent by carefully reasoned argument within a traditional rhetorical framework. A skilled pastor and confessor, he gave the same attention to the prisoners and the sick as to the famous and wealthy, and was a living answer to the attacks of B. Pascal* against the

Jesuits. In addition to being an opponent of the Jansenists,* he was used successfully to convert Montpellier Protestants after the revocation of the Edict of Nantes.* IAN BREWARD

BOURGEOIS, LOUIS (1510-1561). French Protestant musician. Invited to come to Geneva about the time of Calvin's return in 1541, he four years later took the place of Guillaume LeFrenc as chief musician of the city. In 1551, because of some of the tunes he published, he was imprisoned, but was released on the plea of Calvin. In 1557, apparently unable to withstand the rigorous attitude of the Genevan city fathers, he returned to Paris, where he died. In 1547 he produced a harmonization of Clement Marot's* psalms, using some currently popular tunes, some from the psalter of Strasbourg, and some which he himself composed.

BOURIGNON, ANTOINETTE (1616-1680). Enthusiast and visionary. Born at Lille in the Spanish Netherlands, she was brought up as a Roman Catholic. An unusual child, she withdrew from ordinary social life and became convinced that God spoke to her in visions and chose her to be a new Eva, a new Mary, the "woman clothed with the sun" of Revelation 12, with the mission of reforming Christianity. Attacking all established churches, she taught the inner light, direct contact with God, and the destruction of man's sinful ego. By her late forties she began attracting disciples; after a sojourn on Nordstrand in the North Sea, she moved to Hamburg, where the mystic P. Poiret joined her, and died in Friesland on her way to Amsterdam. Her extreme version of Quietism* was spread by her followers and gained support in Scotland (where Presbyterian assemblies denounced the movement in the early 1700s). The sect gradually faded out in the course of the 1700s. Her extensive and rambling writings were published by Poiret in Amsterdam, 1679-84, in twenty-one volumes.

See A.R. MacEwen, *Antoinette Bourignon, Quietist* (1910). DIRK JELLEMA

BOURNE, HUGH (1772-1852). Founder of the Primitive Methodist Church.* Born at Stoke-on-Trent, he joined the Wesleyan Methodists, and in 1802 built a chapel for them at his own expense. By this time a local preacher, he organized with others in 1807 a camp meeting* of the type that had proved so successful in the USA. It was the first of many, despite the disapproval of his denomination, which eventually expelled him. Unwillingly Bourne organized a separate body which held its first conference in 1820, with a name that indicated a desire to restore Methodism to its primitive simplicity. Bourne journeyed to Scotland, Ireland, and the USA, and was everywhere well received, and by the time he died the movement had some 110,000 members. The body was not styled "church" until 1902; in 1932 nearly a quarter of a million Primitive Methodists in Great Britain united with others in the new Methodist Church. Bourne was a man of high principle and a total abstainer; for most of his life he worked as carpenter and builder so he would not be a charge on his church. J.D. DOUGLAS

BOWEN, GEORGE (1816-1888). American missionary called the "White Saint of India." Born in Middlebury, Vermont, his early life was irreligious. In 1844 his dying sweetheart gave him a Bible, which became the means of his conversion and of his decision to become a missionary. After seminary training he was appointed to the Marathi Mission by the American Board.* From 1848 to 1888 he served in the Bombay region, unmarried and without furlough. After a year on the field he determined to live a life of self-denial as a self-supporting missionary on the level of the people. In 1872, influenced by William Taylor,* he joined the Methodist Church in India. By his writings, and particularly through editing the *Bombay Guardian*, he strongly influenced both missionaries and Indians. HAROLD R. COOK

BOWING. A physical gesture used in Christian worship, performed—like genuflecting, kneeling, and prostration—to express reverence. It may be directed towards a person, e.g., the bishop; an object, e.g., the altar; or it may be a response to an utterance, e.g., the name of Jesus. It may also be used by the minister to express thanks to someone for assistance in the service. The origin of the use of bowing is problematical. Bowing to bishops was part of the court ceremonial introduced into the church in the time of Constantine, who gave bishops the rank of imperial officials. Bowing at the name of Jesus is obviously suggested by Philippians 2:10. In the Church of England it was approved by canon 18 of 1604 (cf. canon B9 of 1969), although its observance has been generally restricted to the creed. The further practice of bowing toward the altar was ordered by Archbishop Laud's ill-fated canons of 1640, which were of disputed legality and never took effect.

JOHN TILLER

BOWRING, SIR JOHN (1792-1872). English linguist, writer, traveler, diplomat. He served for some time as a member of Parliament before going to the Far East in the foreign service. After a period as consul in Canton, he eventually became governor of Hong Kong. For his research work he was made a Fellow of the Royal Society. Author of many poems and hymns, he is now chiefly remembered for his "In the cross of Christ I glory," written in 1825 and made famous by Stainer in his oratorio *The Crucifixion* (1887). Also well known is "Watchman! tell us of the night."

BOY BISHOP. A name given in medieval times to the leader of choirboys' revels, elected on St. Nicholas' Day (6 December), who executed the functions of the bishop until Holy Innocents' Day (28 December). Originating in cathedrals, the custom spread to larger monastic and scholastic establishments. In England it proved more popular and enduring than the Feast of Fools*; it was not finally abolished until the reign of Elizabeth I.

BOYCE, WILLIAM (1710-1779). English composer. Trained as a chorister at St. Paul's Cathedral, and a pupil of the distinguished organist Maurice Greene, Boyce was organist of several London churches, and from 1758 to 1769 of the Chapel Royal. He was forced to retire because of loss of hearing and devoted his last years to his great historic anthology, *Cathedral Music* in three volumes. This collection did much to keep alive the great music of such men as Byrd and Gibbons. Boyce was a fine and voluminous composer in his own right, producing music of all kinds. He is certainly the best native English composer of the later eighteenth century, and he was known also for his excellent personal qualities.

J.B. MAC MILLAN

BOYLE, ROBERT (1627-1691). Natural philosopher. Fourteenth child of the great earl of Cork, he was educated at Eton till eleven, then traveled with a tutor. A violent storm in Geneva in 1641 led to his conversion; thereafter his life, talents, possessions were dedicated to Christ. As an Anglican he favored a national church to include Nonconformists* (toward whom he was singularly tolerant). Though ordination was often pressed upon him, with tempting preferment, Boyle refused, believing that the taking of vows led men to suppress their doubts. For the same reason he refused the presidency of the Royal Society, of which he was co-founder. In his day Boyle's reputation was comparable with that of his contemporary and friend Isaac Newton.* He wrote voluminously and exerted a profound influence on subsequent thought in science ("The Father of Chemistry"; originator of chemical analysis; "Boyle's Law"), philosophy (through Locke), and theology. His views on science and theology are surprisingly modern; he identified the "nitre" of the Bible with native sodium carbonate, financed much Christian work and Bible translation, and endowed the Boyle Lectures. He loved science passionately as God's "second book," revealing His power and wisdom. To fight science, he supposed, was to fight God. His mission in life was to convince men that God's two "books" were in harmony. His *Seraphic Love* (1660), a gem of Christian literature, went through many editions.

See *Collected Works and Life* (ed. T. Birch, 6 vols., 1772); and biographies by M.S. Fisher (1945) and R.E.W. Maddison (1969).

R.E.D. CLARK

BOYS' BRIGADE. The pioneer of uniformed, voluntary organizations for boys or girls, it was founded in Scotland by William A. Smith in 1883, for "the advancement of Christ's Kingdom among boys and the promotion of habits of obedience, reverence, discipline, self-respect and all that tends towards a true Christian manliness." In the uniform, in the ordered life of the company, and in the progressive program and award system, it appeals to and meets the needs of a boy's nature. The movement, which provides for boys in the eight-to-nineteen age range, has spread all over the world and currently has more than 3,000 companies (140,000 boys) in the United Kingdom, and nearly 2,000 companies (80,000 boys) in other countries. In 1971 the executive reaffirmed that the Boys' Brigade "must remain a Church-centred organisation, with the purpose of helping boys to achieve a true Christian faith. . . ."

J.D. DOUGLAS

BRABOURNE, THEOPHILUS (1590-c.1661). Controversialist. Frustrated in his plans to enter university and the ministry, he worked for a time in his father's hosiery business in London. After private study he was eventually ordained and preached in Norwich. His idiosyncratic views began to appear with two pamphlets in which he argued that Saturday should be observed as the Christian Sabbath (*Discourse upon the Sabbath Day,* 1629; and *Defence . . . of the Sabbath Day,* 2nd ed., 1632). The latter was more dogmatic and brought him eighteen months' imprisonment. One of the judges at the trial desired to burn him as a heretic. After this Brabourne gave up the ministry, but did not attempt to found any sect and wrote against separation from the national church. His last pamphlet (1661) attacked Quaker scruples over the oath of allegiance and supremacy.

JOHN TILLER

BRADBURY, WILLIAM BATCHELDER (1816-1868). American music teacher and hymntune writer. He studied music in Boston under Summer Hill and Lowell Mason* and became a music instructor and organist in New York City. He also edited over fifty Sunday school and choir music books, and he engaged in piano manufacturing. From 1847 he studied for two years in Leipzig. He was criticized because he treated music as a business and was not essentially a performer. He is best remembered for his hymntunes, two of the most famous being used for "Just as I am" and "He leadeth me." Both represent Bradbury at his best, with simple yet reverent expression.

ROBERT C. NEWMAN

BRADFORD, JOHN (1510-1555). Protestant martyr. Educated at grammar school in Manchester, he saw service at Boulogne and the siege of Montreuil in 1544 under Sir John Harrington, paymaster to the English forces. Admitted to the Inner Temple in 1547, he turned to a study of divinity at the instance of a fellow-student, T. Sampson (later dean of Christ Church). Taking his Cambridge M.A. in 1549, Bradford became a fellow of Pembroke Hall, where John Whitgift* was among his pupils and the continental Reformer Martin Bucer* a personal friend. In the following year he was ordained deacon, and for his ability and sympathy to the Reformation licensed to preach by Nicholas Ridley.* In 1551 he became prebendary of St. Paul's and a royal chaplain. Arrested in 1553, he was imprisoned in the Tower, first with Edwin Sandys,* then with Ridley, Hugh Latimer,* and Thomas Cranmer* for a time. After examination he was burnt at Smithfield. "A bold, intrepid yet sweet and earnest preacher," he was of singularly gentle character and "of more soft and mild nature than many of his fellows" (Parsons). His complete works were published in two volumes by the Parker Society (1848, 1853).

G.S.R. COX

BRADFORD, WILLIAM (1589?-1657). Second governor of Plymouth Colony; writer. Baptized in 1590 in Yorkshire, England, he is referred to by Cotton Mather* as the Moses who brought the Pilgrims out from England to the New Canaan. He governed the colony for most of the years between 1621 and 1656. Bradford wrote the *History of Plymouth Plantation,* a historical masterpiece, describing how God had providentially led the Pilgrims in the creation of their new society between 1620 and 1646. His scholarship and managerial ability were great, but, in Mather's words, "the crown of all was his holy, prayerful, watchful and fruitful walk with God, wherein he was very exemplary." JOHN D. WOODBRIDGE

BRADWARDINE, THOMAS (c.1290-1349). Archbishop of Canterbury in 1349. A native of Chichester, he studied at Merton College, Oxford, achieving great distinction in mathematics and divinity, and earning the nickname "the Profound Doctor." At Oxford he was elected chancellor and professor of divinity. Among other honors he was appointed chaplain and confessor to Edward III, whom he accompanied abroad on his French campaign. He was appointed archbishop of Canterbury in 1349, but died of the Black Death* within forty days of consecration. Through his influence Oxford University was freed from subservience to the bishop of Oxford. In theology, Bradwardine's attack on Pelagianism* in which he stressed God's grace and irresistible will as the ultimate cause of events paved the way for the later development of the doctrine of predestination. He also wrote extensively on mathematical subjects. R.E.D. CLARK

BRADY, NICHOLAS (1659-1726). Irish clergyman. Born in County Cork, he was educated at Westminster School; Christ's College, Oxford; and Trinity College, Dublin. After ordination he became prebendary of Cork in 1688. He was a staunch supporter of the Prince of Orange in 1690, and after James II had ordered the destruction of his hometown of Bandon, Brady was successful on three occasions in preventing its burning. Returning to England, he presented the grievances of the people of Bandon to Parliament. Later he held various livings in London, translated the classics (Virgil, etc.), was chaplain to royalty, and with Nahum Tate* wrote a metrical version of the Psalms which was in constant use until about 1800. Parts of this are still to be found in modern hymnbooks (e.g., "As pants the heart" and "Through all the changing scenes of life"). Many of his sermons were published. R.E.D. CLARK

BRAHMS, JOHANNES (1833-1897). German composer. Born in Hamburg, he showed extraordinary musical gifts at a very early age. Like Beethoven* before him, he later settled in Vienna where he was active as a choral conductor as well as being the foremost composer of symphonic and chamber music of the late Romantic era. Although he was not a church musician, he was grounded in the historic traditions of Lutheran church music, much of which he edited. He was one of the first to seek to perform the cantatas of Bach* in an authentic manner. He composed a small number of beautiful sacred motets after the example of Schütz* and Bach. His great choral masterpiece is *A German Requiem.* Using entirely words from the German Bible, it has no relation to the Requiem Mass of the Roman rite. It might justly be regarded as the greatest major sacred choral work of the nineteenth century. J.B. MAC MILLAN

BRAINERD, DAVID (1718-1747). Pioneer missionary to North American Indians. Born at Haddam, Connecticut, he had a profound conversion experience in 1739 and thereafter went to Yale College. Expelled in 1742 through misdemeanors arising out of "intemperate, indiscreet zeal" (Edwards), he studied divinity privately and was licensed to preach. The Scottish Society for the Propagation of Christian Knowledge appointed him missionary to the Indians, and he labored diligently in eastern Pennsylvania amid severe hardships until overcome by disease. By November 1745 he had ridden over 3,000 miles on horseback, and in the years 1745-46 saw, in his own words, "a remarkable work of grace." By March 1746 more than 130 Indians had been converted. Increasing illness forced his retirement, and his work was taken over by his brother John. He died at the New England home of Jonathan Edwards.* His *Journal* became a devotional classic, influencing hundreds to become missionaries. J.G.G. NORMAN

BRAMHALL, JOHN (1594-1663). Irish archbishop and primate. Born at Pontefract, Yorkshire, he attended a local school, then entered Sidney Sussex College, Cambridge (1608). He came under the influence of the master, Samuel Ward, who represented the Church of England at the Synod of Dort in 1621 and from whom he derived his Anglo-Catholic leanings. After eight brilliant years at Cambridge he had two brief pastorates in Yorkshire. Marriage to a wealthy lady enabled him to devote his time to reading and study. His leadership in the Church of England seemed assured when he was appointed a high commissioner by Charles I. He was invited to undertake the difficult task of reorganizing the Church in Ireland, and was successively archdeacon of Meath (1633), bishop of Derry (1634), and archbishop of Armagh and Primate of All Ireland (1660). Jeremy Taylor* preached his funeral sermon. He used his great gifts as speaker in the Irish House of Lords, in brilliant controversies with Hobbes, Baxter,* and the Puritans generally, and as one of the outstanding preachers of his day. His bitter opposition to Cromwell led to exile from 1648 to 1660. ADAM LOUGHRIDGE

BRAMWELL, WILLIAM (1759-1818). Wesleyan preacher and revivalist. Born in Elswick, Lancashire, of well-to-do, devout Anglican parents, he was the tenth of eleven children. He was apprenticed to a currier but, devoting his spare time to Bible study, he came under deep conviction of sin and found Christ as Savior for himself. Lacking Christian companionship among Anglicans, he joined the unpopular "Wesleyan devils" to the consternation of his parents. After entering the ministry he preached in various Wesleyan circuits, often with great power, for the rest of his life. He is remembered for the hundreds he led to God, for his unfailing generosity, for his self-denial, and his remarkable clairvoyant power by which he came to know intimate details about the lives of those he encountered. Preaching to the last, he died suddenly after declaring that the Lord had told him that he had only a little while to live. R.E.D. CLARK

BRANT, JOSEPH (Indian name: **Thayendanegea**) **(1742-1807).** Christian Mohawk chief. Born near the Ohio River, he became an Anglican convert to Christianity. The English called his stepfather Brant. When Sir William Johnson was Indian commissioner, he sent Joseph, whose sister he was later to marry, to be educated in Connecticut and afterwards made him his assistant. Chosen chief of the Mohawks, Brant visited England in 1775. There he was lionized and promised to lead 3,000 Indians on the royalist side. Appointed captain in the British army, he was involved in the Cherry Valley Massacre of 1778. In 1785 he was presented at court, and the king gave him an estate at the head of Lake Ontario. He built the first church in Upper Canada. HAROLD R. COOK

BRAY, THOMAS (1656-1730). Anglican divine and missionary promoter. Born at Marton, Shropshire, and educated at Oswestry School and Oxford, he became rector of Sheldon, Warwick, in 1690. Five years later he was appointed commissary for Maryland by H. Compton,* bishop of London, to establish the churches there, but found that he could only enlist poor men unable to buy books. He worked at a scheme to form parochial libraries in the colonies and then at home in every deanery in England and Wales, so that by his death more than eighty existed in the United Kingdom and thirty-nine in North America, some of over 1,000 volumes. Out of other educational projects developed the Society for Promoting Christian Knowledge (1698). In 1699 Bray at last reached Maryland, but found that he could develop his projects better at home. He returned to found the Society for the Propagation of the Gospel (1701). Rector of St. Botolph-Without, Aldgate, in 1706, he continued to develop his libraries, but failed in his attempts to provide a bishop for New England. G.S.R. COX

BRAY, WILLIAM ("Billy") (1794-1868). Methodist preacher. Born near Truro, Cornwall, son of a mining convert from Wesley's visit, he entered a mine after his father's early death, was dismissed, and lived a dissipated life as a publican. He was converted after reading Bunyan* and thereafter spent his life preaching, often ill-clad, hungry, and sleeping outdoors. His wonderful joy in Christ, powers of repartee, and wit made him famous throughout Cornwall. "I am the son of a King," he would say; prayer he called "cutting the devil's ould claws." He saw to it that "his wheels were kept nicely oiled and ready for work."

BRAZIL, see LATIN AMERICA

BREAD, see COMMUNION, HOLY

BRÉBEUF, JEAN DE (1593-1649). Jesuit missionary to Canada. He entered the Jesuit novitiate in Rouen, France, at the age of twenty-four, was ordained in 1622, and in 1625 commissioned to go to New France with Charles Lalemant and E. Massé. On his first visit to the Huron country he traveled some 800 miles by canoe, carefully chronicling the event. After the capture of Quebec by the Kirke brothers, he returned briefly to France. On his second visit to the Huron country he founded a permanent mission under the direction of Paul Le Jeune. His missionary ambitions were thwarted by epidemics, low moral standards, and animosity among Indian nations. Brébeuf's description of the missionary endeavors in the *Jesuit Relations* is an invaluable sociopolitical commentary of the Huron milieu of the day. He and others were massacred at the hands of the Iroquois in 1649. EDWARD J. FURCHA

BREECHES BIBLE, see GENEVA BIBLE

BREMOND, HENRI (1865-1933). Roman Catholic scholar. Member of the Society of Jesus from 1882-1904, he left on account of "incompatibility of temperament." Friend of the Catholic Modernists, though not himself one, he was especially influenced by Blondel* and by J.H. Newman,* whom he discovered while in England and discussed in his controversial *The Mystery of Newman* (ET 1907). Among many works, his greatest was *Histoire littéraire du sentiment religieux en France, depuis les guerres de religion jusqu'à nos jours* (11 vols., 1915-33). He used literary rather than theological sources to describe the spiritual revival in French Catholicism in France in the sixteenth and seventeenth centuries, portraying the devout humanism of the saints with psychological penetration. Volumes 7-11 became more of an exposition and defense of their teaching on prayer. Bremond was particularly interested in the close relation between mysticism and poetry. His scholarship, dialectical skill, and human insight were recognized by his election to the *Academie Française* in 1924.

HADDON WILLMER

BRENT, CHARLES HENRY (1862-1929). Canadian bishop and ecumenist. Born in Ontario, he was educated at Trinity College, Toronto, and ordained in 1887. After serving a parish in Boston, he was elected bishop of the missionary district of the Protestant Episcopal Church in the Philippines, and while serving there was a leader in the fight against the opium traffic. He was elected bishop of Western New York in 1917 and from 1926 to 1928 was in charge of the Episcopal churches in Europe. In his later years he gave much time to conferences promoting Christian unity. On returning from the Edinburgh Conference* of 1910 he induced the general convention of the Protestant Episcopal Church to convene a "World Conference on Faith and Order." He was elected president of the conference, held at Lausanne in 1927 (see LAUSANNE CONFERENCE). He held that all Christian groups belonging to such movements had something to contribute to all the others, and that such a great church would thereby become the repository of the great spiritual, intellectual, and moral wealth of the Christian centuries. His works included *The Mind of Christ in the Church of the Living God* (1908), *Presence* (1914), and *The Mount of Vision* (1918).

KEITH J. HARDMAN

BRENZ, JOHANN (1499-1570). German Reformer. He entered the University of Heidelberg in 1514 and came under the influence of Oecolampadius* and Luther. Ordained priest in 1520, he was appointed to the town of Hall in Swabia (1522), but was so strongly attracted to the Reformed position that he ceased celebrating Mass in 1523 and gave himself to biblical exposition. Among his works, published in an incomplete edition in Tübingen (1576-90), are many expository writings. He took a firm stand against the Peasants' Revolt* in 1525, but was compelled to flee from Hall in 1548, when it was captured by the imperial forces in the Smalcaldic War. He found protection under Duke Ulrich of Württemberg, who appointed him as minister of the collegiate church of Stuttgart, where he gave outstanding service to the Reformed cause. He took a prominent part in theological discussions of his time and was one of the members of the conference on doctrine called by Philip of Hesse at Marburg in 1529. He aligned himself with Luther on the doctrine of the Lord's Supper, and in 1525 published his *Syngramma Suevicum*, expounding Lutheran teaching on the presence of Christ in the sacrament.

HUGH J. BLAIR

BRES, GUIDO DE (c.1522-1567). Protestant martyr. Little is known about his early life except that he was born in Mons, Hainaut, and fled to England about 1548, where he joined a refugee congregation in London. Returning home to Belgium in 1552, he played a leading part in establishment and nurture of congregations until once again forced to flee in 1561 from the Spanish authorities who were increasingly concerned about the inroads of Protestantism and the growing Huguenot* influence. Deeply disturbed at the anarchic tendencies of many of his co-religionists, de Bres emphasized the importance of obedience to the magistrate and worked closely with William of Orange. When Valenciennes was besieged by the Spaniards, de Bres failed to persuade the radicals to surrender and was himself executed for rebellion when the city fell. His concern for unity, pastoral zeal, and theological leadership (he helped to draft the Belgic Confession*) made him a notable Reformer.

IAN BREWARD

BRETHREN, PLYMOUTH, see PLYMOUTH BRETHREN

BRETHREN IN CHRIST. This church originated in a society founded between 1775 and 1788 along the Susquehanna River near the present town of Marietta in Lancaster County, Pennsylvania. The initial group name was "Brethren." Jacob and John Engel of Swiss Mennonite* ancestry were two prominent leaders. The Brethren synthesized into a new pattern concepts already present in the religious life of their community. To a personal, conscious experience of the new birth, as stressed by the eighteenth-century Pietistic awakening, they joined concern for discipleship and restitution of the visible church along NT lines, as emphasized by Anabaptist* tradition. Late in the nineteenth century they accepted Wesleyan perfectionism as the third principal element of their eclectic faith. After a century of relative quiescence and slow growth, the group burst into new activity. They began Sunday schools, orphanages, and a home for the aged, supported higher education, founded a church periodical *(Evangelical Visitor),* and launched into formal evangelism and missions at home and overseas. In 1970 more than one-third of their 17,000 members were in mission churches in India, Japan, Nicaragua, Rhodesia, and Zambia.

Soon after their origin, the Brethren in the USA became "River Brethren," and in Canada, to which their faith was carried in 1788, they became "Tunkers." About 1862 the River Brethren changed their name to "Brethren in Christ"; in 1933 the Canadian branch of the movement also adopted the latter name. Three other living churches—Old Order River Brethren (Yorkers), United Zion Children, and Calvary Holiness—share a common heritage with the Brethren in Christ. The first two emerged from divisions within the River Brethren movement in the mid-nineteenth century; the third separated from the parent stock in 1962.

The Brethren in Christ are affiliated with the Mennonite Central Committee, the National Association of Evangelicals, and the National Holiness Association. Their headquarters are at Evangel Press, Nappanee, Indiana, and their archives at Messiah College, Grantham, Pennsylvania.

CARLTON O. WITTLINGER

BRETHREN OF THE COMMON LIFE. During the fourteenth and fifteenth centuries in Germany and the Netherlands, a rising tide of mystical lay piety grew up outside the official church. Under the leadership of Gerhard Groote* (1340-84), an interest in the inner life of the soul and the necessity of imitating the life of Christ by loving one's neighbor as oneself had become popular in the Low Countries. When the church ordered Groote to stop preaching, he retired to Deventer, his hometown, and gathered a commune around him. It was this group that, led by Florentius* Radewijns after Groote's death, founded the association known as the Brethren of the Common Life. The movement spread from one city to another as houses for men and also for women were founded throughout the Netherlands and Germany. These were to continue until the Reformation era. The Brethren did not constitute regular religious orders, but they took informal vows. They were entirely self-supporting, but pooled their money in a common fund from which each drew expenses, the surplus being used for charity. Groote had urged the copying of books as a method of earning a living and also to make reading materials more available. This work led to the founding of schools in many communities. From these emerged many influential religious leaders and humanists, such as Nicholas of Cusa* and Erasmus.* One pupil, Thomas à Kempis,* wrote *The Imitation of Christ,* which gives an understanding of the spirit and the teaching of the movement.

See A. Hyma, *The Christian Renaissance, a History of the Devotio Moderna* (1924); and T.P. Van Zijl, *G. Groote, Ascetic and Reformer* (1963).

ROBERT G. CLOUSE

BREVIARY. The liturgical book of the Roman Catholic Church which contains the material used in the Daily Office. Prior to the late eleventh century, this material was not collected together, but divided into separate books according to its constituent parts—e.g., Antiphonary, etc. In the Breviary the material is arranged under the "common of time" (called the Psalter, since the recitation of the psalter forms the basis of all the daily offices); the "proper of time" (i.e., seasonal variations); the "common of saints"; and the "proper of saints." This cycle of prayer had developed under monastic influence from its primitive Christian original into a system of seven daytime hours (Lauds to Compline) and a night office (Matins). In the Middle Ages, commemoration of the saints obscured the early scheme of Bible reading and weekly coverage of the Psalter. An attempt was made by Cardinal Quignon to reform the Breviary in 1535, and his work was known to Cranmer.* In medieval England the Breviary was often called the "portiforium" or "portuise."

JOHN TILLER

BREWSTER, WILLIAM (1567-1644). Founding member of Plymouth Colony. Born in Scrooby, England, he attended Cambridge University, where he acquired Separatist* ideas. He served the English ambassador to Holland for several years. Returning to Scrooby in 1589, he became a leading member of the small Puritan congregation which separated from the established church in 1606. Because of persecution the Separatists migrated to Holland in 1608 with their leaders John Robinson* as teacher and Brewster as elder. Brewster supported himself by printing Puritan books. He favored immigration to America, and in 1620 sailed on the *Mayflower* and helped establish Plymouth Colony. He became one of the most important members, playing a major role in its civil and financial affairs. He was the only church officer until 1629, but though he led the congregation in praise and prayer and in teaching the Bible and Christian doctrine, he did not preach or administer the sacraments because he was not an ordained minister.

HARRY SKILTON

BRIAND, JEAN OLIVER (1715-1794). Roman Catholic bishop. Native of France, he went to Canada in 1741 to follow Bishop Pontbriand. During the siege of Quebec by the British he directed diocesan affairs in the absence of the bishop. He is generally hailed as the second founder of the Catholic Church in Canada because of the reconciling manner with which he hailed the British under General Murray for their humanity toward the conquered in 1763. Though Roman Catholicism was officially outlawed, the British government gave informal consent to his consecration as "superintendent of the Roman Church in Canada" in 1766. He kept the French passively loyal to the British crown. His *Catechism* (1765) is one of the first books printed in Canada. Due to his efforts, the Quebec Act and the Habeas Corpus Act of 1774 broadened the privileges of Roman Catholics as British subjects. He consecrated his successors before resigning his see.

EDWARD J. FURCHA

BRIDAINE, JACQUES (1701-1767). French Roman Catholic preacher. He conducted missions in towns and villages all across France; these led to many conversions. His style of preaching was extemporaneous in a day when this was extremely rare. Bridaine edited a volume of *Cantiques spirituals* (Spiritual Songs, 1748) which passed through more than fifty editions. A number of books containing his thoughts and sermons were published, including five volumes of sermons.

BRIDE, see BRIDGET OF IRELAND

BRIDGE, WILLIAM (1600-1670). Nonconformist* preacher. Born in Cambridgeshire and educated at Emmanuel College, Cambridge, where he also became a fellow, he served several churches in Essex and Norfolk as lecturer from 1631 to 1636. Deprived in the latter year by the bishop of Norwich, he migrated to Holland, renounced his episcopal ordination, and became the teacher of a gathered church in Rotterdam. Returning to England, he was town preacher at Yarmouth, Norfolk, and pastor of a Congregational church there. He was nominated to serve in the Westminster Assembly* and became one of its five Dissenting Brethren. He attended also the Savoy Assembly in 1658. Though having a great reputation as a preacher, he was deprived in 1661 of his Yarmouth post. At first thereafter he lived in or near London, but in 1667-68 he returned to Yarmouth to preach in conventicles. He died in East Anglia. Most of his printed works were sermons.

PETER TOON

BRIDGES, ROBERT (SEYMOUR) (1844-1930). Poet. Educated at Eton and Oxford, he practiced medicine until 1882, when he retired to Yattendon, Berkshire, to devote himself to literature. He wrote poems, plays, and essays. His concern for hymn-singing and church music led to his *Yattendon Hymnal* (1895-99), for which he wrote, translated, and adapted forty-four hymns, including "Ah, holy Jesu, how hast thou offended" (Heermann); "All my hope on God is founded" (Neander); and "The duteous day now closeth" (Gerhardt). Although the collection of 100 hymns had a restricted circulation, its literary and musical merits were such that it was used extensively for the *English Hymnal* (1906) and the *Oxford Hymn Book* (1908). In 1913 he became Poet Laureate. His long poem, *The Testament of Beauty* (1929), sought to reconcile science and Christianity. He was the first to edit the poems of G.M. Hopkins* (1918).

JOHN S. ANDREWS

BRIDGET (Bride). (c.455-c.523). Patron saint of pity and mercy. She largely inspired the convent system that was to make such an impact on life in Ireland. She was born at Fochart, near Dundalk, where her mother was slave to a Druid. Bridget was possibly the Druid's illegitimate child. On obtaining freedom she settled at Kildare where she built for herself and her female friends a house for refuge and devotion. It is difficult to get a true picture of her life and work on account of the excessive legend surrounding her. Her name is commemorated in places like Kilbride or Kirkbride, and as these place names are found in Scotland as well as Ireland it is possible that she exercised a missionary influence there before the days of Columba.* ADAM LOUGHRIDGE

BRIDGET OF SWEDEN (c.1303-1373). Founder of the Brigittines.* Daughter of the governor of Uppland, she had from an early age spiritual influences that directed the course of her life. She married and had eight children, but on her husband's death in 1344 she retired to a life of penance and prayer. To the prior of the nearby Cistercian monastery at Alvastra she dictated her revelations, and he translated them into Latin. One vision commanded her to found a new religious order (of St. Bridget, or of the Most Holy Savior), and for this she secured the necessary papal permission in 1370. The order flourished in Sweden until the Reformation. In 1350 she went to Rome, and she remained there for the rest of her life, apart from several pilgrimages, one of which was to the Holy Land the year before her death. In Rome she ministered widely to rich and poor, homeless and sinners, giving God's messages in a restless and corrupt age. She was canonized in 1391; one of her daughters was Catherine of Sweden.* J.D. DOUGLAS

BRIDGMAN, ELIJAH COLEMAN (1801-1861). First American missionary to China. Born in Belchertown, Massachusetts, on graduation from seminary in 1829 he was appointed to China by the American Board.* For a year he and David Abeel of the Seamen's Friend Society were supported by D.W.C. Olyphant, a China trader. Bridgman learned the Cantonese dialect and prepared a 730-page manual on it. In 1832 he began publishing in English the influential *Chinese Repository*, which for many years carried general information about China as well as about missionary work there. A faithful preacher of the Gospel, he was also a founder of the Morrison Educational Society and was a member of the North China branch of the Royal Asiatic Society and editor of its journal. After the first Opium War he started work in Shanghai, where he supervised production of the Bible and was pastor of a church. HAROLD R. COOK

BRIGGS, CHARLES AUGUSTUS (1841-1913). Clergyman and scholar. Born in New York City, he was educated at the University of Virginia and Union Theological Seminary and later studied at the University of Berlin. After Presbyterian ordination (1870) he served a church in New Jersey, and in 1874 was appointed professor of Hebrew and cognate languages at Union Seminary. In 1890 he was appointed to a new chair of biblical theology. In his inaugural address he vigorously condemned "the dogma of verbal inspiration." He was tried for heresy before the Presbytery of New York in 1892 and was acquitted, but after the prosecution appealed to the general assembly he was condemned and suspended from the ministry (1893). Union Seminary ignored the decision, and in 1900 Briggs was ordained in the Episcopal Church. Among his many scholarly works was his *Critical and Exegetical Commentary on the Book of Psalms* (1906-7). BRUCE L. SHELLEY

BRIGHT, WILLIAM (1824-1901). Patristics scholar. Educated at Rugby School under Thomas Arnold and at Oxford, he taught at Trinity College, Glenalmond, and in 1868 succeeded H.L. Mansel* as regius professor of ecclesiastical history at Oxford. His main field was patristic history, though he also had a liturgical interest. Of his numerous works, the most important were *History of the Church, 313-451* (1860); *Early English Church History* (1878); and *The Age of the Fathers* (2 vols., 1903). He was a dynamic lecturer; his religious views were High Church; and he wrote a number of hymns, some of which are still sung.

BRIGID, see BRIDGET (OF IRELAND)

BRIGITTINES. Order of the Most Holy Savior, founded by Bridget of Sweden* at Vadstema about 1346, following the Augustinian Rule. For some two centuries it was organized in double communities of men and women. Members were allowed to possess books for study. Bridget's daughter Catherine* became abbess of the Vadstema monastery, but the order was banished from Sweden in 1595. A new branch was brought back to Sweden by Elisabeth Hesselblad in 1923. Autonomous houses following the original rule survive in Bavaria, Holland, and England (Syon Abbey). In Spain, the "Brigittines of the Recollection" follow a modified rule introduced by Marina de Escobar in the seventeenth century. J.G.G. NORMAN

BRITISH AND FOREIGN BIBLE SOCIETY, see BIBLE SOCIETIES

BRITISH CHURCH, see CELTIC CHURCH

BRITISH COUNCIL OF CHURCHES. An associated national council of the World Council of Churches,* sharing the same doctrinal basis, namely a "fellowship of churches in the British Isles which confess the Lord Jesus Christ as God and Savior according to the Scriptures and therefore seek to fulfill together their common calling to the glory of the one God, Father, Son and Holy Spirit." It was formed in 1942 through an amalgamation of existing ecumenical-type bodies. Currently twenty-five bodies in the British Isles belong to the council. The Roman Catholic Church sends observers to the meetings. There are over 700 local councils of churches in association with the BCC. The council itself consists of 131 members, 90 of whom are elected by the various member churches. They meet twice a year for two days. The council works through a departmental structure of which the best known is Christian Aid, but others deal with education, international affairs, mission and unity, social responsibility, and youth. PETER S. DAWES

BRITISH ISRAELITES. Those who hold that the British and American peoples are part of the ten "lost" tribes of Israel. Sometimes known as Anglo-Israelism, the idea probably originated with John Sadler (*Rights of the Kingdom,* 1649), but in its modern form it dates from John Wilson's book *Our Israelitish Origin* (1840). The first Anglo-Saxon Association was founded in England in 1879. British Israelites vary in their views, but the following is typical. The kingdom, though promised to David's seed in perpetuity, failed to survive in Israel so that we must look for its continuance elsewhere. The connection with England is through Zedekiah's daughters (Jer. 41:10), who escaped death in Egypt (Jer. 44:12-14) and took root elsewhere (Isa. 37:31f.), that is, in Ireland, one of the "isles of the sea" (Jer. 31:10), to which they sailed in a ship with Jeremiah. From Ireland they or their progeny reached England and became the royal house. The common people reached England after much continental wandering, being "sifted through many nations" (Amos 9:9), but some remained in western Europe.

Many OT prophecies about Israel are said to have been fulfilled in the history of the British Empire, e.g., that Israel would lend and not borrow, would possess the gates (such as Gibraltar, Singapore, and Hong Kong) of her enemies; that Joseph's branches would run over the wall (Gen. 49:22), which means that the Pilgrim Fathers belonged to the tribe of Manasseh, but left their Ephraimitish relatives behind in England. The Great Pyramid is said to enshrine these truths. British Israelites do not form a separate sect, but belong to many churches.

Critics urge that the evidence for British Israelitism is very slender; that if true, British Israelitism is not important (Col. 3:11); that the promises of God are sometimes conditional (Deut. 28:68; 1 Sam. 2:30) while 2 Sam. 7:16 is messianic; that the "lost" tribes were largely absorbed into Judah, and that there are better claimants than the British.

BIBLIOGRAPHY: H.L. Goudge, *The British Israel Theory* (1933)—against; H.W. Armstrong, *The United States and British Commonwealth in Prophecy* (1967)—for; B.R. Wilson (ed.), *Patterns of Sectarianism* (1967), chap. 10—history.

R.E.D. CLARK

BRITISH NORTH AMERICA ACT (1867). "Foundation charter" of the Dominion of Canada, passed by the British Parliament. The self-governing colonies of Nova Scotia, New Brunswick, Quebec, and Ontario, after a series of conferences, were joined together under a federal government. The Act preserved the monarchy in British North America while providing a bicameral legislature with a House of Commons elected by the people and a Senate appointed from the three basic regions in the nation. The Act also reflected the religious struggle between the two founding peoples, for certain powers were left to the provinces which safeguarded the rights of French Canada. Two key questions left to the provinces were property and education which allowed separate schools to continue and placated the French Canadians. ROBERT WILSON

BROADCASTING, RELIGIOUS. This began in the *British Isles* 24 December 1922, when the Rev. J.A. Mayo gave a ten-minute talk just forty days after the inauguration of "the wireless." In 1924 the first worship service was broadcast from St. Martin-in-the-Fields, London, and since 1928 there has been a daily service on radio without a break. The scope has widened to include talks, hymn-singing, panel discussions, and magazine-style material as well as services of various kinds.

Little was done by the British Broadcasting Corporation to develop religious television programs until the era of competition with Independent (i.e., commercial) Television began in 1955. One of the ITV companies launched "About Religion" in 1956; another began "Sunday Break" in late 1957. Both set out to popularize Christianity, and they used the so-called closed period (6:15-7:25 P.M.) on Sunday evenings. The BBC responded to the challenge with "Meeting Point" and "Songs of Praise"—the latter a hymn-singing program.

No program regularly succeeded in reaching mass audiences until Yorkshire Television's "Stars on Sunday" in 1969. This long-running series of unashamedly sentimental songs, poems, and readings—largely chosen by viewers—has captured audiences of up to fifteen million, but has been heavily criticized for its saccharine style. Nevertheless successive archbishops of Canterbury have appeared on the program. One reason for its popularity may have been the overintellectual style of most BBC religious programs, which have until recently been heavily weighted in favor of high-level debate or formal church events. With the arrival of BBC's "Anno Domini" program, the balance seems to have been partly restored.

Local radio began in Britain in 1967 and from the start offered much wider access for the Christian layman. In practice, local programs have often allowed experimental ideas to be tested before they reach the larger regional and national audiences.

Training centers for radio and television were set up by the late Lord Rank (at Bushey) and the Roman Catholic Church (at Hatch End). The former is now fully ecumenical, and the latter is open to non-Catholic participants. Both centers are independent of the public broadcasting media.

From the beginning, religion has not been advertised, nor could would-be sponsors purchase time. The main churches generally favor this policy, believing that any change opens the public to those for whom money alone is the means of gaining air-time.

Religious broadcasting has inevitably been dominated by the Christian churches as a result of the BBC and a "mainstream" policy. By 1976, however, this was being interpreted much less rigidly. The end of the "closed period" also is in sight.

Some evangelicals have been dissatisfied with Britain's religious broadcasting and have attempted to beam shortwave programs into Britain from Monte Carlo and elsewhere. These programs are much closer in style to American religious broadcasting and have not won mass audiences. In the 1960s similar programs were transmitted by the now-defunct "pirate" radio stations. More recently, attempts have been made to gain access to BBC and ITV by media-trained evangelicals, and this seems a more fruitful method. MICHAEL SAWARD

Predicated on the Bill of Rights of the Constitution, the doctrines of "Freedom of Speech, Press and Religion" have guaranteed that individuals and organizations in the *United States* have access to the airwaves for responsible religious instruction and persuasion. The history of American religious broadcasting has had three phases:

The pioneer phase, 1921-31. This began with the endeavors of individual clergymen, usually broadcasting their own church services. The first of these apparently was from Calvary Episcopal Church, Pittsburgh, on 2 January 1921. Within two years, several ministers had regular programs from various cities.

The first religious station license was granted on 22 December 1921 to the National Presbyterian Church of Washington, D.C. Within five years there were more than sixty licensed religious stations, including KJS (Bible Institute of Los Angeles) in 1922, KFUO (Concordia Seminary, St. Louis) in 1924, and WMBI (Moody Bible Institute, Chicago) in 1926. When the first American network (NBC) was organized in 1926, free time was given to S. Parkes Cadman's program from New York City, and it was co-sponsored by the Federal Council of Churches as the "National Radio Pulpit," the first network religious program. In 1928, D.G. Barnhouse* purchased program time on the CBS network; in 1930 came Walter Maier's "The Lutheran Hour." Roman Catholic network broadcasting began at the same time, when Fulton Sheen* preached the first sermon on "The Catholic Hour." Foreign missionary broadcasting started when the World Radio Missionary Fellowship was founded in 1929. Its station, HCJB, went on the air from Quito, Ecuador, on Christmas Day, 1931.

The development phase, 1932-42. The Great Depression forced the sale of most religious stations to commercial interests; WMBI was one of the few to survive. Because the FCC controlled most of the free time, the independent gospel broadcaster had to buy time on commercial stations. This led to development of many types of religious programs. C.E. Fuller* launched his own independent radio ministry, as did other pastors. This period of development was retarded by two events: America's entry into World War II in 1941, and the FCC's persuading networks to discontinue the sale of time to religious broadcasts and channel all free time through the FCC.

The expansion phase, 1943 onward. The exclusion of most independent gospel broadcasters from the networks soon led to the former purchasing time on hundreds of individual stations as well as organizing their own. Metropolitan, denominational, and ecumenical agencies were formed, such as the Southern Baptist Radio Commission (1938). The National Religious Broadcasters, an evangelical association, was established in 1943, partly to counter the FCC's restrictive policies. The Broadcasting and Film Commission of the National Council of Churches now serves the interests of the mainline denominations.

In 1943 Billy Graham began the popular program "Songs in the Night," which has had many imitators; his even more widely heard "Hour of Decision" was launched in 1950. A year earlier, the radio networks resumed the selling of time to independent religious broadcasters.

Postwar broadcasting saw two significant technical achievements: FM broadcasting and television. FM has meant many new religious stations granted frequencies in the noncommercial portion of the band. The current era has seen much creative energy in programing religious television, beginning with Percy Crawford's "Youth on the March" in 1949, the first network gospel program. Others followed, including the Lutheran "This is the Life" and the telecasts of Billy Graham and Oral Roberts. Network television began to sponsor programs such as "Frontiers of Faith" and "Lamp Unto My Feet." Roman Catholics were well represented by Bishop Sheen, Jews by "The Eternal Light." EDWIN A. HOLLATZ

BROAD CHURCHMEN, see ENGLAND, CHURCH OF

BROOKS, PHILLIPS (1835-1893). Preacher and bishop. Born in Boston and educated at Harvard College and the Protestant Episcopal Seminary at Alexandria, Virginia, he was made deacon in 1859, ministered in Philadelphia, and in 1862 became rector of Holy Trinity Church, also in that city. His eloquence brought a call to Trinity Church, Boston. At first he refused, but after further urging he began his memorable twenty-two years of ministry there in 1869. Early in 1877 he delivered his *Lectures on Preaching* before Yale Divinity School; in these he stressed preaching as "the bringing of truth through personality." In 1880 he preached in Westminster Abbey, and before Queen Victoria in the Royal Chapel at Windsor. He wrote the lovely hymn "O little town of Bethlehem." In 1891 he was consecrated bishop of Boston. He died fifteen months later.

BRUCE L. SHELLEY

BROOKS, THOMAS (1608-1680). Nonconformist* preacher. Born into a Puritan family, he was sent to Emmanuel College, Cambridge. He soon became an advocate of the Congregational way and served as a chaplain in the Civil War. In 1648 he accepted the rectory of St. Margaret's, New Fish Street, London, but only after making his Congregational principles clear to the vestry. On several occasions he preached before Parliament. He was ejected in 1660 and remained in London as a Nonconformist preacher. Government spies reported that he preached at Tower Wharf and in

Moorfields. During the Great Plague and Great Fire he worked in London, and in 1672 was granted a license to preach in Lime Street. He wrote over a dozen books, most of which are devotional in character. He was buried in Bunhill Fields. PETER TOON

BROTHER LAWRENCE (c.1605-1691). Christian mystic. Born Nicholas Herman of Lorraine and of humble background, he had spent many years as a soldier and then as a footman—"a great awkward fellow who broke everything." At over fifty years of age he entered the Carmelite* Order in Paris as a lay brother and served in the kitchens. He became known as Brother Lawrence. After his death his *Conversations* and *Letters* were published and are printed together in *The Practice of the Presence of God.* They stress the need to do everything, including kitchen work which Brother Lawrence naturally disliked, for the love of God, and thus achieve a condition in which the presence of God is as real in work as in prayer. They influenced current mystical thought (cf. the writings of Fénelon*). C. PETER WILLIAMS

BROTHERS, RICHARD (1757-1824). Naval lieutenant, writer on prophecy, and early British Israelite. Born in Newfoundland, he came to England, and in 1771 joined the navy. He saw action under Rodney in 1781 and retired on half-pay in 1783, at which time he became a Christian pacifist. Refusing to take the oath required of pensioners, he was later left without means. From about 1790 he became increasingly eccentric, suffered from megalomania, and was placed until 1806 in a madhouse in Islington. His book *Revealed Knowledge* (1794) created a stir. He predicted that by 1794 the ten lost tribes of Israel—that is, the English—would have returned to Jerusalem where he, the Nephew of the Almighty, would be proclaimed their Prince. For a time he had many influential followers, among them Riebaud the publisher, to whom Michael Faraday was apprenticed.

See C. Roth, *The Nephew of the Almighty* (1933). R.E.D. CLARK

BROTHERS HOSPITALLERS. Mostly a lay order, it was founded by John of God* (d.1550) at Granada, and continued with the support of Philip II with hospitals founded in Madrid, Cordova, and elsewhere in Spain. In 1572 Pius V approved the order, under Austin rule, and it spread through Europe to distant colonies, with other hospitals in Rome, Naples, Milan, then in Paris by 1601. The brothers were expelled from their forty hospitals in France by the Revolution, though others emerged later. The order is governed by a prior general at Rome and divided into provinces, extending to Nazareth, England, and Ireland. Brothers take training and vows, wear habits, and perform duties of the religious life. Beyond personal sanctification, they seek their patients' spiritual and physical well-being.

C.G. THORNE, JR.

BROUGHTON, WILLIAM GRANT (1788-1853). Anglican bishop. Educated at Cambridge, he was ordained in 1818 and spent his years as a country curate in scholarly work. In 1828 he was appointed second archdeacon of New South Wales. Although a High Churchman with sympathy for the Tractarians,* he was regarded by the evangelical chaplains of the colony as not hostile to the Gospel. His ecclesiastical opinions led to conflicts with liberal opinion in the colony. In 1836 he was consecrated as first bishop of Australia. He traveled widely and in the 1840s made the first divisions of his vast diocese. He became the first bishop of Sydney. In 1850 he initiated discussions about self-government for the Church of England in Australia at a conference of six bishops in Sydney. His death delayed progress in this direction. NOEL S. POLLARD

BROWN, JOHN (1800-1859). Controversial leader of part of the abolitionist* movement. He was born in Torrington, Connecticut, into a family with a history of mental illness. He engaged in many different business ventures in several states and was a defendant in numerous legal cases resulting from his failure to meet financial obligations. In 1854 he began organizing guerrilla warfare activities to rid Kansas of the evil of slavery. Believing himself divinely appointed to destroy the supporters of slavery, he led a group of six associates in the infamous Pottawatomie Massacre of 1856; five pro-slavery men were seized in their homes at night and hacked to death. From this incident John Brown gained notoriety and provoked fear and hatred, but he was defended by radical abolitionists. In 1859 he tried to set up a revolutionary government by force when he attacked the arsenal at Harpers Ferry, Virginia, with a band of twenty-one followers. This free state could be a refuge for slaves and encourage slave insurrections. The effort was easily crushed by the U.S. Marines, and Brown was convicted of treason and hanged. His violent acts were glorified by the abolitionist extremists and made the extreme pro-slavery faction stronger in the South by the panic they aroused. HARRY SKILTON

BROWN, JOHN (of Edinburgh) (1784-1858). Scottish minister. Son of the manse, and grandson of John Brown of Haddington, he entered the Burgher* ministry. Ordained at Biggar in 1806 and translated to Edinburgh in 1822, he soon became one of the most influential ministers of any denomination in the Scottish capital, with many friends in all religious bodies, many of whom met regularly for prayer and discussion. Appointed Scotland's first professor of exegetical theology, first in the Secession, then later in the United Presbyterian divinity hall, he became involved in prolonged controversy over his liberal views on the Atonement and other important doctrines. Indicted under twelve counts before the UP synod, he was triumphantly vindicated.

D.P. THOMSON

BROWN, JOHN (of Haddington) (1722-1787). Scottish minister. A self-educated boy from the hills above Abernethy in Perthshire who was accused by his own minister of witchcraft because of his prodigious learning, he taught himself

Latin, Greek, and Hebrew, and after serving as soldier and peddler, studied under Ebenezer Erskine, taking the Burgher* side at the "Breach" in the Secession Church in 1747. Ordained in 1751 to his sole charge at Haddington, East Lothian, he fulfilled a ministry of thirty-six years, during the larger part of which he also trained the Burgher divinity students. From his humble manse went out through the years a long series of volumes which earned him worldwide fame, the best known and most popular being his *Self-Interpreting Bible,* still being reprinted. With an income that never exceeded sixty pounds per annum, he founded a clerical dynasty which has given scores of learned men to all the professions.

D.P. THOMSON

BROWN, JOHN (of Wamphray) (c.1610-1679). Scottish minister. Little is known of him until about his fiftieth year. He was evidently not a member of the famous 1638 Glasgow Assembly (the year of the National Covenant*), nor does he seem to have taken any notable part in church affairs until 1660. Ejected after the Restoration of Charles II, he was one of those charged for opposing prelatical and arbitrary power, was imprisoned in Edinburgh, and then in 1663 banished to Holland, never again to be allowed to enter his native land. In 1665 he published *An Apologetical Relation,* which deals minutely with every aspect of the dispute between Crown and Covenant, and strongly upholds the righteousness of the principles and actions of the Covenanters, justifying their resistance to unconstitutional rulers by armed rebellion and defensive war. One of John Brown's last acts was to participate in the ordination of Richard Cameron* in Rotterdam.

J.D. DOUGLAS

BROWN, WILLIAM ADAMS (1865-1943). Liberal theologian. Born in New York City and educated at Yale, Union Seminary, and the University of Berlin, he afterward joined the faculty at Union, teaching theology there from 1892 until 1930, when he became research professor in applied theology until 1936. Through his widely used textbook *Christian Theology in Outline* (1906) Brown exerted considerable influence upon American theology until the 1930s. Himself influenced by his professor, Harnack,* Brown sought to elaborate a theology not of "dogmas to be received on authority" but of "living convictions, born of experience." He declared that Christ must be the center of Christian theology, but for him this meant the centering of faith upon the life and teachings of the historical Jesus rather than upon the orthodox understanding of the incarnation, objective atonement for sin, or the Resurrection.

HARRY SKILTON

BROWNE, GEORGE (d.1556). Archbishop of Dublin. Educated in an Augustinian friary at Oxford, he later became general of the Mendicant* Orders in England. He had attracted notice for his sermons, in which he ignored the invocation of the saints and exhorted his hearers to address their prayers to Christ alone. Thomas Cromwell* saw his potential usefulness to the king's cause and appointed him archbishop of Dublin, with the particular tasks of having the royal ecclesiastical supremacy recognized in Ireland, as it had been in England, and of suppressing the monasteries. Despite clerical opposition, Browne succeeded in having the Supremacy Bill passed. His suppression of the monasteries seems to have had in it more of a desire for spoil than for radical reformation; his destruction of images and relics failed to win the support of the people, who did not understand his sermons attacking idolatry, delivered in English to a largely Irish-speaking population. In the reign of Edward VI (1547-53) Brown was given the title of Primate of Ireland when the primacy was temporarily withdrawn from Armagh, where Archbishop Dowdall had maintained his allegiance to Rome. When Mary came to the throne, the title was restored to Dowdall, and Browne was set aside from his office as archbishop of Dublin.

HUGH J. BLAIR

BROWNE, ROBERT (c.1553-1633). English Separatist leader. Born in Rutland, he went to Cambridge and there came under the Presbyterian influence of Thomas Cartwright.* In 1573 he graduated, and while he was a school teacher his thinking developed in a Puritan separatist direction. In 1579 he began to preach in Cambridgeshire churches, refusing to accept the bishop's permission on the ground that the calling and authority of bishops was unlawful and that true authority lay in the gathered church. He concluded that the parishes were incapable of reform and that this would have to begin among "the worthiest were they never so few." In 1580 he was forbidden to preach by the council. In Norwich, Browne convinced Robert Harrison of his views, and together they organized separatist churches locally. His ideas effectively undermined the church-state relationship achieved by the Elizabethan Settlement,* and he was imprisoned several times—though his relationship to Lord Burghley helped to secure his release.

Persecuted by the authorities, the new church was persuaded that it was better to settle abroad; thus in 1582 it emigrated to Middlebrugh in Zeeland. While there, Browne set forth his views in the *Treatise of Reformation without Tarrying for Anie* and the *Booke which Sheweth the Life and Manners of all True Christians.* He argued that the Church of England was beyond reformation and that the true church should be established without tarrying for the magistrate. He quarreled with Harrison, however, and was excommunicated by the church in 1583 and eventually returned to England. By 1585 he had made peace with Archibishop Whitgift* and by 1591 was sufficiently orthodox to be ordained to a Northamptonshire living which he occupied for the next forty-three years. In 1633 he died in prison after a fit of aggression against the local constable. He is often called "the Father of English Congregationalism" and the followers of his ideas were known as Brownists (a term of abuse).

BIBLIOGRAPHY: C. Burrage, *The True Story of Robert Browne* (1906); A.L. Peel and L.H. Carlson, *The Writings of Robert Harrison and Robert*

Browne (1953); H.C. Porter, *Puritanism in Tudor England* (1970). C. PETER WILLIAMS

BROWNE, SIR THOMAS (1605-1682). Physician and writer. Graduate of Oxford, he traveled widely before settling in Norwich in 1637. In 1642 he published *Religio Medici*—a highly original attempt to work out a religious outlook in an increasingly scientific age. He asserts the right to examine nature which has much to reveal about God. With regard to the apparent contradictions between revealed faith and scientific truth he oscillates between acceptance by simple faith and explanation of the scriptural narrative by way of allegory. In his *Pseudodoxia Epidemica* he tried to separate scientific truth from the myths which had accumulated over the centuries. He wrote with vividness and feeling and did not allow himself to be confined to any one area of knowledge. In 1658 he produced *Hydriotaphia, or Urn-Burial,* which was a learned and fascinating study of burial in many countries. By now his encyclopedic knowledge had attracted considerable attention in scientific and antiquarian circles, and he was knighted when Charles II visited Norwich in 1671. Published after his death were *Certain Miscellany Tracts* (1683), *A Letter to a Friend upon the Occasion of the Death of an Intimate Friend* (1690), and *Christian Morals* (1716).

C. PETER WILLIAMS

BROWNISTS, see BROWNE, ROBERT

BRUCE, ALEXANDER BALMAIN (1831-1899). Scottish theologian. Son of a Perthshire farmer, he graduated at Edinburgh, attended the Free Church divinity hall there, and in 1859 became minister at Cardross. He was translated to Broughty Ferry in 1868, and in 1875 was appointed professor of apologetics and NT exegesis at his church's Glasgow college. His *Training of the Twelve* (1871) established his reputation as a scholar; then came *The Humiliation of Christ* (1876) and *The Kingdom of God* (1889). The latter was criticized by a general assembly suspicious of new principles of biblical criticism (perhaps the memory of his defense of W.R. Smith some years before was still fresh), but no official action was taken against him. Bruce was active also in helping to compile hymnbooks. His other writings include *St. Paul's Conception of Christianity* (1894) and volumes on the synoptic gospels (1897) and the epistle to the Hebrews (1899).

J.D. DOUGLAS

BRUCE, ROBERT (1554-1631). Scottish minister. Of a family that claimed descent from the royal family of Bruce, he studied law at Paris, but after a deep spiritual experience he took up theology at St. Andrews, was ordained in 1587, and became minister in Edinburgh. A few months later, in striking testimony to the worth of one so young, he was elected moderator of the general assembly. His moderate influence was appreciated at first by James VI, whose queen Bruce anointed in 1590. In 1596, however, having opposed the king's resolve to impose episcopacy on Scotland, he and others were banished from the capital. Though permitted to return, he was ordered away again in 1601 and spent his last thirty years in no settled home. For two periods of four years each he was confined to Inverness by the king's command, but wherever he could preach great crowds attended—and his life was acknowledged by all to be in accord with his preaching. His best-known writing is the *Way to True Peace and Rest* (1617). J.D. DOUGLAS

BRUCKNER, ANTON (1824-1896). Austrian composer. In his earlier years as the cathedral organist at Linz, Austria, he attained fame for his playing. He also wrote a considerable quantity of church music. His masterpieces are his *Te Deum* and three great masses for soloists, chorus, and orchestra. The second of these, for double chorus and wind instruments, is closely allied in spirit to the aims of the *Cecilian Movement* in Catholic music that led up to the *Moto Proprio* of Pope Pius X in 1903. In his later years he taught composition at the conservatory in Vienna. Known most widely for his gigantic symphonies, he is associated with the last flowering of the symphonic type of church music that goes back to Haydn* and his contemporaries. J.B. MAC MILLAN

BRUDERHOF, see HUTTERITES

BRÜGGLERS. A small sect founded at Brügglen, near Berne, Switzerland, about 1745 by two brothers, Christian (b.1710) and Hieronymus Kohler (1714-53). They proclaimed themselves the Holy Trinity with a convert, Elizabeth Kissling, functioning as the Holy Spirit. Because of this and the prediction of the end of the world at Christmas 1748, they were exiled by the Berne government in 1749. In 1753 Hieronymus was tried for heresy and executed in Berne. The group survived for a time in remote mountain regions, and was eventually absorbed by the Antonian* sect.

BRUNNER, (HEINRICH) EMIL (1889-1966). Swiss theologian. One of the most influential scholars of the interwar years, he enjoyed an international reputation through lecture tours and translations of his writings. After a pastorate at Obstalden (1916-24) he was professor of systematic and practical theology at Zurich (1924-53). In retirement he still traveled, spending two years in Japan as professor of Christian philosophy at the International Christian University of Tokyo (1953-55).

Brunner has long been thought of as a less extreme colleague of Barth* who parted company over natural theology in the 1930s. In fact, despite similarities, Brunner's approach was largely independent. He came from a Christian background. His early thought was influenced by the Christian Socialism of H. Kutter and L. Ragaz. But World War I caused him to reappraise his ideas in the light of the message of Christ. This he did independently of Barth. Among his early publications was a critique of Schleiermacher,* *Die Mystik und das Wort* (1924), which stressed the priority of divine revelation over human knowledge, reason, and experience.

The Mediator (1927: ET 1934) was the first presentation of the doctrine of Christ in terms of Dialectical Theology.* Brunner saw the Gospel as an exposition of the First Commandment. Without Christ's fulfillment in the Gospel this commandment would be unreal and unintelligible. Christ comes as the one who has fulfilled the law, as mediator, revealer, and reconciler. Faith is essentially obedience. Christ's mediatorship is the basis of the Christian ethic. Only in the Mediator do we know ourselves as we really are. Only in Him is the will of God known as love. Only in Him is it possible to see and love one's neighbor. Only in Him is our arrogant self-will broken and God honored. Only through faith in justification does the good become a reality instead of a mere postulate. Only through faith in Christ, the Mediator, does man gain a really ethical relation to historical reality.

Brunner's thought was deeply influenced by Kierkegaard's* dialectic and Martin Buber's* I-Thou concept. He saw revelation essentially in terms of personal encounter with God who communicates Himself. Brunner opposed both altruistic, theological liberalism and evangelical orthodoxy with its concept of revealed truth. The ground of his objection to the latter—that God Himself is a personal subject who cannot be reduced to an object—draws attention to a truth. But it causes difficulty when it is asked how Brunner's view compares with Scripture and how one can speak meaningfully of God and revelation on Brunner's premises. He regarded Scripture as somehow normative, though not above criticism. Revelation is always indirect. It is even mythical in form, but this is necessary because of the incommensurability of Creator and creature. Brunner felt no tension between his stress on revelation and a positive attitude to culture and philosophy. Unlike Barth, he believed in an already existing point of contact between the Gospel and non-Christian man. He pleaded for a positive reformed attitude to natural theology, though he failed to state convincingly what would be involved in it.

The church is the fellowship in faith and love of those who believe in Christ and is therefore the presupposition of faith. There is, however, the constant danger of institutionalism. In his treatment of eschatology Brunner has sought to get rid of what he regarded as the inadequate temporal conceptions of orthodoxy. Though he rejected the "realized eschatology" of Dodd,* he insisted that the existence of a hell has no place in the Christian hope. Brunner regarded Communism as "an anti-religion without God" in which all the elements of antichrist are present.

His writings include *The Divine Imperative* (1932; ET 1937); *Man in Revolt* (1937; ET 1939); *Justice and the Social Order* (1943; ET 1945); *Our Faith (1935; ET 1936); Christianity and Civilization* (2 vols., 1948-49); *Revelation and Reason* (1941; ET 1947); *The Divine-Human Encounter* (1938; enlarged as *Truth as Encounter*, ET 1954); *Dogmatics* (3 vols., 1946-60; ET 1949-62); and with Karl Barth, *Natural Theology* (1934; ET 1954).

BIBLIOGRAPHY: *The Theology of Emil Brunner* (ed. C.W. Kegley, 1962) contains an autobiography by Brunner, seventeen studies on him, Brunner's reply, and an exhaustive bibliography. Other studies include P.K. Jewett, *Emil Brunner's Concept of Revelation* (1954); P.G. Schrotenboer, *A New Apologetics, an Analysis and Appraisal of the Eristic Theology of Emil Brunner* (1955); and "Emil Brunner" in P.E. Hughes (ed.), *Creative Minds in Contemporary Theology* (2nd ed., 1969), pp. 99-130. COLIN BROWN

BRUNO (925-965). Archbishop of Cologne. He was educated at the cathedral school in Utrecht, then at the court of his brother Otto I. Abbot of the monasteries of Lorsch near Worms and of Carvei on the Weser from 941, he became archchaplain to Otto in 951. As archbishop from 953 he contributed greatly to every department of Otto's reign. With great personal sanctity and deep concern for clerical and lay education, he was a generous benefactor of churches and monasteries, having established three foundations in Cologne. An exemplary prince-bishop, he demonstrated the successful union of church and state. He participated in the Synod of Verdun (947) and improved relations between France and Germany. He was also a patron of learning. His disciple, Ruotger, wrote his biography shortly after Bruno's death. C.G. THORNE, JR.

BRUNO, GIORDANO (1548-1600). Italian Renaissance philosopher. He joined the Dominican Order* in 1562, but later was accused of heresy and fled, abandoning the Dominican habit in 1576. Thereafter he wandered through Europe, teaching in France and England, and visiting Wittenberg and Prague. In 1592 he returned to Italy and was arrested by the Inquisition.* After eight years' imprisonment at Rome he was sentenced as a heretic and burned. Bruno was a devotee of the Renaissance Hermetic tradition. Based on the writings of Hermes Trismegistus, who was supposed to have been an Egyptian sage who foretold Christianity and inspired Plato, this belief encouraged the use of magic and the worship of the sun (see HERMETIC BOOKS). Although he opposed Aristotle, Bruno did this on Hermetic grounds and not because he cared for Copernicus's mathematical proofs. His two major works are *On the Infinite Universe and Worlds* and *Concerning the Cause, Principle and One.*

ROBERT G. CLOUSE

BRUNO OF QUERFURT, see BONIFACE OF QUERFURT

BRUNO THE CARTHUSIAN (c.1030-1101). Founder of the Order. He was born in Cologne, where he began his studies at St. Cunibert and completed them at Reims. A canon of Reims, he taught arts and theology, becoming master of the schools (1056), then chancellor of the archdiocese (c.1075). Having no secular ambitions, he twice refused a bishopric and sided with Gregory VII against clerical decadence. He left Reims about 1082 for a monastic life, going first to Sèche-Fontaine. He found the cenobitic* vocation not sufficiently solitary and moved on to the Chartreuse where with a few clerics and laymen he lived the eremitic life, without a rule. Unintentionally, the Carthusian Order was founded by 1084 with the aid of the bishop, Hugh of Grenoble. The Chartreuse site, high above sea level with rugged mountains and severe climate, guaranteed silence, poverty, and small numbers. The first Carthusians combined the cenobitic and solitary without reference either to Benedictine or Camaldolese practices. In 1090 Urban II, a former pupil, called him to S Italy for counsel; there he founded the hermitage Santa Maria of La Torre, where he died. His works include two letters, being ascetical treatises, a commentary on the Psalter, and perhaps a commentary on the Pauline epistles. Canonization by Leo X is disputed.

See H. Löbbel, *Der Stifter des Carthäuser-Ordens* (1899); and A. Wilmart, "La Chronique des premiers chartreux," *Revue Mabillon* 16 (1926), pp. 77-142. C.G. THORNE, JR.

BRUYS, PIERRE DE, see PETER DE BRUYS

BRYAN, WILLIAM JENNINGS (1860-1925). Populist, editor, Chautauqua lecturer, secretary of state, and opponent of evolution. He was born and educated in Illinois and was admitted to the bar there in 1883. He practiced law in Illinois and Nebraska (1883-91) and was editor of the Omaha *World-Herald* (1894-96). From 1891 to 1895 he was a congressman from Nebraska. As secretary of state in Wilson's government he negotiated thirty treaties of arbitration with other nations. Three times the unsuccessful candidate of the Democratic Party for the presidency, he was a folk-hero for rural America because of his conservative Protestantism and concern for the agrarian Midwest. Catapulted onto the national political stage by his "Cross of Gold" speech in 1896, Bryan brought the program of the Populists into the Democratic Party. In 1925 he debated with the lawyer Clarence Darrow in the famous "Monkey Trial" at Dayton, Tennessee, when J.T. Scopes was accused of teaching evolution in the classroom (see SCOPES TRIAL). Bryan died five days after the trial finished. JOHN D. WOODBRIDGE

BRYANITES, see BIBLE CHRISTIANS

BRYENNIOS, PHILOTHEOS (1833-1914). Greek ecclesiastic and scholar. Born of poor parents in Constantinople, he was educated theologically at Halki and in 1856 went to Germany where he studied at Leipzig, Berlin, and Munich. In 1861 he became professor in Halki, in 1875 metropolitan of Serrae, and in 1877 of Nicomedia (until 1909). His fame springs from his publication in 1883 of the *Didache** of the Twelve Apostles, from a manuscript of the Metochion of the Holy Sepulchre in Constantinople (residence for the visiting patriarch of Jerusalem) often called "the Jerusalem manuscript" (written in 1056). It was removed to Jerusalem in 1887. From the same manuscript he previously published (1875) *1 and 2 Clement,* both in complete form for the first time. Bryennios was an Orthodox representative at the Bonn Reunion Conference* of 1875 and later wrote a critique of the encyclical *Satis Cognitum.* J.N. BIRDSALL

BUBER, MARTIN (1878-1965). Jewish religious philosopher. Born in Vienna, he was educated there and at German universities, and from 1916 to 1924 edited *Der Jude,* a paper for German-speaking Jews. He taught philosophy and religion at Frankfurt University (1923-33) and religion at the University of Jerusalem (1933-51). He was much influenced by the mysticism of the Hasidim and by Kierkegaard's* Christian existentialism. His own contribution was the "I-Thou relationship," for which he saw Judaism uniquely suited. Religion is essentially the act of holding fast to God; religious truths are dynamic, not dogmatic; and Judaism stressed the encounter of God and man. Buber held that three primary terms—It, I, and Thou—cannot be analyzed as separate concepts, but are purely relational words. "It" is a broad term representing the material world, which permits itself to be experienced. "Without It, man cannot live. But he who lives with It alone is not a man." By "Thou," Buber meant a more unlimited concept of deity than the term "God." The Thou is addressed and not expressed, for to express God is to imply that we can know Him as we can know an object in the realm of It. To maintain a deity worthy of the I-Thou realization, Buber called the God of the man-God relationship the "Eternal Thou," contending that this was the living personal God of the OT.

See his *I and Thou* (2nd ed., 1958), *Moses* (1958), *Mamre* (1946), and several other works. KEITH J. HARDMAN

BUCER (Butzer), MARTIN (1491-1551). Strasbourg Reformer. Born of humble parentage at Sélestat, he was schooled in Alsatian humanism and, as a Dominican,* in Aquinas's Scholasticism, but became an enthusiastic Erasmian and then, after moving to Heidelberg, an ardent "Martinian" through Luther's disputation in 1518. Released from his order in 1521, he was one of the first Reformers to marry (1522), was excommunicated while preaching reform at Wissembourg, and took refuge in Strasbourg (1523). He quickly assumed leadership in Strasbourg's reformation, together with Matthew Zell, Capito,* and Caspar Hedio, and retained it for over two decades. His gifts and industry soon established him as the chief statesman among the Reformers, an ecclesiastical diplomat of European stature, rarely absent from colloquies and diets from Marburg* (1529) onward. A prolific compiler of "Church orders" *(Kirchenordnungen),* he participated in the constitution of several Reformed churches, though unsuccessfully in Cologne with Hermann von Wied* (1542-43). In three formative years at Strasbourg (1538-41), Calvin sat at Bucer's feet, notably in church organization, ecumenism, and perhaps theology (e.g., predestination and Eucharist). Strasbourg's Reformed liturgy likewise shaped Genevan and Scottish patterns. Bucer's ideals attained fullest realization outside Strasbourg, through Geneva and in Hesse (1538-39), whose prince, the Landgrave Philip,* was long his intimate. (Bucer's radical views on divorce and remarriage extended even to justifying Philip's bigamy.) The magistrates restricted his scope in Strasbourg and refused to exercise fully the religious responsibilities he assigned them. Troublesome Anabaptist and spiritualist refugees provoked a tightening of ecclesiastical doctrine and structure (1533 synod), but Bucer also responded positively to the radicals, e.g., in developing discipline and confirmation, with signal success in Hesse.

He assiduously attempted to overcome the Zwinglian-Lutheran divide on the Lord's Supper. Having begun as an unquestioning Lutheran, he adopted the symbolist interpretation of Carlstadt,* Zwingli,* and Oecolampadius* (1542-46), but moderated from 1529 and treated the dispute as largely verbal. The Wittenberg Concord of 1536 climaxed his peace efforts, but it received a limited welcome. Bucer stressed participation in the true or real presence of Christ's body and blood, presented or conveyed *(exhibere)* by and with the signs in a "sacramental union" of earthly and heavenly realities. Thus Bucer's Strasbourg led a middle group of Reformed S German cities (cf. the *Tetrapolitan Confession,* 1530).

In the late 1530s and early 1540s he was the leading Protestant negotiator for agreement with the Catholic Church in Germany, especially at the conferences of Leipzig (1539), Hagenau and Worms (1540), and supremely Regensburg (1541), where a remarkable concord on justification was attained. (Bucer's intellectualist view of faith as conviction or persuasion allowed for justifying faith to be "faith active through love.")

Bucer was exiled for resisting the imperial Interim settlement (1548) and went to England as Cranmer's* guest. He was appointed regius professor at Cambridge; influenced the revision of the 1549 Book of Common Prayer and especially the 1550 Ordinal; wrote for Edward VI *The Kingdom of Christ,* a blueprint for a Christian society; mediated in the vestments controversy; and left his impress on John Bradford,* Matthew Parker,* and later John Whitgift.*

Bucer's distinctive greatness was long eclipsed. He highlighted the importance of love and service in community, an ordered and disciplined church life, and personal holiness. He was a profuse biblical commentator (here too a source for Calvin), a humanist advocate of patristic antiquity, a pastoral theologian, and a zealous, even risqué irenicist.

BIBLIOGRAPHY: Details of edition of *Opera Omnia* in progress (ed. F. Wendel, R. Stupperich et al., 1955ff.) in D.F. Wright, *Common Places of Martin Bucer* (1972), translated selections with introduction. Fuller bibliography in R. Stupperich, *Bibliographia Bucerana,* bound with H. Bornkamm, *Martin Bucers Bedeutung für die Europäische Reformationsgeschichte* (1952). Other selected studies: A. Lang, *Der Evangelienkommentar Martin Butzers und die Grundzüge seiner Theologie* (1900); H. Eells, *Martin Bucer* (1951; rep. 1971); J. Courvoisier, *La Notion d' Église chez Bucer dans son Développement Historique* (1933); C. Hopf, *Martin Bucer and the English Reformation* (1946); G.J. Van de Poll, *Martin Bucer's Liturgical Ideas: The Strasbourg Reformer and His Connection with the Liturgies of the Sixteenth Century* (1954); J.V. Pollet, *Martin Bucer: Études sur la Corresspondance* (2 vols., 1958, 1962); W. Pauck, *The Heritage of the Reformation* (2nd ed., 1961), chap. 5, "Luther and Butzer"; chap. 6, "Calvin and Butzer"; J. Müller, *Martin Bucers Hermeneutik* (1965); W.P. Stephens, *The Holy Spirit in the Theology of Martin Bucer* (1970); H. Vogt, *Martin Bucer und die Kirche von England* (typescript, Münster, 1968).

D.F. WRIGHT

BUCHANAN, CLAUDIUS (1766-1815). Anglican chaplain in India. Born at Cambuslang, Scotland, he was educated at Glasgow and Cambridge universities, was ordained in the Church of England in 1796, then went to India. He was chaplain at Barrackpur for two years before becoming chaplain in Calcutta and vice-principal of Fort William College which had been established by Lord Wellesley. At Cambridge Buchanan had come under evangelical influence from Charles Simeon* and his circle, and though his official connection with the East India Company prevented his ministering to the natives, he encouraged Scripture translation and native education. He wrote *Christian Researches in India* (1811), and on his return to England he helped to establish the first Indian bishopric.

G.E. DUFFIELD

BUCHANAN, GEORGE (1506-1582). Scottish humanist. Born at Killearn, he was educated at the universities of Paris and St. Andrews, began to favor Protestantism, and wrote against the Franciscans, which activity led to his arrest in 1539. He escaped to France, taught at Bordeaux and Paris, then was regent at the Portuguese university of Coimbra. There he was imprisoned, charged with heresy before the Inquisition,* and suffered much restriction before final acquittal. All the time he was furthering his classical studies, which were to enhance his reputation as the most distinguished British humanist of his time. He was friend and tutor to Mary Queen of Scots, but later supported the Protestant lords against her. In 1566 he became principal of St. Leonard's College, St. Andrews, and in 1567 moderator of the general assembly of the (now Reformed) Church of Scotland. In 1570 he became tutor to the child James VI. In 1579 he completed his famous treatise *De Jure Regni apud Scotos* which, though dedicated to James, marked the beginning of that conflict in Scotland that was to end more than a century later with the overthrow of the House of Stuart. Buchanan taught that kings are chosen and continued in office by the people, that they are subject to both human and divine laws, and that Scots had always claimed the right to call wicked rulers to account. Samuel Rutherford* later took up and elaborated these views. In 1683 Buchanan's political works, with those of John Milton,* were publicly burned by the common hangman because they militated against the Stuart belief in the Divine Right of Kings.* Among Buchanan's other writings were *Baptistes* (1642), a dramatic presentation of the life of John the Baptist, and *History of Scotland* (ET 1690).

J.D. DOUGLAS

BUCHANAN, JAMES (1804-1870). Scottish theologian. Born in Paisley, he studied at Glasgow University, was briefly minister of Roslin, then in 1828 was inducted to the large and influential parish of North Leith, where his evangelical preaching attracted great crowds. In 1840 he became minister of the High Church (St. Giles'), Edinburgh. At the 1843 Disruption,* in common with most of his fellow Evangelicals, he left the established church and became minister of St. Stephen's Free Church, Edinburgh. In 1845 he was appointed professor of apologetics at New College, and two years later he succeeded Thomas Chalmers* in the chair of systematic theology. Buchanan's works include *Faith in God and Modern Atheism Compared* (2 vols., 1855) and *The Doctrine of Justification* (1866).

J.D. DOUGLAS

BUCHANITES. This most bizarre of all Scottish religious sects (1783-1846) originated in Irvine with Elspeth Buchan, a fanatical visionary of doubtful antecedents, and Hugh White, a popular young minister there, who became her devoted disciple. Claiming to be "the third person in the Godhead and the woman clothed with the sun of the Book of Revelation," she led a mixed multitude, including White and the town clerk, on a weird pilgrimage to the south of Scotland, where at successive sites over a period of many years the steadily dwindling band awaited their vainly promised translation to heaven. The dramatic history of this venture, with the last lonely disciple keeping vigil over Mrs. Buchan's remains as he awaited her resurrection, has fascinated many thousands of readers.

D.P. THOMSON

BUCHMAN, FRANK NATHAN DANIEL (1878-1961). Founder of Moral Re-Armament, earlier known as the Oxford Group.* Born in Pennsburg, Pennsylvania, and educated at Muhlenberg College and Mt. Airy Seminary, he was ordained as a Lutheran clergyman. After engaging in unsuccessful work with students, he experienced a religious conversion at Keswick, England, in 1908. He soon began a movement to develop new methods of evangelism as a means of fostering world change. In 1921 he founded the First Century Christian Movement, in 1929 the Oxford Group, and finally in 1938 Moral Re-Armament (MRA). His house parties and stress on confession and the four absolutes won many among the upper classes and students.

ROBERT C. NEWMAN

BUCK, DUDLEY (1839-1909). American composer. It is customary to depreciate him today and all he stood for, but he did great service to American church music in the era in which he lived. He was the first influential American-born church musician to be trained in Europe. He held important posts in Hartford, Chicago, Boston, and New York. He was a voluminous composer of cantatas, anthems, and also secular music. Although the musical substance of his works is undistinguished and depends much on facile clichés, he provided his generation with material that was readily attractive and encouraged choral activity. He was also a distinguished organist and by his recitals did much to advance the popularity of that instrument.

J.B. MAC MILLAN

BUDDHISM. Eastern religion. The founder, Siddhartha Gautama (c.566-486 B.C.), grew up as the son of a petty ruler in NE India. Tradition relates that, surrounded by luxury, he saw four sights—a diseased man, an old man, a dead man, and a wandering ascetic. These convinced him of the inevitability of suffering and death, and he set out, abandoning his wife and son, to seek enlightenment and release from inevitable rebirth. After several fruitless attempts to find enlightenment by means of accepted ascetic techniques, he finally reached full enlightenment under a tree at what is now Bodh Gaya. He became "the Buddha" (the Enlightened One). Conscious that he was living out his last existence on earth, he determined not to enter into Nirvana directly, but to proclaim the *dhamma* (law) he had rediscovered, and he did so for some forty years. He founded monasteries for both men and women, and the later spread of Buddhism was linked closely with the fortunes of these institutions. After a couple of centuries during which Buddhism was little more than an unorthodox Hindu school, it began to expand as the result of the work of King Ashoka (third century B.C.)—northward into Tibet, China, and ultimately Japan; southward into Ceylon and SE Asia, where it now has its strongest centers. By the Middle Ages, Buddhism had virtually disappeared from India.

Buddhist teaching rests on four "excellent truths": all existence involves suffering; suffering is caused by desire; suffering can be ended if desire can be conquered; and there is an eightfold path to the conquering of desire. This path consists of right views, intentions, speech, action, livelihood, effort, mindfulness, and concentration. To this is added an elaborate monastic discipline. Classical Buddhist scriptures are called the *Tripitaka* (three baskets) and are written in Pali (a dialect of Sanskrit), though there are important writings also in Sanskrit, Tibetan, Chinese, Japanese, etc.

The great historical division among Buddhists is between the conservative or *Theravāda* (way of the elders) school, and the comprehensive or *Mahāyānā* (great vehicle) school. Theravāda—also called *Hīnayāna* (little vehicle) because its opponents maintained that it offered salvation only to monks—is strongest in the South (Ceylon, etc.), Mahāyānā in the North (formerly Tibet and China, now mainly Japan). Theravāda is atheistic in principle; Mahāyānā has sometimes tended to reckon the Buddha as a savior-god.

Buddhism and Christianity rest on entirely different conceptual foundations. Buddhism acknowledges the reality of neither God nor the soul; all is constant flux, and personality is an illusion. As in Hinduism, the doctrine of rebirth is axiomatic. Perhaps for these reasons, Buddhism in various forms (Zen, etc.) has of late gained ground in the West and is currently regarded by many as an attractive alternative to Christianity.

BIBLIOGRAPHY: E.J. Thomas, *The Life of Buddha* (3rd ed., 1949); T.R.V. Murti, *The Central Philosophy of Buddhism* (1955); E. Conze, *Buddhism: Its Essence and Development* (1959); H. Dumoulin, *A History of Zen Buddhism* (1963); W. Rahula, *What the Buddha Taught* (1965); T.O. Ling, *A History of Religion East and West* (1968); H. von Glasenapp, *Buddhism: a Non-Theistic Religion* (1970).

E.J. SHARPE

BUGENHAGEN, JOHANN (1485-1558). German Reformer. Born at Wollin near Stettin in Pomerania, he studied at the University of Greifswald, and through the writings of Erasmus* and the Humanists came to see the need for a reform of the corruptions of the Roman Catholic Church. In 1504 he became rector of the town school of Treptow, where his outstanding organizing ability became apparent in the school's success. In 1509 he was ordained, and in 1517 became a lecturer on the Bible and the Fathers in the monastery school at Belbuck. It was in 1520 that Luther's *Concerning the Babylonian Captivity of the Church* led him to realize that a much more radical reform of the church was needed and that the root of the corruptions of the Roman Church was its erroneous doctrine. Convinced by Luther's argument, he became an enthusiastic Reformer. In 1521 he went to Wittenberg and became closely associated with Luther and Melanchthon. He became minister of the collegiate church in Wittenberg in 1522 and served as preacher there for the rest of his life. His organizing ability found an outlet in the establishment of Reformed churches in Brunswick, Hamburg, and Lübeck. In 1537 he went to Copenhagen at the invitation of Christian III and remained there in Denmark for five years, reconstituting the Danish Church and reorganizing the country's education. Bugenhagen gave valuable assistance to Luther in his translation of the Bible; his best-known book was a commentary on the Psalms, highly praised by Luther. He also wrote a history of Pomerania, which was not published until 1728.

HUGH J. BLAIR

BULGAKOV, SERGEI NIKOLAEVICH (1871-1944). Russian philosopher, theologian, and economist. Son of a priest, he studied at a seminary in Orel, the University of Moscow where he was graduated in 1894, and then at Berlin, Paris, and London, before receiving his doctorate from Moscow in 1912. He taught at the Kiev Polytechnic Institute (1901-6), at Moscow (1906-18), and was elected a Cadet delegate to the second Duma. In 1918 he was ordained priest and because of Soviet disapproval went to teach at the University of Simferopol in the Crimea. He was expelled from Russia in 1922, and went to Prague, then in 1925 to Paris where he was dean and theology professor at the Orthodox Theological Institute which he helped to found. Bulgakov's ideas were strongly influenced by the philosophy of Soloviev* and Pavel Florensky. His thinking developed from Marxism to idealism and then to mysticism. He believed that the world or cosmos was an organic whole, animated by a world soul. God created the world out of nothing and as an emanation of His own nature. Mediating between God and the cosmos and uniting them is a third being, the Sophia or Divine Wisdom. A prolific writer, his many works include *The Unfading Light* (1917), *Jacob's Ladder* (1929), *Agnus Dei, The God-Manhood* (1933), and *The Comforter* (1936).

BIBLIOGRAPHY: L. Zander, *God and the World, the World Conception of Father S. Bulgakov* (1948); N.O. Lossky, *History of Russian Philosophy* (1951); V.V. Zenkovsky, *A History of Russian Philosophy*, vol. II (1953).

BARBARA L. FAULKNER

BULL, GEORGE STRINGER ("Parson") (1799-1865). An Evangelical and a staunch churchman who resigned from missionary schoolteaching through ill health, he became a clergyman in the industrial West Riding and Birmingham. He worked for temperance, for emancipation of slaves and of factory children (the Ten Hours movement), and opposed the 1834 Poor Law legislation as injurious to workers.

BULL, PAPAL. Derived from the Latin *bulla*, "seal," a bull originally referred to the seal affixed to papal edicts but later was transferred to the edicts themselves. The term is now restricted to the most important papal mandates, which are stamped with an official wax seal. Bulls have been issued to assert major Roman Catholic doctrines. For example, papal supremacy was declared by Pius II in *Execrabilis* (1460); in *Innefabilis* (1854) and *Pastor Aeternus* (1870), Pius IX proclaimed the doctrine of the Immaculate Conception and papal infallibility. Since 1878 only "consistorial bulls," signed by the pope and cardinals, are sealed in a special way. The originals of these bulls are kept at Rome and copies are sent out.

S. TOON

BULLINGER, JOHANN HEINRICH (1504-1575). Swiss Reformer. Son of the parish priest at Bremgarten, Canton Argau, he received his education first at the School of the Brethren of the Common Life* in Emmerich, duchy of Cleves, and then at the University of Cologne, the citadel of the *via antiqua*. At Cologne he also became familiar with the works of Erasmus,* Melanchthon,* and Luther whose writings decisively influenced him. After his return to Switzerland in 1523, he joined those who supported Zwingli's* reformation at Zurich and took part in the Berne Disputation of 1528. The disastrous Second Kappel War in 1531 destroyed his fortune and compelled him to flee from Bremgarten, where he had succeeded his father, and to take refuge at Zurich. After the Zurich Council agreed to guarantee the clergy's freedom to preach on all aspects of life in the city, he consented to become Zwingli's successor in December 1531. As the Zurich *antistes*, Bullinger performed the functions of a "Reformed" bishop presiding over the cantonal synod which he helped to reorganize and mediating between the Zurich Council and the clergy. He was also responsible for the reform of the school system and the creation of a central administration for the income from cloister lands confiscated by the city.

By this time he had already begun his repetitive but massive literary activities which included several important polemical works against the Anabaptists, *The Decades* (fifty sermons on Christian doctrine), the Diary *(Diarium)*, and the History of the Reformation *(Reformationsgeschichte)*. Bullinger used his literary skills to mediate the quarrels which arose within the Reformed churches, but they were of no avail to him in his repeated attempts to seek a theological agreement with the Lutherans. He played an important part in the writing of the *Consensus Tigurinus* and the Helvetic Confessions.*

Though he accommodated his own moderate Augustinian doctrine of predestination to the more rigorous one advanced by Calvin, Bullinger remained a lifelong opponent of Calvin's theory of the two polities within the Christian commonwealth and the Genevan ecclesiastical discipline. He was Thomas Erastus's closest ally in the partially successful struggle to prevent the introduction of a presbyterian polity into the Rhineland Palatinate, and he supported the English bishops against Thomas Cartwright's* Presbyterianism because he viewed it as a new form of papal tyranny. Denying that the punishment of Christians should include exclusion from the Lord's Supper, he delegated all coercive power to the secular magistrate whom he assumed was Christian. It was left to the clergy to fulfill their prophetic function by preaching the Word and administering the sacraments to a Christian people whom Bullinger, a covenant theologian, believed were in a covenant relationship with God.

The Zurich Letters (2 vols., 1542, 1545) reveal his interest in English affairs and bear witness to his hospitality to many Marian exiles. Bullinger viewed the leaders of the Church of England as fellow Reformed churchmen and at their behest wrote his refutation of Pius V's Bull of Excommunication against Elizabeth I. Portions of his *Decades* were dedicated to Edward VI and Lady Jane Grey, and they also provided Whitgift* with an educational tool to protect the clergy against the "prophesyings."

BIBLIOGRAPHY: H. Bullinger, *Reformationsgeschichte* (ed. J.J. Hottinger, H.H. Vögeli; 3 vols., 1839); G.W. Bromiley (ed.), *Zwingli and Bullinger* (1953); T. Harding (ed.), *The Decades of Henry Bullinger* (4 vols., 1849-52); F. Blanke, *Der junge Bullinger*, 1942); A. Bouvier, *Henri Bullinger le successeur de Zwingli* (1940); H. Fast, *Heinrich Bullinger und die Täufer* (1959); W. Hollweg, *Heinrich Bullingers Hausbuch* (1956); H. Kressner, *Schweizer Ursprünge des anglikanischen Staatskirchentums* (1953); C. Pestalozzi, *Heinrich Bullinger Leben und ausgewählte Schriften* (1858); J. Staedtke, *Die Theologie des jungen Bullinger* (1962); D. Keep, *Henry Bullinger and the Elizabethan Church* (1970).

ROBERT C. WALTON

BULTMANN, RUDOLF (1884-1976). German theologian. After studying at the universities of Marburg, Tübingen, and Berlin, he was professor of NT at Marburg (1921-51). One of the major NT scholars of the century, he was during the first twenty years of his professorship known mainly for his pioneering work on the form-criticism of the gospels. Starting at almost the same time as K.L. Schmidt and M. Dibelius, he argued that the gospels could be broken down into smaller units which had grown up in the oral stage of the tradition. He took a largely skeptical view of the authenticity of these units, whether reported sayings or deeds of Jesus. His book *Die Geschichte der synoptischen Tradition* was first published in Germany in 1921 and made an immediate impact, though there was no English translation until 1963. His next major work was simply called *Jesus* (1926; ET 1934). In this there is little emphasis on the deeds and teaching of Jesus apart from His call to decision. This is interpreted by Bultmann in terms closely akin to those of existentialist philosophy, and he has been strongly criticized for failing to give enough objective content to the grounds upon which a decision is to be based. Bultmann contributed a number of major articles to Kittel's *Theologisches Worterbuch zum Neuen Testament* but the next landmark among his publications came in 1941 with publication of his commentary on John's gospel. He suggested in this the dependence of the evangelist upon Gnostic* ideas. In the same year there was published in duplicated form his essay "Neues Testament und Mythologie" (ET in H.W. Bartsch, *Kerygma and Myth*, 1953). His advocacy here of the need to "demythologize" all the concepts of the NT has had a powerful influence upon theological thought since World War II. Having shown himself skeptical of the historical content of the gospels, and having emphasized the need for decision, here he showed that this decision was to be based upon the *kerygma*, which was not to be abandoned (as it had been by the liberals) but reinterpreted with its mythological elements expressed in existential terms. His last great work was *New Testament Theology* (1948-53; ET 1952-55).

BIBLIOGRAPHY: *Festschriften* in honor of his 65th birthday (ed. E. Wolf, 1949) and his 70th birthday (ed. W. Eltester, 1954); G. Miegge, *Gospel and Myth in the Thought of Rudolf Bultmann* (ET 1959); R.H. Fuller, *The New Testament in Current Study* (1962).

R.E. NIXON

BUNSEN, CHRISTIAN KARL JOSIAS VON (Chevalier Bunsen) (1791-1860). Theologian and Prussian diplomat. He founded the German *Evangelische Gemeinde* and prepared its liturgy. He thus came to be associated with King Frederick William III's ecclesiastical policy, and even more with that of Frederick William IV. The latter appointed him Prussian minister in London (1841-54), where his enthusiasm for Anglicanism increased, and he helped to establish the Anglo-Prussian Jerusalem bishopric (1841). He came to know England well, and his significance may be gauged partly by Rowland Williams's writing in *Essays and Reviews* (1860) on Bunsen's biblical researches. In the 1850s Bunsen's approach had become less confessional and conservative. He defended freedom of conscience and sought to present the living meaning of the Bible in accord with its historical sense as then understood in critical scholarship. He interpreted dogma morally and psychologically rather than as metaphysical, and placed emphasis on the personal nature of God who reveals Himself in the course of history and through personality. None of his theological or historical work had permanent value.

HADDON WILLMER

BUNTING, JABEZ (1779-1858). English Wesleyan Methodist. Son of a radical Manchester tailor, he was brought up as a Methodist and entered its ministry in 1799. He gradually achieved an unusual ascendancy over conference, which made him its secretary, president (four times), president of the theological institution, and secretary of the missionary society (which his enthusiasm had done much to form). A great organizer, he more than any other single figure determined the shape of Wesleyan Methodism over against the Church of England and the old Dissent; one effect was the loss of Methodists fearing his tendencies to centralism and ministerial dominance. His influence for political conservatism has probably been exaggerated. His opponents paid tribute to his sincerity, eloquence, and gift of prayer.

A.F. WALLS

BUNYAN, JOHN (1628-1688). Puritan writer and preacher. Born at Elstow, near Bedford, into a poor home, he probably acquired his grasp of the English language from reading the Bible. As a youth he was involved in the Civil War on the Roundhead side. In 1649 he married, and his wife brought him Dent's *Plain Man's Pathway to Heaven* and Bayly's* *Practice of Piety.* In 1653 he joined Pastor Gifford's Independent church at Bedford. A year or two later he began to preach with no little success, except with the magistrate who remanded him in custody for refusing to undertake not to preach. His imprisonment lasted intermittently from 1660 to 1672, but it enabled him to produce his masterpiece *Pilgrim's Progress* and other writings, including some verse. After 1672 he spent most of his time in preaching and evangelism in the Bedford area.

The Bedford tinker's fame rests chiefly on three works: *Pilgrim's Progress* (1678, 1684), *The Holy War* (1682), and *Grace Abounding to the Chief of Sinners* (1666). He proved to be a master of simple, homely English style, narrative, and allegory. The first-named book especially was, with Foxe's* *Martyrs,* read in virtually every Victorian home, and remains a best seller for children and adults alike. Theologically Bunyan was a Puritan in that he held a Calvinist view of grace, but he was a separatist in his views of baptism and the church. He has been much studied by literary experts, and has gathered round his name a number of biographies, chiefly of a devotional nature. The standard biography is still, however, J. Brown's *John Bunyan: His Life, Times and Work* (1885; rev. F.M. Harrison, 1928).

For a major study of Bunyan's theology, see R. Greaves, *John Bunyan* (2 vols., 1969). A complete edition of Bunyan's works is currently in preparation.

G.E. DUFFIELD

BURCHARD (c.965-1025). Bishop and canonist. Of a noble Hesse family, he entered the service of Archbishop Willigis of Mainz, and was ordained. Appointed bishop of Worms in 1000, he built new churches, reconstructed the cathedral, and disciplined the clergy. He took a leading part in ecclesiastical reform in Germany. Between 1007 and 1014 he compiled his *Decretum,* an influential collection of canon law, and between 1023 and 1025 promulgated a celebrated body of laws known as the *Leges et statuta familiae S. Peter Wormatiensis.*

BURGHERS (also known as the Associate Synod). Scottish Presbyterian secessionist group. In 1733 the first secession from the Church of Scotland took place over the question of patronage.* Despite efforts by the general assemblies of 1732 and 1736 to effect reconciliation, Ebenezer Erskine* and three others constituted themselves into a presbytery and were joined by his brother Ralph and four others in 1737. In 1740 they were formally deposed by the Church of Scotland, and the secession was made definitive. In 1747 they split into Burghers and Antiburghers (also called the General Associate Synod), who respectively regarded it as lawful or sinful for members to take the oath required of burgesses of certain cities by which they acknowledged the true religion publicly preached within Scotland and authorized by law. In 1799 the Burghers split into Auld Lichts* and New Lichts; the Antiburghers similarly split in 1806. The New Lichts of both groups formed the United Secession Church* in 1820. The Auld Licht Antiburghers joined the Free Church of Scotland in 1852, though a remnant continued as the Original Secession Church. The Auld Licht Burghers reunited with the Church of Scotland in 1839.

J.W. MEIKLEJOHN

BURGON, JOHN WILLIAM (1813-1888). Anglican scholar and controversialist. Educated at Worcester College, Oxford, and thereafter fellow of Oriel, he became vicar of St. Mary's, Oxford in 1863, and dean of Chichester in 1876. An old-fashioned High Churchman, he was famous on three counts: as author of the very popular *Lives of Twelve Good Men,* a series of pungent sketches of High Churchmen of his age (1888); as an unremitting protagonist for the "Textus Receptus" of the NT, both against higher criticism* (cf. *The Last Twelve Verses of the Gospel according to St. Mark Vindicated,* 1871), and against the then new RV with *The Revision Revised* (1883); and as the loser in at least three major conflicts in which he violently denounced the disestablishment of the Irish Church in 1869, vigorously opposed the appointment of A.P. Stanley* as a Select Preacher to the university in 1872, and severely criticized the Prayer Book Lectionary of 1872.

G.S.R. COX

BURGOS, JOSE (1837-1872). Philippine-born Spaniard, theologian-curate at Manila Cathedral. From 1863 he led the nationalistic clergy in the Philippines,* and wrote a "Manifesto" for Filipino priests' rights. With Fathers Mariano Gomez and Jacinto Zamora he headed the reform committees.

BURIAL SERVICES. In earliest Christianity, actual burial practices followed the customs of Judaism with the exception that (because of the resurrection of Jesus) a more positive note was introduced in the funeral service. Greater care was taken for the body due to the conception of it as the "temple of the Holy Spirit." With persecution and martyrdom, Christian burial practices gave greater emphasis symbolically, liturgically, and spiritually to those who witness *(marturion)* to their faith with their lives.

Our earliest distinctively Christian cemeteries are in the area of Rome, "and as martyrs to the faith multiplied, such cemeteries became consecrated ground, and the tombs of the martyrs were ere long places of pious meditation and devotion" (*HERE,* IV, p. 456). In time, churches were erected either on gravesites or near them, and when the eucharistic celebration became the major focus of Christian worship and liturgy, it was easy to associate this worship with death. Supporting the practice of eucharistic ceremonies (Requiem Mass) were two developments: the Jewish tradition of saying prayers for the dead (*2 Macc.* 12: 40-46); and the ecclesiological developments which came to distinguish the Church Militant and Church Triumphant from the Church Suffering (said to be Christians in purgatory). Early Christian sources do not forbid cremation as a burial practice; however, the custom of burying in the "earth" was stressed as preferred (cf. Tertullian, *De Anima,* 51; Origen, *Contra Celsus,* 5.23, 8.30; Augustine, *De Civitate Dei,* 1.12-13). Partly because of the Christian doctrine of the Resurrection and Second Coming, as well as the positive emphasis upon martyrdom, funeral services in the centuries prior to the eighth were occasions of joy and celebration; however, from the eighth century black rather than white characterized funeral dress, and the liturgy gave great emphasis to prayers for speedy purification (from purgatory) and even deliverance from hell. The ceremonies themselves came to include Vespers, night before funeral; Matins and Lauds, the dirge during the night; and the Requiem Mass with prayers for absolution in the morning. At the graveside special committal prayers are offered. Protestant practice generally became less liturgical.

See also DEATH; E.K. Mitchell, "Death and Disposal of the Dead: Early Christian," *HERE,* IV, pp. 456-58; E. Bendann, *Death Customs: An Analytical Study of Burial Rites* (1930; rep. 1970). DONALD M. LAKE

BURKITT, FRANCIS CRAWFORD (1864-1935). Biblical and patristic scholar. After taking his degree at Cambridge, he settled there and was Norrisian/Norris Hulse professor of divinity from 1905 until his death. He did important work on the Syriac text of the NT and in 1904 published a two-volume edition of the Old Syriac Gospels, entitled *Evangelion da-Mepharreshe.* He contributed an important article on the text and versions of the NT in *Encyclopaedia Biblica* (vol. 4, 1903). His best-known work is *The Gospel History and Its Transmission* (1906). He followed Johannes Weiss* in rejecting the views of liberal Protestantism and making an eschatological interpretation of the mission and teaching of Jesus. In a small book, *The Earliest Sources for the Life of Jesus* (2nd ed. 1922), he referred to "the stormy and mysterious Personage portrayed by the second Gospel." He was influential in arranging for the publication in an English edition of Schweitzer's* famous work *The Quest of the Historical Jesus* (1910). Burkitt was a man of original mind with a gift for clarity and an extraordinary range of academic interests, After his death the *Journal of Theological Studies* set apart a whole issue as a memorial to him, dealing among other things with his work on Gnosticism, on Manichaeism, and on Franciscan studies.

See S.C. Neill, *The Interpretation of the New Testament 1861-1961* (1964), pp. 114f.; E.C. Ratcliff, "Francis Crawford Burkitt," in *JTS* XXXVI (1935), pp. 225-53. R.E. NIXON

BURMA. A land strongly Buddhist, but with various animistic hill tribes who number some five or six millions. It is one of the few countries that have recently excluded all foreign missionaries. Yet the Protestant Christian church is stronger there than in any other country of the southern Asiatic mainland in proportion to the population. It has roughly 800,000 adherents, nearly three per cent of the population, though principally among the tribes.

British Baptists, including William Carey's eldest son Felix, entered Burma first from India, but did not long continue. The most important ongoing work was that of the American Baptists, begun in 1814 by Adoniram Judson* and his wife Ann. Judson was one of the first party of missionaries sent out by the American Board.* Shortly after his arrival in India, he changed his affiliation from Congregational to Baptist and severed his connection with the sending society. Unable to remain as a missionary in East India Company territory, he finally made his way across to Burma. There in spite of great suffering and severe opposition he laid the foundation of a flourishing Baptist work.

Judson worked primarily with the Burmese, who have never responded to the Gospel in large numbers. It was seven years before the first converts were baptized. But when George Dana Boardman was sent to Tavoy, he helped begin there a great movement among the Karen tribes that soon spread to other areas. Some groups of Karens became largely Christian and developed a strong, indigenous church. Later other tribes, such as the Chins, Kachins, and Shans, were also effectively reached. Other works, such as that of the Bible Churchmen's Missionary Society (Anglican), have made a significant contribution, but have been overshadowed by the Baptist achievement.

World War II and the independence of Burma in 1948 failed to open up new opportunities. Internal strife forced the Buddhist government to make a few concessions to the Christian minority, but increasing restrictions on foreign missionary activity culminated in the exclusion of all missionaries in 1966. The church has continued to grow, however. In addition, some tribal Christians, such as the Lisu, have fled from Communist China to N Burma, where conditions are still unsettled.

BIBLIOGRAPHY: A. McLeish, *Christian Progress in Burma* (1929); R.L. Howard, *Baptists in Burma* (1931); G.A. Sword, *Light in the Jungle* (c.1954); C. Anderson, *To the Golden Shore* (1956); Tegenfeldt, *Through Deep Waters* (c.1968); H.R. Cook, *Historic Patterns of Church Growth* (1971). HAROLD R. COOK

BURNET, GILBERT (1643-1715). Bishop of Salisbury. Born in Edinburgh, he entered Marischal College, Aberdeen, in 1653, studying arts, law, and divinity, and seldom working less than fourteen hours a day. After continental travel he was episcopally ordained and became minister in 1665 of Saltoun, East Lothian. While there he published *A Memorial of Diverse Grievances,* which attacked both bishops and clergy for the low moral state of the land—and nearly led to his deposition and excommunication. In 1669 he took the chair of divinity at Glasgow, a post to which he gave characteristic zeal. His moderation in an age of extremes, however, was resented by both Presbyterian and Episcopalian parties and (having twice refused bishoprics offered for political reasons) finally he resigned and settled in London, where he became chaplain of the Rolls Chapel and lecturer of St. Clement's (1675-84). Initially a royal chaplain, Burnet rebuked Charles II for his way of life, was subsequently dispossessed by him, was outlawed by James II (1687), and became adviser and staunch supporter in Holland of William of Orange. Appointed bishop of Salisbury in 1689 under William III, he was a faithful counselor to high and low. He preached the sermon at William's coronation, and attended his deathbed. Burnet's other writings include a *History of the Reformation in England* (3 vols., 1679-1714); *Exposition of the Thirty-Nine Articles* (1699); and *History of His Own Time* (1723-34). J.D. DOUGLAS

BURNEY, CHARLES FOX (1868-1925). Anglican Bible scholar. Educated at Merchant Taylors School and Oxford, he was ordained, lectured in Hebrew at Oxford (1893), and was university librarian (1897-1908). He lectured on the Septuagint at Oxford, where in 1914 he became Oriel professor of the interpretation of Holy Scripture. His many publications include *Outlines of Old Testament Theology* (1899), *Notes on the Hebrew Text of the Books of Kings* (1903), *The Book of Judges* (1918), and *Poetry of our Lord* (1925).

BURNS, WILLIAM CHALMERS (1815-1868). Scottish missionary to China. Son of a Forfarshire minister, he was educated at Aberdeen University and was licensed to preach by the Church of Scotland in 1839. He wished to go to the mission field, but for some years delayed his departure to engage in remarkably fruitful evangelism in Scotland, Ireland, and Canada. Finally in 1846 he went to China as agent of the Presbyterian Church of England. He began the patient study of Chinese, adopted native dress, and endured years that showed little outward response to his labors. Nevertheless he laid the foundations for the main centers of English Presbyterian work in China, both in the south around Amoy and in the north in Manchuria. In the 1850s Burns had been an inspiration and help to the young Hudson Taylor.* He translated into Chinese *Pilgrim's Progress* and some hymns. He died in a remote spot he had chosen to visit because of its destitution.

LESLIE T. LYALL

BURROUGH, EDWARD (1633-1662). Early Quaker. Born near Kendal, England, of godly parents, he heard George Fox* preach in 1652 and after disputing with him soon became one of his followers, even though disowned by his family. He was, with Howgill, Audland, and Camm, one of the pioneer group of Quaker preachers who dispersed in pairs from their native county to spread the light through Commonwealth England. After working in the northeast, he and Howgill went to London in 1654 and also paid short visits to Bristol and the eastern counties. A year later they went to Ireland. Burrough was esteemed by his companions as a controversialist and engaged in a pamphlet war with Bunyan* in 1656-57, although he had little learning or literary polish. At the Restoration he pleaded with Charles II on behalf of persecuted Quakers in New England. He was himself thrown into prison in 1662 for holding an illegal meeting, and he died there. JOHN TILLER

BURROUGHES, JEREMY (1599-1646). English Independent. Educated at Emmanuel, Cambridge, he began his ministry assisting Edmund Calamy* at Bury St. Edmunds, and in 1631 became rector of Tivetshall, Norfolk. He was suspended after Bishop Wren's visitation in 1636, whereupon he went to Rotterdam and became teacher in the English Congregationalist Church there which had William Bridge* for its pastor. Returning in 1641, Burroughes became famous as lecturer at Stepney and Cripplegate. With Bridge he was among the five Dissenting Brethren of the Westminster Assembly,* and presented the *Apologeticall Narration* to Parliament in 1644. He died of consumption before the Confession of Faith had been completed. JOHN TILLER

BURTON, EDWARD (1794-1836). Patristic scholar and church historian. Educated in Christ Church, Oxford, he studied on the Continent (1818-24) and on his return to England gained a reputation for wide and exact learning. In 1829 he was appointed regius professor of divinity at Oxford. His works include *Testimonies of the Ante-Nicene Fathers to the Divinity of Christ* (1826); editions of the works of Bishop George Bull, Pearson's* *Exposition of the Creed,* and Eusebius's *Ecclesiastical History;* and two volumes of lectures. He is chiefly remembered because his early death resulted in R.D. Hampden's succeeding him as regius professor, and the consequent campaigning by the leaders of the Oxford Movement* against Hampden.

JOHN A. SIMPSON

BUSCH, JAN (1399-1480). Member of the Brethren of the Common Life.* He became an Augustinian Canon in 1424. An enthusiastic clerical reformer, he supported Nicholas of Cusa,* who made him visitor of Augustinian houses in Saxony and Thuringia. He ultimately became prior of Windesheim where he had taught in early life. He wrote Windesheim's history (*Chronicon Windesheime*), an invaluable source for the Brethren's early years and for an account of the Dominican Martin Gabow's attack on the Brethren at the Council of Constance* (1414-18), and how they countered it most successfully.

BUSHNELL, HORACE (1802-1876). Congregational minister and theologian. Born in Connecticut, he graduated from Yale in 1827. After a brief experience in journalism he returned to Yale to study law. He passed his examinations and was ready for admission to the bar when during a revival at the college (1831) he suddenly decided to enter the divinity school. Here he encountered the vigor of N.W. Taylor,* champion of the New Haven Theology.* Bushnell's imaginative mind, however, was uneasy even with Taylor's modified Calvinism. He was attracted instead by Coleridge's *Aids to Reflection.* In 1833 he was ordained pastor of the North Church of Hartford, Connecticut. There he remained until 1859 when he was forced to resign because of ill health. Through his writings he sponsored three cardinal propositions, each elaborated in a major work. In *Christian Nurture* (1847) he argued that conversion should be educative rather than spontaneous or sudden. In *Nature and Supernatural* (1895) he contended that these may be harmonized. In *The Vicarious Sacrifice* (1866) he declared that Christ's atonement was an illustration of an eternal principle of love rather than a satisfaction by which God was reconciled to man.

BRUCE L. SHELLEY

BUTLER, ALBAN (1711-1773). Hagiographer. Educated at the Roman seminary at Douai from the age of eight (both his parents had died), he became professor first of philosophy and then of divinity, being ordained priest in 1735. For a time chaplain to the duke of Norfolk, he then went with Edward Howard, the duke's nephew, to Paris, where he completed his four-volumed *Lives of the Fathers, Martyrs, and other principal Saints*...—a monumental and wide-ranging collection of material involving some 1,600 lives. Its weakness lies in the lack of clear differentiation between history and legend. After all the travels to collect material and some time as a mission priest in England, he spent the last seven years of his life as president of the English College at St. Omer in France.

G.S.R. COX

BUTLER, JOSEPH (1692-1752). Bishop and scholar. Born at Wantage, Berkshire, son of a Presbyterian draper, he was sent to a dissenting academy where, inspired by Samuel Clarke's *Boyle Lectures,* he corresponded with the lecturer on philosophy and God's existence. He was intended for the Presbyterian ministry but, deciding not to proceed, he entered Oriel College, Oxford, in 1715, apparently with the aid of borrowed money. Here he read law followed by divinity. Though ordained in 1718, Butler did not become financially independent for eight or ten years. His eventual preferment followed a friendship with Queen Caroline; he became bishop of Bristol in 1738, and of Durham in 1750. He was never politically active, but throughout life maintained his early interest in philosophical questions.

The *Analogy of Religion,* on which Butler had worked for many years, appeared in 1736 at a time when the Deist controversy was at its height. It proved to be the greatest theological book of its age and did more to discredit Deism than any other book. It influenced many later writers, including Hume* and J.H. Newman.* Butler's argument is empirical, stressing *fact* in support of religion. The order we find in nature paralleled by the order we find in revelation, suggesting joint authorship by God, is his theme. The difficulties we encounter in Christianity, he held, bear a close analogy with those we encounter in nature, neither presenting more difficulties than the other. What we cannot understand in both spheres is due to lack of knowledge or limitation of intellect. Later in his *Sermons* he expanded the theme, arguing that man's psychological make-up, on which a rational ethical theory must be built, is consistently interrelated, thus affording an analogy to the constitution of the world at large. In discussing Christian evidences, Butler argues that evidence is based on probability, this being of three kinds: matters of speculation, matters of practice, and matters of great consequence. Religion is placed in the third class.

BIBLIOGRAPHY: There are many editions of the *Analogy,* e.g., H. Morley, 1884 (with biography); the best edition of the collected works is by W.E. Gladstone (3 vols., 1896); see too I. Ramsey, *Joseph Butler* (1969).

R.E.D. CLARK

BUTLER, JOSEPHINE ELIZABETH (1828-1907). Social reformer. She was first concerned with the promotion of educational facilities for women, but after 1866, first in Liverpool and thereafter in other places, notably Winchester, she supported refuges for all kinds of destitute women. Her life soon centered on an attempt to remove all forms of sexual exploitation of women. In general she objected to the hypocrisy of different standards of sexual morality for men and women, and led two major campaigns for reform. The first was for the repeal of the Contagious Diseases Acts of the 1860s which, in attempts to control disease, virtually provided a form of officially recognized prostitution in seaports and garrison towns and subjected many women to ignominious harassment. She formed the Ladies National Association for Appeal in 1869, and while she could not easily be described as an evangelical, her technique of parliamentary and extraparliamentary opposition owed much to the example of some of the earlier evangelical social reformers. Repeal was achieved in the 1880s. Her other great campaign was to promote legislation to raise the age of consent and to stamp out organized prostitution, especially the procurement of young girls. Her campaign received somewhat flamboyant support from W.T. Stead, editor of the

Pall Mall Gazette, and Bramwell Booth. In 1885 a famous procurement was carried out and publicized by Stead. Legislation, including the raising of the age of consent to sixteen, which had long been filibustered, was quickly passed. Criminal prosecutions of Stead and Booth followed, and Stead was imprisoned.

See biography by M.G. Fawcett and E.M. Turner (1927). R.H. CAMPBELL

BUTZER, MARTIN, see BUCER

BUXTEHUDE, DIETRICH (1637-1707). German composer. For many years organist of St. Mary's Church in Lübeck, he gained wide fame for the *Abendmusiken* (performance of church music on the five Sundays before Christmas), which he instituted in 1673. His choral and organ music is of outstanding merit. J.S. Bach* journeyed to Lübeck in his youth to hear Buxtehude's performances and was greatly influenced by him. The seventeenth century produced a great many Lutheran composers of merit, many of whom deserve attention; Buxtehude stands out among them for the vigor and imagination displayed in his chorale-preludes for organ, and for the beauty of the melodic line in many of his vocal compositions. His music is less difficult than Bach's for the average church choir today. J.B. MAC MILLAN

BYE PLOT (1603). William Watson, a Roman Catholic secular priest, supported James I's accession, believing that he had promised withdrawal of the recusancy fines. Disillusioned when they continued, he plotted to capture James, in company with another priest, William Clarke, and two dissatisfied Protestants, George Brooke and Lord Grey of Wilton. The Jesuits, anxious to discredit the seculars, revealed the plot to James who gratefully relieved Catholics of payment of fines for a period. Grey died in prison, the others were executed.

BYRD, WILLIAM (1543-1623). British composer. One of the greatest of all British composers, he remained a staunch Catholic in the England of Elizabeth and James I. He composed in practically all the musical forms of the day, sacred and secular. In spite of his religious affiliation, he was active at court and composed important music for the new Anglican rite. Although the chronology is not wholly clear, he has a good claim to be considered the father of the Anglican anthem. His beautiful "Christ rising again," for two sopranos, chorus, and instrumental accompaniment (1589)—although it may not have been intended for church performance—is a prototype of the *verse* anthem (i.e., with solo parts), which became and remained popular with Anglican composers. His magnificent "Great" Service is a landmark in Anglican music. It includes canticles for Morning Prayer, Evensong, and Holy Communion, set for the traditional English cathedral choir of *decani* and *cantoris* (the two groups of singers who face each other across the chancel). In addition to anthems and services, he wrote a very large quantity of excellent Latin music for the Catholic rite, including three complete Masses and two large volumes of *Gradualia.* A complete edition of his music in twenty volumes has been edited by Canon E.H. Fellowes, who also wrote a biography (1948). J.B. MAC MILLAN

BYROM, JOHN (1692-1763). English poet. Educated at Merchant Taylors School, he became a fellow of Trinity College, Cambridge and subsequently lived in Manchester. He is now known for his hymn "Christians, awake!" but his real claim to fame lies in his devotion to the ideas of the mystical writer, William Law.* Many of Byrom's poems are versifications of Law. Hence it is not surprising to find his exaltation of divine love, especially in the exemplary character of Christ's death, and—what was so uncharacteristic for the eighteenth century—his awareness of the activity of the Holy Spirit; hence also his frequent spiritual application of natural law and his frank avowal of "enthusiasm," the spontaneous response, in preference to the rational. ARTHUR POLLARD

BYZANTINE CHURCH, see EASTERN ORTHODOX CHURCHES

BYZANTINE TEXT. The name given to the form of text of the Greek NT to which the great majority of MSS dating from the Byzantine period and since bear witness. It underlies the "Textus Receptus" and therefore also the earlier English versions of the Bible. It is now thought to be due to a (perhaps lengthy) process of revision by which a generally standard text form arose. It is more closely related to the Syrian family of MSS than to any of the other families. This is marked by a tendency to conflate the shorter readings of earlier MSS, to harmonize differences, and to provide a smoother literary style. Because of this secondary character, the text is a much less reliable witness to the original text of the NT than are some of the other text families. R.E. NIXON

BYZANTIUM, see CONSTANTINOPLE

C

CABALA, see KABBALAH

CABASILAS, NIKOLAOS, see CAVASILAS

CABLE, MILDRED (1877-1952). Missionary to China. Having been delayed by the Boxer Uprising, she went to China in 1901 and with Eva and Francesca French built up a flourishing work at Hochow in Shansi province, including one of the first girls' schools in China. When principles of self-support began to be applied in 1928, they dealt the work a severe blow. Soon afterward, the indomitable trio of women decided to move to the extreme northwest of China, where they established their base in Suchow, the "City of Criminals." From there they made repeated evangelistic journeys through central Asia, traveling from oasis to oasis, and in the main cities of Turkestan (or Sinkiang), preaching and distributing Scriptures. LESLIE T. LYALL

CABRINI, FRANCES-XAVIER (1850-1917). Founder of the Missionary Sisters of the Sacred Heart. Born at S. Angelo Lodigiano, Italy, she was shy and physically delicate, but possessed of an iron will and implacable spirit. She abandoned her elementary-school teaching career to fulfill her lifelong desire to be a missionary to China. Rejected because of poor health, in 1880 she founded her own women's missionary society. Her vast vision and bold faith give ample proof that she thoroughly believed her society's motto: Philippians 4:13. Ordered to New York by Pope Leo XIII, she began her work among Italian immigrants in 1889. Her boundless energy gave birth to schools, charitable organizations, and hospitals throughout North and South America, Asia, and Europe. Mother Cabrini was especially adept at procuring funds and aid from non-Catholic sources. Regarded as patron of emigrants and all displaced persons, she was canonized in 1946.

ROYAL L. PECK

CADBURY, HENRY JOEL (1883-). American NT scholar. Member of the Society of Friends, he taught at various eastern colleges and seminaries, then from 1934 to 1954 served as Hollis professor of divinity at Harvard. He was a member of the RSV translation committee, secretary for many years of the American Schools of Oriental Research, and president of the Society of Biblical Literature and Exegesis. His primary contribution to biblical scholarship has been in the area of Lucan research. His publications include *The Style and Literary Method of Luke* (1919-20), *The Making of Luke-Acts* (1928), *The Book of Acts in History* (1955), and numerous essays in the same area. With K. Lake* he co-authored the commentary volume of *The Beginnings of Christianity: Part I: The Acts of the Apostles* (vol. IV, 1933) and also wrote many of the most valuable essays in the appendix volume of the same work. His books are models of careful scholarship.

W. WARD GASQUE

CAECILIAN. Bishop of Carthage, A.D. 311/12-c.340. As archdeacon he assisted Bishop Mensurius in checking extravagant devotion to confessors and martyrs during the Great Persecution. His rigorist critics, joined by disaffected clergy when he was elected bishop, called in the Numidian bishops who declared his consecration invalid because an alleged *traditor* (see DONATISM), Felix of Apthungi, had participated and because Caecilian had neglected the confessors. They proceeded to create a counter bishop. The dissidents protested against Constantine's assigning his relief measures of 312-13 to Caecilian, but at synods at Rome in 313 and, after further (Donatist) appeals, Arles in 314 and in Constantine's own inquiries in 315-16, Caecilian was completely vindicated. He was the only African bishop at the Council of Nicea in 325. D.F. WRIGHT

CAEDMON (d. c.678). English poet. He was a cowherd who suddenly, according to Bede's *Ecclesiastical History* (IV. 24), was endowed with the gift of poetry and composed the short piece known as Caedmon's Hymn. Taken before Hilda,* abbess of Whitby, he composed other verses, as a result of which she persuaded him to enter the monastic life. He is said to have written on the Genesis story, the Exodus, Incarnation, Passion, Resurrection, and Ascension, as well as other topics. Of surviving Old English poetry, only the Genesis A text of the Junius MS is now thought possibly to have been his, and there are considerable doubts even about this. His hymn is written, in the four-stressed alliterative meter, with any number of unstressed syllables, characteristic of much Old English verse. ARTHUR POLLARD

CAESAREA (Palestine). A city built by Herod the Great on the site of Strato's Tower, about halfway between Joppa and Dor, on the Mediterranean coast. Named in honor of the Roman emperor, Caesar Augustus, it was the Roman metropolis of Judea and the official residence of the Roman procurators and Herodian kings. The city was built between 22 and 10 B.C., on an insignificant site, though an important location. It lay on the coastal route between Egypt and Syria, but the harbor of Strato's Tower was a minor one, unsheltered and never more than an insignificant anchorage. Yet here, unrestrained by earlier constructions or by Jewish sensitivities, a Greco-Roman masterpiece was built, surpassing the engineering feats of Masada. A circular, artificial harbor was created larger than Piraeus and a major seawall built. The city was laid out on a grid, the main streets orientated to the harbor, linking it with a magnificent theater, forum, and amphitheater that overlooked the sea. A major aqueduct was built that brought a copious water supply from springs in the hills several miles away; an intricate drainage system underlay the streets. One hundred fifty years later, Hadrian doubled the size and capacity of the aqueduct, possibly as a result of an earthquake that damaged the city's monuments. Since 1960 air surveys have plotted the outline of the harbor, and excavations of the forum, theater, and aqueduct have been made. But most sections of the city have not yet been excavated, presently preserved beneath an extensive golf course.

The allusions to Caesarea in the Book of Acts are important. Philip the Evangelist brought Christianity to his home city and here later entertained Paul and his companions (21:8). Here also dwelt Cornelius, and this was the locale of his conversion (10:1, 24; 11:11). It was in this cosmopolitan city of Jews and Gentiles that Peter gained his first insight (10:35) of the divine kingdom that has no discrimination of peoples as "clean" or "unclean." Pontius Pilate, the procurator, lived here, and in 1961 Italian archaeologists found a stone inscribed with his name. Paul made Caesarea the port of his landing when he returned from his second and third missionary journeys (18:22; 21:8), and it was to Caesarea he was sent for trial by Felix (23:23-33). He made his defense here before Festus and Agrippa and sailed from here in chains to Rome (25:11). It was difficulties between Jews and Gentiles at Caesarea that led to the Jewish revolt of A.D. 66 which ended in the destruction of Jerusalem in A.D. 70. It subsequently became an important Christian center, and it was the home of the church father Eusebius.

See F.M. Abel, *Géographie de la Palestine,* II (1938), pp. 296f. JAMES M. HOUSTON

CAESAREAN TEXT. The name given by B.H. Streeter* (*The Four Gospels,* 1924) to a family of MSS of the Greek NT related to the text which Origen used at Caesarea. Its leading representatives are the Koridethi* MS and the two families of minuscules—fam. 1 and fam. 13. This text seems to have been a sort of compromise between the two previous "Western" and "Alexandrian" texts, and because of its consequent similarity in certain points to the Syrian text, its separate existence long escaped detection. K. Lake,* R.P. Blake, and S. New suggested that it probably first came into being in Egypt and was brought by Origen to Caesarea. From there it seems to have been taken on to Jerusalem, to the Armenians (who had a colony in Jerusalem at an early date), and then to the Georgians (the Koridethi MS was found in Georgia). It has therefore been questioned whether the Caesarean text should be thought of as something as distinct as the Western or Alexandrian texts. It is certainly more mixed and less homogeneous. But the work of Streeter has helped to open up an important new line of investigation into the history of the NT text.

R.E. NIXON

CAESARIUS OF ARLES (470-543). Bishop of Arles. Born at Chalons and educated in the monastery of Lérins, he epitomized the virtuous monk-bishop for forty crucial years, and by his legislation at his frequent church councils reformed the conduct of both secular and monastic clergy, writing rules for monks and nuns. An indefatigable preacher, he ruled that his clergy should preach in town and country frequently, simply, and briefly (his own practice was the fifteen-minute sermon). He fostered daily attendance at worship, congregational singing and memorization of Scripture, and involving laymen in the administration of funds. At the Second Council of Orange in 529 his representative countered the Semi-Pelagians* with his statement concerning prevenient grace and baptismal regeneration, and his denial of predestination to condemnation. MARY E. ROGERS

CAESARIUS OF HEISTERBACH (c.1170-c.1240). Author and prior of the Cistercian House of Heisterbach, near Cologne. He was educated at Cologne, where he received a good grounding in the classics and the Fathers, and acquired a fluent Latin style. Entering the monastery at Heisterbach in 1198 or 1199, he employed his talent in works prepared for the novices under his care, most notably the *Dialogus miraculorum* (c.1223), a delightful collection combining detail about contemporary life in the monastery and in the empire with credulous tales of witches, *incubi,* and *succubi.* One of the most popular writers of the thirteenth century, he wrote also a second compilation of miracles in eight books, of which only three are extant: a *Catalogus episcoporum Coloniensium;* an admirable biography of Engelbert, bishop of Cologne murdered in 1225; and sermons intended for a monastic audience, with copious scriptural and historical references. With true Cistercian zeal he criticized the ecclesiastical abuses he observed, including those of the confessional system. MARY E. ROGERS

CAESAROPAPISM. A system whereby supreme authority over the church is exercised by a secular ruler, so even doctrine is subject to state control. The term is applied chiefly to the authority exercised by the Byzantine emperors over the Eastern Church during the sixth to tenth centuries. Similar terms are "Byzantism" and "Erastianism."*

CAINITES. A dissolute Ophite* Gnostic sect (c.175-225), the Cainites believed the Creator to be so evil that his laws should be inverted and his recorded enemies (Cain, Esau, Korah) commended. For example, the mastery of Cain over Abel showed the impotence of the Creator. Judas Iscariot was commended, by some for opposing Christ, by others for facilitating His soteriological work. Their writings included a "Gospel of Judas" and an account of the revelations to Paul in the third heaven.

CAIRD, EDWARD (1835-1908). Scottish philosopher. One of the leading representatives of the Neo-Hegelian movement in British philosophy during the latter part of the nineteenth century. Previously professor of moral philosophy at Glasgow, he succeeded Benjamin Jowett* as master of Balliol College, Oxford, in 1893. Caird accepted the then prevailing concept of the progressive evolution of thought, and his own idealistic religious philosophy was expressed in his two important works, *The Evolution of Religion* (1893) and *The Evolution of Theology in the Greek Philosophers* (1904). He produced two important works on Kant* and a monograph on Hegel.* JOHN A. SIMPSON

CAIRD, JOHN (1820-1898). Scottish theologian and philosopher, older brother of Edward.* Born in Greenock and educated at Glasgow University, he was ordained and became minister successively at Newton-on-Ayr (1845), Lady Yester's, Edinburgh (1847), Errol (1849), and Park Church, Glasgow (1857). He was appointed to the chair of theology at Glasgow in 1862 and became principal of the university in 1873. Earlier he had preached before Queen Victoria from Romans 12:11 what Dean Stanley called "the greatest single sermon of the century," which was published and translated into several languages. His writings include *Introduction to the Philosophy of Religion* (1880), which contained, it was said, the essence of Hegelianism (see HEGEL) as applicable to the Christian religion, and *The Fundamental Ideas of Christianity* (2 vols., 1899).
J.D. DOUGLAS

CAIRNS, JOHN (1818-1892). Scots divine. The outstanding leader of the United Presbyterian Church of Scotland of his time. Brought up among the Seceders, the son of a border shepherd, he was ordained at Berwick-on-Tweed in 1845, where for thirty years he sustained a remarkable ministry. Having refused numerous calls to influential city pulpits and to university and college chairs, he moved to Edinburgh in 1876 to become professor of systematic theology and apologetics, and later also principal, in the U.P. divinity hall. He made his mark on the religious and public life of Scotland, alike by personality and character, and "his death called forth a manifestation of public feeling such as does not occur twice in one generation." D.P. THOMSON

CAIUS, see GAIUS

CAJETAN (Gaetano di Tiene) (1480-1547). Founder of the Theatine Order.* A noble of Vicenza, he was made protonotary apostolic by Julius II in 1505, ordained priest in 1516, and became a member of the Oratory of Divine Love. Concerned about the state of the secular clergy, he conceived the idea of a community of secular priests living together under the three vows while engaging in pastoral work. With Pietro Caraffa (later Paul IV) he began the Theatine Order, aimed at reforming the church from within, which played a part in the Counter-Reformation.

CAJETAN, TOMMASO DE VIO (1464-1534). Dominican cardinal and philosopher. Born of noble stock and studious by nature, he defied his parents' wishes and entered the Dominican Order* before he was sixteen. He studied at Naples, Bologna, and Padua and established a reputation through lectures and writing (he is credited with some 115 works). Appointed to the chair of metaphysics at Bologna, he enhanced his reputation through attacking the prevailing humanism. His celebrated work *De Ente et Essentia* was directed against Averroism.* He became a recognized exponent of Thomas Aquinas.* His defense of the power and monarchical supremacy of the pope at the Pseudo-Council of Pisa (1511) increased his favor at Rome. Through Ferdinand of Spain he was successful in sending the first Dominican missionary for the conversion of the natives of America. He was created cardinal in 1517. His greatest disappointment was his failure to persuade Luther to recant when they met on three successive days in Augsburg in 1518. Cajetan was one of Luther's most competent opponents.
GORDON A. CATHERALL

CALAMY, EDMUND (1600-1666). Puritan divine. Educated at Pembroke Hall, Cambridge, he entered the church and was successively bishop's chaplain (at Ely), vicar of St. Mary, Swaffham Prior, and then lecturer at Bury St. Edmunds in 1626. At first he regarded "ceremonial" as neutral, but later strongly opposed Laudian policy. In 1639 he was elected to the perpetual curacy of St. Mary Aldermanbury. During the controversy over divine-right episcopacy in 1640/41 Calamy contributed, as one of the authors who wrote under the pseudonym of Smectymnuus, to the Presbyterian reply. Later he was prominent in the Westminster Assembly.* He opposed the execution of Charles I and remained quiet during the Commonwealth and Protectorate. After welcoming back Charles II, he was a member of the Savoy Conference* (1661), and a leader of those who wanted a broadly based national church. He was ejected in 1662 and imprisoned briefly in 1663 for disobeying the Act of Uniformity.* His last years were spent in quiet retirement. His son, Edmund the Younger, was also an ejected minister; his grandson, also Edmund,* was the historian of early Nonconformity.* PETER TOON

CALAMY, EDMUND (1671-1732). Historian of English Nonconformity*; grandson of Edmund Calamy* (d.1666). After preliminary education in several private homes and schools, he went to the University of Utrecht in 1688. On his return in 1691 he was able to read privately in the Bodleian Library, Oxford. On 22 June 1694, with six ejected ministers taking part, seven young men of whom Calamy was one were publicly ordained in London to the nonconformist ministry. He then assisted Daniel Williams at the meeting house in Hand Alley, Bishopsgate, before he succeeded Vincent Alsop as pastor of the congregation that eventually moved to Princes Street, London. Though he did much traveling and preaching, he is best known for his work as a writer of biographic accounts of early Nonconformist ministers with Richard Baxter* as the central figure. In four volumes published between 1702 and 1727 he continued the work begun by Baxter in his *Reliquiae* (ed. M. Sylvester, 1696). In 1775 Samuel Palmer published an improved edition of Calamy's biographical studies; a further improved edition appeared in 1802-3, and then in 1937 A.G. Matthews published his *Calamy Revised.* Apart from his historical writings, Calamy defended the orthodox doctrine of the Trinity (1722) and the principle of Nonconformity (1703-4).

PETER TOON

CALCED (Lat. *calceus,* "shoe"). A term applied to certain religious orders who wear boots or shoes to distinguish them from other branches of the order who go barefoot or in sandals. Thus the Calced Carmelites, the unreformed branch of the order, wear shoes; the Discalced Carmelites of the Teresian reform wear sandals.

CALDERWOOD, DAVID (1575-1650). Scottish minister and historian. Said to have been born in Midlothian and educated at Edinburgh University, he was ordained minister of Crailing in 1604. He soon showed himself a strong opponent of James VI's policy of imposing episcopacy upon the Church of Scotland, for which stance he was confined to his parish and debarred from attendance at church courts. After standing up to the king in a personal confrontation he was deprived, imprisoned, and in 1619 ordered to leave the country. He spent several years in Holland and there produced *Altare Damascenum* (1623), a superb statement of the church's doctrine, ministry, and worship, which refuted prelacy and greatly encouraged Presbyterians. His name is usually associated, however, with his *History of the Kirk of Scotland* (1678 in folio; 8 vols., 1842-49), to complete which he was granted an annual pension by the general assembly. J.D. DOUGLAS

CALENDAR. Primitive man measured his calendar by the cycle of recurring natural phenomena which he observed, such as the alteration of day and night and the phases of the moon. The calendar in use in NT times was the Julian, based on the Roman republican calendar. By 46 B.C. the republican calendar had grown out of step with the seasons to the extent of three months, and the seasons were no longer in proper relationship with the calendar months. Julius Caesar instituted a four-year cycle, the first three having 365 days and the fourth 366, the additional day being placed in February. The fourth years were known as bissextile years. The modern term "leap year" is derived from the Old Norse *hlaupar.* In the Julian calendar each year was eleven minutes fifteen seconds too long, a fact which was significant only over a long period. After several delays the new Gregorian calendar was promulgated by Gregory XIII in 1582. The leap year rules were altered to deal with the fault in the Julian calendar, and ten days were omitted from 1582 to balance the accumulated error. The calendar is now correct to within one day in 20,000 years. The Gregorian calendar, now used for civil purposes throughout the world, was not generally adopted immediately. Great Britain did not adopt it until 1752, and the Russians did not do so until the rise of the Soviet government in 1917. The Orthodox Church has not adopted it, with the result that its year is now thirteen days behind the Gregorian year. The Gregorian calendar restored New Year's Day to 1 January. Formerly it was 25 March, the supposed anniversary of the Annunciation. The date of Easter is calculated in reference to the epact (the age of the moon at the beginning of the year), and the rules laid down by the Council of Nicea are largely adopted. Other Christian festivals have fixed dates.

JAMES TAYLOR

CALFHILL, JAMES (1530?-1570). Anglican theologian. Educated at Eton and King's College, Cambridge, he became in 1548 one of the first students of Henry VIII's renewed foundation of Christ Church, Oxford. Although an M.A. by 1552, he preferred to wait for the return of Protestantism under Elizabeth before proceeding to his B.D. and ordination in 1559-60. Calfhill was quickly preferred: canon of Christ Church (1560), prebend of St. Paul's (1562), Lady Margaret professor of divinity at Oxford (1564). He was nominated bishop of Worcester in 1570, but died before consecration. He preached two sermons at Bristol in 1568 against the views of Cheyney, bishop of Gloucester and Bristol. His most important published work was *An Answer to the Treatise of the Cross* (1565). JOHN TILLER

CALIXTINES. The moderate party of the Hussites, also called the Utraquists. The name is derived from the Latin word for cup *(calix),* indicating their demand that communicants should also receive the wine in the sacrament. The Calixtine program was formulated in 1420 in the Four Articles of Prague and included as a major tenet Communion for the laity in both kinds. For a time they were united with the Taborites,* the most radical group among the Hussites, but they eventually came to an agreement with Rome in the Compacts of Prague (1433) which conceded their demand for Communion in both kinds. The agreement was later repudiated by the pope, but the Calixtines survived as a semiautonomous Bohemia national church until the Reformation, when they merged with the Protestant movement in Bohemia.

See also UTRAQUISM. RUDOLPH HEINZE

CALIXTUS, GEORGE (1586-1656). Early ecumenist. He suffered the fate of many irenic writers in being suspected and rejected by the factions which he sought to reconcile. Born at Medelby in Schleswig, he studied philology, philosophy, and theology at the University of Helmstedt and elsewhere. He became acquainted with the leading Reformers through his travels in Holland, England, and France. In 1614 he was appointed professor of theology in Helmstedt, and became the most influential representative of the school of Melanchthon.* The great aim of his life was to attempt the reconciliation of divided elements in the church by getting rid of unimportant differences and concentrating on the fundamental articles of belief. His books, including *Epitome Theologiae*, *Theologia Moralis*, and *De Arte Nova Nihusii*, aroused the antagonism of Roman Catholics, who felt they were directed against them; but they were also rejected by orthodox Lutherans, who detected in them leanings towards Romanism. The Conference of Thorn* brought him into further difficulties when he was accused of Calvinist leanings. His dispute with the Lutherans (the "Syncretistic Controversy") lasted for many years. HUGH J. BLAIR

CALLISTUS I (d. c.222). Bishop of Rome from 217. According to his enemy Hippolytus, Callistus had been a slave, deported to Sardinia for fraud and released by Marcia, concubine of Commodus. Zephyrinus, bishop of Rome (198-217), put him in charge of the Roman clergy and the cemetery now called San Callisto. As bishop, Callistus excommunicated Sabellius and maintained a moderate policy on discipline—no sin was unforgivable; married clergy were acceptable; second baptism was permitted. This charitable policy he defended biblically from Romans 14:4 and the parable of the wheat and the tares (Matt. 13:29-30). Further, he allowed women of high rank to live *in contubernium* with slaves or free men. Roman law forbade marriage between certain social classes. Callistus sought to facilitate partnerships which were unrecognized by state law, but which were confined to one partner. Tertullian as a Montanist* reacted with his *De Pudicitia*, and Hippolytus withdrew from communion and set up an antipope. G.T.D. ANGEL

CALLISTUS II (Calixtus) (d.1124). Pope from 1119. Gui or Guido, fifth son of Count William of Burgundy, became archbishop of Vienne (1088), papal legate in France, and cardinal. He used his position and personality to make Pope Paschal II repudiate concessions granted to Emperor Henry V. At a council in Vienne (1112), he denounced lay investiture, rejected Paschal's decree (1111) that clergy should surrender their temporalities in return for Henry's relinquishment of investiture, and excommunicated Henry. On becoming pope he had to face an antipope, Gregory VIII (Burdinus), set up by Henry. After negotiation at the Concordat of Worms (1122), Henry abandoned Gregory VIII and the right to invest prelates with ring and staff. Ecclesiastical elections remained under imperial influence. Callistus convoked in the Lateran the first ecumenical council to be held in the West (1123). This confirmed the Concordat of Worms and issued decrees against clerical marriage and simony. Callistus gave judgment for the independence of York in the dispute between the sees of York and Canterbury. J.G.G. NORMAN

CALLISTUS III (Calixtus) (1378-1458). Pope from 1455. Spanish by birth, Alphonso de Borgia studied and taught law at the university of Lérida, where he was cathedral canon before becoming a jurist to King Alphonso V. For reconciling Alphonso with Pope Eugenius IV he was made bishop of Valencia (1429) and cardinal (1444). Chosen pope as a neutral, he maintained the balance of power between the Colonna and the Orsini. His major achievement was to organize a crusade against the Turks to recover Constantinople. In this objective it failed, though it forced the Turks to lift the siege of Belgrade (1456) and defeated their fleet at Metelino (1457). Callistus instituted the Feast of the Transfiguration to commemorate the Belgrade victory. A man of austere life, he was yet a renowned nepotist, making his nephew Rodrigo de Borgia (later Pope Alexander VI) cardinal and generalissimo of the papal forces.

There was also an antipope Callistus III (John of Struma) (1168-78). J.G.G. NORMAN

CALOV(IUS), ABRAHAM (1612-1686). German theologian. Born at Mohrungen, he was largely self-taught in his youth, but graduated from the University of Königsberg. Here he taught for a time, moving later to Rostock and Danzig, and in 1650, at the invitation of the elector, to Wittenberg, where he spent the rest of his life. A strenuous defender of rigid Lutheran orthodoxy, he attacked the Syncretistic School of Helmstedt and his Königsberg followers, and later wrote against Calixtus* and his school. He assailed Roman Catholicism, Arminianism, Socinianism, Pietism, and Calvinism, even intervening in the internal controversies of the Calvinist divines. He drew up the *Consensus repetitus fidei verae Lutheranae* in an attempt to exclude all Syncretists from the Lutheran Church, but the state authorities were by now tired of controversial divinity and imposed a silence so thorough that Calovius's own account of the Syncretistic Controversy (1682) was published anonymously. He wrote against the liberal critical views of Grotius* in *Biblia Illustrata* (1672-76) and against Boehme* (1684). Calovius always maintained that he deplored controversy and preferred constructive theology—and indeed his great twelve-volume *Systema locorum theologicorum* (1655-77) ranks with the work of Gerhard* as the leading expression of seventeenth-century Lutheran scholasticism. IAN SELLERS

CALVARY. The AV (KJV) rendering at Luke 23: 33 of the Greek *kranion*, "skull." The Vulgate has *locus calvariae.* The other gospels read "Golgotha" from the Aramaic *gulgoltâ.* The name may have been derived from its being a place of execution, or because it was a skull-shaped hill, though no mention is made of a hill in the text. There is division of opinion whether the place is to be identified with the site of the Church of the Holy Sepulchre or with the "Garden Tomb" also known as "Gordon's Calvary."

CALVARY HOLINESS CHURCH, see BRETHREN IN CHRIST

CALVERT, JAMES (1813-1892). Wesleyan missionary to the Fiji (Cannibal) Islands. Born in Pickering, England, he was converted at eighteen while apprenticed to a printer. In 1837 he was accepted for missionary service and sent to Hoxton Theological Institution. After six months there, his studies were interrupted by an urgent appeal for help from Fiji. He, John Hunt,* and Thomas Jaggar arrived there just three years after the work began. A quarter-century of labors witnessed a marvelous transformation of the islands. Returning to England in 1865, the Calverts seven years later were asked to help in the South African diamond fields. They spent more than eight years there before retiring again. In 1886-87 Calvert revisited Fiji and was able to appreciate what fifty years of missionary work had accomplished.

HAROLD R. COOK

CALVERT FAMILY, THE. George Calvert (1580?-1632), secretary of state under James I, resigned that post in 1625 when he became a Roman Catholic. Thereafter the king made him a baron and gave him large estates in Ireland. The baron spent most of his remaining years in colonial pursuits, first in trying to establish a colony in Newfoundland and later in seeking a grant farther south. The charter for Maryland was actually issued to his son Cecilius (1605-75) in 1632. He established the colony in 1634 with the double motive of providing a haven for the persecuted Catholics of England and carving out for himself a lucrative estate in the New World. Cecilius was not successful in attracting Catholics to his settlement, and from the beginning Protestants were in the majority. To prevent Catholic persecution he followed a liberal policy of religious freedom and in 1649 proposed the first toleration act in the New World, which quickly passed the Maryland assembly. Calvert's troubles with Puritans and his support by Cromwell do not warrant mention here. In 1692 the Crown set up a royal government in Maryland, and the Church of England was established, but the Calverts retained their territorial rights. When a Protestant Calvert (Benedict Leonard) became heir to the proprietorship in 1715, proprietary government was reestablished and continued until the Revolution, as did the Anglican establishment.

HOWARD F. VOS

CALVIN, JOHN (1509-1564). French Reformer. He was born 10 July 1509, in Noyon, Picardy, sixty miles NE of Paris, to Gerard Cauvin (Calvinus was the Latinized form of his name) and his wife Jeanne la France of Cambrai. John was the second of five sons. His father, a notary public, was primarily employed in the service of the bishop of Noyon, and as a result while John was still young he obtained for him two ecclesiastical benefices. The young Calvin at an early age became friendly with the sons of one of the local gentry, Joachim de Hangest, sire de Montmor, who suggested when his sons were going to Paris for further education that Calvin should accompany them. Gerard agreed. After spending a few months at the *Collège de la Marche*, John enrolled in the *Collège de Montaigu*. When Gerard, however, came into conflict with the bishop of Noyon he decided that his son should give up all thought of the priesthood. He therefore ordered him to study law at Orléans where Pierre de l'Estoile was teaching, and while there John also took lectures from Andrea Alciati, the humanist legal scholar, at Bourges. When his father died in 1531, however, he returned to Paris to continue his literary studies, although he did go back to Orléans for a term to complete his law course.

Although we know little about Calvin's conversion, we have information to indicate that he had frequent contacts with men of Protestant tendencies while a student. At the *Collège de Montaigu* he may have met John Major, the Scottish conciliarist, and at Orléans and Bourges we do know that he studied Greek under Melchior Wolmar, a humanist with strong Protestant leanings. A number of his friends at Orléans and his cousin Francis Olivetan were also moving in this direction. It may have been as a result of these influences, coupled with attendance in Paris at secret Protestant meetings, that Calvin, despite his later acknowledgement of his "obdurate attachment to papistical superstitions," became a Protestant. As a young man with very considerable ability as well as a reputation for learning, he soon became one of the leaders in the Protestant movement in Paris.

In April 1532 Calvin, typical of young humanist scholars of his day, published his first book, a commentary on Seneca's *De Clementia.* But soon after, he became caught up in the Reformation movement, which led him to concentrate his attention on biblical studies. When his friend Nicholas Cop* was elected rector of the University of Paris, John helped him to prepare his rectoral address delivered on 1 November 1533, which was an attack upon the church and a demand for reform along the lines advocated by Luther. The result was an explosion of anti-Protestant feeling which forced both Cop and Calvin to leave Paris. Although Calvin later returned for a short time, his reputation as one of *les Réformes* soon obliged him to leave again, and for the next three years he spent his time traveling in France, Switzerland, and Italy. During this period he also resigned his ecclesiastical benefices at Noyon.

Despite the necessity of being continually on the move to avoid arrest or persecution, Calvin had begun to use his pen in behalf of the Protestant faith. In 1534 he published his first religious work, *Psychopannychia,* an attack upon the doctrine of soul sleep after death. Shortly afterward, Olivetan's French translation of the Bible appeared with a preface by Calvin. Most important of all, however, in March 1536 he published in Basle a slim volume of seven chapters with the title *Christianae Religionis Insitututio,* prefaced by a letter to Francis I of France defending the Protestants against their calumniators. A short summary of the Christian faith, this work whose author was at the time virtually unknown soon became popular among Protestants as both an able exposition and a forthright apology for the new doctrines.

After spending a few more months wandering, Calvin accompanied by his brother Antoine and his half sister Marie headed for Strasbourg, where Protestantism had been officially accepted and which consequently would provide the peace necessary for his projected literary work. Owing to fighting between France and the empire, however, he had to travel through Switzerland via Geneva, where he planned to stay only one night. That stopover was decisive, for the Protestant preacher Guillaume Farel,* who had brought about a considerable measure of reform in Geneva, heard of the young scholar's presence. Farel thereupon demanded that Calvin stay to assist him in completing the work. Calvin at first refused, but when Farel threatened the curse of God upon him, he consented much against his will. But his residence in the city did not last long. He and Farel sought to introduce into a notoriously profligate society a measure of ecclesiastical discipline which only raised up enemies. When the two reformers refused to obey the civil government's demand that they accept the liturgy of Berne, their opponents used this as an excuse to force them out of the city. Farel moved to Neuchâtel, while Calvin on the invitation of Martin Bucer* set out once again for Strasbourg.

Probably some of Calvin's happiest years were spent in Strasbourg. Although constantly plagued with poverty, he seems to have enjoyed his life. The most important event personally was his marriage to Idelette de Bure, widow of an Anabaptist whom Calvin had converted to the Reformed position. She bore him one son, who lived only a few days. Shortly after his arrival in Strasbourg, Calvin became pastor of the French refugee congregation which he organized along what he believed to be NT lines. Especially important in this was his drawing up of a liturgy and the preparation of a psalm book made up of his own and Clement Marot's French metrical translations. At the same time he was busy preparing his commentary on Romans and taking part as a representative of Strasbourg in colloquies with Lutherans and Roman Catholics at Worms and Regensberg. From these activities his fame as a biblical scholar and theologian gradually spread.

He would probably have spent the rest of his life in Strasbourg had it not been for Cardinal Sadoleto's efforts to bring Geneva back under Roman control. After Calvin and Farel had departed from the city, no one arose to give the needed leadership in the church. The result was confusion and conflict. In this situation the supporters of the old regime felt it was a propitious time to undo the Reformers' work. To this end, in March 1539, Jacopo Cardinal Sadoleto,* a well-known humanist, wrote a letter urging the Genevans to submit to the pope. Since no one in Geneva seemed capable of replying, the letter was sent to Calvin, who dealt with it very effectively. About the same time, a change in Geneva's government put the control of the city in the hands of his friends, who invited him to return. Although he had no desire to do so, under the exhortations once again of Farel he finally agreed to go, reentering the city on 13 September 1541.

Although he realized only too well that Geneva, which long had a Europe-wide reputation for immorality, would be no easy community to reform, Calvin set about his task immediately. One of his first responsibilities was the revision of the city's laws, while at the same time he drew up a form of government for the church and revised his Strasbourg liturgy and psalter. Eventually (in 1559) he even succeeded in persuading the people that an academy, later to become a university, should be founded for training the youth for service in the commonwealth. In all this, his one great aim was to make Geneva a "holy city," conformed to the will of God. This meant a strict and sometimes harsh discipline of which most people, even Calvinists, would not approve today, but it had the effect of changing Geneva's character and of making it a power in the world of the sixteenth century.

Calvin's efforts to reform Geneva and the Genevans naturally led to internal conflict. Not all the inhabitants were Calvinists, and even some of those who agreed with him at times felt he carried his rigorous demands too far. The result was at times riots and disturbances aimed at eliminating him from the city. The final test came when Michael Servetus,* a Spaniard under sentence of death by the Inquisition* for denying the doctrine of the Trinity, came to Geneva, apparently to cause trouble. He was recognized, denounced by Calvin, and with the approval of the other Swiss Protestant cities, as well as the Roman Catholic authorities, burned at the stake (1553). Although during the sixteenth century thousands of Protestants suffered the same fate at the hands of Roman Catholic persecutors, Calvin has been constantly vilified for his part in this single execution.

While he held no government position, nor indeed even became a citizen of the city until invited to in 1559, Calvin undoubtedly dominated the whole community, by moral suasion rather than by any other means. Not only did he play a large part in the devising of a church government with wide powers of oversight over the population and in helping to make the city's laws more humane, but he also exercised a wide influence in other areas. He was largely responsible for establishing a universal system of education for the young, and he took a large part in arranging for the care of the poor and the aged. He sought to make Geneva a Christian commonwealth, in practice as well as in doctrine.

Naturally, from this endeavor Geneva gained a widespread reputation, particularly among persecuted Protestants throughout Europe. Situated at the crossing point of a number of important trade routes between the north and Italy, it had a strategic geographic position. But what was even more important, under Calvin's influence the city authorities threw open the gates to refugees who flocked in from all directions: France, Holland, England, Scotland, Germany, Italy, Spain, Hungary, Poland, and practically every other country of Europe. From Geneva these people often returned home as missionaries to spread the Gospel as they had learned it in Geneva. Out of these contacts, constantly maintained by a voluminous correspondence, Calvin wielded an influence far beyond the borders of the Genevan commune. He became the dominant figure of the Protestant Reformation in the middle of the century.

Equally, if not more, important than his personal contacts and letters were his more formal writings. During his lifetime he wrote commentaries on twenty-three books of the OT, including all the Pentateuch and all of the prophets, and on all of the NT but the Apocalypse. With his background of humanistic studies and his theological knowledge, these works have been influential in the church down to the present time. Besides preparing commentaries, he preached frequently, every day on alternate weeks, many of his sermons being taken down in shorthand notes which he may have revised, and then were published. The notes of others were lost until this century, but are now being published for the first time. Along with these labors he constantly produced pamphlets dealing with current topics affecting both Protestant thought and action.

Most important of all his writings, however, is the *Institutes of the Christian Religion*. Published originally in 1536, a book of six chapters, as a theological handbook for French Protestants, it was revised by Calvin five times, usually translating the original Latin version into French, by which he greatly influenced the development of the modern French language. By the time the definitive edition appeared in 1559, it had been so changed and enlarged that it was in four books with a total of seventy-nine chapters. This work quickly became disseminated widely in many different translations to form, except in countries where Lutheranism dominated, the systematic theology of the Reformation. And it has endured to the present, as is indicated by the numerous scholarly editions which have appeared recently in English, French, Japanese, and other tongues.

Idelette Calvin died in 1549, leaving her husband a sad and lonely man. He did not apparently ever think of remarrying, although he could well have done with the care of a loving wife, for he was not one who would take great care of himself. The result was that he suffered from stomach ulcers and similar troubles to the end of his life. Such weaknesses of the flesh nevertheless did not restrain him from working intensively almost up to the time of his death on 27 May 1564. At the age of fifty-four Calvin literally burned out in the service of God.

To many since his time, Calvin has been the epitome of rigor and cheerlessness in this life. They feel that he was a legalist who would exclude all joy from Christianity and would make it into an unyielding bondage. Yet if one really studies his works and the life of the man himself, this does not appear to be the case. He was a very human individual, as he reveals so frequently in his letters. True, he was intense in the service of God, to whom he offered his heart fully. Using all his undoubted gifts, he laid the groundwork for much of the Protestantism of the next four centuries. But his influence extended far beyond the borders of the church, as it did beyond the confines of Geneva, for many of his ideas in politics, aesthetics, science, and history became so interwoven in Western thought that we must recognize him as one of the great seminal minds, one of the formative factors in the development of Western culture and civilization.

BIBLIOGRAPHY: E. Doumergue, *Jean Calvin. Les hommes et les choses de son temps* (7 vols., 1899-1927; rep. 1969); J. Moura and P. Louvet, *Calvin. A Modern Biography* (1932); *idem*, *Calvin et l'institution Chrétienne* (1935); Imbart de la Tour, *Calvin. Der Mensch, die Kirche, die Zeit* (1936); F. Wendel, *Calvin, sources et évolution de sa pensée religieuse* (1950); E. Stickelberger, *Calvin, A Life* (tr. D.G. Gelzer, 1954); W. Niesel, *The Theology of Calvin* (1956); G. Harkness, *John Calvin, the Man and His Ethics* (1958); A. Bieler, *La Pensée Economique et Sociale de Calvin* (1959); J.T. Hoogstra (ed.), *John Calvin, Contemporary Prophet* (1959); J. Cadier, *The Man God Mastered: A Brief Biography of John Calvin* (1960); D.A. Erichson, *Bibliographia Calviniana* (rep. 1960); W. Niesel, *Calvin Bibliographie 1901-1959* (1961); J. Rilliet, *Calvin* (1963); G.E. Duffield (ed.), *John Calvin* (1966); A. Ganoczy, *Le Jeune Calvin: Génése et évolution de sa vocation reformatrice* (1966); R.W. Collins, *Calvin and the Libertines of Geneva* (1968); J.N. Tylenda, "Calvin Bibliography 1960-1970," *Calvin Theological Journal*, VI (1971), pp. 156ff.

W.S. REID

CALVINISM. This term comes out of the seventeenth century, largely in opposition to the teachings of Arminius* condemned by the Synod of Dort* in 1618. It had been used by Roman Catholics sometimes in the sixteenth century, but always in a pejorative sense. It is therefore a term that has been used in many different ways over the past three centuries, coming to have many different meanings, both good and bad. Consequently, one must understand its true meaning if one is to employ it properly.

The first problem involved in its interpretation is its relation to John Calvin* himself. He would not have accepted it as a good description of his doctrine and on one or two occasions made comments to this effect. He believed the doctrine he set forth was nothing more or less than the teaching of the Scriptures of the OT and NT. In his dedicatory epistle to Francis I of France with which he prefaced the first edition of his *Institutes of the Christian Religion* (1536), he made this quite clear, insisting he was writing to show that the doctrines espoused by the Protestants were entirely biblical. This thought reappears repeatedly in his commentaries and other writings.

Yet Calvinism is largely derived from Calvin's own interpretation and exposition of Scripture. He was a prolific writer who set forth clearly a system of doctrine which he believed he found in the Bible. Employing the most up-to-date techniques of biblical exegesis developed by the humanists of his day, he wrote commentaries on most of the books of the Bible, summing up his findings in succeeding editions of the *Institutes*, which grew from a small handbook of six chapters in 1536 to a large volume of seventy-nine chapters in the definitive edition of 1559. This work has been the textbook of Calvinism since that time, having been translated into many languages and expounded and explained by those calling themselves Calvinists.

What is the essence of Calvinism? Many have tried to answer this question in different ways, usually on the basis of their own particular theological or philosophical presuppositions. As Calvinism is a many-faceted structure of thought that seeks to interpret the whole of reality from a Christian perspective, to attempt to sum it up in a few words is extremely difficult. Nevertheless, in order to obtain something of an understanding of it, one must attempt some form of analysis and synthesis in order to reduce it to comprehensible size.

The formal principle of Calvinism is the Bible, the source of Calvin's doctrine. He and most of the other sixteenth-century Reformers held a very high view of the Bible, insisting it is the Word of God, bringing God's revelation to man in documents written under the inspiration of the Holy Spirit. They did not, however, foresee all the controversies that were later to arise over this doctrine and so did not develop all the various theories of revelation and inspiration formulated by Calvinists in the nineteenth and twentieth centuries. Yet they did hold very firmly to the view that the Bible is man's only infallible rule of faith and practice. In this all the leading Protestant Reformers were at one.

Because of this belief, Calvinism insists the Bible is the only source of man's knowledge of God and of His will and works. Although creation and providence do indeed reveal God's power and divinity, both nature and man have been so corrupted by sin that they cannot be an adequate means of God's self-manifestation. Furthermore, they do not reveal anything concerning God's redeeming love or action. Thus they are inadequate for the full knowledge of God which comes only through His direct revelation to man in the words and actions of the prophets, of the apostles, and above all others, of Jesus Christ, the Living Word Himself, as recorded in the Bible.

The Bible by revealing God also gives the true understanding and interpretation of man. It lays down first of all that man is God's creature, who is to fulfil the duties and responsibilities God has laid upon him. Thus the Bible simultaneously tells man what he is to believe concerning himself and what he is to believe concerning God. At the same time, it insists that because there is an absolute discontinuity between the being of God and man, as between Creator and creature, man's knowledge of God and His ways can never be more than partial, ultimately surrounded in mystery that even the Bible does not remove. Man, in seeking to understand the biblical revelation concerning himself and his relationship to God, must in the last analysis accept it in faith.

This, however, does not mean man then lapses into some form of Quietism or Mysticism. The Bible is the charter for Christian action or activism. First of all, in the matter of worship, the Scriptures are the final authority, for in them God tells man how he must approach Him. Furthermore, the Scriptures also inform man how he must live and conduct himself in this world, in relation both to its material resources and to other persons. Finally, the Bible is the inspired statement of man's ultimate purpose and aim in life as the creature of God. Thus in Calvinism the Bible holds an absolutely central position as the source of both Christian thought and action.

From the biblical teaching comes what we might call the material principle of Calvinism: the sovereignty of God. Some believe this is the real core of Calvinistic thinking, and to a certain extent it is. The Calvinist believes that the central thought in the Scriptures is that the Triune God, one God in three persons, is totally independent of all else and absolutely self-sufficient. Within the interrelation of the three persons of the Godhead, God is completely and fully expressed in every way. Man cannot by any means understand what this means, except that with regard to everything outside Himself, God is completely and fully sovereign. God has no correlates, but rather is completely and totally absolute.

Everything in the space-time universe, including space and time, therefore, exists only by the creative decision and providential action of God. He has made all things, which means everything in existence is different from and subordinate to Him. The Calvinist can never accept any idea that the space-time universe is a divine emanation, or part of God. Nor does he believe that once created, the universe (as Deists maintain) runs automatically by innate natural laws. The continued existence and operation of the universe, including the free actions of man, are sustained and determined from moment to moment by the mysterious and all-powerful providence of God. For a proper and ultimately true understanding of both natural science and history, therefore, the sovereign God must always be the ultimate point of reference and of interpretation. As Calvin would put it, all things must be seen "*sub specie aeternitatis*" (in the perspective of eternity).

In pursuance of His ultimate purpose God allowed man to sin, although man did so according to his own will and desire, alienating himself from God. At the same time, God in His grace purposed to redeem men from their sin and bring them to glory. Therefore, from the beginning of history two opposing principles have existed in conflict: sin and redemption, alienation and reconciliation. These two principles are revealed very clearly in OT history and come to full fruition in the redemptive work of Jesus Christ, the incarnate Son, on Calvary. Since that time the conflict has continued through the ages as God the Holy Spirit has effectually called His people out of the kingdom of this world into the kingdom of God, to be His people upon earth.

These people are those whom God has chosen in Jesus Christ from all eternity, not with any prevision of their faith or righteousness, but solely of His own free grace and love. No man would of himself turn in repentance and faith to God, because of the corruption of his sinful nature, unless God by the Holy Spirit should regenerate him that he might do so. Christ, therefore, died and rose again that His elect should be reconciled to God, who bestows upon them the gift of the Spirit who infallibly brings them to faith in Christ as Savior and Lord. When they have so experienced conversion, He then works constantly within them that they may grow in grace and in the likeness of Jesus Christ, to be more and more conformed to His image in this life. To Calvinism, man's reconciliation to God is all of God and of

His eternal and sovereign grace. Thus the elect can never be lost, but shall persevere until the very end.

For those who accept this position, the biblical principle of the sovereignty of God involves also a basic ethical principle. Because God is sovereign, the Lord and Creator of all, all men are responsible to serve Him in this life in all they do. It is God's sovereignty that makes man truly responsible. Moreover, God has from the beginning committed to man the responsibility of acting as the great prophet, priest, and king of creation. He is to interpret creation, as God's possession, to lead it in the praise and worship of God and to govern it for God. To this end God gave him the mandate to rule over, subdue, and replenish the earth. This involves both the development of its physical resources and the organization of man for this purpose and objective.

Because of his alienation from God, however, man has failed to meet his responsibilities, seeking to use the physical and human resources of creation for his own pleasure, ease, and glory. The result has been both the perversion and pollution of God's good creation. While man has compulsively developed creation and its riches, including his own abilities, he has usually tended to misuse them, even for the destruction of his fellowman. The Christian, on the other hand, recognizing his responsibility to God, should and often does see his duty as lying in the development and use of both the material creation and his own gifts for the benefit of society and for the glory of God. This is his vocation in life.

The final end or ultimate principle of Calvinism is, then, the glory of God. Creation and even redemption are not primarily for the satisfaction and pleasure of man. Evangelism, social service, and similar activities should not be thought of ultimately as being for man's benefit, but to glorify the sovereign Triune God. In the service of God upon this earth the Christian seeks to manifest God's majesty, power, and grace that he may glorify him in all things. He does not look upon the things that he is doing as something required of him merely as earthly activities, but as those which will redound to the praise of God through all eternity.

While this system of thought was made explicit by Calvin in his writings, it was further elaborated (often in a controversial setting) in the latter part of the sixteenth century, and partially summarized in the Canons of the Council of Dort (1618) in what are commonly known as the "Five Points of Calvinism": (1) total depravity of man; (2) unconditional election; (3) limited [particular] atonement; (4) irresistible grace; (5) perseverance of the saints. The Reformed Confessions drawn up after 1618 also express these doctrines, although they set them in the much wider context of God's universal sovereignty. Nevertheless, many noted theologians (J. Ussher, J. Davenant, J. Cameron, etc.) taught a doctrine of general redemption.

Originating in Geneva and France, Calvinism gradually spread along the Rhine Valley to Germany and Holland, along the Danube River Valley to Hungary and Transylvania and across the Alps into France, shaping and forming the Reformation as it took place in those various countries. From France and from Holland Calvinism soon spread to England and to Scotland. It largely dominated the thinking of the Church of England into the seventeenth century, forming the core of Puritan thought which was transplanted to New England. In Scotland, Holland, and France it was the basic doctrine of the Reformed churches, who took it not only to America, but to many other parts of the world, with the result that today Calvinistic churches are a worldwide phenomenon. Because of its all-inclusive nature, Calvinism has wielded a powerful influence on every aspect of Western man's life for the past four hundred years, even although its impact may sometimes have been unrecognized.

As we might expect, Calvinism's contribution has been most obvious in the fields of theology and Christian life and action. One could give a long list of theologians, preachers, and reformers over the past four centuries: John Owen, Thomas Boston, George Whitefield, William Wilberforce, the seventh Earl of Shaftesbury, Abraham Kuyper, Charles Hodge, B.B. Warfield, J. Gresham Machen, and many others who have held to a strongly Calvinistic position. Yet none of them have adopted the point of view that their religion was something separate from their life in the world. They saw Calvinism as something that overarched all of life, influencing every sphere of thought and action.

Calvinism has also from the beginning had a very considerable influence upon the development of natural science. Pierre de la Ramée, Ambroise Paré, Bernard Pallissy, Francis Bacon, John Napier of Merchiston, and others in the earliest days of the Scientific Revolution were Calvinists, and many scientists since the seventeenth century have held this theological position, believing that God by His providence upholds all nature according to its created law-structures, so that man may be able both to understand and to use it in this world.

From the time of John Knox in Scotland and Admiral Coligny in France through the Puritan Revolution in England in the seventeenth century, to Abraham Kuyper and Herman Dooyeweerd of the Netherlands and Emile Doumergue of France in the nineteenth and twentieth, Calvinists have also played a major part in seeking to develop and apply a Christian view of politics and the state. Believing that Christ is "Lord of Lords and King of Kings," they have sought to bring both subjects and rulers to recognize Him as the one to whom they are responsible. At the same time, they have insisted, as did Calvin, that despotism or oligarchy, because of man's sinful nature, leads only to oppression, but that democracy under law provides the only true political organization for freedom and liberty. Because of this point of view, Calvinism has provided much of the basis for modern constitutionalism.

In the arts also, Calvinism has had its effect. Not only did Calvin by use of French do much to establish that language on a firm foundation, but also his employment of Clement Marot, Theodore Beza, and others to prepare vernacular psalms for singing in the church service stimulated Protestant poetic interest. Under this influence vernacular psalms soon appeared in Dutch, English, and Magyar, and, significantly, the writing of poetry in general was encouraged. Milton's early works reflect this stimulus, as do the writings of men such as William Cowper, Willem Bilderdijk, and many others. In the visual arts, the so-called Little Calvinistic Masters of Holland in the seventeenth century and many others who followed them in France, England, and America were also strongly influenced by the Calvinistic viewpoint.

Usually Calvinism has been accused of originating modern exploitive capitalism because of its doctrine of vocation and its insistence upon hard work and moderation in all things. Max Weber, the German sociologist, followed by R.H. Tawney and Ernst Troeltsch and many others, has set forth this particular interpretation. Undoubtedly, there is a certain amount of truth in some of the contentions, i.e., that the Calvinist felt it was his duty to work hard and to live moderately to the glory of God. But the insistence that Calvinist acceptance of the propriety of the taking of interest on a business loan and of the rational approach to economic activity eventually led to exploitation of the worker, and so laid the basis for modern soulless capitalism, lacks historical evidence for its verification. So writers have pointed out that the opponents of Calvinism rather than the Calvinists favored and developed capitalism.

Over the past four hundred years Calvinism has known its ups and downs. Although weakened considerably by the influence of Enlightenment* rationalism, it experienced a considerable revival under the aegis of the evangelical revival in England and the Great Awakening in America in the eighteenth century. In the nineteenth century, however, it was attacked on two fronts. Not only did higher criticism* and scientism from one side oppose it most vigorously, but so too did Wesleyan and Quietistic evangelicalism from the other. As a result, Calvinists tended to become ingrown and frequently defensive. In the past two or three decades, however, they have regained some of their former confidence. With the founding of organizations such as the International Association for Reformed Faith and Action, the founding of journals holding a Calvinistic point of view, and the publication of an increasing number of books written from this perspective, it would seem Calvinism is perhaps experiencing a revival in a world that has lost most of its moorings.

BIBLIOGRAPHY: Calvin's writings have been published in many editions and languages. Some of the most useful in English are: *Commentaries on the Bible* (44 vols., 1948); *Institutes of the Christian Religion* (ed. J.T. McNeill, F.L. Battles, 2 vols., 1960); *Calvin: Theological Treatises* (ed. J.K.S. Reid, 1954); *Tracts and Treatises on the Reformation* (tr. H. Beveridge, add. by T.F. Torrance, 1959).

On descriptions and history of Calvinism, see: A. Dakin, *Calvinism* (1940); A. Ganoczy, *Calvin, Théologien de l'Église et du Ministère* (1964); S. Kistemaker, *Calvinism, Its History, Principles and Perspectives* (1966); J.T. McNeill, *The History and Character of Calvinism* (1954); H.H. Meeter, *Calvinism*, n.d.; D. Nauta, *Het Calvinisme in Nederland* (1949); A.A. Van Schelven, *Het Calvinisme Gedurende zijn Bloeitijd* (2 vols., 1943); C. Van Til, *The Case for Calvinism* (1964); A. Kuyper, *Lectures on Calvinism* (1931).

W.S. REID

CALVINISTIC METHODISM. The title was first given to those who in the eighteenth-century revival adhered to the doctrinal emphases of G. Whitefield.* It developed into a denominational differentiation with reference to the church in Wales which eventually emerged. More than twenty years before the conversion of either Whitefield or the Wesleys, Griffith Jones* of Llandowror had heralded the awakening in the principality with his evangelical preaching. He was soon to be supported by Howel Harris,* Daniel Rowland,* Howell Davies, and the hymnwriter William Williams* of Pant-y-Celyn.

The first Methodist Association in Wales met in 1742, thus anticipating Wesley's earliest Conference (1744). The societies were regarded as belonging to the Church of England, like those of the Wesleyan Methodists. But as opposition grew, separation became virtually inevitable. From 1763 onward Rowland was no longer permitted to exercise his parish ministry at Llangeitho, although there is no evidence his license was actually revoked by the bishop, as has been alleged. He preached in a meeting house erected by his sympathizers. It was largely on account of Harris's unswerving allegiance to the Church of England that steps toward formal secession were deferred until 1795.

The name of Calvinistic Methodists was also attached to other groups which owed their existence to the ministry of Whitefield. Those belonging to the Countess of Huntingdon's Connexion* fall into this category, along with what was known as the Tabernacle Connexion of Whitefield Methodists.

In 1770 the publication of an annotated Welsh Bible by Peter Williams led to a renewal of interest in the Scriptures, and in 1784 the work was extended to N Wales through Thomas Charles of Bala. In 1795 the protection of the Toleration Act (1559) was sought, while in 1811 the body was officially recognized as the Calvinistic Methodist Connexion and regular ordinations began. The *Confession of Faith*, containing forty-four articles based on the Westminster Confession* as "Calvinistically construed," was published in 1823, and the Connexional *Constitutional Deed* was ratified in 1826.

A ministerial training college was opened at Bala in 1837, with another for the south at Trevecka in 1842 (transferred to Aberystwyth in 1905). Until 1840 the Calvinistic Methodists supported the London Missionary Society, but in that year they started their own work in France and India. The constitution of the church combines features of both Presbyterianism and Congregationalism. Its membership in 1969 was 110,155.

BIBLIOGRAPHY: W. Williams, *Welsh Calvinistic Methodism* (1872); D.E. Jenkins, *Calvinistic Methodist Holy Orders* (1911); J. Roberts, *The Calvinistic Methodism of Wales* (1934); M.H.

Jones, *The Trevecka Letters* (1932); *Legal Hand Book for the Calvinistic Methodist Connexion* (1911). A. SKEVINGTON WOOD

CAMALDOLESE. The austere order of Camaldoli was founded by Romuald in 1012 near Florence. Previously abbot of several Benedictine monasteries, he had been expelled by the monks unable to meet his rigorous demands. Members of the Camaldolese observed two Lents in the year, abstained from meat, and lived on bread and water for three days in the week. Rudolph, the fourth general, slightly mitigated the original severity of the rule in its first written constitution in 1102. Gregory XVI belonged to the Camaldolese.

CAMBRIDGE PLATFORM (1648). A statement of Congregational polity for the New England churches. American Puritans wanted to distinguish from the Brownists* or Separatists on one hand, Presbyterians on the other. The General Court of Massachusetts Bay Colony authorized a synod to meet at Cambridge in 1646 to develop statements on doctrine and polity. An epidemic postponed the final statement on polity until 1648. The platform sets forth the theory of the "Catholick Church" as "the whole company of the elect and redeemed." The visible church comprises those who have a personal knowledge of salvation and whose lives are upright, as well as the children of such. It provides for regular church offices of pastors, teachers, ruling elders, and deacons in an autonomous congregation.

DONALD M. LAKE

CAMBRIDGE PLATONISTS. The name given to a group of theologians centered at Emmanuel College, Cambridge, the chief members of which were Benjamin Whichcote,* Ralph Cudworth,* Henry More,* John Smith (1618-52), and Nathanael Culverwell (d. c.1651). In reaction against the dogmatic Calvinism of the Puritans and the materialism of Hobbes,* they sought by a relationship of philosophy and theology to apply the idealism of Plato and particularly of Neoplatonism to religion. In one of his sermons Whichcote declared it "a very profitable work to call upon men to answer the principles of their creation, to fulfil natural light, to answer natural conscience, to be throughout rational in what they do; for these things have a divine foundation. The spirit in man is the candle of the Lord, lighted by God and lighting man to God." This quotation from Proverbs was a favorite with him and gives some idea of his view of man, reason, and conscience.

Reacting against Calvinistic ideas of total human depravity, the Cambridge Platonists saw man as a creature endowed with reason, not as simply a narrow faculty of ratiocination, but as an inner light. Likewise, they regarded right and wrong as part of the eternal nature of things, part of the law of the ideal world, imprinted on the will of man and which even the will of God could not change. "Had there not been a Law written in the Heart of Man, a Law [outside] him could be to no purpose" (Whichcote). It is easy to see how the degeneration of these views of reason and morality could so easily lead to the narrow and complacent views of eighteenth-century Deism.* This is particularly so when, as in this quotation from Whichcote and in another from John Smith to the effect that "God hath provided the truth of divine revelation [as an addition] to the truth of natural inscription" following the decline of reason after man's fall, it appears that the revealed word of God is merely a kind of supplement to existing truth.

The Cambridge Platonists were saved from the arid rationalism of the Deists by their mystical apprehension of God. They recognized the limits of philosophy and realized that some forms of knowledge cannot be apprehended in conceptual forms but are the product of a personal relationship with God. They had the awareness that "nothing can explain the phenomena of religious experience except the sense of the infinite within the heart of man" (G.R. Cragg).

BIBLIOGRAPHY: C.A. Patrides (ed.), *The Cambridge Platonists* (1969); F.J. Powicke, *The Cambridge Platonists* (1926). ARTHUR POLLARD

CAMERARIUS, JOACHIM (1500-1574). German Reformer. Born in Bamberg, he studied at Leipzig, Erfurt, and Wittenberg, and later instituted reforms at the universities of Tübingen and Leipzig. Noted as philologist and humanist, he was an outstanding Greek scholar. He participated in religious colloquies and imperial diets—e.g., Speier (1526 and 1529) and Augsburg (1530). In 1525 he journeyed with Melanchthon* to the Palatinate and visited Erasmus* in Basle. His friendship with Melanchthon stemmed from mutual humanistic, theological, and pedagogical concerns. He translated the Augsburg Confession* into Greek. His Greek catechism is Melanchthonian in its theology. He wrote biographies of Melanchthon (his most famous work), George, Prince of Anhalt, and Eoban Hesse. He gathered letters of prominent religious leaders of the Reformation, which though partially scattered proved valuable for historical research. He discussed the possibility of Lutheran reunion with Roman Catholicism with both Francis II (1535) and Maximilian II (1568).

CARL S. MEYER

CAMERON, JOHN (1579-1625). Scottish theologian and exegete. Born and educated in Glasgow, where he early taught Greek, he was in 1600 appointed professor of philosophy at Sedan, later becoming collegiate minister of the Reformed church in Bordeaux (1608-17) and professor of divinity at Saumur. In the Reformed Church of France, in which he spent so much of his life, he was greatly esteemed, and to it he was devoted. Returning to Scotland in 1622, he was for one year principal of Glasgow University. Back again in France, he lived out the rest of his life as professor of divinity at Montauban, where he achieved a growing reputation as a linguist. Posthumously published, his theological lectures and treatises continued to be reprinted almost up to our own time. D.P. THOMSON

CAMERON, RICHARD (c.1648-1680). Scottish Covenanter* from whom the Cameronians* took their name. Originally an Episcopalian in Falkland, he was converted and found his spiritual home thereafter among Presbyterians. Unable to obtain ordination in a Scotland on which episcopacy had been imposed, he went to Holland in 1679 and was ordained in Rotterdam at a service in which John Brown of Wamphray* and Robert MacWard (exiled Scots ministers) and James Koelman (a Dutch pastor) participated. He soon returned to Scotland and resumed his field preaching. He was one of the chief authors of the Sanquhar Declaration* that purported to depose Charles II, and he constantly prophesied the extinction of the Stuart line "for their treachery, lechery, but especially their usurping the royal prerogatives of King Christ." Finally, in June 1680 the dragoons who had long sought him, surprised him and his little band at Ayrsmoss, and "the Lion of the Covenant" was killed.

J.D. DOUGLAS

CAMERONIANS. Originally a Covenanting group named after Richard Cameron,* they fought for religious liberty in the last persecuting years of the Stuart dynasty. A majority refused thereafter to join the reestablished (Presbyterian) Church of Scotland, holding that some of the most precious rights of Christ for which the Covenanters* had suffered were not recognized by church and state. Known since 1743 as the Reformed Presbyterian Church,* modern Cameronians maintain substantially their former principles, particularly Christ's claim to national obedience.

CAMILLUS OF LELLIS (1550-1614). Founder of the Ministers of the Sick. A soldier's son, he was born at Bucchiano near Naples. After serving in the Venetian army (1571-74), he was reduced to penury because of inveterate gambling and worked for the Capuchins* until dismissed because of an incurable wound. He nursed at the hospital of San Giacomo and eventually became its superintendent. Deeply influenced by Philip Neri,* he was persuaded to study for the priesthood and in 1584 founded his own order, the Ministers of the Sick, who took a fourth vow to care for the sick. Officially approved in 1586, the order followed the Augustinian Rule, and until 1607 Camillus was its superior. He resigned to devote himself personally again to the sick. Canonized in 1746, he was named patron of the sick (1886) and of nurses (1930).

IAN BREWARD

CAMISARDS. French Protestant resistance fighters provoked to revolt by the brutal repression of all public practices of their faith following the revocation of the Edict of Nantes* in 1685. A period of passive resistance (1686-98) was followed by fierce outbreaks of open warfare lasting until 1709. Some 12,000 Protestants were executed in Languedoc alone. The Camisards, inspired by apocalyptic writings of prophets and intellectuals, notably Pierre Jurieu,* rose in revolt in the Cevennes district in 1702. Numbering about 3,000, they organized armed bands, and an army of some 60,000 was needed to put them down. Many suffered unspeakable tortures. In 1704 Camisards were offered pardon and the right to leave the country if they laid down their arms. Of those who accepted, some entered the British army. A few zealots held out, but were finally suppressed. Some Camisards found sanctuary in England and formed a small sect known as the "French Prophets."

J.G.G. NORMAN

CAMPBELL, ALEXANDER (1788-1866). One of the founders of the Disciples of Christ and the Churches of Christ.* Son of Thomas Campbell,* he attended Glasgow University, then emigrated to Pennsylvania, joined his father's Christian association, and was ordained to the ministry (1812). Alexander was very interested in unity among Christians, and he spread his theological views in a series of preaching tours in Kentucky, Ohio, Indiana, West Virginia, and Tennessee. Despite his emphasis on Christian unity, several congregations divided from the Baptist church because of his teaching, and these were united into the Disciples of Christ (1832, nicknamed "Campbellites"). While advocating a return to the simpler theology of the early church, he was opposed to both speculative theology and emotional revivalism. Campbell's desire for an educated clergy led to the founding of Bethany College, West Virginia (1840), which he directed until his death. He engaged in numerous debates on religious topics with such leaders as the secularist Robert Owen and John Purcell, Roman Catholic bishop of Cincinnati. Campbell wrote or edited over sixty volumes, the most important of which, in addition to his published debates, are a book, *The Christian System* (1835), and his periodicals *The Christian Baptist* and *The Millennial Harbinger.* He believed that baptism and confession of Jesus Christ as Savior are the only requirements of Christianity.

See J. Kellems, *Alexander Campbell and the Disciples* (1930).

ROBERT G. CLOUSE

CAMPBELL, ARCHIBALD (1598?-1661). First marquis of Argyll, and Covenanter.* At first adviser and strong supporter of Charles I (who in 1641 was to promote him from earl to marquis), he signed the National Covenant,* fought strongly against the king, but deplored his execution, and was a principal participant in Charles II's coronation at Scone on New Year's Day 1651. Nevertheless, he acquiesced in Cromwell's Commonwealth, and after the Restoration (1660) was arraigned for high treason. Charles II (who resented Argyll's strictures on his immoral life) determined his fate, and the young advocates who had defended the marquis were bullied and harassed. On receiving the sentence, Argyll said: "I had the honour to set the crown upon the king's head, and now he hastens me away to a better crown than his own." With his beheading in Edinburgh the killing of Covenanters began.

J.D. DOUGLAS

CAMPBELL, JOHN McLEOD (1800-1872). Scottish theologian. Son of an Argyll minister, he studied at Glasgow and Edinburgh universities, and in 1825 became minister of Row (Rhu) in Dunbartonshire, where he applied himself with "almost apostolic zeal." He was not to serve there long, however, for in 1830 he was accused of heresy before Dumbarton presbytery and found guilty of preaching "the doctrine of universal atonement and pardon through the death of Christ, and also the doctrine that assurance is of the essence of faith and necessary to salvation." Despite appeals, the general assembly in 1831 overwhelmingly voted to depose him from the ministry—a decision accepted characteristically without bitterness. His views were later incorporated in *The Nature of the Atonement* (1856), which is regarded as a substantial contribution to the development of Scottish theology. From 1833 to 1859 Campbell ministered to an independent congregation in Glasgow. J.D. DOUGLAS

CAMPBELL, THOMAS (1763-1854). One of the founders of the Disciples of Christ. A Scots-Irishman, he was a minister in the Secession church that had broken away from the Church of Scotland. The lack of unity in the seceding body led Campbell to become an enemy of Sectarianism. In 1807 he emigrated to Pennsylvania, where he established a Christian association for people from various professions. Since it seemed a new denomination would grow from this experiment, he tried unsuccessfully to merge with the Baptists. Later these Campbellites did join with a similar group organized by B.W. Stone.* Campbell was a popular preacher who moved constantly and usually made his livelihood by teaching school. Becoming blind in later life, he resided with his son Alexander* at Bethany, West Virginia. Alexander wrote *The Memoir of Elder Thomas Campbell* (1861). ROBERT G. CLOUSE

CAMPBELLITES, see CAMPBELL, ALEXANDER

CAMPEGGIO, LORENZO (c.1472-1539). Archbishop of Bologna. At twenty-six he became a doctor of canon and civil law, and he was ordained in 1510 after the death of his wife. He was a great diplomat and became involved in the political intrigues of his time. In 1512 he was appointed to the bishopric of Feltre and from 1513 to 1517 was nuncio to Maximilian I. Leo X wanted him to bring peace among Christian princes and unite them in a crusade against the Turks, but he failed to gain Henry VIII's support. Campeggio also took a leading part in some of the greatest events of the Reformation. In 1519 he was appointed *Segnatura,* a post of the highest dignity and honor. When Adrian VI was elected pope in 1522, many plans for reform of abuses in the church were submitted to him, but the pontificate was too short to implement them. In 1523 Campeggio was created archbishop of Bologna and in 1524 was made protector of England in the Roman Curia. Henry VIII made him bishop of Salisbury in 1524; in 1528 he was sent to England to form a court with Thomas Wolsey* to try Henry's divorce case. He was deprived of the see of Salisbury by act of Parliament in 1535. In 1537 he was made cardinal. S. TOON

CAMPION, EDMUND (1540-1581). English Jesuit.* Son of a London bookseller, he was a precocious youth, accomplished at giving addresses of welcome to royalty. He was maintained by the Grocers' Company at Christ's Hospital, and afterward at St. John's, Oxford, where he became a junior fellow in 1557. Despite his doubt, Bishop Cheyney persuaded him to be ordained deacon, but he left Oxford in 1569 and went to Ireland. An attempt to resurrect Dublin University failed, and he returned to England in disguise. At Douai in 1571 he entered the Roman Church. Next year he went on a pilgrimage to Rome, where he became a Jesuit. He was sent to Bohemia and ordained by the archbishop of Prague in 1578. When the Jesuits agreed to take part in the English mission, Parsons and Campion were the first two chosen. They reached England in June 1580, but Campion was arrested just over a year later and executed at Tyburn in December 1581. He had time, however, to print and distribute his *Decem Rationes.* JOHN TILLER

CAMP MEETINGS. A distinctive feature of religious life on the American frontier in the early decades of the nineteenth century. The open-air auditorium was always located in the center of the surrounding tents, whether in a rectangular, horseshoe, or circular pattern. The meetings promoted vivid conversion experiences, emotional and even physical activities, such as the "jerks," prostration, and dancing. James McGready* developed the technique in Logan County, Kentucky, in the summer of 1800. Soon other preachers, especially Methodists, adopted the camp meeting. The most famous camp meeting was at Cane Ridge in Bourbon County, Kentucky, in August 1801. Estimates of the crowd range from ten to twenty-five thousand. BRUCE L. SHELLEY

CAMPUS CRUSADE FOR CHRIST. An evangelistic organization working primarily with college students in the USA. Over 2,000 staff members (missionaries), most of them in America, seek in a personal conversation to present the Gospel in the form of "four spiritual laws." Headquarters for the organization are in a converted luxury hotel called Arrowhead Springs, near San Bernardino, California. The 1,800-acre site provides administrative offices and training facilities for nearly 5,000 collegians annually. The organization was created in 1951 shortly after William R. Bright, a young businessman, was converted. He attended Princeton and Fuller theological seminaries, but left school in order to contact student leaders at the University of California at Los Angeles. From UCLA the organization spread to many other campuses. Recently other ministries with laymen, athletes, military men, and high school students have been added. BRUCE L. SHELLEY

CANADA. The Christian religion in Canada was established and maintained by Protestants and Roman Catholics in a dual culture. In personnel and financial support they came originally from France, Great Britain, and the Thirteen Colonies. The ideas of Canadian Christianity largely reflected these outside influences, not only in matters of basic belief, but also in controversies over such subjects as church establishment and clergy reserves.* From the period of origins (c.1600-

1840) to the present, five religious groups have comprised the greater part of the Christian community in Canada: Roman Catholics, Anglicans, Presbyterians, Methodists, and Baptists.

The Church in New France. From the beginnings of French colonization in Canada in the early seventeenth century, the Roman Catholic Church occupied a position of importance. The church gave cohesion and stability to French Canadian society and was associated with French expansion into the interior as well as with the establishment of the church in the settled part of the colony. Samuel de Champlain* (c.1570-1635), the virtual founder of New France, brought Franciscan *Récollets** from France, hoping they would Christianize the Indians. In 1625 the Jesuits joined the *Récollets* in Quebec and soon became the dominant element in mission work among the Indians, particularly in the Huron country south of Georgian Bay. In 1648-49 several Jesuits were martyred at the hands of the Iroquois, the enemies of the Hurons.

While missions were being developed, the church was also established strongly in the French settlements of Quebec. Organization of the domestic church was largely accomplished by F.-X. de Laval.* The Roman Catholic Church was active also in Acadia (Nova Scotia) from the first expedition, authorized in 1604 and led by the Sieur de Monts.* Expansion into western Canada began between 1731 and 1741, when Jesuits accompanied the explorer Pierre de la Vérendrye on his trips into the west.

The Churches Under British Rule, 1760-1866. After the transfer of Nova Scotia to Great Britain in 1713 and of the rest of Canada in 1763, Protestantism and English-speaking Catholicism became established in the former French territory.

Roman Catholicism held its ground in Lower Canada (Quebec) and secured a hold in Upper Canada, chiefly as a result of the immigration of Glengarry Scots and, subsequently, Irish settlers. Scots, settling in the Maritimes, particularly in Cape Breton, were predominantly Catholic. In general, the French Catholic clergy were loyal to the British civil government after the conquest. Bishops J.O. Briand* and J.O. Plessis* were successful in consolidating the relations of the church with the British authorities.

Protestant churches entered Canada from both Britain and the British American colonies. They were largely supported by British missionary societies. The Anglicans drew support from pre-Loyalist New Englanders, United Empire Loyalists,* British garrisons and administrators, and immigrants from the British Isles. The Methodists consisted chiefly of British Wesleyans and American Episcopal Methodists. Presbyterianism, while derived from Britain and the United States, reflected the traditional breach between the Church of Scotland and the various Secession churches. The Baptist Church was pioneered in the Maritimes, in the Eastern Townships of Quebec, and in the Niagara Peninsula from the United States, but also derived support from the Scottish Highlands. The Lutheran Church in Canada kept pace with the immigration of German and Scandinavian peoples. The Lutherans established their first permanent congregation in Nova Scotia about 1750 and entered Upper Canada some twenty-five years later.

Expansion of the churches into the west soon followed their development in eastern Canada. Early missionaries in the west were J.N. Provencher,* John West,* James Evans,* and John Black.

During the first half of the nineteenth century, the position of the churches in regard to education emerged. By 1840 it was clear that secondary education was to be in the hands of the state, but with some provision for religious instruction on a nondenominational basis. Upper Canada (later Ontario) made provision for separate schools for the Roman Catholics; Lower Canada (later Quebec) developed a system divided into Catholic and Protestant sections. For most of the nineteenth century, higher education was largely controlled by the Catholic and Protestant churches which had founded some two dozen church-related colleges by 1867.

The Churches Since 1867. Expansion of the churches continued in the late ninteenth century and resulted also in ecumenical movements of organized reunion and confederation. The Presbyterians in Canada were united in 1875, most of the Methodists in 1884, and the Anglican General Synod was formed in 1893. In 1925 the Methodist and Congregational churches and a large part of the Presbyterian Church united to form the United Church of Canada.*

Toward the end of the nineteenth century, the Protestant churches in Canada felt the impact of new movements of thought among the scientists and the biblical critics. The challenge gave rise to the development of Christian liberalism, a viewpoint which tended to discard belief in the supernatural aspects of Christianity and to concentrate upon the Christian ethic. Early exponents were clergy such as G.M. Grant and Professors George Paxton Young (1819-89) and John Watson (1847-1939). The growth of Christian liberalism continued into the twentieth century. Many of its exponents, notably J.S. Woodsworth* and other Methodists, laid great emphasis on the social implications of Christianity. After 1930 the Protestant church in Canada was influenced by other trends of thought such as Neoorthodoxy, associated with the name of Karl Barth,* and also Christian existentialism,* but Christian liberalism continued to be the dominant theme.

There has always been, however, a strong conservative reaction to Christian liberalism, based on the idea of an inspired, authoritative Bible and on adherence to early Christian creeds, particularly the Apostles' and Nicene. Although the major denominations became largely liberal, there were many Christians within their membership who were orthodox and resisted the onslaughts of liberal theology. Other denominations, some of them fairly new, did not have this problem, among them Pentecostals, Plymouth Brethren, Christian and Missionary Alliance, Mennonites, and the Fellowship of Evangelical Baptist Churches. The Salvation Army first entered Canada in 1882.

Roman Catholic thought in the middle and late nineteenth century reflected the struggle between Ultramontanism* and liberalism which was raging in Europe. The working out of relations between church and state in regard to secondary education controlled by the provinces involved the Catholics in difficulties with civil authorities in Ontario, New Brunswick, and Manitoba. In large measure such problems continued to plague relations between Catholics and provincial governments in the twentieth century.

The Modern Church. For the most part, the groups which were numerically large in the early nineteenth century remained so in the latter part of the twentieth. In 1961 the Roman Catholic, Anglican, United, Presbyterian, and Baptist churches comprised over fifteen million adherents—more than 87 percent of the total population. The Lutheran Church and the Mennonites numbered over 800,000. Immigration from central Europe after 1890 had brought to Canada several churches representing old Christian traditions, notably the Greek Orthodox Church and the Ukrainian Greek Catholic Church. Other Protestant groups were still increasing, notably the Pentecostals (143,000).

Participation of the Canadian churches in missions continued. Domestic missions among the North American Indians and the Eskimos were chiefly maintained by the Roman Catholic, Anglican, and United churches. Protestant missions in the foreign field were overwhelmingly in the hands of missionaries who were evangelical and conservative in theology.

The struggle between liberalism and conservatism was still apparent in the last third of the twentieth century. In Protestantism the struggle cut across denominational lines, with the liberals stressing social justice, permissive morality, and flexibility in doctrine, and the conservatives emphasizing the importance of personal salvation and of adherence to the historic Christian creeds. Roman Catholics were concerned with such issues as birth control, services in the vernacular, and relations with other Christian groups.

BIBLIOGRAPHY: G.F. Playter, *The History of Methodism in Canada* (1862); W. Gregg, *History of the Presbyterian Church in the Dominion of Canada* (1885); A. Sutherland, *Methodism in Canada* (1903); J.E. Sanderson, *The First Century of Methodism in Canada* (2 vols., 1908-10); A.G. Morice, *History of the Catholic Church in Western Canada* (2 vols., 1910); E.R. Fitch, *Baptists of Canada* (1911); R.H. Gosselin, *L'Église du Canada* (4 vols., 1911-17); A. Shortt and A.G. Doughty, *Canada and Its Provinces*, vol. XI (1914); A. Dorland, *A History of the Society of Friends in Canada* (1927); W.S. Reid, *The Church of Scotland in Lower Canada* (1936); V.J. Eylands, *Lutherans in Canada* (1945); J.H. Riddell, *Methodism and the Middle West* (1946); W.E. Mann, *Sect, Cult and Church in Alberta* (1953); S. Ivison and Rosser, *The Baptists in Upper and Lower Canada before 1820* (1956); C.A. Tipp and T. Winter, *The Christian Church in Canada* (1956): H.H. Walsh, *The Christian Church in Canada* (1956) and *The Church in the French Era* (1966); C.B. Sissons, *Church and State in Canadian Education* (1959); G. French, *Parsons and Politics* (1962); P. Carrington, *The Anglican Church in Canada* (1963); D.C. Masters, *Protestant Church Colleges in Canada: A History* (1966) and *The Rise of Liberalism in Canadian Protestant Churches* (Annual Report, Canadian Catholic Historical Association, 1970); J.D. Wilson, *The Church Grows in Canada* (1966); L.K. Shook, *Catholic post-secondary education in English-speaking Canada* (1971). D.C. MASTERS

CANADA, PRESBYTERIAN CHURCH OF, see PRESBYTERIAN CHURCH OF CANADA

CANADA, UNITED CHURCH OF, see UNITED CHURCH OF CANADA

CANADIAN COUNCIL OF CHURCHES. Organized in 1944 "to promote the Church's mission, unity, renewal and obedience." Its formation resulted from interchurch cooperation in Canada and ecumenism abroad. Current members are the Anglican Church of Canada, Armenian Church, Baptist Federation of Canada, Christian Church (Disciples), Greek Orthodox Church, Lutheran Church in America (Canada Section), Presbyterian Church in Canada, Reformed Church, Salvation Army, Society of Friends, and United Church of Canada. The council has a close association with several interdenominational societies, and a working relationship with the Canadian Catholic Conference. The supreme decision-making body is the triennial assembly, with members appointed by the participating churches.

RICHARD N. MUGFORD

CANDLE; CANDLEMAS. The use of candles as ornaments in the Western Church probably had its origin in the lighted candles which were carried in procession before the bishop of Rome and then placed behind the altar. By 1200 two candles were actually put on the altar of the papal chapel and the custom, using more candles, quickly spread. They are now lit in both the East and West during liturgical services. Smaller candles, votive candles, are lit by worshipers and placed in front of statues of saints, especially before the Virgin Mary. In the Church of England the legality of two candles on the holy table was established by the Lincoln Judgment of 1890; the Lambeth Opinion of 1899 condemned the carrying of candles in procession; both these rulings are now widely disobeyed, and many parish churches follow Roman Catholic practice.

Candlemas is the feast which commemorates the purification of the Virgin Mary and the presentation of Jesus in the Temple (Luke 2:22-38). Originally a festival of the church in Jerusalem which began about 350, it became widespread after Emperor Justinian ordered its observance at Constantinople in 542. In the East the festival is called "The Meeting" (i.e., of Jesus and Simeon in the Temple). The major rite of the day in the West is the blessing and distribution of candles of beeswax to the singing of the *Nunc Dimittis* and in commemoration of Christ as the "Light of the World." PETER TOON

CANDLISH, ROBERT SMITH (1806-1873). Scottish minister. Educated at Glasgow University, he became minister of St. George's, Edinburgh (1834), and was one of the most prominent of those who at the Disruption* of 1843 left the establishment to form the Free Church of Scotland. In 1862 he became principal of his church's New College in Edinburgh, which post he exercised concurrently with his ministry at Free St. George's. In his 1861 Cunningham Lectures, Candlish disagreed with the F.D. Maurice* view of the universal fatherhood of God; he denied that unfallen Adam was the possessor of true sonship and argued that in adoption the sonship received by believers is an entirely new relationship. While contemporaries applauded his attempts to widen the scope of systematic theology, some scholars particularly in his own denomination criticized his reasoning. Candlish was one of the founders of the Evangelical Alliance (1845) and in 1861 was moderator of his church's general assembly.

J.D. DOUGLAS

CANISIUS, PETER (Petrus) (1521-1597). Roman Catholic reformer. Born in Nijmegen, he studied theology at Cologne and Louvain (1535-46). Deeply influenced by followers of the *devotio moderna,* he became a Jesuit* novice in 1543. In Cologne he helped to defeat Archbishop Hermann von Wied, and rapidly made a deep impression on Catholic leaders by his ability and dedication. More than any other single person, he established the Catholic reformation in S Germany and Austria. His three *Catechisms* went through hundreds of editions and did for Catholics what Luther's had done for Protestants. Jesuit provincial for upper Germany between 1556 and 1559, he not only attracted many other talented men to the order, but demonstrated singular ability in educational reform, preaching, teaching, apologetics, and pastoral work. Good relations with Catholic rulers were an indispensable part of his achievements. Tending to perfectionism, he was an unsparing critic of abuses, and his letters are an invaluable historical source. He died in Freiburg and was canonized in 1925.

IAN BREWARD

CANO, MELCHIOR (1509-1560). Spanish Roman Catholic theologian. Born at Tarancon, New Castile, he entered the Dominican Order* in 1523. After a professorship in theology at Alcala, he became professor of theology at Salamanca in 1546, and in 1551 was sent by Charles V to play an active part in the deliberations of the Council of Trent.* In the following year the emperor presented him for the bishopric of the Canary Islands, but a month later he resigned. In 1553 he became rector of St. Gregory's College, Valladolid, and in 1557 he was elected provincial of the Dominican Order. The appointment was contested, and a subsequent reelection was not confirmed by Pope Paul IV because of Cano's support for the Spanish crown against the papacy. His election was subsequently ratified by Pius IV. His reputation rests on the twelve books of his *De Locis Theologicis* (1563), an elegantly written inquiry into the sources of theological knowledge, which helped to lay the foundations of theological methodology.

HOWARD SAINSBURY

CANON. The Greek word *kanōn* was the rod of straightness, from which meaning emerged the idea of that which was measured or against which another could be measured; hence, in the derived sense, a rule or order of arrangement, and thereafter the "order of priests" and "clergy" in general. Its English use as a title stems from the early medieval ordered life of cloistered clergy housed within the close of a cathedral or collegiate church, thus *secular* canons. Those partially reformed in the eleventh century in their common living by the renunciation of private property were distinguished as *regular* canons (see AUGUSTINIAN CANONS and PREMONSTRATENSIANS). Since the Reformation, all Church of England canons are secular, the system being retained—often only in an honorary sense—for a cathedral chapter under a dean, originally advisory to a bishop with duties in his church.

CLYDE CURRY SMITH

CANONESS. Since titular "canon" involves all clergy, including minor orders, cloistered within a single house, the feminine usage (in English only since the seventeenth century) covers all members of a corresponding community of women living under a rule, although not necessarily under a perpetual vow. For all practical purposes there is little distinction from the designation "nun."*

CANONIZATION. A papal decree commanding public veneration to be paid to an individual by the universal church. It thus creates a cultus which is both universal and obligatory. Urban VII published in 1634 a bull which reserved to the Holy See exclusively its right of canonization. Papal authority is generally given only after a long legal process. In the primitive church, martyrs were the first to be publicly venerated by the faithful. From the fourth century, a cultus was extended also to confessors. The first historically attested canonization is that of Ulrich of Augsburg by John XV in 993. Canonization is said to confer a seven-fold honor: the name is inscribed in the catalogue of saints; his/her name is invoked in the public prayers of the church; churches may be dedicated to God in his/her memory; the Eucharist is celebrated in his/her honor; his/her festival day is observed; pictures of the saint show him/her surrounded by a halo; and his/her relics are enclosed in precious vessels and publicly honored.

S. TOON

CANON LAW. A body of ecclesiastical rules or laws drawn up and imposed by authority in all matters of faith, morals, and discipline. Such laws stem from the early practice of convening councils of church leaders to settle matters of uncertainty and dispute (cf. Acts 15). The importance of the councils determined the degree of authority attached to the canons—those from the Council of Nicea (325), for example, possessed great significance, and the results of other councils appear to be attached to the Nicean. The African churches held frequent plenary sessions that produced a large collection of canonical material, evidenced by reference in the Council of Chalcedon (451) to the Antiochean canons of 341 (or 330).

The councils were not alone in producing canon law. Their work was supplemented by that of individuals, particularly bishops, men such as Gregory Thaumaturgus,* Basil* of Caesarea, and Amphilochius* of Iconium. There was also the work of anonymous and fictitious authors such as the Apostolic Canons.* Papal letters (Decretals) also gained special authority from as early as the letter of Pope Siricius to Himerius of Tarragona in 385. With the Decretal of Gratian about 1140, scholars have drawn the dividing line between *ius antiquum* and *ius novum:* all canons after the Council of Trent (1545-63) are called *ius novissimum.* Gratian's Decretal was eventually extended into the *Corpus Iuris Canonici* which became the authoritative law for the Roman Church until 1904, when Pius X called for it to be completely overhauled and codified, the standard text now being the *Codex Iuris Canonici* issued in 1917. Local canons have also been used in conjunction with the authoritative law providing the basic canonical guide.

See A.G. Cicognani, *Canon Law* (ET 1934), and R. Metz, *What Is Canon Law?* (1960).

GORDON A. CATHERALL

CANON OF SCRIPTURE. Although the word "canon" came to be used of ecclesiastical pronouncements, it has a wider connotation when applied to Scripture. It was used in the sense of rule of faith and in the sense of a catalogue or list. Both these usages occur before the first decision was made on the subject of Scripture by a church council (at Laodicea in A.D. 363). This at once focuses attention on an important feature in the history of the canon, i.e., the fact that the content of the canon was determined by general usage, not by an authoritarian pronouncement.

The Christian Church took over the OT Scriptures in the Septuagint version, but there is no evidence that the Apocrypha,* which formed part of the Septuagint, was regarded as part of Scripture. Indeed Melito* in the second century thought it necessary to send to Palestine to discover the content of the Hebrew Bible because it was assumed that this and not the Greek canon should be used in the Christian Church. By the time of Jesus there seems to have been general agreement on the contents of the OT canon, despite the fact there was later discussion on a few of the books. The Jewish elders at Jamnia during the period A.D. 70-100 were in general agreement on canonicity, but discussed whether Esther, Proverbs, Ecclesiastes, Canticles, or Ezekiel "soiled the hands" of those who used them. Both Josephus and *2 Esdras* assume the same position in their acceptance of all the books. It should be noted that the OT canon of the Roman Church is wider than that of the Protestant churches because of the former's inclusion of the Apocrypha, which is regarded as of equal inspiration to the OT itself.

The acceptance of the Hebrew canon of Scripture as authoritative by the early church exercised an important influence on the formation of the NT canon. Following the regular reading of the OT in Christian worship after a pattern similar to the Jews', there was at once the need to relate also the teaching of Jesus and of the apostles. During the most formative period of the NT there is little evidence of the precise procedure in Christian worship, but it is certain the teaching of Jesus would have commanded equal respect to the OT. Moreover, literature giving authoritatively the teaching of the apostles would soon have been valued, particularly after the decease of the apostles themselves. In spite of the lack of specific information about church life in the later part of the first century, there is enough to show that considerable respect had emerged for the Gospels, Acts, and Pauline epistles by the early part of the second century. This does not mean that there was an official line on the NT canon, but rather a developing agreement on the use of these books. The concerted testimony of Clement of Rome, Ignatius, Polycarp, and the *Didache* shows the importance attached to these books, although direct citation of them is slight and there is no discussion of their canonicity. It is in fact significant that the earliest canon of the NT was from a heterodox source, Marcion,* who excluded everything except ten Pauline epistles and the gospel of Luke, the latter in a mutilated form. There is no doubt that the emergence of heretical groups claiming secret books to be authoritative promoted vigilance on the part of the orthodox church toward its authorized books.

By the close of the second century there was general acceptance of all the NT books except James, 2 Peter, 2 and 3 John, Jude, Hebrews, and the Apocalypse, which were only partially accepted. It was during the third and fourth centuries that the position of the canon became clarified, although it must not be supposed that lack of use necessarily implied doubt. Nevertheless, Origen mentions the hesitation of some churches over the two Johannine epistles and over 2 Peter, although he himself appears to regard them as Scripture. He questioned the Pauline authorship of Hebrews, although he clearly accepted its canonicity. The same attitude is seen in Dionysius, who rejected the apostolic authorship of the Apocalypse, but accepted the book as Scripture. The church in the East seems to have arrived at the full canon of the NT as it now exists at an earlier date than the Western Church. The canon is set out in detail in Athanasius's Easter Letter (A.D. 367) and contains the twenty-seven books to the exclusion of all others, although certain other books, such as Hermas's *Shepherd* and the *Didache* are allowed for private reading. A similar list was confirmed at the Synod of Carthage in A.D. 397.

In the Western Church there was greater tardiness, and it was not until the time of Jerome* and Augustine* that certain of the NT books were accepted, no doubt mainly under their influence. Those books over which there had been hesitation were Hebrews, James, 2 and 3 John, 2 Peter, and Jude. In the case of Hebrews and Jude, there is earlier evidence of acceptance at the close of the second century, after which they appear to have fallen into disuse for a period. Augustine at first considered Hebrews as Pauline, but in his later works he cited it anonymously. All of the NT books except 2 and 3 John, 2 Peter, Jude, and the Apocalypse were included in the Peshitta, the Bible of the Syriac-speaking church, but these omitted books were included within the next century (the Philoxenian version produced in A.D. 508 included them).

During the Reformation, discussion over the NT canon opened again. Erasmus, Luther, and Calvin, among others, discussed the authenticity of certain of the books. Luther* is most notable for creating what was almost a deuterocanon. He placed Hebrews, James, Jude, and the Apocalypse at the end of his Bible to signify that they were of less value than the rest. But the Reformers were subjective and largely uncritical in their comments.

During the era of critical inquiry, many of the books of Scripture have been considered nonauthentic, but the position of these books within the canon has never been seriously discussed. It has been implicitly assumed that even nonauthentic or pseudonymous works can be regarded as canonical, but such a view finds no support from early Christian testimony.

BIBLIOGRAPHY: H.E. Ryle, *The Canon of the Old Testament* (2nd ed., 1904); H.B. Swete, *An Introduction to the Old Testament in Greek* (rev. R.R. Otley, 1914); W. Bauer, *Rechtgläubigkeit und Ketzerei im ältesten Christentum* (1934); W.O.E. Oesterley, *An Introduction to the Books of the Apocrypha* (1935); J. Knox, *Marcion and the New Testament* (1942); J.N. Sanders, *The Fourth Gospel in the Early Church* (1943); A. Souter, *The Text and Canon of the New Testament* (2nd ed., 1954); B.M. Metzger, *An Introduction to the Apocrypha* (1957); E.J. Young, "The Canon of the Old Testament," in *Revelation and the Bible* (ed. C.F.H. Henry, 1958). DONALD GUTHRIE

CANONS, APOSTOLIC, see APOSTOLIC CANONS

CANONS, BOOK OF. Passed by James I and by Canterbury Convocation in 1604, and by York Convocation in 1606, the book contained 141 canons covering such things as the conduct of services, the duties of church officers, and the discipline of the clergy. They betray at points an anti-Puritan bias, but generally reflect the Elizabethan Settlement.* Down to 1936 there were minor changes, the convocations reforming their own composition (without seeking statutory authority) and thus becoming more representative in order to be joined with the house of laity in the newly empowered church assembly (see GENERAL SYNOD). In 1939 the archbishops appointed a canon law commission which drew up a draft new code (1947). Revision was undertaken by the convocations (consulting the house of laity). Many changes in statute law were required. The new canons were promulgated in 1964 and 1969, but further, continuous reform of the code has gone on, and the new general synod of the Church of England has a standing commission handling such needs. While the present church-state relationship lasts, such amendments themselves sometimes require a parliamentary measure before gaining the royal assent. COLIN BUCHANAN

CANONS REGULAR, see AUGUSTINIAN CANONS

CANTATA, see MUSIC, CHRISTIAN

CANTERBURY. In 597 Pope Gregory the Great sent Augustine to evangelize Britain, with instructions to establish sees at London and York. Augustine was, however, welcomed by Bertha, the Christian wife of Ethelbert, king of Kent, and established his first church in Ethelbert's capital, Canterbury. This became the center of his missionary activity and consequently took the place of London. From the thirteenth century the archbishops were regarded as permanent papal legates, and by the fourteenth century they had established precedence over the archbishops of York as Primate of All England. Many archbishops played significant parts in national history. The diocese of Canterbury consists of most of Kent and part of Surrey, but the province of Canterbury covers England south of Cheshire and Yorkshire, and the archbishop is also regarded as head of the worldwide Anglican Communion.*

Augustine consecrated a Roman-British basilica as his cathedral and founded a Benedictine monastery beside it, which was reorganized by Lanfranc* as the priory of Christ Church. After its destruction by the Danes in 1067, the cathedral was rebuilt in Norman style under archbishops Lanfranc and Anselm* and consecrated in 1130. Archbishop Becket* was murdered in the cathedral in 1170, and after a fire in 1174, the choir was reconstructed in Transitional style, with a magnificent shrine for Becket, dedicated in 1220. Thousands of pilgrims visited it from all over Europe, bringing great wealth to cathedral and city. From 1376 the Norman nave was reconstructed and transepts added. The central tower (Bell Harry Tower) was begun about 1495, also in Perpendicular style. Edward the Black Prince and Henry IV were buried in the cathedral. In 1538 Becket's shrine was destroyed and the priory dissolved, to be replaced by a dean and twelve canons, appointed by the Crown.

BIBLIOGRAPHY: A.P. Stanley, *Memorials of Canterbury* (1855); W.F. Hook, *Lives of the Archbishops of Canterbury* (1860-76); R. Willis, *Architectural History of Canterbury Cathedral* (1945); M.A. Babington, *Canterbury Cathedral* (1948); J. Shirley, *Canterbury Cathedral* (1970).

JOYCE HORN

CANTICLE, see MUSIC, CHRISTIAN

CAPISTRANO, JOHN OF (1386-1456). Franciscan* leader and preacher. Having studied civil and canon law in Perugia, he joined the Observant Franciscans there and preached against the Fraticelli.* He did much to win for the Observants their own provincial vicars (1431) and to gain them holy places in Palestine. Later he became vicar general of the Cismontane community and wrote a course of studies, having founded other convents and monasteries. He defended Bernardino of Siena* before Martin V, and (unsuccessfully) the Angevin cause in Naples as papal legate (1435-36). He obtained for Third Order Franciscans the right to live in common (1436) and spent two years in Milan preaching and writing (1440-42). Nicholas V sent him to Austria to preach against the Hussites (1451), extending his responsibility to Styria, Hungary, and Bohemia. In 1454 Pius II had him preach the crusade which brought about the 22 July victory. He was canonized in 1690. C.G. THORNE, JR.

CAPITO, WOLFGANG FABRICIUS (1478-1541). Protestant Reformer. Born Köpfel, son of a Hagenau blacksmith, he studied medicine at Pforzheim, jurisprudence at Ingolstadt, and theology at Freiburg im Breisgau, and became expert in Hebrew. As chapter preacher at Bruchsal (1512) he met Melanchthon, and as professor of theology and cathedral preacher at Basle (1515) encountered Oecolampadius* and corresponded with Luther and Zwingli, associated closely with Erasmus, and published a Hebrew grammar and a translation of the Psalter. From Mainz, where as chancellor and preacher he mediated uncomfortably between the archbishop and Luther (1519-23), he moved to Strasbourg as provost of St. Thomas's. Drawn from scholarship into reform mainly by Matthew Zell, he was dismissed from the chapter but appointed pastor of New St. Peter's, and married in 1524. The tension between study and pulpit produced lectures on the OT and commentaries in humanist Latin on Habakkuk (1526) and Hosea (1528).

Towards Strasbourg's quieter Anabaptists he was markedly benevolent and hospitable, confident in the outcome of irenical discussion. For years he was unhappy with infant baptism, espousing an illuminist or spiritualist piety which inculcated inwardness and distrusted externals. He was influenced by radicals like Schwenckfeld, Michael Sattler, and especially Martin Cellarius, with whom he cooperated in translation and publication. But progressively unsettled by radical excesses, illness (1529), and his wife's death (1531; in 1532 he married Oecolampadius's widow), he inclined towards closer alignment with Bucer* in securing a firmer ecclesiastical system for Strasbourg (1533-34). Together they compiled the *Tetrapolitan Confession** (1530) and pursued union negotiations (Wittenberg, 1536; Worms, 1540). Capito also drew up church orders for Berne (synod of 1532) and Frankfurt (1535). His later works (*Responsio de Missa* . . ., 1537; *Hexemeron Dei* . . ., 1539) favor even more strongly a magisterial, "Neo-Catholic" ecclesiasticism. Above all a scholar in the Erasmian mould (active in patristic studies and promoting works by Luther and Oecolampadius), he failed to fulfill his early promise even in this sphere.

BIBLIOGRAPHY: J.W. Baum, *Capito und Butzer, Strassburgs Reformatoren* (1860); O.E. Strasser, *La Pensée Théologique de Wolfgang Capiton dans les Dernières Années de sa Vie* (1938); *idem*, "Un Chrétien Humaniste," in *RHPR* 20 (1940), pp. 1-14; B. Stierle, *Capito als Humanist* (1974); J.M. Kittelson, *Wolfgang Capito, From Humanist to Reformer* (1975). D.F. WRIGHT

CAPITULARY. A name applied to legislative decrees issued by the Merovingian or Carolingian kings of France or by the Lombard kings of Italy, covering varied aspects of administration, including ecclesiastical regulations and moral prescriptions along with more general political and economic edicts. The acts, composed in Latin, were often long and discursive, so were divided into *capitula*, or chapters, and were of various types. Some were confirmed by local assemblies or by church councils, while others took effect without such confirmation. Some were effective for a limited time or in a specific area, others were binding and permanent throughout the entire realm. None survives in its original form, but many are available in collections, beginning with that of the Abbot Ansegisus in 827.

MARY E. ROGERS

CAPPADOCIAN FATHERS. In the second half of the fourth century, three theologians from the province of Cappadocia had a profound influence upon the character of Christian theology. They were Basil of Caesarea,* his brother Gregory of Nyssa,* and Basil's close friend Gregory of Nazianzus.* They gave final shape to the Greek doctrine of the Trinity, and through their efforts Arianism was finally defeated.

CAPREOLUS. (fifth century). Bishop of Carthage who probably succeeded to the see shortly before the death of Augustine (430). Principally known for his letter to the Council of Ephesus (431) against the views of Nestorius (see NESTORIANS), he also successfully begged the council not to reopen discussion on the Pelagian heresy. He stated the principle in his letter that old questions, once settled, should not be reopened. The judgment of the Fathers, under the Spirit's guidance, should decide new questions.

CAPREOLUS, JOHN (c.1380-1444). "Prince of the Thomists." A Dominican who taught in French universities, he was foremost in the revival of Thomism. His *Four Books of Defences of the Theology of St. Aquinas* (1409-33) used the sources systematically against critics such as Scotists and Ockhamists.

CAPUCHINS. Reformed branch of the Franciscan* Order. In 1525 Matteo da Bascio (1495-1552), a Friar Observant at Montefalcone, desired to return to the primitive simplicity of the order. He adopted the pointed cowl (*capuce*) which St. Francis wore, together with sandals and a beard. He and his companions cared for plague victims at Camerino and established themselves there with the approval of Pope Clement VII in 1528. They met much opposition from other Franciscans. Their rule (1529) emphasized the ideals of poverty and austerity, and they devoted themselves to charitable work. The movement was nearly suppressed in 1542 when their third general, Bernardino Ochino,* became a Protestant, but they survived through the influence of Cardinal Sanseverino and Vittoria Colonna, duchess of Amalfi. By dint of their missionary zeal and preaching enthusiasm they became a powerful weapon of the Counter-Reformation. In 1619 they became a fully independent branch of the Franciscan family. J.G.G. NORMAN

CARDINAL (Lat. *cardo*, "hinge"). One of the ecclesiastical princes who form the Sacred College of Cardinals in Rome and choose the pope. This title was first applied to the priesthood generally, especially to those with permanent church attachments, but eventually it came to denote specific priests and deacons in Rome who formed a council to advise the bishop of Rome. From the eighth century the consistory included neighboring "cardinal" bishops. The action of Leo IX (1002-54) enhanced the position of the Roman cardinals. The cardinalate was formed into a collegiate body, and its members ranked as Roman princes, who when in consistory became the immediate papal advisers and assumed the government of the Roman Catholic Church during the vacancy of the Holy See. The present function of cardinals is chiefly administrative, and they are appointed by papal nomination. In 1568 the number was fixed at seventy—six bishops, fifty priests, fourteen deacons—but modern appointments generally are made from the ranks of the episcopate. Pope John XXIII increased the number, and there are now over one hundred.

Among their duties they are to reside in Rome, unless excused or bishops of foreign dioceses; to act as heads of curial offices and Roman congregations; to preside over ecclesiastical commissions. Their title is "Eminence," and they are afforded rights in all dioceses, such as the use of a portable altar everywhere. Their insignia include the "red hat"—a flat-crowned, broad-brimmed hat with two clusters of fifteen tassels (which is not worn again after a cardinal's first consistory)—a biretta and skullcap, the sacred purple, a sapphire ring, and a pectoral cross. They meet in conclave for the election of a new pontiff, a privilege they have held since the Third Lateran Council (1179).

GORDON A. CATHERALL

CAREY, WILLIAM (1761-1834). Missionary to India. Born near Northampton, England, he worked as a shoemaker from the age of sixteen to twenty-eight. Following his conversion at eighteen, he became a preacher among the Calvinistic Baptists, working by day and ministering in his spare time; while he worked he studied. In 1785 he became preacher to the Baptist church in the village of Moulton and taught also in the village school; in 1786 he was made pastor. During this period a great burden for the unevangelized in heathen lands was given to him. In 1792 his pamphlet *An Enquiry into the Obligations of Christians to use Means for the Conversion of the Heathens* was published. He proposed the formation of a society to achieve this—the first modern missionary society. In 1789 he became pastor of a run-down Baptist church in Leicester; and in 1792 preached his famous missionary sermon—"Expect Great Things from God; Attempt Great Things for God"—at a ministers' meeting. At Kettering, four months later, the "Particular (Calvinistic) Baptist Society for Propagating the Gospel among the Heathen" (now the Baptist Missionary Society) was founded.

In 1793 Carey, with John Thomas, sailed for Bengal, India. At first destitute in Calcutta, he quickly mastered the language, and in 1794 was made manager of an indigo factory near Madras. Soon he set to work translating the Bible into Bengali, in addition to his business, evangelistic, and pastoral labors. By 1798 he had learned Sanskrit and had translated into Bengali the whole Bible, except Joshua to Job. To print it he set up his own press. He established schools and medical work. In 1800 he moved to the Danish colony of Serampore, which was his base for the remaining thirty-four years of his many-sided missionary labors. During these years he worked untiringly at the comprehensive pattern of missionary service which he had already laid down—Bible translation and production, evangelism, church-planting, education, and medical relief—spreading its influence and activities throughout India and then stimulating missions in other parts of Asia. He himself served as professor of Sanskrit, Bengali, and Marathi at the College of Fort William; he supervised and edited translations of the Scriptures into thirty-six languages; produced a massive Bengali-English dictionary, pioneered social reform, and founded the Agricultural and Horticultural Society of India. Carey has generally been acclaimed as "the Father of Modern Missions."

BIBLIOGRAPHY: J. Taylor, *Biographical and Literary Notices of William Carey* (1886); S.P. Carey, *William Carey* (1923); E.D. Potts, *British Baptist Missionaries in India 1793-1837* (1967).

A. MORGAN DERHAM

CARGILL, DONALD (Daniel) (c.1619-1680). Scottish Covenanter.* Son of a Perthshire notary, he was ordained minister of the Barony Church, Glasgow, in 1655, but was ejected with many other Presbyterian ministers in 1662. He became a field preacher, and generally ascribed to him is the Queensferry Paper, the most advanced of the Covenanting documents which, along with a strong confession of faith, advocated the establishment of a republic, since monarchy was "liable . . . to degenerate into tyranny." In 1680 at a service at Torwood, Cargill excommunicated Charles II and the other chief persecutors of the Covenanters. The sentence on the king was implicitly founded on the same grounds as were afterward used in the British renunciation of the Stuarts as a whole. J.D. DOUGLAS

CARISSIMI, GIACOMO (1603-1674). Italian composer. He is important in the history of sacred music for the role he played in developing *oratorio.* He worked in Rome, and the objective of his works in this form was didactic under the influence of the Jesuits. What came to be called *oratorio* was originally sacred opera and began about 1600 with the work of Cavalieri. Carissimi's works presented in dramatic style episodes mainly from the Bible. They were in Latin, with *dramatis personae* represented by solo voices, but without costumes and scenery. A significant role was given to the chorus, which sang in simple chordal style to give greater audibility to the words. Although living much later, Handel* was indebted to Carissimi. J.B. MAC MILLAN

CARLILE, WILSON (1847-1942). Founder of the Church Army.* His successful business career was virtually ruined by the economic difficulties of 1873, and this led to serious thought and his conversion. Almost immediately he showed great concern for the "roughs" of society and great skill in presenting the Gospel message vividly. After study at the London College of Divinity he was ordained in 1880 to a Kensington parish, where he combined remarkable ability, imagination, and efficiency, with a certain musical talent, in evangelism of the poorest areas, especially through open-air and after-church meetings. He began to train lay preachers, and this led to the foundation of the Church Army in 1882, in which he played a dynamic part until the end of his life.

C. PETER WILLIAMS

CARLSTADT (Karlstadt), ANDREAS BODENSTEIN VON (c.1477-1541). German Protestant Reformer. Born in Bavaria, he was educated at Erfurt, Cologne, and Wittenberg, where he became a member of the theological faculty. In 1511 he traveled to Rome and Siena to receive a doctorate. He was at first a defender of Scholasticism* and an opponent of Luther, but after reading Augustine he became an advocate of grace and divine sovereignty. In 1518 he wrote 380 theses on the supremacy of Scripture and the fallibility of councils in support of Luther's Ninety-Five Theses.* He debated these principles against J. Eck* at Leipzig (1519). Later he gave his interpretation of the debate in a tract, *Against the Dumb Ass and Stupid Little Doctor Eck.* The bull *Exsurge Domine* which condemned Luther and other Reformers included Carlstadt.

There were, however, differences between Luther and Carlstadt, and these became clear in 1521. While Luther was hiding in the Wartburg, Carlstadt made many reforms. Luther allowed a good deal of liberty in the Christian life, but Carlstadt considered some changes as necessary, such as Communion in both kinds, the marriage of the clergy, and ridding the liturgy of music. He also believed infant baptism was unnecessary and Communion was a memorial service. When Luther returned to Wittenberg, Carlstadt left for Orlamünde. Here he became a very popular preacher and renounced his academic degrees. He took an anticlerical attitude, began dressing as a peasant, wearing no shoes, and asked that people call him "Brother Andrew." These actions were based upon his conviction that inner religious experience demanded social equality. Luther visited Orlamünde, and in a debate with him Carlstadt claimed he spoke by direct revelation of the Holy Spirit, rather than with the "papistical" talk of Luther. In 1524 the Saxon authorities asked Carlstadt to leave the city. Eventually he settled in Switzerland, associating for a while with Zwingli* in Zurich and later with Bullinger* in Basle.

A brilliant, often petty, man, Carlstadt in his "turgid and long-winded pamphlets," as Gordon Rupp has pointed out, anticipated much of Puritanism.

See K. Müller, *Luther und Karlstadt* (1907), and E.G. Rupp, "Andrew Karlstadt and Reformation Puritanism," *JTS*, NS, X (1959).

ROBERT G. CLOUSE

CARLYLE, ALEXANDER ("Jupiter") (1722-1805). Scottish minister. Born in a Dumfriesshire manse, he studied at Edinburgh, Glasgow, and Leyden universities, was ordained as minister of Inveresk in 1746, and held that charge until his death. Having earned his nickname because of his imposing appearance, he walked in high social circles, was a brilliant conversationalist, scandalized many by going openly to the theater, and played cards at home "with unlocked doors." He seems, however, to have carried out his pastoral duties zealously, and he championed many good causes in the general assembly, whose moderator he was in 1770. During his many visits south of the border he attended Church of England services as a matter of course. His *Autobiography* (1860) is a fascinating commentary on his times.

J.D. DOUGLAS

CARLYLE, THOMAS (1795-1881). Scottish writer. Son of a Scottish peasant farmer, he was early attracted to German literature. His concern with social conditions (*Chartism*, 1839, and *Past and Present*, 1843) led to his propounding the need for hero-rulers—strong, just men who emerged to leadership by their own innate powers rather than by election (*Heroes and Hero-Worship*, 1841). He produced two massive works on Oliver Cromwell (1845) and Frederick the Great (1858-65) respectively, besides the earlier *French Revolution* (1837). Carlyle believed in order and that order is realized only through power, which evokes a sense of duty from those who obey. There may be something of his own religious background in the stress that he placed on the responsibility of the individual will, as there certainly is in his admiration of the Puritans and Covenanters. In *Sartor Resartus* (1843-44) and *Past and Present* he proclaims his gospel of action. It is a sad, but not surprising, reaction to this that expresses itself in the bitter, disillusioned, and mocking tone of *Latter-Day Pamphlets* (1850). ARTHUR POLLARD

CARMAN, ALBERT (1833-?). General superintendent of Canadian Methodism. After a brief period as a high school teacher, he was appointed in 1858 as principal of Albert College, Belleville, the major educational institution of the Methodist Episcopal Church, and soon secured for it a place within the federated structure of the University of Toronto. Ten years later he became chancellor, and in 1873 he became the bishop of his denomination. Despite his responsibilities in a circuit-riding church, he was also a moving force in the founding of Alma Ladies College in St. Thomas. In 1884, when Canadian Methodism united, Carman was chosen superintendent, and his great gifts of administration and leadership resulted in his being reelected repeatedly and continuing in office until 1915. He was deeply involved in

Christian social action and latterly was a doughty opponent of theological liberalism. In 1899 and 1907 his opposition forced G.B. Workman to be relieved of his OT post in two Methodist institutions, and in 1909 he engaged in controversy with George Jackson, whose views he believed were erroneous and would weaken the doctrinal basis of the burgeoning church union movement.

IAN S. RENNIE

CARMELITES. They emerged in 1593 as the reformed part of the Carmelite Order with its own general and its special emphasis on the contemplative life. This reform was begun in 1562 by Teresa of Avila.* The Carmelite Order was originally founded by Berthold on Mount Carmel in Palestine about 1154. It once claimed to have descended directly from Elijah and the community of prophets who lived there. A new period in the history of the order began with the fall of the Crusader States and migration of the Carmelites to Europe. The sixth general, Simon Stock, obtained from Innocent IV certain modifications of the primitive rule as laid down in 1209 by Albert de Vercelli, Latin patriarch of Jerusalem, who insisted on absolute purity, total abstinence from flesh, and solitude. Although abstinence was not abolished, it became less stringent and silence was restricted to specific times. In 1452 an order of Carmelite nuns was founded.

In the sixteenth century, discipline among monks and nuns deteriorated. Teresa of Avila resolved to revive the primitive rule and follow the contemplative life. This reform movement came to be known as the Discalced.* In the course of fifteen years Teresa founded sixteen more convents of nuns. The ideal of the contemplative life attracted many followers, among whom was John of the Cross,* who extended the reform to the male houses of the order. The Carmelites emphasized special devotion to Mary and the Child Jesus, and not unnaturally Carmelite theologians were among the earliest to defend the Roman Catholic dogma of the Immaculate Conception.

S. TOON

CARMICHAEL, AMY WILSON (1867-1951). Missionary to India. Adopted daughter of Robert Wilson, chairman of the Keswick Convention, she worked in Japan for a time with Barclay Buxton. After a breakdown in health she served with the Church of England Zenana Missionary Society in South India, where she was a colleague of Thomas Walker* of Tinnevelly. In 1903 she wrote *Things as They Are,* a moving account of stark realism which influenced many towards missionary work. In 1901 the Dohnavur Fellowship began, with the purpose of rescuing children devoted to temple service with all its attendant corruption. It was part of the CEZMS until it became independent in 1926. After a fall in 1931, she was crippled by arthritis, but remained at the center of the life of the Fellowship, writing many devotional books and poems which were marked by an intense, almost mystical, spirituality.

J.G.G. NORMAN

CARNELL, EDWARD JOHN (1919-1967). American evangelical theologian. Born in Antigo, Wisconsin, he received his college and seminary education at Wheaton College and Westminster Theological Seminary. He later won doctorates at Harvard and Boston. From 1945 to 1948 he was professor of philosophy and religion at Gordon College and Divinity School. In 1948 he joined the faculty of the recently founded Fuller Theological Seminary and served in several positions during his nineteen years there. He was president of the seminary (1954-59) and at the time of his death was professor of ethics and philosophy of religion. He was one of the leaders in the intellectual awakening of conservative evangelicalism after World War II. He excelled in giving a contemporary and relevant statement of the historic Christian faith. His books included *An Introduction to Christian Apologetics* (1948), *Christian Commitment* (1957), and *The Burden of Sören Kierkegaard* (1965).

BRUCE L. SHELLEY

CAROL. The word came into English from the medieval French for "round dance." In fifteenth-century England the carol developed into an important type of vocal composition. Although no longer danced, it retained structural evidence of its origin. A refrain called "burden" was sung before the first stanza, repeated between each succeeding one, and at the end. The subject matter was not always religious, but many concerned the Nativity or the Virgin Mary. Music has survived for only about a third of the extant texts. A well-known composition employing authentic medieval texts with modern music is *A Ceremony of Carols* by Benjamin Britten (cf. for texts, R.L. Green, *The Early English Carol,* 1935; for music, *Musica Britannica,* vol. IV, 1952).

In later times, anonymous folk ballads and lyrics dealing with aspects of the Nativity and attendant events, real or mythical, became known as carols. In England these correspond to the *noëls* of France and the *Weihnachtslieder* of Germany. From the sixteenth century, such lyrics were frequently circulated in broadsheets. There are carols for seasons other than Christmas. The Romantic era brought an interest in folksong, which led to the preservation of texts and melodies in print that had been transmitted only in oral tradition. Excellent examples from the eighteenth century are "God rest you merry" and "A Virgin unspotted." Today all Christmas hymns are frequently but incorrectly referred to as Christmas carols.

See P. Dearmer, R.V. Williams and M. Shaw (eds.), *The Oxford Book of Carols* (1928); and E. Routley, *The English Carol* (1959).

J.B. MAC MILLAN

CAROLINE BOOKS *(Quattuor Libri Carolini).* A four-volume Frankish commentary written about 790-92 on the place of ecclesiastical images, debating the decisions of the iconoclastic council (753-54) and the Second Council of Nicea (787), which advocated extremes, destruction and adoration, respectively. Its position is neither; rather, images are for instruction alone, while the Cross of Christ, Scripture, sacred vessels, and saints' relics are worthy of adoration only. Probably Alcuin* or a Spanish or Irish theologian at Charlemagne's court is the author. The Nicene position did gain increasing Frankish recognition after its more accurate version by Anastasius Bibliothecarius* appeared under John VIII (872-82). C.G. THORNE, JR.

CAROLINE DIVINES. The title used, often loosely, to refer to a group of Anglicans who flourished in the reigns of Charles I and II. Their theology is referred to, in some ways misleadingly, as "Arminian," and they also had distinctive views on spirituality and ceremonial. The father of the movement was Lancelot Andrewes,* and the effective leader William Laud.* The outstanding theologians, however, were younger men. It is important to correct a modern distortion which sees this school as representing "true Anglicanism" in contrast with the contemporary Calvinists and Latitudinarians.*

CAROLINGIAN RENAISSANCE. The revival of learning in Charlemagne's reign (768-814) which lasted until the Norse invasions. Charlemagne's empire was the first attempt at unified government since the Roman Empire's collapse and was represented as its re-creation. Charlemagne wanted a literate clergy so that it would succeed and endure, hence his attempt to revive the ancient learning. Though functional rather than creative, the renaissance checked ignorance and illiteracy and preserved the classics for future generations. Charlemagne's concern to continue the ancient classical culture had a significant difference—it was to be Christian. As Alcuin* put it in a letter to him, "If your intentions are carried out it may be that a new Athens will arise in 'Francia', and an Athens fairer than of old, for our Athens, ennobled by the teachings of Christ, will surpass the wisdom of the Academy."

Charlemagne drew upon contemporary scholarship. From Italy he brought the grammarian Peter of Pisa and the historian Paul the Deacon. From Spain came the Visigoth Theodulf, poet, man of letters, to be bishop of Orléans. Principally he drew upon the flourishing Anglo-Saxon culture resulting from the fusion of Irish and Benedictine monasticism in Northumbria under Bede, bringing Alcuin from York to become head of the palace school at Aix-la-Chapelle and chief organizer of the renaissance. The latter had three aspects: (1) measures to preserve literacy. Alcuin established a standard spelling throughout the empire, and developed a clear script in the beautiful Carolingian miniscule which was responsible for some of the finest medieval manuscripts; (2) schools established at monasteries (e.g., Tours, Fulda, Fleury) and cathedrals to give wider education; (3) content of education. The old Roman system was mainly rhetorical and literary. Alcuin constituted the curriculum from the seven liberal arts of the *trivium* (grammar, rhetoric, logic) and the *quadrivium* (arithmetic, astronomy, geometry, music) as laid down by Boethius and Cassiodorus.

Capitularies* issued by Charlemagne from 787 gave effect to these measures. Though the movement ended with the empire's collapse, it rescued culture from extinction and set it upon an educational foundation which survived to form the starting-point of the eleventh-century renaissance.

BIBLIOGRAPHY: M. Deanesley, *A History of the Medieval Church, 590-1500* (1925); R.H.C. Davis, *A History of Medieval Europe* (1957); G. Leff, *Medieval Thought* (1958). J.G.G. NORMAN

CARON, JOSEPH LE (1586-1632). Roman Catholic missionary to Canada. One of three members of the *Récollet** Order, a reformed branch of the Franciscans, who responded in 1615 to an appeal by Samuel de Champlain* for missionaries to take the Gospel to the Indians in New France, he moved from Quebec further west to establish the first mission among the Hurons. He was for a year (1617-18) provincial minister of Quebec. In 1618 Caron went to Tadoussac to work among the Montagnais, whom he served intermittently for some years. He compiled dictionaries of the Huron, Algonquin, and Montagnais languages, but these have been lost.

CARPENTER, JOSEPH ESTLIN (1844-1927). Unitarian minister. Educated at University College, London, he served Unitarian churches in Bristol and Leeds (1866-75) before becoming lecturer at Manchester New College. Subsequently he was principal of Manchester College, Oxford (1906-15). Prominent in his denomination, he had also a wide knowledge of Near Eastern studies and translated Ewald's* major work into English. His own books included *The First Three Gospels* (1890), *The Composition of the Hexateuch* (1902), and *The Johannine Writings* (1927).

CARPOCRATES. He taught, about A.D. 135 in Alexandria, a syncretistic Christianity. He believed that God is an unrevealed First Principle; the world was created by subordinate beings; souls transmigrate on the cyclic model of the *Phaedrus*; Jesus, a mere man, perceived eternal truths and rose above world powers; full exploitation of human experience qualifies the soul for direct experience of God without reincarnation. Carpocratians, renowned for licentiousness, revered images of both Christ and philosophers.

CARROLL, JOHN (1735-1815). First Roman Catholic bishop in America. Educated in Jesuit schools in France, he joined the order, was ordained to the priesthood (1769), and taught philosophy at Liège and Bruges. When the Society of Jesus was dissolved in 1773, he returned to America and became a leader of the Catholics in the colonies. In 1776 he accompanied Benjamin Franklin and his cousin Charles Carroll, a signer of the Declaration of Independence, on an unsuccessful mission to Quebec to gain support for the Revolution. John Carroll was appointed by Pius VI as the first Prefect Apostolic of the United States in 1784, and in 1790 was consecrated as the first American Roman Catholic bishop. He urged Catholic Americans to be patriotic defenders of the new nation, and he defended the rights of Catholics to have religious freedom and justice equal to that enjoyed by Protestants. In 1784 he responded to anti-Catholic attacks by publishing an effective defense of Catholicism, *An Address to the Catholics of the United States.* He founded Georgetown College for the training of capable native priests and in 1808 became the first archbishop of Baltimore. HARRY SKILTON

CARSTARES, WILLIAM (1649-1715). Scottish churchman and statesman. His influence with William of Orange did much to ensure that the form of church government of the Church of Scotland was settled as Presbyterian rather than Episcopalian at the time of the 1688 Revolution. His influence was decisive also in the consummation of the Union of Scotland and England, for he secured the support of the Church of Scotland for the cause of union. He spent much of the two decades preceding the Revolution in Holland as a refugee from the tyrannical Scottish government of Charles II, but acted from time to time as a link between William of Orange and dissident elements in both Scotland and England. His discretion when captured and tortured earned William's gratitude, and he had a place of weight during William's reign. After the king's death his power declined, and in 1703 he accepted the principalship of Edinburgh University, where he did much to raise academic standards. As one of the ministers of Edinburgh and moderator on four occasions of the general assembly, he was a main architect of the form of church government which survives to the present day.

HENRY R. SEFTON

CARTHAGE. The ecclesiastical metropolis of Roman Africa, in the West second only to Rome. Councils held here, both provincial—for Proconsular Africa alone—and all-African, fall into five periods:

(1) Cyprian* and the Donatists regularly appealed to the decision of seventy African and Numidian bishops under Agrippinus about 220 to rebaptize heretics. The ninety bishops who condemned their heretical colleague, Privatus of Lambaesis, between 236 and 248 in the episcopate of Donatus, possibly convened in Carthage.

(2) Cyprian initiated the holding of annual synods to channel the power of the corporate episcopate. Those in 251 and 252 dealt with the *lapsi* and the related schismatic disorders at Carthage and Rome. One in 253 reaffirmed baptism for newborn babies, and like that in 254 considered appeals from clerical *lapsi* against disciplinary sanctions. In 255 and twice in 256 the rebaptism controversy was dominant.

(3) In 312, seventy bishops, mostly Numidian, deposed Caecilian* and elected a counter bishop of Carthage, whence issued Donatism.* A protracted assembly of 270 Donatist bishops under Donatus about 330-35 decided that rebaptism of catholics was not invariably obligatory. Tyconius was condemned at a Donatist council about 380, possibly in Carthage, and in 392/3 Maximian's supporters met in synod against Primian, the "regular" Donatist bishop of Carthage. In 348/9 the catholic bishop Gratus presided at the first African council whose canons survive. Like those of Genethlius's council of 389 and 390 they mostly regulate clerical discipline. In 386 a council under Genethlius received Pope Siricius's letter concerning Donatism at Rome.

(4) Bishop Aurelius presided over at least ten provincial and fourteen general councils at Carthage from 394 (or 397) to 424. Donatism, the Pelagians, and Roman jurisdiction in Africa constituted the major extraordinary business. The council of 419 (resumed in 421-22) is credited with the *Code of Canons of the African Church,* mainly canons from earlier councils, which was accepted into the universal code of canons (text, Mansi 3, 699-843; translation, *NPNF* II. 14, 437-510).

(5) As Vandal control of Africa ended, councils met to restore church life, under bishops Boniface in 525 and Reparatus about 534, when the reception of Arian converts was discussed. In the Byzantine era general synods in 549 and 550 adhered to the Three Chapters* and excommunicated Pope Vigilius for complying with Justinian's condemnation, and in 594 and 646 councils ruled against Donatism and Monothelitism respectively.

See E.S. Foulkes in *DCA* 1, pp. 36-39; the *DHGE* surveys of A. Audollent (1, pp. 747-50, 811-22) and Ferron and Lapeyre (11, pp. 1220-26) give references to the volumes of Hefele-Leclercq and the conciliar collections of Hardouin and Mansi. D.F. WRIGHT

CARTHUSIANS. Founded by Bruno of Cologne in 1084 at the Grande Chartreuse (from which derives the name Carthusian), the Carthusian Order lays special emphasis on the contemplative life. At the beginning there was no special rule. In time, however, three collections of the customs of the order were made. One of these was by Guigues de Chatel, prior of the Grande Chartreuse, who compiled the *Consuetudines Carthusiae* in 1127, and this was approved by Innocent II in 1133. The monks lived a rigorous and austere life, eating no meat, drinking only watered wine, fasting frequently, and wearing hair shirts. They vowed to observe silence, lived in individual cells within the monastery, and devoted several hours each day to the discipline of mental prayer. Their physical needs were met by a community of lay brothers, and only on feast days did they meet their brethren for meals. The Carthusians thus combined the solitary life of Egyptian desert ascetics with the discipline of a monastic life.

The abbot of the order was the bishop of Grenoble, and the head was an elected prior. The prior of the Grande Chartreuse was the general who was elected by the monks of his own house and a general chapter consisting of visitors and priors which met annually. The order also has a few houses of nuns who observe a rule similar to the monks'. The rigorous life of the movement prevented it from becoming widespread. In 1104 there were still only thirteen monks at the original foundation. By 1300 there were thirty-nine foundations, most of these being in France. Because of its austere life, however, it was least affected by the monastic decline of the late Middle Ages. The most notable event in the history of the order is the split which occurred between 1378 and 1400, known as the Great Schism, healed only by the resignation of two generals and the election of another to replace them. The order has also been subject to persecution. During the Reformation Henry VIII put a number of English Carthusians to death. Their property was confiscated during the French Revolution (much of it was restored in 1816). They were driven by the anticlerical legislation of 1901 from the Grande Chartreuse and sought refuge in Spain, but returned in 1904. The order has had many mystics and devotional writers, among them Hugh, founder of the first English Charterhouse at Witham in 1175-76.

S. TOON

CARTWRIGHT, PETER (1785-1872). American Methodist pioneer circuit rider. Born in Virginia and reared in Kentucky, he was converted in an 1801 camp meeting in the wake of the Cane Ridge Revival, after intense spiritual struggle over his "delight in horse-racing, card-playing and dancing." Made an exhorter the following year, he was ordained deacon in 1806 by Francis Asbury* and elder in 1808. He served several circuits in Kentucky and adjoining states before he requested transfer in 1824 (because of his distaste for slavery) to Illinois, where he served as presiding elder for forty-five years. He attended twelve general conferences and two Illinois legislatures. He was defeated by Abraham Lincoln in an 1846 race for Congress. Rough, uneducated, and eccentric, he possessed unusual stamina, a quick wit, clear perception of human nature, and profound devotion to the work of God. He wrote *Fifty Years a Presiding Elder* and his *Autobiography* (1857).

DONALD W. DAYTON

CARTWRIGHT, THOMAS (1535-1603). Puritan divine. Educated at Clare Hall, Cambridge (1547), he was then scholar at St. John's (1550), but had to leave on Mary's accession (1553). Returning after her death, he eventually became a major fellow of Trinity (1562). The Reformation doctrines and practice of St. John's—no surplice, etc.—spread to Trinity with Cartwright's encouragement, and in 1569 he was appointed Lady Margaret professor of divinity after a two-year absence as chaplain to the archbishop of Armagh. Particularly in his lectures on the first two chapters of Acts he compared the constitution and hierarchy of the Church of England most unfavorably with that of the early Christians. John Whitgift* as vice-chancellor deprived him of his chair (1570), and as master of Trinity deposed him from his fellowship (1571).

After some time in Geneva with Beza,* possibly as a professor of divinity, he returned to England in 1572, but left again the following year owing to the uproar ensuing on John Field's and Thomas Wilcox's *Admonition to Parliament*—a Presbyterian work with which he was in full agreement. While abroad at Heidelberg and Antwerp (where he ministered to the English congregation) he married the sister of John Stubbs* and produced an answer to the Reims NT (which was suppressed until 1618). On his illicit return in 1585 he was seized by Bishop John Aylmer,* but released and attacked in passing in the Marprelate Tracts.* A somewhat pedantic theorist, Cartwright grew alarmed when his followers tried to put his theories into practice, but showed a contemptuous indifference to John Greenwood and Henry Barrow, and separated himself from the Brownists.* Tried by the Court of High Commission in 1590, he was committed to the Fleet Prison for a time, but released through Burghley's intervention, and spent his last years in Warwick, "a rich and honoured patriarch" helping with the production of the Millenary Petition* (1603), but not surviving until the Hampton Court Conference.

See biography by B. Brook (1845); and A.F.S. Pearson, *Thomas Cartwright and Elizabethan Puritanism, 1535-1695* (1925).

G.S.R. COX

CASAUBON, ISAAC (1559-1614). Classical scholar. Born at Geneva, the son of Huguenot* refugees, when the city was famous as a center of Greek scholarship, he studied under Francis Portus and succeeded him as professor of Greek in 1581. He went to Paris at the turn of the century and received a pension from Henry IV in return for duties in the royal library. On the king's death (1610), he resisted pressure to become a Roman Catholic, crossed to England, received a pension from James I, and was made a prebend of Canterbury (1611). He was employed to answer the historical defense of the papacy by Baronius,* and published various commentaries and critical editions, notably of Suetonius and Polybius.

JOHN TILLER

CASE, SHIRLEY JACKSON (1872-1947). Liberal church historian. Born in New Brunswick, Canada, he was educated at Acadia University, then studied theology at Yale. From 1908 to 1938 he taught NT and early church history at the University of Chicago Divinity School, and became dean there in 1933. He once remarked he was "born a liberal," and he contributed much to the development of the liberal "Chicago School" of theology. He rejected the supernatural element in Christian belief and attempted to explain the development of Christianity solely in terms of natural environmental influences. Among Case's many books are *The Evolution of Early Christianity* (1914), *Jesus—A New Biography* (1927), *Bibliographical Guide to the History of Christianity* (1931), and *The Christian Philosophy of History* (1943).

HARRY SKILTON

CASHEL, SYNOD OF (1171). A meeting of Irish bishops with a representative of Henry II of England to reform the Irish Church. It dealt with such matters as the collection of tithes, the paying of Peter's Pence* to Rome, the regulation of marriages, and baptisms. The archbishop of Armagh was recognized as primate of Ireland directly responsible to the pope, but in fact English dominance was obvious, for although no mention was made of the superiority of the archbishop of Canterbury, all native liturgies were replaced by that of the Church of England. This was another aspect of Henry II's conquest of Ireland.

CASSANDER, GEORG (1513-1566). Roman Catholic theologian who sought to mediate between Catholics and Protestants (especially Anabaptists). Born at Pitthem, near Bruges, he graduated from the *Collège du Chateau,* Louvain (1533), and taught literature at Ghent and Bruges. After a tour of Italy he enrolled (1544) in the theological faculty of Cologne, and in 1549 undertook both the teaching of theology and the direction of the newly formed academy of Duisberg. He joined in the programs of emperors Ferdinand I and Maximilian II to promote unity in the church (1561-66). His writings met with strong opposition from both sides, being accused of excessive tolerance and readiness for compromise. His chief work, *De Officio Pii ac Publicae Tranquillitatis vere amantis viri in hoc Religionis Dissidio* (1561), was submitted to the Colloquy of Poissy.* His writings were placed on the Index in 1617. J.G.G. NORMAN

CASSELS, WILLIAM WHARTON (1858-1925). Missionary bishop. He was one of the four Cambridge undergraduates who formed the nucleus of the "Cambridge Seven," the band of young men who went to China with the China Inland Mission in 1884, causing a sensation in Great Britain and the USA. After gaining experience in Shansi, Cassels went to Szechwan, western province of China (population 68,000,000), where Hudson Taylor* decided to establish a "Church of England" district. Cassels, an ordained Anglican, now took the lead in this new enterprise and in 1895 was consecrated first bishop in Szechwan by Archbishop Benson. The new diocese was under the joint sponsorship of the Church Missionary Society and the China Inland Mission.
LESLIE T. LYALL

CASSIAN, JOHN (d. c.433). Writer on asceticism. Brought up probably at Dobrudja (in modern Romania), he entered a Bethlehem monastery by 392. Leaving there for Egypt, he studied for seven years the aims and practices of ascetics, especially Paphnutius of Scete. On this experience he based the *Collectiones Patrum* (after 420). Made deacon at Constantinople about 402, he admired Chrysostom,* on whose downfall (405) he went to Rome, where he met Leo, who later invited him to refute Nestorius (*De Incarnatione,* 430). At Marseilles, Cassian founded a monastery and a nunnery about 415. In reply to a request from Bishop Castor of Apt for advice on Egyptian asceticism, he wrote the *Institutes* (425-30), which influenced the Benedictine* Rule and was regarded as an ascetic classic for centuries. Following Origen and Evagrius, he made contemplation of God with a pure mind and will his central aim. His methods were Bible study, withdrawal of the soul from ecclesiastical office or sex, striving to defeat the seven temptations and to shun the eight capital sins, and practicing humility and love. Beginners needed the set prayer and routine of a coenobium to stimulate this, but the acme lay with the silent, secluded anchorite.* Fear of idleness arising from omnipresent grace made him stimulate Gallic opposition to some aspects of Augustinianism (*Collectio* 13). G.T.D. ANGEL

CASSIODORUS, FLAVIUS MAGNUS AURELIUS (c.477-c.570). Roman noble, statesman, scholar, and monk. Under Theodoric the Great and the regency of Amalasuntha he was successively quaestor, consul, *magister officiorum,* and praetorian prefect. His official *Variae* (twelve books of imperial edicts and decrees) is an important source for knowledge of late Roman administration and of the Ostrogothic kingdom. After the collapse of Gothic rule he withdrew about 540 from Ravenna to the monastery he had built at Vivarium on his ancestral estate in Calabria on the shores of the Gulf of Squillace. He made his foundation a great center of secular and religious learning; he established the copying of manuscripts and helped to create the monastic tradition of preserving classical culture in the Dark Ages. His own *Institutiones Divinarum et Saecularium Litterarum* (550-60) was a classic, advocating the Augustinian fusion of secular and sacred learning in Christian education. The first part is an introduction to theological study, and the second a manual of studies on the seven liberal arts. He also wrote historical books and biblical commentaries. PETER TOON

CASSOCK. A long close-fitting black tunic worn by clergy under surplice or gown in church, or as ordinary attire, sometimes in the latter case with topcoat. It originated in the *vestis talaris,* or ankle-length dress, laid down for clergy by the Council of Braga in 572, and was retained despite the barbarian pressure for shorter garments in secular life. The Roman Catholic Church used to insist on its invariable use by all clergy, although overcoats were permitted, and the Anglican Canons of 1604 similarly insisted on the use of cassocks in public. In both churches a greater freedom is now implicitly allowed.

CASTELLIO, SEBASTIAN (1515-1563). Protestant theologian. Born in Savoy, he went to Geneva as a teacher, having met Calvin in Strasbourg. His disagreements with Calvin concerned several points (for example, the interpretation of Christ's descent into Hades); but Castellio particularly opposed Calvin's doctrine of predestination. Beza suggests there were personal grounds of disagreement also. When plague broke out in Geneva in 1542, Castellio was one of three volunteers—Calvin himself and Peter Blanchett were the two others—who offered to serve as pastor to a hospital treating plague victims. When lots were cast and Castellio was chosen, he refused to go. Calvin wanted to serve, but was prevented by the Senate from doing so. Castellio, piqued because Calvin had not commended his French translation of the NT, attacked some of the Reformed doctrines and insisted that the Song of Songs should be expunged from the canon as impure and obscene. There was, furthermore, ground for suspicion that Castellio had had a hand in some anonymous tracts against Calvin's teaching on predestination and particularly that he was responsible for a treatise published under the name of Martinus Bellius, with the title *De non Puniendis Gladio Haereticis,* opposing Calvin's view that the state should be responsible for the punishment of heretics. Castellio, who had been forced to leave Geneva and go to Basle, published there elegant French and Latin translations of the Bible.

HUGH J. BLAIR

CASUISTRY. The science of applying the standards of ethics, or moral principles, to bear on particular kinds of cases. The word is generally restricted to an established code, though it can refer to individual judgment. Its true, though difficult, function is to make the larger moral principles clear in the complexities of human situations. In theology and ethics it concerns questions of conduct and conscience. Its history includes the development of "universal private penance," evolving into a system of complete legal digests. The Jesuits of post-Reformation days used casuistry to defend conduct that appeared wrong to the intuition of common sense, developing "Probabilism,"* which held that an action may be justified if supported by the opinion of one Christian doctor.

GORDON A. CATHERALL

CASWALL, EDWARD (1814-1878). Hymnwriter. Son of an Anglican vicar, he was educated at Oxford, and after ordination was incumbent of Stratford-sub-Castle, Salisbury (1840-47). Having become a Roman Catholic, he joined J.H. Newman* at the Oratory, Edgbaston, in 1850. There he wrote most of his poetry. His translations of Latin hymns from the Roman Breviary, etc., are widely used: for example, "Bethlehem! of noblest cities"; "Hark, an awful voice is sounding"; "Jesu, the very thought of Thee"; and "My God, I love Thee—not because." Also much used in Christian worship are "See amid the winter's snow" (an original composition); "Glory be to Jesus" (from Italian); and "When morning gilds the skies" (from German). He published *Lyra Catholica* (1849); *The Masque of Mary* (1858); *A May Pageant* (1865); and *Hymns and Other Poems* (1863).

JOHN S. ANDREWS

CATACOMBS (probably from Gr. *kata kumbas*, "at the ravine"). A term used for the subterranean Christian cemeteries whose origins go back to the first century A.D. They were labyrinths of underground galleries with connecting passages, often at more than one level. The majority of the excavations were carried out in the third and fourth centuries. Bodies were put into each of the spaces *(loculi)* hewn out of the rocky sidewalls. The *loculi* were then sealed by means of large marble slabs or tiles. Though catacombs existed in Paris, Asia Minor, Malta, and North Africa, Rome had the largest number. The latter have been extensively excavated in the nineteenth and twentieth centuries. This period of excavation was inspired by the work of such men as Adami, Raoul-Rochette, and Settele, and the classic work of G.B. de Rossi, *Roma sotterranea cristiana* (3 vols., 1864-77). About forty catacombs survive in an area of three miles around Rome, and they include those named after such saints as Callistus, Praetextatus, Sebastian, Domitilla, Agnes, Pancras, and Commodilla. Most are situated at the sides of the great Roman roads (e.g., the Via Appia) which led out of the city. Their excavation has brought to light many inscriptions, paintings, and sarcophagi. The earliest inscriptions, written in beautiful characters, are distinguished by their sober wording and the use of ancient symbols (e.g., the anchor and cross). Christian symbolism reached its most lofty expression in the third century, as is seen in the symbols of the dove, palm branch, fish, bread and basket. The fourth century saw the development of Christian epigraphy. Since burial grounds were regarded as sacrosanct in Roman law, Christians were able in times of persecution to worship in the catacombs. They also used them for services held on the anniversaries of martyrs. In order to provide sufficient air and light for the crowds who attended the services, shafts to the surface were constructed. Certain popes (e.g., Damasus, 366-84) encouraged the beautification of the catacombs—hence the first examples of Christian art date from this period. From the fifth century they were no longer used for burials, but services continued to be held. During the Middle Ages they seem to have been virtually forgotten.

BIBLIOGRAPHY: G.M. Bevan, *Early Christians of Rome, Their Words and Pictures* (1928); O. Marucchi, *The Evidence of the Catacombs* (1929); see also entries under ARCHAEOLOGY and ART.

PETER TOON

CATAPHRYGIANS, see MONTANISM

CATECHESIS. Derived from Greek, this term described the teaching and instruction given to Christian catechumens who were being prepared for baptism in the early church. Later it was used also of the books containing this teaching—for example, the *Catecheses* of Cyril of Jerusalem.

CATECHISMS (Gr. *katēcheō*, "to teach, to instruct"). Probably from the second century onward, there came into use a "catechumenate," or period of instruction preparatory to baptism, and there grew up well-drilled quasi-liturgical procedures, including responsive material. Some of this was proper to the baptismal service itself (i.e., the interrogations and responses), but other features (e.g., the *redditio symboli*) came at an earlier stage. The universal use of infant baptism after the sixth century meant that basic instruction of this catechetical sort had to be given to young children within the church after their baptism rather than as a preparation for it. Thus, in England popular expositions of the Ten Commandments, the Creed, and the Lord's Prayer are found as far back as Anglo-Saxon times, and these formed the basis for the later, considerably elaborated, medieval manuals of instruction. The term "catechism" itself seems to be without primitive precedent, and evidently not until the Reformation was it used to mean specifically the documents setting out instruction by the responsive method—though it could and did mean simply any manual of instruction. These latter became particularly popular with the rise of printing in the latter half of the fifteenth century.

At the Reformation the term was employed by Luther to describe his *Kleiner Katechismus* (1529), and this meaning attaches to virtually all

Protestant and Reformed use of the word thereafter. Luther's Catechism was based upon the Ten Commandments, the Creed, the Lord's Prayer, and the sacraments, and these became the staple components of Reformed instruction. Catechizing by this responsive method was a great instrument of reformation of the common people throughout Europe.

Three uses of the manuals may be discerned. One was simply *instructional,* applicable to all ages and classes. A second was *preparatory* for confirmation. In the pre-Reformation church the idea had grown up that confirmation should not be administered until "years of discretion." The Reformers grasped the concept as affording opportunity to teach every Christian while still in adolescence and as giving expression to their desire for intelligible theology purged of childish superstition. Thus Calvin cites some wholly mythical church history to make the case (*Inst.* IV.xix.4), and recommends the restoration of the same discipline (xix.13). Catechisms became linked in the Reformers' minds with a Reformed practice of confirmation; the English Reformers ensured that from 1549 to 1662 the catechism was actually included in the Prayer Book confirmation service.

A third use of catechisms is only slowly discernible. As the contents expanded to include detailed discussion of justification and other finer points of theology, the catechetical form tended to conceal a *confessional* purpose to the documents (cf. the *Heidelberg Catechism,* 1563, written by Olevianus and Ursinus, and revised by the Synod of Dort*). For this purpose the form was purely a literary convention, but catechisms, however elaborate, never served this purpose solely. With few books available and with widespread illiteracy, passages of great length and complexity were committed to memory by ordinary worshipers. Nevertheless, the catechisms tended to take their place among the formularies of the respective churches and to declare their public stance. The Church of England Catechism to this day is part of the *Book of Common Prayer,* * its doctrine being cited in detail in the famous heresy hearings of Gorham versus the bishop of Exeter (1846-50), and still included within the doctrine of the Church of England to which all ministers declare assent when they are ordained, instituted, or licensed. This catechism is much less confessional in appearance than other Reformation counterparts on the Continent (e.g., the catechisms of Calvin, Oecolampadius, Bullinger, etc.). The origins of the Prayer Book Catechism of 1549 appear to lie in William Marshall's *Goodly Primer in English* (1534), itself an approved version of a banned (more Lutheran) predecessor.

At the Hampton Court Conference* (1604), the Puritan impatience with the brevity of the Prayer Book provision led to Dr. Reinolds urging that, as Nowell's *Catechism* (1563) was too long, a uniform *via media* should be produced. The idea arose of adding a section on the sacraments to the Prayer Book one, and this was "penned" by Dean Overall. This is usually taken to mean he was the author, but he may have been the amanuensis of a small group working from Nowell's catechism, which obviously lies behind this section. The additions to the Prayer Book were made by royal proclamation (1604). At the 1661 Savoy Conference* this same section was cited by the Puritans as a model for the expansion and augmentation of the rest of the catechism. Their hopes were not fulfilled.

The seventeenth century saw the production of the most famous catechism of history—the Westminster *Larger Catechism* (1647). Closely related to the Westminster *Confession of Faith* and *Shorter Catechism,* this again combines the three uses of catechisms set out above and has become a foundation document of the English-speaking Presbyterian churches. It is obviously largely confessional in purpose, though in fact traditionally learned by heart by some. Less well known, because abortive, is Richard Baxter's own program for instruction and preparation in *Confirmation and Restoration* (1658). He follows Calvin in urging a laying on of hands, completing the course in catechizing and examination.

The use of the Church of England Catechism was the subject of visitation queries by many bishops in the century or more following 1662. In the nineteenth century the revival of a pastoral concept of confirmation led to increased use of the catechism, which persisted until the second quarter of the twentieth century. The Anglo-Catholics of the nineteenth century wrote their own catechisms, giving expression to the new emphases they had brought into the teaching of the Faith. Gradually, in the twentieth century, other teaching methods replaced learning by rote for confirmation preparation, and the catechism became an auxiliary aid to the teacher, not the verbal substance of the teaching, nor the basic procedure for instruction. The Convocations of the Church of England approved in 1962 *A Revised Catechism* (which includes questions on the church and ministry, the Anglican Communion, and, along with the sacraments, five other sacramental "ministries of grace"). This, perhaps because of its catechetical form, has not passed into general use or found widespread favor.

The Church of Rome also employed the printing press, this time to rebut the Reformers and confirm the faithful by way of catechisms. The great sixteenth-century work of this type is Canisius's *Summa Doctrinae Christianae* (1554). This had 211 questions and was in time issued in many different translations. Others have succeeded it to the present day, and catechizing is still used in the Church of Rome. There is, however, no one document in which instruction is mandatory prior to first confession, first Communion, or confirmation. One catechism of historical interest was the Irish *Keenan's Catechism.* This, in the editions prior to 1870, asked "Is then the Pope infallible?" and gave the answer "No, this is a Protestant calumny." The text was predictably changed after Vatican I.* There has been a tendency in the Church of Rome to call other teaching manuals "catechisms" even when not in catechetical form. Thus the *Catechism of the Council of Trent* (1566) is not in this sense a catechism at all, nor indeed is the recent Dutch avant-garde manual *A New Catechism* (1968).

BIBLIOGRAPHY: W.A. Curtis, *A History of Creeds and Confessions of Faith* (1911); T.F. Torrance, *The School of Faith* (1959); B.A. Gerrish, *The Faith of Christendom* (1963).

COLIN BUCHANAN

CATECHIST. In the early church the word was used to describe the person who taught the catechesis* to the catechumens. Clement and Origen were famous Alexandrian catechists. In modern times it has been used (by missionaries) of native pastors, teachers, or preachers, and sometimes in the Western world of those who teach children the Christian faith—especially in the Roman Catholic Church.

CATECHUMENS. Converts to Christianity being prepared for baptism. In the early church there was a very thorough preparation before entrance into the privileges of church membership. According to the *Apostolic Tradition* of Hippolytus, the catechumenate was in two parts: a preliminary (often long) training in doctrine and ethics, followed by an intensive spiritual preparation immediately before baptism at Easter. The latter included fasting, prayer, and exorcism. After the time of Constantine the Great, the number of converts was too great to continue this lengthy preparation. After a brief period of teaching, those approved by the church had special spiritual preparation during Lent; this included prayer, fasting, and exorcism, and learning the Creed. No catechumen could share in the Eucharist until after baptism. The more widespread infant baptism became, the less there was heard of the catechumenate. It was finally compressed into a brief rite to be performed at the church door before the baptism of an infant (see e.g., the Latin Sarum rite and the 1549 English Prayer Book).

PETER TOON

CATENA. The Latin word for "chain," it was originally used to describe biblical commentaries, from about A.D. 400 onward, in which portions of Scripture were explained by means of chains of sentences or paragraphs derived from earlier commentaries, without any personal comment by the compiler. Later it was used of collections of writings by different authors, which related to one topic—e.g., a gospel. Thomas Aquinas made a famous collection of the gospels, the *Catena Aurea*.

CATESBY, ROBERT (1573-1605). Originator of the Gunpowder Plot,* which he devised in 1603-4. Of a recusant* family in Warwickshire, his father having suffered in the second half of Elizabeth's reign for befriending Roman emissaries, Robert was educated at Douai and at Gloucester Hall—later part of Worcester College—Oxford, the recusant parallel to Peterhouse, Cambridge. On the death of his father in 1598 he became very rich, but suffered a heavy fine for his support of Essex's rebellion in 1601. After the failure of the Gunpowder Plot he fled to Holbeche in Staffordshire, where he was killed while allegedly resisting arrest.

CATHARI (Gr. *katharoi*, "pure ones"). An appellation assumed by third-century Novatianists,* but more usually associated with a widespread ascetic sect of medieval times. The latter probably arose in Armenia or the Balkans, possibly resulting from a fusion of Paulician and Euchite doctrines. In Bulgaria they were called Bogomiles,* in France Albigensians.* They spread to western Europe and were known in Orléans by 1017. Despite persecution they survived until the fourteenth century, when they succumbed to the Inquisition.* Their doctrines were akin to Manichaeism* and Gnosticism,* with elements such as dualism, universalism, Docetism, and metempsychosis. They were divided into two classes: *credentes* ("Believers") and *perfecti*. The latter received the baptism of the Spirit by the imposition of hands, called *consolamentum*, which removed original sin and restored immortality. They rejected marriage and sexual intercourse, practicing a rigid asceticism. From them were chosen bishops and priests. The *credentes* only had to promise to become *perfecti* before death, i.e., to receive the *consolamentum*. The *endura*, or ritual suicide, was sometimes permitted or recommended when the recipient of the *consolamentum* was seriously ill. Infant baptism and purgatory were rejected.

J.G.G. NORMAN

CATHARINUS AMBROSIUS (1485-1553). Dominican theologian and archbishop of Conza. Born at Siena with the name of Lancelotto Politi, he studied both canon and civil law before receiving a doctorate. Influenced by Savonarola,* he entered the Dominican House of San Marco at Florence in 1517, taking the name Ambrosius Catharinus (from the two names Ambrosius Sansedoni and St. Catherine). He wrote against Luther in 1520 with *Apologia Pro Veritate Catholicae Ac Apostolicae Fidei*. Luther responded with *Ad Librum A. Catharinii Responsio*, to which Catharinus replied with *Excusatio Disputationis contra Lutherum*. Later he even wrote against a fellow Dominican, Cardinal Cajetan*; this book, *Annotationes*, defended the doctrine of the Immaculate Conception. Another object of his literary skill was Bernardino Ochino, the friend of Peter Martyr. While living in France he published his *Opuscula Magna* (1542), which revealed that he was not afraid to deviate from Dominican orthodoxy on matters such as predestination, original sin, the Virgin Mary, etc. After taking a prominent part in the Council of Trent* he was made bishop of Minori in 1546 by Paul III and archbishop of Conza by Julius III in 1552.

PETER TOON

CATHEDRAL. A church which contains the throne or *cathedra* (Gr. = "chair") of the bishop of the diocese. In the early Christian basilicas the *cathedra* was placed at the center of the back of the apse behind the altar. From there the bishop presided at the celebration of the Eucharist and preached the sermon. The place where the bishop had his teaching chair assumed a special dignity in the eyes of his people, and gradually the "cathedral church" became the mother-church of the diocese.

At first the cathedral was always in the immediate vicinity of the bishop's residence, and its services were maintained by the bishop and his household of chaplains. But as the bishop came to be responsible for a greater area and his pastoral and administrative duties became more exacting, the care of the cathedral and its services were delegated to a separate body of clergy which developed into an ecclesiastical corporation or "chapter" with its own privileges and rights. Notably the chapter came to have the right of electing a new bishop when the see fell vacant. The bishop's relationship with "his" cathedral tended to become formal and distant, and he visited it only on special occasions.

A cathedral is not necessarily the largest or even the finest church in a diocese, but the term has come to connote a building of outstanding size, grandeur, or beauty. This is largely due to the immense amount of cathedral building undertaken in the Middle Ages, described by Jean Gimpel as "the cathedral crusade." Cathedrals were usually situated in towns, and during the Middle Ages such a town gradually increased in prosperity, size, and independence. It was natural that this should be reflected in the enlargement and enrichment, or rebuilding on a grander scale, of the principal church within its walls. In addition to the bishop's throne and the clergy's choir, the cathedral included a nave for the people. Guild business and even buying and selling took place there. At Chartres the transepts of the cathedral served as a kind of labor exchange, and the crypt was always open for the shelter of pilgrims and the sick. The cathedral of Amiens with an area of over 84,000 square feet could accommodate the town's entire population of over 10,000. The legacy of this era is a large number of buildings of great size and grace, and also a feeling that Gothic is the "natural" style of architecture for churches in general and especially for cathedrals. In the USA, recent cathedrals such as St. John the Divine, New York, are built in this style. In England, the new Roman Catholic Cathedral in Liverpool is an attempt to break free from both the style and the arrangement of the Gothic cathedral.

Cathedral worship is often characterized by the beauty and elaborateness of the music. Many cathedrals maintain schools where choirboys are trained in music besides receiving a general education.

See articles under ARCHAEOLOGY, ARCHITECTURE, and ART. HENRY R. SEFTON

CATHERINE DE' MEDICI (1519-1589). French queen and regent. As a relative of Pope Clement VII she became a pawn of the rivalry between the Hapsburg and Valois monarchs, resulting in her marriage to Francis I's son, Henry. In 1547 she became queen of France. Widowed by 1559, Catherine, first as queen mother and after 1560 as regent of France, was deeply involved in the political intrigues associated with the Wars of Religion. Notable was her calling of the Colloquy of Poissy,* and her move toward toleration for the Huguenots* (whose support she even sought). This made the latter more militant and led to civil war. In 1568 she abandoned her policy of toleration and, partially due to her jealousy over Coligny's growing influence over her son, Charles IX, plotted the infamous St. Bartholomew's Day massacre* (1572), for which she accepted full responsibility. She was a typical woman of the Renaissance courts: desiring power, given to intrigue, ruthless, a great patron of the arts and letters, but with little understanding of religious conviction.

See J.E. Neale, *The Age of Catherine de Medici* (1943); and D. Stone, Jr., *France in the Sixteenth Century* (1969). KENNETH R. DAVIS

CATHERINE OF ALEXANDRIA. Virgin and martyr. According to Simeon Megaphrastes in the tenth century, she defended the faith at Alexandria before philosophers and courtiers, and was tortured to death by "Maxentius" the emperor. A corpse disinterred by monks on Mt. Sinai in the eighth or ninth century, the legend continues, had been transported there by angels from Alexandria. In 1063 the Order of the Knights of Mt. Sinai was formed to protect her relics and observe the Rule of St. Basil. Their emblem was a broken spiked wheel. Historical detail forbids her identification with the anonymous Christian lady of Alexandria in Eusebius *HE* 8.15.

CATHERINE OF GENOA (1447-1510). She was born Caterinetta Fieschi, daughter of a distinguished Guelph family. For diplomatic reasons her brother arranged her marriage in 1463 to Giuliano Adorni, the headstrong and pleasure-loving son of an equally distinguished Ghibelline family. In 1474 she was suddenly converted and persuaded her husband to release her from the obligations of their marriage. While Giuliano entered the Third Order of the Franciscans, Catherine worked with the Ladies of Mercy in the care of the hopelessly ill in the hospital of St. Lazarus in Genoa. Her teaching, which grew out of her own remarkable spiritual experiences, was recorded in her *Dialogues on the Soul and the Body* and her *Treatise on Purgatory.* She fasted regularly and received Communion every day, a practice unusual for lay Christians in the fifteenth century. In her later life she took Cattaneo Marabotto as her confessor and spiritual director. She remained in Genoa as the rector of the hospital, nursing her own husband in his last illness (1497-98). She was canonized in 1737.

DAVID C. STEINMETZ

CATHERINE OF SIENA (1347-1380). Dominican tertiary.* She had a vision when she was seven, when she vowed her virginity to Christ. She became a Dominican tertiary in 1364/5, and from 1368 to 1374 lived in Siena, gathering a circle, clerical and lay, around her, who gave rise to her well-known letters on many subjects. In 1375 interest in a crusade took her to Pisa, where she also received the stigmata. In 1376, at the request of the Florentines, she journeyed to Avignon to meet with Gregory XI. While unsuccessful in her attempts to help the church's Babylonian Captivity,* she did figure in Gregory's decision to have the Curia removed to Rome that year. After the Great Schism* in 1378 the rest of her life was spent in Rome working toward unity in support of Urban VI. She went about fearlessly, and her political involvements were always for spiritual ends. Bearing the unmistakable marks of her order, and having always its protection in her worldly associations, she and Francis of Assisi* were named the chief patron saints of Italy by Pius XII in 1939. A *Dialogue* of four treatises was her testament, and Augustine and Bernard were for her as significant sources as Aquinas. She was canonized in 1461. A complete edition of her letters (4 vols., ed. N. Tommasèo), was published in Florence in 1860.

See also biographies by Raymond of Capua (ET 1960); and A. Levasti (ET 1954); and R. Fawtier, *Sainte Catherine de Sienne et la critique des sources* (2 vols., 1921-30). C.G. THORNE, JR.

CATHERINE OF SWEDEN (1331-1381). The daughter of Bridget of Sweden,* she succeeded her mother as head of the Brigittine Order, and under Catherine the order was confirmed. Her attempt to secure the canonization of her mother, however, came to nothing. She was an ally of Catherine of Siena* and supported Urban VI and his followers during the Great Schism.

CATHOLIC (Gr. *katholikos,* "according to the whole," hence "general"). In Ignatius the church was "wherever Jesus Christ is present." Thereafter no definition for the term more completely summarizes both its intent and the inherent problems that emerge therefrom than does the usage of Vincent of Lérins* (fifth century): "that which has been believed everywhere, always, by everyone"—a condition which historically has never been met. By analogy the concept appears also in the Athanasian Creed* applied to the Christian faith itself, but even earlier the church had been referred to as "one, holy, catholic, and apostolic." Subsequent, more limited, understanding of "catholic" has separated the Roman Church from those protesting for its reform, the Western Church from its Eastern counterpart called "ecumenical" or "orthodox," and an assumed orthodoxy from an assumed heterodoxy which, as in Valentinian practice, also dared to use the term.

CLYDE CURRY SMITH

CATHOLIC ACTION. Organized activity by Roman Catholic laity seeking to influence public affairs. It received strong impetus in the pontificate of Pius XI, whose first encyclical *Ubi Arcano* (23 December 1922) encouraged the new lay organizations like *Azione Cattolico* in Italy, Jocists in Belgium and France, the Legion of Mary in Ireland, the Grail Movement in Holland. There were notable confrontations between Catholic Action and Fascism in Italy in 1931.

CATHOLIC APOSTOLIC CHURCH. The first decades of the nineteenth century saw an increasing dissatisfaction with the oversimplified Gospel of the earlier evangelical movement. The quest for a more experimental faith and a fuller biblical exegesis led to greater emphasis on the work of the Holy Spirit, ecclesiology, and prophecy. These subjects were of major interest to such orthodox churchmen as Haldane Stewart, Hugh MacNeil, and William Marsh, who together with Edward Irving* and many others attended at Henry Drummond's* invitation the Conferences for Biblical Study at Albury Park, Surrey, in 1826. The Catholic Apostolic Church grew out of the fact that under Irving's influence many attending these meetings came to believe that the special gifts of the apostolic age were a permanent endowment of the church, restrained only by the faithlessness of later Christians.

In 1830 Mary Campbell (later Mrs. Caird), of Fernicarry in J.M. Campbell's* parish of Row, and James and Margaret Macdonald of Port Glasgow spoke in tongues and experienced miraculous healing. Great interest was shown among Drummond's friends, and in 1831 similar gifts appeared in Irving's congregation in Regent Square, London. The church which Irving founded in Newman Street after his repudiation by the London Presbytery became "a rallying point of millennial expectation," though his own influence on the community declined as he had declared the utterances in his congregation to be the authoritative voice of the Holy Spirit.

The conviction that the Day of the Lord was near dictated the new community's structure. In 1832, under prophetic direction, twelve latter-day apostles (including Drummond, but not Irving) were recognized. These, together with the original Twelve, were expected to occupy the twenty-four thrones of Revelation 4. The "Restored Apostolate's" mission was to warn the church of the impending Second Coming and to heal her schisms. In 1836 their testimony was delivered to King William IV and the Anglican hierarchy, and in 1838 the apostles and their assistants delivered similar memoranda in the different regions of Christendom overseas.

In the face of almost complete apathy, the community developed elaborate ritual involving the Real Presence, Perpetual Reservation, Holy Water, etc., together with the distinctive sacrament of Sealing by which members were numbered among the 144,000 of Revelation 7 and would thus escape the Great Tribulation. The movement spread on the Continent as well as in England, but in theory at least did not withdraw from other ecclesiastical bodies. Rather, they regarded themselves as an order within the church universal. Attempts were made on the Continent to come to terms with the death of the Apostles (see NEW APOSTOLIC CHURCH), but the movement is now almost nonexistent, as the 144,000th witness has been sealed.

The striking similarity between the Catholic Apostolics and the Tractarians* must be noted. Not only is there a common abandonment of traditional Protestant austerity in favor of ritualism and a fuller ecclesiology, but also a response to the experimental and intuitive emphases of the Romantic movement (cf. Irving's connections with Coleridge). Likewise, both had very little use for social reform. There are also several personal links between the movements—e.g., Joseph Wolff, Gladstone's* respect for Irving, H.J. Owen's influence on churchmen (see *The Guardian,* 25 March 1863).

BIBLIOGRAPHY: M.O.W. Oliphant, *Life of Edward Irving* (1862); E. Miller, *The History and Doctrines of Irvingism* (1878); G.W.E. Russell, *The Household of Faith* (1902), pp. 264-74; A.L. Drummond, *Edward Irving and His Circle* (1938); P.E. Shaw, *The Catholic Apostolic Church* (1946); J. Robert, *Catholiques-apostoliques et néo-apostoliques* (1960). TIMOTHY C.F. STUNT

CATHOLIC EPISTLES, see EPISTLES, GENERAL

CATHOLICOS. Basically an adjective denoting "general, universal," it naturally became a title, both secular ("supervisor of accounts") and ecclesiastical ("archbishops," "heads of monastic houses"). Today the title has become limited in use to certain patriarchs: among Orthodox, the head of the Georgian Church; among Monophysite churches, the head of the Armenian (Gregorian) Church, and the Armenian patriarch of Sis; among Nestorians (Assyrians), "catholicos of the East"; and among Roman Catholics, the heads of the Uniate Armenians, and of the Uniate Nestorians (Chaldeans).

CATON, WILLIAM (1636-1665). Quaker. He was convinced in 1652, when George Fox* first visited Swarthmore Hall on the edge of the Lake District, where Caton was companion to Margaret Fell's son George. Caton soon became a Publisher of Truth, traveling at first in the company of John Stubbs.* They went to Kent and converted Samuel Fisher, the notable Baptist pastor at Folkestone. They were imprisoned briefly at Maidstone. In 1655 they made the first important visit by Quakers to the Continent. Caton spent much of his remaining life in Holland and also visited Scotland. His autobiography, edited by Fox, was published in 1689. A collection of early Quaker letters put together by Caton has apparently now been lost. JOHN TILLER

CAUSSADE, JEAN PIERRE DE (1675-1751). French mystical writer and preacher. He joined the Jesuit order in 1693 and worked in Lorraine, Perpignan, Albi, and Toulouse. He was the last of the seventeenth-century mystical school, struggling to bring mysticism out from under the cloud induced by the condemnation of Quietism.* In *Instructions spirituelles en forme de dialogue sur les divers états d'oraison* (1741), he used the authority of J.B. Bossuet,* who had played the leading part in the condemnation of Quietism, in an attempt to rehabilitate the mystical approach. He also wrote *L'Abandon à la Providence divine,* an influential book concerned with "the sacrament of the present moment," consisting first of his teaching and second of a series of letters of spiritual direction addressed to nuns of The Visitation of Mercy. His manuscript notebooks were passed from hand to hand, and the work was first published in 1867 (ET 1921). His work has a continuing influence. HOWARD SAINSBURY

CAVASILAS, NIKOLAOS (c.1320-1371). Greek theological and mystical writer. He was born in Thessalonica of an aristocratic family, and although his father's name was Chamaetos, he is known by the more famous surname of his mother and his uncle. After an early political career linked with the pretender John VI Cantacuzenus, his public activity waned with the latter's abdication; we know little of his life after 1354, although his friends sought his participation in affairs. It is uncertain whether he was ordained. His best-known works are treatises on *Life in Christ* and *The Exposition of the Divine Liturgy* in which he expounds the saving work of Christ mediated through the sacraments, especially the Eucharist, appropriated by the believer steadfast in his adherence. J.N. BIRDSALL

CAVEN, WILLIAM (1830-1904). Canadian Presbyterian leader. He went to Canada from Scotland in 1847, graduated in theology from Knox College, Toronto, was ordained to the Presbyterian ministry (1852), and did parish work before becoming professor of exegetical theology at Knox in 1866. He was principal there, 1873-1904. Strongly ecumenical, president of the Pan-Presbyterian Alliance (1900-1904), leader in the Evangelical Alliance, and active in missions, he was an architect of the future United Church of Canada. When he was moderator of the general assembly in 1875, there came union with the Church of Scotland, forming the Presbyterian Church in Canada. A number of his papers were published posthumously as *Christ's Teaching concerning the Last Things* (1908).

C.G. THORNE, JR.

CAXTON, WILLIAM (c.1422-1491). First English printer. Born in Kent, he was apprenticed to a cloth dealer in the City of London in 1438 and very soon was sent to Bruges for his employer. In 1446 he set up in business on his own in Bruges and became the leader of the English business community there. He began to travel widely in Europe and developed a literary interest which led him to begin translating books into English. In 1471 he finished his first translation and decided to learn the new skill of printing in order to produce his own books. After studying the technique, either in Cologne or Bruges, he set up a printing press in Bruges. In 1476 he returned to Britain and set up his press (probably the same one he had had in Bruges) in the precincts of Westminster. The first book he printed in England was the *Sayings of the Philosophers,* and this reflected his wide literary interest. From 1477 to 1491 he was continuously busy printing and translating at Westminster. Among the early books he printed were Chaucer's *Canterbury Tales* and Malory's *King Arthur* (1485). In fourteen years he printed 18,000 folio pages, making up altogether eighty separate books; he also translated twenty-one volumes, mostly from French and Dutch. During this period he designed and put into use six fonts of type of the Gothic style.

BIBLIOGRAPHY: W. Blades, *The Life and Typography of William Caxton* (rev. ed., 1882); E.G. Duff, *William Caxton* (1905); W.J.B. Crotch (ed.), *The Prologues and Epilogues of William Caxton* (1928). A. MORGAN DERHAM

CECILIA. Virgin and martyr. She is mentioned in several Western breviaries and missals. A church of St. Caecilia housed a council at Rome in 498. According to Fortunatus of Poitiers in the sixth century, she died in Sicily between 176 and 180, but her *Acta* given by Simeon Megaphrastes, the tenth century legendist, sets her death at Rome when Urban was bishop (220-30). Her first clear association with music was in 1584, when Pius V put a new academy of music under her protection.

CEDD (Cedda) (d.664). Bishop of the East Saxons. A Northumbrian, he was with his brother Chad one of the twelve pupils of Aidan* on Lindisfarne. In 653 he was sent by King Oswy of Northumbria to evangelize Mercia by permission of the heathen King Penda. Next year he was sent to Essex and consecrated bishop of the East Saxons. He built a number of churches and founded two monasteries. On a return visit to Northumbria he founded the abbey at Lastingham, Yorkshire, and became its first abbot. He was present as one of the Celtic representatives at the Synod of Whitby in 663/4 and accepted the verdict in favor of Roman customs.

CELESTINE I (d.432). Bishop of Rome from 422. Soon after he succeeded Boniface I, he was in controversy with African bishops. When Apiarius,* a deposed African cleric, requested reinstatement, Celestine complied, sending the unpopular legate Faustinus to execute the decision. A Council of Carthage about 424 dismissed Faustinus, asking Celestine to obey Nicene canons and to support the decisions of local bishops. Celestine opposed heresy and irregularity. In 425 the emperors expelled heretics from Rome. He censured bishops around Marseilles for allowing priests to teach doctrines opposing Augustine. Others, in Vienne and Narbonnensis, he criticized for electing monastic bishops and for ascetic clerical dress (428). He sent Germanus of Auxerre to England (429) to counter Pelagianism,* and he closed Novatianist* churches at Rome. In 430 he empowered Cyril of Alexandria* to execute a ten-day ultimatum on Nestorius. Although his representatives were absent from most of the Council of Ephesus (431), his approval of its decisions later earned Celestine the reputation of having presided with Cyril.

G.T.D. ANGEL

CELESTINE III (c.1106-1198). Pope from 1191. A follower in his youth of Abelard,* he had built up a reputation by the time he became pope as a learned theologian, good administrator, and irenic negotiator. He was anxious to avoid a conflict with the empire for political supremacy. Thus he was willing to crown Henry VI as emperor in 1191, and though relations became strained because of Henry's Italian ambitions, they never completely collapsed. He accepted Henry's clever promise to lead a Crusade in 1195, but made sure it had a wide political base by having it preached in other countries. He prevented a marriage of Alfonso IX of Leon within the prohibited degrees, and he refused to agree with the French bishops' nullification of Philip Augustus's marriage. Administratively he improved the Curia, and the *Liber censuum*—a survey of all property dependent on Rome—was taken. He continued the extension of papal jurisdiction particularly by using delegate-judges to hear disputed cases.

C. PETER WILLIAMS

CELESTINE V (1215-1296). Pope in 1294. A noted hermit, he had founded a community of disciples later called the Celestines.* His election as pope was a conscious attempt to raise the spiritual tone of the papacy, debilitated by political involvement in the War of the Vespers. The inexperienced Celestine, however, soon became the tool of Charles II of Naples, appointing twelve cardinals of Charles's choice, placing his nominees in the papal states, and taking up residence in Naples. Administratively the church slipped into great confusion. Having been made aware of his own failings, Celestine had the spirituality and humility to resign at the end of 1294. He was succeeded by an opposite in every way—Boniface VIII, who, fearing conspiracy around the former pope, had him imprisoned until he died in 1296. Celestine was canonized in 1313.

C. PETER WILLIAMS

CELESTINES (Celestinians). An order of hermits who adopted the Rule of St. Benedict about 1251 (see BENEDICTINES). When their leader Peter of Morrone became Pope Celestine V in 1294, his monks took the name of Celestinians. Their discipline was severe. Celestine V introduced them into Monte Cassino, from which monastery they were removed by Boniface VIII. The order increased rapidly, spreading into France and Germany. At one time there were 150 priories. The German priories perished during the Reformation. In the early eighteenth century there were ninety-six houses in Italy and twenty-one in France, though the French Celestinians were suppressed by a commission in 1766. In 1785 the last surviving house at Calavino, near Trent, was closed. An attempt to revive the order in the nineteenth century met with no success.

JAMES TAYLOR

CELESTIUS (early fifth century). Associate of Pelagius.* Probably an African, he deserted the Roman bar about 400 for Pelagius's reform movement, accompanying him in 409-10 to Sicily and Carthage, where Pelagius left him hoping for ordination. He received instead in 411 the first Catholic condemnation of a "Pelagian," for teaching that Adam's mortality was independent of his sin, which in turn injured only himself, so that the newborn enjoy Adam's condition prior to the Fall. Infants are baptized in order to obtain sanctification or the kingdom, not remission of sins. His denial that transmission of sin was *de fide* challenged African tradition, which was also sensitive about baptism. Withdrawing to Ephesus, Celestius acquired ordination, but here and at Constantinople African denunciation hounded him, and further condemnation followed in absentia at the Palestinian synod of Diospolis in 415, with Pelagius, unconcerned about infant baptism, partly disavowing his opinions. Back in Rome in

417, Celestius appealed personally to the vacillating Pope Zosimus,* but in 418 Emperor Honorius expelled the Pelagian troublemakers and papal reprobation was reaffirmed. In 428/9 he secured refuge in Nestorius's Constantinople, but Marius Mercator's *Memorandum on the Name of Celestius* heralded imperial exile and conciliar condemnation at Ephesus in 431, after which he vanished from sight. Augustine preserves much of his *Definitions on Sinlessness* and *Statement of Faith* (to Zosimus). Lost works included *The Monastic Life* and a *Book against Original Sin.* He was a successful propagandist, tenacious and candid. His theology owed much to Rufinus "the Syrian."*

See PELAGIUS bibliography. D.F. WRIGHT

CELIBACY, CLERICAL. The Roman Catholic practice of requiring its clergy to remain permanently unmarried and devoted to personal purity in thought and deed. The NT seems to be ambivalent on the subject of marriage. On the one hand, some of the apostles were married (Matt. 8:14; 1 Cor. 9:5), and Paul recommended marriage for the leaders of churches (1 Tim. 3:1), but on the other hand, the values of virginity are stressed. Besides the examples of Christ, Mary, and John the Baptist, there are (1) the teaching of our Lord that celibacy is a way of consecrating oneself to God (Matt. 19:12, 19) and (2) the statements of Paul that celibacy is the condition for a more fervent consecration to God because it avoids earthly entanglements and prepares the soul for the coming of Christ (1 Cor. 7:26-35). The idea developed early in the church's history that the unmarried state was preferable. During the fourth century most of the bishops in Greece, Egypt, and western Europe were unmarried or left their wives after consecration. Still, priests and deacons married and no law was passed prohibiting clerical marriage during the first three centuries of the Christian era.

In the East, the sixth and seventh centuries saw laws enacted which forbade the marriage of bishops. (If he were already married before consecration, he had to put his wife away in a distant monastery.) Yet the lower orders of clergy were allowed to marry. Celibacy in the Western Church became a canonical obligation for the clergy through the combined efforts of the popes and regional church councils. The earliest canonical statement, canon 33 of the Council of Elvira (c.305), states: "We decree that all bishops, priests, and deacons, and all clerics engaged in the ministry are forbidden entirely to live with their wives and to beget children: whoever shall do so shall be deposed from the clerical dignity." Later Hosius* of Cordova attempted unsuccessfully to have this decree enacted by the Council of Nicea. If the ecumenical council would not act, the pope would, and the decretals of Damasus I, Siricius, Innocent I, and Leo I enjoin the clergy to celibacy. Other local councils in Africa, France, and Italy issued decrees enforcing this practice.

After the fall of the Carolingian empire there was a movement away from clerical celibacy, but with the Hildebrandine reform of the eleventh century a new ascetic spirit came to the church. Gregory VII, for whom this movement is named, struggled with great zeal to restore sacerdotal celibacy. Even after his time, however, there was a considerable gap between theory and practice with regard to this requirement.

The Protestant Reformers did not value celibacy. Calvin taught that it should not be judged of greater value than the married state, and he protested the despising of marriage by writers such as Jerome. The Council of Trent* (1545-63) reaffirmed the teaching of clerical celibacy, but it stated that this was enjoined on the clergy by the law of the church and not by the law of God. Currently, the Roman Church feels celibacy is useful for ministers as it gives them greater freedom in the service of God, but it also states the church may abrogate this rule if it chooses.

See H.C. Lea, *History of Sacerdotal Celibacy in the Christian Church* (2 vols., 1907); "Celibacy," in the *New Catholic Encyclopedia,* vol. II, pp. 366-74. ROBERT G. CLOUSE

CELL. Since 1945 this word has been increasingly used to describe a small group of Christians in any given locality who meet, usually in private homes, for prayer, Bible study, and fellowship, in order to equip themselves for Christian service in their everyday lives. Traditionally, however, a cell denotes one of three things: first, the private room, usually scantily furnished, of the member (monk or nun) or a religious order; second, the dwelling place of a hermit; third, a small religious house dependent on a larger convent or monastery.

CELSUS (second century). Author of the first known philosophical and religious critique of Christianity, entitled *The True Doctrine.* Writing during a persecution (177-80?) and perhaps reacting against Justin Martyr,* he debunks Judaism, Christ, and Christians, some of whom he met at Rome or Alexandria. Despite an allusion to conservation of matter, there is not enough internal evidence to identify him with the Epicurean friend of Lucian of Samosata. Only Origen's reply, *Contra Celsum,* written just before the Decian persecution (249-51) provides evidence of Celsus. Platonic philosophical monotheism combines in Celsus with Greco-Roman ancestral polytheism to produce an unknown and unmoving supreme God, who has set various demons over human experience. True religion is demonstrated both by concentrating the soul on God and by propitiating the traditional cultic demons on whom depend the empire and the everyday functions of life. Worship and service are therefore due to their agent the emperor, by celebrating public feasts, holding public office, and joining the army. On these presuppositions Celsus makes his main criticism of contemporary Christians. Further, his Platonic conception of soul and body, his pantheism, and comparative study of religions make him reject, sometimes through misconception or misrepresentation, the Christian doctrines of Creation, man, the Incarnation, the *unique* character and ministry of Christ, and the Resurrection.

BIBLIOGRAPHY: H. Chadwick, *Origen contra Celsum* (ET and introduction, 1953); J. Altaner,

Patrology (ET 1960), pp. 115-16; E.R. Dodds, *Pagan and Christian in an Age of Anxiety* (1965).
G.T.D. ANGEL

CELTIC CHURCH. The church which existed in parts of the British Isles before the mission of Augustine (597) and which maintained its independence for some time in competition with the Anglo-Roman Church. Little is known of the introduction of Christianity into Britain, but by the fourth century it was sufficiently organized to send representatives to the Synod of Arles (314) and the Council of Arminum (359). The Pelagian heresy spread to the Celtic Church, and Germanus of Auxerre visited England to try to combat it (429). Monasticism came to the Celts by way of Gaul, illustrating further the contacts the church had with the Continent.

This was changed by the Saxon invasions about 450, which isolated the British Church from continental life and resulted in the extermination of Christianity in England. It survived only in remote areas of the British Isles. When Augustine's mission reestablished contacts with Rome, the Celtic Christians argued with the Romans over such matters as the calculation of the date of Easter and variants of tonsure. These differences were settled at the Synod of Whitby* (663/4) resulting in a victory for the Roman practices.

The Celtic Church under the leadership of missionaries such as Ninian* (c.400) and Patrick* (c.440-61) featured the monastery under the abbot, rather than the bishop's diocese, as the unit of ecclesiastical organization. Each monastery served a single tribe. The abbot was a tribal leader, and the bishop was a subordinate official in the monastery whose duties were wholly spiritual. The leaders and heroes of this movement equaled the ascetic extremes of the early founders of monasticism. A typical act of self-mortification involved standing for long periods of time immersed to the head in an icy stream. Another characteristic of the Celtic monastic life was an emphasis on missionary work. The highest service to Christ was lifelong exile to evangelize foreign lands. Thus the Celts sent out men like Columbanus* (585-615), who preached in France, Switzerland, and N Italy. The Celts also encouraged scholarship and had a rich artistic tradition, particularly in sculpture and the illumination of manuscripts.

BIBLIOGRAPHY: N.K. Chadwick (ed.), *Studies in the Early British Church* (1958); *idem, The Age of the Saints in the Early Celtic Church* (1961); L. Gougaud, *Christianity in Celtic Lands* (ET 1932).
ROBERT G. CLOUSE

CENOBITES (Coenobites). The word is derived through Latin from the Greek meaning "common life." Monks within a community or order are cenobites and thus distinguished from anchorites or hermits.

CENTURIES OF MAGDEBURG, see MAGDEBURG, CENTURIES OF

CERDO. The predecessor of Marcion,* he came to Rome from Syria, according to Epiphanius, about 137-41. Irenaeus continues that he publicly professed orthodoxy, but taught heresy in secret until he separated from the congregation. He taught that the OT God is known and righteous, while the Father of the Lord Jesus Christ is unknown and good. Later writers expand or contradict Irenaeus, making Cerdo a Docetic* believing in the resurrection of the soul and presaging the canon of Marcion.

CERINTHUS (c.100). Heretic who lived in Asia Minor. His theology appears to have been a combination of Ebionite theology with Gnostic speculation. The world, he taught, was not the creation of the Supreme God, but that of an inferior angel who held the world in bondage. According to him, Jesus was a normal man, the son of Joseph and Mary, who differed from ordinary men only in greater wisdom and righteousness. He was chosen by the Supreme God to proclaim Him and release the world from its bondage. For this task the Christ descended upon him at his baptism in the form of a dove, sent from the Father. This Christ departed from Jesus before his crucifixion, and it was only Jesus who suffered and rose again. Cerinthus taught also a carnal doctrine of the Millennium, in that at his coming Jesus would introduce 1,000 years of sensuous pleasure before the consummation. According to the Alogi,* Cerinthus was the author of the gospel of John and the Apocalypse. G.L. CAREY

CERULARIUS, MICHAEL (d.1059). Patriarch of Constantinople. Under his direction the final severing between the Eastern and Western churches took place, often referred to as the Great Schism of 1054. Differences between Rome and Constantinople did not originate under his regime, but went back over many centuries. The patriarch brought to his office the ability to use political power to his end. Constantine Monachus, who was emperor at the time, was negotiating with Pope Leo IX concerning the defense of S Italy against the Normans. As a part of the deal the emperor of Constantinople agreed to give back the southern Italian churches to the jurisdiction of the papacy. When Patriarch Cerularius found out about these negotiations, he decided to flex his power against the weak emperor. He began by trying to force the Latin churches in Constantinople to use Greek language and practices. When they refused in the year 1052, he closed them down.

The ritual divergences which were used as the excuse for this became the focal points of the dispute. The real issues which had been dividing them and are still dividing them, such as the equation of the papacy and the Holy Spirit, remained beneath the surface at the time of the schism. The papacy apparently got the message that Cerularius was reacting to the capitulation of the Byzantine emperor to Rome. This led to an extreme assertion of papal authority to which, of course, Cerularius reacted. When the legates came from Leo, Cerularius refused to see them—so they excommunicated him. The patriarch in turn excommunicated those who had come representing the pope. Thus the division was complete.

GEORGE GIACUMAKIS, JR.

CESARINI, JULIAN (1398-1444). Cardinal-bishop of Tusculum and papal legate. Born of a distinguished Roman family, he entered the service of the papal court and carried out several important missions. In 1419 he was in Bohemia on a mission against the Hussites; in 1425 he represented the pope in France and in 1426 in England. In that year also he was created cardinal. He was present, as papal legate, at Domazlice in 1431 when the Czechs defeated the German princes. He was appointed by Martin V to be president of the Council of Basle. The rout at Domazlice convinced him of the need for a general council to deal with Hussitism. He tried to persuade Eugenius IV to cooperate with the council, but after the council divided and moved to Ferrara, he played a leading part in the negotiations for union between the Roman and Greek churches. Cesarini went to Hungary in 1442 to preach the crusade against the Turks. In 1444, due to his influence, the Peace of Szegedin was repudiated by King Ladislaus. In the renewed war, Cesarini, along with Vladyslav III, king of Poland and Hungary, was killed at the final, fatal battle of Varna in Bulgaria.

JAMES TAYLOR

CHAD (Ceadda) (d.672). Bishop of Lichfield. A Northumbrian, he was with his brother Cedd* one of the twelve pupils of Aidan* on Lindisfarne. In 664 he succeeded Cedd as abbot of Lastingham, Yorkshire. Because of Wilfrid's* dallying in France when he had gone there to be consecrated as bishop of York, King Oswy of Northumbria had Chad appointed to that see. After Wilfrid's return Theodore of Tarsus, archbishop of Canterbury, ruled the appointment was irregular, and in 669 Chad returned to Lastingham. The same year, King Wulfhere of Mercia asked Theodore for a bishop, and Chad was sent. He made his see at Lichfield. Chad was noted, like many others for whom Lindisfarne was a spiritual home, for his humility, his devotion, his teaching ability, and his missionary journeys.

R.E. NIXON

CHADERTON, LAURENCE (1538?-1640). English Puritan leader. Born into a wealthy Catholic family in Lancashire, he studied at Christ's College, Cambridge. When he came under extreme Protestant influences and adopted Puritanism, his father disowned him. He managed, however, to take his degree, in 1567, and to become a fellow of the college. As a tutor he was a success, but his fame rested on his preaching. For nearly fifty years he served as the afternoon lecturer at St. Clement's, and through the influence of his preaching many young men began to study the Bible and practice godliness. When Sir Walter Mildmay decided to found Emmanuel College in 1584, it was Chaderton whom he persuaded to become master. The purpose of the foundation was to train "godly ministers," and this Chaderton did. The college became a center of church Puritanism. Chaderton represented the Puritans at the Hampton Court Conference* and was connected with the translation of the Authorized Version of the Bible (1611). In 1622 he gave up the mastership of Emmanuel, but survived till he was over a hundred. His published works are few, but his influence on Puritanism was great.

PETER TOON

CHAFER, LEWIS SPERRY (1871-1952). American Presbyterian clergyman and educator. Born at Rock Creek, Ohio, he was educated at New Lyme Academy and at Oberlin College and Conservatory, and he entered the ministry as a gospel singer and evangelist. In 1900 he was ordained to the Congregational ministry, but after three years changed his affiliation to Presbyterian when he became professor of music at Mount Hermon School for Boys. When C.I. Scofield* helped to found Philadelphia School of the Bible (1914), Chafer joined the school as professor and remained there until moving to Dallas, Texas, in 1923. He was pastor of Scofield Memorial Church there (1923-27) and in 1924 founded Dallas Theological Seminary, where until his death Chafer served as president and professor of theology. He also edited the journal *Bibliotheca Sacra*. He is best known today for his eight-volume *Systematic Theology* (1947)—a most detailed discussion of the dispensational premillennial system of theology.

BRUCE L. SHELLEY

CHALCEDON, COUNCIL OF (451). Held in the martyry of Euphemia, it was summoned by Marcian and Valentinian and attended by bishops from the civil dioceses of Oriens, Asia, Pontus, Thrace, Egypt, and Illyricum. Legates represented Leo of Rome, who like most Western bishops feared the Huns too much to travel far. Including two refugee Africans and a Persian, no more than 340 attended at one time, although 450 subscriptions were made. The emperors sought a common declaration between the Eastern bishops in accord with the creeds of Nicea and Constantinople, canonical epistles of Cyril, and the Tome of Leo. Thus the imperial commissioners resisted the scruples of bishops who feared to add to the faith and insisted on a new formulary, the *Definitio*, drawn up within the biblical and patristic tradition (see CHALCEDON, DEFINITION OF).

The immediate rift had centered on Eutyches, condemned at Constantinople under Flavian and reinstated at the Ephesian synod (449) dominated by Dioscorus of Alexandria. At Chalcedon, Dioscorus was deposed, ostensibly for uncanonical action, although in effect his teaching was ostracized by conciliar support of the Tome. Theodoret and Ibas of Edessa, tending towards Nestorian teaching, were restored, but the earlier condemnation of Nestorius remained. The fall of the Alexandrian patriarch favored the see of Constantinople. Bishop Anatolius in committee steered preparation of the *Definitio*, and several canons proceeded to boost his authority. Either the bishop of Constantinople or the exarch of a diocese became the final appeal court in disputes involving (Eastern) metropolitans. Canon 28 confirmed Constantinople canon 3 correlating ecclesiastical honor with political importance, to the disgust of Rome and Alexandria. Further, Constantinople received the right to ordain metropolitans in the dioceses of Thrace, Asia, and Pontus as well as bishops among barbarian peoples.

Several canons reflect current affairs. Gang conspiracy, false accusation, and plunder directed against bishops were forbidden. Redundant clergy should not congregate at Constantinople. Clergy should not minister outside their local churches nor bishops outside their sees except as visitors bearing commendatory letters. Monks and clergy should not do state or secular work. Bishops should control the movements of monks and clergy. Other canons simply aim for good order. Clerical appointment is regulated, e.g., by opposing sinecure ordination. Marriage is forbidden after taking vows or orders. Episcopal churches should have short vacancies, a proper sequestration fund, and a financial steward. In new cities parochial division should follow the civil model.

Unity was not achieved. Rome rejected canon 28. Monophysite and Armenian churches still reject the council. But most Christian churches still respect the elusive Definition as the zenith of patristic christological endeavor.

BIBLIOGRAPHY: J. Altaner, *Patrology* (ET 1960), pp. 291-93; K. Sarkissian, *The Council of Chalcedon and the Armenian Church* (1965); essay by S.L. Greenslade in *The Councils and the Ecumenical Movement* (1968). G.T.D. ANGEL

CHALCEDON, DEFINITION OF (451). The majority at the Council of Chalcedon* acknowledged the Nicene Creed, confirmed at Constantinople, together with two Cyrilline letters and the Tome of Leo. Alarmed by reluctant Egyptian bishops and recalcitrant supporters of the condemned Eutyches, the imperial commissioners pressed for a new unifying formula. A committee headed by Anatolius of Constantinople produced a draft containing the Cyrilline "out of two natures." Others preferred the Leonine "in two natures," and the commissioners reconstituted the committee. This produced the extant *Definitio.* Basically it affirms, first, that the Lord Christ is one, His two natures preserved in one *prosopon* and *hypostasis.* This favored the Alexandrine stress on One God. Second, it states that both natures, God and man, are unimpaired, "perfect," consubstantial with God and man, preexistent and born of the Virgin. He is "acknowledged in two natures" (a phrase perhaps taken from Basil of Seleucia) "unconfusedly, unchangeably, indivisibly and inseparably." The commissioners had commended this last Leonine phrase. Third, the Definition affirmed that the distinct natures are fully God and man, thus securing salvation by a saving God and a man identified with men. Confirmed by Emperor Marcian and Pulcheria, the Definition has split the Monophysite* from other Eastern churches.

BIBLIOGRAPHY: T.H. Bindley and F.W. Green, *The Oecumenical Documents of the Faith* (1950 ed.); A. Grillmeier, *Christ in Christian Tradition* (ET 1965); R.V. Sellars, *The Council of Chalcedon* (1953). G.T.D. ANGEL

CHALDEAN CHRISTIANS. This group had its origins in Iraq and is an offshoot of the Nestorian* branch of Christianity. In the last quarter of the seventeenth century the Nestorian patriarch of Diarbekir quarreled with the head of the Nestorian sect of Christianity and then turned to the Roman pope for recognition. The pope promptly established the patriarchy of the Chaldeans, which tended to fix the division between the Nestorian groups. Thus the Chaldeans are the branch of Nestorians who recognize papal authority.

CHALICE (Lat. *calix,* a regular, small drinking cup). For its domestic background, reference can be made to the Greek particular two-handled *kulix,* in contrast with the NT's generalized *potērion,* from which is derived the special liturgical usage for the administration of the liquid portion of the Lord's Supper. Archaeologically interesting, though of doubtful date, is the Chalice of Antioch, discovered in 1910.

CHALLONER, RICHARD (1691-1781). Roman Catholic writer. Born to Protestant parents, he was educated in Roman Catholic households after his father's death and embraced that faith at the age of thirteen. A student at Douai, he became successively professor and vice-president there, returning to England in 1738 when he joined the London Mission. Following a controversy with Conyers Middleton, he returned to Douai in 1738 and was consecrated a bishop *(in partibus)* at Hammersmith in 1741, becoming vicar apostolic in 1758. He produced a modernized version of the Douai Old and New Testaments (1749, 1750), wrote anti-Protestant tracts, martyrologies, and two popular devotional works, *The Garden of the Soul* (1740) and *Meditations for Every Day of the Year* (1753). IAN SELLERS

CHALMERS, JAMES (1841-1901). Scottish missionary. Born at Ardrishaig and brought up at Inverary, Argyllshire, he heard God's call at fifteen and vowed to take the Gospel to cannibals—a vow soon forgotten until his conversion three years later. In 1862 he went to Cheshunt College, Cambridge, to train under the London Missionary Society, and in 1867 he sailed for the Cook Islands of Polynesia. For ten years he continued the work begun at Raratonga by John Williams, but he longed for unevangelized areas, especially New Guinea. The work there was begun in 1872 by six pastors from Raratonga; Chalmers joined them in 1877. During his twenty-four years there, his concern and determination carried the work through despite many setbacks. He opened up a wide area for the Gospel through exploration, and rendered great service to government officials when SE New Guinea was annexed by Great Britain and became a crown colony. He established a training institution at Port Moresby and saw whole areas transformed by the Gospel.

Chalmers was essentially a pioneer. He made up for his imperfect knowledge of native languages by faith, prayer, Christlikeness, and love for the people. His success is attributable to the fact that he never doubted he had a Gospel for them. He was murdered by cannibals during a journey to explore new territory.

See R. Lovett, *James Chalmers* (1902).

J.W. MEIKLEJOHN

CHALMERS, THOMAS (1780-1847). Scottish minister. Born in Anstruther and educated at St. Andrews University, he was inducted to Kilmany parish in 1803 and lectured part-time in mathematics at his university. In 1811 he experienced evangelical conversion, and his ministerial emphasis and activity were completely changed. He began to write for the *Christian Instructor*, publication of which marked the turning of the tide in the Moderate-Evangelical encounter. In 1815 he was inducted to the Tron Church, Glasgow, and his theology, kept within the bounds of the Shorter Catechism, was heard by crowded congregations twice every Sunday. The city's social needs were staggering. Chalmers created a new parish, St. John's, out of three existing overgrown parishes, with 10,000 people in its limits, many of them the poorest in Glasgow. He divided the parish into twenty-five areas with about 400 people in each. An elder was appointed to the spiritual oversight, and a deacon to the social welfare, of each area. Day and Sunday schools were provided for all children. Chalmers and his assistants conducted four services each Sunday. This pastoral experiment made evangelicalism a force to be reckoned with.

Glasgow reeled in shock when Chalmers left in 1823 to teach moral philosophy at St. Andrews, but when he moved on to Edinburgh in 1828 to teach divinity, his friends were sure he had found his sphere. He became moderator in 1832 and thereafter the leader and symbol of the Evangelical party. During the next decade he championed the cause of church extension, building in six years 216 churches, and raising £290,000. After the Disruption,* Chalmers became moderator of the Free Church assembly and also professor of theology in New College. His *Institutes of Theology* was published posthumously in 1849.

See H. Watt, *Thomas Chalmers and the Disruption* (1943).

ARTHUR CLARKE

CHAMBERLAIN, JACOB (1835-1908). Missionary to India from the (Dutch) Reformed Church in America. Born in Sharon, Connecticut, he studied both theology and medicine and in 1859 joined the Arcot Mission in South India. As a doctor he began two hospitals and conducted several dispensaries. But perhaps his most enduring work was literary. From 1873 to 1894 he chaired a committee to revise the Telugu Bible. He prepared a popular Telugu hymnbook and completed the first volume of a Bible dictionary. Repeated illness obliged him to spend altogether ten years in the USA, where he inspired much missionary interest. Concerned for church union, he became the first moderator of the Synod of South India in 1902. In 1878 he was the first missionary ever chosen to be president of the general synod of his denomination.

HAROLD R. COOK

CHAMBERS, OSWALD (1874-1917). Bible teacher and evangelical mystic. Born in Aberdeen where his father was a Baptist pastor, he was converted on hearing C.H. Spurgeon,* studied art in London and Edinburgh, and entered Dunoon College in 1897 to train for the Baptist ministry. During this time he met William Quarrier, founder of the Orphan Homes, and from him learned a simplicity of faith and prayer. In 1907 he visited America and Japan, afterward becoming traveling missioner for the Pentecostal League of Prayer founded by Reader Harris. He was principal of the Bible Training College at Clapham Common, London (1911-15). His last great ministry was to the troops in the desert camps of Egypt, as superintendent of the YMCA huts in Zeitoun and Ismailia. Many of his Bible studies given to students and troops have been published, and his *My Utmost for His Highest* is a devotional classic.

J.G.G. NORMAN

CHAMPLAIN, SAMUEL DE (c.1570-1635). Explorer, cartographer, and colonizer. Son of a French sea captain, he made his first trip to Canada in 1603. Next year he returned to the New World, accompanying Monts to Acadia, where they established a settlement on an island in the St. Croix River. After a very hard winter, the settlers moved across the Bay of Fundy to found Port Royal, whence Champlain explored and mapped the Acadian and New England coasts. In 1608 he established a fur-trading colony at Quebec. The rest of his life was dedicated to the success of the colony as he explored the interior, made alliances with the Indians, and pleaded its cause in frequent trips to France. He strongly emphasized Christian missions, for he dreamed of a Christian Canada ruled by the French king and peopled by a race of Indian-French people who had intermarried. In 1612 he became commandant of New France, and in 1615 encouraged the *Récollet** missionaries to come to Canada. Made governor of New France in 1633, he earned by his contribution to Canadian development the title "Father of Canada."

ROBERT WILSON

CHANCELLOR. In the Middle Ages, the chancellor acted as secretary to the cathedral chapter. In the Church of England now, he usually has jurisdiction (authorized by a patent under the seal of the bishop) over the consistory court, and is the chief representative of the bishop in the administration of the temporal affairs of his diocese. Not normally ordained, the chancellor must be at least twenty-six years old, "learned in the civil and ecclesiastical laws and at least a Master of Arts, or Bachelor of Law"; he must be "zealously bent to religion," must take the oath of allegiance, and assent to the Articles of Religion. He issues marriage licenses through his surrogates, hears applications for faculties, dispensations, etc., and hears complaints against clerics for immorality.

HOWARD SAINSBURY

CHANNING, WILLIAM ELLERY (1780-1842). Unitarian leader and abolitionist. Born in Newport, Rhode Island, he received his B.A. from Harvard in 1798 and was "regent" there from 1802 until in 1803 he was ordained as pastor of the Federal Street Congregational Church in Boston. His sermon at the ordination of Jared Sparks in Baltimore in 1809 set forth the tenets of Unitarianism, such as denial of the Trinity, the deity of Christ, total depravity, and a substitutionary atonement. In 1820 he organized the Berry Street Conference of liberal ministers out of which the American Unitarian Association developed in 1825. He also supported abolition of slavery, the merits of temperance, and peace instead of war. GEORGE MARSDEN

CHANTRY (Lat. *capellania*). The aisle of a church, or part of an aisle, set apart for the offering of Mass for the benefit of the founder or some other holy person. The name was used also for the institution and endowment of such a service. All chantries in Britain were dissolved by Acts in 1545 and 1547.

CHAPEL. The word has a variety of uses, not all of them religious. It is widely used in Britain to describe Nonconformist* places of worship, while in Ireland it is often used to refer to Roman Catholic churches. In Europe, since the Middle Ages, a chapel has described the place of worship (which is not a parish church) belonging to a school, college, hospital, palace, etc.; that part of a cathedral or parish church where there is a separate altar/Communion table (often a "Lady Chapel"); and the chancel (hence also sometimes the choir) of a church or cathedral. Specialized usage of the word includes "Chapel Royal" (a church under the direct control of the sovereign), "Chapel of Ease" (a building subordinate to a parish church erected for the ease of parishioners), and "Proprietary Chapel" (primarily an eighteenth or nineteenth century phenomenon, built by subscription and maintained by private individuals but without parochial rights).
PETER TOON

CHAPLAIN. The term derives from the place ("chapel") wherein the duty is performed. The feudal structure of medieval Christian society had enabled private places of worship for the ranking personages of the society to be established on their land holdings, or in relation to those institutions of military, political, penal, welfare, or educational proportion which they supported. It was the duty of the chaplain to conduct religious services therein, his living being provided thereby; in middle English history he was specifically a chantry-priest, capable of singing those services. From the range of institutional possibilities, already in the eighteenth-century England of Jonathan Swift the specialized chaplain of a regiment was recognized, from which the modern sense of the military chaplaincy is derived.
CLYDE CURRY SMITH

CHAPMAN, JOHN WILBUR (1859-1918). Presbyterian evangelist. Born in Richmond, Indiana, and educated at Oberlin College, Lake Forest University, and Lane Seminary in Cincinnati, he was ordained in 1882 and held pastorates in Ohio, Indiana, New York, and Pennsylvania, but devoted more than half of his ministry to evangelism. Once associated with D.L. Moody,* he worked during his last ten years with C.M. Alexander.* Evangelistic tours took him to a number of foreign countries as well as many large cities in the USA. He was the first director of the Winona Lake Bible Conference and contributed to the progress of other conference centers. In 1917 he was appointed moderator of the Presbyterian general assembly. BRUCE L. SHELLEY

CHAPTER. Members of a religious house used to assemble regularly to hear a chapter (Lat. *capitulum*) of the Rule of St. Benedict or of Scripture read publicly, and the name "chapter" became attached to their assembly and later to the people who met. Meetings of a whole province or order of monks became known as "provincial chapters" or "general chapters."* In particular, the term became used of monks or canons in a cathedral or collegiate church, presided over by the dean and responsible for its administration, fabric, and worship. Chapter-houses were built from the ninth century, almost exclusively in England, often polygonal externally and vaulted within. In some cathedrals, the lesser chapter consists of residentiary canons, and the greater chapter comprises these and the honorary canons. JOYCE HORN

CHARDON, LOUIS (1595-1651). French Dominican mystic and theologian. Born at Clermont (Oise) of a wealthy family, he studied in Paris and entered the Dominican* Order in 1618 at the Annunciation Priory. He became ordinary preacher in 1632 and spent most of his life as spiritual director and novice master. All his works were written during the last four years of his life. His principal work, *The Cross of Jesus*, was written in 1647 and is a speculative and practical theology of Christian suffering, emphasizing the Christian's progress in suffering through grace. He produced French translations of *Dialogue of St. Catherine of Siena* (1648) and *Institutiones divinae* of John Tauler (1650). He wrote *Meditation on the Passion of our Lord Jesus Christ* (1650), *La Vie de St. Samson d'Yore* (1647), *Raccourci de l'art de méditer* (1649), and other minor works. HOWARD SAINSBURY

CHARISMATA. Plural of the Greek word *charisma* which is almost exclusively a NT word and, excepting 1 Peter 4:10, is peculiar to Paul. It means essentially a free, graciously conferred gift. In the NT it is used sometimes in a broad general sense, but more often it refers specifically to the gracious gifts derived from the Holy Spirit (1 Cor. 12:4), manifested by Christians (12:7) according to His will (12:11), in proportion to faith (Rom. 12:6), for the profit of the whole fellowship (1 Cor. 12:7), and for the work of the ministry. Their purpose is edification, not revelation. Two lists of "gifts" occur (Rom. 12:6-8; 1 Cor. 12:8-10); also, 1 Corinthians 7:7 refers to an additional gift of self-control making possible the celibate state, and 1 Peter 4:10 describes generosity in hospitality as a *charisma.* 1 Timothy 4:14 and 2 Timothy 1:6 use *charisma* with reference to the evangelistic gift. The term *charismata* is being used increasingly of religious movements claiming a restoration to the church of the more spectacular "gifts," such as healing, prophecy, and "tongues-speaking." Max Weber has popularized the term "charismatic leadership" to contrast with institutionalized, formal leadership.

See also GLOSSOLALIA. KENNETH R. DAVIS

CHARLEMAGNE (Charles the Great) (742-814). King of the Franks and first medieval Roman emperor. Son of Pepin the Short, he became sole ruler of the Frankish kingdom in 771 and spent the next three decades in warfare. His greatest military achievement was the conquest of the Saxons in a protracted war that drained his resources. Victory was gained through massacres, forced conversions, mass deportations, and the organization of Saxony into counties and dioceses under Franks loyal to Charles. In 774 he defeated the Lombards and annexed N Italy. He conducted several campaigns along the southwestern frontier and founded the Spanish March. In the east he crushed Bavaria, organized the Ostmark as a defense against the Slavs, and later destroyed the Avars, an Asiatic people who had penetrated to the middle Danube. These campaigns brought the heathen Slavic tribes under Charles's influence and opened the way for German colonization of eastern Europe. On 25 December 800, in Rome, Leo III crowned him emperor, an event that has been the object of intense historical controversy. The motives of both Leo and Charles are extremely unclear, and the significance of the action itself is disputed. As this challenged the Byzantine emperor's position, Charles worked to improve relations with the East.

The Carolingian administrative structure contained few innovations, but Charles was more effective in securing respect for his authority and instituting better government than his predecessors. He utilized permanent professional judges and royal envoys *(missi dominici)* to extend his authority in the realm, while promulgating ordinances (capitularies*) to correct abuses. Charles possessed power over both church and state, and practiced a kind of religious paternalism in his church reforms. He intervened in questions of clerical appointments, discipline, and even doctrine. Convinced a better-educated clergy was needed, he fostered a revival of learning by bringing to his palace the intellectual elite of Latin Christendom (e.g., Alcuin of York,* Paul the Deacon, Peter of Pisa, Theodulf, and Einhard) to form the Palace School at Aachen. The Carolingian Renaissance* spread throughout the empire, and by uniting pagan and Christian classical knowledge it reestablished the common culture of the West.

BIBLIOGRAPHY: Einhard, *The Life of Charlemagne* (many translations); F.L. Ganshof, *The Imperial Coronation of Charlemagne* (1949); P. Munz, *The Origin of the Carolingian Empire* (1960); R. Winston, *Charlemagne: From the Hammer to the Cross* (1954). RICHARD V. PIERARD

CHARLES I (1600-1649). King of Great Britain and Ireland from 1625. In that year his marriage to Henrietta Maria, daughter of Henry IV of France, and his inheritance of Buckingham as political adviser, proved serious encumbrances from the start. The queen was a fervent Roman Catholic and headstrong in seeking to obtain special concessions for Catholic worship. Buckingham embarked on rash military expeditions abroad that ended disastrously, incurring serious financial loss. When Parliament refused the king the money he requested, Charles dissolved the body and resorted to forced loans. For eleven years from 1629 he ruled without Parliament. His unpopularity increased as he exemplified the principle of the Divine Right of Kings.* When Buckingham was replaced as royal adviser by Strafford and Laud* with their policy of "thorough," the king's will was enforced through the Courts of High Commission (for clerical offenders) and of the Star Chamber (for laymen). Laud became archbishop in 1633, and began to press forward High Church reforms and to crush Puritan opposition. The Puritans Burton, Bastwick, and Prynne were savagely sentenced by the Star Chamber in 1637. Laudian divines were intruded into high office; one such was Juxon, bishop of London, who became lord treasurer. The High Church party, with its Arminian theology and seemingly Romanist tendency, thus became identified with this policy of acquiring a firm hold over church and state under king and bishops. By his inept handling of Scottish affairs Charles assured his downfall. The attempt in 1637, approved by Laud, to force the Scottish Prayer Book on Calvinistic Scots led to the signing of the Scottish national Covenant* and the abolition of episcopacy. Charles invaded Scotland, but shortage of funds compelled him to withdraw and summon Parliament to vote him the necessary capital. The Long Parliament of 1640 abolished the special courts and severely curtailed the independent power of the monarchy. Charles had to submit, but his attempt to arrest five members for high treason in 1642 played into the extremists' hands and issued in civil war. Presbyterianism ousted a religious establishment that had become so offensive. Even as a prisoner in 1647, however, Charles negotiated a secret treaty with the Scots, which led to a second civil war. It was this perfidy that prompted the army and Independents to bring him to trial and execution. In family life Charles was an exemplary character, but in public affairs he displayed a grave lack of trustworthiness and judgment. After the Restoration he was acclaimed a martyr and officially celebrated as such by royal mandate until 1859.

BIBLIOGRAPHY: Edward Earl of Clarendon, *The History of the Rebellion and Civil Wars in England* (new ed., 1888); F.M.G. Higham, *Charles I* (1932); G. Davies, *The Early Stuarts* (1937); C. Hill, *Puritanism and Revolution* (1958), chap. 1; J.P. Kenyon, *The Stuarts* (1958); R. Lockyer (ed.), *The Trial of Charles I* (1959).

J.W. CHARLEY

CHARLES II (1630-1685). Second son and successor of Charles I* (d.1649); king in exile until his restoration in 1660. Since the High Church party had identified the Anglican and Royalist causes, his religious flexibility for political ends was a source of constant concern. In 1650 he became Presbyterian to enlist Scottish support for the recovery of his throne, but soon reverted after defeat by Cromwell at Worcester (1651). It was a largely Presbyterian Parliament that invited him to return, and in the Declaration of Breda (1660) he promised a "liberty to tender consciences." But the Savoy Conference* and 1662 Act of Uniformity soon showed that the High Church party had no desire for toleration. Subsequent legislation treated Dissenters as a political danger. There followed the Clarendon Code,* the Act Against Conventicles,* and the Five Mile Act.* Charles's court soon became "the center of corruption and good taste." Shortage of finance for his extravagant living led him into secret intrigues with his cousin Louis XIV. In the secret treaty of Dover (1670) he promised in return for money to make war on the Protestant Dutch and to promote Roman Catholicism in England. Such was the real background behind the trumped-up "Popish Plot"* of Titus Oates (1678). For similar reasons he had been compelled by Parliament to retract his Declaration of Indulgence* (1672). He quashed, however, the Whig attempt to exclude the Roman Catholic duke of York from the throne. As final token of his political discretion, not till his deathbed did the king make an open avowal of conversion to the Roman Catholic faith which he had long espoused secretly.

See R.S. Bosher, *The Making of the Restoration Settlement* (1957), *passim;* and D. Ogg, *England in the Reign of Charles II* (1935).

J.W. CHARLEY

CHARLES V (1500-1558). Holy Roman Emperor, king of Spain. The grandson of Maximilian of Habsburg, Mary of Burgundy, Ferdinand of Aragon, and Isabella of Castile, he fell heir to an empire greater than Charlemagne's. Born and reared in Flanders, he inherited the Netherlands and Franche-Comté (1506), and Spain, Naples, Sicily, and the American possessions (1516). Although initially resented by his Spanish subjects, he earned their loyalty by identifying with their national traits, by his religious zeal, and by marriage to a Portuguese princess. Maximilian's death in 1519 brought him the Habsburg duchies in Austria and Germany, together with rights over Bohemia and Hungary. He also obtained the imperial throne through bribery and granting significant concessions to the estates. The medieval ideal of unifying all Christendom under one scepter, however, was a hopeless anachronism. He was a man of perseverance, patience, and intelligence, but the obstacles facing him were insuperable. His empire had no common ties, while three great problems plagued his reign—the religious revolt in Germany, the Turkish threat, and the continuing struggle with France for European hegemony.

Although strongly opposed to Lutheranism, Charles was absent from the empire during the crucial 1520s. His brother Ferdinand, appointed regent over the German lands in 1521-22, was preoccupied with the Turkish menace. Charles gained the upper hand over Francis I of France in two Italian wars (his army sacked Rome in 1527) and was the last emperor to be crowned by the pope (1530). Unable to crush Lutheranism at the Diet of Augsburg (1530), faced by a defensive coalition (Smalcald League*), and needing help against the Turks, Charles made a truce with Protestant princes (Nuremberg Standstill, 1532). While he campaigned in North Africa against Barbary pirates and vassals of the Sultan (1535, 1541) and fought with Francis (1536-38, 1542-44), Protestantism spread rapidly in Germany. But when divisions appeared in the movement, Charles attacked and defeated the Smalcald League in 1546-47. The defeat, along with his unpopular theological settlement (Augsburg Interim, 1548), forced the Protestants into an alliance with Henry II of France in 1552, and into a new round of conflict. Finally recognizing that Lutheranism could not be destroyed, Charles authorized Ferdinand to conclude a religious peace at Augsburg in 1555. Tired and discouraged, Charles divided his possessions between his son Philip (Spain, the Netherlands, and Sicily) and Ferdinand (Habsburg hereditary lands), while retaining the imperial title. His last two years were spent in retirement at the monastery San Yuste in Estremadura, Spain.

BIBLIOGRAPHY: K. Brandi, *The Emperor Charles V* (1939); H. Holborn, *History of Modern Germany*, vol. 1 (1959); W. Robertson and W.H. Prescott, *The History of the Reign of the Emperor Charles V* (3 vols., 1902); G. von Schwarzenfeld, *Charles V: Father of Europe* (1957).

RICHARD V. PIERARD

CHARLES, ROBERT HENRY (1855-1931). Archdeacon of Westminster and biblical scholar. Born in Cookstown, County Tyrone, and educated at Belfast Academy; Queen's College, Belfast; and Trinity College, Dublin, he was ordained and held curacies in London, 1883-89. He subsequently held several teaching posts, notably in Dublin and Oxford. His primary academic interests lay in the field of intertestamental Judaism. He mastered all the requisite languages for the study of the apocryphal and intertestamental literature, and produced basic critical texts for several of them. This activity culminated in the Oxford edition of *The Apocrypha and Pseudepigrapha of the Old Testament in English* (1913), of which Charles was general editor and which incorporates much of his own work. These two volumes remain the standard work. It was inevitable that he should take a special interest in apocalyptic, and he produced exhaustive commentaries on the two apocalyptic books of the Bible: Daniel (1929) and Revelation (2 vols., 1920).

D.F. PAYNE

CHARLES, THOMAS (1755-1814). Welsh Methodist leader. Born at Longmoor in Carmarthenshire, and educated at Carmarthen Academy and Jesus College, Oxford, he underwent an evangelical conversion at Carmarthen under the ministry of Daniel Rowland.* He was ordained deacon at Oxford and served as a curate in Somerset and in Merioneth before casting his lot with the Methodist Society at Bala, Merioneth, in 1784. For the remainder of his life, Bala was to be his home. Charles's influence on Welsh religious and cultural life was immense. In the field of education he set up a system of circulating schools to replace the now defunct system initiated by Griffith Jones.* His pioneering work as organizer of Sunday schools was originally an extension of his day schools. The need to provide guidance for a new generation of readers made him into a writer. His little catechism (*Yr Hyfforddwr*, 1807) ran to some eighty-five editions before the end of the century, and his Bible dictionary which he began in 1805 was at the same time a popular and a scholarly work which became part of the furniture of every Welsh home.

It was the crying need for Welsh Bibles that compelled Charles to draw the attention of the Religious Tract Society to the matter in 1802, and from this initiative sprang the British and Foreign Bible Society (see BIBLE SOCIETIES). He himself was responsible for editing the version of the Welsh Bible for the society, and his work was based on an unsurpassed knowledge of the text of the numerous Welsh versions that had appeared since 1567. But above all he was the greatest leader of the Calvinistic Methodists* of Wales after the deaths of Williams Pantycelyn and Daniel Rowland, and it was he—albeit with the greatest reluctance—who finally led the Methodist Societies to break with the Church of England by ordaining ministers of their own in 1811 and so forming the Calvinistic Methodist Church of Wales. Through his work as educationist, writer, and religious leader, he was the foremost link between the first generation of Welsh Methodists and the men of the nineteenth century.

See D.E. Jenkins, *The Life of the Rev. Thomas Charles, B.A., of Bala* (3 vols., 1908).

R. TUDUR JONES

CHARLES BORROMEO, see BORROMEO, CHARLES

CHARLES MARTEL (689-741). As a member of the Frankish aristocratic family, the Arnulfings, he succeeded to the office of Mayor of the Palace, the real power overshadowing the weak Merovingian kings. As military leader of the Frankish forces, in 732 his prestige grew both by his victory over the invading Saracens near Poitiers, which ended the threat of serious invasion of western Europe by Muslims from Spain, and by his ceaseless campaigns against the Frisians and Saxons in the northeast. Possibly Charles and his military advisers began the shift away from relying on the infantry to emphasizing heavy cavalry in the Frankish army, due to realizing the potential of the stirrup. This change contributed significantly to the development of medieval feudalism. Charles also encouraged and helped finance two great missionary monks, Willibrord* and Boniface,* in their work of Christianizing and pacifying the Germanic peoples north and east of the kingdom. But within the Frankish kingdom itself, he kept close control of both the ecclesiastical and lay magnates alike, confiscating at will church lands for military purposes or endowing churches and monasteries as the situation required. Such lay control may account partially for the spiritual degeneration and clerical indiscipline of the Frankish church.

See L. White, *Medieval Technology and Social Change* (1962); and J.M. Wallace-Hadrill, *The Barbarian West* (rev. ed., 1962).

KENNETH R. DAVIS

CHARNOCK, STEPHEN (1628-1680). Puritan minister. Son of a London solicitor, he was educated at Emmanuel College, Cambridge, and later became a fellow of New College, Oxford. In 1655 he was appointed chaplain to Henry Cromwell, governor of Ireland, and won a reputation for preaching in Dublin. From 1675 he ministered in a London Presbyterian church. His sermons were published mostly after his death; they reflect the characteristic Puritan divine's concern for central Gospel themes; the most important work was entitled *Existence and Attributes of God*.

CHARRON, PIERRE (1541-1603). French Roman Catholic philosopher and theologian. Born in Paris, he studied at the Sorbonne, Orléans, and Bourges, and was called to the Paris bar. He became a priest and won a great reputation for pulpit oratory, serving as preacher in ordinary to the queen of Navarre. He was also theological adviser to several dioceses, and canon of Bordeaux. He wrote *Les Discours Chrêtiens* in 1589. *Les Trois Vérités* followed in 1593, an apologetic work attacking Calvinism. His most important work, however, was *De la Sagesse* (1601), which was strongly influenced by his close friend Montaigne. In the context of a generally skeptical view of religion, it declares that outside of revelation man cannot be sufficiently certain of religious and moral truth and, as a skeptic, should live as conveniently as he can on the basis of what society allows. The book contributed to the separation of ethics from religion and the growth of free-thought and Deism. He was bitterly attacked and vigorously defended within the church. His ultimate intentions remain equally puzzling today.

HOWARD SAINSBURY

CHARTERIS, ARCHIBALD HAMILTON (1835-1908). Scottish minister. Chaplain successively to Queen Victoria and King Edward VII, and moderator of the Church of Scotland general assembly in 1892, he ranks with Norman Macleod as one of the two outstanding leaders of his church in the post-Disruption era. Highly successful in three widely different charges, he was appointed in 1868 to the chair of biblical criticism at Edinburgh University, and one year later became first convenor of the church's new Christian Life and Work Committee. He was the founder of both the Woman's Guild of the church, and the Young Men's Guild, revived the Order of Deaconesses, and launched both the Deaconess Hospital and the church's influential magazine, *Life and Work*.

D.P. THOMSON

CHASUBLE. A sleeveless outer vestment worn by the celebrant at Mass or Eucharist. Originally shaped like a tent or poncho with a hole for the head, and in the course of time reduced in size, it is said to be derived from the *paenula* or *planeta*, the outdoor cloak of Greek and Roman times. At the English Reformation the chasuble was retained by the first Prayer Book of 1549 as an alternative to the cope,* which was, however, regarded as having no doctrinal significance, but was abolished in 1552. It does not seem to have been used again in the Church of England until the nineteenth-century Oxford Movement,* despite one widely publicized interpretation of the Ornaments Rubric. G.S.R. COX

CHATEAUBRIAND, FRANÇOIS RENÉ, VICOMTE DE (1768-1848). An aristocrat who initially supported the French Revolution but emigrated to London in the Terror (1793). Returning in 1800, he gave occasional service to both Napoleon and the restored Bourbons, but he was primarily a Romantic writer, exercising immense influence on the literature of the time. An essay on revolution (1797) showed him to be versed in the skeptical philosophy of the eighteenth century, but also conscious of religious needs. While still in London he was converted. *The Génie du christianisme* (1802), apology for Christianity, made him famous. Its argument went beyond the eighteenth-century appeal to reason, justifying Christianity to the imagination and aesthetic sense, as more poetic and favorable to the arts and letters than any other religion. He dwelt upon the impressiveness of its ritual and interpreted dogmas poetically. Classical paganism was by such methods shown to have *anima naturaliter christiana,* and so to bear witness to the profound humanity of Christianity. HADDON WILLMER

CHAUNCY, CHARLES (1705-1787). American Congregational clergyman and leader of the "Old Light" (those opposed to the Great Awakening*). Born in Boston, he received his education at Harvard and was pastor of the First Church in Boston (1727-87). When revivalism broke out in New England in the early 1740s, Chauncy at first spoke sympathetically of the movement; however, by 1742 he had become one of the most outspoken critics and leaders in the antirevivalist camp. He charged the revivalists with excessive enthusiasm and antinomianism. His *Seasonable Thoughts on the State of Religion in New England* (1743) contains his serious and satirical criticism of revivalism. He also opposed the appointment of any Episcopal bishop in the colonies.
DONALD M. LAKE

CHAUTAUQUA MOVEMENT. Aimed at promoting popular education, this American movement began at Lake Chautauqua, New York, in 1874, when John H. Vincent,* first chairman of the International Sunday School Lesson Committee and later Methodist Episcopal bishop, began a brief summer course for Sunday school teachers. In 1878, as the Chautauqua Literary and Scientific Circle, it began to provide popular education through home reading in literature, science, etc. It holds an annual assembly at Chautauqua during July and August, marked by outstanding lectures and artistic presentations. Home study courses are accredited by New York University. Readers meet each week in local areas for discussion. Four years of work are available, but most complete no more than two. The movement has been so successful that the word "chautauqua" has come to mean any educational assembly on this pattern.
HAROLD R. COOK

CHELCICKY, PETER (c.1390-1460). Founder of the Chelcic Brethren. A layman of earnest piety, he came under the influence of the writings of John Wycliffe* and made known his ideas in Prague from about 1420. He condemned or rejected the worldly power of the church and the use of secular force in spiritual matters. He did not believe the monastic life was valid as a service to God, and he worked for a reconstruction of society based on a mystical doctrine of the Body of Christ. With his followers, known as the Chelcic Brethren, and other likeminded people influenced by Archbishop Rokycana, he helped to sow the seeds from which the later Bohemian Brethren* movement was to grow. Among his books are *Netz des wahren Glaubens* (1455) and *Postilla* (1434-36). He died in Chicice, S Bohemia. PETER TOON

CHEMNITZ, MARTIN (1522-1586). Lutheran theologian. Born of a poor family in Treuenbrietzen, he studied at the *Trivialschule,* Wittenberg, Magdeburg, the University of Frankfurt on the Oder, and at the University of Wittenberg (1545) where he formed a lasting friendship with Melanchthon.* At this time he was interested chiefly in mathematics and astrology. After he was driven out of Wittenberg because of the Smalcaldic War, he came to Königsberg in Prussia, where he was made rector of the Kneiphof School, and received his master's degree from the new university. In 1550 he became librarian at the ducal castle library in Königsberg, primarily because of his reputation as an astrologer. Influenced by Sabinus, Melanchthon's son-in-law, he seriously applied himself to the study of theology. In 1553 he came again to Wittenberg, where he was made a member of the philosophical faculty. After a year and a half he went to Brunswick, where he was made coadjutor of Joachim Mörlin and later (1567) superintendent. He was one of the founders of the University of Helmstedt and was active as a practical churchman in Brunswick, both in city and province, highly regarded by the city council and Duke Julius. He set up a church order and a *corpus* or *forma doctrinae,* consisting of the Scriptures, the Apostolic, Nicene, and Athanasian creeds, the Augsburg Confession, the Apology, the catechisms of Luther, the Smalcald Articles,* and other writings of Luther. Duke William of Lüneburg asked him to draw up the *Corpus Wilhelminum.*

One of the most important theological treatises written by Chemnitz is his *Examen Concilii Tridentini* in four volumes. It has been called "a detailed study of the Council of Trent, illuminated by the Author's penetrating Biblical theology and a profound knowledge of the history of the Christian church and its teachings." His *De Duabus Naturis in Christo* is an extensive analysis of Christology. His *Loci Theologici* was published after his death. He worked successfully with James Andreae* to mitigate the doctrinal

controversies among German Lutherans between about 1550 and 1575. He was one of the co-authors of the Formula of Concord* in 1577. Theologically he took a position between the Gnesio-Lutherans and the Philipists. Biblical, temperate, scholarly, it has been said of him, "If Martin had not come, Martin would hardly have stood," i.e., referring to Chemnitz and Luther respectively.
CARL S. MEYER

CHENEY, CHARLES EDWARD (1836-1916). First bishop of the Reformed Episcopal Church.* Graduate of Hobart College and the Episcopal seminary in Virginia, he was ordained in 1860 and became rector of Christ Church, Chicago, a position he held until his death. As a low church Episcopalian and pronounced evangelical, he favored "the great fundamental principles held by all *evangelical* Christians" and opposed both "Romanism" and "radicalism involved in the destructive criticism of God's word." His practice of omitting the words "regeneration" and "regenerate" from the baptismal service led to his deposition from the ministry, an action which was overruled by the civil court. In 1873, while retaining his rectorship, he with Bishop G.D. Cummins* and others organized the Reformed Episcopal Church.
MILLARD SCHERICH

CHESTER BEATTY PAPYRI. Portions of three NT manuscripts, designated P45, P46, and P47, which comprise 126 leaves, partially mutilated, and afford valuable additional early textual evidence utilized in more recent critical editions of the Greek NT. Said to have been found near Memphis, on the banks of the Nile, these papyri were acquired by Mr. Chester Beatty from a dealer in Egypt about 1930. Most of the material is now housed in Dublin, but one page of P45 is in Vienna, part of P46 is in Michigan. P45 (third century) contains gospel fragments and Acts; P46 (about 200) has most of the Pauline epistles. P47 (late third century) gives the earliest extant text of the Apocalypse, chapters 6-17. These documents, part of a considerable stream of newer textual evidence, must be evaluated in the exegesis of particular passages.
ROY A. STEWART

CHEYNE, THOMAS KELLY (1841-1915). OT scholar. Educated at Oxford and at Göttingen, where he was greatly influenced by H.G.A. Ewald,* he returned to Oxford for the rest of his life, apart from a brief period as rector in Essex, and from 1885 until his death was Oriel professor of the interpretation of Scripture. A pioneer in England of the critical approach to the OT, he participated in the preparation of the Revised Version of the Bible, and was co-editor of the *Encyclopaedia Biblica* (4 vols., 1899-1903). In latter years some of his views were condemned by scholars as wild and unbalanced. Cheyne's writings include *The Origin and Religious Contents of the Psalter* (1891), *Founders of Old Testament Criticism* (1893), *Introduction to the Book of Isaiah* (1895), and *Jewish Religious Life after the Exile* (1898).
J.D. DOUGLAS

CHICAGO LAMBETH ARTICLES. In 1870 William Reed Huntington, a member of the Protestant Episcopal Church, proposed four principles upon which Christian churches might agree in order to bring about unity. In 1886 these principles were adopted by the American Episcopalian bishops meeting in Chicago. Known ever since as the Articles, or the Chicago Quadrilateral, they consist of these commitments: (1) adherence to the Holy Scriptures as the ultimate standard of faith; (2) adherence to the Apostles' Creed and the Nicene Creed; (3) adherence to the two sacraments of baptism and the Lord's Supper; (4) adherence to a belief in the historic episcopate. The fourth article has created a difficulty for other Christian churches whose leaders fear that unity with the Episcopalians might force them to reordain their own clergy. Nonetheless, the influence of the Chicago Lambeth Articles in ecumenism is great. In 1888 the Lambeth Conference also adopted these articles in a revised form (the Lambeth Quadrilateral) as the basis for its own discussion of Christian unity with other churches.
JOHN D. WOODBRIDGE

CHICHELE, HENRY (c.1362-1443). Archbishop of Canterbury from 1414. One of the able lawyer-bishops of the Middle Ages upon whom so much of the administration in church and state depended, he was the son of a Northamptonshire merchant, and furthered his career through the patronage of the house of Lancaster and of William of Wykeham. The latter had him educated at Winchester and his foundation at Oxford. After diplomatic missions abroad, Chichele became bishop of St. David's (1408) and went as a delegate to the Council of Pisa* (1409). He enjoyed the special favor of Henry V and became archbishop of Canterbury in 1414. After Henry's death, however, Pope Martin V and the powerful Bishop Beaufort attacked him for failing to extend papal authority in England against the Statute of Provisors, and his legatine authority was briefly suspended (1427-29). He supported education by promoting graduate clergy and founding All Souls College, Oxford (1438).
JOHN TILLER

CHILDREN'S CRUSADE (1212). This was a mass migration of tens of thousands of children in an effort to reach the Holy Land. A French boy named Stephen told of receiving a visitation of Christ, who gave him a letter for King Philip Augustus. The boy began preaching and attracted large crowds. He excited children from the age of six upward, and they formed bands to go to Jerusalem. They believed God would deliver the city to their innocence, whereas the nobles had been unable to conquer it. Some of them reached Genoa, where they could not get passage to the Holy Land. Part of this group returned home, others went to Rome where they were urged by the pope to go home as their friends had done. One group went to Marseilles, whence they were shipped to North Africa and sold into slavery. The lay participation and social discontent exhibited in this crusade is regarded as a forerunner to the heresies and revolts of the fourteenth century.
ROBERT G. CLOUSE

CHILE, see LATIN AMERICA

CHILIASM, see MILLENARIANISM

CHILLINGWORTH, WILLIAM (1602-1644). Anglican apologist. Godson of William Laud,* scholar and fellow of Trinity College, Oxford, and friend of Gilbert Sheldon and John Hales, Chillingworth had a reputation of "an impartial and well-balanced mind, a large store of learning and a keen power of dialectics." He was converted to Rome by (John) Fisher the Jesuit* (d.1641), through the weakness of the logical basis of Laudian theology. At Douai in 1630 he was set to write against the Church of England, but reweighing the arguments he returned to England, where in 1634 he again declared himself Protestant, though not yet Anglican. Hotly attacked by the Romans, he wrote at Laud's request *The Religion of Protestants a Safe Way to Salvation* (1638). On the basis that "the Bible only is the religion of Protestants," he maintained the rights of reason and free inquiry, and the necessity of personal conviction. At last persuaded into the Anglican ministry by Laud, he became chancellor of Salisbury in 1638 and incurred the wrath of Parliament in 1640. Serving as chaplain or soldier in the Civil War, he was captured at Arundel in December 1643 and died in captivity a month later. G.S.R. COX

CHINA. The first appearance of Christianity in China was the arrival in 635 of the Nestorian missionary Alopen,* who entered Sian, the T'ang dynasty capital. The Nestorians met with imperial favor and some success, but provided no Bible and condoned Buddhist-Christian syncretism. By the end of the tenth century Christianity had disappeared. Nestorian missionaries returned to China in the thirteenth century with the Mongol conquerors, but their adherents were few.

John of Monte Corvino,* a Franciscan, followed Marco Polo to Peking in 1294. He claimed 6,000 converts, but the Franciscans also failed in acculturizing Christianity; their converts were merely alien enclaves amid a hostile population, and with the end of the Yuan (Mongol) dynasty in 1368 Christianity had again disappeared in face of persecution.

In the sixteenth century Christian missionaries spurred by the religious awakening in Europe, pressed eastward with European traders into the Pacific Ocean. Francis Xavier,* the great Jesuit, reached Japan in 1549 and hoped to enter "Cathay," but died near Canton in 1552. The Italian, Alexander Valignano, landed in Macao in 1574; he recognized the importance of knowing the Chinese language and of cultural adaptation. But it was Matteo Ricci* who reached Peking in 1601 who commended Christianity to the Chinese court and intelligentsia by his learning and complete adoption of Chinese culture. He even sought to graft Christianity onto the Confucian system and consequently won over some high officials, but also started the Chinese Rites controversy.* Ricci died in 1610, having made a profound impression on the Chinese, but severe persecutions followed and lasted until the end of the Ming dynasty (1644).

The early Ch'ing (Manchu) emperors proved sympathetic toward Christianity, and missions were opened in most of the provinces. The Jesuits sincerely aimed at a Christian Church composed of Chinese believers. They were joined later by Dominican and Franciscan missionaries, and the number of believers rose to 250,000. In 1717, however, Emperor Kang Hsi ordered the banishment of all missionaries, and persecution continued for a century afterward. The West was not yet ready for a genuine encounter with Chinese culture. In 1840 the Jesuits established their famous base at Zikawei in Shanghai.

China was virtually closed to foreign trade and Protestant missionaries until 1841, but Robert Morrison,* living in Macao from 1807, had by 1819 completed a translation of the Bible into Chinese. The iniquitous "Opium War" (1841) compelled a conservative China to cede Hong Kong to Britain and pay an indemnity; later China was required to open five ports to foreign trade and residents, free from Chinese jurisdiction. What are regarded in Chinese minds as aggressive and humiliating actions of this sort by Western Christian powers have rankled ever since and have implicated Western missions. The Taiping Rebellion (1850-56) was a pseudo-Christian peasant movement ending in a disastrous orgy of death and destruction.

The "opening of China" through the "unequal treaties" was seen by missionary societies as their long-awaited opportunity. Their representatives first moved into the Treaty Ports and then after 1866, led by J. Hudson Taylor* and the China Inland Mission, into all the inland provinces. Protestant Christianity now began to make an impact on Chinese life, through its schools, hospitals, and churches. By the end of the century there were 500,000 Roman Catholics and 75,000 Protestants.

As Western aggression increased, however, so did feeling against foreigners. In 1900 the conservative empress dowager seized power, and nationalist feeling exploded into the Boxer Uprising; 181 missionaries, Catholic and Protestant, and more than 49,000 Chinese Christians were killed. China now entered on twenty-five years of rapid social, political, and economic change for which Christianity had helped prepare the way. A modern educational system introduced in 1905 produced a new student class. The decay of the Manchu dynasty hastened the Nationalist Revolution (1911), the leaders of which, including Dr. Sun Yat-sen, the first president, were mostly the product of Christian schools. 1901-14 were years of unprecedented prosperity for missions, but warlordism rendered the revolution futile and plunged China into chaos. The Versailles Peace Treaty (1919) cold-shouldered China and favored her enemy Japan. China's only friend seemed to be Russia, and in 1921 the Communist Party of China was formed.

Meanwhile, the Protestant Church was striving to achieve its own identity. Education received prominence as a means to influence the whole country through its future leaders. Excellent universities, medical schools, and high schools multiplied—often at the expense of direct evangelism. But phenomenal growth followed, and Protestant church membership in 1915 was 331,000. New life was stirring, and a national conference in Shanghai (1921) brought into prominence able leadership concerned with the future development and unity of the church. The years 1920-27 were the heyday of missionary expansion; in 1927 there were 8,518 Protestant missionaries in China and three million baptized Christians, of whom one-fifth were Protestants. This period also saw the rise of independent church movements which were expressions of a nationalistic desire to shake off missionary domination.

The rapid growth of the Communist movement was now a serious threat. In 1926 Communist-inspired agitation brought about a general evacuation of missionaries from inland China and slashed Protestant missionaries to 3,000. Between 1924 and 1934, twenty-nine Protestant missionaries were killed and a number kidnaped. To survive, the church had clearly to achieve genuine autonomy. A nationwide spiritual awakening in the 1930s facilitated this development, but continuing civil war boded ill for the future.

The war with Japan (1937-45) severely tested the Chinese Church and left China at the mercy of Communism. Though 3,000 missionaries again dispersed throughout the country, their time was short. In 1949 Chairman Mao inaugurated the Peoples' Republic of China. During 1950-51 all Protestant missionaries withdrew from the China mainland, and the church was left to face the future alone—a church strengthened, however, by a remarkable postwar revival among university students. Roman Catholic missionaries were largely expelled one by one over a period of years.

Church membership, Protestant and Catholic combined, had never exceeded one percent of the population, and the modern missionary movement had again failed to penetrate deeply into Chinese cultural structures. Christianity to the intellectuals of China was Western and alien. To the Communists, the church was too closely associated with Western imperialism to escape opprobrium. The Protestant "Manifesto" of 1950 committed the church to shaking off imperialist shackles and pledged total subservience to the Communist party. Christians were in a dilemma as to their loyalties. Mass trials of clergy were held all over China. Among the millions who died at Communist hands were hundreds of Christians. There were numerous acts of heroism both by Roman Catholics and Protestants. By 1958 the church was immobilized. Finally, in 1966 the visible church was destroyed by the Red Guards in the "great proletarian cultural revolution." Thereafter silence fell like a shroud over the church in China, though Christians continued to meet secretly. The church endures, though many leaders remain in prison. The underground church forms a nucleus for future expansion when a new day comes.

BIBLIOGRAPHY: K.S. Latourette, *A History of the Christian Missions in China* (1929); A.C. Moule, *Christians in China Before the Year 1550* (1930); A.H. Rowbotham, *Missionary and Mandarin: the Jesuits at the Court of China* (1942); D.V. Rees, *The "Jesus Family" in Communist China* (1959); L.T. Lyall, *Come Wind, Come Weather* (1961); *International Review of Missions*, vol. LV (January 1966); D.E. MacInnis, *Religious Policy and Practice in Communist China* (1972).

LESLIE T. LYALL

CHINESE RITES CONTROVERSY. A debate about missionary methods in China during the late sixteenth and early seventeenth centuries, associated with the Roman Catholic priest Matteo Ricci.* He arrived at Macao in 1582, established a work at Nanking in 1599, and did so at Peking two years later, where he remained until his death. Ricci made a point of explaining Christianity to the Chinese in their own terms. In this connection he tolerated the practice of indigenous rites in honor of ancestors and deceased relatives, and also (for the more educated) in honor of Confucius, since he believed that the Chinese worshiped the true God in their own way. This action, the forerunner of much modern missionary strategy, provoked a long and heated controversy, chiefly because Ricci forgot the part played by superstition in such rites as generally practiced. Long after Ricci's death, Clement XI issued decrees (1704, 1715) which condemned the rites.

STEPHEN S. SMALLEY

CHOIR. The church choir can be traced back to the establishment of *schola cantorum* in Rome, probably in the fourth century. It served to train singers in liturgical chant, and from it sprang the papal choir. In time, similar *scholae* functioned elsewhere, notably in Paris. Until the Renaissance, most adult singers were clerics, and choirs were found normally only in cathedral churches, monastic institutions, and the chapels of the aristocracy.

Luther promoted the continuance of choral foundations. Such institutions as the *Thomasschule* in Leipzig, under a succession of distinguished musicians—among whom in the eighteenth century was J.S. Bach*—continued to train choristers, and to establish the fame of the Lutheran choral tradition. Calvin restricted music in worship to unison, congregational singing of metrical psalms. Anglican parish churches adopted this custom, but the English Chapel Royal and the cathedrals retained choirs to chant prose psalms and perform anthems, until abolished during the Commonwealth (1649-60). After the Restoration, volunteer parish choirs also sprang up to aid in singing the metrical psalmody. Such groups increasingly added simple anthems to their musical fare, their example being followed in an increasing number of Nonconformist chapels in the eighteenth century.

In America, a parallel development took place, especially in the wake of the *singing school* movement. Volunteer choirs in both America and Britain proved a social and recreational attraction while contributing to public worship, and sometimes to church controversy. In larger cities, professional quartets were employed in the nineteenth century, adding an element of entertainment rather than devotion. Some Episcopal churches emulated the English cathedral tradition.

Most of the historic choral traditions are represented in America: that of the Moravians continuously from about 1740, and more recently the distinctive *a cappella* choralism of the Russian Orthodox Church. Important also is the colorful and improvisatory tradition of the Afro-American churches, which increasingly influences many segments of the current choral practice.

BIBLIOGRAPHY: L. Ellinwood, *The History of American Church Music* (1969); R.T. Daniel, *The Anthem in New England before 1800* (1966); E.A. Wienandt and R.H. Young, *The Anthem in England and America* (1970): this work contains an excellent bibliography; C. Dearnley, *English Church Music, 1650-1750* (1970).

J.B. MAC MILLAN

CHORALE, see MUSIC, CHRISTIAN

CHOREPISCOPUS. Traditionally a bishop of the countryside. The Arabic version of the Nicene canons sets them "in the place of a bishop over villages, monasteries and village-priests." The chorepiscopus exercised limited episcopal functions in the East before 314. Fifteen Asian and Syrian chorepiscopi signed the Nicene decrees (325). The Dedication Council of Antioch (341) confined their responsibilities to issuing letters dimissory; superintending the church in their areas; appointing readers, subdeacons, and exorcists; and ordaining, only by permission of the city-bishop, deacons and presbyters. The Synodal Letter of Antioch refers to "bishops of the adjacent countryside and cities." The mid-fourth-century Canons of Laodicea forbade further appointment of bishops "in villages and country districts," but chorepiscopi appeared in Caesarea about 375, at the Council of Ephesus in 431, at Rome in 449, and at the Council of Chalcedon in 451. Second Nicea (787) allowed them to appoint readers, but with the permission of the bishop. By the late twelfth century they were extinct.

Western chorepiscopi, first mentioned at Riez (439), met opposition in the ninth century. Nicholas I (858-67) confirmed their episcopal acts, but ninth-century councils canceled them and forbade further appointments. By 1048 they were thought to be equivalent to archdeacons.

G.T.D. ANGEL

CHRISM (Gr. *chrisma*, "anointing"). A mixture of oil and balsam is used by the Roman and Byzantine churches in certain liturgical ceremonies—in the conferment of the three sacraments which claim to put a special mark on the soul and strengthen it spiritually (baptism, confirmation, ordination) and in the consecration of churches, chalices, patens, and the blessing of church bells and baptismal water. There is a very detailed and formal procedure for the consecration of chrism, which in the Roman Church is done by a bishop on Maundy Thursday. Christ Himself was supposed to have used chrism, but the evidence is not authentic—although it was used by the early Christians. Allegorically, the olive oil symbolizes strength, and the balsam the fragrance of virtue.

L. FEEHAN

CHRIST. From the Greek equivalent of the Aramaic "Messiah," "Anointed One." Most first-century Jews anticipated the advent of a great King through whom the kingdom of God would come. This eschatological use of the term is rare in the OT (probably in Dan. 9:24-26 only), although the *idea* is much more frequent. Jesus used the term sparingly, probably because of its political and militarist associations for His contemporaries. He welcomed it from Peter (Matt. 16:13-17), however, and accepted it from the high priest when He was, humanly speaking, a helpless prisoner without any possible political future (Mark 14:61f.). As a distinctively Jewish term its use as a title receded somewhat, once the Gentile mission was under way, and it tended to become a name. The view that Jesus did not see Himself in any sense as the Messiah cannot be substantiated unless the gospels are subjected to historical skepticism.

G.W. GROGAN

CHRISTADELPHIANS. A sect founded by a physician, John Thomas (b.1805), who emigrated to America in 1832, and at first associated with the Campbellites and Millerites. He revisited England three times, and founded ecclesias, of which the dominant one has always been in Birmingham, although there have been several splinter groups. The most influential British leader was Robert Roberts, founder of the journal *The Christadelphian* and expounder of Christadelphian doctrines in his standard textbook *Christendom Astray from the Bible.*

Adherents accept the Bible as their sole authority. They reject the immortality of the soul. They are anti-Trinitarian and believe in one personal God the Father. They hold that Jesus Christ had no existence, except in the mind of the Father, before he was born of the Virgin Mary. Although his body was necessarily "unclean," he was personally free from sin, and received the resident divinity from the Father through the Holy Spirit at his baptism. The Holy Spirit is the name for the power of God in action. There is no personal devil, but Satan is a personification of sin in the flesh. The death of Christ on the cross was not expiatory, but in order "to express the love of the Father in a necessary sacrifice for sin." Salvation is through perseverance in good works and through acceptance of Christadelphian doctrines and baptism. Christadelphians reject any teaching of "heaven beyond the skies," but emphasize the promises to Abraham and Israel, and look for the return of Jesus to reign permanently in Jerusalem. The saved will be raised to live in the renewed earth, while the wicked will be annihilated.

They are democratic in organization, and as brethren they have no separate ministry. Each ecclesia is independent, although a member of the wider Christadelphian fellowship. It is estimated there are some 20,000 Christadelphians in Britain and perhaps a similar number in the USA. They spread their views largely by public lectures and Bible exhibitions.

See R. Roberts, *Christendom Astray* (1862, continually reprinted), and B. Wilson, *Sects and Society* (1961): the fullest "neutral" account of the movement. J. STAFFORD WRIGHT

CHRISTIAN. According to Acts 11:26, "the disciples were first called Christians at Antioch." It was apparently Gentiles outside the church who coined the word as a nickname, i.e., "Christ's ones." While it is little used in the NT (thrice only), it soon became established as the obvious title (Acts 26:28). In the NT, it is used only once of themselves by Christians (1 Pet. 4:16), and some scholars feel Peter is here employing it as a term of accusation on the lips of the church's enemies, as the context might indicate. Other titles were used concurrently during the apostolic age (as expressed in KJV): Christianity was "the way" (Acts 9:2; 19:9, etc.), and Christians were "followers" (Eph. 5:1; 1 Thess. 1:6, etc.) and "believers" (Acts 5:14; 1 Tim. 4:10, etc.). The favored designations in the NT are "saints" (Acts 9:13; Rom. 12:13), "the elect" (Rom. 8:33; Col. 3:12), "the brethren" (Acts 9:30; Rom. 16:14), and "the disciples" (Acts 11:26, 29). Until the appellation "Christian" became common, doubtless several others were used, such as "Nazarenes" (Acts 24:5), which would bring little comprehension outside Palestine, but could be used there as a pejorative term (cf. John 1:46) for believers in Jesus of Nazareth.

When the name "Christian" was coined in Antioch, it gained usage quickly and was used widely within the course of a single generation. Outside the NT, the Roman historian Tacitus declares that this name (or "Chrestian") was popularly used in Rome at the time of Nero (c. A.D. 64). By the middle of the second century it had been taken up as one of jubilant testimony by those who might have expected martyrdom if they admitted to being a follower of Jesus Christ. As the church adopted the name, deeper meanings began to be seen in it. The Greek word *christos* ("anointed") suggested the more familiar term *chrēstos* ("gracious, good"). Such terms witnessed to the dignity of the church's Founder and Lord, "the Anointed One," and Peter (1 Pet. 2:3) may be making a play on the word "Christ" when he writes, "Now that you have tasted that the Lord is good." KEITH J. HARDMAN

CHRISTIAN AND MISSIONARY ALLIANCE. The name given in 1887 to a group organized in the USA and Canada, part of a worldwide movement begun six years earlier by A.B. Simpson (1844-1919). Some 1,400 of more than 4,000 fully organized congregations are in North America; the others are separately organized under various names in some thirty-eight other countries on six continents. The movement has over 5,000 indigenous ministers and nearly 1,000 missionaries serving outside their home countries. More than 900 of the latter are from North America. The movement's leadership denies that over the years the emphasis on sanctification as a crisis-experience and on miraculous healing has been modified, but inevitably the transdenominational character has been replaced by a self-conscious, broadly evangelical denomination not classifiable with the Calvinistic, Wesleyan, Baptist, and premillennial streams which flowed into it. DONALD TINDER

CHRISTIAN CATHOLIC CHURCH, see DOWIE, J.A.

CHRISTIAN CHURCHES, see CHURCHES OF CHRIST

CHRISTIAN ENDEAVOR SOCIETY. The Young People's Society of Christian Endeavor was the first widespread nondenominational youth organization in the American churches. Begun by pastor Francis E. Clark* to conserve the results of special meetings in connection with the Week of Prayer, January 1881, it soon spread to other churches and ultimately around the world. Though later many denominations withdrew to form their own youth societies, on its seventy-fifth anniversary CE could still claim about three million members. Basically it is a church-connected society conducted exclusively by and for young people. Through its pledge, its weekly devotional meeting which once a month becomes a consecration meeting, and its various committees, it encourages young people to confess Christ and to serve him in fellowship with other young people in the Church. HAROLD R. COOK

CHRISTIAN ETHICS. A wide variety of views on the relation between the Christian faith and moral decision has been held by Christians, and this is reflected in important differences over concrete issues such as war, race, and social morality, and on its relationship to other major religions and philosophical views. These differences emerge early in the history of the church with the Augustinian emphasis on the need for renewal of the human will and the search for God as man's chief good, rather than on detailed duties (e.g., *Didache*). Perfect moral freedom is to be found in obedience to God. Similarly in Reformation theology and again in Neoorthodox theology, moral obligations arise out of a direct encounter with God and depend on His sovereign will. Moral problems are a matter of the will, not the intellect; hence the need for the regenerative power of God to enable one to do what is right. Luther sees God as giving the believer freedom—to serve so that the Christian is both "subject to none and subject to all." Calvin stressed the subjection of the Christian to the law and the Gospel, though for the unbeliever the law is a reminder of his inability to fulfill God's commands and may therefore lead him to repentance.

Puritanism is indebted to Calvin for its sense of divine sovereignty and, with the teaching on the civil magistrate, attempted to place the whole of society under God's law, though there were differences, e.g., about the degree of religious toleration. Greater emphasis was placed on the Bible as defining specific unchangeable duties to God and the neighbor, and to these many Puritans added further detailed instructions. Examples of this Reformed tradition are found today in the work of John Murray and Carl F.H. Henry.

Neoorthodox theology shares with the Augustinian tradition the view that of ourselves we cannot know the good, the will of God. This is known only through revelation, not to be identified with the Bible. The command to love God and one's neighbor does not vary in intention, but in content according to the conditions with which it deals. Ethics is essentially a matter of free decision, and it is not possible to know beforehand what the requirements of God are for any situation. This is the interpretation of the Christian ethic to be found in the work of H.E. Brunner*; the same motif appears in Bonhoeffer's* call for discipleship, and in Bultmann's* call to "radical obedience." This view has led to the development and popularization of "situation ethics" and *koinonia* ethics.

The Roman Catholic approach to ethics depends on the notion of natural law. Morality is "natural" in the sense that it has its source in rational reflection on the true "end" of human life. The pluralism of modern Western civilization makes this implausible. The extent to which the church is thought necessary to enable people to discern their duty has varied. Theologians such as Karl Rahner have tried to modify the rigidities (as they see it) of the classic natural law position by stressing (in Neoorthodox fashion) the importance of the individual and the concrete. Debates on birth control and abortion show the extent to which the traditional position has been eroded by secular utilitarian morality, or by truly biblical insights, or "situationalism."

OONAGH MC DONALD

CHRISTIAN NURTURE, see BUSHNELL, HORACE

CHRISTIAN RADICALS, see DEATH OF GOD SCHOOL

CHRISTIAN REFORMED CHURCH. An American denomination with its background in Dutch Calvinism. In the Netherlands, conservative dissatisfaction with modernism and doctrinal laxness led to the secession *(Afscheiding)* of 1834. A group of seceders emigrated in 1846, settling in western Michigan under H. Van Raalte. They joined the Dutch Reformed Church; for the ultraconservatives among them, this was an uneasy union from the start, and in 1857 four congregations separated and formed the Christian Reformed Church. For some decades its growth was gradual. In 1886 a second and much larger secession (the *Doleantie*) took place in the Netherlands. Conservative Calvinist emigrants to the United States tended to join the Christian Reformed Church, which now grew rapidly. A seminary was established in the 1890s, a system of Calvinistic day schools started, and by World War I a college established (Calvin College). Membership, including children, was around 100,000. During the 1920s, controversy over "liberalism" and "ultra-Calvinism" erupted; H. Hoeksema, denying God's common grace *(gratia generalis)* to the nonelect, formed the Protestant Reformed Church. After World War II a wave of emigration from the Netherlands to Canada produced additional members for the Christian Reformed Church, which now includes over 275,000 souls. Growth has been primarily through immigration from the Netherlands and through internal growth. The church has associated with it three colleges and a large system of private elementary and high schools. It has been notable for adherence to Calvinistic orthodoxy. Its headquarters are in Grand Rapids, Michigan.

See J.H. Kromminga, *The Christian Reformed Church: A Study in Orthodoxy* (1949).

DIRK JELLEMA

CHRISTIAN SCIENCE. The religion which had its origin in Mary Baker Eddy.* She claimed it came to her by direct revelation, and that Christian Science's definitive book, *Science and Health, With a Key to the Scriptures,* was written by her under divine dictation, though she conceded that a clergyman edited its poor grammar. Mrs. Eddy affirmed the inspiration of the Scriptures, but made the same affirmation for her own book which takes priority over Scripture with its not-to-be-changed-or-doubted key to the Scriptures. There are more than 2,000 Church of Christ, Scientist groups in the USA, and fewer than 1,000 in the rest of the world. The Mother Church is in Boston.

A distinctive feature of Christian Science is that it is never preached. The church has no preachers, no sermons. Instead, each church has a First and Second Reader who are obligated to read a selection of Scripture and a selection from *Science and Health.* No comments, explanations, or interpretive remarks are permitted, decreed the founder. Thus Christian Science received and maintains its final meaning, and protects the special status of Mary Baker Eddy by excluding the possibility that a greater than she should arise. She is the first and the last word about the truth of Christian Science.

Christian Science's one truth is that God as Spirit is All in All. Everything is Mind or Spirit—or, rather, there is no reality except Mind or Spirit. Mind, or Spirit, is Truth, Love, Power, Life, Goodness. Materiality is evil, sin, sickness, death, unreality. Since God is All, man is coexistent with God and his being, therefore, resides eternally in, and is not to be differentiated from, God's being.

Since Jesus of Nazareth is a physical man, he is not to be identified with God; only Christ, as the Principle of Mind, is identified with God. Jesus neither died on the cross nor arose from the grave. Jesus left the grave knowing that he had not died, that no man can die. Each of us must come to that knowledge; from discernment of the illusory character of death, our salvation comes. In the name of this same metaphysical idealism, Christian Science tolerates marriage (Mary was thrice married) and such things as food and money (Mary had great interest in the sale of her book and in the acquisition of money) since in the imperfection of their faith men do not wholly accept the fact that God, or Mind, is All. Everything Christianity posits in terms of the biblical teaching about creation, fall, and redemption, Christian Science declares to be unreality.

It is a mistake to think of Christian Science as a faith-healing religion. It does not claim to *heal* sickness, for it claims sickness is an illusion. Nor does Christian Science claim to *save* men, for it teaches that all which men could be saved from is unreal. It takes neither sickness nor sin seriously. One might add that if Christian Science took *itself* seriously, it could dispense not only with Readers as well as preachers, but also with its key to the Scriptures as well as with Scripture, and no less with Mary Baker Eddy herself. The latter's claim to fame rests on her use of both these writings to overcome what by Christian Science's own claim is really nothing at all. For God is All.

BIBLIOGRAPHY: Official biographies of Mary Baker Eddy by S. Wilbur (1908) and L.P. Powell (1930); others by E.F. Dakin (1929), E.S. Bates and J.V. Dittemore (1932), and H.A. Studdert Kennedy (1947). See also M.C. Sturge, *The Truth and Error of Christian Science* (1903); H.A.L. Fisher, *Our New Religion* (1933); C.S. Braden, *Christian Science Today* (1958); R. Peel, *Christian Science: Its Encounter with American Culture* (1958).

JAMES DAANE

CHRISTIAN SOCIALISM. By definition the term applies to the activities of a group of Anglicans between 1848 and 1854, but their ideas inspired subsequent generations. The group, which formed as a response to the Chartist fiasco of 1848, consisted of F.D. Maurice,* J.M.F. Ludlow,* and Charles Kingsley,* though later they were joined by Tom Hughes, Archie Campbell, Vansittart Neale, and others. They reacted against the dominant utilitarianism of the age, laissez-faire economics, and the indifference of the Anglican Church to social issues. Though not united politically, they were united in believing that Christianity stood for a structure of society which would enable men to live and work as brethren, and that competition is not a universal law. Ludlow was the founder of the movement, but Maurice was its prophet and thinker. Maurice had a dread of societies and hated the prospect of Christian Socialism's becoming a party. He aimed to "Christianise Socialism and to Socialise Christendom, not to Christian-Socialise the universe."

The day following the failure of the Charter, the group brought out a poster introducing the Christian element into socialism. This was followed by the short-lived, much-criticized journal *Politics for the People.* Workers suspected this journal as a middle-class trap, but in 1849 the group began regular meetings with workingmen, which improved relations. Kingsley, meanwhile, wrote his novels *Yeast* and *Alton Locke* in defense of working-class aspirations, and Ludlow produced a program of founding workers' cooperatives. In 1850 associations of tailors, bakers, needlewomen, builders, bootmakers, and printers were formed, together with a Society for the Promotion of Working Men's Associations. Through lack of money, some of the associations foundered, but the group did make a direct contribution to the Industrial and Providential Societies Act (1852), which gave cooperatives their charter. In 1850 a new journal *Christian Socialist* appeared and met with much hostility. The driving force of the group was its Monday evening Bible study, though on Fridays it met to discuss social problems and the action to be taken. There were, however, clashes in the group, and from associations Maurice began to turn his attention to education, founding in 1854 the first workingmen's college, soon to be followed by others throughout the country.

The failure of several associations, the rising prosperity of England, and the indifference of the church at large ended the Christian Socialists, but the movement marked the beginning of modern social concern in the Anglican Church, inspired the later Guild of St. Matthew, the Christian Social Union, and the twentieth-century protests, as well as influenced trades unions, cooperative legislation, and working-class education.

BIBLIOGRAPHY: G.C. Binyon, *The Christian Socialist Movement in England* (1931); C.E. Raven, *Christian Socialism, 1848-1854* (1920); M.B. Reckitt, *Maurice to Temple: A Century of the Social Movement in the Church of England* (1947).

JOHN A. SIMPSON

CHRISTIAN YEAR, THE. The early Christians who were mainly Jews were used not only to keeping one day in the week as separate but also to marking the year with certain religious festivals, notably Passover, Tabernacles, and Pentecost. From early times Christians kept a commemoration of Christ's resurrection. This was held at Passover time and was finally fixed on the Sunday following Passover. Pentecost was then celebrated at the appropriate time; the fifty days between the two were days of joy and rejoicing. The choice of 25 December (in the East, 6 January) for the birth of Christ is almost certainly because that day was the great pagan day of honor to the sun, and in Rome in the fourth century it was transformed into a Christian festival.

From the fourth century the Christian calendar became more historical in character, and Holy Week and Ascension Day appeared. Pentecost became the day of the giving of the Holy Spirit. Lent arose out of the custom of preparing catechumens for baptism at Easter. Saints' days came into the calendar either through the commemoration of a martyrdom or through the date of a dedication of a church in honor of a particular saint. The advantage of the Christian Year is that through the church services, and in particular the choice of Scripture passages to be read, worshipers are regularly reminded of the great events of the Christian faith and a balance is kept between them. In recent years there have been various suggestions for modification of the calendar, particularly in relation to Advent and Lent, and some demand, supported by secular sources, for a fixed date for Easter.

PETER S. DAWES

CHRISTLIEB, THEODOR (1833-1889). German preacher and professor of pastoral theology. Born in Württemberg and educated at Stuttgart and Tübingen, he held a tutorship in Montpellier followed by two German curacies, then worked in Islington, London (1858-65) as minister to the local German population. In 1865 he became pastor at Friedrichaften where he influenced members of the German royal family. From 1868 to the end of his life he held the chair of pastoral theology at Bonn, where he taught many generations of theological students, upheld conservative views against German biblical critics despite much opposition, and organized missionary work. At Bonn he influenced Prince William (later the emperor) in the latter's student days (1877-80). Christlieb's fine and often original sermons still make stimulating reading. His wife wrote his biography (1892).

R.E.D. CLARK

CHRISTMAS. The English name for the Feast of the Nativity of Christ kept on 25 December by the Western Church. There is no evidence of a Feast of the Nativity before the fourth century, except possibly among the Basilidians. The earliest mention of 25 December is in the Philocalian Calendar, compiled in 354, which cites its observance in Rome in 336. It would not appear to have been celebrated in Antioch until approximately 375. By 380 it was being observed in Constantinople, and by 430 in Alexandria. It was still unknown in Jerusalem early in the fifth century—it was not until the sixth century that the Nativity was finally detached from 6 January and celebrated on 25 December. By the middle of the fifth century it was being gradually observed throughout East and West. The Armenians still observe 6 January, the closely related Feast of the Epiphany, as Christmas Day.

There is no authoritative historical evidence as to the day or month of Christ's birth in Jerusalem. 25 December was the date of a Roman pagan festival inaugurated in 274 as the birthday of the unconquered sun which at the winter solstice begins again to show an increase in light. Sometime before 336 the Church in Rome, unable to stamp out this pagan festival, spiritualized it as the Feast of the Nativity of the Sun of Righteousness. Christmas in the Eastern Church celebrates the birth of Christ together with the visit of the shepherds and the adoration of the wise men. In the Western Church the adoration of the Magi is attached to Epiphany on 6 January. In the Roman Catholic Church three masses are usually said to symbolize the birth of Christ eternally in the bosom of the Father, from the womb of Mary and mystically in the soul of the faithful. The traditional customs associated with Christmas have been derived from several sources. The merrymaking and the exchange of presents find their origin in the Roman Saturnalia festival (17-24 December), and the greenery and lights come from the Kalends of January (1 January, the Roman New Year) with its solar associations. The Germano-Celtic Yule rites introduced the tradition of feasting and fellowship. In the USA (and in England during the Commonwealth) Christian celebrations were at first suppressed by the Puritans, who objected to their pagan origins. Since the nineteenth century the celebration of Christmas has become increasingly popular.

JAMES TAYLOR

CHRISTOLOGY. The study of the person of Christ. Our Lord is unambiguously called God by the NT writers (John 1:1,18; 20:28; Col. 2:9; Titus 2:13; Heb. 1:8,10). The truth of His divinity pervades all strata of NT witness and teaching. He is called the Son of God, and while this does refer to His sonship by incarnation (Luke 1:35; John 1:34; Rom. 1:4; Heb. 1:2), it is not limited to the Incarnation because the terms relate Him to the Father as His "own" Son in a special way (Matt. 11:27; John 5:18). In John the terms "Father" and "Son" are not used only temporally but on the footing of eternity (John 3:13; 17:5; 1 John 4:10). "Son of God" is certainly a title of and claim to deity (Matt. 16:16; 26:63-65; Luke 22:70,71; John 19:7). "Only begotten Son" is to be understood in relation to Christ's preincarnate dignity and privilege (Rom. 8:29; Col. 1:15-18; Heb. 1:6) and in the special sense of "begotten from everlasting," begotten from the being not the will of the Father. The begetting is an eternal fact of the divine nature.

Christ is the Word of God. "Logos" in John 1:1-18 is not explained, but is simply used to declare Christ's deity. Omission of the definite article in "the Word was God" means the Word is identified with the essential nature of God (cf. Rom. 9:5). OT titles ascribed to Him are inexplicable unless Christ is being identified with the nature of Yahweh (cf. Matt. 3:3 with Isa. 40:3; Acts 13:33 with Ps. 2:7, etc.). He is honored and worshiped as God (John 20:28; Phil. 2:10,11; Rev. 5:12-14, etc.). His name is associated with the Father and the Spirit on equal terms in the baptismal formula (Matt. 28:19), in the benediction (2 Cor. 13:14), and in the bestowal of eternal life (John 5:23,24). Finally, the whole biblical structure rests on the claim that redemption belongs to God alone (2 Cor. 5:19; 1 Tim. 2:5). The heart of Athanasius's great argument against Arius was that only God could redeem and reconcile.

The pressure of NT witness to the truth of Christ's humanity is intense, including His birth at Bethlehem (Luke 1:35), boyhood and growth at Nazareth (Luke 2:39-52), fasting and temptation (Matt. 4:1-11), weariness (John 4:6), and death (John 19:28-30; Acts 2:23,36). His true humanity is in part the condition of the work of redemption (Acts 2:22; Rom. 5:15; Phil. 2:7; 1 Tim. 2:5). NT Christology is concerned to show the ideal and normative character of Christ's humanity. His uniqueness is variously shown and emphasized including His birth from the Virgin Mary (Luke 1:34,35), His knowledge and foreknowledge (Matt. 11:27), His moral perfection (Luke 1:35; 2 Cor. 5:21), His teaching (Matt. 5—7), and His transfiguration and exaltation (2 Pet. 1:16-18).

In the patristic period Christology developed chiefly under pressure of the fourth-century Arian heresy. The creeds of Nicea (325) and Constantinople (381) reaffirmed Christ's full deity and full humanity. They insisted on the faith that Christ is truly God, not an intermediate being (which safeguarded the biblical doctrine of creation against Greek forms of thought), and that Christology must be adequate to the facts of redemptive experience, i.e., only God can redeem.

At the Council of Chalcedon* (451) the unity of Christ's person was affirmed, influenced by the differing traditions of the Alexandrian* and Antiochene* schools. Chalcedon does not purport to define the mystery, but to set limits outside which believing Christians cannot go: our Lord took on human nature, not an adult personality, the Godhood and Manhood are each whole and perfect, the two natures are united in one person, and we confess the one Christ.

Classical controversies in the early church reflect divergent viewpoints on the divinity and humanity of Christ. Those who started from the Manhood but failed to do justice to the Godhood of Christ included: Ebionites and Cerinthians who said Jesus was a man specially endowed by God for his mission (cf. 1 John 5:6-12); Adoptianist and Dynamic Monarchians who taught the Incarnation as the inspiration of Jesus by the Spirit at His baptism; and Nestorians who kept Christ's natures apart in the union, i.e., they advocated a prosopic rather than real union of the two natures in the one person. Others started from the Godhood but failed to do justice to the Manhood of Christ. They included: Docetists who made of our Lord's humanity merely appearance; Modalistic Monarchians who made of Christ a revelatory mode of the Father; Apollinarians who substituted the divine nature for the human nature; and Eutychians who said the human nature was swallowed up by the divine nature.

Modern christological controversy follows upon the quest for the historical Jesus. While the quest has been largely abandoned, recent NT scholarship nevertheless concludes that the inner witness of faith came to the disciples who knew Jesus of Nazareth in the flesh as a historical personality. It is not possible ultimately to bifurcate Jesus of Nazareth and the Christ of faith. Kenotic Christology has attempted to probe the meaning of Christ's self-emptying (Phil. 2:7). It has been fiercely attacked (W. Temple,* D.M. Baillie*), but most christological formulations attempt to take account of Christ's self-limitation in some way.

Modern Adoptianist Christology (the American theologian John Knox) is widespread. Through Jesus' goodness, the divine broke through into human life historically, which should be paralleled in our lives. This, however, is not the coming of the eternal Second Person of the Trinity into actual human existence. Others deny that revelation implies factual assertions (Paul Tillich*), which means for Christology that it is irrelevant to our faith if Jesus Christ had never actually lived on earth. The Incarnation becomes the projection of the Christ-Spirit into the world within man through the Christ-event. This sets up metaphysical and existential categories of interpretation rather than those of historical fall, incarnation, and redemption.

Christians confess the true and full Godhood and Manhood of Jesus Christ and the indivisible unity of His person. No theological formula is adequate to this greatest of all Christian mysteries. The Incarnation means the Son of God experienced fully the conditions of personal and individual manhood in such a way that as man He was yet one person with the Son of God. Christians confess they do not know the intensity of unity of the two natures necessary to achieve this, but they accept the apostolic witness. In Jesus Christ is revealed the perfection of God for man (Rom. 5:8-21; Heb. 2:14-18) in virtue of whose response to the Father's will men can respond in faith to become like Him by His Spirit.

See also ATHANASIAN CREED and TRINITY.

BIBLIOGRAPHY: B.B. Warfield, *The Lord of Glory* (1907); H.R. Mackintosh, *The Person of Jesus Christ* (1912); A.E.J. Rawlinson (ed.), *Essays on the Trinity and the Incarnation* (1928); L. Hodgson, *And Was Made Man* (1933); D.M. Baillie, *God Was in Christ* (1948); J.K. Mozley, *The Doctrine of the Incarnation* (1949); T.H. Bindley, *The Oecumenical Documents of the Faith* (1950); R.V. Sellers. *Two Ancient Christologies* (1954); J.M. Creed, *The Divinity of Jesus Christ* (1964); D. Jenkins, *The Glory of Man* (1967).

SAMUEL J. MIKOLASKI

CHRISTOPHERS. A movement established in 1945 among American Roman Catholics to promote the common good of society. Under the leadership of its founder, James Keller, it has sought to enlist individuals from all walks of life to become "Christophers"—a title derived from the Greek word meaning "Christ-bearers." They attempt to "overcome evil with good" (Rom. 12:21), and seek to penetrate and transform every social institution by expressing love for all men through concrete constructive actions in the varied spheres of government, labor, and industry, the family, education, entertainment, and the arts. With headquarters in New York City, the Christophers distribute a newsletter and numerous books, and broadcast their message through radio and television. HARRY SKILTON

CHRODEGANG (d.766). Bishop of Metz from 742, he was also appointed chancellor by Charles Martel and exercised an influence almost unique at that time in both church and state. He played a part in the severing of ties which bound Italy and Rome to the Byzantine Empire, and in the various conquests and acts of legislation which prepared for the unity of Christendom under the joint sovereignty of emperor and pope. His "rule" for his clergy was intended to revive discipline and abolish laxity, though they were not bound by vows of poverty or of unquestioning obedience. His canonical clergy lived together under his supervision. He founded monasteries at Gorze, Lorsch, and St. Arda. JAMES TAYLOR

CHROMATIUS (d.407). Bishop of his native Aquileia in N Italy from about 388. He encouraged others to advance biblical knowledge. Ambrose wrote an exposition of the prophecy of Balaam on his behalf, and his efforts helped to stimulate and finance the translation of the OT by Jerome, who dedicated to Chromatius a commentary on Habakkuk. He tried in vain to reconcile Jerome to Rufinus, whom he had baptized, in their dispute over the translation of the *De Principiis* of Origen. He persuaded the latter to translate Eusebius's church history and the Homilies on Joshua by Origen. When Chrysostom shared with him, together with Innocent of Rome and Venerius of Milan, his complaints about deposition, Chromatius wrote on his behalf to the emperor

Honorius. His extant writings are eighteen homilies on Matthew 3:15-17; 5; 6. G.T.D. ANGEL

CHRONICON EDESSENUM. An anonymous Syriac chronicle of events in Edessa from the time of its first king in 133/2 B.C. until A.D. 540. Its notices are derived from contemporary records in the city's archives and from works such as the Chronicle of Joshua the Stylite. There are only a few entries for the period prior to the Flood of Edessa in A.D. 201/2. The account of the flood, and of King Abgar's measures to prevent a repetition of the disaster, is given a disproportionate amount of attention, but supplies incidental information about the city's early history. Written from a basically orthodox standpoint, the majority of the entries concern the bishops of Edessa and their activities, though ecclesiastical matters of wider significance are noted. The chronicle was compiled about A.D. 550. ROBERT P. GORDON

CHRONICON PASCHALE. A chronicle composed shortly after 628, the point at which its record ended. The Vatican MS from which it is known is mutilated, ending in 627. The work takes its usual name from treatises about chronology and the calculation of Easter in its introduction. The birth of Christ is dated in the "year of the world" 5507, and His crucifixion in 5540. The chronicler expands the framework by excerpts from many sources, e.g., the Chronographies of Sextus Julius Africanus, the Bible, works of Eusebius and Epiphanius, the *Fasti consulares,* acts of the martyrs. For Byzantine history he follows John Malalas to the year 532, and after 600 records events contemporary with his own life. His name is unknown. His work was much read and used in the Byzantine period. It is of great value, especially for the later history, and for the history of chronological techniques and theories in the ancient Christian world. J.N. BIRDSALL

CHRONOGRAPHER OF A.D. 354, THE. A modern title given to a state directory compiled from fragments of MSS. Its chronicled entries cease at 354, when it might have been compiled by Dionysius Philocalus, the later calligrapher of Damasus I, for the use of Christians at Rome. It contains an illustrated state calendar (Roman holidays); a register of the consuls from A.U.C. 245 to A.D. 354; a list of city prefects from 254 to 354; a description of the fourteen districts of Rome; Easter tables from 312 to 354 (with an extension to 410); a chronicle of the world, called *Chronica Horosii,* which is a Latin version and extension to A.D. 334 of the chronicle of Hippolytus; a chronicle of the city to the death of Licinius (324/5); anniversaries of Roman bishops from 255 to 352 (= *Depositio Episcoporum*); the oldest existing list of martyrs (= *Depositio Martyrum*); a list of bishops of Rome from Peter to Liberius (352-66), defining the length of their episcopates (= Liberian Catalogue). This differs significantly from the early list in Irenaeus, which distinguishes the apostles Peter and Paul from the first bishop, Linus. G.T.D. ANGEL

CHRYSIPPUS (d.479). Religious writer. He left his home in Cappadocia in search of advice on the religious life at the monastery of Euthymius in Jerusalem. According to Cyril of Scythopolis, Juvenal of Jerusalem made him superior of the monastery and then of the Church of the Resurrection. He was ordained presbyter and later, about 469, appointed "guardian of the Holy Cross" at the Church of the Holy Sepulchre. Although he wrote copiously, his extant works are four panegyrics, on St. Theodor Teron, the Archangel Michael, the Mother of God, and John the Baptist. This last work is written in the rhetorical style of panegyrists, crammed with metaphors and similes and heavily illustrated with allusions to classical mythology. G.T.D. ANGEL

CHRYSOLOGUS, PETER (d. 449/450). Archbishop of Ravenna. Renowned preacher and conscientious administrator, he sought to eradicate pagan customs, to regulate the Lenten fast, and to correct delay of baptism by catechumens. When Eutyches was deposed by Flavian (448) and asked for support, Chrysologus replied sympathetically but referred him to Leo, writing, "We cannot listen to matters of faith without the consent of the Bishop of Rome." Apart from his extant sermons, no evidence of him survives until the ninth-century *Liber Pontificalis* of Agnellus. According to this legendary account, he was born at Imola and consecrated archbishop by Sixtus III (432-40). Recent examination of his works by Alejandro Olivar endorses the authenticity of all but eight of the 176 sermons traditionally attributed to him by Felix of Ravenna. Olivar adds fifteen sermons from other sources. G.T.D. ANGEL

CHRYSOSTOM, JOHN (c.344/354-407). Bishop of Constantinople. Born at Antioch of exalted Christian parents, he studied philosophy and rhetoric (under the celebrated pagan professor Libanius) before adopting the religious life under the direction of Melitius* and Diodore of Tarsus. Responsibility for his widowed mother curbed his monastic aspirations until about 373, when he became a hermit in the nearby mountains. Harsh austerities impaired his health, and he returned to Antioch to be made deacon by Melitius (381) and priest by Flavian (386), entrusted with preaching in the cathedral. Here in the next decade he delivered most of the series of sermons, chiefly on biblical books, which merited him the sixth-century name *chrysostomos,* "golden-mouthed." This most distinguished of Greek patristic preachers excelled in spiritual and moral application in the Antiochene tradition of literal exegesis, largely disinterested, even untutored in speculative and controversial theology. *The Homilies on the Statues* (387) mastered a congregation terrified of imperial retribution for wrecking statues in a tax riot.

In 398 John was unwillingly appointed patriarch of Constantinople, where his uncompromising reforming zeal and political innocence antagonized Empress Eudoxia and sundry clergy, including Theophilus, bishop of rival Alexandria, who contrived to have him condemned on twenty-nine charges, including Origenist heresy, at the Synod of the Oak* near Chalcedon (403). Deposed and exiled but rapidly recalled, he again infuriated Eudoxia, and after disobeying an imperial fiat to relinquish episcopal duties, which led to bloodshed at the Easter baptisms, he was again exiled (404) to Cucusus in Armenia Secunda; there he proved so accessible and influential that he was ordered to migrate to Pityus on the E Black Sea. He died of the rigors of this forced journey at Comana in Pontus (407). Pope Innocent I broke communion with Constantinople, Alexandria, and Antioch over John's deposition and resumed it only after his posthumous vindication. John's remains were honorably interred in Constantinople in 438.

John's writings, in an attractive Attic style, have nearly all survived. Besides hundreds of sermons (among them a set of recently discovered baptismal homilies and eight against the Jews), they comprise 236 letters concerning his second exile, and several practical treatises, including *The Priesthood* discussions on the monastic life, a pamphlet on the nurture of children (ed. A.M. Malingrey in *Sources Chrétiennes,* 1972; ET in M.L.W. Laistner, *Christianity and Pagan Culture in the Later Roman Empire,* 1951), and a polemical anti-Jewish apologia.

For his straightforward, if artless, integrity and his lively and earnest inculcation of Christian mores, John has enjoyed a wider esteem than any other Father. After Augustine, none was so popular with the Reformers.

BIBLIOGRAPHY: J. Quasten, *Patrology* 3 (1960), pp. 424-82; D. Burger, *Complete Bibliography of Scholarship on the Life and Works of St. John Chrysostom* (1964).

Works: *PG* 47-64; several letters and treatises in *Sources Chrétiennes* series, esp. *Baptismal Catcheses* (vol. 50, 1957, ed. A. Wenger). For numerous spuria, J.A. de Aldame, *Repertorium Pseudochrysostumicum* (1965).

ETs: In *Library of the Fathers* (16 vols., 1839-52) and *Nicene and Post-Nicene Fathers* (1st series, vols. 9-14, 1888-93); many ETs of *The Priesthood* (esp. T.A. Moxon, 1907); *Baptismal Catecheses* (tr. P.W. Harkins, 1963); Biographical *Dialogue* by Palladius (c.408), (ed. P.R. Coleman-Norton 1928, ET by H. Moore, 1921).

Lives by W.R.W. Stephens (1872), D. Attwater (1939), B. Vanderberghe (1958), J.C. Baur (2 vols., 1959, 1961).

Other selected studies include F.H. Chase, *Chrysostom: A Study in the History of Biblical Interpretation* (1887); S.C. Neill, *Chrysostom and His Message* (1962); H. von Campenhausen, *Fathers of the Greek Church* (1963); J. Pelikan, *The Preaching of Chrysostom* (1966); T.M. Finn, *The Liturgy of Baptism in the Baptismal Instructions of St. John Chrysostom* (1967). D.F. WRIGHT

CHURCH (Anglo-Saxon *circe;* from Gr. *kuriakon,* meaning "the Lord's," i.e., house or body). The English word renders the NT *ekklēsia,* used in the free city states of Greece to denote the assembly of citizens, called out from their homes to vote on legislation and transact other public business. As taken up into redemptive history, in the Septuagint *ekklēsia* refers to Israel as assembled for religious and cultic purposes (Deut. 31; 1 Kings 8). As such, Israel is the people of God, called by the divine initiative into holy convocation. Because of the essential continuity with Israel of the new people of God (Gal. 6:16), it is not surprising that the NT writers follow the usage of the Septuagint and speak of those who are united by a common confession of Jesus as Lord, as the church *(ekklēsia),* reserving the word synagogue *(sunagōgē)* for the assembly of the Jewish people. James 2:2 is the only exception to the usage.

The NT church was created out of the band of disciples associated with Jesus during His earthly ministry, who on the day of Pentecost received His Spirit poured out upon them, empowering them to witness to Him as the risen Christ (Acts 1:8). This "new humanity" (Eph. 2:15) takes the outward form of congregations of believers meeting in various places regularly for worship (Acts 20:7; Heb. 10:25) and governed by elders, after the analogy of the synagogue, who are chosen from their own numbers (Acts 14:23). With the rise of the office of monarchical bishop early in the second century, the institutional form of the church rapidly became episcopal, the unity of the churches being found in the bishops. Eventually (in the West) the bishop of Rome was recognized as supreme head of the church. Concomitantly with this development in the institutional form of the church, the original NT emphasis on fellowship and brotherhood *(koinōnia)* as the essence of the church recedes into the background. Although the churches emerging from the Reformation have rejected all claims to papal absolutism, yet they have also had to struggle, if to a lesser degree, to keep the church from becoming a mere institution.

In its spiritual essence the church has often been called (following Augustine and especially Luther) the "church invisible" in that its members enjoy the invisible grace of the indwelling Spirit (1 Cor. 12:4f.) who animates them as a body and unites them to Christ the head (Eph. 4:15,16). United thus in Christ and to one another, the church is the "communion of saints" and though sin yet clings to them, its members are sanctified by the Word and sacraments. This one, holy church (Eph. 4:4; 5:27) is apostolic, being built upon the foundation which the apostles laid in their teaching (Eph. 2:20), and embraces all men (Acts 10:13f.) in a catholic fellowship of faith working by love. The fellowship is visibly present wherever the Word is truly preached and the sacraments rightly administered. In the Reformed tradition (following Calvin), discipline, by which the integrity of Word and sacrament is ensured, has also been made a mark of the church as visible.

In recent times the question of the true nature and task of the church has been much discussed in the ecumenical movement. The fundamental article of the World Council of Churches* defines the council as a fellowship of churches which "confess our Lord Jesus Christ as God and Saviour." The churches which make up the council have understood themselves in terms of the following affirmation: the church, established through the will of the living God, is the fellowship of those who believe in Jesus Christ and is the chosen instrument for the mediation of salvation. The marks of the church on earth are the recognition of the word of God in Holy Scripture, the confession of God in Christ, the proclamation of the Word, the use of the sacraments, the existence of the office of the ministry, and the reality of communion in the exercise of faith and love. This statement was first drafted at the Lausanne Conference in 1927 and has been used as a kind of foundation document in subsequent assemblies of the WCC.

BIBLIOGRAPHY: K. Barth, *The Church and the Churches* (1936); F.F. Bruce, *The Spreading Flame* (1953); H.E. Brunner, *The Misunderstanding of the Church* (1953); F.J.A. Hort, *The Christian Ecclesia* (1908); H. Küng, *The Church* (1967). PAUL KING JEWETT

CHURCH, RICHARD WILLIAM (1815-1890). Dean of St. Paul's. Born in Portugal, he spent early years in Italy (1818-28) and in 1833 went to Oxford where he became fellow of Oriel (1838-52). From early evangelicalism he was moved through the influence and friendship of J.H. Newman* and others to High Churchmanship, in which cause he helped to found the *Guardian* in 1846. In 1851 he was appointed to the rectory of Whatley in Somerset, and in 1871 was appointed on Prime Minister Gladstone's recommendation as dean of St. Paul, a post he held until his death. Church's writings include *Anselm* (1870) and a much acclaimed work on *The Oxford Movement, Twelve Years, 1833-1845* (1891). J.D. DOUGLAS

CHURCH ARMY. The conviction on the part of Wilson Carlile* that laymen should be trained to reach others for Christ led to the formation of the Church Army in 1882. While aggressively evangelistic in his outlook, Carlile, led by study of the Franciscan and Wesleyan movements, insisted from the start that the organization of the Church Army should represent a broad spectrum of the Church of England and that the officers should be dependent, not only on their own headquarters, but on the bishops and clergy in whose dioceses and parishes they worked. Despite considerable early opposition from clergy who objected to laymen ministering in consecrated buildings, it soon was playing a vigorous and invaluable part in evangelism through its caravans and missions, and these were allied to a social concern seen especially in its homes and its ministry to prisons. The Church Army also developed a role in the Anglican Church in other countries.

C. PETER WILLIAMS

CHURCH ASSEMBLY, see GENERAL SYNOD

CHURCHES OF CHRIST (Great Britain). Also called Disciples. They began in sporadic reform movements in the eighteenth century, with Glasites, Baptists, Haldaneites, and Sandemanians all contributing to their formation. The first known church was at Dungannon, Ireland (1804), the second at Auchtermuchty, Scotland (1807). They have been organized since 1842, when there were fifty congregations and 1,300 members. They practice weekly breaking of bread and believers' baptism. Church government is congregational. They have an ordained ministry trained since 1920 at Overdale College, Birmingham. Both sexes can be locally ordained as deacons and elders. They have been responsible for missions in India and Africa. The church is a member of the World Council of Churches. Latest figures available show 119 congregations and 7,000 members. Since 1825 they have maintained contact with the Churches of Christ in the USA,* receiving ministerial and financial help from them in recent years. J.W. MEIKLEJOHN

CHURCHES OF CHRIST (USA). These churches share many doctrinal beliefs and a common heritage with the Disciples of Christ. They follow a congregational pattern of church polity which they regard as being in accordance with NT teaching. Indeed, to align all church practice and belief with the Scriptures is a major goal of the Churches of Christ. Nothing is to be accepted as an article of faith or as a condition of communion but "what is expressly taught and enjoined . . . in the Word of God," which is "the perfect constitution for the worship, discipline, and government of the New Testament Church." The Bible is the best source of information about God, far better than creeds or statements of dogma. The Churches of Christ affirm their belief in the Trinity, the Virgin Birth, the vicarious atonement, the necessity of spiritual rebirth, and the need for believer's baptism by immersion.

The histories of the Churches of Christ and the Disciples of Christ were for all practical purposes the same between 1832 and 1906. Earlier, in 1824, Alexander Campbell,* who in 1811 organized the Disciples, and Presbyterian minister Barton Stone* met each other and found they shared many beliefs. In 1832 at Lexington, Kentucky, the two men led their respective groups into a union to form the "Christian Church (Disciples of Christ)," a church designed to imitate as closely as possible the church life described in the Book of Acts. The creation in 1849 of a missionary society and the use of organ music in church services began to bring division within the ranks. Moreover, the Federal Religious Census of 1906 listed the Churches of Christ and the Disciples of Christ separately. The division was based more on differences of practice than on differences in doctrine.

BIBLIOGRAPHY: R. Richardson, *Memoirs of Alexander Campbell* (1868); T.W. Philips, *The Churches of Christ* (1905); W. Robinson, *What the Churches of Christ Stand For* (1926); B.L. Smith, *Alexander Campbell* (1930); C.C. Ware, *Barton Warren Stone* (1932).

JOHN D. WOODBRIDGE

CHURCHES OF GOD. A name designating about 200 various religious bodies in the USA. Adherents feel the term avoids a doctrinal connotation while stressing a scriptural church designation. Historically the name was first adopted by a revival group within the German Reformed Church (American) in 1825. Later they used the title "Churches of God in North America." At present the name serves also to identify several bodies produced by post-Civil War revivals.

Church of God movements divide into five groups, the first being those of Pentecostal persuasion. Emphasis is placed upon the gifts of prophecy, divine healing, and speaking in tongues as evidence of entire sanctification. Two early leaders were R.G. Spurling and A.J. Tomlinson.* The (Original) Church of God, Inc., with headquarters in Chattanooga, Tennessee, was organized in 1856 from a split in Spurling's followers. In 1970 it reported fifty churches and 6,000 members. The first Spurling group (1886), now known as the Church of God (Cleveland, Tenn.), listed 272,276 members in 4,024 churches in 1972. Tomlinson withdrew from this body in 1923, forming what is now the Church of God of Prophecy. His death in 1943 divided the church into two bodies, one following each of his two sons. The Tennessee organization retained the 1923 name, while the other group, moving to New York, chose "Church of God" for a title. In 1971 it listed a membership of 75,290 in 1,933 churches. The Tennessee faction claimed 48,708 in 1,531 churches. A small splinter group left the Church of God of Prophecy under G.R. Kent in 1957, forming the Church of God of All Nations. In 1923 the Church of God by Faith, Inc., was chartered in Florida. Its headquarters in Jacksonville reported 5,300 members in 105 churches in 1970. C.H. Mason and C.P. Jones of Arkansas founded the Church of God in Christ in 1895. Mason and Jones were rejected by Baptist bodies due to Pentecostalism. In 1970 the church listed 419,466 members in 4,150 churches. The Church of God and Saints in Christ came into being in 1896 under W.S. Crowdy, a Negro deacon. Sometimes called "Black Jews," the movement stresses Jewish ritual.

A second group, in the Wesleyan pattern, is essentially "holiness" in its teaching and emphasizes a conversion experience, followed by Holy Spirit baptism which brings entire sanctification. The principal churches in this category are the Church of God (Anderson, Indiana) and the Church of God (Apostolic).

A third division arose from the Second Advent Movement. The Church of God (Seventh Day), which keeps the Sabbath, is headquartered in Denver and listed 5,000 members in seventy-six churches in 1970. A sister body in Salem, West Virginia, claimed seven churches and 2,000 members. The Church of God General Conference (Abrahamic Faith) was organized in Oregon, Illinois, in 1888 and listed 124 churches with 6,700 members in 1971. Like their Seventh-day Adventist counterparts, these churches hold a strong premillennial eschatology.

A fourth division cannot be catalogued, since there are literally thousands of small "storefront" churches which use the title "Church of God" but belong to no specific denomination.

Finally, the organization founded by H.W. Armstrong in 1947 has used the title "Worldwide Church of God." This American version of British Israelitism* has major defects in several doctrines and is classed by most scholars as a cult rather than church or sect.

See PENTECOSTAL CHURCHES.

BIBLIOGRAPHY: F.E. Mayer, *The Religious Bodies of America* (1956); F.S. Mead, *Handbook of Denominations in the United States* (5th ed., 1970); C.H. Jacquet, Jr. (ed.), *Yearbook of American Churches* (39th issue, 1971).

ROBERT C. NEWMAN

CHURCH OF CHRIST, SCIENTIST, see CHRISTIAN SCIENCE

CHURCH OF JESUS CHRIST OF LATTER-DAY SAINTS, see MORMONISM

CHURCH OF THE BRETHREN. One of the three major "peace churches" of the USA, the Brethren originated in 1708 at Schwarzenau, Germany, as part of the Pietist protest against the state church. They emphasized a warm enthusiasm as well as a study of the Bible and holy living. Their leader, Alexander Mack, Sr.* and seven companions were baptized by trine immersion and began to live by the Brethren practices. These included—in addition to believer's baptism by immersion three times forward—the love feast (including a meal, the Eucharist, and the washing of the saints' feet), anointing of the sick with oil, laying on of hands for Christian service; congregational church government; and opposition to war, oaths, secret societies, and "worldly" clothes and habits. Because of persecution many Brethren fled in 1719 to America, after a short period in Holland, and by 1729 Mack himself had come to the New World. During the American Revolution they refused to fight, but aided some of the German mercenaries who fought on the British side. Again they were persecuted and forced from the eastern urban centers to follow the frontier westward. At present the Church of the Brethren, with headquarters in Elgin, Illinois, and a membership of about 200,000, is the largest Brethren body, but because of a division in 1882 there are also Brethren churches with centers at Ashland, Ohio, and Winona Lake, Indiana (the Brethren Church and the National Fellowship of Brethren Churches respectively). In addition there is a group called the Old German Baptist Brethren or Old Order Brethren.

See H.A. Kent, Sr., *250 Years Conquering Frontiers: A History of the Brethren Church* (1958).

ROBERT G. CLOUSE

CHURCHWARDENS. In the Church of England this office, dating from about the fourteenth century, concerns the lay guardians of a parish church. The churchwardens are responsible for the business and financial side of parochial activity, especially the care of church furnishings and fabric, the seating arrangements, the collection and distribution of alms, and the keeping of order during worship. There are usually two wardens who should be chosen jointly each year in Easter week, by the congregation and the incumbent; but in practice the incumbent has frequently chosen one (the vicar's warden) and the congregation the other (the people's warden). They are officers of the parish, but their report to the bishop (or archdeacon) at his annual visitation illustrates their role as officers of the bishop also.

HOWARD SAINSBURY

CIMABUE (fl. c.1240-c.1300). Florentine painter. Little is known of his life, but his fame is attested by Dante in *Divine Comedy* (Purgatory 11: 94). His style is heavily influenced by the Palaeologan painters of Byzantium. He used also the methods of the Western painters of the Florentine Duecento such as Coppo di Marcovaldo and the Florence Baptistry mosaicists. The works of Cimabue include "The Madonna Enthroned With Angels," and frescoes in the lower and upper churches of San Francisco in Assisi. His painting of the Crucifixion is very impressive with its huge sagging body of Christ, weeping angels, and despairing mourners. Because of his creativity and expressiveness, Cimabue has been considered the first outstanding Western painter.

ROBERT G. CLOUSE

CIRCUIT RIDER. Early Methodist preacher who visited regularly on horseback a number of appointments. Typically poor, single, young, and uneducated, he lived with the people he served. Developed by Wesley, the itineracy was perfected on the American frontier, where it greatly contributed to the advance of Methodism under the leadership of Francis Asbury.*

CIRCUMCELLIONS. Fanatical Donatist* fringe. Their origins are obscure, perhaps antedating Donatism, and their name of uncertain meaning, though interpreting the *cellae* around *(circum)* which they assembled as martyrs' shrines (rather than barns, storehouses) harmonizes with their enthusiastic martyr-dominated ethos. They are most convincingly understood in religious terms, as pilgrim "warriors *(agonistici)* of Christ" marked by apocalyptic defiance of worldly powers, and not as a social or economic grouping only accidentally connected with Donatism. Neither their passion for righting injustice nor their support for local anti-Roman revolts justifies casting them in a revolutionary role. Their asceticism and undisciplined wanderings led to their being confused with Eastern monks (cf. the suggested derivation of their name from monastic *cellae*), and their violence, randomly exerted for egalitarian ends, found frequent outlet in intimidating Donatists against joining the Catholic Church.

D.F. WRIGHT

CIRCUMCISION. Practiced in ancient times in many parts of the world, it was especially important for Israel, for whom it was a sign of God's covenant with Abraham. It is still practiced by Jews, Muslims, and other peoples. It was abandoned early by the Christian Church, notably when the Jerusalem Council (Acts 15) decided it was not obligatory on Gentiles. From a remote date it has been in use in the Church of Ethiopia, where it is performed before baptism, between the third and eighth day after birth.

CISTERCIANS. A Benedictine order founded at Cîteaux in 1098 by Robert of Molesme,* it emphasized poverty, simplicity, and eremitical solitude. Approved by Paschal II in 1100, they were unlike other reformed Benedictine monasteries, rejecting all feudal incomes and basing their economy on the monks' labor assisted by lay brothers. Liturgy was simplified together with church vestments and furnishings; their habit was white or gray under a black scapular. Expansion was early and rapid, due mostly to Bernard,* founder and abbot of Clairvaux (1115), who was responsible for organizing sixty-five new houses in France and abroad, making more than 300 by his death (1153); by 1200 there were over 500 throughout Europe. With Bernard, concerns were widened much beyond contemplation to crusades, missionaries, and pastoral concern for surrounding population.

General decline came in the fifteenth century, assisted by the Reformation and civil wars. Reform followed in the sixteenth century, not from Cîteaux but within congregations, notably the Feuillants* and Common and Strict Observance communities which were, however, dissolved entirely by the French Revolution. After the Bourbon restoration, the Strict Observance was revived by former members of La Trappe (Trappists*), insisting on contemplation, while the Common Observance took to teaching and pastoral duties. In 1892 both became completely separate, though fully Cistercian, and each continues widely established.

BIBLIOGRAPHY: A.A. King, *Cîteaux and Her Elder Daughters* (1954); C. Bock, *Les Codifications du droit cistercien* (1955); L. Bouyer, *The Cistercian Heritage* (tr. E. Livingstone, 1958).

C.G. THORNE, JR.

CLANDESTINITY. Celebration of marriage in secret without proper authority. Because of its prevalence in the later Middle Ages, both Protestants and Catholics were anxious for reform: Luther, e.g., regarded as invalid marriages contracted without parental knowledge and consent. Melanchthon, Brenz, Calvin, and Beza followed his views. Clandestinity posed a problem for Roman Catholic canonists with their concept of marriage as a sacrament. The Council of Trent decreed that though clandestine marriages were true and proper, in future all such marriages in places where the decree obtained would be reckoned as null. All marriages were to be before the parish priest. In Britain, publicity is secured by the publication of banns or issue of a license; clandestinity is commonly held not to invalidate marriage. In 1754 civil legislation was introduced to prevent clandestine marriages (Lord Hardwicke's Act). The Marriage Act (1823) required as a minimum of publicity two or more credible witnesses besides the minister.

J.G.G. NORMAN

CLAPHAM SECT. The name given, probably for the first time, by Sir James Stephen in an article in the *Edinburgh Review* (1844) to the group of wealthy Anglican Evangelicals who lived mainly in Clapham. Its characteristics were those of a large intimate family. Its most famous figure was William Wilberforce,* but around him was a very remarkable galaxy of talent including Henry

Thornton, the banker whose home at Battersea Rise was in many ways the center of the "family"; John Venn, rector of Clapham; Charles Grant, a director of the East India Company; Lord Teignmouth, a governor-general of India; James Stephen, a leading barrister; Zachary Macaulay;* and William Smith, the dissenting member of Parliament who can be counted as a member though he was a Unitarian. Other intimates of the circle included some who did not live in Clapham, such as Hannah More,* Grenville Sharp, Isaac Milner,* and Charles Simeon.*

The establishment and support of a colony in Sierra Leone for ex-slaves, the abolition of the slave trade in 1807 and of slavery in the British colonies in 1833 are their most famous achievements, owing much to a skillful ability to mobilize public opinion and thus to bring pressure to bear on Parliament. There were also significant attempts to widen the basis of education, to make the evangelical message known in carefully timed and presented works for the upper classes, as in Wilberforce's *Practical View,* and for the lower classes, as in Hannah More's *Cheap Repository Tracts.* There was a formidable commitment to a great variety of societies and pressure groups for social improvement. Aristocratic, conservative, and upholders of the status quo they were, yet there was a liberalism in their conservatism not common to men of their background. They were also closely connected with the foundation of the Church Missionary Society (1799), the British and Foreign Bible Society (1804), and the successful parliamentary battle of 1813 to legalize the sending of missionaries to India.

BIBLIOGRAPHY: J. Stephen, *Essays in Ecclesiastical Biography,* II (1849); E.M. Forster, *Marianne Thornton* (1956); M. Hennell, *John Venn and the Clapham Sect* (1958); E.M. Howse, *Saints in Politics. The "Clapham Sect" and the Growth of Freedom* (1960); F.K. Brown, *Fathers of the Victorians. The Age of Wilberforce* (1961); S. Meacham, *Henry Thornton of Clapham, 1790-1815* (1964). C. PETER WILLIAMS

CLARE (c.1193-1253). First abbess of the Poor Clares.* Born at Assisi into a wealthy family, she was attracted by the rule and lives of St. Francis and his friars. At the age of sixteen she ran away from home. At Portiuncala in 1212 Francis usurped the bishop's privilege and received her profession as a nun. At first he placed her in a Benedictine convent, but when her sister Agnes and other women came to join her, he established a community at St. Damian, based on Franciscan lines, with Clare as abbess (1215). This community, the Poor Clares (Clarisses), practiced a strict discipline. Clare governed the convent till her death, though she would have preferred to wander like the friars and to look after the poor and ill. She was canonized by Alexander IV in 1255.
JAMES TAYLOR

CLARENDON, CONSTITUTIONS OF, see CONSTITUTIONS OF CLARENDON

CLARENDON CODE. The Cavalier or Pensioner Parliament first met on 8 May 1661. It passed a series of severe statutes known as the Clarendon Code. These were the Corporation Act (1661), the Act of Uniformity* (1662), Conventicle Act* (1664), and Five Mile Act* (1665). Their aim was to remove from the ministry of the church as well as from national and local government those who did not subscribe to the liturgy and doctrines of the Church of England. They are named after Sir Edward Hyde, earl of Clarendon (1609-74), who was the Lord Chancellor under Charles II, but he was not wholly responsible for them.

CLARK, FRANCIS EDWARD (1851-1927). Father of the Christian Endeavor* movement. Born Francis E. Symmes in Aylmer, Quebec, he was orphaned of both parents before he was eight and was adopted by an uncle, the Rev. E.W. Clark. He prepared for the Congregational ministry; his first pastorate was Williston Church, Portland, Maine, where he increased the membership in two years from fifty to 400. In February 1881 he started the first Christian Endeavor Society. In 1883 he went to Phillips Church, South Boston. By 1887 rapid expansion of Christian Endeavor led him to give it full time as president of the United Society and editor of *The Christian Endeavor World.* During the next forty years he made five trips around the world for this youth work. HAROLD R. COOK

CLARK, WILLIAM SMITH (1826-1886). American agriculturalist. President of Massachusetts Agricultural College, he came to the notice of Japanese government emissaries who invited him to establish a similar institution in their country. In July 1876 the dynamic Clark reached Sapporo, capital of Japan's newly colonized northern island, and within eight months set up a college, preparatory school, and experimental farm, introduced crops and trees, agricultural buildings and methods, and converted all sixteen students to Christianity. The young believers, exemplifying his parting words, "Boys, be ambitious," won all the second class who also signed Clark's "Covenant of Believers in Jesus." The zealous group became known as the Sapporo Band, whose most notable member was Kanzo Uchimura.*

See also JAPAN. DAVID MICHELL

CLARKE, ADAM (1762?-1832). Methodist theologian. Born in County Londonderry, he received a local education, supplemented through John Wesley's influence in England at Kingswood School, Bristol. Having become a Methodist in 1778, he was appointed circuit preacher in Wiltshire in 1782, later traveling all over the British Isles as his fame grew. From 1805 he lived chiefly in London, and was three times president of the Methodist conference. Awarded an Aberdeen LL.D., his scholarship was impressive, encompassing classics, patristics, oriental languages and literature, history, geology and natural science, and even the occult. Theologically orthodox in most things, Clarke nevertheless denied Christ's eternal sonship while maintaining His divinity; held that Judas repented and was saved; and rejected Calvin's view of predestination. His great achievement was an eight-volume Bible commentary (1810-26). Selected to edit Rymer's *Foedera,* he was forced by ill-health to relinquish the task at volume 2. Clarke's miscellaneous works in thirteen volumes were printed in 1836.

J.D. DOUGLAS

CLARKE, JOHN (1609-1676). One of the founders of Rhode Island. Born near London and trained as a physician, he left England in 1637 for Boston. Following a dispute there with the Puritans when he supported the banished Anne Hutchinson,* Clarke and other colonists were driven out and bought from the Indians an island they called Rhodes. In April 1639 he was one of those who settled Newport, where he practiced medicine, participated in local government, and became also pastor of the Baptist church. Having collaborated with Roger Williams,* Clarke's unifying zeal was further seen in his accompanying Williams to England in 1651 to seek a new charter. Williams went home three years later, but Clarke remained as sole agent for the colony until 1663, when he secured from Charles II the royal charter that determined Rhode Island law until 1842. He resumed his pastorate at Newport and was three times elected deputy governor of the colony.

J.D. DOUGLAS

CLARKE, SAMUEL (1675-1729). Philosophical theologian. Son of a British member of Parliament, he was educated at the Norwich Free School and at Caius College, Cambridge. There Rohault's *System of Natural Philosophy* (in indifferent Latin), a Cartesian work, was in use. While still a student Clarke retranslated this (published 1697) and in footnotes explained how much better Newton's physics explained the facts. As a result, the views of Newton, who became his friend, were quickly accepted in the university. After ordination Clarke wrote several theological texts, including paraphrases of the gospels, and later threw himself, on the Christian side, into the controversies of the day—the immortality of the soul, ethics, the Trinity, natural theology, Locke's empiricism, the materialism of Hobbes, Spinoza's pantheism, etc. He is best known for his Boyle Lectures of 1704 *(Being and Attributes of God)* and 1705 *(Discourse on Natural and Revealed Religion),* in which he argued for belief in God and in the Christian religion, in the form of propositions in mathematical style. Learning that Newtonianism was being blamed for the decline of natural theology, Clarke corresponded with Leibnitz,* defending Newtonianism and arguing that it supported rather than endangered the principles of religion. Clarke is said to have had a happy, playful disposition. The level of his thinking was above that of his contemporaries, who occasionally suspected him of heresy; but for this, it is said, he might have become archbishop of Canterbury. His *Works* (4 vols.) were published in 1738-42.

See also W. Whiston, *Historical Memoirs* (1730).

R.E.D. CLARK

CLARKE, WILLIAM NEWTON (1841-1912). Baptist minister and theologian. Born in Cazenovia, New York, he was educated at Madison (now Colgate) University and Theological Seminary. Particularly influential was his pastorate at Newton, Massachusetts (1869-80), where he was drawn to the liberalism of Andover and Newton theologians. From 1890 to 1908 he taught Christian theology at Colgate. Sensitive and cosmopolitan, he stressed the evolution of divine revelation and human goodness. His combination of Christian theology and evolutionary thinking is best expressed in his major work, *An Outline of Christian Theology* (1898), the first systematic theology of liberalism in America. By 1914 this work, described by Lyman Abbott as the most religious theology ever written, was in its twentieth printing.

DARREL BIGHAM

CLARKSON, THOMAS (1760-1846). Abolitionist. While at St. John's College, Cambridge, he wrote in 1785 a prize-winning essay on slavery, which influenced him to devote his life to the cause of abolition. Leaving Cambridge, he was ordained deacon in the Church of England, though he rarely exercised a ministry. In 1787 he joined with William Wilberforce* and others to campaign in the country and in the House of Commons for abolition, and thenceforth his life is the history of the antislavery struggle, in which he spent most of his fortune. His main work was collecting information for the campaign, which achieved success with the ending of the slave trade in 1807 and emancipation in the British dominions in 1833. After the outbreak of the French Revolution he propagated ideas of abolition in France, and after 1815 he presented his views at the European Congresses. Among his written works are an account of the abolition of the African slave trade and books on Quakerism, religious origins, war, and baptism.

JOHN A. SIMPSON

CLASS MEETING. Dating from the early organization of Methodism, this was part of John Wesley's policy to increase understanding of the faith and to ensure growth in holy living. Members of each local society are divided into groups that meet weekly under a lay leader for "fellowship in Christian experience."

CLAUDE, JEAN (1619-1687). French Huguenot* preacher. After studying theology at Montauban, he began his work as a pastor at La Treisse. Next he moved to St. Affrique before settling at Nîmes in 1654; here he also lectured in the Protestant Academy. In 1661, the year when a synod over which he presided refused the idea of reunion with Rome, he was prohibited from preaching. He moved to a professorship at Montauban (1662-66), but again fell foul of the government. Between 1666 and the revocation of the Edict of Nantes* (1685) he was in Charenton, Paris, where as a respected Huguenot leader he engaged in controversy with leading Roman Catholics, the Jansenist Arnauld* and the celebrated J.B. Bossuet,* among others. About 1686 he went to live at The Hague, where he died. He wrote many books; two of them were widely read in English translations: *An Account of the Persecutions of the Protestants in France* and *On the Composition of a Sermon.*

PETER TOON

CLAUDIANUS MAMERTUS (d. c.474). Priest and poet. Friend of Sidonius Apollinaris* and brother of Mamertus, archbishop of Vienne, he was trained for monastic life and educated in Greek, Roman, and Christian literature. As assistant to his brother he trained clergy, organized services, and arranged the local psalter and lectionary. When Faustus of Riez published anonymously a treatise affirming that the soul is corporeal, Claudianus was urged by his friends to reply, and produced *De Statu Animae* in three books, which alone among his extant writings has theological value. Written between 467 and 472, it argues that a soul is incorporeal because it is made in the image of God and does not conform to the categories of place and quantity.

G.T.D. ANGEL

CLAUDIUS, MATTHIAS (1740-1815). German poet. Descended from Lutheran pastors, he studied at Jena and later moved to Wandsbeck, near Hamburg, where in 1771 he became literary editor of *Der Wandsbecker Bote.* For this he wrote, under the pseudonym Asmus, prose and poems in a popular "calendar" style. In 1776 he moved to Darmstadt, where he was influenced by freethinkers, but in 1777 after an illness he regained his faith. He returned to Wandsbeck to redirect the *Bote* on Christian lines and provide ammunition for a revolt against rationalism. While many of his poems were Christian in spirit, he wrote nothing for church use. Miss J.M. Campbell's translation "We plough the fields and scatter" (1861), reportedly a paraphrase of part of a village sketch, met a then new demand for harvest festival hymns.

JOHN S. ANDREWS

CLAUDIUS APOLLINARIUS, see APOLLINARIUS, CLAUDIUS

CLAUDIUS OF TURIN (d.827). Bishop of Turin. Of Spanish birth, he was trained in theology at the school of Leidradus at Lyons, and also under Felix,* bishop of Urgel, an Adoptianist. He was a priest in the court of Louis the Pious in Aquitaine and followed him to Aachen when Louis became emperor. Made bishop of Turin (817/8), Claudius attacked the cult of images in his *Liber de imaginibus,* wrongly attributed to Agobard* of Lyons, on which Dungal and Jonas of Orléans based their attacks on him. He wrote commentaries on some of the OT books and the gospels, and probably all the Pauline epistles, using the *catenae* form which obtained in the Carolingian and later medieval period.

C.G. THORNE, JR.

CLAVER, PETER (1580-1654). Jesuit missionary. Born in Verdu, Spain, he entered the order in 1602. Three years later at Mallorca, Alphonso Rodriquez encouraged him to become a missioner. Until 1610 Claver studied at Barcelona, whence he went to Cartagena, Colombia. There Alonso de Sandoval deeply impressed on him the misery of African slaves. In 1616 Claver went to Bogota, where he was ordained, and on his return to Cartagena did much to alleviate the misery of those in the disease-ridden slave ships. He befriended them as a doctor and teacher. Valtierra considers that by 1651 more than 300,000 had been converted. Pope Urban VIII condemned slavery in a bull of 1639.

Claver died at Cartagena. He was part of a small band like Sandoval and Bartolome de Las Casas who protested the inhumanity of European expansion in the New World. Claver was canonized in 1888, and in 1896 became the patron of all Catholic missionary activity among Negroes.

MARVIN W. ANDERSON

CLAVIUM POTESTAS, see KEYS, POWER OF THE

CLEMENS, JAKOB (d.1558?). Flemish musical composer. He was called "Clemens non Papa" to distinguish him from a priest of that name in Flanders or from Pope Clement VII (1523-34). He may have been imperial choirmaster under the Holy Roman Emperor Charles V (1519-21). His sacred compositions, many published posthumously, include metrical psalms, Masses, motets, and chansons, and there are indications works were interrupted by an untimely death.

CLEMENT V (1264-1314). Pope from 1305. Born Bertrand de Got (or Gouth) into an influential French family, he studied at Toulouse, Orléans, and Bologna. He was appointed bishop of Cominges in 1295, and archbishop of Bordeaux in 1299. A Roman conclave met in 1304 after the short reign of Benedict XI (1303-4) to select a new pope; after eleven months (during which envoys from France attended) Bertrand was proclaimed Pope Clement V. Being under the influence of Philip the Fair of France, who seemed able to exploit the weaknesses of the new pope, Clement was crowned at Lyons and in 1309 finally installed himself and his Curia at Avignon. Thus began the seventy years of the "Babylonian Captivity" of the papacy.

To serve his own personal interests, Philip IV made two major demands of Clement. The first was to have Boniface VIII condemned as a heretic. Clement obliged insofar as he annulled Boniface's excommunications and interdicts, especially the bull *Unam Sanctam* (bull of 27 April 1311). The second was to dissolve the Order of Knights Templar.* Again Clement obliged, since the Council of Vienne (1311) did suppress the order. It is not unfair to say that Clement was in a state of servility to Philip. However, he did attempt to salvage what he could for the Knights Templar in 1311.

Despite his political problems, which related to England and Scotland as well as France and the Holy Roman Empire, he was a competent scholar and is remembered as the founder of two universities, Orléans and Perugia, and for his collection of decretals in *Liber Septimus* (usually known as *Clementina*).

PETER TOON

CLEMENT VI (1291-1352). Second Avignon* pope from 1347. Born Pierre Roger, he entered the Benedictine monastery at La Chaise-Dieu, later moving to Paris where he became a student, then a teacher. Ecclesiastical promotion came quickly: abbot of Fécamp in 1326, then after two bishoprics, archbishop of Rouen in 1330. He was made cardinal by Benedict XII in 1338, and pope at Avignon nine years later.

Clement was a brilliant orator and a great and ostentatious nobleman; thus he has been described as the first of the Renaissance popes. Like Clement V, his attachment to France was strong, and he was an ardent admirer of Philip the Fair, which prejudiced his attempts to bring peace between France and England. Also, following French policy, he carried on the struggle with Louis the Bavarian until 1347. He received the submission of William of Ockham who had advocated the separation of church and state together with the denial of temporal authority to the pope. The Franciscans also came under his displeasure as schismatics. That same year (1347), Cola di Rienzo* was imprisoned for similar reasons as Ockham—the encroachment of the pope's temporal power.

Clement seemed more concerned with Avignon than Italy. The sovereignty of Avignon he bought from Joanna I of Naples in 1348 for 80,000 golden florins, and surrounding himself with French cardinals, he turned Avignon into one of the most resplendent courts in Europe, from where also he directed the political campaigning in Italy. All this was paid for by extending taxation to ecclesiastical benefices, claiming for himself the overall right of church property and the right to intervene in the presentation of benefices, which brought opposition from many quarters, especially from Edward III of England (who in 1351 claimed monarchial right of presentation in all papal appointments to benefices). In spite of his tendency to nepotism, Clement did aid the poor. He helped those who were caught in the Black Death* at Avignon (1348-49). He defended the Mendicant against the secular orders. He was kindly disposed toward the Jews, who generally were safer in papal states than elsewhere in Europe. In 1351 he established a bishopric for the Canary Islands. His contribution to theology is best seen in his jubilee bull of 1350 concerning indulgences,* a practice which had received his approval in 1343.

See W. Ullman, *The Origins of the Great Schism* (1948); and G. Mollat, *Les Papes d'Avignon* (9th ed., 1949). GORDON A. CATHERALL

CLEMENT VII (1478-1534). Pope from 1523. Giulio de' Medici, illegitimate son of Giuliano de' Medici and cousin of Pope Leo X,* became archbishop of Florence, being granted a special dispensation on account of his birth. Made cardinal in 1513, he practically controlled papal policy in Leo's pontificate and succeeded Hadrian VI as pope. Indecisive and timid, he failed to deal with the problems raised by the Reformation, preferring to live as a prince and to forward Medici interests.

He patronized the arts, encouraging Cellini, Raphael,* and Michelangelo.* He vacillated between support of Francis I of France and emperor Charles V, siding with Francis at the League of Cognac (1526), then, after his capture by imperial troops at the sack of Rome (1527), with Charles at the treaty of Barcelona (1529). He was equally irresolute over the divorce of Henry VIII of England. Eventually yielding to imperial pressure (1529), he resisted Henry's subsequent break with Rome.

There was also an antipope Clement VII (d.1394): Robert, son of Count Amadeus III of Geneva, who became archbishop of Cambrai in 1368 and was elected pope when the French cardinals seceded in 1378, so beginning the Great Schism.* J.G.G. NORMAN

CLEMENT VIII (1536-1605). Pope from 1592. Born Ippolito Aldobrandini, son of an Italian lawyer, he was the first of the series of "restorer" popes, though in 1594 his confessor Philip Neri* refused him absolution unless he recognized the coronation of Henry IV (who was received back into the church in 1595). In 1598 Clement helped to prepare the Treaty of Vervins between France and Spain, and three years later he negotiated the Treaty of Lyons between France and Savoy. In 1599 he created a preliminary sketch, as it were, of the future Congregation of Propaganda, under the presidency of Cardinal Santonio; its main function was to vitalize and control the Apostolate in every part of the world. In 1600 Clement refused the request of a number of orders to break the monopoly of the Jesuits to work in Japan, which had been granted them in 1585 by Gregory XIII. His attempts at restoring the Roman Empire failed, particularly in the case of James I (of England) and VI (of Scotland), whom he attempted to woo back to Roman Catholicism.

In ecclesiastical matters Clement made his presence felt soon after his accession. He ordered a revision of the Vulgate,* which had been done previously in haste, known as the *Sixtus Vulgate;* the new edition, generally known as Clement's edition, appeared in 1592 and contained some 3,000 corrections. Later research reveals that the Clementine edition departs from the text of Jerome at many points. His interest in the revision of the service books of the church extended to the Missal, Breviary, *Caeremoniale episcoporum*, and Pontifical. He also built a monumental altar within St. Peter's, over the site believed to contain the apostle's body, at which only the pope can celebrate Mass. His popularity is indicated by the fact that in 1600 he was acclaimed by three million pilgrims to Rome. GORDON A. CATHERALL

CLEMENT XI (1649-1721). Pope from 1700. Born Giovanni Francesco Albani, native of Urbino, he was educated at Rome, gaining a doctorate in civil and canon law. He entered the papal administration, success at which brought him the appointment of secretary of papal briefs in 1687. Three years later he was made cardinal, and ten years later he reluctantly accepted election as pope. His reign occurred during the period of the War of Spanish Succession, and he was set between the interests of the Hapsburgs and the Bourbons. Not surprisingly his political actions were unsuccessful. In 1701 he vainly protested at the elector of Brandenburg's use of the title "king of Prussia." In the same year he was forced to recognize Philip of Anjou, but in 1709 he had to abandon him in favor of the Archduke Charles.

The result was the Treaty of Utrecht (1713), with Clement's rights in Sicily, Sardinia, Parma, and Piacenza being ignored. Also in 1709 Clement quarreled with the duke of Savoy over his rights of investiture in Sicily. The unhappy outcome was that Clement issued an interdict, from which his clergy suffered most, because all who accepted it were banned from Sicily until it was conquered by Philip V of Spain in 1718. He further failed to gain the support of the princes when the Turks declared war on Venice in 1714. On the pastoral side, Clement had to deal with Jansenism,* which he condemned in a bull entitled *Vineam Domini Sabaoth* (1705). In 1708 he condemned the work of P. Quesnel* usually called *Réflexions morales.* In 1713 he issued the famous bull *Unigenitus Dei Filius,* in which he attacked Quesnel's thesis that grace is irresistible and without it man is incapable of spiritual good. This produced a pessimistic theology and subsequent harshness and moral rigidity. Clement had also to judge between Franciscans and Jesuits concerning Chinese Rites,* and he upheld the Dominican position. He took a similarly conservative decision about the Malabar* Rite in India. In 1708 he made obligatory the Feast of the Immaculate Conception.* He secured for the Vatican Library one of its most valuable collections of manuscripts from the East, made by J.S. Assemani. When he died, his reign had lasted longer than any other pope for five centuries. GORDON A. CATHERALL

CLEMENT XII (1652-1740). Pope from 1730. Lorenzo Corsini, born at Florence, studied at the Roman College and Pisa University. On his father's death he entered the church, and became titular bishop of Nicomedia and nuncio to Vienna (1691), governor of Castel Sant' Angelo (1696), and cardinal (1706). He became blind in 1732, but despite ill health tried to halt the decline of papal power. He defended church rights threatened by encroachments of the Catholic powers, and though forced to allow the papacy to lose its feudal rights in Parma and Piacenza, he made significant agreements with Spain and Portugal. He took action against the Jansenists. In 1738 he condemned Freemasonry,* forbidding Catholics to belong to Masonic lodges under pain of excommunication. He vigorously supported missionary activity, founding a seminary for training priests of the Greek rite at Ullano, S Italy, helping the Lebanese Maronites, and sending Franciscans to Ethiopia. J.G.G. NORMAN

CLEMENT XIII (1693-1769). Pope from 1758. Carlo della Torre Rezzonico, born in Venice, studied at the Jesuit college at Bologna, and became doctor of theology and canon law at Padua. He was ordained and appointed governor of Rieti (1716) and of Fano (1721). Benedict XIII called him to Rome (1725) and made him auditor of the Rota for Venice (1729). Created cardinal by Clement XII (1737), he became bishop of Padua (1743). He was elected pope at a time when the papacy's prestige was declining and the Jesuits were under attack. He took up the cause of the Jesuits, to whom he owed his election. Portugal expelled them and severed relations with Rome (1760), while France demanded drastic alterations in the Jesuit constitution (1761). Clement refused this demand in the famous words "Let them be as they are or not be." Louis XV abolished the order in France (1764), and Clement responded by the bull *Apostolicum pascendi munus* (1765). The Jesuits were expelled from Spain, Naples, and Malta; Parma ordered a commission to investigate monastic charters. As traditional suzerain of the duchy of Parma, Clement attempted to reassert temporal power there (1768), but found himself faced with the seizure of Avignon, Benevento, and Ponte Corvo, and an almost universal call for the suppression of the Jesuits. He consented to call a consistory, but died almost immediately of apoplexy. The suppression was effected by his successor, Clement XIV,* in 1773. J.G.G. NORMAN

CLEMENT XIV (1705-1774). Pope from 1769. Born Giovanni Vincenzo Antonio Ganganelli near Rimini, he joined the Franciscan Order in 1723, at which time he took the name Lorenzo. He was named cardinal in 1759. His election to the papacy came at a time when Catholic powers were pressing for the suppression of the Jesuits. It was thought that his coming to a secret understanding with the former (though this has never been proved) led to his choice as pope after many of the other cardinals had been rejected. Whatever the truth of the accusation, Clement did in fact suppress the Jesuits in 1773; he may have thought it a price worth paying in order to appease potential allies in an age when the papacy had to cope with anticlericalism and growing secularism.
J.D. DOUGLAS

CLEMENTINE LITERATURE, see CLEMENT OF ROME

CLEMENT OF ALEXANDRIA (c.155-c.220). The first known Christian scholar. Titus Flavius Clement, probably an Athenian, succeeded his teacher Pantaenus as head of the Catechetical School at Alexandria in 190. Clement's greatest literary activity was displayed while he held that post (190-202). His principal works extant are the *Protrepticos (Exhortation to Conversion),* the *Paidagogos (The Tutor),* and *Stromateis (Miscellanies).* The three works constitute a trilogy. The Logos,* Clement says, first of all "converts" us, then "disciplines" us, and finally "instructs" us.

Clement engaged in a perpetual debate with the Gnostics who disparaged faith as an inferior possession of ordinary Christians while they rejoiced in their possession of esoteric *gnosis.* He maintained that faith, instead of being the "prop of the ignorant," was the means by which mankind arrived at true *gnosis* ("knowledge"). The doctrine of the Logos is the mainspring of Clement's whole system of theology. The Logos is conceived of as eternally with the Father and the principal cause of all things that are. Some of Clement's statements concerning the person of Christ have a docetic echo, but he defended vigorously the reality of the Incarnation, even though the humanity of Jesus has little importance in his theology. The work of the Logos, or Christ, is considered as the redemption from the bondage of sin and error which has left mankind blind and helpless. Clement's most characteristic thought is that Christ is the true "teacher" who gives men the true *gnosis* which leads to freedom from sin, to immortality, and to righteousness. By contemplation of the Logos man is deified. Thus Clement's soteriology is a Christ-mysticism in which the Lord's passion and death have little or no redemptive part to play.

Clement's other surviving works include the *Hypotyposes*, a commentary on the Scriptures, and *Quis Dives Salvetur? (Who is the Rich Man that shall be Saved?)*, the theme of which is the stewardship of wealth. This last attractive homily concludes with the well-known story of the aged Apostle John rescuing and restoring a young Christian who had become a bandit.

BIBLIOGRAPHY: C. Bigg, *The Christian Platonists of Alexandria* (2nd ed., 1913); G.W. Butterworth, *Clement of Alexandria* (1953); J.E.L. Oulton (ed.), *Alexandrian Christianity* (1954); J.N.D. Kelly, *Early Christian Doctrines* (1958); F. L. Cross, *The Early Christian Fathers* (1960); H. Chadwick, *Early Christian Thought and the Classical Tradition* (1966). G.L. CAREY

CLEMENT OF ROME (fl. c.90-100). Prominent early Roman presbyter-bishop. Perhaps already mentioned in Philippians 4:3 (if from Rome) and often identified or connected (as slave or freedman?) with the Titus Flavius Clemens executed by Domitian, he is most probably the Clement in Hermas's *Shepherd* whose duty was to write to other churches, which accords with the traditional authorship of *1 Clement,* a letter in the name of the Roman to the Corinthian Church. (The ascription to Clement is first attested about 170 by Dionysius of Corinth, but is absent from Irenaeus.)

The letter was almost certainly written soon after the supposed persecution under Domitian, i.e., about 96. It attempts to heal a division which had led, perhaps through the ascendancy of a Gnostic party, to the deposition of senior Corinthian presbyters (also called bishops). The letter betrays no knowledge of monepiscopacy, but appeals to a simple form of apostolic succession. It argues incessantly for the preservation of divine order, by cataloguing moralistically the entail of jealousy or strife in the OT and in the Christian times, by rehearsing the virtues of humility, and by natural and military analogies. It adheres to traditions of Hellenistic rhetoric and Stoic political philosophy, but also follows Jewish precedents. As probably the oldest extra-NT Christian writing, it is invaluable for plotting the development of "early catholicism" *(Frühkatholizismus).* Though it appeals to the authority of works like 1 Corinthians and Hebrews, it reveals a marked change of outlook from Paul's epistles. In the later second century it enjoyed almost scriptural status in several churches.

For editions, etc., see APOSTOLIC FATHERS, and J. Quasten, *Patrology* 1 (1950), pp. 42-53; also ed. A. Jaubert (*Sources Chrétiennes* 167, 1971); ET by W.K.L. Clarke (1937). See also L. Sanders, *L'Héllenisme de S. Clément de Rome et le Paulinisme* (1943); K. Beyschlag, *Clemens Romanus und der Frühkatholizismus: Untersuchungen zu I Clemens 1-7* (1966).

Later tradition promoted Clement to a monarchical bishop, the third pope after Peter or even his successor, and spawned legends about his travels and death. The following writings were wrongly ascribed to him (see also APOSTOLIC CONSTITUTIONS):

(1) *2 Clement:* this *Second Epistle to the Corinthians,* read as Scripture in Syria but recognized as spurious by Eusebius, is actually a sermon, probably the oldest in postapostolic literature. It appears to have been delivered about 140-50, in Rome (by Bishop Hyginus, c.138-42?) or Corinth (hence its MS connections with *1 Clement*), or perhaps Alexandria. It touches on several aspects of Christian doctrine, especially the church and repentance, revealing Pauline and gnosticizing influences, but is mainly a moral exhortation.

For literature see APOSTOLIC FATHERS and J. Quasten, op.cit. 1, pp. 53-58; also K.P. Donfried, *The Setting of Second Clement in Early Christianity* (1974).

(2) *Two Letters to Virgins:* an important source for early Christian asceticism, from the first half of the third century, probably Palestine or Syria. It extols the angelic and Christlike quality of ascetic life, and deplores the irregularities of the *syneisaktoi* (see AGAPETAE).

ET in *Ante-Nicene Christian Library* 14 *(The Writings of Methodius),* pp. 365-95; bibliography in Quasten, op.cit. 1, pp. 58-59.

(3) *Clementine Homilies* and *Recognitions:* didactic novels built around Clement's career mostly as Peter's companion. After a religious-philosophical pilgrimage, Clement is converted by Peter at Caesarea. The twenty Greek *Homilies* (before 380) are Clement's account of Peter's missionary preaching and contest with Simon Magus, preceded by letters to Jesus' brother, James of Jerusalem, from Clement and Peter, and Clement's instructions for the use of the work. Two Greek epitomes preserve mostly narrative sections, with additions, e.g., regarding Clement's martyrdom. The *Recognitions* (before c.360-80) get their title from Clement's successive reunions, through Peter's intervention, with long-lost members of his family. The ten books survive only in Rufinus's Latin abridgment.

The two collections possess a common narrative core (more developed in the *Recognitions*), which results from common dependence on a Jewish-Christian *Grundschrift* from early third-century Syria. Scholars still debate whether the *Recognitions* are independent of the earlier *Homilies,* whether they have suffered interpolation (e.g., the *Recognitions* by a Eunomian), how extensive were the *Grundschrift* and its presumed sources, the *Sermons (Kerygmata) of Peter* and the *Acts* (or *Travels, Periodoi*) *of Peter,* and how deviantly Jewish-Christian their theology. The *Homilies* display marked Ebionite or Elkesaite features (Christianity as reformed Judaism; Christ as merely the true prophet; Paul as hostile to Peter's message), and reflect a Jewish-Christian Gnosticism, but in the *Recognitions* Judaistic elements are largely obscured by orthodoxy, presumably partly through Rufinus. For all their difficulty, the *Clementines* are an invaluable source for the fortunes of Jewish Christianity.

BIBLIOGRAPHY: Text: *Homilies*, ed. B. Rehm, J. Irmscher, and F. Paschke, in *Die griechischen christlichen Schriftseller der ersten drei Jahrhunderte* 42 (2nd ed., 1969); *Recognitions*, ed. Rehm and Paschke, ibid. 51 (1965).

ETs: in *Ante-Nicene Christian Library* 3 *(Recognitions)* and 17 *(Homilies)*; extracts with introduction in E. Hennecke, W. Schneemelcher, R. McL. Wilson (eds.), *New Testament Apocrypha*, vol. 2 (1965), pp. 102-27 (G. Strecker) and pp. 532-70 (J. Irmscher).

Other studies: O. Cullmann, *Le problème Littéraire et Historique du Roman Pseudo-Clémentin* (1930); H.J. Schoeps, *Theologie und Geschichte des Judenchristentums* (1949); G. Strecker, *Das Judenchristentum in den Pseudo-Klementinen* (1958), and on the *Sermons of Peter* source in W. Bauer, *Orthodoxy and Heresy in Earliest Christianity* (1972), pp. 257-71. D.F. WRIGHT

CLERICALISM. A term denoting that which relates to clerics and the clergy. During the nineteenth century in Italy and France it was used to attack the efforts of the Roman Catholic Church to influence secular affairs unduly. Thus in 1877 Léon Gambetta coined the phrase *Le cléricalisme, voilà l'ennemi!* The term has also been used opprobriously to denigrate an excessive professionalism on the part of clergymen or their imitators.

CLERICUS, see LECLERC, JEAN

CLERGY RESERVES. The term applied to that portion (one-seventh) of the land which was set apart in the Canadas by the Canada Act of 1791 for the "support and maintenance of a Protestant clergy." Because the proceeds were used in the interests of the Church of England, the other denominations soon began to urge that the funds be used for the support of all denominations. The Durham Report cited the Clergy Reserves as one of the major causes of unrest preceding the Upper Canada rebellion in 1837. A compromise was therefore included in the Imperial Act of 1840 by which the monies realized from the sale of the reserves were divided among the denominations. The reserves remained a political sore spot until they were secularized in 1854, with payments being made to the various denominations, and with the Presbyterians and Anglicans retaining their former grants of land.

See CANADA. ROBERT WILSON

CLERK MAXWELL, JAMES, see MAXWELL, JAMES CLERK

CLERKS REGULAR. Roman Catholic clergy who have taken solemn vows for the purpose of more effective pastoral work. The term applies particularly to orders like the Theatines* (1524) and Jesuits* (1534) founded in the sixteenth century. Their numbers have not increased since then, but their example has had a striking influence on later religious communities which have lived under rule, whether strictly monastic or not, though the great majority of these communities take simple vows.

CLERMONT, COUNCIL OF (1095). Called at the instigation of Urban II, it was well attended by both clergy and laity. As a result of an embassy received from Emperor Alexius Comnenus at the Council of Piacenza (earlier that year), Urban called a crusade, urging the assembly to gain paradise fighting in God's cause, instead of losing their souls in fratricidal wars. There was an enthusiastic reception for Urban's proclamation. Influential support was given the council by the presence of Raymond of Toulouse, whose vassal Bishop Adhemar Le Puy was chosen to be papal representative during the crusade. Thirty-two canons were issued. It was decreed, *inter alia*, that no king or prince should grant investiture; no bishop or priest should pay homage to a layman; no flesh should be eaten between Ash Wednesday and Easter; and communicants should receive in both kinds separately whenever possible. The council also excommunicated Philip I for adultery, and confirmed the primacy of the see of Lyons.

JAMES TAYLOR

CLIMACUS, JOHN, see JOHN CLIMACUS

CLOISTER (Lat. *claustrum*, "clergy-house, closed-off place"). The verbal concept "to cloister" referred originally to any enclosure, but it came in practice to apply to monasticism in fourteenth-century England, when "cloister" and "convent" had become parallel terms, one stressing the seclusion from other life, the other the communal character of those secluded. In the derived sense, meaning is transferred from the religious process to the space in which provision is made for that process to take place. By the fifteenth century the concept identified the covered walkway connecting the buildings of the institution, often forming a quadrangular perimeter with an open colonnade toward the inner court. CLYDE CURRY SMITH

CLOSE, FRANCIS (1797-1882). Dean of Carlisle. Son of an Anglican rector, he was born at Frome, graduated from Cambridge, and became curate (1824), then incumbent (1826) at Cheltenham. For the next thirty years, while the population of this fashionable watering-place doubled, he exercised a powerful Evangelical ministry, basing his sermons on those of Charles Simeon,* denouncing sundry vices, promoting five new parish churches with schools attached, and assisting in the founding of Cheltenham College. Now one of the best-known Evangelical ministers, he was in 1856 appointed dean of Carlisle. There his ministry was especially directed to the poor, whose moral and physical condition he strove to improve. He was a stern critic of Roman Catholicism, Tractarianism,* and the English Church Union, and saw the revival of Gothic architecture as a tool of papal aggression. He resigned in 1881. Among voluminous works, his sermons were popular in the Evangelical world of his day.

IAN SELLERS

CLOUD OF UNKNOWING, THE. An English mystical treatise from the latter fourteenth century, of East Midland origin. The anonymous author was probably a solitary, possibly a Dominican. The sources of *The Cloud of Unknowing* may be identifiable as Dionysius the Areopagite's *Theologia Mystica* and the Rhineland mystics, especially Tauler, who were known in England through the Dominicans. Having written up to six other pieces, two being translations, the author addresses that specific person who seeks the solitary, contemplative life, but finally includes the Marthas as well as the Marys. The instructions remind of those given by St. John of the Cross,* and the Thomist position is apparent throughout; grace is central, bodily travail discounted. The aim is to put all created things under a cloud of forgetting, and in love press on to the cloud of unknowing, Dionysius's *caligo ignorantiae,* which lies forever between the prayerful soul and God. C.G. THORNE, JR.

CLOUGH, JOHN EVERETT (1836-1910). Baptist missionary to India and supervisor of a mass movement to Christianity among the Telugus. Born in W New York, frontier life in Illinois and Iowa prepared him for strenuous labors in India. As a converted skeptic he joined the Baptist church and offered himself for missionary service. In 1864 he and his wife sailed for India to a mission that three times the Baptists had been about to give up in discouragement. At Ongole he saw the beginning of a movement among the outcaste Madigas that he wisely let continue along Indian lines. The flow of converts became a flood after the famine of 1876-78, in the relief of which he played a significant role. In six weeks 8,691 were baptized. When he left India in 1910 the Baptist Telugu Mission had reached 60,000 members.

HAROLD R. COOK

CLOVIS (d.511). Son of Childeric I, he became king of the Salian Franks in 481. Almost at once he extended his territories from the Somme to the Seine and the Loire at the expense of Roman power. About 492 he married the Christian Clotilda, but he became a Christian only after being convinced of intervention by Christ in a battle. His baptism on Christmas Day 496 led to the support of the Catholic bishops and Roman officials in governing the country. Subsequent conquests laid the foundations of the modern French nation. In waging war against the Visigoths he invoked God's help and killed their king, Alaric II, with his own hands at the battle of Vouille in 507. He made Paris the capital of his kingdom.

JAMES TAYLOR

CLOWES, WILLIAM (1780-1851). Co-founder of the Primitive Methodist Connexion.* Son of a drunken potter in Staffordshire, he early followed his father's trade and habits. Converted after a Methodist "love feast" (where he received the bread and water "under the idea of a sacrament") in 1805, he became an active Methodist, participating in the first Mow Cop camp meeting. Like Hugh Bourne* earlier, but independently, he was expelled in 1810 through Methodist alarm at such developments. Many local Methodists supported him, calling him (to his own sacrifice) as a fulltime preacher paid from their meager wages. Primitive Methodism emerges from the coalescence of this movement with Bourne's, though the relations of the founders were often strained. Clowes became a hard-living traveling evangelist, especially in the industrial Midlands and North, with Hull as his center later. Many record the remarkable power of his preaching.

A.F. WALLS

CLUNIACS. Those forces which were part of the monastic foundation at Cluny, established in 909 as a reform movement based on the Benedictine Rule. Although Cluny came to an end legally in 1790, after a long decline (despite reforms in the seventeenth century), it was the center of monastic reform and experiment for two centuries, having a vast network of communities (more than 1,100 at its height) attached to the central hierarchy. Their reforms extended to older monasteries, and their adaptability to numerous situations brought great success, extending also into political life, for Cluniacs were much involved over the investiture struggle. Many of its monks became popes, cardinals, and bishops, and among its notable abbots were Odo, Odilo, Hugh, and Peter the Venerable.

Combating ecclesiastical decadence and withstanding the empire were not its chief concerns. Even though it grew weak because of its vastness, its mission was admirably accomplished. Cluny renewed the ideals of its predecessors as Cîteaux (see CISTERCIANS) was to do for Cluny (though without the same cultural achievement). The order had centers in Germany, Italy, Spain, and England, in addition to France. It emphasized biblical scholarship and liturgical renewal, and at its height manifested the fruits of simplicity.

See J. Wollasch et al., *Neue Forschungen über Cluny und die Cluniacenser* (ed. G. Tellenbach, 1959); and J. Evans, *Monastic Life at Cluny, 910-1157* (1931). C.G. THORNE, JR.

COADJUTOR BISHOP (Lat. *adjutare,* "to help"). Ecclesiastically, the bishop appointed as assistant, usually with right of succession. Such appointment is found most frequently in the Roman Catholic Church, particularly where a bishop is old or infirm. Modern usage distinguishes coadjutor from suffragan,* but seventeenth-century English treated the terms interchangeably.

COCCEIUS, JOHANNES (1603-1669). German theologian. Educated at Hamburg and Franeker, he taught at Bremen, Franeker, and Leyden. He studied under Maccovius and William Ames.* In his works, based upon considerable knowledge of oriental languages, he tried to present theology on a purely biblical basis, and although a Calvinist, he objected to the Calvinist orthodoxy of his day. His major work, *Summa doctrinae de Foedere et Testamento Dei* (1648), presents an outline of the scriptural teaching of salvation. He pictures the relationship between God and man, both before and after the Fall, in the form of a covenant. In Eden there was a covenant of works which promised salvation for obedience, but when man sinned it was no longer valid. Then the covenant of works was replaced by a covenant of grace which offered salvation as a gift of God. This covenant originated in an agreement between the Father and Son and is realized in a succession of historical steps culminating in the kingdom of God. In this way Cocceius was able to introduce the ideas of the history of salvation and of millennialism into scholastic Reformed theology.

ROBERT G. CLOUSE

COCHLAEUS, JOHANNES (1479-1552). Roman Catholic controversialist. Born near Nuremberg of peasant origin, he studied there and at Cologne (1504), where he developed a distaste for Scholasticism and a sympathy for Platonism and Renaissance humanism. He became rector of the Latin school of St. Lawrence, Nuremberg. He studied law at Bologna, graduated in scholastic theology at Ferrara (1517), and was ordained at Rome. From 1521 he engaged in bitter controversy with Luther, and in 1525 endeavored to prevent the printing of Tyndale's English NT at Cologne. He was canon at Mainz (1526), Meissen (about 1534), and Breslau (1539), and attended many conferences between Catholics and Protestants. His best-known works were *Historiae Hussitarum Libri XII* and *Commentaria de Actis et Scriptis M. Lutheri, 1517-1546.* J.G.G. NORMAN

CODEX, see MANUSCRIPTS OF THE BIBLE

CODEX IURIS CANONICI (CIC), see CANON LAW

COENOBITES, see CENOBITES

COFFIN, HENRY SLOAN (1877-1954). Theological educator. Born in New York City and educated at Yale, Edinburgh, and Union Theological Seminary, New York, he was ordained in 1900 and established a Presbyterian mission congregation in the Bronx. From 1905 to 1926 he was pastor of the Madison Avenue Presbyterian Church, where he became known as one of the most eloquent preachers in the USA. During that period he also taught practical theology at Union Theological Seminary, and in 1926 he became Brown professor of homiletics and president of the seminary, continuing in that capacity until 1945.

Coffin was influential in American Protestantism as a preacher, a theological educator, a proponent of the Social Gospel,* a liturgist and hymnologist, and a participant in the ecumenical movement. He was a leader of the liberal faction in the Presbyterian Church in the USA, who worked hard to promote theological inclusivism in that denomination. He called himself an "evangelical liberal" and sought to moderate the antagonism between conservatives and extreme liberals. Among his many books are *In a Day of Social Rebuilding* (1918) and *The Meaning of the Cross* (1933). HARRY SKILTON

COILLARD, FRANÇOIS (1834-1904). Missionary to South Africa. Born at Asnières, France, he was sent to Lesotho in 1857 by the Paris Evangelical Mission and stationed at Leribe. In 1877 he led a party of Basuto to the Banyai, but their Matabele overlord, Lobengula, refused to allow a mission. Coillard then pressed on to Barotseland (1878-79). In 1880-82 he visited Europe to enlist support for a Barotse mission, which was launched in 1886. Coillard's difficulties were immense. At first political conditions were unsettled. Chief Lewanika was friendly, but stubbornly refused conversion. Many colleagues died, and critics suggested withdrawal. Toward the end, Ethiopianism threatened his work, but developments after his death proved he had not labored in vain. D.G.L. CRAGG

COKE, THOMAS (1747-1814). Methodist preacher, superintendent, and missionary enthusiast. Born at Brecon and educated at Oxford, Coke was made deacon in 1770 and served as curate at South Petherton, Somerset (1772-76). In 1775 he became doctor of civil law. Associated with Wesley from 1777, he was indefatigable as an itinerant preacher and increasingly served as Wesley's right-hand man. He was set apart by Wesley as superintendent for America (1784) and presided at the Christmas Conference which constituted the Methodist Episcopal Church of America. He maintained links with both sides of the Atlantic (which he crossed eighteen times). Relationships with Asbury, whom he had set apart as fellow superintendent in 1784, were not always easy.

In England, Coke presided over conference several times and served the connection assiduously. He had a strong link with Ireland, where he repeatedly presided over conference. He was a staunch opponent of slavery and a vigorous promoter of overseas mission. Tirelessly he raised funds, sent out missionaries, and opened up new areas. He organized the Negro Mission in the West Indies, developed missionary activity in Gibraltar, Sierra Leone, and Cape of Good Hope, and was recognized by conference as general superintendent of the missions (home as well as foreign). He died at sea, on his way to Ceylon with a party of missionaries.

See J. Vickers, *Thomas Coke* (1969).

HAROLD H. ROWDON

COLENSO, JOHN WILLIAM (1814-1883). First Anglican bishop of Natal. Born at St. Austell, Cornwall, he read mathematics at Cambridge, held a living in Norfolk, and was consecrated in 1853. His theology wedded the conclusions of F.D. Maurice* to liberal Protestant presuppositions and aroused much controversy. Unlike most missionaries, he favored the baptism of polygamists and respected African beliefs and customs. His missionary theology was expounded in a commentary on Romans (1861): the Atonement is entirely objective; men are redeemed from birth, and baptism merely proclaims this fact; the heathen must be shown the pattern of Christ. Colenso had already quarreled with James Green, Tractarian* dean of Pietermaritzburg, about church government and the Eucharist. Green now reported the commentary to Bishop Gray of Cape

Town, who referred it to the archbishop of Canterbury.

Before Colenso's orthodoxy could be tested, he caused a further storm by publishing the first part of a work which questioned the historicity and authorship of the Pentateuch and Joshua. Although Colenso's critical views were moderate by German standards, they scandalized most Englishmen and dominated the controversy which culminated in his deposition by Gray in December 1863. Colenso denied Gray's jurisdiction and was supported by the judicial committee of the Privy Council, which confirmed his position as bishop. Further legal action secured his stipend and the control of church property. Many laymen supported him, but with insufficient clergy and finance he could do little.

In his closing years, Colenso championed the African people against injustice. This cost the friendship of his principal supporter, Theophilus Shepstone, but earned undying African affection. Colenso was a tragic figure whose positive contributions to missionary policy and biblical scholarship were vitiated by an inadequate theology, an impetuous manner, a difficult personality, and the intense conservatism of his contemporaries.

See P.B. Hinchliff, *John William Colenso* (1964), and *The Anglican Church in South Africa* (1963).

D.G.L. CRAGG

COLET, JOHN (c.1466-1519). Dean of St. Paul's. Born in London where his father was lord mayor, he was educated at Oxford, and in Paris and Italy. Returning to Oxford, he delivered in 1497 a series of lectures on Paul's epistles marking the spirit of the S European Renaissance which was increasingly to prepare the way for the Reformation as it moved northward. The lectures are marked by critical comment and a concern to get back to the early sources behind all the medieval glosses and allegorizings.

Colet also shared the Renaissance humanist concern for reforming the clergy and church institutions, and also for furthering enlightened education. He attacked many clerical abuses, and though he did not advocate doctrinal reform, the suspicion of heresy was never far from him. Yet he was listened to by a wide circle; among the contemporaries whose thinking he influenced were Erasmus* and Sir Thomas More.* Colet founded St. Paul's School, London (still in existence), where he laid stress on the teaching of the classics. He was appointed dean of St. Paul's in 1505 and held the post until his death.

See J.H. Lupton, *Life of John Colet* (1909); E.W. Hunt, *Dean Colet and His Theology* (1956).

G.E. DUFFIELD

COLIGNY, GASPARD DE (1519-1572). Huguenot* leader and French statesman. Scion of one of the most powerful families in sixteenth-century France, he became the great hero of the first generation of French Protestants and a formidable Calvinist political chieftain. As a member of the noble Châtillon family and nephew of the powerful Constable Anne de Montmorency, he spent most of his early life in military and public service and eventually was named an admiral of France. Sometime between 1555 and 1560 he was converted and joined the Reformed Church. Historians have debated whether this was basically a religious or a political decision, but he became a leading Calvinist in the wars of religion which began in 1562. After most of the other Huguenot captains were killed, he became the unchallenged political head of the movement. Gaining the favor of Charles IX, he arranged an acceptable peace between Calvinists and Catholics in France in 1570, but the chance of a permanent settlement vanished when he was murdered in Paris with several thousand other Huguenots in the infamous St. Bartholomew's Day massacre.*

ROBERT D. LINDER

COLIGNY, ODET DE (c.1517-1571). French cardinal, later a Protestant leader. Created cardinal in 1533 and bishop-count of Beauvais shortly afterward, he became the national grand inquisitor in 1560, but did little in that capacity. In 1561 he created a sensation by embracing the Reformed faith. Endeavoring with his brother Gaspard de Coligny* to alleviate the persecution of the Huguenots,* he was excommunicated, and in 1568 fled to London, where Queen Elizabeth welcomed him warmly. Poisoned—it is said, by his valet—at the instigation of Catherine de' Medici, he is buried in Canterbury Cathedral.

COLLECT. A short form of prayer peculiar to Western church liturgies. It consists of an invocation of God, a petition, a pleading of Christ's merits, and an ascription of praise to God. The term itself *(collecta)* is found originally in the Gallican Rite,* and there it appears to mean the "collecting" of the individuals' silent petitions into a public and corporate form. Another possibility is to connect the term with the assembly *(collecta)* of people which gathered at each of the churches in Rome and said a prayer *(oratio ad collectam)* before proceeding to the stational church for the papal Mass. Most of the Sunday collects in the Church of England's Book of Common Prayer* are freely translated from the Latin (largely from the Sarum service books and the Gelasian Sacramentary); the remainder were composed by Thomas Cranmer* and others, or were added later in 1662.

HOWARD SAINSBURY

COLLEGIALISM. A theory of church-state relationship associated with H. Grotius* and S. Pufendorf.* Church and state are both purely voluntary associations *(collegia)* in which supreme authority rests with the body of the members. A civil magistrate has no relations with the church other than those which he enjoys with every other voluntary association within the territory. C.M. Pfaff (1686-1760), chancellor of Württemberg, defending collegialism against territorialism, claimed that the rights which princes possess in ecclesiastical matters are conferred on them by the church.

COLLEGIANTS. Dutch religious group, originating in the controversies surrounding the Synod of Dort* (1619). The triumph of the orthodox Calvinists there, and the ousting of the Remonstrant ministers from the Reformed Church, left many congregations without pastors. At Warmond, near Leyden, Gijsbert VanderKodde and his brothers began leading such a congregation in informal services, stressing baptism by immersion, withdrawal from the world, and a minimal creed. The style was closer to Mennonite* than to Remonstrant; and the group refused to join the Remonstrant Brotherhood. Meeting as *collegia* rather than regularly organized churches, the group moved to the neighboring village of Rhynsberg (hence the names "Collegiants" and "Rhynsbergers"). The movement spread, with the "congregations" joining once a year at Rhynsberg. Spinoza* lived in Rhynsberg for some years (1661-64) and found the group impressive, though his advanced ideas (denial of miracles, etc.) caused a split which took some time to heal. During the 1700s, membership dwindled, with many turning to Socinianism* or Deism*; the last meeting was held in 1787. DIRK JELLEMA

COLLIER, JEREMY (1650-1726). English nonjuring divine. Born at Stow-by-Quy, Cambridgeshire, he was educated at Ipswich and Caius College, Cambridge, and in 1679 became rector of Ampton, near Bury St. Edmunds, and lecturer of Gray's Inn in 1685. He was imprisoned for provocative political writing supporting James II. Released without trial, he was again imprisoned briefly in 1692 on suspicion of treasonable correspondence with James. In 1696 he publicly absolved on the scaffold two would-be assassins of William III. He was outlawed, but returned to London later. He wrote a controversial pamphlet, *A Short View of the Immorality and Profaneness of the English Stage* (1698), which led to a furious literary debate. In 1713 he was consecrated as a "bishop of the Nonjurors"* and became primus in 1716. He was in favor of union with the Eastern Church. He preferred the 1549 Book of Common Prayer, and his romanizing views on certain "usages" in the Communion service led to the fatal split in the nonjuring community. Collier wrote voluminously. His most important work is *An Ecclesiastical History of Great Britain* (2 vols., 1708-14). HOWARD SAINSBURY

COLLOQUY OF MARBURG, see MARBURG, COLLOQUY OF

COLLUTHUS (fourth century). Presbyter of Alexandria. The first to sign the deposition of Arius in an encyclical of Alexander* of Alexandria about 321, Colluthus was accused by the latter of indiscretion by 324. A synod of Alexandria in the same year under Hosius deposed Ischyras, a presbyter ordained by Colluthus, and received him back into communion as a layman. By 339 an Egyptian synod affirmed that Colluthus had died as a presbyter and that his ordinations were invalid. No reason is apparent for his unauthorized ordinations. Later Philastrius (d. c.397), followed by Augustine, charged Colluthus with the heresy of denying that God made evil.

COLLYRIDIANS. The *collyris* was a small cake, distributed by King David at a sacrifice (2 Sam. 6:19 LXX). According to Epiphanius (fourth century), women had imported from Thrace to Arabia a cult-sacrifice to the Virgin Mary, at which a *collyris* was offered and then eaten by the devotees. Epiphanius objected to women offering sacrifice, holding that while honor is due to Mary, God alone is to be worshiped. The term *Theotokos** might have stimulated the cult.

COLMAN (d.676). Bishop of Lindisfarne. An Irishman, he was sent from Iona, where he was a monk, to succeed Finan, Aidan's* successor, as third bishop of Lindisfarne. He supported King Oswy in maintaining Celtic customs, particularly over the date of Easter. At the Synod of Whitby* in 663/4 he argued for the retention of this practice, and when the decision went in favor of the Roman customs he resigned his see. He took some of the monks of Lindisfarne with him to a monastery at Innisboffin, County Mayo, Ireland, where he remained until his death.

COLOMBINI, GIOVANNI (c.1300-1367). A wealthy Sienese merchant and magistrate converted to a life of service after reading a life of St. Mary of Egypt,* he cared for the poor and sick in his own house for almost a decade. When his son died and his daughter became a nun, he settled an annuity on his wife, disposed of his remaining property, and formed an association dedicated to poverty and service about 1360. Alarmed at their following, the magistrates banished the group, until an outbreak of plague led to their recall. Urban V constituted the group as *Clerici apostolici S. Hieronymi* (or Gesuati*) in 1367, and Gregory XIII beatified Colombini.

COLORED METHODIST EPISCOPAL CHURCH, see AMERICAN NEGRO CHURCHES

COLOSSAE. A city in the Roman province of Asia, in the valley of the Lycus, a tributary of the Maeander. The Lycus was the ancient thoroughfare eastward from the west coast of Asia Minor through Phrygia. Of the three cities of the valley, including Hierapolis and Laodicea, it was the oldest, mentioned by Herodotus and Xenophon in the fifth century B.C. At Colossae the Lycus valley narrows, to be dominated by Mt. Cadmus at 8,013 feet, and the site itself is 1,150 feet above sea level. Here the road to Sardis and Pergamum branched off. It was a wool center, known for its sheep-raising. But the rise of Hierapolis and especially Laodicea brought crippling competition, and Colossae was eventually abandoned between 600 and 700.

The site has never been excavated, though it was first identified by W.J. Hamilton in 1835. Epaphras, a member of Paul's missionary team and native of Colossae, seems to have been largely responsible for its evangelization (Col. 1:7; 4:12,13). Paul had apparently not visited it when he wrote his epistle, though he may have gone subsequently (Col. 1:4; 2:1; Philem. 22). It is clear from the Colossian epistle that Christianity had severe contests there with various aspects of paganism and heresies.

See D. Magie, *Roman Rule in Asia Minor* (1950), pp. 126f., 985f.; and W.M. Ramsay, *The Cities and Bishopries of Phrygia* (1895), pp. 208-13.

JAMES M. HOUSTON

COLOSSEUM (Coliseum). Bede's eighth-century name for the Flavian Amphitheater in Rome. Built from A.D. 72 to 82 under the Flavian emperors, its axes are 188 meters and 156 meters. It is 48.5 meters high. Seated in three tiers, about 50,000 spectators viewed animal hunts *(venationes)*, combats of hundreds of animals at once, gladiatorial fights, and seafights *(naumachiae)*, with the arena flooded. The tradition that Christians were martyred there remains a possibility but has no ancient basis, dating from Benedict XIV's consecration of the structure to the martyrs in 1750. His dedicatory cross was removed in 1874, replaced in 1927. It was frequently damaged by lightning and otherwise; restorations and additions spanned the period from Nerva to Theodoric the Ostrogoth (d.526). It served as a fortress in medieval times, and its Renaissance use as a travertine quarry—Palazzo Farnese was built from its stones—ceased under Benedict XIV. Its efficient passageways, protective canopy for spectators, aromatic refreshment of the air, and subterranean elevators to the arena are remarkable.

DANIEL C. SCAVONE

COLOSSIANS, see EPISTLES, PAULINE

COLUMBA (Columkille) (521-597). The most illustrious Irish churchman of the sixth century, he was born at Gartan, County Donegal; his parents were both of royal lineage. He came to Derry about 546 and built a church, and some years later founded a notable monastery at Durrow. A keen student, his passion for books caused him trouble when his claim to a copy that he had made from a psalter borrowed from Finnian of Movilla was overruled by King Diarmid. His hasty Irish temper led to frequent involvement in quarrels, and his departure from Ireland in 563 is shrouded in mystery. It may have followed excommunication from the Irish Church for his part in civil war, or he may have been seeking both release from a sense of frustration and freedom to propagate the Gospel.

After a perilous journey to Iona Columba found scope for his great talents in establishing a college for the training of young men for the evangelization of the N Picts. He visited Ireland in 574 to attend a convention at Drumceatt, near Limavady in County Derry. He survived for a further twenty-three years and was buried with the kings in his beloved Iona, off the Scottish coast. A man of outstanding gifts as scholar and preacher, Columba commanded attention for courageous leadership. Though his temper was hot, his indignation was often righteous and his spirit just and generous. He set a noble example in prayerfulness, self-discipline, and pastoral concern.

ADAM LOUGHRIDGE

COLUMBANUS (c.543-615). Irish saint and scholar. A disciple of Comgall of Bangor, he was a native of Leinster. He had a passion for learning, and had the distinction, unusual in his day, of being able to study the Scriptures in both Hebrew and Greek. When almost fifty he felt the call to evangelism on a wider scale, and after an abortive attempt to found a monastery in the south of England, he went to France where he established a noted school of learning at Luxorum in Burgundy. The discipline of his school was extremely severe, and while his movement was approved by the people, it was bitterly opposed by the clergy. He showed great courage in denouncing the vices of the Burgundian court, and of King Thierry in particular. Expelled from France, he worked for a time at Lake Constance in Switzerland before forming a notable monastery at Bobbio in N Italy, where he died.

Probably the sharpest and ablest controversialist of his time, Columbanus revived an interest in the findings of the General Council of Constantinople in 553, and pleaded with Pope Boniface IV not to condone Eutychianism or encourage those who believed there was only one nature in Christ. His correspondence with Boniface and later with Gregory the Great makes striking claims for the purity and independence of the Celtic Church, and challenges their claim to papal supremacy. He was a poet of singular gifts, and an able expositor of Scripture; his commentary is in the Ambrosian library in Milan.

ADAM LOUGHRIDGE

COLVILL, JOHN, see COVEL, JOHN

COMBA, EMILIO (1839-1904). Waldensian* historian and theologian. Born at San Germano Chisone, he graduated at the theological school in Geneva, where he had some of the most distinguished professors of the time, including Merle d'Aubigné.* A man of great intelligence, strong personality, and deep spirituality, he was ordained pastor in 1863, and after a short period of evangelical work in Brescia was sent (1867) to Venice (then just freed and annexed to the kingdom), where with his warm eloquence he attracted great numbers, thus founding a prosperous Waldensian church. In 1872 he was called to Florence to succeed G.P. Revel, one of the founders of the theological faculty; there he taught for thirty-two years, devoting all his time to his students, his academic studies, and for some time caring for the Waldensian community in that city as well as editing a journal, *La Rivista Cristiana.* His works include two short volumes in English, *Who are the Waldenses?* (1879) and *Waldo and the Waldenses before the Reformation* (1880). Later came a more substantial history of the Waldensians in French (1887) and Italian (1893), and two volumes on the Reformation in Italy (1881-95).

DAISY RONCO

COMBER, THOMAS JAMES (1852-1887). Baptist pioneer missionary. Born at Camberwell, London, he was baptized in 1868 and studied at Spurgeon's College evening classes and Regent's Park College. Accepted by the Baptist Missionary Society, he sailed for the Cameroons (1876), meeting Alfred Saker* en route. With George Grenfell* he was appointed to the proposed Congo

Mission which followed H.M. Stanley's explorations. They established a base at San Salvador (1878). His wife died one month after arriving in Africa. After several attempts, Stanley Pool was reached, and mission stations established at Wathen and, on Stanley's advice, at Leopoldville. Comber and Grenfell explored the Congo River as far as Liboko on the mission steamer *Peace*. Many missionaries died, including Comber's brother and sister, and eventually he himself succumbed to fever in his thirty-fifth year.

J.G.G. NORMAN

COMENIUS (Komensky), JAN AMOS (1592-1670). Bohemian educational reformer. Born in E Moravia, Comenius was educated at the Latin school in Prerov and at the Herborn Academy in Nassau, and later studied theology at Heidelberg. On his return he was ordained priest of the Unity of Brethren (Unitas Fratrum), and in 1618 became pastor at Fulnek. On the outbreak of the Thirty Years' War, Fulnek was invaded (1621) and Comenius sought refuge with his coreligionist Lord Charles the Elder of Zerotín in E Bohemia. The wife and two children of Comenius died in a plague.

When imperial laws proscribed all non-Catholic clergy from Bohemia-Moravia, he left the country, never to return. He settled at Leszno in Poland, where he became rector of the gymnasium. His *Janua linguarum reserata* (1631), written to make the study of Latin easier, was translated into eleven European and four Asiatic languages. When soon after he produced his *Didactica magna*, he was acclaimed as the outstanding educational reformer, and is so considered to this day.

Comenius was invited to England, and while in London during the winter of 1641-42 he outlined in his *Via lucis* a plan of cultural reform calling for the establishment of a "pansophic college." Education was defined as aiming at learning all that was necessary for this and the future life. It included not only scientific study of nature with deductive reasoning, but also the Bible as integrated into universal education of both sexes and as securing universal peace and religious harmony. In 1645 Comenius began the monumental work *De rerum humanarum emendatione Consultatio*. Only two of the seven volumes were published in his lifetime; the rest were unfinished. Lost early in the eighteenth century, the text was found in 1935 and 1940, and all seven volumes were published in 1966 by the Czechoslovak Academy in Prague.

The first volume asserts that all men are capable of being educated. The second *(Panaugia)* says God has provided men with three sources of knowledge: nature, reason, and revelation (Scripture). All three are necessary for good life, for the whole truth consists in the synthesis of all three. In the fourth volume *(Pampaedia)* Comenius presents his final reworking of educational theories, extending the training of all men literally from the cradle to the grave. Volume 6 *(Panorthosia)*, the most important of the whole series, advocates reform of culture, politics and religion. To achieve this, the Council of Light is to deal with worldwide educational reforms; the Court of Justice is to govern political reorganization and to serve as the supreme judicial tribunal; the Ecumenical Consistory is to supervise and regulate the universal spread of Christianity.

Since his hopes for establishing the pansophic college in London were disappointed, he accepted an invitation to reform the Swedish educational system, and in 1650 undertook a similar reform in Hungary. When in 1656 Leszno was burned by the Poles, Comenius lost all his property including many valuable manuscripts. He was then invited to Amsterdam, which became his home for the rest of his life and where he published many of his books (which total more than 150).

Though known primarily as an educational reformer, he was equally prominent as a religious leader (he was the last bishop of the Czech branch of the Unity of Brethren), and as an ecumenical pioneer who strove throughout his life for the unification of Christendom..

BIBLIOGRAPHY: M. Spinka, *John Amos Comenius, that Incomparable Moravian* (1943, 1967); M. Spinka (tr.), Comenius's *Labyrinth of the World* and *The Bequest of the Unity* (1942, 1940); J.E. Sadler, *J.A. Comenius and the Concept of Universal Education* (1966); G.H. Trumbull, *Hartlib, Dury and Comenius* (1947).

MATTHEW SPINKA

COMGALL (517-c.601). Irish monk. Son of Setna, a Pictish warrior, he was born at Magheramorne, County Amtrim, and was educated under Fintan at Clonehagh, Clairenach at Glasnevin, and Finnian at Clonard. Bishop Lugidius ordained him, possibly at Connor. In 558 he founded at Bangor, County Down, a monastery which became an outstanding seat of learning with at one period a reputed enrollment of 3,000 students. He and Columba* were close friends, and Columba had his support when he went to Inverness to seek permission from King Brude to settle at Iona.

COMMANDMENTS, THE TEN. These are said to have been written by the finger of God (Deut. 9:10) and spoken directly by God to the people. They are thus distinguished from other collections of laws. They form in themselves a significant unity. They had a special place in the continuing life of Israel (cf. Deut. 6; Jer. 7:9). There are slight differences between the two biblical statements of the Decalogue (Exod. 20:1-17; Deut. 5:1-21). In Exodus, Sabbath observance is based on God's rest in His creative activity, whereas in Deuteronomy it is based on the rest required by a servant as a sign that God has redeemed him. In Deuteronomy, moreover, a man's wife is clearly distinguished from his servants and is accorded special status and protection.

The commandments were intended to be divided into two tables, the first teaching duty to God, the second teaching duty to the neighbor. Christ Himself referred to the commandments in His teaching, and summed up their meaning by combining the texts Deuteronomy 6:4 and Leviticus 19:18, each of which sums up one of the two tables (Matt. 22:34-40). In the NT the commandments are sometimes directly quoted (Matt. 5:17, 19; Mark 10:19; Luke 18:20; Rom. 7:7,8; 13:9; 1 Tim. 1:9,10). It has sometimes been supposed that the number ten had some religious significance, since it occurs in the description of the details of the Temple and in other similar connections. Goethe found ten commandments in Exodus 34:-14-26, and it was later suggested that this formed a cultic decalogue corresponding to, and possibly rivaling, the ten moral commandments. It has recently been pointed out that in passages such as Psalm 15:2-5 commandments are grouped in tens.

The dividing and numbering of the commandments has created controversy. Philo divided the two tables equally, the first pental ending with the commandment to honor parents. Subsequent scholars felt the first table should end with the Sabbath commandment. The Talmudic tradition held that the commandments against idolatry and the forbidding of images formed one long, indivisible unit. Augustine, who was followed by the Roman and Lutheran traditions, accepted this suggestion and found two commandments under the rubric "thou shalt not covet." A further tradition, following the lead of Origen, separated the commandment against images from that against idolatry; this is the view of Calvin and the Reformed tradition.

The more the Decalogue is listened to with faith, the more clearly will it impress itself upon us as a Word through which the commandment and grace of the living God reach us today. The more it is studied, the more it will be recognized that in spite of its brevity—and indeed perhaps because of its brevity—it says what is sufficient to establish the claim of God upon every aspect of our lives, each in its due place and proportion. Its significance has been recognized throughout church history. Irenaeus recognized it as a universal law common to Jews and Gentiles, and receiving new sanction rather than abrogation from Jesus Himself. It came to be commonly used in the instruction of catechumens and was identified by the Schoolmen with natural law. Luther gave fresh prominence to it through his use and exposition of it in his catechisms, and along with the Lord's Prayer it was used as the basis of instruction in the Christian life in all the Reformed tradition. Our interpretation of the commandments must take account of the fact that the law was fulfilled in Christ, and that its true meaning is found in His words and life.

BIBLIOGRAPHY: R.W. Dale, *The Ten Commandments* (1895); J. Davidman, *Smoke on the Mountain* (1954); R.S. Wallace, *The Ten Commandments* (1965); J.J. Stamm, *The Ten Commandments in Recent Research* (1967); E. Nielsen, *The Ten Commandments in New Perspective* (1968). RONALD S. WALLACE

COMMON ORDER, BOOK OF. Frequently referred to as "John Knox's Liturgy," this was *The Form of Prayers and Ministration of the Sacraments, etc. used in the English Congregation at Geneva,* with the addition of Calvin's second catechism and a collection of metrical psalms. Although some have tried to show that the original form of worship employed by Scottish Protestants was the second Edwardian *Book of Common Prayer,* it is doubtful that it was ever in common use. In the *First Book of Discipline,* the Genevan book is referred to as "our Book of Common Ordour," and in 1562 the general assembly of the church ordained that it should be used for the administration of the sacraments and the solemnization of marriages and burials. It was not, however, a prayer book, but a guide to the conduct of services. As David Calderwood pointed out in the next century, no minister was tied to a specific form of prayer or even of service. It quickly became the standard directory of worship for the Scottish Reformed church, republished numerous times. The attempt by Charles I and Laud* to do away with it led to the signing of the National Covenant* in 1638 and the subsequent "Bishops' Wars."* It was replaced in 1645 by the Westminster Assembly's *Directory of Public Worship,* but has continued to influence Presbyterian worship down to the present. W.S. REID

COMMON PRAYER, BOOK OF. The official servicebook of the Church of England. Services in English replaced those in Latin somewhat gradually at the Reformation. An English Litany was introduced in 1544; and in 1548, after the accession of Edward VI, an *Order of the Communion* was issued with which the people were to prepare for and receive Communion in both kinds (i.e., bread and wine). This order was inserted into the Latin Mass immediately after the priest's own Communion. A commission, meeting at Chertsey and Windsor in 1548, considered Archbishop Cranmer's* drafts for a complete English liturgy and submitted a book to Parliament at the end of the year. *The Book of Common Prayer and Administration of the Sacraments, and other Rites and Ceremonies of the Church, after the Use of the Church of England* came into use on Whitsunday 1549. The title indicates that it replaced three of the five main Latin servicebooks: the breviary, the missal, and the manual. The processional had been abolished in 1547; and the Church of England never produced an equivalent of the pontifical. An ordinal was added to the Book of Common Prayer in 1550, but common forms for the induction of incumbents, consecration of churches, etc., have not been fixed.

The new services were conservative in form and ceremonial, keeping close to the structure of the medieval Sarum Rite. Priestly vestments were retained and most liturgical gestures permitted except "any elevation, or shewing of the Sacrament to the people." This book already contained, however, most of Cranmer's creative liturgical writing, which gave the services a revolutionary content. In the Holy Communion, a propitiatory sacrifice of Christ was replaced by a self-offering of the worshipers who here received the benefits of Christ's "full, perfect, and sufficient sacrifice" once offered upon the cross. Other major changes included a systematic scheme of Bible-reading, a drastic reduction of saints' days, and increased vocal participation by the congregation.

In the light of criticisms made by continental Reformers then living in England, especially Martin Bucer* and Peter Martyr,* prayer for the dead, reservation of the sacrament, vestments, and various ceremonial features were eliminated from a revised book which appeared in 1552. New material included the form of confession and absolution at the start of Morning and Evening Prayer. But the most revolutionary changes affected the structure of the Communion service. Intercessions were removed from the canon, administration followed immediately the words of institution, and the oblation of the worshipers became a response to Communion.

A few months later this revised Book of Common Prayer was abolished by Queen Mary, but Elizabeth I restored it in 1559 with a very few changes which did not prove controversial. A rubric did enjoin the vestments, but it remained a dead letter; indeed there was a struggle to retain even the surplice of the 1552 book. The Puritans opposed it and other ceremonial features; and there was a deeper opposition from some who had been at Geneva and disliked a "stinted, prescript liturgy" which did not leave scope for the minister's gift of extempore prayer. The book survived, with further small alterations in 1604, until it was again abolished in 1645. At the Restoration a new Act of Uniformity (1662) enforced what was essentially the same book, to the exclusion of dissenters. At the same time many detailed changes were made, and forms provided for adult baptism and prayers at sea. Since then the lectionary has been revised; and an Act of 1872 allowed certain abbreviations in the services. There was an abortive attempt at revision in 1927-28. After ten years of limited experiment with alternative services, the Worship and Doctrine Measure of September 1975 finally replaced the 1662 Act and gave general synod the right to approve new forms without recourse to Parliament. The 1662 Book is safeguarded where still in use, and it remains a standard of doctrine in the Church of England.

BIBLIOGRAPHY: G. Harford and M. Stevenson, *The Prayer Book Dictionary* (1912, rev. 1925); F.E. Brightman, *The English Rite* (2 vols., 1915); B.J. Wigan, *The Liturgy in English* (1962, rev. 1964); C.O. Buchanan, *Modern Anglican Liturgies, 1958-1968* (1968); G.J. Cuming, *A History of Anglican Liturgy* (1969); P.J. Jagger, *Christian Initiation, 1552-1969* (1970). JOHN TILLER

COMMUNION, HOLY. After Pentecost the disciples "devoted themselves to the apostles' teaching and to the fellowship, to the breaking of bread and to prayer" (Acts 2:42). The "breaking of bread" has since then been called the Eucharist, Lord's Supper, or Holy Communion, has been celebrated in a wide variety of ways and with widely differing interpretations, but has never been neglected.

At first there may have been a full fellowship meal at which the action of Jesus with the bread at the Last Supper was repeated at the beginning, and that with the cup at the end. When abuses crept into the meal (sometimes called a "love feast"), or when the church came to a better understanding of the mind of Christ (cf. 1 Cor. 11:20-34), the separated actions were brought together, and the meal as a replenishment was gradually stopped. In Christian worship in the second century, according to Justin Martyr, the Word of God was first expounded, in a manner corresponding to that of the Jewish synagogue, then the bread and wine were brought, blessed and thus consecrated, distributed, and partaken of by the Church in an action corresponding as closely as possible to that of Jesus at the Last Supper.

In interpreting the meaning of the rite, the church has always tried to understand what Jesus intended at His last supper. Recently it has been suggested that the breaking of bread in the earliest church life was dominated by the belief that Jesus willed the continuation of His fellowship and resurrection meals with His disciples on a similar pattern so that the church could continue after the Ascension to realize His presence and have joyful communion with Him. However accurate this may be, the words "this is my body" and "this is my blood" may indicate the wish of Jesus that these elements should be regarded as such signs of His own real presence in the midst as He would never fail to honor. He established the Lord's Supper in the continuity of a well-known tradition that when the messianic age came, the Messiah would feast with those who had expected and waited for Him (cf. Isa. 25:6). The vow by which Jesus pledged His own abstinence from participation in the feast, while making His disciples partake, is a vivid reminder that the freedom and joy which the Cross brings involve Him in the inevitable, in a shame and agony from which we are spared.

From the earliest time the church felt the celebration of the Supper was much more than a mere act enabling it to recall mentally the significance of the life and death of Jesus. Just as the celebration of the Passover as a "memorial" involved the Jews not only in memory but in a real participation in the continuing power of the past redemptive events of their history, so in the Lord's Supper "remembrance" was held to involve becoming vitally affected today in the whole of existence, through the re-presentation of a unique event in the past which already had in it more significance than that of an ordinary historical occurrence. The Supper, therefore, was a means through which the efficacy and power of the death and resurrection of Jesus (the unique eschatological event) is applied to successive generations throughout the course of history. Jesus no doubt had this in mind in identifying Himself with the paschal lamb, and His death with the sacrifice of the lamb for deliverance from bondage. He had found foreshadowed in the redemptive history of Israel His own work of delivering man from sin and death. Moreover, since He had come to establish the New Covenant in which sin would be fully forgiven, and the law would be written on the hearts of men through the Spirit, He also referred to this at the Last Supper, and desired that the Supper should be understood in this light.

The church has always found significance in the fact that Jesus, with words which identified the bread and wine with Himself, actually gave them to be eaten and drunk rather than to be merely looked at and adored. This deliberately planned giving and receiving implies some kind of real participation in Christ Himself—that Christ in the Supper is seeking to impart to men in some real ways the actual life which dwelt in His own flesh and blood, and this imparting of life is in some way connected with the giving and receiving at the Lord's Table. The Early Fathers interpreted Jesus' teaching in John 6 about eating the flesh and drinking the blood of the Son of Man to refer not only to a participation in Jesus' humanity by faith through His Word, but also to a participation in Jesus' humanity through the Eucharist. Paul's interpretation of the communion given in the Supper (e.g., in 1 Cor. 10:14-22) can be interpreted in the same light.

Indeed, so realistically did Paul interpret this communion with Christ that he uttered a warning about the damnation which might occur to those who ate and drank "unworthily, . . . not discerning the Lord's body" (1 Cor. 11:29 KJV). It does not seem adequate to interpret the presence of the Lord's body in the Supper as simply the presence of the church. There is no doubt, however, that the Supper is a meal in which, through communion with Christ, the members of the church understand and experience—in a way not otherwise so possible—the reality and depth of their unity in Christ, and find strength in each other through a mutual sharing of gifts and burdens.

The Fathers of the ancient Catholic Church tried to give expression in their theology of the Eucharist to their belief that the union with Christ given and confirmed in the Supper was as real as that which took place in the incarnation of the Word in human flesh. Origen and others insisted this fellowship with Christ was spiritual and indeed nothing more than that which comes by believing in the Word. Yet there arose the belief that somehow or other, through consecration in the prayer of thanksgiving, the elements themselves became a sacramental food enabling men to assimilate the essence of deity. This also led to the belief that in the sacrament a sacrifice was made, placating God and repeating the sacrifice of Calvary. This development was encouraged by the doctrine of transubstantiation, made official in 1215, which supposed that the whole substance of the bread and wine was changed into the substance of the body and blood of Christ. The church controlled the grace given through the rite, and the laity were allowed to partake only of the bread, since it was believed that the whole Christ was given under one species.

The Reformers returned to a more biblical view of God's presence among men in word and sign, and were greatly influenced by the best elements in Augustine's teaching. They carefully distinguished between the sign and the thing signified, and insisted that faith alone could receive, for salvation, the reality present in the Supper. They condemned transubstantiation. Luther vigorously denied any sacrificial implications in the rite. He believed the humanity of Christ at the Ascension took on the attributes of deity, including that of omnipresence, and could be given "in, with, and under" (to use a later Lutheran phrase) the elements. The bread remained bread, and the wine remained wine. He insisted the body and blood of Christ are partaken of orally by all, by believers to their blessing, by unbelievers to their judgment. In some Reformed circles it was taught that the elements were simply bare signs encouraging mental remembrance and stimulating faith and brotherly love. This view is sometimes attributed to Zwingli. Calvin insisted on a much more realistic, indeed a more substantial, reception of the humanity of Christ than this latter doctrine would allow. He believed that, though the body of Christ had ascended to heaven retaining its true human properties, nevertheless the Holy Spirit could unite things in heaven and things on earth and could by divine mystery raise the soul to heaven there to feed by divine mystery on Christ Himself. Calvin, following Augustine, noted that at the heart of the Supper there takes place a sacrifice of thanksgiving, a true offering of Christ in His body, the Church, to the Father.

BIBLIOGRAPHY: A.J.B. Higgins, *The Lord's Supper in the New Testament* (1952); R.S. Wallace, *Calvin's Doctrine of the Word and Sacrament* (1953); G. Aulén, *Eucharist and Sacrifice* (tr. E.H. Wahlstrom, 1956); F.J. Leenhardt and O. Cullmann, *Essays in the Lord's Supper* (1958); R. Bruce, *The Mystery of the Lord's Supper* (ed. T.F. Torrance, 1958); E.J. Kilmartin, *The Eucharist in the Primitive Church* (1965); K. MacDonnell, *John Calvin, the Church, and the Eucharist* (1967); J.J. Von Allmen, *The Lord's Supper* (1969).

RONALD S. WALLACE

COMMUNION OF SAINTS, THE. This clause was one of the last to find a place within the Apostles' Creed,* first appearing in the Creed of Niceta* about 375. The clause may, however, have originated in the third century. Because in Latin the genitive in the phrase *communionem sanctorum* may be either masculine or neuter, it has sometimes been thought that the reference is to a communion in holy things, in which case it would be a reference to the sacraments. Despite the fact that this would then provide a parallel with the Nicene Creed—where a reference to the sacraments follows a reference to the church—and despite the fact that in some medieval explanations the reference is taken to be the sacraments, this is unlikely. The clause more probably refers to people: to the fellowship, first, which we enjoy with the saints, saints here being synonymous with Christians. This fellowship was one of the marks of the early church as we see in Acts 2:42f. It would then be an expansion of one aspect of the church of God referred to immediately preceding.

Nor is fellowship to be limited simply to the living. This clause emphasizes we are one with those who have died in Christ. We are in His keeping, and so are they. The clause has sometimes been pressed beyond its proper scriptural meaning by suggesting that we may pray for the dead or even that prayer is warranted to the saints and that the saints in turn remember us. Both the present and future aspects of its true meaning are well summarized in an early English service for the Visitation of the Sick, where the clergyman says to the sick man, "Dearest brother, dost thou believe in . . . the communion of saints, that is that all men who live in charity are partakers of all the gifts of grace which are dispensed in the church and that all who have fellowship with the just here in the life of grace have fellowship with them in glory?" PETER S. DAWES

COMMUNION TOKENS. Vouchers of fitness to receive Communion, usually stamped pieces of metal, but also written tickets. "Houselling" tokens, given after confession to admit penitents to Communion, are noted in England in 1534, and from about 1561 *mereaux* (or *marreaux*) were used in the French Reformed Church. From the Reformation in Scotland the distribution of Communion tokens after catechizing of the recipients was strictly controlled by the kirk session. French tokens were round, but various shapes were used in Scotland. Early Scottish tokens bear only the initial of the parish and "K" (for "Kirk"), but later ones bear also the date and the minister's initials. The Communion card now used is more a check on attendance than a token.

HENRY R. SEFTON

COMMUNISM. A name given to any economic scheme which advocates common ownership of property to the exclusion of private ownership of property. Obviously, in actual life situations neither course can be pursued absolutely, but any structure of communal living which accents the sharing of goods by all alike may be considered a manifestation of communism. Communism is not merely an economic theory; it has significant philosophical and even religious implications. The myth of a "golden age" when men had all things in common is found in many religions, and the thought that all men should be treated as equals has inspired some of the noblest philosophical ideals; and it is with such myths and ideals that communistic schemes have been associated throughout the ages of human history. Even the Marxist-Leninist form of Russian communism, though political in character, rests on assumptions which have religious overtones.

As for Christianity, in its earliest manifestations of fellowship and devotion, provision for the poor, not as an economic program but as an expression of worship, was a central concern. The report of a free-will sharing of goods in the mother church at Jerusalem (Acts 2:42f.; 4:32f.) is understood by the sacred historian as a manifestation of the work of the Holy Spirit (Acts 5:3), though scholars have suggested that such imprudent enthusiasm contributed to the indigence of the Jerusalem community which Paul later sought to alleviate with an offering from the Gentile churches (1 Cor. 16).

Many times this example in the early church of "having all things common" has been appealed to, especially when the church has been tempted by temporal wealth and material possessions to identify with the world. This appeal has been variously united in the different ages of Christian history with a Greek depreciation of matter, a Stoic emphasis on natural law, and an apocalyptic enthusiasm which sits lightly by all worldly possessions. While the church as a whole has condemned luxury and protested social inequality, it is only in the inner circle of the ascetic elite, beginning with the ancient anchorites* and later in the monastic orders, that the common possession of goods and the renunciation of private property have been carried through with consistency. The monastic lifestyle has always been for the Roman Catholic Church the ideal; held in a kind of critical balance with the so-called compromise ethic by which the devout layman is allowed to live.

In the late Middle Ages, the heirs of the Franciscan spiritualists within the church, and the heretical sectarians outside the church, formed protest groups against the wealthy institutional church. These parachurch groups often embraced laymen (e.g., the Brethren of the Common Life*) who practiced a mutual sharing of goods, not from commitment to any economic theory, but by a common devotion to the ideal of apostolic poverty. Alongside such essentially religious developments, Christian history also reflects sociological movements of a more revolutionary nature. In the late Middle Ages and at the time of the Reformation there were protests by the peasants against feudal class distinctions and the rigorous exactions of ecclesiastical law, especially the law of the double tithe.

Negotiating Peter's Pence* and other matters of papal finance on behalf of the British crown, John Wycliffe became convinced of the scriptural warrant for a communistic society. Such a society could be achieved only by grace, according to Wycliffe; but in Bohemia, where his thought was influential, the more radical Hussites (Taborites) advocated revolutionary change of the economic order (early 1400s). No permanent new structures were attained, however, because of the military reverses which they suffered.

At the time of the Reformation the communism practiced by the radical Anabaptists* of Münster (1534/35) proved short-lived, but that of the Hutter Brethren (Jacob Hutter was burned in 1536) has endured even to the present in the Hutterite* colonies of the New World, making them the oldest communistic societies in history.

The best-known romantic apology for a communistic state is Thomas More's* *Utopia* (1516), comparable in many ways to Plato's *Republic*, though More rejected Plato's community of wives. The government of Utopia is democratic in form; a community of goods prevails, the magistrates distribute the instruments of production among the inhabitants, and dispense the wealth resulting from their labors to all citizens equally. All wealth and ostentation is proscribed in this ideal Christian state. More expresses a keen sympathy for the poor in their helpless misery (though, anomalously, slaves are found in Utopia to perform the toilsome, dangerous, and offensive forms of labor). "The rich," complains More, "desire every means by which they may, in the first place, secure to themselves what they have amassed by wrong; and then take to their own use and profit, at the lowest possible price, the work and labor of the poor; and as soon as the rich decide on adopting these devices in the name of the public, then they become law." In this stricture More strikes a note found in virtually all subsequent forms of communistic social agitation.

Of the various communistic experiments in the English-speaking world, arising out of some form of the Christian vision (e.g., Diggers* of Cromwellian England), the most numerous and enduring have been those that moved to America from the end of the eighteenth century (Shakers,* Harmony Society,* the Amana* community, the Owenite communities, the Fourierist communities, the Icarian communities). There have also been indigenous societies in the United States such as the Perfectionists of Oneida, New York. In the 1970s there is a recrudescence of interest in communal living throughout the free world, especially among the youth. While traditional Christian values of fellowship and equality inform the ideals of many who participate in these youthful communes, there is the new element of disenchantment with the depersonalized world of technology and a renunciation of the materialism which the profit motive in Western capitalism has tended to encourage.

In conclusion, it should be observed that the vision of a society where the needs of all are met by the sharing of goods held in common has been too persistent to be rejected as incompatible with Christian teaching. On the other hand, because of the moral flaw in human nature, the communistic espousal of communal property can never have more than marginal significance in ordering human society; and this is true even for the church insofar as the church is a social organism. Failure to take into account the doctrine of original sin accounts for much of the romantic and even irrelevant character of religious schemes for the sharing of wealth.

In contemporary usage the word "communism" generally refers, not to the religiously motivated renunciation of private property, but rather to the political policy and program of the party which has controlled the USSR since the days of the Bolshevik revolution in Russia (October 1917) and has since gained the ascendancy in other nations of the world as well, notably mainland China. This manifestation of communism, unlike Christian types, has achieved its ends, not by a voluntary devotion to common ideals of brotherhood (though many of the party members have evidenced a commitment that shames the followers of Christ), but by overthrowing the established order through violent revolution employing military means. Its master theoreticians—Marx, Engels, and Lenin—were avowed atheists, denouncing all religion as the "opiate of the masses." In Russia, where the Orthodox Church had aligned itself with the tsarist regime, there has been an intense effort by the "militantly godless" to banish religion from the life of the people.

Russian communism has represented itself as the movement of the proletariat, i.e., the wage-earning laboring class, and has openly espoused a world view known as dialectical materialism. By the process of revolutionary change the dialectic of history is to be achieved, and in the place of the antithesis of rich and poor will emerge the synthesis of a classless society. Material reality is the only reality, and the laborer who has been defrauded of his rightful share in material riches is summoned to shake off his chains and join in the struggle to establish the "dictatorship of the Proletariat." This program has not as yet been realized. Contrariwise, the Communist Party, wherever it has achieved the ascendancy, has spawned dictatorships, depriving the masses of their freedom. It has used its political powers (especially under Stalin) in such a way as to give the state the demonic form of a "total state"—i.e., a state from whose decision there is no recourse or appeal. This tyranny over the minds of men has more than offset the advantages of economic reform which have been introduced. It also explains why the traditional affinity between Christian idealism and communism has changed into a radical antinomy, so much so that today communism and Christianity are regarded by many as the two major world views competing for the allegiance of mankind.

BIBLIOGRAPHY: K. Marx and F. Engels, *The Communist Manifesto* (1848); N. Berdyaev, *The Origin of Russian Communism* (1937); J.C. Bennett, *Christianity and Communism* (1949); M. Holloway, *Heavens on Earth: Utopian Colonies in America* (1951); W. Hordern, *Christianity, Communism and History* (1954); R. Lowenthal, *World Communism: the Disintegration of a Secular Faith* (1964). PAUL KING JEWETT

COMMUNITY OF THE RESURRECTION. In 1868 B.F. Westcott, then a master at Harrow, preached a sermon that planted in the mind of the future Bishop Gore the idea of a religious order suited to the age. In 1886 the first stage in the creation of such an order was the founding of the Society of the Resurrection in England by Gore and the Brotherhood of the Epiphany in Calcutta. After lengthy study of various orders, in 1892 six men in Pusey House Chapel made their profession. The distinctive feature of the rule was to be a communal life lived in simplicity. At first each member of the community followed the line of work in which he had been engaged before his profession, but the members soon found themselves in request as missioners, as conductors of retreats, and as wardens to communities of women. After a period in Radley, the community moved in 1898 to Mirfield, and after four years there it began the training of ordinands, a work it has carried on ever since. The community consists of lay as well as ordained brethren, undertakes mission work in South Africa and the West Indies, and has been active in ecumenical affairs, while continuing to serve the church in missions and retreats. PETER S. DAWES

COMNENA, ANNA, see ANNA COMNENA

COMPANY OF JESUS, see JESUITS

COMPARATIVE RELIGION. Otherwise known as "the history of religions," *Religionswissenschaft,* etc., the term originated in the last quarter of the nineteenth century to denote one of the sciences based on the evolutionary theories of Darwin and Spencer. In its first phase it attempted to compare religions with a view to assigning each a place on an evolutionary ladder, and was especially concerned with the problem of the origins of religion, usually independent of any doctrine of revelation. Early representatives included Max Müller, C.P. Tiele, J.G. Frazer, W.R. Smith, and A. Lang, who were as a rule liberal Christians, though some were agnostics, and comparative religion tended as a result to be classified by evangelicals as a species of "modernism." After World War I, as the idea of unilinear evolution began to lose ground, and as material accumulated, comparative religion split into specialist fields—the history of religion, the psychology of religion, the sociology of religion, and the phenomenology of religion among them—and the synthetic approach was less and less cultivated. The present situation is that scholars in comparative religion agree that all religions are worthy of dispassionate study as phenomena in their own right, but there is otherwise little agreement concerning precise methods. A broad division may, however, be observed between "pure historians," who study phenomena in sequence within given traditions, and "phenomenologists," who prefer to take a cross section of religious phenomena from a variety of traditions. Strictly speaking, it is an axiom with comparative religion that no value judgments may be imposed on the material; the student's first duty is to describe and (hopefully) to understand. To pass beyond this to theological or philosophical evaluation, however legitimate, is not comparative religion.

BIBLIOGRAPHY: L.H. Jordan, *Comparative Religion: Its Genesis and Growth* (1905); J.N.D. Anderson, *Christianity and Comparative Religion* (1970); S.G.F. Brandon (ed.), *A Dictionary of Comparative Religion* (1970); E.J. Sharpe, *Fifty Key Words: Comparative Religion* (1971).

E.J. SHARPE

COMPLINE (Lat. *completorium*). The last hour of prayer in the Daily Office. It was established in the West by Benedict in his monastic rule, providing a retiring office for religious communities later than Vespers (Evensong), the public evening service of the church. It included various psalms appropriate to the time of day, and a later addition outside the monasteries was the canticle *Nunc dimittis,* or Song of Simeon (Luke 2:29-32), which had been used at Vespers *(Hesperinon)* in the East since the fourth century. In compiling his evening office for the Book of Common Prayer, Cranmer made use of this, and other parts of Compline from the Sarum breviary, including the collect, "Lighten our darkness." The Eastern Compline is the *Apodeipnon.* JOHN TILLER

COMPLUTENSIAN POLYGLOT. The first complete Bible printed in the original languages, published in 1521 or (more probably) 1522. It was printed between 1514 and 1517, in six folio-size volumes. The OT volumes (1-4) incorporate the Septuagint, Vulgate, and Hebrew texts among other things; volume 5 (the first volume printed) contains the NT in Greek, and was the first Greek NT printed, though that of Erasmus was published before it (1516); volume 6 offers lexical and other aids to Bible study. The Polyglot was produced at the newly founded University of Alcalá de Henares (near Madrid), which was called in Latin *Complutum,* under the direction of Cardinal Jiménes (Ximénez) de Cisneros (1436-1517), and dedicated to Pope Leo X. Some 600 copies were printed. D.F. PAYNE

COMPTON, HENRY (1632-1713). Bishop of London. Son of the earl of Northampton, he was educated at Queen's College, Oxford, served in the army, and then was ordained and became rector of Cottenham, near Cambridge. He was consecrated as bishop of Oxford in 1674 and was transferred to the see of London the following year. He was appointed tutor to the princesses Mary and Anne. Compton was a leading advocate of tolerance and comprehension toward Protestant nonconformists, but his hostility to Rome led James II to find an excuse to suspend him. He signed the invitation to William of Orange and officiated at his coronation (his old position having been restored to him). He was selected to help revise the liturgy, was one of the commissioners to arrange the union between England and Scotland, was a privy councillor under Queen Anne, but was twice passed over for the primacy. He wrote several theological works and supported the newly founded Society for the Propagation of the Gospel. G.E. DUFFIELD

CONCELEBRATION. In the Roman Catholic Church the joint recitation of the Canon of the Mass by a number of priests simultaneously with the principal celebrant. It is claimed this was the practice in the primitive church, and is still the regular practice of the Orthodox communion. It only survived in the Roman Catholic Church at ordinations where the newly ordained priests concelebrated with the bishop, but has revived in recent years in that church and in occasional ecumenical celebrations.

CONCLAVE. Used to describe either the closed rooms in which the cardinals of the Roman Catholic Church are shut as they choose a new pope, or the actual meeting of the cardinals themselves. The origin of the enclosure of the cardinals goes back to 1271 when, after three years of indecision, they were locked in a room until they finally came to an agreement over who should be the next pope.

CONCOMITANCE. The doctrine held in the Roman Catholic Church that the body and blood of Christ are received even when the communicant receives only one of the elements. Although it is a logical extension of the theory of transubstantiation, the practical pressure for this doctrine of concomitance was provided by the withdrawal of the cup from the laity within the Roman Church.

CONCORD, BOOK OF (1580). Also known as the Concordia, this contains the confessions or symbols of the Lutheran Church: the three ecumenical creeds—the Apostolic, Nicene, and Athanasian; the Augsburg Confession (1530) and its Apology (1531); the Smalcald Articles and the "Tract concerning the Power and Primacy of the Pope" (1537); Martin Luther's Small and Large Catechisms (1529); and the Epitome and Thorough Declaration of the Formula of Concord (1577). It supplanted the various *corpora doctrinae* among German Lutherans. Subscription to these confessions among Lutherans varies. Among some it is made *quia* (because) they are believed to be in conformity with Scripture; among others, *quatenus* insofar as they are.

Although Luther's Small Catechism (in 1548) and the Augsburg Confession and its Apology (in 1536 by Richard Taverner) were translated into English already in the sixteenth century, no English translation of the entire Book of Concord appeared before 1851, when one was issued in Virginia. H.E. Jacob's translation appeared in 1882. In 1921 the translation by F. Bente and W.H.T. Daw, known as the *Triglot Concordia,* was published in St. Louis, Missouri. In 1959 T.G. Tappert, with the assistance of A.C. Piepkorn, J.J. Pelikan, and R.H. Fischer, produced the *Book of Concord* (printed in Philadelphia). The texts in the original languages can best be found in *Die Bekenntnissschriften der evangelisch-lutherischen Kirche* (1952 ff.). CARL S. MEYER

CONCORD, FORMULA OF (1577). A Lutheran confession which settled the doctrinal controversies within German Lutheranism after Luther's death, primarily those between Flacians and Philippists. The controversies dealt mainly with: Adiaphorism, which arose because of the subscription on the part of some Lutherans to the Leipzig Interim (1548); Majorism, the place of good works in salvation; synergism, free will, and conversion; original sin; Antinomianism, the distinction between Law and Gospel; the Lord's Supper; Christ's descent into hell; and predestination. Martin Chemnitz* and James Andreae* were the chief architects of the Formula, which grew out of six sermons by the latter, the Swabian-Saxon Concordia, the Torgau Book, and the Bergen Book. It is made up of an Epitome and a Thorough Declaration, each with twelve articles. The Epitome states briefly the *status controversiae,* the affirmative theses, and the antitheses; the second part has the title: "Thorough, Pure, Correct, and Final [Solid, Plain, and Clear] Repetition and Declaration of Some Articles of the Augsburg Confession concerning which, for Some Time, there has been Controversy among Some Theologians who Subscribe Thereto, Decided and Settled according to the Analogy of God's Word and the Summary Contents of Our Christian Doctrine." CARL S. MEYER

CONCORDANCE. A reference book to the Bible (or other sacred books and standard texts), especially an alphabetically arranged word list. The term (originally a Latin plural, *concordantiae*) dates from the thirteenth century, as do the first examples of concordances. Ideally each language in which the Bible is to be found requires its own concordance, and so does each individual Bible version. The earliest concordance to the Hebrew OT dates from 1523, by Rabbi Isaac Nathan. The Buxtorfs produced one in 1632. Available today are those by S. Mandelkern and G. Lisowsky, and for English readers, *The Englishman's Hebrew and Chaldee Concordance.* The earliest concordance for the Greek NT was that of Betulius (1546); those by W.F. Moulton and A.S. Geden, and J.B. Smith, are today current. For the Septuagint there is the work of E. Hatch and H.A. Redpath. For the Latin Bible, the concordance of R. Estienne (1555) is famous; the most used is that of F.F. Dutripon (1838, with many reprints).

The sixteenth century saw the first concordance to the English Bible, by T. Gybson (1535) for the NT and by J. Merbecke (1550) for the whole Bible. The most famous is that of A. Cruden* (1737) on the Authorized Version; it has been revised more than once (most recently, 1954) and remains in print. R. Young's* concordance (1873) has the additional advantage of offering the reader the Hebrew and Greek equivalents of English words. Nelson's published their *Complete Concordance of the Revised Standard Version* in 1957. In 1968 Zondervan published their *Expanded Concordance,* which covers several English versions, including the New English Bible NT. D.F. PAYNE

CONCORDAT OF 1801. An agreement made between Napoleon Bonaparte and Pius VII in 1801 by which the Roman Catholic Church was formally restored in France. By its terms the French government recognized the Roman Catholic religion as the national faith. The pope, in concert with the government, was to make a new division of dioceses, requiring if necessary the resignation of existing bishops. The state was given the right of nominating new bishops. Bishops were required to swear allegiance to the government and to offer ritual prayers for the consuls. Alienated church property was to remain with those who had acquired it, but the government agreed to provide fittingly for prelates and clergy. These provisions were considerably modified by Napoleon's "Organic Articles" (1802). At the same time, Protestants were accorded full religious rights. The Concordat was to govern the relations between France and the papacy until the separation of church and state in 1905.

J.G.G. NORMAN

CONCORD OF WITTENBERG, see WITTENBERG, CONCORD OF

CONDITIONAL IMMORTALITY. This doctrine is an attempt to steer a middle course between what is felt to be the harshness of the doctrine of eternal punishment and the sentimental tendencies of restorationism. It is held to be realistic without being repugnant. It is the view that the soul is immortal by grace and not by nature, and is a form of annihilationism, for the soul that persists in sin is ultimately destroyed. It is defended partly on philosophical and partly on biblical grounds. Philosophically it would seem to owe something to privative ideas of evil, with their ultimate origin in Plato or even in oriental religious philosophy. The biblical evidence cited is largely lexical and involves a discussion of the sense in which the Bible uses such terms as "death," "destruction," and "perdition." In fact none of these terms is properly the equivalent of "annihilation," but refer to a condition of great loss and of continued existence. Conditional Immortality is maintained by some of the sects (e.g., Jehovah's Witnesses, Christadelphians), but also by some who, in other doctrines, stand within historic evangelicalism. G.W. GROGAN

CONDREN, CHARLES DE (1588-1641). French preacher and theologian. Born in Vaubuin, he entered the priesthood in 1614, instead of following the expected military career, and became an Oratorian* in 1617. Already noted as a preacher and spiritual director, he was called to Paris in 1624 after founding houses at Nevers, Langres, and Poitiers. As second superior-general from 1629, he gave the order clearer aims and better constitution, and especially channeled its efforts into seminary education. Of great humility, he unsuccessfully tried to resign his office, concealed his strict asceticism, and refused to publish any writings. His teaching on sacrificial Christian life is of considerable interest, and some Tractarians* were influenced by posthumously edited works. IAN BREWARD

CONFESSING CHURCH. The Confessing Church in Germany grew from such movements as Martin Niemöller's* Pastors' Emergency League and the free confessing synods, which sought to oppose the theology of the German-Christians* and the Nazi-supported church government of Ludwig Müller, elected *Reichsbischof,* 27 September 1933. Its theological basis was set out in the Declaration of the Barmen* Synod, 29-30 May 1934, which with the Synod of Dahlem (October 1934) provided for the direction of the Confessing Church through Councils of Brethren and a Provisional Church Administration.

The existence and tenacity of the Confessing Church helped to discredit Müller's regime, since despite coercive measures it was unable to secure the docile unity which Hitler looked for in the churches. But from the beginning the Confessing Church suffered near-crippling internal differences. Its member churches had varying legal positions. In "destroyed" churches (i.e., where the church government had been taken over by German-Christians, in, for example, the Old Prussian Union, Hesse, Nassau, Saxony), the Councils of Brethren sought to exercise totally independent emergency government; in "intact" churches (e.g., Bavaria, Württemberg, Hanover, Baden), they worked with the existing leadership. There as elsewhere, however, serious tensions developed between Lutheran and Reformed wings of the Confessing Church. Much support for the Confessing Church came from the tradition of the nineteenth-century confessional revival which had been concerned about the preservation of pure doctrine as defined by the *historic confessional statements* of the churches. But among Confessing Church theologians, in the line of Barth and Bonhoeffer, a new concept of confession stressed the *act* of confessing Christ in which the church is realized ever anew as the church of Christ.

In December 1935 the Provisional Administration under Bishop Marahrens of Hanover decided to cooperate with the Minister for Church Affairs, H. Kerrl, and with his National Commission under W. Zöllner. This decision implied that the Confessing Church had been unable to establish its legal claim to be the true Evangelical Church. The decision was opposed by Niemöller, who held to the Barmen-Dahlem line that the Confessing Church was not a movement within the church, but the true church itself. The split became obvious in the Synod of Bad Oeynhausen, February 1936; the Confessing Church lost the episcopally led, intact churches, and a new provisional leadership was appointed; many men drifted into cooperation with Zöllner's church committees as apparently the only way of exercising effective ministry. Harassed by the Gestapo, dependent on hard-pressed congregations for finance, the Confessing Church lived precariously in the years before and during the war, unable to play a large part on the public stage.

The Confessing Church was never a political protest movement against Nazism, though its mere existence embarrassed the regime; its witness to Christ's lordship over the world challenged Hitler's totalitarianism in principle. The wing led by Niemöller, to which D. Bonhoeffer* belonged, was especially aware of political responsibility, though often inhibited by its conservatism and nationalism from explicit opposition. On only a few occasions did the Confessing Church try to criticize the government for its general policies, rather than for attacks on the rights of the church; and then their political ineptness and the power of the police state prevented their orations from having significant effect. After World War II the Confessing Church structures were merged into the re-formed Evangelical Church in Germany. Its traditions have been kept alive by both radical and conservative movements in the church. HADDON WILLMER

CONFESSION. In the NT, the noun *homologia* can be related in the sense of "confession" to the Gospel or the Faith; the verb *homologeō* already has as object both faith and sin. Under the impact of Roman persecution, patristic needs created the "confessor." While the oldest examples of the Apostles' Creed, still in interrogative form, had "confessional" usage at baptisms in the second century, the insertion of a declarative creed into the liturgy seems first to have occurred in 473 when Peter the Fuller* brought in the Creed of Nicea to confound the requirements of the Council of Chalcedon. Confession in this declarative sense includes also the orthodox Protestant formulations, of which that of Augsburg (1530) is the first example (see AUGSBURG CONFESSION).

The verb *exomologeō* came to be adapted to signify the confession of sins to the congregation (cf. *Didache* 4.14: *1 Clement* 51.3). By the late fourth century Chrysostom indicated the need for confession before baptism or Communion, and others specified it regularly, though the form still seems to be communal rather than private.

Confession of sin, especially in the sense *ad auriculam* ("into the ear" of the priest), was a medieval development which also came to play a role in worship. After the passing of frequent lay communion, and under the transformed character of penance, at the Fourth Lateran Council* in 1215 confession was made minimally an annual obligation, as was receipt of the host for which it was prerequisite. As aid to the accomplishment of this requirement, the confessional stall dates from the sixteenth century. Litanies of ancient usage also serve as vehicles for communal or private confession in the several senses.

CLYDE CURRY SMITH

CONFESSION OF 1967. As part of the merger of the United Presbyterian Church of North America and the Presbyterian Church in the USA, in May 1958 a special committee was appointed with Edward A. Dowey, Jr. of Princeton Seminary as chairman. Though the proposed confession with its Book of Confessions, including eight historic creeds and confessions, aroused some opposition after being published in 1965, it was approved in May 1967 with only 19 of 133 presbyteries against.

The Confession starts with the biblical idea of 2 Corinthians 5, God's reconciling work in Jesus Christ. It places a heavy stress on the true humanity of Jesus, "a Palestinian Jew." On the doctrine of Scripture "the confession carefully avoids saying either that Scripture 'is' God's word or that Scripture 'is' unique and authoritative as such in its own right" (Dowey). Its statement on reconciliation in society includes exhortations for the church to act in international conflicts, pleading for a search for peace "even at risk to national security;" about sex; about racial discrimination and poverty. The original committee included Arnold Come, Addison Leitch, David H.C. Read, John MacKay, John Wick Bowman, Leonard Trinterud, G. Ernest Wright, and Samuel Thompson.

See E.A. Dowey, Jr., *A Commentary on the Confession of 1967* (1968).

ROBERT B. IVES

CONFESSOR. In the early church the term was applied to one who had suffered in times of persecution for confessing the faith, but without actually being martyred. Later the word was loosely applied to those who had not suffered so, but were markedly holy men, and still later to those pronounced so by the pope. Edward the Confessor was so named by Pope Alexander III ninety-five years after his death. "Confessor" denotes also the priest who hears confessions (usually private) in the Roman Catholic Church.

CONFIRMATION. This rite is often traced back in the NT to Acts 8:14-17; 19:1-7. On both occasions, in Samaria and in Ephesus, the action of manual imposition is associated with baptism and the gift of the Holy Spirit; but in neither case is a second stage of Christian experience, separate from commitment to Christ in baptism, implied. And even if it were, this would not add up to confirmation as we now know it. It is impossible, in fact, to deduce from the laying on of hands in Acts 8 or 19, or anywhere else in the NT, a clear precedent for the later rite of confirmation.

This first appears distinguished from baptism in the third century. From the third to the fifth centuries a complex initiation rite was practised by the church, which included baptism followed by the laying on of hands, or anointing with oil, or both. Gradually, by the twelfth century, confirmation was separated from baptism as a "sacrament" in its own right. It is still such in the Roman Catholic Church. At the Reformation the practice of confirmation continued, but it was associated less directly than in the patristic and medieval periods with the gift of the Spirit.

Confirmation, as Anglican prayer book reformers now insist, has greatest meaning when it is associated with (infant) baptism.* Then it can be regarded as primarily an opportunity to assume personal responsibility for baptismal vows; and this may have been the view taken of it in the early church. In any case, as in current Baptist teaching, such a rite (whether or not it is called "confirmation") denotes reception into full church membership, with or without the imposition of episcopal hands. The theology of confirmation, however, and the nature of the gift (if any) it confers, is disputed. Those who associate the rite with the reception of the Spirit see it as in this sense the completion of baptism.

BIBLIOGRAPHY: G.C. Richards, *Baptism and Confirmation* (1942); G. Dix, *The Theology of Confirmation in Relation to Baptism* (1946); L.S. Thornton, *Confirmation: Its Place in the Baptismal Mystery* (1954); J. Hickinbotham et al., articles on confirmation from the Islington Clerical Conference of 1963, in *The Churchman*, LXXVII (1963), pp. 84ff.; G.W.H. Lampe, *The Seal of the Spirit* (2nd ed., 1967).

STEPHEN S. SMALLEY

CONGO, BRAZZAVILLE, see WEST AFRICA

CONGO, KINSHASA, see ZAIRE

CONGREGATION, THE. A name adopted by leaders of the Protestant party in Scotland who in December 1557 entered into a "Band" which bound the signatories to strive for "the most blessed Word of God and his congregation . . . unto which holy word and congregation we do join us." Cautious though they were initially, this covenanting together of the "Congregation (of the Lord)" was a major factor in that build-up that gained momentum with John Knox's return in 1559, and was climaxed when in 1560 the Scottish Reformation became a reality.

CONGREGATIONALISM. This may be traced back to the reign of Queen Elizabeth I, whose objective for the church in England was an enforced uniformity. There were those, however, who thought otherwise. Puritans* wanted to see the national church reorganized on presbyterian rather than episcopal lines. A few others repudiated the whole concept of a state church and favored the "gathered church" principle. These became known as "Separatists"* and were the forerunners of those who later were termed

"Congregationalists." They contended that the church should consist only of those who had responded to the call of Christ and who had covenanted with Him and with each other to live together as His disciples.

A leading figure among the Separatists was Robert Browne,* who in 1582 published in Holland his famous treatise, "Reformation without tarrying for any," in which he set forth his congregationalist principles. He asserted that "the Church planted or gathered is a company or number of Christians or believers, which, by a willing covenant made with their God, are under the government of God and Christ and keep His laws in one holy communion." Such churches, he claimed, are subject neither to bishops nor magistrates. Ordination is not vested in elders, but is in the hands of the whole church. In East Anglia and in London, around Gainsborough and in the West country, companies of men and women put Browne's teaching into practice. Rather than submit to ecclesiastical regimentation, many sought religious freedom in Holland, and some of these later crossed the Atlantic where churches of the congregational pattern became one of the formative influences in the New World.

It was from John Robinson's church at Leyden that the Pilgrim Fathers set off in 1620 in the *Mayflower.* Congregationalism became the established order in Connecticut and Massachusetts until 1818 and 1824. Meanwhile, in England the pattern of church life taught by Robert Browne spread with the growth of Congregational and Baptist churches throughout the country.

Following the civil wars of 1642-46 and 1647-48, Oliver Cromwell* assumed power, and the direction of the religious life of the country was shared between Congregationalists, Presbyterians, and Baptists. Insistence upon the independence of the local Christian community was never regarded as precluding a loose fellowship of independent local churches for purposes of mutual consultation and edification, and in 1832 the Congregational Union of England and Wales was formed. The local church is entitled to manage its own affairs, determine its own forms of worship, and call its own minister.

The officers of a Congregational church are usually a minister, a diaconate, and a church secretary and treasurer. Membership of a Congregational church is on profession of personal faith in Christ as Savior and Lord, and new members are normally welcomed by the giving of the right hand of fellowship at a Communion service. The church meeting is the assembly of the members of the church, gathered under the guidance of the Holy Spirit to discuss and decide matters affecting the life of the church. The call to a minister to assume the pastorate of a local church is issued by the church meeting. Deacons are elected by the membership to assist the minister in the administration of the church and also to share with him pastoral responsibilities. The concept of the "gathered church" as set forth by Robert Browne has had a profound effect upon the British and American ways of life. The spiritual principles upon which this doctrine of the church rests have been a source of great strength and inspiration, particularly in times of persecution.

In Britain, Congregationalism took a decisive step in 1966 when local Congregational churches were invited to covenant together to form the Congregational Church. This step has since been followed by the union with the Presbyterian Church of England to form the United Reformed Church.* With this new development, Congregationalism, as traditionally understood, disappears from the scene in Britain, although it is still the basis of church order in Baptist churches, in a growing number of independent evangelical churches, in those Congregational churches which voted not to enter into the scheme of union, and in a number of smaller denominations.

Missionary interest in Congregational churches has historically found expression through the London Missionary Society founded in 1795, which in 1966 became the Congregational Council for World Mission. Former LMS missionaries have included such illustrious names as James Chalmers,* David Livingstone, and John Williams.*

At the world level, Congregationalism has been closely involved with the ecumenical movement, and this fact no doubt explains to some extent the underlying reasons for the several mergers which have taken place in various countries between Congregational, Presbyterian and, in some cases, Methodist churches. The general trend in Congregationalism worldwide has been away from independency.

In 1949 the International Congregational Council* came into being following a conference in Boston in the USA. At that time it was reckoned that Congregational Church members throughout the world numbered just under 2.5 million, of whom well over half were in the USA. Both Scotland and Ireland have had separate Congregational unions, and in Wales the Union of Welsh Independents has existed as a distinct body. Most of the Commonwealth countries have had their own Congregational unions. On the continent of Europe, Congregationalism has had close links with the Mission Covenant churches of Scandinavia, the Dutch Remonstrant Church, and the United Protestant Church of the Palatinate. At Nairobi in 1970 the International Congregational Council was dissolved to make way for the World Alliance of Reformed Churches* (Presbyterian and Congregational).

In 1919 moderators were introduced into Congregationalism—men charged with the spiritual oversight of churches in different geographical areas, with no legal authority over the churches, but available to give advice. Each year Congregationalists have met together in an annual assembly to which each church has sent its representatives. A chairman, either minister or lay, has been elected annually. Inevitably the newly formed United Reformed Church involves a number of modifications in these procedures.

Congregationalists have, generally speaking, set themselves against credal tests for church membership. They have prided themselves on their breadth of understanding and tolerant spirit. While from some points of view this has been their strength, it has also been a weakness in that as a denomination Congregationalism has been particularly open to liberal and modernistic teaching; so much so that some Congregationalists have at times approximated to a Unitarian position theologically.

Erik Routley in his book, *The Story of Congregationalism,* makes this observation: "When Higher Criticism came, Congregationalists drank more deeply of it than did any of the others. Congregationalism freed by its new federation from the bondage of parochialism, and freed traditionally by its intellectual ethos from any risk of becoming mentally stagnant, offered enthusiastic hospitality to the new critical teachings. . . ." In spite of these tendencies, there has always remained a small but closely knit group of evangelical Christians. Since 1947 the Congregational Evangelical Revival Fellowship has served as a rallying point for most of the more evangelically inclined ministers and members of Congregational churches.

The Congregationalist system has often wrongly been described as democratic, whereas a more correct, even if somewhat idealistic, description would be Christocentric. It is not surprising that, for the most part, Congregational churches have tended to be relatively small in membership—a large congregation finds it less practicable to work out the principle of Christocracy as expressed through the medium of the church meeting. There have, however, been a number of outstanding Congregational churches which have been served by such eminent preachers as R.W. Dale,* J.H. Jowett,* G. Campbell Morgan,* and Joseph Parker.* Among distinguished British theologians one could mention Sydney Cave, A.E. Garvie, P.T. Forsyth,* J.S. Whale, and Nathaniel Micklem.

Undoubtedly the greatest contribution which Congregationalism has made to the church generally is its whole concept of the local church as a Christ-ruled fellowship. Many would feel that the decline of Congregationalism in modern times is largely due to the fact that a humanistic liberalism affected the denomination to such a large extent.

BIBLIOGRAPHY: H.M. Dexter, *The Congregationalism of the Last Three Hundred Years* (1880); R.W. Dale, *Manual of Congregational Principles* (1884), and *History of English Congregationalism* (ed. A.W. Dale, 1907); W. Walker, *The Creeds and Platforms of Congregationalism* (1893); W.E. Barton, *The Law of Congregational Usage* (rev. ed., 1923); W.B. Selbie, *Congregationalism* (1927); R.P. Stearns, *Congregationalism in the Dutch Netherlands* (1940); G.G. Atkins and F.L. Fagley, *History of American Congregationalism* (1942); A. Peel and D. Horton, *International Congregationalism* (1949); G.F. Nuttall, *Congregationalism* (1951); D. Horton, *Congregationalism* (1952); D. Jenkins, *Congregationalism: A Restatement* (1954); P. Miller, *The New England Mind: The Seventeenth Century* (2 vols., 1961); R.T. Jones, *Congregationalism in England, 1622-1962* (1962); W.W. Sweet, *The Congregationalists (Religion on the American Frontier,* vol. 3, 1964).

GILBERT W. KIRBY

CONGREGATIONS, RELIGIOUS. The term applied to communities which have taken simple vows. They are regulated by the Roman Congregation for Religious, and governed by their own superiors and chapters. Pontifical communities are exempt from episcopal jurisdiction, but superiors of diocesan communities are responsible to their diocesan bishop. Papal permission is required for their establishment, but the rule requiring episcopal permission for their establishment within particular dioceses is now generally suspended.

CONNOR, RALPH, see GORDON, C.W.

CONRAD OF GELNHAUSEN (c.1320-1390). Theologian. He studied and taught at the University of Paris, and became a canon of Mainz (1539), procurator of the German nation at Bologna University (1369), and provost of Worms (c.1380). One of the earliest advocates of the conciliar movement, he urged on Charles V of France that the circumstances of the Great Schism* provided sufficient reason for calling a general council without papal convocation. Since the position of the papacy was doubtful, authority to convoke a council lay in the universal church, as in apostolic times. He expounded this in his principal work *Epistola Concordiae* (1380). Charles VI's adherence to the antipope Clement VII caused Conrad to leave France in 1386 to lecture at Heidelberg University, where he became the first chancellor.

J.G.G. NORMAN

CONRAD OF MARBURG (c.1180-1233). Inquisitor-general of Germany. Learned and ascetic, and either a Dominican or Franciscan, Conrad gained initial fame by his enthusiastic support for and preaching of the crusade of Pope Innocent III. This led to his being used as a reformer and visitor of religious houses in Germany. Subsequently Ludwig IV of Thuringia, impressed by his zeal and competence, entrusted him with great ecclesiastical powers. In 1225 he became the confessor and spiritual director of Elizabeth, daughter of King Andrew II of Hungary. Conrad is reputed to have treated her with excessive severity. After showing himself a devoted opponent of every kind of heresy, he was nominated by Pope Gregory IX as the first papal inquisitor of Germany. During 1232-33 he used his absolute authority ruthlessly, handing over people on debatable evidence to the secular power for punishment. After accusing Count Von Sayn of heresy, he was condemned by a court of bishops and princes at Mainz in 1233, and then murdered by certain Hessian knights on his way home to Marburg.

PETER TOON

CONSALVI, ERCOLE (1757-1824). Italian statesman. Born in Rome of a noble family, he entered papal service early. After the French invasion of the Papal States (1798) following General Duphot's murder, he was imprisoned and later exiled. Gaining his freedom, he worked for the election of Cardinal Chiarmonti as Pope Pius VII, who made him secretary of state. He was chiefly responsible for the Concordat of 1801* with Napoleon. When Napoleon seized Rome in 1809, Consalvi was compelled to go to Paris, becoming leader of the "black cardinals" until forcibly retired to Rheims. Following Napoleon's abdication, he was reappointed secretary and represented the pope at the Congress of Vienna (1815), where he secured the restoration of the Papal States. He spent his remaining years reorganizing them. J.G.G. NORMAN

CONSCIENCE. It has come to be recognized that there is in man some faculty by which he can form a habit of asking what he ought to do, and of making judgments about the moral quality and value of his actions and thoughts. This faculty is regarded as seated in the depths of his personality. The habit of making such judgments is called "conscience." It seems to be innate in man, but it appears to be more developed in certain historical circumstances than in others. The Greek word for it *(syneidesis)* appears prominently in use for the first time in late Hellenistic and Stoic philosophy. Especially with Seneca it was regarded as a holy spirit within us, an observer of the good and evil.

In the OT there is no special word for "conscience," but the phenomena that gave rise to the word are described and attributed to the heart (1 Sam. 24:5; 2 Sam. 24:10; Job 27:6). In the NT, however, the word *syneidesis* is adopted from the Stoics and used in many different contexts, especially in Pauline theology. Christ describes what is meant by conscience by the phrase "the light within you" (Matt. 6:23). In Paul's thought, the voice of conscience can anticipate the last judgment, can become a true answer to the Gospel. Though conscience can become bad and weak, it can nevertheless become pure and strong if it is properly bound to God.

In the Middle Ages, conscience was regarded as a volitional ability of the soul *(synteresis)*, unimpaired after the Fall and related to natural law, to which the immediate voice of God could speak. Luther and Calvin stressed that conscience was the sphere of battle between God and evil, that it had no autonomy or independent justification, and that it required to be liberated by faith before God. Kant exalted the role of conscience in life. More recently it has been increasingly pointed out that conscience has been a fallible guide in the development of morals, that it could be the product of instinctive habit, as in animals, of the gulf between the ego and the superego (Freud), or of society's attempt to kill vital instincts (Nietzsche). In spite of such analyses, it still remains an aspect of man which requires sanctification and education, and on which God's Word can take hold with compelling authority. The problems relating to conscience arise mainly within the problem of giving it a proper direction.

BIBLIOGRAPHY: H. Rashdall, *Conscience and Christ* (1916); K.E. Kirk, *Conscience and Its Problems* (1927); O. Hallesby, *Conscience* (1950); C.A. Pierce, *Conscience in the New Testament* (1955); J.N. Sevenster, *Paul and Seneca* (1961); P. Delhaye, *The Christian Conscience* (1968); E. Mount, *Conscience and Responsibility* (1969).

RONALD S. WALLACE

CONSECRATION. An act by which a person or a thing is separated from a secular or profane use to the service and worship of God by prayer, rites, and ceremonies. Common in the OT and existing analogously in paganism, in Christian usage the term usually refers to the setting apart of the euchristic elements, the ordination of bishops, the dedication of churches, and the consecration of various sacred objects such as eucharistic vessels and altars. There are wide divergencies in the rites themselves, in their scope and in the significance attached to them by Roman Catholic, Orthodox, and Protestant churches.

CONSENSUS TIGURINUS, see ZURICH AGREEMENT

CONSERVATIVE BAPTIST ASSOCIATION OF AMERICA. An association of about 1,150 Baptist churches chiefly in the N United States, organized at Atlantic City, New Jersey, in 1947. It is closely allied with the Conservative Baptist Foreign Mission Society, the Conservative Baptist Home Mission Society, and four educational institutions. These seven organizations constitute the Conservative Baptist "movement." Churches within the association cooperate with other denominations through the National Association of Evangelicals. The association arose from the fundamentalist-modernist struggle in the Northern (now American) Baptist Convention. In 1943 several hundred conservative churches, impatient with the Northern Baptist Convention's refusal to adopt doctrinal standards for its missionary program, formed the Conservative Baptist Foreign Mission Society. The association of churches was organized when it became apparent that the Northern Convention would not tolerate a competing missionary agency within its organization. BRUCE L. SHELLEY

CONSISTORY (Lat. *consistorium*). Originally that part of the imperial palace in Rome where the emperor and his council administered justice. In the Western Church it came to refer to the assembly of the clergy of the city of Rome under the presidency of the bishop; later it achieved its current meaning of the college of cardinals. There are three types of consistory today: the public one (e.g., when a cardinal is given his red had); a semipublic one (when Italian bishops attend); and a private one (wherein the normal work of the college is conducted). In the Church of England each diocesan bishop still has a consistory court to administer ecclesiastical law under the judge (chancellor). Its work has been much reduced in modern times. In Presbyterian churches the term has been applied to the meeting of the parish minister and lay elders (kirk session*), which is the lowest court. Calvin's consistorial court in Geneva was the forerunner of a wider usage: that of a court of presbyters. In Lutheranism the term has been used to describe a board of clerical officers (provincial or national) set up to oversee ecclesiastical affairs (e.g., those in Germany in 1587). PETER TOON

CONSTANCE, COUNCIL OF (1414-1418). The medieval papacy suffered a series of reverses during the fourteenth century, including the removal of the popes from Rome to Avignon and the Great Schism,* when for a time there were two and even three claimants to the office. There had been other instances of antipopes, but they were not nearly so serious as the fourteenth century schism, because in the latter instance the nations of Europe lined up behind their rival popes. There were many suggestions for ending the schism, and it is a high tribute to the office of pope that few people thought of abolishing the institution. The solution finally accepted by leaders such as Jean Gerson* and Pierre d'Ailly* was that of calling a council representing the entire church to settle the matter.

One of the popes, John XXIII, who was in military trouble, agreed to call the Council of Constance to secure the help of the Holy Roman Emperor, Sigismund. The emperor wished to call a council to enhance his prestige, reform the church, and suppress heresy. The council was attended by representatives of the lay rulers, bishops, and abbots or their representatives, and representatives of other ecclesiastical corporations and the universities. Several months passed before the council reached its full strength, and instead of voting by head, the group was organized into five nations after the pattern of medieval universities. Each nation—Italian, German, English, French, and Spanish—had one vote in the formal casting of ballots. This minimized the importance of the numerous Italian clergy present.

Among its achievements, the council ended the papal schism by deposing all three popes and electing Martin V. The problem of heresy was dealt with in the case of John Hus.* Hus came to Constance under safe conduct, firmly believing he could convince the council that his views were not heretical. The safe conduct was not honored, and he was imprisoned, tried, and burnt for heresy. The execution of Hus did not extinguish his teaching, but rather led to the Hussite wars. In the matter of reform, the council won two impressive victories, but lost on most of the practical issues. It passed the decrees *Sacrosancta*, affirming the authority of councils over the church, and *Frequens*, which set the intervals at which councils should meet. On specific matters, however, not much progress was made. Various committees of the council worked on reform of the abuses connected with papal revenues and provisions. Each party was ready to reform whatever did not affect its own selfish interests. The result was that Martin V was presented with a bill of particulars which he accepted in principle, but did not bother to enforce.

ROBERT G. CLOUSE

CONSTANTINE, DONATION OF, see DONATION OF CONSTANTINE

CONSTANTINE THE GREAT (c.274/280-337). First Christian emperor of Rome. Son of Constantius Chlorus, the future Western emperor, Constantine spent the years 293-305 as an apprentice-cum-hostage under the Eastern emperors Diocletian and Galerius, the instigators of the Great Persecution, but at York in 306 was proclaimed emperor *(Augustus)* by his father's troops on his death. From the contest for supremacy in the West he emerged triumphant by defeating Maxentius in 312 at the Milvian Bridge, north of Rome. According to Eusebius and Lactantius,* before the battle he adopted the emblem of the labarum in obedience to a vision and dream assuring him of victory from the Christians' God, whose worship he may have confused with his family's monotheistic reverence for the Unconquered Sun or Apollo.

In 313 he and Licinius, soon to control the Eastern empire, decreed full legal toleration for Christianity (Edict of Milan*), and the church enjoyed increasing favor—restitution of confiscated property, financial aid for Catholics, clerical exemption from hereditary offices, civil jurisdiction for bishops. The Donatists* protested against exclusion from his benefactions, and his determination to let churchmen resolve the dispute foundered on repeated Donatist appeals until in frustration he adjudicated personally in 316. His abortive coercion of the schismatics in 321 set another precedent. As sole emperor after conquering Licinius in 324, Constantine tackled the Arian* conflict. After a fruitless mission by Ossius, bishop of Cordoba, his adviser since about 312, he summoned the Council of Nicea (325), presided, at least at first, and influenced the inclusion of *homoousios* in its creed. The unity thus achieved soon disintegrated. Constantine, pursuing harmony without theological insight, and guided increasingly by the Arianizing Eusebiuses of Caesarea and Nicomedia, exiled obstructionist orthodox bishops like Athanasius.* Civil sanctions now regularly enforced ecclesiastical censures.

Constantine in 330 inaugurated his new foundation of Constantinople, located for strategic and Christian reasons far from old Rome, symbol of the pagan past and citadel of paganism's continuing vigor. Constantine spurned Rome, but his career magnified the Western Church's appreciation of its bishop's function as arbiter. He was baptized shortly before he died, and was buried amid the apostles in the basilica he founded in their honor in Constantinople. Constantine and his mother Helena* were keen patrons of church building, especially in the Holy Land, thus promoting the revived importance of Jerusalem.

The genuineness of Constantine's adoption of Christianity has been hotly debated, especially since Jacob Burckhardt's portrayal of the megalomaniac motivated solely by political expediency (*The Age of Constantine the Great,* 1853; ET 1949). The authenticity of Eusebius's panegyric *Life of Constantine* is now almost universally accepted. If Constantine often appears to stand outside the church, it is because he bore a heavenly commission to ensure its welfare and unity in the interests of imperial peace and prosperity which are divine blessings. He conceived of himself as "the servant of God," alongside or above the bishops, the *pontifex maximus* in Christianized dress. His religious ideas were dominated by an almighty Supreme Divinity who assigns responsibilities and distributes temporal and eternal rewards and punishments. Worship, the observance of the religious law, must be correct and united. Lactantian influence here is discernible.

The ambiguity of much of Constantine's public life hinges on the circumstance that, though a Christian, he was an absolutist monarch ruling an empire still largely pagan. His elevation of Christianity into virtually the imperial religion, though fraught with baleful consequences in the short and long term, was inevitable once an emperor became Christian, for the secular or religiously neutral state was unknown in antiquity.

BIBLIOGRAPHY: N.H. Baynes, *Constantine the Great and the Christian Church* (1931; new ed. with full bibliography, 1972); N.H. Baynes in *CAH* XII (1939); A. Alföldi, *The Conversion of Constantine and Pagan Rome* (1948); A.H.M. Jones, *Constantine and the Conversion of Europe* (1948); J. Stevenson, *A New Eusebius* (1957); R. MacMullen, *Constantine* (1970). D.F. WRIGHT

CONSTANTINOPLE. Many centuries before Constantine established the city of Constantinople, the Megarians had founded a colony at the same place on the European side of the Bosporus. When they established the colony in the seventh century B.C. they were fully aware of the strategic and commercial advantage in locating the community on the border of two continents and at the entrance to two seas, the Black and the Mediterranean. Even the Greek historian Polybius saw the importance of this location, and wrote in the second century B.C. that the inhabitants of Byzantium controlled all commercial vessels entering or leaving the Black Sea, thus placing them in a very powerful position.

When Constantine became emperor at the beginning of the fourth century A.D., he immediately recognized that Rome was too far away to deal with the eastern problems of the Roman Empire. At first he had planned to locate his eastern capital at the site of ancient Troy, but he soon changed his mind for the site at Byzantium. On 11 May 330, Constantine dedicated the "New Rome" and called his new capital "Constantinople." Historians mark this event as the beginning of the Byzantine Empire, even though this officially took place under the Roman Empire.

Constantine was responsible for an added dimension to the imperial throne. He considered himself the representative of God on earth and so brought a sacred character to his sovereign power. When he made Christianity the state religion of the Roman Empire, it was only natural he should take his role among the bishops of the church as if he were one of them. The emperor fixed this into the physical structure of the city by making the church building the center of the city, which is still to be found in many eastern towns and cities. Thus church and state began to operate in a more uniform way in Constantinople.

The fourth and fifth centuries A.D. were the period of the great heresies in the Christian Church. As these theological battles were fought, Constantinople began to emerge as the second most important religious center of power in Christianity, next to Rome. The bishop of Constantinople began to compete with the bishop of Rome for the primacy of the Christian Church. The new power which came with the New Rome provided the basis for the Greek East to consider itself equal to the Latin West. As more Germanic influence penetrated Rome and since paganism still dominated much of Roman life, the Eastern church leaders felt less and less loyal to the authority of the Roman papacy. Constantinople, on the other hand, was a capital with all the power and prestige that went with this role. The eastern city was, moreover, Greek and tended therefore to separate from Latin Rome so far as ecclesiastical authority was concerned.

After the Turks captured the city in 1453, the patriarch of Constantinople continued to keep his office in the city. The Turks tended to treat their Christian subjects with generosity. In fact, with the establishment of the Ottoman Turkish *millet* system (religious nations within an empire), it made possible the survival of the Greek Orthodox Church during the four centuries of Turkish rule. To this day the patriarch of Constantinople is the head of the Greek Orthodox Church, though the Greek Orthodox Church does not include the other branches of Eastern Orthodoxy.

GEORGE GIACUMAKIS, JR.

CONSTANTINOPLE, FIRST COUNCIL OF (381). Summoned by Theodosius I, it was attended by bishops from the civil dioceses of Oriens, Asia, Pontus, and Thrace at first, and later by Timothy of Alexandria and Ascholius of Thessalonica. Mainly it confirmed earlier decisions: the appointment of Melitius* to Antioch and Gregory Nazianzus* to Constantinople; acceptance of the Creed of Nicea and the *homoousia* of the Holy Spirit by Damasus of Rome (372), by the Council of Antioch under Melitius (379), and by Theodosius (380), whose condemnation of heretics is reflected in canon 1. The authority of this local Eastern council, entitled "ecumenical" in 382, was recognized at Chalcedon.

When Melitius died and Gregory retired, attacked for his appointment to a second see (Constantinople), the council appointed Flavian to Antioch and Nectarius, a government official, to Constantinople. The council presented six canons (the seventh is spurious) to Theodosius for ratification. Canon 1 confirms the Nicene faith and anathematizes all heretics, mentioning Anomoeans, Eudoxians ("Homoeans"), Pneumatomachi, Sabellians, Photinians, and Apollinarians. Canon 2 forbids bishops functioning outside the civil diocese of their see, although founding churches may still regulate mission churches among barbarians. The controversial third canon gives Constantinople primacy of honor next to Rome "because Constantinople is new Rome." Alexandria resented this relegation, and Rome rejected political prominence as a ground of her ecclesiastical supremacy. Canon 4 cancels the *Acta* of Maximus, while 5 and 6 (probably promulgated in 382) accepted Roman and Antiochene "tomes" and regulated accusations against bishops. The *Acta* of Chalcedon (451) attribute the "Nicene" creed to this council. Possibly known to Epiphanius of Salamis in 374, it was based probably on a local creed used at the baptism of Nectarius.

BIBLIOGRAPHY: J.D. Mansi, *Sacrorum Conciliorum Nova et Amplissima Collectio,* III (1759), cols. 521-600; W. Bright, *Notes on the Canons of the First Four General Councils* (2nd ed., 1892); J.N.D. Kelly, *Early Christian Creeds* (1960).

G.T.D. ANGEL

CONSTANTINOPLE, SECOND COUNCIL OF (553). The fifth ecumenical council. In order to understand its decisions and decrees it is necessary to investigate some of the previous events. The Council of Chalcedon* (451), which attempted to present both sides of the incarnation question about the nature of Christ, did not produce the unity hoped for by its leaders. Instead, the decisions at Chalcedon began the Monophysite controversy which plagued the church for centuries. Egypt and other centers in the Middle East were the main centers of Monophysitism* and were in constant conflict with Constantinople and Rome. The emperors Justin and Justinian tried to use force to impose orthodoxy by appointing orthodox bishops to seats in Monophysite areas. These bishops could, however, only occupy their seats with the aid of the police, for the masses were clearly on the side of the Monophysites.

Justinian in 531 decided to change his policy and follow one of compromise. This was probably due to the influence of his wife Theodora, who was a secret Monophysite believer. She was able to mellow her husband's orthodoxy. Thus the emperor began to enforce a somewhat double policy. On the one hand he opposed Monophysitism, as is evidenced in an edict against them in 536; on the other, he wanted to please his wife, so he allowed the Monophysites to reestablish their church in Constantinople in 543. When the Second Council of Constantinople took place, he wanted to please the Monophysites by having the council and the papacy condemn the Three Chapters,* involving Theodore of Mopsuestia, Theodoret of Cyrrhus, and Ibas of Edessa, who were all linked with Nestorianism,* the direct opposite of Monophysitism. The earlier council had cleared these men of heresy charges. Justinian, prior to 553, condemned these three leaders, and the Eastern bishops for the most part submitted to his edict. The Western bishops, however, including the Africans and Pope Vigilius, refused to sign it.

The council in 553, attended by 165 bishops, met and condemned the Three Chapters and thus moved closer to the Monophysite position. At the same time it condemned also the other extreme—that of Origenism.* Another decision of the council was to add a title to Mary, the mother of Jesus. Besides the already existent *Theotokos* (Mother of God), the council bestowed also the title *Aeiparthenos* (Ever-Virgin). They thus fixed in church dogma the perpetual virginity of Mary and considered the brothers of Jesus mentioned in Mark 3:31,32 as half brothers, cousins, or near relatives.

From the time of the Second Council one can begin speaking about the distinctiveness of Byzantine Orthodoxy in contrast with the Western Church. One of the distinctives is this particular interest to synthesize the two extremes of Christianity concerning the nature of Christ.

GEORGE GIACUMAKIS, JR.

CONSTANTINOPLE, THIRD COUNCIL OF (680). The sixth ecumenical council. With the rise of Islam in the middle of the seventh century, there was a marked attempt by the Eastern emperor to bring about unity and to induce the Monophysites back into the Orthodox Church. One of the ways by which this was attempted was through the emphasis called Monothelitism.* This was put forth by Emperor Heraclius in 638 along with the support of Patriarch Sergius. It stated simply that in Christ there were two natures, but not two wills. Christ was a single person and therefore acted with one will.

A few of the leaders in Egypt favored this new interpretation, but most of the Eastern Christians rejected the compromise. This council wanted to bring religious unity between Rome and Constantinople. Thus it rejected Monothelitism and restated the Chalcedonian definition, adding the interpretation that Christ had two wills as well as two natures. The council went on to anathematize the leaders of Monothelitism and reestablish the Orthodox faith. It also condemned the then dead Pope Honorius because he had sanctioned Monothelitism. Because of these decisions and the fact that Islam had taken over much of the Middle East, the Nestorian* and Monophysite* wings of the church became permanently separated from the Orthodox Church.

GEORGE GIACUMAKIS, JR.

CONSTANTINOPLE, FOURTH COUNCIL OF. It is uncertain which fourth Council of Constantinople was in fact the eighth ecumenical council. The Eastern Church recognizes only seven ecumenical councils, but they do sometimes recognize an eighth council as that which took place in 879 under the leadership of Patriarch Photius* of Constantinople. The Latin West, however, recognizes the eighth ecumenical council as being the one which took place in 869 under Patriarch Ignatius. The latter had decided to rebuke the immorality of Emperor Caesar Bardas. Ignatius was arrested and the layman Photius was promptly nominated in his place. Ignatius was exiled, then voluntarily resigned. The supporters of Ignatius, however, who considered Photius to be illegitimate, were able to convince Pope Nicholas I that he should intervene. He convoked a synod at Rome which at once voted to excommunicate Photius and to reinstate Ignatius. Photius in turn held a counter synod and deposed the Roman pope.

Photius's greatest support came from Emperor Michael III. But the emperor was murdered in 867, and his successor, Basil, decided to restore Ignatius as patriarch. Basil convened the 869 council which voted against Photius and in favor of Ignatius. Another action of the council was to condemn the Monothelites and the Iconoclasts as had been done in past councils. Photius, however, was reappointed to the patriarchal see shortly after the death of Ignatius in 877. He then convened his own Council of Constantinople in 879.

It annulled the council of 869, branding it as fraudulent, and it readopted the Nicene Creed. The council specifically spoke against the *Filioque*, the phrase "and from the Son" which was added to the Nicene Creed by the Western Church to show that the Holy Spirit came from the Father *and from the Son*. The council ended by praising the virtues of Photius.

GEORGE GIACUMAKIS, JR.

CONSTANTIUS II (337-361). Eastern emperor from 337; sole emperor from 350. Acceding after defeating Magnentius in the West, he continued the benevolent policy of Constantine I toward the church, which freed clergy from taxation and public service, although his laws in 360/1 made such privileges conditional. Some of his pro-Christian legislation was directed against Jews. Doctrinal disputes threatened the cohesion of the empire. To alleviate Western discontent with Eastern bishops, Constans induced Constantius to recall Athanasius* to Alexandria (346). In 350 he reassured the latter of his support, but gradually, influenced by the Arian Valens* of Mursa, he sought to unite the church with a vague creed which excluded the unscriptural term "substance" and stressed "likeness." At Milan in 355 he exiled bishops who refused to depose Athanasius, and, despite a brief period of favor towards Homoiousions, imposed the "Dated Creed" on the Councils of Rimini and Seleucia (359).

G.T.D. ANGEL

CONSTITUTIONAL ACT. The commonly accepted name for the Canada Act of 1791, which divided the old province of Quebec into Upper and Lower Canada, with the French of Lower Canada maintaining their own religion, customs, and laws. The people received representative but not responsible government. The legislatures consisted of an appointed legislative council and an elected legislative assembly. The executive powers remained in the hands of the governor, who acted in conjunction with the Crown-appointed executive council. While the assembly voted funds for roads, bridges, and other public projects it was the executive council that administered both the collection and the spending of tax and custom monies. A ruling élite emerged that in Upper Canada was called the "Family Compact" and in Lower Canada was dubbed the *Château Clique*.

ROBERT WILSON

CONSTITUTIONAL CHURCH. This was established in France at the Revolution by the "Civil Constitution of the Clergy" (1790). It was organized in 1791 under the protection of the National Assembly, which had passed a law requiring all bishops, pastors, and functionary priests to take an oath of fidelity to the Civil Constitution under pain of deposition. About one-third of the clergy obeyed. Those refusing to take the Constitutional Oath became the "Refractory Church" siding with the papacy, to which most loyal Roman Catholics adhered, especially when the old discipline was abandoned, priests and bishops married, and divorce was permitted. The church established itself for three years, but after the royalist and Catholic uprising in La Vendée (1793) and the triumph of Jacobinism, it was itself persecuted and many of its priests apostatized, though Constitutional bishops like Henri Baptiste Grégoire showed dauntless courage and integrity. After Robespierre's fall (1794), a measure of toleration was granted, but the Thermidorian Convention adopted a regime (1795) separating the state from the churches, thereby abandoning the Civil Constitution and refusing to pay Constitutional priests. When Napoleon concluded the Concordat of 1801,* Pope Pius VII had little difficulty in obtaining the church's abolition.

J.G.G. NORMAN

CONSTITUTIONS OF CLARENDON. Articles laid down by Henry II of England at a council called to the royal hunting lodge at Clarendon, near Salisbury, in 1164. They purported to record established customs in relations between church and state, but in fact incorporated some new regulations designed to reduce the power of the church over against the Crown. For example, guilty "criminous clerks" were to be handed over to the king's officers for punishment, and royal permission was required for the imposition of excommunication and interdict on the king's tenants-in-chief and his officers, for clergy to leave the realm, and for appeals to Rome. After initial opposition, the bishops led by Thomas Becket* gave way, but Becket later withdrew his consent and fled to France.

HOWARD SAINSBURY

CONSUBSTANTIAL. A term occasionally used in connection with the fourth-century Arian* controversy. The orthodox party, led by Athanasius,* defined the Son as *homoousios*—the same identical essence or substance as the Father. Arian groups taught that He was *homoiousios* (like essence) or *homois* (like) or *anomoios* (unlike). The Greek *ousia* (essence) had an imprecise Latin equivalent in *substantia*, and *consubstantialis* therefore became the equivalent of the orthodox *homoousios*.

CONSUBSTANTIATION. A term used to describe Luther's view of how Christ is present in the Eucharist, although the term does not appear in his writings. Luther's belief was that Christ is present "under (or with) the species of bread and wine"—a position William of Ockham* had earlier held. Luther nevertheless insisted that we know Christ is present only through the Word, whose pledge guarantees it. A union of the earthly and the heavenly occurs in the sacrament, but the bread and wine are not changed in substance, an explanation Luther found too reasonable. The question that intrigued him, rather, was why Christ was really present. His answer: for our salvation. "Of course," he held, "it is a miracle that Christ's body and blood should be in the sacrament and not be visible; yet we are content to know through the Word and by faith that they are there."

ROBERT B. IVES

CONTARINI, GASPAR (1483-1542). Venetian ambassador and cardinal. After an education at Padua appropriate to his noble birth, he became in 1521 ambassador of the republic of Venice to Charles V. Later he also served in Spain, England, and Rome. A fruit of this activity was his *De Magistratibus et Republica Venetorum.* His general integrity (in part the result of his adoption of the New Learning) led to Pope Paul III's making him a cardinal in 1535, even though he was a layman. He had previously, however, shown himself to be a competent theologian in his defense of the immortality of the soul (1516) and in his treatise against Luther (1530). He favored reform within the church and for this reason was put on the commission set up by Paul III to suggest reforms. His proposals made in 1537, entitled *Consilium de emendanda ecclesia,* were too radical for many, and the book was put on the Index* in 1539. Prior to this he became bishop of Belluno. In 1541 he attended the Diet and Conference of Ratisbon where he made a valiant attempt to bring back the Lutheran movement into the church. Also he composed a treatise on Justification which, in the opinion of other Roman Catholics, went too far towards the Protestant doctrine. He died while serving as a legate at Bologna, at a time when the Inquisition* was beginning to make life difficult for reforming Catholics.

PETER TOON

CONTRA-REMONSTRANTS. A name given to the defenders of Calvinist orthodoxy in the Dutch controversies of the early 1600s aroused by the teachings of Arminius and his followers. A theologian at Leyden, Arminius tried to soften the doctrine of predestination so as to preserve something of human free will. His colleague Francis Gomar* (Gomarus) attacked his views as dangerous innovations, and the controversy spread rapidly. After Arminius's death in 1609, his followers issued the Remonstrance* of 1610, stating the Arminian* position. In 1611 the Counter-Remonstrance appeared, reiterating the orthodox position as understood by scholastic Calvinism. In general it stressed that predestination did not depend in any way on man's actions and thus salvation was assured for the elect. More particularly, it held that predestination is not based on God's foreknowledge of man's choice; children, though unable to make a mature choice, may be among the elect; election is due to grace alone, and man does not cooperate; Christ died for the elect, whom He willed to save; the elect are assured of salvation; and these doctrines lead to a virtuous life, not carelessness.

During the next few years, the controversy aroused heated debate among the people as well as among theologians; for those who favored the Contra-Remonstrant position, the Remonstrants appeared to take away the assurance of salvation and make it dependent on man's will rather than divine grace. Political issues also became involved. The Remonstrants were strongest in the province of Holland and were supported by its leader, Jan van Oldenbarneveldt, who favored provincial autonomy rather than centralization, and peace with Spain; the Contra-Remonstrants thus turned to the *stadhouder* Maurice of Orange, who favored continuation of the war by a centralized government. The political struggle between Maurice and Oldenbarneveldt resulted in the latter's imprisonment (and later execution on charges of treason). Thus, at the Synod of Dort* which followed (1618), the Contra-Remonstrants were in control. The Remonstrant positions were condemned, Remonstrant ministers ousted from their pulpits, and Remonstrant leaders exiled as disturbers of the public peace. The Canons of Dort became one of the official standards of the Dutch Reformed Church.

See D. Nobbs, *Theocracy and Toleration: A Study of the Disputes in Dutch Calvinism, 1600-1650* (1938).

DIRK JELLEMA

CONVENT (Lat. *con* and *venire,* "to come together"). To convent is to assemble persons for some common purpose; thus the noun can designate any general or specific gathering or company. Already in thirteenth-century English, "convent" designated particularly men or women living under a disciplined religious order with a single superior. From the institutional phenomenon the term was applied by the sixteenth century to the set of buildings thus occupied, and since the late eighteenth century, without historical warrant, popularly restricted to that of women only.

See also CLOISTER.

CONVENTICLES, ACT AGAINST (1670). Legislation passed by the Scottish Parliament against field preaching* and illegal house services conducted generally by Presbyterian ministers who had been ejected for nonconformity after the Restoration of Charles II. The punishment for infringing the act was death and confiscation of goods. It was required of everyone on oath to give information regarding conventicles, and those who had had their children baptized by the nonconforming ministers could be punished with exile. Gilbert Burnet* claims Charles himself said that bloody laws did no good and that he would not have sanctioned the act if he had known of it beforehand, but despite this disclaimer the repressive policy was maintained.

J.D. DOUGLAS

CONVERSION. A radical change, a transformation, a turning around. The term applies to nonreligious responses to stimuli, or to reorientation of mental attitudes and behavior, but usually religious conversion is intended. The term does not have a prominent place in the NT, but the idea of conversion is abundantly present in both testaments, particularly with regard to the apostolic preaching of the Gospel through which men are converted to Jesus Christ. Repentance (turning from) and faith (turning to) are usually seen as the two sides of conversion; they figure more prominently in the biblical language.

Conversion is a conscious act on the part of the subject, not an event passively experienced. For the Christian, the changed life of the converted man is the outward expression of a changed heart. Biblical examples are Paul's conversion (Acts 9), the Ethiopian eunuch (Acts 8:26-40), the Prodigal Son (Luke 15:11-32), and Zaccheus (Luke 19:2-10). While conversion is usually thought of in relation to individuals, societies and nations also have been profoundly affected by religious awakenings. These include Israel under Moses' leadership and during Hezekiah's reign, Nineveh as a result of Jonah's preaching, and more recent events like the English revival under John Wesley, and the Welsh revival.

The need of sinful men to be converted is declared by Jesus (Matt. 18:3) and the apostles (Acts 3:19; 15:3). In Acts, conversion is also presented under the figure of the two ways and choosing the Way of the Lord (9:2; 19:9,23; 22:4; cf. James 5:19,20). The new Way involves a new kind of life (Eph. 5:2; Col. 1:10; 2:10-12). Bunyan's *Pilgrim's Progress* is a classic which presents conversion as entrance upon the pilgrimage from the City of Destruction to the Celestial City.

Conversion entails intellectual, emotional, and volitional elements, including a doctrinal relationship to or affirmation of Jesus Christ's lordship, acceptance of His redemptive work, devotion to Him personally, commitment of fellowship to the community of Christians, and the ethical transformation of life.

Many psychological explanations of religious conversion have been attempted. Most of these, following William James,* see conversion as a conscious unification or reunification of a hitherto divided self, with a sense of wholeness, being right and happy, resulting. Conversion is thus seen as a profound step in the creation of a self. The biblical language concerning the Prodigal ("he came to himself" KJV) is distinctly parallel. Other explanations include such terms as these: integration of personality, new being, freedom, reorientation, and brainwashing. William Sargant's thesis, while interpreted as a critique of brainwashing techniques in religious conversions, usually draws attention to the dangers of religious manipulation.

In its biblical sense, conversion is the soul's turning to Christ and union with Him in His death and resurrection, which baptism signifies as entering by faith upon a new life (Rom. 6:1-14).

BIBLIOGRAPHY: Augustine, *Confessions* (fifth century; many editions); W. James, *The Varieties of Religious Experience* (1907); A.D. Nock, *Conversion* (1933); B. Citron, *New Birth* (1951); O. Hallesby, *Religious or Christian* (1954); R.E.O. White, *Into the Same Image* (1957); W. Sargant, *Battle for the Mind* (1959); O. Brandon, *The Battle for the Soul: Aspects of Religious Conversion* (1960); E. Routley, *Conversion* (1960); E.F. Kevan, *Salvation* (1963).

SAMUEL J. MIKOLASKI

CONVOCATION. The name given to each of the two provincial gatherings of clergy in the Church of England: those of Canterbury and York. The origins of these two provincial convocations are lost in the mists of medieval antiquity, but they certainly antedate Parliament. Until 1665 they taxed the clergy instead of Parliament, but in 1717 the two convocations were suspended until 1852 (Canterbury) and 1861 (York). The suspension largely arose out of clashes between Whig Upper Houses (bishops) and Tory Lower Houses (other clergy). It is sometimes alleged that the Church of England was paralyzed by this suspension, but being a church established by the law of the land, church government took place through Parliament, which in England was a Christian body. More recently, powers over the Church of England's doctrine and liturgy were claimed by the convocations, but this was largely a myth since changes in basic church law had to come through Parliament. In 1920 the national assembly of the Church of England (usually called the Church Assembly) began to function alongside the convocations. In 1969 the Synodical Government Measure more or less coalesced Church Assembly and convocations into a new General Synod,* though formally the convocations still remain.

G.E. DUFFIELD

CONVULSIONARIES. The tomb of a young Jansenist,* François de Paris, was the scene of a series of supposed miracles after 1727. Jansenists regarded them as a vindication of their cause, and thousands visited the cemetery of St. Medard. Ecstatic behavior led to the nickname "Convulsionaries," and even after the cemetery was closed by the authorities, portions of earth from the graveyard induced the same results. The movement ended in discredit.

CONYBEARE, FREDERICK CORNWALLIS (1856-1924). Armenian scholar. Elected fellow of University College, Oxford, in 1880, he resigned in 1887 to devote himself to research, particularly in Armenian, in order to obtain material for the textual criticism of the Greek classics. Because of the religious nature of the MSS he became interested in church history and in the textual criticism of the Septuagint and the NT. His discoveries included a MS with the ascription of the Last twelve verses (16:9-20) of Mark's gospel to the "Presbyter Aristion." He was generally skeptical toward Christianity as is shown in his *Myth, Magic and Morals, a Study of Christian Origins* (1909), though he refuted those who denied the historicity of Christ in *The Historical Christ* (1914). He also entered into controversy on a number of political issues of his day.

R.E. NIXON

CONYBEARE, WILLIAM JOHN (1815-1857). Biblical scholar. He was ordained in 1841 and the following year became the first principal of the Liverpool Collegiate Institution. In 1845 he was joined there by J.S. Howson,* later dean of Chester. Three years later he had to resign on account of his health, and became vicar of Axminster, Devon, until 1854. While he was there he combined with Howson to write *The Life and Epistles of St. Paul* (2 vols., 1852). This was an important book for introducing to the English-speaking world new ways of understanding the apostle's writings in the context of their time. He also wrote *Essays Ecclesiastical and Social* (1856) and several articles for the *Edinburgh Review.*

R.E. NIXON

COOK, DAVID CALEB (1850-1927). American Sunday school leader and publisher. Born in East Worcester, New York, he gained a lasting impression from his first contact with a Sunday school teacher in Wheaton, Illinois. He entered business in Chicago as a sewing-machine salesman and later developed a thriving mail-order business. Involvement in several Sunday schools led him to give up business and dedicate himself to the Sunday school, particularly in the field of publications. In 1882 rapid expansion obliged his publishing house to move to nearby Elgin, where it still is. While recuperating from a nervous breakdown, Cook built up a flourishing estate in California, but sold out to return to Illinois and his beloved Sunday school work, to which he made valuable contributions. HAROLD R. COOK

COOK, STANLEY ARTHUR (1873-1949). Biblical and Semitic scholar. His wide interests within the Semitic field included epigraphy, archaeology, and religion. In 1925 he gave the Schweich Lectures at the British Academy under the title *The Religion of Ancient Palestine in the Light of Archaeology* (published in 1930). He admired the work of W.R. Smith, and especially his use of anthropological evidence in the study of the early religion of the Semites. In his own work he explored the close cultural links between Israel and her neighbors. From 1902 to 1932 he served as editor of the *Quarterly Statement* of the Palestine Exploration Fund, and was a co-editor of the *Cambridge Ancient History*, to which he contributed several chapters. He was regius professor of Hebrew at Cambridge (1932-38). Apart from numerous articles in journals, Cook's writings include *The Study of Religions* (1914); *The Old Testament: A Reinterpretation* (1936); and *The Rebirth of Christianity* (1942).

ROBERT P. GORDON

COONEN CROSS. The cross outside the church at Mattancherry in Cochin—*coonen* meaning "crooked" or "bent"—at which a crowd of Syrian Christians took an oath not to be subject to the Jesuit Archbishop Francis Garcia on 3 January 1653 and thereby initiated a revolt which affected almost all the Syrians in the Roman Catholic Church in Kerala, India. Discontent with the rule of the Jesuit archbishops of Cranganore was nothing new, and led the Syrians to look again to Mesopotamia for a bishop of the old church, although the Synod of Diamper* had apparently terminated that connection and bound them to Rome.

In 1652 a Bishop Ahatalla did in fact come to India, but was detained by the Portuguese at Mylapore (Madras). He managed to get a letter to the Syrians in Kerala, and a great agitation was stirred when the Syrians learned that he was being taken to ship and that the ship was now at Cochin. The Portuguese shut the fort and manned the walls until the ship sailed. A rumor was that the bishop had been thrown overboard and drowned; in fact he probably reached Goa and may have been shipped to Europe after trial as a heretic. The fury of the Syrians left hardly any of them under Roman authority, but it is evident that the revolt was not against Romanism as such, for efforts led by the Carmelites to win back the lost Syrians were very successful. The Coonen Cross incident, however, marked the renewed independence of a considerable section of the Syrians, and tempered Roman policy in Kerala.

ROBERT J. MC MAHON

COORNHEERT, DIRCK VOLKERTSZOON (1522-1590). Dutch humanist and evangelical. Born in Amsterdam, he settled in Haarlem as a skilled engraver. In his twenties he read widely on religious matters, in Erasmus, Luther, Calvin, Menno, Franck, and others, and settled on a sort of humanistic evangelicalism, stressing the role of the Bible as ethical teacher. Expert in Latin, he became town secretary, and during his forties became involved with William of Orange and the growing opposition to Spanish rule. In 1566 he helped draw up William's manifesto against Spanish misrule, and in 1568, as the revolt broke out, he went over the border to Cleves; there he worked as Orange's agent. A brief return to Haarlem (1572) brought controversy with militant Calvinists there, and he went back to his work in Cleves; a second return (1577) brought more of the same, and in his sixties he moved to Emden (1585) and finally to Gouda.

Coornheert wrote extensively on religious matters, and his views led him into controversies with Catholics, Mennonites, with Calvin (on original sin; Calvin replied briefly in 1562) and Beza, and many others. For Coornheert, man has free will and can choose for Christ, who is divine but whose role is to lead men to moral worthiness. Neither church nor state should interfere with the exercise of this free will. His views influenced Arminius (who, appointed to refute Coornheert, instead came to agreement with many of his positions). His main religious work (1568) is *Zedekunst: dat is Wellevenskunst.* Coornheert was also a figure of some importance in literature; he translated from Latin (Seneca and Boethius and part of a Latin version of the *Odyssey*) and French (a French version of the *Decameron*) and wrote a series of allegorical "comedies" on biblical themes.

See B. Bicker, *Bronnen tot de kennis van het leven en de werken van Dirck V. Coornheert* (1928); and G. Güldner, *Das Toleranz-Problem in den Niederlanden in Ausgang des 16. Jahrhunderts* (1968). DIRK JELLEMA

COP, NICHOLAS (sixteenth century). German-French scholar. Son of Guillaume Cop, former royal physician, he was professor of medicine in the *College Sante-Barbe* and one of a circle of humanists in Paris with Roussel and Calvin. In 1533 he was elected rector of the University of Paris. This involved a university sermon on All Saints' Day. Choosing as his text Matthew 5:3, he contrasted the slavery of the law which man cannot fulfil with the saving merits of Christ. He minimized the value of good works and reviled the "Sophists" of the Sorbonne for their intolerance. The address was largely made up of citations from Erasmus and Luther. The theologians proceeded against him for heresy, and he fled to Basle. Nothing more is known of him. J.G.G. NORMAN

COPE. A full semicircular cloak used as an alternative to the chasuble* in the 1549 Prayer Book, but retained when the latter was forbidden in 1552, and prescribed for use at the Holy Communion in cathedral and collegiate churches in the 1604 canons of the Church of England. While there is little doubt that its origin was either the Roman *paenula* or *pluviale,* scholars disagree as to the date of its use as a specifically clerical vestment, from as late as the mid-ninth century to as early as Gregory of Tours in the sixth. Revived with other ceremonial robes by the Oxford Movement* in the nineteenth century, it is highly embroidered with a vestigial hood as a triangular or semicircular ornament on the back. G.S.R. COX

COPERNICUS, NICOLAS (1473-1543). Polish doctor and astronomer. Copernicus's interest in astronomy was aroused in his early years at Cracow (1491-94); thereafter he maintained his interest, but was better known as a compassionate physician in his lifetime rather than as an astronomer. His great work, *De Revolutionibus,* which marks the beginning of the modern scientific era, did not appear until the year of his death. In this work he set forth the modern heliocentric theory: "In the middle of all sits the sun on his throne, as upon a royal dais ruling his children the planets which circle about him."

Prior to Copernicus it was known that the *idea* of a heliocentric planetary system was simpler than a geocentric one, but as heavenly bodies were then supposed to move in circles (rather than ellipses) this resulted in no great simplification. With lack of observational evidence either way, the geocentric scheme was favored, admittedly on slender biblical grounds. Copernicus urged that planetary movements were best explained by the heliocentric theory. His researches, though not proscribed by the church, were not encouraged. He lectured in Rome in 1533, but Clement VII feared reaction. In 1541 Copernican views were ridiculed in a comedy. *De Revolutionibus* was on the papal Index,* 1616-1758.

After publication of his book, Copernicus's views spread slowly, undermining not religion but astrology. The effect on religion was mainly indirect: in geocentric astronomy the fixed stars lie just outside Saturn's orbit; in heliocentric astronomy they are very far away, and an infinite universe is possible. Theologians argued that God could not create infinity, therefore an infinite universe was an atheistic conception.

BIBLIOGRAPHY: C. Singer, *A Short History of Science* (1941); C.C. Gillespie, *The Edge of Objectivity* (1960); P. Duhem, *To Save the Phenomena* (1969); biographies include those by H.S. Jones (1943); J. Rudnicki (1943); H. Kesten (1946); and A. Armitage, *Sun, Stand Thou Still* (1947).

R.E.D. CLARK

COPTIC. The latest form of the ancient Egyptian language; the language of the important Nag Hammadi* Gnostic texts. Its alphabet comprises the twenty-four letters of the Greek alphabet, in uncial form, and seven Demotic characters expressing consonantal sounds not represented in Greek. The transcription of Egyptian into Greek letters first appears in pagan texts of the first or second century A.D.; translation of biblical books began in the third century. Coptic remained a spoken and literary language until the seventh century, thereafter being gradually replaced by Arabic. It is still the ecclesiastical language of the Coptic Church.* Greek influence upon vocabulary and syntax is considerable, while many characteristics of Egyptian proper (e.g., the suffix conjugations) had disappeared, or had all but disappeared, by the time of the Copts. Several dialects of Coptic existed in the early Christian period; the Bible translations in Sahidic (Upper Egypt) and Bohairic (Lower Egypt) are of particular value. The earliest Bible translation was probably into Sahidic. Coptic was instrumental (with Demotic and Greek) in the decipherment of hieroglyphic Egyptian; the only form of Egyptian with a regular system of vocalization, it has also thrown light on the pronunciation of the ancient language.

BIBLIOGRAPHY: W.H. Worrell, *Coptic Sounds* (1934); W.E. Crum, *A Coptic Dictionary* (1939); J.M. Plumley, *An Introductory Coptic Grammar (Sahidic Dialect)* (1948); W.C. Till, *Koptische Grammatik (Saïdischer Dialekt)* (2nd ed., 1961). ROBERT P. GORDON

COPTIC CHURCH. The Christian Church in Egypt traces its birth to St. Mark, and counts as theological ancestors Dionysius, Clement, Origen, and Athanasius. Spearheading the struggle which culminated at the Council of Chalcedon (451) was Cyril of Alexandria* (d.444). At stake in the debate over Nestorianism* were the unity of the incarnate Christ and the eternal preexistence of the Logos,* both of which Cyril deemed threatened by Nestorius's emphasis on two natures. To Cyril this implied two persons. At the council, Dioscorus* (d.454) led the Monophysite cause in an effort that was defeated by Roman Pope Leo's doctrine of two natures in one person and Byzantine Emperor Marcian's political ambitions. Dioscorus died in exile, a hero-martyr to most Egyptians, who rejected the Chalcedonian symbols and the puppet patriarch, Proterius.

Egyptian opposition to Chalcedon was more than theological. Politically Dioscorus's defeat meant the triumph of the younger see of Constantinople over the ancient throne of Mark. Culturally Chalcedon seemed the triumph of Greek language over the indigenous culture of Egypt. The very name "Coptic" Church and the persistent use of the Coptic language in liturgy and literature show the vigor of the Egyptian fight for national identity.

In Egypt, the century after Chalcedon was scarred by religious civil war. Possession of the throne of St. Mark became a game of musical chairs, alternately occupied by Melchites* (Chalcedonian Christians still loyal to the Byzantine emperor) and Monophysites,* depending on the emperor's ability to support his man against the hostile Copts. Emperor Zeno's attempts at compromise to bring unity to the Christians of the East were a failure (482). So were the efforts of Justin II to recognize two patriarchs of Alexandria

—one Melchite, one Monophysite. Virtually the whole population was Monophysite by this time (567).

The Muslim invasion of 642 nearly broke the battered church, sorely weakened by decades of religious strife. Though the Copts found temporary relief in freedom from Byzantine pressures, the cure proved worse than the disease. The Arab conquerors used heavy taxations and other threats to persuade Copts to become Muslims—tactics that induced mass conversions.

Under the Caliphs, especially El-Haken biamr Allah (c.1000), the destruction of churches and monasteries and the massacre of Christians helped to spark the Crusades. By about 1100 Arabic had replaced Coptic as the common language. To maintain identity, the Copts instructed the faithful in their traditional canon and civil law (cf. the major compilation of Patriarch Cyril III, c.1236).

The coming of the Turks in 1517 expanded the history of persecution. By 1700 Monophysites had been reduced to five per cent of the population. Of the once influential monastic community, only a handful of monasteries survived. Even before Chalcedon, Egyptian Christianity exhibited a strong ascetic side which has persisted to the present. The reforms of Muhammad Ali (c. 1840) and additional concessions from the Turks in 1911 allowed some Coptic participation in government and permitted the establishment of schools and printing presses. During the recent decades of independence, Egypt has been officially free from religious intolerance. Yet Copts tend to feel that opportunities for political and economic progress are denied them by the Muslim majority.

Copts still worship and observe sacraments according to the Alexandrian Rite, with some monastic and Syrian modification. The Eucharist is usually offered in one kind, and baptism is combined with confirmation as a prominent sacrament. Masses are frequently two hours in length, scented with lavish use of incense and punctuated by triangles and cymbals which set the rhythm for the chanting.

Current figures estimate a Coptic population of about four million in twenty-four dioceses and a Roman Catholic population of 85,000 including over 100 priests. The Coptic patriarch, called Pope of Alexandria, Pentapolis, and Ethiopia, is elected by the people through a religious tribunal, subject to confirmation by the government. His dominion over Ethiopia has been *sine cura* since 1959 when the Ethiopian Church declared its autonomy by consecrating its own bishop or *abuna*.

BIBLIOGRAPHY: E.L. Butcher, *The Story of the Church of Egypt* (2 vols., 1897); W.A. Wigram, *The Separation of the Monophysites* (1923); E.R. Hardy, *Christian Egypt: Church and People* (1952); R.V. Sellers, *The Council of Chalcedon* (1953); D. Attwater, *The Christian Churches of the East* (2 vols., rev. ed., 1961); E. Wakin, *A Lonely Minority: The Modern Story of Egypt's Copts* (1963); A.S. Atiya, *A History of Eastern Christianity* (1968).

DAVID A. HUBBARD

CORINTH. One of the great seaports of ancient Greece, it was situated at the western end of the isthmus linking central Greece and the Peloponnesus. It was sited some five miles southwest of the modern canal that cuts the isthmus. Its important location thus enabled Corinth to control the trade between N Greece and the Peloponnesus and across the isthmus. It had two harbors: its eastern harbor, Cenchreae, was on the Saronic Gulf, an arm of the Aegean Sea; its western harbor, Lechaeum, was on the Gulf of Corinth, an arm of the Ionian Sea. The Corinthian canal conceived by Nero to cut through the isthmus, and so avoid a lengthy and dangerous sea voyage around the Peloponnesus, was completed only in 1893. So the two harbors of Corinth, with ox cart transference of goods between the two, had to suffice in classical times.

Corinth had a checkered history, being twice destroyed by earthquakes; this old city was about two miles inland on an elevated plateau at the foot of the Acro Corinth (about 1,886 feet above sea level). Depopulated and destroyed after a cruel siege in 146 B.C., it was rebuilt by Julius Caesar in 46 B.C., and settled with freedmen from Italy, Orientals, and Jews. The mixed population, scorned by the proud Athenians, was licentious and tumultuous, as Paul learned to his sorrow. His two epistles to the Corinthians are vivid evidence of this unruly church. Paul stayed eighteen months in Corinth during his second missionary journey (Acts 18:1-18). This has been dated by an inscription from Delphi which shows that Gallio came to Corinth as proconsul in A.D. 51 or 52 (Acts 18:12-17). His *bēma*, judgment seat, has been excavated (Acts 18:12), as also the meat market (1 Cor. 10:25). An aedile Erastus is mentioned on an inscription near the theater, who has been identified with the treasurer of Roman 16:23. It was from Cenchreae, seven miles from Corinth, that Paul sailed to Ephesus, after he had written 1 Thessalonians, the oldest preserved Pauline letter.

BIBLIOGRAPHY: J.G. O'Neill, *Ancient Corinth*, (1930); O. Broneer, "Corinth, Center of Paul's Missionary Works in Greece," *BA* XIV (1951), pp. 77-96; *Corinth I-VIII* (1951 ff.).

JAMES M. HOUSTON

CORINTHIANS, EPISTLES TO THE, see EPISTLES, PAULINE

CORINTHIANS, THIRD EPISTLE TO THE, see APOCRYPHAL NEW TESTAMENT

CORNELIUS (d.253). Pope from 251. His election, which came after the persecution initiated by Decius had died down, ironically stirred up trouble within the church itself. A minority declared in favor of Novatian, a fierce rigorist in church discipline who refused to admit the *lapsi** to the sacrament of penance. Cornelius convened a synod which, attended by some sixty bishops, excommunicated Novatian and his adherents. In this Cornelius had the weighty support of Cyprian of Carthage* who, although taking a serious view of the lapsed, held more strongly that the unity of the visible church should be maintained and was prepared to see them restored on evidence of true sorrow and penance. Among letters from the hand of Cornelius that have survived are several addressed to Cyprian. With the renewal of persecution in 253, Cornelius was exiled to Centumcellae (Civitavecchia) by Emperor Gallus, and died that same year.

See also NOVATIANISM. J.D. DOUGLAS

CORNELIUS À LAPIDE (1567-1637). Flemish biblical exegete. Born Cornelis Cornelissen van den Steen at Bolcholt, near Liège, he was educated at the Jesuit colleges of Maastricht and Cologne, and at the age of twenty-five entered the Jesuit order. Four years later he became professor of exegesis at Louvain, and in 1616 was translated to Rome, to lecture in the same subject. There also he completed his commentary on all the canonical books except for Job and Psalms. His works had an enduring quality and were especially attractive to preachers because of their clarity and deep spirituality, backed by an erudition enabling the writer to draw on the works of the Fathers and the medieval theologians. The list of his writings is impressive, including a number of posthumously published works. GORDON A. CATHERALL

CORPUS CATHOLICORUM. Leonhard von Eck as Bavarian chancellor shrewdly opposed the national council at Speyer. In the presence of papal legate Compeggio, Eck saw that the Regensberg Union was organized in June 1524 to enforce the Edict of Worms. Archduke Ferdinand, the two dukes of Bavaria, the cardinal-archbishop of Salzburg, and eleven bishops of S Germany endorsed the Union to stamp out heresy. N Germany formed a similar league in 1524 with Duke George of Saxony and Archbishop Albert of Mainz. This League of Dessau led to a similar 1525 Protestant League of Torgau under Elector John of Saxony and the Landgrave Philip of Hesse. MARVIN W. ANDERSON

CORPUS CHRISTI, FEAST OF. One of the greatest festivals of the Roman Catholic Church, honoring the presence of Christ in the sacrament of Communion, celebrated on the Thursday after Trinity Sunday by carrying the host in procession, accompanied by hymns. In the wake of the enunciation of the dogma of transubstantiation at the Fourth Lateran Council in 1215, Juliana of Mont-Cornillon, the Belgian mystic, persuaded Bishop Robert of Liège to establish a special celebration in honor of the Eucharist in 1246, and the practice was made universal by Pope Urban IV in his bull *Transiturus* in 1264, which provided indulgences for those observing it. The rejection of transubstantiation as well as the theory of indulgences* by the Reformers led them to suppress this festival, which was the first prohibited by Luther. MARY E. ROGERS

CORPUS EVANGELICORUM. The delegates from the Protestant states in the Holy Roman Empire charged with the protection of Protestant interests in the imperial diet. The organization of these delegates into a structured party was the product of a slow evolution, and it was not until the Diet of Ratisbon (1653) following the Thirty Years' War that a clear structure emerged. Saxony was chosen to serve as the permanent president of the Corpus, which now consisted of the thirty-nine Protestant states represented in the diet. The Corpus Evangelicorum was opposed by the Corpus Catholicorum,* and both existed until the dissolution of the Holy Roman Empire in 1806.

CORPUS IURIS CANONICI, see CANON LAW

CORTESE, GREGORIO (1483-1548). Benedictine prior and cardinal. Baptized Joannes Andrea, he shared in many important Catholic reforming attempts. Having been secretary to the future Leo X and then canon of the Modenese cathedral after 1503, he entered the monastery of Polirone near Mantua in 1507. His vast patristic learning was used for personal reform. In 1513 he urged Leo X to restore Christian morality. At the island monastery of Lérins (1516-24) he also studied Protestant writings, since Sadoleto* mentions on a visit in 1518 and again in 1523 the Lutheran works in Cortese's possession. In 1522 he used all this to refute the Bohemian writer Ulrich Velenus, who in 1519 published a treatise denying that Peter was ever in Rome. Cortese's response, dedicated to Adrian VI, defended papal authority by answering Velenus's eighteen "persuasions." In so doing he cited Epiphanius's *Against Heresies* long before the manuscript in Melanchthon's possession became the *editio princeps.* Conrad Gesner listed Cortese's treatise as a manuscript in a Roman library before its first printing in 1573.

As abbot of Lérins and then at Modena and Perugia (1529), Cortese became a monastic examiner for the Cassinese congregations. In 1532 as abbot of San Giorgio Maggiore, Venice, he took part in a brilliant study circle with Pole,* Contarini, and other scholars. In 1537 he joined these and six others on Paul III's reform commission. After drafting the famous *Consilium de Emendanda Ecclesia* he returned to Polirone as abbot. In 1542 Paul III made him a cardinal, partly to help stem heresy in Modena. As an intimate of Catholic reforming cardinals, staunch defender of the papacy, and learned scholar, he looked for "a beautiful and new form of the holy Church." He died before his monastic simplicity could be corrupted by the Baroque papacy.

MARVIN W. ANDERSON

COSIN, JOHN (1594-1672). Bishop of Durham. Born at Norwich, he studied at Caius College, Cambridge, took holy orders, and became prebendary of Durham (1625) and rector of Elwick and Brancepeth (1626). At the request of Charles I he compiled his *Collection of Private Devotions* (1627), which included his fine translation of *Veni Creator Spiritus* ("Come, Holy Ghost"). He became master of Peterhouse (1635) and dean of Peterborough (1640), but was deprived of all his benefices by the Long Parliament. He became chaplain of the Anglican royalists in Paris (1642), and befriended Huguenots and disputed with Roman Catholics, writing against transubstantiation. He became bishop of Durham (1660) and attend-

ed the Savoy Conference* (1661).
J.G.G. NORMAN

COSMOLOGICAL ARGUMENT. Broadly, any argument for the existence of God that proceeds from some feature or features of the world to God. More narrowly, the term "cosmological" is reserved for a group of arguments postulating the existence of God as the explanation of an otherwise inexplicable universe, of "why there should be any world rather than none, and why it should be such as it is" (Leibnitz). The first three of Thomas Aquinas's* Five Ways are versions of the cosmological argument. God is thought of as the Unconditioned, the Unmoved Mover, the Sufficient Reason for all that is. Apart from general theological or philosophical objections to proofs for God's existence, the cosmological argument requires a number of doubtful premises—e.g., that for something to change it must be changed, and that there cannot be an infinite regress of such changes. PAUL HELM

COSTA, ISAAK DA, see DA COSTA, ISAAK

COTTON, JOHN (1584-1652). Puritan minister and author. Graduate of Cambridge University, he was ordained in 1610, and from 1612 to 1633 he was vicar of Boston, Lincolnshire, before migrating to the Massachusetts Bay Colony where he assumed the pastorate of a church in Boston. A gifted theologian, he quickly became a dominant figure in the colony, where theological prowess often brought political influence. One of his opponents noted that some people in Massachusetts "could hardly believe that God would suffer Mr. Cotton to err." He engaged in stormy controversies with Roger Williams* and Anne Hutchinson,* both of whom were driven out of the colony. Moreover, although personally not democratic, he helped create the pattern of church government for New England by his book, *The Way of the Churches of Christ in New England* (1645). JOHN D. WOODBRIDGE

COUGHLAN, LAWRENCE (? - ?). Pioneer preacher in Newfoundland. An itinerant Wesleyan preacher in Ireland (1755-65), he went to Newfoundland in 1765 under the auspices of the Society for the Propagation of the Gospel. Arriving in Harbour Grace, he proceeded to establish a congregation. On the recommendation of John Wesley and the Countess of Huntingdon, Coughlan was ordained by the bishop of London in 1767. A revival broke out after his return to Newfoundland, but he was soon in trouble with the more influential people of the area because of his preaching against conditions in the colony. Incessant opposition, long difficult journeys, and an aversion to sea travel broke Coughlan's health. He returned to England in 1773, leaving the work in the hands of two merchant converts.
ROBERT WILSON

COUNTER-REFORMATION. The movement for reform and missionary expansion within the Roman Catholic Church in the sixteenth and seventeenth centuries, that was quickened, if not caused, by the Protestant Reformation.* It found expression in a variety of forms. One time-honored way to reform the church was to renew or reform monastic orders. Recognized by the pope in 1528, the Capuchins,* who sought to recover the ideals of Francis of Assisi, devoted themselves to charitable work and evangelism. Entirely new creations, indicative of the changing times, were the Theatines* (1524), Somaschi (1532), Barnabites* (1533), Ursulines* (1535), and Oratorians* founded by Philip Neri (1575). They sought to show that the old ideals of celibacy, chastity, self-sacrifice, and compassionate service were still practicable in the sixteenth century. The most important order to be founded, however, was the Jesuits,* established by a papal bull in 1540 but formed in Rome eighteen months earlier by Ignatius Loyola,* Francis Xavier,* and others. The "Company of Jesus" was intended to be a society of priests who ministered to the poor, educated boys, and evangelized the heathen. It certainly did these things, but it also proved to be a most powerful anti-Protestant force, counting among its theologians Robert Bellarmine* and Peter Canisius.*

Another traditional approach to reform was through a general council. Emperor Charles V wanted such a council in Germany, but the Vatican opposed this. Some Catholics, led by such men as Cardinal Contarini, wished to conciliate and win Protestants through dialogue (cf. Colloquy of Ratisbon,* 1541) and an ecumenical council, but it was the conservative element in the church that triumphed. This bore fruit in the Roman Inquisition from July 1542 and the Council of Trent* (1545-63). The latter was indirectly under the control of the papacy and had no intention of making concessions to Protestantism in its doctrinal declarations. Its disciplinary decrees were intended to reform the structure of the church, and included the establishment of seminaries in every diocese in order to improve the standard of the clergy.

With Pius V (1565-72), a period of internal reform began within the Roman Curia and Vatican. Militantly anti-Protestant, Pius also issued edicts against simony, blasphemy, sodomy, and concubinage in his own church. In 1568 he reformed the breviary, restoring the reading of Scripture to a dominant place. The devotional power of the Reformation was, however, reflected more in personal religion than in liturgical reform. Apart from the general increase in personal confessions and communions, this was the age not only of the mystics—Teresa of Avila,* the Carmelite, and Juan de Yepes known as John of the Cross,* but also of Francis of Sales,* author of *Introduction to the Devout Life* (1609), which related piety to real life situations outside the monastery.

The term "counter-reformation" is applied also in a political sense to the revival of the Catholic powers of Europe. This lasted from about 1562 to 1629, at a time when France was internally weak and the Hapsburg powers had a free hand for their foreign policies. Encouraged by Pius V, a league of Catholic princes which included Philip II of Spain existed to defend the church and destroy Protestantism. Though the Spanish Armada failed to capture England, success was registered in Europe—in Poland, for example. The Thirty Years' War* was from 1618 to 1635 a religious war, with Calvinists fighting Protestants. It was in this war that the Catholics achieved a great success in driving Hussitism and Protestantism from Bohemia and forcibly making that land adopt the old religion.

Some scholars would see the Counter-Reformation as beginning with such men as Jiménes* and Savonarola,* gaining impetus in the sixteenth century and continuing, with different degrees of success, through the centuries until today. Vatican II* is thus seen as a part of the whole movement for renewal.

BIBLIOGRAPHY: A.W. Ward, *The Counter-Reformation* (1889); B.J. Kidd, *The Counter-Reformation, 1550-1600* (1933); P. Janelle, *The Catholic Reformation* (1949); H. Daniel-Rops, *The Catholic Reformation* (1962); see also *Cambridge Modern History*, vol. II, chap. 9, etc.

PETER TOON

COUNTESS OF HUNTINGDON'S CONNEXION. A body founded by Selina Hastings, countess of Huntingdon (1707-91). After her conversion she was briefly a Moravian before joining the Methodists in 1739. An early supporter of itinerant lay preaching, she became on her husband's death (1746) a prominent figure in the Evangelical Revival. More Calvinist than John Wesley, she retained her Anglican links and appointed Evangelical Anglicans as her chaplains (among them was George Whitefield*). Aiming to reach the upper classes particularly, she opened chapels at Brighton, Tunbridge Wells, Bath, and London, where Evangelicalism was combined with a liturgical form of service. In 1768 she established at her own expense a college in South Wales for the training of Evangelical clergy. Her ecumenical tendencies received a setback when a legal decision in 1779 forced her to register her chapels as dissenting meetinghouses and led to the resignation of her Anglican chaplains.

The countess helped to sponsor Whitefield's orphanage in Georgia, took an interest in the American Indians (about whom she corresponded with George Washington, a distant relative), and encouraged the beginnings of Dartmouth College and Princeton University in the USA. Her college was moved to Hertfordshire in 1792 and to Cambridge in 1904. Her chapels, which organize an annual conference, hold membership in the British Council of Churches. There are some thirty-six chapels at present, but it is doubtful if membership exceeds one thousand.

See *Life and Times of the Countess of Huntingdon* (2 vols., 1844); and S.C. Carpenter, *Eighteenth-Century Church and People* (1959).

J.D. DOUGLAS

COURT, ANTOINE (1696-1760). Minister of the French Reformed Church. He succeeded in reorganizing his fellow Protestants after their church had been broken and scattered by the revocation of the Edict of Nantes* (1685). Fanaticism, disorder, and apostasy were strong, and Court began to preach and organize, insisting on order and restraint. In 1715 he called together at Monobet the first provincial synod of the Reformed Church of France since 1685, whose wise and responsible decisions did much to stabilize Protestant attitudes. In 1718 he was ordained. Other synods were arranged and attracted increased persecution. Court continued to counsel restraint and withdrew several times to Switzerland where, in Lausanne, he founded and directed a seminary for the training of ministers. He wrote *Histoire des troubles des Cevennes ou de la guerredes Camisards* (1760) and left considerable material in manuscript now in the public library in Geneva.

HOWARD SAINSBURY

COURTENAY, WILLIAM (c.1342-1396). Archbishop of Canterbury from 1381. Son of the earl of Devon, he studied law at Oxford and in 1367 became university chancellor. In 1370 he became bishop of Hereford, in 1375 bishop of London, before becoming primate. He consistently opposed John of Gaunt's policies, proceeded against John Wycliffe* for heresy (1377), held a synod in 1382 to condemn Lollardy, and urged bishops to imprison heretics. During 1381 he was briefly chancellor of England, and one of the commissioners appointed to reform the kingdom and the royal household in 1386. He opposed the Statute of Provisors* (1390), and succeeded in slightly modifying the Statute of Praemunire* (1393).

J.G.G. NORMAN

COUSTANT, PIERRE (1654-1721). Roman Catholic scholar. Educated by the Jesuits at Compiègne, his birthplace, he studied at Maurist* houses in Reims and Soissons before going to Paris. After assisting with the great Maurist edition of Augustine, he was entrusted with an edition of Hilary of Poitiers* in 1687, after which he became prior at Nogent-sous-Courcy in 1693. Returning to Paris in 1696, he did his main work, an edition of papal letters, of which only those up to 440 had appeared by his death.

COVEL, JOHN (1638-1722). Also, "John Colvill." Anglican scholar. Educated at the grammar school at Bury St. Edmunds and graduated from Cambridge in 1658, he was for the next decade chaplain at the British Embassy in Constantinople. At the request of some of his colleagues he investigated the Greek Church, especially to see whether it held the doctrine of transubstantiation. In 1681 he was appointed chaplain to the Princess of Orange, but the discovery of a letter criticizing the prince led to his dismissal. In 1687 he was elected master of Christ's College, Cambridge. Though William III was friendly when he came to Cambridge and Covel was vice-chancellor, saying that he could distinguish "between Dr. Covel and the Vice-Chancellor," nevertheless the earlier incident probably prevented further preferment. His account of the Greek Church was not published until 1722, and by that time interest in the subject had waned.

PETER S. DAWES

COVENANT, NATIONAL, see NATIONAL COVENANT

COVENANTERS. A name applied particularly to Scottish Presbyterians who signed the National Covenant* of 1638 and the Solemn League and Covenant* of 1643, and to their successors who during the reigns of Charles II and James VII resisted the episcopal system forced upon Scotland. Initially resistance came when Charles I* and William Laud* tried to impose a new liturgy on the Scots (1637), continued with a Scots Covenanter/English parliamentary alliance, and ended with the defeat and subsequent execution of the king. This outcome horrified the Covenanters, who had a high view of the kingship, but did not prevent their compelling the young Charles II to assent to both covenants in order to obtain the Scottish crown. George Gillespie* and Samuel Rutherford* were contemporary Scottish writers who opposed the theory of the Divine Right of Kings,* holding that limitless sovereignty pertains to God alone.

Such Covenanting views brought trouble when, after the eight-year interlude of Commonwealth and Protectorate, the Stuart dynasty was restored to the united kingdom. Presbyterianism was outlawed and replaced by Episcopacy, the Covenants were denounced as illegal, and with the execution of James Guthrie* and Archibald Campbell,* marquis of Argyle, began a savage repression of the dissentients. Seeing the issue as obedience to God or to king, Covenanters became rebels. Many resorted to field preaching,* and were hunted, jailed, killed (sometimes without trial), or banished to Holland or America. Even moderate Covenanters like Robert Baillie* were eventually driven to admit that their more extreme colleagues had been right in their distrust of Charles II. In desperation at the merciless persecution, some of the Covenanters were led from justifying rebellion to justifying assassination, and this was the fate of Archbishop James Sharp* (1679), one of their chief adversaries. Others grew weary of strife and accepted the governmental Declarations of Indulgence.* Perhaps nothing justifies the Covenanters more than that their Sanquhar Declaration* (1680) included reasoning remarkably similar to that used a decade later when the country as a whole rejected the royal House of Stuart. Presbyterianism was restored to Scotland by William III in 1690, but a remnant, the Cameronians,* objected to him as an uncovenanted king and refused to rejoin the national church of Scotland. Their successors are still to be found not only in Scotland, but in Northern Ireland and in North America.

BIBLIOGRAPHY: W.L. Mathieson, *Politics and Religion in Scotland . . . from the Reformation to the Revolution* (2 vols., 1902); J.K. Hewison, *The Covenanters* (2 vols., 1908); A. Smellie, *Men of the Covenant* (1908); H. Macpherson, *The Covenanters Under Persecution* (1923); J.D. Douglas, *Light in the North: The Story of the Scottish Covenanters* (1964).

J.D. DOUGLAS

COVENANT THEOLOGY. Sometimes called "Federal Theology," this system describes the relationship between God and man in the form of covenants. One of the features in the development of Calvinism, it was especially popular with Puritans and the Reformed theologians of Germany and Holland in the latter sixteenth and during the seventeenth century. It holds that God entered into an agreement with Adam at creation, promising him eternal life if he would obey the divine commands. Adam failed by eating the forbidden fruit and thus plunged himself and his descendants into eternal death. To remedy this, God (eternally) entered into a second agreement with Christ on behalf of the elect, promising them forgiveness and eternal life on the basis of Christ's sacrifice. The elect may have assurance of salvation with all its attendant blessing because of their faith in Christ.

This teaching helped Calvinists reconcile the sovereignty of God with man's desire for assurance. The covenant, or federal, theologians believed that man as a sinner has no right before a holy, sovereign, omnipotent God. Man ought to be perfectly obedient to the will of God, but even then there is no reward to be earned. Fellowship with God must come through a voluntary divine agreement establishing a relationship which is not necessarily according to nature. This was done by the covenant, which caused God to act in a kindly way, thus removing the uncertainty from dealing with the Almighty.

Covenant theology in a strict sense began in Germany when a number of Calvinists such as Kaspar Olevianus* and Zacharias Ursinus* emphasized the idea of the covenant of God with man and the believer's mystic union with Christ. Parallel with this German movement was the British development of covenant theology which was sometimes related to political thought (see NATIONAL COVENANT and SOLEMN LEAGUE AND COVENANT). William Ames* became the leading British exponent of federal theology, which in a moderate form appears in both the Westminster Confession* and the Savoy Declaration.* Debtor to both British and German schools, John Cocceius* published a book, *Summa Doctrinae de Foedere et Testamento Dei* (1648), which has the most elaborate explanation of the covenant principle produced to that time.

Interest in covenant theology continued into the eighteenth and nineteenth centuries, but in a much diminished form. The Princeton theologians Charles and A.A. Hodge* gave it much attention, and federal theology still occupies a central position in reformed doctrine.

BIBLIOGRAPHY: W.A. Brown, "Covenant Theology," *Encyclopedia of Religion and Ethics* (ed. J. Hastings), vol. IV, pp. 216ff.; P.Y. DeJong, *The Covenant Idea in New England Theology* (1945); P. Miller, *The New England Mind, The Seventeenth Century* (1961), pp. 365ff.; P. Toon, *The Emergence of Hyper-Calvinism in English Nonconformity* (1967), chap. 1.

ROBERT G. CLOUSE

COVERDALE, MILES (1488-1569). Bible translator. A native of York and graduate of Cambridge, he was ordained in 1514 and became an Augustinian Friar. Under the influence of his prior, Robert Barnes, he embraced Lutheran teaching. Abandoning his order (1528), he preached against the Mass, images, and confession until forced to flee the country. As a translator, Coverdale served his apprenticeship helping William Tyndale* revise his translation of the Pentateuch. In 1535 Coverdale's version of the Bible appeared, the first printed English Bible, for which he utilized Tyndale's work, supplemented by Latin and German versions (including Luther's). He enjoyed the patronage of Thomas Cromwell who commissioned him to revise Matthew's Bible,* a task which was completed in 1539 with the publication of the Great Bible. From 1540 he was again in exile, engaged in literary work in Strasbourg, taking a doctorate in divinity at Tübingen, and serving as pastor and schoolmaster in Bergzaben. Once more in England under Edward VI, he excelled as a preacher, and became bishop of Exeter (1551). Though at first imprisoned under Mary, he was allowed to go into exile, during which he spent a year in Geneva, where he probably worked on the Geneva Bible.* He returned to England in 1559, but played little part in public life.

See J.F. Mozley, *Coverdale and his Bibles* (1953). HAROLD H. ROWDON

COWL. A garment with a hood, *vestis caputiata,* traditionally worn by monks (whence the title "Capuchins"* of one of the orders of Franciscan Friars). While two orders of monks wear only a cowl unattached, the others from the time of Benedict take the cowl to include the cloak and mantle together with the hood.

COWPER, WILLIAM (1731-1800). English poet. Educated at Westminster School and thereafter articled to a solicitor, he suffered throughout his life from fits of depression which more than once developed into mania. From 1765 he was cared for by Mary Unwin, who until her death in 1796 did much to keep Cowper serene and happy. At her suggestion he wrote his generally mild satiric poems ("Table Talk," "The Progress of Error," "Truth," "Expostulation," "Hope," "Charity," "Conversation," and "Retirement"), published in 1782. This work was followed by the discursive poem, *The Task* (1784), and Cowper later translated Homer (1791). His earliest publication, however, was the collection of hymns written in collaboration with John Newton,* under the title *Olney Hymns* (1779). Cowper's contributions were written mainly around 1771-72 and included "O for a closer walk with God," "God moves in a mysterious way," and "Hark, my soul! it is the Lord," as well as the controversial but at the same time deeply moving "There is a fountain fill'd with blood."

The 1782 poems are more didactic than satiric and, allowing for a certain narrowness, they provide a good poetical survey of Evangelical doctrines. Lines 258-278 of "Truth," for instance, give a masterly summary of the experience of conviction. The narrowness expresses itself in a negative asceticism, what Norman Nicholson has called an "instinctive distrust of certain types of pleasure" and an occasional but also virulent anti-intellectualism. The most moving expressions of his faith, however, are those which relate to his intense feeling of his own predetermined damnation, whether it be in the passage in *The Task* (III. 108ff.), describing himself as "a stricken deer," or in his last terrible poem "The Castaway."

See Norman Nicholson, *William Cowper* (1951). ARTHUR POLLARD

COX, FRANCES ELIZABETH (1812-1897). Translator of German hymns. Little is known about her life. Daughter of George V. Cox, she apparently spent all her days in Oxford. She ranks next to John Wesley* and Catherine Winkworth* in the quality of her translations of German hymns and in the extent to which they are still sung. They first appeared in her *Sacred Hymns from the German* (1841), a collection of about fifty items. In her second edition, entitled *Hymns from the German* (1864), she dropped twenty-two, modified others, and added twenty-nine new items. Her most popular translations are "Jesus lives! no longer now" (Gellert); "Sing praise to God who reigns above" (Schütz); and "Who are these like stars appearing" (Schenck).

JOHN S. ANDREWS

COX, RICHARD (c.1500-1581). Bishop of Ely. Becoming convinced of Reformed views while at Cambridge, he was appointed in 1540 to the commission which composed *The Necessary Doctrine and Erudition of a Christian Man,* and from 1544 was tutor to Prince Edward. He became the first dean of Christ Church, Oxford (1547), and dean of Westminster (1549). He helped to compile the "Order of the Communion" of 1548 and the Prayer Books of 1549 and 1552. As chancellor of Oxford University (1547-52) he was responsible for introducing Peter Martyr, Stumphius, and John ab Ulmis to the university. On Mary's accession he was imprisoned and deprived, going into exile at Frankfurt in 1555. In the English congregation there, Cox wanted to maintain Edward VI's Prayer Book, while John Knox* wanted a more thorough reformation. The disputes between them gave rise to the names Coxians and Knoxians. Under Elizabeth, Cox was appointed bishop of Ely (1559-80), and translated Acts and Romans for the Bishops' Bible.* He was severe toward both Roman Catholics and Puritans, but refused to minister in the queen's chapel because of its candles and crucifix, and becoming disgusted with the court he asked to resign his see. He died a year later.

See various letters in *The Zurich Letters* (ed. H. Robinson, 1842); and J. Strype, *Annals of the Reformation* (1824). JOYCE HORN

CRABBE, GEORGE (1754-1832). English poet. Born at Aldeburgh in Suffolk, much of his poetry (e.g., *The Borough*) is set in this area. He was successively rector of Muston (Leicestershire) and Trowbridge (Wiltshire). His first major poem, *The Village* (1783), is an exposure of social conditions in a small seaside community. *The Parish Register* (1807) is a series of character sketches based on his experience at Muston and elsewhere. *The Borough* (1810), *Tales* (1812), and *Tales of the Hall* (1819) are virtually short stories in verse. There is a strong underlying moral comment in his skillful analyses of character and action, and this is found also in his sermons (1850).

See *Life* by his son (1834), and L. Haddakin, *The Poetry of Crabbe* (1955).

ARTHUR POLLARD

CRAIG, JOHN (1512-1600). Scottish Reformer. He was a Dominican Friar who became one of the outstanding personalities of the Reformation in Scotland, doing much to shape the future policy of the national church. Imprisoned for heresy in 1536, he escaped to become rector of the Dominican convent in Bologne, where his conversion to Protestantism had taken place. Condemned to death by the Inquisition in Rome, he reached Vienna after a series of dramatic escapes to become the favorite preacher of Emperor Maximilian II. Returning to Scotland in 1560, he joined John Knox* as collegiate minister of St. Giles', Edinburgh, two years later. In 1570 he became chaplain to James VI, drafting the first Scots Catechism and being largely responsible for the King's Confession, or National Covenant,* of 1581. D.P. THOMSON

CRAKANTHORPE, RICHARD (1567-1624). Anglican scholar. Born in Westmoreland, he was educated at Queen's College, Oxford, where he also became a fellow. Coming under the influence of John Reynolds (Rainolds*), he became a zealous Puritan. He was, however, known as a learned man and was chosen to accompany, as chaplain, Lord Evers, the ambassador of James I to the court of Emperor Rudolph II. Later he served as a chaplain both to the bishop of London and to the king. In 1605 he became rector of Black Notley, Essex, and in 1617 rector of Paglesham, Essex. He engaged in controversy with Mark Antonio de Dominis, archbishop of Spalato, who claimed to be a convert to the Protestant Church of England. Crakanthorpe's most famous book, *Defensio Ecclesiae Anglicanae* (1625), was an answer to the defense of the retraction *(Consilium Reditus)* of de Dominis after he had returned to Roman Catholicism. PETER TOON

CRAMP, JOHN MOCKETT (1791-1881). Baptist pastor and scholar. Educated at Stepney College, London, he was in 1818 ordained pastor of Dean Street Baptist Chapel, Southwark. From 1827 to 1842 he assisted his father at the Baptist Church in St. Peter's, Isle of Thanet, and then served in Hastings. In 1844 he went to Canada to become president of Montreal Baptist College. In 1851 he became president of Acadia College, Nova Scotia. He resigned in 1869 to devote himself to literary work, which included writing theological and historical books, and editing magazines and a newspaper. Among his many books the most noteworthy was his *Baptist History from the Foundation of the Christian Church to the Present Time* (1868). J.G.G. NORMAN

CRANACH, LUCAS (1472-1553). German painter. For more than forty years he was the prosperous, leading official painter at the court of Saxony. Several times mayor of Wittenberg, patronized by the powerful, worldly Cardinal Albrecht von Hohenzollern, simultaneously a friend of Luther and godfather of Luther's first son—Cranach set up a workshop of a dozen or so journeymen artists who could paint in the wavy, serpentine Cranach manner to keep up with commissions to decorate with "scenes from antiquity" the town hall ceiling, a duke's hunting lodge, new wedding quarters in some castle, or, later on, mass-produce portraits of Luther and Melanchthon. Before he became painterly spokesman for courtly love taste, Lucas Cranach the Elder's style showed a late Gothic, precious, curlicue landscape fascination, alive with the innovation of a virtuoso—Christ's cross is not center but on the right side in a *Crucifixion* of 1503. As established pacesetter at the court, however, especially after Dürer's death, he painted the nudes which became his trademark. So close to the Reformation, Cranach's art did not become captivated by its reforming spirit. CALVIN SEERVELD

CRANMER, THOMAS (1489-1556). Archbishop of Canterbury from 1533. Born in Nottinghamshire, he was educated at Jesus College, Cambridge and became a fellow there. He was influenced by humanist and Lutheran opinions, and became strongly antipapalist. In 1529 Henry VIII heard Cranmer had suggested consulting the theologians at the universities on his "divorce," and employed him for this purpose as an ambassador in Europe. While in Germany in 1532 Cranmer married Margaret, niece of the Lutheran Reformer Osiander.* On the death of Warham he was consecrated archbishop of Canterbury with papal approval. He pronounced Katherine of Aragon's marriage null and void in the same year. In the years that followed Cranmer was able to bring about a moderate doctrinal reform, mirrored in the Ten Articles* and the Bishops' Book.* He supported Thomas Cromwell in securing an official English translation of the Bible, for which he wrote a preface. While not directly involved in the dissolution of the monasteries, he approved of it, but protested against the financial abuse involved.

A Catholic reaction which led to the fall of Cromwell was opposed by Cranmer with great courage. In the last years of Henry's reign and under his protection, Cranmer began the task of liturgical revision. In 1544 he produced the first of his vernacular services, the English litany. From 1547 onward under Edward VI he had a greater freedom to reform the liturgy.

In 1549 the Communion service of his first Book of Common Prayer* embodied his recently adopted receptionist view in the framework of the Latin Mass. The reactions of English Catholics like Gardiner and of continental Protestants such as Bucer* and Martyr were the occasion for producing a second Book of Common Prayer in 1552. The Communion service in this broke away from the Latin Mass entirely. The climax of the service was now the receiving of the bread and wine, while kneeling around a table. Cranmer defended his new understanding of the Eucharist in his major theological work, *The True and Catholic Doctrine of the Lord's Supper.* He published a book of Homilies, a confession of faith in forty-two articles, and a revision of canon law, *Refor-*

matio Legum Ecclesiasticarum. He fought a losing battle against the increasing inroads on the finances of the church made in the name of reform. He was involved in the plot to make Lady Jane Grey queen only after strong persuasion from the dying Edward VI.

When Mary came to the throne in 1553, Cranmer was condemned to death for treason, but the sentence was not carried out. Under the renewed heresy laws of 1555 he was tried at Oxford and was convicted and degraded. He was forced to watch the burning of Latimer* and Ridley.* After much pressure he signed a number of recantations through fear of suffering and through loyalty to the royal supremacy. On the eve of his execution his courage returned, and he went to the stake on 21 March 1556, denying his recantations and suffering for his faith.

BIBLIOGRAPHY: C.H. Smyth, *Cranmer and the Reformation Under Edward VI* (1926); T.M. Parker, *The English Reformation to 1558* (1950); J. Ridley, *Thomas Cranmer* (1962); G.E. Duffield (ed.), *The Work of Thomas Cranmer* (1964).

NOEL S. POLLARD

CRASHAW, RICHARD (1612-1649). English poet. After some years as fellow of the then High Church college, Peterhouse, Cambridge, he entered the Roman Catholic Church and spent his last years at the court of Charles I's queen, Henrietta Maria, in Paris and then as subcanon of Loretto where he died. His *Steps to the Temple* (1646) links him with the group known as the Metaphysical poets (see DONNE and HERBERT), but his work differs from theirs in its more exotic and emotional coloring. Crashaw was influenced by the Italian Marino in poetry and the cult of St. Teresa in religion. His work is expressive of intense devotion, but it is often marred by extravagance of language and imagery.

ARTHUR POLLARD

CRAWAR, PAUL (c.1390-1433). Hussite martyr. Probably born in Kravare in Moravia, he studied medicine at Montpellier, then entered the University of Paris (M.A., 1415). He then returned to Prague where he taught medicine, becoming about 1421 physician to Wladislaw Jagiello, king of Poland. By this time he had become a radical Hussite, writing two books against the papacy: *Sermones de Antichristi* and *De Anatomia Antichristi.* After ten years in the service of the Polish king, he returned to Prague, perhaps as a result of persecution for his religious beliefs. He then left, possibly having met Scots traders in Poland, to work with the Lollards* in Scotland. He was arrested, however, and burned at the stake in St. Andrews.

W.S. REID

CRAWFORD, DANIEL (1870-1926). Missionary to central Africa. Born in the Scottish town of Gourock, he joined the Plymouth Brethren in 1887 and two years later accompanied F.S. Arnot* to central Africa. Traveling alone, he reached Katanga in 1890. After the murder of Chief Msiri a year later, he itinerated widely, settling in 1895 at Luanza on Lake Mweru. Apart from a visit overseas (1911-15) he remained there until his death. Crawford was a strong individualist who relied for support upon unsolicited gifts and preferred to work alone. In his view, preaching and individual conversion were all-important. He itinerated frequently with little regard for health and safety, and was averse from institutionalized mission stations. Education in his village Bible schools concentrated upon the Scriptures, which he himself translated into Luba (NT in 1904; OT in 1926). His converts were encouraged to participate in teaching, preaching, and church affairs.

D.G.L. CRAGG

CREATIONISM. Theologically the term has been variously interpreted:

(1) The doctrine that the universe was created by God out of nothing, or (in recent years) that the world was created a few thousand years ago, in opposition to the view that it is several thousand million years old.

(2) The doctrine that species were created, in opposition to the evolutionary view.

(3) In Catholic theology, the doctrine (held by Jerome, Hilary, Aquinas, and others, as well as Calvin), that God creates from nothing each new soul that comes into the world, in contrast with the Traducian view (held by Tertullian, Luther, and others) that souls are formed naturally as the body develops, or that they are reincarnated after previous existences on earth (metempsychosis). Augustine suspected that Creationism and original sin were incompatible, for a new creation by God cannot be tainted with sin. Nevertheless, orthodox Catholicism holds to Creationism, Aquinas insisting that its denial is heretical. In medieval theology the soul's creation occurred on the fortieth day after conception for a male, on the eightieth for a female.

R.E.D. CLARK

CREED. "A concise formal and authorized statement of important points of basic Christian doctrine." The word comes from the Latin *Credo* ("I believe"), since the statement of faith involves not merely acceptance of truth, but personal commitment—*Credo in Deum,* "I believe in God." The creed was also termed the *regula fidei* as equivalent of the Eastern *kanōn tēs alētheias,* the standard of faith, *kanōn* being a builder's square. *Symbolum,* the military "pass," or *tessara* was also used, whereby the faithful would be known to each other throughout the world as against the heretics. Creeds were both the earliest development of the formal faith of the church, and the first and most authentic form of oral tradition, and are likely to have grown out of the rudimentary forms of confession we find in the NT—e.g., Rom. 10:9; 1 Cor. 12:3—not only at baptism but also in worship and instruction. (Both Trinitarian and purely christological forms are in use in the NT.) By the fourth century these confessions had become uniform, and the Apostles' Creed* in the West and the Nicene Creed* in the East became the only baptismal confessions in use.

G.S.R. COX

CREIGHTON, MANDELL (1843-1901). Bishop of London, and historian. Born at Carlisle, and educated at the universities of Durham and Oxford, he taught history at Oxford, was ordained in 1873, and in 1875 became vicar of Embleton in Northumberland. Nine years later he was appointed Dixie professor of ecclesiastical history at Cambridge, and in 1891 was consecrated bishop of Peterborough. In 1897 he was translated to the see of London, where the popular lecturer and erudite scholar exercised a valuable moderating influence in controversies of the day, particularly in connection with the vexed question of ritualism. His many writings include a biography of Cardinal Wolsey (1888), but he is best known for his (unfinished) *History of the Papacy* (5 vols., 1882-94). J.D. DOUGLAS

CREMATION. The reduction of human remains to ash. It was widely practiced in the ancient world, except in Egypt and China. Cremation among the ancient Israelites was exceptional (e.g., 1 Sam 31:12). The Etruscans and the Greeks passed on the practice to the Romans, with whom it became the fashionable means of disposal among the aristocracy. Though the early Christians did not fear cremation, they preferred to follow the burial customs of the Jews. The growth of the Christian doctrine of the resurrection of the physical body was largely responsible for its lack of popularity in Europe. Its modern development in Britain can be traced to 1874 (the founding of the Cremation Society), and in the USA to 1876. Since then it has grown in popularity. Statistics for 1960 reveal that one-third of those who died in the United Kingdom were cremated, and the proportion is increasing. The Roman Catholic Church still bans it as a regular means of disposal. JAMES TAYLOR

CREMER, HERMANN (1834-1903). German Protestant theologian. He became professor of theology at the University of Greifswald (1870), combining this with a city pastorate which he held until his death, refusing further preferment. He strongly resisted the liberalizing movement in theology. In his *Die paulinische Rechtfertigungslehre im Zusammenhange ihrer geschichtlichen Voraussetzungen* (1899), he reaffirmed a traditional interpretation of Paul's soteriology. In this work also, with what W. Eichrodt called "the insight of genius," he described "righteousness" as a "term of relation," being apparently among the earliest to do so. His best-known work was *Biblisch-theologisches Wörterbuch* (ET *Biblico-Theological Lexicon of New Testament Greek*, 1878), in which he theologically defined Greek words and terms used in the NT. J.G.G. NORMAN

CRISIS THEOLOGY, see DIALECTICAL THEOLOGY

CRITOPULOS, METROPHANES, see METROPHANES CRITOPULOS

CROMWELL, OLIVER (1599-1658). Lord Protector. Born near Huntingdon of a lesser English landowning family, he was educated at the local Free School and at Sidney Sussex College, Cambridge. Elected to represent Huntingdon in the 1628 parliament, he spent the following eleven years (when Charles I* ruled without Parliament) occupied with East Anglian concerns and came to a strong personal Christian faith. He championed the rights of commoners against enclosures and Fen drainage, supported Puritan lecturers, and may have contemplated emigrating to New England. As member for Cambridge in the 1640 Parliament, he moved the second reading of the bill for annual parliaments, spoke for the abolition of episcopacy, and served on numerous committees.

On the outbreak of war between Parliament and king, he raised a cavalry troop of "godly, honest men" to fight under Essex at Edgehill (1642). Though aged forty-three at his first battle, he became one of the great cavalry leaders of history. In 1643 he enlarged his troop to a regiment and secured the Eastern counties for Parliament. He commanded the left wing in the victory at Marston Moor (1644). With Fairfax, he remodeled the parliamentary forces into the New Model Army, a disciplined, professional force of men "who know what they fight for and love what they know," which won at Naseby (1645). Cromwell supported the soldiers in their grievances against Parliament over arrears of pay.

He tried to make terms with Charles, who endeavored to play off army, Parliament, and Scots against each other. After defeating the Scots at Preston (1648), Cromwell supported the execution of Charles I as a "cruel necessity." He subdued the Royalist rebellion in Ireland, 1649-50, justifying his ruthlessness on the grounds that it would "prevent the effusion of blood" in the future, but he was clement to noncombatants. Made commander-in-chief of the army, he defeated the Scots at Dunbar (1650) and Worcester (1651)—his "crowning mercy." Anxious for Parliament to rule, he nevertheless dissolved the "Rump" of the Long Parliament in 1652, when members tried corruptly to perpetuate their own tenure of office. A new Parliament also proved incompetent.

Cromwell was made Lord Protector in 1653 and ruled by ordinances confirmed later by Parliament. He reorganized the Church of England, trying to provide faithful preachers in every church. He protected Quakers and Jews. His foreign policy raised England's standing in Europe, he acquired Dunkirk, and he championed the persecuted Vaudois Protestants. He refused the title of king, was buried in Westminster Abbey, but was disinterred in 1661. Without personal ambition, and motivated by Christian belief, he ensured that England would be ruled by Parliament and not absolute kings, though much of his work seemed overthrown at the Restoration.*

BIBLIOGRAPHY: W.C. Abbott (ed.), *The Writings and Speeches of Oliver Cromwell* (4 vols., 1937-47); C. Hill, *God's Englishman* (1970); biographies by C. Firth (1900) and C.V. Wedgwood (1973). JOYCE HORN

CROMWELL, THOMAS (c.1485-1540). English statesman. In early life he traveled widely, and his varied adventures produced a mind essentially functional and secular, and with considerable business and legal acumen. From 1520 he served under Wolsey* and survived his master's disgrace to enter Parliament in 1529. There he quickly made his mark and, though debate surrounds his precise contribution to Henry VIII's policies, it is clear that from 1532, when first he held high office, there is an ordered, creative intelligence not previously evident. The Henrician Statutes bear his stamp and especially the Act in Restraint of Appeals (1533), which is a concise statement of Erastian* principles held by Cromwell and culled from Marsiglio of Padua. By 1535 Cromwell was vicar-general and thus the effective controller of the church. The *Valor Ecclesiasticus*, which attempts to estimate the income of every cleric, is a typical example of his administrative genius. He maneuvered the Dissolution of the Monasteries. He was created earl of Essex in 1540.

Cromwell's religious policies demonstrate a sympathy for Lutheran ideas. In particular he encouraged the translation of the Bible leading to the publishing of the Great Bible in 1539. His Injunctions of 1536 and 1538 reveal his dislike of superstitious practices and the importance of providing a Bible in every church. He stirred up conservative opposition, especially because of his Protestant leanings. His attempt to establish an alliance with Lutheran princes, by way of marriage, was unsatisfactory to the king in terms of foreign policy—the more so because of the unattractiveness of Anne of Cleves. The result was Cromwell's condemnation in June 1540 under the Act of Attainder for heresy and treason. He was beheaded the following month.

BIBLIOGRAPHY: G.R. Elton, "Thomas Cromwell's Decline and Fall," *Cambridge Historical Journal*, X (1951); *idem*, "King or Minister? The Man Behind the Henrician Reformation," *History* 39 (1956); *idem*, "The Political Creed of Thomas Cromwell," *Transactions of the Royal Historical Society*, 5th series, VI (1956); A.G. Dickens, *Thomas Cromwell and the English Reformation* (1959).

C. PETER WILLIAMS

CROSBY, FANNY (Mrs. F.J. Van Alstyne) (1823-1915). American hymnwriter. Born in Putnam County, New York, she lost her sight when six weeks old. At the age of eleven she entered the New York City Institution for the Blind, remaining there for twenty-three years as pupil and teacher. While at school she began writing poetry (published first in 1831). She married in 1858 a blind musician, Alexander Van Alstyne, and together they produced a number of hymns. From 1864 she published more than 2,000, of which sixty are still commonly used. Her association with Ira D. Sankey* and W.H. Doane yielded many of her most popular compositions, notably "Safe in the arms of Jesus," and "To God be the glory."

J.G.G. NORMAN

CROSBY, HOWARD (1826-1891). Presbyterian scholar. Graduating from New York University, he held the chair of Greek there from 1851, and in Rutgers University (named for his great-uncle) from 1859. He then read theology, was ordained, and later became minister of Fourth Avenue Presbyterian Church, New York. He was a member of the New Testament Committee for the American edition of the Revised Version (1872-81), was moderator of the general assembly (1873) and its delegate to the Pan-Presbyterian council, Edinburgh (1877), and founded a society for crime prevention (1877). His writings include commentaries on Joshua, Nehemiah, and the entire NT; Yale lectures; other books and numerous pamphlets.

C.G. THORNE, JR.

CROSIER; CROZIER. The history of the use of these words is complicated. In current usage they refer to the crook-shaped staff carried primarily by bishops. The origin of the staff is not clear. Some think it derives from the rod of Roman augurs, but its symbolism is derived from the shepherd's crook. In the Eastern churches, the staff is surmounted by a cross between two serpents. Popular nineteenth-century uses of the words included "one who bears the cross or staff of a bishop or abbot" and "the cross of an archbishop."

CROSS, FRANK LESLIE (1900-1968). Anglican scholar and editor. Born in Devon and educated at Bournemouth School, he was elected to a scholarship at Balliol College, Oxford. After serving in World War I, he studied at Oxford and London and in Germany, part of the time under Rudolf Otto. He was appointed to the staff of Ripon Hall, Oxford, and was ordained (1925-26), his affiliations being modernist and Anglo-Catholic. By 1934-35 he was an established university lecturer in the philosophy of religion and in comparative religion. His overriding interests were the relations of science to religion, and the history of Christian thought, especially patristics. He was elected to the Lady Margaret professorship of divinity at Oxford in 1944. Cross is remembered for the breadth and accuracy of his scholarship. He was editor from 1941 of the *Lexicon of Patristic Greek*, and in later years conceived and edited the monumental *Oxford Dictionary of the Christian Church* (1957).

R.E.D. CLARK

CROWTHER, SAMUEL AJAYI (c.1806-1891). Anglican bishop. Born at Oshogun in Yorubaland, he was enslaved in 1821 and taken to Sierra Leone after liberation by the British navy. He was baptized in 1825 and entered the African Institution (later Fourah Bay College) in 1827. In 1841 he was a Church Missionary Society representative on the Niger expedition. His report so impressed the society that he was summoned to London for further training and ordained in 1843. For some ten years after 1846 he worked at Abeokuta, giving special attention to the Yoruba language. His most significant work was in the Niger Mission which he initiated in 1857 and led for thirty years.

In 1864 he was consecrated bishop of Western Africa beyond colonial limits, but as the European missionaries in Yorubaland would not accept his jurisdiction, he was virtually bishop of the Niger. The mission depended upon local financial support and could not attract well-qualified staff. Furthermore, poor communications made supervision difficult. Crowther's achievement nevertheless led his former European opponents to recommend an African bishop for Yorubaland in 1875. This confidence was not shared by younger European missionaries, who attacked the policy of African leadership. Crowther's personal integrity was never questioned, but many charges were brought against his subordinates, and financial administration was taken out of his hands. His position was progressively undermined, and by 1889 he was ready to resign. His discipline was probably too gentle and his administration faulty, but this did not justify the arrogant behavior of his critics. The conflict led to the secession of the Niger Delta Pastorate in 1891, the formation of the United Native African Church in Lagos in 1892, and the appointment of a white successor after Crowther's death. It helped to discredit Henry Venn's* "three-self" policy in pursuance of which Crowther had been appointed.

See J.F.A. Ajayi, *Christian Missions in Nigeria 1841-1891* (1965); J. Page, *The Black Bishop* (1908). D.G.L. CRAGG

CRUCIFIX (Lat. *crux,* "cross"; *figo,* "I fix"). A figure of Christ attached to a cross. The earliest known crucifixes appear to date from the sixth century, not in the form of a suffering Christ, but as a victorious Christ, reigning from the tree. He is shown alive, standing, head erect, clothed, and crowned. The Lamb may be taken to be the earliest visual representation of Christ's humanity and sufferings. Realism in Christian art began to replace symbolism from about the tenth century, culminating in a stress on the realistic aspects of Christ's suffering and death from the thirteenth century. In the Eastern Church, the Iconoclastic Controversy represented the aversion from sculptural portrayals in the round and restricted itself to the icon.* In the pre-Reformation West, the crucifix became an object of personal and public devotion, sometimes idolatrous. It was the central feature of the rood (alternate English name for crucifix) screen. Protestants in general, with the exception of Lutherans, make little use of it, but now tend to view it with greater tolerance.

HOWARD SAINSBURY

CRUCIFIXION. A form of execution practiced in the ancient world which involved fixing the victim to a wooden cross and leaving him to die. It seems to have been invented by the Phoenicians and to have been taken up by a number of other peoples. The Romans made considerable use of it for the execution of slaves, foreigners, or the lowest criminal classes. It was used in Palestine by Antiochus Epiphanes and by Alexander Jannaeus, as well as by the Romans. Josephus describes the crucifixion of 2,000 Jews by the Roman general Varus in 4 B.C.

Three shapes of cross were used: **T**-shaped and **X**-shaped as well as in the shape familiar to us through representations of the crucifixion of Christ. The victim was normally brutally flogged and then made to carry the cross-beam *(patibulum)* to the place of execution outside the city. The *titulus* (tablet of execution) would usually be hung round his neck. His clothes would be removed, and his hands were fastened by nails or cords to the cross-beam. That was then raised and fixed to the upright which would have been already erected. There was a peg on which the victim had to sit so that the weight of his body did not pull him down. His feet were attached to the upright, and he was left to die in intense suffering and often in the face of jeers and insults. On occasion, the process of dying took several days. Sometimes a drugged drink was given to relieve pain, and a victim's legs might be broken to hasten his death.

Jesus was put to death by crucifixion, as all the gospels narrate, and what was accounted as the most degrading form of punishment became a symbol for Christians of forgiveness and dedication (Gal. 3:13; 6:14; Phil. 2:8). R.E. NIXON

CRUCIGER, KASPAR (1504-1548). Reformed scholar. Born at Leipzig, he served as secretary at many of the theological discussions of his time, notably at Worms and Ratisbon in 1541. One of his outstanding services was to preserve in shorthand many of the lectures and sermons of Luther, with whom also he collaborated in Bible translation. In his youth Cruciger studied theology under Luther and Melanchthon and became proficient also in Hebrew, natural history, and medicine. For a few years he served as a teacher in Magdeburg, but went to Wittenberg in 1528 as preacher and professor, remaining there until his death. Beza records that after a private conference with Calvin on the doctrine of the Lord's Supper, Cruciger became attached to the Calvinist position on that question. HUGH J. BLAIR

CRUDEN, ALEXANDER (1699-1770). Scottish author of the famous *Concordance* which is still the standard reference guide to the King James Version. Shortly after graduating from Aberdeen he suffered a mental breakdown and was confined to an asylum (1721-22). Upon his release he went to London, where at first he was a tutor, but later opened a bookstore and did proofreading. The first edition of his concordance appeared in 1737. It was not a financial success, and the pressure of the work caused him to suffer another attack of insanity (1738). Confined again, he escaped after a few weeks and issued an indignant account of his experience entitled *The London Citizen Exceedingly Injured, or a British Inquisition Displayed.* He then sued for damages, but lost the case. Confined a third time in 1753, he was released by the following year and once more set to work on his concordance.

The appearance of the second and third editions won Cruden considerable recognition and profit. Influenced by the preaching of Wesley, he came to think of himself as the public guardian of the nation's morals. Calling himself "Alexander the Corrector," he was especially concerned with swearing and Sabbath-keeping. He ran for Parliament in 1754, and unsuccessfully sought to marry the daughter of the lord mayor of London. He published also a *Scripture Dictionary* and an index to the works of John Milton. A great deal of autobiographical information is in his three pamphlets, *The Adventures of Alexander, the Corrector* (1754-55).

See E. Oliver, *The Eccentric Life of Alexander Cruden* (1934). ROBERT G. CLOUSE

CRUSADES, THE. A series of seven major and numerous minor campaigns into the Levant by West Europeans between 1095 and 1291. Christians had gone on pilgrimages to the Holy Land during much of the medieval period, but with the arrival of the Seljuk Turks their travels were hampered. After wresting Jerusalem from their fellow Muslims, the Seljuks pushed north and defeated the Byzantine forces at the Battle of Manzikert (1071). Within the next few years, Asia Minor, the chief recruiting ground for Byzantine soldiers, was lost, and the emperor was writing to Western princes and to the pope, seeking mercenaries with which to regain the lost territories. Pope Urban II responded to this appeal by proclaiming the *First Crusade* in a sermon at Clermont (1095). At the conclusion of this address, the crowd shouted, "God wills it!" and this became the battle cry of the movement. The primary reason for the Crusades was religious, for they constituted a holy war, and following Urban's appeal there was an outpouring of religious enthusiasm. In addition, the pope saw in the Crusades an outlet for the energies of the warring nobles of Europe.

The First Crusade, consisting of about 5,000 fighting men, proceeded overland to Constantinople. Alexius Comnenus, the eastern emperor, was frightened by the group, but he provisioned them well, surrounded them with guards, and got them into Asia Minor. Antioch and Jerusalem were among the places that fell to the Crusaders; victory in the Holy City was followed by frightful slaughter of their enemies.

The Crusaders did not free all the Middle East from Muslim control, but by the establishment of several states in the Levant they maintained a balance of power between Byzantines and Muslims. The fortunes of these states varied, and when Jerusalem was endangered, Bernard of Clairvaux* organized the *Second Crusade* in 1147, which ended in defeat at Damascus. By 1187 Saladin had united the Muslims and conquered Jerusalem. This provoked the *Third Crusade*, called the Crusade of Kings because its leaders were Frederick I, Richard I, and Philip II. Frederick was drowned; Philip and Richard quarreled until Philip returned to France, leaving Richard in command against Saladin. This campaign resulted in a three-year truce and the granting of free access to Jerusalem for Christian pilgrims.

The crusader states were protected by the semi-monastic orders of Templars* and Hospitallers.* Combining monasticism and militarism, they were to protect pilgrims and wage perpetual war against Muslims. They were unable, however, to stop the more numerous onslaughts of their enemies, hence further crusades were necessary. The few knights who answered Innocent III's call to the *Fourth Crusade* were unable to pay the passage charges demanded by the Venetians. This led the two groups to strike a bargain and agree to attack Constantinople. After conquering and sacking the city, the Crusaders set up the Latin Empire of Constantinople and forgot about recovering the Holy Land.

During the thirteenth century there were more crusades such as the Children's Crusade* (1212), the Fifth Crusade against Egypt (1219), the Sixth Crusade led by the excommunicated Frederick II, and the Seventh Crusade of Louis IX. Each of these failed in its efforts to shore up the Latin crusading kingdom, and in 1291 Acre, the last stronghold of the Christians in the Holy Land, fell to the Muslims, thus ending the eras of the crusades. Despite this failure to achieve their main objective, the Crusaders led Europe to have more contacts with the East. This experience stimulated Western trade and thought, thus bringing to an end Western isolation.

BIBLIOGRAPHY: S. Runciman, *A History of the Crusades* (3 vols., 1951-54); J.A. Brundage, *The Crusades, Motives and Achievements* (1964); K.M. Setton (ed.), *History of the Crusades* (5 vols., two published to date). ROBERT G. CLOUSE

CRUSIUS, CHRISTIAN AUGUST (1715-1775). German theologian. Born at Leuna and educated at Leipzig University where he became successively professor of philosophy (1744) and of theology (1750), he attacked in a series of important works the determinism of Leibnitz, the perfectionism of Wolff, and the biblical criticism of his colleague Ernesti as dangerously anti-Christian. Founding all spiritual knowledge on divine revelation which he tried to prove harmonizes completely with reason, Crusius constructed in his three-volume *Hypomnemata ad theologiam propheticam* (1764-78) a theological system heavily dependent on typology and special theories of prophetic interpretation. For a number of years the University of Leipzig was divided into Ernestians and Crusians. His "prophetic theology" was rediscovered and popularized in the nineteenth century by Hengstenberg* and Delitzsch.*

IAN SELLERS

CRUTCHED FRIARS, An order of Mendicants said to have been founded by Gerrard, prior of St. Mary of Morello at Bologna, and confirmed by Alexander III in 1169, being brought by him into the Augustinian Rule. Initially they carried a cross fixed to a staff and were therefore called *fratres cruciferi* (sometimes translated as "Crossed Friars"). Later they wore a cross of red on their backs or chests. They appeared in England in 1244, and similar orders existed in France and the Low Countries from the thirteenth century. They were suppressed by Alexander VII in 1656.

CRYPT. An underground cell, chamber or vault, especially one beneath the main floor of a church, used as a burial place, chapel, or oratory. Some crypts containing the remains of saints are designed as the focus of pilgrimages. Modern uses of crypts, in addition to the traditional ones, include museums, refuges for the "down and out," and youth clubs.

CUDWORTH, RALPH (1617-1688). English divine. Son of an Anglican rector, he became regius professor of Hebrew at Cambridge (1645), to which was later (1654) added the mastership of Christ's College. He was perhaps the most distinguished exponent of the Cambridge Platonists.* Cudworth formulated the theory of a "plastic nature" and declared, "Things are what they are, not by Will but by Nature." He asserted the need for revealed religion, was a staunch advocate of the reality of moral freedom and responsibility, and was convinced that the "impossibility" of atheism was demonstrable. He was imperfectly understood, however, partly because of his habit of introducing much extraneous lore into his arguments, partly because of his scrupulously fair delineation of his adversaries' views (a policy to which his controversial age was not notably drawn). An excellent if somewhat diffuse scholar, he was consulted in 1657 by a parliamentary committee exploring the need for a new translation of the Bible. His own works included *The True Intellectual System of the Universe* (1678); and a *Treatise concerning Eternal and Immutable Morality* (published posthumously in 1731), directed in part against Calvinism. J.D. DOUGLAS

CULDEES. The word is of doubtful origin and may be derived from *keledei*, "friends of God," or *cultores Dei*, "worshipers of God." The name appeared first in the tenth century, and continued in use, particularly at St. Andrews, until the middle of the fourteenth century. It was given to a somewhat enigmatic but highly influential group in the Scottish Church who have sometimes been referred to as the "evangelicals" in the pre-Reformation church. They were in the first instance monks with high ideals and a sense of spiritual values who maintained a good degree of spiritual life in the Scottish Church when it was at a low ebb elsewhere. Their influence enabled the church to survive the troubled period of the Norse invasions. They never owed allegiance to Rome, and maintained a separate and independent existence until the Reformation. About 1100 they abandoned their regular monastic habits and became a college of secular priests. Their resistance of King David's attempts to have them incorporated in the new Augustinian priory that he had founded at St. Andrews in 1144 was an evidence of their strength and independence. At Brechin they fulfilled the function of a cathedral chapter, but their status then was largely secular, and their contribution to the work of the church came from their material wealth as landowners and their capacity for administration.

ADAM LOUGHRIDGE

CUMBERLAND PRESBYTERIANS. The Cumberland Presbyterian Church emerged as a new denomination in 1810 during the Second Great Awakening.* Three ministers—Finis Ewing, Samuel King, and Samuel McAdow—organized the independent Cumberland Presbytery which formed the nucleus of the Cumberland Presbyterian Church. Refusing to adhere to the licensing requirements for the education of the clergy as set forth by the Presbyterian Church, these men maintained that ministers in the frontier regions of Kentucky and Tennessee did not necessarily need a formal education as a prerequisite for their religious vocations. Moreover, they modified Calvinistic doctrines to suit the revivalistic preaching on the frontier, and this move set them at odds with the mother denomination. In 1906 an attempted union with the Presbyterian Church met with only partial success. The continuing Cumberland Presbyterian Church has some 90,000 members. JOHN D. WOODBRIDGE

CUMMINS, GEORGE DAVID (1822-1875). Founder of the Reformed Episcopal Church.* Born in Delaware, he was educated at Dickinson College and ordained in the Protestant Episcopal Church after a time as a Methodist preacher. He served parishes in Norfolk, Richmond, Washington, Baltimore and Chicago, and in 1866 became assistant bishop of Kentucky. As a low churchman, however, he noticed that even his colleagues were using the Book of Common Prayer to provide for a growing "ritualism and sacerdotalism." He tried to have the Prayer Book of 1785 approved to change this, but failed. As he could no longer obey the church's mandates—its increasing "popishness," and strictures against celebrating Holy Communion with non-Anglicans (for which Cummins got into trouble with the church)—he resigned his orders (1873), and a few weeks later he and others of like mind, clerical and lay, founded the Reformed Episcopal Church, using the 1785 Prayer Book and with himself as presiding bishop. C.G. THORNE, JR.

CUNNINGHAM, WILLIAM (1805-1861). Scottish theologian. One of the leaders of the Disruption* in 1843, and later Thomas Chalmers's successor as principal of New College, Edinburgh, he was a man of massive theological learning and an able controversialist both in theology and in the ecclesiastical affairs of his day. Judged by *Historical Theology*, a posthumously published set of lectures (2 vols., 1862), Cunningham's gifts as a teacher lay in his judicious summings-up of the chief features of theological systems, from one who had an unswerving commitment to the Westminster Standards. He was a theological heir both of the Westminster Divines and of the continental Reformed theologians of the seventeenth century, in whom he was widely read. Cunningham was a close friend of Charles Hodge* of Princeton. PAUL HELM

CURATE. One who has the care of souls in a particular area, hence, properly, a vicar or rector is a curate. But a curate in the Church of England is popularly understood to be an assistant or unbeneficed clergyman—although strictly speaking he is an assistant curate. He is nominated by the incumbent or the bishop, and licensed by the bishop. His license may be summarily revoked by the bishop, although he may appeal to the archbishop of the province. The employing incumbent may dismiss him on six months' notice.

CURÉ D'ARS, THE (1786-1859). French priest. Born Jean-Baptiste Marie Vianney at Dardilly, near Lyons, he had little formal education because of the unsettled times of the Revolution in France. His desire for the priesthood was long frustrated by an inability to study Latin, and also hindered by conscription to the army (1809) and subsequent desertion. Dismissed from seminaries in Varrieres (1811) and Lyons (1814), he eventually with the persistent private tuition of Abbé Balley, pastor of Écully, was ordained at Grenoble in 1815. He became Balley's assistant, and in 1818 began his famous ministry at Ars-en-Dombes, a village of 230 inhabitants. In eight years its religious tone was transformed. His fame as confessor spread, and from 1827 penitents in thousands came to him. During his last years he spent sixteen to eighteen hours daily in the confessional. He was created the patron of parish priests in 1929. J.G.G. NORMAN

CURIA, THE. Used in English since the first half of the nineteenth century as a description of the papal court, including all its authorities and functionaries. In medieval Latin the word was used to signify "court" generally, but the original meaning of the word denoted an ancient Roman political division.

CUSANUS, NICOLAUS, see NICHOLAS OF CUSA

CUTHBERT (c.634-687). Bishop of Lindisfarne. After a vision connected with the death of Aidan* he entered the monastery at Melrose in 651. Some years later he went as guestmaster to the monastery at Ripon founded by Eata. In 661 he returned with Eata to Melrose and soon became prior. In 664 he moved to Lindisfarne as prior, and introduced Roman customs there, at first against opposition. He continued with missionary journeys in the tradition of Aidan. In 676 he withdrew for greater solitude and closeness to God through nature to one of the neighboring Farne islands. In 684 he declined the see of Hexham, but the next year he consented to become bishop of Lindisfarne. After his death on Farne in 687 his body was buried at Lindisfarne. Due to the Danish raids, it was moved in 875 by monks, who wandered with it all over Northumbria. In 883 it came to rest at Chester-le-Street in County Durham. Because of further danger it was moved again in 995 to Ripon, and on its return north in 999 was finally buried at Durham, where a church was built as a shrine. R.E. NIXON

CYNEWULF (d.783). Bishop of Lindisfarne. He was the tenth bishop (Aidan being the first), and his episcopate from 740 to 780 was easily the longest of that see. He incurred the displeasure of King Eadfrid because of the murder of a member of the royal family who had taken sanctuary on the island. He was imprisoned for two years, but was later reconciled to the king. He is probably to be identified with the Anglo-Saxon poet Cynewulf (see following entry), who flourished in the second half of the eighth century.

CYNEWULF (fl. 750-780). Old English poet. Little is known of him, but a runic signature enables us to identify four poems—the *Juliana*, *Ascension*, *Elene*, and *Fates of the Apostles*—as his. He may also have written other poems surviving from the Old English period such as *The Dream of the Cross* and *The Last Judgment*, while *The Phoenix* and *The Mission of St. Andrew* (or *Andreas*) are probably the work of his followers. His work shows Northumbrian characteristics, and it has been conjectured that he may have been Bishop Cynewulf* of Lindisfarne (d.783). The *Juliana* and *Ascension* stress Trinitarian doctrine and support an ecclesiastical authorship. Modern versions of the poems are found in C.W. Kennedy, *Early English Christian Poetry*, 1952. See also K. Sisam, *Cynewulf and His Poetry*, 1933.
ARTHUR POLLARD

CYPRIAN (c. 476-549). Bishop of Toulon. He was a disciple of Caesarius of Arles,* whose life he wrote in 530. As bishop he took part in the various councils of the bishops of Provence held by Caesarius: Arles in 524, Carpentras in 527, Orange and Vaison in 529, Marseille in 533. When after the Council of Orange the bishops of the neighboring province of Vienne met to criticize the decisions of the council concerning grace, Cyprian was sent to plead the cause of orthodoxy.

CYPRIAN (Thascius Caecilius Cyprianus) (c.200/10-258). Bishop of Carthage. Son of wealthy and cultured pagan parents, Cyprian was a prominent Carthaginian master of rhetoric (and perhaps advocate) before his total conversion to Christianity about 246. His dedication (to celibacy, poverty, and the Scriptures—he repudiated pagan literature) and native ability led quickly to the presbyterate and episcopate (c.248), to the displeasure of some senior presbyters.

During the Decian persecution (250-51), he went into hiding, a move that predictably incurred suspicion in Rome and criticism in Carthage. In his absence some presbyters and confessors assumed the initiative in reconciling those who had "lapsed" in the persecution. Cyprian resisted what he viewed as innovation and usurpation, and reserved the issue for an episcopal synod. On his return he delivered an address on *The Lapsed* and presided over a council which stipulated that the gravity of the lapse should determine the duration of penance (251). The *libellatici** were readmitted forthwith. The disaffected clergy, led by the deacon Felicissimus and the presbyter Novatus, though demanding milder terms for the *lapsi*,* aligned themselves against Cyprian with the rigorist Roman schism of Novatian, and soon both parties appointed rival bishops of Carthage, undeterred by Cyprian's strictures on schism. In 252 a serious plague (which evoked widespread anti-Christian animus, despite Christian relief measures) and fears of renewed persecution dictated the prompt readmission of all penitents.

Cyprian was next in conflict with Bishop Stephen of Rome over schismatic baptism, and probably revised *The Unity of the Catholic Church* to prevent misappropriation by his opponents. African (and Asian) tradition, confirmed by three councils at Carthage in 255-56, repudiated sacraments outside the Catholic Church, and so practiced "rebaptism." The point of divergence was not the nature of the church (as in the Novatianist protest) but the relation of the gift of the Spirit to water baptism and the laying-on of hands and chrism. Stephen's threat of excommunication was averted in 257 by his death and Valerian's resumption of persecution. Cyprian was banished to Curubis and beheaded a year later outside Carthage.

Cyprian's letters and treatises reveal a pastor and administrator rather than a theologian. His rigid correlation of Church and episcopate and his application of OT sacrificial and priestly categories to Christian ministers and sacraments constituted an unhappy legacy. Crisp in thought and action, one of the first great magisterial bishops, strongly but never slavishly indebted to Tertullian, he has subsequently suffered at the hands of ecclesiastical controversy.

BIBLIOGRAPHY: Writings: J. Quasten, *Patrology* 2 (1953), pp. 340-83; *Clavis Patrum Latinorum* (2nd ed.), nos. 38-67; ET of selections in F.A. Wright, *Fathers of the Church* (1928), S.L. Greenslade, *Early Latin Theology* (1956), and J. Stevenson, *A New Eusebius* (1957); ET of *The Lapsed* and *The Unity of the Catholic Church* by M. Bévenot (1957; and with text, 1971). Full ET in *Ante-Nicene Christian Library.*

Life by his deacon Pontius, and official *Proconsular Record (Acta)* of trial and death.

Select Studies: E.W. Benson, *Cyprian, His Life, His Times and His Work* (1897); P. Monceaux, *Histoire Littéraire de l'Afrique Chrétienne* 2 (1902); A. d'Alès, *La Théologie de S. Cyprien* (1922); H. Koch, *Cyprianische Untersuchungen* (1926); A. Ehrhardt in *Church Quarterly Review* 133 (1941-42), pp. 178-96; M. Bévenot, *The Tradition of Manuscripts: A Study in the Transmission of St. Cyprian's Treatises* (1961); H. von Campenhausen, *Fathers of the Latin Church* (1964) and *Ecclesiastical Authority and Spiritual Power in the Church of the First Three Centuries* (1969); G.S.M. Walker, *The Churchmanship of St. Cyprian* (1968); M.A. Fahey, *Cyprian and the Bible: A Study in Third-Century Exegesis* (1971); P. Hinchcliff, *Cyprian of Carthage and the Unity of the Christian Church* (1974). D.F. WRIGHT

CYPRUS. The Christian Church in Cyprus claims as its founders Paul and Barnabas, who ministered there during missionary journeys. Because of its island geography and its apostolic founder, the early church on Cyprus strove to proclaim its independence from any patriarchal see. The Third Ecumenical Council which met in Ephesus in 431 decided in favor of the independence of the Cypriot Church. This decision went against the wishes of the patriarch of Antioch, who desired to dominate the island church. The council declared the Cypriot Orthodox Church to be autocephalous, or self-governing—and this is its present status.

The island came under the control of Islam when it was captured in 647 by the Arabs. The Cypriot Christians struggled and maintained their identity. Cyprus freed itself from the control of Islam from the middle of the eighth century to the early part of the ninth, but it was then recaptured by the Arabs under the Abbasid dynasty, and remained under Islamic rule until the middle of the tenth century. During the Crusader period the island passed from one controlling power to another. Richard I of England who had taken the island sold it to the Knights Templars*; they did not remain long; they were soon replaced by the Knights of St. John. Once Constantinople was captured by the participants of the Fourth Crusade (1204), Latin ecclesiastical superiors attempted to dominate the Cypriot scene. The Greek Christians of Cyprus found it difficult to determine where they should direct their loyalty.

When the Ottoman Turks took over in the last part of the sixteenth century, Cypriot Christianity came under the millet system of the Turks. Millets were religious nations under the Ottoman Turkish Empire. Each organized branch of Christianity which won a millet status was given power to exercise legal authority over its followers. This involved collection of certain taxes, marriage, and divorce courts, and certain aspects of local laws. Thus the head of the church became legal authority with close connections to the state. This tradition of the closeness of church and state was familiar from the days of the Byzantine Empire.

In light of the millet structure, it is not surprising that the Cypriot people elected the head of the church to become their first president on gaining independence from the British in 1960. During the eighty-eight year rule by the British the Greek Orthodox Church was entirely free to carry on its activities. Protestant missionary societies are found on the island, but their numbers and their effect on Cypriot life tend to be negligible.

BIBLIOGRAPHY: H.T.F. Duckworth, *The Church of Cyprus* (1900); J. Hackett, *A History of the Orthodox Church of Cyprus . . . A.D. 45-1878* (1901); G. Hill, *A History of Cyprus* (4 vols., 1940-52). GEORGE GIACUMAKIS, JR.

CYRIL (d.444). Patriarch of Alexandria. He was probably destined early for an ecclesiastical career and perhaps learned theology among desert monks. He assisted his uncle, Patriarch Theophilus, at the Synod of the Oak* (403) and contrived to succeed him in 412. With familial bellicosity he assailed Novatianists, Jews, Origenists, and pagans, and most strenuously Nestorius of Constantinople, who heeded complaints of refugee Egyptian monks against Cyril and disapproved of designating Mary *Theotokos*, "God-bearer" (428-29). Cyril's vehemence against Nestorius (see NESTORIANISM) vented traditional Alexandrian jealousy of Constantinople as well as horror of his typically Antiochene Christology. Armed with synodal condemnations from Rome (Celestine I) and Alexandria, Cyril sent Nestorius twelve anathematisms and demanded his prompt recantation (430), and at Ephesus in 431 secured his deposition (and the canonizing of his second letter to Nestorius, but not formally his third, with the anathematisms) by initiating proceedings before the Antiochenes arrived. The emperor recognized Cyril's assembly, not the Antiochenes' (which deposed Cyril), as the general council, and exiled Nestorius.

Such extremism demanded redress. With imperial prompting and after tortuous negotiations, Cyril approved an Antiochene "Formula of Union," John of Antioch* accepted Nestorius's excommunication, and Cyril's anathematisms quietly lapsed (433), although he campaigned still against unbudging defenders of Nestorius, especially Theodoret of Cyrrhus* (the compiler of the "Formula of Union"), and even Nestorius's old teacher, Theodore of Mopsuestia.*

Cyril's writings voluminously reveal the man—the ruthless theologian, forceful, acute, profuse, but inelegant, pompous, and myopic. Extensive exegetical works survive (allegorical commentaries on Isaiah, the Minor Prophets, and parts of the Pentateuch, more literal ones on John and Luke), plus a third of his massive reply to Julian's *Against the Galileans,* about twenty sermons, and many letters of dogmatic interest. Of his anti-Nestorian treatises, *That Christ is One* is a mature example. The early anti-Arian *Treasury* and *Dialogues on the Holy and Consubstantial Trinity* present an undeveloped Christology. Their heavy dependence on Athanasius* highlights Cyril's influential practice of arguing from "the holy Fathers."

Against Nestorius he vindicated the "Hypostatic Union"* of divine and human in Christ which guaranteed the eucharistic reception in His divinized flesh and blood, but gave the impression, by his mistaken appeal to Apollinarian phrases (especially "the one incarnate nature of the Word") as Athanasian, by his analogies, and by his use of *physis* to mean both "nature" and "person" of teaching Apollinarianism* and Monophysitism.* A major forerunner of Chalcedon (which also canonized his 433 letters to John of Antioch), Cyril nevertheless failed to develop a fully fledged appreciation of the role of Christ's humanity.

BIBLIOGRAPHY: Works: J. Quasten, *Patrology* 3 (1960), pp. 116-42; *PG* 68-77; several works ed. P.E. Pusey (1868-77) and E. Schwartz, *Acta Concil. Oecum* I, 1-5 (1927-30).

Selected Studies: B.J. Kidd, *A History of the Church to A.D. 461* (1922); R.V. Sellers, *Two Ancient Christologies* (1940); G.L. Prestige, *Fathers and Heretics* (1948); H. von Campenhausen, *Fathers of the Greek Church* (1963); A. Grillmeier, *Christ in Christian Tradition* (2nd ed., 1975); R.W. Wilken, *Judaism and the Early Christian Mind: a Study of Cyril of Alexandria's Exegesis and Theology* (1971). D.F. WRIGHT

CYRIL (826-869) and METHODIUS (c.815-885). Two brothers from Thessalonica known as "the Apostles of the Southern Slavs." After ordination they went to Constantinople where Cyril (whose real name until 868 was Constantine) was librarian at the famous church of St. Sophia. In 860 they appear to have participated in a mission to the Khazars. Two years later Emperor Michael III and Patriarch Photius sent them to Moravia to evangelize and organize the Slav church. The legendary *Life* of Cyril credits him with the invention of an alphabet known as Glagolitic.* The brothers certainly used the vernacular in the liturgy, and eventually gained the pope's approval for this when they visited Rome after encountering opposition from German clergy.

Cyril entered a monastery in Rome in 868, but soon died and was buried in the church of San Clemente. Methodius was consecrated bishop by Adrian II in order to return to the Slavs. In Moravia he was opposed by the German bishops and imprisoned for a period by Prince Sviatopolk. John VIII procured his release and later consecrated him archbishop of Pannonia. He died probably at Velehrad in modern Czechoslovakia; his followers took refuge in Bulgaria. The memory of Cyril and Methodius is still treasured by Czechs, Croats, Serbs, and Bulgars. The details of their lives are, however, shrouded in some mystery. PETER TOON

CYRILLIC. Under the impact of missionary expansion from Byzantium into regions dominated by Slavic language groups, there were devised modifications of Greek uncials by which those languages, and hence texts useful for missionary enterprise both biblical and liturgical, could be rendered into written form. Tradition ascribes the creation to ninth-century activities of SS. Cyril and Methodius, from the former of whom this particular script (still used for Slavonic languages) receives its name.

CYRIL LUCAR, see LUCAR, CYRIL

CYRIL OF JERUSALEM (c.310-386). Bishop of Jerusalem. According to the *Synaxary* he was born "of pious parents professing the orthodox faith and was bred up in the same in the reign of Constantine." For Cyril's early clerical career in Jerusalem we depend upon Jerome's *Chronicle,* which is unfair to Cyril by representing his activities as part of the squabbles within Arianism. It seems he was made deacon about 330 by Macarius and priest by Maximus about 343. He renounced his orders received through the latter for canonical reasons (not because Maximus supported Athanasian doctrine) and thereby gained the temporary favor of Acacius* of Caesarea. About 350 he became bishop, but soon was in conflict with Acacius because, during a famine, he sold church property (donated by the emperor) to feed the poor.

After being deposed by a provincial synod in 357, he appealed to the emperor. Though cleared by the synod of Seleucia in 357, he was exiled by Constantius but was able to return in 361 when Julian the Apostate began to reign. From his fifteenth catechetical lecture we gather that Cyril opposed the move of Julian to help the Jews rebuild their Temple. After Acacius's death in 366, Cyril consecrated his nephew Gelasius as the new bishop of Caesarea. When Valens adopted the ecclesiastical policy of Constantius in 367, Cyril was again banished. In the eleven years of exile he cooperated with the body of bishops of northern Syria and eastern Asia Minor, which ranged itself around Melitius of Antioch, to restore the Nicene faith. In 378 he was able to return to his see; the synod of Jerusalem (381-82) referred to Cyril as rightful bishop of "the mother of all the churches" and one who had striven to preserve the true faith against Arianism. Ten years after his death, Aetheria, the famous pilgrim, visited Jerusalem and wrote in her journal of the beautiful liturgy, especially for Eastertide.

Cyril's twenty-four *Catecheses* are his chief surviving work. See English edition by W. Telfer (1955). PETER TOON

CZECHOSLOVAK CHURCH. National church of Czechoslovakia, founded in 1920. Against the background of Hapsburg central government repression, which sought to curb Czech nationalism by appointing bishops favoring centralization and Germanization, an association of Catholic priests called "Jednota" was formed in 1890. They sought the introduction of the Czech language into the liturgy, the abolition of priestly celibacy, and lay participation in church government. Their demands were rejected by Rome in 1919, whereupon they formed an independent religious body, forty priests meeting in the national house in Prague-Smichov in January 1920. They won many adherents and were soon recognized by the secular government. Constituted on Presbyterian lines, the church elected but did not consecrate four bishops, there being no doctrine of apostolic succession. Since 1946 bishops are appointed for seven years only. A strongly rationalistic church—largely because of the influence of the first patriarch, Karl Farsky—it rejected the doctrines of original sin, purgatory, and veneration of saints; christological and eucharistic doctrines were liberally interpreted.

A more conservative group under Matthias Pavlik, who was ordained an Orthodox bishop under the name Gorazd (1921), entered into relations with the Serbian Orthodox Church. The majority followed Farsky, but after his death (1927) they too became more conservative. Patriarchs succeeding Farsky were Gustav Prochazka (1928-42), Francis Kovar (1946-61), Miroslav Novak (from 1961). In 1963 there were some 750,000 members in 345 parishes and five dioceses. Candidates for the ministry are trained in the Hus Czechoslovak theological faculty in Prague.

J.G.G. NORMAN

DABNEY, ROBERT (1820-1898). Young contemporary of J.H. Thornwell,* he is generally regarded as the second great theologian of the Southern Presbyterian Church. He graduated from the University of Virginia in 1842, and in 1844 entered Union Seminary, was licensed to preach in 1846, and became minister of the Tinkling Spring Presbyterian Church (1847-53). Thereafter he was appointed to the chair of ecclesiastical history and polity at Union Seminary, transferring in 1859 to the chair of systematic theology. In 1881 he took a prominent part in the formation of the Southern Presbyterian Church. From 1861 he served as a chaplain, then as chief of staff to Stonewall Jackson. In 1866 he returned to Union, remaining there until 1883 when for health reasons he moved to Austin where until 1884 he was professor of mental and moral philosophy at the recently established University of Texas, and played a prominent role in the founding of Austin Theological Seminary. In 1870 he published his *Syllabus and Notes of the Course of Systematic and Polemic Theology Taught in the Union Seminary in Virginia* which, revised and reissued in 1878, went through six editions until 1927. C. GREGG SINGER

DA COSTA, ISAAK (1798-1860). Dutch poet and theologian. Born into a wealthy Jewish family in Amsterdam, he was through the persistent witness of Willem Bilderdijk* converted from admiration of Voltaire to fervent Christianity. Thereafter his life was devoted to the Christian cause; his lectures and writings had a profound influence on the subsequent religious history of the Netherlands. He edited the poems of Bilderdijk in sixteen volumes. His own collected works were published in Holland (1861-63). Of those translated into English, *Israel and the Gentiles* appeared in 1850 and *The Four Witnesses* in 1851. The last named, compiled over many years, was first published after Strauss's *Life of Jesus* had begun to make an impact in Holland. R.E.D. CLARK

DAILLÉ (Dallaeus), JEAN (1594-1670). French Protestant theologian. Born at Châtellerault and educated at Poitiers and Saumur, he was tutor to two grandsons of Philippe de Mornay* from 1612 to 1621. Ordained in 1623, he became private chaplain to Du Plessis-Mornay, whose memoirs he subsequently wrote. Appointed pastor at Saumur in 1625 and at Charenton (1626-70), he gained a reputation for oratory and theological controversy. In *Traité de l'employ des saints pères* (1632) he rejected the authority of the Fathers, for which he was assailed by Roman Catholic and Anglican scholars. In *La Foi fondée sur les saintes écritures* (1634) he held that all Christian doctrines are either explicitly stated in Scripture or directly deducible from it. He attacked the genuineness of the Ignatian literature in *De Scriptis Quae sub Dionysii Areopagitae et Ignatii Antiochenii Nominibus circumferuntur* (1666), adducing no fewer than sixty-six objections. His *Sermons* on Philippians and Colossians vindicate his claim to rank as a great preacher as well as an able controversialist. He was president of the last national synod held in France before the revocation of the Edict of Nantes, held at Loudun (1659), when he defended the universalism of Moise Amyraut. He also wrote *Apologie pour les Églises Reformées.* J.G.G. NORMAN

D'AILLY, PIERRE (1350-1420). French cardinal and theologian. Born in Compiègne, he entered the College of Navarre, Paris, in 1363, graduating in arts in 1368 and becoming doctor of theology in 1381. He had many interests and wrote on scientific, philosophical, geographical, and astronomical, as well as theological, subjects. His work shows the influence of Bacon and William of Ockham.* He was made canon at Noyon in 1318, rector of his college in 1384, and chancellor of the University of Paris in 1389. He had close contact with the French court, as confessor and almoner to Charles VI. Among the many benefices he held in plurality was the archdeaconry of Cambrai, and on the accession of Benedict XIII (1395) he was appointed bishop of Puy, which office he never fulfilled, for in 1397 he was translated to the see of Cambrai. Deeply concerned to end the Western Schism which had divided Western Christendom since 1378, he supported the Council of Pisa* which had been convoked by the cardinals in 1409. This ended in the existence of three popes, with d'Ailly supporting the third, Alexander V. On the accession of John XXIII, d'Ailly was created cardinal in the hope that his support would be given to Rome. At the Council of Constance* (1414) he supported the theory of the supremacy of the general council over the pope. In 1416 he published his *Tractatus super Reformatione Ecclesia,* which was the third part of a much larger work, *De Materia Concilii Generalis.* Some of his suggested reforms were adopted by the Council of Trent,* and the *Tractatus* was well received in England and Germany. His theology was greatly influenced by that of Ockham, as seen in his belief that God could not be proved by reason, only faith; he argued also that the pope was not essential to the church. It is not surprising that these views found acceptance with Luther and other Reformers.

See J. McGowan, *Pierre d'Ailly* (1936).

GORDON A. CATHERALL

DALE, ROBERT WILLIAM (1829-1895). Congregational preacher. Of Nonconformist background, he was graduated at London with a first class degree in philosophy (1853). Dale started his career as a schoolmaster, but soon turned to the ministry, becoming in 1853 pastor at Carr's Lane Chapel, Birmingham, where for several decades he exerted a powerful influence upon the religious, educational, and social life of the city. In politics a Liberal of forthright views, he pioneered social work, especially among the young, opposed the imposition of oaths upon members of Parliament, fought religious education in state schools, and joined effectively in every controversy of the day. Dale published many theological books and tracts, and is remembered particularly for his fine work on *The Atonement* (1875) in which, though critical of the legalistic views of some theologians, he saw in the death of our Lord the sole ground of man's reconciliation with God. Though he upheld conditional immortality, Dale never minimized the seriousness of sin and so the need for God's forgiveness. He strenuously opposed extreme Calvinism and the High Church movement, but cooperated in revival meetings with D.L. Moody.* His books remained in wide use long after his death.

See A.W.W. Dale (his son), *The Life of R.W. Dale of Birmingham* (1898); and L.H. Hough, *Dr. Dale after Twenty-Five Years* (1922).

R.E.D. CLARK

DALMAN, GUSTAF HERMANN (1855-1941). Biblical scholar. Of Moravian background, he served as professor of the Institutum Delitzschianum in Leipzig (1895-1902), director of the German Protestant Institute for Archaeology in Palestine (1902-17), professor at Greifswald (1917-25), and again in Jerusalem as director of his own Institute for the Study of Antiquity in the Holy Land (1925-41). His many writings are a mine of information for the student of the NT, though some of his conclusions are in need of revision in the light of more recent research. His works include a Grammar of Jewish-Palestinian Aramaic (in German, 1894); *The Words of Jesus* (1898; ET 1902); *Sacred Sites and Ways* (3rd ed., 1924; ET 1935); *Jesus-Jeschua* (1922; ET 1929); and *Arbeit und Sitte in Palastina* (7 vols., 1928-42).

W. WARD GASQUE

DAMASUS I (c.304-384). Pope from 366. Born in Rome of Spanish parentage, he succeeded Liberius in the papacy, but was established only after Emperor Valentinian I came to his support and forcibly suppressed supporters of Ursinus, set up as a rival pope. Under the pontificate of Damasus, Apollinarianism and Macedonianism were condemned (368, 369), the famous Tome of Damasus* was issued (382), an imperial decree against clerical worldliness was enforced, and Jerome (then his private secretary) was commissioned by Damasus to revise the Latin translations of the Bible, the outcome of which was the Vulgate.

DAMASUS, TOME OF, see TOME OF DAMASUS

DAMIAN, PETER (1007-1072). Roman Catholic reformer. Born in Ravenna and raised amid hardship, he entered the Benedictine hermitage at Fonte Avellana in 1035, and was made prior eight years later. Having founded new monasteries and reformed old ones, gaining notice from Henry III's court and the papal Curia, he was made cardinal bishop of Ostia against his will. He attacked clerical decadence including marriage *(Liber gomorrhianus),* viewed church reform as a joint effort by papacy and empire, and gave himself to synodal work, diplomatic missions to France and Germany, and matters of highest ecclesiastical policy and principle. He defended the validity of orders conferred gratis by simonists *(Liber gratissimus),* upheld Alexander II during the schism of antipope Honorius II *(Disceptatio synodalis),* and benefited Cluny by supporting Abbot Hugh against Bishop Drogo of Macon *(Iter Gallicum).* Yet church and state beyond Italy did not concern him; he showed no interest in the open struggle between the Greek and Latin churches. His writings, inspired by antiquity as well as by the Carolingian Renaissance, ordered severe mortification and reveal his tension between the active and contemplative life, whether letters, sermons, treatises, or minor works of prose and verse. Though he was never canonized, Leo XII in 1828 made him "Doctor of the Church."

See O.J. Blum, *St. Peter Damian: His Teaching on the Spiritual Life* (1947). C.G. THORNE, JR.

DAMIEN, FATHER (Joseph de Veuster) (1840-1889). Roman Catholic missionary priest. Born in Tremelo, Belgium, he was trained for the priesthood by the Fathers of the Sacred Heart (Picpus Fathers) at Louvain and Paris, and took the religious name of Damien. He was sent in 1864 to Honolulu to be ordained as a missionary in the Sandwich Islands. He served at Puna and Kohala on the island of Hawaii. In the 1860s the government decided to use the island of Molokai as an isolation settlement for lepers, but made no provision for permanent medical staff. In 1873 Damien heard of the lepers' plight and was allowed to join them. The colony thereafter increased to over 1,000. He undertook the duties of nurse, builder, superintendent, as well as priest. By his own labors and by pressure on the authorities he created a well-ordered community. By 1885 he knew that he had leprosy, yet continued to serve the lepers right up to his death. His growing fame in the outside world made him many admirers, but also created hostility among local officials and the Catholic hierarchy on the islands. The most famous defense of his character came from Robert Louis Stevenson in 1905. NOEL S. POLLARD

DAMNATION. The King James Version renders three Greek words as "damnation": *krisis* (Matt. 23:33), *krima* (Matt. 23:14), and *apōleia* (2 Pet. 2:3). Newer translations have usually preferred "condemnation," "judgment," or "destruction" to "damnation." In Luke 16:19-26 Jesus spoke of a division between saved and lost, and the NT always declares that the saved go immediately into the Lord's presence (Luke 23:43; 2 Cor. 5:8; Phil. 1:23), while the lost will remain in Hades until the judgment, after which they will pass into the lake of fire (Rev. 20:11-15), "the second death" (Rev. 2:11; 21:8). During the nineteenth century particularly, theological liberals attempted to teach that Christ's gospel of love was perverted and oriented toward punishment, especially by Paul but precisely the opposite is apparent. While Paul does indicate that Christ-opposing powers will come to final ruin (1 Cor. 15:24-26; 2 Thess. 2:8-10), the NT emphasis is rather on the Lord's infinite love which prompts repeated warnings of impending damnation for those who defiantly resist the truth (Matt. 5:22, 29-30; 10:28; 18:9, etc.).

See also HELL and ESCHATOLOGY.

KEITH J. HARDMAN

DANIEL THE STYLITE (409-493). Syrian ascetic. Born near Samosata, he began his religious life as a monk not far from his home. Twice he visited Simeon the Stylite* on his pillar at Telanissus. When Simeon died, Daniel decided to emulate him. Upon two conjoined pillars he built a small platform on which he lived for the rest of his life. Only once in thirty-three years did he descend, and that was to reprimand an emperor. He was consulted by the patriarch of Constantinople and the emperors Leo I and Zeno; thousands flocked to see and hear him, and he prayed for many sick people. He died at the age of eighty-four and was buried at the foot of his pillars.

DANIEL-ROPS, HENRI (1901-1965). French Roman Catholic historian, essayist, and novelist. Graduate in history of the University of Grenoble (1922) and high school teacher until 1946, he changed from the family name of Petiot to Daniel-Rops (a character in one of his short stories) as a pseudonym so as to avoid the difficulty of obtaining permission from the ministry of education to publish. His literary output was enormous, including numerous short stories, essays, and nearly seventy books (twenty of them novels). For a time he was a leader among a group of young intellectuals called "The New Order" (1932ff.). At first merely a nominal Roman Catholic, from 1934 he began to assert himself as a man of faith and zeal. His novel *Death, Where is thy Victory?* (1934; ET 1946) had a widespread impact in France and was made into a film. His twelve-volume biblical and ecclesiastical history (1943-56) has been translated into English. In 1955 he was elected to the French Academy.

W. WARD GASQUE

DANTE ALIGHIERI (1265-1321). Italian poet. Born in Florence of a family that had a slight claim to nobility, Dante (a contraction of Durante) began early to write poetry and to take part in the political life of the city. While on an embassy to Rome as one of the priors of Florence, one of the opposing factions that supported Boniface VIII (whose secular ambitions Dante opposed) took control of the city. Several charges were lodged against Dante, and he was fined and exiled from the city. He never saw his wife or Florence again, though later he was given permission to return. He died at the court of Guido da Polenta in Ravenna, where he is buried.

Dante's literary achievement is universally acclaimed. His works include *La Vita Nuova; La Divina Commedia* (*Commedia* originally; *Divina* was added later); *Convivio,* an allegorical poem and numerous treatises. He wrote also several epistles and eclogues. A Latin treatise, *De Monarchia,* sets forth his political views on the conflict between the church and the Holy Roman Empire.

The poet's idealized love for Beatrice Portinari was first told in *The New Life,* but Dante is remembered best for *The Divine Comedy.* Composed of 100 cantos divided into *The Inferno, The Purgatory* and *The Paradise,* written in *terza rima,* this "cathedral in words" represents the poet astray in a dark wood (sin) at Eastertide. Virgil (Philosophy) appears and offers to act as guide through Hell and up to Mount Purgatory. Then Beatrice (Theology, Revealed Religion) conducts him to the heights of Paradise, where he contemplates the glory ineffable. The account is an elaborate Christian allegory of the soul's vision of sin, its purging from guilt and stain, and its rising in newness of life. This profound vision is imaginatively expressed in simple but vigorous style.

BIBLIOGRAPHY: C. Williams, *The Figure of Beatrice* (1943); M. Barbi, *Life of Dante* (tr. and ed. by Paul Ruggiers, 1954); D. Sayers (ed.), *The Divine Comedy* (1955); T. Bergin, *Dante* (1965).

ERWIN RUDOLPH

DARBOY, GEORGES (1813-1871). Archbishop of Paris. From humble origins in Fays-Billot, he was curate at St. Dizier, taught at Langres (1839-45), then became chaplain at the *Lycée Henri IV* where he occupied increasingly important diocesan posts. As bishop of Nancy, his writings showed a deep interest in education. Archbishop from 1863, he was a strong defender of episcopal rights and a determined Gallican. So strongly did he oppose the definition of papal infallibility that he tried to persuade Napoleon III to intervene at Vatican Council I.* He eventually submitted and exemplified important pastoral instructions by his distinguished care for the needy during the siege of Paris (1870-71). He was executed by the Commune while blessing his executioners.

IAN BREWARD

DARBY, J(OHN) N(ELSON) (1800-1882). Plymouth Brethren* leader. Born into a distinguished Anglo-Irish family (his middle name came from the famous admiral), Darby had a distinguished career at Trinity College, Dublin, and after graduation (1819) was called to the Irish Chancery Bar (1822). Ordained three years later, he served with tireless devotion a Church of Ireland parish in County Wicklow. Uneasy about church establishment, however, he startled the country in 1827 with his tract *On the Nature and Unity of the Church of Christ.* Resigning his curacy that same year, he associated with a group meeting in Dublin for breaking of bread and prayer, and quickly assumed leadership because of his teaching gifts and his powerful and attractive personality. From 1830 he made frequent preaching journeys to the Continent. In 1845 Darby broke with B.W. Newton* in Plymouth through disagreement on prophecy and ecclesiology, and later led the attack on Newton on heresy charges that forced the 1848 division of the Brethren into Open and Exclusive groups. Darby's later travels included North America, the West Indies and New Zealand. His literary style was often obscure, but his hymns breathe a strain of mystical devotion. He translated the Bible into German, French, and English. Passionate in controversy and at times blinded by prejudice, he was kind and generous to children and poor people. Though not the founder of the so-called Plymouth Brethren, he was undoubtedly their most gifted teacher at the beginning.

See W.G. Turner, *John Nelson Darby* (1944); and books listed under PLYMOUTH BRETHREN.

G.C.D. HOWLEY

DARK AGES, THE. Originally an alternative description of the Middle Ages,* the term was later restricted to the six centuries ending about 1100. The "Dark" has been taken to signify depreciation of morality, scholarship, and achievements, but modern research has increasingly tended to interpret the adjective as denoting rather the paucity of our knowledge of the period.

DARWIN, CHARLES ROBERT (1809-1882). English naturalist. Graduate of Christ College, Cambridge (1831), his theory of evolution* was formulated during a five-year voyage around South America. In 1837 he "opened his first notebook on the Transmutation of Species," but he hesitated until 1859 before publishing his *Origin of Species by Means of Natural Selection.* Lifelong vacillation between agnosticism and faith was accompanied by psychosomatic pains. Originally intending to become a clergyman, by 1850 he declared himself agnostic. Hesitation continued until A.R. Wallace's MS of a similar theory expedited Darwin's *Origins.*

In 1857 at the Oxford meeting of the British Association, T.H. Huxley as "Darwin's Bulldog" attacked Bishop Samuel Wilberforce,* whose ridicule and lack of science was an open target for Huxley. Thus evolution and religion were set at variance, counteracting the accepted *Bridgewater Treatises,* by renowned scientists on *The Power, Wisdom and Goodness of God as Manifested in the Creation.* Fears that the new theory would brutalize humanity were not unfounded. Herbert Spencer opposed the betterment of the unfortunate because it might hinder selection by survival of the fittest. Marx, Nietzsche, and Hitler justified war on the same grounds. Darwin's inner conflict continued into old age, according to the Duke of Argyle unrelieved by his wife's prayers and Bible reading. Some credence is given to his nurse's record, however, that the epistle to the Hebrews brought him final consolation. He was buried in 1882 in Westminster Abbey, a few feet from Isaac Newton. Darwin's other works included *Descent of Man* (1871), *Different Forms of Flowers* (1877), and a theory of reef formation (1842).

E.K. VICTOR PEARCE

D'AUBIGNÉ, JEAN HENRI MERLE (1794-1872). Protestant historian. Born at Eaux-Vives near Geneva, son of French Protestant refugees, his initial studies were in Geneva, where he was influenced by Robert Haldane and the current evangelical awakening. Later he studied in Berlin and became a friend of J.A.W. Neander.* In 1817 he was ordained, and a year later took the pastorate of the French congregation at Hamburg. Then in 1823 he was appointed court preacher at Brussels, but after the Revolution of 1830, he declined the post of tutor to the Prince of Orange and returned to Geneva. Here he deeply involved himself in the work of the Evangelical Society of Geneva and was appointed to a professorship in its theological school. His primary interest was church history, and his most popular work, a mixture of rhetoric, apologetics, and history, was *Histoire de la Reformation du XVIe siècle* (5 vols., 1835-53; ET 1846-53). A more scholarly production was *Histoire de la Reformation en Europe au temps de Calvin* (8 vols., 1863-78). He was also a founder of the Evangelical Church of Switzerland.

PETER TOON

DAVENANT, JOHN (1576-1641). Bishop of Salisbury. Graduate of Cambridge, by the age of thirty-three he had become D.D. and divinity professor there. In 1618, at the height of the Calvinist controversy, King James I (himself not without scholarly pretensions) sent him and three other churchmen to represent the Church of England at the Synod of Dort. Despite his colleagues' uneasiness, Davenant on their behalf read to the synod a paper advocating the doctrine of universal redemption. His king evidently approved, for Davenant was appointed bishop of Salisbury in 1621. Even his very moderate Calvinism was, however, unacceptable under the new king, Charles I, and the growing power of William Laud.* In 1631 Davenant was summoned before the council after preaching at court a sermon on predestination and election which provoked Charles's displeasure. Thereafter Davenant in his west country diocese meekly carried out Archbishop Laud's High Church commands in ecclesiastical matters. Among his works was a much-praised commentary on Colossians.

J.D. DOUGLAS

DAVENPORT, JOHN (1597-1670). New England pastor. Born at Coventry and educated at Oxford, he became vicar of St. Stephen's in Coleman Street, London. He was of Puritan sympathies and helped to procure the charter of the Massachusetts Company in 1629. He was also one of the twelve London feoffees who formed a committee to buy up lay impropriations and use the income to finance a Puritan preaching program. In 1632 he entertained John Cotton who was on his way to America. Becoming a Nonconformist, Davenport fled to Holland in 1633, and after a brief return to England, sailed with Theophilus Eaton to New England in 1637, where finally he became pastor to an independent colony at what is now New Haven, where Eaton was governor. In 1661 they sheltered the English regicides Edmund Whalley and William Goffe. Davenport opposed the Half-Way Covenant* and the absorption of the New Haven colony by Connecticut. In 1667, against the wishes of his church, he accepted a call to become minister of the First Church of Boston, which could not get the services of John Owen. PETER TOON

DAVID (c.520-589). Ascetic monk and patron saint of Wales. Of a southern Welsh princely family, he was a great founder of monasteries, including one at Mynyw in Pembrokeshire at which he lived and built up a fine library destroyed later by Scandinavian raiders. The Council of Brefi chose him as primate of Wales, but he accepted only on condition that the seat was moved from Caerleon to St. Davids. In order to strengthen his argument that the see of St. Davids was independent of Canterbury's authority, an eleventh century biographer, Rhygyfarch, fabricated the story that David then went to Jerusalem to receive episcopal consecration from the patriarch. In religious art David is depicted standing on a mound with a dove on his shoulder, a reference to the tradition that when he was speaking at the Council of Brefi a white dove rested on his shoulder and the ground below his feet rose to form a hill so that everyone could hear him speak. His connection with the leek is unknown. He was a popular saint in South Wales, Devon, Cornwall, and Brittany. L. FEEHAN

DAVID, CHRISTIAN (1691-1751). Moravian Brethren leader. Born in Moravia and by trade a carpenter, he was converted in 1717 and became a lay evangelist. In 1722 he met Count Zinzendorf,* helped him to found the Christian community of Herrnhut on his estate, and recruited settlers from the exiled Brethren (Unitas Fratrum) in Moravia. In 1733 David led a group of missionaries to Greenland to assist the Norwegian pioneer Hans Egede,* but sharp differences developed between them. David, strongly self-willed and at times intolerant, criticized Egede's Lutheran orthodoxy, while the latter resented the sentimental nature of the Moravian message. When Egede departed in 1736 they had been reconciled, and the work in Greenland prospered. David traveled extensively in Europe, returned to Greenland in 1747 with more missionaries, and built a church and mission residence. He journeyed to Pennsylvania, then returned to Greenland to construct a storehouse before dying in Herrnhut. RICHARD V. PIERARD

DAVID OF AUGSBURG (c.1200-1272). Mystic. Joining the Franciscans at Regensburg (Ratisbon) where he became novice master, he moved in 1243 to the order's new foundation at Augsburg. An Official Visitor of the Franciscan friaries, and missionary and inquisitor of Waldensian* heretics, he also wrote works on mysticism and contemplation in Latin and German which influenced the *Devotio Moderna.* The main ones were *Die Sieben Vorregeln der Tugend, Der Spiegel der Tugend, Formula Novitiorum, Expositio Regulae.* His authorship of *Christi Leben unser Vorbild, Von der Anschauung Gottes* and *De Inqisitione Haereticorum* is doubtful.

DAVIDSON, ANDREW BRUCE (1831-1902). Scottish OT scholar. Born into a poor Aberdeenshire family which made considerable sacrifices for his education, he graduated in Arts (1849) at Aberdeen. After three years' schoolteaching, during which time he mastered Hebrew and some modern languages, he entered New College, Edinburgh, the theological training school of the Free Church of Scotland (then only nine years old). Licensed as a preacher in 1856, he was subsequently assistant (1858) and then (1863) successor to the famous "Rabbi" Duncan as professor of Hebrew and oriental languages. A bachelor of diffident disposition, Davidson was a superb teacher of whom it was said, "Easy mastery of his subject, lucid and attractive discourse, the faculty of training men in scientific method, the power of making them think out things for themselves, were united in him with the capacity of holding their minds, quickening their ideas, and commanding their imagination." For his listeners the OT prophets came alive. He was also an influential member of the OT revision committee (1870-84). A pioneer in introducing historical methods of OT study in Scotland, who taught his students to read the Bible with grammar and dictionary in hand, he was the author of a number of Bible commentaries and theological works, but is now best remembered for his *Introductory Hebrew Grammar* (1874), known as a textbook to many generations of students. He also contributed extensively to publications such as the *Encyclopaedia Britannica* and Hastings's *Bible Dictionary.* Although his own orthodoxy was never called into question and he disliked controversy, Davidson's methods as furthered by W. Robertson Smith provoked the latter's trial for heresy.

See J. Strahan, *Andrew Bruce Davidson* (1917).

J.D. DOUGLAS

DAVIDSON, RANDALL THOMAS (1848-1930). Archbishop of Canterbury, 1903-28. The son of Scottish Presbyterians, Davidson was educated at Harrow and Oxford, was ordained priest in the Church of England, and rose rapidly to become dean of Windsor (1883), bishop of Rochester (1891), and bishop of Winchester (1895) before going to Canterbury. In the primacy Davidson presided over his church at a particularly difficult time. An administrator rather than a spiritual leader, he strove for compromise, though on some issues—especially the Balfour Education Bill and Welsh Disestablishment—his Anglicanism was unbending. In social matters his attempts to mitigate the horrors of war and of the ensuing industrial depression were widely approved, as were his ecumenical endeavors, the Lambeth Quadrilateral and the Malines Conversations, gestures respectively to the Free Churches and the Roman Catholics, and both fruitless. Evangelicals were not happy under his archiepiscopate, convinced that his concern for moderation and comprehension was a cloak for the advancement of ritualists or advanced liberals like Hensley Henson. Davidson promoted no evangelicals, on the grounds that they had no outstanding men, and confirmed their worst fears when he lent his weight to the new Prayer Book, rejected by Parliament to his disgust in 1928. He retired the same year. His writings include a biography of Archbishop Tait (2 vols., 1891).

See G.K.A. Bell, *Randall Davidson* (2 vols., 1935). IAN SELLERS

DAVIES, RICHARD (1501?-1581). Bishop and translator. Born a curate's son at Gyffin, Conway, he was educated at New Inn Hall, Oxford, and after serving two Buckinghamshire livings, was deprived on the accession of Mary Tudor and withdrew to Frankfurt to join other Protestant exiles. He was consecrated bishop of St. Asaph in 1560 and translated to St. David's the following year. As an administrator he was the foremost protagonist of the Elizabethan settlement in Wales, and profoundly concerned about the sorry condition of his diocese and the low spiritual condition of both clergy and people.

He participated in the production of the Bishops' Bible* (1568) and was responsible for the section Joshua to 2 Samuel. He applied his scholarly gifts to the task of translating Scripture into Welsh. An act of 1563 (which owed much to him) had put the responsibility for preparing a Welsh version of the Bible upon the four bishops of Wales together with the bishop of Hereford. In order to comply with this command, he invited William Salesbury* to join him at Abergwili, the episcopal residence, to cooperate in the venture. In 1567 their joint efforts saw the light of day in a Welsh translation of both the Book of Common Prayer and the NT. Davies's contribution to the NT consisted of an interesting preface, *Epistol at y Cembru*, together with 1 Timothy and Hebrews-2 Peter. Although the Book of Common Prayer has traditionally been attributed to Davies, this is doubtful. He combined with his excellent knowledge of the classical languages a solid familiarity with the classical Welsh literary tradition and was capable of producing poetry in the strict meters.

See excellent biography by G. Williams (1953); and I. Thomas, *William Salesbury and his Testament* (1967). R. TUDUR JONES

DAVIES, SAMUEL (1723-1761). Founder of Southern Presbyterianism. After theological training in Samuel Blair's school at Foggs Manor, Pennsylvania, he was ordained as an evangelist in 1747 and sent to Hanover County in Virginia to preach to Presbyterian converts of the Great Awakening.* He preached in licensed meeting houses in Hanover, and evangelized in other counties of Virginia and North Carolina. He went to England with Gilbert Tennent in 1753, and raised £3,000, mostly in Scotland, for the College of New Jersey (now Princeton University), preached with great acceptance, and won by personal argument before the king in council legal status for dissenters in Virginia. He became president of the College of New Jersey in 1759 and improved its standards. He led in organizing Hanover Presbytery (the first in the South) in 1775. His always feeble health gave way in 1761 after he contracted pneumonia and blood poisoning from being bled. EARLE E. CAIRNS

DEACON. The Greek work *diakonos* means "servant," with an emphasis on usefulness rather than inferiority. Such service was expected of all members of the early church, but there soon evolved a special auxiliary ministry with this title. Deacons were apparently assistants to the bishops or overseers (Phil. 1:1). Their origin was early attributed to the appointment of the Seven in Acts 6:3,4, though they are not actually called deacons. They were, however, to "deacon" tables, and certainly such administrative service characterized the later work of deacons (note the qualification "not greedy for gain" in 1 Tim. 3:8 RSV).

By the time of Ignatius in the early second century, their flexible beginnings had evolved into a specific office in the threefold ministry of bishops, presbyters, and deacons. The office soon gained in significance, with a combination of administrative, pastoral, and liturgical functions. The *Apostolic Tradition* of Hippolytus in the beginning of the third century regards the deacon as the bishop's link with the church. Collecting and distributing alms was a major responsibility and liable to create an undue sense of their importance. From the Council of Arles in 314 onward, their temptation to arrogance is alluded to and their subordinate position to that of presbyters stressed. In the Eastern Church a man often remained a deacon for life while continuing in a secular occupation; his diaconal role was almost entirely liturgical. In the West the diaconate steadily declined in importance during the Middle Ages until it became simply a stepping-stone to the priesthood.

In many of the Reformation churches the liturgical function of deacons was discarded. Among the Lutherans pastoral care and witness are their special responsibility. Consequently they are not ordained, but on the European continent are consecrated as members of a community (*Bruderhaus*). Calvin traced two kinds of deacon in the NT, one to dispense alms and the other to care for the poor and sick (*Institutes* IV 3.9), a dual role still exercised by this "lay" ministry in the Reformed churches. Among Baptists, deacons approximate to the executive committee of the church, being normally elected by the church meeting for a limited period.

Experiments in permanent diaconate and the advocacy of fresh study of the subject by both Vatican II and the Lambeth Conference of 1968 indicate widespread uncertainty concerning their role.

BIBLIOGRAPHY: F.J.A. Hort, *The Christian Ecclesia* (1897), pp. 198ff.; K.E. Kirk (ed.), *The Apostolic Ministry* (1946); J.G. Davies, "Deacons, Deaconesses and the Minor Orders in the Patristic Period," *JEH* XIV (1963), pp. 1-15; *The Ministry of Deacons*, World Council Studies No. 2 (1965). J.W. CHARLEY

DEACONESS. In the early church, a woman involved in the pastoral ministry of the church. Exactly what her function or official status may have been is a matter of dispute. Paul mentions Phoebe, "deaconess" (*ousan diakonon*) of the church at Cenchreae (Rom. 16:1) and there is also an ambiguous reference in 1 Timothy 3:11. The "widows" mentioned in 1 Timothy 5:3-10 may also be connected with the role. Not until the end of the fourth century is much known about the office of deaconess (Gr. *diakonissa*). The "Didascalia" and the "Apostolic Constitutions" describe their functions as assistants to the clergy in the baptizing of women, ministers to the poor and sick among women, instructors of women catechumens, and in general intermediaries between the clergy and women of the congregation. Fears of the usurping of priestly functions and other considerations led to the extinction of the office in the church at large by the eleventh century.

The modern deaconess movement began clearly in 1836 when the Lutheran pastor T. Fliedner* founded a Protestant community at Kaiserswerth,* near Dusseldorf, devoted chiefly to nursing. The movement spread rapidly through the Protestant world. In the second half of the nineteenth century, deaconesses were established in the Church of England, the Methodist Church, and the Church of Scotland. They act as pastoral assistants to the minister.

HOWARD SAINSBURY

DEAD SEA SCROLLS. Discovered in 1947 and subsequently in caves near the northwestern end of the Dead Sea, the scrolls are the outstanding find of the century. They have posed and continue to pose many problems, particularly in regard to date and connection with the Essenes* already known from the writings of Philo, Josephus, and the elder Pliny. But intensive work on the scrolls and on the site of their discovery has made certain conclusions reasonably assured.

The scrolls are the work of a community of Jews who after a violent quarrel with the priesthood of Jerusalem made a home for themselves in the torrid region adjoining the Dead Sea. The settlement began probably at the time Jonathan, brother and successor to Judas Maccabaeus, became high priest in 153 B.C., and continued (save for a break of about thirty or forty years in the time of Herod the Great) until the war with Rome of A.D. 66-73. The scrolls are probably from between 20 B.C. and A.D. 70.

All the scrolls have suffered damage to some extent, and many are reduced to small fragments. The story of how the copper scroll was unrolled illustrates the drama which surrounds the scrolls generally. Those scrolls so far discovered and deciphered can be classified thus:

(1) *Interpretation.* There is a whole series of commentaries on the books of the OT. They interpret the text as fulfilled in the events surrounding the founding of the community. One of the two Isaiah scrolls found in cave 1 contains the complete Hebrew text of Isaiah. It indicates that the Massoretic text was substantially fixed by Christian times, which few scholars would previously have dared to assert with any confidence. The work on Habakkuk, useful for its information on the history of the sect, follows a phrase-by-phrase type of commentary *(pesher)* peculiar to Qumran. The fact that this scroll has nothing on the third chapter of Habakkuk is taken to support the long-held view of some scholars that this chapter was part of the original Habakkuk. Fragments of other *pesherim* on Nahum and Psalm 37 have been found. A looser sort of commentary is the *midrashim.* The best example is the Genesis Apocryphon. It eulogizes the patriarchs as examples to be followed, emphasizing the beauty of Noah and Sarah. A new literary genre found only in Qumran is the *florilegia,* or collections of biblical texts. Some of these are interesting for their Messianic allusions.

(2) *Discipline.* Very influential in the life of the community was the so-called Manual of Discipline. Fragments of no less than eleven MSS have been found. This scroll contains very valuable information on rules for membership of the community, beliefs held, directions on admission, discipline, the holy life, and the praise of God.

(3) *Praise.* The scrolls known as the Psalms of Thanksgiving are moving compositions. Written on the lines of the OT Psalms, they are intended as individual prayers more than for community worship. Some of these compositions appear to reflect the experiences of the teacher of the sect and may in fact have been written by him. Fragments of hymns, prayers, and blessings for public worship have also been discovered.

(4) *Hope.* In all the scrolls there is a strong eschatological hope. These Jews believed they would one day resume the worship of God at Jerusalem. This explains the "New Jerusalem" texts which are based on the vision of Ezekiel 40—48. A dramatic description of the battle whereby the Children of Light will overcome their enemies and establish themselves at Jerusalem is found in the War Scroll.

The scrolls are invaluable for the study of the text of the OT and its transmission. The fragments of the Pseudepigrapha and Apocrypha provide specimens of what these texts were like in the original, and thus enable scholars for the first time to assess the fidelity of our translated versions. Most important of all is the invaluable information which the scrolls have to give on Jewish life and thought at the time of Jesus Christ and the beginning of the Christian Church. Compared with the rigidity and ultraconservatism of Qumran, Orthodox Judaism is seen to be flexible and progressive, and compared with both, Christianity is unique and inspired.

BIBLIOGRAPHY: Material on the subject is considerable, and includes M. Burrows, *The Dead Sea Scrolls* (1955); T.H. Gaster, *The Scriptures of the Dead Sea Sect in English Translation* (1956); E. Wilson, *The Scrolls from the Dead Sea* (1957).

R.J. MC KELVEY

DEAN. The Latin title *decanus* (from *decem,* "ten") probably entered ecclesiastical usage as the title of a monk supervising ten novices. Its principal use is to describe the head of a cathedral or collegiate church, who is responsible with the chapter for the services, fabric, and property. Rural deans assist the bishop's administration by superintending and representing the clergy of a subdivision of a diocese. Other usages include "the Dean of the Province of Canterbury," who is the bishop of London; "the Dean of the Arches," who is a lay judge in the archbishop of Canterbury's Court of Arches; "the Dean of the Sacred College," who is the cardinal bishop of Ostia. University or college deans are usually lay officials. JOYCE HORN

DEARMER, PERCY (1867-1936). Anglican scholar. Educated at Oxford, he was influenced by Charles Gore* in the direction of Christian Socialism.* He was vicar of St. Mary's, Primrose Hill, London (1901-15), where he adapted medieval English liturgy to the Prayer Book, as advocated in his *Parson's Handbook* (1899; rev. 1965). He was professor of ecclesiastical art at King's College, London (1919-36), and from 1931 canon of Westminster. His books treated such topics as ecclesiology, topography, and English church history. He edited the Anglo-Catholic *English Hymnal* (1906, with R. Vaughan Williams* as musical editor; new musical ed. 1933); the modernist *Songs of Praise* (1925, with Martin Shaw and Vaughan Williams as musical editors; enlarged ed. 1931); *Songs of Praise Discussed* (1933); and *The Oxford Book of Carols* (1928). He wrote original hymns and translations, and he bowdlerized older hymns. JOHN S. ANDREWS

DEATH. The Bible implies that man was made to enjoy the present life, not as a final end in itself, but as a transitory stage leading to another life. Man was therefore meant to experience at some stage a transition to a higher life. As a consequence of the Fall, this transition has taken the form of death, and the hope of life beyond has become faint and obscured. Christ came to reverse this sentence, to restore death to its true meaning, and to be the first to ascend to the glory man was originally meant to inherit by his nature. The Bible therefore seeks to understand death both as a natural aspect of man's created life and as a curse laid on man. Thus, though two men are "taken up" (Gen. 5:24; 2 Kings 2:11) and though several references are made to death as if it were a satisfactory climax to life, a "gathering to the fathers," and as a merciful and natural end to life, death is nevertheless more often regarded as a sign of God's wrath and judgment on man as the result of sin, and as a realm of unhappiness to which all men are given over at the end.

Yet there is also in the OT an obvious movement toward a more cheerful and positive view. Death and Sheol are at times regarded as if only limited duration. The earth is not able to cover her slain for ever, and God will raise them. Death cannot separate from the love and fellowship of God.

In the NT there are passages that imply death is part of the original order of creation and of the intrinsic nature of man (John 12:24; 1 Cor. 15:36; 1 Tim. 6:16), and passages that recognize that death can be used positively in the service of God (John 15:13; Phil. 2:17; 2 Tim. 4:6), and as something which brings gain and liberation (Rom. 6:3ff.; 2 Cor. 5:6ff.; Phil. 1:21ff.). Yet it is by the grace of God alone that death can play such a role. As men know it, death is the "last enemy" (1 Cor. 15:26,55). Its intrusion into life is the effect of sin (Rom. 5:12; James 1:15). Its shadow tends to blight all human life (Matt. 4:16), and it needs to be destroyed by Christ (Heb. 2:14f.). The destruction of death by Christ is linked with His conflict and destruction of the powers of evil (Rev. 20:14). He breaks the connection between death and sin, tastes its pangs (Acts 2:24; Heb. 2:9), becomes the firstborn from the dead (Rev. 1:5,18), in order that men might be saved from even experiencing its sting (John 6:49; 8:52; 1 Cor. 15:55f.). The final destruction of death at His second coming is now certain. (See also BURIAL SERVICES.)

The apostles believed that in some cases they had the authority to seek signs of Christ's power in raising the dead. The early church in both East and West proclaimed that Christ had come to give deliverance from death and corruption as well as from sin. Therefore it stressed the saving effect of the union of the incorruptible Word with our mortal flesh in the Incarnation—a theological emphasis Calvin sought to revive.

BIBLIOGRAPHY: L. Morris, *The Wages of Sin* (1955); M. Achard, *From Death to Life* (1960); J. Pelikan, *The Shape of Death* (1961); E. Duckett, *Death and Life in the Tenth Century* (1967); N. Tromp, *Primitive Conceptions of Death and Life in the Old Testament* (1969); A. Toynbee et al., *Man's Concern With Death* (1969); H. Thielicke, *Death and Life* (1970). RONALD S. WALLACE

DEATH OF GOD SCHOOL. The thought of the death of God dates from before the nineteenth century. The German poet, Jean Paul, wrote a "Discourse of the dead Christ from atop the cosmos: there is no God" (*Siebenkäs*, 1796-97). Hegel's *Phenomenology of Mind* (1806) spoke of the death of God in Christ, signifying his belief that through him God ceased to be pure, abstract spirit by becoming immanent in the profane. Whereas Hegel* retained the thought of divine immanence, Nietzsche rejected God altogether. For him the death of God meant cessation of belief in God, and hence meant that man is free to be master of his own destiny (*The Joyful Wisdom*, 1882). In *Existentialism and Humanism* (1946), Jean-Paul Sartre took as the starting point for existentialism* the remark of Dostoevsky: "If God did not exist, everything would be permitted." Since God does not exist, man is the author of his own existence, creating his own values and making his own decisions unaided. He may not, like some humanists, abandon belief in God and yet try to retain Christian values and morals.

In the 1960s, the thought of the death of God became a rallying cry for certain radical theologians, especially in the USA. It is doubtful, however, whether they could be called a school in view of their sharp differences. In *The Death of God* (1961) and *Wait without Idols* (1964), Gabriel Vahanian has attempted to analyze why belief in God has become culturally irrelevant. God dies as soon as He becomes a cultural accessory or a human ideal. Vahanian himself believes in a God who is wholly other, the transcendent God who can never be objectified. In *The Secular City* (1965), Harvey Cox argued that, in the light of biblical faith, secularization and urbanization are not curses to be escaped but opportunities to be embraced. Through art, social change, and teamwork relationships, the transcendent may eventually reveal a new name, for the word "God" has perhaps outlived its usefulness. We must come to terms with the hiddenness of God, even in Jesus. Nevertheless, Cox believes in a transcendent God.

T.J.J. Altizer's *The Gospel of Christian Atheism* (1967) rejected divine transcendence in favor of an immanent dialectic based on Hegel, but influenced also by Nietzsche and William Blake. William Hamilton and Paul van Buren both changed from a more orthodox position comparatively late in their careers. At the age of forty, Hamilton began to feel that his grounds for belief in God were giving way, though he still retained a belief in Jesus as the one to whom he may repair. Van Buren, on the other hand, abandoned his Barthian position under the influence of linguistic analysis, feeling the word "God" is no longer meaningful. His *Secular Meaning of the Gospel* (1963) urged a purely secular restatement of Christianity in which the doctrine of creation expresses an affirmative view of the world, and the mission of the church the practice of the liberty for which it has been set free.

Such views make it difficult to see what advantages they have to offer over atheism and agnosticism. They also raise the question whether the God who is allegedly dead was not a figment of the scholar's imagination all along. Nevertheless, the views of these writers pose the valid question of the basis and content of Christian belief in God and the form in which it is presented.

BIBLIOGRAPHY: E.L. Mascall, *The Secularisation of Christianity* (1965); T.J.J. Altizer and W. Hamilton, *Radical Theology and the Death of God* (1966); C.I. Glicksberg, *Modern Literature and the Death of God* (1966); T.W. Ogletree, *The "Death of God" Controversy* (1966); Ved Mehta, *The New Theologian* (1966); J.W. Montgomery, *The "Is God Dead?" Controversy* (1966) and *The Altizer-Montgomery Dialogue* (1967); R. Bultmann, "The Idea of God and Modern Man" in R.G. Smith (ed.), *World Come of Age* (1967), pp. 256-73; D. Peerman (ed.), *Frontline Theology* (1967); C. Brown, *Philosophy and the Christian Faith* (2nd ed., 1971); A. Kee, *The Way of Transcendence: Christian Faith Without Belief in God* (1971).

COLIN BROWN

DE BARRI, GERALD, see GERALD DE BARRI

DE BRUYS, PIERRE, see PETER DE BRUYS

DECALOGUE, see COMMANDMENTS, THE TEN

DE CAUSSADE, J.P., see CAUSSADE

DECIUS (d.251). Roman emperor from 249. A native of Pannonia, he was proclaimed emperor by the troops after restoring discipline in the Danubian armies. After Emperor Philip had been killed near Verone, Decius was accepted by the Senate in 249 only to be killed two years later in an attempt to defeat the Goths. He was a staunch advocate of the old Roman traditions, and his persecution of Christians proceeded from his belief that the restoration of state cults was essential to the preservation of the empire. The persecution of the church began with the execution of Fabian, bishop of Rome, in January 250, and it led to thousands of deaths. These resulted from Christian disobedience of the imperial order that all citizens should offer sacrifice to the name of the emperor. Those Christians who did sacrifice and later repented caused many problems for the church.

PETER TOON

DECLARATIONS OF INDULGENCE. Five English and four Scottish royal declarations may be so called.

England. (1) In 1660 Charles II issued a Declaration on Ecclesiastical Affairs, wherein it was announced that differences of opinion in regard to ceremonies were to be left to the determination of a national synod. Its provisions lasted only briefly, and its principles were not made law by the Convention Parliament.

(2) In 1662 Charles issued what is usually called "the first Declaration of Indulgence," announcing that he intended to ask the Cavalier Parliament to pass a measure to "enable him to exercise with a more universal satisfaction that power of dispensing which he conceived to be inherent in him." No such measure, however, was passed.

(3) In 1672 Charles issued a Declaration of Indulgence for those with tender consciences and suspended the penal laws in ecclesiastical matters against Nonconformists. Both Protestants and Catholics made good use of this indulgence, which remained in force until March 1673.

(4) In 1687 James II issued a Declaration of Indulgence more thoroughgoing than his brother's of 1672. It suspended the Test Act* and all penal laws against Protestant and Catholic Nonconformists.

(5) In 1688 James issued his second Declaration of Indulgence. This repeated the substance of that of 1687 and promised that Parliament should meet by November that year. An order in council required it to be read on two successive Sundays in all parish churches. Seven bishops were imprisoned because of their opposition to it.

Scotland. (1) In 1669 Charles II offered reinstatement to ejected ministers where their parish was still vacant, or alternatively institution to another charge, if they would undertake to be orderly, receive collation, and attend church courts. Lesser inducements were held out to those with reservations but of orderly behavior.

(2) In 1672 opportunity was given Nonconformist ministers to be appointed to parishes in pairs; each was to receive half the stipend; pastoral duty was to be confined to parishioners; preaching was to be in churches only; and ministers could not leave their parishes without the bishop's sanction.

(3) In 1679 "A Proclamation suspending laws against Conventicles" authorized the remission of fines and other disabilities on condition of good behavior; hitherto dissident ministers were offered appointment to parishes, with authority to

dispense the sacraments, provided they had had no part in the last rebellion.

(4) In 1687 James VII announced toleration of moderate Presbyterians, with permission to worship in private houses and to hear indulged ministers; to Quakers with licensed meeting houses; and to Roman Catholics, against whom penal laws were abrogated and who were given political equality. PETER TOON
J.D. DOUGLAS

DECLARATORY ACTS. Ecclesiastical legislation passed in order to allow two Scottish Presbyterian bodies to modify their attitudes toward the Westminster Confession of Faith.* The United Presbyterian Church* took this course in 1879; the Free Church of Scotland* followed in 1892. In the latter it caused a secession leading to the formation of the Free Presbyterian Church.* In the Church of Scotland,* this form of legislation was considered to be beyond the powers of the general assembly, but in 1910 the same end was achieved with authorization of a new formula by which ministers and office-bearers subscribed the Westminster Confession. While the modifications differed in all three cases, a common motive was the softening of that rigorous Calvinism acceptable to an earlier age. J.D. DOUGLAS

DE CONDREN, CHARLES, see CONDREN

DECRETALS. Papal letters, themselves having the force of law and specifically written to an individual or to a group in answer to a question on some point of canon law. The first such definitely established decretal was from Pope Siricus to Bishop Himerius of Tarragona (A.D. 385), though Damasus (d.384) might have issued one. From the start, popes insisted that decretals had the force of law. Dionysius Exiguus* compiled the earliest influential decretal collection. Various collections were made in the following centuries, but Gratian's *Decretum* (c.1140) set a new standard in the grouping of the decretals according to the different points of law they dealt with. Though his collection (which included some forgeries) ranged over many centuries, it concentrated on more recent pontificates and this became the accepted trend of subsequent decretal collections. The definitive medieval collections were the Decretals of Gregory IX (1234), Boniface VIII's *Liber Sextus* (1298), and Clement V's *Constitutiones Clementiae.* Later John XXII's *Extravagantes* (1325) and the *Extravagantes Communes* (c.1500) were added unofficially. L. FEEHAN

DECRETALS, FALSE, see FALSE DECRETALS

DECRETUM GELASIANUM *(Decretum de libris recipiendis et non recipiendis).* A document purportedly from the pontificate of Gelasius (492-96), although copies exist claiming provenance from the reigns of Damasus (366-84) and Hormisdas (514-23). The five divisions of the text as it stands are: Christ and the Holy Spirit; the books of the biblical canon, including those now placed in the Apocrypha; a statement of the bases of the claims to supremacy of the Roman see; a list of writings from the Fathers and the Councils approved by the Roman Church; and a list of noncanonical biblical and patristic books (from which the *Decretum* takes its customary title), the first appearance of any Index of Forbidden Books. Many copies of the text omit one or more parts. On the bases of both internal and external evidence compiled by E. von Dobschütz, who edited the published text, the Gelasian origin of the document is generally rejected, as is its claim to be an official document of the Roman see. The opening section and the list of the books of the Bible are attributed to a Roman synod of 382 under Damasus, and the rest of the text to an unknown private person, a native of Italy or Gaul, writing in the sixth century. MARY E. ROGERS

DE DOMINIS, MARCANTONIO (1566-1624). Scholar and controversialist. Born in what is now Rab, Yugoslavia, he became a Jesuit but left in 1596 on appointment as bishop of Segni. He was subsequently archbishop of Spalato (1602) and primate of Dalmatia until 1616. He had also been a professor of mathematics at Padua (propounding a scientific explanation of the rainbow which anticipated Newton) and of rhetoric and logic at Brescia. Involved in the quarrel between the papacy and Venice, he went to England (1616) and was appointed dean of Windsor and master of the Savoy by James I (1617). There, having written *De Republica Ecclesiastica,* attacking Rome and defending national churches, he attacked the English Church just as violently *(Sui reditus ex Anglia consilium).* He died in the Castel Sant' Angelo as a relapsed heretic, having been seized by the Inquisition. C.G. THORNE, JR.

DEER, BOOK OF, see BOOK OF DEER

DEFENDER OF THE FAITH, see FIDEI DEFENSOR

DE FOUCAULD, CHARLES EUGÈNE (1858-1916). Roman Catholic missionary and ascetic. Born in Strasbourg of a distinguished and devout family, he had a somewhat dissolute army career, but was honored by the Paris Geographical Society for his expeditionary work in North Africa. Impressed by Muslim religiosity, he sought faith under the direction of Abbé Henri Huvelin and gave himself to prayer and asceticism, followed by residence in Trappist* monasteries, in search of increased poverty and self-sacrifice (1890-96). He read theology in Rome (1897), lived thereafter until 1900 with the Poor Clares* at Nazareth, and in 1901 returned to France for ordination. Thereafter he went to the Sahara, established a hermitage at Beni-Abbès, and in 1905 went on to the Ahaggar mountains near Tamanrasset in Algeria. He learned the language and engaged in dictionary and translation work, but gave himself chiefly to prayer and charitable works, winning much native affection. He was murdered in circumstances that remain obscure. Publication of his personal papers led to the founding of Little Brothers of Jesus (1933) and Little Sisters of Jesus (1936). C.G. THORNE, JR.

DEISM. This designates a rationalistic mode of explaining God's relationship to the world which evolved in the late seventeenth and early eighteenth centuries. Following the classical illustration of God as a clock-maker, used by Nicolaus of Oresmes in the fourteenth century, the Deists stated that God gave the world its initial impetus and then left it to run its course. Consequently divine providence, revelation, and a supernatural scheme of salvation were called in question. Samuel Clarke* in his *Demonstration of the Being and Attributes of God* (1704-6) could distinguish four separate classes of Deists, for there was no single school of thought.

The religious wars and the beginnings of modern science in the seventeenth century created an atmosphere where reasonableness would replace intolerance and authoritarianism. Isaac Newton* had begun to unlock the secrets of the universe and John Locke* those of the human mind. Locke's *Reasonableness of Christianity* (1695) was a spur to the rationalization of the Christian faith, though he disavowed claims of Deists to be following his lead. As early as 1624 Lord Herbert of Cherbury had taught that all religions had five basic ideas in common and denied the need for revelation. One strand of Deism was not hostile to Christianity, but the tendency was toward a natural religion. In 1696 John Toland* published *Christianity not Mysterious,* and Matthew Tindal* produced the most competent exposition of this natural religion in *Christianity as old as the Creation* (1730).

Opposition by the orthodox provoked hostile attacks on Christian "evidences" cited in defense, especially fulfilled prophecy and miracles. Anthony Collins disputed the authority of the Bible; Thomas Woolston* questioned Christ's miracles and resurrection. Several ended in pantheism or atheism. Such attacks called forth a multitude of replies, the most famous being Joseph Butler's* *Analogy of Religion* (1736). While Deism never won a substantial following in England, it was widely acclaimed on the Continent in Germany and France, where Voltaire* and J.-J. Rousseau* became its chief advocates.

The Deists lacked any historical sense in assessing the biblical revelation. Faith was dethroned and man's rationality viewed with a false optimism. The initial revelation in nature was sufficient.

BIBLIOGRAPHY: C.J. Abbey and J.H. Overton, *The English Church in the Eighteenth Century* (1896); L. Stephen, *English Thought in the Eighteenth Century,* vol. I (1902); W.R. Sorley, *A History of English Philosophy* (1937); F. Copleston, *A History of Philosophy,* vol. V (1964).

J.W. CHARLEY

DEISSMANN, ADOLF (1866-1937). German NT scholar. He taught in Heidelberg (1895-1908) and Berlin (1908-34). In the light of secular Greek inscriptions and papyri, first intensively studied in the late nineteenth century, he showed NT language to be popular rather than classical, and held that the writings, especially the epistles, were unliterary. He gave a picture of primitive Christianity as a "popular cult," growing from mystical personal reaction to Jesus; it was not therefore to be explained in terms of the coherence of the development of "doctrine." For him, Paul was not a theologian but a man of the people responding to the impact of the Damascus road encounter. Deissmann focused scholarly attention on the Pauline concept "in Christ" (1892), by suggesting that Christ and Spirit were interchangeable and by using the analogy of man's being in air, as the air also was in man. Deissmann was both scholar and popularizer. He was deeply concerned for the relation between the church and contemporary culture, and was active in the early ecumenical movement.

HADDON WILLMER

DE LA CRUZ, APOLINARIO (Hermano Pule) (1815-1841). Filipino religious leader. Born in Quezon Province, he went to Manila to join a monastic order, but was rejected because he was a native. He worked instead in a religious hospital and studied theology in his free time. In 1840 he returned to Quezon and founded a religious brotherhood for Filipinos called *La Confradia de San Jose.* The movement grew widely among the masses, but was refused recognition by both ecclesiastical and colonial authorities. Although the movement was banned, meetings continued in secret. De la Cruz went into hiding after government troops were used to break up his meetings. In 1841 he proclaimed war in the name of religious freedom, and took the title "King of the Tagalogs." During a skirmish, a provincial government official was killed. De la Cruz was later captured, executed, and quartered—and the brotherhood was disbanded. He is now honored as "the first martyr to the cause of religious liberty and Filipino priestly equality." RICHARD DOWSETT

DE LAGARDE, PAUL ANTON (1827-1891). German oriental and OT scholar. He studied theology and philology at Berlin and Halle, but rejected the Pietistic faith of his father and teachers and became a lifelong foe of organized religion. He entered into a productive scholarly career in ancient oriental literature, especially the Septuagint, but his unpleasant personality made it difficult to secure a university chair. Even after receiving an appointment at Göttingen in 1869 he remained an unrestrained polemicist. Although his published works were fragmentary, he was regarded by contemporaries as an outstanding orientalist. A fervent nationalist, Lagarde was a leading critic of German society, and he saw a national religion, a Germanic-Christian faith, as the means to effect a spiritual regeneration.

RICHARD V. PIERARD

DELITZSCH, F.J. (1813-1890). Lutheran OT scholar. Born and educated at Leipzig where he taught for some years, he later held chairs at Rostock (1846-50), Erlangen (1850-67), and Leipzig (1867-90). From a pietistic background and of Jewish descent, he sought to combat both the extremes of anti-Semitism and of Zionism, and to aid in the conversion of the Jews to Christianity. To this end he edited a periodical, *Saat aut Hoffnung,* from 1863; founded a Jewish missionary college; translated the NT into Hebrew (1877); and established at Leipzig an *Institutum Judaicum* (1886). He published a number of OT commentaries of a conservative character. He examined carefully the critical theories of Wellhausen* and cautiously and without abandoning his concern for evangelical truth came to uphold the different literary strands in the Pentateuch and the dual authorship of Isaiah. During his lifetime his moderately critical views were probably more widely accepted in the English-speaking world than those of Wellhausen himself. He wrote extensively on rabbinical subjects, published with S. Baer an edition of the Hebrew text of most books of the OT, and wrote a few essays on dogmatic theology. But these do not impress, and it is as an exegete that he is chiefly remembered.

IAN SELLERS

DELLA ROBBIA, ANDREA (1435-1528). Nephew and adopted son of Luca della Robbia.* His works are principally in terracotta, and progressively demonstrate a departure from the refined simplicity of his uncle. Although greatly impressed by Luca's training, Andrea yielded to many outside influences, chiefly Verrochio's. His *Madonna and Child* is in the National Museum of Florence, and his series of medallions with infants *(i Puttini)*, done for the Foundling Hospital in Florence, is one of his best works.

See M. Cruttwell, *Luca and Andrea Della Robbia and Their Successors* (1902).

DELLA ROBBIA, LUCA (c.1399-1482). Master of sculpture and terracotta *(majolica)*. His works reveal a profound spirit of religiosity. His masterpiece, *la Cantoria di S. Maria del Fiore* (the singing and dancing boys), is presently in Museo del Duomo in Florence; it took seven years to complete (1831-38). Few Florentine churches of this period are lacking the touch of his deft hand and Gothic flair. The bronze doors of the sacristy of the Florence cathedral took him thirty years to finish and not only illustrate his serene vision of life, but attest that it is not without merit he is called the "most classic" of the 1400s.

ROYAL L. PECK

DE LUGO, JUAN (1583-1660). A founder of modern Scholasticism. Born in Madrid, he studied law at Salamanca until he entered the Jesuits in 1603. After teaching philosophy at Medina del Campo (1611) and theology at Valladolid (1616), he was called to Rome in 1621, where he further enhanced his reputation by important dogmatic and moral works, in addition to becoming a cardinal in 1643. A profound and independent thinker, he was inclined to over-subtlety, but was an important influence on Alphonsus Liguori,* and combined dogmatic and moral theology more satisfactorily than many of his contemporaries.

DE MAISTRE, JOSEPH MARIE (1754-1821). Catholic philosopher, and one of the initial proponents of Ultramontanism.* He experienced the French Revolution as terror and anarchy when the revolutionary armies invaded Savoy (1792). He linked such action to the rationalism of the *philosophes*, many of whom he knew well. His *Lettres d'un royaliste savoisien* (1793) first stated the argument which he elaborated in *Essai sur le principe générateur des constitutions politiques* (1814) and his foremost work *Du pape* (2 vols., 1819). Society, he affirmed, is dependent on authority to cohere. It is God who establishes authority by divine sovereignty which is then reflected in the sovereignty of the popes, who are infallible in spiritual things, and of monarchs, infallible in temporal things. Divine Providence in history creates the traditions of society which are a revelation of the order necessary to men's existence (cf. TRADITIONALISM). De Maistre's emphasis on faith, rather than rationalism, and on an organic, rather than mechanistic, view of history, found wide acceptance among Catholics in France. He formulated most of his ideas while serving as Savoy's ambassador to Russia (1802-16). After an education in law at Turin, he entered the Savoy civil service (1774), the senate (1788), and was appointed twice as regent (1799, 1817).

BIBLIOGRAPHY: J. de Maistre, *Oeuvres complè tes* (14 vols., 1884-86); F. Bayle, *Les idées politiques de Joseph de Maistre* (1945); A. Caponigri, *Some aspects of the philosophy of Joseph de Maistre* (1945).

C.T. MC INTIRE

DEMETRIUS (d. c.231). Bishop of Alexandria. He is said to have sent Pantaenus, head of the catechetical school at Alexandria, to preach to the Indians, but this is conjectural. He did, however, have a lively interest in the catechetical school, and about 203 appointed Origen as its head. The breach that occurred between Demetrius and Origen, when Origen preached to the congregations of Theoctitus of Caesarea and of Alexander of Aelia Capitolina (Jerusalem) in 216 was reopened in 228 when Alexander and Theoctitus ordained Origen a presbyter. Demetrius brought Origen to trial and deposed him. His attitude toward Origen's ordination may have been prompted by Alexandrian usage, which later explicitly prohibited the ordination of a eunuch, but Eusebius suggests that it was prompted rather by jealousy. There is some evidence that Demetrius changed the system of appointing bishops in Egypt, and that he wrote letters on the keeping of Easter, maintaining the view adopted at Nicea.

DAVID JOHN WILLIAMS

DEMIURGE. from a Greek word denoting a skilled worker, it was in the Platonic system the name for the maker of the world, and so used by Greek Christian writers to refer to God's creative activity. It was, however, taken by the Gnostics to refer to a god who, as creator of the material world, thinks himself to be the Supreme God, but who in reality is incapable of knowing spiritual things. He deceives men by declaring in the Bible that he is the true God.

DEMYTHOLOGIZATION. The English translation of the German *entmythologisierung*, used originally by R. Bultmann* to describe a particular aspect of his hermeneutic. In 1941 he circulated an essay in duplicated form called *Neues Testament und Mythologie*. In this he argues that the whole thought-world of the NT is mythological. Previous critics had argued for the presence of myth in the NT, as in the OT. "Myth" was taken to mean the pictorial expression in narrative form of some great philosophical or theological truth. Bultmann's argument was that the "three-decker universe," together with the preexistence, virgin birth, deity, resurrection and ascension, and parousia of Jesus, and also the doctrines of the Trinity, sin, and the atonement, needed to be translated out of the mythical form in which they were cast. He believed that modern man found the *kerygma* incredible because he was convinced the mythical view of the world was obsolete.

Bultmann also stated that the mythology of the NT contained contradictions. His direct onslaught upon the historicity of the gospel narratives, while having skeptical features similar to those of the older liberalism, was distinct from it in trying still to see a meaning for everything. This meaning was found in terms of existentialist philosophy. Bultmann laid great stress on the *kerygma* as a proclamation calling men to authentic decision, but this could only be meaningful when it had been demythologized. Many other leading scholars, including K. Barth* and E. Brunner,* have worked in a more moderate way on the same lines.

Bultmann's essay is included in H.W. Bartsch, *Kerygma and Myth* (ET 1953). R.E. NIXON

DENCK, HANS (c.1495-1527). Anabaptist leader. Born in Heybach, Bavaria, he studied in Ingolstadt. His training was along humanistic lines, and he became acquainted with humanists in Augsburg. He went to Basle, where he became the friend and student of Oecolampadius.* In 1523 he went to Nuremberg to teach at St. Sebald's. There he became involved in the judicial trial of the "three impious painters," Sebald, Barthel Behaim, and George Peng. Here Spiritualism together with the ideas of Thomas Münzer and Andreas Karlstadt influenced him greatly. About October 1525 he was forced to leave Nuremberg, and he became a wanderer throughout S Germany. He spent about a year in Augsburg, where he was rebaptized by Hubmaier; there he also attended the "martyr synod," and there established himself as a leader of the Anabaptists.* In his writings he opposed the doctrines of predestination, the bound will, justification by faith, the sufficiency of Christ's atonement, the authority of the Scriptures, the necessity of baptism and the Lord's Supper, and the ministry. He returned to Basle in 1527, where he succumbed to the plague. CARL S. MEYER

DENMARK. Christianity developed in Denmark from about 735 to 1060, partly through influences from the Anglo-Saxon world, partly through direct Christian mission from the south (see ANSKAR). The baptism of King Harold Blue Tooth about 960 gave impetus to the movement and gave the Christian Church an officially recognized status in the kingdom. The mission period was followed by a great church-building era. Before the end of the twelfth century, about 1,800 parish churches were built throughout the country. In 1104 a Scandinavian archbishopric was established in Lund. Generally, the subsequent history of the medieval church in Denmark advanced along the same lines as in other European countries. Archbishop Eskil of Lund (1137-71), who was inspired by Gregorianism, succeeded in realizing several of his ideas of church policy. At the same time he laid the foundation of a durable and harmonious cooperation of church and crown. There were indeed quarrels between 1231 and 1340, and during the late Middle Ages the church frequently had to endure the interference of king and councils, but in many respects the cooperation between church and state was continued for the good of both.

The Reformation in Denmark (1500-1560), as in other countries, was a complicated process, closely bound up with social and political conditions. The way was prepared by a biblical humanism of the Erasmian type. During the period 1523-36, the preaching of some Lutheran ministers brought about a spiritual revival that finally led to the official accomplishment of the Reformation in 1536 (see ODENSE, DIET OF; TAUSEN, H.; PALLADIUS, P.). During the "Age of Orthodoxy" (1560-1700) nothing but pure Lutheranism was tolerated in the kingdom. About 1660 Denmark became a hereditary and absolute monarchy. In fact the church had no kind of independence, but was simply the institution through which the absolute monarch ruled the religious affairs of his country. Nevertheless, throughout this age of political and religious absolutism, true spiritual life was upheld and nourished by the prevalence of serious biblical preaching and through the reading of devotional works such as J. Brochmand's book of family sermons. This period fostered also Thomas Kingo, one of the Danish Church's greatest hymnwriters.

About 1700 the influence from German Pietism was felt in Denmark; soon the revival developed into two different main streams: a church-pietism, adopted and supported by the clerical authorities and the king; and a conventicle-revivalism, which was decidedly critical of the church and developed separatistic tendencies. The church-pietism was victorious, but died out shortly after the middle of the century. The conventicles were put under the strict control of the clergy by a special decree *(Konventikelplakaten)* in 1741. After this, only a few small groups survived here and there. More lasting effects of the Pietist revival were achieved through the beginning of a foreign mission in 1705, a number of philanthropic and educational initiatives, and through the hymns of H.A. Brorson, the greatest hymnwriter of the Pietist movement. For some years after 1750 supranaturalism was predominant, but soon it was superseded by various shades of rationalism. During the "Age of Enlightenment and Rationalism" (1750-1800), much energy was spent on useful social reforms. Otherwise it was a period characterized by widespread religious indifferentism and serious spiritual decline.

The first half of the nineteenth century was a period of transition. Gradually the old rationalism was overcome through the influence of men such as N.F.S. Grundtvig,* J.P. Mynster,* and H.L. Martensen,* and through the rising tide of revivalism, originating in the conventicles of pietistically minded laymen. In 1849 the common people's fight for freedom and independence, religiously as well as politically, finally led to the overthrow of the absolute monarchy and the spiritual coercion of the state-church system. The principle of religious liberty and freedom of conscience was legally established in the basic law of the new constitution. From about the middle of the century, many from "the awakened circles" joined the Grundtvigian movement. Others founded the more pietistic Indre Mission.* Smaller groups joined the Lutheran Mission (Bornholmians*) or various free churches. Although he never became the founder of a movement, S. Kierkegaard* exerted a deep and lasting influence.

Twentieth-century church life in Denmark has developed mostly as a continuation along the lines established in the nineteenth century. None of the old movements, however, has escaped the influence of modern trends: liberal theology, Neoorthodoxy, existential and secular theology. The most important influence has been exerted by the *Tidehverv* (i.e., epochal turning point) movement, started in the late 1920s by a group of theologians inspired by Kierkegaard, the early Karl Barth, Luther, and afterward more profoundly by Bultmann. The post-World War II years have been characterized by religious indifferentism and by moral and spiritual decline, but there have also been some signs of an approaching evangelical renewal and awakening.

So far the great majority of the population maintain membership of the Evangelical Lutheran Church, recognized and supported by the state as the national church. The free churches have not gained much following in Denmark: Roman Catholics number 26,000; Baptists no more than 8,000; Salvation Army 5,500; Methodists 3,500; Pentecostals 4,000; Seventh-Day Adventists 4,000; and Mission Covenanters 2,500.

BIBLIOGRAPHY: Most of the material is in Danish, but books in English include J.C. Kjaer, *History of the Church of Denmark* (1945); E.H. Dunkley, *The Reformation in Denmark* (1948); P. Hartling (ed.), *The Danish Church* (1964).

N.O. RASMUSSEN

DENNEY, JAMES (1856-1917). Scottish theologian. Born in Paisley, he graduated in arts with great distinction at Glasgow University, was trained in theology at the Free Church College in that city, and in 1886 became minister of the East Church, Broughty Ferry. During his eleven years there he published commentaries on *Thessalonians* and *Second Corinthians*, and *Studies in Theology*. In 1897 he took the chair of systematic theology at his former college, transferring to the NT department in 1900. He was college principal for the last two years of his life. In 1900 appeared his monumental work on *Romans*, followed in 1903 by *The Death of Christ*. He once declared that the only theology he was interested in was the theology that could be preached. He became a leader in his church's courts after his church united with the United Presbyterians in 1900 to form the United Free Church,* and contributed to the distant (1929) union with the national church of Scotland. His other writings include *Jesus and the Gospel* (1908), his book of sermons *The Way Everlasting* (1911), and the posthumous *Christian Doctrine of Reconciliation*. Denney was also a strong protagonist of temperance and civic righteousness.

J.D. DOUGLAS

DE NOBILI, ROBERT (1577-1656). Jesuit missionary to India. Born into a wealthy family at Montepulciano, he became a Jesuit in 1596 despite family opposition. Arriving in India in 1605, he served on the Fisher Coast before going to Madura for thirty-six years. His senior companion aimed to turn converts into Portuguese. De Nobili rejected this and met bitter criticism because he dressed as a *sannyasi* and lived in the Brahmin quarter, so that he could be met without defilement. He became the first European to have firsthand knowledge of Sanskrit, the Vedas, and Vedanta. His first convert, Sivadarma, was baptized in 1609 and permitted to retain Brahmin insignia. Fellow priests regarded this as a betrayal of Christianity, and he was inhibited from ecclesiastical functions by the local authorities. After a long delay, which de Nobili used to write a number of books in the local vernaculars, Gregory XV upheld de Nobili's appeal in *Romanae sedis antistes*. Resuming missionary activity, he traveled widely in new areas till 1654, when he was retired to Myalpore and spent his two remaining years revising his books, despite increasing blindness. By the time of his death there were several thousand converts, including some from high castes. His attempt to distinguish between Christianity and its cultural trappings has proved to be of permanent importance in Christian missions.

BIBLIOGRAPHY: J. Bertrand (ed.), *La Mission du Madure* (4 vols., 1847-54); P. Dahmen (ed.), *Première Apologie* (1931); M. De Crisenoy, *Robert De Nobili* (1939); V. Cronin, *A Pearl to India* (1959).

IAN BREWARD

DENYS, see DIONYSIUS OF PARIS

DE RANCÉ, ARMAND JEAN LE BOUTHILLIER (1626-1700). Founder of the Trappists.* He was presented with lucrative benefices (including the abbacy of Cistercian La Trappe) at an early age, and for over thirty years lived the life of a noble. Ordination in 1651 made little difference, but the death of his close friends the duchess of Montbazon and the duke of Orléans made him much more seriously religious. Renouncing his possessions, he became a Cistercian at Perseigne in 1663 and regular abbot of La Trappe. An unsuccessful series of attempts to gain autonomy for Cistercians of Strict Observance from Cîteaux was followed by a life devoted to the reform of La Trappe, which soon became both admired and criticized for its austerity. In *De la sainteté et des devoirs de la vie monastique* (1683), de Rancé argued for the strictly penitential nature of monastic life, excluding even study. Taken vigorously to task by Jean Mabillon, he engaged in lively controversy. He resigned in 1695, his health broken by austerity. IAN BREWARD

DE' RICCI, SCIPIO, see RICCI

DESANCTIS, LUIGI (1808-1869). Parish priest in Rome, Protestant pastor, and theologian. Perturbed by the moral and doctrinal corruption of the church, he went through a long spiritual crisis and finally left the Roman Catholic Church, taking refuge in Malta (1847). There he revived the *Chiesa evangelica italiana,* collaborated in the first Protestant journal, and published *Il cattolico cristiano,* another of deeper theological content. After a stay in Geneva (1850-52) as evangelist to the Italian exiles, he was ordained pastor in the Waldensian* Church and sent to Turin, where his presence gave a great impulse to the community. Disagreements with the Waldensians caused him to leave them and work for ten years among the free Italian churches, mainly in Liguria and Piedmont. He joined A. Gavazzi,* attracted by his dream of founding a national Italian church, but finally returned to the Waldensians, moving to Florence as pastor and lecturing in their faculty of theology. He was a prolific author, and his *Roma papale* remains a document of great value for its detailed account of the religious life in Rome at the time of Pope Gregory XVI.

See also RISORGIMENTO and GUICCIARDINI.

DAISY RONCO

DES BRISAY, THEOPHILUS (1755-1825). First Anglican clergyman on St. John's (Prince Edward) Island, he went there in 1775, and although the only Protestant clergyman there, he received no stipend and so became the chaplain for an army regiment stationed there. After he provided intermittent services for some time, the government set apart £150 for his support, and so he began to provide regular services, although he visited his parish of Charlottetown only on Sundays. A tolerant churchman, he even welcomed the Methodist evangelist William Black during the latter's preaching tour of 1783.

DESCARTES, RENÉ (1596-1650.). French philosopher. Born at La Haye in Touraine, he studied at the Jesuit College of La Flèche (which he came later to regard as a waste of time). After studying law and attempting a military career in Holland, he became absorbed in the idea of extending the "geometrical method" to all branches of learning, thus freeing them from doubt and disorder. After some years in France and Italy, he lived in Holland until 1649. He visited Sweden at the invitation of Queen Christiana, and he died there.

His philosophy must be understood as an attempt to establish absolute certainty by means of an appeal to the "clear and distinct ideas" of human thought. By a process of methodical doubt (made famous by the *Discourse of Method* and the *Meditations*), Descartes arrives at an indubitable first principle, *cogito ergo sum.* By means of *a priori* arguments, he establishes the existence of God who "secures" the reality of the external world despite the uncertainty of sense perception. Implicit in the methodical doubt is Descartes's dualism—man is essentially a "thinking thing," a spiritual substance, having free will, inhabiting a body, a material substance. Between the two there is a mysterious interaction.

The sort of impact that his views had may be seen in the controversy in the University of Utrecht between the distinguished Reformed theologian Voetius and Leroy of the medical faculty. Descartes was accused by Voetius of atheism and defended himself in a long letter in 1643. But Descartes's influence has been pervasive, in the development of continental rationalism (Spinoza, Leibnitz) and through Locke on the kinds of questions that empiricism has been concerned with. OONAGH MC DONALD

DESCENT INTO HELL. This should more properly be called "the descent into Hades," for the Greek word *Hades* renders the Hebrew *Sheol* and means "the place of the departed." This is to be distinguished from the word *Gehenna,* which refers to "the place of punishment." Peter on the day of Pentecost quotes Psalm 16 and then refers it to Jesus, saying that "he was not abandoned to Hades" (Acts 2:31 RSV), and Paul tells his readers not to ask who will go into the abyss to bring up Christ from the dead (Rom. 10:7). These references are reinforcing NT statements about the reality of the death of Christ and therefore the reality of His resurrection. He shared to the full the human experience of death.

The descent was asserted in some early Arian creeds. It was not in the Old Roman Creed, but reached the Apostles' Creed via the Aquileian Creed of Rufinus about 400. It is not mentioned in the Nicene Creed. As to what Christ did in descending to Hades, beyond sharing human experience, we have the two very difficult passages in 1 Peter 3:18-20; 4:6. The proclamation of Christ's triumph seems to be involved, but there is no clear agreement about the nature of the message or the type of people who heard it.

R.E. NIXON

DES PREZ, JOSQUIN (c.1440-1521). Musical composer. Born in what is now southern Belgium in the twilight of the Middle Ages, Josquin lived to exemplify in his own works almost every aspect of the fully developed musical style of the Renaissance. He served the Sforzas in Milan, worked at Ferrara at two periods of his life, sang for a time in the papal choir, graced the court of Louis XII, and returned a revered artist to end his days at Condé near his birthplace. He was greatly admired by Luther, who mentions him several times. His works were known and performed from Seville to Warsaw. His greatest music is in his motets, of which almost one hundred are preserved, as well as twenty Masses. In his music, the customary four-part texture of soprano, alto, tenor, and bass becomes standard. He also used consistently the imitative contrapuntal texture that became the norm in choral music throughout the sixteenth century and beyond. His music continued to be admired long after his time, and an Italian writer in 1567 places him on an artistic plane with Michelangelo. J.B. MAC MILLAN

DETERMINISM (Lat. *de* + *terminus,* "end"). The general philosophical thesis which states that all events are subject to a rigid law of cause and effect. Given any set of conditions, only one outcome is possible. This rules out any concept of free will. There are five basic approaches to determinism:

(1) *Ethical:* knowledge determines choices. Therefore, if one knows the good, he automatically follows it (cf. Socrates, Plato, Descartes, Aquinas, Leibnitz).

(2) *Logical:* men's minds are fettered and that nothing can be altered by them. This is equal to fatalism (cf. Stoics).

(3) *Theological:* the universe and everything in it are absolutely dependent upon God. The absolute goodness of God predisposes some to say that this means all things are good. The omniscience and omnipotence of God leads others to the ideas of foreordination and predestination. God foreknows, purposes, and does everything according to His eternal, changeless, and infallible will. Man without God is determined to sin, and with God he is determined to salvation. The latter statements are indicative of the theology of Luther and Calvin.

(4) *Physical:* all things in nature (men included) behave according to inviolable and unchanging natural laws. Thomas Hobbes expressed this philosophy.

(5) *Psychological:* all human behavior is precipitated by causal factors. B.F. Skinner is the best modern exponent of this view.

JOHN P. DEVER

DEVIL. The devil is referred to in the OT sometimes by the name Satan, which originally means "an opponent" (cf. Num. 22:22ff.; 1 Sam. 29:4; 1 Kings 11:25). In Genesis 3, while there is no direct identification made of the serpent and the devil, it is at least implied that what has gone wrong with human life cannot be explained except on the presupposition of the invasion and infection of human life by some malignant power whose hatred of God goes beyond anything of which man is capable on his own initiative (cf. Matt. 13:28). At the climax of the Bible story we see Christ struggling, not simply to reform and repair the evil wills of men and call them back to God, but against some force of titanic proportions whose challenge to what is good demands an agonizing and total response from God Himself (1 John 3:8). Jesus' life is a struggle against one who is mighty (Mark 3:27); the hour of his agony is the climax of this struggle (Luke 22:53). Yet there is no question about his triumph (Luke 10:18). The devil must therefore be regarded as an alien personal force in the universe, which seeks to annihilate what God has created, to bring chaos where there is order, darkness where there is light; which defies God with superb hatred and pride; which manifests itself in human sin, especially in the opposition which Jesus met and overcame in the cross. The intrusion of this power into human life does not absolve man from responsibility. Man's consent to its work rather increases his guilt, yet the powers of evil, even in the temptation of man, are under the control of God.

In the OT the figure of Satan is seldom depicted in such a sinister and powerful role as in the NT, but a development toward such a view can be traced (1 Chron. 21:1; Isa. 14:12-21; Zech. 3:1,2). The NT repeats the doctrine of fallen angels (Jude 6-8), and Christ's words in John 8:44 have to be interpreted in this light. There are OT references to various kinds of demons (Isa. 34:14; Lev. 17:7), who under the decree of God can incite men to folly and sin (Judg. 9:23; 1 Sam. 16:14; 1 Kings 22:22ff.), and it is recognized that those who practice communion with departed spirits are in danger of becoming allied with, and possessed by, a whole world of evil (Lev. 19:31; Deut. 18:10; Exod. 22:18). In apocalyptic literature (e.g., Daniel) there are suggestions of a kingdom of evil, a continuing war in heavenly places, and of satanic powers involved in earthly political movements and conflicts.

The NT recognizes the unity of this kingdom (Mark 3:22-27) under one head (Rev. 12:7,8). Because the devil is doomed and his time is short, he is the more full of anger. Therefore Christians are warned to watch and resist (1 Pet. 5:6-9; 2 Pet. 2:4; Jude 6), recognizing the greatness of his power to possess and trouble men (Luke 22:31f.; Eph. 6:11,12). The NT suggests a connection between idols, witchcraft,* and demons (1 Cor. 10:20; Gal. 5:20; Rev. 9:21).

Theology has tried to do justice to the fact that all creation is basically good. It has stressed the absurd, accidental, and voluntary nature of the fall of the angels. It has been careful to avoid any idea that the devil could be a dualistic counterpart of God—an eternal being involved in an eternal struggle between good and evil. "The devil," said Luther, "is God's devil." Christian piety has always sought to do justice to the fact that all evil is made infinitely serious in its guilt and consequences in the light of the cross of Christ, and that nevertheless all evil has been overcome and exposed in its meaninglessness and futility.

BIBLIOGRAPHY: E. Langton, *Essentials of Demonology: A Study in Jewish and Christian Doctrine* (1949); G.B. Caird, *Principalities and Powers: A Study in Pauline Theology* (1956); M.F. Unger, *Biblical Demonology* (1963); J.G. Kallas, *The Satanward View* (1966); R.S. Kluger, *Satan in the Old Testament* (1967).

RONALD S. WALLACE

DEVIL'S ADVOCATE *(advocatus diaboli).* The person appointed by the Roman Catholic Congregation of Rites to contest the claims of a candidate proposed for beatification or canonization before a papal court. Officially called *promotor fidei* ("promoter of the faith"), he is opposed by God's advocate *(advocatus Dei),* who supports the proposal.

DE WETTE, WILHELM MARTIN LEBRECHT (1780-1849). German biblical scholar; one of the most influential theologians of the nineteenth century. After studying with J.G. von Herder at Weimar and J.J. Griesbach, J.P. Gabler, and H.E.G. Paulus at Jena, he taught at the universities of Heidelberg (1807-10), Berlin (1810-19), and Basle (1822-49). He wrote many books, including OT and NT introductions (1817, 1826), a monumental work on Christian ethics (1819-23), numerous commentaries and historical works, and a translation of the Bible (with J.C.W. Augusti, 1809-13; reworked and published under his own name in 1831-32). Although he was himself a thoroughgoing nonsupernaturalist, he was a continual critic of the theories of F.C. Baur* and his disciples.

W. WARD GASQUE

DIADOCHUS. Bishop of Photike about the middle of the fifth century. He belonged to the generation immediately after the great Greek fathers Basil and Gregory of Nazianzus, and wrote against the Arians and on the Ascension. But he is best known for his work on Christian perfection, *peri gnōseōs pneumatikēs.* This comprises a foreword with ten definitions of virtue and one hundred *Capita Gnostika,* under which he attempts to analyze what it means to be Christlike and how to live such a life. He finds its basis in the three theological virtues, especially love. Diadochus was evidently a man of considerable culture which he combined with a deep faith in Christ. His work on Christian perfection enjoyed great popularity in succeeding generations and is quoted by Maximus the Confessor, in the *Doctrina Patrum,* and by Photius. It is still of practical value and deserves to be better known than it is.

DAVID JOHN WILLIAMS

DIALECTICAL THEOLOGY. The title used to characterize the theological methodology of Karl Barth,* to distinguish his dogmatic principles from those of the liberal traditions which had reached their climax in F.D.E. Schleiermacher.* The first phase of the Barthian movement was termed the "theology of crisis." It appeared superficially as a "desperation-theology" born out of disillusionment with World War I, and expressing theologically the historical and cultural pessimism given vogue in Spengler's *Decline of the West.*

The Dialectical Theology was not, however, just an echo of a tragic historical situation. It was concerned rather with the judgment of God, not in a particular concrete situation, but with the Divine "No" to all human efforts, and especially to the religious search for righteousness. Yet the "No" is not God's only and final word, for it is the very occasion for His "Yes." This is the true dialectic: God's "No" finding an answer in His "Yes." God's judgment is, so to speak, the darker side of His grace: the "No" is overcome by the "Yes." The dialectic finds its resolution in God alone, for the "No" of God's judgment cannot be met by a balancing "Yes," having its origin and impetus on man's side. "There is no way from us to God, no via negativa, no via dialectica, no via paradoxa" (Barth). Thus almost paradoxically the dialectical apprehension of God transcends every dialectical method.

Dialectical Theology sought its method and principles in the theology of the Reformation, and especially in John Calvin.* Its general thesis was expounded by, among others, H.E. Brunner* and F. Gogarten.* The influence of Dialectical Theology has been extensive, especially in the Reformed Churches in Europe and in the Church of Scotland. The Church of England and the Free Churches of Britain have also been affected.

See bibliography under BARTH, KARL.

H.D. MC DONALD

DIAMPER, SYNOD OF (1599). An unofficial synod held on the Malabar coast of SW India, uniting the so-called Thomas Christians with the Church of Rome. Allegedly dating back to the missionary work of Thomas the Apostle, these native Christians followed the Syriac liturgy of Addai and Mari.* The Portuguese discovered them in 1498, and for a time they lived in harmony with the Portuguese-sponsored churches. By the end of the sixteenth century, however, their Nestorianism* and their refusal to conform to a Latin liturgy led the Portuguese archbishop of Goa, Alexis de Menezes, to convoke the synod. It rooted out the Nestorian heresy and began the Westernization of the church.

BRIAN G. ARMSTRONG

DIASPORA. A technical term for Jewish communities outside Palestine from about 100 B.C. to A.D. 100; the Greek word thus transliterated means "dispersion." Settlements of Hebrews outside the Holy Land began with the deportations to Assyria and Babylonia in the eighth-to-sixth centuries B.C., and later migrations to Egypt from about 525 B.C. The Jewish communities in Babylon and Alexandria became large and prosperous, and further migration led to the establishment of settlements in major cities throughout the Roman Empire as well as the East. During NT times, it is estimated, Jews represented eight to ten percent of the population of the Roman Empire. Synagogues of the Dispersion served as centers of the early Christian mission. The word occurs three times in the NT: once in the technical sense (John 7:35) and twice as a symbolic description of Christians (James 1:1; 1 Pet. 1:1).

W. WARD GASQUE

DIATESSARON. A Greek musical term, meaning "a harmony of four parts," which was the title given to a harmony of the four gospels composed by the Assyrian Christian Tatian* soon after A.D. 150. This was, so far as is known, the first time such a task was attempted. Tatian spent some time in Rome as a disciple of Justin Martyr before returning to Assyria about 172. He founded a sect called the "Encratites"* who had certain ascetic practices including vegetarianism, so that in the *Diatessaron* John the Baptist is made to feed not upon locusts but upon milk and honey. The *Diatessaron* was very popular among Syriac-speaking Christians, but it is not clear whether it was first composed in Syriac. The title is Greek, and F.C. Burkitt even suggested that it might have been composed originally in Latin.

There are no extant MSS of the whole of the *Diatessaron,* though a fragment in Greek of Joseph of Arimathaea's request for the body of Jesus was discovered in 1933 at a Roman fort at Dura-Europos on the Euphrates. In the fourth century, Ephraem Syrus wrote a commentary on the *Diatessaron* and an Armenian translation of this was discovered in 1836. A considerable portion of the

Syriac original of this came to light in 1957. There are extant in addition two late Arabic MSS of the *Diatessaron,* a medieval Dutch harmony of the gospels which is dependent upon it, and the Latin Codex Fuldensis which has the order of the *Diatessaron,* though the text has been assimilated to the Vulgate. The *Diatessaron* was very popular among Syriac-speaking Christians, and steps had to be taken in the fifth century to abolish its use.
R.E. NIXON

DIBELIUS, MARTIN (1883-1947). German NT scholar and theologian. He taught in Berlin (1910-15) and Heidelberg (1915-47), where he succeeded J. Weiss. When his early interest in Semitic languages and comparative religion gave way to NT studies, he soon made the important transition from problems of literary criticism to concern with oral tradition in the gospels. Thus, along with K.L. Schmidt, H. Gunkel, and R. Bultmann,* he helped to found Form-Criticism.* *Die Formgeschichte des Evangeliums* (1919; ET, *From Tradition to Gospel,* 1934) presented his view that the church's preaching was the medium of the transmission of the words of Jesus. The gospels are thus to be regarded as popular rather than "high" literature, and the Evangelists as compilers of traditional material rather than independent authors. Dibelius tended to be more restrained in his judgments than others in the Form-Critical school. Dibelius also did important work on the origin and history of ethical instruction in the NT and the relation of eschatology and ethics. As a leader of the Faith and Order* movement he worked for an adequate theological basis for the ecumenical movement as a whole. HADDON WILLMER

DIBELIUS, OTTO (1880-1967). Bishop of the Berlin-Brandenburg Church. Son of a civil servant, he studied theology at Wittenberg, became a pastor of the Reformed Church, and for some years ministered to what he called "quite unpretentious congregations." He came to prominence in 1933 when he was removed from his post as Lutheran superintendent in Berlin on refusal to recognize the church overseer appointed by Adolf Hitler. In 1934 he was a strong supporter of the Barmen Declaration* that asserted the primacy of Christ and opposed nationalization of the churches. Three times he was arrested by the National Socialist authorities, and though acquitted was forbidden to speak or publish. At the close of World War II he became bishop of a Berlin divided between East and West, a division later symbolized by the notorious Wall, and was instrumental in uniting several Protestant bodies into the German Evangelical Church. He was as fearless in resisting Communist demands as he had been under the Nazis; nevertheless, his resignation in 1961 was not accepted because East and West could not agree on a successor. A staunch supporter of the ecumenical movement, he had attended the 1910 Edinburgh Conference* and later became a president of the World Council of Churches (1954-61). Shortly before his death he participated in the 1966 Berlin World Congress on Evangelism. J.D. DOUGLAS

DICKINSON, JONATHAN (1688-1747). Presbyterian theologian, educator, and revivalist. Born in Hatfield, Massachusetts, he graduated from Yale (1706-7) and became pastor of Elizabethtown Presbyterian Church, New Jersey (1708-47). Said to have been "the most powerful mind in his generation of American divines," he was a reconciling influence in the controversy over subscription to the Westminster Confession,* and authored the 1729 Adopting Act which brought about a compromise. With Francis Makemie, Dickinson led the opposition to Anglican establishment attempts in the middle and southern colonies, stressing in numerous pamphlets "the unalienable rights of mankind" and preparing an increasing revolutionary mood, although his intent was not political but strictly theological. In 1739 the Great Awakening* divided Dissenters; the Presbyterians split in 1741. At first Dickinson kept a mediating position, but seeing that the "Old Side" (antirevivalists) were bent on schism, he abandoned reconciliation efforts in 1741 and joined George Whitefield and Gilbert Tennent's* "New Side," providing theological works to curb excesses and align revivalism with moderate Calvinism. One of the institutional embodiments of the Great Awakening was Dickinson's leadership in founding the College of New Jersey (1746), for men preparing for the New Side ministry. Dickinson served as first president until his death. A prolific and profound theologian, he was regarded as the most efficient champion of Calvinism in the colonies, with the exception of Jonathan Edwards.* KEITH J. HARDMAN

DIDACHE, THE. A Greek handbook of instruction in morals and church order, of which the full title is "The Teaching of the Lord to the Gentiles through the Twelve Apostles." It was first discovered in modern times by Philotheos Bryennios,* Greek metropolitan of Nicomedia, who found it in a library at Constantinople in a manuscript dated 1056 which contained also the epistles of *Barnabas* and *Clement.* Its discovery immediately provoked a flood of literature about it, but it has not been possible to give it a precise setting with any confidence. It was probably known to Clement of Alexandria and was considered by Eusebius to be almost a canonical NT book. It was used as a basis for part of the fourth-century *Apostolic Constitutions** as well as other "Church Orders." There is a close relationship between the *Didache* and the *Epistle of Barnabas,* normally now accounted for by their use of a common source.

Various views about its dating have been put forward, ranging from the idea that it is a first-century document to the suggestion that the author was writing in the third century and consciously archaizing. On the whole, a date in the earlier part of that range seems likely, but it is impossible to pin down either its date or place of origin with certainty.

The first six chapters present a Christian moral code under the headings "The Way of Life" and "The Way of Death." While there are references to the Sermon on the Mount, this section may be dependent upon a Jewish source. Chapters 7-10 deal with baptism, fasting, and the Eucharist.

They specify immersion in the threefold name in running water, but other water and affusion are allowed if this is not possible. Fasting was not to be done "with the hypocrites; for they fast on the second and fifth days of the week, but you must fast on the fourth and on the Preparation." The Eucharistic prayer is strongly eschatological, using the phrase "Let grace come, and let this world pass away" and the Aramaic *Marana tha.* Chapters 11-15 are largely concerned with the ministry, special emphasis being placed upon prophets as well as bishops and deacons. Tests are given to discover who are false prophets. The final chapter deals with the Second Coming and the end of the world.

BIBLIOGRAPHY: Ed. with facsimile by J.R. Harris (1887); numerous other editions and translations; F.E. Vokes, *The Riddle of the Didache* (1938); C.C. Richardson (ed.), *Early Christian Fathers,* I (1953), pp. 161-79; J.-P. Audet, *La Didachè, Instructions des Apôtres* (1958).

R.E. NIXON

DIDASCALIA APOSTOLORUM. A book of church order probably of Syrian origin. Apart from fragments of the original Greek, it has survived complete only in Syriac, partially in Latin. Various oriental adaptations are known, and it was included in the later *Apostolic Constitutions** in a revised form. It is to be dated in the third century. It deals with six main topics: (a) standards of the Christian life: the Decalogue binds the Christian, but the ritual commands are a second law *(deuterosis)* imposed as punishment; (b) the bishop, who is supreme leader and teacher of his church: usually he should be over fifty, husband of one wife, generous and merciful to his flock; presbyters are mentioned only incidentally; (c) widows, evidently numerous and troublesome, who ought to attend to duties of mercy; (d) orphans, who should if possible be adopted by other childless Christians; (e) martyrs and confessors, who should be cared for by the church, and who are upheld by the resurrection hope and the example of Christ; (f) heresy and schism, both Judaizing and Gnostic interpretations. The book is significant for the history of penitential discipline. Apart from canonical Scriptures, widely used, the unknown author uses some apocryphal Christian writings and works of the Apostolic Fathers.

J.N. BIRDSALL

DIDEROT, DENIS (1713-1784). French encyclopedist and key figure of the Enlightenment* philosophy. Born in Langres, son of a master-cutler, he studied in local Jesuit schools and at the College of Louis le Grand, Paris, receiving his master's degree (1732). He married his secretary in 1743. After translating English works, he published a defense of natural religion, *Pensées philosophiques* (1746). He became editor-in-chief of the *Encyclopédie,* which monumental enterprise was his chief occupation until its completion in 1772. In 1749 he was imprisoned briefly because of his *Lettre sur les aveugles* which questioned the existence of purpose in the universe. The encyclopedia was officially suspended in 1759 because of its advanced ideas, but was published clandestinely. He traveled in Russia (1773-74), meeting Catherine the Great, who purchased his library, paying him in advance to provide his daughter's dowry. His closing years were spent in semiretirement in France. Many of his writings were published posthumously.

J.G.G. NORMAN

DIDYMUS THE BLIND (309/314 to 398). Alexandrian theologian. Although blinded at the age of four, he acquired a considerable learning and was appointed by Athanasius head of the catechetical school at Alexandria. His ability was widely acknowledged, and he numbered Antony, Rufinus, Palladius, and Jerome* among his pupils. Jerome was greatly indebted to him, and it is from Jerome we learn most about him. He translated into Latin Didymus's *On the Holy Spirit,* to which he prefixed a preface referring to the author as having "eyes like the spouse in the Song of Songs," and as "unskilled in speech but not in knowledge, exhibiting in his very speech the character of an apostolic man, as well by luminous thought as by simplicity of words." His extant works are *On the Holy Spirit, On the Trinity,* and *Against the Manichaeans.* Some of his comments on the Catholic Epistles survive, and the work *Against Arius and Sabellius,* usually ascribed to Gregory of Nyssa, may also be his. His extant works show little evidence of Origenism,* nor was he charged with this by Epiphanius or Theophilus. But he was charged with Origenism by Jerome, and may have been also by the Council of Constantinople in 553 which condemned Origen.

DAVID JOHN WILLIAMS

DIETRICH OF NIEHEIM (Niem) (c.1340-1418). Historian and papal lawyer. Papal notary in Avignon, he served in the chancery office of several popes. In 1395 he was appointed bishop of Verden by Boniface IX, but was deprived of the office four years later, having never taken possession of his see. Dietrich wrote tracts to help end the Great Schism* and declared his allegiance to Alexander V and John XXIII, whom he later attacked in tracts and in his history of the Council of Constance. Among his works also are an attack on John Wycliffe, guides for curial administration, a history of the papal chancery from 1380, and a work in which he asserted the plenary power of a general council, including its right to depose a pope.

JAMES TAYLOR

DIGAMY. One of the early manifestations of asceticism in the church was the objection to second marriage as a lower state for the laity (although it was widely conceded, even by Tertullian, to be lawful) and as a forbidden state for the clergy. 1 Timothy 3:2 was often deemed to disqualify a digamist from ordination. Athenagoras opposed digamy on the grounds that the relationship between husband and wife was an eternal one which not even death could annul. Second marriage after the death of a husband or wife was therefore "a specious adultery." Second marriage in Tertullian's view was a concession to "fleshly concupiscence," but his arguments against it tell equally as well against first marriage.

Marriage is permitted, but "what is permitted is not absolutely good." It is better not to marry and not to have the care of children. Tertullian's attitude hardened even more against digamy once he became a Montanist. This attitude was most characteristic of groups such as the Montanists* and Novatianists.* But the imposition of a small penance on digamists is presupposed in the seventh canon of the orthodox council of Neocaesarea in 314, although the Council of Nicea in 325, in providing for the reconciliation of Novatianists, insisted in its eighth canon that those who had married twice should not be excluded from Christian fellowship. The Eastern Church has always been more severe on digamists than the Western Church. But even in the West second marriage was held to be a disqualification for ordination, and this remains so in Roman Catholic canon law to this day. DAVID JOHN WILLIAMS

DIGGERS. English communistic movement. Led by Gerrard Winstanley and William Everard, a group of about twenty men began in 1649-50 to cultivate common land in Surrey, and to plant vegetables to feed the needy. Local gentry and other conservatives incited mobs to harass them; later the government dispersed the group. The Diggers would have been forgotten but for Winstanley's writings which made new followers. Winstanley held that the Civil War had destroyed the claims of landholders and caused the land to revert to a "common treasury" which presumably had existed before the Norman Conquest. Holding land in common would be accompanied by complete social equality, abolition of trade, universal suffrage, education for all, and the arrival of the Millennium. Some have tried to find in the Digger movement an agreement with modern materialistic communists; others have stressed their basic adherence to a millennial interpretation. The weight of evidence suggests that the digging was not the beginning of a worldly revolution, but the outward manifestation of an inner confidence that the time had come for God's intervention in history. ROBERT G. CLOUSE

DILLMANN, CHRISTIAN FRIEDRICH AUGUST (1823-1894). Lutheran biblical scholar and orientalist. Born at Illingen, he studied at Schönthal and Tübingen under Ewald* and Baur.* He became interested in the neglected field of Ethiopic studies, and worked on Ethiopic manuscripts in Paris, London, and Oxford, producing catalogues of the collections at the British Museum and the Bodleian. In 1848 he returned to Tübingen, where he became a professor in 1853. He later held chairs at Kiel (1854), Giessen (1864), and Berlin (1869). He produced a grammar and lexicon of Ethiopic, and editions of Ethiopic texts of the OT. In later life he wrote commentaries on some OT books, one of which (Genesis) was translated into English in 1897. More critical than Delitzsch,* he did not lose the friendship and admiration of that great commentator. Dillmann's *Handbook of Old Testament Theology* was published posthumously in 1895. IAN SELLERS

DINSMORE, CHARLES ALLEN (1860-1941). Congregational clergyman and scholar. Born in New York City and educated at Dartmouth and Yale Divinity School, he served Congregational pastorates in Connecticut and Massachusetts (1888-1920), then was appointed lecturer on the Bible as literature at Yale Divinity School. He became an authority on Dante and was widely known in the USA and abroad for his scholarly work on the Italian poet. Five of his major works are on Dante. Dinsmore reflects the changing mode regarding biblical studies during the first quarter of the twentieth century, by which the Bible became primarily a piece of religious literature. His major works are *The Teachings of Dante* (1901); *Atonement in Literature and Life* (1906); *Religious Certitude in an Age of Science* (1924); and *The English Bible as Literature* (1931).

DONALD M. LAKE

DIOCESE. In ecclesiastical usage, the sphere of jurisdiction of a bishop. The word is of secular origin, having been employed to describe an administrative division in the Roman Empire. The pattern was adapted by the Christian Church, with province and dioceses controlled by metropolitan and bishops. In the Western Church, nevertheless, there was a certain flexibility: up to the Middle Ages it was not unknown for a metropolitan to refer to the province as his diocese, and to his own bishopric as his parish. The Eastern Orthodox Church still retains a similar practice, using "diocese" to denote the patriarch's territorial sphere, and "parish" to indicate the area for which a bishop is responsible.

J.D. DOUGLAS

DIOCLETIAN (245-313). Roman emperor 284-305. A Dalmatian of low birth, originally named Diocles, he rose to be commander of Emperor Numerian's bodyguard, and later the Augustus after defeating Numerian's brother, Carinus. In 293 he established his famous tetrarchy. Galerius was his Caesar in the East, while Maximian and Constantius Cholorus were in the West. His genius was as an organizer, and many of his administrative measures lasted for centuries. He believed that the old Roman religion, tradition, and discipline would help to reinforce imperial unity. This policy formed the background to the persecution of Christians undertaken in 303, possibly at the insistence of Galerius. Four edicts were issued before Diocletian's abdication in 305, and they were enforced with varying degrees of severity, most harshly in Palestine and Egypt. The first edict required the destruction of churches and books, the next two were aimed at the clergy, and the fourth included both laity and clergy.

PETER TOON

DIODATI, GIOVANNI (1576-1649). Calvinist theologian. Born in Geneva of an Italian Protestant family, he became professor of Hebrew (1597) and pastor (1608) there before succeeding Beza* as professor of theology (1609), a post he retained until his death. He was Genevese representative at the Synod of Dort.* His Italian translation of the Bible appeared in 1607, a revised edition with notes in 1641. This version, which established his reputation, is still commonly used by Italian Protestants. Highly acclaimed for accuracy and lucidity, it shows his theological interests and tendencies, leading some to suggest that he was more theologian than critic. He produced also a French translation of the Bible in 1644.

J.D. DOUGLAS

DIODORE OF TARSUS (d. before 394). Bishop of Tarsus. After a thorough secular and religious education in his native Antioch and in Athens, he became a Christian monk and a persuasive, influential teacher in Antioch. His two most distinguished disciples were Theodore of Mopsuestia* and John Chrysostom.* In 372 he was banished from Antioch to Armenia by Emperor Valens, but in 378 he became bishop of Tarsus. An opponent of paganism and generally regarded as orthodox, he was nevertheless condemned by a synod at Antioch in 499 as the author of Nestorianism.* This condemnation meant that copies of his writings were doomed to destruction. Some scholars have assigned various extant treatises to him, but only *De fato* seems sure. Large fragments are also extant in catenae on the Octateuch (Genesis to Ruth) and on Romans. His original writings, however, covered a wide range of topics. He expounded cosmology, defended the Faith against heresies of many kinds, and wrote many commentaries on Scripture. In the latter he followed the historical-grammatical method of exegesis in opposition to the allegorical method of Alexandria.

PETER TOON

DIOGNETUS, EPISTLE TO. An anonymous work, sometimes attributed to Quadratus or Pantaenus. Though written in the late second century, it is never mentioned in antiquity or in the Middle Ages; it was preserved with the works of Justin Martyr in a single manuscript which was burned at Strasbourg in 1870 in the Franco-Prussian War. The epistle invites a certain Diognetus to consider the superiority of Christianity to paganism and Judaism. It seeks to answer three questions: What is the nature of Christian worship and how does it differ from other forms of worship? What is the nature of Christian charity? Why has Christianity appeared so late in human history? While the style is good, the argumentation and apologetic are not very profound. The last two chapters (11 and 12) almost certainly come from a later hand—perhaps that of Hippolytus.

PETER TOON

DIONYSIUS (of Corinth) (c.180). Bishop of Corinth. A man of considerable influence, he was credited by Eusebius with seven epistles to various churches. These are of encouragement and admonition, and include an attack on Marcion, the information that Clement's epistle was read in the Corinthian church, and that Dionysius the Areopagite was the first bishop of Athens.

DIONYSIUS (of Rome) (d. 269). Bishop of Rome from 259. A Greek by birth, he succeeded Xystus. His only surviving doctrinal work is directed against Dionysius of Alexandria, who was under suspicion of Sabellianism.* In a letter to Alexandria he refuted the Sabellian doctrines, emphasizing instead the divine "monarchy." In a separate letter he invited his namesake to explain himself, which he did to Dionysius's satisfaction. Dionysius sent help to the church of Caesarea when that city was invaded by barbarians (perhaps in 264), and in 268 joined with Dionysius of Alexandria and the Council of Antioch in condemning Paul of Samosata.

DIONYSIUS EXIGUUS (d. between 525-544). Writer and chronologer. Called "Exiguus" (the less) because of his humility, he is best known for his early-sixth-century edition of the first influential decretal* collection. It comprised forty-one decretals from Siricius (384-98) to Anastatius II (496-98), and certain conciliar and synodal canons which he translated from the Greek because previous translations were not readily available to the clergy. Only his *Apostolic Canons** were apocryphal, and he had reservations about them. He also translated into Latin the works of early Greek Fathers, especially those concerning the Nestorian heresy. Dionysius helped to establish the accepted Roman method of dating Easter. Christians dated Easter according to the Jewish Passover (calculated on a lunar calendar, which was shorter than the Julian calendar, which did not always fall on a Sunday). The Council of Nicea* established how the date was to be calculated, and Cyril of Alexandria produced a series of tables which Dionysius continued up to 626—with two differences. He worked on a nineteen-year cycle, unlike the existing eighty-four-year cycle, and he took as his base year not 284 when Diocletian became emperor, but the year of Christ's incarnation—in his estimate 755 years after Rome's foundation. Though a few years off, it was the basis of the present *Anno Domini* system.

L. FEEHAN

DIONYSIUS OF PARIS (d. c.250?). Also known as Denys, patron saint of Paris. Traditionally it has been held that he was one of seven bishops sent to convert Gaul and that later he became bishop of Paris before becoming a martyr at Montmartre (Martyrs' Hill). In 626 his remains were translated to King Dagobert's foundation of St.-Denis, near Paris (the famous Benedictine abbey). Often he has been confused with other men of the same name (e.g., Dionysius the Areopagite).

DIONYSIUS THE AREOPAGITE. A member of the council of the Areopagus, he came to believe during Paul's visit to Athens (Acts 17:34). One early source states he became bishop of Athens, and later writers claim his martyrdom. The tradition that he became bishop of Paris has been rejected.

DIONYSIUS THE CARTHUSIAN (1402/3-1471). Mystic and writer. Born in Ryckel, Belgium, he was educated at Cologne University and joined the Carthusians at Roermond in 1423. From 1465 to 1469 he was in charge of the order at Bois-le-Duc. He wrote commentaries on Scripture, Pseudo-Dionysius (who influenced him greatly), Peter Lombard, Boethius, and John Climacus; produced also twenty-one treatises on the reformation of the church and Christian society, and letters for a crusade against the Turks. He had premonitions of calamities threatening the Christian world. Keen to lead souls to contemplation, he wrote *De contemplatione.* In the *Opuscula* he dealt with devout recitation of the Psalms, mortification, inconstancy of heart, and reformation of the inner man. He assisted Nicholas of Cusa* in his reform visitations in the Rhineland (1451-52). Though an eclectic, he was no mere compiler in his writings; Ignatius of Loyola and Francis de Sales read him. Dionysius published also a compendium of Aquinas's *Summa.* His mystical experiences made him known as "Doctor Ecstaticus." C.G.THORNE, JR.

DIONYSIUS THE GREAT (d. c.264). Bishop of Alexandria. After having been a pupil of Origen, he became head of the famous Alexandrian Catechetical School for about fourteen years. In 247 he was elected bishop of Alexandria. During the Decian persecution he was arrested but managed to escape to the Libyan desert, where he remained until the death of Decius. On his return he was faced with the problem of how to treat those church members who had apostatized. He advocated lenient treatment of them. He also sought to mediate in the heated dispute over heretical baptism between Cyprian and Pope Stephen. In the persecution under Valerian (257-58) he was again banished, but returned to his church in 260. During his last years he was much involved in combating Sabellianism.* Though he wrote much, his writings exist only in fragments, mainly in the extracts preserved by Eusebius, Athanasius, and others. Some of his writings date from his period in the Catechetical School, while others—e.g., his Easter Letters (the earliest Paschal Letters)—come from his period as bishop. Of his theological writings the most important is his letter to Pope Dionysius of Rome on the doctrine of the Godhead, prompted by the Sabellian controversy. In some aspects his theology anticipated that of Arius, but much later Athanasius expressed the opinion that Dionysius had an orthodox doctrine of God. PETER TOON

DIONYSIUS THE PSEUDO-AREOPAGITE. The name identifies an author who probably lived in Syria in the fifth or early sixth century A.D. His writings were originally held in high honor, being initially attributed to Dionysius of Athens. His works made a significant impact on medieval theology. Gregory the Great, Martin I, and the Lateran Council of 649 all approved his writings, and in the Western Church they exerted considerable influence toward mysticism. Hugh of St.-Victor, Albertus Magnus, Thomas Aquinas, and Dionysius the Carthusian all drew inspiration from him. So too did Platonists of the Italian Renaissance, John Colet, Dante, and John Milton, among others.

His extant writings include *The Celestial Hierarchy* (describing the mediation by angels of God to man), *The Divine Names* (on the attributes of God), *The Ecclesiastical Hierarchy* (which describes the sacraments and the three "ways" of spiritual life), and a work on mystical theology which describes the ascent of the soul toward union with God. He wrote also ten letters to monks, priests, and deacons on the points raised in his treatises. His works reveal a knowledge of Plotinus, Proclus, and other Neoplatonists, and considerable familiarity with Scripture and the Apocrypha. His writings attempt a synthesis between Christian truth and Neoplatonist thought. His central emphases are the union between man and God, and the progressive deification of man in which the soul abandons both the perceptions of the senses and the reasoning of the mind. The soul is consequently illuminated and carried ultimately to a knowledge of the ineffable Being. The Pseudo-Areopagite also taught that God is related to the world by a graded series of beings or angels corresponding to the hierarchy of the church (bishops, priests, and deacons). These hierarchies are intended to lead man to deification, a goal which is reached through the purgative, illuminative, and unitive stages.

In the sixteenth century the Reformers and Roman Catholic scholars doubted the authenticity of these writings, a doubt intensified by the development of literary criticism. Comparison of the writings of the Pseudo-Areopagite with those of the Neoplatonists is now held to establish their genuineness and a single, actual author.

See bibliography in B. Altaner, *Patrologie* (5th ed., 1958). JAMES TAYLOR

DIOSCORUS (d. 454). Patriarch of Alexandria from 444 to 451. The successor of Cyril, he became a leading figure in the Monophysite controversy. In 444 he had accused Theodoret* of Nestorianism, and when Eutyches was accused by Theodoret and others of the opposite error, he came to his aid. With the help of Chrysaphius, Dioscorus persuaded Theodosius II to call a council at Ephesus in 449 which, with Dioscorus presiding, declared Eutyches orthodox and deposed Theodoret and others, including Flavian of Constantinople. Following the death of Theodosius, the new rulers, Pulcheria and Marcian, leaned to the other side. Another synod was held at Constantinople (450) under Flavian's successor, Anatolius. Leo's Tome was read and received with acclamation and many of Dioscorus's victims were rehabilitated. But when a new council was called for Nicea, Dioscorus, supported by ten Egyptian bishops, excommunicated Leo. The site for this council was subsequently changed by Marcian to Chalcedon, where it assembled in 451. Presided over by the four papal legates and with imperial officers acting as secretaries, it deposed Dioscorus and exiled him. Many of the charges brought against him at Chalcedon were prompted by vindictiveness, but probably not all of them. His great ability was perverted by his great power. There is little of what remains of his writings that cannot be interpreted as orthodox. His deposition permanently divided Egyptian Christians. The majority continued to venerate Dioscorus and to repudiate Chalcedon. This remains the position of the Coptic Church* today. DAVID JOHN WILLIAMS

DIPPEL, JOHANN KONRAD (1673-1734). German Pietist. Born at Frankenstein and educated at Giessen University, he at first sought academic preferment by professing orthodoxy, but was disappointed and removed to Strasbourg where he taught astrology and palmistry and professed Pietism. He then fled from his creditors to the court of Darmstadt, where his Pietism received a better hearing and was deepened into intense conviction under the influence of G. Arnold. There followed the satirical *Orthodoxia Orthodoxorum* (1697) and the vigorously apologetical *Papismus Protestantium Vapulans* (1698). For the rest of his life Dippel upheld his controversial form of Pietism, repeatedly enraged the Lutheran authorities, and interfered, usually to his loss, in political affairs. Forbidden to publish in 1702, he fled to Berlin where he dabbled in alchemy and accidentally discovered the Prussian blue. In 1707 he removed to Köstritz and then to Holland, where he took a medical degree (1711). Later he appeared in Denmark and was sentenced to exile on the isle of Bornholm. Freed and expelled in 1726, he went to Sweden and became physician to King Frederick I, but was again expelled and returned to Germany, finding refuge first at Liebenberg and then at Berleberg, where he died, and where his *Works* (3 vols.) were published in 1747. IAN SELLERS

DIRECTORY FOR THE PUBLIC WORSHIP OF GOD, see WORSHIP OF GOD

DISCALCED. The act of going barefoot. First introduced by John of Guadalupe about 1500 among the Friars Minors of the Strict Observance, the austerity is still practiced by certain orders of friars.

DISCIPLES OF CHRIST, see CHURCHES OF CHRIST

DISCIPLINE. In the context of church life, this term is used to describe the practical methods and rules by which Christ, through the influence of the whole community, seeks to help each member to be healthy in his own Christian growth and discipleship, and to make his best contribution to the life and witness of the whole body. From the beginning some form of discipline was accepted as an aspect of the Gospel. Christ was regarded as the Master whose teaching and example contained patterns for such discipline (cf. Matt. 11: 29; 28:19).

The most acute problem relating to discipline was ensuring that members whose conduct brought offense to the community were challenged about their behavior and convicted so that they could be restored. Repentance had to be ensured. This problem takes up some space in the NT because the church felt it had definite guidance from Jesus on the matter. The offending brother was to be approached privately, and only if he refused to respond was the matter to be brought before the church. If he then remained impenitent, he was to be excluded from the fellowship in the hope of his ultimate return and repentance (Matt. 18:15-17). The church claimed that Christ had given it the power to exercise such "binding and loosing" of sin in his name (Matt. 18:18-20; John 20:23).

We have some indication in the NT of how moral advice was given and discipline was effected, in early church life. The case of Ananias and Sapphira was exceptional (Acts 5:1-11). Paul gives various instructions (e.g., 1 Cor. 4:21; 5:1-12; 2 Cor. 2:1-11; Titus 3:10ff., etc.). Von Campenhausen points out that discipline in the first and second centuries seems to have had the forgiveness and winning back of the erring, rather than their punishment, as its aim. It was directed toward the individual. It was not regarded as annulling baptism, and was exercised only with purely spiritual authority.

From the fourth century, discipline began to show undesirable features. More concern came to be shown for the sanctity of the congregation as a whole than for the expelled individual. The authority to exercise discipline was taken from the congregation and was regarded as residing in the clergy, and often the monarchical bishop alone. The system of penitence began to be concerned too much with trivial offenses. There was partiality in its exercise. Private confession was made compulsory for all. The church began to enforce its discipline by use of civil power.

The Reformation saw sincere attempts by Luther to deliver men from priestly ecclesiastical tyranny in discipline, and by Calvin to restore in its integrity the discipline of the NT church. Unhappily, in the seventeenth century the pursuit of discipline became in some quarters more important than the pastoral care of the individual. Severity in certain areas of life was exercised at the expense of slackness in other areas. Discipline tended to stifle growth. Today, in reaction, it is asserted that it is impossible and undesirable within pluralistic society to set standards to which church members should conform. Our attitude to such an assertion will be determined by our understanding of the Gospel. Christ, in fulfilling the New Covenant on our behalf, presented to God a definite pattern of response into which He seeks to conform up by the Spirit. The church cannot decide to ignore this pattern. Moreover, while repentance is not a prior condition of forgiveness, it is inseparable from forgiveness, and it produces signs and fruits that man must look for and encourage. RONALD S. WALLACE

DISCIPLINE, BOOKS OF. The doctrine, worship, and government of the Reformed Church of Scotland in 1560 are clearly set out in the Scots Confession, the *Book of Common Order*, and the *First Book of Discipline*. They were the work of a committee of six members appointed by Parliament, and John Knox was their acknowledged leader.

The First Book of Discipline aimed at "a total Reformation of Religion in the whole Realm." It provided for the government of the church by kirk sessions, synods, and assemblies. It recognized the office-bearers in the church to be ministers, teachers, elders, deacons, superintendents, and readers. The outlined form of government clearly followed the pattern at Geneva and was influenced by the Ecclesiastical Ordinances of the Reformed Church in France. The most revolutionary of the recommendations were the offices of superintendent and reader. The duties of a superintendent were the supervision of the work of the church in defined areas, and the submission of reports to the general assembly. The appointment of readers was due to the scarcity of trained ministers and the lack of funds to support a pastor in every parish.

The sixteen chapters in the Book of Discipline require the Word to be preached and the sacraments administered in every parish, the abolishing of idolatry, the provision of suitable stipends in money and in kind, the recommendation of a school for every parish, the enforcement of scriptural discipline, and regulations for marriages, funerals, and the repairing of church buildings. Though approved by the general assembly, the Book never received the legal sanction given to the Confession, but it helped considerably in consolidating the work of Reformation in Scotland.

In 1581 the *Second Book of Discipline* was published. It was virtually a demand for the reversal of the Erastian policies of the Regent Morton. It defined the relationship between church and state, outlined more precisely the functions of office-bearers in the church and the constitution of her courts, and strengthened the position of the ruling elder. Despite a remarkable omission of any reference to the function of the presbytery as a court, even though the importance of this court was recognized by the assembly of 1580, the Second Book of Discipline became the charter of Presbyterianism in the Church of Scotland.

ADAM LOUGHRIDGE

DISPENSATIONALISM. The term "dispensation" occurs in the KJV in 1 Corinthians 9:17; Ephesians 1:10; 3:2; Colossians 1:25. In each case it translates the Greek *oikonomia,* from which the English "economy" is derived, and so is concerned with administration. Dispensationalism is the view that there is much variety in the divine economy in the Bible, that God has dealt differently with men during different eras of biblical history. A dispensation is "a period of time during which man is tested in respect of obedience to some *specific* revelation of the will of God," according to C.I. Scofield.*

Some variety exists among dispensationalists, but Scofield's scheme of seven dispensations is widely accepted. These are Innocence (before the Fall), Conscience (from the Fall to Noah), Human Government (from Noah to Abraham), Promise (from Abraham to Moses), Law (from Moses to Christ), Grace (the Church age), the Kingdom (the Millennium). The close of the Millennium ushers in the Eternal State. J.N. Darby* is usually regarded as the founder of Dispensationalism, although some of its elements are found in Augustine. All dispensationalists are necessarily premillenialists, but the reverse is not always the case. Dispensationalism was greatly popularized through the Scofield Bible, and its fullest theological expression is in the works of L.S. Chafer.* An extreme version of it ("Ultra-Dispensationalism") was due to the work of E.W. Bullinger. The term "dispensation" (as the virtual equivalent of "covenant") is often used by those who do not accept Dispensationalism as defined above.

G.W. GROGAN

DISRUPTION, THE (1843). The withdrawal of 474 ministers from the Church of Scotland, which brought the Free Church of Scotland* into existence. The Church of Scotland had been debilitated spiritually for about a century by various secessions. The seceders were usually earnest men, deeply concerned about what they felt to be the wrongs of patronage, and the national church could ill afford to lose them. The growth of Moderatism further weakened the spiritual quality of its ministry. The French Revolution, however, seemed to many to be a direct challenge to the complacent rationalism of the Moderates,* and the Calvinism they so much despised seemed much more in touch with reality. Rationalism had never really affected the pew as it had the pulpit. People flocked to the Secession churches, but at the same time the tides of evangelical life and power began to flow more strongly in the state church itself.

The Evangelical party found an outstanding leader in Thomas Chalmers,* and from 1815 (the year of his settlement in Glasgow) it grew rapidly both in numbers and influence. There was a breath of genuine revival in the air. Chalmers had once been a Moderate himself, but now the fire of the gospel of Christ burned in his heart, and he became increasingly aware of the social and ecclesiastical implications of the new faith to which God had brought him. The twin enemies of the Evangelical were Moderatism and patronage, and the matter of state interference in the affairs of the church came more and more to the fore.

The general assembly of 1842, by a large majority, declared that the Church of Scotland must be free to govern itself, and it protested against any attempt by Parliament or the courts to interfere in matters spiritual and ecclesiastical. Parliament rejected this claim of right; thus, as the general assembly of 1843 opened, Chalmers and some two hundred other ministers, mostly Evangelicals, walked out. They and nearly three hundred others founded the Free Church of Scotland, with Chalmers as its first moderator. They claimed to stand for "the confession of faith and standards of the Church of Scotland as heretofore understood." It was one of the most dramatic moments in the history of the Scottish Church. The movement which led to the Disruption was undoubtedly spiritual at its heart, but it was also an expression of the nineteenth-century swing toward greater democracy.

See also TEN YEARS' CONFLICT.

BIBLIOGRAPHY: *Disruption Worthies, A Memorial of 1843* (1876); A.J. Campbell, *Two Centuries of the Church of Scotland, 1707-1929* (1930); H. Watt, *Thomas Chalmers and the Disruption* (1943); J.H.S. Burleigh, *A Church History of Scotland* (1960), pp.334ff. G.W. GROGAN

DISSENTING ACADEMIES. After the 1662 Act of Uniformity* some English Nonconformist ministers opened small academies to prepare young men for the ministry. The first to do so was probably Jarvis Bryan of Coventry sometime after 1663. By 1690 some twenty-three academies existed, the most famous being that at Newington Green run by Charles Morton (later of Harvard College). Daniel Defoe was a student at Newington. After the 1689 Toleration Act,* academies grew in number and importance. In the eighteenth century there were large ones in many cities and towns. Their teaching methods and curricula—especially in the new science subjects—were often superior to those of the ancient universities. The campaign against them early in the eighteenth century was led by Henry Sacheverell, who believed they were hotbeds of fanaticism, and reached its climax in the 1714 Schism Act (later repealed by the House of Hanover). Philip Doddridge and Joseph Priestley were among those educated in academies. In the nineteenth century most academies ceased to exist or became theological colleges as the sons of Nonconformists were now able to attend the universities. PETER TOON

DISSENTING BRETHREN, see WESTMINSTER ASSEMBLY

DISSOLUTION OF THE MONASTERIES. In the early sixteenth century, one in every 375 people in England was in a religious order. Monasteries were great landowners, and thirty abbots were lords in Parliament. The widespread view that they were wholly corrupt is probably exaggerated. However, the supposed corruption allowed the official case against the monasteries to rest on moral grounds, even though the motive was to gain finance for the Crown. Henry VIII needed money, and Thomas Cromwell* saw in the dissolution a good way to enrich the royal purse. A subsidiary motive of Cromwell was possibly the desire to stamp out the veneration of relics.

The dissolution began when Parliament gave its approval in early 1536 to an Act for the suppression of smaller monasteries. This meant in practice that 243 were actually closed, which was three of every ten religious houses. Many of the dispossessed monks and nuns transferred to larger houses. Despite the violent language of its preamble, this Act of 1536 should not be regarded as the first stage of a carefully planned attack upon English monasticism, but rather as a moderate measure of reorganization to release surplus property for secular use. In the north of England the dissolution provided rebels with a popular cause and rallying cry. The Pilgrimage of Grace* was, however, soon put down. The fact that monks from the larger abbeys had taken part in the rebellion resulted in the establishment of important precedents for surrender and forfeiture when once the government decided later to suppress all the religious orders.

Between 1537 and 1540 the larger abbeys and houses of the friars were gradually taken over through a process of surrender. By March 1540 the religious orders in England were no more. The Act of 1539, sometimes called "the Second Dissolution Act," did not transfer to the Crown the property of any abbeys. Rather, its purpose was to set at rest the doubts concerning the validity of deeds of surrender and to legalize all surrenders that had occurred and would occur.

See G.W.C. Woodward, *The Dissolution of the Monasteries* (1966). PETER TOON

DISTLER, HUGO (1908-1942). German composer, perhaps the most influential composer of church music in Germany between the two world wars. In 1931 he began a series of musical vespers in Lübeck reminiscent of those of Buxtehude* in the seventeenth century. In Stuttgart he came under the influence of Orff and Hindemith. He strove in his compositions to create functional church music in a contemporary idiom that would carry on the great musical tradition of the Lutheran past. He was honored with professorial rank and made the conductor of the *Staats und Domchor* in Berlin. He took his own life rather than be conscripted into the Nazi army.

DIVINE, MAJOR J. ("Father Divine") (1865?-1965). Founder of the Peace Mission Movement. Born George Baker into a poor Negro family on Hutchinson's Island, Georgia, he—a former Baptist—was by 1907 claiming to be "God in the sonship degree." He traveled awhile, then settled in New York City. About 1919, as Major J. Divine, he moved to Sayville, Long Island, and later to Harlem. His socioreligious movement grew rapidly in the 1930s and 1940s as he spoke across the country and published his magazine *The New Day.* As Father Divine, he ran a massive cooperative agency and employment service, providing low-cost meals and lodgings in his "heavens." After a court judgment against him, headquarters were moved to Philadelphia. He equated Americanism, Christianity, and democracy, and imposed an exceedingly strict moral code. The interracial movement regards him as God, has over a million members in several states and foreign countries, and is continued by his white widow and heir, Mother Divine.

ALBERT H. FREUNDT, JR.

DIVINE RIGHT OF KINGS. In the Middle Ages it was widely held that royal authority was divinely ordained, but not the person of the king. The Puritans of both England and New England held that all government was of divine origin and received its just powers from God alone. On the other hand, the divine right theory, as practiced by the Stuarts in seventeenth-century England and by Louis XIV and his successors in France, was founded on the belief that the king possessed an absolute grant of authority from God Himself. The king was, therefore, above the law of the land, but at the same time he was directly responsible to God for the welfare of his people, as a father to his family. Thus disobedience to the king was disobedience to God, and therefore sin. This theory of government virtually disappeared in England after the Revolution of 1688-89.

C. GREGG SINGER

DIX, DOM GREGORY (1901-1952). Anglican scholar. Born George Eglinton Alston, son of a clergyman, he graduated from Oxford and held a lectureship in modern history there, 1923-26, afterward entering monastic life among the Anglican Benedictines at Nashdom Abbey. He received his final vows in 1940 and became prior in 1948. While well known for his religious broadcasts, his most important contribution lay in his comprehensive study of Christian worship, *The Shape of the Liturgy* (1945). Earlier, his edition of the *Apostolic Tradition* of Hippolytus had provided pointers about how in the Roman Church theological thinking is moved to confirm what the worshiping Christian already espouses. Several posthumous publications reflect his thinking even in the midst of terminal illness. The two cautions expressed in *Jew and Greek* (1953) bear out the essence of his theological thought: "religions pray" and "history happens through men and women, not through abstractions."

CLYDE CURRY SMITH

DIXON, A(MZI) C(LARENCE) (1854-1925). Baptist pastor and author. Born in North Carolina, he studied at Wake Forest College and Southern Baptist Theological Seminary. Among his many pastorates the most important were at Moody Memorial Church, Chicago (1906-11), and the Metropolitan Tabernacle, London (1911-19). He was active in the conservative Bible conference, prophetic movement and in evangelistic crusades from about 1875. He joined R.A. Torrey* in the publication, beginning in 1909, of *The Fundamentals,** a twelve-volume paperback series dedicated to the defense of the fundamental doctrines of the Christian faith. His more important publications include *Evangelism Old and New* (1905): *Destructive Criticism Vs. Christianity* (1912); *The Birth of Christ: The Incarnation of God* (1919); and *Higher Critic Myths and Moths* (1921).

DONALD M. LAKE

DOBER, JOHANN LEONHARD (1706-1766). Moravian Brethren* leader. Born in Swabia and a potter by trade, he came to Herrnhut in 1725 where he soon became one of the most significant spiritual figures. In 1732 Dober and David Nitschmann* volunteered to go to St. Thomas, Virgin Islands, as the first Moravian missionaries. Unable to practice his trade there, Dober supported himself by working as a plantation watchman. He tried to evangelize Negro slaves on the island, but his black congregation numbered only four when he left in 1734. Summoned to Herrnhut to become superintending elder, he was replaced by a new group of fourteen missionaries. After 1738 he served for three years as a missionary to Jews in Amsterdam. Dober spent his remaining years in constant travel, visiting the various Moravian congregations in Europe, for which purpose he was consecrated bishop in 1747.

RICHARD V. PIERARD

DOCETISM (Gr. *dokein*, "to seem," "appear to be"). In the history of Christian theology this is the view that Jesus Christ was not a real man, but simply appeared so. This undermines not only the Incarnation, but also the Atonement and Resurrection. Through Eusebius we know about Cerinthus,* the Docetist opponent of the Apostle John at Ephesus. The presence of such a heresy there probably accounts for the strong emphasis upon the "flesh" of Christ and His "blood" (although this word has sacrificial overtones also) in the Johannine group of writings (e.g., 1 John 4:2; 5:6-8). There was a docetic element in the Gnostic group of heresies, and this accounts for the exceptional emphasis upon the reality of our Lord's humanity in some early Christian writers, beginning with Ignatius.

The origins of Docetism are not biblical, but Hellenistic and oriental, and are due to the idea that matter is essentially evil and to a particular construction of the doctrine of divine impassibility. Alexandria was a melting-pot of Hellenistic and oriental ideas, and the home of some of the greatest Gnostic teachers. It is therefore not surprising to find that there are docetic tendencies even in some of the more "orthodox" Alexandrian Christian writers, such as Clement and Origen. Later christological heresies emanating from the Alexandrian school (such as Apollinarianism, Eutychianism, and Monophysitism) all have something of a docetic flavor. Although modern theology normally takes the humanity of Jesus very seriously (sometimes to the neglect of His deity), those theologians who tend to drive a wedge between faith and history are confronted with the charge of opening the door to a new Docetism.

BIBLIOGRAPHY: See under GNOSTICISM.

G.W. GROGAN

DOCTORS OF THE CHURCH. A term that came to be applied to eight Early Fathers conspicuous for learning, sound doctrine, and saintliness. These comprised Ambrose, Augustine of Hippo, Gregory the Great, Jerome, John Chrysostom, Basil the Great, Gregory of Nazianzus, and Athanasius. During the past four centuries the Roman Catholic Church has proclaimed twenty-four more "Doctors," beginning with Thomas Aquinas (1567). The most recently proclaimed were Teresa of Avila and Catherine of Siena (both 1970).

DODD, CHARLES HAROLD (1884-1973). British Congregational minister and NT scholar. Educated at Oxford, he taught NT at Mansfield College (1915-30) and lectured also in the university (1927-31). In 1930 he succeeded A.S. Peake as Rylands professor of biblical criticism and exegesis at Manchester, and from there went to Cambridge as Norris-Hulse professor of divinity—the first non-Anglican to hold a chair of divinity there since 1660.

Dodd was the most influential figure in British NT scholarship during the middle decades of the twentieth century. His emphasis on "realized eschatology" in the teaching of Jesus (*The Parables of the Kingdom,* 1934) and his isolation of an outline of early Christian preaching *(kerygma)* common to all apostolic writings (*The Apostolic Preaching and Its Development,* 1935) have proved to be important in the development of contemporary NT theology. In *According to the Scriptures* (1952) Dodd stressed the unity of approach in the use of the OT in the New, and suggested that the key was Jesus' interpretation of the Old. In his commentary on Romans (1932) and elsewhere he argued that the biblical concept of God's wrath should be understood as an impersonal process of retribution in human history rather than as the divine reaction to the sin of man; similarly, he argued against the concept of "propitiation" as a biblical idea.

After retirement from formal academic teaching in 1940 Dodd wrote more than a dozen books, including two important works on John (1953, 1963), served as the general director of the New English Bible translation, and lectured extensively in various parts of the world.

See F.F. Bruce, in *Creative Minds in Contemporary Theology* (ed. P.E. Hughes, 1966), pp. 239-66 (with bibliography). W. WARD GASQUE

DODDRIDGE, PHILIP (1702-1751). Nonconformist* divine. Born in London in 1702, Doddridge was educated at Kibworth Academy, was minister of Kibworth (1723-29), began his academy at Market Harborough (1729), and then moved to Northampton (1729-51). Here he accomplished his life-work, training generations of students for the ministry, letting each one decide controverted theological points for himself, encouraging village preaching, and promoting unity among the Nonconformist bodies. Theologically he occupies a curious position. He adhered to the modified Calvinism of Richard Baxter* and was the leader of the "Middle Way" men after the death of Edmund Calamy,* but he also inclined to Sabellianism, though his alleged heresies are probably due to lack of necessary mental equipment to articulate his thoughts clearly. At the same time he was deeply influenced by the warmth of the Methodist revival, and he regarded Dissent as the religion of the common people, not as a political prop for the Hanoverian dynasty. *The Rise and Progress of Religion in the Soul* may be the last great Puritan spiritual autobiography, but it is shot through with evangelical fervor; so too are his hymns, particularly "Hark the glad sound." To Arians and thoroughgoing Calvinists alike he appeared a trimmer. Most of his students became liberal Presbyterians, though a few embraced an earnest evangelicalism. Doddridge died at Lisbon. IAN SELLERS

DODS, MARCUS (1834-1909). Scottish biblical scholar. Born at Belford, Northumberland, where the elder Marcus Dods was Presbyterian minister, he graduated at Edinburgh and in 1864 became minister of Renfield Free Church, Glasgow, a charge he held until 1889. In that year he was called to the chair of NT criticism in New College, Edinburgh. In 1890 a complaint was brought against him in the general assembly that he had denied the inerrancy of Scripture, but it was dismissed, and a more liberal view in his church was consolidated with the passing of the 1892 Declaratory Act.* In 1907 Dods was appointed principal of his college. His works include commentaries on Genesis (1888) and 1 Corinthians (1889), and *The Bible, Its Nature and Origin* (1905). J.D. DOUGLAS

DOGMA (Gr. *dokein,* "to seem"). The word ranged in meaning from "thinking" or "having an opinion" to "appearing best" or "being determined." The noun formation *dogma* is first found in early fourth-century B.C. writings of Xenophon and Plato, with an application comprehending legal or military decrees or commands, and philosophical or religious tenets or understandings. Patristic citation shows the process over three or four centuries of Christian confrontation with Judaism and with its own deviations, by which the legal weight of commandment was carried into the philosophical dimension, so that dogma came to identify fixed doctrines or the total system of creedal religion. That which had expressed opinion became the determined or right opinion *(orthodoxia).* Collectively dogma is the intellectual side of the Christian faith. The Nicene church reviewed it historically; the nineteenth century subjected it to critical analysis (cf. F.C. Baur and Adolf Harnack). CLYDE CURRY SMITH

DÖLLINGER, JOHANN JOSEPH IGNAZ VON (1799-1890). Roman Catholic church historian and theologian. Ordained in 1822, he was professor of church history in Munich from 1826. Influenced by both romantic Catholicism's feeling for the past and by developing scientific history, he practiced and encouraged church history as a discipline based on critical study of sources, which involved setting the subject within world history and interpreting it in the light of Catholic truths. Lord Acton studied under Döllinger. A friend of Gladstone* and of Lamennais,* he like the latter blended liberalism in theology and politics (he represented his university in the Frankfurt national assembly, 1848) with an Ultramontanist loyalty to the papacy, expressed clearly as late as 1860 in *The First Age of the Church* (ET 1866). Nevertheless, from the 1850s his liberalism was eroding his Ultramontanism. He had a growing distaste for the modern institutions of the papacy and Curia, for the Papal States and the influence of the Jesuits. He disliked the decree on Immaculate Conception (1854), wanted the scholastic method replaced by the historical, and saw the Syllabus of Errors as an outright attack on the modern world and some of his own positions.

When Vatican I was announced and it was thought that papal infallibility would be defined, Döllinger immediately wrote adverse newspaper articles, later published as *The Pope and the Council* (1869-70). He distinguished between the primacy which the papacy had always had by divine appointment, and the papacy which had developed since the ninth century. The evils of the political papal monarchy were devastatingly exposed, but he was probably wrong to fear that the definition of papal infallibility presaged its revival. He was excommunicated when he refused to accept the decree (1871), and shared in the founding of the Old Catholic Church,* taking part in its discussions with Anglicans and Orthodox (1874-75) out of concern for Christian reunion. He defended Anglican ordinations, but became estranged from the Old Catholics when they discarded some traditions (clerical celibacy, auricular confession). There is some doubt whether he died an Old Catholic or an isolated, excommunicate Roman Catholic.

See J. Friedrich, *Ignaz von Döllinger* (3 vols., 1899-1901). HADDON WILLMER

DOM; DON. This shortened form of the Latin *dominus,* which itself conveyed a range of titular meanings from "Lord" in the abstract to "owner, possessor" in the common—all based on etymological derivation from Indo-European words—was a result of linguistic process among the Romance family of languages. Within general usage it identifies a nobleman; within the church it entitles ordered monks or canons regular.

DOMINIC (1170-1221). Founder of the Order of Preachers, generally known as Dominicans* or Black Friars. Born at Calaruegan in Old Castile, he was educated at the University of Palencia which was later moved to Salamanca. In 1199 he was appointed a canon by the bishop of Osma, who strenuously enforced the Rule of St. Augustine on his canons. Dominic rose quickly to a position of subprior. In 1203 he accompanied his bishop, as chaplain, on a royal embassy to the south of France which had been greatly infected by the Albigensian heresy. Challenged by the need to combat heresy, they obtained papal permission to stay in Languedoc to preach. They went barefoot, practicing great abstinence. In 1206, with the support of Fulk, bishop of Toulouse, Dominic opened a house at Prouille where girls and women might be taught under strict supervision. This was the first Dominican convent.

During the seven years' crusade against the Albigensians,* launched by Innocent III in 1208, Dominic worked with great zeal to bring the heretics back into the church. He felt he must bind his helpers to him to give stability and unity of purpose to his work. At the Fourth Lateran Council in 1215, therefore, Dominic laid before Innocent his scheme for an order of well-educated preaching friars, directly subject to the papacy. Innocent approved, but the council refused. Dominic then placed his sixteen brethren under the Rule of St. Augustine. Their first monastery was built at Toulouse.

In 1216 Honorius III granted a bull legalizing the order. It was to be a mendicant order, devoted to preaching and the conversion of heretics. Dominic met Francis of Assisi* in Rome in 1218. Adept at organization, Dominic traveled tirelessly in Italy, France, and Spain for the rest of his life, establishing and consolidating. The first general chapter of the order was held in Bologna in 1220. In the same year he became ill on his way to preach in Hungary and, returning to Bologna, died there, after encouraging his brethren "to show charity, maintain humility, and accept poverty." He was canonized in 1234. Dominic was a determined leader of men, devout in life, of firm faith, and with a passion for winning souls. His refusal three times to accept a bishopric was typical of his humility and austere self-negation.

See B. Jarrett, *Life of Saint Dominic* (1924); and M.H. Vicaire, *Saint Dominic and His Times* (2 vols., ET 1964). JAMES TAYLOR

DOMINICANS. A preaching order founded in 1216 by Dominic,* a Castilian who became a canon in the diocese of Osma, where the bishop had adopted the Augustinian Rule for his canons. Dominic became head of this community and remained there until 1203. Having spent several unsuccessful years trying to convert the Albigenses, he applied for papal authority to found a new monastic order devoted to defense of the Faith. It was granted on condition that he choose an established Rule. Dominic chose the Augustinian, and the order was officially established by Honorius III in 1216. In order to devote itself to study and preaching, the order abolished manual labor and had its divine office shortened. Easy movement of the preachers was ensured by requiring members to swear allegiance to the order and not to a particular house. At the two chapter meetings at Bologna in 1220-21, the order decided to live by voluntary alms and relinquished ownership of property and fixed incomes. The general chapter assigned authority to a master general chosen for life, and members were required to vow obedience directly to him.

Each house was ruled by a prior chosen by its members and sent its prior together with one elected member to an annual provincial chapter which in turn elected a provincial prior for four years. The provinces sent representatives to the general chapter—the supreme legislative authority—which chose the master general. There is a second and third order attached to the Dominican order. The second order consists of nuns who observe a similar rule to that of the men but who live an enclosed and contemplative life somewhat mitigated later by their undertaking to educate girls. The third order is not enclosed, and a majority of its members live active lives in the world.

Well-organized and having preaching at the center of their activities, the Dominicans were particularly useful to the pope for preaching crusades, collecting monetary levies, and the execution of various diplomatic missions. Their zeal for missionary work led them to seize the opportunities for such activity provided by the Spanish and Portuguese explorations in the West and East. They were interested in establishing their order in the centers of intellectual life such as Rome, Paris, and Bologna. This concern was furthered by Dominic's successors—with the result that by the middle of the thirteenth century each province had its own Dominican university. Many of the leaders of European thought in the Middle Ages were Dominicans. They were innovators in the teaching of languages such as Hebrew, Greek and Arabic among the religious. The order is noted for its impressive literary and scholastic output, the works of Thomas Aquinas and Albertus Magnus being famous examples. True to their role as opponents of heresy, the members have produced outstanding works in apologetics. The activity which detracted from their popularity and aroused the hostility of other orders was their involvement in the work of the Inquisition,* which was often staffed by Dominican members. In addition, the rise of new orders, especially the Jesuits, pushed them into the background, but they remain champions of learning and orthodoxy.

BIBLIOGRAPHY: B. Jarrett, *The English Dominicans* (2nd ed., 1927); R.F. Bennett, *The Early Dominicans* (1937); W.A. Hinnebusch, *History of the Dominican Order* (1966). S. TOON

DOMINIS, M.A. DE, see DE DOMINIS

DOMITIAN, TITUS FLAVIUS (A.D. 51-96). Roman emperor from A.D. 81. Son of Vespasian, he succeeded his brother Titus and at first ruled well. After the failure of his campaigns (87) against the Dacians and Marcomanni, however, he became cruel and, among other things, demanded that he be worshiped as *Dominus et Deus* (Lord and God). Christians and Jews were persecuted for their refusal to give him divine honors. Tradition declares that one who was banished for his faith was the Apostle John, who then wrote the Apocalypse on the island of Patmos as a message to his suffering brethren. Domitian's behavior became so intolerable even to pagans that a conspiracy was formed against him, and he was assassinated.

PETER TOON

DOMITILLA FLAVIA (c. A.D. 100). A Christian Roman matron of the imperial family, she was married to Titus Flavius Clemens, a first cousin of the emperor Domitian. Her grandmother and mother were the wife and daughter of the emperor Vespasian. According to some early historians, her husband was also a Christian. Possibly for the confession of Christianity, Clemens was put to death and Domitilla banished to the island of Pandateria. Property of Domitilla on the Via Ardeatina was used from the first century onward as a Christian cemetery, being known as the *Coemeterrium Domitillae.* Domitian's treatment of Clemens and Domitilla is most striking since he had designated their two sons as his heirs.

PETER TOON

DONATION OF CONSTANTINE. A document supposedly written by Constantine giving Rome and the western region of the empire to Pope Sylvester. The pope was said to have cured the emperor of leprosy, and Constantine decided to withdraw to a new city, Constantinople, feeling unworthy to live in the same city as the pope. Although this work was a forgery (probably written in the eighth century), not until the Renaissance did scholars such as Nicholas of Cusa and Lorenzo Valla demonstrate the fact. Valla's arguments, indeed, helped to establish the science of textual criticism. He held that Constantine was not the sort of ruler to give away his empire, and that Sylvester would not have accepted such a gift as he was most concerned with his spiritual office as the shepherd of souls. Through philological and critical reasoning Valla showed that the document could not have been written in the fourth century, as it refers to satraps, the stockings of the Roman senators, and the papal crown. None of these terms or items was in use during Constantine's time. Valla's scholarship was so thorough that even the pope accepted his conclusions.

ROBERT G. CLOUSE

DONATISM. African separatist church. After Caecilian became bishop of Carthage in 312, objectors alleged that one of his consecrators, Felix of Apthungi, had committed *traditio,* the "surrender, betrayal," of the Scriptures in the recent Great Persecution. Motivated partly by personal grievances, the opposition—including the Numidian bishops (not all guiltless of *traditio*), affronted at their primate's exclusion from Caecilian's irregular and precipitate consecration—elected as bishop Majorinus, whose successor was Donatus (313).

When Constantine granted compensation and exemptions only to the Caecilianists, the dissenters appealed to him to arrange adjudication of the dispute. Ecclesiastical and imperial inquiries cleared Felix (and Caecilian), and Constantine, with Catholic connivance, vainly attempted a coerced reunification (317-21). The Donatists rapidly multiplied under the able leadership, unmatched among the Catholics (but cf. Optatus of Milevis), of Donatus (d. c.355), his successor Parmenian (c.355-91/2), and others (cf. the "unorthodox" Tyconius), and enjoyed the ascendancy throughout the fourth century except following the "Macarian persecution" under Emperor Constans (347-48). The effects of this imposed "unity" persisted until Julian removed restrictions and repatriated exiles (361).

Only in the era of Augustine and Aurelius did the Catholics begin to prevail, but not without imperial coercion in the Edict of Unity (405) and the decrees proscribing Donatism after a great confrontation of the two episcopates at Carthage in 411 under Marcellinus, the imperial commissioner. Though repressed, Donatism survived until the Moorish conquests eclipsed African Christianity. Under the Vandals, Catholics and Donatists probably suffered alike. Increasing mutual toleration heralded a resurgence of Donatism in the Byzantine era, especially in Numidia, and perhaps even an ecumenical rapprochement. Gregory the Great* repeatedly rebuked the African bishops' complacency towards the Donatists, but the African Church as a whole had now recovered its traditional independence of Rome.

Donatism professed authentic African beliefs. Its rebaptisms enjoyed Cyprian's authority, and its rigorism, puritan ecclesiology, adulation of martyrdom, and apocalyptic rejection of state and society bore an African stamp as old as Tertullian. Its fundamentally religious inspiration is all-pervasive. It was "nationalist" only in its hostility to the ruling (and often persecuting) power. The embarrassing violence of Circumcellions* and Donatists' support for the revolts of Firmus (372-75) and Gildo (397-98) hardly betoken political motivation, although such excesses, e.g., under Primian, Parmenian's successor at Carthage, provoked splinter groups like the Maximianists. Cultural particularism, such as the fostering of a Libyan (Berber) language, was unimportant as cause or consequence. The stronger concentration of Donatism in the Numidian countryside by the fifth century was largely due to more effective imperial and episcopal repression in the cities, but the movement naturally gathered up economic and social discontents. Donatism provoked in fourth- and fifth-century Catholicism an uncharacteristically African alignment with ecclesiastical and imperial Rome.

BIBLIOGRAPHY: P. Monceaux, *Histoire Littéraire de l'Afrique Chrétienne* 4-7 (1912-23); H. von Soden, *Urkunden zur Entstehungsgeschichte des Donatismus* (2nd ed., 1950); W.H.C. Frend, *The Donatist Church* (1952); J.P. Brisson, *Autonomisme et Christianisme dans l'Afrique Romaine* (1958); G. Bonner, *St. Augustine of Hippo* (1963); E. Tengström, *Donatisten und Katholiken: soziale, wirtschaftliche und politische Aspekte einer Nordafrikanischen Kirchenspaltung* (1964); E.L. Grasmück, *Coercitio: Staat und Kirche im Donatistenstreit* (1964); R.A. Markus, "Donatism: the Last Phase," in *Studies in Church History* 1 (1964), ed. C.W. Dugmore and C. Duggan, pp. 118-26; R. Crespin, *Ministère et Sainteté: Pastorale du Clergé et Solution de la Crise Donatiste dans la Vie et la Doctrine de S. Augustin* (1965); P.R.L. Brown, *Augustine of Hippo* (1967) and *Religion and Society in the Age of St. Augustine* (1972), part III.

D.F. WRIGHT

DONNE, JOHN (1573-1631). English poet and dean of St. Paul's. Born a Catholic and with Jesuit relatives, he did not go abroad to Douai but went instead to Hart Hall, Oxford, then to Trinity College, Cambridge, and afterward studied law at Lincoln's Inn. He joined the household of Egerton, the lord chancellor, with whose niece, Anne More, he eloped in 1601. Their marriage resulted in Donne's dismissal from Egerton's service and his imprisonment. After failure to secure advancement in other directions, he acceded to the king's desire, took orders in 1615, and became dean of St. Paul's six years later.

Donne is the first and greatest of the group known as the Metaphysical poets. His *Songs and Sonnets* and *Elegies* are variations on the theme of love, but they are not the conventional outpourings of the usual Elizabethan love-poet. They are surprising, even outrageous, in the range of experience they treat and the language used to describe it. Whether autobiographical or not, both they and the *Divine Poems* are marked by an unsurpassed intensity of passion, which issues in most elaborate and original "conceits"—lovers compared to compasses or the body to a map, for example.

Donne abounds in paradox, and the imagery of religion is cited in the love-poems, that of sexual experience in the religious. His religious poetry is permeated by his deep sense of sin and his awareness of judgment. As T.S. Eliot has said, he "was much possessed by death," and some of his most powerful sermons, powerful alike in vision and in argument, vividly illustrate Donne's almost medieval awareness of dissolution: "Ask not for whom the bell tolls; it tolls for thee."

BIBLIOGRAPHY: *Sermons*, sel. and ed. T. Gill (1958); W.R. Mueller, *John Donne, Preacher* (1962); R.C. Bald, *John Donne* (1970).

ARTHUR POLLARD

DORDRECHT, see DORT, SYNOD OF

DORNER, ISAAC AUGUST (1809-1884). German Lutheran theologian. Born at Neuhausen ob Eck, son of a pastor, Dorner was educated at Tübingen under Baur, became professor of theology there in 1838, and a year later replied to the *Life of Jesus* of his rationalist colleague Strauss* in a treatise which was expanded into a multivolume work a few years later and translated into English between 1861 and 1865. He became successively professor at Kiel (1829), Königsberg (1843), Bonn (1847), Göttingen (1853), and Berlin (1862). Deeply influenced by Schleiermacher, Hegel, and Kant, he brought these philosophic insights to the study of doctrine, which he interpreted in a traditional Evangelical and historical sense. Among the most distinguished of German christological scholars whose work is still significant, Dorner was the founder and editor of the *Jahrbücher für deutsche Theologie* from 1856.

IAN SELLERS

DOROTHEUS (sixth century). Ascetic and writer. Initally influenced by Barsanuphius, he founded a monastery near Gaza in Palestine about 540 and became its archimandrite. For the members of the community he wrote *Didaskaliai Psychopheleis,* instructions on the life of discipline and asceticism. He made use of earlier collections of rules and added to them his own insights. The later standard edition of the *Didaskaliai* contains further additions to Dorotheus's original rules. He placed great emphasis on humility, maintaining its supremacy over love and its basic relationship to all other virtues.

DOROTHY (d. c.313). Legendary martyr. The earliest mention of her is in the Hieronymian Martyrology. Her emblem is a laden basket. She is reputed to have been a Christian maiden of Caesarea in Cappadocia who was arrested during Diocletian's persecution of the church. On the eve of her arrest she reconverted to Christianity two apostate women who were sent to pervert her. As she was being led to execution she was mocked by a lawyer named Theophilus, who asked her for a basket of flowers and fruit. Soon afterward a child came to Dorothy with a basket of roses and apples. She sent this to Theophilus, who was converted to the faith and later died as a martyr. Gottfried Keller describes her in *Sieben Legenden* (1872).

PETER TOON

DORT, SYNOD OF. Held in 1618-19 in the Netherlands town of Dort (Dordrecht), the synod produced the Canons of Dort, one of the doctrinal standards of the Dutch Reformed Church. It affirms the orthodox Calvinist position on predestination and related issues, and was directed against the Remonstrants or Arminians, who wanted a statement which allowed some role for the human will. Arminius died in 1609; in 1610 his followers issued the Remonstrance* against the orthodox insistence on unconditioned predestination; in 1611 a Counter-Remonstrance reiterated the orthodox stand; and bitter controversy flared.

Apart from predestination, other issues became involved: the Remonstrants wanted a tolerant church, but one under state supervision, which the Contra-Remonstrants saw as an attack on the independence of the church. Even worse, political issues became entangled with theological passions. After the assassination of William the Silent (1584), two leaders emerged to carry on the fight against Spain: William's son Maurice, the *stadhouder* and military leader, and Jan van Oldenbarneveldt, the statesman. By 1609 a truce with Spain was arranged, and the two leaders drifted into disagreement. Maurice favored a strong centralized government to carry on the war of liberation; Oldenbarneveldt, controlling the province of Holland, wanted provincial autonomy and peace. Oldenbarneveldt supported the Remonstrants, and Maurice the Contra-Remonstrants. The political struggle escalated, and when Oldenbarneveldt raised a provincial militia under his control, Maurice sent in the army and arrested him (he was later executed for treason). It was in this situation that Dort convened. Maurice's victory meant, among other things, that the churches elected Contra-Remonstrant delegates; the Remonstrants, who had hoped for Oldenbarneveldt's powerful support, faced a synod packed against them.

Called by the Estates-General, the synod included delegates elected by the synods of the various provinces. Also present as advisers were delegates from Calvinist churches in England and Scotland (James I was strongly anti-Remonstrant), and in the German states; French Calvinists were invited, but were forbidden (by Louis XIII) to attend. The Estates-General chose five theological professors and eighteen commissioners, also to give advice. The regular delegates numbered fifty-six. The synod took the position that it was convened to judge whether the Remonstrant position was in accord with the Calvinist confessions, and cited Episcopius and other Remonstrant leaders to appear before it. Despite Remonstrant protests that the issue was whether the confessions should be revised, synod proceeded. Episcopius denounced the synod as unqualified and unrepresentative, and he refused to cooperate.

Judging the Remonstrants by their writings, then, the synod not surprisingly concluded that they were not orthodox. The Canons were written to summarize the orthodox position against the Remonstrants, and affirmed total depravity (i.e., man, after the Fall, cannot choose to serve God), unconditional election (God's choice of the elect is not conditioned on any action by them), limited atonement (Christ died for the elect only, since those He died for are saved), irresistible grace (divine grace cannot be rejected by the elect), and perseverance of the saints (once elect, always elect). The Canons were adopted as one of the standards of the Dutch Reformed Church.

Remonstrant ministers were ousted from their pulpits, and Remonstrant leaders ousted from the country (by the Estates-General, as disturbers of the peace). The synod ended with a banquet 9 May 1619), celebrating the triumph of Calvinist orthodoxy. Oldenbarneveldt was executed shortly afterward.

See D. Nobbs, *Theocracy and Toleration: A Study of the Disputes in Dutch Calvinism, 1600-1650* (1938). DIRK JELLEMA

DOSITHEOS (1641-1707). Patriarch of Jerusalem. Placed in a monastery at the age of eight and educated at Athens, he entered the service of the patriarch of Jerusalem in 1657. Soon he became archdeacon of Jerusalem (1661) and archbishop of Caesarea (1666). When Nectar Pelopides resigned in 1669, a synod at Constantinople appointed Dositheos in his place as patriarch of Jerusalem. In his new position he showed himself to be a great defender of traditional Greek theology and an opponent of Western theology, both Roman and Protestant. In 1680 he established a printing press at Jassy in order to aid this defense.

He is particularly well known for his presidency of the Synod of Jerusalem in 1672. Its decrees were intended to root out all Protestant influence from the Greek Church. He also sought to reform the monasteries and general administrative structure of the Greek Church; in particular he tried to limit the rights of Western religious orders (e.g., Franciscans) in the Holy Places of Palestine. Further, he tried to extend the influence of the Greek Church into the Russian Orthodox Church when the latter was embroiled in controversy surrounding Patriarch Nikon. While his chief literary work was the posthumous *History of the Patriarchs in Jerusalem* (2 vols., 1715), he wrote also on many other topics, many of them controversial. Though not an original thinker, he displayed wide erudition and learning. PETER TOON

DOSITHEUS. Founder of a Samaritan sect that exercised influence at Nablus until opposed by the priestly school (according to Bowman). He is dated in the second century B.C. by Josephus, the first century A.D. by Origen and the Clementine Recognitions, and the fourth century A.D. (under "Dusis") in the Samaritan Chronicles 3,6,7). His proto-Gnostic sect in the first century A.D. (Hegesippus) supported his claim to be the Christ foretold by Moses, kept the Sabbath strictly, read his books, and claimed he was still alive (Origen). They practiced circumcision, vegetarianism, and possibly chastity (Epiphanius). Epiphanius thought Dositheus taught resurrection but that Sadducean influence made the sect deny it. From this J. Montgomery argues for two Dosithean sects. Dositheans survived to the twelfth century, according to Arabic sources. G.T.D. ANGEL

DOSTOEVSKY, FYODOR (1821-1881). Russian writer. Born in Moscow, son of a doctor, he was educated as an engineer, but early turned to writing. His first novel, *Poor Folk* (1846), was highly acclaimed by the critics for its penetrating psychological study of the poor. Shortly thereafter, Dostoevsky became involved with an antigovernment socialist group. For this he was arrested and sentenced to death, but at the public execution he was given a last-minute reprieve. He was forced to spend ten years in Siberia in prison and military service instead. Returning to St. Petersburg in 1859, he began to write again. *The House of the Dead* (1861) offers a realistic account of his prison experiences. *Notes from Underground* (1864) is an extraordinary picture of a mentally disturbed and alienated man. For a time Dostoevsky was nearly overwhelmed by gambling debts, emotional tensions, and epileptic seizures. In 1866 he won wide acclaim for his superb novel *Crime and Punishment*, a tale of deep spiritual insights. For years he wandered over Germany, Switzerland, and Italy, often in abject poverty. *The Idiot* (1868) and *The Possessed* (1871) added to his stature as one of the greatest Russian novelists. His masterpiece is *The Brothers Karamazov* (1880), finished the year before his death. Dostoevsky's profound psychological insights have made him in the twentieth century the most influential and most widely read of the older novelists. His works are novels of ideas in which there is brilliant characterization, tense dramatic situations, and a struggle between good and evil. His Russian orthodoxy is represented by characters who seek salvation through suffering. PAUL M. BECHTEL

DOUAI-REIMS BIBLE. Roman Catholic translation of the Bible into English, so called because it was the production of the English College founded by Roman Catholic refugees in the Elizabethan period at Douai, later removing to Reims, and returning to Douai in 1593. The work was begun in 1578 at the instigation of William Allen,* not to promote Bible reading, but "with the object of healthfully counteracting the corruptions whereby the heretics have so long lamentably deluded almost the whole of our countrymen" (i.e., in the Protestant versions).

The chief translator was Gregory Martin, an Oxford scholar, and his daily stint of two chapters was revised by Allen and Richard Bristow. It was a translation of the Latin Vulgate, because of its antiquity and freedom from discrepancies visible in the Greek manuscripts, and because the Council of Trent defined it as exclusively authentic. Martin, however, did use the Greek text, and also the Protestant versions, notably Coverdale's "Diglott" version (1538). The style of the Douai-Reims was strongly Latinate, and it deliberately retained many technical terms in their original form: e.g., neophyte, Paraclete, sancta sanctorum, archysynagogue. The NT translation appeared in 1582 and was extensively used in the preparation of the KJV. The OT was ready at the same time, but did not appear until 1609-10 because of lack of funds. A revision of the Bible was made by Bishop Richard Challoner in 1749-50 and again in 1763-64, and in 1941 it was revised in accordance with Hebrew and Greek sources, and completely modernized. J.G.G. NORMAN

DOUBLE MONASTERIES. Houses used by nuns and canons of the Order of Sempringham, founded by Gilbert of Sempringham.* On his death in 1189 he left nine double houses. The only common portion was the church in which the nuns and canons could neither see nor hear each other. All other buildings were well apart.

DOUBLE PROCESSION, see PROCESSION OF THE SPIRIT

DOUKHOBORS. A group of so-called spiritual or rationalist Christians which arose in Russia some time before the late eighteenth century, when its members first appear as the objects of persecution. Their name "spirit-wrestlers," originally intended by enemies to suggest strife against the Holy Spirit, was taken by them to designate striving by means of the Spirit. Christian doctrines are interpreted by them as manifested in the nature of man. The Trinity is Light, Life, and Peace, with which each man may be linked by Memory, Understanding, and Will. The story of Jesus symbolizes a spiritual development which anyone may undergo. Death is insignificant since the soul migrates. Advance beyond the revelation through Jesus is possible, and others may be called "son of God." Ritual acts of all kinds are rejected. They are pacifist, agrarian, acknowledge no earthly government, and will not own property. Doukhobors were fiercely persecuted from the start, although their upright life was acknowledged. They were exiled first to Siberia, later to "Milky Waters" in Taurida, and again to Georgia. In each case they came into conflict with the authorities. Under the influence of Tolstoy they refused army service. Tolstoy and English Quakers (who felt some affinity) publicized their plight and arranged emigration to Canada in 1898. There also conflict with authority arose over landownership, and registration of births, deaths, and marriages. In protest, extremist Doukhobors (Sons of Freedom) have resorted to parading naked and to arson and dynamiting. The majority have come to compromise with the authorities, buying their land, in the prairies and British Columbia. Many Doukhobors remaining in the Soviet Union were liquidated in the Stalinist era, but some villages remain in Georgia.

BIBLIOGRAPHY: F.C. Conybeare, *Russian Dissenters* (1921); W. Kolarz, *Religion in the Soviet Union* (1961), pp. 353-56; G. Woodcock and I. Avakumovic, *The Doukhobors* (1968).

J.N. BIRDSALL

DOWIE, JOHN ALEXANDER (1847-1907). Faith healer and founder of the Christian Catholic Church. Born in Edinburgh and taken to Australia in 1860 by his parents, he was in business in Adelaide for seven years, then studied at Edinburgh University. Ordained as a Congregational minister in 1870, he served as pastor of churches in Alma and Sydney. In 1878 he resigned to become an evangelist and faith healer. By 1888 he had built a large independent tabernacle in Melbourne and organized the International Divine Healing Association. He then organized churches along the American Pacific coast for two years. He moved to Chicago and in 1896 founded the Christian Catholic Church, a theocracy headed by himself as "Elijah III, the Restorer." By 1901 he moved his community to what became Zion, Illinois. As First Apostle after 1904, he ruled the theocratic community strictly and banned pork, alcohol, tobacco, and drugs. Failure in a New York Madison Square Garden campaign and fiscal problems led to his deposition in 1906 by his successor W.G. Voliva, a year after Dowie had been stricken with paralysis. EARLE E. CAIRNS

DOWSING, WILLIAM (1596-1679?). Often described as iconoclast. Born of Yeoman parents in Suffolk, he was appointed by the earl of Manchester in 1643 as an official visitor of East Anglian churches in order to remove papistical ornaments such as images and pictures. He did his work with great relish, keeping careful journals of what he and his colleagues accomplished. Robert Loder printed in 1786 part of the journal which described the work in Suffolk; a transcript of the journals describing work in Cambridgeshire survives in Cambridge University Library, and this was printed in 1928, edited by A.C. Moule. The edition of the Cambridgeshire journal published by Zachary Grey in 1739 seems to have been made from a copy now lost. PETER TOON

DOXOLOGY. From the same root as "dogma," the ancient formation *doxa*, "that which seemed to one," came by the fourth century B.C. to identify the "reputation" or "fame" of another. The extension to gods or God ("His fame") was made in Septuagint translation and in magical Greek papyri, and from this came in the Greek of the patristic period both verbal and nominal forms associated with the uttering of praise in general *(doxologia)*. More specifically, the *Gloria in Excelsis* (adapted from Luke 2:14, in the *Apostolic Constitutions**) and the *Gloria Patria* (a Trinitarian liturgical conclusion for Psalms, enlarged with an anti-Arian counter-clause) are called greater and lesser, while in English Reformation circles the Doxology refers to a refrain which closed three hymns by Thomas Ken.

CLYDE CURRY SMITH

DRAMA, CHRISTIAN. It is customary to ascribe the beginnings of postclassical drama to the action of the Mass, and even beyond this the first identifiable dramatist is Hroswitha, the tenth-century nun of Gandersheim in Saxony, whose attempts at comedy owe much to Terence but add strict moral and religious teaching in a manner foreign not only to the Roman writer but even to the dramatic mode itself.

The church relied upon visual means—paintings, stained glass, mime, etc.—to convey its message to illiterate congregations, and gradually there grew the practice of dramatizing the events associated with the major festivals, and especially Easter. In due course these representations outgrew the places where they were performed, and so they moved out of the church and into the churchyard. In addition, the range of representation was extended. The churchyard in its turn proved inadequate, and so a further move was made into the open spaces of the town, while in some places there grew up the practice of playing the several episodes on different carts, which might move from one location to another and thus present the possibility of multiple performance.

The Mysteries or Miracle Plays,* as the several sequences came to be known, covered the whole of biblical history and even included nonbiblical material. They were mounted by the several trade guilds and often performed on the Feast of Corpus Christi, especially after 1311 when the Council of Vienne ordered the strict observance of this feast. It is known that these cycles of plays were given in at least a hundred English towns. Those which survive include, in part or whole, the plays performed at Chester (from the early fourteenth century until at least as late as 1600), York (with forty-eight plays—from 1360 to 1579 and revived over the last twenty years), Wakefield (the so-called Towneley cycle of thirty-two plays), and Coventry. Of these the last is least dramatic. The others, however, contain a variety of effort and achievement ranging from such solemn and moving scenes as the Crucifixion to the farcical of Noah's wife railing at what she considers her half-crazy husband, and the thoroughly English sheep-stealing and concealment of the Second Shepherd's play in the Towneley cycle. The whole shows the capacity of medieval man to regard the sacred and the secular in a coalescence which he did not find either irreverent or inappropriate.

The Morality Plays,* by contrast, tend to be more what they say they are. One of the earliest is *The Castell of Perseverance* (c.1405), tracing the history of Humanum Genus from birth to judgment, assailed by Mundus, Belyal, and Caro and protected by his Good Angel. This is a prolix and tedious play and does not compare with what is doubtless the greatest example of the genre, *Everyman* (early sixteenth century), which, though presenting basically the same story, does so with effects of psychological tension quite remarkable for its type and time.

The sixteenth century saw the full flowering of the Renaissance with its stress on humanistic learning at the inevitable expense of the religious. There *were* plays on sacred themes, such as John Bale's *God's Promises* (1538) with its strong Calvinistic propagandist intent. Later there was George Peele's *David and Bethsabe* (c.1594), a play first of high erotic sensuality and then of war, revenge, and retribution. The plot is close to the Bible, but the tone owes much to Ovid and Seneca, those models who inspired the Elizabethans to their equivalents of modern Hollywood spectaculars of blood and lust.

It is not plays like these, but rather one like Marlowe's *Faustus* (c.1590) which best illustrates the secular and religious tension in Renaissance man. Like others of Marlowe's heroes, Faustus is an overreaching egotist; nothing is beyond man's aim and, with Lucifer's help, his attainment; but with figures such as the Good Angel to balance Mephostophilis there is an inescapable morality element which reaches the joys and agonies of high tragedy first in Faustus's intercourse with Helen of Troy and then in his dying speech, a piece of writing unmatched in English for its evocation of torture, remorse, and final damnation.

Hamlet also presents something of the tension —"What a piece of work is a man!" and

Imperial Caesar, dead and turn'd to clay
Might stop a hole to keep the wind away—

but in general Shakespeare is concerned not with religious, but metaphysical questionings. His most explicit treatment of religious problems and attitudes is in *Measure for Measure* (c.1602), where questions of law and liberty, charity and chastity, justice and mercy are examined with that fulness and subtlety of which Shakespeare alone is capable.

His great successor in English poetry, John Milton* wrote two dramatic pieces, but neither is actable drama. *Comus* (1634) is a masque and, like *Measure for Measure,* much concerned with the demands of chastity and the lure of license, seeking to define the nature of responsible freedom. *Samson Agonistes* (1671) traces the history of fallen Samson and of his restoration to fulfil God's will, even though it be at the cost of his own destruction. Shaped in the Greek tragic manner, *Samson* lacks dramatic urgency and tension. Dryden dramatized *Paradise Lost* in his *State of Innocence* (1677).

The eighteenth century is the age of the opera and the oratorio, and notably of Handel.* When drama reemerges in the nineteenth century, it is represented biblically by Byron with *Cain* and *Heaven and Earth,* but now in the age of Romantic individualism the hero is Cain, puzzled and "Satanic," set against an unjust and capricious God—and yet the end with Cain is full of remorse. It shows, in fact, the contrasting polarities of Byron's own disturbed personality. Like most dramas of its century, *Cain* is all but unactable. In addition, the nineteenth century by its rigid censorship practically annihilated the possibility of religious drama.

At the end of the period a work like Wilde's *Salome* (1893) is a deliberate flouting of this rigidity. He wrote it in French, and performance was prohibited in England. Despite continuing restraints, writers like Yeats with his own symbolic *Calvary* (1920) and *Resurrection* (1931), presenting a strangely dehumanized Christ, and D.H. Lawrence with *David* (1926), aiming to bring passion into religion, pursued biblical themes. (Lawrence reverently retells the Resurrection story in *The Man Who Died.*)

The developing freedom allowed Laurence Housman to present his *Old Testament Plays* (1950) with their bitter hostility to the biblical accounts. A more orthodox approach is represented by Norman Nicholson's *The Old Man of the Mountains* (1946), with Elijah and Ahab in modern Cumberland. Its producer was E. Martin Browne, who did much for T.S. Eliot* including *Murder in the Cathedral* (1935), where the dilemma of self and service is explored unto death in Thomas à Becket. Charles Williams* is a lesser figure, but his *Seed of Adam* (1936) and *Thomas Cranmer of Canterbury* (1936) should be mentioned. The latter is a story much like that of Becket, while the former is an attempt to put human history into a single act. Finally there is Dorothy Sayers* with *The Man Born to be King* (1943), a BBC series of twelve plays, and Christopher Fry's treatment of the mystery of human existence through Moses in *The Firstborn* (1948) and his psychological reinterpretation of biblical story in *A Sleep of Prisoners* (1951).

See M. Roston, *Biblical Drama in England* (1968).

ARTHUR POLLARD

DREXEL(IUS), JEREMIAS (1581-1638). German spiritual writer. Born at Augsburg in 1581 of Lutheran parents, he was converted to Catholicism in his youth and educated by the Jesuits, which order he joined in 1598. He was later professor of humanities at Munich and Augsburg, taught at the Jesuit seminary at Dillinger, and was court preacher to the elector of Bavaria. Between 1620 and 1638 he wrote a series of twenty works, mainly of a devotional nature, which were eagerly read and translated into many languages. Among the most popular were *Considerations on Eternity,* four separate English translations of which had appeared by 1710 and which discusses man's apprehension of the eternal dimension, and the *Heliotropium* (1627; ET 1682), on the nature of revelation.

IAN SELLERS

DRIVER, SAMUEL ROLLES (1846-1914). OT scholar. Born at Southampton, he was educated at Winchester and Oxford, with which university he was connected all his working life, and where he succeeded E.B. Pusey in the regius chair of Hebrew (1883-1914). Influenced by the critical approach to the OT of German scholars, he did much to publicize their views in his teaching and writings. Apart from many commentaries on OT books, his works include *Introduction to the Literature of the Old Testament* (9th ed., 1913), *Notes on the Hebrew Text and the Topography of the Book of Samuel* (2nd ed., 1913), and co-editorship of the *Hebrew and English Lexicon of the Old Testament* (1906). He was a member of the Old Testament Revision Company (1875-84).

J.D. DOUGLAS

DROSTE-VISCHERING, CLEMENT AUGUST VON (1773-1845). Archbishop of Cologne. Ordained to the Roman Catholic priesthood in 1798, he became curate to the chapter of Münster and auxiliary bishop of Münster in 1827. When Von Spiegel died in 1835, he was elected archbishop of Cologne at the suggestion of the Prussian government. Soon he came into conflict with the government by refusing to sanction the teachings of the Bonn professor George Hermes, which had been condemned by Gregory XVI in 1835. When he further refused to approve the Prussian policy on mixed marriages between Protestants and Catholics he was imprisoned by Frederick William III in the fortress of Minden in 1837. J.J. Von Görres* wrote his *Athanasius* (1838) in defense of Droste-Vischering, and the archbishop was restored to his former honor, but the government of the diocese was left to a coadjutor more favorable to the Crown. Among his writings are *Über die Religionsfreiheit der Katholiken* (1817) and *Über den Frieden der Kirche und der Staaten* (1843).

WAYNE DETZLER

DRUMMOND, HENRY (1786-1860). Politician, writer, and a founder of the Catholic Apostolic Church.* Educated at Harrow and Oxford, he entered the banking profession, was elected to Parliament (1810), where his vote on major issues was uninfluenced by party considerations, and founded a chair of political economy at Oxford (1825). His individualism was carried into his religious activities. Going to Switzerland, he contended strongly against Socinian tendencies in Genevan Protestantism. In later years (during which time he was again in Parliament) he was closely associated with the origin and spread of the Catholic Apostolic Church. Meetings of those in sympathy with the views of Edward Irving* were held for the study of prophecy at his home in Surrey. He became the new body's "angel" for Scotland. His many writings include *Social Duties and Christian Principles* (1830), *The Fate of Christendom* (3rd ed., 1854), and *Discourses on the True Definition of the Church* (1858).

J.W. MEIKLEJOHN

DRUMMOND, HENRY (1851-1897). Scottish writer and evangelist. Born at Stirling and educated at Edinburgh University, he was persuaded by D.L. Moody* to suspend his theological course and to work with him in evangelistic campaigns during the American's first visit to Britain (1873-75). From 1877 he taught natural science at the Free Church College, Glasgow, and in 1883 published the best seller *Natural Law in the Spiritual World.* In 1884 he was ordained and became professor of theology in the college. He influenced many generations of students through his evangelistic work, visiting also Australia (1887) and the USA (1890). His lectures at Boston were published as *The Ascent of Man* (1894). His best-known work was, however, *The Greatest Thing in the World,* a meditation on 1 Corinthians 13. During the last fourteen years of his life he was involved in controversy about the relation of science and religion, and about the authority of the Bible. He might have been a great scientist had not evangelism been the master passion of his life. He died after two years of crippling illness.

J.W. MEIKLEJOHN

DRYDEN, JOHN (1631-1700). Essayist, playwright, and poet. Born in Northamptonshire and educated at Cambridge, he dominated the English literary scene throughout the latter part of the seventeenth century. Johnson considered him "the father of English criticism," and of his influence on English prose said that "he found it brick and left it marble." His satire vigorously supported Charles II's cause against the Whigs. This typifies the chief strain in his thinking, namely, his profound regard for authority, perhaps a legacy from growing up amid the upheaval and uncertainty of the Civil Wars. This search for authority also governed his poems on religion. *Religio Laici* (1682) is a defense of rational Anglicanism as a *via media* between extremes, but it is no surprise in Dryden's progress that he eventually became a Roman Catholic and that *The Hind and the Panther* (1686) is a satiric allegory in which the milk-white hind of Romanism triumphs over the spotted, Anglican panther. Dryden's is an intellectual rather than a spiritual faith.

See his *Poems* (ed. J. Kinsley, 1962); and C.E. Ward, *John Dryden* (1962).

ARTHUR POLLARD

DUALISM. When used of a religion, this word implies the doctrine of two divine powers or principles in opposition within the universe. Such is found in Zoroastrianism and Manichaeism. In the field of philosophy the term describes the existence of two essentially different constituents in the world—e.g., the Cartesian bifurcation of reality into material substance and mental substance. Within Christian theology, Nestorians were accused of dualism because they supposedly taught the doctrine that Jesus was two Persons linked together and not truly one Person. In general the word dualism describes any system of belief or thought which contains two opposing principles.

PETER TOON

DUBOURG, ANNE (1520/21-1559). French Protestant martyr. Born at Riom and trained as a lawyer, he became professor of law at Orleans (1547), received his doctorate (1550), and held important posts at the university. In 1558 he became a member of the Parlement de Paris, but in 1559 fell into disfavor with Henry II because, having become a Protestant, he made a violent attack upon the Roman Catholic Church and advocated reform. Arrested for his bold declaration of his position, he was tried for heresy, but used every legal means of escaping punishment through appeals to different courts. Finally his appeal was rejected, and he was condemned, strangled, and burned. His martyrdom caused widespread horror, particularly among university students, some of whom had been responsible for his conversion.

W.S. REID

DUCHESNE, LOUIS MARIE OLIVIER (1843-1922). French archaeologist and church historian. Born in Saint-Servan, he read theology in Rome and developed archaeological and patristic interests under G.B. de Rossi, with whom he edited the *Martyrology of St. Jerome.* Ordained priest in 1867, he lectured in schools for six years, then served as a member of the French archaeological school of Rome (1874-76), overseeing work in Epirus, Thessaly, Mt. Athos, and Asia Minor. He was to return to the school as director in 1895, a post he held till his death. Holding the chair of church history at the *Institut Catholique* in Paris (1877-85), he resigned because of opposition to his views on pre-Nicene doctrine and the founding of the French Church. Member of the French Academy from 1910, Duchesne's publications include *Histoire ancienne de l'Église chrétienne* (3 vols., 1906-10); studies on Macarius Magnes and *Liber Pontificalis; Les Fastes épiscopaux de l'ancienne Gaule* (3 vols., 1894-1915); and *L'Église au sixième siècle* (1925).

C.G. THORNE, JR.

DUFAY, GUILLAUME (c.1400-1474). Netherlandish composer. He was the first of a series of great composers coming from the Low Countries (especially from the duchy of Burgundy) who gave leadership to the development of polyphonic music until well into the sixteenth century. As with most composers of the period, he was a singer and took holy orders. Like many of his compatriots, he was drawn to Italy, where he spent two significant periods of his life and sang for a time in the papal choir. He wrote a highly interesting motet for the consecration of the *duomo* in Florence, *Santa Maria del Fiore* in 1436. A large amount of his music survives, much secular as well as sacred. He composed a cycle of three-part settings of the office hymns for the church year, as well as motets and other liturgical pieces.

Although earlier examples are known, Dufay was the first great master of the unified setting of the ordinary of the Mass, employing a preexistent melody *(cantus firmus)* as the tenor part, repeated in each section as a unifying device. He seems to have begun the custom of often using a secular melody for this purpose. Especially famous is his *L'Homme armé* Mass, using a secular melody of disputed origin. Almost every composer of Masses employed this melody in at least one work until it was banned by the Council of Trent. Beginning with the generation of Dufay, the polyphonic Mass became the most important major form of composition until the end of the sixteenth century. J.B. MAC MILLAN

DUFF, ALEXANDER (1806-1878). Scottish missionary to India. Born in Perthshire and educated at St. Andrews University, he became in 1830 the first church of Scotland missionary in India (he and his wife were twice shipwrecked en route to Calcutta). Realizing the value of a strong educational policy, he opened an English school in which the Bible was the central textbook, but which offered a variety of subjects to university standard. There was some opposition from both Hindus and fellow missionaries, but he had a powerful ally in the (British) governor general, and the school developed notably. Poor health compelled his return home in 1834, but he had recovered sufficiently to see India again in 1840.

When the Disruption* came in 1843, he like most missionaries left the establishment to form the Free Church of Scotland.* Consequently the Indian property had to be relinquished and the building of a new institution begun. In 1844 Duff helped to found the *Calcutta Review,* and was from 1845 its editor until he left for Scotland again in 1849. Earlier he had declined an invitation to succeed his old teacher Thomas Chalmers as principal and theology professor at New College, Edinburgh. In 1851 he was moderator of his church's general assembly, and in 1854 he impressed his concern for missions on American and Canadian listeners. Another address, delivered at an Evangelical Alliance gathering in 1855 was, said a friend, "like a trump of doom uttered over the worldliness of existing churches and a call to assembled Christendom to turn from luxury and pomp and to remember the perishing nations." A further spell in India (1856-64) was concerned with the advancement of higher education in the country, and with the foundation of the University of Calcutta. Ill health forced him to leave India, but he labored in the missionary cause until his death. From 1867 he occupied the first chair of evangelical theology at New College, Edinburgh.

See biographies by G. Smith (2 vols., 1879) and W. Paton (1923). J.D. DOUGLAS

DUKHOBORS, see DOUKHOBORS

DU MOULIN (Molinaeus), PIERRE (1568-1658). French Protestant pastor. Born in France and educated at Sedan, he was sent to England in 1588 and acted as tutor to the duke of Rutland. He studied at Cambridge under William Whitaker. In 1593 he became professor at Leyden, lodging with Scaliger,* and having Grotius* as a pupil. Ordained in 1599, he became pastor in the French Reformed Church at Charenton near Paris. His house was the resort of leading Protestants. On a visit to England he was made D.D. by Cambridge and given a prebend at Canterbury by James I. In the latter part of his life he preached and lectured at Sedan. He wrote many books and treatises, mostly of a controversial nature. PETER TOON

DUNCAN, GEORGE SIMPSON (1884-1965). Scottish biblical scholar, one of the translators of the NT in the New English Bible. Duncan is associated with the theory that all Paul's "imprisonment" epistles were written from Ephesus (cf. his *St. Paul's Ephesian Ministry,* 1929) and with the view that Galatians is the earliest of Paul's extant letters (cf. his *Commentary on Galatians,* 1934). After acting as chaplain to Earl Haig (1915-19), he spent his teaching life at the University of St. Andrews where he was professor of biblical criticism (1919-54), principal of St. Mary's College (1940-54), and vice-chancellor (1952-53). In 1920 he was one of the founders of the St. Andrews Summer School of Theology which still provides refresher courses for clergy. In 1949 he was moderator of the general assembly of the Church of Scotland. HENRY R. SEFTON

DUNKARDS, DUNKERS, see CHURCH OF THE BRETHREN

DUNS SCOTUS, JOHN (1266-1308). Scholastic theologian. Born in Scotland, he entered the Franciscan Order at the age of fifteen and was ordained priest in 1291. After studying at Paris (1293-96) he returned to England to lecture on the *Sentences* of Peter Lombard* at Oxford. Later he taught at Paris and in 1303 was banished by Philip IV (the Fair) because he supported Pope Boniface VIII. In 1304 he again lectured at Paris, but was transferred to Cologne (1307) where he died. Although venerated as a saint in his order, his cult is not universally recognized in the Roman Church. Duns Scotus's thought is so intricate that he has been given the title "the Subtile Doctor" by Roman Catholics, and Protestant Reformers called anyone whose ideas seemed obscure a "duns," hence "dunce." He wrote commentaries on the *Sentences* of Lombard, explanations of Aristotle, and explanations of Holy Scripture.

Critical of the philosophy of Thomas Aquinas* which attempted to harmonize Aristotle with Christianity, he argued that faith was a matter of will and could not be supported by logical proofs. This division between philosophy and faith was to have far-reaching effects. Although arguing for the existence of God from efficiency, finality, and the degrees of perfection, he taught that all other knowledge of the divine, including the Resurrection and immortality, must be accepted by sheer belief. Creation he believed was the effect of God's love as He extends His goodness to creatures so that they will love Him freely. Grace is identical with love and has its origin in the will. Because of his idea of the superiority of the will over the intellect, Duns Scotus believed that heaven consists of sharing the love of God. Divine love can best be seen in Jesus Christ who would have come, Duns Scotus taught, even if man had not sinned. Thus the incarnation as the center and end of the universe was not determined by original sin. Although much of Duns Scotus's teaching gained wide recognition among theologians, he is especially remembered for championing belief in the Immaculate Conception. Scholars in the Franciscan school, Scotists, who followed him, moved ever further in the separation of faith and reason, leading to the eventual decline of Scholasticism.* His works were edited by Luke Wadding (12 vols., 1639) and reprinted in Paris (26 vols., 1891-95). Recent studies have demonstrated that some of the writings attributed to him are spurious; thus a new edition of his works is now appearing with the title *Opera omnia, studio et Cura Commissionis scotisticae ad fidem cadicum edita* (Vatican City, 1950-).

BIBLIOGRAPHY F. Copleston, *A History of Philosophy*, vol. II (1950); E. Bettoni, *Duns Scotus: The Basic Principles of His Philosophy* (tr. B.M. Bonansea) (1961); J. Weiberg, *A Short History of Medieval Philosophy* (1964); J.K. Ryan and B.M. Bonansea (eds.), *John Duns Scotus, 1265-1965* (1965), vol. III. ROBERT G. CLOUSE

DUNSTABLE, JOHN (d.1453). English composer. Details of his life are meager. He seems to have gone to the Continent with the duke of Bedford during the Hundred Years' War, and to have spent much of his career there. He was known also as a mathematician and astronomer. Several European authorities of his century testify to his great influence and reputation. About sixty of his works survive, mostly sections of the Mass and motets, and are found chiefly in Italian manuscripts. It was through him that the English richness of triadic harmony was communicated to the early masters of the Renaissance, such as Dufay.*

DUNSTAN (c.909-988). Archbishop of Canterbury from 959. After serving at King Aethelstane's court, he became a monk and then abbot (c.943) at Glastonbury, which he made famous for asceticism and learning. In 959 King Edgar of Mercia and Northumbria became king of all England, and appointed Dunstan to Canterbury. Together the two carried out a complete reform of church and state, continued under Edward the Martyr who succeeded to the throne on Edgar's death in 975. When Edward was murdered three years later, Dunstan's star waned. A versatile man, he is remembered for having revived monastic life in England and for making it an influence in the country's affairs. One of his illuminated manuscripts is in the British Museum.

J.D. DOUGLAS

DUNSTER, HENRY (1609-1659). Congregational minister and educator. Born in Bury, England, he was educated at Cambridge, taught school, and served as curate in his hometown. To escape High Church tyranny he fled to Massachusetts in 1640 and was immediately appointed first president of the newly established Harvard College. His reputation and administration gave Harvard a standing and character which persisted throughout the colonial period. Although poorly paid, he was a benefactor to the college. He lost his position in 1654 for advocating anti-paedobaptist views and refusing to have his child baptized. After public admonition he retired to the pastorate in Scituate, where he labored until his death. His revision of Eliot's Bay Psalm Book was used for many years. ALBERT H. FREUNDT, JR.

DUPANLOUP, FÉLIX ANTOINE PHILIBERT (1802-1878). Bishop of Orléans. Born in Savoy, he was educated in Paris and ordained in 1825. He was curate of the Madeleine Church and later became superior of the minor seminary of St. Nicholas (1837-45). A leading educationist, he favored freedom for Catholic secondary schools and was the chief architect of the Falloux Law (1850) as bishop of Orléans. Amid the major quarrels besetting the church of France—Ultramontanism and modern liberties—he initiated many diocesan charities and inspired his subordinates. With the Italian war he moved to the forefront of the European politico-religious scene, writing brochures defending papal temporal power, but he won disfavor at Vatican Council I.* Elected to the French Academy (1854), French National Assembly (1871), and the Senate (1875), he made his way even among unbelievers in French society, and vigorously promoted women's education. Apart from sermons, speeches, and catechetical pieces, his major works are *De l'Education* (6 vols., 1850-66) and *La Femme studieuse* (1869).

C.G. THORNE, JR.

DUPERRON, JACQUES DAVY (1556-1618). Archbishop of Sens. Son of a Reformed pastor who had fled to Bern, Duperron went to Paris in 1573 and was converted to Roman Catholicism by his study of Aquinas and Bellarmine. His considerable gifts soon won Henry III's favor. In 1591 he became bishop of Évreux and played an important role in the conversion of Henry IV and his reconciliation with Rome. Though not an original scholar, he was a formidable controversialist, and humiliated P. Duplessis-Mornay* by demonstrating misuse of patristic texts on the Eucharist. Appointed cardinal in 1604, he used his great diplomatic talents to reconcile Venice to the papacy. Returning to France, he became archbishop of Sens and was an important opponent of Gallicanism, also engaging in theological controversy with James I of England. Duperron's great gifts were given unreservedly to the papacy and the French crown. IAN BREWARD

DUPIN, LOUIS ELLIES (1657-1719). French church historian and theologian, he produced an immense history and criticism of theologians and their writings since the first century. It appeared at Paris in forty-seven volumes (1686-1714) as *Nouvelle Bibliothèque des auteurs ecclésiastiques*, and was placed on the Index.* He was severely attacked by both Rome and Gallicans, especially

Bossuet,* and variously accused of Jansenist, Gallican, and Romanist doctrines, although he fits precisely into none of these categories. He sought reunion of the Catholic, Greek, and Anglican churches, and wrote a sympathetic *Histoire des Juifs* (7 vols., 1710). He earned a doctorate in theology at the Sorbonne and was professor at the Royal College. C.T. MC INTIRE

DURANDUS OF ST. POURÇAIN (c.1270-1332). Dominican theologian, and bishop successively of Limoux (the only cleric ever to hold this title) in 1317, Le Puy-en-Velay (1318), and Meaux (1326). Known as the *Doctor Modernus* and the *Doctor Resolutissimus,* he lectured at Paris until 1313, when he was called to Avignon and entrusted with a diplomatic mission by Pope John XXII. Though a Dominican at a time when Thomas Aquinas was already recognized as the official doctor of the order, Durandus was not a Thomist but a kind of nominalist, maintaining that the universal comes after the thing and that intellection is a psychological rather than a metaphysical operation. His partiality for nominalist solutions did not prevent his serving on the papal commission which condemned fifty-one propositions taken from William Ockham. Under heavy pressure from his own order for his anti-Thomist ideas, Durandus revised his *Commentary on the Sentences* (before 1308) twice (1310-12, 1317-27), removing some of his more offensive theses. He never repudiated his position, however, and eleven articles from his treatise *De Visione Dei* were censured by a papal commission in 1333.
DAVID C. STEINMETZ

DÜRER, ALBRECHT (1471-1528). Painter, engraver, and woodcut designer. Son of a Nuremberg goldsmith, he study-lived across Europe during his formative years. During his forties he worked for Emperor Maximilian I. Late in life he consorted with the rich, cultured literati of the day; he was also a friend of Philip Melanchthon.* Deeply taken by the exacting perspective of Mantegna and the Italian search for perfect body proportion, he strove to mate those concerns with the strange, landscape wildness indigenous to his native Gothic training. In technique this led to an astounding, powerful refinement of the woodcut, because Dürer used the detailed subtleties possible with engraving to modulate rough woodcut simplicity. The unnerving line, monumental complexity, yet classic motifs in this *Apocalypse* series (c.1497-98) transformed woodcut art in Europe. His most famous engravings also juxtaposed erudite humanist motifs and composed forms with curious, moody grotesqueries. Both *Fall of Man* (1504) and *Melancholia* (1514) show the unresolved yet compelling hybrid: studied hieroglyphic elements and Reformation directness bursting with Renaissance energy. Dürer spent years probing theoretically the criterion for artistic action, fussing with the concept of qualitative harmony. While his formulation of such a foundational aesthetic was inconclusive, the muscular soldier-angel figures and massive, severe grandeur in the *Four Apostles* painting (1523-26) indicate his direction. He consciously chose Luther's side; his art, however, has deep affinity with the piety and Christian humanism of Erasmus.
CALVIN SEERVELD

DURIE, JOHN (1596-1680). Scottish ecumenist who devoted much of his life to "ecclesiastical pacification." Both his father and grandfather were militant Presbyterians who incurred the displeasure of James VI*, and Durie accompanied his father into exile at the age of ten. This early experience of controversy helped to form the peacemaker he became soon after his 1624 settlement as minister of a congregation of English and Scottish Presbyterians at Elbing in West Prussia. When Elbing came under Swedish rule, he petitioned Gustavus Adolphus* "for the obtaining of aid and assistance in this seasonable time to seek for and reestablish an ecclesiastical peace among the Evangelical Churches."

In 1634 Durie accepted ordination in the Church of England hoping that this would give support to his schemes for the union of all Protestant churches. He never seems to have doubted the possibility of its early realization but "was always too ready to mistake his own dreams for solid realities of the near future, whenever he was entertained with kindness and friendly words" (Westin). His journeys all over Europe included a short visit to Scotland where the Aberdeen Doctors* warmly supported him. He tried to mediate in the English Civil War and took part in the Westminster Assembly.* After the Restoration he settled at Cassel where he continued his efforts toward church union.

See G. Westin, *Negotiations about Church Unity 1628-34* (1932); and J.M. Batten, *John Dury* (1944). HENRY R. SEFTON

DUTCH REFORMED CHURCH *(Hervormde Kerk).* The major Protestant church in the Netherlands, Calvinist in theology, presbyterian in church government, organized during the revolt of the Low Countries against Spanish rule in the sixteenth century. The Lowlands (the Netherlands and Belgium), after partial unification under the Burgundian dukes in the 1400s, passed to Hapsburg rule: Charles V, being Luther's sovereign, reigned also over the Lowlands. Anabaptism and Lutheranism spread during the 1520s and 1530s. Thereafter the dominant version of Protestantism was Calvinism.* Under the Spanish son of Charles, Philip II, the Inquisition* was stepped up, and martyrs soon abounded. The "seventeen provinces" revolted against Philip under the leadership of William of Orange (1568), with the Calvinists playing the role of a militant and influential minority. The Belgic Confession* (1561), with the Heidelberg Catechism,* were accepted generally as standards of the Reformed Church. In the "liberated" areas, Calvinism was the religion favored by the state. The first national synod was held in 1578. As the revolt went on, the N Lowlands gradually drove out the Spanish, while the revolt was slowly crushed in the south, which thus remained Catholic. By 1609, when a Twelve Year Truce recognized for all practical purposes the independence of the north, the Calvinists were free to turn to difficulties within their own ranks.

The controversy on the teachings of Arminius* and his followers, the Remonstrants, ended with the triumph of orthodox Calvinism at the Synod of Dort* (1618-19); the Remonstrants were ousted from the Reformed Church.

During the Dutch "golden age" of the 1600s, when the Netherlands was a major power, the Reformed Church, as the established church, played an important role in Dutch life. Its efforts in theology became increasingly defensive. The effort to preserve the orthodoxy of Dort resulted in controversy on doctrinal detail (notably the confrontation between Voetius* and Cocceius,* which caused a great uproar), as well as protest against overstructuralization (Labadie). By the 1700s the great days of scholastic Calvinism were over. Intellectuals turned to the new ideas of the Enlightenment* rather than to theology, and Deism* made some inroads in the church itself. By the 1780s, as the rhetoric of the "Patriot" movement showed, many regarded it as a bulwark of privilege. The storms of the French Revolution affected the Netherlands as well as the rest of Europe. French troops, greeted by many as liberators, occupied the country (1792). The privileges of the Reformed Church were taken away, and full religious freedom granted. The Napoleonic reorganization of the Revolution resulted in a modification: existing churches were recognized by the state, and supported by it, at the price of submitting to some regulation. After Napoleon's defeat and the end of the Revolution (1815), the Dutch Republic was replaced by a kingdom (which for a while, until 1830, included Belgium). William I retained the Napoleonic approach in matters of religion.

The Reformed Church was by now given to a good deal of tolerance in matters of religious dogma. Partly in reaction to this tolerance, a conservative wing emerged, as the "Awakening" *(Réveil*)* called for a revival of heartfelt religion; Bilderdijk,* Da Costa,* Groen Van Prinsterer, and others opposed the increasing "modernism" of the church. Some conservatives left the church (1834, the Separation or *Afscheiding*). The tension between evangelical and modernist helped the efforts of the Groningen School* to bridge the gap by stressing way of life rather than dogma; it controlled most of the theological faculties around midcentury. In the 1880s, the revival of a dogmatic Calvinism (notably by Abraham Kuyper*) produced another and larger exodus of conservatives (1886, the *Doleantie;* they soon joined with the earlier separatist group to form the *Gereformeerde Kerk*). The Reformed Church remained as it had been, with evangelical and modernist in the same communion, stressing heritage of three centuries, united in desiring a Christian way of life. It remains today by far the largest Protestant church in the Netherlands, with some three million members.

The Reformed Church spread also wherever the Dutch colonized or emigrated: thus, in the 1600s, to the East Indies, the West Indies, Ceylon, South Africa (see following entry), and New Amsterdam (New York). Mission efforts had some success in the Indies and in South Africa. In North America, the Reformed Church* grew out of the early Dutch settlement; emigration to the USA in the 1800s increased its numbers (and also produced the more conservative Christian Reformed Church*).

DIRK JELLEMA

DUTCH REFORMED CHURCH IN SOUTH AFRICA. With the first Dutch settlement on the Cape (1652, Jan Van Riebeck), the Reformed Church appeared in South Africa. The Cape Colony grew only slowly; it was primarily a way station on the Dutch East India Company route to the Indies. The company paid the pastors, who were under the jurisdiction of the classis of Amsterdam. During the French Revolution, the cape came under British control. The South African Dutch, the "Boers," moved northward, formed their own independent states (Transvaal, Orange Free State), and organized their own Reformed churches. By 1859 these churches had their own seminary at Stellenbosch. Events in the Netherlands had echoes in South Africa, so that the Separation of 1834 produced similar small conservative breakaways, which in turn started a more orthodox seminary, at Pochefstroom (1869). The Boer War at the turn of the century brought the independent *trekker* states under British control, and the churches united (1909) in the Reformed Church of South Africa, with the conservative splinter churches remaining separate. Characterized by relative orthodoxy in dogma and adherence to traditional morality, the Reformed Church has had some difficulty in defining its role in relation to the mission churches among the natives. It has viewed the problem as one similar to that of relations between whites and Indians in America, and has supported "apartheid," or the independent cultural growth of the two groups. In practice, apartheid has proved a cloak for white supremacy and has thus come under attack from within the church. Total membership is around 1.4 million, with some 150,000 in mission churches.

DIRK JELLEMA

DWANE, JAMES MATA (1848-1916). South African independent church leader. Born of heathen parents, he entered the Wesleyan Methodist ministry in 1875 and seceded in 1895 to the independent Ethiopian Church. In 1896 he visited the USA to arrange a union with the African Methodist Episcopal Church; a visiting AME bishop consecrated him vicar-bishop in 1898. This action was disputed in the USA, and Dwane began to resent American Negro control. He was also convinced that his orders were invalid, and requested Anglican ordination for his clergy. In 1900 the Anglican bishops agreed to constitute the Order of Ethiopia within the Church of the Province, and Dwane left the AME with some Xhosa followers. He was ordained deacon in 1900, priested in 1911, and held office as provincial of the order, with one break, till his death.

D.G.L. CRAGG

DWIGHT, TIMOTHY (1752-1817). Congregational theologian and educator. Born in Northampton, Massachusetts, he graduated from Yale and taught for some years before ordination as a Congregational pastor in Fairfield, Connecticut (1783-95). There he became famous as an educator, endeavored to establish an American literary tradition in poetry, and was a recognized leader in Connecticut Congregationalism. The College of New Jersey and Harvard both conferred honorary doctorates on him. From 1795 until his death he was president and professor of divinity at Yale, reforming administration and curriculum and tripling enrollment. A religious revival took place under his preaching, which by 1802 converted a third of the students. His chapel sermons, constituting a moderately Calvinistic or Edwardsean system of theology, were posthumously published as *Theology, Explained and Defended* (five vols., 1818-19). He was a leading conservative force in New England and exerted powerful influence in the Second Great Awakening.* ALBERT H. FREUNDT, JR.

DYER, MARY (d.1660). Quaker martyr. Wife of William Dyer of Somerset, with whom she came to Massachusetts in 1635, she sympathized with Anne Hutchinson* and others in the Antinomian controversy. Alienated from their orthodox neighbors, the Dyers left Boston in 1638 and helped to found Portsmouth, Rhode Island. During a stay in England (1650-57) Mrs. Dyer became a Quaker. She was arrested in Boston on her return journey, but was soon released. She was expelled from New Haven in 1658 for preaching Quakerism. She was jailed on each of three trips to Boston in 1659-60 to visit imprisoned Quakers and bear witness to her faith. Twice reprieved when entreaty was made for her, the last time she was hanged when she would not promise never to return.

ALBERT H. FREUNDT, JR.

DYKES, JOHN BACCHUS (1823-1876). English composer. While at Cambridge he studied composition under Walmisley. In 1849 he became precentor of Durham Cathedral, and from 1862 he was vicar of St. Oswald's in Durham. He is important for his hymntunes, a large number of which became extremely popular, especially through their inclusion in the influential *Hymns, Ancient and Modern.* His "Nicaea" for "Holy, holy, Lord God Almighty" and "Lux benigna" for "Lead, Kindly Light" are examples of the many tunes that enjoyed unsurpassed popularity. His style found numerous imitators, and while he has been much criticized, he possessed a remarkable gift for memorable and readily singable melody.

J.B. MAC MILLAN

DYOPHYSITES. In patristic literature *diphysites* was used by Monophysites* like Timothy Aelurus of Alexandria to describe adherents of the Chalcedonian *definitio.* The description lampooned the clause "in two natures" for contradicting the Monophysite understanding of the oneness of Christ after the union of the Word and man. Modern writers have used "dyophysite" either in the patristic sense (e.g., Dorner) or to denote the Antiochene tradition of two natures in Christ (e.g., Loofs).

DYOTHELETES. The Greek translates as "two-willers" and signifies those who hold the view that Christ had two wills, a divine and a human. The opposite doctrine was held by the Monothelites.*

EADIE, JOHN (1810-1872). Scottish Secession and United Presbyterian Church minister and NT scholar. Born in Alva, Clackmannanshire (where today one of the two parish churches bears his name), he was the son of an elderly Relief Kirk father and a youthful but pious Antiburgher mother. In 1843 he was appointed professor of biblical literature in the United Presbyterian Divinity Hall; in 1857 he was moderator of his church's general assembly. His *Analytical Concordance, Family Bible,* and *Biblical Cyclopaedia* proved very popular, and his widely acclaimed commentaries on several of the Pauline epistles helped to secure for him a place as one of the New Testament Committee engaged in preparing the Revised Version of the Bible in English (1870).

D.P. THOMSON

EADMER (b. c.1055). Anglo-Saxon biographer and theologian. Placed as a boy in the monastery of Christchurch, he grew up there and ultimately became precentor. Meeting Anselm* because the archbishop of Canterbury was also *ipso facto* abbot of Christchurch, he became Anselm's secretary, chaplain, and constant companion during the latter's dispute with William II and Henry I, even sharing his exile. Eadmer's two books about Anselm deal respectively with his private life *(Vita Anselmi)* and the disputes *(Vita Novorum in Anglia).* The books naturally have an eyewitness character about them, and understandably Eadmer presents the issues at dispute in a light favorable to Anselm. The dispute was soon forgotten after the compromise settlement of 1107. Eadmer's writings include also biographies of Wilfred and Dunstan.

L. FEEHAN

EAST AFRICA. The Christian gospel reached Ethiopia* at a very early date (Acts 8:27-39), but took many centuries to reach East Africa. The first contacts were probably through Nestorian and Jacobite merchants from India, followed by Roman Catholic priests who came with the Portuguese in the sixteenth century. All traces of such contacts have now disappeared, and the effective missionary penetration of the area began with the arrival of Protestant missionaries from Europe in the mid-nineteenth century.

(1) *Kenya, formerly British East Africa.* Effective preparation for the evangelization of Kenya began in 1844 with the arrival of J.L. Krapf,* a German Lutheran sent out by the Anglican Church Missionary Society. He began to work in Mombasa and then moved inland to Rabai, thirty miles from the sea, when he was joined by a colleague John (Johannes) Rebmann* in 1846. By journeys of exploration inland and down the coast, and by study of the local languages, these two prepared the way for missionary occupation. Krapf was forced by ill health to return to Europe in 1853, but he continued his linguistic work. Meanwhile the CMS reinforced its staff at Rabai, but several died from malaria. In 1861 the United Methodist Church from Britain opened a mission station at Ribe, a few miles north of Mombasa, and later extended its work up to the Tana River. In 1873 slavery was legally abolished within the Sultanate of Zanzibar, and the CMS established a colony for freed slaves at Freretown on the mainland just north of Mombasa. This became the main base of the CMS on the coast.

With the consolidation of missionary work at the coast, the next stage was that of penetration of the hinterland which began in 1891. In 1889 the directors of the Imperial British East Africa Company, who were in virtual administrative control of the area now known as Kenya and Uganda, invited the churches of Scotland to send out missionaries to Kenya. An exploratory party arrived in 1891 under James Stewart of Lovedale in South Africa and chose Kibwezi as the first station about 200 miles inland from Mombasa. The choice proved to be a bad one, and in 1898 the Rev. Thomas Watson, sole survivor of the original party, moved the station to Kikuyu near Nairobi. Meanwhile the British government had taken over administrative control of the area from the company in 1895. In that year the Africa Inland Mission was formed by Peter Scott, entered Kenya, and began to work among the Kamba people at Machakos in 1902. Other missions such as the Friends Africa Mission and the Seventh-Day Adventist Church moved up to stations in W Kenya, then called Kavirondo and under the administration of Uganda. Anglican missionaries entered W Kenya from Uganda, and began church, educational, and medical work in that area.

The modern history of Roman Catholic missionary work in Kenya begins with the arrival of the Holy Ghost Fathers, who were French in origin, at Mombasa in 1892. They moved up to Nairobi in 1899 and were at first mainly interested in the Goanese immigrants from India. In 1902 the Consolata Fathers from Turin settled in the Kenya Highlands with their main center at Nyeri, about 100 miles north of Nairobi. The Mill Hill Fathers of London came in 1904, and soon became the largest Roman Catholic missionary agency in the country.

With so many Protestant missions at work in the country, the problem of cooperation arose early and gave rise to a series of joint mission conferences. The main centers and dates at which these were held were Nyanza (1908), Nairobi (1909), and Kikuyu (1913, 1918, 1919, 1922, 1926). The most famous of these was in 1913 (see KIKUYU CONTROVERSY), but much good work was done at all of them. This included agreement on spheres of influence for the different missions, the production of common versions of the decalogue, the creed, and the Lord's Prayer, and the setting up of the Alliance of Missionary Societies as the permanent means of cooperation. In 1924 the Kenya Missionary Council was established, and finally the Christian Council of Kenya in 1943 to which almost all non-Roman churches and missions belong.

The missions established church work and to a varying degree educational, agricultural, and medical work, and pioneered much of the modern development of the country. Most of the missions have now handed over to the indigenous African churches, which have arisen out of their work and have now assumed control of the work formerly carried on by the missions.

(2) *Tanzania, formerly German East Africa and then Tanganyika.* Tanzania is the political union of Tanganyika and Zanzibar achieved in 1964. The island of Zanzibar was the center of Arab control of the coast and trade routes of East Africa, and was the base of all expeditions to the mainland. In 1864 Bishop Tozer moved the headquarters of the Anglo-Catholic Universities Mission to Central Africa from Malawi to Zanzibar, and in 1868 the Holy Ghost Fathers arrived from the island of Reunion. Soon the various missionary societies began to penetrate the mainland. The UMCA landed at Tanga and worked inland as well as working up the north bank of the Rovuma River in the south.

Before the German occupation of the country, all the missionary societies were British and included the Anglican CMS in the central area and the London Missionary Society along Lake Tanganyika. With declaration of a German protectorate in 1885, Lutheran and Moravian missionaries began to arrive. The Bethel Mission started work in Dar es Salaam in 1887, and in 1891 the Moravians took over part of the work of the LMS south of Lake Victoria. In 1893 the Leipzig Mission took over the work of the CMS among the Chagga people at the foot of Mount Kilimanjaro where the latter society had run into difficulties with the German administration. On the Roman Catholic side, the Holy Ghost Fathers were followed by the White Fathers in 1879, and the Benedictines in 1888.

By 1914 most of the country had been occupied by missionary societies, but under the British occupation most of the German missionaries were interned in World War I, and in 1920 they were all repatriated when the country came under British mandate. Replacements came from America and Scandinavia, and in 1925 the German missionaries were allowed to return. In 1940 they were again interned during World War II, but this time the work was less severely hampered as the local church was more organized, and the German mission stations were leased to the Augustana Synod of the American Lutheran Church. In the postwar period all the churches consolidated their work, and local autonomous churches were established. The largest Protestant body was the Lutheran Church which formed in 1958 the Federation of Lutheran Churches in Tanganyika and then in 1963 the Evangelical Lutheran Church of Tanzania. The Anglican Church set up several dioceses and in 1970 became a separate province of the Anglican Communion. A Christian Council of Tanzania was established to promote cooperation between the Protestant churches.

(3) *Uganda.* The pioneer of missionary work in Uganda may be said to have been the explorer H.M. Stanley. In April 1875 he had several interviews with Mutesa I, the *Kabaka* (king) of the Baganda, in which he found the king to be very interested in the Christian faith. The result was Stanley's famous letter to the *Daily Telegraph* and *New York Herald,* in which he appealed for "some pious practical missionary" to come to the kingdom of Buganda. The Anglican CMS took up the challenge, and in 1876 a party of eight missionaries led by Lieut. Shergold Smith set out from Britain. Only three reached Buganda, and of these, two were killed in a local dispute, leaving only the Rev. C.T. Wilson, who was alone for the next year or so. In November 1878 Alexander Mackay,* a Scottish Presbyterian, arrived.

Mutesa welcomed the Anglican missionaries and showed more interest in the Gospel than in Islam. Stanley's letter, however, had been read also by Charles Lavigerie, head of the White Fathers (founded in 1874 in North Africa), and in 1878 he sent a party of missionaries to Buganda. He did this in spite of a personal request from the CMS secretary not to do so in order to avoid competition and consequent confusion in the minds of the Baganda. Much unhappiness and even warfare would have been avoided if this request had been heeded. In 1884 Mutesa I died unbaptized, although he had asked for baptism from both Anglican and Roman missionaries. He was succeeded by his eighteen-year-old son Mwanga. Mwanga was a cruel and treacherous ruler, and the infant Christian Church was subjected to a persecution which produced many martyrs both Anglican and Roman. The first Anglican bishop of Eastern Equatorial Africa, James Hannington,* never reached Buganda, but was murdered at Busoga on Lake Victoria in 1885 by Mwanga's orders.

Mwanga's reign was marked by religious war and disorder until finally in 1894 Uganda was declared a British protectorate and its administration taken over by the British government from the East Africa Company. That government also built a railway from Mombasa to Lake Victoria which was a major factor in opening up Uganda to the world. Meanwhile, more Anglican and Roman Catholic missionaries had arrived. Outstanding among the Anglicans was Alfred Tucker, third bishop of Eastern Equatorial Africa, who arrived in 1890. He was the first Anglican bishop to reach Uganda, and he proved an active and able administrator. He had firm views on the need to establish an indigenous church. In 1898 he was installed as the first Anglican bishop of Uganda, but because of ill health was forced to resign in 1911.

On the Roman Catholic side the leader was Father Livinhac, who arrived in 1879 and later became superintendent general of the White Fathers. In 1894 the Mill Hill Fathers came from Britain to work in E Uganda. After the establishment of law and order under the British administration, there was a mass movement among the Baganda into both the Anglican and Roman Catholic churches. From early 1892 the work of evangelization began to extend out from Buganda into the other Uganda kingdoms of Bunyoro, Ankole, and Toro. Much of this work was done by the Christian Baganda. The Protestant influence in Uganda has remained predominantly evangelical and Anglican, and few other Protestant missions have entered the country; thus there has been no need for a Christian Council as in Kenya and Tanzania. In addition there have been few separatist movements in the Uganda Church. The most important was that led by Reuben Spartas in 1929 when he broke away from the Anglican Church to establish the African Orthodox Church, which in 1946 was recognized by the Greek Orthodox Patriarch of Alexandria. The church in Uganda has grown more rapidly than any other church on the African continent, and both the Anglican and Roman Catholic branches have established their own local hierarchies with African archbishops.

(4) *Significant movements in the East African Church.* In reaction to the spiritual decline in the church in Uganda came the movement of spiritual awakening and renewal known as the "East African Revival." The history of this movement has still to be written, but its origins have been traced to 1929 when an Anglican missionary doctor and a Buganda Christian came together in a newfound spiritual fellowship. Another source was the preaching of Blasio Kigozi, a young Anglican teacher in Ruanda. Other influences may be traced to the Keswick Convention* and the Oxford Group.* The Revival stressed the need for spiritual renewal and refused to admit to its fellowship those whom it did not regard as renewed. This led to the possibility of separation from the church in the early years of the movement, but this fortunately did not occur and the movement became a great source of spiritual strength in all the churches. The Revival soon spread to Kenya and Tanzania, and beyond to Central Africa and to the Sudan and Ethiopia. In Kenya it was a great source of inspiration to the church during the Mau Mau troubles in the 1950s. It continues to be a powerful stimulus in the church life of East Africa.

East Africa was the last region of Africa to become involved in the "Independent Church" movement which first appeared there about 1914. It has been most widespread and numerous in Kenya, and least so in Tanzania. The movement has been represented as an indigenous reaction to the European domination and paternalism of the Christian missions, and as expressing a desire to combine Christianity with features of African traditional religion. It has often been the vehicle of political nationalism. The movement probably has both religious and nonreligious causative factors. It is an important feature of the impact of Christianity on East Africa.

The "Church Union" movement began early in Kenya, but received a setback as a result of the Kikuyu Controversy.* Conversations were resumed and have been actively promoted in recent years between the main Protestant denominations of Kenya and Tanzania, but union seems unlikely in the near future.

BIBLIOGRAPHY: J.D. Richter, *Tanganyika and Its Future* (1934); H.R.A. Philp, *A New Day in Kenya* (1936); C.P. Groves, *The Planting of Christianity in Africa* (1948-58), vols. 2-4, *passim;* R. Oliver, *The Missionary Factor in East Africa* (1952); M. Warren, *Revival: An Enquiry* (1954); J.V. Taylor, *The Growth of the Church in Buganda* (1958); F.B. Welbourn, *East African Rebels* (1961); *The New Catholic Encyclopedia* (1967), *passim;* D.B. Barrett, *Schism and Renewal in Africa* (1968); R. Macpherson, *The Presbyterian Church in Kenya* (1970); G. Hewitt, *The Problems of Success: History of the Church Missionary Society, 1910-1942* (1971).

JOHN WILKINSON

EASTER. The celebration of Christ's resurrection. Although the Scriptures make no provision for the observance of Easter as the day of resurrection, all the evidence suggests that the celebration of the death and resurrection of Christ began at a very early date in the history of the church, probably as early as the apostolic age. It would seem also that the Christians of the first century consciously sought to create a Christian parallel to the Jewish Passover, since the close relationship between the significance of that event in the OT and the crucifixion in the NT made a transformation of that Jewish feast into Easter both logical and easy.

After A.D. 100, Easter, Pentecost, and Epiphany became the final parts of the church year. The time of the celebration in those early years is obscure, but during the second and third centuries serious controversies arose between some Catholic churches and the church in Rome concerning the proper time for the celebration of Christ's resurrection from the dead. This eastern group, known as the Quartodecimani,* insisted that Easter be celebrated on the fourteenth of Nisan. Basically the controversy was concerned with the question of whether the Jewish Paschal day or the Christian Sabbath should determine the time for the celebration, and whether the day of crucifixion or the day of resurrection should be the focal point of the celebration. It was a prolonged struggle, and toward the close of the second century it became so bitter that Bishop Victor of Rome denounced the Quartodecimans as heretics. The controversy was finally settled by the Council of Nicea in 325; it was decreed that Easter should be celebrated on the first Sunday after the vernal full moon and never on the fourteenth of Nisan. Because of different calculations, the time of the Eastern Orthodox Churches' celebration varies in relation to that of the Western Churches, and can be as much as five weeks later.

C. GREGG SINGER

EASTERN ORTHODOX CHURCH. A federation of several self-governing or autocephalous churches. "Orthodox" comes from the Greek words meaning "right believing." Included in the church are the four ancient patriarchates of Constantinople, Alexandria, Antioch, and Jerusalem. Because of their historical significance they rank highest in honor. The heads of these churches are given the title "patriarch." The other autocephalous churches are Russian, Romanian, Serbian, Greek, Bulgarian, Georgian, Cypriot, Czechoslovakian, Polish, Albanian, and Sinaian. The heads of the Russian, Serbian, Bulgarian, and Romanian churches are called "patriarch." The head of the Georgian Church is called "catholicos-patriarch," and the heads of the other churches are referred to as either "archbishop" or "metropolitan."

Besides the churches mentioned above, there are several other churches which are self-governing in many ways, but do not yet have full independence. They are called autonomous, but not autocephalous. These are the churches of Finland, China, Japan, and three administrations among Russians who live outside of Russia. Then there are ecclesiastical provinces which depend on one of the autocephalous churches or on one of the Russian jurisdictions in emigration. These provinces are located in W Europe, North and South America, and Australia.

The major area of distribution of Orthodox Christians is in E Europe, in Russia, and along the coasts of the E Mediterranean. Many of the autocephalous churches are located in countries where Orthodoxy is the predominant Christian faith. Most of the churches are in lands that are either Greek or Slavonic. It is estimated that about one-sixth of all Christians are of the Orthodox faith. Because so many live in Communist-dominated countries, exact statistics on membership are not available. It is usually estimated, however, that Orthodox Christians number about 150 million.

The Orthodox Church claims to be a family of self-governing churches held together, not by a centralized organization or a single prelate, but by a bond of unity in the faith and communion in the sacraments. The patriarch of Constantinople is known as the Ecumenical or Universal Patriarch. He has a position of special honor, but not the right to interfere in the internal affairs of other churches.

Orthodoxy claims to be the unbroken continuation of the Christian Church established by Christ and His apostles. Timothy Ware finds three main stages of fragmentation of Christendom. The first occurred in the fifth and sixth centuries when the Nestorian Church of Persia and the five Monophysite churches of Armenia, Syria, Egypt, Ethiopia, and India divided from the main body of Christianity. The second stage happened in 1054 when the Great Schism* divided the Roman Catholic Church of the West from the Orthodox Church of the East. Thus between the Semitic Eastern churches and the Western Latin Church there was the Greek-speaking world with its Orthodox faith. The third stage in separation came with the Protestant Reformation in the sixteenth century.

In 313 the persecuted Christian Church received its first toleration in Constantine's Edict of Milan. Constantine in 324 decided to move the capital of the Roman Empire to the site of the Greek city, Byzantium, which was now renamed Constantinople. Constantine also presided at the first general council of the Christian Church, held at Nicea in 325. Constantinople grew in wealth and power as Rome declined. It became the center of Greek culture and a center of the Christian Church.

The Orthodox Church often calls itself the Church of the Seven Councils. These councils, held between 325 and 787, clarified the organization and teachings of the Christian Church. They were Nicea* (325), Constantinople* (381), Ephesus* (431), Chalcedon* (451), Constantinople* (553), Constantinople* (680-81), Nicea* (787). These councils condemned as heresy Arianism* and Monophysitism,* and clearly defined the doctrine of the Holy Trinity and the person of Christ. The Nicene Creed* and the Chalcedonian Definition* carefully described Christian doctrine on these vital issues. The councils also decided on the order of priority among the five patriarchal sees. Rome was given the primacy of honor, Constantinople second, and Alexandria, Antioch, and Jerusalem in that order.

In the eighth and ninth centuries the iconoclastic dispute occurred within the Byzantine Empire. Some of the emperors supported iconoclasm and saw the use of icons as a form of idolatry to be suppressed. A bitter struggle arose over this issue, but in the end the Iconodules (venerators of icons) successfully defended the place of icons in church life. The struggle lasted 120 years. The Orthodox consider this far more than a minor dispute over Christian art. They dismiss the charge of idolatry which the Iconoclasts brought against them by saying that the icon is not an idol but a symbol, and that the veneration is not directed toward the object itself but toward the person depicted. The Iconodules then argued the necessity of icons to safeguard the correct doctrine of the Incarnation. Material images can be made of the One who took a material body. With the ending of the Iconoclastic Controversy* and the meeting of the seventh council, the age of the ecumenical councils came to an end. This was the great age of theology and definition of the Christian faith.

Byzantium has often been called "the icon of the heavenly Jerusalem." Religion permeated all aspects of life. Monasticism was a significant form of religious life in the East. Early monasticism took different forms, and we still find these in the Orthodox Church today. First of all, there are the hermits who lead a solitary life. Then there is community life where hermits live together in a monastery under a common rule. Finally there is the semi-eremitic life or middle way where a loosely knit group lives together in a settlement under the guidance of an elder. The elder or *starets* in Russia is characteristic of Orthodox monasticism. Antony was the most famous of the monastic *startsi.*

The Eastern Church followed a policy of converting Slavs to Christianity. In the ninth century the patriarch Photius* sent Cyril and Methodius* as missionaries to the Slavs. They not only gave the Slavs a system of Christian doctrine, but also created their written language. The Bulgarians and Serbs were converted to Christianity in the ninth century, and the Russians in the tenth. Greek civilization and culture followed the faith into Slavic lands.

In Byzantium there was no separation of church and state. Although the emperor participated widely in church affairs, Orthodox historians object to the term "Caesaropapism," as they do not believe the church was subordinated to the state, but that they worked in harmony with neither having absolute control over the other.

In 1054 occurred the Great Schism which marked the separation of the Orthodox Church in the East from the Roman Church in the West. The East and West had been growing further apart economically, politically, and culturally, but at the end when the split came doctrinal issues were given as the cause. One of these was the matter of papal claims. The pope was claiming absolute power in the East as well as the West. Greeks were willing to accord honor to the pope, but not universal supremacy. They felt that matters of faith were finally decided by a council with all the bishops of the church, not by papal authority.

The other doctrinal issue was the *Filioque*. Originally the Nicene-Constantinopolitan Creed read: "I believe ... in the Holy Spirit, the Lord, the Giver of Life, who proceeds from the Father, who with the Father and the Son together is worshipped and together glorified." The West inserted a phrase, so that the creed now read "who proceeds from the Father and the Son." The Greeks objected to this change because they believed the ecumenical councils forbade any changes in the creed, and if a change was to be made, only another ecumenical council could make it. The Greeks also believed the change was doctrinally wrong because it destroyed the balance between the three persons of the Trinity and could lead to an incorrect doctrine of the Spirit and the Church. Besides these major differences between Greeks and Latins there were minor differences such as priestly celibacy in the West (the Greeks allowed married clergy), different rules of fasting, and the use by Greeks of leavened bread in the Eucharist (the Latins used unleavened).

Even after 1054 there were friendly relations between East and West. In 1204, however, Constantinople was captured during the Fourth Crusade. The destruction and sacrilege of the Crusaders shocked the Greeks, and the division between East and West was thereupon final.

In 1453 the Turks attacked Constantinople by land and sea. The Byzantines put up a brave defense, but were hopelessly outnumbered. After seven weeks the city fell and the Church of the Holy Wisdom became a mosque. The Byzantine Empire had come to an end, but not the Orthodox faith. Moscow was becoming increasingly strong in this period, and the claim of a Third Rome was asserted. The marriage of Ivan III and the niece of the last Byzantine emperor helped enhance this claim. The Turks did not treat the Byzantines with undue cruelty and were more tolerant than many Christian groups were toward each other during the Reformation and seventeenth century. Christians under Islam, however, had to pay heavy taxes, were not allowed to serve in the army, and were forbidden to undertake missionary work.

The Orthodox Church did not undergo either a Reformation or Counter-Reformation, but these movements did have some influence upon the East. Through the Poles there were contacts with Roman Catholicism. The Uniat Church* was formed in Poland, recognizing the supremacy of the pope, but keeping many of the traditions of the Orthodox Church, including married clergy. Cyril Lucar,* patriarch of Constantinople, combated Catholicism and turned to Protestant embassies at Constantinople for help. He fell under the influence of Calvinism in matters of theology.

The Orthodox Church of the twentieth century is divided by the Iron Curtain. The four ancient patriarchates and Greece are on the one side, the Slavonic Churches and Romania on the other. It is estimated that eighty-five per cent of Orthodox people live in Communist countries.

In the midst of the many changes that have occurred in the world, the Orthodox claim a living continuity with the church of the past and a strict adherence to its traditions. The three greatest sources of its traditions are the Bible, the ecumenical councils, and the creed. The statements of faith issued by the seven ecumenical councils, used along with the Bible, serve as the basis for the traditions. The Nicene-Constantinopolitan Creed is considered the most important of the ecumenical statements of faith. Other sources of tradition which are also accepted, but not with the same authority as the above, are the statements of later councils, the writings of the Church Fathers, the liturgy, canon law, and icons. The Orthodox believe the traditions of the church are expressed not only in words, but in actions, gestures, and art used in worship. An icon is considered one of the ways whereby God is revealed to man.

Central to the Orthodox faith is the belief in the Holy Trinity. This is best defined as "one essence in three persons." God is described as transcendent, but not cut off from the world which He created. Man was created in the image and likeness of God which indicates rationality, freedom, and assimilation to God through virtue. Included in this is the belief in the free will of man. Although man fell through Adam's sin, the Orthodox do not believe that man is entirely deprived of God's grace, thus the picture of fallen man is not the total depravity of Augustine or Calvin. Jesus Christ is seen as true God and true man. An overwhelming sense of Christ's glory is seen especially in His transfiguration and resurrection. Christ's humanity is not overlooked, however, and is seen in the love for the Holy Land where the incarnate Christ lived and in the veneration of the cross on which he died. The Holy Spirit's work, of sanctification, is emphasized. The true aim of the Christian life is the acquisition of the Holy Spirit of God. This involves the process of deification. The church sees this as something intended for all believers, and that which involves a social process and leads to practical results.

Deification is achieved through the church and the sacraments.

Orthodoxy believes in the hierarchical structure of the church, apostolic succession, the episcopate, and the priesthood. It believes in prayers to the saints and prayers for the dead. In this it agrees with Roman Catholicism, but differs in that it rejects papal infallibility. The church is pictured as the image of the Holy Trinity, the body of Christ, and a continued Pentecost. "The Church is a single reality, earthly and heavenly, visible and invisible, human and divine." The Orthodox view, according to Timothy Ware, is that there is unity in the church, and although there can be schisms from the church, there will be no schisms within the church. The church is held together by the act of communion in the sacraments. The church is infallible, and this is expressed through ecumenical councils.

Religion is approached by the Orthodox through liturgy. Because of this, even the smallest points of ritual are extremely important. The whole basic pattern of worship is the same as in the Roman Catholic Church—the Holy Liturgy, the Divine Office, and the Occasional Offices. Besides these the Orthodox Church has a number of lesser blessings. In the services of the church the language of the people is used. All services are sung or chanted. In most Orthodox churches singing is unaccompanied, and instrumental music is not used. Normally the worshiper stands during the church services, although there are occasions to kneel and sit. In the Orthodox Church the sanctuary is separated from the remainder of the interior by a solid screen known as the *iconostasis.* There are three doors in the *iconostasis*—the center one is the Holy Door which gives a view of the altar, the left door leads into the chapel of preparation, and the right door leads into the *Diakonikon* which is used as a vestry. Orthodox churches are filled with icons which are venerated by the worshipers.

The Orthodox Church accepts seven sacraments: baptism, chrismation (similar to confirmation in the West), the Eucharist, repentance or confession, holy orders, marriage, and anointing of the sick. Of the seven, the Eucharist and baptism have a special position. Baptism is accomplished by threefold immersion. Although both married and unmarried may receive holy orders, bishops are chosen from the unmarried clergy. The Christian year consists of Easter as the central event, twelve great feasts, and a number of other festivals and fasts. In relation to the ecumenical movement of the twentieth century, most Orthodox believe that there must be full agreement in the Faith before there can be reunion among Christians.

BIBLIOGRAPHY: S.N. Bulgakov, *The Orthodox Church* (1935); R.M. French, *The Eastern Orthodox Church* (1951); N. Zernov, *Eastern Christianity* (1961); J. Meyendorf, *The Orthodox Church: Its Past and Its Role in the World Today* (1962); E. Benz, *The Eastern Orthodox Church, Its Thought and Life* (1963); A. Schmemann, *The Historical Road of Eastern Orthodoxy* (1963); T. Ware, *The Orthodox Church* (1963).

BARBARA L. FAULKNER

EASTON, BURTON SCOTT (1877-1950). American Episcopalian theologian and NT scholar. Educated at the universities of Pennsylvania and Göttingen, and at Philadelphia Divinity School, he taught NT at Nashotah House (1905-11), Western Theological Seminary in Chicago (1911-19), and General Theological Seminary of New York (1919-48). His works include commentaries on Luke (1926) and the Pastorals (1947), and an English translation with notes of *The Apostolic Tradition of Hippolytus* (1934). A collection of his essays was edited by F.C. Grant and published under the title *Early Christianity: The Purpose of Acts and Other Papers* (1954); this contains also a memoir and discussion of his work.

W. WARD GASQUE

EBEDJESUS (d.1318). Nestorian theologian. He became bishop of Sigar and Bet Arabaje in 1284/5, and metropolitan of Nisibis and Armenia in 1298. A prolific author, his Syriac writings included treatises on philosophy and science, a Bible, commentary, a polemic against heresy, hymns of praise, and anthems. As bibliographer he cataloged nearly 150 Syrian authors and their works, beginning with Simeon bar Sabbae (third century). He compiled the *Nomocanon,* the most complete collection of Nestorian canon laws, from three sources: (1) "Western Synods," i.e., before the Schism (e.g., Antioch, 341; Ancyra, 358); (2) "Eastern Synods," i.e., those held by Nestorian *Katholikoi* down to eighth century; and (3) laws made since the eighth century. Also extant is a theological work, *Margaritha (The Pearl)*, and a series of fifty poems, *Paradisus-Eden.*

J.G.G. NORMAN

EBERHARD, JOHAN AUGUST (1739-1809). German philosopher and theologian. Born in Helberstedt, he studied philosophy, theology, and classical philology under such teachers as A.G. Baumgarten and J.S. Sember. After 1766 he moved into the Berlin circle of F. Nicolai and M. Mandelssohn. His *Neue Apologie des Sokrates* (2 vols., 1772-78), a criticism of Kantian philosophy and a critique of such ideas as original sin, caused a controversy, and he was attacked by G.E. Lessing* and others. Eberhard was appointed a professor of philosophy at Halle where he was recognized as being in the tradition of Leibnitz* and Wolff. Later he turned his attention to aesthetics, publishing *Handbuch der Ästhetik* (4 vols., 1803-5).

PETER TOON

EBERLINN, JOHANN (1470-1533). Reformation preacher and writer. Born at Günsburg-on-the-Danube, he studied at Basle (1490) and Freiburg (1493), and became a Franciscan in Heilsbronn. Thereafter he lived in Tübingen, Ulm, and Freiburg, where he encountered Luther's writings and became a Reformation supporter. Returning to Ulm, he was expelled from his order. He spent a year in Wittenberg, then traveled as an evangelist to Basle, Rheinfelden, Rottenburg, and Ulm. He married, and after some years in Erfurt, and Wertheim (1525-30), he ended his days in Leutershausen, near Ansbach. In 1521 he published a series of fifteen pamphlets, *Bunds-*

gennossen, describing a utopian state called Wolfaria, and attacking, *inter alia*, the Lenten fast, priestly celibacy, and friars. He also sought to foster radical school changes. The extremities of some radicals and the moral laxity of some German Protestants led him to moderate his views in later writings. As a relaxation he translated into German Tacitus's *Germania*. J.G.G. NORMAN

EBIONITES. The name is derived from a Hebrew word meaning "poor" (cf. Luke 6:20). It seems clear that after the fall of Jerusalem many of the survivors from Qumran joined the Jewish Christian Church. Their influence caused a split. Some remained orthodox, being distinguished from Gentile Christians by their observance of the Sabbath* and circumcision. The Ebionites exalted the Law, though they considered it contained false pericopes, rejected the Pauline epistles, and regarded Jesus as the son of Joseph and Mary, but elected Son of God at his baptism when he was united with the eternal Christ, who is higher than the archangels, but not divine. This Christ had appeared in various figures from Adam on. His work was that of a teacher rather than savior. From Qumran they learned dualism, vegetarianism, and hatred of sacrifices. They had their own gospel, now called the "Gospel of the Ebionites"; it has survived mainly in quotations in Epiphanius. Apparently it was a developed form of the "Gospel according to the Hebrews," i.e., essentially Matthew. Our main information about their theology is derived from the "Journeys of Peter," which has been recognized in the pseudo-Clementine *Homilies* and *Recognitions*. Along with other Jewish Christians they suffered heavily during the Bar-Kochba revolt (132-35), because they would not accept him as Messiah. They then gradually dwindled away, their last remnants being swept away by the Muslim conquest of Syria.

BIBLIOGRAPHY: Quotations from "Gospel of the Ebionites" will be found in collections of Apocryphal NT literature, e.g., Hilgenfeld, M.R. James; the fullest modern treatment of Ebionites is in H.J. Schoeps, *Theologie und Geschichte des Judenchristentums*, (1949), *Urgemeinde, Judenchristentum, Gnosis*, (1956), *Jewish Christianity (1969; German original, Das Judenchristentum*, 1964); J. Daniélou, *The Theology of Jewish Christianity* (1964), *passim*. H.L. ELLISON

ECCHELLENSIS, see ABRAHAM ECCHELLENSIS

ECCLESIASTICAL TEXT OF THE NEW TESTAMENT, see BYZANTINE TEXT

ECK, JOHANN (1486-1543). Roman Catholic scholar and orator. Born in Eck on the Günz in Swabia, his proper name was Johann Mayr or Mai(e)r. He is best known for his opposition to Martin Luther's theological position. Eck attended the universities of Heidelberg, Tübingen, and Freiburg, and was professor in Ingolstadt from 1510 until his death. Scholastic, canonist, dogmatist, humanist, endowed with an excellent memory and oratorical ability, he upheld Roman Catholicism in his writings, disputations, and negotiations. His defense of a 5 percent interest rate on loans in 1514, in opposition to medieval prohibitions, gained him the favor of capitalists such as the Fuggers. In 1519 he opposed Andreas Carlstadt and then Martin Luther in the famous Leipzig Debate. He was largely responsible for procuring the bull *Exurge, Domini* against Luther (1520). In 1530 he presented 404 propositions against Luther and composed the *Confutatio* of the Augsburg Confession. He participated in the colloquies at Hagenau (1540), Worms (1541), and Ratisbon (1541). His *Enchiridion locorum communium adversus Lutherum et alias hostes ecclesiae* (which appeared in forty-six printings between 1525 and 1576) was directed against Luther, Melanchthon, and Zwingli. His earlier *De primatu Petri adversus Ludderdum* (1520) was a defense of the papacy. His translation of the Bible into German (1537) lacked originality.

See T. Wiedermann, *Dr. Johann Eck* (1865).

CARL S. MEYER

ECKHART VON HOCHHEIM (Meister Eckhart) (c.1260-1327). German mystic. Born in Hochheim near Gotha, he entered the Dominican Order and was prior in Erfurt before becoming vicar of Thuringia. In 1300 he went to Paris, where he graduated in theology. He was made provincial of Saxony, and in 1307 vicar general of Bohemia. The next period of his life is obscure, but he probably spent part of this time in Strasbourg and Cologne. In 1326 he was accused of heresy, tried by the archbishop of Cologne, and convicted. He appealed to Rome, where some of his teachings were judged heretical by John XXII in 1329, two years after his death. He was noted as a preacher; many of his sermons were delivered in convents of nuns.

Eckhart's German writings were classified by Franz Pfeiffer (1857) into 110 sermons, 18 tracts, and 60 *Sprüche* (brief notices). Since that time other pieces have been found and the authenticity of some of the earlier ones disputed. Questions of higher and lower criticism of the corpus of his works remain. For instance, his "tracts" are probably portions of sermons. His Latin writings are largely expositions of some of the OT books. His writings reveal him as a Scholastic* and as a mystic.

His principles have been interpreted variously. It was said he was pantheistic, but this charge is generally discounted. That he was influenced greatly by Thomism and Neoplatonism can be asserted with considerable confidence. He differentiated between God and the Godhead, and this caused discussion. In man he found a divine remnant which he called *Funck* or *Füncklein* or *Füncklein der Seele*. His incarnation theology makes Christ in the first instance the focal point of mankind and also its Redeemer. In his ethical teaching he reached a lofty plane. His followers included Henry Suso* and John Tauler.*

BIBLIOGRAPHY: O. Karrer, *Meister Eckhart* (1926); J.M. Clark, *Meister Eckhart, An Introduction to the Study of His Works* (1957); J.M. Clark and J.V. Skinner, *Meister Eckhart, Selected Treatises and Sermons* (1958). CARL S. MEYER

ECLECTICISM. An intellectual procedure involving the taking from various systems of thought of whatever appears to be true or striking or apt, without much regard to method or systematic consistency. Origen's use of classical and pagan sources is an example of eclecticism in Christian theology.

ECSTASY. The term refers to the supernatural state of being beyond reason and self-control, as when obsessed by emotion or overpowering feeling such as joy or rapture. External patterns of behavior such as incoherent speech, insensibility to pain, wild leaping contortions, jerking, and glossolalia (or "speaking in tongues") are examples of the ecstatic state. In the OT there are several instances of ecstasy recorded, usually in connection with the prophets. In the Christian Church, ecstasy has been held to be one of the normal stages in mystic life, although never regarded as normative for the Christian life. The ecstatic state is not sought for its own sake, but rather as an external indication of the union of the human will with the divine. A central characteristic of the ecstatic state is the alienation of the senses, with the individual perhaps becoming immovable and inoperative. WATSON E. MILLS

ECTHESIS (Gr. = "a statement of faith"). A theological formula drafted by Sergius, patriarch of Constantinople, and issued by Emperor Heraclius in 638. It forbade the mention or teaching of one or more principles of energy or modes of activity in the person of Christ, and it claimed that there was only one will in Him. Two councils held at Constantinople in 638 and 639 accepted the Ecthesis, but the Western Church did not. It was finally withdrawn by Emperor Constans II in 648. Eventually the church agreed that there were two wills in Christ, the divine and the human.

See also MONOTHELITES.

ECUMENICAL COUNCILS. Church councils representing the whole church, hence called ecumenical from the Greek word *oikoumene* (inhabited world). The Roman Catholic Church recognizes twenty-one councils as having been ecumenical. According to canon law, an ecumenical council must be convoked by the pope, and all diocesan bishops of the church must be invited. Its decrees are binding only upon papal ratification, and the rulings of the papacy cannot be appealed to a council. Although in modern Roman Catholic theology councils are held to be subordinate to the papacy, this was not always the case.

Ecumenical councils originated in the Christian Roman Empire, and the early councils were convoked by emperors who summoned the bishops, paid their expenses, and gave their decisions binding force. Whether or not a council was finally accepted as ecumenical was, in fact, based on later recognition by the church rather than on its actual characteristics. Some councils which believed themselves to be ecumenical were later not included in the list of ecumenical councils. Others, such as the Council of Constantinople (381) which was held without the pope's knowledge and which included only Eastern bishops, were later accepted as ecumenical although they did not conform to the modern definition.

There is little agreement among Christians on the number of ecumenical councils. Some churches accept only the first three (Coptic, Armenian, and Syrian). The Eastern Orthodox Church and many Protestants accept the first seven, while Luther regarded only the first four as ecumenical. Luther believed the decisions of councils were not infallible since they were subordinate to the Word of God; however, if those decisions were in harmony with God's Word, they deserved respect since they were the expression of the community of believers guided by the Holy Spirit. He therefore respected the decisions of the early councils, but rejected the medieval ones because he felt they had introduced superstitions and errors into Christian teaching. The councils considered by the Roman Church as ecumenical can be divided into four groups:

(1) The first eight, which were convoked by emperors and normally had representation from both Eastern and Western clergy: Nicea I (325); Constantinople I (381); Ephesus (431); Chalcedon (451); Constantinople II (553); Constantinople III (680-81); Nicea II (787); Constantinople IV (869-70).

(2) The seven medieval councils which were convoked and controlled by the papacy: Lateran I (1123); Lateran II (1139); Lateran III (1179); Lateran V (1215); Lyon I (1245); Lyon II (1274); Vienne (1311-12).

(3) Three late medieval councils which were held during the period when the conciliar movement was challenging the power of the papacy and which witnessed both the initial success of the movement and its final defeat: Constance (1414-18); Basle-Ferrara-Florence (1431-37); Lateran IV (1512-17).

(4) The last three councils which were all convoked by popes and which best fit the characteristics described by modern Roman Catholic theology: Trent (1545-63); Vatican I (1869-70); Vatican II (1962-65).

See entries under individual councils.

RUDOLPH HEINZE

ECUMENICAL MOVEMENT. The original Greek word *oikoumene* meant "the inhabited world" (as in the decree of Caesar Augustus, Luke 2:1), being derived from the verb *oikeo*, "I dwell." From this concept of the whole world it was but a short step to the idea of ecumenical councils* such as those of the fourth and fifth centuries. The distinctive feature of these councils was specifically that the bishops of the "whole world" were present, whereas provincial or other councils would involve only the bishops of a small area. In the sixteenth century, when the Roman Catholic Church had convened the "Ecumenical" Council of Trent,* Cranmer wrote to Calvin that the Protestant Churches should arrange their own council to meet (and if necessary oppose) the claims of that council. The appeal to a future council, bringing together representative leaders of all the churches, has been a minor part of the Anglican atmosphere ever since.

It is but a further short step to the twentieth-century use of the word "ecumenical." Any such gathering must bring together not only those who are scattered geographically, but also those who belong to different churches or denominations. The non-Roman churches would be unhappy to allow the title "ecumenical" to the general councils of the Roman Catholic Church just because such councils do not include representatives of all Christians. "Ecumenical" has come to mean "uniting." Indeed, by an etymological paradox, it is possible to have "ecumenical" trends in one small country or area, and in Britain today the British Council of Churches* sponsors "areas of ecumenical experiment."

The ecumenical movement itself is normally dated from the Edinburgh Missionary Conference* of 1910. This was the first really international conference of a multidenominational character, and, although its theme was "mission," it was inevitable that the degree to which the various bodies represented could cooperate, converge, or even merge was never far off the agenda. From this conference sprang further international organizations, which eventually merged into the World Council of Churches* in 1948.

The whole concept of "ecumenism" has been a source of theological division, some of which remains. Those bodies which have made exclusive claims to truth have been unable to meet with others in any such way as to suggest that they recognize the others as holding the truth. This has meant that the Eastern Orthodox, the Roman Catholics, and a large part of the evangelical churches stood aside from the movement initially. The Orthodox are now involved fully, the Roman Catholics are participating in various ways short of actual membership (even the latter had seemed likely in Britain in 1973), and evangelicals remain divided on the issue. It has been a very live issue in English evangelicalism, with Anglican evangelicals working out a theology of involvement without compromise, and of dialogue without sell-out, and expressing this in the Keele Congress Statement of 1967 and in later writings. This stance has seemed frankly incredible to non-Anglican evangelicals and has led to a thorough polarizing. A similar tension has been felt among evangelicals in the Presbyterian churches in Wales and Ireland, and in other historic, more mixed denominations in America. At the same time, some of the bodies which have been traditionally most separatist, e.g., the Salvation Army and some Pentecostalists, are becoming involved with the ecumenical movement at every level from the World Council downward.

BIBLIOGRAPHY: W.A. Visser 't Hooft, *The Meaning of Ecumenical* (1954); J.D. Murch, *Cooperation Without Compromise* (1956); M. Villain, *Introduction à l'oecumenisme* (1958); P.A. Crow, *The Ecumenical Movement in Bibliographical Outline* (1965); N. Goodall, *The Ecumenical Movement* (2nd ed., 1964); R. Rouse and S.C. Neill (eds.), *A History of the Ecumenical Movement 1517-1948* (1967); S.C. Neill, *The Church and Christian Union* (1968).

COLIN BUCHANAN

ECUMENICAL PATRIARCH. The patriarch of Constantinople which since the Council of Chalcedon in 451 has been regarded as the ranking see of the Eastern Orthodox Church.* Its sphere of jurisdiction in its own geographical area has shrunk considerably, and in post-World War I days the Turks all but abolished it, alleging that it was a center of pro-Greek intrigue—an accusation reiterated in more recent times in connection with Greek treatment of the Turkish minority in Cyprus. Formerly the partriarchate had civil authority over the Greek community in Constantinople (modern Istanbul), but its function is now restricted to ecclesiastical matters.

EDDY, MARY BAKER (1821-1910). Founder of Christian Science.* Born into a Congregational family on a farm near Concord, New Hampshire, she was from infancy subject to attacks of convulsive hysteria; even as a grown child she had to be rocked to sleep in a cradle made by her father. Her first husband married her "cradle and all." She was highly sensitive, intensely religious, seeing God everywhere. Reared on the Westminster Confession, she was accepted on confession of faith at the age of twelve by her father's church, despite her rejection of the Westminster's predestinarianism. At twenty-two she married a bricklayer who died a year later. Ten years afterward she married a roving dentist, who left her because of her "fits." In 1862 Mary visited "Doctor" Phineas Quimby, an ignorant, nonreligious blacksmith who practiced hypnotism and set her free from years of suffering. Impressed by his healings through the use of mind, Mary combined Quimbyism with her understanding of Christianity and gave birth to her Divine Science of healing, which she claimed came by direct revelation from God. Her book *Science and Health,* with a Key to the Scripture, she claimed was dictated by God, though she hired a clergyman to edit out the bad grammar. She was so filled with the Spirit, she explained, that her "grammar was eclipsed."

In 1877 she married Asa Gilbert Eddy, a man of poor health whom she cured. Eddy also left her a widow in due course, but she kept his name. She summoned a doctor before her husband died, later explaining that his death was from "arsenic mentally administered." After his death this remarkable widow of sixty-one went on to fame, wealth, and the founding of her own church. She died at eighty-nine, after years of loneliness and mortal terror that her enemies were projecting some mental arsenic into her mind.

See bibliography under CHRISTIAN SCIENCE.

JAMES DAANE

EDERSHEIM, ALFRED (1825-1889). Biblical scholar. Born in Vienna of Jewish parentage, he entered the University of Vienna in 1841 but was forced to leave after a few months by the illness of his father. Shortly after, he came under the influence of the Scottish Presbyterian John Duncan, who was chaplain to workmen on the Danube bridge at Pesth. Edersheim accompanied Duncan to Scotland and was enrolled as a student at New College, Edinburgh, and later at the University of Berlin. In 1846 he entered the Presbyterian ministry and became a missionary to the Jews at Jassy, Rumania. Three years later he was inducted as minister of the Free Church in Old Aberdeen. In 1861 the church of St. Andrew (Presbyterian) was built for him at Torquay. In 1875 he took orders in the Church of England, and was subsequently vicar of Loders, Dorset (1876-82). Of his writings the most widely read was his *Life and Times of Jesus the Messiah* (1883-90). WAYNE DETZLER

EDICT OF MILAN, see MILAN, EDICT OF

EDICT OF NANTES, see NANTES, EDICT OF

EDINBURGH CONFERENCE (1937). Convened as the second Conference on Faith and Order following the first conference in Lausanne in 1927, it comprised 504 delegates representing 123 churches. The discussions between the conferences became the basis for four main reports studied at Edinburgh: the Doctrine of Grace, the Ministry and Sacraments, the Church of Christ and the Word of God, the Church's Unity in Life and Worship. While there were wide areas of agreement among the delegates, no attempt was made to conceal disagreements, and further studies were instituted on them. So wide a measure of agreement was reached in the discussion on "the grace of our Lord" that the report stated: "There is in connexion with this subject no ground for maintaining division between the Churches." Agreement could not be reached on the means of grace and their right ordering and the Communion of the Saints. The authority of the Church to interpret Scripture revealed differences of conviction, and the report on the ministry proved the most contentious of all, with the episcopacy as the center of the disagreement. The root of the differences lay in conflicting views regarding the nature of the church. The proposal of the report on the church's unity that a World Council of Churches be formed was approved by the conference. An affirmation was issued, speaking of the unity of those who confess allegiance to Christ as Head of the Church.

See L. Hodgson (ed.), *The Second World Conference on Faith and Order* (1938).

JAMES TAYLOR

EDINBURGH MISSIONARY CONFERENCE (1910). This ten-day gathering for discussion was significant for its representative character, its leadership, the range of its discussions, and its outcome. Previous conferences on the missionary task of the church had been undenominational in character; this was interdenominational. All churches, with the exception of the Roman Catholic, were represented. The Anglo-Catholic representatives insisted that South America, which they regarded as being Roman Catholic, be omitted from the agenda, and that matters of doctrine and church polity should not be considered, as they were the business of the churches. There were 1,355 delegates, the places being allocated on the basis of missionary society incomes. Less than a score were from the "younger churches." The chairman was John R. Mott,* the secretary J.H. Oldham.*

Discussion ranged over the reports of the eight preparatory commissions: (1) conveying the Gospel to all the non-Christian world; (2) the Church in the mission field; (3) education in relation to the Christianization of national life; (4) the missionary message in religion to non-Christian religion; (5) the preparation of missionaries; (6) the home base of missions; (7) missions and governments; (8) cooperation and the promotion of unity.

During the discussions the need became apparent for a permanent representative body, able to coordinate missionary cooperation and to speak to governments. The only resolution of the conference, that a continuation committee with a fulltime executive staff be appointed, was carried unanimously. The continuation committee was the first-ever representative, interdenominational organization to be formed and, with its originating conference, is regarded as the beginning of the modern ecumenical movement.* J.R. Mott began his closing address with the words, "The end of the Conference is the beginning of the Conquest. The end of the Planning is the beginning of the Doing."

See W.H.T. Gairdner, *Edinburgh 1910* (1910).

JAMES TAYLOR

EDISON, THOMAS ALVA (1847-1931). Inventor. Born in Ohio, he was a newspaper boy at eleven and later became a telegraph operator. In 1868 he purchased Faraday's *Experimental Researches in Electricity* which inspired his life's work. Of his 1,100 inventions, the best known are the phonograph, the electric lamp (Edison's used a carbon filament), and the alkaline storage battery. He discovered the "Edison effect," later utilized in Fleming's thermionic valve, and was also responsible for the first industrial research laboratory. Edison was a strong believer in God: "The existence of an intelligent Creator, a personal God, can to my mind almost be proved from chemistry." He was motivated by a firm faith that, where man faced technological problems, God had in nature supplied materials necessary to solve them. Thus despite early losses, no difficulty daunted him. R.E.D. CLARK

EDKINS, JOSEPH (1823-1905). Missionary to China. When Shanghai was declared a "Treaty Port" in the settlement of the Opium War (1841), it soon became a center of missions. Edkins, sent by the London Missionary Society, arrived there in 1848. In 1860, at the invitation of Hung-ren, a convert of Hung Hsiu-ch'üan, the Taiping leader, he twice visited the Taiping rebels in Suchow and Nanking to instruct them in the Christian faith. That same year Edkins moved to Chefoo, and 1861 to Tientsin, whence he visited Peking in 1862 and baptized the first three Protestant converts in that city, which then became a new LMS base for evangelism. An eminent philologist, Edkins also wrote extensively about China's religions. LESLIE T. LYALL

EDMAN, V(ICTOR) RAYMOND (1900-1967). American college president. Born in Chicago

Heights, Illinois, he served as army medical corpsman in Europe (1918-19), then after two years at the University of Illinois and a year of biblical studies at Nyack Missionary Training Institute, he graduated from Boston University in 1923. From then until 1928 as a missionary in Ecuador, he helped found a Bible institute for training national workers. Forced by illness to return home, he became a pastor in Worcester, Massachusetts, and earned a Ph.D. degree from Clark University between 1929 and 1935. He taught for a year at Nyack and taught political science from 1936 to 1940 at Wheaton College, where he became president from 1940 until appointment as its first chancellor in 1965. Wide travel, love of the Bible, and a deep sense of God's presence made him a valued counselor and the writer of over twenty devotional books. EARLE E. CAIRNS

EDMUND (c.840-870). King and martyr. He succeeded to the throne of East Anglia at fifteen. For fourteen more years nothing is known for certain about his life, though he seems to have been an acceptable monarch. Then the invading Danes, who had gained ground in other parts of the country but had for some reason until then left his territories alone, came south. Edmund engaged them in battle, possibly at Hoxne on the Suffolk-Norfolk border. He was defeated. It is difficult to reconcile different versions of what happened on that occasion. According to some accounts, he was slain in the fighting; others say he was captured, refused to renounce his faith or to hold his kingdom as vassal of heathen overlords, was thereupon killed by the Danish archers, and then beheaded. Many legends sprang up around his memory, and his resting place at Bury St. Edmunds became one of the most famous shrines in England. J.D. DOUGLAS

EDMUND OF ABINGDON (Edmund Rich) (c.1175-1240). Archbishop of Canterbury from 1233. After graduating at Paris, he taught liberal arts there and at Oxford (where he was the first to lecture on Aristotle). In 1227 he was appointed to preach the crusade in England, and six years later became archbishop of Canterbury at the pope's behest, although the monks of Canterbury had suggested three others. Edmund unsuccessfully challenged papal exactions, and rebuked King Henry III for following the advice of foreign favorites, and for other misdemeanors threatened him with excommunication. The king submitted, but asked the pope to send him a legate. When Cardinal Otto arrived in 1237, Edmund's influence declined, for the legate took precedence over him on public occasions. Edmund protested in vain, and on further papal encroachments withdrew to Pontigny. Though he was somewhat ineffectual when challenged by great national issues, he was one of the most saintly and attractive figures of the English Church. St. Edmund Hall, Oxford, was dedicated to him in 1682

J.D. DOUGLAS

EDUCATION, CHRISTIAN. Christian education is rooted in Scripture. From its beginning the religion of the Bible has gone hand in hand with teaching. Parental responsibility for youth, the supreme worth of persons, the obligation to develop personal capabilities, the motivating power of love, the necessity of literacy, the unity of all truth in God—these and other principles basic to Christian education have biblical sources. Christianity is par excellence a teaching religion, and the story of its growth is largely an educational one.

In the *Old Testament,* education begins with God (Exod. 4:12), who continues to teach His people (Ps. 32:8; Isa. 48:17; Jer. 32:33). And early in His dealing with Israel, God makes parents responsible for teaching their children about Him, as the great educational principle the Jews call the Shema (Deut. 6:4-9) shows. Until the Exile, home and school were one. Parents were the teachers, except for special cases such as tutors for the royal family (2 Kings 10:1-5). Adults learned from priests and Levites, and also from the prophets. Along with their religious training, sons were taught a trade by their fathers; daughters learned household arts from their mothers. Both sexes were taught to read. Thus Hebrew education combined the two essentials of learning and doing.

The rise of the synagogue during the Exile and the growing importance of the scribes as teachers after Ezra had redirected attention to the Law (Neh. 8—10) led to expansion of Hebrew education. The synagogue was primarily a center for instructing the people in the Scriptures, and the Scribes became the professional students of the Law. The instruction of younger children, however, remained in parental hands until about 75 B.C., when elementary education, given either in the teacher's house or in the synagogue, became compulsory. The ordinary name for the elementary school was "house of the book," for the Scriptures were the only textbook. It is in its devotion to the Word of God and its relation to life that ancient Hebrew education has relevance for Christian education in modern times. Not only are the Torah (Pentateuch) and the Book of Proverbs the oldest educational handbooks, but the entire OT stands along with the NT as the chief sourcebook for an authentically Christian education.

First century. In keeping with its OT roots, Christianity is a teaching religion. Its Founder is by common consent acknowledged as the greatest of all teachers. In His ministry, teaching occupied a place second only to His work of redemption. His great commission (Matt. 28:18-20) obligates His followers to "teach all nations," and it was through the apostolic ministry, especially that of Paul and his colleagues, that the infant church grew.

The first Christian churches met in homes (Rom. 16:3-5; 1 Cor. 16:19; Col. 4:15; Philem. 2). Christian parents undoubtedly taught their children, and the church meetings fulfilled a teaching as well as worship function, as shown in the reference to "pastors and teachers" (Eph. 4:11) in Paul's list of the gifts of the Spirit, and the many allusions in his epistles (e.g., Rom. 12:7; Col. 3:16; 1 Tim. 3:2; 2 Tim. 2:2). Of great and continuing significance are the two elements in the NT *kerygma* (proclamation of the gospel of Jesus

Christ) and *didachē* (moral and social teaching based on the proclamation). These are organically related, the *kerygma* providing dynamic motivation for the *didachē* and being itself a form of teaching. It is evident that Christian education, though not carried on in separate schools, went on in the first-century church, which, without its unremitting faithfulness in proclaiming the Gospel and teaching the Word, would not have grown.

Patristic age. As Christianity spread, patterns of more formal education developed. Early in the second century, the catechumenate (instruction in the Scriptures, worship, Christian conduct, etc.) began as preparation of adults for baptism and membership in the Christian fellowship. Depsite the persecutions prior to the reign of Constantine, it continued—in some places until the fifth or sixth century. One of the earliest Christian schools was founded in Alexandria about A.D. 190 (see ALEXANDRIAN THEOLOGY); others developed at Caesarea, Antioch, Edessa, and Nisibis. At the catechetical schools not only was thorough instruction in the Scriptures given, but Greek philosophy (aside from Epicureanism), literature, grammar, rhetoric, science, and other subjects could be studied. Thus a relationship in education between Christianity and classical learning began. Augustine* indeed wrote: "Every good and true Christian should understand that wherever he may find truth, it is his Lord's." Others, notably Tertullian,* who said, "What indeed has Athens to do with Jerusalem . . . we want no curious disputation after possessing Jesus," had repudiated classical learning; and Jerome* wrote of Latin literature, "How can Horace go with the psalter, Virgil with the gospels, Cicero with the apostle?"

During the *Dark Ages,* the alliance between Christianity and classical learning was drastically obscured but not obliterated. The barbarian invasions were largely responsible for the decline of the public schools of grammar and rhetoric, which, though formerly prevalent throughout the Roman Empire, were by the sixth century practically gone. The church, in which distrust of pagan learning was strong, began to step into the gap and, for the next thousand years and more, dominated education. Education for the people was generally in eclipse during the Dark Ages. Learning flourished, however, in some monasteries, especially in Ireland, where from the sixth to the eighth centuries there was a genuine intellectual resurgence of learning. Also, after the disappearance of the public schools of grammar and rhetoric, the bishops established schools for training clergy. These began to teach grammar as well as theology, and as time passed reached out to some of the laity. Probably the first such schools were in England. Still, education for the people was almost nonexistent. About this time, certain kings began to foster education. Most notable of them was Charlemagne* (742-814), who made Alcuin,* a former head of the school at York in England, his education minister.

Scholasticism. In the period of the Schoolmen (ninth century to the end of the fourteenth century) there came a rediscovery of Greek philosophy, particularly that of Aristotle.* During the peak of Scholasticism, Thomas Aquinas* in his *Summa Theologiae* reconciled Aristotelian philosophy with the historic Christian faith. And with the rise of Scholasticism and the concurrent development toward the end of the twelfth century of the universities—Bologna, Paris, Oxford, and (somewhat later) Cambridge—Europe emerged from the Dark Ages.

The Reformation brought a new day for education. Two of its principles—the full authority of Scripture and the priesthood of the believer—served as a catalyst for developments that changed the face of education. The former principle made education mandatory, so all might read the Word of God (a motivation akin to that of ancient Hebrew education); the latter shifted the responsibility for education from the priestly hierarchy to the people. Luther, Calvin, Melanchthon, and other leading Reformers were scholars of the first rank and saw the strategic importance of Christian education. Luther had a medieval rather than a humanistic (in the Renaissance sense) education. But he worked closely with the Christian humanist Melanchthon, who not only "provided the foundation for the evangelical school system of Germany" but also "put into the curricula of his schools, especially the higher schools, those subjects which would contribute most to an understanding of the Scriptures" (C.L. Manschreck). His concept of correlating the curriculum with the Scriptures points to the effort to integrate biblical faith and learning which has become a major concern for evangelical educators in our times.

The effects of the Reformation upon education reached beyond Germany to Switzerland, Scandinavia, England, and other lands. From Geneva, Calvin's powerful influence led to a burgeoning of Christian schools in France, the Netherlands, and Scotland. In England the Christian humanism of men like Grocyn, Linacre, Erasmus, Colet, and Ascham had profoundly affected education there, and when England and its schools and universities became Protestant this influence continued. In seventeenth-century England, the Reformed faith affected education through Puritanism.

In Comenius,* "the founder of modern educational theory," the Reformation bore some of its most enduring educational fruits. This Moravian bishop, who was an evangelical Christian, "stands in education in the direct line of succession from Luther" (William Boyd). Thus for him the Bible was the supreme authority and norm for all knowledge. There are elements in the teaching of Comenius that relate to the present-day emphasis in Christian educational philosophy upon the unity of all truth in God. Although this article deals primarily with Protestant Christian education, it should be noted that through the Council of Trent* (with its reaction to the Reformation), the Reformation affected Roman Catholic education, as Loyola and his followers took into the Jesuit schools ideas borrowed from such places as Geneva and Strasbourg.

The modern period. With the founding of the colonies in the seventeenth century, the impetus given education by the Reformation came to

America. Here the influence of Calvin through the Puritans in New England and through the Dutch colonists was strong, though not exclusive (e.g., the Church of England also had its effect on education, though it too reflected Calvinistic influences). Up to the end of the eighteenth century, all schools and colleges in America (with the exception of the University of Pennsylvania, which though nonsectarian was not hostile to Christianity) had roots in evangelical Christianity, and the same was true for most of them until the beginning of public education in the nineteenth century.

Broadly speaking, it may be said that almost the whole of education in the Western world from the first until the nineteenth century was in one way or another Christian. But with the rise of rationalism and the French Enlightenment at the close of the eighteenth century, a shift toward secularism began. In America the roots of democracy were not only Calvinistic but Deistic, as in such leaders as Franklin and Jefferson. Here the principle of separation of church and state, laid down in the First Amendment to the Constitution, has gradually led to the exclusion of religious training and practices from public education through various decisions of the Supreme Court.* On the other hand, private schools and private colleges and universities flourish in the United States and are permitted to have complete programs of Christian education, provided that educational standards are maintained. Whether the wall of separation between church and state will in all respects remain intact is questionable, as in a time of rising costs pressures mount for federal aid to private (especially Roman Catholic) education. In England and Scotland, however, and in other countries where there is an established church, some Christian teaching continues in state schools. Yet the winds of secularism are blowing there also.

Among evangelicals in the United States, Christian elementary, secondary, and higher education has had a remarkable resurgence, particularly since about 1920. During this time, existing institutions have been strengthened and many new ones founded. Christian liberal arts colleges, theological seminaries, and both parent-controlled and parish-controlled day schools (elementary as well as secondary) have multiplied. Noteworthy has been the development of Bible institutes and Bible colleges, about two hundred being founded since the 1880s (of these, many have begun during the last five decades). These Christian institutions constitute nothing less than a new educational genre and represent one of America's distinctive contributions to Christian education.

There are, however, aspects of Christian education other than those having to do with school and college. Some of these, such as the relation of Christian education to the home and to the local church, have already been touched upon in the discussion of education in OT times and in the first-century church. Nothing that has happened in the long history of education has excused Christian parents from their primary responsibility, grounded in Scripture, for Christian training in their homes. Yet it must be said that in a day of pervasive secularism when homes are invaded by television and the mass media, their effectiveness as essential agents of Christian nurture is being eroded. Not even the development of strong evangelical schools and colleges or the renewal of the Sunday school or church school (as it is sometimes called) can make up for parental defection from their educational responsibility.

The Sunday school. In the historical development of Christian education the Sunday school is a comparative newcomer. It began with Robert Raikes,* and the movement rapidly gained ground in Britain and within a few years spread to America. Until about 1815, Sunday schools in the United States were attended mostly by children of the poor, and centered, along with Christian teaching, on dispelling illiteracy. After that date, the Sunday school became an educational arm of the evangelical Protestant churches. The length of the sessions was shortened, teaching was voluntary rather than paid, and pupils represented all social backgrounds and ages. The aim became more exclusively that of conversion and Bible teaching, and Sunday schools served as feeders for the churches. Growth was widespread not only in Britain and the United States but elsewhere, and in 1889 the First World Sunday School Convention was held in London. Subsequent developments, such as the establishment of Uniform Lessons, the shift to separate denominational and independent curricula, the relation of the movement to the International Council of Religious Education, which became part of the National Council of Christian Churches in the USA, need not be detailed here. It is, however, important to note that during the past five decades tensions respecting the Sunday schools have developed between evangelicals and more liberal Protestants. These led to the establishment in 1945 of the evangelically oriented National Sunday School Association and have led also to the development of certain independent and theologically conservative curricula.

Through the years, Sunday schools have grown until their pupils in the United States have totalled annually well up in the tens of millions, yet growth has not been without fluctuations. From 1926 to 1947 there was a decline, followed by a definite recovery which went on until about 1960, when a loss of enrollment in the American Sunday school set in. This has chiefly affected the Sunday schools of larger, mainline denominations and has reached drastic proportions. Ironically these denominations have invested millions of dollars in new Sunday school curricula, but the downward trend has not been reversed. Independent evangelical publishers have also been active in publishing new curricula (generally more biblical and Gospel-centered than the mainline denominational materials). For the Sunday schools of conservative evangelical churches—either those affiliated with larger denominations or with smaller bodies and independent churches—the enrollment situation is different. Here, while in some areas there has been decline, in others there has been marked growth. On the whole, the evangelical Sunday school has been holding its own and even showing a slight gain. Nevertheless, it is evident that in the period of revolutionary social changes during the latter part of the twentieth century the Sunday school is in serious trouble. This is true on the British as well as the American scene, so much so that the yearbooks of some of the major denominations no longer carry statistics on Sunday school work. The Church of Scotland, where the work has been strong in the past, has seen Sunday school numbers decline from close to half-a-million in 1901 to less than half that figure in 1971.

For many years, students of Christian education have recognized such problems of the Protestant Sunday school as the inadequacy of the weekly teaching period of an hour or less as compared with the time spent on secular education, the difficulties of teacher recruitment and preparation, the frequent ineffectiveness of teaching, the lack of adult Christian education in the churches, and the general failure of the Sunday school to communicate a coherent knowledge of the Bible and the elements of Christian truth. "The typical Christian of our time," says J.D. Smart, "however noble his character is, is unable to speak one intelligent word on behalf of his faith."

Other agencies of Christian education. The dramatic slippage of the Sunday school in numbers and influence has led some to question the continuing usefulness of this form of Christian education, which in the past has contributed so much to the church and society. In a time of radical change in attitudes of youth that often carries with it a reaction against organized religion, churches are venturing into new ways of ministering to young people. These include coffee houses, social action groups that work among the underprivileged, and contemporary music as a means of communicating with youth. Such para-church agencies as Daily Vacation Bible Schools, Christian camps, Youth for Christ, Young Life, Child Evangelism, Scripture Union (for college and university students), Inter-Varsity Christian Fellowship, Campus Crusade for Christ, and International Students continue to do effective work and are adapting to changing conditions without blunting their evangelical thrust. The triennial conventions of the Inter-Varsity Christian Fellowship at Urbana, Illinois (17,000 students attended the 1976 convention), Explo-'72, at which 85,000 youth met at Dallas, Texas, under the sponsorship of Campus Crusade for Christ, and the Jesus Movement (despite its vagaries) show that young people today respond to the evangelical presentation of Christ with a remarkable openness and readiness. This must be met by more effective Christian nurture through new forms of Christian education together with renewal and revision of older forms such as the Sunday school.

An aspect of education relating to the Bible concerns the American public school, from which formal worship (devotional Bible reading and prayer) have been excluded by judicial decision. Yet the same decision that did this approved the study of the Bible as literature in public schools. Accordingly Christian groups are promoting such Bible study on the ground that, while it can be neither doctrinal nor sectarian, the reading and study of portions of Scripture as great literature are not futile. The Word of God does not return to Him void, and when it is studied even under secular restrictions its spiritual power cannot be bound.

The centrality of the Bible. Whatever methods are employed in Christian education, it remains indissolubly united with the Bible. In the continuing concern for Christian education, the vital place of the pastor is too often overlooked. Pastors are called to be teachers as well as preachers (cf. Paul's reference in Eph. 4:11). The pastor who never expounds the Word of God from the pulpit has a truncated ministry. Biblically illiterate laity reflect the lack of expository preaching. No amount of topical preaching, no matter how inspirational, can build up the people of God in the knowledge of the Bible essential to spiritual growth. The incomparable educational resource of the church is the Word of God. When the Word has been truly and faithfully taught, Christian education has flourished; when it has been lost sight of and obscured, Christian education has waned; when it has been rediscovered, as in the Reformation, Christian education has been revived. This is the lesson of history. Therefore one of the most hopeful developments in the latter part of this twentieth century is the outburst of new and contemporary Bible translations and the opening of the Roman Catholic Church to Bible reading and study.

Nowhere has the educating power of the Bible been more potent than in missions.* Through the translation of the Scriptures in whole or in part into over 1,300 different languages and dialects, the door to literacy has been opened for millions who would otherwise have remained in intellectual and spiritual ignorance. No survey of Christian education can be complete without recognition of the missionary movement in which the teaching of the Bible and the establishment of schools has had a major part on an ecumenical scale reaching beyond the Western world to every continent and the farthest islands of the seas.

BIBLIOGRAPHY: F.P. Graves, *A Student's History of Education* (1921); S. Leeson, *Christian Education* (1947); E.P. Cubberly, *The History of Education* (1948); L. Cole, *A History of Education* (1950); J.D. Smart, *The Teaching Ministry of the Church* (1954); F.E. Gaebelein, *The Pattern of God's Truth* (1954); P. Le Fevre, *The Christian Teacher* (1958); C.L. Manschrek, *Melanchthon: The Quiet Reformer* (1958); M.J. Taylor, *Religious Education* (1960); K.B. Cully, *The Westminster Dictionary of Christian Education* (1963); J.E. Hakes, *An Introduction to Evangelical Christian Education* (1964); C.B. Eavey, *A History of Christian Education* (1964). FRANK E. GAEBELEIN

EDWARD VI (1537-1553). King of England. Son of Henry VIII and Jane Seymour, he reigned for six years and was only sixteen when he died. Physically frail, intelligent, and sincere, he was naive and inevitably became the tool of counselors, notably the earl of Northumberland, whose motivation was by no means religious. With a serious ferocity now difficult to imagine, the Europe of his time argued with words and tortures the mystery of the sacrament. England under Henry VIII had rejected papal authority, but retained medieval dogma. Edward's reign saw decisive moves in a Protestant direction. Most legislation against heresy was repealed, and England became a sanctuary for the persecuted. English Bibles were freely printed. The 1552 Prayer Book, going further than the already strongly Protestant version of 1549, set forth the sacrament as essentially an act of remembrance. The Forty-Two Articles* of 1553 codified in irenic terms these and other changes.

Despite his inevitable limitations, Edward was sincerely Christian. His reign saw few executions. Mary, his half sister, said her Mass and had her chaplain. The Roman Catholic bishop, Stephen Gardiner, though imprisoned and deprived of his see, was able to write six volumes of theological controversy.

See C.R. Markham, *King Edward VI* (1907); J.D. Mackie, *The Earlier Tudors, 1485-1558* (1952). P.W. PETTY

EDWARDS, JOHN (1637-1716). Calvinist divine. Cambridge graduate, he ministered at Trinity Church, Cambridge, from 1664 and stuck to his task even when plague struck the area. Later he held a fellowship at St. John's College, where his position became untenable because of his Calvinist views. After two further brief pastorates he retired because of declining health and the anti-Calvinist temper of the times. He lived for thirty years more and published more than forty works, notably the *Socinians' Creed,* intended as an answer to John Locke.* J.D. DOUGLAS

EDWARDS, JONATHAN (1703-1758). "The greatest philosopher-theologian yet to grace the American scene" (Perry Miller). After a precocious childhood (before he was thirteen he had a good knowledge of Latin, Greek, and Hebrew and was writing papers on philosophy) he entered Yale in 1716. It appears that it was during his time at college that he "began to have a new kind of apprehensions and ideas of Christ, and the work of redemption, and the glorious way of salvation by him." After a short pastorate in New York, he was appointed a tutor at Yale. In 1724 he became pastor of the church at Northampton, Massachusetts, a colleague of his grandfather Samuel Stoddard until the latter's death in 1729. Under the influence of Edwards's powerful preaching, the Great Awakening* occurred in 1734-35, and a geographically more extensive revival in 1740-41. Edwards became a firm friend of George Whitefield,* then itinerating in America.

After various differences with prominent families in his congregation, and a prolonged controversy over the question of the admission of the unconverted to the Lord's Supper, he was dismissed as pastor in 1750 (though, curiously, still preached until a suitable replacement could be found) and became, in 1751, pastor of the church in the frontier town of Stockbridge, and a missionary to the Indians. He was elected president of Princeton in 1757, but was reluctant to accept because of his desire to continue writing. Finally yielding to pressure, he was inaugurated in February 1758. One month later he died of the effects of a smallpox injection.

Edwards was, and was content to be, firmly in the tradition of New England Calvinism and the Westminster Divines. Efforts to demonstrate that he consciously shifted away from this position do not carry conviction. The influence of the "new way of ideas" of John Locke was mainly confined to his anthropology and is clearest in Edwards's classic *Freedom of the Will.* Because of his commitment to salvation by sovereign grace, Edwards was agitated by what he considered to be the religiously destructive developments in New England, particularly incipient Arminianism and Socinianism, and revivalistic excess. The first concern prompted the *Freedom of the Will* and, later, *Original Sin.* The second inspired a group of writings, notably the *Religious Affections.*

In Edwards, as in Augustine, there is a union of a highly intellectual and speculative spirit and an often ecstatic devotion to God-in-Christ. The same mind deployed the relentless logic of the *Freedom of the Will* and resolved "to cast and venture my whole soul on the Lord Jesus Christ, to trust and confide in him, and consecrate myself wholly to him." Edwards was a complete stranger to that separation of "heart" and "head" that has often plagued evangelical religion. Edwards's influence has been widespread. Some of his successors in America, such as Emmons, Hopkins, and Nathaniel Taylor, while appealing to Edwards, developed the "New England Theology"* in directions that he would surely have disapproved of. He had a wide circle of correspondents, compensating somewhat for his cultural isolation. His writings greatly influenced Thomas Chalmers, Andrew Fuller, and Robert Hall, among others.

BIBLIOGRAPHY: O.E. Winslow, *Jonathan Edwards* (1941); P. Miller, *Johathan Edwards* (1949); P. Miller (ed.), *The Works of Jonathan Edwards* (1957-). PAUL HELM

EDWARDS, LEWIS (1809-1887). Welsh Calvinistic Methodist minister. Born at Pen-llwyn, Cardiganshire, he had a patchy education in local schools, engaged in a little teaching himself, and went to Edinburgh University in 1833. In 1837 he and his brother-in-law, David Charles (1812-78) opened a school at Bala which was eventually adopted by the Calvinistic Methodists* as the institution for training its ministers. He spent the remaining fifty years of his life as principal there. Edwards was a powerful personality and became the undisputed leader of his denomination, particularly in intellectual matters. He was an eager protagonist of institutional efficiency and led the Methodists to adopt a modified form of Presbyterianism. His most substantial work, however, was done in literary and intellectual circles. He was the founder in 1845 of the quarterly magazine *Y Traethodydd,* in which he introduced Welsh readers to a much wider range of international ideas than had previously been available to them. He wrote books on the person of Christ and the doctrine of the Atonement, and a brief history of theology. As a theologian he sought to evade controversy while maintaining a somewhat moderate Calvinism. His son, Thomas Charles Edwards (1837-1900), first principal of the University College of Wales, Aberystwyth, is better known to English readers through his commentaries on 1 Corinthians and Hebrews. He also wrote his father's biography (1901). R. TUDUR JONES

EDWARD THE CONFESSOR (1003-1066). Son of Ethelred II ("the Unready"), he was taken into exile by his father and resided at the Norman court until recalled by his half-brother Hardicanute in 1041. Thereafter he was evidently regarded as heir to the English throne, largely through the influence of Earl Godwin, whose daughter he married in 1045. He preferred his Norman advisers to the Saxons, however, and one of his favorites, Robert of Jumieges, became archbishop of Canterbury in 1051. Civil war seemed imminent, but the king's chief adversary, Godwin, regarded as representing the cause of the nationalists, fled into exile. Soon a reconciliation was effected, the foreigners fled, and the influence of Godwin followed by that of his son was complete. The king suffered further blows to his pride, his failing health caused his absence from the consecration of his new abbey of Westminster late in 1065, and just after the new year the charming, mild-mannered ascetic died. He was canonized in 1161.
J.D. DOUGLAS

EGBERT (639-729). English monk from Lindisfarne. He lived and taught in Connaught. In 664, when afflicted by the plague, he vowed that if spared, he would never return to his native land. He was fired with missionary zeal and was largely responsible for organizing the evangelization of Germany. He spent the last thirteen years of his life on Iona,* persuading the monks to accept the Roman date for Easter and the crownlike tonsure. He died on the very day they first celebrated Easter in common with the rest of Europe.

EGBERT (d.766). Archbishop of York. Pupil of the Venerable Bede* and cousin of Ceolwulf, king of Northumbria, Egbert was ordained in Rome and appointed to the diocese of York in 732. Bede continued his interest in Egbert by giving him some good advice on that occasion. Three years later, Pope Gregory III made Egbert archbishop—only the second to be so styled at York. His position was consolidated and enhanced when his brother Eadberht succeeded to the Northumbrian throne in 738. Egbert wrote an epitome of ecclesiastical law and was correspondent and consultant of the English Boniface,* the "Apostle of Germany." As befits a churchman who added wisdom and rectitude to learning, Egbert was known for the care he took to ordain as priests only men worthy of the sacred charge. He is, however, remembered most as the founder of the cathedral school, where he himself taught theology. Numbered among his pupils was Alcuin,* who himself became master of the school in the year Egbert died. J.D. DOUGLAS

EGEDE, HANS (1686-1758). "Apostle of Greenland." Born in Norway, he became a pastor in Vagan in N Norway. Deeply concerned about the descendants of Norwegian settlers on Greenland, of whom nothing was known for about two centuries, and aware of the fact that the inhabitants were pagans who spoke an unknown language, he decided to go there as a missionary. He eventually got the king's permission to go to Greenland and arrived there in 1721. He learned the difficult language of the Eskimos, fought the witch doctors, and evangelized the people. From 1736 he lived in Copenhagen, supervising the mission work. From 1740 he was titular bishop of Greenland. Egede's publications are important contributions to the theory of missions. Influenced by H. Müller, P.J. Spener, and C. Gerber, he believed in the responsibility of the church to bring the Gospel to the pagans. His sons Paul and Hans carried on the missionary work of their father. Paul wrote a dictionary and a grammar of the language of the Greenland Eskimos and translated the NT into that language.
CARL-FRIEDRICH WISLOFF

EGERTON PAPYRUS. Among a collection of papyri purchased in the summer of 1934 and included in the Egerton collection, there were found to be some fragments from a codex of a life of Christ written in Greek. The four pieces surviving show clear affinities to a number of passages from the synoptic gospels and John's gospel. Two of them deal with questions of the relationship between Jesus and the Law, and one with the healing of a leper, while the detail of the last is uncertain. On palaeographical grounds they must be assigned to a date not later than A.D. 150, and it is possible that they may have been written some years earlier. The provenance of the fragments is unknown, though there are some similarities to the Oxyrhynchus* fragments. The fragments are of particular importance in helping to provide a limit to the possible late dating of the four gospels. They are clearly dependent upon the canonical works, and time must be allowed for the circulation and acceptance of those after writing, before another work such as this, of which we may not have the first copy, could be written drawing upon them.

See H.I. Bell and T.C. Skeat, *Fragments of an Unknown Gospel and Other Early Christian Papyri* (1935) and *The New Gospel Fragments* (1935). R. E. NIXON

EGYPT, see COPTIC CHURCH

EGYPTIANS, GOSPEL ACCORDING TO THE, see APOCRYPHAL NEW TESTAMENT

EGYPTIAN VERSIONS OF THE BIBLE. Only grudgingly have Egyptian sands and monastery ruins yielded information about the history of the Coptic (Egyptian) Bible. Of the translators we know nothing; of the time, very little. We do know that Greek was the theological language of the Egyptian Church until after the Council of Chalcedon (451), when the see of Alexandria severed its connections with Rome and Constantinople and pursued its own Monophysite paths. Although Coptic literature flowered most fully after Chalcedon, as Greek was rejected and the vernacular Coptic (the last stage of the Egyptian language) became the common tongue of cleric, monk, and layman, the beginnings of translation probably go back to the third century.

Whether the Sahidic translation from Upper (southern) Egypt or the Bohairic from Lower (northern) Egypt came first we have not yet discovered. Nor have scholars decided whether one translation was influenced by the other or whether they were made independently from the Greek.

The *Sahidic* version can only be pieced together from fragments and codices that range in date from the fourth to the fourteenth century. No single manuscript has been discovered covering the NT let alone the whole Bible. Though for liturgical purposes the gospels and Psalms may have been given priority by the translators, it is evident from the fragments discovered that the whole Greek Bible, as well as many apocryphal books, was available to the Sahidic-speaking Copts. The text seems to have followed the Neutral tradition prevalent in Egypt, with many affinities to the great Vaticanus manuscript (B), although with some Western readings.

The *Bohairic* version exists in more substantial manuscripts than the Sahidic and can be more readily reconstructed. The extant manuscripts, however, are generally younger; the earliest of the major ones (Curzon-Catena) dates from 889. The scarcity of earlier manuscripts (a fragment of Philippians from a fourth or fifth-century codex is a notable exception) is due both to the more moist climate of Lower Egypt and the frequent raids which sacked the library of the monastery of Wadi 'n-Natrun, the most productive center of Bohairic literature. The text of the NT follows closely the Alexandrian tradition, particularly Codex L, though it includes Aristion's ending of Mark.

Of the *Akhmîmic* version (Akhmîm in Upper Egypt) we have only fragments and scattered sections of both Testaments. Though the whole Bible may have been translated into Akhmîmic, little effort was made to preserve it once Sahidic replaced it as the language of the region.

A handful of verses from the gospels and a few fragments of Pauline epistles are our only legacy of the *Faiyumic* (Nile Valley, west bank) version, while a copy of the gospel of John is the chief biblical vestige of the *Sub-Akhmîmic* or *Asyutic* (Asyut, Upper Egypt) dialect.

The OT, extant mainly in Sahidic and Bohairic manuscripts, probably stems from the third century. In general it follows the so-called Hesychian recension of the Septuagint. The main value of the OT fragments is for Septuagint studies, while the various Coptic versions of the NT play a not insignificant role in helping to fill in the details of the history of textual transmission of the NT. The growing horde of fragments and codices is eloquent testimony to the vitality of Egyptian Christianity in the centuries before the Muslim invasion.

BIBLIOGRAPHY: For OT bibliography, see O. Eissfeldt, *The Old Testament, An Introduction* (tr. P.R. Ackroyd, 1965), pp. 713-14; for NT bibliography, see A. Vöobus, *Early Versions of the New Testament* (1954), pp. 211-41.

DAVID A. HUBBARD

EICHHORN, JOHANN GOTTFRIED (1752-1827). German biblical scholar. Born at Dörrenzimmern, he became professor at Jena (1775) and at Göttingen (1788). He dismissed as spurious many of the OT books and was a pioneer of "Higher Criticism."* His three-volumed introduction to the OT was influential for many years after its publication in 1780-83. Though purportedly based upon scientific study, the accuracy of some of his work has been questioned by scholars. He was one of the early advocates of the so-called primitive gospel hypothesis, which holds that all three synoptic gospels* are based on a lost Aramaic gospel.

EIGHTEENTH AMENDMENT. This amendment to the U.S. Constitution, ratified in 1919, prohibited "the manufacture, sale, or transportation of intoxicating liquors." Late nineteenth and early twentieth-century Protestantism, whether theologically evangelical or liberal, was devoted to the moral reform of society. The increase in drunkenness in growing urban industrial centers aroused reform-minded people. The temperance movement gradually grew in strength and shifted its emphasis from moderation to total abstinence enforced by law. There was widespread popular resentment against the Eighteenth Amendment, and it was difficult to enforce, especially in urban areas. Many people disregarded the law, and bootlegging became a lucrative business, bringing with it the establishment of violent, organized crime syndicates. The passage in 1933 of the Twenty-First Amendment repealed the Eighteenth Amendment, but left local or state authorities to deal with the right to make their own alcoholic beverages. HARRY SKILTON

EKTHESIS, see ECTHESIS

ELDER, see PRESBYTER

ELECTION. In Christian theology this normally refers to the divine choice of persons to salvation. There are differences of approach to this, however. Augustine, Luther, and Calvin all held a doctrine of unconditional election, in which the choice is sovereign and in no way dependent upon anything in man. Arminius and Wesley held that it was conditional and was dependent upon the individual's faith, foreseen by God. Karl Barth* held that election applies primarily to Christ and so to mankind as seen in him. In the Barthian type of theology, therefore, election is not inconsistent with universalism, although by no means are all Barthians universalists.

In the OT, election terminology is applied to Abraham and to Israel, the election of which latter is a mystery of the divine love. The Son of God is seen in the NT to be God's Elect One (Matt. 12:18; Luke 9:35; 1 Pet. 2:4,6). Believers are chosen in Christ (Eph. 1:4), for only in him do we know God. This is a pretemporal election (Eph. 1:4; 2 Thess. 2:13). It is also according to divine foreknowledge, a term which has been understood historically in several ways. Christians are exhorted to make their calling and election sure, in a context which lays great emphasis on the moral qualities of Christian living (2 Pet. 1:10).

G.W. GROGAN

ELIAS, JOHN (1774-1841). Regarded by many as the greatest of all Welsh preachers, he was born John Jones, but adopted his grandfather's name out of respect for the spiritual nurture he had received from him. His only formal education was at a private school in his native Caernarvonshire. He was ordained a Calvinistic Methodist* minister in 1811. Although his published sermons are somewhat laborious, his power as a preacher was extraordinary. He had an overwhelming conviction of the truth and efficacy of the Gospel as a means of salvation, and of the inerrancy of Scripture. There was an intense seriousness in his preaching, and never a suggestion of humor. His voice was haunting and powerful; his finger stretched in accusation or raised in warning brought mockers to their knees. In theology he was an unreserved Calvinist and opposed with great determination the tendency to flirt with "Modern Calvinism," still less with Arminianism.

After the death of Thomas Jones* he was the unchallenged leader of the Calvinistic Methodists. He was a man of indomitable will and unyielding principle, and this produced a kind of autocracy which many younger men—and some of his contemporaries—resented. In political matters too he was a bitter critic of the radicalism that was emerging among Welsh nonconformists. He was a constitutionalist of the old school, with a decided bias in favor of Toryism. On the other hand, he was energetic in his promotion of moral virtue and social betterment. This, however, did not prevent his being nicknamed "the Anglesey Pope" because of his hostility to democratic agitation and Catholic emancipation. Even so, his influence as a preacher far exceeded that of any of his contemporaries, and the legend of his miraculous eloquence in the cause of the Gospel has not died away even yet in Wales.

See E. Morgan, *Memoir* (1844), and *Letters, Essays ... of ... John Elias* (1847).

R. TUDUR JONES

ELIAS OF CORTONA (c.1180-1253). Minister-general of the Franciscan Order. A lay brother, he and Francis of Assisi* had great mutual affection. Elias was provincial-minister of Syria (1217-20) and Francis's spokesman at the 1221 general chapter. Gregory IX entrusted him with the construction of the basilica at Assisi where Francis's body was to rest. Elias is the most controversial early Franciscan figure, gaining in his own lifetime an evil reputation among the friars which his declining years, spent with the excommunicated emperor Frederick II, could only confirm. Francis's death apparently deprived him of vital personal inspiration. His resolution weakened, Elias did display un-Franciscan pride and ostentation, but it was unjust and unhistorical of his enemies, in the light of his later deviations, to reinterpret the events of Francis's lifetime so that Elias even then appeared in a bad light. The official element in the order detested him because he favored lay brothers equally with clerics (as had Francis), did not obtain many privileges, and generally did not consult them in his decision-making. They had him deposed as minister-general by Gregory IX in 1239.

L. FEEHAN

ELIGIUS (c.588-660). Bishop and metalworker. A goldsmith in the Merovingian royal mint and master of the royal mint at Marseilles under Chlothar II, king of the Franks (613-28), Eligius became one of Dagobert I's chief advisers. He was a noted builder of abbeys and churches, including the abbey of St.-Denis, Paris. He also commissioned, or perhaps made, a number of reliquaries, though none survive. A renowned philanthropist, he was ordained in 640, consecrated bishop of Noyon and Tournai in 641, and proselytized among the many pagans in his diocese. Of the sixteen writings credited to him, only two—on Superstition and the Last Judgment—seem genuine. The *Vita Eligii*, supposedly written by St. Ouen, a contemporary and fellow adviser of Dagobert, is apparently a later work.

L. FEEHAN

ELIJAH OF MOQAN (fl. c.800-820). Eastern missionary bishop. As a monk of Bait Abe in the Arbil province of N Mesopotamia, he was an ascetic of simple faith and something of a mystic, achieving concentration by repeating "Hallelujah! Glory be to God!" before reciting each verse of a psalm. Appointed as missionary bishop to Moqan, on the SW shores of the Caspian, he insisted on being consecrated on the day of Pentecost, and Timothy I* agreed to this. Elijah's mission was among tree-worshipers, and culminated in his going single-handed and felling a great oak. His methods included preaching, teaching and healing, and distributing copies of prayers and hymns. He went to Bait Abe about 820 to report on his work, but died there before he could return.

WILLIAM G. YOUNG

ELIOT, JOHN (1604-1690). "Apostle to the Indians." Born in England, he graduated at Cambridge in 1622, was ordained in the Church of England, and taught for a time at an Essex school run by Thomas Hooker,* whose views were decidedly Nonconformist. Eliot came to share them, and this led to his leaving for the New World in 1631. In 1632 he became teacher to the church at Roxbury, which connection he retained until his death. Having perfected himself in the Indian dialects, he began his work among them in 1646. Having soon discovered that they preferred to live by themselves, he had by 1674 gathered his "praying Indians" (numbered at 3,600) into fourteen self-governing communities. The work at Natick, where the first Indian church was founded in 1660, was to continue until the death of the last native pastor in 1716. Eliot arranged for them to have jobs, housing, land, clothes. Teetotaler and nonsmoker himself, he did not forbid alcohol and tobacco to his converts. His fellow ministers, among whom he was held in high respect, approved his work, and money for the founding of schools and other purposes came in, even from England.

The medicine men were hostile, and when the war associated with King Philip's name broke out, a severe blow was struck to the mission cause. It was some years before enough support could again be enlisted. Eliot never despaired, and in 1689 gave seventy-five acres of land in Roxbury for the teaching of Indians and Negroes (he may have been the first to champion the latter also).

His literary zeal was not abandoned despite the incessant demands of his ministry. With Richard Mather and Thomas Welch he prepared for printing in 1640 an English metrical version of the Psalms. This *Bay Psalm Book** was the first book printed in New England. He translated many works into the Massachusetts dialect of the Algonkian language; here his crowning achievement was the Bible (1661-63), the first time the Scriptures had been printed in North America. With the help of his sons he produced also an Indian grammar (1666). His translation of the Larger Catechism followed in 1669. He published also *The Christian Commonwealth* (1659), which curious work on government was suppressed by the authorities for its republican sympathies. Finally, his *Harmony of the Gospels* (1678) was a life of Jesus Christ. Renowned for learning, piety, evangelistic zeal, and practical wisdom, Eliot lived to a great age. "He that writes of Eliot," said Cotton Mather, "must write of charity, or say nothing."

See C. Mather, *Magnalia* (2 vols., 1820); and W. Walker, *Ten New England Leaders* (1901).

J.D. DOUGLAS

ELIOT, T(HOMAS) S(TEARNS) (1888-1965). Poet, dramatist, and critic. Born at St. Louis, Missouri, and educated at Harvard, Oxford, and the Sorbonne, he settled in England and worked first as a teacher and a bank clerk, afterward turning to journalism and ultimately to publishing. He edited *The Criterion* throughout its existence from 1922 to 1939, but his work as a critic began before this with *The Sacred Wood* (1920). In his later years his main creative work went into drama, but his chief fame rests upon his poetry. His early work such as "Prufrock" and "Gerontion" prepared the way for his statement of modern man's loneliness and isolation in *The Waste Land* (1922).

Some years later in *For Lancelot Andrewes* Eliot enumerated his now famous definition of himself as "classicist in literature, royalist in politics and Anglo-Catholic in religion." All his subsequent work, like that of Dryden whom he much admired, has sought to support ideas of order and authority. In particular he was concerned by the ways in which man fell short of a high ideal. Thus in *Ash Wednesday* he stresses appropriately the need for repentance, while in *The Four Quartets* he explores more subtly and more extensively man's hapless search without God, concluding impressively in the last, *Little Gidding,* that prayer is man's occupation even in a place of defeat. The need for sacrifice marks Eliot's drama right from the first, *Murder in the Cathedral* (1935), though perhaps it is most powerfully expressed in *The Cocktail Party* (1950)—but, as he emphasizes in the former, it must be sacrifice for the right reasons. Eliot is rigorous, and in *Notes towards the Definition of Culture* (1948) he argues boldly for an élitist ideal in modern mediocre and egalitarian society.

See *Complete Poems and Plays* (1969).

ARTHUR POLLARD

ELIPANDUS (c.718-802). Originator and exponent of Adoptianism* in Spain. In reaction against the teaching of Migetius that Jesus was one of the divine persons of the Trinity, he drew a very sharp distinction between the eternal Son, the second person of the Trinity, and the human nature of Christ. The Logos,* eternal Son of God, had adopted the humanity—not the person—with the result that Christ became the adoptive Son. Such views were condemned by the councils of Regensburg (792), Frankfort (794), and Aix-la-Chapelle (798), and by popes Adrian I and Leo III. Elipandus was metropolitan bishop of Toledo, and though his views were decisively condemned, his position was so strong—possibly because of the Muslim presence in Spain—that he was able to retain his see until his death.

HAROLD H. ROWDON

ELIZABETH I (1533-1603). Queen of England and Ireland from 1558. Daughter of Henry VIII and Anne Boleyn, she succeeded her Roman Catholic sister Mary at a time when the country was divided and nearly defenseless. Three decades later, the Armada's bid for a Spanish and Catholic England had been defeated, the flag of St. George flew in all the oceans, and the country was experiencing an extraordinary cultural flowering. It was a remarkable achievement for one who early had lost her mother (executed by her father), and had at twenty-one been imprisoned by her sister. She had learned to screen her thoughts from others and to circumvent difficulties with a resourcefulness which was just what England needed. Of rival religious dogmatisms she was impatient, but of the providence of God she had no doubt.

To the problems of state she brought a fine intellect. French, Latin, and Italian she spoke fluently, and she read Cyprian and Greek. Setting about securing a religious settlement (see ELIZABETHAN SETTLEMENT), she sensed that her subjects were now basically Protestant and prepared to move in that direction. Initially she dropped the title "Supreme Head of the Church," but left "Etc." at the end of her other titles to leave the needed loophole. In due course came the Act of Supremacy* and the Act of Uniformity.* Persecution there was, for this was an age when religion and politics could not be separated; but compared with the fury under Mary and the horrors on the Continent, it was comparatively little. It was a golden age of literature and exploration. The awakening power of the Commons was controlled by the queen with a dexterity the Stuarts could not produce. The Scottish Reformers' need for military help was met at precisely the right moment, and James's succession was prepared for. Possessed of extraordinary powers of rapid decision, she could temporize when it seemed prudent; only over the vexed and complex case of Mary Stuart did she long hesitate before consigning her to the executioner.

BIBLIOGRAPHY: A.L. Rowse, *The England of Elizabeth* (2 vols., 1951-55); J.E. Neale, *Queen Elizabeth I* (rep. 1952), and *Elizabeth I and Her Parliaments* (2 vols., 1953-57); J.B. Black, *The Reign of Elizabeth, 1558-1603* (2nd ed., 1959).

P.W. PETTY

ELIZABETHAN SETTLEMENT (1559). The changes effected in English ecclesiastical affairs after the Protestant Elizabeth I* had succeeded the Roman Catholic Mary in 1558. Briefly, these involved: the abolition once more of papal power in England, and the restoration of Henry VIII's* ecclesiastical legislation, with penalties for recusants; an Act of Supremacy* that declared the queen to be "supreme of all persons and causes, ecclesiastical as well as civil"; an Act of Uniformity,* accepting (in the main) Edward VI's Second Prayer Book, making orders about vestments and ornaments, and reenacting Edward VI's "Articles of Religion," reduced from forty-two to thirty-nine; and the dissolution of those monasteries that had been restored by Mary.

The Settlement only gained partial success; it was opposed by papists and Puritans. Pius V excommunicated Elizabeth in 1570, but the short shrift given to priests who infiltrated the country about 1579 showed where the majority opinion lay. Protestants for their part chafed at the moderate nature of the Settlement, and some left the national church to form separate congregations; hence the origin of Independency. The outcome of the queen's policies, nevertheless, was the emergence of an essentially Protestant religion identified in English minds with patriotism, and the rejection of Spain and other foreign elements. Elizabeth never asserted the Divine Right* theory that was to prove the downfall of the Stuart dynasty; she had a sure touch in stirring up and maintaining loyalty, and was generally discriminating in furthering the Protestant cause in her kingdom. While not notably possessed of strong personal beliefs, she showed great wisdom in choosing as archbishop of Canterbury Matthew Parker* who for sixteen years from 1559 firmly resolved ecclesiastical disorder.

BIBLIOGRAPHY: C.S. Meyer, *Elizabeth I and the Religious Settlement of 1559* (1960); see also bibliography for previous entry. J.D. DOUGLAS

ELIZABETH OF HUNGARY (1207-1231). Ascetic. The daughter of King Andrew of Hungary, she was betrothed to Louis IV of Thuringia at the age of four and married him ten years later when he was twenty-one. During her husband's short lifetime she gained a reputation for prayer and charitable works; on one occasion in 1225 she gave away food and money despite her officials' protests. In 1227 Louis died of plague while on crusade, leaving her widowed with three children. Louis's body was brought home in 1228 and buried at Reinhardsbrunn. That same year she became a member of the Third Order of St. Francis. From then until her premature death she led a life of rigorous self-mortification and service to Marburg's poor and sick. Her spiritual adviser, and renowned Franciscan inquisitor Conrad of Marburg,* exercised some restraint over her enthusiasm, but in other ways his methods were very extreme. Germany's first Gothic cathedral, the Elizabethskirche, was built for her body by her brother-in-law Conrad, and it was translated there in 1236 with Emperor Frederick II in the audience. In 1539 the Protestant Philip of Hesse moved her body, and its subsequent fate is unknown. L. FEEHAN

ELKESAITES. Elkesai lived about 100. It is not clear whether he was an Ebionite* who developed particular views, or whether he came from a common background. He stressed the Law, though cutting out the false pericopes, rejected sacrifices and Paul, and taught vegetarianism. His Christology seems to have been Ebionite. In addition he claimed a special revelation given him by an angel (the Son of God) and a feminine being (the Holy Spirit). There is a common background for many of his concepts and the *Shepherd of Hermas.* Though strongly ascetic, there was an insistence on marriage and a great stress on baptism. Because his teaching was somewhat more orthodox than that of the Ebionites and showed more Gnostic tendencies, it spread to Alexandria and Rome. We know details largely through quotations in Hippolytus and Epiphanius.

See fragments of the "Book of Elkesai" in Hilgenfeld; its main points are given by J. Daniélou, *The Theology of Jewish Christianity* (1964).

H.L. ELLISON

ELLERTON, JOHN (1826-1893). English hymnwriter. Educated at Cambridge where he was influenced by F.D. Maurice* but did not identify himself with any party in the Church of England, he held several parochial appointments from 1850. His last living was at White Roding, Essex. Among the hymnals he co-edited were *Church Hymns* (1871, with W.W. How*); *The Children's Hymn Book* (1881, again with How); and *Ancient and Modern* (1875 and 1889). His own eighty-six compositions, which he refused to copyright, were included mostly in his *Hymns Original and Translated* (from the Latin, 1888). Many are still sung, e.g., "Behold us, Lord, a little space"; "Saviour, again to Thy dear name we raise"; "The day Thou gavest, Lord, is ended"; and "Throned upon the awful tree." JOHN S. ANDREWS

ELLICOTT, CHARLES JOHN (1819-1905). Bishop of Gloucester. Educated at grammar schools and at St. John's College, Cambridge (elected fellow in 1845), he held an incumbency in Rutland from 1848, and was thereafter professor of divinity at King's College, London (1858-61), and Hulsean professor of divinity at Cambridge (1860-61). Dean of Exeter in 1861, he became bishop of Gloucester and Bristol in 1863. On the separation of the sees in 1897, he was bishop of Gloucester until his resignation in 1905. He was a notably energetic diocesan, chairman of the New Testament Revision Company for eleven years, wrote a series of highly acclaimed commentaries on most of the Pauline epistles, and a number of other religious and theological books.

HOWARD SAINSBURY

ELMO (c.1190-1246). Dominican preacher. The popular name for Peter Gonzalez. Of a noble Castilian family, he was educated by his uncle, the bishop of Astorga, and was appointed to a canonry while still under age. He renounced his easy life, however, and became a Dominican. As chaplain to Ferdinand III of Leon, he helped him in his crusade against the Moors, but urged generosity after their defeat at Cordova. Subsequently he left the court and began to preach among the poor and sailors of Galicia and the Spanish coast. A very popular preacher, he later came to be regarded as the patron saint of Spanish and Portuguese sailors. They called the electrical discharge sometimes seen on the decks of ships "Elmo's fire," and regarded it as a sign of his protection.

C. PETER WILLIAMS

ELVIRA, COUNCIL OF (c.305). Held in Spain after a period of persecution, it was attended by nineteen bishops and a number of presbyters. The council reveals a community facing the problems created by rapid conversion followed by something of a decline from the original fervor. The problems basically resolve themselves around the degree of compromise acceptable in a mixed pagan-Christian society. Thus the eighty-one canons deal with matters such as continued nominal attachment to the pagan priesthood, remarriage, adultery, and celibacy amongst the clergy. The penalties are severe, including lifelong excommunication, without the possibility of reconciliation, for offenses such as sacrificing and bigamy. There is also evidence of a developing asceticism in a regulation for virgins "who have dedicated themselves to God."

C. PETER WILLIAMS

EMBER DAYS. Fast days on the Wednesdays, Fridays, and Saturdays after St. Lucy's Day (13 December), the first Sunday in Lent, Whitsunday, and Holy Cross Day (14 September), which are peculiar to the Western Church. The origin of these days is obscure, but probably three of the four groups were a Christianization of the pagan religious fasts connected with seedtime, harvest, and autumn vintage. As Christian observances, they were well established at Rome by the pontificate of Leo I (440-61), but not until the eleventh century was the practice finally fixed. During the Middle Ages these fasts became associated with ordination. The Anglican Church retains this association, appoints two special collects for Ember Weeks, and normally ordains ministers on the following Sunday. It is suggested that the name "ember" is a corruption of the Latin *quatuor tempora*—"four seasons."

JOHN A. SIMPSON

EMBURY, PHILIP (1728-1773). Probably the first Methodist minister in America. Born in Ireland of a German Palatinate refugee family, he was a carpenter by trade. John Wesley's preaching led to his conversion in 1752, and he became a local preacher in Ireland six years later. He migrated to New York City in 1760. Barbara Heck* encouraged him to preach to migrants who were spiritually careless, and he did this in a meeting in his own home in 1766. He erected a chapel in New York in 1768. He later moved to Camden, New York, where he worked as a carpenter and preached on Sundays until his early death on the farm.

EMERSON, RALPH WALDO (1803-1882). American "Transcendentalist" minister. Descended from nine successive generations of ministers, he graduated from Harvard College and attended the divinity school there before accepting a pastorate in 1829 at Second Church of Boston, then Congregationalist and now Unitarian. For years he struggled over his faith and his vocation. Except for preaching, he disliked his work in the ministry. His sermons increasingly complained about "historical Christianity," denied the distinction between natural and supernatural, and stressed the immanence of God. In 1832, with his refusal to administer Communion as the immediate reason, he resigned his pastorate. His first book, *Nature,* which became a kind of Transcendentalist bible, appeared in 1836, but it was his address before the Harvard Divinity School in 1838 which clearly drew the lines of the Unitarian controversy. Emerson's Christ was strictly human; he advocated a "faith in man," not "in Christ" but "like Christ's." The battle over Christology and miracles was in the open.

Emerson's mature religious thought was essentially pantheistic and syncretistic. His essays were more suggestive than closely reasoned, and in pieces like "Self-Reliance" he advocated a religion of self. His rebellion against Lockean epistemology was an intuitionist stance strongly influenced by German Romanticism via Coleridge and Carlyle. His extreme optimism about man's moral nature and potential was tempered somewhat in his later writings. He was a successful lecturer and essayist during the 1840s and 1850s. Despite his reformist philosophy he kept aloof from the slavery controversy until the 1850s. His most famous writings were the *Essays* of 1841 and a second series in 1844. Other writings include *Poems* (1847); *Representative Men* (1850); *English Traits* (1856); and *The Conduct of Life* (1860).

BIBLIOGRAPHY: The twelve-volume "Centenary Edition" (1903-4) of his *Works* edited by his son Edward Waldo Emerson is considered standard, though it has been supplemented by several later volumes of uncollected lectures, sermons, and letters. The definitive biography is that of R.L. Rusk (1949). For the history of Emerson's role in the Unitarian controversy see W.R. Hutchison, *The Transcendentalist Ministers: Church Reform in the New England Renaissance* (1959).

JOAN OSTLING

EMINENCE. From a root *(eminere)* expressive of loftiness or prominence in a physical sense, an ecclesiastical title of honor was derived. It was limited by Urban VIII in 1630 to the grand master of the order of Malta, the three archiepiscopal electors of the Holy Roman Empire, and the cardinals in general—which latter usage alone has survived.

EMMANUEL, see IMMANUEL

EMMONS, NATHANAEL (1745-1840). Congregational theologian. Born in rural Connecticut and educated at Yale College, he entered the ministry of the Congregational Church. He served as pastor at Franklin, Massachusetts, from 1773 to 1827, during which period he published more than two hundred articles in periodicals, and personally instructed about one hundred young men in theology and preaching, many of whom attained positions of leadership in the church and in theological education. He generally followed the teaching of Jonathan Edwards* as developed by Samuel Hopkins,* but Emmons elaborated the Hopkinsian theology further into a system called "Consistent Calvinism." He affirmed that "holiness and sin consist in free, voluntary exercises"; consequently only Adam, and not mankind generally, was guilty of original sin. Yet God in His sovereignty determined to treat Adam's posterity as sinners. God executes His decision that all men must "choose evil before they choose good" by "directly operating on the hearts of children when they first become moral agents." In fact God Himself placed within Adam the first inclination to evil, which resulted in the Fall. Emmons believed it is consistent with God's righteousness to implant either sinful or holy exercises within man. Men act freely at the same time they are being determined by divine agency. Although God determines that men sin, he has the "right to require them to turn from sin to holiness." Therefore, preachers should "exhort sinners to love God, repent of sin, and believe in Christ immediately." He helped to found the Massachusetts Missionary Society, favored the abolition of slavery, was a zealous patriot during the American Revolution, and became a Federalist thereafter. HARRY SKILTON

EMS, CONGRESS OF (1786). A meeting at Ems in Hesse-Nassau of the representatives of the three elector-archbishops of Mainz, Cologne, and Trier, and the prince-archbishop of Salzburg. It sought to prevent a papal move to establish a new nunciature at Munich that would have enabled the Bavarian crown to communicate directly with Rome instead of through the archbishops as before. Another nunciature had been created at Cologne. The congress adopted the twenty-three "Points of Ems," embracing both Febronian and Josephinian principles, asserting the political and ecclesiastical claims of the elector- or prince-archbishops. They would: accept only a limited primacy of the pope, require episcopal assent to papal communications and decrees, discontinue appeals to Rome and payment of annates to the Roman Curia, hold authority over members of religious orders themselves. Joseph II approved the points, but many bishops and German princes opposed them as unwarranted usurpation of authority by the archbishops. The effect of the French Revolution in Germany terminated the controversy. C.T. MC INTIRE

EMSER, HIERONYMOUS (1477-1527). German editor and essayist, and Luther's bitterest opponent. After study at Tübingen (1493) and Basle (1497), he taught classics at Leipzig. As George of Saxony's secretary at Dresden, he was present at the 1519 Leipzig debate with Duke George. Emser broke with Luther and exposed him in a defense of papal primacy. Luther responded in the famous retort, "On the stinking Goat Emser," burning Emser's writings with the canon law and papal bull of excommunication in 1520. When Luther's "December Bible" of 1522 appeared, Emser prepared a German version identical in appearance, even using Cranach* woodcuts. Luther scorned this "correction" in *On Translating: An Open Letter.* In 1523 Emser wrote a *Defense of the Canon of the Mass Against Huldreich Zwingli.* Zwingli answered in 1524 with the *Antibolon.* Emser's polemics ended only with his death at Dresden.

MARVIN D. ANDERSON

ENCRATITES. The name is derived from the Greek *enkrateia,* "self-control," and was applied to various groups by Irenaeus, Clement of Alexandria, and Hippolytus. It was never used precisely, but included all those movements given to extremer ascetic practices. Their origins go back to Jewish Christianity, especially inasmuch as it was influenced by Qumran, to Gnosticism, and to those Docetic sects influenced by Greek philosophy. They tended to reject the use of wine (which, as among the Ebionites, would influence the celebration of the Lord's Supper) and of meat. Often marriage was repudiated. These groups were not necessarily heretical, but they were always in danger of going too far. Two of their leading figures were Tatian* (according to Jerome) and Julius Cassian, who expressed the ideals of the movement in his book *Peri Eunouchias.*

H.L. ELLISON

ENCYCLICAL. That which "circles around." The term is applied especially to epistolary literature: Paul's letter to the Ephesians represents the earliest example within Christian documents: his statement in Colossians 4:16 provides explication of both intention and operational mode. In the subsequent generation several of the documents now grouped under Apostolic Fathers were similarly intended, and thereafter any of the letters of a bishop to his church. Modern terminological usage distinguishes churchwide teachings of the bishop of Rome.

ENCYCLOPEDISTS. A term applied to the *philosophes* of the Enlightenment* who contributed articles to the *Encyclopedia.* Published between 1751 and 1772, and edited by Denis Diderot* and Jean d'Alembert, this was the most prodigious intellectual undertaking of the eighteenth century. The encyclopedists intended that it be a source of information and means of education in the crafts, sciences, and every area of learning, and serve as a clearinghouse for new ideas on religion, politics, and society. Prominent themes were the autonomy of man, the secularization of knowledge and thought, the natural goodness and perfectibility of human nature, and a belief in reason, science, and progress. Because it exposed abuses in the French government and Roman Catholic Church, it was officially suppressed in 1759, but continued to circulate freely.

RICHARD V. PIERARD

ENDECOTT (Endicott), JOHN (c.1589-1665). Puritan colonial governor of Massachusetts. Born in England, he emigrated to Salem, Massachusetts, in 1628, moved to Boston, and was chosen governor in 1629. After John Winthrop,* the already-appointed governor, arrived in the colony in 1630, Endecott's life was a stormy one. His hot temper and impetuous acts led to his removal from public office, fines for assault and battery, and a generally controversial career. His ineffective expeditions against the Indians may have led to the noted Pequot War in 1637. Following Winthrop's death, he served as governor for thirteen years; he was also active earlier in the founding of Harvard College. Always a zealous Puritan, he opposed the Quakers even to the extent of having some of them publicly executed. Some historians regard him as the real "Father of Massachusetts."

DONALD M. LAKE

ENERGUMENS (Gr. *energoumenoi*). In general ecclesiastical usage, those whose bodies were possessed by an evil spirit. The early church made special provision for such and placed them in the oversight of exorcists. Baptism was denied them until a complete cure had been effected. In cases where church members were afflicted, they were debarred from the sanctuary and restricted to the outer porch. A partial cure permitted them to rejoin public worship, but not to communicate until wholly restored. Rules regarding baptism and Communion, however, could be waived where death appeared to be imminent.

See also EXORCISM.

ENGLAND, CHURCH OF. The origins of English Christianity are unknown, but the presence of British bishops at the Council of Arles (314) indicates the existence of an organized church. Following the Roman withdrawal and the Teutonic invasions, Christianity retreated to the Celtic lands, but in the late sixth and early seventh centuries, a Roman mission under Augustine* and a Celtic mission under Aidan* began the reconversion of England. Celtic and Roman Christians disagreed over several minor customs, but the Synod of Whitby (663/4) secured the observance of Roman forms. Theodore of Tarsus,* archbishop of Canterbury (668-90), united and organized the church on a diocesan basis, but though continental monastic reform influences were felt during the tenth century, under Dunstan* the English church was largely isolated from continental ecclesiastical affairs until the Norman invasion of 1066. William I and his archbishop, Lanfranc,* brought the church into line with the main features of the Hildebrandine reform, though William himself managed to avoid a complete subservience to the papacy. The Investiture Controversy* had repercussions in England in the conflict between Anselm* and first William II and then Henry I; the church-state struggle for supremacy produced its most dramatic example in England with the quarrel between Henry II and Thomas Becket* over the trial of criminous clerks, resulting in Becket's martyrdom and consequent victory. The triumph of papalism was clearly seen in King John's recognition of the kingdom as a papal fief in 1213, after the lifting of the papal interdict, and during the thirteenth century the extension of canon law gave the papacy wide influence in England. However, distance from Rome, the conflict between England and France (which in the fourteenth century controlled the papacy), and also papal decline made English submission more nominal than real in the later Middle Ages.

By the sixteenth century the situation was such that it was an easy matter for Henry VIII* to use his divorce from Catherine of Aragon as grounds for detaching England from the papal obedience. The parliament of 1532-36 created Henry "Supreme Head on earth of the Church of England" and severed the financial, judicial, and administrative bonds between England and Rome—a move supported by the majority in the church. For financial more than religious reasons, the monasteries were dissolved (1536-39), but otherwise the church retained a Catholic position. Under Edward VI* (1547-53), the church underwent a liturgical and doctrinal reformation, linked particularly with the two Prayer Books of 1549 and 1552, the latter being distinctly Protestant in character. The accession of Mary Tudor* (1553-58) inaugurated a period of Roman reaction, during which many of the Edwardine reformers were martyred, including Thomas Cranmer,* Nicholas Ridley,* and Hugh Latimer,* as well as many ordinary people. Elizabeth I* (1558-1603) restored a Protestant settlement, but her aim was a comprehensive, national, episcopal church, with the monarch as Supreme Governor. The Elizabethan Prayer Book was based on the 1552 Book, but with significant changes to assist comprehension, and the same moderate Protestantism was reflected in the church's doctrinal basis, the Thirty-Nine Articles of Religion.* A Puritan party, wanting to influence Anglicanism in a Calvinistic direction, emerged under Elizabeth, but the queen would allow no interference in the Elizabethan Settlement,* and in the writings of John Jewel* and Richard Hooker,* Anglicanism gained its classic *Via Media* statements. Puritan hopes were dashed when James I* maintained Elizabeth's policy, and further conflict resulted from the emergence of a High Church "Laudian" party, with Arminian emphases, and a stress on the Church of the Fathers, worship, and ceremonial. The dispute between Laudians and Puritans was the religious counterpart of the conflict between Charles I* and Parliament, and with the Civil War, Parliament abolished episcopacy and the Prayer Book, and executed Archbishop William Laud* (1645). Though proscribed during the Commonwealth and Protectorate, Anglicanism survived and with the Restoration of Charles II* in 1660 the Church of England was restored to its position as the national church. The Prayer Book, with Laudian alterations, came into force again in 1662, and the Clarendon Code* imposed legal disabilities on all who would not conform to the national church.

The post-Restoration church had its High and Low wings, the High Churchmen maintaining Laudian emphases, and Low Churchmen (or Latitudinarians), inspired by the Cambridge Platonists,* stressing the place of reason in religion. Like most Protestant denominations, the Anglican Church was affected by Deism* in the eighteenth century, but the key movement of this period was the Evangelical Revival (see REVIVALISM), with its emphasis on justification by faith, personal conversion, and the Bible. Though the Wesleys and Whitefield increasingly worked outside the Anglican system, a sizable Evangelical party emerged in the church, valuing the Prayer Book and the parish system, gaining its leadership from laymen such as William Wilberforce and the members of the Clapham Sect.* Eighteenth-century Anglicanism also produced important philosophers in George Berkeley* and William Paley.*

The early nineteenth century with its movements for Catholic Emancipation, the removal of Nonconformist disabilities, utilitarian reform of the church, and parliamentary reorganization of the Irish Church, saw the position of the Establishment threatened. A financial, administrative, and diocesan reform of the church, bringing it into line with a modern world, was undertaken by Parliament and the newly created Ecclesiastical Commissioners, under Bishop Blomfield, but spirituality was revived by the Oxford Movement,* led by J.H. Newman,* John Keble,* and E.B. Pusey,* with an emphasis on the church, apostolic succession, sacramental grace, and ascetic holiness. By many the movement was seen as a Romanizing tendency, a suspicion which seemed to be confirmed by the secession of Newman and others to Rome in 1845; but the majority in the movement were loyal in their Anglicanism and weathered persecution within the church, though until the mid-twentieth century party conflict has been rife.

In 1854 the convocations of the clergy were revived, thereby giving the church a forum for debate. But though laymen began to be more involved in church affairs in the later part of the nineteenth century, it was not until the passing of the Enabling Act (1919) and the creation of the church assembly* and parish church councils that laymen gained an official place in church government; and not until the introduction of synodical government in 1970 have clergy and laity achieved an equal footing in the councils of the church.

Since the mid-nineteenth century, due to the activity of the Christian Socialists,* the church has become increasingly aware of its social responsibilities, and a number of Anglicans, particularly Archbishop William Temple,* have played a prominent role in this sphere.

Having both Catholic and Protestant features, the Anglican Church has had an important function within the ecumenical movement, but to date her attempts at union negotiations with other churches have not achieved success. Like other denominations, the Anglican Church has been affected by the Liturgical Movement,* and despite the abortive attempt to revise the Prayer Book in 1928, liturgical reform has gone ahead since 1965, by the use of alternative services for experimental periods. Theologically, neo-biblicalism dominated the church from the thirties until the early sixties, but radicalism has had a growing influence since, as has conservative evangelicalism, which has grown numerically since World War II.

BIBLIOGRAPHY: H. Gee and W. Hardy, *Documents Illustrative of English Church History* (1896); W.R.W. Stephens and W. Hunt, *A History of the English Church* (9 vols., 1899-1910); J.R.H. Moorman, *A History of the Church in England*, (3rd ed., 1973). JOHN A. SIMPSON

ENHYPOSTASIA. While the term itself does not appear in patristic citation, and even the English lexicon knows only the rare verbal form "to enhypostatize" as late nineteenth-century encyclopedic jargon, the doctrinal concept is descriptive of the effort by the sixth-century theologian Leontius of Byzantium* to recast the formula of Chalcedon in an Aristotelian framework under the political pressure from Justinian's demand to settle the Monophysite question. Leontius's term ("intrahypostatic"), from which derives the descriptive heading, is meant to define the unity of substance *(hypostasis)* one nature may achieve with another, so that its own peculiarity *(eidos)* is retained. To Leontius and his time the analogy of soul and body, or fire and torch, provided examples, which then permitted the divine-human life of Christ to be described in terms made famous by Cyril of Alexandria. CLYDE CURRY SMITH

ENLIGHTENMENT (Aufklärung), THE. A movement seen in particularly clear-cut form in eighteenth-century Germany. Karl Barth characterized it as "a system founded upon the presupposition of faith in the omnipotence of human ability." Immanuel Kant defined it in his *Religion Within the Bounds of Reason Only* (1793): "The Enlightenment represents man's emergence from a self-inflicted state of minority. A minor is one who is incapable of making use of his understanding without guidance from someone else ... *Sapere aude!* Have the courage to make use of your own understanding, is therefore the watchword of the Enlightenment." According to F.A.G. Tholuck, the theological and philosophical pacemaker of the *Aufklärung* was Christian Wolff (1679-1754), who sought the path to absolute truth through "pure reason." A more radical form is seen in H.S. Reimarus.* Influenced by English Deists, he rejected supernatural revelation and expressed this in his *Wolfenbüttel Fragments*, published after the author's death by G.E. Lessing.* The latter held that man had developed beyond the need for Christianity (*Education of the Human Race*, 1780). In Lessing's best-known work, *Nathan the Wise* (1779), he argued that truth was found in Christianity, Islam, and Judaism, therefore toleration was imperative.

Predicated upon the reliability of reason, the *Aufklärung* rejected both supernatural revelation and man's sinfulness. God, the all-wise creator, had implanted in man a natural religion which taught both morality and immortality.

WAYNE DETZLER

ENNODIUS, MAGNUS FELIX (c.473-521). Bishop of Pavia. He was born at Arles, but with the invasion of the Visigoths was brought up either in Pavia or Milan. He married in 489, and this saved him from poverty, but by mutual consent the marriage was subsequently renounced when he was ordained deacon by Epiphanius, bishop of Pavia, about 493. In the years that followed he taught rhetoric at Milan. In 494 he accompanied Epiphanius on an embassy to Gundebaud, king of the Burgundians, to procure the ransom of prisoners. In the dispute over the succession to Pope Anastasius II (d.498), he defended Symmachus and the synod of 501 acquitted him. Somewhat later he composed a panegyric to Theodoric who had confirmed Symmachus's election.

Ennodius succeeded Maximus II in the see of Pavia about 514, and in 515 and 517 he was sent by Pope Hormisdas on an embassy to the emperor Anastasius I to oppose the Monophysite heresy and to effect a reconciliation between Rome and the East necessitated by the Acacian Schism.* Neither embassy was successful, and the schism continued until the succession of Justin in 518. The abundant writings of Ennodius are strongly imbued with pagan culture and are unattractive in style. They throw valuable light, however, on the age in which he lived.

DAVID JOHN WILLIAMS

EPARCHY. While Greek of the fifth century B.C. (Aeschylus) could identify a commanding officer as *eparchos,* significant usage began with observation of the expansion of Roman jurisdiction. From Polybius the term equates the Latin *praefectus* (1 *Clem.* 37.3) and *eparchia,* the province under his jurisdiction (Acts 23:34). Patristic citation, while noting the variations of third century A.D. Roman imperial reform, retained the basic meaning, adding only to it the awareness of Christian ecclesiastical presence directly within the imperial framework. Hence "eparchy" became also the jurisdiction of the metropolitan,* that bishop of a province being referred to as *eparchiotēs.*

CLYDE CURRY SMITH

EPHESIANS, EPISTLE TO THE, see EPISTLES, PAULINE

EPHESUS. The chief city of the Roman province of Asia, it stood at the crossroads of the coast route between Smyrna and Cyzicus and the interior route up the Maeander and Lycus valleys. It has a long history traceable to one of the twelve cities of the Ionean Confederation, first founded by early Greek colonists about 1044 B.C. About 560 B.C., the city was relocated to low ground and henceforth the great temple of Artemis became a focal feature of Ephesus. In 287 B.C. the city site was changed again, and this Hellenic foundation lasted for another thousand years. The plan was shaped like a bent bow between Pion and the hill of Astyages. Many fine buildings were added by the Romans, especially in the time of Augustus when a destructive earthquake in A.D. 17 had occurred. At its peak Ephesus possibly reached 500,000 inhabitants. By the tenth century A.D., however, the city had become completely deserted.

Christianity possibly first reached Ephesus with the visit of Paul, Aquila, and Priscilla (Acts 18:18,19). On his third missionary journey Paul stayed there for two years (Acts 19:8,10), attracted no doubt by its strategic importance, its large Jewish colony, its nodality and wealth. During this second visit, Christianity spread to the other churches of the Lycus valley (Col. 1:7; 2:1). At Ephesus the cult of Artemis was extremely popular. It is a cult known from archaeological evidence to have been widespread in over thirty places of the ancient world, but her temple in Ephesus was one of the Seven Wonders of the World. The temple's riches were such that it was the biggest bank of Asia (cf. Acts 19:27). It is no wonder that the economic effects of Paul's preaching there are highlighted (Acts 19:10,26), and a large church established (1 Cor. 16:9).

BIBLIOGRAPHY: W.M. Ramsay, *The Historical Geography of Asia Minor* (1890) and *The Letters to the Seven Churches* (1908); D.G. Hogarth, *Excavations at Ephesus. The Archaic Artemisia* (1908); E.F. Campbell and D.N. Freedman (eds.), *The Biblical Archaeologist Reader,* vol. 2 (1964), pp. 331-52.

JAMES M. HOUSTON

EPHESUS, COUNCIL OF (431). Summoned by Emperor Theodosius II to solve the problems raised by the Nestorian controversy, it has become known as the Third General Council. With sixty bishops present, it was opened by Cyril,* bishop of Alexandria. Neither the Syrian bishops (who were expected to support Nestorius) nor the representatives of the bishop of Rome were present. Nestorius was deposed from his see and excommunicated; his theology was condemned, the faith of Nicea reaffirmed. *Theotokos** was approved as a right title for the Virgin Mary; Pelagianism, the Western heresy that had been gaining ground in the East, was anathematized; and Chiliasm was condemned. When the Syrian bishops arrived they held a rival meeting where Cyril and Memnon, bishop of Ephesus, were excommunicated. But it was Cyril's assembly that was eventually endorsed by the papal legates when they arrived. John of Antioch (the Syrian leader) and Cyril were reconciled two years later, but the Nestorian schism gained momentum and led to the separate Nestorian Church, although the emperor tried to prevent this.

PETER TOON

EPHESUS, ROBBER SYNOD OF (449). It is often called "Latrocinium," a name derived from an expression in Pope Leo's letter to Empress Pulcheria where he described the synod as being *non iudicium sed latrocinium.* (*Latrocinium* =

robber or band of robbers.) The purpose of this council, called by Theodosius II, was to consider the implications of the condemnation of Eutyches* at the Synod of Constantinople in 448. In fact, under the influence of the Monophysite Dioscorus, patriarch of Alexandria, the synod reinstated Eutyches and deposed his opponents. The papal legates who carried the famous Tome of Pope Leo were insulted. The decisions were reversed at the Council of Chalcedon in 451.

PETER TOON

EPHOR. From a verbal root *ephoraō* already in Homer conveying general vision and specific supervision, there emerged the designation for a specific office of overseer, *ephoros.* By the time of Herodotus the five annual eponymous ephors at Sparta were understood to constitute its government. General usage is also known, since it is but a rarer, alternate Greek derivation with little essential distinction from *episkopos.* Hence, by the fifth century A.D. Philostorgius can make the equation, apply a verbal form to the bishop's rule, and identify by nominal derivative the diocese of Tyre. "Ephor" takes on a restricted use after the tenth century, when it is applied to the lay overseer of Byzantine monastic property.

CLYDE CURRY SMITH

EPHRAEM THE SYRIAN (c.306-373). The great classical writer of the Syrian Church, he was born at Nisibis. It is uncertain whether his parents were Christian. After baptism in early manhood he was made deacon about 338. At some time he probably lived as a monk, but apparently never entered the priesthood. After the Persian occupation of Nisibis, he fled to Edessa where his life was spent in teaching, preaching, and literary activities. Details of his life are few: there is no contemporary biography, and much legendary accretion. His writings are many, covering differing aspects of theology and church life. In exegesis, commentaries on Genesis, Exodus, the "concordant gospel" (i.e., harmony, viz. the *Diatessaron*), Paul, and Acts have survived, with fragments of other work in the catenae and elsewhere. His dogmatic works are all in polemical form, against Bar-Daisan, Marcion, and Mani, against Julian the Apostate, and other topics. The ascetic life also is his theme, both in spiritual teaching and in the praise of famous ascetics, while he also composed many hymns and poems, among the latter the Nisibene hymns which reflect contemporary conditions and events up to 363. Many polemical and ascetic works are in the metrical form "memre" (the poetical form is called "madrash").

Ephraem's teaching is orthodox, but conveyed in flowery rhetoric, and a definitive study of his theology is still lacking. His poetic gifts were much prized among the Syrians, however, and they named him "the lyre of the Holy Spirit." To this is due the wide popularity of his works, of which there is a rich and complex tradition; in Syriac much has been lost, but early translation into Armenian has preserved much. The Greek tradition is also fairly early, and from it stems translation into Latin and Christian Oriental languages. Much spurious matter has also been attributed to him, and the whole tradition still represents an important area of patristic research.

BIBLIOGRAPHY: F.C. Burkitt, *Early Eastern Christianity* (1904), pp. 95-110; O. Bardenhewer, *Geschichte der altkirchlichen Literatur,* vol. IV (1924), pp. 342-75; I. Ortiz de Urbina, *Patrologia syriaca* (2nd ed., 1965), chap. 3.

J.N. BIRDSALL

EPHRATA SOCIETY. A cloistered, Protestant commune founded at Ephrata, Pennsylvania, by German Pietist mystic J.K. Beissel* and his Dunker disciples. By 1750 some 300 Brethren and Sisters lived in monastic austerity within the cloister, practicing celibacy and pacifism, keeping Saturday as the Sabbath, sharing agricultural and trade labor, and holding all property and profit in common. The society printed about 200 books from 1745 to 1800, most notably *Martyr's Mirror* by Mennonite J.V.T. Braght and the first American edition of *Pilgrim's Progress.* The first music printed in America was published at Ephrata, often embellished in the European monastic tradition by the Sisters. Led by scholar and linguist Peter Miller after Beissel's death (1768), the society, with its monastic features deteriorating, incorporated as the German Religious Society of the Seventh-Day Baptists in 1814, finally dissolving in 1934.

D.E. PITZER

EPICTETUS (c.50-c.130). Stoic philosopher. Born a slave, probably at Hierapolis in Phrygia, he was permitted to sit under Rufus, a teacher of Stoicism, and later was given his freedom. As a young teacher in Rome, Epictetus met with some success until expelled by Domitian about A.D. 90. He wrote nothing, but some of his lectures were taken down by a disciple and are entitled *Discourses of Epictetus* and *Encheiridion* (the latter is shorter and more popular). While his conception of God is more akin to that posited in Christian theology than in Stoic pantheism, his discourses reveal only one reference to the Christians. Thus it is difficult to argue that the Christian viewpoint was a significant factor in the formulation of his philosophy, although there is a close similarity in statements relating to morality.

WATSON E. MILLS

EPICUREANISM. Popularly this is taken as equivalent to hedonism. More strictly, however, Epicureanism regarded some pleasures as unnatural, and some unnecessary, emphasizing the need for practical wisdom to secure pleasure. This ethical emphasis was part of an antiskeptical philosophy in which knowledge was derived from sense-experience which gave rise to skepticism-defeating "anticipations." Epicureans advocated the avoidance of political and public responsibilities, and the cultivation of friendship. Epicureanism was revived by Gassendi and became popular with the English Deists and French encyclopedists. Paul encountered Epicureans at Mars Hill (Acts 17). The teaching of Epicurus (341-270 B.C.) can be found summarized in Diogenes Laertius's *Life of Epicurus* (Book X of the *Lives*).

PAUL HELM

EPIPHANIUS (c.315-403). Bishop of Salamis. After a brief visit to meet Egyptian monks, he founded at Eleutheropolis in Judea (c.335) a monastery. In 367 he was elected by the bishops of Cyprus as bishop of Constantia (Salamis) and metropolitan of the island. His qualities included orthodoxy, scholarship, linguistic ability, and austerity. His weaknesses included an unenlightened zeal for orthodoxy and an inability to understand the points of view of others. He died at sea after a visit to Constantinople on behalf of Theophilus bishop of Alexandria. In contrast with the Cappadocian Fathers,* Epiphanius denied any right or place in the church to Greek learning, theological speculation, and historical criticism. Despite his dogmatism, however, his works have importance, for in them are found extracts from earlier sources now lost. The *Ancoratus* is a compendium of the doctrine of the church, and includes several baptismal creeds. The *Medicine Box (Panarion)* was intended to heal those Christians who had been bitten by poisonous snakes (heresies). This work contains many extracts from earlier authors, Christian and pagan. Other works included a Bible encyclopedia—*De mensuris et ponderibus.*

PETER TOON

EPIPHANY. The feast of the epiphany is celebrated on 6 January, to commemorate (in the West) the visit of the wise men to Jesus (Matt. 2) and (in the East) the baptism of Christ. The name derives from the Greek *epiphaneia* (manifestation), and it recalls the spiritual significance of the occasion when Gentile magi came from the East (Matt. 2:1) to adore the infant Messiah. The birth of Christ concerns the whole world. The (Eastern) origin of the festival is clear; for Clement of Alexandria, in the third century, refers to a Gnostic commemoration of the baptism of Jesus on 6 January. The date was probably chosen under Egyptian influence. The object of the feast is less easy to determine, since by the fourth century Epiphany celebrated the birth of Jesus, His baptism, His adoration by the wise men, and the miracle at Cana (John 2:1-11). In the East, baptismal water was blessed on that day; and this is still the custom in the Eastern Church. The River Jordan itself is blessed in Palestine at Epiphany. In the Roman liturgy, from the fourth century onward, the feast became primarily a recollection of the manifestation of Christ to the world after his birth; although subordinate themes are included to commemorate any disclosure of His divine power.

See H. Usener, *Das Weihnachtsfest* (*RU* I, 1889), esp. pp. 18-213. STEPHEN S. SMALLEY

EPISCOPACY. From the Greek *episkopos* (overseer), this denotes the system of church government in which a bishop possesses the chief ecclesiastical authority. The NT evidence appears inconclusive. *Episkopos* and *presbuteros* (presbyter) are used interchangeably (cf. Acts 20:17, 28). The threefold order of bishop, priest, and deacon seems likely to have emerged in the second century. The subject has provoked wide divergence of opinion and bitter controversy.

EPISCOPALIAN. Strictly, a member of any church governed by bishops *(episcopi),* but popularly applied to churches when contrasted with other nonepiscopal bodies. The existence of the Episcopal Church in Scotland over against the presbyterian government of the Church of Scotland, and that of the Methodist Episcopal Church in the USA, are examples of this usage.

EPISCOPIUS, SIMON (Simon Biscop) (1583-1643). Dutch Protestant theologian. Educated at Leyden under Arminius,* who taught a softened version of predestination, he and Uyntenbogaert were prominent among his former teacher's followers who issued the Remonstrance* of 1610. After taking part in an abortive conference at the Hague with Contra-Remonstrant leaders, Episcopius at twenty-nine took the place of F. Gomar(us),* a Contra-Remonstrant leader, at Leyden. Gomarus had resigned in protest against Remonstrant influence at the university. The Remonstrant controversy became entangled with political issues, and passions ran high: Episcopius at one point was mobbed in the streets and accused of plotting with Catholics.

In 1618, Oldenbarneveldt, the political protector of the Remonstrants, was arrested, by the stadhouder Maurice of Orange, and the Synod of Dort,* convening to discuss the Remonstrant controversy, was thus filled with Contra-Remonstrants. Episcopius, though one of the few Remonstrant delegates, was cited to defend the Remonstrant position; cast as a defendant, his procedural protests were to no avail, and the synod condemned the Remonstrants, adopted the Canons of Dort as a statement of the orthodox view on the disputed issues, ousted Remonstrant ministers, and arranged for the exile of Remonstrant leaders.

After stays in Antwerp, where he helped organize the Remonstrants in exile, and France (Paris and Rouen), Episcopius returned to Rotterdam in 1625 at the death of Maurice. He worked as a Remonstrant preacher, and later as professor at the Remonstrant college in Amsterdam. His own theological views went considerably beyond those of Arminius, but in the next generation became common among the Remonstrants. He denied predestination and election, interpreted the Trinity in a symbolical sense, viewed Christ as basically an ethical model for man, and stressed right conduct rather than dogma. His main works are collected in his *Opera theologica* (2 vols., 1650-55) and *Institutiones theologica* (1650).

See A.H.W. Harrison, *Arminianism* (1937).

DIRK JELLEMA

EPISCOPI VAGANTES. In the second century A.D. all concepts of episcopacy seem to have related the person of the bishop closely to the structure of the church. In the third century Cyprian held this so strongly that he denounced as not being bishops at all those who, although they had been duly consecrated, were not "in communion" with him. On this view, a bishop who was excommunicated ceased to be a bishop. Cyprian was opposed by Stephen of Rome, and in the fourth and fifth centuries Stephen's view prevailed, being adopted by Augustine in his dealings with the Donatists.* Augustine held that, although from the Catholic standpoint the Donatists were no church (there being only one church, and it fully in communion with itself), yet their orders were validly transmitted and their ministers did not need to be ordained if they were received into the Catholic Church. In some sense they remained true ministers, able to perform true ministerial actions (such as bishops ordaining), even though outside the fold of the church.

Augustine's doctrine paved the way in time for bishops to take actions contrary to the discipline of Rome, and it is on the basis of Augustine's doctrine that Anglicans have always claimed that *even from the Roman Catholic standpoint* the consecrations and ordinations performed by Cranmer and Parker and their successors ought to be recognized as true ministerial actions, though outside the communion of Rome. The Roman reply has always conceded the Augustinian doctrine, but declared Anglican orders null and void on the grounds that the rites used did not have the intention of conferring true orders.

In the case of the Old Catholics,* originating in a hostility of Rome toward the see of Utrecht in the early eighteenth century, not only were orders initially conveyed by a Roman bishop using the Roman rite, but even with a century-and-a-half of "single-bishop" consecrations true orders have been conferred, and recognized as such by the Church of Rome. The Church of England had a similar case with the Nonjurors* in the eighteenth century, and the Scottish succession springs from bishops who were simply a "college" without dioceses or parishes to oversee in the days of the repressing of Jacobites.

It is but a short step from these cases to the thoroughgoing "Episcopi Vagantes." Once validity is suspended solely upon pedigree, without regard to ecclesial context, then a line of bishops may arise without any real church connections at all. In the twentieth century such lines of succession have in fact sprung from clandestine or ill-advised consecrations, the most famous, that of A.H. Mathew, being by the Old Catholic bishops in Holland; and others, notably of J.R. Vilatte and Vernon Herford, being by schismatic bishops of the East in Ceylon. In each instance the recipient of episcopal orders seems to have deceived his consecrators to greater or lesser degree.

The final states are both tragic and comic. The recipients of episcopal orders of this sort have suffered from megalomania, due to the validity of their orders even in the eyes of Rome. They have themselves conferred "valid" orders recklessly. Very often the recipients have in turn quarreled with the donors of orders, and have started new lines of succession. Congregations, or other reasons for being pastors, have been absent. The "valid" orders have been their own justification. Some such "bishops" have clandestinely given "valid" orders to Anglican clergy doubtful of their own. Others have lapsed into infidelity (without giving up ordaining and consecrating). Most have adopted grandiose titles (e.g., Mar Georgius, patriarch of Glastonbury, whose titles run to ten lines of print in Anson's book). Brandreth reckoned there were 200 or more alive in 1961.

See H.R.T. Brandreth, *Episcopi Vagantes and the Anglican Church* (1961); and P.F. Anson, *Bishops at Large* (1964). COLIN BUCHANAN

EPISTLES, GENERAL. This name is given to seven short epistles of the NT, namely James; 1 and 2 Peter; 1, 2, and 3 John; and Jude. They are also known as the Catholic epistles. It is usually held that "general" means much the same as "catholic" and that the reason for the name is that these letters, unlike the Pauline epistles, are not addressed to specific churches or individuals. This is not true of 2 and 3 John, which carry specific addresses, nor is it quite true of 1 Peter, though the address there covers quite an area. It may well be that writings like 1 John, which is in fact general and not addressed to anyone at all, were first given the name and it afterward attached to the group as a whole. Some seek another explanation of the name, and maintain that "catholic" was originally equivalent to "canonical"—i.e., it signified epistles received in the Catholic Church. Against this is the fact that this would apply equally to the Pauline epistles. Further, some of the seven had not come to be regarded as canonical at the time the expression was first used. It seems that "general" is the way we should understand the term.

The church was slow to accept all the letters in this group. The Muratorian Fragment, regarded as giving us the canon accepted at Rome in the second half of the second century, lists two epistles of John, and also Jude, but none of the others. 1 Peter appears to have been accepted in Africa at this period. In the East there was a greater readiness to welcome these writings, and Origen uses all seven (though with doubts about James, 2 Peter, and 2 and 3 John). His attitude was not universal, for in the early fourth century Eusebius of Caesarea regarded only 1 Peter and 1 John as canonical. He put the five others into the category of disputed books. However, they steadily made their way and by the end of the century seem to have been accepted in most places, East and West alike, except in the Syrian Church. The Peshiṭta of that church included James, 1 Peter, and 1 John. But it never did include the four others. They were added only in the Philoxenian revision of A.D. 508.

James is usually held to have been written by James the brother of the Lord, but the attribution is uncertain. So is the date, though the epistle seems early. It is concerned with the way the Christian faith is to be lived out in daily life, perhaps the most important section being that in which James opposes a corruption of the Pauline teaching that a man is justified by faith alone. While he does not espouse a doctrine of justification by works, he denies that faith without works is viable. Faith shows its presence by works (2:18).

Traditionally the Apostle Peter is held to have written 1 Peter, and while this has sometimes been denied it seems the preferable view. This is a letter written to Christians facing suffering, and it encourages them to a steadfast and joyful endurance of what confronts them, together with a steady insistence on the importance of living out the Christian faith in innocence and purity.

In recent times many have been prepared to deny the apostolic authorship of 2 Peter, partly on the grounds of style, partly on those of the nature of the teaching given and opposed. This has been met by pointing to the possibility of the use of an amanuensis in one or both epistles and to our ignorance of the kind of teachers that arose in the early church. This letter addresses itself to refuting heretics of unsound doctrine and immoral life and to pointing its readers to the coming of Christ as the hope of the church.

2 and 3 John are written by "the elder" but he remains unnamed and there is no author mentioned in 1 John. The style of all three, however, is much that of the fourth gospel, and this has led to the general acceptance of all three of these letters as from the Apostle John. This is denied by some, but is widely accepted among conservative Christians. 1 John is concerned to insist on the reality of the Incarnation. It is important to recognize that "Jesus Christ has come in the flesh" (4:2). The epistle insists just as strongly on the importance of the Christian virtues, especially love. The same concern for sound doctrine and upright living runs through 2 and 3 John.

Jude is written by the "brother of James" (v. 1), but unless we can identify James this does not help. It is usually taken that this is the James who wrote the epistle which bears his name. Jude sternly denounces heretics for their false teaching and for their immoral lives. He speaks of the divine punishment which awaits such men and urges believers to build themselves up on their most holy faith.

BIBLIOGRAPHY: B.F. Westcott, *The Epistles of St. John* (1883); J.H. Ropes, *A Critical and Exegetical Commentary on the Epistle of St. James* (1916); E.G. Selwyn, *The First Epistle of St. Peter* (1946); D. Guthrie, *New Testament Introduction: Hebrews to Revelation* (1962); E.F. Harrison, *Introduction to the New Testament* (1964); Bo Reicke, *The Epistles of James, Peter and Jude* (1964); W.G. Kümmel, *Introduction to the New Testament* (1966); J.N.D. Kelly, *A Commentary on the Epistles of Peter and of Jude* (1969).

LEON MORRIS

EPISTLES, PAULINE. Of the twenty-seven documents making up the NT canon, thirteen are epistles or letters bearing the name of Paul. Possibly the majority of these letters are the oldest writings in the NT. Since Paul was called from the day of his conversion to be Christ's Apostle to the Gentiles, his letters form our primary sources of information about primitive Gentile Christianity, and they shed some incidental but welcome light on the early Jewish mission too. Yet the earliest of them dates from a time when Paul had been a Christian and herald of the Gospel for fifteen years; his extant correspondence comes from the second half of his apostolic career.

In the traditional order of his letters, those to churches precede those to individuals, and within these two groups they are arranged in (approximately) descending order of length. Here it is more convenient to adopt a grouping which is more nearly chronological.

(1) *The Thessalonian Correspondence.* The two letters to the Thessalonians were written in the early stages of Paul's Aegean ministry, soon after the Council of Jerusalem (Acts 15) which decreed that circumcision should not be imposed on Gentile Christians. Rioting in Thessalonica compelled Paul to leave that city (c. A.D. 50) before he had given the newly formed church there all the teaching it required; the two letters, sent from Corinth, were designed largely to supply what was lacking in this regard. In 1 *Thessalonians* he (with his colleagues Silvanus and Timothy who had shared in the evangelization of Thessalonica) congratulates the church on remaining steadfast in face of opposition and propagating the Gospel (1:2—3:13), he reminds them of the ethical standards of Christianity, especially in sexual relations (4:1-12), and clears up some difficulties concerning the Parousia, particularly concerning the status of those believers who had died before that event (4:13-18).

In 2 *Thessalonians,* written very soon afterward—some (e.g., T.W. Manson) have argued it was written earlier—he clears up further eschatological difficulties. There was a tendency to imagine that the Parousia was so imminent that there was no point in going on working. Those who took this attitude are reminded that certain developments associated with the rise of Antichrist must precede the Parousia (2:1-12) and that for able-bodied men to give up working and live at the expense of others is quite inconsistent with the demands of the Gospel and the example set by the apostle and his companions (3:6-12).

(2) *The Capital Letters.* This designation is commonly applied to the epistles to the Galatians, Corinthians and Romans, which are our most important sources for Paul's teaching.

Galatians. While in subject matter Galatians goes closely with 2 Corinthians and Romans, there are features which suggest it may be eight or nine years earlier, possibly even the earliest of Paul's extant epistles. The life-setting—an attempt to persuade the churches of Galatia that circumcision is not essential to the Gospel—could be anterior to the Council of Jerusalem, which settled that question, especially if the churches addressed are those of Pisidian Antioch, Iconium, and Lystra (cities of S Galatia), planted by Paul and Barnabas before the council (Acts 13:14—14:23). News of this situation impels Paul to make an uncompromising defense of the gospel of justification by faith, as opposed to legal works, and incidentally to vindicate the independence of his apostleship and Gentile mission in relation to the leaders of the Jerusalem church. Occasional as his defense of justification is, this is no subsidiary or accidental element in his gospel, but its very pith and core, and however vigorously he asserts his independence of Jerusalem, he plainly attaches great importance to maintaining fellowship with the mother-church and its leaders.

1 and 2 Corinthians. These two epistles are the surviving parts of a larger correspondence revealing Paul's pastoral concern for the church which he had planted during his eighteen months' stay in Corinth (A.D. 50-52). The correspondence belongs to the later part of his Ephesian ministry and the months immediately following (55-56).

1 Corinthians, which followed a "previous letter" (1 Cor. 5:9) warning the recipients against the proverbial sexual laxity of Corinth, begins with an admonition deprecating a tendency to party-spirit (chaps. 1-4) and goes on to deal with ethical problems (chaps. 5-6) and then to answer questions raised in a letter sent by the church to Paul, concerning marriage and divorce, food consecrated to idols, the exercise of spiritual gifts in their meetings, etc. (chaps. 7-16).

After the dispatch of 1 Corinthians, Paul appears to have paid the church a painful visit (2 Cor. 2:1; 13:2), which was followed by a letter of such severity that the church was stung into disciplinary action against the leadership of the anti-Pauline faction and into a desire for full reconciliation with Paul (2 Cor. 2:3ff., 7:8ff.). *2 Corinthians* (at least chaps. 1-9) is Paul's response to the news of this welcome change of heart—welcome because it reached him at a time when to his anxiety over Corinth had been added some especially deadly peril in proconsular Asia. In his relief Paul pours out his heart to the Corinthians and enlarges on the glory of the ministry of reconciliation committed to him and his fellow-preachers of the Gospel (chaps. 1-7). He judges the atmosphere favorable for an exhortation to participate generously in the gift being collected in the Gentile churches for the relief of their brethren in Jerusalem (chaps. 8-9). That the mood of reconciliation did not last long is indicated by chapters 10-13—unless (which is unlikely) they are a displaced fragment of the earlier severe letter—for the denigration of Paul is more vigorous than ever, fomented (it appears) by visitors from Judaea claiming the authority of the Jerusalem leaders.

Romans was sent by Paul to the Christians of the capital at the beginning of A.D. 57, when his plans were maturing for launching in Spain a missionary program such as he had just completed in the Aegean world, and for visiting Rome on the way and making it (as he hoped) his base for the evangelization of Spain. He writes to prepare the Roman Christians for his visit and takes the opportunity of setting before them a statement of the Gospel as he understood and proclaimed it, in its bearing on Israel and the Gentiles alike. Many of the themes of his earlier epistles are repeated here, especially of Galatians, but more dispassionately and systematically, Galatians being related to Romans "as the rough model to the finished statue" (J.B. Lightfoot).

(3) *The Captivity Letters.* In Romans 15:25ff. Paul says that before setting out for Rome he must visit Jerusalem with the Gentile churches' gift for the Christians there. Events in Jerusalem led to his detention for two years at Caesarea (Acts 24: 27), after which he was sent under armed guard to Rome, where he remained under house arrest for two years more (Acts 28:30). The four "captivity letters"—Philippians, Philemon, Colossians, and Ephesians—are traditionally assigned to these two years at Rome, but arguments have been put forward for assigning them to his Caesarean detention (e.g., by E. Lohmeyer, J.J. Gunther) and even to an earlier undocumented but probable imprisonment in Ephesus (e.g., by G.S. Duncan). All four captivity letters need not come from the same period: Philemon, Colossians, and Ephesians are closely interrelated, and the development of Pauline themes in the last two bespeaks a later date than the "capital letters"; but certain affinities between 2 Corinthians and Philippians might suggest an Ephesian provenance for the latter, if it is a literary unity.

Philippians (on the assumption that it is a unity and not an editorial construct of two or three letters sent by Paul to Philippi) was in intention written to thank the Philippian church for a gift sent to Paul in prison by the hand of Epaphroditus (4:10-20). But first he expresses pleasure at the progress of the church (1:3-11), tells them how his imprisonment has turned out for the furtherance of the Gospel (1:12-18), invites their prayers (1: 19-26), and urges them to maintain a spirit of concord among themselves, following the example of humility shown by Christ (1:27—2:5). This exhortation is reinforced by the quotation of what is commonly regarded as a pre-Pauline hymn or confession celebrating Jesus' self-denial and His consequent exaltation by God (2:6-11). Further personal news (2:12-29) is followed by a warning against troublemakers such as had threatened to disrupt his churches elsewhere, Judaizers at one extreme and libertines at the other (3:2-16), and renewed injunctions to rejoice and be of one mind in the Lord (3:1; 4:1-9).

Philemon is a charming personal letter to a Christian of that name in Colossae, a city of the Lycus valley in Asian Phrygia. Onesimus, a former slave of Philemon, whom Paul had befriended and won for Christ, is sent back to be reconciled with his master, to be received "no longer as a slave, but . . . as a dear brother" (v. 16), and (as Paul strongly hints) to be sent back in order to continue making himself useful to the imprisoned apostle as he had already begun to do.

Colossians was sent at the same time to the whole church of Colossae to put it on its guard against a form of Judeo-pagan syncretism which was flexible enough to take some elements of Christianity into its system, but in effect undermined the Gospel by robbing Christ of His uniqueness as the one who embodied the fulness of deity, and by its ascetic demands imposed a yoke of bondage on those who should enjoy the emancipation which was theirs by faith-union with the crucified and risen Lord. In his reply to this false teaching Paul develops more fully than in his earlier epistles his conception of the church as the body of which Christ is the head, together with the doctrine of the cosmic sovereignty of Christ, and draws out their practical implications for Christian life.

But if the church is the body of One who wields cosmic sovereignty, what is the cosmic significance of the church? This question, arising out of the argument of Colossians, is dealt with in *Ephesians,* where Paul unfolds his vision of the church as being not only God's present masterpiece of reconciliation but also God's pilot scheme and agency for the reconciled universe of the future, when He has achieved His purpose of uniting all things in Christ. This letter was sent by the hand of Tychicus (the bearer of Colossians) to Ephesus and other churches in proconsular Asia to serve in some way as Paul's testament to them.

(4) *The Pastoral Letters. 1 and 2 Timothy* and *Titus* have been known since 1703 as the "Pastoral epistles" because so much of their contents consists of directions to Paul's colleagues and lieutenants, Timothy and Titus, for the organization of church life and ministry in Ephesus and Crete respectively. In their present form they have been widely, but not universally, regarded as post-Pauline, partly because of the difficulty of finding an appropriate setting for them in Paul's career, but mainly because of deviations in style and vocabulary. Some have attempted to account for these deviations in terms of freedom granted to a confidential amanuensis (such as Luke); others have thought of a posthumous editor collecting *disiecta membra* of Paul's correspondence and instructions. 2 Timothy 4:6-18 envisages his death as imminent.

Pauline Corpus. Even in Paul's lifetime, and occasionally (it appears) at his own instance, some of his letters circulated outside the territories to which they were primarily addressed. This process was well advanced by the end of the first century, when Clement of Rome so readily quotes 1 Corinthians. Early in the second century Paul's letters were gathered together into one corpus, perhaps in two stages—the first comprising ten letters, the second adding the three Pastorals. The extant textual tradition of the letters is almost entirely derived from the collected corpus; only rarely does it bear independent witness to the pre-corpus phase.

BIBLIOGRAPHY: K. Lake, *The Earlier Epistles of St. Paul* (1911); H.A.A. Kennedy, *The Theology of the Epistles* (1919); P.N. Harrison, *The Problem of the Pastoral Epistles* (1921) and *Paulines and Pastorals* (1964); H.N. Bate, *A Guide to the Epistles of St. Paul* (1926); G.S. Duncan, *St. Paul's Ephesian Ministry* (1929); J. Knox, *Chapters in a Life of Paul* (1950) and *Philemon Among the Letters of Paul* (2nd ed., 1959); C.H. Dodd, "The Mind of Paul" in *New Testament Studies* (1953), pp. 67-128; G. Zuntz, *The Text of the Epistles* (1953); C.L. Mitton, *The Formation of the Pauline Corpus of Letters* (1955); D. Guthrie, *The Pastoral Epistles and the Mind of Paul* (1956) and *New Testament Introduction: The Pauline Epistles* (1961); J. Munck, *Paul and the Salvation of Mankind* (ET 1959); A.M. Hunter, *Paul and His Predecessors* (2nd ed., 1961) and *The Gospel According to St. Paul* (1966); F.W. Beare, *St. Paul and His Letters* (1962); T.W. Manson, *Studies in the Gospels and Epistles* (1962); A.T. Hanson, *Studies in the Pastoral Epistles* (1968); B. Rigaux, *The Letters of St. Paul* (ET 1968); J.J. Gunther, *St. Paul: Messenger and Exile* (1972).

F.F. BRUCE

EPISTULA APOSTOLORUM, see TESTAMENT OF OUR LORD

ERASMUS (c.1466-1536). The leading Christian humanist, who wished to reform the church through scholarship and instructions in the teachings of Christ. Born the illegitimate son of a Dutch priest, he was educated by the Brethren of the Common Life* at Deventer (1475-84). When his father died, Erasmus transferred to another school and eventually became a monk. Later he secured the position of secretary to the bishop of Cambrai, thus escaping the secluded life. An opportunity arose for him to study at the *Collège de Montaigu* in Paris, and after this experience he visited England. Here he met John Colet,* who influenced him to apply his humanistic interests to biblical scholarship and the revival of primitive Christianity. After a visit to Italy and another trip to England Erasmus settled in Basle (1514-29) where, except for some short excursions, he was to live and work for many years. When the reform in the city became too radical for him, he moved to Freiburg-im-Breisgau, but returned to Basle to die.

Erasmus was the first best-selling author in the history of printing. Some examples of his popularity include *The Praise of Folly,* which has appeared in more than 600 editions, and the *Colloquies,* more than 300 editions. Among his publications, in addition to these satirical works, are a critical edition of the NT based on Greek manuscripts; a paraphrase of the NT (except for the Book of Revelation); editions of the Greek and Latin Fathers; *Adages* (a collection of sayings taken from the Greek and Latin classics); the *Enchiridion Militis Christiani (Handbook* or *Weapon of a Christian Knight);* and *De Libero arbitrio* (on the freedom of the will, an attack on Luther's ideas).

There are many interpretations of the career of Erasmus. Some say he was weak—a Lutheran at heart, but for fear of the church a conforming Catholic. Others have pictured him as a devotee of reason, a precursor of the eighteenth-century Enlightenment.* Another interpretation makes him the forerunner of Luther. "Erasmus laid the egg that Luther hatched," it has been said. According to this view, Erasmus with his critical work, his emphasis on the original texts of Scripture, and on the teachings of Christ took the first step toward the Reformation. Luther, with his stress on Paul's presentation of the Gospel, took the second and left Erasmus behind.

There is truth in each of these positions, yet another view comes nearer to an understanding of the man. Erasmus had his own reform program, partly critical but for the most constructive. He believed it was necessary for reform to use the tools of scholarship and the materials provided by Christian antiquity. Philology, a critical sense, and diligent labor would enable the scholar to reveal the truth in the Bible and in the Church Fathers. The philosophy of Christ thus recovered when taught to the learned and to the simple would infuse new spiritual life into all Christendom. As he stated in his most famous lines, "I would to God that the plowman would sing a text of the Scripture at his plow and that the weaver would hum them to the tune of his shuttle . . . I wish that the traveler would expel the weariness of his journey with this pastime. And, to be brief, I wish that all communication of the Christian would be of the Scriptures" (*Opera,* V, 140). It is the tragedy of Erasmus that history passed him by, leaving him doggedly defending his position against Reformers and Counter-Reformers.

BIBLIOGRAPHY: P.S. Allen, *The Age of Erasmus* (1914); J. Huizinga, *Erasmus* (1924); M.M. Phillips, *Erasmus and the Northern Renaissance* (1950); P. Smith, *Erasmus* (1962); W. Kaiser, *Praisers of Folly: Erasmus, Rabelais, Shakespeare* (1963); R. Bainton, *Erasmus of Christendom* (1969). ROBERT G. CLOUSE

ERASTIANISM. The doctrine that the state has the right to intervene and overrule in church affairs; it takes its name from Thomas Erastus. Born in 1524 in Switzerland, Erastus studied theology at Basle and then medicine and philosophy at Bologna and Padua. In 1558 he became physician to the elector Palatine and professor of medicine at Heidelberg. In the city there was a strong Calvinist party led by Kaspar Olevianus* which wanted to introduce the Presbyterian polity and discipline in the church. Erastus, a Zwinglian in theology, opposed this and eventually had to leave the city. Six years after Erastus's death, G. Castelvetro, who married Erastus's widow, published a work found among his papers and entitled, *Explicatio gravissimae quaestionis utrum excommunicatio* (1589). In this Erastus argued against excommunication being practiced by the church and for the rights of the state in ecclesiastical matters. An English translation appeared in 1659 as *The Nullity of Church Censures.* Its teaching was far from new in England. Richard Hooker* had given supremacy to the secular power in his *Ecclesiastical Polity* (1594), and there were both in the Long Parliament and in the Westminster Assembly of Divines those (e.g., Selden, Lightfoot, Coleman) who claimed the right of the civil magistrate to control to a large extent the administrative and disciplinary machinery of the church. The Church of England is sometimes described as Erastian in that bishops are appointed by the Crown and major liturgical changes must have the agreement of Parliament. PETER TOON

ERDMAN, CHARLES ROSENBURY (1866-1960). Presbyterian minister. Graduate of Princeton, he was ordained in 1891 and held two pastorates in Pennsylvania (1890-1905). He was then professor of practical theology in Princeton Seminary, simultaneously pastor of First Church (1924-34), and was moderator of assembly in 1925, and president of the Board of Foreign Missions (1928-40). Balancing academic with keen pastoral concern, Erdman was a popular preacher who aided the Bible Conference Movement. He published thirty-five books, and all of his NT expositions were translated into Korean and other languages.

EREMITE, see HERMIT

ERIGENA, JOHN SCOTUS (c.810-c.877). Irish scholar. Noted chiefly as an interpreter of Greek thought in the West, he entered also into the religious controversies of his day, notably on predestination and the Eucharist. He translated the Neoplatonist author known as Pseudo-Dionysius and also Greek theologians such as Gregory of Nyssa. Erigena is of importance in the period between Augustine and Anselm. His work *De Divisione Naturae* (produced about 862) is markedly ambiguous and was exploited by various parties. It was condemned by Pope Honorius III in 1225. He makes no distinction between theology and philosophy, and he attempts a rational demonstration of the substance of Christian truth. This seems, however, to lead him in the direction of pantheism. The opposed tendencies in his work can be seen in the fact that he insisted on both a sharp distinction between God and the creation and on the emanation of the created order from God. Similarly he wishes to deny that creatures are a part of God, but claims, in Neoplatonist fashion, that God is the only true reality. Though he was not himself a mystic, there is thus a strong mystical strain in his writings. PAUL HELM

ERNESTI, JOHANN AUGUST (1707-1781). German Lutheran theologian. Born in Thuringia and educated at Wittenberg and Leipzig, his early career was given to classics; he was later to produce famous editions of Homer, Polybius, Aristotle, Xenophon, Tacitus, Suetonius, Cicero, and others. In 1742 he became professor of ancient literature at Leipzig, to which post was added in 1758 a chair in theology, which dual role he sustained until 1770. As a grammarian and philologist he discarded dogma for historical evidence in theological studies, and he tried to reconcile traditional Lutheran beliefs with biblical scholarship. Ernesti's most important work is *Institutio interpretis Novi Testamenti* (8 vols., 1761).
C.G. THORNE, JR.

ERSKINE, EBENEZER (1680-1754). Founder of the Secession Church* in Scotland. Son of a minister ejected in 1662 for nonconformity, he graduated at Edinburgh University in 1697, and in 1703 was ordained to Portmoak, where for twenty-eight years he ministered faithfully and imaginatively. His preaching was such that regular adjournment to the open air became necessary when the church could not contain the congregation. He was one of those who protested his general assembly's condemnation of Edward Fisher's *Marrow of Modern Divinity.*

Just after he moved to a Stirling charge in 1731, Erskine as synod moderator preached against assembly legislation on patronage, convinced that it took away the right of Christian people to elect and call their minister. Rebuked by synod and assembly, Erskine with three others handed in a formal protest. This led in 1733 to the suspension of the four and to their constituting the "Associate Presbytery." They nevertheless continued their parish work. The 1734 assembly admitted that its 1732 predecessor had acted illegally, but the breach had widened too far. In 1740 Erskine and seven other ministers were deposed. Within five years the Seceders were ministering to more than forty congregations in Scotland. When they themselves split over a Burgess Oath imposed by the state, Erskine adhered to the Burgher* majority.
J.D. DOUGLAS

ERSKINE, THOMAS (of Linlathen) (1788-1870). A qualified advocate who had never practiced law, this landed proprietor of distinguished ancestry and ample means lives in the history of Scotland as the most outstanding lay theologian that country has ever produced. Alike by his personality and writings he influenced profoundly some of the ablest men of his day—among those who came regularly to sit at his feet in Linlathen, his beautiful Forfarshire home, being Dean Stanley, Thomas Carlyle, Benjamin Jowett, and Charles Kingsley. Well known also in western Europe, his *Internal Evidence for the Truth of the Christian Religion* passed rapidly through nine editions, while books like *The Brazen Serpent* and *The Unconditional Freeness of the Gospel* introduced to thousands a daring thinker of deep spiritual insight with something new and compelling to say about the Fatherhood of God, the nature of the Atonement, and the doctrine of election. His *Letters* (2 vols.) have long since taken their place as a minor religious classic and have been used as a textbook on practical Christianity in many college classrooms. D.P. THOMSON

ERSKINE OF DUN, SIR JOHN (1509-1591). Scottish Reformer. It has been said that while there might have been a Protestant Reformation in Scotland without John Knox, it could not have happened without Sir John Erskine of Dun, its distinguished lay leader. Destined to become after 1560 one of the church's seven area superintendents, and five times elected moderator of its general assembly, this well-known Scotsman introduced the teaching of Greek into the country, befriended George Wishart,* and later gathered the lairds of Angus and Mearns to his home at Dun to sit at Knox's feet. Championing the Protestant cause on many dramatic occasions, and living to become the "Grand Old Man" of the movement, he won the respect even of Mary Queen of Scots. D.P. THOMSON

ESCHATOLOGY. The doctrine concerning the "Last Things," among them being the resurrection of the dead,* the Second Coming* of Christ, the final judgment,* and the creation of the new heaven and the new earth. Traditional Christian theology continues to apply passages like those in Daniel, Isaiah, Zechariah, and Christ's teaching in parables—such as that of the Tares among the Wheat, along with Mark 13 and Matthew 24—as well as 1 and 2 Thessalonians and the Revelation to these coming events. Liberal Protestants, however, led by Albert Schweitzer,* have given eschatology a new meaning. Schweitzer argued that the eschatological teachings of Jesus were central, and that He believed that by sending out the twelve apostles He would bring history to an end. When this failed, He felt He must focus in His own person the troubles of man and offer Himself as a ransom to God so that the new age could begin. He went to Jerusalem with this conviction, but the statements made from the cross made Schweitzer doubt whether He maintained this conviction to the end.

This thesis has had a powerful effect upon twentieth-century scholarship. Critics have arisen not only from the traditional groups, but also among the liberals. It has been pointed out that Schweitzer overstressed Jewish apocalyptic sources, did not accept rabbinical teachings, and that since the discovery of the Qumran documents scholars realize that Messianic expectations were much more complex than Schweitzer pictured them. As early as the 1930s C.H. Dodd* introduced the idea of "realized eschatology," i.e., that in Jesus' ministry the kingdom in all its essentials had already come. Christ, according to his interpretation, was not particularly interested in the future, and the apocalyptic prophecies are in reality additions made to His statements by the early church. Dodd's outlook has been accepted by some scholars, but other more radical critics such as R. Bultmann* followed Schweitzer in believing that Jesus felt there was to be no interval between His death and the start of the New Age.

BIBLIOGRAPHY: S.D.F. Salmond, *The Christian Doctrine of Immortality* (1895); T.F. Glasson, *His Appearing and His Kingdon* (1933); J. Baillie, *And the Life Everlasting* (1934); C.H. Dodd, *The Apostolic Preaching and Its Development* (1936); J.A.T. Robinson, *In the End God* (1950); R.H. Fuller, *The Mission and Achievement of Jesus* (1954). ROBERT G. CLOUSE

ESSENES. According to Josephus (*War*, II.viii.2) the Essenes were the third of the main Jewish philosophies, but unlike the Sadducees and Pharisees they kept their main tenets secret among their adherents. Hence the details given by him and Philo of Alexandria are of necessity suspect. They are mentioned by the elder Pliny in a way that links them unmistakably with Qumran.* Philo connects them rather dubiously, with the Therapeutae,* a contemplative Jewish group in Egypt. From these sources it appears that they were marked out by asceticism, communism, and rejection of animal sacrifices, but the more exaggerated forms suggested by Philo seem to have been derived from the Therapeutae. Josephus acknowledges that some of them married and seems to suggest they were prepared to bring sacrifices, if they could be kept separate from those they regarded as polluted.

Their name is probably derived from *hasidim* (the loyal ones); they probably claimed to be the true representatives of the pious in the time of Antiochus IV Epiphanes. If that is so, their main peculiarities will have resulted from a conviction that they were living in an end-time condition of virtually complete apostasy. After A.D. 70 many probably joined the Palestinian church and helped to produce most of its characteristic heresies. There seems virtually no doubt that we must equate them with the Qumran sect, though it may be that by the time of Christ there were a number of groups bearing the name Essene. The differences between the Qumran documents and our other information are reconcilable, if we remember that our informants based their statements on hearsay knowledge.

BIBLIOGRAPHY: C.D. Ginsburg, *The Essenes* (1864); J.B. Lightfoot, *St. Paul's Epistles to the Colossians and to Philemon* (1875); E. Schürer, *Geschichte des jüdischen Volkes im Zeitalter Jesu*

Christi (4th ed., 1907); M. Dupont-Sommer, *The Jewish Sect of Qumran and the Essenes* (1954); H. Kosmala, *Hebräer-Essener-Christen* (1959); M. Black, *The Scrolls and Christian Origins* (1961): relevant material from Josephus and Philo quoted in appendix. H.L. ELLISON

ESTIENNE, ROBERT ("Stephanus") (1503-1559). Scholar-printer. He was appointed in 1539 as printer in Latin, Greek, and Hebrew to Francis I. Becoming a Protestant, he provoked severe attacks from the Sorbonne because of his Bible annotations. In 1551 he fled to Geneva, embracing the Reformed faith. In his Latin Bibles (1527-28, 1532, 1540) he followed as closely as possible the text of Jerome. He produced editions of the Hebrew OT (1539, 1544-46). As royal printer he published first printed editions of Eusebius (1544), Alexander of Tralles (1548), Justin Martyr (1551), and others. In 1544 he began to print Greek and, helped by his son Henri, published the NT in two tiny volumes (1546). The text was taken chiefly from the fifth edition of Erasmus (1535), although the Alcalá edition of Ximenez was also used. His large edition (1550) was the first to contain a critical apparatus, which is in general the source of the Textus Receptus. He was responsible also for the verse division of the NT, first printed in his fourth edition in Geneva in 1551. According to Henri, he did most of the work on a horseback journey between Paris and Lyons. Estienne subsequently published several of Calvin's works. J.G.G. NORMAN

ESTIUS (Willem Hessels van Est) (1542-1613). Dutch biblical exegete. Born in Gorcum he studied at Utrecht and Louvain, where he was deeply influenced by Michel Baius.* Seminary professor at Douai from 1582 and chancellor of the university there from 1595 till his death, he was respected by colleagues and students alike for his learning, discernment, and saintliness. He played an important part in the predestinarian controversies which led to the *Congregatio de auxiliis* (1597-1607). Wide theological reading and fine judgment made his commentaries widely used.

ETHERIA, PILGRIMAGE OF, see PILGRIMAGE OF ETHERIA

ETHICAL MOVEMENT. A quasi-religious movement having as its motto "Need, not Creed," and as its goal a society embracing the ideals of love, loyalty, brotherhood, and peace. Instigated in 1876 by Felix Adler as the New York Society for Ethical Culture, its central purpose, as stated in the constitution adopted in 1906 at an international conference at Eisenach, was "to assert the supreme importance of the ethical factor in all the relations of life—personal, social, national and international, apart from theological considerations." An outgrowth of the movement was the International Humanist and Ethical Union formed in 1952 "to promote an alternative to the religions which claim to be based on revelation on the one hand and to totalitarian systems on the other." MILLARD SCHERICH

ETHICS, see CHRISTIAN ETHICS

ETHIOPIA. An East African empire that dates its acceptance of Christianity from the fourth century, and its dynasty from Solomon. It was said that Frumentius* and Edesius of Tyre were taken prisoners to Abyssinia, but on gaining favor with Emperor Ezana were set free and began to evangelize the country. About 340 Frumentius was consecrated bishop of Ethiopia by Athanasius in Alexandria. At the close of the fifth century, nine monks reportedly came from Syria, and the Ethiopian Church was confirmed in the Monophysitism that had characterized the original link with Alexandria. The Christian influence declined as Islamic influence spread in Africa, and the church was cut off from contact with other Christians, except for the Coptic Church.* Both the Coptic connection and the isolation are significant factors in the history of a land long shrouded in mystery, beset even today by paganism in the interior, and with a church overlaid by superstition and syncretism, one in which Judaism is still a potent feature.

In 1268 the old dynasty was restored; the church took new life, but excesses of zeal led to the forcible baptism of conquered tribes. Attempts to bring the church into communion with Rome ended with the martyrdom of Dominican missionaries. Only the Abyssinian monastery in Jerusalem retained relations with the West.

When the Muslim onslaught was renewed early in the sixteenth century, an appeal to Rome brought further attempts at reunion in exchange for Portuguese aid, and during the pontificate of Julius III (1550-55), Portuguese Jesuits entered the country. They impressed the court, but alienated the clergy. In 1614 belief in Christ's two natures was imposed on pain of death. The Monophysites resisted but were defeated, and the emperor Susenyos became a Roman Catholic. In 1626 this was proclaimed the official religion, but in 1632 Susenyos's son succeeded him, the old religion was restored, the Jesuits ousted.

In 1634 Peter Heyling* introduced Protestantism into the country, but he too was finally expelled. Later Franciscan efforts proved to be not only unrewarding but dangerous, and the indigenous church was to reach its nadir from the mid-eighteenth to the mid-nineteenth century because of doctrinal difficulties and isolation. The church was suspicious of change as interference with God's established order, and with education controlled by the clergy they wielded immense power. The isolation continued until 1935 when Ethiopia was opened up, not by missionaries, but by the military might of Mussolini. Many clergy, including two bishops, suffered martyrdom, and almost all non-Italian missionaries were expelled.

After World War II, the Ethiopian Church broke with the tradition that its *abuna** should be a Copt sent from Egypt. In 1951 the patriarch of Alexandria consecrated an Ethiopian catholicos-patriarch, and in 1959 the church became independent of Egypt. It is distinctive in several ways. Its canon includes some of the apocryphal books; it observes the Sabbath, circumcision, and the difference between clean and unclean meats. The ark is to be found in every church and at every outdoor festival. The church holds that Christ has one nature, but insists He is perfectly human as well as perfectly divine (though it has known divisions on this point). There are two kinds of clergy: the somewhat illiterate priests responsible for administering the sacraments; and the educated lay clerks who chant the church offices in the long-dead Ge'ez tongue, and teach in the schools. Monasticism is widespread. Each church has its school, and until about 1900, church schools were the sole source of education. A translation of the liturgy has now been made into Amharic, in which language a revised version of the Scriptures was authorized in 1960. More than one-third of Ethiopia's twenty-four-million population belongs to the Ethiopian Orthodox Church. Priests number some 170,000, parishes more than 11,000.

The country is now open to foreign missionaries, albeit with some restrictions in this very tradition-conscious land. Emperor Haile Selassie moved resolutely in advancing national education. He participated in the 1966 Berlin World Congress on Evangelism and was host to the meeting of the World Council of Churches central committee in Addis Ababa in 1971. Deposed and imprisoned in 1974, he died in 1975.

BIBLIOGRAPHY: J.M. Harden, *Introduction to Ethiopian Christian Literature* (1926); H.M. Hyatt, *The Church of Abyssinia* (1928); DeL. O'-Leary, *The Ethiopic Church* (1936); A.F. Matthew, *The Teachings of the Ethiopian Church* (1936); D. O'Hanlon, *Features of the Abyssinian Church* (1946); J.S. Trimingham, *The Christian Church and Missions to Ethiopia* (1950); M. Daoud (tr.), *The Liturgy of the Ethiopian Church* (1954); R. Crummey, "Foreign Missions in Ethiopia," *Bulletin of the Society for African Church History*, II, 1 (1965); E. Isaac, *The Ethiopian Church* (1967); M. Abir, *Ethiopia: the Era of the Princes* (1968); E. Ullendorff, *Ethiopia and the Bible* (1968); M. Geddes, *Church History of Ethiopia* (1969). J.D. DOUGLAS

ETHIOPIC VERSIONS OF THE BIBLE. Although the process of translation into Ethiopic (i.e., Ge'ez, the classical Semitic tongue) may have begun in the late fourth century soon after the conversion of King Ezana, the major impetus probably began in the middle of the fifth century with the flight to Ethiopia of numbers of Syrian monks (particularly the Nine Saints) who sought haven for their Monophysite faith which had been banned by the Council of Chalcedon* (451).

The process of translation probably continued into the seventh century and included both Testaments. The literary revivals of the Solomonic dynasty (1270) produced the only biblical manuscripts now extant. Not only are these manuscripts late, but they show the marks of considerable revision that took place about the fourteenth century under the influence of Arabic versions from Alexandria. It is impossible to tell when the canon of the Ethiopian Orthodox Church was expanded to include not only all the books of the Septuagint, including the Apocrypha, but pseudepigraphical works like *Enoch, Jubilees, Ascension of Isaiah, Paralipomena of Baruch,* and the *Apocalypse of Esdras.* The long history of translation and revision produced manuscripts so varied that Ludolf, the great seventeenth-century pioneer in Ethiopian studies, posited several independent translations.

Doubting this, the majority of scholars, following the lead of A. Dillmann, have voted for a Greek original similar to the text used in Antioch. According to A. Vööbus, the Ethiopic NT is based on a Syriac original (akin to the Sinaitic and Curetonian texts as well as quotations from Syrian fathers) which was later revised with an almost slavish dependence on Greek and then further modified by Arabic and Coptic intrusions and wildly free readings.

Though there is general agreement that the OT was based largely on a Lucianic recension of the Septuagint, the verdict, then, is not in on the NT, virtually a virgin field for textual critics. A thorough investigation would enrich our knowledge of textual development and, more, would enhance our understanding of the variegated and unique structure of Ethiopian Christianity.

BIBLIOGRAPHY: F. daBassano (ed.), *Ethiopic Old Testament* (4 vols., 1926); *Ethiopic New Testament* (1949; rep. from 1899 Leipzig edition); A. Vööbus, *Early Versions of the New Testament* (1954), pp. 243-69; E. Cerulli, *Storia Della Letteratura Etiopica* (1956). DAVID A. HUBBARD

EUCHARIST, see COMMUNION, HOLY

EUCHERIUS (d. c.449). Bishop of Lyons. Engaged at first in public administration and married with two sons, he retired in middle life to Lerins, where he pursued an ascetic life of study and worship. His reputation for sanctity became widespread, and about 434 he was elected bishop of Lyons. Very little is known of his episcopate. In 441 he presided over the Synod of Orange jointly with Hilary of Arles.* His writings include two ascetic treatises and two exegetical works which display an extensive biblical knowledge and anticipate many favorite usages of medieval mystics and hymnwriters.

EUCHITES, see MESSALIANS

EUDES, JEAN (1601-1680). French missioner and pastor. Educated by the Jesuits in Caen, he became an Oratorian* in 1623. Appointed superior of the Caen congregation in 1639, his pastoral concern was demonstrated by the foundation of the Order of Our Lady of Charity of the Refuge (1641), dedicated to rehabilitation of reformed prostitutes. In 1643 he left the Oratory to found the Congregation of Jesus and Mary (or Eudists). Despite their work as missioners and seminary teachers, the congregation never gained papal approval. Eudes was also an influential pioneer in devotion to the Sacred Heart of Jesus and the Sacred Heart of Mary, to further which he wrote offices and devotional and theological works, and actively fostered lay confraternities. He was canonized in 1925. IAN BREWARD

EUDOXIUS (d.370). Bishop of Constantinople. Native of Cappadocia and an Arian, he became bishop of Germanicia and in 358 of Antioch (Constantius later denied his nomination of Eudoxius). His extreme views led to his deposition by the Council of Seleucia in 359, but with the adjustment of his declared views he was elected bishop of Constantinople in 360. Eudoxius declared then that the Father was impious and the Son pious, explaining to the angry assembly that the Son reverences the Father, but the Father has no one to reverence.

EUGENIUS III (d.1153). Pope from 1145. Born Bernardo Pignatelli of Pisa, he was a Cistercian monk at Clairvaux and subsequently abbot of SS. Vincent and Anastasius at Rome. Bernard of Clairvaux dedicated *De Consideratione* to him and preached the Second Crusade over his *Bulla cruciata* (1145/6). Though dejected over the crusade's failure he would not participate in the wave of anti-Byzantine bias sponsored by Roger of Sicily and Louis VII of France, wanting good relations with Conrad III and his successor Frederick Barbarossa, which the Treaty of Constance (1153) secured for the church. In England Eugenius deposed William of York, supported Theobald of Canterbury even to the extent of not banning King Stephen, and named Nicholas Breakspear (later Adrian IV) cardinal bishop and legate to Scandinavia. He held synods at Paris, Trier (1147), and Cremona (1148), and in Ireland, and a council at Reims (1148) on Gilbert de la Porrée's heresy and Hildegard's visions. Reform canons were issued to strengthen the Lateran decrees (1139), and an examination of papal revenues foreshadowed the *Liber censuum* (1192).

C.G. THORNE, JR.

EUGENIUS IV (1383-1447). Pope from 1431. Born Gabriele Condulmaro, of a wealthy Venetian family, he entered an Augustinian monastery at an early age. Brought to the papal court by his uncle, Gregory XII, he was appointed cardinal priest of San Clemente in 1408. Under Martin V he governed the March of Ancona and Bologna for a time. On his election as pope he dismissed the Council of Basle convoked by Martin which had sought to limit papal power. The council refused to dissolve, and reasserted and extended the principle of conciliarism enunciated at the Council of Constance. Eugenius was forced to withdraw his dissolution in 1433. A popular insurrection in Rome led by the Colonnas caused him to flee to Florence. His relations with the council worsened. It tried to destroy papal authority completely and in 1439 elected Amadeus VII, duke of Savoy, as antipope (Felix V). Eugenius had, however, called a council at Ferrara, transferred later to Florence. He concluded a short-lived reunion of Eastern and Western churches (1439) and excommunicated the bishops at Basle. In 1443 he returned to Rome, and in 1444 promoted a crusade against the Turks which ended in defeat at Varna.

J.G.G. NORMAN

EUNAN, see ADAMNAN

EUNOMIANISM. A theological heresy propagated by Eunomius (d.395). Born in Cappadocia, he went to Alexandria where he became a follower of Aetius,* the Anomoean, who carried the principles of Arianism to their logical limit, asserting the complete unlikeness of the Son and Father. The extreme Arianism of Eunomius remained latent until he became bishop of Cyzicus in Mysia. Here, after openly expounding his heresy, he was forced to resign his see and return to Cappadocia. But later, after the death of Aetius (370), he became the leader of the Anomoeans.* Through lecture tours and by means of books he ardently propagated his views. His chief work was an *Apology* which was answered by Basil the Great. He also wrote a commentary on the epistle to the Romans.

For Eunomius, God was the ungenerated Being, the single, supreme, ultimate, and simple Substance. He held that the "Son of God" was actually created by the Father, and though possessing creative power was not of His essence; further, the "Holy Spirit" was created by the Son in order to be the Sanctifier of souls. Putting great emphasis on doctrine, he depreciated the value of the sacraments and the ascetic life. His teaching had no permanent success, and it was refuted by Gregory of Nyssa in *Contra Eunomium* (c.382).

PETER TOON

EUSEBIAN CANONS. A system devised by Eusebius of Caesarea to facilitate the location of parallel passages in the gospels. Each gospel is divided into sections, numbered consecutively (355 in Matthew, 233 in Mark, 342 in Luke, 232 in John), and the sections, indicated by these numbers, are listed in parallel with the corresponding sections of other gospels. There are ten lists or canons. The system may have been suggested by the work of Ammonius of Alexandria (c.200)—they are sometimes called the Ammonian Sections*—and is explained by Eusebius in a letter to Carpianus.

EUSEBIUS (d.341/2). Bishop of Nicomedia; latterly patriarch of Constantinople. As a young man he studied with Arius under Lucian of Antioch. After ordination he was bishop of Berytus and later of Nicomedia. When Arius was deposed in 320, Eusebius decided to support and defend his friend. Though he signed the creed at the Council of Nicea* (325), where he was a prominent figure, he later led a widespread reaction against its teaching. His personal contacts with the imperial household, situated at Nicomedia, allowed him to engineer the deposition and exile of the principal opponents of Arianism*—Athanasius,* Eustathius, and Marcellus—and to propagate Arian views. He had the honor of baptizing Constantine just before the latter's death in 337 and then exercised great influence over Constantius. His leadership of the anti-Nicene party was so well recognized that his followers were called Eusebians. In 339 he became patriarch of Constantinople.

PETER TOON

EUSEBIUS (d. c.359). Arian bishop of Emesa. Of a noble Edessan family, he studied theology under Eusebius of Caesarea and, subsequently, philosophy at Alexandria. He then settled in Antioch and became intimately acquainted with the bishop, Flacillus. As he was a man of considerable learning, high personal character, with a theology acceptable to the Eusebian party (see previous entry) and with a knowledge of Alexandria, it is not surprising that the Council of Antioch (341) offered Eusebius the bishopric of Alexandria, made vacant through the exile of Athanasius.* He refused, having no great desire for preferment, and certainly not for the unpopularity the usurper of Athanasius was certain to encounter. He was, however, persuaded to become bishop of Emesa. There he successfully overcame the opposition of the people who had heard he was a sorcerer. He was a friend and adviser of Emperor Constantius. Only fragments of his writings have survived, the most important being seventeen sermons recently discovered. C. PETER WILLIAMS

EUSEBIUS (d.371). First bishop of Vercilli. A strong upholder of the Nicene orthodoxy, he was respected for his holiness. As the leading spokesman for Pope Liberius at the Council of Milan (355), he courageously resisted efforts of the powerful pro-Arian bishops, who were supported by Emperor Constantius, to join in a condemnation of Athanasius.* He requested that before anything was decided there should be an acceptance of the Nicene Creed. He was defeated and exiled in the East. When released, he was one of the instigators of the Council of Alexandria (362). With Athanasius he was responsible for its conciliatory tone, and was the bearer of its letter attempting to end the schism in Antioch between the Eustathians and the Melitians. He was, however, frustrated by the impetuosity of Lucifer of Cagliari in consecrating Paulinus as bishop, thus perpetuating the conflict. On returning to the West, Eusebius joined with Hilary of Poitiers in an unsuccessful attempt to oust the Arian bishop of Milan, Auxentius. After this he devoted himself to the needs of his large diocese. Three of his letters survive, unlike his Latin translation of Eusebius of Caesarea's commentary on the Psalms. He may have written the "Codex Vercellensis"—a pre-Jerome text of the gospels.
C. PETER WILLIAMS

EUSEBIUS (d.380). Champion of the Nicene faith; bishop of Samosata from 361. He was a member of the synod held under Melitius of Antioch in 363 which accepted the formula *homoousios* ("of the same substance") as describing the relationship of the Son and Father. He was associated with Basil of Ancyra and Gregory of Nazianzus in opposing Arianism, but in 374 was banished first to Cappadocia and then to Thrace. He has the reputation of being one of the few bishops of the fourth century of whom nothing evil is known. He is reckoned a saint because he was killed by a missile thrown at him by a woman supporter of Arianism at Doliche in Syria.

EUSEBIUS (fifth century). Bishop of Doryleum from 448. A Constantinople lawyer, he posted there in 429 a document comparing excerpts from the sermons of Nestorius (see NESTORIANISM) with the utterances of the third-century heretic Paul of Samosata,* thus directly accusing Nestorius of denying the divinity of Christ. After becoming bishop he presented formal charges of heresy against Eutyches* before a synod at Constantinople presided over by Flavian. In 449, along with Flavian, he was deposed and exiled by the Robber Synod held in Ephesus. He was restored to his see by the pope and by the emperor Marcion in 451, and subsequently took a prominent part in the Council of Chalcedon* that year, principally in helping to draft its Definition of the Faith. JAMES TAYLOR

EUSEBIUS OF CAESAREA (c.265-c.339). "The Father of Church History." Born probably in Palestine, of humble parentage, in early youth he became associated with Pamphilus,* founder of the theological school of Caesarea, assisting him in preparing an apology for Origen's teaching. After Pamphilus's martyrdom (310), he withdrew to Tyre, naming himself "Eusebius Pamphili" in honor of his master. Later he went to Egypt, where he was apparently imprisoned for a short time. He was subsequently accused by Potammon at the Synod of Tyre with having escaped martyrdom by sacrificing, but this seems unlikely, unless he had been forced by the soldiers to go through the motions of burning incense (as J.W.C. Wand suggests).

Eusebius was unanimously elected bishop of Caesarea about 314, and in 331 declined the patriarchate of Antioch. At the Council of Nicea* in 325 he led the large moderate party, submitting the first draft of the creed which was eventually accepted after important modifications (notably the *homoousios* clause). He seems to have discovered during the council that Arius's subordinationism was more radical than he had supposed, and he veered toward the Alexandrian position, though he never accepted the extreme views of the Athanasian party which, he believed, tended to Sabellianism.* He presided over the Council of Caesarea in 334 which endeavored to draw Athanasius into negotiation, and took part in Athanasius's condemnation at Tyre (335). On the occasion of Constantine's thirtieth anniversary (335) he delivered at Constantinople an encomium setting forth the political theory which came to be embodied in the Byzantine Empire. He was chief prosecutor of Marcellus of Ancyra* at a synod in Constantinople (336). He was the ecclesiastical and spiritual voice of the Constantine era, and the heir and master of the Origen tradition in that age.

A diverse author, his histories are most notable. First to appear was *Chronicon*, a history of the world to 303 (later to 328); in this he "liberated Christian chronography from the bonds of apocalypticism ... basing it on purely logical foundations" (H. Lietzmann). Best known of all his works is his *Historia Ecclesiastica*, the most important church history of ancient times, invaluable for its wealth of material, much of it preserved here only. The definitive edition in ten books appeared in 325. Apologetic books include *Contra Hieroclem* (against a pagan governor of Bithynia); *Prae-*

paratio evangelica (explaining why Christians accept the Hebrew tradition); *Demonstratio evangelica* (trying to prove Christianity by the OT); and *Theophania* (on the Incarnation). Among other writings are a collection of Origen's letters; a biography of Pamphilus; a *Life of Constantine; De Martyribus Palestinae* (an account of the Diocletianic persecution); *Eclogae Propheticae* (a general elementary introduction); *Contra Marcellum* (against Marcellus of Ancyra); *Onomasticon* (a biblical topography); and commentaries on Psalms and Jeremiah.

BIBLIOGRAPHY: H.J. Lawlor, *Eusebiana* (1912); D.S. Wallace-Hadrill, *Eusebius of Caesarea* (1960); H. Lietzmann, *A History of the Early Church,* vol. III (1961). J.G.G. NORMAN

EUSTATHIUS. Bishop of Antioch 324-c.326. Prior to his elevation to Antioch he had been bishop of Berea and also a confessor. A prominent and eloquent opponent of Arianism at the Council of Nicea* (325), he attracted the opposition of the Eusebians, the more so because of his strong criticisms of Origen* and because after his return to Antioch he refused to accept Arian clergy and entered into a bitter correspondence with Eusebius of Caesarea.* Perhaps as early as 326 his opponents were able to depose him at a synod in Antioch. Various reasons are given for the deposition. Theodoret's suggestion that it was because of immorality with a prostitute seems very unlikely. His sharp tongue, however, makes more probable Athanasius's statement that he was accused of insulting the emperor's mother, Helena. He may also have been charged with Sabellianism.* Constantine banished him to Thrace, and this further suggests nontheological factors. His followers formed the Eustathian sect which survived for some eighty years. His developed Christology is an anticipation of Nestorianism.* His only complete surviving work is a sermon on the witch of Endor (an attack on Origen).

C. PETER WILLIAMS

EUSTATHIUS (c.300-c.377). Bishop of Sebaste. The ascetic practices of Eustathius and his followers were extreme enough to earn the condemnation of several synods, but he was nevertheless sufficiently respected to be elected bishop of Sebaste in Armenia Minor in 356. Famous as an exponent of asceticism, he attracted and became a formative influence in the development of Basil the Great.* He was a prominent member of the Synod of Ancyra (358)—which stood for the Homoiousion* position in the Arian controversy—and was consequently deposed in 360. He was one of the delegates from the East who appealed to the Western emperor Valentinian and Pope Liberius in 366. Later he became a leader of the Macedonian heresy, and his relations with Basil the Great were completely severed.

C. PETER WILLIAMS

EUSTOCHIUM, JULIA (c.370-418). First Roman lady of noble birth to take the vow of perpetual virginity. She was the third of five children to her mother Paula,* the friend of Jerome.* It is from the latter's writings that we gather all that is known of Eustochium. It was to her that Jerome addressed his famous letter in which he discusses the motives that ought to actuate those who devote themselves to a life of virginity, and the rules by which they ought to live. The animosity aroused by this and other letters in which he satirizes Roman society was largely responsible for his departure from Rome in 385. He was followed by Paula and Eustochium, and after joining company in Antioch, they traveled together through Palestine to Egypt, visiting the monks of Nitria and Didymus the Blind.* They returned to Palestine in the autumn of 386 and settled at Bethlehem. A monastery was built of which Jerome was head, and a convent for women of which Eustochium was head from the time of her mother's death in 404. Jerome speaks glowingly of Eustochium's devotion to the ascetic life, to the study of the Scriptures, and to the training of the virgins. He attributes the writing of many of his commentaries to her thirst for knowledge of the Scriptures. DAVID JOHN WILLIAMS

EUTHALIUS. He appears to have been a deacon, devoted to the study of the NT, and is now best known as the supposed author of a collection of editorial material on the NT. This consists of an arrangement of the text in short lines to facilitate its reading aloud; a division of the books into chapters with summary headings of their contents, extended over the Pauline epistles, the Acts, and the Catholic epistles; a table of OT quotations in the epistles; a list of place names at which the epistles were thought to be written; and a list of names associated with Paul's in the headings to the epistles. There is also a lengthy sketch of Paul's life, writings, and chronology, and a brief statement of his martyrdom. It has been argued that Euthalius lived in the seventh century and is identifiable with a bishop of Sulca of the same name. The name "Evagrius" also occurs in some Euthalian MSS. More commonly he is thought to have been a deacon of Alexandria about the mid-fifth century (J.A. Robinson thinks a century earlier). The system is not due entirely to one hand. The Euthalian apparatus seems to have been known fairly early in the library at Caesarea. DAVID JOHN WILLIAMS

EUTHYMIUS ZIGABENUS (eleventh/twelfth century). Byzantine monk, exegete, and theologian. Emperor Alexis Comnenus commissioned him to write a work against heresies; the result was *Panoplia Dogmatica.* This work contains twenty-eight chapters, of which the last six are devoted to the contemporary heretical movements; these chapters are our only sources for some such as the Bogomiles.* In addition, Euthymius wrote extensive commentaries on the Psalms, the four gospels, and the Pauline epistles. Although he depends heavily upon patristic sources, especially Chrysostom, these exegetical works are noteworthy for their hermeneutical approach, especially in the emphasis that Euthymius gives to the literal meaning of the text—and this in an age when allegorical exegesis dominated most commentaries. DONALD M. LAKE

EUTYCHES (c.378-454). Early Monophysite.* After the third Ecumenical Council at Ephesus in 430, Cyril* of Alexandria, worked out a compromise between the theologians advocating two natures of Christ and those holding to one nature. On Cyril's death in 444, however, open opposition broke out against the compromise, with Cyril's successor leading the opposition along with Eutyches, who was an archimandrite of a monastery in Constantinople. Eutyches had come out of retirement to contest the error of Nestorianism toward which he felt the compromise leaned, but went to such an extreme in stressing the single nature of Christ that the supporters of orthodoxy in Constantinople became uneasy. The obstinacy of Eutyches in refusing to recognize the two natures of Christ brought the condemnation of Patriarch Flavian, who declared Eutyches's views unorthodox. Eutyches would not accept this condemnation and maneuvered the bringing about of the scandalous Robber Synod of Ephesus* (449) to support his views.

GEORGE GIACUMAKIS, JR.

EUTYCHIANISM, see MONOPHYSITISM

EVAGRIUS (c.536-600). Eastern church historian. Born probably in Syria, he seems to have been a lawyer by profession. His history of the church begins where Eusebius's *Ecclesiastical History* stops, with the Council of Ephesus (431), and brings the account of the church to his own lifetime, about 590. His six books contain both valuable and now nonexistent source materials as well as worthless legends. He generally reflects the speculative theology of the period and especially the curious interest in the miraculous. His ecclesiastical history can be found in England in Bohn's Ecclesiastical Library (1854), pp. 251-467; and in the Greek text in J.P. Migne, *Patrologia Graeca*, vol. 86:2, pp. 2415-2906.

DONALD M. LAKE

EVAGRIUS PONTICUS (345-399). Eastern writer. Born at Ibora in Pontus Galaticus, he was ordained reader by Basil and deacon by Gregory of Nazianzus at Constantinople. He attended the council in 381, and on Gregory's departure from Constantinople remained to assist the new bishop, Nectarius, in dealing with theological questions. Because of a developing relationship between himself and a married woman, Evagrius left Constantinople for Jerusalem. There he was influenced by Melania to adopt the ascetic life. At her instigation he went to Egypt, where he practiced and taught the ascetic life in Nitria and Cellia, north of Nitria, until his death. He numbered among his pupils Palladius, Rufinus, and Heraclides of Cyprus, later bishop of Ephesus.

It seems probable that John Cassian* met Evagrius during his visit to Egypt, and certainly Cassian's own writings on monasticism reveal the influence of Evagrian ideas. Although his extant works show little evidence of Origenism,* he was condemned by Jerome for this and for his association with Melania and Rufinus. Jerome speaks contemptuously of his writings, especially his book *Peri apatheias*, when combating the tenet ascribed to the Origenists that a man can rise above temptation and live without sin. In the same context Jerome refers to another work by Evagrius on monks, but as this book is not referred to by anyone else, Jerome may have wrongly attributed Palladius's *Historia Lausiaca* to Evagrius. The latter's works are largely lost or extant only in Latin or Syriac translations. They include works on the ascetic and spiritual life and commentaries on the Psalms and Proverbs.

For a list of Evagrius's known writings see H. Wace and W.C. Piercy, *A Dictionary of Christian Biography and Literature* (1911). His extant works are given by J.P. Migne, *PG*, XL, pp. 1213-86, also LXXIX, pp. 1165-1200 (*De Oratione* of "Nilus of Ancyra"). DAVID JOHN WILLIAMS

EVANGELICAL. The term means pertaining to the Gospel (as expounded by the four gospels) or conforming to the basic doctrines of the Gospel (as enunciated by the NT as a whole). By extension it signifies one who is devoted to the Good News—or "Evangel"—of God's redemptive grace in Jesus Christ. The Apostle Paul summarizes the Christian evangel in 1 Corinthians 15:1-4. There he affirms, as the central preaching-content of the primitive missionary churches, that Jesus Christ died for our sins and was buried and rose the third day, and was seen, and that all this eventuated in fulfillment of the prophetic-scriptural disclosure of God's gracious salvational purpose to provide redemption for sinful man.

In its secular Greek sense the word *euaggelion* could refer not simply to news or ordinary events, but could be used even of a false story of victory fabricated in wartime to boost military morale. But the Word-Event Jesus Christ—His incarnation, teaching, death, resurrection, and exaltation—particularized *euaggelion* as "good news." Related terms depict the messenger or bearer *(euaggelos)* of these good tidings, and the evangelist, one who proclaims the good news, designated by the rare word *euaggelistēs* which occurs three times in the NT (Acts 21:8; Eph. 4:11; 2 Tim. 4:5).

In subsequent Christian history a distinction evolves between "evangelical" and "evangelistic," the former stipulating conformity to the fundamental facts and truth of Christianity, the latter designating a sense of missionary compassion and urgency. But primitive Christianity had no category of believers who were not at the same time missionary-minded. Nor was Christian evangelism compatible with defection from the truth of revelation. To deny the vicarious death and historical resurrection of Jesus Christ is to forfeit the Gospel and the central theme of Christian faith and preaching, the exclusive sufficiency of Christ and His work for our salvation.

The term "evangelical" therefore categorizes a commitment, not a negation or divisive attitude. Its original content is supplied by the apostolic preaching, at first in vocal and then in written form, so that the substance of the Good News is conveyed by the gospels and in the NT as a whole. Evangelical Christians are thus marked by their devotion to the sure Word of the Bible; they are committed to the inspired Scriptures as the divine rule of faith and practice. They affirm the fundamental doctrines of the Gospel, including the incarnation and virgin birth of Christ, His sinless life, substitutionary atonement, and bodily resurrection as the ground of God's forgiveness of sinners, justification by faith alone, and the spiritual regeneration of all who trust in the redemptive work of Jesus Christ.

CARL F.H. HENRY

EVANGELICAL ALLIANCE. Formed in 1846 after Christian leaders had felt the need to present a more united front in the face of political upheaval in Europe. It was stressed at the inaugural conference in London that those present had met "not to create Christian union, but to confess the unity which the Church of Christ possessed as His body." Those who formed the Alliance declared they were "deeply convinced of the desirableness of forming a Confederation on the basis of great evangelical principles held in common by them, which may afford opportunity to members of the Church of Christ of cultivating brotherly love, enjoying Christian intercourse and promoting such other objectives as they may hereafter agree to prosecute together; and they hereby proceed to form such a Confederation under the name of the Evangelical Alliance." They drew up a basis of faith expressing their convictions as evangelical Christians.

One of the first difficulties encountered was a difference of opinion within the ranks regarding the rights and wrongs of slavery. The progress of the Evangelical Alliance during the nineteenth century was significant. Great and inspiring conferences were held in most of the capital cities of Europe and in America, and the Alliance quickly established itself as a body worthy of respect in the religious world. In the course of the first century of its existence the Alliance concentrated its attention on a number of different projects, including the relief of persecuted Protestant minorities, the promotion of a united week of prayer throughout the world during the first full week of January, the defense of biblical Christianity, and the promotion of missionary work.

Between the wars the Alliance went through a somewhat quiescent period, but blossomed into new life after World War II. The first notable postwar project it sponsored was the United Evangelistic Exhibition in the Central Hall, Westminster, in 1951, which coincided with the Festival of Britain and which enjoyed the support of 180 different societies. In 1952 the Alliance opened its first hostel for overseas students in central London; another was opened in 1963. It sponsored also the crusades led by Dr. Billy Graham in 1954-55 and again in 1966-67. Thereafter the Alliance embarked on the work of film evangelism, which brought the Gospel message to, among others, numerous prisoners and to members of the armed forces. Another outcome of the crusades was the Alliance's organizing of ministers' conferences in which ministers of different denominations have been encouraged in the work of evangelism in their own locality. Yet another Alliance project was the launching of the religious monthly *Crusade.* This has proved widely acceptable to the Christian public, particularly to young people.

Other notable developments during the postwar years include the formation of the Evangelical Missionary Alliance (1958) which links together almost all the evangelical missionary societies, whether denominational or interdenominational, and the holding of two united Communion services in London's Royal Albert Hall. There have also been several National Assemblies of Evangelicals, when delegates from churches and societies affiliated with the Alliance have met together to discuss matters of vital concern. Reports on such subjects as evangelism, the missionary task of the church, and church extension in new housing areas have been produced for consideration at these assemblies. Probably one of the most publicized of the Alliance's recent activities has been the launching of the Relief Fund (TEAR), which provides a channel whereby evangelical Christians are able to send gifts for relief work in particularly needy areas of the world.

Though the Alliance appears to have embarked on numerous projects, its real *raison d'être* has remained unaltered—fellowship in the Gospel. The Alliance has always stressed that evangelical Christians should enjoy such fellowship regardless of any denominational allegiances which they may have. When the Alliance was founded, membership was on an individual basis and it remained so for many years. Until 1912, prospective members were required to assent to the full doctrinal basis of the Alliance as agreed upon at its inception. In that year, however, the council opted for a simplified form as follows: "All are welcomed as members of the Evangelical Alliance (British Organization) who acknowledging the divine inspiration, authority and sufficiency of the Holy Scriptures, believe in One God—the Father, the Son, the Lord Jesus Christ our God and Saviour who died for our sins and rose again; and the Holy Spirit by whom they desire to have fellowship with all who form the One Body of Christ." In 1970 the doctrinal basis was revised and expressed in terminology more appropriate to the times without in any sense departing from its traditionally conservative evangelical position. Membership of the Alliance is now open to local evangelical fellowships, societies, denominations, and individual churches in agreement with the basis of faith and with the aims and objects of the Alliance. Those who attend national assemblies do so as delegates from different societies and churches.

The Evangelical Alliance was one of the founder members of the World Evangelical Fellowship* formed in 1951. Through this means it enjoys fellowship with similar bodies across the world such as the National Association of Evangelicals* in the United States, the Evangelical Fellowship of India,* and the various Alliances in Europe. The revitalization of the Alliance after World War II coincided with the rise of the ecumenical movement and the formation in 1948 of the World Council of Churches. From the outset, evangelical Christians have not spoken with one voice about their attitudes toward the WCC, and this fact has been reflected in tensions within the Alliance at various times over this issue. The Alliance has consistently adopted the policy of seeking to unite all evangelical Christians regardless of their denominational affiliations. This policy has not proved acceptable to some, with the result that the British Evangelical Council has tended to attract those who wish to have no connections, directly or indirectly, with the WCC or with denominations affiliated with it, while the Alliance covers a somewhat wider spectrum.

BIBLIOGRAPHY: J.W. Ewing, *Goodly Fellowship* (1946); J.E. Orr, *The Second Evangelical Awakening in Britain* (1949); J.B.A. Kessler, *A Study of the Evangelical Alliance in Great Britain* (1968). GILBERT W. KIRBY

EVANGELICAL AND REFORMED CHURCH. An American Protestant denomination formed by the merger of the Evangelical Synod of North America with the Reformed Church in the United States in 1934. The union brought together Lutheran and Reformed Christians of predominantly German, Swiss, and Hungarian ancestry. At the time of the merger the denomination had 2,648 pastors, 2,929 congregations, and 631,271 communicant members. The new church adopted as its doctrinal standards the Augsburg Confession, the Heidelberg Catechism, and Luther's Catechism, allowing freedom of interpretation where those standards differed, and subjecting every theological judgment to the norm of the Word of God. Because of its own history and theological heritage, the church was concerned to promote ecumenical relations and the reunion of separated churches. It supported hospitals and homes, educational institutions, and missionary work in Africa, China, Honduras, India, Iraq, Japan, and South America. In 1940 the body began merger negotiations with the Congregational Christian Church, which led in 1957 to the creation of the United Church of Christ. DAVID C. STEINMETZ

EVANGELICAL ASSOCIATION, see EVANGELICAL CHURCH (ALBRIGHT BRETHREN)

EVANGELICAL CHURCH (Albright Brethren). An American Protestant denomination founded by Jacob Albright (1759-1808), a Pennsylvania tilemaker and farmer. Following his conversion to evangelical Christianity in 1791, Albright, though raised as a Lutheran, associated himself with a class meeting of the Methodist Episcopal Church and was licensed as a lay preacher. In 1796 he undertook a preaching mission in German throughout E Pennsylvania. Though he and his followers were on friendly terms with the English-speaking Methodists led by Francis Asbury,* the language barrier made it necessary for the Evangelicals to create their own independent organization.

Stressing a personal and experiential relationship with God, the Evangelicals held their first council in 1803. The first annual conference of preachers was in 1807, and a book of *Discipline* was adopted in 1809. In 1816, eight years after the death of its founder, the first general conference of the "so-called Albright People" named its new denomination the Evangelical Association. In 1891 controversies led to a schism and the birth of the United Evangelical Church (1894). In 1922 the two groups were reunited in the Evangelical Church. Negotiations with another Wesleyan denomination of predominantly German background, the United Brethren in Christ,* led in 1946 to the creation of the Evangelical United Brethren Church.* In 1968 this body merged with the Methodist Church to form the United Methodist Church,* healing the division caused by the old language barrier and bringing together into one body the church of Francis Asbury and the church of Jacob Albright.

BIBLIOGRAPHY: W.W. Orwig, *History of the Evangelical Association* (1858); R. Yaekel, *History of the Evangelical Association* (2 vols., 1892-95); R.W. Albright, *A History of the Evangelical Church* (1942). DAVID C. STEINMETZ

EVANGELICAL COVENANT CHURCH OF AMERICA. Founded in Chicago in 1885, it traces its origins to the Reformation, biblical instruction in the Lutheran State Church of Sweden, and the awakenings of the nineteenth century. It took the name Evangelical Mission Covenant Church of America until 1957. It has traditionally cherished the historic confessions and creeds, but recognizes the sovereignty of the Word of God over their interpretations. The constitution states that "the principle of personal freedom, so highly esteemed by the Covenant, is to be distinguished from the individualism that disregards the centrality of the Word of God and the mutual responsibilities and disciplines of the spiritual community." It allows divergent views of baptism, though traditionally it is paedobaptist. Its chief institutions are North Park College and Seminary in Chicago, and it has an extensive missions outreach. Membership in the mid-1970s has been about 70,000 in more than 500 churches.

EVANGELICAL FELLOWSHIP OF INDIA, see INDIA, EVANGELICAL FELLOWSHIP OF

EVANGELICAL FOREIGN MISSIONS ASSOCIATION. Both denominational and nondenominational, this American association has sixty-four members representing nearly 7,000 missionaries in 120 fields. It was organized in Chicago in 1945 "to provide a medium for voluntary united action among the evangelical foreign missionary agencies." It represents the missions before governments through its Washington, D.C. office and conducts annual retreats for mission executives where subjects of common interest are discussed. In cooperation with the Interdenominational Foreign Mission Association* it has sponsored the Wheaton Congress of 1966, the Summer Institute of Missions at Wheaton College, the Evangelical Missions Information service with news releases and the *Evangelical Missions* quarterly, the Committee to Assist Missionary Education Overseas, and area committees such as those on Latin America, with its congresses on communications and evangelism, and Africa, which initiated the Association of Evangelicals of Africa and Madagascar. HAROLD R. COOK

EVANGELICAL FREE CHURCH OF AMERICA. This body developed from the seventeenth-century Pietistic free church movement in Scandinavia which was carried to the United States by immigrants from 1870 to 1914. The Midwest churches coalesced to form the Evangelical Free Church (Congregational) and the eastern churches to form the Eastern Association in 1891 for fellowship and mutual aid. These two Norwegian-Danish groups merged to form the Evangelical Free Church in 1909. The Swedish Evangelical Free Church, organized at Boone, Iowa, in 1884, merged with the Norwegian-Danish group at Medicine Lake, Minn. in June 1950 to form the Evangelical Free Church of America. Trinity College and Seminary at Deerfield, Illinois, are its major educational centers, and it carries on an extensive missions program. In 1970 it had a membership of 59,041 in 539 churches with 762 ordained clergymen. EARLE E. CAIRNS

EVANGELICALISM. A term in common use only in the twentieth century, used to describe the international movement which is committed to the historic Protestant understanding of the Evangel. Its adherents should be distinguished from those of three other broad groupings within professing Christianity: nonevangelical Protestantism; Catholicism; and the so-called sects and cults. Evangelicalism has become the defender of the historically orthodox Protestant theologies (and their subsequent variations) and the underlying biblical exegesis; as a result, some have labeled the movement "Conservative Evangelicalism."

Because of its emphasis on personal commitment (rather than comprehension of all of a given population) and acceptance of the Bible as the basis for its authority (rather than institutional bishops in supposed apostolic succession), Evangelicalism has remained clearly distinct from Catholicism, both Roman and Orthodox, despite their common Trinitarian supernaturalism in the face of naturalistic trends in Protestantism. Evangelicalism's acceptance of historic Trinitarianism, however, distinguished it from various non-Protestant sectarian movements—Mormonism, Christian Science, Jehovah's Witnesses, etc.

Since it is usually missionary-minded, Evangelicalism is found almost everywhere in the world. Its manifestation is primarily to be found within the histories of the various Protestant denominational families, chiefly Lutheran, Anglican, Reformed (Presbyterian and Congregational), Mennonite (Anabaptist), Baptist, Quaker, Moravian, Dunker Brethren, Wesleyan (including parallel movements among non-English-speaking people), Plymouth Brethren, Campbellite, Adventist, Pentecostal, Bible Churches, and some of the Third World denominations rising indigenously or resulting from transdenominational missions. The diversity is best accounted for by the differences in time, place, and context of independently originating evangelical awakenings which become institutionally self-perpetuating. Amid all the organizational confusion, evangelicals recognize each other by the common message of eternal salvation which they proclaim. They work also in many nondenominational enterprises—faith missions, Christian education, Evangelical Alliance, world congresses on evangelism, etc.

See A.S. Wood, "Evangelicalism: a Historical Perspective," *Theological Students' Fellowship Bulletin*, 60 (Summer 1971), pp. 11-20.

DONALD TINDER

EVANGELICAL UNITED BRETHREN CHURCH. This American Protestant denomination was formed in 1946 by the union of the Church of the United Brethren in Christ* with the Evangelical Church.* Both these churches originated among the German-speaking people of Pennsylvania and Maryland during the Second Great Awakening.* Both groups had their roots in German Evangelical Pietism, were Arminian in theology, and had an episcopal form of polity similar to that of Methodism. The Church of the United Brethren in Christ arose in 1800 from the impact of the revivalist preaching of P.W. Otterbein* and Martin Boehm.* The Evangelical Church was founded in 1807 by Jacob Albright.* Only their usage of the German language kept these two groups from joining with the Methodist Church during the nineteenth century. In 1968 the Evangelical United Brethren Church merged with the Methodist Church to form the United Methodist Church.* HARRY SKILTON

EVANGELISCHE BUND, DER. This "evangelical league" was an alliance of German Protestants founded in 1886-87 by Willibald Beyschlag* and others. Its purpose was to defend Protestant interests against the growing power of Roman Catholicism. It issued its own literature and furthered its purpose by other methods such as making good use of contacts with the secular press. In the years preceding World War I the movement took on a strongly nationalistic emphasis. By 1914 it had gained 540,000 members. Since that time, however, its influence has declined considerably.

EVANGELISM-IN-DEPTH. A cooperative mass evangelistic effort. It originated in Latin America and later extended in different forms to other parts of the world. It was started by R.K. Strachan* of the Latin America Mission. After some years of study and observation of apparently successful movements, he developed the theorem: "The growth of any movement is in direct proportion to the success of that movement in mobilizing its total membership in the constant propagation of its beliefs." Basically E/D attempts to make a Christian impact on a single country in an all-out, yearlong campaign that mobilizes the whole Christian constituency of that country under local leadership. Local committees make preparations, organizing multitudes of neighborhood prayer-cells and programs of visitation, and setting up training classes for witnesses. Communitywide evangelistic services culminate in a great rally and parade in a principal city for maximum impact on the country. The first campaign in Nicaragua (1960) was followed by numerous others.

HAROLD R. COOK

EVANGELIST. One who proclaims the Gospel ("good news") or Evangel. The task of proclamation was committed by Christ to the apostles as representatives of the church throughout the entire Christian era (Matt. 28:18-20; Mark 16:15). The NT references to evangelists (Acts 21:8; Eph. 4:11; 2 Tim. 4:5) are therefore to be understood of those who are divinely gifted specialists in the work to which the entire church is called. The term has perhaps wrongly been employed sometimes of those who do virtually the work of a pastor but with a lower status, usually on educational grounds. In fact, in the NT the roles of evangelist and pastor seem to be distinct but related, one being a "fisher of men," the other a shepherd of Christ's flock. The use of the term in reference to the gospel writers dates from the close of the second century. Modern emphasis on the Gospel or *kerygma* as basic to the whole NT and on the four gospels as detailed expansions of this Gospel has served to show the appropriateness of this derivative use of the term. G.W. GROGAN

EVANS, CHRISTMAS (1766-1838). Welsh Baptist preacher. Born at Esgairwen, Llandysul, Cardiganshire, on Christmas Day, son of a cobbler, he served as a farmhand in the neighborhood, and the only education he enjoyed was informal instruction by the well-known schoolmaster David Davis, Castellhywel (1745-1827), and a short period at his school. Evans joined the Baptist church at Aberduar. He went as a missioner to Caernarvonshire and was ordained in 1789. From 1791 to 1826 he ministered in Anglesey, and after two brief periods at Caerphilly and Cardiff he returned north to Caernarvon in 1832 and spent his remaining years there. He died in Swansea while on a journey, and is buried there in Bethesda churchyard.

Together with John Elias* and William Williams* of Wern, he is enshrined in Welsh tradition as one of the three greatest figures in the history of the nation's preaching. Like Elias, his formal education did not amount to very much, and he was virtually self-educated. This is not without significance for the historian seeking to demonstrate how Nonconformity* developed into a large-scale working-class movement in Victorian Wales. These men were exceptionally able communicators with the largely uneducated public of the period.

Evans's great strength as a preacher lay in his oratorical imagination. To call it baroque would be no exaggeration. His sermons on such themes as the Prodigal Son or the Last Judgment became existential dramas of the most poignant kind at his hands. This method of preaching he had learned from Robert Roberts of Clynnog (1762-1802), and Roberts was (indirectly) indebted to George Whitefield. Evans's preaching was inspired by a profound personal godliness and a "passion for souls." Although he was a busy writer of theological pamphlets, he tended to be somewhat wayward in his theological opinions.

R. TUDUR JONES

EVANS, JAMES (1801-1846). Missionary and linguist. He entered Canada as a Methodist missionary from England in 1823, and in 1828 began teaching at the Rice Lake Indian Mission School in Upper Canada. Ordained a Methodist minister in 1833, he went to the Ojibwa Indians, and in 1837 published an Ojibwa grammar and translated biblical extracts and some hymns into Ojibwa. In 1840 he became general secretary of all the Wesleyan Missionary Society's Indian Missions in the northwest. Extensive travel from Norway House made him realize the need for a written Cree language. After inventing a Cree syllabic alphabet in 1840, he published a *Cree Syllabic Hymnbook* in 1841 and organized a group of translators who by 1861 had translated the Bible into Cree. Opposition from the Hudson's Bay Company and false charges led to his recall to England and death in 1846.

ROBERT WILSON

EVANS, OWEN (1829-1920). Welsh divine. Born in Penybontfawr, Montgomeryshire, of the same family stock as Ann Griffiths,* his only education was at elementary school. He served as minister at several Congregational churches, and finally at King's Cross (Welsh) Tabernacle (1881-1906). He was chairman of the Union of Welsh Independents in 1887. Apart from being a successful pastor, he was a prolific author. He specialized in books of a popular nature dealing with biblical themes and written in a clear and interesting style. Some of these had a substantial circulation. Throughout his life he was an assiduous defender of the Reformed faith against the attacks of modernism and liberalism, yet was his stand in no way weakened by the tendency of liberals in theology to poke fun at him as "an old fossil." He represents one of the connecting links in Wales between the Calvinism that was in decline by the end of the last century and the biblical theology of our own day.

R. TUDUR JONES

EVANSTON ASSEMBLY (1954). The second international meeting of the World Council of Churches, held at Northwestern University, Evanston, Illinois; 132 member denominations were represented at the gathering, the theme of which was "Christ—The Hope of the World." The program was divided into four parts: (1) the theme was presented and debate focused on varying views of eschatology, the part that evangelism of the Jews plays in hastening the return of Christ, and the relationship of the other-worldly to this world's problems; (2) the assembly was split up to study the six subthemes, the most controversial of which were those on the evangelizing church and on racial and ethnic tensions. The group on the evangelizing church recommended day schools as a means of providing Christian nurture, and that on racial tensions condemned segregation, urging member churches to renounce it (it also condemned anti-Semitism); (3) structural changes in the WCC were considered; (4) routine business was undertaken.

If many agreed that there ought to have been fewer plenary sessions and more time for personal interaction, it was because the machinery was so complex that some felt the Holy Spirit was hampered by seemingly insurmountable restrictions.

Many also agreed that the concerns of the assembly were generated from the top down and that some delegates, especially non-English-speaking lay people, had difficulty sharing the concerns. A hint of the machinery may be seen in the fact that six-and-a-half tons of mimeograph paper were used to print official news releases and to record assembly speeches. Optimism about the assembly lay in the fact that Christians of a variety of backgrounds were listening to one another and interacting over differences, including theological ones, and that even without agreement on theological issues they could carry out a program of world relief and refugee help.

As council presidents for the ensuing seven-year period the assembly elected Dr. John Baillie of Scotland, Bishop Sante Barbieri of Argentina, Bishop Otto Dibelius of Germany, Metropolitan Juhanon of the Mar Thoma Syrian Church, Archbishop Michael (Greek Orthodox) of New York City, and Bishop Henry Knox Sherrill of the USA.

See J.H. Nichols, *Evanston. An Interpretation* (1954).

ROBERT B. IVES

EVELYN, JOHN (1620-1706). Diarist and author. Fourth son of a landed gentleman in Surrey, he remained a pious Christian, loyal Anglican, and Royalist all his life. In the Commonwealth period he refused office, often traveling abroad. After the Restoration he enjoyed Charles II's confidence, but was assigned difficult and unremunerative appointments such as the care of prisoners. Evelyn consistently sought the good of the realm, suggesting remedies for air pollution, dirt in streets, traffic congestion, deforestation of the land, and other things. He held it is man's duty to study God's handiwork in nature. He was a co-founder of the Royal Society. "Whatever you love besides God only, pray you may not love too much" summarizes his attitude to life. His *Diary* (6 vols., ed. E.S. de Beer) was published in 1955. His own books number more than forty.

See biography by F. Higham (1968).

R.E.D. CLARK

EVENING PRAYER; EVENSONG. Evensong was the medieval English name for Vespers,* one of the two evening hours of prayer. Cranmer retained the title for his reformed daily office in the 1549 Prayer Book, changing it to Evening Prayer in 1552. His service made use of elements from both Vespers and Compline,* including recitation of the psalter, and the canticles *Magnificat* and *Nunc dimittis,* which in the Prayer Book follow readings from OT and NT. A penitential introduction was added in 1552, and further prayers (known as the State Prayers) at the end in 1662, following an anthem "in quires and places where they sing." It is now the practice on Sundays to supplement the service with additional hymns and intercessions and a sermon.

JOHN TILLER

EVERARD, JOHN (c.1575-c.1650). English clergyman. Cambridge D.D. and excellent preacher, his preaching on the unlawfulness of matching with idolaters was taken as criticism of the projected Spanish marriage, and he several times went to prison (once for six months) until he apologized. Meanwhile, as reader in St. Martin's-in-the-Fields, he attracted fashionable congregations, although he averred that his sermons were aimed at his less sophisticated hearers. By 1636 he had a parish in Essex, but later summoned before the Court of High Commission for heresy, he was found not guilty, yet was later deprived of his benefice, and in 1639 was fined £1,000. He was no favorite of Archbishop Laud,* who "threatened to bring him to a morsel of bread because he could not make him stoop or bow before him." Those of Everard's sermons that escaped confiscation were published in 1653 as *Some Gospel Treasures opened.* Their mystical flavor suggests the influence of John Tauler.*

J.D. DOUGLAS

EVOLUTION. A theory of organic development by natural processes of descent, in which modifications are selected by environment. Ancient Greece speculated upon the origin of man and animals, and Aristotle (382 B.C.) made a classification. Some in the early church, influenced by Genesis 1, speculated on the progression of living things. Julian (A.D. 331) held that man had been modified by soil and climate. Augustine (354) believed creativity operated within matter over long ages. Speculation was renewed in the seventeenth century. Linnaeus, son of a Swedish pastor, introduced modern taxonomy (1735) and believed he was cataloging God's creation. The problem of the mechanism of transformation occupied the eighteenth century. Buffon thought it was by environment perpetuated by heredity. Erasmus Darwin (grandfather of Charles) and Lamarck postulated the inheritance of acquired characteristics.

Many distinguished clergy were evolutionists. Charles Darwin* felt indebted to Thomas R. Malthus, who wrote on the struggle for existence. It was a clergyman, also, who recommended Darwin for two appointments. The agnostic T.H. Huxley saw the advantage of championing Darwin's *Origin of Species* (1859). Cuvier opposed this, arguing that the fossil record showed catastrophes followed by re-creations.

Christian reactions varied. Orthogenesis received wide support, seeing evolution as directed by an internal force. Later Ambrose Fleming founded the Evolution Protest Movement, which became a useful critique of plausibility. The genetics of G.J. Mendel, the Austrian Augustinian monk, ignored until 1900, showed that acquired characteristics could not be inherited. Evolution was soon extended to culture, morals, sociology, and religion. Bible Christianity suffered harm by Julius Wellhausen's* reshuffle of Scripture to present an evolution of religion from animism to polytheism to monotheism, instead of one of God's revealing Himself to man. Anthropologists and theologians have long since abandoned these concepts of Wellhausen, J.G. Frazer, and Herbert Spencer.

Evidence adduced for evolution includes comparative anatomy, fossil succession, ramifications of like organisms, and geographical distribution. Contrary to evolution are the inability of mutations to produce higher orders of life, and the lack of successive bridging fossils between the main orders (they have major organic differences).

Attention has recently shifted to molecular biology and life's origins. Previously the cell was regarded as simple (Teilhard de Chardin thought this), but the smallest viable life unit is more complex than any manmade machinery.

E.K. VICTOR PEARCE

EWALD, GEORG HEINRICH AUGUST VON (1803-1875). German biblical scholar. Born in Göttingen, he studied there under J.G. Eichhorn* and thus was put in touch with the first generation of modern critical studies of the OT. His interest in the poetic and prophetic materials of the OT reveal strong reactions to the romantic interpretations of J.G. Herder* and Eichhorn, just as his conservative historical bent made him an opponent of F.C. Baur* and D.F. Strauss.* Ewald succeeded Eichhorn at Göttingen, and among his pupils were C.F.A. Dillmann* and Julius Wellhausen.* Ewald's place in the history of biblical scholarship is fixed by his initiating role in two of its major dimensions: Semitic linguistics in a historical vein (from 1827) and the history of Israel as *Volk* (from 1843). Political views forced him to leave Göttingen in 1837, and for ten years before he could return he taught at Tübingen. In later life he was deeply involved in political affairs.

CLYDE CURRY SMITH

EXARCH. In Homeric Greek, the one who takes the lead or makes a beginning, presumably militarily, could be called *exarchos,* but until the Christian era the term seems to have disappeared. Plutarch called the *pontifex maximus* by the term *exarchos tōn hiereōn.* Synodical canons from the fourth century used the term interchangeably with "metropolitan,"* or provincial bishop (see EPARCHY), and from the fifth century the term could be applied to the supervisor of a council or of a monastery (see ARCHIMANDRITE). In the Byzantine ecclesiastical state, "exarch" can designate also a viceroy with civil and military powers, or an archdeacon, as well as the founder of heresy.

EX CATHEDRA. The Greek noun *hedra* from Homer, *kathedra* from Thucydides, designates a chair or seat. In the Roman architectural form—the basilica, built with or without permanent seats, *exedras,* and used for the operation of government or court—a moveable chair could be brought in for the presiding officer. The chair, like the more specialized "throne," came to represent the authority of the office. In the church secular it is the seat of the bishop, in the church regular that of the abbot, who are understood when seated to preside; the cathedral is simply the place of the chair. The notion *ex cathedra,* "out of the chair," conveys the voice of authority or its codified or written pronouncements. In the Roman Catholic Church at Vatican Council I the status of the bishop of Rome was confirmed in that, when speaking as the successor of Peter, his words concerning faith and morals had the infallible character of apostolic doctrine.

CLYDE CURRY SMITH

EXCEPTIONS, THE. The objections to the existing Prayer Book (that of 1604) made by the Presbyterian Puritans (e.g., R. Baxter) in 1661 after the restoration of Charles II. They were made at the Savoy Conference* which met that year. The Presbyterian ministers composed a document, *The Exceptions against the Book of Common Prayer,* to which the bishops replied in *The Answer of the Bishops.* The *Exceptions* falls into two parts: a tabulation of the historic Puritan objections to the Prayer Book and religious ceremonies (e.g., no lessons to be read from the Apocrypha, and saints' days to be abolished), and a catena of critical comments and suggested amendments to parts of the Book. Among the latter was the request that the famous Black Rubric (Declaration on Kneeling of 1552) be restored, and that modifications be made in the Communion and Baptismal Services. Very few of the exceptions were in fact taken notice of and incorporated in the 1662 Book of Common Prayer.

PETER TOON

EXCOMMUNICATION. This involves varying degrees of exclusion from the community of the faithful because of error in doctrine or lapse in morals. The term *excommunicatus* first appears in ecclesiastical documents in the fourth century. Discipline in the primitive church followed the Jewish model; cf. the threefold warning recommended for an offending brother in Matthew 18:15-17 (privately, before two or three witnesses, before the whole assembly), which conforms to Jewish practice.

The origin of excommunication in Christian terms is normally traced to the saying of Jesus about "binding and loosing" in Matthew 16:19 (to Peter) and 18:18 (to the disciples; cf. John 20:23). Even if such legislation were relevant at the time when the Evangelists wrote, there is no need to regard it as post-Easter invention. Paul advocates degrees of sanction to deal with offenders in the church, ranging from social deprivation (2 Thess. 3:10,14f.) to full exclusion from the community (1 Cor. 5:13; cf. v. 5 and 2 Cor. 2:5-11). The punishment in this case was the responsibility of the whole assembly (1 Cor. 5:4) and intended for the good of both the offender and the church (vv. 5-7; cf. 1 Tim. 1:19f.). With the growth of the church, the problem of the authority to excommunicate also arose (cf. 3 John 9f.).

In the primitive Christian community, excommunication as such ("hand over to Satan," 1 Cor. 5:5) implied complete isolation from the faithful. By the fifteenth century, a distinction had been introduced between excommunicates who were to be shunned for gross error (the *vitandi*) and those to be tolerated (the *tolerati,* who were rigidly excluded only from the sacraments). This distinction still operates in the Roman Catholic Church. In modern Protestant circles, despite the Anglican canons, formal excommunication is rarely imposed.

See also ANATHEMA, DISCIPLINE, and HERESY.

STEPHEN S. SMALLEY

EXCOUNTIANS, see ANOMOEANS

EXEMPLARISM (Lat. *exemplum,* "pattern or example"). The theological doctrine that finite things are copies of originals existing in the divine mind. Plato's archetypal ideas or forms were located by later Greek Platonist philosophers in the divine mind. Christian philosophers (Augustine and his school) made this a central point of their metaphysics and used the theory to distinguish God's free and intelligent creation from that of purely spontaneous production. The theory was prominent until its peak in the thirteenth century with Bonaventura.* The term "exemplarism" is sometimes used also to describe the "moral influence" view of the Atonement, originating with Abelard and developed by Hastings Rashdall.*

HOWARD SAINSBURY

EXISTENTIALISM. The term has been defined as an attempt to philosophize from the standpoint of the actor, rather than, as in classical philosophy, from that of the detached spectator. The word derives from the German *Existenzphilosophie.* The movement grew in Germany after World War I and flourished in France from the time of World War II. It is best regarded as indicating an approach rather than a fixed body of philosophical doctrines. Its leading exponents have tended to coin their own vocabulary of technical terms and to develop their thought in their individual ways. Some are atheists, others profess Protestant or Catholic faith. The existentialist protest against philosophical systems has in the hands of some of its advocates been transformed into highly elaborate systems.

The origins of existentialism have been traced back to S. Kierkegaard* with his attack on absolute idealism and concern for individual existence, to the atheism of F.W. Nietzsche,* and to the disenchantment of F.M. Dostoevsky* with rationalistic humanism and his saying: if God did not exist, everything would be permitted. The premise of atheistic existentialism is that God does not exist, therefore man must fend for himself. He must work out his own values and create his own existence. At the same time he has a sense of the absurdity of it all. The choices he has to make are often impossible, giving rise to profound anxiety.

Existentialism represents a revolt against external authority, ready-made world views, authoritarian and conventional moral values and codes of conduct. Man has been dumped into the world whether he likes it or not. He has to make his own way in it, creating his own values and determining his existence as he goes along. It is this which distinguishes man from things and animals. But if he refuses he relapses into the kind of existence that things have, instead of living an authentic human existence.

In Germany, Karl Jaspers* and Martin Heidegger are the best-known existentialist thinkers. The latter's *Being and Time* (1927; ET 1962) was a seminal work which set out what was virtually an existentialist metaphysics. Leading French existentialists have not only written philosophical treatises, but have expressed their ideas in plays and novels. Among them are the communist Jean-Paul Sartre, the atheist Albert Camus, and the Catholic Gabriel Marcel. Sartre's philosophical studies include *Being and Nothingness* (1943; ET 1957) and *Existentialism and Humanism* (1946; ET 1948).

In his demythologizing program, Rudolf Bultmann* has made use of Heidegger's existentialism, interpreting the gospel of the death and resurrection of Christ as a challenge to men to choose between authentic and inauthentic existence. In their different ways Paul Tillich* and John Macquarrie have combined existential analysis with ontological speculation in an attempt to provide a new metaphysical basis for interpreting the Christian faith.

Although existential analysis has yielded rewarding insights into aspects of human existence, the speculative systems have been sharply criticized from the standpoint of linguistic analysis for category mistakes in the use of language. The attempt to restate Christian belief in existentialist terms has tended to eliminate the transcendent and divine personal element of biblical theism.

BIBLIOGRAPHY: H.J. Blackham, *Six Existentialist Thinkers* (1952); J.M. Spier, *Christianity and Existentialism* (1953); F.H. Heinemann, *Existentialism and the Modern Predicament* (1954); J. Macquarrie, *An Existentialist Theology* (1955), *The Scope of Demythologizing* (1960), *Studies in Christian Existentialism* (1966), and *Existentialism* (1971); W. Kaufmann (ed.), *Existentialism from Dostoevsky to Sartre* (1956); F. Copleston, *Contemporary Philosophy* (1956); W. Barrett, *Irrational Man* (1961); M. Warnock, *The Philosophy of Sartre* (1965); S. Keen, *Gabriel Marcel* (1966); C. Brown, *Philosophy and the Christian Faith* (2nd ed., 1971).

COLIN BROWN

EXORCISM. The practice of expelling evil spirits by means of prayer, divination, or magic. There is one example of this in the Apocrypha (Tobit's expulsion of a demon), but in the NT the casting out of evil spirits by Christ and His apostles is common (cf. Mark 1:25; Acts 16:18). Since that time exorcism has been practiced by the church until the present day. In the early church it became common to exorcise catechumens from pagan and Jewish backgrounds before baptism. This practice was mentioned at the Council of Carthage in 255.

In the Middle Ages exorcism formed part of infant baptism; the service included the *exsufflatio*—the thrice-repeated breathing on the face of the infant with the accompanying words, "Depart from him, thou unclean spirit, and give place to the Holy Spirit." This was condensed in the *Rituals Romanum* of 1614. Early Lutheran baptismal services, however, as well as the First Prayer Book of Edward VI (1549), contained a brief exorcism. The title of "exorcist" described in the early church a minor order* of the ministry, whose office included laying hands on the insane, exorcising catechumens, and helping at Holy Communion. Today in the Roman Catholic Church the order is retained as a stepping-stone to the priesthood, but it has no real significance. The Eastern Church has no order of exorcists. Within the Pentecostalist movement the casting out of devils is often practiced by charismatics, and the Church of England has several licensed exorcists.

See D. Omand, *Experiences of a Present Day Exorcist* (1970). PETER TOON

EXPIATION. This term employed in connection with atonement and sacrifice means to render satisfaction for sin. It has replaced "propitiation" in some modern translations of the Bible. It is an unexceptionable term in itself, yet is not adequate to translate the Greek *Hilaskomai* word-group or its Hebrew equivalents (see L. Morris, *The Atonement in New Testament Teaching* [1955], pp. 125-185). Expiation treats sin as something to be dealt with, while propitiation lays stress on the fact that it merits the wrath of God. Expiation is impersonal, propitiation is personal. The commentaries and translations of C.H. Dodd (chairman of the New English Bible translators) reflect modern opposition to ideas of wrath and propitiation, and modern preference for the weaker term "expiation." G.W. GROGAN

EXTRAVAGANTES. Originally applied to papal decretals* not included in Gratian's collection, and then to those excluded successively from collections by Gregory IX (1234) and Boniface VIII (1298), the term strictly applies to all decretals after 1298, including the Clementinae (1317). Normally, however, it applies to decretals of John XXII and those covering the pontificates of Urban IV to Sixtus IV, known respectively as the *extravagantes* and *extravagantes communes.*

EXTREME UNCTION, see UNCTION, EXTREME

EYCK, see VAN EYCK

EZNICK (fifth century). An Armenian writer and bishop, he was born at Kolb and was a disciple of the patriarch Isaac and of Mesrob. Besides Armenian, he understood Persian, Greek, and Syriac and was familiar with the theological literature of those languages. As bishop of Pakrewand in the province of Airerat he took part in the synod of Artashast in 450 which rejected the Persian demand that the Armenians adopt Zoroastrianism. He also took part in translating the Armenian version of the Bible and according to tradition wrote a number of homilies, all of which are now lost. His best-known work, however, is the *Refutation of the Sects,* which is still preserved in the Armenian original. It consists of four books, the first dealing with the pagan idea of the eternity of matter, the second with Zoroastrianism, the third with Greek philosophy, and the fourth with the Gnostic sect of the Marcionites. The latter book is particularly interesting in that it reveals a number of later developments in Marcionite ideas.

DAVID JOHN WILLIAMS

FABER, FREDERICK WILLIAM (1814-1863). English hymnwriter. Although his upbringing was Calvinist, Faber was influenced at Oxford by J.H. Newman,* and collaborated in work on *The Library of the Fathers.* He took orders in the Church of England in 1837, became rector of Elton, Huntingdonshire, in 1843, but seceded to Rome in 1845. With others he formed a community in Birmingham, which was merged in 1848 with Newman's Oratory of St. Philip Neri. In 1849 he started a branch of the order in London, which developed into Brompton Oratory. He wrote many devotional books and several collections of verse and was an ardent propagandist for the Roman Catholic Church. His enthusiasm for Italian styles of devotion sometimes lapsed into sentimentality. His 150 hymns, collected and published as *Hymns* (1861), were intended to have the same popular appeal as those of Newton and Cowper. A number appear in the Roman Catholic *Westminster Hymnal* (1940). Among those sung also by Protestants are "Hark, hark, my soul"; "My God, how wonderful Thou art"; "O come and mourn with me awhile"; and "Souls of men, why will ye scatter." JOHN S. ANDREWS

FABER, JACOBUS (c.1455-1536). French humanist. Known also as Jacques Lefèvre d'Étaples, or Faber Stapulensis, he was a native of Picardy. He spent some years in the schools of Italy, where he was influenced by Pico della Mirandola, and in Paris, where he studied classics. This awakened in him the importance of the language for the study of the Bible, which in turn helped him to create an interest for others and thus blaze a trail for Christian humanism. His Latin translation of Paul's epistles (1512), published with a commentary, resulted in his being branded heretic, especially to theologians of medieval outlook, on the questions of transubstantiation, justification, and the merit of good works. The Pauline translations and his earlier work on Psalms (1509) had some influence on Luther. Faber was subsequently denounced by the Sorbonne (1517-18) and by the government (1525), and had to flee to Strasbourg. Later he was given the protection of the queen of Navarre. He never accepted the Reformed doctrines on grace, justification, and predestination, but there is a link between him and the Anabaptists; a volume of mystical and prophetic writings he edited (1513), later bound with sermons by Tauler, are said to have been read by Thomas Mü nzer translation of the NT (1525) was ordered to be burnt with the French translations of the treatise of Erasmus. He had an interview with Calvin about the time of Calvin's conversion (1534) which may have proved of significance in the latter's break with Rome.

GORDON A. CATHERALL

FABER, JOHANN (1478-1541). Bishop of Vienna. German by birth, he became vicar general of the diocese of Constance in 1518, and chaplain and confessor to Ferdinand I in 1524. He conducted several important missions for Ferdinand, including one to engage the assistance of Henry VIII against the Turks. He was appointed to the see of Vienna in 1530. He belonged originally to the humanistic and liberal party, and his friendship with Erasmus* led to an initial sympathy with Reformation leaders, including Zwingli and Melanchthon. In 1521 he returned from a visit to Rome to become a zealous adversary of the Reformation and its leaders, earning the title "the Hammer of the Heretics." He defended Catholic orthodoxy in conferences and disputations, and burned heretics in Austria and Hungary. He wrote against Luther and in defense of celibacy and papal infallibility. JAMES TAYLOR

FABIAN (d.250). Bishop of Rome from 235. Ruling with vigor, he furthered the tendency to a hierarchical structure by dividing Rome into seven ecclesiastical areas with a deacon over each, and a subdeacon to assist. He increased the amount of catacomb property held. He made his opinion known to the African Church with respect to the heresy of Privatus, bishop of Lambesis, and he received correspondence from Origen. He enjoyed considerable favor under the pro-Christian emperor Philip, but when Decius reversed the policy of his predecessor and seized the church leaders, he was the first to be martyred.

FABIOLA (d.399). Patrician Roman lady. The unfaithfulness of her first husband led her to divorce him. A second marriage while he was still alive, however, cut her off from the fellowship of the church. On the death of her second husband she went through extreme penitential discipline and decided to give away her great wealth. Her many acts of benefaction included building a hospital in Rome in which she herself worked. In 395 she went to the Holy Land and became a disciple of Jerome. Fear of the Huns and her own gregarious nature led her to return to Rome and continue her good works.

FABRI, FRIEDRICH (1824-1891). German mission executive. A pastor in Bavaria, he was appointed inspector of the Rhine Mission in 1857, largely because he occupied a neutral position in the confessional struggle of the time. Although he did little to change the internal structure of the mission, he did stress better training of candidates and some expansion of the work in South Africa. On his initiative the highly successful Sumatra field was opened in 1860. As an apologist he published tracts against materialism and Darwinism, and during the *Kulturkampf** supported the concept of a church free from state control. Fabri also fostered German imperialism in a popular book, *Bedarf Deutschland der Kolonien* (1879), and founded the West German Association for Colonization and Export in 1880 to pressure for a colonial policy. In 1884 he retired from the Rhine Mission, accepted an honorary professorship at Bonn, and spent his last years in colonial agitation. RICHARD V. PIERARD

FACUNDUS. Bishop of Hermiane in Byzacena c.546-571. When Emperor Justinian published an edict in 543-44 condemning the Three Chapters,* Pope Vigilius at first refused to approve, but later in Constantinople published his *Judicatum* (548) in which he too condemned the Three Chapters, but explicitly upheld the Council of Chalcedon.* The Western bishops, particularly those of North Africa, were opposed to the *Judicatum.* A North African council excommunicated Vigilius until he should withdraw it, and Facundus, who was also in Constantinople, refused communion with the pope and published his own *Pro defensione trium capitulorum,* in which he defended the orthodoxy of the three men concerned and argued that to deny this was to deny the orthodoxy of the Chalcedon Christology. He laid the main blame for the present controversy on the emperor. When the Three Chapters were condemned by the Council of Constantinople in 553, Vigilius excommunicated Facundus. Facundus wrote two other works: *Contra Mocianum Scholasticum* and *Epistola fidei catholicae in defensione trium capitulorum.*
DAVID JOHN WILLIAMS

FAIRBAIRN, ANDREW MARTIN (1838-1912). Congregationalist divine. Born at Inverkeithing, Scotland, of Covenanting stock, he had little regular schooling and was earning his living before he was ten. In his spare time he read omnivorously and eventually studied at Edinburgh University. He entered the Evangelical Union* theological college in Edinburgh in 1857 and thereafter ministered at Bathgate and Aberdeen before becoming principal of Airedale Theological College in 1877. Nine years later he transferred to Mansfield College, Oxford, of which he was the first principal—a post he held for twenty-three years. An original and refreshing teacher whose theological liberalism reflected the views of German scholars, Fairbairn was much in demand as preacher and lecturer, and paid several visits to the USA. His writings include *Studies in the Philosophy of Religion and History* (1876); *Studies in the Life of Christ* (1880); *Christ in Modern Theology* (1893); and *The Philosophy of the Christian Religion* (1902). He was active also in religious and political controversy, notably in connection with the significant Education Act of 1902, and helped develop theological education in English nonconformity. J.D. DOUGLAS

FAIRBAIRN, PATRICK (1805-1874). Scottish theologian. Born in Berwickshire, he studied at Edinburgh University, and after ordination served parishes in Orkney, Glasgow, and Salton (East Lothian) where he continued as minister of the free church after the 1843 Disruption.* In 1853 he went to his church's Aberdeen college as professor of divinity, and three years later he became principal of the Glasgow college, which post he held until his death. A scholar who became widely known outside his own denomination, he was a member of the Old Testament Revision Company. He wrote a definitive work, *The Typology of Scripture* (rep. 1953), and his other publications included commentaries on Ezekiel and the Pastoral epistles, and the editorship of the *Imperial Bible Dictionary.* J.D. DOUGLAS

FAITH, DEFENDER OF THE, see FIDEI DEFENSOR

FAITH AND ORDER, see LAUSANNE CONFERENCE; EDINBURGH CONFERENCE (1937); WORLD COUNCIL OF CHURCHES

FALSE DECRETALS. These were allegedly discovered in Spain by Archbishop Riculf of Mainz. They were really mid-ninth century Frankish forgeries interpolated into an earlier genuine collection of Conciliar Acts edited by Isidore of Seville (d.636). The forgeries (called also the Pseudo-Isidorian Decretals) reflect current mid-ninth century church reformers' views on the status of the church in general. In particular, the forgeries were most frequently used in the ninth, tenth, and eleventh centuries to establish historical grounds for contemporary papalist views on papal monarchy. For this the Donation of Constantine* and the letter from Clement I to James (the brother of Christ), both included in the False Decretals, were key documents. The forgeries also defended bishops' rights against metropolitans, which in turn effectively increased papal powers. Certain passages from the Donation of Constantine were incorporated into canon law by Gregory VII,* although the document's dangerous implication that Emperor Constantine had given power to the pope meant that the Donation was outdated by the eleventh century. The forged nature of these "Decretals" was completely demonstrated in 1558, but some like Nicholas I,* the first to use them (865), apparently knew much earlier that they were spurious. Their authenticity was, however, accepted throughout the Middle Ages.
L. FEEHAN

FAMILISTS. Alternatively named "Family of Love," the sect was founded in Emden about 1540 by a prosperous businessman, Henry Nicholas (1501-80), as a result of a series of visions. Nicholas, who never left the Roman Catholic Church into which he was born, is said to have claimed to be an incarnation of the deity and to have taught a mystic pantheism. The Familists were much persecuted on the Continent, Nicholas himself being frequently imprisoned. The group later took root in England, but laws were enacted against it under Elizabeth I in 1580, though persecution was not intense. After the Restoration it ceased to exist, the remaining members joining the Quakers and Congregationalists.

R.E.D. CLARK

FAMILY COMPACT. A term originally derived from alliances between the crowned heads of Europe during the fifteenth and sixteenth centuries. After 1828 it was applied popularly to the governing elite in Upper Canada. The Constitutional Act* of 1791 (more accurately, the "Canada Act") had created a governmental system dominated by the executive and legislative councils. The members of these were appointed by the crown from the wealthy and powerful segments of society who were usually Anglicans and Tories. Because the majority of the population had little political power, resentment against the Family Compact mounted until it erupted in the Rebellion of 1837.

ROBERT WILSON

FARADAY, MICHAEL (1791-1867). English scientist. Born in London, son of a blacksmith who had shortly before removed from Yorkshire, Faraday became a laboratory assistant to Sir Humphrey Davy at the Royal Institution (1813) and later succeeded him as professor of chemistry (1827). His discoveries in physical science were numerous and outstanding: he made, for example, the first electric motor, the first dynamo, and the first transformer. In religion he was of Sandemanian* ancestry; his grandfather Robert Faraday (1724-86) had been an Inghamite, but was converted to Sandemanianism around 1759. Faraday's parents, like himself and his wife, were lifelong adherents of the group. His outlook in science was deeply influenced by religion; in his lectures he often used science as evidence of God's power and wisdom. Believing that the universe was in some way a manifestation of the one and only God revealed through Christ, Faraday looked for and found unity in natural phenomena. His life was devoted to Christian work and to science; he was a brilliant lecturer who made science popular in his day. On Christian grounds he rejected wealth, and on retirement was very poor, but a government pension was granted and later (1858) a house in Hampton Court was provided for him by Queen Victoria.

BIBLIOGRAPHY: Biographies by J.H. Gladstone (2 vols., 1872); S.P. Thompson (1898); and L.P. Williams (1963); for background, see J.F. Riley, *The Hammer and the Anvil* (1954); for Faraday's works, see A.E. Jeffreys, *Michael Faraday, A List of his Lectures and Published Writings* (1960).

R.E.D. CLARK

FAREL, GUILLAUME (1489-1565). French Reformer. Born at Gap in Dauphiné, he went in 1509 to Paris where he studied under Jacques Lefèvre d'Étaples (J. Faber Stapulensis). By about 1520 his humanist training led him to adopt reforming ideas. Soon after this he began to assist in the reform of the diocese of Meaux under Bishop Briçonnet. In 1523 he was influenced by radical ideas on the Eucharist stemming from the treatise of Cornelius Hoen, which Farel may have translated into French. In the same year growing intolerance and persecution led to his expulsion from France. In 1524 he was involved with Oecolampadius* in a disputation in Basle, but his fiery attacks on the theological faculty soon led to his expulsion. He also visited Strasbourg, Montbéliard, and Neuchâtel.

From 1526 he became the leader of a peripatetic band of evangelists preaching mainly in French-speaking Switzerland. His own fiery preaching often led to rough handling by mobs of opponents. He took part in the disputation which won the city of Berne to the Reformation in 1528. Thereafter he received support from Berne in his preaching activities in the Pays de Vaud. In 1532 he began to evangelize Geneva, and in 1535 that city accepted the Reformation. Farel was instrumental in persuading John Calvin to serve the church in that city in 1536. With Calvin, he was expelled from Geneva in 1538. Farel now made Neuchâtel his base and spent many years there working in close harmony with Calvin. In 1558 Farel married a young girl, and for a time there was coolness between Calvin and himself, but in 1564 the rift was healed on Calvin's deathbed. Throughout the long years in Neuchâtel, Farel continued to undertake evangelistic work in France, especially at Metz where he died.

NOEL S. POLLARD

FARRAR, FREDERIC WILLIAM (1831-1903). Dean of Canterbury. Born in India of missionary parents, he went to school at King William's College, Isle of Man. There the religious teaching was strongly evangelical, and Farrar was to write about it in his best-selling school story *Eric; or Little by Little* (1858). He studied at King's College, London, where he was influenced by F.D. Maurice,* and after graduation and ordination was a schoolmaster until his mid-forties. He had a tremendous influence on the Victorian middle classes in both religious and cultural matters. His *Life of Christ* (1874) went through twelve editions in a year, and his *Life and Works of St. Paul* (1879) also had a great vogue.

Much controversy was aroused by his *Eternal Hope* (1878), a collection of sermons in which he questioned the doctrine of eternal punishment for the wicked. A particular adversary here was E.B. Pusey.* Farrar was to modify his position somewhat in *Mercy and Judgment* (1881). In 1882 he preached at Charles Darwin's funeral. It was held that Farrar's broad outlook long hindered his ecclesiastical promotion, but eventually, after having been a royal chaplain and canon of Westminster, he was appointed dean of Canterbury, which post he held for the last eight years of his life. Farrar was a pioneer in introducing into modern education some of the results of philological research, for which he was, on Darwin's nomination, elected in 1866 as a Fellow of the Royal Society—an honor not accorded to many modern churchmen.

J.D. DOUGLAS

FASTING. Abstinence from eating and drinking. It may be complete or partial, for a set length of time or intermittently, and for religious or other purposes. It has been practiced in numerous societies for reasons as varied as to produce evidence of virility, to coerce or to appease a supposed god or spirit, and to prepare for ceremonial observance. Among Jews, the Day of Atonement was the most prominent occasion for a public fast (Lev. 16:29,31; 23:27,29,36; Num. 29:7), though the OT refers to many special fasts, both individual and public (Judg. 20:26; 1 Sam. 14:24; 31:13; 2 Sam. 1:12; 12:16-23; 1 Kings 21:27; 2 Chron. 20:3).

Jesus engaged in a prolonged fast following His baptism, but He seems neither to have expressed strong approval nor disapproval of fasting as such. He urged that, if fasting is to be practiced, it should be to the glory of God rather than to gain the acclaim of men (Matt. 6:16-18). But when He was asked expressly about fasting, He said it would be appropriate for His hearers to fast after He left them (Matt. 9:14,15). There is some evidence of fasting in the early church (Acts 13:2,3; 14:23), but it seems not to have had so much emphasis then as, in certain branches of the church, it received in postbiblical times.

MILLARD SCHERICH

FATHER (Lat. *pater*). A title now used in the English-speaking world to describe or to address Roman Catholic or Anglo-Catholic clergy, whether secular or religious. "Pater" was originally a description of bishops as "fathers in God" or later of confessors. It seems that due to Irish influence in the nineteenth century the term took on a much wider meaning. The pope is still called "the Holy Father," but certain classes of monks prefer the title "Dom." On the continent of Europe the term "Father" is not often used of the secular clergy.

FATHER DIVINE, see DIVINE, M.J.

FATHERS, APOSTOLIC, see APOSTOLIC FATHERS

FAULHABER, MICHAEL VON (1869-1952). Archbishop of Munich. Born at Klosterheidenfeld, Bavaria, he studied at Schweinfurt and Würzburg where he later became lecturer (1899-1903). Ordained in 1892 and thereafter in Rome (1895-98), he was professor of OT in Strasbourg until becoming bishop of Speyer (1911). He was appointed archbishop of Munich in 1917 and cardinal in 1921. He was a monarchist who favored the Wittelsbach kings, and condemned paganism, racism, and totalitarianism. In World War I he ministered to Bavarian forces in the field and later became the leader of the right-wing German Catholics. He emphasized the Jewish background in Christianity and attacked anti-Semitism. Earlier he had contributed to the patristics field, but afterward wrote much on contemporary issues.

C.G. THORNE, JR.

FAUNCE, WILLIAM HERBERT PERRY (1859-1930). Baptist minister and educator. Son of a Baptist pastor, he graduated from Brown University and Newton Theological Institution and in 1884 entered the Baptist ministry. After fifteen years as pastor, he was appointed president of Brown University, where he served until his retirement in 1929. Esteemed as lecturer and preacher, he was the greatest builder in Brown's history. Active in many social causes, he was at various times president of the World Peace Foundation, of the National Education Association, and of the Religious Education Association. A liberalizer, he moderated between modernists and fundamentalists and opened Brown to non-Baptist leadership. He was also noted for his writings, particularly *Social Aspects of Christian Missions* (1914) and *Religion and War* (1918). Contemporaries knew him as a poetic, thoughtful, and soft-spoken man.

DARREL BIGHAM

FAUSSET, ANDREW ROBERT (1821-1910). Anglican scholar. Born in County Fermanagh, Ireland, he graduated from Trinity College, Dublin, in 1843. He went to England, was curate of Bishop Middleham, County Durham (1848-59), and rector of St. Cuthbert's, York, from 1859 until his death. Evangelical and premillennialist, he was a prolific writer and editor. His works include *Scripture and the Prayer Book in Harmony* (1854); the editorship of the first ET of Bengel's *Gnomon of the New Testament* (1886); *The Englishman's Critical and Expository Bible Cyclopedia* (1878); and (with Robert Jamieson and David Brown) *Critical, Experimental and Practical Commentary on the Old and New Testaments* (1871).

H. CROSBY ENGLIZIAN

FAUSTUS OF RIEZ (d. c.490/500). Semi-Pelagian* bishop, probably of British origin. Abbot of Lérins from 437, he became involved in a dispute with the bishop of Arles, the solution of which helped to lay down the lines of demarcation between dioceses and monasteries for the future. He then became bishop of Riez and—as an ascetic, Bible scholar, and preacher of very considerable oratorical powers—he was clearly the most outstanding prelate in Roman Gaul. Resistance to the Arian king of the Visigoths, Euric, led to his exile (477-85). In *De gratia Dei* he argued, more strongly than Cassian,* for the operation of man's free will in obtaining salvation.

FAWCETT, JOHN (1740-1817). English Baptist theologian. After some years' engagement in a secular occupation, his early impressions of George Whitefield's preaching prevailed and he became a Baptist pastor. His ministry was spent entirely in the Halifax area of Yorkshire, where he also taught school for most of his active life. He was a vigorous preacher, zealous and much respected among his people, who might have held high office in his denomination had he been so inclined. He is best known, however, for his *Devotional Commentary on the Holy Scriptures* (1811). His method was to follow each expository section with a paragraph of "aspirations" to guide the feelings of his readers. Though his health had long been indifferent, he survived until his seventy-seventh year.

J.D. DOUGLAS

FAWKES, GUY (1570-1606). Prominent participant in the 1605 Gunpowder Plot.* Born into an Anglican family, he was converted to Roman Catholicism after his father's death (1579) and his mother's remarriage to a Roman Catholic. He served in the Spanish army for some years, and on his return to England in 1604 was persuaded to join Robert Catesby* and others who were conspiring to blow up the Parliament building while James I and many governmental leaders were inside. Fawkes duly planted an impressive amount of explosives, but the authorities learned of the plot, arrested Fawkes on the night before the gunpowder was due to go off, and forced him under torture to identify his accomplices. The abortive plot is still celebrated by fireworks in England on 5 November (Guy Fawkes Day). J.D. DOUGLAS

FEAST OF FOOLS/ASSES. Mock-religious festivals, often identified, which were celebrated about New Year time, notably in France (e.g., Rouen, Beauvais) but also in England (e.g., Lincoln, Salisbury), sometimes associated with Balaam's Ass or the Flight into Egypt. The festivities were controlled by subdeacons and included much buffoonery and burlesque of religious services. They flourished during the twelfth, thirteenth, and fourteenth centuries, but severe penalties for their observance were imposed by the Council of Basle (1435).

FEATLEY (Fairclough), DANIEL (1582-1645). Anglican controversialist. Oxford graduate, he became chaplain to Archbishop Abbot and, in 1619, rector of Lambeth. His anti-Roman views and his preaching impressed James I, but during an outbreak of plague Featley forsook controversy to produce *Ancilla Pietatis* (1626), a popular manual of devotion later a favorite of Charles I in his troubles. Featley produced also an exposition of Paul's epistles which with his marginal annotations was printed in the Bible issued in 1645 by the Westminster Assembly, from which he was the last episcopal member to withdraw. Imprisoned as a spy by Cromwell, he continued his writing against Roman Catholics and his defense of the Church of England against Puritan divines. After eighteen months he was released for health reasons, but soon died. "A most smart scourge of the church of Rome," it was said at his funeral, "a compendium of the learned tongues, and of all the liberal arts and sciences." J.D. DOUGLAS

FEBRONIANISM. A German and Austrian movement in the late eighteenth century to limit papal authority in the church. Its chief doctrines were defined by J.N. von Hontheim,* writing under the pseudonym "Justinius Febronius," *The State of the Church and the Legitimate Authority of the Roman Pontiff, a Book Composed for the Purpose of Uniting in Religion Dissident Christians* (in Latin, 1763). Hontheim, learning from Gallicanism,* argued with evident Catholic, not secular, devotion, that the "keys of the kingdom" (Matt. 16:19) were not given to the papacy, but to the whole church, which acts through general councils composed of all the bishops, who hold office from God, not the pope. The bishop of Rome should be understood as *primus inter pares* to establish the unity of the universal church and preserve its canons, while bishops should exercise most of the authority which wrongly accrued to the pope. The doctrine, although similar to some Gallicanism, differs significantly from it in its universality and in not advocating royal supremacy. It remained for proponents of Josephinism* to make Hontheim's ideas serve their own secular statist centralism; Kaunitz, Austrian chief minister, found them useful and ordered them taught in the universities. The Synod of Pistoia* (1786) and the Congress of Ems* (1786) adopted Febronian principles. I.H. von Wessenberg* succeeded Hontheim as leading advocate of Febronianism, and hoped to build a nearly independent national German church. Clement XIII condemned Hontheim's book in 1764, Hontheim later recanted (1778), and the Syllabus of Errors* (1864) and the dogma of papal infallibility (1870) made it dogmatically inconsistent with Catholicism. The Old Catholic* movement continued the doctrines after the Vatican Council (1869-70).

BIBLIOGRAPHY: O. Meyer, *Febronius, Weihbischof Johann Nicolaus von Hontheim und sein Widerruf* (1880); L. Just, "Febronianismus," in *Lex. Theol. Kirche*, IV (1960), pp. 46-47; M. O'-Callaghan, "Febronianism," in *New Catholic Encyclopedia*, V, pp. 868-69; C.B. Moss, *The Old Catholic Movements, Its Origins and History* (1964). C.T. MC INTIRE

FEDERAL THEOLOGY, see COVENANT THEOLOGY

FEGAN, JAMES WILLIAM CONDELL (1852-1925). English philanthropist. His father was a leader of the Christian Brethren in Southampton. Converted at seventeen as he read in the epistle to the Romans, he was soon involved in Ragged School work in London. At twenty-one he gave up a commercial career to rescue and care for children. His first home for boys was at Deptford. His marriage to Mary Pope in 1889 brought him a loyal partner in this enterprise. Boys were trained in farming and encouraged to go to Canada, and by 1900 nearly 3,000 had emigrated. "Mr. Fegan's Homes" were established at Stony Stratford in 1900 and Goudhurst in 1912, and at present there are family homes for boys and girls in Surrey, Sussex, and Kent. J.G.G. NORMAN

FELIX OF URGEL (d.818). Bishop of Urgel in Spain. An exponent of Adoptianism,* he defended his views in the presence of Charlemagne at the Council of Regensburg (792) where he was induced to recant. Sent to Rome by Charlemagne, he was compelled to sign an orthodox confession which he subsequently repudiated. Alcuin* wrote extensively against him, opposing his use of the phrase "adopted son" with regard to Christ in His human nature. At the Council of Aix-la-Chapelle (798) Felix again acknowledged himself defeated, wrote a recantation, and called on the clergy of Urgel to follow his example. He was placed under the supervision of the archbishop of Lyons till his death. It is doubtful whether his recantation was sincere, for a treatise discovered after his death contained evidence of his former views.

HAROLD H. ROWDON

FELLER, HENRIETTA (1800-1868). Missionary and educator. A Protestant from Switzerland, she arrived in St. Johns, Quebec, in 1835, and began spreading the Gospel from house to house. In Grande Ligne she and her colleague Louis Roussy found a home to receive them, and Madame Feller began a school in the attic for local children. During the Rebellion of 1837 the two missionaries and their sixty-three converts were persecuted for being Protestants and forced to flee to New York. When peace was restored they returned, and in 1840 they erected the first building of what was to be Feller College. Henrietta was instrumental in founding the Grande Ligne Mission in 1836, the first Canadian French Baptist Church.

ROBERT WILSON

FELLOWSHIP OF INDEPENDENT EVANGELICAL CHURCHES (UK). This body, founded by the Rev. E.J. Poole-Connor in 1922 and originally named the Fellowship of Undenominational and Unattached Churches and Missions, brings together in evangelical unity isolated ministers and churches, pastors desiring legal recognition as ministers of religion and professional qualifications, and others disturbed by the spread of liberalism within their own denominations. Its Declaration of Faith is broad enough to embrace evangelicals of Calvinist, conservative Methodist, and other traditions; questions of church organization and administration of the sacraments are carefully avoided. The Fellowship has drawn up lists of churches and personal members, has an annual assembly, a permanent council and provincial auxiliaries, issues publications, and acts as a trustee of church property. At present it embraces more than 400 churches, but the number is increasing.

IAN SELLERS

FELLOWSHIP OF RECONCILIATION. Probably the most significant of the peace organizations created during World War I. It was founded in England in December 1914 by Henry Hodgkin, a Quaker, and rapidly spread to the United States in 1915. It is now an international organization and the leading American pacifist organization. It is based upon the essential unity of mankind and the power of truth and love to resolve human conflict. It has been active in nonviolent intervention on behalf of victims of injustice and exploitation, working for the abolition of war, the redemption and rehabilitation of public offenders, encouraging reverence for personality, and the maintenance of a spirit of self-giving love while engaged in the efforts to achieve these purposes.

JOHN P. DEVER

FÉNELON, FRANÇOIS DE SALIGNAC DE LA MOTHE (1651-1715). French ecclesiastic. Educated at the University of Cahors, a Jesuit college in Paris and the seminary of St. Sulpice, he was ordained in 1675, and for thirteen years conducted a mission to the Huguenots, whom he endeavored to convert with a mixture of oratory, threats, and open bribery. From 1689 to 1697 he was tutor to Louis XIV's grandson, the duke of Burgundy, for whom he wrote his famous educational novel, *Télémaque.* In 1695 he was at the height of his influence and was made archbishop of Cambrai, but he fell suddenly because of his association with the Quietist followers of Madame Guyon.* For a time he was banished to his see city, but on the pope's condemnation of his treatise on true and false mysticism (1699), he issued a submission which on his later testimony was insincere. In the Jansenist controversy Fénelon defended the infallibility of the church and supported the bull *Unigenitus* in sermons and letters. He displayed throughout a baffling mixture of ecclesiastical authoritarianism and broad humanitarian ideals: his optimism, historicism, and instinctive belief in progress make him a forerunner of the Enlightenment, while his idea of God leads directly to the Deism of the next century.

IAN SELLERS

FERDINAND II (1578-1637). Archduke of Austria, he became king of Bohemia (1617), king of Hungary (1618), and emperor of the Holy Roman Empire (1619). Zealous for the cause of Roman Catholicism, he favored the Jesuits in their aggressive endeavors to regain territories lost to Protestantism. In Austria he banished the Utraquists, the Reformed, the Lutherans, and the Anabaptists, and in 1624 required adherence to Roman Catholicism. He carried through the Counter-Reformation in Tirol. Peter Canisius* was his court preacher between 1571 and 1577. In the Thirty Years' War, Wallenstein brought him a succession of victories, so that in 1629 he considered himself strong enough to issue the Edict of Restitution, which required that properties confiscated from the Roman Catholic Church since 1555 be returned by the Protestants. The intervention of Gustavus Adolphus checked the emperor's successes, although the Treaty of Prague in 1635 resolved matters in his favor.

CARL S. MEYER

FERDINAND V (1452-1516). King of Aragon. Son of John II of Aragon, he married his cousin, Isabella of Castile in 1469, in order to unite his claims to the crown of Castile with hers. In practice, however, she asserted her own claims to authority in the government of Castile. A faithful Roman Catholic, Ferdinand was very ambitious; he helped to establish royal authority in Spain, he carried on a long struggle against France in Italy, and he supported the voyages of Columbus. He achieved acclaim in Christendom after the capture of Granada in 1491 and the subsequent expulsion of the Moors (Muslims) from Europe. Further honor came his way after the expulsion of the Jews from Spain in 1492. For his enthusiasm for the Inquisition he earned the title "the Catholic." He was succeeded by his grandson Charles.

PETER TOON

FERGUSON, JAMES (1621-1667). Scottish minister and writer. Of aristocratic lineage, he graduated at Glasgow University in 1638 and five years later became minister of Kilwinning in Ayrshire. There he was to stay for the twenty-four years of life left to him, declining invitations to richer parishes and the divinity chair of his own university. A man of notable piety, he was reckoned by his biographer to have been one of the wisest men in the land. Living in troubled times, he maintained the moderate (Resolutioner) side against the more fervent Covenanters* (Protesters), but admitted later that he had been wrong. He wrote various excellent commentaries on Paul's epistles (1656-75), commended by C.H. Spurgeon as being those of "a grand, gracious, savory divine." Ferguson produced also a *Refutation of the Errors of Toleration, Erastianism, Independency, and Separation.* J.D. DOUGLAS

FERRAR, NICHOLAS (1592-1637). Founder of the Little Gidding community. A brilliant academic career at Cambridge, a period of continental travel, work with the Virginia Company, and a year in Parliament all preceded his establishment at Little Gidding in 1625 of a small religious community based on biblical and Anglican principles. In 1626 William Laud* ordained him deacon, but he was never priested. At Little Gidding, Huntingdonshire, the Anglican offices were said in church, and the other canonical hours in the manor oratory. Vigils were kept throughout the night, and life was ordered by rule. Everyone learned a trade, and the community specialized in bookbinding. There was a free school for local children, and many charitable works were done in the locality. In 1633 Charles I visited the community and was greatly impressed. Puritans, however, were hostile, and the institution was attacked in a pamphlet called *The Arminian Nunnery* (1641). In 1647 the community was sacked by the parliamentary army. JOHN A. SIMPSON

FERRAR, ROBERT (c.1500-1555). Bishop of St. David's, Wales. Reportedly a graduate of both Oxford and Cambridge, he joined the Augustinians. Influenced by Lutheran literature, he was compelled to recant in 1528. Records are conflicting, but he became bishop of St. David's in 1548, an appointment savagely resisted by a "greedy and turbulent chapter" who opposed him on legal technicalities and absurd charges. Nevertheless, he was imprisoned and on the accession of the Roman Catholic Queen Mary was transferred to a London jail where John Bradford and other Protestants renewed his Reformation principles. Deprived of his bishopric in 1554, Ferrar was arraigned before his successor at St. David's, found guilty of heresy, and burnt at Carmarthen. He told a spectator that "if he saw him once to stir in the pains of his burning he should then give no credit to his doctrine." J.D. DOUGLAS

FERRETTI, SALVATORE (1817-1874). Organizer of the Evangelical Italian church in London. A man of great faith and deep humanity, in 1846 he founded a school in London to save poor Italian children from the exploitation of other Italians who used them to beg in the city and cruelly ill-treated them. The school, which he supported, giving private lessons in Italian, was put under the auspices of the "Society for the religious care and instruction of foreigners," founded by Lord Shaftesbury. Ferretti then edited *L'Eco di savonarola*, a bilingual journal intended to spread the Gospel among the Italian exiles and to expose the errors of Roman Catholicism. Returning to Italy in 1861, he founded an orphanage for girls in Florence, which is still in existence. DAISY RONCO

FEUARDENT, FRANÇOIS (1539-1610). Franciscan preacher and scholar. A Norman by birth, he studied the humanities at Bayeux, became a Franciscan, was ordained in 1561, and later lectured at the University of Paris. He took a leading part in France's political and religious struggles, and was a prominent preacher for the cause of the Catholic League. He retired to the convent of Bayeux, where he provided a library. His works include biblical commentaries, writings against Calvinism, patristics, including editions of Ildefonsus of Toledo, Irenaeus, Ephraem; and editions also of Arnobius, Michael Psellus, and Nicholas of Lyra.

FEUDALISM. Although at varying periods its incidence was international, it was most notably present in Western Europe in medieval times, arising amid decaying central authority, civil war, invasion, and economic stagnation. Feudalism was a way of governing by the strong (lords) over the weak (vassals), from the nobility through the peasantry; property (fiefs) determined rank. Churchmen were excluded from but dependent on this stratification, and its development varied within Europe and chronologically: England, for example, did not experience it until the Normans introduced it in 1066. Whatever the degree of feudalization, there never was anarchy, and government was always local; personalness and proximity were paramount, and interdependency reigned. To be both lord and vassal was common, and there was even hope for the serf's progress. At first the vassal was the lord's fighting man, but as property was passed down, more distributed, duties and standing changed. Civilization began to replace war, and the erstwhile military class were becoming country gentlemen. Feudal institutions lasted to the *Ancien Régime*, and some elements still survive; but feudalism ceased to be important by 1300 when the bourgeoisie were acquiring fiefs alongside the nobility. Manors and serfdom were bound to shift, but the ideal engendered by them—chivalry—bore a timeless universal truth. A deep sense of law emerged, and the later courts of justice and the parliamentary system were but natural feudal outgrowths. Feudalism was founded on, and developed and bequeathed a commitment to aristocracy, believing that to be a law of nature.

BIBLIOGRAPHY: C. Stephenson, *Medieval Feudalism* (1942); F.L. Ganshof, *Feudalism* (tr. P. Grierson, 1952); M. Bloch, *Feudal Society* (tr. L.A. Manyon, 1961); F. Stenton, *The First Century of English Feudalism, 1066-1166* (1961).

C.G. THORNE, JR.

FEUILLANTS. Originally a branch of the Cistercian Order, so named because they had their home at Les Feuillans. Although they were directly under Pitaea, they became an independent order under the leadership of Jean de la Barrière. He reestablished the early strict discipline of the order. In 1595 the pope forbade Cîteaux to claim any jurisdiction over the monastery, and confirmed its independence.

FICINO, MARSILIO (1433-1499). Florentine humanist. He was the son of Cosimo de' Medici's physician, and his early life is little known, but by 1456 he began the study of Greek which resulted in his translation of the complete works of Plato (1463-73), Plotinus (1482-92), and Pseudo-Dionysius (1492). Cosimo had given him the use of a villa (1462), and here the Platonic Academy was founded. Later, while teaching at this famous school Ficino wrote his major work, *Theologia platonia* (1469-72). In 1473 he became a priest, later writing *De Christiana religioni* (1476). When the Medici were forced from Florence, he retired to the country.

Ficino believed that Neoplatonism could be used to win intellectuals to Christ. His outlook presupposed that truth was found only in poetry and faith and was transmitted through a long line of ancient philosophers, the most important of whom were Plato and his followers. He thought there was no difference between divine revelation and the teachings of the ancient philosophers. In fact, the Platonic works contained all that man could know of truth, beauty, and goodness. The world of the Platonist was a hierarchy of emanations from the original essence. In this stepladder of bodies, qualities, souls, and intelligences, man occupied an intermediary role, related to the world of matter by his body and to the world of the spirit by his soul. Christ is identified as the mediator who leads man to love God and to emulate his perfection. Ficino exercised an enormous influence not only on the Renaissance, but on later European thought also. His editions of Plato were standard for several centuries, and scholars such as Colet Spenser and the Cambridge Platonists* owe much to him.

BIBLIOGRAPHY: P.O. Kristeller, *The Philosophy of Marsilio Ficino* (tr. V. Conant, 1943); E. Cassirer et al. (eds.), *The Renaissance Philosophy of Man* (1948), pp. 193-212; J.C. Nelson, *Renaissance Theory of Love* (1958).

ROBERT G. CLOUSE

FIDEI DEFENSOR. The title granted to Henry VIII by Leo I on 11 October 1521 as a result of his *Defence of the Seven Sacraments* against Luther and his persistent pursuit of a special papal title to parallel those of the "Catholic King" of Spain had "the Most Christian" King of France. Although Henry had not written the reply to Luther's *Babylonian Captivity of the Church* with his own hand, he had taken a considerable part in its composition and all of the credit for its authorship It tipped the scales in favor of his demand for a special papal title, despite its misrepresentation of Luther's position and its general theological weakness. Although not intended to be hereditary, Henry accepted the title for his successors, and it is still borne by the British monarch, despite the change in the "faith."

W.S. REID

FIDEISM. A view which assumes that knowledge originates in a fundamental act of faith, independent of rational presuppositions. Though the term in this form dates only from about 1885 when it was associated with and adapted by French theologians, the standpoint represented by it had several times been officially condemned during the pontificate of Gregory XVI (1831-46).

FIDES DAMASI, see TOME OF ST. DAMASUS

FIELD, FREDERICK (1801-1885). Anglican scholar. Directly descended from Oliver Cromwell, Field was a Cambridge graduate who ministered briefly in Suffolk before becoming rector of Reepham, Norfolk, in 1842. He being partly deaf from an early age, the afflication worsened, and in 1863 the scholarly bachelor retired to Norwich where he devoted himself to his books. In 1870 he was appointed a member of the Old Testament Revision Company. Age and infirmity prevented his attendance at meetings, but his meticulous notes were always welcomed by his colleagues. He was an erudite patristics scholar, specializing in the works of Chrysostom and of Origen, whose *Hexapla* he edited (1875). Bishop Christopher Wordsworth called him "the Jerome of the Church." In the field of NT linguistics he made a significant contribution with his *Otium Norviceuse* (1864-81). In theology Field described himself as holding the catholic faith of the reformed Church of England, avoiding all extremes and party labels.

J.D. DOUGLAS

FIELD, RICHARD (1561-1616). Anglican divine. Graduate of Oxford, Field continued to lecture and study there until 1592. Soon after, he became rector of Burghclere in Hampshire. He was appointed a royal chaplain, and he took part in the Hampton Court Conference* (1603). Six years later he was made dean of Gloucester, where his occasional preaching attracted great crowds. He preferred, however, to reside at Burghclere or at Windsor, of which he was a prebendary. Field's chief work, *Of the Church* (1606), was intended as an apology for the Church of England against Rome and was hailed as one of the best works of polemical divinity, unsurpassed even by his friend Richard Hooker.* Field drew a parallel between the Roman Catholic Church of his day and fourth-century Donatism.* He argued also that the continental Protestant bodies were part of the church of Christ.

J.D. DOUGLAS

FIELD PREACHING. Generally associated with Scotland, field preaching has passed through a number of distinctive phases in the last 450 years. Beginning with George Wishart,* in the early stages of the Reformation struggle, when churches were either not available or inadequate to hold the thronging crowds, it came into its own in the second half of the Covenanting era. It was after the restoration of Charles II, with the subsequent legislation which drove so many ministers from their parishes, that men like John Welsh and John Blackadder, Richard Cameron* and Donald Cargill,* took to the moors and the mountains where great conventicles were held, ten to fifteen thousand people attending, and where the sacraments were celebrated. With the coming of George Whitefield* and the Evangelical Revival, field preaching took on a new dimension, as many as 30,000 people gathering at Cambuslang in 1742 to hear the famous evangelist. Later came the advent of the great Highland open-air communions, which were to remain a characteristic feature of religious life in the north down to our own day. Among recent developments have been Covenanting commemoration conventicles, Easter Sunday hilltop services, and evangelistic rallies. Closely associated with these has been the growth of summer seaside missions, with their services on beach and promenade involving thousands of workers. D.P. THOMSON

FIFTH MONARCHY MEN. An apocalyptic movement that came to prominence in England during the Commonwealth and Protectorate. Its adherents hoped to see the prophecy of Daniel (cf. 2:44) fulfilled by the establishment of the rule of Christ and His saints upon earth as the successors to the Assyrian, Persian, Macedonian, and Roman empires. This was to be done by destroying all anti-Christian forms such as the established church. After the fall of the Commonwealth they supported Oliver Cromwell. The nominated or Barebones Parliament of 1653 raised their hopes of bringing in the Millennium, but the establishment of the Protectorate destroyed their hopes and they turned against Cromwell. One of their leaders, Christopher Feake, even called him "the most dissembling and perjured villain in the world." This agitation led to the arrest of their leaders, including Maj.-Gen. Thomas Harrison, Maj.-Gen. Robert Overton, Feake, and John Rogers. A Fifth Monarchist, Thomas Venner, attempted two uprisings (1657, 1661), both of which were easily suppressed, and the movement died out.

See B.S. Capp, *The Fifth Monarch Men* (1972). ROBERT G. CLOUSE

FIGGIS, JOHN NEVILLE (1866-1919). Anglican historian. Son of a minister in the Countess of Huntingdon's Connexion* in Brighton, Figgis in his youth reacted against his father's evangelical religion. He had a brilliant Cambridge career as student and teacher of history, being a pupil and friend of Mandell Creighton, Maitland, and Acton. He won lasting fame through *The Divine Right of Kings* (1892) and *Studies of Political Thought from Gerson to Grotius 1414-1625* (1907), in which he pioneered interpretation of the transition from medieval to modern periods, especially in the struggle between absolutism and constitutionalism, and emphasized the theological matrices of modern secular political theory. He came to oppose the omnicompetent state, supporting Guild Socialism and, using Gierke's idea of the real personality of groups, he argued that the freedom of the church was a bulwark of the freedom of all groups within the state (*Churches in the Modern State,* 1913).

Figgis surprised his friends by deciding to be ordained to the Church of England's ministry in 1892, for he had not been noticeably religious. He was not very successful as a parish priest, but as vicar of Marnhull (1902-7) he had a "middle age conversion" from a humanitarian and moralist religion, entered the Community of the Resurrection, and became a prophetic preacher of the gospel of supernatural, sacramental, and disciplined redemption in Christ, in a world in crisis where indifference and hostility—he was an early acute English interpreter of Nietzsche—threatened the faith and the Christian civilization that rested on it. More of an evangelist and apologist than a theologian, Figgis's influence hardly survived World War I and his death.

See M.G. Tucker, *John Neville Figgis* (1950). HADDON WILLMER

FILASTER (Philaster) (d. c.397). Writer of the famous treatise *Liber de Haeresibus,* which sought to refute 128 Christian and 28 Jewish heresies. It met a real contemporary need and was used by Augustine of Hippo. Its weakness was to place side by side major heresies such as Arianism and minor aberrations such as astronomical speculation. Before writing this book Filaster traveled widely, preaching against Arianism and other heresies, and thereby ensuring personal acquaintance with heretics. Toward the end of his life he became bishop of Brescia in N Italy, and he was succeeded in that see by Gaudentius.

FILLMORE, CHARLES, see UNITY SCHOOL

FINAN (d.661). Second bishop of Lindisfarne, where he succeeded Aidan* in 652. He worked with Oswy, king of Northumbria, for the conversion of the country to Christianity of the Celtic or Irish type. Intensely missionary in his outlook, he sought to bring the Gospel to people beyond the borders of Northumbria, and did so with such success that he baptized both the kings of the East Saxons and Mercia. In the controversy with the Roman Church over the date of Easter he took the side of the Celtic tradition, but died before the matter was finally settled in favor of Rome by Oswy at the Synod of Whitby (663/4).

FINDLATER, SARAH, see BORTHWICK, JANE

FINDLAY, GEORGE GILLANDERS (1849-1919). Methodist biblical scholar. Born in Montgomeryshire, Wales, he studied at Wesley College, Sheffield, and London University. He served his denomination's theological colleges at Headingley and Richmond from 1870 until his retirement in 1917. Findlay contributed to *The Expositor's Bible, The Expositor's Greek Testament, The Cambridge Bible,* and *The Pulpit Commentary.* He also wrote *Epistles of the Apostle Paul* (1892); *Church of Christ in the New Testament* (1893); *The Books of the Prophets in their Historical Succession* (3 vols., 1896-1907); and *Fellowship in the Life Eternal* (1909).

FINLAND. Christianity was first introduced to Finland mainly through trade relations—from Novgorod in the East and from Birka in the West. In addition, the bishopric of Hamburg-Bremen did missionary work in Scandinavia. The position of the Western Church and Swedish rule were secured in 1249 through the crusade to Tavastland (Häme) led by Birger Jarl of Sweden. Åbo (Turku) was soon made an episcopal seat, and a cathedral was built there during the thirteenth century. From the beginning of the fourteenth century until the end of the Middle Ages all the bishops were graduates of foreign universities. The Reformation came peacefully to Finland when Peter Särkilax,* Mikael Agricola,* and others came from Wittenberg and became church leaders at home.

The Pietistic revivals were most important for the inner development of the church and continued to enrich it from the end of the seventeenth century until the present day (see, e.g., P. RUOTSALAINEN, H. RENQVIST, F.G. HEDBERG). During the present century new groups have arisen, such as the "Fifth Movement," and these carry on the old revival inheritance as well as receiving new impulses from revivals of the English-speaking world. These groups emphasize faithfulness to the Bible and the Lutheran Confession, in reaction to liberal theology and higher criticism. These movements enjoy much freedom within the church and have influenced it to a great extent.

The Church of Finland is a state church, but has considerable liberty. The church assembly meets every five years; its enactments must be ratified by Parliament. In 1889 a church law was passed, giving everyone the right of choosing his religion; in 1923 this was enlarged to include the right to freedom from religion. The Orthodox Church is also regarded as a state church; active mostly in the eastern parts of the country before World War II, its members were spread out over almost the entire country after the evacuation of Karelia.

The Evangelical-Lutheran Church of Finland, which still claims more than ninety per cent of the population, has rather more than four-and-a-half million members; the Orthodox Church numbers about 68,000; none of the other registered bodies has a membership in excess of 10,000, according to available statistics. Roman Catholics number only about 3,000. The so-called Civil Register (which curiously includes Pentecostals and other groups unregistered as religious bodies) numbers some 250,000.

BIBLIOGRAPHY: E. Jutikkala, *A History of Finland* (1962); M. Juva, *The Finnish Evangelical Church* (1962); K. Pirinen, *L'Église de Finlande* (1962); L. Pinomaa, *Finnish Theology Past and Present* (1963); G. Sentzke, *Finland, Its People and Its Church* (1963) and *Reformation und Pietismus in Finnland* (1963).

STIG-OLOF FERNSTROM

FINLEY, SAMUEL (1715-1766). Presbyterian minister and educator. Born in Ireland, he emigrated to America after 1730 and studied under William Tennent, Sr.* at the "Log College" in Neshaminy, Pennsylvania, for the Presbyterian ministry. He was licensed in 1739 by the pro-revivalist Presbytery of New Brunswick and sent on itinerations as the Great Awakening* grew in intensity. A fiery preacher of repentance and divine grace and forgiveness, he defended the "New Side" (revivalist) position in a pamphlet war with John Thomson. Finley was aligned with the positions on excesses held by Jonathan Dickinson* and Jonathan Edwards.* He founded an academy on the Maryland-Pennsylvania border which helped to feed the College of New Jersey with youth aspiring to the New Side ministry. He became one of the early trustees of the new college at Princeton in 1746, was elected its fifth president in 1761, and served in that post until his death five years later.

KEITH J. HARDMAN

FINNEY, CHARLES G(RANDISON) (1792-1875). American revivalist. Born in Warren, Connecticut, and reared in Oneida County, New York, he entered a law office in Adams, New York, after limited formal education and was later admitted to the bar. At Adams he began attending church services conducted by a friend, George W. Gale. Although at first critical of religious dogmas, Finney after studying the Bible for himself was converted in 1821. This involved, as he said, "a retainer from the Lord to plead his cause." Turning from law, he began to preach and in 1824 received Presbyterian ordination. For the next eight years he conducted revivals in the eastern states with unusual results. In 1832 he became pastor of the Second Presbyterian Church in New York City, but dissatisfaction with the disciplinary system in Presbyterian churches soon led to his withdrawal from the presbytery. He also delivered a series of lectures on revivals during these years in New York. These were soon published (1835) and widely read.

In 1835 he became professor of theology at a new college in Oberlin, Ohio. During the remainder of his life he was linked with the school, serving as president from 1851 to 1866. Through most of these years, however, he remained active in evangelistic work, devoting a part of each year to revivals. Finney fits no theological pattern. In general he was a New School Calvinist, but he laid heavy stress on man's ability to repent, and he made perfectionism* a trademark of Oberlin teaching.

See also OBERLIN THEOLOGY.

BRUCE L. SHELLEY

FINNIAN OF MOVILLE (c.495-579). Irish monk. Probably trained at the monastery founded by Ninian in Galloway, he supposedly went to Rome where he was ordained priest. Returning to his native Ulster, he formed a religious community at Moville, about 540. From Italy he brought back to Ireland a copy of the Vulgate NT and Pentateuch. He is sometimes identified with St. Frigidian (or Frediana) who is venerated at Lucca in Italy, and sometimes confused with Finnian of Clonard (d.549), the traditional initiator of true monasticism in Ireland. Many legends surround him.

FIRMILIAN (d.268). Bishop of Caesarea in Cappadocia from about 230. A great admirer of Origen, they exchanged visits and he was able to study under him. Firmilian was clearly a man of great influence in the East, but his only surviving work is a letter to Cyprian (Cypr., Ep.75). In this he supports Cyprian's contention that baptism belongs to the church alone and is therefore invalid outside its confines. He rejects the opposing view of Stephen, bishop of Rome, with all the considerable power of scorn at his command. He presided over at least one synod to deal with the heretical Paul of Samosata, bishop of Antioch, in 264, and died at Tarsus on his way to Antioch to a further synod to decide this issue. C. PETER WILLIAMS

FIRST AMENDMENT, THE. The First Amendment to the Consitution of the United States states: "Congress shall make no law respecting an establishment of religion, or prohibiting the free exercise thereof; or abridging the freedom of speech, or of the press; or the right of the people peaceably to assemble and to petition the Government of a redress of grievances." The Supreme Court has held that the first phrase is not applicable when the free exercise takes a form of activity thought to be antisocial or self-destructive (e.g., polygamy, snake-handling, or refusal to be vaccinated). In the 1947 *Everson* v. *Board of Education* decision, the court interpreted the law as an intention to erect "a wall of separation between church and State." Since then, the court has invalidated religious instruction, Bible reading, and prescribed prayers in the public schools; however, the payment of public funds for the busing of children to parochial schools was ruled acceptable. JOHN P. DEVER

FISH. There are various biblical references to fish: the OT uses the word as a figure to point to man's helplessness (Eccl. 9:12; Hab. 1:14); Matthew (13:47f.) lets the variety of fish caught characterize the kingdom of heaven. In biblical material, however, the fish is not a primary symbol. In Christian art and literature, on the other hand, *ichthus* ("fish" in Greek), has basically been a symbol of Christ. The Greek letters form an acrostic (Jesus Christ, Son of God, Savior), but it is not known whether the acrostic preceded the symbol or vice versa. Neither is it known how early this usage developed. In addition, the term has been used to refer to neophytes (Tertullian, *De Baptismo*) and to the Eucharist (as in the paintings made inside the catacombs). From early times fish has been substituted for meat on days when fasting is observed. WATSON E. MILLS

FISHER, EDWARD (c.1601-1655). Anglican theological writer. He graduated in Arts at Oxford (1630), where he showed an excellent command of church history and the classics. While records are not clear, he is known to have been a royalist supporter and strongly anti-Puritan. Monetary difficulties set him traveling, and for a time he taught school at Carmarthen in Wales. His creditors found him, however, and he fled to Ireland, where he died. Fisher was the author of *The Scriptures Harmony* and *An Appeale to thy Conscience,* both published in 1643. Parts of another work of his, *A Christian Caveat to the old and new Sabbatarians,* in which he declined to regard Sunday as a Sabbath (a purely ecclesiastical device), were reprinted in New York two centuries after his death by the Seventh-Day Baptists of America. An early eighteenth-century school of thought identified Fisher with the "E.F." who wrote the *Marrow of Modern Divinity,* but Fisher's obviously superior education makes this highly improbable. J.D. DOUGLAS

FISHER, GEOFFREY FRANCIS (1887-1972). Archbishop of Canterbury, 1945-1961. Born in a Midlands rectory, he was educated at Marlborough and Oxford, taught at Marlborough for three years, then in 1914 was appointed headmaster of Repton at an astonishingly young age. In 1932 he became bishop of Chester; in 1939 he was translated to London. He was appointed to Canterbury in 1945 after the untimely death of William Temple. Deeply rooted in the church-and-community of nineteenth-century rural England, Fisher always combined deep devotion to the cause of Christ, an understanding of widely different viewpoints, and a strong sense of duty and of the need for discipline. As bishop he could appreciate Evangelical and Anglo-Catholic, those two wings of his church that between them have nearly all its "life," but who do not always appreciate each other. Bishop of London in the war years, he was successful in bringing about financial reorganization, but not in bringing order in place of the eccentricities of Anglo-Catholic worship. As primate he was largely responsible for the planning by which different provinces of the Anglican Communion acquired independence. Though deeply involved in interchurch relations, he became increasingly suspicious of organic union, and in his retirement staunchly opposed the Anglican-Methodist merger scheme. During his last ten years at Canterbury he traveled extensively, and made history by meeting Pope John XXIII* in Rome, in addition to visiting Jerusalem and Constantinople. P.W. PETTY

FISHER, GEORGE PARK (1827-1909). Church historian. Born in Wrentham, Massachusetts, he graduated from Brown University in 1847, continuing his studies at the Yale Divinity School, at Andover Seminary, and in Germany. From 1854 to 1861 he was a pastor, then he joined the faculty of Yale Divinity School, where he became professor of ecclesiastical history. He was a prolific writer whose first major work, *The History of the Reformation* (1873), was followed by others dealing with early Christianity and the problems of apologetics.

FISHER, JOHN (1469-1535). Roman Catholic martyr, sometimes known as St. John of Rochester. Born of a Yorkshire family, he was educated at Michaelhouse, Cambridge, of which college he became master in 1497. As the chaplain of Lady Margaret Beaufort he enjoyed her support for raising academic standards. In 1504 he became bishop of Rochester and chancellor of Cambridge. Against Lutheran ideas for reform he defended traditional Roman Catholic doctrine. He was scholarly, an admirer of Erasmus,* and genuinely interested in moderate reforms. As confessor to Catherine of Aragon he strongly protested against Henry VIII's determination to divorce her. His property was confiscated, and in 1534 he was put in the Tower for refusing to take the oath relating to the Act of Succession. Then Pope Paul III created him cardinal. This infuriated the king, and within a month Fisher was brought to trial in Westminster Hall, charged with treason. Found guilty, he was executed on 22 June 1535. With Thomas More he was canonized by Pius XI in 1936. PETER TOON

FISHER, JOHN (1569-1641). Otherwise known as Fisher the Jesuit. A native of County Durham whose original surname was Percy, he was early converted to Roman Catholicism and was educated at the English colleges at Reims and Rome. After admission into the Jesuit Order in 1594 he returned to England, was imprisoned on several occasions, and banished. He persisted in his mission and disputations, however, and was persuasive enough to bring William Chillingworth* into the Roman Catholic Church for a time.

FISKE, FIDELIA (1816-1864). First single woman missionary to Persia. Born in Shelburne, Massachusetts, she was a niece of Pliny Fisk, one of the first two missionaries of the American Board to the Near East. She was converted at fifteen and joined the Congregational Church. A graduate of Mount Holyoke Seminary (1842), she was deeply influenced by Mary Lyon. The American Board appointed her to Persia in 1843. For fifteen years she served in that land, principally among Nestorian women and girls near Lake Urmia. She directed the first boarding school for girls. Ill health forced her return to the USA in 1858. She was offered the principalship of Mt. Holyoke, but refused in the hope, never realized, of being able to return to Persia.

HAROLD R. COOK

FISKE, JOHN (1842-1901). American philosopher, historian, and lecturer. A precocious child, he was graduated from Harvard and admitted to the bar without formal legal training. He turned, however, to propagating Herbert Spencer's evolutionary philosophy that had replaced the Calvinism of his youth. Claiming "Evolution is God's way of doing things," he infused evolutionism with religious values in such works as *Outline of Cosmic Philosophy* (1874), *The Destiny of Man* (1884), *The Idea of God* (1885), and *Through Nature to God* (1899). In his last twenty years he turned more to the study of American history. Essentially a popularizer, he was nevertheless one of the most important intellectual influences in America at the end of the nineteenth century.

DONALD W. DAYTON

FITZRALPH, RICHARD (d.1360). Archbishop of Armagh. Born at the end of the thirteenth century, he studied at Oxford, and about 1333 was chancellor of the university. In 1334 he became chancellor of Lincoln Cathedral. He was in favor with the Avignon papacy, was advanced to the deanery of Lichfield in 1337, and ten years later was consecrated archbishop of Armagh. In 1349 he visited Avignon and came into conflict with the mendicant orders. Between 1353 and 1356 he wrote *De Pauperie Salvatoris* in seven books. He was spokesman for the secular clergy, claiming that poverty was neither of apostolic observance nor of present obligation, and that mendicancy was without warrant in Scripture or primitive tradition. In a series of sermons at St. Paul's Cross, London, he preached against the mendicants. He was hotly opposed by the English friars, notably by Roger Conway, and was cited to Avignon where he preached a famous sermon, "Defensio Curatorum," before Innocent VI, but died before judgment was given. In his opposition to the friars, and in the doctrine of "Dominion" in *De Pauperie Salvatoris* he was a forerunner of John Wycliffe. J.G.G. NORMAN

FIVE MILE ACT (1665). One of the statutes that comprised the Clarendon Code* by which Parliament sought to penalize those who did not subscribe to the liturgy and doctrine of the Church of England. The Five Mile Act forbade Nonconformist ministers and teachers to come within five miles of any city, town, or parliamentary borough. Offenders were subject to a severe fine.

FLACIUS, MATTHIAS ILLYRICUS (1520-1575). German Lutheran theologian. Born in the Adriatic peninsula of Istria, his Croatian name was Vlacic (Latinized as Flacius) and Illyricus was added later to refer to his coastal homeland. An orphan, he worked his way through school and studied with the humanist Baptista Egnatius, a friend of Erasmus, in Venice (1536-39). Through the influence of his uncle he was dissuaded from becoming a monk and persuaded to attend the university. Consequently he studied at the universities of Basle, Tübingen, and Wittenberg. At the latter school he underwent a deep spiritual crisis which ended only when he was converted to evangelical doctrine through contact with Luther. He became a professor of Hebrew (1544) and lectured on Aristotle and the Bible. He subsequently differed with Melanchthon over the compromise Augsburg Interim (1548) and then wandered from Jena to Regensburg, Antwerp, Strasbourg, and finally to Frankfurt, where he died. He was almost Manichaean in his view of sin and evil in man. His fame rests upon his *Clavis* or key to the Scriptures, a monument in the history of hermeneutics, and the *Magdeburg Centuries*, an interpretation of church history which in its extremely antipapal emphasis had a strong effect on later Protestant thought.

ROBERT G. CLOUSE

FLAD, JOHANN MARTIN (1838-1915). German missionary to Ethiopia. A saddler by trade, as a boy he enrolled in the St. Chrischona school. Bishop Gobat asked the school for four lay craftsmen to staff a new work in Ethiopia, and Flad was among those selected. In 1855 he went to Ethiopia and initially was cordially received. He became a victim of persecution by Emperor Theodorus II in 1864, and two years later, with his family held hostage, the emperor compelled him to undertake a diplomatic mission to England. After the British had forced the downfall of Theodorus in 1868, Flad was able to pursue an undisturbed ministry of literature distribution and evangelism among the Falascha (Ethiopian Jews) until his death. RICHARD V. PIERARD

FLAGELLANTS. In the Middle Ages these were groups of people, under the influence of a form of religious hysteria, who often went barefooted and inflicted beatings on their bare shoulders by scourges as an act of penance. They first appeared in Bologna in 1260 following a period of famine and strife. The prophecies of the impending end of the world by Joachim of Flora combined with the state of the times to create a mass hysteria at the possibility of divine displeasure. The orthodox belief in the efficacy of the scourge *(flagella)* as a sign of repentance degenerated into a depraved delight in self-torture and the conviction that flagellation was the only effective sacrament. The ecclesiastical approval given to the bands on their first appearance was later withdrawn. Clement VI repressed them, and they were condemned by the Council of Constance.

The most spectacular appearance of the Flagellants took place in N Europe in 1349 and was associated with the outbreak of the Black Death.* Their bloodletting was an attempt to stanch God's anger by sacrificial propitiation. The Flagellants believed that, because of their self-inflicted tortures, they would all be saved, that they bore on their bodies the stigmata of Christ, and that their blood mingled with his blood. They also called for the killing of the Jews, whom they believed were the enemies of God and responsible for the plague. Their language and customs were those of the commonly approved piety of their day, adapted to their desire to make a sacrifice to God of their own bodies and those of their enemies. They revealed all the signs of a mass religious reaction during a period of great popular stress.

See W.M. Cooper, *Flagellation and the Flagellants* (1908). JAMES TAYLOR

FLAVEL, JOHN (d.1691). English Puritan and Nonconformist divine. Educated at University College, Oxford, he was ordained by a presbytery at Salisbury and became curate of Diptford, Devon. Later he was lecturer of St. Saviour's, Dartmouth. Ejected in 1662, he became a Nonconformist. In 1672 he was licensed as a Congregational to preach in his own house at Dartmouth. Persecuted in Devon, he removed in 1682 to London, but returned to Devon where he was instrumental in promoting the "Happy Union" of Presbyterians and Congregationalists in 1690-91. He is best known for his practical writings, which were usefully collected in six volumes in the nineteenth century and have now been reprinted as *The Works of John Flavel* (1968). PETER TOON

FLAVIA, DOMITILLA, see DOMITILLA

FLAVIAN. Bishop of Antioch, 381-404. He was leader of Melitius's* supporters during the latter's banishment under Constantius (360) and Valens (370), and was largely responsible for Melitius's recognition as the rightful claimant to the see of Antioch under Gratian. The rival claimant, Paulinus,* continued as bishop of the orthodox Eustathians who had separated from the Antiochene church on the deposition of Eustathius. Flavian accompanied Melitius to the Council of Constantinople in 381. Melitius died at the council, and against the wishes of Gregory of Nazianzus, the council chose Flavian as his successor, although his actual election must have taken place in Antioch. Like Melitius, Flavian was supported by the Eastern bishops, while the Egyptian and Western bishops continued to recognize Paulinus. A council was held at Rome to decide this and other matters in 382, but the Eastern bishops did not attend, holding a synod of their own at Constantinople in which they ratified Flavian's election. Theophilus of Alexandria gave his recognition to Flavian in 394, and Rome followed some time before 398. DAVID JOHN WILLIAMS

FLAVIAN (d.449). Patriarch of Constantinople. He took a prominent part in the movement for the condemnation of Eutyches and presided at a synod held in Constantinople for this purpose in 441. Eutyches was found guilty of denying the two natures in the person of Christ, but appealed from Flavian to Pope Leo I at Rome. The result of this was the meeting of the Council of Ephesus, better known as the Robber Synod,* in 449. This council then absolved Eutyches from all charges of heresy in regard to the person of Christ. It also deposed Flavian from office, and the latter was killed by his opponents there.

FLÉCHÉ, JESSE (d.1611?). Missionary priest. He was chosen by Jean de Biencourt de Poutrincourt to accompany him to Acadia in 1610. After landing in Port Royal in late May of 1610, Fléché, under pressure from Poutrincourt, baptized the Micmac chief Membertou and twenty of his tribe on June 24. The Indians were not ready for baptism, but Poutrincourt wanted to be able to present a favorable report on the progress of the mission in order to keep the support of wealthy and pious people in France. Within a year Fléché baptized over 100 Indians, but when the Jesuits arrived in 1611 they were shocked that these new converts were ignorant of the rudiments of faith. Fléché returned to France in 1611.
ROBERT WILSON

FLÉCHIER, ESPRIT (1632-1710). Bishop of Nîmes. Born at Pernes, France, he studied at Tarascon, was ordained, and went to Paris (1660) where he gained royal favor for a poem on Louis XIV, becoming tutor to the Dauphin. His preaching became famous, his panegyrics being compared with Bossuet,* and his work in history and literature flowered. Elected to the French Academy in 1673, Fléchier was made bishop of Lavaur in 1685 and of Nîmes two years later. At Nîmes he dealt with the consequences of the Edict of Nantes and won many Huguenots to Catholicism.

He was an elegant man of letters and a pretentious orator, but neither a moralist nor a humble and spiritual preacher. His works include *Oraisons Funèbres, Sermons,* and *Panégyriques,* and several histories. C.G. THORNE, JR.

FLEMING, PAUL WILLIAM (1911-1950). American missionary and founder of the New Tribes Mission. Through his mother's prayers and the ministry of Paul Rader he went to Malaya in 1937 to reach inland tribes. Repeated attacks of malaria obliged him to return to the USA, where he tried to stimulate enthusiasm for missions to unreached peoples. In 1942 he started the New Tribes Mission with headquarters in Chicago and sent out a first party to eastern Bolivia. The disappearance of the first five men in the jungle in 1943 did not dampen his ardor. Neither did the fatal crash of the first mission transport plane in Colombia in 1950. Later that year, however, came a second fatal crash on Mount Moran in Wyoming, with Fleming on board. HAROLD R. COOK

FLETCHER, JOHN WILLIAM (1729-1785). English clergyman. Born in Switzerland (his original name was de la Fléchère), he came to England with a distinguished university record from Geneva, and was appointed as a private tutor in 1752. Converted under the influence of the Methodists, he was ordained by the bishop of Bangor in 1757. After assisting John Wesley in London, Fletcher took the living of Madeley, Shropshire, in 1760, preferring it to one double its value previously offered to him. For a time he superintended the Countess of Huntingdon's ministerial training college at Trevecca. During the Calvinistic controversy Fletcher was the chief defender of evangelical Arminianism against the objections of Shirley and others. His five *Checks to Antinomianism* (1771-75) have been compared with Pascal's *Provinçiales* as models of polite controversial irony.

In his personal relationships with theological opponents Fletcher was a model of Christian reconciliation. Above all, he exemplified in his own character the holiness he preached. Herein lay the secret of his influence over the rough colliers in his parish. Robert Southey said of him: "No church has ever possessed a more apostolic minister." That Wesley recognized his worth can be seen in the fact that he designated Fletcher as his successor, had he consented, as the leader of Methodism.

BIBLIOGRAPHY: *Works* (9 vols, 1800-4); L. Tyerman, *Wesley's Designated Successor* (1882); F.W. Macdonald, *Fletcher of Madeley* (1885); J. Marrat, *The Vicar of Madeley, John Fletcher* (1902). A. SKEVINGTON WOOD

FLEURY, CLAUDE (1640-1723). French church historian. Born in Paris, son of a Normandy lawyer, he was educated in the Jesuit College of Clermont and began practicing law in 1658. Bossuet and Fénelon brought him into contact with leading French personages, and his own studies in civil and canon law, history, literature, and archeology advanced. Ordained in 1669, he won positions in the French court as tutor to the Princes de Conti (1672) and then to Louis XIV's grandsons (1689), was abbot of the Cistercian Loc-Dieu (1684), elected to the French Academy in 1696, and was made prior of Notre-Dame-d'Argenteuil in 1706. Chosen confessor to Louis XV, he was averse to disputes and remained aloof from Jansenism and other movements. His great work was *Histoire ecclésiastique* (20 vols., 1690-1720), which he ended at 1414 and others continued. This, together with *Les Moeurs des chrétiens* (1682) and *Grand Catechisme historique* (1683) and other works, was placed on the Index* for Gallican tendencies. C.G. THORNE, JR.

FLIEDNER, THEODOR (1800-1864). Founder of the deaconess organization in the German Lutheran church. Born in Eppstein/Taunus, he was educated for the ministry at Giessen and Göttingen. As a young pastor in Kaiserswerth near Düsseldorf he became acquainted with the Mennonite practice of appointing deaconesses, and in 1833 he established a home for ex-convicts which he placed in charge of a woman. In 1835-36 he set up a school for children which also trained women teachers, and a hospital in 1836 which gave nursing instruction. By reviving the office of deaconess, Fliedner provided opportunities for unmarried women to be active in public life. He founded other mother houses in Germany, modeled after the Kaiserswerth Sisters, and introduced the idea into the United States and Palestine. RICHARD V. PIERARD

FLORENCE, COUNCIL OF (1438-45). An ecclesiastical assembly noteworthy for its attempt to unite the Greek and Latin churches. While the Council of Basle* was in session, the Greek Empire was under threat from the Turks. The emperor, John VIII Palaeologus, decided to propose to the pope, Eugenius IV, that the Greek and Latin churches unite and thereby offer effective resistance to the infidels. From Basle the council was transferred to Italy by the pope in order to bring it under his control. The sessions began on 8 January 1438 in Ferrara, and three months later the Greek representatives arrived as guests of the pope. They included the emperor, the archbishop of Nicea (John Bessarion*), and the metropolitan of Ephesus (Mark Eugenikos). The latter was an antiunionist.

When the cost of the council became too much for the pope, he accepted the offer of the city of Florence to pay for it, and it was moved there in February 1439. Here the most important discussions and agreements took place. Difficulties were encountered in four areas: the Double Procession of the Spirit, the use of unleavened bread in the Eucharist, the doctrine of purgatory, and the primacy of the bishop of Rome. Of these, the first and the last gave the most problems and were subjects discussed by commissions appointed in the council.

The famous conciliatory discourse by Bessarion on the doctrine of the Spirit, promises of help against the Turks, and the death of the patriarch of Constantinople on 10 June 1439 all helped to make an agreement possible. The union document was prepared by Ambrose Traversari, and the decree of union, beginning with the words *Laetentur Coeli,* was signed on 5 July 1439. A few, led by Mark Eugenikos, did not sign. On face value it seemed the Latins had won on all points of doctrine, but the Greeks did not believe they had conceded any important points. On 6 July, in the cathedral of Florence, divine service was held to celebrate the union. Cardinal Cesarini read the decree in Latin, and Bessarion in Greek; then the pope celebrated Mass.

In August the Greek emperor left. The fall of his capital was not prevented, however, and in addition the Greek Church renounced the union made at Florence. With the Greeks gone, the council dealt with the continuing irregular Council of Basle and excommunicated its members; also it sought union with other Eastern churches (Mesopotamian, Chaldean, and Maronite). The pope's ascendancy over councils was affirmed in the bull *Etsi non dubitemus* of 20 April 1441. In 1443 the council was moved to Rome, where it concluded its sessions in 1445. It is regarded as either the sixteenth or seventeenth ecumenical council, due to the fact that the status of Basle (1431-49) is debated.

BIBLIOGRAPHY: J.D. Mansi, *Concilia* (1789), vol. XXXI and Supplement; Hefele-Leclercq, *Histoire des conciles d'après les documents originaux* (1916), vol. VII, part 2; J. Gill, *The Council of Florence* (1959) and *Eugenius IV* (1961).

PETER TOON

FLORENTIUS RADEWIJNS (1350-1400). Follower of Gerard Groote,* he helped found the Brethren of the Common Life.* He studied at Prague, became a canon at Utrecht, and was converted to a serious Christian life by Groote's fiery preaching. At his suggestion, and with his help, Groote founded the Brethren, to be devoted to the cultivation of practical piety. After Groote's death in 1387, Florentius became leader of the movement and formed the Congregation of Windesheim, with regular canons, associated with the Brethren. Under his leadership the Brethren expanded rapidly. If Groote founded the movement, Florentius was its organizer. During the 1400s it was to become a major devotional and educational force in the Low Countries. Florentius also wrote several brief devotional works (e.g., *Tractatus devotus de extirpatione vitiorum*).

DIRK JELLEMA

FLORILEGIA (Lat. *flores legere,* "to gather flowers"). Collections of quotations, particularly from patristic sources or from Scripture. Such anthologies were common in early Christian times when books were not readily available. Reformation collections of "proof-texts" (ranging from dogmatics in outline to oratorio librettos) and their related precedent medieval collections of passages from the Church Fathers in addition to Scripture belong to this class of literature. They show the degree to which subsequent thought in the history of the tradition included as an essential element the direct commentary upon its predecessors.

FLORUS (d. c.860). Scholar and controversialist. Nothing is known of his early life before he became a deacon of Lyons during the period when Agobard* was its bishop (816-40). After Agobard's deposition (he was later reinstated) in 835 because of his opposition to the schemes of Empress Judith, Florus defended the rights and independence of the Church of Gaul in *De iniusta vexatione ecclesia Lugdonesis.* His other writings included a defense of moderate predestination against the extreme views of Gottschalk, three treatises on liturgy, a commentary on the epistles of Paul, some additions to the Martyrology of Bede, and a collection of poems.

FLUE, NIKOLAUS VON, see NICHOLAS OF FLUE

FLÜGEL, OTTO (1842-1914). German Herbartian philosopher and theologian. Born at Lützen, he studied at Halle and was strongly influenced by C.S. Cornelius, a disciple of J.F. Herbart, and so became himself an Herbatian and opponent of monism. He served as a Lutheran minister at Laucha, Schochwitz, and Wansleben, but ended his life as a free-lance writer. He was involved in the publication of the journal *Zeitschrift für exokte Philosophie* which later was entitled *Zeitschrift für Philosophie und Pedagogik* in 1894.

FOAKES-JACKSON, F.J., see JACKSON, F.J.F.

FONT. A container of baptismal water, usually made of stone but sometimes of metal. Early fonts were basically pits in which the baptismal candidates could actually stand and be immersed. Later, in the Middle Ages, when infant baptism was common custom, fonts were raised above the ground so that babies could be immersed. Later still, when pouring or sprinkling replaced immersion, fonts were made smaller and raised on higher pedestals. Often wooden or metal lids covered fonts to preserve the purity of the water. Groups that practice only believer's baptism prefer the term "baptistry," associating "font" with infant baptism.

FONTANA, DOMENICO (1543-1607). Italian architect. Born at Melide near the Swiss-Italian border, he worked in Rome before his appointment as chief architect by Sixtus V in 1585. During the latter's five-year pontificate, Fontana's many works included the Laleian Palace and the Quirinal, and a significant contribution to the completion of the dome of St. Peter's. Dismissed for alleged misconduct be Clement VIII (1592), he later built the royal palace at Naples, in which city he died.

FOOLS, FEAST OF, see FEAST OF FOOLS

FORBES, ALEXANDER PENROSE (1817-1875). Bishop of Brechin. Of Scottish aristocratic descent, he was educated at Glasgow University, then after three years with the East India Company in Madras, he returned to Britain and in 1844 graduated from Oxford. Ordained in the Church of England, he served a Leeds parish briefly before consecration in 1848 to the Scottish see of Brechin. Close friend of E.B. Pusey,* he advocated the doctrine of the Real Presence, a stand which brought censure from his episcopal colleagues. It was at Pusey's instigation also that he wrote *An Explanation of the Thirty-Nine Articles* (2 vols., 1867-68). His other works include *Kalendars of Scottish Saints* (1872) and an edition of *Lives of St. Ninian, St. Kentigern, and St. Columba* (1875). J.D. DOUGLAS

FORBES, JOHN (1593-1648). Most prominent of the Aberdeen Doctors.* After studying at continental universities he was appointed professor of divinity at King's College, Aberdeen, in 1620. A man of great integrity, he was a defender of episcopacy and a notable theologian. He was deposed in 1639 after disagreeing with the policies of the Covenanters,* and was evicted from the residence which, formerly his own, he had made over to the university. Worse was to follow. Being reluctant later to sign the Solemn League and Covenant,* he felt compelled in 1644 to leave the country for Holland, where he spent two years. Even after his return to Scotland the more extreme among the Presbyterians had not forgiven his opposition: it was Forbes's wish to be buried beside his wife and father in St. Machar's Cathedral, Aberdeen, but this was refused, and he rests in the kirkyard at Leochel. Among his works was *Instructiones Historico-Theologicae de Doctrina Christiana* (1645). J.D. DOUGLAS

FORGED DECRETALS, see FALSE DECRETALS

FORM-CRITICISM (Ger. *Formgeschichte,* "Form-history"). This is a method of literary study, applied both to secular and religious literature, which seeks to classify the forms which underlie written documents, and to reconstruct the process by which they reached their present shape. The pioneer in using this method for study of the Bible was Hermann Gunkel,* who first applied it to the narratives of Genesis. Among the more significant forms which were found to be present in the OT were "legend" and "myth." A "legend" was a story with a historical base which was recounted for an instructional or devotional purpose. A "myth" was a story to explain in pictorial form some supernatural truth. The presuppositions of some form-critics and the use of the words "myth" and "legend" (which were generally taken to indicate a lack of historical reliability) tended to give form-criticism of the OT a negative bias. But much useful work has been done, particularly in the classification of various kinds of poetic and prophetic literature.

The application of form-criticism to the NT, though dependent upon the work of Gunkel and that of J. Wellhausen* and E. Norden, was specifically made almost simultaneously between 1919 and 1921 by K.L. Schmidt, M. Dibelius,* and R. Bultmann.* The particular area of study was the synoptic gospels and, as source-criticism had suggested Mark to be the earliest of these, form-criticism, seeking to go a step further back, concentrated particularly on Mark. Schmidt sought to show that the paragraphs of Mark were units on their own, and that the gospel was "a heap of unstrung pearls." The main classifications involve those stories which are told principally for a saying of Jesus ("apophthegms," "paradigms," or "pronouncement stories") and those which are told principally for an action of His ("miracle stories" or *Novellen*).

Form-critics have tried to find the *Sitz im Leben* ("setting in life") of the various units, and so a great deal of study has been directed to the understanding of the everyday life of the early church, including its liturgical and evangelistic activity. There has been a tendency among many form-critics to suggest that the early church created the gospel material to serve its own needs and thus to find its original *Sitz im Leben* after the ministry of Jesus. But this is due, not to the method itself, but to philosophical presuppositions held about the nature of the gospel history. The method may equally well be used to suggest that incidents from the life of Christ were not created, but preserved by the church because of their usefulness, and thus provide a genuine double *Sitz im Leben* for the gospel material.

BIBLIOGRAPHY: V. Taylor, *The Formation of the Gospel Tradition* (1933); M. Dibelius, *From Tradition to Gospel* (ET 1934); C.F.D. Moule, *The Birth of the New Testament* (2nd ed., 1966); R. Bultmann, *History of the Synoptic Tradition* (ET 1968). R.E. NIXON

FORMOSUS (d.896). Pope from 891. Probably born in Rome, he became cardinal bishop of Porto in 864 and was entrusted with papal missions by Nicholas I and Hadrian II. Papal legate to Bulgaria (866-76) to promote the country's conversion, and emissary to France (869) and Trent (872, to deal with the question of Louis II's successor), Formosus was dismissed as prince of Boris in 876 for opposing the policies of John VIII after serving ten years. He lived at Sens until 882 when Pope Marinus restored him to the diocese. Elected pope, he asked the German king Arnulf to protect Italy against Guido, former duke of Spoleto, and named him emperor in 896. His pontificate witnessed strife over Photius of Constantinople, the suffragan of Bremen, and the successor to the French crown (he supported Charles the Simple). Beholden to the Spoletanian party, the new pope Stephen VI exhumed Formosus's body, conducted a posthumous trial, declared him deposed, and allowed his decomposed body to be thrown into the Tiber. JOHN GROH

FORMULA OF CONCORD, see CONCORD, FORMULA OF

FORSYTH, PETER TAYLOR (1848-1921). Congregationalist theologian. A postman's son in Aberdeen, he was educated at the university there, then studied at Göttingen under A. Ritschl.* After serving various Congregational churches in England, he became in 1901 principal of Hackney College, London, a post he retained until his death. Keenly interested in historical critical theology, and concerned to open the way to a "better, freer, larger Church," he took part in the Leicester Conference which had such aims in 1877, and was suspected in his denomination of heterodoxy.

Although he never went back to the earlier conservative scholastic theology, Forsyth gave increasing emphasis to the need for using the new theological critical freedom to live by and for the evangelical realities, not to supplant them. Writing with learning, passion, and an idiosyncratic style, he argued that man must not take God's central place in theology; that God's love, being holy, was necessarily wrathful against sin, which could not be explained away. He stressed that atonement was by the cross, in which God as well as man was reconciled at cost. In his greatest work, *The Person and Place of Jesus Christ* (1909), he made a creative contribution to Christology in suggesting that *kenosis* (self-emptying) and *plerosis* (fulfilling) are the two movements from God to man and from man to God which savingly occur in Christ.

In this later period he was increasingly respected as a Congregationalist leader. He developed a high doctrine of the church, ministry, and sacraments (including preaching), and was a sharp critic of the laymindedness, individualism, and tendency to nondoctrinal religion prevalent in the Free churches. He opposed the "New Theology" of R.J. Campbell. His churchmanship was not an imitation of Anglican or Roman Catholic forms, but a disciplined working out of Christian truth according to basic Congregationalist principles.

See G.O. Griffith, *The Theology of P.T. Forsyth* (1948); and R.M. Brown, *P.T. Forsyth: Prophet for Today* (1952). HADDON WILLMER

FORTY MARTYRS OF SEBASTE, see SEBASTE

FORTY-TWO ARTICLES ACT (1553). The accession of Edward VI to the throne in 1547 marked an important step in the development of the Reformation in England. Under the influence of Thomas Cranmer the young king began to exercise a direct influence on the church. As a result Cranmer drew up the Forty-Two Articles Act, which became the first truly Protestant confession of faith for the Church of England. Promulgated in 1553, they reflected the position of the Augsburg Confession in their statement of the doctrines of the Trinity and justification, but in their position on predestination and the Lord's Supper they were clearly Calvinistic. They were revised in 1562 by the Convocation of the Anglican clergy, and in 1563 they were promulgated by Elizabeth as the Thirty-Nine Articles Act.

C. GREGG SINGER

FOSDICK, HARRY EMERSON (1878-1969). American Baptist minister. He was a pastor in Montclair, New Jersey (1904-15) and taught practical theology (especially homiletics) at Union Theological Seminary (1908-46). From 1918 he served as guest minister at First Presbyterian Church, New York City, where he played a prominent role in the fundamentalist-modernist controversy when his 1922 sermon "Shall the Fundamentalists Win?" led to his 1925 resignation. Soon thereafter he became minister of the influential Riverside (then Park Avenue Baptist) Church, where he remained until retirement (1946). A popularizer of evangelical liberalism, biblical criticism, psychology of religion, and psychologically oriented "personal religion," Fosdick greatly influenced American preaching through his "problem-centered" homiletical style. Among his thirty widely read books were *The Modern Use of the Bible* (1924), *A Guide to Understanding the Bible* (1938), and devotional books *The Manhood of the Master* (1913), *The Meaning of Prayer* (1915), and *On Being a Real Person* (1943). His autobiography, *The Living of These Days,* was published in 1956. DONALD W. DAYTON

FOSSORS (Fossarians) (Lat. *fodere,* "to dig"). Gravediggers; officers of the early church charged with the burial of the dead. They were initially regarded as inferior clergy and in late fourth or early fifth centuries became powerful corporations, with the management of the catacombs in their hands. They had the power to sell grave-spaces, and numerous inscriptions recording such sales survive. Included in the corporations were the artists who adorned Christian tombs. After the fall of Rome (410), burial in the catacombs becoming impossible, inscriptions of the fossors virtually cease, but a later chronicle (possibly sixth century) includes them among the clergy. Other names for them were *lecticarii* (from *lectica,* "bier") and *copiatae* (from Gr. *kopos,* "labor"). J.G.G. NORMAN

FOSTER, GEORGE BURMAN (1858-1918). Baptist scholar. Born in West Virginia, he graduated from West Virginia University and Rochester Theological Seminary, he taught at McMaster University before becoming one of the early members of the "Chicago School of Theology," where he was named professor of systematic theology in 1897, and of the philosophy of religion in 1905. For his views on the relation between Christianity and such subjects as Darwinian evolution, comparative religion, and relativistic physics, he was "excommunicated" by fellow Baptists in the fundamentalist-modernist controversy of the early twentieth century. He even wrestled with the "death of God" as voiced in the thought of F.W. Nietzsche,* despite criticism from conservatives. CLYDE CURRY SMITH

FOUCAULD, CHARLES DE, see DE FOUCAULD

FOURSQUARE GOSPEL, see MCPHERSON, AIMEE SEMPLE

FOX, GEORGE (1624-1691). Founder of the Society of Friends. Born in Leicestershire and apprenticed to a shoemaker, he apparently had no formal schooling. In 1643 he parted from family and friends and traveled in search of enlightenment. After long, painful struggles he came in 1646 to rely on the "Inner Light of the Living Christ." He forsook church attendance, dismissed contemporary religious controversies as trivial, and in 1647 began to preach that truth is to be found in God's voice speaking to the soul—hence, "Friends of Truth," later abbreviated to

"Friends." In 1649 he was jailed for interrupting a Nottingham church service with an impassioned appeal from the Scriptures to the Holy Spirit as the authority and guide. In 1650 at Derby he was imprisoned as a blasphemer, and there a judge nicknamed the group "Quakers," after Fox had exhorted the magistrates to "tremble at the word of the Lord."

The prospect of a new government more sympathetic to his views did not attract Fox, for he declined a captaincy in the parliamentary army. NW England he found specially responsive, and it was there at Swarthmore Hall, near Ulverston, that he established his headquarters. His irenic spirit was more highly developed than that of many of his associates, and his discipline of religious silence had a sobering influence. Fox spent six years in different prisons, sometimes under terrible conditions. He campaigned against the latter and against other social evils. His later years were spent in the London area, working to the end in helping others, promoting schools, and campaigning for greater toleration—and all this despite poor health caused by prision severities. His famous journal was published posthumously in 1694.

See also FRIENDS, SOCIETY OF.

BIBLIOGRAPHY: J. Smith (ed.), *Descriptive Catalogue of Friends' Books* (1867); T. Hodgkin, *George Fox* (1896); A.N. Brayshaw, *The Personality of George Fox* (3rd ed., 1933); H.J. Cadbury, *George Fox's Book of Miracles* (1948); editions of Fox's *Journal* by N. Penney (1911) and J.L. Nickalls (1952). J.D. DOUGLAS

FOXE, JOHN (1516-1587). Protestant historian and martyrologist. Born at Boston, Lincolnshire, he studied at Brasenose College, Oxford, and was fellow of Magdalen (1539-45). He became tutor to Thomas Lucy of Charlecote, and later to the earl of Surrey's children, when he met John Bale,* who stimulated his interest in history. He was ordained deacon by Nicholas Ridley in 1550.

On Mary's accession (1554) Foxe fled to the Continent, taking a manuscript designed to be the first part of a history of the movement for the reformation of the church, published eventually in Strasbourg with the title *Commentarii Rerum in Ecclesia Gestarum....* By September 1555 he was at Frankfurt, where he met other refugees, including Edmund Grindal,* who was recording the stories of the martyrs. Foxe joined Bale at Basle, where he found employment at the printing establishment of Oporinus; he also turned Grindal's martyr stories into Latin. At Mary's death, Grindal and his associates hurried to England, but Foxe remained to complete his book, bringing his history up to date and making use of Grindal's materials. Oporinus published it in 1559.

Returning to England, Foxe was ordained priest by Grindal, now bishop of London. He joined forces with John Day, the printer, who published the first English edition of his book in 1563 as *Actes and Monumentes* (popularly known as *Foxe's Book of Martyrs*). Four editions appeared in his lifetime, profoundly influencing Elizabethan England. More recent editions have often been truncated, unrepresentative versions. Foxe spent the rest of his life in London with Day, and was buried in St. Giles', Cripplegate.

See J.F. Mozley, *John Foxe and His Book* (1940); and W. Haller, *Foxe's Book of Martyrs and the Elect Nation* (1967). J.G.G. NORMAN

FRA ANGELICO, see ANGELICO, FRA

FRANCE. The Christian faith made its appearance in Gaul at an early date, probably in the first century. Missionaries and merchants from the East brought the Gospel to Marseille from which town it spread up the Rhone River valley to Vienne and Lyons. The greatest impact was made initially in the cities, among the Roman and Greek populations. Progress was much slower in rural areas among the native Celts. Martyrdom became the lot of many Christians in Gaul in the second and third centuries. Under the vigorous leadership of Irenaeus* and others, however, the faith spread northward, reaching Paris in 250. By the time religious toleration was granted throughout the Roman Empire in the early fourth century, Christianity was established in the cities of Gaul, but had only begun to penetrate the countryside.

When the Germans overran the empire beginning in the fifth century, the Christians of Gaul were faced with the task of converting and civilizing them. Since the church was the strongest surviving institution in the West, the invaders sensed its importance. A turning point in the history of French Christianity came in 496 when Clovis,* king of the Franks, was baptized. Genuine or not, his conversion made the task of the further evangelization of his people much easier for Christian missionaries.

During the early centuries, Gallic Christianity produced a number of illustrious saints such as Martin of Tours,* Hilary of Poitiers,* the talented scholar-poet Paulinus of Bordeaux (d.431), Germanus of Auxerre (d.448), and Genevieve.* The work of Gregory of Tours* also adds luster to the history of Christianity in Gaul in this period.

By about 500, Frankish Gaul had been divided into dioceses. Although the history of the Christian faith in this period is obscure, one important service which the Franks rendered to the Western Church was their firm stand against the Muslim invasion from the south which in the eighth century threatened all of Christendom. Charles Martel* stemmed the tide of Muslim advance with his victory over them near Tours in 732.

It was, however, the grandson of Charles Martel, Charlemagne,* who proved the major benefactor of the Christian Church in Frankland in the early Middle Ages, doing all he could to further it, even using force to convert the heathen when necessary. But by 843 his former empire was divided into three parts, and this date marks the beginning of the modern kingdom of France. After a protracted power struggle, the Capetians, with the support of the French clergy, were enthroned to replace the ineffectual Carolingians as the kings of France.

The history of Christianity in medieval France was marked by great vitality and achievement in the realms of piety, reform, learning, and politics.

No other kingdom of Europe surpassed the French in medieval times in their leadership and enthusiasm for piety and reform. The Capetians supported the French Church and, in turn, the church greatly influenced the affairs of state to an extent that it has never attained either before or since. One of the most devout of all Christian kings in history was Louis IX,* and under his reign the alliance of church and state in France reached its highest point of development.

Medieval France was also the home of the Cluny reform movement (see CLUNIACS), which began in 910 and was to contribute a host of reformers to the church, including several popes. Bernard of Clairvaux* was another whose influence extended to every part of the Christian world. It was he who preached the Second Crusade so effectively that he himself later noted it had reduced the ratio of women to men in France to seven to one. Significantly, the First Crusade had been preached and organized at Clermont in 1095. The history of the Crusades, indeed, was so inexorably linked with French leadership that in the Holy Land the Crusaders as a whole were known simply as "Franks," and the Christians of those lands until very recently looked to France as their protector.

In addition, medieval France was the home of many popular medieval heresies. A majority of the people of S France may have been "heretics" during the twelfth and thirteenth centuries. The major heretical groups were the Albigensians* and Waldensians (see WALDENSES). The former outnumbered the latter, but their exact beliefs were not known because both they and their records were destroyed with ruthless thoroughness. The Waldensians survived and flourish today in Italy and other parts of the world. Both groups represented vigorous protest movements against the lax Christianity which they felt existed in their day.

Christian piety in medieval France expressed itself also in the building of magnificent Gothic cathedrals in the later Middle Ages. Illustrative of this great outburst of church construction in the period was the celebrated Notre Dame cathedral of Paris which was begun in 1163. Medieval Paris, moreover, was the home of the first and greatest of the universities of N Europe. Sponsored and protected by the church, the University of Paris was chartered in 1200. Even before the official founding of the university, Paris had been the site of a celebrated cathedral school. Many of the most famous teachers of the day were either French or taught in the French schools: Anselm,* Peter Abelard,* and Thomas Aquinas.*

Finally, the church in France in the Middle Ages played a major role in the development of both French and papal politics. The part of Joan of Arc* in restoring the credibility of the French monarchy, the challenge of Philip IV* to the papacy, and the subsequent removal of the latter to Avignon* for two generations all testify to this fact. But perhaps the most important of all was the development of what later would be called the "Gallican Church" (see GALLICAN ARTICLES; GALLICANISM).

The history of French Christianity in the modern period begins with the Renaissance,* the Reformation,* and the monumental figure of John Calvin.* Christian humanism and sympathy for reform permeated France in the early sixteenth century. For reasons not yet fully understood, however, Protestantism never captured the allegiance of the majority of Frenchmen. Calvinism came close. At one time as many as one-tenth of the population had embraced the Calvinist doctrines, with perhaps as many as twice that number in sympathy with their cause. Popularly known as Huguenots,* political complications deflected their original aims, plunging the country into a long and bitter period of civil war. In the end the Calvinists lost, but managed to salvage a certain amount of toleration when their political leader, Henry of Navarre, converted to Roman Catholicism in order to receive the crown as Henry IV.* His Edict of Nantes* in 1598 gave Huguenots a measure of religious freedom for more than two generations. The scars of the civil wars, however, lingered for a long time, and the Protestant churches of France have never recovered fully from their impact.

The Catholic Reformation was successful in seventeenth-century France, producing an era of Catholic piety and a battery of saints: Francis of Sales,* Vincent de Paul,* Quietism,* Jansenism,* and the expulsion of many Huguenots with the revocation of the Nantes edict in 1685.

The eighteenth century brought a dramatic reaction to the growing power of the church in the life of the nation. Voltaire,* Diderot,* and other men of the Enlightenment* flourished where once the seventeenth-century French saints had walked. Hostility to organized religion, Deism, naturalism, and materialism spread along with other ideas spawned by the Enlightenment. These religious trends culminated with the French Revolution of 1789. When the Revolution triumphed, it tried to abolish the church in France as well as the dynasty. From 1793 vigorous attempts were made to remove all traces of the Christian past from France: Notre Dame cathedral became the Temple of Reason, and Fanny Aubry danced there, "natural religion" was encouraged by the state, and the Roman Catholic Church was outlawed. The church persisted, however, and regained its freedom and some of its former privileges under Napoleon I (d.1821). The French emperor and the pope reached an understanding in the Concordat of 1801.*

Nevertheless, the reverberations of the bitter struggle between the revolution and the church echoed throughout the century. Anticlericalism* became widespread, and the Catholic Church in France found itself constantly on the defensive. A phenomenon known as "Catholic atheism" made its appearance among those who remained within the church, but who had lost their faith. Leo XIII* tried to accommodate Catholicism to the increasingly liberal and secular mood of the nation, without much success. The anticlericalism of the period climaxed with the anti-Roman Catholic legislation of the early twentieth century, including the law of 1905 which decreed complete separation of church and state.

France today is deeply secular, but no longer as hostile to Christianity as it was previously. Roman Catholic Christianity continues to flourish among the peasantry, the Reformed Church (Calvinist) still claims more than a million adherents, and various other non-Roman Catholic groups have grown recently despite certain disabilities still attached to religious nonconformists. Further, many nonpracticing Catholics maintain their ties with the church despite their anticlericalism.

Modern French Christianity is virile enough to produce such first-rate thinkers as Teilhard de Chardin* and Jacques Ellul (b.1912), and the worker-priest movement. It is still powerful enough to help mold international politics, as when great progress was made in Franco-German relations after Chancellor Konrad Adenauer and President Charles de Gaulle attended Mass together.

Thus, despite the curious paradox of a France divided into various shades of belief and unbelief, into practicing and nonpracticing Catholics, into a growing interest in Christianity on one hand and a widespread and historical hostility toward organized religion on the other, the Christian faith remains an important ingredient in French civilization. As it has been central to French history in the past, so it continues to be a significant force even in the secular twentieth century.

BIBLIOGRAPHY: There is no satisfactory general history of Christianity in France available in either French or English. Treatment of various periods can be found in the following generally reliable works: C.S. Phillips, *The Church in France, 1848-1907* (1907) and *The Church in France, 1789-1848* (1929); T.S. Holmes, *The Origin and Development of the Christian Church in Gaul During the First Six Centuries of the Christian Era* (1911); H. Bremond, *Histoire littéraire du sentiment religieux en France* (8 vols., 1916-33); F.V.A. Aulard, *Christianity and the French Revolution* (1927); R.R. Palmer, *Catholics and Unbelievers in Eighteenth Century France* (1939); E. Amann and A. Dumas, *L'Église au pouvoir des laïques* (1948); S.W. Herman, *Report from Christian Europe* (1953); J.W. Thompson, *The Wars of Religion in France* (1956); G.R. Cragg, *The Church in the Age of Reason, 1648-1789* (1960); A.R. Vidler, *The Church in an Age of Revolution, 1789 to the Present* (1961); G. Mollat, *The Popes at Avignon* (1963).

ROBERT D. LINDER

FRANCIS BORGIA, see BORGIA, FRANCIS

FRANCISCANS. An order founded by Francis of Assisi* in 1209, when he gave his followers a simple rule advocating a life of apostolic poverty, preaching, and penance. The order was officially recognized when Francis and eleven others went to Rome in 1209 where Innocent III gave his oral approval and the brothers became known as the Order of Friars Minor. At first the primitive rule and the life of Francis provided enough cohesiveness for the order, but as the numbers increased with astonishing rapidity problems of administration became apparent. In 1221 Francis composed a second rule which reflected more the needs of a larger community, but was still not workable. In 1223 the third rule written by Francis and Cesarius of Speyer was confirmed by a papal bull of Honorius III and became known as *Regula bullata.* This rule maintained some of the spirit of the early rule but was more concerned with the official character of the order and its organization in accordance with church hierarchy. Francis expressed his regret for the loss of the freedom of the early rule in his *Testament,* which he wrote in 1226 reaffirming the need for a life of apostolic poverty and the imitation of Christ.

The conflict within the order which lasted for almost four centuries arose between those who wished to adhere strictly to the admonitions of the saint and those who felt modification was necessary in the practical administration of a large organization. Elias of Cortona* was instrumental in expanding the order, but was disliked for his worldliness and overbearing methods. During these formative years two schools of thought arose: the *zelanti,* or Spirituals, wished to follow the rule and *Testament* to the letter, while the Conventuals advocated moderation. Eventually John XXII decided in favor of the less strict interpretation in 1317-18, causing many Spirituals to rebel and form the schismatic Fraticelli.*

As the Franciscans grew in material wealth, laxity increased and a general decline ensued in the fourteenth century, aided by the Black Death* and the Great Schism.* Reform movements were in force, however, and a group of friars known as the Observants, who wished to live austere lives, were granted ecclesiastical recognition in 1415 and became a separate order in 1517. The Capuchins* also adhered to a doctrine of absolute poverty, adding an eremitical element. Altercations and divisions within the branches continued throughout the next three centuries, and culminated in the decree of Leo XIII which united all the different factions under a uniform constitution in 1897. Today the Franciscans consist of three orders: the Conventuals, the Observants, and the Capuchins.

The Franciscans have contributed much to the development of the Roman Catholic Church. Five of their members have been pope (Nicholas IV, Sixtus IV, Julius II, Sixtus V, Clement XIV). Franciscan scholarship has had great significance (Bonaventure, Duns Scotus, William of Ockham, and numerous educational institutions). The missionary and social work carried out has been outstanding.

See also POOR CLARES and TERTIARY.

BIBLIOGRAPHY: A.G. Little, *A Guide to Franciscan Studies* (1920); E. Hutton, *The Franciscans in England 1224-1538* (1926); R.M. Huber, *A Documented History of the Franciscan Order* (1944); H. Holzapfel, *History of the Franciscan Order* (tr. A. Tibesar; 1948); I.C. Brady (ed. and tr.), *The Marrow of the Gospel: A Study of the Rule of St. Francis of Assisi* (1958).

ROBERT G. CLOUSE

FRANCIS OF ASSISI (1182-1226). Founder of the Franciscan Order. Son of Pietro de Bernardone, a wealthy textile merchant, he was christened Giovanni, but supposedly nicknamed Francesco by his father upon returning from a trip to France. He received the usual education for his time and enjoyed a carefree life as a popular youth of Assisi. While taking part in a feud with the nearby city of Perugia he was imprisoned for a year in 1202. He joined the army upon his release, but could not complete a campaign against Apulia because of illness.

Francis's thoughts gradually began to turn to serious reflections, and in 1205 he made a pilgrimage to Rome, after which he had a vision wherein God told him to rebuild the church of St. Damian near Assisi. Selling his horse together with some of his father's cloth goods, he gave the proceeds to the priest for that purpose. His father disowned him, and Francis renounced his worldly possessions, taking up begging to provide for the reconstruction of more churches. In 1209 a sermon on Matthew 10:7-10 impressed him as being a personal admonition to take up a life of apostolic poverty, and he began preaching brotherly love and repentance. Attracting a number of followers, he composed a short rule in 1209 and succeeded in gaining the approval of Innocent III in 1212. Since Innocent required that the brothers receive minor orders, they called themselves the Friars Minor.

Embarking on a course of preaching and caring for the sick and the poor, the Friars came together each year at Pentecost for a meeting at Portiuncula in Assisi. In 1212 the Second Order was founded when an heiress of Assisi, Clare, was invested by Francis and formed the order for women, the Poor Clares.* Francis began missions to Syria (1212) and Morocco (1213-14), but was unable to complete them due either to illness or other misfortune, and traveled to the Middle East in 1219 in an unsuccessful attempt to convert the sultan Kameel. Since the order was growing out of the bounds of the early simple rule, Francis requested Pope Honorius to name Cardinal Ugolino as protector of the order. A new rule was approved by Honorius in 1223, and the character of the order began to move away from the simple ideal of Francis, especially when an ambitious, politically minded brother, Elias of Cortona,* was appointed vicar-general in 1221.

After his abdication of leadership in 1223, Francis spent the remaining years of his life in solitude and prayer, retiring to a hermitage on Monte Alverno in 1224. During this period he composed his "Canticle to the Sun," his *Admonitions,* and his *Testament.* In 1224 he allegedly received the stigmata. He was canonized by Gregory IX two years after his death. Revered by Protestants and Catholics alike, the ideal of St. Francis is still a vibrant force on the current religious scene, as the professions of the "Jesus freaks" and recent biographies of him indicate.

BIBLIOGRAPHY; P. Sabatier, *Life of St. Francis of Assisi* (1917); O. Englebert, *St. Francis of Assisi* (1966); L. Cunningham (ed.), *An Anthology of Writings by and About St. Francis of Assisi* (1973); J.H. Smith, *Francis of Assisi* (1973); M. Habig (ed.), *St. Francis of Assisi, Writings and Early Biographies* (1973). ROBERT G. CLOUSE

FRANCIS OF PAOLA (1416-1507). Founder of the Minims* Order. After a year at the Franciscan friary of San Marco, a pilgrimage to Rome and Assisi, and living as a hermit at Paola, he started the mendicant order in 1435; a church and house were built for them in 1453, and further foundations came later. Sixtus IV urged him to leave Italy for France, the court of Valois, where he ministered to Louis XI and became tutor to his son Charles VIII. He helped restore peace between France and Brittany by advising the marriage of the Dauphin and Anne of Brittany, and between France and Spain by urging Louis to return the counties of Rousillon and Cerdagne. Many miracles were attributed to Francis; he was declared patron of seafarers by Pius XII (1943); was honored in Latin countries by the devotion of the "Thirteen Fridays"; was a subject for paintings by Murillo, Velasquez, and Goya. He was canonized in 1519. His letters have been preserved.

C.G. THORNE, JR.

FRANCIS OF SALES (1567-1622). Counter-Reformation leader. Born in Savoy of aristocratic family, he read law at Paris and Padua, but soon abandoned legal studies for theology, and became priest in 1593. As missioner among the Calvinists of the Chablais (an arduous and dangerous project), he is credited with 8,000 conversions in two years. In 1599 he was appointed bishop-coadjutor of Geneva, succeeding to the see three years later. With Jane Frances de Chantal* he founded the Visitation Order in 1610. Francis achieved considerable success in his campaign to win the erring Swiss back to the Roman fold. An indefatigable worker, he organized clergy conferences and there insisted on simple teaching and preaching. "Love alone," he declared, "will shake the walls of Geneva." He established a seminary at Annecy, near his birthplace. Among his writings is the classic *Introduction to the Devout Life* (1608). Canonized in the mid-seventeenth century and declared a Doctor of the Church in the nineteenth, Francis was made patron saint of Roman Catholic journalists in 1923.

J.D. DOUGLAS

FRANCIS XAVIER (1506-1552). Jesuit missionary to the East Indies and Japan. Son of an aristocratic Spanish-Basque family, he was born at the castle of Xavier in Navarre. He studied law and theology at the University of Paris, where he met and befriended Pierre Favre and Ignatius Loyola.* Together with five others, Favre and Xavier became Loyola's associates in the founding of the Society of Jesus in 1534. The society vowed to follow Jesus in poverty and chastity, and to evangelize the heathen. It was in the latter activity that Xavier excelled and earned fame as an outstanding missionary pioneer and organizer.

He was ordained in Venice in 1537, and in 1539, at the request of John III of Portugal, he was appointed papal legate and sent to evangelize the East Indies. He arrived in Goa in 1542 and spent three years preaching to and serving the sick. He was very successful in evangelizing the pearl fishermen of SW India, who were baptized in thousands. He extended his missionary activity to

Japan, where he arrived in 1549, accompanied by Hachiro whom he met at Malacca and converted. He studied the Japanese language and within two years established a flourishing Christian community of 2,000 but he was driven out by Buddhist monks while his community endured great persecutions. He paid a short visit to China, but returned to Goa in 1552 and worked at the Goa college. During the same year he left for China, was refused entry, and died on the island of Sancian. His body was brought back to Goa and lies enshrined in the Church of Jesus the Good.

The success of Xavier's evangelization has not exempted his methods from criticism. He has often been accused of lack of understanding of oriental religions—a situation which he did little to remedy. His use also of the Inquisition has detracted from the glory of numerous conversions. He seems also to have made use of the government of Goa in proselytizing. Nevertheless, his outstanding missionary work aroused much interest in overseas missions in Europe. More than 700,000 conversions have been attributed to him by the Jesuits; Pius X conferred upon him the name "Patron of Foreign Missions." He was canonized in 1622.

See J. Brodrick, *St. Francis Xavier, 1506-1552* (1952). S. TOON

FRANCK, SEBASTIAN (1499-1542). Humanist and Spiritual Anabaptist. Born at Donauwörth, he studied at Ingolstadt and Heidelberg, was ordained priest (1524), became a Lutheran (1525), and married Ottilie Behaim (1528), whose brothers had Anabaptist leanings. He mildly opposed Johannes Denck, translating into German a Latin work directed against him, but subsequently left the Lutheran ministry and moved to Strasbourg (1529), where he began his friendship with Caspar Schwenkfeld. He now advocated complete freedom of thought and undogmatic Christianity, views which were expressed in his most important book, *Chronica, Zeitbuch und Geschichtsbibel* (1531). He also wrote *A Letter to John Campanus,* expressing his spiritual conception of the church, which marked him off from normative evangelical Anabaptism. Expelled from Strasbourg for his views, he eventually settled in Basle as a printer in 1539. J.G.G. NORMAN

FRANCKE, AUGUST HERMANN (1663-1727). German Lutheran minister, professor and early advocate of Pietism. Born in Lubeck, he studied at Erfurt and Kiel and became professor of Hebrew at Leipzig in 1684. Converted in 1687, he began to conduct Bible classes at Leipzig which led to a revival among both students and townspeople. When the theological faculty objected to his religious endeavors, he left and eventually became a minister at Glaucha and professor of oriental languages at the nearby University of Halle (1692). By 1698 he was professor of theology and in this post made important contributions in the scientific study of philology. He was concerned also for poor children, founding an orphanage, common school, teacher training school, and high school. In the course of time he added a drugstore, bookstore, bindery, and other industries, to train his wards and to help finance his work. He helped make Halle a center of piety and missionary enthusiasm. ROBERT G. CLOUSE

FRANK, FRANZ HERMANN REINHOLD VON (1827-1894). German Lutheran theologian. A professor at Erlangen from 1857, he was the systematizer of the Erlangen School. In his basic work *System der christlichen Gewissheit* (1870-73) he proceeds from the experience of rebirth as a certainty in self-consciousness. There are certain necessary objects of belief in this experience, namely, the immanent (sin, righteousness, certainty of perfection), transcendent (church, means of grace, revelation, inspiration). By working from the subject to the necessary reality, the Christian's experience is validated. In *System der christlicher Wahrheit* (1878-80), Frank moves from the ultimate principles of the spiritual world to the subject. In both his approach was apologetic, i.e., to make the certainty of faith secure in face of modern doubt. RICHARD V. PIERARD

FRANK, JACOB (1726-1791). Founder of the sect called "Frankists." Born Jankiev Lebowicz, son of a rabbi in Poland, he early came under the influence of the Sabbatarians (the Shabbetai-Tzevi) and later also, when visiting Turkey, of the Jewish Donmeh sect. He declared himself to be an embodiment of the Messiah and adopted a doctrine of the Holy Trinity in which he was the holy king. Returning to his homeland, he was attacked by the Talmudists and eventually he and his followers became Catholics, professing conversion to Christianity. After Frank allowed himself to be worshiped as the Messiah, however, he was imprisoned and released only when the Russians partitioned Poland in 1773. He then resided in Vienna and other places before his death in Offenbach. His daughter Eve continued to lead his followers. PETER TOON

FRANKFURT, COUNCILS OF. Frankfurt was the scene of a number of councils during Carolingian times. The most famous of them was held in 794 to condemn the Spanish Adoptianist heresy. The two foremost Adoptianists were Felix,* bishop of Urgel, and Elipandus,* archbishop of Toledo. Felix was ultimately imprisoned until his death, while Elipandus remained free but stubbornly recalcitrant. The council in 794 issued fifty-five other canons. One repudiated the Second Council (787) of Nicea's decree on icon worship, and the others dealt with a variety of matters, including metropolitan jurisdiction and monastic discipline.

FRANKLIN, BENJAMIN (1706-1790). Inventor, author, and diplomat. Born in Boston, he left school at the age of ten to help with his father's candle and soap business. Two years later he was apprenticed to a printer, his half-brother. He achieved great success in several fields—as publisher, author, businessman, philanthropist, moralist, inventor, scientist, civil servant, and statesman. He influenced American religious thought and popular morality through his writings, especially through the widely read *Poor Richard's Al-*

manac (1732-57) which extolled the virtues of hard work, thrift, moderation, and common sense in a humorously homespun way.

Though he contributed money to many religious institutions and valued the churches insofar as they promoted individual and social morality, he personally rejected the distinctive doctrines of orthodox Christianity in favor of an optimistic and undogmatic natural religion. He was a Deist who believed that nature rather than Scripture is the place where human reason recognizes God. He admired Jesus and His teachings, but doubted His divinity, and believed the essence of religion is to do good to men. He advocated separation of church and state and helped write the Declaration of Independence, which he also signed, and the U.S. Constitution. He helped to found the University of Pennsylvania, the first circulating library in America, and the American postal system. HARRY SKILTON

FRANSON, FREDRIK (1852-1908). Evangelist and founder of The Evangelical Alliance Mission. Born in Sweden, he migrated in 1869. He was influenced by Moody to become an evangelist to Swedish immigrants in north-central and western United States. He was ordained in 1881 and engaged in successful evangelism in Scandinavia and Germany from 1881 to 1890. In the latter year he led in the founding of the Scandinavian Alliance Mission, known as The Evangelical Alliance Mission since 1949, and was its general director from 1896 to 1908. In his evangelism on every continent he stressed Christ's second coming.

FRANZELIN, JOHANNES BAPTIST (1816-1886). Roman Catholic scholar. Born at Aldein, Tyrol, he received his early training in the Franciscan college in Bolzano and entered the Jesuit novitiate at Graz in 1834, teaching for six years in Austrian Poland. He studied theology at Rome and Louvain, was ordained in 1849, and was prefect of studies and confessor at the German College, Rome. He lectured in oriental languages, then held the chair in dogmatic theology (1857) in the Gregorian University. Papal theologian at Vatican I and prefect of the Sacred Congregation of Rites, he was made cardinal (1876). He wrote a number of theological works and, in addition, his *Examen doctrinae Macarii Bulgakov* (1876) arose from his participation with Greek Orthodox and Protestants on the work of the Holy Spirit. C.G. THORNE, JR.

FRATICELLI. A group within the Franciscan Order during the Middle Ages which insisted on a very strict observance of the rule of poverty and vigorously opposed the decrees of John XXII which held that Christ and His apostles owned property. They were quite active in Italy and S France. Wherever they appeared they were hunted down by the Inquisition and were regarded as heretical.

FREDERICK I (Barbarossa) (c.1122-1190). German king and Roman emperor. Nephew of the weak Conrad III, he was elected king in 1152. Although a Hohenstaufen, his mother was a Guelph, and it was hoped he could mediate between the two parties. Frederick's endeavor to restore the rights of the German monarchy and expand his territorial control while reviving the imperial authority made him a controversial historical personage. He conciliated the Guelphs by recognizing Duke Henry the Lion's position in Saxony and granting him the duchy of Bavaria. Frederick issued a proclamation of peace *(Landfriede)* in Germany (1152) and built up an efficient royal government based upon the non-noble *ministeriales.* He gained control over the German church, utilized feudal obligations to strengthen the monarchy, and enlarged his own family domains.

The wealth of the Italian cities and the political disorder there tempted the ambitious Frederick, who decided to embark upon restoring the empire. He invaded Italy in 1154-55, repressed the Lombard communes, allied with Pope Adrian IV to oust Arnold of Brescia, and was crowned emperor. The uneasy alliance collapsed at the Diet of Besançon (1157) when Frederick firmly rejected the concept of papal feudal overlordship. Imperial documents began referring to the Holy Empire *(sacrum imperium),* an indication that secular authority was divinely sanctioned, not bestowed by papal coronation.

In four subsequent Italian campaigns Frederick had only limited success. Pope Alexander III (1159-81) turned against the emperor, who then supported an antipope, and the schism lasted until Frederick's defeat at Legnano in 1176 by the Lombard League and reconciliation with Alexander. After the Peace of Constance in 1183 (the communes recognized the emperor's suzerainty but were granted self-government), Frederick arranged a marriage alliance with the kingdom of Sicily, thus depriving the papacy of secular allies in Italy. In 1180 he exploited the legal technicalities of feudal practice and the hostility of Henry the Lion's many enemies to eliminate the Saxon rival and diminish Guelph power in Germany. At the pinnacle of success he died while taking part in the Third Crusade.

BIBLIOGRAPHY: J.W. Thompson, *Feudal Germany* (1928); G. Barraclough, *The Origins of Modern Germany* (1947); Otto of Freising, *The Deeds of Frederick Barbarossa* (ET 1953).

RICHARD V. PIERARD

FREDERICK II (1194-1250). King of Germany and Sicily, and Roman emperor. The Sicilian-born grandson of Frederick Barbarossa,* his father had him elected German king in 1196. When, however, Henry VI died the next year, the princes refused to accept the youthful Hohenstaufen heir. The ensuing struggles in Germany and Italy, abetted by French and English pressures and Innocent III's endeavors to restore papal power, resulted in a decline of imperial authority. In 1212 Frederick was again named king through the contrivance of Innocent and Philip Augustus. The French victory at Bouvines (1214), followed by the deposition of the Guelph emperor Otto IV (1215) and Frederick's own imperial coronation (1220), placed him in a dominant position.

In the next three decades he was involved in a continuous struggle with the papacy. His primary interest lay in Italy, and Germany (under the regency of his sons) occupied a clearly subordinate role. In 1213 he relinquished authority over German church personnel and recognized those rights acquired by the nobles since 1197, while his privilege of 1220 eliminated all royal power over the internal administration of ecclesiastical principalities. In 1231 these concessions were extended to all secular princes and included control over local courts and coinage. This signified the victory of princely particularism over the monarchical ideal in Germany. In Sicily, Frederick had by 1224 restored the power of the Norman monarchy, and after a crusading interlude in 1228-30 sought to extend his absolutistic rule to northern and then central Italy, actions which incurred papal opposition. In 1245 Innocent IV excommunicated and preached a crusade against Frederick which had little effect. The execution of Frederick's grandson Conradin in 1268 ended the Hohenstaufen dynasty.

Frederick was a patron and student of mathematics, astronomy, medicine, zoology, and poetry. His court at Palermo, noted for its oriental splendor, was the leading cultural center of southern Europe. A skeptic in religion, he was tolerant of Jews and Muslims. In his dealings with Christian and Muslim leaders alike, he proved to be a brilliant diplomat, administrator, and general. Frederick's achievements and interests were so many that some called him the "wonder of the world."

BIBLIOGRAPHY: E. Kantotowicz, *Frederick II* (1931); G. Barraclough, *The Origins of Modern Germany* (1947); D.P. Waley, *The Papal States in the Thirteenth Century* (1961).

RICHARD V. PIERARD

FREDERICK III (the Wise) (1463-1525). Elector of Saxony. Born in Torgau, the eldest son of the elector Ernst and Elizabeth (daughter of Albert, duke of Bavaria), he was later called "the Wise" because of his reputation for fair play and justice. He succeeded his father as elector in 1486. Before this his education had been influenced by Renaissance ideals. His court at Wittenberg was a center of artistic and musical activity: Albrecht Dürer and Lucas Cranach were patronized. Nevertheless, he was a devout Catholic, interested in the cult of relics: the catalog produced by Cranach in 1509 revealed that he had 5,005 particles.

Always insisting on the need for constitutional reform in the empire, Frederick became president of the newly formed council of regency (Reichsregiment) in 1500, but later (1519) refused to stand as a candidate in the imperial election. In 1502 he founded the University of Wittenberg. To it in 1511 came Martin Luther and in 1518 Philip Melanchthon. When Luther was required by the pope to go to Rome in 1518, Frederick intervened and had the trial transferred to Augsburg on German soil. Two years later the elector refused to execute the bull *Exsurge Domine* against Luther. After the Diet of Worms (1521) had imposed the imperial ban, he provided a refuge for Luther at Wartburg. There is no firm evidence that he totally accepted the Lutheran faith, but just before his death at Annaberg he received Communion in both kinds from George Spalatin,* who had acted on so many occasions as an intermediary between the elector and Luther. The latter preached at Frederick's funeral, and Melanchthon gave an oration in which he highly commended the elector's work in promoting the Gospel.

PETER TOON

FREDERICK III (the Pious) of the Palatinate (1515-1576). Eldest son of Duke John II of Palatinate Simmern, he received his princely education and various administrative experiences before succeeding his father in 1557. In 1537 he married Mary, daughter of Margrave Casimir of Brandenburg, who had been reared a Lutheran. Eleven years later he announced his conversion to the Evangelical cause. He opposed the Augsburg Interim (1548). In 1559 he became heir to the electorate of the Palatinate. Here not only Lutheranism but also Calvinism had found a strong foothold, and under the leadership of the Lutheran Tileman Hesshusius a controversy raged about the correct doctrine of the Lord's Supper. Frederick and his wife plunged into a thorough theological study of the question and in 1541 came to the conclusion that Article X of the Augsburg Confession* was popish. With the help of various divines Frederick supported Calvinism in Heidelberg and commissioned Ursinus and Olevianus to write the "Heidelberg Catechism"* (1563). His support of Calvinism brought opposition from Frederick of Saxony and others. A request from the princes at Augsburg in 1566 that he abide by the Peace of Augsburg (which recognized only Lutheranism or Roman Catholicism) did not turn him from his convictions. In 1570 the presbyterian form of church government was introduced in the Palatinate. Frederick aided the French Huguenots and the Dutch Calvinists. His son, Louis VI (elector, 1576-83), returned to Lutheranism.

CARL S. MEYER

FREDERICK IV (1671-1730). King of Denmark and Norway. From his accession in 1699 he was continually at war, but he did introduce reforms in the treatment of peasants, the administration of justice and public finances, military organization, and commercial relations. While crown prince, he was greatly influenced by Pietism and decided to emulate Roman Catholic rulers by caring for the spiritual welfare of his subjects in the South Indian dependency of Tranquebar. Finding no suitable candidate in Denmark, he secured Bartholomaeus Ziegenbalg* and Heinrich Plütschau* Over the objections of several Danish theologians Frederick obtained their ordination as missionaries in November 1705, and they immediately sailed for India. In 1718 he approved the mission of Hans Egede* to Danish-owned Greenland.

RICHARD V. PIERARD

FREE CHURCH OF ENGLAND (otherwise called the Reformed Episcopal Church). A Reformed and Protestant church established in 1844 as a reaction to the doctrines and development of the Oxford Movement* in the Church of England. Its constitution was formally registered in 1863. It is pledged to the Thirty-Nine Articles, and its Prayer Book is, for all practical purposes, the 1689 revision of the 1662 Book of Common Prayer—a revision acceptable at the time to the Puritans, but not adopted in the Anglican Church. The Free Church of England recognizes and adheres to episcopacy "not as of Divine right, but as a very ancient and desirable form of Church polity," but its ministry is presbyterian. A similar organization called the Reformed Episcopal Church was formed in the USA in 1873 and in England from 1877. The two English bodies maintained close relations and were finally united in 1927. There are two dioceses, north and south, each with its own bishop. The existence of a vigorous evangelical wing in the Church of England has militated against the development of the Free Church of England, but it has persisted in the face of many difficulties. In 1971 there were thirty-three congregations and thirty-nine clergy.

HOWARD SAINSBURY

FREE CHURCH OF SCOTLAND. Popularly known as the "Wee Free" Church, it represents the minority of the former Free Church of Scotland who in 1900 refused to enter the union with the United Presbyterian Church to form the United Free Church of Scotland. The United Presbyterian Church when constituted in 1847 had accepted the principle of voluntaryism* and the minority felt that union with it would compromise the Free Church belief in the national recognition of the Christian religion. The original Free Church was constituted in 1843 after the Disruption* when about one-third of the ministers and members seceded from the Church of Scotland rather than submit to what they regarded as state control of the church. But their leader, Thomas Chalmers,* declared: "We quit a vitiated establishment and would rejoice in returning to a pure one."

After the majority of this Free Church entered the union of 1900, the dissenting minority laid claim in the civil courts to the entire property of the Free Church on the grounds that they alone were true to the Disruption principle of a free established church. After losing in the Scottish courts they appealed to the House of Lords, which decided in their favor in 1904. This judgment caused a sensation, and a parliamentary commission was appointed to distribute the property in accordance with the relative strengths of the two parties. The present Free Church, though not established, holds to the principle of establishment. Conservative in theology, it affirms its loyalty to the whole of the Westminster Confession of Faith.* Strongly Sabbatarian, it has no instrumental music and uses only the metrical psalms in congregational praise. It is strongest in N and NW Scotland, with a total of just under 6,000 members and more than 17,000 adherents.

HENRY R. SEFTON

FREE METHODISTS, see METHODIST CHURCHES, AMERICAN

FREE PRESBYTERIAN CHURCH OF SCOTLAND. This originated in a group in the Free Church of Scotland which objected to the Declaratory Act of 1892. The act stated that the Free Church disclaimed intolerant or persecuting principles and did not consider her office-bearers, in signing the Westminster Confession of Faith, committed to any principles inconsistent with liberty of conscience and the right of private judgment. This Scottish secession is unique in that it turned primarily on a doctrinal issue and not on relationships between church and state. The post-1900 Free Church repealed the Declaratory Act, but in such a way that the Free Presbyterians have felt constrained to continue a separate witness, although the two churches are similar in most other respects. Their numerical strength is officially estimated at 10,000 members and adherents.

HENRY R. SEFTON

FREEMAN, THOMAS BIRCH (1809-1890). First Wesleyan Methodist missionary in Ghana to survive for more than a short period. Born near Winchester, son of a West Indian father and an English mother, he worked as a gardener and botanist, offering for overseas service in 1837. He arrived at Cape Coast at the beginning of the following year and started work single-handed after the death of all his predecessors. His interests extended to architecture, botany, agriculture, and education, but his dominant concern was the geographical expansion of Christianity. He visited Kumasi, capital of Ashanti, in 1839, and Badagry and Abeokuta, now in Nigeria, in 1841-42, calling at Abomey, the capital of Dahomey, during his return journey. By 1856 he had built up a strong church and an educational system which included thirty-five schools, four of them in Nigeria and Dahomey. He had also, however, incurred cumulative over-expenditure of more than £10,000, and in 1857 he resigned from the ministry. For some time he worked in government service in Ghana, returning to the ministry in his later years.

PAUL ELLINGWORTH

FREEMASONRY. An international organization, claiming adherents of all faiths, whose principles are embodied in symbols and allegories connected with the art of building and involving an oath of secrecy. The origins of Freemasonry probably lie in the twelfth century, when English masons founded a fraternity to guard the secrets of their craft. The "lodge," the name given to the meeting place of modern masons, was not only a workshop, but a place to exchange views, air grievances, and discuss craft matters. Hence their secrecy.

There are two elements in the masonic tradition: (1) *the Old Charges.* Two manuscripts, now in the British Museum, dating from 1390 and about 1400, detail the customs and the rules of the craft. Rules apply to the master in charge, the journeymen, and the apprentice who is learning the trade; (2) *the Masonic Word.* This is probably a Scottish institution and is somewhat obscure in origin and development. It is a distinguishing secret sign, either a word, a handshake, or both.

The development of Freemasonry falls into three periods. In the first, all members were operative masons. During the age of "Accepted Masonry," nonoperative masons either joined existing lodges or formed new ones. From the eighteenth century there developed "Speculative Masonry," Freemasonry as it is known today. The Grand Lodge was founded in 1717 principally to maintain communication and harmony among lodges. Following 1721, many of the highest offices were filled by members of the aristocracy. The origins of the modern masonic ceremonies are obscure, though they probably derive from seventeenth-century practices. The influence of speculative masonry on these practices has almost obscured their operative origins. There are ceremonies for entry into each grade—entered apprentice, fellow of the craft, and master mason. These grades, and their associated secrets and rituals, are fundamental to modern Freemasonry.

Freemasonry places considerable emphasis on social and welfare activities. It is to be found throughout the world, though it is proscribed in Communist countries. Freemasonry claims to be based on the fundamentals of all religions held in common by all men. Among many reasons for its criticism by Christian bodies are the following. Freemasonry was closely connected with the upsurge of Deism in eighteenth-century England, and this outlook continues to prevail. Freemasonry calls for a "common denominator" God who incorporates Assyrian and Egyptian elements. The name of God in masonic rituals veils the doctrine of a blind force governing the universe. In its elaborate ritual Freemasonry omits the name of Jesus Christ, Savior and Lord. The masonic vows involve a depth of commitment which Christians should give only to Jesus Christ. The masonic initiation is to an unknown course of action and is often for this reason held in grave suspicion. The bloodcurdling vows, if taken seriously, are at best rash, and if not taken seriously, are frivolous.

Because of its invitation to men of all faiths, Freemasonry does not hold to the uniqueness of Jesus Christ. It does not teach the necessity of salvation through Christ alone. Good works, it believes, will cause a man to ascend to "the Grand Lodge Above." It accords no preeminent place to the Bible and claims that masonic initiation gives a measure of illumination unattainable elsewhere. The Roman Catholic Church has frequently condemned Freemasonry, mainly for its masonic secret. Six papal bulls have been directed against it—by Clement XII in 1738, Benedict XIV in 1751, Pius VII in 1821, Leo XII in 1826, Pius IX in 1864, and Leo XIII in 1884.

BIBLIOGRAPHY: A.G. Mackey, *Encyclopedia of Freemasonry* (rev. ed., 3 vols., 1946); W. Hannah, *Darkness Visible* (5th ed., 1953), and *Christian by Degrees* (1954).

JAMES TAYLOR

FREER LOGION. The saying ascribed to Christ in an addition to the text after Mark 16:14 in Codex W. This is a late fourth or early fifth century MS of the four gospels discovered in 1906 by C.L. Freer and now in the Freer Museum in Washington. It runs: "And they excused themselves, saying, 'This age of lawlessness and unbelief is under Satan, who does not allow the truth and power of God to prevail over the unclean things of the spirits. Therefore reveal thy righteousness now'—thus they spoke to Christ. And Christ replied to them, 'The term of years for Satan's power has been fulfilled, but other terrible things draw near. And for those who have sinned I was delivered over to death, that they may return to the truth and sin no more; that they may inherit the spiritual and incorruptible glory of Righteousness which is in heaven.' " It has some affinity with the general style and content of the longer ending of Mark's gospel as a whole.

See V. Taylor, *The Gospel according to St. Mark* (1952).

R.E. NIXON

FREE THINKERS. Those who refuse to submit reason to the control of authority in questions of religious belief. The term which seems to have appeared first in 1692, was used by Deists and other opponents of orthodox Christianity in the early eighteenth century in their emphasis on reason above all else. In *A Discourse on Freethinking* published in 1713, Anthony Collins assailed ministers of all denominations and asserted that free inquiry was the only means of attaining truth—and that this procedure was, indeed, commanded by Scripture. It can be seen from this that in its earlier manifestations the term was not chiefly involved with a direct attack on religion as such, but rather on the exclusive claims of, and the stress on revelation by, the Christian religion. Coupled with this was an attempt to throw doubt on the authority of the Bible. Freethinking has now come, however, to be a general description of any agnostic or atheist whose rejection of theism is conscious and real rather than stemming from apathy or indifference. Its weakness as a description comes from its fallacious assumption that freedom of thought must inevitably involve rejection of the supernatural.

It has been associated with a great number of movements. Modern secularism in its militantly atheistic form claims the term brings together many different strands of thought. Among these would be listed modern Unitarianism (1825); Mexican secularism (1833); the German free-religious movement (1848); organized positivism (1854); New Zealand rationalism (1856); Australian secularism (1862); the Belgian Ligue de l'Enseignement (1864); the English religion (later the ethical) society (1864); Italian anticlericalism (1869); Vosey's theistic church (1871); American free thought (based on the *Truth Seeker*, 1873); American ethical culture (1876); the Dutch *Dageraad* (dawn) as a national movement (1881); Argentinian secularism (1883), and Austrian secularism (1887). One could legitimately add to these the more radical groups within the major denominations and extend the list even to include political revolutionaries.

J.D. DOUGLAS

FRELINGHUYSEN, THEODORE JACOBUS (1691-1747). Dutch Calvinist and pietist. Born at Lingen in East Friesland, steeped in the pietistic emphases then current in Dutch Calvinism, he served as a pastor in the Netherlands. He became aware of the need for trained ministers among the Dutch congregations in America and emigrated to New Jersey in his late twenties. From 1720 he served in the Raritan Valley area. An eloquent preacher, he stressed the need for spiritual revival, and he found in the Great Awakening* that swept the colonies a similar emphasis. He was in active touch with Gilbert Tennent* and other revivalistic leaders. Frelinghuysen aroused some opponents among the Dutch settlers, who appealed to Amsterdam; he, meanwhile, was working for a separate organization for the American Dutch Calvinist churches. He organized an assembly *(coetus)* which asked approval from Amsterdam; in 1747, the year he died, this approval was given. He is thus an important figure in the history of the Dutch Reformed in America.

DIRK JELLEMA

FRIAR (from Old French *frere*, "brother"). Title of a member of one of the Mendicant ("Begging") Orders founded during the Middle Ages. They are distinguished from monks in that, though they have a local headquarters in a "friary," their work is an active ministry in the world. Part of a highly organized, widespread body with a central authority, they also are often distinguished, in England, by the color of their habits, e.g., "Grey Friars" (Franciscans), "Black Friars" (Dominicans), "White Friars" (Carmelites).

FRIEDRICH, JOHANNES (1836-1917). Church historian. Born in Upper Franconia, Germany, he was educated at Bamburg and Munich and ordained in the Roman Catholic Church. He lectured in the theological faculty at Munich, first in ecclesiastical history, then in philosophy, until his retirement in 1905. Secretary to Cardinal Gustav von Hohenlohe at Vatican I, he considered papal infallibility historically indefensible and joined in opposing such a dogma. Leaving Rome before the council ended, he refused to accept the decrees, and in 1871 was excommunicated. The Bavarian government gave him protection in respect of his university appointment at Munich. He continued as a priest with the Old Catholics,* whom he influenced profoundly, but left them because they did not uphold clerical celibacy. Among his many writings were *Johann Wessel* (1862); *Kirchengeschichte Deutschlands* (2 vols., 1867-69), completed only to the Merovingian period; *Geschichte des Vatikanischen Konzils* (3 vols., 1877-87); *Beiträge zur Geschichte des Jesuitenordens* (1881); and *Ignaz von Döllinger* (3 vols., 1899-1901), his teacher and intimate friend, for whose *Letters from Rome* (1869-70), published under the pseudonym "Quirinus," he was a major informant.

C.G. THORNE, JR.

FRIENDS, SOCIETY OF (Quakers). A religious group whose origins are traced to the radical wing of English Puritanism of the 1640s. The term "Quaker" was used from 1650, partly because people were expected to tremble before the Word of God, partly because a sect of women in Southwark had previously been so called. The first leader was George Fox,* who in the 1650s preached the message of the New Age of the Spirit. The Seekers of Westmorland were converted, and with their help Fox and others moved south in their aggressive evangelism, opposed by Puritans and Anglicans alike. In Scotland many espoused Quaker views introduced into the country during the Protectorate, and they, like the Covenanters,* suffered persecution for their beliefs.

From their emphasis on realized eschatology and the presence of the Spirit emerged the typical Quaker meeting wherein people waited for the Spirit to speak in and through them. The "Inner Light" was as important as Scripture; sacraments, ceremonies, and clergy were abandoned. Persecuted at home, they evangelized North America which they reached in the mid-seventeenth century. Two decades later, another of their leaders, William Penn,* established the colony of Pennsylvania. In 1796 they opened the first asylum in England, where also Elizabeth Fry* began notable work in prison reform. In America in 1827 a schism developed under the influence of the Hicksites.* During the nineteenth century there was a steady move westward in which Quakers participated; the first yearly meeting in Canada was established in 1867.

Their theology was given classic expression by Robert Barclay* in *Apology for the True Christian Divinity* (1678), and their meetings were regulated by Fox's "Rule for the Management of Meetings" (1668). The body believes in the priesthood of all believers and holds that women equally with men have a share and responsibility in worship and organization. Many in modern times have abandoned the traditional form of worship in favor of a service led by a pastor. Called upon to act toward others in the way most likely to lead to a response of goodness, the Friends have obeyed with great consistency. Early in the eighteenth century they began to oppose slavery, and their efforts contributed much to Wilberforce's ultimate success. Their well-known opposition to war is not based primarily on Scripture, but on the conviction that warlike feelings are a sign something is wrong in men's thinking and attitude toward one another. Although refusing combatant duties, Quakers have a notable record of valiant service on and off the battlefield. "Walking in the light" means speaking the truth, so Quakers refused to take oaths. To them is due credit also for our system of fixed price trading—they held it wrong to ask a higher price than one was willing to take.

There are now estimated to be some 200,000 Friends throughout the world, of which more than 60 percent are in the USA, 11 percent in the British Isles. They maintain missions and international centers in several countries.

BIBLIOGRAPHY: R.M. Jones et al., *The Quakers in the American Colonies* (1911); W.C. Braithwaite, *The Beginnings of Quakerism* (1912, rev. 1955), and *The Second Period of Quakerism* (1919, rev. 1961); A.N. Brayshaw, *The Quakers, Their Story and Message* (1921, rev. 1953); R.M. Jones, *The Faith and Practice of the Quakers* (1927; 7th ed., 1949); H.H. Brinton, *Friends for 300 Years* (1952); D.E. Trueblood, *The People Called Quakers* (1966). P.W. PETTY

FRIENDS OF GOD *(Gottesfreunde).* A term used in Scripture, by the Fathers, and in medieval writings, which refers also to a group of German mystics and other Christians in the fourteenth century. They exchanged visits, letters, and writings for their own spiritual growth and service. Some lived alone, others in groups, and several were nuns in the convents to whom the mystics preached and ministered. Profoundly influenced by Meister Eckhart's* works as well as the ideals of earlier German prophetesses, they cultivated intense prayer, austerity, and self-renunciation. Fully supporting the church, they were concentrated in Bavaria, the Rhineland, Switzerland, and the Low Countries, with Basle, Strasbourg, and Cologne as chief centers. Their number included Dominicans, Franciscans, and lay people of every estate. Associated with them were John Tauler,* Henry Suso,* Jan van Ruysbroeck,* and the author of the *Theologia Germanica,** and there were links also with the Brethren of the Common Life.* The Friends are to be distinguished from the heretical Beguines* who took their name. The decline of mysticism brought the end of their association, but their influence long continued. C.G. THORNE, JR.

FRITH, JOHN (c.1503-1533). Protestant martyr. Born at Westerham and educated at Eton and King's College, Cambridge, he was made junior canon by Wolsey of his newly founded Cardinal College (Christ Church), Oxford. He was briefly imprisoned in 1528 for his Lutheran views. In Marburg he helped Tyndale* in his translation work. He returned to England in 1532 where his writings against the doctrines of purgatory (*A Disputation of Purgatory,* 1531, combating Sir Thomas More and Bishop Fisher) and transubstantiation precipitated his arrest on the orders of More on a charge of heresy. He refused to recant, was condemned to death, and burned at Smithfield. Some ten works are credited to him, including one of the first antipapistical books in English, *An Epistle to the Christian Reader: Antithesis wherein are compared together Christ's Acts and our Holy Father the Pope's* (1529). Several were written in the Tower or in Newgate Prison in defense of his views. HOWARD SAINSBURY

FROMENT, ANTOINE (1510-1584). Reformer of Geneva. Born at Tries, near Grenoble, he was educated at Paris, met Lefèvre d'Étaples,* and received a canonry on an estate of the queen of Navarre. He accompanied Guillaume Farel* on his evangelistic tours through Switzerland. He went to Geneva (1532), opened an elementary school to teach French, but turned his lessons into sermons. His followers daily increased. In 1533 he protested publicly after an attack on evangelical doctrine by Guy Furbiti, a Sorbonne theologian; he was forced into hiding, but returned with Farel and Pierre Viret.* He became pastor of St. Gervais Church (1537) and was engaged by Bonivard, the republic's historian, to help in his *Chronicle* (1549). He renounced his ministry and became public notary (1553) and a member of the "Council of the Two Hundred" (1559). He had domestic troubles, was banished after adultery (1562), but was permitted to return in view of past services (1572), and reinstated as notary (1574). J.G.G. NORMAN

FRONTIER RELIGION. A reference to the character of Christianity on the American frontier during the century of western expansion (1790-1890). Pioneer religion tended to encourage an individualistic faith, emotion-filled meetings, and democratic church government. Employing revivals and camp meetings* freely, Baptists and Methodists proved to be the most effective denominations in winning frontiersmen to the Christian faith. Baptist ministers, usually farmers during the week and preachers of the "simple gospel" on Sunday, readily identified with the homesteader; while the Methodist circuit-riding preachers with their message of free will and free grace seemed to offer the right combination of method and message for the scattered, democratically minded frontiersmen. BRUCE L. SHELLEY

FROST, HENRY WESTON (1858-1945). Mission director. Born in Detroit, reared in New York, and educated at Princeton, Frost joined his father in oil production. Originally his was a Danish family who moved to Cambridge, England, his mother being descended from a long line of English and Flemish knights. To his immediate family and a quickening evangelistic and missionary experience he owed his Christian vocation, being much encouraged by his wife. He founded the North American branch of the China Inland Mission and served as its director in Philadelphia for over forty years. Presbyterian, a premillennialist, and active in the Bible Conference Movement, he wrote numerous pamphlets, three books of poetry, and a dozen volumes on devotional, doctrinal, and missionary subjects. C.G. THORNE, JR.

FROUDE, RICHARD HURRELL (1803-1836). Tractarian* leader. Son of the archdeacon of Totnes and elder brother of J.A. Froude the historian, he was educated at Eton and Oriel College, Oxford, served as tutor at the latter, and was ordained priest in 1829. In 1831 the first signs of consumption appeared, and he traveled widely in search of healthier climes. On one of these journeys, to Italy (1832-33), he was accompanied by John Newman* whom he influenced greatly, being responsible for bringing Newman and John Keble* together—hence he is sometimes known as the "third man" of the Oxford Movement.* He died in obscurity. In 1838-39 his friends published his *Remains,* consisting of extracts from his essays, sermons, and letters, in the mistaken belief that readers would admire the exemplary High Church piety of the deceased. In effect the book revealed that Froude, beneath a debonair and cavalier exterior, was melancholy, self-torturing, cruel, arrogant, and somewhat schizophrenic. He bitterly hated the Reformation, and he was devoted to clerical celibacy and the cult of the Virgin. The *Remains* convinced many that Tractarianism's real goal was reunion with Rome. IAN SELLERS

FRUCTUOSUS (d.259). Bishop of Tarragona in Spain, Fructuosus with two deacons was arraigned before Roman officers for refusing to offer sacrifice to Roman state gods. They contravened the edicts of Valerian and Gallienus in A.D. 257-58, requiring nonpagans to join in Roman religious ceremonies. He was burnt to death.

FRUMENTIUS (fourth century). First bishop of Axumis. According to Rufinus, the brothers Frumentius and Aedesius accompanied their teacher Meropus on a voyage from Tyre to Ethiopia. Here, probably at Adoulis, Meropus was killed by the Ethiopians and the brothers taken captive to the king at Axumis. He made Frumentius his secretary and Aedesius his cupbearer. The brothers were Christians, and when they were persuaded by the queen mother to remain after the death of the king and to help with the education of the prince and with the government during the prince's minority, they were able to promote the Christian faith. On the prince's majority they were permitted to return to their own country. Frumentius visited Alexandria to report on his work and was consecrated bishop of Axumis there by Athanasius, either shortly before 339 or shortly after 346. He took the title "Our Father" *(abuna)* which was retained by primates of the Ethiopian Church. He was apparently opposed to Arianism.

DAVID JOHN WILLIAMS

FRY, ELIZABETH (1780-1845). Quaker prison reformer. Born in Norwich, daughter of John Gurney, a Quaker banker, she married a London merchant in 1800 and had a large family. Her religious upbringing created in her a deep concern over social issues, and in 1808 she was in a position to found a Girls' School at Plashet, East Ham. In 1811 she was admitted as a Quaker "minister." It was not until 1813 that she became interested in prison work and began her welfare work at Newgate Prison among the women prisoners, visiting them daily, teaching them to sew, and reading the Bible to them. In 1817 she began her campaign for the separation of the sexes in prisons, classification of criminals, women warders to supervise women prisoners, and provision of both secular and religious instruction. In 1818 she gave evidence before a select committee of the House of Commons on the subject of prisons, and her views played a significant part in the design of subsequent legislation.

Later, in 1839, realizing the necessity for the care and rehabilitation of discharged criminals, she formed a society with this as its prime concern. She did much to foster prison reform on the Continent by frequent visits. Other philanthropic causes also occupied her time and attention. In an attempt to deal with mendicancy she sponsored the "Nightly Shelter for the Homeless in London" (1820), as well as visiting societies in Brighton and other places. In 1827, with her brother, she produced a report on social conditions in Ireland, and in 1836 she secured the provision of libraries at coastguard stations and certain naval hospitals. Her husband went bankrupt in 1828, however, and this curtailed her work. Throughout her life she combined an evangelistic zeal with her social work, and her *Texts for Every Day in the Year* (1831) had a very wide circulation. Her maxim was "Charity to the soul is the soul of charity."

BIBLIOGRAPHY: Her two daughters published a two-volume *Memoir* in 1847; her numerous biographers include G.K. Lewis (1910) and J.P. Witney (1937).

JOHN A. SIMPSON

FULGENTIUS (468-533). Bishop of Ruspe. Born at Telepte in Byzacene, he later left Hunneric's court for the monastic life, first under the exiled bishop Faustus and then under Abbot Felix. He visited Rome in 500 and on his return to Byzacene founded his own monastery, from which he soon retired to practice a more ascetic life on an island. At the instigation of Felix he was ordained presbyter by Faustus, and in 508 was consecrated bishop of Ruspe by Victor of Byzacene. His first task was to build a monastery of which Felix became abbot. Soon after, Fulgentius was banished by Thrasimund to Sardinia with sixty other Catholic bishops, but was recalled to Carthage in 515 to answer objections to the Catholic faith. He returned into exile in 517 and remained in Sardinia until Thrasimund's death in 523. Hilderic allowed him to return to Ruspe, where he died. Fulgentius was a devotee of Augustine, as seen in his letters, sermons, and treatises against the Arians and the Pelagians.

DAVID JOHN WILLIAMS

FULKE, WILLIAM (1538-1589). Puritan divine. Born in London, he studied at St. John's College, Cambridge, and Clifford's Inn. He became a friend of Thomas Cartwright and took a prominent part in the Vestiarian Controversy* in the university. For his extremism he was deprived of his fellowship and expelled. Later he was readmitted. He became chaplain to the earl of Leicester and through his help received the livings of Warley and Dennington in Essex-Suffolk. He became head of Pembroke Hall in 1578 and vice-chancellor in 1580. The last decade of his life was taken up with literary activity in defense of Protestantism and against Roman Catholicism, especially against Cardinal Allen and other leaders of the Counter-Reformation. His defense of the English translation of the Bible ("Bishops' Bible"*) against the attacks of Gregory Martin of Reims revealed Fulke's wide learning and ability. It also helped to make known the Douai-Reims NT in England, so that its language influenced the AV of 1611.

PETER TOON

FULLER, ANDREW (1754-1815). Baptist theologian. Son of a Cambridge farmer and a powerful wrestler in his youth, Fuller was ordained as minister of Soham Baptist church in 1775 and inducted to the Kettering church in 1783. Entirely self-taught and possessed of a blunt, incisive style, Fuller was the greatest original theologian among eighteenth-century Baptists. Reared in an atmosphere of deadening hyper-Calvinism, he was led through vigorous independent study of Scripture, the encouragement of Robert Hall, John Ryland, and others, and his reading of Jonathan Edwards to evolve, or rather revive, an evangelical Calvinism which was the substance of his greatest work, *The Gospel Worthy of All Acceptation* (1785).

This involved him in various controversies: with hyper-Calvinists like John Martin and William Button who denounced "Fullerism" as "Duty-faith" and led their churches apart from the evangelical Baptists (and thus created the Strict and Particular Baptist denomination); with Dan Taylor the Arminian Baptist; with Archibald McLean's Scotch Baptists (whose eccentricities Fuller deplored and whom he countered by encouraging the Haldane brothers and R.C. Anderson to establish orthodox Baptist churches in Scotland); with William Vidler the Universalist; and with various Unitarian apologists. As the Baptist churches of Britain responded increasingly to his evangelicalism, Fuller's role in denominational affairs grew more important: his was a profound influence on William Carey* and the Baptist Missionary Society (of which he was secretary, 1792-1815), and he was a loyal servant of the infant Baptist Union and the Baptist Irish Society.

IAN SELLERS

FULLER, CHARLES E. (1887-1968). American Baptist radio evangelist. He graduated from Pomona College in 1910 and engaged in orange-growing. Following his conversion under Paul Rader, Fuller for three years studied at the Bible Institute of Los Angeles where he came under R.A. Torrey's* influence. He was ordained at Calvary Church, Placentia, California, in 1925, where he was pastor until 1932. Two of his early radio programs were "The Pilgrim Hour" and "Heart to Heart Talks." His "Old Fashioned Revival Hour" was aired nationwide over the Mutual Broadcasting System beginning in 1937 and later switched to CBS. The program reached its coast-to-coast peak in the 1940s with live broadcasts over 625 stations from the Municipal Auditorium in Long Beach where Fuller spoke to several thousands every Sunday. His wife pioneered the technique of reading excerpts from listeners' letters on the air. He was a co-founder of Fuller Theological Seminary in 1947. His "folksy" style produced great numbers of converts.

ROBERT C. NEWMAN

FULLER, THOMAS (1608-1661). Divine and historian. Son of the rector of Aldwincle St. Peters in Northamptonshire, and nephew of Bishop Davenant, he studied at Queens' College, Cambridge. Through his uncle's help he had access to good livings at Broadwinter, Waltham Abbey, and Cranford. As a preacher he was very popular in the 1630s and 1640s. During the Civil War he spent most of his time at Oxford, although he was only a moderate Royalist. His fame rests on his books. In 1650 he published *A Pisgah Sight of Palestine,* in 1655 *A Church History of Britain.* A year after his death his *Worthies of England* was published. Into these three books went years of research, and they have often been reprinted. A less famous and more popular work was his *The Holy and Profane State* (1642), a book on the Christian life. He is reputed to have been one of the first authors to make an income by his pen.

PETER TOON

FUNDAMENTALISM. A conservative theological movement in American Protestantism, which arose to national prominence in the 1920s in opposition to "modernism."* Most interpretations of the movement try to explain it in socioeconomic or psychological terms, but the movement was rooted in genuine theological concern for apostolic and Reformation doctrine growing out of American revivalism.* Further confusion has arisen from repeated reference to five basic doctrines (or "five points") of fundamentalism, supposedly springing from the Niagara Bible Conference of 1895.

Fundamentalism should be understood primarily as an attempt to protect the essential doctrines or elements (fundamentals) of the Christian faith from the eroding effects of modern thought. Such doctrines include the Virgin Birth, the resurrection and deity of Christ, His substitutionary atonement, the Second Coming, and the authority and inerrancy of the Bible.

The roots of fundamentalism go back into the nineteenth century when evolution, biblical criticism, and the study of comparative religions began to challenge old assumptions about the authority of the biblical revelation. At the same time new ethical problems accompanied the emerging urban-industrial society in America. Men such as William H. Carwardine and Washington Gladden appealed to the Christian conscience and advocated what came to be called a "social gospel."* The so-called higher criticism* (historical and literary, in contrast with textual) of the Bible entered the mainstream of American Protestantism following the Civil War. By World War I higher criticism was generally accepted in seminaries and colleges. This success came, however, only after strong resistance. Heated debates took place in scholarly journals. Baptists dismissed professors such as C.H. Toy and E.P. Gould, and Presbyterians held heresy trials of C.A. Briggs and A.C. McGiffert. By the turn of the century, major conflict between progressives and conservatives appeared certain.

A significant offensive against modernism was launched in 1910 with the publication of the first of *The Fundamentals.** By 1918 the term "fundamentals" had become common usage, but "fundamentalist" and "fundamentalism" were coined in 1920 by Curtis Lee Laws, Baptist editor of the *Watchman-Examiner.* Laws proposed that a group within the Northern Baptist Convention adopt the name "fundamentalist." During a conference in Buffalo, New York, in 1920, Laws and his associates accepted the title. This group, popularly called "The Fundamentalist Fellowship," were moderate conservatives, who believed that the modernists were surrendering the "fundamentals" of the Gospel, namely, the sinful nature of man, his inability to be saved apart from God's grace, the indispensability of Jesus' death for the regeneration of the individual and the renewal of society, and the authoritative revelation of the Bible. This group, the first to apply the name "fundamentalist" to itself, was identified neither with dispensationalism nor with a crusade against evolutionary teaching. They asserted repeatedly that they were concerned only about the preservation of the central affirmations of the Christian faith.

Historians have often portrayed fundamentalists as "losers." While it is true that the conservatives were unable to gain the adoption of a confession of faith in any of the northern denominations, Laws and his associates did not consider their cause a lost one. Laws wrote in 1924 that certain schools of his denomination had checked the inroads of liberalism and that the investigation of the mission societies, as advocated by the fundamentalists, resulted in certain changes which made the creation of a new mission unnecessary.

A more militantly conservative voice had been raised in 1923 with the formation of the Baptist Bible Union. Composed of Baptists from the South and Canada, as well as the North, the union broadened the fundamentalist cause to include the struggle against evolutionary teaching.

Among Presbyterians, the conservative position was championed by J.G. Machen* of Princeton Theological Seminary. When he refused to break his ties with the Independent Board of Presbyterian Foreign Missions, he was tried and found guilty of rebellion against superiors. Thus evolved the Orthodox Presbyterian and Bible Presbyterian churches.

Gradually "fundamentalism" came to be used loosely for all theological conservatism, including militants, moderates of the Laws type, and a scholarly type represented by Machen. Due to the tactics of certain leaders, the fundamentalist image eventually became stereotyped as close-minded, belligerent, and separatistic.

In the 1950s a growing number of conservatives attempted to set aside the fundamentalist label. Harold John Ockenga was one of the first to propose "new evangelical" as an alternative. He called for a conservative Christianity which held to the central beliefs of the Christian faith, but which was also intellectually respectable, socially concerned, and cooperative in spirit. Since the late fifties this perspective has deepened and broadened. Carl F.H. Henry, Edward John Carnell, the periodical *Christianity Today*, the Billy Graham Evangelistic Association, and other individuals and groups have been identified with the new evangelicalism, which considers itself the heir of the spirit and purpose of the original fundamentalists.

BIBLIOGRAPHY: S.G. Cole, *A History of Fundamentalism* (1931); N.F. Furniss, *The Fundamentalist Controversy* (1954); J.I. Packer, *Fundamentalism and the Word of God* (1958); E.R. Sandeen, *The Roots of Fundamentalism* (1970).

BRUCE L. SHELLEY

FUNDAMENTALS, THE. A series of twelve small books published from 1910 to 1915, containing articles and essays designed to defend fundamental Christian truths. Three million copies of the books were sent free to every theological student and Christian worker whose address was obtainable. The project arose in the thinking of Lyman Stewart, a wealthy oilman in Southern California, who was convinced that something was needed to reaffirm Christian truths in the face of biblical criticism and modern theology. After listening to A.C. Dixon* preach in 1909, Stewart secured Dixon's help in publishing *The Fundamentals.* Stewart then enlisted the financial support of his brother, Milton, and Dixon chose a committee, which included the evangelist R.A. Torrey,* to assist in the editorial work.

Sixty-four authors were eventually chosen. The American premillennial movement and the English Keswick Convention were well represented. Other conservatives such as E.Y. Mullins* of Southern Baptist Seminary and B.B. Warfield* of Princeton Seminary were also among the contributors.

BRUCE L. SHELLEY

FUNERALS, see BURIAL SERVICES

FUX, JOHANN JOSEPH (1660-1741). Musical composer. Although chiefly remembered today for his *Gradus ad Parnassum,* the most influential eighteenth-century treatise on counterpoint, he was also perhaps the greatest Catholic composer of church music in the Germanic cultural sphere of his day. He strove to keep alive the tradition of church style stemming from Palestrina and the Counter-Reformation. He was active at court in Vienna, and for a time at St. Stephen's Cathedral. His influence continued to be felt by composers down into the nineteenth century.

GABRIEL SEVERUS, (1541-1616). Greek theologian. Born in Morea and educated at Padua, he later lived in Crete and Venice. In 1577 he was consecrated metropolitan of Philadelphia (Ala-Shehr) in Asia Minor (Turkey). Much of his time was spent in Venice ministering to the Greek-speaking people there. His proximity to the Roman Catholic Church led him to feel the need to expound and defend the principles of his own church. Two of his major works were explanations and defenses of the Orthodox custom of venerating the elements of the Eucharist before the actual consecration had taken place. Another treatise, *An Exposition against those who . . . teach that the members of the Eastern Church are schismatics . . .*, defended the validity of his church against the critical remarks of leading Jesuits such as Bellarmine.* He was known to scholars in England and helped in the edition of Chrysostom's works prepared by Henry Savile in 1610-13.

PETER TOON

GABRIELI, GIOVANNI (1557-1611). Italian composer. Giovanni is the most important of several composers named Gabrieli. He was the last great composer to cultivate the Renaissance polychoral style at St. Mark's Cathedral in Venice. He also was a key figure in developing some aspects of music that are thought of as distinctly Baroque. He almost completely abandoned the composition of Masses in favor of motets, these often for two or three choirs and in up to nineteen voice parts. He gave instruments a new role, not only combining them on equal terms with voices in many of his motets, but also writing magnificent *canzoni* for wind instruments to be played in the great cathedral. He was a pioneer in the new "concertato" style that dominated the Baroque era and had a direct influence on almost all the leading figures among his younger contemporaries.

J.B. MAC MILLAN

GAEBELEIN, ARNO C(LEMENS) (1861-1945). Writer on prophecy. Born in Thuringia, Germany, he went to America at eighteen and was later ordained in the Methodist Episcopal Church, holding pastorates in Baltimore, Hoboken (New Jersey), and New York City where he began a remarkable ministry to the Jews. He founded and edited *Our Hope* magazine (published 1894-1958), and a press by that name, to provide literature for Jews, chiefly on prophecy and biblical exposition. Its distribution was worldwide. A student of biblical and major national, including Middle Eastern, languages, he wrote nearly fifty books and numerous pamphlets, mostly on prophecy. He lectured and preached widely and was active in the Bible Conference Movement.

C.G. THORNE, JR.

GAIRDNER, JAMES (1828-1912). Historian and records scholar. Born in Edinburgh, he worked as a clerk and an editor in the Public Record Office, London (1846-93). From 1856 he first collaborated with J.S. Brewer in the preparation of *The Letters and Papers of Henry VIII,* then became editor after Brewer's death. He also edited collections of documents for the Rolls Series and the Camden Society and prepared the definitive edition of the *Paston Letters.* He wrote the volume for the period 1509-59 for Stephens and Hunt's *History of the Church of England,* closing his career with the four-volume work *Lollardy and the Reformation in England.* Always strongly Protestant in his outlook, he also sought to be objective in his editing and writing.

W.S. REID

GAIRDNER, WILLIAM HENRY TEMPLE (1873-1928). Anglican missionary and scholar. Born in Ardrossan, Scotland, he was educated at Trinity College, Oxford, and was associated with J.R. Mott* in work among British students. He went with the Church Missionary Society to Cairo in 1898, with a "special view to work among students and others of the educated classes of Moslems." He was ordained in 1901. A gifted linguist, he broke new ground by teaching missionaries and native teachers colloquial Arabic, produced a handbook on phonetics and two textbooks on the subject, and wrote hymns, poems, plays, and popular biblical literature in Arabic. He founded an English and Arabic Christian magazine, *Orient and Occident,* in 1904. He collected some three hundred Near Eastern tunes for use in Christian worship. His deep study of Islamics and Arabic literature is revealed in *The Reproach of Islam* (1909) and *The Values of Christianity and Islam* (with W.A. Eddy, 1927). He worked to make the Arabic Anglican Church into a welded group of believers, to train indigenous leaders, and to improve relations with the Coptic Church.* He believed that Islam could be won by a living exemplification of Christian brotherhood. His deep awareness of beauty is seen in his love of music, poetry, and the world of nature. Zest characterized his life and his faith; he seemed to have the ability to enjoy everything intensely.

HOWARD SAINSBURY

GAIUS (Caius) (third century). Roman presbyter and author of a *Dialogue* in which he maintained a debate with the Montanist Proclus* during the pontificate of Zephyrinus. Proclus defended the prophesying of his sect by referring to Philip's prophesying daughters (Acts 21:9), who were buried with Philip at Hierapolis. Gaius defended the authority of Rome by referring to the tombs of the apostles in the Vatican and on the Via Ostia. Gaius accepted thirteen epistles of Paul, but denied the Pauline authorship of Hebrews. It appears also that he rejected the fourth gospel and the Apocalypse as the work of Cerinthus. Two later Syriac writers, Dionysius Bar-Salibi (twelfth century) and Ebedjesus (fourteenth century), mention a treatise of Hippolytus in which he defends the apostolic authorship of these works against Gaius. Eusebius may not have been aware of Gaius's attitude toward these works, as he calls him a "churchman," a title usually reserved for the orthodox. DAVID JOHN WILLIAMS

GALATIA. (1) A region and Roman province in central Asia Minor, named after a Celtic tribe that migrated into the valley of the Halys River in the third century B.C. from central Europe. Although never the majority, these "Gauls" ruled the indigenous tribes of Phrygians and Cappadocians. Three different tribes were involved, each of whose Celtic tribes were divided into four classes, called "tetrarchies" by the Greeks. Their military prowess made the Galatians desirable as mercenaries.

(2) Hellenistic Galatia was the central plateau of Asia Minor that is bounded by the upper Sangarius and middle basin of the Halys River, limited to the north by the kingdoms of Bithynia and Pontus. After the fall of the Hittite Empire, this area was called Phrygia, and later Galatia. The Galatians did not dwell in towns, but lived tribally, until finally crushed by the Romans in 25 B.C.

(3) Roman Galatia was created from Galatia proper and major extensions to include Lycaonia, Isauria, Pisidia, as well as the cities of Iconium, Lystra, Derbe, Apollonia, and Antioch. A variety of peoples was thus added, and the city of Ancyra* was selected as the capital. Further extensions were added to this vast territory in 6-5 B.C., A.D. 64 and 72, to a size approaching the former Hittite Empire.

(4) NT usage, A subject of dispute has been Paul's usage of the term "Galatia" (Gal. 1:2). Does he refer to the original ethnic sense or to the Roman province? The latter is now more favored. Acts 16:6 seems to imply that Paul visited those ports of Phrygia which had been incorporated into the Roman province of Galatia. Likewise, in Acts 18:23 it is doubtful if Paul ever visited the northern area of Galatia. Two other references to Galatia likewise imply the territory of the Roman province: 2 Timothy 4:10; 1 Peter 1:1, while 1 Corinthians 16:1 will be interpreted according to one's view of the "Galatia" meant in the other passages. JAMES M. HOUSTON

GALATIANS, EPISTLE TO THE, see EPISTLES, PAULINE

GALERIUS, VALERIUS MAXIMIANUS (d.311). Roman emperor. A native of Illyricum and of humble origin, he was invested with the title of Caesar in 293 by Diocletian, whose daughter he married, and was given responsibility for the Danube frontier. Here he won several victories over the Germans (293-95). When Diocletian decided on measures against the Christians, the severity with which this decision was implemented in the series of edicts after 303 was due largely to the influence of Galerius. As the emperor's health failed, Galerius's power increased, and in 305 he persuaded both Augustuses to abdicate and himself became the Augustus of the East while Constantius became that of the West. The church in the West enjoyed comparative peace after this, but Galerius and his Caesar, Maximin, continued their policy of persecution. After 307 there was some remission, but it was not until his own health failed and he was under the threat of an alliance between Constantine and Maxentius that Galerius published his Edict of Toleration from Nicomedia in the same year that he died. DAVID JOHN WILLIAMS

GALESBURG RULE. "Lutheran pulpits for Lutheran ministers only; Lutheran altars for Lutheran communicants only" was adopted by the Lutheran General Council meeting at Galesburg, Illinois, in 1875. It was intended to preserve confessional distinctiveness threatened by practices of some Lutherans that seemed to promote unionism and Americanization. The rule, suggested by president C.P. Krauth, was enacted at Akron, Ohio, in 1872, with carefully worded provisions for exceptional cases. One party within the council demanded an exclusive interpretation of the Galesburg Rule, without exceptions; another insisted on the understanding reached at Akron. It was finally decided (Pittsburgh, 1889) that the Galesburg Rule had not annulled the Akron statement. ALBERT H. FREUNDT, JR.

GALILEO GALILEI (1564-1642). Italian astronomer and physicist. He studied at the University of Pisa, and after teaching at Siena and Florence returned to Pisa as professor of mathematics (1589). Two years later, because of his opposition to Aristotelianism, he moved to the University of Padua. Here he conducted mechanical research, made mathematical instruments for sale, and wrote several articles that were circulated in manuscript to his pupils and friends. In 1610, with the aid of his newly invented telescope, he discovered four moons that revolve around Jupiter. By analogy he reasoned that the planets revolve about the sun. This led him to support the Copernican explanation of the solar system. He also noted in his book *The Starry Messenger* many other observations which could not be accounted for with the Ptolemaic view of the universe. His publication of these ideas gained him Europe-wide fame, and appointment as philosopher and mathematician to the duke of Tuscany.

He also came into conflict with the Inquisition,* and when the Copernican theory was condemned Galileo was forbidden to teach it (1616). In 1624 he visited Rome and obtained permission to write on the Copernican and Ptolemaic systems provided that the treatment was impartial. The book which resulted from his work, *Dialogue*

Concerning the Two Chief Systems of the World (1632), caused him to be brought to trial by the Inquisition. The work was condemned, Galileo recanted, and he was sentenced to life imprisonment. He was, however, permitted to live under house arrest till his death.

BIBLIOGRAPHY: G. de Santillana, *The Crime of Galileo* (1955); L. Geymonat, *Galileo Galilei* (tr. S. Drake, 1965); C.L. Golino (ed.), *Galileo Reappraised* (1966). ROBERT G. CLOUSE

GALL (c.550-640). Irish monk and missionary. He was one of the twelve monks who accompanied Columbanus from Ireland to Gaul, remaining there with him until 612. Then he settled with a few friends in a waste place to the west of Bregenz, in Austria near Lake Constance. Many legends surround him—e.g., that he was the founder of the Benedictine monastery at St. Gallen. In fact he died a century before its foundation.

GALLA PLACIDIA (c.390-450). Roman empress. Daughter of Theodosius I by his second wife Galla, she was captured by Alaric in 410 and married Alaric's successor, Ataulf, at Narbonne in 414. On Ataulf's death she returned to Italy, and in 417 married Constantius. When he died in 421, she retired to Constantinople, but when her brother Honorius died in 423, Theodosius II recognized her son Valentinian III as Honorius's successor and she returned to the West, establishing her court at Ravenna. An uncompromising Catholic, her influence may be seen in the edicts against all "Manicheans, heretics, and schismatics, and every sect opposed to the Catholic faith." She also supported Leo against the Council of Ephesus in 449. She died shortly after this council and was buried at Ravenna.

DAVID JOHN WILLIAMS

GALLICAN ARTICLES, THE FOUR (1682). A declaration concerning the respective authorities of the crown, papacy, and French bishops, adopted at Paris by a special assembly of the French clergy. The immediate occasion was a conflict over the *regalia,* involving opposing claims by Louis XIV and Innocent XI to fill vacant French bishoprics and to control their revenues. The declaration, drafted by Bishop Bossuet, was intended to avoid outright break with Rome while acknowledging the supremacy Louis XIV wanted. The crucial first article asserted that the king was not subject "in temporal things" to any ecclesiastical power, he could not be deposed, nor could his subjects be relieved from obedience to him by papal authority. The second claimed that while the pope enjoyed full spiritual authority, he was subject to general councils as decreed by the Council of Constance (1414-18). The next added that the exercise of papal authority was further subject to the canons and constitutions of the French kingdom and church. The fourth allowed the pope "the principal part in questions of faith," but claimed that his judgments were not above correction. Louis XIV later denounced the declaration (1693), although its principles remained the core of Gallicanism* throughout the eighteenth century. C.T. MC INTIRE

GALLICANISM. A movement, triumphant in the seventeenth century, defining the authorities of, and relations among, the French king, the French Church, the papacy, and indirectly the French Parlements. The central event was the French bishops' declaration of the Four Gallican Articles* of 1682, at the insistence of Louis XIV. Common to the varieties of Gallican theories and practices are three assertions, as shown by Victor Martin: the sovereignty of the crown in temporal things, the authority of general councils over the pope, and the authority of crown and bishops to regulate papal interference in France.

Gallicanists professed to recognize the universal spiritual authority of the pope, but with these qualifications. The secular absolutist statism of Louis XIV *de facto* subjugated the French Church to the Crown, and completed the reversal of the pre-1300 relations between papacy and Crown. The progress of this reversal and the assertion of Gallicanism was marked by the resistance of Philip the Fair to Boniface VIII,* the Pragmatic Sanction of Bourges* (1438), the Concordat of Bologna* (1516), the nonreception of decrees of the Council of Trent in France, and similar events, whereby the Crown claimed rights in control of episcopal elections, liturgy, canon law, education, and in many other ecclesiastical matters. The Crown made such successful claims often in conflict with contrary claims by the bishops and the Parlements, as the modern self-sufficient sovereign state emerged by the late eighteenth century. Pierre Pithou's *Les Libertés de l'Église gallicane* (1594) served as the standard handbook until the nineteenth century.

The French Revolution and the Civil Constitution of the Clergy (1790) effected an even more radically secularist Gallicanism, only somewhat modified by the Napoleonic Concordat of 1801* as unilaterally amended in 1802 by Napoleon's Organic Articles. The Ultramontane Catholic revival aroused among the French faithful and clergy a devotion to the unity and teachings of the church under the pope. This provided a core support for the reception of the Syllabus of Errors* and the dogma of papal infallibility (1870), which effectively rendered Gallicanism an unacceptable doctrine.

BIBLIOGRAPHY: V. Martin, *Les Origines du gallicanisme* (2 vols., 1939), *Le Gallicanisme et la réforme catholique* (1919), and *Le Gallicanisme politique et le clergé de France* (1929); F. Mourret, *History of the Catholic Church,* VI (1947); C.B. du Chesnay, "Gallicanism," in the *New Catholic Encyclopedia,* VI, pp. 262-67; A.-G. Martimort, *Le gallicanisme de Bossuet* (1953).

C.T. MC INTIRE

GALLICAN PSALTER. At different times Jerome made three versions of the Psalter, known respectively as the Roman, the Gallican, and the Hebrew. The first, produced about 383 at the request of Pope Damasus, was a revision of the Old Latin version, in which Jerome made use of the Septuagint. This version remained in use in the Roman Church until the time of Pius V (1566-72). Outside of Italy, however, the Roman Psalter was superseded by the Gallican, a version which

Jerome made in Bethlehem about 389, using the Septuagint, Theodotion's Greek version, and the Hebrew, by means of Origen's Hexapla. This version was taken to Gaul by Gregory of Tours. Later it was removed to England, where it gradually replaced the Old Latin version and ultimately became the basis for the Prayer Book version of the Psalms. DAVID JOHN WILLIAMS

GALLICAN RITE. Although there was no uniformity of worship in the churches of Merovingian Gaul, there was a basic pattern to the liturgies. Thus it was possible to distinguish a Gallican rite from the Roman Rite. The Gallican forms for the Mass, baptism, and ordination were less austere and more oratorical than the Roman, and had important differences of order and of content. For example, in the baptismal service the confession of faith preceded immersion or affusion, and part of the ceremony was the washing of feet. Also in the Mass the Trisagion ("Holy, Holy, Holy") was sung in Greek and Latin before the *Kyries* ("Lord, have mercy") as well as before and after the Gospel, and a Trinitarian hymn (the *Trecanum*) was sung during the actual Communion. Various theories have been put forward to explain these differences. The most ancient of these is that the Gallican Rite came from Ephesus through the influence of Irenaeus. A more recent view (e.g., that of Louis Duchesne) is that it came from Milan. The Gallican Rite was formally abolished by Emperor Charlemagne; however, the present Roman Rite shows signs of being influenced by or conflated with that of Gaul. PETER TOON

GALLIC CONFESSION (1559). The French Calvinist Confession of Faith. Its history begins with the statement of faith sent by the Reformed churches of France to Calvin in 1557 during a period of persecution. Working from this, and probably with the help of Beza and Pierre Viret, Calvin wrote a confession for them. This took the form of thirty-five articles. When persecution subsided, twenty delegates representing seventy-two churches met secretly in Paris from 23 to 27 May 1559. With François de Morel as moderator, the brethren produced a Constitution of Ecclesiastical Discipline and a Confession of Faith. Calvin's thirty-five articles were all used in the confession, apart from the first two which were expanded into six. Thus the Gallic Confession had forty articles.

Scholars point out that the revisions of the delegates in the first part caused the introduction of natural theology into Reformed creedal statements. Article II speaks of God revealing Himself firstly in creation and only secondly through His Word. In 1560 the Gallic Confession was presented to Francis II with a preface requesting that persecution should cease. The confession was confirmed at the seventh national synod of the French churches at La Rochelle in 1571, and recognized by German synods at Wesel in 1568 and Emden in 1571. The original draft of Calvin's articles is in the Genevan Archives; for the confession, see P. Schaff, *Creeds of Christendom* (4th ed., 1905), vol. I; and A. Cochrane *Reformed Confessions of the Sixteenth Century* (1966). PETER TOON

GALLITZIN, DEMETRIUS AUGUSTINE (1770-1840). Roman Catholic priest and missionary to the Alleghenies (also known as Smith or Schmet). Son of a freethinking Russian scientist and ambassador to Holland, he was, after his mother's return to the faith, converted to Catholicism. He came to America in 1792, and after attending Baltimore Seminary he was ordained. Refusing offers of prestige, both in the church and by the Russian government, he devoted his life to upbuilding the church in the Alleghenies. In this effort he exhausted his personal fortune. He was honored by the church for his success in establishing a large Catholic settlement in the area, also for his writing *A Defense of Catholic Principles.* MILLARD SCHERICH

GAMBLING. This may be defined as an agreement between two or more parties in which the transfer of something of value from one party to another is made to depend solely on an event the outcome of which is unknown, and perhaps unknowable, to the parties. Covered by the definition are betting, lotteries, and financial speculation. Insurance is excluded, being understood as a way of minimizing insecurity in the face of what are regarded as inevitable risks.

The dominant view in the church has been that though it is not wrong to make decisions depend on the outcome of a "chance" event such as a lottery, gambling is wrong because it involves covetousness, the seeking of gain at another's expense, and financial irresponsibility. Objections to gambling based on the view that all events are, or ought to be, within the control of "rational" human decisions are obviously incompatible with a serious recognition of divine sovereignty and so are unsound from a Christian standpoint.

On the above definition, gambling is not a greater evil than other abuses that the church has been more reluctant to condemn, for example, financial and economic exploitation. Further, where the amounts of money or objects of value are small, where the practice is carefully regulated, and where it has the unconstrained agreement of all parties, the evils of gambling are sometimes regarded as at a minimum. Many Christians, however, regard gambling as intrinsically evil.

Where gambling is uncontrolled and large amounts of money become involved, or where poor people are pressed into parting with what they have in the belief they may easily win large amounts of money, it can interfere with family life and with work. It can attract crime and lead in some cases to addiction and compulsive neuroses. In evaluating gambling from a Christian point of view, attention should be focused not only on gambling as an activity that is socially evil, but also on the conditions, moral and social, that give rise to it. OONAGH MC DONALD

GANSFORT, W.H., see WESSEL OF GANSFORT

GARDINER, STEPHEN (c.1490-1555). Bishop of Winchester. After graduating in canon and civil law, he became master of Trinity Hall, Cambridge, in 1525, and private secretary first to Wolsey and then to the king. He was employed in legal proceedings against heretics and in negotiations with Rome for annulling the king's marriage. In 1531 he was made bishop of Winchester. In a famous oration, *De Vera Obedientia,* in 1535 he argued that the pope has no legitimate jurisdiction over other national churches and that kings and princes are entitled to supremacy in their respective churches. Unsympathetic with Protestant doctrines, he was generally regarded as responsible for the Six Articles* of 1539. In Edward VI's reign he was deprived of his offices and imprisoned, but Mary restored him, making him lord high chancellor. With Bonner* he organized vigorous proceedings against Protestants, securing the reenactment of the statute *de heretico comburendo* and taking a leading part in the trials of John Bradford and John Rogers. Though jealous of Archbishop Pole, he approved the submission of the English Church to Rome, despite his earlier views, and aimed at the reestablishment of ecclesiastical courts. He died a wealthy man, and was buried in Winchester Cathedral. His *Letters* (ed. J.A. Muller) were published in 1933.

See also J.A. Muller, *Stephen Gardiner and the Tudor Reaction* (1926). JOYCE HORN

GARNIER, JEAN (1612-1681). Jesuit* scholar. Native of Paris, he entered the Jesuit Order at Rouen in 1628 and taught philosophy at Clermont-Ferrand for ten years from 1643, and theology thereafter at Bourges until his death. A church historian, patristics scholar, and moral theologian, he published in 1648 *Libellus fidei,* arising from the Pelagian controversy, providing critical and historical notes. At Paris he edited all the works of Marius Mercator, with important treatises on Pelagianism (see PELAGIUS) and Nestorianism* (1673), and the *Breviarium causae Nestorianorum et Eutychianorum* with his own reflections on the Fifth Council (1675). He wrote on scholastic philosophy and moral and doctrinal theology, as well as a library manual, *Systema bibliothecae collegii Parisiensis S.J.* (1678), and made a critical edition of the papal *Liber Diurnus* (1680). C.G. THORNE, JR.

GASCOIGNE, THOMAS (1403-1458). English theologian. Son of the lord of the manor of Hunslet, he was educated at Oriel College, Oxford, where he spent the rest of his life, refusing outside appointments and often appearing as either the chancellor or vice-chancellor of the university. Wealthy and somewhat conceited, Gascoigne was a strong defender of the established order in church and state, and a foe of Lollardy. He was, however, very concerned about current abuses in the church, especially pluralism and monastic decline, and his *Dictionarium Theologicum* (written between 1431 and 1457) is, theological interest apart, a highly personal guide to the affairs of the English Church and government and to the Oxford of his day. IAN SELLERS

GASQUET, FRANCIS NEIL AIDAN (1846-1929). Benedictine* scholar. Born in London of French and Scots parentage, he was educated at Downside, was made prior there in 1878, but resigned in 1885 because of ill health. Convalescence led him to historical research. The result was his *Henry VIII and the English Monasteries* (2 vols., 1888-89), which won acclaim for its vindication of English monastics at the Reformation, and *Edward VI and the Book of Common Prayer* (1891), which led to his nomination to the Commission on Anglican Orders (1896). Abbot-president of the English Benedictine Congregation (1900-1914), he was also the first president of the International Commission for the Revision of the Vulgate (1907). He was created cardinal in 1914, negotiated for a British minister to the Vatican, and became prefect of the Vatican Archives in 1917 and librarian in 1919. His other works included *A History of the Catholic Church in England* (2 vols., 1897) and *Monastic Life in the Middle Ages* (1922). C.G. THORNE, JR.

GAUDENTIUS (fourth/fifth centuries). Bishop of Brescia in Italy. While on pilgrimage to the Holy Land he was elected about 387 to succeed Philaster as bishop of Brescia and was persuaded by Ambrose and some Eastern bishops to accept the election. Little is known of his episcopate, but in 404-5 he was sent to Constantinople by Honorius and Innocent I to plead the cause of Chrysostom to Emperor Arcadius. He achieved nothing, but nevertheless was thanked by Chrysostom. Rufinus dedicated his Latin translation of the *Clementine Recognitions* to Gaudentius and refers in particular to the latter's knowledge of Greek. A number of Gaudentius's sermons survive which reflect this knowledge and show a propensity to allegorization.

DAVID JOHN WILLIAMS

GAUL, see FRANCE

GAUSSEN, FRANÇOIS SAMUEL LOUIS (1790-1863). Swiss Reformed pastor. While studying theology in Geneva, he found personal faith through the student group influenced by the orthodox Scot, Robert Haldane.* After becoming minister of the Satigny church, Gaussen underwent a long period of conflict with the heterodox *Vénérable compagnie des pasteurs,* which finally dismissed him in 1831. He republished the Second Helvetic Confession of 1566, helped found the Evangelical Society of Geneva (1831), and with Merle d'Aubigné* and others formed an independent and orthodox Reformed theological seminary in Geneva (1834). The author of many generally Calvinistic works, Gaussen is best known for his widely circulated *Theopneustia,* a statement of verbal biblical inspiration which drew fierce attacks. ROBERT P. EVANS

GAVAZZI, ALESSANDRO (1809-1889). Italian patriot and religious reformer. Born at Bologna of a large and very religious family, he joined the Barnabite* Order and taught in their schools in various Italian cities. Endowed with great oratorical ability, he soon began to preach sermons, mainly political, championing the cause of liberalism and Italian freedom against ecclesiastical authorities and the Jesuits, who tried vainly to silence him. Threatened and enclosed in a convent, he was liberated at the election of Pope Pius IX, who sent him as chaplain with the papal volunteers fighting with Charles Albert against Austria in the first war of independence (1848). Influenced by Vincenzo Gioberti* and deeply disappointed after the pope's *volte-face,* he took an active part in the Roman republic of 1849, organizing hospital assistance during the siege. After the defeat, pursued by papal police, he fled to Britain, joining the many Italian exiles, some of whom had founded an Italian evangelical church in London.

It is impossible to determine when Gavazzi left Roman Catholicism and whether he was genuinely converted. Soon, however, he became known as an eloquent speaker, and Lord Palmerson suggested hiring a hall in Oxford Street where crowds went to hear him denounce papal abuses and Jesuit politics. During the next ten years (spent in Britain, apart from ten months in North America) he traveled widely in England, Scotland, and Ireland, making many friends. In 1859 he returned to Italy to join Garibaldi in the wars of independence, the expedition of the Thousands, and the various attempts to conquer Rome, his main object now being the destruction of the papacy and the foundation of one great reformed Italian church.

Disagreeing with the organization and policy of the Waldensians, he tried to join the Free Italian Church (see GUICCIARDINI), meeting with Mazzarella* and Desanctis,* preaching in Genoa and Florence, but his political approach and his lack of spirituality made him suspect and led to a division in 1863. Aided by J.R. McDougall, minister of the Free Church of Scotland in Florence, he devoted himself thereafter to the establishment of a new Free Italian Church, which by 1870 comprised twenty-two congregations and a theological school.

Gavazzi's last years were spent in frequent journeys to Britain to collect funds for his church and in repeated attempts to join with the Waldensians, a move much opposed by McDougall. (At the turn of the century the various congregations were absorbed by the Methodists.) Disillusioned and embittered, rejected by all his friends, Gavazzi died in Rome. Author of innumerable and often contradictory pamphlets (some in English) written to propagate his views and attack his enemies, Gavazzi was, in the words of his most objective biographer, "the greatest intruder of Italian evangelism."

See L. Santini, *Alessandro Gavazzi* (1955).

DAISY RONCO

GEDDES, JENNY. The supposed instigator in 1637 of a riot in St. Giles', Edinburgh, against the imposition of "Laud's Liturgy." A little investigation might suggest that the only historically authenticated Jenny figured in the High Street of Edinburgh at the Restoration of Charles II more than twenty years later, as a staunch Royalist. It may, of course, be that the lady, older and wiser, had had a change of heart, but it is hard to avoid the conclusion that despite the memorial in St. Giles', the monumental Jenny is a myth.

J.D. DOUGLAS

GEDDIE, JOHN (1815-1872). Pioneer Presbyterian missionary to the New Hebrides Islands. Born in Banff, Scotland, his family emigrated to Nova Scotia in the following year. He became interested in missions at a very early age. After ordination he was instrumental in getting his small denomination to undertake a mission of its own. He became its first missionary, sailing from the USA in January 1847. As founder of the New Hebrides Mission, he labored amid great difficulties on the island of Aneityum. After his death in 1872, a memorial was placed in the mission church with these words: "When he landed in 1848 there were no Christians here, and when he left in 1872 there were no heathen."

HAROLD R. COOK

GEILER VON KAYSERSBERG, JOHANNES (1445-1510). Roman Catholic preacher and reformer. Born in Schaffhausen, he studied at the universities of Freiburg im Breisgau and Basle, taught at Freiburg, then became people's priest in Strasbourg. In theology he was a nominalist, but highly oriented his theology to his pastoral concerns. According to the nominalists, in the covenant relationship between God and man, God is regarded as a God who is faithful to His commitments. His grace is given to those who turn to Him, disposing themselves by their natural capacities to His gracious gifts. God gives His grace to those who do their very best. God's justice, human responsibility, and the act of creation are emphasized in nominalism. Geiler is noted especially as a preacher, effective in reaching the common people through his German sermons. Although not a humanist, he stressed the need for reform and influenced Bishop Wilhelm von Honstein in his reformatory endeavors. Geiler has been called "the prince of the pulpit in the late fifteenth century."

CARL S. MEYER

GELASIAN DECREE *(Decretum Gelasianum)*. An early Latin document referred to also by the title of its last section *Decretum de Libris Recipiendis.* It comprises five sections, dealing with: Christ and the Holy Spirit; the canonical books of Scripture; the Roman Church; the orthodox councils and fathers; and the works of the fathers to be accepted and those to be rejected. Hincmar of Reims in the seventh century was the first to assign the work to Gelasius (492-96), but it may include earlier material. Some MSS assign it to Damasus (366-84), and its statement concerning the Roman Church that it "has not been set above the rest by any synodical decisions" may express the reaction of Damasus's council of 382 to the third canon of the Council of Constantinople in 381, which implied that the ecclesiastical prestige of a city was directly related to that city's political power. Some MSS assign the *Decretum* to Hormisdas (514-23). In its present form it belongs to the end of the fifth or beginning of the sixth century, but earlier material could have been used by either Gelasius or Hormisdas to produce the *Decretum* as it now stands.

DAVID JOHN WILLIAMS

GELASIUS (d.394/5). Bishop of Caesarea from about 367. Nephew of Cyril of Jerusalem and a convinced Nicene, he was removed from his see during the reign of Valens, but restored on the accession of Theodosius in 378. In 381 he was present at the Council of Constantinople. According to Theodoret he "was renowned alike for lore and life." He wrote, according to Jerome, "more or less in carefully polished style, but not to publish his works." Those works include a continuation of Eusebius's *Ecclesiastical History,* on which both Rufinus and Socrates may have drawn for their own histories. He wrote also a treatise against the Anomoeans* and an *Expositio Symboli,* of which fragments survive.

DAVID JOHN WILLIAMS

GELASIUS I (d.496). Pope from 492. Although his birthplace and nationality are uncertain, it seems probable that he was a Roman citizen in Africa. It is clear he was one of the most able in a century of outstanding pontiffs. Coming to the office a decade after the Acacian Schism* began, he was constantly engaged in controversy in upholding the Roman primacy against Constantinople. His writings include treatises and letters on the two natures in Christ, Arianism, Pelagianism, and Manichaeism. His views on the relative places of church and state predated much later thinking on the subject. "There are two powers by which this world is chiefly ruled," he wrote to the emperor Anastasius I, "the sacred authority of the priesthood and the authority of kings." Each he held to be of divine origin and to be independent in its own sphere. The *Decretum Gelasianum* and the Gelasian Sacramentary have been wrongly attributed to him, although some scholars find traces of his thinking in the latter.

J.D. DOUGLAS

GELASIUS OF CYZICUS (latter fifth century). Church historian. Our only knowledge of him is derived from Gelasius himself. He was the son of a presbyter of Cyzicus in Asia Minor, and having found in his father's house a document which had belonged to Dalmatius, bishop of Cyzicus, containing an account of the proceedings of the Council of Nicea, he was prompted to write a history of the council. He appears to have used other sources—such as Eusebius, Rufinus, Socrates, and Theodoret—in compiling it, and adds little of value to his earlier sources. The history contains a number of errors and anachronisms. Thus a number of chapters are devoted to disputations on the divinity of the Holy Spirit, a matter not yet in question. It is sometimes assumed that the history contains a complete collection of the synodal acts of the council, but there is no evidence of the existence of such a collection.

DAVID JOHN WILLIAMS

GELLERT, CHRISTIAN FÜRCHTEGOTT (1715-1769). German poet. Born in Saxony, son of a Lutheran pastor, he studied theology at the University of Leipzig, but was temperamentally unsuited to preaching. From 1751 he taught philosophy, poetry, and rhetoric at the university. He was greatly esteemed by his students, including Goethe* and Lessing,* for his piety and generosity. Although he wrote a novel and plays, his fame rests chiefly on his moral tales in verse, *Fabeln und Erzählungen* (1746-48), and his devotional poems, *Geistliche Oden und Lieder* (1757), which provided Beethoven with the text of six of his songs, e.g., "Die Himmel rühmen des Ewigen Ehre." Contemporary rationalism made the assertion of faith in his Easter hymn all the more impressive: "Jesus lebt, mit ihm auch ich," well known in English as "Jesus lives! no longer now." His hymns, often too didactic, were at their best full of rational piety.

JOHN S. ANDREWS

GENERAL ASSOCIATION OF REGULAR BAPTISTS. Founded by twenty-two churches withdrawing from the Northern Baptist Convention in protest against modernism in 1932, the association adopted the New Hampshire Confession of Faith (1832) with a premillennial interpretation to the final article. Strictly congregational in church government, the association has grown to well over 200,000 members and is served by a Council of Eighteen. Various missions agencies and six institutions of higher learning are approved.

See AMERICAN BAPTIST CHURCHES.

GENERAL BAPTISTS, see BAPTISTS

GENERAL CHAPTER. The meeting of a religious order to determine policy for the organization and especially to elect new leaders. Although the orders vary widely in the frequency, functions, and powers accorded to the general chapter, it is generally composed of heads and representatives of the constituent communities, convoked every three or four years, and constituted as the highest authority in the order under the pope, although in some cases the head is not bound by its decrees. The institution only became standard with the Cistercians—who established an annual general chapter of all abbots as the ultimate authority within the order—and was made mandatory for all orders by the Fourth Lateran Council.

MARY E. ROGERS

GENERAL CONFESSION. A set form of confession suitable for all persons and all occasions and in the Book of Common Prayer* used at the beginning of Morning Prayer and Evening Prayer. The rubric directs that it is to be said by the whole congregation, after the minister, kneeling. It is based on Romans 7:8-25 and consists of a confession of sin to God, a prayer for forgiveness, and a prayer for grace to live rightly. It dates from the revised Prayer Book of 1552 and may have been suggested by the Confession in the Strasbourg Liturgy.

GENERAL COUNCILS, see ECUMENICAL COUNCILS

GENERAL SUPERINTENDENT. Formerly the highest ecclesiastical office in many German Protestant churches, exercising authority in conjunction with the provincial consistory and synod. It was purely ecclesiastical, not being confirmed by the state. Numbers would vary in different provinces, e.g., four in Brandenburg, three in Saxony. In the west the title was replaced by that of "Präses," and in the east the bishop of Berlin has been set over them. It has also been an office among British Baptists since 1916, having pastoral and administrative care of areas in the country in an advisory capacity. It recalls the "Messenger" of the seventeenth-century General Baptists.

J.G.G. NORMAN

GENERAL SYNOD (of the Church of England). In 1969 the Synodical Government Measure replaced the Church Assembly with the General Synod, and transferred to the latter some of the Convocation* responsibilities. Like its predecessor, the synod consists of three houses: bishops (the two archbishops and all diocesan bishops), clergy (some *ex officio*, some elected by clergy on a diocesan basis), and laity (almost entirely elected by laity on a diocesan basis, with a few coopted). Under Parliament (so long as the Church of England remains by law established) the synod is the Church of England's central legislative body, meeting normally three times a year for week long sessions, and occasionally meeting separately by houses. It is elected (except for bishops who are all *ex officio*) by proportional representation on a diocesan basis, though the option of subdivision into smaller units has led to great diversity between dioceses and in some cases to the total annulment in practice of the proportional system which was designed to protect minorities.

Church assembly was set up by Parliament in 1919 by the Enabling Act, with elections every five years. Church assembly prepared measures for parliamentary approval, after which they became part of English law. The assembly rarely clashed with Parliament except on the famous 1928 draft Prayer Book, which Parliament twice rejected on doctrinal grounds. This action infuriated many bishops and a large number of High Churchmen. Evangelicals and some other High Churchmen were, on the other hand, relieved that doctrinal innovations had not been forced on the church. Later it was clear that Parliament saved the Church of England from the folly of most of its then leaders, and the 1954 Church and State Report recognized that many clergy and laity still believe that Parliament is a more reliable and impartial judge than the assembly, and perhaps even the new synod.

The general synod is much smaller than the church assembly and has been widely advocated as "bringing the laity in more." Those laymen who have observed it closely are more inclined to regard it in terms of bureaucratic streamlining designed to minimize opposition from powerful minorities (such as defeated the Anglican-Methodist union scheme) and to concentrate effective power and control in the hands of a few people.

G.E. DUFFIELD

GENEVA BIBLE. A 1560 translation of the Bible into English popularly known as the "Breeches Bible" from its rendering of Genesis 3:7 ("They made themselves breeches"—AV "aprons"). It was translated at Geneva by a committee of Protestant exiles which probably included William Whittingham and John Knox. Its Calvinistic annotations greatly irritated James I of England, but delighted and instructed his increasingly Puritan subjects. Used widely for two generations, it became the official version of the Scottish Kirk and the household Bible of English-speaking Protestants everywhere. Gradually replaced by the Authorized Version of 1611, the last Geneva Bible was published in 1644.

ROBERT D. LINDER

GENEVAN ACADEMY. This was founded in 1559 under the influence of John Calvin,* who believed that one of the great needs of the Reformed church was an educational institution that would prepare not only ministers to preach the Gospel, but men who could take their place as Christians in every walk of life. Influenced by the example of Martin Bucer and John Sturm in Strasbourg, he desired to extend the public school established in 1537 to cover the whole course of education. This he succeeded in doing in 1559, at which time Theodore Beza* became the first rector, along with three other professors. Joint control over the institution was exercised by ecclesiastical and civil authorities, although the basic law emanated from the mind of Calvin. One important characteristic of the academy was the emphasis upon the use of French as well as Latin. Physical science and mathematics also became part of the standard curriculum. In the faculty of theology, biblical studies were fundamental although much attention was also paid to preaching. From the academy in the years following its foundation many leaders of the Reformation in other countries such as France, Hungary, Holland, England, and Scotland went out to carry on the Calvinistic tradition.

See C. Borgeaud, *Histoire de l'Université de Genève*, (1900), I; W.S. Reid, "Calvin and the Founding of the Academy of Geneva," *Westminster Theological Journal*, XVIII (1955), pp. 1ff.

W.S. REID

GENEVAN CATECHISM. The first Genevan Catechism was drawn up by John Calvin in 1537, originally French and then translated into Latin. It would seem, however, to have been verbose and not easily memorized. Consequently, after his return from exile in Strasbourg in 1541, at the request of a number of people including some of the ministers of East Friesland, Calvin produced another which had a more catechetical form with crisper phrases more easily learned. Again written in French and then translated into Latin, this work sought to set forth the basic doctrines of the Christian faith. At the same time a schedule was prepared indicating how it could be learned and recited over a period of fifty-five weeks. This catechism was adopted by the French Reformed Church and in translation was used as the Scottish Church's catechism until it accepted the Westminster Catechisms in 1648. The best modern edition is in *Calvin, Theological Treatises* (ed. J.K.S. Reid, 1954). W.S. REID

GENEVIÈVE (Genovefa) (c.422-500). A virgin of Paris, and the city's patron saint. As a child she is said to have attracted the attention of Germanus of Auxerre and at his instigation devoted herself to a life of asceticism. From her fifteenth until her fiftieth year she ate only twice a week, and then only barley bread. Thereafter, at the command of the bishop, she added fish and milk to her diet. The diversion from Paris of the Huns under Attila in 451 was attributed to her prayers.

GENNADIUS I (d.471). Patriarch of Constantinople. While a member of a monastery in Constantinople, he wrote against the anathemas of Cyril of Alexandria during the Nestorian controversy. On the death of Anatolius in 458, Leo the Thracian made him patriarch. After a synod in Constantinople in 459 he tried to heal the schism which followed the Council of Chalcedon (451) by action as well as by sending an encyclical to his bishops and clergy. He wrote a number of biblical commentaries, e.g., on Genesis and Daniel, but parts of these only survive within *catenae.* A letter from Pope Leo to him is extant (*ep.* 170). In the East he is regarded as a saint.

GENNADIUS II, see GEORGE SCHOLARIUS

GENNADIUS OF MARSEILLES (d.496). Ecclesiastical historian. He is best known for his work *de Viris Illustribus,* which contains 101 short but useful and generally accurate biographies of ecclesiastics between 392 and 495, and thus provides a continuation of Jerome's work of the same name. In its commonly accepted form there is evidence of a second hand. The laudatory account of Jerome at the beginning of the book, for example, seems inconsistent with the hostile references to him in the biography of Rufinus. A presbyter, Gennadius seems to have been semi-Pelagian, as he censures Augustine and Prosper and praises Faustus. On the other hand, in his other work, *Epistola de Fide mea* or *de Ecclesiasticus Dogmatibus Liber,* while the freedom of man's will is strongly asserted, the beginning of goodness is assigned to divine grace. The work was long included among those of Augustine.
DAVID JOHN WILLIAMS

GENUFLEXION. In Roman Catholicism the momentary bending of the right knee so as to touch the ground. The body is held erect and the sign of the cross is made. In the Western Church this has largely superseded the profound bow which was general before the sixteenth century and which is still performed by some religious orders, and is almost universal in the Eastern Orthodox Churches. Genuflexion is used frequently in Roman Catholic Church ritual and is found also in the Anglo-Catholic wing of Anglicanism. A double genuflexion consists in kneeling on both knees, bowing the head and rising, and is made "before the Blessed Sacrament when it is exposed," according to a Roman Catholic source. Some early Christians who bent the knee only in sign of penance were known as "Genuflectentes."
J.D. DOUGLAS

GEORGE, BISHOP (c.640-724). Prominent Monophysite* church writer. After 676 he was chief bishop of Arab tribes in the Tigris and Euphrates area that were still Christian. Among his writings and works were a translation of Aristotle's *Organon* into Syriac, rhymed treatises on the sacraments, and treatments of patristic, dogmatic, exegetical, and astronomical questions in tracts and letters. He also completed the *Hexameron* begun by Jacob of Edessa.

GEORGE, MARGRAVE OF BRANDENBURG-ANSBACH (1484-1543). German prince and patron of the Reformation, also known as "George the Pious." A younger son in the Franconian branch of the Hohenzollern family, he had little prospect of succession and spent a long period of involvement in military ventures and family squabbles. In 1523 he gained possession of the principality of Jägerndorf in Silesia, where he brought in settlers and encouraged economic development. George also introduced the Reformation into both Jägerndorf and Ansbach, which he inherited in 1527 because of the unexpected death of his elder brother. He resisted the efforts of the Catholic king, Ferdinand I of Bohemia, to expel him from Silesia, and he stood firmly for Protestantism at the Diets of Speyer (1529) and Augsburg (1530). RICHARD V. PIERARD

GEORGE HAMARTOLOS (ninth century). Byzantine monk. He wrote a world chronicle *(Chronicon Syntomon)* from the Creation to the death of the Eastern emperor Theophilus in 842. It was not an original work, but an eclectic compilation whose main source was Theophanes Confessor (d.817). In turn, later writers borrowed from Hamartolos's chronicle, which had been continued up to 948 by another scribe, perhaps Simeon Metaphrastes. The work gives useful information on ninth-century monasticism, and though nothing is known about George's personal life, his chronicle reveals him as strongly anti-Iconoclast.

GEORGE OF CAPPADOCIA (d.361). Bishop of Alexandria. An Arian of the Acacian school (see ACACIUS), he was intruded into the see of Alexandria from 357, following Athanasius's flight the previous year. His arrival was accompanied by soldiers, and the cruelty that had earlier marked the eviction of the orthodox from their churches in favor of the Arians was now resumed by George. Nor was he disliked any less by the pagans of Alexandria, against whom his measures were equally violent. His tyranny ultimately brought about his death. In 358 he was rescued only with difficulty when a mob seized him. He was forced to flee and probably did not return for about a year. But when in 361 news of Julian's accession reached the city, a mob again seized him—and killed him some weeks later. Described by his enemies as unlearned, George nevertheless had an extensive library, according to Julian. De-

spite his cruelty, even his enemies conceded that he was a man of resolution and action.
DAVID JOHN WILLIAMS

GEORGE OF ENGLAND. Though nothing is known of his life, his authenticity is generally accepted. He is believed to have been martyred in the third century, probably at Lydda in Palestine. Mythical tales about George abound, and the best known of these is the story of "George and the Dragon." This later medieval tale from the "Golden Legend" of James of Voraigne really derives from Perseus's killing of the sea monster which threatened Andromeda. This event had supposedly occurred at Jaffa, near Lydda, and so George, through his connection with Lydda, inherited the tale. It has not been established why he became so popular in England. He was already known there in Anglo-Saxon times, and by 1222 his feast day was being celebrated. Edward III's famous Order of the Garter adopted George as its patron, and during that reign he became patron of England as well.
L. FEEHAN

GEORGE OF LAODICEA (d. probably after 360). A native of Alexandria, he was ordained presbyter by Alexander, but was later deposed by him for false doctrine and misconduct following an unsuccessful attempt to reconcile the Arians and the Catholics in Antioch. Eustathius would not receive him after this in Antioch, but when Eustathius was himself deposed, George was received there by the Arians. Later he succeeded the Arian Theodotus of Laodicea and as bishop took part in a number of councils opposed to Athanasius, who described George as one of the worst of the Arians. He was deposed by the Council of Sardica in 347 on the grounds that he had previously been deposed by Alexander, but he remained unaffected by the council's decision. A leading Arian, George played an important role in the Council of Seleucia in 359 which deposed the Anomoeans* Acacius and Eudoxius. He also wrote against heresies, particularly that of the Manichaeans.
DAVID JOHN WILLIAMS

GEORGE SCHOLARIUS (c.1400-c.1468). Patriarch of Constantinople. At first a teacher of philosophy, he became a civil court judge in Constantinople. At the Council of Florence (1439) he favored the scheme for reunion with Rome, but later became a bitter opponent, writing numerous works on the subject. He entered the Monastery of the Almighty and took the name of "Gennadius." In 1453 the Turks under Sultan Mohammed II took Constantinople; and Gennadius, now leader of the anti-union party, was appointed (as "Gennadius") patriarch with full confirmation of his rights over the Orthodox community in return for their political obedience. After two (or possibly five) years in office, he resigned and lived in the monastery of St. John Baptist at Seres in Macedonia until his death. Over 100 books are credited to him, including speeches, anti-Latin polemical works, translation of works by Aquinas, philosophical treatises, and many theological pieces, especially his *Confession*, an apologetic dialogue with Mohammed II.
HOWARD SAINSBURY

GEORGE SYNCELLUS (eighth/ninth century). Byzantine church historian. Almost nothing is known of his personal life except that he may have lived for a time in Palestine. He should not be confused with Syncellus of Tarasius, who was patriarch of Constantinople (784-806). He wrote a *Chronicle* which traces human history from Creation to the time of Diocletian (245-313). But the form in which the *Chronicle* is now preserved also contains an additional section probably by his friend Theophanes Confessor. The appendix covers the period 313-813. Although it does contain some original material, its value lies more as an illustrative piece of historiography than factual accuracy.
DONALD M. LAKE

GEORGIAN VERSION. The Georgians (called "Iberians" in antiquity) of the Caucasus were converted in the fourth century. Closely linked with Armenia at first, they at length separated in A.D. 608/9, adhering to the Chalcedonian orthodoxy of the Greek Church. We have evidence that at least parts of the Scriptures were known in Georgian in the fifth century, although our earliest MSS date from the sixth or seventh only. The earliest form of the gospels is known in the Adysh MS of 897, whose text shows clear signs of translation from Armenian. As with the Armenian version, this has earlier links with Syriac traditions and the *Diatessaron*.

Earlier fragmentary MSS in the archaic forms of Georgian known as *han-meti* and *hae-meti* contain many parts of the Old and New Testaments: where comparison is possible we find that these reveal textually a revised form, to some extent corrected to a Greek standard. In the gospels this is related to the so-called Caesarean Text* and the quotations of Origen and Eusebius. In the earliest forms of the OT, a Lucianic text is seen; some traces of the versions of Aquila and Symmachus are also known in marginal notes. In the tenth century and later, Georgian monks on Mt. Athos (Euthymius, George the Hagiorite, Ephrem Mcire) produced revised versions of various parts of the Bible upon which the editions current later depend. In the nineteenth century, however, Russian influence debased the version, which has only recently been freed from this corruption. Revelation was not translated until the tenth century, and Maccabees until the eighteenth.

BIBLIOGRAPHY: M. Tarchnishvili, *Geschichte der kirkchlichen georgischen Literatur* (1955); L. Leloir, "(Versions) Orientales de la Bible," *Dictionnaire de la Bible*, Supplement VI (1960); J.N. Birdsall, "A Georgian Palimpsest in Vienna," *Oriens Christianus* 53 (1969), pp. 108-113.
J.N. BIRDSALL

GERALD DE BARRI (1147-1223). Archdeacon of Brecon. Born Giraldus Cambrensis of the Welsh royal family, he originated in Pembrokeshire, and studied and lectured at Paris before becoming archdeacon in 1175. He went again to Paris to read theology, civil and canon law, was made royal chaplain in 1184, and preached the

Third Crusade in Wales in 1188. He made expeditions to Ireland, accompanying Henry II's son John. He produced his *Topographia Hibernica* and *Expugnatio Hibernica* and then, having obtained no preferment, retired in 1189 to pray and study in Lincoln. He wrote an autobiography, and his complete works were published in eight volumes in 1861-91. He was buried in St. David's Cathedral. C.G. THORNE, JR.

GERARD OF ZUTPHEN (Gerard Zerbolt) (1367-1398). Associated with the early history of the Brethren of the Common Life,* he was born in Zutphen, the Netherlands, and as a youth was a pupil of the Brethren's founder, G. Groote (d.1384). An able scholar, he studied at the University of Paris, but then returned to Deventer, the headquarters of the movement. There, though in his twenties, he was highly valued as a spiritual adviser. He wrote several treatises on religion and ethics during the 1390s, stressing the value of love, the spiritual ascent of the soul to God, and the imitation of Christ. Best known are *De reformatione virium animae* and *De spiritualibus ascensionibus.* Gerard's writings were doubtless known by and an influence on Thomas à Kempis* and his *Imitation of Christ.* Gerard died of plague at an early age. DIRK JELLEMA

GERBILLON, JEAN-FRANÇOIS (1654-1707). Jesuit* missionary to China. Born in Verdun, France, he became a Jesuit in 1670 and taught humanities for seven years before going to China as a missionary. Sent to found a French mission there, he and a colleague, Bouvet, found favor with the emperor, Kang-Hi, with whom they shared Western scientific achievements, and who advanced in turn the mission's aims. Gerbillon was used for many scientific and diplomatic services and made extensive journeyings. He was in charge of the French college in Peking, then became superior-general of the mission. In 1692 he received an edict granting freedom for Christianity, and the emperor presented a site for a chapel and residence in gratitude for the personal kindness shown toward him by Gerbillon and Bouvet. Gerbillon was the author of several mathematical works; he wrote also accounts of his travels, and on philosophy and linguistics. He died in Peking. C.G. THORNE, JR.

GERHARD, JOHANN (1582-1637). Lutheran theologian. Born at Quedlinburg, he studied at the universities of Wittenberg, Jena, and Marburg, entered the service of Duke Casimir of Coburg whose churches he was commissioned to reorder, but was released in 1616 to become a professor at Jena. Here he joined with Johann Major and Johann Himmel; the three distinguished teachers became known as the "Trias Johannea." Though afterwards employed on numerous ecclesiastical, political, and diplomatic matters by a number of German princes, he devoted most of his time to scholarship. His *Confessio Catholica* (1634-37) was a strong defense of the evangelical faith, and his *Loci Theologici* (1610-22) are regarded as the climax of the Lutheran dogmatic theology. His fifty-one devotional *Meditations* written in 1606 were deservedly popular and were translated into English in 1627, a selection of fourteen of the same being reprinted in English as late as 1846. IAN SELLERS

GERHARDT, PAUL(US) (1607-1676). German hymnwriter. Born in Saxony, he studied at Wittenberg and spent some years as a tutor in Berlin. In 1651, already middle-aged, he became a Lutheran pastor at Mittenwalde. In 1657 he was appointed to the St. Nicolaikirche in Berlin. Here he won esteem as a preacher. An uncompromising Lutheran, he refused to accept even tacitly Friedrich Wilhelm I's edict restricting freedom of speech on disputed points between Lutherans and Calvinists. In 1668 he became archdeacon of Lübben where, as a widower with one surviving child out of five, he remained until his death. Among German hymnwriters he ranks second only to Luther. About one-third of his 133 hymns, first published in the collections of J. Crüger, are still sung in Germany. They mark the transition from the confessional to the devotional type of hymnody: cf. Luther's "Ein feste Burg" with Gerhardt's "Befiehl du deine Wege," which John Wesley translated as "Commit thou all thy griefs." Other translations include "All my heart this night rejoices" (Winkworth); "Jesu, Thy boundless love to me" (Wesley); "O Sacred Head! now wounded" (J.W. Alexander), a paraphrase of "Salve caput cruentatum"; and "The duteous day now closeth" (Bridges). JOHN S. ANDREWS

GERHART, EMANUEL VOGEL (1817-1904). American Reformed theologian. He studied at Marshall College (1836-38) and the Theological Seminary of the Reformed Church in Mercersburg. Ordained in 1842, he served parishes in Pennsylvania and Ohio. He was elected president of Heidelberg College, and professor of systematic theology at the Reformed Seminary in Tiffin, Ohio (1851-55). He became the first president of Franklin and Marshall College (1855-68), resigning in 1868 to become the successor of Henry Harbaugh as professor of systematic theology at Mercersburg. Editor and contributor to the *Reformed Church Review* (from 1854), his most important works were the Triglot edition of the Heidelberg Catechism (1863) and a general systematic theology, *The Institutes of the Christian Religion* (1891, 1894).
DAVID C. STEINMETZ

GERHOH OF REICHERSBERG (1093-1169). Prominent Augustinian and a leading advocate of Gregorian reforms in Germany. Born in Bavaria, he was *scholasticus* of the cathedral school in Augsburg (1119), but came into conflict with his simoniacal bishop, Hermann, and was forced in 1121 to surrender his post. Later reconciled to his bishop, he advised him at the First Lateran Council (1123), summoned by Callistus II to confirm the Concordat of Worms. Gerhoh attempted unsuccessfully to persuade the council to adopt his program for the reform of the secular clergy through the introduction of communal life. He returned in 1124 to Germany and entered the cloister of the Augustinian Canons* Regular in

Rottenbuch. He reformed the rule of the Canons and explained his ideas in *Liber de aedificio Dei* (1130). As provost of the Canons Regular (Reichersberg, 1132), he traveled widely and established friendly relations with Bernard of Clairvaux.* Always a champion of theological orthodoxy—though libeled by many opponents of his reforms as a heretic—he attacked the Christology of Peter Abelard and Gilbert de la Porrée. His views on the relation of imperial and papal power were summarized in his treatise, *De Investigatione Antichristi* (1161), an essay which did not endear him to Frederick I, whom he further alienated in 1166 because of his unwillingness to support an imperialist antipope.

DAVID C. STEINMETZ

GERMAN BAPTISTS, see CHURCH OF THE BRETHREN

GERMAN-CHRISTIANS *(Deutsche Christen).* The so-called German Christian Church which at first reflected the ecclesiastical policies of the Nazi dictatorship. After World War I, nationalist and racist traditions of the nineteenth-century German Protestantism expressed themselves in a number of movements. With the rise of Adolf Hitler they achieved great influence in the church, often by force. At one extreme there were those who made a pagan religion of German blood and destiny, hostile to traditional Christianity (German Faith Movement); while on the other side the Faith Movement of the German-Christians, supported by theologians of the stature of Emanuel Hirsch and (for a time) Friedrich Gogarten,* believed that God was calling the church through the contemporary German situation to be again the church of the German people, with a living faith freed from a dead and alien past and with the organizational unity of a *Reichskirche* under one *Reichsbischof.* In this way it might be seen as a nationalist exploitation of liberal theology; its rejection of the OT and the Jewish element in Christianity was perhaps not unrelated. Nazi support for the German-Christians waned after 1934, though it was sufficient to keep many in their official positions. They had failed to realize that Hitler did not wish the Nazi state to be complemented by a nazified but still influential Christian church.

HADDON WILLMER

GERMANUS (Germain) (c.496-576). Bishop of Paris. Born at Autun, he was ordained priest in 530 and later became abbot of the monastery of St. Symphorian at Autun. In 555 he succeeded Eusebius as bishop of Paris, and from this position seems to have exercised considerable influence over Childebert, whose edict against pagan revelries on holy days is probably due to him. Germanus's influence may also be seen in Childebert's building of the Church of St. Vincent, in which Germanus was later buried. It then became the Church of St.-Germain-des-Prés. Germanus was present at the third council of Paris (555), the second council of Tours (566), and the fourth council of Paris (573). A treatise on the old Gallic Liturgy is attributed to him, as is a privilege exempting the monastery of St. Symphorian from episcopal jurisdiction, but in both cases his authorship has been disputed.

DAVID JOHN WILLIAMS

GERMANUS (c.634-c.733). Patriarch of Constantinople. Son of noble Byzantine parents, he became associated about 668 with the church of St. Sophia; later he became primate of the church and patriarch from 715 to 730. Some mystery surrounds his position regarding the Monothelite controversy and the Sixth Ecumenical Council (680-81). Under pressure he may have rejected the declarations against the heresy, but if he did, there is good indication that he returned to the Chalcedonian christological formula later, since one of his first acts as patriarch was to condemn the Monothelites.* When the Iconoclastic Controversy* began to surface about 725 (the emperor Leo III issued his edict against the veneration of icons) Germanus opposed the emperor's decree and was finally forced to leave his office. He then went to Platonium, where he spent the remaining three years probably writing.

Due to his conflicts with the emperors, most of his writings have been destroyed; however, his *De Haeresibus et Synodis* is extant. He may also be the author of the *Historia Mystica Ecclesiae Catholicae,* a liturgical work which includes several poems. His keen interest in the Virgin Mary can be seen in his seven homilies on her virtues and role in salvation. His part in the Iconoclastic Controversy is indicated by four surviving letters. Germanus is unusual as he reflects the influence of Western Christianity upon the Eastern Church at a time when the two sections of the church were being driven apart by ecclesiastical, political, and cultural issues.

For his extant works see J.P. Migne, *Patrologia Graeca,* vol. 98, pp. 9-454.

DONALD M. LAKE

GERSON, JEAN CHARLIER DE (1363-1429). French theologian and church leader. He entered the University of Paris in 1377 and succeeded his friend and teacher Pierre d'Ailly* as chancellor in 1395. Schooled in the Nominalism of William of Ockham, he yet resisted its potential speculative and skeptical excesses. Strongly interested in practical Christian living, he sought to curb academic intellectualism by lecturing on mysticism and the spiritual life in the university; at the same time, he wrote against popular superstition and irrational enthusiasm. Such a man was naturally pained by the Great Schism.* At first he sought to moderate the hostility of factions in the church, but eventually he came to throw the authority of the leading university in Christendom behind a more radical cure for the schism. Disillusioned by the failure of the papal leadership, he held that in an emergency canon law could be set aside; in particular he followed Henry Langenstein in insisting that the pope was not absolute, but must be understood as the head, and so as part, of the body, the church, i.e., the totality of the faithful, and existing for the sake of the church. From this followed the revolutionary step, that the body has the right to call a failing head to account. So he took part in the Council of Pisa (1409) and the Council of Constance (1415-17) which burnt Hus and deposed three popes to end the schism. In all this, Gerson acted as the man of traditional order; the law could only be set aside in order to maintain the spirit of the law. His views were set out in *De potestate ecclesiastica,* (1416-17).

His concern for social order and morality was to be seen in his hostility to the priest John Petit, who wrote a tract on tyrannicide in order to justify the action of the Burgundian faction in France. Gerson secured the condemnation of Petit at Constance, but since the Burgundian party gained dominance at that time, his return to Paris after the council was impossible, and he ended his days in exile. As one of the chief theorists of the Conciliar Movement, Gerson has considerable importance in the history of the doctrine of the church and of Christian political thinking.

HADDON WILLMER

GERTRUDE THE GREAT (1256-c.1302). German mystic. Born at Eisleben, she was brought up from the age of five and educated in the Black Benedictine nunnery of Heltfa, Thuringia. She had her first mystical experience in 1281 and from then on led a life of contemplation. Her *Legatus Divinae Pietatis* is a classic of Christian mysticism. She was one of the first exponents of the devotion to the Sacred Heart. She also wrote *Exercitia Spiritualia,* a collection of prayers. She was never formally canonized, yet her cult was first authorized in 1606 and extended to the entire Roman Catholic Church by Clement XIII (1738). She is patroness of the West Indies.

GERVASIUS and PROTASIUS. Two brothers and protomartyrs of Milan of whom nothing certain is known. When Ambrose was about to consecrate his new church, he discovered (386) the burial place of the two in the Church of Sts. Felix and Nabor. This greatly encouraged the orthodox and discomfited the Arians, for whom Empress Justina was demanding a church in Milan at this time. The remains of the two brothers were intact, but with their heads severed. The protomartyrs were reburied two days later under the altar of the new church. A number of miracles were attributed to them. How Ambrose discovered their burial place is not clear. He says he acted on "a presentiment" *(cuiusdam ardor praesagii),* but as custodian of the church records his presentiment may have had some basis of knowledge.

DAVID JOHN WILLIAMS

GESENIUS, HEINRICH FRIEDRICH WILHELM (1786-1842). German Orientalist and biblical scholar. Born at Nordhausen, Hanover, he received theological training at Helmstedt and Göttingen, and was professor of theology at Halle from 1811. He concentrated on problems of Semitic philology, becoming the most outstanding Hebraist of his generation. His chief work was *Hebräisches und Chäldaisches Handwörterbuch* (1810-12), which passed through several editions and was the basis of the Hebrew lexicon of Brown, Driver, and Briggs (1906). In 1813 he published the first edition of his Hebrew grammar, edited and enlarged by E. Kautzsch (1899 onward; ET by A.E. Cowley, 1910). He wrote a commentary on Isaiah (1820-21), and his monumental *Thesaurus philologico-criticus linguae Hebraeae et Chaldaeae Veteris Testamenti* (1829-58) was completed after his death by his pupil E. Rödiger.

J.G.G. NORMAN

GESS, WOLFGANG FRIEDRICH (1819-1891). German theologian. He studied at Tübingen where he came under the influence of F.C. Baur* and J.T. Beck.* After serving as assistant minister to his father, he taught at the missionary college, Basle (1850). He became a professor at Göttingen (1864), where he was a colleague of Ritschl,* then moved to Breslau (1871). He became general superintendent of the province of Posen (1880), but retired early for health reasons. Gess came from the Württemberg pietism which stressed biblical theology and Christian experience, but which adopted a looser attitude toward biblical inspiration. He was a leading exponent of kenotic Christology which sought to explain the union of Christ's full humanity with true divinity by positing a self-emptying of the latter (Gr. *kenosis*) in the Incarnation. The theory was based on an interpretation of Philippians 2:5ff., as a key to explaining the orthodox view of Christ's divinity in the light of His evident humanity in the gospels and modern criticism.

Gess's chief works were *Lehre von der Person Christi* (1856), *Christi Selbstzeugnis* (1870), *Das apostolische Zeugnis von Christi Person und Werk* (1878-79), *Das Dogma von Christi Person und Werke* (1887), and *Die Inspiration der Helden der Bibel und der Schriften der Bibel* (1891).

COLIN BROWN

GESUATI. A congregation of laymen founded by Giovanni Colombini* about 1360, devoted to prayer, mortification, and charitable works. Officially *Clerici apostolici S. Hieronymi,* their popular name arose from their frequent ejaculations of "Praised be Jesus" or "Hail Jesus" in preaching. Approved by Urban V (1367), they established monasteries and adopted a white tunic and grayish-brown cloak as habit. They spread throughout Italy and established houses in Toulouse (1425), but were dissolved by Clement IX in 1668 as having lost the spirit of their order. Their female counterpart, the "Jesuatesses" (Sisters of the Visitation of Mary), was founded about 1367 by Colombini's cousin, Catherine Colombini, and existed until 1872.

J.G.G. NORMAN

GEULINCX, ARNOLD (1624-1669). Philosopher. Born at Antwerp, he studied philosophy at Louvain, and taught there (1646-52) until dismissed for his sympathy for Jansenist teachings. He moved to Leyden in the Netherlands, and after a decade of poverty, during which he turned to Calvinism, became a lecturer at the university, and (in his early forties) then professor. He died not long after from the plague. His main works were *Metaphysica* (1651) and *Ethica* (1655). As a philosopher, starting with the sovereignty of God, he stressed submission to God's will as the main virtue, and in his metaphysics tried to reconcile sovereignty with man's apparent free will. Descartes* had not fully clarified the relation of soul and body (how can an immaterial soul influence material things?), and Geulincx elaborated on this question in an almost paradoxical manner. Since God is the sole cause of all events, second causes do not exist. God is thus the immediate cause of every event, and what we take to be causal relationships are illusory. Man's free will is also thus illusory, for if God is the immediate cause of every event, man cannot will events: he can only resolve to conform to God's causal will.

See *Opera Philosophica* (ed. J.P.N. Land, 3 vols., 1891-93). DIRK JELLEMA

GEYMONAT, PAOLO (1827-1907). Waldensian* evangelist and scholar. Born at Villar Pellice (Waldensian Valleys), he graduated from the theological school in Geneva and was ordained pastor in 1850. His earnest desire to spread the Gospel in Italy led him to work as an evangelist in Rome, Florence, Turin, and Genoa. In 1855 he was invited with G.P. Revel to start a Waldensian school of theology at Torre Pellice, a task which he carried out with success amid many difficulties. After the unification of most of Italy in 1860, he supported strongly and successfully the removal of the school to Florence, where he lectured until 1902. He then retired and resumed his work as evangelist, founding an autonomous church in Florence. Author of several treatises and articles, his theology, missionary zeal, and piety are typical of Le Réveil,* by which he was greatly influenced. He had a more conciliatory attitude than most Waldensians; he tried throughout his life to solve the conflict between the Waldensian Church with its rigid organization and the other Italian denominations. DAISY RONCO

GHÉON, HENRI (1875-1944). French Roman Catholic writer. Born at Bray-sur-Seine, son of a chemist, his real name was Henri Léon Vangeon. Educated at Sens and Paris, he became a doctor. He lost his faith in adolescence, regaining it on the death of his niece and friends (1914). He published a collection of verse, *Chanson d'aube* (1897), and was a founder of the *Nouvelle Revue Française.* His first play, *Le Fils de M. Sage—Le Pain,* a popular tragedy in verse, was performed in 1912. He devoted himself to developing a Christian theater, producing his own plays and founding a company of young Catholics, "Les Compagnons de Notre-Dame" (1924), and their successors, "Les Compagnons de Jeux" (1931). Many of his works had kinship with fifteenth-century miracle plays, having as subjects the lives of saints and other sacred themes, deliberately aimed at reproducing the atmosphere of medieval hagiography. He wrote biographies, novels, and a film script. J.G.G. NORMAN

GHETTOS. Streets or sections of a city where Jewish people were compelled to reside. In more modern parlance the compulsory character of the situation may be implicit. The church continually agitated for segregation of Jews, but this was not officially done until the Third and Fourth Lateran Councils (1179 and 1215). In 1555 Pope Paul IV insisted upon the enforcement of the principles of segregation. Ghettos were established in Italy, North Africa, the Germanic countries, E Europe, and most of W Europe. The Muslim countries also enforced a rigid ghetto system. Within the ghettos the Jews experienced considerable autonomy and religious freedom, and developed some local pride. Some ghettos were enclosed with walls, and the gates were kept locked at night and during church festivals. These conditions were abolished by the end of the nineteenth century in Europe and by 1917 in Russia with the fall of the czarist regime. They continued in N Europe until the founding of Israel in 1948.

The ghettos denoted a way of life which led the Jews to a paradoxical desire for freedom and the fear of being exposed to discrimination. A type of ethnic and racial mentality has existed and still does exist to some extent in the United States. Legislation has been passed, however, which discourages these patterns of living.

See also ANTI-SEMITISM.

BIBLIOGRAPHY: I. Zangwill, *Children of the Ghetto* (1894); L. Wirth, *The Ghetto* (1928); P. Freedman, "The Jewish Ghettos of the Nazi Era," in *Jewish Social Studies,* vol. XVI, no. 1 (1954). JOHN P. DEVER

GHIBELLINES, see GUELFS AND GHIBELLINES

GIBBON, EDWARD (1737-1794). English historian. Born in Surrey, the son of a parliamentarian, he was of independent means throughout his life. His *Decline and Fall of the Roman Empire* (7 vols., 1766-88; best edition by J.B. Bury) helped to make church history a critical discipline. In some ways his work is still unsurpassed. Friend of Voltaire, Diderot, and d'Alembert, Gibbon surveyed Roman history from the second to the fifteenth centuries from the point of view of the ironic humanism of the eighteenth century. He saw it as the story of the fall, through the progressive "triumph of religion and barbarism," from the intellectual freedom evidenced in classical literature. This theme was more fitting to the Western than to the Eastern Roman Empire, and Gibbon dealt less adequately with the latter. He did not believe in the supernatural and sought to explain the growth of Christianity naturalistically, on the principle that the religious is at least a phenomenon of human experience. He was always sharply aware how religious claims could cloak ambition, incredulity, and inhumanity, though he could respect genuine piety. HADDON WILLMER

GIBBONS, JAMES (1834-1921). Archbishop of Baltimore. Son of Irish immigrants, he rose from simple surroundings in Baltimore to become "the American Cardinal," the dominant Roman Catholic prelate in United States history. Appointed archbishop of Baltimore in 1877, and named cardinal in 1886, he led the nation's first archdiocese, and thus much of the American church, until his death. Although untalented as a writer and a thinker, he extended Catholic influence in an age of intense anti-Catholicism. Faced with a church that was glutted with non-English-speaking immigrants and a nation that feared aliens, he tried to prove that loyalty to Rome actually improved

American Catholic citizenship. His leadership, which fostered such institutions as the Catholic University of America and the National Catholic Welfare Conference, also created an "Americanism" which supported the established order and ignored some church traditions like the just war.
DARREL BIGHAM

GIBBONS, ORLANDO (1583-1625). English composer. Of a family distinguished in English music for several generations, he served as organist both of the Chapel Royal and of Westminster Abbey. He was the first great composer to write church music only for the Protestant rite in England. He wrote services in the cathedral style, preces, festival psalms, and both full and verse anthems. Like Byrd,* his secular works are characterized by a notable seriousness. "Hosanna to the Son of David" and "Almighty and everlasting God" are among his best-known and most performed anthems today.

GIBSON, EDMUND (1669-1748). Bishop of London. Educated at Bampton Grammar School and Oxford, he was successively domestic chaplain to Archbishop Tenison, archdeacon of Surrey, and bishop of Lincoln before moving to London. Offered the see of Canterbury, he declined. Gibson was distinguished in several fields; his huge work on English ecclesiastical law (1713) became a standard work of reference and earned for him the nickname "Dr. Codex." He was an enemy of ecclesiastical abuses, and though a Whig was a moderate High Churchman who strove to reconcile the Tory Power Clergy and the universities to the Hanoverian Succession. He was chief adviser to Walpole in church matters till they quarreled in 1736. He wrote fiercely against Deists and freethinkers, Catholics, intemperance, and Sabbath-breaking. Toward the Methodist revival he was at first sympathetic, but later he became a hostile critic. A patron of the arts, he tried to improve the academic standards of the universities and began the cataloging of the Lambeth Library.
IAN SELLERS

GICHTEL, JOHANN GEORGE (1638-1710). German mystic and theosophist. Born in Regensberg, he studied at Strasbourg, then in his twenties was attracted by the writings of Jakob Boehme,* broke with traditional Lutheranism, attacked the established churches, got into trouble with the authorities, and settled in Amsterdam (1668). Developing his own blend of mysticism, pietism, and theosophy, he stressed the "heavenly marriage" between the spiritual man and divine Wisdom, and founded the "Angelic Brethren" who renounced earthly marriage. He gained some following in N Germany. In 1682 he edited Boehme's complete writings (11 vols.). He had some influence on Gottfried Arnold,* also a critic of orthodoxy.
DIRK JELLEMA

GIDEONS INTERNATIONAL. An association of Christian business and professional men that grew out of a meeting between John Nicholson and Samuel Hill in Central Hotel, Boscobel, Wisconsin (1898). Participating in evening devotions, they discovered they shared a common Christian faith. The next year, with W.J. Knights, they organized an association of Christian traveling men, named the "Gideons," after the OT figure who led a small band of Israelites to victory over the Midianites (Judges 6—7). From this beginning the membership has grown to more than 42,000 in ninety countries. Their primary purpose is to win individuals to faith in Christ, particularly through the free distribution of Scripture. Distribution began in 1908; copies of the Bible and NT were placed without cost in public places such as hotel rooms, and in the hands of school children, prisoners, nurses, soldiers, and others. The work is supported primarily by voluntary offerings received in local churches. By 1971 more than eleven million Bibles and ninety-one million New Testaments in some thirty-two languages had been distributed. A monthly organ, *The Gideon*, is published.
ALBERT H. FREUNDT, JR.

GIESELER, JOHANN KARL LUDWIG (1792-1854). German Protestant church historian. Born in Petershagen in Westphal, he was educated at Halle, and in 1819 was called to a chair of theology at the University of Bonn, where he became a colleague of K.I. Nitzsch.* His earliest book, *Historisch-kritischer Versuch über die Entstehung der Evangelien* (1818), attempted to examine the oral tradition on which the synoptic gospels are based. In 1831 he was appointed professor of church history and doctrine at the University of Göttingen, succeeding G.J. Planck. Gieseler was co-founder of the journal of the *Vermittlungstheologie*, the *Theologische Studien und Kritiken* (1828). His most important publication was his *Lehrbuch der Kirchengeschichte* (5 vols., 1824-57), valued by Hase and Bauer for its rich documentation.
DAVID C. STEINMETZ

GILBERT OF SEMPRINGHAM (c.1083-1189). Founder of the Gilbertine Order. Son of a Norman knight who had come over with the Conqueror, Gilbert as parish priest of Sempringham encouraged seven women to adopt the Cistercian Rule and with the support of the king and many nobles formed a number of houses which, however, the authorities at Cîteaux refused in 1148 to incorporate. Accordingly he arranged for them to be supervised by Augustinian Canons, and thus were born the curious "mixed" Gilbertine communities, the only purely English order. There were nine houses at the time of Gilbert's death, and twenty-five at the Dissolution of the Monasteries under Henry VIII.
IAN SELLERS

GILES OF ROME (c.1243/47-1316). Theologian and philosopher. After studies at the Paris house of the Augustinian Hermits, he read theology under Aquinas in the University of Paris. From 1285 to 1291 he was the first Augustinian master in theology there. He was a prodigious writer. In 1287 the order prescribed that his teaching be followed in its schools. He became its general in 1292. As tutor to Philip IV of France he wrote *De Regimine principum;* as archbishop of Bourges from 1295 his *De Renuntiatione papae* upheld the validity of Celestine V's abdication and Boniface VIII's election. He later sided with Boniface in a quarrel with Philip, writing *De ecclesiastica potestate* (1301/2) which inspired Boniface's *Unam Sanctam* (1302). In papal theocracy he saw fulfillment of Augustine's City of God ideal, the theology of which he defended. A witness for Aquinas's thought, though differing appreciably on some issues, Giles also commented on Aristotle's and Peter Lombard's works, and produced exegetical writings on Paul's letters and John's gospel. C.G. THORNE, JR.

GILL, JOHN (1697-1771). Baptist minister and biblical scholar. Born in Northamptonshire and educated at Kettering grammar school, he was ordained in 1718, and in 1719 entered upon a pastorate at Horsleydown, Southwark, that was to last over fifty years. A Wednesday evening lectureship was founded for him in Great Eastcheap by his admirers in 1729, and this he held until 1756. A profound scholar and voluminous writer, his works include *The Doctrine of the Trinity Stated and Vindicated* (1731); *An Exposition of the New Testament* (3 vols., 1746-48) which with his *Exposition of the Old Testament* (6 vols. 1748-63) forms his major work; *A Dissertation on the Antiquity of the Hebrew Language* (1767); *A Body of Doctrinal Divinity* (1767); and *A Body of Practical Divinity* (1770). A hyper-Calvinist, he was so zealous to maintain the sovereignty of God that he denied that preachers had the right "to offer Christ" to unregenerate sinners.

ROBERT G. CLOUSE

GILLESPIE, GEORGE (1613-1649). Scottish minister. Son of the manse, he was ordained in 1638 by Kirkcaldy Presbytery to the parish of Wemyss, despite the disapproval of the archbishop of St. Andrews (who was that year ejected with other bishops when thoroughgoing Presbyterianism was restored to Scotland). Gillespie became one of the ministers of Edinburgh in 1641, a chief apologist for the National Covenant,* a participant in the Westminster Assembly, and the champion against English opposition of the place of the elder in the kirk and of the Presbyterian system of church courts. Though a victim of chronic ill health culminating in his early death, he was one of the most learned and prolific of the Covenanter writers. His chief work, *Aaron's Rod Blossoming* (1646), a comprehensive study of the Erastian controversy in the light of Scripture, so stung the Episcopalians that when they regained ascendancy in 1661 they had his tombstone "solemnly broken" by the public hangman at Kirkcaldy.

J.D. DOUGLAS

GILLESPIE, THOMAS (1708-1774). Scottish minister. Trained under Philip Doddridge at Northampton, friend of Jonathan Edwards, and minister of the country parish of Carnock in Fife, he was deposed by the Church of Scotland general assembly in 1752 for refusing to take part in a "forced settlement" at Inverkeithing. Supported by the large congregation that built his Dunfermline church, he stood alone for nine years, being then joined by two other ministers, one of them Thomas Boston's son. Passing into history as the founder of the Relief denomination, Gillespie sponsored the first Presbyterian body in Scotland to espouse the cause of foreign missions, and to open its pulpits to all ministers of Christ and its communion table to all believers. "The Relief" joined with "the Secession" in 1847 to form the United Presbyterian Church of Scotland.

D.P. THOMSON

GILMOUR, JAMES (1843-1891). Scottish missionary to Mongolia. Born in Glasgow, he studied at Glasgow University and in Congregationalist theological colleges, and left in 1870 for Mongolia to reopen under the London Missionary Society work that had long been in abeyance. After language study in Peking he went on to Krechta, where in order to learn the Mongol language and customs he went to live in a tent on the plains, preferring to reach the nomads rather than the settled Chinese-speaking agriculturists whom his seniors advised him to evangelize. With indomitable perseverance, and despite an almost total lack of response from all except the Chinese, Gilmour persisted in his task for fifteen years in the face of adverse criticism. His custom was to winter in Peking, where also he sought to reach the Mongols. His last years were spent among the agriculturists. Promised colleagues failed to materialize, and his was a lonely, hard, self-sacrificing task, with seemingly little effect made on Mongolian Buddhism. LESLIE T. LYALL

GILPIN, BERNARD (1517-1583). Anglican divine. Born in Westmorland, he was educated at Queen's College, Oxford, and became thereafter a student at Christ Church. At first he had no inclination toward the Reformed faith, and disputed against John Hooper and Peter Martyr, but during the reign of Edward VI, while continuing to attack changes in and neglect of the externals of worship, he proceeded slowly toward the Reformed position. From 1552 to 1556 he resided in France, but returned home when he was presented by his great-uncle, the easy-going Bishop Tunstall of Durham, to the living first of Easington and then of Houghton-le-Spring. Now began his lifelong series of missionary tours and his social and educational work among the neglected masses which earned him the title "Apostle of the North." Arrested in 1558, he would undoubtedly have perished in the flames but for Queen Mary's timely death. Gilpin accepted Elizabeth's religious settlement, albeit with some hesitation, his most significant example being followed by most of the clergy of the north. In the 1570s the growing Puritan party made approaches to him, but his innate conservatism precluded active support. To the end he continued his tours, denouncing ecclesiastical scandals and arousing hostility.

IAN SELLERS

GILSON, ÉTIENNE HENRY (1884-). Roman Catholic philosopher. Son of a Paris merchant, he studied at the Sorbonne and the Collège de France, and taught philosophy at Lille (1913), Strasbourg (1919), the Sorbonne (1921), and the Collège de France (1932). He lectured at Harvard (1926-28) and in 1929 became director of the

Pontifical Institute of Medieval Studies, University of Toronto. He retired in 1951. An outstanding medievalist, he produced works on Descartes, Thomas Aquinas, Bonaventure, Francis, Augustine, Bernard, and Duns Scotus, but he produced also broad interpretative essays concerned with the whole of the medieval mind. These included *L'Esprit de la philosophie médiévale* (1931-32), *Reason and Revelation in the Middle Ages* (1939), *God and Philosophy* (1941), and *History of Christian Philosophy in the Middle Ages* (1955). In addition to later philosophical studies, his retirement years have seen also publications that reflect his interest in and knowledge of the fine arts. CLYDE CURRY SMITH

GIOBERTI, VINCENZO (1801-1852). Italian philosopher and statesman. Born in Turin, he was ordained priest at the age of twenty-four and was soon known for his great scholarship, leading to his appointment as professor in the theological school of Turin University. Suspected and hated for his liberal ideas, he was imprisoned and exiled to Paris in 1833. In exile his published work praised Italy and its civilization, and at the same time exhorted Italians to strive for unity and join in a confederation under papal leadership ("Neo-Guelphism"). The impact and success of his writings were great, and Pius IX's liberal policies seemed to confirm Gioberti's expectations. He returned to Turin, was elected member of parliament, and became prime minister of Piedmont during the first war of independence. Back in Paris after the Piedmontese defeat, disillusioned by the pope's *volte-face* and the waning of his utopian federation, he wrote a work that condemned the pope's temporal power and advocated the unity of Italy under the new Piedmontese king, Victor Emmanuel II, and his minister Cavour. His latter writings suggest the necessity of suppressing the papacy in order to renew the church.
DAISY RONCO

GIOTTO (c.1266-1337). Italian painter. A Florentine, he replaced the Byzantine cool distance of a majestic Christ with a Lord solemnly moving among men who experienced His presence. The frescoes in Padua confront an observer at real-life eye level. The saints wear golden halos but are crying, gesturing, intensely responsive; and their crowded compositional groupings bespeak daily life. The soft watercolor tones and limited range of color areas natural to fresco murals reinforce the quiet, tender simplicity of Giotto's interacting figures. Everyone about Christ assumes a kind of "sacred" (separated from the ordinary) solidity that is very "spiritually" real. Giotto antedates Boccaccio's secularizing spirit and critical developments like *devotio moderna;* he painted while the *Divine Comedy* was being written. But unlike Dante with his hell, purgatory, and heaven locations, Giotto presented Christ on firm earth. When painter Giotto was made head of the Florence Cathedral school in 1334, the appointment signaled the rising, formative influence of painting in a period (Gothic) previously dominated by stained-glass architecture and stone sculpture.
CALVIN SEERVELD

GIRALDUS CAMBRENSIS, see GERALD DE BARRI

GIRGENSOHN, KARL (1875-1925). Protestant religious psychologist. Born at Carmel auf Ösel, he was successively professor at Dorpat (1907), Greifenwald (1919), and Leipzig (1922). Concerned with the challenge of modern psychoanalysis to the Christian faith, he strove to establish the study of religious psychology on a firmly empirical basis, and in his most important work, *The Spiritual Structure of Religious Experience* (1921), he argued that religious experience is a synthesis of intuitive awareness of God and consciousness of self. The importance of his work was at once recognized in Germany, but far less readily in England and America. One of the leading architects of modern post-Freudian religious psychology, Girgensohn also wrote *Foundations of Dogma* (1924) and *The Inspiration of Holy Scripture* (1925) and founded the periodical *Christentum und Wissenschaft* (1925f.). IAN SELLERS

GLABRIO, MANIUS ACILIUS. A consul in 91, he was ordered by Emperor Domitian to fight with lions in the amphitheater at Albano, but having emerged successfully from this he was finally put to death. His execution may have been part of Domitian's general proceedings against people of senatorial rank whom he suspected of conspiring against him; but his implication in the charge of "Judaism and atheism" and the presence of his family crypt in the Cemetery of Priscilla, an early Christian burial place on the Via Salaria, suggest that he may have been put to death as a Christian.

GLADDEN, WASHINGTON (1836-1918). Liberal theologian and exponent of the Social Gospel. He received his college degree from Williams College and served churches in New York and Massachusetts from 1860 to 1882, except for some years as religious editor of the *Independent.* His main pastorate was at First Congregational, Columbus, Ohio, from 1882 to 1914. Here he applied Christ's teaching to social problems, upheld the rights of unions, favored profit sharing and industrial arbitration, and tried to have his denomination turn down a large gift from Standard Oil as "tainted money." He wrote the hymn "O Master, let me walk with Thee."

GLADSTONE, WILLIAM EWART (1809-1898). British prime minister. Son of a Liverpool merchant, he was educated at Eton and Oxford, distinguished himself in classics and mathematics, and would have entered the ministry had not his father, himself a member of Parliament, planned a political career for his son. He entered Parliament in 1832 and continued a member (with one brief interruption) until 1895. He was throughout a man of principle and humanity: his first speech was mainly an attack on slavery, and on many occasions he championed oppressed minorities, always aiming at a decision on moral grounds. He supported Catholic Emancipation, not because of religious indifference, but from principle. The Oxford Movement* greatly influenced him, and he opposed Archbishop Tait,* who wanted to abolish ritual. Gladstone nevertheless opposed the claims of Roman Catholicism. He knew his Bible and called one of his books *The Impregnable Rock of Holy Scripture.* P.W. PETTY

GLAGOLITIC. A system of forty alphabetic signs invented presumably by Cyril about A.D. 863 for writing Slavonic. Though reminiscent of some other alphabets, it is largely original. It was supplanted probably in the tenth century by the Cyrillic alphabet, the basis of modern alphabets for Russian, Bulgarian, Ukrainian, etc. Several important manuscripts survive written in these characters, two gospel manuscripts *(Zographensis, Marianus),* a lectionary *(Assemanianus),* psalter *(Sinaiticum),* prayer-book *(Euchologium Sinaiticum),* and fragments of homilies *(Clozianus).*

GLAS, JOHN (1695-1773). Founder of the Glasites, or Sandemanians.* Born in Scotland, he graduated from St. Andrews, pursued theological studies at Edinburgh, and became minister of the Angus parish of Tealing in 1719. There his able preaching rapidly increased his congregation. Later, while lecturing on the Shorter Catechism, it struck him that since Christ is king of the church, power cannot be exercised over it by the state or magistrates. These views led to his final deposition in 1730; on returning to his church he found it locked against him. He continued to preach in the nearby fields; most of his congregation remained loyal to him. Soon he moved to Dundee, and later to Perth where he met Robert Sandeman, who became his son-in-law, and where he was to minister in a church built by his followers.

Glas's writings are scholarly, kindly in spirit, and devout. He took the Bible more literally than most of his day: Christians were forbidden to eat blood or to store wealth; they should practice foot-washing and the holy kiss. Creeds and catechisms he regarded as useless: it was easier to learn from the Bible itself. Attempts to enforce uniformity in the church, or to permit domination by single individuals, he looked upon as utterly wrong. Glas's meek and gentle spirit ill befitted him to become leader of the sect he had unwittingly founded, and the role was taken over by the contentious and vigorous Sandeman. In old age Glas himself became bitter and controversial, as extant correspondence plainly shows.

See H. Escott, *A History of Scottish Congregationalism* (1960). R.E.D. CLARK

GLEGG, ALEXANDER LINDSAY (1882-1975). British lay evangelist. Born in London of Scottish parents, he trained as an electrical engineer at London University and subsequently became director of several companies. But while he worked hard in business, his great love was evangelism. Converted while at the Keswick Convention* as a young man, he soon became involved in mission work in Wandsworth, and for nearly fifty years was responsible for the ministry at Down Lodge Hall. He was a regular speaker at campaigns and conventions, and many thousands throughout the British Isles became Christians through his lucid and winsome presentation of the Gospel. He was known particularly for his Albert Hall meetings in the 1940s, his leading part in beginning the Christian holiday camp which annually draws thousands to Filey, and for his active support of numerous missionary societies. He encouraged many young evangelists, both spiritually and financially—and counseled them to play golf for their soul's good. Billy Graham is among those who acknowledge a great debt to Glegg who in his nineties was still preaching. His published works, which went all over the world, include *Life with a Capital "L"* and *Four Score and More.*

J.D. DOUGLAS

GLOSSOLALIA. The term as such is not found in the Bible, though its hybrid form is built from *glōssa* and *lalein* which occur, e.g., in Acts 2:4. They are usually translated "speaking in tongues" (1 Cor. 12—14; Acts 2:3ff.; 10:46; 19:6; Mark 16:17). Apparently glossolalia is the spontaneous utterance of uncomprehended and seemingly random vocal sounds. It appears to describe a form of spiritually affected speaking which is of particular value to the individual. It has been (and is) a feature of religious, especially revivalist, activities at many periods of church history. It was not, however, until the late seventeenth century that the phenomenon occurred among numerous people of one locality. In S France the Cevenols, who lived in constant fear of death, had ecstatic experiences which included speaking in tongues. In the nineteenth century a second major outburst of tongue-speaking occurred in England, among the followers of Edward Irving* who himself strangely never "received" the gift of glossolalia. Aside from sporadic instances during these same centuries among the several revival movements in England and America, glossolalia was relatively infrequent until its phenomenal rise in connection with Pentecostalism.

Several Pentecostal revivals sprang up in the United States just after the turn of the twentieth century. The earliest recorded instance of glossolalia in this century was in Topeka, Kansas, in 1901, when the "baptism of the spirit" fell upon Agnes N. Ozman, a student at the Bethel Bible College. From Kansas the movement spread to Missouri and Texas. By 1906 tongue-speaking was being practiced in Los Angeles, and "the movement began to take on international proportions" with twenty-six contemporary church bodies (two million plus in membership) tracing their origin to Los Angeles.

In the last decade, glossolalia scored great gains among the non-Pentecostal groups. Laymen also, in large numbers, have become involved. The movement has been given a considerable boost by the Full Gospel Business Men's Fellowship International, founded in 1953 by a group of Pentecostals. Today, however, its membership includes virtually all denominations. In 1959 this organization began publishing a periodical entitled *View,* which is by far the intellectual superior to the other publications, *Voice* and *Vision.*

Another organization particularly interested in glossolalia is the Blessed Trinity Society of Van Nuys, California, which was started by Episcopalians. Between 1962 and 1966 it published a handsome, slick magazine entitled *Trinity*, which was distributed among the historic denominations. It carried testimonials which were aimed at "proving" tongues to be more than a mere religious fad.

By far the most common association in which tongue-speaking flourishes is the small Holy Spirit Fellowships which have sprung up throughout Christendom. Usually these groups are small and relatively unstructured. They often have prayer, testimonials, and singing as well as speaking in tongues.

BIBLIOGRAPHY: M. Barnett, *The Living Flame* (1953); A.A. Hoekema, *What About Tongue Speaking?* (1960); W.H. Horton (ed.), *The Glossolalia Phenomenon* (1966); I.J. Martin III, *Glossolalia: A Bibliography* (1970).

WATSON E. MILLS

GLOVER, TERROT REAVELEY (1869-1943). English Baptist scholar. Son of a Baptist minister in Bristol, he graduated from Cambridge, became fellow of St. John's College there, then in 1896 went to Ontario as professor of Latin in Queen's University, Kingston. He returned to teach at Cambridge in 1901 and later became university lecturer in ancient history (1911-39). He traveled widely, particularly in the United States. His classical scholarship refused to segregate the emergent Christian tradition from its environment. Thus he was at home in *The World of the New Testament* (1931) as much as in the Greek world (*From Pericles to Philip*, 1917). As a teacher who loved the classics he sought their instructive discipline, yet he willingly assisted in the editing of Scripture for little children (1924). Among his many other works were *The Conflict of Religions in the Early Roman Empire* (1909) and *The Jesus of History* (1917). CLYDE CURRY SMITH

GNOSTICISM. The term designates a variety of religious movements in the early Christian centuries which stressed salvation through a secret *gnōsis* or "knowledge." These movements are most clearly attested in the writings of the church fathers of the second century. They viewed the various Gnostic movements as heretical perversions of Christianity. Modern scholars conceive of Gnosticism as a religious phenomenon which was more independent of Christianity. There is no clear consensus, however, as to how it originated. German scholars, who define Gnosticism rather loosely, are able to find Gnostic traces wherever there is an emphasis upon "knowledge" for salvation, as in the Dead Sea Scrolls.* Other scholars who define Gnosticism more strictly would require the presence of a cosmological dualism before conceding that a document is Gnostic.

Sources. (1) *Patristic.* Until the nineteenth century we were entirely dependent for our knowledge of the Gnostics upon the writings of the church fathers of the second and third centuries: Justin Martyr, Irenaeus, Hippolytus, Origen, and Tertullian, together with the later descriptions of Epiphanius (d.403). Some of the church fathers preserved extracts of primary Gnostic documents, but for the most part their accounts are highly polemical. Scholars were thus not sure how accurate a picture of the Gnostics they had in the patristic accounts. E. de Faye, writing early in this century, was extremely skeptical. He viewed any information relating to movements earlier than Justin's writing, the lost *Syntagma* (c.150), as completely legendary. Scholars have recently gained more confidence in the patristic sources as the Nag Hammadi* treatises have confirmed some of their materials.

Although the NT itself in Acts 8 does not describe Simon Magus* as a Gnostic, the patristic accounts are unanimous in regarding Simon as the fount of all heresies. Unlike the later Gnostics, Simon claimed to be divine and taught that salvation involved knowledge of himself rather than any self-knowledge. Simon was followed by a fellow Samaritan, Menander, who taught at Antioch toward the end of the first century. He taught his followers that those who believed in him would not die. In Justin's time (c.150) it seemed that almost all the Samaritans had become followers of Simon. But by the year 178 Celsus no longer attributed any importance to the Simonians. Teaching in Antioch at the beginning of the second century was Saturninus,* who unlike Simon and Menander held that Christ was the redeemer. In Asia Minor, Cerinthus* was a contemporary of Polycarp of Smyrna. Somewhat of an atypical Gnostic was Marcion* of Pontus, who taught at Rome from 137 to 144. Other Gnostic teachers include Basilides* and his son Isidore, and Carpocrates* and his son Epiphanes—all of whom taught at Alexandria. The most famous Gnostic teacher was Valentinus,* who taught at Alexandria and who came to Rome about 140. He had a number of able disciples, including Ptolemy and Heracleon* in the West, and Theodotus in the East.

(2) *Coptic.* In the nineteenth century, two original Gnostic codices in Coptic were translated: the Codex Askewianus containing the Pistis Sophia, and the Codex Brucianus containing the Books of Jeu. A third codex, Codex Berolinensis 8502, though acquired late in the nineteenth century, was not fully published until 1955. It contains a *Gospel of Mary* (Magdalene), a *Sophia of Jesus, Acts of Peter,* and an *Apocryphon of John,* a work mentioned by Irenaeus (A.D. 180).

In 1946 a cache of thirteen Coptic codices was discovered near Nag Hammadi in Upper Egypt. These contain some fifty-three treatises, of which more than one-third have now been published. The cache was deposited about A.D. 400. Among the works which have been published are: (a) *The Gospel of Truth,* which some have ascribed to Valentinus; (b) *The Epistle of Rheginos,* which is a discourse on the resurrection as a nonphysical phenomenon; (c) *The Gospel According to Thomas,* which contains sayings attributed to Jesus; (d) *The Gospel of Philip,* which reflects a Valentinianism similar to that of the Marcosians of the late second century; (e) *The Apocryphon of John,* which gives a cosmogony similar to that ascribed to the Sethians and Ophites by the church fathers; (f) *The Hypostasis of the Archons,* which gives a cosmogony similar to that of the *Apocryphon of John;* (g) *The Apocalypse of Adam,* which the editor A. Böhlig considers to be a document representing non-Christian Gnosticism. There are no explicit references to Christianity in the *Apocalypse,* but according to other scholars there are some clear allusions to Christianity.

(3) *Mandaic.* The Mandaean* communities in Iraq and in Iran are the sole surviving remnants of ancient Gnosticism. Their texts, although quite late, have been used by German scholars such as R. Reitzenstein and R. Bultmann* to reconstruct presumably earlier Gnostic traditions. Three major Mandaic texts were translated by M. Lidzbarski early in the twentieth century: (a) The *Ginza,* which presents a detailed cosmology; (b) the *Johannesbuch,* which contains some late traditions about John the Baptist, whom the Mandaeans revere; (c) the *Qolasta,* which is a collection of Mandaic liturgies. More recently E.S. Drower has published a number of other manuscripts, including the *Haran Gawaita,* which is a legendary account of the sect's migration from Palestine. In addition to these late manuscripts (sixteenth to nineteenth century) there are earlier Mandaic magic bowl texts (c. A.D. 600) and some lead strips which have been dated as early as the third/second century by R. Macuch. Macuch, Drower, and K. Rudolph have argued for a pre-Christian origin of the Mandaeans. The present writer has suggested an origin in the second century A.D. for the Mandaeans.

(4) *Other sources.* Mani (A.D. 216-75) was born near Seleucia-Ctesiphon in Babylonia. He established a highly syncretistic form of Gnosticism called Manichaeism,* which became widespread and which even included Augustine among its converts. R. Reitzenstein and G. Widengren have assumed that the late Manichaean texts preserve early Gnostic Iranian elements. A new codex from Cologne has now shown that the baptist sect to which Mani and his father belonged was not the Mandaean sect but that of the Elchasaites, a Jewish-Christian group. Other texts which have been adduced as evidence for early forms of Gnosticism, but whose Gnostic character has been disputed, included the "Hermetica," the "Syriac Odes of Solomon," and the "Hymn of the Pearl" (in the *Acts of Thomas*), the writings of Philo of Alexandria, the references to the *minim* in rabbinical sources, Jewish Merkabah mysticism, and the Dead Sea Scrolls.

Teachings of the Gnostics. In Gnostic systems there is an ontological dualism—an opposition between an ineffable, transcendent God and an ignorant, obtuse demiurge (often a caricature of the OT Jehovah), who is the creator of the cosmos. In some systems the creation of the material world results from the fall of Sophia. The material creation is viewed as evil. Sparks of divinity, however, have been encapsuled in the bodies of certain pneumatics destined for salvation. These pneumatics are ignorant of their celestial origins. God sends down to them a redeemer, often a docetic Christ, who brings them salvation in the form of secret *gnōsis*. Thus awakened, the pneumatics escape from their fleshly bodies at death and traverse the planetary spheres of hostile demons and are reunited with the deity. Since salvation is not dependent upon behavior but upon the knowledge of an innate pneumatic nature, some Gnostics manifested extremely libertine behavior. They held that they were "pearls" who could not be sullied by any external "mud." On the other hand, many Gnostics took a radically ascetic attitude toward marriage, deeming the creation of woman the origin of evil and the procreation of children but the multiplication of souls in bondage to the powers of darkness.

Gnostic Origins. There is no unanimity as to how, where, and when Gnosticism originated. S. Pétrement follows the church fathers in holding to a post-Christian and an inner-Christian development of Gnosticism. Many German scholars assume a pre-Christian origin of Gnosticism, though the evidences which they adduce to support this position are all either early texts which are not clearly Gnostic or very late Gnostic texts. An increasing number of scholars, including G. Quispel and G. MacRae, assume an important Jewish role in the origin of Gnosticism, though the Gnostic texts themselves are openly anti-Jewish. R. Grant has suggested that the disappointment of Jewish apocalypticism at the fall of Jerusalem in A.D. 70 may have resulted in Gnosticism.

Gnosticism and the New Testament. German scholars such as R. Bultmann and his disciples assume a pre-Christian origin of Gnosticism. They believe they can detect both direct and indirect references to Gnosticism in the NT, especially in the writings of John and of Paul. But the primary evidence which is used are NT passages themselves, which can be interpreted in a non-Gnostic sense. Bultmann has held that the NT was dependent upon a pre-Christian Gnostic myth of a "redeemed redeemer." C. Colpe has made some devastating criticisms of the work of the history-of-religions scholars that was responsible for the formulation of this myth. Most scholars today are now convinced that such a Gnostic redeemer myth is a post-Christian development patterned after the person of Christ. It seems safest to agree with the judgment of R. McL. Wilson, who accepts the existence of a rudimentary Gnosticism at the end of the first century, combated in the later NT books, and who warns against reading back traces of the fully developed Gnosticism of the second century into earlier texts.

BIBLIOGRAPHY: I. General: F. Burkitt, *Church and Gnosis* (1932); G. Quispel, *Gnosis als Weltreligion* (1951); R. Wilson, *The Gnostic Problem* (1958); H. Jonas, *The Gnostic Religion* (2nd ed., 1963); R.M. Grant, *Gnosticism and Early Christianity* (2nd ed., 1966); J.M. Robinson and H. Koester, *Trajectories Through Early Christianity* (1971).

II. Patristic Sources: E. de Faye, *Gnostiques et gnosticisme* (2nd ed., 1925); R.M. Grant, "The Earliest Christian Gnosticism," *Church History,* XXII (1953), pp. 81-98; V. Corwin, *St. Ignatius and Christianity in Antioch* (1960); R.M. Grant, *Gnosticism, A Sourcebook* ... (1961).

III. Coptic Sources: J. Doresse, *The Secret Books of the Egyptian Gnostics* (n.d.); A. Helmbold, *The Nag Hammadi Gnostic Texts and the Bible* (1967); J.M. Robinson, "The Coptic Gnostic Library Today," *NTS,* XIV (1968), pp. 356-401; D.M. Scholer, *A Classified Bibliography of the*

Coptic Gnostic Library and of Gnostic Studies 1948-1969 (1971).

IV. Mandaic Sources: M. Lidzbarski, *Das Johannesbuch der Mandäer* (rep. 1966) and *Ginza: Das Grosse Buch der Mandäer* (1925); E.S. Drower, *The Haran Gawaita...* (1953) and *The Canonical Prayerbook of the Mandaeans* (1959); *idem, The Secret Adam* (1960); K. Rudolph, *Die Mandäer* I and II (1960-61); E. Yamauchi, *Gnostic Ethics and Mandaean Origins* (1970).

V. Gnostic Origins: U. Bianchi (ed.), *The Origins of Gnosticism* (1967).

VI. Gnosticism and the New Testament: J. Dupont, *Gnosis: la connaissance religieuse dans les epîtres de St. Paul* (1949); C.H. Dodd, *The Interpretation of the Fourth Gospel* (1953; rep. 1968); R. Bultmann, *Theology of the New Testament* I and II (1952, 1955); G. Quispel, "Gnosticism and the NT," *The Bible in Modern Scholarship,* (ed. J.P. Hyatt, 1965), pp. 252-71; R. Wilson, *Gnosis and the New Testament* (1968). E. YAMAUCHI

GOBAT, SAMUEL (1799-1879). Bishop of Jerusalem. A French-speaking Protestant, he entered the Basel Mission Society school in 1821 where he showed considerable linguistic aptitude. After studying Arabic in Paris, he transferred to the English Church Missionary Society for service in Ethiopia. He spent two terms there in the 1830s and then went to Malta to do translation work. In 1845 the Lutheran Gobat was ordained in the Anglican Church and a year later King Frederick William IV appointed him to the joint English-Prussian bishopric of Jerusalem. In Palestine he founded hospitals, schools, and orphanages, and brought in German workers from the Kaiserswerth Sisters and the St. Chrischona Mission to assist in his ministry. Gobat's proselytizing activities among the Eastern churches aroused so much controversy that in 1853 several Anglican bishops publicly affirmed their confidence in him. After his death Prussia withdrew its support from the bishopric, leaving it a purely Anglican post. RICHARD V. PIERARD

GOD. For the *ancient Greeks,* the divine plenitude of life was reflected in a full Pantheon of gods, and it was to these "gods many" (1 Cor. 8:5) that the ordinary man looked for the supply of his religious needs. Though they were beyond the reach of death, the common lot of mortal men, the gods could not alter the dark counsels of fate, for they were themselves of the same order of being as men. Inasmuch as the Greeks conceived of the divine majesty in terms of man—the highest creature in the hierarchy of being—the noblest qualities with which they endowed their gods were human also.

For the Greek philosophers, the word "God" *(theos)* was a general term gathering up into itself all the impersonal, metaphysical forces and powers whereby order struggles out of chaos. God was the great sustaining Reality, the final necessary and adequate Condition of the existence of world order. In this process of rationalizing and ethicizing the Olympian deities, the philosophers did not necessarily deny the presence of the divine in the world; rather, in many instances they affirmed it. But they thought of God in terms of the regularity of being and immanent righteousness, more than in terms of the personal categories which dominated the Homeric world.

There was no possibility, therefore, in Greek philosophic thought, that man should enjoy a personal relationship with God, much less that God should assume our human mode of existence. That God should be a God-who-is-for-man is a view alien to Greek thought. By the same token it was unthinkable that man should address God in prayer as a loving Father. Man, indeed, may be moved by erotic attraction to the perfection residing in the diving Being, but this experience is not prayer in any sense of personal communion. Rather, since the thought of the philosophers about God and the cosmos tended toward identity, religious experience became the inner freedom and blessedness of self-fulfillment through striving toward higher forms of existence.

In contrast with Greek religious speculation, where the movement of thought is from the world to God, *the biblical view of God* is the other way. The Bible always speaks of God as a personal God who comes to man in his self-revelation.

In the OT, the general word for God is "El," denoting a personal object of religious perception and pious awe having to do with power, a power which man cannot master, but which fills his religious consciousness. Hence "El" is contrasted with "man" in passages like Ezekiel 28:2 and Hosea 11:9. Significantly the term is interchangeable with God's personal name "Jahweh" (commonly translated "LORD" in our English versions) in a way that makes it clear that the two are synonymous. Note the expression, found from time to time in the OT, "Jahweh Elohim," where Elohim is in explanatory apposition: "Jahweh, that is, God." The use of the plural when referring to Jahweh can have no numerical significance, since "Elim" occurs as the plain plural of "El," while "Elohim" is used of other individual gods in the OT such as Baal of Sidon. We may therefore regard Elohim as a "plural of amplitude in addressing God"; to speak of Jahweh as Elohim is to confess that the God whose name is Jahweh possesses the quality of El in the fullest measure.

Though her faith in Jahweh was confronted by a plurality of deities in the surrounding world, through many a crisis Israel came to leave behind all thought of a tribal, national God and to recognize that God by His power fashioned and rules the whole world. Faith in this one, personal, all-powerful God was pledged in the covenant established between Jahweh and His people by the hand of Moses. This sense of Jahweh's absolute uniqueness was not obtained by rational argument, but by the impelling experience of the divine reality. Jahweh is He who helps, delivers, judges, and consoles His people. Hence He must be taken seriously in His transcendent power and steadfast love. Israel's monotheism, in other words, is not the end product of polytheism, driven by some inner motif of unity to a more satisfying concept, but a confession of God's overpowering reality in the lives of his people. "I am the LORD your God, who brought you out of the land of Egypt, out of the house of bondage. You shall have no other gods before me" (Exod. 20:2,3 RSV).

Since faith in Jahweh, the God of Israel, is the only proper response to His mighty acts of redemption, the prophets attacked the power of heathen piety by noting the obvious fact that the gods of the heathen can neither hear nor speak. Fashioned in wood and stone, the symbols of nature's forces numinously conceived, they are of no avail to help in time of trouble. They are a silent mystery and their devotees deluded fools (Isa. 44).

This monotheism (confessed in Deut. 6:4) comes to its finest expression in the second part of Isaiah. The overthrow of the nation of Israel, in the natural course of things, would have marked the demise of the worship of Jahweh. But Israel's history is not natural, as is evidenced by the fact that her national tragedy was the occasion for the prophet to affirm Jahweh's sovereign lordship over all the nations of the earth. Jahweh is God alone, unique and incomparable in His power and wisdom, the almighty Creator of heaven and earth, the Lord of all human history (cf. Isa. 41:1-5; 43:10,11; 44:7,8; chaps. 45,46).

The OT doctrine of God is the presupposition of all that the NT writers teach about Him. The God of Abraham, Isaac, and Jacob, almighty, holy, living, and faithful, is the God who is uniquely revealed in Jesus the Christ. Only now man's reverence and awe as a worshiper of the true God is informed by a heightened intimacy of relationship. This is because Jesus taught His disciples to call God "Father" (Matt. 6:9; Luke 11:2) and gave them His Spirit, by whose inner witness they were enabled to cry, "*Abba*, Father" (Rom. 8:15, 16). Since Jesus is uniquely God's Son (Luke 10:22), and since His Spirit indwells His followers, Christians could hardly conceive God's unity, so stressed in the OT and reiterated in the New, in terms of an undifferentiated monad. The great events of redemptive history, incarnation, and Pentecost which precede the writing of the NT explain why the doctrine of the Trinity, though not expressly elaborated in the NT, is nonetheless there in primordial form. Hence the spontaneous use of threefold expressions when speaking of God on the part of the apostolic community (Matt. 28:19; 2 Cor. 13:13; Eph. 4:4-6; 2 Thess. 2:13,14; 1 Pet. 1:1,2).

In the early Christian mission to the Gentiles, biblical religion confronted Greek philosophy, and of this encounter *Christian theology* was born. So far as the Christian doctrine of God is concerned, we may say that the OT revelation of God, culminating in the work of Jesus Christ, provided the church with the substance of its faith, while Greek philosophy supplied the intellectual categories and concepts for the systematic articulation of this faith. The marriage of Greek and biblical thought is a fact to be accepted, but not with uncritical approval. On the one hand, "the fullness of time" may be seen in that Greek philosophy had prepared men's minds for the theological task confronting the church. This task was to work out the implications of God's final self-revelation in the person and work of Jesus Christ, so as to meet the need of the catechumen and the challenge of the heretic and pagan. On the other hand, there was always the threat of distortion, since Greek thought is not only other than, but in a sense alien to, the thought world of the Bible.

As we have seen, the accent in Greek philosophy was on the impersonal Ideal, whereas the Bible is concerned with a personal God who speaks to His people and acts in history to redeem them. Because of this essential difference, philosophy can never be more than the handmaid of theology. Some Christian thinkers have sought to deny even this modest place to abstract, rational thought about God, but this is patently an overreaction. It is a fact that philosophy has provided the critical categories of thought with which Christian theology has gone about its task, and indeed, if one accepts the providence of God and the ultimate unity of truth, it is difficult to see how the theological task of the church could have been pursued in any other way.

There have been *persistent theological problems;* these concern the being and existence of God and his relationship to the world, especially His relationship to man. In an article of such limited scope one can do little more than offer the briefest survey of the answers Christian thinkers have suggested to these problems. It must always be remembered that these Christian answers which take the form of doctrinal formulations of the church are the extension of faith. Dogmatic pronouncements about God define truths concerning which there is a broad consensus and concerning which the church has an obligation to preach and teach as it upholds the faith once for all delivered to the saints. The central confession of the church is that Jesus is Lord, and the elaborated, explicit form of this confession is the doctrine of the Trinity.*

The church answers the question of God's being by the fundamental affirmation that God is a Trinity. Although the dogma of the Trinity contains no definition or classification of the attributes nor inference of specific attributes from the nature of the divine Being, yet when discussed, the attributes describe the Trinity in Unity which is the Godhead.

The above position is stated as preferable to that of the medieval Thomists and Protestant scholastics who discussed the doctrine of the Trinity only after they had defined God's essence as pure actuality *(actus purus)* and established His existence and attributes on general principles of reason. Such a metaphysical approach has its merits and is certainly to be preferred to the antimetaphysical stance of German liberalism whereby pronouncements about God are turned into pronouncements of religious experience (Schleiermacher) or statements of the ethical values of the kingdom of God (Ritschl). Such a theological method can hardly escape the criticism of Feuerbach that all theology is anthropology. On the other hand, metaphysics tends to alter pronouncements about God from confessional statements to be used in worship, to general statements of reason, uniting God and the world in a rational system in which the doctrine of God is more philosophic than biblical (see NATURAL

THEOLOGY). Because of our reservations about such an approach, in this article we shall first discuss the *divine attributes* and only then raise the question of the proofs of God's being and existence.

The theologians' difficulty in classifying and ordering the divine attributes confirms one in the opinion that our understanding of God is inadequate to comprehend His essence, though we may postulate a genuine analogy between His being and the properties which we ascribe to Him. The more metaphysical attributes are God's aseity or independence, which means that God is unlike anything He has made; God's infinity, which includes both His eternity (He is beyond temporal limitations) and His immensity (He is beyond spatial limitations, i.e., omnipresent); and finally, God's impassibility (He is pure actuality, devoid of mere potentiality). As Augustine has said: "To God it is not one thing to be, and another to live, as though he could be, not living; nor is it to him one thing to live, and another to understand, as though he could live, not understanding; nor is it to him one thing to understand, another to be blessed, as though he could understand and not be blessed. But to him to live, to understand, to be blessed, are to *be.*"

The more religious attributes describe God in the perfection of His intelligence, will, and holy love. His perfect intelligence we call omniscience; His perfect will omnipotence; while holy love refers to His justice (wrath) and His grace (mercy). Since these attributes are based on God's self-revelation in history, they should always be understood in the light of that revelation. God's infinity, for example, is not simply His independence from time and space, but His lordship over them as displayed in the Incarnation, wherein He freely reveals Himself in time and space.

Of equal importance to faith with the question of God's attributes is the question of how we should conceive and speak of God's relationship to the world of which man is a part. This is the question of *God's immanence and transcendence.* Pantheism tends to press the divine immanence to the point of identity between God and the world. Whatever is, is God, and nothing can be conceived apart from God. Deism takes the opposite tack and stresses the divine transcendence. God is so apart from the world as to be an absentee Lord, the laws of nature being sufficient unto themselves. The biblical view is described by the word "providence," which is the doctrine that God preserves and governs all His creatures and all their actions, by a personal exercise of His power, freely, according to the counsel of His will and for His own glory. Classically, this providential rule of God has been construed in terms of causality. But to affirm that God is the ultimate "cause" of everything that comes to pass has raised difficult questions, so far as evil is concerned, for no theologian will say that God is the author of evil. In contemporary thought, moreover, causality is an impersonal category of science, which makes it especially problematical as describing the divine agency in the area of the free and responsible acts of men. It is best, therefore, to think of God's providence in such personal categories as are suggested by the biblical titles of Ruler, King, Lord, and Father.

Traditional theology has been much concerned to prove the existence of God. While such an effort is entirely understandable, it has no express biblical warrant. For the writers of Scripture, God's presence and power in the world were as self-evident as the axioms on which the traditional demonstrations of His existence were supposed to rest. It is possible, therefore, to view the traditional arguments for God's existence as clarifications of mental concepts derived from revelation.

Contemporary theology is marked by efforts, not to prove God's existence as an eternal and absolute Being, but to reconstruct our thought about God in terms of evolutionary process. Such "Process Theology" stresses the thought that God is not only eternal, but also eminently temporal, affected in His being by all that transpires in the creation. Since the universe is a changing, dynamic, living reality, God must also be conceived as changing, dynamic, and living; open to the possiblities of creation, actualizing his potentialities. Many modern existentialist theologians have gone further than this in their departure from the traditional thought of the church about God (see EXISTENTIALISM). If God is transcendent in any sense, it is a hidden transcendence in the depth of Existence. To speak about God is to speak about man's existence. His transcendence is the transcendence of man's inner life. Hence the complaint of God's "silence," "absence," "concealment," "eclipse," even "nonbeing" and "death" on the part of many religious existentialists. Much of the disarray of contemporary theology could be due to the fact that modern man does not want a sovereign God to rule over him. As Augustine once pointed out, however, man's freedom is not preserved by banishing God, but by serving Him.

BIBLIOGRAPHY: J.S. Candlish, *The Christian Doctrine of God* (1888); H.M. Gwatkin, *The Knowledge of God and its Historical Development* (1908); G.S. Hendry, *God the Creator* (1937); J. Baillie, *Our Knowledge of God* (1939); E. Gilson, *God and Philosophy* (1941); E. Brunner, *Dogmatics I: The Christian Doctrine of God* (1949); H. Bavinck, *The Doctrine of God* (ET 1951); G.E. Wright, *God Who Acts* (1952); K. Barth, *Church Dogmatics II: The Doctrine of God,* I (1957); W. Eichrodt, *Theology of the Old Testament* (ET 1967). PAUL KING JEWETT

GODET, FRÉDÉRIC LOUIS (1812-1900). Swiss Reformed theologian and exegete. Educated at the universities of Neuchâtel, Bonn, and Berlin, he became chaplain to the king of Prussia and tutor of Prince (later Kaiser) Frederick Wilhelm (1838-44). He was pastor (1851-66) and professor of biblical exegesis (1851-73) in Neuchâtel and then professor of NT exegesis in the Free Evangelical Faculty of the same city (1873-87). He was one of the most influential Reformed scholars of his day, his works being translated into various languages. He defended the orthodox Christian position against the growing theological liberalism in academic Protestant theology, and combined a deep, Christian piety with positive biblical and historical criticism. He is best known in the English-speaking world for his commentaries on John (1864-65; ET 1877), Luke (1871; ET 1875), Romans (1879-80; ET 1880-81), and 1 Corinthians (1886; ET 1886-87).

W. WARD GASQUE

GODFREY OF BOUILLON (c.1060-1100). French Crusader. Member of the French nobility, he led a German contingent in the First Crusade of Urban II in 1096. Three years later, after Raymond of Toulouse had marched on Jerusalem, he took a leading part in the siege and capture of the city. When Raymond refused the offer to rule Jerusalem, Godfrey was chosen and took the title "Protector of the Holy Sepulcher." After his death his successor and brother Baldwin established the Latin Kingdom of Jerusalem. In later legend Godfrey was often depicted as the personification of the ideal Christian knight. The *Assizes of Jerusalem,* a law-book claiming to contain the laws of the kingdom of Jerusalem laid down by Godfrey, was in fact a fifteenth-century work by John of Ibelin.

PETER TOON

GODPARENTS. Also called "sponsors," they act on behalf of infants at baptism, and in their stead pronounce the renunciation of Satan, the confession of faith, and assist during the baptism itself. The sponsor either touches or holds the person being baptized and receives him from the hands of the minister following the ceremony. Sponsors act as witnesses on behalf of adults too. For solemn baptism there can be no more than two sponsors, a man and a woman, who must not be closely related to the person baptized. Sponsors themselves must have been baptized, have reached the age of reason, must intend to execute the office of sponsor, must not be an excommunicate, a heretic, father, mother, or spouse of the subject, and must be designated to the role by the recipient, his parents, his guardian, or the minister.

HAROLD LINDSELL

GOETHE, JOHANN WOLFGANG VON (1749-1832). German poet, novelist, and scientist. As a student of law at Leipzig and Strasbourg he became interested in occult philosophy and mysticism. In 1775 he was appointed to the court at Weimar and had at this time an increasing interest in scientific questions. In 1794 he became friendly with Schiller, which friendship lasted until Schiller's death in 1804. Goethe died in Weimar and was buried beside his friend. To summarize his career and work briefly is impossible. The central philosophical influences on him were Spinoza, Jacobi, and Kant. He was attracted by Spinoza's pantheism and his ethics; like Kant, he held that God is unknowable. Man, a part of nature, has a natural impulse to develop and fulfil ideals inherent in himself. Goethe's religious views were ambiguous, for he was a pantheist when studying nature, a polytheist when poetizing, and a monotheist in morality.

OONAGH MC DONALD

GOFORTH, JONATHAN (1859-1936). Canadian Presbyterian missionary to China. Educated at Knox College, Toronto, he was ordained in 1886 and with his wife went to China (1887) where they were pioneers in the Canadian Presbyterian work in Honan Province. Theologically conservative and a firm believer in preaching, he became famous for his leadership and participation in the revivals which swept over China in the early twentieth century. Though he lost his eyesight during his final years in China, he continued to minister there until 1934. He wrote *By My Spirit,* which told of the remarkable revivals in China and which inspired much support for missions.

GOGARTEN, FRIEDRICH (1887-1967). German Protestant theologian. A native of Dortmund, he established his reputation with an essay on Fichte as a religious thinker (1914), was pastor at Stelzendorf and Dorndorf, and from 1927 taught systematic theology at Jena. Reacting like Barth from religious liberalism, but unlike others of the new orthodoxy from a Lutheran rather than a Calvinist background, he believed he had recaptured the true insight of Luther by upholding an existential interpretation of sacred history which sees it not as an objective series of happenings to be accepted from the outside, but as dynamically to be apprehended by ourselves who are within the same historical process. Two of his numerous theological writings to be translated into English are *Demythologizing and History* (1955) and *Christ and Crisis* (1970).

IAN SELLERS

GOMAR, FRANCIS (also "Gomarus") (1563-1641). Low Countries Calvinist theologian. Born in Bruges as the revolt of the Lowlands against Spanish rule was imminent, he studied as a boy at Strasbourg under Johann Sturm, went on in theology under Zanchius at Neustadt, continued at Oxford and Cambridge, and received his doctorate at Heidelberg in 1593. In his thirties he became professor of theology at Leyden. An ardent and skilled defender of Calvinist orthodoxy, he protested against the teachings of Arminius (see ARMINIANISM), from 1603 his colleague at Leyden, seeing these as effectively denying the doctrine of election. The growing controversy soon spread through the Dutch Calvinist churches, with "Gomarist" and "Arminian" factions in increasingly bitter debate. Arminius's death (1609) was followed by the Arminian Remonstrance of 1610 and by the Contra-Remonstrance of 1611.

Meanwhile, when the Remonstrant Conrad Vorst (Vorstius) was appointed at Leyden to replace Arminius, Gomar resigned in disgust. As the Remonstrant controversy raged, increasingly mixed with political factionalism, Gomar taught in the Huguenot seminary at Saumar (1614-18) and was then called back to be professor of theology at Groningen. Known for his Contra-Remonstrant* views, he was chosen as a delegate to the Synod of Dort,* played a prominent role there, and rejoiced at the condemnation of the Remonstrants. The rest of his career was spent at Groningen. He held to a scholastic version of Calvinism, stressed the importance of doctrine, and took a supralapsarian position regarding predestination (Dort left the question open). His *Opera theologi-*

ca omnia were published soon after his death (2 vols., 1645).

See G.P. Van Itterzon, *Francisciscus Gomar* (1930). DIRK JELLEMA

GOMEZ, MARIANO (c.1788-1872). Chinese-Filipino priest and martyr. He had long defended his fellow priests in their just grievances against the friars. When in January 1872 Filipino soldiers and workers at the Cavite arsenal mutinied and killed their Spanish officers, the government and the friars used this as an opportunity to suppress all dissent. In this connection Gomez, with his colleagues Jose Burgos and Jacinto Zamora, was arrested, and following a secret trial with perjured witnesses, the three were garrotted before a large crowd in Manila. The governor-general requested Archbishop Martinez to unfrock them first, but he refused. The three martyred priests rapidly became a symbol of united nationalism. In 1904 they were canonized as saints of the Philippine Independent Church.

RICHARD DOWSETT

GOOD FRIDAY. The title used in many English-speaking countries for the day on which Christ's death is particularly remembered. In other countries it was known as Long Friday, Day of Preparation, Day of the Lord's Passion, and the Passion of the Cross. It is called "Good" because of the benefits which flow from what the day commemorates. It came to be observed as a result of the development of the calendar in the fourth century. In the Pilgrimage of Etheria* is a first-hand account of the ceremonies practiced in Jerusalem at the end of the fourth century, with a description of the veneration of the Cross which still continues in the Roman rite. Popularly known as "creeping to the Cross," this was condemned by the Reformers. Holy Communion was not usually celebrated on Good Friday. When weekday masses began in the sixth and seventh centuries, Friday was already a special fast day with Bible readings and prayers, and this tradition was left undisturbed.

The Reformers provided in the Book of Common Prayer an Epistle and Gospel for Good Friday, and in England there is some evidence that up to and including Queen Victoria's reign some churches held Communion on Good Friday, though generally it has dropped out. In its place normally is a devotional service based on the Seven Words from the Cross. PETER S. DAWES

GOODELL, WILLIAM (1792-1867). Pioneer American Congregational missionary to the Near East. Born in Templeton, Massachusetts, he became one of a succession of notable scholarly missionaries in the Near East. He served there forty years, not counting one furlough. Appointed by the American Board in 1823, he helped establish the work in Beirut, which became the center of the Syrian Mission. In 1828 the mission was obliged to move to Malta, where for three years he supervised the press and worked on his Armeno-Turkish translation of the Bible. Sent with H.G.O. Dwight to Constantinople in 1831, he helped found the work in Turkey. His major accomplishment was the translation of the Bible into Armeno-Turkish from the Hebrew and Greek. HAROLD R. COOK

GOODRICH, CHAUNCEY ALLEN (1790-1860). American Congregational clergyman, educator, and lexicographer. Graduate of Yale (1810) and student of Timothy Dwight,* Goodrich at the former's suggestion published *Elements of Greek Grammar* (1814). He was ordained in that year, engaged in pastoral work, then in 1817 became professor of rhetoric at Yale, where he helped to establish the theological department in 1822. He was professor of preaching and pastoral work from 1838 until his death. As editor of the quarterly *Christian Spectator* (1828-36) he promoted Nathaniel Taylor's "New Haven Theology."* He published also several other writings, the most important of these being *Select British Eloquence* (1852). He also worked on an abridgment of the *Webster Dictionary.*

ROBERT C. NEWMAN

GOODSPEED, EDGAR JOHNSON (1871-1962). NT scholar. Born at Quincy, Illinois, he studied at Denison University, Yale, and the universities of Chicago and Berlin. He taught biblical and patristic Greek at Chicago (c.1900-1937), and pioneered in collating NT manuscripts and in the study of Greek papyri in America. He translated into American idiom *The New Testament* (1923), *The Apocrypha* (1938), and *The Apostolic Fathers* (1950), and was an original member of the committee that produced the RSV New Testament (1946). He taught history at the University of California, Los Angeles (1938-51). He wrote over sixty books, some of the more important being *An Introduction to the New Testament* (1937), *History of Early Christian Literature* (1942), *How to Read the Bible* (1946), *A Life of Jesus* (1950), and an autobiography, *As I Remember* (1953). ALBERT H. FREUNDT, JR.

GOODWIN, JOHN (1594-1665). Puritan divine. Educated at Queens' College, Cambridge, he became a fellow there. He was rector of East Rainham, Norfolk (1625-33), then moved to London as vicar of St. Stephens, Coleman Street. Though nominated as a member of the Westminster Assembly, he did not attend. He was an ardent supporter of Parliament against the king in the Civil War, and became known as a leading republican thinker and defender of religious liberty. In the 1640s he attacked the Presbyterians as a persecuting party. He formed a gathered church in his parish (c.1644) and for a brief period was removed from his vicarage. Goodwin was one of the few Puritans who were also Arminian. This fact involved him in much controversy. He opposed Cromwell's National Church, and printed his opinions about the Triers in *Basaoistai, or The Triers (Or Tormenters) Tried* (1657). He was one of those exempted from the Act of General Pardon of Charles II in 1660, but he was not executed. PETER TOON

GOODWIN, THOMAS (1600-1680). Congregational divine. Born in Norfolk and educated at

Cambridge, he became a fellow of St. Catherine's and vicar of Holy Trinity Church, Cambridge. On becoming a Congregationalist in 1634 he resigned and moved to London. In 1639 persecution drove him to Holland, where he was a pastor of a church at Arnheim. He returned to London when the Long Parliament began to sit and formed a gathered church in London. Nominated as a member of the Westminster Assembly,* he became the leader of the Dissenting Brethren in it. In 1649 he was appointed a chaplain to the Council of State, and in 1650 president of Magdalen College, Oxford. Goodwin was a leading member of both the Board of Visitors in the university and the Cromwellian Triers. From 1656 he enjoyed the confidence of Oliver Cromwell. He was a prominent member of the Savoy Assembly of Congregational elders in 1658 and was much esteemed among the gathered churches of the nation. After the Restoration he moved from Oxford to London and was pastor of a gathered church in the City. His works were published in five folio volumes between 1682 and 1704 and have often been reprinted. They include devotional, expository, doctrinal, and ecclesiastical studies and are Calvinistic in outlook.

PETER TOON

GORDON, ADONIRAM JUDSON (1836-1895). Baptist minister, educator, and author. Born in Hampton, New Hampshire, he decided for the ministry when entering preparatory school. He graduated from Brown University (1860) and Newton Theological Seminary (1863), and for six years was minister of Jamaica Plain (Massachusetts) Baptist Church. In 1869 he went to Clarendon Street Baptist Church, Boston, a center of evangelistic and philanthropic work. He founded a school for training missionaries for home and foreign service, and for pastors' assistants, from which came Gordon College and its divinity school. A student of prophecy, for a time he edited *Watchword,* a monthly given to biblical exposition. His writings include *The Ministry of Healing* (1882), *Ecce Venit* (1890), *The Ministry of the Spirit* (1894), and *When Christ Came to Church* (1895). He also helped to compile two hymnals: *Service of Song* and *Coronation Hymnal.*

C.G. THORNE, JR.

GORDON, CHARLES WILLIAM (1860-1937). Presbyterian minister and writer, better known by the pseudonym of "Ralph Connor." Born in an Ontario manse, his *The Man from Glengarry* gives an interesting picture of his boyhood Highland community, including the intense spirit of revival that gripped many evangelical Presbyterian congregations during the 1860s. After ordination in 1890 he carried out mission work in the lumber camps and mines of W Canada, then accepted a call to St. Stephen's Church, Winnipeg, where he remained, apart from a period of chaplaincy service, until retirement in1929. Among his other novels were *Sky Pilot* (1899) and *Glengarry School Days* (1902). With considerable powers of description and an understanding of certain types of men, his stories found a large readership. His avowed reason for writing he stated thus: "Not wealth, not enterprise, not energy, can build a nation into true greatness, but men and only men with the fear of God in their hearts."

IAN S. RENNIE

GORDON, GEORGE ANGIER (1853-1929). Congregational minister and writer. Scottish-born, he graduated from Bangor Theological Seminary and Harvard University, proving himself a brilliant student of philosophy. A preacher of great power, during a long and influential pastorate at the Old South Church in Boston (1884-1927), and through numerous books, and lectures in all the leading American universities, he was a leader in introducing liberalism into Congregationalism. He called Calvinism "the ultimate blasphemy of thought" and held that moral progress was the key to history. He nevertheless, with other liberals, rejected the contemporary drift toward a merely human view of Jesus. His works include *Ultimate Conceptions of Faith* (1903), *Religion and Miracle* (1909), *Through Man to God* (1906), and *Aspects of the Infinite Mystery* (1916).

KEITH J. HARDMAN

GORDON, SAMUEL DICKEY (1859-1936). American devotional writer. Born and educated in Philadelphia, he was for ten years an assistant secretary and later state secretary of the Ohio YMCA. He then began to preach and lecture on religious subjects in America, and traveled for four years in the Orient and Europe holding Bible conferences and missionary conventions. He wrote more than twenty devotional books under the title "Quiet Talks," such as *Quiet Talks on Power* (1901), . . . *on Prayer* (1904), and . . . *on Jesus* (1906). These books were well received and widely used.

GORDON RIOTS (1780). These broke out in London on 2 June, when Lord George Gordon led a mob to the House of Commons with a petition for the repeal of the Catholic Relief Act of 1778. Lord George, an eccentric Scot and fanatical anti-Papist, had become president of the Protestant Association in 1779. The demonstrators soon became violent. Roman Catholic chapels were destroyed. On 6 June Newgate and other prisons were burned down and the following day attacks made upon the Bank. While the magistrates were acting feebly, the crowd had been swollen by released criminals and resorted to wholesale looting. George III personally ordered in the troops to quell the riots, in which nearly three hundred people died. Dickens graphically described the events in *Barnaby Rudge.* Many of the rioters were convicted and twenty-five executed. Arrested on a charge of high treason, Gordon was nevertheless acquitted. Later he became a Jew, was convicted for libel, and died in Newgate Prison in 1793.

J.W. CHARLEY

GORE, CHARLES (1853-1932). Anglican bishop. Educated at Oxford where he proved to be a brilliant scholar, he was ordained in 1875 and was elected a fellow of Trinity College. He was vice-principal of Cuddesdon Theological College from 1880, and three years later became the first principal of Pusey House. During those years in the Oxford area he exercised a strong influence on the religious life of the university, mainly through personal relationships. A lifelong Anglo-Catholic, he nevertheless brought a more conciliatory and liberal spirit to the Oxford Movement.* He was active on behalf of the Christian Social Union and was the founder in 1892 of the Community of the Resurrection. He upset some of his friends by inferring from Philippians 2:7 that Christ's humanity involved certain limitations.

Gore became canon of Westminster (1894) and a royal chaplain (1898); bishop of Worcester (1902); and, when the new diocese of Birmingham was formed largely through his efforts, he became its first bishop in 1905. In the latter place he formed excellent relations with civil authorities, non-Anglicans, and evangelicals (this despite his unyielding views on the episcopal system). He supported also the Workers' Educational Association. In 1911 he became bishop of Oxford, but found it more resistant to his masterful personality. It may be that some there had not forgotten his views expressed in the symposium *Lux Mundi** (a volume he edited also), which created a sensation and caused the High Church movement increasingly to take account of modern developments in scholarship. In 1919 he resigned and settled in London. Gore's many works include *The Sermon on the Mount* (1896), *The Body of Christ* (1901), *The Ministry of the Christian Church* (new ed., 1919), *The Holy Spirit and the Church* (1924), and *Christ and Society* (1928). He was the most versatile, and probably the most influential, churchman of his generation.

J.D. DOUGLAS

GÖRRES, JOHANN JOSEPH VON (1776-1848). German Roman Catholic publicist and lay theologian. Born in Coblenz, he was deeply influenced by the Enlightenment, hostile to religion, and an enthusiast for the French Revolution. As he became increasingly disillusioned, he moved in the direction of Catholic mysticism and German romanticism. He was first a teacher, but then became a scholar of religious history and a journalist. His early works included *Glauben und Wissen* (1805) and *Mythengeschichte der asiatischen Welt* (1810).

During the Napoleonic Wars Görres was a vocal supporter of German nationalism. Now a popular figure, in 1814 he started the *Rheinische Merkur,* the first important German newspaper. Although suppressed by the Prussian government in 1816, it established his reputation as a founder of modern political journalism. He published controversial tracts in 1819 and 1837, demanding more freedom for the Catholic Church in public life. He accepted a professorship at Munich University in 1817, and there dominated a circle of noted scholars who promoted a Catholic renewal emphasizing romanticism and mysticism. His major work, *Die Christliche Mystik* (4 vols., 1836-42), was a vague, sentimental, and even fanciful treatment of mysticism. The Görres-Gesellschaft, founded in 1876, is a leading German Catholic society devoted to scholarly research and publication.

RICHARD V. PIERARD

GORTON, SAMUEL (1592/3-1677). Early American colonist. Born in England, he migrated in 1637 to Massachusetts. His opposition to the union of church and state, an ordained ministry, and the sacraments, his denial of heaven and hell, and his advocacy of unorthodox views of the Trinity and inner illumination of the Holy Spirit soon forced him to flee to Rhode Island in the winter of 1637-38. After some years of wandering in Rhode Island and Massachusetts, in 1643 he founded Shawomet which he later renamed Warwick. He was in England from 1644 to 1648 and secured religious liberty for his colony from the earl of Warwick. He served several times as a member of the Massachusetts Assembly between 1649 and 1666, and was a member of the Warwick town council in 1677, the year of his death.

EARLE E. CAIRNS

GOSPEL (OE *godspel,* "good tidings"; Gr. *euangelion,* "good news"). The message of God's redemption in Jesus Christ, which lies at the heart of the NT and the church's faith. In the NT it is, first, the proclamation by Jesus that the kingdom has drawn near and, then, the proclamation by His disciples that in His life, death, and resurrection the kingdom has been established and that salvation and forgiveness are offered to all who believe. At a later date the term came to be used of those early Christian writings which tell the story of that unique manifestation of the "good news" in the person and work of Jesus Christ (cf., Justin Martyr, *First Apology* 66.3; Clement of Alexandria, *Stromata* iii.13). Strictly speaking, there is only one Gospel: the four writings called "gospels" are really only variations on a single theme. It would be more accurate to speak of the "fourfold gospel" (Irenaeus) than of the "four gospels": the gospel *according to* Matthew . . . Mark . . . Luke . . . John.

The background of the use of the noun *euangelion* and the related verb *euangelizomai* in the NT is the Greek translation of the second part of the prophecy of Isaiah (40:9; 52:7; 60:6; 61:1), which is quoted or alluded to many times in the NT (e.g., Mark 1:3; Rom. 10:15; Luke 4:17-21; Matt. 11:5/Luke 7:22).

In Lutheran theology the term is used to represent the NT revelation as contrasted with "law" (the old dispensation).

W. WARD GASQUE

GOSPELS, see SYNOPTIC GOSPELS; JOHN, GOSPEL OF

GOSSE, PHILIP HENRY (1810-1888). Son of an itinerant painter of miniatures, he was born in Worcester and worked as a clerk and a farmer in North America (1827-38). On returning to England he was poverty-stricken until his *Canadian Naturalist* (1840) was accepted by a publisher. After a brief spell with Methodists he associated himself with Brethren. He visited Jamaica in 1847 and soon became a prolific writer and lecturer on natural history. He invented and popularized the aquarium and was a sensitive, accurate illustrator. His *Actinologia Britannica* (1858-60) is a standard history of sea anemones and corals. After his first wife's death in 1857 he retired to St. Marychurch, remarried, and shepherded a Brethren assembly on individualistic lines. His attempt in *Omphalos* (1857) to reconcile Genesis with geology satisfied no one. His failings were those of an intellectual recluse, not a bigot.

JOHN S. ANDREWS

GOSSNER, JOHANNES EVANGELISTA (1773-1858). Founder of the Gossner Missionary Society. Born at Hausen near Augsburg, he studied at the University of Dillingen and then enrolled in the seminary at Ingolstadt. In 1796 he was ordained priest and assigned to a curacy at Neuburg, where he became an evangelical. From 1797 to 1804 he officiated at Augsburg and then was parish priest at Dirlewang (1804-11). Thereafter he accepted a benefice and engaged in literary pursuits at Munich. In 1819 he followed his friend Martin Boos to Düsseldorf. Gossner served a German congregation at St. Petersburg, Russia, from 1820 to 1824, until doubts regarding the celibacy of the clergy forced him to resign. In 1826 he joined the Lutheran Church and was appointed in 1829 to the pastorate of the Bethlehem Church in Berlin, where he remained seventeen years. During his tenure at Berlin he founded schools and asylums and a missionary society bearing his name in 1836. Missionaries with it served mainly among the Khols of East India. After resigning from the Bethlehem Church in 1846 he devoted the remainder of his life to the hospital which he founded. WAYNE DETZLER

GOTHIC VERSION. The Goths, a Germanic people of the Balkans who later moved westward, already had Christians among them in the third century, and in the fourth their countryman Bishop Ulfila is said to have been instrumental in gaining their adherence to the Arian heresy and in translating the Scriptures into Gothic, for which he had formed an alphabet. The main witness is the gospel Codex Argenteus, Stockholm; and there are about nine other fragmentary MSS, eight of them palimpsest. The gospels and Pauline epistles are fairly well represented, but of the OT only words of Genesis 5:11; 2 Esdras 15—17 (viz., Neh. 5—7); and two verses of Psalm 52 survive. The basic textual complexion in the NT is early Byzantine; the westward wanderings of the Goths, however, brought Old Latin influence to act upon the text, while Gothic influence upon the Old Latin is also known. J.N. BIRDSALL

GOTTESFREUNDE, see FRIENDS OF GOD

GOTTSCHALK (Godescalus) (c.805-869). Theologian and monk. He was compelled by his father, the Saxon Count Bruno, to enter the Benedictine abbey of Fulda. The Synod of Mainz (829) released him from his vows, but this dispensation was canceled on an objection by Rabanus Maurus,* the newly elected abbot, and he was moved to the Franciscan monastery of Orbais. He devoted himself to theological study, particularly to the teaching of Augustine and Fulgentius on predestination. He appears to have been the first to teach "double predestination," i.e., the elect are predestined "freely" to bliss, while the wicked are predestined "justly" to condemnation, on foreknowledge of their guilt.

Gottschalk was opposed by Rabanus Maurus, whom he charged with Semi-Pelagianism, and by Hincmar, archbishop of Reims, who accused him of denying the universal saving will of God as well as human free will. He was defended by Walafrid Strabo, Prudentius of Troyes, Servatus Lupus, Ratramnus, and others, but was condemned by the Synods of Mainz (848) and Quiercy (849), deprived of priesthood, flogged, and imprisoned for life at the monastery of Hautvilliers. There he continued his controversy with Hincmar, accusing him of Sabellianism, and expounded his own views in two confessions. He died unreconciled in a disturbed mental state as a result of his privations. He was also a lyric poet, and is accepted as the author of the *The Eclogue of Theodolus,* a colloquy between Truth and Falsehood, with Reason as umpire. Falsehood cites incidents from pagan mythology, giving a quatrain to each. Truth caps every incident with a contrast from Scripture. The work survived as a school book into the Renaissance period.

See K. Vielhaber, *Gottschalk der Sachse* (1956).

J.G.G. NORMAN

GOUDIMEL, CLAUDE (c.1510-1572). French composer. Little is known of his early life, and much misinformation is found in older accounts. He was apparently converted to the Huguenot faith before 1565, but had already shown great interest in the metrical psalms and their tunes being compiled under Calvin's influence at Geneva. Starting in 1551, he published at intervals eight books, each containing a selection of eight of these Genevan psalms in extended motet settings. As part-singing was not permitted in church by Calvin, these ambitious works must be regarded as Huguenot madrigals. Many other composers, Catholic as well as Protestant, published arrangements of these evidently popular psalms, until Catholic reaction forbade them. Goudimel published also the 150 Genevan psalms in 1564 in very simple four-part settings, and again in more elaborate versions in 1568. These had a wide circulation in the Netherlands and Germany as well as in France and Switzerland. Goudimel was also the composer of Masses and secular chansons in his earlier days. He perished in the St. Bartholomew's Day massacre in Lyons.

J.B. MAC MILLAN

GOUNOD, CHARLES FRANCOIS (1818-1893). French composer. While he is remembered today almost exclusively for his popular opera, *Faust,* Gounod maintained throughout his career a great interest in sacred music. He spent several years in England, where he attained much popularity as a choral conductor and composer. He wrote a number of anthems and sacred songs, some of which still enjoy a measure of popularity, and he had a very strong influence, like Spohr and Mendelssohn before him, on the lesser Victorian church composers, much to the distaste of later critics. His best-known oratorio, *Redemption,* has passed out of the repertory, although parts of his *St. Cecilia Mass* are still heard. While he often exemplified what is least favored today in music of the Victorian era, he was a highly gifted composer and had a fine gift of melody. He was at his best in the operatic and orchestral mediums.

J.B. MAC MILLAN

GOVETT, ROBERT (1813-1901). Theological writer (pseudonym "Mathetees"). Reared in Staines, Middlesex, he entered Worcester College, Oxford, in 1830, and after graduation was awarded a life fellowship in 1835. Ordained (1836-37), he became curate at St. Stephen's Church, Norwich, where his preaching attracted great crowds until in 1844 he confessed that he had forced his conscience on the matter of infant baptism and forthwith resigned his curacy and his fellowship. Most of the congregation left the Church of England and made Govett their pastor; services were held in Victoria Hall, Norwich, and by 1848 he had baptized 300-400 former Anglicans. Surrey Chapel, Norwich, was opened in 1854, and Govett ministered there to the end of the century. This nondenominational church still flourishes.

Govett's writings are extensive, of varying quality, and often marked by a high level of scholarship, a superbly logical approach, extraordinary originality, and complete faithfulness to biblical revelation. Much concerned with eschatology (*Apocalypse*, 1864, and other works), he held that much of the Book of Revelation is to be understood literally.

R.E.D. CLARK

GRABE, JOHANNES ERNST (1666-1711). Anglican scholar. Born and educated at Königsberg, where he was appointed privatdocent in 1685, he questioned the validity of Lutheran orders and contemplated becoming a Roman Catholic. On P.J. Spener's recommendation, he went to England in 1697, seeking a church possessing apostolic succession. Ordained into the Anglican priesthood, he had close links with the Nonjurors.* On receiving a pension from William III, he gave himself to biblical and patristic research. He published *Spicelegium S.S. Patrum ut et haereticum seculi post Christum natum I, II et III* (1698-99), editions of Justin Martyr's *First Apology* (1700) and of Irenaeus (1702), and a transcript of Codex Alexandrinus (Septuagint) with numerous emendations (1707-9).

J.G.G. NORMAN

GRACE. The favor shown by the Sovereign Creator to human sinners. In the OT, two words are basically used to convey the idea of God's mercy and free favor: *chesed* (e.g., Lam. 3:22) and, more importantly, *chen* (Gen. 33:8,10,15; Jer. 31:2). Grace is revealed in God's choice of and care for Israel. In the NT, the two equivalent Greek words are *eleos* (e.g., Rom. 9:15-18) and *charis* (e.g., 1 Cor 1:4). The divine love and initiative reached its greatest manifestation in the person and work of Jesus Christ (2 Cor. 8:9; Phil. 2:6ff.). Sinners, having transgressed God's law, cannot expect anything from God. In that He freely moves toward them and offers to them reconciliation, fellowship, and salvation, God is said to be the "God of grace" and Christianity to be "a religion of grace."

Since grace is so fundamental and many-sided a concept, it is to be expected that Christians will have had partial or unbalanced understanding of it. In church history there have been important controversies over the nature of grace. Of these we may note those between Augustine and the Pelagians and between Roman Catholicism and Protestantism. For Augustine, grace was absolutely necessary in order to begin, continue, and complete the salvation of an individual sinner. God must give the desire, the faith, and the perseverance. The Pelagians understood grace not as a supernatural power at work in the human soul, but as the normal functioning of the human faculties. So a man could freely accept salvation and later, if he wished, renounce his salvation.

Within Roman Catholicism, grace has usually been portrayed as a power conveyed through the priestly ministry and sacraments by which justification and sanctification are achieved. So personal faith and works go hand in hand. For Protestants, the connection between grace and faith has been central. As the sinner believes in God through Christ, the grace of God is active in that his sins are forgiven, a declaration of justification is made on his behalf, and he is reconciled with God. Works follow as the believer, accepting God's help through the means of grace (prayer, worship), continues to trust in his Lord.

In dogmatics, various adjectives are sometimes added to "grace" in order to describe aspects of it: e.g., *Actual Grace* is used by Roman Catholics to describe any supernatural help given in order to avoid sin or do a good work. *Habitual (Sanctifying) Grace* is used by Roman Catholics to describe the divine power which assists men to perform righteous acts; for Protestants it describes the sanctifying work of the Holy Spirit in the heart of the justified believer. *Irresistible Grace* is used by Calvinistic Protestants to describe the sovereign activity of God in regeneration and conversion. *Prevenient Grace* is used by Roman Catholics of God's work in the heart of the infant who is baptized, and by Protestants of God's secret, preparatory work in the heart of a sinner before he actually believes. *Sufficient Grace* is used by Roman Catholics to describe God's offer of help made to all Christians; when used it becomes *Efficacious Grace.*

BIBLIOGRAPHY: N.P. Williams, *The Grace of God* (1930); J. Moffatt, *Grace in the New Testament* (1931); D. Hardman, *The Christian Doctrine of Grace* (1947); H.D. Gray, *The Christian Doctrine of Grace* (1949); C.R. Smith, *The Bible Doctrine of Grace* (1956); J. Daujat, *The Theology of Grace* (1959); P.S. Watson, *The Concept of Grace* (1960).

PETER TOON

GRACE AT MEALS. The custom of giving thanks at meals. Not an exclusively Christian practice, it has its roots in the religious instincts of humanity, witnessing to the solemnity which attaches to every meal as an act of maintaining life. The Jews consecrated their meals with the "blessing," the Mishna describing the procedure (cf. Deut. 8:10; 1 Sam. 9:13). Our Lord followed the custom (e.g., Matt. 14:19; 15:36), as did the early Christians (e.g., 1 Cor. 10:30; 1 Tim. 4:3-5). In the second century Clement of Alexandria (*Paedagogus* 2.4) and Tertullian (*Apologeticum* 39) witness to the practice, and the church generally has kept the custom. W.T. Brooke (Julian's *Dictionary of Hymnology*) gives examples of metrical graces.

J.G.G. NORMAN

GRAFFITI. Inscriptions and drawings that often give graphic expression to thought or feeling, made upon whatever convenient surface is available. Some of the most ancient examples appear to be those of mercenaries or forced labor serving in foreign lands; subsequent usage reflected all groups and occupations of society. Ossuary graffiti from Jerusalem, claimed as mid-first century, pray to/for some Jesus, among others, for help, and petition that the bones contained may rise from the dead. Also noteworthy is the crudely drawn anti-Christian sketch with label from the plaster of a wall of the Paedagogium on the Palatine hill of the third century, deriding one Alexamenos about the worship of a crucified figure with an ass's head—the oldest representation of the centrality of *the* crucifixion. CLYDE CURRY SMITH

GRAFTON, RICHARD (d.1572). Chronicler and printer. A prosperous London merchant and a member of the Grocers' Company, Grafton, a convinced Protestant, arranged in association with Edward Whitchurch for the printing of "Matthew's Bible," produced under royal license at Antwerp in 1537. The following year Grafton was in Paris supervising the printing of the "Great Bible"—Coverdale's revision of Matthew's Bible without the extensive and overtechnical critical aids. Halted in his efforts by the Inquisition, he escaped to England where a year later the Great Bible was published. In its revised edition of 1540 this remained the official Bible of the English Church until the "Bishops' Bible" of 1568. In disfavor after the fall of Thomas Cromwell, Grafton prospered under Edward VI, receiving appointment as official printer of statutes and acts of Parliament as well as of Bibles and Service Books. Briefly imprisoned by Queen Mary, he retired from business but was subsequently a member of Parliament for London, and later for Coventry. In the 1560s he produced a rather pedestrian and controversial chronicle. Grafton's love for the Reformed faith, his zeal for the printing of the Bible in English, and his heroism in Paris in 1538-39 are undoubted. His rather sharp and occasionally dishonest business methods are a reproach.

IAN SELLERS

GRAHAM, WILLIAM FRANKLIN ("Billy") (1918-). Evangelist. Born in North Carolina, he was educated at Bob Jones University, Florida Bible Institute, and Wheaton College, Illinois. After ordination (Southern Baptist) and a brief pastorate, he became in 1943 the first evangelist of the newly founded Youth for Christ.* In 1949, while president of Northwestern College, Minneapolis, he acquired national fame through his Los Angeles Crusade. He founded the Billy Graham Evangelistic Association with headquarters at Minneapolis. In 1954 he grew world famous through his first Greater London Crusade, extending for three months and creating a more profound impression on the United Kingdom than any mission since those of Moody and Sankey seventy years before. After 1954 he crusaded, with songleader Cliff Barrows, singer George Beverly Shea, and a team of associates, in most parts of the world.

Graham crusades have always been cooperative evangelism between his team and staff and the churches. Planned on a big scale, they involve great numbers of laity and ministers in an invigorating enterprise over a long period before and after. Despite the huge attendance figures, the emphasis has been on the decision of individuals, carefully counseled and followed up by specially trained local people in the context of the church. The number of lasting converts probably runs into millions.

Graham has developed also subsidiary evangelistic ministries. The weekly "Hour of Decision" radio broadcast began in 1950. His films have included *Souls in Conflict, Two A Penny,* and *His Land.* He has a weekly syndicated column, *My Answer;* was a co-founder of the fortnightly *Christianity Today;* and in 1960 founded the monthly magazine *Decision.* Television has greatly extended his evangelism, especially in the USA. His books, *Peace with God* (1952) and *World Aflame* (1965), have been worldwide best sellers.

In 1966 he inspired the World Congress on Evangelism* in Berlin, which in turn inspired similar congresses in various parts of the world. Each crusade includes a school of evangelism. He has always had a strong appeal to youth. He has also been the close friend and counselor of several heads of state, including U.S. presidents.

His personal character, his power to preach a fully biblical Christ-centered message in terms that simultaneously reach the most varied hearers, and his skill at using modern techniques of mass communication have made him one of the twentieth century's best-known religious figures.

JOHN C. POLLOCK

GRANT, GEORGE MONRO (1835-1902). Canadian minister and educator. Born in Nova Scotia and educated at Pictou Academy, he studied afterward at West River Seminary and Glasgow University. Ordained a minister in the Church of Scotland in 1860, he became pastor of St. Matthew's Church, Halifax, in 1863. In 1877 he was appointed principal of Queen's University, Kingston, Ontario, and held that office until his death. Emphasizing very strongly the practical social and political aspects of Christianity, he became a national figure because of his involvement in national and imperial interests, and because of his book *Ocean to Ocean* (1873). In 1899 he was moderator of his denomination's general assembly, and in 1901 became president of the Royal Society of Canada. Described as "the most influential churchman in Canada of his day," Grant was known as author, educator, politician, and minister, and was regarded almost as a national institution. ROBERT WILSON

GRATIAN (twelfth century). Called the father of canon law, he composed the *Concordia discor-*

dantium canonum, better known as the *Decretum.** Very little is known about him except that he was born in Chiusi, Italy, probably became a Camaldolese monk, and lived in the monastery of SS. Felix and Nabor.

GRATRY, AUGUSTE JOSEPH ALPHONSE (1805-1872). French Roman Catholic scholar. Born in Lille, he was irreligious until in 1822 he awoke to the folly of worldly ambitions. He studied theology at Strasbourg, was ordained (1834), was a college director and chaplain, then in 1863 was appointed professor of moral theology in the Sorbonne. He was elected to the French Academy four years later. He first opposed papal infallibility, but submitted to the decrees, being much concerned for a renewal in French church life which led to his work in restoring the Oratory.* His works are apologetic, even to including a proof for the existence of God with the aid of mathematics: *De la connaissance de Dieu* (1855), *Les Sources* (1862), *Les Sophistes et la critique* (1864), and *La Morale et la loi de l'histoire* (1868).

C.G. THORNE, JR.

GRAVEDIGGERS, see FOSSORS

GRAY, GEORGE BUCHANAN 1865-1922). English OT scholar. Born at Blandford, Dorset, son of a Congregational minister, he was educated there and at Exeter, was a schoolmaster, then matriculated at London University (1882). He went on to Mansfield College, Oxford, and completed his studies at Marburg. He became tutor at Mansfield (1891), was ordained a Congregational minister (1893), and from 1900 was professor of Hebrew and OT exegesis at his college. His books include commentaries on Numbers, Isaiah 1—27, and Job (with S.R. Driver). In *Sacrifice in the Old Testament* (published posthumously in 1925), he made a notable contribution to OT theology. An independent and original thinker, though accepting the views of the Wellhausen* school of biblical criticism, he had a positive, constructive approach, and his preaching and teaching were permeated by the devotional spirit of the OT.

J.G.G. NORMAN

GRAY, JAMES MARTIN (1851-1935). Author and Bible teacher. Born in New York City, he was educated in New England schools, served as rector of First Reformed Episcopal Church, Boston (1879-94), and lectured at Reformed Episcopal Seminary, Philadelphia, and A.J. Gordon's Boston Missionary Training School (Gordon College). He lectured also at Moody Bible Institute summer sessions from 1893 until he became permanently associated with it as dean in 1904. He was president for nine years from 1925. Guiding the institute through a period of growth, he developed and popularized the synthetic approach to Bible study which greatly influenced Bible institute and Bible college curricula. He wrote many books, including *How to Master the English Bible* (1909) *Synthetic Bible Study* (1920), and *Prophecy and the Lord's Return* (1917), was one of the editors of the Scofield Reference Bible, and produced a number of popular hymns.

HOWARD A. WHALEY

GRAY, ROBERT (1809-1872). First Anglican bishop of Cape Town. Born near Sunderland and educated at Oxford, he was consecrated in 1847. He found South African Anglicanism weak and disorganized. Under his leadership it developed into the Church of the Province of South Africa—an independent, disestablished province of the Anglican Communion with five synodically governed dioceses (1870). This was not achieved without difficulty. His High Churchmanship aroused antipathy. Some local Anglicans resented their loss of independence and opposed his introduction of synodical government. This conflict, and the case of J.W. Colenso,* led to costly and confusing litigation which overshadows his more positive achievements. Despite delicate health Gray traveled widely in his diocese and overseas, organizing the church, promoting missions, enlisting recruits, and raising money. He favored the appointment of missionary bishops to unevangelized areas and inspired the formation of the Universities' Mission to Central Africa. The tragic failure of Bishop Mackenzie's Zambesi mission was a great disappointment.

D.G.L. CRAGG

GREAT AWAKENING, THE. A series of revivals in the American colonies between 1725 and 1760. The earliest stirrings occurred among the Dutch Reformed in the Raritan Valley of New Jersey, through the fervent preaching of T.J. Frelinghuysen.* This early revival reached a peak in 1726 when, encouraged by Frelinghuysen, Gilbert Tennent,* a Presbyterian pastor in New Brunswick, began to preach for "conviction." Within a short time several Scotch-Irish Presbyterian churches around New Brunswick were experiencing conversions and fresh excitement. Simultaneously in New England in 1734-35 a recovery of heartfelt religion appeared in Northampton, Massachusetts, through the preaching of the able theologian and preacher Jonathan Edwards.*

The one man more than any other, however, who linked these regional awakenings into a "Great Awakening" was the English evangelist George Whitefield.* By traveling throughout the colonies and calling men to repentance and faith in Christ, Whitefield after 1740 helped to plant evangelical Christianity on American shores and to prepare the colonies religiously for the trials of the revolutionary age.

Through the "Reading houses" of Samuel Morris, and the preaching of William Robinson and Samuel Davies,* Presbyterians experienced revival in the South. Methodism with the preaching of Devereux Jarratt and the Baptist movement through the work of Daniel Marshall and Shubal Stearns grew rapidly in the era of the Great Awakening.

The revivalists soon met resistance. The established clergy, led by Charles Chauncy* in New England, criticized the revivalists' preaching and practices. The encouraging of lay preaching or "exhorting," the criticisms of revivalists who charged the established clergymen with "spiritual darkness," the uninhibited "enthusiasm," and the divisions within churches after the revivalists had swept through them—these taken together erected a sizable barrier to the spread of the revival.

In Edwards, however, the awakening had a vigorous defender. In his work *Some Thoughts Concerning the Present Revival* and in his later treatise on *Religious Affections,* Edwards discriminates between revivals' beneficial and detrimental effects and argues that nothing deserves the name religion that falls short of a remarkable change of disposition, created in the heart by the Holy Spirit, and showing itself in unselfish love for the things of God and in a burning desire for Christian conduct in other men.

In New England those who followed Edwards and other defenders of the Great Awakening were known as New Lights, and became proponents of the New England Theology*; those who opposed it were known as Old Lights. The Presbyterians also split into New Side and Old Side groups between 1741 and 1758, and the Baptists into Separate and Regular Baptists.

In spite of its faults, the Great Awakening made its impact upon the American colonies. Dissenting groups growing from the revival in Virginia helped to overthrow the established Anglican Church in that colony. Early antislavery sentiment was fostered, and increased missionary activity among the Indians arose from the movement, as represented by the work of David Brainerd,* Eleazar Wheelock,* and Samuel Kirkland.

The movement made also a great contribution to education. Princeton University, the University of Pennsylvania, Rutgers, Brown, and Dartmouth were some of the more significant schools created as a result of the awakening. Of equal importance was the mood of tolerance that cut across denominational lines. This attitude not only contributed to a national spirit of religious tolerance that helped to make the First Amendment to the U.S. Constitution a workable arrangement; it also provided for an evangelical consensus that is traceable to the present.

See M. Gewehr, *Great Awakening in Virginia, 1740-1760* (1936); and E.S. Gaustad, *Great Awakening in New England* (1957).

BRUCE L. SHELLEY

GREAT BIBLE, see BIBLE, ENGLISH VERSIONS

GREAT SCHISM, THE. This may designate either one of two unrelated events which disunited Christendom. The schism of 1054 ("Eastern Schism") formally ruptured communion between the churches under the pope at Rome, Leo IX, and those under the patriarch at Constantinople, Michael Cerularius, until then considered second in the hierarchy. Already two very different churches, both culturally and theologically, they have since developed their separate ways: a Byzantine-Greek church became Eastern, Greek, and Russian Orthodoxy, while a Roman-Latin church became Roman Catholicism.

The schism of 1378-1417 ("Western Schism"), following the pope's Avignon residency (1309-77), divided allegiances in disintegrating Western Christendom between first two, then three, simultaneous popes, each excommunicating the other. A line with seat in Rome began with Urban VI, backed by the German Empire, England, Hungary, Scandinavia, and most of Italy. An Avignon line began with Clement VII, backed by France, Naples, Savoy, Scotland, Spain, and Sicily. Attempts (1409) to end the schism brought a third Pisan line. Martin V's election (1417) ended the schism. The crisis partially arose out of tensions between the authority of the papacy and feudal monarchies, and of the papacy and cardinals.

C.T. MC INTIRE

GREBEL, CONRAD (1498?-1526). Leader of the Swiss Brethren movement (commonly called Anabaptism*). He was educated at Basle, Vienna, and Paris, where he encountered humanism. Returning to his home in Zürich he made contact with Zwingli and other humanists, and studied Greek with them. About 1522 he was converted to biblical Christianity and began to work for reform in Switzerland. He became dissatisfied, however, with the incomplete reformation advocated by Zwingli. With friends he diligently studied the Bible, searching for the true doctrine of the church. On 21 January 1525 the Anabaptist movement was born when Grebel baptized Georg Blaurock* and then Blaurock baptized others present, making thereby a gathered church. This action provoked the wrath of the city council and led to persecution of the Brethren. Grebel himself, weakened by imprisonment, died at Maienfeld.

PETER TOON

GREECE. When the good news of God's love to man began to be spread, Greece was under Roman domination. But Greeks were to be found everywhere; in a sense Achaia in Paul's time was "Greece"—but the Greek world was dominant throughout the Mediterranean countries and beyond.

The colonies of the Jews, on the other hand, were scattered throughout the old country from the very north to the extreme south. Thus when Paul with his companions crossed the sea from Troas and put his foot on the soil of Europe there were already bridges prepared for the transfer of the Good News. From Philippi—where the first convert was won in the person of Lydia—to Thessalonica, Berea, Athens, and Corinth, there were synagogues where not only the Jews but large numbers of devout persons, the proselytes, offered to the apostles a most receptive soil for the good seed.

The old religion of the Greeks was on the decline. The efforts of the Neoplatonists to revive paganism were made in vain. The temples remained magnificent, but the wealth that belonged to them had become private. Christianity gained a victory, though not without a long struggle, against paganism. Such terms as *"ecclesia"* and "liturgy" were not unfamiliar to the popular mind, and this undoubtedly was a contributing factor.

When Paul stood in the midst of the Areopagus (A.D. 51/52) proclaiming "Jesus and the resurrection" in the face of the Epicureans and the Stoics, he got a favorable response at least from a few individuals (Acts 17). It is significant that no serious opposition is reported on the part of the Greeks; the opposition was always raised by the intolerant Jews. This was the case in Thessalonica, Berea, and Corinth. Very shortly after Paul's preaching, churches were organized so that the apostle addressed himself not only to the principal church at Corinth, but to "all the saints throughout Achaia" (2 Cor. 1:1).

According to tradition, Andrew the apostle came to Achaia and suffered martyrdom there. In the second century two Athenian philosophers, Aristides and Athenagoras, became apologists of the Christian faith. Origen, the great Alexandrian teacher, visiting Athens in the middle of the third century, found the church flourishing there.

By the sixth century all opposition to the Christian faith ceased. Only the mountainous tribes of Mane insisted on the old heathen religion. They were converted in the ninth century and then perhaps only nominally.

When Constantine the Great removed the capital of the Roman Empire to Constantinople (330), Greece proper continued in oblivion, but the Greeks were so dominant that the Byzantine Empire became eventually "greater" Greece. Christianity became the religion of the state, the Greek language was the language of the empire, and Greek philosophy and dialectic came to contribute to the shaping of Christian doctrine and teaching. This had such extended implications that the question was raised whether Christianity converted Hellenism or Hellenism absorbed the Christian faith, covering under the Christian mantle much of the old heathen practice. It was out of this situation that reformation movements appeared during the eighth and ninth centuries, known as "iconoclastic," which after a long struggle culminated in the prevalence of icon worship and the subsequent shaping of Orthodox Christianity.

The Crusades of the thirteenth and fourteenth centuries tried hard to convert the Eastern Church to Rome, but in vain. The only remnant of the Crusades in Greece was a small Roman Catholic element and a bitter animosity toward the Western invaders.

The great Reformation of the church in the West in the sixteenth century found Greece and the Greek Church struggling under the Muslims who had swept away the Byzantine state and captured Constantinople in 1453. With the Reformation, the Greek Orthodox Church remained untouched—though not entirely so. The Greek patriarch of Constantinople, Cyril Lucar,* embraced the doctrines of Calvin and attempted to introduce the Reformation in his own Greek Church. He was strongly opposed both by the majority of his clergy and by the Jesuits. He suffered martyrdom by the Turks, and eventually was formally anathematized by the Synods of Constantinople and Jerusalem.

In the nineteenth century part of Greece was liberated from the Ottoman Empire (1827). About the same time, Protestant missionaries from the West and from America came to Greece for relief, educational, and evangelistic work. The Protestant minority that exists today in the country has been the direct and indirect fruit of the activities of the missionaries. But the Greek Orthodox Church, claiming over 95 percent of the people and following the Byzantine pattern, is the state church. One of the interesting facts is that the first two monarchs of modern Greece who undertook officially to protect the Orthodox Church were not themselves Orthodox. The first, the Bavarian Otto (1832-186?), was Roman Catholic; the second, George I from Denmark (1863-1912), was Protestant.

See also EASTERN ORTHODOX CHURCHES.

BIBLIOGRAPHY: Much of the literature is in Greek only, but see W. Smith, *A History of Greece* (1857); G. Hadjiantoniou, *The Protestant Patriarch* (1961); C.M. Woodhouse, *The Story of Modern Greece* (1968). MICHAEL KYRIAKAKIS

GREEK, HELLENISTIC. The spoken and written language of Hellenistic times, a period that begins with the conquests of Alexander the Great, covers NT times, and reaches the time of Constantine the Great. Within this period, the Hebrew text of the OT was translated into the Greek of the Septuagint, and the books of the NT were written. Hellenistic Greek became the language used by Greeks and non-Greeks, including Jews of the diaspora of pre-Christian and NT times. It was also the common language in Palestine at the beginning of the Christian era. The seven deacons mentioned in Acts (6:5) belonged to the Hellenistic party. Saul of Tarsus was a Hellenist, and it is most probable that the Lord had been familiar with some words and sentences of Hellenistic Greek.

In ancient times the Greek language was not a single, uniform tongue. Each of the divided Greek city-states developed its own dialect according to its progress and achievements. There were numerous dialects; four of them were prominent: Attic, Ionian, Doric, and Aeolic. Little by little the variety of dialects gave way to a "common" dialect. The great classical writers on one hand and the pan-Hellenic athletic games and festivals (Olympic, Delphi, Corinth) on the other contributed to this development.

This common Greek was based mainly on the Attic dialect. Athens being the great center of letters and arts, it became natural that the language of Athens became the universal language of the Greeks. In time the Attic dialect was no longer the pure language of the past. Elements from the other Greek dialects were mixed to form the common Attic.

The common Attic Greek was the language adopted by the Macedonian kings. It became the official language of the court and subsequently was brought with the conquests of Alexander to the conquered lands and peoples of the East. Thus it became the language of the Egyptians, Syrians, and Jews as well as of the Greeks who moved with the military forces and as merchants, educators, etc. From this the language termed Hellenistic or *Koine* emerged. In the new cosmopolitan centers such as Alexandria, Pergamos, and Antioch the new international language was molded.

In the course of time, beside the changes that are inevitable in every living language, expressions and words were added to Hellenistic Greek, not only from the variety of Greek dialects, but also from the languages and dialects of the "barbarians." In addition, during the Roman domination, Latin elements were introduced. It is a universal law that those who learn and use a language not only acquire but also give elements of their own modes of expression: idioms, local words. The Hellenistic Greek of the Septuagint and of the NT books is a demonstration of this fact. The variations of the language in the sacred books are easily explained when we consider the ethnic and cultural backgrounds of translators and authors.

The OT translation of the Septuagint was made under one of Alexander's successors, King Ptolemae (third century B.C.). Philo, Josephus, and early church fathers support the idea that the translation was made in Greek understood by the king and by Greeks in general. Becoming international, Hellenistic Greek was gradually simplified in grammar and syntax while the distinction between long and short vowels tended to disappear. The refined and highly cultured philologists despised as "barbarian" the Septuagint version of both OT and NT books.

The Septuagint, with apparent Semitic elements, might be said to belong to the Alexandrian version of Hellenistic Greek, while the NT language is the Palestinian version of the same. The Septuagint translators in their attempt to render the text as accurately as possible could not avoid hebraisms, while the NT writers, as original authors, were freer from such elements.

MICHAEL KYRIAKAKIS

GREEK EVANGELICAL CHURCH. This body has been from the first a national movement. Its first leader, Michael Kalopothakes, a native of Areopolis, near Sparta, had come under the influence of Protestant missionaries, having attended a missionary school run by two missionaries of the Southern Presbyterian Church in the USA. As a student in Athens he attended the meetings of Jonas King,* but the missionaries had no intention of establishing a Protestant church in Greece—and only strong opposition compelled Kalopothakes and other Greeks to organize an Evangelical church. After graduating in medicine he studied at Union Theological Seminary, New York, and in 1858 organized the First Church in Athens, opened the first Sunday school, and in 1871 erected the first Evangelical building at the foot of the Acropolis. He became the first agent of the British and Foreign Bible Society, and editor of the weekly paper *Astir Tis Anatolis* which is still being published by the church as a monthly magazine. Within a few years several churches were organized in other parts of the country.

Parallel to this movement was a similar Evangelical effort among the Greeks in Asia Minor, Turkey, where a number of churches were organized. When in 1922-23, as a result of war, the Greeks had to evacuate Asia Minor, the Evangelicals in Turkey went to Greece as refugees, joined the local church, and also formed new congregations in different areas. Today under the general synod there are some thirty congregations with a membership of 12,000. There are seventeen ordained pastors and a few lay workers. There is an Evangelical orphanage in Katerini, Macedonia; a Bible school; two summer camps for the church's children (in Attica and Macedonia); and a family camp in Thrace.

MICHAEL KYRIAKAKIS

GREEK ORTHODOX CHURCH, see EASTERN ORTHODOX CHURCHES

GREENHILL, WILLIAM (1591-1671). English Nonconformist* minister. Matriculating at Oxford when only thirteen, he graduated four years later, and for eighteen years from 1613 held the living of New Shoreham in Sussex. John Howe and others were later to speak highly of his pastoral gifts and dedication. He became afternoon preacher to the congregation ministered to in the morning by Jeremy Burroughes*; they were known as the "Morning Star" and the "Evening Star" of Stepney. As a member of the Westminster Assembly he opposed the Presbyterian party, and in 1644 became the first pastor of a Congregational church in Stepney. His first volume of a commentary on Ezekiel was dedicated to a daughter of Charles I (after the king's execution Greenhill became chaplain to three of his children). He found favor with Cromwell, however, and was made one of the "Triers" for the approbation of public preachers. Ejected in 1660 from his vicarage of St. Dunstan's-in-the-East, he retained the congregational pastorate till his death. The four other volumes of his Ezekiel had been published by 1662, a work described as being full of erudition and practical wisdom.

J.D. DOUGLAS

GREENWOOD, JOHN (d.1593). English separatist. Educated at Corpus Christi College, Cambridge, he became a zealous Puritan and a chaplain in the Essex home of Lord Robert Rich. Moving to London, he associated himself with those who were taking Puritanism to its logical conclusion, and becoming separatists and critics of the idea of the state church. He was arrested in 1586 for holding an illegal conventicle and was examined before Archbishop Whitgift. Other men imprisoned with him were Henry Barrow* and John Penry; with them he composed tracts defending separatism and some of these were printed in Holland. Released in 1592, he formed a church with Francis Johnson; this church later migrated to Holland and was known as the "ancient church." But Greenwood was arrested in 1592 and charged with Barrow for writing and publishing seditious books. At Tyburn in April 1593 they were hanged.

PETER TOON

GRÉGOIRE, HENRI (1750-1831). Bishop of Blois. Born in Lorraine, he first won a reputation as a scholar, but became especially known for his role as a Roman Catholic bishop of the constitutional church during the French Revolution. He led the marathon session of the Third Estate during the attack on the Bastille (1789), and was elected by Nancy to the states-general. Grégoire was the first priest to sign the loyalty oath of the civil constitution of the clergy demanded by the constituent assembly (1790). As bishop of Blois (1790-1801) he ruled the Loire-et-Cher diocese and was elected president of the national assembly (1792). At the height of the Terror of 1793 Grégoire refused to abjure his faith or doff his robes. His Gallican opposition to Napoleon's conciliation with the Vatican stirred Ultramontane reaction and caused his resignation as bishop in 1801. ROBERT P. EVANS

GREGORAS, NICEPHORUS, see NICEPHORUS GREGORAS

GREGORIAN CALENDAR, see CALENDAR

GREGORIAN SACRAMENTARY. Early Roman liturgy. The liturgy itself is probably based on a liturgy from the papacy of Gregory the Great (590-604). Pope Hadrian I sent the Sacramentary to Emperor Charlemagne probably about 790. It was widely used in the Frankish Empire during and after the Carolingian period. A ninth-century manuscript of the Sacramentary is extant as well as several later editions and translations.

GREGORY I (the Great) (540-604). Pope from 590. Born in Rome, he was brought up in a household that encouraged piety and enabled him to receive a thorough education in grammar and rhetoric. His outstanding performance as a student of law led to his appointment as prefect of the city about 570. Later he decided to renounce worldly things and provided for the founding of seven monasteries, including one in his family home which he dedicated to St. Andrew and entered about 575. The experience in business affairs gained by his service as prefect and his predilection for the contemplative life were valuable in shaping the policies of his pontificate. Gregory was brought back into public life by Benedict I, who ordained him a Roman deacon. He was active as a papal representative to Constantinople and was successful in some instances, but failed to obtain aid for Rome against the Lombards. He reluctantly accepted his election as pope and was consecrated in 590.

His term in office had important and far-reaching consequences for the future of the papacy. In an effort to secure Rome against invasion by the Lombards he entered into a factional dispute with the church at Ravenna and the imperial exarch. Unable to reach an agreement which would unify Italian peacemaking efforts, Gregory sent his own troops against Lombard forces and made a truce with the Lombard duke, Aruilf of Spoleto, in 592. When the Lombard king entered Rome in 594, Gregory moved to save Rome by paying a large ransom and committing himself to an annual tribute. Gregory continued to work for peace throughout Italy, but this effort was not fruitful until 598. Revenues from the papal patrimony (lands in Italy, Sicily, Corsica, Sardinia, Gaul, North Africa, and Illyricum) were administered by Gregory to care for poor families, ransom captives, and pay for the campaigns against and peace settlements with the Lombards. Since it was Gregory and not the emperor who undertook these duties usually assumed by the civil government, this was an important step in the formation of the Papal States,* thus making the pope a temporal ruler.

In ecclesiastical affairs, Gregory strengthened the position of the Roman pontificate through his handling of the church in both East and West. While recognizing the jurisdictional rights which the other churches had over their own territories, he maintained that the See of Peter had been entrusted with the care of the entire church and therefore had universal jurisdiction. He reversed a decision against two priests made by the patriarch of Constantinople (John IV the Faster) and strongly objected to the patriarch's use of the title "ecumenical (universal) bishop." Gregory also asserted his position in the Western Church by seeing that the bishops were elected according to correct canonical procedure and by working to heal the Donatist* schism. He was not always successful in his attempts to enforce Roman primacy, especially in Aquilia where a previous schism remained unhealed until after his death. Gregory was able to link the independent Frankish Church to Rome by restoring the vicariate. He rejoiced over the conversion of the Arian Visigoths in 589 and was able to place the Spanish Church in the care of his friend Bishop Leander of Seville. Missionary work began in England under Augustine of Canterbury in 597 and succeeded in converting the Anglo-Saxons.

Gregory's importance is that of a transmitter of the wisdom of the ancient world to the medieval world. He is considered one of the four great doctors of the Roman Catholic Church in moral theology, not so much for the originality of his thought as for his didactic method. His works include forty *Homilies on the Gospel* (590-91), aimed at preparing his subjects for the Judgment; twenty-two *Homilies on Ezekiel* (593), profound and masterful pieces on many aspects of Christian life, including historically important accounts of Italy and the Lombards; the *Book of Morals,* a commentary on the Book of Job, his longest work and highly valued in the study of ethics during the Middle Ages; *Pastoral Care,* an exposition on the duties and qualities of the bishops of the church; fourteen books of *Letters,* which contain valuable information on his pontificate; and *The Four Books of Dialogues on the Life and Miracles of the Italian Fathers and on the Immortality of Souls* (593-94). The *Dialogues* are especially significant in that they simplified the doctrines expressed in Augustine's *The City of God,* and were thus very influential during the Middle Ages. Gregory was also active in the reform of the liturgy of the Roman Rite.

BIBLIOGRAPHY: F.H. Dudden, *Gregory the Great* (2 vols., 1905); C. Butler, *Western Mysticism* (2nd ed., 1927); P. Batiffol, *St. Gregory the Great* (tr. J. Stoddard, 1929); N. Sharkey, *St. Gregory the Great's Concept of Papal Power* (1956).
ROBERT G. CLOUSE

GREGORY II (c.669-731). Pope from 715. Born to a noble Roman family, he served as the first papal librarian known to us by name, during the pontificate of Sergius I. His first task as pope was to repair the walls of Rome destroyed by the Lombards. Then among other duties he commissioned Boniface to convert the Bavarians, consecrating him a bishop in 722. This positive mission bore fruit as Gregory interested Charles Martel the Frankish leader in the mission. The defense of Rome against Muslim advance and Lombard intrigue, the reception of important pilgrims, and the encouragement of the Bavarian mission, as well as the growing alienation of the papacy from Byzantium—all mark the importance of Gregory's eighth-century pontificate. The rift with the Eastern Church opened when Gregory condemned the policy of Emperor Leo III in two famous letters about 726, and in a council at Rome (727) proclaimed that images should be maintained. The two letters are now considered authentic, apart from errors of translation and interpolation. G. Ostrogorsky concludes that since the letter to Patriarch Germanus is unquestioned, Gregory's attitude of opposition is clear. Since Leo III acted with caution by not promulgating any iconoclastic laws until 730, perhaps Gregory reacted as strongly to Leo's tax policies in Italy. Gregory II was known in the West as Gregory the Younger.

ROBERT G. CLOUSE

GREGORY VII (Hildebrand) (c.1023-1085). Pope from 1073. Born in Saona, Tuscany, he was educated at a school of the Lateran in Rome. When Gregory VI was exiled to Germany after the Synod of Sutri, Hildebrand accompanied him and came in contact with many proponents of church reform. At some time he became a monk, either before or after his stay in Germany. In 1049 Leo IX brought Hildebrand back to Rome, ordained him a subdeacon, and appointed him administrator of the monastery St. Paul-Outside-the-Walls. Increasingly active in the Curia, he was very influential in the pontificates preceding his own and was elected pope by popular acclamation in 1073. Although this method of the election was in violation of the law of 1059, its validity was not questioned until after 1076.

Gregory believed that the foremost function of the papacy was to serve as a governmental institution, and that to fulfill this capacity the law must be given an important role. In order to achieve his goals for the papacy he felt it necessary to purify the higher clergy, and Gregory renewed decrees against simony and clerical marriages. The execution of such a reform program necessitated action against lay investiture, since this practice lessened the legal allegiance which bishops owed the Roman See. In 1075 Gregory entered into a dispute with the German king, Henry IV, over the question of lay investiture which was to last throughout his pontificate. When Gregory threatened Henry with excommunication for violation of the reform decrees in 1075, Henry retaliated by having the diet at Worms depose Gregory in 1076. Gregory then excommunicated Henry and released his subjects from their oaths of allegiance to the king. At Canossa (1077) Gregory received Henry as a submissive penitent, but did not reinstate his royal powers.

A short while later, when a group of German princes elected Rudolph of Swabia as king, Gregory did not oppose the move. Civil war ensued in Germany, and Gregory excommunicated Henry for the second time. On this occasion, however, popular opinion was against the pope. At the Synod of Brixen in 1080, Henry encouraged the election of a counter pope, Clement III, and formally deposed Gregory in 1084. In that year a Norman prince, Robert Guiscard, carried Gregory into exile at Salerno.

In addition to the conflict with Henry, Gregory was concerned with other matters. He felt he was the vicar of Peter and in this position thus had the responsibility of governing the church, whose officers included both bishops and kings. He attempted to maintain this position of governor in his relations with the temporal powers of Christendom, such as Hungary, Russia, and England. He wished to lead a crusade to the Holy Land which would result in the unification of Western and Eastern Christians, but was unable to accomplish this goal. He also introduced several liturgical reforms. His program of papal reform as set down in his *Register* consists of twenty-seven short sentences known as the *Dictatus papae* which contained both traditional and innovative applications of the doctrine of papal authority. Gregory's important contribution to the development of the papacy was in the influence he exerted on canon law which shaped both ecclesiastical and political policy for many years.

BIBLIOGRAPHY: A.J. MacDonald, *Hildebrand: A Life of Gregory VII* (1932); G. Tellenbach, *Church, State and Christian Society at the Time of the Investiture Contest* (1940); W. Ullmann, *Growth of Papal Government in the Middle Ages* (2nd ed., 1962).

ROBERT G. CLOUSE

GREGORY IX (c.1170?-1241). Pope from 1227. Born at Anagni of the noble house of Segni, he studied at Paris and Bologna and in 1198 was made cardinal deacon by his uncle, Innocent III, and cardinal bishop of Ostia in 1206. Made papal legate for a series of diplomatic missions to Germany, he was commissioned in 1217 to preach a crusade in northern and later central Italy, with a vow of assistance from Frederick II. As pope (1227) there was constant difficulty between him and the emperor, with two excommunications because of not carrying through the crusade (1227) and invading Lombardy and usurping the rights of the church in Sicily (1239). The election of an anti-king was plotted, and a general council summoned to Rome in 1241 which Frederick prevented from convening.

Heresy, too, preoccupied Gregory because of Albigensian* activity in France, Italy, and Spain. In a Paris treaty (1229) Raymond VII of Toulouse pledged assistance in suppressing Waldenses* and Cathari,* and the punishments included death. As the heresy spread to Italy, with a Cathar bishop in Florence and Rome, then in Spain, an inquisition was established. Gregory made special use of the Dominicans for this work, having canonized St. Dominic (1234), and supported the Camaldolese, Cistercians, and the followers of Joachim of Fiore. A close friend of Francis of Assisi, whom he canonized (1228), Gregory protected the order and fostered the Third Order and the Poor Clares. He sent both Franciscans and Dominicans as missionaries from Finland to Rumania. He labored at length, but unsuccessfully, to unite the Greek and Latin churches.

C.G. THORNE, JR.

GREGORY X (1210-1276). Pope from 1271. Born Teobaldo Visconti of Piacenza, he succeeded Clement IV (d.1268) after a vacancy of three years which ended when Bonaventura, Franciscan minister general, forced the cardinals into action. Canon of Lyons, then archdeacon of Liège, the future pope studied in Paris, then went to England in 1270 before setting out for the Holy Land. Unlike his predecessors, he concentrated less on secular concern than on spiritual revival and reunion. Interested in the affairs of the Latin East, the Kingdom of Jerusalem, he also established monarchical authority in Germany by inviting electors to designate a king of the Romans, persuading Alfonso of Castile to resign his claims. In 1273 at Lausanne he confirmed their election of Rudolph I of Hapsburg. He convoked the Second Council of Lyons to resume talks with the Greek Church, asking Michael VIII Palaeologus to send his ambassadors; agreement was reached, but the reunion was short-lived. A new crusade was discussed and financial preparations made, but nothing happened. Knowing the difficulty of his own election, Gregory had the council establish rules for papal election, for which he constitutionally created the conclave with *Ubi periculum*, 1274.

C.G. THORNE, JR.

GREGORY XI (1329-1378). Last of the Avignon popes. Born Pierre Roger de Beaufort, he was created a cardinal in 1348 by his uncle, Clement VI, and studied law at Perugia. Faced with arbitration between the houses of Anjou and Aragon concerning territorial rights and papal homage, he was compelled to wage a painful war against Florence and the Visconti which unleashed a general revolt in the Papal States (1375); Florence ended it by negotiation in 1377. He was concerned with reform in the religious orders, especially the Dominicans and Hospitallers, and with heresy. The Inquisition was reactivated, principally against the Waldensians, and certain of Wycliffe's* theses were condemned. As with Urban V before him, he wanted to take the papacy back to Rome, where he arrived in January 1377. Unable to settle the disturbances, he considered returning to Avignon, but death intervened—and the Great Schism* was to follow.

C.G. THORNE, JR.

GREGORY XII (c.1326-1417). Pope, 1406-1415. Angelo Correr was born in Venice, and was named Latin patriarch of Constantinople in 1390. From secretary and cardinal (1405) under Innocent VII he became pope in 1406. He promised to resign if elected so that his dual resignation with the antipope at Avignon could end the Western Schism caused by the double papal election of 1378. Gregory's envoys had reached Paris, where at Notre Dame in 1407 a solemn service of thanksgiving took place. Benedict XIII agreed to meet, but Gregory lost interest when Benedict changed his mind. Angry cardinals met at Pisa in 1409 to depose both popes and elect a third, Alexander V. Gregory though forsaken was still true pope and was supported by King Ladislaus of Naples, among others. When the Council of Constance recognized Gregory as true pope in 1415, he resigned his office. Benedict XIII refused to accept the conciliar decision, and in 1417 Constance named him heretical. Gregory became cardinal bishop of Porto and legate of the March of Ancona until his death.

MARVIN W. ANDERSON

GREGORY XIII (1502-1585). Pope from 1572. Born at Bologna, Ugo Buoncompagni became a professor of law at Bologna from 1531 to 1539. When Cardinal Parisio brought him to Rome in 1539, Paul III made him first a judge, then responsible to the Council of Trent, and finally vice-chancellor in the Compagna. He was ordained at forty, and under Paul III became bishop of Viesti. After the Council of Trent, Pius IV named him cardinal-priest in 1565. The papal election after Pius's death quickly resulted in Buoncompagni's elevation. Gregory XIII took as his motto the words, "Confirm, O God, what thou hast wrought in us." He was acceptable to Philip II, for in 1565 he had been sent to Spain for the Inquisition's trial of Carranza, archbishop of Toledo.

Barely three months after his election, Gregory celebrated the massacre of St. Bartholomew's Day* with a *Te Deum* at Rome. His election in 1572 and political events conspired to make him a restorer of Catholicism. Though he failed against England, the Turks, Sweden, and Russia, his pontificate is known as a high point of the Catholic revival. Charles Borromeo* inspired him, while Jesuit support from abroad and founding of the Gregorian University at Rome were solid achievements. The Quirinal Palace and the fountain in the Piazza Navona speak of the Baroque splendor which implemented the ideals of Trent. By 1585 all five points announced at Gregory's first papal consistory of 1572 had been started. Consolidation of the League against the Turks, fight against heresy by use of the Inquisition, and internal reform by employing Tridentine legislation were realities. In addition there were now friendly relations with the Catholic princes and improved supervision of papal states.

MARVIN W. ANDERSON

GREGORY XVI (1765-1846). Pope from 1831. Born Bartolomeo Alberto Cappellari, he devoted his reign to the consolidation of the papacy as the locus of authority in the church and as definer of religious principles for society. The Revolution of 1831 at Rome faced him immediately with revolutionary principles; he called in Austrian troops to put it down. He determined to implement the ideas he published earlier in *Il trionfo della Sante Sede e della Chiesa* (1799), which claimed that the church was divinely ordained with an independent and unchanging constitution with the pope the infallible head; the Papal States* were an unchanging patrimony to ensure spiritual independence from all states. Gregory's two secretaries of state, Cardinals Bernetti and Lambruschini, helped him hold his own, with help from Austrian troops, against intervention by the powers of revolution. The holocaust only broke in 1848 under Pius IX.* In numerous encyclicals he tried to pinpoint the religious errors animating movements against, or at least unsympathetic with, his own religious-cultural ideal. *Mirari vos* (1832) and *Singular nos* (1834) were his most significant, occasioned by the troubles in the Papal States and the writings of Lamennais.* He condemned revolution, liberalism, traditionalism, and separation of church and state, and mandated support of the "alliance between Throne and Altar," and his Temporal Power. In promotion of the Catholic Church he stimulated enormous missionary activity worldwide, especially in Asia and Latin America. He named nearly 200 missionary bishops, as he managed to centralize Catholic missions directly under the papacy. Before election as pope he became a monk (1783), procurator-general (1807) and vicar general (1823) of the Camaldolese Order, prefect of the Propaganda Fide (1826), and cardinal (1826).

See J. Leflon, *La crise revolutionnaire, 1789-1846* (1949), and E.E.Y. Hales, *Revolution and Papacy, 1769-1846* (1966). C.T. MC INTIRE

GREGORY OF AGRIGENTUM (c.559-c.638). Byzantine prelate. Born near Agrigentum, Sicily, he traveled in North Africa and the Near East when he was only eighteen. He was ordained deacon by the patriarch of Jerusalem and was consecrated bishop in Rome at the age of thirty-one. Accused and imprisoned by enemies, on appeal to Pope Gregory I he was declared innocent and received with honor by Emperor Maurice. His ten-volume commentary on Ecclesiastes has been preserved, and he had an influence on the development of Byzantine ecclesiastical and literary styles. His life was written by Leontius, hegumen of the St. Sabas monastery in Rome.

GREGORY OF ELVIRA (d. after 392). Bishop of Elvira (Eliberis) near Granada, and greatly esteemed as a defender of Nicene orthodoxy and an opponent of Arianism. After the death of Lucifer of Calaris (whose decision not to pardon those who became Arians at the Council of Arminum [359] he approved), he became head of the followers (Luciferians) of Lucifer. Later he also attacked Priscillianism.* Recent research has restored his claim to literary fame by ascribing to him books that were thought to have been written by others: e.g., *De fide orthodoxa,* a defense of the use of *homoousios; Tractatus Origensis,* twenty essays (homilies) each arising from a text of Scripture; and *Tractatus de epithalamio,* homilies on the Song of Solomon. Gregory's exegesis of the Bible was allegorical. PETER TOON

GREGORY OF NAZIANZUS (330-389). Cappadocian Father.* Brought up on the family estate near the town in Cappadocia where his father, also named Gregory, was bishop and whence he derived his title, he was educated at Caesarea where he met Basil* and eventually the two friends, about 350, went on to the University of Athens. Gregory returned home about 358, and after a short career as a teacher of rhetoric he spent some time helping his aged father at Nazianzum and the remainder at Basil's monastic retreat. In 362 against his will his father had him ordained priest. Ten years later he reluctantly agreed with Basil's wish that he be bishop of Sasima, a position he never in fact fulfilled, and a place he never visited, preferring to assist his father at home. After the latter's death in 374 he retired to Seleucia in the province of Isauria.

Gregory was summoned out of his monastic peace to Constantinople to defend the Nicene faith against Arianism.* His ministry at the "Church of the Resurrection" in Constantinople made a significant contribution to the final establishment of the orthodox faith. During the council he was appointed bishop of Constantinople, but characteristically resigned the see when his election was disputed. After the council he went back to Nazianzum, where he took charge of the church, but from 384 he retired to his family estate where he finally died.

Although of unimpressive personal appearance and bearing, Gregory had an outstanding power of oratory which was used to great effect in his ministry at Constantinople. Most worthy of note are the famous five *Theological Addresses* against the Arians. After dealing with the Eunomians in the first oration and the nature of God in the second, he develops in the third and fourth the doctrine of God the Son. He shows that the orthodox teaching concerning the equality of Father and Son is much more Christian and more logical than the Arian concept of the Godhead. In the fifth oration Gregory treats the doctrine of the Holy Spirit and argues for the consubstantiality of the Spirit with the Father and the Son. Other writings include the *Philocalia,* a selection from the works of Origen which he compiled with Basil; several writings against Apollinarianism; and 242 letters and poems.

BIBLIOGRAPHY: P. Gallay, *La Vie de Saint Grégoire de Nazianze* (1943); J.H. Newman, *Essays and Sketches,* vol. III (1948); S. Plagnieux, *Saint Grégoire de Nazianze théologien* (1952).

G.L. CAREY

GREGORY OF NYSSA (330-c.395). Bishop of Nyssa, Cappadocian Father,* and younger brother of Basil* of Caesarea. A shy, gentle man of studious disposition, Gregory was totally dominated by his forceful brother whom he sometimes called "the Master." After a brief spell as reader in the church, he became a teacher of rhetoric and thereby incurred Basil's great displeasure at entering upon a secular life. In penitence he entered a monastery founded by Basil. In 371 he accepted Basil's invitation, although rather unwillingly, to become bishop of Nyssa.

Because he supported the Nicene faith Gregory was deposed by a synod of Arian bishops in 376, but regained his see in 378 when Emperor Valens died. Gradually his fame spread; about 379 he was asked to visit the Church of Syria to help solve the problem of schism in that see, and at the Council of Constantinople he took a leading part, not only delivering the inaugural address which is not extant, but also the funeral oration of Melitius of Antioch, the first president of the council. Very little is known of the later years of Gregory's life, but he appears to have traveled extensively.

In some respects Gregory was the most gifted member of a distinguished family. Although deficient in practical ability so clearly marked in the career of Basil, in originality and intellectual ability he was not only superior to his brother but an outstanding thinker of the fourth century. His theological views were more profoundly influenced by Origen than by any other teacher. His idealism, allegorical interpretation of Scripture, and doctrine of human freedom and the final hope indicate the extent of Origen's influence. But Gregory was no mere plagiarist; each subject was worked out carefully. In general his theology turned on the assumption that the world was ruined by the Fall which was a consequence of man's free will. Redemption is made possible by a remedial process both human and divine in the incarnation of Christ, the beneficient results of which are communicated through the sacraments. He was the first theologian to interweave firmly the doctrine of the sacraments into a systematic theology of the Incarnation.

His chief apologetic work was the *Sermo Catecheticus*, a manual of theology in which he deals at length with Christology and eschatology. The latter doctrine is based upon the views of Plato and Origen which Gregory believed to be consonant with Scripture. He took Paul's statement literally that God will eventually be "all in all," and saw hell as a process of ultimate purification rather than a place of eternal suffering. Gregory was a staunch supporter of the Nicene faith and was among the first to distinguish between *ousia* and *hypostasis*. The former he used to express essence, and the latter the distinctive peculiarity which was equivalent to *prosōpon*, "person." His supposed marriage to Theosobeia based upon allusions in his treatise on *Virginity* cannot be proved and must remain only a conjecture until fuller evidence comes to light. Such was Gregory's fame that at the Seventh General Council of the church he was entitled "Father of Fathers."

BIBLIOGRAPHY: S.M. Shea, *The Church According to St. Gregory of Nyssa's Homilies on the Canticle of Canticles* (1966-67); *Gregorii Nysseni Opera* (1967); R. Staats, *Gregor von Nyssa und die Messialianer* (1968); C.W. Macleod, "Allegory and Mysticism in Origen and Gregory of Nyssa," *JTS*, XXIII (October 1971). G.L. CAREY

GREGORY OF RIMINI (d.1358). Augustinian philosopher. Born at Rimini, he joined the Augustinian Hermits, studied in Italy, Paris, and England, and subsequently taught at Paris, Bologna, Padua, and Perugia. In 1340 he lectured on *The Sentences* in Paris, and in 1345 was made a doctor of the Sorbonne by Clement VI. He was elected vicar general of his order in 1357 and spent the last eighteen months of his life in Vienna. Considered by his contemporaries as one of the most subtle of philosophers, he furthered the Nominalist teaching of William of Ockham, though he was less skeptical. He held it was possible to demonstrate philosophically the spirituality of the soul, and he rebutted the Ockhamist assertion that God could cause a man to sin. He defended Augustinianism vigorously, teaching that works done without grace are sinful, and that unbaptized infants are damned. This last earned him the nickname *tortor infantium* ("infant torturer").

J.G.G. NORMAN

GREGORY OF TOURS (c.538-594). Frankish bishop and historian. Born Georgius Florentius Gregorius of a noble Roman family at Arverna (now Clermont-Gerrand), he was in 573 appointed bishop of Tours, and carried out his immense tasks with zeal and devotion. He saw to the administration of an important diocese, disciplined an unruly clergy and members of religious orders, defended Catholicism against Arianism, kept order in Tours (the site of a pilgrimage center), and attended to secular judicial duties. Except for a short period of antagonism by King Chilperic (576-84), Gregory was on amiable terms with all four rulers of Tours during the time he was bishop, and often advised them on matters of state.

Gregory's writings consist of ten books of history, seven of miracles, a book on the lives of the Fathers, a commentary on the Psalms, and a treatise on offices of the church. His best-known work, *Historia Francorum*, treats the history of the world to 511 in the first two books, and the history of the Franks to 591 in the remaining eight books. Some of the latter give an almost exhaustive account of Gregory's activities around Tours. The dominant theme of his history is concerned with the spread of Christianity through the exploits of Catholic kings and the work of missionaries and martyrs. Although he wrote in crude Latin and his historical methods were questionable, his works provide an invaluable knowledge of sixth-century Gaul. His writings and life reveal him to be a sincere and eloquent spokesman for the developing early church.

BIBLIOGRAPHY: O.M. Dalton (ed. and tr.), *The History of the Franks* (2 vols., 1927); W.C. McDermott (tr.), *Selections from the Minor Works* (1949); J.M. Wallace-Hadrill, *The Long-Haired Kings* (1962). ROBERT G. CLOUSE

GREGORY OF UTRECHT (c.707-c.775). Missionary and abbot. Son of noble Frankish parents, he was educated at the abbey of Pfalzel where his widowed mother was abbess. Hearing Boniface speak on the apostolic life in 722, he immediately joined him and remained associated with him as a fellow-laborer for more than thirty years, accompanying him to Rome in 738, where Gregory acquired valuable manuscripts. He became abbot of St. Martin's at Utrecht in 750, establishing there a kind of missionary college to which students flocked from almost all the German tribes and even England. He was appointed administrator of the diocese of Utrecht, but was never consecrated bishop, and following the death of Boniface was commissioned by Pope Stephen III to convert the Frisians. His biography as recorded by his pupil Liudger, first bishop of Munster, stresses his contempt of riches, seriousness, forgiveness, and charity. MARY E. ROGERS

GREGORY PALAMAS (c.1296-1359). Greek theologian. After a broad education in Constantinople, at twenty years of age he became a monk on Mt. Athos in Greece. It was during his stay at Athos that Palamas developed much of his thinking on the subject of the mystical communion with God. He is reported to have excelled all his fellow monks in the area of asceticism. He was not able to isolate himself completely from the world, for he was called to be archbishop of Thessalonica after he became well known for his defense of Hesychasm.* The Hesychasts were attacked both for their mysticism and for their physical positions in relationship to prayer. Their opponents felt that God could not be known through mystical communion with God in intense meditation. They considered this as bordering on the deification of man. Some even thought the doctrine of Uncreated Light held by the Hesychasts was close to a doctrine of pantheism.

Gregory Palamas strongly defended the Hesychasts. He affirmed the theology of experience through meditation with God. God's *essence* could not be known, but His *energies* could. Palamas was not presenting something totally new, for Basil and the Cappadocian Fathers had earlier proposed this. By coming into contact with God's energies, man could have a direct relationship with God. Since God is light, the experience of God's energies takes the form of light. It is the same uncreated light of the Godhead which appeared on Mt. Tabor at the Transfiguration, according to Palamas. A council was held on this subject in Constantinople in 1341, and it sanctioned the doctrine of Uncreated Light relating it to the divine energy. It thus upheld the ideas of Gregory Palamas. GEORGE GIACUMAKIS, JR.

GREGORY THAUMATURGUS (c.213-c.270). Bishop of Neo-Caesarea. One of Origen's pupils during the first half of the third century, Gregory had been a pagan lawyer from Neo-Caesarea in Pontus when he came under the influence of Origen. He was converted and became an enthusiastic supporter of Origen's emphasis: that the church should attempt to use all wisdom and literature for its own use. The good of heathen learning should be used, the evil cast away. Gregory, like all of Origen's students, was led through various academic disciplines, culminating in theology. Soon after leaving Origen, who was in Caesarea of Palestine, Gregory was consecrated bishop of Neo-Caesarea. As bishop he made attempts to draw the Christian believers away from their pagan festivals by instituting martyrs' festivals which could substitute as times of celebration. His ministry appears to have been successful on a numerical scale, for there was a marked increase in the number of Christians on the completion of his ministry in Neo-Caesarea. His biography was written in the fourth century by Gregory of Nyssa. GEORGE GIACUMAKIS, JR.

GREGORY THE ILLUMINATOR (or "Enlightener") (c.240-332). Known as the "Apostle of Armenia." He is reported to have been the son of a Parthian who murdered King Khosrov I of Armenia. As a baby Gregory was removed to Caesarea in Cappadocia where he eventually became a Christian. He married and had two sons before returning to Armenia where he converted King Tiridates III to Christianity. He was consecrated bishop *(catholicos)* of the Armenians by the metropolitan of Caesarea. For several generations the episcopate remained in his family. His son and successor, Aristakes, attended the Council of Nicea (325). Agathangelus,* the first historian of Armenia, wrote a biography of him.

PETER TOON

GRELLET, STEPHEN (1773-1855). Quaker missionary. Étienne de Grellet du Mabillier was born in Limoges, France, and educated at the College the Oratorians, Lyons, becoming skeptical of Roman Catholic dogmas, however. During the revolution he joined the royal army, but was taken prisoner. Escaping to Amsterdam, he sailed for Demerara in Guyana, South America, and in 1795 went to New York. By this time a disciple of Voltaire, he was moved by William Penn's book *No Cross, No Crown* and later converted while hearing Deborah Derby, a Friend from Coalbrookdale, England. He joined the Friends in 1796. Speaking and "travelling in ministry" followed in America and Europe. In England (1813) he visited Newgate Prison and introduced Elizabeth Fry* to her life-work among prisoners.

J.G.G. NORMAN

GRENFELL, GEORGE (1849-1906). Baptist missionary. Born in Cornwall, he grew up in Birmingham, and after a short apprenticeship in the Cameroons (1875-78) he led a pioneer party to the Congo in 1878. By inclination an explorer, he traveled 15,000 miles on the Congo and its tributaries (1884-86), winning recognition from the Royal Geographical Society. For over twenty years he supervised Baptist Missionary Society work, and continued exploring in two steamers he himself assembled. His base after 1889 was Bolobo, where he engaged in conventional missionary work. He considered European rule preferable to intertribal conflict and Arab slave-raiding. He therefore welcomed the Congo Free State and was at first highly regarded by its authorities. King Leopold consulted him in 1887 and appointed him to an international boundary commission in 1892-93. Later Grenfell condemned official atrocities and was treated with marked disfavor. This prevented him from completing a chain of stations linking up with the Church Missionary Society in East Africa. D.G.L. CRAGG

GRENFELL, SIR WILFRED THOMASON (1865-1940). Medical missionary and author. Born near Chester, England, his life of Christian service began in 1885 in response to a challenge presented by D.L. Moody at an East London tent meeting. After graduating in medicine he joined the Mission to Deep Sea Fishermen, became its superintendent in 1890, and cruised from the Bay of Biscay to Iceland, ministering to the physical and spiritual needs of the fishermen. In 1892 he

went to Labrador and devoted the rest of his life to the welfare of its inhabitants. Before he retired in 1935 he had founded five hospitals, seven nursing stations, three orphanage boarding schools, cooperative stores, industrial centers, agricultural stations and, in 1912, the King George V Seaman's Institute in St. John's, Newfoundland. He annually cruised along the Labrador and Newfoundland shores with a hospital ship to minister in remote communities. Honored by numerous medical societies, geographical societies, and universities, he was knighted in 1927.
ROBERT WILSON

GREY FRIARS, see FRANCISCANS

GREY NUNS. A name given to "Sisters of Charity." The North American "Grey Nuns of Charity" were founded by Madame d'Youville in Montreal in 1737 as a small community of women devoted to the care of the sick. Their Rule (1745), besides the usual three vows, included the promise to devote their lives to the relief of suffering. They persisted in their dedication despite hostility and invective. They were called *les soeurs grises* ("the drunken sisters"), so they deliberately chose grey *(gris)* as the color of their habit (1755). They spread to other parts of North America, forming separate congregations, e.g., "Grey Nuns of the Cross" at Ottawa (1845), "Grey Nuns of Quebec" (1849), "Grey Nuns of the Immaculate Conception" (1926). The name is given also to Sisters of Charity in France, and to the "Grey Sisters of St. Elizabeth" in Germany. J.G.G. NORMAN

GRIESBACH, JOHANN JAKOB (1745-1812). NT scholar. Born at Butzbach, he was educated at Frankfurt-am-Main, Tübingen, Leipzig, and Halle, and became professor at Halle in 1773, and professor of NT at Jena in 1775. He was the first critic to make systematic application of literary analysis to the gospels, maintaining that Mark was the latest synoptic gospel, and basing his work on Matthew and Luke (the "dependence theory"). His major work lay in NT textual criticism. He published a critical edition of the Greek NT based on Elzevir (1774-77). He collated a great number of MSS, and developed Bengel's "family" theory, classifying the authorities into three classes or "families," Alexandrian, Western, and Byzantine (or Constantinopolitan). Subsequent NT criticism has built on his work.
J.G.G. NORMAN

GRIFFITHS, ANN (1776-1805). Welsh hymnwriter. Born Ann Thomas, she joined the Methodist society at Pontrobert in 1797 after experiencing evangelical conversion. She married John Griffiths of Meifod in 1804 and died the following year after the birth of a child. She was in the habit of composing hymns which she recited to her servant, Ruth Evans, who later married Methodist minister John Hughes. Between them they preserved and published the hymns of Ann Griffiths, comprising a remarkable literary phenomenon characterized by a masculine strength, complex biblical allusions, and a profound Christocentric mysticism.

GRIGNION DE MONTFORT, LOUIS-MARIE (1673-1716). Missionary, trained at the Jesuit College, Rennes. After a life devoted to prayer and poverty, he was ordained priest in 1700. From 1701 to 1703 he was chaplain to the hospital at Poitiers where he founded the "Daughters of Wisdom," a congregation devoted to the nursing of the sick and the education of poor children. In 1704, however, he regarded his true vocation as the missionary in W France. During the initial period he suffered from the jealousies of the Jansenists, who envied him his influence. Several years before his death he founded a second congregation, the "Company of Mary," a congregation of missionaries. Both foundations suffered severe numerical losses during 1715, but they have since been revived. His most famous work on devotion and mariology, *Traité de la vraie dévotion à la Sainte Vierge,* first printed in 1842, was popular with some English Catholics and was translated in 1863 by F.W. Faber. It has since been reprinted several times. Pius XII canonized him on 20 July 1947. GORDON A. CATHERALL

GRIMSHAW, WILLIAM (1708-1763). Anglican clergyman. Born in Lancashire of obscure parentage, he was educated at Christ's College, Cambridge, and became a typcial curate of his times until a long spiritual struggle culminated in a conversion experience in 1742. He knew nothing then of Whitefield or Wesley, but like the former was converted through reading Scripture and seventeenth-century books. He was incumbent of Haworth, Yorkshire (afterward famous for the Brontes), in a wild country with rough, illiterate people. His uncouth, racy preaching with plenty of humor; his athletic prowess that won their respect; his affection for sinner and saint; and his passionate sense of Christ as Savior made him a powerful evangelist. He transformed the whole place. Before sermon he would go out and round up shirkers with a riding crop, and his preaching brought many hearers from a distance. He took particular pains with the very poor, the isolated, and the sick. Because neighboring parishes never heard the Gospel, he went around preaching, and when their own slack clergy protested, his archbishop supported him. Grimshaw was an ally of both Whitefield and Wesley, but disapproved of the Wesleys' movement toward separating the Methodists from the Church of England. He trained many curates, and was a fine example of the Evangelical Revival in parish life.
JOHN C. POLLOCK

GRINDAL, EDMUND (1519?-1583). Archbishop of Canterbury from 1575. Born in Cumberland and educated at Cambridge, he adopted the Protestant views which were widespread in the university, and became vice-master of Pembroke Hall in 1549. Two years later he became chaplain to Ridley, then to Edward VI, and a canon of Westminster in 1552. During Mary's reign he was in exile, chiefly at Strasbourg, although he visited Frankfurt and was involved in the liturgical disputes there. Elizabeth made him bishop of London in 1559, whence he proceeded to York in 1570 and Canterbury five years later. In 1576 he rebuked the queen for ordering him to suppress the meetings of clergy known as "prophesyings," which he believed were an important means of improving the standard of preaching in the church. For his disobedience he was sequestered from his jurisdiction. Despite efforts at mediation, no real reconciliation with the queen was achieved before Grindal died, a blind and pathetic figure. Though his primacy has often been judged a disastrous failure, it is now coming to be viewed as an interesting and important attempt to establish a Reformed type of episcopacy in which the bishop sought a much closer working relationship with his brother clergy.

BIBLIOGRAPHY: J. Strype, *The History of the Life and Acts of ... Edmund Grindal* (1821); E. Grindal, *Remains* (ed. W. Nicholson, 1843); S.E. Lehmberg, "Archbishop Grindal and the Prophesyings," *Historical Magazine of the Protestant Episcopal Church*, xxxiv (1965), pp. 93-97; P. Collinson, "Episcopacy and Reform in England in the Later Sixteenth Century," *Studies in Church History*, vol. III, ed. G.J. Cuming (1966), pp. 91-125.

JOHN TILLER

GRONINGEN SCHOOL. A theological movement which flourished in the Dutch Reformed Church in the middle third of the nineteenth century. Dissatisfied with the formalistic religious establishment of the day, but opposed both to the conservative and pietistic approach of the "Awakening" *(Réveil*)* and to the rising influence of dogmatic liberalism, it tried to revive the humanistic and evangelical emphasis of earlier figures such as Coornheert* and Erasmus.* P.W. Van Heusde, who taught philosophy from a Christian Platonist standpoint at Utrecht, inspired most of the leaders of the movement, who gathered at the theological faculty at Groningen. Petrus Hofstede De Groot was the leader of the group; other prominent figures included W. Muurling, H. Muntinghe, L.G. Pareau, and J.F. Van Oordt.

After 1835, when the synod refused to censure them for their rejection of traditional Calvinism, their influence steadily widened. In its heyday around mid-century, the "Groningen School" was perhaps the dominant influence in the church. Its teachings emphasize walk of life rather than dogma. Man is characterized by an innate spiritual feeling which relates him to the divine. God has revealed Himself in Christ, who has taught us that our spiritual nature is fulfilled in love. The Gospel calls us to follow Christ, which we can do if we so will. God is to be seen as loving Father. Love of neighbor is more important than dogmatic system (the Trinity, e.g., is interpreted as a valuable symbolic insight rather than a statement of fact).

Such teachings were expounded in the movement's periodicals, *Waarheid in Liefde* ("Truth in Love") from 1837, and *Geloof en Vrijheid* ("Faith in Love") from 1867. The Groningen School was reluctant to deal in dogma and metaphysics, and this left it exposed to attacks, during the last third of the century, from "scientific" modernism, and from the Kuyperian revival of Calvinism, both of which stressed world-view as well as personal ethics.

See J.H. Mackay, *Religious Thought in Holland During the 19th Century* (1911); and Th. L. Haitjema, *De richtingen in de Nederlandse Hervormde Kerk* (2nd ed., 1953).

DIRK JELLEMA

GROOTE, GERARD (Geert) (1340-1384). Founder of the Brethren of the Common Life.* Born of a wealthy family at Deventer, he studied law and theology at Paris. During his twenties, holding benefices at Aachen and then Utrecht, he led a worldly life, which did not satisfy him. He was attracted by the ideas of the mystic J. Ruysbroeck,* whom he met and conversed with. After a serious illness, Groote about 1374 was influenced by the Carthusian Hendrik Van Calkar and turned to a devout Christian life. Monastic life at the Carthusian house near Arnhem did not meet his needs, and he gained permission to preach in the diocese of Utrecht. He immediately gained wide popularity. His attacks on clerical abuses aroused some opposition. In 1380 Groote with his younger friend Florentius Radewijns* decided to form a group in Deventer for the cultivation of piety; this was the nucleus of the Brethren. He turned his own house over to a similar gathering of devout women, for whom he wrote a Rule. In 1383 his enemies were able to have permission to preach withdrawn from him. Groote started an appeal to Rome, but died in 1384 of plague. As an admirer of Ruysbroeck, Groote translated the mystic's *Horarium* into Dutch, and his *Brulocht* ("Marriage") into Latin.

See Th. Van Zijl, *Geert Groote, Ascetic and Reformer* (1963).

DIRK JELLEMA

GROPPER, JOHANN (1503-1559). Roman Catholic theologian. Born in Soest, Westphalia, he studied in Cologne, and his chief activities were in the service of Hermann von Wied,* archbishop of Cologne. At the provincial synod in 1536, Gropper's reform program did not receive approval. In 1538, together with the canons of the Cologne council, he published a handbook of Christian doctrine which contained an exposition of the Decalogue, the Creed, and the seven sacraments. At the Colloquies of Hagenau and Worms (1540-41) he took a mediating position between Roman Catholics and Lutherans, setting forth the doctrine of double justification (justness by faith and justness by love), in which he evidenced the influence of Erasmus. His *Liber Ratisbonesis* became the basis of negotiations at the Colloquy of Ratisbon* (1541). However, he prevented Hermann von Wied from carrying through the Protestantization of Cologne as advocated in the *Consultatio* by Martin Bucer and Philip Melanchthon. Gropper participated in the Council of Trent, especially the third and fifth sessions (1546) and the thirteenth and fifteenth sessions (1551-52). He declined the appointment as cardinal by Paul IV.

CARL S. MEYER

GROSSETESTE, ROBERT (c.1168-1253). Bishop of Lincoln and initiator of the English scientific tradition. Little is known of his life, but he was born of poor parents and studied at either Oxford or Paris. He became a member of the Arts faculty at Oxford and was made chancellor sometime between 1214 and 1221. He then became lecturer to the Oxford Franciscans (1229), leaving this post to take the bishopric of Lincoln (1235), England's largest diocese, where he remained till his death. He was a zealous bishop, deposing many abbots and priors because they neglected to staff adequately the parish churches in their care. He attended the Council of Lyons (1245) and in 1250 visited Rome, where he delivered a sermon in which he declared that the papal court was the origin of all the evils in the church; he objected also to the appointment of Italian friends and relatives of the pope to rich English benefices. The last years of his life were spent in a struggle to stop one of these appointments.

Grosseteste was just as independent in his dealings with the English monarch. He believed churchmen should not hold civil office and asserted that a bishop did not in any way derive his authority from the civil power. At times he refused to carry out royal orders in his diocese and threatened the king with excommunication.

He combined the churchman's active life with a variety of scholarly interests. He lived at a crucial period in the intellectual history of W Europe when the philosophic and scientific works of Aristotle were being recovered from the Muslims. As a teacher, commentator, and translator he took an active part in this movement. Although basically Augustinian in outlook and relying on standard authors, he was heavily influenced by Muslim, Jewish, and Aristotelian works. He never wrote a comprehensive philosophical work or devised a system, but his views had a profound effect upon later scientific thought. The most important of his many works are *De Luce* ("Light"), *De Motu Corporali et Luce* ("Corporal Motion and Light"), *Hexameron*, and commentaries on Aristotle's *Posterior Analytics* and *Physics*. Basic to Grosseteste's view of the universe is his metaphysics of light. He believed light was the first form to be created in prime matter, and from it all else developed. He also taught that God's existence could be proved from the argument of motion. Twentieth-century scholars have been interested in his recovery and elaboration of scientific method.

BIBLIOGRAPHY: S.H. Thomson, *The Writings of Robert Grosseteste, Bishop of Lincoln, 1235-1253* (1940); A.C. Crombie, *Robert Grosseteste and the Origins of Experimental Science* (1953); D.A. Callus (ed.), *Robert Grosseteste, Scholar and Bishop* (1955). ROBERT G. CLOUSE

GROTIUS, HUGO (1583-1645). Dutch jurist and statesman. He entered the University of Leyden at the age of eleven and was practicing at the bar at The Hague at sixteen. In 1612/13 he became pensionary of Rotterdam and worked with Oldenbarnevelt in his struggle with Prince Maurice and the Calvinist party. In 1618 he was imprisoned for life by Maurice, but escaped to Paris in 1621, where in relative poverty he produced *De jure belli et pacis* (1625), the fruit of twenty years' thought, on which his fame largely rests. He served as Swedish ambassador in Paris, and was disappointed that because of religious prejudices he was never recalled to the service of Holland.

Grotius was a man of undeniable piety and prodigious learning, yet in the history of Christianity he has the ambiguous significance of a transitional figure, a humanist placed between Scholasticism and Enlightenment. He sought to interpret the Bible by the rules of grammar without dogmatic assumptions, but he had inadequate philological resources for the task. As a Christian and a statesman he sought to moderate the dogmatic controversies then rife in Europe, praying God in his last testament "to unite the Christians in one church under a holy reformation." He had tried to get ecclesiastical peace in Holland by preventing preaching on disputed points in the Calvinistic controversy, and he was often suspected —unfairly—of tendencies to Roman Catholicism.

Grotius did not look for a return to the Christian Middle Ages. The truth about him can be seen rather in his *De veritate religionis Christianae* (1627), a defense of basic Christianity for sailors meeting other religions. It is a simplification of parts of the Scholastic theological tradition, presenting Christianity as the true religion in harmony with God's rationally ordered world. Grotius thus points forward to the writers of Christian evidences in the eighteenth century. His faith in the orderliness of the world is basic to his work in theology as in jurisprudence. He sought amid disorder to realize and extend this order. He believed there was a law of nature deriving from God's will and known by reason. It was both to guide and to be upheld by the processes of law; where there was no judge, as in war, conflict was to be seen as a form of litigation. Thus human strife properly understood was at once limited by law and directed toward its realization. Grotius is regarded as a father of international law. He believed that the law of nature is intrinsic to the social being of man, and that God cannot alter it, any more than laws of number; the skeptical question, whether God therefore is unnecessary to the law of nature, was not pressing at this time and Grotius did not tackle it.

This problem lies near the heart of his defense of the Catholic doctrine of the Atonement against Socinus. He argues that God is free to relax the law that death follows sin, but not in such a way that the fundamental order of the universe, for which he is responsible as Moral Governor, is subverted. The sufferings of Christ are a penal example by which God upholds this order while remitting sin. This theory had considerable influence in Protestant theology into the nineteenth century.

See W.S.M. Knight, *The Life and Works of Hugo Grotius* (1925). HADDON WILLMER

GROVES, ANTHONY NORRIS (1795-1853). Plymouth Brethren* leader. After studying chemistry, dentistry, and surgery in London, Groves settled in dental practice first in Plymouth (1813), then in Exeter (1816). In 1826 he entered Trinity College, Dublin, to prepare for ordination, but came to see that "ordination of any kind to preach the gospel is no requirement of Scripture." In Dublin he associated with the group that included J.G. Bellett and J.N. Darby.* He influenced Bellett to the view that the principle of union among Christians was "the love of Jesus, instead of *oneness* of judgment in minor things...." In *Christian Devotedness* (1825) he advocated complete dependence on God for temporal needs. This influenced George Müller toward his lifelong principle of faith, and through him a host of others.

Groves sailed with his party for Baghdad in 1829, remaining there three years, during which time his wife died of plague. He remarried in 1835. For nineteen years from 1833 he labored in India and was latterly joined by others. Watching with concern Darby's tendency to domination, Grove's letter to Darby in 1836 struck a prophetic note of the results of setting more store on correctness than love. Unwell in 1852, he returned to England and died in George Müller's house in Bristol (Müller's wife was Groves's sister). Groves's views strongly influenced early Brethren; he was probably the pioneer of simpler, apostolic missionary principles. His eldest son Henry, a gifted Bible teacher, was also a leader among early Brethren.

BIBLIOGRAPHY: H. Groves (his widow), *Memoir of the late Anthony Norris Groves* (1869); G.H. Lang, *Anthony Norris Groves* (1949); see also books listed under article PLYMOUTH BRETHREN.

G.C.D. HOWLEY

GRUBER, F.X., see MOHR, JOSEPH

GRUNDTVIG, NIKOLAI FREDERIK SEVERIN (1783-1872). Danish bishop and hymnwriter. In his youth he experienced a personal crisis that led him first to the Romantic view of nature and religion, later to a personal Christian faith and more clarified biblical convictions. Except for short periods of service as a pastor, he lived as an independent writer (1810-25), struggling for the reintroduction of an orthodox Lutheran Christianity. About 1824 he once more experienced a religious crisis, caused by biblical criticism, which seemed to him to make his former orthodox view of the Bible untenable. It was during this crisis that he made his "unique discovery." This he published in 1825 in a pamphlet called *Kirkens Genmaele* ("The Church's Reply"), which argued that the sure foundation of faith is not to be found in the Bible, but in the living Word of God in his living congregation, i.e., in the risen Christ Himself, who lives and works in His congregation, when it gathers around the sacraments.

About 1830 Grundtvig three times visited England and was strongly impressed by the spirit of liberty and activity which he found characteristic of English society. This made him in later years an indefatigable advocate of liberty in both church and society. It is also reflected in his fairly optimistic view of man and in his educational ideals which made him one of the fathers of the Danish folk-high-school movement. From 1825 he was the leader of an ever-increasing following. In 1839 he was appointed clergyman of the Vartov Foundation in Copenhagen, where he worked until his death. He was given the rank of bishop in 1861. Through his views of church and sacraments, through his educational and liberal ideals, and as an unsurpassed writer of innumerable hymns, he left a lasting mark upon the Danish Church as a whole and on Danish society.

BIBLIOGRAPHY: N. Davies, *Grundtvig of Denmark* (1944); E.L. Allen, *Bishop Grundtvig: a Prophet of the North* (1947); P.G. Lindhardt, *Grundtvig: an Introduction* (1951).

N.O. RASMUSSEN

GUARANTEES, LAW OF. This defined the relationship between the Roman Catholic Church and the Italian kingdom after the annexation of the States of the Church. Passed by the Chamber on 13 May 1871, it determined the papal rights and prerogatives. The law declared the pope's person was inviolable, he would be accorded sovereign honors, he would receive an annual state grant, the Vatican, Lateran, and Castel Gandolfo properties would remain in his possession and be tax-exempt, the freedom of conclaves and general councils was assured, the seminaries in Rome and the Suburbicum were solely under papal control, foreign envoys to the Holy See were to enjoy the usual rights under international law, and the pope was guaranteed freedom of communication with the Catholic world. Pius IX categorically rejected the law, refused the financial offer, and withdrew into the Vatican as a voluntary "prisoner." It was formally abrogated by the Lateran Treaty (1929).

RICHARD V. PIERARD

GUELFS and GHIBELLINES. The two main party groups in medieval Italian politics. "Guelf" is derived from the name Count Welf (d.825), father-in-law of Louis the Pious and founder of the great German family of Welf; "Ghibelline" comes from Waiblingen, seat of the Hohenstaufen in Swabia, and from the battle cry *Hie Weibling.* Thus in the struggle between Frederick II and the papacy, the imperialists were termed Ghibellines and the pope and papal party Guelfs. From use in Tuscany in the thirteenth century the names spread throughout Italy and were used of opposing parties in many cities and towns. Hence diverse social, political, and religious factors contributed to the creation of the parties of Guelfs and Ghibellines. The exact meaning of each term thus differed somewhat from city to city—so much so, that in the sixteenth century the French kings and their supporters in Italy were called Guelfs, while the supporters of Charles V were called Ghibellines.

PETER TOON

GUEUX. A nickname used during the revolt of the Low Countries against Spanish rule. As revolt neared (1566), a group of nobles including William of Orange and Henry of Brederode presented a list of grievances to the Spanish regent, Mary of Parma; one of her advisers jeered at the petitioners as *ces gueux* ("those beggars"). Brederode adopted the name proudly, and it soon was widely used for the rebels. It was applied more particularly to the "Sea Beggars," hit-and-run sea raiders, to whom Orange granted letters of marque (1569). Bitterly anti-Catholic (cf. their motto, "Sooner the Turk than the Pope"), they gained a reputation for desperate courage and cold-blooded plundering. When refused refuge in English ports, they took the port of Brill in Zeeland from the Spanish; it was thus the first "liberated" territory. As the revolt proceeded, the Sea Beggars were gradually merged into the growing rebel fleet. DIRK JELLEMA

GUICCIARDINI, FRANCESCO (1483-1540). Italian historian and statesman. Born into an aristocratic Florentine family, he studied law at Florence, Ferrara, and Padua, established a legal practice, then in 1511 became ambassador to King Ferdinand of Aragon. From 1516, three years after Leo X became pope, he served as governor of part of the Papal States, carrying out this and subsequent public positions with distinction. The changing face of the political scene and the differing attitudes of successive popes saw fluctuations in Guicciardini's fortunes thereafter, and finally he retired into private life, spending his last years in completing his *History of Italy*, regarded as a most significant source, especially for contemporary events. J.D. DOUGLAS

GUICCIARDINI, PIERO (Count) (1806-1886). Italian Protestant leader. Born in Florence of a noble and ancient family which included in its history Francesco Guicciardini,* he received a good general education and was at once attracted by the spiritual revival spreading in Tuscany. There, under Leopold II, there was a certain religious toleration. Invited to participate in the educational project directed by Lambruschini,* he founded the first kindergarten in Florence, showing marked organizing ability, intelligence, and common sense. The contacts with the Swiss Protestant Church in Florence and the study of the Bible led to his conversion in 1836, a date which he desired to be remembered on his tomb. Actively involved in preparing a religious reform, in the reaction and repression after 1848 (see RISORGIMENTO), he was imprisoned and exiled with many others, and took refuge in Britain. There he was warmly received by many noble families and came in close touch with the Open Brethren. At the invitation of the Society for Promoting Christian Knowledge he collaborated in a revision of the Italian Bible which had been translated by Diodati. The Guicciardini Bible of 1853 remained for a long time the best Italian translation.

A year later he returned to Italy at Nice, then part of the Sardinian kingdom, followed in 1857 by his great friend and collaborator Pietrocola-Rossetti,* with the purpose of preaching the Gospel to their compatriots. Guicciardini was the organizer of the movement, administering the gifts which came from Britain and adding much of his fortune. The communities formed were called Free Italian Churches, so anxious were the leaders to vindicate their antidenominational character and their Italian origin. The spreading of the movement occurred simultaneously with the liberation, and by 1870 there were more than thirty churches scattered throughout Italy. Guicciardini spent his last years in Florence. He had gathered a rich collection of religious works from Savonarola to the Italian reformers of the sixteenth century, which he bequeathed to the National Library in Florence. DAISY RONCO

GUILLAUME DE PARIS; GUILLAUME D'AUVERGNE, see WILLIAM OF AUVERGNE

GUINNESS, HENRY GRATTAN (1835-1910). Evangelist and writer. Born near Dublin and educated at New College, London, he was ordained an evangelist (1857), preached in Europe and America (1857-72), and had a part in the conversion of Dr. Barnardo in Dublin (1866). He founded the East London Institute for training missionaries (1873), the Livingstone Inland Mission in the Congo (1878), and other missions in South America and India, and all of these societies were in 1899 amalgamated into the Regions Beyond Missionary Union, which supported nearly one hundred and sent out more than one thousand missionaries. In 1903 he made a missionary tour of the world. His books include *The Divine Programme of the World's History, Romanism and Reformation, History Unveiling Prophecy,* and grammars of the Congo language. C.G. THORNE, JR.

GUNKEL, HERMANN (1862-1932). German Protestant biblical scholar. Born at Springe (Hanover), he taught NT exegesis at Göttingen (1888), OT exegesis and history of Israelite literature at Halle (1889-93), Berlin (1894-1907), Giessen (1907-20), and Halle again (1920-27). He was a leading member of the *Religionsgeschichtliche Schule* (i.e., comparative religion school), and one of the first to develop the Form-Critical method in relation to the OT. His writings include *Schöpfung und Chaos in Urzeit und Endzeit* (1895), studying popular mythology underlying biblical ideas of the beginning and end of the present world order, commentaries on Genesis (1901) and 1 Peter (1907), and the influential *Die Psalmen* (1926-28), in which the Psalms are dated and interpreted on the basis of classification according to literary form. J.G.G. NORMAN

GUNPOWDER PLOT (1605). A conspiracy to blow up James I and the Parliament at the palace of Westminster, evidently with the aim of restoring Roman Catholic supremacy in England. Chief mover in the affair was Robert Catesby,* a zealous Catholic disappointed that his co-religionists had not received the greater toleration promised by James before his accession in 1603. Having gained access to the palace from a neighboring cellar, Guy Fawkes planted a considerable quantity of gunpowder some months before Parliament was due to assemble. Ten days before the opening on 5 November, the plot was revealed through an anonymous letter. Of the thirteen conspirators, four were killed resisting arrest, one died in prison, and the others were in due course executed. J.D. DOUGLAS

GUNTHER, ANTON (1783-1863). German religious philosopher. Born at Lindenau in Bohemia, he studied law and philosophy at Prague, where his faith was shaken by his study of Kant, Fichte, and Schelling; but his appointment as tutor in the household of Prince Bretzenheim brought him under the influence in particular of C.M. Hofbauer* and his Christian convictions were restored.

He then began the study of theology and in 1822 entered the Jesuit novitiate at Starawicz, Galicia, but left in 1824. He lived at Vienna for the rest of his life, propagating his system of philosophy and speculative theology. He refused chairs at Munich, Bonn, Breslau, and Tübingen in the vain hope of a professorship at Vienna.

In the interests of apologetics he tried to combat the contemporary pantheistic idealism of Schelling and Hegel. His approach was rationalistic. He argued that the fundamentals of the Christian faith could be established by reason alone, that revelation was not an absolute necessity, and hence that faith should be changed into knowledge. Although his work was supported by a number of influential scholars and clerics, his writings were condemned by the Index in 1857 for their basic rationalism and its application to Christian doctrine. Gunther's system is implicit rather than explicit in his main writings: eight works in a collected edition, *Gesammelte Schriften* (9 vols., 1882); with J.E. Veith, *Lydia, Philosophisches Jahrbuch* (5 vols., 1849-54); *Anti-Savarese,* published posthumously (1883). After the definition of papal infallibility at the Vatican Council in 1870, many of his followers joined the Old Catholics.* HOWARD SAINSBURY

GURNEYITES. An American Quaker group named after Joseph John Gurney (1788-1847), a Quaker philanthropist. Born near Norwich, England, he briefly attended Oxford University, and became a Quaker minister in 1818. Many American Friends had by that time become intrigued by evangelical concepts and the revivals sweeping the country, when most of the Protestant denominations were engaged in the Second Great Awakening.* The Friends experienced a schism in 1827 when Elias Hicks of Long Island rebelled against a thoroughly evangelical statement of faith adopted by most of the Philadelphia Quakers (see HICKSITES). During 1837-40 Gurney toured America and the West Indies, preaching widely and becoming a rallying point in conforming to the revivalistic pattern, and eventually giving his name to the movement. His followers in time took on the characteristics of normative Protestantism, using the sacraments and having a minister preach at worship services. On his return to England, Gurney helped his sister, Elizabeth Fry,* in her work, and collaborated with Thomas Clarkson* and others for the abolition of the slave trade. He also wrote tracts on temperance and other subjects. KEITH J. HARDMAN

GUSTAV-ADOLF-WERK. This project of the German Evangelical Church, known until 1946 as *Gustav-Adolf-Verein,* originated in 1832 with G. Grossman of Leipzig who envisaged a living memorial to King Gustavus Adolphus of Sweden. It was popularized by R. Zimmermann in Darmstadt from 1841. Its purpose was to help Protestants in predominantly Roman Catholic areas like Bohemia, providing resources for all kinds of church buildings and activities. Since it was not exclusively Lutheran, stricter Lutherans preferred their own *Lutherischen Gotteskasten.* It was not a movement directed against Roman Catholicism, though it has been so interpreted. Its inspiration has been found in Galatians 6:10, and the application of the concept of the Diaspora (the dispersed people of God) to isolated groups of evangelical Christians in various kinds of alien societies, first made in 1855 by H. Rendtorff, has been influential. Since World War II, the work has been extended to cover the refugees from East Germany. A new concept of diaspora recognizes that in East Germany congregations live in the midst of Marxist atheism, and elsewhere they are scattered amid the godlessness of the world.

HADDON WILLMER

GUSTAVUS ADOLPHUS (Gustav II) (1594-1632). King of Sweden (1611-32) and one of the most influential leaders of the seventeenth century. Soon after his accession he was forced to lead his armies into battle, defeating in turn Denmark (1611-13), Russia (1613-17), and King Sigismund of Poland (1621-29). His goal was the establishment of a Swedish empire in the Baltic area. When the Thirty Years' War* was going badly for the Protestants, Gustavus decided to intervene (1630). At first the German Protestant princes were frightened by the Swedish armies, but after the sack of Magdeburg (1631) they rallied to his cause. With their support Gustavus defeated Tilly, the imperial commander, at the Battle of Breitenfeld (1631) and at the Lech River (1632). Following these victories he restored the freedom of the Protestants in S and SW Germany. The worried Holy Roman Emperor reinstated Wallenstein, the mercenary commander, who met the Swedish menace at the Battle of Lützen. Although victorious, Gustavus was mortally wounded and died on the battlefield.

He is considered among the greatest military leaders in history. His field tactics, including the use of small, mobile, well-disciplined units skilled in musketry, were revolutionary. Within Sweden itself, with the support of his chancellor, Oxenstierna, he reformed the judiciary, established schools, encouraged industry, and built a strong economy. His intervention in Germany has been described by some as an attempt to become emperor, while others have felt that the "Lion of the North" wished only to save Protestantism.

BIBLIOGRAPHY: G.F. Macmunn, *Gustavus Adolphus: The Lion of the North* (1931); N.G. Ahnlund, *Gustav Adolph the Great* (tr. M. Roberts, 1940); M. Roberts, *Gustavus Adolphus: A History of Sweden 1611-1632* (2 vols., 1953-58).

ROBERT G. CLOUSE

GUSTAVUS VASA (Gustav I) (1496-1560). King of Sweden from 1523. He not only led Sweden to independence from Denmark, but established the Lutheran state church in his domain. During the Swedish war of independence the Roman Catholic Church made several ill-advised moves and stirred great opposition to Catholicism among Swedes at a particularly sensitive period in their history. For instance, Gustavus Trolle, archbishop of Uppsala, placed himself firmly on the side of Denmark; for his political activities he became known as the "Swedish Judas Iscariot." Though Gustavus Vasa (who led the independence movement after 1520) had some leanings toward Lutheranism, the issue that especially led him to break with Rome was his great need for money. His main support was the poor peasants; most of the nobles had been massacred by Denmark in 1520. The Roman Church controlled a great percentage of Sweden's wealth; some claim it owned as much as two-thirds of the land. The Diet and Ordinances of Westeras* (1527) confiscated most church property, ordered teaching of the Gospel in the schools, and provided for royal confirmation of the higher clergy. Lutheranism henceforth gained rapidly, especially under the influence of Lars Petersson, professor of theology at Uppsala and translator of the NT into Swedish (1526), and Lars Andersson, archdeacon of Uppsala, royal chancellor, and publisher of the entire Bible in Swedish (1540-41). HOWARD F. VOS

GUTENBERG, JOHANN (1398?-1468). Regarded as the inventor of printing in Europe. Born in Mainz, he moved to Strasbourg in the 1430s. Evidence given by witnesses in a lawsuit there seems to indicate he had constructed a printing press with movable type, although there are no printed specimens to support the claim. Having returned to Mainz by 1448, he borrowed money from Johann Fust who later became his partner. In 1455 Fust foreclosed on his loans and took over the printing operation in association with Peter Schoeffer. Although their appearance postdates this foreclosure, Gutenberg probably played an important part in the production of the 42-line Bible (the "Gutenberg Bible") which appeared in 1456 and is frequently considered to be the first book printed in Europe, and in the production of the Psalter of 1457 which apparently was the first dated book to appear in print in Europe.
T.L. UNDERWOOD

GUTHRIE, JAMES (c.1612-1661). Scottish minister. Son of an Angus landowner, he at first favored episcopacy, but at St. Andrews University he became a Presbyterian, and in 1642 was ordained at Lauder in Berwickshire. In 1646 he was one of the Scots commissioners sent to press upon Charles I the claims of Presbyterianism and the Solemn League and Covenant.* In 1649 he became minister of Stirling, where he confirmed his stance as a strong Covenanter* by declining Charles II's judgment in matters of doctrine. Later during the Cromwellian regime, however, he upheld the principle of the kingly office. In 1653 he published *Causes of God's Wrath against Scotland* which governed all conduct by "the duty of preserving and defending true religion." After the Restoration in 1660, Guthrie with others reminded Charles of his covenant obligation, but the petitioners were seized and imprisoned. Guthrie continued to deny the king's authority in ecclesiastical affairs, and was hanged in Edinburgh. J.D. DOUGLAS

GUTHRIE, THOMAS (1803-1873). Scottish minister and social reformer. Born in Brechin, he had an extensive university education at Edinburgh (1815-25), then, refusing to renounce evangelical principles in order to obtain a parish, studied medicine and social conditions in Paris. In 1830 he became minister of Arbirlot and in 1837 collegiate minister at Old Greyfriars, Edinburgh. His concern for the poor and for improving environment was expressed in *The City: Its Sins and Sorrows* (1857). He felt that the 1843 Disruption* was the inevitable response to the law that enslaved the church of Christ, and he became minister of Free St. John's until 1864. His "Manse Fund" within a year gathered £116,370 for those who had sacrificed home for conscience in joining the Free Church of Scotland.* His social reform proposals were numerous, and he championed especially the cause of Ragged Schools. He was one of the earliest supporters of the Evangelical Alliance.* His preaching attracted rich and poor in great numbers. His writings include *The Gospel in Ezekiel* (1855) which sold 50,000 copies. Guthrie's funeral in Edinburgh brought out some 30,000 mourners. J.D. DOUGLAS

GUTHRIE, WILLIAM (1620-1665). Scots Covenanting divine. He graduated in 1638 from St. Andrews, where he studied under his cousin James and in divinity under Samuel Rutherford.* Licensed to preach in 1642, he was a tutor before being called in 1644 to the Ayrshire parish of Fenwick. His preaching filled the church, and in pastoral visitation he was most diligent. In 1651, with Scotland divided between Resolutioners* and Protesters,* he supported the latter group. Under Cromwell he was one of the Triers. Such affiliations were inevitably suspect at the Restoration, yet when Charles II forced episcopacy on Scotland, Guthrie was overlooked until 1664, perhaps because of influential friends. Even then he was not hustled summarily out of his parish, like so many, but left Fenwick in 1665, only to die that year in Brechin of the kidney disease that had long afflicted him. Guthrie is best known for his little book *The Christian's Great Interest* (1658) which has passed into many libraries and languages. John Owen called him "one of the greatest divines that ever wrote." J.D. DOUGLAS

GUTZLAFF, KARL FRIEDRICH AUGUST (1803-1851). Missionary to China. Born in Pomerania, he was sent by the Netherlands Missionary Society to Singapore in 1823, and some three years later went on to Batavia, where he met W.H. Medhurst* and began the study of Chinese. In the 1830s he traveled along the Chinese coast distributing Christian literature before succeeding Robert Morrison* as Chinese secretary to the East India Company at Canton. He helped negotiate the Treaty of Nanking, and in Hong Kong elaborated a plan for the evangelization of China, writing voluminously to Germany urging support for his scheme. Unhappily, he became a victim of dishonest Chinese assistants. He died at forty-eight in Hong Kong, but not before inspiring others to form missions for China's evangelization. The Chinese Evangelistic Society, under which Timothy Richard* and J. Hudson Taylor* originally went to China, was one of those which owed its beginnings to Gutzlaff. LESLIE T. LYALL

GUYARD, MARIE (1599-1672). Roman Catholic missionary to Quebec. Born in Tours, France, she married to please her parents, though her real desire was to become a nun. After four years, during which she had a son, her husband died. In 1631 she decided to enter the Ursuline Convent in Tours. Eight years later she agreed to the request of her archbishop to go with two other nuns to the Jesuit mission in Quebec and there found a community of nuns. She became the first superior. The work was primarily educational and social among both Indians and French. She showed great fortitude, determination, and bravery. Throughout her life, both in France and Quebec, she received visions which on the instruction of her confessors she recorded. Her letters have been published in several French editions. In 1911 Pius X pronounced her Venerable.

PETER TOON

GUYON, MADAME (1648-1717). French Quietist.* Born Jeanne Marie Bouvier de la Mothe, she was an introspective and deeply religious girl who after a conventual education desired to enter a religious order, but was forced by her mother to marry in 1664 Jacques Guyon, a middle-aged invalid dominated by a tyrannical mother. Mme. Guyon's response was to retreat more deeply into a life of private contemplation, aided by excruciating forms of mortification and a mystical espousal to Christ.

Guyon died in 1676, but his widow now came under the spell of Molinos's writings and of a neurotic Barnabite friar, Lacombe. In 1680 she achieved a "unitive state" with the divine: "God-me" had supplanted "self-me," and a year later she began to receive visions and revelations. During her "Lacombe period" (1681-88) she wandered from place to place, often with the friar, seeking to found an "interior church" and give birth to "spiritual children," and writing some of her best-known mystical essays. She was arrested in 1688, but was freed on the intervention of Mme. de Maintenon. She began to exchange a series of spiritual letters with Fénelon,* who admired and later defended her, and she became prominent in court circles, lecturing in the Girls' School at St. Cyr. Bossuet,* alarmed by the nature and effects of her teaching and rumors of her private life, examined her writings in 1694. Thirty of her propositions were condemned, but though she recanted, a year later she was again imprisoned and spent six years first at Vincennes and then in the Bastille. Released in 1701, she spent the rest of her life in Blois.

IAN SELLERS

HABIT. Distinctive dress of religious orders. The practice of wearing a special habit goes back to the earliest days of monasticism in Egypt, possibly having its origin in the rough clothes of the anchorites, or even in the rough cloaks of the wandering Cynics (first century). It normally consists of tunic, belt or girdle, scapular, hood (men) or veil (women), and a mantle for use in choir or outdoors. Colors are usually white, brown, or black. Worn by all the old orders (monks, friars, nuns), it is dispensed with by some modern orders.

HADRIAN (76-138). Roman emperor from 117. At his accession as adopted heir to Trajan, Publius Aelius Hadrianus was already in his forties, a mature man with considerable humanistic concern for things Greek, for which he was to be known and ultimately judged. The decade of the 120s saw him tour almost all the empire. The Jewish Revolt of 132-35 was the sole severe strain within the state during his administration; one result was the distinguishing of Christianity from its roots. Hadrian's awareness of Christianity seems negligible, and he was certainly no persecutor; at most he maintained the policy enunciated by Trajan, while his reign saw the first Christian apologist Quadratus,* and the expansion of Gnosticism with Basilides. The story that he built temples without images for Christ is a fallacious product of later syncretistic concerns.

CLYDE CURRY SMITH

HADRIAN, for popes of that name see ADRIAN

HADRIAN THE AFRICAN (d.709). Monk and scholar. Native of Roman Africa, he became abbot of Niridan, a Benedictine house near Naples. Proficient in Greek and Latin learning and adept in canon law, he resisted his appointment by Pope Vitalian, a personal friend, to the vacant archbishopric of Canterbury. He did agree, however, to accompany his suggested nominee, Theodore of Tarsus, to England as adviser and defender of Roman orthodoxy. As abbot of SS. Peter and Paul's monastery and head of the school in Canterbury, he introduced a variety of disciplines, taught pagan and patristic literature, founded other schools, and educated students—notably Aldhelm—from as far as Ireland and the Continent. His contributions helped assure the dominance of Roman Christianity in Britain.

JAMES DE JONG

HAGENAU, COLLOQUY OF (1540). Called by the emperor Charles V in an attempted reconciliation between his Lutheran and Roman Catholic subjects. Hermann von Wied, Johann Gropper, and Martin Bucer were among the participants. The Colloquy had no permanent results. It was adjourned to meet in Worms in November 1540.

HAGIOGRAPHY. That special category of biographical literature centering upon those regarded as "holy." Such hagiography ranges from the life of one saint to the complex collections of lives of saints encyclopedically arranged. Hagiographic literature within the Hebrew-Christian tradition belongs already to the literature of the OT. It is further embellished in the Apocrypha and Pseudepigrapha, plays little part in the NT (unless one includes in it certain aspects of the gospels or the Acts of the Apostles), but is of increasing importance from the second century on, including much of the Apocryphal New Testament,* with Eusebius of Caesarea representing the first major collector and illustrating the close alliance with "martyrology."

CLYDE CURRY SMITH

HAGUE, DYSON (1857-1935). Clergyman, educator, and author. Graduate of arts and divinity at the University of Toronto, he was ordained in 1883, was curate at St. James' Cathedral, Toronto, rector of St. Paul's, Brockville, Ontario, and of St. Paul's, Halifax, Nova Scotia. In 1897 he became professor of apologetics, liturgics, homiletics, and pastoral theology in Wycliffe College. In 1901 he returned to pastoral work. He wrote numerous pamphlets, and several books on Anglican liturgy.

HALDANE, JAMES ALEXANDER (1768-1851). Scottish evangelist. Born in Dundee two weeks after his father's death, he lost his mother when he was six; he and his brother Robert* were placed under the guidance of two kind and understanding uncles. After schooling in Dundee and Edinburgh, he studied at Edinburgh University, then in 1785 joined the navy, and at an early age attained command of an East Indiaman, achieving fame as an officer of coolness and resource. After spiritual self-questioning he left the sea in 1794 and settled in Scotland. Later he was converted, and this led eventually to a series of remarkable itinerancies during which he preached the Gospel in every part of Scotland. In 1797 he founded the Society for Propagating the Gospel at Home, after discovering that the Church of Scotland was as little interested in home as in foreign missions. In 1799 he became the first Congregational minister in Scotland, and two years later was installed in the new "Tabernacle" in Edinburgh—an impressive building with seating for more than 3,000, where he was to minister for almost fifty years. Like his brother Robert, he embraced Baptist principles. Concerned with restoring the life and conditions of the apostolic church, he was an advocate of the kind of church fellowship which in the next generation was developed by the Christian Brethren.

See A. Haldane, *The Lives of Robert ... and James Alexander Haldane* (1856).

J.D. DOUGLAS

HALDANE, ROBERT (1764-1842). Scottish evangelist, writer, and philanthropist. Born in London of an ancient Scottish family, he lost both parents by the time he was ten years old. Educated in schools at Dundee and Edinburgh, he joined the navy in 1780, but soon abandoned that career to return to his Stirlingshire estate, where he gained a reputation as a farmer and landscape gardener. Converted in 1795, he resolved to devote life, talents and fortune to the Christian cause. He sold his estate, determined to finance and participate in missionary work in India, but that door was closed through opposition from the East India Company. The 1796 Church of Scotland general assembly, controlled by Moderates,* also decided against foreign mission work, and much of Robert's money went into establishing preaching "tabernacles" and theological seminaries.

Although not best known for his preaching, he greatly furthered the work of evangelism, in cooperation with his brother James.* He was the moving spirit behind the bringing of twenty-four children from Sierra Leone to be educated in Britain for five years, and was himself prepared to assume complete financial responsibility for the project. As an active friend of the Bible Society, he challenged its circulation of the Apocrypha with the Bible in continental Europe, thus beginning a controversy that lasted for many years. His written works include *Evidences and Authority of Divine Revelation* (1816), and a commentary on the epistle to the Romans, based on lectures given to students in Geneva during a period of evangelistic work in Switzerland and France (1816-19).

J.D. DOUGLAS

HALES, ALEXANDER OF, see ALEXANDER OF HALES

HALF-WAY COVENANT (1662). The admission to New England church membership of more than the dedicated elite. The Massachusetts Synod of 1662 asserted that baptized adults who professed faith and lived uprightly, but who had had no conversion experience, might be accepted as church members. Their children, baptized as "half-way" members, could not receive the Lord's Supper or participate in church elections. This dual conception of membership, forced on churches by declining power and widely modified by 1700, opened the churches to a cross-section of New Englanders. That practice prompted attempts, notably that of Jonathan Edwards, to restate Calvinist orthodoxy. DARREL BIGHAM

HALL, JOSEPH (1574-1656). Bishop of Norwich. He gained early distinction by publishing satirical verse, meditations, and (in 1610) a controversial work against the Brownists, John Robinson and John Smyth. James I made him dean of Worcester and in 1618 sent him as his representative to the Synod of Dort, where he advocated moderation and mutual charity. Though brought up a Calvinist, he regarded the Church of Rome as corrupt but still catholic. As bishop of Exeter from 1627, he secured conformity by a conciliatory policy toward the Puritans which was regarded with suspicion by Archbishop Laud.* In 1640 Hall published his *Divine Right of Episcopacy*, and his *Humble Remonstrance to the High Court of Parliament* (1640-41) brought a reply from five Puritans whose initials made the name "Smectymnuus." In 1641 he was translated to Norwich, but the revenues of his see were sequestrated by Parliament. His *Hard Measure* relates his subsequent poverty and sufferings. JOYCE HORN

HALL, ROBERT (1764-1831). English Baptist minister. Born in Arnesby, Leicestershire, where his father was a Baptist pastor, he was a precocious boy, writing hymns before he was nine and preaching his first sermon at eleven. He was educated at the famous Nonconformist academy run by John Ryland, baptized in 1778, studied at the Baptist Academy, and graduated in 1785 from Aberdeen University. Hall began his ministry as an assistant in Bristol and very quickly established a reputation as an eloquent preacher and a shrewd apologist for Christianity. Theologically he moved his position from an early Calvinism to a basically Arminian system. He was deeply interested in the rapid progress of scientific research and was friendly with Joseph Priestley, the leading Unitarian. In 1791 he moved to Cambridge, where he succeeded Robert Robinson in the pastorate, and continued to cultivate his scientific acquaintances. A period of ill health included two periods of mental breakdown, and at this time, like many others, he sought alleviation in drugs. Then during a period of prescribed rest he had a profound religious experience which he himself described as his "conversion," and in 1807 moved to become minister of Harvey Lane Baptist Church, Leicester, where his ministry was wide and influential.

Like many evangelicals of the time, Nonconformist and Anglican, he took a close interest in social need. He published a pamphlet appealing for help for a fund to provide relief for the distressed stocking-makers of Leicester during periods of unemployment: the germ of the trade union movement could be seen in this development. In 1825 he returned to Broadmead Baptist Church, Bristol. His influence on the Baptist denomination, particularly in moving it away from the at-times sterile Calvinism of the eighteenth century, left its permanent mark. His *Works*, extending to six volumes, were published with a biography by O.G. Gregory a year after his death.

A. MORGAN DERHAM

HALLBECK, HANS PETER (1784-1840). Moravian missionary to South Africa. Born in

Sweden, he studied theology at Lund before joining the Moravian Brethren.* From 1817 until his death he was superintendent of their mission in the Cape Colony. This was a period of consolidation and expansion. Five new missions were established, two of them among Africans, and pastoral work was extended from the closed settlements to neighboring farms. Hallbeck laid stress upon Christian education and established a training school at Genadendal to provide indigenous helpers for the mission (1838). He was a gifted administrator. The regulations governing mission settlements were improved, giving communicant householders a share in the administration. After 1838 many emancipated slaves were successfully integrated into the communities. Hallbeck maintained the good reputation of the Moravians and cooperated wherever possible with the colonial government. He detested the pass laws which restricted Hottentot freedom, but did not oppose them publicly. In 1836 he attended the Moravian Synod, at which he was elected and consecrated bishop. D.G.L. CRAGG

HALLELUJAH. From the Hebrew word meaning "praise ye Yah(weh)"; the Greek and Latin versions transliterated it as *alleluia,* whence the alternative English spelling. In the Bible it occurs only in the latter part of the Psalter and in Revelations 19:1-6. The precise phrase is found only in hymnic context, and only as the beginning or conclusion of a cultic acclamation (with a single exception). All the Psalms containing the phrase appear to be relatively late (none is ascribed to David), and it seems evident that the term became a fixed part of the later temple liturgy. The preponderance of occurrences is at the beginning of individual psalms, and it may be inferred from Psalms 135:19ff. that the Levites had special responsibility for uttering it, probably as a summons to praise. 1 Chronicles 16:36 suggests that the congregation uttered it in response, at the close of hymns of praise. Revelation 19 (as also *Tobit* 13:18) uses the term in eschatological context. The Hallelujah Psalms had considerable use in the synagogue liturgy by NT times, and the Christian Church also adopted the term (in transliteration) from the earliest times. It has played its part throughout the history of Christian liturgy (for which see s.v. "Alleluia" in the *New Catholic Encyclopedia*) and in Christian hymnody.

D.F. PAYNE

HALLER, BERCHTOLD (1492-1536). Swiss Reformer. Born at Aldingen, he studied theology at Cologne, taught for a time, became a spiritual notary in 1517, and assistant and then successor to Thomas Wyttenbach at the Church of St. Vincent, Berne. Here he became imbued with Zwinglian ideas. One of a circle of Evangelical clergy in Berne, he was left alone when the rest were forced to flee in 1523, he himself being tried but acquitted of heresy. With popular backing he gradually won the town council to his Reformed views, defending his actions at the conferences of Baden and Berne (1526 and 1528), compiling with Kolb a Protestant liturgy, and issuing a reformatory edict in 1528. By now the acknowledged spiritual leader of his town, he spent his later years in a round of preaching, visiting and catechizing, in efforts to strengthen the Reformed cause diplomatically, and in controversy with the Anabaptists. IAN SELLERS

HALLESBY, OLE KRISTIAN (1879-1961). Norwegian theologian. Son of a farmer and originally deeply rooted in the Lutheran piety of the "Haugean" tradition (see HAUGE, H.N.), he studied theology and adopted the outlook of the liberal school. In 1902 he experienced a conversion and reverted to the biblical faith and piety of his fathers. For some years he worked as an itinerant lay preacher and was the means of awakenings in several places. He was called to the chair of dogmatics at the Free Faculty of Theology and took the post after acquiring his doctorate in Berlin. From 1909 to 1952 he lectured on dogmatics and in a sense became the teacher of a whole generation of Norwegian ministers. As chairman of the Norwegian Lutheran Home Mission he exerted a great influence on lay Christians all over the country.

Hallesby was the leading light in the opposition of conservative pastors and laymen to the liberal theology. In 1920, on his initiative, a meeting of representatives of home and foreign mission societies decided to cease cooperation with liberal theologians. During the years of German occupation in World War II (1940-45) he was one of the leaders in the church's resistance to the Nazi government. He was arrested in 1943 and lived in a concentration camp until liberation came in 1945. Hallesby wrote textbooks on dogmatics and ethics, and many devotional books, including *Prayer* (ET 1948) and *Why I am a Christian* (ET 1950), some of which were translated into many languages. He was also the first president of the International Fellowship of Evangelical Students (1947). CARL FR. WISLOFF

HALO (Gr. *halōs,* "nimbus"). A circle or disc of light with which the head of Christ, the Virgin Mary, or a saint is surrounded in Christian art. This usage was taken over from religious symbolism within Hellenism and the Roman Empire, wherein gods and some emperors were represented with halos around their bodies or heads. Since the third century there has been a gradual development of usage within Christendom. At first it was deemed proper only to use the halo, whose usual color was blue, for Christ, but from the time of Leo I it was extended to include the Virgin and the saints. During the Middle Ages several types were in currency, but the color now was usually yellow or gold. A plain, round halo was used for angels and saints; a round one with a suitable distinguishing characteristic (e.g., a cross or monogram) for Christ; and a rectangular one for living dignitaries—e.g., that of Gregory the Great in the monastery of Clivus Scauri at Rome. Within contemporary Catholicism the halo is used only for the saints and those of the "blessed" who are venerated. PETER TOON

HALYBURTON, THOMAS (1674-1712). Scottish theologian. Born near Perth, son of a Presbyterian minister ejected in 1662 for nonconformity, he was educated at Rotterdam and at St. Andrews University, was minister of Ceres (1700-10), then became professor of divinity at St. Andrews two years before his early death. Halyburton was a champion of Reformed theology in the controversy over Deism, and this prompted his chief work which was reissued in 1865 as *Essay on the Ground of Formal Reason of a Saving Faith.* His other writings include *Memoirs* which have been frequently reprinted. Archibald Alexander of Princeton spoke most highly of Halyburton, who was regarded by others as one of Scotland's greatest theologians. J.D. DOUGLAS

HAMANN, JOHANN GEORG (1730-1788). German religious thinker. Born in Königsberg, he had an irregular education, became a private tutor, and underwent a religious experience in 1758 during a business trip to London. Returning to Königsberg, he secured a minor customs post and in his spare time began to study as extensively as possible. Soon, despite his eccentric, angular style of writing, he was an acknowledged leader of the literary *Sturm und Drang* movement. The most evangelical of this school, Hamann rediscovered in Luther's work a spontaneous personal faith, a universal concern, and a vastly widened field of religious experience which rose superior to Protestant scholasticism, pietistic subjectivism, and rationalistic philosophy, and which in a series of notable works, particularly *Golgotha und Scheblimini* (1784), he strove to commend to his countrymen. As a self-appointed rejuvenator of German Christianity he exercised an important influence on Herder, Schleiermacher, and Kierkegaard. IAN SELLERS

HAMARTOLOS, GEORGE, see GEORGE HAMARTOLOS

HAMILTON, JOHN (1512-1571). Archbishop of St. Andrews; the last Roman Catholic to hold that office. Born in Edinburgh, he became abbot of Paisley when only fourteen and held the post until his death. Few details of his early career are known for certain. He matriculated at St. Andrews University (1528), later spent some time studying in France, and through influential family connections was appointed in 1543 as privy seal, to which was later added the lord treasurership. After protracted controversy he was in 1546 consecrated as bishop of Dunkeld, but was soon translated to the primacy. In 1552 there was published *Archbishop Hamilton's Catechism,* a highly regarded product of St. Andrews University, the precise authorship of which cannot be determined. Even after the Reformation in 1560 he maintained his opposition, continued to be known as archbishop of St. Andrews though latterly a political figure only, retained his seat in Parliament, and baptized the future James VI after the Roman form. A supporter of Mary, Queen of Scots, in her trials, he was finally indicted as a traitor and hanged at Stirling. J.D. DOUGLAS

HAMILTON, PATRICK (1503-1528). Generally regarded as the proto-martyr of the Scottish Reformation. Of aristocratic lineage, he was sent to Paris University about 1515 and graduated five years later, soon after Luther had posted his theses. He matriculated at St. Andrews University in 1523, fired up by the Lutheran opinions expressed by those for whom the Word of God had become a living force. Hamilton probably taught in St. Andrews, but in 1527 he fell foul of Archbishop Beaton and was compelled to flee to the land of Luther. Later that year he returned to Scotland, intent on preaching the Gospel. Early in 1528 Beaton summoned him to St. Andrews, ostensibly to debate, but in reality to put him to death hurriedly before influential friends could muster support. Sentence was passed by the ecclesiastical court, and Hamilton was burnt in St. Andrews, His murderers expected by summary treatment of one of high rank to intimidate others and suppress the rising tide of Reformation. The opposite effect was produced; great discussion ensued, and as one bystander said, "The reek [smoke] of Mr. Patrick Hamilton has infected as many as it did blow upon." J.D. DOUGLAS

HAMLIN, CYRUS (1811-1900). American Congregational missionary and educator. Born near Waterford, Maine, he graduated from Bowdoin College (1831) and Bangor Theological Seminary (1837). He went to Turkey under the American Board of Commissioners for Foreign Missions. In 1840 he founded a seminary at Bebek. He resigned from the American Board and in 1863 opened Robert College, later moving it to Constantinople. Roberts, the school's benefactor, was persuaded because of a misunderstanding to dismiss Hamlin as college president in 1877. Hamlin then taught at Bangor Seminary for three years, became president of Middlebury College, Vermont, in 1881, and retired in 1885. He published *Among the Turks* (1878) and *My Life and Times* (1893). ROBERT C. NEWMAN

HAMMOND, HENRY (1605-1660). Anglican divine. Born at Chertsey, Surrey, and educated at Eton and Magdalen College, Oxford, he was ordained in 1629 and in 1633 was appointed rector of Penhurst, Kent. His ten years of parish ministry helped to prepare him to defend and explain the doctrine and practice of the Church of England in the revolutionary period, 1640-60. Returning to Oxford during the Civil War, he was appointed by the king, whose faithful servant he became, to a canonry at Christ Church in 1645. During the Parliamentary Visitation he lost this, and eventually settled down at Westwood in Worcestershire. He excelled as a writer and is chiefly remembered for his *Practical Catechism* (1644) and *Paraphrases and Annotations on the New Testament* (1653). Most of his numerous controversial tracts are printed in his *Miscellaneous Theological Works* (4 vols., 1847). PETER TOON

HAMPTON COURT CONFERENCE (1604). Following the presentation of the Millenary Petition* by the Puritans, James I agreed to the suggestion of a conference between representatives of the bishops and the Puritans. He decided to be the chairman. The conference, lasting four days, took place in January 1604 at Hampton Court Palace; the leading participants were Bishops Bancroft and Bilson on one side, J. Reynolds and L. Chaderton on the other. Though many of the Puritan demands were dismissed, the king did admit the justice of some of them. He agreed to allow minor changes in the Book of Common Prayer, to attenuate the power of the High Commission, to improve parish livings and eliminate plurality, to change the methods of suspension and excommunication, and to gain a new translation of the Bible. In reality only the latter—the famous Authorized Version—ever came to fruition. PETER TOON

HANDEL, GEORGE FRIDERIC (1685-1759). Music composer. Unlike Bach, who was born in the same year, Handel was not of a musical family, and his father only grudgingly acknowledged his musical talent and destined him for law. He received his early musical training from the distinguished organist and composer Zachow, in Halle. The greater part of his career was concerned with dramatic music, opera, and oratorio. His only music written for the church consists of his early German passions, the Latin psalms written during his sojourn in Italy, the cantata-like anthems composed for the British duke of Chandos, and occasional festal works for coronations and national celebrations. In the last category are the magnificent "Utrecht" and "Dettingen" *Te Deums.* The failure of his Italian opera enterprises in London led him to turn more and more to oratorios based on biblical themes. These appealed to a wider public in England because of the vernacular text and familiar plots. With the exception of *Messiah* and *Israel in Egypt,* which draw entirely upon direct biblical texts, the oratorios employed versified librettos, not always of great poetical merit.

Handel completely overshadowed his English contemporaries with the dramatic grandeur of his style and his instinct for excellent choral effect. The nobility of his melody, together with these other attributes, places him among the greatest composers of all time. The classical masters, Haydn, Mozart, and Beethoven, admired and drew inspiration from his choral style. Oratorio was conceived as edifying Lenten entertainment, but numerous extracts from Handel's works of this sort found their way into the repertory of church choirs. *Messiah* (written 1741, first performed 1742), became the most performed major choral work in history, and continues to be. Handel was also an outstanding composer of chamber music and concertos.

BIBLIOGRAPHY: R.M. Myers, *Handel's Messiah: A Touchstone of Taste* (1948); W. Dean, *Handel's Dramatic Oratorios and Masks* (1959); P.H. Lang, *George Frideric Handel* (1966).

J.B. MAC MILLAN

HANNINGTON, JAMES (1847-1885). Anglican missionary to East Africa. Born at Hurstpierpoint, Sussex, he was educated at Brighton and then entered business and the army. In 1868 he enrolled at St. Mary Hall, Oxford, to train for the Anglican ministry, to which he was ordained in 1874. In 1882 he offered to the Church Missionary Society and was appointed to Uganda. He reached Lake Victoria on his way out, but was forced to return to England suffering from malaria and dysentery. On his recovery he again offered to the CMS and was consecrated the first bishop of Eastern Equatorial Africa at Lambeth in 1884. In 1885 he reached Mombasa, and after superintending the work at Freretown near Mombasa he set off on foot with porters for Uganda. He reached Busoga on Lake Victoria, and there he was arrested and later killed on the orders of Mwanga, the treacherous *Kabaka* (ruler) of the Baganda. JOHN WILKINSON

HARDENBERG, ALBERT RIZAEUS (1510-1574). German Reformer. Born at Emden in East Friesland, his early education was at the school of the Brethren of the Common Life in Groningen, where he became an admirer of Wessel Gansfort.* As a student at Louvain he aroused suspicion by his humanistic-evangelical leanings. While gaining a doctorate at Mainz he came into contact with the Polish Calvinist Jan Laski* (John à Lasco). After briefly teaching at Louvain, he returned to the Groningen area, keeping in touch with Laski and Melanchthon. After a visit to Wittenberg he became actively involved in the Reformation. In 1544 he went to Cologne, aiding Archbishop Hermann von Wied in his short-lived attempt to reform the Catholic Church there. After appearing at the Diets of Speyer and Worms (1544, 1545) he became military chaplain to the count of Oldenburg, and was wounded in battle. The count made him the first evangelical preacher at the cathedral in Bremen. Expelled from Bremen for Calvinistic views of the Eucharist, Hardenberg later became pastor of a Calvinist church in his birthplace of Emden. He edited for publication the works of Wessel Gansfort.

DIRK JELLEMA

HARDENBERG, F.L.F. VON, see NOVALIS

HARDING, STEPHEN (d.1134). Abbot of Cîteaux. Harding was born at Sherborne, Dorset, and as a young man traveled widely, visiting Scotland, Paris, and Rome. He joined the community at Molême in Burgundy where he failed to secure acceptance of the Rule of St. Benedict. Departing in 1098 with twenty others, including the abbot and the prior, he established a strict and austere religious house at Cîteaux, a very desolate spot. Here he was successively subprior, prior, and third abbot. The community flourished, and another thirteen houses of this new Cistercian* Order were founded largely by Stephen himself. He drew up the rule, instituted a general chapter, and introduced the famous white habit for his communities, obtained papal support and a large measure of freedom from episcopal control, and must along with Abbot Robert and Alberic the Prior be regarded as the founder of the Cistercian Order. IAN SELLERS

HARDOUIN, JEAN (1646-1729). Classicist and polemicist. Born at Quimper in Brittany, son of a bookseller, he joined the Jesuits, studied theology at Paris, and became librarian of the Jesuit *Col-*

lège Louis-le-Grand in 1683. He was a gifted numismatist, philologist, classical scholar, and editor, but his wild assertions earned him a certain notoriety. He claimed that, with a few salient exceptions, the Greek and Latin classics were the productions of thirteenth-century monks, and he asserted that some of the works of the Fathers were likewise spurious. Though he declared that all the church councils before Trent were fabrications, he nonetheless prepared careful transcripts of the texts for his excellent editions of the church councils from the year 34 to 1714, including a number never before published. This definitive work, *Conciliorum collectio regia maxima*, cosponsored by Louis XIV and his clergy, was suppressed by the French clergy for ten years until 1725 because it countered their Gallican pretensions. Hardouin rejected the Greek NT, insisting that Jesus had preached, and the NT had been written, originally in Latin. MARY E. ROGERS

HÄRING, THEODOR (1848-1928). German theologian. Born at Stuttgart and educated at Urach and Tübingen, he became professor at Zurich (1886), Göttingen (1887), and Tübingen (1895). In a series of theological treatises, particularly *The Righteousness of God in Paul* (1896), *The Ethics of the Christian Life* (ET 1909), and *The Christian Faith* (ET 1913), he showed himself with his Pietistic background among the more conservative of Ritschl's* followers, reaffirming traditional Christian doctrine and trying to save the system from immanentism and mere moralism. At the same time he strongly emphasized the kingdom of God and an ethical system closely patterned on doctrine, and was one of the first German theologians to apply in detail Christian ethics to political and social questions.
IAN SELLERS

HARKLEAN VERSION, see SYRIAC VERSIONS OF THE BIBLE

HARLESS, GOTTLIEB CHRISTOPH ADOLPH VON (1806-1879). German Lutheran theologian. Born in Nuremberg, he studied philology and law at Erlangen and Halle, where he was influenced by F.A.G. Tholuck* toward theology and particularly Luther's doctrine of justification. He was professor of NT exegesis and university preacher at Erlangen (1829-45), where he considerably raised the standard of theological teaching and founded the *Zeitschrift für Protestantismus und Kirche* (1838-76). He was professor at Leipzig (1845-50) and court preacher at Dresden (1850). Called to Munich (1852) to become president of the supreme consistory of Bavaria, he reorganized the Lutheran state church and gave it a new hymnbook and a new order of services. He was one of the most influential representatives of Lutheran orthodoxy of his generation. Among his writings was *Christliche Ethik* (1842), translated as *System of Christian Ethics* (1865). J.G.G. NORMAN

HARMONIES. Efforts to make explicit the close proximity or exact relationship of parallel pieces of literature. Within the NT, the special relationships of the "synoptic" gospels provide not only the basic and best-known example for harmony, but the classic instance of the historical problem related to priority. While Tatian harmonized, and Augustine understood the need for viewing the gospels synoptically, the ancient church produced no "harmony" in the modern literary sense. The earliest modern usage of the analogous concept "synoptic" in the context of printing a "harmony" was that resulting from the textual studies of J.J. Griesbach.*

HARMONY SOCIETY. A Protestant communal society established in 1805 north of Pittsburgh, Pennsylvania by 500 Pietist dissenters from Württemberg, Germany, seeking religious freedom. Led by George Rapp (1757-1847) and his adopted son Frederick, the group moved to 30,000 acres on the Wabash River at New Harmony, Indiana, in 1815. In 1825 the society sold New Harmony to British socialist Robert Owen for his communal experiment, and moved to Economy (now Ambridge), Pennsylvania, where it survived until 1916. Practicing first-century Christian communism, Harmonists labored cooperatively as farmers, brewers, millers, and spinners, making their communities showpieces of economic growth and security. They pioneered in prefabricated buildings, oil refining, and underwriting railroad construction. Well-ordered lives, uniform dress, and simple, nonceremonial religious observances characterized "Rappites." Believing that Father Rapp would present them personally to Christ on His imminent second coming, they became perfectionists, by 1807 adopting celibacy, a factor in the society's ultimate demise.
D.E. PITZER

HARMS, CLAUS (1778-1855). Lutheran preacher and theologian. After helping his miller father at Fahrstadt/Holstein, he went to the university of Kiel in 1799 where he became an evangelical through reading Schleiermacher's *Monologues*. Elected deacon at Lunden/Holstein in 1806, Harms rapidly gained fame as a preacher. Ten years later he became archdeacon at St. Nicolai's Church in Kiel and was elevated to the position of provost in 1835. In 1834 he had declined an offer to succeed Schleiermacher* as pastor of the Church of the Holy Trinity in Berlin. On the occasion of the tercentenary of Luther's Ninety-Five Theses, Harms wrote his *Ninety-Five Theses*. These attacked rationalism with its attendant Pelagianism. The union of Reformed and Lutheran churches proposed by the Prussian monarch Frederick William III was also criticized by him. His writings contributed significantly to the advancement of Lutheran piety. His further writings included a *Pastoraltheologie* and collections of sermons. WAYNE DETZLER

HARMS, LUDWIG (1808-1865). Mission organizer in Germany. Son of a pastor in the Lüneburger Heath, he was at first influenced by rationalism, but in 1830 was converted to a strongly biblical Christianity. After serving for some years as a teacher at Lauenburg, he succeeded his father in the tiny village of Hermannsburg. A natural outgrowth of his deeply pietistic orientation was an interest in foreign missions, and he assisted in forming the North German Mission in 1836. Although his parishioners were simple peasants, they founded a missionary training school in 1849 and sent a group of missionary colonists to Ethiopia in 1853. Forbidden to land there, they located in Natal and established a settlement named Hermannsburg. They stressed a strongly confessional Lutheranism and trained the Africans in agricultural techniques. Harms himself never left Germany, but continually fostered the work at home and dispatched more agricultural missionaries to open new stations elsewhere in South Africa. RICHARD V. PIERARD

HARNACK, ADOLF (1851-1930). German scholar. Son of the Lutheran scholar Theodosius Harnack (1817-89), he taught at Leipzig (1874) before becoming professor at Giessen (1879), Marburg (1886), and Berlin (1889-1921). The last appointment was challenged by the church because of Harnack's doubts about the authorship of the fourth gospel and other NT books, his unorthodox interpretations of biblical miracles including the Resurrection and his denial of Christ's institution of baptism (see his *History of Dogma*, 7 vols., 1894-99). The appointment was, however, upheld by the Prussian cabinet and the emperor. But the dispute cast a shadow over the rest of his career, and he was denied all official recognition by the church, including the right to examine his own pupils in church examinations. Nevertheless, Harnack was perhaps the most influential church historian and theologian until World War I.

Harnack's main field was patristic thought, on which he published numerous monographs. His standpoint was a form of Ritschlianism that regarded metaphysics in early Christian thought as an alien intrusion ("Hellenization"). In the winter of 1899-1900 he delivered a course of public lectures assessing Christianity in the light of modern scholarship. They were taken down in shorthand and published as *Das Wesen des Christentums* (ET, *What is Christianity?*, 1901). Jesus was depicted as a man who had rest and peace for his soul and was able to give life and strength to others. The gospel that he preached was not about himself, but about the Father. It concerned the kingdom, the fatherhood of God, the infinite value of the human soul, the higher righteousness and the command to love. The work was a best seller and the center of much controversy.

In many ways Harnack was positive. Though liberal in theology (later clashing with his former pupil Barth*), he was conservative and perceptive in his studies on the NT. He held that Acts was written by Luke while Paul was a prisoner in Rome, assigning an early date to "Q," the synoptic gospels, and Acts. Such views would undermine much contemporary liberal and radical scholarship. His studies were published in English as *Luke the Physician* (1907); *The Sayings of Jesus* (1908); *The Acts of the Apostles* (1909); and *The Date of the Acts and of the Synoptic Gospels* (1911). English translations of other works include *The Mission and Expansion of Christianity in the First Three Centuries* (2 vols., 1904-05), and *The Consitution and Law of the Church in the First Two Centuries* (1910).

In 1906 Harnack was appointed director of the Prussian Royal Library (the largest in Germany), and he became also president of the Kaiser Wilhelm *Gesellschaft* for learning and science. His numerous honors included the title "von Harnack" in 1914. He was interested in social questions and published with W. Hermann *Essays on the Social Gospel* (ET 1907). He declined the post of German ambassador to Washington (1921).

BIBLIOGRAPHY: Agnes von Zahn-Harnack (his daughter), *Adolf von Harnack* (1936); G.W. Glick, *The Reality of Christianity: A Study of Adolf von Harnack as Historian and Theologian* (1967); W. Pauck, *Harnack and Troeltsch: Two Historical Theologians* (1968). COLIN BROWN

HARPER, WILLIAM RAINEY (1856-1906). Semitics scholar. Born in Ohio of Scotch-Irish ancestry, he held a Ph.D. from Yale by 1875, and in 1879 became professor of Hebrew at the new Baptist Union Theological Seminary at Morgan Park, Illinois. He returned in 1886 to teach Semitics at Yale, to which was added in 1889 the chair of biblical literature. Two years later he became head of the new University of Chicago, into which the Morgan Park seminary was integrated with Harper assuming responsibility also for the Semitics department. Over the next fourteen years his enormous energies burnt out his life, but not before he had created a great graduate university. He produced a major commentary on Amos and Hosea (1905), and from 1884 edited the periodical *Hebraica*. CLYDE CURRY SMITH

HARRIS, HOWEL (1714-1773). Welsh preacher. Born at Talgarth, Brecon, of humble parents, Harris intended at an early age to enter the ministry of the established church; for a time he supported himself as a schoolmaster, undergoing a vital conversion experience in 1735. He went to Oxford but spent only a week there, returning to Wales to begin a campaign of tireless evangelism. He aroused first the south by his stately appearance, powerful voice, and overwhelming passion, and though often threatened by mobs and magistrates he extended his activities with equal success to the north in 1739.

Though he must be regarded as the principal founder of Welsh Calvinistic Methodism and the greatest spiritual force in the principality of his day, Harris was shy and awkward in the presence of other evangelical leaders, and quarreled with both Rowland and Whitefield.* Many influences —Wesleyan, Moravian, and even Antinomian and Universalist—molded his thought, but he always remained loyal to the Church of England, and deplored any tendency to break away from it. But his excursions into theology are not impressive—he was a revivalist, not a systematizer, and scenes of wild enthusiasm accompanied his preaching. In 1752 he retired to a house at Trevecca Fach which he built up as a center for revivalist activity. He was supported by the Countess of Huntingdon who after 1768 sent her own students to train at Trevecca Isaf. Harris died in 1773, leaving behind a small number of very popular Welsh hymns and numerous letters and journals, some of which are still unpublished. IAN SELLERS

HARRIS, JAMES RENDEL (1852-1941). Biblical scholar and Orientalist. Born in Plymouth, he was educated at Clare College, Cambridge, and taught mathematics at the university until 1882, when he migrated to the USA and taught at Johns Hopkins University. After denouncing vivisection, he was compelled to leave. He then joined the staff of Haverford College in Pennsylvania. On returning to England he gained the reputation of a brilliant but unorthodox scholar who specialized in textual problems. His final post was that of curator of MSS at the John Rylands Library, Manchester. He edited and published many ancient texts, but is best known for his discovery in 1889 of the long lost *Apology of Aristides* (published in 1891). In early life Harris was a Congregationalist, but in 1880 he joined the Society of Friends. In 1896 he organized relief for Armenians at the time of the massacres. Throughout his life his work output was enormous. Theologically he was a "liberal" Christian, scornful of "fundamentalism." R.E.D. CLARK

HARVARD, JOHN (1607-1638). Benefactor of Harvard University. Born in Southwark, he studied at Emmanuel College, Cambridge, and because of the religious situation in England under Archbishop Laud he joined the Puritan emigration to the New World. He settled in Massachusetts and was admitted a member and teaching elder of the Congregational church at Charlestown; he was also a freeman of the colony. In his will he left half his estate with a library of about 400 volumes to the new college, recently founded by the colony in 1636. The General Court of Massachusetts named the college after him in 1638/39, and with the aid of his legacy buildings were erected. PETER TOON

HASSLER, HANS LEO (1564-1612). Probably the ablest German composer of the Renaissance. He studied in Venice with Andrea Gabrieli and was much influenced by the latter's distinguished nephew, Giovanni.* His music is notable for its melodic charm. Hassler was active in Augsburg and Nuremberg. He wrote music for both Lutherans and Catholics, some of which might have been used in either form of worship. Notable are his settings of great Lutheran chorales in both simple and elaborate motet style. The well-known *Passion* chorale (sung now to "O Sacred Head") is his melody, originally composed for secular words, but adapted by him as a church tune.

HASTINGS, JAMES (1852-1922). Scottish minister and editor. Born in Huntly and educated in arts and divinity at Aberdeen, he held charges in the Free Church and (after the 1901 union) the United Free Church (1884-1911) before retiring to engage in editorial work. In 1889 he had founded the monthly *Expository Times,* which he edited until his death. His many other works as an editor and writer include a five-volume *Dictionary of the Bible* (1898-1904), a two-volume *Dictionary of Christ and the Gospels* (1906-8), and a two-volume *Dictionary of the Apostolic Church* (1915-18). He is probably best known, however, for successfully undertaking a daunting project: the *Encyclopaedia of Religion and Ethics* (12 vols., 1908-21; an index volume appeared later). Hastings was also a magnificent preacher, a man whose message was always unmistakably evangelical, spoken without the aid of notes and with that eloquent simplicity which is not infrequently associated with a wide range of knowledge. J.D. DOUGLAS

HATCH, EDWIN (1835-1889). Anglican divine. Born at Derby, he graduated from Oxford and in 1859 was appointed professor of classics at Trinity College, Toronto. He became rector of Quebec High School (1862), vice-principal of St. Mary's Hall, Oxford (1867-85), rector of Purleigh, Essex (1883), and reader in ecclesiastical history at Oxford (1884). His most important work was his Bampton Lectures on *The Organization of the Early Christian Churches* (1880), which aroused considerable controversy, especially in High Church circles. They argued that the Christian episcopate derived from the financial administrators *(episkopoi)* of Greek religious associations. Hatch continued the subject with *The Growth of Church Institutions* (1887). He produced also *Essays in Biblical Greek* (1889) and *Concordance to the Septuagint* (with H.A. Redpath, published posthumously in 1897). His Hibbert Lectures, *The Influence of Greek Ideas and Usages on the Christian Church* (1888), reflect his philosophical interests. HOWARD SAINSBURY

HATFIELD, COUNCIL OF (679). This provincial synod or meeting of the bishops and teachers of the English Church at Hatfield (or Heathfield) was summoned and presided over by Archbishop Theodore. It met at the wish of Pope Agatho, who hoped for and secured the church's condemnation of the Monothelite* heresy, its acceptance of the decrees of the first five general councils of the church, and a profession of faith in the Double Procession of the Holy Spirit. At a council in Rome in the following year, Wilfrid* attested this decision of the Church in England as bishop of York and legate of the synod of Britain.

HAUGE, HANS NIELSEN (1771-1824). Norwegian lay preacher. A farmer's son, he was brought up in a pious Lutheran home, and in 1796 had a religious experience in which he felt called by God to exhort the people of Norway to repentance. He traveled throughout the country (1796-1804), usually on foot, preaching his message and gathering followers wherever he went. At the same time he started factories and other industrial enterprises. Itinerant preaching was not lawful, and his economic efforts were looked on with suspicion. Arrested ten times, he was in prison from 1804 to 1811. After a prolonged trial he was sentenced in 1814 to pay a fine for unlawful preaching and strong criticism of the clergy. Helped by friends, who came to be called "Haugeans," he bought a farm near Oslo. During his last years relations with the authorities were friendly. Hauge wrote many books which had a large circulation. His preaching was pietistic inasmuch as it stressed personal holiness. Hauge is generally regarded as the initiator of the powerful Christian laymen's movement in Norway.

CARL FR. WISLOFF

HAUSRATH, ADOLF (1837-1909). German Lutheran theologian. A leading liberal scholar, he served as a pastor and later professor at Heidelberg and belonged to the consistory of the Baden state church. Stimulated by the ideas of the Tübingen School,* he transformed and popularized its historical picture of primitive Christianity. Hausrath was a founder and the secretary of the German *Protestantenverein,* the group which represented a distinctly liberal position in the Lutheran Church. He was also noted for his biographical and epic writings, and believed the novel was an effective means of making history live.

HAVERGAL, FRANCES RIDLEY (1838-1879). Hymnwriter. Born in a Worcestershire rectory, she early gave evidence of possessing great gifts. She became very proficient in Latin, Greek and Hebrew, but her main interest was the writing of poetry, which she began when she was only seven. Converted at fifteen, she spent the rest of her life in various Christian activities. Her first accepted poem was "I gave my life for thee," but even more popular is the well-known "Take my life and let it be." She published several volumes of poems and hymns, the best known of which is *Kept for the Master's Use.*

HAWEIS, THOMAS (1734-1820). Co-founder of the London Missionary Society and trustee-executor of Lady Huntingdon. Son of a Redruth solicitor, he attended Truro Grammar School during the mastership of George Conon, through whom he was first introduced to the doctrines of the evangelical revival. He was converted and called to the ministry under Samuel Walker, curate of St. Mary's. At Oxford he started a second holy club among the undergraduates, and later served as curate to Joseph Jane at St. Mary Magdalene. After assisting Martin Madan, chaplain to the Lock Hospital in London, Haweis took the living of All Saints, Aldwincle, Northamptonshire, in 1764. His church quickly became a center of evangelical influence throughout the area. In 1774 he was appointed chaplain to Lady Huntingdon. When the London Missionary Society was formed in 1795 he was instrumental in ensuring that Tahiti was the first field to be evangelized.

A. SKEVINGTON WOOD

HAWKER, ROBERT STEPHEN (1803-1875). English poet. Educated at Cheltenham and Oxford, he spent most of his life as vicar of Morwenstow on the north coast of Cornwall. Although his Anglo-Catholicism was marked by his own eccentricities, he was undoubtedly sincerely attracted to the lore of the Celtic saints, and "A Rapture on the Cornish Hills" seems to record a genuine mystical experience. His most ambitious work is *The Quest of the Sangraal* (1864), in which he brings out the explicit Christian associations of the Arthurian legend and records the quest as one full of spiritual vitality. There is some controversy as to whether or not he was accepted into the Roman Catholic Church on his deathbed.

ARTHUR POLLARD

HAWTHORNE, NATHANIEL (1804-1864). American writer. Born in Salem, Massachusetts, he attended Bowdoin College, where one of his classmates was the poet H.W. Longfellow.* After college he returned to Salem for a period of reading, reflection, and writing—training in the craftmanship that was to make him the first major American novelist. In 1837, the same year in which he married Sophia Peabody, he published his first collection of stories, *Twice Told Tales.* He spent two years in a Boston customs house and seven months in the utopian community of Brook Farm. Neither was a pleasant experience. Nor did he feel any enthusiasm for the transcendentalism of his friends Emerson and Thoreau. His first novel, *The Scarlet Letter* (1850), acclaimed by many as the greatest American novel, won him literary success for its masterful structure, beauty of style, and penetrating assessment of the Puritan moral conscience. *The House of Seven Gables* (1851) examines the decadence of Puritanism. From 1853 to 1857 he served as U.S. consul in Liverpool. He spent the next two years in Italy, the setting of his last complete novel, *The Marble Faun* (1860). Among his best tales are "Young Goodman Brown," "The Birthmark," "Rappaccini's Daughter," and "My Kingsman, Major Molineaux." Together with Poe, Hawthorne did much to shape the short story as a distinctive American form. His art tended toward the projection of moral ideas through symbol and allegory. In American literature he is the classic interpreter of Puritanism.

PAUL M. BECHTEL

HAYDN, FRANZ JOSEPH (1732-1809). Music composer. Son of a humble wheelwright in lower Austria, he rose to be the *Kapellmeister* of the dazzling princely court at Esterhaz, Hungary, and one of the most sought-after composers in Europe. In his later years he wrote his greatest symphonies for London audiences, and received an honorary doctorate at Oxford. He spent his last years in Vienna. Although his greatest energies were spent in the realm of the symphony and the string quartet, he wrote at least a dozen Masses, the last six postdating his symphonic output and considered by some critics his crowning achievement. He also wrote a variety of works for the Catholic rite in the classical, symphonic style, a setting of *The Seven Last Words,* and his magnificent oratorio, *The Creation,* inspired by his experiences with Handel's* music in England. His church music has been frequently criticized unjustly as being too lighthearted for the sanctuary; it simply represents the taste of the classical era. His younger brother Michael was also a distinguished and voluminous composer of Catholic church music in the classical vein.

J.B. MAC MILLAN

HAYMO OF FAVERSHAM (d.1244). English Franciscan. Born at Faversham, Kent, he became a Master of Divinity at Paris, and entered the Order of Friars Minor (c.1226). Returning to England, he lectured at Oxford before 1229. He was sent as a deputy by the general chapter of his order at Assisi to Pope Gregory IX to seek official explanation of the Rule (1230). Gregory sent him on a mission to Constantinople (1233) to negotiate reunion with the Eastern Church. He took a leading part in the deposition of Elias of Cortona,* and was himself elected provincial of the English province, succeeding Albert of Pisa. In 1240 he was elected fourth general of the Franciscan Order, the only Englishman ever to hold the position. He was called *Speculum honestatis.* At the request of Innocent IV he revised the ordinals for the *Breviarum Romanum* (1243-44). He died at Anagni. J.G.G. NORMAN

HEADLAM, ARTHUR CAYLEY (1862-1947). Anglican bishop and theologian. Educated at Winchester and New College, Oxford, he was successively fellow of All Souls (1885-96), parish priest at Welwyn (1896-1903), principal and professor of theology at King's College, London (1903-18), regius professor of divinity at Oxford (1918-23), and bishop of Gloucester (1923-45). His early work was concentrated in the area of NT, where he is best known for his collaboration with W. Sanday* in a classic commentary on Romans (1895). His theological position was that of a moderate conservative; one of his chief concerns was Christian unity. In addition to important essays in Hastings's *Dictionary of the Bible* (1898-1902), Headlam wrote *St. Paul and Christianity* (1913), *The Miracles of the New Testament* (1914), *The Doctrine of the Church and Christian Reunion* (1920), *The Fourth Gospel as History* (1948), and numerous other books and essays. W. WARD GASQUE

HEALING, see SPIRITUAL HEALING

HEAVEN. The word is used in the Bible in a twofold sense: as the visible heaven over our heads, including all that is apart from the earth (Gen. 1:1; 2:1), and as the invisible heaven, the dwelling place of God, the holy angels, and the redeemed of all ages. Some Jews held that there are a number of heavens—as many as seven—although this is nowhere taught in Scripture. Paul, however, speaks of being caught up into the third heaven (2 Cor. 12:2-4), but he probably meant by this nothing more than the invisible heaven. The invisible heavens are the abode of God (Deut. 26:15; 1 Kings 8:30; Job 22:12; Matt. 12:50), although they cannot contain God, who is omnipresent (Ps. 139:8-10). By metonomy the word "heaven" comes to be used for God Himself (Matt. 16:19; 18:18). The Jews had a scruple against the use of the divine name, and therefore used substitutes, one of which was "heaven" (Mark 11:30; Luke 15:18,21). In the incarnation Christ descended from heaven (John 3:13; 6:38), and at His ascension He returned to heaven (Mark 16:19; Acts 1:11), in which He prepares a place for His own (John 14:2-4) and from which He shall come to judge the living and the dead (Matt. 24:30).

The invisible heavens are also the dwelling place of the holy angels (Matt. 22:30; 24:36) and of the righteous dead. The souls of the latter enter heaven directly after death (Phil. 1:23; 2 Cor. 5:6-8), but at the second coming of Christ they will receive new spiritual bodies, adapted for existence in their new environment. In heaven there is an end of death, pain, tears, sin (Rev. 21:4,27; 22:3,5). The redeemed sing songs of redemption (Rev. 14:3) and they serve God (Rev. 7:14). At the end of time God will create new heavens and a new earth (Isa. 65:17; 66:22, 2 Pet. 3:13; Rev. 21:1). Some theologians regard heaven as a state of the soul rather than a place, but there is no warrant for this opinion in the teaching of the Bible.

BIBLIOGRAPHY: R. Baxter, *The Saints' Everlasting Rest* (1650); F.E. Marsh, *What is Heaven?* (n.d.); U.E. Simon, *Heaven in the Christian Tradition* (1958); W.M. Smith, *The Biblical Doctrine of Heaven* (1968). STEVEN BARABAS

HEBER, REGINALD (1783-1826). Bishop of Calcutta and hymnwriter. Educated at Oxford, he was appointed to the family living at Hodnet, Shropshire, in 1807. He published an edition of Jeremy Taylor* in 1822, and the next year he became bishop of Calcutta, a see that then included all of British India. He worked tirelessly to spread Christianity there, but after three years of travel and administration, during which he ordained the first Indian, he died suddenly. His fifty-seven hymns, all written at Hodnet, were collected and published in 1827 as *Hymns written and adapted to the Weekly Church Service of the Year,* the title indicating pioneer work. Heber led a movement toward a literary type of hymn and helped popularize the use of hymns in the Church of England. His compositions include "Bread of the world," "Brightest and best," "From Greenland's icy mountains," "Holy, Holy, Holy," and "The Son of God goes forth to war." JOHN S. ANDREWS

HEBICH, SAMUEL (1803-1868). A founder of the Basel Mission work in India. Born near Ulm in Württemberg, Germany, son of a pastor, he went to Mangalore in 1834, but later moved south to Malabar. Hebich had a remarkable ministry to British soldiers, making converts despite his poor English and eccentric manners. One regiment was termed "Hebich's Own." At the same time he was truly a missionary to the people of India. He left in poor health in 1859 and died at Stuttgart.

HEBREW. Semitic language in the Canaanite branch and the language of the OT, except for short Aramaic portions chiefly in Daniel and Ezra. It is called "the language of Canaan" (Isa. 19:18) and "Jewish"/"Judean" (2 Kings 18:26, etc.). Canaanite glosses in the fourteenth-century B.C. Amarna letters as well as the Ugaritic texts from the same period have thrown light on the early history of the language adopted by the Israelites after the Exodus. In common with the other Semitic languages the majority of Hebrew roots are triconsonantal, with a certain amount of evidence to suggest the priority of a biconsonantal theme.

The alphabet has twenty-two letters; in the absence of a written vowel-system, certain consonants were used to represent pure long vowels. The ancient Phoenician script was replaced by the Aramaic square script about 250-150 B.C. By this time Aramaic had become the vernacular language in Palestine. From about the fifth century A.D. the Massoretes set about providing the consonantal text of the Hebrew Bible with a written vowel-system. Much of the early rabbinical literature (Mishnah, Midrash, etc.) is written in Hebrew, and the chain continues through the medieval era to Modern Hebrew, the language of the state of Israel.

BIBLIOGRAPHY: W.J. Martin, "The Genius of the Language of the Old Testament," *Journal of the Transactions of the Victoria Institute,* LXXIV (1942); W.L. Moran, "The Hebrew Language in its Northwest Semitic Background," in *The Bible and the Ancient Near East* (ed. G.E. Wright, 1961); D.W. Thomas, "The Textual Criticism of the Old Testament," in *The Old Testament and Modern Study* (ed. H.H. Rowley) 1951, 1961.

ROBERT P. GORDON

HEBREWS, EPISTLE TO THE. For long the early church doubted the canonicity of Hebrews, largely because of uncertainty regarding its author. Although it was not written by Paul, nobody today would deny its authenticity of content and spiritual worth. Its author remains unknown, the best guess still being Apollos or his spiritual twin. Its readers, also unknown, who may have lived in Italy, were in danger of losing their earlier Christian zeal, or even perhaps of giving up their faith, partly through the pressure of persecution. Possibly they were being tempted to retreat into Judaism. Their spiritual *ennui* occasioned this unique document, which is more of a written sermon than a letter (except in ch. 13), a brilliant piece of rhetoric with a carefully wrought theological argument, mingling doctrine and exhortation in alternating sections.

The basic theme is the finality of the Christian revelation, God's last word, to turn aside from which is spiritual suicide. Christians are called to a persevering faith, similar to that of OT saints, but with the added incentive of already enjoying the partial fulfillment of the promises. The writer offers an elaborate proof of the superiority of the Christian revelation to the OT revelation, and in particular of Jesus to the angels, Moses, and the Jewish high priests. The OT system of worship, valid for its time, was inherently defective, only a reflection of the true spiritual reality; it has now been superseded by the coming of Jesus, the Son of God, as high priest to offer Himself once-for-all as the perfect sacrifice in the heavenly tabernacle. Stressing that the old has given place to the new, the writer brings out the continuity between the old and new covenants and the oneness of God's pilgrim people in all ages.

This argument could scarcely have been addressed to other than "Hebrew" (Jewish) Christians, although the case for a Gentile Christian audience is not lightly dismissed. The author's developing theology suggests a late rather than a very early writing; it antedates *1 Clement* (c. A.D. 96), and the absence of indication that the Jerusalem temple no longer exists points to a date before A.D. 70. The author's theology here leaves some gaps, such as the lack of teaching on faith-union between the believer and the Lord, but nothing can diminish the worth of this noble witness to the Jesus who is the same unchanging Savior, yesterday, today, and forever.

BIBLIOGRAPHY: W. Manson, *The Epistle to the Hebrews* (1951); R. Williamson, *Philo and the Epistle to the Hebrews* (1970); commentaries by B.F. Westcott (2nd ed., 1892), J. Moffatt (1924), F.F. Bruce (1964), H.W. Montefiore (1964), J. Héring (1970), G.W. Buchanan (1972).

I. HOWARD MARSHALL

HEBREWS, GOSPEL ACCORDING TO THE, see APOCRYPHAL NEW TESTAMENT

HEBRONITES, see HEPBURN, JOHN

HECK, BARBARA (1744-1804). "Mother of American Methodism." Born in Ireland of a German Palatinate refugee family, she migrated to New York with her husband in 1760. She encouraged her cousin Philip Embury* to hold the first Methodist meeting in America in his home, and encouraged him further in the building of the first Methodist chapel in America. The family moved to Canada early in the Revolutionary War because of their Tory views.

HECKER, ISAAC THOMAS (1819-1888). Founder of the Paulist Order Born in New York City of Protestant German parentage, he worked in a bakery with his brothers until Transcendentalist ideas led him to the communal Brook Farm and Fruitland in 1843. He became a Roman Catholic in 1844 and entered the Redemptorist Order. After studying in Belgium, Holland, and England, he was ordained in 1845 and worked with Roman Catholic German immigrants after returning to the USA in 1851. Because of an unauthorized trip to Rome, he was excluded from his order, but freed from his vows by Pius IX to found in 1858 the Missionary Priests of St. Paul the Apostle to convert Protestants. Hecker was superior of the order until 1888. He also founded and edited *The Catholic World* (1865) and the *Young Catholic* (1870).

EARLE E. CAIRNS

HEDBERG, FREDRIK GABRIEL (1811-1893). Finnish pastor; founder of the "Evangelical movement." As a schoolboy he experienced his first spiritual revival through influence from Herrnhut groups. He studied the Bible together with pietistic revival literature, which books through their strictness extinguished his spiritual life. In 1834 he was ordained as pastor of the Church of Finland, at which time he was completely a theologian of the Enlightenment and mainly tried to improve people's ability to read. He soon found this foundation inadequate, for it had nothing to offer souls in need. He came into contact with Pietism* and this influenced him decisively, but in 1844 he published *The Doctrine of Faith Unto Salvation,* indicating a complete break with the Pietists. He then founded and became leader of the Evangelical movement, based on the writings of Luther. In the center are the grace and forgiveness of God, the redemption of Christ, and the appropriation of it through the means of grace. Hedberg speaks less about sanctification and prayer, but with more boldness about the universal grace of God, which grace is given to man already in (infant) baptism.

STIG-OLOF FERNSTROM

HEERMANN, JOHANN (1585-1647). Silesian hymnwriter. Son of a furrier, he was dedicated by his mother to the ministry in early childhood. After studying at Fraustadt, Breslau, and Brieg and holding various teaching posts, he became in 1611 pastor of Köben an der Oder. He suffered during the Thirty Years' War, and in 1634 had to give up preaching, at which he excelled. His health was always poor. His 400 hymns, many inspired by suffering and many still sung in Germany, are ranked by some as second only to those of Gerhardt.* Although many exist in English, only two are commonly used: "Ah, holy Jesu, how hast thou offended" (Bridges), and "O Christ, our true and only Light" (Winkworth).

JOHN S. ANDREWS

HEFELE, KARL JOSEPH (1809-1893). Roman Catholic bishop and historian. Born in Unterkochen bei Aalen, he was ordained priest in 1833. After serving in minor academic posts he was called in 1840 to succeed his own teacher, J.A. Möhler,* as professor of church history at Tübingen. His most famous work as a church historian was his monumental *Conciliengeschichte* (1855-74) in seven volumes. His study of the councils and his joint editorship of the *Theologische Quartalschrift* (from 1839) established him as one of the most important Roman Catholic scholars of his day. He was appointed a consultant to the preparatory commission for Vatican Council I* (1868) and, following his consecration as bishop of Rottenburg (1869), returned to Rome to take his seat as a council father. He was a leader of the minority opposed to the doctrine of papal infallibility, though he submitted eventually to the decision of the council. His last years were spent principally in pastoral work in his own diocese.

DAVID C. STEINMETZ

HEGEL, GEORG WILHELM FRIEDRICH (1770-1831). The dominant figure in German idealism, and one of the great philosophical system-builders. He studied at Tübingen (1788-93), and after holding teaching positions at various universities, including Jena and Bern, was professor of philosophy at Berlin (1818-30). Hegel rejected both realism (the view that reality exists independently of the mind) and subjective idealism (that reality is the product of individual consciousness) because, in his view, they involved unavoidable contradictions. Rather, true knowledge is only possible of ultimate reality, the product of the Spirit which, in a dynamic development, reconciles the self-contradictions that permeate every aspect of human experience. The ideas of the unity and comprehensiveness of thought, and of its dynamic development, are dominant.

In claiming that realism and subjective idealism embodied fundamental contradictions, Hegel held that all experience presupposes the unity of the knower and the known. This unity, however, is not achieved, but is in the process of fulfillment in human experience, becoming explicit in aesthetic and religious experience, and fully developed in truly philosophical thinking. The dialectical method is the only true philosophical method, since it alone corresponds to the process of nature and history and of all reality. Error lies in partiality and incompleteness. This basic view underlies his diverse philosophical productions. Some of the most notable of these are *The Phenomenology of Mind* (1897), an account of various stages of human consciousness from sense awareness to absolute knowledge; *Logic* (1812-16), the analysis of categories basic to all discourse; *Philosophy of Right* (1821), in which Hegel's view of the state, as the synthesis of the family and civil society, is given.

Hegel is important in any account of the development of Christian thought, with which his philosophy is fundamentally incompatible. Religion for Hegel is simply an imaginative, pictorial way of representing philosophical truth. His overall position obviously has strong affinities with pantheism. Hegel's system was the inspiration behind the destructive biblical criticism of D.F. Strauss,* and, in more complicated ways, Hegel influenced both Feuerbach and Marx.

OONAGH MC DONALD

HEGESIPPUS (second century). Church historian. A *terminus a quo* is provided for him by his reference to Hadrian (117-38) establishing certain games in his day; and a *terminus ad quem* by his addition of the names of Soter and Eleutherus (175-89) to a succession list of the bishops of Rome which he had drawn up in Rome in the time of Anicetus (156-67). Jerome corroborates these dates when he says that Hegesippus lived near the time of the apostles. Eusebius draws the conclusion that Hegesippus was a Jew and says his work comprised five books of "Memoirs." These appear to have been directed against the Gnostics and to have ranged over the whole of church history to his day in a random fashion (James is dealt with in the last book) and an unpretentious style. The "Memoirs" survive now only in fragments, nearly all in Eusebius. One fragment in Photius has been taken as an attack by Hegesippus on Paul's words in 1 Corinthians 2:9. It is more likely, however, to be an attack on the misuse of Paul's words by the Gnostics.

DAVID JOHN WILLIAMS

HEIDELBERG CATECHISMS, see CATECHISMS

HEILER, FRIEDRICH (1892-1967). German theologian. A Roman Catholic, he studied theology, philosophy, and oriental languages at Munich. Under the influence of N. Söderblom* he became a Protestant, joining the Lutheran Church at Uppsala (1919). He was appointed professor of comparative history of religions at Marburg (1922). Influenced by the writings of Friedrich Von Hügel,* he took a more Catholic line and became a leader of the German High Church Union from 1929. He founded an Evangelical order of Franciscan Tertiaries. He edited the *Hochkirche* from 1930. His finest work was *Das Gebet* (1918; ET *Prayer*, 1932), a study of prayer from its most primitive forms to mystical contemplation. Other works include *Der Katholizismus* (on Roman Catholicism, 1923); *Evangelische Katholizität* (1926), and *Die Wahreit Sundar Singha* (1927). J.G.G. NORMAN

HEILSGESCHICHTE. German term meaning "salvation history." It was coined in the mid-eighteenth century and employed by J.T. Beck* who combined Hegel's philosophy with the notion that God's dealings with mankind required a logical connection between the various events composing that revelation. *Heilsgeschichte* emphasized the importance of each stage of the process because each became a part of the whole.

As new approaches to history developed, the Lutheran theologian J.C.K. Hofmann* offered some revisions in his use of the term. Against Beck, he noted that in a teleological view of history the earlier elements could not serve the same function as the later ones; nonetheless, Hofmann stopped short of rejecting the OT as inferior to the New. Rather, he maintained that superiority of the New is lost when it is studied in isolation.

More recently the idea of *Heilsgeschichte* has served to help theologians out of the corner into which historicism had forced them. Refusing to relegate the Bible to a purely human and therefore completely relative phenomenon, theologians such as Oscar Cullmann have once again turned to the *Heilsgeschichte* approach. According to this view, the events in the biblical narratives point to an increasing awareness of God's saving work in history, and confront the believer in the present with divine challenge.

BIBLIOGRAPHY: E.C. Rust, *Salvation History: A Biblical Interpretation* (1962); *idem, Towards a Theological Understanding of History* (1967); O. Cullmann, *Christ and Time* (1964); *idem, Salvation in History* (1967). WATSON E. MILLS

HEIM, KARL (1874-1959). Lutheran theologian. Native of Württemberg and of Pietistic background, he studied at Tübingen, was for some years pastor and schoolmaster, taught at Halle (from 1907), at Münster (1914), and returned in 1920 to Tübingen as professor of theology. While fully appreciating the achievements of recent scientific civilization, he was anxious to restate faith in a transcendent God in a manner intelligible to modern minds. In his early writings he stressed the Ritschlian contrast between faith and reason, but later under the influence of existentialist thinkers, and especially of Martin Buber,* he developed his notion of spaces: impersonal relations of an I-It character and personal ones of the I-Thou sort can only subsist within an archetypal or suprapolar space where the very presence of God is to be found. Recognized as one of Germany's leading postwar theologians, he defended his theological system against both scientific secularists and Nazi perversions of the Christian faith. His monumental work is *Der evangelische Glaube und das Denken der Gegenwart*, two sections of which were translated into English as *God Transcendent* (1935) and *Christian Faith and Natural Science* 1953). IAN SELLERS

HELENA (c.248-c.327). First wife of Constantius and mother of Constantine.* Of humble origin, her relationship with Constantius may have been that of *concubinatus* and not wife. In any case Maximian required that she should be divorced in favor of his stepdaughter Theodora. Nothing is known of Helena's life during the subsequent reign of Constantius, but after Constantine's accession in 306 she was at his court, where she was greatly honored. Through Constantine she became a Christian and aided by his bounty did great works of charity. In her old age she visited the Holy Land, where her name is associated with the erection of churches on sites connected with Jesus. There is no basis, however, in the tradition either that the Cross was discovered or that it was discovered by her. The place of her death is unknown, but she was probably buried at Constantinople. DAVID JOHN WILLIAMS

HELL (Gr. *geena*, from Aramaic *ge-hinnam* and Heb. *ge-hinnom*). The Hebrew derivation is an abbreviation of the full title "valley of the son of Hinnom," probably after the original Jebusite owner of the property which divided ancient Jerusalem from the hills to the south and west. Today the area is known as Wadi er Rababi and joins the valley of the Kidron at the S extremity of the hill of Zion. During the monarchy this ravine was the location of an idolatrous cult which practices human sacrifice (2 Kings 23:10; Jer. 32:35) and the passing of children through fire 2 Chr. 28:3; 33:6; Jer. 7:31). In the first century "the valley of Hinnom" was used as a metaphor to denote a fiery punishment which awaited the wicked after death or ultimately after the last judgment (Rev. 19:20ff.; 2 Pet. 2:4; Jude 6). The general idea of a raging destructive fire may be found in the earlier portions of the OT, but it is only during the Graeco-Roman period of Jewish history that the specific concept of a lake or abyss of fire begins to emerge (cf. Dan. 7:10). The ancient biblical toponym—"Gehenna"—was first made in the gospels, possibly in the light of Jeremiah's prophecy against the notorious valley.

In Christian theology, virtually without exception, "hell" signifies the state to which the unrighteous pass at death. Its character is inferred from certain biblical teachings, especially the words of Jesus concerning those who reject the kingdom of God. He notes these will be cast "into the darkness" (Matt. 25:30) or into "eternal fire prepared for the devil and his angels" (Matt. 25: 41). The punishment of fire is mentioned elsewhere in Matthew and Luke. Also, Isaiah 66:24 is applied with its suggestion that what is already corrupt will be destroyed: that God is able to "destroy both soul and body in hell" (Matt. 10:28). Paul suggests that the fate of the unrighteous is "death" (Rom. 6:21) or simply destruction (2 Thess. 1:9; Phil. 3:19). WATSON E. MILLS

HELLENISTIC GREEK, see GREEK, HELLENISTIC

HELVETIC CONFESSIONS. Two creedal standards of the Swiss Reformed churches. The *First Helvetic Confession* (1536) is remembered primarily as an attempt to reconcile Lutheran and Zwinglian views, before the spread of Calvinism. Aimed at the German-speaking Swiss cantons, the confession was drawn up by the young H. Bullinger, M. Bucer, and L. Jud. Also taking part were Megander, Myconcius, and other theologians. The first draft of the confession was modified by Jud after complaints that it was too Lutheran. The statement on the Eucharist, however, made it unacceptable to the Lutherans. The confession was accepted by the Swiss Zwinglian churches, which soon merged with the Calvinist movement.

The *Second Helvetic Confession* (1566) was a major Calvinistic or Reformed confession, accepted as a standard not only in Switzerland, but also in the Palatinate, France, Scotland, Hungary, and Poland, and well received in the Netherlands and England. The Elector Palatine, Friedrich III, who had recently turned Protestant and published the Heidelberg Catechism (1563), important as a Calvinistic statement, desired a confession of his personal beliefs to aid him against charges of fomenting religious dissension which were to be made at the upcoming diet, and turned to Heinrich Bullinger for help. Bullinger had drawn up a lengthy statement of his own personal beliefs which, with slight modification, became the Confession. It had an immediate and warm reception.

A product of Bullinger's mature thought, this second confession presents Calvinism as evangelical Christianity, in conformity with the teachings of the ancient church. Though scholastic and lengthy, it is moderate in tone. Harmony with the teachings of the ancient church is important; variety in nonessentials is allowable. The teachings of the Greek and Latin theologians of early days are valuable, though tradition must always be subordinated to Scripture. The ecumenical creeds of the early undivided (pre-Roman) church are scriptural. The Roman claim to be the true successor of the early church is vigorously assailed. The doctrine of election from eternity is affirmed, as befitted a Calvinistic confession. Against the Anabaptists, the Confession defends baptism of children, participation in civil life, and taking up arms under certain conditions (only in self-defense and only as a last resort). DIRK JELLEMA

HELVIDIUS. A Western writer of whom nothing is known except that he was in Rome at the same time as Jerome, during the papacy of Damasus (366-84). He wrote a tract in which he asserted that, after the birth of Jesus, Mary had other children by Joseph, who are referred to in the Scripture as Jesus' brothers and sisters. "And why not? Are virgins in any way superior to Abraham, Isaac, and Jacob, who were married men?" He sought the authority of Tertullian and Victorinus for this attack on the ascetic ideal and in favor of marriage. He was not known to Jerome, but his tract was strongly opposed by the latter, who maintained that Joseph was not really Mary's husband, that those whom Helvidius regarded as brothers and sisters were in fact cousins, and that virginity is a better state than marriage.

DAVID JOHN WILLIAMS

HELWYS, THOMAS (c.1550-c.1616). Founder and pastor of what was probably the first General Baptist church in England. He joined the English Independent ("Brownist") Church in Amsterdam, founded by John Smyth* in 1606. In 1609 he and Smyth, probably influenced by the Mennonites, were expelled because they advocated believer's baptism, and they were Arminian in theology. Smyth became pastor of a Baptist church in Amsterdam; when he died in 1610, Helwys succeeded in the pastorate. In 1611 the church issued a "Declaration of Faith," notable for its definition of baptism: "the outward manifestation of dying with Christ and walking in newness of life; and therefore in nowise appertaineth to infants"; and its declaration—perhaps the first ever—of the right of full individual freedom of conscience: "the magistrate not to meddle with religion or matters of conscience, nor compel men to this or that form of religion." In 1611 Helwys and his flock returned to England and established their church in Newgate Street. Although practicing believer's baptism, they did not normally immerse candidates, but used a Mennonite-style affusion. Helwys was a powerful preacher and the church grew rapidly. In 1615 he published a treatise against persecution. A. MORGAN DERHAM

HENDERSON, ALEXANDER (1583-1646). Scottish minister. Born in Fife, he matriculated at St. Andrews when he was sixteen, and shortly after graduation (1603) was appointed teacher of philosophy there. Having found favor with the archbishop, he became minister of nearby Leuchars contrary to the wishes of parishioners who secured the church doors against him. A few years later he was converted through hearing Robert Bruce preach on John 10:1, and threw in his lot with the Presbyterian party that was resisting James VI's ritualism. When that royal policy was intensified under Charles I, Henderson was prominent among those who defied the king and became co-author of the National Covenant.* Elected moderator (1638) of the first general assembly for two decades, Henderson became leader of the Covenanters,* was appointed minister of the High Kirk of Edinburgh (1639), and largely drafted the Solemn League and Covenant.* He did much to further the cause of education in Scotland, introduced Hebrew into the regular curriculum at Edinburgh University (of which he was rector during the last six years of his life), and was highly respected even by the Episcopal party. J.D. DOUGLAS

HENGSTENBERG, ERNST WILHELM (1802-1869). Lutheran scholar. Born at Fröndenberg near Hamm, he studied theology at the University of Bonn and became a private-dozent at Berlin in 1824, subsequently becoming professor there.

During his early years at Berlin he was associated with Evangelicals such as August Neander, Frederick Strauss, and Theremin, but after 1840 he developed into an outstanding spokesman of Lutheran orthodoxy. Hengstenberg's influence was enhanced by the *Evangelische Kirchenzeitung* which he founded in 1827 and edited until his death. This organ combated rationalism and defended confessional Lutheranism with equal vigor. He also wrote several significant works in the field of OT studies.

WAYNE DETZLER

HENOTICON. A decree of union issued by Zeno in 482. Monophysite bishops had succeeded to the sees of Alexandria (Timothy, 457) and Antioch (Peter, 470), and with Basiliscus's usurpation in 475 there was a Monophysite emperor on the throne. Basiliscus's encyclical anathematizing the Council of Chalcedon, however, had so enraged the Greeks that he had been forced to withdraw it. When Zeno regained the throne in 476 his policy, therefore, was one of conciliation between the orthodox and the Monophysites.* With the aid of Acacius he issued the "Henoticon." It took the form of a letter addressed by the emperor "to the bishops, clergy, monks, and faithful of Alexandria, Libya and Pentapolis," declaring the sufficiency of the creeds of Nicea and Constantinople (381) and the Twelve Anathemas of Cyril. It denounced any contrary doctrine to these "whether taught at Chalcedon or elsewhere," and in particular denounced the doctrines of Nestorius and Eutyches. But by tacitly setting aside Leo's Tome* and the Chalcedon Definition,* the Henoticon had made an important concession to the Monophysites which the Western Church could not accept. After an angry controversy Pope Simplicius excommunicated Acacius, Peter of Alexandria (Timothy's successor), and Zeno himself. Thus began the first ecclesiastical schism between East and West.

DAVID JOHN WILLIAMS

HENRY II (973-1024). German king and Holy Roman Emperor. Duke of Bavaria and direct descendant of Otto I, Henry was elected king in 1002 and crowned emperor in 1014. He stressed the consolidation of his position in Germany and paid minimal attention to Italy, but did regain some territory lost to the Slavs. He depended heavily on the church, and appointed churchmen to most important administrative positions. Henry made large endowments of crown lands to churches and monasteries, and founded the see of Bamberg in 1007. He put churchmen in charge of vacant counties while freeing other clerics from noble control. By zealously encouraging ecclesiastical reform, he unwittingly paved the way for the destruction of the empire, because a reformed church would not be able to reconcile its spiritual ideals with the political duties imposed on it. Canonized in 1146, many legends have grown up around Henry as the model Christian ruler. His pious wife, Kunigunde, was canonized in 1200.

RICHARD V. PIERARD

HENRY III (c.1017-1056). Holy Roman Emperor from 1039. He was responsible for "the cleansing of the papacy." The tenth and eleventh centuries were for the papacy a period of degeneracy; it had become the tool of violent Roman nobles. In 1046 there were three popes. A deeply religious man, Henry was grieved at the situation and responded to an appeal by marching on Rome, summoning the Synod of Sutri which deposed all three popes and installed a German, Clement II, and forcibly subdued the nobles. He was opposed by "high sacerdotalists," but many supported him, especially those desiring reform such as Peter Damian* and Cardinal Humbert. As a result, the papacy began the task of reforming itself and the church generally. Henry appointed the next three popes, all German, and all zealous reformers. During his life, emperor and pope worked in amicable partnership, but subsequently the papacy began to assert its independence.

J.G.G. NORMAN

HENRY IV (1050-1106). German king and Holy Roman Emperor. He succeeded Henry III in 1056 and endured a regency marked by civil strife that instilled in him a resolve to strengthen the monarchy. After reaching majority in 1065, his overly hasty actions in extending royal power in Saxony resulted in war, and he sought church backing in the struggle. Just as victory was gained in 1075, Pope Gregory VII* forbade lay investiture, thus denying Henry a voice in the selection of German church officials. He indignantly deposed Gregory, and the pope responded by excommunicating Henry in 1076. Because most German nobles supported Gregory, Henry sought to forestall his imminent deposition by going to Canossa in 1077 to obtain papal absolution just before a council was to meet at Augsburg.

Civil war followed as the nobles elected another king and Gregory supported him. In 1080 Henry again deposed Gregory (who died in exile, 1085), set up Clement III as his antipope, and was crowned by him in 1084. Henry's last years were filled with insuperable difficulties as his sons Conrad and Henry revolted and the imperial government collapsed. Resulting from these wars was the growth of feudalism and princely sovereignty in Germany and powerful urban communes in Italy.

BIBLIOGRAPHY: G. Barraclough, *The Origins of Modern Germany* (1947); G. Tellenbach, *Church, State and Christian Society at the Time of the Investiture Contest* (1959).

RICHARD V. PIERARD

HENRY IV OF FRANCE (1553-1610). First Bourbon king of France. Reared as a Protestant by his staunchly Calvinist mother Jeanne d'Albrêt, Henry inherited from her the throne of Navarre. He was related to the ruling Valois dynasty through both his father, Anthony of Bourbon, and his mother, a niece of King Francis I (d.1547). With the coming of the wars of religion in 1562, Henry's family became leaders of the Huguenot forces. In 1572 a peace marriage was arranged between Henry and Margaret of Valois, sister of Charles IX (d.1574). Four days later, on 22 August, the St. Bartholomew's Day massacre occurred, Henry was captured, forced to convert to Catholicism, and held prisoner for three and one-half years. He finally escaped, returned to his Protestant faith, and assumed leadership of the Huguenot cause.

In 1589 his cousin, Henry III of France, died without issue, and Henry of Bourbon was heir apparent. Most Frenchmen refused, however, to accept him as king because of his Calvinism. Finally in 1593, with the country on the verge of total collapse, Henry once more embraced the Roman faith and marched triumphantly into Paris. Historians have debated the real motives for his reconversion: personal advancement or the survival of France. Whatever the case, he inaugurated an era of toleration for his former Huguenot compatriots with the Edict of Nantes* in 1598. Henry was assassinated by the Catholic fanatic Francis Ravaillac. ROBERT D. LINDER

HENRY VI (1421-1471). King of England. Son of Henry V and Catherine of Valois, he became king in 1422. He was crowned twice: as king of England in 1429 and as king of France in 1431. In 1445 he married Margaret of Anjou, a woman of forceful character. Henry was an extremely devout and kindly person. He was generous to the poor and abhorred cruelty and immorality. He prayed and meditated frequently and exhorted his barons to do likewise. His interest in education led to his two foundations: at Eton College (1440) and King's College, Cambridge (1441). Yet he was not without personal courage. In 1450, during Jack Cade's rebellion, he rode openly through the streets and refused to fight against his subjects. Unfortunately his character and temperament and bouts of mental disorder made him ill-suited for the task of ruling a politically turbulent country.

When civil war came, Richard Duke of York by 1460 had imprisoned Henry and forced him to recognize Richard (rather than his own son Prince Edward) as his heir, but Richard died that same year, and Edward of York seized the throne while Henry went into exile. Captured in 1465, Henry was held in the Tower until 1470 when, although now completely mad, he was nominally reinstated, only to be sent back to the Tower on Edward's triumphant return to power in May 1471. Henry died probably that same month, reportedly murdered. Pilgrims soon started to visit his tomb in Chertsey, Surrey, and continued to do so after its removal to St. George's Chapel, Windsor. Miracles were reported, and Henry's fellow Lancastrian king, Henry VII, tried unsuccessfully to obtain his canonization. L. FEEHAN

HENRY VIII (1491-1547). King of England. Second son of Henry VII and Elizabeth of York, he was an intelligent boy who received a Renaissance education. On the issuance of a papal dispensation, in 1509 he married Catherine of Aragon, widow of his elder brother Arthur, thus continuing an alliance between the Tudors and the Spanish throne. He became king that same year, with Thomas Wolsey* managing the realm for him.

Shortly after the appearance of Luther's tracts of 1520, Henry VIII with some help replied in 1521 with *Defence of the Seven Sacraments*, which resulted in the papal grant of the title "Defender of the Faith." Toward the end of the decade Henry became increasingly concerned with his role as king in the spiritual welfare of his people, and with his inability to produce a legitimate male heir which could result in a civil war. The only surviving child of his marriage to Catherine was Mary Tudor. Wolsey thought he could arrange for a divorce and settle the "great matter," but the special legatine court of 1529 presided over by Wolsey and Campeggio* failed to resolve the problem.

In 1529 Wolsey was removed from office and Henry began his assault upon papal control in England. With the death of Archbishop Warham and the resignation of Lord Chancellor Sir Thomas More,* Henry moved quickly. Cranmer was named archbishop; the divorce was granted; Henry married Anne Boleyn. Parliament with the guidance of Thomas Cromwell* proceeded to pass a series of laws that placed England outside the sphere of Rome's control. Appeals to Rome were forbidden, annates and Peter's Pence were stopped, dissolution of monastic property was begun, and the clergy were required to submit to the throne. Protests arose in the Pilgrimage of Grace,* which was crushed, and in the objections of Thomas More and John Fisher, who were executed as a result.

With the birth of Elizabeth the succession question was still unresolved. Three years later Anne Boleyn was accused of adultery and beheaded. Next day, Henry married Jane Seymour who did produce a son, the future Edward VI, but twelve days later the queen died. In 1540 Henry was enticed into marrying Anne of Cleves, but upon her arrival he was so displeased with her that the marriage was not consummated and was dissolved. He next married Catherine Howard, later charged with adultery and beheaded in 1542. Finally he married Catherine Parr, who survived him.

Henry apparently remained basically Catholic, unwilling to subscribe to many Protestant doctrines. The Six Articles* of 1539 mark a return to Catholic doctrine, as perhaps did his marriage to Catherine Howard. The last years of his reign did involve some effort to reform the church while maintaining the exterior of Catholicism. His reign not only started the Reformation in England, but through the use of the Star Chamber, the employment of parliamentary law to work the reforms, the establishment of a national church under direction of the Crown, and the restructuring of the councils of the north and west, he greatly strengthened the Tudor throne in England. Yet as Scarisbrick says: "Few kings have had it in their power to do greater good than Henry, and few have done less."

BIBLIOGRAPHY: J.S. Brewer et al. (eds.), *Letters and Papers ... of the Reign of Henry VIII* (22 vols., 1862-1932); A.F. Pollard, *Henry VIII* (1951); C. Read (ed.), *Bibliography of British History: Tudor Period, 1485-1603* (2nd ed., 1959); E. Doernberg, *Henry VIII and Luther* (1961); H.M.

Smith, *Henry VIII and the Reformation* (1962); J.J. Scarisbrick, *Henry VIII* (1968); L.B. Smith, *Henry the Eighth: The Mask of Loyalty* (1973).

ROBERT SCHNUCKER

HENRY, MATTHEW (1662-1714). Biblical expositor. Son of an evangelical Church of England minister, he was born shortly after his father had been ejected from his living as a result of the Act of Uniformity.* A studious boy, he dated his conversion in 1672. He studied at a Nonconformist academy in London, and then read law at Gray's Inn. He considered becoming an Episcopalian minister, but decided to be a Nonconformist and was privately ordained as a Presbyterian. His first pastorate was in Chester (1687-1712), followed by Hackney (1712-14). Greatly influenced by the Puritans, he made exposition of Scripture the central concern of his ministry. Beginning work at four or five o'clock each day, he aimed to use time to the full. In 1704 he began the seven-volume *Commentary on the Bible* for which he is remembered. He finished up to the end of Acts; ministerial friends completed the NT from his notes and writings. It set a style in detailed, often highly spiritualized, exposition of Scripture which has shaped evangelical ministry ever since; C.H. Spurgeon* acknowledged his debt to Henry; many others have neglected this courtesy. Critical textual problems were not within his purview. Suffice to say that he could write 190 words of comment, including a three-part sermon outline, on Genesis 26:34.

A. MORGAN DERHAM

HENRY, PHILIP (1631-1696). Puritan divine. Born in London of Welsh parents, his father being one of the king's gardeners at Whitehall, he was educated at Westminster and Christ Church, Oxford, and joined the Presbyterians. He became tutor to Judge Puleston's children at Emral in Maelor, and was ordained as a Presbyterian minister in 1657. He was ejected from Worthenbury chapel in 1660. Under the 1672 Declaration of Indulgence* his friends secured him a license to hold services at his home at Broad Oak. His diaries (ed. M.H. Lee, 1882) provide a vivid picture of life for the persecuted under the Penal Code. They also cast a revealing light on the piety and strictness of a Puritan home. He was the father of Matthew Henry.*

R. TUDUR JONES

HENRY OF BLOIS (d.1171). Bishop of Winchester. Trained in Cluny, Henry in 1126 became abbot of the wealthy Glastonbury abbey, which post he continued to hold by special dispensation after he became bishop of Winchester in 1129. Proud and ambitious, he was instrumental in his brother Stephen's accession to the throne in 1135, and was therefore aggrieved because he failed to become archbishop of Canterbury in 1139. He became papal legate, however, and as such was very powerful. Far from being an instrument of Stephen, he represented the interests of the church and papacy, and even of Matilda, against those of his brother. The appointment as legate was not renewed after the death of Innocent II (1143), and because of this Henry was prepared to support Stephen's resistance to the papal will, especially in 1148. Under Henry II, age lost him Canterbury to Becket* (1162), but in the ensuing controversy with the Crown, Henry was a steady opponent of excessive royal authority within the church. He is famous also as a builder of churches and castles.

C. PETER WILLIAMS

HENRY OF GHENT (d.1293). Theologian and philosopher. Born at Ghent, he became archdeacon successively of Bruges and Tournai, and taught in Paris where he became the most outstanding secular master for many years. Involved in the condemnation of aspects of Thomist teaching in Paris in 1277, he also opposed the privileges of the mendicant orders in 1282. His most famous philosophical and theological writings are *Quodlibeta* and his *Summa Theologica.* These are critical of Aristotle and produce a significant synthesis of Augustinian teaching and the new learning. Henry was an important catalyst to Duns Scotus,* who took much from him as well as criticizing him freely. Fundamental is his idea of being, and he holds that from this, rather than from the sense perceptions as Aquinas argued, God's existence can be proved.

C. PETER WILLIAMS

HENRY OF LANGENSTEIN (c.1325-1397). German philosopher and theologian. Born in Hesse, he was educated at the University of Paris, where he taught Nominalist philosophy and, later, theology. As the school's leading theologian and vice-chancellor, he led her four faculties to a public position advocating a general church council to heal the Great Schism.* When the French court rejected the idea, some half of the faculty and student body returned to their native Germany. Henry withdrew to the monastery of Eberbach. In 1384 he helped found a theological faculty at the University of Vienna, where he became vice-chancellor and, after 1393, rector. As a competent scholar and writer he produced works on astronomy, the Great Schism, the Immaculate Conception, asceticism, and the errors of astrology.

JAMES DE JONG

HENRY OF LAUSANNE (d. mid-twelfth century). Monk and theologian, he later lapsed into heresy. He rejected the objective efficacy of both the priesthood and the sacraments. His message was the evangelical life of poverty and penance which he himself lived. In 1101 he went from Lausanne to Le Mans, but his views on the priesthood and the sacraments led to his expulsion by Bishop Hildebert. He then preached in various parts of S France, and was condemned by the Council of Toulouse in 1119. In 1135, after arrest by the bishop of Arles, Henry recanted temporarily but soon relapsed and continued his preaching. After 1135 he was influenced by Pierre de Bruys, whose doctrines had many similarities with Henry's. In 1145 Bernard of Clairvaux was sent to combat Henry's preaching. Henry was arrested and died at Toulouse shortly afterward. Although perhaps a precursor of the Waldensians, Henry was not a Manichaean.

L. FEEHAN

HENRY SUSO (c.1300-1366). Swabian mystic. Well-born, he entered the Dominican friary at Constance when he was thirteen and had a deep conversion five years later. Completing his studies at Cologne under the influence of Meister Eckhart, he returned to Constance as lector in the friary school, then became prior. At forty he abandoned his extreme asceticism to preach and be a pastor. As an itinerant preacher, teacher, adviser, and confessor, he visited regularly the Dominican convents about Constance. His writings are essentially devotional; an important one is *The Life of the Servant,* which records his mystical experiences. His speculative book, *The Little Book of Truth,* and the more practical, *The Little Book of Eternal Wisdom,* discuss mysticism in detail. He settled in 1348 in the Dominican convent at Ulm.

C.G. THORNE, JR.

HENSON, HERBERT HENSLEY (1863-1947). Bishop of Durham. After becoming a Fellow of All Souls College, Oxford, in 1884, he was ordained and in 1888 became vicar of Barking. In 1895 he was appointed chaplain of Ilford Hospital, and in 1900 rector of St. Margaret's, Westminster, and canon of Westminster Abbey. In 1912 he became dean of Durham. Six years later he was consecrated bishop of Hereford, but was translated to Durham in 1920, retiring in 1939. A man of liberal churchmanship, his appointment to Hereford was strongly opposed by Anglo-Catholics. A supporter of the establishment, he changed his views after the rejection of the revised Prayer Books by Parliament in 1927 and 1928. A man of courage and wit, he was noted for his pungent utterances on a variety of subjects. His publications include *Anglicanism* (1921), *Christian Morality* (1936), *Ad Clerum* (1937), and his three-volume *Retrospect of an Unimportant Life* (1942-50).

R.E. NIXON

HEPBURN, JAMES CURTIS (1815-1911). Missionary to Japan. Born at Milton, Pennsylvania, he was converted at Princeton and decided to become a medical missionary. In 1840 he and his wife joined the Presbyterian Board, but were invalided home after five arduous years in Java, Singapore, and Amoy. In 1859 they were among the Protestant pioneers to Japan. Though preaching was forbidden, Hepburn diligently applied himself to learning Japanese while living in a Buddhist temple. A lifetime of industrious and devoted service included the opening of the first dispensary, initiating classes for medical students, inventing a system of romanizing Japanese sounds, compiling the first Japanese-English dictionary, helping to found Meiji Gakuin University, and a major part in the Japanese translation of the Bible which was completed in 1888.

DAVID MICHELL

HEPBURN, JOHN (1649-1723). Scottish minister. Brought up an Episcopalian, this turbulent son of a Morayshire farmer became one of the most contentious ministers Scottish Presbyterianism has known. Ultimately hailed as "The Morning Star of the Secession," and ordained while an exile in London, he was for some thirty-six years minister of the parish of Urr in Galloway without ever having been formally elected or inducted, and despite admonition, suspension, banishment, imprisonment, and deposition at the hands of the ecclesiastical authorities. Ranging over wide tracts of country—preaching, marrying, baptizing—he gathered several thousand followers. These Hebronites, as they came to be known, after taking part in Scotland's first agrarian rebellion, formed the nucleus of many Secession churches once that movement had taken shape under Ebenezer Erskine.*

D.P. THOMSON

HERACLAS (d.247). Alexandrian scholar and bishop. He had a pagan background, and with his brother Plutarch met Origen* at the lectures of the philosopher Ammonius Saccas, and they became Origen's disciples. Heraclas was appointed Origen's colleague in the catechetical school, took over from him the training of catechumens, and succeeded him as head when Origen went to Caesarea. He became bishop of Alexandria in succession to Demetrius in 232. During his episcopate the number of bishops in Egypt increased from four to twenty-four. Eusebius called him "an outstanding exponent of philosophy and other secular studies." He was succeeded as head of the catechetical school and later as bishop by Dionysius, another of Origen's pupils.

J.G.G. NORMAN

HERACLEAN VERSION, see SYRIAC VERSIONS OF THE BIBLE

HERACLEON (fl. c.170-180). A Gnostic teacher who is described as the most esteemed *(dokimōtatos)* of the school of Valentinus and had known Valentinus personally, but who differed in some points from other writers of Valentinus's school. He is the first-known commentator on the Gospel according to John, and parts of his highly allegorizing commentary *(hypomnēmata)* are preserved in quotations in Origen, while Clement refers to his exposition of Luke 12:8, which suggests that he had commented on that gospel also. It is not known where he taught.

HERACLIUS (575-641). Byzantine emperor from 610. His reign marked the revival of the Eastern Empire. In 611 Persian attackers captured Edessa, Apamea, and Antioch. Heraclius fought the Persians throughout his reign, reorganizing his army for efficiency and establishing the "theme" as the empire's basic military and administrative unit. Antioch was his headquarters until about 636; the center of the empire survived despite the loss of Syria and Egypt to the Arabs. Fearing that Monophysitism in Syria, Armenia, and Egypt would bring support to the Persians by alienating the indigenous population from the central government, he tried to reconcile Monophysite and Chalcedonian views on Christology by proposing a Monothelite solution in 633: Christ had one divine human will. Sergius of Constantinople was his chief religious counselor (drawing on Cyril and Dionysius the Areopagite). This effort began as early as 628, when occupied territories were freed. Negotiations pivoted on

Athenasius, Jacobite patriarch of Antioch, but he died in 631. Strong opposition to the plan focused in a monk named Sophronius, later patriarch of Jerusalem, and in Honorius of Rome.

See ECTHESIS. JOHN GROH

HERBERT, GEORGE (1593-1633). English poet. Educated at Westminster and Cambridge (where he was in due course to become public orator), it seemed that he was set for a distinguished public career, but the deaths of patrons and of James I himself dashed such possibilities. These deaths and that of his mother in 1626 may have influenced Herbert's always serious mind into deciding to enter Holy Orders. He became rector of Bemerton, near Salisbury, in 1629, but died of consumption four years later. Herbert was an exemplary pastor, and *A Priest to the Temple* is a fine analysis of what is desirable in clerical character and care. An early Anglo-Catholic, associated with the Little Gidding community of Nicholas Ferrar,* he expresses in the poems contained in *The Temple* a personal piety as deep as, if less demonstrative than, that of his self-confessed mentor, John Donne,* and wider than that of his master. In much of Herbert's poetry there is that subdued, but nonetheless sincere, sense of devotion that is the very essence of Anglican worship. ARTHUR POLLARD

HERDER, JOHANN GOTTFRIED VON (1744-1803). Lutheran scholar. Born in East Prussia, he studied at the University of Königsberg (1762-64) where he came under the divergent influences of Immanuel Kant* and J.G. Hamann.* He was successively Lutheran pastor at Riga (1764-69), court pastor at Bückeberg (1771-76), and general superintendent and court preacher at Weimar (1776), where he lived, uncomfortably opposed by the official clergy, for the rest of his life. In addition to extensive literary and philosophical publications, Herder produced a study of the synoptics (1796) which recognized Mark as no epitome, and to be in parallel passages longer, and perhaps older, than the others. His study of John (1797) indicated that that gospel could not be harmonized with the synoptics, and that while a life of Jesus could come from either John or the synoptics, it could not be derived from an artificially constituted harmony of them all. Herder's collected works were critically edited by B. Suphan (33 vols., 1877-1913).

CLYDE CURRY SMITH

HERESY. In Hellenistic Greek the term *hairesis* meant a philosophical school or teaching (e.g., Stoicism). Its use in Judaism was similar (e.g., the "party" of the Pharisees or the Essenes). The term appears in the NT colored by this background, and is at first used neutrally (Acts 24:5; 26:5; cf. Acts 5:17; 15:5; 28:22). But the term is also used in the NT in a specifically Christian context with a pejorative sense, to mean divisions within the church which threaten its unity (1 Cor. 11:19; Gal. 5:20; cf. Titus 3:10). The problem of heresy as it was to be later defined, over against orthodoxy, shows itself in the NT at 2 Peter 2:1, referring to false teachers who will "introduce destructive heresies" in their denial of Christ. Although the term *hairesis,* however, is not used in this connection, the letters of Paul and John reveal early pressure on the Christian Church to resist doctrinal error within its ranks (pre-Gnosticism), as well as persecution from outside (cf. Col. 2:8-23; 1 John 2:22; 4:2f.; 2 John 7ff.).

In the early church the concept of *hairesis* as theological error predominated, although at first (as with Cyprian on the Novatians), "heresy" and "schism" were not always distinguished (cf. 1 Cor. 11:18f.). From the late second century, however, "heresy" usually meant doctrinal error, departure from accepted rules of faith; while "schism" implied dissent from the church for any reason whatever. The existence of heterodoxy in the early church encouraged the definition of the faith by the councils, in the creeds, and in the canon of the NT. Walter Bauer holds neverthless that diversity of belief at the local levels of the early church (in the second century, at least) was such that "orthodoxy" and "heresy" (as these came to be described eventually) originally coexisted.

The early Fathers regarded heterodoxy as sinful, because of the inflexibility of will from which (they claimed) it derived. This view of the moral aspect of heresy strongly influenced medieval Scholastic thought on the subject; although the terms "faith" and "heresy" acquired at the same time a wider meaning, related generally to Christian life and conduct, and not only to the denial of revealed truth as taught by the church (so Aquinas). In more recent times heresy has come again to denote a strictly doctrinal heterodoxy which deserves censure.

See also CANON; MARCION; EXCOMMUNICATION.

BIBLIOGRAPHY: H.E.W. Turner, *The Pattern of Christian Truth* (1954); A. Ehrhardt, "Christianity Before the Apostles' Creed," *Harvard Theological Review,* LV (1962), pp. 73-119; W. Bauer, *Rechtgläubigkeit und Ketzerei im ältesten Christentum* (2nd ed., 1964); S.S. Smalley, "Diversity and Development in John," *NTS* XVII (1970-71), pp. 276-92. STEPHEN S. SMALLEY

HERGENRÖTHER, JOSEPH (1824-1890). Roman Catholic scholar. He studied in his native Würzburg, the German College in Rome, and at Munich where he took his doctorate in theology and lectured (1850-52) before appointment as professor of canon law and church history at Würzburg. He published Photius's *Liber de Spiritus Sancti mystagogia* (1857), followed later by his own massive work on Photius, and the *Handbuch der allgemeinen Kirchengeschichte* (3 vols., 1876-80). He then became consultant in the preparations for Vatican I, cardinal (1879), and the first prefect of the Vatican archives. For the sake of his studies he declined the bishopric of Limburg. Defending papal infallibility, he refuted Döllinger in *Anti-Janus* (1870), having earlier attacked his liberalism with *Der Zeitgeist* (1861). Hergenröther also edited the register of Leo X to 1515, and wrote *Der Kirchenstaat seit der französischen Revolution* (1860). C.G. THORNE, JR.

HERMANN OF REICHENAU (Hermanus Contractus) (1013-1054). Author of the earliest extant universal chronicle. His father, Count Wolverad II of Altshausen in Swabia, entrusted him at the age of seven to Abbot Berno of Reichenau in Lake Constance, where he took vows in 1043. Although severely handicapped physically (hence the nickname), Hermann gained the reputation of the most scholarly man in eleventh-century Germany. He became proficient in theology, Latin, Greek, and Arabic, and achieved fame as poet, mathematician, astronomer, and musician. He was a faithful monk and a genial teacher, and students flocked to him. His greatest achievement is his chronicle, which is more interpretive and less strictly chronological in organization than its antecedents. His narrative, remarkable for accuracy, objectivity, and careful chronology, begins with the birth of Christ and ends with the year of Hermann's death. He constructed timepieces and musical and astronomical instruments, wrote mathematical treatises, poems and hymns, and has often been credited with the famous hymns *Salve Regina* and *Alma Redemptoris Mater.* MARY E. ROGERS

HERMANN VON WIED (1477-1552). Church reformer. Trained in law, he became archbishop-elector of Cologne while only a subdeacon (1515), and later also administrator of Paderborn diocese (1532). Ardent for reform but hostile to Protestantism, especially Anabaptism, he convened a provincial council in Cologne in 1536 which, under Gropper's* guidance, enacted disciplinary and reforming canons. Disappointed at the outcome, the theologically unsophisticated Hermann welcomed the Regensburg Recess instruction to "institute and establish a Christian order and reformation" (1541), and invited Bucer, Melanchthon, and other Protestants to promote renewal and compile a new church order for his territory (1542-43). To the *Cologne Ordinances* or *Didagma* (*Einfaltigs Bedencken einer Christlichen Reformation,* 1543-44) Melanchthon contributed most of the doctrine and Bucer the institutional and ceremonial. The revised Latin version (*Simplex ac Pia Deliberatio,* 1545) greatly influenced Cranmer's 1549 Prayer Book, and an English translation followed (*A Simple and Religious Consultation,* 1547-48). In Cologne the proposals met increasing opposition from councillors, university and chapter (including the formerly favorable Gropper, who published an *Antididagma,* 1544), although the temporal estates backed Hermann. After a pivotal Catholic-Protestant struggle, he was excommunicated by Paul III (1546) and deposed by Charles V (1547). At last completely a Protestant, he died in the principality of Wied.

See C. Varrentrapp, *Hermann von Wied und sein Reformationsversuch in Köln* (1878); and M. Köhn, *Martin Bucers Entwurf einer Reformation der Erzstiftes Köln* (1966). D.F. WRIGHT

HERMAS. Traditionally one of the Apostolic Fathers, known almost exclusively from his work *The Shepherd.* Formerly a (Jewish?) slave emancipated at Rome, he farmed and prospered, but lost his property and saw his sons apostatize in persecution. *The Shepherd* reveals a prophet of mediocre intellect, narrow concerns, and simple, sometimes unstable, piety. He was a contemporary of Clement, yet the Muratorian Canon says he wrote while his brother Pius was bishop of Rome (i.e., 140-54). Internal evidence confirms that *The Shepherd* was composed in stages c.90-140/150, perhaps by three different authors.

The work consists of five *Visions,* twelve *Mandates,* and ten *Similitudes.* Hermas receives revelations from a woman whose age turns to youthful beauty (*Vis.* 1-4), who is the Church (also depicted as a tower under construction), and from the "angel of repentance" in a shepherd's guise, whose appearance in *Vis.* 5 introduces the remaining sections. Inconsistencies, the apocalyptic and allegorical genres, and colorful imagery greatly complicate interpretation. The major themes are ethical—purity and repentance. Moral instruction largely occupies *Simil.* 1-5 and the *Mandates,* which embody a "two ways" pattern widely attested in Jewish and early Christian literature.

Debate has surrounded *The Shepherd's* teaching on postbaptismal repentance. The view that it was first generally countenanced by Hermas is now being overtaken by the interpretation that he *assumes* it from the outset but limits it, because of the approaching end, to sins committed up to the present. A rudimentary penitential system is already operative (cf. *Simil.* 7-10).

The Shepherd's chief importance lies in the light it throws on the beliefs of Jewish Christianity, whose literary forms it employs, and on the "vulgar catholicism" of a Christian congregation in Hellenistic Roman society. The work enjoyed high regard in the early centuries, especially in the East. It was widely included among the Scriptures until the third century and was still used for catechetical purposes in Athanasius's day. Nevertheless, it survives in a poor textual tradition.

See also APOSTOLIC FATHERS.

BIBLIOGRAPHY: W.J. Wilson, "The Career of the Prophet Hermas," *HTR* 20 (1927), pp. 21-62; B. Poschmann, *Paenitentia Secunda* (1939), pp. 134-205, and *Penance and the Anointing of the Sick* (1964), pp. 26-35; J. Quasten, *Patrology* 1 (1950), pp. 92-105; S. Giet, *Hermas et les Pasteurs* (1963); L. Pernveden, *The Concept of the Church in the Shepherd of Hermas* (1966); J. Reiling, *Hermas and Christian Prophecy* (1973).

D.F. WRIGHT

HERMESIANISM. A philosophical and theological system propounded by Georg Hermes (1775-1831), German Roman Catholic theologian, professor of theology at Münster. He had studied philosophy at Münster and had been deeply influenced by the rationalism and idealism of Kant and Fichte. He tried to establish the truth of Christianity by reason alone. Theology, he argued, must begin with positive doubt. Kant had held that God's existence was a postulate of man's reason in its practical or moral use. Hermes thought God's existence could be demonstrated by theoretical reason, which determines and categorizes the data supplied by sense intuition.

Hence the consciousness that "I know" and the thought that "something is there" involve variations that require a sufficient and absolute reason for their origin. From this point Hermes argued for the possibility of divine revelation. The dualism of theoretical reason and practical reason runs throughout his work. The "belief of the reason" is brought about by demonstration. The "belief of the heart" is the accepting of revealed truths by a free surrender of the will. Hermes's principal writings were *Einleitung in die christkatholische Theologie* (1819-29) and *Positive Einleitung* (1829). His followers were influential in the universities, but in 1835 Gregory XVI condemned the system, largely for its basic rationalism and tendency to skepticism. There was strong opposition to the decision, but in 1870 it was confirmed by Vatican Council I.* HOWARD SAINSBURY

HERMETIC BOOKS. This collection of writings deals with religious and philosophical subjects and reflects a degree of syncretism with reference to Platonic, Stoic, Neo-Pythagorean, and Eastern religious thought. The collection dates from the second or third century and is ascribed to Hermes Trismegistus which represents a later designation for the Egyptian god Thoth, who was said to be the source and protector of all knowledge. The literary form of the Hermetic Books is basically that of the Platonic dialogue. The single most significant of the several writings is "Poimandres," which tells of the soul's ascent to God through the various spheres of the planets.

HERMIAS. Author of the *Irrisio Gentilium Philosophorum.* Nothing is known of him, and his work has been variously dated from the second to the sixth century. It is an attempt to show that the opinions of the Greek philosophers are contradictory, but it is "disfigured by bold caricature and over-simplification" (H.E.W. Turner). He holds that their contradictions are due to the influence of demons and extend to such basic matters as the being and attributes of God and the nature of Providence. Although Hermias is himself called a philosopher in the title of the book, his attitude to philosophy "seems to rest upon anti-intellectualist premises" (Turner). As Neander suggests, he may once have worn the philosopher's mantle, but his enthusiasm for philosophy had turned to abhorrence. DAVID JOHN WILLIAMS

HERMIT (Gr. *erēmitēs* from *erēmia* = desert). A person seeking to please God who voluntarily adopts the solitary religious life. Within the early Church of Egypt, Christian hermits first appeared in the third century—e.g., Antony.* They lived in desert areas. Their fame was widespread, and many came both to see and to emulate them, so that in the next few centuries the number of hermits vastly increased. Some lived alone, others maintained their solitude in a community of hermits—e.g., the famous Augustinian Hermits.* Since the Counter-Reformation in the sixteenth century, hermits have disappeared from the Western Church, though their tradition is still partly retained in such religious orders as the Carthusians. The Eastern Orthodox Churches still have hermits. PETER TOON

HERODS. A Jewish family of Idumean descent prominent in the government of Palestine in the NT period, after their ancestor Antipater had been appointed procurator of Judea by Julius Caesar in 47 B.C.

(1) Herod the Great (c.73-4 B.C.) was the son of Antipater and a man of ruthless ability. He married Mariamne, who was the heiress of the Hasmonaean dynasty, so hoping to win the favor of the Jews. He rebuilt the Temple at Jerusalem, but he was never fully accepted by the Jews. He worked faithfully for the Roman authorities and was given by them the title "King of the Jews" in 37 B.C. He promoted Hellenism in Palestine. His suspicion of plots led to the murders of his wife and members of his family, and of the children in the area of Bethlehem (Matt. 2). After his death the Jews petitioned for direct Roman rule. In his will the territory was allotted to his sons—Judea, Samaria, and Idumea to Archelaus; Galilee and Perea to Antipas; and Batanea, Trachonitis, Iturea, and Auranitis to Philip.

(2) Archelaus was ethnarch in Judea from 4 B.C. to A.D. 6, but after a deputation went to Rome, he was deposed and exiled.

(3) Herod Antipas, known as "the Tetrarch," features most prominently in the gospels, which show him as a man of ability and cunning ("that fox," Luke 13:32), responsible for the execution of John the Baptist (Mark 6:14-28). Jesus was sent to him by Pilate for judgment (Luke 23:7-12). Herod was defeated in battle by Aretas IV of Nabatea in A.D. 36, and three years later was deposed and exiled by the Romans as a plotter.

(4) Philip seems to have ruled well until his death in A.D. 34.

(5) Herod Agrippa I was given by the emperor Gaius (Caligula) the tetrarchy of Philip after his death and that of Antipas after the latter's banishment. In A.D. 41 Claudius gave him the Roman province of Judea and Samaria. He was in favor with the Jews, but became extremely arrogant and was struck by sudden fatal illness in A.D. 44.

(6) Herod Agrippa II was his son, but was given less territory than his father. He was involved in the examination of Paul in Acts 25 and 26. He tried to dissuade the Jews from revolting in A.D. 66.

BIBLIOGRAPHY: in addition to traditional sources, A.H.M. Jones, *The Herods of Judaea* (1938); S. Perowne, *Life and Times of Herod the Great* (1956) and *The Later Herods* (1958); H.W. Hoehner, *Herod Antipas* (1972). R.E. NIXON

HERRNHUT, see MORAVIAN BRETHREN

HERTFORD, COUNCIL OF (673). A council of bishops summoned by Theodore of Tarsus, archbishop of Canterbury, to promote the reorganization of the English Church. Among its ten canons it reaffirmed the Roman calculation of Easter (canon 1), prohibited bishops intruding in the affairs of neighboring dioceses (2), forbade monks and clergymen from leaving their places without permission (4/5), provided for future episcopal

synods twice a year (7—later amended to an annual meeting at Clovesho), established precedence of bishops according to dates of ordination (8), and recognized adultery as the only ground for divorce (10). It was the first occasion on which the English Church deliberated as a unity and has been called "the first constitutional measure of the English race" (Stubbs), representing a landmark in the development of the English constitution.
J.G.G. NORMAN

HERZOG, JOHANN JAKOB (1805-82). Swiss-German Reformed theologian. Born in Basle, he studied theology there and later in Berlin, where he was a pupil of F. Schleiermacher* and J.A.W. Neander.* Appointed to Lausanne in 1835, he served as professor of historical theology from 1838 to 1846 and authored several works on the Zwinglian and Calvinist Reformation. He assumed the chair of church history at Halle in 1847 where he published two major studies of the Waldensians (*De origine et pristino statu Waldensium,* 1848, and *Die romanischen Waldenser,* 1835), and in 1854 was named professor of Reformed theology at Erlangen. In 1848 he was invited to undertake the editorship of a comprehensive religious encyclopedia from the Protestant perspective to counter a Catholic work then being published. The editing of this twenty-two volume, *Realencyklopädie für protestantische Theologie and Kirche* (1853-68), was his most significant endeavor and he himself contributed 529 articles to it. He began a second edition with his colleagues G.L. Plitt and Albert Hauck which the latter completed after Herzog's death. The work was modified and condensed into an American edition by Philip Schaff in 1882-84, and subsequent editions of this were known as the *Schaff-Herzog Encyclopedia of Religious Knowledge.*
RICHARD V. PIERARD

HESYCHASM. The Hesychast movement made its appearance in Byzantium in the first part of the fourteenth century. The name comes from the Greek word *hesychia,* "quiet" or "silence." It was applied to those individuals who devoted themselves in silence to mystical meditation, attempting to come into a full unity with God. This movement in the Byzantine Church illustrates the difference between official theology and the theology of experience. The ascetic monks of Mt. Athos attempted to meditate so intensely as to isolate themselves completely from the world and so attain the Divine Light. A monk of this persuasion would usually press his chin on his chest while focusing his eyes on his navel and holding his breath until his vision became dim. He would soon enter an ecstatic trance which would be the ultimate in union with God.

An extensive controversy broke out over this issue, primarily started by one Barlaam who in the West had been involved in an attempt to reunite the Eastern and Western churches. On his return to Constantinople he immediately began to ridicule the Hesychasts for their ecstatic experience of God. Barlaam and those who took similar views against the Hesychasts felt that it was wrong to have such an experience, for God could only be known indirectly. The Divine and Uncreated Light which came through the Hesychast experience was not authentic but simply an illusion. The various physical positions of prayer carried on by the Hesychasts were also part of the attack.

Gregory Palamas,* a former monk of Athos and then archbishop of Thessalonica, took up the defense of the Hesychasts. His articulate exposition of the position that man can know God even though God is by nature unknowable won the day. This was explained by differentiating between the *energies* of God which are knowable and the *essence* of God which is unknowable. Proper meditation involves the whole body since man is a unified being and must use that total being in his communion with God. A council was summoned on this issue in 1341; it sanctioned the doctrine of Uncreated Light and thus declared in favor of the Mt. Athos monks and Orthodox mysticism. To illustrate the extremes to which the victory was taken, Nicephorus Gregoras who was against the Hesychasts was reportedly dragged along the streets of the city after his death.

This was a difficult time for the Eastern Christian world because the Byzantine Empire being greatly weakened after the Crusades was having its very existence threatened by the advancing Muslims. The Muslim Turks were constantly creeping closer to Constantinople. The Slavs were stirring and causing Byzantium difficulty in the north. In the struggle for the identity of Byzantine theology, the mystical emphasis of the theology reigned supreme at this particular time.

See J. Meyendorff, *St. Grégoire Palamas et la mystique orthodoxe* (1959).
GEORGE GIACUMAKIS, JR.

HESYCHIUS (third century). Egyptian bishop. He is associated with Phileas, Theodorus, and Pachumis in addressing a letter to Melitius of Lycopolis. The letter, a Latin version of which is still extant, remonstrates with Melitius on his irregular ordinations. The bishops were in prison when the letter was written and were martyred under Galerius. This Hesychius is usually identified with the reviser of the text of the Septuagint and NT, or at least of the gospels, which was extensively used in Egypt. Jerome refers to it more than once, charging Hesychius with making apocryphal additions to the text, a charge that is later repeated in the Gelasian Decree.

HESYCHIUS OF JERUSALEM (fifth century). Greek writer. Born and educated in Jerusalem, he early became a monk and was subsequently ordained presbyter by the bishop of Jerusalem. His knowledge and eloquence were held in high esteem. He wrote against the Manichaeans, Arians, Apollinarians, and others. He is said to have written a history of the Council of Ephesus (431) and to have commented on the whole of the Bible. What survives of his writing suggests that he adopted the Alexandrian style of exegesis. Cyril of Scythopolis in his *Vita Euthymii* mentions Hesychius as having accompanied Juvenal of Jerusalem to the consecration of the Church of the Laura of Euthymius about 428, and he is said by Allatius to have been *chartophylax* of the

Church of the Anastasius at Jerusalem. Hesychius was a friend of Eutyches and opposed to the Council of Chalcedon (451). The date of his death is unknown. DAVID JOHN WILLIAMS

HETZER, LUDWIG (c.1500-1529). Anabaptist reformer, translator, and hymnwriter. Born at Thurgau, he matriculated at Basle (1517). From the chaplaincy at Wädenswil he came to Zurich, and wrote advocating an iconoclasm like that of Carlstadt. Disillusioned with Zwingli's caution, Hetzer, Grebel, and Manz established their own conventicles. Hetzer was expelled from Zurich (1525), led a group of Anabaptists at Augsburg, was banished to Basle, then stayed with Capito in Strasbourg (1526), where he was joined by Hans Denck. The three were gifted Hebraists, and Hetzer busied himself translating the Prophets. Expelled again, he went to Worms, and there published with Denck's help *Alle Propheten verdeutscht* (1527), the earliest Protestant version of the Prophets in German. By this time tending to anti-Trinitarian spiritualism, he was accused of adultery (1528) and beheaded at Constance. He composed hymns which were highly prized by the Hutterite* tradition. J.G.G. NORMAN

HEUMANN, CHRISTOPH AUGUST (1681-1764). German Protestant theologian. He studied at Jena, where he taught philosophy for some years, was director of the theological seminary at Eisenach (1709-17), and then of the gymnasium at Göttingen (1717-34). When the latter became the new University of Göttingen, Heumann became professor of the history of literature and associate professor of theology (1734-45) and then full professor of theology (1745-58). Resigning in 1758 when he came to reject the Lutheran understanding of the Eucharist, he devoted the remainder of his life to writing. He wrote extensively in theology, philosophy, linguistics, history, and literary criticism. He translated the NT (1748), wrote a twelve-volume commentary on the whole NT (1750-63), and published numerous controversial papers on the Lord's Supper.
W. WARD GASQUE

HEYLING, PETER (1607/8-1652). First German Protestant missionary. Born in Lübeck, he was from childhood noted for piety. When he began legal studies at Paris in 1628 he came under the influence of the Dutch legal scholar Hugo Grotius,* who resided there. His tract, *On the Truth of the Christian Religion*, was a handbook for missionaries. A member of a band of pious German students concerned about the church in the Near East, Heyling volunteered for missionary service without ecclesiastical support or connection. He studied Arabic on Malta and then went to Egypt where he encountered opposition from Orthodox and Catholic clerics. The Coptic Abuna (bishop) invited him to come to Ethiopia in 1634 where Heyling tutored children of prominent families and even gained the king's favor. He translated the gospel of John into Amharic and assisted in preparing a compendium of Roman law for use in Ethiopia. In 1652 he was martyred by a Muslim fanatic. RICHARD V. PIERARD

HEYLYN, PETER (1600-1662). Anglican polemicist and historian. Born at Burford, Oxfordshire, and educated at Magdalen College, Oxford, where he held a fellowship from 1618-29, he was ordained in 1624 and early showed strong High Church sympathies. A series of anti-Puritan treatises and debates gained him notoriety, and in 1630 he was made a royal chaplain. He basked in the favor of Charles I and Laud, receiving several livings, including a prebend of Westminster (where eventually he became subdean). He came into conflict with the dean there, Bishop J. Williams, continually slandering Williams until the latter was suspended by the Star Chamber (1637). Heylyn's inveterate anti-Puritanism brought him to grief during the Puritan revolution. W. Prynne,* perhaps in revenge for Heylyn's part in the condemnation of his *Historiomastix* and the subsequent personal ignominies, brought him before a Long Parliament committee. In the event Heylyn was heavily fined, his goods and library confiscated, and his life jeopardized. After 1648 he settled in Oxfordshire and, though continuing his rancorous treatises against Puritanism and Presbyterianism, lived in relative peace until the Restoration. He regained his influence in the church, but died soon after in London. His main works include *Ecclesia restaurata, or the History of the Reformation of the Church of England* (1661; new ed. 1849); *Cyprianus Anglicus, or ... the Life and Death of ... William Laud* (1668); *Aerius redivivus, or the History of the Presbyterians* (1670; 1672 ed. reprinted 1969).
BRIAN G. ARMSTRONG

HICKES, GEORGE (1642-1715). Nonjuror.* Graduate of Oxford where he later taught, he was ordained in 1666 and was briefly rector of St. Ebbe's, Oxford. While chaplain to the duke of Lauderdale he participated in Scottish church affairs. He was prebendary of Worcester where he later (1683) became dean, but he declined the bishopric of Bristol. Refusing allegiance to William and Mary, he was deprived of his deanery in 1690. He was in 1694 consecrated titular bishop of Thetford by the Nonjurors. His scholarship ranged from a specialized grammar and thesaurus to monographs on baptism, priesthood, and church order, editions of Thomas à Kempis and Fénelon, and a posthumously-published work, *Constitution of the Catholic Church* ... (1716).
C.G. THORNE, JR.

HICKSITES. In 1827-28 a number of American Quakers, following the preaching of Elias Hicks (1748-1830), withdrew from the orthodox Society of Friends and established their own yearly meetings. Hicks, an eloquent preacher and social crusader who had attacked such institutions as slavery, contended that man was capable of saving himself, and described the Bible and church dogma as functional but not authoritative. This group included those who had been influenced by Unitarianism, those who wished to resist the attempt by evangelical Quakers to unite all yearly meetings and create written doctrine, and those who believed inner experience was primary. Out of the schism, which generally included urban, progressive Quakers, came increased social activism. In 1902 the seven yearly meetings claiming Hicksite loyalties formed their own confederation, the Friends General Conference, to provide mutual help, but not coercion, for the "liberal, silent" meetings. In recent years they have cooperated with orthodox Friends.

DARREL BIGHAM

HIERARCHY (Gr. *hierarchia,* "the administration of sacred things"). The term has been used by Christians since the Church Fathers to denote the body of persons participating in church rule. To Roman Catholics it means collectively the organization of clerics into rank and order of position. More specifically, the Roman Catholic Church gives a twofold meaning to "hierarchy," with several subdivisions under each. In the hierarchy of *order,* those deriving authority directly from God comprise bishops, priests and deacons. Non-divine-right functions in this hierarchy are the subdiaconate and minor orders. In the hierarchy of *jurisdiction,* all grades derive authority from ecclesiastic sources, except the papacy and the episcopate, which are divinely ordained. The former grades of this hierarchy exercise authority conferred either from the pope or from the bishop.

ROYAL L. PECK

HIERONYMIAN MARTYROLOGY. A compilation comprising as its chief elements the calendars of Rome, Carthage, and Syria. The nucleus of the work is the Roman calendar, with which were incorporated in part or in whole the calendars of other Italian cities and the calendar of Carthage. To the Western calendar thus formed, a later editor added the first part of the Syrian festival list. It is noteworthy that he has taken over with the Syrian calendar its Arian coloring, the commemoration of the two bishops Eusebius being retained and even that of Arius himself, his name appearing in the corrupted forms "Arthoci," "Artotes," or "Ari Thoti" in different MSS. The preface takes the form of a letter addressed by Chromatius of Aquileia and Heliodorus of Altinum to Jerome, asking him to send them the festal calendar of Eusebius, and a reply from Jerome in which he says he is sending a shortened form of the calendar with the names of the most notable martyrs arranged according to the months and days of the year. It is from this apocryphal correspondence that the martyrology takes its name. The preface is first cited by Cassiodorus in 544. Its final compilation probably dates therefore from between the late fifth and early sixth centuries, and was probably made in N Italy. DAVID JOHN WILLIAMS

HIERONYMITES. In the fourth century certain Roman ladies at Bethlehem placed themselves under Jerome's direction. Though he founded a monastery there, it is in the fourteenth century that one looks to Fernando Pecha, who founded the Hieronymites in Spain. Gregory XI confirmed the order in 1373. In 1389 the monastery of Our Lady of Guadalupe passed into their hands. The Palace Monastery of San Lorenzo del Escorial, erected by Philip II outside Madrid, contains the richest library in Spain. A third important monastery is Belem, where the Portuguese kings are buried. Generous almsgiving marks the order, though its original strict observance of an Augustinian Rule relaxed so that in 1780 Charles III received special papal permission to solve disciplinary problems. In 1837 the Hieronymite Order as reorganized in 1585 was suspended.

MARVIN W. ANDERSON

HIGHER CRITICISM. The older term for what today is more generally known as "literary criticism." Though used earlier by students of the classics, it seems to have been first applied to biblical literature by J.G. Eichhorn* in the preface to the second edition of his *Old Testament Introduction* (1787). Higher criticism is so designated to distinguish it from "lower" or textual criticism. The image is that of a building: the first task in the study of any ancient document is to determine the true text insofar as that is possible; discussion of such matters as literary form, date, authorship, and purpose is "higher" in that it builds on the foundation of textual (lower) criticism. Although higher criticism is an essentially positive term, it is sometimes used by conservative Christians in a pejorative sense. W. WARD GASQUE

HILARION (c.291-371). Eastern ascetic. Born of pagan parents at Thabatha, near Gaza, he was educated at Alexandria and there converted to Christianity. For a time he was a disciple of Antony in the Egyptian desert, but soon returned to Palestine, to the desert south of Majoma, where he continued to practice the ascetic life which he had adopted in Egypt. Jerome traces the origin of the practice of the ascetic life in Palestine to Hilarion. The fame of his sanctity soon spread. He gathered disciples and organized them into societies. He also exercised an influence over the nomadic Arab tribes that came into contact with him. But his fame interfered with his life as a hermit and he returned to Egypt about 356. Some years later Julian's police forced him to flee further afield. He stayed in Sicily and Dalmatia and finally at Paphos in Cyprus. There he enjoyed the company of his disciples Hesychius and Epiphanius. He died there, and his body was taken by Hesychius back to Majoma.

DAVID JOHN WILLIAMS

HILARY OF ARLES (401-449). Bishop of Arles. Born of noble family and educated in philosophy and rhetoric, he was persuaded to renounce secular society for the solitude of Lérins by its founder, Honoratus, his kinsman. When Honoratus became bishop of Arles in 426, Hilary accompanied him there and succeeded him two years later. He presided over the councils of Riez (439), Orange (441), and Vaison (442). The canons of Riez and Orange are concerned mainly with discipline. The seventh canon of Riez is concerned with the rights of the bishop of Arles which Hilary was most energetic to further, howbeit from no selfish motives. He remained an ascetic throughout his episcopate, but he came into conflict with Leo, who was equally energetic in furthering the rights of the bishop of Rome.

At a council at Vienne in 444, Hilary deposed Chelidonius, bishop of Besançon. When the latter appealed to Rome, Hilary went there to defend his decision. Leo, however, reversed that decision, depriving Hilary of his metropolitical rights. A rescript was obtained against him from Valentinian III which also ordered provincial governors to enforce obedience to the bishop of Rome. Of his remaining years little is known. It is evident from the letters of Prosper and another Hilary that while he was a great admirer of Augustine, Hilary did not accept Augustine's teaching on predestination. From this, and from his respect for Faustus of Riez, we must regard Hilary as a semi-Pelagian.* Fragments of his works were collected in editions of Leo's works by P. Quesnel (1675) and P. and H. Ballerini (1753-57; rep. J.P. Migne, *Patrologia Latina,* 1, pp. 1213-92, with additions). DAVID JOHN WILLIAMS

HILARY OF POITIERS (c.315-368). Bishop of Poitiers. Born of good family, he was educated in the Latin classics, converted about 350 to Christianity, and some three years later was made by popular choice bishop of Poitiers, his birthplace. He became a leader of the orthodox in Gaul, although he confessed that he only discovered the creed of the Council of Nicea on the eve of his exile, but had held the teaching it contained on the basis of his study of the Bible. After the Council of Milan in 355 he led the protest against the banishment of those bishops who refused to condemn Athanasius and against the intervention of the civil power in questions of faith. As a result of this he was himself condemned by the council at Beziers in 356, and banished by Emperor Constantius to Phrygia. In this enforced leisure he was able to pursue his study of theology, and his *De Trinitate* belongs to this period. In 359 the councils of Arminum in the West and Seleucia in the East were held. Hilary was obliged to attend that of Seleucia, at which he defended the cause of orthodoxy. From there he went to Constantinople, only to find the delegates from the Western council betraying the orthodoxy which he had upheld. He appealed to the emperor for an audience, but was refused. Constantius sent him back to Gaul without annulling his banishment. The emperor's attitude provoked a bitter attack by Hilary in his *Contra Constantium.* Meanwhile, the emperor had forced the orthodox bishops at Arminum to subscribe to an Arian creed. On his return to Gaul, Hilary set about counteracting this Arian victory. In 362 he traveled to N Italy and Illyria for the same purpose, but was ordered back to Gaul by Valentinian after a dispute between Hilary and Auxentius, bishop of Milan.
DAVID JOHN WILLIAMS

HILDA (Hild) (614-680). Abbess of Whitby. She was the daughter of Hereric, nephew of King Edwin of Northumbria, who was converted through the preaching of Paulinus and baptized by him in 627. She served God faithfully in the secular world for a number of years, being influenced by both the Roman and the Celtic streams of Christianity. She then decided to become a nun, and was on her way to France to join a religious community when she was recalled from East Anglia by Aidan* in 649. Aidan appointed her abbess of the convent at Hartlepool in County Durham. In 659 she became the founder and abbess of the double monastery for men and women set strikingly on the cliff top at Streanshalch (Whitby) in Yorkshire. This community became famous as a school of theology and literature, nurturing five future bishops and Caedmon, the earliest known English poet. At the Synod of Whitby* in 663/4 she defended the Celtic customs, but when the decision went in favor of the Roman usage she accepted that. R.E. NIXON

HILDEBRAND, see GREGORY VII

HILDEGARD (1098-1179). German abbess, mystic, and writer, who became the leader of a convent near Bingen. She experienced visions which increased in frequency as she grew older. An investigation by the archbishop of Mainz gave a favorable verdict on the authenticity of her experiences, and he assigned a monk, Volmar, to act as her secretary. Pope Eugenius III also investigated her activities, and again a favorable report followed. Her principal work, *Scivias,* is an account of twenty-six visions with an apocalyptic emphasis dealing with creation, redemption, and the church. She also wrote saints' lives, two books of medicine and natural history, hymns, homilies, and a language of her own consisting of 900 words and an alphabet of twenty-three letters. Her influence extended beyond her convent through her extensive correspondence and travels in Germany and France. She spoke to people of all classes and called them to repent and obey the warnings God had given to her. Although miracles have been attributed to her and canonization procedures have been started, they have never been completed. ROBERT G. CLOUSE

HILGENFELD, ADOLPH BERNARD CHRISTOPH (1823-1907). German Protestant scholar. He taught in Jena University from 1847 until his death, and was editor of *Zeitschrift für wissenschaftliche Theologie* from 1858. He adopted the principles of F.C. Baur and the Tübingen School,* though he was less radical, accepting, for example, the genuineness of 1 Thessalonians, Philippians, and Philemon. He was a pioneer of research into apocalyptic literature and wrote extensively on later Judaism. He was the author of an edition of extracanonical NT books, *Novum Testamentum extra Canonem receptum* (4 vols., 1866). In his *Die jüdische Apokalyptik in ihrer geschichtlichen Entwickelung* (1857), he attempted to show, *inter alia,* that the *Similitudes of Enoch* were Christian in origin—an idea generally abandoned, and refuted by E. Sjöberg and others.
J.G.G. NORMAN

HILL, ROWLAND (1744-1833). Preacher. Educated at Eton, he entered St. John's College, Cambridge (1764), at a time when evangelical views were unpopular. He believed at first that he was the only evangelical Christian there, except for the shoeblack at the gate, but soon he led several students to Christ. He continued to visit the poor and sick, and to preach as opportunity offered, even after six of his friends were expelled from Oxford for so doing. Following ordination he was appointed to Kingston and preached to great crowds, often in the open air, for ten years. He then inherited money and built Surrey Chapel, Blackfriars, where he exerted a powerful London ministry. He welcomed advances in science, himself vaccinating the children of his congregation. He was instrumental in founding the Religious Truth Society, the British and Foreign Bible Society, and the London Missionary Society. Spurgeon described him as full of fun in the pulpit—"a childlike man in whom nothing was repressed." Sir Rowland Hill of the penny post was named after him. R.E.D. CLARK

HILLIS, NEWELL DWIGHT (1858-1929). Presbyterian clergyman and author. Born into a Quaker home in Magnolia, Iowa, he graduated in arts from Lake Forest College (1884) and in divinity from McCormick Seminary (1887). He organized Sunday schools in the west each summer for the American Sunday School Association. After serving Presbyterian churches in Illinois (1887-99) he became pastor of the Plymouth Congregational Church in Brooklyn. He organized Plymouth Institute in 1914 as a social service agency to provide educational opportunities for young people. From 1914 to 1917 he lectured in 250 cities on U.S. obligation to enter World War I, and later reportedly sold $100 million in "Liberty Bonds." He resigned his pastorate in 1924 because of ill health. EARLE E. CAIRNS

HILTON, WALTER (d.1396). English mystic. Although long considered a Carthusian, he was an Augustinian canon of the priory of Thurgarton, Nottinghamshire. Events of his earlier career and dates of his writings are still uncertain. *De Imagine Peccati* and *Epistola Aurea* he wrote as a solitary during a period between university career and his religious association. *The Scale of Perfection* is his most famous work, originally written for the guidance of a religious friend, an anchoress. In two parts or books, the second is addressed to no one and is more advanced. The first book has been reckoned austere and theocentric, the second of warmer christocentric piety. He distinguishes between the life of "faith" (active, ascetic) and "feeling" (contemplative, mystical), and makes wide use of Augustine, Gregory, Bonaventura, the Victorines, Rolle, and *The Cloud of Unknowing.* C.G. THORNE, JR.

HINCMAR OF REIMS (c.806-882). Archbishop of Reims. He was educated at the abbey of St.-Denis, Paris, under Abbot Hilduin, who in 822 introduced him to the court of Louis the Pious. Officially entering the king's service in 834, he attached himself on Louis's death to Charles the Bald, thus incurring the hostility of Lothair I. Hincmar administered the abbeys at Compiègne and St.-Germer-de-Flay. Elected archbishop of Reims in 845, he faced imperial opposition but avoided his own deposition at the Synod of Soissons (853).

In opposing the king of Lorraine (Lothair I's second son) who wanted to divorce his wife, Hincmar produced his *De divortio Lotharii,* which displayed a great knowledge of canon law. He was unsuccessful in his attempt to depose Rothad II, bishop of Soissons, who had long attacked his rights, but did manage to quiet his own nephew, Hincmar of Laon, who refused to recognize his authority. This occasioned his *Opusculum LV Capitulorum,* wherein he defended the rights of a metropolitan over his bishops. Hincmar protested episcopal appointments at Cambrai, Noyon, and Beauvais, and was able to place his own appointments in these places. When the Council of Mainz (848) condemned Gottschalk's errors on predestination, he published a refutation of the monk, *Ad Reclusos et Simplices,* which brought attack on himself, and later on his colleague John Scotus Erigena for the latter's *De Divina Praedestinatione.* The controversy continued at the synods of Quiercy (853) and Valence (855), whereupon Hincmar wrote his defense, *De Praedestinatione Dei et Libero Arbitrio,* arguing that if God predestines the wicked to hell then He is the author of sin. Reconciliation was finally achieved at the Council of Thuzey (860). With Lothair's death in 869, Hincmar no longer feared to support Charles the Bald, and he proceeded to crown him despite papal objection.
C.G. THORNE, JR.

HINDUISM. This is not a religion, if by that we mean a single closed system of beliefs and practices observed by all Hindus. It is rather an infinitely complex aggregate of beliefs and practices bound together by their common location on the Indian subcontinent, and by their links with the social system of caste. The word "Hindu" is derived, through the Persian, from the name of the River Indus; "Hinduism" is the European blanket term which covers all forms of Indian ethnic religion that acknowledge, directly or indirectly, the authority of those scriptures called *Veda* (dating from c.1200 to c.600 B.C.), and that acknowledge the *dharma* (law) of caste. The sacred language of Hinduism is Sanskrit.

Most Hindus would accept (1) the belief in transmigration, i.e., that every person lives many times on earth, in human or other form; (2) the belief that one's status, or caste, in any given existence, depends upon one's conduct in a previous life; (3) that man's ultimate goal is release *(moksha)* from rebirth, and from the phenomenal world; (4) that the priestly (Brahmin) class is worthy of special reverence; (5) that the cow should be cared for and revered as a symbol of the earth's bounty. Beyond this point it is difficult to generalize. Very many Hindus are theists, and believe in a personal God under such names as Vishnu or Shiva, who should be worshiped with love and devotion *(bhakti).* Others, following the philosopher Shamkara, hold the Supreme Reality to be impersonal. A few are theoretically atheists. Most would believe God to be immanent in all creation and would now consider all religions to be equally valid as means of access to God. This particular view has been expressed strongly by such prominent leaders of Hindu thought as Ramakrishna, Gandhi, and Radhakrishnan.

The Hindu scriptures fall into two broad classes: *shruti* (revelation) and *smriti* (tradition). The former comprises the Vedic hymns, commentaries *(Brahmanas)* and speculative writings (Upanishads); the latter includes the two great epics (*Mahabharata* and *Ramayana*), the *Bhagavad Gita* (part of the *Mbh.*), the law books, the later mythological writings, and the documents of the sects. All in all, the Hindu scriptures are of immense size and staggering diversity. Hindu worship takes place in the home and the temple, the latter being thought of as a dwellingplace of a god or goddess, and not as a place of assembly. Daily and seasonal patterns of worship are followed.

The main point at issue between Hindus and Christians is the uniqueness of Christ. Many Hindus can accept Jesus as a divine teacher *(Yesuswami)*, but not as sole Savior. Christians for their part are not able to accept the basic Hindu belief in transmigration and rebirth, and insist that God is one and personal (a view held by some, but not all, Hindus).

BIBLIOGRAPHY: Books on Christianity and Hinduism include A.G. Hogg, *The Christian Message to the Hindu* (1947); E.J. Sharpe, *Not to Destroy but to Fulfil* (1965); K. Klostermaier, *Hindu and Christian in Vrindaban* (1970).

E.J. SHARPE

HIPPO, COUNCIL OF. Held in 393 with Aurelius of Carthage presiding and Augustine present as a presbyter in the entourage of Valerius of Hippo, it is chiefly important for its conciliatory measures towards Donatism.* It resolved to accept as clergy those Donatists who came over to the Catholic communion *cum suis plebibus*, but this should not be done before the *transmarina ecclesia* had been consulted. A *breviarium* of the canons of Hippo was read at the Council of Carthage in 397 and passed ultimately into general canon law.

HIPPOLYTUS (d. c.236). Presbyter and teacher in the Church of Rome. Origen heard him preach there in 212. Very little is known about his early life, but he was a presbyter under Bishop Zephyrinus whom he accused of compromise with the views of Sabellius. Perhaps his theological judgment was affected by his opposition to Callistus, the archdeacon, who himself became pope in 217. Hippolytus then set himself up as an antipope and continued as such until deported in 235 by Emperor Maximin during a period of persecution. In exile he was reconciled to the pope, and after his martyrdom his body was brought to Rome with honor by the church.

In the centuries following his death his identity was confused and he was equated with various people—e.g., in the Roman Breviary he is identified as a soldier converted by St. Lawrence. After many years of oblivion he was given prominence again by the discovery near his tomb in Rome of a (headless) statue of him enthroned as a bishop (erected by his followers who later merged with the Novatians?). Inscribed on the statue were a table for computing the date of Easter and a list of his writings. Among those which survive in translation are the *Philosophoumena* (the title given to parts 4-10 of his longer *Refutation of all Heresies*) which was thought to be by Origen until J.J.I. Döllinger in 1859 showed it to be by Hippolytus. And there is the *Apostolic Tradition* which E. Schwartz in 1910 and R.H. Connolly in 1916 demonstrated was also by Hippolytus. The *Philosophoumena* is of value for its description of the Gnostic sects, and the *Apostolic Tradition* preserves for us a conservative picture of Roman church order and worship at the end of the second century. Mention may also be made of his *Commentary on Daniel* which is the oldest Christian Bible commentary to survive in its entirety.

Theologically, Hippolytus taught a Logos doctrine inherited from Justin Martyr. He distinguished two states of the Logos, the one eternal and immanent, the other exterior and temporal. By his opponents he was, with some justice, called a ditheist. In disciplinary matters he was a rigorist who strenuously opposed the mitigation of the penitential system in order to cope with the entry into the church of large numbers of converts. Also he seems to have been the first scholar to construct an Easter table that was independent of contemporary Judaism.

BIBLIOGRAPHY: C. Wordsworth, *Saint Hippolytus and the Church of Rome* (1853); works of Hippolytus in J.P. Migne, *Patrologia Graeca*, vol. X (1857); translation of *The Apostolic Tradition* by G. Dix (1937).

PETER TOON

HIPPOLYTUS, CANONS OF. An early sixth-century collection of canons, originally written in Greek, and relating to liturgical and disciplinary matters. The Greek text is lost, and they survive in Ethiopic and Arabic MSS of the thirteenth century. These were made from a Coptic translation. They are wrongly attributed to Hippolytus,* whose *Apostolic Tradition* was one of the sources for them. Until this century the canons were regarded as a genuine production of Hippolytus and thus thought very valuable as a source for the early history of the church. Louis Duchesne made much use of them in his influential *Origines du culte chrétien* (1889). Scholars now regard them as having only minor importance.

PETER TOON

HISPANA CANONS. The lengthiest and most significant of several recensions of early conciliar and papal decisions. The material from sixty-six Eastern, African, French, and Spanish councils is arranged geographically-chronologically. The 103 papal decretals extend from the reigns of Damasus I (d.384) to Gregory I (d.604). Initially appearing in eighth-century Gaul, the collection was, according to questionable tradition, ascribed to Isidore of Seville* as compiler; hence its other name, *Isidoriana*. After the papacy of Alexander III (d.1181) it was recognized as the official body of Spanish canon law. It constitutes a major portion of the authentic material in the False Decretals.*

JAMES DE JONG

HOADLY, BENJAMIN (1676-1761). Anglican theologian. One of the most scandalous of the eighteenth-century bishops, he owed his ecclesiastical career and translation to the rich dioceses of Bangor (1715), Hereford (1721), Salisbury (1723), and Winchester (1734) entirely to his championship of the Whig party, and his authorship of skillful pamphlets against Tories and High Churchmen. As the leader of the "Whig" or "Low Church" Anglicans, he was a notorious Latitudinarian,* writing down all mysteries and dogma, and justifying the most generous inclusion of all groups within his church, including Arians. In 1716 a sermon which denied that there was a visible Church of Christ at all and defined Christianity as merely "sincerity" provoked the Bangorian Controversy* which led to an outcry and the government's suspension of Convocation, which did not meet for 150 years thereafter. A later essay (1735) on the merely memorialist nature of the Lord's Supper led to an accusation of Socinianism. This minimizing and controversial prelate survived for another twenty-six years.

IAN SELLERS

HOBBES, THOMAS (1588-1679). British political philosopher. Trained at Oxford as a classicist, he was a tutor to the Cavendish family, but when the civil war came to England he went into exile in France (1640-51). In Paris he was tutor to the Prince of Wales (later Charles II). By 1651 he returned to England, submitted to the Commonwealth, and published his great work *Leviathan.* This work propounds an absolutist government based, not on divine right, but on an analysis of human psychology. All men, according to Hobbes, possess instinctive feelings of fear and self-preservation. These instincts provide the motivation for social organization. If there were no government and if all men were equal, life would be unendurable ("solitary, poor, nasty, brutish and short," as he put it). Driven by self-preservation, men contracted with each other to transfer all their power to an absolute sovereign, who would use his unlimited power to enforce obedience and unity. The contract between ruler and people was unbreakable. Hobbes's ideas offended both the divine-right theorists and those who held to the historic rights of Englishmen. After the Restoration, however, he was granted a pension and free access to the king.

ROBERT G. CLOUSE

HOCHMANN VON HOCHENAU, ERNST CHRISTOPH (1670-1721). German Pietist mystic. Born to a customs collector in Lauenburg/Elbe, he was raised a Lutheran in Nuremberg. While studying law in Halle he was influenced by A. H. Francke* and converted in 1693. The initial radicalism of his beliefs was intensified by contacts with Gottfried Arnold* at Giessen in 1697. Hochmann taught that the church was primarily spiritual in character and minimized the importance of structures, creeds, and sacraments. Regarding the conversion of the Jews as the sign of Christ's impending return, he engaged briefly in Jewish missionary work. From 1701 to 1711 he wandered about Germany preaching the necessity of revival and frequently suffered persecution and imprisonment. He worked with Alexander Mack* in the Palatinate in 1706 and played a major role in founding the Brethren movement. While jailed in Detmold in 1702 he was compelled to prepare a statement of his beliefs which the Brethren highly regarded and had reprinted in Pennsylvania in 1743. He eventually broke with Mack, feeling that the Brethren were too sectarian. In Schwarzenau/Eder he spent his last years as an ascetic in a small hut he named Friedensburg. Besides being the spiritual father of the Church of the Brethren* and the most important separatist mystic of the early eighteenth century, he was also on friendly terms with the Mennonites.

RICHARD V. PIERARD

HODGE, ARCHIBALD ALEXANDER (1823-1886). Presbyterian theologian. Son of Charles Hodge* and Princeton-trained, he held several teaching posts before succeeding his father as Princeton's systematic theologian in 1877. Explainer of his father's ideas rather than creator of new concepts, and less prolific and scholarly, he was noted for his *Outlines of Theology* (1860; rev. 1878), *The Life of Charles Hodge* (1880), and *Popular Lectures on Theological Themes* (1887). Although defending the Princeton fundamentals of divine sovereignty and human depravity, he attempted to enliven those ideas through social application. He believed that America, if it respected "the Fatherhood of God, the Elder Brotherhood and redeeming blood of Christ, and the universal brotherhood of men," was placed "at the crisis of the battles on which the fate of the Kingdom for ages turns."

DARREL BIGHAM

HODGE, CHARLES (1797-1878). Leading American theologian of the nineteenth century. Born in Philadelphia, son of an army surgeon, he was educated at Princeton, graduating from the college in 1815 and from the seminary in 1819. His theological studies under Archibald Alexander* determined his life-work. He became an instructor at Princeton Seminary in 1820, and remained there for the rest of his life, except for two years' study in France and Germany (1826-28). He was professor of oriental and biblical literature (1822-40), then professor of theology. His own theology was mainly that of the Westminster Confession with obvious traces of scholastic Calvinism, notably from Turretine. His thought was governed by a high view of verbal inspiration and infallibility. While orthodox Calvinism was declining in American thought generally, and the evolutionary idea was beginning to exert unusual power, Hodge unswervingly defended a supernaturally inspired Bible and thereby placed his stamp upon what came to be called "Princeton theology." This had a powerful influence, not only in his own Old School Presbyterian circles, but in other churches as well.

His writings carried his influence beyond the 3,000 students he taught during a half-century. He started the *Biblical Repertory* in 1825 (later called the *Biblical Repertory and Theological Review,* and after 1836 the *Biblical Repertory and Princeton Review*) and edited it for more than forty years. His first book, *A Commentary on the Epistle to the Romans* (1835; 19th ed., 1880), established his scholarship. Among his later works none exerted greater influence than his *Systematic Theology* (3 vols., 1872-73).

He also held a commanding position in the Presbyterian Church. He was moderator of the general assembly (Old School) in 1846, and a prominent member of the missionary and educational boards. In the controversy of 1837 he opposed the New School views of doctrine and polity. When division came, he supported it.

See A.A. Hodge. *The Life of Charles Hodge* (1880); and C.A. Salmond, *Princetonia: Charles and A.A. Hodge* (1888). BRUCE L. SHELLEY

HOFBAUER, CLEMENT MARY (1751-1820). Redemptorist* priest. Born John Dvorvák, youngest of twelve children of a Moravian grazier and butcher, his desire for the priesthood had to be postponed until 1780. He worked as a baker from his father's death in 1757, except for a period of living as a hermit and as a servant in the Premonstratensian* monastery of Buck (1771-75). The generosity of three Viennese ladies saw him through Vienna University, after which he went to Rome and joined the recently founded Redemptorists (C.SS.R.). He was ordained in 1785. Returning to Vienna, where he was unable to establish a house because of Josephinism,* he went to Warsaw (1787-1808) where he did much pastoral work and opened schools; he founded several houses in Poland, working mainly among the German-speaking population. He also introduced the order into Switzerland and S Germany, acting as vicar-general for the regions north of the Alps. Driven from Warsaw by Napoleon, he returned to Vienna, serving first in a Franciscan church, then as chaplain to the Ursulines and as pastor of St. Ursula's church from 1813. His influence ranged from the emperor Francis throughout the populace, especially with leading Romanticists, winning many converts and causing spiritual renewal. To him more than any other the extinction of Josephinism is due. He established the Redemptorists there in 1819, and at his death Pius VII said, "Religion in Austria has lost its chief support." He was canonized in 1909, named patron saint of Vienna in 1914.

BIBLIOGRAPHY: Lives by J. Hofer (tr. J.B. Haas, 1926) and J. Carr, C.SS.R. (1939).

C.G. THORNE, JR.

HOFMANN, JOHANN CHRISTIAN KONRAD VON (1810-1877). German theologian. He is widely regarded as the most significant of the Erlangen School of theologians which represented a modified form of Lutheran orthodoxy. He taught at Erlangen from 1845. His writings include *Weissagung und Erfüllung* (1841-44), *Der Schriftbeweis* (1852-56), and *Die heilige Schrift* (1862-78). Hofmann laid great emphasis on biblical exegesis which he coupled with stress on Christian experience. The latter is expressed historically in salvation history, of which Scripture is the record prior and subsequent to the coming of Christ. Hofmann interpreted this within the framework of an orthodox theism which saw world history rooted in the Trinitarian character of God and His purposes for man. The ultimate goal of the historical process is the union of God and man in Christ. Hofmann's rejection of the penal satisfaction view of atonement caused considerable controversy. Satan, he held, is defeated because Jesus maintained His oneness with God even in His greatest extremity on the cross.

COLIN BROWN

HOLCOT, ROBERT (c.1290-1349). Biblical expositor. As a Dominican theologian he commented on a wide range of theological topics, though he asserted free will contrary to his contemporary Thomas Bradwardine.* There is no clear evidence that Holcot studied or taught at Cambridge. His commentary on the *Sentences* (Oriel MS 15) uses the term *potentia absoluta* of God's complete freedom to will all things. God's grace bears no necessary relation to His love, for God's will can dispense with anything. This philosophical Pelagianism is precisely what Bradwardine and Wycliffe attacked in fourteenth-century Oxford. By loving God less than another man loves Him, one can gain the greater reward. The sixteenth century knew Holcot for his questions on Lombard's *Sentences*, published in 1497, 1510, and 1518. The *Commentary on Wisdom* passed through several editions after 1480. The 1586 Basle edition amended his extreme views on the Immaculate Conception. Holcot died in the black plague of 1349. MARVIN W. ANDERSON

HOLINESS MOVEMENT, AMERICAN. A religious movement dating from the mid-nineteenth century that tried to preserve the original thrust of the Methodist teachings on entire sanctification and Christian perfection as taught by John Wesley in such writings as the *Plain Account of Christian Perfection.* This teaching expects that entire sanctification normally takes place instantaneously in an emotional experience similar to conversion. At this point one is cleansed from inbred sin and enabled to live without conscious or deliberate sin. In the American revivalistic context and under the influence of the camp meeting* there was a subtle mutation of these concepts in the direction of individualism, emotionalism, and emphasis on the crisis experience.

Early in the nineteenth century, groups began to emerge from Methodism in protest against the decline of discipline. In the 1840s abolitionist Orange Scott led the Wesleyan Methodists out because Methodism had grown comfortable with slavery. In 1860 B.T. Roberts and the Free Methodists were expelled from the Genesee Conference because of controversy over similar issues and the decline of the holiness emphasis. Both groups added statements on Christian perfection to their articles of religion and gradually identified themselves as holiness bodies.

At the same time there was a movement within Methodism to reemphasize holiness. In the 1830s two sisters, Sarah Lankford and Phoebe Palmer, organized a weekly prayer meeting known as the "Tuesday Meeting" which along with similar meetings became a major force in this movement. In the late 1860s was founded a "National Camp

Meeting Association for the Promotion of Holiness," which evolved over the years into the National Holiness Association (NHA), renamed in 1971 the Christian Holiness Association (CHA), the present ecumenical body representing non-Pentecostal holiness bodies. By the end of the century this movement had spawned innumerable holiness camp meetings, periodicals, and state and local holiness associations. Increasing conflict with Methodist leaders and the decline of national holiness leadership resulted in a period of fragmentation into a myriad of small groups. Many of these clustered to form such holiness denominations as the Church of the Nazarene, the largest independent holiness body, and the Pilgrim Holiness Church, which in 1968 merged with the Wesleyan Methodists to form the Wesleyan Church.

Also founded at this time were a number of denominations taking the name "Church of God."* Many of these moved into Pentecostalism, but the group centered at Anderson, Indiana, remained closely identified with the holiness movement. The impact of the holiness movement extended far beyond the bounds of Methodism. Two Mennonite bodies, the Missionary Church and the Brethren in Christ,* adopted Wesleyan views and identified with the movement. Other denominations such as the Christian and Missionary Alliance,* reveal holiness influence, but have not completely identified with the movement. Salvation Army founder William Booth was converted in England under an American holiness evangelist. When the Army came to the USA in the 1880s it had a strong holiness orientation and later identified with the CHA.

The twentieth century has produced other holiness groups. The Evangelical Methodist Church withdrew from Methodism in the wake of the fundamentalist–modernist controversy, and the Evangelical Church of North America was formed after the merger of the Methodist Church and the Evangelical United Brethren in 1968. Other related groups include the Holiness Christian Church, the Churches of Christ in Christian Union, the Methodist Protestants, the Primitive Methodist Church, the Congregational Methodist Church.

Many of these denominations developed in the wake of the revival movements associated with C.G. Finney,* with whose Oberlin Theology* holiness theology has many affinities. Another parallel movement in the mid-nineteenth century was the British Keswick* movement, whose teachings on the victorious life are distinguished from holiness thought primarily by their context in Reformed theology.

Many interpreters fail to distinguish between the holiness movement and Pentecostalism. There are many similarities and historical connections. In the late nineteenth century, holiness writers began to speak of entire sanctification as a "baptism of the Holy Spirit" on the model of Pentecost. It was in this milieu and thought pattern that Pentecostalism was born in America. Some holiness bodies, such as the Pentecostal Holiness Church, moved in this direction, but most dropped the term "Pentecostal" and reaffirmed non-Pentecostal Wesleyan doctrine.

In the twentieth century the holiness movement has shed some of the trappings of revivalism* and is better viewed as conservative Methodism. This development has produced a conservative reaction leading to a number of very small groups such as the Allegheny Wesleyan Methodist Connection, the Bible Missionary Church (originally Nazarene), the Wesleyan Holiness Association (originally Bible Missionary), the United Holiness Church, and the Evangelical Wesleyan Church (both originally Free Methodist), loosely grouped together in the Inter-Denominational Holiness Convention.

At present the holiness movement would claim a constituency of at least two million, from fifty to a hundred schools (including three theological seminaries), two interdenominational missionary societies (Oriental Missionary Society and World Gospel Mission), innumerable local associations, camp meetings, etc., and denominational agencies.

BIBLIOGRAPHY: T. Smith, *Called Unto Holiness* (1962); D. Rose, *A Theology of Christian Experience* (1965); K. Geiger, *The Word and the Doctrine* (1965); D.W. Dayton, *The American Holiness Movement: A Bibliographic Introduction* (1971); H.V. Synan, *The Holiness-Pentecostal Movement* (1971). DONALD W. DAYTON

HOLL, KARL (1866-1926). Patrologist. Born in Tübingen, he was a pastor before returning to Tübingen to lecture. He became professor of church history there in 1901 and from 1906 to 1926 held a similar chair in Berlin. Having collaborated with Harnack* on an edition of Greek patrology (the Berlin Corpus), he also investigated the origins of Epiphany and the Easter feast and fast, and wrote *Enthusiasmus und Bussgewalt beim griechischen Mönchtum.* One of Germany's most influential church historians, for both Eastern and Western churches, his studies on Luther, including *The Cultural Significance of the Reformation,* created a renascence in the Evangelical Church. He studied Russian and brought attention to Tolstoy, but avoided the theological disputes of the turn of the century. C.G. THORNE, JR.

HOLLAND, see LOW COUNTRIES

HOLLAND, HENRY SCOTT (1847-1918). Anglican preacher and theologian. Educated at Eton and Balliol, he became in turn senior student of Christ Church (1870-84) and canon of St. Paul's (1884-1910) before returning to Oxford as regius professor of divinity. A witty and prominent member of the *Lux Mundi** group (he contributed the article on Faith), he combined a ritualistic High Churchmanship with a vague liberal theology formulated under the influence of T.H. Green and F.D. Maurice. An advanced member of the Christian Social Union of which he was sometime vice-president, he popularized a view of Christ as "the solution of all human problems." His religious and political optimism was shattered by the experiences of World War I. IAN SELLERS

HOLTZMANN, HEINRICH JULIUS (1832-1910). German theologian and NT scholar. Born in Karlsruhe and educated at Berlin, he held a pastorate at Baden (1854) before beginning his academic career. He taught in Heidelberg (1858), then in Strasbourg (1874) until retirement in 1904. In his study of the synoptic gospels (1863) he developed the two-source theory, with its dual necessity of accepting a "teachings" source for Matthew and Luke, and the priority of Mark, which yielded the kind of portrait of Jesus desirable to the liberal-psychological scheme, and one in which the eschatological interpretation of the kingdom of God in Jesus' preaching could be avoided. As a consequence Holtzmann was involved in church-political squabbles of the German pastors. At Strasbourg his work was mainly directed toward the preparation of textbooks in NT introduction (1885) and theology (2 vols., 1896-97), and a *Lexicon für Theologie und Kirchenwesen,* with Richard Zoepffel (1882). By his contribution on the synoptics to the *Hand-Kommentar zum Neuen Testament* (1889) and his *New Testament Theology,* he laid the foundations for the NT research of the twentieth century.
CLYDE CURRY SMITH

HOLY ALLIANCE. A declaration in treaty form signed on 26 September 1815 in Paris, by the Orthodox Czar Alexander I of Russia, Catholic Emperor Francis I of Austria, and Protestant King Frederick William III of Prussia after the final allied victory over Napoleon. It proclaimed that international relations would henceforth be based on "the sublime truths which the Holy Religion teaches" and that the rulers of Europe would abide by the principle that they were brothers and whenever necessary would "lend each other aid and assistance." They would recognize no other sovereign than "God our Divine Saviour, Jesus Christ." It was once believed that the pietistic Baroness Von Krüdener had inspired Alexander to advance such a compact, but recent scholarship holds that either he had for some time considered breaking with the old system of relations based on power politics or he wished to establish an international concert to counterbalance English seapower. Only the British government, the Sultan, and the pope refused to accede to it. Although it had no practical binding power, for liberals and revolutionaries the term "Holy Alliance" took on a sinister connotation as a conspiracy of reactionary powers to maintain the status quo in E Europe.

See J.H. Pirenne, *La Sainte Alliance* (2 vols., 1946-49). RICHARD V. PIERARD

HOLY CLUB. This name was derisively given to the group of earnest "Methodists" which in the early 1730s met in John Wesley's rooms at Lincoln College, Oxford, and included Charles Wesley, Benjamin Ingham, and George Whitefield. Its members, in addition to spiritual exercises, visited prisoners, relieved the poor, and maintained a school for neglected children. Membership was never more than twenty-five, and when John Wesley left Oxford in 1735 the group disintegrated. The Club owed much to Moravian example and to the earlier religious societies which flourished in the Anglican Church. The practices and discipline of the Holy Club became the model for the later bands, classes, and societies of the Methodist revival, and the inspiration for the movement's social concern. IAN SELLERS

HOLY COMMUNION, see COMMUNION, HOLY

HOLY GHOST, see HOLY SPIRIT

HOLY MOUNTAIN, see ATHOS, MOUNT

HOLY OFFICE. Organized in 1542 by Paul III, the Holy Office or *Sacra Congregatio Romanae et Universalis Inquisitionis seu Sancti Officii* was established to serve as the last court of appeal in heresy cases brought before the Inquisition.* Sixtus V in 1587 increased the number of cardinals designated to serve on it from six to thirteen. While the pope may preside over the deliberations of this body, it customarily functions without him. In 1908 the Holy Office was reorganized as the court to which final decisions in faith and morals were referred, and its name was shortened to the *Congregatio Sancti Officii.* It was subjected to further reorganization as a result of Vatican Council II.* DAVID C. STEINMETZ

HOLY ORDERS, see ORDINATION

HOLY ROMAN EMPIRE. Political entity in medieval Europe. The empire's founding may be dated to Otto I (962), although the precise term "holy" was first used in 1157. A long struggle between the emperors and popes beginning with the Investiture Controversy* in 1076 undermined the empire, while the growing power of kings elsewhere in Europe and a deteriorating situation in Italy destroyed it as a supranational institution. The Golden Bull (1356) which delineated the rights of the electors revealed clearly that it had become essentially a German institution. The failure of imperial reform under Maximilian I and the religious cleavage of the Reformation left the German princes autonomous and entrenched in their rights. Although Hapsburgs occupied the throne continuously from 1438 (except 1742-45), their dynastic interests took preference over the empire. After 1648 it was merely a loose federation, and responding to Napoleon's expansionist pressures, Francis II dissolved it on 6 August 1806.

See F. Heer, *The Holy Roman Empire* (1968).
RICHARD V. PIERARD

HOLY SEPULCHRE. The name of the church which today houses the traditional sites of both the crucifixion and the tomb of Christ. It is a dilapidated, heterogeneous, and unlovely building inside the Old City of Jerusalem. The site has been inside the city walls since A.D. 41-44, when Herod Agrippa had a third wall built; recent archaeological investigation has at last clarified the fact that the site lay outside the city wall before then (cf. Heb. 13:12). The Roman emperor Constantine gave orders in A.D. 326 to build the church, which in its earliest phase was a complex in three parts (on different levels), the *Anastasis* (Resurrection grotto), the *Martyrium* (the basilica), and the elevated site of Calvary. Under the basilica was also a chapel dedicated to St. Helena, Constantine's mother, whom legend soon credited with the discovery of "the true cross."

The original church was burnt down by the Persians in 614, but was soon restored. The caliph Hakim then destroyed or damaged much of the edifice in 1009. Restoration was effected in 1048; and then the Crusaders did major rebuilding work between 1099 and 1149. On his conquest of Jerusalem, Saladin did no more than destroy the bells (1187). Since then, the church has suffered occasionally from earthquakes, fire, and well-intentioned alterations. Several Christian confessions have rights in different parts of the church. The authenticity of the site can be argued on the grounds that in Constantine's reign there was already a strong and unrivaled tradition in support of it, and that during the preceding two centuries the site was inaccessible to pilgrims.

See A. Parrot, *Golgotha and the Church of the Holy Sepulchre* (ET 1957); see also bibliography for JERUSALEM. D.F. PAYNE

HOLY SPIRIT. In the OT, the expression "Holy Spirit" is rare, but there are references to "the spirit of the LORD," which is used of God in action, God doing something. God is, of course, to be discerned in quietness (1 Kings 19:11,12), but it is not this that is meant when "the spirit of the LORD" is used. Then it is rather the irresistible God who is in mind (e.g., 2 Kings 2:16; Ezek. 3:14). The Spirit is active in the Creation (Gen. 1:2; cf. Job 33:4). The Spirit gives life (Ezek. 37:-14).

The Spirit is at work in men in a variety of ways. He may give strength to Samson (Judg. 14:6) or skill to Bezaleel (Exod. 31:3). It was when the Spirit of the Lord "came upon" men like Othniel or Jephthah that they were able to do their work as judges of the people (Judg. 3:10; 11:29). The Spirit "came mightily upon" David (1 Sam. 16:13 RSV). Nehemiah (9:20) saw the knowledge that took the Israelites through the wilderness as coming from God's Spirit.

The Spirit gave Ezekiel his message (11:5; cf. 2:2; 3:24, etc.), and other prophets too, such as Balaam (Num. 24:2), Amasai (1 Chr. 12:18), Zechariah son of Jehoiada (2 Chr. 24:20). Isaiah and Micah report similar experiences (Isa. 61:1; Mic. 3:8). And the Spirit may be expected to help others than the prophets as they seek to serve the Lord (Psa. 51:11; 143:10; Ezek. 36:27).

In all this there is nothing which compels us to see the Spirit as a hypostasis in the NT manner. The full flowering of Christian teaching on the Spirit is future in the prophetic writings (Joel 2:28f.). There are hints that the Spirit may be understood as in some sense different from the Father (e.g., Isa. 48:16), but no more. But in the NT there is a very great advance.

In the early chapters of the gospels, it is true, the Spirit appears to be used in much the OT manner (e.g., Luke 1:41, 67). Throughout the lifetime of Jesus there is not much more. John explains this by saying, "Up to that time the Spirit had not been given, since Jesus had not yet been glorified" (John 7:39). The coming of the Spirit in all His fulness was something that would follow, not precede, the passion and resurrection. But after this had occurred, the Spirit came on the infant church in a striking manifestation of enlightenment and power (Acts 2). From that time on, the presence of the Spirit is the characteristic thing about the Christian Church. It is a Spirit-filled body.

Two things are especially noteworthy about NT teaching on the Spirit: His universality among Christians, and His bringing of power for ethical achievement. First-century religions often held that a divine spirit would from time to time come upon men. But it was thought he would come only upon a few outstanding people. To be possessed by the spirit was a mark of outstanding distinction. But among the Christians the possession of the Spirit was the distinguishing characteristic. "Those who are led by the Spirit of God are the sons of God," wrote Paul, and again, "If one does not have the Spirit of Christ, he does not belong to Christ" (Rom. 8:14, 9). This is made clear by such incidents as that in Acts 19 where, when Paul met some men who claimed to be Christian, his first question was, "Did you receive the Holy Spirit when you believed?" (v. 2). It was apparently unthinkable that anyone should be a Christian and not have the Spirit. This seems implied also throughout the epistles of the NT. The church is plainly regarded as a community indwelt by the Spirit of God. He is expected to be at work in believers constantly.

The second unusual thing about NT teaching on the Spirit is that "the fruit of the Spirit is love, joy, peace, patience, kindness, goodness, faithfulness, gentleness and self-control" (Gal. 5:22,23). In the religions of antiquity generally, the divine spirit made his presence known by causing those in whom he came to engage in unusual behavior of an ecstatic kind. It was in the "whirling dervish" type of activity that the spirit's presence was to be discerned. It was something new and important when his presence was revealed rather by the manifestation of ethical qualities. The NT, it is true, does know of ecstatic gifts, such as the gift of "speaking in tongues." But such activities are subordinated to love and the like, which represent the "most excellent way" (1 Cor. 12:31).

Apart from these two points it is most important to see the Spirit in the NT as personal, not a force or an influence. Personal words are used of Him (like *parakletos,* "advocate"), and the activities ascribed to the Spirit are those which are normally fulfilled by persons. He gives gifts as He wills (1 Cor. 12:11), He leads believers and bears witness in them (Rom. 8:14, 16). He has knowledge (1 Cor. 2:11) and mind (Rom. 8:27). He loves (Rom. 15:30), grieves (Eph. 4:30), intercedes (Rom. 8:26f.), and cries out (Gal. 4:6).

The first Christians lived exultantly in the joy of the Spirit. But in succeeding generations the enthusiasm tended to wane, and the doctrine of the Holy Spirit was accepted formally as taught in Scripture rather than seen as a basis for living. Not surprisingly, in time there came a reaction. Montanus, a native of Phrygia in Asia Minor, who lived in the second part of the second century, put great emphasis on the Holy Spirit. He thought that revelation did not cease with the end of the NT period, and held that he himself was the source of important new revelations. As Jesus had been the incarnation of the Second Person of the Trinity, Montanus saw himself as the incarnation of the Holy Spirit. He was supported by others, notably by two prophetesses, Priscilla and Maximilla. They thought the new Jerusalem would come down from heaven to a spot in Phrygia, and they prepared for this happy event with a strenuous asceticism. Their protest against the clericalism of the church of their day and against the lax morality of many professing Christians was important, and won them many adherents. But they were in serious error in their teaching of the new dispensation of the Spirit inaugurated by His "incarnation" in Montanus, and the church had no alternative but to condemn them.

The only other important heresy in the doctrine of the Spirit is that associated with Macedonius, bishop of Constantinople at the end of the Arian period. He accepted the full deity of the Son, but held that the Spirit was a created being, not unlike the angels. In a day when men were coming to see that the full Arian position was impossible, it seemed to many that it was a useful compromise to accept the deity of the Son (with the orthodox) but to deny that of the Spirit (with the Arians). But sound doctrine is not built up on political compromises of this sort, and Macedonianism was soon rejected.

The precise relationship of the Spirit to the Father and the Son is nowhere stated in Scripture, and it has caused discussion and even division in the church. The only passage which even appears to bear on the subject is that in which Jesus speaks of "the Spirit of truth, who proceeds from the Father" (John 15:26 RSV). This passage does not deal with the eternal interrelationships between the persons of the Trinity, but it has given us the terminology. It has become customary to speak of the "procession" of the Spirit or of the Spirit as "proceeding."

In the earliest statements it was customary simply to take up the passage in John's gospel and speak of the Spirit as proceeding from the Father. But the Nicene Creed came to be transmitted in the West in the form "proceeding from the Father and the Son." It seems that this arose in the first instance from a copyist's mistake. But it became common to recite the creed in the West in this form. Not unnaturally, the Easterners demand that the creed be recited in its original form and that the double procession (i.e., procession from the Son as well as the Father) be renounced. The West's refusal was the formal cause of the break between the Eastern and Western churches.

The West has resisted the demand that it surrender the doctrine of the double procession because, however the disputed words originally got into the creed, they point to something true. The NT may not speak of the Spirit as "proceeding from" the Son, but it does link the two closely. The important point is that the Spirit is the "Spirit of Christ" (Rom. 8:9; 1 Pet. 1:11; cf. Acts 16:7; Phil. 1:19). Jesus baptized with the Holy Spirit (Matt. 3:11; Mark 1:8; Luke 3:16; John 1:33), and sent the Spirit (John 20:22; Acts 2:33).

More important than the citing of any individual texts is the general thrust of NT teaching that the Spirit comes upon men as a result of what Christ has done. We know and receive the Spirit only because we have been saved by Christ's atoning death and brought into newness of life. It is in this new life that Christ brings that we know the Spirit. The doctrine of the double procession safeguards this as the single procession does not. See also PROCESSION OF THE SPIRIT.

BIBLIOGRAPHY: A. Kuyper, *The Work of the Holy Spirit* (1900); H.B. Swete, *The Holy Spirit in the New Testament* (1910); W.H.G. Thomas, *The Holy Spirit of God* (1913); H.W. Robinson, *The Christian Experience of the Holy Spirit* (1928); F.W. Dillistone, *The Holy Spirit in the Life of To-day* (1946); C.K. Barrett, *The Holy Spirit and the Gospel Tradition* (1947); J.E. Fison, *The Blessing of the Holy Spirit* (1950); E.F. Kevan, *The Saving Work of the Holy Spirit* (1953); R. Pache, *The Person and Work of the Holy Spirit* (1956); N.Q. Hamilton, *The Holy Spirit and Eschatology in Paul* (1957); G.S. Hendry, *The Holy Spirit in Christian Theology* (1957); E.H. Palmer, *The Holy Spirit* (1958); G. Smeaton, *The Doctrine of the Holy Spirit* (1958); L. Morris, *Spirit of the Living God* (1960); J.R.W. Stott, *The Baptism and Fullness of the Holy Spirit* (1964); A.M. Stibbs and J.I. Packer, *The Spirit Within You* (1967); J.D.G. Dunn, *Baptism in the Holy Spirit* (1970).

LEON MORRIS

HOLY SYNOD. The Russian emperor Peter the Great (1682-1725) created the Holy Synod as part of his efforts to reorganize the Russian Church, thus replacing the old patriarchate of Moscow in 1721. The synod, a committee of the higher clergy, became the supreme authority in the Russian Orthodox Church second only to the emperor. As a result the clergy became a kind of auxiliary police force in the subordination of the church to the state, and helped to create an Erastian relationship between church and state which endured until the Communist revolution of 1918. The synod was charged with the responsibility of maintaining schools for the training of young men for the priesthood. C. GREGG SINGER

HOLY WATER. Ordinary water which has been blessed by a priest in order to be used for religious purposes—e.g., blessings, exorcisms, burials, and the Asperges of the Mass. In the West the actual blessing of the water is accompanied by exorcism and by the addition of exorcized salt. Holy water is usually found in the stoups, situated near the main door of churches; this practice goes back to Norman times, and the use of holy water in the church began in the fourth century, but gained momentum in the Middle Ages.

HOLY WEEK. The week beginning with Palm Sunday and ending with Holy Saturday, observed as a solemn fast, commemorating the Passion. Its origin lies in the two-day pre-paschal fast of the ante-Nicene Church, but with the historicization of festivals in the fourth century it became a week, and its liturgical observances were modeled on developments in Jerusalem. By the Middle Ages a complex of services had been adopted, including the Palm procession, the *Pedilavium,* the Veneration of the Cross, the burial of the Cross and Host in the Easter Sepulcher, and ceremonies with the new fire and Paschal Candle. The Anglo-Catholic movement revived some of these customs in Anglican worship. In 1957 the Roman Catholic Church completed a reform of Holy Week. JOHN A. SIMPSON

HOLY YEAR, see JUBILEE, YEAR OF

HOMBERG, SYNOD OF (1526). Called by Philip of Hesse to reorganize the church in his territory. At the synod a Frenchman, Francis Lambert,* formerly of the Franciscans, was mainly responsible for producing an ambitious scheme of church reform. This developed the logic of the priesthood of all believers, creating a democratic church in which the local congregation had the right of appointing pastors and excommunicating, and the general supervision was given to an annual synod, which appointed three visitors to examine local churches. The scheme was abandoned when Luther rejected the proposals on the grounds that the time had not yet come for such definitive legislation, but also because, since the Peasants' War, he had become more skeptical of the mass of the laity and more dependent on princes. C. PETER WILLIAMS

HOMILETICS. The discussion of the art and theology of preaching. The earliest Christian sermon was called a "homily," a term deriving from the Latin *homilia,* "a conversation." From the earliest times preaching played a basic part in the religious life of the OT. Moses, Joshua, and Elijah all appealed by the spoken word to the listening congregation. The great literary prophets were preachers as well as writers. The prophetic oracles may have been used as proclamation in the cultic life of the community during the Exile. In the Jewish synagogue service on the Sabbath there developed the custom of an address delivered on the portion of Scripture which had been read in the congregation (cf. Luke 4:16-21; Acts 15:21).

The NT opens with the preaching of John the Baptist which reechoed the prophetic message of the coming Messiah and the kingdom of God. Jesus Himself devoted a great part of His ministry to preaching, and His word had the unique personal power and authority of God Himself. He also sent out His disciples to preach in His name (Matt. 10:7), promising that their word would have the same power and authority as His own personal utterance (Luke 10:16), and would cause the spread of His kingdom till the end of the world (Matt. 24:14). After the resurrection, the apostles found that Jesus continued His ministry through their proclamation (i.e., His former ministry was only a *beginning*—Acts 1:1). God Himself, they believed, appealed to men and stretched out His hand to heal as they preached. They had no hesitation in claiming that what they preached was the Word of God (1 Thess. 2:13; Acts 4:31; 1 Pet. 1:23) which saved men (1 Cor. 1:21). Moreover, they believed that through their preaching the powers of salvation and of the new age were being implanted within human life and history to hasten the fulfillment of the hitherto hidden purpose of God to unite all things in Christ and fully to perfect the new creation (cf. Col. 1:22-29; Eph. 1:9, 10; 3:4-13). The accounts of the apostolic church indicate that preaching in the form of the exposition of Scripture took place not only as a missionary activity, but also in the context of the assembled congregation of believers, especially on the Lord's Day (Rom. 15:4; Acts 18:24-28; 20:7ff.). The continuance of this custom is reflected in Justin's account of worship in the second century.

The earliest Christian preaching took the form of a simple conversational, practical, and pastoral homily, based on the text which had been read, and often following the varied topics suggested by the text, in the order in which they arose within the text, with little concern to attain a satisfying rhetorical structure. The sermon began when and where the text began, and ended when the text ended. It was often delivered extempore, though arising out of careful preparation. It was regarded as the primary duty of the bishop to preach, and he often did so seated, while the congregation stood. The great preachers of the third and fourth centuries—Basil, the Gregories, Chrysostom, and Augustine—were conscious that they lived in a world in which the normal method of communication involved the use of traditional rhetoric; and while they recognized the deep difference between their task as Christian preachers and that of pagan rhetoricians whose aim was merely to impress an audience with a great speech, they felt that the church must accept the help which the study of rhetoric could bring to preaching. The first important discussion of homiletics in this light was made in Augustine's *On Christian Doctrine,* a work that has never ceased to be important.

In the early Middle Ages, arrangements and excerpts from the sermons of Augustine, Caesarius of Arles, and other Fathers were circulated to help preachers who could not produce their own sermons. In the revival of preaching with the Dominican and Franciscan friars, a great variety of homiletic helps, with sermon suggestions for every possible occasion, was published. The art of illustrating sermons with fantastical allegorical fables was greatly developed, giving rise to many collections of *exempla* for this purpose. Treatises on preaching (e.g., by Humbert of Romans) were issued. In the thirteenth century a "modern" new form of thematic preaching from a short text with careful introductions, transitions, conclusion, and, of course, three headings, appeared in university circles. It was called "modern" in contrast with the older form of homily. Calvin and Luther at the Reformation tended to return to the older form of preaching, but the medieval forms prevailed with succeeding generations. Influential works on homiletics were produced, e.g., by Hyperius and Keckermann. The most important Puritan work, recently republished, was William Perkins's *Art of Prophesying.* Later there appeared Simeon's

Horae Homileticae and Vinet's *Homiletics*, immensely popular in their day.

What is usually said on the subject of homiletics has been best discussed by following the topics traditionally used in works on rhetoric: invention (finding out what to say), disposition (arranging the material), style (clothing it in suitable language), memory (the task of fixing for the mind what is to be delivered), and delivery.

BIBLIOGRAPHY: In addition to books cited above, J. Bingham, *The Antiquities of the Christian Church* 2 vols., (1878); J.A. Broadus, *Lectures on the History of Preaching* (1893); R.G. Owst, *Preaching in Mediaeval England* (1926); C.H. Dodd, *The Apostolic Preaching and Its Development* (1936); R.F. Bennett, *The Early Dominicans* (1937); J.J. von Allmen, *Preaching and Congregation* (1962); J.W. Blench, *Preaching in England* (1964); St. Francis de Sales, *On the Preacher and Preaching* (1964); Y. Brilioth, *A Brief History of Preaching* (1965).

RONALD S. WALLACE

HOMILIES, BOOKS OF. Authorized sermons issued in two books by the Church of England in the reigns of Edward VI and Elizabeth I. They were to provide for Protestantism sermonic models for the new simplified style of topical preaching as well as a proper theological base. Thomas Cranmer broached the idea as early as 1539, it was authorized by Convocation in 1542, and within a year the twelve homilies of the first book were collected and edited by Cranmer, who also wrote at least five of them. They were not published, however, until 1547. The first six homilies present distinctive Protestant theology, namely the perspicuity and sufficiency of Scripture, the radical sinfulness of man, justification by faith alone (entitled "Of the Salvation of all Mankind"), evangelical faith, and sanctification. The homilies were revoked by Mary, but reinstituted by Elizabeth. In 1562-63 the second book was issued, though only published with the full twenty-one homilies in 1571. Bishop John Jewel wrote all but two of these; they are more practical and liturgical in content than the first book. The two books were united into one volume in 1632.

BIBLIOGRAPHY: G.E. Corrie (ed.), *Certain Sermons Appointed by the Queen's Majesty* ... (1850); J.T. Tomlinson, *The Prayer Book Articles and Homilies* (1887); M. Donovan and A.R. Vidler, "The Homilies," *Theology* (1941), pp. 284-95; H. Davies, *Worship and Theology in England From Cranmer to Hooker* (1970).

BRIAN G. ARMSTRONG

HOMILIES, CLEMENTINE, see CLEMENT OF ALEXANDRIA

HOMOEANS. In the controversy stirred by the insertion of *homoousios* into the creed of Nicea* (325), one group sought to counter the objectionable term by using *homoios* ("like"). They are usually called "Homoeans," though distinctions among them ranged from the unqualified use of the term (Valens of Mursa and Ursacius of Singidunum) through a qualified "like in all things" (Acacius of Caesarea) to a full "like in substance" (Basil of Ancyra). The compromise saw its chief, but short-lived, success under the patronage of the Arian-sympathizing emperor Valens (364-78) at the synodal sessions of Nice (359) and Constantinople (360), whose creeds incorporate the Homoean formula qualified by "according to the Scriptures."

CLYDE CURRY SMITH

HOMOOUSIONS, see HOMOOUSIOS

HOMOOUSIOS (Gr. = "of the same substance"). This technical term from late Greek philosophical tradition was first used in a Christian setting by Gnostics. Both Origen in Alexandria and Sabellius in Rome played some role in its adoption in Christianity. Tertullian was ultimately responsible for its reappropriation, though as late as the Synod of Antioch (268) *homoousios* was severely criticized. That Arius himself opposed the term is evident in his letter to Alexander, bishop of Alexandria, before the Council of Nicea* (325). Evidently it was the emperor Constantine as presiding officer there, presumably on the advice of Hosius* of Cordova, who introduced the concept into creedal discussions, with reference to the relationship between Father and Son.

CLYDE CURRY SMITH

HONORIUS I (d.638). Pope from 625. A native of Campania in Italy, he was interested in the Christianization of the Anglo-Saxons, administered the financial affairs of the papacy wisely, but is chiefly remembered for his involvement in the Monothelite* controversy. Sergius, patriarch of Constantinople, wrote to Honorius seeking support for a formula that would reconcile the Monophysites. While confessing two natures in Christ, this formula attributed to Him "one theandric operation," i.e., one mode of activity—that of the Divine Word. This had been strongly opposed by Sophronius of Jerusalem. In his reply, Honorius supported the forbidding of further discussion of either one or two operations, adding that such questions should be left to the grammarians. He went on to write, "Whence also we confess one will of the Lord Jesus Christ since plainly our nature was taken by the Godhead, and that nature sinless, as it was before the fall." In a second letter, of which only fragments remain, he again repudiated as inexpedient the formula "two operations," though he confessed two natures in Christ, "unmixed, undivided, unchanged," operating what is characteristic of each.

He died in the year that the Ecthesis,* the charter of Monothelitism, was published, making use of his formula of "one will." His successors condemned Monothelitism, and at the Council of Constantinople in 681 Honorius was formally anathematized with Sergius and Cyrus of Alexandria. This anathema has created difficulties for the supporters of papal infallibility, particularly in the Gallican controversies of the seventeenth and eighteenth centuries, and before the Vatican Council definition of 1870.

See J. Chapman, *The Condemnation of Pope Honorius* (1907).

J.G.G. NORMAN

HONORIUS III (d.1227). Pope from 1216. Born Cencio Savelli at Rome and well educated, he succeeded Innocent III after long experience in ecclesiastical administration, and his pontificate carried on an inherited policy with some changes. At once he found himself amid plans set by the Fourth Lateran Council, as initiated by Innocent, while his chief concern was a crusade to recover the Kingdom of Jerusalem. This was to be led by Emperor Frederick II, but it was not, and seeming political stability was attempted to launch it: aid given to Henry III (England's nine-year-old monarch) and the attention of King Philip Augustus and his son, Louis VIII, redirected from England to Toulouse. A major concern was always the relationship with Frederick, who wanted for himself and his son Henry VII control of both the imperial and Sicilian crowns. This brought immense papal opposition, leading to a major struggle and finally a victory which ended his crusading interest.

The Fifth Crusade did proceed, but not happily. The crusade against the Albigenses* continued, and heresy was unrelentingly pursued; here were unmistakable shades of the Inquisition. Honorius formally approved and used the new mendicant orders, Dominicans and Franciscans. Among his many writings are *Liber Censuum* (1192); *Compilatio quinta* (1226); his decretals which have been considered the first official book of canon law; a life of Gregory VII; and a continuation of the *Liber Pontificalis.*

C.G. THORNE, JR.

HONORIUS OF AUTUN (c.1090-c.1156). Monk and writer. Describing himself as a priest and teacher ("Scholasticus"), he was a well-known writer—over 500 manuscripts of his works have survived. Although a monk with solitary inclinations, he showed a lifelong interest in the outside world. He wrote books on liberal education, he defended the right of ordained monks to preach and to exercise sacramental functions, and his popular *Imago Mundi* dealt with astrology, astronomy, geography, and history. Honorius was a Christian Platonist and a great admirer of John the Scot. He wrote widely on religious and theological matters. His *Elucidarium,* in its three books on God, man, and paradise, ranged from evil and free will to a variety of contemporary problems which he dealt with very sensibly. He wrote also on the Virgin Mary's intercessionary powers, the Liturgy, the Psalter, the Song of Songs, and papal supremacy. Various other writings condemned ecclesiastical abuses—simony, clerical marriage, misuse of ecclesiastical offices, and the ordination of illiterates. It appears now that his Latin name, *Augustodensis,* links him not with Autun in Burgundy, but with Regensburg in S Germany.

L. FEEHAN

HONTHEIM, JOHANN NIKOLAUS VON (1701-1790). Suffragan bishop of Trier, and known pseudonymously as "Justinius Febronius," was the formulator of Febronianism.* After eighteen years' work, he published his doctrines in *The State of the Church and the Legitimate Authority of the Roman Pontiff, a Book Composed for the Purpose of Uniting in Religion Dissident Christians* (in Latin, 1763). It reflected Gallican and Protestant motifs, although remaining devoutly Catholic, and not secular. Clement XIII condemned it (1764), but intensive public debate continued throughout Europe. Hontheim later unconvincingly recanted (1778), while others, including Austrian chief minister Kaunitz, used his ideas to support a more secular Josephinism.* He studied law at Louvain, Leyden, and Rome, winning a doctorate in jurisprudence (1724); then became priest (1728), professor (1732), university pro-chancellor (1746), and suffragan bishop (1748), all at Trier.

C.T. MC INTIRE

HOOK, WALTER FARQUHAR (1798-1875). Dean of Chichester. Born in London and educated at Winchester and Oxford, he served in parishes in Birmingham and Coventry, and was from 1837 to 1859 a remarkably successful vicar of Leeds, frustrating Dissenters, increasing the number of parish churches from fifteen to thirty-six, befriending the poor, and adapting Anglicanism to the challenge of the new urban areas. A High Churchman, Hook helped the Tractarian* party to consecrate St. Saviour's, Leeds (1845), but later quarreled with Pusey over the ritualism practiced there. He wrote extensively, producing an eight-volume *Dictionary of Ecclesiastical Biography* and twelve volumes of lives of archbishops of Canterbury. Hook's immense physical and spiritual energies, together with his belligerent High Church Toryism, have suggested a comparison with Dr. Samuel Johnson.

IAN SELLERS

HOOKER, RICHARD (c.1554-1600). English theologian and apologist. Educated at Exeter Grammar School and Corpus Christi College, Oxford, where he was fellow from 1577 to 1584, he lectured in Hebrew and logic and was rector of Drayton Beauchamp (1584), master of the Temple in London (1585-91), and rector of Bishopbourne from 1595 until his death. Though an able preacher and sensitive pastor, he is primarily remembered as one of the greatest apologists for a Church of England which was not obliged slavishly to copy sister churches. His famous encounter with Walter Travers* at the Temple showed an independent Reformed position on matters like predestination, assurance, and judgment of Rome, in addition to a shrewd insight into the doctrinal and psychological weaknesses of militant Puritanism. Released from duties at the Temple, he produced the first four books of his *Laws of Ecclesiastical Polity* in 1593, followed by the fifth in 1597. The history of the remaining books is obscure, and they were not fully published until 1662. As well as being a classic of English prose, his work was a profound contribution to the English theological tradition. His skillful restatement of Thomism, combined with a careful discussion of the relation between reason and revelation, enabled him to meet Puritan criticisms of the Elizabethan Church at a far more creative level than apologists like Whitgift or Bancroft. He showed that a church could justifiably be ordered without either claiming divine institution for every detail or falling into Roman error by continuing medieval practices; his exposition of law showed an appreciation of historical continuity which was lacking in much Protestant ecclesiology. The defense he offered for the role of redeemed reason helped to inspire the flowering of Caroline theology and has since provided many members of the Church of England with a theological method which has combined the claims of revelation, reason, and history. Though this account of the relationship between church and state was unduly optimistic, it has continued to be influential. He is one of the most important English theologians of the sixteenth century.

BIBLIOGRAPHY: The best edition of his *Works* is the 7th revised edition of J. Keble (1888). For his life, C.J. Sisson, *The Judicious Marriage of Mr. Hooker* (1940). Recent studies are P. Munz, *The Place of Hooker in the History of Thought* (1952); G. Hillerdal, *Reason and Revelation in Richard Hooker* (1962); J.S. Marshall, *Hooker and the Anglican Tradition* (1963); W.S. Hill, *Studies in Richard Hooker* (1974). IAN BREWARD

HOOKER, THOMAS (1586-1647). Puritan clergyman and founder of Connecticut. Born in Leicestershire, he was educated at Cambridge, where he was converted to Puritanism. His popularity as lecturer at St. Mary's, Chelmsford, forced authorities to retire him, and in 1630 he fled from England to Holland. Persuaded to emigrate to Massachusetts, he became pastor at Newtown (Cambridge) in 1634. Restiveness led the congregation to move to Connecticut in 1636, against the wishes of the Massachusetts officials. In 1638 he was the primary figure in the creating of the Frame of Government for communities around Hartford. Virtual dictator of Connecticut thereafter, and perhaps the most powerful pulpit orator of his day, Hooker was an expositor who dealt eloquently with Puritan fundamentals of religious experience and moral duty. His writings were chiefly sermonic, but his *Survey of the Summe of Church Discipline* (1648) was a notable work on Congregational polity and social theory. DARREL BIGHAM

HOOPER, JOHN (d.1555). Protestant martyr and Anglican bishop. Graduate of Oxford and later a monk, he moved to London after the dissolution of the monasteries. After reading some reformational writings he was converted to Protestantism and then sought to spread his views at Oxford. As a result of his activities he had twice to flee from England. In 1546 he married a woman from Antwerp. For a brief period they settled in Zurich, where he enjoyed the friendship of H. Bullinger* and a correspondence with M. Bucer* and J. à Lasco.* Returning to England in 1549, he became chaplain to Protector Somerset.

Hooper gained fame as a supporter of the principles of the Swiss reformation. His preaching was very popular, being devoted to biblical exposition and exposure of imperfect reformation in his own land. Following the fall of Somerset he was Northumberland's chaplain. In 1550 he was nominated to the see of Gloucester, but his consecration was delayed until 1551 due to his opposition to vestments. During 1552 the sees of Gloucester and Worcester were amalgamated, and he was made bishop of both. With the accession of the Catholic Mary he was imprisoned, deprived, degraded, and publicly burned. His record as a bishop was praiseworthy. He preached several times each day, visited all the parishes of his dioceses, was generous to the poor, denounced ruthless landlords, and sought to persuade his clergy and people to read the Bible.

For his writings, see the two-volume collection by the Parker Society (1843 and 1852). There are short biographies by J.C. Ryle (1868) and by W.M.S. West (1955). See also L.B. Smith, *Tudor Prelates and Politics* (1953). PETER TOON

HOPKINS, GERARD MANLEY (1844-1889). English poet. Born at Stratford (Essex) and educated at Baliol College, Oxford, under Jowett and Pater, he was influenced in art by the Pre-Raphaelites and in religion by the later Tractarians.* In 1866 he seceded to Rome. He joined the Jesuits and held several teaching and pastoral posts, including Stonyhurst, St. Helens, and the chair of Greek at Dublin. On entering the order he destroyed the poetry he had written up to that date and only returned to the art at the request of his superior to write a poem on *The Wreck of the Deutschland,* celebrating the heroic death of four exiled nuns from Bismarck's Germany. The oddity of the poem's language resulted in its rejection by *The Month,* to which it was submitted.

This oddity of expression, now recognized as the source of Hopkins's strength—"All things counter, original, spare, strange," to use one of his own lines—is consistent with his emphasis on individuality—"What I do is me: for that I came." In this Hopkins revealed his allegiance to Duns Scotus, rather than to the official theologian of the Jesuits, Thomas Aquinas. Consistently with this emphasis, he developed his theories of inscape (or individually distinctive form in things) and instress (the force that determines the form). This gives to his vision of things, particularly natural phenomena, a sharp, unusual, and special vividness and to his account of his own experience a penetrating sensitiveness that in the so-called Terrible Sonnets and in "Spelt from Sibyl's Leaves" becomes almost unbearably painful—"selfwrung, selfstrung, sheathe- and shelterless, thoughts against thoughts in groans grind."

Hopkins's poems were not published in his lifetime and indeed did not appear until his friend and literary executor, Robert Bridges, issued them in 1918. It was then clear how he was indeed a poet born out of due time, and the subsequent publication of his letters has revealed the acuteness of his criticism and dissatisfaction with the poetic mode of his own contemporaries. Hopkins remains a difficult, but immensely rewarding poet, a man who agonized before God.

See *Poems* (ed. W.H. Gardner and N.H. MacKenzie, 1967); and N.H. MacKenzie, *Hopkins* (1968). ARTHUR POLLARD

HOPKINS, SAMUEL (1721-1803). Congregationalist theologian of the New England Theology* or "Hopkinsianism." Born in Waterbury, Connecticut, he received his A.B. from Yale in

1741, and two years later was ordained as pastor of Great Barrington, Massachusetts. In 1770 he became pastor of the Congregational church in Newport, Rhode Island. He was an early exponent of the abolitionist cause, seeing slavery as a moral evil. He is even better known for his modification of Calvinism. To him sin was essentially self-love but without legal imputation of Adam's sin to us. Man was responsible to seek a change of heart which would lead to "disinterested benevolence" in the life of the regenerate.

GEORGE MARSDEN

HOPKINSON, FRANCIS (1737-1791). American composer. Although a lawyer by profession and a musical amateur, Hopkinson was very active in the early musical life of Philadelphia. He was one of the first American-born composers, and took a great interest in church music as well as secular. He published *A Collection of Psalm Tunes with a Few Anthems.* He was a signer of the Declaration of Independence.

HORMISDAS (d.523). Pope from 514. Born at Frosinone in Latium, he succeeded Symmachus as bishop of Rome in 514. Eastern and Western churches had been divided since 484 over the *Henoticon,** but Hormisdas negotiated with the emperor Anastasius I in 515 to hold a council. His extravagant demands were rejected by the emperor. In 519 further negotiations induced the emperor Justin I and the patriarch John to sign a dogmatic statement ("Formula Hormisdae"), accepting the Chalcedonian Definition* and the Tome of Leo. The names of Acacius, Zeno, and Anastasius were removed from the "diptychs," and the authority of the Roman see was emphasized (based on Matt. 16:18). Hormisdas conferred with the Arian Goths in ecclesiastical matters and maintained good relations with Theodoric, the Ostrogoth ruler of Italy.

J.G.G. NORMAN

HORNE, GEORGE (1730-1792). Bishop of Norwich. Born in Kent, he graduated from University College, Oxford (1749), became president of Magdalen College (1768), and later was dean of Canterbury (1781) before going to Norwich in 1790. A High Churchman, he nevertheless adopted some of the views of John Hutchinson and sympathized with Methodism's spiritual earnestness. He strongly disapproved of the expulsion of Methodist students from St. Edmund's Hall, Oxford, and refused to forbid John Wesley to preach in his diocese. He actively promoted the Naval and Military Bible Society (founded in 1780), and espoused the cause of Scottish bishops who petitioned Parliament (1789). He wrote a commentary on Psalms, interpreting them messianically.

J.G.G. NORMAN

HORNE, THOMAS HARTWELL (1780-1862). Librarian and Protestant biblical commentator. Born in London and educated at Christ's Hospital (where he was a contemporary of Samuel Taylor Coleridge), he became clerk to a barrister, undertaking varying literary work in his spare time. At first a Wesleyan, he was ordained in the Church of England (1819), later joining the staff of the British Museum (1824), where he worked on the compilation of the catalog for many years. Horne was the author of more than forty books, many of them on Christian apologetics and bibliography. In his earlier days he cataloged the Harleian MSS for the British Museum; later he edited Charles Simeon's *Horae Homileticae,* a twenty-one volume commentary on the Bible. He is remembered chiefly for his *Critical Study of the Holy Scriptures* (3 vols., 1818), which was widely used for half a century by students.

R.E.D. CLARK

HORNER, RALPH (1853-1921). Founder of the Holiness Movement Church in Canada. Born in Shawville, Quebec, he was converted at a Methodist meeting in 1876 and began almost immediately to preach to his neighbors. Realizing he had the ability to elicit strong emotional responses from his hearers, he decided to enter the Methodist ministry. He attended Victoria College (1883-85), during which time he continued to organize holiness meetings. Refusing a regular circuit in 1886, he embarked on an independent evangelistic tent ministry. Because of mounting protests against the speaking in tongues which accompanied his preaching, the Montreal Annual Conference deposed him from its ministry in 1895. He organized his followers into a Holiness Movement Church, and in 1895 at a convention in Ottawa attended by evangelists from Ontario, the Western Provinces, and Quebec, he was elected bishop. He held this position until he withdrew from the Holiness Movement Church in 1916 over the true interpretation of sanctification and formed the Standard Church of America. After his death both churches declined rapidly.

ROBERT WILSON

HORSLEY, SAMUEL (1733-1806). Anglican bishop. Son of a clergyman, Horsley was educated at Cambridge and held various livings and domestic chaplaincies, was made a Fellow of the Royal Society in 1767 and later served as its secretary, was consecrated bishop of St. David's in 1788, of Rochester in 1793, and of St. Asaph in 1802. He was a man of scientific bent, and in the course of a twelve-year controversy with Joseph Priestley over the latter's philosophical and historical methods strongly defended the Trinitarian and christological beliefs of the early church. In religion a High Churchman and in politics a strong Tory, he upheld the establishment and opposed all innovations, particularly Sunday schools. He was a genuine, if ostentatious, friend of the poor, and his social teaching is a novel blend of Utilitarianism and traditional Christian philanthropy. Under the impact of the French Revolution he adopted an extravagant millenarianism and was mentally unbalanced when he died.

IAN SELLERS

HORT, FENTON JOHN ANTHONY (1828-1892). NT critic and biblical scholar. With B.F. Westcott* he edited an edition of the Greek NT (1881) which formed the basis for the English Revised Version and which set the pattern for nearly all future editions of the Greek text. The fifty-seven-page introduction by Hort sets out the basic elements of the science of textual criticism which remain, in all essentials, valid to the present. With his friends Westcott and J.B. Lightfoot,* he planned to write a complete commentary on the NT. Hort was to be responsible for the synoptic gospels, Acts, the general epistles, and the Apocalypse. Due to a tendency towards perfectionism he was able to publish very little, though it is generally regarded that he was the greatest of the three Cambridge scholars. His written legacy consists of a few fairly fragmentary works, most of them published posthumously, including *Two Dissertations* (on John 1:18 and on the Eastern Creeds, both 1876); *The Way, The Truth, the Life* (1893); *Judaistic Christianity* (1894); *The Christian Ecclesia* (1897); *I Peter i.1-ii.17* (1898). W. WARD GASQUE

HOSANNA. The NT Greek transliteration of a Hebrew term meaning "save, we pray." The phrase in a slightly different form occurs in Psalm 118:25, which was used in the Jewish Passover rites of NT times. There is evidence that by then "hosanna" was a ritual exclamation (of praise as much as supplication), associated with Messianic hopes, which was uttered in the context of several Jewish festivals. The NT use of the word in the context of Christ's final entry into Jerusalem (e.g., Matt. 21:9) brought it into the early Christian eucharistic liturgy, and also into the Palm Sunday liturgy somewhat later.

HOSIUS (Ossius) (c.256-357). Bishop of Cordova. Born in Spain, he was probably already bishop when he suffered persecution under Maximianus Herculius, before the edicts of Diocletian in 303. He was present at the Synod of Elvira (c.300) and was later in attendance on Constantine. The Donatists* blamed their condemnation by the emperor in the Council of Milan (316) to his being advised by Hosius. As sole master of the Roman Empire by 323, Constantine sent Hosius as his commissioner to Alexandria to settle the dispute between Alexander and Arius. Apart from refuting the dogmas of Sabellius it is not clear what he achieved; probably on his advice the emperor called the Council of Nicea to settle the still unsolved Arian problem.

The role of Hosius in the council has been much discussed. He appears to have presided, but not as papal legate. His influence over the emperor can probably be seen in the latter's explanation of the *Homoousion* and letter to the churches concerning the council. After Nicea, Hosius appears to have returned to his diocese, but reemerged again in 345 at the Council of Sardica. Constans, at the instigation of certain bishops, persuaded Constantius to call the council chiefly to settle the question of the orthodoxy of Athanasius. Athanasius traveled to Sardica with Hosius, who was to preside. When the Eastern bishops arrived, they refused to attend because of Athanasius's presence, and withdrew to issue an encyclical letter condemning Hosius and Julius of Rome and others for holding communion with Athanasius.

In 355 Hosius was summoned to Milan by Constantius to condemn Athanasius and to hold communion with the Arians. He refused, and persisted despite imperial persuasion. His only extant letter, a reply to Constantius, belongs to this period. Finally at Sirmium he was forced to sign the second (the "Blasphemy") of the three creeds that seem to have emanated from here over this period (351, 357, 359). This was an Arian creed which Hosius probably repudiated before his death soon afterward in Spain. He has been described as dictatorial, harsh, and inflexible, and some of the blame for the failure of the Council of Sardica must rest on him. On the other hand, he was held by Athanasius and by Liberius of Rome in the highest honor. DAVID JOHN WILLIAMS

HOSIUS (Hos, Hosz), STANISLAUS (1504-1579). Polish Counter-Reformation controversialist and cardinal. Born at Cracow, he received humanistic training there and, sponsored by Tomicki, bishop of Cracow and vice-chancellor of Poland, studied law at Padua and Bologna, earning a Bologna doctorate in canon and civil law. On his return home he served as secretary to Tomicki, eventually becoming royal secretary. He was invested with several benefices, ordained priest (1543), and became bishop first of Culm (1549), then of Ermland (1551). His burning desire was the extirpation of the Protestant heresy. His consummate polemical skill against reformers such as Johann Brenz* and Jan à Lasco,* his educational programs, and his violent repressive measures made him perhaps the greatest of the Polish counter-reformers, winning him both the sobriquet "hammer of the heretics" and the favor of Rome. His *Confessio fidei catholicae christianiae* (1553, 1557) was one of the most successful polemic pieces of the Catholic Reformation. In 1558 Paul IV called him to Rome, where he became a leading voice in the Curia, playing an important role in the final sessions of the Council of Trent.* His two-volume works were published at Cologne in 1584.

See also L. Bernacki, *La Doctrine de l'Église chez le Cardinal Hosius* (1936); and F.J. Zdrodowski, *The Concept of Heresy According to Cardinal Hosius* (1947). BRIAN G. ARMSTRONG

HOSKYNS, SIR EDWYN CLEMENT (1884-1937). Anglican clergyman and theologian. He was educated at Jesus College, Cambridge; Wells Theological College; and the University of Berlin. From 1919 he was a fellow of Corpus Christi, Cambridge. A pioneer of the "biblical theology" movement in England, Hoskyns was the translator of Karl Barth's famous commentary on Romans into English (1933). His most important book was *The Riddle of the New Testament* (1931), with his pupil F.N. Davey, concerning the relation of Jesus of Nazareth to the primitive Christian Church. He argued that no interpretation of the person and teaching of Jesus which fails to explain this relationship can be true to history. In his discussion it was argued that no matter where one looks, or how far one goes back into the tradition (Mark, Q, M, L, etc.), one never finds a "simply human" Jesus, and that the best explanation of the historical data is that the so-called Christ of faith is a historical "given" rather than the creation of the church. His other works include a massive theological commentary on the fourth gospel (ed. F.N. Davey, 1940) and a number of important essays: "The Christ of the Synoptic Gospels" in *Essays Catholic and Critical* (1926) and "Jesus the Messiah" in *Mysterium Christi* (1930).

W. WARD GASQUE

HOSMER, FREDERICK LUCIAN (1840-1929). American Unitarian clergyman. Born in Framingham, Massachusetts, he graduated from Harvard in Arts and Divinity and subsequently was pastor of several Unitarian churches, including First Unitarian, Berkeley, California, from which he retired in 1904. Considered a liberal even in his denomination, Hosmer devoted his remaining years to hymn and verse writing. Among his publications is *Unity Hymns and Chorals* (1880).

HOSPITALLERS. Medieval men and women given to the care of the infirm, with both themselves and their patients usually observing religious vows. By 800 the Muslim world had medical hospitals, but W Europe did not in any proper sense until 1200. There were both hospices for the permanently poor, insane, and incurable, and hospitals for temporary medical treatment. One foundation could be both, and often hospitals developed from hospices. Monasteries also became hospitals, while St. Bartholomew's, London, was both. The Rules when observed were Austin, Benedictine, Franciscan, or that of the Knights Hospitallers of St. John of Jerusalem. Diocesan bishops had some control, and the popes bestowed many favors: chapels, cemeteries, indulgences. Laymen also shared in the work, and while the master was sometimes a layman more often he was a religious. The attendant brothers and sisters were mostly nurses, and in larger hospitals were assisted by clerks in minor orders and lay servants.

The eleventh and twelfth centuries saw a huge increase in both hospitals and nursing orders: Antonines, Order of the Holy Spirit, Order of St. William of the Desert, Bethlehemites, Order of St. Catherine, and Beguines and Beghards. The Knights of St. John of Jerusalem, after 1310 known as the Knights of Rhodes, and from 1530 as the Knights of Malta, were exemplary hospitallers. Founded not later than 1108 to care for the sick and provide for pilgrims and crusaders in Jerusalem, they subsequently established and managed hundreds of hospitals and hospices across Europe and the Levant, together with their military efforts. Having conquered Rhodes in 1309, they moved their center there, with the emphasis being more military and their wealth and power much increased after suppressing the Knights Templars in 1312. In 1530 their center became Malta until 1798, and they were reestablished in Rome in 1834.

Hospitallers also served the leper communities and had a great influence on medical progress. Leprosy appeared in Europe about 500 and reached its height in the thirteenth century, declining by 1350 and very rare by 1500. The Hospitallers of St. Lazarus who began to treat leprosy in twelfth-century Jerusalem are but one of many such examples. Most medieval hospitals were very small, fewer than thirty beds, and every country had its great infirmaries: S. Spirito, Rome; Holy Ghost, Lübeck; St. Leonard's, York. By 1200 also, medieval hospitallers were working under physicians trained at Salerno, Montpellier, and elsewhere, and much documentation testifies to the careful standard of their nursing. The orders flourished well into the modern period, when drastic changes occurred.

BIBLIOGRAPHY: R.M. Clay, *The Mediaeval Hospitals of England* (1909); D. Riesman, *The Story of Medicine in the Middle Ages* (1935); G. Bottarelli, *Storia politica e militare del sovrano ordine di S. Giovanni di Gerusalemme detto di Malta* (2 vols., 1940); E.E. Hume, *Medical Work of the Knights Hospitallers of St. John of Jerusalem* (1940).

C.G. THORNE, JR.

HOTMAN, FRANÇOIS (1524-1590). French jurist, scholar, and Reformer. Born in Paris of a family originally from Silesia, Hotman became one of the foremost legal experts and law professors of his day. His conversion to Reformation doctrines in 1547 led to a life of uncertainty and periodic exile. In his most influential book, *Franco-Gallia* (1573), Hotman argued for a limited constitutional monarchy in France. Actually written before the St. Bartholomew's Day massacre of 1572 and thus not merely a *livre de circonstance* prompted by that notorious event, it became a landmark in Western political thought because it signaled a clear transition from medieval to modern constitutionalist principles. Hotman died in poverty at Basle.

ROBERT D. LINDER

HOUGHTON, WILLIAM HENRY (1887-1947). Baptist minister and president of Moody Bible Institute. Born in South Boston, he was educated in New England schools. An early career on the stage was cut short for theological training after a deepening religious experience, and subsequently he joined R.A. Torrey* as songleader. Ordained to the Baptist ministry in 1915, he held several pastorates, including Baptist Tabernacle, Atlanta (1925-30), and Calvary Baptist Church, New York City (1930-34), in each of which membership doubled. He conducted a successful evangelistic campaign in Ireland in 1924. Ten years later he became president of Moody Bible Institute, Chicago, a post he held until his death. During his tenure he strengthened the faculty; doubled the circulation of *Moody Monthly* of which he was editor (1934-46); encouraged Irwin Moon to join the institute extension staff (1938) to begin what became Moody Institute of Science; and provided the stimulus for the founding of the American Scientific Affiliation (1941). He was well known for his gospel songs.

HOWARD A. WHALEY

HOW, WILLIAM WALSHAM (1823-1897). Bishop and hymnwriter. A solicitor's son, educated at Shrewsbury and Oxford, he held several parochial posts before his consecration in 1879 as bishop of Bedford, where he became known as "the poor man's bishop." In 1888 he became the first bishop of Wakefield. Although he published a commentary on *The Four Gospels* (1870) and numerous sermons, he is remembered chiefly for several of his fifty-four hymns. These were contributed to a collection of *Psalms and Hymns,* compiled with T.B. Morrell (1854), and to *Church Hymns* (1871) which he edited jointly with J. Ellerton.* B.L. Manning thought him possibly the greatest nineteenth-century hymnwriter. He wrote, e.g., "For all the saints," "O Jesu, Thou art standing," "O Word of God Incarnate," "It is a thing most wonderful," and "We give Thee but Thine own." JOHN S. ANDREWS

HOWARD, JOHN (1726-1790). Prison reformer. Born in Hackney, after a short period as apprentice to a grocer he came into a modest inheritance in 1742 and traveled in Europe. After one trip to Portugal his boat was captured by a privateer on the return journey, and he was imprisoned in France. This may have set his mind moving in the direction which was to be his life-work. In 1758 he settled in Bedfordshire, where he built model cottages, promoted educational experiments, and developed rural industry. Following the death of his second wife, he traveled again through Europe, and on his return in 1773 was made high sheriff of Bedfordshire. Thereafter he devoted his time and strength and a good part of his fortune to the reform of conditions in prisons in England and in Europe.

There were many abuses—for example, jailers in England received fees rather than a salary, and this led to extortion and corruption. Howard visited all the county jails, and he promoted parliamentary bills designed to reform conditions. He then toured jails throughout England and published his book, *State of the Prisons* (1777), which caused great concern. Further travels in Europe were followed by another act of Parliament in 1779. In 1782 he made a third general inspection of English prisons, and then a tour of Europe, giving special attention to the *lazarettos,* particularly in Italy and the Near East, which were designed for the control of infectious diseases. He himself was put in one when he was in quarantine in Venice. He publicized the terrible conditions in such institutions and agitated for reform. In 1789 he took his last journey, which led him to Prussia and Poland and on into Russia; there he caught camp fever from a woman in the course of his researches, and died. Howard was a very earnest evangelical Christian, a teetotaler and a vegetarian, whose life was devoted to the cause of prison reform.

See biographies by D.L. Howard (1958) and M. Southwood (1959). A. MORGAN DERHAM

HOWARD, PETER (1908-1965). Oxford Group* movement leader. Born at Maidenhead, son of a schoolmaster, he was educated at Mill Hill and Oxford, and worked for the Beaverbrook Press till his conversion to Moral Re-Armament ("Oxford Group") in 1941 provoked his resignation. Thereafter he farmed in Suffolk, undertaking worldwide journeys for the movement, and after Frank Buchman's* death in 1961 he became its foremost representative. A forthright personality with many admirers and some bitter critics, Howard wrote a large number of books and plays which were performed at the Westminster Theatre, London. Though frequently invoking the names of the three persons of the Trinity to support his crusade for the "moral absolutes" or "ideology" of the movement, no consistent Christian theology can be said to underlie his work. IAN SELLERS

HOWE, JOHN (1630-1705). English Nonconformist minister. Born at Loughborough, Leicester, he studied at Christ's College, Cambridge, and then Brasenose College, Oxford. In 1652 he became a fellow of Magdalen and was also ordained by Charles Herle and other ministers at Winwick, Lancashire. In 1654 he was given the "perpetual curacy" of Great Torrington, Devon, by the dean and canons of Christ Church, Oxford. At Torrington he labored to unite Presbyterians and Independents, but his work was halted when Oliver Cromwell called him to court in 1657 as a chaplain. Once again he tried to heal divisions among the various groups who frequented Whitehall. He also served Richard Cromwell. After the latter's resignation of the Protectorate, he returned to Torrington.

In 1662 Howe was ejected and for the next eight years he had his share of harassment under the Clarendon Code* while he preached from time to time in the homes of local gentry. Moving to Ireland in 1670 he became the chaplain to Lord Massareene in Antrim Castle. During his six years there he engaged in various schemes to educate Presbyterian clergy and also wrote *The Living Temple of God* (1675). In 1676 he returned to London as co-pastor of the Presbyterian congregation at Haberdashers' Hall. During 1685-87 he lived abroad, mostly at Utrecht. After the Toleration Act (1689) he labored to unite the Presbyterians and Independents, but the "Happy Union" he helped to forge was but a brief one. A six-volume edition of his works was published in 1862-63. PETER TOON

HOWE, JULIA WARD (1819-1910). Writer and reformer Born in New York City, she was reared an Episcopalian and was privately educated. Her marriage to Samuel G. Howe (1843), a humanitarian and teacher of the blind, placed her in the company of prominent Bostonian intellectuals, poets, and social reformers. She belonged to the Radical Club and assisted her husband in editing the abolitionist paper *The Commonwealth* (1851-53). Becoming a Unitarian, she occasionally preached from Unitarian pulpits. After her husband's death (1876) she gave herself unceasingly to public service, a leader in every humanitarian movement or cause. She advocated woman suffrage, prison reform, international peace, and children's welfare. She also wrote travel books, essays, drama, and verse. Her most famous piece of poetry is "The Battle Hymn of the Republic."
ALBERT H. FREUNDT, JR.

HOWSON, JOHN SAUL (1816-85). NT scholar. Graduate of Trinity College, Cambridge, he served as teacher (1845-49) and then headmaster (1849-66) of Liverpool Collegiate Institute. From 1867 he was dean of Chester. Howson is remembered chiefly as co-author with W.J. Conybeare*

of the influential work, *The Life and Epistles of St. Paul* (2 vols., 1852), which remains in print. Howson was chiefly responsible for the historical, geographical, and archaeological aspects of the work. He published also several other works on Paul and on NT subjects generally.

HROMADKA, JOSEF LUKI (1889-1969). Czech Reformed theologian. Born in Hodslavice, Moravia, he was educated at several universities and in 1912 became pastor of the Evangelical Church of Czech Brethren. From 1920 to 1939 he was professor of systematic theology at the Jan Hus Theological Faculty in Prague, and during the war he taught at Princeton Theological Seminary. On his return home in 1947, he became a controversial figure for urging reconciliation between Christians and Communists. He was a founder of the World Council of Churches and served on its central committee. He was also founder and chairman of the Christian Peace Conference, an organization which served as a vehicle for Christian-Marxist dialogue and communication between Christians in the East and West, and he received the Lenin Peace Prize in 1958. Because of his protest against the Soviet invasion of Czechoslovakia, he was forced to resign as chairman of the CPC in November 1969. He died six weeks later. RICHARD V. PIERARD

HROSVIT. Tenth-century nun and poetess who lived at Gandersheim in Saxony. She took as her model the literary style of Terence, but she was very familiar with all the leading Latin poets and with Holy Scripture. One of her main aims was to oppose Terence's descriptions of the frailty of women by portraying the courage and chastity of Christian virgins. After her death she was quickly forgotten, but Conrad Celtes, the humanist, discovered a Latin manuscript of her writings at Ratisbon which was printed in 1501. One of her poems, "Passio Sancti Pelagii," claims to be based on the report of an eyewitness account of the martyrdom. There is an English translation of her plays by H.J.W. Tillyard (1923) and of her other works by Sister Gonsalva Wiegand (1936).

PETER TOON

HUBER, SAMUEL (1547-1624). Protestant scholar. Born at Burgdorf, near Bern, he was active in religious controversy, usually defending Lutheran doctrines against the Calvinism of the Swiss Reformed Church. He caused special offense by his assertion of Christ's universal atonement *("Christum Jesum esse mortuum pro peccatis totuis generis humani")*, and was banished from Switzerland in 1588. He joined the Lutheran church, signed the Formula of Concord,* and became first a pastor near Tübingen, then professor in Wittenberg University. His assertion that God has from eternity elected all men to salvation gave offense to other Lutherans, and he was opposed by Polycarp Lyser and Aegidius Hunnius (1593), whom he in turn charged with Calvinism. According to Albrecht Ritschl, Huber was "a very poor theologian," but noteworthy as representing the protest against the doctrine of twofold predestination. The title "Huberianism" was widely given to expositions of a universal atonement such as he advocated.

J.G.G. NORMAN

HUBERT WALTER (d.1205). Archbishop of Canterbury from 1193. He faithfully served three kings, was essentially a royal administrator, and was rewarded for his services by promotion in the church. He was successively dean of York, bishop of Salisbury from 1189, and then primate. He was also papal legate from 1195. He studied law at Bologna and had been a member of the famous Angevin lawyer Glanvill's household. He played a major role in collecting Richard I's ransom money, and was Richard's justiciar from 1193 until 1198. The newly ransomed Richard waged war in France from 1194 and never returned to England, so Walter as justiciar was in virtual charge of the kingdom. His achievements as justiciar were many. He systematized the organization and recording of judicial and legal matters and insisted on a more thorough enforcement of law and order. After Richard's death in 1199, Walter helped to rally support for King John. As the latter's chancellor from 1199 he gained equally high praise for his reorganization of chancery practice and the systematic recording of its transactions.

PETER TOON

HUBMAIER, BALTHASAR, see ANABAPTISTS

HUC, ABBÉ (Régis Evaniste) (1813-1860). Missionary to China. The period after 1840 and the Opium Wars was one of persecution for Roman Catholics, but Abbé Huc with his companion, Father Gabet, both French Lazarists, undertook a remarkable journey from Peking through Mongolia and Tibet to Lhasa and back to Canton in 1844-46. They were protected by wearing mandarin dress and the crimson sash that signified kinship to the emperor. Everywhere they found terror-stricken little groups of Christians, but nearly succeeded in establishing a mission in Lhasa itself. The famous account of his journeyings *(Travels in Tartary and Tibet)* by Abbé Huc stimulated fresh interest abroad in the Christian mission in China. LESLIE T. LYALL

HUGHES, HUGH PRICE (1847-1902). Wesleyan divine. Born at Carmarthen and educated there and at Swansea, he trained for the Wesleyan ministry at Richmond College and graduated from London University. He served in various circuits, founding new churches wherever he went, and in 1885 while superintendent of the Brixton (London) circuit he launched the Wesleyan Forward Movement, a campaign of evangelism and social service characterized by the erection of central halls and encouraged by the *Methodist Times* (founded 1885) which he edited. Other related causes were the Wesleyan Twentieth Century Fund which, as president of conference in 1898, Hughes instigated to raise a million pounds for church extension, Methodist reunion, and social reform—his bitter attacks on Parnell after the divorce crisis of 1890 led him to formulate the term "the Nonconformist Conscience." His self-styled "Christian Imperialism" caused him to champion British imperial expansion overseas and a campaign of social and ecclesiastical aggression against the Anglican Church at home. Despite his missionary enthusiasm, Hughes's theology was reductionist, and appears particularly so in his teaching on sin and on the Last Things. The central halls and big city missions of contemporary British Methodism remain his most permanent memorial. IAN SELLERS

HUGHES, JOHN JOSEPH (1797-1864). Roman Catholic archbishop of New York. Born in Annalogham, Ireland, he went to the USA in 1817, was educated in Mount St. Mary Catholic College in Maryland, and was ordained in 1826. He served two parishes in Philadelphia until 1837, when he became coadjutor bishop of New York. He founded St. John's College (now Fordham University) in 1841. He became bishop of New York in 1842 and archbishop in 1850. He organized the parochial school system of New York and freed it from public and lay control. He began the building of St. Patrick's Cathedral in 1858 and led in the establishment of the North American College in Rome. Early in the Civil War he was successful as an envoy of the United States government in winning sympathy for the Union cause in France, Ireland, and Italy. EARLE E. CAIRNS

HUGHES, STEPHEN (1622-1688). Welsh Puritan. He was son of a Carmarthen silk merchant, but little is known of his youth and education. He was appointed Puritan minister of the parish of Meidrum, Carmarthenshire, in 1654. He contributed largely to the work of publishing "good books" in Welsh. He began by issuing the first part of the work of Vicar Rhys Prichard in 1659. During the era of persecution his home was at Swansea, and the work of publishing was in abeyance. It recommenced in 1670. He secured the cooperation of such men as Thomas Gouge (1605?-1681), Bishop William Thomas (1613-1689), Charles Edwards (1628-1691), Richard Jones (1603-1673), and William Jones (d.1679). Between them they published a series of books culminating in a translation of *Pilgrim's Progress* in 1688. The significance of this work for Welsh Christianity was immense in that it ensured that Puritanism would make wide use of the Welsh language in literature and education. Hughes was also a preacher of great influence and was the founder of the strong Congregationalist tradition in Carmarthenshire. R. TUDUR JONES

HUGH OF CLUNY (1024-1109). Abbot of Cluny. Descended from Burgundian nobility. He rejected the knightly life for academic training under Bishop Hugh of Auxerre, his great-uncle, and entered the Cluny novitiate when fourteen. He took vows a year later, was ordained to the priesthood in 1044, appointed prior in 1048, and named successor to Abbot Odilo* in 1049. At the Council of Reims (1049) he eloquently advocated reforms before Leo IX. Thereafter enjoying the confidence of nine popes, several from Cluny ranks, he served them as personal adviser, diplomatic emissary, and executor of Vatican policy. His presence at numerous councils and synods contributed to significant decisions: condemnation of Berengar's heresy (Lateran, 1050), decree on papal elections (Lateran, 1059), implementation of reforms (Avignon and Vienne, 1060), defense of Cluniac privileges (Lateran, 1063), organization of the First Crusade (Clermont, 1095).

This capable disciplinarian governed Cluny for sixty prosperous years. Houses were added to the order in France, Italy, Spain, Germany, and England. Civil and ecclesiastical privileges were gained. The magnificent abbey church of Cluny was erected and in 1095 its altar personally dedicated by Urban II. Concubinage, simony, and investiture were attacked. Hugh feared secular domination of the church. Diplomatic in his sympathy with Gregory VII against Emperor Henry IV, he remained a wise irreproachable mediator in a factional age, winning the tributes and friendship of both civil and ecclesiastical leaders. Most of his voluminous correspondence has been lost, as has his *Life of the Blessed Virgin*.

JAMES DE JONG

HUGH OF LINCOLN (1135-1200). Bishop of Lincoln. Born in Burgundy and educated by Regular Canons at Villard-Benoit, he was professed there when fifteen years old, and subsequently became head of one of that House's dependencies at Maximum. Later he joined the Carthusian Order at Chartreuse where he remained for seventeen years. At Henry II* of England's request, he was sent to England as first prior of the Carthusian House at Witham, Somerset, founded by Henry II as part expiation for Becket's murder. Hugh built the house at Witham. He quickly impressed everyone by his personal holiness and integrity. He became a close friend of Henry, but could be fearlessly critical of his policies toward the English Church. Henry respected his attitude, and overriding Hugh's objections, he appointed him bishop of Lincoln in 1186.

Hugh applied himself tirelessly to the improvement of the see, which had been vacant for eighteen years. He introduced a program of clerical reform and started the rebuilding of Lincoln Cathedral, occasionally carrying hods of stones and mortar himself. As an important bishop he was involved also in political matters. In 1197 he was one of the feudatories who denied Richard I's right to insist that his barons should serve personally on the continent. In John's reign Hugh visited France on the king's behalf. He was a much-loved campaigner for justice. He insisted that Henry II compensate those evicted to make room for the House at Witham, and during the popular persecutions of the Jews in England he did all he could to protect them. He died in London and was buried at Lincoln. He was canonized in 1220.

L. FEEHAN

HUGH OF ST.-VICTOR (c.1096-1141). Exegete and theologian. Descended from the courts of Blankenburg in Saxony, he early joined the Austin Canons Regular at Hamersleven, and settled finally about 1115 in the new monastery of St.-Victor in Paris. Conflicting accounts of these years exist for lack of information; he himself said: "Since my childhood I have been an exile." From 1120 until death he was the leading master in the school of St.-Victor, where he was prior of the abbey for a time after 1133. A recognized scholar, he was concerned about the task of the *trivium* and *quadrivium*, distinctions between natural reason and divine faith and the objects of each, the nature of philosophy, scientific classification, the importance of the literal interpretation of Scripture, and rules for exegesis.

As a pure philosopher his contribution was limited. His forte was exegesis, and together with expounding Scripture he was a student of the science of interpretation—seen in his notes on the first books of the OT. He developed into a theologian (cf. his *Summa Sententiarum* and *De sacramentis christianae fidei*). He was given to studying the Fathers, and was called "the second Augustine."

In the belief that original sin is a corruption contracted at birth, his theological system begins with Adam and goes through Advent and the final consummation, defining faith as "a certainty about things absent, above opinion and below science." He was indebted to his contemporaries, Anselm of Canterbury, Anselm of Laon, and William of Champeaux. A mystic, Hugh wrote on mystical union and believed that as the soul ascends to God it acquires the gift of wisdom or contemplation which original sin canceled; he distinguished sharply between contemplation and Beatific Vision. His writings on all these subjects are many.

BIBLIOGRAPHY: J.P. Kleinz, *The Theory of Knowledge of Hugh of St. Victor* (1945); R. Baron, *Science et Sagesse chez Hugues de Saint-Victor* (1957); J. Taylor, *The Origin and Early Life of Hugh of St. Victor* (1957). C.G. THORNE, JR.

HUGUENOTS. A nickname for the French Calvinists (the origin is uncertain: perhaps a corruption of the German *Eidgenossen*, "confederates"). Under Francis I (d.1547), persecution of Protestants was sporadic; his sister Margaret, indeed, made Navarre a center for reform-minded humanists. By the 1540s Calvinism spread rapidly in France, bringing increased repression. Under Henry II's reign (1547-59), special courts were set up to try heretics, who were often burned at the stake. As martyrs multiplied, so also did Calvinism spread, aided by massive mission efforts from Geneva. Powerful noble clans adopted the new faith, notably the Bourbons, led by Antoine of Navarre. A national synod was held in 1559. With Henry's death, the political situation began to disintegrate rapidly. The princely family of the Guises, militant Catholics, opposed any toleration of the heretics. An extremist Huguenot attempt to kidnap the new king (the weak Francis II) failed; a Catholic-Calvinist colloquy at Passy (1560) achieved nothing; attempts at compromise by allowing limited toleration produced militant Catholic protest, climaxed by a Guise march on Paris (1562).

Civil war broke out. It was to last for a generation. The main parties were three: the Huguenots, the militant Catholics, and the *politiques*, who wanted above all the restoration of order. Religious differences were entangled with political ambitions. Both Huguenots and militant Catholics proved ready to intrigue for foreign support. The wars were marked by political assassinations and even by mass "executions" (the Massacre of St. Bartholomew's Day,* 1572), an attempt to wipe out the Huguenot leadership).

Huguenot political theorists developed justifications for revolt against tyrants (e.g., *Vindiciae contra tyrannos,* 1579). Given the conventional wisdom of the age—namely, that a state could not survive if its citizens were divided in ideology (religion)—the situation seemed insoluble. The wars were ended in ironic fashion: the assassination of Henry III (by a fanatic Catholic) made Henry of Navarre, the Huguenot leader, heir to the throne. To gain it, he turned Catholic ("Paris is well worth a Mass"). He quickly ended the civil war and in 1598 issued the Edict of Nantes,* granting the Huguenots full toleration, civil rights, and the right to their own fortified towns. To some extent, thus, the Huguenots remained a "state within a state."

During Henry IV's* reign (1598-1610) the Huguenots felt secure. After his assassination their position slowly worsened. Huguenot militant revolts (1615, 1625) merely led to the loss of the fortified towns. Though entrenched enough under Louis XIII (1610-43) to engage in internal controversy over the attempts by Amyraut (Amyraldus) of Saumur to soften the orthodox idea of predestination, the Huguenots' days were numbered. Louis XIV (reigned 1643-1715) was determined to make France the most powerful state in Europe, and this involved ruling a state committed to one religion. Repressive measures were instituted (e.g., the *dragonnades,* or quartering of soldiers on Huguenot families), persecution followed, and in 1685 the Edict of Nantes was revoked. Calvinism was now illegal. Hundreds of thousands of Huguenots left in a mass exodus from the lands of the "Sun King."

Those who remained, mostly the poor, suffered sentences to the galleys, hangings, and other punishments. The Calvinist peasants of the Cevennes rose in desperate revolt (1702); though the "Camisards"* were gradually hunted down, by Louis's death in 1715 a regular "underground" church had been organized, led by Antoine Court* and later by Paul Rabaut.* By the later 1700s, with the spread of Enlightenment ideas (Voltaire, etc.), persecution for religious reasons seemed increasingly antiquated, and by 1787 the Huguenot remnant gained limited civil rights.

The French Revolution brought full toleration and civil rights. The Napoleonic regime recognized Calvinism as an established religion, along with others, at the cost of some degree of state regulation. This was continued in the post-Revolutionary era. The Huguenots, though a small minority, produced many noted figures (e.g., the political leader Guizot). As the revolutionary storms subsided, new influences affected the Calvinists: higher criticism and "modernism" on the one hand, and the "Awakening" *(Réveil*)* on the other, the latter a conservative and Pietist return to traditional orthodoxy. By 1848 a conservative group led by Adolphe Monod* split off; another conservative schism followed in 1872, despite the efforts of the aging Guizot to reconcile evangelicals and modernists.

By 1905 anticlerical liberalism brought an end to all ties between state and religious groups. Among the Calvinists four separate bodies resulted. After World War I they cooperated increasingly, and by 1938 most Calvinists united in the Reformed Church of France.

BIBLIOGRAPHY: A.G. Grant, *The Huguenots* (1943); E. Leonard, *Histoire du Protestantisme français* (1961); B.G. Armstrong, *Calvinism and the Amyraut Heresy* (1969). DIRK JELLEMA

HUMANISM, RELIGIOUS. The Renaissance may be said to be the source of religious humanism, at least in modern times. In its early phases it expressed itself in the revival of "human" learning, the rebirth of classicism, as against the "sacred" learning of the Middle Ages. This in turn involved both the revival of classical languages (incidentally benefiting biblical studies) and the development of a historical perspective made necessary by the rejection of medievalism. In its later phases, religious humanism showed itself in the repudiation of the Augustinianism of the Reformers by Erasmus and, later, Arminius. Thus as a "movement," if it can be called such, it embraced parts of the Roman Catholic Church (in such individuals as Colet, More, and Erasmus) and Protestantism (Arminius, Socinus, Locke) as well as independent thinkers such as Spinoza. The Moderates* in the Church of Scotland and the "Broad Church" school of Anglicanism, as well as certain themes in German Pietism and in the philosophy of Kant, may be said to have carried many of the emphases of religious humanism into the eighteenth and nineteenth centuries.

These emphases were: (1) a confidence in human nature, coupled with a belief in the power of education. This expressed itself characteristically in a repudiation of the Augustinian (and biblical) teaching on the bondage of the will (see the Erasmus-Luther debate), and in an anthropocentric religiosity. This confidence in human nature was tempered by skepticism, particularly in theological matters; (2) a belief in toleration, due less to conviction about fundamental human rights than to theological indifferentism and skepticism, combined with the belief that what was right in Christianity was but republication of ancient wisdom or, later, a restatement of "natural religion." Though some individuals such as Colet and More were religiously earnest, for many religious humanists the church was treated in a thoroughly secular way, or thought of as having simply a "civic" function to fulfill. PAUL HELM

HUMBERT (d.1061). Cardinal bishop of Silva Candida. A Burgundian by birth, he became a monk at the monastery of Moyenmoutier, where he showed himself to be a good scholar and keen reformer. Leo IX called him to Rome in 1049 and a year later made him cardinal bishop. In this position he was a principal adviser to Leo IX, Victor II, and Stephen IX with regard to the reform of the church and relations with the East. He strongly denounced simony in *Libri tres adversus Simoniacos.* At the time of the Great Schism* he was a leading member of the mission which Leo IX sent to Byzantium to the patriarch, Michael Cerularius, in 1054. Humbert is to be classed with Peter Damian* as a leading reformer of the eleventh century. PETER TOON

HUME, DAVID (1711-1776). Scottish philosopher, historian, and man of letters. His philosophical program, first outlined in *A Treatise of Human Nature* (1739-40), involved the application of Newtonian scientific method to human nature. The ultimate data of investigation are "impressions," those sensations directly presented to the mind of which "ideas" are the copies. The philosopher must discover from what impression or impressions ideas are derived. This amounts, in essence, to an early application, in a psychological idiom, of the logical positivists' verification principle. Hume's analyses of memory, personal identity, and (most famously and successfully) causation are attempts to fulfill this program. In morals, Hume argued that moral judgments were the product of the passions, not of the reason.

In religion, Hume is chiefly noteworthy for his skeptical attacks on miracles and on the argument from design. Miracles are denied on *a posteriori* grounds; it is always more reasonable to reject someone's testimony about a miracle than to accept it. This view has implications for historiography. His attack on the argument from design involved showing the ambiguity of the evidence. Hume was a skeptic about metaphysical claims and theories, in religion and elsewhere, though not about the "natural beliefs" men have about, for instance, the external world. He set many of the problems currently discussed by analytic philosophers. His work in religion can be regarded as one of the most fundamental attacks on natural theology in modern times. PAUL HELM

HUMILIATI. An order of penitents, probably founded by Johannes Oldratus (d.1159), which was partly suppressed in the sixteenth century. Following the Benedictine Rule, they cared for the poor and mortified their bodies. The order had three types of members: those who lived ascetically in their own homes, those who were nuns, and those who were monks. During the late Middle Ages their discipline and devotion deteriorated, and when Charles Borromeo,* who had sought to reform the order, was assaulted in 1571 by one of the monks, the pope suppressed the monasteries but allowed the Humiliate Nuns to continue. The Humiliati are sometimes called the "Barettines of Penitence." PETER TOON

HUNG, HSIU-CH'ÜAN (1813-1864). Leader of the Taiping Rebellion. A Hakka by birth, he was a disillusioned scholar in Canton when handed a treatise on Christianity written by Liang A-fah,* a colleague of Robert Morrison in Malacca. Thereafter he gave Christian teaching serious attention. Not having received any Christian instruction, he began to teach a syncretistic faith combining Chinese beliefs with the doctrines of God and of Jesus, His Son, and adopted much Christian teaching and practice. Being also eccentric and with delusions of grandeur, he eventually assumed the imperial title in 1851 and led a peasant uprising against the Manchu rule. His declared intention was to set up "the Heavenly Kingdom of Great Peace" (Taiping). The rebellion at first met with dramatic success, but degenerated into arson and massacre and was finally suppressed by foreign troops led by Lt.-Col. C.G. ("Chinese") Gordon. Hung committed suicide. This pseudo-Christian movement which, wisely directed, might have given Christianity a strong start in China became the most desolating insurrection of the nineteenth century. LESLIE T. LYALL

HUNGARY. In the latter years of the ninth century the pagan Magyars made their first permanent settlements in Hungary. Using their new home as a base, they raided large portions of W Europe. In 955 Otto I, Holy Roman Emperor, won a great victory over them at Unstrut. This victory checked the advances of the Magyars, and their conversion to Christianity followed this major defeat. It took place in the closing years of the tenth century under Stephen (see STEPHEN OF HUNGARY) who preached to his subjects, urging them to accept Christianity. From Pope Sylvester II Stephen received the royal crown and title as king of Hungary. When Stephen died in 1038, a sharp reaction set in against Christianity. Later in the eleventh century, however, powerful monarchs gave new support to the Christian Church, and Christianity gained a stronghold in Hungary. The situation changed again when the fall of Constantinople to the Turks in 1453 opened the way for their conquest of Hungary. Christianity faced increasing difficulties and lost its privileged position.

In the latter half of the sixteenth century Protestantism made its way into the country. It had been adumbrated by the Hussite movement of the fifteenth century and the resultant translation of the Scriptures into the Hungarian language. Lutheranism made great headway in Hungary after 1525, and many Hungarian students went to Wittenberg for their training in theology. Calvinism later had an impact, particularly among the Magyars, while Lutheranism had the greater appeal for the German and Slavonic peoples of the kingdom. On the other hand, the upper classes, particularly the landed aristocracy, remained loyal to the Roman Catholic Church, and the activities of the Jesuits also helped to keep the Protestants as a minority in Hungary. Not until 1787, in fact, did the Hungarian Protestants gain a degree of freedom, when the Hapsburg Edict of that year either eased or removed entirely the earlier restrictions. Protestants were thereafter given the same civic rights as Roman Catholics.

During the nineteenth century, Protestants in Hungary felt the effects of the evangelical movements which were so influential in Britain, Switzerland, and the USA. With the restoration of the monarchy in 1920 after a short period of communistic control, the religious situation of the nineteenth century was restored, but in 1944 a Nazi regime was created which brought great hardship to both Protestants and Catholics. When Nazis gave way to Communists after World War II, the situation became even worse, and all branches of the Christian Church have suffered severely since 1949.

BIBLIOGRAPHY: E. Horn, *Christianisme en Hongrie* (1906); E. Revesz, S. Kovats, and L. Ravasz, *Hungarian Protestantism, Its Past, Present and Future* (1927); V. Gsovski (ed.), *Church and State Behind the Iron Curtain* (1955); R. Tobias, *Communist-Christian Encounter in East Europe* (1956). C. GREGG SINGER

HUNT, JOHN (1812-1848). English missionary to Fiji. Born into the family of a Lincolnshire farmer, he had little formal schooling. Converted in a Methodist meeting at seventeen, he educated himself and preached in various chapels. He entered a Wesleyan theological college in 1835, and after ordination in 1838 sailed for Fiji as a missionary where he stressed Bible translation and the training of indigenous pastors. As he was a strong enthusiast for the doctrine of entire sanctification, his preaching sparked a revival in 1845. The rigors of extensive traveling in visiting the scattered Fijian congregations led to his death. His translations of the Scriptures were published posthumously, the NT in 1853 and the whole Bible in 1864. RICHARD V. PIERARD

HUNT, WILLIAM HOLMAN (1827-1910). English painter. He led the "Pre-Raphaelite Brotherhood" grouping of young British artists (and literary figures) who in the mid-nineteenth century wanted to return to simple painting technique, direct study of outdoor events with detailed depiction of exactly what is there, avoiding the academistic rules of chiaroscuro lighting and the coloring virtuosity Raphael had made the trend. In search of serious subject matter, Hunt himself practiced what he preached by going to Egypt and Palestine to paint biblical scenes with authentic local settings and types of people. Although Charles Dickens attacked the Pre-Raphaelite program as arrogant presumption, Ruskin defended their work, insuring its influence in England.

Holman Hunt helped break the conventional, iconographic picturing of Christ, and paintings like his well-known *The Light of the World* (1854) appealed to the new middle-class patrons of the arts. But the artistic skill and sincerity of principles behind the bright colors, exactly detailed foregrounds, and choices of "elevated" topics suffer from a somewhat maudlin spirit and anecdotal bent which keeps the art tied to being period pieces, illustrational tracts for the times.

CALVIN SEERVELD

HUNTING. The Bible calls Nimrod (Gen. 9:9) a mighty hunter, and later in the patriarchal period it cites Esau (Gen. 25:27) as an ideal outdoorsman. Although the cultural period of hunting had disappeared before the Hebrews settled in the Promised Land, there were still numerous references to the activity. A wide variety of game animals, beast of prey, and game birds were available in Palestine. Although "quiet" hunting was sanctioned for the clergy by the church for centuries, there are many today who would condemn it altogether on moral and humanitarian grounds.

There is little direct guidance available in the biblical material concerning the modern-day ethics of hunting. It is made clear in Genesis 1:28 that man is to have dominion over the fish of the sea, the birds of the air, and animals. It is also clear that the killing of animals was not forbidden. Cruelty to animals, however, appears to be contrary to the nature of God's creation. It seems consistent to add here that when hunting threatens the existence of a species, this could be termed exploitation of God's creation and in direct disregard for the God-given laws of nature. Theologians agree that conservation laws directed toward the preservation of the species of game are binding in conscience, for they promote the common good. JOHN P. DEVER

HUPFELD, HERMANN CHRISTIAN KARL FRIEDRICH (1796-1866). German OT scholar. Born at Marburg, where he began his education, he continued at Halle under H.F.W. Gesenius,* staying on as instructor before going back to Marburg as professor of theology in 1825. In 1838 he returned to Halle and remained there until his death. In addition to more technical works, Hupfeld sought to define the methods of biblical interpretation* (1844). In contrast with rationalists of his day, he preserved more of the sense of revelation in the OT. His work on the sources of Genesis (1853) rediscovered K.D. Ilgen's theory (1798) and cautiously divided into two distinct documents the sources in which *Elohim* had been used. His four-volume work on the Psalms (1855-61) was the first modern commentary on that book. He had a significant influence on H. Gunkel.* CLYDE CURRY SMITH

HUS, JAN (1373-1415). Bohemian Reformer. He was born of poor parents at Husinec in S Bohemia, the name of which place he assumed. His mother greatly desired that her son become a priest, and Jan at about thirteen years old entered the elementary school in the nearby Prachatice. In 1390 he matriculated in the university of Prague and four years later received his A.B. degree, ranked sixth of twenty-two. Going on to obtain the master's degree in 1396, he began teaching in the faculty of arts.

In 1402, after receiving priestly ordination, he was appointed rector and preacher of the Bethlehem Chapel, the center of Czech preaching in the spirit of the previous Czech reform movement. He thus became its most outstanding popular exponent. Nevertheless he continued his teaching in the arts faculty and enrolled in the faculty of theology for the doctor's degree. When the theological works of John Wycliffe* were brought to Prague about 1401, Hus became acquainted with them; prior to that he knew only Wycliffe's philosophical realism, with which he agreed. In 1403 a German university master, Johann Hübner, selected forty-five theses from Wycliffe's writings and secured their condemnation as heretical by the university, where the Germans, largely Nominalists, had three votes against the one Czech vote. This caused a rupture between the German and Czech masters, for the latter generally defended Wycliffe. Hus, however, did not share Wycliffe's radical theological views, such as Remanence, although some members of his party did.

The new archbishop, the young nobleman Zbynek Zajíc of Hasmburk, knew but little theology. Fortunately for Hus and the reform party, he favored its ecclesiastical reforms. This benevolent attitude lasted five years, during which the reform party grew in strength. Finally in 1408 the opponents of reform, mostly the higher clergy, won the archbishop to their side. The final break came in 1409 over the deposition of Pope Gregory XII and the election of Alexander V at the Council of Pisa. King Wenceslas and the Czech university masters, including Hus, sided with Alexander, while Zbynek and the German masters remained faithful to Gregory. When the king forced the archbishop to acknowledge the new pope, Zbynek secured from Alexander prohibition of preaching in chapels, including the Bethlehem Chapel. Hus refused to obey and was excommunicated by Zbynek, and the case was then turned over to the Curia. Hus was cited to appear in Rome, but sent procurators instead. Thereupon he was excommunicated by Cardinal de Colonna for contumacy. The king, angered by the opposition of the German masters to his ecclesiastical policy, changed the constitution of the university by depriving the Germans of their three votes and granting them to the Czechs. The Germans left in a body, and Hus was elected rector of the now Czech university.

But even greater conflict arose in 1411, when Pope John XXIII issued his "crusading" bull against King Ladislas of Naples. Shortly after, he appointed a commission for the sale of indulgences. Hus vehemently denounced this "trafficking in sacred things" as heresy. The Prague populace rose in revolt and burned a simulated papal bull. During the uprising three young men were beheaded for opposing the sale of indulgences. The process against Hus at the Curia was renewed in 1412, and he was declared under major excommunication by Cardinal Peter degli Stephaneschi. Prague was placed under interdict because of Hus's presence; thereupon he left for exile.

Hus found refuge chiefly in S Bohemia, and during the next two years he engaged in literary conflict with his adversaries, particularly Stanislav of Znojmo and Stephen Pálec. He also preached far and wide. Among the most important Czech works he now wrote were the *Expositions of the Faith, of the Decalogue, and of the Lord's Prayer,* as well as *Postil.* In 1414 he accepted the invitation of the leading Czech noble, Henry Lefl of Lazany, the chamberlain of King Wenceslas, to his castle of Krakovec.

Because of the Great Schism,* aggravated at the Council of Pisa* by the division of the West among the three popes, it was decided to hold still another council for the final settlement of the controversy. Emperor Sigismund* was the leading promoter, though Pope John XXIII unwillingly cooperated. The council was to be held in Constance* on 1 November 1414. Sigismund invited Hus to attend and promised him safe conduct for the journey both ways, even if the charges against him were not lifted. After much hesitation, and upon the urging of even King Wenceslas, Hus consented to go. He left Krakovec on 11 October and arrived at Constance on 3 November. At first he was left unmolested; but within less than a month he was treacherously lured into the papal residence and then imprisoned in a dungeon in the Dominican monastery. A panel of judges was thereupon appointed, and he was subjected to what amounted to a continuation of the previous trial for heresy. The judges endeavored to convict him of adhering to Wycliffism; but when he successfully repudiated most of the charges, Pálec extracted forty-two articles from Hus's chief work, *De ecclesia.* When the Parisian chancellor, Jean Gerson, arrived at the council, he brought with him an additional twenty charges of heresy and error.

When Pope John, who presided over the council, found himself in danger of losing his bid for confirmation in the papal office, he fled from Constance on 21 March 1415. He was seized, however, and brought back a prisoner. He was finally condemned and deposed on the basis of fifty-four charges. The council meanwhile reorganized itself. Hus, who had been transferred to the castle of Gottlieben, was now being judged by a new commission, the head of which was Peter Cardinal d'Ailly.* He was finally permitted a public hearing before the council on 5, 7, and 8 June, but was not permitted to present and defend his own views, but only to answer charges falsely formulated against him by his enemies or testified to by false witnesses. Finally d'Ailly demanded that Hus abjure the articles charged against him. In vain Hus protested that to abjure what he did not hold would be to perjure himself. He was willing to abjure if instructed from Scripture in what way his teaching was wrong. This the council refused to do. Even so, it would have preferred to secure Hus's retraction. It gave him a final formula which likewise proved unacceptable, since he still would have to admit having taught heresy and error.

The last session was held on 6 July in the cathedral before the general congregation. The final thirty articles, none of which correctly stated his own teaching, were read. Since he still refused to recant on the ground that they ascribed to him views he did not hold, he was declared an obstinate heretic, a disciple of Wycliffe, deposed and degraded from the priesthood, and turned over to the secular arm for execution. He was burned at the stake the same day on the outskirts of the city.

BIBLIOGRAPHY: M. Spinka, *John Hus and the Czech Reform* (1941), *John Hus at the Council of Constance* (1965), and *John Hus, a Biography* (1968); P. de Vooght, *L'hérésie de Jean Huss* (1960).

MATTHEW SPINKA

HUTCHINSON, ANNE (1591-1643). Early American colonist. Born Anne Marbury, she married William Hutchinson and emigrated to Massachusetts in 1634. Her particular way of expressing her Calvinistic doctrines (and her criticism of the monopoly of preaching, education, and administration by one social class) brought her into conflict with the leaders of the young colony. At first she was supported by John Cotton* and Henry Vane,* but a synod of Congregational churches denounced her supposedly antinomian views. After this the General Court of the Colony sentenced her to banishment after a travesty of a trial. In 1638 she moved with her family to Aquidneck (now Rhode Island). After her husband's death in 1642 she moved first to Long Island, then to what is now Pelham Bay, New York, where she was killed by Indians.

PETER TOON

HUTTEN, ULRICH VON (1488-1523). German Reformer. Born at Steckelberg, he was in 1499 placed in a monastery with a view to a religious vocation, but fled in 1505 and wandered from university to university, studying the classics and humanist writings. In 1515 he made a bitter attack on Duke Ulrich of Württemberg who had murdered the head of his family, Hans von Hutten, and in 1517 he settled permanently in Germany in the service of the archbishop-elector of Mainz. Hitherto a humanist scholar, he was suddenly caught up in enthusiasm for the Reformation and the freeing of Germany from papal control. Bitter ironical attacks on the papacy led to an order of arrest from Rome in 1520 and his dismissal from the elector's service. He fled at first to the castle of Franz von Sickingen, but was forced later to remove to Schlettstadt, Basle, and Mühlhausen, all of which towns refused to receive him. In 1522, afflicted by disease and poverty, he approached Zwingli,* who secured him refuge on an island till his death.

Hutten is a puzzling figure whose precise influence on the course of the Reformation has been hotly debated by historians. Undeniably he sought the political emancipation of Germany rather than her spiritual renewal, advocating what is often called the "Knights' Reformation," i.e., an alliance of the German nobility and free cities against the princes, an impossible ideal rendered quite abortive as early as 1520. But he was not without spirituality and derived from Luther not only inspiration to address his German audience in its native tongue, but also those evangelical sentiments which characterize his later works.

See H. Holborn, *Ulrich von Hutten* (1929; rev. ET 1937) and T.W. Best, *Humanist Ulrich von Hutten* (1969).

IAN SELLERS

HUTTERITES. Anabaptist* sect. They first emerge in Moravia in 1529; reorganized by Jacob Hutter in 1533, they were able despite their leader's martyrdom in 1536 to develop their distinctive ideas, in particular their pacifism and Christian communism, in the comparative peace and security of Moravia. Until 1599, in fact, they enjoyed their "golden period," expanding into Slovakia and building up about a hundred *bruderhofs*, or farm colonies, with a membership of about 25,000. The Counter-Reformation at last caught up with them in the person of the persecuting Cardinal Franz von Dietrichstein, and their discomfiture was completed by the Catholic victory at the battle of the White Mountain (1620).

Moravia being lost to them, they retreated to Slovakia and Transylvania where despite Turkish invasions and Jesuit harassment they held out for

150 years, producing a rich devotional literature which is still the basis of their worship and witness. Renewed and vicious persecution fell upon them during the reign of Maria Theresa (1740-80), but in 1767 the rump of the sect, now confined to Transylvania, crossed the mountains into Walachia and in 1770 removed again to the Ukraine. In Russia they flourished under such leaders as Johannes Waldner (1794-1824), but the introduction of military conscription in 1870 determined them to emigrate to the USA, where they settled mainly in South Dakota, Some again emigrated in 1917 when their pacifism proved unpopular, this time to Canada. They now number about 7,500 in the USA, still practice community of goods, learn German, cherish their ancient manuscripts, and maintain hostility to most forms of modern culture.

BIBLIOGRAPHY: R. Friedman, *Hutterite Studies* (1961); P.K. Conkin, *Paths to Utopia* (1964); J.A. Hostetler and G.E. Huntington, *Hutterites in North America* (1966); V. Peters, *All Things Common* (1966); J.W. Bennett, *Hutterian Brethren* (1967). IAN SELLERS

HYACINTH (1185-1257). "Apostle of the North." Nobly born Jacek Odrawaz (the former name was corrupted by a later writer), he studied at Cracow and probably Bologna. As a priest he entered the Order of Preachers at Rome in 1217/18. Having accompanied his uncle who was to be consecrated in Rome as bishop of Cracow, he is said to have witnessed St. Dominic's miracle of raising the dead. The earliest biography was written a century after Hyacinth's death by Stanislaus of Cracow. Later accounts are not reliable. Hyacinth had an important part in the extensive Dominican missionary efforts, having founded several houses of the order—e.g., at Cracow, Danzig, and Kiev—as well as making journeys of thousands of miles on foot to preach in Scandinavia, Lithuania, Bohemia, Greece, Russia, Tibet, and his native Poland. Miracles of crossing rivers dryshod, restoring the blind, and raising the dead have been attributed to him. Hyacinth was canonized in 1594. C.G. THORNE, JR.

HYMNS. A Christian hymn is a song, normally metrical and strophic, used in worship. Augustine's requirement that it be "with praise of God" would outlaw hymns of meditation, description, exhortation, or teaching. An ideal hymn has something definite to communicate, is scriptural, poetic yet simple and singable, theistic, preferably christocentric, orthodox, and truly ecumenical. Although a good tune is important, this should carry the words, and not vice versa. This article treats mostly hymns popular in Britain and America.

The NT shows that the apostlic church sang hymns. The Psalter was soon supplemented by the canticles (*Magnificat,** etc.) of Luke 1, 2, and the doxologies, e.g., Luke 2:14. The younger Pliny* (C.A.D. 115) reported in a letter to Trajan that Christians sang "a song to Christ as a god." This might, however, have been a liturgical recitation. The earliest hymn whose full text has survived is one used at a lamplighting ceremony (C.A.D. 200 or earlier) and translated from the Greek by John Keble*: "Hail, gladdening Light." Another Greek hymn (fourth century or earlier) is known to us, via Latin, as the *Gloria in excelsis*. Most other Greek hymns in our collections were translated by J.M. Neale.* Latin hymnody can be traced back to the fourth century. The *Te Deum,** possibly by Niceta, bishop of Remesiana (d.c. 414), used to be ascribed to Ambrose (d.397), the anti-Arian bishop of Milan, to whom the Western Church owes the recognition of hymns as an integral part of public worship and indirectly the invention of "Long Meter." From the fourth to the eleventh centuries there were many Latin hymns, mostly translated by Neale. The authorship is not known of the originals of "Jesu, the very thought of Thee" (Caswall) and "Come, Holy Ghost, our souls inspire" (J. Cosin). There are over 150 versions of *Dies Irae* (Thomas of Celana, thirteenth century), the greatest of the medievel Sequences (hymns sung before the Gospel at Mass).

During the two centuries after the Reformation there was, largely owing to Calvin,* no book of hymns for use in the Church of England. Their place was largely taken by the metrical Psalms, notably the *Whole Booke of Psalmes Collected into English Metre* (1562), by Thomas Sternhold* (d.1549) and John Hopkins (d.1570), and printed by John Daye. Only one hymn associated with this is now used, the "Old Hundredth": "All people that on earth do dwell" (ascribed to William Kethe, d.1594?). This "Old Version" of the Psalms held sway until 1696 when *A New Version* by Nahum Tate* and Nicolas Brady* was published. To this we owe "As pants the heart" and "Through all the changing scenes." The books co-existed until c.1870. The *Scottish Psalter,* still used by Scottish Presbyterians, dates from 1650.

In 1623 there appeared George Wither's *Hymnes and Songs of the Church,* the first attempt at a comprehensive English hymnbook; it had little success. Hymns have been taken or adapted from Herbert* ("Let all the world"), Milton* ("Let us with a gladsome mind"), Baxter* ("Ye holy angels bright"), Bunyan* ("Who would true valour see"), and Addison* ("When all Thy mercies"). Samuel Crossman (1623-84) gave us "My song is love unknown," and Thomas Ken,* "Awake my soul" and "Glory to Thee, my God, this night." Late in the seventeenth century hymns began to be freely written, and Dissenters began to use them in congregational worship. In 1671 hymns were sung at Broadmead Baptist Church, Bristol, and in 1673 Benjamin Keach published a collection of Communion hymns.

The Independent, Isaac Watts,* threw open the door by publishing in 1707 his *Hymns and Spiritual Songs* and in 1719 *The Psalms of David.* He made David speak "like a Christian" (cf. Ps. 72 with his paraphrase, "Jesus shall reign"). In over 600 hymns, many still in use (e.g., "O[ur] God, our help in ages past" and "When I survey the wondrous cross"), he expressed wonder, praise, and adoration at all aspects of Christian experience. His co-religionist, Philip Doddridge,* composed about 370 hymns, notably "Hark, the glad sound," "My God, and is Thy table spread?" and "O God of Bethel."

The Collection of Psalms and Hymns, compiled by John Wesley* (1737), was for use in the Church of England. John edited many subsequent collections, mostly consisting of some of the 6,000 or so compositions by his brother Charles Wesley* and his own thirty-three paraphrases from, e.g., Gerhardt,* Scheffler,* Tersteegen,* and Zinzendorf.* The definitive *Collection of Hymns, for the Use of the People called Methodists* (1780, with supplements 1831 and 1876) contain such hymns as Charles's "And can it be," "Hark! the herald angels sing," "Jesu, Lover of my soul," "Love divine," and "O for a thousand tongues." Whereas Watts and Doddridge freely paraphrased Scripture, Charles Wesley also paraphrased the Prayer Book and versified Christian doctrine and experience.

The Wesleys were not the only hymnwriters of the Evangelical Revival. John Byrom* is remembered for "Christians, awake," A.M. Toplady* for "Rock of ages." In 1779 John Newton* and William Cowper* produced the *Olney Hymns*, including "Glorious things of Thee are spoken" and "There is a fountain fill'd with blood."

Thomas Cotterill's *Selection of Psalms and Hymns* (8th ed., 1819), which led to the quasi-legalization of hymn-singing in the Church of England, contained many hymns by James Montgomery.* In 1820 Heber* failed to get his MS collection authorized for use in the Church of England. When published (1827), his *Hymns* [etc.], which introduced us to H.H. Milman* ("Ride on! ride on in majesty!"), led a movement toward a literary type of hymn. H.F. Lyte* is remembered for "Abide with me."

The Oxford Movement (1833 onward) revived an interest in Latin and Greek hymns, which Neale,* Caswall,* and John Chandler (1806-76) sought to satisfy. Many Tractarians wrote original hymns: notably, Keble, whose *Christian Year* appeared in 1827; J.H. Newman,* whose "Lead, kindly Light" and "Praise to the Holiest" are well known; F.W. Faber,* whose compositions tended toward sentimentality; and Mrs. C.F. Alexander,* who expounded the Creed to children in "All things bright and beautiful," "Once in royal David's city," and "There is a green hill."

The 600 or so hymns of the Presbyterian Horatius Bonar* included "Fill Thou my life," "Here, O my Lord," "I hear the words of love," and "I heard the voice of Jesus say." Thomas Kelly,* the Church of Ireland minister whose evangelical preaching cost him his living, wrote some 760 hymns, such as "Look, ye saints, the sight is glorious," "The Head that once was crowned with thorns," and "We sing the praise of Him who died." Brethren writers should be better known: E.F. Bevan*; J.G. Deck (1802-84), "O Lamb of God, still keep me"; Sir Edward Denny (1796-1889), "Light of the lonely pilgrim's heart"; and Alexander Stewart (1843-1923), "Lord Jesus Christ, we seek Thy face."

Among many other nineteenth-century British hymnwriters were Henry Alford,* "Come, ye thankful people"; Thomas Binney*; John Ernest Bode (1816-74), "O Jesus, I have promised"; Matthew Bridges (1800-94) and Godfrey Thring (1823-1903), "Crown Him with many crowns"; W.C. Dix (1837-98), "As with gladness"; James Edmeston (1791-1867), "Lead us, Heavenly Father"; Charlotte Elliott (1789-1871), "Just as I am"; Richard Mant (1776-1848), "Bright the vision"; George Matheson*; J.S.B. Monsell (1811-75), "Fight the good fight"; E.H. Plumptre (1821-91), "Thy hand, O God, has guided"; C.G. Rossetti*; W.C. Smith, "Immortal, invisible, God only wise"; S.J. Stone (1839-1900), "The Church's one foundation"; H.K. White (1783-1806), "Oft in danger"; William Whiting (1825-78), "Eternal Father, strong to save"; and C. Wordsworth.*

Some compilers of Anglican hymnals contributed hymns. E.H. Bickersteth,* whose *Hymnal Companion to the Book of Common Prayer* (1870) is still used by some Evangelicals, wrote "Peace, perfect peace" and "Till He come." The 1871 edition of *Church Hymns* was edited by John Ellerton* and W. W. How*; the 1903 edition is still used. An extreme product of the Oxford Movement was *The Hymnal Noted* (1852-4) by J.M. Neale.

A more moderate product was *Hymns Ancient and Modern* (1861). The comprehensive quality of the book eventually won it favor with most Anglicans. Although not an official collection, the total sales of all editions were by 1960 about 150 million.

In the 20th century excellent hymns have been written by G.W. Briggs (1874-1960), "God, my Father, loving me"; Cyril Alington (1872-1955), "Good Christian men, rejoice and sing"; Timothy Rees (1874-1939), "O crucified Redeemer"; Frank Houghton (1894-1972), "Thou who wast rich beyond all splendour"; and Timothy Dudley-Smith (1926-), "Tell out, my soul, the greatness of the Lord."

More typical of the times has been the study of hymns (see bibliography) and compiling of hymnals (by, e.g., Bridges and Percy Dearmer* and most denominations). Recent interdenominational collections are *The B.B.C. Hymn Book* (1951); *Christian Praise* (1957); *The Cambridge Hymnal* (1968); *Hymns for Church and School* (1964) and *Youth Praise* (1966-69).

The earliest American Psalter was *The Bay Psalm Book* (1640). The earliest American hymns were Wesleyan or Calvinist, notably "Great God of wonders" (Samuel Davies*) and "I love Thy kingdom, Lord" (Timothy Dwight*). The Kentucky revival (1797-1805) inspired "negro spirituals" full of longing for release from slavery, e.g., "Swing low, sweet chariot."

Most American nineteenth-century hymnody reflected a highly literary, but liberal or Unitarian outlook: J.W. Chadwick (1840-1904), "Eternal Ruler"; Oliver Wendell Holmes (1809-94), "Lord of all being"; F.L. Hosmer,* "Thy kingdom come!"; Julia Ward Howe,* "Mine eyes have seen the glory"; Samuel Johnson (1822-82), "City of God," the friend of the hymnodist, Samuel Longfellow (1819-92); W.P. Merrill (1867-1954), "Rise up, O men of God"; E.H. Sears (1810-76), "It came upon the midnight clear"; and J.G. Whittier,* "Immortal love" and "Dear Lord and Father."

There were exceptions: G.W. Doane (1799-1859), "Thou art the Way"; George Duffield

(1818-88), "Stand up, stand up for Jesus"; Phillips Brooks,* "O little town of Bethlehem"; and, especially, Ray Palmer (1808-87), "Jesus, these eyes have never seen," "Jesu, Thou joy of loving hearts," and "My faith looks up to Thee."

From 1870 onward D.L. Moody* and Ira D. Sankey* inspired many new compositions. Although their *Sacred Songs and Solos* grew from a sixpenny pamphlet (1873) to a book of 1,200 pieces (1903), its restricted range (ideal for the original revival meetings) disqualifies it from general congregational use. Some of the better hymns have been taken into standard collections. Even the Anglo-Catholic *English Hymnal* contains five examples, including "There were ninety and nine" by E.C. Clephane (1830-69). The most extensive contributors to what Americans know as the *Moody and Sankey Hymn Book* or *Gospel Songs* are Fanny Crosby*; P.P. Bliss*; F.R. Havergal*; and D.W. Whittle (1840-1901).

Among more recent American hymnals are those for the Congregational (2nd ed., 1958), Methodist (1935), Episcopal (1940), Presbyterian, etc. (1955), Lutheran (2nd ed., 1958), and the Evangelical and Reformed (1941) churches.

Although German hymns had been translated at the Reformation (Miles Coverdale's *Goostly Psalmes and Spirituale Songes*, c.1536) and later by the Moravians and J. Wesley, the nineteenth century produced the most translators: J.W. Alexander (1804-59) (of Gerhardt); E.F. Bevan; J.L. Borthwick* (and Sarah Findlater); S.A. Brooke (1832-1916) (of Joseph Mohr*); Miss J.M. Campbell (1817-78) (of Claudius); Carlyle and F.H. Hedge (1805-90) (of Luther's *Ein' feste Burg*); F.E. Cox*; Richard Massie (1800-87) (of Luther* and K.J.P. Spitta*); William Mercer (1811-73) (*Church Psalter and Hymn Book*, 1854); and Catherine Winkworth.*

Other modern languages represented in hymnals are (translators only given): Danish ("Through the night of doubt and sorrow," Sabine Baring-Gould*); French ("Thine be the glory," R.B. Hoyle, 1875-1939); Indian (Marathi) ("One who is all unfit to count," Nicol Macnicol, 1870-1952); Irish ("I bind unto myself," C.F. Alexander); Italian (Caswall); Russian ("O Lord my God!" chorus: "How great Thou art!" Stuart K. Hine, 1899-); and Welsh ("Guide me, O Thou great Jehovah," Peter Williams, 1722-96).

Missionary hymnals include *Hymns of Universal Praise* (Shanghai, 1948); London Missionary Society, *Dihela tsa Tihelo ea Modimo* (Bechuanaland, 16th ed., 1951); the *E[ast] A[sia] C[hristian] C[onference] Hymnal* (Tokyo, 1964); and the *Treasury of Praise* (Taiwan, 2nd ed., 1967), for Chinese-speaking Brethren assemblies in the Far East. *Cantate Domino* (World Student Christian Federation) contains (Geneva, 3rd ed., 1951) 120 multilingual hymns.

See also MUSIC, CHRISTIAN; CAROL; and entries under individual composers and hymnwriters.

BIBLIOGRAPHY: See bibliographies in A. Pollard, *English Hymns* (1960), and E.R. Routley, "Hymn," *Encyclopaedia Brittanica* (1973). Other works: H.W. Foote, *Three Centuries of American Hymnody* (1940 new ed., 1968); G. Sampson, "Century of Divine Songs," *Proceedings of the British Academy*, 29 (1946), pp. 37-64; A.E. Bailey, *Gospel in Hymns* (1950); W.J. Reynolds, *Survey of Christian Hymnody* (1963); C. Northcott, *Hymns in Christian Worship* (1964); E.R. Routley, *Hymns Today and Tomorrow* (1964); T.B. McDormand and F.S. Crossman, *Judson Concordance to Hymns* (1965); C.S. Phillips and L.H. Bunn, "Hymn," *Chambers' Encyclopaedia* (1966); C.J. Allen, *Hymns and the Christian Faith* (1966).

JOHN S. ANDREWS

HYPOSTATIC UNION. The doctrine of the substantial union of the divine and human natures in the one person *(hypostasis)* of Jesus Christ, formulated by Cyril of Alexandria* in opposition to Nestorius. He described the union as "natural" *(kata physin)* or "hypostatic" *(kath' hypostasin)*. The doctrine was formally accepted at the Council of Chalcedon (451), though the phrase "hypostatic union" was not used, in the words "the property of each nature being preserved and coalescing in one 'prosopon' and 'hypostasis.'"

See INCARNATION.

HYPSISTARIANS (Hypsistians). An obscure sect of the fourth century, probably confined to Cappadocia. Nothing is known of them apart from what we know from Gregory of Nazianzus* and Gregory of Nyssa.* The sect appears to have held a syncretistic doctrine, containing elements derived from heathen, Christian, and Jewish sources. It was strictly monotheistic, rejecting both polytheism and the doctrine of the Trinity, and worshiping God who was symbolized by fire and light, and referred to as the Almighty and the Most High. The sect rejected sacrifice and every outward form of worship, holding that adoration was a purely spiritual act. But paradoxically, it adopted the observance of the Jewish Sabbath and the Levitical prohibition of certain foods. Members of the sect were few in number and of little influence even in Cappadocia. The father of Gregory of Nazianzus (also called Gregory) had been a member, but was converted through the influence of his wife Nonna.

DAVID JOHN WILLIAMS

I

I AM MOVEMENT. An American organization founded in 1934. Its views include an eclectic assortment of beliefs and practices drawn from Hinduism, Theosophy, Unity, Spiritualism, and the American fascist Silver Shirts, in addition to the special revelations supposedly given to Guy W. Ballard.* I AM teaches that human beings remain subject to continual transmigrations of the soul within history unless they are cleansed perfectly by heavenly light and ascend from this world in full harmony with the eternal I AM Power. I AM is also an intensely patriotic movement which views America as under the special guidance and protection of St. Germain. It is extremely conservative on social and economic issues and sponsors an auxiliary organization called "Minute Men of St. Germain" which crusades against Communism, spy activities, labor unions, and everything believed to threaten the American way of life. Since the federal trials of I AM leaders in 1940-41 on fraud charges, the movement has become secretive. Serious inquirers are enrolled in study groups and cautioned not to disclose information to outsiders. Fully committed members are expected to abstain from certain foods and drinks, tobacco, card playing, and all sexual activities even within marriage in order to purify themselves. HARRY SKILTON

IBAS (d.457). Bishop of Edessa from 435, he is best known for the letter he sent in 433 to Bishop Mari(s) in Persia. This reveals that he took a mediating position between Nestorianism and the views of Cyril of Alexandria. He also helped to translate into Syriac the writings of Theodore of Mopsuestia, the Antiochene theologian. Because of supposed Nestorianism, he was deposed by the "Robber Synod" of Ephesus in 449. Two years later his orthodoxy was vindicated by the Council of Chalcedon. His epistle to Mari was condemned by Emperor Justinian and the Fifth Synod of Constantinople (553) (see THREE CHAPTERS CONTROVERSY). Only a Greek translation of the letter has survived; all his other writings are lost.

PETER TOON

IBN GABIROL, SOLOMON BEN JUDAH, see AVICEBRON

ICELAND. The first Christians in Iceland were Celtic monks who arrived about 800, after having been introduced to Christianity in Britain. Early missionary activity is attributed to Thorvaldr, an Icelandic viking, and the Saxon bishop Frederick, in the late tenth century. Although their efforts proved abortive, this preparatory work laid the groundwork for further missionary labor under the Norwegian king, Olaf Tryggvason, who introduced Christianity as the national religion in the year 1000.

In 1056 Isleifur Gizurarson became the first native bishop, and his ancestral estate at Skaholt thereafter became the episcopal residence. Isleifur's son, Gizur, as the succeeding bishop (1082-1118), established a second see at Holar. Due to the subservience of these sees to the civil powers, the state of the priesthood declined morally and intellectually. Because of political pressures and the attempts by Norwegian and other foreign bishops to centralize the organization of the church and its estates, the Icelandic Church experienced considerable conflict and suffering. Monastic activity declined; intellectual pursuits went into eclipse; the level of popular piety concerned neither people nor priest. Only the reviving breezes of the Lutheran Reformation were to bring new life to this disjointed religious situation.

Since Iceland had come under Danish control in the late fourteenth century, it fell to the Lutheran King Christian III to introduce the new doctrine to the Icelandic Church. Aided by the biblical scholar Oddur Gottskalksson, who translated the NT into the native language, Christian III declared to the nation through its legislature that the Lutheran system should be adopted. His most determined opposition came from Bishop Jon Aresson* of Holar, who, resorting to violence to save Skaholt for the pope, was himself imprisoned and beheaded.

A succession of able Lutheran bishops appeared after 1540, and the Reformation entered a constructive period. The sees of Holar and Skaholt were more happily united under the energetic bishop Gudbrandur Thorlaksson (1570-1627); the next two centuries saw the zenith of Lutheran preaching, hymnody, and devotional literature. The first complete translation of the Bible appeared in 1584.

In 1801, due to rationalistic inroads, and led by Magnus Stephensen, Holar and Skaholt were merged into one diocese located at Reykjavik; the Lutheran hymnal and service were altered to reflect the new thought. This liberalism has been perpetuated in the current century by the church's bishops and theological faculty.

See J. Helgason, "Die Kirche in Island," in *Ekklesia* (ed. F. Siegmund-Schultze, 1937), and J.C.F. Hood, *Icelandic Church Saga* (1946).

H. CROSBY ENGLIZIAN

ICON; IKON (Gk. *eikōn* "image"). Traditionally very popular in the public and private worship of members of Orthodox Churches (e.g., in Greece and Russia), icons are flat images of Christ, the Virgin Mary, or a saint. They usually take the form of wooden pictures painted in oils; some have ornate decoration and some are made of ivory or in mosaic. Their usage may be traced back to the fifth century; during and after the Iconoclastic Controversy* in the eighth and ninth centuries it was much intensified. To the icons is given full veneration—genuflexions, incense, etc. They are believed to be the channel through which the divine blessing or healing comes to the faithful; for those who use them this view is confirmed by the stories of miracles associated with them—e.g., that of the *Theotokos** in the monastery of the Abramites at Constantinople. The Western Church has never widely used icons, but there is a famous one, that of Our Lady of Perpetual Succor at Rome. PETER TOON

ICONOCLASTIC CONTROVERSY (Gr. *eikonoklastēs,* "image-breaker"). The dispute involving church and state over the presence of paintings, mosaics, and statues in churches, in the period from 717 to 843. Though early councils (e.g., that of Elvira) had prohibited pictures in churches, their usage became widespread between 400 and 600. It was claimed that pictures of the martyrs would teach the illiterate to follow their good examples. The veneration of pictures, however, had its opponents such as Epiphanius.*

In 717 Leo III,* the Isaurian, acceded to the imperial throne and in 725 legislated against image-worship. His motivation is not clear, but it was possibly affected by his knowledge of Muslim opposition to images and by a desire to gain greater control over the church. His legislation was rejected in Rome by Gregory II* as heretical, and the latter's successor, Gregory III, called a council of ninety-five bishops in 731 to confirm this position. John of Damascus* also wrote against Iconoclasm. This action did not stop Leo and his successor Constantine V in their crusade against images. In 753 Constantine summoned a council to meet at Hieria, near Chalcedon; this resulted in a full condemnation of images by the 338 bishops present. Iconodules were accused of circumscribing the divinity of Christ and of confusing his two natures by the veneration of pictures of Him.

The Seventh Ecumenical Council of Nicea (787), however, guided by the empress Irene and the patriarch Tarasius* (both iconodules) reversed the decisions of Hieria. Icons were justified by reference to the tradition of the church through quotations from the Fathers. When Leo the Armenian became emperor in 813, though opposed by the patriarch Nicephorus,* he reverted to the policy of Leo III, and at an assembly of bishops in Sancta Sophia in 815 had the decrees of Hieria restored. His successors, Michael and Theophilus, continued the policy of Iconoclasm, but after the latter's death his widow, Theodora, restored the use of icons. She caused a "feast of orthodoxy" to be instituted on the first Sunday in Lent in 843, and arranged the return of the exiled iconodules. This marked the end of imperial support for Iconoclasm.

The writings and records of the councils of the Iconoclasts were destroyed, thus our knowledge of them is drawn from what their opponents said. The theological significance of the controversy was threefold: it caused a development of thinking about the use of icons and of sacramental theology; it emphasized the importance of tradition in the church; and (in the West) it strengthened the papacy. Iconoclasm has often reappeared in European history, with Carlstadt,* Luther's colleague, an example of a fervent Iconoclast.

BIBLIOGRAPHY: E.J. Martin, *A History of the Iconoclastic Controversy* (1930); E. Bevan, *Holy Images* (1940); P.J. Alexander, *The Patriarch Nicephorus of Constantinople* (1958); E. Gilson, *The Arts and the Beautiful* (1965). PETER TOON

ICONOGRAPHY, see ART, CHRISTIAN

IGNATIUS (d.98/117). Bishop of Antioch. He is known of almost exclusively through seven letters whose authenticity was established in the seventeenth century, largely by James Ussher,* and vindicated in the nineteenth, chiefly by J.B. Lightfoot.* While traveling under armed guard to be executed in Rome, preceded by other Syrian Christians, he was welcomed by Polycarp* and delegates from other churches at Smyrna, whence he wrote to the churches at Ephesus, Magnesia, Tralles, and Rome. Later from Troas he wrote to the Philadelphian and Smyrnaean congregations and to Polycarp. His death at Rome is asserted about 135 by Polycarp, who had earlier collected his letters for the Philippian church.

In six of his letters Ignatius attacks a heresy compounded of Docetic, Judaistic, and perhaps gnosticizing features, and advances the antidote of adhesion to the bishop, presbyters, and deacons. Probably still the only monarchical bishop in Syria, and the earliest witness to the threefold ministry, he magnifies the bishop's unifying authority as representing God (apostolic succession is unmentioned).

Ignatius displays prophetic qualities (he calls himself Theophoros, "God-bearer" or "God-borne," perhaps his baptismal name), and is a colorful and vigorous writer, influenced by Judeo-Christian and Gnostic conceptions. The flattering letter to the Romans (silent on monepiscopacy) pleads with them to do nothing to thwart his passion for martyrdom, by which, in language indebted to Maccabean and *imitatio Christi* ideals, he will "attain to God," become at last a disciple, and offer a ransom for the church.

Ignatius falls heir to apostolic tradition (explicitly Pauline rather than Johannine), but is led by personal and ecclesiastical circumstances into dramatic, even bizarre emphases. Against Docetism* he stresses Christ's true humanity and identifies it with the healing food of the Eucharist, a further focus of congregational unity.

BIBLIOGRAPHY: J.B. Lightfoot, *Apostolic Fathers,* Part II (3 vols., 2nd ed., 1889); C.C. Richardson, *The Christianity of Ignatius of Antioch* (1935); study by J. Moffatt in *Harvard Theological Review* 29 (1936), pp. 1-38; F.A. Schilling, *The Mysticism of St. Ignatius of Antioch* (1932); H.W. Bartsch, *Gnostisches Gut und Gemeindetradition bei Ignatius von Antiochien* (1940); V. Corwin, *St. Ignatius and Christianity in Antioch* (1960); M.P. Brown, *The Authentic Writings of Ignatius* (1963); T. Camelot in *Sources chrétiennes* 10 (4th ed., 1969).

D.F. WRIGHT

IGNATIUS, FATHER, see LYNE, J.L.

IGNATIUS OF LOYOLA (Iñigo López de Loyola) (1491-1556). Spanish ecclesiastic reformer and mystic, founder and first general of the Society of Jesus.* He was born in the Basque province of Guipúzcoa in NW Spain. Little is known of his youth. His father died when he was about fourteen, whereupon he attached himself to the court of King Ferdinand, pursuing a military career. In 1521, defending Spain's claim to Navarre against France, while at the fortress of Pamplona, Loyola was struck by a cannon ball. One leg was badly mangled, ending his military career. While recuperating at the castle of Loyola he chanced to read Ludolph of Saxony's *Life of Christ.* Inspired to become a soldier for Christ, he vowed lifelong chastity and soon entered the monastery at Manresa. Here he spent nearly a year in ascetic practices, experienced several mystical visions, and composed the essence of that great manual of spiritual warfare and conquest, the *Spiritual Exercises.* After a pilgrimage to Jerusalem in 1523, he commenced his schooling, culminating it with the M.A. at Paris University (1535).

At Paris he gathered around him a band of associates who worked through the *Exercises* and became fired with Loyola's ideal. After graduation, Loyola and six dedicated colleagues (Nicolas de Bobadilla, Peter Faber, Diego Laynez, Simon Rodriguez, Alfonso Salmeron, and Francis Xavier), at St. Mary's Church in Montmartre, vowed together a life of poverty, chastity, and a career of service in the Holy Land or, failing that, of unreserved service to the pope. They met the following year in Venice, found their way to Jerusalem blocked by war, and finally won a favorable response from Paul III, leading to their approval as an Order of the church. This was officially confirmed by the bull *Regimini militantis ecclesiae* (1540).

In early 1548 Loyola was unanimously chosen "general" for the society. He provided the organization of the group by the famous *Constitutions,* which outlined a paramilitary structure with obedience, discipline, and efficiency the key ideas. Heavy stress was laid also on education and preparation, and Loyola founded in 1551 the Roman College as a model. Based on these ideals the Jesuits* took the lead in the Catholic reform movement. Loyola was beatified in 1609, canonized in 1622.

BIBLIOGRAPHY: *Autobiography* (1900 and 1956); P. Dudon, *St. Ignatius of Loyola* (1950); J. Brodrick, *St. Ignatius Loyola: The Pilgrim Years* (1956); *Letters to Women* (1960); R. Gleason (ed.), *Spiritual Exercises* (1964).

BRIAN G. ARMSTRONG

IHMELS, LUDWIG HEINRICH (1858-1933). German Lutheran theologian and churchman. Born in East Frisia, he served in pastorates there after completing his education. Appointed professor of systematic theology at Erlangen in 1894, he was the last major representative of the Erlangen theology. His most important work, *Die christliche Wahrheitsgewissenheit* (1901), dealt with assurance of salvation. He moved to Leipzig University in 1902 and was named bishop of the church in Saxony in 1922. He took a leading part in promoting the Lutheran union movement and presided over the first Lutheran World Congress at Eisenach in 1923. In his speeches and writings he stressed the ecumenical character of Lutheranism.

RICHARD V. PIERARD

IHS. This monogram is built by using the first three uncial (capital) letters of Jesus' name as it is written in Greek. It was overlooked that the second letter was a Greek *eta,* not a Latin *H.* (The monogram is also sometimes found as IH or IHV.) This explains why it has occasionally been incorrectly expanded to "Ihesus." There have been several other attempts to explain the etymology of this abbreviation. A popular theory regards each of the letters in the IHS monogram as the first letter of separate words instead of the first three letters in the Greek term for Jesus. Among others, the letters have been held to refer to *Iesus Hominum Salvator* ("Jesus, Savior of Men") or *Iesum Habemus Socium* ("We have Jesus as our companion").

WATSON E. MILLS

ILDEFONSUS (c.607-667). Archbishop of Toledo. From a distinguished family, he studied apparently under Isidore of Seville.* Against his father's wishes he entered the monastery at Agalia near Toledo, later becoming its abbot and founding a nunnery nearby. As abbot he served at the eighth and ninth Councils of Toledo (653, 655). From 657 he was archbishop of Toledo. Ildefonsus contributed much to Roman Catholic veneration of Mary through his *De virginitate sanctae Mariae.* His legendary encounters with the Virgin, recounted by his early biographers, became the subject of medieval art and poetry. Three other major works and two letters to Quiricus, bishop of Barcelona, survive. *De viris illustribus* demonstrates the seventh-century Spanish Church's debt to fourteen notables. Valuable insight into the medieval catechumenate and baptism is derived from *Annotationes de cognitione baptismi. De progressu spiritualis deserti* describes the believer's pilgrimage from baptism to heaven.

JAMES DE JONG

ILLUMINATI (Alumbrados). Members of a Spanish group of mystical tendencies. At the dawn of the sixteenth century Spain was touched by a religious movement which brought about renewal. The common folk, thirsty for a personal understanding of the Gospel, met in small groups to study the Bible. Bataillon contrasts this simplistic search for freedom with the intellectual aspirations of Cardinal Jiménes,* whose movement at Alcala had limited appeal. The Alumbrados, or "enlightened ones," were mystical in nature, drawing from Neoplatonism certain concepts also used by the Sadilies or Islamic mystics. A Franciscan first used the word "Illuminati" in 1494 in a letter to Cardinal Jiménes. Its doctrines seem to be known from the Inquisition,* which in 1623 condemned a wide range of Illuminist opinions. More recent scholars point to the positive leadership of Pedro Ruiz de Alcarez rather than repeat the general condemnation of an amorphous Spanish religious phenomenon of 129 years. Alcarez in his letter of 22 June 1524, written to the Inquisitors after four months' imprisonment, identifies the movement with the Franciscan Order. The key term was "love of God."

Alcarez may be called a member of the *dexados* movement, since Isabel de la Cruz initiated him into it about 1510. When Alcarez took Isabel's views on the freedom in God's love, thereby opposing merits by grace, a definite doctrinal content was added to the amorphous Franciscan illuminative way. The propositions condemned from his writings were made public on 25 September 1525. The perfection which Alcarez taught is the submission to God's will rather than an eradication of evil from the soul. Alcarez fought against the *recogidos* movement of Francisco De Osuna. This form of Spanish mysticism in the Franciscan movement sought by a method of prayer (*recogimento*) to recollect sufficient data through the senses to achieve illumination and union with God. The allegorical use of Scripture here in contrast to Alcarez assumes a natural divine light in which grace builds upon nature. This form of Illuminati thought influenced Loyola and St. Teresa. In Juan de Valdés* and Peter Martyr* such views were conveyed to Little Gidding and New England.

BIBLIOGRAPHY: A. Sánchez Barbudo, "Algunos aspectos de la vida religiosa en la Espana del siglo XVI: Los Alumbrados de Toledo" (unpublished Ph.D. dissertation, University of Wisconsin, 1953); J.E. Longhurst, "The Alumbrados of Toledo," *Archiv für Reformationsgeschichte,* XLV (1954), pp. 233-52; *idem, Luther's Ghost in Spain* (1969).

MARVIN W. ANDERSON

IMAGES. The use of images to represent Yahweh was absolutely prohibited by the Mosiac Law (Exod. 20:4, 5; Deut. 5:8, 9; Lev. 26:1) and reinforced by the prophets (Amos 5:26; Hos. 13:2; Isa. 2:8; 40:18-26). Many objects such as the ark of the covenant, the oxen that supported the bronze sea, and the cherubim in the Holy of Holies were employed to assist worship, but they were never considered objects of worship. The bronze serpent (Num. 21:8, 9) was venerated with incense for a period, but this was abolished by Hezekiah. The illegitimate use of images is epitomized in the use of the golden calves in Israel by Jeroboam I. Although they were intended as symbols for the presence of God, they soon deteriorated into objects of worship. The syncretism of Baal worship and Yahwehism brought a flood of images into both the northern and southern kingdoms. Even with the reforms of Hezekiah and Josiah, this idolatry continued until the exilic period (587-537 B.C.). After this period, the prohibition was intensified.

During the Roman occupation, the observers of the law, at the risk of death, elicited promises from the procurators and governors not to bring the standard adorned with the emporer's image into the holy city or through Jewish territory. Soon after the establishment of the early church, the use of images arose and was later justified by such leaders as Augustine and Ambrose. Opposition did arise and precipitated the Iconoclastic Controversy* of the eighth and ninth centuries. The Second Nicean Council ruled that honor paid to an image passes on to its prototype and that he who worships an image worships the reality of him who is painted in it.

Icons became an integral part of worship in the Orthodox Churches, and veneration of images was later strengthened by Thomas Aquinas in the Western Church. The use of images was strongly opposed by some of the Reformers, especially Calvin, Zwingli, and the Puritans. Following this lead, most Protestant churches continue to oppose the use of images.

See also ICON.

BIBLIOGRAPHY: Thomas Aquinas, *Summa Theologica,* III, i; W. Palmer, *An Introduction to Early Christian Symbolism* (1885); E. Bevan, *Holy Images* (1940); A.D. Lee and C.H. Pickar, *New Catholic Encyclopedia,* vol. 7, pp. 370-72 (1967).

JOHN P. DEVER

IMMACULATE CONCEPTION. In 1854 Pius IX in the papal bull *Ineffabilis Deus* stated that "from the first moment of her conception, the Blessed Virgin Mary was, by the singular grace and privilege of Almighty God, and in view of the merits of Jesus Christ, Saviour of Mankind, kept free from all stain of Original Sin." This dogma is based on a particular view of conception. It is believed that a person is truly conceived when the soul is created and infused into the body. At the moment of her animation Mary was given sanctifying grace which excluded her from the stain of original sin. Mary was redeemed at conception by Christ in anticipation of His atoning death. At the same time the state of original sanctity, innocence, and justice was conferred upon her. Thus Mary was sinless from the moment of her conception, although this did not exempt her from sorrow, sickness, and death, consequent upon Adam's sin.

No direct or categorical proof of the dogma can be brought forward from Holy Scripture, though basis is sought in such texts as Genesis 3:15; Psalm 45:12ff.; Luke 1:28, 41, 48. This basis is strengthened by the designation of Mary as the "new Eve" by the Christian Fathers such as Justin Martyr and Irenaeus. Augustine exempted the Virgin Mary from actual but not original sin. Thomas Aquinas argued against Augustine because he believed that Mary's conception was a natural one and in every natural conception original sin is transmitted. The Council of Basle in 1439 affirmed that the belief was in accordance with the Catholic faith, with reason, and with Holy Scripture. The universities followed the lead of the Sorbonne which in 1449 required its candidates to make an oath to defend the dogma. The Franciscans, Carmelites, and especially the Jesuits were staunch defenders of it.

As early as the seventh century a Feast of the Conception of Mary originated in the monasteries of Palestine. In 1476 Sixtus IV approved the feast with its own mass and office, and adopted it for the Roman Church. In 1708 Clement XI imposed it on the whole Western Church. "Immaculate" was added to the title after its promulgation in 1854. Since 1854 Eastern Orthodox theologians have rejected the doctrine as detracting from the merits of Mary's actual sinlessness. Protestants have always rejected this dogma since it appears to have no direct scriptural basis. S. TOON

IMMANUEL. A biblical name (Isa. 7:14, 8:8; Matt. 1:23), meaning "God (is) with us" in Hebrew. (The KJV spelling "Emmanuel" in Matt. 1:23 is due to the Greek form of the name.) The name is unambiguously applied to Christ in Matthew, but its function in Isaiah 7f. has been much debated; Jews traditionally have related it to Hezekiah, the crown prince of Judah, and many modern Christian scholars have similarly argued that some contemporary of the prophet Isaiah was intended. The traditional Christian view that it was a title of the Messiah is still by no means untenable.

IMMERSION, see BAPTISM

IMPANATION (Lat. *impanare,* "to impane, to embody in bread"). A description of certain theories of the Eucharist propounded in the Middle Ages and at the Reformation which sought to safeguard the Real Presence of Christ in the elements without positing the change of the natural bread and wine. Originally, in the eleventh and twelfth century, the word described the "heretical" doctrines of such people as the followers of Berengar who taught that the relation between Christ and the bread and wine was similar to that of the divine and human natures of Jesus Christ. During the sixteenth century Roman Catholics accused Luther, and Carlstadt accused Osiander, of teaching impanation. The first use of it as a technical term seems to have been by Guitmund of Aversa (d. c.1090); Alger of Liége (d.1131) also used the term when writing against transubstantiation. PETER TOON

IMPRIMATUR. The term employed to signify that the Roman Catholic authorities have approved of a book's publication. This approval must be given to all works of clergy or laity dealing with the Bible, doctrine, church history, church law, and ethics. It indicates that the censor has found in them nothing that is in conflict with the church's teachings. It is thus more negative than positive. The approval is given through the bishop of the diocese. All works—whatever the subject—written by clergy are supposed to receive this imprimatur. This practice grew up as a result of decrees of the Council of Trent (1545-63) which established a list of forbidden and expurgated books.

See also INDEX. W.S. REID

INCARNATION (Lat. *in carne,* "in flesh"). Although this term does not occur in Scripture, it is scriptural in the sense that it expresses the meaning of Scripture by teaching that Jesus of Nazareth was the eternal Word who became flesh (John 1:14). Jesus is the Son of God who, being sent by the Father, comes into this world "in the likeness of sinful flesh" (Rom. 8:3 KJV). At the heart of the Christian faith is the confession of the "mystery of godliness: He appeared in a body" (1 Tim. 3:16).

Having declared in the doctrine of the Trinity (A.D. 325) that the Father and the Son are co-eternal and consubstantial, the fathers of the church could not avoid the question: How could the eternal Son, who is equally God with the Father, so partake of our flesh as to become a man as we are men? Some (e.g., Apollinarius) suggested that the Son assumed a true body and soul, but in place of the human spirit had, or rather was, the divine Logos. Realizing that this impugned our Lord's full humanity, others (e.g., Nestorius) affirmed this humanity, but spoke of Jesus in a way that made Him virtually a distinct person from the divine Logos ("He who was formed in Mary's womb was not himself God, but God assumed him . . .").

Reacting against suggestion that the divine Son and Jesus were two persons, Cyril of Alexandria and his followers argued that, as a result of incarnation, the human and the divine were fused into one nature (Monophysitism*). After much controversy, following the lead of Pope Leo I, the church came to define the orthodox doctrine of the incarnation at the Council of Chalcedon in A.D. 451 by declaring that our Lord Jesus Christ is true God and true man *(vere Deus, vere homo)*, consubstantial with the Father in all things as to His divinity, yet in His humanity like unto us in all things, sin excepted. This one and the same Jesus Christ is known in two natures "without confusion, without conversion, without severance and without division, the distinction of natures being in no wise abolished by their union, but the peculiarity of each nature being maintained, and both concurring in one person and subsistence." This union of the human and the divine natures in one person (known technically as the "hypostatic union," from the Greek *hypostasis,* "person") is the common confession of the church, Eastern Orthodox, Roman Catholic, and Protestant.

It is not that the Definition of Chalcedon* removes the mystery of the Incarnation—one might say that it rather heightens the mystery—but it has proven remarkably effective in marking out the proper boundaries of believing thought about the person of Jesus Christ the only Mediator between God and man.

As for the terms of Chalcedonian Christology, the following should be noted. The word "nature" (Gr. *physis,* Lat. *natura*), as used by the Fathers, does not refer to the physical order which is the object of investigation in the "natural" sciences. "Nature" rather designates "being" or "reality" in distinction to "appearance." To say that Jesus Christ has a "divine nature" is to say that all the qualities, properties, or attributes by which one describes the divine order of being pertain to Him. In short, He is God Himself, not *like* God, but just *God.* So also with the affirmation that Jesus Christ has a "human nature." He is not God *appearing* as a man; He *is* a man. He is not *only* a man or *only* God; He is the God who *became* a man. He did not cease to be God when He became a man, He did not *exchange* divinity for humanity; rather He *assumed* humanity so that, as a result of the incarnation, He is both human and divine, the God-man.

As for "person" (Gr. *hypostasis*, Lat. *persona*), this term was used by the Fathers to describe our Lord as a self-conscious, self-determined *Subject*, one who designates Himself by the word "I" over against a "thou." The word *hypostasis* literally means "that which stands under," i.e., what is there, in each individual case, at the deepest level. While we must ascribe to Jesus as the Christ all the qualities which belong to the human order of being (including bodily, physical, objective being—the Word "became flesh," John 1:14), we cannot say that at the most ultimate level of His being He is a human person. He is a divine person, with a human nature. The Son of God did not assume a man's person to His own nature, but a man's nature to His own person. He is, then, a divine person who has assumed our humanity. (Were he not a divine person, He would not be the object of Christian worship, for Christians worship God only, never the creature.)

As for the personal qualities of Christ's humanity, the position generally—though not universally—held by theologians subscribing to Chalcedonian Christology is one which speaks of the "impersonal humanity" of our Lord (anahypostasy, enhypostasy). Not that there is no manifestation in Jesus of personality at the human level; rather, the thought is that this humanity, of itself, has no existence independently of the divine Person. That which is human, in Him, exists in and through the Word which is God Himself. There is, to be sure, a sense in which God is present to *all* created reality, especially in a *gracious* way in His word and sacraments. But howsoever we may conceive of this divine presence of power (creation-providence) and of grace (word-sacrament), there can be no thought of identity between God and the creature. But of the man, Jesus Christ, something absolutely unique is affirmed; this man is declared to be identical with God Himself, because He, the Person, is the "Word made flesh and dwelling among us" (cf. John 1:14). Therefore we can never think of Him as man, without at the same time thinking of Him as God.

If the incarnate Son of God unites true deity and true humanity in a personal self, then there is a communion of attributes in the person of the Mediator whereby we may speak of Him in any way that is proper to speak of God or of man (see, e.g., Acts 20:28, where the best text speaks of God's having purchased the church with His own blood). Lutheran theologians, in distinction to the Reformed, have gone further, arguing not only for a communion, but also for a "communication" of attributes (Article VIII, *Formula of Concord*). To suppose, however, as the doctrine of the *communicatio* does, that the human nature of our Lord possessed of the attributes of the divine by virtue of the hypostatic union, is to draw a conclusion that has not commended itself widely outside Lutheran circles.

Likewise, the church as a whole has rejected the Lutheran doctrine of kenosis* (from the verb *kenoō*, "to empty," Phil. 2:7) though others besides Lutherans, especially in the Anglican Communion, have accepted some form of kenoticism.

In all of these views the fundamental assumption of Chalcedonian Christology is not challenged; the Incarnation is understood to mean that a preexisting divine person, the eternal Son of God, has revealed Himself in history as the man Jesus of Nazareth. A significant shift, however, begins with Schleiermacher,* who observed that it is unfortunate that the church should speak of the union of the human with the divine in Jesus the Christ as the *act* of the person himself, rather than making it *constitutive* of the person. The "Son of God" is the subject of the union of the human and divine in Jesus, according to Schleiermacher, not the divine person who exists before the union took place.

Ultimately, for Schleiermacher, Jesus is a man "who was uniquely endowed with God consciousness"; for Ritschl, "Jesus taught a lofty morality, but in the exercise of this vocation never transgressed the limits of a purely human estimate of himself"; for Harnack, "Jesus knew God in a way no one had ever known him before." In this tradition of German liberalism there is such a radical shift away from Chalcedon toward a view of Jesus as only a man, albeit an exceptional one, that it is hardly possible to speak of "incarnation" any longer. For this reason Tillich* has put the word "incarnation" in quotation marks and declared that its traditional meaning that "God has become a man" is "not a paradoxical statement but a nonsensical one." For contemporary theologians who concur in this judgment, Jesus was a first-century Palestinian Jew, no different from other men, save that his ethical integrity and religious genius have made us all his debtors.

In contrast with this "theology of the horizontal," the Christian doctrine of incarnation affirms that in the Jesus-event God, like a "perpendicular from above," visited our planet and became a part of our history. To understand this difference, in the light of the crucial question, "Whom say ye that I am?" is to perceive how important lucid theological thought can be for Christian faith, as that thought is preserved in the historic symbols of the church.

See also CHRISTOLOGY; JESUS CHRIST.

BIBLIOGRAPHY: A. Ritschl, *Justification and Reconciliation* (1902); F.E.D. Schleiermacher, *The Christian Faith* (1928); H.E. Brunner, *The Mediator* (1947); D.M. Baillie, *God Was in Christ* (1948); P. Tillich, *Systematic Theology*, vol. II (1951); K. Barth, *Church Dogmatics*, part I, vol. 2 (1956); K. Heim, *Jesus the World's Perfector* (1961); K. Rahner, *Theological Investigations*, vol. 1, chap. 5, "Current Problems in Christology" (1961).

PAUL KING JEWETT

INCENSE. Usually regarded as a symbol of prayer ascending to God, incense has been, and is, widely used in many religious ceremonies, both Christian and pagan. The word describes both the substance used for burning and the aroma. In the Jewish temple it accompanied all sacrifices (except the sin-offering of the poor and the meat-offering of the leper). On the Day of Atonement it was solemnly burned by the high priest in the Holy of Holies. For the ingredients of incense and its use see Exodus 30:34-38; Leviticus 16:12ff.; and the Talmud (cf. Keritot 6a). Unless there is a reference in Revelation 8:3-5, there is no sure evidence of its use in Christian worship until the sixth century. This use may have risen in imitation of the custom of carrying incense in a *thuribulum* (thurible) before Roman magistrates. By the ninth century its use was widespread in both West and East. Today it is only used in solemn sung services in the West, but in the East it is used rather more frequently. Within the Church of England its use is technically illegal, but it is nevertheless used by Anglo-Catholic clergy in the service of Holy Communion. PETER TOON

INCUBATION (Lat. *incubare,* "to brood"). The practice of sleeping in a temple or sacred place for oracular purposes was common in many ancient cultures, e.g., in the worship of Ascelpius, the patron of medicine. After the fall of paganism in the Roman Empire the custom was introduced into Christianity, and so the term came to refer to the practice of sleeping in or near churches or holy places in order to receive healing or visions. Certain centers, with which a saint was connected, were believed to have special possibilities for healing—e.g., St. Michael in the church of Anaplous, near Istanbul.

INDEPENDENTS; INDEPENDENCY. In Britain in the seventeenth and eighteenth centuries Independency was a synonym for Congregationalism.* The word came into general usage in the revolutionary decade of the 1640s and was used at first in both a political and an ecclesiastical sense. In the former sense it came to represent those who believed in the provision of some form of religion toleration inside or alongside the state church; thus some members of the Long Parliament and many army officers were termed Independents. In the second sense it described those who believed in or practiced a Congregational form of church government; such people could be separatists* or ministers within the parish system who both gathered a church and preached in the parish "church."

Famous examples of conservative Independents are the Five Dissenting Brethren of the Westminster Assembly: Thomas Goodwin,* Philip Nye,* Sidrach Simpson, William Bridge,* and Jeremy Burroughes.* They taught a doctrine of the church which was halfway between separatism (Brownism) and Presbyterianism* (see further *An Apologeticall Narration,* ed. R.S. Paul, 1963). An example of a separatist Independent is Vavasor Powell,* the Welsh evangelist. After the Restoration in 1660 those who practiced the Congregational way and paedobaptism were, in the main, those who were deemed Independents. The term was not much used in the nineteenth century and has never been popular in the USA. However, the growth of the Fellowship of Independent Evangelical Churches* in Britain since 1945 has again brought the term in general use. PETER TOON

INDEX OF FORBIDDEN BOOKS. A list of books which are prohibited reading for members of the Roman Catholic Church. This practice is in keeping with a tradition which regards certain works as dangerous to the faith and morals of Catholics. Under Pope Gelasius I (492-96) a decree was issued which divided books into three categories: authentic scriptural ones, recommended works, and heretical works. Many specific works were denounced until the formal codification of forbidden books took place under Paul IV (1555-59). One of the tasks of the Congregation of the Inquisition was to compile a catalog of forbidden books, and this was published with papal approval in 1559, the first such list to bear the official title of "Index."

Due to dissatisfaction with this Index the leaders of the Council of Trent (1545-63) began a revision, but were unable to complete it and appointed a commission for this purpose. As a result the Tridentine Index or Index of Pope Pius IV (1559-65) was issued in 1564. In addition to the list of forbidden books, it contained ten guidelines for regulating censorship. Revisions of the Index include those of Sixtus V (1590), Clement VIII (1596), Alexander VII (1664), Benedict XIV (1751), and Leo XIII (1897 and 1900). The last revision of the Leonine Index took place in 1948. In 1966 at Vatican Council II,* it was declared that no further editions of the Index would be issued, and its chief value today is historical. Catholics are still, however, bound to abide by certain guidelines prohibiting the reading of books which constitute a possible spiritual danger.

See IMPRIMATUR.

See also R.A. Burke, *What Is the Index?* (1952); and H.C. Gardiner, *Catholic Viewpoint on Censorship* (1958; rev. ed., 1961).

ROBERT G. CLOUSE

INDIA. The tradition that the Apostle Thomas came to India grows no weaker in the Indian Church. There are two separate considerations. The first is the tradition held particularly by the Syrian Christians of Kerala: that Thomas came to Cranganore in A.D. 52, founded churches in seven places in Kerala, later proceeded to the east coast and indeed beyond India, and finally was martyred at Mylapore in 72. Mylapore is within the modern city of Madras, and the reputed burial place is within the present Roman Catholic cathedral at Mylapore. There is, however, no written evidence for this South India tradition earlier than a Portuguese account about 1600.

The second consideration—whether Thomas was ever in India at all, leaving aside the South India details—is different. There are early references to Thomas's being in India. But the difficulty lies in demonstrating beyond a doubt that "India" in these references corresponds with India as we know it. One example is the Syriac *Doctrine of the Apostles* (c.250), which states that "India and all its countries, and those bordering on it, even to the farthest sea, received the Apostles' hand of priesthood from Judas Thomas, who was Guide and Ruler in the Church which he built there, and ministered there." Where the reference is to "India," it perhaps refers to somewhere around S Arabia, as when Eusebius writes of Pantaenus in India (about 180), and incidentally associates Bartholomew and not Thomas with the Church there.

Regarding the *Acts of Thomas** with its allusions to the apostle, even if it indicated that Thomas was in NW India, this would in no way substantiate the South India theory, but it has a grain of credibility which contributes to modern hesitancy before dismissing the traditions. There is no inherent impossibility in the theory that Thomas came to India. The sea route, using the monsoon, was known then. Both proof and disproof are lacking.

The existence of Syriac liturgy in the Keralan churches (giving them the name "Syrian," for they are not racially distinct from the rest of the population) is sufficient proof of a link between these churches and the Near East. But in trying to trace the antiquity of the link, the same difficulty, of defining the word "India" in documents, arises. For example, there is mention of a bishop of Basra, Dudi, leaving his see and evangelizing in "India" (c.295-300), and there is the signature of John the Persian to the Creed of Nicea (325) on behalf of churches "in the whole of Persia and in the great India." Keralan tradition is of an immigration of Syrian Christians in 345, to strengthen an already existing if ailing church, and such an exodus would be likely from the Sassanid Empire at that period of persecution of Christians. Another tradition speaks of an immigration in 823; but long before that Cosmas the "Indian Sailor" (c.552) writes of Christians "in the land called Male [Malabar] where the pepper grows" and of a bishop appointed from Persia at a place called "Kalliana," which was probably equivalent to "Kalyan" (beside modern Bombay).

This link, whenever it began, determined certain things of importance about the Keralan Church. There was a distinct foreignness about having worship in Syriac and depending on a supply of foreign bishops; there was an indirect link with the patriarch of Antioch, via the Church of the Sassanid Empire, but that link would be severed when the Eastern Church stressed its autonomy; and when the Eastern Church adopted a Nestorian confession, the Indian Church would be Nestorian also, from the seventh century until the Portuguese period. Further, when the mother church suffered losses, and in parts vanished altogether, during the Muslim era, its dependent church in India became weak and neglected. It was, in extent, a church in Kerala with an outpost on the east coast at Mylapore, the Thomas shrine. Whatever had existed further north in India would hardly survive Muslim invasions.

Marco Polo was in India in 1288 and 1292 and was shown a tomb said to be that of St. Thomas. From about this time, friars were calling at India on their way to China, and one of these, John of Monte Corvino,* made converts and recommended missions. In 1321 four Franciscans who landed at Thana, near modern Bombay, were martyred; but their companion, a French Dominican called Jourdain de Severac, who was evidently assigned to work in India, remained in the area and baptized many people. The friendly Nestorian Christians whom Jourdain met were a neglected, ignorant people; and, of course, they were sadly heretical. In 1330 Jourdain was sent back to India as bishop of Quilon (in Kerala), and he carried a letter from the pope urging the Syrian Christians to submit to Rome. This was the first outright papal claim to authority, and although we know little of Jourdain's subsequent ministry, a Latin Rite church was established in Kerala.

The Portuguese discovery of the Cape route to India (1498) and settlement at Goa (1510) changed everything. These were men with a religious mission, for the pope had granted the kings of Portugal perpetual right of church patronage in the east, the *Padroado.* By 1534 a bishop was at Goa, head of a powerful church, but not a church whose life could in many aspects have pleased Francis Xavier,* founder of Jesuit missions in the East, when he arrived in 1542. Xavier's character contrasted with the general picture of that Portuguese period, earning him the respect of almost all (despite his recommending the Inquisition* in India). At Cape Comorin he instructed and established the neglected converts who had been won from the pearl-fisher community, and he laid foundations for Jesuit work in India.

To a degree, Roman Catholic success corresponded with Jesuit fortunes. They were agents in establishing authority among the Syrians of Kerala, becoming somewhat less than popular in the process (see MALABAR CHRISTIANS and DIAMPER, SYNOD OF). They responded to an invitation from Akbar, the Mogul emperor, and conducted a mission at his court which actually proved fruitless but might have had great consequences. In the south an Italian Jesuit, Robert de Nobili,* led the way in trying to overcome the foreign appearance of Christianity by means of "accommodation" to Hinduism. He dressed and lived like a Brahmin *sannyasi,* dissociated himself from the existing church at Madurai, and when he made converts permitted them to remain separate from the other Christians and to keep their outward signs of Hinduism such as the sacred thread. They thus preserved caste, which to Nobili was simply social custom. A bitter church controversy arose over these methods, and Nobili was withdrawn from Madurai in 1645. Eventually the method of "accommodation" was condemned. The Jesuits themselves were suppressed as an order. Their lot was to be succeeded by others; but, again, it might be said that the Roman recovery from a period of decline coincides with the restoration of the Jesuits in the nineteenth century.

If a difficulty exists about defining India, one might sometimes have pause in determining what was Christianity. From a Protestant viewpoint, it was the arrival of Protestant missions which gave to India the Bible; and it was the belief of many of these early missionaries that they were introducing biblical Christianity, in contrast with what stood for Christianity in the existing churches. The first Protestant missionaries were the German Lutherans, B. Ziegenbalg* and H. Pluetschau,* sent personally by Frederick IV of Denmark to his trading territory at Tranquebar in South India. They arrived in 1706, royal missionaries about to enjoy a far from royal welcome at the hands of the Danish authorities, not even wanted by the chaplains who cared for the European community; but it was nevertheless a great day in the story of Christianity in India and, to single out what is basic to the Protestant period, by 1714 Ziegenbalg had translated and printed the Tamil NT, the first in an Indian language.

Tranquebar was a tiny territory, but the mission had reverberations to the end of the earth. A new missionary conscience was stirred, as news was spread in the annual letters distributed by the Pietists at Halle, Germany. Although Germans led the way, impetus was to shift to Britain—as political power in India was to be won by the British—in that eighteenth century. Even before there were British missionaries, Anglicans were so inspired by news of Tranquebar that they gave financial support to "English" missions at other places in South India, staffed by German Lutherans. The East India Company countenanced these missions and even lauded a man of the caliber of C.F. Schwartz,* but it became alarmed at mounting pressure in Britain that it should itself make provision for mission work. Talk of missions was a threat to good trading conditions, and the Company for two decades pursued a policy of strictly forbidding the entry of missionaries.

Thus when William Carey* and Dr. John Thomas of the Baptist Missionary Society came to Calcutta in 1793, they were undesirable, illegal immigrants. They found employment as managers of indigo plantations, and Carey prepared himself in Bengali and Sanskrit for his real mission. The next Baptist families, in 1799, had to bypass the Company and make for Danish territory at Serampore, sixteen miles from Calcutta, so that Denmark had again a special place in this history; Serampore proved to be the birthplace of "the modern missionary movement." The Serampore trio of Carey, Joshua Marshman,* and William Ward attempted great things for God, and their college (1819) and their translations of the Bible (Bengali, Oriya, Assamase, Sanskrit, Hindi, and Marathi, by the time of Carey's death in 1834) indicate the foundations they laid for later work. Even in the period before 1813 when missionaries were forbidden, the Serampore men were not really alone, for there were evangelical chaplains to the Company who shared the missionary vision. Most famous of these was Henry Martyn,* who translated the Urdu NT and whose holy "burning out for God" inspired subsequent generations and is a symbol of the cost of mission to India.

As doors opened to missionaries, an Anglican Church organization was established in India, paid from Indian revenues; early bishops at Calcutta included such men as Reginald Heber* and Daniel Wilson.* Church of England missionary societies had important roles, inheriting work previously under the Danish-Halle German missionaries, helping the Syrian Church in Kerala, and sharing in the rise of a strong church at Tinnevelly.

In 1833 restrictions on non-British missions were lifted and the whole process of covering the map of India was accelerated. In the coming century churches were to rise which would approximately equal the number of Roman Catholics in India. Sometimes the converts were won singly and slowly, and sometimes in the rush of mass movements, as in Bihar and the Telugu country. It was an enterprise of foreign mission, but some of the converts were themselves leaders for the young churches. They included Pandita Ramabai,* Narayan Vaman Tilak,* and Sadhu Sundar Singh.*

Christian missions led the way in education. Alexander Duff* used higher education as a means of evangelism, but the belief that Western education would necessarily erode Hinduism and win over higher castes (Duff himself had some notable converts) was belied. Christianity undoubtedly influenced the nineteenth-century reform movements within Hinduism, and education did change belief. There was, however, a new factor of revived nationalism, with a revulsion toward the West, and if some of the great Indians of the twentieth century were to be professedly unreligious (like Jawaharlal Nehru and his daughter, Mrs. Indira Gandhi), there was also a noticeable revival of orthodox Hinduism. Some of the reform movements within Hinduism had been close to Christianity, but others, such as the Arya Somaj (1875), were militantly anti-Christian.

India was large enough to absorb Protestant missions without much friction. The rules of "Comity" regulated boundaries. More positively there was active cooperation in such union institutions as Madras Christian College (1887); and the 1910 World Missionary Conference in Edinburgh,* which gave birth to the ecumenical movement, led directly in India to the formation of the National Missionary Council (1914), which evolved into the National Christian Council. Missions integrated with national churches, and the NCC restricted full membership to churches.

Church union was another process. The strength of feeling among Indians was detected in a manifesto issued by a meeting of ministers at Tranquebar in 1919, deploring denominational disunity as something foreign and a brake on evangelism. V.S. Azariah* was a leader on that occasion. An important point conceded at Tranquebar was that acceptance of the "historic episcopate" was necessary if Anglicans were to be in union, and when the Church of South India (1947) and the Church of North India (1970) were formed, it was on this basis.

The Roman Catholic Church was not affected by this. Fully recovered from its period of decline, it weathered a sore controversy in the nineteenth century as it began to shed the vestiges of the Portuguese *Padroado,* established an Indian hierarchy (1886), formed a satisfactory separate organization for the Keralan Syrians in its fold, and displayed its progress by creating an Indian cardinal (1952) and holding the 38th International Eucharistic Congress at Bombay (1964).

Generally this period of readjustment in churches was not marked by expansion. Of course there were always external checks on that. Hinduism's tolerance seemed embodied in Gandhi, who had favorite Christian hymns but remained firmly a Hindu; by contrast, evangelism seemed a variant of Western imperialism, and that was in full retreat. Independence for India in 1947 did not inhibit evangelism, for the constitution guaranteed freedom to propagate one's religion, and some princely states now within the Indian Union were actually opened to missionaries for the first time. But it is also true that two states, Orissa and Madhya Pradesh, passed acts in the late 1960s which were designed to make conversions more difficult; and the missionary force declined sharply as government virtually stopped inflow. The greatest deterrent to evangelism, however, was not from outside the church, but within. Evangelism was sometimes out of favor in the church leadership and replaced by a social gospel. Some theologies minimized the uniqueness of Christ, and (as elsewhere in the world) spiritual decline meant decline in evangelism.

Fewer than 3 per cent of India's population of 547 million is even nominally Christian, but pessimistic signs are matched by the vitality of such an indigenous movement as the assemblies led by Bakht Singh, a converted Sikh, and by the quality of leadership appearing in the larger and older denominations.

BIBLIOGRAPHY: E. Chatterton, *A History of the Church of England in India* (1924); A. Mingana, *The Early Spread of Christianity in India* (1926); J.W. Pickett, *Christian Mass Movements in India* (1933); D. Ferroli, *The Jesuits in Malabar* (1939); E.G.K. Hewat, *Christ and Western India* (1950); C. Dawson, *The Mongol Mission* (1955); L.W. Brown, *The Indian Christians of St. Thomas* (1956); C.B. Firth, *An Introduction to Indian Church History* (1961); J.S.M. Hooper (rev. W.J. Culshaw), *Bible Translation in India, Pakistan and Ceylon* (1963); M.E. Gibbs, *From Jerusalem to New Delhi* (1964); K. Baago, *A History of the National Christian Council of India, 1914-1964* (1965); J.E. Orr, *The Light of the Nations* (1965) and *Evangelical Awakenings in India* (1970); C.P. Mathew and M.M. Thomas, *The Indian Churches of Saint Thomas* (1967); W.G. Young, *Handbook of Source Materials for Students of Church History* (1969). See also *The Christian Handbook of India* (25th and last large-scale ed., 1959) and *The United Church of North India Survey, 1968* (report).

ROBERT J. MC MAHON

INDIA, EVANGELICAL FELLOWSHIP OF. An association of individuals and groups (such as churches and missions) formed in 1951, affiliated later to the World Evangelical Fellowship,* and aiming at revival in the Indian Church, evangelism, and "effective witness to, and safeguard of, the evangelical faith in the Church." The EFI has become known as the main meeting ground of evangelicals ("ecumenically linked" and otherwise), holds an annual conference in January, and has furthered its aims by activities like a national congress on evangelism, national prayer assembly, and city "penetration plans," and by permanent departments for Christian literature, relief aid, and Christian education (this last producing Sunday school lessons in the various languages). At first, EFI was mainly a missionary movement, owing much to the inspiration of the National Association of Evangelicals* in the USA, but leadership is now entirely Indian. In 1970 individual membership was about 1,400; group membership represents a significant segment of the church.

ROBERT J. MC MAHON

INDIANS, AMERICAN, see AMERICAN INDIANS

INDONESIA. There are slight traces of Christianity in Indonesia from the seventh century, but little is known. The first Franciscan mission reached the Spice Islands with the Portuguese in 1522 and saw mass conversions in Halmahera (1534) and other places. Francis Xavier* spent a short period on Indonesian soil. An anti-Portuguese reaction, however, severely reduced the number of Indonesian converts: the first martyrdoms on Halmahera also occurred in 1534. When the Dutch ousted the Portuguese in 1605, some 30,000 Indonesian Christians became Protestants, following the faith of their new masters. The Dutch East India Company church, however, catered chiefly for its Dutch employees, and local evangelization was scant and superficial. Most converts "turned" in social groups, their Christian knowledge was small and their zeal less—Sanghir Christians, it was reported, thought Christ had died on Christmas Day! The Bible was nevertheless translated into Malay, the trade language; a manuscript of the Lord's Prayer dates to 1627. The NT was published in 1688, and the whole Bible translated by Leydekker in 1733. Some 55,000 baptized Indonesian Christians (although only about a thousand were communicants) were counted in 1727. By the end of the company regime in 1799 there were hardly any more.

The evangelical (Pietistic) movement in Europe transformed the picture. The first efforts to evangelize Java were made, separately, by Coolen (1770-1863) and Emde (1774-1859), a planter and a watchmaker respectively, in the east of the island. Raffles, the British governor (1811-15), was the first to instigate missionary work, and thereafter Dutch and German missionaries gave themselves to the Indies. The Dutch government kept strict control over them and the churches they planted, and prohibited work in politically sensitive areas such as Atjeh and Bali. Churches were planted all over the country: in nineteen ethnic areas the whole people turned to Christianity, and at least as many "gathered" churches were planted elsewhere. L.I. Nommensen* evangelized the animistic Bataks of Sumatra; Kam (1772-1833) earned by his labors in the nominally Christian eastern islands the title of "Apostle of the Moluccas"; Bruckner (1783-1847) pioneered in Central Java. These are only a few among many. Portions of Scripture were produced in a score of local languages. Most of the schools and hospitals of the colonial era were provided by missions, education in particular being fruitful as a method of evangelization.

Nowhere else in the world was so large a church established in the midst of Islam. Its weakness was that it was totally under missionary control and financed from Europe. Shocked by the obvious weakness of the state church, Schuurman opened the Depok Seminary in 1878 to train indigenous evangelists. Depok was closed in 1926, but its work was continued by the Djakarta Theological Seminary, founded in 1934. In Dutch days, however, missionaries were supreme—only a Dutch pastor could baptize, for instance. The credit for altering the situation belongs largely to H. Kraemer,* who after many study-tours (1926-35) recommended that the churches should be freed from foreign control and that missionaries "turn from chiefs into teachers of independence." In the following decade many churches (differentiated largely by ethnic, linguistic, and geographical factors rather than by theology) received independence: the Batak Church in 1930, the East Java Church in 1931, the National Protestant Church in 1935. Even after independence, however, most of the church finances originated from Europe. Similar progress was being made in other areas when the Japanese occupation in 1942 abruptly brought the period of tutelage to an end. With their missionaries in concentration camps, Indonesian Christians were compelled to take responsibility for their own church life.

Since independence, Indonesian Christians have lost the state protection they formerly enjoyed, but have shared the full religious liberty granted by the constitution. A great effort began to equip the whole church with Indonesian literature, with theological colleges, with a trained ministry, and an Indonesian pattern of life and worship.

The two distinctive features during this period have perhaps been the ecumenical movement and the mass turnings to Christianity. The Indonesian Council of Churches (DGI) was formed in 1951 and is supported by thirty-seven member bodies. It has always hoped to unite all the Christian bodies of the country, but this remains an ideal. In the twentieth century there has been a great deal of Anglo-Saxon missionary effort in Indonesia, and the churches which have resulted from this effort (including the majority of the Pentecostal groups) do not support the national council, largely because of their suspicions of the ecumenical movement as a whole.

Mass turnings have always been a feature of the Indonesian church scene. Recently, especially after the abortive Communist coup d'état in 1965, there has been a flood of "new converts" seeking to enter the Christian Church, particularly in North Sumatra, Central and East Java, and other places. There are good sociological explanations for this phenomenon, but in some areas, particularly the island of Timor, numerous miraculous events have been reported. Sadly, many of the existing churches were not prepared for an influx of converts seeking instruction. In some places little teaching was or could be given, and by 1970 the mass movements in many places had slowed down.

Apart from the mainstream of Indonesian Protestantism, Roman Catholicism has made steady progress since it was permitted by the Dutch government to enter Indonesia in the nineteenth century. Flores and N Timor in the Lesser Sunda Islands have become Catholic strongholds. With a membership of about two million, the Roman Catholic Church was granted an independent hierarchy in 1962 and has in recent years experienced the same rapid growth as the Protestant community.

The Chinese in Indonesia, too, have not gone unevangelized. The most notable events in their history were the visits of John Sung the evangelist, just before the Japanese war. He began a revival movement of which the effects can still be felt. There is a large Indonesian-Chinese community with its own churches, and several Chinese-language churches, mostly independent of each other.

Both Protestant and Catholic communities have their own political party, and individual Christians have held high office in the government. In a society subject to disintegrating pressures the Indonesian Church, still predominantly rural, is one of the factors making for national unity; although numbering only one-tenth of the total population, it is truly part of the life of the nation.

BIBLIOGRAPHY: In addition to general books on missions and an extensive literature in Dutch and German, volumes in English include the following: H. Kraemer, *From Mission Field to Independent Church* (1958); N. de Waard, *Pioneer in Sumatra* (1960); S. Houliston, *Borneo Breakthrough* (1963); D. Bentley-Taylor, *The Weathercock's Reward* (1967); F.L. Cooley, *Indonesia: Church and Society* (1968); R. Peterson, *Storm over Borneo* (1968); P. van Akkeren, *Sri and Christ* (1970); K. Koch, *The Revival in Indonesia* (1970); P.B. Pedeusen, *Batak Blood and Protestant Soul* (1970); [Christian name unknown], *Christian Opportunity in Indonesia* (1970); E.C. Smith, *God's Miracles in Indonesian Church Growth* (1970); J. Warneck, *The Living Christ and Dying Heathenism* (n.d.).

MARTIN B. DAINTON

INDRE MISSION. The popular name of the *Kirkelig Forening for Indre Mission i Danmark* (the Danish Church Home Mission Society), an evangelical movement within the Danish national church. It was formed in 1861 by some Pietistic clergymen and laymen from "the awakened circles." Toward the end of the nineteenth and the beginning of the twentieth centuries, the Mission spread as a dynamic revival movement all over Denmark; many clergymen joined it, numerous laymen were employed as colporteurs or lay-preachers, and gradually meeting-houses were built all over the country. Originally the Indre Mission was characterized by the preaching of a straightforward and simple message of doom and salvation, always stressing the necessity of conversion and sanctification, and drawing a sharp line between believers and unbelievers. In spite of their critical attitude to the national church, the Indre Mission people stayed within it, considering it their divinely given mission to function as its salt. Today the Mission has become a recognized and established party within the church, but at the same time it has lost much of its original zeal and spiritual power, and to some extent even drifted away from its original biblical and evangelical position. Nevertheless, it still exerts a stronger influence than any other movement upon Danish church life. N.O. RASMUSSEN

INDULGENCE, DECLARATIONS OF, see DECLARATIONS OF INDULGENCE

INDULGENCES. In the Roman Catholic Church this is the remission of all or part of the debt of temporal punishment owed to God due to sin after the guilt has been forgiven. This grant is based on the principle of vicarious satisfaction, which means that since the sinner is unable to do sufficient penance to expiate all his sins, he is able to draw on the spiritual treasury formed by the surplus merits of Christ, the Virgin Mary, and the saints. The authority for granting indulgences rests with the pope, although he may designate others (e.g., cardinals, bishops) to have this power, with the exception of indulgences for the dead.

Most indulgences granted by the pope are applicable to the souls in purgatory, Apostolic indulgences are those attached to religious articles (such as crucifixes, statues, medals, and rosaries), or to the performance of certain works on special feast days, or the recitation of certain holy names at the hour of death. Other indulgences may be gained by fulfilling the prescribed conditions and completing a designated act (usually the recitation of certain prayers) in the manner specified by the granting authority. To gain an indulgence it is necessary to be a member of the Catholic Church, in a state of grace, and to have the intention of gaining the indulgences. Other conditions, such as confession and Communion, may be required in some instances. A plenary indulgence remits the entire payment of punishment due up to the point when it is gained, while a partial indulgence remits only part of the punishment.

Although instances of forms of indulgences, such as commutations of penance and absolution grants, can be found in the early church, it was not until the eleventh century that indulgence grants appeared which relaxed penitential acts on the condition that contributions be made to a church or monastery. The practice of granting indulgences became more widespread with the advent of the Crusades, beginning with the First Crusade in 1095 when Urban II promised the remission of all penance to those who set out to liberate the Holy Land. Later this was extended to a plenary indulgence and came to include those who contributed to the support of the Crusades. The abuse of the granting of indulgences in return for financial support was considerable during the Middle Ages, and it eventually touched off the Protestant Reformation when Martin Luther attacked the doctrine itself in his "Ninety-five Theses."

BIBLIOGRAPHY: E.J. Ross (tr.), *Indulgences as a Social Factor in the Middle Ages* (1922); W. Herbst, *Indulgences* (1955); P.F. Palmer (ed.), *Sacraments and Forgiveness*, vol. 2 of *Sources of Christian Theology* (2 vols., 1955-60).

ROBERT G. CLOUSE

INFALLIBILITY. The Roman Catholic doctrine that ecumenical councils of bishops and the pope speaking *ex cathedra* are immune from error when teaching concerning faith and morals. The need for infallibility in the church has been argued in recent times from two standpoints. First, since the Holy Spirit indwells the church, it is to be expected that He will ensure that the shepherds of the flock will understand and teach aright the divine message of salvation. Second, since eternal punishment is threatened to those who disobey the Gospel (Mark 16:16), it is to be expected that God will provide a correct understanding of the Gospel in the world.

The infallibility of the pope was first defined by Vatican I* on the basis of such passages as Matthew 16:18; Luke 22:31; John 21:15—understood in the light of their interpretation in the West from early times. Cardinal Cullen composed the definition, making much use of Reginald Pole's *De Summo Pontifice* (1569). Infallibility is understood as a *charisma*, given by the Holy Spirit for the preserving and expounding of divine truth. The teaching of previous councils (Constantinople IV, Lyons, and Florence) is referred to as pointing to this conclusion. Vatican II* reaffirmed the doctrine of papal infallibility, but set it in a larger context. The *charisma* belongs to the pope insofar as he is the head of the college of bishops; when he speaks *ex cathedra* he has this authority in a special way *(singulariter)*. The extent of the infallibility of bishops in council, and the pope pronouncing *ex cathedra*, is declared to be "as wide as the divine deposit of faith, which is to be kept as a sacred trust and faithfully expounded." Roman Catholic theologians also speak of the infallibility of the body of the faithful insofar as it maintains its faith and practice.

Protestants have always opposed this doctrine, arguing that only God and His Word are infallible.

BIBLIOGRAPHY: G. Salmon, *The Infallibility of the Church* (1888); T.G. Jalland, *The Church and the Papacy* (1944); B.C. Butler, *The Church and Infallibility* (1954); O. Rousseau, *L'Infallibilité de l'Église* (1962); O. Karrer, *Peter and the Church* (1963).

PETER TOON

INFANT BAPTISM, see BAPTISM

INFRALAPSARIANISM, see SUBLAPSARIANISM

INFUSION, see BAPTISM

INGE, WILLIAM RALPH (1860-1954). Dean of St. Paul's, London. Born in Yorkshire and educated at Eton and Cambridge, he taught at Eton and elsewhere (1884-88), lectured in Hertford College, Oxford (1889-1904), was vicar of All Saints', Knightsbridge (1904-7), and for four years thereafter was Lady Margaret professor of divinity at Cambridge. He went to St. Paul's in 1911 and held the post until 1934, despite differences with a theologically conservative and mainly Anglo-Catholic chapter. Of great ability and wide-ranging interests, Inge wrote on many subjects, including mysticism and especially Plotinus. He held that Platonic philosophy and Christianity belong together. For many years he contributed a column to a London daily, expressing himself frankly and not always cheerfully. Theologically he is regarded as a liberal, but he strongly opposed the Catholic modernists Tyrrell and Loisy, holding strongly to the importance of the historical in Christianity. P.W. PETTY

INGERSOLL, ROBERT GREEN (1833-1899). American politician and agnostic. Born in New York and educated in Illinois, he was admitted to the bar in 1854 and served in the Civil War from 1861 to 1863. His unsuccessful attempt in 1860 to become the Democratic candidate for Congress led him to change party tickets; as a Republican he was appointed attorney general (1867-69) for the state of Illinois. He moved his residence to Washington, D.C., in 1879 and to New York in 1885. Ingersoll received as much as $5,000 for some of his famous antireligious speeches. His main attacks were directed against the authority of the Bible and its alleged inaccuracies. He gave titles to his famous speeches: "The Gods," "Ghosts," "Skulls," and "Some Mistakes of Moses." Few have rivaled his eloquence. Some historians feel that his agnostic beliefs kept him from becoming the Republican nominee for the presidency. DONALD M. LAKE

INGLIS, CHARLES (1734-1816). First Anglican bishop of Nova Scotia. Born and educated in Ireland, he was ordained in 1758 in London before being sent as a Church of England missionary to Dover, Delaware. In 1765 he went to New York to become assistant to the rector of Trinity Church. The American Revolution disrupted his life, for his church was burnt, his congregation scattered, and his property confiscated. In 1783 he sailed with a group of Loyalists to Nova Scotia. Four years later he was consecrated the first bishop of Nova Scotia. A devoted churchman and an incessant traveler, he did much to organize the Anglican cause in his diocese. Because of failing health he retired to his farm near Annapolis in 1796, but remained active, being named to the Council of Nova Scotia in 1809.

ROBERT WILSON

INNER LIGHT, see FRIENDS, SOCIETY OF

INNERE MISSION. Proclaimed by J.H. Wichern* at a *Kirchentag* in 1848, this mission within Christendom aimed to show the love of Christ meeting practical human needs as the realization of the Kingdom of God. The central committee and local associations of the Innere Mission flourished despite considerable opposition in the church. They coordinated many already existing evangelistic and welfare activities in the church, and encouraged many fresh enterprises. Wichern saw all these as parts of the one mission of the church, and he looked for an Evangelical *Volkskirche,* a people's church, though he was conservative in church and politics, working with the Prussian government and hostile to socialism. Later some leaders of the Mission adopted a more radical socialist approach to winning the alienated working classes, but with little success. Meanwhile the many activities of the Mission continued up to the present as part of the German Evangelical Church's attempt to minister to human wholeness in the Spirit of Jesus.

HADDON WILLMER

INNOCENT I (d.417). Pope from 401. In him great talent, upright character, vigorous determination, and a high view of the papacy met at a time which, because of the rapid collapse of the Roman power, was ripe for the extension of papal influence. His thirty-six surviving letters reveal his relationships with other churches and are important because of the doctrinal positions of the church to which they point. He argued that the Western bishops had an obligation to follow the Roman Church because they belonged to churches formed through the agency of Peter. When he granted the African bishops' request to condemn Pelagius, he took the opportunity of asserting that nothing should be completed, even in the most distant provinces, "until it comes to the knowledge of this see." He was more restrained in his dealings with the East. He ensured that Illyricum remained under his jurisdiction even though it was partly in the East. When dealing with an appeal from John Chrysostom* for support, however, he first called for a council in the East, and only when this failed did he break off communion with Chrysostom's enemies. His secular influence also increased, especially after Alaric sacked Rome (410). C. PETER WILLIAMS

INNOCENT III (1160-1216). Pope from 1198. One of the greatest popes of the Middle Ages, Giovanni Lotario de' Conti was unanimously elected at the age of thirty-seven, following the death of Celestine III. A member of one of the noble families of Rome, he studied theology at Paris and canon law at Bologna and was made a cardinal deacon by Clement III (1190). Although not an outstanding cardinal, he did actively participate in Curial matters and wrote a book, *On the Contempt of the World,* an exhortation to asceticism and contemplation which became popular throughout Europe. He was able to marshal traditional ideas of the papacy and thus make it the pivot of W and E Europe. He had a keen intellect and knew exactly what had to be done for the Roman Church to dominate all human relationships. In addition to mastering the tasks of the Curia, he was also an able statesman and his dealings with European monarchs are evidence of his desire to be viewed as the supreme arbiter in all cases and the preserver of unity throughout Christendom.

Innocent began his reign by reorganizing the administration of Rome. He obtained the right to nominate the senator who ruled Rome and to receive an oath of fidelity from him. Following the collapse of German rule in Italy after the death of Emperor Henry VI, he was able to restore and expand the Papal States, thus strengthening the position of the papacy and preventing the unification of Italy. Under Innocent, the Fourth Crusade

(1202-4) was launched, resulting in the formation of the Latin Empire of Constantinople. It failed, however, to unite the Eastern and Western churches. Also the great Fourth Lateran Council (1215) was summoned by Innocent. Some of the more important of its seventy decrees included the official approval of the term "transubstantiation," the suppression of heresy, the role of the church in secular justice, the necessity of paying the tithe, and several other actions which shaped church policy for centuries. The council also fixed 1217 as the year for a crusade against Islam.

Innocent was heavily involved in European politics. In the controversy over whether the Hohenstaufen, Philip of Swabia, or the Guelf, Otto of Brunswick, should become Holy Roman Emperor, he decided in favor of the purportedly pro-papacy Otto. But when Emperor Otto IV invaded Sicily, the pope excommunicated him and supported the election of his ward, Frederick II. In England, Innocent put the land under interdict and excommunicated King John for his refusal to allow the papal appointee to the archbishopric of Canterbury, Stephen Langton,* to enter the country. John submitted in 1213. Innocent also intervened in the matrimonial affairs of such monarchs as Peter II of Aragon, Alphonso IX of Léon, and Philip II of France.

BIBLIOGRAPHY: S.R. Packard, *Europe and the Church under Innocent III* (1927); L. Elliott-Binns, *Innocent III* (1931); C.E. Smith, *Innocent III, Church Defender* (1931); J. Clayton, *Pope Innocent III and His Times* (1941).

ROBERT G. CLOUSE

INNOCENT IV (c.1200-1254). Pope from 1243. Born of a family that produced another pope (Adrian V) and several cardinals, he was trained in canon and Roman law at Bologna where he also taught briefly. Confidant of Gregory IX and made cardinal (1227), he succeeded Celestine IV after an eighteen-month vacancy owing to the pressure of Emperor Frederick II. Innocent excommunicated him as had Gregory, and at the Council of Lyons* (1245) charged him with perjury and heresy. The intrigue between empire and papacy was alleviated with his death, but papal interference continued with his successors until an agreement was reached in 1254. Innocent's *Commentaria* on canon law speaks defensively to the many extreme worldly entanglements he initiated and inherited.

C.G. THORNE, JR.

INNOCENT X (1574-1655). Pope from 1644. Born Giovanni Battista Pamfili, he rose rapidly as a papal advocate, auditor, and nuncio, became a cardinal in 1629, and was elected pope as the nominee of the Spanish party. He began his pontificate by trying to recover money purloined by the Barberini, the relatives of the late pope who had the support of Mazarin, while not neglecting to advance his own family. He waged a cruel war against Parma, promised considerable help for Venice against the Turks but sent little, encouraged Spain to revolt against Naples, and was at first hostile to the independence of Portugal. He protested strongly against the Peace of Westphalia* (1648) and issued a bull against it as a violation of the laws of the church. In 1653 he condemned five Propositions from Jansen's *Augustinus.* In his later years Innocent fell under the dominance of the arrogant Donna Olimpia Maidalchini, his brother's wife, who mulcted the papal treasury and contributed to the spread of corruption.

IAN SELLERS

INNOCENT XI (1611-1689). Pope from 1676. Born Benedetto Odescalchi at Como, he was educated there at the Jesuit College, then gained the doctorate in civil and canon law from the University of Naples (1639). He was appointed apostolic protonotary by Urban VIII and became a powerful influence in the Curia because of his model Christian life. He was made cardinal in 1645 and cardinal legate to Ferrara in 1648. In 1650 he took holy orders and was consecrated bishop of Novara, distinguishing himself here as at Ferrara by charitable works. He returned to Rome in 1656, and he served in the Curia till his election as pope.

His pontificate is one of continuous reform, continuous struggle against vested interests, and concerted effort against the forces of Islam. Perhaps his most famous work was his defense of the traditional rights and freedoms of the church against the Gallicanism* of Louis XIV of France. He opposed Louis's use of the *régale* (right to use the revenues of a bishopric when vacant), of the *franchise* (right of ambassadors to the papal court to grant asylum to anyone in their compounds), of the Gallican Articles of 1682, and also Louis's effort to secure the archbishopric of Cologne for a pro-French candidate. Innocent eventually excommunicated Louis and firmly followed policies he felt were best for the church. He also decreed against laxity in moral theology, seeming to favor the Jansenists over the Jesuits, and he decreed against Quietism.* He was beatified in 1956.

BIBLIOGRAPHY: E. Michaud, *Louis XIV et Innocent XI* (4 vols., 1882-89); *Epistolae, ad principles* (2 vols., 1891-95); J. Orcibal, *Louis XIV contre Innocent XI* (1949).

BRIAN G. ARMSTRONG

INNOCENT XII (1615-1700). Pope from 1691. Born Antonio Pignatelli outside of Naples, he was educated at the Jesuit Roman College whence he entered the Roman Curia in 1635. He received several advancements under succeeding popes, especially under Innocent XI who created him cardinal in 1681, bishop of Faenza in 1682, and archbishop of Naples in 1687. He was elected pope as a compromise candidate in the Hapsburg-Bourbon struggle, and he patterned his pontificate after Innocent XI except that he sided with the Bourbons rather than the Austrian Hapsburgs. He proved to be a reformer of iron will as well as being successful in healing divisions within the church. He issued a bull against nepotism, abolished sinecures, established the *Curia Innocenziana* (lately known as the *Camera dei Deputati*) for the forceful and fair administration of justice, and instituted many charitable and educational works. He resolved the long-standing Gallican problem, condemned Quietism without alienating the great Fénelon,* quieted the Jansenist storm, and helped contribute to the War of the Spanish Succession by convincing Charles II of Spain to name Philip of Anjou as his successor.

BRIAN G. ARMSTRONG

INQUISITION, THE. A special tribunal established by the medieval church for the purpose of combating heresy. In the Middle Ages the growing threat of heretical groups, particularly the Cathari,* led to the acceptance by the church of the use of the secular authority, of physical penalties, and of an inquisitorial method as means for their suppression. Alexander III* at the Council of Tours (1163) urged secular princes to prosecute heretics, to imprison them, and to confiscate their property. He also directed the bishops to search out heretics, encouraging them to replace the older method of trial by accusation, which depended upon the initiative of an accuser, with the more vigorous and effective method of inquest in which the judge took the initiative.

It was Gregory IX,* however, who in a series of actions from 1231 to 1235 imposed on such activities a formal organization and set of procedures whereby the apprehension and trial of heretics was reserved to the church and the major responsibility for such work was given to papal inquisitors. He is therefore often credited with having established the Inquisition. In 1478 the Spanish Inquisition was authorized by Sixtus IV.* It followed procedures similar to those described below but was characterized by its subservience to the state, with its appointments officially made by the secular authority.

Dominicans and Franciscans were most often chosen as papal inquisitors. Assisted by numerous aides, the inquisitor would begin his work in a town by calling the clergy and people to a solemn assembly at which those who knew themselves to be guilty of heretical views were urged to confess within a period of grace frequently ranging between two and six weeks. Those who did so were normally given light penalties. At the expiration of this period the inquisitor began a systematic search for suspects who would be summoned before the tribunal for interrogation. The suspect, who was not allowed legal defense but could have a counselor, was encouraged to confess his errors, and toward this end Innocent IV in 1252 allowed the use of torture.

When confessions were not forthcoming, the testimony of two witnesses, if it could not be refuted, was considered sufficient for conviction. The practice of withholding from the accused the names of the witnesses was not modified until the time of Boniface VIII (1294-1303). Some safeguards were provided, however, for the accused, such as the opportunity to discredit as witnesses one's enemies, the punishment of false witnesses, and the various restrictions placed on the use of torture. Nevertheless, frequently the accused found himself in the position in which he was assumed to be guilty; his safest mode of escape was to confess, since by persistently asserting his innocence he ran the risk of being judged an obdurate heretic, for which the punishment was death. If he were so judged, he would be abandoned to the secular authority for burning, for the church did not participate officially in the shedding of blood. Milder punishments included imprisonment, confiscation of property, wearing a yellow cross, prayer, fasting, almsgiving, flagellation, and pilgrimage.

In the early modern period the Roman Inquisition, established by Paul III in 1542, was used to combat witchcraft and the Protestant Reformation. Some Protestants also were to be found employing inquisitorial procedures against those suspected of sorcery and incorrect doctrine. The increasing secularization of Western society, however, was accompanied by the decline of such activities by religious institutions. To modern minds the Inquisition rightly seems abhorrent, but to be fair one must recall the *Zeitgeist* in which it operated and remind oneself of the activities of modern institutions which also, but sometimes more subtly, suppress those who hold unorthodox views.

BIBLIOGRAPHY: H.C. Lea, *A History of the Inquisition of the Middle Ages* (3 vols., 1888); A.S. Tuberville, *Medieval Heresy and the Inquisition* (1920); H. Kamen, *The Spanish Inquisition* (1966).

T.L. UNDERWOOD

INSCRIPTIONS, CHRISTIAN. Understanding of the NT and early church has been revolutionized by the many written documents recovered from the past. The term "inscription" may be used in several senses: to describe any written document, including those written upon stone, clay, papyrus, or other material; to include all inscribed artifacts, except the parchment and papyri; or, more commonly, to refer to the larger monuments only. The excessively large number of artifacts makes study very complicated—e.g., more than 11,000 Christian inscriptions earlier than the seventh century have been found in Rome alone. Archaeologists usually limit the texts of significant value to the first seven centuries of the Christian era.

An inscription is said to be "Christian" if it bears evidence of the Christian faith. Many inscriptions were no doubt the work of the Christian community, yet bear no specific reference to the faith. Some Christian inscriptions are composed with correctness and some even with elegance, both in form and content; others, however, are written in barbarous style. Christian inscriptions often give few, if any, personal details: epitaphs, for example, usually give only name, age, and date of death. On the other hand, the total number of Christian inscriptions gives valid evidence about the nature and essence of a community of believers, and helps in an analysis of the expansion of the Christian faith. Occasionally an inscription has been found which records the building of a specific church or perhaps the death of an early Christian martyr. More rarely, archaeologists have unearthed inscriptions which relate to doctrine, but these are generally of little value when compared with the many more or less fully preserved literary sources.

Second-century inscriptions are fairly common, especially in the famous Roman catacombs. By the middle fourth century inscriptions were common in Rome, North Africa, and Asia Minor, although some of these are from heretical groups such as the Montanists or the Donatists. Most of these inscriptions are in either Latin or Greek, and most are original, although the texts of others survive only in copies. Dating the inscriptions is complicated, since many of the dates which the original inscriptions bear refer to consular years. Often the archaeologist must infer the date of the inscription from either the site where it was found, or from some kind of internal evidence such as content or style of writing.

See A. Parrot, *Le Musée du Louvre et la Bible* (1957), pp. 142-44; D.J. Wiseman, *Illustrations from Biblical Archaeology* (1958).

WATSON E. MILLS

INSPIRATION. By the term "inspiration" Christian theology designates a particular activity of the Spirit of God whereby specially chosen prophets and apostles spoke and wrote the veritable Word of the Living God. God's spiration, or breath, is associated in biblical theology with the primal divine gift of human life, with the regeneration of sinners, and with the production of sacred scripture. In view of God's agency in inspiration, historic Christianity distinguished the Bible from all other literature as a unique canon of written revelation.

Inspiration as a spiritual phenomenon is not common to all believers, but is divinely reserved for specially authorized and authoritative bearers of God's message. This does not, however, imply that the message transmitted by chosen prophets and apostles is a product of mechanical divine dictation. Inspiration neither suppresses the personalities of the writers, nor puts an end to their human fallibility. Although prophets and apostles remained fallible men who shared the culture of their times, God nonetheless revealed to them information beyond their natural resources, and what they taught as doctrine has its basis in the Holy Spirit as ultimate author of their message.

Nor is inspiration understandable biblically in terms of mantic ecstaticism. The obscure and sporadic pronouncements of mantic seers contrast with prophetic-apostolic declaration which is emphatically moral and centers in the historic redemptive purpose of God. Nor is scriptural inspiration to be confused with internal psychic excitement. The biblical view declares, not simply the writers, but their very writings to be inspired. The Apostle Paul in 2 Timothy 3:16 uses the term *theopneustia,* which not only emphasizes that God is the original author, but also affirms Scripture itself to be God-breathed.

Inspiration is nonetheless consistent with, and does not violate, the human personality of the prophets and apostles through whom God communicates the truth about Himself and His purposes. The modernist notion that men could not have told the truth about divine things unless they transcended their humanity—that is, ceased to be human—is self-refuting. Neo-Protestant theologians have long endeavored, within the Bible itself, to distinguish inspired from supposedly uninspired strata. But such efforts turn serious biblical study into a shambles. Without the reliability of Bible history, scriptural theology cannot be credited, since the two are intertwined. And the Bible view of creation and miracle (centrally Christ's incarnation and resurrection) has clear implications for nature as well as history. Scripture implies that the sacred writings are plenarily inspired. No theologian has adduced objective criteria for discriminating that which he presumes to be errant in Scripture from that which he contends to be trustworthy. If one assumes that the biblical writers are to be trusted only where their assertions can be presently validated, he distrusts the writers, finding them credible on grounds other than their supposed divine inspiration.

Karl Barth acknowledged the futility of the modernist attempt to divide Scripture into trustworthy Word of God and fallible word of man. He affirmed that none of Scripture is objectively Word of God, yet held that any of it can *become* Word of God through a personal divine confrontation. But this alternative forfeits the inspiredness of Scripture on which Scripture itself insists, and it obscures both the authority and truth of the Bible in its propositional form.

BIBLIOGRAPHY: B.B. Warfield, *The Inspiration and Authority of the Bible* (1948); R. Preus, *The Inspiration of Scripture* (1955); C.F.H. Henry (ed.), *Revelation and the Bible* (1958); K. Runia, *Karl Barth's Doctrine of Holy Scripture* (1962).

CARL F.H. HENRY

INSTANTIUS (late fourth century). Spanish bishop. A follower of Priscillian,* he may have been deposed by a synod at Saragossa in 380. When Idacius, bishop of Emerita, obtained a rescript from Gratian that he could use against them, Instantius and Salvianus (another bishop and follower of Priscillian) accompanied Priscillian, whom they had made bishop of Avila, to Italy to seek the aid of Damasus and Ambrose. Unsuccessful in this, they nevertheless gained the aid of Macedonius, the *magister officiorum,* who used his influence to obtain their protection. Gratian's assassination, however, brought another change of fortune. Priscillian and Instantius were summoned before a synod at Bordeaux in 385. Instantius was deposed. When they appeared before Maximus at Treves, Priscillian was put to death on a charge of witchcraft, and Instantius banished. He may be the author of eleven treatises attributed by G. Schepps to Priscillian and published from a Würzburg MS in 1886.

DAVID JOHN WILLIAMS

INSTITUTES, THE, see CALVIN; CALVINISM

INSTITUTIONAL CHURCH. Gymnasiums, libraries, handcraft centers, and medical services were all part of the work of an institutional church. A former president of Dartmouth, William J. Tucker, probably developed this description of a church that served the life of the whole man in the inner city each day of the week. Thomas Beecher, in the latter half of the nineteenth century, built the first full-fledged institutional church, in Elmira, New York, with facilities for social services as well as worship. In 1882 William S. Rainsford, with the financial help of his vestryman, J.P. Morgan, developed the Episcopal Church of St. George's on the East Side of New York City into an institutional church. Temple University in Philadelphia grew out of the night school for workers in Russell Conwell's Baptist Temple. Over 150 of these churches united under the leadership of Josiah Strong and William E. Dodge to form the Open and Institutional Church League, which helped to develop the Federal Council of Churches. Many churches as a result give attention to human need in all aspects of life.

EARLE E. CAIRNS

INTERDENOMINATIONAL FOREIGN MISSION ASSOCIATION. Composed of interdenominational faith missionary societies, this was founded in 1917 by seven societies to provide spiritual fellowship and cooperation. The organization has a conservative creedal statement, provides member societies with relevant information through *IFMA News*, helps churches to set up missionary programs, and promotes cooperation among evangelical missionaries who rely mainly on prayer and faith in money-raising. This organization in 1969 embraced forty-four societies sending out nearly 6,000 missionaries with an income of nearly $33 million.

INTERDICT. In ancient Roman law this was a negative command by the *praetor* forbidding certain actions. In the Roman Catholic Church it is a command which prohibits or denies to the faithful participation in certain sacred acts, even though they remain in communion with the church. An interdict may be directed toward two purposes: its general use is a censure to bring about a desired cessation of an act or condition, but it may also be used as a vindicative penalty to bring about atonement for an offense.

There are several types of interdict. A personal interdict affects the person against whom it is directed wherever he goes, while a local interdict is concerned only with a certain locality and does not affect any who leave the area. In a particular personal interdict, the recipient or recipients of the ban are explicitly named; but in a general personal interdict, all those in a certain group are affected. A particular local interdict is directed against one sacred place, and a general local interdict affects a larger area, such as a diocese, province, state, or nation. A bishop may impose an interdict on a particular parish or the people of a parish, but only the Apostolic See may impose interdicts affecting larger areas or groups. The sacred acts which are prohibited by an interdict include the celebration of the Mass, Benediction of the Blessed Sacrament, and burial rites. Exceptions to the prohibitions of the interdict are granted to the dying and on some of the greater feasts (e.g., Christmas and Easter).

Although the interdict was used in the early church in the sixth century, it was not part of ecclesiastical law until the eleventh, and was not completely institutionalized until later. It was a frequent weapon of the medieval papacy in dealing with obstinate monarchs, as in the case of Innocent III's* interdict on England when King John (1199-1216) refused to allow the papal appointee (Stephen Langton) to the archbishopric of Canterbury entry into the country. Theoretically the interdict is still applicable, but is no longer held to be an active instrument of the Roman Catholic Church.

ROBERT G. CLOUSE

INTERIM OF AUGSBURG, see AUGSBURG, INTERIM OF

INTERNATIONAL BIBLE READING ASSOCIATION. A movement to encourage personal Bible study. It was founded in 1882 by the National Sunday School Union under the inspiration of Charles Waters, a bank manager and a member of C.H. Spurgeon's congregation at the Metropolitan Tabernacle, who became the first secretary. It first issued a scheme of Bible readings related to the International Sunday School Lessons. In three years the membership rose to 100,000 and by 1900 reached nearly 750,000. Weekly comments were introduced in 1886, followed in 1887 by a monthly leaflet, "Hints on the Daily Readings." A badge was adopted in 1895.

The movement spread to Australia (1882), New Zealand and Canada (1883), the USA (1885), and many other countries. The first foreign language used was French (1884), followed soon by German and Swedish (1886). Members supported a Sunday school missionary to India, Dr. J.L. Phillips (1890), thus beginning the IBRA Missionary fund. Fuller daily Bible notes began in 1909, written by Dr. Alexander Smellie for 14 years, continued by his widow editing his material after his death until 1944. Other authors have since maintained these notes, and further Bible reading aids have developed for children, young people, and more advanced students. Expansion overseas has continued, both in association with missionary societies and independently, e.g., Nigeria (1971/2).

J.G.G. NORMAN

INTERNATIONAL BIBLE STUDENTS ASSOCIATION, see JEHOVAH'S WITNESSES

INTERNATIONAL CONGREGATIONAL COUNCIL. This was formed in London in 1891 for the purpose of obtaining greater cooperation among the Congregational churches over the world. The Congregational response to the emerging ecumenical movement of the latter part of the nineteenth century, it was designed not only to bring greater unity among Congregationalists, but also to bring Congregationalism into greater cooperation with other Protestant churches in the task of evangelism. It proved, however, to be a suitable vehicle for the promotion of the Social Gospel* and was increasingly used for this purpose by liberal theologians.

INTERNATIONAL COUNCIL OF CHRISTIAN CHURCHES. An interdenominational council of churches of strong fundamentalist beliefs characterized by militant opposition to the World Council of Churches, to Communism, and to defections from orthodox Christianity. The principal founder of the movement was Carl McIntire, an American Presbyterian minister who led also in the founding of the Bible Presbyterian Church, Faith Theological Seminary, Shelton College, and the American Council of Christian Churches.* The ICCC was founded at Amsterdam in 1948 and includes constituent bodies (national church groups), consultative bodies (local churches), and associated bodies (such as missionary societies and Bible leagues). In 1972 there were 155 "Bible believing Protestant denominations" in ICCC membership, most of them small. Throughout its history there have been some defections, usually stemming from disagreements with McIntire.

The missionary arm of the council is known as The Associated Missions (TAM). The ICCC operates also International Christian Relief, and its youth movement called International Christian Youth offers a separatist alternative to ecumenical programs. The ICCC holds regional and national council meetings yearly, and in 1973 held its eighth world congress in Cape May, New Jersey. The international headquarters is in Amsterdam; national offices are maintained in Africa, the Middle East, Latin America, the Far East, Canada, the USA, and Europe. The main organ of the movement is McIntire's *Christian Beacon*, a weekly publication from his headquarters in the Bible Presbyterian Church of Collingswood, New Jersey, of which he is pastor. McIntire is also widely known for his controversial "Twentieth Century Reformation Hour" radio broadcast.

ROBERT C. NEWMAN

INTERNATIONAL MISSIONARY COUNCIL. Founded at Lake Mohonk, New York, in 1921, the IMC was designed to be a council of councils. It was an outgrowth of the great World Missionary Conference at Edinburgh* in 1910, responding to the increasing desire for more cooperation among Protestant Christian missions. Its membership was restricted to regional cooperative agencies, such as the national missionary councils in Europe and America and the national Christian councils in what were considered the mission lands. The total membership ultimately included thirty-eight agencies. Mission boards or societies were not members, except as they were represented in these regional agencies.

The IMC was not an administrative, but a consultative and advisory body. Its chief function was to stimulate cooperation, arrange for joint conferences, make careful surveys or studies of the missionary enterprise, and recommend procedures. Its major conferences were those of Jerusalem (1928), Madras (1938), Whitby (1947), Willingen (1952), Ghana (1957), and New Delhi (1961). Its official organ became *The International Review of Missions*, which had started publication in 1912. It also published a number of missionary studies of permanent value. During World War II and until 1955, it assisted the "orphaned" missions whom the war had cut off from support from their homelands. At the Ghana assembly it created a theological education fund to assist theological schools overseas.

After the formation of the World Council of Churches* at Amsterdam in 1948, the IMC came under increasing pressure to become the missionary arm of that body. At the cost of losing a few of its members, in 1961 it took the final step at New Delhi of integration with the WCC as its Commission/Division of World Mission and Evangelism.

HAROLD R. COOK

INTER-VARSITY FELLOWSHIP. This movement was founded in 1927 to further the cooperation between evangelical Christian unions in the universities and colleges of Great Britain. Four Christian unions in the teaching hospitals had united in 1873 to form the Medical Prayer Union. The Cambridge Inter-Collegiate Christian Union began in 1877 and later became affiliated to the Student Christian Movement.* In 1910 the link was broken because of the broadening outlook of the SCM. In 1919 the first annual inter-varsity conference was organized to promote evangelical and missionary activity in the colleges; the growth of this work led to the formation of the IVF.

Typical of the aims of membership of a Christian union are "to present the claims of the Lord Jesus Christ to the members of the university; to unite those who desire to serve him; and to promote the work of Home and Foreign Missions." Groups are found also in polytechnics, colleges of education, and technical colleges. Missionary conferences are organized, and much is done among overseas students. On completing their studies, students are encouraged to join graduate fellowship groups which are located in many areas. Special interests are catered for by organizations such as the Christian Medical Fellowship and the Theological Students' Fellowship. A wide range of Christian literature is produced. The IVF is affiliated to the International Fellowship of Evangelical Students.

J.W. CHARLEY

INVESTITURE CONTROVERSY. This developed into a fifty-year struggle after Gregory VII* (pope from 1073) charged the Salian emperor Henry IV* of Germany with making ecclesiastical appointments through lay investiture, a practice condemned by Nicholas II in 1059. Henry claimed that an imperial divine right for a century had been withdrawn; he sought Gregory's dethronement. Excommunication followed; the imperial ecclesiastics feared for their own security and hesitated to support their king. Henry found himself isolated and sought reinstatement to the extent of humiliating himself before the pope at Canossa (1077). After the intervention of Matilda of Tuscany and the abbot Hugh of Cluny, Gregory heard Henry's plea and absolution ensued. But conniving and fighting followed as a result of earlier resentments: the two protagonists set up an antipope and an anti-king, but with no effect.

After Gregory died in 1085, Urban II turned instead to the crusade without German support: both sides wanted to save face. Paschal II renewed the struggle fruitlessly; the new leaders in Rome agreed with Gregory's aims, but not his means. The principle used to terminate the German investiture controversy (1103-7) was embodied in the Concordat of Worms (1122) and reasserted by the Second Lateran Council (1123), between Callistus II and Henry V: the emperor abandoned lay investiture with ring and staff, but could still demand homage of bishops and abbots in his domains before their ecclesiastical investiture. The German king did have the right of veto over ecclesiastical appointments. This struggle kept German churchmen from cultural advances while they attended to political affairs, and Germany fell behind in its intellectual leadership of W Europe. C.G. THORNE, JR.

IONA. A small Scottish island off the SW coast of Mull from which Columba,* who arrived there in 563, evangelized W Scotland and N England. His monastery became a famous ecclesiastical center, but was repeatedly ravaged by Norsemen. A Benedictine house was established on Iona in the early thirteenth century, but the buildings were pulled down at the Reformation in 1561. In 1900 the duke of Argyll gave the monastic ruins into the care of the Church of Scotland. The abbey church of St. Mary, dating from the twelfth century, has been restored and is now used in connection with the work and worship of the Iona Community.* Forty-six Scottish kings are said to have been buried on the island. Many ancient remains have been discovered, including a number of Celtic crosses. Iona forms part of a Church of Scotland parish which embraces also a section of Mull. J.D. DOUGLAS

IONA COMMUNITY. Founded by Scottish minister-baronet George MacLeod in 1938 on the island of Iona,* this imaginative experiment brought together ministers and laymen sharing the fellowship of work and worship. On the wall (where an extensive rebuilding project was planned) and in the abbey they would seek to carry out what they regarded as the task of the church: "to find a new community for men in the world today." Three obligations are involved in full membership of the Iona Community: ministers and craftsmen are expected to spend some time on the island, particularly during the initial stages of their membership; members accept a threefold rule concerned with prayer, Bible reading, and tithing; and they are expected to attend the monthly meetings on the mainland in winter and the annual regathering on Iona in June. There are thousands of associate members throughout the world. In 1951 the Iona Community was brought under Church of Scotland auspices; a special church committee reports annually to the general assembly on the Community's affairs.

J.D. DOUGLAS

IRELAND. Pre-Christian Ireland was a Celtic land of tribal institutions, Druidic influences, and pagan worship, especially of the oak, the ash, and the yew tree. The island had escaped the ravages and the benefits of Roman invasion. There were in Ireland Christians from an early age, such as Kieran of Cape Clear Island. Palladius* was sent in 431 to minister to the Irish who were believers in Christ; but the establishing and development of the Christian Church was largely the work of Patrick,* and the church he founded developed in isolation from the Western Church (see IRELAND, CHURCH OF). Its monasteries were the main centers, and its abbots exercised a wide influence in learning and art. Saints like Finnian of Clonard (d. c.589) and Comgall* of Bangor at home, and Columba* and Columbanus* abroad, made famous the name of the Irish Church.

This period of brilliance ended disastrously with the Danish invasions that ravaged the island for more than three centuries before their defeat by Brian Boru at Clontarf in 1014. But the land still disunited was poor soil for Christian progress, and the days were dark during Anglo-Norman invasions that began in 1170. Though the conquest by the Anglo-Normans was never complete, the pope's recognition of the sovereignty of Henry II in 1172 was the death knell of the independent Celtic Church,* and the dividing of the church into a majority section that accepted oversight from Rome and a minority section which was largely the church of the ruling classes and continued as the Church of Ireland.*

The policies of reformation for the Church of England adopted and enforced by Henry VIII were applied to Ireland with equal tactlessness and with the same disastrous results. In 1537 the king was declared to be head of the church in Ireland, and the submission to Roman authority was forbidden. The gap between native Irish and Anglo-Norman widened. Reformation was identified with English law, and the people were drawn to Roman supervision as never before. Indeed, Romanism and patriotism became almost synonymous from that time.

A new element was introduced in the early seventeenth century when English and Scottish settlers were planted in Ulster to replace the native Irish who had been hostile to the English rule, and to develop the land by their industry. For some years their religious life was directed by godly Anglican bishops like Ussher of Armagh, Echlin of Down, and Knox of Raphoe. But the Scots in particular looked to their homeland for ministers, and eminent preachers like Brice, Cunningham, Blair, and Livingstone served them well. The coming of a Scots army to Carrickfergus in 1642 to quell a bitter rebellion led to the organizing of a presbytery by the army chaplains, and from this there has grown the strong, virile Presbyterian Church in Ireland. This church submitted to a period of stern testing after the Restoration in 1660. It had been largely committed to the Scottish Covenants of 1638 and 1643. Following the Revolution Settlement of 1690, a small remnant adhered to the Scottish Covenants and still exists as the Reformed Presbyterian Church* of Ireland.

The Secession from the established church in Scotland had its effect in Ireland. For about a century the original Presbyterians, the synod of Ulster, and the Secession Synod worked side by side in Ireland, but a felicitous union in 1840 formed the general assembly of the Presbyterian Church in Ireland. The influence of Arianism affected Presbyterianism in Ireland. The synod of Ulster in 1721 by a large majority affirmed its belief in the essential deity of Christ, and called for a voluntary subscribing of the Westminster Confession of Faith.* The minority who were unwilling to subscribe subsequently became the synod of the Non-Subscribing Presbyterian Church.

Two other vital factors affected Christianity in Ireland. The first was the vigorous impact of the ministry of John Wesley.* He preached in many parts of Ireland, and while his influence was greatest where English settlers were most numerous, the cause he fostered has been permanently established throughout Ireland, Methodism today makes a big contribution to the spiritual, educational, and cultural life in both Northern Ireland and Eire. The second and very important factor was the revival of religion in 1859. There had been earlier evidences of the special working of the Holy Spirit, particularly among the Presbyterian settlers in County Antrim in 1625, but the 1859 revival was wider in its scope and deeper in its effects. Most of the branches of the Protestant Church received benefit from the movement.

The onward progress of Christianity in Ireland has been closely identified with its political life. The outstanding politicians in every age have been closely connected with some branch of the church. As a general rule, Roman Catholics have been nationalist and republican in their outlook, while the Protestant population has sought to maintain the link with Great Britain. This strong division in political sympathies has produced much bitterness and tension, but it has also led to a deeper involvement in political life by the churches.

Ireland has made a significant contribution to missionary work throughout the world. From the days of Columba in the sixth century until the present time, thousands of missionaries of every type of ecclesiastical attachment have taken the Gospel to many countries.

BIBLIOGRAPHY: J.S. Reid, *History of the Presbyterian Church in Ireland* (3 vols., 1833); J. Godkin, *The Religious History of Ireland* (1873); W.D. Killen, *Ecclesiastical History of Ireland* (1875); T. Olden, *The Church of Ireland* (1892); W.A. Phillips (ed.), *A History of the Church of Ireland* (3 vols., 1933-34); R.P. McDermott and D.A. Webb, *Irish Protestantism Today and Tomorrow* (1945); T.J. Johnston, J.L. Robinson, and R.W. Jackson, *A History of the Church of Ireland* (1953).

ADAM LOUGHRIDGE

IRELAND, CHURCH OF. This owes its origin to Patrick,* who was loyal to the church order of his times and particularly to the customs of the church in Gaul where he had been trained and ordained. The latter included a preference for the monastic system, which at first was concerned with the education of men and boys of high social rank and later developed a zeal for evangelism. In the century after Patrick's death, Brigid introduced a convent system at Kildare with a special concern for the poor.

The Church of Ireland retained her independence from the see of Rome, though surrendering her practice in favor of the Roman observance of Easter in 704. Roman influence was seen in the Irish Prayer Book at the end of the seventh century. Extreme sufferings were endured during the Viking invasions from 795 to 1014. Monasteries and churches were plundered and ruined, the people and their pastors were led captive or put to death, and standards of culture and of religion inevitably suffered. The defeat of the invaders brought quieter times, and in the twelfth century the constitution of the church was reformed.

The period from 1200 to 1500 saw the transfer of authority to English government, the building of some fine cathedrals, and the development of a distinctive spiritual character. The breach with Rome in the sixteenth century heralded a return to a measure of independence, and while the Reformation was inadequate in origin, the Irish Articles* were strongly Calvinistic, her bishops were commendable for their soundness and piety, and the founding of the university in Dublin in 1591 was a mark of progress.

The Church of Ireland bridged the gap between the people of the land and the ruling classes. She survived the tragedy of the 1641 rebellion, the pressures of Cromwell—and there was no gap in the succession of her bishops. The progress of the eighteenth and nineteenth centuries was steady, but the census of 1861 showed that the church had claim on only one-eighth of the total population. The Church of Ireland (disestablished in 1869-70) has made an outstanding contribution to culture and government and is well geared today for work especially in education and in the industrial areas of the country.

See IRELAND for bibliography.

ADAM LOUGHRIDGE

IRENAEUS (fl. c.175-c.195). Bishop of Lyons. Probably a native of Smyrna, where as a boy he listened to Polycarp,* he perhaps studied and taught at Rome before moving to Lyons. As presbyter in 177/8 he mediated in his church's behalf with Bishop Eleutherus of Rome over the Montanists.* On his return he succeeded Bishop Pothinus (who had died in the persecution of 177/8), probably without episcopal consecration. He represented an important link between East and West, corresponded widely, and protested against Pope Victor's excommunication of the Asian Quartodecimans.*

His diocese included also Vienne and possibly congregations further afield (he was Gaul's sole monarchical bishop) and involved him in speaking Gallic (Celtic). He encountered Gnostic activity and devoted five books to the *Detection and Overthrow of Falsely-named Knowledge (Gnosis)*, usually called *Against Heresies (Adversus Haereses)*, which are invaluable in recording Gnostic teachings, especially of the Valentinians. He drew on Gnostic works and earlier refutations, mostly lost, and was himself heavily used by later antiheretical writers. Irenaeus's Greek survives only in extensive extracts, but can be reasonably reconstituted from a close Latin translation produced before 421 (perhaps as early as about 200), an Armenian version of Books 4-5, and several Syriac fragments. His *Demonstration of the Apostolic Preaching*, rediscovered in an Armenian translation in 1904, is both catechesis and apologia, expounding Christian theology and christological proofs from OT prophecy.

Irenaeus almost belongs to the Apostolic Fathers. Through Polycarp he claimed contact with the apostolic generation and the traditions of the Elders, and in his time the Spirit still dispensed *charismata* and the bishop was still a presbyter. Yet his apostolic tradition, embodied in the Rule of Faith and transmitted by successions of teachers in churches of apostolic foundation, was a developed ecclesiastical tradition. The NT writings are paralleled with the OT as Scripture, and the four-gospel canon stoutly defended. Against Gnostic scriptures, traditions, and successions he erects the apostolic pillars of catholic orthodoxy. The unity of Father, Son, and Spirit in both creation and redemption (including the millennial resurrection of the flesh) is strongly emphasized, and in his key concept of "recapitulation" he develops Paul's Adam-Christ parallel (extending it to Eve-Mary) and views the Incarnation as the climactic summation of God's dealings with mankind in creation, education, and salvation, and the unification of the whole human race.

BIBLIOGRAPHY: J. Quasten, *Patrology* 1 (1950), pp. 287-313. *Against Heresies:* best complete edition to date, W.W. Harvey (2 vols., 1857). Extensive new ed. in progress in *Sources Chrétiennes* series (1952ff.). *Demonstration:* ETs by J.A. Robinson (1920) and J.P. Smith (1952).

Selected studies: F.M.R. Hitchcock, *Irenaeus of Lugdunum, A Study of His Teaching* (1914); J. Lawson, *The Biblical Theology of St. Irenaeus* (1948); G. Wingren, *Man and the Incarnation: A Study in the Biblical Theology of Irenaeus* (1959); A. Benoit, *S. Irénée: Introduction à l'Étude de sa Théologie* (1960); H. von Campenhausen, *Fathers of the Greek Church* (1963), Chap. 2; R.A. Norris, *God and World in Early Christian Thought* (1965), Chap. 3; J. Daniélou, *Gospel Message and Hellenistic Culture* (1973), pp. 144-53, 166-83, 221-34, 357-64, 398-408. D.F. WRIGHT

IRISH ARTICLES (1615). The Church of Ireland* had drawn up twelve short articles of religion in 1566. A convocation of the Irish Church held between 1613 and 1615, and moved by a spirit of independence, decided to prepare a set of articles that would reflect their particular beliefs. The new articles, 104 in number, were largely the work of James Ussher.* Though in general agreement with the Lambeth Articles of 1595, they reflected Ussher's Calvinism and the spirit of Puritanism which then prevailed in Trinity College, Dublin. They had also a Presbyterian flavor, for they made no reference to the prelatic orders of bishop, priest, and deacon. Their strong emphasis on predestination and reprobation was to many a stumbling block. Though approved by the convocation and ratified by the lord deputy in 1615, they were replaced twenty years later by the English Thirty-Nine Articles,* largely due to the efforts of Wentworth and John Bramhall, bishop of Derry. Though never used since then, they were never repealed. ADAM LOUGHRIDGE

IRONSIDE, HENRY ALLEN ("Harry") (1876-1951). Bible teacher and author. Born in Toronto, he moved to California in 1886. Although never ordained, he began preaching when fourteen years old. For a time a Salvation Army officer, he later joined the Plymouth Brethren (1896). For over fifty years he traveled widely as a home missionary, evangelist, and Bible teacher. After 1924 he held meetings under the auspices of Moody Bible Institute, was visiting professor at Dallas Theological Seminary (1925-43), pastor of Moody Memorial Church, Chicago (1930-48), and author of over sixty books and pamphlets, mostly biblical and popular, including *Things Seen and Heard in Bible Lands* (1936), *In the Heavenlies* (1937), *Lamp of Prophecy* (1939), *The Way of Peace* (1940), and *The Great Parenthesis* (1943). He died on a preaching tour in New Zealand.

ALBERT H. FREUNDT, JR.

IRVING, EDWARD (1792-1834). Scottish minister. Born at Annan, he graduated in arts from Edinburgh, where he also studied divinity, taught school at Haddington and Kirkcaldy (1810-19), then became assistant to Thomas Chalmers* at St. John's, Glasgow. In 1822 he went to the Caledonian Chapel, London, which proved so inadequate for the hundreds who wanted to hear him that a new church was built in Regent Square in 1827. Many famous people were among his listeners, entranced by one who so eloquently attacked the spirit of the age and the callous indifference of the rich to the poor man at his gate and in his factories.

Irving's friends included Carlyle, Coleridge, and Henry Drummond. Gradually, however, the novelty wore off, and Irving found it difficult to settle to the ordinary pastoral round. Unbalanced emphases crept into his preaching, and many were alienated by his treatment of prophecy, eschatology, his high view of the sacraments, and his encouragement of speaking in tongues during public worship. A sad process of deterioration set in, and latterly he inveighed against political reform, Catholic emancipation, and the University of London ("the synagogue of Satan"). Predictably there was a split in his congregation. Six hundred followers went with him into the wilderness and a succession of temporary meeting-houses. Many of these erstwhile Presbyterians were to join the Catholic Apostolic Church,* the founding of which is often wrongly attributed to Irving.

His writings include *For the Oracles of God* (1823) (*The Times* published daily extracts), *The Doctrine of the Incarnation Opened* (1828), and *The Orthodox and Catholic Doctrine of Our Lord's Human Nature* (1830). The latter particularly led to his arraignment before the London presbytery, charged with holding the sinfulness of Christ's humanity. Though he claimed his words had been misunderstood, he was excommunicated, and in 1833 deposed from the Church of Scotland ministry by the presbytery of Annan. He became an itinerant preacher, was given (and humbly accepted) the modest position of deacon in the emerging Catholic Apostolic Church, died in Glasgow, and was buried in the cathedral there.

See A.L. Drummond, *Edward Irving and His Circle* (1937), and H.C. Whitley, *Blinded Eagle* (1955).

J.D. DOUGLAS

ISAAC, HEINRICH (c.1450-1517). Musical composer. He was one of the most voluminous and versatile of the Netherlandish composers who helped to create the Renaissance style during the later fifteenth century. He was active in Germany and Austria as well as in Italy, where he died. He created a gigantic cycle of polyphonic compositions for the *Proper* of the Mass for the whole church year; commissioned for the diocese of Constance, it is known as the *Choralis constantinus.* This monumental work was published in the middle of the sixteenth century, showing the continuing esteem in which his work was held. Among his Masses is the attractive *Missa carminum,* based entirely on secular songs adapted to the traditional sacred text. One of these tunes, his own "Innsbruck," was later adopted as a Lutheran chorale and is among those included by Bach in his *St. Matthew Passion.*

J.B. MAC MILLAN

ISAAC THE GREAT (c.350-440). Armenian patriarch (catholicos). Son of Nerses, the sixth catholicos of the Armenian Church,* Isaac received a good education at Constantinople. He married, but after the death of his young wife he became a monk. In 390 he became the tenth catholicos and negotiated the severance of the Armenian Church from submission to the see of Caesarea in Cappadocia. He also encouraged the creation of an Armenian literature, including the translation of the Bible and theological works; he himself helped with the latter. Some hymns are attributed to him. His name is highly venerated in the Armenian Church, and two days in the church calendar are set aside for remembrance of him.

PETER TOON

ISAAC OF NINEVEH (d. c.700). Nestorian* bishop of Nineveh; sometimes called "Isaac Syrus." Originally a monk in Kurdistan, he was made bishop by the patriarch George, but after only about five months in the episcopate he retired to a monastery at Rabban Shapur. During his later life he was suspected of departing from Nestorian tenets. Although he wrote in Syriac, many of his extensive writings (mostly on asceticism and related topics) were translated into Greek, Arabic, and Ethiopic. The Greek translation was undertaken by two monks, Patricius and Abraham, of the monastery of Mar Saba, near Jerusalem. Part of this was eventually published at Leipzig in 1870 by Nicephorus Theotokios. Most of the Syriac text remains unpublished, but there are Latin versions in J.P. Migne, *Patrologia Graeca,* LXXXVI (1) pp. 811-86. PETER TOON

ISABELLA OF CASTILE (1451-1504). Daughter of John II of Castile, she married Ferdinand of Aragon in 1469. On the death of her half-brother Henry IV in 1474, she ascended the Castilian throne, and in 1479 Ferdinand became king of Aragon (see FERDINAND V). The result was a union, not of countries, but of crowns—whose bearers came to be known as the "Catholic Sovereigns." Isabella was successful in increasing royal power at the expense of the independence of the nobles, the towns, and the church. She developed a regular army and played a personal role in its successful campaign to recapture Granada (1492). She was the friend of exploration and learning, and the enemy of heretics and infidels, supporting the endeavors of Columbus and the universities while instigating the work of the Spanish Inquisition (1478) and developing the policy for the expulsion of Jews (1492) and Muslims (1502). Isabella's five children included Joanna "the Mad," who gave birth to the future emperor Charles V, and Catherine who married Henry VIII.

T.L. UNDERWOOD

ISHO'DAD OF MERV (ninth century). Nestorian* bishop of Hedatta on the Tigris. Facts concerning his life have been gathered largely from Arab sources. At one point he was considered as a candidate for a patriarchal see. His fame rests on his commentaries, which were written in Syriac and which demonstrate unusually rich acquaintance with earlier exegetical writings. Because of his attempt to reconcile in his own work the allegorical method of exposition current among Monophysites with the more scientific approach espoused by Theodore of Mopsuestia and the Nestorians, he ranks as a key figure for understanding biblical exposition in Eastern Christianity.

ISHU'-YAB III (before 600-658). Patriarch of the East from 650. Son of a Persian Christian, he was a monk at Bait Abe in N Mesopotamia, then became bishop of Nineveh before 627, and metropolitan of Arbil about 637. He fled from his diocese during the Byzantine invasion of 627, was sent as a member of a peace embassy to Constantinople in 630, and on his way back stole some relics from a church in Antioch. He vigorously opposed the teachings of Sadhona of Ariwan. After appointment as patriarch he set up a monastic school, and encouraged high standards of singing. By this time the Arabs were masters of Mesopotamia. His long-winded, rhetorical letters are important first-hand evidence for the history of the time. He succeeded, probably with Muslim help, in bringing Shim'un, metropolitan of Fars, into obedience, but not before the latter had lost the Christians of Oman to the Muslims. Ishu'-Yab probably created four new metropolitan provinces—Samarqand, China, India, and Qatar. He had a high, mechanical view of apostolic succession; his views on schism were similar to those of Cyprian of Carthage.* WILLIAM G. YOUNG

ISIDORE OF PELUSIUM (c.360-c.440). Ascetic and theologian. Born probably at Alexandria where he received a systematic theological education, he was for forty years abbot of a monastery near Pelusium on the eastern estuary of the Nile. He took part in fifth-century controversies, supporting the memory of Chrysostom, whose exegesis he followed, and warning Cyril of Alexandria to be moderate in his dealings with Nestorius. He also seems to have opposed Eutyches.* Isidore left some 2,000 letters which contain much of doctrinal, exegetical, and moral interest. He followed Athanasius in Christology and appears to have anticipated the terminology of the Council of Chalcedon. He held that the Holy Spirit was consubstantial with the Father and the Son. He defined the church as "the assembly of saints knit together by correct faith and excellent manner of life," adding that it should abound in spiritual gifts. J.G.G. NORMAN

ISIDORE OF SEVILLE (c.560-636). Archbishop of Seville and encyclopedist. His parents fled to Seville from Cartagena when the city was destroyed by Arian Goths. He was born in Seville and educated in a monastery primarily by his elder brother, Leander, who became archbishop of Seville. From his earliest years he showed great aptitude in learning, and his studies covered virtually all areas of contemporary knowledge. About 600 he became archbishop and as such founded schools, laid plans for the conversion of the Jews, and also presided over church councils —e.g., at Seville (619) and Toledo (633).

Without any doubt his chief importance lies in his writings. His *Sententiarum libri tres* was the first manual of Christian doctrine in the Latin Church: the first book dealt with dogma, the second and third with ethics. The *Etymologiarum sive originum libri viginti* was an encyclopedia in twenty books distilling all the knowledge of his time in all fields; by means of it he virtually became "the schoolmaster of the Middle Ages." Topics covered included grammar, rhetoric, mathematics, music, jurisprudence, history, theology, heresies, geography, geology, clothing, agriculture, and anthropology. Many of his earlier writings were used in this massive work. On the Bible he wrote a general introduction, *Prooemiorum liber unus;* biographical sketches of biblical characters, *De vita et morte sanctorum utriusque Testamenti;* and an allegorical interpretation of the OT, *Quaestium in Vetus Testamentum libri duo.* His *Historia de Regibus Gothorum, Vandalorum et Suevorum* is the principal source for the history of the Visigoths.

He died at Seville and became the national hero of the Spanish Church. He was canonized in 1598 and formally accepted as a "Doctor of the Church" in 1722.

BIBLIOGRAPHY: His works are in J.P. Migne, *PL* LXXXI-LXXXIV. For a modern edition of his *Etymologiae* see the two-volume edition by W.M. Lindsay (1911). For his life and work see E. Bréhaut, *An Encyclopaedist of the Dark Ages* (1912), and P. Séjourne, *Le Dernier Père de l'Église, Saint Isidore de Séville* (1929); J. Fontaine, *Isidore de Séville et la Culture Classique* (2 vols., 1959).

PETER TOON

ISLAM (Arabic = "obedience": one who is "obedient" is a Muslim). The youngest of the world's great religions, founded in Arabia by Muhammad (c.570-632). It contains material drawn from both Judaism and Christianity, but regards Muhammad as the final revealer of the unity and the will of God (Allah). The basic confession of Islam ("There is no god but Allah, and Muhammad is his prophet") is simple, but the implication is total obedience, as the name Islam indicates.

Muhammad was active in two Arabian cities: Medina and Mecca, the latter of which is the holy city of Islam. It was there he received the revelation of the *Qur'an** (Koran), and it is to Mecca that every Muslim is expected to make a pilgrimage at least once in his life. Within a century of Muhammad's death the influence of Islam had been extended from Spain in the west to India in the east, and this process of expansion continued for several centuries. Today expansion by conquest has ceased, but Muslim missionary activity is intensive in Africa south of the Sahara, and increasingly in the West. Centers of Muslim influence are still in North Africa and the E Mediterranean, Asia Minor, Iran, Pakistan, Malaysia, and Indonesia.

The doctrinal history of Islam is complex. The main historical division is between *Sunnite* (from *sunna,* "accepted practice") and *Shi'ite* (from *shi'a,* "party") groups, which disagreed over the leadership question. *Sufism* is the main Islamic form of mysticism. But theological distinctions are slight: accepted by all are such points of doctrine as monotheism, the prophetic office of Muhammad, and the infallibility of the Qur'an. Worship is closely regulated. Five times a day the faithful should pray—if possible in a mosque, but if not, on a prayer-mat—and always in the direction of Mecca. On Fridays special services are held, and during the month of Ramadan fasting takes place between sunrise and sunset. Islam forbids the making of images, and the ornamentation of the mosque is restricted to patterns and elaborately carved texts from the Qur'an.

Since Islam is post-Christian, maintains its own finality as a divine revelation, and holds Jesus to have been a forerunner of Muhammad, relations between Christianity and Islam have always been strained, and Islam has always been passionately resistant to the Christian message. So far there has been practically no accommodation on either side: proclamation has been met with counter-proclamation, and in Muslim countries (theocracies) the Christian Church has as a rule a minority position and little direct influence.

BIBLIOGRAPHY: *Encyclopaedia of Islam* (1913-38; new ed. in progress); T. Andrae, *Die Person Muhammeds in Lehre und Glauben seiner Gemeinde* (1917); W.M. Watt, *Muhammad at Mecca* (1952), and *Muhammad at Medina* (1955); J.W. Sweetman, *Islam and Christian Theology* (1945-55); K. Cragg, *The Call of the Minaret* (1956) and *Sandals at the Mosque* (1959); W.C. Smith, *Islam in Modern History* (1957).

E.J. SHARPE

ISSY, ARTICLES OF (1695). These thirty-four articles issued at Issy near Paris were the result of an ecclesiastical investigation of the works of Madame Guyon* in 1694-95. She chose three bishops to sit in judgment on her doctrine, and their verdict (the Articles) condemned for Quietism*: e.g., indifference to one's own salvation or past sins, desire to suppress explicit acts of faith, secret mystical doctrine, or any claim that extraordinary states of prayer are the only way to perfection. The Articles were a compromise; she signed them herself, retracting her Quietism, as did Archbishop Fénelon* of Cambrai. Consequently Bishop Bossuet* of Meaux, who had been one of her judges, was determined to remove any traces of her influence, while Fénelon wanted her principles to stand. Sympathetically he published forty-five articles (1697) on true and false mysticism which Bossuet attacked and the Holy See condemned. C.G. THORNE, JR.

ITALY, see PAPAL STATES, RISORGIMENTO

IVO OF CHARTRES (1040-1116) Bishop of Chartres. A great canonist, he had studied first at Paris and then at Bec under Lanfranc.* In 1090, having become a celebrated teacher, he was appointed to the see of Chartres which was already famous for its school. As bishop he showed courage in opposing Philip I's proposals to desert his wife and remarry. As a result he was imprisoned in 1092. He was a moderate in his involvement in the Investiture Controversy,* suggesting that a king could not grant the spiritual office but might bestow the temporalities. In this solution he prefigured the Concordat of Worms (1122). Of his works, *Panormia* (seventeen books) and the *Decretum* (eight books) are of greatest significance. In these he brought together ecclesiastical rules from a wide background in an orderly collection, suggested ways of discovering underlying unity in the face of apparent incompatibility between authorities, and also provided good patristic documentation. In so doing he paved the way for the synthesis of the canons which Gratian was to complete, provided theological texts for many later theologians, and by his technique inspired Abelard's* *Sic et Non.* Under his guidance Chartres flourished, as a school both of theology and of canon law. Ivo's surviving 288 letters provide a good insight into the political and religious life of the period. C. PETER WILLIAMS

J

JABLONSKI, DANIEL ERNST (1660-1741). Theologian, hebraist, and bishop of the Unitas Fratrum. Born at Nassenhuben, near Danzig, he studied at Frankfurt-an-der-Oder and Oxford. On the death of his father Petrus Figulus Jablonski (1670), he was chosen to succeed his maternal grandfather Comenius* as leader of the Unitas Fratrum, and he was consecrated bishop in 1699. He was appointed preacher at Magdeburg (1683), head of the United Brethren College at Lissa (1686-91), court preacher at Königsberg (1691) and Berlin (1693). When the revived Moravian Brethren sought an episcopal link with the older Brethren Church, Jablonski, assisted by Christian Sitkovius, bishop of the Polish Brethren, consecrated David Nitschmann as bishop, in Berlin (1735). He worked for the union of Lutherans and Calvinists and later sought to reform the church of Prussia by introducing the episcopate and liturgy of the Church of England. He was one of the founders of the Berlin Academy of Science, and its president in 1733. J.G.G. NORMAN

JACKSON, FREDERICK JOHN FOAKES (1855-1941). Anglican theologian. Educated at Trinity College, Cambridge, he was chaplain and lecturer (1882-95) and then dean (1895-1916) at Jesus College. In 1916 he became Briggs graduate professor at Union Theological Seminary of New York, a position held until 1934. He wrote numerous books, primarily in the area of church history, including *History of the Christian Church* (1891), *Christian Difficulties in the Second and Twentieth Centuries* (Hulsean Lectures, 1903), *English Society, 1750-1850* (1916), *Studies in the Life of the Early Church* (1924), *Eusebius* (1933), and many others. He edited *The Parting of the Roads* (1911), a collection of essays of some of his former pupils at Jesus, and *The Beginnings of Christianity: Part I: The Acts of the Apostles* (5 vols., 1919-33) with Kirsopp Lake. W. WARD GASQUE

JACKSON, SAMUEL MACAULEY (1851-1912). American Presbyterian church historian, educator, and philanthropist. Educated at College of the City of New York, Princeton and Union seminaries, and the universities of Leipzig and Berlin, he engaged in pastoral work from 1876 to 1880 and thereafter devoted much time and money to philanthropic causes. From 1895 to 1912 he taught church history at New York University for which he refused any salary. He published widely in that field, and from 1908 edited the *New Schaff-Herzog Encyclopaedia of Religious Knowledge.* He produced several volumes on Zwingli, including *Hulderich Zwingli, the Reformer of German Switzerland* (1901).

ROBERT C. NEWMAN

JACKSON, SHELDON (1834-1909). Presbyterian missionary to the West and Alaska. Born in New York State, he earned his degrees at Union College and Princeton Seminary, afterward serving Presbyterian churches in Minnesota (1859-69). He was then put in charge of his church's western missions (1870-82) and there pioneered the use of prefabricated church buildings. After two years in New York he supervised Alaskan Presbyterian missions from 1884 until 1907. He set up a public school system in Alaska for the government, and in 1892 introduced reindeer into mainland Alaska to help the natives. When elected moderator of his denomination in 1897 he was described as a man who was "by inside measurement a giant." EARLE E. CAIRNS

JACOB BARADAEUS (d.578). The Monophysite bishop from whom Jacobites* take their name. Born at Tella, east of Edessa, he was educated at the monastery of Phasilta near Nisibis. Then, after a visit to Constantinople to plead the cause of Monophysitism* with the sympathetic Theodora, he stayed in the city as a monk. About 542 he was consecrated bishop of Edessa. The nickname "Baradai" was given him because, in order to avoid arrest by the imperial forces he traveled around "clad in rags" (*baradai*). For nearly forty years he moved from place to place in the area from the Nile to the Euphrates, but especially in Syria. He preached, established monasteries, ordained clergy, consecrated bishops, and even created patriarchs. He died in the monastery of Cassianus. A Syriac *Life of Jacob* has been preserved and was edited by J.P.N. Land in *Anecdota Syriaca,* vol. II (1875). PETER TOON

JACOBINS. A name originally applied to French Dominicans whose first house in Paris was under the patronage of St. James and was located on Rue St. Jacques. As an intellectual center of the order it attracted many students, who undoubtedly promoted the term's popularization. When the radical Society of Friends of the Constitution began using these Dominican facilities early during the French Revolution, her members soon found the old term applied to them.

JACOBITES. The Monophysites* of Syria who rejected the doctrine of the two natures in Christ and who have been traditionally named after Jacob Baradaeus.* After the Council of Chalcedon (451) the Syrian patriarch withdrew his church from communion with other Eastern churches because he did not accept the christological doctrine set forth by the council. Often persecuted, this Monophysite church experienced strengthening through the labors of Jacob Baradaeus and his supporters. Also, Empress Theodora treated it sympathetically during the mid-sixth century. It was at the Second Council of Nicea (787) that it was described as "Jacobite" in the anathemas against the Monophysite doctrine.

Though suffering numerical losses through the Muslim conquests and through internal schisms and losses to the Roman Catholic Church in the seventeenth century, the Jacobite Church still exists, but with a small membership. Its patriarch, while taking his title from Antioch, lives elsewhere. The bread for the Eucharist is made of leavened dough mixed with salt and oil; in the liturgy to the *trisagion* is added the words "who was crucified on your account"; and the sign of the cross is made with one finger (perhaps to emphasize the doctrine of the one nature of Christ). Among the theologians of the Jacobites are reckoned Isaac of Antioch, Jacob of Edessa,* and Jacob of Sarug.*

See D. Attwater, *The Dissident Eastern Churches* (1937). PETER TOON

JACOB OF EDESSA (c.640-708). Jacobite* scholar. Born near Antioch, he was elected bishop of Edessa in 684, but soon yielded his position to spend most of his time in monasteries. His presence often led to open hostility, especially at the monastery in Tell 'Addā. He was one of the most important writers of the Monophysite Church in W Syria; Greek theology influenced the area through him. His works included Syriac homilies in verse and prose, liturgies, commentaries on the OT and NT, and a chronicle that took up where Eusebius left off. Guided by Greek and Syriac versions, he revised the Peshitta OT. Among the Greek writers he translated into Syriac was Severus of Antioch. He wrote the earliest known Syriac Grammar, and introduced the use of Greek letters for Syriac vowels. JOHN GROH

JACOB OF NISIBIS (fourth century). Jacob (or James), a solitary ascetic, was made bishop of Nisibis by popular acclaim. A confessor in the persecution of Maximus, he later took a leading part at the Council of Nicea and was classed favorably by Athanasius with Hosius of Cordova and Alexander of Alexandria. Apparently he subscribed to the decrees of the Dedication Council of Antioch (341). He baptized and befriended Ephraem the Syrian,* with whom tradition sometimes confused him. Theodoret and Gennadius record legends of his austerity and wonder-working. Nicknamed "the Moses of Mesopotamia," he organized a week of public prayer which was answered by the death of Arius, and he so defended Nisibis against the Persian Sapor II in 338, 346, and 350 that after he died, his bones were treated as guarantees of civic protection. Christians removed them when Nisibis fell in 363. The Chronicle of Edessa dates his death in 338, but the legendary defense of Nisibis in 350 rules this out. Modern scholars reject the claim that eighteen tracts discovered in Venice and published by Antonelli in 1756 are part of twenty-six Syriac treatises on faith and practice attributed to Jacob by Gennadius. The Roman, Syrian, Greek, Mennonite, and Coptic churches commemorate him. G.T.D. ANGEL

JACOB OF SARUG (451-521). Bishop of Batnae and Syriac theological writer. Born at Kurtnam on the Euphrates and educated at Edessa (now Urfa), the center of Syriac theology and Bible translation, he became a presbyter. During Persian dominance of his country he achieved fame for his care of his people. In 519 he became bishop of Batnae, the main town of Sarug in Osrhoene, but died two years later. He wrote many letters, sermons, funeral orations, hymns, and edifying biographies. He also translated the six *Centuria* of Evagrius Ponticus. Perhaps his principal work was a long series of metrical homilies (3,300 lines, for example, on the Passion of Christ). Several other works, such as liturgies, were ascribed to him. Some scholars regard him as a Monophysite, but he was considered orthodox by his contemporaries. Many of his writings remain unpublished. PETER TOON

JACOPONE DA TODI (c.1230-1306). Franciscan monk and poet. Born into nobility, he studied law at Bologna and became a wealthy lawyer in his hometown. After the tragic death of his wife, he experienced spiritual conversion and repentance, donated his wealth to the poor, and became a lay brother. He supported the "Spirituals," with the cardinals Peter and Jacob Colonna and Angelo Clareno, in their opposition to papal opulence and political machinations. Boniface VIII excommunicated and imprisoned him in 1298. He was released in 1303. Jacopone is the mystic who gave to Italian poetry its sharpest notes of religious experience. His two great works are the hymns *Laude* ("hymns of praise") and *Stabat mater dolorosa* ("the sorrowing mother stood"). He has long been called "blessed" in Todi, but the Congregation of Sacred Rites refuses to consider his beatification, probably due to his sarcastic treatment of Boniface VIII in *Laude*. ROYAL L. PECK

JAMES. Apostle, son of Zebedee, sometimes known as "James the Great." This James was the brother of the Apostle John. He was called from his occupation as a fisherman to become one of the twelve apostles (Mark 1:19,20). It is possible that their mother, Salome, was a sister of Mary and that James was therefore a cousin of Jesus (cf. Matt. 27:56; Mark 15:40; John 19:25). As he is normally mentioned first, he was probably older than his brother. In the lists of the apostles their names are associated with Peter and Andrew. Mark records that Jesus gave the brothers the nickname "Boanerges" ("Sons of Thunder," Mark 3:17). Luke records their asking Jesus whether they should call down fire on a Samaritan village (Luke 9:51-56). Mark states that they themselves, and Matthew that their mother, asked that they be given places at the right and left hand of Jesus in His glory. The two brothers are associated with

Peter in a sort of inner circle of Jesus' disciples at the home of Jairus (Mark 5:37, etc.), on the Mount of Transfiguration (Mark 9:2, etc), and in the Garden of Gethsemane (Mark 14:33, etc.). James is the only one of the apostles (apart from Judas) whose death is recorded in the NT. He was martyred by Herod Agrippa in A.D. 44 (Acts 12:1,2). There was apparently no attempt to replace him.

R.E. NIXON

JAMES. Apostle, son of Alphaeus. The name comes from the Greek *Jacobos* which renders the Hebrew Jacob, and it was a common one. He heads the third group of four disciples in each list of the twelve. His name is not found elsewhere in connection with any incident during or after the ministry of Christ. As Levi is said to be the "son of Alphaeus" (Mark 2:14), it has been suggested that James and Levi were brothers, but there is nothing else to link them together. It is not clear whether James, son of Alphaeus, is the same as "the younger James," or the "the smaller James" (AV "James the less") of Mark 15:40. Some have tried, implausibly, to identify him with James the brother of Jesus, thus giving the latter a place among the twelve.

R.E. NIXON

JAMES II (of England) and VII (of Scotland) (1633-1701). King of Great Britain, 1685-88. Coming to the throne because his brother Charles II was officially childless, James was a Roman Catholic who had been abroad for many years. He might have been passed over in favor of his daughters Mary and Anne, especially as Mary's husband, William of Orange, had strong claims in his own right. Respect for monarchy was strong, however, and James would have retained his position had he not pushed Catholic policies and ensured a Catholic succession. Though Parliament initially gave tangible sign of its goodwill and support, James overreacted to uprisings in his kingdom, alienated Parliament and influential Anglicans, and appointed his coreligionists to high office. As a political expedient he eventually courted the Nonconformists, but he failed to win them or to allay Anglican fears. In Scotland he intensified the persecution of the Covenanters.* The birth of a son in June 1688 precipitated the issue. William of Orange was ready, and the country ready to receive him. James fled and heard Mass in France as his people celebrated a Protestant Christmas.

See F.C. Turner, *James II* (1948), and J.P. Kenyon, *The Stuarts* (1958).

P.W. PETTY

JAMES VI (of Scotland) and I (of England) (1566-1625). King of Scots from 1567, his descent from Henry VII made him the nearest heir when Queen Elizabeth of England died in 1603. Although he assumed the title James I of Great Britain, the kingdoms were not united until 1707.

Son of Mary, Queen of Scots* and Lord Darnley, he was proclaimed king by the nobles who forced his mother's abdication and placed him under the tutorship of George Buchanan.* Four regents followed in quick succession, and even after James' coming of age in 1578 rival groups made bids for power by seizing his person. With this background James was determined to become a "universal king"—king of the whole nation and beyond the power of factions. This view conflicted with the "Two Kingdoms" theory of Andrew Melville* and the Presbyterians which meant that the secular kingdom of the state should not interfere with the spiritual kingdom of the church. Despite Buchanan's constitutional teaching, James was unwilling to have any area excluded from his jursidiction and aimed at the "One Kingdom" which he would rule under God alone. As he was a notable exponent, therefore, of the Divine Right of Kings,* his political ideas found expression in his *Trew Law of Free Monarchies* (1598) and *Basilikon Doron* (1599).

Astute use of opportunities and nicely calculated gifts of the church temporalities which had been annexed to the Crown in 1587 enabled him to break the power of the Scottish nobility and to impose various forms of episcopacy over the Presbyterian structure of the Scottish Church. Thus, when he went to England he boasted that he was able to rule Scotland by his pen which others had not done by the sword.

James seemed much less able to assess the political and ecclesiastical situation in England. His policy of peace with Spain and his extravagant expenditure on his favorites led to quarrels with Parliament. His attempts to improve the lot of Roman Catholics were unpopular and ineffectual enough to provoke the Catholic Gunpowder Plot* (1605). At the Hampton Court Conference* (1604) James astonished the English divines with his theological learning, but failed to understand the Puritans' viewpoint. Confusing them with Presbyterians, he sternly ordered them to conform in ceremonial matters. A positive result of the conference was the planning of the Authorized (King James) Version of the Bible, which was published in 1611.

In Ireland James was responsible for the settlement of Protestant English and Scots in Ulster and indirectly for its modern divisions.

See D.H. Willson, *James VI and I* (1956), and D. Mathew, *James I* (1967).

HENRY R. SEFTON

JAMES, BOOK OF ("Protevangelium"). An apocryphal infancy gospel of the late second century professing to have been written by James the Lord's brother. The original Greek text has survived, and also various oriental versions are in existence, the oldest being the Syriac, though there is no Latin version. It describes the miraculous birth and infancy of the Virgin Mary, whose parents are here for the first time given the names of Joachim and Anne. It then goes on to deal with her relationship to Joseph and the birth of Jesus (using the early chapters of Matthew and Luke as a basis). Here Joseph is presented as an old man who already had sons. The story of the death of Zechariah seems not to belong to the original. The "Protevangelium," as it has been called since the sixteenth century, may have been known by Clement of Alexandria. It was certainly known by Origen, for he refers to it when asserting that the brothers of Jesus were sons of Joseph by a former wife.

R.E. NIXON

JAMES, EPISTLE OF, see EPISTLES, GENERAL

JAMES, LITURGY OF. A very ancient liturgy which is extant in Greek and Syriac. Early tradition ascribes it to James, Jesus' brother, mentioned in Matthew 27:56 and Galatians 1:19, etc. Tradition also makes him the first bishop of Jerusalem. Modern scholarship generally views the *terminus a quo* of its origin as A.D. 450, since this liturgy was used by both the Syrian Jacobites* and the main stream of orthodoxy after the Council of Chalcedon. The liturgy has some similarity to the one associated with Cyril, a fourth-century bishop of Jerusalem, and seems to contain a reference to the discovery of the "true cross" at Jerusalem in the fourth century.

JAMES, WILLIAM (1842-1910). American psychologist and philosopher. After a career first as an artist, then as a medical student, James developed an interest in experimental psychology (1867), and taught physiology, psychology, and philosophy at Harvard. Although plagued with ill-health after 1865, he was very active in lecturing both in America and Europe, and in writing what were to become classics of American philosophy. Among his most important books are *The Varieties of Religious Experience* (1902), *Pragmatism* (1907), and *A Pluralistic Universe* (1909). James wrote for a wide, popular audience. It is partly for this reason that his work seems often unguarded and difficult to summarize. James was a pragmatist in the sense that for him truth is that which we must take account of if we are not to perish. The mind is not simply a passive recipient of sense-data, as in classical empiricism, but is characteristically active. He was anti-reductionist in temper in that he stressed the richness, the "pluralism" of experience, including religious experience, against what he took to be the rigidities of scientific or religious orthodoxy. Religious experience is a well-nigh universal phenomenon; it endures; there must therefore be "truth" in religion. James's brilliance as a descriptive psychologist is apparent in his accounts of religious experience. His brother was Henry James the novelist.

PAUL HELM

JANE FRANCES DE CHANTAL (1572-1641). Founder of the Congregation of the Visitation of Our Lady. Born Jeanne Françoise Frémiot at Dijon, she was the daughter of the president of the *parlement* of Burgundy. She married Baron de Chantal (1592), who was killed in a hunting accident nine years later, leaving her with four young children. She heard Francis de Sales* preach, and she placed herself under his direction in 1604. After her eldest daughter married, and as her fourteen-year-old son was provided for, she took her two remaining daughters to Annecy, and there founded the Visitation Order (1610). At her death there were eighty-six houses, and by 1767 when she was canonized there were 164.

J.G.G. NORMAN

JÄNICKE, JOHANNES (1748-1827). Founder of the first German missionary training school. A Bohemian-born weaver, he was influenced by the Moravians and eventually became pastor of the Bethlehem Bohemian-Lutheran congregation in Berlin. Encouraged by the pious Saxon official, Von Schirnding, Jänicke opened a school in 1800 to train young men for missionary service. It being a faith venture, he gave the instruction himself and received assistance from English missionary societies, individual German Christians, and even the Prussian king. Its eighty graduates, among them Karl Rhenius* and Karl Gutzlaff,* served under various societies in Africa and Asia. Because his son-in-law and successor refused to merge the school with the Berlin Mission Society (founded 1824), it rapidly declined after his death and was dissolved in 1849. In 1805 Jänicke had formed a Bible Society and in 1811 a tract society which were forerunners of the later Prussian Bible (1814) and Tract (1816) societies.

RICHARD V. PIERARD

JANSEN, CORNELIUS OTTO (1585-1638). Roman Catholic bishop of Ypres (1636). He studied at Louvain and Paris, where he met Jean Duvergier de Hauranne, later abbot of Saint-Cyran. At Bayonne and Champre (1612-17) in the company of Saint-Cyran, Jansen immersed himself in the writings of Augustine. Against the Jesuits and the theologians of the Counter-Reformation,* Jansen and Saint-Cyran wished to reshape Catholicism with the teaching of Augustine and defeat Protestantism with its own weapons. Jansen became the director of a college at Louvain (1617), a public opponent of the Jesuits (Madrid, 1626-27), and eventually bishop of Ypres. His masterpiece was the *Augustinus of the Doctrine of St. Augustine on the Health, the Sickness and the Cure of Human Nature: against the Pelagians and those of Marseilles* (1640). This treatise was condemned as heretical by the Sorbonne in 1649 and by Innocent X in 1653.

See JANSENISM. DAVID C. STEINMETZ

JANSENISM. A radically Augustinian movement in the Roman Catholic Church in the seventeenth and eighteenth centuries, whose teaching was summed up in five propositions condemned by Pope Innocent X in 1653: (1) that it is impossible to fulfill the commands of God without special grace; (2) that grace is irresistible; (3) that only freedom from compulsion is needed for merit, not freedom from necessity; (4) that it is semi-Pelagian to teach that grace can be resisted or complied with by free will; and (5) that it is semi-Pelagian to teach that Christ died for all men.

While such ideas had been found in the writings of strongly Augustinian theologians throughout the history of the church, the Jansenists (who took their name from C.O. Jansen*) drew practical conclusions from these ideas which undercut the sacramental and hierarchical claims of the church of the Counter-Reformation.* The sacraments of the church were only efficacious when God had already transformed the inner disposition of the recipient by His grace. Because the grace of God was strictly limited to the elect, the church need not preoccupy itself with the conversion of men still outside the visible institution, but should rather purify itself by severe discipline and rigorous asceticism. The sacraments were restricted in their use to those who by their moral discipline had qualified themselves to receive them. Everything in the church which did not have divine sanction should be mercilessly excised.

The Jansenists were antipapal in their sentiments, admitting the right of the pope to condemn the five propositions taken from the *Augustinus*, while rejecting the condemnation itself. Jansenist views of free will, predestination, stringent moral asceticism, the sacraments, the hierarchy, and the mission of the church brought them into inevitable conflict with the Jesuits.

The first Jansenists, including the convent of Port-Royal, were generally known as Cyranists after the abbot of Saint-Cyran (Duvergier), the friend and colleague of C.O. Jansen. The Jansenists had already assumed a definable shape by 1638. After the death of Jansen and Saint-Cyran, Antoine Arnauld* became the acknowledged leader of the movement (1643), whose most illustrious member was Blaise Pascal.

While Arnauld and his generation died in communion with Rome, Jansenist ideas were repeatedly condemned, most vigorously in the decree *Unigenitus* issued by Clement XI in 1713 against the teaching of Pasquier Quesnel.* The headquarters of the Jansenists at Port-Royal was destroyed, and the movement was subject to persecution in France. In Holland, however, Jansenism was tolerated, and in 1723 the Jansenists created the schismatic bishop of Utrecht. Jansenism also flourished in Tuscany, giving articulation to its views in the Synod of Pistoia (1786). Thousands of volumes were produced in the seventeenth and eighteenth centuries by and about Jansenism, many of which were written in the vernacular and sold to the public at large.

BIBLIOGRAPHY: J. Carreyre, "Jansenisme," in *Dictionnaire de Theéologie Catholique*, vol. VIII (1924); N.J. Abercrombie, *The Origins of Jansenism* (1936); M. Escholier, *Port-Royal* (1968).

DAVID C. STEINMETZ

JAPAN, CHRISTIANITY IN. Although some claim evidence of Nestorian influence on early Japanese Buddhism, Christianity's introduction to Japan is generally held to have been the arrival of the Spanish Jesuit, Francis Xavier,* with two Japanese converts on 15 August 1549. Under the patronage of the ruling warlord, Nobunaga, and his successor, Hideyoshi, the Roman Catholic faith spread rapidly. In 1587, however, Hideyoshi issued an edict banning all missionaries. Fierce persecution set in, with thousands dying for their faith, including twenty-six Christians who were publicly crucified. By 1640 only 150,000 secret Christians remained. For the next 250 years the Tokugawa line cut off Japan from the outside world until 1853, when the isolation was forcibly broken. Christianity was still forbidden to the Japanese, but the trade treaties drawn up with the West opened the door for missionaries.

The first Protestant missionaries, the Rev. John Liggins and Bishop Channing-Williams of the Episcopal Church, arrived in May 1859. Later that year two other American denominations sent missionaries: Dr. James Hepburn* (Presbyterian) and Drs. Guido Verbeck and S.R. Brown (Reformed Church). In 1860 the Baptists sent their first. Despite the edicts outlawing Christianity, the first convert was baptized in 1864, and the first church was organized at Yokohama in 1872. Catholicism's renaissance began with Father Girard's arrival in September 1859. The first church was established in 1862, and in 1865 thousands of secret Christians, descendants of the seventeenth-century Catholic believers, revealed themselves. Intense persecution followed, but was eased in 1873 when the edicts banning Christianity were removed. Ivan Kasatkin, later Bishop Nicolai,* founded the Eastern Orthodox Church and saw it grow to 30,000 members.

The main characteristic of the Protestant movement in the 1870s was the emergence of Christian bands: at Yokohama under the Rev. John Ballagh, at Kumamoto under Captain L.L. Janes, at Sapporo under Dr. W.S. Clark,* at Nagasaki under Dr. G. Verbeck,* and at Hirosaki, etc. From these bands came many leaders in the period of rapid growth from 1880 to 1889, notable Yuzuru Neeshima,* Kanzo Uchimura,* Masahisa Uemura, and Yoichi Honda. Mergers between denominational missions and their churches resulted in the Japan Episcopal, Presbyterian, Congregational, and Methodist denominations, and numerous interdenominational bodies were founded such as the Japan YMCA (1880), Scripture Union (1884), Christian Endeavour (1886), and the Bible Society (1890).

Although religious freedom was granted with the promulgation of the national constitution in 1889, the 1890s were marked by a reaction against Christianity, two contributory factors being the Imperial Rescript on Education and the realization that modernization did not require Christianization. With the rise of capitalism in the mid-1890s came modern social and labor problems. Newly arrived Salvation Army officers found in Gunpei Yamamura a leader for the work of freeing the indentured prostitutes and helping the poor in the slums. The catastrophic influence of liberal theology which had made its first appearance in 1885 was being increasingly felt in the church. A touch of revival came at the turn of the century and with it development of cooperative evangelism and inspirational conferences arranged by Barclay Buxton and Paget Wilkes* who in 1904 formed the Japan Evangelistic Band.

From 1900 to 1920, as Japan's industrial revolution was intensified, Christians became more involved in social work, the contribution of Toyohiko Kagawa* being the most outstanding. Through the preaching of Charles Cowman, Ernest Kilbourne, and Juji Nakada who had organized the Oriental Missionary Society in 1898, the Japanese Holiness Church came to birth in 1917. In Hokkaido the work of the Rev. John Batchelor, the "apostle to the Ainu," was at its height, and on the main island the Central Japan Pioneer Mission was begun in 1925. The rise of militarism in the 1930s led to increasing curtailment of religious freedom until 1941 when thirty-two major Protestant groups were forcibly amalgamated into the United Church of Christ in Japan (called the

"Kyodan"). In the war years many Christians suffered greatly for their faith, but many compromised.

In the postwar period, with the dissolution of State-Shintoism and emperor-worship, Christians began to reorganize, spurred on by a new wave of missionaries in a climate of unprecedented religious freedom. The Anglicans, Presbyterians, Lutherans, and some others soon seceded from the Kyodan, which set up a partnership with eight denominations for channeling financial aid and missionary personnel to churches and schools, the main ones being Meiji Gakuin, Aoyama Gakuin, Doshisha, St. Paul's (Anglican), Kanto Gakuin (Baptist), and the International Christian University. Other Kyodan-related ministries are the KyobunKwan (publishing), Avaco (audio-visual), and the Japan Union Theological Seminary.

The majority of the postwar Protestant Missionary force comprises the denominational missions of the Southern Baptist, American Baptist, Reformed, Nazarene, Christian and Missionary Alliance, Lutheran, Pentecostal churches, and numerous interdenominational faith missions, The Evangelical Alliance Mission (outgrowth of the Scandinavian Alliance Mission started by F. Franson* in 1891 and associated with the Domei denomination), the Far Eastern Gospel Crusade, and the Overseas Missionary Fellowship being the three largest. All engage in evangelistic work and founding churches, most of which have either a denominational link or membership in the Federation of Independent Churches. The prewar Holiness churches reorganized, the largest group forming the Immanuel Church under Dr. David Tsutada.

Cooperative ministries include Pacific Broadcasting Association under Dr. Akira Hatori, Japan Inter-Varsity Christian Fellowship, Japan Sunday School Union, and Word of Life Press. The main evangelical seminaries are the Japan Bible Seminary, Japan Christian Theological Seminary, Kobe Theological Seminary, Tokyo Christian College (liberal arts), and more than a score of Bible institutes.

The Roman Catholic Church has grown conspicuously, aided by many foreign personnel. Churches, hospitals, schools, and universities (e.g., Sophia, Seishin, Nanzan) are their main emphases, and about 50 per cent of Japan's total baptized church membership is in the Catholic Church. These, with Greek Orthodox, nonchurch Christians, conservative, liberal, and neoorthodox, and exogenous and indigenous sects (e.g., Tejimakyo, Spirit of Jesus Church) make up the Christian streams. Evangelicals nationwide have formed the Japan Evangelical Association, while ecumenically minded Protestants and Catholics have the Japan Ecumenical Association.

The Christian Church, with less than one-half of 1 per cent of the population, looks insignificant set against the vast majority whose customs and beliefs are deeply rooted in a Shintoism and Buddhism allied with materialism, and the resurgent religions such as the Soka Gakkai, but its influence is widely felt. With capable Japanese leaders at the helm, with a spirit of partnership between missionaries and the national church, with urban area evangelism expanding, and with an increasing number of Japanese missionaries going out, the Christian cause should continue to advance steadily.

BIBLIOGRAPHY: O. Cary, *A History of Christianity in Japan* (2 vols., 1907); C.R. Boxer, *The Christian Century in Japan, 1549-1650* (1951); J. Natori, *Historical Stories of Christianity in Japan* (1957); T. Yanagita, *A Short History of Christianity in Japan* (1957); C.W. Inglehart, *A Century of Protestant Christianity in Japan* (1959); D. Pape, *Captives of the Mighty* (1959); J.M.L. Young, *The Two Empires in Japan* (1961); J.J. Spae, *Christian Corridors to Japan* (1965) and *Christianity Encounters Japan* (1968); M. Griffiths, *Take Off Your Shoes* (1971).

DAVID MICHELL

JASPER, JOHN (1812-1901). American Negro Baptist preacher. Born into slavery on a plantation in Fluvanna County, Virginia, he was raised by a pious mother. He was employed as a slave in the Richmond tobacco factory of a prominent Baptist layman where he experienced a dramatic conversion in 1837. At his master's urging he began preaching, and after learning to read he immersed himself in the Bible. His fame as a lay preacher and pulpit orator spread rapidly throughout Virginia, and after emancipation he took over a regular congregation, eventually becoming pastor of the Sixth Mount Zion Church in Richmond. Thousands thronged to hear him, and one sermon in particular—"De Sun Do Move" (a defense of the literal movement of the sun around the earth), which he repeated over 250 times—gave him a national reputation. Not a sensationalist, he was known for his simple biblical faith, remarkable eloquence, and ability to relate tenderly to his parishioners.

RICHARD V. PIERARD

JASPERS, KARL (1883-1969). German existentialist* philosopher. He read medicine at Heidelberg, where he became a lecturer in psychology before promotion to the chair of philosophy there in 1921. Jaspers was relieved of his duties by the Nazis in 1937, but reinstituted in 1945. His position under the Nazis was made all the more acute by the fact that his wife was Jewish. From 1948 he taught at Basle. Already in his medical studies he was influenced by philosophy, and he used Husserl's phenomenology and Dilthey's descriptive analytical psychology.

As a philosopher Jaspers developed an independent approach, though he paid considerable respect to the classical philosophers of the past. He early rejected the view that philosophy is a branch of science. He was equally opposed to the idea of the omnicompetence of science. Though science has its proper place, it does not disclose the meaning of life. Philosophy is a type of thinking which is not compelling and does not have the universal validity of the natural sciences, but which nevertheless leads the thinker to himself. It arises out of his inner activity and awakens sources within him which give ultimate meaning. For Jaspers, *Existenz*-philosophy is "the way of thought by which man seeks to become himself."

It does not cognize objects, but elucidates the being of the thinker.

Jaspers held that there is no law of nature or history which determines the way of things as a whole. The future depends upon the decisions and deeds of men, in the last analysis of the individual among the billions of men. He stood apart from institutional religion and regarded the concrete forms of religion as symbols or ciphers. He spoke of "the Encompassing" *(das Umgreifende)* to denote the Being that surrounds us and the Being that we are. It is neither subject nor object, but contains both. The transcendent denotes both the source and the goal of our existence, out of whose depths alone we become authentically human. Although often difficult, Jaspers regarded philosophy not as a specialist study, but as a way of thinking for all who seek illumination of "the Ground within us and beyond us, where we can find meaning and guidance." It is virtually an alternative to religion.

Among the English translations of Jaspers' writings are *Nietzsche and Christianity* (1961), *Truth and Symbol* (1959), *Philosophical Faith and Revelation* (1967), and (with R. Bultmann*) *Myth and Christianity* (1958).

BIBLIOGRAPHY: P. Koestenbaum in *The Encyclopedia of Philosophy,* IV, pp. 254-58; P.A. Schilpp (ed.), *The Philosophy of Karl Jaspers* (1957): containing an autobiography, twenty-four studies, a reply by Jaspers, and a bibliography; C.F. Wallraff, *Karl Jaspers: An Introduction to his Philosophy* (1970). See also bibliography under EXISTENTIALISM.

COLIN BROWN

JEFFERSON, THOMAS (1743-1826). Third president of the USA and political philosopher. Born at Shadwell, Virginia, he was educated at the College of William and Mary (1760-62) and admitted to the bar in 1767. He was a member of the House of Burgesses (1769-76), the Continental Congress (1775-76, 1783-84), and the Virginia House of Delegates (1776-79), governor (1779-81), minister to France (1785-89), secretary of state (1790-93), vice president of the USA (1797-1801), and president (1801-9). His administration saw the Louisiana Purchase, the Lewis and Clark Expedition, and war with Algerian pirates. He retired to his home at Monticello in 1809. He was a founder of the Democratic-Republican party and advocated democratic simplicity, agrarianism, state rights, and separation of church and state. He considered the Declaration of Independence (1776), the Virginia Bill for Establishing Religious Freedom (1786), and the founding of the University of Virginia (1819) to be his greatest achievements. A Deist, he deleted the miraculous from his edition of the gospels, *The Life and Morals of Jesus of Nazareth.*

ALBERT H. FREUNDT, JR.

JEHOVAH. The traditional English spelling, introduced by Tyndale, of the Hebrew name for the God of Israel; it is generally agreed today that the correct spelling should be "Yahweh." The earliest MSS of the books of the Hebrew Bible contained no vowels, so the sacred name appeared simply as *YHWH* ("the Tetragrammaton"); for knowledge of the pronunciation we are indebted to the Greek writers Clement of Alexandria and Theodoret. Even before the Christian era, the Jews refused to pronounce the name at all and substituted the word "my Lord" (in Hebrew *'adonay*); in later (Massoretic) MSS the vowels of this word were therefore attached to the consonants of Yahweh, as a guide to synagogue readers to substitute *'adonay.* The erroneous reading of this hybrid form as "Jehovah" dates from the medieval period. The KJV and its successors usually prefer to translate the Tetragrammaton as "the LORD" (in capitals). The original meaning of the name is uncertain; suggestions include "He who is," "He who is present," and "He who causes to be."

D.F. PAYNE

JEHOVAH'S WITNESSES. This movement originated with C.T. Russell* as the Watch Tower Bible and Tract Society and the International Bible Students Association—not to be confused with the International Bible Readers Association. The title of Jehovah's Witnesses was assumed in 1931 under Russell's successor, J.F. Rutherford.* On the death of the latter in 1942 the leadership passed to Nathan H. Knorr (b.1905).

Theologically Jehovah's Witnesses resemble Arians in their view that the Son was the first and highest created being. He is identified with Michael the Archangel. When he became man, he became only man, and although at his resurrection he was exalted above the angels as a spirit being, his body remained dead, although it was removed from sight by Jehovah. Christ's appearances were "in materialized bodies." The holy spirit is the active force of God. As a perfect man Jesus died to ransom all the descendants of Adam from the physical death which Adam's sin had inflicted on them. "As a childless man his unborn human offspring counterbalanced all the race that Adam has reproduced." Since Jehovah's Witnesses do not believe in a soul that can live apart from the body (see CONDITIONAL IMMORTALITY), the primary purpose of Christ's ransom is to give the right either not to die physically or to be restored by resurrection. Salvation is through faith in the ransom, through baptism by Jehovah's Witnesses, and through proclamation of their message, together with a moral life. There is virtually no interest in the devotional life. The Lord's Supper is celebrated once a year only at the Passover, and only those who have the inner witness that they are members of the 144,000 elite may partake.

The sect has been continually expecting Armageddon and the setting up of the Kingdom. This kingdom will be governed by Jesus Christ through the 144,000 in heaven, and on earth through an indefinite number of "men of goodwill," "other sheep," or "Jonadabs." This extra class was discovered when it was obvious that Jehovah's Witnesses numbered more than the expected 144,000. The rest of mankind will be raised at intervals, except presumably those who fought against God at Armageddon, and will be judged for life or destruction according to their behavior during the Millennium. In *Life Everlasting in Freedom of the Sons of God* it is thought likely that 1975 will see the start of the millennial kingdom.

The movement is probably the most authority-ridden religious body in the world. Members are told by the central government what they must find in the Bible, and may not deviate. Thus blood transfusion must be rejected as though it were banned by Scripture. Dogmas may be changed. In 1929 "the clear light broke forth" in the *Watchtower* that the higher powers of Romans 13 were not earthly rulers, but Jehovah and Christ Jesus, and Rutherford's books took this up. Recent books (e.g., *Life Everlasting*) have returned to the orthodox interpretation. It used to be stated that the 144,000 were raised in 1918 (e.g., *Let God be True*, p. 192), but now they will be raised in the Millennium (e.g., *Things in which it is Impossible for God to Lie*, p. 350f.).

The sect has over one million active members who have the title of "publishers" (of the good news of the kingdom), and a number of full-time "pioneers." A further million-plus are interested followers. Some 30 per cent are found in America, 8 per cent in West Germany, and 5 per cent in Britain. Many Witnesses have stood firm under shocking persecution, especially under Communist regimes. Since the theology turns on the assertion that Jesus Christ is not Jehovah, it is worth noting that John 12:39-41 says that He was the one whom Isaiah saw in the Temple, and Isaiah 6:5 says this was Jehovah. In Revelation 1:17 Christ describes himself as "the First and the Last," which is the unique title of Jehovah in Isaiah 44:6. Note also similar equations in 1 Peter 2:8 (Isa. 8:13,14) and Revelation 2:23 (Jer. 17:9, 10).

BIBLIOGRAPHY: W.R. Martin and N.H. Klann, *Jehovah of the Watchtower* (1953); M. Cole, *Jehovah's Witnesses* (1956); J.K. Van Baalen, *The Chaos of Cults* (1956); W.J. Schnell, *Thirty Years a Watchtower Slave* (1957) and *Into the Light of Christianity: Basic Doctrines of Jehovah's Witnesses* (1959); G.D. McKinney, *The Theology of Jehovah's Witnesses* (1962); A. Hoekema, *The Four Major Cults* (1963); T. Dencher, *Why I Left Jehovah's Witnesses* (1966).

J. STAFFORD WRIGHT

JEROME (Eusebius Hieronymus) (c.345-c.419). Biblical scholar and translator, he was born of Christian parents in Stridon in NE Italy. Around the age of twelve he went to Rome and studied Greek, Latin, rhetoric, and philosophy under Aelius Donatus. While in Rome he met Rufinus of Aquileia. He allegedly spent his Sundays in the catacombs translating the inscriptions. At the age of nineteen he was baptized. He journeyed to Gaul, became acquainted with monasticism at Treves, and on his return joined a small group of ascetics including Rufinus. About 373 he left the group and went to the East and spent some time living as an ascetic in the desert near Chalcis.

During this time he began to master the Hebrew language, perfected his Greek, and had his famous dream in which he was accused of being a Ciceronian rather than a Christian. He left his ascetic existence and went to Antioch, where he heard the lectures of Apollinaris of Laodicea on Scripture and was ordained without pastoral responsibility by Bishop Paulinus, recognized by Rome as an orthodox bishop. Jerome then went to Constantinople and studied with Gregory Nazianzus* and perhaps Gregory of Nyssa. While there he translated some of the works of Eusebius, Origen, and others. In 382 he journeyed to Rome with Bishop Paulinus and became involved in the dispute surrounding the Melitian Schism.* He became the friend and secretary of Pope Damasus. While in Rome he praised the ascetic life of monasticism and decried the lax moral life of the Christians in the city. He was most successful in winning the female sex to his views of ascetic living, but due to rumors about his relationship with them and the accusation that his harsh asceticism caused the death of one of them, he left Rome after the death of Pope Damasus and in 386 made his home in Bethlehem for the rest of his life. There he oversaw a men's monastery and continued to serve as the spiritual adviser to some of the women who followed him from Rome to establish a convent.

He engaged in theological controversy with Vigilantius, Origen, Pelagius, Jovinian, his good friend Rufinus, and even Augustine of Hippo. In these controversies he used irony, personal attacks, sarcasm, and bitter invective. Yet his service to the church was invaluable and should not be obscured because of the flaws in his complex personality. His scholarship and grasp of languages was unsurpassed in the early church. He engaged in a voluminous correspondence, compiled a bibliography of ecclesiastical writers, wrote *De Viris Illustribus*, wrote commentaries on virtually all the books of the Bible, and perhaps most important of all, upon the urging of Pope Damasus used his great linguistic skills and erudition to translate the Bible into the common tongue of that day.

In the process of producing the Vulgate, Jerome apparently used Origen's *Hexapla* and consulted local rabbis in order to perfect the OT section. He questioned the inclusion of the Apocrypha section although he did use it for edification. His translation is important in that he set the example of working from the original languages. The Vulgate has left a tremendous imprint upon the development of the church, and thus Jerome's scholarship extends its influence into our own day.

BIBLIOGRAPHY: Works in J.P. Migne (ed.), *Patrologia Latina* (1844-64), vols. XXII-XXX; L. Huizinga, *Hieronymus* (1946); P. Antin, *Essai sur Saint Jérôme* (1951); F.X. Murphy (ed.), *A Monument to Saint Jerome* (1952); E. Arns, *La technique du livre d'après saint Jérôme* (1953); J.G. Nolan, *Jerome and Jovinian* (1956); C.C. Mierow, *Saint Jerome: The Sage of Bethlehem* (1959); J.N.D. Kelly, *St. Jerome* (1975).

ROBERT SCHNUCKER

JEROME EMILIANI (Girolamo Miani) (1481-1537). Founder of the Somaschi, a Counter-Reformation* order of clerks regular, specializing in the care of orphans. Born in Venice of noble parentage, he fought in the Venetian army against the League of Cambrai (1508). He was ordained

(1518), returning to Venice to devote himself to relieving suffering following the invasion of N Italy. He opened a hospital at Verona (1518) and the house in Somasca near Bergamo, from which his order was named. He founded at Bergamo Italy's first home for prostitutes. He died at Somasca from typhus contracted while tending sufferers. He was made patron of orphans and abandoned children in 1928. The Somaschi, founded in 1528, began a community life under Augustinian Rule in 1532, were approved by Paul III in 1540, and raised to the rank of order by Pius V in 1568. J.G.G. NORMAN

JEROME OF PRAGUE (c.1371-1416). Bohemian Reformer. A layman, he was a brilliant orator and debater, and a close friend and disciple of John Hus.* Intellectually he was a Realist, following closely upon the Wycliffism which he ardently propounded in Bohemia and especially at Prague, but he always remained orthodox on transubstantiation. After graduating from Prague University in 1398, he greatly contributed to the spreading of Wycliffe's works in Bohemia. Later he received the M.A. degree from Paris where he also lectured, but adverse reaction to his Realism and Wycliffism led to flight successively from Paris and Heidelberg.

In 1407 he helped to lead the nationalist-Wycliffite campaign which granted the Czechs equality of powers in the hitherto German-dominated Prague University. The archbishop of Prague apparently excommunicated him in 1409. In 1410 his preaching at the court of Sigismund of Hungary led to his dismissal. He appeared in Vienna, where the Inquisition arrested him, but despite an oath to the contrary he fled. From 1410 he became much more radical and activist in his opposition to the church. In Prague he was a leader of popular demonstrations against his friend Hus's excommunication, against indulgences, and against religious relics. In 1413 the local episcopate expelled him from Cracow and he went to support the Ruthenian schismatics in White Russia. In 1414 he went to the Council of Constance, honoring a promise to help his friend Hus. He quickly left to avoid arrest, but was brought back and jailed. In June 1415, after Hus's condemnation, Emperor Sigismund demanded that Jerome be dealt with, and he was burnt as a heretic at Constance in 1416.

See "Jerome of Prague," *University of Birmingham Historical Journal,* vol. I (1947).

L. FEEHAN

JERUSALEM. An important Palestinian city, sacred because of its association with David and Jesus to Jews and Christians respectively; it is also sacred to Islam because of its traditional associations with Abraham and Muhammad. The city and the name Jerusalem (? "Foundation of Salim") go back at least to 1800 B.C. Its pre-Israelite kings included Melchizedek (cf. Gen. 14). Conquered by David about 1000 B.C., it became the capital and also the central shrine of the united monarchy of Israel, and subsequently of the kingdom of Judah. During the monarchy the city grew, especially to the north, where the Temple was erected by Solomon.

Both city and Temple suffered destruction by the Babylonians in 587/6 B.C., but were partially rebuilt by the end of the century, and Jerusalem was refortified by Nehemiah (445-433 B.C.). The city expanded greatly in the Maccabean period; and Herod the Great (37-4 B.C.) carried out a major building program. His grandson Agrippa I further added to the northern fortifications (A.D. 41-44), but in A.D. 70 the city and Temple suffered a major destruction by the Roman armies under Titus. Yet further damage was done in the Second Jewish Revolt (A.D. 132-35), after which Hadrian rebuilt Jerusalem as a pagan city named Aelia Capitolina. It became a Christian city in Constantine's reign, with the construction of important churches.

The Persian invasion of A.D. 614 caused some destruction. In 638 the Muslims under the caliph Omar captured Jerusalem, and both the El-Aqsa mosque and the Dome of the Rock ("Mosque of Omar") were erected in the seventh century. The Muslims proved tolerant until the caliph Hakim in 1009 ordered the destruction of many Christian buildings. The First Crusade soon afterward resulted in the creation of the Latin Kingdom of Jerusalem (1099-1187); the Crusaders were again great builders. Saladin drove them out, but did no harm to the Christian shrines. The Muslim rulers in the following centuries were responsible for much of the Muslim architecture still to be seen; the Turkish sultan Suleiman the Magnificent built the existing city walls about 1535.

Gen. E.H.H. Allenby captured Jerusalem in 1917, and a British administration followed until 1948, with Jerusalem as the capital of Palestine. Then the city was divided, the W suburbs held by Israel, the Old City and the E sector in Jordanian hands. The Israelis captured the whole city in 1967. Even in its divided days, the Israelis made it their capital.

Archaeological exploration has been limited and difficult because of the dense population of the city. It is clear, however, that the city's earliest expansion was northward, then westward in the postexilic era. Nehemiah's rebuilding retracted the E walls somewhat, while Hadrian's city retracted the S wall to a marked degree. Modern expansion has been chiefly to the west and north of old Jerusalem.

Solomon's temple, symbolizing the presence of God, gave the city a powerful religious appeal as "the holy city"; the names "Jerusalem" and "Zion" early came to symbolize the people of God. As a theological symbol Jerusalem exercised a profound influence on Christian writing (as early as Galatians 4 and Revelation 21) and hymnology (e.g., "Jerusalem the golden").

BIBLIOGRAPHY: J. Simons, *Jerusalem in the Old Testament* (1952); S. Perowne, *Jerusalem and Bethlehem* (1965); J. Boudet (ed.), *Jerusalem: A History* (ET 1967); K.M. Kenyon, *Jerusalem: Excavating 3000 Years of History* (1967); N.W. Porteous, "Jerusalem-Zion: The Growth of a Symbol," in *Living the Mystery* (1967), chap. 7; J. Jeremias, *Jerusalem in the Time of Jesus* (ET 1969); C. Gulston, *Jerusalem: The Tragedy and the*

Triumph (1978). See also Bible atlases for plans and maps. D.F. PAYNE

JERUSALEM, KNIGHTS OF, see HOSPITALLERS

JERUSALEM, PATRIARCHATE OF. From the account in the Acts of the Apostles, enlarged in Eusebius, it is evident that the first organized Christian ecclesiastical structure was brought into being in Jerusalem, with James, the Lord's brother, as presiding officer. That leadership remained in Judeo-Christian and dynastic hands until the devastation of the city under Hadrian, whence the episcopal list becomes Gentile in name, and the relevance of the city is eclipsed by the primary churches in the major cities of the Greco-Roman world. The honor of birthplace of Christianity was however retained, and intensified when Constantine's mother, Helena, made the city a place of pilgrimage which restored prestige to its diminished jurisdiction.

By canon law the Council of Nicea (325) accorded place of honor to its episcopal structure next after Alexandria, Rome, and Antioch—under the last of whose territorial jurisdiction it might actually be assumed to have come. While the Council of Chalcedon (451) raised the see to patriarchal rank, the Islamic conquest reduced the significance of that rank. The Crusaders disrupted the residential continuity of the office, so that its holders intermittently until 1845 often were to be found in Constantinople instead; they also created in 1099 a Latin patriarchate frequently resident at Acre which lasted until 1291 (nominally 1374), but was reconstituted in 1847. The non-Roman Armenians also have a patriarch of this title, while the Melchites include Jerusalem in the titulary of their patriarch of Antioch.

CLYDE CURRY SMITH

JERUSALEM, SYNOD OF (1672). A generation after the violent death of Cyril Lucar,* who had brought Calvinism into the Eastern Church, the occasion of the consecration of the restored Church of the Holy Nativity in Bethlehem provided an opportunity to exterminate the lingering effects of that influence. Dositheus,* patriarch of Jerusalem (since 1669), with his retired predecessor Nectarius,* convened a synod in Jerusalem in March 1672, in which six other metropolitans among sixty-eight Eastern bishops and ecclesiastics from as far as Russia participated. Their signed decisions serve Eastern Orthodoxy as an equivalent to the Roman Catholic Council of Trent.*

The acts of the synod are in two parts. The "Six Chapters" attack the so-called Confession of Cyril Lucar, declaring it to be forgery, so as to refute the Calvinist impact in a context which shows the patriarchate to have been free from error even while the writings of a patriarch are anathematized. The "Eighteen Articles," to which are appended four questions catechetical style, serve as a renewed declaration of faith; these were drafted chiefly by Dositheus as his "Confession" and have come to be called "the shield of orthodoxy." Article I states the Trinitarian formula with single procession of the Spirit. Article II and the first three questions deal with the relation of Scripture to an authoritative Tradition, the questions making it clear that Scripture, defined as including the Apocrypha (3), being full of difficulties (2), ought not be read indiscriminately by all (1), and hence can receive proper interpretation only within that church free from error. Others equally emphasized traditional "Catholic" doctrine over against the "new" doctrines of the Reformation.

CLYDE CURRY SMITH

JERUSALEM CONFERENCE (1928). World missionary conference gathered on Mount of Olives at Easter 1928. It was the first conference held since the formation of the International Missionary Council,* which was itself the outgrowth of the World Missionary Conference at Edinburgh,* 1910. Its purpose was to reexamine the Christian mission in the light of the spread of secularism. The first globally representative assembly of non-Roman Christians, nearly one-quarter of the 231 members represented in full equality the churches of Asia, Africa, and Latin America. The agenda included urbanization and industrialization in Asia and Africa, rural problems, race relations, war, medical work, religious education, relations between younger and older churches. Fears were expressed, especially by European representatives, that this agenda signified the triumph of the Social Gospel and might lead to syncretistic compromise. Some interpretative reports were given to keep the balance, but some evangelical societies withdrew from their respective national conferences, e.g., the China Inland Mission. The "Message" was drafted by William Temple and incorporated part of the statement prepared by the Lausanne Conference,* 1927. After acknowledging elements of truth in other religions, it affirmed that "Christ is our motive and Christ our end. We must give nothing less, and we can give nothing more."

See *Reports of the Jerusalem Meeting of the International Missionary Council* (8 vols., 1928).

J.G.G. NORMAN

JERUSALEM CONFERENCE ON BIBLICAL PROPHECY (1971). Most Protestants traveling to Bible lands are evangelical believers whose faith embraces a belief, not only in Jesus of Nazareth as the Messiah of OT promise, but also in His return to judge the world and fully establish the kingdom of God. The Jerusalem Conference attracted some 1,500 such evangelicals from thirty-two nations from 15-18 June 1971, to hear speakers expound eschatological themes. The program concentrated mainly on widely shared evangelical views and reflected differences only secondarily. The resurrection of Jesus Christ was affirmed to be the hinge of human history, and decision for or against Christ the final determinant of man's destiny. The conference focused interest on eschatology in a scriptural context at a time when end-time concerns gained wide secular emphasis due to possibilities of nuclear annihilation, global ecological pollution, and world famine due to human overpopulation. That God, not man, determines the outcome of history, and that men and nations are destined for final moral judgment, was a central conference emphasis. The purpose of God in creation, reasserted in the redemption of fallen man, was affirmed to be the ultimate conformity of regenerate mankind to the image of Jesus Christ.

The Jerusalem Conference was significant for the fact that its prophetic interest was not correlated either with social withdrawal or neglect of evangelism. The lively expectation of Christ's return was said to require more earnest missionary engagement in fulfillment of the entrusted Great Commission, and also a more vigorous quest for social justice and widespread repentance and moral renewal since the Risen Lord will judge men and nations for their misdeeds. Most participants saw in the return of modern Jewry to Palestine a fulfillment of OT prophecies. Conference leaders emphasized, however, that Israel as a nation is answerable to the requirements of divine justice no less than her Arab neighbor-nations.

CARL F.H. HENRY

JESUITS (Society of Jesus). The name given in 1540 to a brotherhood founded six years earlier by Ignatius Loyola.* He had been joined by six others: Francis Xavier,* Pierre Le Favre (Faber), James Laynez,* Alphonsus Salmeron, Nicholas Bobadilla, and Simon Rodriguez, and they vowed to go to Palestine or anywhere the pope would send them. They soon gained a reputation in Italy as preachers, leaders of retreats, and hospital chaplains. In 1539 they formed a "Company of Jesus" in Rome, dedicated to instructing children and illiterates in the law of God. In 1540 the Society of Jesus was established by a bull entitled *Regimini militantis ecclesiae.* During the period 1540-55 it grew rapidly, requiring an autocratic structure, which was provided by Loyola's military training and the exercises he had worked out in *The Spiritual Exercises.* These were not novel, for there were parallels in the Rules of Francis and Benedict. Nevertheless, they provided the atmosphere of religious obedience so essential to such a disciplined constitution, culminating in the special promise of obedience to the pope, demanded in full commitment to the Society. Loyola refused to turn it into a contemplative order, convinced that its task was to minister to society; thus he removed the obligation of the religious to say the offices in choir. It was this readiness of Loyola to adjust the old ideals of the monks to the new demands of the age that prepared the way for their success.

They established orphanages, houses for reclaiming prostitutes, schools, centers of poor relief, and even a system of banking for the destitute peasants. Their missionary work expanded; one of their most famous missionaries was Francis Xavier. By the time of Loyola's death in 1556, the Society was one thousand strong and its direction of ministry had changed in that its influence was felt more acutely among the aristocracy than among the poor. This change was accomplished mainly through Loyola's wisdom in adopting modern methods of education. The first Jesuit secondary school was established at Messina in 1548. Colleges were founded in university settings, and the Society became a teaching order and the leading movement in Catholic higher education, providing the most effective teaching methods in contemporary Europe.

The Jesuits were the pope's strong supporters at the Council of Trent* and also found themselves spearheading the intellectual attack on the Reformation* and becoming the foremost Catholic apologists. They arrived in England in 1578 and were much feared. The eighteenth century saw them expelled from Portugal (1759), France (1764), and Spain (1767). Pressure from various states forced Clement XIV in 1773 to issue the bull *Dominus ac Redemptor,* suppressing the Society. Not until 1814 were they restored by Pius VII's *Sollicitudo omnium ecclesiarum.* The Society is today still a powerful force in the world of education, responsible for the Gregorian University in Rome, and nine others in their eastern missions, in addition to their many schools and academies throughout the world.

BIBLIOGRAPHY: T. Hughes, *The History of the Society of Jesus in North America* (4 vols., 1907-17); T.J. Campbell, *The Jesuits, 1534-1921* (1921); J. Brodrick, *The Origin of the Jesuits* (1940); M.P. Harney, *The Jesuits in History* (1941; rep. 1962); L. Polgar, *Bibliography of the History of the Society of Jesus* (1967); B. Basset, *The English Jesuits* (1968).

GORDON A. CATHERALL

JESUS CHRIST (c. 5 B.C.-c. A.D. 30). The Founder of Christianity bore "Jesus" (the Greek form of Joshua or Jeshua) as His personal name; "Christ"* (Gk. *christos,* "anointed") is the title given Him by His followers, who acknowledged Him to be the expected Messiah of Israel. Since the theological aspects of His person and work are treated under other entries, this article confines itself to the historical Jesus. By this is meant not simply Jesus as He is accessible to the scientific methods of the historian who views history as a closed continuum of cause and effect, from which any divine initiative is excluded, but Jesus as He actually was —a Jesus who is adequate to account for the Christian faith and life which have found their basis in Him.

Early days. According to the only two evangelists who give us any record of His birth and infancy, Jesus was born in Bethlehem, about six miles south of Jerusalem, toward the end of Herod the Great's reign (37-4 B.C.), but spent his boyhood and youth in Nazareth, a town in Galilee, where he was brought up, with four brothers and some sisters, in the household of Joseph, the carpenter or builder (who probably died before Jesus emerged into public view), and his wife Mary (who lived on to become, with the rest of the family, a member of the primitive church of Jerusalem). It was generally understood that the family was descended from King David, but Jesus, although He did not repudiate the designation "son of David" when it was given to Him, laid no weight on such descent.

His life up to the age of thirty is unchronicled apart from Luke's record of a visit which He paid to Jerusalem with Joseph and Mary when He was twelve years old. His early years, however, cannot have been uneventful. His home overlooked the mighty highway between Syria and Egypt, and news of the world outside Galilee would reach Nazareth quickly, losing nothing in the telling. At the age of about ten He would hear of the revolt led by Judas the Galilean against Quirinius's census in Judea, and of the severity with which it was crushed (A.D. 6). Sepphoris, four miles to the north, was involved in an earlier revolt during Jesus' infancy and was destroyed by the Romans, but it was subsequently rebuilt by Herod Antipas, who ruled from there as tetrarch of Galilee and Perea until he moved (c. A.D. 22) to his new capital, Tiberias; reports of happenings at his court would readily be carried to Nazareth. Scenes of Israel's ancient history, such as Mt. Tabor, where Deborah and Barak mustered their forces for victory over the Canaanites, or Mt. Gilboa, where Saul fell in battle against the Philistines, were familiar to Jesus from early days; even more pressingly familiar were the realities of Israel's present plight, dominated as she was by the Romans, either indirectly through their clients the Herod family, in the north, or directly farther south in Judea.

Beginnings of Jesus' ministry. Jesus' first public appearance was His receiving baptism in the Jordan near Jericho at the hands of John the Baptist (c. A.D. 27-28). This experience was the inauguration of His ministry. The descent of the dove in the baptismal narrative marked Him out as the one anointed with the Spirit of God to be the Servant-Messiah (cf. Isa. 11:2; 42:1; 61:1); the simultaneous voice from heaven addressed Him in terms which indicated He was Israel's king, acclaimed as Son of God in the oracle of Psalm 2:7, but a king who was to fulfil His royalty as the Servant of Isaiah 42:1 devoting His life to the will of God and the blessing of men. The conviction of His destiny thus impressed on Jesus' mind was confirmed by His temptation experience immediately following.

An early ministry in south and central Palestine, concurrent with the later period of John's activity, is attested for Jesus by the fourth evangelist (John 3:22ff.). But the main phase of His public ministry began in Galilee after John's imprisonment by Herod Antipas (Mark 1:14). In this ministry, the proclamation of the kingdom of God was accompanied by works of mercy and power, in which—especially in the cure of people who were demon-possessed—the advancing forces of that kingdom were manifested. All this aroused great popular enthusiasm throughout Galilee. When Jesus' refusal to be bound by religious convention, together with His insistence on interpreting and applying the law of God in the light of its original intention and not according to the tradition of the elders, meant that the synagogues were no longer available for His preaching, He found larger congregations on the hillside and by the lake shore.

From among His disciples He selected twelve men whom He commissioned to share His ministry, and sent them out two by two throughout the Jewish districts of Galilee to proclaim the kingdom of God. Their activity increased the popular enthusiasm and stimulated the unfriendly suspicion of Herod Antipas who, having recently executed John, felt that he had now a potentially more serious situation on his hands (Mark 6:7-16). On the return of the Twelve, Jesus took them out of Antipas's territory, but was pursued across the lake of Galilee by an excited crowd, prepared for militant action, who hoped that Jesus would put Himself at their head. They had to be dissuaded by "hard sayings" in which Jesus made it plain He had no intention of being the kind of leader they had in mind.

From then on, Jesus' Galilean following largely fell away, although a minority, more appreciative of the inwardness of His mission, remained loyal to Him. The Twelve, in particular, adhered to Him, and Jesus devoted a good part of the next few months to preparing them for the crisis which they would have to face when they went up with Him to Jerusalem for His fateful confrontation with the national and imperial authorities there.

Early preaching. His early Galilean preaching is summed up by Mark (1:15) in the announcement: "The appointed time has fully come, and the kingdom of God has drawn near; repent, and believe in the good news" (own trans.). In its setting this announcement could only mean that the time had come, in accordance with Daniel's apocalyptic visions, for Gentile dominion over Israel to give way to the everlasting and indestructible regime which God had promised to establish, in which dominion and judgment would be exercised by the "saints of the Most High." But that in itself differed little from the program of the Zealots.* The distinctiveness of Jesus' message appears in the content with which He filled this framework.

In this content the principal place is occupied by Jesus' teaching about the God to whom the kingdom belonged. Although He had so much to say about the kingdom of God, He did not speak of God as King (except indirectly, as in some of His parables), but as Father (not that there was any incompatibility between the concept of King and that of Father). When He spoke of God (or to God) as Father, He apparently used the form *Abba,* not a formal liturgical term but the designation which children gave to their father in the affectionate intimacy of the family circle. By His use of this designation Jesus expressed His awareness of loving nearness to God and implicit trust in Him, and—while He spoke on occasion of "the Father" and "the Son" on a plane of their own (e.g., Matt. 11:27; Luke 10:22)—He taught His disciples to use the same designation and think of God in the same way as He did. In His ethical teaching He insisted that the children of God should display the same qualities of grace and generosity as their Father did: if He did not withhold sunshine and rain from sinners, they must show compassion and a forgiving spirit to uncongenial and malicious persons, and be merciful as their Father was merciful (Luke 6:36). Nor should this attitude be a matter only of outward deed and word; it should, like all other ethical attitudes, be rooted in the inward thought and desire.

Those who listened to such teaching, with its emphasis on the supremacy of love, must have marked its fundamental divergence from that of the Zealots and their sympathizers, who inculcated an attitude of mortal hostility toward the Romans and those who collaborated with them. With His advocacy of nonretaliation and nonresistance Jesus coupled an active ministry which did not satisfy the messianic expectations of those who expected the restoration of David's kingdom or the execution of judgment upon the ungodly. When challenged on this point, Jesus emphasized He was indeed fulfilling the prophetic hope; as He said in the message which He bade the two disciples of John the Baptist deliver to their incarcerated master: "the blind receive their sight and the lame walk, lepers are cleansed and the deaf hear, and the dead are raised up, and the poor have good news preached to them" (Matt. 11:5 RSV; Luke 7:22). These were things which the greatest prophets of Israel had associated with the advent of the new age (cf. Isa. 35:5f.); in particular, the emphasis on the preaching of the good news of liberation to the poor marked Jesus out as the Spirit-anointed speaker of Isaiah 61:1f., who is commissioned to engage in precisely such a ministry. No wonder that in His programmatic sermon in the Nazareth synagogue, recorded in Luke 4:16ff., He began his exposition of Isaiah 61:1f. with the words: "Today this scripture is fulfilled in your hearing."

The kingdom of God and the Son of Man. The teaching of Jesus envisages two phases in the coming of the kingdom. The earlier, preparatory phase in which it was subject to limitations was present in His ministry—"If it is by the finger of God that I cast out demons, then the kingdom of God has come upon you" (Luke 11:20 RSV; cf. Matt. 12:28)—and should be manifested in the lives of His followers. But one day soon it would come "with power" (Mark 9:1), the limitations having been removed; and for this day Jesus taught His disciples to pray "Thy kingdom come" (Matt. 6:10 RSV; Luke 11:2). The frontier between the two phases is marked by the passion of the Son of Man.

The close association of the Son of Man with the kingdom of God in Jesus' teaching reflects Daniel 7:13f., where eternal and universal dominion is bestowed by God on "one like a son of man," whose counterpart in the angelic interpretation of the vision takes the form of "the saints of the Most High" (Dan. 7:18, 22, 27). The expression "the Son of Man" (with an exception in Acts 7:56 which proves the rule) is peculiar to Jesus in the NT and denotes His own relation (and also, in part, that of His followers) to the kingdom of God. Two main phases in the career of the Son of Man are distinguished in Jesus' teaching: He will indeed (as Daniel saw) receive dominion and appear in glory; "but first he must suffer many things and be rejected by this generation" (Luke 17:25 RSV; cf. Mark 8:31, etc.). The suffering of the Son of Man, as much as his investiture, is something that is "written" (Mark 9:12); and if it be asked where it is "written," a more promising answer is offered by the fourth Servant Song of Isaiah 52:13—53:12 than by the visions of Daniel.

If Jesus at His baptism was hailed from heaven in terms of Isaiah 42:1, then the other oracle which, like that, begins with the introduction "Behold my servant" would naturally have presented itself to Him as something which He was called upon to fulfil. His last words to the high priest at His trial confirm that His passion would be the prelude to vindication and manifestation in power: "From now on the Son of man shall be seated at the right hand of the power of God" (Luke 22:69 RSV; cf. Mark 14:62). Or, as His words are reported by John in another context: "When you have lifted up the Son of man, then you will know that I am he" (John 8:28). At times He spoke of His impending passion as the cup which He had to drink or the baptism which He had to undergo, and made it plain to His disciples that only by sharing His cup and baptism could they hope to share His glory—a glory which was not the external reward for sharing His ministry as the Servant-Messiah, but consisted in their sharing that ministry.

Jesus and the Ruling Powers. Jesus' teaching was revolutionary enough in its inner essence, but not directly revolutionary in the political sense which would attract the immediate hostility of the civil and military power. In Galilee there was little occasion for Him to refer to the Roman occupation of Judea. When news was brought to Him of a massacre of Galilean pilgrims in the temple area by the soldiery of Pilate, a prefect of Judea, Jesus made the report the basis of a general warning but expressed no judgment on the perpetrators. He is credited with one unflattering personal reference to the ruler of Galilee—"that fox" (Luke 13:32)—and may have indirectly criticized Herodias's remarriage (Mark 10:12). Antipas's hostility was perhaps aroused against Him because of indiscretions of the Twelve during their Galilean mission, but Jesus refused to countenance any seditious talk or action on their part, just as, in the sequel, He sharply discouraged the leaderless multitude which hailed Him as a second Moses and tried to compel Him to be their king (John 6:14f.; cf. Mark 6:34ff.). The proclamation of the kingdom of God and the inculcation of the Golden Rule as the way of life had incalculable implications for the Roman Empire and the Herodian dynasty alike, but Jesus deplored the Zealots' suicidal policy of armed revolt, which could lead only to disaster.

So, when He went to Jerusalem at the end of His ministry, His lament over the city and prediction of its overthrow were due to its refusal of the way of peace which He offered. The spirit of revolt was in the air; a recent abortive insurrection (Mark 15:7) had aroused widespread sympathy for its doomed leaders. When the Zealots' test-question, if God's chosen people should pay tribute to a pagan emperor, was put to Him, He answered that Caesar's coinage was best given back to Caesar, and that the important issue was to discharge one's duty to God. While His answer was bound to disappoint those who followed the Zealot line, it could be misrepresented (as in the event it was) to suggest a disavowal of allegiance to Caesar.

Death of Jesus. It was the enmity, in the first instance, of the temple authorities, not of the Roman administration, that He incurred in Jerusalem. His "cleansing" of the outer court of the Temple was not a violent demonstration such as would have led quickly to the intervention of Roman auxiliaries from the Antonia fortress; it was a "prophetic action" in the OT tradition, whose message was not lost on the leaders of the sacerdotal establishment: the restoration of the area to be "a house of prayer for all people" (Isa. 56:7) in terms of the prophetic ideal would prejudice their privileges. His action was as unacceptable to them as Jeremiah's temple speech had been to their predecessors six centuries before (Jer. 7:1ff., 26:1ff.). In addition, such attacks on an unpopular hierarchy, unless they were checked, could lead to a popular movement which would attract Roman reprisals. It was therefore decided to arrest Jesus unobtrusively, if possible. With the aid of Judas Iscariot, the Jewish authorities found an earlier opportunity of doing this than they had expected.

Jesus was arrested on Passover Eve. According to John (18:3,12), members of the Antonia garrison took part in the arrest; this suggests that the chief priests had already made preliminary arrangements with the Roman authorities. Jesus was brought first before a Jewish court of inquiry. After an unsuccessful attempt to convict Him of a threat to the Temple (the one department in which the Romans still allowed the Sanhedrin to exercise capital jurisdiction), His unexpected acceptance of the designation "Messiah" (albeit in His own preferred understanding of the term) gave His inquisitors the occasion they sought to hand Him over to Pilate as a leader of sedition, who claimed to be "the king of the Jews." This was the charge recorded in the "title" affixed above His head on the cross, to which Pilate, after some temporizing, condemned Him.

The abiding significance and saving efficacy of the death of Jesus belong to theological rather than historical study (see ATONEMENT). But the study of the historical Jesus includes what can be known about the way in which He viewed His death. When He spoke of giving His life "a ransom for many" as the crown of His service (Mark 10:45), He used language which was not unfamiliar in Israel. The martyrs of Maccabean times, for example, offered up their lives as an atonement for their people and land. In offering up His life as a ransom "for many," Jesus had in His mind the Servant of Yahweh who in His suffering bears the sin of "many" and procures their justification (Isa. 53:11f.). To the same effect His words at the institution of the cup interpret it as His "covenant [blood], which is poured out for many" (Mark 14: 24), to which Matthew supplies the epexegesis, "for the forgiveness of sins" (Matt. 26:28). This is no peculiarly Markan *theologumenon*, of Pauline or comparable origin; it underlies the Johannine presentation of Jesus as "the Lamb of God, who takes away the sin of the world" (John 1:29) and His portrayal in the Apocalypse as the one who by His blood has ransomed men for God "from every tribe and language and people and nation" (Rev. 5:9). These and similar words convey the testimony of the early Christians that Jesus' own understanding of His death validated itself in their experience of its purifying and redemptive power.

With His resurrection on the third day after His crucifixion, the Jesus of history becomes the exalted Lord—in pre-Pauline Christianity as well as in Pauline; among Aramaic-speaking as well as among Greek-speaking believers (see CHRISTOLOGY). But the value of His exaltation to be Lord of all depends wholly on the continuity and identity of the exalted Lord with the Jesus of history, in whom already the Word had become flesh.

BIBLIOGRAPHY: A. Schweitzer, *The Quest of the Historical Jesus* (ET 1910); J. Klausner, *Jesus of Nazareth* (ET 1929); T.W. Manson, *The Teaching of Jesus* (2nd ed., 1935) and *The Servant-Messiah* (1953); V. Taylor, *The Life and Ministry of Jesus* (1954); J.M. Robinson, *A New Quest of the Historical Jesus* (1959); G. Bornkamm, *Jesus of Nazareth* (ET 1960); E. Stauffer, *Jesus and His Story* (ET 1960); N. Perrin, *The Kingdom of God in the Teaching of Jesus* (1963) and *Rediscovering the Teaching of Jesus* (1967); H. Zahrnt, *The Historical Jesus* (ET 1963); H. Anderson, *Jesus and Christian Origins* (1964); M. Kähler, *The So-Called Historical Jesus and the Historic Biblical Christ* (ET 1964); G.E. Ladd, *Jesus and the Kingdom* (1964); J. Peter, *Finding the Historical Jesus* (1965); O. Betz, *What do we know about Jesus?* (ET 1968); X. Léon-Dufour, *The Gospels and the Jesus of History* (ET 1968); D. Flusser, *Jesus* (ET 1969); C.H. Dodd, *The Founder of Christianity* (1971); H.J. Schultz (ed.), *Jesus in His Time* (ET 1971); E. Schweizer, *Jesus* (ET 1971).

F.F. BRUCE

JEWEL, JOHN (1522-1571). Bishop of Salisbury. Educated at Merton and Corpus Christi Colleges, Oxford, he became a fellow of Corpus in 1542. In 1547 the continental Reformer Peter Martyr* came to Oxford and greatly influenced Jewel, who became one of the leading thinkers in the Reforming party. On Mary's accession, Jewel agreed to sign anti-Protestant articles, but was still obliged to flee in danger of his life, and reached Frankfurt in 1555. There he was regarded suspiciously by John Knox* because of the articles he had signed, and he publicly expressed sorrow for his cowardice. With Richard Cox he defended the 1552 Prayer Book against Knox and the more advanced Reformers. Later he joined Peter Martyr in Strasbourg and accompanied him to Zurich. After his return to England in 1559 he corresponded with Martyr on the religious state of the country. In 1560 Elizabeth made him bishop of Salisbury, which he administered conscientiously and vigorously. He carried out several visitations, and on account of the lack of capable preachers, engaged himself in preaching and literary work.

In 1562 he published his *Apologia pro ecclesia Anglicana*, the first systematic defense of the Church of England against the Church of Rome. Written in Latin for circulation abroad, it was translated into English by Lady Bacon in 1564. It described the beliefs and practices of the Church of England, defending their deviation from the Roman ones, and sought to demonstrate both that a reformation had been needed and that local churches had the right to legislate for reform in provincial synods. The treatise was distinguished by weighty learning (especially of the early

Church Fathers), logical reasoning, and absence of emotional appeal. It was given official approval in James I's reign by Archbishop Bancroft.

Lengthy and bitter controversy followed with the Roman Catholic Thomas Harding. A convinced Anglican, Jewel took his stand on the Elizabethan Settlement* and also opposed the Puritans with their desire for further reforms such as the abolition of the surplice. At Salisbury, Jewel built the cathedral library and educated and supported a number of poor boys, among them Richard Hooker,* whose *Ecclesiastical Polity* shows the strong influence of Jewel.

BIBLIOGRAPHY: *Collected Works* (ed. J. Ayre, 1845-50, and R.W. Jelf, 1848); N.M. Southgate, *John Jewel and the Problem of Doctrinal Authority* (1962); J.E. Booty, *John Jewel as apologist of the Church of England* (1963). JOYCE HORN

JEWS, MISSIONS TO THE. Except under Hadrian, the Jews in the Roman Empire down to the reign of Constantine retained their position as a tolerated cult and were therefore more favorably placed than the Christians. Much the same was true in the East under the Parthian and Sassanid dynasties. Hence the Church tended to be on the defensive against the Synagogue, except for efforts by Hebrew Christians, of which we know little. The only surviving evidence from this period of a genuine attempt to reach the Jews is Justin Martyr's *Dialogue with Trypho.* Once Christianity had triumphed in the Roman Empire the only methods of dealing with Jews known to most in authority were discrimination, persecution, and sometimes forced baptism. During the Reformation and Counter-Reformation some little interest was shown by the Church in the conversion of the Jews, but not until the Moravians (1738) and Pietists did it become important. The work of Ezra Edzard (1629-1708), J.H. Callenberg (1694-1760) and A.H. Franke (1663-1723) led to the founding of the *Institutum Judaicum* at Halle in 1728.

A new era began when J.S.C.F. Frey (1771-1851), a Hebrew Christian from Germany, came to London. His work led to the founding of the London Society for Promoting Christianity among the Jews (later Church Missions to Jews, now The Church's Ministry among the Jews) in 1809 as an interdenominational society. Christian opinion was not yet prepared for such an experiment, and it soon had to be reformed as a purely Anglican one. In 1842 Free Church supporters of Jewish missions founded the British Society for Promoting Christianity among the Jews (British Jews' Society, now International Jews' Society).

The first acceptance of responsibility by a church as such was in 1840 by the Church of Scotland, and its example was followed by the Presbyterian Church of Ireland the following year. Beginning with Norway in 1844, the new concern spread to Scandinavia and Finland and gradually to all the main Protestant churches of Europe, though in varying degree. Especially important was the founding in 1886 of the *Institutum Judaicum Delitzschianum* in Leipzig by F. J. Delitzsch.* This provided the necessary intellectual tools for the missionary and apologist. The Swedes set up a similar institute in Jerusalem in 1951. In the Roman Church the most important development was the founding of the Sisters of Zion by Father Marie Ratisbonne, a Jew converted in 1842 by a vision.

A second generation of missionary societies came into existence in the second half of the century, largely as an answer to the westward surge of E European Jews. England gave the lead, most of the societies being nondenominational. The most important were the Mildmay Mission to the Jews (1876), the Barbican Mission to the Jews (1889, Anglican in leadership), and the Hebrew Christian Testimony to Israel (1893). In America most missionary work was purely local in its nature, but in 1894 Leopold Cohn founded the American Board of Missions to the Jews, interdenominational and fundamentalist in nature, which has grown to be the largest Jewish mission in the world, both in resources and missionaries. The only other mission of this kind that needs to be named is The Friends of Israel (1938). Many of the larger denominations support some form of missionary activity, normally depending on local response; the Southern Baptists are the only one active outside America.

The International Missionary Council* had from the first envisaged including Jewish missions in its scope, but it was not able to set up its Committee on the Christian Approach to the Jew (IMCCAJ) until after its Jerusalem meeting (1928), and it was scarcely functioning before the Nazi holocaust changed the whole position of Jewry. Its chief contribution was its insistence on the "Parish approach," i.e., in most countries the emergence of the Jew into the mainstream of life placed responsibility for Christian witness on the local church.

The effect of the holocaust on both Jews and Christians and of the coming into being of Israel led to an increasing stress on dialogue, which Vatican Council II declared to be the policy of the Roman Church. With the entry of IMC into the framework of the World Council of Churches, IMCCAJ became the Committee on the Church and the Jewish People (CCJP), and its main purpose has become the furthering of dialogue. In turn this has meant that active missionary work among Jews is being increasingly confined to conservative evangelical circles, which have all along been its mainstay. The list of converts is in every way an imposing one, but above all missions have succeeded in making Jesus a reality to His people.

BIBLIOGRAPHY: R. Allen, *Arnold Frank of Hamburg* (n.d.); J. Wilkinson, *Israel My Glory* (1894, 1921); W.T. Gidney, *The History of the London Society for Promoting Christianity amongst the Jews from 1809 to 1908* (1908); D. McDougall, *In Search of Israel: A Chronicle of the Jewish Mission of the Church of Scotland* (1941); G.H. Stevens, *"Go, Tell My Brethren": A Short Popular History of Church Missions to Jews (1809-1959)* (1959). H.L. ELLISON

JIMÉNES (Ximenez) DE CISNEROS, FRANCISCO (c.1436-1517). Spanish cardinal and inquisitor. Born Gonzales Jiménes at Torrelaguna in Castile, he took law degrees at Salamanca (1456)

and after some years in Rome served as secular priest and administrator in the Spanish Church. In 1484 he unexpectedly entered the strict Observantine Franciscan Order. For nearly ten years he lived an austere penitential life, gaining fame as a man of great spirituality. His secluded life changed when in 1492 Isabella made him her confessor. She consulted him often for political as well as spiritual advice. In 1495 he became archbishop of Toledo, head of the Spanish Church, and chancellor of Castile. Remaining an ascetic, he used his power and wealth to effect rigorous reform of the church and to convert or extinguish the Moors and Jews of the kingdom. After Isabella's death (1504) he became increasingly active politically. He financed and led para-Crusade military expeditions in North Africa and supported the monarchy against uprisings within Castile. For his services he was created cardinal in 1507.

Jiménes was a great patron of all types of philanthropic work, especially education. He founded and funded the University of Alcalá (1500) and the College of San Ildefonso. Alcalá he made a center of humanism, and financed the Complutensian Polyglot* Bible. He died at Roa, having effectively renewed the church and strengthened the monarchy of Spain.

BIBLIOGRAPHY: Primary works are in Spanish, but among secondary sources are R. Merton, *Cardinal Ximenes and the Making of Spain* (1934); W. Starkie, *Grand Inquisitor* (1940); J.C. Niceto, *Juan de Valdes and the Italian Reformation* (1970). BRIAN G. ARMSTRONG

JOACHIM OF FIORE (c.1135-1202). Mystic philosopher of history. He lived in Calabria, Italy, where he became a Cistercian monk. After being abbot of Curazzo, he retired to a more remote region and founded the order of San Giovanni in Fiore (1192). He recorded two mystical experiences which gave him the gift of spiritual intelligence enabling him to understand the inner meaning of history. At times he prophesied on contemporary events and the advent of Antichrist. He also meditated deeply on the two great menaces to Christianity, the infidel and the heretic.

With papal encouragement, Joachim explained his beliefs in three major works: the *Exposition of the Apocalypse,* the *Concordance of the Old and New Testaments,* and *Psalterium of Ten Strings.* His explanation involves interwoven patterns of twos and threes. The two testaments represent two eras of history culminating in the first and second advents. These periods are marked by other agreements such as twelve tribes and twelve churches and seven seals and seven openings. History is also trinitarian, with the first age being that of the Father when mankind lived under the law as recorded in the OT. The second age, that of the Son, is the period of grace and covers the NT dispensation which Joachim believed would last for forty-two generations of thirty years each. The third age was to be that of the Spirit during which the liberty of spiritual intelligence would prevail. This new age was to begin about A.D. 1260 and would be characterized by the rise of new religious orders that would convert the world.

Joachim's teaching was not meant to undermine ecclesiastical authority, but it inspired groups such as the Spiritual Franciscans and the Fraticelli, who carried his ideas to revolutionary conclusions to claim to be Joachim's spiritual men ready to usher in the third age.

See H. Bett, *Joachim of Fiore* (1931).

ROBERT G. CLOUSE

JOAN, POPE. The story of an alleged female pope was widely believed from the thirteenth to the seventeenth centuries. So much so, that when David Blondel argued in 1647 that it was a legend he was criticized by fellow Protestants. The story has it that a scholarly woman, disguised as a man, succeeded to the chair of Peter about 1100 (later versions state 855). After about two years in office she gave birth to a child as she was taking part in a procession to the Lateran, and then died. It seems that a thirteenth-century Dominican chronicler, Jean de Mailly, first gave respectability to the legend and that it gained wide currency in the Middle Ages, partly due to the influence of Martinus Polonus (d.1278) and the use of the story in fifteenth-century controversies over the extent of papal power. Some scholars interpret the story as a modification of a Roman folk-story, the original possibly relating to a priest of Mithra and a child. PETER TOON

JOAN OF ARC (1412-1431). "The Maid of Orléans," national heroine of France. An illiterate though devout peasant girl from Domrémy, Champagne (called Jeanne la Pucelle), she began at the age of thirteen to experience inward promptings, voices accompanied by light, which urged her to save France from the aggressors. As these voices increased, she could even distinguish those of SS. Michael, Catherine, Margaret, and others. By this time the dauphin, Charles, was at war with the joint forces of England and Burgundy. Although she was unsuccessful at persuading the French commander at Vaucouleurs in 1428 of the reality of her visions, she was sent to Charles who became convinced when she recognized him in disguise. On close examination by theologians at Poitiers, she was given armor and attendants, then joined the army at Blois, later to rout the English besieging Orléans.

After another victory in the Loire she persuaded Charles to be crowned at Reims (1429). Going in 1430 to relieve Compiègne, she was there taken prisoner by the Burgundians and sold to the English without intervention from Charles VII. Appearing before the court of the bishop of Beauvais, she was charged with witchcraft and heresy; fearlessly enduring the long trial, she refused to betray her inward leading. Found guilty, with the verdict confirmed by the University of Paris, her visions were declared "false and diabolical." Facing death, she recanted slightly, only to stand firm once again. She was burnt as a heretic in the marketplace of Rouen. Charles VII twice sought a changed verdict, but not until Callistus III (1456) was the case declared fraudulent and her innocence acknowledged. Her death proved her influence; her banner carried the symbol of the Trinity and the words "Jesus, Maria." Canonized in 1920, she is the second patron of France.

BIBLIOGRAPHY: P. Champion (ed.), *Procés de condamnation de Jeanne d'Arc* (2 vols., 1920-21); W.F. Barrett, *The Trial of Jeanne d'Arc* (1931); L. Fabre, *Joan of Arc* (tr. G. Hopkins, 1954); R. Pernoud, *Joan of Arc* (1965). C.G. THORNE, JR.

JOGUES, ISAAC (1607-1646). Jesuit missionary and martyr. Born in Orléans, France, and educated at the local Jesuit college, he entered the Society of Jesus in 1624 and in 1636 was ordained and began missionary work among the Huron Indians in Canada. He made numerous journeys into the interior of North America. Returning once from Quebec, the canoes were attacked by the Iroquois and Jogues was taken prisoner. He was flogged, bitten, stripped, mutilated, and insulted by the Indians. He was rescued by the Dutch and sent to France, only to return to Quebec in 1644. He immediately sought permission to go as a missionary to the Iroquois, but his request was denied because of unsettled conditions. In 1646, while acting as part of a peace mission, he was again taken prisoner by the Iroquois and died of a hatchet blow. Jogues's martyrdom emphasized his great piety already apparent in his writings, and his self-sacrifice provided inspiration and impetus to Canadian missions. ROBERT WILSON

JOHANSSON, GUSTAF (1844-1930). Archbishop of Finland. Successively professor of dogmatics and ethics at the University of Helsinki (1877-85), bishop of Kuopio (1885-96), and bishop of Savonlinna (Nyslott) (1896-99), he became primate thereafter. His literary production was large and included a volume on dogmatics and publications on justification and on the Church of Finland. Although a conservative theologian, he struggled for many ecclesiastical reforms, the abolition of obligatory participation in Communion, and the advancement of *diakonia*. Chairman of the committee for a new translation of the Bible, he worked also toward a new church hymnal and catechism. He participated in politics as a clerical representative and boldly maintained the special position of Finland within the Russian Empire, even in the presence of the czar and his officials. In spite of this his policy of loyalty has been much criticized.

Johansson opposed the ecumenical movement, and especially the 1925 Stockholm Conference,* chaired by Archbishop Nathan Söderblom of Sweden. According to Johansson, the ecumenical trend was a grave danger to the church because it opened the door to syncretism and cooperation with liberal theologians. He felt also that ecumenicity could not be united with Christian eschatology. He was a disciple of J.T. Beck* of Tübingen, whose biblicism he tried to follow. Johansson retains a central position in the modern church history of his country, and his influence is still felt in the present generation of Finnish theologians. STIG-OLOF FERNSTROM

JOHN I (d.526). Pope from 523. He was Tuscan. His pontificate was marked by much friction with Theodoric the Ostrogothic ruler at Rome who adopted a very severe policy against the orthodox Catholics in Italy as his answer to an equally stringent policy by Justin against the Arians of the Eastern Empire. John went to Constantinople to secure a reversal of this policy, and he was partially successful. But on his return to Italy he was imprisoned by Theodoric and died in Ravenna.

JOHN VIII (d.882). Pope from 872. He devoted much of his pontificate to defending Italy against the Saracens. The various regnal and imperial coronation ceremonies which he conducted were designed to persuade the recipients to accept responsibility for Italy's defense. He had supported the emperor Louis the Stammerer who had overrun the duchy of Benevento in 873, but Louis died in 875. John then turned to his friend Charles the Bald and crowned him emperor on Christmas Day 875, but Charles's subsequent visits to Italy were fleeting and unsuccessful. John's great difficulty was an inability to detach the princes of Palermo, Naples, and Capua and the maritime power of Amalfi from their alliances with the Saracens. The hope of obtaining the Eastern emperor Basil I's support against the Saracens led John, at Basil's request, to recognize the previously deposed Photius* as patriarch of Constantinople, and to conclude peace between Rome and Byzantium at the Photian Council (879-80). In his final years John was forced to pay annual tribute to the Saracens. He was assassinated by conspirators at the Lateran Palace.

L. FEEHAN

JOHN XII (c.936-964). Pope from 955. He was the son of Alberic II of Spoleto, ruler of Rome, before whose death in 954 the nobles had to swear to elect his only son, Octavian, who was only eighteen. Called John, he was much given to vicious living, according to the historian Liutprand, bishop of Cremona. In 959 John tried to recover former papal lands from King Berengarius II and had to appeal for help from Otto I, who obligingly sent his army to Italy in 961. In return, Otto was crowned emperor in 962, and the archbishopric of Magdeburg and the bishopric of Merseburg were erected. Otto then issued the *Privilegium Ottonianum*, promising to make the pope temporal ruler of most of Italy, but with the provision that he recognize imperial suzerainty over the Papal States. Future popes were to take an oath of fealty to the emperor before consecration. Bitter strife between pope and emperor ensued; with increasing support Otto called a synod at Rome to accuse John of immorality. Deposition followed (963), and a Roman lay official was elected and consecrated Leo VIII. With Otto's departure from Rome in 964, however, John returned and took revenge. Leo was deposed and all his actions canceled. Liutprand records that John's death was sudden and scandalously mysterious.

C.G. THORNE, JR.

JOHN XXII (1244?-1334). Pope from 1316. Born Jacques d'Eudes, he studied law at Paris and elsewhere in France, and was successively bishop of Frejus (1300), bishop of Avignon (1310), and cardinal bishop of Porto (1312). At one time he had been chancellor of the Angevin kingdom of Naples, and his Angevin connection contributed to his election as pope. The first "Avignon" pope who really resided at Avignon,* he was an outstanding administrator. He was mainly responsible for that perfecting of the machinery of papal government which was one of the positive contributions of the Avignon papacy. He failed in his repeated attempts to break the Ghibelline (imperialist or antipapal) power in Italy and to reestablish a papal state there.

In pursuit of these aims he tried to assert the papacy's ultimate theocratic authority during imperial vacancies and over disputed imperial elections. This policy led to a long drawn-out quarrel with Emperor Lewis of Bavaria, whom he excommunicated in 1324 in the unsuccessful hope of fomenting civil war in Germany. Lewis counterattacked strongly. He organized a Ghibelline League which made great inroads into Italy. He encouraged Marsilius of Padua and John of Jandun, whose book *Defensor Pacis* was a fundamental challenge to John XXII's views on papal theocracy. Lewis also supported those Franciscan Spirituals who had protested against John's vesting in the Order of Friars Minor all property which had been given to it (1322-24). John had added, in justification of his decision, that although Christ and His apostles had lived in poverty, they had exercised the right of ownership both in common and individually (*Cum inter nonnullos*, 1323). The Spirituals claimed that this was heresy, but those who refused to accept the ruling were persecuted. Their leaders were excommunicated and a few Spirituals were burnt. When Lewis captured Rome, he had an antipope, Nicholas V, crowned and had John XXII condemned for heresy (1328). Lewis was, however, soon driven permanently from Italy, and his antipope submitted. L. FEEHAN

JOHN XXIII (c.1370-1419). Antipope. Born in Naples, Baldassare Cossa studied law at Bologna after an early military career. In 1402 he served as cardinal deacon under Boniface IX, then became cardinal and legate in Bologna (1403-8). In 1408 he collaborated in convening the Council of Pisa* to terminate the schism between Roman and Avignon popes by withdrawing support from Gregory XII, the Roman pope. He presided at the council in 1409. Both popes were deposed, and Alexander V was elected, but his sudden death led to Cossa's election to the papacy at Bologna in 1410.

With John XXIII in his company, Louis of Anjou recaptured Rome and defeated Ladislaus of Durazzo, an ally of Gregory XII. A new council in Rome condemned the works of Hus and Wycliffe in 1412, after Hus revolted against the emissaries John sent to Prague to sell indulgences and defame his political enemy, the king of Naples. In 1413 Ladislaus drove John from Rome to Florence. Out of constraint he complied with Emperor Sigismund's proposal for a new council at Constance.* The council, which opened in 1414, summarily deposed John in 1415, and the schism was ended with the election of Martin V. John's attempted escape failed when Sigismund placed him in custody. After submitting to the new pope in 1418, John was pardoned and soon named cardinal bishop of Tusculum (Frascati). The papal insignia graces his tomb in the baptistery at Florence, but after John XXIII ascended the throne in 1958 Cossa was no longer officially numbered among the popes. JOHN GROH

JOHN XXIII (1881-1963). Pope from 1958. Elected following a conclave of three days with eleven ballots, Angelo Giuseppe Roncalli had previously served as a secretary to the bishop of Bergamo in N Italy, directed the Congregation for the Propagation of the Faith in Italy, and then been apostolic delegate until in 1953 he became a cardinal and patriarch of Venice. It was thought at first that he would be only a "transitional" pope.

Of medium height, sturdy and stout, he looked very different from his predecessor, Pius XII. His whole approach was also different from other modern popes. He chose the name "John," a title not used for five and a half centuries; at the Christmastide following his election he made lengthy visits to two hospitals, caused sick and crippled children to visit him, and also visited the Regina Coeli prison. Soon he gained the love of Christians everywhere; beneath the vestments of the pope they felt there still existed the soul and heart of a country priest. An all-important day in his pontificate was 25 January 1959. To a stunned extraordinary congregation of seventeen cardinals who were in Rome he announced he would do three things: (1) call a synod of the church in the city and diocese of Rome; (2) summon an ecumenical council to promote Christian unity; (3) promote the reform of canon law. There had not been a Roman synod since medieval times, and the last council had met in 1869-70. If there was opposition to his ecumenical ideals within the Vatican, leaders in other churches were most gratified. Athenagoras, Orthodox patriarch of Constantinople, was delighted; and Archbishop Fisher of Canterbury even called upon the pope, an event that caused a great stir in England.

Pope John made two famous speeches in 1962, one on the relationship between the church and the world, and another at the opening of the council. The latter urged the church to respond to the twentieth century and make the *depositum fidei* relate to the world and its needs. As the council progressed in its first session, he took a lively interest in it and prodded often in order to facilitate progress. Both during and after the first session John XXIII was a sick man, but he continued his work unabated. His visits to hospitals and parishes in Rome continued. Statesmen—e.g., Harold Macmillan of Britain—and church leaders—e.g., the prior of the Protestant community of Taizé—were given audiences.

On 29 March 1963 he set up the Commission of Cardinals to revise the Code of Canon Law. During this same period he showed great concern for a happy relationship of the church with Communist governments. He saw the release of the archbishop of Lvov in the Ukraine from prison, and he received at the Vatican the son-in-law of Nikita

Kruschchev. Furthermore, bishops from Poland, Hungary, and other countries behind the Iron Curtain were allowed to attend the council. His encyclicals included: *Ad Petri Cathedram*, which related to the council; *Princeps Pastorum*, which dealt with missions; *Mater et Magistra*, which looked at social questions; and *Pacem in terris*, which dealt with peace on earth. Part V of the latter was concerned with the relationship of the church to Communist governments.

The last week of his life was followed intensely by the press of the free world, for his warm, irradiating humanity had endeared him to many who did not share his religious views.

BIBLIOGRAPHY: L. Algisi, *Giovanni XXIII* (1959); F.X. Murphy, *Pope John XXIII Comes to the Vatican* (1959); E. Balduci, *Papa Giovanni* (1964); E.E.Y. Hales, *Pope John and His Revolution* (1965); G. Lercaro and G. De Rosa, *John XXIII, Simpleton or Saint* (1967); M. Trevor, *Pope John* (1967). See also John XXIII, *Journal of a Soul* (ET 1965), and bibliography under VATICAN II.

PETER TOON

JOHN, ACTS OF. A legendary document belonging to the early third century, purporting to give information about the life and work and death of John the Apostle. The original in Greek consisted of 2,500 lines. Large sections of this survive, although the beginning is lost. We also have a Latin version of some otherwise unknown incidents involving John, and several fragmentary MSS. The work probably originated from Ephesus, which is the setting for many of the events and traditionally the place where John died. The work is remarkable as much for the words attributed to John as for the works. It contains also a docetic account of the death of Jesus. The *Acta Ioannis* came eventually to be regarded as the work of the Valentinian Leucius, a supposed disciple of the apostle. In the fifth century an orthodox revision of the *Acts of John* was ascribed to one Prochorus.

STEPHEN S. SMALLEY

JOHN, EPISTLES OF, see EPISTLES, GENERAL

JOHN, GOSPEL OF. The history of the interpretation of the gospel of John shows that until the middle of the twentieth century the major concern of critics has usually been with the identity of the fourth evangelist. During the nineteenth century, for example, radicals like A.F. Loisy* regarded this gospel as an unhistorical, theological reconstruction of the first three gospels, having no connection with John the Apostle; whereas conservative scholars like B.F. Westcott* maintained that the fourth gospel was mainly if not entirely an apostolic work. At the turn of the twentieth century, and under the influence of the religio-historical method of biblical exegesis, the gospel of John was viewed as representative of a late stage in the process of hellenizing Christianity (O. Pfleiderer; cf. R. Bultmann, who claims that John used ideas current in Gnostic circles to give expression to Christian truth). Later, scholars such as A. Schlatter* and C.F. Burney,* taking serious account of the Semitic background to the language and ethos of the gospel, began to assert its Jewishness.

Tradition. Johannine criticism since 1950 has tended to find the key to the problem of the fourth gospel (namely, the apparently large differences between John's gospel and the synoptic gospels) in the issue of tradition rather than authorship. The "old look," represented by the work already mentioned, assumed that the fourth evangelist's tradition was derived directly from the synoptic gospels and was either John's own work of history (in which case the supposed conflict between the two versions presented a difficulty) or a later, theological reshaping of the so-called historical tradition of the synoptists by someone other than the apostle. But the more recent "new look" on the gospel of John recognizes that a sharp division of this kind between history and theology in the gospels is invalid; and also that there probably lies behind the fourth gospel a reliable and primitive Christian tradition, parallel to that behind the synoptic gospels and independent of it.

The likelihood that the synoptic and Johannine traditions are related in this way (at the substructural level, rather than by direct literary dependence) has been confirmed in three directions. First, it is suggested by a straight literary comparison between John and the other gospels (P. Gardner-Smith). Secondly, the evidence of the Dead Sea Scrolls makes it likely that a religious setting which could have influenced the writing of John's gospel existed before the Christian tradition began. In this milieu, Jewish and Greek (even pre-Gnostic) ideas and terms were combined in a way that was previously known outside the fourth gospel only from late, Hellenistic literature. Thirdly, archaeological excavation in and around Jerusalem has established the probability that John's gospel drew on a genuinely historical tradition which was originally transmitted in a S Palestinian setting.

Composition. The "new look" on John's tradition, which is accepted by many but not all contemporary writers on the fourth gospel, clearly affects the question of the composition of this gospel, and the extent to which the John the Apostle (traditionally regarded since Irenaeus as both "the disciple whom Jesus loved" and the Fourth Evangelist) is associated with it. The beloved disciple plays the role in the fourth gospel that John the Apostle does in the synoptic gospels, and also appears as a witness to the tradition being recorded (cf. John 19:35; 21:24). It is possible therefore that John was responsible for the tradition behind the fourth gospel, which explains why the church eventually accepted the gospel as his; but that he did not write the gospel in its final form, which explains why there was hesitation at first to accept it as coming from an apostolic hand. In this case three basic stages may have been involved in the composition of the gospel. At the first stage, the tradition about Jesus, mostly in oral form, would have taken shape in Palestine in association with John the Apostle himself. Secondly, this tradition may have been recorded by a disciple-friend of John in later life, at Ephesus (traditionally the place where the apostle ended his days, and where his gospel was published). This writer may be called the fourth evangelist. Finally, after John's death the fourth evangelist, in the context of a Johannine school perhaps, may have edited the gospel material so far written down, adding the epilogue (John 21) and prologue (1:1-18) in that order, and drawing out the implications of its distinctive theology already present in seminal form at the second stage (R.V.G. Tasker, R.E. Brown; cf. R. Bultmann, R.T. Fortna). Despite the early background to the fourth gospel just suggested, it was probably still the fourth to be written (c. A.D. 85); cf. its developed theology, and especially its high Christology.

Purpose. Like all the evangelists, John wrote his gospel to proclaim Jesus as Lord and Christ. He expresses his intention directly in John 20:31. The background and character of the gospel suggest that he is in the first place addressing a group of Hellenistic Jews, probably in the diaspora. This accounts for the shaping of a basically Palestinian tradition in a Greek direction, and suggests that the Hellenism of the fourth gospel belongs to the environment in which it was published rather than to its original tradition; although the Hellenistic elements in John (concepts such as Logos, light, and knowledge) also have their own Hebraic background which a Jew could appreciate. But the fact that the writer explains Jewish terms (such as "Messiah," John 1:41) and customs (such as the Passover, 6:4), which cannot have been unfamiliar to the most hellenized Jew, indicates that ultimately the scope of John's audience was probably without limit. He is anxious that *any* reader of the gospel should "see" who Jesus is, and receive the life He came to give (20:29-31).

For this purpose the writer selects seven "signs," beginning with the Incarnation (1:14), which reveal the true nature of Jesus as the Savior of the world (4:42), and point toward the fulfillment of the salvation in His glorification (13:31). Thus John 1—12 deals with the revelation of the Word to the world, and 13—21 with the glorification of the Word for the world. A theological development of this kind makes better sense of the structure of the fourth gospel than sacramental (O. Cullmann) or liturgical-lectionary (A. Guilding) theories.

The gospel in the early church. The fourth gospel seems to have been used by Gnostics, especially the Alexandrians, before it was used by the orthodox; and it was the Valentinians (e.g., Ptolemaeus and Heracleon, the first commentator on the gospel) who first ascribed it to "John," presumably wishing to secure apostolic authority for their teaching. Evidently the heretics found that the doctrinal diversity in the gospel favored their dualist and docetic theology. The gospel probably influenced as well the Valentinian *Gospel of Truth* and the apocryphal *Gospel of Peter* (c.150). The Apologists, beginning with Justin Martyr (mid-second century), are the first orthodox writers to show any likely knowledge of the gospel, although there are possible traces in Ignatius (c.115), and Tatian used the gospel in his *Diatessaron* (c.150); but it is not until Theophilus of Antioch (late second century) that any certainty about its authenticity is reached. Theophilus quotes the opening phrases of the prologue to the fourth gospel as the words of John, implying but not stating that this John was the apostle.

Thereafter the gospel is attributed directly to John the Apostle, the beloved disciple, who died at Ephesus where the work was published (cf. the Muratorian Canon, Irenaeus, Polycrates, and Clement of Alexandria, who gives us the famous description of John as the "spiritual gospel"; also the anti-Marcionite prologue to the gospel of John, although its date is uncertain). Papias, described by Irenaeus as a disciple of John, and Polycarp his companion, are strangely silent about the gospel; so also is the third century apocryphal document, the *Acts of John,** where allusion to the gospel would have been natural. From the beginning of the third century the apostolic authorship of the fourth gospel was denied by the church, probably because of its use by Gnostics (cf. the Alogi in the late second century, who went so far as to ascribe the gospel and Revelation of John to the Gnostic heretic Cerinthus); and it was not until much later that the fourth gospel was accepted back into the NT canon as genuinely apostolic. The history of the gospel of John in the early church accords with its origin as suggested above.

BIBLIOGRAPHY: Ancient commentaries: Cyril of Alexandria (ed. P.E. Pusey, 1872); Chrysostom (ed. J.P. Migne, *PG* LIX); Heracleon (fragmentary; see *Texts and Studies* I [1891], pp. 50-103); Origen (ed. A.E. Brooke, 1896); Theodore of Mopsuestia (Syriac in *Corpus Scriptorum Christianorum Orientalium: Scriptores Syri,* IV.3, 1940); Augustine (*Corpus Christianorum, Series Latina* XXXVI, 1954); also J. Calvin (1553).

Modern commentaries: B.F. Westcott (1882); B. Weiss (4th ed., 1902); A. Loisy (2nd ed., 1921); J.H. Bernard (1928); E.C. Hoskyns (2nd ed., 1947); C.K. Barrett (1955); R.H. Lightfoot (1956); R. Bultmann (4th ed., 1959); R.V.G. Tasker (1960); R. Schnackenburg (1965); R.E. Brown (1966-70); J. Marsh (1968); J.N. Sanders (1968); L. Morris (1971); B. Lindars (1972).

Special studies: W. Sanday, *The Criticism of the Fourth Gospel* (1905); C.F. Burney, *The Aramaic Origin of the Fourth Gospel* (1922); P. Gardner-Smith, *St. John and the Synoptic Gospels* (1938); R.H. Strachan, *The Fourth Gospel: Its Significance and Environment* (3rd ed., 1941); J.N. Sanders, *The Fourth Gospel in the Early Church* (1943); O. Cullmann, *Early Christian Worship* (ET 1953); C.H. Dodd, *The Interpretation of the Fourth Gospel* (1953); W.F. Howard, *The Fourth Gospel in Recent Criticism and Interpretation* (4th ed., 1955); A.E. Guilding, *The Fourth Gospel and Jewish Worship* (1960); M.F. Wiles, *The Spiritual Gospel* (1960); C.H. Dodd, *Historical Tradition in the Fourth Gospel* (1963); A.M. Hunter, *According to John* (1968); J.L. Martyn, *History and Theology in the Fourth Gospel* (1968); L. Morris, *Studies in the Fourth Gospel* (1969); R.T. Fortna, *The Gospel of Signs* (1970); T.E. Pollard, *Johannine Christology and the Early Church* (1970); B. Lindars, *Behind the Fourth Gospel* (1971); C.K. Barrett, *The Gospel of John and Judaism* (1975).

STEPHEN S. SMALLEY

JOHN, GRIFFITHS (1831-1912). Co-founder with Robert Wilson of the first Protestant mission in Inland China. Born in Wales, he was appointed to China by the London Missionary Society in 1855. The first mission was established in Hankow six years later. John traveled widely from Hankow through Hupeh and Hunan and up the Yangtze River into Szechwan, though no work could be started there until 1888, when a Chinese evangelist took up residence in Chungking. Land was purchased in Wuchang in 1863 despite the opposition of officials. John also prepared a translation of the NT in Mandarin, and started on the OT for the Bible Societies. He undertook too a translation into easy Wen-li. In 1901 the LMS under John's leadership established several stations in the long-resistant province of Hunan. John was undoubtedly one of the five most prominent missionaries in China in the nineteenth century. LESLIE T. LYALL

JOHN BAPTIST DE LA SALLE, see LA SALLE

JOHN CHRYSOSTOM, see CHRYSOSTOM

JOHN CLIMACUS (579-649). Ascetic and mystic; also known as "Sinaites" and "Scholasticus" (though not to be confused with Patriarch John III). A monk of Sinai, he became an anchorite and later abbot of the monastery. His life was written by Daniel, a monk of Raithu on the Red Sea. His name was derived from his book *Ladder of Paradise (Klimax tou Paradeisou).* There are thirty "steps of the ladder," corresponding with the age of Christ at His baptism. Each step is a chapter describing a particular monastic virtue or vice, showing the way it may be acquired or eliminated. Step 30 is entitled "Faith, Hope and Charity," where the monk receives the crown of glory from Christ. J.G.G. NORMAN

JOHN FREDERICK, ELECTOR (1503-1554). German Lutheran prince, called "the Magnanimous." Born at Torgau, he was the son of Elector John the Constant. Well educated under the Lutheran Spalatin,* he strongly supported Luther and was instrumental in the publication of early editions of his works. His reign as elector was crucial in the consolidation of Lutheranism, and particularly in the establishment of a Lutheran state church. A bulwark of the Smalcald League, he was defeated by Charles V's forces at Mühlberg in 1547, taken prisoner, and lost Wittenberg and his electoral title. Released in 1552, he established his government at Weimar and founded the University of Jena to replace Wittenberg, but he died at Weimar before his programs developed. BRIAN G. ARMSTRONG

JOHN GUALBERT (c.990-1073). Founder of the Vallumbrosan Order. Originally he was a member of the Benedictine monastery at San Minato, near Florence. After four years there he joined the monastery at Camaldoli, and some time later moved to Vallumbrosa, a city probably near Florence, where he formed his own monastic settlement. The Rule of the Vallumbrosan Order was a modified Benedictine, in some ways more austere, and it particularly excluded manual labor. The order eventually included the institution of lay brothers. Traditional date for the founding of the order is about 1300. John Gualbert has been canonized by the Roman Catholic Church.

JOHN OF ANTIOCH (d.441). Patriarch of Antioch. Former student of Theodore of Mopsuestia, he is known chiefly through the writings of Cyril* of Alexandria and the records of the Council of Ephesus* of 431. John arrived late at the latter; Cyril had already illegally proceeded without him. John's arrival reversed the sides and the decisions, except for the condemnation of Nestorius (see NESTORIANISM), and added a procedural condemnation of Cyril. A compromise was effected in 433: Cyril kept his *Theotokos,* but in the context of an Antiochene union of one person in two natures. Measured in terms of the condemnation of Nestorius, victory lay with Cyril; but that a moderating reconciliation should have been required and accomplished suggests the importance of John. The whole situation was complicated by the fact that in the midst of theological debate there were also nontheological factors: imperial concern, patriarchal rivalry, and human personality. CLYDE CURRY SMITH

JOHN OF AVILA (1500-1569). Spanish missionary and scholar. Born near Toledo, he studied law at Salamanca only to abandon it (1515) for philosophy and theology at Alcalá under Domingo de Soto. He dispensed the family fortune to the poor after his ordination in 1525 and hoped to go to America as a missionary. Persuaded to work in Spain instead, he was a missionary in Andalusia for nine years. A great preacher and counselor, he pleaded strongly for reform and denunciation of vice in high places, which brought him before the Inquisition, where he was declared innocent (1533). From Seville he went to Cordova and then to Granada in 1537 where he helped found the university. His greatest work was in reforming Spanish clerical life, with a large circle of disciples around him who taught in colleges he founded. The University of Baeza became a model for seminaries and schools of the Jesuits, who revered him and whose work he widely encouraged. John of God, Francis Borgia, Teresa of Avila, and Louis of Granada (his biographer) benefited from his friendship and counsel. His sermons and letters of spiritual direction are literary classics, but *Audi Filia* (c.1530) on Christian perfection is his best work. C.G. THORNE, JR.

JOHN OF BEVERLEY (d.721). Bishop of York. He was educated at the school of Canterbury and later studied at Hilda's double abbey at Whitby. He was one of the five pupils of hers there who later became bishops. John was appointed to the see of Hexham in 686. One of those whom he ordained while he was at Hexham was Bede.* In 705 he was translated to York. During his episcopate there he founded an abbey at Inderawood (the modern Beverley), to which he retired on resigning his see in 718. He had a considerable reputation for personal holiness and seems to have possessed gifts of spiritual healing. An extensive cultus grew up around him in medieval England. R.E. NIXON

JOHN OF CAPISTRANO, see CAPISTRANO

JOHN OF DAMASCUS (c.675-c.749). Greek theologian and last of the great Eastern fathers. After serving as chief representative of the Christians in the court of the caliph of Damascus, he left (or was compelled to leave) and entered the monastery of St. Sabas near Jerusalem where he was ordained priest. In the Iconoclastic Controversy* he defended in three treatises the use of icons. His fame, however, is particularly associated with his *Fount of Wisdom* (or *Sources of Knowledge*) which achieved lasting fame in both East and West. It is divided into three parts, covering philosophy, heresies, and the orthodox faith. The last part presents the teaching of the Greek fathers on important doctrines and has always been used as a textbook in the Orthodox Churches.

He was, however, unknown in the West, apart from references in florilegia or Catenae of patristic quotations, until the twelfth century when his *Exposition of the Catholic Faith* was translated into Latin as *De Fide Orthodoxa.* A century later the other two parts of the *Fount of Wisdom* were likewise translated. Peter Lombard appealed to the authority of John of Damascus twenty-seven times in his *Sentences* (1150). In the thirteenth century the *De Fide* was divided into four books on the model of the *Sentences.* Also, the discovery of two versions of a concordance to the *De Fide,* dating from the mid-thirteenth century, would seem to go far in supporting the view that the *De Fide* was important in the creation of Western medieval theology. In 1890 Pope Leo XIII declared John to be a "Doctor of the Church." John also wrote a treatise on the ascetic life, a commentary on the Pauline epistles, and various poems (hymns). Some of his sermons also are extant.

BIBLIOGRAPHY: *Works* in J.P. Migne (ed.), *PG* XCIV-XCVI; J. Nasrallah, *Saint Jean de Damas* (1950); B. Studer, *Die theologische Arbeitsweise des J. von Damaskos* (1956); B. Kotter, *Die Überlieferung der pege Gnoseos des Johannes Damaskenos* (1959) and *Die Schriften des J. von Damaskos* (1969). PETER TOON

JOHN OF GOD (1495-1550). Founder of Brothers Hospitallers.* Born John Ciudad in Portugal, he was a soldier until forty, when he returned to Spain as a shepherd lamenting his sinful life. Hoping for martyrdom in Africa, he returned to Granada (1538) to sell books and religious pictures. Ministered to by John of Avila, he began his ministry to the sick poor and founded a hospital there. Many were attracted to the work, which had archiepiscopal approval, and Bishop Tuy named him "John of God," prescribing a habit for him and his companions. The order, Brothers Hospitallers, received papal approval in 1572, and its Rule was devised after his death. Canonized in 1690, he was declared patron of hospitals and the sick (1886) and of nurses (1930). C.G. THORNE, JR.

JOHN OF LEYDEN (Jan Beukelszoon) (1509-1536). Militant Anabaptist.* When Anabaptism was spreading rapidly in the northern Low Countries in the late 1520s, Jan Beukelszoon was rebaptized by Jan Mattheys of Haarlem, a fiery chiliast preacher. Militant "apostles" gathered in Westphalia, where Münster seemed ready to be the Anabaptist "city of refuge." Many of the lower classes there had turned Anabaptist, taken control, banned unbelievers, and incurred a siege by the outraged bishop of Münster. Mattheys, followed by Beukelszoon and others, went there, and after Mattheys was killed in battle, the tailor of Leyden was crowned king of the "New Zion." He instituted community of goods and polygamy, and executed his opponents. Protestant nobles joined Catholic forces against the city, and after its fall the defenders were slaughtered (1536), with their leader numbered among the dead. The memory of Münster helped shape the stereotype of the Anabaptists as disturbers of the peace and as erratic revolutionaries. DIRK JELLEMA

JOHN OF MATHA (Jean de Matha) (1160?-1213). One of the founders of the Order of Trinitarians. A native of Provence, France, he became a priest and consecrated himself to the redemption of Christian captives in the hands of the Turks. To further this purpose, with Felix of Valois (d.1212) he founded the Trinitarian Order at Cerfroid in the diocese of Meaux (1198). He was canonized in 1679.

JOHN OF MONTE CORVINO (d. c.1330). Founder of the first Franciscan mission in China. Born at Monte Corvino (Salerno), he was commissioned by Pope Nicholas IV in 1291, with letters for eastern kings, and journeyed through Persia, spent a year in India where he made about one hundred converts, and finally reached Khanbalik (Peking) in 1294. Khan Timor Olcheitu (Chentsung) was receptive to John and his traveling companion, a merchant, Peter Lucalonga. John was most successful in establishing a mission at Tenduk, northwest of Khanbalik, where the ruling prince, George, already a Nestorian Christian, converted to Western orthodoxy. Lucalonga became a successful merchant in China and donated land for the establishment of three churches. John was named archbishop in 1307 by Pope Clement V. Besides making 6,000 converts, he translated the NT and Psalms and established a native boys' choir that was highly popular with the people and the khan. DONALD M. LAKE

JOHN OF PARIS (c.1250-1306). Preacher and early conciliarist. Also known as John Quidort, he was born and educated in Paris. Joining the Jacobins,* he established a reputation for brilliance as an apologist for Thomism and respondent to William de la Mare's *Correctorium fratris Thomae.* A popular preacher and lecturer on Peter Lombard's *Sentences,* he was stripped of these offices near the end of his life for his writings on the Eucharist and the papacy. *De potestate regia et papali* (1302) anticipated conciliarist positions by advocating Christ as head of the church, collegiate authority, and deposition of the pope under certain circumstances. His interpretation of Christ's presence at the sacrament approximated consubstantiation and influenced Lutheran thought. He died in Bordeaux while appealing his suspension to the pope. JAMES DE JONG

JOHN OF PARMA (1209-1289). Franciscan leader. He taught logic at Parma, became a Franciscan in 1233, and later went to Paris for further study. He taught theology and was a master in logic and an eloquent preacher. As minister-general of his order from 1247 he sought to restore its asceticism and original standards. To this end he traveled throughout Europe, trying also to unite the Eastern and Western churches. His austerity and adherence to the doctrines of Joachim of Fiore,* together with a suspect treatise, brought his resignation, at Alexander IV's insistence, in 1257. He suggested Bonaventura* as his successor. The rest of John's life was spent in penance and contemplation as a solitary at the hermitage of Greccio. In 1289 he set out for Greece again to further reunion, but died en route at Camerino. The Franciscan Salimbene (1221-88) described John in his *Chronicle* as handsome, charming, learned, musical, strong, and energetic.

C.G. THORNE, JR.

JOHN OF RAGUSA (c.1380-c.1443). Dominican theologian. Born John Stoikovic, he became a considerable scholastic figure. In 1422, as a representative of Paris University, he successfully urged Martin V to hold a council, and was present at its sessions in Pavia and Siena. He became procurator general of the Dominicans in 1426, residing in Rome, and was influential with Martin V, helping to persuade him to call the Council of Basle* (1431). He attended its opening stages as papal theologian, preaching the opening sermon, and debating with the Hussites. In 1435 the council sent him to Constantinople as part of an embassy to attempt to achieve the union of Eastern and Western churches. He spent some years there persuading the emperor John Palaeologus to send representatives to the council, studying Greek, and when in 1437 there were envoys of the pope and council in Constantinople at the same time, preventing open hostility between them. He supported Felix V and became a cardinal, though some argue that he was faithful to Eugenius IV. Of his extant writings, his accounts of the councils of Siena and the early stages of Basle, together with his condemnation of Hussite teaching, are the most important. C. PETER WILLIAMS

JOHN OF ST. THOMAS (1589-1644). Dominican scholar. Portuguese and named from his devotion to Thomas Aquinas, he read arts and theology at Coimbra, continuing at Louvain, where he studied under a Dominican and entered the order. In 1620 he began lecturing in theology, at Piacenza and Madrid, then held the chair at Alcalá, simultaneously being a qualificator of the Supreme Council of the Spanish Inquisition. Confessor and adviser to Philip IV, he refused on insufficient evidence to condemn Louvain professors brought before him. *Cursus theologicus* was his chief work (1637-67), a commentary on Aquinas's *Summa Theologica*, the first four volumes of which were published during his lifetime. *Cursus philosophicus* (1631) also followed Thomist forms, and the *Compendium* (1640) on doctrine was published in seven Spanish editions and several languages. C.G. THORNE, JR.

JOHN OF SALISBURY (c.1115-1180). Medieval philosopher and classical scholar. Born of a poor family at Salisbury, Wiltshire, he studied under Abelard* at Paris and at Chartres. He became a papal clerk and was employed on various missions. In 1153 he returned to England to become chief minister and secretary to Theobald and Thomas Becket,* archbishops of Canterbury. He supported Becket against Henry II and was present at the archbishop's death in 1170. He became bishop of Chartres in 1176. His principal writings are the *Policraticus* (1159), a political treatise, and the *Metalogicon* (1159), which defends the study of logic and metaphysics and argues fittingly, as the best Latinist of his age and a notable man of letters, that this should be related to literary education. The latter work demonstrates his acquaintance with Aristotle's logical writings. He also wrote *Historia Pontificalis* and a number of *Letters*. HOWARD SAINSBURY

JOHN OF THE CROSS (1542-1591). Spanish mystic. Born Juan de Yepez y Alvarez in Old Castile, of a poor family of noble origin, he entered the Carmelite monastery at Medina del Campo in 1563. After studying theology at Salamanca, he was ordained (1567). Dissuaded by Teresa of Avila* from becoming a Carthusian, he introduced her reforms among the friars, joining the first of their reformed houses (Discalced*) at Duruelo. He then became master of the Carmelite College at Alcalá de Henares (1570-72) and confessor of the Convent of the Incarnation at Avila (1572-77).

Imprisoned in 1577 in the Carmelite monastery at Toledo for his reforms (his general favored a more relaxed rule: Calced*), he wrote the beginning of *The Spiritual Canticle*, but the entire work derives from this experience. With his escape after nine months to the monastery of Calvario came the separation between Calced and Discalced. He and Teresa were joint founders of the Discalced Carmelites. He was rector of the college at Baeza (1579-81) and prior at Segovia from 1588. In 1581 he went to Granada and became acquainted with the Arabian mystics. Writing out of personal experience and as a student of Scripture and Thomism, he wrote three poems, with commentaries: *The Dark Night of the Soul* (on which *The Ascent of Mount Carmel* is a second commentary), *The Spiritual Canticle*, and *The Living Flame of Love*. In this order is spiritual progression, classically conceived, with his own mystical experience bursting out in the third poem. Distrusted by his superiors, he was removed from the friary at La Penuela to Ubeda, where he died from inhumane treatment. He was canonized in 1726 and made a "Doctor of the Church" in 1926.

See B. Frost, *Saint John of the Cross* (1937), and E.A. Peers, *Handbook of the Life and Times of*

Saint Teresa and Saint John of the Cross (1954).
C.G. THORNE, JR.

JOHN OF WESEL (c.1400-1481). Roman Catholic reformer. John Ruchrath or Rucherat was born in Oberwesel am Rhein, studied at Erfurt (where he later served as rector of the university), and after a brief period as professor in Basle (1461), he became a cathedral preacher in Worms (1463). He fought for the reform of theology, even when it led him to defend disturbing and unpopular positions. He rejected the *Filioque* clause of the Nicene Creed as unbiblical, and virtually denied the Augustinian doctrine of original sin on the same grounds. Canon law was binding only inasmuch as it accorded with Scripture. Fasting, clerical celibacy, distinctions between bishops and priest were all human institutions and held no authority over the conscience of the faithful. Indulgences were a pious fraud, because only God could remit the penalties for sin. Indicted for suspected Hussite doctrines, he was deposed in 1477 and brought before the Inquisition in 1479. He recanted, but his books were burnt and he was sentenced to confinement for the rest of his life in the Augustinian cloister at Mainz. He died soon after, a broken and dispirited man.
DAVID C. STEINMETZ

JOHN SCHOLASTICUS (d.577). Born at Sirimis, near Antioch, the son of a cleric, he was a lawyer at Antioch, where he made a famous collection ("Synagoge") of canons, which he later re-edited and enriched from the *Novellae* of Justinian, so that the work became one of the primary sources for subsequent Eastern canon law. He was appointed patriarch of Constantinople by Justinian in 565, succeeding Eutychius who had been exiled for opposing Aphthardocetism, which Justinian had espoused. John became a close friend of Emperor Justin II and took an active part in trying to win over the Monophysites, but later carried out Justin's policy of suppression. He is said to have written a "Catechetical Oration" on the Trinity. He may be identical with John Malalas, the Byzantine chronicler, whose *Chronography* woven around the fortunes of Antioch, is a source of religious and secular history, written in the common Greek of the age. J.G.G. NORMAN

JOHNSON, RICHARD (1753-1827). Anglican clergyman. Educated at Cambridge, he was appointed first chaplain to the convict colony at Botany Bay, New South Wales, Australia, in 1786. Both the Eclectic Society and members of the Clapham Sect* helped to secure his appointment. He conducted the first service in Australia on 3 February 1788. As the only chaplain his task was made more difficult by opposition from the military junta in the colony. He built a church at his own expense. His emphasis on personal salvation brought charges of "methodism." Yet independent observers recognized his unsparing help towards the sick among the convicts, the aborigines, and the orphans. Like most government officials he acted as farmer and magistrate, but his main concern was for his clerical duties. He left the colony in 1800 officially on leave, but in 1810 he took a London parish. NOEL S. POLLARD

JOHNSON, SAMUEL (1709-1784). Moralist, essayist, and lexicographer. Born at Lichfield and educated at Pembroke College, Oxford, Johnson unsuccessfully attempted schoolmastering before he went to London with the actor Garrick. He had a harsh struggle, in the course of which he wrote for *The Gentleman's Magazine.* His satiric poems *London* and *The Vanity of Human Wishes* appeared in 1739 and 1749, and in 1747 he began work on his great *Dictionary* which appeared in 1755. He also ran two periodicals, *The Rambler* (1750-52) and *The Idler* (1758-60), while in 1759 he published the moral tale, *Rasselas.* The last twenty years of his life are the Age of Johnson, the great conversationalist, as recorded by Boswell, but they also include his edition of Shakespeare (1765) and his *Lives of the Poets* (1779-81). Johnson, Tory and Anglican, held deeply sincere religious views, though these are expressed more often in moral than in spiritual terms. Nowhere is this better shown than in the last lines of *The Vanity of Human Wishes,* where a fine reticence controls profound beliefs. In this regard Johnson ranks among the most reverent of English writers.
ARTHUR POLLARD

JOHN THE APOSTLE. Traditionally the author of the NT documents that bear the name of John: the fourth gospel, the epistles of John, and Revelation. He was the son of Zebedee, and the (probably younger) brother of James (Matt. 4:21). His mother was possibly the Salome who was present at the crucifixion of Jesus (Matt. 27:56; cf. Mark 15:40). The two fishermen brothers James and John were called Boanerges ("Sons of Thunder") by Jesus (Mark 3:17), presumably because of their headstrong character. With Peter they formed an "inner group" of the Twelve, and were the only disciples present with Jesus on three important "revelatory" occasions during His ministry: the raising of Jairus's daughter (Mark 5:37), the Transfiguration (Mark 9:2), and the agony in Gethsemane (Mark 14:33). Luke also tells that Peter and John were the two disciples sent to prepare the final Passover meal for Jesus (22:8; cf. Mark 14:13). The Apostle John is not mentioned by name in the fourth gospel, although "the sons of Zebedee" appear in 21:2; but he may have been present at the call of the first disciples (John 1:35-41) and is usually but not always identified with the beloved disciple. In that case, we have further evidence about him (see John 13:23; 19:26; 20:2-10, 21:7; 21:20-23). The beloved disciple in the gospel of John* has the special function of witnessing to the reliability of the Johannine tradition (John 19:35; 21:24).

The Apostle John is mentioned in three passages of Acts, each time with Peter (1:13f.; 3:1—4:31; 8:14-25). He seems to have played an important part in the life of the early church (Gal. 2:9), despite his subordination to Peter, and was presumably present at the Jerusalem council (Acts 15). He does not appear in the later part of Acts, and we do not know when he left Jerusalem (cf. Acts 12:17).

Later tradition connects John the apostle with Ephesus, where he is reported to have died a natural death in old age. The testimony of Irenaeus affirms also that John lived at Ephesus until the time of Trajan (A.D. 98-117). Although this tradition has been challenged, its chronology is plausible. Eusebius follows Irenaeus, but says that John was exiled to Patmos during the reign of Domitian (81-96) and afterward returned to Ephesus. Thus John the apostle could be the author of the Revelation. Eusebius, however, assigns this to John the Elder,* whom he mentions in association with the confused witness of Papias to the origins of John's tradition.

Various stories support the link between John and Asia Minor, but other accounts are less straightforward. Tertullian claims John was exiled from Rome, after he had been "plunged, unhurt, into boiling oil." The Muratorian Canon says John was with the other apostles when he wrote his gospel, perhaps suggesting a Palestinian rather than Asian provenance. This evidence, however, is probably unhistorical, like the tradition that John was martyred early in life, possibly at the same time as James his brother (Acts 12:2; cf. Mark 10:39). The case for John's association with Ephesus in old age appears strong, although this does not by itself establish his responsibility for the final edition of the Johannine corpus. Its diversities of style and thought rather suggest origin in a school, behind which (no doubt) stood the Apostle John himself.

BIBLIOGRAPHY: See under JOHN, GOSPEL OF. Also, A. Harnack, *Die Geschichte der altchristlichen Litteratur bis Eusebius* (1893); P. Parker, "John and John Mark," *JBL* LXXIX (1960), pp. 97-110; J.N. Sanders, "St. John on Patmos," *NTS* IX (1962-63), pp. 75-85. STEPHEN S. SMALLEY

JOHN THE BAPTIST. Son of the priest Zechariah and Mary's kinswoman Elizabeth, John was born in circumstances suggesting divine intervention (Luke 1:5-79). He grew up in the desert (Luke 1:80). His association with the Qumran* community in his early days is possible, and in this case he may have been responsible for the Qumranic influences in the fourth gospel; but this suggestion should not be exaggerated. After his call, John was too independent a figure to be committed to another movement, whatever his background.

John's ministry began in the region of the Jordan; although if W.F. Albright is right to identify "Aenon near Salim" (John 3:23) with the Samaritan territory near Nablus, it had significantly wider scope. John appeared as an ascetic preacher (cf. Luke 1:15), clothed like the OT prophets with camel's hair, and eating locusts and wild honey (Mark 1:6). He preached "a baptism of repentance for the forgiveness of sins" (Mark 1:4 RSV), which included a reference to the dawning kingdom of God (Matt. 3:2). The eschatological content of his message, with its stress on judgment, is given by Matthew and Luke (Matthew 3:7-12). The baptism which John administered had its background in Jewish lustration and initiation rites; but in John's hands it acquired a new dimension, both because it was no longer self-administered, and because it was accompanied by the demand for repentance in the light of Israel's need for renewal, and in the face of the coming messianic age (Mark 1:5; contrast Josephus, *Antiquities* 18.5.2).

John's relation to Jesus in the NT is of special and obvious importance. He was similar to Jesus in many ways (cf. the parallelism of Luke 1 and 2); yet after him the line was drawn across history (Matt. 11). For he was above all the forerunner of the Messiah he recognized and confessed (Mark 1:2f., 7f.), even if he possibly had doubts on this score eventually (Matt. 11). The Johannine estimate of his own identity is probably correct. He regarded himself as neither the Christ nor Elijah, the expected forerunner of the new age; but simply as "the voice of one crying in the wilderness" (John 1:23 KJV). Probably, however, he modeled himself on Elijah. John is thus an important bridge-figure between Judaism and Christianity; his popular preaching marked a new beginning in Israel's history. His ministry was, as a result, regarded in the early church as the beginning of the Gospel period (Acts 10:36f.; cf. 1:22). John was put to death by Herod Antipas in the fortress of Machaerus, after the hostility of both Herod and Herodias, his wife, had been aroused (Mark 6:14-29; differing in some respects from the account in Josephus, *Antiquities* 18.5.2). The disciples of John the Baptist evidently remained together after his death (cf. Mark 6:29); but the *Clementine Recognitions* 1.60 is the only source to suggest that their existence constituted a threat to the Christian Church.

BIBLIOGRAPHY: C.H. Kraeling, *John the Baptist* (1951); P. Winter, "The Proto-Source of Luke 1," *Novum Testamentum* I (1956), pp. 184-99; J.A.T. Robinson, "The Baptism of John and the Qumran Community," *HTR* L (1957), pp. 175-91; P. Benoit, "Qumran et le Nouveau Testament," *NTS* VII (1960-61), pp. 279-88; C.H.H. Scobie, *John the Baptist* (1964); W. Wink, *John the Baptist in the Gospel Tradition* (1968); F.F. Bruce, *New Testament History* (1969/70), pp. 145-54; E. Bammel, "The Baptist in Early Christian Tradition," *NTS* XVIII (1971-2), pp. 95-128.

STEPHEN S. SMALLEY

JOHN THE CONSTANT (1468-1532). Saxon prince and brother of Frederick the Wise. Born at Meissen, John became an early supporter of the Lutheran cause. He urged his brother to protect Luther from the ban of the Empire, and welcomed Luther when he preached at his own court in Weimar in 1522. John attempted in his practice of statecraft to follow the principles outlined by Luther in *Von Weltlicher Obrigkeit.* He was reluctant to suppress Müntzer and Carlstadt at first and tolerated their radical reform. When on the death of his brother he became sole ruler (1525), he confessed himself a Protestant and entered into a treaty with Philip of Hesse (1526). He established visitations, reorganized the University of Wittenberg, and stopped the appropriation of church property by the Saxon nobility. He defended the Protestant interests at the Diet of Speyer and accepted the Schwabach Articles.* At the Diet of Augsburg he signed the Augsburg

Confession as leader of the Protestant party. Opposing the election of Ferdinand as Roman king, he agreed to the formation of a Protestant league of defense against the emperor, leading to the peace of Nuremberg (1532).

DAVID C. STEINMETZ

JOHN THE ELDER. Probably a mythical figure, invented by Eusebius of Caesarea to account for the authorship of the Apocalypse. Those who could not accept the book as the work of John the Apostle, particularly because of its millenarianism, and yet hesitated to regard it as pseudonymous, looked around for another John who could have been its author. Dionysius of Alexandria (d. c.264) appealed to the existence of two tombs at Ephesus, both purporting to be the burial place of "John," as evidence that such a second John existed. Eusebius followed Dionysius, mistakenly discovered the other John in the tradition of Papias, called him "John the Presbyter," and ascribed the Revelation to him. But Papias was no doubt referring to John the *Apostle* in two different contexts, among the disciples (also called apostles and elders who had died, and those who were still alive). Although the "Presbyter" has been claimed as the author of all or part of the Johannine corpus, there is no need to postulate his existence in order to account for its particular character.

STEPHEN S. SMALLEY

JOHN THE SCOT, see ERIGENA

JOINVILLE, JEAN (c.1225-1319). French nobleman and chronicler. Becoming Seneschal of Champagne after his father's death in 1233, he was subsequently active in French court life, and established contacts which enlisted his participation in the Sixth Crusade (1248-54). He joined the assaults on Egypt and Palestine and became a friend of Louis IX during their imprisonment. While at Acre, he composed a credo which reflects medieval piety. Returning to his estates, he followed Louis's career with interest, though he declined the royal invitation to participate in the Crusade of 1270. At the suggestion of Queen Jeanne, wife of Philip the Fair, he composed a discursive but graphic and engaging biography of Louis. It chronicles the Sixth Crusade and captures the spirit of the age. JAMES DE JONG

JONAS, JUSTUS (1493-1555). Protestant Reformer and scholar. Born at Nordhausen, he studied at Wittenberg and Erfurt, became provost of All Saints' and professor at Wittenberg, and sided with Luther in his movement. He attended the Marburg Colloquy* (1529) and the Diet of Augsburg (1530) and participated in the Concord of Wittenberg* (1536). As pastor in Halle he contributed to the progress of the reform movement there. He opposed the Augsburg Interim* (1548). He is known for translating the Apology of the Augsburg Confession from Latin into German. His Latin translation of the Brandenburg-Nuremberg Catechism was translated into English under the direction of Archbishop Cranmer. It was also translated into Icelandic. His contributions to the Lutheran Reformation include hymnwriting, church visitation, organizational ability, the drafting of church orders, and preaching.

CARL S. MEYER

JONES, BOB (1883-1968). American evangelist. Born in Alabama, son of a Confederate Army veteran, he held his first evangelistic meeting at the age of fourteen, and a year later was licensed to preach by the Methodist Church. He attended Southern University in Greensboro, Alabama. In 1924 he decided to start his own college to promote unflinching fundamentalism.* At the time of his death, Bob Jones University, located in Greenville, South Carolina, had an enrollment of 4,000 and a modern campus valued at $50 million. Jones's evangelistic preaching took him to every state of the union and into thirty foreign countries. During the height of his career he preached an estimated 12,000 folksy sermons to more than fifteen million people. The school which bears his name came to be identified with fundamentalist theology and Southern conservative politics, including segregation of the races.

BRUCE L. SHELLEY

JONES, E(LI) STANLEY (1884-1973). Missionary to India. Born in Maryland, he was ordained as a Methodist minister and appointed first to the Lal Bagh English-speaking church in Lucknow. Soon, however, he was released to a wider ministry among English-speaking Indians. Possessing a deep understanding of Indian culture and religion, he sought to interpret the Christian faith to the educated, not as a Western import, but as the fulfillment of their own spiritual longings. A popular preacher, he was also a spokesman for peace, racial brotherhood, and social justice. His encouragement of Indian independence led to his being barred from the country for a period by the British authorities. In 1930 he founded a Christian ashram for study and meditation in Sal Tal, which many continue to attend. During his extensive world travels he was responsible for the foundation of similar centers in America and several European countries. He also established the Nur Manzil psychiatric center in Lucknow. He was the author of twenty-nine books, the best known being *Christ of the Indian Road* and *Abundant Living*. JAMES TAYLOR

JONES, GRIFFITH (1683-1761). Welsh preacher. Born a Nonconformist in Carmarthenshire, he was ordained in the established church in 1709, served two curacies, and was rector of Llandilo Abercowyn from 1711 and of Llanddowror from 1716. A traveling preacher and keen to improve religious and social conditions in Wales, he established in 1730 his first charity school, where children and adults were taught by day and night to read the Bible in Welsh; education grew despite ecclesiastical opposition. Teachers traveled a circuit and schools multiplied: before his death over 3,000 schools were opened and 150,000 taught. Jones had strong Calvinist leanings and wrote many theological books in Welsh and English. Daniel Rowland* of Llangeitho, founder of Welsh Methodism, was supposedly converted under his ministry. C.G. THORNE, JR.

JONES, JOHN CYNDDYLAN (1840-1930). Welsh expositor and theologian. Born at Capel Dewi, Cardiganshire, he was educated as a candidate for the Welsh Calvinistic ministry at Bala and Trevecca colleges. After serving as minister at the English Calvinistic Methodist church at Pontypool (1867-69), he became a Congregational minister in London, and in 1875 returned to the Methodist fold as minister of Frederick Street, Cardiff. His eagerness to introduce an element of Anglican ceremonial into the services ended in his resignation in 1888. He then joined the staff of the British and Foreign Bible Society in South Wales.

He was an assiduous writer. His books, *Studies in Matthew, Luke, John, Acts* and *The Epistles of Peter,* became popular, not least among American readers. He published similar expositions in Welsh on the gospel of John and the epistle to the Philippians. He was a vigorous opponent of liberal trends in biblical scholarship; some of his views on this topic can be seen in his *Primeval Revelation.* His systematic theology in four volumes, *Cysondeb y Ffydd,* expounds a moderate Calvinism much influenced by such American theologians as W.G.T. Shedd.* At his best he was a very good writer, and there was a rare power and unction in his preaching. He was passionately concerned to maintain the evangelical position in theology and biblical exegesis, but lacked the scholarly equipment to do full justice to his intentions.

R. TUDUR JONES

JONES, RUFUS MATTHEW (1863-1948). Quaker scholar and professor. Born in South China, he was educated at Haverford College, Harvard, and in continental universities. Having taught in Friends schools and been principal of Oak Grove Seminary (1889-1893), he returned to Haverford to lecture in philosophy. A birthright Quaker, at twenty-four he had a mystical experience which reshaped his life entirely. By mysticism he meant a kind of Pauline experience and understanding of Christianity. Philosophy and theology were his life, together with a conspicuous concern for others. To Kant's philosophy of ethics he became increasingly indebted. Considered a prophet of Quakerism and hailed as a saint, he was as prominent in Britain as in America, and lectured more widely still. He helped to found the American Friends Service Committee (1917) and served as its chairman for over twenty years. He wrote over fifty books and hundreds of articles, most of which concerned mystical experience and action.

C.G. THORNE, JR.

JONES, SAMUEL PORTER (1847-1906). Evangelist and Prohibitionist. Born in Alabama, he grew up and graduated from high school in Cartersville, Georgia. Chronic nervous indigestion kept him from college, and alcoholism soon ended a law career he began in 1868. After promising his dying father to reform, he was converted under his Methodist grandfather's preaching and became a circuit rider for the North Georgia Conference of the Methodist Episcopal Church, South, in 1872. Eminently successful in winning converts in Georgia before 1880 and in raising funds as agent for the Methodist Orphanage Home, Decatur, Georgia, "Sam" was thereafter invited to conduct revival meetings in major southern cities and became after 1885 an evangelist of national prominence. His meetings, characterized by controversial, vernacular oratory, often produced organized efforts to enforce local blue laws and helped stimulate the national Prohibition movement. The Sam P. Jones Lectureship at Emory University perpetuates his memory.

D.E. PITZER

JONES, THOMAS ("of Denbigh") (1756-1820). Welsh Calvinistic Methodist* theologian. He was born at Penucha, Flintshire, of a fairly well-to-do family. His only education was at schools in Caerwys and Holywell, but it gave him excellent grounding in the classics. He began to preach among the Calvinistic Methodists in 1783 and ministered to their societies at Mold, Ruthin, and finally Denbigh (1806-20). He was closely associated with Thomas Charles after whose death in 1814 Jones was the most learned and distinguished leader of the Calvinistic Methodists.

Jones was a man of wide culture. His massive volume *Hanes Diwygwyr, Merthyron, a Chyffeswyr Eglwys Loegr* (1813) is at the same time a plea for toleration in matters of religion and a sustained defense of the thesis that Evangelicalism is a continuation of the Augustinian understanding of Christianity which is the true catholicism. Jones's autobiography is a moving document describing his spiritual pilgrimage. In the theological controversies that dominated Welsh intellectual life at the beginning of the nineteenth century he took a firm but moderate Calvinist position. He was firmly opposed to Arminianism on the one hand, and although he was critical of hyper-Calvinism he was not able to embrace fully the "Modern Calvinism" of his friend, Edward Williams of Rotherham. He was also well versed in the Welsh literary tradition. His prose style is forceful and elegant, and as a poet and hymnwriter he combined the classical heritage of the strict meters with the newer use of free meters in the tradition of Williams Pantycelyn. The Welsh Calvinistic Methodists had no wiser or more scholarly and godly leader at the beginning of the last century than Thomas Jones.

R. TUDUR JONES

JORDANES (Jornandes) (fl. c.550). Latin historian. A native Alan, he served as *notarius* to an Alan king, later becoming an orthodox bishop in Italy. His *Getica,* written about 551, an epitome of Cassiodorus's lost *Gothic Histories,* relates the earliest migrations of the Goths to their defeat by Belisarius in 541, then carries events beyond Cassiodorus in hopes of reconciling Gothic and Roman royal families. Uncritical and profuse with errors, it is nevertheless much cited as principal extant source for our knowledge of the Goths, especially their settlement on the Black Sea under Hermanric (mid-fourth century) and of Attila's second invasion of Gaul. He also authored a short compendium of universal history from Creation to 552, *de regnorum ac temporum successione,* also known as the *Romana.*

DANIEL C. SCAVONE

JOSEPH. Husband of the Virgin Mary. He is scarcely mentioned in the NT outside the birth and infancy narratives in the gospels of Matthew and Luke. There are indirect references to him in John 1:45; 6:42. It is likely that he had died before the ministry of Jesus began. Matthew and Luke both show him as a man of Davidic descent (Matt. 1:19f.; Luke 2:4) who was betrothed to Mary at the time of the conception of Jesus, but had no intercourse with her until after His birth (Matt. 1:18; Luke 1:27, 35). It is natural to assume that the brothers of Jesus referred to in the gospels were subsequent children of Joseph and Mary. While the account of the Annunciation* in Luke's gospel is told from Mary's point of view, that in Matthew is told from Joseph's. Through the appearance of angels in dreams he was able to take Mary as his wife rather than divorcing her (Matt. 1:18-25), and to escape from Herod to Egypt (Matt. 2:13-15) and return again to Israel (Matt. 2:19-23). The veneration of Joseph goes back to early times in the Eastern Church, but seems to have started in the West only in the late Middle Ages. R.E. NIXON

JOSEPH CALASANCTIUS (1556-1648). Founder of the Piarists.* Born near Petralta de la Sal, he studied law and theology at Lérida, Valencia, and Alcalá; was ordained in 1583; and went to Rome in 1592, where he was patronized by the Colonna family and became active in charitable works. Convinced of the need to provide religious and secular education for the children of the poor, he opened in 1597 the first free public school in Europe (in Rome) and established in 1602 the Piarist Order for those teaching, giving full privileges of a religious order. Dissension arose: Galileo's associations were suspected, and fears were expressed that the educated poor might unbalance society. By 1643 crises led to Joseph's own trial, and in 1646 destruction came when Innocent X reduced the order to a federation of independent religious houses, not to be fully restored until Clement IX (1669). Joseph was canonized in 1767; Pius XII declared him "the heavenly patron of all Christian schools."

C.G. THORNE, JR.

JOSEPHINISM. The Austrian Hapsburg policy of secular state control of the church implemented in the eighteenth century under Empress Maria Theresa and culminating with intensity under Joseph II (1780-90). Its motivation was secular and rationalist, aimed at "rationalizing" the organization of the whole of society through an "enlightened" program of statist centralism. Instructions (1767, 1768) by the chief minister, Kaunitz, initiated the program.

Then under Joseph II came a rush of projects, including the following. The law of toleration (1781) ended the Catholic monopoly, allowing Protestants and Jews a certain freedom to worship—a step permitted, Joseph argued, because any church could be made obedient to the state. Traditional censorship, including the Index, was abolished, but a new rationalist censorship was practiced against, *inter alia*, "superstitious" works, all ecclesiastical publications, and most public discussion of religion. Monasteries were either dissolved or their members reduced in number (beginning 1781) on grounds that many were useless, or wasteful; their properties were confiscated, and the revenues used to fund a state reorganization of parishes, and state-controlled schools, shops, or factories. All links between the papacy and the Hapsburg Church were abolished or controlled, since the pope was viewed principally as a foreign political power. Bishops were obliged to swear loyalty to the state and forbidden corporately to oppose Joseph's decrees, while priests were made *de facto* state officials. The system thus erected continued without major alterations until 1850.

BIBLIOGRAPHY: M.C. Goodwin, *The Papal Conflict With Josephinism* (1938); F. Maass, *Der Josephinismus* (5 vols., 1951f.), "Josephinism" in *New Catholic Encyclopedia* VII, pp. 1118-19, and "Josephinism" in *Sacramentum Mundi* III, pp. 209-10, and "Josephinismus" in *Lex Theol Kirche* V (1960), pp. 1137-39; S.K. Padover, *The Revolutionary Emperor: Joseph II of Austria* (rev. ed., 1967); C.A. Macartney, *The Hapsburg Empire, 1790-1918* (1968). C.T. MC INTIRE

JOSEPH OF ARIMATHEA. Although coming only from a small village in Judea, Joseph was a respected member of the Sanhedrin (Mark 15:43). The fact that he had a private tomb and was able to provide the linen for it is evidence of his wealth (Matt. 27:57-60). He is described by Luke as "a good and upright man, who had not consented to their decision and action [i.e., the Sanhedrin's plot against Jesus] . . . and he was waiting for the kingdom of God" (23:50f.), and by John as "a disciple of Jesus, but secretly, because he feared the Jews" (John 19:38). The crucifixion seems to have emboldened him, and he used his position to go to Pilate and obtain from him permission to take down the body of Jesus from the cross (Mark 15:-43-46, etc.). This may have been partly the act of a Jew not wishing to see the land defiled by a corpse (Deut. 21:23), but is more closely bound up with his devotion to Jesus. A number of legends were associated with him, especially that of a visit to England with the Holy Grail and the building of a church at Glastonbury. R.E. NIXON

JOSEPH OF COPERTINO (1603-1663). Franciscan ascetic. The son of a poor carpenter in a small town of SE Italy, Joseph in his youth suffered much ill health aggravated by his asceticism. He entered the Franciscan Order, became a priest in 1628, and thereafter lived a life of extreme austerity, punctuated by many loathsome self-tortures. In his devotions he frequently swooned in ecstasy and rose in the air. His superiors, doubtless moved by notoriety or envy, were determined to end this "miracle." He was made to attend chapel alone, and even charged before the Inquistion; but in paying homage to Urban VIII by kissing the papal feet, Joseph astonished Urban by forthwith levitating. There are numerous eyewitness accounts of this achievement, including that of the duke of Brunswick (d.1679), patron of the German philosopher Leibnitz. The duke, though a Lutheran, was so impressed by the "miracle" (which he saw twice) that he forthwith became a Roman Catholic. Levitation seems now to be unknown or extremely rare: the last reasonably well authenticated cases were connected with the medium D.D. Home (1833-86).

R.E.D. CLARK

JOSEPHUS FLAVIUS (A.D. 37-post 100). The Roman name of Joseph ben-Matthias, a Jew of aristocratic family and Pharisaic adherence, author in Greek. Religious ascetic for a time in youth, he was later a priest, and in A.D. 64 member of a mission to Nero. After the outbreak of the Jewish revolt in A.D. 66 he became commander in Galilee, eventually capitulating to Vespasian at Jotapata. Prophesying his captor's imperial destiny, he became his protégé and attempted to urge surrender upon his compatriots. He accompanied Titus to Rome where he lived in literary activity, having taken Roman citizenship and the Flavian name.

Devious and self-centered, he wrote from a mixture of motives: sycophancy, self-defense, and patriotism. Between A.D. 75 and 79 appeared his *Jewish War*, addressed first in Aramaic to races prone to trouble Rome on her borders (still extant in Greek). In six books it draws upon personal reminiscence, that of Herod Agrippa II also, and official records, for the events of A.D. 66-70, for which period it is a valuable historical source. The *Archaeology* (or *Antiquities*) *of the Jews*, in twenty books, appeared in A.D. 93-94: an apologia for his people, it follows the Septuagint order, and later draws upon Nicolas of Damascus, Herod's secretary, for recent events. In book XVIII is found the renowned passage ("Testimonium Flavianum") about Jesus, which modern scholarship confirms as basically authentic, though some alteration of Josephus's style and/or interpolation by Christians is generally admitted. Other passages deal with John the Baptist and James the Just. The Life was an appendix to the *Antiquities; Contra Apionem* is written against a contemporary anti-Semite.

BIBLIOGRAPHY: Edition: Loeb Classical Library in nine volumes with translation and notes (1926-65), by H. St. J. Thackeray et al.; H. St. J. Thackeray, *Josephus the Man and the Historian* (1929); L.H. Feldman, *Studies on Philo and Josephus (1937-62)* (1963); H. Schreckenberg, *Bibliographie zu Flavius Josephus* (1968).

J.N. BIRDSALL

JOSQUIN DESPREZ, see DESPREZ

JOURNALISM, CHRISTIAN. In current usage of the term, Christian journalism relates most commonly to Christian periodicals. Only rarely is it connected with other media. It is not so closely tied to news in the narrow sense as is secular journalism. In the pre-Reformation church, "journalism" was for the most part confined to official announcements and newsletters. The Roman government used *acta diurna*, bulletins posted daily in public places. Beginning with Constantine, these may have taken on something of the character of Christian journals. The newsletters were sent out regularly by Roman scribes to businessmen and politicians in distant cities to keep them abreast of doings in Rome. Such newsletter service continued from European capitals into the 1700s, after printed papers became common.

With the development of printing at the time of the Reformation came the rise of Christian journalism in the form of pamphlets. It is doubtful whether the Reformation would have had nearly the impact without them. Luther himself wrote that "there has never been a great revelation of the Word of God unless He has first prepared the way by the rise and prosperity of languages and letters, as though they were John the Baptists." His own tracts were the forerunners of modern periodicals, and they were highly successful. Four thousand copies of his address "To the Christian Nobility" were sold in five days. Roman Catholics also began to use printed materials, but they have been dogged to the present day by theological doctrines which fail to recognize the freedom of the press.

There was little effort to develop Christian journalism in the seventeenth and eighteenth centuries. The Puritans argued extensively in books and pamphlets, and recorded extensive history there, but despite many attempts to produce regular news journals in 1640-60, they never became seriously interested in periodicals. State churches saw no good reason to enlighten the masses.

Christian periodicals did not become established until the nineteenth century. In North America they took the form of highly partisan newspapers which bred much strife among Christians, but which nonetheless often enjoyed as much respect and readership as their secular counterparts. Religious newspapers were sold on the streets along with the others. During the latter part of the nineteenth century and the early part of the twentieth, most Christian daily newspapers died out, to be replaced by weekly papers and magazines. Eventually many of these changed into or gave way to monthlies, and independent periodicals were replaced by denominational or organizational house organs. These have been the mainstay of Christian journalism for most of the twentieth century, and their proliferation has been little short of phenomenal, although virtually all require subsidies on top of circulation and advertising income.

Modern Christian publications tend to equate progress with improvement in physical appearance rather than rhetoric, and investment of resources has reflected this emphasis. The popularity of offset printing with its much cheaper picture reproduction has encouraged the trend. Articles, reviews, and editorials generally espouse strongly held theological positions, but little great literature has been produced. Emergence of the Jesus Movement, which immediately started dozens of new papers on inexpensive newsprint, may have been the start of a reversal away from the slick, multicolored, profusely illustrated pattern toward more concentration on quality editorial content, although the first issues carried rather standardized devotional and evangelistic material.

The scope of Christian journalism is seen in the size of the several "trade associations" in the field. In *North America*, these are the Catholic Press

Association (435 member publications in 1971 with an aggregate circulation of 24,346,826), Associated Church Press (195 and 21,958,111), and Evangelical Press Association (195 and 11,400,000). There is some overlapping membership among these. Canadian editors, for example, have their own group in addition to belonging to one or more of the others. There are hundreds of other religious publications, mostly small, which belong to none of these.

All these periodicals rely upon mail promotion and delivery. Very few have been available on public newsstands. Many churches subscribe for their members. The prospect of sharp increases in postal rates during the 1970s spelled possible reconsideration of this distribution method. Most Christian publications have been selling large amounts of advertising space. In 1971 the top fifteen consumer-type Protestant periodicals received an estimated total of more than $3 million in advertising revenue. The advertisers generally are publishers of books and educational materials, organizations appealing for donations, insurance companies, schools, and travel agencies.

The liberal *Christian Century* was undoubtedly the most influential religious journal until *Christianity Today* was founded in 1956 to challenge its stature. As of 1972, both were being indexed in the standard reference work, *Reader's Guide to Periodical Literature,* along with *America,* a Jesuit weekly, the liberal lay Catholic weekly *Commonweal,* and the monthly *Catholic World. Christian Herald* has also long been a leader. *Decision,* published by the Billy Graham Evangelistic Association, captured top circulation honors in going from an initial press run of 286,000 in 1960 to more than 4,500,000 in twelve years. *The Christian Science Monitor* is highly reputed as a daily newspaper for secular coverage; it devotes one page a day to Christian Science dogma. The African Methodist Episcopal Church prints the largest continuously published Negro paper in America, the *Christian Recorder.* The sprightly *United Church Observer* is the most prominent in Canada.

Most religious weeklies that survived the two world wars subsequently established sound footing and maintained wide respect. Most notable among these have been the Southern Baptist state papers, one of which, the *Baptist Standard* of Texas, has had a larger circulation than most national and international religious publications. Among weeklies that do remarkably well despite very small staffs are *The Mennonite* and *Presbyterian Journal.* Among the few churchmen who still recognize the incomparable value of a weekly is Carl McIntire, who edits and publishes the polemical but very timely *Christian Beacon.*

In many African countries, mission-spawned publications are among the most widely read and are sold on newsstands because they serve secular news as well as religious purposes. There are some daily Christian newspapers published in Scandinavia and the Netherlands. *L'Osservatore Romana,* voice of the Vatican, is the world's best-known religious daily. In Indonesia Christians have their own daily newspaper, *Sinar Harapan* ("Ray of Hope"), begun in 1961. It distributes 65,000 copies daily, is twelve pages in length, and in size is the largest newspaper in the country.

The most-respected training ground for Christian journalists has been Syracuse (New York) University, which has a department of religious journalism in its school of communications and which offers bachelors, masters, and doctoral degrees in communications. Wheaton (Illinois) College and Oklahoma Baptist University offer graduate programs in Christian communications which embrace journalism.

Evangelicals have seized opportunities in radio and television to a much greater extent than liberals. Thousands of Christian programs took to the air, and hundreds of Christian stations have been commercially successful. Few broadcasts and telecasts, however, have been concerned enough with timely affairs and current developments to warrant the designation of journalism. Most have simply followed the pattern of church services. The coming generation will probably have more sophisticated media such as cassettes, electronic video recordings, and facsimile at its creative disposal.

A definitive work on Christian journalism has been long overdue. Indeed, an authentic philosophy of Christian journalism relating reportage to evangelism, service, education, worship, and fellowship has been desperately needed. Under the influence of empiricism, secular journalism long championed a theory of verifiability wherein biblical truth must be labeled "opinion" and subordinated to the "facts" of observable data. In recent years, a new art form of journalism that is basically existential has risen to compete with the objectivity school. To do justice to its scriptural origin an evangelical view must be developed which transcends both these theories of meaning.

DAVID KUCHARSKY

In the *British Isles,* the nineteenth century could aptly be called "a continuing communication revolution" in the field of print and journalism. Technical developments with faster transportation made possible the distribution of periodicals on a national scale. Modern Christian journalism in Britain can, however, be dated from the Wesleyan Revival of the eighteenth century. Before the foundation of the Religious Tract Society in 1799, John Wesley* and his friends were using tracts on a wide scale. The first Christian magazines, it should be noted, were either collections of sermons aimed at the elect, or tracts—stories or homilies—designed to reach the unconverted. The Religious Tract Society really shaped the magazine revolution of the nineteenth century. Originally founded as a means of developing new styles of print evangelism, the RTS became a worldwide enterprise, producing literature for distribution in many languages and countries.

As the century proceeded and reform movements developed, Christians were faced with the implications of wider literacy. The Society for the Propagation of Christian Knowledge, like the RTS, became a foremost publisher of "wholesome Christian literature." Magazines such as *The Sunday at Home* and *The Leisure Hour* offered superb combinations of articles on travel, popular science, the arts, literature, etc., combined with very readable articles on Scripture. Later *The Quiver* and *Cassell's Family Magazine* mingled popular appeal with spiritual content; *The Sunday Strand* had evangelist George Clarke as editor.

Denominational newspapers also flourished in the later nineteenth and early twentieth centuries, particularly as so many evangelists were deeply involved in journalism. The firebrand Hugh Price Hughes, with his campaigning *Methodist Times,* was obviously encouraged in literature evangelism by D.L. Moody.* Former missionary Michael Paget Baxter was another great crusader in print with his *Christian Herald* (still the best-selling Protestant weekly in Britain) and *The Signal.* World War I saw for the first time free church evangelists working as chaplains on an equal footing with Anglicans and Roman Catholics. Reports sent home from the front, published in Christian papers at home, helped prepare the way for interchurch evangelism and ultimately the ecumenical movement as well as the cause of Christian journalism.

Between the wars, some of the finest campaign journalism was seen in papers like *The Methodist Recorder* and *Joyful News.* The 1930s tended to be years of retrenchment, however, and radio, movies, and television (the latter in its infancy) held out alternative attractions to reading. World War II closed down many publications, and the bombing of London destroyed the premises of many publishing houses. *Crusade* (monthly) magazine, launched in the 1950s by the Evangelical Alliance, was probably the most noteworthy feature of the decade, but twelve issues a year does not give scope for much "hard" news. Where such is found it is generally in denominational publications such as *Church Times, Church of England Newspaper,* and *Baptist Times.* Of the interdenominational weeklies, *The Christian* tried to blaze new trails in offering a wide range of news coupled with a professionalism of presentation, but it fell a financial casualty in 1969. The *British Weekly,* launched by W. Robertson Nicoll in 1886, still maintains a precarious existence, having changed hands several times in the past two decades. The *Life of Faith* (founded in 1874) maintains its traditionally strong links with the Keswick Convention.* The Roman Catholic weeklies—the *Catholic Herald, The Universe,* and *The Tablet*—while their ecclesiastical attachment is never in any doubt, offer an imposing treatment of international affairs as they overlap the religious scene. DAVID LAZELL

JOVIANUS, FLAVIUS CLAUDIUS (c.332-364). Roman emperor. A Christian from Pannonia and senior member of the imperial bodyguard, Jovian was a compromise choice as emperor in 363. His brief reign was hailed as guaranteeing the church a place in the state from which no subsequent emperor again sought its removal, though it appears that his official policy was one of broad toleration for all religious parties. To get the army safely out of Mesopotamia and to gain his own security, he agreed to a treaty with Shapur II with unfavorable provisions including the loss, not only of Transtigridian territory, but the frontier cities of Nisibis and Singara also. He did not have the chance to establish himself, for he died suddenly and mysteriously at Ancyra before he reached Constantinople. CLYDE CURRY SMITH

JOVINIAN (d. c.405). Writer and monk. His treatise entitled *Commentarioli* has not survived, and its thought can at most be minimally reconstructed from Jerome's refutation, *Adversus Jovinianum,* in two books written in 392 at the request of Pammachius.* By that date Jovinian, presbyter and formerly strict monk, and his followers had already been synodically condemned twice: in Rome under Siricius (390) and in Milan under Ambrose (391). Jerome's first book deals entirely with Jovinian's notion that, provided persons did not differ in other respects, marriage is of equal esteem to virginity—which struck at the heart of monasticism. In the second book three further notions are covered: those receiving baptism in full faith cannot again be led into sin; fasting has no greater merit than eating with thankfulness; and there is no inequality based on status in life in heavenly reward. It appears also that Jovinian was associated with Helvidius as one rejecting the perpetual virginity of Mary and affirming that Jesus had full brothers by Joseph.

CLYDE CURRY SMITH

JOWETT, BENJAMIN (1817-1893). English classicist and theologian. Born at Camberwell, London, he was educated at Oxford where he was elected fellow while still an undergraduate (1838). After a succession of college offices and the regius chair of Greek (1855), he became master of Balliol (1870). He was ordained in 1842 and was vice-chancellor of Oxford, 1882-86. He wrote commentaries on the epistles of Paul (1855), and through an essay on scriptural interpretation contributed to *Essays and Reviews* (1860), he fell under suspicion of unorthodoxy and ceased to write on theological subjects. He worked to secure the abolition of theological tests for university degrees and offices. His classical learning was almost unsurpassed in his day, and he was known as "the Great Tutor." J.G.G. NORMAN

JOWETT, JOHN HENRY (1864-1923). English Congregationalist preacher. Born near Halifax, Yorkshire, he studied at Airedale College, Bradford, at Edinburgh University, and Mansfield College, Oxford. He was pastor at St. James' Congregational Church, Newcastle-upon-Tyne (1889), and succeeded R.W. Dale at Carr's Lane Chapel, Birmingham (1895). He became chairman of the Congregational Union (1906-7) and president of the National Free Church Council (1910-11). His reputation as a preacher grew, and he became pastor of Fifth Avenue Presbyterian Church, New York City (1911), returning to Westminster Chapel, London (1918-22), in succession to Campbell Morgan. He delivered the Yale Lectures on Preaching in 1912, wrote many devotional books, and was made Companion of Honour in 1922. J.G.G. NORMAN

JUBILEE, YEAR OF (Holy Year). A Roman Catholic institution, based only indirectly on the

Levitical Year of Jubilee (Lev. 25), but rather an extension of the pilgrimage movement and the system of indulgences,* first observed in 1300 under Boniface VIII, apparently by spontaneous popular demand. Boniface in the bull *Antiquorum fida relatio* offered plenary indulgences to all pilgrims to Rome who met the conditions within the year. Intended to be celebrated each century, the next jubilee was proclaimed for 1350 by the Avignonese Pope Clement VI at the request of the Romans. Urban VI in 1389 reduced the interval to thirty-three years, and in 1470 Paul II further diminished it to twenty-five—the present interval. In 1500 Alexander VI extended the indulgence to all churches in the year following the jubilee, until Easter Saturday, though subsequent practice has varied. A jubilee indulgence was proclaimed throughout Christendom by Paul VI for 1 January to 29 May 1966, later extended to 8 December, to mark the completion of Vatican II. MARY E. ROGERS

JUD, LEO (1482-1542). Reformer in Switzerland. Born at Gemar, Alsace, after studies at Basle and Freiburg-im-Breisgau (1499-1512) he became pastor at St. Hippolyte, Alsace (1512-19), and then at Einsiedeln, Switzerland (1519-22), where he succeeded Zwingli.* One of the earliest followers of Zwingli, he was a staunch supporter of the Zwinglian Reformation. In 1523 he became pastor of St. Peter's, Zurich; on 1 September he preached against images and set off a wave of iconoclasm. He played a part in the suppression of convents. He helped Zwingli in his conflict with the Anabaptists, although at one time he had been "quite taken by some of Schwenkfeld's ideas on tolerance and some of the democratic principles of the Anabaptists" (G.H. Williams). He introduced a baptismal liturgy in 1523, which was in German, though it retained several Catholic features. He translated into German works of Augustine, Thomas à Kempis, and Erasmus; made a Latin translation of the Hebrew OT; and was author of the Swiss-German version of the Prophets (1525) which was incorporated in the complete Zurich Bible of 1529, preceding Luther's version by five years. His Protestant catechism (1538) was long in use. J.G.G. NORMAN

JUDAISM. The religion of the Jews in contrast with that of the OT, from which it was derived. The two focal points in its development were the two destructions of the Jerusalem temple in 586 B.C. and A.D. 70, which ended the centrality of sacrifice found in the OT. It was encouraged above all by the widespread dispersion of Jewry both East and West, which made the Law the center around which Jewish life and religion had perforce to revolve outside Palestine. During the intertestamental period, in which Judaism was developing, various directions became obvious—e.g., Pharisees, Sadducees, Essenes, Zealots, Hellenists—but the situation created by the destruction of the Jewish state in A.D. 70 and confirmed by the crushing of the Bar-Kochba revolt in 135 left the Pharisaic interpretation of Judaism without rivals. It reached its full development by 500, its authoritative documents being the Talmud, composed of the Mishnah and Gemara, and the Midrashim (official interpretations of the OT books).

An outstanding feature of Judaism has been its ability to adapt itself to pressure and persecution from Christianity and Islam, to minority social status, and to changing cultural circumstances without altering its essential nature. These adaptations have periodically been codified, the most important being Maimonides's *Mishneh Torah* (1180) and Karo's *Shulcan Aruch* (1565). Once Judaism had been able to eliminate the Jewish Christian, there was only one movement, until modern times, which it was not able to assimilate, viz., the Karaites, who emerged in the eighth century and reached their climax in the twelfth; today they claim only some two thousand adherents. In the nineteenth century, under the influence of modern thought, a reform, or liberal, movement began and has steadily increased. Other Jews have turned to materialism, Marxism, or a religionless nationalism.

Though Judaism had its great philosophical thinkers, it is essentially a historical religion based on God's election of Israel, shown above all in the Exodus, giving of the Law, and conquest of Canaan. Though Judaism recognizes the existence of the righteous among the nations, who will have "a share in the world to come," a full knowledge of God's will and the possibility of carrying it out are confined to Jewry. It is the possession of the Law that gives real meaning to God's election, and any Gentile who is prepared to accept "the yoke of the Law" is welcomed into the community of Israel. Since under Muslim and medieval Christian rule successful missionary attempts by Jews involved the death penalty for converter and converted alike, the zeal for converts shown in an earlier period vanished and is only rarely met with today.

The Jewish doctrine of God is not merely monotheistic, but strongly and deliberately anti-Trinitarian. God's unity is declared to be unique, "like no other unity." In its more philosophical forms Judaism declares that no human attribute may be postulated of God; where they are used in the Bible, it is merely an accommodation to human weakness. The transcendence of God is stressed in a way that makes any concept of incarnation impossible. Various concepts to bridge the gulf between God and His creation are found, but none are authoritative or universally received.

The main stress in normative Judaism is laid on the Torah. Though under Septuagint and Christian influence this is very generally rendered "Law," it is realized that "Instruction" is a more accurate rendering. The Torah is perfect, written in letters of fire in heaven antedating the creation of the world. Israel was chosen for the sake of the Torah and apart from it has no reason for existing. The Torah consists of two parts, the written and the oral. The written Torah contains 613 precepts, 365 being negative and 248 positive; the oral Torah is the extension of these precepts to cover all life and all its contingencies. Except for adaptations to more recent conditions, the oral Torah has found its definitive expression in the Talmud, to which modern developments must conform. Much scorn has been poured on the methods used by the rabbis, but once we grant the basic concept of Torah, it is difficult to see what other results could have been reached. The Talmudic system is more humane than that of the Samaritans or Karaites and is in many ways comparable with the casuistic methods of the Schoolmen and Jesuits.

Legalistic such a system must be, but this is mitigated by stress on *kawwanah* and *lishmah*, i.e., that in the carrying out of the commandment the heart must be directed *(kawwanah)* to God and that it should be done for its own sake *(lishmah)*, out of the love to God, and not purely for reward. If it is asked how this can be reconciled with a statement like Acts 15:10, it must be remembered that it was said by Peter, who came from a mixed population of Galilee. The real keeping of the Law virtually demands a thoroughly Jewish environment.

The outstanding weakness of the system from the Christian standpoint is that the Torah is conceived of as something given to man for his interpretation and application. On the basis of passages like Leviticus 18:5 it is stressed that the Torah was given that men might *live* by it; hence commandments that weighed too heavily on the community have been mitigated or circumvented. In addition, a genuine threat to life releases the Jew from all commandments except those prohibiting idolatry, murder, and adultery—Christian baptism is regarded as idolatry. This outlook, the tendency to place all commandments on the same level, and the vanishing of sacrifice have in the course of time diminished the sense of sin. Traditional Judaism recognizes that there are two impulses in man, one good, one bad, by virtue of his creation, *not fall;* the concept is played down today. The evil impulse can be checked by study of Torah.

The messianic hope never took on a fixed and official form. It is universally accepted that God will yet set up His perfect rule on earth, and it was generally agreed that this would be achieved through the Messiah.* Mainly due to repeated disappointments, he has become for many the personification of the hope of the kingdom of God. With this was linked the expectation of the resurrection of the body. Under Greek influence the concept of the immortality of the soul was gradually accepted. The virtual incompatibility of the two ideas has led to a blurring of the hope of a future life and bodily resurrection. For the Liberal, future life is purely spiritual.

There is judgment to come for all, yet only the exceptionally wicked Jew need fear punishment; for most it will mean rewards. Various views have been held about *Gehinnom*, or hell. For some it is a limited period of punishment, which purifies or annihilates; others consider it eternal.

Theology, beyond the unique oneness of God, the divine origin and primacy of the Torah, the coming kingdom of God, and the election of Israel, plays little part in Judaism. Controversies, until recently, have been about doing. A Jew who does the right things is presumed to believe the right things unless he expressly denies them. Hence it is usual to speak of an "observant" rather than of an "orthodox" Jew. "Orthopraxy" would be a better term than "orthodoxy."

Judaism by its nature has always stressed the community rather than the individual, this world rather than the next. The ghetto system, by shutting the community in on itself, made charity and justice paramount virtues necessary for survival, and the Torah scholar, using his knowledge for the good of the community, was its outstanding member. In Judaism there are no sacraments, and the rabbi occupies his position solely by virtue of his knowledge of the Torah.

Mysticism has played a major role in the synagogue. At first it was confined to small circles "whose intellectual and religious background fortified them against the dangers of straying into the paths of heresy." Its influence widened over the centuries until in the late eighteenth century it became a popular mass-movement in E Europe known as Hasidism. Stress on the transcendence of God and the claims of the Torah prevented mysticism from degenerating into pantheism and antinomianism. All estimates of Judaism in practice must allow for the influence of the mystic element.

With few exceptions Jewry missed the impact of the Renaissance. It was the growth of humanism and the politically liberating effects of the French Revolution that exposed Jewry, especially in W Europe, to the impact of modern thought. Religiously the result has been the rise of the Liberal Synagogue (Reform Synagogue in America), which in turn has deeply influenced the thought, if not the practice, of much traditional Judaism. In it the center of gravity has moved from the Law to prophetic ethics, and but for its historical coloring, it is hardly distinguishable from Unitarianism. The advent of the State of Israel, especially since the war of 1967, has normally added a strong nationalistic coloring.

In Israel itself, Judaism is under the control of the stricter traditionalists. The tendency is for the more obvious demands of the Law to be treated as national customs without religious significance.

BIBLIOGRAPHY: S. Singer (ed.), *The Authorised Daily Prayer Book of the United Hebrew Congregations of the British Empire* (1891): with commentary by J.H. Hertz (2 vols., 1941); *Jewish Encyclopaedia* (12 vols., 1901-6); C.G. Montefiore, *Liberal Judaism* (1903); L. Baeck, *Das Wesen des Judentums* (1905—ET *The Essence of Judaism*, 1936); W.O.E. Oesterley and G.H. Box, *The Religion and Worship of the Synagogue* (1907) and *A Short Survey of the Literature of Rabbinical and Mediaeval Judaism (1920);* M. Friedländer, *The Jewish Religon* (1909, 1921); S. Schechter, *Aspects of Rabbinic Theology* (1909); H.L. Strack and P. Billerbeck, *Kommentar zum Neuen Testament aus Talmud und Midrasch* (4 vols., 1922-28); G.F. Moore, *Judaism in the First Centuries of the Christian Era* (3 vols., 1927, 1930); E.R. Bevan and C. Singer (eds.), *The Legacy of Israel* (1927); H.L. Strack, *Introduction to the Talmud and Midrash* (ET 1931); A. Cohen, *Everyman's Talmud* (1932); H. Danby (tr.), *The Mishnah* (1933); I. Epstein (ed.), *The Talmud* (ET, 35 vols., 1935-52); C. Roth, *A Short History of the Jewish People* (1935); C.G. Montefiore and H. Loewe, *A*

Rabbinic Anthology (1938); G.C. Scholem, *Major Trends in Jewish Mysticism* (1946); J.H. Hertz, *The Pentateuch and Haftorahs* (1947); S.S. Cohon, *Judaism, A Way of Life* (1948); J. Jocz, *The Jewish People and Jesus Christ* (1949); M. Waxman, *Judaism, Religion and Ethics* (1953); I. Epstein, *The Faith of Judaism* (1954) and *Judaism* (1959); L. Roth, *Judaism, A Portrait* (1960); J. Parkes, *The Foundations of Judaism and Christianity* (1960); R.A. Stewart, *Rabbinic Theology* (1961); L. Jacobs, *Principles of the Jewish Faith* (1964), R.J. Zwi Werblowsky and D.G. Wigoder, *The Encyclopedia of the Jewish Religion* (1966).

H.L. ELLISON

JUDAIZERS. A party of Christians in the early church who thought it was necessary that Gentile converts to Christianity should be circumcised and observe the Jewish law—in fact that they should become Jews in order to become Christians. The epistle to the Galatians was written by Paul to demonstrate that such an attitude contradicted the gospel of the grace of God. Acts 15 records the taking up of the issue at the Council of Jerusalem and the decision that such requirements should not be made of the Gentiles, but that they should be asked to observe as requirements for table fellowship certain food and marriage regulations.

JUDAS ISCARIOT. One of the twelve apostles; the betrayer of Jesus. The meaning of the name "Iscariot" is uncertain. It is most likely to signify "man of Kerioth" (a place in S Palestine), which would make him probably the only Judean among the Twelve as well as being the bearer of the name of Judah. Others have suggested links with Sychar, Issachar, or Jericho, or a derivation from *sicarios*, "assassin." He held a position of importance as treasurer of the group (John 12:6; 13:29) and was close to Jesus at the Last Supper (John 13:21-26). He went to the chief priests to offer to betray Jesus (Mark 14:10f., etc.) and took the opportunity to do so in the Garden of Gethsemane (Mark 14:43-50, etc.). Accounts of his remorse and suicide are found in Matthew 27:3-10 and Acts 1:15-20. The motives which lay behind Judas's action have long fascinated students of the NT. John makes explicit his avarice (12:4-6), but it is likely that he was in one way or another disillusioned with how Jesus was working out His messianic vocation—and acted to put down someone he considered dangerous, or to force His hand to bring matters to a head.

R.E. NIXON

JUDAS OF JAMES. He may have been the son or the brother of James (whose identity is unknown). He is probably to be identified with Thaddaeus.*

JUDE, EPISTLE OF, see EPISTLES, GENERAL

JUDGMENT. An important doctrine of Scripture as Christ Himself stated: "For judgment I have come into this world" (John 9:39). This NT idea grew from the OT teaching of the Day of the Lord. That New Era was thought to be the final crisis of history when God would judge men and reward them for their works. Both testaments in describing God picture Him as a holy and righteous being who must judge sin. Many leaders of the early church focused on the idea of a general judgment at the end of history. Irenaeus, Hippolytus, and Lactantius show the profound effect of this teaching as they argue for the resurrection of the body, the completion of the ministry of Christ, and the significance of the present work of God in history, all on the basis of the doctrine of God's judgment. After the Reformation and the development of Protestantism there was a great diversity of opinion on the precise number as well as the time of the judgment. Some continue to speak of a general judgment whereas others distinguish as many as seven actions in this work.

Evangelicals who believe in a number of judgments usually mention those at the cross, at the *Bema* seat of Christ (2 Cor. 5:10), of the living nations at the end time (Matt. 25), of the angels (Jude 6 and 2 Peter 2:4), and of the wicked dead at the Great White Throne (Rev. 20:5,7). Advocates of these elaborate schemes of judgment are usually premillennialists and dispensationalists who hold to a more complicated and detailed plan for the second coming of Christ. They agree with other evangelicals, however, that the final judgment is the climax of a process which was inaugurated by the coming of Christ, who claimed that a rejection of Him caused a person to be "condemned already because he has not believed in the name of God's one and only Son" (John 3:18). When the consummation comes, those who have not followed Christ will be eternally condemned to hell (Matt. 25:31, 46; 2 Thess. 1:7-10; Rev. 20:14,15).

The judgment of God is to be universal, and even though Christians will not be judged for their salvation, they still will be examined as to their works (Rom. 14:10). The achievements of some will be superficial ("wood, hay, or straw"), while others will have done worthwhile acts for Christ ("gold, silver, costly stones," 1 Cor. 3:12-15). In this judgment Christ's words of praise or blame will be the reward or punishment (Matt. 25:21-23; Luke 19:17).

BIBLIOGRAPHY: J. Baillie, *And the Life Everlasting* (1934); O. Cullmann, *Christ and Time* (1951); L. Morris, *The Wages of Sin* (1955; rep. 1957) and *The Biblical Doctrine of Judgment* (1960); L. Boettner, *Immortality* (1956); A.T. Hansen, *The Wrath of the Lamb* (1957).

ROBERT G. CLOUSE

JUDSON, ADONIRAM (1788-1850). Missionary, lexicographer, and Bible translator. Born at Malden, Massachusetts, son of a Congregational minister, he was graduated as valedictorian from Brown University (1807). After teaching for a year at Plymouth, he studied divinity in Andover Theological Seminary. He was a leader in the founding of the American Board of Commissioners for Foreign Missions.* In 1812 he was ordained, and he and his wife embarked for Burma as Congregational missionaries. During that voyage they reexamined their views on baptism and both were baptized in Calcutta (1812). That cut off their support, but this was taken over by the Baptist Triennial Convention, organized in 1814.

Reaching Rangoon, Judson learned Burmese to preach and translate the Bible, and worked on an English-Burmese dictionary. The war with England (1824) brought him seventeen months' imprisonment, but the peace saw him work as an interpreter. He continued missionary work at Ava, but by 1826 he had lost his wife and two children. He then went to Maulmain. His second wife died in 1845, and he remarried again. In 1849 the Burmese-English half of his dictionary was published. He died at sea.

C.G. THORNE, JR.

JULIAN CALENDAR, see CALENDAR

JULIAN OF ECLANUM (380-c.455). Pelagian theologian. Son of Memorius, bishop of Eclanum, he was ordained on the death of his wife and succeeded his father as bishop. He was very learned in Latin. Greek, logic, and theology. He became a supporter of Pelagius* and in 418 attacked the *Epistola Tractoria* in which Pope Zosimus had condemned Pelagius and Celestius. He was deposed and expelled from Italy. He traveled in the East and was received by Theodore of Mopsuestia and Nestorius. Returning to Sicily, he taught there until his death. Julian was Pelagianism's "most systematic exponent;" he defined freedom of the will as "the possibility of committing sin or abstaining from it." He reduced grace to simple, protective divine assistance and denied the solidarity of the human race in Adam's sin. He defended marriage against asceticism, and the innocence of the sexual impulse. Augustine* answered him in three works. J.G.G. NORMAN

JULIAN OF NORWICH (c.1342-after 1413). English mystic. About her very little is known, except that she was an anchoress who lived in a cell built to the wall of the Norman Church of St. Julian in Norwich. Her only testament is *Revelations of Divine Love.* A series of sixteen revelations, or shewings, which took place on 8 May 1373, gave rise to this work, and that record is then extended with the fruit of her meditation thereon for the next twenty years. Her life was total solitude but for occasional counseling, and her thinking was without other influences except for possible knowledge of *The Cloud of Unknowing** and the writings of Walter Hilton.* The *Revelations* contain no formal theological system nor any elaboration because of the ineffable nature of mystical experience, though great theological questions are wrestled with more simply. God's love is defined often in terms of pain: the sufferings of Christ and what believers must be willing to suffer for Him, which must be understood before joy can be had. Julian makes a distinction between accidental element—sickness, and shewings, which are only props to help the soul advance to God—and essential ones—prayer and contemplation, which establish the true union between God and man.

BIBLIOGRAPHY: R.H. Thouless, *The Lady Julian* (1924); P. Molinari, *Julian of Norwich: The Teaching of a Fourteenth Century English Mystic* (1958); G. Warrack (ed.), *Revelations of Divine Love* (1958). C.G. THORNE, JR.

JULIAN THE APOSTATE (Flavius Claudius Julianus) (c. 331-363). Roman emperor who endeavored to restore pagan religion. Born in Constantinople, he was the son of Julius Constantius, brother of Constantine I. His mother, Baslina, died soon after his birth. In 337 all his family apart from a half-brother, Gallus, were murdered by soldiers to ensure undisputed succession of Constantine's sons. Educated under Mardonius and Eusebius, bishop of Nicomedia, in 341 he was sent with Gallus to Macellum, a remote castle in Cappadocia. Returning to Constantinople (347), he studied grammar and rhetoric until exiled again to Nicomedia, where, listening to Libanius the philosopher, he was awakened to the glories of classical Greece. Hitherto a Christian by conviction, now he was won over to the old gods; hence his nickname "the Apostate."

When Gallus was executed (354), he escaped, thanks to the intercession of Empress Eusebia, and he was allowed to continue his studies in Athens. Trouble in Gaul forced Constantius to make him Caesar (355). He married Julian to his sister Helena and sent him to govern Gaul, where he achieved some notable military victories. Proclaimed Augustus by his troops (360) when Constantius died unexpectedly, he was everywhere acknowledged sole ruler (361).

Immediately Julian set about restoring the old religions. He issued an edict of universal toleration, swept away the court sycophants, became *Pontifex Maximus* in more than name, and ordered the restoration of the old cultus, reopening the temples and reviving the sacrifices. He even attempted to rebuild the Temple of the Jews at Jerusalem. At first tolerant of Christians—indeed, exiled Nicenes like Athanasius were permitted to return—later he began a persecuting policy. His efforts were doomed to failure, as he himself discovered at Antioch. The old religions were moribund, and his philosophic version of them never existed. The revival ended with his death on campaign in Persia. J.G.G. NORMAN

JÜLICHER, ADOLF (1867-1938). German NT scholar. Professor of theology at Marburg from 1889 to 1923, he wrote a very influential work on *The Parables of Jesus* (2 vols., 1888-89), in which he argued that Jesus' parables were originally intended to illustrate one truth only (i.e., as true similes and not as allegories) and that all allegorical features are therefore secondary. His *Introduction to the New Testament* (1894; ET 1904) offers a good summary of critical opinion in German Protestant circles at the turn of the century and was a standard work for many years. His other writings include a study of gospel criticism (1906), *Paulus und Jesus* (1907), and a posthumously published edition of parts of the Old Latin Versions of the NT (1938-54).

W. WARD GASQUE

JULIUS I (d. 352). Pope from 337. Elected to the Roman see in the year that Constantine died, he had the task of presiding over the Western Church in the difficult years of the Nicene theological crisis, which cannot be separated from the equally trying conflicts by which the church was ensnared in the imperial division among the sons of Constantine. That the West had from the outset found the *homoousion* position more acceptable than the East helps to define Julius' stance. He provided refuge for Athanasius* during the latter's second deposition (339-46), presiding over the synod in Rome (341) which went on record supporting that Nicene position represented, not only by Athanasius, but also by the more extreme Marcellus of Ancyra who had also been deposed from his see. Letters to Julius from both these men survive, as does Julius' letter of support for them, which arose out of the synod and was sent to the Eastern bishops. When the Arians responded by their own synod at Antioch (341), Julius prevailed upon the Western emperor Constans to summon a general council at Sardica (343). But the session split between West and East over the seating of Athanasius, and the two halves met separately—the Eastern within Constantius's jurisdiction—and produced opposing decisions. Within Rome, Julius was responsible for building two new churches. CLYDE CURRY SMITH

JULIUS II (1443-1513). Pope from 1503. Born Giuliano Della Rovere at Albisola, near Savona, Giuliano was influenced by his uncle, Francesco, to enter the Franciscan Order where he studied with them in Perugia and was ordained. When his uncle became Pope Sixtus IV, Giuliano became a cardinal in 1471 and served as a legate to the French king Louis XI in 1480-82. He exerted some control over the papacy when Sixtus ruled, but in 1484 when his uncle died, Giuliano secured the election of Innocent VIII and determined papal policy until Innocent's death in 1492. When Rodrigo Borgia (Alexander VI) was elected pope, Giuliano was forced to flee from Rome due to the animosity between the two powerful churchmen. Although a brief reconciliation was effected for political reasons, it was not until 1503, the year of Alexander's death, that Giuliano was able to return to Rome without fear. After the one-month pontificate of Pius III, Giuliano was able to secure his own election through means of bribery and extensive promises.

He had promised not to make war to recapture the losses which the papacy had suffered under the Borgias, but nevertheless demanded that Cesare Borgia return the dukedom of Romagna in the Papal States. Julius incurred the enmity of the Venetians with his thrust to strengthen the landholdings of the Papal States, and entered into alliances with the great European powers, including the Holy Roman Empire and France, to gain the desired territories from Venice. When this was accomplished, he strove to drive the French out of Italy and formed new alliances, keeping European statecraft in a constant turmoil. The emperor and Louis XII of France encouraged the calling of an antipapal council at Pisa, and Julius retaliated by calling the Fifth Lateran Council in 1511. Julius also entered into the Holy League with Spain, Venice, and later England, against France and finally succeeded in expelling French forces from Italy.

Julius brought about administrative reforms in the Curia and was an avid patron of the arts (notably Michelangelo), but his reputation is that of the warrior-pope. ROBERT G. CLOUSE

JULIUS III (1487-1555). Pope from 1550. Born Giovanni Maria Ciocchi del Monte in Rome, of a Tuscan family of lawyers, he completed humanist studies under Raffaelo Lippo Brandolino, then took to jurisprudence in Perugia and Bologna. Following theological training he became chamberlain to Julius II. Succeeding his uncle as archbishop of Siponto (1511), he held administrative posts under Clement VII and Paul III. He was taken hostage by imperial forces after the sack of Rome (1527). As one of three papal legates he opened the Council of Trent* as first president. The first assembly ended in 1549, and in the following year he was elected pope. He ordered the resumption of the council (1551), but it had to be suspended because of the opposition of Henry II of France. On the accession of Mary Tudor (1553), Julius sent Reginald Pole as legate to England. Julius fostered reform in the church, encouraged the Jesuits, and was a generous patron of Renaissance humanism. J.G.G. NORMAN

JULIUS AFRICANUS, SEXTUS (d. after 240). Christian scholar. Brought up in Palestine, Julius traveled widely and inquisitively, both during and after his military service. He listened to Heraclas in Alexandria, met Origen, and hunted with Bardesanes in Edessa. After settling in Emmaus he went as its representative to the emperor Elagabalus in Rome to ask that it be rebuilt. He was successful enough to be given charge of the task. He later impressed the emperor Severus (222-35) so much that he was commissioned to organize his public library in Rome. In his five-volume *Chronographia* he attempted to synchronize sacred and profane history and predicted that the world would last 6,000 years and that Christ had been born in the year 5,500. His twenty-four-volume *Cesti* was an encyclopedic work on subjects ranging from natural science to military tactics, but revealing a belief in superstition and magic. Only fragments of either work have survived. Two letters also are extant: one to Origen arguing that the Septuagint story of Susanna cannot be regarded as canonical because the evidence is against its having a Hebrew original, and one to an unknown Aristides on the differences in the genealogies of Christ in Matthew and Luke. C. PETER WILLIAMS

JUNG CODEX, see NAG HAMMADI

JURIEU, PIERRE (1637-1713). Prominent theologian and apologist of the French Reformed Church. Descended from notable families, he studied philosophy at Saumur and theology at Sedan (1656-58) before traveling to England and the Netherlands. He succeeded his father as minister at Mer (1671) and remained there until he was appointed professor of Hebrew at the Sedan academy (1674-79). During his years at Sedan, Jurieu vigorously defended the Reformed faith against the attacks of clerics like Bossuet.* When the academy was dissolved by Louis XIV, Jurieu went to the Netherlands, where on the recommendation of Pierre Bayle* he became minister of the Walloon Church at Rotterdam (1681). There he fostered French Calvinism through his writings and by caring for exiled French Reformed pastors. He came to believe that the Calvinists would soon be restored to France, because of his interpretation of the prophecies of the Apocalypse (as seen in his work: *L'Accomplissement des prophéties ou la délivrance de l'église,* 2 vols., 1686; ET 1687).

Jurieu gradually adopted a view that distinguished between temporal and spiritual power. He demanded full liberty of conscience for the citizen. These ideas, developed in *Histoire du calvinisme et du papisme* (2 vols., 1683) were accepted by many Protestants in 1685 when the Edict of Nantes* was revoked by Louis XIV. Jurieu then continued during the years 1686-89 to fashion an ideology of revolution. He stated that the "right of princes to use the sword does not extend to matters of conscience." Since Louis's use of the sword to coerce men's consciences had put himself outside the pale of law, revolt was thus lawful. Violence was to be repaid with violence.

A prolific writer, Jurieu's style is marked by impressive erudition and polemical bitterness. In addition to his books against Louis and Bossuet, he wrote against the Jansenists (Pierre Nicole and Antoine Arnauld) as well as against the indifference of Bayle. His other important works include *Histoire critique des dogmes et des cultes* (1704-5; ET 2 vols., 1705), *Unité de l'église et points fondamentaux* (1688), and *Traité de l'amour divin* (1700).

BIBLIOGRAPHY: C. Van Oordt, *Pierre Jurieu* (1879); H.M. Baird, *Huguenots and the Revocation of the Edict of Nantes* (2 vols., 1895); G.H. Dodge, *The Political Theory of the Huguenots With Special Reference to Pierre Jurieu* (1947); P. Hazard, *The European Mind, the Critical Years* (1680-1715) (tr. J.L. May, 1953).

ROBERT G. CLOUSE

JUSTIFICATION (Lat. *justificatio,* Gr. *dikaiosis*). Any consideration of this term in the end resolves itself into an argument over etymology: the verb *justificare* undoubtedly has the "forensic" connotation of pronouncing a guilty person acquitted. *Dikaioun,* on the other hand, while clearly having this meaning in the majority of cases, is understood by some to imply making or becoming actually righteous.

The doctrine of Justification is clearly adumbrated in the gospels, but is brought to full realization by Paul, particularly in the Roman and Galatian epistles. Here it is presented as the result and completion of the redemptive work of Christ when man in faith responds to Him, and God in His mercy treats him as though he were righteous.

The clarity of Paul's teaching was obscured in the patristic period: Augustine at first sight seems to reaffirm the Pauline position, but really conflates the immediacy of the act of justification with the later process of sanctification. This became the accepted medieval view, reaffirmed by Thomas Aquinas for whom justifying grace was a supernatural quality infused like hope or love into the human soul, with faith its preliminary rather than its channel. Justification is thus no longer the acquirement of a status, but the production of a state, dependent especially on loyal observance of the sacraments. When the Renaissance threw men back onto the original Greek text of the NT and highlighted once more the significance of individual personality, the way was opened for Martin Luther's most vital contribution to Reformation theology: his rediscovery after agonizing search of the Pauline emphasis that in justification Christ's righteousness becomes our righteousness, or is imputed to us, by faith through grace.

This central belief of the Reformers, reiterated by Melanchthon and later by Calvin, Wesley, and Spurgeon, the Council of Trent anathematized in favor of the medieval view. Justification in post-Tridentine Catholicism became again an imparted gift, not a pronouncement of acquittal; a gradually realized psychological condition, not a once-for-event in the believer's experience. The way was opened, as before, to salvation by merit.

In later Protestant theology the theme of justification was variously handled. The basic Protestant emphasis was never entirely obscured, but Calvinists, particularly under the influence of Federal theology, dwelt on the highly contentious doctrine of eternal justification, others depressed faith at the expense of grace or vice versa, still more treated justification as progressively realized through different stages, others subsumed it under the general idea of reconciliation, while Ritschl's teaching that the community of believers is the object of justification raised the issues of the interdependence of justification, church, baptism, and the Holy Spirit. More recently Hans Küng has argued impressively that the differences between the Catholic and Protestant views are largely imaginary and capable of reconciliation, a theory more deserving of study by evangelicals than recent and impatient radical dismissals of justification as an "archaic term" (Macquarrie), its significance having been "vastly exaggerated" in previous debate.

BIBLIOGRAPHY: W. Cunningham, *Historical Theology* (1863); C. Hodge, *Systematic Theology,* vol. 3 (1873); A. Harnack, *History of Dogma* (ET 1894-99); A. Ritschl, *Critical History of the Christian Doctrines of Justification and Reconciliation* (ET 1900); J. Denney, *The Christian Doctrine of Reconciliation* (1917); K. Barth, *Church Dogmatics,* I, 2 (ET 1956); F. Gogarten, *The Reality of Faith* (ET 1959); G. Ebeling, *Word and Faith* (ET 1963) and *The Nature of Faith* (ET 1966); H. Küng, *Justification: The Doctrine of Karl Barth and a Catholic Reflection* (1964); J. Jeremias, *The Central Message of the New Testament* (ET 1965).

IAN SELLERS

JUSTINIAN I (483-565). Greatest of the Byzantine Roman emperors. Illyrian by birth and adopted by his uncle, Justin I, whom he succeeded in 527 as emperor, he changed his name to Flavius Justinianus. In 523 he married Theodora, who until her death in 547 had a similar influence over him as that of Livia over Augustus. A new legal code, the *Corpus Juris Civilis,* was made during his reign by the jurist, Trebonianus (see JUSTINI-

AN CODE). Justinian built on a grand scale, the church of Hagia Sophia being his greatest monument. During 533-4 his armies reconquered North Africa, defeating the Vandals, and a year later he drove the Goths from Italy. A champion of Nicene orthodoxy, he closed philosophical schools in Athens, forced pagans to accept Christian baptism, and persecuted the sectarian Montanists. He failed, however, in his anti-Nestorian zeal, to win the Monophysites over to his viewpoint, and this failure led to the condemnation of Origenism* and the Three Chapters Controversy.* PETER TOON

JUSTINIAN CODE. Consolidation of Roman law promulgated by Emperor Justinian* in 529 *(Codex Constitutionum).* Justinian found the laws of the empire in great disarray and appointed a commission of ten legal experts, including the great jurist Trebonianus, to enlarge and rearrange the existing laws, eliminating contradicting and useless constitutions. Encouraged by the success of this endeavor, Justinian turned to the simplification of more difficult works—the writings of the jurists. After issuing the "Fifty Decisions" which settled certain important legal questions on which there had been disagreement among the older jurists, Justinian appointed a new commission under Trebonianus whose task it was to condense the writings of the jurists. This *Digest* was promulgated in 533, shortly after Trebonianus completed and published a revised edition of the *Institutes* of Gaius, which was to be used as a manual for law students. Justinian then appointed another commission headed by Trebonianus to revise the *Codex* and incorporate the new constitutions of the "Fifty Decisions" into it. This revised edition of twelve books was promulgated in 534 *(Codex repetitae praelectionis)* and is what has survived to the present. From 534 to the end of this reign Justinian continued to enact new ordinances *(Novellae constitutiones post codicem).* The *Codex, Digest, Institutes,* and *Novels* constituted the *Corpus Juris Civilis* which became the basic collection of Roman law. The Code was influential in the development of canon law in the West and today is valuable for its historical and legal interest. ROBERT G. CLOUSE

JUSTIN MARTYR (c.100-165). Christian apologist. Born of pagan parents at Flavius Neapolis, formerly Shechem, in Samaria, he appears from his youth to have been intent on finding intellectual peace and satisfaction. He studied the leading philosophies of his day: Stocism, Aristotelianism, Pythagoreanism, and Platonism. At last through a conversation with an old man he discovered that Christianity was the "one sure worthy Philosophy." From his conversion (c.132) he sought to proclaim his newfound faith, and he taught in many of the leading cities of the ancient world. He seems to have spent considerable time in Rome, where Tatian was one of his pupils. Justin was one of a number of Christian apologists who set themselves to defend the Christian faith against misrepresentation and ridicule. Justin, especially, attempted to show that Christianity was the embodiment of the noblest concepts of Greek philosophy and was the Truth *par excellence.*

In his *First Apology* (c.152) addressed to Emperor Antoninus Pius and to his son Verissimus and the philosopher Lucius, he argued that the teaching of Christ and the prophets is alone true and older than all other writings. He asserts that the divine Logos had been in the world from the beginning, and that those who lived according to "reason," whatever their race, were Christians. Justin emphasizes, however, that the whole Logos resided in Jesus Christ. Christianity was not therefore a *new* revelation, but supremely the *full* revelation of truth because Christ was Himself the incarnation of the whole divine Logos. The purpose of His coming was to save men from the power of demons and to teach the truth. More than any other second-century apologist, Justin states frequently that Christ saves mankind by His death on the cross and by His resurrection. Although he speaks of Father, Son, and Spirit, it is clear that his emphasis upon the transcendence of God led him to subordinate the Son and Spirit. In the later chapters of the *First Apology* he gives an account of the sacraments of baptism and the Eucharist which are of great value to students of the early liturgy.

The *Second Apology* (c.153) is much shorter than the first and was called into existence by Justin's indignation at the unjust persecution of Christians. The *Dialogue with Trypho* comes from a different background: it narrates Justin's conversation with a learned Jew, Trypho, and certain of his friends. This writing shows Justin's desire to win Jews for Christ as well as the Gentiles. The book closes with an eloquent appeal to Trypho to accept the truth and "enter upon the greatest of all the contests for your own salvation, and to endeavour to prefer to your own teaching the Christ of Almighty God."

The main significance of Justin, indeed, is that he is the first Christian thinker after Paul to grasp the universalistic implications of Christianity. With his own distinctive understanding of the Logos concept, he sums up in one bold stroke the whole history of mankind as finding its consummation in Christ.

BIBLIOGRAPHY: *Works* in J.P. Migne (ed.), *PG* VI (1857); E. Goodenough, *The Theology of Justin Martyr* (1923); H. Chadwick, "Justin Martyr's defence of Christianity," *BJRL* XLVII (1965), pp. 275-97, and *Early Christian Thought and the Classical Tradition* (1966); W.A. Shotwell, *Biblical Exegesis in Justin Martyr* (1965); L.W. Barnard, *Justin Martyr: His Life and Thought* (1967).
G.L. CAREY

JUVENAL (d.458). Bishop of Jerusalem. Some ten years after his appointment he came to prominence at the Council of Ephesus* (431) where he supported the anti-Nestorian side. At the Robber Synod of Ephesus* (449) his voice was raised for Dioscorus,* but the political opportunism so characteristic of Juvenal's career led to his voting for the former's condemnation at the Council of Chalcedon* (451), which body confirmed the diocese of Jerusalem's jurisdiction over all Palestine. Some monkish followers of Dioscorus so threatened Juvenal's position that imperial help was needed for him to regain control of the see in 453.

JUVENCUS, GAIUS VETTIUS AQUILINUS (fl. c. 330). Latin Christian poet. The fame of Juvencus, Spanish priest of a noble family, rests on his epic of Christ's life in about 3,200 hexameters, the earliest important Christian counterpart to pagan epic. Drawn from Matthew primarily, it is called *Historiae evangelicae libri IV.* He was well read in the pagan classics, especially Virgil, whose influence is seen in Juvencus's genre, meter, vocabulary, and mythological allusions. However, the nature of Juvencus's work, closely paraphrasing the gospels, especially Christ's words, precluded any flights of poetic imagination. Still, the versification is generally pleasing and Juvencus was highly respected in medieval times.
DANIEL C. SCAVONE

JUXON, WILLIAM (1582-1663). Archbishop of Canterbury. Born at Chichester and educated at Oxford where he studied law, Juxon in 1609 became vicar of St. Giles, Oxford, and in 1621 succeeded William Laud* as president of St. John's College. He was vice-chancellor in 1626-27 and aided the Laudian reform of the university statutes. Laud was his friend and patron, and in 1633 Juxon succeeded him as bishop of London. In 1636 he became Lord High Treasurer of England, and after Laud's imprisonment he was Charles I's constant adviser, ministering to him up to his execution in 1649. Though a thoroughgoing Laudian, he was trusted by men of other persuasions as a man of integrity and tolerance. During the Commonwealth he was left in seclusion. At the Restoration he was made archbishop of Canterbury, but because of infirmity exercised little influence over the church. JOHN A. SIMPSON

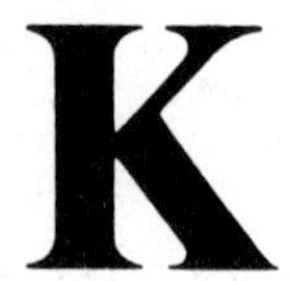

KABBALAH (Cabala). The broad stream of Jewish mysticism, and more especially those schools which flourished in parts of Europe and in Palestine between the twelfth and seventeenth centuries. The Hebrew word *qabbalah* denotes "tradition," and by it the mystics meant both Jewish oral tradition in general and also their own esoteric lore. The roots of such mysticism lay in intertestamental developments, evidenced in the Qumran scrolls and the apocalyptic literature; in the rabbinical period, moreover, there was a considerable amount of speculative and mystical thought, influenced by Gnosticism and Neoplatonism. The *Sepher Yetsirah* (seventh or eighth century) already testified to a speculative interest in the nature of God and of the universe, and posited ten mediating emanations from God (called *sephiroth*). In due course a distinction was made between theoretical and practical kabbalah, the latter approximating to white magic. The movement, though tending to heterodoxy, was nevertheless deeply attached to the Hebrew Bible, which it interpreted both literally and cryptically (by ciphers, numerology, etc.); its authoritative work, the thirteenth-century *Zohar* (by Moses de Léon of Granada), purports to be a commentary on the Pentateuch. The movement became popular in the late Middle Ages, due partly to the historical stresses to which European Jewry was subjected. Hasidism is the linear descendant of the kabbalah.

The chief figures of the movement included Eleazar of Worms in Germany (thirteenth century) and Isaac Luria and Hayim Vital in Safed in upper Galilee (sixteenth century). Nahmanides and Joseph Caro were much influenced by the kabbalah, and so were some Christian thinkers, e.g., J. Reuchlin and Paracelsus.

BIBLIOGRAPHY: J.L. Blau, *The Christian interpretation of the Cabala in the Renaissance* (1944); C.D. Ginsburg, *The Essenes ... the Kabbalah* (1863-64; rep. 1955); A.J. Heschel, "The mystical element in Judaism," in L. Finkelstein (ed.), *The Jews: Their History, Culture, and Religion* (3rd ed., 1960), ii, pp. 932-53; G.G. Scholem, *Major Trends in Jewish Mysticism* (ET, 3rd ed., 1961) and *On the Kabbalah and Its Symbolism* (ET 1965).

D.F. PAYNE

KAEHLER, MARTIN (1835-1912). German Protestant theologian. He was born near Königsberg, Prussia, and except for three years at Bonn (1864-67) his entire academic career from 1860 to his death was spent at the University of Halle. He was strongly influenced in his theological development by Rothe, Tholuck, Müller, Beck, and von Hofmann. Though his lectures on Protestant theology, the *Geschichte der protestantischen Dogmatik im 19. Jahrhundert,* were published posthumously (1962), Kaehler was best known for his penetrating study, *Der sogenannte historische Jesus und der geschichtliche biblische Christus* (1892). In this book he opposed the tendency of biblical scholars to drive a wedge between the historical Jesus and the proclamation of the apostles. The real Jesus is not the portrait of Jesus of Nazareth which historians are able to reconstruct, but the Christ of faith who is experienced again and again by the Christian community ("the real Christ is the preached Christ").

DAVID C. STEINMETZ

KAFTAN, JULIUS WILHELM MARTIN (1848-1926). German Protestant theologian. Educated at Erlangen, Berlin, and Kiel, he later taught at the universities of Basle (1873-83) and Berlin (from 1883). Strongly influenced by Ritschl, Kaftan stressed personal religious experience and the historical revelation in Christ. The Atonement was interpreted in mystical and ethical categories, rejecting any idea of satisfaction or need for God's reconciliation with man (rather than the reverse).

KAFTAN, THEODOR (1847-1932). German Lutheran churchman. Elder brother of Julius Kaftan,* he was born in Loit, Schleswig, and educated at Erlangen, Berlin, and Kiel. He occupied various ecclesiastical posts in Schleswig, including general superintendent for the province (1886-1917). A strongly confessional Lutheran, he clashed with spokesmen for both the orthodox and neo-Protestant positions. He supported Lutheran ecumenism on the national and international levels and was a leading figure in the General Evangelical Lutheran Conference. He opposed Prussian attempts to Germanize North Schleswig, and sought to protect the Danish character in church life there. He disliked the church's growing dependence upon the state, and advocated the formation of an episcopalian territorial church. He was the author of numerous works in the field of practical theology.

RICHARD V. PIERARD

KAGAWA, TOYOHIKO (1888-1960). Japanese Christian leader. Born at Kobe, he was an illegitimate child of a wealthy cabinet minister and a geisha; yet his father took a liking to him and formally adopted him, but both his parents died before he was five. His childhood in the ancestral home was filled with bitter loneliness and tragedy, but when he was at school in Shikoku the first ray of hope and love entered his life through the friendship of a Japanese Christian teacher and two missionaries. His conversion at fifteen brought disinheritance from his family, but an overpowering experience of the love of Christ moved him to dedicate his life to serve the destitute in the slums. In his second year at the Presbyterian College in Tokyo he was stricken with tuberculosis, and all but succumbed. The appalling conditions of prostitution, poverty, and exploitation impelled him to his God-given mission, and despite poor health he entered Kobe Theological Seminary, soon to exchange his living quarters for the city's slums, where 10,000 people lived in cell-like houses six feet square.

For fifteen years from 1919 he toiled in the slums, striving to improve labor conditions and the laborers themselves. In 1921 he became the leader of the nascent labor movement, and he formed also the first peasant union. Fired by a passion for social righteousness, Kagawa preached, wrote, and worked unceasingly for the cause of Christian socialism; in 1925 trade unions were given legal right to organize, and in 1926 legislation was finally passed for abolishing the slums. Kagawa made his mark as a mystic, ascetic, and pacifist, but it was as a soldier of movements that his influence was greatest. He once stated as his aim "the salvation of 100,000 poor, the emancipation of 9,430,000 labourers, and the liberation of twenty million tenant-farmers." Prominent as a church leader and patriot, he continued as Japan's apostle of love to the end. Among his many books are *Before the Dawn* (ET 1925), *Christ and Japan* (1934), and *Love, the Law of Life* (1930).

See W. Axling, *Kagawa* (1932), and J.M. Trout, *Kagawa, Japanese Prophet* (1959).

DAVID MICHELL

KAISERSWERTH. Rhineland town where Theodor Fliedner,* pastor of the small Protestant community, founded an institution in 1836 to train deaconesses for nursing, educational, and social work. At Kaiserswerth itself, hospitals, homes for the old and needy, and schools formed a considerable estate. Fliedner's work was helped by Frederick William IV of Prussia and was linked with the Innere Mission.* Florence Nightingale trained there. Deaconesses from Kaiserswerth work throughout Germany and in several other parts of the world, and Kaiserswerth is the mother house of an association of about 28,000 sisters in seventy-two houses (1958).

KANT, IMMANUEL (1724-1804). German philosopher. Born in Königsberg, Prussia, into a Pietist* family, he lived there all his life. He was professor of logic and metaphysics in the university from 1770. His contact with the ideas of David Hume* "awoke him from his dogmatic slumbers" and turned him into the "critical" philosopher of the *Critique of Pure Reason* and later works. Thenceforth his aim was to show how reason functions in the acquisition of knowledge, and how the *a priori* knowledge which (according to Kant) the mind has, in logic, mathematics, and physics, can be justified; to preserve the notion of human freedom; and to give an account of the true nature of morality. The only actions that are moral are those which are in accordance with the Categorical Imperative and which are performed from a sense of duty alone. Nonmoral reasoning is by contrast hypothetical, not categorical in character.

His inquiry into the limits of knowledge demonstrates that metaphysical knowledge (including knowledge of God) is impossible; for all our knowledge arises from sense experience, although it does not end there, for the general structure of knowledge is given by the combinatory power of the human mind: neither reason nor sense experience can provide knowledge by itself. The classical rational proofs of the existence of God have to be abandoned. This denial of the knowledge of God makes room for faith. God, though unknowable, is still required (postulated) by practical reason, since the moral law demands that we should promote the highest good (happiness commensurate with virtue) which only God can bring about. The way is thus open for "rational faith," that is, the viewing of all one's duties as divine commands.

Kant's view that any knowledge of God is impossible has been extremely influential in Protestantism. Theology has become anthropology. On it has been based the romanticism of Schleiermacher* and the ethical religion of Ritschl.* Less obviously but just as surely, Karl Barth's* "wholly other" God has connections with Kant's unknowable God. With the view that God is unknowable has gone a recasting of classic dogmatic theology, from the notion of revelation onward.

OONAGH MC DONALD

KARLSTADT, see CARLSTADT

KASATKIN, IVAN, see NICOLAI

KATTENBUSCH, FERDINAND (1851-1935). German church historian and theologian. He was born in Kettwig and studied under A. Ritschl. He became professor of systematic theology at Geissen (1878), Göttingen (1904), and Halle (1906). His principal work was a history of the Apostle's Creed, *Das apostolische Symbol* (two vols., 1894, 1900), based on the researches of C.P. Caspari. He dated the Old Roman Creed about 100. He also contributed to the history of the Reformation confessions, the theology of Luther, and systematic theology. He "has the merit of being the first to have applied the idea of the Son of Man = people of the saints to the Christian concept of the Church" (Cullmann). J.G.G. NORMAN

KEBLE, JOHN (1792-1866). Hymnwriter and Tractarian. Educated at Oxford, where he won a double first and became a fellow of Oriel (1811-23), he assisted his vicar-father before accepting the country living of Hursley, Hampshire, where he remained until his death. In 1827 he published anonymously *The Christian Year.* According to A. Fox, its influence (over 100 editions were published by 1867) has been overrated and has in any case waned. Parts of it and Keble's other collections are still sung, e.g., "Blest are the pure in heart," "New every morning," "Sun of my soul," "There is a book," and "When God of old." In 1833 Keble preached his Oxford Assize sermon in which he denounced contemporary Erastianism as "national apostasy." Newman regarded the sermon as the start of the Oxford Movement.* Keble also contributed to *Tracts for the Times* (1833-41). After Newman's secession to Rome, Keble and Pusey led the movement. Keble's standard edition of the *Works* of Hooker appeared in 1836. Keble is remembered less for his many books than for his hymns and as a devoted parish priest. In 1870 Keble College, Oxford, was founded in his memory. JOHN S. ANDREWS

KEIL, JOHANN KARL FRIEDRICH (1807-1888). Lutheran scholar and exegete. Born in Lauterbach, Saxony, he studied theology at the universities of Dorpat and Berlin, and was professor of Old and New Testament exegesis and oriental languages at Dorpat (1833-58). Influenced by Hengstenberg toward conservative orthodoxy, Keil helped to shape Lutheran ministerial thought in the Baltic provinces for twenty-five years. From 1859 until his death he lived in Leipzig and Rodlitz, engaging in literary work and in serving the Lutheran Church. A vigorous advocate of conservative theology, he rejected the rationalistic, critical views of Scripture. His chief work is his biblical commentary on the OT begun in 1861 in collaboration with Franz Delitzsch. Later expositions appeared on the gospels, Peter, Hebrews, and Jude. H. CROSBY ENGLIZIAN

KEIM, KARL THEODOR (1825-78). German Protestant theologian and church historian. After studying under F.C. Baur* at Tübingen, he was professor at Zurich (1860-73) and then at Giessen (from 1873). He concentrated on the history of primitive Christianity and the Protestant Reformation. His liberal-critical life of Christ, *Die Geschichte Jesu von Nazara* (3 vols., 1867-72; ET in 6 vols., 1873-82), is his best-known work; here he attempted to disengage the "facts" of the life of Jesus from the miraculous elements of the gospels. He rejected the historicity of the fourth gospel altogether, though (following Baur) he argued for the primitive nature and primacy of Matthew. Luke and Mark were understood to have adapted material from Matthew and other sources in the direction of a mediating Paulinism and world-embracing universalism respectively.

W. WARD GASQUE

KEITH-FALCONER, ION GRANT (1856-1887). Missionary and Arabic scholar. Born in Edinburgh, son of the eighth Earl of Kintore (who was also an evangelist), he was educated at Harrow, Cambridge, and Leipzig. Very tall, of attractive personality, he was a Hebrew and Arabic scholar, and also one of the earliest bicycle champions (on "penny-farthings" and heavy boneshakers). He helped found the Cambridge Inter-Collegiate Christian Union in 1877 and engaged in evangelism among the poor. In 1885, partly through the influence of the Cambridge Seven, he went as a missionary to Arabs, though appointed Lord Almoner's professor in Arabic at Cambridge (nonresident). He and his young wife founded the Sheikh Othman hospital (Free Church of Scotland, afterward Church of Scotland) near Aden, but he died of fever within a few months. His social, academic, and athletic standing, together with his early death, made him a great Christian influence on his contemporaries.

JOHN C. POLLOCK

KELLS, BOOK OF, see BOOK OF KELLS

KELLY, THOMAS (1769-1854). Irish hymnwriter. Son of an Irish judge, he was born in Dublin and educated at Trinity College there. He was originally intended for the law, but was converted in 1792 and took holy orders. He remained in the Church of Ireland only a short time, and for most of his life conducted a vigorous evangelical ministry in unconsecrated buildings in Dublin, Wexford, and elsewhere. As a hymnwriter his name is often linked with those of Watts and Newton. He is credited with 765 hymns in all, of which the best known are "Look, ye saints! the sight is glorious," "The Head that once was crowned with thorns," and "We sing the praise of Him who died." Kelly excels in hymns of praise, handles difficult meters dexterously, and often succeeds in compressing the whole evangelical economy within the compass of a few verses.

IAN SELLERS

KELLY, WILLIAM (1821-1906). Plymouth brother and biblical critic. Only son of an Ulster squire, he was educated at Downpatrick and Trinity College, Dublin. He was in 1841 converted to the principles of the Brethren, and began to write extensively on behalf of the Darbyite section of that body. He edited two periodicals, *The Prospect* from 1848 to 1850, and the *Bible Treasury* from 1856 till his death; wrote a long series of devotional works and commentaries; and edited the collected works of J.N. Darby* (34 vols., 1867-83). In 1879 the excommunication of Dr. Edward Cronin led to the "Kelly schism" in the Exclusive Brethren, Kelly heading the more moderate and understanding faction till his death. The original Kelly assemblies are now much reduced, being entirely separate from the Exclusives and inclining more to the independent Brethren. Spurgeon said of Kelly that he was "born for the universe" but "narrowed his mind by Darbyism." IAN SELLERS

KELVIN, LORD, see THOMSON, WILLIAM

KEMPE, MARGERY (c.1373-d. after 1433). English mystic. Born at Bishop's Lynn in Norfolk and daughter of its many-times mayor, John Burnham, she married a local official, John Kempe, and they had fourteen children. She is best remembered for her *Book of Margery Kempe*, a detailed but far from complete account of her life over forty years. She suffered mental disorder, vowed chastity in 1413, and made pilgrimages to Palestine and Europe. Although her autobiography has been criticized as devoid of spiritual understanding, she counted among her advisers and supporters theologians and religious of repute, including Julian of Norwich.* Recipient of visions and revelations, she lived in a period remarkable for lady saints, and in an area which was the scene of preaching and other activities by several distinguished friars of the time. C.G. THORNE, JR.

KEMPIS, THOMAS À, see THOMAS À KEMPIS

KEN, THOMAS (1637-1711). Bishop of Bath and Wells. He became a fellow of New College, Oxford, in 1657 and taught at Winchester College from 1672. Here he probably wrote his two famous hymns, "Awake, my soul, and with the sun" and "Glory to Thee, my God, this night." Although a king's chaplain, he refused to allow his house to be used by the king's mistress, Nell Gwynne, on a royal visit to Winchester. Charles II greatly respected him, made him bishop of Bath and Wells in 1684, and received absolution from him on his deathbed. Ken refused to read James II's Declaration of Indulgence, but also declined to take the oath of allegiance to William and Mary in 1689 and was deprived of his see. Despite an offer of reinstatement in 1703, he lived an ascetic life in retirement. He wrote in the Laudian tradition a manual of devotion for boys, and *The Practic of Divine Love* (1685). JOYCE HORN

KENOSIS. This Greek term is formed from a reflexive verb meaning "he emptied himself" (Phil. 2:7). As a christological statement it has been appealed to as a scriptural warrant for a highly distinctive understanding of the Incarnation. In fact, P. Henry calls this theory "the fourth great attempt at a theological explanation of Christ's being." The essence of the original kenotic view which goes back to Thomasius of Erlangen is stated by J.M. Creed: "The Divine Logos by His Incarnation divested Himself of His divine attributes of omniscience and omnipotence, so that in His incarnate life the Divine Person is revealed and solely revealed through a human consciousness." F. Loofs demonstrates that nothing approaching an acceptance of this kenotic idea is to be found in the Church Fathers before the modern period. As a christological theory it is an innovation inspired by liberal theology.

Kenoticism falls into two categories corresponding to the two main presuppositions which underlie the theory. The notion of a surrender of divine attributes took its rise in Lutheran theology which, starting from the premise of *communicatio idiomatum,* so divinized the human nature of Christ as to produce a type of Monophysitism. But this raised problems for the nineteenth-century Lutherans who, led by Thomasius, proceeded to invert the *communicatio idiomatum* formula and assert the communication of Christ's human attributes to His deity. In this way they sought to safeguard the reality of His humanity—but at the expense of abolishing a continuance of His deity into His incarnate existence.

A more attractive and reasonable version of the principle that the incarnation of the divine Logos required a self-limitation was offered by the British theologians, H.R. Mackintosh and P.T. Forsyth, both of whom operated with categories of consciousness rather than metaphysics. For them, consciousness became the essence of personality; and they argued that it is a monstrous thought that the human Jesus could have sustained a divine consciousness. Instead, the eclipse of that divine awareness was the price He paid to become man, and it is that surrender which constituted the kenosis. But it remains doubtful if this modified kenoticism is any improvement, and both forms of the theory have to face the irreducible facts that God is unchangeable and the Atonement must be a work of God.

See R.P. Martin, *Carmen Christi: Philippians ii. 5-11 in Recent Interpretation and in the setting of early Christian worship* (1967), pp. 165-96.

RALPH P. MARTIN

KENRICK, FRANCIS PATRICK (1796-1863). Roman Catholic archbishop and educator. Born in Dublin and educated in local schools, he graduated from the College of Propaganda in Rome and was ordained there in 1821. He taught at the college and seminary at Bardstown, Kentucky (1821-30), then in 1831 was consecrated bishop and became coadjutor in the Philadelphia diocese. He succeeded to the see in 1842. In 1851 he was named archbishop of Baltimore, which post he held until his death. Kenrick was a fine scholar, brilliant teacher, and able administrator. He published several erudite books on biblical, theological, and apologetic topics.

KENSIT, JOHN (1853-1902). Protestant preacher and controversialist. Born in London of working-class parents, he was successively draper's assistant, stationer, and subpostmaster. From his youth an ardent Protestant, he was deeply incensed by the romanizing trends within the Anglican Church, and founded successively a Protestant book depot in 1885, the *Churchman's Magazine,* and the Protestant Truth Society (1890). The strong antiritualist agitation of 1898-1900 led him to establish the Wycliffe Preachers to bear staunch witness to Protestant principles. He fought against the consecration of liberal and ritualistic bishops, and was charged by his enemies, especially Bishop Creighton of London, with fanaticism. While conducting a Protestant crusade in Liverpool and Birkenhead in 1902, he was assaulted by a Catholic mob and died in hospital a few days later. He is generally regarded as the founder of what is known in Britain as "Political Protestantism." IAN SELLERS

KENTIGERN (Mungo) (d.603). One of the early Christian leaders and missionaries in Scotland. He came of Strathclyde Briton stock. Brought up in Culross and trained by Servanus, he moved to the Glasgow area where he became the leader or bishop of the local church. With the rise of an anti-Christian party in the kingdom of Strathclyde, however, he was forced to retire to Wales, where he is reported to have founded the monastery of Llanelwy. Later he returned to Glasgow to continue his work and, from church dedications in the north of Scotland, would seem to have carried on evangelization there also. About 590 he and Columba* met, probably in the Tay Valley, where they agreed to delimit the areas of their work. St. Mungo's Cathedral, Glasgow, is named after him. W.S. REID

KENYON, SIR FREDERIC GEORGE (1863-1952). Greek manuscript scholar. Educated at Winchester and New College, Oxford, he was appointed to the staff of the British Museum in 1889. From 1898 until 1909 he was assistant keeper of manuscripts, and from 1909 until 1930 director and principal librarian. His scholarly interests covered a good range, and he published a number of books concerned with Robert and Elizabeth Barrett Browning. His main work was done with Greek papyri and particularly with NT manuscripts. His publications included *Our Bible and the Ancient Manuscripts* (1895; rev. 1939); *A Handbook to the Textual Criticism of the New Testament* (1901; rev. 1912); *Recent Developments in the Textual Criticism of the Greek Bible* (1933); *The Text of the Greek Bible* (1937); *The Bible and Archaeology* (1940); and *The Bible and Modern Scholarship* (1948). He used his considerable knowledge of MSS in the ancient world to demonstrate the substantial reliability of the NT text and its closeness to the events which it records. R.E. NIXON

KEPLER, JOHN (1571-1630). One of the founders of modern astronomy. Born near Stuttgart, he became a theological student, teacher of astronomy and mathematics, assistant to Tycho Brahe, imperial mathematical aide to Rudolph II, and astrologer to Wallenstein. His principal scientific discoveries were the three laws of motion which bear his name, the principle of continuity in geometry, and the Keplerian telescope. He worked also on the theory of optics and on the calculus and coined a number of scientific terms such as "satellite" and "camera obscura." He was led to his discovery of the three laws of planetary motion by his belief in Neoplatonic mysticism. Although he accepted the Bible and the Christian religion, his understanding of nature was pantheistic. He thought the universe was an expression of the being of God Himself, and that the sun was the image of the Father.

ROBERT G. CLOUSE

KERR, ALEXANDER (1885-1970). First principal of Fort Hare, South Africa. Born near Kilmarnock, Scotland, he took charge in 1915 of Fort Hare, the first institution of higher learning for Africans in Bantu-speaking Africa; it resulted from cooperation between the churches, the state, and the African people. Initially it pioneered African secondary education, but was soon preparing students from South, Central, and East Africa for degrees of the University of South Africa. There were 720 graduates between 1923 and Kerr's retirement in 1948. He served on various official commissions, including those which recommended the establishment of Makerere College, Uganda, and the University of Rhodesia. Although a layman, he was moderator of the general assembly of the Presbyterian Church of South Africa in 1942. He opposed the transfer of Fort Hare to government control in 1960.

D.G.L. CRAGG

KESWICK CONVENTION. This annual summer gathering of evangelicals at Keswick in the English Lake District originated in the Moody–Sankey revival of 1875 through the efforts of the then vicar of Keswick, Canon Harford-Battersby. The keynotes of Keswick are prayer—especially invocation of the Holy Spirit to indwell the gatherings—reverent Bible study, addresses, and a marked enthusiasm for foreign missions. The movement aims to promote "practical holiness," and its motto is "All One in Christ Jesus." A quantity of literature appears annually, most notably the yearly report *Keswick Week*, the journal *The Life of Faith* (from 1879), and the volumes of the Keswick Library (from 1894). Local "Keswicks" or conventions are held in various cities. Supporters come mainly from Christians of the Reformed tradition, especially evangelical Anglicans. Unlike Wesleyan–Arminian concepts of holiness, Keswick maintains that the Christian's tendency to sin is not extinguished but merely counteracted by victorious living in the Spirit. IAN SELLERS

KETTLEWELL, JOHN (1653-1695). Writer and Nonjuror.* Born in Yorkshire, he was educated at Oxford, and ordained in 1678. His first book, *The Measures of Christian Obedience* (1681), brought him the chaplaincy to the countess of Bedford and the vicarage of Coleshill (1682). He established several charities, preached as a Nonjuror against the rebellions of 1689, and was deprived of his living in 1690. He removed to London and there wrote many devotional books and some controversial tracts. He founded also a fund for deprived clergy, which was declared illegal after his death.

KEYS, POWER OF THE (Lat. *Clavium potestas*). A term symbolic of the authority of Christ and of church leaders. In the Book of Revelation (1:18; 3:7,8), "key(s)" is used as a symbol of the Lord's authority over His church, or that of one of His messengers to whom is given power over those in the "Abyss" (20:1). In Matthew's gospel (16:19) it is the symbol of that authority given to Peter as the leader of the apostolic band. The Roman Church has traditionally understood this authority as belonging to Peter alone, and thus to the bishop of Rome as the head of the Church Universal. Protestants understand this authority either as having been given to Peter as representative of the whole band of apostles (cf. Matt. 28:18), or as having been fulfilled by Peter as an individual when he "opened the door of faith to the Gentiles" (Acts 14:27) by preaching to and baptizing the household of Cornelius, as he had done earlier for the Jews on the day of Pentecost (Acts 2).

W. WARD GASQUE

KGAMA III (c.1828-1923). African Christian chief. Baptized in 1862 by H.C. Schulenbourg of the Hermansburg Mission, he soon proved himself an uncompromising Christian. Confirmed in 1875 as chief of the Ngwato, he tried to apply Christian standards of government despite strong opposition. He opposed many tribal customs, banned the liquor trade, dispensed evenhanded justice, and refused to alienate tribal lands to white men. Although jealous of his position and somewhat intolerant, he treated his opponents (particularly his father and brother) with astounding patience and magnanimity. He consistently supported the London Mission Society which replaced the Hermansburg Mission, but sometimes had strained relations with individual missionaries. He accepted a British Protectorate in 1885 and visited England in 1895 in a successful attempt to avoid control by the Chartered Company. His grandson, Seretse Khama, became first president of Botswana. D.G.L. CRAGG

KHOMYAKOV, ALEKSEI STEPANOVICH (1804-1860). Russian philosopher and theologian. A member of the landed gentry, he was graduated from the University of Moscow in 1822. He was an officer in a cavalry regiment before traveling to France, Italy, Switzerland, and Austria. From 1828 to 1829 he served in a hussar regiment during the Russo-Turkish War. In 1830 he retired to his estates of Bogucharov and Lipitzy where he tried to improve the conditions of his serfs and eventually advocated the abolition of serfdom. He spent his winters in Moscow where he was active in the intellectual life of the salons. In 1847 he traveled to Germany and Bohemia. His writings cover a wide range of subjects from tragedy and poetry to philosophy and theology.

Khomyakov's philosophy emphasized the concreteness and wholeness of reality. A leader in the Slavophile movement, he believed that the Slavs were destined to take over the leadership of the world from the decadent Western civilization which was characterized by reason, logical necessity, and materialism, in contrast with the spiritual and moral freedom of Russian thought. Although a layman, he was well read in theology and believed the Orthodox Church as a mystical body was the guiding light of true Christianity. He criticized both Roman Catholicism and Protestantism for destroying the unity of Christianity. At the heart of his theology is his doctrine of *Sobornost,* or commonality, which portrays the church as the divinely inspired fellowship of truth and love. Truth, then, comes not from the decisions of a hierarchy or a council, but from the whole Christian community; not from reason, but from the illumination in the depth of man's soul by faith. Khomyakov was a religious populist who saw the Russian peasant commune as that which preserved Christianity in its pure form and would lead the nations into a new Christian era.

BIBLIOGRAPHY: V.Z. Zavitnevich, *Aleksei Stepanovich Khomiakov* (2 vols., 1902-13); A. Gratieux, *A.S. Khomiakov et le mouvement slavophile* (2 vols., 1939); N. Zernov, *Three Russian Prophets* (1944); S. Bolshakoff, *The Doctrine of the Unity of the Church in the Works of Khomyakov and Moehler* (1946).

BARBARA L. FAULKNER

KIDD, BERESFORD JAMES (1864-1948). Church historian. Son of a Church of England clergyman, he was educated at Oxford, was ordained (1887), served as assistant curate in Oxford (1887-1900), and lecturer in theology at Pembroke College (1902-11). He was vicar of St. Paul's, Oxford (1904-20), and in 1920 became warden of Keble College, where he remained until retirement (1939). Kidd's publications were concerned with the history of Christianity, beginning with the Thirty-Nine Articles (1899) and the English (1901) and Continental (1902) Reformation, pursuing in detail the church to 461 (1922), adding the Eastern Churches from 451 (1927), and the Counter-Reformation (1933). His work in the history of Christianity was based on careful study of its documents, illustrative collections of which he also published. Throughout his career he pursued the matter of Anglican catholicity, and later wrote on Roman primacy (1936) and Validity (1937). CLYDE CURRY SMITH

KIERKEGAARD, SØREN AABY (1813-1855). Danish philosopher. Born in Copenhagen, son of a wealthy Lutheran who retired early to devote his life to piety, Søren's melancholy disposition, inherited from his father, may have influenced his highly individual and introspective writing. Attempts have been made to explain his thought in psychological terms. He took ten years to take his degree; his engagement was broken off and he never married; he prepared for ordination in the Danish Lutheran Church, but was never ordained.

His writings have been divided into two main groups, though the division is only a rough one. Those written between 1841 and 1845 are largely philosophical and aesthetic. Some were pseudonymously attributed to John Climacus and contain numerous pseudonymous characters who express indirectly the writer's viewpoint. The works of this period include his thesis on *The Concept of Irony with Constant Reference to Socrates* (1841), *Either-Or* (1843), *Fear and Trembling* (1843), *The Concept of Dread* (1844), *Stages on Life's Way* (1844), *Philosophical Fragments* (1844), *Concluding Unscientific Postcript to the Philosophical Fragments* (1846), and numerous *Edifying Discourses.* The works of Kierkegaard's later period are sometimes described as his Christian writings, though they might with equal accuracy be described as works attacking formal Christianity. In fact, in both periods he wrote from a Christian standpoint. Among his later writings are *Works of Love* (1847), *Christian Discourses* (1848), and *Training in Christianity* (1850). He kept a *Journal* to the end of his life. A study of its contents and his various other papers is an invaluable supplement to his other writings.

In Holy Week 1848, Kierkegaard underwent a second conversion experience, after which he largely abandoned his pseudonymous writing in favor of direct communication and Christian witness. When Bishop Mynster died in 1854, his successor, H.L. Martensen, delivered an oration celebrating his predecessor's witness to the truth. Although Mynster had been a lifelong friend of the family, Kierkagaard could not forbear writing a withering series of attacks on the man who had come to symbolize for him the formal, conformist, indifferent Christianity into which Protestantism had now fallen. Kierkegaard died while the controversy was at its height.

His thought was shaped by his reaction to Hegel and German idealism in general; his debt to Greek thought, especially to Socratic irony; his sense of the otherness of God; and an overpowering awareness of the personal demands of NT Christianity as contrasted with the lukewarm, official Christianity of the day.

At the heart of Kierkegaard's thought lies the distinction between time and eternity, finite and infinite, immanent and transcendent. Man and his world belong to the former; God to the latter. There is no continuity between the two, for God is wholly other. The gulf can be bridged only from God's side. This is done in the Incarnation. But even here the divinity of Christ is hidden. Christ comes to men incognito. It cannot be otherwise, for to be known directly is the mark of an idol. Christ can be known only by faith. It is by faith that man becomes a true contemporary of Christ, transcending the limits of time and space. The Christian life is one of personal commitment in faith.

Consequently Kierkegaard's pronouncements on history appear disparaging; for merely historical knowledge without faith does not lead to Christ. Conversely, a minimal historical knowledge is enough to afford an occasion for faith. But unlike some of his twentieth-century followers, Kierkegaard did not favor radical, biblical criticism. His writings take Scripture at its face value and show no interest in the criticism of his day. The key to his attitude was the conviction that the finite cannot contain or express the infinite. The temporal is merely the occasion for encountering the eternal.

Through translation into German, English, and other languages, Kierkegaard is more influential today than in his lifetime. He is widely regarded as a forerunner of existentialism.* But although he was deeply concerned with human existence, his thought has more in common with the dialectical theology of the early Barth than with later radical existentialism. It worked within a theistic framework that was concerned above all with the transcendence of God. He preserved this transcendence by making it a hidden one. Consequently Kierkegaard has been criticized for irrationalism. Others have dispensed with his theistic framework and made his approach the basis of a nontheistic existentialism.

BIBLIOGRAPHY: There is no collected edition of Kierkegaard's writings in English, but all the most important works have been translated. A comprehensive list, with descriptive analysis, is given by G.E. and G.B. Arbaugh, *Kierkegaard's Authorship* (1968). A. Dru has edited his *Journals* (1938). This is supplemented by R.G. Smith (ed.), *The Last Years: Journals 1853-1855* (1965). His *Journals and Papers* are being edited by H.V. and E.H. Hong (1967-).

Studies: W. Lowrie, *Kierkegaard* (2 vols., 1938) and *A Short Life of Kierkegaard* (1942); H. Diem, *Kierkegaard's Dialectic of Existence* (1959); L. Dupré, *Kierkegaard as Theologian* (1963); E.J. Carnell, *The Burden of Søren Kierkegaard* (1965); P. Sponheim, *Kierkegaard on Christ and Christian Coherence* (1968). COLIN BROWN

KIKUYU CONTROVERSY. A dispute within the Anglican Church about the nature of the church and its ministry, which arose from the proceedings of a conference of missionary bodies working in Kenya, held on the Church of Scotland mission station at Kikuyu near Nairobi in 1913. The conference discussed a scheme of federation under which all Christian missionary work in Kenya would be brought together. Opposition to the scheme came from missionaries holding Baptist views who wished to rebaptize those baptized in infancy, and from the Anglican delegates who wished to insist on episcopal confirmation. Both parties, however, withdrew their opposition, and the scheme was approved for transmission to the overseas parent bodies of the missions involved. The conference closed with a communion service according to the Anglican Rite conducted by the bishop of Mombasa (William Peel), in which all members took Communion except the Friends.

When the Anglo-Catholic bishop of Zanzibar, Frank Weston, heard of the proceedings he wrote a letter of protest to the archbishop of Canterbury, Randall Davidson. He objected to what he regarded as the inadequate view of the church and its ministry reflected in the scheme of federation, especially the lack of emphasis on the historic episcopate, and the admission of nonepiscopally confirmed Christians to a Communion service conducted by an Anglican bishop. A month later he sent the archbishop a formal indictment of the bishops of Mombasa and Uganda (J.J. Willis, who chaired the conference) and asked him to arraign them on charges of "propagating heresy and committing schism." The archbishop refused to do this, and referred the matter to the Central Consultative Body of the Lambeth Conference.

Meanwhile the controversy spread throughout the English-speaking world and was even the subject of a *Punch* cartoon. After some delay due to the outbreak of war, the archbishop delivered his findings at Easter 1915. He was obviously sympathetic to the two bishops, but advised caution in the matter of intercommunion in future, and felt that he could not advise Anglican acceptance of the scheme of federation in the form proposed.

BIBLIOGRAPHY: H.M. Smith, *Frank, Bishop of Zanzibar* (1926); G.K.A. Bell, *Randall Davidson* (1935); vol. 1, chap. 42; J.W. Arthur and J.J. Willis in the symposium *Towards a United Church* (1947); R. Macpherson, *The Presbyterian Church in Kenya* (1970). JOHN WILKINSON

KILHAM, ALEXANDER (1762-1798). Founder of the Methodist New Connexion.* He was born in Epworth, Wesley's birthplace, son of a Methodist weaver. Converted while young, he entered the service of Robert Carr Brackenbury, the Methodist gentleman–preacher, and assisted him in pioneer preaching in the Channel Islands. He became an itinerant preacher in 1785 and was soon involved in controversies (arising first from the expectation and then the event of Wesley's death) about the relation of Methodism to the Church of England. His pamphlets, under his own name or pseudonyms, defended administration of the sacrament to Methodists by preachers who were not in Anglican orders, called for lay representation in circuit and connexional government, and opposed a project for Methodist bishops. An attack on conference abuses brought his expulsion in 1796, and the Methodist New Connexion was formed, embodying his principles. He died, still only 36, worn out by extreme toil. One of Kilham's stations was Aberdeen, and Scottish Presbyterian practice may have affected him. Certainly he represents the tradition always present in Methodism which stood closer to English Dissent than to Anglicanism. His widow Hannah (1774-1832) joined the Society of Friends and became a pioneer of African linguistics and education. A.F. WALLS

KILIAN (c.640-c.689). "The Apostle of Franconia." A native of Ireland, he was probably already a bishop when he left with eleven companions to evangelize the Franks. Having reached Würzburg, he reputedly traveled to Rome for papal approval of his mission. He converted many in Franconia and Thuringia, including Duke Gozbert, whom he persuaded to separate from Geilana, his brother's widow. In revenge she had him murdered with two of his fellow-missionaries, Coloman and Totman. Their relics were solemnly transferred by Burchard, first bishop of Würzburg, to the new cathedral in 752, and are now enshrined in the Neumünster, traditionally the site of the martyrdom. J.G.G. NORMAN

KILWARDBY, ROBERT (c.1210-1279). Archbishop of Canterbury from 1273. One of the foremost theologians of his time, he studied at Oxford and Paris and taught in both universities. In 1240 he became a Dominican and in 1261 prior provincial for England. Vigorous and respected as a theologian, his tenure of the former post was marked by controversy with the Franciscans as to whether interior poverty was not more spiritual than material poverty, and by increasing confrontation with the Aristotelian influences in his own order. In 1273 he became archbishop of Canterbury and was energetic in visitations. In 1277, after an Oxford visitation, he condemned several Thomist doctrines, but Thomist support was powerful at the papal court, and Kilwardby was summoned to Rome and "promoted" to the cardinalate. He resigned as archbishop and died shortly afterward. C. PETER WILLIAMS

KING, EDWARD (c.1735-1807). Archaeologist and writer on science and religion. Born in Norwich and educated at Cambridge and Lincoln's Inn, King (who had private means) practiced law, occasionally writing and studying for most of his life. His interests were many, but the study of sacred Scripture as related to secular knowledge predominated. His writings are characterized by Christian devotion, a pellucid style, and heavy documentation, but also, not infrequently, by lack of a balanced judgment. His extreme originality shocked his contemporaries, arousing strong opposition, but many of his ideas, suitably modernized, would be more acceptable today. In a brilliant book (1796) he argued on biblical and observational grounds for the reality of meteorites at a time when this view was commonly ridiculed. His discussions and speculations on the use of "heaven" and "heavens" in the NT, the meaning of Genesis 1-3, and the possibility of a multipopulated universe, etc., are of abiding interest.

R.E.D. CLARK

KING, JONAS (1792-1869). American Congregational author and missionary to Greece. Born in Hawley, Massachusetts, and educated at Williams College (1816) and Andover Seminary (1819), he was ordained in 1819, studied Arabic for a year, then went to Palestine under the American Board of Commissioners for Foreign Missions (1822-25). Returning home, he remained in Greece for a time, marrying Anetta Mengous in 1829. Assigned to Greece in 1830, he began a career of distinguished missionary and consular service. Besides founding a Greek Protestant church, King also started a school and wrote Christian literature in several languages. His *Farewell Letter* (1825) with his reasons for not becoming a Roman Catholic was translated into several languages. He died in Athens.

ROBERT C. NEWMAN

KING, MARTIN LUTHER, JR. (1929-1968). American civil rights leader. Born in Atlanta, Georgia, and educated at Morehouse College, Crozer Theological Seminary, and Boston University (Ph.D., 1955), he became pastor of Drexler Avenue Baptist Church, Montgomery, Alabama (1954), and co-pastor with his father of Ebenezer Baptist Church, Atlanta (1959). He rose to national prominence as leader of the movement to secure equal rights for Negroes through nonviolent, mass demonstrations, beginning with the Montgomery bus boycott (1956). He organized the Southern Christian Leadership Conference,* was the leading figure in the March on Washington (1963) that led to the 1964-65 Civil Rights Acts, and was active in voter registration drives. He received the Nobel Peace Prize (1964). King urged settlement of the Vietnam conflict and admission of Communist China to the U.N. Strongly criticized by segregationists and militant blacks, he was assassinated in Memphis, Tennessee, by a white man. He wrote *Stride Toward Freedom* (1958), *Strength to Love* (1963), *Why We Can't Wait* (1964), and *Where Do We Go from Here: Chaos or Community?* (1967).

ALBERT H. FREUNDT, JR.

KINGDOM OF GOD. This phrase (and the phrase "the kingdom of heaven" which is used often in Matthew's gospel with the same meaning) does not occur in the OT or in Jewish literature, but its source is to be found there. The idea of God reigning is prominent especially in the Psalms. The messianic promises were concerned with a king who should reign. The apocalyptic writers stressed God's reign as something which would break into the present world order and establish a new one, while the rabbis saw the kingdom as being connected with obedience to the Law. In the NT, the kingdom of God plays a very important part, particularly in the synoptic gospels, of which it is the central theme. The Greek word *basileia,* like its English translation, suggests too strongly a "realm" rather than a "reign," but it is the latter concept which, drawing its significance from the dynamic Hebrew work *malkuth,* is the more prominent in the NT.

A proper understanding of "the kingdom of God" in the teaching of Jesus has been bedeviled by the tendency of scholars to read into the gospels their own views of what Christianity is about. The liberal Protestants of the nineteenth century reduced the conception to a set of humanistic values ("the kingdom of self-respect," etc.). Albert Schweitzer, in reaction to this, found a revolutionary content in Jesus' message and the expectation of an imminent consummation of the kingdom. More recent scholars have tried to do more justice to the various sides of the teaching. Many of the parables of Jesus deal with the crisis caused by the coming of the kingdom, and if this in many cases is meant to have its primary reference to the response of the Jewish people to the ministry of Jesus, it may have a legitimate extension to the situation of the Christian Church as it waits for the final consummation.

There has been divergence of opinion as to whether the "drawing near" of the kingdom proclaimed by Jesus (Mark 1:14f.) means that it had actually arrived when He began his ministry. There are passages which seem unquestionably to imply that the presence of Jesus meant the presence of the kingdom (Mark 14:25; Matt. 25:34; Luke 22:29f.). A number of other passages may refer to the Parousia or to some event such as the Resurrection or Pentecost which could be described as a coming of the kingdom (e.g., Mark 9:1). Whether present or future, the reign of God demands a response of commitment from men who are called frequently to enter the kingdom. This is also found in John's gospel (3:3-5). The term is comparatively rare in the rest of the NT. Since the time of Augustine there has been a tendency to institutionalize the concept of the kingdom by identifying it with the church.

BIBLIOGRAPHY: A. Schweitzer, *The Quest of the Historical Jesus* (1910); T.W. Manson, *The Teaching of Jesus* (1931); C.H. Dodd, *The Parables of the Kingdom* (1935); R. Otto, *The Kingdom of God and the Son of Man* (1943); R.H. Fuller, *The Mission and Achievement of Jesus* (1954); N. Perrin, *The Kingdom of God in the Teaching of Jesus* (1963); G.E. Ladd, *Jesus and the Kingdom* (1966). R.E. NIXON

KING JAMES VERSION, see BIBLE, ENGLISH VERSIONS

KING'S CONFESSION, see SCOTS CONFESSION

KING'S EVIL, THE. Scrofula, a tubercular disorder afflicting the skin with draining sores and ugly scars. French and English tradition regarded the disease curable by the monarch's touch, hence the term "the king's evil." Presenting those afflicted to the king may date from the reign of Edward the Confessor (d.1066). The practice certainly existed in the court of Louis IX (d.1270) in France. His curative power was supposedly passed to the Valois monarchs of France and to English royalty through Edward III (d.1377). The latter embellished the practice by presenting the sufferer with a touchpiece or coin serving as a talisman. At times the touching was accompanied by ablutions and litanies. The last to practice the rite were the English Stuarts. JAMES DE JONG

KINGSLEY, CHARLES (1819-1875). English novelist and Christian Socialist. Born at Holne, Devonshire, and educated at King's College, London, and Magdalene College, Cambridge, he was ordained in 1842, and from 1844 was vicar of Eversley, Hants. Though ill-qualified, he was professor of modern history at Cambridge (1860-69) and he subsequently held canonries at Chester and Westminster. From 1869 he was prominent in the Educational League. He was precocious, athletic, romantic, and interested in the sciences, particularly botany. He was influenced by Thomas Carlyle and F.D. Maurice, and as "Parson Lot" was the pamphleteer of the Christian Socialist* movement. His concerns were educational and sanitary reform, and the extension of the co-operative principle. He was a critic of Tractarianism;* "Muscular Christianity" is associated with him; and his insinuation in 1863 that Newman had little respect for truth led Newman to write his *Apologia.* Kingsley's novels generally had some bearing on social issues. *Westward Ho, Hereward the Wake,* and *The Water Babies* are still read. JOHN A. SIMPSON

KIRK; KIRK SESSION. In Scotland the word "kirk" is still used interchangeably with "church." Originally the allusion was with particular reference to the national Church of Scotland (hence, "Auld Kirk"), but it has come to be employed generally to denote the denomination or the place of worship of any of the mainline Scottish churches other than Roman Catholics and Episcopalians. The kirk session is the lowest court in any Presbyterian Church, concerned with the oversight of a single congregation. In the case of the Church of Scotland, the kirk session's interest technically extends to the whole parochial area.

KIRK, KENNETH ESCOTT (1886-1954). Bishop of Oxford. Educated at Oxford, he saw chaplaincy service in World War I before returning to Oxford (1922-37) where finally he held the chair of moral and pastoral theology. Consecrated bishop in 1937, he continued his writing, particularly on moral theology, and is probably best known for *The Vision of God* (1931) and *The Apostolic Ministry* (1946).

KISS OF PEACE. A salutation as a token of Christian brotherhood named in 1 Peter 5:14 and often referred to in the NT as the "holy kiss" (Rom. 16:16; 1 Cor. 16:20; 1 Thess. 5:26) and later, among patristic writers, as the "kiss of peace." It was so mentioned first by Justin Martyr in the second century; he described it as a mutual greeting of the faithful. No limitation of its use is stated or implied; the Christians were simply bidden "to greet one another." A sign of love and union, the kiss of peace was a part of the Eucharistic liturgy and was maintained in the Western Church until after the thirteenth century. Once actually a kiss, the symbol has been modified so that the persons exchanging it face each other, and each places his hands upon the other's shoulders and each bows his head. WATSON E. MILLS

KITTEL, GERHARD (1888-1948). German biblical scholar. Born in Breslau, youngest son of Rudolph Kittel,* he taught at Kiel and Leipzig before assuming the chair of NT at Griefswald in 1921. Five years later he took up a similar post at Tübingen and held it nominally until his death. Originally interested in rabbinical studies, Kittel in 1931 launched a major composite project—*Theologisches Wörterbuch zum Neuen Testament*—wherein he insisted that a lexicon of the NT must trace the history of each word with reference to its secular usage in classical and *koine* Greek as well as its religious connotations derived from the Septuagint and the Hebrew background. By World War II four massive volumes (A-N) had been completed. While Kittel held back from full Nazi demands to suppress Christianity, *Das antike Weltjudentum* (1943), written with Eugen Fischer, shows his propagandistic usefulness.

CLYDE CURRY SMITH

KITTEL, RUDOLPH (1853-1929). German OT scholar. Born of a Swabian family, he studied at Tübingen where in 1881 his prize criticism of J. Wellhausen* impressed C.F.A. Dillmann of Berlin, whose OT handbook he was later to edit, and whose commentary on Isaiah he revised. In 1888 he became professor of OT at Breslau, transferring in 1898 to the chair of biblical theology at Leipzig, whence he retired in 1924. He prepared a critical edition of the Hebrew text, *Biblia Hebraica* (3rd ed., with Paul Kahle, 1929-37), and prepared commentaries on Kings (1900), Chronicles (1902), Psalms (1913), Judges (1922), and Samuel (1922). He wrote *Geschichte der Hebräer* (2 vols., 1888-92)—which later became *Geschichte des Volkes Israel, Die hellenistische Mysterienreligion und das Alte Testament* (1924)—and *Die Völker des vorderen Orients* (1931). He did not live long enough to see the Ras Shamra Tablets* which his work anticipated.

CLYDE CURRY SMITH

KITTO, JOHN (1804-1854). English biblical scholar. Born in Plymouth, Kitto at the age of twelve sustained an accident while assisting his father, a drunken stonemason, which left him permanently deaf. Dragging out a miserable existence first as a workhouse inmate and then a shoemaker's apprentice, he showed a great interest in books and, having been converted in 1824, was rescued by A.N. Groves,* who sent him to Islington Missionary College to train as a printer for the Church Missionary Society. This body found his services both in London and Malta (1827-29) unsatisfactory, and in 1829 he traveled to Muslim lands as one of Groves's party of Brethren missionaries. In Baghdad he set up a missionary school which was destroyed in 1832 when he returned to England. He now broke with the Brethren and began to write for the wider evangelical world. His *Pictorial Bible* (1836f.), *History of Palestine* (1843f.), *Cyclopedia of Biblical Literature* (1845f.), *Journal of Sacred Literature* (1848f.), and the *Daily Bible Illustrations* (1850f.) were once well-known works of devotion and popular scholarship. Latterly academically honored, he struggled against severe physical and pecuniary hardships till his death.

IAN SELLERS

KLOPSTOCK, FRIEDRICH GOTTLIEB (1724-1803). German poet. Born at Quedlinburg, eldest son of a lawyer, he attended the classical school of Schulpforta, near Naumburg (1739), and studied theology at Jena (1745) and Leipzig (1746). While still at school he drafted the plan of a religious epic, *Der Messias,* inspired by Milton's *Paradise Lost.* He was a private tutor in Langensalza (1748). Frederick V of Denmark, on the advice of his prime minister, Count von Bernstorff, invited him to his court to complete *Der Messias* (1751). He married Margareta (Meta) Moller of Hamburg (1754), who died four years later. Apart from a year at Karlsruhe from 1770, he lived in Hamburg, completing *Der Messias* (1773) and marrying the niece of his first wife. He wrote religious odes, hymns, and lyrical and epic poems, and made important contributions to philology and the history of German poetry.

J.G.G. NORMAN

KNEELING. A posture used in Christian worship, generally for prayer, which in the early church signified penitence. It was actually forbidden on Sundays and during the Easter festival. Standing was the normal posture for prayer adopted from Jewish practice (but cf. Acts 20:36). It remains customary in the East, apart from penitential devotions. In the West kneeling has become more normal, apart from some Protestant churches where sitting is usual. At the Reformation in England there was some dispute over the direction to kneel when receiving the Communion. An explanation (known as the Black Rubric) that this signified no adoration of the elements, but simply humility and gratitude, was inserted in the 1552 Prayer Book and restored in revised form in 1662. JOHN TILLER

KNIBB, WILLIAM (1803-1845). Perhaps the best known of the early Baptist missionaries in Jamaica, he arrived in the island in 1824 to manage the Kingston school. In 1830 he went as minister to Falmouth, near Montego Bay, where he remained until his death. His ministry in Falmouth spans momentous years: the slave revolt of 1831-32, the persecution of evangelicals which followed it, emancipation, the shift from plantation to freehold residence—all crowd into the fifteen years of Knibb's ministry. His principal contribution in these events was as propagandist. A vigorous and flamboyant orator, he could always be counted on to flay the devil. In 1832 and again in 1841, when the Baptists badly needed an advocate in England, they sent Knibb. He was a tireless champion of the blacks, in slavery, in apprenticeship, and in freedom, when he risked his personal credit to settle the slaves on their own land. He was also a prime mover in the decision to declare the Jamaica churches independent of the Baptist Missionary Society, in the formation of Calabar College for training ministers, and in organizing the first West Indian mission to Africa.

See H.J. Hinton, *Memoir of William Knibb, Missionary in Jamaica* (1847) and G.A. Catherall, *William Knibb: Freedom Fighter* (1972).

GEOFFREY JOHNSTON

KNIGHTS HOSPITALLERS, see HOSPITALLERS

KNIGHTS OF COLUMBUS. A fraternal benefit society of Roman Catholic men, founded by Michael J. McGivney of New Haven, Connecticut, and chartered by the state of Connecticut in 1882. The organization is represented in every state of the USA and in Canada, Mexico, Puerto Rico, the Philippines, and the Canal Zone. With more than 1.5 million members, affiliated through 5,000 subordinate councils and sixty-one state councils with its supreme council, it seeks to further charity, brotherhood, and patriotism. It has promoted war relief, disaster relief, parochial education, veterans' benefits, parish assistance, and historical studies, and it has lobbied for aid to parochial schools. It is also notable for its provision of insurance to protect the wife and children of each of its members. DARREL BIGHAM

KNIGHTS OF MALTA/RHODES, see HOSPITALLERS

KNIGHTS TEMPLAR, see TEMPLARS

KNOX, EDMUND ARBUTHNOTT (1847-1937). Bishop of Manchester. Educated at Oxford, where he became fellow and tutor (1869-84), he subsequently engaged in parish work before consecration as suffragan bishop of Coventry in 1894, whence he was translated to Manchester in 1903. He retired in 1921. Knox was a prominent evangelical; his opposition contributed much to the rejection of the revised Prayer Book (1927-28), and he strove vigorously against liberal and High Church tendencies. His writings include *Sacrifice or Sacrament* (1914), *On What Authority?* (1922), and *The Tractarian Movement, 1833-1845* (1933).

KNOX, JOHN (c.1514-1572). Scottish Reformer. Born at Haddington and educated at St. Andrews, probably under the conciliarist and scholastic John Major, Knox was ordained by the bishop of Dunblane (1536) and later served as a notary (by 1540) and a private tutor (by 1543). Thomas Gilyem (Gwilliam) converted him to Protestantism. He was subsequently influenced by the more zealous principles of John Rough and George Wishart,* a disciple of Lutheran and Swiss theology. Knox was indebted to Wishart for his sense of prophetic vocation, his tendency toward theological eclecticism, and his adherence to Bucer's doctrine of the Lord's Supper. Following Wishart's martyrdom (1546), Knox contemplated going to Germany, but renewed action against heretics caused him to go to St. Andrews Castle, where he was called as preacher. When the castle fell, he was taken to France and made a galley slave. While thus detained, he wrote a précis of Henry Balnaves's compendium of Protestant thought, which drew heavily on Luther's commentary on Galatians. In it Knox embraced Luther's doctrine of justification.

After being freed early in 1549, Knox went to England and was appointed preacher at Berwick. His sermons attacked the Mass as idolatrous, and he was summoned to answer for his views before the Council of the North at Newcastle (1550). Through the influence of Northumberland, Knox preached before the royal court in 1552. At Windsor he criticized the provision in the forthcoming Second Book of Common Prayer calling for kneeling during Communion; his efforts were largely responsible for the inclusion of the Black Rubric. Sensing trouble, he refused the bishopric of Rochester, but not because he opposed episcopacy. As one of the preachers of the 1553 Lenten sermons, he warned of the dangers of secret Catholics in political offices. Following Mary's accession, he fled to the Continent. He met with Calvin* in Geneva, Bullinger* in Zurich, and other Swiss leaders, posing questions on rebellion against idolatrous monarchs and female sovereigns.

At Calvin's urging Knox became pastor of the English congregation at Frankfurt in 1554. A dispute over the Book of Common Prayer led to his ouster and return to Geneva in 1555. The same year he went back to Scotland and openly preached Protestant doctrine. He was summoned to appear in Edinburgh in May 1556 on a charge of heresy, but the regent's intervention resulted in a quashing of the summons. He left Scotland that year to become pastor of the English congregation in Geneva. There he wrote *The First Blast of the Trumpet against the Monstrous Regiment of Women* (1558), arguing that female sovereignty contravened natural and divine law. The *Blast* was aimed primarily at Mary Tudor, but shortly after its appearance Elizabeth was crowned, making Knox's name odious in Elizabeth's court. Even Calvin was displeased, prompting Knox to write a treatise against an "Anabaptist" in defense of the Calvinist doctrine of predestination. The treatise is untypical of Knox's style and basic theological concerns. In the summer of 1558 Knox wrote three more tracts setting forth his theory of lawful rebellion against idolatrous princes, including rebellion by commoners.

The Protestant lords in Scotland sought Knox's return, and he arrived in May 1559. In addition to preaching he negotiated with the English for troops and money. With John Willock and others, he played a major role in drafting the Scots Confession,* which Parliament approved in August 1560. With Willock, John Douglas, and three others, he also drafted the *Book of Discipline.* After Mary Stuart's return in 1561, Knox denounced her masses and court life at Holyroodhouse. During her reign Knox had three interviews with Mary in which he defended his opposition to idolatry. In 1561-62 he engaged in a controversy on ordination with Ninian Winzet, a Catholic priest and educator. Knox claimed that he, like Amos and John the Baptist, had extraordinary calling, but lacked the miraculous power to demonstrate it. Knox also disputed with Quintin Kennedy, abbot of Crossraguel, on the Mass. In 1567 he visited England, afterward refusing to sanction separation from the Church of England. Following Darnley's murder, he returned to Scotland the month (June) Mary was captured. He demanded her execution. After her abdication he preached at the coronation of her son James (see JAMES VI AND I).

Knox was a man of conviction and courage, whose declamations against idolatry overshadowed the warmer side of his nature. His most notable work was the *History of the Reformation of Religion within the Realm of Scotland,* the first complete edition of which was published in 1644. He gave to the Kirk of Scotland an eclectic theology and polity, helped draw up its *Book of Common Order,* and planted the seeds for the later development of Covenant thought in Scotland. His broad view of ecumenical fellowship with Protestant churches in England and on the Continent tempered the spirit of Scottish nationalism during his lifetime.

BIBLIOGRAPHY: Knox's *Works* (ed. David Laing, 6 vols., 1846-64); his *History* (ed. W.C. Dickinson, 2 vols., 1949): major biographies by P.H. Brown (1895), E. Percy (1937), J. Ridley (1968), and W.S. Reid, *Trumpeter of God* (1974); on Knox's thought, see J.S. McEwen, *The Faith of John Knox* (1961). RICHARD L. GREAVES

KNOX, RONALD ARBUTHNOTT (1888-1957). Roman Catholic scholar. Born at Kibworth, son of the (Anglican) bishop of Manchester, E.A. Knox,* he was educated at Eton and Balliol, and became a strong Anglo-Catholic and a bitter critic and satirist of modernists within the church. He joined the Church of Rome in 1917, was ordained priest in 1919, taught at St. Edmund's College, Ware, and was Catholic chaplain at Oxford from 1926 to 1939. Thereafter he devoted himself to his translation of the Bible into English (1944) and to a study of religious enthusiasm, which appeared in 1950. Knox dealt learnedly and half-admiringly with Protestant religious leaders, especially Anabaptists, Quakers, and Methodists, while disapproving of their beliefs. His collected sermons are valued highly by English Catholics.

IAN SELLERS

KNOX, WILFRED LAWRENCE (1886-1950). Anglican churchman and NT scholar. He was the son of a famous evangelical bishop, E.A. Knox,* and the brother of an even more famous Roman Catholic cleric and writer (Ronald A.*). Wilfred was the scholar of the family and was elected to membership in the British Academy. Educated at Trinity College, Oxford, he was ordained priest in 1915; he moved to Cambridge in 1920, where he remained, except for two years in London, until his death. His early writings were primarily apologetic in nature and aimed at a defense of "liberal" Anglo-Catholicism. His major scholarly works include *St Paul and the Church of Jerusalem* (1925), *St Paul and the Church of the Gentiles* (1939), *Some Early Hellenistic Elements in Primitive Christianity* (1944), and many articles in the *Journal of Theological Studies.* W. WARD GASQUE

KOCH, JOHANN, see COCCEIUS

KOHLER, C. and H., see BRÜGGLERS

KORAN, see QUR'AN

KOREA. Christianity was planted in Korea by Koreans, not by foreigners. Prior missionary contacts were only peripheral—the first Catholic, de Cespedes, in 1593 as chaplain to invading Japanese troops, and the first Protestant, Karl Gutzlaff,* in 1832 exploring the coast. Not until Lee Sung-hun in 1784 returned, baptized, from a visit to the ex-Jesuit mission in Peking did Catholicism begin to spread among Koreans. In the next one hundred years, despite great persecutions (1801, 1839, 1846, and 1866), the Catholic Church, though still a hidden movement, grew to some 17,500 members. The first foreign missionary was Chinese, Father James Chou in 1794, followed in 1835 by Father Pierre Maubant of the Paris Missionary Society.

Protestantism similarly was introduced by Koreans. A year before the arrival in 1884 of the first resident Protestant missionary, a Korean convert of Scots missionaries in Manchuria, Suh Sang-yun, brought Korean Scripture portions into forbidden Korea and secretly gathered together the country's first group of worshiping Protestants. The missionaries followed—first a Presbyterian physician, Dr. Horace Allen, and in 1885 two clergymen, H.G. Underwood* (Presbyterian) and H.G. Appenzeller (Methodist). The pioneer in opening N Korea was S.A. Moffett. It was in the north that church growth was greatest, particularly after 1895, later reinforced and vitalized by the great revival of 1906-7. By 1910 Protestants had outstripped Catholics 167,000 to 73,000. Methodists and Presbyterians cooperated in a comity agreement, but growth was greatest in Presbyterian areas which followed the "Nevius Plan," a strategy stressing Bible classes for all Christians, lay witness, self-support, and self-government. The Korean Presbyterian Church was organized as an independent body in 1907; the Korean Methodist Church in 1930. Other major denominations are the Anglicans (1890), Seventh-Day Adventists (1903), the Holiness Church of the Oriental Missionary Society (1907), and the Salvation Army (1907). Southern Baptists revived an earlier, independent work after World War II.

Japanese annexation in 1910 brought harassment to the church, culminating in open persecution when Christians in the 1930s refused government demands to participate in Shinto ceremonies. But at the same time Christian identification with the Korean independence movement won nationwide respect, and a spreading network of Christian hospitals and colleges (Ewha, Soongsil, Yonsei, and later Keimyung, Taejon, and Seoul Women's) broadened the Christian witness. The end of World War II ushered in a second period of church growth which not even the disastrous church schisms of the 1950s, the division of the country, or the Communist invasion could block. Despite the loss of all North Korea to organized Christianity, the number of Protestant adherents has almost doubled in every decade since 1940, and since 1960 Catholic growth has been even more rapid. The largest groups are Presbyterians (1,438,000), Catholics (839,000), Methodists (300,000), and Holiness (217,000).

Officially Christianity, with just under four million adherents including marginal sects, is smaller than Buddhism (5.5 million) and Confucianism (4.42 million), but more realistic surveys suggest that the claims of the older, weakening religions are exaggerated and that Christianity, with 13 per cent of the population, is now the largest organized religion in Korea.

BIBLIOGRAPHY: A.D. Clark, *History of the Korean Church* (1961); S.H. Moffett, *The Christians of Korea* (1962); J.C. Kim and J.J. Chung, *Catholic Korea, Yesterday and Now* (1964); R. Shearer, *Wildfire: Church Growth in Korea* (1966); L.G. Paik, *The History of Protestant Missions in Korea 1832-1910* (2nd ed., 1971).

SAMUEL HUGH MOFFETT

KORIDETHI CODEX. A manuscript of the gospels which once belonged to the Church of St. Kerykos and St. Jolitta at Koridethi in the Caucasus Mountains near the Caspian Sea. It is written in a rough uncial hand by a scribe who knew little or no Greek. Edited in modern times by G. Beermann and C.R. Gregory, who dated it in the ninth century, it contains a Byzantine-type text in Matthew, Luke, and John, but one akin to the type of text used by Origen and Eusebius at Caesarea in Mark. The MS is identified by the Greek letter *theta* or the number 038 in critical texts of the NT. It is now located in the museum at Tiflis, the capital of the Soviet Socialist Republic of Georgia.

W. WARD GASQUE

KRAEMER, HENDRIK (1888-1965). Educator, ecumenist, and writer on missions. After specializing in oriental languages and cultures in his native Netherlands, the Bible Society of that land sent him to serve the Dutch Reformed Church in Indonesia from 1922 to 1937, as a linguistics and Bible translation consultant. His most famous book, *The Christian Message in a Non-Christian World,* written as a study guide for the third World Missionary Conference at Tambaram* in 1938, emphasized the uniqueness of the biblical message in missions. From 1937 he was professor of religion in the University of Leyden, until in 1948 he became the first director of the Ecumenical Institute of the World Council of Churches at Bossey, near Geneva. He was interned by the Nazis from 1942 to 1943. From 1955 to 1957 he was visiting professor at Union Seminary, New York, and thereafter lived in Holland until his death.

EARLE E. CAIRNS

KRAPF, JOHANN LUDWIG (1810-1881). Pioneer missionary to Kenya. Born at Derendigen, near Tübingen where he received his education, he spent a short time in the Lutheran parish ministry before offering to the Anglican Church Missionary Society in 1838. After a frustrating six years in Abyssinia from which he was finally expelled, he transferred to Mombasa in 1844. Here he laid his wife and newly born child in "a lonely missionary grave." He took a house in Mombasa and studied Swahili, into which he translated the NT, and produced a standard dictionary and grammar. In 1846 he was joined by a fellow German Lutheran, Johannes Rebmann,* and they moved to Rabai, about ten miles inland from Mombasa, to work among the Wanyika. From here he and Rebmann went on several important journeys of exploration inland. In 1850 he was in Europe on furlough and returned to Kenya in 1851, but in 1853 had to return to Europe due to ill-health. He maintained his interest in East Africa* and continued important linguistic work. He was twice back in East Africa, once conducting a pioneer party of British Methodists to their new station of Ribe to the north of Mombasa in 1861, and again as interpreter to the Napier expedition to Abyssinia in 1867. He died at Kornthal near Stuttgart.

BIBLIOGRAPHY: W. Claus, *Dr Ludwig Krapf, weil Missionar in Ostafrika* (1882); J.L. Krapf, *Travels, Researches and Missionary Labours in Eastern Africa,* (1860); E. Stock, *History of the Church Missionary Society,* (1899).

JOHN WILKINSON

KRAUTH, CHARLES PORTERFIELD (1823-1883). Lutheran theologian and editor. Born in Martinsburg, Virginia, he studied at Gettysburg College and Seminary, was ordained into the Lutheran ministry in 1842, and served congregations in Maryland, Virginia, and Pennsylvania as a member of the Evangelical Lutheran General Synod. In 1861 he became editor of the *Lutheran and Missionary;* in 1864 he was elected professor at Mt. Airy Lutheran Seminary in Philadelphia and, after 1868, served also as professor of philosophy at the University of Pennsylvania. He was the leading spirit in the establishment of the General Council (1867), a federation of Lutheran synods opposed to "American Lutheranism," which sought to compromise Lutheranism with Puritanism. He authored "Fundamental Articles of Faith and Church Polity," adopted in the meeting preliminary to the founding of the General Council in Reading, Pennsylvania, in 1866. His "Theses on Altar and Pulpit Fellowship" reiterated the stand he advocated in the Akron and Galesburg Rules, "Lutheran altars for Lutherans only; Lutheran pulpits for Lutherans only." He was editor of *The Lutheran* and *The Lutheran Church Review.* His most important work is *The Conservative Reformation and Its Theology* (1871).

See A. Spaeth, *Dr. Charles Porterfield Krauth* (2 vols., 1909).

CARL S. MEYER

KRÜDENER, BARBARA JULIANA VON (1764-1824). Russian-born Pietist. Unfaithful to her husband, a Russian minister of state, she formed an attachment with a young French officer which was described in her idealized autobiographical novel, *Valérie.* A few months later (1804) during a sojourn at Riga she experienced a sudden conversion, after which she held Pietistic conventicles throughout Württemberg. As a confidante of Czar Alexander I she was regarded by some as the prime mover of the Holy Alliance with Prussia and Austria. Alexander, however, contended that he had conceived the treaty while at Vienna in 1815. Further conventicles sprang up in 1816-18 in N Switzerland and S Germany. She died during a visit to the Crimea. WAYNE DETZLER

KRUMMACHER, FRIEDRICH WILHELM (1796-1868). German Reformed pastor. Son of Friedrich Adolf Krummacher, he was born in Mörs, near Düsseldorf, and studied at the universities of Halle and Jena. He became pastor at Frankfurt in 1819, Ruhrort (1823), Gemarke in the Wuppertal (1825), Elberfeld (1834), and the Trinity Church in Berlin (1847). In 1853 he was appointed court chaplain at Potsdam. A powerful preacher, he strenuously opposed rationalism and was an influential leader of the Evangelical Alliance in Germany. His most important and enduring work is the classic, *Elias der Thisbiter* (1826; ET *Elijah the Tishbite,* 1836), and his other works include *Der Prophet Elisa* (1837; ET *Elisha,* 1838), and *David, der König von Israel* (1867; ET *David, The King of Israel,* 1867).

KEITH J. HARDMAN

KUENEN, ABRAHAM (1828-1891). Netherlands Protestant theologian. Born at Haarlem, he was professor of NT, ethics, and OT interpretation at Leyden from 1855. Although Julius Wellhausen is generally acclaimed as chief exponent of the so-called "literary-historical school," the earliest scientific exposition of its essential themes must be ascribed to Kuenen and K.H. Graf. Kuenen's first book, showing G.H.A. Ewald's* influence, was *Historisch-Kritisch Onderzoek* (1861-65; ET *The Hexateuch,* 1886). With Graf he came to hold that "P" (the Priestly code) was the latest element in the Pentateuch. His later views were promulgated in *De Godsdient van Israel* (1869-70; ET *The Religion of Israel,* 1873-75) and *De profeten en de profetee onder Israel* (1875). J.G.G. NORMAN

KULTURKAMPF (German "struggle for civilization"). Church-state conflict in Prussia and elsewhere in the 1870s. It was so called first by Richard Virchow, an atheistic scientist, in 1873. It was mainly inspired by Otto von Bismarck, who feared that Catholic influence would endanger German unity. Its antecedents included the mixed-marriage dispute in Cologne in the 1830s, Protestant resistance to Catholic demands for increased liberties, hostility by German liberalism, and the 1870 decree of papal infallibility. It began with the abolition of the Catholic bureau in the Prussian Ministry of Education and Public Worship (1871). Bismarck appointed Adalbert Falk as Minister of Public Worship (1872). He expelled the Jesuits, brought education under state control, and passed the famous May Laws.* When Pius IX protested, Bismarck severed diplomatic relations with the Vatican. In 1875 the Roman Catholic Church was deprived of all financial assistance from the state, and religious orders were compelled to leave the country. Catholic resistance remained firm, and several bishops and priests were imprisoned.

A change of policy came later when Emperor William I favored a more moderate approach; the rise of the Socialists as a new political enemy, coupled with the election of a more conciliatory pope (Leo XIII), convinced Bismarck that a concordat with the Vatican was a better solution. Falk was dismissed (1879), diplomatic relations were restored with the Vatican (1882), and the May Laws modified (1886-87). In other German states, in Austria, and in Switzerland, similar but less extreme legislation prevailed for a time, but religious peace was eventually restored.

See J.B. Kissling, *Geschichte des Kulturkampfes im Deutschen Reich* (3 vols., 1911-16); and H.W.L. Freudenthal, "Kulturkampf" in *New Catholic Encyclopedia* (1967), vol. 8.

J.G.G. NORMAN

KÜNG, HANS (1928-). Roman Catholic theologian. Born in Switzerland, he studied at the German College in Rome, the Gregorian University, the *Institut Catholique,* the Sorbonne, and later in universities in Berlin, London, Amsterdam, and Madrid. Ordained in 1955, he served for several years as a parish priest. In 1960 he was appointed professor of fundamental theology in the Roman Catholic faculty of the University of Tübingen. He attended Vatican Council II* and gained fame as a progressive but not radical thinker. His name became known in English-speaking lands after the publication of his *Council and Reunion* (1961). Other titles that have been well received in English include *The Church, Justification, Infallible?* and *On Being a Christian.* His name is now associated with moderate, progressive, ecumenical Roman Catholic theology.

PETER TOON

KUYPER, ABRAHAM (1837-1920). Dutch Calvinist theologian and political leader, a major figure in recent Dutch history. Born in Maassluis, his father a minister in the Reformed Church, Kuyper was brought up in a perhaps naïve version of orthodoxy. As a student at Leyden, he rebelled against it and turned to the prevalent "modern" theology. A brilliant student, he went on in theology, studying under the "modernist" Scholten and others. This proved not to satisfy him emotionally. As a young preacher at Beesd, he was attracted by the deep-seated Calvinistic pietism of the villagers; this along with other influences led him to embrace orthodox Calvinism. At thirty a rising preacher, he moved to Utrecht, and soon to Amsterdam. Attracted by the "anti-revolutionary" political views of the Calvinist theorist Groen Van Prinsterer, whom he finally met in 1869, Kuyper's thoughts turned to politics. The aging Groen's protégé, as an "Anti-Revolutionary Party" made its first appearance, he ran for parliament.

Groen's death (1867) left Kuyper as the Anti-Revolutionary leader. With ferocious activity he moved toward making the orthodox Calvinists a political force. A daily newspaper was started, Kuyper elected to parliament (1874), party chapters organized, and a specific political program drafted. Codified in Kuyper's *Ons Program* (1878), it called for state aid to religious schools, extension of the suffrage, recognition of the rights of labor, reforms in colonial policy, and a revitalization of national life. Its theoretical basis was the idea of the autonomy of the various social spheres, each of which had its own God-given rights. The school issue gave Kuyper the opportunity to organize a massive petition campaign (1878), which provided a mass base for the party, and paved the way for the "Monstrous Coalition" with the Catholics, who also wished state aid for their schools.

By 1880 Kuyper started an orthodox Calvinist "Free University" (free from church and state control), and taught in its seminary. By 1886 he led an exodus of over 100,000 orthodox from the Reformed Church (the *Doleantie*: joining with an earlier separatist group, they formed the *Gereformeerde Kerk,* the second largest Protestant group in the Netherlands). By 1888, after the extension of suffrage to many of the middle class, the Coalition won brief control of the government, to the dismay of the Liberals, who saw Kuyper as a potential Cromwell. By 1892, as left-wing liberal proposals for major suffrage extension split all major parties, the conservative wing of the Anti-Revolutionary left (to form the Christian Historical Party). Kuyper, calling for "Christian democracy," drove on, a whirlwind of activity. Further suffrage extension brought a notable Coalition victory in 1900, and Kuyper was made prime minister (1901). This was in many ways the high point of his career. As prime minister he encountered difficulties (notably the railroad strike of 1902) and was ousted after the heated campaign of 1905. He was now sixty-eight. He lived on for a decade and a half, the "grand old man" of the Anti-Revolutionary Party and still a political force to reckon with. He lived to see the granting of full financial equality to religious schools, and the extension of suffrage to all (1917). Though the Coalition broke up (1925), Kuyper's Anti-Revolutionaries have remained a major party. Kuyper's achievement was to give the long-submerged "common people," the lower-middle-class orthodox Calvinistic group, a religious and political voice. He contributed to the development of the Netherlands' present "plural society" (ideological groupings each having their own political parties, trade unions, etc.). As a theologian he revived a systematic, orthodox Calvinism, marked by an emphasis on "common grace."

See P. Kasteel, *Abraham Kuyper* (1938), and F. Vandenberg, *Abraham Kuyper* (1960).

DIRK JELLEMA

L

LABADISTS. Followers of Jean de la Badie (1610-74), born near Bordeaux, the son of the governor of Guienne. He possessed a strong bent to mysticism, becoming a Roman Catholic priest, then about 1650 embracing Protestantism. He held pastorates in Geneva (1659-66), then in Holland (until 1670). There under his leadership his congregation at Middelburg became a religious community dedicated to simple living, holding children and property in common. Enthusiastic disciples flocked to him, among them Pierre Dulignon (d.1679), Pierre Yvon (d.1707), Theodor Untereyk (d.1693), and more important than any, Anna Maria von Schürman (d.1678), whose book *Eucleria* set forth the above principles and others such as the continuance of prophecy and the continuous Sabbath. The Dutch authorities found them too independent of the religious community in Holland, and in 1670 they moved to Westphalia, then to Bremen in 1672, and later to Altona, where they were dispersed on the death of the leaders. By 1730 the movement was dead, although settlements had been made in Maryland and New York. KEITH J. HARDMAN

LABARUM. The first Christian military standard, designed by Constantine from his celestial vision and dream on the eve of his victory at the Milvian Bridge (313). From 324 it was the official standard of the Roman Empire. Fashioned after legionary standards, it substituted for the old pagan symbols the form of a cross, surmounted by a jeweled wreath containing the monogram of Christ, intersecting *chi* (X) and *rho* (P), on which hung a purple banner inscribed *hoc signo victor eris* or a Greek or Latin variant. As a new focal point for Roman unity the monogram appeared on coins, shields, and later public buildings and churches.

LACHMANN, KARL KONRAD FRIEDRICH WILHELM (1793-1851). German philologist, founder of modern textual criticism. Born at Brunswick, he studied at Leipzig and Göttingen, and joined the Prussian army in 1815. He was professor of philology at Berlin from 1825 until his death. His life was spent in the study of philology, especially of Old and Middle High German, and he was one of the finest classical scholars of his day. He applied to the text of the NT the same critical principles as he applied to the texts of classical authors, and he was the first scholar to produce an edition of the Greek NT in which the Textus Receptus was abandoned in favor of older MSS. He aimed at presenting the text current in the latter part of the fourth century, and he gave the impulse to later scholars such as Tischendorf, Westcott, and Hort. He published his smaller edition of the NT in 1831, and his larger in two volumes in 1842-50.

J.G.G. NORMAN

LACORDAIRE, JEAN-BAPTISTE HENRI (1802-1861). Celebrated French Roman Catholic orator, he was at first a Deist of the Rousseau type. After a legal education at Dijon, he practiced law in Paris until Lammenais's* *Essai sur l'indifférence* convinced him of the credibility of Christianity. Following theological study, he was ordained in 1827, but immediately became a revolutionary. An attempt to open a progressive school in Paris after the Revolution of 1830 failed, and the pope condemned and terminated his periodical *L'Avenir.* Chastened, Lacordaire proclaimed ultramontanism at Notre Dame in a series of fiery sermons which electrified Paris. He succeeded in reviving the Dominican Order, forbidden since the Revolution, but experienced continuous conflicts with Rome over periodicals he launched to air republican principles.

ROBERT P. EVANS

LACTANTIUS (c.240-c.320). Latin rhetorician, Christian apologist, and historian. Taught by Arnobius in North Africa, his accomplishments attracted the emperor Diocletian to appoint him as teacher of Latin oratory in Nicomedia. Having become a Christian, he felt it necessary to resign when persecution started in 303 and consequently knew real poverty. In this period he turned to writing Christian apologetics for the educated pagan and for Christians disturbed by the challenges of the accepted intellectual wisdom. Feeling that technical Christian terminology had obscured the effectiveness of previous apologists, he shunned its use whenever possible. His masterly Ciceronian style has earned him the title "the Christian Cicero." His *Divine Institutes* argues that pagan religion and philosophy are absurdly inadequate. Truth lies in God's revelation, and the ethical change which the teaching of Christ brings points conclusively to its accuracy. Lactantius draws on a wide range of pagan sources at some cost to theological orthodoxy. He later became tutor of Constantine's eldest son, Crispus. He wrote *The Death of the Persecutors* —an account of the recent persecutions arguing passionately, though with good historical documentation, that persecuting emperors suffer and that virtuous and just emperors prosper. This has become a major primary source for the persecutions of the period. C. PETER WILLIAMS

LAGARDE, P.A. DE, see DE LAGARDE

LAGRANGE, MARIE JOSEPH (1855-1938). French Roman Catholic scholar. Born in Bourg-en-Bresse, he studied at the minor seminary, Autun; at Paris where he took his doctorate in law; at Salamanca; and at Vienna. He was ordained at Zamora (1883), lectured in history and philosophy at Salamanca and Toulouse, and founded in Jerusalem the *École Pratique d'Études Bibliques* (1890), and the *Revue Biblique* (1892). Engaged in biblical criticism, he supported Leo XIII's efforts in that direction, was appointed to the biblical commission (1902), and came as close to higher criticism as Catholic orthodoxy would permit. Among his innumerable works are studies on the OT, NT commentaries, *Études sur les religions sémitiques* (1903), *Synopsis evangelica graece* (1926), *Le Judaïsme avant Jesus Christ* (1931), and the popular *Gospel of Jesus Christ.*

C.G. THORNE, JR.

LAINEZ, DIEGO, see LAYNEZ

LAITY (Gr. *laos,* "people"). In the Roman Catholic and Orthodox churches and in High Anglicanism a sharp distinction is drawn between clergy and laity who owe them allegiance and depend on them for guidance and help. Evangelicals find this division unknown in the primitive church and, as they regard the ultimate authoritative ministry as that of the risen Lord and all Christian people as dependent on and sharing in the same, they would include the church's ministry within the *laos,* assigning to it, with varying degrees of emphasis, a functional and representative role. During the last twenty years, theologians within the mainstream churches have been much concerned to elevate the position of the laity and have paid some heed to the evangelicals' contention.

IAN SELLERS

LAKE, KIRSOPP (1872-1946). British biblical scholar. Lake was educated at Lincoln College, Oxford. Following two brief curacies in England, he held professorships at the universities of Leyden (1904-14) and Harvard (1914-37, various subjects). He was controversial in both his academic and theological views. His book *Historical Evidence for the Resurrection of Christ* (1907) was an attempt to cast doubt on the story of the empty tomb. Probably his best book was *The Earlier Epistles of St. Paul* (1913), in which he emphasized the influence of Hellenistic religion on early Christianity. Also of great value was the epochal, multivolumed work which he edited with F.J. Foakes Jackson, *The Beginnings of Christianity: Part I. The Acts of the Apostles* (1920-33), which is still a standard reference tool for all students of the NT. It should be noted that Lake's views regarding the historical value of Acts were much more skeptical than the majority of the other contributors. He also published extensively in the area of Greek paleography and textual criticism.

W. WARD GASQUE

LALEMANT, JEROME (1593-1673). Jesuit missionary to Canada. He entered the Jesuit novitiate in 1610, and after probation held several positions at Jesuit colleges in France. In 1638 he went to Canada as a missionary to the Hurons, and in the same year was named superior of the Huron mission. He began building Sainte-Marie-des-Hurons in 1639, with much of the work being done by the *donnés.* These were laymen who dedicated their lives for religious service without taking religious orders, and it was Lalemant who created the *donnés.* In 1644 he became superior of the Jesuits in New France, and in 1658 was made rector of the Jesuit college at La Flèche, France. He returned with Laval* to Quebec in 1659 and was again until 1665 superior of the Jesuits. Lalemant was known for his spirituality and was held in esteem by both French and Indian Canadians.

ROBERT WILSON

LAMBERT (c.635-c.700). Bishop, monk, and missionary. Some uncertainty surrounds the exact dates of his life, but sometime around 670 he became bishop of Maestricht. Although bishop until his death, he was for political reasons exiled from the monastery of Stavelot which served as the headquarters for his see. During the exile he traveled as a missionary. He is sometimes known under the name "Landebertus the Martyr," since some sources claim that he met a violent and bloody death. The exact causes are unknown; one conjecture involves the rebuke of Pepin of Heristal, mayor of the palace, for adultery. After his death the see was transferred to Liège.

LAMBERT, FRANCIS (1486-1530). Reformer of Hesse. Born at Avignon of noble parents, he entered the Franciscan Order and became a noted preacher. Influenced by Luther's writings from 1520 and by Zwingli from 1522, he left the order and traveled under an alias to Wittenberg to study the Reformation in its original setting. Here he received a state pension and translated Protestant works into French and Italian. After a short and troubled stay in Metz and Strasbourg he was called to Hesse by the Landgrave Philip* in 1526. Though distrusted by the Lutherans as a Frenchman and a supporter of Zwingli's sacramentalism, he took a leading part in the promotion of the Reformation in Hesse, and from 1527 was professor of exegesis at the new University of Marburg. He wrote a number of commentaries on the Prophets, the Writings, and the Apocalypse, and some controversial pamphlets, notably an attack on Erasmus.

IAN SELLERS

LAMBERT OF HERSFELD (c.1025-c.1085). Benedictine abbot. He took Benedictine orders at the abbey of Hersfeld in 1058. Sympathy with the Cluny reforms and opposition to Emperor Henry IV precipitated his move to the abbey of Hasungen, where he became abbot in 1081. His reputation rests on a number of historical writings, more notable for their polished Latin style than for their accuracy. *Annales,* the most significant, chronicles world history from Creation to 1077. Only Lambert's extensive treatment of the period 1069-77 is original work. Until the mid-nineteenth century, when von Ranke tempered this judgment, Lambert was widely regarded as a highly credible medieval historian.

LAMBETH ARTICLES. Nine theological propositions drawn up in 1595 at Lambeth Palace, London, by Archbishop Whitgift and his advisers. Their purpose was to clarify the doctrine of predestination which was mildly stated in Article Seventeen of the Thirty-Nine Articles.* At Cambridge there had been a controversy over predestination, due to the advocacy of a "weak" doctrine of divine election (similar to that of the later Arminians) by Peter Baro and William Barrett. William Whitaker and others defended the doctrine of double predestination based wholly on God's good pleasure. A series of articles were drawn up and these, modified slightly by Whitgift, became the Lambeth Articles. Queen Elizabeth did not like them, so they were not officially authorized; they were, however, incorporated into the 1615 Irish Articles.* PETER TOON

LAMBETH CONFERENCES. The origin of these gatherings was fortuitous. A synod of the Anglican Church in Canada conceived the idea of a general council to deal with the Colenso* Affair and the effects of *Essays and Reviews.* This scheme posed great problems, but the archbishop of Canterbury, C.T. Longley, proposed an informal gathering of bishops which would meet at his personal invitation to discuss Anglican problems, though having no legislative powers. In 1867 the first conference of seventy-six bishops met, and its success ensured the calling of future conferences, which have occurred every ten years, with the majority of Anglican bishops attending. The 1888 conference was important for its endorsement of the Lambeth Quadrilateral*; the 1920 conference for the "Appeal to all Christian People"—a plea for reunion addressed to the heads of all Christian communities; and the 1958 conference for its progressive statements on race relations and family planning. The 1968 conference discussed the Christian ministry, as well as current world and reunion issues. The plan of each conference is the same: a theological issue; domestic issues; church unity; and current social issues. Though the conference lasts a month, large themes are frequently treated superficially.

JOHN A. SIMPSON

LAMBETH QUADRILATERAL, see CHICAGO-LAMBETH ARTICLES

LAMB OF GOD. A description of Jesus by John the Baptist found only in John 1:29, 36. The Greek word *amnos* refers to Jesus also in Acts 8:32 and 1 Peter 1:19, and the word *arnion* is found frequently in Revelation. As John 1:36 seems to allude to the lamb of the Passover, it may be that this is also in mind in the Baptist's words. But as reference is made to removing the sin of the world this is probably not an adequate explanation. The lamb of the sin offering was meant to signify the forgiveness of sin, and this concept may have some bearing on the saying. It is possible also that some contribution to the understanding of it may come from Isaiah 53 and the Suffering Servant who is led like a lamb to the slaughter. Likewise the lamb God was to provide for Abraham (Gen. 22:8) may give some comparison. John's gospel is so rich in imagery that there could be allusions to several of these concepts.

R.E. NIXON

LAMBRUSCHINI, RAFFAELLO (1788-1873). Italian educational, social, and religious reformer. Born in Genova, ordained a Roman Catholic priest and in charge of the Orvieto diocese, he opposed Napoleonic policies and was imprisoned and sent to Corsica. Freed in 1814, he withdrew to Figline near Florence where he possessed an estate, and devoted himself to improve social conditions by means of schools, educational publications such as *La Guida per l'educatore* (first review on education), political journals, and other treatises. He also took an active part in government affairs before and after the annexation of Tuscany to the Kingdom of Italy. A man of high ideals and fine intellect, in close touch with all the leading men of Tuscan and Swiss evangelism, he championed the cause of liberty in all its aspects: economic, political, moral, and religious. He opposed the temporal power of the pope and advocated a reform of the Roman Church from within, based on a deepening of the spiritual life of the individual and a return to the simplicity of the Gospel. DAISY RONCO

LAMENNAIS, FÉLICITÉ ROBERT DE (1782-1854). French Roman Catholic writer who personally epitomized the spiritual conflict between Catholic faith and the new democratic ideal. From budding rationalism in his youth, he converted (1804) to devout Catholic faith, became a priest (1816), and early supported Joseph de Maistre* and Louis de Bonald. He came to believe that the church, as the supreme guardian of the truth, ought to be independent of any state control or alliance. His concern promoted some dimensions of Ultramontanism,* but worked against even a favorable alliance between throne and altar. By 1830 he founded a newspaper *L'Avenir* to promote liberty for the church from the state. Even though he did not then accept the secularist basis for liberalism, his ideas were condemned by Gregory XVI (1832, 1834). *Paroles d'un croyant* (1834) summarized his commitments. Gradually he left the church to advocate the new liberal democratic ideal and to extol the common man, not the church, as the hope for societal regeneration. He joined the republican government (1848-52) which was formed out of the Revolution of 1848. Liberal Catholicism counted him among its founders. C.T. MC INTIRE

LANFRANC (c. 1005-1089). Archbishop of Canterbury from 1070. Born in Pavia, he studied and practiced law before becoming a pupil of Berengar of Tours (1035). An excellent student of logic, he opened a school at Avranches (1039), but gave up his work and entered the Benedictine abbey of Bec in 1042. There he started another school that became famous throughout Europe, numbering among its alumni Anselm of Canterbury and Ivo of Chartres. Lanfranc also became an adviser to the future William I (the Conqueror) of England while the king was duke of Normandy. Against his inclinations and only because of papal orders, Lanfranc was consecrated archbishop of Canterbury. With the support of William he reformed the church by enforcing clerical celibacy, purifying the cathedral chapters, and introducing Norman personnel committed to the Hildebrandine reform program into England. Although he enjoyed the full confidence of William I, the reform movement alienated him from William II (Rufus). Lanfranc's work as a theologian includes glosses on the epistles of Paul and participation in controversies over the nature of Holy Communion. He developed the teaching of transubstantiation in opposition to Berengar of Tours at the Council of Rome and Vercelli (1050), at Tours (1059), and in his *Liber de Corpore et Sanguine Domini* (1059-1066).

See A.J. MacDonald, *Lanfranc: A Study of His Life, Work and Writing* (2nd ed., 1944).

ROBERT G. CLOUSE

LANGE, JOHANN PETER (1802-84). German Protestant theologian and biblical scholar. Born near Elberfeld, Prussia, he attended the University of Bonn (1822-25), where he was brought under the influence of K.I. Nitzsch, one of the chief advocates of the *Vermittlungstheologie.* While serving as a parish minister in Duisburg (1832-41), he published a sharp attack on D.F. Strauss* under the title, *Über den geschichtlichen Charakter der kanonischen Evangelien* (1836). When Strauss was prevented in 1841 from assuming the chair at the University of Zurich to which he had been elected, Lange was called to occupy the chair in his place. At Zurich Lange elaborated further his own alternative to Strauss's *Leben Jesu* in his multivolume *Leben Jesu nach den Evangelien* (1844-47). In 1854 he was appointed to the chair of dogmatic theology at the University of Bonn previously held by I.A. Dorner.

DAVID C. STEINMETZ

LANGTON, STEPHEN (d.1228). Archbishop of Canterbury from 1207. Born in England but educated at Paris, where he won a reputation as an OT commentator, he was created a cardinal by Innocent III in 1206, and in 1207 consecrated archbishop of Canterbury. King John refused to accept him, and until 1213 England was under papal interdict. In that year John yielded to Innocent, and Langton arrived in England. He strongly sympathized with the baronial opposition, and though he seems to have been the first to suggest a Charter, he tried to hold a mediating position. On the Charter itself, his name heads the list of counselors. Late in 1215 he was suspended by papal commissioners for not excommunicating the barons (the pope supported John). Langton went to Rome to plead before Innocent, but though the sentence was revoked, he was not allowed to return to England until 1218. He supported the regency against baronial attacks and papal claims, and in 1225 obtained from Henry III the final edition of Magna Carta. From Honorius III he secured the right for the archbishop of Canterbury to be the pope's *legatus natus,* and at the Synod of Oseney (1222) he issued the decrees of the Fourth Lateran Council and special constitutions for the English Church.

See F.M. Powicke, *Stephen Langton* (1928).

JOHN A. SIMPSON

LAODICEA. A city in SW Phrygia (Asia Minor), near the juncture of the Lycus with the main Maeander valley. Built on a spur (c.850 feet above sea level), it commanded the great coast road that passed from Ephesus 100 miles away on the coast, to the interior of Asia Minor. The city's origin is unknown, but it was refounded by Antiochus II (261-246 B.C.) and named after his wife Laodice. When the Pergamene kingdom was willed to the Roman state in 133 B.C., Laodicea became part of the province of Asia. A textile and banking center, it also had a celebrated medical school. Commentators of John's condemnation of Laodicea (Rev. 3:14-19) have related these features to the geographical background of the city, including allusion to hot spring water brought six miles by aqueduct and cooled to lukewarm temperatures en route. The Christian church there may have been founded by Epaphras (Col. 4:12, 13).

JAMES M. HOUSTON

LAODICEA, CANONS OF. In the second canon of the Quinisextum Council held at Constantinople (692), reflecting an Eastern stance over against Rome, the sources for canon law were specified in a list of synods. Standing between that of Antioch (341) and Constantinople (381) is the otherwise unidentifiable synod at Laodicea in Phrygia from which come sixty canons, about one-third of all those datable before 381, and sufficiently repetitious to suggest a compilation of even older collections. The twelfth-century Gratian* speaks of thirty-two bishops present, and a Theodosius as their chief author. These canons of Laodicea are concerned with relationships of Christians to non-Christians, Jews, and heretics; with conditions and requirements for the clergy; and with worship practices in general, but including in addition preparation for baptism, fasting before Easter, and penance. Regarding bishops, the canons declare against the widespread practice of rural appointments, and against the mob making the choice. There are references also to female presbyters, and the necessity to prohibit clergy from being "magicians, enchanters, mathematicians, or astrologers." The concluding canon, which may be an appendage, provides one of several lists of Scripture known from the fourth century; it omits most notably Revelation, and the Apocrypha of the OT.

See English translation in *A Select Library of the Nicene and Post-Nicene Fathers of the Christian Church* (ed. P. Schaff and H. Wace, rep. 1956).

CLYDE CURRY SMITH

LAODICEANS, EPISTLE TO THE, see APOCRYPHAL NEW TESTAMENT

LAPIDE, CORNELIUS À, see CORNELIUS À LAPIDE

LAPSI. Latin word for the "lapsed" who abandoned the faith in persecution. Some early Christians escaped by bribery, others by flight, but apostates were few until Decius's persecution (250-51), when especially in Africa many offered sacrifice or incense *(sacrificati, thurificati)*, procured false certificates of conformity, or volunteered professions of compliance, apostasy, or paganism *(libellatici*)*. Their reconciliation was complicated by letters of recommendation *(libelli pacis)* issued by prestigious confessors. Cyprian* wrote numerous letters and a treatise on *The Lapsed*, and steered the Council of Carthage* (251) into decreeing that penance be proportioned to the gravity of the offense, a policy followed also at Rome, where it occasioned Novatian's rigorist protest. The African *traditores* were among the lapsed dealt with in councils following the Great Persecution. Lapsed clergy were normally readmitted only as laity or merely titular clergy. D.F. WRIGHT

LARDNER, NATHANIEL (1684-1768). Nonconformist* scholar. Educated at Deal, Hoxton Academy, Utrecht, and Leyden, Lardner was both an Independent minister and a domestic chaplain between 1709 and 1729, and assistant preacher at a Presbyterian chapel from 1729 to 1751. He was a nonsubscriber in the Salter's Hall controversy of 1719. His theological opinions developed from Baxterian Calvinism through Arminianism to a modified Arianism. Between 1727 and 1757 he published his remarkable *Credibility of the Gospel History* in which with disarming candor and immense learning he strove to defend the facts of the NT against Deist critics; in effect by detaching the OT from the New and developing novel ideas concerning the Logos* he was unconsciously adopting the rational methods of the Socinians and was responsible for Joseph Priestley's conversion to that system. Lardner's works were translated into several languages and were three times reprinted. IAN SELLERS

LA SALLE, JEAN BAPTISTE DE (1651-1719). French educational reformer. Born in Reims of a noble family, he became a priest in 1678. His parish work brought to his attention the lack of education among the poor, and he helped to set up charity schools for them. He set about improving the standards, status, and morale of his teachers, forming them into a religious order—Brothers of the Christian Schools (1684). He established what was in effect the first training college for secular teachers in conditions of great hardship and poverty, centering first on Vaugirard, near Paris, and then on St. You, near Rouen. He met much opposition from the vested interests of the Writing Masters and the Little Schools. Reformatories and boarding schools were also founded. At his death, the schools were to be found in twenty-two French towns and they are now spread widely through the world.

As a practical necessity he pioneered the simultaneous method (class teaching) in his primary schools in contrast to the individual methods common in primary schools of the time. He also substituted French for Latin learning to read. In the classroom, great stress was laid on silence, and signs were used instead of words wherever possible. "The Brothers should be careful to punish their pupils but rarely," says his Rule. He was canonized in 1900. His educational works include *Les Règles de la bienséance et de la civilité chrétiennes* (1703); *La Conduite des écoles chrétiennes* (1720); and various school manuals. He produced also a number of spiritual works.

HOWARD SAINSBURY

LAS CASAS, BARTOLOMÉ DE (1474-1566). Spanish missionary to the West Indies, sometimes called "the Apostle of the Indies." Born in Seville, son of a merchant who had gone with Columbus on his second voyage, he received a law degree from the University of Salamanca. In 1502 he went with Governor Ovando to Hispaniola as a legal adviser, and was ordained priest there in 1510. He became concerned about the harsh treatment of the natives under the colonial system, and spent the years 1515-22 traveling between Spain and America, obtaining the power from Charles V to set up various projects and enforcing them. In 1521 the Indians revolted against the church-controlled Indian settlement he had established, and blaming his fellow Spaniards for its failure, he entered the Dominican Order in 1523.

Emerging from a long retreat, Las Casas again campaigned for humane treatment of the Indians, working for their conversion also. He gained acceptance from them, but alienated many of his colonial countrymen by his fanaticism. He played a key role in Indian-Spanish relations in the New World, and while in Spain succeeded in getting the New Laws of 1542-43 promulgated. Although these laws would have improved the lot of the Indians, they were not completely successful, due to opposition from the colonists. As bishop of Chiapa in Mexico from 1544-47, Las Casas could not even enforce the New Laws in his own diocese. He spent his last years in Spain working tirelessly for improved conditions for the Indians. He entered into controversy with the historian Sepulveda (1550) over the question of wars of conquest in the New World. His major works include *De unico vocationis modo; Apologética historia;* and *Historia de las Indias;* along with his best-known work, *Brevísima relacion de la destrucción de las Indias,* which denounced the evils of colonialism. Although Las Casas has been accused of promoting Negro slavery in America, this practice had already long existed and was not considered morally wrong in his time.

See L. Hanke, *The Spanish Struggle for Justice in the Conquest of America* (1949); and J. Friede and B. Keen (eds.), *Bartholomew de Las Casas in History* (1972). ROBERT G. CLOUSE

LASKI (à Lasco), JAN (1499-1560). Nephew of Jan Laski, archbishop of Gniezno and primate of Poland, the younger Jan benefited from his uncle's aid especially in ecclesiastical preferments and his treatment by distinguished people abroad. In 1521 he was ordained priest and appointed dean of Gniezno. With his elder brother Jerome he went on many diplomatic missions for his uncle and for Poland and thereby met (and even bought the library of) Erasmus in Basle. Soon after this he had made contacts with both Zwingli and Oecolampadius. Exactly when his conversion to Protestantism occurred is not known, but it was probably completed by 1538 when, being offered a Polish bishopric, he abruptly departed for Frankfurt-am-Main. Thence he moved to Louvain, where he married. Next he was in Emden in Frisia where in 1543 he became superintendent of the churches in the territory of Countess Anna of Oldenburg.

A gifted organizer, Laski used the type of discipline favored by Oecolampadius. As the major link in this order he established the Coetus, which was composed of all the clergy and which met each Monday in Emden from Easter to Michaelmas. The Interim of 1548 necessitated his departure, and in 1550 he arrived in England. He was granted the use of the church of the Austin Friars for the German, Dutch, Belgian, and French Protestants and as superintendent of the "churches of strangers" was given a free hand in preaching, teaching, worship, and discipline—a unique concession. He published an influential book on church discipline and also a confession of faith and a catechism. His only real friend among the bishops was John Hooper. Following the death of his protector, Edward VI, he had to leave England. He ended his days as superintendent of the Reformed churches of S Poland.

See O. Bartel, *Jan Laski* (1955), and B. Hall, *John à Lasco* (1971). PETER TOON

LASSUS, ORLANDUS (c.1532-1594). Musical composer. Probably born at Mons as Roland de Lattre, he was also known by his Italianized name, Orlando di Lasso. As a boy he was thrice kidnaped for his beautiful voice. Active in Italy and France, he spent most of his later life in Munich. No composer of the sixteenth century was more widely sought after, and his music was printed by almost every European publisher of the day. Over 1,200 of his works are known, the greater part sacred, but he also wrote French *chansons*, Italian madrigals, and German *lieder*. This latter category contains some beautiful pieces based on Lutheran chorales and Genevan Psalms, showing the appeal these had to many Catholics as well as Protestants. He composed over fifty Masses, but it is his motets that show his true greatness. Notable are his settings of the penitential Psalms. The *Magnum opus musicum* is a collection of 516 motets published by his sons after his death. J.B. MAC MILLAN

LAST JUDGMENT, see JUDGMENT; ESCHATOLOGY

LAST SUPPER. There are slight differences in the four accounts of the Last Supper in the NT (Matt. 26:26-29); Mark 14:22-25; Luke 22:14-20; 1 Cor. 11:23-26). Luke and Paul introduce the words "do this in remembrance of me." Matthew and Mark have "this is my blood of the covenant" instead of "this cup is the new covenant in my blood." The synoptic accounts each preserve a pledge by Jesus to abstain from drinking the fruit of the vine till the kingdom has come. Paul, instead, gives the rubric: "whenever you eat . . . and drink . . . you proclaim the Lord's death until he comes." In certain texts of Luke, two cups are mentioned, one before and one after the bread, and scholars differ about what could be the original text. The varied nature of the accounts enriches our understanding of the incident. The large area of consensus no doubt arises from our Lord's careful impression of Himself, His words and actions on His immediate followers at this time.

The synoptic accounts indicate that the supper was a Passover meal eaten on the Passover night, but according to John, Jesus was slain on the cross when the Passover lambs were slain in the Temple. There may therefore have been two different current calendars for calculating the feast date, each followed by a rival group. Some suggest that Jesus deliberately ate a Passover meal earlier than on the official date. Others suggest that the meal was a farewell festive meal of a type common among friends, or a Jewish Kiddush—a simple meal of preparation either for a Sabbath or a festival.

See also COMMUNION, HOLY.

RONALD S. WALLACE

LAST THINGS, see ESCHATOLOGY

LATERAN COUNCILS. Ecumenical church councils held in the Church of St. John Lateran, one of the major Roman basilicas. There were five such councils, the *first* of which was summoned by Callistus II in 1123 to signal the end of the Investiture Controversy,* at which time the Concordat of Worms was confirmed. The council promulgated a number of canons, chiefly restatements of previous decrees dealing with ecclesiastical ordinations and offices, in keeping with the Gregorian Reform, and with crusading indulgences and the Peace and Truce of God.

At the *Second Lateran Council*, summoned by Innocent II following the death in 1138 of his rival Anacletus II who had challenged his rule from the double election of 1130, Innocent announced the deposition of all supporters of Anacletus, the excommunication of Roger II of Sicily, and the condemnation of the adherents of Pierre de Bruys and Arnold of Brescia. The canons followed the reforming lines of the First Lateran Council, prohibiting payment for such priestly services as extreme unction and burial, the study of civil law or medicine by religious, marriage of the clergy, usury, tournaments, use of the crossbow, and incendiarism, among others. They provided that monks and canons were to be consulted in episcopal elections, and confirmed the Peace and Truce of God.

The *Third Lateran Council*, like the Second, marked the end of a schism. It met in 1179 at the behest of Alexander III, following the discomfiture of his rivals, whose support from Frederick Barbarossa had ended with the agreement at Venice in 1177. Alexander III* was the first great canonist pope of the period of the revived study of law, and issued at the council a series of important decretals, the first of which stipulated a two-thirds majority of the cardinals for a papal election, and another required majority decisions in ecclesiastical communities. Other reforming decrees set up cathedral schools with free instruction, and attacked simony, pluralism, and clerical vices. There were sanctions against usurers, Cathari, Jews, and Saracens, and against those aiding the latter or pirates. The Truce and Peace of God were reconfirmed.

The *Fourth Lateran Council,* summoned in 1215 by Innocent III,* marks the pinnacle of the achievements of the most powerful of medieval popes. It confirmed the election of Frederick II; denounced Magna Carta as an affront to the pope and his vassal, King John of England; enunciated the dogma of transubstantiation; made yearly confession and Communion mandatory; confirmed the new Franciscan Order; and required distinctive dress for Jews and Saracens. Condemnations were directed against the Cathari and the Waldensians, though they were not named, and against the teachings of Joachim of Fiore* and Amalric* of Bena. Reforming canons stipulated, among others, that no new orders were to be founded, and that general chapters were to be held in existing orders. The abuses surrounding indulgences were to be curbed. Clerks were enjoined against participating in judicial ordeals. Provisions were made for the forthcoming (Fifth) crusade. The council rejected a proposal that regular payments be collected from the entire church to support the papal administration.

Julius II* responded to the antipapal Council of Pisa (1511-12) by summoning the *Fifth Lateran Council* in 1512. No canons were issued, only pontifical constitutions. The chief concern of Julius was to achieve the condemnation of conciliar theory in general, and of the decrees of the councils of Constance* and Basle* and the recent Council of Pisa,* plus the Pragmatic Sanction of Bourges,* in particular. Both Maximilian and Louis XII were persuaded to disavow their previous support of the Council of Pisa. There were futile gestures toward reform of abuses surrounding commendations, pluralism, and clerical immunities, and a recognition of the need for church reform generally. A new crusade was projected, to be supported by a three-year tax on all benefices. The failure of the Fifth Lateran Council to deal decisively with the issues confronting it led directly to Luther's reform.

BIBLIOGRAPHY: G.D. Mansi, *Sacrorum Conciliorum nova et amplissima collectio* (31 vols., 1759-98); K.J. von Hefele, *Histoire des conciles d'après les documents originaux* (ed. H. Leclercq, 1907ff.); G. Tangl, *Die Tielnehmer an den allgemeinen Konzilien des Mittelalters* (1922); P. Hughes, *The Church in Crisis* (1961); R. Foreville, *Histoire des counciles oecuméniques,* vol. VI (1966); H.J. Margull (ed.), *The Councils of the Church: history and analysis* (1966).

MARY E. ROGERS

LATERAN TREATY. Concluded between the Vatican and the Italian Kingdom in 1929, this restored the relations ruptured at the seizure of Rome in 1870. It acknowledged the Holy See's independence and sovereign jurisdiction over the Vatican City, and proprietary rights over several churches and buildings elsewhere. Italy affirmed that Catholicism was the sole religion of the state, while the papacy formally recognized the Italian Kingdom and Rome as its capital. Italy agreed to compensate the Apostolic See for the loss of the Papal States. The attached concordat guaranteed to the Church the free exercise of its spiritual powers, but bishops were to take an oath of loyalty to the state. Mussolini thereby enhanced his reputation by neutralizing the papacy.

RICHARD V. PIERARD

LATIMER, HUGH (1485-1555). English Reformer and martyr. Born in Leicestershire and educated at Cambridge, he was at first a staunch defender of the unreformed faith, but was convinced by Bilney* of his error and thereafter was foremost as a reformer. Appointed bishop of Worcester in 1535, he was twice imprisoned for his beliefs during the reaction of Henry VIII's later years; and ultimately with Cranmer* and Ridley* he was to become one of the most celebrated victims of the Marian persecution, being burnt at Oxford in October 1555. The earliest controversies of the English Reformation were concerned with a strange mixture of papal pretension, clerical corruption, and doctrinal error. Much, for instance, was made of pilgrimages, purgatory, and the view of the Virgin Mary. It is to Latimer's credit that as early as 1533, in a letter to Morice about the accusations leveled against him by Powell, he recognized the central and necessary doctrine of justification, quoting Romans 5:1, adding, "If I see the blood of Christ with the eye of my soul, that is true faith that his blood was shed for me." After that he could say, with Luther, right to the end of his life: "Here I stand; I can do no other."

This being the case, it was essential that men should read and understand for themselves. Hence therefore the emphasis he placed on the need for acquaintance with the English Bible. Hence also his part in the composition of the First Book of Homilies, the twelfth of which—"A Faithful Exhortation to the Reading of Holy Scripture"—is possibly from his pen.

It is, however, as a preacher that Latimer still lives—and that says much for the force and vividness of his style. During the encouraging years of Edward VI's reign Latimer preached his series at St. Paul's Cross "On the Plough" (1547) and before the court in Lent 1548-50. In the first three sermons of the first series he concentrated on the doctrine to be taught; in the last, which has come down to us in detail, he sought to define "what men should be the teachers and preachers of it." Latimer was unsparing of clerical and especially episcopal shortcomings ("Since lording and loitering hath come up, preaching hath come down . . . For they that be lords will ill go to plough . . . They hawk, they hunt, they card, they dice; they pastime in their prelacies with gallant gentlemen, with their dancing minions and with their fresh companions"). And he could go further than this with his racy references to "pampering of their paunches . . . munching in their mangers, and moiling in their gay manors and mansions."

Latimer saw the preacher's office as to teach truth and "to reprehend, to convince, to confute gainsayers, and spurners against the truth," but he always asserted that the way a man lives will be the clue to what he believes. He therefore emphasized conduct, and never more vigorously than in the seven discourses in Lent 1549 with his exposures of public misdemeanors—bribery, exploitation, and the suborning of justice among them. Latimer was not only forthright; he was also gifted with a fine colloquial turn of phrase, a fund of arresting anecdotes and a capacity for vivid narrative. He was a popular preacher in the best sense of the phrase.

See *Sermons* (ed. H.C. Beeching, 1906); and H.S. Darby, *Hugh Latimer* (1953).

ARTHUR POLLARD

LATIN, ECCLESIASTICAL. Latin succeeded Greek as the official language of the Western Church during the third century. Ecclesiastical Latin originated in the popular speech and was popularized and formalized by the Vulgate of Jerome. It served as the *lingua franca* until late in the Middle Ages. By the time of the Renaissance, some humanists tried to purify the ecclesiastical Latin by returning it to its classical structure and phonetics. Comparatively, ecclesiastical Latin stands to classical Latin as Koine Greek relates to classical Greek. The phonetics and pronunciation of ecclesiastical Latin is known as the Italian pronunciation, and for liturgical purposes none but this pronunciation is permissible.

DONALD M.LAKE

LATIN AMERICA. This is the commonly accepted term for the twenty-one republics located to the south of the United States together with assorted territories and islands of the Caribbean Sea. The original inhabitants of Latin America were American Indians, many of whom later mixed with colonizers from Spain and Portugal to form the large ethnic group called *mestizos.* The population of Latin America is approximately 300 million with a high annual growth rate of 2.8 percent. Major languages spoken are Spanish, Portuguese, Quechua, and French. The Latin American nations gained their independence from the Iberian conquerors in the early nineteenth century. Economic progress has been slow, with agriculture predominating and a wide gap between rich and poor in many areas. The two major rival political options are so-called developmentalism and Marxist-type socialism.

The Roman Catholic religion entered Latin America with the Iberian conquest in the sixteenth century. Spain at that time had just won back her territory from the Moors after almost 800 years of struggling, and the idea of imposing religion upon a people through military conquest was still strong in Hispanic mentality. In the New World, Catholic missionaries found three major civilizations, the Aztecs of Mexico, the Mayas of Central America, and the Incas of South America, each with a form of animistic religion. The success of the Franciscans, Dominicans, Augustinians, Jesuits, and other missionaries in bringing the Indian peoples to true Christianity was spotty. In many cases the religious institutions of Spain were merely imposed alongside the political institutions, and a nominal Christianity resulted which in many cases could be described as "Christo-paganism," or "folk Catholicism." Outstanding among early missionaries were Bartolomé de Las Casas* and José de Acosta. Separation of church and state was unknown. The wars of independence in the early 1800s caused severe difficulties for the church until the Vatican was finally able to come to terms with the new independent governments.

Abortive attempts to plant Protestantism in Latin America were made by the French Huguenots in 1555 and the Dutch Reformed in 1624, both in Brazil and both effectively crushed by the Catholic Portuguese. The German Moravians settled permanently in British Guiana in 1735 and carried on successful evangelistic work among the Arawak Indians in Dutch Guiana. Argentina, Uruguay, Brazil, and Chile received European immigrants in the nineteenth century, among whom were colonies of Lutherans, Scotch Presbyterians, Anglicans, and Italian Waldensians. Their religious influence was typically confined to their own ethnic communities. Most courageous of the pioneers of Protestantism in this period were the colporteurs of the British and Foreign Bible Society and the American Bible Society, such as James (Diego) Thompson (1788-1854), his convert, Francisco Penzotti (1851-1925), and many more.

A milestone for Protestant missions was the founding of the Patagonian Missionary Society (later South American Missionary Society) by an Anglican sea captain, Allen Gardiner, in 1844. This was the first missionary of the aggressively evangelistic type. Under SAMS, Barbrooke Grubb (1865-1930) successfully planted churches among the Aracua Indians in S Chile. Presbyterian churches were planted by missionaries David Trumbull (1819-89) in Chile in 1868 and H.B. Pratt in Colombia in 1865. William Taylor (1821-1902), stressing self-supporting missions, was an outstanding pioneer of Methodist work particularly in Chile, Peru, and Central America. In 1882 J.H.L. Ewen of Great Britain, traveling through Argentina in a horse-drawn "Bible coach," planted Plymouth Brethren assemblies there. Interdenominational work began when the American Board of Commissioners for Foreign Missions sent workers to Mexico between 1860 and 1880, and when C.I. Scofield founded the Central American Mission in 1890, pioneering the work in almost every one of the Central American republics. Lack of religious liberty and harsh persecution from Roman Catholics severely hindered Protestant missions during this period. By the year 1900, only approximately 50,000 Protestants could be located in Latin America.

The turning point in the Protestant advance in Latin America came with the convening of the Conference on Christian work in Latin America in Panama in 1916. The World Missionary Conference which met at Edinburgh in 1910 did not regard Latin America as a mission field, but this was corrected at Panama. Following the Panama Conference, the Committee on Cooperation in Latin America was organized under the leadership of Robert E. Speer (1867-1947) and Samuel Guy Inman. The CCLA convened subsequent conferences in Montevideo in 1925 and Havana in 1929. Latin Americans then assumed the initiative and convened the First Latin American Evangelical Congress in Buenos Aires in 1949, the second in Lima in 1961, and the third in Buenos Aires in 1969. In 1965 the CCLA phased out and became the Latin America Department of the Division of Overseas Ministries of the National Council of Churches, USA.

The end of World War II brought a great wave of conservative missionaries to Latin America, both from interdenominational missions and from the newer denominations. Easing of Catholic persecution of Protestants during the 1960s, combined with aggressive evangelism, has expanded the Protestant movement from the 50,000 of 1900 to over twenty million today, with some projections anticipating one hundred million by A.D. 2000. The Second Vatican Council (1962-65) convened by Pope John XXIII produced a radical change in the Latin American Roman Catholic Church. An ecumenical spirit now prevails, Bible reading is encouraged, a social conscience has been awakened, the Mass is said in the language of the people, and new emphasis is being placed upon the laity. Modern innovations, however, have produced divisions, three of which have become quite sharply defined. A large number of Catholic leaders remain conservative, attempting to preserve the traditions of the past. Among the progressives, some have chosen to emphasize the social implications of Christianity, casting their lot with a form of radical Marxism, while other progressives have taken a more spiritual line, stressing a return to the Bible and biblical Christianity.

Ecumenical overtures on the part of the Catholic Church and the World Council of Churches have not found wide acceptance among Latin American Protestants, and Latin America remains for ecumenists the most difficult continent. Evangelicals (as Latin American Protestants call themselves) at the grass roots level are still largely converts from nominal Catholicism, and they tend to identify the Roman Church with spiritual emptiness and even idolatry. A decision taken by the United Bible Societies at their first Regional Conference of the Americas in Oaxtepec, Mexico, in 1968, to produce a "common Bible" approved by both Protestants and Catholics met with such widespread opposition that the plan was later dropped. The WCC has encouraged the development of the Committee for Latin American Evangelical Unity (UNELAM) as its chief Latin American arm.

Social conditions in Latin America have strongly influenced the development of Christian theology. Over four centuries of a semifeudal socioeconomic system, in which the church played an important supportive role, have, particularly within the past twenty years, been repudiated by a growing body of Latin Americans. The unjust distribution of wealth and the submission of the economy of many republics to foreign economic interests seem to many to be the central issue of life in Latin America. A group of theologians, both Catholic and Protestant, has rallied to the cause and is in the process of developing a "theology of liberation," stressing the horizontal man-to-man or man-to-society dimensions of Christianity. Vatican II, papal encyclicals such as Paul VI's *Populorum Progressio*, the establishment of research centers in Chile (Centro Belarmino) and Mexico (Center of Intercultural Formation), the martyrdom of Marxist priest Camilo Torres, and the impact of outspoken advocates of liberation such as Archbishop Helder Camara of Brazil and theologian Gustavo Gutierrez of Peru are all symbolic of the direction taken by a significant number of Catholic thinkers.

The Second Assembly of the Latin American Roman Catholic Episcopate (CELAM), held in Medellín, Colombia, in 1968 perhaps marked the beginning of Roman Catholic theology of liberation. On the Protestant side, a vocal organization called Church and Society in Latin America (ISAL) was established in 1962 after a consultation in Huampaní, Peru, in 1961. It has become the rallying point for Protestant theologians of liberation, and its journal, *Cristianismo y Sociedad*, their principal mouthpiece. The two streams converged in 1972 in the first Latin American Congress of Christians for Socialism, held in Santiago, Chile. The close similarity between Catholic and Protestant approaches to the theology of liberation became apparent in Santiago.

Another significant group of Protestant theologians has been critical of the theology of liberation, at least in its more radical expressions. Disturbed by the questionable hermeneutics of the radical theologians and their failure to maintain the biblical emphasis on vertical reconciliation (man-God) and on the centrality of conversion and the pious life, these theologians met in Cochabamba, Bolivia, in 1970 to form the Latin American Theological Fraternity. This has given visibility to thinkers who are as aware of the social problem of Latin America as the ISAL group, but who have set to work to develop what they feel is a more biblical and balanced approach to the theology of liberation. The Fraternity is more representative of grassroots Latin America Protestantism than the ISAL theologians, although ISAL may be shedding some of its former élitism.

Until recently, theological education in Latin America has followed traditional North American and European patterns. Missionaries from the historic denominations tended to establish theological seminaries on the post-secondary school level, whereas missionaries from the interdenominational groups and newer denominations usually set up Bible institutes on lower academic levels. Outstanding among the seminaries taking more liberal positions are the *Instituto Superior Evangélico de Estudios Teológicos* (formerly Union Seminary) of Buenos Aires and the Evangelical Seminary of Rìo Piedras, Puerto Rico. Equally high-level training with a more evangelical and evangelistic emphasis is given at the Latin American Biblical Seminary and evangelistic emphasis is given at the Latin American Biblical Seminary of San José, Costa Rica. The total number of institutions for training the Protestant ministry is estimated at 360.

Roman Catholic theological training has likewise followed traditional lines through a program of major and minor seminaries, although the present trend is to reduce the number of minor seminaries as such. The dual problems of reduction in the number of vocations on the part of Latin Americans themselves and the subsequent disproportionate number of foreign priests and nuns ministering in Latin America continue to plague the church. Between 1955 and 1969 the population of Latin America increased from 202,000,000 to 270,000,000 while the number of students in major seminaries increased only from 6,385 to 7,013.

An innovative type of ministerial training called theological education by extension originated in the Presbyterian Seminary of Guatemala in 1962. Faced with the inability of traditional approaches to theological education to keep up with the extremely rapid multiplication of Protestant congregations in Latin America, it was decided to restructure seminaries in order better to meet the needs of church leaders. Through the use of programmed instructional materials, seminary training is taken out to the students (typically mature people) rather than requiring residence, and it is adjusted to several different academic levels. After a consultation in Armenia, Colombia, in 1967, the method began to spread and by 1972 over 10,000 were taking theological training by extension in Latin America. Institutions in Africa and Asia were also adapting the Latin American system to their own conditions.

Evangelism in Latin America was given new impetus in 1960 when Kenneth Strachan of the Latin America Mission initiated the nationwide program of Evangelism-in-Depth* in Nicaragua. This year-long program of intensive evangelism has since been repeated in over half of Latin American republics. It became the prototype of what is now called "saturation evangelism," which is being applied in principle in a number of countries in Asia and Africa as well as in the United States. Evangelists under the Billy Graham Association and Overseas Crusades (SEPAL) have also enjoyed wide international and interdenominational ministries. The First Latin American Congress on Evangelization was held in Bogotá, Colombia, in 1969, drawing together 1,000 delegates from the entire continent.

The various branches of the Pentecostal church have been outstanding in their ability to bring large numbers of Latin Americans to a commitment to Christ. Approximately two-thirds of Latin American Protestants are Pentecostals. Particularly large Pentecostal movements have sprung up in Chile and Brazil, characterized by their indigenous qualities. Church buildings seating 16,000 and 25,000 respectively have recently been constructed. Studies showed that this growth is aided by a culturally relevant liturgy, concentration on the receptive working classes, the apprenticeship system of leadership training, the mother-daughter church planting scheme, and the encouragement of lay ministry.

A charismatic movement, not directly related to denominational Pentecostalism, has been growing in such countries as Brazil, Argentina, and Costa Rica among the more traditional churches over the past decades, and is currently gaining momentum. It has penetrated Catholic as well as Protestant churches, bringing fresh winds of renewal. Although the movement is unstructured, it has adopted a name, *el movimiento de renovación,* and some eighty of its advocates from Argentina, Brazil, Paraguay, Ecuador, Colombia, and Costa Rica held a consultation in Buenos Aires in 1972. The close fellowship of Catholics and Protestants in this kind of meeting is extraordinary in the Latin American historical context.

Although the Protestant educational and medical ministries which were important earlier in the century are being phased out as governments become better able to care for these needs, other specialized ministries continue to be significant. Radio station HCJB in Quito, Ecuador, pioneered missionary radio in 1931 and missionary telecasting in 1961. Now almost without exception, every Latin American republic enjoys an evangelical radio broadcasting station. These stations, as well as other related ministries, are coordinated by Difusiones Interamericanas (DIA) of San José, Costa Rica. Several substantial Protestant publishing houses such as Editorial Caribe, Editorial Vida, Editorial Moody, Casa Bautista de Publicaciones, Editorial Aurora, Editorial Libertador, and many others keep a steady supply of Christian literature flowing in Spanish and Portuguese. Major Protestant periodicals include *Pensamiento Cristiano* and *Certeza,* both from Argentina, and *La Estrella de la Mañana* from Venezuela. Both Campus Crusade for Christ and the International Fellowship of Evangelical Students have active staff workers throughout Latin America for ministry on the university campuses. Wycliffe Bible Translators are working in some 200 Latin American tribes, reducing their languages to writing, and translating the NT.

BIBLIOGRAPHY: J. Mackay, *The Other Spanish Christ* (1932); W.S. Rycroft, *Religion and Faith in Latin America* (1958); S.U. Barbiere, *The Land of Eldorado* (1961); W. Scopes (ed.), *The Christian Ministry in Latin America and the Caribbean* (1962); W.M. Nelson, *A History of Protestantism in Costa Rica* (1963); R. Wood, *Missionary Crisis and Challenge in Latin America* (1964); J. Bishop, *Latin America and Revolution* (1965); F. Houtart and E. Pin, *The Church and the Latin American Revolution* (1965); W.R. Read, *New Patterns of Church Growth in Brazil* (1965); J.L. Mecham, *Church and State in Latin America* (1966); J.B.A. Kessler, *A Study of the Older Protestant Missions and Churches in Peru and Chile* (1967); S. Shapiro (ed.), *Integration of Man and Society in Latin America* (1967); E. Willems, *Followers of the New Faith* (1967); C. Bennett, *Tinder in Tabasco* (1968); W.R. Read, V. Monterosso, and H. Johnson, *Latin American Church Growth* (1969); R.D. Winter (ed.), *Theological Education by Extension* (1969); M. Bradshaw, *Church Growth Through Evangelism in Depth* (1969); C.L. d'Epinay, *Haven of the Masses* (1969); J. Lara-Braud (ed.), *Social Justice and the Latin Churches* (1969); C.P. Wagner, *Latin American Theology* (1970) and *The Protestant Movement in*

Bolivia (1970); T.E. Quigley (ed.), *Freedom and Unfreedom in the Americas* (1971); I. Illich, *Deschooling Society* (1971); A.W. Enns, *Man, Milieu and Mission in Argentina* (1971); R.R. Covell and C.P. Wagner, *An Extension Seminary Primer* (1971); L. Colonnese (ed.), *Conscientization for Liberation* (1971); Q. Nordyke, *Animistic Aymaras and Church Growth* (1972).

C. PETER WAGNER

LATITUDINARIANISM. A term applied both to those seventeenth-century Anglican divines who appealed to reason as a "source" of religious authority besides Scripture and church practice (e.g., Chillingworth,* Stillingfleet*), and to later Broad Churchmen such as the contributors to *Essays and Reviews* (1860), and men such as Whateley, S.T. Coleridge, and Kingsley.* The term signifies more a theological temper and method than a set of doctrines. Latitudinarians have often been distinguishable more by what they oppose—e.g., Puritanism, Deism, Tractarianism—than by any peculiar tenets of their own.

LATOURETTE, KENNETH SCOTT (1884-1968). Church historian. Born and raised in a devout Baptist family in Oregon, he studied at Yale (1904-9), traveled for the Student Volunteer Movement,* then taught in China until invalided home in 1912. After recovering he began his life's work of teaching and writing church history, especially the history of missions, and Far Eastern history. He returned to Yale in 1921, continuing to live on campus in his bachelor quarters after his retirement from full-time teaching in 1953. He served as president of the American Baptist Convention, the American Historial Association, and the Association for Asian Studies. He served on numerous editorial boards and was actively associated both with the ecumenical movement and with various evangelical organizations. He perennially hosted informal student discussion groups. His works include *The Development of China* (1917); *The Development of Japan* (1918); *History of Christian Missions in China* (1929); *The Chinese: Their History and Culture* (1934); *History of the Expansion of Christianity* (7 vols., 1937-45); *A History of Christianity* (1953); *Christianity in a Revolutionary Age* (5 vols., 1958-62); and *Beyond the Ranges: An Autobiography* (1967). He was honored with a *festschrift* edited by W.C. Harr, *Frontiers of the Christian World Mission* (1962).

DONALD TINDER

LATROCINIUM, see EPHESUS, ROBBER SYNOD OF

LATTER-DAY SAINTS, see MORMONS

LAUBACH, FRANK CHARLES (1884-1970). American Congregational missionary and linguist. Born in Benton, Pennsylvania, he was educated at Princeton, Columbia, and Union Seminary, and following ordination in 1914 he did literacy work in the Philippines. In 1929 he began his famous educational project of teaching reading by phonetic symbols and pictures, eventually developing literacy primers for some 300 languages and dialects in over 100 countries in Asia, Africa, and Latin America. Laubach came in contact with several world leaders, including Gandhi, who became an advocate of his literacy work in India. The Laubach Method, now world famous, is essentially "each one teach one" in which each new literate teaches another the language. Out of Laubach's efforts grew the Committee on World Literacy and Christian Literature of the Foreign Missions Conference of North America. As an author he wrote widely, often moving beyond linguistics. Among his main works are *India Shall Be Literate* (1940); *Teaching the World to Read* (1947); *Prayer, the Mightiest Force in the World* (1946); and *Making Everybody's World Safe* (1947).

ROBERT C. NEWMAN

LAUD, WILLIAM (1573-1645). Archbishop of Canterbury from 1633. Educated at St. John's College, Oxford, he reacted, under the influence of the president, John Buckeridge, against the dominant Calvinism and became convinced of the importance of the episcopal organization of the Anglican Church, and of the observance of external order. Himself elected president in 1611, he tried to reintroduce pre-Reformation liturgical practices. In 1616 he became dean of Gloucester, where he moved the Communion table from the nave to the east end of the choir. Made bishop of St. Davids in 1621, he engaged in a conference with "Fisher the Jesuit" during which he admitted that the Church of Rome was a true church "because it received the Scriptures as a rule of faith . . . and both the sacraments." His views commended themselves to Charles I, who translated him to Bath and Wells in 1626, to London in 1628, and to Canterbury in 1633.

He encouraged the reintroduction into churches of stained-glass windows, crosses, even crucifixes, and railed altars, and of practices such as bowing whenever the name of Jesus was mentioned, and making the sign of the cross in baptism. During his provincial visitation of 1634-36 he tried to secure uniformity without regard for conscientiously held objections, and used the Star Chamber to enforce this. Charles I's Declaration of Sports allowed on Sundays, published in 1637, was probably instigated by Laud in opposition to Puritan views of the Sabbath. He supported the king's new Prayer Book for Scotland, which led to the eruption in St. Giles' in 1637. In 1640 he secured the passing by Convocation of canons maintaining the divine right of kings,* but was obliged to suspend the oath binding men never to alter the government of the church. Imprisoned by Parliament in 1641, he was executed for treason in January 1645.

BIBLIOGRAPHY: Collected works (ed. W. Scott and J. Bliss, 1847-60); E.C.E. Bourne, *The Anglicanism of William Laud* (1947); H.R. Trevor-Roper, *Archbishop Laud, 1573-1645* (1962).

JOYCE HORN

LAUDS. Although this is the first service of the day hours in the Daily Office, in practice it is said together with the night office (Matins). Derived from the ancient morning prayer of the church, the service always includes Psalms 148-150, the laudes, or praises sung to God at daybreak. The canticle is the *Benedictus*, or Song of Zechariah (Luke 1:68-79). The Eastern Lauds is the Orthros.

LAURENCE (Lawrence, Laurentius) (d.258). Martyr. Possibly born at Huesca, Aragon, he was one of seven deacons at Rome during the pontificate of Sixtus II. He suffered martyrdom in Valerian's persecution. According to traditions preserved by Ambrose, Prudentius, and others, when required by the Roman praetor to deliver up the church's treasure, Laurence assembled the poor who were his special charge, saying, "These are the treasures of the Church," for which action he was roasted to death on a gridiron. This story has been widely rejected by modern scholars who hold that he was beheaded like Sixtus and other contemporary martyrs. He was buried on the Via Tiburtina, at Campus Veranus, and during Constantine's reign a basilica was built over his tomb, later enlarged to the present San Lorenzo fuori le Mura. His name occurs in the canon of the Roman Mass and the litanies. J.G.G. NORMAN

LAUSANNE CONFERENCE (1927). First conference of the "Faith and Order" movement, held at Lausanne, largely through the initiative of Bishop C.H. Brent and Robert H. Gardiner. Over 400 delegates from about ninety churches participated, notable absentees being the Roman Catholic and Russian Orthodox churches and several Baptist groups. Roman Catholic nonparticipation was explained by Pius XI in the encyclical *Mortalium Animos* (6 January 1928), in which he forbade Roman Catholic involvment in the ecumenical movement which he called "panchristian." The conference was concerned with the doctrinal questions which divided the churches. Subjects discussed were: the call to unity, the message of the church to the world, the essence of the church, episcopacy and apostolic succession, and the sacraments.

The Greek archbishop Germanos declared that union was impossible without the acceptance of the seven Ecumenical Councils, and some delegates came away more conscious of differences than when they went. Even so, the conference did much to stimulate interest in reunion and encourage theological cooperation. The concluding statement, called "The Lausanne Message," was an admirable exposition of the essential Gospel, defined as "the joyful message of redemption, both here and hereafter, the gift of God to sinful man in Christ." It won the delegates' unanimous assent, and part was incorporated in the message of the Jerusalem Conference, 1928.

See H.N. Bate (ed.), *Faith and Order: Proceedings of the World Conference, Lausanne* (1937). J.G.G. NORMAN

LAUSANNE CONGRESS ON WORLD EVANGELIZATION (1974). Convened by an international group of 142 evangelical leaders under the honorary chairmanship of Dr. Billy Graham, this congress aimed: to proclaim the biblical basis of true evangelism; to relate biblical truth to contemporary issues; to share and strengthen unity and love in Christ; to identify those yet unreached with the Gospel; to learn from each other the patterns of evangelism the Holy Spirit is using today; to awaken Christian consciences to the implications of expressing Christ's love in attitude and action; to develop cooperative strategies toward partnership in the work; to pray together that the congress might notably further world evangelization; and to be God's people, available for His purposes in the world.

There were nearly 3,000 official participants from 150 countries. All had previously received, and most had responded in writing to, the major papers; at Lausanne the main work was done in seminars and study groups. Regional and national groups met to consider local implications of the insights gained. The congress produced the widely acclaimed Lausanne Covenant and set up a continuation committee "to further the total biblical mission of the Church," with special reference to the 2.7 billion of the world's people yet unreached.

Lausanne was the culmination of the 1966 World Congress on Evangelism* and a series of national and regional congresses.

See *Let the Earth Hear His Voice* (ed. J.D. Douglas, 1975). GOTTFRIED OSEI-MENSAH

LAVAL, FRANÇOIS XAVIER (1623-1708). First bishop of Quebec. Born in Montigny, France, he was educated by the Jesuits at La Flèche and Paris. Ordained in 1647, he prepared himself for missionary service, and in 1658, just before leaving for New France, was appointed apostolic vicar by the pope. Laval played a significant role in the affairs of New France, for upon his arrival in 1659 he became a member of the Quebec Council and, in 1663, of its replacement the Sovereign Council which governed the colony. As spiritual leader of the colony he sought to maintain high moral standards which, in his refusal to allow the sale of liquor to the Indians, earned him the enmity of the fur traders. Because of his strong will he was often at odds with the governor and other leaders of the colony. He was also responsible for organizing the parochial system of New France and was a key supporter of missions to the Indians. Named the first bishop of Quebec in 1674, he held the post until 1688 when he retired to the Quebec Seminary which he had founded, and which became Laval University in 1852. ROBERT WILSON

LAW, WILLIAM (1686-1761). English devotional writer. Born at King's Cliffe, Northants, he was educated at Emmanuel College, Cambridge, of which he was elected a fellow in 1711, the year of his ordination. He declined to take the oath of allegiance to George I in 1714, was deprived of his fellowship, and became a Nonjuror* for the rest of his life. In 1727 he first became associated with the Gibbon family at Putney, on his appointment as tutor to Edward Gibbon, father of the historian. Here he remained as a valued friend and family adviser until Gibbon's death in 1740 when, with the break up of the household, he returned to King's Cliffe for the rest of his life. He became recognized as a notable controversialist with his *Three Letters to the Bishop of Bangor*

(1717), which refuted Bishop Hoadly's attempt to "dissolve the Church as a society." He ridiculed the bishop's theory that sincerity alone should be the test of religious profession, though it might testify to moral integrity, and instead he built up a constructive apologetic for orthodox Christianity. His *Case of Reason* (1732) was an answer to Tindal's *Christianity as old as Creation,* and in part anticipates Bishop Butler's argument in his *Analogy.*

But Law's most influential work was his *Serious Call to a Devout and Holy Life* (1728), which influenced the lives of many early Evangelicals, including Whitefield, the Wesleys, Henry Venn, Thomas Scott, and Henry Martyn, and others such as Samuel Johnson and Gibbon. In this book Law strongly commends the Christian faith for its moral and ethical teaching, especially for its advocacy of self-denial, humility, and self-control. All life must be lived for the glory of God. But the book has no strong doctrine of the Atonement, and lacks any joy in the good news of the gospel message. A master of logical argument, Law also wrote *The Spirit of Prayer* (1749, 1752), and *The Spirit of Love* (1752, 1754). In association with two ladies, Mrs. Hutcheson and Miss Hester Gibbon, his closing years were spent in founding schools and almshouses and in other practical works of piety. G.C.B. DAVIES

LAWRENCE, BROTHER, see BROTHER LAWRENCE

LAWS, ROBERT (1851-1934). Scottish medical missionary. Born in Aberdeen, he qualified in arts, theology, and medicine by part-time study, and joined the Livingstonia Mission (1875) as medical officer and second-in-command. As leader after 1877, he founded stations at Bandawe (1881) and Livingstonia (1894) and helped to develop extensive work west of Lake Nyasa. He regarded evangelism, education, industrial training, and medical work as complementary aspects of the Christian mission and gave practical expression to this belief at Livingstonia, which he superintended from 1894 to 1927. His work here provided trained leaders for the autonomous African Church which he hoped to create. Laws was elected moderator of the United Free Church of Scotland General Assembly in 1908, and was an unofficial member of the Nyasaland Legislative Assembly from 1912 to 1916. D.G.L. CRAGG

LAXISM. Not a theoretical system of moral theology as such, but an interpretation of Probabilism* which its critics thought to be excessively lenient. Laxism maintained that if the less safe opinion (favoring liberty) were slightly probable, it could be followed with a safe conscience. It came to the fore in the seventeenth century through Juan Sanchez (d.1620), Bauny (d.1649), Leander (d.1663), Diana (d.1663), Tamburini (d.1675), Caramuel (d.1682), and Moya (d.1684), and was bitterly attacked from a Jansenist viewpoint by Blaise Pascal in 1657 in his *Lettres provinciales.* Laxism was condemned in 1665 and 1666 by Alexander VII and by Innocent XI in 1679.

HOWARD SAINSBURY

LAY BROTHER. A member of a religious order employed primarily in manual labor who, unlike a priest, is not obliged to recite the Divine Office daily. He has, however, to attend daily Mass and say a short, regular office (varying from order to order). The custom of unordained monks originated in the eleventh century to fill the gap created in monasteries by the fact that priests were excused many manual chores. A lay brother has a special habit and serves a novitiate. There are also lay sisters.

LAYNEZ, JAMES (Iago or Diego Lainez) (1512-1565). Spanish Jesuit theologian and leader in the Catholic Reformation. Born at Almazán in Castile, he graduated in philosophy at Alcalá in 1532, then after studying theology for a year moved to the University of Paris, joining Ignatius of Loyola.* He was one of the six who took vows at Montmartre in 1534, thus forming the nascent Society of Jesus. The group moved to Italy, where Laynez spent his remaining life as a powerful counter-Protestant preacher and philosophy–theology teacher. He was a leader of the Jesuits, becoming provincial in Italy (1552), vicar general when Ignatius died (1556), and second general (1558). The leading papal theologian at the Council of Trent, he decisively shaped the uncompromising canons on justification, the sacraments, purgatory, and papal absolutism. He died in Rome.

BRIAN G. ARMSTRONG

LAY READER. A nonordained Anglican licensed to read the lessons, conduct Morning and Evening Prayer (except the Absolution), preach at non-Eucharistic services, and, with special episcopal permission, read the Epistle and administer the chalice at Holy Communion. The office is a modern attempt to revive the ancient office of reader, which dates from 1866 and received its earliest wide development in the USA—where the number of clergy was inadequate to the needs of expansion, and where the services of the Episcopal Church were conducted in many areas and for long periods solely by lay readers. In England, Convocation issued regulations on a lay reader's work in 1905, and these regulations have been revised subsequently on a number of occasions. A bishop admits a reader to his office, granting him either a parochial or a diocesan license. Each diocese has a readers' board, which stipulates a minimum training, and coordinates their work.

JOHN A. SIMPSON

LAZARISTS (Vincentians). Popular name of the Congregation of the Priests of the Mission which originated in the successful mission to the ordinary people on the Gondi family estates conducted by Vincent de Paul* and five others in 1625. Approved by papal bull (1632), the society was constituted a congregation with Vincent as superior. They received the priory of St. Lazarus, Paris (formerly a lazar-house), hence the popular name. The society was confirmed by Alexander VII (1655), and rules framed on the Jesuit model were published in Paris in 1668. Their special concerns were the evangelization of the poorer classes, the training of clergy, and foreign missions. Suppressed during the French Revolution, they were restored by Napoleon.

J.G.G. NORMAN

LAZARUS. The name of a beggar in a parable told by Jesus (Luke 16:19-31) and also of the brother of Mary and Martha of Bethany, who was raised from the dead by Jesus (John 11). It has been suggested that the former was a real person, since it was not Jesus' usual practice to name the characters in His parables, but this is questionable. Nothing is known of the latter outside of what is contained in John 11 and 12.

The hypothesis has been suggested by O. Cullmann and F. V. Filson that Lazarus is the author of the fourth gospel, or at least the one on whose testimony the narrative depends, because of the words of John 11:3, 36 (cf. 13:23; 19:26; 20:2; 21:7,20), but this view has not found widespread acceptance. Various valueless traditions and legends connect him with Cyprus, Constantinople, and Marseilles.

W. WARD GASQUE

LEANDER (c.550-c.600). Bishop of Seville. Born in Cartagena, Leander came from a family which eventually included four saints—himself, his sister Florentina, and his brothers Fulgentius and Isidore. The latter succeeded him as bishop. Leander assumed the episcopal see about 577, and as Spain's leading churchman led the country's Visigothic rulers from Arianism to Catholicism. By converting Prince Hermenegild, Leander initially incurred the wrath of King Leovigild and was forced to flee to Constantinople. Here he befriended Gregory the Great, several of whose letters to Leander are extant. Returning to Seville, he presided at the Synod of Toledo in 589. *De triumpho ecclesiae ob conversionem Gothorum*, his closing address to that assembly, along with a rule for nuns, constitute his two preserved writings.

JAMES DE JONG

LE CARON, JOSEPH, see CARON, JOSEPH

LECLERC, JEAN (1657-1736). French Protestant theologian. Born in Geneva, he studied philosophy and theology, was ordained in Geneva, and then went to Saumur where he published his *Liberi de Sancto Amore Epistolae Theologicae* in which he dealt with the doctrine of the Trinity, original sin, and the problem of the two natures in the person of Christ. After meeting John Locke and Philip Lumbach in Amsterdam, he became a Remonstrant* in theology. After a brief return to Geneva he settled in Amsterdam where he became a professor of Hebrew in the Remonstrant Seminary. Between 1684 and 1712 he held the chair of church history at that school. He was a prolific writer and exercised good influence in Arminian circles, particularly through the reviews he edited.

C. GREGG SINGER

LECTIONARY. A book containing portions (pericopes) of Scripture appointed to be read at public worship on particular days of the year, or one listing such lessons. The practice of reading extracts from the Scriptures is found in the synagogue and in the early church. Systems of lessons began to appear from the third century, and appropriate readings for differing churches' ecclesiastical calendars followed. Western Protestant churches followed the emphasis of the Roman Church upon Advent, but Lutheran and Anglican lectionaries now differ widely from present Roman Catholic usage. The Anglican lectionary of 1871 governs the lessons read at Morning and Evening Prayer. There is an alternative revised Table (1922). Further revisions following in 1944, 1946, and 1956 illustrate a general determination to make the lectionary more meaningful for the present day.

HOWARD SAINSBURY

LECTOR, see MINOR ORDERS

LECTURER. A term used of a clergyman of the Church of England who preached a sermon on a specified day in a given parish church or cathedral. The lectureship was much used by Puritans in the sixteenth and seventeenth centuries as a means of propagating Protestant theology and gaining direct access to many people whose parish clergy were not committed preachers. The lecturer was not usually the incumbent of the parish, and his support came from other than normal ecclesiastical funds—e.g., directly from parishioners, a corporation, or a nobleman.

LEE, ANN (c.1736-1784). Founder of the Shakers.* Originally "Shaking Quakers," her movement took form near Manchester, England, in 1758-72. After unfortunate experiences in marriage, childbirth, and the loss of four infants, Ann withdrew from her husband in 1766 and announced her "complete conversion." Assuming leadership of the local Shakers shortly thereafter, she enunciated her cardinal doctrines: confession was the door to the regenerate life, celibacy its rule and cross. Failure to progress and increasing persecution led "Mother Ann, the Word," as she was now regarded, and seven followers to emigrate to Watervliet, New York, in 1774. In America the movement grew rapidly under Ann's energetic leadership. She was largely responsible for the formulation of the characteristic beliefs of the Shakers: celibacy, communism, pacifism, millennialism, élitism, and spiritual manifestations through barking, dancing, and shaking.

ROBERT D. LINDER

LEE, JESSE (1758-1816). "The Apostle of Methodism in New England." Born in Virginia, he early became a Methodist preacher, but was a pacifist during the Revolutionary War when Methodists were regarded as unpatriotic, due to the statements of John Wesley. He was appointed to the first circuit in New England from 1789 to 1798, achieving much success over a wide area. After serving as assistant to Bishop Asbury* (1797-1800), he was appointed as presiding elder of the South District of Virginia in 1801. He served three terms as chaplain in the House of Representatives, and one in the Senate of the United States. During all his other activities he attempted to chronicle the progress of Methodism in America, and published his *Short History of*

Methodism in America, the first such account, in 1810. KEITH J. HARDMAN

LEFÈVRE D'ÉTAPLES, JACQUES, see FABER, JACOBUS

LEGATE, BARTHOLOMEW (1575?-1612). Last heretic burnt in London. Born in Essex, he became a cloth merchant, and when business took him to Zealand he became a preacher among the Seekers.* Expecting a new revelation, he held that meanwhile there was no true church or true baptism, nor any "visible Christian." He rejected the Mennonite tenet of the celestial origin of Christ's body as an "execrable heresy." By 1604, though believing in propitiatory sacrifice, he had concluded that Christ was only a man, but born free from sin. In 1611 with his brother Thomas, Bartholomew was imprisoned, charged with heresy. King James I took a personal interest and tried to convince him of error, but found Legate incorrigible. In February 1612 Legate appeared before a formidable consistory of episcopal, clerical, and legal assessors. Thirteen articles of heresy were cited. Found guilty and handed over to the secular arm for execution, he refused to recant and was publicly burnt. Legate was reportedly of good appearance, articulate, and of excellent character. J.D. DOUGLAS

LEGATE, PAPAL, see NUNCIO AND LEGATE

LEGGE, JAMES (1815-1897). Missionary and Chinese scholar. Born in Scotland, he graduated in arts at Aberdeen (1835) and early revealed scholarly gifts. In 1839 he went to Malacca under the London Missionary Society to become principal of the Anglo-Chinese College, and three years later he founded a theological college in Hong Kong where he also revised part of the NT. He returned to England in 1873 and two years later became the first professor of Chinese at Oxford University, where he enhanced his reputation by monumental translations of classical Chinese literature.

LEIBNITZ, GOTTFRIED WILHELM (1646-1716). German philosopher. Born at Leipzig, he studied jurisprudence, mathematics, and philosophy at Leipzig and Jena universities, entered the service of the elector of Mainz in 1666, and in 1676 at the invitation of the duke of Brunswick became the ducal librarian and historiographer at Hanover. There he spent the rest of his life, working on his massive history of the House of Brunswick. As a philosopher Leibnitz was dissatisfied with Descartes's dualism and the mechanistic views of man and society propounded by Newton and Locke. In his *Monadology* (1714) and other works he asserted the dynamic and spiritual nature of the world which he saw in terms of motion, but whether, as he believed, his idea of God is essential as the first link in his great schematic chain of causation must be a moot point. His outlook is buoyant and hopeful, and leads on to the optimism of the Enlightenment.

As a Protestant theologian Leibnitz's optimism in conspicuous, and his *Théodicée* (1710), written in reply to Bayle's* *Dictionary*, demonstrates the harmony of faith and reason. But his intellectualist view of Christianity as the summation of all religions, and even more his idea of evil as merely the unfortunate consequence of the necessary limitation of all things created mark a serious departure from Lutheran orthodoxy. Repelled by his knowledge of the excesses of the Thirty Years' War, Leibnitz arranged during the years 1686-91 a number of fruitless negotiations between Protestant and Roman Catholic theologians, and strove to promote union between the Protestant churches, helping to establish the *Collegium Irenicum* in Berlin in 1703. A profound if sometimes abstruse metaphysician, he is rightly regarded as the real founder of the German philosophic tradition.

BIBLIOGRAPHY: R.W. Meyer, *Leibnitz and the Seventeenth century Revolution* (1952); N. Rescher, *The Philosophy of Leibnitz* (1966); K. Muller and W. Totok, *Studia Leibnitiana* (1969f.). IAN SELLERS

LEIGHTON, ROBERT (1611-1684). Archbishop of Glasgow. Graduate of Edinburgh University, he was ordained in 1641 when he became minister of Newbattle, which post he relinquished on appointment in 1653 as principal of Edinburgh University. On the reimposition of the episcopal system he was "reordained" and consecrated bishop of Dunblane in 1661. A man of ascetic habits, he gave his entire income, apart from his own frugal expenses, to the poor, but his passive acquiescence in the government's persecution of the Covenanters* has tarnished his reputation—though he described persecution as "scaling heaven with ladders fetched out of hell." His ruling passion was to achieve the unity of the Church in Scotland, to further which aim he reluctantly became archbishop of Glasgow in 1670. Disappointed in his quest, he retired to Sussex and engaged in works of charity. Among his devotional writings is a notable commentary on 1 Peter. J.D. DOUGLAS

LEIPZIG, DISPUTATION OF (1519). A debate arranged by Johann Eck,* pro-chancellor of Ingoldstadt and Luther's former friend but later chief adversary, in an attempt to discredit Luther's and Carlstadt's recently discovered Augustinian theology and force them into dangerous antipapal admissions. The choice of Leipzig was Eck's, as here Wittenbergers were known to be unpopular. The first and third phases of the debate between Eck and Carlstadt were arid affairs, with the Reformer, anxious to plead for the open Bible and the witness of the Fathers, forced by his opponent to retreat into tortuous Scholasticism. The second phase, however, when Luther debated with Eck, came to life as the former was led to affirm that church councils may not only err, but have in fact erred; that the "power of the keys" had been given to the church (i.e., the congregation of the faithful) rather than to the pope; and that belief in the preeminence of the Roman Church was not necessary to salvation. Luther left the disputation depressed by the levity and insincerity of the proceedings and the hostility of the

Leipzigers, while Eck boasted that he had triumphed over the heretic. The debate had in fact cleared the air, furnished Luther's enemies with a case against him, and prepared the way for his condemnation by the Diet of Worms* the following year.
IAN SELLERS

LE JEUNE, CLAUDE (c.1530-1600). French composer. Much of his life was spent close to court circles or in the service of noblemen friendly to the Huguenot* cause. At some time before 1564 he seems to have become a Protestant. He had connections with the family of William of Orange and the Duc de Bouillon, and finally became court composer to Henry of Navarre. Le Jeune was not a church composer, although he did write one Mass and a few motets. Like Goudimel,* he was strongly attracted to the texts and melodies of the Genevan Psalter. Since Calvin forbade the use of part-singing in the Reformed service, Le Jeune's settings must have been conceived for social purposes, even though he described some of them as *"en forme de motets."* He was one of the most talented and versatile French composers of his time. There are over 300 settings by him of Genevan psalms, as well as settings of moralistic Huguenot poems. His simpler, four-part settings are superior to those of Goudimel and were reprinted well into the seventeenth century, also with Dutch and German text. He wrote, in addition, much important secular music.
J.B. MAC MILLAN

LENT (Old English *Lencten,* German *Lenz,* "spring," Lat. *Quadragesima*). The period of forty days' fasting before Easter. One or two days of fasting in preparation for Easter is attested by Irenaeus in the third century, but the earliest reference to a period of forty days (Greek) as the name for Lent occurs in the fifth canon of the Council of Nicea (325). "Forty" no doubt was suggested by the forty days' fast of Jesus, while the fast itself may have been originally part of the preparation of candidates for baptism on Easter night. The length of the fast differed. In Rome in the fifth century, for example, it was three weeks, and in the Eastern churches seven. Not until the seventh century was a period of forty days determined in the West.

Originally the fast was rigorous. One meal a day was allowed, and all flesh and "white meats" forbidden. Gradually the fast was relaxed in the West from the eighth/ninth centuries. By the sixteenth century the evening office of Vespers was advanced to before midday so that the rule of not eating before Vespers could be maintained. A light meal (collation) was also allowed.

In the Roman Catholic Church the Lenten Masses reflect the baptismal associations of the fast by their references to water, raising from the dead, and light. Penitence is another ancient aspect of Lent, deriving from the practice of publicly excluding penitents from Communion at the beginning of Lent and their public reconciliation on Maundy Thursday. The Passion theme dominated the fast. Eastern Church liturgies reflect the same themes. Roman Catholics now usually keep only Ash Wednesday and Good Friday as fast days, but Lent remains a time of penitence.

The Book of Common Prayer prescribes the observance of Lent with fasting. The Tractarians revived the observance in the nineteenth century after a period of comparative disuse, and it is now widespread in the Anglican Church with the emphasis on penitential practice and private devotion at the discretion of the individual. Lent forms part of the Lutheran church year and is observed in some measure in other Protestant churches.
HOWARD SAINSBURY

LEO I (the Great), (St. Leo) (c.400-461). Pope from 440. Born in Tuscany, he was a deacon under Pope Celestine I (422-32) and was active before his election as bishop of Rome, succeeding Sixtus III. One of the greatest administrators of the early church, Leo is known for combining Roman Law with ecclesiastical procedure and for strengthening the primacy of the Roman see in church structure.

As pope, Leo was dedicated to the duty of preaching, and he wrote many sermons for the liturgical cycle. He vigorously enforced uniformity in church government and doctrine, both locally (i.e., in the ten surrounding bishoprics) and universally. When the co-emperor Marcian convened the Council of Chalcedon in 451, Leo sent representatives and his *Tome to Flavian* (patriarch of Constantinople), part of which concerned a doctrine of Christ that was adopted by the council. He maintained peaceful relations with Marcian's successor, Emperor Leo the Thracian, in spite of some friction over the support of Chalcedon. Leo's belief in the use of moderation in the wielding of power is illustrated by his dealings with the African Church and his violent reaction to the discovery that one of his vicars (Anastasius of Thessalonica) had acted hastily in dealing with his assistants. He defended the faith against such heretical groups as the Manichaeans, Monophysites, and Pelagians. He was also instrumental in preventing the destruction of Rome in 452, when he persuaded Attila to withdraw, and in 455, when he persuaded the Vandals to refrain from murdering the populace.

Leo did not write any treatises, but his preserved sermons indicate his beliefs in matters of doctrine. The essence of his teaching lies in his awe for the mystery of Christ and the church. He believed that he acted in the place of Peter, and that Christ actively participated in the governing of the church.

See W.J. Halliwell, *The Style of Pope St. Leo the Great* (1939), and T.G. Jalland, *The Life and Times of St. Leo the Great* (1941).
ROBERT G. CLOUSE

LEO III (c.680-741). Byzantine emperor from 717, known as "the North Syrian" or "Isaurian." After revolting against Theodosius III, he marched to Constantinople and was elected emperor. In a series of battles he repelled and defeated the Arabs. Also he introduced important administrative reforms—e.g., in the *Ecloga Legum,* based in part on the Justinian Code* and on canon and customary law. His ecclesiastical policy caused the outbreak of the long Iconoclastic Controversy.* In 726 he issued an edict against the use of images in worship, but when he tried to remove an honored icon of Christ in Constantinople he was strenuously opposed by the local people, the monks, the patriarch Germanus, and John of Damascus. A revolt broke out in Hellas. Undeterred, he banned in 730 the use of icons and ordered their destruction. Germanus refused to cooperate and was deposed; Pope Gregory III held two synods at Rome in 731, condemning Leo's supporters. In part a punishment to Rome, Leo transferred to the patriarchate of Constantinople lands (e.g., S Italy and Greece) previously under the papacy and also appropriated some papal patrimonies. His policy was continued by his son, Constantine V. PETER TOON

LEO III (d.816). Pope from 795. He was a native Roman, and his first significant act on election was to send to Charlemagne* the standard of Rome and the keys of Peter's sepulcher and of the city. In 799, having been accused by enemies of serious misbehavior, he barely escaped with his life. He appealed to Charlemagne, who came to Rome in 800 and before whom Leo purged himself on oath of the accusations. Two days later, on Christmas Day, he crowned Charlemagne emperor, perhaps in order to suggest that this was the prerogative of the pope alone. This put a strain on the improving relations with the East. Leo tried to correct this in 809 when, confirming the correctness of the *Filioque* clause introduced into the Nicene Creed,* he urged that the creed should not be chanted in the public liturgy. Leo was canonized in 1673. J.D. DOUGLAS

LEO VI (866-912). Byzantine emperor from 886; known also as "the Wise" and "the Philosopher." Eldest son of Basil I, he was educated under Photius,* became co-emperor in 870, and after ascending the throne renounced his former teacher to gain support with intransigents in the church. Facing threats from Arabs and Bulgarians, he staved them off with diplomacy rather than a military campaign. He reformed the state's legal and administrative functions, and in 888 completed the *Basilica,* a corpus of Roman law still valid in the empire, by enlarging the number of Greek books from forty to sixty. Leo married four times to secure an heir and was severely censured by churchmen for this action, although his son was legitimatized in time to inherit the throne. Leo constructed a church and a monastery in Constantinople to honor the biblical person Lazarus. His writings included laws and decrees regarding secular and religious affairs; a treatise on military strategy that incorporated much of Aelian's text on military affairs and urged feigned retreats without apology; homilies; religious verse; orations; secular poetry; and a funeral oration on his father's accomplishments.

JOHN GROH

LEO IX (1002-1054). Pope from 1048. Born in Alsace, he did military service in Lombardy during the reign of his relation, Emperor Conrad II, to whom he owed his appointment in 1027 as bishop of Toul. Having been influenced by the work of Cluny and Lorraine, he reformed various monasteries, which reforming zeal was apparent to the church at large after his election as pope. He traveled the Continent fostering a new ideal of the papacy. The councils at Bari, Mainz, Pavia, and Reims issued decrees against simony, clerical marriage, and other abuses. He forcefully opposed the Norman devastation of S Italy which antagonized both the German court and the Eastern Church. Defeated by the Normans in 1053, he was ready to embark on a triple alliance of papacy, empire, and Byzantium, but he died that year. Hildebrand, later Gregory VII, began his career in Rome under Leo's pontificate.

C.G. THORNE, JR.

LEO X (1475-1521). Pope from 1513. The second son of Lorenzo the Magnificent, Giovanni de'Medici received the tonsure before the age of eight and became a cardinal deacon at age thirteen. His tutors in his father's court included such men as Marsilio Ficino, Angelo Poliziano, and Giovanni Pico della Mirandola. He studied theology and canon law at Pisa and became a member of the college of cardinals in 1492. During the intervening period until his election to the papacy, Giovanni took part in the election of Alexander VI (1492), went into exile from Florence during Savonarola's* reign (1494), became head of the Medici family (1503), participated in the successive elections of Pius III and Julius II, and was able to regain power in Florence through a bloodless revolution by the Florentines (1512).

On 15 March 1513, he received holy orders, on the seventeenth he was consecrated bishop, and he was crowned pope on the nineteenth. He was chosen because of his peace-loving qualities as opposed to the warriorlike tendencies of Julius II. Leo personified the Renaissance—he loved art, music, and the theater, and was the patron of many humanists. His piety was probably sincere, but his lavish spending impoverished the papacy. He managed the culminating work of the Fifth Lateran Council (1512-17) and negotiated a settlement with Francis I of France which clarified the duties of king and pope (1515). In his constant search for sources of revenue Leo renewed an indulgence to support the building of St. Peter's, an act which led the Protestant Reformation.

See W. Roscoe, *Life and Pontificate of Leo X* (2 vols., 1853). ROBERT G. CLOUSE

LEO XIII (1810-1903). Pope from 1878. Vincenzo Gioacchino Pecci was a native of Carpineto, he was educated by the Jesuits of Viterbo, and studied at Rome. He was ordained priest in 1837, and made apostolic delegate of Benevento. In 1841 he became delegate of Perugia, where he gained a reputation as a social reformer. He was appointed nuncio to Brussels (1843) and consecrated archbishop of Damiato, and during his three years' residence he mediated in an educational controversy between the Jesuits and the Catholic University of Louvain. He became bishop of Perugia (1846) and in 1853 was created cardinal priest of St. Crisogono. His long episcopate found him building and restoring churches, and encouraging learning and social reform. Though not altogether *persona grata* to Pius IX, he yet protested against the loss of the pope's temporal power in 1870.

Elected to the papacy, he ought to have come to terms with the civilization of the day. By conciliatory methods he overcame the anticlericalism in Germany which followed the decree of papal infallibility in 1870. In Belgium he saw the Catholic party return to power (1884), while in 1892 he established an apostolic delegation in Washington. He renewed contacts with Russia and Japan and improved relations with Britain. In France, however, he tried with little success to dissociate the Catholic clergy from the royalist party, and during his last years relations between church and state there deteriorated into a period of triumphant anticlericalism. In Italy, too, he failed to recover the lost temporal powers of the papacy, and the pope remained "the Prisoner of the Vatican."

Leo did much by way of improving social attitudes, and attempted to stem the drift of the working classes into irreligion. His encyclical *Rerum Novarum* (1891) emphasized that labor should receive its just reward, and approved of social legislation and trade unionism. He encouraged the study of the Bible, and in 1883 he opened the Vatican archives to historical research. He made some approaches to other churches, notably to the Church of England in his apostolic letter *Ad Anglos* (1895), even though a commission he appointed rejected Anglican orders as invalid (1896). He also promoted the spiritual life of his own church.

BIBLIOGRAPHY: M. Spahn, *Leo XIII* (1912); H. Somerville, *Studies in the Catholic Social Movement* (1933); E. Soderini, *The Pontificate of Leo XIII* (ET 1934-35). J.G.G. NORMAN

LEONARDO DA VINCI (1452-1519). Florentine artist, scientist, and inventor. Born the natural son of a notary, he studied under Verrochio and in 1482 left Florence for Milan, where he served the duke until 1499. He wandered for several years, but finally settled in France under the patronage of the king where he remained until his death. Although a universal man, Leonardo placed painting above the other arts, for he felt it was the best method by which to present the work of nature to the senses. Thus it extends to the surfaces, colors, and forms of natural objects which science studies in their intrinsic forms. The beauty that painting seeks is the proportion of the things themselves. Proportion is found not only in numbers, but in sounds, weights, time, space, and any natural force. These principles he applied in such works as the *Virgin of the Rocks*, the *Last Supper*, and the *Mona Lisa*.

But Leonardo was also quite interested in science, which he felt should be based on experience and mathematical calculation. He had contempt for those who spent their time studying Aristotle and his commentators. In these attitudes he demonstrates his dependence upon Marsilio Ficino and the Renaissance Platonists of Florence. Da Vinci left a sizable collection of notebooks which have been scattered in various libraries of Europe. They consist of notes and sketches on various topics: mechanics, anatomy, physics, physiology, and philosophy. They also contain suggestions for machines of all types, such as airplanes, tanks, automatic guns, gears, and parachutes. In addition there are methodological notations on the procedures of scientific inquiry and the processes of nature.

See K. Clark, *Leonardo da Vinci* (1939), and I.B. Hart, *The World of Leonardo da Vinci* (1961). ROBERT G. CLOUSE

LEONTIUS OF BYZANTIUM (sixth century). Anti-Monophysite theologian. In his early years he lived among Nestorians. Probably a Palestinian monk, he is known from the *Vita Sabae* of Cyril of Scythopolis. He entered a monastery in Palestine called "New Laura" about 520 and accompanied Sabas to Constantinople in 530, where he defended Chalcedon against the Monophysites (531-36). Back in Palestine in 537, he defended Origenism against orthodox attacks, then returned to Constantinople about 540. He with Boethius in the West and John Philoponus in the East helped make Aristotelian philosophy available for use in theology. His works show him to be well versed in Aristotelian logic and the psychology of the Platonists. He forcefully opposed Monophysitism* and Severus* of Antioch in his works, which included *Three Books against the Nestorians and Eutychians; A Resolution of the Arguments Advanced by Severus; Thirty Chapters against Severus;* and perhaps *Against the Fraud of the Apollinarians.* His Christology appeared to be closer to Theodore of Mopsuestia's* than to Cyril of Alexandria's. He argued that Christ's two natures were permanently distinct, but the existence of the humanity was concretely manifested in the one hypostasis of the divine Word. He used the term *enupostatos* (later developed by Maximus the Confessor and John Damascene) to argue that if a nature had its subsistence in another hypostasis it need not thereby become an accident. His favorite term was used earlier; the idea came from the Neoplatonists. In sum, in his Christology Leontius seemed to be trying to reformulate the Christology of Evagrius Ponticus* into Chalcedonian terms. JOHN GROH

LEPORIUS (fl. 425-430). Latin theologian of Gaul. Probably born in Trier, he was a monk in Marseilles. Cassian spoke of him as a Pelagian, and his christological teaching shows points of contact with Theodore of Mopsuestia* and early Nestorianism.* He issued a letter in which he taught the essential similarity between Christ's moral experience and our own. When rebuked by several bishops of S Gaul, he and his followers fled to Africa, where he met Augustine and publicly recanted his error at Carthage in a *Libellus Emendationis* which was subscribed by Aurelian of Carthage and Augustine.

LEPSIUS, JOHANNES (1858-1926). Founder of the German Orient Mission. Son of a noted Egyptologist, his acquaintance with the East began as a child. A Lutheran pastor, he was deeply moved by the Armenian massacres of 1894-95, and his

1896 tract indicting Turkish policies received European-wide notice. In 1895 he established a German relief organization for Armenia which soon consisted of orphanages, medical clinics, and a carpet factory in Urfa, Turkey, to provide employment. In 1900 Lepsius transformed this into the German Orient Mission and endeavored to evangelize Muslims, but with little success. Most of its properties were lost in World War I, but it did carry on a relief work among the Armenian diaspora and a small Muslim mission which the Berlin Mission took over in 1937.

RICHARD V. PIERARD

LE QUIEN, MICHEL (1661-1733). French Dominican scholar. Librarian of Saint-Honoré, Paris, he wrote *Défense du texte hébreu et de la version Vulgate* (1690) and *Panoplia contra schisma Graecorum* (1718), taking issue with Patriarch Nectarius of Jerusalem on papal primacy. His posthumous *Oriens Christianus* (3 vols., 1740) treats Eastern Church history. He was responsible also for the still standard though incomplete edition of John Damascene's *Opera omnia* (2 vols., 1712). Unfinished too was an edition of *Opera omnia Leontii Byzantini.* He wrote two major works concerning the validity of Anglican orders.

LESSING, GOTTHOLD EPHRAIM (1729-1781). German writer and dramatist. Son of a Saxon pastor, he became librarian to the duke of Brunswick (1774-78), during which period he published a series of *Fragments of an Unknown Writer.* He claimed to have discovered them in the ducal library at Wolfenbüttel (hence their popular name of *Wolfenbüttel Fragments**). In reality they were extracts from a massive unpublished manuscript by H.S. Reimarus.* The work was a defense and restatement of skeptical Deism. The last of the Fragments was entitled *The Goal of Jesus and His Disciples,* and claimed to expose the gospel accounts of Jesus as a piece of fraud on account of their alleged unfulfilled eschatological predictions. Jesus had promised the imminent coming of the kingdom of God, and it had not come. On His death the disciples had cunningly postponed it indefinitely, claiming that Jesus had risen from the dead and had gone to heaven. In the meantime, people have failed to notice that Christianity is built on unfulfilled fraudulent claims.

The ensuing controversy raged fiercely. Among those who replied were J.M. Goeze, J.C. Döderlein, J.D. Schumann, and J.S. Semler. Lessing did not commit himself, feigning ignorance of the author's identity and replying to critics in a series of pamphlets ostensibly trying to put the debate into perspective, while adding fuel to the fire. He adopted an enlightened attitude to religion, maintaining ambiguously that religion is not true because the apostles taught it; they taught it because it is true. Historical evidence is insufficient basis for religious belief, for the accidental truths of history can never become proof for the necessary truths of reason. The truth and value of a religion are to be apprehended in experience. Those who live right will show that they have true religion. This was the message of the dramatic poem *Nathan the Wise* (1779) and the essay on *The Education of the Human Race* (1780). Lessing also wrote an essay in gospel criticism, *New Hypothesis on the Evangelists considered as merely human historical Writers* (1788), which posited a single Hebrew or Aramaic source behind the gospels, portraying Jesus as a purely human messiah.

See H. Chadwick, *Lessing's Theological Writings* (1956), and G. Pons, *Gotthold Ephraïm Lessing et le Christianisme* (1964). COLIN BROWN

LEVELLERS. A democratic party in England during the Puritan Revolution and Commonwealth period. The name was given by enemies of the movement to suggest that it was aimed at "levelling men's estates." The party developed in 1645-46 among radical supporters of Parliament in and around London. The war had been waged in the name of Parliament and the people, and the Levellers demanded that sovereignty be transferred to the House of Commons elected by universal suffrage, that there be a redistribution of seats and annual (or biennial) parliaments. They also advocated equality before the law and freedom of religion. Since their reforms were not inaugurated by Parliament, the Levellers turned to agitation within the New Model Army. A debate held at Putney (October 1647) between Gen. Henry Ireton, the representative of the prevailing view that only property owners should have the franchise, and the Levellers ended in deadlock. Discipline was restored by the generals with force, and by 1650 the leading Levellers such as John Lilburne were imprisoned and the movement crushed. ROBERT G. CLOUSE

LEWIS, C(LIVE) S(TAPLES) (1898-1963). Novelist, poet, and apologist. Anglican layman who taught at both Oxford and Cambridge, Belfast-born Lewis attracted wide readership during and after World War II because he had, as C.E.M. Joad wrote, "the rare gift of making righteousness readable." He burst on the scene in 1941 with the clever satire *Screwtape Letters*—instructions from a senior devil to a junior devil on how to snatch a new Christian from the snares of heaven. Soon after, Lewis delivered a series of twenty-nine widely popular broadcast talks on basic Christian doctrine. They were, like Lewis's later writings, traditionally orthodox; he avoided denominational distinctives and called it "mere Christianity." The stamp of his style was wit, urbanity, clarity, an effortless elegance, and disciplined logic.

In scholarly circles Lewis was a respected literary critic before Screwtape surfaced. His *Allegory of Love* (1936), a study of the courtly love tradition in literature, was considered a landmark in medieval scholarship. *A Preface to 'Paradise Lost'* (1941) was at the center of controversy over Milton's theology and Romantic interpretations of Satan.

Books flowed from his pen at the rate of one or more a year, and in astonishing variety: novels, children's books, theology, philosophical apologetics, poetry, literary criticism. Many of his books seemed especially designed to remove obstacles facing the Christian in an agnostic age of scientific materialism. *The Abolition of Man* (1943), on the rational and social necessity of a normative ethic, and *Miracles* (1947), a philosophical defense of the possibility of miracles, were closely reasoned philosophical treatises. *The Problem of Pain* (1940) dealt with the ancient difficulty of justifying the ways of a good God to suffering man. *Reflections on the Psalms* (1958) and *Letters to Malcolm: Chiefly on Prayer* (1963) discussed problems in the Psalms, prayer, and the private devotional life. His three most popular novels, *Out of the Silent Planet* (1938), *Perelandra* (1943), and *That Hideous Strength* (1945), wove Christianity into a hauntingly beautiful fictionalized cosmic myth. The seven Chronicles of Narnia became modern children's classics.

C.S. Lewis disliked cant, positivism, and "chronological snobbery," the notion that newest is best. His lively apprehension of man's evil was balanced by his vision of joy—he called it *Sehnsucht*—man's longing for his eternal home. His was a world of clarity shaped by a commitment to the use of sound reasoning, furnished by emphasis on the changeless and detachment from passing fashions. He believed reason was the organ for truth, and imagination the vehicle for understanding. A certain dash of Platonic philosophy and philological expertise lay behind his concept of image and myth. He believed myth contained universal truth, and that Christianity was the archetypal myth. For Lewis, the Christian myth had an objective correlative.

Lewis was raised Anglican but became an atheist as a teenage schoolboy. His education at Oxford was interrupted by military service in World War I. After recovery from a shell wound he returned to Oxford to read philosophy and English literature. His first published work, a slim volume of lyric poetry, *Spirits in Bondage*, appeared in 1919, and a narrative poem, *Dymer*, followed in 1926: both were published under the pseudonym Clive Hamilton. After nearly thirty years as fellow of Magdalen College, he left Oxford in 1954 for the newly created chair of medieval and Renaissance English at Cambridge. Lewis moved slowly from atheism through Yeatsian romanticism to absolute idealism and finally theism, returning to worship in the Church of England in 1929. His conversion journey is traced in *The Pilgrim's Regress* (1933), an allegory, and *Surprised by Joy* (1955), his spiritual autobiography. Most of his life was spent in quiet bachelorhood, but in 1956 he married Joy Davidman Gresham, an American Jewish Christian convert, when she was critically ill with cancer. After her death in 1960 he wrote the poignant *A Grief Observed*, initially published under the pseudonym "N.W. Clerk"; until his own death three years later he served as guardian of her two sons.

Other writings include *The Great Divorce* (1946); *Mere Christianity* (1952); *Till We Have Faces* (1956); *Christian Reflections* (1967); and works on literature and criticism.

See biography by R.L. Green and W. Hooper (1974) and A. Arnott, *The Secret Country of C.S. Lewis* (1975). JOAN OSTLING

LIANG A-FAH (1789-1855). First ordained Chinese Protestant evangelist. When Robert Morrison could not gain access to China, he established a base in 1814 among the 4,000-strong Chinese community in Malacca, where his colleague, William Milne,* set up a printing press. Among Milne's converts and assistants was Liang A-fah, who became a Bible Society colporteur and wrote a long treatise on Christianity entitled *Good Words Exhorting the Age*, which he eventually distributed among the civil service examinees in Canton. One set of these books fell into the hands of Hung Hsiu-ch'uan* and sparked the Taiping Rebellion.

LIBELLATICI. The name given to those who during the Decian persecution (249-51) purchased certificates from the civil authorities which stated they had sacrificed to idols when in fact they had not. The practice was condemned by church leaders, but those guilty of the circumvention were treated more liberally than those who had actually sacrificed. A council at Carthage in 251 decided that penitent *libellatici* might be restored, but that *sacrificati* must submit to lifelong penance.

See LAPSI.

LIBERAL CATHOLIC CHURCH. A body that may be dated from 1918 when it appeared in London as a synthesis of theosophical and Old Catholic* doctrines and practices. Four members of the English Theosophical Society were ordained (1913-14) priests in the miniscule British Old Catholic Church. A lapse in the Old Catholic bishopric led to the recognition (1916) of J.I. Wedgwood as bishop, who in turn ordained C.W. Leadbeater (the leading London Theosophist since 1895) as bishop of the Old Catholic Church for Australasia. Wedgwood and Leadbeater compiled a new liturgy and renamed the body the Liberal Catholic Church (1918) to distinguish it from Old Catholicism. An American branch, begun in 1917, established its headquarters at Los Angeles, California, and built a procathedral (1924). The church does not hold to a firm doctrine, believing that there are "many paths to truth." Instead it stresses liturgy—whereby the "living Christ" is experienced—theosophy,* and reincarnation. It maintains a hierarchy of regional bishops, selected by a general episcopal synod, with a presiding bishop. Membership numbered 10,000 (1964), including 2,500 in the United States. Headquarters are in London.

BIBLIOGRAPHY: E.E. Beauregard, "Liberal Catholic Church" in *New Catholic Encyclopedia* XIII, pp. 699-700; R.K. McMaster, "Theosophy" in *New Catholic Encyclopedia* XIV, pp. 74-75; F.W. Pigott, *The parting of the ways* (1927); C.W. Leadbeater, *The science of the sacraments* (1920; rep. 1957). C.T. MC INTIRE

LIBERIAN CATALOGUE. A list of Roman bishops from Peter down to Liberius (352-66), one part in a collection of documents made by a compiler known as the Chronographer of A.D. 354.* the name given by T. Mommsen in his studies. Probably the earliest version of the *Liber pontifi-*

*calis,** it contains two sections: (a) dating the pontificates from Peter to 231 (Pontianus 230-35) based on authentic but faulty traditions as with a list in the *Chronicle* of Hippolytus; (b) the period 231-352 which seems to evidence official documentation. The twenty-five years given Peter's reign is substantiated in their chronicle by Eusebius and Jerome. Probably prepared in 336 under Pope Mark before its publication later, the Liberian Catalogue was first edited by the Jesuit, A. Bucherius, at Antwerp (1636), and is sometimes called the Bucerian Catalogue.

C.G. THORNE, JR.

LIBERIUS (d.366). Pope from 352. Shortly after his accession he was faced with the Eastern demand to condemn Athanasius. Not present at the Council of Milan (355) which succumbed to anti-Athanasian pressure, he was summoned before Constantius, but refused to condemn Athanasius and asked that the Nicene Creed be confirmed. He was exiled to Berea, and after two years he signed, under pressure, an agreement to the exile of Athanasius and to a heretical formula. The evidence for this is contained in four letters from exile, the authenticity of which is challenged, but cannot be doubted. There is, however, doubt as to which formula he signed and hence how heretical it was. He returned to Rome, and in 366 received into communion representatives of the Eastern Church, fleeing from the pro-Arian emperor Valens. He died that same year. Other items of his correspondence have been preserved, including three letters to Eusebius, bishop of Vercelli.

C. PETER WILLIAMS

LIBER PONTIFICALIS. The "Book of Popes" contains biographies of popes from the Apostle Peter's day to the mid-fifteenth century. The entries give such basic information as each man's birthplace, date of pontificate, and contributions to Catholicism. They tend to increase considerably in size with later pontiffs and are of uneven reliability, often reflecting the biases of the era in which composed. The first edition, based on a fourth-century list of Roman bishops, *Catalogus Liberianus,* appeared in the sixth or perhaps seventh century. Thereafter *Liber pontificalis* was periodically updated. It is least adequate on tenth- and eleventh-century popes. Yet it remains a basic source of information and is indispensable as a reflection of medieval attitudes towards the papacy.

JAMES DE JONG

LIBERTINES. Sometimes this refers to a Jewish synagogue group mentioned in Acts 6:9 KJV, but it is more commonly associated with two groups of opposition to Calvin in Geneva. The first—not known as Libertines in their century—were Genevan patriots and influential families (the Perrins, Favres, Vandels, Bertheliers, etc.) who led the republic to independence and the Reformation. They resented the dominant influence of Calvin and "foreigners" in Genevan affairs. A bitter struggle with Calvin ended in their complete disgrace in 1555. The second group, called Libertines by Calvin, were spiritualists, professing a pantheistic, antinomian creed, denying evil, and rejecting all formal Christianity. They came to Geneva from France, but Calvin traces their beginnings to one Coppin of the Netherlands. They were suppressed in Geneva in 1555.

BRIAN G. ARMSTRONG

LICENTIATE. A term common in Scottish Presbyterianism to signify one who, having completed his theological studies, has received authority to preach the Word, but has not yet been by ordination given authority to administer the sacraments. Sometimes the word "probationer" is used to describe one having this status; in previous times in Scotland the word "preacher of the gospel" or "expectant" was used.

LIDDELL, ERIC (1902-1945). Athlete and missionary to China. While a student at Edinburgh University, he became the most popular and most widely known athlete Scotland had produced. Rugby football internationalist, Olympic champion, and world record-breaker, he established a national reputation also as an evangelist before returning to his birthplace at Tientsin to join the staff of the Anglo-Chinese Christian College there in 1925. Interned by the Japanese at Weihsien in China in 1942, he was highly regarded by people of all ages and races because of his Christlike life and unremitting, self-denying service in the internment camp, where he died of a brain tumor just before he would have been released. "Scotland," said a leading national newspaper, "has lost a son who did her proud every hour of his life."

D.P. THOMSON

LIDDON, HENRY PARRY (1829-1890). Anglican preacher. Born in London and educated at Oxford, he became a member of the Tractarian* group after the secessions to Rome. He exercised great influence both as a professor and administrator in Oxford, but his real fame rests on his preaching from the pulpit of St. Paul's as one of its canons over the last two decades of his life. Steeped though he was in the ethos of Oxford, Liddon nonetheless succeeded in the requirements of a popular preacher. This is not to say he was shallow or flashy. He marshaled his vast learning, strong logic, lucidity, and sense of order, and his sustained grasp of ideas within strictly controlled and methodically developed structures and expressed them in prose that was brilliantly expressive and finely modulated. He had, in particular, a fine ear for the effects of inversion and a falling close, and a ready memory for apt and succinctly narrated illustrations. In his Oxford sermons, sometimes taking as long as eighty minutes to deliver, he provided exhaustive treatments of theological topics in a closely argued and deeply scholarly manner.

ARTHUR POLLARD

LIETZMANN, HANS (1875-1942). German church historian. Having studied at the universities of Jena and Bonn, he became professor first at Jena (1905-24) and then at Berlin (1924-42) as the successor of A. Harnack.* He served as the editor of the *Zeitschrift für Neutestamentliche Wissenschaft* from 1920 until his death and also, with others, *Handbuch zum Neuen Testament*

(1906ff.), to which he contributed commentaries on Romans, 1 and 2 Corinthians, and Galatians. Lietzmann authored numerous important monographs, including an early study of Apollinarius of Laodicea (1904), another on Peter and Paul in Rome (1915; rev. ed. 1927), and an epochal work on the Mass and the Lord's Supper (1926). His four-volumed *History of the Early Church* (1932-44; ET 1937-51) remains a standard work. His academic interests bridged all the disciplines related to early Christian history—NT exegesis, classical archaeology and philology, papyrology, Hellenistic religion, canon law, and the like.

W. WARD GASQUE

LIFE AND WORK, see OXFORD CONFERENCE; STOCKHOLM CONFERENCE; WORLD COUNCIL OF CHURCHES

LIGHTFOOT, JOHN (1602-1675). English biblical scholar. Born at Stoke-on-Trent, he graduated from Cambridge, spent two years teaching at Repton, and after ordination was engaged in pastoral work in Shropshire and Staffordshire (1630-42). He supported Parliament in the Civil War. In 1643 he became rector of St. Bartholomew's, London, and was a member of the Westminster Assembly* in which he was on the Erastian side against the extreme Presbyterians. He was later rector of Great Munden and master of Catherine Hall, Cambridge (1650), and was vice-chancellor of Cambridge University (1654). He took part in the Savoy Conference* (1661), this time supporting the Presbyterians. His works include *Erubhin; or Miscellanies, Christian and Judaical* (1629), a book on Genesis (1642), and *Horae Hebraicae et Talmudicae* (6 vols., 1658-78), showing the bearing of Jewish studies on NT interpretation. He also assisted Brian Walton with the Polyglot Bible (1657).

J.G.G. NORMAN

LIGHTFOOT, J(OSEPH) B(ARBER) (1828-1889). Bishop of Durham. A sickly child, he was at first educated by tutors at his Liverpool home, but in 1844 moved to King Edward's School, Birmingham, where he was much influenced by the headmaster, Dr. J.P. Lee, and formed a friendship with E.W. Benson, later archbishop of Canterbury. In 1847 he went to Cambridge, studied under B.F. Westcott,* became a fellow of Trinity and, after ordination, a tutor at the college, lecturing on the classics and the Greek NT. In 1861 he became Hulsean professor of divinity, and in 1875 Lady Margaret professor. In 1879 he was appointed bishop of Durham and devoted himself with great energy to his episcopal duties. He arranged for the division of the diocese into two and saw to the building of many new churches in the expanding industrial areas. He made benefactions to the University of Durham, and he used to have living with him at Auckland Castle six or eight young graduates who were training for the ministry.

Lightfoot was one of the most learned men of his time. Fluent in seven languages, he read very widely, had an accurate memory, wrote lucidly and powerfully, and was at his best when dealing with and assessing facts rather than ideas. This made him a good foil to both Westcott and F.J.A. Hort,* with both of whom he intended to write a commentary on the complete NT. Lightfoot was to do the Pauline epistles, and he completed three definitive commentaries: Galatians (1865), Philippians (1868), and Colossians and Philemon (1875). His work on the Apostolic Fathers (1869 and 1885) was epoch making. In nine articles in the *Contemporary Review* he completely demolished the arguments advanced in J.A. Cassels's anonymous *Supernatural Religion* and magnificently defended the historicity of the Christian faith. His historical gifts are displayed in a collection of sermons in the diocese of Durham entitled *Leaders of the Northern Church* (1890).

See G.R. Eden and F.C. Macdonald (eds.), *Lightfoot of Durham* (1931).

R.E. NIXON

LIGHTFOOT, ROBERT HENRY (1883-1953). English biblical scholar. Born in Wellingborough, Northamptonshire, son and maternal grandson of Church of England clergy, he was an Eton scholar and exhibitioner of Worcester College, Oxford, from which he received first class theological honors (1907) and the senior Greek and Septuagint prizes (1908). Subsequent to ordination he was curate of Haslemere (1909-12), and then in succession at Wells Theological College bursar (1912), vice principal (1913), and principal (1916). He served as examining chaplain to the archbishop of Canterbury (1913-53). He was named fellow and chaplain of Lincoln College, Oxford (1919), and then fellow of New College (1921) which he held through retirement.

He delivered the Bampton lectures (1934) in which he presented his basic understanding of the problem of *History and Interpretation in the Gospels* (1935), championing the form-critical method which had been developed among German scholars of the interwar period, and which he introduced into England, though mellowed by his own dependence upon F.C. Burkitt.* In 1935 he was named Ireland professor of exegesis of Holy Scripture at Oxford, receiving emeritus status 1949. In 1938 he published his study of *Locality and Doctrine in the Gospels.* He was editor of the *Journal of Theological Studies* from 1941 until his death. Among his mature studies were volumes on the gospels of Mark (1950) and John (published posthumously in 1956).

CLYDE CURRY SMITH

LIGUORI, ALPHONSUS (1696-1787). Roman Catholic moral theologian. Born of noble Neapolitan parents near Naples, he became a successful barrister, but retired from the profession following a crucial oversight in a case, and became a priest in 1726. His simple oratory stood in direct contrast with the pompous rhetoric of the day, and he became a mission preacher around Naples. In 1731 he founded the Congregation of the Redemptoristines for women, and in 1732 the Congregation of the Most Holy Redeemer (Redemptorists*) for men, communities particularly dedicated to mission work among the poor in rural areas.

His moral theology reacted against the gloomy rigorism prevalent under the influence of Jansenism,* which he attacked fiercely. His articles were later summarized in his *Moral Theology.* He eschewed equally the rigorism of Tutiorism and the possible laxity of Probabilism, maintaining a middle-of-the-road teaching to be identified as "Equiprobabilism." This view has gained the approbation of the Roman Church, although at the time it was bitterly attacked. Alphonsus can be considered the father of moral theology.

In 1762 he became bishop of St. Agatha of the Goths, a small diocese near Naples, but ill health caused him to submit in 1770 a resignation which Clement XIV would not accept. In 1775, however, Pius VI allowed him to retire. He remained major superior of his congregation and was involved in persistent controversy within his order and outside it. He was beatified in 1816 and canonized in 1839. In 1959 he was proposed as heavenly patron of confessors and moralists. He produced popular devotional and mystical writings, often on the themes of the Sacred Heart of Jesus and Mary as semi-divine mediatrix, and wrote many hymns. He also wrote apologetic and theological works.

See Lives by A. M. Tannoia (ET 4 vols., 1848-49) and A. Berthe (ET 2 vols., 1906).

HOWARD SAINSBURY

LIMBO. From a Teutonic word denoting the hem or border of a garment, limbo (or *Limbus infantum*) is the place between heaven and hell to which unbaptized babes are assigned at death. In the time of Augustine and for many centuries after, limbo was held to be a place of torture, but the torture was later replaced by natural bliss. The Jansenists in 1786 revived the torture theory, but Pius VI (1794) permitted Catholics to think that such children feel a pain of loss but not of the senses in the life to come. More liberal views (e.g., that unbaptized children may after all be saved) were first heard about 1900, but are still regarded as heterodox by most Catholics. In 1958 the Holy Office reiterated the urgent necessity for child baptism since the church still teaches the "absolute necessity of Baptism for eternal salvation." Fear that children might suffer in limbo led Catholic gynecologists to invent syringes so that baptism might be administered (under condition) before birth.

R.E.D. CLARK

LINCOLN, ABRAHAM (1809-1865). Sixteenth president of the United States. Raised on frontier farms, he was a self-educated lawyer who served in the Illinois legislature and the House of Representatives before becoming president in 1861. A Republican, he attracted support through his simplicity of manner, his defense of established authority, his fusion of farmers and industrialists, his equation of slavery's expansion with threats to Northern prosperity, and his recognition of the inferior condition of black Americans. His election prompted the lower South to secede, and his intransigence on secession, in turn, led the upper South to rebel and sparked civil war. Political and military concerns forced him into decisions many thought dictatorial, and guided him to interpret the war, which was initially to preserve the Union, as a crusade to free the slaves and finally as a national tragedy, a shedding of blood from which would come a new nation. Assassinated by a Southern sympathizer within days after his forces gained the victory, he epitomized the nation's ideals of self-reliance, opportunism, and churchless religion.

DARREL BIGHAM

LINCOLN, WILLIAM (1825-1888). English preacher. Born in East London in 1825, and converted at seventeen partly through reading Doddridge's *Rise and Progress of Religion,* he studied with missionary service in mind, but abandoned the plan owing to poor health. After further study at King's College, London, he was ordained in 1849. Following a curacy in Southwark he was appointed in 1859 to be minister of Beresford Chapel in Walworth, where his preaching was very popular. Even at this stage he was doubtful about his position in the Established Church, and he announced his intention to abandon it in 1862, shortly after which he published his *Javelin of Phinehas,* a lengthy condemnation of the union of church and state. Many of his congregation chose to stay with him, and as he had acquired the lease of the chapel he continued to minister there and encouraged congregational participation to such an extent that it was eventually regarded as a Brethren assembly. After his death a number of Brethren continued the ministry there.

TIMOTHY C.F. STUNT

LINDLEY, DANIEL (1801-1880). American missionary to South Africa. Born in W Pennsylvania and ordained a Presbyterian minister in 1832, he reached South Africa as an American Board missionary in 1835. A mission to the Matabele was abandoned after a Boer attack in 1837, and he joined colleagues among the Zulu in Natal. In 1840-46 he served as minister to the Boers (or *Voortrekkers*), believing that this would ultimately benefit the Zulu Mission. His devotion won their lasting regard. In 1847 he returned to mission work in the Inanda Location, NW of Durban, where he remained until retirement in 1873. The Inanda Girls' Seminary was opened in 1869. Lindley's attitude to tribal custom mellowed with the years, and he welcomed the ordination of native pastors in 1869, in which respect he proved more liberal than younger colleagues.

D.G.L. CRAGG

LINDSAY, THOMAS MARTIN (1843-1914). Scottish church historian. Born in Lanarkshire, he was educated at the universities of Glasgow and Edinburgh, and was ordained in the Free Church of Scotland* in 1869. Three years later he was appointed to the chair of church history at his church's Glasgow college, where he was later (1902) to become principal. Lindsay was a defender of W.R. Smith in the heresy trial (1877-81) that led to the latter's deposition. Among Lindsay's better-known works are *Luther and the German Reformation* (1900), and *A History of the Reformation in Europe* (2 vols., 1906-7). He made substantial contributions also to such projects as the *Encyclopaedia Britannica* and the *Cambridge Modern History.*

J.D. DOUGLAS

LINGARD, JOHN (1771-1851). English Roman Catholic historian. Member of an old Roman Catholic family, Lingard was trained at Douai, ordained priest in 1795, taught at Crook Hall seminary and later at Ushaw, and became parish priest of Hornby near Lancaster in 1811. A man of genial and kindly temperament, he produced between 1819 and 1830 an eight-volume *History of England* with the intention of disarming Protestant critics by his candor and scholarship. Many were, and still are, bemused by this approach, but his essential premise was long ago exposed by T.B. Macaulay: no Protestant opinion on any subject can possibly be correct. In 1836 he produced a new version of the four gospels which relied on the Greek rather than the Vulgate text. This novel approach, combined with the tenor of his historical writing and the well-known Gallicanism* of his youth aroused the wrath of the hierarchy and the Ultramontane party. IAN SELLERS

LINUS (first century). Beginning with Irenaeus and further documented by Eusebius, Linus is identified as the first appointed bishop of Rome "after the martyrdom of Paul and Peter" and taken to be the same as that one named in 2 Timothy 4:21 as companion in Rome with Paul. His term of office was put at twelve years terminating in the second of the emperor Titus—thus approximately 68-80. The language of this documentation would indicate him to be successor to "Paul and Peter."

LIPPI, FRA FILIPPO (c.1406-1469). Italian painter. He painted his commissioned pieces in the incomparable Florence contemporary with Fra Angelico, Masaccio, Donatello, and Ghiberti. His set pieces, molding together both Flemish detail and the current Italian, classic sense of bodies, were distinguished by a more pronounced linear movement—turning robes of the Madonna, birds in flight. The painting conventions he used (e.g., the Virgin Mary seated on a throne) seem somewhat at odds with items of everyday normalcy accompanying them, as if the self-conscious artist is no longer content working as a hireling within prescribed genres. One of Filippo Lippi's Renaissance students was Botticelli.*

CALVIN SEERVELD

LIPSIUS, RICHARD ADELBERT (1830-1892). German Protestant theologian. Graduate of the University of Leipzig, after various appointments he became professor of systematic theology at Jena, where he wrote extensively on dogmatics and the history of early Christianity. His writing showed the influence, first, of Hegel, later of Kant and Schleiermacher. His mature thought was close to Ritschl.* He attempted to harmonize scientific principles and methods with those of religion. In this he was opposed by the Lutheran Church for an alleged theological liberalism. He found the basis of the harmonization he attempted in the unity of the personal ego, arguing that though one may come to an objective knowledge of the world through science, a real understanding of the world and, thus, of value required subjective experience. Through such experience, God is revealed as the ultimate unity. Lipsius was co-founder of the Evangelical Alliance and the Evangelical Protestant Missionary Union.

MILLARD SCHERICH

LISZT, FRANZ (1811-1886). Hungarian composer and pianist. One of the most influential figures of the Romantic era in music, he is famed not only as a superlative pianist and composer, but as one of the great innovators of the nineteenth century. His interest and involvement in religious music is not so generally recognized. He wrote three large-scale settings of the Mass, two oratorios, and an assortment of other sacred works.

LITANY (Gr. = "supplication"). A form of alternating prayer in which the biddings or petitions are spoken (or sung) by the minister, and the people make the same response at short intervals, e.g., in: *Kyrie eleison:* "Lord, have mercy" "We beseech Thee, hear us." It may derive from the prayers and psalms of the synagogue, and it has analogies in pagan worship. The earliest known Christian litanies are found in fourth-century Antioch and from there spread through the Eastern Church. They play a prominent part in the worship of Eastern Christians today. In the Roman Catholic Church, litanies indicate special services such as Rogationtide processions. Here the invocation of Christ and the Holy Trinity is followed by that of numerous saints. The Litany of the Book of Common Prayer omits the invocations of saints. Appointed to be sung or said after Morning Prayer on Sundays, Wednesdays, and Fridays, it is used also at ordinations.

HOWARD SAINSBURY

LITTLE LABYRINTH, THE. A lost third-century treatise directed against the Adoptianists—e.g., Theodotus, Artemon—assigned with some probability to Hippolytus of Rome. It was referred to by Theodoret, who ascribed it to Origen. Eusebius quoted two passages from it. There are grounds for thinking that a fragment may be preserved under the name *Contra Noetum,* in which Hippolytus expounds his rich and somewhat mystical doctrine of the Incarnation. It should be distinguished from *The Labyrinth* mentioned by Photius.

LITURGICAL MOVEMENT. There have been many liturgical movements in the history of the church, and the modern one had its immediate precursor in the revival of Roman Catholic Benedictine worship associated with Guéranger at Solesmes in nineteenth-century France. This stimulated scientific liturgical research and a concern for the correct performance of the liturgy. In the present century, this rather reactionary and limited movement in the Roman Catholic Church has been transformed into a powerful and remarkably ecumenical force through a revolution in theology. The rediscovery of the corporate nature of the church has resulted in a widespread attempt to restore to the laity a full and active participation in worship, in place of a solo performance by Catholic priest or Protestant minister. In the Roman Catholic Church this may be traced from the congress at Malines (1909), through the encyclical *Mediator Dei* (1947), to the constitution on the sacred liturgy of Vatican II* (1963) and the radical changes in the services which have followed. In England the modern movement did not appear until the 1930s, but since the war it has had a profound effect on the design of church buildings, and the adoption of a parish Communion as the main Sunday service, house meetings, and various attempts to integrate church services with daily life.

BIBLIOGRAPHY: E.B. Koenker, *The Liturgical Renaissance in the Roman Catholic Church* (1954); J.H. Srawley, *The Liturgical Movement: its Origin and Growth* (1954); A.R. Shands, *The Liturgical Movement in the Local Church* (1959; rev. 1965); M. Thurian, "The present aims of the Liturgical Movement" in *Studia Liturgica*, iii (1964), pp. 107-114. JOHN TILLER

LITURGY. From the Greek word *leitourgia*, which originally meant a public duty of any kind. In Jewish usage the word was specially applied to temple services (e.g., 2 Chr. 8:14, Septuagint). In the NT it is used for Christian service of God, though no distinction is made between worship and other kinds of service (Phil. 2:17). In English the word denotes a service of public, or corporate worship, and hence also a prescribed form used on such occasions, which may be a baptismal, eucharistic liturgy, etc. In churches with a liturgical tradition, however, the word on its own is normally understood to refer to the Eucharist. Services of daily prayer, and those (e.g., burial) which are less than full assemblies of the church, are referred to as "offices." JOHN TILLER

LIUDGER, see LUDGER

LIUTPRAND (c.920-c.972). Bishop and church historian. Of noble Lombard stock and educated at the court of Pavia, he was ordained deacon in that city. In 949 King Berengar II appointed him envoy to Constantinople. Subsequent disaffection with the king led him to the court of Emperor Otto I, who made him bishop of Cremona in 961. On Otto's behalf, Liutprand paid diplomatic visits to Rome and Constantinople, the latter to contract the marriage of the future Emperor Otto II with a Byzantine princess. Liutprand's reputation as a historian rests on three highly rhetorical though biased works. *Antapodosis* treats Italy, Rome, and Byzantium between 887 and 949. *Historia Ottonis* describes Otto I's vicissitudes with the papacy from 960-64. *Relatio de legatione Constantinopolitana* recounts Liutprand's second visit to Constantinople. JAMES DE JONG

LIVINGSTONE, DAVID (1813-1873). Scottish missionary and explorer. Born in Blantyre, he left school at ten years of age, worked incredibly long hours in the mill, but kept a book beside him while he worked. He attended evening classes and often studied until midnight. When he was about seventeen he experienced Christian conversion and dedicated his life to spreading the Gospel in other lands. He saved in order to study medicine and theology at Anderson's College, Glasgow, and heard God's call to go to Africa through Robert Moffat* who had labored there for twenty-three years under the London Missionary Society, and whose daughter Mary he was later to marry.

Arriving in 1841, he soon moved north from Kuruman into unexplored and unevangelized territory, thus beginning the travels which were to take him some 30,000 miles over the African continent.

His first great discovery was Lake Ngami (1849). Four years later he began "the greatest journey of exploration ever made by one man"—north from Cape Town to the Zambesi River, west to the Atlantic Ocean, then right across the continent to the Indian Ocean. In the course of it he discovered the falls on the Zambesi which he named after Queen Victoria. He went home in 1856 to find himself famous. In 1857 he published *Missionary Travels and Researches in South Africa.* He returned in 1858 as a consul to explore the Zambesi River and find whether it was navigable; the expedition was recalled in 1863. After a second trip home he planned to do something to expose and possibly end the Arab slave trade, and also to discover the sources of the Nile.

During his subsequent journeyings he dropped out of sight, and rumors reached home that he was dead. The *New York Herald* sent out H.M. Stanley to find him, and they met at Ujiji in November 1871. Stanley tried to persuade the doctor to return home, but he refused, convinced that God still had work for him to do. He died about the beginning of May 1873, and was subsequently buried in Westminster Abbey. Many came to do homage to one who was missionary, writer, poet, linguist, scientist, doctor, and geographer. Africa owes more to him than to any other. It has been said that he not only discovered Africa, but the African too. Largely due to his reports, it was not long before slavery was made illegal throughout the civilized world.

BIBLIOGRAPHY: H.M. Stanley, *How I Found Livingstone* (1872); W.G. Blaikie, *Personal Life of Livingstone* (1880 and numerous later editions); R. Coupland. *Livingstone's Last Journey* (1945); J.I. Macnair, *Livingstone's Travels* (1954); G. Seaver, *David Livingstone: His Life and Letters* (1957). J.W. MEIKLEJOHN

LLWYD, MORGAN (1619-59). Welsh Puritan author. Born at Cynfal in Merioneth, he was educated at Wrexham where he experienced conversion under the ministry of Walter Cradock (1610?-59), whom he then followed south to join the Puritan group centered on Brampton Bryan, Shropshire. Llwyd settled in Wrexham in 1647 after participating on the parliamentary side in the Civil Wars. He was an approver under the Propagation Act (1650) and joined the protest movement led by Vavasor Powell against Oliver Cromwell's Protectorate; but unlike Powell, he soon made his peace with the Protector and continued to minister at Wrexham until his death at forty years of age. There is a memorial to him at Rhos-ddu cemetery, where he was buried.

In his theology Llwyd veered toward the Quaker position, although for a time he adopted the views of the Fifth Monarchy Men*. He was much impressed by the thought of Jacob Boehme* and translated some pieces by him into Welsh. He expounded his views in a number of books and poems, but his little volume, *Llyfr y Tri Aderyn* ("The Three Birds") of 1653 is considered one of the prose classics of the Welsh. It expresses in memorable language Llwyd's concern for the spiritual life and his burning desire to see his countrymen embrace the Gospel. He left behind him a considerable amount of poetry. Two volumes of his works were published by the University of Wales in 1899 and 1908 while another substantial volume still remains in manuscript.

See G.F. Nuttall, *The Welsh Saints, 1640-1660* (1957). R. TUDUR JONES

LOBSTEIN, PAUL (1850-1928). Protestant theologian. Born at Epinal in France, he was educated in the universities of Strasbourg, Tübingen, and Göttingen. From 1876 he taught at Göttingen as a professor of theology. Essentially a systematic theologian with a Ritschlian emphasis, he wrote mostly in French, but a few books were in German—e.g., his books on Calvin's ethics and Peter Ramus's theology, both published at Strasbourg (1877-78). His later studies were in historical theology and dogmatics: aspects of Calvin's thought, studies of the doctrines of God, the person and work of Christ, baptism, etc. Two of his major books, *The Virgin Birth of Christ* (1890) and *An Introduction to Christian Dogmatics* (1902), appeared in English translations. He is probably best known as an associate editor of the works of Calvin in the huge *Corpus Reformatorum.*

PETER TOON

LOCKE, JOHN (1632-1704). English philosopher. He was the first major British empiricist, conceiving his role as a philosopher as an "underlabourer" to the "natural philosophy" of the Royal Society. In *An Essay Concerning Human Understanding* (1690), Locke rejected the innate ideas of the Cambridge Platonists* and Descartes* and claimed that the mind is a *tabula rasa;* all knowledge is the product of ideas, which are in turn derived either from sense-experience or self-awareness. Knowledge of the external world is the product of the ideas of the qualities of things. Some of these qualities ("primary qualities" such as "solidity, extension, figure, motion or rest, and number") are in the world; others ("secondary qualities" such as sounds and colors), are in the perceiver. Primary qualities inhere in in-principle unknowable substances. Qualities must have substances, but it is impossible to say what these substances are. This doctrine of substance, together with the unclarity of Locke's central notion "idea," is damaging. In his writings, "idea" stands both for a quality of mind and for a quality of the external world. If the former, then (as Berkeley* showed) Locke is committed to a version of idealism or, at best, to the existence of a world forever veiled in the unknown. If the latter, Locke is committed to a naïve realism.

In religion Locke is known chiefly as the opponent of "enthusiasm" (in the *Essay*) and as the proponent of an undogmatic rationalism (in *The Reasonableness of Christianity,* 1695). These views paved the way for the Deism of the following century. Locke denied the deistic implications of his remarks on religion in the *Essay,* while in controversy with Edward Stillingfleet,* bishop of Worcester, but at the cost of some loss of consistency. It is undeniable that Deists such as Toland* appealed to Locke's "new way of ideas."

In his *Second Treatise of Government* (1690), Locke argued that a civil society, with true rights and liberties for its members, is produced out of a "state of nature" by means of a "contract" between participating individuals. He is noteworthy as an advocate of toleration, though this did not extend to Roman Catholics. PAUL HELM

LOEHE, JOHANNES KONRAD WILHELM (1808-1872). Born in Fuerth, Loehe attended the *Gymnasium* in Nuremburg and the universities of Erlangen and Berlin. He was ordained a Lutheran pastor on 25 July 1831. He served as vicar in Fuerth and in Kirchenlamitz before going to St. Giles in Nuremburg as second pastor and from there briefly to Altdorf, Gertholdsdorf, and Merkendorf before becoming pastor in Neuendettelsau in 1837 where he served the remainder of his life. He married Helene Andreae the same year; she died six years later. Loehe sponsored foreign missions. He was responsible for sending missionaries *(Sendlinge)* to North America. Some of them were instrumental in founding The Lutheran Church—Missouri Synod. After the break with the Missouri Synod (1853), some of Loehe's followers organized the Evangelical Lutheran Synod of Iowa. Loehe also supported mission work in New Guinea. He was a strong supporter of the Innere Mission* movement and founded a deaconess home in Neuendettelsau which was still in 1972 (a hundred years after his death) an important center. Loehe emphasized private confession and made significant contributions to liturgics. His *Three Books About the Church* was valuable for the discussions in Germany about the nature of the church. He published an *Agenda,* a book of church forms and order of services for Lutheran congregations. The periodical he issued, *Kirchliche Mittheilungen aus und ueber Nord-Amerika,* tells about the work of his missionaries in North America. Loehe was known as one of the foremost preachers of his time. He published books of sermons, devotional writings, treatise on liturgical practices, and the like.

CARL S. MEYER

LOGIA (Gr. "sayings"). The term generally used to denote the supposed collection of the sayings of Christ which circulated independently of the gospels in the early church. Some scholars have interpreted Papias's statement that "Matthew compiled the 'logia' in the Hebrew language" to mean that Matthew was responsible for the source "Q." The word "logia" is used also to describe the numerous sayings of Jesus discovered at Nag Hammadi* and Oxyrhynchus.*

LOGOS. The term is used in the NT and in Christian doctrine for Jesus Christ. It derives mainly from the prologue of the gospel of John. *Logos* meant both "word" and the thought or reason expressed in a word. Heraclitus (c. 500 B.C.) conceived it pantheistically as the universal reason penetrating everything, and the Stoics took it over and popularized it as the rational principle inhabiting and governing the universe. Under the influence of Plato's teaching on the eternal forms the idea of the Logos as an immanent power underwent development. It was thought of by Philo as an intermediary agent between God and the world. At the same time, other Jewish thinkers, working from the dynamic conception of the Word in Hebrew thought (Isa. 55:11; Ps. 33:6) and using Greek ideas, aimed at a very similar doctrine of the divine Wisdom (Prov. 8:22-31). Later Jews, writing in Greek, combined the two conceptions, using by preference the term "Logos" (*Wisdom of Sol.* 9:1f.), which is now personified (18:15). Philo took the significant step of making the Logos the intermediary between the transcendent God and the created order.

In the NT Paul calls Christ "the wisdom of God" (1 Cor. 1:24), "the firstborn over all creation," in whom "all things were created" (Col. 1:15ff.). John takes the further step of identifying Christ with the Logos of contemporary Greek and Jewish thought (1:1, 14). The identification is telling; but the fact that the term *logos* is not again used in the fourth gospel in the same sense raises questions as to its importance for the evangelist's doctrine of Christ. These questions are confirmed if the prologue of the gospel was not originally an integral part of the gospel and, moreover, if the evangelist used a hymn which may not have come originally from Christian sources. The extent to which the author of the fourth gospel was indebted to Greek and biblical thought is debated and raises questions which go far beyond his use of *logos.* In no doubt is the Christian stamp which John gives the term. Augustine, commenting on the statement that the Word was made flesh, declared, "This I never read of the *Logos* in the Neoplatonists" (*Confessions* 6:9).

See C.H. Dodd, *The Interpretation of the Fourth Gospel* (1958), pp. 263-85.

R.J. MC KELVEY

LOISY, ALFRED FIRMIN (1857-1940). Founder of Roman Catholic modernism in France. Born in Ambrières, he studied at Châlons-sur-Marne seminary (1874-79) and the Institut Catholique, Paris, under Louis Duchesne, remaining there as professor of Hebrew and exegesis (1884-93) until dismissed for his views on biblical inerrancy. Much shaken in his faith by 1886, he rejected all traditional dogmas and turned to pantheism. Excommunication came in 1908 when he publicly renounced his faith, as he had done his priestly functions in 1906. He was professor of the history of religions in the *College de France* (1909-30) and in the *École des Hautes Études* (1924-27). He never recanted his position, and died without reconciliation to the church.

His *Choses Passées* (1913) and *Mémoires pour servir à l'Histoire religieuse de notre temps, 1860-1931* (1930-31) provide much autobiographical comment: the tortures of his thought, difficulties of conscience, and relationships with colleagues. The lack of intellectual honesty in the church disturbed him profoundly. In 1902 he published *L'Évangile et l'Église* in answer to Harnack's *Wesen des Christentums* (1900), holding that Christianity had developed in a way Christ had not prepared for. This was immediately condemned, as was *Les Évangiles synoptiques* (1908) and *Le Quatrième Évangile* (1903) which respectively defied authorized interpretation and John as author. *Simples Réflexions* was based on the decree *Lamentabili* and encyclical *Pascendi* (1907) as an attack on the authorities in Rome, while *Naissance du christianisme* (1933) sums up his final NT views. A student of biblical criticism with some extraordinary insights, he shifted too often in his views to have any permanently solid conclusions.

See A.R. Vidler, *The Modernist Movement in the Roman Catholic Church* (1934), and M.D. Petre, *Alfred Loisy: His Religious Significance* (1944).

C.G. THORNE, JR.

LOLLARDS. A term applied to the English followers of John Wycliffe.* Although the derivation of the word is not clear it seems to have meant "a mumbler" or "mutterer." The original group of Lollards was composed of Oxford scholars led by Nicholas of Hereford,* the translator of the first Lollard Bible. These students spread their ideas to Leicester, where laymen were won to the cause. From this center William Swinderby led preaching missions to nearby towns. Although the academic followers of Wycliffe's teachings were forced to recant, the movement continued among other classes under the leadership of John Purvey,* Wycliffe's secretary. By 1395 the Lollards had become an organized sect with specially ordained ministers, spokesmen in Parliament, and considerable strength among the middle and artisan classes.

Lollard beliefs are summarized in a document, the *Twelve Conclusions,* drawn up for presentation to the Parliament of 1395. This manifesto expressed disapproval of the hierarchy in the church, transubstantiation, clerical celibacy, the church's temporal power, prayers for the dead, pilgrimages, images, war, and art in the church. Though not mentioned in the *Twelve Conclusions,* the Lollards also felt that the main purpose of priests was to preach and that the Bible should be available in the vernacular for all believers. Due to persecution and the loss of the leadership of scholars such as Wycliffe, the movement came to include many strange extremists.

In 1401 Parliament passed a statute, *De heretico comburendo* ("On the Burning of a Heretic"), aimed specifically at Lollards. This law stated that a heretic convicted by the spiritual court who did not recant, or relapsed, should be turned over to the civil power and burned. Despite this legislation and the measures taken against them by Archbishop Thomas Arundel,* the Lollards remained strong and in 1410 found a leader in Sir John Oldcastle.* He succeeded in identifying Wycliffe's reform of the church with middle-class dissatisfaction with the wealth and conduct of the clergy. Arrested in 1413 for maintaining Lollard preachers and opinions, he was examined and condemned. However, he escaped imprisonment and organized a great Lollard march on London (1414). Henry V and his soldiers dispersed the group, but Oldcastle escaped once again. Later he was caught and hanged. The abortive uprising shattered the power of Lollardy and henceforth it existed as an underground movement. In 1431 another Lollard plot aimed at the overthrow of the government and the disendowment of the church was brought to light.

The continued popularity of the movement may be attested by the appearance of the *Repressor of Overmuch Blaming of the Clergy* (1455) by Reginald Pecock,* a strong attack on Lollard beliefs. There was a Lollard revival in the early sixteenth century in London, East Anglia, and the Chiltern hills. By 1530 this movement began to merge with Protestantism and amplified the undercurrents of dissent and anticlericalism that were present during the reign of Henry VIII. Lollardy facilitated the spread of Lutheranism, helped to make the king's anticlerical legislation popular with the people, and may have created the base for popular nonconformity.

BIBLIOGRAPHY: W.H. Summers, *Our Lollard Ancestors* (1904); J. Gairdner, *Lollardy and the Reformation* (1908); M. Deanesly, *The Lollard Bible* (1920); A.G. Dickens, *Lollards and Protestants in the Diocese of York, 1509-1558* (1959).

ROBERT G. CLOUSE

LOMBARD, see PETER LOMBARD

LONGFELLOW, HENRY WADSWORTH (1807-1882). American poet. Born in Portland, Maine, and educated at Bowdoin College, he became the most popular American poet of his day. After study and travel abroad, he was appointed Smith professor of literature at Harvard University in 1836. His most popular longer works are *Evangeline* (1847), which aroused national interest for its narrative power; *The Song of Hiawatha* (1855), based on Indian legends and sometimes regarded as the American epic; and *The Courtship of Miles Standish* (1858), which popularized the legend of Plymouth Colony. Longfellow also wrote energetic ballads, beautifully reflective lyrics, and many sonnets, some of them among the best written by an American. The tragic death of his second wife Fanny in a fire in 1861 encouraged him to undertake as a source of solace one of his greatest works, a translation of the *Divine Comedy.* Longfellow's poetry reflects the optimistic sentiment and humanitarianism of the day. He was the first American poet to receive wide recognition abroad. There is a bust of him in Westminster Abbey's Poet's Corner.

PAUL M. BECHTEL

LOOFS, FRIEDRICH ARMIN (1858-1928). Lutheran theologian. Born at Hildersheim and educated at Tübingen under Harnack, at Göttingen under Ritschl, and at Leipzig, he became a devoted Ritschlian and was appointed professor of church history at Leipzig in 1886 and at Halle in 1888 where he remained until his death. He played a leading role in Lutheran affairs and became a member of the Saxon Consistory in 1910. He wrote several important monographs on the Fathers and the patristic period, of which *Paulus von Samosata* (1924) is probably the best known, and a notable work *Guidelines to the Study of Dogma* in 1890. A book severely critical of the materialistic philosophy of Haeckel was published in English in 1903; in 1913 there came a christological study, *What is the truth about Jesus Christ?* and in 1914 a series of lectures given in London entitled *Nestorius and his place in the history of Christian Doctrine.*

IAN SELLERS

LOPEZ, GREGORY (A-lou) (1615-1691). First Chinese bishop. Baptized by the Franciscan missionary Antonio de Santa Maria, A-lou studied in Manila, was admitted to the Dominican Order, and in 1656 became the first Chinese to be ordained. Accounts differ about subsequent events, but it seems clear that A-lou, who had assumed the name Gregory Lopez, shrank from ecclesiastical preferment and the offer of a titular bishopric. In 1690, however, a year before his death, he became bishop of Nanking. Thereafter no Chinese bishop was appointed until the twentieth century. Lopez had been involved in the renewed Chinese Rites Controversy,* and favored Matteo Ricci's* sympathetic attitude toward Confucian ritual. LESLIE T. LYALL

LORD HIGH COMMISSIONER. The representative of the sovereign at the general assembly of the Church of Scotland. After the assembly has been constituted and has appointed its moderator, the Lord High Commissioner presents his commission in evidence that he is to be regarded as the royal representative. He is invited to address the assembly, which he customarily assures of the monarch's intention to maintain Presbyterian church government in Scotland. The Lord High Commissioner is not a member of assembly in his official capacity, but may be a member in his private capacity if he holds also a valid commission from a presbytery.

LORD OF MISRULE, see ABBOT OF UNREASON/MISRULE

LORD'S DAY, see SUNDAY

LORD'S PRAYER. The prayer which Jesus taught His disciples, recorded in different forms in Matthew 6:9-13 and Luke 11:1-4. The two versions are introduced differently. Matthew has, "This is how you should pray: ..." Luke has, "When you pray, say: ..." It may be that the prayer was given to His disciples in slightly different forms on two occasions. If, on the other hand, it is the same utterance of Jesus which has reached us in these two forms, most scholars would hold that Luke's version is nearer the original and that Matthew's is an expansion, perhaps in the form which it had already acquired in the worship of the church for which he wrote. The doxology, "For yours is the kingdom and the power and the glory forever," is found only in later MSS of Matthew, and may have been added by a scribe because he was used to saying the prayer with that ending.

The prayer contains much which can be paralleled from pre-Christian Jewish sources, but its originality lies in its compact arrangement and its emphasis on the Fatherhood and the reign of God. The address in Luke is simply "Father"; this would be *Abba* in Aramaic and is a characteristic way in which Jesus addressed God and taught His disciples to do so (cf. Rom. 8:15; Gal. 4:6). Matthew has the more explanatory "Our Father in heaven." The first three petitions deal with God's will and glory. Both gospels have "hallowed be your name." The name implies the character and nature of God which must be honored. The next petition is "your kingdom come," which may be primarily an eschatological prayer for God's final reign to be brought in, but which has present significance as well, as is shown in "your will be done, on earth as it is in heaven" (in Matthew only). The prayer then turns to human need. The petition for daily bread (Matthew "today," Luke "each day") may not be purely material in view of the picture of the messianic banquet and the symbolism of the Lord's Supper. The unusual word *epiousion* probably means "for tomorrow," which may have eschatological overtones. Next comes the request for the forgiveness of "debts" (Matthew) or "sins" (Luke) in the same measure as the petitioner forgives others. Finally there is a petition not to be led into "temptation," this being continued in Matthew with "but deliver us from evil" (or "the evil one"). The thought seems to be of the eschatological trial when faith might give way.

A vast amount of theology and devotion is packed into these few short phrases, and the Lord's Prayer (Lat. *Paternoster* from its first two words) has been used in instruction and worship in almost all sections of the Christian Church from the earliest times.

BIBLIOGRAPHY: F.H. Chase, *The Lord's Prayer in the Early Church* (1891); J. Jeremias, *The Lord's Prayer* (ET 1964); E. Lohmeyer, *The Lord's Prayer* (ET 1965); H. Thielicke, *The Prayer that Spans the World* (ET 1965). R.E. NIXON

LORD'S SUPPER, see COMMUNION, HOLY

LOSEE, WILLIAM (b. c.1764). A Methodist Episcopal Church itinerant preacher from the Lake Champlain area of New York, he organized the first Methodist circuit in Upper Canada. While visiting relatives in the Bay of Quinté area of Upper Canada he preached a number of sermons, and after a few months the people in the district urged him to ask the New York Conference for a permanent minister. Losee himself was sent, and he set up a circuit by 1791. In 1792 the first Methodist church in Upper Canada was built at Adolphustown, and in that year another preacher from New York joined Losee and the circuit was divided, with Kingston marking the boundary. Losee was only the first of a number of itinerant preachers who came to Canada as a result of the second great American frontier revival.

ROBERT WILSON

LOUIS I ("the Pious," also "the Weakhearted") (778-840). Frankish emperor from 814. King of Aquitaine in 781 and co-emperor in 813, Louis was the youngest and the sole surviving son of Charles the Great. In 817 he divided his empire among his three sons, Lothair, Pepin, and Louis. His later attempt to include a fourth son born in his second marriage (819) was thwarted by the other sons. Louis was greatly interested in missions and monastic reform. He wanted mission work to proceed apart from territorial conquest, and as a result he stimulated the creation of a large Scandinavian mission. Anskar* was his chief missionary. For leadership in monastic reform he turned to Benedict of Aniane.* In 815 Louis built a model abbey (Inden) for him near Aachen. In 816, 817, and 818 he gathered all the abbots of the empire to Aachen for meetings with Benedict. The council in 817 endorsed a set of interpretations of monastic rule, but this uniformity and reform were soon shattered by Scandinavian raiders. Louis's reform scheme was too advanced for his age; Benedict's early death left the abbeys without a visible leader. JOHN GROH

LOUIS IX (St. Louis) (1214-1270). King of France. Grandson of Philip II (Augustus) of the house of Capet he inherited the throne as a child (1226) and was long dominated by his mother, Blanche of Castille. His lifestyle was characterized by a devotion to justice that led the French to consider him the ideal Christian king. He dressed modestly, avoiding luxury and ostentation, was deeply pious and ascetic, and delighted in building churches and hospitals. Louis's axiom was when in doubt to favor others above himself, and the poor over the rich. An astute politician and administrator, he systematically extended royal control over the barons, the cities, and the church. He put down several noble rebellions, forbade the construction of private castles, appointed royal manors for the cities and inquisitors to keep watch over the *baillis.*

Louis's commitment to fairness even led him to negotiate a treaty with England unfavorable to himself because he felt that the lands had been unjustly gained by his predecessors. Although he attempted to establish peace in Europe, Louis was no pacifist, for he wished to free the resources of Christendom to make war on the infidel. In 1264 he launched the Seventh Crusade against Damietta in Egypt. The campaign was unsuccessful and Louis was captured, but later freed by the payment of a ransom. Later (1270) he led the Eighth Crusade directed against Tunis in North Africa, when he died of fever. The prestige of the medieval French monarchy reached its zenith under Louis, who was canonized in 1297.

ROBERT G. CLOUSE

LOUIS XIV (1638-1715). King of France. When he began his personal reign (1661), France was the strongest nation in Europe. Within the land, the nobility was subdued and forced to attend the king at his new palace of Versailles (the court moved there in 1682, although the buildings were not completed until 1710). The French administrative structure, the most highly developed in Europe, was centralized in the king. During the seventeenth century, French culture reached its highest point and was consciously and slavishly imitated by other lands. Louis's religious policy emphasized the autonomy of the French Church (Gallicanism*) and he persecuted the Jansenists with papal approval. He also issued the revocation of the Edict of Nantes* (1685) which rejected religious toleration in France and forced the Huguenots to convert to Roman Catholicism or else emigrate. Louis's foreign policy was based upon the desire to extend France to what was considered her natural boundaries. This led to the building of a large war machine and to four major wars. His reign can be divided into two parts at the year 1685. Until that time he was very successful, but in his last twenty-five years when the nations of Europe united against him, the resulting wars drained the strength of France.

BIBLIOGRAPHY: M. Ashley, *Louis XIV and the Greatness of France* (1946); W.H. Lewis, *The Splendid Century* (1953); W.F. Church (ed.), *The Greatness of Louis XIV: Myth or Reality* (1959); J.B. Wolf, *Louis XIV* (1968).

ROBERT G. CLOUSE

LOUIS (Luis) OF GRANADA (1504-1588). Spanish Dominican mystic, preacher, and writer. Born at Granada into the impoverished de Sarría family, he received schooling thanks to the aid of the marquis of Mondéjar and the Dominican priory of the Holy Cross, which he entered in 1524. He distinguished himself in studies and in 1529 was sent to the College of St. Gregory in Valladolid. Here he took the name "Louis of Granada" and was introduced to Christian humanism and perhaps mysticism. Louis was drawn to preaching and refused the offer of a professorship at Valladolid. About 1536 he was sent to restore the dilapidated monastery at Escala Coeli near Cordova. About 1548 he became prior at Badajoz, bringing him into contact with Portugal. A renowned preacher, in 1553 he became confessor for Queen Catherine of Portugal, in 1556 was elected provincial of his order for Portugal and spent the last three decades of his life there. His lasting contributions are his ascetical treatises *On Prayer and Meditation* (1544) and *Sinners' Guide* (1567), through which he decisively influenced, among others, Francis of Sales and Vincent de Paul.

BRIAN G. ARMSTRONG

LOURDES. A famous town on the Gave de Pau River in SW France. It gained fame in 1858 when a fourteen-year-old resident, Bernadette Subirous,* reported eighteen visions of the Virgin Mary between 11 February and 16 July. After the first vision, crowds began to accompany her to the grotto of Massabielle on the riverside, but only Bernadette saw the visions. In one of the visions she was instructed to dig for a spring, which gushed forth as she dug, and it now flows at the rate of 32,000 gallons a day. The spring water is used for sacramental baths by pilgrims. In other visions the Virgin told her that she, the Virgin, was the Immaculate Conception* and instructed Bernadette to have a chapel built and to encourage pilgrims to attend. Bernadette has since been beatified (1925) and canonized (1933) by the Roman Catholic Church.

After a period of opposition, the pilgrimage to Lourdes and the public cult of Our Lady of Lourdes were given official sanction. A Gothic church was constructed immediately, and the magnificent Rosary Basilica was added between 1883 and 1901. In 1891 Leo XIII approved an Office and a Mass of Lourdes for a local province, and Pius X extended it to the universal church in 1907. The pilgrimages reached a peak of six million in the centenary year of 1958, and they still continue at a yearly average of two million. Thousands of cures have been reported at Lourdes; after very careful check by the International Medical Commission in Paris, fifty-eight had been officially designated miracles by 1959.

BIBLIOGRAPHY: R.P. Cros, *Historie de Notre Dame de Lourdes d'après les documents et les témoins* (3 vols., 1925-27); D.C. Sharkey, *After Bernadette* (1945); R. Laurentin, *Lourdes: Histoire authentique des apparitions de Lourdes* (6 vols., 1961-66).

JOHN P. DEVER

LOVE, CHRISTOPHER (1618-1651). Presbyterian divine. Born in Wales and influenced in a Puritan direction by William Erbury, the Independent, he became a "poor scholar" at New Inn Hall, Oxford, graduating in 1639. Disapproving of Laudianism and prelacy, he would not accept episcopal ordination and so sought it from a Scottish presbytery, but without success. On his return he was imprisoned at Newcastle-on-Tyne for his extreme criticisms of the Book of Common Prayer. After his release he became a chaplain in the regiment of Colonel Venn within the army of Parliament in the Civil War. Eventually in 1645 he was ordained by a London presbytery in Aldermanbury Church. A zealous Presbyterian, he had little patience with the Independents and was accused in 1651 of plotting against the Commonwealth—the plot being commonly known as "Love's Plot." One charge against him was that he had corresponded with the young Charles Stuart and his mother, Henrietta Maria. He was condemned to death and executed on Tower Hill. In his lifetime he published several books, but his executors published many more after his death.

PETER TOON

LOVEJOY, ELIJAH PARISH (1802-1837). Presbyterian editor and abolitionist. Born at Albion, Maine, he graduated from Waterville College, taught school in Maine and Missouri, and was converted in 1832 through Presbyterian abolitionist preacher David Nelson. Lovejoy edited the Presbyterian weekly *St. Louis Observer* after attending Princeton Seminary and licensing by the Philadelphia Presbytery in 1833. Gradually adopting abolitionist views and staunchly defending freedom of speech, press, and petition, he became the focus of controversy in pro-slavery St. Louis where, in 1836, his establishment was assaulted because he denounced the lynching of a black man. Moving his press to free soil in Alton, Illinois, and protesting the 1836 Presbyterian general assembly's failure to endorse abolition petitions, Lovejoy made the *Observer* a principal abolitionist periodical with circulation reaching 1,700. Anti-abolitionists, however, unable to drive him from Alton, destroyed his first two presses and shot him to death to destroy a third he was protecting. D.E. PITZER

LOW COUNTRIES. The modern Netherlands and Belgium, the Low Countries, are roughly divided by the Rhine estuary. In Roman times the S Lowlands were a frontier province along the Rhine boundary. Christianity spread there during the third century, and Tongeren and Cambrai were centers of bishoprics (the Armenian St. Servatius, from the E regions of the far-flung empire, was bishop at Tongeren in mid-century). During the fourth century, imperial rule in the West crumbled, and there were only scattered survivals of Christianity in the S Lowlands, now under barbarian rule. The conversion of Clovis,* chief of the Franks, in 496 opened the way to the expansion of the faith; the Merovingian kings encouraged missions, and during the fifth and sixth centuries the S Lowlands were gradually converted (Vaast, Falco, Herebert, Lambert were among the better-known missionaries), the effort culminating in St. Amandus, the "Apostle to the Belgians," around 650. The region north of the Rhine, then known as Frisia, was a strong pagan kingdom, suspicious of Frankish expansion. Mission effort there came from the Anglo-Saxon Church. The isolated effort of Wilfred of York (678) was soon followed by the mission of Willibrord* (690), the "Apostle to the Frisians," who from 695 was bishop of Utrecht. His successor Boniface,* working with the expanding Frankish power, was martyred by the N Frisians (still independent) at Dokkum (754). Mission efforts by Wilehad,* Lebuinus, Ludger,* and others, and the conquest of most of the independent Frisians by Charlemagne, completed the Christianization of the Lowlands by around 800.

The temporary stability furnished by Charlemagne's tribal "empire" was soon followed by the invasions of the pagan Northmen (Vikings), and much of the N Lowlands came under Viking control (Rorik, mid-800s). The Viking storm passed, and the Northmen too were converted. Feudalism emerged from the wreckage of the Carolingian state. The N Lowlands were part of the German feudal "empire" while the south was part of the kingdom of Frankland, or France. The cathedral school at Liège in the ninth century gained fame as the center of learning. In the north the bishop of Utrecht became an important political figure. By around 1050 feudalism had evolved into a system furnishing a relative degree of political stability, and the religious history of the Lowlands merged with the general religious history of feudal Europe. The S Lowlands became a center of commerce and industry; the counts of Flanders played important roles in the Crusades*; the Lowlands produced philosophers (Henry of Ghent,* Siger of Brabant*) and church reformers (Norbert* of Xanten) and in general took part in the religious life of the twelfth and thirteenth centuries.

By the 1300s medieval institutions were in disarray. The church suffered through the papal "Babylonian Captivity"* at Avignon, followed by the Great Schism.* Attempts at reform were many. The Lowlands produced a strong mystical movement (particularly Jan Van Ruysbroeck*), and the "new piety" associated with Gerard Groot* and the Brethren of the Common Life.* This movement, during the 1400s, stressed not only piety (Thomas à Kempis,* *The Imitation of Christ*) but also education; the young Erasmus* was among those trained at its schools. Meanwhile, theologians such as Wessel of Gansfort* and Cornelis Hoen worked out positions which in some ways anticipated Luther. Politically the Lowlands were to some extent united, under the dukes of Burgundy (Philip the Good, Charles the Bold); but towns and nobles alike strove to retain their feudal "liberties." The Burgundian heritage passed to the Hapsburgs, so that Charles V* (as well as being Luther's sovereign) ruled the Lowlands in the early 1500s.

The Lutheran "heresy" found fertile soil in the Low Countries. By the 1520s not only Lutheranism but Anabaptism found adherents. Dutch Anabaptist militants (John of Leyden*) took part in the Anabaptist "New Zion" at Münster: more significantly, Menno Simons* reorganized the peaceful wing of the movement. The "heretics" were sporadically persecuted under Charles V,* with the Anabaptists in particular suffering. In the next generation Calvinism spread rapidly and overshadowed other versions of Protestantism in both north and south. In 1555 the Spanish-born Philip II,* fervently Catholic, striving to centralize the administration of his domains, became the ruler of the Low Countries. The persecution of the Calvinists and other "heretics" was escalated; the Spanish attempts to ride roughshod over the cherished "feudal liberties" of towns and nobles aroused opposition; and by 1568 the Low Countries were in revolt under William of Orange. During the next forty years the fortunes of war changed many times. By the Twelve Year Truce of 1609, the north had gained its independence and become Protestant; the south remained Catholic and under Spanish rule. Renewal of the war (1621-48) did not change the situation.

During the seventeenth century the Netherlands (the N provinces) became a major European power. The Calvinists secured at the Synod of Dort* (1618-19) a victory for scholastic orthodoxy against the followers of Arminius*: The Remonstrants, who wished a Reformed church with a good deal of tolerance in matters of dogma, were expelled from the church. The great universities (e.g., Leyden, Utrecht) became internationally known centers of Calvinist learning, drawing students not only from the English Puritans,* but from all over Europe. But orthodoxy became increasingly defensive, and the church was soon factionalized by controversies over minor doctrinal matters (Voetius* against Cocceius*). In the "Spanish Netherlands," the S provinces, the Counter-Reformation made great strides, and the south became a center for Catholic orthodoxy. It, also, had its internal problems: Jansenism* spread to France, where Port-Royal became its headquarters and Pascal* its most famous defender. It eventually resulted in schism (1713: the Old Catholic Church* of Utrecht). For Catholicism, the N provinces were mission territory; the normal hierarchical organization was replaced by direct control from Rome, through vicar-generals or other special officials. In the north many remained Catholic, but the Reformed church was for all practical purposes the established church.

The "Enlightenment Era" of the 1700s found religion generally on the defensive against new currents of thought, secular in emphasis, stressing reason rather than divine revelation. In the Netherlands the Reformed church settled into what seemed to many to be a petrified orthodoxy, a bulwark of privilege, increasingly infiltrated as the century wore on by Deism.* The common people turned to an experiential Pietism,* while the educated classes lost interest in theology. In the S Lowlands, now under Austrian rule, the Catholic Church likewise took a defensive posture. No longer a center of intellectual development, it could no longer attract the ablest talents. Internally the main controversy of the era arose over the antipapal emphases of "Febronius" (J.N. von Hontheim*).

This era of religious torpidity was not to last. In 1789 the French Revolution broke out, soon to become a "European civil war." Revolutionary troops occupied the Lowlands in the 1790s, set up revolutionary regimes, and took away the privileges of Catholic priest (in the south) and Reformed minister (in the north). Napoleon, the "organizer of the Revolution," viewed religion as a useful ally and was willing (in return for support of revolutionary regimes) to grant recognition to existing religious bodies. After Napoleon's defeat at Waterloo in the S Lowlands (1815), the anti-Revolutionary allies planned to set up a strong state on the northern borders of France. The Lowlands were briefly again united, as a kingdom, under the Dutch leader William I.

Following Napoleon's general approach to religious matters, the king encountered problems. The Belgian Catholic hierarchy opposed his constitution as too liberal; the Belgian "Liberals" (who wanted a share in government for parliament) opposed it as being too conservative. Both groups, temporarily in uneasy alliance, opposed rule by the Dutch. Belgium became independent in 1830-31. The temporary allies soon fell out. The Belgian constitution of 1832 was, for the Catholic hierarchy, far too liberal, and Catholics consequently found it difficult to engage in politics. Agitation on the "school question"—state subsidies for religious schools—grew after mid-century, and (after Leo XIII, in 1879, approved participation in politics in a religiously neutral state) a Catholic political party emerged in the 1880s. It soon became the dominant party, held a majority until World War I, and pushed through legislation giving state aid to religious schools. The introduction of universal suffrage weakened its hold, but it remains a major party. Catholic trade unions and numerous other organizations were formed, making Belgium (like the Netherlands) in some ways a "split society," with religion playing an important role in social and economic life.

In the Netherlands the Reformed church was the major recognized religious body. Partly due to this state recognition, it was tolerant of dogmatic differences. A conservative-pietist wing emerged with the "Awakening" *(Réveil)* of the 1820s (Da Costa,* Groen Van Prinsterer et al.); in 1834 a small group of orthodox split off (the *Afscheiding*, or "Separation"). The main emphasis of the Reformed church as stressed by the "Groningen School"* at mid-century, was on way of life rather than on dogma. Thus the growing tensions between evangelical and modernist could be held in check. In 1886 A. Kuyper* led a second separation (the *Doleantie:* the two separatist groups soon united in the *Gereformeerde Kerk*, the second largest Protestant group in the Netherlands). Catholicism, meanwhile, had allied uneasily with the Liberals in politics. Both opposed the Reformed "establishment." In 1853 the Liberal leader Thorbecke arranged for the restoration of the traditional hierarchy in the Netherlands. But growing agitation on the "school question," among orthodox Calvinists as well as among Catholics, raised problems for the Liberals. Kuyper's charismatic leadership produced a "Monstrous Coalition" between Geneva and Rome in politics, and the coalition of Catholic and orthodox Calvinists parties gained political control from the Liberals and gained state support for religious schools. Catholic and orthodox Calvinist trade unions and other organizations were formed also, thus producing the "splintered society," or "plural society," of the present-day Netherlands.

BIBLIOGRAPHY: J.J. Altmeyer, *Les Précurseurs de la réforme aux Pays-Bas* (2 vols., 1886); P.H. Ditchfield, *The Church in the Netherlands* (1893); H. Pirenne, *Histoire de Belgique* (7 vols., 1902-32); T.M. Lindsay, *A History of the Reformation*, vol. II (1907); J.H. Mackay, *Religious Thought in Holland during the Nineteenth Century* (1911); E.C. Vanderlaan, *Protestant Modernism in Holland* (1924); E. de Moreau, *Histoire de l'Église en Belgique* (2 vols., 1940-48); A. Keller, *Christian Europe Today* (1942).

DIRK JELLEMA

LOYALISTS, see UNITED EMPIRE LOYALISTS

LOYOLA, see IGNATIUS LOYOLA

LUBBERTUS, SIBRANDUS (c.1556-1625). Calvinist theologian, best known as a Contra-Remonstrant* leader in the controversy leading up to the Synod of Dort.* Born at Langweer in East Friesland, Lubbertus studied at Wittenberg, Marburg, and Geneva, where he became an enthusiastic disciple of Beza. After stays in Basle and Heidelberg, he returned (now in his late twenties) to his home territory. He preached in Emden (1583), and at twenty-nine became a professor at the new university at Franeker. As a mature scholar he became increasingly alarmed by the teachings of Arminius* and engaged in active controversy against Vorstius,* Grotius,* and other leaders of the Arminian or "Remonstrant" (after the Remonstrance* of 1610) party. He was an ardent defender of orthodox (or scholastic) Calvinism, an able controversialist, a voluminous pamphleteer. The Remonstrant controversy became entangled with politics, increased in intensity, and led to the Synod of Dort; there Lubbertus, in his sixties, was a leading delegate, and took part in the deliberations which led to the condemnation of the Remonstrants and the triumph of the dogmatic Calvinism he had so long affirmed. He died soon after. DIRK JELLEMA

LUCAR, CYRIL (1572-1638). Patriarch of Constantinople and theologian. He knew much of Western ways, for he studied in Venice and Padua, could read and write Italian with ease, and served the Orthodox Church in Poland as rector of the Vilna Academy. This service in Poland transformed his outlook, for in controversies with the Roman Catholics he found allies in the Protestants. Appointed patriarch of Alexandria in 1602, he became patriarch of Constantinople in 1612. Although deposed several times by the Muslims, he continued as patriarch until his murder at the hands of the troops of Sultan Murad.

Lucar tried to bring the Orthodox Church closer to a Calvinist theological position. He entered into cordial relations with the archbishop of Canterbury and other important Protestant leaders. As a sign of this friendship he gave the Codex Alexandrinus to Thomas Roe, the English ambassador to Constantinople, who presented it to Charles I (it is now in the British Museum). He sent some of his ablest young priests to study at Oxford, Helmstedt, and Geneva. He also allowed his *Confession of Faith* to be printed in Geneva. This thoroughly Calvinist document taught that the church was subject to Scripture and could err; predestination to eternal life irrespective of good works; justification by faith; two sacraments; and a Reformed doctrine of the Eucharist. This statement caused a reaction in Europe. The confession's effect upon the Orthodox Church was limited, however, since it was repudiated shortly after Cyril's death. Finally, in 1672 the great Orthodox Synod of Jerusalem formally condemned the "error" of Protestantism.

See G.A. Hadjiantoniou, *Protestant Patriarch* (1961). ROBERT G. CLOUSE

LUCIANIC TEXT. A revision of the text of the Greek Bible by Lucian of Antioch (c.240-312), which became the standard text of the Eastern Church. It lies behind the so-called Syrian (Westcott and Hort) or Byzantine *(=Koinē)* text of the NT and is thus the ultimate authority for the Textus Receptus, which lies behind the AV (KJV) and other early Protestant translations. This text is characterized by smoothness of language, which is achieved by the removal of barbarisms, obscurities, and awkward grammatical constructions, and by the conflation of variant readings.

LUCIAN OF ANTIOCH (c.240-312). Teacher and martyr. Born at Samosata of distinguished parents, he completed his education at Antioch. Though for a time under the censure of the church because of his theological views, he became the head of the theological school at Antioch and there made his impression. As an exegete he encouraged a literalistic interpretation of Scripture, and so opposed the allegorical methods of Origen. He accepted the preexistence of Christ, but insisted that this had not been from all eternity. Many of his students, who included Arius and Eusebius of Nicomedia, came to occupy the most important sees in the East, and as fellow-disciples of Lucian were sympathetic to Arius. Lucian is often called the father of Arianism.* He produced a very distinguished revision of the Septuagint. The Second Creed of Antioch (341) is reputed to have been written by him and, though this is probably not so, he may have had some connection with it. He was martyred at Nicomedia. C. PETER WILLIAMS

LUCIAN OF SAMOSATA (c.125-c.190). Pagan satirist. Originally a lawyer in Antioch, he turned to that literary creativity in which he was unrivaled by any but Aristophanes. He became well known as a traveling lecturer through Greece, to Rome and beyond, and is a significant witness to human affairs in that urban society of the Greco-Roman world wherein Christianity found its place and began its growth. His satires penetrate into mystery cults, expose religious frauds, and reveal the skepticism of traditional modes of life and thought. Twice he pointed to Christianity, and while his view was that of an outsider, it cannot be said that he was the blasphemer which a later Christian age made him out to be. Christians, with Epicureans, are identified as opponents of the fraudulent prophet Alexander. In order to account for their victimization at the hands of the unscrupulous Peregrinus, he summarizes their behavior in a fashion not far from the Acts portrait of their communal life. CLYDE CURRY SMITH

LUCIFER (d.370/1). Bishop of Cagliari and strong opponent of Arianism.* He was one of the envoys of Pope Liberius to Emperor Constantius in 354, requesting a council to confirm adherence to the Nicene position and to settle the question of Athanasius. The council met at Milan (355) and there, with a few others, Lucifer fiercely resisted the emperor's wishes and refused to sign the condemnation of Athanasius. He was sent into exile, but had to be moved several times in an unsuccessful attempt to silence him. During this period he wrote five aggressive pamphlets to Constantius, revealing a dualistic concept of church and state and arguing that the emperor should be subordinate to the church. On his release, by Julius, he impetuously bypassed the deliberations of the Council of Alexandria (362) and traveled to Antioch to try to deal with the schism. There he consecrated Paulinus as bishop and thus effectively undermined the conciliatory plans of the council and its emissary to Antioch, Eusebius of Vercilli. Incensed by their compromise, he separated himself from them, and a small band of Luciferians sprang up. Lucifer returned to Sardinia and may have died excommunicate. His writings are also important because of the evidence they give as to the pre-Jerome biblical text. C. PETER WILLIAMS

LUDGER (Liudger) (c.744-809). Missionary to the Saxons. Born near Utrecht, he studied under Gregory (of Utrecht) before visiting England and studying under Alcuin at York.* In 775 he was sent to continue the missionary work of Lebuin at

Deventer. Afterward he lived at Dockum and evangelized the Frieslanders. His work was not without opposition. Charlemagne sent him to preach to the Saxons of Westphalia, and about 803 he became bishop of Münster (Mimigernaford). Before this he had established a monastery in the same area. He died on a preaching tour and was buried in the Benedictine monastery at Werden.

LUDLOW, JOHN MALCOLM FORBES (1821-1911). Christian Socialist. Educated in France, where he was influenced by socialists and social Catholics, he came to London in 1838 to study law and was called to the bar in 1843. At Lincoln's Inn he came into contact with F.D. Maurice,* and from Paris during the 1848 Revolution he wrote his famous letter to Maurice, insisting that "the new Socialism must be Christianized." The Chartist fiasco of 1848 united him with Maurice and Charles Kingsley* into the Christian Socialist movement, but it was Ludlow who was the real leader and who supplied the social ideas—cooperative associations being one of his main contributions. With Maurice he edited the short-lived journal *Politics for the People,* and in 1850 edited alone a new journal *Christian Socialist,* which contained the first attempt to state coherently the Christian view of a socialist society. He had wide contacts with trades union and workers' leaders; he had a large part in the Industrial and Provident Societies Act (1852); and he conceived the scheme for the Working Men's College, which he and Maurice opened in 1854, and in which he taught for many years. Ludlow profoundly believed that religious as well as intellectual education must accompany political and industrial emancipation, and this led him to concentrate on educational work in later life, though in no way abandoning the official organs of Christian Socialism.*

See C.E. Raven, *Christian Socialism* (1920), and N.C. Masterman, *John Malcolm Ludlow* (1963). JOHN A. SIMPSON

LUDOLF OF SAXONY (c.1300-1378). Carthusian spiritual writer, author of the most widely read life of Christ written during the Middle Ages. Little is known of his life. He entered the Carthusian Order in 1340 and became prior at Coblenz in 1343. He retired to the charter house at Mainz in 1348 and died at Strasbourg thirty years later. He is best known for his book, *The Life of our Lord Jesus Christ.* It was first printed in 1474 and since that time has been translated into numerous languages and has had over sixty editions. The book was the most extensive life of Christ written until that time. It dealt with the events and teachings recorded in the gospels and included commentaries by church fathers and medieval writers as well as prayers and moral instructions. The work later influenced Ignatius Loyola, who utilized it in writing his *Spiritual Exercises.* Ludolf also wrote a commentary on the Psalms which was first printed in 1491.

RUDOLPH HEINZE

LUGO, see DE LUGO, JUAN

LUIS OF GRANADA, see LOUIS OF GRANADA

LUKE, GOSPEL OF, see SYNOPTIC GOSPELS

LUKE THE EVANGELIST. Luke, a physician and companion of Paul on his missionary journeys (Col. 4:14; Philem. 24; 2 Tim. 4:11), is generally though not universally agreed to have been the author of the third gospel and the Acts of the Apostles. It is widely thought that he joined the Pauline party where the "we" passages begin in Acts (16:10), and he remained as Paul's only companion at the end of the apostle's life (2 Tim. 4:11). Attempts have been made to identify him with either Lucius of Cyrene (Acts 13:1) or Lucius a kinsman of Paul at Corinth (Rom. 16:21). These were Jews, and Luke has usually been held to be a Gentile in view of the universalistic outlook of his gospel and his being distinguished from the circumcision party in Colossians 4:10-14. But the Jewish nature of the early chapters of the gospel has been stressed, as has been the emphasis Luke puts on Jerusalem.

Some have also challenged the view that "the circumcision party" in Colossians 4 refers to Jewish Christians as a whole, and have suggested that it means those Jewish Christians who were strict in the observance of the Law, but did not (like the Judaizers) try to force their practice upon others. If these views are correct, Luke may have been a Hellenistic Jew. He was not an eyewitness of the ministry (Luke 1:2), and is therefore unlikely to have been a member of the Seventy (Luke 10:1) or the companion of Cleopas on the road to Emmaus (Luke 24:18). The anti-Marcionite prologue to his gospel says he died unmarried in Boeotia, aged eighty-four. He is the patron saint of doctors. R.E. NIXON

LULL, RAYMOND (c. 1232-1316). Franciscan missionary, mystic, and scholar. From Palma (on Majorca in the Balearic Islands), he was educated as a knight and then converted from a life of dissipation (1263). Resolving to dedicate himself to winning Muslims to Christ, Lull learned Arabic. He also had a vision which revealed to him a method of approaching Muslims with the Christian message. Then he persuaded James II of Majorca to found a monastery at Miramir where Franciscans could study Arabic and the art in his method in order to prepare for missionary work among the Moslems. Lull taught at Miramir, Montpellier, and Paris. He took missionary journeys to Tunis and Algeria and tried unsuccessfully to enlist the rulers of Europe in his projects. The traditional account of his martyrdom in North Africa does not seem to be true; he probably died on Marjorca.

Lull was the first Christian theologian of the Middle Ages to use a language other than Latin for his major works. He wrote in Catalan and Arabic, in addition to Latin, and produced 290 books of which 240 survive. His writings center about his art, a method to demonstrate the unity of all truth. He attempted to work out a system by which all possible knowledge could be reduced to, or derived from, certain first principles. The art, he believed, would lead to a unification of the

Greek and Latin churches and to the reunification of all mankind through Christianity. In addition to refuting Islamic teaching, Lull also struggled against the "Averroists" such as Siger of Brabant.*

BIBLIOGRAPHY: S.M. Zwemer, *Raymond Lull, First Missionary to the Moslems* (1902); E.A. Peers, *Ramon Lull* (1929); F.A. Yates, *The Art of Memory* (1966). ROBERT G. CLOUSE

LULLY, JEAN-BAPTISTE (1632-1687). Italian composer. Of humble origin, he rose to a position of extraordinary influence at the court of Louis XIV. He was the virtual creator of French opera, successfully adapting recitative, which had been developed in Italy, to the French language. He also wrote lengthy motets for the royal chapel, in which he employed soloists, large choral forces, and orchestra. His followers, LaLande and Charpentier, wrote more extensively in the sacred categories, using his developments; the latter distinguished himself particularly in oratorio. Charles II of England sent Pelham Humfrey to study with Lully and apply his style to music for the English court. Thus the church music of the Restoration period was strongly influenced by French models, an influence which extended to the church music of Henry Purcell.*

J.B. MAC MILLAN

LUND CONFERENCE (1952). Conference of the "Faith and Order" commission of the World Council of Churches,* held at Lund, Sweden. There were 225 delegates from 114 churches; Roman Catholic observers were present. Documents issued following the 1937 Edinburgh Faith and Order conference formed the basis for study. Doctrinal differences of the churches were listed under the following heads: definition and limits of the church; church continuity and unity; the goal of the reunion movement; the number and nature of the sacraments and their relation to church membership; Scripture and tradition; priesthood and sacrifice. After stating the diverse views on these topics, the conference concluded that "comparative ecclesiology" (i.e., comparing and contrasting different convictions about the church) had been pursued to the limits and offered no prospect of reconciliation. "We need, therefore, to penetrate behind our divisions to a deeper and richer understanding of the mystery of the God-given union of Christ with His Church." The conference selected four major points for study for at least ten years: union of Christ and the church; tradition and traditions; ways of worship; institutionalism (the church as a sociological entity with its laws and customs).

See O.S. Tomkins (ed.), *The Third World Conference on Faith and Order, Lund, 1952* (1953).

J.G.G. NORMAN

LUTHARDT, CRISTOPH ERNST (1823-1902). Lutheran scholar. Born at Maroldsweisack, he was educated at Erlangen and Berlin and became professor of systematic theology at Marburg in 1854 and at Leipzig in 1856. His *Apologetic Lectures on Christianity* (1864f.), *Compendium of Dogma* (1865), and *Compendium of Religious Ethics* (1896) reveal him as a conservative Lutheran scholar, and in translation brought him acknowledgment in the English-speaking world, where his Johannine studies were also popular. In 1864 he attacked the skeptical portraits of Jesus by Strauss and Renan, but during the next few years, when German unification was being accomplished, he as a loyal Saxon and president of the Leipzig Mission Society led the opposition to the Prussianization of the religious institutions of his kingdom. He helped found the General Evangelical-Lutheran Conference in 1868 and edited its *Kirchenzeitung.* He also edited the *Zeitschrift für Kirchliche Wissenschaft* (1879-89) and the *Theologische Literaturblatt* (1890-1902).

IAN SELLERS

LUTHER, MARTIN (1483-1546). Born in Eisleben, Luther attended the *Ratsschule* (city school) in Mansfeld, came under the influence of the Brethren of the Common Life* while in Magdeburg, and continued his preparatory training in the *Georgenschule* in Eisenach, where he was a member of the Cotta and Schalbe circle, before enrolling at the University of Leipzig (1501). At Leipzig Jodocus Trutvetter, a Nominalist *(via moderna),* seems to have influenced him most. Luther received his B.A. in 1502 and the M.A. in 1505. In July of that year Luther entered the chapter house of the Hermits of St. Augustine in Erfurt as a novice, due to a vow made in "a moment of terror," when thrown to the ground by a bolt of lightning during a thunderstorm. However, he was troubled about his salvation before this, and other incidents probably led to the decision to become a monk. In the monastery he pursued some theological studies and was ordained priest in 1507.

In 1508 Luther was transferred to the University of Wittenberg, where he earned the *Baccalaureus Biblicus* degree in 1509 and the doctor of theology degree in 1512. During these years he lectured on moral theology, the *Sentences* of Peter Lombard, and the Bible. Between November 1510 and March 1511 he was on a journey to Rome as a companion of a fellow friar on business for his order. With the doctor's degree Luther received the permanent appointment to the chair of *lectura in Biblia* at Wittenberg.

During these years before he became a doctor of theology, Luther was wrestling with the problem of his personal salvation. While in the monastery and as friar in Wittenberg he assiduously performed the required tasks and offices, frequently went to confession, and fulfilled the imposed penances. The problem of the dating of his "Tower Experience," when he came to a full realization of the meaning of justification by grace alone, has occupied numerous scholars and has not been fully solved. Some have placed it in 1514; others have put it as late as 1518. The continued study of Scripture, the influence of Augustine, the writings of John Tauler* and other mystics, the *Psalterium Quintuplex* of Jacques Lefèvre d'Étaples,* and the advice of his superior Johann Staupitz* were determinative probably in that order in clarifying his thoughts and convictions, and no one date or moment can be predicated. In 1518 his theology of the Cross was thoroughly

Pauline, and Luther was the champion of *sola fide* (faith alone), *sola gratia* (grace alone), *sola Scriptura* (the Bible alone).

In his writings between 1516 and 1518 Luther evidences his Augustinianism. In *Two Kinds of Righteousness* (1518), Luther clearly speaks of Christ and His work from His birth to His death and resurrection as constituting the righteousness of believers; the promise is the assurance of faith for them. These thoughts are present in his lectures on the Psalms and on Hebrews (1518). By this time Luther had issued his famous protest against the scandals of the indulgence traffic, the Ninety-Five Theses* of 31 October 1517. The most noteworthy of these is number 62: "The true treasure of the church is the most holy Gospel of the glory and grace of God." His *Explanations of the Ninety-Five Theses* (1518) states, "The merits of Christ perform an alien work." The alien righteousness is the righteousness which he defines in his sermon on the *Two Kinds of Righteousness* as "the righteousness of another, instilled from without, the righteousness of Christ by which He justifies through faith."

The controversy brought about polemics from Rome, tracts and treatises, blunt and subtle attempts to silence him, a debate with Johann Eck* at Leipzig (1519), and an interview with Tommaso de Vio (Cardinal Cajetan*) in Augsburg (1519). In 1519 Charles V was elected emperor of the Holy Roman Empire and soon became aware of the magnitude of the religious problem in Germany, because of Luther's writings and the anticlericalism of the German people.

The year 1520 marked the appearance of some of Luther's most important reformatory writings. The *Treatise on Good Works* (May 1520) used the Decalogue as a basis for showing how faith is implemented in the life of the believer. In his *Sermon on the Mass* (April 1520) he taught that every Christian is a priest. *On the Papacy at Rome* (June 1520) branded the pope as "the real Anti-Christ of whom all the Scripture speaks." In *The Address to the German Nobility* (August 1520), Luther disallowed the authority of the pope over temporal rulers, denied that the pope was the final interpreter of Scripture, decried the corruption of the Curia, affirmed again the universal priesthood of the believers, and spelled out a program of the church reforms. *Concerning the Babylonian Captivity of the Church* (October 1520) reduced the number of sacraments from seven to two and aroused the ire of Henry VIII* of England. *The Freedom of a Christian* defended two propositions: "A Christian is a perfectly free lord of all, subject to none. A Christian is a perfectly dutiful servant of all, subject to all."

In April of the next year (1521) Luther stood before the emperor and the estates of the empire in the Diet of Worms,* declining to recant unless overcome by Scripture. Taken to Wartburg Castle by order of Frederick the Wise,* elector of Saxony, Luther had the opportunity to continue his writings and especially to translate the NT from Greek into German (the "September Testament" of 1522). The translation of the entire Bible was not completed until 1534, perhaps the greatest single achievement of the great Reformer.

Although condemned by the Edict of Worms and declared an outlaw, Luther returned to Wittenburg in March 1522 to cope with the Wittenburg Movement, braking its radical direction, assuring the essentially conservative character of his reformation. This conservative character is evident in Luther's revised *Order of Mass and Communion* (1523), the *Order of Baptism* (1523), and the *German Mass and Order of Service* (1526), collects and chants, orders for occasional services, a *Litany*, a German *Te Deum* and a *Magnificat*, and hymns for congregational singing. Among these *Ein' feste Burg* must be singled out. The conservative nature of the Lutheran Reformation was affirmed, too, by Luther's not depending on the knights, the humanists, or the peasants.

Luther's break with Erasmus* resulted from Erasmus's *Diatribe on Free Will* (1524). In his answer, *The Bondage of the Will* (1525), Luther affirmed that man cannot will to turn to God or play any part in the process leading to his own salvation. He granted that man has freedom regarding "things below him."

Luther's position in the Peasants' Revolts* (1524-25) upheld authority, denied the right to rebel, called for social justice, and urged consideration for the economic welfare of the lower classes. His language was intemperate in urging the princes to put down revolt, and alienated some of the lower class.

Between 1525 and 1529 Luther carried on a controversy with Ulrich Zwingli* of Zurich and others regarding the Lord's Supper. He took the words of institution, "This is my body" and "This is my blood," in a literal sense and opposed all attempts to interpret them figuratively. *The Sacrament of the Body and Blood of Christ—Against the Fanatics* (1526) and *That These Words of Christ, "This is My Body," etc. Still Stand Firm Against the Fanatics* (1527) define his position, often erroneously labeled consubstantiation. Philip of Hesse* attempted to bring about a reconciliation by calling the Colloquy of Marburg* (1529). Fourteen articles of the Christian faith were agreed on by the participants (Luther, Melanchthon,* Zwingli, Oecolampadius,* Bucer,* and others), but no agreement was reached on the fifteenth, regarding the Lord's Supper. However, in 1536 Bucer and Luther agreed on the Wittenberg Concord.*

Luther showed his concern for education by writing his appeal *To the Councilmen of All Cities in Germany, that they Establish and Maintain Christian Schools* (1524) and his *Sermon on the Duty of Sending Children to School* (1530). His preface to Melanchthon's *Instructions for the Visitors of Parish Pastors in Electoral Saxony* (1528) is evidence, not only of his churchmanship, but also of his genuine desire to foster schools. Both his "Large Catechism" (1529) and his "Small Catechism" (1529) came in response to the findings of the visitation. They are his efforts to raise the level of the understanding of Christian doctrine. His Small Catechism has been dubbed the "Layman's Bible." Moreover, Luther contributed to the revision of the theological curriculum of the University of Wittenberg; he was continually occupied in training pastors and preachers for the Lutheran churches; and he expended many efforts on behalf of non-German students who came to Wittenberg.

The history of Luther's life is not simply the history of the professor of Wittenberg. He was the leader of the movement of the professor that spread through much of Germany and the Scandinavian countries. Although he could not attend any of the diets of the Holy Roman Empire, because of the Edict of Worms, he was frequently consulted at his quarters in Castle Coburg during the Diet of Augsburg (1530). He prepared the Smalcald Articles* (1537) for the Smalcald League* in preparation for the council which was summoned, but did not meet, at Mantua. In 1535 his *Lectures on Galatians* were published; they are regarded by some scholars as his most profound theological treatise. In that year he began his *Lectures on Genesis*, which he completed ten years later, shortly before his death.

In 1525 Luther married Katharina von Bora, an ex-nun. His family life was a happy one. His home was a gathering place of friends and students, and the voluminous *Table Talk* was recorded by at least ten students between 1531 and 1544. Luther died in the town in which he was born (Eisleben) while on a mission to reconcile the princes of Anhaldt. His greatness can be gauged from the fact that during the four-hundred-plus years since his death, more books have been written about him than about any other figure in history, except Jesus of Nazareth.

BIBLIOGRAPHY: J. Besizing, *Lutherbibliographie: Verzeichnis der gedruckten Schriften Martin Luthers bis zu dessen Tod* (1966).

D. Martin Luthers Werke: Kritische Gesamt-Ausgabe (1883-), known as the Weimar Ausgabe (WA) is still being published, and includes his letters, German Bible, and *Table Talk.* The most comprehensive English edition (55 vols., 1957-) is still in production.

W. Pauck (ed.), *Lectures on Romans* (1961); T.G. Tappert (ed.), *Luther: Letters of Spiritual Counsel* (1955); J. Atkinson (ed.), *Luther: Early Theological Works* (1962); G. Rupp and P. Watson (eds.), *Luther and Erasmus: Free Will and Salvation* (1969); B.L. Woolf (ed.), *Reformation Writings of Martin Luther* (2 vols., 1952, 1956); P. Smith and C.M. Jacobs (eds.), *Luther's Correspondence* (2 vols., 1913, 1918).

Among biographies of Luther in English the following may be singled out: R. Bainton, *Here I Stand* (1950); E.G. Schwiebert, *Luther and His Times* (1950); U. Saarnivaara, *Luther Discovers the Gospel* (1951); H. Lilje, *Luther Now* (ET 1952); G. Rupp, *The Righteousness of God* (1953); R.H. Fife, *The Revolt of Martin Luther* (1957); H. Bornkamm, *Luther's World of Thought* (ET 1958); F. Lau, *Luther* (ET 1963); J.J. Pelikan, *Obedient Rebels* (1964); J.M. Todd, *Martin Luther* (1964); A.G. Dickens, *Martin Luther and the Reformation* (1967).

CARL S. MEYER

LUTHERAN CHURCH BODIES IN THE USA. There were three major Lutheran churches in the USA as of 1974 and nine minor bodies.

(1) *The Lutheran Church in America* was organized in 1962, a merger of the Augustana Synod (founded in 1860), the American Evangelical Lutheran Church or the "Danish Evangelical Lutheran Church" (founded in 1878), the Suomi Synod or the Finnish Evangelical Lutheran Church (founded in 1890), and the United Lutheran Church in America (organized in 1918).

The United Lutheran Church was the largest and most influential of the church bodies that merged in 1962. Its roots go back to colonial America and the Pennsylvania Ministerium, formed in 1748 by H.M. Muhlenberg.* Lutherans settled early in New York, Delaware, New Jersey, Maryland, Virginia, the Carolinas, and Georgia (the Salzburgers). By 1820 some of these were organized into state synods; they federated in the General Synod in 1820. In 1867 the General Council was organized, made up in part of synods previously belonging to the General Synod. The General Synod South was composed of synods that had been formed in the South and had separated from the General Synod because of the slavery issue. These three groups (the General Synod, the General Synod South, and the General Council) united in 1918 to form the United Lutheran Church in America.

The Augustana Evangelical Lutheran Church in America was organized in 1860 among Scandinavian immigrants in America's Midwest. In 1870 the Norwegian constituents withdrew to form their own synod. The Augustana Synod expanded rapidly between 1871 and 1910. The period of Americanization followed between 1910 and 1930. The Augustana Synod was a charter member of the Lutheran World Federation and also of the American Lutheran Conference formed in 1930. The other two synods that formed the Lutheran Church in America were relatively small church bodies. In 1962 the United Lutheran Church numbered about 2,495,000 members; the Augustana Synod, 618,000; the Suomi Synod, 35,500; and the AELC, or Danish Synod, 24,000.

The Lutheran Church in America has a highly centralized administration. It has thirty-one constituent synods. In 1971 the LCA numbered about 3,229,000. Foreign missions, theological education, social ministry, and publications are the responsibility of the LCA as a whole, and not of its synods. Its theology has been within the framework of Lutheranism's tenets, but it has generally been regarded as more receptive to advanced theological thinking than the other two large church bodies within American Lutheranism.

(2) *The American Lutheran Church* is also the result of a merger of previously formed synods. Organized in 1960 it merged (the) American Lutheran Church, the Evangelical Lutheran Church, the Lutheran Free Church, and the United Evangelical Lutheran Church. The Lutheran Free Church was a small group of Norwegian Lutherans, the followers of Georg Sverdrup* and Sven Oftedal, organized in 1890. The United Evangelical Lutheran Church was a Danish group, organized in 1896. American Lutheran

Church (as distinguished from The American Lutheran Church) was organized in 1930, the merger of the Buffalo synod (1818), the Iowa Synod (1854), and the Ohio Synod (1818). These were made up of German immigrants and Germans who had moved westward into the Ohio region in the "great crossing" of the early nineteenth century.

The Buffalo Synod, under the leadership of J.A.A. Grabau, held a "high church" view of the ministry. It early came into conflict with the Missouri Synod, a conflict which resulted in the breakaway of some of its members as a result of the Missouri-Buffalo Colloquy (1866). The Iowa Synod was organized under the leadership of Georg Grossmann and Johannes Deindoerfer as a result of the differences between the Missouri Synod leaders and Wilhelm Loehe.* It emphasized "open questions" in doctrinal formulations. Michael Reu was its outstanding leader in the twentieth century.

The Ohio Synod was organized partly because of the geographical barriers with the Lutherans in Pennsylvania under the leadership of Paul Henkel. In 1868 it reached an accord with the Missouri Synod and became a member of the Evangelical Lutheran Synodical Conference when organized in 1872. It severed its connections with that group in 1881 as a result of the Predestinarian Controversy. Its attempts to draw closer to the Iowa Synod culminated in 1930 with the formation of (The) American Lutheran Church.

The Evangelical Lutheran Church reaches back into the history of Norwegian immigrations and the formation of various Norwegian church associations in the nineteenth century. The Norwegian Synod (1853) was the largest of these. The 1890 merger of four Norwegian bodies resulted in the United Norwegian Lutheran Church in America. In 1917 the Norwegian Synod, the United Church, and the Hauge's Synod joined to form the Norwegian Lutheran Church in America, which changed its name to Evangelical Lutheran Church in 1946. Among the various Norwegian church bodies over the years the Norwegian Synod alone joined the Synodical Conference, of which it was a member from 1872 to 1883. In 1930 (the) American Lutheran Church and the Evangelical Lutheran Church joined with the Augustana Synod to form the American Lutheran Conference, which was dissolved in 1954. In 1971 the ALC numbered about 2,540,000 members.

(3) *The Lutheran Church—Missouri Synod,* organized in 1847, has assimilated other Lutheran synods, but has not merged with any large Lutheran bodies. Its founding fathers were made up of emmissaries sent over by Wilhelm Loehe as missionaries to the German Lutherans in America and followers of Martin Stephan who settled in Perry County and St. Louis, Missouri. C.F.W. Walther* was its first president and acknowledged theological leader until his death in 1887. The Evangelical Lutheran Synodical Conference (1872) united like-minded, confessional Lutheran synods into a loose federation, which existed until 1970. The Missouri Synod has been rigidly confessional in its theology, subscribing to the Lutheran Confessions because (not in so far as) they were in conformity with the Scriptures. In its church polity it has been congregational, the synod having only advisory jurisdiction over the congregations. Between 1910 and 1930 the process of Americanization produced a language transition from German to English and altered attitudes within the synod.

The Missouri Synod has supported a strong system of parish or parochial schools. It has maintained a system of preparatory schools for training future professional workers within the synod, two theological seminaries, and two teacher training institutions. Especially since 1894 it has been active in foreign missions and since World War II has been increasingly concerned about social questions. The "International Lutheran Hour," a radio ministry, has been under the sponsorship of the Lutheran Layman's League, an organization within the Missouri Synod. The Missouri Synod numbered about 2,877,000 members in 1971.

(4) *Independent Lutheran church bodies* in America in 1972 are the Wisconsin Evangelical Lutheran Synod (1853); the Evangelical Lutheran Synod of Canada; the Evangelical Lutheran Synod (1918); the Church of the Lutheran Confession; the Apostolic Lutheran Church (1961); the Church of the Lutheran Brethren; the Eielson Synod; the Association of Free Lutheran Congregations; the Fellowship of Authentic Lutherans (1971). The total number of Lutherans in the USA in 1971 was 8,872,000 members.

(5) *The Lutheran Council in the USA* was organized in 1967 as a federation of The Lutheran Church—Missouri Synod, The American Lutheran Church, the Lutheran Church in America, and the small Synod of Lutheran Churches which merged with the Missouri Synod in 1970. LCUSA continues the work of the National Lutheran Council, organized in 1918. The Lutheran Council has a theological commission to which all three members must belong for the purpose of carrying on theological discussions. Membership in its other commissions, e.g., publicity, armed services, student welfare, is voluntary. Public relations and welfare have been major concerns. However, the Lutheran Council has fostered dialogues with the Roman Catholic, the Presbyterian, and the Episcopalian churches. It operates independently of, but in coordination with, the Lutheran World Federation.

See J. Bodensieck (ed.), *The Encyclopedia of the Lutheran Church* (3 vols., 1965) as the best reference work for American Lutheranism.

CARL S. MEYER

LUTHERANISM. The system of religious beliefs ascribed to the followers of Martin Luther* is generally called "Lutheranism." The term may also be used in reference to the activities of the churches calling themselves "Lutheran." Both of these aspects are discussed in this article.

Lutheranism's doctrinal position is embodied in the Book of Concord* (1580), consisting of the three ecumenical creeds, the Augsburg Confession* and its Apology, Luther's Small and Large Catechisms, the Smalcald Articles,* and the Formula of Concord.* Justification by grace alone through faith in Jesus Christ is the primary doctrine accented by Lutheranism. Because of original sin man is in need of reconciliation with God. Reconciliation and the forgiveness of sins are the essence of justification; the righteousness of Christ is imputed to the believer who accepts it by the action of the Holy Spirit. It is not man's merit or works, but solely the grace of God that makes him justified before God. *Sola gratia* and *sola fide* are the phrases used to summarize this doctrine, most carefully explained in Article IV of the Apology of the Augsburg Confession. Good works are the fruit of faith. A good tree bears good fruit, Luther said, so a believer does good works. These good works, too, are the fruit of the Spirit. The believer, *simul justus et peccator* ("justified but still a sinner"), strives against evil and labors to do good. His spiritual life, according to Lutheran doctrine, is endangered and sustained by the means of grace. These are baptism, the Word, and the Lord's Supper.

The Word of God in Lutheranism is often equated with the canonical writings of the Old and New Testaments. These have been given by divine inspiration and therefore are authentic, reliable, and able to accomplish their divine purposes. Their purpose is, first of all, "to make wise unto salvation through faith which is in Christ Jesus." They are also meant, according to Lutheranism, to instruct men in questions of morality, to comfort him in tribulation, to refute those who repudiate the Christian religion, and to teach divine doctrine. By the Word the Holy Spirit calls men, enlightens and instructs them, sanctifies them, and gathers them into His church. The Scriptures are the sole source, rule, and norm of faith, *sola Scriptura.*

The sacraments—baptism and the Lord's Supper— are more than cultic rites. Baptism is regarded as the water of regeneration, a means by which the new birth is effected, especially in infants. The Lord's Supper is not a mere memorial meal, but was instituted by Christ for the forgiveness of sins, the strengthening of faith, and as an expression of union with Him and with fellow believers. In the bread and wine of the Holy Supper the body and blood of Christ Himself are present, Lutherans teach. They believe in the Real Presence (not to be termed "consubstantiation").

Lutheran theology is strongly christocentric. The message of Christ's redemptive act is the central message of the Scripture; being reborn in Christ and rising to newness of life are the essence of baptism; Communion with Christ and partaking of His body and blood are the essential of the sacrament of the altar. *Solus Christus* is the heart of Lutheran theology.

Lutheranism emphasizes the differences between Law and Gospel. The Law condemns; the Gospel saves. The Law terrorizes; the Gospel comforts. The Law reveals God's wrath; the Gospel reveals God's grace. Regarding predestination, Lutheranism teaches that God has elected certain men for salvation in Christ Jesus before the beginning of the world. This doctrine is given for the believer's comfort, to assure him of his salvation. Lutheranism does not teach an election to reprobation.

Christ Jesus, in Lutheran theology, is true God and true man. With the Father and Holy Spirit He is a member of the Holy Trinity, of the one Godhead. He became incarnate, born of the Virgin Mary, in order to fulfill the law, suffer, die, and rise again for the atonement of mankind. In him the two natures, the human and the divine, are united in one Person. This union is true and real, a personal one, and a perpetually enduring one. Lutheranism adheres to the formula of the Council of Chalcedon*: "We confess one and the same Jesus Christ, the Son and Lord only-begotten, in two natures, without mixture, change, division, or separation."

To Him is given the Headship of the church. The church is made up of all those, but only those, who trust in Christ Jesus as their Savior, Redeemer, and Mediator. They have entered into a saving fellowship with Christ. The church is holy, because its members are sanctified by the Holy Spirit. It is one, because it has one Lord and is united to Him; it is apostolic, because it is founded on the proclamation of the apostles, the gospel of Christ; it is catholic, or universal, because it is not restricted to one people, nation, or time. Lutheranism, too, speaks of the church invisible and the church visible. The church invisible is not discernible in a structure; the church visible is made up a structure. The marks of the true church, according to Lutheranism, are the pure preaching of the Word of God and the correct administration of the sacraments according to Christ's institution.

Lutherans do not insist on a uniform church polity. Some of the Lutheran churches are episcopal in character; some are congregational; others tend toward a presbyterial form of organization. Some are supported by the state; others are free churches or voluntary ecclesiastical societies. In its worship services Lutheranism tends to be ritualistic. Luther's "conservative reformation" retained much of the liturgy of the Western Catholic Church. There are Lutheran churches in the twentieth century, however, that have plain orders of service. The rites and ceremonies are regarded as *adiaphora* ("things indifferent") so long as the Gospel is not vitiated or nullified by them. The holy ministry has been instituted by Christ. Ordination is a good ecclesiastical custom, stemming from the ancient church, but not absolutely necessary. Believers in Christ are a royal priesthood.

The largest concentration of Lutherans is found in Germany and Scandinavia. In the USA, Lutheran membership was 8,872,000 in 1971. Lutheranism is found in Africa, South and Central America, Canada, Japan, India, Korea, the USSR, and possibly even in China. The largest of the territorial churches in Germany is the Evangelical Lutheran Church of Hanover, with approximately 4,-000,000 members. The Evangelical Lutheran Church of Saxony numbers 3,600,000 members. In Württemberg, Schleswig-Holstein, and Bavaria there are 2,500,000 Lutherans each; in Mecklenburg and Thuringia, a million Lutherans each. The Scandinavian countries are almost entirely Lutheran. The Evangelical Lutheran Church in

Denmark numbers 4,300,000 members; that in Finland, 4,375,000 members. The Church of Norway is somewhat smaller, numbering about 3,500,000 members. The Church of Sweden is the largest of the Lutheran Scandinavian churches with 7,000,000 members.

The Lutheran World Federation is the ecumenical voice of Lutheranism. It was organized at Lund, Sweden, in 1947, having been preceded by four Lutheran World Conventions (Eisenach, 1923; Copenhagen, 1929; Paris, 1935; Lund, 1947). Assemblies of the Lutheran World Federation have been held in Hanover (1952), Minneapolis (1957), Helsinki (1963), and Evian, France (1970). The constitution of the LWF gives its doctrinal basis in Article II: "The Lutheran World Federation acknowledges the Holy Scriptures of the Old and New Testaments as the only source and the infallible norm of all church doctrine and practice, and sees in the three Ecumenical Creeds and in the Confessions of the Lutheran Church, especially in the Unaltered Augsburg Confession and Luther's Small Catechism, a pure exposition of the Word of God."

The functions of the Federation are to further a united witness to the gospel of Jesus Christ; to cultivate unity among the Lutherans of the world; to foster Lutheran participation in ecumenical movements; to provide a channel for meeting the physical needs of the destitute; most importantly, to "support Lutheran Churches and groups as they endeavor to meet the spiritual needs of other Lutherans and to extend the Gospel." The Lutheran World Federation consists of an assembly which meets normally every six years, an executive committee which meets annually, national committees, and commissions. Its headquarters are at Geneva, Switzerland. There, too, the executive staff, headed by a general secretary, is located.

The department of theology of the LWF has made some basic studies into the problems facing Lutheranism in the twentieth century. Its studies have centered on the unity of the church, freedom and unity in Christ, justification, and the church and her confessions. *The Lutheran World* is a quarterly journal published by the LWF.

Lutheran doctrine, polity, church structures, and federations cannot be treated as completely united. Even in doctrine Lutherans are not totally at one. Some Lutherans and Lutheran church bodies have been influenced greatly by modern biblical criticism in recent decades and have repudiated strongly held Lutheran beliefs. In the age of the Reformation there were the so-called Crypto-Calvinists, Lutherans who held to the Calvinistic doctrines of the Lord's Supper. In the seventeenth and eighteenth centuries Pietism* gained a strong foothold in Lutheranism, both in Germany and in Scandinavian countries. Rationalism muffled Lutheranism's doctrinal accents, which were heard again only as a result of the revival of Lutheran Confessionalism in the nineteenth century. Lutheranism in North America, in general, has been more conservative than European Lutheranism.

Lutheranism's involvement in the questions of society and social welfare have varied from time to time and country to country. Lutheranism did not fail to emphasize the need to "love thy neighbor as thyself" and to be helpful in meeting his bodily needs. Due to the state control practiced by the Lutheran countries of Europe, this was often made a matter for the state rather than the church. In North America, Lutheran quietism resulted in a "hands off" policy in political and societal problems that persisted into the twentieth century.

Lutheranism's contributions to music, literature, the arts, and sciences cannot be recounted here. Its contributions are not confined to the "Lutheran" countries, but are evident in many parts of the world. Lutheranism's outreach has been an extensive one in the nineteenth and twentieth centuries, and its influence pervasive. Lutheranism has emphasized the educational aspects of the church's total task.

BIBLIOGRAPHY: H. Schmid, *The Doctrinal Theology of the Evangelical Lutheran Church* (ET 1899; rep. 1961); E.L. Lueker (ed.), *The Lutheran Cyclopedia* (1954); C. Lund-Quist (ed.), *Lutheran Churches of the World* (1957); T.G. Tappert (ed.), *The Book of Concord* (1959); J. Bodensieck (ed.), *The Encylcopedia of the Lutheran Church* (3 vols., 1965). CARL S. MEYER

LUTHERAN WORLD FEDERATION, see LUTHERANISM

LUX MUNDI. The title of a book containing "A Series of Studies in the Religion of the Incarnation," edited by Charles Gore,* bishop of Oxford, published in 1889. Dissatisfied with the superficial level of the Anglo-Catholic movement during the latter half of the nineteenth century, the contributors to this volume pressed for a more liberal and socially informed Catholicism in the Church of England. Bishop Gore, in his preface, maintained that the old faith and the new scientific hermeneutics were not necessarily incompatible; changing times, he said, required "new points of view." Nevertheless, these Oxford essays were violently attacked by the conservative members of the Church of England and assailed publicly in Convocation. H. CROSBY ENGLIZIAN

LYNE, JOSEPH LEYCESTER (1837-1908). Religious community leader. As "Father Ignatius" he revived Benedictine monasticism in the Church of England. Ordained deacon in 1860, he started communities at Claydon (1863) and Norwich (1864). In 1870 he began building a monastery at Capel-y-ffin, near Llanthony. As abbot he adapted the Benedictine Rule, although perforce remaining a deacon until irregularly ordained priest in 1898 by an *Episcopus vagans*, Mar-Timotheus. He was often away on missions and in fund-raising, including an extended visit to America (1890-91). During these times the life of the abbey became unsettled. Despite his ritualism, his theology was soundly evangelical after his conversion in 1866. He rejected penance and purgatory as detracting from Christ's finished work. Evangelical hymnbooks often include his hymn, "Let me come closer to Thee, Jesus." Crowds flocked to his simple preaching of Christ as Savior. Gladstone put him among the first of contemporary orators. In Wales he was admitted to the *gorsedd* of the bards. J. STAFFORD WRIGHT

LYONS, COUNCILS OF. Two general church councils were held at Lyons. The First Council of Lyons, or Thirteenth Ecumenical Council, met in 1245 and held three sessions. It was summoned by Innocent IV, who told the assembled prelates that five problems tormented him: the sins of the priests; the loss of Jerusalem; the dangerous situation in the Byzantine Empire; the Mongol attacks on Europe; and Emperor Frederick II's* persecution of the church and pope. Although the council recommended a new crusade, nothing actually happened. Its main concern was with the dispute between Frederick II and the pope, a conflict which had its origins in Gregory IX's* excommunication of Frederick because he had not gone on crusade as promised. Frederick naturally forbade the imperial prelates to attend the council and blocked the routes to Lyons, and only 150 bishops, mainly French and Spanish, participated. The formal deposition was announced on the grounds of perjury, sacrilege, heresy, and felony, and the Franciscans and Dominicans were deputed to promulgate the decision throughout Europe.

The Second Council of Lyons, or Fourteenth Ecumenical Council, met in 1274. It held six sessions and voted seventeen canons. Some 1,600 ecclesiastics attended, including 500 bishops. Its convener, Gregory X,* wanted to organize a general crusade, but only one king came—and he soon left, convinced like many other European rulers that Gregory's plans were impractical in the current political climate. Gregory's efforts to obtain general peace in Europe, an obvious precondition for a crusade, were not entirely successful either. All he really achieved was a six-year tithe for financing such a future crusade. The council also strove for reunion between the Roman and Byzantine churches. Agreement would reopen a crusading route across Anatolia, guarantee the Eastern emperor Michael VIII against attack from Charles of Anjou, and make possible joint action against the Saracens. The council also decreed that in future, on the deaths of popes, the cardinals were to wait only ten days for absent colleagues before going into conclave, and in an effort to avoid long interregnums, they were to receive no salaries or emoluments until after the election. Further, the council suppressed a number of recent Mendicant orders, but specially commended the Franciscans and Dominicans.

L. FEEHAN

LYRA, NICOLAUS DE, see NICHOLAS OF LYRA

LYTE, HENRY FRANCIS (1793-1847). Born near Kelso in Scotland, he was educated at Enniskillen and Trinity College, Dublin, where he three times won the prize for an English poem. In 1815 he became curate at Taghmon, near Wexford. In 1817 he moved to Marazion, Cornwall, where he underwent a great spiritual change after attending a friend's deathbed. His whole outlook was altered, and his preaching revitalized. After various curacies he went as perpetual curate to Lower Brixham, Devon, in 1823 and remained until his delicate health broke down. He died at Nice. His main works were *Tales on the Lord's Prayer in Verse* (1826); *Poems, chiefly Religious* (1833; 2nd ed., 1845); *The Spirit of the Psalms* (1834), consisting of paraphrases, e.g., "God of mercy, God of grace," and "Pleasant are Thy courts above"; also an edition of *The Poems of Henry Vaughan* (1846). Among his many hymns were "Abide with me," and "Praise, my soul, the King of Heaven." JOHN S. ANDREWS

MABILLE, ADOLPHE (1836-1894). Swiss missionary to South Africa. Born in Switzerland, he went to Lesotho in 1860 with the Paris Evangelical Mission. Apart from enforced absences and an expedition to the E Transvaal in 1873, he spent his entire ministry at Morija, where he operated a printing press and established normal, Bible, and theological schools. He initiated a local synod but aroused resentment by his negative attitude to tribal custom. He was the trusted adviser of Chief Moshweshwe and his successor Letsie. The British annexation of Lesotho in 1868, which prevented Boer domination, owed much to his advocacy. When the Cape Colony, which took over the administration, tried to disarm the Basuto in 1879, Mabille opposed the measure but vainly urged the chiefs to obey. During the subsequent rebellion he presented the Basuto case in England and helped to obtain the transfer of Lesotho to direct British rule in 1884.

D.G.L. CRAGG

MACARIUS (d.333). Bishop of Jerusalem from c.313. He was probably the Macarius whom the heretical Arius labeled an "uneducated heretic" in his letter to Eusebius of Nicomedia, since we have the tradition from several sources that Macarius attended the Council of Nicea (325) and may have actively debated the Arians and helped draft the Creed. Further, his differences with his metropolitan, Eusebius of Caesarea,* the church historian, stemmed from the latter's soft stand against Arianism,* though a more basic cause was Macarius's attitude that Jerusalem, birthplace of Christianity, ought not to be subordinate to Caesarea, provincial capital. Indeed Jerusalem later became a patriarchate. Constantine's letter assigning construction of the Church of the Holy Sepulchre to Macarius is extant in Eusebius. The church complex, enormous with porticoed courtyard and sumptuous in gold coffered ceilings and marble from all parts of the empire, took nearly ten years to build. Macarius may also have helped identify the true cross discovered by St. Helena.

DANIEL C. SCAVONE

MACARIUS MAGNES (fourth/fifth century). Christian apologist. Nothing is known for certain of his life. Some have identified him with the bishop of Magnesia who accused Heraclides at the Synod of the Oak* (403). He was the author of the *Apocriticus,* in which the objections formulated against the Christian message by a learned Neoplatonist (possibly Porphyry) were attacked. In the ninth century the treatise was used by the Iconoclasts in defense of their doctrines. It was quoted by Francisco Torres in the sixteenth century from a copy then in the Marciana Library, Venice, but later this was lost. In 1867 a defective MS was found at Athens, but this also is missing. There do survive fragments of a spurious series of *Homilies on Genesis* which have been ascribed to him.

J.G.G. NORMAN

MACARIUS OF ALEXANDRIA (c.320-c.404). Desert Father. He is also surnamed "Politicus" (i.e. "of the city") or "Junior" to distinguish him from Macarius of Egypt, his older contemporary and rival in asceticism, prophecy, and healing, with whom he was banished to a Nilotic island by Lucius, Arian patriarch of Alexandria, during Valens's persecution of orthodoxy (364-78). An extant monastic rule containing thirty regulations for his 5,000 monks of the Nitrian desert and a sermon on the eschatology of souls are among the writings ascribed to him.

MACARIUS OF EGYPT (d. c.390). Ascetic writer. Also called "the Elder" or "the Great," he was a native of Upper Egypt and lived for some sixty years in the wilderness of Scete, the center of Egyptian monasticism. He gained a reputation as an ascetic and therefore became a highly respected holy man. Palladius and Rufinus give accounts of him, although without mentioning that he was a writer. From the sixteenth century, however, he has been credited as the author of fifty homilies. These are of great importance in mystical theology, and scholars point out that they have similarities to the Messalian* heresy. It is possible these writings were by an anonymous writer who was called "blessed" *(makarios).* Seven further homilies were discovered in 1918, but it is not now considered that these can be attributed to Macarius.

C. PETER WILLIAMS

MACARIUS OF MOSCOW (1810-1882). Metropolitan of Moscow. Educated at the seminary at Kursk, then at Kiev, he became a leading theologian in the Russian Orthodox Church, holding the professorship of dogmatic theology at the Academy of St. Petersburg. In 1857 he was made bishop of Tamlov, and in 1879 metropolitan of Moscow. In this position he became an authority on the official theology of the Roman Catholic Church. His best-known works are his *Introduction à la Théologie dogmatique orthodoxe* (1845), *Théologie dogmatique orthodoxe* (5 vols., 1845-53), and *History of the Russian Church* (12 vols., 1857-82).

MACAULAY, ZACHARY (1768-1838). Evangelical leader. Son of a Scots Presbyterian minister, he went to Jamaica at the age of sixteen as a bookkeeper on an estate which used slave labor. Deeply impressed with the evils of slavery, he returned to England in 1792 and became a member of the Sierra Leone Company. From 1793 to 1799 he was governor of the colony and ruined his health with overwork. Thereafter he was secretary of the company till the colony was transferred to the Crown in 1808, and editor of the *Christian Observer* (1802-16). He resided in Clapham with other prominent evangelicals and played a leading role in the abolition of the slave trade in 1807 and in the renewal of the antislavery agitation in 1823. He took a great part also in the affairs of the Bible Society and the Church Missionary Society. He failed in business in 1823 and was thereafter dogged by ill-health till his death. His son was Thomas Babington Macaulay, historian and essayist. IAN SELLERS

M'CHEYNE, ROBERT MURRAY (1813-1843). Church of Scotland minister. Educated at Edinburgh University, he was moved by the death of an elder brother to seek "a brother who cannot die." In 1836 he was ordained to the charge of St. Peter's, Dundee, where the fruitfulness of his ministry and his own spiritual growth were the outcome of a strict daily program of Bible study, prayer, meditation, visiting his people, and preparation of sermons—and all this despite frequent illness. His missionary interest involved a visit to Europe and Palestine in 1839 to study the possibility of a mission to the Jews—the beginning of a notable Church of Scotland work. Few ministers have so greatly influenced their own and succeeding generations in so short a life. He became known as a preacher throughout Scotland and even more widely known through Andrew Bonar's *Memoir and Remains of Robert Murray McCheyne* (1862). McCheyne's other writings include the moving hymn, "When this passing world is done." J.W. MEIKLEJOHN

McCULLOCH, THOMAS (1777-1843). Presbyterian minister in Nova Scotia. Born in Scotland and educated at Glasgow University, he was ordained a minister of the General Associate Synod. He migrated to Pictou, Nova Scotia, where he founded a church and (in 1808) Pictou Academy, which was incorporated in 1816 with McCulloch as principal. In 1838 he became president of Dalhousie College in Halifax, a post he held until his death. He devoted his life to the cause of educational freedom and labored in Nova Scotia to break the Anglican monopoly of higher education and of access to the professions.

MacDONALD, GEORGE (1824-1905). Scottish novelist and poet. Born at Huntly, Aberdeenshire, he studied at King's College, Aberdeen, and Highbury Theological College and became minister of a Congregational church at Arundel, Sussex, in 1850. For expressing views on final judgment that left some hope for the heathen, he was opposed by his deacons and had his salary reduced from £150. By 1853 the situation had become intolerable, and he resigned, thereafter supporting himself and his wife by lecturing, tutoring, writing, and occasionally preaching. Though his health was poor and his poverty great, his writings show little trace of this, and much of a deep faith in God. He reacted against the Calvinism of his day, but not violently, and never became liberal in theology. C.S. Lewis,* who owed much to him, rated *Phantastes* (1858), the Curdie books, and *Lilith* (1895) among his great works. MacDonald's novels, while containing many quotable sayings, are too verbose to be good. He was at his best as a myth-maker, and it was the quality of cheerful goodness in his work that captured Lewis's imagination and convinced him that real righteousness is not dull.

P.W. PETTY

McGIFFERT, ARTHUR CUSHMAN (1861-1933). American church historian and educator. Born at Sauquoit, New York, he graduated from Western Reserve College and Union Theological Seminary, afterward pursuing studies in Paris, Rome, Berlin, and Marburg (Ph.D., 1888). Ordained to the Presbyterian ministry, he taught church history in Lane Theological Seminary (1888-93) before succeeding Philip Schaff* as professor of church history at Union in 1893, where he was subsequently also president (1917-26). Under the influence of Harnack, he wrote *A History of Christianity in the Apostolic Age* (1897). His ideas aroused such opposition that he withdrew from the Presbyterian Church and became a Congregationalist (1899). Other major works include a translation of Eusebius's *Church History* (1890); *The Apostles' Creed* (1902); *Protestant Thought Before Kant* (1911); *Martin Luther, The Man and His Work* (1911); *The Rise of Modern Religious Ideas* (1915); *The God of the Early Christians* (1924); and *A History of Christian Thought* (2 vols., 1931-33).

ALBERT H. FREUNDT, JR.

McGREADY, JAMES (c.1758-1817). Presbyterian revivalist. Born in W Pennsylvania and raised in North Carolina, he studied theology between 1785 and 1788 in Pennsylvania and was thereafter licensed to preach. His revivalistic zeal appears to have been kindled during a visit to Hampden-Sydney College in Virginia. His early ministry in North Carolina resulted in the conversion of some twelve young men who entered the ministry, among them B.W. Stone.* In 1796 McGready moved to pastor three churches in Kentucky. Revival began in 1797 and was climaxed during the summer of 1800 at a great outdoor camp meeting to celebrate Communion and admission of church members. The revival spread throughout the W and S states (see SECOND GREAT AWAKENING). McGready is credited with originating the camp meeting* in 1800 at Gaspar River, and with helping in the origin of the Cumberland Presbyterian Church.* He finished his life as a pioneer missionary in S Indiana.

HOWARD A. WHALEY

MacGREGOR, JAMES (1759-1820). Presbyterian minister in Nova Scotia. Born in Perthshire, Scotland, and educated in Edinburgh, he was ordained a Presbyterian minister in 1786. In 1787

he was sent to Pictou, Nova Scotia, by the General Associate Synod. He became concerned about the decline of the moral and cultural life of the Nova Scotian Scottish communities, which he felt was caused by the lack of clerical leadership. For this reason he acted as a part-time itinerant minister while appealing for more workers from Scotland. He was the first Presbyterian minister to preach in New Brunswick and on the Island of St. John (Prince Edward Island). He also aided the establishment of the Pictou Academy and in 1817 became the first moderator of the Presbyterian Church assembly in Nova Scotia.

ROBERT WILSON

MACHEN, JOHN GRESHAM (1881-1937). American Presbyterian scholar and apologist. Born in Baltimore, he was educated at Johns Hopkins, Princeton University and Theological Seminary, Marburg, and Göttingen. He was ordained in 1914. He taught NT at Princeton Seminary from 1906 to 1929, apart from a brief period of YMCA service in France. As a defender of the classic Reformed position, he was influenced by his teacher B.B. Warfield.* When Warfield died in 1921, the mantle of leadership for the "Princeton Theology" fell upon Machen. He resigned in 1929 due to the Liberal realignment of the seminary. Machen was a principal founder of Westminster Theological Seminary (1929) and what is now the Orthodox Presbyterian Church* (1936). He served as president and professor of NT at Westminster from 1929 to 1937.

In 1935 he was tried and found guilty of insubordination by a presbytery convened at Trenton, New Jersey, on charges brought by the general assembly of the Presbyterian Church in the USA. It condemned him for activities in connection with an independent mission board. He was forbidden to defend himself and was suspended from the Presbyterian (PCUSA) ministry. Machen is regarded by friend and foe as a leading conservative apologist in the modernist-fundamentalist era. Among his most significant publications are *The Origin of Paul's Religion* (1927); *Christianity and Liberalism* (1923): most definitive of his thought; *New Testament for Beginners* (1923); and *The Virgin Birth of Christ* (1930).

ROBERT C. NEWMAN

MACHRAY, ROBERT (1831-1904). Anglican primate of Canada. Born in Scotland, he was educated at the universities of Aberdeen and Cambridge, and after Anglican ordination and three years of travel returned to Sidney Sussex, Cambridge, as a dean, to which was added four years later the vicarage of Madingley (1862). In 1865 he was consecrated as the second bishop of Rupert's Land. His revival of St. John's (Anglican) College in Winnipeg (1866) was but a prelude to the establishment of the University of Manitoba (1877), of which he was first chancellor. In 1875 he became metropolitan of Western Canada, and in 1893 primate of All Canada, at which time, despite English reluctance, each of the two Canadian metropolitans received the designation of archbishop. These changes reflect the strong sense in which Machray helped the church in Canada pass from missionary status dependent on England to a self-sustaining independence indicative of the growing needs of the dominion.

CLYDE CURRY SMITH

MACK, ALEXANDER (1679-1735). Organizer and first leader of the New Baptists, or Brethren. This group, founded in 1708, in 1871 adopted the name German Baptist Brethren, but its main body today is known as the Church of the Brethren.* Born in Schriesheim, Germany, of Reformed parents, and a miller by trade, Mack was attracted to radical Pietism, and became a close associate of one of the leaders of that movement, E.C. Hochmann* von Hochenau. Mack accompanied Hochmann on some of his preaching missions. Due to governmental persecution of Pietists, Mack sold his mill and moved to Schwarzenau in the county of Wittgenstein. There eight Pietists under the leadership of Mack and because of Anabaptist influence became convinced that complete fidelity to the NT required the actualization of the gathered community of believers. The Brethren were born. Mack shepherded the body through persecution, resettlement, and finally to Pennsylvania. The first party reached the New World in 1719; the second, including Mack, arrived in 1729. Some 250 reached William Penn's land. Mack provided leadership until his death.

Mack's leadership role is ably exemplified in his two writings. The shortened titles of the works are *Basic Questions* (1713) and *Rights and Ordinance* (1715). Total obedience to Jesus Christ, including the constituting of the visible brotherhood community, was the key concept. English translations of Mack's two writings are in D.F. Durnbaugh (ed.), *European Origins of the Brethren* (1959).

Alexander Mack, Jr. (1715-1803) followed in his father's footsteps, guiding the Brethren in his adult life.

MARTIN H. SCHRAG

MACKAY, ALEXANDER MURDOCH (1849-1890). Missionary to Uganda. Born in Aberdeenshire, he studied at the Free Church training college for teachers, and in 1873 went to Berlin to acquire qualifications with an engineering firm. In 1875, having read H.M. Stanley's book on David Livingstone,* he changed his original intention about working in Madagascar and applied to the Church Missionary Society for work in Uganda. He arrived in East Africa* and began work on making a road through to Lake Victoria Nyanza, 230 miles inland. It took two years. He arrived on the lake shortly after the murder of two CMS colleagues, and after all others had left because of ill-health. The boat, intended for the lake and brought up in sections, had suffered severe damage, but Mackay's engineering knowledge and resourcefulness resulted in its completion and made an enormous impression. While work was going on, Mackay resolved to visit Lkonge, whose warriors had murdered his colleagues, and he succeeded in reaching an agreement with him. The boat completed, the party set out for Entebbe, headquarters of King Mtesa. Mackay's skill with metal won him acceptance, and on Sundays he was free to read and expound the NT. Arab traders opposed him, and French Catholic priests introduced a divisive element. Mackay protested at appalling cruelties without entirely losing favor.

James Hannington* had been consecrated bishop and approached Uganda from the East. There was an old prophecy that the country would be conquered by invaders who came that way, and though Mackay assured the king no harm was intended, warriors were dispatched to kill the party. There followed a general persecution of Christians; the lives of the missionaries at court were saved by one of the ministers who interceded for them. Mackay finally withdrew to the south of the lake. There he taught and translated, and there he met Stanley. Mackay died from malaria, but not before he saw the first copies of Matthew's gospel printed.

See J.W. Harrison, *A.M. Mackay, Pioneer Missionary of the C.M.S. in Uganda* (1890).

P.W. PETTY

MACKAY, GEORGE LESLIE (1844-1901). Scots-Canadian Presbyterian missionary. Born in Zerra, Canada, he was educated at the universities of Toronto and Edinburgh and Princeton Theological Seminary, and ordained in 1871. He spent the rest of his life in missionary service in Formosa, an island then scarcely touched by Protestant missionary societies. He made converts among both the Chinese and the aboriginal inhabitants, built up a strong church, and trained indigenous leaders. He compiled a Chinese Romanized dictionary of the Formosan vernacular and wrote an important account of his work, *From Far Formosa* (1896). In 1894 he became moderator of the general assembly of the Presbyterian Church in Canada. He died on the mission field.

IAN SELLERS

McKENDREE, WILLIAM (1757-1835). First American-born bishop of the Methodist Episcopal Church. He was converted at the age of twenty-nine and began preaching two years later with but scant formal education (he had previously been a small planter and an officer in the Revolution). He was first a helper on Mecklenburg's circuit, and in 1790 Bishop Francis Asbury* ordained him deacon, and in 1791 elder. Finally working closely with Asbury after nearly siding with the Republican Methodist schism, McKendree served on circuits for twenty years in Virginia, Ohio, Kentucky, and parts of Illinois, Tennessee, and Mississippi. He was a leader in the Great Revival in the West. He was elected bishop in 1808 and served ably until his death in Tennessee.

ROBERT C. NEWMAN

MACKENZIE, JOHN (1835-1899). Scottish missionary to South Africa. Born in Morayshire, he went to South Africa under the London Missionary Society in 1858. His wife's health prevented his joining the disastrous Makalolo mission. He eventually settled among the Ngwato at Shoshong (1864-76) and gained the confidence of Kgama III. From 1871 to 1882 he superintended the Moffat Institution, first at Shoshong, later at Kuruman. Mackenzie was convinced that the protection of the Africans demanded the extension of British rule to the Zambesi. He therefore became politically involved, first as a government representative in Bechuanaland, and in 1885-91 as a propagandist of imperial expansion in Britain. His efforts were repeatedly frustrated, principally by his fellow-imperialist C.J. Rhodes, whose motives and methods differed fundamentally from Mackenzie's humanitarianism. Mackenzie's closing years (1891-99) were spent as missionary at Hankey, Cape Colony.

D.G.L. CRAGG

MACKINTOSH, HUGH ROSS (1870-1936). Scottish theologian. Born in Paisley, he was educated at the universities of Edinburgh, Freiburg, Halle, and Marburg, and ordained to the ministry of the Free Church of Scotland* in 1897. He served parishes in Tayport and Aberdeen before appointment as professor of systematic theology at New College, Edinburgh (1904-36). He was moderator of the general assembly of the Church of Scotland in 1932. Mackintosh, who had a wide grasp of the teaching of German theologians, was regarded as a liberal evangelical. His works include *The Doctrine of the Person of Jesus Christ* (1912), *The Christian Experience of Forgiveness* (1927), and *Types of Modern Theology* (1937).

J.D. DOUGLAS

MACLAREN, ALEXANDER (1826-1910). Baptist minister. Born in Glasgow to Baptist parents, he was baptized in 1840 and trained at Stepney College. He ministered successfully at Portland Chapel, Southampton (1846-58), and Union Chapel, Manchester (1858-1903), where he acquired the reputation of "the prince of expository preachers." His sermons drew vast congregations and his methods of subdivision and analogies drawn from nature and life have been widely imitated ever since. In the pulpit he expounded evangelical certainties, yet his writings and private conversations show him prepared to accept a critical position. His attitudes are thus ambiguous, though Spurgeon* excepted him from the "Downgraders." Maclaren was twice president of the Baptist Union and chairman of its Twentieth Century Fund and the first president of the Baptist World Alliance* (1905). He strove unsuccessfully to unite the Baptist and Congregational denominations, but saw the establishment of many "Union" churches at a local level.

IAN SELLERS

MACLEOD, NORMAN (1812-1872). Scottish minister. Son and grandson of famous West Highland ministers, and grandfather of the founder of the Iona Community, he belonged to a celebrated clerical dynasty that gave six moderators to the Church of Scotland general assembly. Queen Victoria's favorite chaplain, he was also her personal friend and spiritual adviser. Minister successively of Loudoun Parish, Ayrshire, of Dalkeith, and of the celebrated Barony Church in Glasgow, in the latter city he made his name as pastor, evangelist, author, editor, churchman, and social reformer. He became widely known as editor of the immensely popular *Good Words*, as a champion of the working man, and as a foreign missionary en-

thusiast. He wrote the hymn "Courage, brother."
D.P. THOMSON

McNICOL, JOHN (1869-1956). Bible college principal. Born in Ottawa, he read classics at Toronto University and divinity at Knox College. For two years he was secretary of the University YMCA and a leader in the Student Volunteer Movement.* He was minister of the Presbyterian church in Aylmer, Quebec (1896-1900), then in 1902 began lecturing in Toronto Bible College, where he was principal from 1906 to 1946. He wrote much on the Holy Spirit; other books were *Thinking Through the Bible* (4 vols.) and *The Bible's Philosophy of History.*

McPHERSON, AIMEE SEMPLE (1890-1944). Canadian-born evangelist. She was converted through the ministry of Robert J. Semple, later married him, then went with him to China in 1908 as a missionary. He died of malaria after three months, and she came home with their baby. A second marriage to Harold McPherson brought the birth of a son, Rolf. Ill-health with two major operations ended with the adoption of an evangelistic career and a later divorce. She and her mother between 1918 and 1923 crossed the continent eight times. By 1922 she had built the 5,000-seat Angelus Temple in Los Angeles and developed her "foursquare gospel" of Christ as Savior and Healer, the baptism of the Holy Spirit with speaking in tongues, and the Second Coming. She broadcast what was claimed as the first radio sermon (1922). In 1927 she incorporated her International Church of the Foursquare Gospel (see PENTECOSTAL CHURCHES) and sent students from her Bible school to preach that gospel. She disappeared for a time in 1926. The Los Angeles grand jury challenged her story of kidnaping, but finally dismissed the case. She died of a heart attack shortly after trips to the Holy Land and the British Isles. EARLE E. CAIRNS

MACRINA THE YOUNGER (c.328-379/380). Eastern ascetic. Granddaughter of Macrina the Elder, she was born in Neocaesarea, Cappadocia, sister of Basil* of Caesarea and Gregory of Nyssa.* She won her brothers Basil and Peter (of Sebaste) to religious vocations. While Basil was choosing the ascetic life, she was establishing one of the earliest communities of women ascetics on the family estate in Pontus. Returning from the Synod of Antioch in 379. Gregory visited her and at the request of the monk Olympus wrote her life *Vita Macrinae Junioris,* also providing details about her brothers. This book, together with Gregory's *De Anima ac Resurrectione,* gives a vivid account of their meeting on her deathbed. Manuscripts of this biography suggest that her cult spread through the Eastern churches; it came much later in the West. C.G. THORNE, JR.

MADAURAN MARTYRS (second century). First reputed martyrs in Africa. Their names are given as Namphano, Miggin, Lucitas, and Sammae, all of which are Punic names. They are said to have suffered at Madaura in Numidia in 180. The record of them comes from the writings of Maximus of Madaura, a pagan grammarian of the late fourth century, who vigorously protested against the popular practice of visiting the tombs of such uncultured barbarians, deserting the pagan cults for the new religion. It has been argued by some scholars (e.g., J.H. Baxter) that the evidence for the martyrs is very uncertain.

MADSEN, PEDER (1843-1911). Danish bishop and theologian. In his youth he studied theology at Erlangen and was strongly influenced by F.H.R. Frank* and "the theology of experience." In 1874 he became lecturer, and in 1875 professor in Christian dogmatics and NT exegesis at Copenhagen University. In 1909 he was appointed bishop of Zealand. Madsen was not distinguished by any special originality, but combined influences from various directions into a full-blown doctrinal system, mainly biblical conservative and orthodox Lutheran, but seriously weakened by tendencies of subjectivism, a mild synergism, a kenotic view of Christology, and by some unnecessary concessions to biblical criticism. He exercised an extensive influence on many students preparing for the ministry. His personal seriousness and piety left a lasting mark on many, and he was the most influential Danish theologian of his time. His principal literary works are a commentary on the Book of Revelation (1887) and a posthumously published textbook on Christian dogmatics (1912-13). N.O. RASMUSSEN

MAGDALENES. A name for convents or communities of penitent females founded under the patronage of Mary Magdalene, the converted prostitute. The monastic revival influenced by Francis of Assisi much helped in the move to form communities in the thirteenth century—e.g., at Goslar, Worms, and Strasbourg. Bulls confirming the privileges of such communities were issued between 1227 and 1251 by Gregory IX and Innocent IV. In the main, the Augustinian Rule was followed, but in a few German houses it was the Franciscan or Dominican Rule. Refuges for fallen women were also established, e.g., at Marseilles in 1272 and Naples in 1324. In 1640 the Magdalenes adopted a new rule after a long period of weak discipline, but they are now virtually extinct. PETER TOON

MAGDEBURG CENTURIES (1559-74). A major Protestant reinterpretation of the history of the Christian Church to 1308, originally in thirteen volumes as *Centuriae Magdeburgenses.* It was written by Matthias Flacius,* in collaboration with six other Lutherans. The work is highly polemical, severely antipapal, and based on selected sources. But it did make some contribution to the development of historical criticism. It also provoked an equally polemical response, the *Annales ecclesiastici* (1588-1607) by Cesare Baronius,* based on sources in the Vatican Library.

MAGI. According to the gospel of Matthew, *magoi,* guided by a mysterious star, came from the East to Bethlehem with gifts for the infant Jesus (2:1-12). The word *magoi* can mean either "wise men" or "magicians." The only other use in the NT is Acts 13:6ff., where it clearly means "magician." Whether Matthew also intended the word to be understood in this sense is not certain. Ignatius of Antioch did. Commenting on the appearance of the star, he says that from that time magic lost its power, since God had appeared in human form. At any rate, it is reasonable to infer that the evangelist looked upon the Magi as representatives of the Gentiles. Tradition has embellished the story in various ways. Origen stated there were three wise men, probably on account of their three gifts. In the sixth century they were named as Gaspar, Melchior, and Balthasar. The Adoration of the Magi soon became a popular subject in art. By the Middle Ages they were venerated as saints. What were claimed to be their relics were taken to Germany by Frederick Barbarossa in 1162 and are now enshrined in Cologne cathedral.

R.J. MC KELVEY

MAGNIFICAT. The song of praise sung by Mary (Luke 1:46-55) when she visited Elizabeth to tell her about the forthcoming birth of Jesus. The name is derived from the first word of the hymn in the Latin version. There are considerable similarities between the *Magnificat* and the Song of Hannah (1 Sam. 2:1-10) and there is no doubt that the memory of this formed the background. The situation of Elizabeth makes a closer parallel in some ways to that of Hannah, and there is some slight MS evidence for ascribing the hymn to Elizabeth. It has been suggested that the original text may have included no name and that scribes have inserted one or the other. The hymn is one of praise for the gracious action of God on behalf of His people and in particular of the singer herself. It is fully appropriate to the occasion and shows no sign of having Christian theology read back into it. It has been used as an evening canticle in the worship of the Western Church since at least the time of Benedict.

R.E. NIXON

MAIER, WALTER ARTHUR (1893-1950). American Lutheran scholar and radio preacher. After graduating from Boston University (1913), he went to Concordia Theological Seminary, was ordained in 1917, and received his Ph.D. from Harvard in 1929. He taught OT language and studies at Concordia from 1922 and established several churches in St. Louis. From 1920 to 1945 he edited the *Walther League Messenger.* He gained international prominence as the regular speaker of the "Lutheran Hour" from 1935 until the late forties. At his death the broadcast went to over 1,200 stations worldwide. Maier's blend of scholarship and practical devotion enabled him to reach both scholar and common man.

ROBERT C. NEWMAN

MAIMONIDES, MOSES (1135-1204). A leading Jewish philosopher of the Middle Ages. His *Guide to the Perplexed* is of central importance. "Perplexity" is what was aroused by the supposed opposition of the Greek sciences to Jewish religious belief. Maimonides, who was concerned with the socially divisive effects of popularized Greek learning, responded by attempting to make Aristotle as *un*systematic and contradictory as possible. This makes the *Guide* something of an enigma. On one view Maimonides seeks to show the limitations of philosophy; on another, he is showing that the practical worth of religion is undiminished by the existence of supposedly alien philosophies. On his view, God cannot be known, but can be appreciated and loved through an acquaintance with His workings in the natural order. Revelation has the role of educating the believer to know God in knowing nature, and not of imparting distinctive truths. The *Guide* not only became a fundamental text of medieval Jewish thought and the subject of much debate, but exercised an influence on medieval discussions of the relation between faith and reason, and later it had a somewhat different influence on Spinoza.*

PAUL HELM

MAISTRE, JOSEPH, see JOSEPH DE MAISTRE

MAITLAND, SAMUEL ROFFEY (1792-1866). Anglican historian and writer. Born in London, son of a Scottish merchant, he was educated at Cambridge but left without a degree. A man of very wide intellectual tastes which ranged from mesmerism to music, he had at first intended to pursue a legal career, but in 1821, his religious views having changed, he was ordained deacon in the Church of England and from 1823-27 was perpetual curate of Christ Church, Gloucester. He made significant contributions to the study of contemporary Judaism and in 1832 produced a masterly account of the Albigenses and Waldenses. Associated with the Clapton sect of High Churchmen, he contributed notable historical essays to the *British Magazine,* which he later edited. In 1838 he was made librarian at Lambeth Palace. Maitland was equally unpopular with the Tractarians,* whose ritualism he deplored, and with the evangelicals, for his attacks on Foxe the martyrologist and on Milner's *Church History.*

IAN SELLERS

MAJORISTIC CONTROVERSY. This was occasioned by George Major (1502-74), professor at the University of Wittenberg. Major taught that "good works are necessary to salvation" and that "it is impossible for a man to be saved without good works." He was attacked especially by Matthias Flacius.* Before this, Flacius had attacked Major because he had subscribed to the Augsburg Interim* (1548) in which the *sola* had been omitted in the phrase "sola fide justificamur." In the controversy Justus Menius sided with Major, both of them contending that faith alone justifies, but that faith is not present without confessing and persevering. In the seven propositions of the Synod of Eisenach (1556), Menius repudiated the proposition "good works are necessary to salvation." Major maintained that "good works are necessary." Nicholas von Amsdorf* opposed Major, saying "good works are harmful to salvation." Article IV of the Formula of Concord* (1577) repudiated both Amsdorf and Major, teaching that good works should be excluded from the question concerning salvation and the article about justification, but that regenerate man is bound to do good works.

CARL S. MEYER

MAKEMIE, FRANCIS (1658-1708). Regarded as the founder of American Presbyterianism, he was born of Scotch-Irish parents in County Donegal, but received his education at the University of Glasgow. He was ordained in 1682 for missionary work in America and arrived there in 1683. He labored as an itinerant evangelist in North Carolina, Maryland, the Barbados, and Virginia. He was the moving spirit in the organization of the Presbytery of Philadelphia in 1706. When he preached in New York colony in that year, Governor Cornbury had him arrested for preaching without a proper license. Makemie ably defended his right to free speech and was acquitted, but Cornbury forced him to pay over £83 for the costs of the trial.

EARLE E. CAIRNS

MALABAR CHRISTIANS. The "Syrian" or "Saint Thomas" Christians of SW India,* the word "Malabar" here being broadly equivalent to the present state of Kerala. There are now three main groups of them: Roman Catholics (who are in two separate blocs); the Syrian Orthodox Church; and the Mar Thoma Church. All trace a common origin to early Christian centuries and generally hold the tradition that Thomas* landed at Cranganore in A.D. 52 and founded churches in seven places in Kerala. A separate tradition concerns immigration of Syrian Christians at Cranganore in 345, after the king and the bishop of Edessa had resolved to reinforce the churches in Kerala. The leader was Thomas of Cana. A section of the Syrian Christians today, known as Southists, comprises those said to be descended from the Cranganore colonists. In the Roman Catholic Church the diocese of Kottyam is explicitly for Southists. The earliest account of the detailed tradition about Thomas in Kerala comes from the Portuguese writer, Antonio de Gouveau.

A subsequent immigration of Syrians, at Quilon, is traditionally dated 823. Five copper plates exist, recording grants of lands and privileges to the Tarisa Church, but are of uncertain date. Five carved stone crosses, with inscriptions in Pahlavi (language of the Sassanid Persian Empire), exist, one at the supposed St. Thomas shrine in Mylapore (Madras) and four in Kerala. They are of the seventh or eighth centuries. The Alexandrian writer Cosmas (c.525) had found Christians in Malabar and some at Kalliana, with a bishop appointed from Persia. Not much is known about the Syrian Christians in the Middle Ages, but the general picture is clear: an established church in Kerala with an outpost at Mylapore; a church dependent on the church of the Persian Empire; and consequently a church with a Syriac liturgy and ultimately with a creed reflecting Nestorianism.*

Roman Catholic influence began with Franciscans calling en route to China in the fourteenth and fifteenth centuries, by which time the Syrians' link with the mother church in Mesopotamia was weakened by the circumstances of the Muslim era. In 1330 a French Dominican, Jourdain de Severac, became bishop of Quilon, the initial papal claim to jurisdiction in India. In 1503 the Nestorians were reinforced by five bishops of their own, but simultaneously the Portuguese had come to India, and Jesuit pressure proved too much. The Syrians' metropolitan submitted to Rome, and finally the Synod of Diamper* (1599) severed the Syrian Church from its past and from its Mesopotamian patriarch. Syriac continued to be used in the Romanized liturgy. Resentment at Jesuit rule, however, boiled over in a revolt at the Coonen Cross* in 1653. For a time it seemed that most of the Syrians had seceded from Rome, but within a decade Rome had won back much of the lost ground, using Carmelites instead of Jesuits and appointing one of them as titular bishop in Malabar, thus bypassing the Portuguese hierarchy.

The Syrians who did not return to Rome were free to look abroad for a bishop, especially when Dutch power replaced Portuguese in Kerala and the foreign Roman clergy were cleared out. But the bishop who came to them in 1665 was not from the Nestorian patriarch, but from the Jacobite one at Antioch—hence the subsequent description of the Syrians as "Jacobite." The Syrians were now in two groups: Roman Catholic and Jacobite. Through the initiative of the British authorities in South India, and partly because of the findings of Claudius Buchanan,* work among the Syrians was begun by the Church Missionary Society in 1816. The purpose was not to make Anglicans of the Syrians, but to seek the renewal of the ancient church. Unhappily, after two decades the mission was unacceptable to the Syrian authorities, and the missionaries turned to other work in Kerala. Some of the Syrians at this point seceded and became Anglicans; the outcome is the Central Travancore diocese of the Church of South India* today.

Others remained within the Syrian Church as a reform party. One of its leaders, Abraham Malpan, was excommunicated because of his reforms, but sought to change things at the top by having his nephew sent to the patriarch of Antioch. The nephew returned to India as Bishop Mar Athanasios and with a claim to be made metropolitan of the Syrian Church. After lengthy legal battles, the reform party lost all claim to property and in 1889 had to begin separate existence as the Mar Thoma Church, which proved to have an evangelistic concern.

The Jacobite Church had to endure continued legal struggles in the twentieth century until a 1958 decision of the supreme court of India led to reconciliation between parties of the metropolitan (catholicos) and of the patriarch of Antioch. The title "Orthodox Church" was now preferred to "Jacobite." The Roman Catholics were the largest single group of Syrians, most of them belonging to the Syro-Malabar Rite, but a smaller number to the Syro-Malankara Rite which owes its origin to an accession of Jacobites in 1930. There are, in addition, Roman Catholics of the Latin Rite in Kerala, but fewer in number than the Syrians. There are other smaller groups, including a Nestorian Church which derives from a split from the Roman Catholic Syrians.

See C.B. Firth, *An Introduction to Indian Church History* (1961), and D.P. Matthew and

M.M. Thomas, *The Indian Churches of Saint Thomas* (1967). ROBERT J. MC MAHON

MALACHY (1094-1148). Archbishop of Armagh. A zealous reformer, he was born at Armagh, son of a clergyman, was educated at Armagh and Lismore, and became successively abbot of Bangor, bishop of Connor, and archbishop of Armagh. He was greatly attracted to the monastic system in Gaul. With four Irish clergy he visited Bernard of Clairvaux* and studied the system with the result that a Cistercian abbey, the first in Ireland, was established at Mellifont, County Louth, in 1142. It was his burning desire to bring the Celtic Church under the supervision of Rome. With this end in view he summoned a synod to Inishpatrick in 1148. The synod appointed him delegate to visit Rome and to receive the pallium, the vestment that indicated the holding of office of papal authority. On his journey he visited his friend Bernard, and on the day he was to have continued his journey to Rome he died. His biography, written by Bernard, gives rather a sad picture of the state of affairs in Ireland. Malachy was a man of ardent temperament, sincere piety, and outstanding influence. He resisted the attempts of Nigellus and Maurice, two laymen who sought to usurp the see of Armagh. ADAM LOUGHRIDGE

MALAN, CÉSAR HENRI ABRAHAM (1787-1864). Swiss preacher. Born in Geneva, he studied theology there and was ordained in the Reformed Church. Previously under the influence of Voltaire and Rousseau, he had been converted in 1817. His conversion brought him into conflict with the ecclesiastical power in Geneva, and he was forbidden to preach on original sin, election, and related doctrines. When Malan disregarded this order he was expelled from his pulpit. Although he may have never formally left the established church, he gathered a group first in his own home, and then in the *Chapelle du Témoignage,* and may have joined the Scottish Church. After 1830 he engaged in missionary tours to other parts of Switzerland, Germany, France, the Netherlands, and Scotland. C. GREGG SINGER

MALCHION OF ANTIOCH (fl. c.270). Christian rhetorician. As presbyter and leading teacher of rhetoric in Antioch he opposed his own bishop, Paul of Samosata,* at the Synod of Antioch in 269, where the latter's view that Christ was by nature merely an ordinary man was condemned. Paul, bishop since 260, was deposed and excommunicated. Eusebius records portions of the actual debate. Jerome further reports that Malchion also drafted the long encyclical announcing the synod's decision.

MALDONALDO, JUAN (1534-1583). Spanish exegete. Otherwise known as Johannes Maldonatus, he was born at Casar de la Rema, Estremadura, and after studies at Salamanca (1547-58) and Rome (1558-62) he became a Jesuit and was ordained. He taught philosophy at Rome (1563) and Paris (1564-65) and became theology professor at the *Collège de Clermont* (1565-74). At first his lectures were traditional, but in 1570 he initiated his own theological course. The Sorbonne accused him of heresy (1574). He was vindicated by the bishop (1576), but withdrew to Bourges where he wrote commentaries on the gospels (1596) which became famous. In 1581 Gregory XIII called him to Rome to work on a critical edition of the Septuagint.

J.G.G. NORMAN

MALEBRANCHE, NICHOLAS (1638-1715). French Catholic philosopher. He entered the Oratorians* in Paris (1660) where he remained throughout his life. He was ordained priest (1664). His philosophical work concentrated on the relation between faith, reason, and empirical observation, seeking therein to find an accommodation between Catholicism and Cartesian philosophy. *De la recherche de la vérité* (2 vols., 1674-75), *Traité de la nature et de la grace* (1680), and *Entretiens sur la métaphysique et sur la religion* (1688) are his greatest works and embroiled him in constant polemics, with Bossuet, Leibnitz, Fontenelle, and many others. As a scientist he studied insects, mathematics, and color; simultaneously he was a *méditatif* in the Oratory.

He agreed with Descartes that we do not perceive the actual physical objects, such as the sun; they are mediated to us via *idées* in our minds. Against Descartes, however, he claimed such *idées* are archetypes of objects in the mind of God. We are assured that the actual objects do exist by supernatural revelation which we accept on faith. This notion is termed the "vision in God." Concerning causation in the physical world, he postulated "occasionalism" by affirming that God, acting through his general laws of motion, is the true cause of all motion (i.e., causing a ball to move), while particular or occasional causes (i.e., a ball striking another ball causing it to move) exist.

C.T. MC INTIRE

MALINES CONVERSATIONS. A series of meetings between Anglican and Roman Catholic theologians at Malines (Mechelen), Belgium, extending over five years from 1921. The prime movers were D.J. Cardinal Mercier* and Lord Halifax, a High Church Anglican. Two such groups inevitably found substantial areas of agreement after the manner of ecumenical exchanges. Mercier did, however, raise the possibility of the Church of England's uniting with Rome as a uniate body with a patriarch. The conversations petered out with Mercier's death in 1926. It was he who had engendered much of the warmth on the Roman side, and in 1928 indications from the Vatican confirmed that the talks had been quietly dropped. Low Church Anglicans had always entertained grave suspicions about the exchanges. J.D. DOUGLAS

MALTA, KNIGHTS OF, see HOSPITALLERS

MANALO, FELIX (1886-1963). Founder of the *Iglesia ni Kristo* in the Philippines. Born near Manila in a devout Catholic home, he was converted to Protestantism in 1902. After study with the Methodists and Presbyterians, he became first a Disciples and then a Seventh-day Adventist preacher. In 1912 he turned to agnosticism, but after special revelations during Bible study in 1913 he began to preach his own doctrine, and in 1914 founded the *Iglesia ni Kristo.* He denied the divinity of Christ and justification by faith, argued that the church did not exist between A.D. 70 and 1914, and therefore claimed that salvation is to be found only in the *Iglesia ni Kristo.* In 1922 he declared himself to be the angel from the East mentioned in Revelation 7:1-3. He built many splendid chapels of distinctive architecture, personally appointed all his clergy (and wrote their sermons), and dictated to his members how they should vote in state elections. He appealed to contemporary nationalism by teaching that there were many specific references to the Philippines in Bible prophecy. The INK doubtfully claimed 3,500,000 members in 1963. Felix's son Erano succeeded him as "Executive Minister."

RICHARD DOWSETT

MANDAEANS. The only-known Gnostic* sect surviving into the present century. Numbering only a few thousand, they are located in S Iraq and Khuzistan, and no longer use the language of their sacred books, speaking instead dialectal Arabic. Mandaic thought from the beginning (i.e., A.D. second century, at latest) was assimilated less to Western concepts than was Hellenism, and remained characterized by a syncretism of highly complex elements which defies systematization. These elements include biblical and other West Semitic materials, late Babylonian ingredients (especially astrological), Iranian dualism, and a peculiar concern for a highly legendized John the Baptist (the source of their designation as "St. John's Christians").

Their significant literature includes doctrinal treatises, liturgical materials, and a variety of popular works including phylactery rolls and incantation bowls. There are three ranks of cult officials; assistants, ordinary priests, and overseers. These preside over two fundamental rites: a frequent ritual washing in the name of Life and the "knowledge of life," and a "deathbed" washing. The former includes anointing with oil and a sacramental meal of bread and holy water.

While known to Westerners since the sixteenth century, publication of Mandaean texts began in the late nineteenth century. Mandaean studies are valuable for the history of religions, and gained further significance with R. Bultmann* and his school in their work on John's gospel.

See E.S. Drower, *The Mandaeans of Iraq and Iran* (1937; rep. 1962), and G. Widengren, "Die Mandäer," in *Religionsgeschichte des Orients in der Zeit der Weltreligionen* (1961), pp. 83-101.

CLYDE CURRY SMITH

MANDE, HENDRIK (c. 1360-1431). Dutch mystic, associated with the Brethren of the Common Life.* Born in Dordrecht, well educated, for a time in the service of Count Willem VI of Holland, he was influenced by the fiery preaching of G. Groote,* and in his thirties joined the Brethren as a lay brother in their cloister at Windesheim. This remained his base for the remainder of his life, spent in conducting various business affairs for the Brethren, and also in compiling several treatises on mysticism. These are taken largely from the thoughts and writings of the famed Flemish mystic, Jan Van Ruysbroeck,* and are thus theologically orthodox.

MANEGOLD OF LAUTENBACH (c.1030-c.1103). Roman Catholic scholar. Born at Lautenbach, he entered the monastery there, and when followers of Henry IV ravaged it in 1086 he fled to the cloister of Raitenbuch, Bavaria, where he became dean. Later he assumed priorship of the monastery at Marbach, Alsace. His fame and consequent difficulty with the imperial party stems from his staunch defense of papal policy in the Investiture Controversy.* In *Manegoldi ad Gebhardum liber* he defined kingship as based on a contract between monarch and subjects which becomes void by royal breach. Hence the pope legitimately dissolved the German oath to Henry IV, argued Manegold. *Opusculum contra Wolfelmum Coloniensem* indicates the danger of studying pagan classics. In 1096 Manegold was delegate to the Synod of Tours, and in 1098 he was an imperial prisoner.

JAMES DE JONG

MANICHAEISM. Mani (Gr. Coptic *Manichaios*), of aristocratic Parthian family and religious father, was born in 216, and grew up in S Babylonia probably among adherents of Elchasai. Revelations at the ages of twelve and twenty-four led him to leave the community of his youth and after study and meditation to embark in 240/41 upon his mission of proclamation of the revealed truth. After the conversion of members of his family, and a time in India, he returned to the center of the Persian Empire, by this time having royal princes among his followers. King Shapur I (241-72) received him, gave him leave to preach his message, and made him one of his entourage from admiration or political expediency. Under this patronage he was able to write the six books and the letters that made up the Manichaean Canon, and to travel widely through the empire spreading his message. Under Hormizd I royal protection continued, but after one year Bahram I succeeded him and the climate changed: Mani was first denied traveling rights, then summoned to the presence, and at the instance of Karter, Zoroastrian high priest, imprisoned in chains. After suffering for a month, he died in February of either 276 or 277.

His religion, which had already reached Egypt in his lifetime, spread throughout the Roman Empire and eastward beyond Persia to Sogdia, by the eighth century reached China, and in the late eighth century was the state religion of the Turkic Uigurs. Almost everywhere bitterly persecuted, it eventually died out. Manichaean writings have come to light only this century, mainly in Coptic, Sogdian, and Uigur, but also in Chinese, Greek, and Persian. The doctrine, though claimed as a system, is highly mythological. Two principles, Light and Dark, God and Matter, are eternal. The invasion of the Light led to the saving expedition of the Primeval Man, some of whose substance remained imprisoned in matter after his return. The creation of sun, moon, stars, and plant life was part of a plan to redeem this imprisoned Light. The appearance of Adam was a counterplot to retain Light imprisoned, through reproduction; "Jesus the Brilliant Light" redeemed him by a vision. The Jesus of the gospels is but an instance of the suffering of imprisoned Light in matter.

The religious practice of Mani's followers (among whom the "elect" or "righteous" ate no meat and abstained from sexual life) was an ascetic means of continuing the process of gradual liberation. At length, after the second coming of Jesus and a millenial reign, the end comes, when the elect are reunited with Light, and this creation is destroyed. The precise relation of Iranian, Mandaean* Gnostic, Buddhist, and Christian elements in the teaching of Mani (who called himself "apostle of Jesus Christ") remains a matter for debate and research. The relation to the medieval "Manichee" is problematic.

BIBLIOGRAPHY: F.C. Burkitt, *The Religion of the Manichees* (1925); S. Runciman, *The Medieval Manichee* (1947); H-C. Puech, *Le Manichéisme. Son Fondateur. Sa Doctrine* (1949); G. Widengren, *Mani and Manichaeism* (ET 1965); A. Adam, *Texte zum Manichaismus* (2nd ed., 1969); A. Henrichs and L. Koenen, "Ein griechischer Mani-Codex," in *Zeitschrift für Papyrologie und Epigraphie* 5 (1970), pp. 97-216; A. Henrichs, "Mani and the Babylonian Baptists," in *Harvard Studies in Classical Philology* 77 (1973), pp. 23-59.

J.N. BIRDSALL

MANNING, BERNARD LORD (1892-1941). English church historian. Son of a Congregational minister, he was educated at Jesus College, Cambridge, where he was successively fellow, bursar, and senior tutor (1933). His chief historical publications, in which he argued that Dissent stood inside the common tradition of Latin Christendom, were *The People's Faith in the Time of Wyclif* (1919); *The Making of Modern English Religion* (1929); *Essays in Orthodox Dissent* (1939); and *The Protestant Dissenting Deputies* (1954). His scholarship, wit, and piety are seen in *The Hymns of Wesley and Watts* (1942), to which all hymnologists are indebted. Similar qualities pervade two volumes of sermons (1942-44). Of delicate health, he was mourned at his death by Christians of all persuasions.

JOHN S. ANDREWS

MANNING, ERNEST C. (1908-). Premier of Alberta and radio preacher. He grew up in W Saskatchewan. Hearing William Aberhart broadcasting from Calgary, he proceeded to the Prophetic Bible Institute and was its first and most distinguished graduate. In the 1935 Social Credit landslide in the Alberta provincial election, Manning was not only returned to the legislature, but was made a cabinet minister. On the death of Aberhart in 1943, Manning assumed the office of premier. Early showing signs of competence and integrity, Manning was the beneficiary of the Alberta oil strikes of the late forties and early fifties. He managed the oil boom well, and as provincial coffers overflowed he channeled much of the money into an enlightened program of social welfare. People so often did not vote for Social Credit; they voted for him. During all these years Manning conducted the weekly nationwide "Back to the Bible" broadcast. In his sepulchral tones he preached the Gospel, expounding prophecy with decreasing emphasis on its predictive element and increasingly using the prophetic passages as a basis for calling the nation to repentance and revival. In all of this Manning was an expression of the western prairies revival movement of the hungry thirties, with its rather sectarian attitude and compelling worldwide missionary vision.

When he retired from politics, Manning quickly became a member of Canada's economic establishment, his prestige readily opening directorships in the country's major corporations. But when he retired an era ended. Although his successor, Harry Strom, was equally evangelical, represented Scandinavian Pietism, and was lay moderator of the Evangelical Free Church, the exciting days of the mid-thirties had run their course. Social Credit was roundly defeated eighteen months after Manning retired.

IAN S. RENNIE

MANNING, HENRY EDWARD (1808-1892). English cardinal, archbishop of Westminster. Born in Hertfordshire, son of a member of Parliament, he was educated at Harrow and Oxford, then followed a career in two parts. He was an Anglican until his forties (1851), then as a Roman Catholic he helped consolidate the Catholic revival in England. He began as an Anglican Evangelical, ordained (1832), then archdeacon at Chichester (1841). Gradually his interest in the Oxford Movement* became commitment to its principles; he wrote Tract 78 and was considered a leader of the movement after Newman's conversion (1845). But Manning himself converted (1851). N.P.S. Cardinal Wiseman,* amid opposition from older Catholic families, took a personal interest in him, ordaining him a priest (1851), encouraging him to found in England the Oblates of St. Charles (1851), then appointing him inspector of schools in the Westminster diocese (1856) and provost of the metropolitan chapter (1857). In 1860 Manning became the chief English defender of papal temporal power.

Upon Wiseman's death (1865), Pius IX made Manning the new primate, an appointment he held for twenty-seven years, meanwhile becoming cardinal (1875). At the Vatican Council (1869-70) he was a leader of the Ultramontane move to define the dogma of papal infallibility as necessary to defend the principle of authority. He consistently promoted Catholic social consciousness, especially concern for the condition of the poor and working men. Leo XIII's social encyclicals were to him the supreme definition of Catholic social principles. He generally supported Gladstone's policies.

BIBLIOGRAPHY: E.S. Purcell, *Life and Times of Cardinal Manning* (2 vols., 1896); P. Thureau-Dangin, *La renaissance catholique en Angleterre au XIXe siècle* (3 vols., 1923); J. Fitzsimons (ed.), *Manning, Anglican and Catholic* (1951); V.A. McClelland, *Cardinal Manning: His Public Life and Influence, 1865-1892* (1962).

C.T. MC INTIRE

MANNING, JAMES (1738-1791). Founder and first president of Brown University. Graduating from the College of New Jersey (Princeton University) in 1762, he had by 1765 finally secured a charter for Rhode Island College and founded a Baptist church in Warren, Rhode Island. He was appointed president of the college in 1765 and held the position of professor of language until 1791. The school was moved from Warren to Providence, where he also served as pastor of the First Baptist Church from 1771 to 1791. In 1786 he represented Rhode Island in the Congress on the Confederation, and in the summer of 1791 he wrote a report that suggested the creation of the state's present free public school system.

EARLE E. CAIRNS

MANSEL, HENRY LONGUEVILLE (1820-1871). Dean of St. Paul's. Born at Cosgrove, Northamptonshire, he was a scholar, then tutor, of St. John's College, Oxford, and was ordained in 1844. In 1859 he was appointed first Waynflete professor of moral and metaphysical philosophy at Oxford; in 1866 he succeeded to the regius chair of ecclesiastical history; and in 1868 was made dean of St. Paul's. He achieved eminence as a teacher of logic, though his real interest lay in the field of metaphysics, which he approached by way of psychology. As a metaphysician, however, he gained little distinction. His Bampton Lectures of 1858 brought him into conflict with F.D. Maurice* and John Stuart Mill. Mansel maintained that man acquires knowledge of the nature of God only from supernatural revelation. In *What is Revelation?* Maurice replied by challenging both Mansel's concept of revelation and his concept of Christianity. The conflict dragged on with a heatedness which did little credit to either man.

JOHN A. SIMPSON

MANSON, THOMAS WALTER (1893-1958). British biblical scholar. Educated at Glasgow and Cambridge universities, he was Yates professor of NT Greek and exegesis, Mansfield College, Oxford (1932-36); Rylands professor of biblical criticism and exegesis, Manchester University (1936-58). In his greatest contribution to NT scholarship, *The Teaching of Jesus* (1931), written while he was Presbyterian minister at Falstone, Northumberland, Manson propounded the thesis with which his name is distinctively associated—that the title "Son of Man" on the lips of Jesus had corporate significance until the end of His ministry, when it became evident that His disciples were not yet ready to take their share in enduring the destiny of suffering appointed for the Son of Man, and Jesus endured it alone. While he published much of lasting value on the NT epistles and on ministry and priesthood in the church, his most important work was concerned with Jesus and the gospels.

F.F. BRUCE

MANTEGNA, ANDREA (1431-1506). First N Italian Renaissance painter. Born near Vicenza, son of a carpenter, he profited from the tutelage of the painter Francesco Squarcione of Padua, an avid antiquarian who adopted him and from whom he may have derived his classical and archaeological interests. A precocious and arrogant genius, member of the guild of painters at eleven, he strengthened a fruitful association with these artists in his marriage in 1454 to Nicolosia Bellini, daughter of Jacopo and sister of Giovanni and Gentile. Mantegna's austere classicism is evidenced in the monumentality of his figures, his concern with perspective, his zeal for historical precision, his use of architectural detail, and his choice of much of his subject matter, particularly the triumphs. He was more preoccupied with fidelity to nature than with idealized beauty, especially in his later works, which proclaimed his sense of tragedy, as shown in his foreshortened *Dead Christ.* He decorated the Belvedere Chapel (now destroyed) in the Vatican for Innocent VIII. His influence on his contemporaries and successors in Italy and beyond, including Dürer,* was tremendous.

MARY E. ROGERS

MANUSCRIPTS OF THE BIBLE. Manuscripts are written texts copied individually by hand. Until the invention of printing in W Europe about 1450, practically all written texts were manuscripts. This was so with biblical texts, in whatsoever language they were written. Here we limit ourselves to biblical manuscripts in Hebrew, Aramaic, and Greek, the three original languages of Scripture. These were written with carbon ink and a reed pen on skin (parchment, vellum, etc.) or papyrus, in scroll or codex form.

Old Testament (Jewish Bible). The earliest known biblical manuscripts are those discovered in 1947 and the following years in the caves of Qumran and other places west of the Dead Sea. These go back to the closing centuries B.C.: they include fragments of all the books of the Hebrew Bible except Esther, and a few portions of the pre-Christian Greek version of the OT, commonly called the Septuagint. But the great majority of manuscripts of the Hebrew Bible exhibit the Massoretic Text, the text edited by the Jewish scholars called Massoretes (i.e., custodians of *massorah,* "tradition") from the sixth century A.D. onward. The oldest Massoretic fragments came to light toward the end of the nineteenth century in the "Ezra Synagogue" of Fustat (Old Cairo). Of complete Massoretic manuscripts, the oldest known is a codex of the Prophets belonging to the Qaraite synagogue in Cairo, dated A.D. 895. Others of comparable date are a codex of the Pentateuch in the British Museum (Or 4445), only a few years younger; an early tenth-century codex of the whole Hebrew Bible, formerly (until 1948) belonging to the synagogue in Aleppo and now, unfortunately mutilated at the beginning, preserved in Israel; a Leningrad codex of the Latter Prophets (P) dated A.D. 916; a Leningrad codex of the whole Hebrew Bible (L) completed in 1008, on which the text of Kittel's *Biblia Hebraica* (3rd ed.) is based; and an almost complete codex of the whole Hebrew Bible in the Bodleian Library, Oxford, a few years younger. Some of these codices exhibit the specially pure form of text edited by the Ben Asher family of Massoretes, of Tiberias.

Other manuscripts which exhibit the Ben Asher text are the British Museum Or 2626-2628 and 2375, together with the "Shem Tob" Bible belonging to the Sassoon family; these were used by N.H. Snaith for the Hebrew Bible published by the British and Foreign Bible Society in 1958.

Samaritan Bible. The Hebrew Pentateuch has been preserved independently by the Samaritan community, which recognizes no other part of the OT as canonical. The Samaritan Bible is based on a popular Palestinian text (of which some samples have been identified among the Qumran manuscripts), with the addition of some sectarian readings upholding distinctive Samaritan beliefs, such as that Mt. Gerizim and not Jerusalem is the dwelling-place of the name of God, where He desires to be worshiped (cf. John 4:20). The oldest known Samaritan codex (in Cambridge) contains a note indicating that it was sold in A.D. 1149-50; it must have been written some decades before that. The Abisha scroll, which is shown to visitors at the Samaritan synagogue in Nablus as the oldest book in the world, is actually composite: the oldest part (Num. 35—Deut. 34) may go back to the eleventh century A.D., but when the remainder of the scroll (Gen. 1—Num. 34) was accidentally lost or destroyed in the fourteenth century, it was replaced by a new copy.

Septuagint. The "Septuagint" (abbreviated LXX) is the name traditionally, though imprecisely, given to the pre-Christian Greek version of the Hebrew Bible and some associated documents. Some fragments of this version have been identified among the manuscripts from the west shore of the Dead Sea. The John Rylands Library, Manchester, possesses a papyrus fragment of Deuteronomy 25–28 in Greek, of date not later than c.150 B.C. (P. Ryl. 458), and there is another pre-Christian portion of the Greek Deuteronomy in Cairo (P. Fouad 266), containing a few verses of chapters 31 and 32. In this last papyrus the consonants of the divine name (YHWH) are left in square Hebrew characters, instead of being turned into Greek.

All our other Septuagint manuscripts are of Christian origin. The Chester Beatty biblical papyri, now housed in Dublin, include seven codices of various parts of the Septuagint, written in the second and third centuries A.D. With the NT papyri in the same collection these probably belonged to the multivolumed Bible of a Greek-speaking church in Egypt which could not afford more expensive or durable copies. One of these codices, containing parts of Ezekiel, Daniel, and Esther, is of special interest because it is one of the very few witnesses to the original "Septuagint" version of Daniel—a version which is so free a paraphrase that in nearly all manuscripts of the Greek OT it is replaced by a later and more accurate version ascribed to Theodotion (late second century A.D.). The principal witness to the original "Septuagint" text of Daniel is a codex in the Chigi collection in Rome (eleventh century A.D.).

Most manuscripts of the Septuagint form the OT part of a complete Greek Bible; they are thus witnesses also to the Greek text of the NT, and can be conveniently treated under that heading.

New Testament. The oldest surviving manuscripts of the Greek NT are written on papyrus. Such are the three NT codices in the Chester Beatty collection, containing the gospels and Acts (P 45), the Pauline letters and Hebrews (P 46), and Revelation (P 47), and dating from the late second-to-mid-third century. The oldest piece of any part of the NT is the papyrus fragment of John 18 in the John Rylands Library (P. Ryl. 457 or P 52), dated before the middle of the second century. The Bodmer Library, Geneva, houses another important collection of NT papyrus codices, including a copy of John's gospel dated c.A.D. 200 (P 66), an incomplete copy of Luke and John perhaps a decade or two earlier (P 75), and a third-century copy of 1 and 2 Peter and Jude with a number of other early Christian documents (P 72).

The most important manuscripts of the NT are the great uncials of the fourth and fifth centuries —so called because they are written in uncial letters, which were based on lapidary capitals. The Vatican and Sinaitic codices (B and Aleph respectively) are the best-known examples of these; they are fourth-century copies of the whole Greek Bible, beautifully produced on vellum, exhibiting a text characteristic of Alexandria. These two manuscripts form the chief biblical treasures respectively of the Vatican Library and the British Museum. The British Museum also houses the Alexandrine Codex (A), which was presented to King Charles I in 1627 by Cyril Lucar, who had recently been patriarch of Alexandria and thus legal owner of the manuscript. The Ephraem Codex (C) in the Louvre, Paris, is also a fifth-century manuscript of the Greek Bible; it owes its name to the fact that in the twelfth century its biblical text was scraped off to make room for some writings of the fourth-century Syriac father Ephraem. The original writing was later made visible again by the use of chemical reagents. A manuscript which has received this treatment is called a palimpsest (from a Greek adjective meaning "scraped again"). In Cambridge University Library the Codex of Beza (D) is preserved; this is the best-known example of a group of bilingual (Greek and Latin) NT manuscripts. This codex, containing the gospels and Acts, was written in the fifth or sixth century; it came into the possession of the Geneva Reformer Theodore Beza, who presented it to Cambridge University in 1581. It exhibits the "Western" text, which is marked by a number of peculiar deviations from other types of NT text, mainly amplifications and additions.

Two important gospel uncials are the Washington Codex (W), in the Library of the United States Congress, and the Koridethi Codex (Theta), in the Georgian State Library, Tiflis. They are important especially for the evidence they provide of the conventionally called "Caesarean" text of Mark; in addition, the Washington Codex includes a substantial expansion of the unauthentic ending of that gospel after Mark 16:14.

While the best-known uncials are traditionally designated by capital letters, all the uncials are officially listed in a series beginning 01 (Sinaitic).

In addition to nearly 270 uncials of the NT, there are about 2,800 minuscules, written (that is to say) in smaller letters approximating to ordinary cursive script. Whereas the official list of uncials is distinguished by 0 preceding each serial number, the official list of minuscules is numbered 1-2800. In addition to the regular uncial and minuscule manuscripts, well over 2,000 lectionaries containing the Greek text of the NT have been listed (their serial numbers are preceded by *l*); in them the text has been arranged in selections for reading in church. In all, the Greek NT text in whole or in part is preserved in some 5,250 extant manuscripts, covering a range of nearly 1,400 years—a wealth of attestation such as no other body of ancient literature can approach.

The majority of later uncials, minuscules, and lectionaries exhibit the Byzantine Text—a revised form of the NT text standardized throughout Greek-speaking Christendom after the peace of the church, from the later years of the fourth century onward. This text was in its main essentials taken over into the earliest printed editions of the Greek Testament (the "Received Text") and is represented in the English AV and other early printed vernacular versions.

See BIBLE (ENGLISH VERSIONS).

BIBLIOGRAPHY: B.J. Roberts, *The Old Testament Text and Versions* (1951); E. Würthwein, *The Text of the Old Testament* (ET 1957); F.M. Cross, *The Ancient Library of Qumran and Modern Biblical Studies* (1958); F.G. Kenyon, *Our Bible and the Ancient Manuscripts* (revised by A.W. Adams, 5th ed., 1958); P.E. Kahle, *The Cairo Geniza* (2nd ed., 1959); F.F. Bruce, *The Books and the Parchments* (3rd ed., 1963); B.M. Metzger, *The Text of the New Testament* (2nd ed., 1968).

F.F. BRUCE

MANZ, FELIX (c.1498-1527). Anabaptist* Reformer. Son of a Zurich canon, he acquired a thorough knowledge of Latin, Greek, and Hebrew, joined Zwingli in 1519, but was alienated by his caution in reform, and with Grebel and Hetzer formed the original Swiss Brethren congregation. He distributed some of Carlstadt's* eucharistic tracts (1524). In 1525 Grebel, Manz, and others faced Zwingli in what was the first baptismal disputation. The council proclaimed Zwingli victorious and decreed that all children were to be baptized on pain of banishment. The brethren promptly performed "believer's baptism" in Manz's house, thus breaking with Zwingli. Manz endured several imprisonments, but his quiet, steadfast witness encouraged many to become Anabaptists. After further baptismal disputations, severer measures were introduced in 1526, including capital punishment by drowning for those rebaptizing. Manz and Blaurock were arrested later that year. On 5 January 1527 Manz was drowned in the River Limmat, the first Protestant martyr at the hands of Protestants.

J.G.G. NORMAN

MANZONI, ALESSANDRO (1785-1873). Romantic poet and Italian novelist. Though educated in the Somaschi schools, he was influenced by the theories of the Encyclopedists,* Voltaire, and the Revolution. By 1810, however, having come in touch with the Jansenist circle in Paris (led by the Abbé Degola), he had returned to the Christian faith, thence devoting his literary talent to the writings of works which had as collateral aim the proclamation of Christianity. Between 1812 and 1832 he published *Inni sacri,* sacred lyrics in which he exalts the great events of Christendom (Christmas, Easter, Pentecost, etc.) and their significant influence on humanity, while in his tragedies *(Conte di Carmagnola, Adelchi)* he develops the theme of justice and sovereignty of God as opposed to the oppression of the rulers. But it is in his great novel *I promessi sposi* ("The Betrothed," 2nd ed., 1820-42) that the author concentrates all his favorite Christian themes, i.e., the absolute control of Providence over men's lives and actions, the beauty and comfort of the simple faith of two humble peasants and other poor people who lived in the seventeenth century—one of the most difficult times in Italian history—and the sanctity of priestly vocation as opposed to the interests and ambitions of many Roman Catholic clergy.

Though Manzoni's conversion took place in Jansenist circles, it is difficult to assess to what extent he adhered to their doctrines. Throughout the novel he attacks with Jansenist rigorism any aspect of *morale facile* such as the end justifying the means; but in the conversion of one of his major characters he compromises, and the event is an act of grace completed by good works. Of interest here is also a treatise *Osservazioni sulla morale cattolica* (1819) in which Manzoni attempted to refute the attacks made by the historian Sismondi in his *Histoire des republiques italiennes.*

See RISORGIMENTO. DAISY RONCO

MARANO, see ANUSIM

MARBECKE, JOHN, see MERBECKE

MARBURG COLLOQUY (1529). A meeting of Protestant theologians to try to form a united front against the Roman Catholic threat. The efforts at harmony originated with Martin Bucer* and the Strasbourg theologians, but were frustrated by Luther's firmness. It was in response to political pressures that Landgrave Philip of Hesse brought Zwingli, Oecolampadius, Bucer, Capito, and John Sturm and other Swiss and Strasbourg theologians together with Luther, Melanchthon, Jonas, Brenz, Cruciger, and Osiander, the Lutherans at Marburg. The main question of debate was the meaning of the Lord's Supper. The S German group followed the teaching of Zwingli* that Communion was a sign or seal of divine grace already bestowed on the believer; the bread and wine were symbols of the body and blood of Christ, who was locally present in His own body in heaven and not on earth. Luther adhered to the interpretation that Christ's words "This is my body" meant a real presence of Christ and were not to be interpreted metaphorically.

As a result of their discussion, fifteen articles were issued expressing general agreement on doctrines such as the Trinity, the person of Christ, justification by faith, baptism, good works, confession, and secular authority. The fifteenth article, which dealt with Communion, rejected transubstantiation and the idea of the Mass as a sacrifice, insisting on the laity receiving both the bread and the wine as the spiritual partaking of the body and blood of Christ. Despite these articles the colloquy served to divide rather than unite the Protestants, setting the pattern of church splits which has continued into the twentieth century.

See W. Köhler, *Das Marburger Religions-Gespräch* (1929). ROBERT G. CLOUSE

MARCELLA (325-410). Christian ascetic. Of a noble Roman family, after her husband's early death she devoted herself to charitable works and ascetic practices. Her palace on the Aventine Hill became a center of Christian influence, and a retreat for Christian patrician women. Jerome was her guest for three years, and under his direction she gave herself to Bible study, prayers, and almsdeeds. When Alaric sacked Rome (410), the Goths ill-treated her, thinking she was concealing her wealth, and she died as a result.

MARCELLINA (c.330-c.398). Sister of Satyrus* and Ambrose* of Milan. Born in Rome, she helped her mother in Ambrose's education after her father's death. She was consecrated a virgin by Pope Liberius (353). Later she lived at Milan with Ambrose, who tried to dissuade her from excessive austerities. He dedicated his *De Virginibus* to her.

MARCELLUS OF ANCYRA (d. c.374). Bishop of Ancyra in central Anatolia. We know of him through his letter to Pope Julius I* (337-52), preserved by Epiphanius, which includes the oldest Greek text of the Old Roman Creed* and 115 fragments of a treatise of his embodied with but one exception in Eusebius of Caesarea's *Contra Marcellum* and *De ecclesiastica theologia.* These same fragments serve also as a principal source for the teachings of Sabellius. Marcellus had been a supporter with Athanasius* of the *homoousion* position at the Council of Nicea* (325), though his Christology seems to have been based on the notion that the Word of God became the Son only at the Incarnation. This fragmented treatise written about 335, which led to his deposition at the synod in Constantinople in 336, was his continuing effort in the *homoousion* cause, and was specifically directed against Asterius,* Eusebius of Nicomedia,* and Eusebius of Caesarea.*

In exile Marcellus found refuge with Julius in Rome, and subsequent to the arrival of Athanasius (339) participated in those council sessions (Rome, 341; Sardica, 343) which cleared him of all charges: namely, "the falsehood of Sabellius, the malice of Paul of Samosata, and the blasphemies of Montanus." While Marcellus was temporarily restored to his see (344), the emperor Constantius again removed him upon dissent of the Eastern bishops (347). He died in exile, only to have his position condemned as heretical at the Council of Constantinople in 381. Jerome indicated that Marcellus was the author of other works; it has been argued that some of them are confused among the writings of Athanasius.

CLYDE CURRY SMITH

MARCIAN (c.396-457). Eastern emperor from 450. An obscure retired tribune who became part of the dynasty through a marriage in form only with the virgin Pulcheria,* an ardently Nicene Christian and sister of the deceased Theodosius II. Marcian was persuaded by her and Leo I of Rome into calling the 451 Council of Chalcedon.* One result was the imperial espousal of the orthodox *Definition.* Marcian's reign saw also the termination of the raiding activities of the Huns under Attila.*

MARCION (second century). Prominent heretic. A wealthy shipowner from Sinope in Pontus (NE Asia Minor), he came to Rome shortly before 140. He was active for a time as a member of the orthodox community, but was excommunicated c.144. He organized his followers into a rival movement to orthodox Christianity; his churches were established in many parts of the empire and were both numerous and influential for nearly two centuries (though the movement did not die out until later).

Marcion stressed the radical nature of Christianity *vis-à-vis* Judaism.* In his theology there existed a total discontinuity between the OT and the NT, between Israel and the church, and even between the god of the OT and the Father of Jesus. Jesus came to reveal the true God, who was totally unknown up to the Incarnation. The god of the OT, the *demiurge,** an inferior being who created the material world and ruled over it, was not exactly an evil being, but he was not good in the same sense as the God and Father of Jesus, a God of love and grace.

Paul was Marcion's hero and the one from whom (he thought) he derived his doctrine. His canon of sacred writings consisted of ten Pauline epistles (minus the Pastorals and and Hebrews) and the third gospel, both appropriately edited to suit his teaching (e.g., all passages were deleted from the letters of Paul which spoke of the Father as Creator, and the birth narratives were absent from his edition of Luke). His theology consisted of a series of antitheses (the title of his major work) —primarily between law (the principle of the demiurge and of the Jews) and gospel (the principle of the God of love and of redemption in Jesus), and between flesh (that which marks the material order and is evil) and spirit (the characteristic of the eternal realm). The law stresses rewards and punishments, and justification by works; the gospel features faith, freedom, and grace.

Scholars debate whether it is right to classify Marcion as a Gnostic.* He is certainly gnostic in his emphasis, especially in his negative attitude toward the body and the physical world; and his docetic Christology (see DOCETISM) and asceticism* also link him with the Gnostics. He does not, however, reproduce their fantastic mythology of redemption.

None of his writings have survived, though we can reconstruct large portions of his *Antitheses* from the extensive quotations in Tertullian's* *Against Marcion,* as well as from the refutations by other Church Fathers (notably Irenaeus*). His prologues or brief introductions to the epistles of Paul found their way into Latin biblical MSS of orthodox origin and have been preserved in this way.

The significance of Marcion lies in the fact that he compelled representatives of orthodox Christianity to deal seriously with the problem of evil, to think deeply about the biblical teaching concerning creation and redemption, to reexamine the Pauline writings, and to decide upon the question of the canon.

BIBLIOGRAPHY: A. Harnack, "Marcion," *TU* 45 (1921; rev. 1924), and *Neue Studien zu Marcion* (1923); J. Knox, *Marcion and the New Testament* (1942); E.C. Blackman, *Marcion and His Influence* (1948). W. WARD GASQUE

MARCIONITE PROLOGUES. These short introductions to the various Pauline epistles are so called because it is generally believed that they originated in Marcionite circles. They are to be found in some of the manuscripts of the Vulgate. Modern scholarship generally admits Marcionite origin (Hans Lietzmann explains their appearance in the Vulgate as the result of Marcionite teachers at Rome in the second century who translated the Greek epistles of Paul into Latin), although there is some disagreement (cf. M.J. Lagrange, *Revue Biblique* XXXV [1926], pp. 161-73).

MARCUS (second century). Gnostic.* Perhaps an Alexandrian, he taught in Asia Minor. Irenaeus and Hippolytus described him as a charlatan who deceived women by magical devices to become his prophetesses. He developed the ideas of Valentinus* in a one-sided fashion. A characteristic feature is "number-symbolism"; from the numerical values of divine names he sought to discover the nature and order of the aeons, and the mode by which the world came into being. His followers, Marcosians, flourished in the Rhone valley in the mid-second century. They stood right outside the church with their own institutions and special baptismal rites. As scriptures they used the *Acts of Thomas** and other apocryphal books. They lingered into the fourth century.

J.G.G. NORMAN

MARCUS AURELIUS (121-180). Roman emperor from 161. A Stoic philosopher, he in his *Meditations* expressed a high sense of duty toward mankind. Many of the Apologists*—including Justin, Athenagoras, Miltiades, Apollinaris, and Melito—felt that he might view the position of Christianity without prejudice and thus addressed their works to him. In fact, his particular intellectual stance made him less flexible. Faced with great military challenges on, and plagues within, his frontiers, he saw a successful outcome in loyalty to the old state religion. His mentors, Cornelius Fronto and Junius Rusticus, believed that Christianity was a dangerously revolutionary force, preaching gross immoralities and with an obstinate longing for death. Marcus Aurelius was sufficiently convinced to allow anti-Christian informers to proceed more easily, and the result was several outbreaks of severe local persecution, notably in Lyons in 177. During his reign the climate of educational opinion regarded Christianity as a sufficient challenge for anti-Christian intellectuals to thrive, most notably Celsus.* C. PETER WILLIAMS

MARGARET (d. 1093). Queen of Scotland. This gifted and masterful woman, famous alike for her austerities and her charities, became in 1070 the second wife of Malcolm Canmore, king of Scotland. The wedding took place a few months after she and her brother and sister, of the English royal line, had landed on the coast of Fife after fleeing from the Norman invasion. Over Malcolm and his subjects she exercised a remarkable influence, bringing Scotland fully under the Roman obedience and sweeping away most of what remained of Celtic Church* life and practice. Guardian of the orphan and succourer of the prisoners in her husband's dungeons, she cleansed the sores of the leper and washed the feet of the beggar with her own hands. Her work was carried to completion by her son, David I. Margaret lives today in the pages of Bishop Turgot's remarkable biography, enshrined in history as a saint and commemorated by the lovely Norman chapel in Edinburgh Castle which bears her name. She was canonized in 1250. D.P. THOMSON

MARGARET OF ANTIOCH. While there is no positive evidence of her existence, she was probably a martyr under Diocletian (303). Ambrose and John Chrysostom knew a Margaret, or Pelagia, who at fifteen preserved her chastity from violation by jumping off a building; others so named have suffered fabulous afflictions. Margaret became the subject of a medieval cult (she was one of the voices heard by Joan of Arc), though she was honored even earlier in the East. More recently she has been included among the fourteen auxiliary saints as a patron of childbearing. In art she is usually represented as a shepherdess or with a dragon.

MARGARET OF NAVARRE (1492-1549). Champion of the reform movement in France. Sister of Francis I, she first (1509) married Charles, duke of Alençon, and after his death Henry d'Albret, king of Navarre. She early came under the influence of the French Reformers Lefèvre d'Étaples, Guillaume Briçonnet, and others, giving them refuge when persecuted, first at Angoulême and later in Navarre. She also sought to reform the churches under her control and to influence her brother Francis to favor the reform movement which was growing at the time in France. She entered into correspondence with a number of the prominent Reformers, eventually becoming a Calvinist. She wrote several books of poetry and prose, the two best known being *Miroir de l'âme pécheresse* and *l'Heptaméron.* She died shortly after being forced by Henry II to marry her daughter to Antoine de Bourbon, who became the parents of Henry of Navarre, later Henry IV of France. W.S. REID

MARHEINEKE, PHILIPP KONRAD (1780-1846). German Protestant theologian and historian. He studied at Göttingen and taught there and at Erlangen and Heidelberg. In 1811 he became professor at the University of Berlin and from

1820 preached in the influential *Dreifaltigkeitskirche* as an associate of Schleiermacher.* He was an ardent admirer of Hegel's philosophy, and he attempted to use the Hegelian dialectic to support and interpret Christianity. His major work lay in the history of doctrine and in the study of symbolics. Among many published lectures and other writings, his most important works are *Christliche Symbolik* (3 vols., 1810-14), *Geschichte der deutschen Reformation* (2 vols., 1817), and *Vorlesungen über die Bedeutung der hegelschen Philosophie in der christlichen Theologie* (1842). A warm admirer of Luther, he espoused a liberal Pietism in his later years, claiming that because a thing was true it was in the Bible, not that because a thing was in the Bible it was true. HOWARD SAINSBURY

MARIANA, JUAN DE (1536-1623/24). Spanish Jesuit. Born at Talavera, he became a Jesuit in 1554 and read philology, history and theology at Alcalá, after which he lectured in theology at Rome (1561-64), Sicily (1564-69), Paris (1569-74), and Toledo, as well as preaching widely. His *De Rege et Regis Institutione* (1599), written at the request of Philip II, made him famous. It advocated the people's right to tyrannicide and made the Jesuits responsible for the assassination of Henry IV of France* and the Gunpowder Plot* in England. This view was condemned by the order in 1610. In *De monetae mutatione* he openly accused the Spanish officials of fraud and when imprisoned wrote a diatribe against the Jesuits. He wrote also several volumes on Spanish history and *Scholia in Vetus et Novum Testamentum* (1613). C.G. THORNE, JR.

MARIANISTS. Founded in 1817 in Bordeaux by William Joseph Chaminade (1761-1850), known as the Society of Mary, and distinguished from the Marists.* The Marianist Order introduced an original note in that priests and lay members have equal rights and privileges, except for those relating to administration of the sacraments. Members consecrate themselves irrevocably to the Blessed Virgin and wear a gold ring on the right hand as a token of this fact. Chaminade established the order as a cooperative effort between clergy and laity in fighting religious indifference, especially by means of educational works. The order was recognized by the pope in 1865, and in 1963 had 2,900 members throughout Europe, America, and Asia. ROYAL L. PECK

MARIAVITES. A Polish sect founded in Warsaw in 1906 by Jan Kowalski and Maria Felicja Kozlowska, who had both founded communities under the Franciscan Rule, but because of their suspected mysticism had been excommunicated from the Roman Church. The new union took its name from devotion to the Virgin Mary and was recognized by St. Petersburg and the Duma. Having negotiated with members of the Utrecht Schism, they ultimately joined the Old Catholics* (1909) with Kowalski their bishop. With Kozlowska's death, fanaticism within the sect increased noticeably and took the form of "mystic marriages" between priests and nuns, whose children were considered to be without original sin, destined to found a new, sinless race. In 1924 the Old Catholics renounced them. Kowalski, dethroned in 1935, was held by the Nazis and a split occurred. The majority followed bishops C.P. Feldmann (1935-45) and M. Sitek (1957-), with a minority at the Felicjanów monastery. By 1962 their total number was but a fraction of the 200,000 it had been in 1911. C.G. THORNE, JR.

MARIOLATRY. The worship of Mary. The term is used critically by Protestants, but strictly speaking the Roman Catholic Church does not encourage *latria* (worship which is due to God alone) but *hyperdulia* (special veneration). The references to Mary in the NT do nothing to encourage such a cult, and it seems to have come unofficially into the church in the fourth century. Pressure arising from popular devotion led to the definition of the dogma of the Immaculate Conception of Mary in 1854, and that of her Assumption in 1950. She was thus officially provided in the Roman Catholic Church with a beginning and end of life parallel to those of Jesus, and terms like "our Lady" and the idea of her as mediatrix of redemption have helped to increase her importance as an object of devotion. R.E. NIXON

MARIOLOGY. The study of doctrine concerning the Virgin Mary connected with her person as such and her role in the plan of redemption, with special reference to the Incarnation. It is thus parallel to Christology, ecclesiology, pneumatology, etc. It has been pursued far more in the unreformed churches than in the Reformed because of different attitudes to Mariolatry.*

MARISTS (Society of Mary). Founded in 1816 by Jean Claude Courveille and Jean Claude Marie Colin, the order held that Mary desired to aid the church through a namesake congregation. Courveille joined the Benedictines at Solesmes (1826) and Colin carried on, drafting the constitutions with Rome's final approval in 1836. Comprising priests and lay brothers, the society sent missionaries to Oceania and spread rapidly to Europe, North America, and the Antipodes. Based on the Jesuit Rule, the Marists did parish work, taught school and seminary, and held home missions and chaplaincies. Four independent congregations constitute the Marists: fathers (with the third order attached), brothers, sisters, and missionary sisters. C.G. THORNE, JR.

MARITAIN, JACQUES (1882-1973). French philosopher. He is noteworthy both as an interpreter of Thomas Aquinas and as an independent thinker. Reared in liberal Protestantism, he was converted to the Roman Catholic Church in 1906, and in 1914 was appointed to the chair of modern philosophy at the Institute Catholique in Paris. From 1945 until 1948 he was French ambassador to the Vatican, and followed this by teaching at Princeton until his retirement in 1956. He lectured at many other places and was the author of over fifty books.

In his philosophy Maritain used not only Aristotle and Aquinas, but insights from other philosophical sources, and has taken account of modern empirical research in anthropology, sociology, and psychology. Maritain claimed there were different ways of knowing reality, as "mobile being" (nature), quantity (mathematics), and being as being (metaphysics), the last involving the use of a metaphysical intuition. Maritain refashioned the Five Ways of Thomas Aquinas and added his own sixth way. This proof involves coming to see, through intuition and reflection, that the "I" who thinks has some pre-personal existence in God, and hence that God exists.

In his moral philosophy, Maritain claimed that account must be taken of the data of revelation, amd allowed that though this made his work not strictly moral *philosophy,* he claimed that philosophical method is appropriate to it. He expended considerable effort in working out a truly Christian politics. He made a sharp contrast between sacral and secular civilization. Man is necessarily social, as is seen by his needs and his possession of gifts, though he is of greater value than society because he is destined for union with God. PAUL HELM

MARIUS MERCATOR (fl. c.418-452). Latin Christian polemicist. Augustine's* grateful response to two anti-Pelagian writings (now lost) submitted to him by Marius places this layman in Rome about 418. Probably originally from Africa, he resided in Constantinople from about 429 to about 451, perhaps as Pope Celestine's legate. His treatise of 429 against Celestius, presented to Theodosius II, and his rebuttal of Julian of Eclanum* helped secure the imperial banishment of these Pelagians from Constantinople. His works on the nature of Christ against Theodore of Mopsuestia* and Nestorianism,* supporting the position of Cyril* of Alexandria, aided in the condemnation of both heresies at the Council of Ephesus* in 431. Marius's extant writing consists chiefly of Latin translations of heretical treatises and orthodox refutations, rendering him a major source for the history of the Nestorian and Pelagian controversies. DANIEL C. SCAVONE

MARK, GOSPEL OF, see SYNOPTIC GOSPELS

MARK THE EVANGELIST. The widespread ancient tradition that the author of the second gospel was Mark is generally accepted today, but it is not so generally accepted that all the references to "Mark" in the NT refer to the same person, particularly as the Roman name "Marcus" must have been so common. John Mark, mentioned in Acts, was the son of Mary who played an important part in the earliest days of the church in Jerusalem and whose house was used for prayer (Acts 12:12,25). There is no mention of his father. Saul and Barnabas chose him as their companion on their first missionary journey in some subordinate role (Acts 13:5). He later left them and returned home (Acts 13:13), and after a quarrel with Paul, Barnabas took Mark with him on a tour of Cyprus (Acts 15:36-40). The Mark mentioned in the Pauline epistles was a cousin of Barnabas (Col. 4:10) who was a useful and faithful companion of Paul (Col. 4:11; Philem. 24; 2 Tim. 4:11). It does not seem hard to imagine that this is the same Mark, now reconciled to Paul. A Mark is also found with Peter at "Babylon" (almost certainly Rome) when 1 Peter was written. John Mark had dealings, with Peter as well as with Paul in Acts, so that this further identification seems natural enough and particularly as Papias connected Peter with the writing of Mark's gospel. Eusebius said Mark later became bishop of Alexandria. R.E. NIXON

MARK THE HERMIT (d. after 430). Eremite. He had been abbot of a monastery in Galatia, a contemporary and probably a disciple of John Chrysostom.* His extant works, written for the edification of the monks in his care, deal with theological questions in a manner which indicates Mark's ethical and practical rather than mystical approach, his independence of tradition, and his intention to base his arguments upon Scripture. He wrote opposing those who expect to gain grace by works, insisting like Paul that grace and justification are free gifts and that all good works are evidences of a prior work of grace. Both the Roman Catholic Bellarmine and the Protestant Ficker claimed to see a Protestant tone to his concept of justification, such that the former charged the Protestants with interpolating the text. Like Chrysostom, Mark accepted the doctrine of original sin, but denied that it utterly destroyed free will, which he felt was perfectly restored at baptism. MARY E. ROGERS

MARNIX, PHILIP (Philip van Marnix van St. Aldegonde) (1540-1598). Calvinist diplomat and religious writer, active in the revolt of the Low Countries* against Spanish rule. Born at Brussels, he was an able student who studied law at Louvain, Paris, and Padua, then went to Geneva where he studied under Calvin and Beza. As a young Calvinist nobleman in the Low Countries, he joined the rising opposition to Spain, defended the anti-Catholic rioting of 1566, fled over the German border to escape reprisal, and rejoiced as revolt broke out (1568). At twenty-nine Marnix wrote his famed anti-Catholic satire, the *Beehive (Den Byencorf),* which gained wide popularity, going through some twenty-five editions. He met the rebel leader William of Orange in Germany, helped influence his conversion to Calvinism, and became a trusted assistant. As the first territory was "liberated" (Brill, 1572), Marnix represented Orange at the Estates of Holland and Zeeland. As military governor of the Rotterdam area, he was captured by the Spanish, but freed in an exchange of prisoners (1574).

As the revolt spread to the S Lowlands, he had a major role in drafting the Pacification of Ghent (1576), demanding religious toleration and traditional liberties. Orange appointed him to his Council of State, and Marnix had a hand in the negotiations leading to the brief exercise of rule by Anjou, in an attempt to gain French aid. The military situation worsened; in 1583 Orange named him to defend Antwerp, in danger of capture. Marnix failed badly, lost Antwerp, and was called before the rebel Estates-General to defend his conduct. Orange's assassination in 1584 meant the end of Marnix's political career. He retired to his estate at Souberg in Walcheren, and spent the next decade in writing, working on a rhymed version of the Psalms, a translation of the Bible, and a lengthy treatise on religious differences. He died in Leyden. Marnix is sometimes credited with the anonymous rebel anthem "*Wilhelmus van Nassua,*" but its authorship remains uncertain.

See A. Gerlo, *Marnix van St. Aldegonde* (1960).
DIRK JELLEMA

MARONITES. The only fully Roman Catholic Uniate church in the East. The name is derived from Maron, a Syrian solitary who died around 423. According to their tradition, Maronites have always been orthodox and in union with Rome, but in fact they were originally Monothelites* who followed Sergius* of Constantinople. In the last quarter of the seventh century they formed their own hierarchy. After associating with Crusaders they entered a rather unstable union with the Roman see in the thirteenth century. This union was strengthened at the Council of Florence in 1445 and at later synods, particularly one in 1736. A Jesuit named John Eliano worked hard at cementing the union, and due to his effort Rome has had a Maronite college since 1584. At a council in 1616 the Maronites moved against abuses in their fellowship; and again at a council between 1733 and 1742 under Patriarch Joseph IV they attacked abuses and inserted the *Filioque* in the creed.

Maronites have communed under one species since 1736. They follow a W Syriac liturgy. Their patriarch, with his auxiliary bishops living at his side, was known as "Patriarch of Antioch and All the East." Two cloisters, Bkirki and Kannobin near Beirut, serve as his residence. Dioceses were first established in 1736; diocesan clergy usually are married. Since 1926 the Maronite faith has been the chief confession of the Lebanese state, although Maronites make up only 30 per cent of the population (Muslims number 50 per cent). No census has been taken since 1932, to avoid trouble. The Lebanese president is always a Maronite. About 470,000 Maronites reside in Lebanon, and 380,000 in North and South America, the majority in the south. Other Maronites are found in Palestine, Syria, and Egypt. Congregations in the diaspora are subject to local Latin Rite bishops, but the Maronite patriarch remains their true head.
JOHN GROH

MAROT, CLÉMENT (1497-1544). French Protestant hymnist and poet. Born at Cahors, son of the court poet to Anne of Brittany, Marot in 1514 presented the future King Francis I with the *Judgment of Minos* and in 1518 entered the circle of Margaret of Navarre,* whose Protestant teachings influenced him. After his capture at Pavia in 1525 Marot wrote a poem on the NT for Francis I. Following the Placard affair, Marot fled to Ferrara. In 1536 at Lyons, Marot rejected Protestantism, receiving a house in the Paris suburbs from Francis I in 1539. Here he completed the metrical version of the Psalms from Vatable's Latin Version. This work appeared as twelve of the eighteen psalms in the first Calvinist hymnbook published at Strasbourg in 1539. In 1542 thirty such psalms were published with a royal dedication. Fleeing to Geneva for a year, Marot translated twenty more psalms at Calvin's urging for the 1542 *Cinquante pseaumes.* In 1562 Beza completed the hymnbook with 101 of his psalms added to 49 by Marot. Sixty-two editions appeared in three years.
MARVIN W. ANDERSON

MARPRELATE TRACTS. The violent tone of these pamphlets which appeared in 1588-89, their vigorous and often crude humor at the expense of the bishops, the mystery of who wrote them and the fascinating circumstances of their printing—all these have made them seem to be more important than in fact they were within the Puritan reforming program. Written by a fictitious "Martin Marprelate," there are eight extant: *The Epistle, The Epitome, Certain Mineral and Metaphysical Schoolpoints, Hay any worke for Cooper* (all printed by Robert Waldegrave); and *Martin Junior, Martin Senior, More worke for the Cooper,* and *The Protestatyon of Martin Marprelat* (all printed by John Hodgkins). All these titles are shortened forms. Their main importance lies in the fact that they spread alarm in official circles concerning secret printing presses and led to more repression of Puritans. There have been various suggestions as to the authorship of the pamphlets, e.g., John Udall, Thomas Cartwright, Job Throckmorton, John Penry, and Michael Hicks (Burghley's secretary). Whoever it was, he caused not a few leading ecclesiastics (e.g., Richard Bancroft) to write in reply.
PETER TOON

MARQUETTE, JACQUES (1637-1675). French Roman Catholic missionary and explorer. Born in Laon, France, he entered the Jesuit Order in 1654 and after twelve years of study and teaching went to Canada, where he was assigned to mission outpost work. He served among the Ottawa and Huron Indian tribes in the Lake Superior and Lake Michigan area. In 1673 he and Louis Joliet explored toward the west, discovering the Mississippi River. Marquette wanted to found a mission among the Illinois Indians, and died after spending some time preaching among them. He is known today mostly for his explorations, yet his missionary labors have not been forgotten.

MARRANO, see ANUSIM

MARRIAGE. The Christian concept of marriage is the union between a man and a woman that is recognized by society and has intended permanency. This ideal has not been, nor is now, practiced by all societies. The number, rights, and duties of married persons and the dissolubility of the marriage are sources of variance from society to society.

The OT presents a definite practice of polygyny with a marked tendency toward monogamy in the latter sections. Celibacy was foreign to the Hebrew culture. Marriage was patriarchal in form and consummated for the purposes of procreation (maintenance of family line and name were important), joy, fellowship, and companionship.

There was an ontological basis for the marriage, and the initial intercourse established the ontic bond. There tended to be a positive attitude toward sex and a general repugnance for its illicit use. Divorce was primarily the prerogative of the man and became much more frequent in the post-exilic age.

The Greeks and Romans saw marriage as a divine institution of importance to the state and to the family. Monogamy became the model for the institution, but divorce and remarriage were common. Since the primary reason for marriage was procreation, the wives often became bearers of children while the husbands amused themselves with concubines and prostitutes. Even with these discrepancies the Roman form of marriage had its effect upon the Christian institution.

There is a basic kerygmatic core in the biblical view of marriage determined primarily by the fact that marriage rests upon a primeval order of creation and is at the same time symbolically or parabolically incorporated in the order of salvation. Jesus declares that marriage as an original order of creation is indissoluble (Mark 10:2-12; Matt. 5:31,32 does give the possible exception of fornication), but marriage was not compulsory; for some the demands of the kingdom might involve a celibate life. Monogamy, based upon the *henosis* concept of one flesh, was rapidly becoming the ideal.

The NT exalts marriage to a new height of sanctity. "Marriage should be honored by all, and the marriage bed kept pure, for God will judge the adulterer and all the sexually immoral" (Heb. 13:4). The epistle to the Ephesians presents a very high view of marriage and even compares it to Christ's relationship to His church (5:22ff.).

Paul advocated celibacy for himself and for others who wished to serve the kingdom without hindrances. He recognized that family responsibilities encumbered a man's ability to serve (1 Cor. 7:32-34), and he wished that all the Corinthians might share his celibacy; but, like Jesus, he recognized that every man's marital status must be determined by God's gift of continency (1 Cor. 7:7). He encourages the unmarried and the widows to abide in the celibate state, but if they do not have this gift then by all means marry rather than burn in passion (1 Cor. 7:8,9). Paul's entire view of celibacy and marriage is greatly colored and even dictated by his eschatology. Paul looked for the imminent return of Christ, and he deemed it advisable not to change one's marital status during this time. It was time to prepare for the world where "those . . . will neither marry nor be given in marriage" (1 Cor. 7:29-31; cf. Luke 20:34,35). Although the celibacy passages are strong, Paul had a high view of marriage (1 Cor. 7:7, 10-11, 17,28,36).

The early Church Fathers, Clement of Alexandria* and Tertullian,* began to emphasize that all sexual desire was evil and that for most, marriage was an escape from sin. Clement thought marriage was superior to celibacy because a married man must practice self-denial and is less selfish; yet at the same time he speaks of the higher spiritual perfection of widows and virgins. Spritual marriages are advocated.

Augustine* advocated celibacy. Virginity was not necessarily higher than marriage, but those who lived it experienced a higher type of life. Marriage was given a sacramental position in the church, but the basic reasons for matrimony were procreation and the curbing of lust. Augustine came close to equating venereal emotion with original sin. Evil accompanied all acts, especially the coital act. Sex, even in marriage, was evil.

The basic teachings of the Roman Catholic Church were propounded by Thomas Aquinas.* Aquinas followed Augustine in exalting celibacy and still maintained that marriage was for procreation, for the curbing of lust, and to experience a sacrament. Coition was not an integral part of marriage, and the very act transmitted original sin. Continence should be practiced as much as possible. Divorce was not possible if the marriage was lawful, consummated, and Christian; Aquinas did recognize that the wife was more than an instrument of sexual gratification and thus exalted her place in marriage.

The monastic ideal, which Luther had once embraced, became the object of his bitter criticism. Even Catholic historians agree that the abuse of monasticism in Luther's time was rife. Clerical concubinage was an accepted institution, and homosexuality was lightly condoned. However, Luther condemned not only these desecrations, but the very institutions of celibacy. He felt that God had ordained marriage for all men as a remedy for sin. Only a very few men were given the ability to lead chaste lives outside the bonds of marriage. These few were to be considered angels on earth.

The rite of marriage, for Luther, was a "worldly" (civil) act for which the church had no constitutive importance. In the very order of creation, marriage was constituted for all men and not only for Christians; the church can only give her blessing to the marriage that has already been contracted. The marriage becomes the business of the church only when a matter of conscience is involved. For Luther to designate marriage as "worldly" did not mean that it had not spiritual relevance; on the contrary, the worldly and spiritual poles of marriage are not antithetical, but complementary. Luther denied the sacramental nature of marriage, but held tenaciously to its permanency. The Reformer John Calvin* gave new and refreshing interpretations of marriage. First, he elevated woman to a position of mutual responsibility within the marriage. He still saw marriage as an institution for procreation and as a remedy for sin, but he went much further. He taught that the primary purpose of marriage is social and not generative and is the highest relationship known to man. No bond in human relations is more sacred than that by which husband and wife unite to become one body and one soul. He did not condemn virginity, but did disapprove of celibate vows. He also was willing to grant grounds for divorce for adultery, impotency, desertion, or religious incompatibility.

The Puritans* built upon the teachings of Calvin and gave greater equality and independence to women; discouraged celibacy; liberalized divorce laws (but few were granted); proclaimed marriage a civil ordinance; and encouraged the companionship concept of marriage. The sexual ethics of Calvin were, however, reinterpreted in a more legalistic, narrow way. Coitus was just for procreation, but was not to be refused if the other partner felt it necessary. Along with the early Church Fathers they saw all sexual relationships, even in marriage, as bad or evil.

In more recent decades the attitudes of Christians toward the sexual aspects of marriage have changed considerably. The deeper understanding of personality and human behavior made possible by modern psychology has caused sexual activities to be viewed as a more positive element within marriage—not merely a "remedy against sin" or a means of procreation. This new understanding has brought about a wide acceptance of contraceptive methods. (Officially the Roman Catholic Church is opposed to these, but many Catholics use them.)

Much has been made of the psychological understanding of the sexual consummation of marriage. The psychological unity produced is akin to the biblical *henosis* and affects the vital wills of these persons. It also produces an intuitive self-awareness and self-understanding that is unavailable in any other context. This too is amazingly kin to the biblical concept of "to know."

Contextual ethics with respect to divorce, remarriage, and all borderline cases have become acceptable to many. They maintain that men must, as Paul did, make their standards relative but never forget that all relative achievement must fall under the judgment of God. As far as the order of creation is concerned there can be no divorce and hence no remarriage, but the "order of necessity" is real and the allowance for divorce is in keeping with reality. Many reject this as being too arbitrary.

Patterns of marriage continue to change in modern society—group marriage, monogamy, polygyny, exogamy, and polyandry are all found to some degree. However, monogamic marriage with premarital chastity and marital fidelity seem to be the ideals established by the Christian community. The ideal of premarital chastity has been heavily challenged by the younger generation, but the church has stood firm on its NT interpretations.

BIBLIOGRAPHY: O.D. Watkins, *Holy Matrimony: A Treatise on the Divine Laws of Marriage* (1895); G.H. Joyce, *Christian Marriage: An Historical and Doctrinal Study* (1948); R.H. Bainton, *What Christianity Says about Sex, Love and Marriage* (1957); W.M. Capper and H.M. Williams, *Towards Christian Marriage* (4th ed., 1958); D.S. Bailey, *The Man-Woman Relation in Christian Thought* (1959); H.A. Bowman, *A Christian Interpretation of Marriage* (1959); P.G. Hansen et al., *Engagement and Marriage: A Sociological, Historical and Theological Investigation* . . . (1959); O.A. Piper, *The Biblical View of Sex and Marriage* (1960); H. Thielicke, *The Ethics of Sex* (1964).

JOHN P. DEVER

MARROW CONTROVERSY, THE. One of the most significant controversies the Church of Scotland has ever known, it began in England in 1645 with publication of a work entitled *The Marrow of Modern Divinity*. Authorship was traditionally attributed to Edward Fisher,* but this seems improbable. An exposition of Federal Theology, the book largely comprises extracts from Reformers, including Luther and Calvin, and from the English Puritans. It had gone through seven editions by 1648, when a second part was published: an exposition of the Ten Commandments which, like the first part, contrived a middle course between antinomianism and legalism.

About 1700 Thomas Boston* purchased the first part from a Berwickshire parishioner, and it greatly influenced his preaching. His recommendation led finally to the book's reprinting in 1718, with a preface by James Hog of Carnock. It proved anathema to the legalism of the Moderates,* and in 1720 the general assembly, condemning the book as heretical and antinomian, passed an act prohibiting ministers from commending it, and enjoining them to warn their parishioners against it. Despite a document signed by Boston, Ebenezer Erskine,* and ten other "Marrowmen," who saw an attack on evangelical truth, the protesters were formally admonished and rebuked by the 1722 assembly. Many Moderates had urged a more severe sentence, and embarked on a systematic persecution of Marrowmen, whose preaching nevertheless attracted great numbers. The controversy gradually subsided, and when Boston produced in 1726 a new edition of *The Marrow* with extensive notes, the establishment found it prudent not to pursue the matter further.

See D. Beaton, "The 'Marrow of Modern Divinity' and the Marrow Controversy," *Records of the Scottish Church History Society*, vol. I, part III (c.1925), pp. 112-34. J.D. DOUGLAS

MARSDEN, SAMUEL (1764-1838). Anglican chaplain to the convict colony of New South Wales. He was educated at Magdalene College, Cambridge, but at the suggestion of William Wilberforce* left for Australia without taking his degree. He arrived in Sydney in 1794 and was stationed at Parramatta, where he remained until his death. On the departure of Richard Johnson* in 1800 he was the only chaplain in the colony, and in 1810 he became senior chaplain after a visit to England to recruit others. Marsden's activities have been the subject of much controversy. Like most officials in the colony, he took up farming, and his very success occasioned comment. On appointment as a magistrate he gained a reputation for severity scarcely excused by the character of the colony. There is no evidence that he neglected parish duties and church affairs. He is famous as the founder of the mission to the Maoris of New Zealand under the Church Missionary Society. He preached the first sermon in New Zealand in 1814 and made seven journeys in support of the infant mission, often at his own expense. He did much too to establish the Evangelical character of the Church of England in Sydney.

NOEL S. POLLARD

MARSHALL, PETER (1902-1949). Presbyterian minister. Born in Coatbridge, Scotland, he studied at technical school and mining college, but in 1927 went to the USA, read theology at Columbia Seminary, Georgia, and was naturalized in 1938. Ordained to the Presbyterian ministry in 1931, he held pastorates in Georgia before becoming minister of New York Avenue Presbyterian Church, Washington, D.C., in 1937. He was chaplain to the U.S. Senate from 1947, where he was known for remarkably pithy prayers. He wrote *The Mystery of the Ages* (a study in Ephesians) and *Mr. Jones, Meet the Master.*

MARSHALL, STEPHEN (1594?-1655). Puritan divine. Born in Huntingdonshire and educated at Emmanuel College, Cambridge, he became lecturer at Wethersfield and then vicar of Finchingfield, Essex. During Archbishop Laud's supremacy he was often in trouble; his Puritan and Presbyterian influence in Essex was far-reaching from 1630 to 1655. In the electioneering for the Short and Long Parliaments of 1640 he was active on behalf of Puritans. He preached before the Long Parliament many times. He was a member of the Westminster Assembly* and one of the commissioners sent by it to Scotland. Despite his earlier Presbyterianism he cooperated with the Independents when they came to power in 1649. In Cromwell's state church he was a "Trier." His most famous publications were sermons preached before Parliament—e.g., *Meroz Cursed* (1641). He died of consumption and was buried in Westminster Abbey, but his body was removed in 1661. PETER TOON

MARSHMAN, JOSHUA (1768-1837). Baptist missionary. Born at Westbury Leigh, Wiltshire, he had a scanty education but read avidly while working with his weaver father. He married Hannah Shepherd (1791), and they had twelve children. He became master of a Baptist school at Broadmead, Bristol (1794). He offered for the Baptist Missionary Society and sailed for India with William Ward to join William Carey* in 1799. Forbidden to land by the East India Company, they settled at Danish Serampore, where Carey joined them. There they preached, taught, itinerated, and translated. Joshua and Hannah opened boarding schools to help pay for printing the Scriptures. An able Orientalist, he published the works of Confucius, Chinese grammars, and a Chinese version of the Bible. J.G.G. NORMAN

MARSIGLIO (Marsilius) OF PADUA (c.1275-1342). Political philosopher. Born in Padua, Italy, he began his academic career in medicine there. Later he went to Paris where he became rector of the university in 1313. In 1324 he completed the work for which he is famous, *Defensor pacis* ("Defender of the peace"). Because of the very strong antipapal tone of the work, when his authorship was discovered in 1326 he was forced to leave Paris and went to the court of King Louis IV of Bavaria, who was excommunicated as the result of a dispute with Pope John XXII. In 1328 Louis IV seized Rome, and Marsiglio was named the imperial vicar of the city. Some of his political theories were put into practice, but the people of Rome turned against Louis and he left, taking Marsiglio with him. Marsiglio went back to Bavaria and spent the rest of his life there. Toward the end of his life he wrote another work, *Defensor minor,* basically a restatement of his earlier and more important work.

Defensor pacis is divided into three books: the first deals with a philosophy of the state; the second with the theology of the church; and the third is a summary. Marsiglio argued that the unifying element in society is the state and not the church. The chief function of the secular ruler is to maintain peace. He gave power to the people to create law, to govern the common welfare, and to choose a ruler whom they could overthrow if he violated their laws. In book two he chastised the papacy for causing dissension in the world as it improperly attempted to control the temporal world. Papal claims for such control were invalid since Christ supported submission to temporal power, the church's hierarchy was not divine but human, and the temporal prerogatives claimed dated to the Donation* of Constantine and thus were ultimately secular in origin and not inherent right of papacy. He stated the only power possessed by the church was spiritual in nature. He was in favor of the secularly called general council as supreme in the church. His ideas ran counter to political theory of the papacy and help to explain why he was condemned by the papacy and why he is often seen as one of the forerunners of the Reformation period political thought. *Defensor pacis* was published in 1517, placed on the Index in 1559, but studied carefully by many Reformers.

BIBLIOGRAPHY: E. Emerton, *Defensor Pacis of Marsiglio of Padua* (1920); C.W. Previté-Orton (ed.), *Defensor Pacis* (1928); A.P. D'Entrèves, *Medieval Contribution to Political Thought* (1939); A. Gerwith, *Marsilius of Padua: The Defender of the Peace* (2 vols., 1951-56); C. Pincin, *Marsilio* (1967); J. Quillet, *La philosophie politique de Marsile de Padoue* (1970).

ROBERT SCHNUCKER

MARTEL, CHARLES, see CHARLES MARTEL

MARTENSEN, HANS LASSEN (1808-1884). Danish bishop and theologian. As a student he was impressed by N.F.S. Grundtvig*; later on a study tour to Germany he was greatly influenced by Hegel and the Roman Catholic philosopher of religion, Father Baader. In 1838 he was appointed lecturer and in 1840 professor of systematic theology at Copenhagen University; from 1854 until his death he was bishop of Zealand. As a teacher he exerted extraordinary influence upon his students. He held also a preaching ministry through which many were reached. His two principal works are *Den christelige Dogmatik* (1849; ET and other languages) and *Den christelige Ethik* (1871-78). He makes a strong attempt to bring about a harmonious synthesis between faith and thought, theology and philosophy, Christianity and culture. He insists upon the principle "*credo, ut intelligam.*" Nevertheless, his theology is stamped by mystical and speculative elements. As a bishop he carried on the "centripetal" policy of his predecessor, J.P. Mynster.* Unlike most other church people of his time, Martensen displayed some understanding of socialism and the legitimacy of the claims of the workers. This finds expression in his book *Socialisme og Kristendom* (1874). He wrote also *Katholicisme og Protestantisme* (1874), provoked by the 1870 Vatican Council. N.O. RASMUSSEN

MAR THOMA CHURCH, see MALABAR CHRISTIANS

MARTIN I (d. 655). Pope from 649. A native of Tuscany, he was elected successor to Pope Theodore I, but before his election had been confirmed by the emperor Constans II, he had condemned Monothelitism* at the Lateran Synod (649). In 648 Constans had issued the "Typos," a mandate forbidding further discussion of the matter. The Lateran Synod condemned the "Typos" and the earlier "Ecthesis,"* and affirmed its adherance to the doctrine of two wills and two energies corresponding to the two natures of Christ. Constans tried vainly to induce Olympius, the exarch of Ravenna, who was friendly to Martin, to arrest the pope. However, the next exarch, Theodore Calliopas, did seize him. After a year's captivity at Naxos, Martin was brought to Constantinople in December 654. He was charged with treason, publicly stripped of his episcopal robes, and banished to the Chersonesus (Crimea). He was treated with great cruelty and seems to have died en route. He was the last of the popes to be venerated as a martyr. J.G.G. NORMAN

MARTIN IV (c.1210-1285). Pope from 1281. Simon de Brie, a native of Touraine, was appointed chancellor of France by Louis IX in 1260, and in 1261 he was created cardinal of St. Cecilia and papal legate by Urban IV. As legate he negotiated the advancement to the Sicilian throne of Louis's brother, Charles of Anjou. In 1281 he was elected pope through pressure exerted by Charles, whose tool he remained. He restored Charles to the position of Roman senator of which he had been deprived by Pope Nicholas III. With a view toward aiding Charles's projected attack on the Eastern Empire, Martin excommunicated Emperor Michael VIII Palaeologus, thus destroying the union of the Latin and Byzantine churches achieved at the Council of Lyons in 1274. The "Sicilian Vespers,"* the anti-French rebellion in Sicily in 1282, forced Charles to abandon his plans to reconquer Constantinople. Pleas by the Sicilians for papal suzerainty were refused by Martin, so they turned to Peter of Aragon. Martin thereupon excommunicated the Sicilians and organized a crusade against Peter under Philip III of France, but it failed dismally. Deeply interested in the Franciscan Order, he extended their privileges in the bull *Ad fructus uberes* in 1281. Martin's pontificate marked the decline of papal political power. J.G.G. NORMAN

MARTIN V (1368-1431). Pope from 1417. Elected pope at the Council of Constance, he ended the Great Schism,* winning general recognition in W Europe except in Aragon, where the former Avignonese antipopes maintained their positions until 1429. Although the French offered Avignon as his residence, he decided to return to Rome. He reached Florence in 1419, but stayed there until 1420 because Rome was in the hands of Joan of Naples. When he got to Rome, the city was in ruins, and other parts of the Papal States were either in revolt or in the hands of usurpers. The pope reestablished papal control, not only over central Italy, but also in the entire Western Church. He corresponded with the sovereigns of Europe and sent peace missions to England and France, who were involved in the Hundred Years' War.

The pope also devoted attention to the Hussites who, reacting violently to the martyrdom of Jan Hus* at Constance, rebelled against the Holy Roman Emperor, Sigismund,* spread terror in neighboring Catholic countries, and necessitated several crusades in an attempt to crush them. In 1423, five years after Constance, Martin summoned the Council of Pavia-Siena, but the attendance was poor and he quickly dissolved it. Despite his previous commitment to conciliarism, he successfully opposed limitation of the papal monarchy. Martin also reorganized the Roman Curia, uniting the bureaucracies of Rome and Avignon and establishing a model administration. A rebuilding program was initiated in Rome, and efforts were made to end the schism between the Eastern and Western churches.

See P. Partner, *The Papal States Under Martin V* (1958). ROBERT SCHNUCKER

MARTINEAU, JAMES (1805-1900). English Unitarian minister and teacher. Born at Norwich, he held various teaching posts before becoming in 1832 minister of a church in Liverpool and simultaneously professor of philosophy in Manchester New College. He became principal there in 1869. He began his career as a follower of Joseph Priestley,* holding the characteristic doctrines of Unitarianism because of their allegedly biblical character. Later, in his *Rationale of Religious Enquiry* (1836) he adopted a more rationalistic position, and came finally to advocate a philosophical theism grounded on the moral consciousness, opposing the materialism of Spencer and Tyndall. Martineau had a great literary output (e.g., *Types of Ethical Theory,* 1885) and became a popular figure in his attempts to harmonize religion and the "modern thought" of the Victorian era. He was also active in the temperance movement. PAUL HELM

MARTIN OF BRAGA (Bracara) (c.510/520-579). Archbishop and writer. Born in Pannonia (modern Hungary), he became a monk in Palestine and moved to Gallaecia in the NW part of Spain around 550. He established a quasi-eremitical monastery in Dumio and became its abbot, then was named bishop of Dumio in 561. Sometime before 572 he was elevated to be metropolitan of Bracara, the Suevian royal city, where he worked successfully to convert Arians, including the king whom he baptized in 556. A competent translator well trained in theology, Martin read and referred to the writings of Seneca, Augustine, Cassian, and Caesarius of Arles. One of his treatises, dedicated to King Miro (570-83), examined the four cardinal virtues on the basis of one of Seneca's lost works. A sermon titled *De correctione rusticorum*, composed for Bishop Polemius of Astorga, contributes much to the history of culture with its description of peasant superstitions. Well respected outside of Spain, he also composed works on canon law, liturgy, and the church calendar. JOHN GROH

MARTIN OF TOURS (c.335 - c.400). Pioneer of monasticism in Gaul. Born at Sabaria, Pannonia (modern Hungary), the son of a pagan soldier, he became a catechumen at the age of ten. His father enlisted him in the Roman army at fifteen, and three years later came the famous incident when he divided his military cloak with a beggar at Amiens and subsequently had a vision of Christ wearing the half-cloak. About two years later he was baptized. Obtaining his discharge from the army in 358, he visited Pannonia seeking his parents' conversion. In 361 he joined Hilary of Poitiers,* adopted the monastic life, and founded a monastic community at Ligugé. His disciples lived as hermits at first, meeting occasionally for common exercises. He wrote no rule; they simply followed the general traditions of the ascetic life. Later he moved his monastery to Marmoutier. In 372, by popular acclaim he was unwillingly made bishop of Tours. He engaged in active missionary work in Touraine, introduced a rudimentary parochial system, and encouraged monasticism. In 386 he protested to Emperor Maximus against the first execution for heresy, that of the Spaniard, Priscillian. His life was written by his friend Sulpicius Severus. J.G.G. NORMAN

MARTYN, HENRY (1781-1812). Anglican missionary to India. Born in Cornwall, he received his early education there before embarking on a brilliant academic career at Cambridge. The sudden death of his father led eventually to a spiritual awakening, and to his ordination in 1803 as curate to Charles Simeon* at Holy Trinity Church, Cambridge. He was the first Englishman to offer to the newly formed Church Missionary Society, but for reasons beyond his control he was not accepted. Further disappointment and intense unhappiness came when after a protracted period his proposal of marriage was not accepted. In 1805 Martyn sailed for India as a chaplain to the East India Company, and arrived at Calcutta where he enjoyed fellowship with two other Evangelical chaplains, Daniel Corrie and David Brown.

His outstanding linguistic gifts led to his great life-work, the translation of the NT and the Book of Common Prayer into Hindustani. His forthright preaching to British congregations caused offense, as did his constant attempts to make contact with native Indians, both Hindus and Muslims. Posted at first to Calcutta, he was appointed to Dinapore in 1806, and to Cawnpore in 1809, where incipient tuberculosis and the intense summer heat almost caused his death. In 1810 he was advised to take a sea voyage, and being anxious to complete an Arabic and Persian translation of the NT, he traveled to Shiraz, where he talked and worked for long hours with Persian scholars, gaining their respect and confidence in argument and debate, and finishing his task in February 1812. He set out for home, but hard traveling and constant fever brought about his death at Tokat in Armenia in October of that year. He was buried there. His *Journals* were brought home after his death and remain among the classics of English devotional literature, revealing the intensity of his dedication to the service of Christ and the cause of Christian missions.

See C.E. Padwick, *Henry Martyn, Confessor of the Faith* (1953). G.C.B. DAVIES

MARTYR; MARTYROLOGY. A basic definition of the term "martyr" is provided by Origen: "One who of his own free choice chooses to die for the sake of religion." The Greek word *(martys)* means simply "witness" in the legal sense, and it carries this neutral connotation in several NT passages (e.g., Mark 14:56,59,63; Luke 22:71). The ground, however, is already prepared in the NT for the later development of the term, when it becomes the equivalent of a blood witness, i.e., one who dies for his faith and "prefers to die rather than deny his religion and live" (Origen). Stephen (cf. Acts 22:20) is appealed to in the later church as the "perfect martyr" as well as the protomartyr. An otherwise unknown figure, Antipas (Rev. 2:13) is mentioned as both a martyr and a Christian who was killed for his faith. In the Apocalypse the term receives its full technical sense (cf. 6:9; 17:6; 20:4).

The background of the idea, if not the precise term itself, lies in Jewish history, especially in the prophets' fidelity to their mission and consequent suffering. In particular, at the time of the Maccabean struggle against the Syrians, the main traits of Jewish martyrdom which later were to influence the Christian martyrs were fixed. These are seen in such features as the expiatory element in human suffering and an apocalyptic dimension that provided the necessary fanaticism to overcome first the Syrian dictator and then the Roman power. The belief in resurrection, clearly articulated in the Maccabean age, obviously was a needed conviction to sustain the Christian heroes and heroines.

In retrospect, both Paul's and Peter's deaths are hailed as acts of martyrdom. With Ignatius of Antioch (c.115) the thought of the martyr's conscious imitation of his Lord's passion appears (Phil. 2:7; Rom. 6:3), and this formed a powerful motif in the later martyrologies.

One early martyrdom became the model for all subsequent resistance unto death. Polycarp* of Smyrna died after interrogation in the amphitheater; and his hagiographer has graphically recorded his last days in a way which became standard for later *Acta.* The veneration of the martyrs' bones was a practice which began with Polycarp's remains, and an annual event was observed as a "celebrating of the birthday of his martyrdom." This is the origin of the idea of martyrology by which is meant the commemoration of the martyrs' sacrifice. Later, intercessory powers were attributed to the martyr.

The fullest description of early Christian martyrs is that of those who died in Lyons and Vienne in Gaul. The later persecutions under Decius (whose edict in 250 initiated the first universal and systematic persecution of the church) and Diocletian (303) produced their crop of martyrs both in the strict sense of those who chose to die rather than recant and those who were confessors and were tortured for their faith.

The standard work with its full bibliography is W.H.C. Frend, *Martyrdom and Persecution in the Early Church* (1965). RALPH P. MARTIN

MARVELL, ANDREW (1621-1678). English poet. Born near Hull and educated at Cambridge, he traveled on the Continent and acted as tutor to the daughter of Lord Fairfax at Nunappleton House, which provided the subject for a poem and may also have inspired "The Garden." He was subsequently assistant Latin secretary to Milton under Cromwell and after the Restoration was member of Parliament for Hull. In this latter period he wrote much controversial prose and satiric verse, most of it now forgotten. His poems were not published until 1681. His fame rests on a slender but nonetheless firm base. "To His Coy Mistress" is a fine love-poem in the *carpe diem* witty manner, and "The Coronet" is a subtle, Metaphysical exploration of the sin that inevitably mars man's noblest efforts at worship. It is "The Garden," however, that marks his supreme achievement with its quasi-mystical treatment of the retirement theme, the mind rediscovering Paradise in its own creative self-sufficiency.

ARTHUR POLLARD

MARY (the Virgin Mary). Mother of Jesus. The Greek name *Maria* or *Mariam* renders the Hebrew *Miriam.* There is comparatively little reference to Mary in the NT outside the birth and infancy stories of Matthew and Luke. Matthew records these largely from the point of view of Joseph. He is referred to as "the husband of Mary, of whom was born Jesus" (Matt. 1:16), but it is made plain that the conception of Jesus took place when they were betrothed but before they had intercourse (Matt. 1:18-25). The Holy Spirit was stated to be the one through whom she had conceived. The statement that he had no intercourse with her until she had borne a son (Matt. 1:25) implies that they had normal marital relations afterward.

Luke's first two chapters are centered around Mary. The angel Gabriel announces to her that she is to bear a son called Jesus (Luke 1:26-38). Here also it is emphasized that she is betrothed to Joseph, that she has had no intercourse with him, and that it is the Holy Spirit who has brought about the conception. We see also in this passage Mary's willing dedication of herself to this unique role of being the mother of the Son of God, and the angel's address to her as "highly favored" (the recipient, not the giver of grace). Luke then records the visit of Mary to her cousin Elizabeth (Luke 1:39-56). Elizabeth describes her as "blessed among women" and "the mother of my Lord." Mary then gives voice to her song, based upon the song of Hannah (1 Sam. 2:1-10), which is known as the *Magnificat.* In this she praises God for His gracious action on behalf of His people and in particular of herself. She exults that all generations shall call her blessed because of what God has done for her. Luke then records the birth of Jesus of Mary at Bethlehem and the visit of the shepherds at the angel's command (2:1-20).

After reference to the circumcision of Jesus, there follows the account of the presentation in the Temple (Luke 2:21-40). Simeon tells Mary that a sword will pierce her soul because of Jesus. Finally in these chapters, there is the visit to Jerusalem for the Passover when Jesus is twelve (Luke 2:41-51). In this Mary is lovingly rebuked by Jesus for not understanding that He must be in His Father's house. A similar sort of rebuke is found in the story of the wedding at Cana (John 2:1-11). Otherwise, references to Mary in the ministry are almost entirely incidental, and it is stressed that obedience to the will of God is more important than blood relationship (Mark 3:31-35; Luke 11:27f.). John records Mary's presence at the crucifixion (John 19:25-27) and her commendation to the care of the beloved disciple. She is mentioned elsewhere in the NT only in Acts 1:14.

Mary clearly holds in the gospels a considerable place of honor because of her unique role, but there is no justification in history or in theology for the cultus which was to grow up around her figure in the church. In the fourth century Epiphanius had to rebuke heretics who worshiped her, but the Nestorian* controversy of the fifth century led to catholic Christians describing her as *Theotokos,* "bearer of God." Together with this went the idea of her perpetual virginity, and these led on to the idea of her immaculate conception. The antithesis between Eve—the cause of the fall of the human race—and Mary—the cause of its redemption—was developed into the idea of her having a mediatorial role in the economy of salvation. This was particularly stressed in popular medieval devotion because of the apparent remoteness of Christ. The idea of the assumption of Mary to heaven was also developed. In 1854 Pius IX* proclaimed the dogma of the Immaculate Conception* and in 1950 Pius XII* that of the Assumption.* Many liturgical observances are associated with Mary, but Reformed churches have at most observed the Annunciation (25 March) and the Purification (or Presentation of Christ in the Temple—2 February).

See G. Miegge, *The Virgin Mary* (ET 1955), and M. Thurian, *Mary, Mother of the Lord, Figure of the Church* (ET 1963); J. McHugh, *The Mother of Jesus in the New Testament* (1975).

R.E. NIXON

MARY, GOSPEL OF. The Gnostic work which goes under this title is found in the Coptic Papyrus Berolinensis 8502, found in Egypt and brought to the British Museum in 1896. It is preserved only in part, with the text of the final pages preserved in Greek (its original form) in Papyrus 463 (Oxyrhynchus). As the text is transmitted, it contains two separate parts. The first is a fragmentary conversation between the risen Christ and His disciples in which the discussion about matter and sin and preaching the Gospel to the heathen is a source of bewilderment to them. Mary (Magdalene) intervenes to comfort them, and she is then besought (in the second part) to reveal secrets of Gnostic redemption. The exchange between Mary and Peter is interesting since it departs from the normal antifeminist Gnostic line in defense of Mary, as Levi is made to rebuke Peter. The document is dated from the second or third century. RALPH P. MARTIN

MARY, GOSPEL OF THE BIRTH OF. The existence of this treatise is known only from the reference to it by Epiphanius.* He ascribes the work to a Gnostic source which twisted the record of the canonical gospels (here Luke 1) into an anti-Judaic polemic. The document tells how Zechariah, father of John, was killed in the Temple. The title *(Genna Marias)* evidently refers to the birth or genealogy of Mary, the mother of Jesus.

MARY, QUEEN OF SCOTS (1542-1587). Daughter of James V and Marie de Guise-Lorraine, she became queen when six days old. During her minority the pro-English and Protestant interests gained at the expense of the pro-French Catholic group, and in 1560 the Estates of Parliament abolished the authority of the pope in Scotland, forbade the celebration of Mass, and adopted a Reformed Confession of Faith (the Scots Confession*). Mary's upbringing was French and Catholic, and she was consort of Francis II of France, 1559-60. She returned to Scotland in 1561 after thirteen years in France. Regarded by Catholics as the rightful queen of England because of her descent from Henry VII and Elizabeth's alleged illegitimacy, Mary was for the rest of her life the focus of international intrigue. Her personal rule in Scotland was remarkably successful at its beginning. While she incurred the opposition of John Knox* over Mass in her private chapel, she conciliated moderate opinion by acquiescing in the division of church revenues whereby a third was shared by the Protestant ministers and the Crown, and in legislation implying the recognition of the Reformed Church.

Her downfall was caused by the English succession question and her marriages to Henry (Lord Darnley) in 1565 and James Hepburn, earl of Bothwell, in 1567. Darnley's claim to the English throne was almost as good as her own, but the marriage was disastrous personally and politically. Her marriage to Darnley's supposed murderer, Bothwell, completed her ruin. In 1567 she was deposed by a coalition of nobles who proclaimed her son by Darnley as James VI. After an unsuccessful bid to regain power, she fled to England where she was imprisoned as an alleged accomplice to Darnley's murder. A series of Catholic plots to place her on the English throne resulted in her execution at Fotheringay.

See A. Fraser, *Mary, Queen of Scots* (1969), and I.B. Cowan (ed.), *The Enigma of Mary Stuart* (1971). HENRY R. SEFTON

MARY MAGDALENE. The name probably refers to her being an inhabitant of Magdala on the western side of the Sea of Galilee. She is first mentioned in Luke 8:2 as someone out of whom seven demons had been cast. The exact nature of her complaint—physical, emotional, or spiritual—is unknown. It is unlikely that she is to be identified with the sinful woman who anointed Jesus (Luke 7:37ff.), as Luke fails to make any explicit connection. She was a witness to the crucifixion along with the other women who had accompanied Jesus on His last journey to Jerusalem (Mark 15:40, etc.). She was particularly prominent at the time of the resurrection. She went with the others to the tomb (Mark 16:1, etc.), but John mentions her alone as running to tell the disciples that the stone had been removed (John 20:1f.). John also records the moving scene where she meets Jesus but does not recognize Him until He addresses her by name, then she seeks to cling to Him instead of accepting a new relationship with Him on a different plane (John 20:11ff.).

R.E. NIXON

MARY OF EGYPT (c.344-c.421). Said to have had an infamous early life in Alexandria, she was supposedly converted in the precincts of the Holy Sepulcher, Jerusalem, whereupon she spent the remaining forty-seven years beyond Jordan in isolation doing penance for her sins. Zosimus, a priest, met her there, giving her Communion before she died that evening. Sources include Cyril of Scythopolis's *Life of Cyriacus* and later Byzantine hagiographers who used details from Jerome's *Life of Paul the Hermit.*

MARY TUDOR (1516-1558). Queen of England. Mary was the third and only surviving child of Henry VIII* and Catherine of Aragon. Early in life she was a pawn on the chessboard of international politics. At the age of two she was betrothed to the dauphin of France and at six promised to Emperor Charles V. In 1526 she was sent as Princess of Wales to Ludlow. The divorce of her parents greatly troubled her, and after 1531 she never saw her mother again. In 1533 she was declared a bastard and cut out of the succession to the throne. Between 1534 and 1536 her father tried to break her "Spanish pride" by petty persecution; after her mother's death in 1536 she even acknowledged under duress that the marriage of her parents was "by God's and man's law incestuous and unlawful." But during 1536-47 her life was fairly easy and carefree. For six years after the death of her father her problems were chiefly religious. She liked Edward VI,* but disliked his Protestantism. She conceded nothing and remained faithful to Catholicism.

On 19 July 1553 she was proclaimed queen in London, and on 3 August entered her capital in triumph. Parliament annulled the divorce of Catherine of Aragon, established Mary's legitimacy, and restored the church to what it was at the end of Henry VIII's reign. But within weeks Mary's popularity had gone. The most sincerely religious and moral of the Tudors was opposed by most of her people. This was because she was a Spaniard first, a Tudor second, and an English Tudor last. She insisted on restoring papal Catholicism and seeking a husband in Spain (Philip II in July 1554). Reginald Pole arrived as papal legate and archbishop of Canterbury, and in 1555 the statute *de heretico comburendo* was re-enacted, giving power to ecclesiastical courts to deal with "heresy." In 1555-56 T. Cranmer,* J. Hooper,* H. Latimer,* and N. Ridley,* with others, were burned as heretics. Mary's actions ensured that England would be a Protestant country after her death.

See J.M. Stone, *History of Mary I, Queen of England* (1901), and H.F.M. Prescott, *Mary Tudor* (1952). PETER TOON

MASHTOTZ, see MESROB

MASON, LOWELL (1792-1872). American composer. Beginning as an amateur church musician, he became perhaps the greatest single influence in Protestant church music in the United States during the nineteenth century. He edited and published a great number of collections of hymntunes and simple anthems, and devoted much of his energy to music education and the betterment of church music. He wrote a great many hymntunes, of which "Missionary Hymn," "Olivet," "Boylston," and "Bethany" are still among those widely used. He was active in the Handel and Haydn Society of Boston, and he founded the Boston Academy of Music. Unhappily, he discouraged the early folklike and often modal tunes that were part of the distinctive American heritage of the late eighteenth and early nineteenth centuries, considering them inferior to those based on European and particularly German models. J.B. MAC MILLAN

MASS, THE. A term used mainly in the Roman Catholic Church for Holy Communion.* The word is derived from Latin *missio,* "dismiss," referring to the dismissal of catechumens before the Eucharist was celebrated. J.A. Jungmann suggested that *missa* came to be synonymous with "blessing" and so linked with the "consecration" of the elements. It was first used strictly for the Eucharist by Ambrose.* The Roman Catholic Mass still closes with *"Ite, missa est."*

Two main ideas are involved in the doctrine of the Mass: (1) the change whereby the bread and the wine become the actual body and blood of Christ, i.e., transubstantiation*; (2) the conception of the Mass as a sacrifice. According to the Council of Trent,* "in this divine Sacrifice which is performed in the Mass, that same Christ is contained in a bloodless sacrifice who on the altar of the cross once offered himself with the shedding of his blood: the holy Synod teaches that this sacrifice is truly propitiatory."

This particular sacrificial interpretation begins with Cyprian. Earlier writers had used the term "sacrifice," but with the ideas of self-offering and commemoration of Christ's passion, e.g., Justin and Irenaeus. The process continued in the following centuries, e.g., Gregory the Great in the sixth century, Paschasius Radbertus in 844. A tendency in the later Middle Ages associated with each Mass the idea of a distinct offering for sin, leading to the multiplication of Masses. The cup was also denied the laity (Council of Constance,* 1415). The Reformers' rejection of these concepts led to the Tridentine definition which established subsequent Roman Catholic doctrine.

See also under MUSIC, CHRISTIAN.

BIBLIOGRAPHY: B.J. Kidd, *The Later Mediaeval Doctrine of the Eucharistic Sacrifice* (1898); D. Stone, *A History of the Doctrine of the Holy Eucharist* (1909); C. de L. Shortt, *The Mass* (1936); C.A. Scott, *Romanism and the Gospel* (1937); W. Barclay, *The Lord's Supper* (1967). J.G.G. NORMAN

MASSILLON, JEAN BAPTISTE (1663-1742). French preacher and bishop. Native of Provence, he studied in the Oratorian colleges in Hyères and Marseilles, and in 1681 entered their congregation at Aix, later lecturing in their colleges at Pé zenas, Marseilles, Montbrison, and Vienne. He was ordained in 1691, and as directory of the Seminary of Saint-Magloire, Paris, from 1696, he gained a great reputation for preaching. He was said to have been the one court preacher to have made Louis XIV dissatisfied with himself. Consecrated bishop of Clermont (1718) and elected to the French Academy (1719), he assisted at Louis XV's coronation and gave Louis XIV's funeral oration. Respected by Voltaire and others, and militantly anti-Jansenist, he was a moralist and brilliant panegyrist, described as the Racine of the pulpit, combating impiety and incredulity. C.G. THORNE, JR.

MATHER, COTTON (1663-1728). Puritan minister. Born in Boston, Massachusetts, the eldest son of Increase Mather, he was educated at Harvard College. Entering the ministry, he served the Second Church of Boston, first as his father's colleague and then as senior pastor. In 1690 he was elected a fellow of Harvard. The greater part of his 400 publications was published after 1692. They reveal that he was primarily a theologian and historian, with amateur interests in a wide variety of subjects. His *Magnalia Christi Americana* (1702) was and is much used. His *Bonifacius* (1710), later called *Essays to do Good,* was widely read in America. Through his books and voluminous correspondence he enjoyed a European reputation. He was also well known as a philanthropist, supporting, for example, a school for slaves. Though his influence in politics diminished when Joseph Dudley became governor in 1702, he exercised influences in the churches of Massachusetts all his life. The theology he expounded in his later life (e.g., in *Christian Philosopher,* 1721) suggests a move away from orthodox Calvinism. PETER TOON

MATHESON, GEORGE (1842-1906). Scottish minister and hymnwriter. A merchant's son, practically blind by his eighteenth year, he was a brilliant student of philosophy at Glasgow University. Licensed to preach in 1866, he was minister in Glasgow and Innellan (Argyllshire) before moving to St. Bernard's, Edinburgh, where he stayed until 1899. He was an influential preacher. He published books on theology which tended toward Neo-Hegelianism (e.g., *Aids to the Study of German Theology,* 1875); on apologetics (e.g., *Can the Old Faith Live with the New?,* 1885); and of devotion (e.g., *Studies of the Portrait of Christ,* 2 vols., 1899-1900). His collection, *Sacred Songs* (1890), included "O Love that wilt not let me go" (which Tyler argues did *not* arise from a disappointed love affair) and "Make me a captive, Lord." JOHN S. ANDREWS

MATHEWS, SHAILER (1863-1941). American theologian. Born in Portland, Maine, he graduated from Colby College and afterward served on the faculty there (1887-94). He then joined the theological faculty at the University of Chicago, first teaching NT, then theology until 1933. He was the leading voice of the "Chicago School of Theology." Mathews's own thought was a functionalism of extreme clarity and simplicity, which he used not only in biblical and historical theology, but in the problems of the school and its role as champion of the modernist cause against American fundamentalism. His role in the formation of the Federal Council of Churches and the Northern Baptist Convention reflects his view that the work of the school should be carried into the church. CLYDE CURRY SMITH

MATINS (Mattins). The Breviary office for the night, derived from the practice of Vigils in the early church. Designed to be said at midnight in the Roman Church, it is now usually said on the preceding afternoon or evening. The term is used in the Church of England for the service of Morning Prayer.

MATTHEW. Apostle. The name means "gift of Yahweh." He is mentioned in the lists of the Twelve (Matt. 10:3; Mark 3:18; Luke 6:15; Acts 1:13), but only Matthew's gospel gives us any further information. Matthew 10:3 describes him as a tax collector, and the name "Matthew" occurs in the story of the call of the tax collector who is named Levi and, in Mark, "the son of Alphaeus" (Matt. 9:9; Mark 2:14; Luke 5:27). There are some difficulties about the identification of this Levi with the Apostle Matthew, but it should probably be made and it would then be likely that he had two names, "Matthew" being a sort of "Christian name." He seems to have been an official of Herod Antipas collecting dues near Capernaum on goods passing along the road from Damascus to the Mediterranean ports. He appears to have been a man of wealth, as he provided a banquet in his house (Luke 5:29) with a large number of tax collectors present. The first gospel has been traditionally associated with the Apostle Matthew, and, whether he was directly the author or not, it is likely that he lies close behind it. The particular skills which he would have had in his profession would equip him well for recording and arranging systematically material about the ministry and teaching of Jesus, and the keen numerical interest in the gospel has sometimes been connected with his previous way of life. R.E. NIXON

MATTHEW, GOSPEL OF, see SYNOPTIC GOSPELS

MATTHEW OF AQUASPARTA (c.1240-1302). Franciscan philosopher. Born in Aquasparta in Umbria, he studied at Todi and Paris and became Bonaventure's* most important pupil. *Lector sacri palatii* in Rome in 1281, he became general of the Franciscans in 1287. In 1288 he was made cardinal and in 1291 cardinal bishop of Porto and Reufina. His *Commentary on the Sentences* and *Questions* underline Bonaventure's position and put him in the Augustinian tradition. He was opposed to the Aristotelianism of Aquinas by which faith and reason were divided. He posited a parallelism between the essence in things and the ideas of these essences; hence, knowing was less intellection than recognition, and the source of knowledge was supernatural illumination. He produced many sermons and biblical commentaries. HOWARD SAINSBURY

MATTHEW OF JANOV (c.1355-1393). Czech Reformer. He studied at Paris (1373-81), gaining first a master's degree, and then spent six years studying Scripture. He returned to Prague through Rome and was received as titular canon of the cathedral, but prevented from fulfilling the office. Appointed a confessor in 1381, he held this office for seven years until granted a poor parish. Greatly influenced by Milic*—"father of the Czech Reformation"—he followed his pattern of biblical preaching, which brought him many enemies, but this served to make him study Scripture more diligently as the exclusive source of doctrine and preaching. This resulted in his profound work in five volumes, *Regulae veteris et novi testamenti,* which served as a source of inspiration to subsequent leaders of the Czech reform movement. Either Matthew himself or his followers translated the whole Bible into Czech. GORDON A. CATHERALL

MATTHEW PARIS (c.1200-1259). English Benedictine historian and artist. A monk at St. Albans in England after 1217, be became chief of the epistolary section in 1236. An unsuccessful attempt to reform the abbey of St. Benet Holm in Norway in 1248-49 led to a friendship with King Haakon IV, whom he invited to join in a crusade with St. Louis of France. As a writer Matthew decorated his manuscripts with illustrations, including coats of arms and the events he was describing in the text. Among his works were *Flores historiarum; Historia Anglorum; Abbreviatio chronicorum;* and his major work, *Chronica majora,* which summarized the chronicle of Roger of Wendover up to 1235, but then relied on firsthand experiences and documents down to 1259. This work pointed up royal foibles, attacked mendicants, and decried the avarice of the papal court. Matthew replaced mere chronicling with a sense of history; he recognized the responsibility of the historian. Some of his information he secured from royal figures who visited the abbey. His account of contemporary times repeatedly mentioned the Beguines,* a new women's movement without a rule that originated in the vicinity of Liège but centered in Cologne. His chronicles blamed Rome for the schism with the Greeks. Wycliffe later took this line of reasoning to its extremes. JOHN GROH

MATTHEW'S BIBLE. First English authorized version of the Bible, and basis of later versions. Its editor was John Rogers,* chaplain to the English merchants at Antwerp, friend of Tyndale, and the first Marian martyr. It is made up of Tyndale's Pentateuch, Tyndale's 1535 New Testament, Coverdale's Ezra-Malachi and Apocrypha, and Joshua-Chronicles in a version later shown to be Tyndale's. Carefully edited by Rogers, it appeared in 1537, probably printed in Antwerp, and was commended by Cranmer to Cromwell. Dedicated to King Henry and Queen Anne under the pseudonym "Thomas Matthew," its use in churches was prohibited, despite its royal license.

MATTHIAS. Apostle. The name is probably an abbreviation of *Mattathias,* meaning "gift of Yahweh." He was chosen to take the place of Judas Iscariot among the twelve apostles (Acts 1:15-26). The qualifications for the office were, to have been a companion of the Twelve during the ministry of Jesus until the ascension and to have met the risen Christ so as to be able to witness to the resurrection. This in effect meant he must be able to preach the *kerygma* from firsthand experience. The choice was by lot, as in the Urim and Thummim of the OT, a system not apparently used after Pentecost. The idea that the choice was a mistake and that the place was meant for Paul is based on a misunderstanding of the nature of Paul's apostleship. Nothing else is known with certainty about Matthias, but Eusebius suggests that he was one of the seventy (Luke 10:1).

R.E. NIXON

MATTHIAS, GOSPEL OF. An apocryphal gospel according to Matthias (Acts 1:23) is reported by Eusebius as part of a corpus known among the heretics. Other Fathers, such as Origen, Ambrose, and Jerome, also mention it. Cognate with this lost work is a document containing "Traditions" ascribed to Matthias. Sayings from this are preserved by Clement of Alexandria; and there are affinities uniting this document with the Gnostic *Gospel of Thomas.**

MAUNDY THURSDAY. The Thursday before Easter, so called from Christ's command (Lat. *mandatum*) that His disciples should love one another (John 13:34). In fourth-century Jerusalem there were special services at the Mount of Olives and Gethsemane, and in North Africa an evening Eucharist commemorating the Last Supper. By the sixth century in the West, the Blessing of Oils and the Reconciliation of Penitents took place on this day, and by the Middle Ages the stripping and washing of altars and the *pedilavium* (footwashing) were added. The latter ceremony was sometimes performed by sovereigns, and the modern English Royal Maundy Service is a modified survival of this. At Mass, the Kiss of Peace was omitted in commemoration of Judas's kiss.

JOHN A. SIMPSON

MAURIAC, FRANÇOIS (1885-1970). French Roman Catholic writer. Born in Bordeaux, he lived as a child a somewhat idyllic life on a beautiful estate controlled by his family. He was early put under the rigors of a Jesuit education. A serious-minded, voracious reader, he studied at the University of Bordeaux and in Paris. His earliest published work in 1909 was a collection of verse. The first of his many novels, *A Kiss for the Leper,* appeared in 1922, followed by *Genetrix* (1923) and *The Desert of Love* (1925). By 1930, besides his novels, he had written literary criticism, biography, plays, and articles on a variety of subjects. During the French Resistance movement in World War II, Mauriac wrote anti-Nazi papers at the risk of death. His best novels are *Viper's Tangle* (1933) and *A Woman of the Pharisees* (1946).

All his novels come intensely to grips with man's sinful nature, his greed, lust, hatred, pride. Critics have noted traces of Jansenism in his outlook on life. In an age of naturalism and realism, he wrote about the unlovely under the restraint of his own artistic ideal: "To dare to say everything but to say everything chastely. Not to divorce ardor from purity." He is a moralist whose conception of his artistic responsibility is suggested in this statement: "If there is a reason for the existence of the novelist on the earth it is this: to show the element that holds out against God in the highest and noblest characters—the innermost evils and dissimulations; and also to light up the secret sources of sanctity in creatures who seem to us to have failed." He was awarded the Nobel Prize in 1952.

PAUL M. BECHTEL

MAURICE. Leader of the Theban Legion. Evidence derived from the account written by Eucherius of Lyons about 445. Emperor Maximian (Caesar from 286 to 305), joint emperor with Diocletian, had occupying forces in Gaul, among which was a Legion of Egyptians from Thebeid, consisting of Christians led by Maurice. The emperors outlawed Christianity, demanding the destruction of churches and the sacrifice to the old pagan gods. The Theban Legion led by Maurice mutinied at Agaunum (St. Maurice en Valais, Switzerland), for which they were massacred. The accuracy of the account is doubted, with the possible source the action of some Christian soldiers during Maximian's campaign against Bagaudae (286). There exists a sixth-century monastery at the place of the martyrdom.

GORDON A. CATHERALL

MAURICE, FREDERICK DENISON (1805-1872). Christian Socialist. He was son of a Unitarian minister; the religious conflicts of the home partly explain his later preoccupation with a search for unity. In 1823 he entered Trinity College, Cambridge, but being a Nonconformist was unable to take a degree. He moved to London and wrote in criticism of Benthamite materialism and developed an interest in social reform. Influenced by Coleridge's writings, he accepted Anglicanism and, deciding to be ordained, went to Exeter College, Oxford, where he was attracted to Tractarianism.* In 1834 he was ordained to a country curacy in Worcestershire and in 1836 became chaplain at Guy's Hospital, London, by which time he had broken with the Tractarians over baptism, which he saw as assuring every man that he is a child of God.

In 1838 he published his most enduring work, *The Kingdom of Christ,* in which most of his fundamental beliefs are expressed—the basic tenets of incarnational theology and, in particular, his belief in Christ as the head of every man, and universal fellowship and unity being possible in Christ alone. While at Guy's, he was doing practical work in the cause of education, and later in 1848 he was associated in the founding of Queen's College, London, the first higher educational establishment for women. In 1840 he was appointed professor of English literature and history at King's College, London, and in his Warburton Lectures (1846) he replied to J.H. Newman's* theory of development. In 1846 Maurice was appointed chaplain at Lincoln's Inn and also professor of theology at King's.

The political events of 1848 restirred his concern in the application of Christian principles to social reform, and with J.M.F. Ludlow* and Charles Kingsley,* he formed the Christian Socialists,* aiming at a Christian reform of the social bases of society, not just charity to the sufferers in society. At King's his orthodoxy was being questioned, and in 1853 the publication of his *Theological Essays,* in which he attacked the popular view of eternal punishment, resulted in his expulsion from the college. In 1854 he started the first Working Men's College in London. He came to prominence again in 1859 with his book *What is Revelation?,* a reply to H.L. Mansel's* Bampton Lectures of 1858. In 1860 Maurice was appointed to St. Peter's, Vere Street, London; in 1866, Knightsbridge professor of moral philosophy at Cambridge; and in 1870 also incumbent of St. Edward's, Cambridge. Throughout his ordained life he was unwilling to attach himself to any church party, yet remarkably he represented the unity those parties lacked. He was a prolific writer and one of the seminal, though much misunderstood, thinkers of the nineteenth century.

BIBLIOGRAPHY: F. Maurice, *Life and Letters of F.D. Maurice* (2 vols., 1884); F. Higham, *Frederick Denison Maurice* (1947); A.M. Ramsey, *F.D. Maurice and the Conflicts of Modern Theology* (1951); W.M. Davies, *An Introduction to F.D. Maurice's Theology* (1964); A.R. Vidler, *F.D. Maurice and Company* (1966).

JOHN A. SIMPSON

MAURISTS. French Benedictine monks of the Congregation of St. Maur, named after St. Maurus (d. 565). Founded in 1621 to represent in France the reform initiated in the abbey of St. Vanne near Verdun, they eventually numbered nearly 200 houses. From 1672 the Maurists devoted themselves to historical and literary works, producing an extraordinary number of large works of erudition. The full Maurist bibliography names some 200 writers and more than 700 works. The congregation was affected by the ecclesiastical controversies that distracted the French Church in the seventeenth and eighteenth centuries, and some members supported Jansenism.* Suppressed by the Revolutionary government (1790), it was finally dissolved by Pius VII in 1818.

J.G.G. NORMAN

MAXIMILLA (d. c.179). Ecstatic prophetess. With Montanus and Priscilla, another prophetess, she formed the leadership of the rigorous Montanist* sect in Asia Minor. The three proclaimed the imminent return of Christ to establish a New Jerusalem in Pepuza in Phrygia, where they made their headquarters. Charged like Priscilla with forsaking her husband to follow Montanus, she seems to have survived the others, and claimed that after her there would be no more prophecy, only the end of the world. She complained of persecution: "I am driven away like a wolf from the sheep. I am not a wolf; I am Word and Spirit and Power." Two bishops, Zoticus of Comane and Julian of Apamea, endeavored unsuccessfully to confute the spirit which prompted Maximilla's utterances, identified by the Montanists as the Holy Spirit. The wars and revolutions she predicted failed to transpire in the period immediately following her death, characterized by contemporaries as thirteen years of peace.

MARY E. ROGERS

MAXIMUS (c.380-c.468). Bishop of Turin. Little is known about his life, though 116 of his sermons and 118 homilies have been preserved and show him to have been influenced by Ambrose,* with whose works many of Maximus's sermons are included. There are also six tractates on baptism ascribed to Maximus, though three are fragmentary and his authorship is doubtful. The only certain dates in his life include his attendance at the Synod of Milan in 451, where he subscribed to the Tome of Leo,* and that of Rome in 465, where his signature stands immediately after that of Hilarius on the decision regulating nonappointment by bishops of their successors. Maximus was probably born near Tridentum (Trento) in the Raetian Alps. Several of his homilies include significant data: the creed of Turin, which is daughter to the Apostles' Creed; the destruction of Milan by Attila. Others are valuable for the religious life of the populace of his area, still greatly influenced by older agrarian fertility rites. While trying to comfort his people in the face of the raiding Huns (though his exhortation comparing David and Goliath failed to save Turin), he also rebuked their readiness to profit from plunder or slaves as the Huns withdrew.

CLYDE CURRY SMITH

MAXIMUS THE CONFESSOR (c.580-662). Byzantine theologian and writer. As a young, well-educated aristocrat he was appointed chief secretary to Emperor Heraclius I. About 615 he renounced his civil career for monasticism, eventually becoming abbot of the monastery of Chrysopolis. Having fled to North Africa during the Persian invasion (626), Maximus contended with the Monothelites,* notably Pyrrhus, temporarily exiled patriarch of Constantinople, whom he bettered in a famed disputation at Carthage in 645. His victory was instrumental in the triumph of orthodox, Chalcedonian Christology at several local African synods and at the Lateran Council of 649, in which he participated at the invitation of Pope Martin I. For his theology, sympathy with Roman hierarchical claims, and opposition to Emperor Constans II, he was tried for treason and exiled to Thrace. His tongue and right hand were subsequently cut off for his recalcitrance, and he was banished to Lazica on the Black Sea, where he died eighteen years later. His approximately ninety remaining works include commentaries on Scripture and the Fathers, doctrinal and polemical writings, and literature on ascetics, ethics, and liturgics. Known for interpreting the mysticism of Pseudo-Dionysius* for the West, Maximus was studied in Byzantium as a doctrinal theologian in his own right.

JAMES DE JONG

MAXIMUS THE CYNIC. Egyptian Cynic of the late fourth century. He was installed as bishop of Constantinople in an election so irregular that it received canonical denunciation at the second ecumenical council (381). Maximus's subsequent attempts to gain Western support, while initially successful, were short-lived, terminating with the Synod of Rome (382). Nothing by him survives, but literary productivity was not a Cynic trait. Maximus is known solely from historical accounts by others with whom he was involved—unless he is identified with the pagan philosopher and miracle-worker under whom the emperor Julian (361-63) had studied.

MAX MÜLLER, see MULLER, FRIEDRICH

MAXWELL, JAMES CLERK (1831-1879). Scottish physicist. Born in Edinburgh, he was educated at the universities of Edinburgh and Cambridge, and became professor successively at Aberdeen, London, and Cambridge. Maxwell was the Newton of his century. His investigations and discoveries cover a wide field, but he is best known as the creator of the electromagnetic theory. Throughout his life he was a committed Christian, whose philosophic consideration of his faith led him to some of his greatest scientific discoveries. He read widely, and "all he read helped only to strengthen that firm faith in the fundamentals of Christianity in which he lived and died" (Sir Richard Glazebrooke).

R.E.D. CLARK

MAYFLOWER COMPACT (1620). With the Virginia Assembly of 1619, this stands as the foundation stone of American institutions. The *Mayflower* passengers, before disembarking on Cape Cod, signed a compact which provided a "civil body politic" for their full protection. Some of these original Massachusetts settlers had been exiled in Leyden for ten years because of religious intolerance in England, and their English instinct for self-government demanded "equal laws, ordinances, acts, constitutions, and offices." While remaining loyal subjects of the king, their freedom was assured by this compact, which remained Plymouth Colony's principal governmental charter until 1691, when that colony was absorbed by Massachusetts.

C.G. THORNE, JR.

MAY LAWS. Legislation associated with Bismarck's *Kulturkampf** directed against German Catholicism. Passed in May 1873 under the direction of Adalbert Frank in the Prussian *Landtag*, they were based on the theory of the absolute supremacy of the state. They limited the extent of episcopal powers of excommunication and discipline, instituted a supreme ecclesiastical court whose members were appointed by the emperor and directly under state control, placed priestly training under close governmental supervision, and required all ordinands to pass through a state university and submit to state examinations in literature, history, and philosophy, and subjected clerical appointments by bishops to government veto. The laws were condemned by Pius IX in the encyclical *Quod nunquam* (1875) and were opposed also by many German Protestants. One effect of the laws was to unify Catholics and strengthen their resistance. They were eventually modified in 1886-87 after agreement between Bismarck and Leo XIII.

J.G.G. NORMAN

MAYNE, CUTHBERT (1544-1577). English Roman Catholic martyr. Born in Devon, he was educated at Oxford, and after Anglican ordination was appointed chaplain of St. John's College there, where Edmund Campion was a colleague. Under the latter's influence Mayne was converted to Roman Catholicism, and in 1573 went to the newly founded English seminary at Douai. Ordained priest in 1575, he returned to England in the following year and settled, ostensibly as steward, at Francis Tregian's estate in Cornwall. He secretly carried out his priestly functions until his arrest in 1577. Despite some disagreement among judges, he was condemned for refusing to take the oath of royal supremacy and for celebrating Mass, and was executed at Launceston. He was beatified in 1888.

J.D. DOUGLAS

MAZARIN, JULES (1602-61). French statesman. Born at Piscina, Abruzzi, he was educated by Jesuits in Rome and spent three years at Alcalá University in Spain. Returning to Rome, he graduated as doctor of laws and became captain in the army of Colonna. Turning to diplomacy, he averted war between France and Spain at Casal. He entered the church and was vice-legate at Avignon and papal nuncio in Paris. He became a naturalized Frenchman (1639) and was created cardinal (1641). He succeeded Richelieu as prime minister (1642) and rapidly became all-powerful through the favor of Anne of Austria, who perhaps married him secretly. At the Peace of Westphalia (1648) he increased France's possessions, but could not control the deteriorating financial situation which led to the civil wars of the Fronde (1648-53). He was banished (1651-52), but returned to be as powerful as ever, using the young Louis XIV to break any opposition. He raised France to the first rank in Europe, maintained its influence in the Baltic, concluded a trade treaty with Cromwell, and fought Spain successfully, eventually securing Louis's marriage to Maria Theresa after the treaty of the Pyrenees (1659). Toward the Huguenots he pursued a policy of reconciliation. He held the see of Metz and numerous abbeys, enriching himself from their revenues, out of which he founded the *Collège Mazarin*.

See A. Hassall, *Mazarin* (1903), and K. Federn, *Mazarin* (1922).

J.G.G. NORMAN

MAZARIN BIBLE. The name given to a rare and beautiful Bible printed by J. Gutenberg* of Mainz about 1455, probably the first book to be printed in Europe. The copy which first attracted the attention of bibliographers was discovered in Cardinal Mazarin's great library, hence the name. It was issued in two volumes, in double columns with forty-two lines to the column. Of the forty copies said to survive, only two (at Munich and Vienna) are known to be complete.

MAZZARELLA, BONAVENTURA (1818-1882). Italian patriot, preacher, and philosopher. Born in Gallipoli (Puglie), he studied law at the University of Naples. As a barrister and judge he fought against poverty, ignorance, and social injustice. Actively involved in the liberation movements and the war of 1848, he was persecuted and took refuge first in Greece, then in Geneva, afterward at Genoa. During this time he came into touch with the Gospel and was converted. He first joined the Waldensian Church, but soon his independent spirit found a more congenial atmosphere in the rising Free Italian Church (see GUICCIARDINI) caring for a large community of believers in Genoa. In 1860 he became professor of pedagogy at Bologna, and then at Genoa, an extraordinary achievement for a non-Roman Catholic. Finally he entered parliament, where he championed the cause of religious liberty and the interests of the much neglected southern regions, and was much respected for his moral integrity and deep humanity. One of the finest intellects of Italian evangelism, he wrote two remarkable philosophical works, *La Critica della Scienza* and *Della Critica.* DAISY RONCO

MECHITARISTS. Congregation of Armenian monks in communion with the Church of Rome. The founder was Mechitar, born at Sebaste in 1676 who formally joined the Roman Catholic Church, and in 1701 with sixteen companions formed a religious institute with himself as superior. Their Uniate propaganda incurred Armenian opposition, and they were forced to move to Morea which then belonged to Venice, where they built a monastery in 1706. On the outbreak of hostilities between Turks and Venetians, they migrated to Venice where they were granted the island of St. Lazzaro in 1717. Mechitar died there in 1749, and it has remained the headquarters of the congregation. J.G.G. NORMAN

MECHTHILD OF MAGDEBURG (c.1212-c.1280). German mystic. From about the age of twenty-three, for the greater part of her life, she lived in Magdeburg as a Beguine, virtually as a hermit. She claimed to have visions and revelations, which she wrote down in a book of six parts with the help of a Dominican, Heinrich von Halle. Later she added a seventh part. Her work is known as *Das fliessende Licht der Gottheit.* During the last twelve years of her life she lived as a nun in Helfta. About 1345 her book was transferred into High German by Heinrich von Nördlingen. A Latin translation was made of the first six parts. Mechthild's strong individualism and her poetic qualities distinguish her from most mystics. She has been called a *Minnesängerin* of God's grace. CARL S. MEYER

MEDALS, RELIGIOUS. A piece of metal (gold, silver, copper, etc.) fashioned in the form of a coin but not circulated as money. Struck to commemorate a religious event, person, or idea, Christian medals are probably to be traced back to the custom, practiced in the Roman Empire, of issuing special coins at temples on which was an inscription of a god, and to the custom of wearing a talisman. John Chrysostom, for example, reproached Christians in his day for wearing medals of Alexander the Great. Only by the ninth century did medals effectively enter the church. In the British Museum there is a gold coin of Wigmund, archbishop of York (b.837), which has on the reverse side a cross in a wreath and the legend *MVNVS DIVINUM.* At the Renaissance the medallic art was quickly revived and from the sixteenth century a whole series of medals depicting popes, the Virgin Mary, saints, and even Christ have been issued by the Vatican, by heads of religious houses, and at shrines. PETER TOON

MEDE, JOSEPH (1586-1638). English biblical scholar. Born in Essex, he graduated from Cambridge, where he subsequently became professor of Greek and taught several Cambridge Platonists, including B. Whichcote,* Henry More,* and R. Cudworth.* In addition to being one of the greatest biblical scholars the English Church has ever produced, Mede demonstrated his universal interests by being a philosopher, botanist, astronomer, and a pioneer Orientalist. He twice refused the provostship of Trinity College, Dublin, preferring to teach. His expository fame rests upon the *Apocalyptica* (*Key of the Revelation,* 1627, 1643). He attempted to construct an outline of the Apocalypse based solely upon internal considerations. In this interpretation he advocated premillennialism in such a scholarly way that this work continued to influence eschatological interpretation for centuries. ROBERT G. CLOUSE

MEDHURST, WALTER HENRY (1796-1857). Missionary to China. Born in London and educated at St. Paul's School and Hackney College, he was a trained printer who went first to Penang, then to Batavia, with the London Missionary Society. In 1843 he went to the newly opened Treaty Port of Shanghai, which was to be his base until his death. In addition to evangelistic work, he had already earned a reputation as a Chinese scholar, and became a member of the Bible Society committee commissioned to produce the first union version of the Bible. Medhurst was a prominent figure among early Protestant missionaries to China.

MELANCHTHON, PHILIP (1497-1560). German Reformer. Born in Bretten, Baden, the son of George Scharzerd, he was given the Greek-derived name "Melanchthon" by his great-uncle, Johannes Reuchlin,* because of his aptitude in languages and humanistic interests. He attended the grammar school at Pforzheim and went on to graduate from the universities of Heidelberg, Tübingen, and Wittenberg. In Tübingen he was a member of a humanistically oriented circle of friends and came to the attention of Erasmus.*

Reuchlin recommended him to the elector Frederick the Wise* as professor of Greek at the University of Wittenberg, where he soon embraced Luther's* cause. In his inaugural address at Wittenberg in 1518 he made a strong plea for the classics and the reform of studies. In his B.D. thesis of the following year he defended the proposition that the Scriptures alone are authoritative, not the decrees of popes and councils. In that same year he accompanied Luther to the Leipzig Disputation.* He published Luther's early commentaries on Galatians and the Psalms and under the pseudonym "Didymus Faventinus" defended Luther against the Parisian theologians. In 1519 Melanchthon lectured on the epistle to the Romans. In that year his *Rhetoric* appeared and in the following year his *Dialetics.* Melanchthon occupied a position on the faculty of liberal arts and on the theological faculty at Wittenberg. By 1519 his concept of justification, the forgiveness of sins, and reconciliation was already fashioned, a concept to which Melanchthon clung throughout his life.

During Luther's stay at Wartburg Castle after the Diet of Worms* (1521) Melanchthon was called on for theological leadership in the Lutheran movement. He and Andreas Carlstadt were not always in agreement, and Melanchthon was not positive what steps were to be taken in regard to the Zwickau Prophets.* Luther's return to Wittenberg (March 1522) arrested the radical movement. The first edition of Melanchthon's *Loci Communes,* the first systematic treatment of Lutheran theology, was published in 1521. In it Melanchthon treated the doctrines of the bondage of the will, the Law-Gospel dichotomy, justification by grace through faith. Scholasticism* was repudiated.

Melanchthon drew up the *Visitation Articles* (1528), the Augsburg Confession* (1530) and the *Apology of the Augsburg Confession* (1531), the *Confessio Saxonica* (1551), and the *Responsio to the Questions of the Bavarians* (1558). He has been criticized for altering the Augsburg Confession, in the so-called *Variata* (1540), but Luther did not fault him. His *Wittenberg Articles* were the basis of the discussion with the English theologians in 1536. He formulated the Wittenburg Concord* (1536) in which Luther and Martin Bucer* reconciled their views on the Lord's Supper and which was embodied in the Formula of Concord* (1577). He participated widely in the negotiations between the Lutherans and Roman Catholics, notably in the Ratisbon Colloquy* (1541).

He was the foremost humanist among the Lutheran Reformers. He became the center of religious controversies because of his stand on the Interim in 1548 and the more extreme statements of some of his followers on free will, conversion, and the Lord's Supper. Recent scholarship has asserted Melanchthon's integrity as a Lutheran theologian against those who fault him for deviations.

BIBLIOGRAPHY: Melanchthon's *Opera* (ed. K. Bretschneider and E. Bindseil, 28 vols., 1834-60); *Supplementa Melanchthonis* (6 vols., 1910-26); *Melanchthons Werke* (ed. R. Stupperich, 1951-).

Bibliography of works by and about Melanchthon in W. Hammer, *Melanchthon im Wandel der Jahrhunderte* (3 vols., 1967-).

See also K. Hartfelder, *Melanchthon als praeceptor Germaniae* (1889); C.L. Manschreck, *Melanchthon: The Quiet Reformer* (1958); V. Vatja (ed.), *Luther and Melanchthon* (1961); R. Stupperich, *Melanchthon* (1965); M. Rogness, *Melanchthon: Reformer without Honor* (1969).

CARL S. MEYER

MELCHIORITES, see RADICAL REFORMATION

MELCHITES (Melkites). A name given to those Christians who adhered to the creed supported by the authority of the Byzantine emperor. Derived from the Greek form of a Syriac adjective, it means "royalists, emperor's men." It was coined by the Jacobites in the tenth century and implies they could only stand with the emperor's support. It was applied to Christians of Syria and Egypt who rejected Monophysitism* and Nestorianism,* accepted the decrees of Ephesus and Chalcedon, and remained in communion with the imperial see of Constantinople.

The term applied also to Arabic-speaking Catholics of the Byzantine Rite in Syria, Palestine, Egypt, etc. They were organized from 1724 when Cyril Taras, a Catholic, became patriarch of Antioch.

J.G.G. NORMAN

MELITIAN SCHISMS. Fourth-century controversies.

(1) *Egypt.* Melitius, bishop of Lycopolis, traveled around consecrating new presbyters and deacons while Peter of Alexandria was imprisoned during the Diocletian persecution (305). Objections were made, and eventually Peter excommunicated him. During further persecution Peter was martyred and Melitius banished to the mines in Arabia Petraea. On returning he formed a schismatic church. He appears to have ordained Arius, thus sharing in the making of the Arian controversy. The Council of Nicea* (325) decreed that the Melitian clergy should be permitted to function under Alexander, Peter's successor, and their bishops, if legally elected, could succeed the orthodox bishops when they died. Melitius himself was to retain his title without a see. When Athanasius* acceded, the arrangement broke down, and the Melitians, encouraged by Eusebius of Nicomedia,* again went into schism. Melitius was succeeded by John Arcaphos of Memphis, who strongly opposed Athanasius. The sect apparently survived till the eighth century.

(2) *Antioch.* Melitius of Sebaste became bishop of Antioch (360), immediately fell foul of the Arians there by preaching an orthodox sermon on Proverbs 8:22, and was sent back to Armenia. The orthodox separated from the new bishop, Euzoïos, an Arian sympathizer, holding their own services. Already there was another orthodox congregation, the Eustathians, led by Paulinus, and soon there was an Apollinarian group. An attempt by Athanasius to unite the Melitians and Eustathians failed. Allowed back under Julian (362), Melitius was twice banished under Valens (365-66; 371-78). Basil of Caesarea worked for his rein-

statement, though Rome and Alexandria opposed him. He finally returned through Gratian's edict of tolerance (379). He presided over the Council of Constantinople (381), during which he died. The council ignored the opportunity to heal the schism by consecrating the aged Paulinus, and elected instead Flavian as the new bishop.

J.G.G. NORMAN

MELITIUS (d.381). Bishop of Antioch. Born in Melitene of Armenia Secunda (Armenia Minor), he is counted among the fourth-century members of the Antiochene* School of exegetical literalism. Of a wealthy and noble family, he first appears among the *homoean* supporters of Acacius* of Caesarea in 357. It appears that he was not old enough to have been a student of Lucian of Antioch, the school's founder, but should have been a contemporary of both Apollinaris of Laodicea and the school's fourth-century master, Diodorus of Tarsus (d.390). It was Melitius who discovered John Chrysostom,* sent him as a youth to Diodorus, and later ordained him deacon (381). Although venerated for his holy and ascetic life, he was caught up in the controversy of the times (see MELITIAN SCHISMS). Held at first to be too Nicene, he was hotly opposed as bishop of Sebaste, then (from 360) as bishop of Antioch where he had both Arian and ultra-Nicene rivals. When Theodosius I became emperor (379), Melitius was not only established in his see, but was designated to preside over the Council of Constantinople* (381), at which he died. Both the funeral oration by Gregory of Nyssa and the panegyric by John Chrysostom survive, as does a homily by Melitius preserved in Epiphanius.

CLYDE CURRY SMITH

MELITO (second century). Bishop of Sardis. Among the known prolific second-century Christian authors preserved solely in testimony and fragment, Melito has fared well in the twentieth-century recovery of papyrus copies. Eusebius placed him as bishop of Sardis in the reign of Marcus Aurelius (161-80) and gave a catalog of his works. He quoted Melito's "petition to the emperor," an early apology dated shortly after the announcement of Commodus as heir in 175, in which Melito initiates the thesis that the church and the imperial state are two conjoint works of God for the benefit of mankind. Another fragment, known already to Clement of Alexandria, to which Eusebius adds reference, indicates that Melito was fully involved in the problem of the date of Easter. Completely recovered texts demonstrate Melito's rhetorical style and his use of the typology of the slain Pascal Lamb to account for Christ's death and resurrection. He had evidently traveled to the places important to Christian origins, making him the first known Christian pilgrim. An analysis of the recovered fragment of his treatise *On Baptism* indicates his pioneering importation of Stoic Homeric exegesis into Christian thought; his theology falls within the Logos Christology of other apologists.

CLYDE CURRY SMITH

MELKITE CHURCHES, see EASTERN ORTHODOX CHURCHES

MELLITUS OF CANTERBURY (d.624). Third archbishop of Canterbury (619). Probably a Benedictine at St. Andrew's monastery in Rome, he was part of the second contingent of missionaries sent by Gregory the Great in 601 to help Augustine in Britain. After three years' work in Kent he was consecrated bishop of the East Saxons by Augustine (604). He established his see in London where St. Paul's was built as his cathedral. In 610 he returned to Rome on business. When the East Saxon king Sigebert died, he was banished to France (617), but he returned the next year and was soon elevated to Canterbury.

MELVILLE, ANDREW (1545-1622). Scottish Reformer. He was born near Montrose and was well educated in Scotland and France. He came under the influence of Beza* in Geneva and was appointed there to the chair of humanity. He returned to Scotland in 1574, soon becoming principal of Glasgow University and later of St. Mary's College, St. Andrews. His academic career was most distinguished, but of greater importance was his influence on the Scottish Church. He had returned to Scotland just two years after the death of John Knox, when Regent Morton was forcing the so-called Tulchan* bishops on an unwilling Kirk. Melville was strongly Presbyterian in conviction and rejected all attempts to "buy" him for episcopacy, including the offer of the archbishopric of St. Andrews.

He was regarded for many years as the leader of the Scottish Presbyterians and in 1582 was moderator of the general assembly. He led the assembly in its ratification of the Second Book of Discipline,* written in 1578, which has been described as the "Magna Charta of Presbyterianism." King James (VI of Scotland and I of England) was a staunch Episcopalian and was keen to impose this system of church polity on the Scots. Because of a sermon, Melville was summoned to appear before the Privy Council, but refused to accept its jurisdiction and fled to Berwick, across the border in England. In less than two years he was back in Scotland continuing the fight. James tried both conciliation and opposition in the attempt to deal with him. Eventually in 1606 he and seven other Scottish ministers were summoned by the king to Hampton Court. Melville's frankness at this interview led to his detention in the Tower of London for four years. He became professor of divinity at the university of Sedan, where he died. His nephew James undoubtedly expressed the feelings of many of Andrew Melville's Presbyterian contemporaries when he wrote, "Scotland never received a greater benefit at the hands of God than this man."

G.W. GROGAN

MELVILLE, HERMAN (1819-1891). American novelist. Born in New York City, he was one of eight children whose father was a cultivated gentleman (he died when Herman was twelve). In 1837 he shipped as a cabin boy to Liverpool, and in 1841 sailed on a whaler to the South Seas. The latter trip provided the material for his first books—*Typee* (1846), *Omoo* (1847), *Redburn* (1849), *White-Jacket* (1850)—and his greatest work, *Moby Dick* (1851), one of the finest of all American novels. In 1849 he made a trip to England and in the following year moved to a farm in Massachusetts, where he established a firm friendship with his neighbor, Nathaniel Hawthorne,* to whom he was to dedicate *Moby-Dick*. *Pierre* was the tale of an intellectually troubled and overwrought young man. *Billy Budd* was one of his best tales, written shortly before his death, though not published until 1924. Melville's last years were spent in New York City where he died in poverty and obscurity. His rediscovery around 1920 by literary scholars brought him the acclaim which was his due as one of America's greatest authors. Like Hawthorne's, Melville's work is rich in allegory and symbol. His style has a graceful, lyrical beauty rarely matched in American literature. He ceaselessly sought some absolute in the world and the lives of men, but he never acknowledged the sovereignty of the Christian God.

PAUL M. BECHTEL

MEMPHITIC, see EGYPTIAN VERSIONS

MENAS (sixth century). Patriarch of Constantinople from 536 to 552, succeeding the deposed Anthimos. Menas unwaveringly condemned the heresy of the Monophysite* leaders and subscribed to Justinian's edict against the Origenists in 543. In 547 and again in 551 Pope Vigilius deposed him, but he remained firmly committed to Justinian's religious program. He conducted a council at Constantinople on assuming office in 536. The council's acts show that by this time eucharistic services in Constantinople customarily included the Constantinopolitan creed. Paradoxically, Monophysite initiative was responsible for this innovation.

MENDELSSOHN-BARTHOLDY, FELIX (1809-1847). Musical composer. He came of a wealthy and cultured Jewish family that had turned to the Lutheran Church. He was extraordinarily precocious, not only in music but in intellect, and was also talented in art. Many of his finest works were produced while he was in his teens. His presentation of Bach's *St. Matthew Passion* in 1829, for the first time since the composer's death, was a historic musical event of importance, and marked the beginning of the revival of Bach's great legacy of choral music. Mendelssohn spent much time in England, where he was enthusiastically received. His *Elijah* was written for the Birmingham Festival in 1846 (although to a German text and then translated) and is one of the greatest nineteenth-century oratorios. He wrote choral music with particular facility. While little of it was specifically for the church, extracts from his oratorios and cantatas are often used as service music. His setting of Psalm 43 for unaccompanied chorus is a fine piece of its kind. He exerted a perhaps too powerful influence on the lesser choral composers of his day.

J.B. MAC MILLAN

MENDICANT ORDERS. The early Mendicant orders developed during the thirteenth century when both the contrast between the church's wealth and the poverty of the primitive church, and the church's weakness in meeting the pastoral requirements of many people were being underscored by groups outside the church such as the Albigenses and the Waldenses. Orders such as the Franciscans* (1210), the Dominicans* (1216), the Carmelites* (1247), the Augustinians* (1256), and the Servites* (1256) provided a response to these challenges by uniting the concepts of apostolic poverty and obedience to the church, and often by proving quite effective in their ministry, particularly to the poor and destitute in the towns. At their founding the characteristic emphasis on poverty among the Mendicant orders was evident in their commitment to the renunciation of all common as well as individual possessions. Difficulties encountered in attempting to function in society while adhering to such a renunciation, however, induced a relaxation of this commitment. Furthermore, although Francis at least intended the members of his order to support themselves normally by work rather than by begging, growth in numbers and specialization led to an increasing reliance upon the latter method, from which arose the appellation "Mendicants" (*mendicare*, "to beg").

T.L. UNDERWOOD

MENNONITES. A body of conservative and evangelical Christians descended from the Anabaptists* of the sixteenth century. The founder was a disciple of Zwingli* named Conrad Grebel.* A free-lance reformer, Melchior Hofmann, carried the basic ideas of Anabaptism to the Low Countries* where Melchiorites were for a time the dominant reformation body. Melchior was imprisoned in 1533, after which his followers broke into a revolutionary movement led by Jan Matthijs (Münster, 1534-35), and a peaceful wing led by Obbe and Dirk Philips.* Menno Simons* united with the Obbenites in 1536 and his name came to be applied to the movement, first as Mennists, now as Mennonites. Dutch Mennonites began settling in the Danzig area in the 1540s and from there went to Russia beginning in 1788. There were 5,000 martyrs by 1600 (see T. J. van Braght, *Martyrs Mirror* [Dutch, 1660], now available in English and German).

Mennonites began to settle in the New World as early as the 1640s, but the first permanent settlement was Germantown near Philadelphia, 1683. About 1700, Mennonites, largely of Swiss ethnic origin, began to locate in Pennsylvania, from which they later migrated to Virginia, to Ontario, and to Ohio and states farther west. In the 1870s, the 1920s, and after World War II, three waves of Mennonites from Russia migrated to the Americas, the three groups settling respectively in Manitoba, Kansas, and other prairie states; in Canada; and in Canada and South America, mostly Paraguay and Brazil. The Mennonites of North America are in three major conferences: (1) the Mennonite Church representing those who came to the USA before the Civil War of the 1860s; (2) the General Conference Mennonites comprising many of the three waves of immigration from Russia; and (3) the Mennonite Brethren who originated as a revivalist movement in Russia in 1860. Mennonites operate a number of colleges and seminaries.

From the beginning, Mennonites have stressed the Free Church principle as well as believer's baptism and biblical nonresistance (pacifism). Discipleship to Christ is stressed, as is church discipline. They emphasize a life of prayer and of holiness and are concerned not to allow the "world" to weaken their strict New Testament values. Infants are regarded as saved, although often a dedicatory service is held to invoke the blessings of heaven on parents and infants. Children are generally baptized (by pouring) in their teen years, but the Mennonite Brethren baptize by immersion. The Lord's Supper is solemnly yet joyfully observed, often semiannually. The more conservative groups practice footwashing. A vigorous program of missions is carried on, as well as the humanitarian service of the Mennonite Central Committee (1920), ministering to the needy in many lands, including the care of the emotionally ill, helping racial minorities, and the like. The Mennonite World Conference (1925) meets every five years. Around the globe Mennonites number 560,000, of which 300,000 are in North America.

BIBLIOGRAPHY: C.J. Dyck (ed.), *Introduction to Mennonite History* (1968); H.S. Bender (ed.), *The Mennonite Encyclopedia* (4 vols., 1955-59); *Complete Writings of Menno Simons* (tr. L. Verduin, ed. J.C. Wenger, 1956). J.C. WENGER

MENNO SIMONS (1496-1561). Founder of the Mennonites.* Born in Friesland, in his twenties a Catholic priest there, he began reading Luther and other Reformers and was attracted to Anabaptism, though opposed to its militant wing (writing, in 1535, a pamphlet against John of Leyden*). In 1536 he joined the group around Obbe Philips* and soon became a leading figure in the Anabaptist* movement. For the next twenty-five years he traveled through the Netherlands and the German North Sea coasts, spreading the Gospel as he saw it, organizing congregations, disputing with other Protestants (e.g., 1544, against the Calvinist Jan à Lasco*), often forced to move on, continually writing. His last years were spent in Holstein; he died in Wüstenfelde, near Lubeck. Menno stressed the idea of the community of believers, committed to a new life, sealed by adult baptism, tightly knit (e.g., no marriage outside the community), withdrawing from the secular world and its follies (thus, distrust of learning, refusal to take part in politics or bear arms). Suspicious of dogmatic theology, Menno relied on Scripture taken literally. He thus refused to use terms and concepts not clearly scriptural, such as (in his view) the Trinity. At some points, curiously, he seemed to question Christ's full humanity. He held to a speedy second coming of Christ.

See I.B. Horst, *A Bibliography of Menno Simons* (1962), and *Complete Writings of Menno Simons* (tr. L. Verduin; ed. J.C. Wenger, 1956).

DIRK JELLEMA

MENTAL RESERVATION. The "reserving" of some clause within one's own mind which makes one's expressed statement true. It arises out of conflicts between the obligation to tell the truth and the obligation to keep a secret. A politician facing an embarrassing question may reply, "I don't know"; his mental reservation is "in my capacity as a private citizen." The doctrine of mental reservation is found in Roman Catholic moral theology, which holds that lying is always and necessarily sinful. A lie is the intentional assertion of what is contrary to a man's inward thought. It is also an offense against justice, truth being a debt which we owe to others. Roman Catholic teaching has distinguished two types of mental reservation. Strict mental reservation involves the speaker's mentally adding some qualification to what he has said so that the sense is different and the hearer deceived. Wide mental reservation, on the other hand, arises from the permitted ambiguity of the words or the circumstances in which they are spoken. Strict mental reservation was condemned by Innocent XI in 1679. Wide mental reservations, however, are not lies, provided there is a good reason. In this case the speaker is permitting the hearer to remain in his misunderstanding of words susceptible to a different interpretation. HOWARD SAINSBURY

MERBECKE, JOHN (d. c.1585). English musician and theologian. He was organist of St. George's Chapel, Windsor, from 1541 until at least 1565. A Calvinist, he was sentenced to the stake in 1544 for his writings against the Six Articles* and for his compilation of the first biblical concordance in English (thus threatening the use of Latin for worship). He was pardoned evidently because Bishop Stephen Gardiner had a high regard for his musicianship. Merbecke dedicated that concordance later to Edward VI. That same year (1550) came also his *Book of Common Prayer Noted*, an adaptation of the plain chant to Edward VI's first liturgy (1549), issued to prevent diversity in "saying and singing." He wrote many theological works, a hymn for three voices, and Mass for five. C.G. THORNE, JR.

MERCATOR, MARIUS, see MARIUS MERCATOR

MERCEDARIANS. The Order of Our Lady of Mercy, derived from the Spanish word *merced* ("mercy") and known also by other names, was founded in 1218 by Peter Nolasco to attend the sick and rescue Christian captives from the Moors. Their white habit facilitated entrance into Muslim territories; and following the Austin Rule, they took a fourth vow pledging themselves as hostages when needed, thus liberating some 70,000. They spread through Europe to the Americas, changing from a military to a clerical order (1319) and becoming mendicant. A post-Reformation discalced* group arose, and though major setbacks came during the nineteenth century, the order was revived under Valenzuela in the 1880s. Parishes, charities, schools, and chaplaincies occupy them, and there are congregations of women. C.G. THORNE, JR.

MERCERSBURG THEOLOGY. The creation of Philip Schaff and J.W. Nevin,* professors at tiny Mercersburg Theological Seminary in south central Pennsylvania, the Mercersburg Theology was a mediation between head and heart, or objective and subjective, in an age of extremes typified by reckless westward expansion and revivalism. Born in an obscure German Reformed Church seminary, it reflected Nevin's desire to embrace "not the notion of supernatural things simply, but the very power and presence of the things themselves." Both men were converts—Nevin from Princeton Presbyterianism and Schaff from the United Church of Prussia—who praised the sacramentalism of the tradition. Nevin the theologian and Schaff the historian found satisfaction in the centrality of the Eucharist, for only it—not the Bible or individual experience—gave the believer true spiritual knowledge. Celebration of the Lord's Supper, in which the believer received the "spiritual real presence," united the believer with a historical and organic church, and kept the church from being a mere aggregate of individuals. That also symbolized divine governance of all society.

The Mercersburg Theology, which had little influence in its day, was strongest from 1840 until the departure of Nevin and Schaff—the former retired because of ill-health in 1853, and the latter went to Union Theological Seminary after the Civil War had virtually destroyed the seminary. The theology survived at the church's new seminary in Lancaster, Pennsylvania, and was revived in twentieth-century American Christianity because of its Christocentrism, organicism, and ecumenism. The first attempt to reconcile German Idealism and American Protestantism, it was also the first substantial critique of American Calvinism.

BIBLIOGRAPHY: T. Appel, *The Life of John Williamson Nevin* (1889); D.S. Schaff, *The Life of Philip Schaff* (1897); J.H. Nichols, *Romanticism in American Theology: Nevin and Schaff at Mercersburg* (1961). DARREL BIGHAM

MERCIER, DÉSIRÉ JOSEPH (1851-1926). Belgian cardinal and philosopher. Educated at Malines and Louvain, he was ordained in 1874. In 1877 he became professor of philosophy at the *Petit Seminaire* at Malines, and from 1882 to 1906 was professor of Thomist philosophy at Louvain. An ascetic, deeply pious, and an able organizer, he was appointed archbishop of Malines and primate of Belgium in 1906. He was made cardinal in 1907. Believing that in the long run reason will lead to the same conclusions as faith, he worked to create a synthesis between Thomist philosophy and the experimental sciences. He made Louvain a major center of Neo-Thomist philosophy by founding the Higher Institute of Philosophy there. He attacked modernism in general and G.H. Tyrrell in particular in a pastoral letter in 1908. He was a fearless spokesman for the Belgian people during the German occupation of World War I. He was the Roman Church's chief representative in conversations (see MALINES CONVERSATIONS) with the Anglicans at Malines (1921-26). His writings include *Psychologie* (1892), *Logique* (1894), *Métaphysique* (1894), and *Critériologie* (1899). HOWARD SAINSBURY

MERIT. Medieval Catholic theology distinguished between *bonitas* (the ethical value of human acts) and *dignitas* (the religious value of those same acts). Merit in this sense has to do with the dignity or religious significance of human acts rather than with their goodness or inherent moral value. A merit is an act which is rewarded by God because it has met certain conditions, only one of which is inherent moral goodness. What those conditions are differ from theologian to theologian, though a representative list would include such items as the use of free will, the assistance of actual or habitual grace, and the promise of God to reward such an act. The exegetical basis for the promise of God to reward works done in His name and for His glory includes such texts as Exodus 23:20-22; Deuteronomy 5:28-33; Matthew 5:3-12; 6:4, 19ff.; 7:21.

The Scholastic doctors distinguished between *meritum de condigno,* a good work which God is obligated to reward because of His own promise, and a *meritum de congruo,* which God is not obliged to reward because it does not meet the usual conditions, but which, nevertheless, it is fitting that God reward in view of His own liberality and merciful goodness. While Franciscan theologians admitted the possibility of *merita de congruo* for someone still in a state of sin, they restricted *merita de condigno* to works performed in a state of grace. Thomas Aquinas,* on the other hand, denied the possibility of merit, though not of moral goodness, prior to entry into a state of grace, while radical Augustinian theologians such as Gregory of Rimini* denied to the sinner the possibility of both. Protestant theology rejected the doctrine of merit, though in certain forms of Reformed, Anglican, and Free Church theology the notion of reward was not consistently excluded. DAVID C. STEINMETZ

MERLE D'AUBIGNÉ, J.H., see D'AUBIGNE

MESROB (Mashtotz) (c.361-440). Important figure in the history of Armenian culture and spirituality. According to the best sources "Mashtotz" is the more correct version of his name. His life is known from the biography by his pupil Korium (mid-fifth century). Mashtotz was a pupil of Nerses the Great (353-73). He entered the royal service well versed in languages, became a monk in 390 or shortly afterward, and after a brief time in evangelism was the associate of Sahak III, whom he succeeded as patriarch. He died the following year. The main activity of his association with Sahak was in the field of translation of biblical and patristic writings; he was acquainted with Persian, Syriac, and Greek. The earliest tradition asserts that biblical translation was the basis of the Greek, but internal evidence shows strong Syriac influence. Similarly both traditions appear active in early patristic translation. For his work Mashtotz devised the Armenian alphabet, since Armenian had previously had no satisfactory means of transcription. Tradition asserts also that he was the inventor of the Georgian alphabet and of the alphabet of the Caucasian Albanians (of which few traces remain). Although these traditions have been contested, they rest on early evidence before the Armenian-Georgian schism and the destruction of the Albanians. Even the Armenian alphabet alone indicates the knowledge and acumen of Mashtotz. J.N. BIRDSALL

MESSALIANS. From an Aramaic word meaning "praying folk." Also known as Euchites and by other names, they were a heretical sect originating in Mesopotamia about 360, spreading to Syria, Asia Minor, and Egypt. The so-called Christian Messalians were vagrant Quietists, ignored the sacraments, and wandered about sleeping in the streets. They feigned orthodox practice to avoid persecution; prayer was their only occupation, and they claimed to see the Trinity as well as evil spirits. They emphasized the indwelling of the Holy Spirit, saying that every man including Christ was possessed of demons. Although they survived until the seventh century, attempts to suppress them were many—by Flavian of Antioch, the Synod of Side (388-90), Nestorians in Syria, decrees in Armenia (mid-fifth century), and the councils of Constantinople (426) and Ephesus (431) where their *Asceticus* was called a filthy book of heresy. They were accused of immorality and their monasteries were burned. They were scarcely known in the West. The later Bogomiles* are a derivative. C.G. THORNE, JR.

MESSIAH. From the Hebrew word meaning "anointed" (cf. Gr. *christos*). Because kings, priests, and perhaps prophets (1 Kings 19:16) were anointed, the term came to be used of God's representatives (e.g., Ps. 89:38; Isa. 45:1). A fundamental tenet of Israel's religion was that God would set up His perfect rule on earth, through an act of decisive divine intervention, the Day of the Lord, a term first found in Amos 5:18 (if a late date is attributed to Joel), but clearly much older. Though in the majority of passages concerned with the setting up and continuance of the rule of God human agency is not mentioned, it is clearly presupposed, for the Mosaic-prophetic tradition throughout sees God working through human representatives, hence the divine attributes given to kings (Isa. 9:6; Ps. 45:6). Hence too, the frequently met distinction between prophetic pictures of the messianic age and of the Messiah has little practical validity.

It is questionable also whether anything is to be gained by the modern tendency to confine Messiah to the royal office. At Qumran two Messiahs were awaited, "the anointed ones of Aaron and Israel," i.e., a priest and king, the former taking precedence. In *4Q Testimonia* the promise of a prophet like Moses is linked with the hope of a king and priest. Thus it seems clear that at Qumran at least, and almost certainly in wider circles, the Jewish messianic expectation included all three offices, as in Christian interpretation, though normally the delivering king predominated.

The messianic hope is essentially eschatological, i.e., it emerges from despair at conditions as they are. So far as the king was concerned, it must have become a reality with the failure of Zerubbabel to reestablish the Davidic dynasty after the Exile, but prophetically it emerges with Isaiah's prophecy of the cutting down of the royal tree (11:1) and the implicit rejection of Ahaz and his descendants in the Immanuel prophecies, which become explicit in Jeremiah 22:30. The concept of a coming prophet will hardly have occupied the popular mind until the gift disappeared about the time of Ezra, and it may be seen in *1 Maccabees* 4:46; 14:41. There was felt to be a lack in the priestly office as early as the return from exile (Ezra 2:63), but the question did not become acute until the ousting of the Zadokite high priests in the time of Antiochus IV Epiphanes, and brought to a head by the recognition of Simon the Hasmonean as high priest in 142 B.C. (*1 Macc.* 14:41)—this was one of the basic causes of the Qumran movement. The messianic hope became a burning necessity with the Roman conquest and its imposition of either the Herodian dynasty or direct rule. The rise of the Zealots* with their doctrine that all authority belonged to God made the end of the second Jewish commonwealth inevitable.

The church's recognition of Jesus as Messiah (Christ) has always been central to its theology. It implies that He is the fulfiller of every promise and hope of the OT revelation and the basis of its interpretation. Indeed, without it, the Gentile Christians might well not have retained the OT as a sacred book.

BIBLIOGRAPHY: A. Neubauer and S.R. Driver, *The Fifty-Third Chapter of Isaiah According to the Jewish Interpreters*, (1876/7); J. Drummond, *The Jewish Messiah* (1877); F. Delitzsch, *Messianic Prophecies* (1880); V.H. Stanton, *The Jewish and Christian Messiah* (1886); C.A. Briggs, *Messianic Prophecy* (1886); J.H. Greenstone, *The Messiah Idea in Jewish History* (1906); A.L. Williams, *The Hebrew-Christian Messiah* (1916); E. Koenig, *Die messianischen Weissagungen des Alten Testaments* (1923); A.G. Hebert, *The Throne of David* (1941); C.H. Dodd, *According to the Scriptures* (1952); H.L. Ellison, *The Centrality of the Messianic Idea for the Old Testament* (1953); A. Bentzen, *King and Messiah* (1955); J. Klausner, *The Messianic Idea in Israel* (1956); S. Mowinckel, *He That Cometh* (1956); H. Ringgren, *The Messiah in the Old Testament* (1956); F.F. Bruce, *This is That* (1968); H.L. Ellison, *The Corner Stone* (1973). H.L. ELLISON

METHODISM. A movement which originated in a search for an effective method to lead Christians toward the goal of scriptural holiness. The epithet was applied to members of the Wesleys'* Holy Club* at Oxford in 1729. Their disciplined, methodical practices gave rise to what Charles Wesley called "the harmless nickname of Methodist." It was readily accepted by his brother John, who provided his own definition: "A Methodist is one who lives according to the method laid down in the Bible." The term had previously been employed in an ecclesiastical context in the sixteenth century with reference to Amyraldists* or Semi-Arminians. Despite the theological similarity, there is no evidence of any direct derivation.

When the subsequent revival got under way, after the evangelical conversion of George Whitefield* and the Wesleys, the title "Methodist" was attached to all who were influenced by it, whether within the Church of England or beyond. Only at a later stage were Methodists distinguished from Anglican Evangelicals. In its strictest connotation "Methodism" refers only to the adherents of Wesley, although it is extended to include the followers of Whitefield and Lady Huntingdon who subscribed to the doctrines of Calvin rather than of Arminius.

Methodism is now accepted as a general term to cover the worldwide family of Methodist churches, stemming from Wesley's societies, most of which are affiliated to the World Methodist Council.

See also METHODIST CHURCHES and CALVINISTIC METHODISM. A. SKEVINGTON WOOD

METHODISM, CALVINISTIC, see CALVINISTIC METHODISM

METHODIST CHURCHES. The World Methodist Council, to which the majority of Methodist Churches is affiliated, represents a recorded membership of over 18 million, and a community of some 40 million. Approximately 750,000 are to be found in Great Britain and Ireland, while the United Methodist Church of America numbers more than 11 million, quite apart from other Methodist groups. In the countries of the British Commonwealth and in Europe there are autonomous Methodist conferences; in many places elsewhere missionary extension continues.

All these churches derive from the ministry of John Wesley* and the religious societies he founded in the wake of his evangelistic missions in the United Kingdom during the eighteenth century. In 1739 Wesley started a society in London in a former cannon foundry and from 1741 utilized lay preachers. In 1742 the first classes were formed, and in 1743 the rules of society drawn up. The first conference assembled in 1744 with six clergymen and four laymen present. From 1746 the societies were arranged in circuits under the superintendency of Wesley's helpers and, after his death, were grouped in districts.

It was not until 1784, however, that the Wesleyan Connexion was fully established in law. In that year Wesley lodged in the Court of Chancery a Deed of Declaration naming one hundred preachers as constituting the "Conference of the people called Methodists," with provisions for its maintenance. On his death the membership of the conference was extended beyond the Legal Hundred to include all preachers in full connection. Laymen were not added until 1878 and women only in 1911. In 1787 Wesley's chapels were registered as dissenting meeting houses under the Toleration Act of 1559. Separation from the Church of England was made yet more explicit by the Plan of Pacification (1795) which allowed the administration of the sacraments as well as the holding of marriage and funeral services in those Methodist chapels where a majority of officials approved. At the same time it was recognized that reception into full connection with the conference sufficiently validated ministerial orders. Only in 1836 was ordination by the imposition of hands adopted as standard practice.

Within six years of Wesley's death the first secession took place when in 1797 the Methodist New Connexion* was formed. In 1805 a group in Manchester was expelled for holding irregular meetings and became the short-lived Band Room Methodists. In the following year the Independent Methodists appeared at Warrington under Peter Phillips, although they did not officially assume the name until 1898. They still exist. The Camp Meeting Methodists led by Hugh Bourne* joined with the Clowesites to form the Primitive Methodist* Connexion in 1811. William O'Bryan's Bible Christians* emerged in 1815. The Tent Methodists, with George Pocock and John Pyer as leaders, were organized in 1822, and next year the Church Methodists made a move in the direction of reunion with the Anglicans. The Protestant Methodists (1827) joined the Wesleyan Methodist Association (1836), who in turn united the main body of Wesleyan Reformers (1850) to constitute the United Methodist Free Churches (1857). The remainder of the Wesleyan Reformers established an autonomous church in 1859 and continue to the present day. The Arminian or Faith Methodists of Derby seceded in 1832, eventually joining the Wesleyan Methodist Association.

The cause of these divisions was governmental rather than doctrinal, and some of the resultant bodies were comparatively small. Despite the defections, the Wesleyan Church expanded. During the second half of the nineteenth century a period of consolidation set in, and before its end plans for reunification were set afoot. These were to come to fruition in the following century. In 1907 the Methodist New Connexion, Bible Christians, and United Methodist Free Churches combined in the United Methodist Church.* In 1932 the Wesleyan, Primitive, and United Methodists came together to constitute the Methodist Church of Great Britain and Ireland with a membership of 859,652.

The missionary enterprise of Methodism may be said to date from 1769, when Richard Boardman and Joseph Pilmoor volunteered to serve in America (see METHODIST CHURCHES, AMERICAN). In 1785 appointments were made also in Nova Scotia, Newfoundland, and Antigua. The prime mover after Wesley's death was Thomas Coke,* acting under the direction of the conference which from 1786 assumed immediate responsibility for missions overseas. He it was who pioneered the work in the West Indies and whose vision embraced Africa and Asia as well as America. In 1813 the Wesleyan Methodist Missionary Society was formally organized.

Today there are autonomous Methodist Churches in Australasia, New Zealand, South Africa, Italy, Ghana, Nigeria, Zambia, Sierra Leone, Kenya, Ceylon, Upper Burma, and the Caribbean—all derived from the British Conference, with others stemming from American Methodism. Methodists have entered the United Church of Canada, the Churches of South and North India, and the United Church of Japan. There are minority Methodist Churches in Switzerland, Scandinavia, Portugal, Austria, Poland, and Germany—the latter being the strongest. In Belgium, France, and Spain the Methodists have united with other Protestant churches.

BIBLIOGRAPHY: A. Stevens, *The History of the Religious Movement of the Eighteenth Century called Methodism* (3 vols., n.d.); G. Smith, *History of Wesleyan Methodism* (3 vols., 1857-61); C.H. Crookshank, *History of the Methodist Church in Ireland* (3 vols., 1885); W.J. Townsend, H.B. Workman, and G. Eayrs (eds.), *A New History of Methodism* (2 vols., 1909); G.G. Findlay and W.W. Holdsworth, *The History of the Wesleyan Methodist Missionary Society* (5 vols, 1921-24); A.H. Williams, *Welsh Wesleyan Methodism* (1935); H. Bett, *The Spirit of Methodism* (1937); W.F. Swift, *Methodism in Scotland* (1947); H. Carter, *The Methodist Heritage* (1951); C.J. Davey, *The March of Methodism* (1951); E.G. Rupp, *Methodism in Relation to the Protestant Tradition* (1952); I.L. Holt and E.T. Clark, *The World Methodist Movement* (1956); R.E. Davies, *Methodism* (1963); R.E. Davies and E.G. Rupp (eds.), *A History of the Methodist Church in Great Britain*, vol. I (1965); H.D. Rack, *The Future of John Wesley's Methodism* (1965).

A. SKEVINGTON WOOD

METHODIST CHURCHES, AMERICAN. Though John Wesley* had served as a missionary in Georgia (1736-38) before his Aldersgate experience, and George Whitefield* had after 1740 visited America several times during the Great Awakening,* Methodism as such was brought to America in the 1760s by unofficial lay preachers. Among these were Irishman Robert Strawbridge who worked in Maryland and surrounding areas, Philip Embury,* and Captain Thomas Webb, a British officer who reinforced the society in New York and planted Methodism in Pennsylvania, Delaware, and New Jersey. In 1768 Wesley sent the first two official missionaries, Joseph Pilmoor and Richard Boardman. In 1771 came Francis Asbury,* later the greatest leader of American Methodism, and Richard Wright. They were followed in 1773 by Thomas Rankin and George Shadford. That year saw the first American conference at St. George's Church, Philadelphia. The South proved particularly receptive to Methodism, and in Virginia, where Methodists worked with Anglican Devereux Jarratt, a major revival broke out as the Revolutionary War started.

The Revolution precipitated a major crisis. All the British Methodist missionaries except Asbury returned to England. After great internal struggle and a period of forced inactivity, Asbury finally identified with the emerging nation. Anglicanism, with which Methodism had identified, was devastated. To meet the chaotic situation Wesley took the step—a very difficult one for him—of ordaining two elders and Thomas Coke* as general superintendent to establish in America an adequate system of church government. Soon after their arrival in America, the Christmas Conference of 1784 was called to found the Methodist Episcopal Church as an autonomous denomination. Asbury was ordained deacon, elder, and joint superintendent on successive days, though on his insistence only after his appointment had been confirmed by unanimous vote of the conference. According to Wesley's recommendations, a ritual, twenty-five Articles of Religion (abridged from the Anglican thirty-nine), and a discipline were adopted. The church thus founded consisted of 18,000 members, 104 traveling preachers (plus as many local preachers and exhorters), 60 chapels, and 800 recognized preaching places.

The next six decades were a period of phenomenal growth. By 1844 the church had grown to about 4,000 preachers and over a million communicants. After 1792, annual conferences became regional and a general conference met quadrennially. By 1812 the general conference was reorganized on a delegated basis. Leadership fell more and more upon Asbury, who began against Wesley's wishes to use the title "bishop." The itineracy, firmly administered by Asbury until his death in 1816, was admirably suited to the American frontier. Annual conferences and presiding elders supervised the work of circuit riders who followed American pioneers in the westward expansion. Methodism rode the crest of the Second Great Awakening* at the turn of the century and picked up and perfected the institution of the camp meeting.*

But growth was not without tensions. In 1792 Asbury clashed with James O'Kelly over the appointment of preachers. O'Kelly wished the right of appeal to the conference if a preacher should be dissatisfied with his appointment. When O'Kelly's resolution lost, he withdrew with a few other preachers to found the Republican Methodist Church. Perhaps 8,000 members defected, but the group eventually withered away. Similar issues led to the founding of the Methodist Protestant Church in 1830. From 1820 some had advocated an elective presiding eldership, lay representation to conference, and major alteration of the episcopacy. Agitation over these issues continued and led to the 1827 expulsion of several persons in Baltimore and final split in 1830. In 1858 about one-half of this group seceded over the issue of slavery to form the antislavery Methodist Church, but after the Civil War the two parts reunited.

It was over the slavery issue that Methodism really floundered. As early as 1786 friction developed between blacks and whites worshipping together in Philadelphia and Baltimore. Under the leadership of Richard Allen,* a remarkable man who had managed to purchase his freedom and become the first Negro ordained by Asbury, and Daniel Coker, a freed mulatto who became a teacher and preacher, the African Methodist Episcopal Church was founded in 1816 (see AMERICAN NEGRO CHURCHES). Similar problems led to the founding of the African Methodist Episcopal Zion Church in 1821 in New York City. But the white church split also over the issues of slavery and race. For Wesley, slavery was "that execrable sum of all villanies." Asbury seems to have held similar opinions, and slavery was regularly condemned by early conferences, but as Methodism became a national church, pressures toward accommodation increased. Asbury acquiesced, and conference resolutions became tamer. Yet the controversy which split the nation remained in the church and was fired by the rise of abolitionism.

The leading Methodist abolitionist was Orange Scott, who addressed the general conference on the subject in 1836. When abolitionism was consistently repressed in 1840, Scott joined Lucius Matlack, LaRoy Sunderland, and others in founding in 1845 a Wesleyan Methodist Connection which opposed the episcopacy as well as slavery. Within a year the group numbered 15,000, but after the Civil War resolved the slavery issue, many returned to the mother church. The Wesleyan Methodists continued as a separate body and identified themselves as a "holiness" church. In 1968 the Wesleyans merged with the Pilgrim Holiness Church, an early twentieth-century product of the American Holiness Movement,* to form the Wesleyan Church of America.

In 1844 the issue of slavery came to a crisis over slave-owning bishop James Andrew. The general conference deadlocked over this and the polity issues involved, and eventually agreed to a separation of the church into the Methodist Episcopal Church and the Methodist Episcopal Church, South, officially founded in Louisville, Kentucky, in 1845. Controversy continued to rage between the two churches until the Civil War. In 1870 the Colored Methodist Episcopal Church in America was founded in the South for blacks who had been members of the Southern Church but now wished to organize separately. In 1956 this denomination was renamed the Christian Methodist Episcopal Church.

In the latter half of the nineteenth century Methodism came into its own. It was populous, affluent, well respected, even imitated. Theology and theological education flourished, the camp meeting declined, discipline slipped, and holiness preaching began to vanish. The Free Methodist Church was born in protest against these tendencies. A party of "Nazarites" emerged in the Genesee Conference to recall the church to holiness, decry organs and choirs, denounce pew rentals needed to support elaborate churches, etc. B.T. Roberts, leader of the Nazarites, was finally expelled and led in the founding of the Free Methodist Church in 1860. The "free" in the name testified that the new church was delivered from secret societies, slavery, rented pews, outward ornaments, and structured worship.

A similar protest may be discerned in the rise of the Holiness Movement within Methodism. In 1839 Timothy Merritt in Boston launched a periodical entitled *Guide to Christian Perfection*, later renamed *Guide to Holiness.* Phoebe Palmer of New York City, founder of the "Tuesday Meeting for the Promotion of Holiness," later edited the *Guide* and had great influence on those who established the National Camp Meeting Association for the Promotion of Holiness in the late 1860s. At that time nearly all agreed that Methodism had been raised up to spread the message of Christian holiness, but by the end of the century the church had polarized. Most of the holiness advocates drifted into such groups as the Piligrim Holiness Church, the Church of the Nazarene, and the Church of God (Anderson, Indiana). Holiness partisans who remained within Methodism rallied around Asbury College and Asbury Seminary in Kentucky.

Before the turn of the century, some had raised the question of reunion within Methodism. Efforts to bring together the black churches failed. Significant steps were taken toward reuniting the Northern and Southern churches in 1876, but it was not until 1939 that reunion was finally achieved when the Methodist Episcopal Church, the Methodist Episcopal Church, South, and the Methodist Protestant Church merged to form the Methodist Church, which remained the largest American denomination until the mid-1960s. In 1968 the Methodist Church merged with the Evangelical United Brethren to form the United Methodist Church.

In the twentieth century the Methodist Church has distinguished itself by continued emphasis on world mission, social reform, and ecumenism. American Methodists have been active in the National Council of Churches, whose predecessor the Federal Council appropriated in 1908 the Methodist Social Creed, and more recently in the Consultation on Church Union. Theologically Methodism has followed the dominant schools—liberalism earlier in the century and to a lesser extent Neoorthodoxy later. In response to these theological and ecumenical currents the Evangelical Methodist Church was founded in 1948 and the Evangelical Church of North America after the 1968 merger.

Membership of the more than a score of American Methodist bodies numbers some 15 million. Methodists vie with Baptists as the largest American Protestant religious movement. But whatever the statistics, Methodism can claim to be the "most American of Churches"—or in the judgment of Lutheran Jaroslav Pelikan, "Methodism, equally with Puritanism, constitutes the mainstream of American religious history."

BIBLIOGRAPHY: W.W. Sweet, *Methodism in American History* (1954); J.L. Peters, *Christian Perfection in American Methodism* (1956); E.S. Bucke (ed.), *The History of American Methodism* (3 vols., 1964); R. Chiles, *Theological Transition in American Methodism 1790-1935* (1965); C.W. Ferguson, *Organizing to Beat the Devil: Methodists and the Making of America* (1971).

DONALD W. DAYTON

METHODIST NEW CONNEXION. The death of John Wesley* in 1791 forced constitutional changes in Methodism. An existence over against, and not simply within, the Church of England could hardly be denied: it was soon clear that the conference of preachers would hold both power and authority in the movement. Alexander Kilham* and others argued for a more radical recognition of separation from the Church of England, and for lay participation in Methodist government. The Methodist New Connexion, formed after Kilham's expulsion, embodied these ideas: its second conference (1798) had fifteen preachers and seventeen laymen. William Thorn, a respected Scottish preacher of Wesley's, was the first president.

The new body made slow headway—it was twenty-five years before it doubled its numbers. In that period its image of "democracy" and consequent association with revolutionary principles was probably a disability. "Rational liberty" became its watchword, as against tendencies reflected in the type of Wesleyan Methodism represented by Jabez Bunting.* The ferment of the 1840s saw the departure of a radical element associated with Joseph Barker. Barker's chief opponent, William Cooke, was perhaps the Connexion's most considerable mid-century figure. By the end of the century there were some 30,000 members, with a strong movement in Ireland and missions in China. Union with other Methodist bodies was often discussed, and less often sought; but in 1907 the MNC joined the United Methodist Free Churches and the Bible Christians to form the United Methodist Church.* The influence of the MNC was disproportionate to its numbers; it is noticeable that MNC principles were eventually adopted by the main Methodist bodies.

See S. Hulme, *Memoir of the Rev. William Cooke, D.D.* (1886), and G. Packer (ed.), *The Centenary of the Methodist New Connexion* (1897).

A.F. WALLS

METHODIUS, see CYRIL AND METHODIUS

METHODIUS OF OLYMPUS (d. c.311). Ecclesiastical writer. Few biographical details about him are certainly known, but he was probably bishop of Olympus, Lycia, and was martyred in Chalcis. He has variously been attributed the sees of Tyre, Patara, and Philippi. He wrote extensively, mostly in dialogue form. His only complete work extant in Greek is *Symposium; or Banquet of the Ten Virgins,* modeled on Plato's *Symposium,* in which Methodius extols the excellence of virginity (as Plato commended "Eros"), ending with a hymn to Christ as the church's Bridegroom. Portions of two other works survive in Greek. *Aglaophon; or On the Resurrection* attacks Origen's* doctrine of the soul's preexistence and maintains the identity of the resurrection body with the earthly body. *On Free Will* attacks the dualism and determinism of Valentinian Gnosticism.* Other works survive only in Slavonic.

J.G.G. NORMAN

MÉTIS. The Métis (from a French word meaning "to mix") are people of mixed European and usually Cree or Ojibwa Indian blood and were originally the offspring of the NW Canadian fur trade. Those of French origin have survived because the French Roman Catholic missionaries encouraged them to keep the language and religion of the fathers while the Métis held to the nomadic way of life of their Indian mothers. During the nineteenth century the Métis began to think of themselves as a "new nation" and sought a way of life which was part Indian and part European, but accepted by neither. They have suffered much during the twentieth century because of the loss of their way of life.

ROBERT WILSON

METRICAL PSALTERS. One of the fruits of the Protestant Reformation. Both Luther and Calvin insisted that the people as a whole, not just groups of professional singers, should participate in singing in the public service of worship. Luther was the first to compose metrical versions of the Psalms, but his example was soon followed by others. In France, Clement Marot; in England, Miles Coverdale; in Scotland, two Wedderburn brothers; and in the Netherlands, Souter Liedekens—all produced metrical psalters, or portions of the psalter in the vaious vernaculars. Calvin, however, probably played the most important role in the popularization of Psalm-singing. Not only did he prepare a collection of Psalms while in Strasbourg, but he was responsible for the publication in Geneva of a complete psalter versified by Marot and Theodore Beza. This became the standard work for most Reformed churches, being used either in French or in translation. In England and Scotland the metrical Psalms of Sternhold and Hopkins, Kethe and others eventually became the versions used. The music for these works was often adapted from popular tunes, although sometimes they were composed specifically for a psalm, probably as in the case of Old Hundredth. These psalms often became the battle hymns of the Calvinists in their resistance to oppressive governments, and have formed the basis for most Protestant psalmody and hymnology ever since.

W.S. REID

METROPOLITAN. Used legally for the first time in the canons of the Council of Nicea (325), the term denoted the bishop of the principal city of a province. Insofar as institutional Christianity achieved a shape analogous to Roman imperial administration, and had a membership somewhat proportional to the population distribution of the empire, the centers of one served as centers of the other. The concept goes back to the first millennium B.C., by the end of which "metropolis" was applied to a major center such as Rome, or to a minor location such as Laodicea. By the sixth century, according to Evagrius, the metropolitan was under the exarch, reflecting the regrouping of provinces into dioceses; by the tenth century he ranked above other archbishops.

CLYDE CURRY SMITH

MEXICO, see LATIN AMERICA

MEYER, EDUARD (1855-1930). German historian. One of the greatest authorities on the ancient world, he was professor of ancient history at the University of Berlin (1902-23). Next to Mommsen, he probably contributed more than any other person to the foundation of the modern critical study of antiquity. His multivolumed *Geschichte des Altertums* (8 vols., 1884-1902) is an amazing feat of scholarship. He wrote a number of works in the area of Jewish and early Christian studies, the more important being *Die Entstehung des Judenthums* (1896); *Der Papyrusfund von Elephantine* (1912); *Ursprung and Anfänge des Christentums* (3 vols. 1921-23). In the last work he argued for the historical value of Luke–Acts and for its early date.

W. WARD GASQUE

MEYER, F(REDERICK) B(ROTHERTON) (1847-1929). Born in London of a wealthy family of German ancestry, he graduated from London University in 1869 and completed theological training at Regent's Park (Baptist) College. He was pastor of Pembroke Chapel, Liverpool (1870-72), then moved to Priory Street Chapel, York, where he helped launch on his memorable campaigns the then-unknown D.L. Moody.* Meyer moved in 1874 to Leicester, and in 1881 opened Melbourne Hall, a center of social and evangelistic activity which was Meyer's abiding memorial. In 1888 he went to London and was pastor successively of Regent's Park Church and Christ Church, Westminster Bridge Road. He was president of the Free Church Council in 1904 and retired in 1921. Meyer was a popular convention speaker at Northfield, Keswick, and Portstewart. He was a man of dignified appearance, compassionate heart, ceaseless industry, prolific pen, and graceful style. His devotional studies on biblical characters are still widely read.

ARTHUR CLARKE

MEYER, HEINRICH AUGUST WILHELM (1800-1873). German Protestant clergyman and NT scholar. He studied theology at the University of Jena. His chief contribution to scholarship was the internationally famous commentary series which he founded, *Kritischexegetischer Kommentar über das Neuen Testament* (1829ff.), which has been steadily revised and kept up to date to the present and is still the most important academic commentary of the NT. In addition to contributing the first two volumes of the series on the Greek text and German translation, he wrote commentaries on the four gospels, Acts, and eight Pauline epistles (excluding 1 and 2 Thessalonians and the Pastorals). In the introduction to the volume on the Greek text, Meyer outlined the principles of historico-grammatical exegesis as he understood them. Although it is customary to award the laurels to F.C. Baur* for founding the modern critical approach to the NT, an equal case could be made for granting the honor to Meyer. In his work one finds the application of historical criticism quite apart from the Hegelian presuppositions and antiorthodox bias of Baur and his school.

W. WARD GASQUE

MIALL, EDWARD (1809-1881). Congregationalist minister. The acknowledged leader of the movement in the nineteenth century to disestablish the Church of England, Miall for this reason left a pastorate in Leicester in 1840 to found and edit a newspaper, *The Nonconformist.* In 1844 he arranged for a large conference of Nonconformists in London at which was organized the British Anti-State-Church Association (later renamed the Society for the Liberation of Religion from State Patronage and Control). Under Miall's leadership the Liberation Society, as it was popularly called, became a highly organized and vigorous extra-parliamentary pressure group. He also espoused a number of radical causes which he considered to be related to the disestablishment issue—universal suffrage, the ballot, repeal of the Corn Laws, and programs for improving the living conditions of the working classes. He frequently contested parliamentary elections and had two terms in Parliament—as member for Rochdale (1852-57) and for Bradford (1869-74). The climax in his career came when in 1871, greatly disturbed by what he considered to be the too favorable treatment of the Church of England in the Education Act of 1870, he moved, although unsuccessfully, for a committee on church disestablishment. Because of poor health he retired from Parliament in 1874 and shortly afterward from public life.

E. MORRIS SIDER

MICHAEL CERULARIUS, see CERULARIUS

MICHAEL OF CESENA (b. c.1270-1342). General of the Franciscan Order. Born in Cesena, Italy, he came to the leadership of the Franciscans at a delicate moment when the order was torn by strife between factions known as the "Spirituals" and the "Community." The former adhered strictly to the fundamentals of poverty and non-ownership of property laid down by Francis, while the latter faction opted for a more liberal interpretation recognizing papal rulings on the "poor use" of possessions. Michael's support of William of Ockham,* theologian and apologist for the Spirituals, and his association with Emperor Louis of Bavaria brought about his excommunication by John XXII in 1328. Ockham and Emperor Louis opposed not only the papal rulings concerning possession of property, but advocated separation of civil and ecclesiastical authority. Michael disputed the excommunication and until his death was recognized by his faction as general of the order.

ROYAL L. PECK

MICHAELIS, JOHANN DAVID (1717-1791). Protestant Orientalist and biblical critic. He was professor of philosophy (1746-50) and oriental languages (1750-91) at the University of Göttingen. A prolific writer, he authored an entire journal (*Orientalische und exegetische Bibliothek,* 1771-91) as well as important studies in Hebrew lexicography and Aramaic and Syriac grammar. His multivolumed study of the Mosiac Law (1770-75; ET 1814) was very influential in the early days of biblical criticism in Germany, as were his other critical and exegetical writings on the OT (Messianic Psalms, 1759; Ecclesiastes, 1762; *1 Maccabbees,* 1776). In a serialized translation of the Bible with annotations (OT, 1769-83; NT 1790-92) he introduced the educated German layman to the results of contemporary criticism. His *Introduction to the New Testament* (1750; ET 1793-1801) laid the foundation for further works of similar nature; he was also author of an *Introduction to the Old Testament* (1787, incomplete).

W. WARD GASQUE

MICHAEL THE ARCHANGEL. An archangel mentioned in four passages of Scripture (Dan. 10:13-21; 12:1; Jude 9; Rev. 12:7) and perhaps alluded to elsewhere (e.g., Acts 7:38). In Daniel, Michael is "the great prince" responsible for the guardianship of God's people Israel; in Revelation he leads the angelic host in the struggle against the dragon (Satan) and his angels. The brief passage in Jude also depicts Michael as in conflict with Satan, but on this occasion over the body of Moses; the allusion is thought to be from a work no longer extant, *The Assumption of Moses.* The OT Pseudepigrapha attribute a major role to Michael (so especially *Enoch* and the *Ascension of Isaiah*); here he is given the additional functions of recording angel and intermediary at the giving of the Law at Sinai.

In the course of Christian history, Michael came to be revered as the guardian of Christian armies in their battles with the heathen, and as the conductor of individual Christian souls to God at death (on the basis of the Daniel and Jude passages respectively). In Phrygia he came to be revered as a healer too, a cult which spread widely through Christendom. A fifth-century legend further credited him with an appearance on Mt. Gargano in Apulia, whence a feast of the "Appearing of St. Michael" came into being. He was canonized. In Christian art Michael has usually been depicted in battle with, or victory over, the dragon (cf. Rev. 12:7).

For bibliography see under ANGEL.

D.F. PAYNE

MICHELANGELO BUONARROTI (1475-1564). One of the greatest artists of Western culture, he was a genius in nearly every medium of art. His work marked the culmination of the Renaissance and a transition to the new age of baroque and mannerism. Son of a minor Italian nobleman, he studied under the Ghirlandaio brothers and at the Medici palace, where he was influenced by Neoplatonic thought. He worked in several Italian cities before settling permanently in Rome (1534). He was an example of the Renaissance universal man, for in addition to his sculpture, painting, architecture, and drawing he also wrote lyric poetry. Deeply affected by Savonarola,* he lived an austere life and was given to melancholy, brooding and apprehensive presentiments.

Despite widespread acclaim for his painting, he preferred to think of himself as a sculptor and he produced the *Madonna Seated on a Step; Battle of the Centaurs; St. John in the Wilderness; Pietà; Christ the Risen Savior; David;* and *Moses.* His paintings include the great cartoon of *The Battle of Pisa* He worked for Popes Julius II, Leo X, Clement VIII, and Paul III. The frescoes in the Sistine Chapel took many years to complete (1508-12, 1535-41) and depict the story of the coming of Christ as envisioned by prophets and sibyls in scenes from Genesis, including the ancestors of Christ and the Last Judgment. The work on the ceiling alone covered over 10,000 square feet of surface and included hundreds of figures, some of them twelve feet high. As an architect he completed the memorial chapel for the Medici family and the façade of San Lorenzo and was put in charge of the fortification of Florence (1629). His greatest structural achievement was St. Peter's in Rome. He reworked all the designs of his predecessors and supervised the construction of the supports and the lower sections of the giant dome, although he did not live to see the work finished.

BIBLIOGRAPHY: E. Steinman and R. Wittkower, *Michelangelo Bibliographie* (1927); C. de Tolnay, *Michelangelo* (6 vols., 1943-); G. Vasari, *The Lives of the Artists* (ed. B. Burroughs, 1946).

ROBERT G. CLOUSE

MICHIGAN PAPYRUS OF ACTS. A fragment containing Acts 18:27–19:6, 12-16, which dates c. A.D. 300. It was discovered in Egypt and now belongs to the University of Michigan Library (inventory no. 1571). It represents the so-called Western text-type of the Greek NT.

MIDDLE AGES. A term used generally to refer to that period in European history between the fall of Rome and the Protestant Reformation. The center of life in this millennium was the church, so much so that the medieval world was a church-state. Emperors and kings received their privilege from the church, and feudal society descended accordingly with an endless round of homage between lords and vassals—all of them ultimately vassals of the church. Cities arose around the bishop's seat (cathedral), and monasteries were writing theology and praying for the souls of men everywhere. Not only was government theocratic, however hypocritically, but also men's interests, whether art, music, education, or economics, were rooted in the church. Architecture was ecclesiastical and achieved an unsurpassed perfection, and the universities began with theology and civil and canon law.

For those who first coined the term, the Middle Ages unjustly meant an age of barbarism, ignorance, and superstition, but historians have long divided this period into workable thirds: early, high, and late. "Early" meant the actual collapse, 476-700, when barbarians again were finding their way. In the high or Gothic period, the rebuilding occurred, and a new civilization flowered; but by 1300, evening had come, and the late period witnessed change that was again preparatory.

C.G. THORNE, JR.

MIGETIUS (eighth century). Spanish theologian. Little is known of him apart from opponents' letters. His obscure teaching contains lurking remnants of Priscillianism,* rejecting all distinction between the Second Person of the Trinity and the Incarnate Christ. He taught that God was revealed successively in David (as Father), Jesus (as Son), and Paul (as Holy Spirit), basing it on an extreme literal exegesis. He was opposed by Elipandus,* whose Adoptianism probably arose out of this controversy. Migetius was condemned at synods at Seville in 782 and 785.

MIGNE, JACQUES PAUL (1800-1875). French Roman Catholic priest, patrologist, and publisher. Because of a controversy with his bishop concerning the revolution of 1830, he left his diocese and went to Paris. He turned to journalism, and after several unsuccessful attempts in newspaper work he decided to publish a universal library for the clergy. In 2,000 volumes he hoped to publish at a moderate price all the Catholic literature to his own day. His press employed 300 people, and he showed himself expert at managing the enterprise. In addition to hundreds of volumes of theology, sermons, church history, apologetics, theological encyclopedias, and works on the Virgin Mary he published editions of the Latin Fathers (221 vols., 1844-64 and the Greek fathers (162 vols. in Greek 1857-66; 81 vols. in Latin 1856-67). His work remains valuable because despite errors it is still the one uniform collection of the Church Fathers which even approaches completion.

ROBERT G. CLOUSE

MILAN, EDICT OF. Generally understood to be that passage quoted by Lactantius and Eusebius in which the two *augusti,* Constantine and Licinius, meeting at Milan in January 313, redressed the two-century-old policy of the Roman government toward the Christian Church, so that Christians not only were free to worship as they wished, but were to receive compensation and return of confiscated property—in exchange for the divine favor upon the state in its precarious hour. The passage cited, however, cannot be the actual edict, but at best the rescript of Licinius, dated 13 June 313 at Nicomedia, reporting to the governor of Bithynia the action taken.

MILES, JOHN (1621-1683). Welsh Baptist pioneer. Born at Newton Clifford in Herefordshire and educated at Brasenose College, Oxford, he was baptized on profession of faith at the Glass House Baptist Church, Broad Street, London, and migrated to Ilston in the Gower peninsula to begin his career as the main founder of the Particular Baptists* in Wales. Through his diligence Baptist congregations were gathered over a wide area extending from Carmarthen in the west to the English border in the east. Although Miles did not possess the eloquence of some of his Puritan contemporaries, he had a rare gift of organization and a firm grasp of Baptist principles. He linked the congregations together in a quasi-Presbyterian system under his own firm control, with general meetings of representatives to make decisions on matters of common concern. In theology he was a strong Calvinist and firmly opposed any departure from this norm, whether by the Arminian Baptists of central Wales, or Quakers, or millenarians. His position among Welsh Puritans is revealed by his appointment as an approver under the Propagation Act of 1650. He was Puritan minister of Ilston from 1657 to 1660, when he was ejected under the terms of the September Act of the latter year. He emigrated to New England about 1663 and founded a Baptist Church at Rehoboth, Massachusetts, but in 1667 he moved again and founded a new settlement, Swansea, Massachusetts. His life in New England was a somewhat tempestuous one. He died at Swansea.

R. TUDUR JONES

MILIC, JOHN (Jan, of Kromeriz) (d.1374). Reformer. Born in Kremsier, Moravia, he served in the chancery of Charles IV before becoming a priest, canon, and finally archdeacon in Prague. As a wealthy prelate he embraced poverty to preach the simplicity of the early church, openly attacking the laxity of laity and clergy in 1363. The Inquisition in Rome ordered his imprisonment for preaching that the Antichrist had arrived in 1367, but Urban V ordered his release and he worked with outcast women in Prague, founding a home for them in 1372. In 1373 the clergy of Prague denounced his preaching, but Gregory XI cleared him of all charges in Avignon, where he died.

JOHN GROH

MILLENARIANISM. Sometimes known as "Chiliasm," this is the belief that there will be a 1,000-year period at the end of this age when Christ will reign on earth over a perfect world order. The primary biblical support for this belief is a literal interpretation of Revelation 20:1-10. Some Jewish thought envisaged a messianic kingdom of limited duration; others, a "sabbath" of 1,000 years before the final perfect state. Revelation 20 links these two concepts together for the first time. By linking the 1,000-year period with prophetic visions such as in Isaiah 55–66, the concept of a time of peace, justice, and righteousness on earth is built up.

The timing and nature of the "millennium" are disputed. "Premillennialists" hold that at Christ's return the dead will be raised, believers still living will be "caught up" to meet him in the air, and they will then reign on earth with Christ for 1,000 years. Then Satan will be allowed to be active again, but the judgment of the Great White Throne will follow. "Postmillennialists" see the return of Christ as taking place after the millennium, which may be a literal "golden age" on earth, or which may be symbolic of the final triumph of the Gospel, in this age. "Amillennialists" hold that there is no literal millennium; instead, they see the Revelation teaching as standing for the present age, the whole period between the ministry of Jesus on earth and His second coming. Each school of thought has possible ways of explaining the "two resurrections" and other concepts in Revelation 20.

The 1,000-year period is an important element in the doctrinal systems of the various Adventist* groups and of Jehovah's Witnesses.*

BIBLIOGRAPHY: S.J. Case, *The Millennial Hope* (1918); C.N. Kraus, *Dispensationalism in America* (1958); N.R.C. Cohn, *The Pursuit of the Millennium* (1961); S.G.F. Brandon, *History, Time and Deity* (1964).

A. MORGAN DERHAM

MILLENARY PETITION. So called because about 1,000 ministers were said to support it, this petition was presented in April 1603 to James as he was traveling from Scotland to London to begin his reign as James I of England. The men who composed the petition spoke "neither as factious men . . . nor as schismatics;" they requested not a full program of Presbyterianism but moderate reforms within the diocesan structure of the church. These reforms could be decided, they suggested, by "a conference among the learned." They did, however, ask for the removal of certain grievances. These were the sign of the cross in baptism, less liturgical music, no bowing at the name of Jesus, no profanation of the Lord's Day, the reform of church courts, and other matters. James agreed to the conference, which was held at

Hampton Court* in January 1604.

PETER TOON

MILLENNIALISM, see MILLENARIANISM

MILLER, WILLIAM (1782-1849). Founder of Adventism.* Born in Pittsfield, Massachusetts, he educated himself by reading, and farmed in his wife's hometown, Poultny, Vermont, where he served as deputy sheriff and justice of the peace. He gained the rank of captain in the War of 1812 and thereafter settled on a farm in Low Hampton, New York. He was converted from Deism in 1816 and after fourteen years of Bible study decided Christ would return in 1843. In 1833 he was licensed as a Baptist preacher. His book, *Evidence from Scripture and History of the Second Coming of Christ, About the Year 1843,* published in 1836, and the publicity work of Joshua Himes from 1839 won many to his views. Disenchanted, he dropped out of the Adventist movement in 1845. He died after loss of sight in old age in Low Hampton.

EARLE E. CAIRNS

MILLIGAN, GEORGE (1860-1934). Scottish NT scholar. Son of William Milligan* and born while his father was minister at Kilconquhar, he was educated at Scottish and German universities before ministering at Edinburgh and Caputh (1883-1910). He then held the chair of divinity and biblical criticism at Glasgow until retirement in 1932. He interested himself in the promotion of Christian education and was moderator of the general assembly in 1923. Among his many writings on the NT, which were known outside his own country, was *The Vocabulary of the Greek Testament* (1914-29).

MILLIGAN, WILLIAM (1821-1893). Scottish NT scholar. Born in Edinburgh and educated at the universities of St. Andrews, Edinburgh, and Halle, he ministered in two Fife parishes before becoming in 1860 professor of biblical criticism at Aberdeen, a post he held until his death. He was moderator of the general assembly of the Church of Scotland in 1882 and became that body's principal clerk in 1886. He was also a member of the company formed for the revision of the NT in English (1870), and the first president of the newly founded Scottish Church Society (1892). A liberal in theology, he published a number of books on biblical and theological subjects.

MILLS, BENJAMIN FAY (1857-1916). Evangelist and Christian Socialist. Born in Rahway, New Jersey, educated at Phillips Academy, Hamilton College, and Lake Forest College, he received Congregational ordination in 1878. He served Minnesota, New York, and Vermont pastorates before entering itinerant evangelism in 1886. Using his District Combination Plan, he conducted the most highly organized citywide revivals of the nineteenth century, perfecting many presently used methods. Believing social and economic problems could be solved only by effecting God's kingdom on earth, Mills became the only major evangelist attempting to unite revivalism with the Social Gospel. Finding this impossible, he terminated his itinerancy in 1895 to preach Christian Socialism* in New York and Boston. In 1899, despairing of an evangelical awakening, he became minister to First Unitarian Church, Oakland, California. He founded and led the Los Angeles Fellowship (1904-11) and Chicago Fellowship (1911-14). Repenting of his heterodoxy, he returned to itinerant evangelism in 1915.

D.E. PITZER

MILMAN, HENRY HART (1791-1868). Anglican historian. Born in London, son of a noted physician, he was educated at Eton and Oxford and ordained priest in 1816, later becoming a canon of Westminster and dean of St. Paul's. At first he achieved acclaim as a poet and translator from the Sanskrit, and was professor of poetry at Oxford from 1821 to 1831, but his *History of the Jews* (1830) was hailed by liberal divines as a masterly application of German critical methods of OT study. Next he wrote the life of Gibbon,* whose *Decline and Fall* he had also edited. In 1855 appeared his *History of Latin Christianity,* which despite mistakes of fact is a work of immense erudition, candor, and balanced judgment. A member of the Broad Church school, Milman was never as extreme as Dean Stanley,* shunned public controversy, and deplored the writings of the more radical German critics, particularly Strauss.

IAN SELLERS

MILNE, WILLIAM (1785-1822). Missionary to China. Born near Aberdeen, he went from Scotland to study at the London Missionary Society's college, where he was ordained in 1812. In the following year he joined Robert Morrison* in Macao, but having been ordered out, he distributed literature in Canton and the East Indies and then made his base in Malacca where he assisted Morrison in the translation and printing of the Chinese Bible. In 1815 he cut the first fonts of Chinese type made by a European, and wrote Christian pamphlets. He ordained his convert Liang A-fah,* and became principal of the Anglo-Chinese College in Malacca founded by Morrison in 1818. Milne died four years later. His son later served in China.

LESLIE T. LYALL

MILNER, ISAAC (1750-1820). Evangelical clergyman. After his father's death when he was ten, he became a weaver but managed to teach himself Latin, Greek, and mathematics. When he was eighteen his brother Joseph* became headmaster of Hull Grammar School and appointed Isaac to the staff. Joseph then paid for him to go to Queens' College, Cambridge. He was ordained in 1775 and the next year became a fellow of his college. In 1784 he went to France with William Wilberforce* and their reading of the NT together led to the latter's conversion. In 1788 he was appointed president of Queens', and his ambitious and forceful personality was asserted in spreading Evangelical influence in Cambridge. In 1791 he was made dean of Carlisle, but never spent more than three or four months a year in residence there. Learned in many fields of science, mathematics, and philosophy, he was a large and jovial man and has been described as "an Evangelical Dr. Johnson." He wrote a number of books, including a life of his brother.

R.E. NIXON

MILNER, JOSEPH (1744-1797). Evangelical clergyman. After education at Cambridge he was ordained and served a curacy at Thorp Arch, near Tadcaster in Yorkshire. In 1768 he became headmaster of Hull Grammar School, where one of his pupils was William Wilberforce.* He was also lecturer at Holy Trinity Church. He employed his brother Isaac* at the school for a short while before paying for his education at Cambridge. In 1770 he became an ardent Evangelical, and his preaching in Hull and at North Ferriby nearby (where he was successively curate and vicar) was very popular among the poor and resented by the more respectable. In due course the opposition died out, and in 1797 he was appointed vicar of Holy Trinity Church, Hull, through the influence of Wilberforce, but died before he could be instituted. His best-known work was his *History of the Church of Christ* (3 vols., 1794-97). He was not a professional historian, and the work has been subjected to a good deal of criticism, but it was marked by a determination to record the bright side of church history and not just controversies. Some of his essays and sermons were also published.

R.E. NIXON

MILTIADES (second century). Associated with his supposed apologetic writings. His works are mentioned in Tertullian and Eusebius, though none are extant. These writers suggest he wrote against the pagans, Jews, Montanists, and Valentinians. He wrote in Asia Minor.

MILTIADES (Melchiades) (d.314). Pope from about 311. Mentioned in the *Liber Pontificalis* ("the Papal Book"), he was by birth an African. During his pontificate Constantine defeated Maxentius (312) and issued the so-called Edict of Milan* (313) which marked the triumph of Christianity over persecution. He commissioned a council at the Lateran in 313 to investigate the Donatist Schism. The council decided against the schismatics and condemned them.

MILTITZ, CARL (Charles) VON (1490-1529). Papal secretary and subnuncio to Germany. Born at Rabenau near Dresden of the lesser Saxon nobility, he studied law at Cologne and Bologna. In Rome from 1513 to 1518, through unprincipled maneuvering he advanced rapidly within the Curia until appointed papal secretary in 1518. He is best known for his meetings with Luther, 1518-19. Sent to Saxony with the Golden Rose for Elector Frederick, and commissioned to act only with the approval of Cajetan* (the papal legate), Miltitz precipitously decided to attempt a reconciliation of Luther with the church. By boast, misrepresentation, and denigration of Cajetan and Tetzel,* he convinced Frederick a conference with Luther would be successful. They met at Altenburg early in January 1519. Though theologically ignorant, Miltitz was a clever diplomat and secured Luther's promise of silence-unless-attacked until his complaints were heard by a German bishop—a settlement too shallow to last. Miltitz wrote to the pope, indicating that Luther was ready to recant, then journeyed to Leipzig to further embarrass Tetzel. The accord failed; Miltitz met twice again with Luther without result. His last years were spent as canon of Mainz and Meissen. He died by accidental drowning.

BRIAN G. ARMSTRONG

MILTON, JOHN (1608-1674). English poet. Born in London and educated at St. Paul's School and Christ's College, Cambridge, he was always of a serious turn of mind and felt himself even in his youth called to a high vocation in the service of God. His earliest important poem, *On the Morning of Christ's Nativity,* (1629) is really a poem, not of Christmas only, but of the Incarnation, its power and effects.

In the next decade he wrote a number of short poems, of which the most significant is *Lycidas* (1637), and the masque *Comus* (1634). The former is a pastoral lament for a college acquaintance, Edward King, but it goes far beyond the occasion to question the whole purpose of life and especially of the dedicated life. In addition it contains Milton's scathing lines of the "hireling shepherds" of the church, concluding almost prophetically with the threat that the Civil War would actualize of that "two-handed engine at the door [which] stands ready to strike once and strike no more." *Comus* presents in a tableau the conflict between chastity and vice *(luxuria),* personified in the persons of the Lady and Comus. The latter enchains her body, but is powerless against her free spirit.

After the outbreak of the Civil War, Milton was occupied first with pamphleteering in the parliamentary cause and then in the service of government as Latin Secretary under the Commonwealth. His pamphlets cover controversy against episcopacy (e.g., *Reformation of Church Discipline in England,* 1641), about divorce which he supported more liberally than most of his contemporaries, and on political and miscellaneous questions such as the *Tractate on Education* (1644) and the immortal plea for freedom of printing in *Areopagitica* (1644). It is said, though with what degree of truth is conjectural, that his unhappy experience of marriage to Mary Powell led to the first divorce pamphlet. Perhaps it need only be added that, despite the known incompatibilities of this first alliance, Milton later married a second and a third time.

With the Restoration, Milton as a regicide and the most eloquent apologist for regicide stood in danger of his life. By this time he was also blind. The intercession of friends gained him inclusion in the general amnesty, and his last years were spent in a return to the poetry which he had forsaken two decades before. In them he was enabled to fulfill the high vocation he had always felt himself called to perform. *Paradise Lost* appeared in 1667, followed four years later by *Paradise Regained* and *Samson Agonistes.* In the first of these he sought no less than "to justify the ways of God to man" in an extensive epic treatment of the Fall. In verse uniquely sonorous and impressive he characterizes the main protagonists in that event and ranges through heaven, hell, and earth in his examination of motive, conflict, and responsibility.

The later poems are less magnificent, and *Paradise Regained* in particular seems to lack that rich humanism of which its predecessor is the last and finest flower in English literature. There is also an occasional harsh note that is paralleled too in the flashes of savagery that mark, and possibly mar, *Samson Agonistes*. So much like Milton himself, Samson, "eyeless in Gaza, at the mill with slaves," wreaks his God's revenge on the Philistine enemy. Yet in the face of suffering and death, Milton's final word on the subject, both stoic and Christian, is: "All is best,/Though we oft doubt what th'unsearchable dispose/Of highest wisdom brings about."

See *Poems* (ed. J. Carey and A. Fowler, 1968), and W.R. Parker, *Milton*, (1968).

ARTHUR POLLARD

MIMICRY, RELIGIOUS. The act of ridiculing or making sport of a religious rite. This entry will limit itself to mimicry of the early Christian rites (see also ABBOT OF UNREASON).

Much of the dramatic entertainment for people in the Hellenistic and imperial eras consisted of scenes parodied from daily life. Gestures and facial expressions played an important role in these productions. On rare occasions the strolling companies of mimes approached the level of drama by concentrating on a person's character instead of a plot. A single individual at times performed all the roles in a mime. Mimes frequently buffooned a character in a novel situation, such as a poor man with sudden riches. Simple plots, abrupt endings, and vile language added spice, since the only fixed prop was a movable curtain. Most of the mimes dealt with sordid themes; in imperial times a stock theme was adultery, often performed on stage. Spectators witnessed an actual execution when a condemned criminal took the actor's place at a critical point in the play. Most actors were not known for their high moral character or social station.

The mime's flexibility and adaptability to current tastes gave it perpetual vitality. Popular songs and dances were introduced as needed. But since it lacked the dramatic art of tragedy and comedy, the mime was a drama of escape rather than interpretation. It had deserted the religious basis of earlier classical forms.

The "Christian" had probably become a stock figure in the mime by the second century. The church's rites, especially baptism, were parodied as the baptismal candidate, accompanied by a number of clerics, was led on stage for the ceremony. Tradition has it that St. Genesius, the patron of actors, was converted while performing his parody of baptism in which an emotional "fit" preceded the rite. Updating earlier interests in execution, the mimes also ridiculed martyrdom.

Since the mimes were among the last popular strongholds of paganism, their continuing concern with the old gods and their mimicry of Christian rituals prompted vehement attacks from Christian spokesmen. Such attacks were leveled in the writings of Minucius Felix, Tatian, Arnobius, Lactantius, and others. Augustine distinguished between comedies or tragedies and the mimes with their filthy language.

Special works against the theater were composed by Tertullian, Cyprian, and Chrysostom. Tertullian asked how a Christian could pray "amen" and still praise the mime with the same lips. Cyprian judged it improper for a Christian to act or instruct others in the trade; the community should support such a person with its poor chest. According to Chrysostom, God spoke through monks while the mimes were the devil's spokesmen. Their songs, dances, and shows were his litany and sacrament, and their guiding principle, like his, was disguise and imitation. People who attended the shows were the devil's children. Seductive actresses with curled hair and painted cheeks, singing their "ballads of the brothel," were nothing less than contemptuous. Chrysostom called the mime an incurable plague, a snare of death, a theater of concupiscence. These Christian writers apparently held that it was unlawful to witness what was unlawful to do.

The Council of Illiberis required that a pantomimist renounce his trade before baptism, while the Third Council of Carthage was more moderate. The Council of Trullo denounced both pantomimists and their theaters. The pagan Zosimos reproached the Christian emperor Constance for patronizing the mimes. A decree at Elvira forbidding Christians to be charioteers or pantomimists was reiterated at Arles (452), although no mention was there made of attending plays. Leo the Great contended that the theater attracted greater crowds than martyrs' festivals.

Christian attacks were largely unsuccessful, since a large body of nominal Christians viewed the mime as harmless. Interest in this entertainment never faltered in the Eastern Empire; some of Eastern hymnody bore witness to the power of the mimes' songs. As the church's power increased, it got the upper hand. All performers of mime were excommunicated in the fifth century, and in the sixth Justinian closed all theaters. The mime remained unacceptable to most churchmen even after it was forced to drop its mimicry of sacraments and rites, but it lived on as a form of popular entertainment. JOHN GROH

MINIMS *(Ordo Fratrum Minimorum)*. An order of friars who regarded themselves as the least of all the religious, below the Friars Minor. Founded initially and informally as a group of hermits in 1435 by Francis of Paola,* and confirmed by the pope in 1474, the order had no written rule until 1493. This rule was similar to that of the Franciscans. A second rule (1501) was less obviously Franciscan, being more austere. Abstinence from animal meats, fish, eggs, cheese, butter, and milk was required; bread, vegetables, fruit, and oil were the staple diet. The brothers dressed in a black wool habit, cord girdle, with cape and hood. The order enjoyed great initial popularity, and by 1550 there were over 400 houses in Europe. Today only a few survive in Italy and Spain.

PETER TOON

MINOR ORDERS. The lowest offices of the ministry: porters, lectors, exorcists, and acolytes in the West, and lectors and cantors in the East. Subdeacons are classed as a Major Order (with bishops, priests, and deacons) in the West, but as a Minor Order in the East. The giving of Minor Orders is still basically governed in the West by the *Statuta Ecclesiae Antiqua* (c.500). All four (which are first mentioned in a letter of Bishop Cornelius of Rome to Fabian of Antioch in 252) are now usually conferred at the same time by a bishop or abbot upon students who are intended for the priesthood. Though each office did have a specific function, this has now been abandoned since the function has been taken over by the laity or priesthood (e.g., lighting of candles, once done by an acolyte, is now done by a layman). The Minor Orders are now only a step toward full ordination. PETER TOON

MINSTER. Properly a monastery or a monastery church, but applied in England to certain large churches (e.g., Beverley and Wimborne) and to certain cathedrals (e.g., York, Lincoln, Ripon, Southwell, Lichfield). The usage in England derives from the term *monasterium* (OE *mynster*) used in the Middle Ages to denote not only monastic foundations but also colleges of secular canons, many of whom lived under the Rule of Augustine and were hardly distinguishable from regular clergy. The term *münster* is used in this way of certain large churches and cathedrals in Europe (e.g., Ulm, Strasbourg, and Zurich).

MINUCIUS FELIX (second or third century). African author of the apology *Octavius*, an elegant, attractive defense of Christianity, which takes the form of a dialogue between a pagan, Caecilius, and a Christian, Octavius, Caecilius echoes the general calumnies against Christians, and Octavius corrects these views, stressing at the same time the virtues of Christianity. The relationship between this work and Tertullian's apology is very close. There are such striking affinities that some interdependence is certain, but it is impossible to say which work is prior.

MIRACLE PLAY, see MYSTERY PLAY

MIRACLES. In the ancient world, miracles were regarded as a normal, though extraordinary, feature of life. Supernatural powers were looked upon as intervening in the human situation. Remarkable cures were reported, e.g., from temples and centers of healing dedicated to Aesculapsius. Certain individuals such as Apollonius of Tyana* (almost contemporary with our Lord) were credited with outstanding gifts in this respect. Emperors and kings were regarded as endowed with the same kind of gift.

A miracle, in the Bible, is an event which, when we are confronted with it, forces us to say, "This is the LORD'S doing; it is marvellous in our eyes" (Ps. 118:23). The hand of God can be clearly seen at work, and His purpose in working is understood. Various words for "miracle" are used in Hebrew and Greek. Some remind us of the fact that miracles can be either extraordinary or ordinary events. Some are used to stress the extraordinary nature of the phenomenon. Some illustrate that the omnipotence of God is at work. Some underline that the miracle attests the word and mission of the teacher or preacher.

In the OT, miracles are regarded as due to the direct intervention of Yahweh in human affairs and are linked up with His purpose of redemption. In the NT, they are aspects of the proclamation of the good news that the nearness of the kingdom of God has made available within this world's life here and now the kind of power that He will ultimately use to restore all things to their proper order, to banish earth's corruption, to destroy death, and to restore human life to its true integrity. The life, death, and resurrection of Jesus Himself is the supreme miracle of this nature, and the miracles which accompany this life draw their significance entirely from this connection. Jesus' physical miracles are signs that redemption involves body as well as soul. They are acts of His compassion and of His kingly power. They are also signs which have to be interpreted by the word which accompanies them and by the whole purpose of His life. They thus belong to the priestly, kingly, and prophetic ministry of Jesus. They are themselves (as D.S. Cairns has said) integral parts of the revelation, and not adjuncts to it.

The early church for some time experienced power to continue in every respect, through prayer and the laying on of hands, the whole miraculous ministry of Jesus. Such miracles seemed especially to accompany missionary preaching. It came to be understood, however, that the ultimate restoration of the physical realm to order would take place at the second coming of Jesus, and that Christian faith was strong enough to wait for this event. It was realized that the greatest work of the church in this interim period was to reconcile men to God and bring them to repentance. When miracles occurred in the pursuit of this central task, they took place as remarkable answers to prayer, and signs confirming the Word especially in the missionary situation. They were regarded as events to be received thankfully by faith, rather than as aspects of the present program of the church.

Gradually, however, the view arose that the seeking of such miracles should be given a more regular and central place in the church's ministry. Pagan views of supernatural powers inherent in persons, shrines, and streams, etc., replaced the belief that men have direct access to a God who hears and answers prayer. At the Reformation it was emphasized that though all things were possible with God, and though the way of prayer was always open, nevertheless after the Apostolic Age the power magisterially to work miracles had been withdrawn, and the place of outward miracles as seals of the Word had been taken by the sacraments, in which the miraculous element was much more hidden.

In early days God was thought of as so related to the natural realm that every occurrence was due to His direct intervention. Miracle was simply a sign that He now willed in this instance to act differently than He had otherwise hitherto acted. When the idea of laws of nature arose, nature could still be viewed as a plastic medium in the hand of God. But when nature was viewed as a closed and rigid system, miracle had to be regarded as violating the laws of nature. Some theologians stress the fact that miracles are not contrary to nature, but contrary to nature as it is known by us. Explanations have been sought of miracles, stressing that most miracles are due to remarkable coincidence or insight. In this case, what would be to one person an explainable event would be to another a miracle. But there is no doubt that in the thought of the ordinary Christian the idea of what *can* take place will arise out of what has already taken place in the incarnation, life, death, and resurrection of Jesus. The Christian experience of forgiveness, providence, and prayer is often so full of miracle that it is inconsistent to be theologically skeptical on the subject.

BIBLIOGRAPHY: J. Wendland, *Miracles and Christianity* (1911); C.S. Lewis, *Miracles* (1947); J.S. Lawton, *Miracles and Revelation* (1960); J. Kallas, *The Significance of the Synoptic Miracles* (1961); A. Richardson, *The Miracle Stories of the Gospel* (1961); R.H. Fuller, *Interpreting the Miracles* (1963); C.F.D. Moule, *Miracles* (1965); H. Van der Loos, *The Miracles of Jesus* (1965); H.H. Farmer, *Are Miracles Possible?* (1966); E. and M. L. Keller, *Miracles in Dispute—A Continuing Debate* (1969).

RONALD S. WALLACE

MISSIONS, CHRISTIAN. The term has usually applied to foreign missionary activity, but development of the world church has led to fresh appreciation of mission as the task of the church wherever it is found. Mission is the joyous and loving response of the Christian community to the universal and exclusive claims of the triune God who has revealed Himself definitively in Jesus Christ. It involves crossing all human boundaries, by Christians who are called individually and corporately to proclaim God's purposes. By their witness and service they summon fellow-sinners to turn to God and share in His promised kingdom, for right response to God is inseparable from the calling of the nations and offer of new life to all who will hear.

The NT concentrates on Paul's missionary activity, but he was only one of many who traveled the Roman Empire witnessing to the risen and coming Lord. The ministry of Jesus and Paul provide classic examples of the exacting nature of proclamation of the Gospel, the varied methods used, and the historic content of the Christian message. There are fragile boundaries between loyalty to what is historically revealed and cultural exclusiveness. Accommodation can lead to syncretism or conservatism and, despite the presence of the Spirit, the risk of misrepresenting Christ demands constant scrutiny of the message.

During the first three centuries the church faced and partly resolved issues which have continued to test her missionary vocation: disagreement about God's nature, definition of the unique historicity of Christ, relation to the state, exposition of the Christian ethic, relationship to other religions, refutation of misunderstanding and slander, development of a pattern of authority which allowed local adaptation without destroying unity, initiation into mission, and worship.

By the end of the second century the work of the Apologists, the triumph of Catholic Christianity over Gnosticism,* and the development of written Scriptures and creeds had given Christians a defined and readily communicable message, which was greatly assisted by a common political framework and the popularity of *koine* Greek. Judging by the strictures of Celsus, even ordinary Christians developed successful methods for communicating their faith. Initially Christianity mainly appealed to urban groups, especially those already interested in monotheism and accessible to trade routes. Pliny's* letters suggest more widespread impact in Bithynia, and there were major movements in Egypt and North Africa (see AFRICA, ROMAN) by the third century. In Rome, Callistus's pastoral problems suggest that the church included a cross-section of society. Though there were wide regional variations in church growth, there was a Christian presence in most imperial provinces and in Edessa, Armenia, and Ethiopia.* In Asia Minor and Egypt, Christians were too strong to ignore and too numerous to eliminate.

Spasmodic persecution contributed to church growth, giving powerful testimony to the manner in which Christianity freed men and women from the fear of death, demons, and fate. The joyous certainty and vitality of Christian literature contrasted strikingly with the pessimism of much pagan writing, while strong traditions of mutual help on an empire-wide scale enabled believers to meet illness and misfortune far more effectively than pagans who saw little connection between religion and responsibility. Official recognition by Constantine* and his successors posed fresh problems, because of the political and cultural overtones conversion acquired after 313, as barbarians like the Goths moved inside Roman frontiers; but the attempted revival of paganism under Julian the Apostate* showed that Christianity did not depend on official support for its growth. The task of translating Jesus' message into Greek and Latin cultures was almost complete by the end of the fifth century, but the church was seriously weakened by barbarian invasions, the slow collapse of political order in the West, and bitter theological divisions in the East.

By the seventh century large Christian communities were dominated by Islam, though the Nestorians spread Christianity through central Asia and as far as China.* An even more significant movement was under way in the West. Following the conversion of Ireland* by Patrick,* Celtic missionaries (Columba,* Columbanus,* Aidan*) moved into Scotland and the N English kingdoms, throughout and beyond Frankish territory, where Clovis* had become a Catholic Christian in 496. The Gregorian mission to Kent in 597 gradually expanded into other kingdoms, and Anglo-Saxon Christians sent a number of notable missionaries like Wilfrid* and Winfrith (or Boniface*) to their kin among the German tribes and Scandinavia, with markedly successful results. The correspondence of Boniface is a missionary classic, and the pattern of tribal conversion with a minimum of cultural disturbance resulted in the development of strong churches within a Roman framework. The sharp division between clergy, religious, and laity meant that the work of conversion was largely regarded as the responsibility of clerics, and it was long before Christianity penetrated isolated rural areas. Illiteracy, lack of vernacular worship, and ineffective pastoral care meant considerable confusion between paganism and Christianity. The work of conversion continued steadily eastward, and official paganism ended with the baptism of Jagellio in Poland* (1386), though groups like the Lapps and the Romany remained largely untouched, as did the Jews.

Orthodox missionaries, Cyril* and Methodius, penetrated Moravia and, by their translations of Scripture and liturgy, played a formative part in the entry of Slavs into Christendom. Further east, the baptism of Vladimir* of Kiev about 988 was a turning point in Christian history, for Russian Orthodoxy has expanded steadily eastward with the extension of Russian territory, and remained in continuous contact with other religions until the present century (see RUSSIA). Great missionaries like J. Veniaminov (1797-1879) and N. Kasatkin* are too little known in the West.

One of the most powerful inspirations for missionary activity has been the rediscovery of apostolic Christianity and the message of Jesus. In the twelfth century, groups like the Waldenses* traveled widely to communicate this, but were proscribed. It was not until Innocent III recognized the followers of Francis of Assisi* that zeal to convert infidels and heretics, inspired by the Crusades, was briefly redeemed by a truly Christlike spirit, which rejected force as a means of conversion. Raymond Lull,* a Franciscan tertiary and a pioneer theologian of mission as Christians' basic responsibility, saw conversion as a work of love, demanding careful intellectual preparation. Kublai Khan's request for teachers in 1260 was not taken seriously in Rome, but John of Montecorvino* reached Peking in 1294 and was sent other Franciscans as bishops. Some converts were made prior to a change of dynasty in 1368, but these Roman missionaries remained a royal chaplaincy, dependent on imperial favor for survival.

European colonial expansion and the renewal of the church during the sixteenth and seventeenth centuries underlay the next major phase of Christian expansion, dominated by Roman Catholic orders. Mission was clerical, inseparable from political goals, increasingly under Roman control after the foundation of the Propaganda* (1622), and only partly successful in dealing with a new missionary problem—preaching the Gospel and establishing churches in primitive cultures in Africa and parts of Latin America. Brutal exploitation of natives by colonists and administrators was partly redeemed by the struggle for the human dignity of the Indians fought by B. Las Casas* and others against official apathy and the theologians who taught that the Indians had no souls. A more permanent solution was found in the Jesuit* "reductions," which were a noble attempt to create Indian Christian communities; but they collapsed with the dissolution of the Jesuit Order (1773). Paternalism has remained one of the most serious missionary problems. In Japan,* F. Xavier* adopted a different approach, attempting to build on local culture, and was followed elsewhere by M. Ricci,* R. de Nobili,* and A. de Rhodes (1591-1660).

Significant gains were made in Japan, China, and India* until political changes led to the virtual extermination of Christianity in Japan by 1650 and its proscription in China in 1723. European colonies in India enabled Roman Christianity to take root, but local adaptation was proscribed by *Omnium sollicitudinum* (1744). Until 1938 all missionary priests took an oath of submission, with the result that there was excessive Romanization in Roman Catholic missions. They received another setback with the abolition of the Jesuits, but revived after the French Revolution. Many new missionary orders, like the Marists,* were founded to take advantage of French imperialist expansion in Asia, Africa, and the Pacific. Great missionary statesmen like Cardinal Lavigerie (1825-92), with his White Fathers* (1868), were remorseless opponents of slavery, intrepid explorers, and contributors of African education and technical advancement. Catholic missions were particularly strong on institutional work, and their doctrine of the church made it easier for them, than for some Protestants, to deal with African tribalism, though they did not escape the dangers of paternalism and over-identification with colonialism which characterized the nineteenth-century missions.

Protestant churches displayed little interest in the heathen during the Reformation,* though their rediscovery of the Gospel, stress on the vocation of all Christians, recovery of vernacular Scriptures and liturgy, and emphasis on a literate and responsible laity were to prove profoundly important for the development of Christian missions once Protestant countries acquired colonies and came into contact with other religions. Initially visionaries like A. Saravia* and J. von Welz (1621-68) won no official support. Such initiatives as were taken by J. Eliot* and T. Bray* were personal and based on the voluntary principle. The Danish Tranquebar Mission (1706) was the pioneer Protestant foreign mission, though the Moravians were the first church to undertake foreign missions (1732).

Evangelical revivals during the eighteenth and nineteenth centuries combined with European imperialist expansion to open up vast new areas to the Christian message. European Protestants concerned about missions formed voluntary societies, beginning with the Baptist Missionary Society (1792), the London Missionary Society (1795), the Netherlands Mission Society (1797), the Basel Mission (1815), and many others. The entry of churches like the Wesleyans (1818) into missionary activity marked a fresh development, as did the founding of the China Inland Mission by J.H. Taylor,* for it repudiated connection with any one church, was resolutely nondenominational, and was the forerunner of a host of "faith missions." Many missions placed great stress on civilization as a partner of evangelism, and the godly artisan or teacher sent overseas was a new missionary phenomenon. W. Carey* exemplified this approach and set a sterling example by translation work, study of local religion and culture, and determination to develop an educated local ministry and people.

This pattern of Protestant mission was repeated throughout Africa, Asia, and Oceania (see SOUTH SEAS; AUSTRALIA; NEW ZEALAND). Usually it led to strict preparation for baptism, probation before admission to the Lord's Table, and restriction of local ministry to teaching and catechizing, with the unspoken assumption that European standards of literate faith were the norm. Often there was a sharp rejection of local culture wherever it impinged on religion, but critics of the Westernness of Christianity like N.V. Tilak* were not taken seriously. The Boxer Rising in China (1900) and the Nyasaland Rising (1915) led by J. Chilembwe showed the depth of resentment at the confusion of colonialism and Christianity. In Africa* the desire for authentically local Christianity has led to the formation of rapidly growing Independent churches. The No-Church movement in Japan has similar roots. H. Venn* and R. Anderson* looked forward to self-governing indigenous churches, but many missionaries did not. The consecration of S.A. Crowther* in 1864 reflected Venn's goals, but inadequate CMS assistance led many to consider the experiment a failure, and there were no more consecrations until the twentieth century. The Nevius Plan (1890) in Korea was more successful in allowing room for local initiative, but unity has suffered considerably (see NEVIUS, J.L.).

By the beginning of the twentieth century, Christianity was a genuinely international religion and had decisively broken out of its Western boundaries. Missionaries had played an important part in the legal abolition of slavery; had championed natives against white exploitation (J. Philip,* G. Scott); pioneered medical services through groups like the Edinburgh Medical Mission Society (1841), or individuals like P. Parker* and I.S. Scudder*; founded schools, colleges like Serampore (1818) and universities like Doshisha (1875); while the personal influence of educators like A. Duff* and T. Richard* was profound. D. Livingstone* was one of many explorers, and notable contributions were made to linguistics (H. Martyn*), ethnography (H.A. Junod, 1863-1934), and comparative religion (J. Legge*), to mention only a few of the missionary contributions to scholarship.

New perspectives were given on family life and the role of women, economies revolutionized through introduction of products like cocoa into Ghana (1857) by the Basel Mission, cannibalism and infanticide checked, and above all, countless lives transformed by the power of Christ and vigorous Christian communities established, especially in animist societies, but also in ancient Asian cultures. Islam* alone remained largely resistant to Christianity.

The very success of Christian missions raised important questions about the nature of Christian faith. Men and women who wished only to preach the Gospel found it necessary to come to terms with the oversight of churches and ancillary institutions. Mass movements in India, Africa, and Oceania created great problems for missionaries, though H. Whitehead (1853-1947) and B. Guttmann (1877-1966) led the way in suggesting solutions which have raised important issues about the relation of mission, church, and society. This also emerged with local criticism of denominationalism, which seemed superfluous in a largely pagan context where European comity agreements had already established one church in particular regions. In addition, practical matters like preparation of missionaries, relations with governments, and need for biblical translations encouraged cooperation on the field and at home.

Between 1860 and 1963 there were ten major international conferences and a host of local and regional ones. Edinburgh (1910)* was a new beginning, for it institutionalized and internationalized cooperation in national missionary councils and by the formation of the International Missionary Council* (1921) which has played a vital role in the missionary and ecumenical movements. The integration of the IMC and the World Council of Churches at New Delhi (1961) was a symbol of the growing recognition that mission is more than a dedicated Christian minority crossing geographical frontiers, but a task for every Christian and the whole church. Increasingly this partnership has extended to leaders of the "younger" churches; men like V.S. Azariah,* D.T. Niles (1908-1971) and T. Kagawa* have helped Western Christians to realize anew the implications of the universality of Christ.

During the nineteenth century, Protestant theologians paid little attention to the theology of missions, or the relation of Christianity to other religions and ideologies. The erosion of Western Christianity, the resurgence of other religions, the growth of anti-Christian ideologies like Marxism, and the development of liberal Protestantism, which denied the uniqueness of Christianity and the need for conversion, caused a great deal of heart-searching about the real motives for missions. W.E. Hocking (1873-1966) and a team of lay investigators produced *Re-thinking Missions* (1932), which was a persuasive statement of the new views. H. Kraemer* produced a powerful statement of the traditional views in *The Christian Message in a non-Christian World* (1938), informed by his own experience in Indonesia* and the inspiration of Karl Barth,* but the Uppsala* statement on mission and the response of the Frankfurt Declaration (1970) show that there is deep disagreement about the nature of mission in churches associated with the World Council of Churches.*

Since 1945 an increasing proportion of missionaries have come from North America, and many of these regard the WCC and its agencies with deep suspicion, though the Interdenominational Foreign Mission Association* (1917) and the Evangelical Foreign Missions Association* (1945) joined forces for a notable conference at Wheaton in 1966. A series of Congresses on Evangelism (from 1966) sponsored by the Billy Graham Evangelistic Association has also initiated important evangelical cooperative ventures and examined the relation between the historic Gospel and the need for indigenous, but catholic, response to the risen Lord.

There are still many nations closed to Christianity, other areas where contact has been slight. Massive population growth, urbanization and rapid social change, and rival religions and ideologies are problems common to all Christians. Political independence in former colonial territories and the closure of China (1949) have brought rapid localization of authority. As never before, Christianity is feeling the strains of its historicity and universality, but the decisive feature of Christian missions is not only conversion of men and nations, but obedient witness everywhere to the Lord who makes all things new, for ultimately the goal of missions is the glory of God and confession of His sovereignty.

BIBLIOGRAPHY: L.E. Browne, *The Eclipse of Christianity in Asia* (1933); K.S. Latourette, *History of the Expansion of Christianity* (1937-45); J. Glazik, *Die russische-orthodox Heidenmission* (1954); K.M. Panikkar, *Asia and Western Dominance* (1954); O.G. Myklebust, *The Study of Mission in Theological Education* (1955); J. Van Den Berg, *Constrained by Jesus' Love* (1956); A. Mulders, *Missiegeshiedenis* (1957); T. Ohm, *Asia Looks at Western Christianity* (1959); P. Maury (ed.), *History's Lessons for Tomorrow's Missions* (1960); S. Neill, *History of Christian Missions* (1965); K. Baago, "The Post-colonial crisis of Missions," *International Review of Missions* (1967); G.S. Parsonson, "The Literate Revolution in Polynesia," *Journal of Pacific History* (1967); E.D. Potts, *British Baptist Missionaries in India* (1967); D.B. Barrett, *Schism and Renewal in Africa* (1968); R.C. Bush, *Religion in Communist China* (1970); E.M.B. Green, *Evangelism in the Early Church* (1970); A.P. Vlasto, *The Entry of the Slavs into Christendom* (1970); J. H. Kane, *The Global View of Christian Missions* (1971); S. Neill et al., *Concise Dictionary of the Christian World Mission* (1971); M. Jarrett-Kerr, *Patterns of Christian Acceptance* (1972); D. McGavran, *The Eye of the Storm* (1972). IAN BREWARD

MISSOURI SYNOD, see LUTHERAN CHURCH BODIES

MITER (Gr. *mitra,* "turban"). A form of hat or headdress, made of embroidered satin and worn by bishops and some abbots in the Western Church. In the East they wear metal crowns. The usage goes back to the eleventh century and to the *camelaucum,* the papal tiara. Worn at all solemn services and occasions, but taken off during prayers and the canon of the Mass, it is found in three types. First, the Precious Miter, worn on Feasts and ordinary Sundays and adorned with precious stones and/or gold. Second, the Golden Miter, used in penitential seasons and made of golden cloth. Third, the Simple Miter, worn at funerals and on Good Friday and made of plain white silk or linen. Since the nineteenth century, Anglican bishops have also used them. They are shaped like a shield. PETER TOON

MITHRAISM, see MYSTERY RELIGIONS

MIXED MARRIAGE. A marriage between Christians and non-Christians, or between members of different Christian denominations. Generally it is understood as indicating marriages between a Roman Catholic and a baptized non-Roman Catholic. The Roman Catholic Church still continues to discourage such marriages, arguing that although they are a consequence of the division among Christians, they do not, except in some cases, help in reestablishing unity among Christians. This view is reciprocated in Protestant denominations, but not unanimously accepted.

Until 1966 it was maintained that mixed marriages could be acceptable only after episcopal dispensation, conditional on the promise of both Catholic and non-Catholic party—normally in writing—that the children of such a union would be baptized and instructed in the Catholic faith, based on *Codex Iuris Canonici.* In that year the "Sacred Congregation for the Doctrine of the Faith" issued an instruction, *Matrimoni Sacramentum,* which sought to find a more realistic norm for a worthy reappraisal of mixed marriages. One of the significant changes was that "excommunication for attempting marriage before a non-Catholic minister is abrogated."

In 1970 Paul VI issued an Apostolic Letter, the *Motu Proprio,* determining norms for mixed marriages, in which he confronted two major issues: (1) that no human authority has the right to dispense a Catholic from the duty of keeping the faith and handing it on to his children; (2) that the church has no right to require a non-Catholic to make a promise against his conscience. The letter indicates certain changes while still holding as firm a position as possible. Marriage between a Catholic and a baptized non-Catholic still requires an episcopal dispensation, without which the marriage is deemed invalid. The dispensation is granted on condition that the Catholic partner "as far as possible" ensures Catholic instruction for children. This is incumbent now on the Catholic party only, but the non-Catholic party must be informed of it. It is now also possible that at the wedding service there is some form of cooperation between priest and minister, subject to previous episcopal sanction, and that the ceremony conducted in the church of one partner be followed by a service of blessing or thanksgiving in the church of the other, so long as there is no second exchange of marriage vows.

GORDON A. CATHERALL

MOBERLY, ROBERT CAMPBELL (1845-1903). Anglican theologian. Son of a bishop, he was educated at Winchester and Oxford where, after service in a parish and two theological colleges, he returned as professor of pastoral theology from 1892 till his death. One of the *Lux Mundi** school of liberal Anglo-Catholics, he wrote *Ministerial Priesthood* (1897), a study of the Christian ministry from a strongly Anglo-Catholic standpoint which criticizes the views of Bishop Lightfoot as too favorable to Protestantism. His other best-known work is *Atonement and Personality* (1901), a difficult book which defines personality in Hegelian terms and veers towards pantheism, develops a moral satisfactionist theory of the Atonement out of an inadequate treatment of sin, and advocates a high doctrine of the church and sacraments. Moberly also defended Anglican orders against Roman Catholic critics and church courts, and the dual system in education against the Nonconformists. IAN SELLERS

MODALISM, see SABELLIANISM

MODERATES. The name given to various groups of Scottish divines in the eighteenth and early nineteenth centuries. Their basic position was that because of "our present happy constitution in Church and State" secured by the Revolution Settlement of 1690, hardships such as the presentation of ministers to parishes by lay patrons and the necessity of subscribing the Westminster Confession* of Faith could be endured. Their opponents regarded lay patronage as a serious infringement of the rights of the church, and many seceded from the "prevailing party" in the Church of Scotland in 1733, in 1761, and most notably in the Disruption* of 1843. The earlier Moderates were very critical of "man-made creeds and confessions" and tolerant of "infidels" such as David Hume and Lord Kames, but did not press for the removal of the legal requirement of subscription to the Confession. A later Moderate, George Hill, wrote a classic textbook of Calvinist doctrine, but the Moderates were more interested in science, history, and philosophy. They helped to found the Royal Society of Edinburgh. William Robertson's histories gained a European reputation, while Thomas Reid and the "Common sense" school of philosophers were influential in America. HENRY R. SEFTON

MODERATOR. The title given to the presiding officer of the various courts in Reformed churches. The term was used occasionally by John Calvin, but with the formation of the Reformed Church of France in 1559 it became the formal title of the chairman of official church gatherings. It seems to have been adopted in order to emphasize the equality of all presbyters. It was adopted by the Scottish Reformed Church in 1563 and has been generally employed by all Presbyterian bodies since. The attempts at different times to have permanent moderators, sometimes bearing the title of "bishops," has usually been successfully resisted as being contrary to the belief that Jesus Christ alone is head of the Church. W.S. REID

MODERNISM. The term was used of a movement within the Roman Catholic Church at the beginning of the century which accepted biblical "higher criticism" and reacted against Scholasticism and traditional Roman Catholic dogmatics, regarding dogmas only as symbols of high moral value. The leading figures of the movement were A.F. Loisy,* who advocated the Wellhausen "reconstruction" of the OT, F. Von Hügel,* and G.H. Tyrrell.* The movement was condemned by the encyclical *Pascendi* of Pius X in 1907. At about the same time a similar movement was formed in the Church of England, centered around the Modern Churchman's Union.

Used more loosely, "modernism" has been used in a derogatory sense to characterize the varieties of post-Kantian theology that have become popular in Protestant churches during the last century or more. These have uniformly adopted a "higher critical" attitude toward Scripture and toward the very idea of revelation as providing men with knowledge of God. The historic Christian faith, embracing Creation, Fall, and gracious redemption through Jesus Christ, was abandoned. In its place successive attempts have been made to reconstruct the Christian faith along largely ethical lines in accordance with "modern findings" of science and history, and to understand the progress of the kingdom of God simply in terms of social and political amelioration. In NT studies, "modernism" expressed itself in the quest for a de-supernaturalized historical Jesus in the gospels. The term "liberalism" is often used interchangeably with "modernism" even though "liberal" attitudes in Protestantism considerably antedate the rise of modernism.

Theologically the source of modernism is largely to be found in the work of F.D.E. Schleiermacher* and A. Ritschl* who followed Kant's* strictures on traditional metaphysics and were in turn followed by a host of popularizers such as R.J. Campbell in Britain and H.E. Fosdick* in the USA. Other sources lay in S.T. Coleridge, T.H. Green and the Broad Church Anglicans (England), the Cairds (Scotland), and later New England Theology* (USA).

Antimodernist attitudes and arguments are represented (in various phases) by Tractarianism,* the "Downgrade" controversy among British Baptists, and the publication and wide distribution in the USA of *The Fundamentals** (1909-15). American "fundamentalism," though antimodernist in its stance, is not to be identified with historic Protestantism due to its anti-intellectualism and its willing cultural isolation. The most brilliant analysis and indictment of modernism from the standpoint of the historic Reformed faith is probably J.G. Machen's* *Christianity and Liberalism* (1923).

World War I, Karl Barth,* and the rise of the "biblical theology" movement with its more constructive attitude toward Scripture brought about the decline of modernism in its "classic" form, though some would regard Barthianism as proceeding on essential modernist presuppositions. Many, such as Reinhold Niebuhr,* working on "modernist" assumptions about Scripture and Christian theology, have adopted less optimistic views of human nature and culture. Since the decline of modernism, those who attempt modern reconstructions of the Christian faith in accord with recognizably post-Kantian premises, such as P. Tillich* and J.A.T. Robinson, prefer to think of themselves as "radical" theologians.

PAUL HELM

MOFFAT, ROBERT (1795-1883). Scottish missionary to Africa. Born in East Lothian, he had scanty education, but after conversion he was, after some hesitation, accepted by the London Missionary Society for work in Africa. There he went in 1816, and in 1825 settled at Kuruman,

Bechuanaland, which became the headquarters of all his activities for forty-five years.

Moffat saw his work as fourfold. (1) *Evangelization,* which he strongly believed must always precede civilization. Acting on this he made Kuruman a center from which Christian influence radiated over a wide area. When he left in 1870, a whole region had been Christianized and civilized, and many African Christian congregations, ministered to by trained African ministers, had been formed; (2) *Exploration,* in order to extend missionary work. In 1816 only the relatively small Cape Colony was known. The Orange River was the northern limit of partially known territory; the Kuruman River, on which Moffat's headquarters were established, was beyond that. By 1870 Africa was largely explored as far as and beyond the Zambesi, much of it by Moffat and his son-in-law David Livingstone*; (3) *Literature.* Through his complete mastery of Sechuana, he translated the whole Bible, composed hymns, and wrote books, providing the Bechuana Africans with a basis for education, tools for worship and study, and the beginnings of a literature; (4) *Civilization,* especially in agriculture. He introduced irrigation, the use of natural fertilizers, forest preservation, and new crops. In this as in other ways his work was largely preparing the way for others.

Complete consecration, perfect disinterestedness, shrewdness, simplicity of character, and unwavering faith in the power of the Gospel—these were some of the qualities which made Moffat a man of God and an outstanding Christian leader. Failing health forced him to leave Africa in 1870; he died in Kent, England, thirteen years later.

See his *Missionary Labours and Scenes in South Africa* (1842), and biographies by J.S. Moffat (1885) and W.C. Northcott (1961).

J.W. MEIKLEJOHN

MOFFATT, JAMES (1870-1944). Bible translator. Born and educated in Glasgow, he was ordained in the Free Church of Scotland in 1896. In 1911 he left parish work and became professor of Greek and NT exegesis at Mansfield College, Oxford. In 1915 he transferred to the United Free Church college in Glasgow to teach church history, and after twelve years went to Union Theological Seminary, New York, as Washburn professor of church history, where he took a leading part in the preparation of the Revised Standard Version. He was a prolific writer of books reflecting biblical criticism.

His fame rests on his single-handed translation of the entire Bible. His NT was published in 1913, the Old in 1924, and the whole revised and reissued in 1935, known popularly as the "Moffatt Bible." It was the first unofficial translation to acquire widespread readership, although regarded as somewhat literary. His OT relied overmuch on critical theories which were subsequently disproved by archaeological or philological discovery; this weakness prevented acceptance by evangelicals on either side of the Atlantic. He was also prone to alter the order of verses or chapters when the Hebrew seemed to him unintelligible; here too he jumped to hasty conclusions which are no longer tenable. His translation was nevertheless a great achievement which still led its field at the time of his death, and was not finally superseded until the New English Bible was completed.

JOHN C. POLLOCK

MOGILA, PETER (1596-1646). Metropolitan of Kiev. From a noble Moldavian family, he studied in Poland and perhaps at Paris, returned to take monastic vows, became abbot of a Kiev monastery in 1627, and was elected metropolitan in 1633. To him the Russian Orthodox Church owes a more progressive attitude toward education of both clergy and laity. He came under criticism because his policies involved Western emphases, including the teaching of Thomist thought. On the other hand, it was he who produced the *Orthodox Confession of the Catholic and Apostolic Eastern Church* which, accepted by Orthodox patriarchs and endorsed by the 1672 Synod of Jerusalem, outlines Eastern Orthodox doctrine against Roman Catholic and Protestant claims.

J.D. DOUGLAS

MOHAMMED, see ISLAM

MÖHLER, JOHANN ADAM (1796-1838). German scholar. Born at Igersheim, he read philosophy and theology at the Catholic Academy, Ellwangen, and then at Tübingen. He was ordained in 1819 and later was professor of church history at Tübingen from 1828 and at Munich from 1835. He became dean of Würzburg Cathedral just before he died from cholera, pneumonia, and general exhaustion. His principal works are *Die Einheit in der Kirche* (1825), *Athanasius der Grosse* (1827), *Symbolik* (1832), and *Neue Untersuchungen* (1834). He took seriously Schleiermacher, Hegel, and Schelling, which offended his more conservative colleagues. His efforts to understand and be understood by Protestants manifested his deep ecumenical interests. He was falsely accused of heterodoxy on infallibility and of being a precursor of modernism.

C.G. THORNE, JR

MOHR, JOSEPH (1792-1848). Composer of *"Stille Nacht."* Born at Salzburg, he was a chorister in the cathedral there, was ordained to the Roman Catholic priesthood in 1815, and held several parish posts near Salzburg. In 1828 he became vicar at Hintersee and in 1837 at Wagrein. His *"Stille Nacht"* was composed in 1818 for a Christmas Eve serve in Oberndorf, near Salzburg, and set for guitar accompaniment by the organist and schoolmaster, Franz Grüber (1787-1863). The carol became popular before publication when sung by wandering Tyrolese singers. At least five English versions of the words, each with three of the five original stanzas, are current. The most widespread in Great Britain, by S.A. Brooke, begins "Still the Night." The translation commonly used in the USA begins "Silent Night! Holy Night!"

JOHN S. ANDREWS

MOKITIMI, SETH MOLEFI (1904-1971). South African Methodist minister. Born near Quthing, Lesotho, he entered the ministry in 1931. He served as chaplain of the educational institution at Healdtown (1936-51) and warden of the institutions at Osborn (1952-61) and Bensonvale (1962-65). In 1941 he became an official member of the Methodist conference and was its first African president in 1964. His balanced leadership and quiet dignity won wide respect. He attended many ecumenical gatherings in South Africa and abroad, was the first African president of the Christian Council of South Africa, and a vice-president of the All-Africa Council of Churches. In a period of growing racial animosity he consistently advocated reconciliation and interracial cooperation. He always remained an effective preacher and a zealous evangelist, and he was a staunch upholder of ministerial standards.

D.G.L. CRAGG

MOLDAVIA, see ROMANIA

MOLINOS, MIGUEL DE (1640-1697). Spanish Quietist. Born at Muniesa of noble parents, he was educated at Coimbra and settled at Rome in 1663 where he became a noted priest and confessor and won the friendship of prominent ecclesiastics, including the future Pope Innocent XI. In 1675 he produced his famous *Guida Spirituale.* This work, deeply influenced by Neoplatonism and medieval mysticism, traces a path to perfection, the annihilation of the will, and oneness with God, to which all external observances, even the overcoming of temptation, are an obstacle. At once the word was attacked by the Jesuits as Jansenist and Quietist in character, but was approved by the Inquisition. In the early 1680s, however, its fame spread over the Christian world and its teachings were applied, with devastating results, in some religious houses. In 1685, at the instigation of Louis XIV, Innocent XI—who felt himself threatened because of his friendship with Molinos—was urged to arrest the authors, who in 1687 was tried and condemned and, although forced to recant, was immured for the rest of his life on a charge of immorality. Molinos's fate aroused deep sympathy in the Protestant world and intense anti-Jesuit feeling among Catholics.

IAN SELLERS

MOMMSEN CATALOGUE. An early canon of biblical books, known also as "the Cheltenham List." Discovered by Theodor Mommsen in 1885 at Cheltenham, England, in a tenth-century Latin manuscript, it was first published in *Hermes* XXI (1886), pp. 142-56. It dates from 359 and appears to come from Africa. The OT list counts the *Wisdom of Solomon* and *Ecclesiasticus* among Solomon's books, and includes *Maccabees, Tobit, Esther,* and *Judith.* It has 151 Psalms. In the NT, four gospels are accepted in the order Matthew, Mark, John, Luke. Hebrews, James, and Jude are omitted, but Revelation is included. The scribe seems to demur at the assignment of three epistles to John and two to Peter, adding after the enumeration which he found "*una sola*" ("one only"). According to the number of verses indicated, three Johannine epistles and two Petrine were in the original list. The canon is thought to be of Western origin.

J.G.G. NORMAN

MONARCHIAN PROLOGUES. Short introductory statements which are prefixed to each of the four gospels in many manuscripts of the Vulgate. Also known as the *Arguments,* they contain brief accounts of each evangelist and his reasons for writing his account. These Latin introductions are involved, and their meaning is often remote and vague. The fact that these documents are known from antiquity as the Monarchian Prologues indicates that they have been held to date from the second or third centuries. But more recent critics have shown them to belong to the fourth century, and that perhaps they are in fact dependent upon the Anti-Marcionite Prologues.

MONARCHIANISM. The name is applied to a second- and third-century theological movement centered chiefly in Asia Minor and Rome, but also common elsewhere. The term "Monarchians" was coined by Tertullian in the third century. While the word can sustain an orthodox view of the Trinity, it usually described those who opted for a unipersonal rather than trinitarian view of the divine nature in order to preserve the unity of God.

Two forms of the doctrine are discernible. First, Adoptianist or Dynamic Monarchianism, which centers on the problems raised by Christology in early Christian times. In this view Jesus is regarded as a unique man who was divinely energized by the Holy Spirit (usually thought of as occurring at his baptism) and called to be the Son of God. Theodotus of Byzantium expounded such a view at Rome, about A.D. 210. Similar views were held by Paul of Samosata.* Much earlier the Ebionites* and Cerinthus* (a contemporary of the Apostle John at Ephesus) maintained that Jesus was a divinely energized Galilean. 1 John condemns this viewpoint (cf. 5:6).

Second, Modalistic Monarchianism, Patripassianism, or Sabellianism.* The incarnation of God the Father was put forward in an effort to maintain both the divinity of the Son and the unity of God. This view was influential at Rome about A.D. 200 through Noetus,* Praxeas,* and Sabellius. It was vigorously opposed by Tertullian* in North Africa and Hippolytus* at Rome. The Patripassian nickname relates to Tertullian's gibe that by his teaching Praxeas "put the Paraclete to flight and crucified the Father." The Modalist appellation concerns their representation of God as revealed at one time under the mode of Father, at another under the mode of Son, and at another under the mode of the Holy Spirit. According to Hippolytus, Noetus taught that if Christ is God, he is surely the Father, or else not God; therefore, if Christ suffered, then God suffered.

Dynamic and Modalistic Monarchianism represented erroneous early attempts to assimilate the empirical facts of the Christian faith associated with the person of Jesus Christ and the Pentecostal descent of the Spirit to an unrevised notion of unity. The facts of the biblical revelation demanded recognition to the full personhood of the Father, the Son, and the Holy Spirit. Only gradually did Christians acquire categories and a language adequate to the new revelation.

See INCARNATION; TRINITY; SUBORDINATIONISM.

BIBLIOGRAPHY: G. L. Prestige, *Fathers and Heretics* (1940); H. Bettenson, *Documents of the Christian Church* (1946); H.E.W. Turner, *The Pattern of Christian Truth* (1954); J.N.D. Kelly, *Early Christian Doctrines* (1958); B. Altaner, *Patrology* (1958).

SAMUEL J. MIKOLASKI

MONASTERY. The abode of a community of persons living secluded from the world, dedicated to a life of asceticism and prayer in pursuit of personal sanctification, generally united under a superior to obey a common rule by vows of poverty, chastity, and obedience. The term covers both eremitic (hermit) and cenobitic (communal) foundations for men or for women. Following its appearance in the fourth century, the Christian monastery became an important civilizing and evangelizing force—sometimes despite the intention of the monks—especially through the encouragement given to study and the copying of manuscripts. Traditionally the monastery has sought to be self-sufficient, incorporating all the buildings necessary for community life, including its own chapel and the cloister, from which outsiders are excluded, within a walled enclosure.

MARY E. ROGERS

MONASTICISM. The Greek term *monachos* at first probably meant "celibate, single," rather than "alone, solitary" (see ASCETICISM). Ascetics, especially women, tended to separate themselves from congregation as well as society long before monasticism proper began. Total withdrawal from the world, following Jewish and Christian traditions of wilderness spiritually and pagan conventions of "dropping-out" to escape social burdens, emerged in the East in the late third century with the hermits (eremites) or anchorites like Antony* of Egypt—who was not the first (even disallowing Jerome's Paul of Thebes) but the most influential. Retreat in pursuit of perfection was stimulated by growing laxity within the church—Hellenized, at peace, and imperially patronized—and by lay ambitions for the heroism of the martyr in face of increasing episcopal domination. The hermits, in Egypt mainly Coptic *fellahin*, abandoned both civilization and church, but as admirers and imitators sought them out, informal colonies developed, especially in the deserts of Nitria and Scete SW of the Nile delta, where in the fourth and fifth centuries a rudimentary corporate life was observed by the solitaries. Manual tasks predominated, learning remained minimal.

Cenobitic ("common life") monasticism was pioneered in Egypt by Pachomius* (d.346), who subjected his several communities to an elementary common "rule." Monks multiplied in these and in independent monasteries and around centers like Oxyrhynchus, and were forward in the eradication of heresy and paganism from rural Egypt. The spirituality of the Desert Fathers, rooted in Origen's* teaching, was preserved in collections of *Sayings of the Fathers (Apophthegmata Patrum)* and systematized supremely by Evagrius Ponticus* (d.399), who influenced Palladius* (the historian of monastic origins), John Cassian* (who transmitted anchoritic piety to the West), and later ascetic-mystical Byzantine theologians.

In Palestine, Antony's disciple Hilarion* (d.371) propagated anchoritism near Gaza, and Epiphanius,* future bishop of Salamis (d.403), founded the first cenobitic establishment nearby. Biblical sites attracted hermits and communities, and Jerome and Rufinus of Aquileia were associated with Roman matrons in nunneries at Bethlehem and Jerusalem late in the fourth century. A major Palestinian development was the *laura*, combining a chiefly eremitic regimen with a common subjection to one "Father." Euthymius the Great (d.473) and Sabas* (d.532) led famous *lauras.*

Monastic origins in Syria were independent of Egyptian models. Jacob of Nisibis (d.338) and Juliana Saba near Edessa (d.366/7) were prominent early anchorites. Syria's inveterate asceticism, latterly vitiated by the Manichaean-type dualism observable in the Messalians,* bred extreme, even suicidal, manifestations in eremitic stylites,* "browsers" (who lived like animals), and vagrant exiles. Cenobitism, resisted by Ephraem the Syrian,* eventually emerged through Egyptian and Manichaean influences. Syria's primary significance in monastic development, evident in Egyptian use of the Syriac *abba*, "father," extends to the missionary monasticism of the Persian Church and beginnings in E Asia Minor. Here the exaggerations of Eustathius* of Sebaste condemned at the Council of Gangra (c.343) as well as Messalian aberrations fostered suspicions largely overcome by Basil* of Caesarea (d.379). Abjuring solitude and ascetic athletics in favor of a "brotherhood" of love and service modeled on primitive ascetic groups, Basil guaranteed the ecclesiastical acceptance of monasticism in Asia Minor, e.g., in Constantinople in the 380s. Basil's informal Rules influenced most subsequent Eastern developments.

The independence and individualism of early Eastern monasticism were progressively eliminated through the discipline of rules and subjection to the church's hierarchy, notably by the canons of Chalcedon. The monks were prominent in the Origenist controversies, intervened tempestuously in the fifth-century christological disputes, and became Byzantine Church's "democratic front."

Monasticism came to the West from the East with travelers like the exiled Athanasius* and Jerome* and through accounts of Egyptian happenings. From the first its impact was felt in clerical and cultured circles as nowhere in the East. The clerical groups in N Italy around Eusebius of Vercelli* (d.371) and later Jerome and Rufinus* were followed by the championship of monasticism by Ambrose* in Milan and Augustine* in North Africa. (In North Africa the Circumcellions* had earlier largely cornered asceticism for the Donatists.* Similarly in Spain the heretical Priscillianists* discredited asceticism.) Jerome's ascetic propaganda attracted a following among the Roman artistocracy, though also ecclesiastical disfavor. Anchoritic ideals had most influence in Gaul through the early efforts of Martin of Tours* (d.397) and later John Cassian (d.435), and in

Celtic Ireland where in the sixth century through influences deriving in part directly from the E Mediterranean, the whole church assumed a monastic mold in which penitential rigors and (missionary) exile were prominent.

Monastic rules multiplied in the fifth and sixth centuries (cf. Augustine, Caesarius of Arles,* Columban), only to be overshadowed in due course by the Rule of Benedict* (c.540), now agreed to be largely based on the *Rule of the Master* (c.530). The Benedictine pattern, delivered from Benedict's isolationism, dominated developments in medieval Europe. In the Byzantine Church Basil was revered as the patriarch of monasticism, while Theodore of Studium* (d.826) was a significant later monastic organizer.

BIBLIOGRAPHY: H.B. Workman, *The Evolution of the Monastic Ideal* (1913); J. Ryan, *Irish Monasticism, Its Origins and Early Development* (1931); H. Waddell, *The Desert Fathers* (1936); K. Heussi, *Der Ursprung des Mönchtums* (1936); A. Vööbus, *History of Asceticism in the Syrian Orient* (2 vols., 1958, 1960); O. Chadwick, *Western Asceticism* (1958); G.B. Ladner, *The Idea of Reform: Its Impact on Christian Thought and Action in the Age of the Fathers* (1959), part 3; J. Leclercq, *The Love of Learning and the Desire for God: A Study of Monastic Culture* (1961); L. Bouyer, *The Spirituality of the New Testament and the Fathers* (1963); D.J. Chitty, *The Desert a City* (1966); M.D. Knowles, *Christian Monasticism* (1969); J. Ryan and P.J. Corish, *The Monastic Institute: the Christian Mission (A History of Irish Catholicism* I:2,3, 1972).

D.F. WRIGHT

MONICA (331/2-387). More correctly "Monnica," a Berber name. Mother of Augustine of Hippo.* The child of Christian parents, strongly influenced by her nurse, she married Patricius of Tagaste (in Numidia), a pagan with civic (curial) responsibilities, limited means and a disorderly temper, who became a Christian shortly before he died in 372. Their children included Navigius (converted with Augustine); a daughter who as a widow headed a convent in Hippo for which Augustine later wrote the basis of the "Rule of St. Augustine"; Augustine (probably the youngest); and possibly another son and daughter. Augustine's *Confessions* depict Monica as his spiritual mother who pursued him with prayers, tears, and admonitions to Carthage and Milan, *une femme formidable* of strong but simple piety, whose designs for his career sometimes conflicted with her purposes for his religious advancement. In Augustine's Cassiciacum writings she possesses an uncomplicated oracular wisdom, and she died at Ostia after sharing with Augustine a vision of (Neoplatonic?) mystic ecstasy. Part of her original epitaph inscription was rediscovered there in 1945. Her cult was promoted by the translation of her relics to near Arras (1162) and to Rome (1430). She remains a hagiographers' favorite.

D.F. WRIGHT

MONK. A word of uncertain origin, probably from Greek *monos* ("alone") through Latin *monicus* and Old English *munuc.* It denotes a member of a religious community living under the vows of poverty, chastity, and obedience. It has never acquired a clear-cut technical sense, but its use is properly confined to groups in which community life is an integral element, and not extended to later developments, e.g., the Mendicant orders.

MONOD, ADOLPHE (1802-1856). The greatest French Protestant preacher of the nineteenth century, he was the leader of *Le Réveil,** a powerful orthodox movement within the Reformed Church and other churches. At first Monod shunned the Geneva revival espoused by his brother Frédéric.* After theological study (1820-24), he was pastor of a French congregation in Naples (1825-27). Contacts with the Scot, Thomas Erskine,* led him to conversion and the acceptance of orthodox theology. Although he briefly held to a schismatic position while a pastor at Lyons, after serving as a theological professor (1836-47) at Montauban he defended the Reformed Church. His greatest influence was reached while pastor of the prestigious Oratoire Church in Paris (1847-56). His published sermons were very popular in France and abroad.

ROBERT P. EVANS

MONOD, FRÉDÉRIC (1794-1863). French Protestant pastor. Brother of Adolphe,* he contributed heavily to the success of the French Protestant *Réveil** as editor, pastor, and leader. His own faith was kindled under the teaching of Robert Haldane* while a theological student at Geneva. After briefly assisting his father, Pastor Jean Monod, in Paris, he became pastor of Oratoire Church in 1832. As editor for forty-three years of the periodical *Archives du christianisme au dix-neuvieme siècle,* Monod helped to formulate the orthodox views of an enlarging element of French Protestantism. He fought church subservience to the state and demanded a conservative Reformed creed. In 1849 Monod led dissident Reformed congregations into a union of free churches, based on strict orthodoxy, which still exists today.

ROBERT P. EVANS

MONOPHYSITISM. Monophysitism was a controversial issue in the Eastern Church causing lasting divisions. Included in this controversy are not just religious factors, but political ones also.

The fourth ecumenical council, at Chalcedon* (451), was called into session in order to pacify the spirit of conflict which arose in regard to the nature and person of Jesus Christ. Pope Leo of Rome had written his *Dogmatic Tome* for the Ephesus meeting held two years prior to the Chalcedon Council. Since the Ephesus conference turned out to be such a disgrace, often referred to as the Council of Robbers (see EPHESUS, ROBBER SYNOD OF), the Chalcedon session condemned the action in Ephesus and accepted Pope Leo's document as the rule of faith. The council went on to approve and to present the famous text of the dogma of Chalcedon.

This text issued at Chalcedon presented both sides of the Incarnation clearly, without getting involved in a philosophical explanation as to how the two natures of Christ are united. They proclaimed him "truly God and truly man." At the same time the council was careful to point out that part of the uniqueness of Christ was that He was one in person and substance, not divided into two persons.

Unfortunately, unity did not proceed from the Council of Chalcedon. Instead of ending the controversy, it was but the beginning of a dispute which would have an immediate effect on the Christian church over the next two centuries as well as a lasting effect. The opposition to the two natures of Christ became known as "Monophysitism." The name comes from the two Greek words *monos* ("only") and *fusis* ("nature"). The main emphasis of this movement was that there is but one nature in the Incarnation and not two. This, they felt, was the only way to protect the teaching of the unity of Christ's person. To ascribe two natures to Christ was a denial that man could gain ultimate oneness with God which was the goal of salvation. The result of this emphasis is to play down the manhood of Christ and relegate it to the realm of unimportance.

This reaction to orthodoxy which seems to suddenly emerge after Chalcedon in reality goes back to previous aspects of Christian history. Part of its roots can be traced to Christian monasticism as practiced in the Syro-Palestinian region and in Egypt. The monks were in constant battle against their own human weakness and sinfulness. To overcome one's humanity was to gain Christian victory. That which was identified as human had to be destroyed within one's character. For Christ to have a similar human nature as their own would be unthinkable to the Eastern monk.

Monophysitism was also a reaction to Nestorianism.* Nestorius, who became patriarch of Constantinople in 428, was opposed to the expression applied to Mary as "the mother of God." Mary, he felt, was the mother of Christ, but not the mother of the eternal Logos.* Nestorianism took the views of its founder a bit further than he intended. It pressed the distinction of the two natures of Christ to the extent of a double personality. Jesus was not the God-man, but instead, the God-bearing man. This led to a definite duality in the person of Christ.

Monophysitism was extremely popular among the laity of the Eastern churches. This mob popularity often found expression in many outbursts of violence such as in Alexandria, Antioch, and other church centers in the Middle East. Even to this day this issue on the nature of Christ is one of the main theological divisions between several of the Eastern churches.

See W.H.C. Frend, *The Rise of the Monophysite Movement* (1972). GEORGE GIACUMAKIS, JR.

MONOTHEISM. The belief that there is one, personal, transcendent God, who is the creator and ruler of the universe. Monotheism is apparently solely the outcome of revelation and is represented by Judaism, Christianity, and Islam. The claim for a primitive monotheism finds support among certain primitive tribes, but if true it has not influenced the world religions. There is no evidence for any development in nature religions from animism through polytheism to monotheism; the final goal of development seems always to have been pantheism, or exceptionally dualism. Equally, philosophical speculation and mysticism seem incapable of reaching the concept of a single, personal God, separate from matter.

Though the OT ridicules polytheism, it does not seek to establish monotheism by intellectual argument, but rather bases it on the experience of God's acts for those who trust Him alone. This attitude, displayed first by the patriarchs and Moses, is called "ethical monotheism," for it shows itself primarily in a life of trust; its rejection of polytheism bases itself primarily on the impotence of the heathen, nature deities. While popular Christianity has often compromised its claims to monotheism, the NT, the Christian creeds, and the standard theologians have throughout maintained a strictly monotheistic position, the reconciliation of it with the doctrine of the Trinity being regarded as a divine mystery.

BIBLIOGRAPHY: W. Schmidt, *The Origin and Growth of Religion* (1931); W.F. Albright, *From the Stone Age to Christianity* (1940); K. Barth, *Church Dogmatics,* part II (ET 1957), especially vol. I, pp. 440-61. H.L. ELLISON

MONOTHELITES. The question whether there were two wills (Dyothelitism) or one will (Monothelitism) in the Word made flesh was inevitable. The Chalcedon* settlement had insisted on the doctrine of the two natures in the incarnate Son. But some of the followers of Cyril* of Alexandria, believing that any suggestion of a duality must lead straight back to Nestorianism* were dissatisfied. These opponents of Chalcedon became known as Monophysites. Holding that the term "nature" and "person" are synonymous, they sought to secure the Cyrillian formula, "one nature in the Word made flesh." Cyril had accepted the Chalcedon terms, but in his endeavor to overcome any idea of a Nestorian juxtaposition of natures, had allowed himself to use the phrase "one nature." His Monophysite followers were less concerned to maintain the two-natures doctrine. Under the influence of Leontius of Byzantium* (sixth century), however, an interpretation of Chalcedon in a Cyrillian sense was effected. But dissatisfaction with the settlement continued and found expression in a renewed conflict over whether there were two wills or one possessed by the "two-natured" Christ.

Clearly the ethical complement of Monophysitism* is Monothelitism. Thus to allow one will would be to sanction Monothelitism. In order to clear up the issue—which, like the preceding controversy over Dyophysitism and Monophysitism, nearly tore the empire asunder—the emperor Heraclius instructed Sergius* (d.638), patriarch of Constantinople, to find a formula of mediation, which would pacify the Monophysites. Sergius thereupon advanced the thesis that the Word-made-flesh did all things through the action of a single divine-human energy.

The formula was however, opposed by Sophronius,* later to become patriarch of Jerusalem, and this compelled Sergius to restate his position. He then set aside the idea of "energy" and affirmed the existence of one will in the divine-human Christ. But to the upholders of the Chalcedon doctrine, two natures implied two centers of volition. A heated controversy over the issue followed until the Council of Constantinople* (681), the sixth ecumenical council of the church, ruled out—though probably wrongly—Monothelitism, and settled for Dyothelitism—for two wills—as being more in harmony with that of two uncompounded natures in Christ for which Chalcedon had declared. H.D. MC DONALD

MONTANISM. Shortly after the middle of the second century Montanus proclaimed the imminent advent of the New Jerusalem, the signal for which was to be a new outpouring of the Holy Spirit. The movement that followed had its chief strength in Phrygia.

Montanus, a new convert to Christianity, believed himself to be the appointed prophet of God, and his followers were encouraged to regard themselves as an élite of "spiritual" Christians. Preparation for the advent was to be preceded by withdrawal from the world. Special fast-days were called, and persecution was to be expected, even encouraged, so that the church would be a purified and fit Bride for the coming Christ.

Opposition to the movement was initiated by Pope Eleutherus and taken up by writers such as Miltiades* and Apollinarius.* In 230, the group was virtually excommunicated: the Synod of Iconium refused to recognize the validity of Montanist baptism. Montanism continued as an underground movement, chiefly as a protest against growing formalism and worldliness in the official church. The most illustrious product of Montanism was Tertullian.* The movement bears resemblance to the many illuminist and millenarian sects that flourished at the time of the Reformation and subsequently. H.D. MC DONALD

MONTEVERDI, CLAUDIO (1567-1643). "Creator of modern music." Perhaps the greatest musical genius of his generation, he excelled in all the musical forms of his day. If not an innovator, he had the ability to seize on new ideas and modes of musical expression and bring them to fruition. He composed church music both in the older Renaissance style *(prima prattica)* and in the new Baroque style *(secunda prattica)* with its freer use of dissonance and chromaticism, and he defended his methods eloquently in print against his critics. He succeeded Giovanni Gabrieli at St. Mark's in Venice, where Heinrich Schütz returned, a mature composer, to learn further from him. Much of his exciting church music had just recently been explored. In the secular field, he was the last great madrigalist and the first great writer of opera. J.B. MAC MILLAN

MONTFAUCON, BERNARD DE, see BERNARD DE MONTFAUCON

MONTGOMERY, JAMES (1771-1854). Hymnwriter. Born at Irvine, Scotland, son of a Moravian missionary, he worked in Yorkshire stores before joining the *Sheffield Register* as an assistant in 1792. Four years later he became editor, and for thirty-one years he continued to edit the renamed *Sheffield Iris.* Twice he was imprisoned for radical political opinions. He advocated foreign missions, the Bible Society, and the abolition of slavery. At various times he associated with Moravians, Wesleyans, and Anglicans. Over fifty of his 400 hymns were contributed to Thomas Cotterill's *Selection of Psalms and Hymns* (1819 ed.), whose publication led to the quasi-legalization of hymn-singing in the Church of England. Over thirty of Montgomery's hymns are still sung, including "Angels, from the realms of glory," "For ever with the Lord," "Hail to the Lord's Anointed," "Prayer is the soul's sincere desire," and "Stand up and bless the Lord." JOHN S. ANDREWS

MOODY, DWIGHT L(YMAN) (1837-1899). American evangelist. Born at Northfield, Massachusetts, he attended school there until he was thirteen, when he went to work. At seventeen he left Northfield for Boston where he secured employment in a shoestore. Though baptized by a Unitarian minister in Northfield, Moody began attending the Mount Vernon Congregational Church in Boston. Through the influence of his Sunday school teacher, Edward Kimball, he was converted to faith in Christ. Because of his ignorance of church doctrine, he was refused full membership in the church for a year, but was finally received in 1856.

Dissatisfied in Boston, he left for Chicago (1856) where as a traveling salesman he became a successful businessman. He joined Plymouth Church and soon rented four pews for men invited from the hotels and street corners. In 1858 Moody organized the North Market Sabbath School and induced John V. Farwell, a prominent businessman, to serve as superintendent. Two years later he decided to give up business and spend his full time in Sunday school and YMCA work.

During the Civil War he threw himself into work among soldiers while continuing his Chicago Sunday school. He soon established the undenominational Illinois Street Church and traveled often to national Sunday School conventions. At one of these he met Ira D. Sankey,* whom he enlisted as a musical associate.

In 1873 Moody sailed for the British Isles, his third visit. This two-year tour was destined to make him a national figure. Beginning inauspiciously at York, he and Sankey met minor successes in N England and then sudden major victories at Edinburgh and Glasgow. When they invaded London for a four-month period, total attendance at their meetings reached more than 2-1/2 million. Moody returned to the United States in triumph. After a brief time in Northfield he undertook campaigns in Brooklyn, Philadelphia, and New York. Careful preparation, cooperation of the churches, and generous publicity became marks of a Moody meeting. After success in these cities he moved on to preach to throngs in Chicago, Boston, Baltimore, St. Louis, Cleveland, and San Francisco.

Moody was not merely a preacher; he was a doer. In 1879 he established a school for girls, Northfield Seminary; two years later a school for boys, Mount Hermon School, followed. In the summer of 1880 he began a summer conference ministry on the grounds of the Northfield Seminary, and in 1886 he started the Chicago Evangelization Society, later to be known as the Moody Bible Institute.

Throughout his life, however, Moody's greatest contribution was evangelism. Some have established that he traveled more than a million miles and addressed more than 100 million people. In the midst of his last evangelistic campaign in Kansas City he became ill. A few days later, in December 1899, death overtook him.

See J.F. Findlay, Jr., *Dwight L. Moody: American Evangelist, 1837-1899* (1969).

BRUCE L. SHELLEY

MOORE, GEORGE FOOT (1851-1931). American OT scholar. Born in West Chester, Pennsylvania, he was educated at Yale College, then pursued graduate study at Union Theological Seminary, New York, and Tübingen, Germany. In 1878 he became minister of the Putnam Presbyterian Church of Zanesville, Ohio. He was named Hitchcock professor of Hebrew language and literature (1883) and lecturer in the history of religions (1893) at Andover Theological Seminary, of which he became president in 1899. He moved to Harvard University as professor of theology (1902) and was named Fotheringham professor of the history of religions (1904), which post he held until retirement. Firmly committed to the documentary hypothesis of the Hexateuch, he initially extended these notions to a detailed analysis of Judges (1895; "Polychrome Bible," 1898; Hebrew text, 1900). A transition in his work is marked by publication in 1914 of his *Literature of the Old Testament* and the first volume of his *History of Religions* (vol. 2, 1919). Thereafter he studied Judaism and its normative development, culminating in his *Judaism in the Christian Era* (2 vols., 1927). CLYDE CURRY SMITH

MOOREHEAD, WILLIAM GALLOGLY (1836-1914). American biblical scholar. Born in Rix Mills, Ohio, he was educated at Muskingum College and Xenia Seminary, was ordained to the Presbyterian ministry in 1862, and held pastorates in Ohio and Pennsylvania. For some eight years from 1862 he was a missionary to Italy, and on his return was pastor, 1870-85. From 1873 to 1914 he was professor of Greek exegesis and biblical literature in Xenia Seminary, where he was president also from 1899. An editor of the "Scofield Bible," leader in the Bible Conference and Student Volunteer movements, he was visiting lecturer at several Bible schools. He was also a dispensationalist and wrote many tracts and essays as well as biblical commentaries.

C.G. THORNE, JR.

MORALITY PLAY. This, like the mystery and miracle plays, was a distinct genre of religious drama which developed in the early fifteenth century with particular popularity in England, France, and the Netherlands. The morality plays perhaps developed from the sermon or homily and taught practical truth at a popular level. Their distinctive element is the portrayal of the cosmic struggle for the soul of man in his lifelong quest for salvation through the allegorical dramatization which personifies virtues (or the forces of good) and vices (the forces of evil).

Perhaps the earliest morality play is *The Castle of Perseverance* (c.1400-1420) which is characteristic of the early plays since nearly all the themes, later singly presented, are combined. *The Castle* presents the various stages of the life of Mankind (Humanum Genus) from birth and baptism through a life of sin till about forty, a period of about twenty years of sanctity, and finally old age and the dominance of the vice of covetousness, to death and a post death debate for Mankind's soul before God's throne. Other fifteenth-century morality plays are *The Pride of Life; Wisdom; Mankind* (which introduces comedy into this genre); *Mundus et Infans;* and the very popular Dutch play, *Everyman.* In the sixteenth century the themes are fewer, the plays more elaborate, and the messages more adapted to propaganda and secular motifs, as in Skelton's *Magnyfycence* (1533) portraying a political prince; John Bale's Protestant polemic *King Johan* (about 1538); or its Catholic counterpart *Respublica* (1553). The morality plays heavily influenced later English drama. For example, Marlowe's *Doctor Faustus* or Dekker's *Old Fortunatus.*

BIBLIOGRAPHY: A.W. Pollard, *English Miracle Plays . . .* (1890; many later eds.); W. Creizenach, *Geschichte des neuren dramas* (5 vols., 1893-1916) and "Miracle plays and moralities" in A.W. Ward and A.R. Waller (eds.), *The Cambridge History of English Literature,* vol. 5 (1910); E.N.S. Thompson *The English Moral Plays* (1910; rep. 1970); W.R. Mackenzie, *The English Moralities from the point of view of Allegory* (1914); E. Hartl, *Das Drama des Mittelalters* (1937); G. Frank, *The Medieval French Drama* (1954); F.P. Wilson, *The English Drama 1485-1585* (1969).

BRIAN G. ARMSTRONG

MORAL RE-ARMAMENT, see OXFORD GROUP

MORAL THEOLOGY. The application of the Christian teachings of natural law and revelation to ethical issues. It has largely been associated with the Roman Catholic Church as the infallible guide and interpreter of revelation and (sometimes) understood as casuistry in the pejorative sense of elaborate and petty legalisms. Moral theology can best be regarded, not as an attempt to provide detailed answers in advance, but as the provision of a cumulative wisdom derived from its sources which can be useful to Christians facing moral problems. These sources are Scripture, reason illuminated by faith, and the teaching of the church (specially as found in Thomas Aquinas* and Alphonsus Liguori*), but the weight attached to these sources varies, the Catholic tradition putting more emphasis than the Protestant on the last.

Some Protestants have written works of moral theology, notably the Anglican Jeremy Taylor* and the Puritan Richard Baxter,* while twentieth-century Anglicans in the Catholic tradition

—such as K.E. Kirk,* R.C. Mortimer, and Herbert Waddams have sought to revive interest in the subject. In general, Protestant theologians use the term "Christian ethics," which concerns itself largely with general principles, over against "moral theology," which applies general principles to particular cases. The Protestant tradition, reacting against the danger of legalism in moral theology, stresses the individual and subjective aspects of the moral life—personal devotion and obedience to God's will.

The history of moral theology suggests there is a recurrent danger of a type of legalistic casuistry which is divorced from the wholeness of Christian living and which, at its worst, appears concerned to discover how much moral responsibility can be evaded. The connection of moral theology with the confessional and the inevitable links with canon law regulations tended to foster the legalistic and external aspects in the Roman tradition, but recent writing has shown a trend toward a wholeness of approach in moral theology involving the related subjects of ascetic, sacramental, and pastoral theology. HOWARD SAINSBURY

MORAVIAN BRETHREN (Unitas Fratrum). Church of the United Brethren, reborn following the decline of the Bohemian Brethren* after the Thirty Years' War. Fugitives from Moravia found refuge on the estates of N.L. Count von Zinzendorf* in Saxony (1722). Joined by others from Bohemia, in association with some German Pietists, they worshiped at Bertholdsdorf Lutheran church under Pastor J.A. Rothe. About 1724 they decided to set up a church constituted according to the old Unitas Fratrum; an agreement permitted them to manage their own spiritual affairs while still worshiping at Bertholdsdorf (1727). Zinzendorf, gradually drawn into their affairs, became superintendent, and succeeding months saw a great spiritual awakening. Their orders of ministry were restored when David Nitschmann* was consecrated bishop by Daniel Jablonski*, bishop of the sole remaining branch of the Bohemian Brethren in Poland (1735). After Zinzendorf's death (1760) the movement was reorganized under a governing body, the "Unity Elders' Conference," whose influential president for many years was A.G. Spangenberg.* The Moravians led eighteenth-century German evangelicalism in the controversy with rationalism.

It was essentially a missionary movement. As early as 1732, Nitschmann and J.L. Dober* went to St. Thomas, Virgin Islands; work followed in Greenland (1733), North America (1734), Lapland and South America (1735), South Africa (1736), Labrador (1771), among Australian aborigines (1850), and on the Tibetan border (c.1856). The proportion of missionaries to home communicants has been estimated as 1:60 compared with 1:5000 in the rest of Protestantism. Moravian influence was a major factor in the Evangelical Revival in Britain. John Wesley owed his conversion largely to the Moravian Peter Boehler.* The movement grew in Britain and was legally recognized as a church in 1749. Concern for education had been inherited from the Bohemian Brethren, and numerous boarding schools were established in Germany, Holland, England, Switzerland, and America.

The Brethren did not always encourage the establishment of local churches, preferring to remain as "a Church within a Church"—e.g., in Lutheranism and Anglicanism. This "Diaspora concept" militated against the survival of the movement in many areas. It is an episcopal church with presbyterian government; the Unity Elders' Conference is appointed by the General Synod. It is divided into autonomous "Home" Provinces (Continental, British, and American), and "Mission" Provinces in transition to autonomy. Each congregation manages its own affairs subject to the general laws of the province. Worship combines liturgy with freedom in extempore prayer. In the threefold ministry of bishop, presbyter, and deacon, the bishop is the minister of ordination. Infant and believer's baptism are both provided, followed by confirmation. The Moravians had the earliest Protestant hymnbook. Doctrine is basically that of the Augsburg Confession,* though liberty of view is permitted in nonessentials. Strongly evangelical, the movement considers Scripture to be the only rule of faith and conduct.

BIBLIOGRAPHY: A. Bost, *History of the Bohemian and Moravian Brethren* (ET 1834); J.T. Hamilton, *History of the Moravian Church* (1900); J.E. Hutton, *A History of Moravian Missions* (1922); see also bibliography under BOHEMIAN BRETHREN. J.G.G. NORMAN

MORE, HANNAH (1745-1833). English writer and philanthropist. She was born near Bristol and spent the early part of her life in the city, where she and her sisters had a successful school. An unexpected settlement gave her financial independence and enabled her to exploit a remarkable range of literary and artistic gifts, which were combined with marked administrative ability. The abilities were exploited differently in two phases of her life. In the first phase she was part of the London literary scene, in which she was much admired, and an associate of Samuel Johnson, Horace Walpole, Sir Joshua Reynolds, and above all of David Garrick, who aided the production of her plays. In the second phase of her life, from the 1780s, John Newton* became a strong influence and she was brought into close contact with the entire evangelical community centered on Clapham.*

Though her later philanthropic and evangelistic activities were centered on the Mendip hills, her influence was widespread. She was much inspired by W. Wilberforce,* who together with Henry Thornton financed many of her activities. The local action was based on a Sunday school at Cheddar, to which was attached a school of industry, with training first of all in spinning and later in domestic service. The effort was extended throughout the Mendips. Hannah More's wider influence came through the use of her literary gifts in producing religious tracts, notably from about 1788 when she aimed at producing cheap tracts for a wide range of readers. The result, financed by Thornton, was the series of Cheap Repository Tracts. Though the connection is not clear, some of the inspiration of the Religious Tract Society can be traced to the success of Hannah More's work.

It is not surprising that a woman of Hannah More's ability encountered controversy. Partly her troubles arose from personal determination, for she was not easily thwarted once she embarked on a course of action. Hence the significance of William Cobbett's description of her as the "Old Bishop in Petticoats." Subsequent commentators have been as critical, especially because she failed to denounce many injustices and in particular assumed that the existing social structure was divinely ordained. She held that, though the condition of the poor should be relieved as far as possible, they had to accept their position and be comforted with the thought of future recompense. But that was an attitude shown by many, especially as fears of revolution became widespread at the end of the eighteenth century.

Hannah More never married but, following the custom of the time, assumed the designation of "Mrs."

See W. Roberts (ed.), *Memoirs of the Life and Correspondence of Mrs. Hannah More* (1834); and biographies by C.M. Yonge (1888) and M.G. Jones (1952). R.H. CAMPBELL

MORE, HENRY (1614-1687). English philosopher and poet. Educated at Eton and Cambridge, he was one of the more famous members of a seventeenth-century group of English ministers, moralists, and scholars known as the Cambridge Platonists.* At Cambridge he was elected to a fellowship at Christ's College, where he remained until his death. Heavily influenced by Joseph Mede,* More rejected a rigorous Calvinism and read widely in Aristotle and the Scholastics. However, he felt they did not give a satisfactory explanation of the relationship of the soul to God, so he turned to mysticism and Neoplatonism. He came to believe that knowledge of the eternal was dependent on moral perfection achieved through subduing egoism. On this basis he tried to defend Christianity against its greatest "enemies": atheism, Roman Catholicism, and "enthusiasm." His writings include *An Antidote Against Atheism* (1653); *Conjectura Cabbalistica* (1653); *A Brief Discourse of the Nature, Causes, Kinds and Cure of Enthusiasm* (1656); *The Grand Mystery of Godliness* (1660); and *An Antidote Against Idolatry* (1674).

See R.L. Colie, *Light and Enlightenment: A Study of the Cambridge Platonists and the Dutch Arminians* (1957). ROBERT G. CLOUSE

MORE, SIR THOMAS (1478-1535). English lord chancellor. Of a prominent London burgess family, he was educated at Oxford, at one time thought of becoming a priest, but eventually turned to law, although at all times he sought to live a very ascetic life. In 1504 he entered Parliament and subsequently rose to the position of chancellor after the fall of Wolsey in 1529. He was knighted in 1521. Although a devout Roman Catholic, he was very much taken with the humanism of the time, as indicated by his large circle of friends such as Dean Colet, Erasmus, Holbein, and others who were prominent in literary and artistic circles.

Noted for his fairness and clemency as a judge, More also became greatly interested in social reform. Out of this concern came his book *Utopia*, in which he sought to describe an ideal state where there was no private property or money, but all things were in common. Religious freedom, with a few exceptions, was also maintained. Yet, his advocacy of religious toleration and his close connection with many of the Renaissance humanists notwithstanding, he was no Protestant, and he wrote a number of books against William Tyndale and Martin Luther. He may also have helped Henry VIII* write his defense of the seven sacraments. Because of his strong Roman Catholic beliefs he came into conflict with Henry over the latter's desire to have his marriage to Catherine of Aragon annulled—and because of his refusal to take the oath renouncing the authority of the pope. As a result he was executed. He was beatified by Leo XIII in 1886. W.S. REID

MORGAN, G(EORGE) CAMPBELL (1863-1945). Bible teacher and preacher. Born in a Gloucestershire village, son of a Baptist minister who had resigned his living to start a faith mission in a hired hall, he was reared in a religious atmosphere, preaching his first sermon at thirteen. Without academic training, he joined the staff of a Jewish school, learning much from the headmaster, a rabbi. After rejection by the Salvation Army and the Methodists, he was accepted by the Congregationalists as a fulltime minister and was pastor of many churches, including Westminster Chapel, London (1904-17 and 1933-45), and president of Cheshunt College, Cambridge (1911-14). Morgan traveled much, especially during 1919-32; his preaching and Bible expositions attracted great crowds with numerous conversions. His literary output of Bible notes, sermons, and commentaries was immense. R.E.D. CLARK

MORGAN, WILLIAM (1541?-1604). Welsh bishop and Bible translator. Born at Ty Mawr, Wybrnant, in the parish of Penmachno, Caernarvonshire, about 1541, he was the son of a tenant on the Gwydir estates. Graduate of Cambridge, he probably began his career as a clergyman in 1572 and served in a number of livings, including Llanrhaeadr-ym-Mochnant (1578-95?), the place with which his name is most closely associated. He was harassed by critics and personal enemies with the result that he was implicated in suits and countersuits in the Court of Star Chamber and before the Council of the Marches, 1589-91. In 1595 he was consecrated bishop of Llandaff and was translated in 1601 to St. Asaph, where his zeal to defend the privileges of his diocese brought him into bitter conflict with local magnates.

Although he had a troubled career, the evidence seems to reveal him as a man of principle and a conscientious promoter of the good of his flock, not least in his emphasis on the need for preaching. His greatest title to the gratitude of Welsh people, however, was his work as translator of the Bible. John Whitgift* warmly patronized the work, and it was printed in London and published in 1588. By any standard it is a superb piece of work. Although subsequently revised, it is substantially still the Bible used by Welsh readers.

Morgan was an ideal man for the task of translating. His training at Cambridge under such scholars as Immanuel Tremellius, and his familiarity with the work of translating already done on the Continent and in England, provided him with the necessary academic equipment. On the other hand, his deep roots in the classical tradition of Welsh literature give his prose a rare dignity. His Bible is the virtual basis of modern Welsh prose-writing and the foundation for modern Welsh Protestantism. In that way William Morgan's Bible has had a more profound influence on the creation of modern Wales than any other single book in the nation's history.

R. TUDUR JONES

MORISON, JAMES (1816-1893). Founder of the Evangelical Union. Born at Bathgate, son of a Secession minister, he was educated at Edinburgh, licensed to preach by the United Secession Church* in 1839, and in the following year became minister of Clerk's Lane Church, Kilmarnock. His views on the universal nature of the Atonement soon led to suspension from the ministry of his denomination, but joined by three others, including his own father, he founded the Evangelical Union in 1843 (popularly known as Morisonians). His remarkable gifts as a preacher drew huge congregations, and this continued after he moved to Glasgow to become the first minister of Dundas Street Church, with which post he combined the principalship of the theological hall he established.

Initially regarded as a heretic, he became one of the most trusted and outstanding theologians of his time, a fact acknowledged by Glasgow University, which made him a D.D. in 1883. When he visited America he was warmly welcomed by the Cumberland Presbyterians,* who shared his dislike of strict Calvinism. His ministry is summarized on the monument erected to him in Glasgow. Among his published works was a remarkable exposition of Romans 9, a new edition of which was called for forty years later, and commentaries on Matthew (1870) and Mark (1873).

Early on, the Evangelical Union was joined by others leaving the Scottish Congregational Union. Four years after its jubilee, the Evangelical Union brought to its union with Scottish Congregationalists some ninety congregations, ranging from Orkney to Dumfriesshire.

See H. Escott, *A History of Scottish Congregationalism* (1960).

J.D. DOUGLAS

MORMONISM. On 6 April 1830, the "Church of Jesus Christ of Latter-day Saints," more commonly known as the "Mormon Church," was organized at Fayette, New York. Soon the group moved to Kirtland, Ohio, not far from present-day Cleveland. Under the leadership of Joseph Smith,* the community now moved to Jackson County, Missouri. Because of opposition encountered there, the group went on to Nauvoo, Illinois. After Smith was killed by a mob, most members of the Mormon Community followed Brigham Young,* the new leader, and settled in what is now Salt Lake City, Utah, where the church still has its headquarters. Independence, Missouri, is the headquarters of the largest of the splinter groups: the "Reorganized Church of Jesus Christ of Latter Day Saints," the current membership of which is about 200,000. The world membership of the Mormon Church at present is approaching the three-million mark, of which rather more than two-thirds reside in the USA. Mormon temples are found, not only in the USA, but in four foreign countries.

The Mormon Church uses, in addition to the King James Version of the Bible, the following sacred books as its main sources of authority: *The Book of Mormon; Doctrine and Covenants;* and *The Pearl of Great Price.* It is also believed that the president of the church may receive revelations for the guidance of the church as a whole. By thus adding to the Bible their own additional sacred books, Mormons have placed themselves outside of historic Christianity, which recognizes the Bible alone as the final source of authority.

An examination of the doctrines taught by the Mormon Church will reveal that they deny most of the cardinal teachings of the Christian faith. Mormonism rejects the spirituality of God, claiming that God the Father has a body of flesh and bones as tangible as man's. Further, it is taught that there are a great many gods in addition to the Father, the Son, and the Holy Spirit; these gods are in an order of progression, some being in a more advanced stage than others. Mormonism also teaches that the gods were once men, and that men may become gods. If a man faithfully observes all the precepts of the Mormon religion, he may advance to godhood in the life to come. One of the early presidents of the Mormon Church, Lorenzo Snow, summed it up: "As man is, God once was; as God is, man may become."

In the area of the doctrine of man, Mormonism teaches man's preexistence. All men existed as spirits before coming to this earth. This preexistent life was a period of probation. Those who were less faithful or less valiant than others during this period of probation are born on this earth with black skins. In Mormon teaching, further, the fall of man is considered a fall upward! If Adam had not eaten the forbidden fruit, he would have had no children; because he ate the fruit, man is now able to propagate the race. "Adam fell that men might be; and men are, that they might have joy" (*2 Nephi* 2:22-25).

Though Christ is called divine in Mormon teaching, his divinity is not unique, since it is the same as that which any man may attain. Christ's incarnation, too, is not unique, for all the gods, after having first existed as spirits, came to an earth to receive bodies before they advanced to godhood. Christ is said to have made atonement for our sins; what this means, however, is that Christ earned for all men the right to be raised from the dead. On the doctrine of salvation, Mormon teaching says that justification by faith alone is a pernicious doctrine which has exercised an influence for evil. One is saved through faith in

Christ (plus faith in Joseph Smith), but especially through works. By the works one does in this life he merits his salvation.

The entire Christian Church is said to have been apostate until 1830, at which time it was restored under the leadership of Joseph Smith. The Mormon Church, therefore, claims to be the only true church. Baptism is said to be absolutely necessary for salvation; it must be done by immersion. Though infant baptism is rejected, Mormon children are usually baptized when they are eight years old. The Lord's Supper is administered weekly, though water is substituted for wine.

Mormons believe in a literal millennium during which Christ will reign over the earth from two capitals: Jerusalem and Independence, Missouri. In the final state the devil, his angels, and a small portion of the human race will be consigned to hell. Most human beings, however, will be assigned a place in one of three heavenly kingdoms: the celestial, the terrestrial, or the telestial.

BIBLIOGRAPHY: Apart from the books mentioned above, works from Mormon sources include J.E. Talmage, *A Study of the Article of Faith* (1899); B.R. McConkie, *Mormon Doctrine* (1958); J.F. Smith, *Answers to Gospel Questions* (3 vols., 1958) and *Doctrines of Salvation* (3 vols., 1960).

Books by non-Mormon authors include J. and T. Tanner, *The Case Against Mormonism* (3 vols., n.d.); A.A. Hoekema, *The Four Major Cults* (1963); W.J. Whalen, *The Latter-Day Saints in the Modern World* (1964); W.R. Martin, *Mormonism* (1968).

ANTHONY A. HOEKEMA

MORNAY, PHILIPPE DE (Seigneur du Plessis-Marly; Duplessis-Mornay) (1549-1623). French Huguenot* leader. Born at Buhy in Normandy and originally intended for the priesthood, he adopted Protestantism upon his father's death (1559), largely through his mother's influence. He excelled in classical studies at the University of Paris (1560-67), then was a traveling scholar at the universities of Geneva, Basle, Heidelberg, Padua et al. (1567-72). This wide experience may explain his characteristic tolerant and broad-minded spirit. Returning to France convinced that the nation's foreign policy must be anti-Hapsburg, he associated with Coligny* and narrowly escaped the St. Bartholomew's Day Massacre.* He fled to England, but returned in 1573 counseling moderation. In 1576 he married the remarkable Charlotte Arbaleste, shortly thereafter entering the serving of Henry IV* (of Navarre) as soldier, diplomat, and adviser. He served Henry so brilliantly he became known as the "Huguenot Pope." In 1589 he was appointed governor of the Huguenot stronghold of Saumur where he founded the greatest of the Huguenot Academies (1603). Bitterly disappointed by Henry IV's conversion (1593), he served him faithfully till callously disgraced in 1600. He was instrumental in the drafting of the Edict of Nantes* (1598). In 1600 he retired to Saumur, but continued to exert a powerful, moderating voice in the turbulent Huguenot affairs till he died at his castle of La Forest-sur-Sèvre.

BIBLIOGRAPHY: *Mémoires et Correspondance* ... (4 vols., 1624-25); J. Ambert, *Duplessis-Mornay, études historiques et politiques* ... (2nd ed., 1848); N. Weiss, *Du Plessis-Mornay comme théologien et comme caractère politique* (1867); H.M. Baird, *The Huguenots and Henry of Navarre* (2 vols., 1886); R. Patry, *Philippe du Plessis-Mornay: Un Huguenot homme d'État (1549-1623)* (1933): good bibliography.

BRIAN G. ARMSTRONG

MORONE, GIOVANNI (1509-1580). Bishop of Modena. While bishop of Modena from 1529 Morone became close friends with the group of Catholic reformers gathered around Pole,* Contarini,* Cortese,* and Flaminio. Pope Paul III sent Morone to Germany as a nuncio in 1536. After attending Ratisbon with Contarini in 1541, Morone submitted an important memorandum to Rome. When he became a cardinal in 1542, heresy broke out in Modena. The academy signed a confession of faith prepared by Cortese and Contarini which involved Morone as bishop. Unpublished letters from his vicar in Modena prove that Morone knew what was happening and approved of its progress. As cardinal he became papal governor in Bologna at Contarini's death. From 1553-55 he studied the English problem and attended the Diet of Augsburg (1555). Paul IV seized Morone and cast him into the Castel Sant' Angelo during 1557-59. As president at the Council of Trent* during 1563, Morone made possible the effective reforms of the church. His latter work in Germany deserves a full-scale study. Among other tributes in his founding of the German College in Rome (1552) and his work at the 1576 Regensburg Diet.*

MARVIN W. ANDERSON

MORRIS, GEORGE FREDERICK BINGLEY (1884-1965). Anglican bishop. Born in Edinburgh and graduate of Cambridge, where he was active in student evangelistic work, he led a pioneer party of the Africa Inland Mission into the NE Belgian Congo in 1913, later becoming field director of the AIM in the Congo and West Nile Uganda. From 1932 he served the Bible Churchmen's Missionary Society in Morocco, followed by some years of parish work in England before consecration by William Temple* in 1943 as bishop in North Africa. He resigned in 1954 to become rector of Christ Church, Hillbrow, Johannesburg, and the following year accepted election as bishop of the Church of England in South Africa which had long been without episcopal ministrations. His action was condemned by the archbishop of Canterbury, who claimed that he had put himself out of communion with the Church of England (which body was later to insist on reordination of a clergyman ordained by Morris). This did not modify Morris's stand; in 1959, acting alone, he consecrated Stephen Bradley, who continued as bishop in South Africa after Morris's death.

D.G.L. CRAGG

MORRISON, ROBERT (1782-1834). Missionary to China. Born in Northumberland of Scottish Presbyterian artisan parents, he was converted when an apprentice at Newcastle. He educated himself and in 1802 went to a Dissenting academy near London, became a Congregationalist, and offered to the London Missionary Society, then looking for someone to translate and distribute the Bible in the almost totally closed Chinese Empire. Morrison learned the rudiments of a language nearly unknown in England, was ordained, and in 1807 sailed to Canton via the USA and Cape Horn, because the controlling East India Company refused to transport missionaries.

He could never get farther than the (trading) "Factories" at Canton. He saw scarcely any converts; he could remain only because he learned Chinese so well that he swiftly became the official Company interpreter. But his single-minded, rather dour devotion was eminently suitable for the lonely, discouraging task of laying the necessary foundation on which others could build when China opened to the West. He completed a translation of the whole Bible by 1818. His dictionary (1821) was the standard work until long after China opened fully. He wrote tracts and hymns.

When he obtained an assistant, William Milne,* in 1813, Morrison sent him to found an Anglo-Chinese college in Malacca, an important element in the eventual growth of missions in China. He encouraged work among expatriate Chinese and dreamed of opening Japan. On his one return to Britain, in 1824, now famous, he promoted understanding of China and concern for its evangelization. Dying alone in Canton eight years before missionaries were admitted anywhere else, this austere Scot is the father of Protestant missions in China.

See M. Broomhall, *Robert Morrison* (1924).

JOHN C. POLLOCK

MORTMAIN. The third reissue of Magna Carta (1217) forbade subtenants to donate lands fraudulently to "any religious house" and then receive them back at a rent. Such donations enabled subtenants to evade paying their superior lords feudal incidents like wardships and reliefs, legally due from the lands. These donations were known as gifts into mortmain. In 1279 Edward I's famous statute of mortmain forbade any gifts of land to ecclesiastical corporations, regardless of whether they were to be received back or not, or whether the superior lords had consented. In practice a license permitting such gifts was obtainable from the king for a fee, and thousands of licenses were granted. Mortmain Acts were also passed in 1290, 1391, 1531, 1736, and 1888. The 1960 Charities Act abolished mortmain. PETER TOON

MORTON, JOHN (d.1912). Founder of the Canadian Presbyterian Mission to the East Indians in Trinidad, brought in as indentured labor for the sugar industry after emancipation. Social and religious factors inhibited their integration into Trinidadian Christianity, and Morton was sent to open a special mission to them. From 1868 until his death he was the leader of the mission which included, besides Trinidad, work in Guyana, St. Lucia, Grenada, and Jamaica. His approach was based on both education and evangelism. Although the Indian churches are small, his emphasis on schools enabled the Indians for the first time to make their way in the West Indies.* Considerable assistance came almost inadvertently from the Trinidad government which in Morton's time began to encourage the Indians to stay in the island. On the question of indentured immigration, Morton was conservative, accepting the system as he knew it, and was particularly enthusiastic about independent Indian settlement.

GEOFFREY JOHNSTON

MOSHEIM, JOHANN LORENTZ VON (1694-1755). Lutheran church historian. Born in Lubeck of a Roman Catholic father and a Lutheran mother, he became a Lutheran, was educated in Lubeck and at the University of Kiel, where he became a faculty member. Later he moved to Helmstadt (1723), thence to Göttingen (1747) where he became chancellor. Theologically mediating between the Pietists and the Deists, he opposed both groups. Although he contributed to most fields of theology, his principal works were in church history, which he endeavored to make more scientific and objective. His interests extended even to Chinese church history. His most important work was *Institutiones historiae ecclesiasticae antiquae et recentioris* (1755), of which a number of English translations were published in the nineteenth century. W.S. REID

MOSLEM, see ISLAM

MOTET, see MUSIC, CHRISTIAN

MOTT, JOHN RALEIGH (1865-1955). Pioneer of the twentieth-century ecumenical movement. Born in New York, he was converted through the ministry of J.E.K. Studd at Cornell. He became in 1888 general secretary of the Student YMCA and chairman of the Student Volunteer Movement* for Foreign Missions. Thereafter he toured the world ceaselessly, promoting Christian missions, and (what he saw as strategically the same question) Christian ecumenism. He was instrumental in the convening of the 1910 Edinburgh* Missionary Conference, and he presided at most of the sessions and chaired the continuation committee. He thus became in turn chairman of the International Missionary Council* (1921), chairman of the second Life and Work* Conference at Oxford (1937), and vice-chairman of the provisional committee of the World Council of Churches* (1938). Finally in 1948 he became a co-president of the World Council itself.

Methodist and a layman, he possessed both vision and energy in enormous quantities. He helped into being not only world institutions such as those mentioned above, but also a host of national councils of churches, particularly in Asia and Africa, specialist studies in Christian mission (e.g., on the confrontation with Islam), and a whole series of initiatives in relation to Life and Work. His international ecumenical career covered over seventy years, and more than any other man he *was* the international ecumenical movement in the formative period from 1910 to 1948. His *Addresses and Papers* (6 vols.) was published in 1946-47.

See R. Rouse and S.C. Neill (eds.), *A History of the Ecumenical Movement 1517-1948* (2nd ed., 1967). COLIN BUCHANAN

MOULE, HANDLEY CARR GLYN (1841-1920). Bishop of Durham. Youngest of eight sons of the vicar of Fordington, Dorchester, he like the others was educated at home. After a brilliant career at Cambridge he taught at Marlborough (1865), was ordained (1867), and was curate at Fordington until going to Trinity as junior dean (1873). He became the first principal of Ridley Hall Theological College, Cambridge, in 1881, and Norrisian professor of divinity in 1899. In 1901 he succeeded Westcott as bishop of Durham. Moule was a convinced evangelical, but was able to understand other views. He represented evangelicals at the Round Table Conference on the Holy Communion (1900) and in 1908 chaired the missionary section of the Pan-Anglican Congress. He was closely associated with the Keswick Convention.* A profound scholar, he could speak and write for ordinary people. He wrote many hymns and poems, and his works include expositions and commentaries on nearly all the Epistles, as well as books on devotion and a down-to-earth work, *Outlines of Christian Doctrine.*

See biographies by J.B. Harford and F.C. Macdonald (1922), and J. Baird (*Spiritual Unfolding,* 1926). J. STAFFORD WRIGHT

MOULIN, PIERRE DU, see DU MOULIN

MOULTON, JAMES HOPE (1863-1917). Greek and Iranian scholar. Elder son of W.F. Moulton,* he entered the Methodist ministry in 1886 and in 1902 was appointed NT tutor at the Wesleyan College, Disbury, Manchester. In 1908 he became Greenwood professor of Hellenistic Greek and Indo-European philology in the University of Manchester. He gave the Hibbert Lectures for 1912 which were published as *Early Zoroastrianism* (1913), but he is best known for his work on the language of the NT. He was to have undertaken a NT grammar with his father, who died before anything was done. He himself managed to produce only the first volume, the Prolegomena to the *Grammar of New Testament Greek* (1906), but had written much of the second volume. The work was eventually finished by W.F. Howard and N. Turner. He also saw the publication of the first two fascicles of his *Vocabulary of the Greek Testament, illustrated from the Papyri and other non-literary Sources* (1914-15), which he undertook with G. Milligan (the whole being completed in 1930). His work was especially important for showing the kinship between the Greek of the NT and that of the recently discovered papyri, even if his conclusions were at times overstated. Moulton died from exhaustion after the ship on which he was traveling was torpedoed in the Mediterranean. R.E. NIXON

MOULTON, WILLIAM FIDDIAN (1835-98). Headmaster and biblical scholar. He was born at Leek, Staffordshire, of a strongly Methodist family. In 1858 he entered the Wesleyan ministry and became tutor in classics at Wesley College, Richmond, Surrey. In 1870 he translated from German Winer's *Grammar of New Testament Greek,* with corrections and a number of additional notes which revealed his competence as a grammarian. There was a new edition of this in 1876, and he had begun a complete revision of it at the time of his death. In 1870 he was appointed to the NT committee working on the Revised Version of the Bible, and was by far its youngest member. In 1875 he was appointed headmaster of the newly founded Leys School, Cambridge, and he stayed there for the rest of his life. In 1897 there was published *A Concordance of the Greek Testament* which he had undertaken with A.S. Geden, but his own part was restricted through illness and Geden received some help from his colleague's son, J.H. Moulton.* He was president of the Wesleyan Conference in 1890.

R.E. NIXON

MOUNTAIN, JACOB (1749-1825). First Anglican bishop of Quebec. He was educated at the Cathedral Grammar School in Norwich and at Caius College, Cambridge, and after being ordained in 1780 held several livings in England. Appointed bishop of the new diocese of Quebec in 1793, he quickly became the center of controversy, for he considered the Anglican Church to be the established church and held that the proceeds from the clergy reserves* were therefore at his disposal—which thinking was deeply resented by the other denominations. His jurisdiction extended over Upper and Lower Canada, and for over thirty years he faithfully administered his large diocese, traveling from one end to the other by sleigh, carriage, canoe, and foot.

ROBERT WILSON

MOWINCKEL, SIGMUND OLAF PLYTT (1884-1965). Norwegian biblical scholar. Son of a pastor, he became after theological studies (1902-8) junior curate at Egersund before studying OT theology and Assyriology at Marburg and Giessen (1911-13). His doctor's thesis (1916) was on Nehemiah. He taught OT theology at Oslo University from 1917, having professorial status, 1922-54. Some of his most influential work was done on the Psalms. *The Royal Psalms in the Bible* (1916) was followed by *Psalmenstudien* (1921-24), which placed the psalms in their cultic context, interpreting them in the light of this background. His *Offersang og Sangoffer* (1951) translated into English as *The Psalms in Israel's Worship* (1963). J.G.G. NORMAN

MOWLL, HOWARD WEST KILVINTON (1890-1958). Archbishop of Sydney. After graduation from Cambridge (1912) and ordination by Bishop E.A. Knox, he became a tutor at Wycliffe College, Toronto, where he taught, apart from one year as an army chaplain in France, until 1922. In that year he was consecrated as an assistant bishop for the diocese of West China, and succeeded William Cassels* as diocesan four years later. His ten years in China were full of travel, adventure, excitement, and development. He was captured and held by brigands; he was attacked and wounded by river pirates. His great contribution to West China was that he saw the need for a strong Chinese Church, and he planned accordingly. He was the bridge from the paternal rule of Bishop Cassels to the progressive aims of Chinese bishops, and he transferred the real control from the missionary conference to the diocesan synod. He went as far as the times would permit; but he looked still further ahead and planned for a division of the diocese so that Chinese bishops might be diocesans in their own right.

In 1933 Mowll was elected archbishop of Sydney, and enthroned in the cathedral in 1934. He entered at once upon a life of such unremitting activity that it left his clergy breathless. He proved himself a great administrator, with a flair for imaginative leadership and organization. The war years brought out all his latent strength in directing church enterprise to the spiritual, moral, and social welfare of men in uniform. He was never afraid to accept and shoulder responsibility, and his aptitude for leadership grew with experience. In 1947 he was elected primate of the Church of England in Australia, and this broadened his whole field of action. Always inspired by fresh thought and distant vision, his creative ministry brought a tremendous stimulus to all kinds of work both in Sydney and in Australia as a whole. He was a convinced and devoted evangelical in faith and churchmanship, but his remarkable gift for friendship enabled him to establish cordial relations with all kinds of people. His death marked the close of the greatest episcopate the Sydney diocese has ever seen.

MARCUS LOANE

MOZART, WOLFGANG AMADEUS (1756-1791). Austrian composer. The meteoric and tragic career of this unique musician began in Salzburg, where his father Leopold was court composer to the ruling archbishop. Taught by his father, and briefly by several outstanding musicians encountered during his childhood travels to the leading musical centers of Europe, Mozart learned to compose with a rapidity and sureness of technique that are almost beyond comprehension. Much of his childhood and youth was spent in musical tours planned by his ambitious father. In his later teens and early twenties he wrote much church music in the classical symphonic style that has since been looked upon as frivolous. As with the similar music of Haydn,* recent criticism has tended to judge it on its true musical merit rather than according to nineteenth-century canons of churchly propriety. In 1782 Mozart's relations with the archbishop, Count von Colloredo, reached an impasse, and he moved to Vienna, where his inability to obtain a secure income and the attendant worry and debt contributed to shorten his amazingly productive life.

A number of works written before leaving Salzburg, i.e., the "Coronation" Mass, the two vesper services, the *Miseracordias Domini*, and the "Munich" *Kyrie*, exhibit a depth and richness seemingly beyond a youth of twenty. Only three sacred works were written after he left Salzburg: the unfinished Mass in C minor, the exquisite miniature *Ave Verum*, and the *Requiem*, which others finished after his death. It can only be wished that circumstances had offered him an inducement to write more church music in his full maturity, when these few examples suggest what riches in this form he might have left to posterity along with his incomparable instrumental works and operas.

See H.C. Robbins Landon and D. Mitchell (eds.), *The Mozart Companion* (1956).

J.B. MAC MILLAN

MOZLEY, JAMES BOWLING (1813-1878). Anglican theologian. Educated at Grantham Grammar School and Oriel College, Oxford, he early acquired the reputation of a formidable if ponderous academic theologian. He was closely associated with the Tractarians* and was joint editor of the *Christian Remembrancer.* The Gorham case, however, caused him to reexamine baptismal theology more profoundly, and in *A Treatise on the Augustinian Doctrine of Predestination* (1855) and *A Review of the Baptismal Controversy* (1862) he conceded the validity of the Evangelicals' case, and was thereby estranged from many of his former colleagues. By this time he was agitated by the dire effects of unorthodox thinking within his church and, as a reply to Dean Stanley and his followers, gave the Bampton Lectures on Miracles in 1865, which defended miracles in a traditional manner reminiscent of Bishop Butler. Mozley was made a regius professor of divinity at Oxford in 1871 and died, a rather isolated figure, seven years later.

IAN SELLERS

MUGGLETONIANS. A sect taking its name from Ludowicke Muggleton (1609-98) who came under Puritan religious influence while working in London before the Civil War. Between 1640 and 1650 he came under a variety of further influences—e.g., the theology of J. Boehme* and the views of the Ranters.* He also formed an association with his cousin, John Reeve (1608-58), who shared similar views. They saw themselves as the "two witnesses" of Revelation 11, who received visions from heaven; they were to seal the elect in preparation for the forthcoming judgment of God on the world. Their views were set forth in a series of tracts. Their theology was antitrinitarian and dualistic (i.e., teaching the eternity of matter). The unforgivable sin was to disbelieve the "two witnesses." After Reeve's death, Muggleton and his followers continued to propound their views and to engage in controversy, especially with Quakers. Small groups survived into the nineteenth century.

PETER TOON

MÜHLENBERG, HENRY MELCHIOR (1711-1787). Father of American Lutheranism. Born at Einbeck, Hanover, son of a master shoemaker, he had a hard upbringing, especially after his father's death in 1723, but graduated from Göttingen in 1738, studied briefly at Jena, and was then appointed to the Weisenhaus at Halle. In 1739 he was ordained and appointed co-pastor and inspector of an orphanage at Grosshennersdorf. He had at first contemplated going as a missionary to the East Indies, but in 1741 was called to serve the United Lutheran Congregations of Pennsylvania (Philadelphia, New Providence, and New Hanover). He hastily learned as much English and American geography as he could.

Landing in America in 1742, he found the churches in a sorry plight and about to surrender to Count Zinzendorf's* far-reaching ecumenical schemes. These he frustrated after a short but inevitable struggle, then set himself to build up the three isolated congregations into a church. An expert linguist, a tireless traveler in almost impossible terrain, urbane yet full of evangelistic zeal, he summoned the first Lutheran Synod in America in 1748—the Evangelical Lutheran Ministerium of Pennsylvania, which supervised the growing number of churches throughout the middle colonies, many of these planted as a result of Mühlenberg's own missionary labors. A man of immense physical strength and organizing ability, he spent the Revolutionary years in semi-retirement, but managed to edit the Ministerium's *Gesangbuch* in 1787, just before his death.

IAN SELLERS

MÜLLER, FRIEDRICH MAXIMILIAN (1823-1900). Comparative philologist and Orientalist. Son of a lyric poet of the Romantic movement who took inspiration from the Greek War of Independence, Max was born at Dessau, matriculated at Leipzig (1841) where he studied Sanskrit, and thereafter went on to Berlin (1844) and the study of philology and metaphysics. At Paris in 1845 Eugene Burnouf introduced Müller to Zen Buddhism and recommended him to edit the *Rig Veda.* This was published at Oxford (6 vols., 1849-74), where he taught modern languages from 1850 and became professor in 1854. There he was to remain for the rest of his life. In 1856 he produced his *Essay on Comparative Mythology,* the first of the studies which sought origins of myth in natural, especially solar, phenomena, and correspondingly, the origins of "gods." He failed to receive the chair of Sanskrit (1860) because of foreign birth and liberal connections, but his reputation was such that there was created for him a chair of comparative philology (1868). His voice was of considerable importance in the further development of oriental studies and of comparative religion in England. Müller created the science of religion, giving Hibbert (1878) and four series of Gifford (1888-92) lectures. His *Life and Letters* was edited by his widow in 1902.

CLYDE CURRY SMITH

MÜLLER, GEORGE (1805-1898). Pastor, philanthropist, and leader in the Christian Brethren movement. He was born at Kroppenstadt, Prussia, and trained for the Lutheran ministry. After a dissolute early life, in 1825 he was converted during a prayer meeting in a private house. He came to London in 1829 to train for missionary service among the Jews. During a period of convalescence in Teignmouth, he met Henry Craik, a gentle, scholarly Scot of the same age who had been tutor to the children of A.N. Groves,* and through him made acquaintance with Groves's teaching, eventually sharing Groves's views on ordination and establishment. As a result, he amicably severed his connection with the Jews' Society and accepted a call to minister at Ebenezer Chapel, Teignmouth, and married Groves's sister, Mary. In 1832, Müller and Craik began a united ministry, first at Gideon Chapel and then at Bethesda Chapel, Bristol, where he was to remain until his death.

In 1834 he formed "The Scriptural Knowledge Institution for Home and Abroad," to stimulate education "upon scriptural principles," to circulate Bibles, and to help missionary work. Early in his life he had observed the orphan work of Auguste Francke* in Halle, and in 1835 he began in Bristol the orphanage for which he is chiefly remembered. This grew from a rented house to a great complex of buildings on Ashley Down, Bristol. In later years he traveled widely.

With Groves, Craik, and Robert Chapman of Barnstaple, Müller was a leading representative of the more moderate tendencies which grew into what is known as the Open or Independent Brethren, in contrast with the views of J.N. Darby* which developed into Exclusivism. In his early ministry he adopted believer's baptism, the weekly celebration of the Lord's Supper, and the principle of freedom to speak at meetings of the church. He renounced a regular salary and refused throughout the rest of his life to make any requests for financial support either for himself or for his philanthropic projects, even though sometimes he was penniless. His attitude was generally adopted in the movement. Through the wide support given by the Scriptural Knowledge Institution to work at home and overseas, he can justly be claimed as "the architect of the growth of independent Brethren" (F.R. Coad).

See A.T. Pierson, *George Müller of Bristol* (1905), and N. Garton, *George Müller and His Orphans* (1963). See also bibliography under PLYMOUTH BRETHREN.

J.G.G. NORMAN

MÜLLER, JULIUS (1801-1878). German Protestant theologian. Born in Brieg, Silesia, he studied law at the universities of Breslau and Göttingen, but changed from law to theology at Göttingen in 1821. Influenced by his friend and later colleague, F.A.G. Tholuck,* Müller began his career as a pastor. Ordained in Breslau in 1825, he served a parish in Schönbrunn in his native Silesia before accepting appointment as university preacher at Göttingen (1831-35). He was called as professor of dogmatics at Marburg (1835) and later at Halle (1839). Müller opposed both the theology of Schleiermacher and the Hegelianism of the Tübingen School. His largest and most important work, which he earned for him the nickname "Sünden-Müller," was a two-volume study, *Die christliche Lehre von der Sünde* (1839, 1844).

Müller described the essence of sin as selfishness *(Selbstsucht)* against God, occasioned by personal decision, and attempted to explain the origin of sin by positing an extratemporal fall. He supported the Prussian Evangelical Union of Lutheran and Reformed Churches and in his book, *Die evangelische Union, ihr Wesen and ihr göttliches Recht* (1854), attempted to draw up a formula of consensus to serve as the doctrinal basis for the Evangelical Church of Prussia. This formula united elements from the Old Lutheran and Reformed confessions, especially where those symbols were in agreement. Together with A. Neander and K.I. Nitzsch, he founded in 1850 the *Deutsche Zeitschrift für christliche Wissenschaft und christliches Leben.*

DAVID C. STEINMETZ

MULLINS, EDGAR YOUNG (1860-1928). Southern Baptist theologian and educator. Born the son of a Mississippi minister, he graduated in 1879 from Texas Agricultural and Mechanical College and in 1885 from Southern Baptist Theological Seminary. He intended to become a foreign missionary, but poor health altered his plans. He held pastorates in Kentucky, Massachusetts, and Maryland until 1899, then joined the faculty of his own seminary and subsequently became its president. He was president also of the Southern Baptist Convention from 1921 to 1924, the stormy years of the fundamentalist-modernist controversy that divided other Baptist groups, and of the financial crisis. For five years from 1923 he was president of the Baptist World Alliance.* His theology was clearly conservative, but he avoided the polemics of fundamentalism. His main interest was in apologetics. Paradoxically his standard work on systematic theology contains no section on ecclesiology. His major works include *Why Is Christianity True?* (1905), *The Axioms of Religion* (1908), *Baptist Beliefs* (1912), *The Christian Religion in its Doctrinal Expression* (1920), and *Christianity at the Crossroads* (1924).

DONALD M. LAKE

MUNGO, see KENTIGERN

MÜNSTER, SEBASTIAN (1489-1552). German biblical scholar. After studying in Heidelberg, Tübingen, and Vienna, Münster took vows as a Franciscan monk, but in 1529(?) he joined the Evangelical cause and assumed a teaching position at the University of Basle. He was an outstanding scholar of the Semitic languages, writing Hebrew and Chaldean grammars. His two-volume edition of the Hebrew Bible was printed in Basle in 1534 and 1535. It was supplied with a literal Latin translation and notes. Miles Coverdale is known to have used this edition for his English translation of the OT. Münster is remembered too for his German geography, *Cosmographia Universalis* (1544).

CARL S. MEYER

MÜNZER, THOMAS (before 1490-1525). A leader in the "Radical Reformation."* He was born at Stolberg in the Harz Mountains, but little is known of his early life. He attended the universities of Leipzig and Frankfurt an der Oder, then lived at Leipzig. He was there when Luther's disputation took place and seems to have made a favorable impression on the great Reformer. Münzer's interest in Protestantism led him to study Eusebius, Jerome, Augustine, the Acts of the Councils of Constance and Basle, and the works of mystics like Tauler.*

With Luther's approval he received a call to preach at Zwickau, where Egranus had introduced the Reformation. Münzer became acquainted with a radical group called the Zwickau Prophets,* and he preached in a violent way against the clergy. He emphasized the importance of the Holy Spirit's guidance and the need for lay involvement in the work of the church. This led to conflict with Egranus, and Münzer was deposed. Next he appeared at Prague, where he issued a statement calling upon the people of the land of Hus to help him bring in a new age. When little interest was expressed in his message at Prague, he left for Allstedt, Germany, where he became parish priest. Here he introduced a series of liturgical reforms which attracted a good deal of attention. He also organized his followers into bands ready to take up arms for the cause of the Gospel. In May 1524 some of these disciples destroyed a shrine near the city. This action coupled with the warnings of Luther caused Duke John and Duke Frederick of Saxony to order Münzer to preach before them. In his sermon based on Daniel 2, he demanded that the rulers use force to establish the true Gospel. The rulers ordered him to a hearing at Weimar, and when the city council joined the opposition against him, he left Allstedt. After some months in S Germany he appeared at Muhlhausen, where he preached to the townsmen and helped to involve them in the Peasants' Revolt.* This led to the defeat of the rebel forces and Münzer's execution.

Although his revolutionary ideas were discredited, his teaching against infant baptism and his emphasis on the inspiration of the Holy Spirit influenced other Anabaptists. Originally a follower of Luther, he later turned against him when he failed to support a social revolution. He called Luther such names as "Brother Fattened Swine," "Dr. Liar," "Brother Soft Life," and "Pope of the Lutheran Scripture Perverters." Marxist historians emphasize Münzer because he anticipated later social revolutionaries.

BIBLIOGRAPHY: F. Engels, *The Peasant War in Germany* (ET 1926); G. Rupp, "Thomas Münzer, Prophet of Radical Christianity," *BJRL* 48 (1966), pp. 467-87; E. Gritsch, *Reformer Without a Church* (1967); H.J. Hillerbrand, *A Fellowship of Discontent* (1967), pp. 1-30.

ROBERT G. CLOUSE

MURATORIAN CANON. A fragmentary list of NT books known at Rome about 200 is called after its discoverer, L.A. Muratori, who in 1740 published the document in Milan. The text is in Latin, but with sure signs that it is a translation, most likely from Greek. The sense is not always clear, and the list is broken off certainly at the beginning and probably at the close.

The fragment attests the books which were received in the Catholic Church in the West and were authorized to be read out in public. Books to be excluded from the canon are also identified. The four gospels are present, Mark at least by implication at the (lost) head of the list. Differences in the gospels are admitted, but declared to be of no importance since they derive ultimately from the "one guiding Spirit." The Acts of the Apostles is from Luke, whose presence at most of the events he records is attested. Paul's letters are addressed to seven churches which, like the sevenfold church of the Apocalypse, signify the totality of Christendom. Letters sent to individuals (Philemon, 1 and 2 Timothy, Titus) deal with church discipline and are so to be accepted. On the other hand, spurious letters (like a letter to Laodiceans, recognized by Marcion) are to be refused. The Catholic epistles are accepted without discussion, as is the *Wisdom of Solomon.* The Apocalypse of John is recognized as acceptable, but the *Apocalypse of Peter* is not to be read in church. The *Shepherd of Hermas* may be read, but not in church worship, since it was of recent origin.

Attempts to identify an author of the canon—Hippolytus of Rome is the most promising candidate (so Zahn)—are not proven.

See R. McL. Wilson (ed.), *New Testament Apocrypha,* 1 (1963), pp. 42ff. RALPH P. MARTIN

MURILLO, BARTOLOMÉ ESTEBAN (1617-1688). Spanish painter. Born in Seville and an orphan at the age of ten, he first studied under a mediocre painter and earned a living by *sargas,* a cheap painting on a canvas, sold in country fairs. After a visit to Cadiz in 1640, he set out to study the great masters, getting no further than Madrid where he was befriended by Velázquez, through whom he was able to study Titian, Veronese, Tintoretto, Rubens, and Velázquez himself. He spent three years in Madrid, returning to Seville in 1644. He revisited Cadiz once only, to paint an altar for the Capuchins, but an accident caused him to go back to Seville, where he later died. As a member of an austere brotherhood, dedicated to serving the dying, pain and misery were for Murillo objects of pity and not curiosity; unlike other painters of his race, his genius is tenderness and affection. Like Rembrandt, he saw the language of the Gospel as the language of the people, and sought to portray the Gospel in human terms. Most of his work was done for convents in Seville, with a number of devotional paintings for individuals. His preeminence lies in his painting of the Immaculate Conception, which theme he treated over twenty times without repetition. These are among the most markedly feminine paintings in Spain. GORDON A. CATHERALL

MUROMA, URHO RAFAEL (1890-1966). Finnish pastor. His early career included work among seamen in New York and participation in the labors of the Finnish Missionary Society. Gradually he realized that evangelization needed an organization of its own. Thus in 1940 the Evangelical Lutheran Inner Mission Foundation was started. Muroma held that those who became Christians needed to be grounded in God's Word, and therefore instruction became an important part of the foundation's work. Muroma was leader of the foundation and the Bible school and also published a large number of books, including some in which he boldly criticized the modern trends both in theology and in the church.

STIG-OLOF FERNSTROM

MURRAY, ANDREW (1828-1917). South African Dutch Reformed leader. Born at Graaff Reinet, he was educated in Scotland and Holland, then ordained in 1848. He served in Bloemfontein (1850-60), Worcester (1860-64), Cape Town (1864-71), and Wellington (1871-1906), and was six times moderator of the DRC in the Cape Colony. Theologically conservative, he led the opposition to liberalism in the DRC during the 1860s. Mystically inclined, he was greatly influenced by William Law* and led a profound devotional life. He undertook frequent evangelistic tours in South Africa and addressed the Keswick and Northfield conventions in 1895. He wrote much for the spiritual guidance of converts, and initiated the Bible and Prayer Union, the holiness conventions, and the Student Christian Association. His interest in education issued *inter alia* in the Huguenot Seminary (1874) and the Mission Institute at Wellington (1877). He was the moving spirit in the missionary awakening which led to DRC missions in the Transvaal and Malawi, and he also supported the South Africa General Mission. The most famous of his 250 publications was *Abide in Christ,* which appeared in numerous translations. Murray was the most influential leader of his own church in the nineteenth century, and an evangelical Christian of international stature.

See J. du Plessis, *The Life of Andrew Murray* (1919). D.G.L. CRAGG

MURRAY, JOHN (1741-1815). Founder of American Universalism. He was born into a well-off, Calvinistic family in Hampshire, England. His parents moved to Cork, Ireland, where this emotional man became a Calvinistic Methodist* shortly before he rejected Calvinism for the Universalism preached by John Relly. He was excommunicated by the Methodists; his wife and child died. He migrated in 1770 to America and preached his Universalist ideas throughout New England. After serving as a chaplain in the Revolutionary War, he settled in Gloucester, Massachusetts, and organized a Universalist congregation. After founding another church in Oxford, Massachusetts, he served as pastor from 1793 to 1809 of the Universalist Church in Boston. Paralysis in 1809 stopped his preaching that all men would ultimately be saved. EARLE E. CAIRNS

MUSCULUS (Mäuslein), WOLFGANG (1497-1563. Reformer. Born in Dieuze, Lothringen, he studied among other places in the humanists' school in Schlettstadt, where he met M. Bucer.* In 1512 his family urged him to enter the Benedictine monastery near Lixheim. In 1518, while still there, he was sent a packet of Luther's books, perhaps by Bucer, that made him "the Lutheran preacher." He left the monastery in 1527 and came to Strasbourg, where he became Bucer's secretary, and deacon in the cathedral church under Matthew Zell. In 1531, through Bucer's recommendation, he went as preacher to Augsburg.

Here Musculus was part of the struggle between the Roman Catholics, the Lutherans, the Anabaptists, and the Bucerian version of the Reformed faith. Two of the major issues were the interpretation of the Lord's Supper and the question of the relationship of the magistrates to the church. Musculus stood with Bucer, and since he could not be reconciled to the Interim of Charles V (1547), he left Augsburg in 1548. Through H. Bullinger's* influence he was appointed professor of theology at the old Franciscan college in Bern in 1549 after preparing an acceptable doctrinal statement on the Supper question. He began his theology lectures where he had left off preaching in Augsburg—on Psalm 104. His extensive commentary on the Psalms was published in 1550.

Bern was in continual conflict with Geneva over the relationship of the church to the magistrates: who was to control church appointments, who was to exercise church discipline, and who had control of church goods. Musculus supported the magistrates' right to control the churches and to order discipline. Of the major Reformation centers in Switzerland, Bern was the only one that did not sign the Zurich Agreement* of 1551. On many issues, however, Musculus insisted upon the principle of *adiaphora:* "let us be tolerant where nothing is unsuitable to the glory of God, the purity of our faith or the salvation of souls." During the Bern years Musculus published a number of commentaries, translations of various patristic sources, and his important *Common Places* of 1560 (ET 1562). The influence of these works were felt in the Low Countries, Hungary, and England, as well as in Germany and Switzerland.

ROBERT B. IVES

MUSIC, CHRISTIAN. Music is a form of communication, albeit nonverbal and hence without exact meanings. It has aspects analogous to spelling, grammatical structure, and in its more spacious forms syntax. It consists of sounds explicable by the natural phenomena of acoustics, which are no more inherently moral than mathematics. Nonetheless the power of music to call forth strong emotional response has many witnesses from antiquity to the present. Ancient mythologies credited it with supernatural powers, often therapeutic. The classical Greek concept of *ethos* clung to Western thinking to a degree that the rise of rationalism failed to erase except in part, and it still colors the thinking of many today.

Western music today exists in three "dimensions". The first of these is *rhythm,* the most essential to music of the three and the one that provides the strongest and least intellectual stimulus to the hearer. *Melody,* the second dimension, can hardly be said to exist without some recurrent stress patterns of a rhythmical nature. *Harmony,* as we know it in Western music since the Middle Ages, arose in the context of Christendom and until late Renaissance times found its theoreticians almost exclusively among the clergy. It was primarily the clergy who evolved our system of notation to preserve the revered music of the *Gregorian* tradition.

Gregorian chant, which is identical with *plainsong,* consists of a vast repertory of monophonic (i.e., in pure, unharmonized melody) music with Latin text. Every aspect of the Roman liturgy for both Mass and office is provided for. Contrary to tradition, Gregory I had little connection with any of this music, although he had much to do with the liturgical reforms of his day. The music comes from many sources, as we know today, from the simple chant formulas (psalm tones) that the early church inherited from the synagogues of the Middle East in Roman times; the music of the Ancient Byzantine Church; the Old Roman chant, the Gallican chant of pre-Carolingian times; and the elaborate and beautiful creations of unknown monks at such centers as St. Gall in the eighth and ninth centuries. Indeed, the repertory continued to grow even after this "golden age" was passed.

Although the Greeks employed a form of musical notation, the musicians of the church had no exact pitch symbols at their command until the advent of staff notation about the eleventh century. Much elaborate melody of earlier times was lost or greatly modified, since its retention depended upon the memory of the singers, with at best a few rhetorical accent symbols (staffless neumes) above the words to refresh the executant's mind. It seems reasonably clear to scholars now that the bulk of the melodies that come down to our day do so in the form they assumed in the Frankish domains of Charlemagne and his heirs, rather than in the Rome of Gregory's time. Plainsong is preserved in nonmetrical neumes upon a four-lined staff, a system referred to as "Roman choral notation."

Gregorian chant is essentially sacerdotal music. It was evolved for the use of trained singers: monks in the canonical hours of the abbeys; choirs and celebrants in princely chapels, larger city churches, and cathedrals. It exhibits a great variety of forms: simple recitation tones for prayers, Epistles, and Gospels, with their slight terminal inflections; the chants for the Psalms and canticles; the slightly more elaborate *antiphons* that relate the Psalms to the ecclesiastical calendar; the largely syllabic tunes of the *hymns* of the office and the *sequences* of the Mass; the luxuriantly melismatic melodies that abound in *graduals* and *alleluias* of the *proper;* and in many settings of the *Kyrie* and *Agnus Dei* in the *ordinary.* There are also the *responsaries, processionals,* and *tracts*—all with characteristics of their own.

Other systems of Christian chant must be mentioned. In Milan there persisted the Ambrosian chant, which is somewhat similar to but distinct from the Gregorian. The Byzantine chant of the Greek Orthodox Church presents another great system of considerable antiquity. In Russia there developed the Znamenny chant. The Syrian, Armenian, and Coptic churches likewise possessed other types. All represent priestly liturgical repertories. The Roman Church developed and encouraged *scholae cantorum* from before the time of Gregory I to train singers for the performance of the music of the liturgy.

The beginnings of polyphony, the simultaneous performance of more than one melodic line, came at least by the ninth century. At first the practice consisted of singing a plainsong with one or more voices duplicating the melody four or five scale-steps below or above. Soon the added part took on a melodic independence, and still later would sing several notes in the time space allotted to each note of the original melody. With such music, notation indicating exact pitches became indispensable. Such music was known as *organum* and seems to have been limited to the participation of a few skilled singers performing those sections of *graduals* and certain other parts of the *proper* of the Mass that were normally sung by a single voice. Organum developed primarily in France at Limoges and came to its highest development at Notre Dame in Paris about 1200, where a third, and even a fourth, part was added above the original plainsong.

In the thirteenth and fourteenth centuries the new art of polyphony made great strides. From organum grew the motet, in which each voice-part (usually three) had its own text. The lowest, or *tenor,* which was almost always a rhythmatized passage of plainsong, was probably performed instrumentally. Much of this music would seem to have filled a social or ceremonial function away from the church, but it must not be overlooked that a seemingly secular text could have had for that day a religious symbolism not apparent to a modern observer. Nevertheless, the fourteenth century has left us little church music. The thirteenth century also produced the polyphonic *conductus.* Unlike the motet, these pieces usually do not make use of plainsong tenors, and require all voice-parts to sing the same text as in a modern chorale. The texts are usually moralizing or ceremonial, and *conductus* seems to have been intended for religious or other ceremonial processions.

More popular religious music must surely have existed, but little that did not have ecclesiastical or political status was committed to parchment. Many princes and not a few clerics read not at all, and fewer could read music. Religious lyrics that are clearly intended to be sung spring from Francis of Assisi and his followers. These *laude* show to some extent the influence of the love-lyrics of the Troubadours. Similar songs of flagellants and pilgrims have survived in some number both in Italy and Germany, and in their wake such songs as would be called carols* today. Some such songs, often with mixed Latin and vernacular, of slightly later vintage, influenced Luther (e.g., *In dulce jubilo*) and were adapted into the vast Lutheran hymnody. In areas of Calvinistic influence they were not welcome.

In the fourteenth century, polyphonic settings of the choral sections of the ordinary of the Mass begin to appear, most of them anonymous. About 1360 Guillaume de Machaut, most widely known as a literary figure, composed an extraordinary and elaborate Mass in four-part counterpoint, containing the *Kyrie, Gloria, Credo, Sanctus,* and *Agnus Dei.* Curiously, no other such works are known until the beginning of the next century.

If the thirteenth and fourteenth centuries produced little music for the church, in the fifteenth century there began a great tide that continued to rise and swell on into the seventeenth and beyond. The second quarter of the fifteenth century saw a great change in the texture of music. The Englishmen, notably John Dunstable,* who were drawn to France in the wake of Agincourt brought to the Franco-Burgundian composers a new richness of sound. The musical intervals of the third and sixth increasingly replaced the austerity of late medieval sound, and the triad became the basis of harmony. Though cultural historians have generally written little of music, it no less than literature and the visual arts entered upon a period of unsurpassed creative vitality. With few exceptions, its composers of genius were churchmen who devoted to the music of the liturgy their highest powers, though secular music of merit was also abundant.

Composers from the French-speaking Netherlands almost completely dominated the European scene from about 1420 until the middle of the sixteenth century. The Vatican choir and the great churches and courts of the Italian states were full of them. Leonel Power, an Englishman, has left us the oldest known *cantus-firmus* Mass, in which the famous "*Alma Redemptorus Mater*" melody forms the tenor of each of its sections. It was the Burgundian Guillaume Dufay* (d.1474) who made the cantus-firmus Mass the touchstone of musical creative achievement for nearly a century and anomalously often employed a secular melody as his tenor! In his numerous works the techniques of the late Middle Ages are fused with the new spirit and manner of the Renaissance. The motet becomes a vehicle of high religious devotion, as in his dedicatory work for the Florentine Duomo, or in the touching simplicity of his office hymns. Ockeghem,* his junior by some thirty years, in his few carefully wrought masterpieces achieved a richer texture and deeper sonority. One of his Masses is in five voice parts of extraordinarily low range.

Of the many distinguished Franco-Netherlandish composers of the later fifteenth century, Josquin Desprez* more than any other individual displayed leadership and mastery in the composition of both Masses and motets, exhibiting to perfection all the technical procedures of the mature Renaissance. The four pitch-ranges of soprano, alto, tenor, and bass become normative, with imitative counterpoint becoming a pervasive structural principle. The sixteenth-century motet is a choral setting of a text, usually liturgical, but not from the ordinary of the Mass. Contrary to widely held opinion, unaccompanied performance was by no means the norm in the sixteenth century. Instruments as often as not supported the voices or substituted for missing ones.

A great multitude of talented composers were active at courts and in larger churches during the sixteenth century. The invention by Petrucci in 1501 of music printing from movable type, and the introduction of copper engraving later in the century, vastly accelerated the dissemination of music. The works by such geniuses as Lassus* and Palestrina* were known from Stockholm to Naples and from Warsaw to Lisbon, and were even carried to the Spanish colonies of the New World. Such music required the presence of skilled choirs of boys and men, and were largely beyond the possibilities of the ordinary parish, where plainsong prevailed if indeed any music was performed. The influence of humanism caused composers to strive for careful accentuation of the text, and the Counter-Reformation called for more homophonic part-writing in the interest of making the words more audible to the listeners. The Spaniard, Victoria, and the Italian, Palestrina, were among the leading figures influenced profoundly by the latter movement. Palestrina composed over one hundred Masses, and his music continued to be regarded as representative of the best sacred style long after his time.

The advent of the Reformation brought great changes in those countries where it triumphed. Luther introduced the chorale for the congregation to sing after the precedent of the Hussites in Bohemia. He adapted and translated from the traditional Latin hymnody and from the popular nonliturgical songs mentioned earlier. He also wrote hymns himself that were set to tunes composed or adapted by Johann Walther* and other musicians among his supporters. From this beginning developed a vast hymnic literature which reached its zenith with the work of Johann Crüger about 1650. Luther also encouraged the musical training of the clergy and of choirboys in parochial schools. Thus the Lutheran church developed a great musical tradition of congregational song and choral repertory. Along with these grew up an equally great literature of organ music, much of it based on the chorale melodies, and reaching its height of attainment with the creative genius of J.S. Bach* in the eighteenth century.

Very different was the course of music in those areas dominated by Calvin's teachings. Polyphonic composition together with instruments and "hymns of human composure" were totally rejected. The metrical versions of the 150 Psalms by Marot* and Beza* completed in 1562, with its store of tunes edited by Bourgeois* and others, was translated into German and Dutch and long remained the sole church song of the Reformed churches in Europe. The metrical psalms were sung only in unison and without the support of "popish" instruments. Only in Holland did the organ continue in use, and there amid much controversy. There the Genevan Psalter is still used, albeit with a recent redaction of the text—the "*Nieuwe Berijming*" of 1967.

Both England and Scotland followed the lead of Geneva with complete metrical psalters. The Scottish version of 1564 was replaced by a simpler version in 1650 which still survives. In England the "Old Version" (Sternhold and Hopkins) of 1562 was partially superseded by the "New Version" (Tate and Brady) at the beginning of the eighteenth century, but was fated to be eclipsed by the new hymnody of Isaac Watts* and his successors. Although the British psalters and their American counterparts had some fine tunes, they were much inferior in both text and music to the Genevan. In the sixteenth and early seventeenth centuries Huguenot and even some Catholic composers made a great number of polyphonic arrangements of the Genevan psalms. These were not for church use, but provided music-loving citizens with highly artistic music of an edifying character for their recreation. In the English-speaking countries the performance of the Psalms sank to an abysmal level, in which all were "lined out" most unmusically to a handful of well-worn tunes, until the era of Watts and the Wesleys brought new life and reform.

After the Reformation in England, while metrical psalmody prevailed in the parish churches, the chapel royal and the cathedrals continued to share in the great age of Elizabethan polyphony until they were suppressed by the Puritan revolution. The Anglican liturgy found a place for the *anthem.* At first adaptations of Latin motets were used, but a generation of great composers headed by Tallis* and Byrd* evolved a new genre which is among the glories of the English choral heritage—the *cathedral service,* consisting of the canticles for Matins and Evensong, and the choral parts of the Communion service inaugurated a new tradition, to which Byrd, Gibbons,* and Tomkins contributed fine examples.

After the restoration of the monarchy, cathedral music was too dependent upon the vagaries of royal taste and suffered from ecclesiastical indifference during the Age of Reason. Apart from Purcell* and the Chandos anthems of Handel,* little of great artistic significance appeared in the English choral repertory for the church until comparatively recent times. The rise of Methodism in the eighteenth century brought congregational hymn-singing to England. The Wesleys (see especially WESLEY, CHARLES) received their impetus from their contact with the Moravians, whose tradition went back to the Hussites and was refreshed by German Pietism. They lacked, however, an outstanding composer, and their tunes were borrowed or adapted from many sources; many tended to be very florid, with many passing notes and verbal repetitions. Hymn-singing spread rapidly throughout Nonconformist groups, and by the early nineteenth century the Anglican Church could no longer resist the demand for hymns among its rank and file. Then an era of avid hymntune composition began, bringing in a simpler, largely syllabic type of tune.

As the century progressed, many tunes tended to depend more upon their harmonic part-writing and seductive chordal progressions than upon the virility of their melodies. These aspects of many of the tunes of Dykes* and Barnby* and much of the music of the influential *Hymns, Ancient and Modern* (1861) brought a strong reaction at the end of the century. The *Yattendon Hymnal* (1899) and even more the *English Hymnal* (1906; rev. 1933) exerted a wide and continuing influence upon hymnbook editors of the major denominations throughout the English-speaking world. Strong tunes with active melodies, often specified for unison singing, have continued to appear—notably those of R. Vaughan Williams.* In fact the hymntune is by far the most important type of church music in Protestant worship.

In the seventeenth century, both Lutheran and Catholic composers continued to produce a great deal for the church. Much of this was in the new *concertato* manner, with independent and important parts for instruments that both combined and contrasted with the voices. This style had its first important flowering in Venice. Schütz,* the greatest Lutheran composer before Bach, studied there with its greatest exponents at St. Mark's, Gabrieli* and Monteverdi.* From this style developed the Lutheran church cantata. Blended later with the *recitative* and *aria* of Italian opera, it reached its fulfilment in the church works of J.S. Bach. A by-product of the new Italian "drama per musica," or opera, was the *oratorio,* beginning its existence as religious opera, but developing into a type of work with a dramatic theme, usually of biblical origin, performed without the appurtenances of the stage. Carissimi* developed the form in a definitive manner toward the middle of the century. Schütz cultivated it in his old age in his *Story of Christmas* and other works. In the following century, Handel wrote his great succession of English oratorios, which led to the later works of Haydn,* Mendelssohn,* and others. Elgar, Vaughan Williams, Walton, and others bring it down to our own day. Oratorio is not church music, but rather belongs in the realm of edifying entertainment.

In the seventeenth and eighteenth centuries, as the influence of the church decreased and the rationalistic tide of humanism rose higher, fewer composers of first magnitude devoted a significant portion of their creative activity to church music. Indeed, J.S. Bach was the last truly great genius to whom music for the sanctuary was his major concern. The later eighteenth century, the Age of Reason, saw only muddled illogic in polyphony, which Rousseau likened to the simultaneous delivery of several speeches. The affective styles and symbolism of the past were repudiated. Clarity and natural simplicity were the paramount virtues of music. In this "classical" era much church music was written, particularly in the Catholic countries. The aristocratic chapels and metropolitan churches of the Austrian Empire resounded to symphonic Masses and vespers, with skilled soloists, choruses, and orchestras. The music was joyful and spritely, and as such was attacked by the Cecilian movement of the nineteenth century as indecorous if not blasphemous. Pope Pius X expressly forbade their use. Today the Victorian vogue for disparaging as secular and flippant the best music of this era is gone, especially in view of the current trends in religious music. During the Romantic era, religious music of all types continued to be written, but little of it aspired to the artistic significance of former times. Some of it was chaste, some theatrical as with Berlioz, much dolorous and sentimental.

The religious music of the United States largely parallels that of Great Britain, but has enjoyed a degree of freedom from official restraint. Here the folk tradition of the Baptists surfaced and reached print in the shaped-note publications of the South, and the singing-school movement of post-Revolutionary times produced simple congregational music with a distinctive flavor. The "Gospel song" carried to Britain by such figures as Ira D. Sankey* adapted the popular idiom of such writers as Stephen Foster to capture the ear of the unchurched multitudes, and became the normal style of large segments of the less institutionalized churches of Protestantism. This music possesses optimistic rhythms and exceedingly simple harmonies, and lends itself readily to highly improvisatory performance.

An account of the religious music since World War II is almost impossible at this proximity. It remains to be seen whether the present trend toward popularization in religious music will continue, with the acceptance of a closer and closer identification with pseudo-folk and "rock" idioms, not only among younger Christians of evangelical persuasion, but in the larger historic denominations and in the Roman Catholic Church.

While much functional church music which goes little beyond the idioms of the late nineteenth century continues to be written, many professional church musicians are composing in styles closer to the early twentieth century. That these are difficult for any but highly trained musicians must be conceded. "Serious" professional composers of secular art music have never before been so far removed from the general listening public. There is beyond question a need for devoted and gifted composers to provide leadership in this as in all eras, so that the scriptural mandate of singing "psalms, hymns and spiritual songs" may be fittingly fulfilled.

See HYMNS.

BIBLIOGRAPHY: W. Davies and H. Ley (eds.), *The Church Anthem Book* (1933); P.H. Lang, *Music in Western Civilization* (1941); G. Reese, *Music in the Middle Ages* (1946) and *Music in the Renaissance* (1959); M. Bukofzer, *Music in the Baroque Era* (1947); E.H. Fellowes, *English Cathedral Music from Edward VI to Edward VII* (1948); L. Ellinwood, *The History of American Church Music* (rev. ed., 1953); E. Routley, *The Music of Christian Hymnody* (1957), *Twentieth Century Church Music* (1964), and *The Church and Music* (1967); W. Apel, *Gregorian Chant* (1958); Baker's *Biographical Dictionary of Musicians* (5th ed., ed. N. Slonimsky, 1958; suppl. 1965); F.L. Harrison, *Music in Medieval Britain* (1958); E. Werner, *The Sacred Bridge* (1959); D. Stevens, *Tudor Church Music* (1961); W. Douglas, *Church Music in History and Practice* (rev. ed., 1962); I. Lowens, *Music and Musicians in Early America* (1964); E. Wienandt, *The Choral Music of the Church* (1965); R. Stevenson, *Protestant Church Music in America* (1966); *The Treasury of English Church Music* (5 vols., various eds., 1966); A.J.B. Hutchings, *Church Music in the Nineteenth Century* (1967); P. Le Huray, *Music and the Reformation in England* (1967); W. Apel and A. Davison, *Harvard Dictionary of Music* (rev. ed., 1969); E. Wienandt and R.H. Young, *The Anthem in England and America* (1970).

J.B. MAC MILLAN

MUSLIM, see ISLAM

MYCONIUS, FRIEDRICH (1490-1546). German Reformer. Born in Lichtenfels, he entered the Franciscan Order in Annaberg in 1510. He was transferred to the monastery in Leipzig and then to Weimar. He was an assiduous student of theology and the Scripture; the doctrine of predestination troubled him greatly. In 1516 he was ordained priest and with the onset of the Lutheran movement found himself greatly in sympathy with it. Feeling himself threatened, he fled the monastery (1524) and went to Zwickau. From there, on the invitation of Duke John he went to Gotha. Here he furthered ecclesiastical and educational reforms, winning the friendship of Luther and Melanchthon, and later of Justus Menius.

He won the respect of Elector John Frederick of Saxony. He participated in visitations in Thuringia, the Marburg Colloquy* (1529), the Wittenberg Concord* (1536), the Smalcald* synod (1537), and the Colloquy of Hagenau* (1540). In 1538 he went with Francis Burkhardt and George von Boyneburg to England for the dialogues with the English theologians, but was disappointed with Henry VIII's attitude. He was especially gratified to be instrumental in aiding the establishment of the Reformation in Ducal Saxony, particularly in Annaberg. He wrote a *Historia Reformationis, 1517-42*—a valuable account by a contemporary—and German tracts, among them *Wie man die Einfältigen, und sonderlich die Kranken, im Christenthum unterrichten soll.* His own spiritual struggles, the integrity of his character, and his irenic spirit make him one of the most appealing figures of the Reformation era.

CARL S. MEYER

MYCONIUS, OSWALD (1488-1552). Swiss humanist, Reformed minister, and theologian. Born Oswald Geisshäusler at Lucerne, he studied at Basle where he became a humanist friend of Erasmus* (who gave him the name "Myconius"). He taught classics at the canons' school in Zurich from 1516 and was the decisive voice in the call of Ulrich Zwingli* to Zurich as people's priest in 1519. From 1520 to 1523 he taught in Lucerne and Einsiedeln, then returned to Zurich. He quietly but solidly supported Zwingli in the work of reform. Though neither ordained nor possessing an academic degree, when Oecolampadius* died in 1531 he moved to Basle, where in 1532 he became *Antistes* (chief pastor) and professor of NT at the university, two crucial posts he held until his death. Myconius wrote the first biography of Zwingli and is the principal author of the First Basle Confession of Faith (1534), a firm but broadminded Reformed theological statement. He followed Oecolampadius on the separation of church and state, and on discipline, and took a middle position between Zwingli and Luther on the Eucharist problem, believing the two positions could and should easily be reconciled.

BRIAN G. ARMSTRONG

MYERS, FREDERIC WILLIAM HENRY (1843-1901). English writer, classicist, and psychical researcher. Born at Keswick and educated at Cheltenham and Trinity College, Cambridge, he taught classics at that university from 1865 until he became a school inspector in 1872. In 1882 he became a co-founder and president of the Society for Psychical Research. In its early days he helped collect a mass of alleged evidence for survival, much of it published in the *Phantasms of the Living* (1886). He made the first deep studies of the relations between hallucination, hypnotism, and mediumship, published posthumously as *Human Personality and Its Survival of Bodily Death* (2 vols., 1903). As a result of his great desire to put religion on an empirical basis, Myers drifted slowly away from his early faith in Christ as exemplified by his beautiful poem *Saint Paul* (1867).

R.E.D. CLARK

MYNSTER, JAKOB PIER (1775-1854). Danish bishop. From Pietistic circles, he turned his back on them in his youth, adopting the theological and political radicalism of the so-called Enlightenment. Afterward the influence of Kant and German Romanticism made him skeptical of rationalism. In 1803 he had a spiritual experience that led to personal conversion and acceptance of the Christian faith. He developed into an eminent preacher, attracting a large section of the "cultural élite," and through his works influencing still greater numbers. He was rector of Copenhagen Cathedral (1811-28), personal chaplain to the king from 1828, and bishop of Zealand from 1834 until his death.

By most of his contemporaries Mynster was regarded as the great central figure in Danish church life, standing between rationalists on one side, revivalists on the other. Though individualistic, he was a conservative, authoritarian champion of the state church in opposition to N.F.S. Grundtvig* and the Pietistic conventicle Christians. He even introduced compulsory baptism of Baptist children. He with H.L. Martensen* and the national church as a whole was later attacked fiercely by S. Kierkegaard.*

N.O. RASMUSSEN

MYSOS DEMETRIUS (1719-1750). Born in either Montenegro or Thessaloniki, he was probably a deacon sent from Constantinople to Wittenberg in 1559 by Patriarch Joasaph II to study at first hand the Reformers and their teaching. He stayed six months, studying under Philip Melanchthon, attending services and lectures, and studying the Reformers' creedal documents. Leaving Wittenberg for Constantinople in 1559, he carried with him a letter from Melanchthon* to the patriarch and also a Greek translation of the Augsburg Confession. Mysos Demetrius wrote a letter to Melanchthon in October 1559, revealing the effect his contact with the Reformers had upon him. Some authorities suggest that he spent three years trying to introduce Lutheranism into the villages of Transylvania; whatever the case, he was one of the first of the Orthodox clergy to have any real contact with the Reformers.

GORDON A. CATHERALL

MYSTERY PLAY. Early medieval religious dramas based upon some part of the biblical narrative were called "mystery plays." In England the term is often used synonymously with "miracle plays," but the latter term should be restricted to drama based upon the life or miraculous deeds of Christian saints. Perhaps the term "mystery" is derived from the Latin *ministerium,* in the sense of service or function. The earliest record of medieval drama is a simple Easter playlet consisting of four characters—one priest symbolizing the angel at Christ's tomb while three other priests similarly symbolize the Marys—and the dialogue consists of four Latin lines. The popularity of the plays grew and their use in church services at special occasions was expanded until they assumed a form and existence of their own. The plays were banished from within the church in the thirteenth century. The laity immediately seized the opportunity and began organizing performances and presenting them in the vernacular. Eventually the performances were given on movable stages that were drawn from place to place.

The plays, centering around various narratives, were popular all over Europe and especially in England. Soon they were molded into elaborate cycles consisting of thirty or more separate pieces based on principal biblical themes from the Creation to the Day of Judgment. These cycles were presented at the Feast of Corpus Christi, at Whitsuntide, and in other special seasons. Only four of these cycles are extant: The York Cycle (forty-eight plays), the Towneley (thirty-two plays), the Chester (twenty-five plays), and the Coventry (forty-two plays). The use of mystery plays gradually disappeared by the end of the sixteenth century. There has been a modern revival of this type of play, especially on the Continent. Probably the most famous of these is the Passion Play at Oberammergau.

BIBLIOGRAPHY: C. Davidson, *Studies in the English Mystery Play* (1892); E.K. Chambers, *The Medieval Stage* (2 vols., 1903); K. Young, *The Drama of the Medieval Church* (2 vols., 1933); H. Craig, *English Religious Drama of the Middle Ages* (1955).

JOHN P. DEVER

MYSTERY RELIGIONS. In the last few centuries B.C. a plethora of cults, mostly from the Middle East, began to spread throughout the Greco-Roman world by means of migration, trade, and military service abroad. Their popularity continued well into the Christian era. Occasionally assimilated into the official religion of their new locations, they more often stood apart as minority "clubs" of individual initiates, representing personal rather than civic religion. The name "mystery religion" derives from the secret symbols and rites revealed to members only, but initiation is probably the more significant factor: it promised salvation now or bliss hereafter, and it certainly gave the security and identity of belonging to an "in-group." Unlike Judaism and Christianity, adherents were not required to give up their traditional religion; initiation into several cults was even possible.

Adaptations of national religions of the Middle East, they only became mystery cults when transplanted. Most were originally fertility religions with a death-resurrection mythology representing the annual cycle of nature. Their underlying similarities encouraged a syncretistic tendency, particularly in the fourth- and fifth-century struggles against Christianity (e.g., the Mithraeum discovered in London in 1954 contains statues relating to Isis, Dionysus, and the Olympian gods); borrowing of the practices of another cult (e.g., *taurobolium*, see "Cybele" below) occurred even earlier.

The most important cults are treated individually here. Others include the Eleusinian mysteries (an important but local Athenian cult) and the Kabiri (Phrygian deities worshiped extensively by sailors from the fourth century B.C.). The worship of Mâ (Cappadocia), Atargatis ("the Syrian goddess"), and Hadad (a Syrian Baal) came to Rome not as mystery-cults (Roman citizens were forbidden to take part), but as bizarre and frenzied public spectacles. For other parallels see HERMETIC BOOKS and GNOSTICISM.

(1) *Dionysus.* An ecstatic cult from Thrace (European Turkey), which flourished among women in classical Greece, had become a mystery-cult involving initiation of both sexes when it appeared in Rome early in the second century B.C. It was suppressed as corrupt, but reintroduced in mid-first century B.C. There were hierarchical grades of initiation and the promise of an afterlife.

(2) *Cybele* ("The Great Mother"). Introduced officially to Rome from Phrygia in 205 B.C. during a national emergency, the splendid but wild processions to Cybele were originally restricted to non-Romans yet were very popular as spectacles. Later the cult, merged with Mâ, offered initiation to all. Priesthood involved castration. At Rome the *taurobolium* seems to have been associated first with Cybele, later with Mithra: candidates were sprinkled with bulls' blood for either national or personal salvation. Some texts refer to rebirth "for ever" or "for twenty years."

(3) *Isis.* A major Egyptian goddess, whose cult became extraordinarily popular throughout the Roman Empire, often taking the form of a mystery religion. Initiation (e.g., see Apuleius, *Metamorphoses* 11) involved abstinence, and promised salvation from disease, fate, and fear of death.

(4) *Mithra.* Mithraism is puzzling, an apparently artificial though extraordinarily successful creation, and perhaps Christianity's most serious rival in antiquity. Mithra was an old Persian god of light, but the cult also draws on the Zoroastrian dualism of Good and Evil, between which Mithra mediates. The first evidence of mysteries or of the bull-slaying myth depicted prominently in the shrines comes from a group of pirates in first-century B.C. Cilicia. Shrines are common from the late first century A.D., particularly in military camps, even on Hadrian's Wall, but they are tiny, often underground. Men only were initiated, to seven grades of membership, in rites involving ordeal, the *taurobolium*, a communal meal, etc. The cult demanded high ethical standards.

BIBLIOGRAPHY: A.D. Nock, *Conversion* (1933); J. Campbell (ed.), *The Mysteries*, Eranos Yearbooks (1955); G.E. Mylonas, *Eleusis and the Elusinian Mysteries* (1961); J. Ferguson, *The Religions of the Roman Empire* (1970).

I: M.P. Nilsson, *The Dionysiac Mysteries of the Hellenistic and Roman Age* (1957).

II: E.O. James, *The Cult of the Mother Goddess* (1959); E. Neumann, *The Great Mother* (1963); R. Duthoy, *The Taurobolium* (1969).

III: R.E. Witt, *Isis in the Graeco-Roman World* (1970).

IV: R.C. Zaehner, *The Dawn and Twilight of Zoroastrianism* (1961); M.J. Vermaseren, *Mithras, The Secret God* (ET 1963); A.L. Campbell, *Mithraic Iconography and Ideology* (1968).

GORDON C. NEAL

MYSTICISM. The term is difficult to define because of its often confessedly ineffable and inexplicable nature. It has no limits historically or geographically, nor can it be contained philosophically or theologically. It concerns the interior life of the spirit, that pilgrimage with the divine which begins outside its awareness and proceeds to the highest stages of personal development possible.

Immediate relation with the ultimate is the essence of mysticism. This may be a psychological or an epistemological experience in which the mystic, apart from a religious institution or sacred book, has religious knowledge directly from the divine. Quakerism stressed this approach; other mystics believed that the contemplative experience led to temporary union of essence now or permanent at death with ultimate reality or God. The Hindus, Buddhists, Neoplatonists, and to some extent Meister Eckhart* illustrate this view. Prayer, contemplation, and ascetic acts promote this experience.

Christian and biblical mysticism usually stresses the personal reality of Christ as compared with the impersonal approach of Hinduism. It subordinates nature to the Creator rather than linking Him with nature as pantheistic mystics do. The union is not one of merging essence which destroys personality, but the biblical one of union of human love and will with God, which does not lose the subject-object relationship. Such mysticism was contemplative, personal, and practical: action on the plain followed retreat to the mountain.

Mysticism is simply a life of prayer, even from the outset when personal confession is paramount because of realization that one stands before God and must beg forgiveness before any growth in Him can begin. Once begun, and life's purpose shifted from self to God, the "scale of perfection" or "steps leading to the mind of God," has also begun. Many stages exist in mystical experience, however, and they are individually determined. Three are common: awareness and confession before God, life lived totally under God, and a most personal experience of God. The last is not often achieved, nor is it expected: referred to sometimes as a mystical marriage or Beatific Vision, it is the most intimate of divine relationships and therefore is usually not expressed in words.

Mysticism usually arose in an era (e.g., in the Middle Ages) when religion became too much institutionalized, and it sought a more individualistic and personal relationship with God. This helps to explain the rise of medieval mystics—such as Meister Eckhart, Julian of Norwich,* and Thomas à Kempis*—and later, of Madame Guyon* and Fénelon.* They did not leave the church, but had these experiences in the fold of the church as they sought to recall it to a more personal and individual approach to God. It also often appeared when theology was overemphasized at the expense of experience and practice.

Mysticism often led to heresy because of ignoring the biblical norm, or to social passivity that concentrated on personal salvation without any idea of service to God in society. Bernard of Clairvaux* seemed to link biblical truths, mystical experience, and practical service.

BIBLIOGRAPHY: W.R. Inge, *Christian Mysticism* (1899); E. Underhill, *Mysticism* (1911); C. Butler, *Western Mysticism* (1922); R. Otto, *Mysticism East and West* (ET 1932); A. Goodier, *Ascetical and Mystical Theology* (1957); R.C. Petry (ed.), *Late Medieval Mysticism* (1957); S. Spencer, *Mysticism in World Religion* (1959); F.C. Happold, *Mysticism* (1963); H. Graef, *The Story of Mysticism* (1965).

C.G. THORNE, JR.

N

NAASSENES. The name given to a Christian-Gnostic sect, derived from their supposed object of worship—the "serpent" of Genesis 3:1ff., the alleged divine seducer and impregnator of Eve. The materials concerning this group, however, give little evidence of their own self-image as being expressible in such terms, and the title may well be their opponent's deprecation. What is known is given in Hippolytus's *Refutation*, where one of their hymns is cited. This shows that remnant of awareness of a more ancient but exterminated Near Eastern religiousness undergoing the Gnosticizing interpretation of quasi-philosophical allegorization, with the assistance of scriptural data. *The Gospel of Thomas** may originate from within this circle.

See also OPHITES. CLYDE CURRY SMITH

NAG HAMMADI. A town in central Egypt on the western bank of the Nile about forty miles north of Luxor, which has given its name to a Coptic Gnostic library unearthed a few miles away in 1945-46 in a jar in a Greco-Roman cemetery. The scene, so far uninvestigated archaeologically, is near the ancient sites of Chenoboskion, an important fourth-century center of Pachomian monasticism (which originated not far distant at Tabennisi), and the Roman regional capital of Diospolis Parva. The documents, probably interred in the later fourth century, perhaps when Pachomian monks were establishing Catholic orthodoxy in the area, consist of thirteen papyrus codices, one (XII) very fragmentary and another (XIII) now quite short, totaling over 1,100 pages. Ten manuscripts are in Sahidic Coptic and three in a sub-Akmîmic dialect. Palaeographic and other evidence suggests they were written c.330-50. They contain in whole or part fifty-three works, most if not all translated from Greek originals. Three were already known: *The Sophia of Jesus Christ, The Sentences of Sextus,* and *The Apocryphon of John.* Of this typical Gnostic "summa," W.C. Till published in *Texte und Untersuchungen* 60 one (short) recension, which Nag Hammadi has duplicated (III) along with two texts of a longer version (II, IV). The collection includes doublets of *The Letter of Eugnostos the Blessed* (III, IV), *The Egyptian Gospel* (III, IV), an untitled treatise "On the Origin of the World" (II, XII), and *The Gospel of Truth* (I, fragments in XII). Thus we have forty-four completely new treatises, some of whose titles were recognized from patristic and other sources, and some identical with, or similar to, those of other extant Christian apocrypha.

The manuscripts reside in the Coptic Museum in Cairo, except for part of I—known as the Jung Codex because most of it was acquired in 1952 for the Carl Jung Institute in Zurich—which awaits return to Egypt. Its accessibility led to early publication of *The Gospel of Truth,* a Valentinian meditation known to Irenaeus. Publication of the Cairo manuscripts was delayed by political vicissitudes, but has quickened since 1960. To date, about half of the treatises have been printed. A complete facsimile edition under UNESCO auspices was initiated in 1972, and a complete ET is planned by the Institute for Antiquity and Christianity of Claremont, California.

This massive accession of new material is bound to prove enormously significant for the study of Gnosticism (hitherto documented chiefly by its opponents), heterodox Judaism, and primitive Christianity and their interrelations. The texts span a wide spectrum of Gnostic thought—Iranian, Hermetic, Jewish, as well as brands of Christian Gnosticism, Valentinian, Basilidian, and Barbelo-Gnostic or Sethian. If vindicated, the claim that *The Sophia of Jesus Christ* is a secondary Christianized version of *The Letter of Eugnostos,* and that other works—e.g., *The Apocryphon of John, The Book of Thomas the Athlete* (II), and *The Hypostasis of the Archons*—are Christian recensions of non-Christian originals, may illumine the much-debated existence of a pre-Christian Gnosticism. Scholars disagree whether *The Apocalypse of Adam* (V) is one such pre-Christian (syncretistic Jewish) Gnostic text. *The Gospel of Thomas* (II), whose markedly less Gnostic character is not unparalleled in the collection, is important for the development of the gospel tradition and early Syrian Christianity.

BIBLIOGRAPHY: J. Doresse, *The Secret Books of the Egyptian Gnostics* (1960); W.C. van Unnik, *Newly Discovered Gnostic Writings* (1960); U. Bianchi (ed.), *Le Origini dello Gnosticismo* (1967); A.K. Helmbold, *The Nag Hammadi Gnostic Texts and the Bible* (1967); J.M. Robinson, "The Coptic Gnostic Library Today," *NTS* 14 (1967-68), pp. 356-401. D.M. Scholer, *Nag Hammadi Bibliography 1948-1969* (1971, initiating the series *Nag Hammadi Studies*), supplemented annually in *Novum Testamentum;* W. Foerster (ed.), *Gnosis: A Selection of Gnostic Texts,* vol. 2 (1973); J.E. Ménard (ed.), *Les Textes de Nag-Hammadi* (1975). D.F. WRIGHT

NANTES, EDICT OF (1598). Agreement signed between Henry IV of France* and the Huguenots,* after Henry (formerly the Protestant ruler of Navarre) had become a Catholic in order to bring to an end the Wars of Religion. It codified and enlarged rights granted to French Protestants by previous measures—e.g., Edict of Poitiers (1577), Convention of Nerac (1578)—permitting them free exercise of their religion in certain areas, civil equality, and fair administration of justice, and granting them a state subsidy for the support of their troops and pastors. By it they remained in complete control of two hundred towns, including La Rochelle, Montaubon, and Montpellier. Elsewhere in Europe, rulers chose and maintained one religion and one only for their subjects, but this edict introduced a new principle of toleration, establishing freedom for two religions to exist side by side. It was revoked by Louis XIV (1685). J.G.G. NORMAN

NARSAI (Narses) (399-502). Nestorian* theologian and poet. A monk who lived to a ripe age, Narsai headed the theological school in Edessa after 437. Exiled about 457, he established a new school at Nisibis on plans laid by Bishop Barsumas* and at his request. The school became the teaching center for Nestorian theology. Among his writings were dialogue songs, liturgical hymns (including a poem on baptism), rhyming and metrical sermons, and OT commentaries which are no longer extant. His poems are still used in Nestorian services.

NASH PAPYRUS. This single-leaf papyrus containing the entire Decalogue (part from Exod. 20:2-17, part from Deut. 5:6-21) and the Jewish confessional statement known as the *Shema'* (Deut. 6:4,5) was published by Stanley A. Cook in 1903. An early report on the find which was purchased by W.L. Nash from an Egyptian dealer was given in *Revue biblique* 1 (n.s.) 1904, pp. 242-50, which contains also an untouched photocopy of the manuscript. A convenient summary of the evidence, including a translation into English, is given by R.H. Charles, *The Decalogue* (1923).

From its extant form it is clear that it did not come from a roll of the Pentateuch, but was written on a separate sheet and used probably for liturgical or teaching purposes. The ancestor of the leaf is clearly a synagogue scroll from which it was copied. It is therefore a valuable witness to the ancient text of Scripture and was hailed in the day of its publication as "the oldest Hebrew biblical papyrus," a description which has now been antiquated by the discovery of the Qumran manuscripts.

Arguments which maintain that it reflects a knowledge of the Deuteronomic Decalogue (in Codex Vaticanus) and that its origin is to be placed, on epigraphical grounds, in the second half of the second century B.C. (c.165-137) are offered by W.F. Albright. R.H. Charles and F.C. Burkitt incline to a first-century A.D. dating, while S.A. Cook argued for a second-century A.D. date. RALPH P. MARTIN

NATALIS ALEX (Noel Alexandre) (1639-1724). French historian and theologian. Born in Rouen, he entered the Dominican Order in 1655 and studied philosophy and theology at the Convent of St. Jacques in Paris, gaining a doctorate from the Sorbonne in 1675. Persuaded by Jean Baptiste Colbert, he entered the Society of Savants, lecturing on historical subjects of which twenty-four volumes were published between 1677 and 1688. The first volumes were welcomed, but later ones gave offense to Rome because of the author's Gallicanism,* causing the works to be placed on the Index, until later revised. His work included six octavo volumes on OT history, ten volumes on the *Catechisma Romanus*, including sermons and instructions to preachers. In 1704 he fell into Jansenism* by signing the *Cas de Conscience*, but soon recanted. He debated the Dominican and Jesuit doctrines of grace and predestination. His literary work ended through blindness. GORDON A. CATHERALL

NATIONAL ASSOCIATION OF EVANGELICALS. A lineal descendant of the creedal Evangelical Alliance* of 1867. Evangelical leaders, called together by Ralph T. Davis and J. Elwin Wright, met at Moody Bible Institute, Chicago, in 1941. They planned another meeting of nearly 150 leaders at St. Louis in 1942, at which the National Association of Evangelicals was organized with a creedal statement. The first convention was held in May 1943. The organization serves over two million members and claims a constituency of ten million. Subsidiary groups such as the Evangelical Foreign Missions Association* and the National Sunday School Association promote missions and Sunday schools. Other groups fulfill other service functions. Members of the Association worked with evangelicals from other lands to organize the World Evangelical Fellowship* in Holland in 1951 to coordinate worldwide efforts of evangelical service. Denominations, churches, and individuals which are willing to sign the statement of faith are admitted to the organizations. EARLE E. CAIRNS

NATIONAL BAPTIST CONVENTION, see AMERICAN BAPTIST CHURCHES

NATIONAL COVENANT (1638). A legal bond of association drawn up by Scottish Presbyterians against Charles I.* It began by repeating the Negative or King's Confession* of 1581 which had condemned Roman Catholic errors and "the usurped authority of that Roman antichrist upon the Scriptures of God, upon the Kirk, the Civil magistrate, and consciences of men; all his tyrannous laws made upon indifferent things against our Christian liberty." The Covenant went on to detail numerous Acts of Parliament which had established the Reformed faith and church government. Thereafter, more specifically, the subscribers bound themselves to maintain the freedom of the church, to defend the Presbyterian religion, "and the King's majesty . . . in the preservation of the foresaid true religion, liberties, and laws of the kingdom." There was no explicit condemnation of Episcopacy. Charles I was nevertheless incensed, but his law officers advised that the Covenant was not a contravention of statute law. Some 300,000 were estimated to have subscribed to this document, which gave its name to the Covenanters.* J.D. DOUGLAS

NATURAL THEOLOGY. This asserts a knowledge of God from the creation outside the bounds of revealed theology in the Bible. Psalm 19; Romans 1:19-21; Acts 14:15-17 and 17:24-29 imply a natural revelation of God available from nature and creation. Usually the question is not whether there exists some form of natural revelation, but rather what theology, if any, can be developed from it. Augustine* asserted that all knowledge of God was revealed. Anselm* developed the ontological argument for God's existence which was essentially a rationalistic argument divorced from revelation. Thomas Aquinas* modified Augustine and others considerably with his Scholastic view of reason and revelation, borrowing heavily from Aristotle.* Between Tertullian* and Aquinas a change had developed, with reason displacing revelation as the starting point of theological discovery.

The Reformers generally emphasized a strong view of special revelation with a natural theology limited by man's fallen condition. David Hume* assaulted the theistic arguments of Aquinas, and since then the debate has shifted back and forth. Karl Barth* and Emil Brunner* carried on a bitter dialogue, with Barth refuting natural theology by stressing special revelation within the context of his Crisis Theology. Brunner countered with a limited natural theology available from creation. Among conservative theologians the trend has been toward a limited natural theology stemming from natural revelation, sufficient to reveal God but insufficient to redeem man, hence the need for special revelation—especially as depraved man has perverted this natural knowledge (cf. Rom. 1). The conservative would also stress a strong view regarding the infallible character of this special revelation. ROBERT C. NEWMAN

NAUMBURG CONVENTION (1561). A conference of twelve Protestant princes and other German Protestant representatives held at Naumburg, Saxony, which sought to secure doctrinal unity, particularly on eucharistic doctrine. It agreed to recognize afresh the Augsburg Confession in both the *Invariata* (unaltered) edition of 1531 and the *Variata* (altered) edition of 1540, together with Melanchthon's *Apology* of 1531. Strict Lutheran divines such as M. Flacius* and T. Heshusius attacked the *Invariata* as heretical. A violent theological war ensued, ending in the triumph of strict Lutheranism in the Formula of Concord.* The convention also received and rejected an invitation to send delegates to the third session of the Council of Trent.*

J.G.G. NORMAN

NAVIGATORS, THE. An organization that fosters Christian fellowship, witness, and systematic Bible study and memorization. It began informally in 1933, when Dawson Trotman discipled a converted sailor, with 2 Timothy 2:2 as his guiding principle. He organized the Navigators as a nonprofit organization in California in 1943. Members worked primarily with servicemen until 1949, when work was begun in the Far East, and a year later in Europe. Their program fosters person-to-person recruitment training and guidance of servicemen, college students, and businessmen into effective Christian witnesses, especially through systematic Bible study and memorization. Headquarters are in Colorado Springs, Colorado. EARLE E. CAIRNS

NAYLOR, JAMES (c.1617-1660). English Quaker. Born near Wakefield of yeoman stock, he joined the parliamentary army in 1642 and served under both Fairfax and Cromwell. By 1650 he was respected in the ranks as an able preacher. Returning home after the Battle of Dunbar, he took up farming and joined a Congregational church. When George Fox* visited his area, he was much impressed with the doctrine of the inner light and became a Quaker. Later he felt a call to the itinerant ministry and began this in Westmorland, a Quaker stronghold, but later moved southward. Often in prison and much occupied in preaching, traveling, and writing tracts, he seems to have been particularly attractive to women. In 1656 he allowed his followers to treat him as though he were the Messiah, making the Bristol authorities think he was reenacting Christ's entry into Jerusalem. He was arrested. His case aroused national interest and was discussed in Parliament. After being pilloried and whipped, he was eventually released. He returned to the Quakers, but died soon afterward. PETER TOON

NAZARENE, CHURCH OF THE. An international denomination largely the result of the merger of approximately fifteen religious groups originating from the nineteenth-century Wesleyan Holiness Movement and whose organization, within the USA, took place at Pilot Point, Texas, in 1908. Originally called (in 1907) the Pentecostal Church of the Nazarene, the term "Pentecostal" was dropped from the title in 1919 due to its association with "speaking in tongues," a practice not in favor with its members.

The Church of the Nazarene began in the British Isles in 1906 through the ministry in Glasgow, Scotland, of George Sharpe, a native of Lanarkshire, who had been profoundly influenced by the Holiness Movement while in the United States. Originally called the Pentecostal Church of Scotland, it united with the Church of the Nazarene in the USA in 1914 to give birth to the vision of an international holiness communion. Congregations were founded throughout the USA by 1933 in a program of extension, and in the following thirty years 2,812 churches were founded worldwide. Where Methodism had flourished in the nineteenth century in the USA, the Church of the Nazarene flourished in the twentieth. A vigorous missionary program involved forty-two overseas fields. The Church of the Nazarene places great emphasis on Christian education in local churches, operates a publishing house, a theological seminary, several theological colleges, a number of liberal arts colleges, and numerous mission schools and hospitals.

The Church of the Nazarene combines congregational autonomy with superintendency in a representational system. Its governing body is the church assembly which meets every fourth year in the USA. In its major emphasis of entire sanctification as a work of grace following conversion, it stands firmly in the Wesleyan tradition. An emphasis is placed on tithed giving, and its members are bound by a *Manual of General Rules* which binds them to renounce alcohol, tobacco, the theater, the cinema, the ballroom, the circus, and also lotteries and games of chance. The members are also required to renounce "the profanity of the Lord's Day, either by unnecessary labor or business or . . . by the reading of Sunday papers or by holding diversions."

See T.L. Smith, *Called Unto Holiness* (1962) and J. Ford, *In the Steps of John Wesley* (1968).

JAMES TAYLOR

NEAL, DANIEL (1678-1743). Best known as the historian of the Puritans. Educated at Merchant Taylor's School and in a Dissenting academy, he later studied at Utrecht and Leyden. In 1704 he became assistant minister of a Congregational church in Aldersgate Street, London. Two years later he himself became pastor and the church flourished. For his *History of New England* (1720) Harvard College gave him the honorary M.A. degree. His interests were wide and included the commendation of the practice of inoculation against smallpox. The first volume of his *History of the Puritans* was published in 1732; other volumes followed in 1733, 1736, and 1738, covering the period from the Reformation to 1689. It was severely criticized by some Anglicans—e.g., Zachary Grey. Nevertheless it has often been reprinted and is best used in the revised edition prepared by Dr. Joshua Toulmin. Neal was buried in Bunhill Fields.

PETER TOON

NEALE, JOHN MASON (1818-1866). Anglican scholar and hymnwriter. Born in London, he was educated at Sherborne and at Trinity College, Cambridge. He became a fellow of Downing College and was eleven times Seatonian Prize winner. Although his parents were Evangelicals, he adopted High Church ideals. To further these he helped in 1839 to found the Cambridge Camden (later Ecclesiological) Society, which greatly affected the style of Anglican buildings and worship. Lifelong ill-health prevented him in 1843 from accepting the living of Crawley, Sussex, but in 1846 he assumed the wardenship of Sackville College, East Grinstead, a refuge for indigent old men. There he remained for the rest of his life, declining the offer of the provostship of St. Ninian's Cathedral, Perth. His ritualism caused the bishop of Chichester to inhibit him for many years. Despite bitter Protestant opposition, Neale founded for the education of girls and the care of the sick the Sisterhood of St. Margaret at Rotherfield in 1854; he transferred it to East Grinstead in 1856.

He was a voluminous writer, with works including *A Commentary on the Psalms* (with R.F. Littledale, 4 vols., 1860-74); *The History of the Holy Eastern Church* (5 vols., 1847-73); and *Essays on Liturgiology and Church History* (1863). He wrote numerous sermons and children's stories. His fame rests chiefly on his many hymns and carols, some original, some paraphrases or translations mainly from Greek and Latin (though he knew about twenty languages). Examples are translations from Bernard of Cluny*; "All glory, laud, and honor" (Theodulf of Orléans,* d.821); "Art thou weary"; "Good King Wenceslas"; "O come, O come, Immanuel" (Latin, eighteenth century); "O happy band of pilgrims"; "Of the Father's love begotten" (Aurelius Prudentius* Clemens).

He translated 94 of the 105 hymns in *The Hymnal Noted* (1852-54). Among his other collections were *Hymns for Children* (1842-46); *Mediaeval Hymns and Sequences* (1851); and *Hymns of the Eastern Church* (1862), in which he broke entirely new ground.

See biographies and assessments by E.A. Towle (1906) and A.G. Lough (1962); also M. Donovan, "John Mason Neale," *Church Quarterly Review* 167 (1966), pp. 317-22, and J.E. Holroyd, "Victorian Hymn-Writer's Gothic Zeal," *Country Life* 140 (1966), pp. 1518-20.

JOHN S. ANDREWS

NEANDER, JOACHIM (1650-1680). German hymnwriter. Born at Bremen, he was converted there through a Pietist preacher and in 1671 became a tutor at Frankfurt. There he was influenced by P.J. Spener.* When appointed in 1674 to the headship of the Düsseldorf *Lateinschule*, a Reformed grammar school, he organized unofficial gatherings for instruction and preaching. This led to his suspension. In 1679 he returned to Bremen as a preacher, and the renewed opposition to his preaching was cut short by his premature death. Accomplished in literature, music, and theology, he was the first important poet of the German Reformed Church. He wrote some sixty hymns with tunes in a volume published in 1680. Although many are still sung in Germany, only two have become well known in Britain: "All my hope on God is founded" (R.S. Bridges*) and "Praise to the Lord, the Almighty" (C. Winkworth*).

JOHN S. ANDREWS

NEANDER, JOHANN AUGUST WILHELM (1789-1850). German Protestant church historian. Born David Mendel, he changed his name after his conversion to Christianity in 1806. He studied under Schleiermacher and afterward was professor of church history at Berlin for nearly four decades (from 1813), where he was a determined opponent of the rationalistic views of F.C. Baur,* D.F. Strauss,* and others. He is generally regarded as the founder of modern Protestant historiography. His two-volumed *Geschichte der Pflanzung und Leitung der christlichen Kirche durch die Apostle* (1832-33; ET of 2nd ed., *History of the Planting and Training of the Christian Church by the Apostles*, 2 vols., 1887-88) was a model for subsequent histories of the apostolic age. He authored many church-historical monographs, including works on Julian the Apostate (1812), Bernard of Clairvaux (1813), Gnosticism (1818), Chrysostom (1822), and Tertullian (1824). His multivolumed church history (6 vols., 1826-52) concentrated on personalities rather than institutions and set the tone for subsequent evangelical historical work (e.g., P. Schaff).

W. WARD GASQUE

NECTARIUS (d.397). Bishop of Constantinople from 381. Born at Tarsus, Cilicia, he was a jurist who rose to the office of praetor at Constantinople. When Gregory of Nazianzus resigned under pressure from the see of Constantinople, Emperor Theodosius I nominated Nectarius to succeed, despite the fact that he was entirely remote from church affairs and had not been baptized. He was baptized forthwith and unanimously elected and installed, and he took over from Gregory the presidency of the closing stages of the second ecumenical council (381). J. Kunze suggested that his formal profession of faith at the council in the words of the Niceno-Constantinopolitan Creed accounts for the association of that creed with the council. J.G.G. NORMAN

NECTARIUS (1605-c.1680). Patriarch of Jerusalem, 1661-69. Educated by the monks of Sinai, he became a monk and later studied at Athens under the Neo-Artistotelian, Theophilus Corydalleus. He vigorously opposed all Western theology, attacking both the claims of Roman Catholicism and the Calvinism of Cyril Lucar,* patriarch of Constantinople (1621-38). In 1662 he approved the "Confession" of Peter Mogila.* He took a prominent part in the Synod of Jerusalem (1672), which repudiated Lucar's doctrines and approved Mogila's "Confession." He wrote a treatise against the papacy published in 1682 by his successor, Dositheus.

NEESHIMA, YUZURU (1843-1890). Japanese Christian leader. Born of Samurai stock in Edo (now Tokyo), he was determined to bring the learning of the West to Japan and secretly fled his country in 1864. He finally reached Boston, where the ship's owner befriended him. Schooling and seminary followed his conversion, and in 1874 the sensitive and frail Neeshima, fired with a desire to evangelize his own people, was commissioned as a missionary by the Congregational Church. In 1875 he founded in Kyoto, the stronghold of Buddhism, the first Christian school in Japan, calling it the *Doshisha* ("one purpose society"). Undaunted by broken health, he worked passionately to give his students an education that united sound biblical teaching with the highest academic standards. He died from overwork when only forty-six. DAVID MICHELL

NEGATIVE CONFESSION, see KING'S CONFESSION

NEGRO CHURCHES, see AMERICAN NEGRO CHURCHES

NEMESIUS (fl. c.400). Christian philosopher and bishop of Emesa in Syria. Despite chronological coincidence, this Nemesius, author of a remarkable Christian philosophical work *On the Nature of Man,* is probably not identical with the pagan governor of Cappadocia (386) and friend of Gregory of Nazianzus. Beyond his treatise nothing is known of his life. Eclectically incorporating or rejecting Platonic, Aristotelian, Stoic, and Neoplatonic thought including Porphyry, and relying on Galen's medical theories, the work of this highly cultured man embodied that Helleno-Christian intellectual synthesis of the fourth century and was a source for medieval knowledge of classical philosophies. Discounting his belief in the soul's preexistence, his treatise provided an early articulation of Scholastic views on the soul's nature, its relation to the body, and free will as a natural concomitant of reason and basis of human acts. DANIEL C. SCAVONE

NEOORTHODOXY. A loose term used to designate certain forms of twentieth-century Protestant theology which have sought to recover the distinctive insights and themes of the Reformation. The latter are seen as relevant to our modern predicament and as an essential part of the church's witness. Nevertheless, they require some restatement in the light of modern knowledge. The term is generally used by those who would not identify themselves with such a theology, either because it seems to deviate too much from the orthodoxy of the Reformation theologians and the classical Protestant confessions of faith, or because it is too narrowly orthodox.

The term indicates a reaction against the liberalism of the nineteenth and early twentieth centuries, with its reduction of Christian faith to general human and religious truths and moral values, and its relativization of Christianity through historical criticism and theories of the history of religions. By contrast, Neoorthodoxy represents an attempt to recover biblical perspectives. Stress is laid (in varying degrees) on the transcendence of God, man's responsibility as a creature, sin and guilt, the uniqueness of Christ as mediator of revelation and grace, and personal encounter with God in revelation.

These themes were sounded by the Dialectical Theology* or Theology of Crises of the twenties and thirties. They were given almost classic expression in Barth's* commentary on Romans (1919). God is seen as the Wholly Other who is not to be identified with anything in the world. He breaks into our world like a vertical line intersecting a horizontal plane in the person of Jesus Christ. But even so He remains incognito, for to encounter Jesus on a merely human level is to know only the man. God is hidden in Him even in the act of revelation. Full revelation occurs only in the risen Christ. Its truth is not perceived on the level of historical investigation, but through encounter by faith. Christ's coming is also the crisis of judgment of the world. It is both the revelation of God and the revelation of man's sin. This act of judgment is also the means of grace.

A Catholic theologian described Barth's work as a bomb falling on the happy playground of the theologians. The liberal historian Harnack* regarded Barth's teaching as unscientific theology. Nevertheless, Barth found himself at the head of a theological revival in Europe. He soon, however, modified his position and eventually abandoned Dialectical Theology. He spoke of the Kantian-Platonic crust which had encased his teaching. After various revisions, he felt his views of the 1920s were still too much influenced by Kierkegaard* and existentialism.* His stress on the difference between God and man was replaced by a doctrine of analogy, albeit one that could only be known by faith through revelation. At the same time Barth continued to distinguish his view of revelation from that of Protestant orthodoxy. He felt that the latter stressed revealed truth and the verbal inspiration of Scripture, whereas he wished to stress that revelation is essentially God revealing Himself in Christ, even though this is human only through the witness of the biblical writers. In his later teaching, especially in the *Church Dogmatics*, Barth paid particular attention to the exegesis of Scripture and the great theologians of the church.

The teaching of Emil Brunner* tended in a similar direction, though their latent differences came into the open through their dispute over natural theology in 1934. Brunner accused Barth of going too far in denying that man had no knowledge of God apart from that mediated by Christ. He urged that man must have some knowledge which would serve as a ready-made point of contact for the Gospel. He saw grounds for this in the image of God in man and man's awareness of such divine institutions as the state and marriage. Brunner pleaded for a new, reformed natural theology, but his case remained unconvincing in view of the concessions he was willing to make to Barth. Brunner's teaching on revelation focused on the element of divine, personal encounter and attacked even more strongly than Barth the concept of objective, revealed truth.

Also associated with Dialectical Theology were Rudolf Bultmann* and Friedrich Gogarten.* But whereas Barth and Brunner developed theologies which had a framework of biblical theism, Bultmann and Gogarten sought to reinterpret biblical themes in terms of an existential philosophy. The former were primarily concerned with exegesis, the latter with a radical demythologizing hermeneutic. Paul Tillich* has also been considered to be Neoorthodox. His sermons, in particular, are often concerned with biblical themes. But his *Systematic Theology* makes it clear that the basis of his thought is his existential ontology. In the USA, Reinhold Niebuhr* has been regarded as Neoorthodox in view of his use of biblical categories in his moral philosophy and interpretation of history. But in their different ways both Tillich and Niebuhr are more concerned with what they conceive to be the underlying principles of Protestantism than with a modern restatement of a corpus of doctrine.

BIBLIOGRAPHY: E. Brunner, *The Theology of Crisis* (1929); J. Baillie, *Our Knowledge of God* (1939); H. Bouillard, *Karl Barth* (3 vols., 1957); P.K. Jewett, "Neo-Orthodoxy" in *Baker's Dictionary of Theology* (ed. E.F. Harrison, 1960); T.F. Torrance, *Karl Barth: An Introduction to His Early Theology 1910-31* (1962); J. Macquarrie, *Twentieth-Century Religious Thought* (1963); H. Gollwitzer, *The Existence of God as Confessed by Faith* (1965); J. Moltmann (ed.), *Anfänge der dialektischen Theologie* (2 vols., 1966-67); P.E. Hughes, *Creative Minds in Comtemporary Theology* (2nd ed., 1969); W. Nicholls, *The Pelican Guide to Modern Theology*, vol. 1 (1969); J. Pelikan (ed.), *Twentieth Century Theology in the Making* (3 vols., 1969-70); C. Brown, *Philosophy and the Christian Faith* (2nd ed., 1971).

COLIN BROWN

NEOPLATONISM. This comprised probably the most important intellectual vehicle of the ancient world after the third century, though unlike Gnosticism* it never acquired a comprehensive religious guise. Its roots lie in the prolific Platonic culture of Alexandria, which had displaced Athens as the intellectual center of the world. Its founder, Plotinus,* was influenced by the unknown philosopher Ammonius Saccas.* There followed an outstanding philosophical progeny, including Porphyry* and Boethius.*

Neoplatonic influences on Christian thought were more as a catalyst and vehicle of thought than as a religion. Christian writers who employ Neoplatonic methods include Basil the Great,* Nemesius* of Emesa, Synesius* of Cyrene, Nestorius (see NESTORIANISM), Augustine,* and the treatises of Dionysius the Pseudo-Areopagite.* In Neoplatonism the ultimate divine principle is above being. The divine light streams from the superabundance of the divine perfections and fades into the inexhaustible void. Existence is like a ladder with the top near to the light, but the bottom mired in the realm of the irrational and lifeless. By abstracting the particulars of existence or by sheer mystical illumination (a form of transcendental meditation) the mind can overcome the hindrances of the psyche to experience the sublime.

Neoplatonism aimed to overcome the duality between thought and ultimate reality by direct union of the soul with God. It maintained an infinite qualitative distinction and distance between the material world (including the flesh) and divine goodness; hence the ascription to Christ of a phantasmal body by some Neoplatonists because a real incarnation was unthinkable. Religious questions were of the utmost importance, based on a dualistic view of reality. Man should turn his face upward; science turns man's face to what is below him. They refused totally to see in the world the manifestation of a spiritual or divine principle. By contrast, Christianity brought the divine goodness down into the world in discrete personal, bodily form by the Incarnation. Salvation is by redemption through the Cross, based upon the creation of the world by God and His personal coming into it in human life, not by aspiration.

BIBLIOGRAPHY: H.E.W. Turner, *The Pattern of Christian Truth* (1954); L. Hodgson, *For Faith and Freedom* (1957); B. Altaner, *Patrology* (1958); C.C.J. Webb, *A History of Philosophy* (1964); J. Quasten, *Patrology* (1966); A.H. Armstrong (ed.), *The Cambridge History of Later Greek and Early Medieval Philosophy* (1967); R.T. Wallis, *Neoplatonism* (1972).

SAMUEL J. MIKOLASKI

NEREUS and ACHILLEUS (first century). Roman martyrs. Their remains are in the cemetery of St. Domitilla on the Via Ardeatina. The church built over their tomb dates from the fourth century. According to the inscription on the tomb by

Pope Damasus, they were soldiers. According to their legendary *Acta,* however, they were eunuchs in Domitilla's household, and with Domitilla* were transported to the island of Terracina, where Domitilla was burnt, and Nereus and Achilleus beheaded. The *Acta* also purport to tell the story of Petronilla, the Apostle Peter's daughter.

NERI, PHILIP, see PHILIP NERI

NERO CLAUDIUS CAESAR (37-68). Roman emperor from 54. He was an enthusiast for the arts and sports. The early part of his reign was stable enough, for he was under the influence of Burrus and Seneca and had the service of able governors in the provinces; but he was soon free from the restraints of more astute men and, as the result of numerous blunders, became extremely unpopular with the nobility and populace alike. His vanity and lust for power seemed limitless, and his suspicions led him to have his closest friends and relatives executed. The fire during July of 64 that destroyed one-half of Rome increased his unpopularity. In response to rumors that he had started the fire and recited his own poetry over the burning city, he tried to pass the blame on to the Christians, many of whom were arrested and executed in a most horrible manner.

Nero was the "Caesar" to whom Paul appealed for justice (Acts 25:10) and whose God-given authority he had carefully supported (Rom. 13:1-7). The details are uncertain, but it is probable that Paul was acquitted by or even released before the trial at the end of his two years in Rome (Acts 28) and that he was arrested again a few years later and was executed (c. A.D. 66-67). The Apostle Peter was probably executed at about the same time or a little later.

The unofficial policy of opposition to Christianity instigated by Nero was later to become the official policy of the empire.

See B.H. Warmington, *Nero* (1969).

W. WARD GASQUE

NERSES (Narses the Great) (c.326-373). Armenian catholicos. Born perhaps of royal stock, he married a princess and after her death was ordained a priest. In 353 he became catholicos (or patriarch) of Armenia, which marked a new era in Armenian history. Until then the church had been identified with the royal family and the nobility; Nerses brought it closer to the people, promulgating numerous laws on marriage, fast-days, and divine worship. He built schools and hospitals, and sent monks throughout the land preaching the Gospel. Some of his reforms brought him into conflict with King Arshak III, who exiled him to Edessa. On the king's death he returned at the command of King Pap (or Bad) in 369, only to repeat his offense. Tradition says he was poisoned at the instigation of the king.

GORDON A. CATHERALL

NESTLE, EBERHARD (1851-1913). German biblical scholar and textual critic. He held professorships at Ulm, Tübingen, and the Protestant seminary at Maulbronn (from 1898). Nestle did a considerable amount of work on the text of the Septuagint, but it is for his edition of the Greek NT that he is best known today. First published in 1898 by the Württemberg Bible Society, it has gone through twenty-five editions and is the standard text used by the majority of theological students and teachers, especially in Germany (though the new Bible Societies edition, edited by K. Aland et al., is also very popular). Subsequent editors have been Erwin Nestle, his son, and Kurt Aland.

W. WARD GASQUE

NESTORIANISM; NESTORIUS. Nestorianism is usually regarded as the heresy, taught originally by Nestorius, which split Jesus Christ, the God-man, into two distinct persons, one human, one divine. Born of Persian parents, Nestorius was probably a pupil of Theodore of Mopsuestia before becoming a monk and presbyter at Antioch. Because of the fame he achieved as a preacher, Theodosius II* elevated him in 428 to the patriarchal see of Constantinople. Soon after, he was called upon to pronounce on the suitability of *Theotokos* ("God-bearing") as a title for the Virgin Mary. He ruled that it would be best not to use the title unless it was balanced with *anthrōpotokos* ("man-bearing"); however, the best title for her was *Christotokos* ("Christ-bearing"). His doctrine, and the vehement way in which he expressed it, led Cyril* of Alexandria to oppose him and the Council of Ephesus* (431) to anathematize him as a heretic and to declare him deposed. The emperor exiled him to his monastery in Antioch and later to the Great Oasis in Egypt, where he died about 451.

Naturally Nestorius claimed that he was no heretic. As his writings were thought only to exist in fragments, it has been difficult to judge his claim. But the discovery in 1910 of *The Book (Bazaar) of Heracleides* in a Syriac translation has provided us with greater understanding of his views. Nevertheless, modern scholars are not in agreement in their assessment of his doctrine. For some he was the unfortunate victim of ecclesiastical politics; for others he remains guilty of the theological errors charged against him by Cyril and others.

What was his teaching? This can only be understood against the background of the traditional Antiochene Christology which stresses the fact that Jesus Christ was truly a man. First of all, he taught that the human and divine natures remained unaltered and distinct in their union within Jesus of Nazareth. He could not conceive of the divine Logos* being involved in human suffering or change, and so he wanted to hold the natures apart. Secondly, he emphasized that Jesus Christ lived a truly human life which involved growth, temptation, and suffering. This would have been impossible, he argued, if the human nature had been fused and overcome by the divine nature. He believed that the Alexandrian Christology overstressed the divinity of Jesus Christ.

To solve the problem of the union of the two natures of Christ—and to emphasize that he taught the doctrine of the one Person, who combined in Himself two distinct elements, Godhead and manhood—Nestorius explained that Jesus Christ, the person described in the gospels, was the "common *prosōpon,*" the *prosōpon* of union. The humanity had the form of Godhead bestowed upon it, and the divinity took upon itself the form of a servant: the result was the *prosōpon* of Jesus of Nazareth, the Son of God, one Person but with two natures. In view of this, Mary his mother was best described as *Christotokos.* While the strong point of Nestorianism is its attempt to do full justice to the manhood of Christ (a true Savior of men), its weak point is that it places the two natures alongside each other with little more than a moral and sympathetic union between them.

After the Council of Ephesus those Eastern bishops who could not accept the views of the majority gradually formed themselves into a separate Nestorian Church. Its center was in Persia, and a school of Nestorian theology developed under Ibas,* a friend of Nestorius, at Edessa. Later the center of Nestorian theology moved to Nisibis in the school founded there by Barsumas,* the pupil of Ibas. The ecclesiastical center and see of the (Nestorian) "Patriarch of the East" was at Seleucia-Ctesiphon on the Tigris until about 775, when it moved to Baghdad. Nestorians were active missionaries and founded communities in Arabia, India (Malabar Christians*), and Turkestan. During the thirteenth and fourteenth centuries the Nestorian churches suffered badly in the Mongol invasions, but a remnant lived on in the mountains of Kurdistan. Today there still exist the "Assyrian Christians" who claim to be the continuation of the Nestorian Church. They still forbid the use of *Theotokos,* and they treat Nestorius as a saint.

BIBLIOGRAPHY: J.F. Bethune-Baker, *Nestorius and His Teaching* (1908); F. Loofs, *Nestorius and His Place in the History of Christian Doctrine* 1914); A.R. Vine, *The Nestorian Churches* (1937); R.V. Sellers, *Two Ancient Christologies* (1940); A. Grant, *History of the Nestorians* (1955).

PETER TOON

NETHERLANDS, see LOW COUNTRIES

NETTER, THOMAS (c.1372-c.1430). Carmelite theologian. Known also as Thomas Walden from his birth at Saffron Walden, he is best known for his confutation of Wycliffe and the Lollards.* From his ordination in 1396, study at Oxford where Netter received the theological doctorate, and attendance in 1409 at the Council of Pisa, he gained a love for royalty and a hatred for the Lollards. As provincial of the Carmelite Order, Netter asked prayers for the young king Henry VI. His love made him thus bold, for Henry V had died in his arms and Netter preached the funeral sermon at Westminster. At Henry V's wish, Netter commenced his *Doctrinale Fidei Ecclesiae Catholicae contra Wiclevistas et Hussitas* in 1421. This major work approved by Martin V at Rome demolished the Lollard theses point by point. Martin requested a second book on the sacraments and a third on rites. Netter also prepared with help the *Fasciculus zizaniorum,* a fully documented source book of Lollardy, the only contemporary account to survive. He died in Rouen. Netter was a learned defender of Catholicism and a worthy opponent of Wycliffe.

MARVIN W. ANDERSON

NEUTRAL TEXT. The name given by Westcott and Hort to a type of text of the Greek NT which they felt to be near the original autograph. They distinguished the Western text as a clearly separate stream and saw the Neutral Text as sharing a common ancestor with the Alexandrian, but being free from later corruption and mixture. The two leading representatives were Codex Vaticanus (B) and Codex Sinaiticus (ℵ), and when these two agreed Westcott and Hort thought themselves to be very near the original reading. They never rejected such readings unless there were strong reasons for preferring something else. Most textual critics today think their dependence upon these two MSS to have been excessive.

R.E. NIXON

NEVIN, JOHN WILLIAMSON (1803-1886). German Reformed theologian. Descendant of wealthy Scotch-Irish farmers, he prepared for the Presbyterian ministry under Princeton's Charles Hodge.* While teaching at Western Theological Seminary, he developed an interest in church history, particularly through the writings of J.A.W. Neander.* In 1840 he was called to teach at the German Reformed Seminary at Mercersburg, Pennsylvania. With Philip Schaff,* another convert to the German Reformed Church, he created the Mercersburg Theology.* His major works were *The Anxious Bench* (1843), which attacked the superficiality of revivalism, and *The Mystical Presence* (1846), which presented the Eucharist, the "spiritual real presence," as central to the church's life. Plagued by illness, he retired in 1853, but later taught at his church college, Franklin and Marshall, where he was president (1866-76). His anti-individualistic thinking had much influence on later American Christianity.

DARREL BIGHAM

NEVIUS, JOHN LIVINGSTON (1829-1893). American missionary to China. Educated at Princeton Seminary, he went to China in 1854 under the Presbyterian Mission Board and served mainly in the Shantung area. He is best known for the "Nevius method" of self-support and propagation. Its principles were: (1) each Christian should support himself by his own work and be a witness for Christ by life and word in his own neighborhood; (2) church methods and machinery should be developed only so far as the indigenous Christians could take responsibility for these; (3) the church should select for fulltime work those who seemed best qualified and whom it was able to support; (4) churches were to be built in native style and by the Christians from their own resources. The Korean missionaries adopted this approach, and a vigorous church rapidly developed there which maintained an independent spirit virtually unmatched in the non-Western world.

RICHARD V. PIERARD

NEW APOSTOLIC CHURCH. In 1863 the senior apostle of the Catholic Apostolic Church,* F.V. Woodhouse, excommunicated the movement's German prophet, Heinrich Geyer, for recognizing new apostles to replace those who had died. In consequence the New-Apostolic Church was founded in Germany. It laid less emphasis on the Second Coming and was distinguished from its predecessor by the establishment of a *successional* apostolate subject to a senior apostle or patriarch with quasi-papal powers, regarded as "the visible incarnation of Christ on earth." The New Apostolic Church continued to flourish in Germany—even under Hitler, whom Johann Bischoff (patriarch 1932-60) claimed to be God's special emissary. Since 1925 the German branch of the community has nearly trebled in size and Neoapostolics are to be found also in Switzerland, France, South Africa, Java, and the Americas. TIMOTHY C.F. STUNT

NEW CHURCH, see NEW JERUSALEM, CHURCH OF THE

NEW DELHI ASSEMBLY (1961). The third assembly of the World Council of Churches,* held in India with its theme "Jesus Christ the Light of the World." Decisions and features of the assembly included a merger of the WCC with the International Missionary Council*; approval given to membership applications from twenty-three churches, including the Russian Orthodox Church and two Pentecostal churches from Chile, adding seventy-one million members to the movement; and the admission of five official Roman Catholic observers.

Some 577 delegates and 1,006 participants were present. Presidents elected were Archbishop A.M. Ramsey,* Sir Francis Ibiam of Eastern Nigeria, Archbishop Iakovos of the Greek archdiocese of North and South America, Dr. Martin Niemöller,* Dr. David Moses of the United Church of North India, and Mr. Charles Parlin, Methodist layman from the USA. Overwhelmingly accepted was a required Trinitarian formula: "The WCC is a fellowship of churches which confess the Lord Jesus Christ as God and Saviour according to the Scriptures, and therefore seek to fulfil together their common calling to the glory of the one God, Father, Son and Holy Spirit."

A variety of problems was encountered, among them the language barrier, time pressures, the subordination of delegates to preparation for the assembly, verbose papers, and distinctions between clergy and laity. Anti-Semitism, proselytism, religious liberty, and concern for refugees were discussed. Reports of the three study groups, concerned respectively with witness, service, and unity, were "approved in substance and commended to the churches for study and appropriate action."

Witness proceeded on these bases: Jesus Christ is the light of the world; the peoples of the world are interdependent; evangelism must proceed in new ways. Proclamation of Christ as Lord and Savior has "deep implications," and differences among WCC members must be studied. Any evangelism must take specific cognizance of youth, the worker, the intellectual. New ways might include dialogue, small groups, listening, mass media, use of laymen, examination of church structures to see if they help or hinder evangelism, and the embodiment of the message in lives.

Service was concerned about technology, social change, and political order. Since government gives a necessary order to society, Christians must work for political institutions which protect individual freedom and oppose governments which deny rights for racial and other reasons. The group was concerned to promote racial equality; international trust, especially between Russia and America; international institutions which promote peace; disarmament; integrity and honesty in political life.

In the *unity* section it was concluded that "the unity which is both God's will and his gift to his church is being made visible" in various ways "in the fully committed fellowship" which yet leaves many questions unanswered, including the inability to have intercommunion and one baptism.

See W.A. Visser 't Hooft (ed.), *The New Delhi Report* (1962). ROBERT B. IVES

NEW DIVINITY, see EDWARDS, JONATHAN

NEW ENGLAND THEOLOGY (c.1750-c.1850). Calvinist movement begun under Jonathan Edwards.* Regarded as one of America's greatest thinkers, Edwards set out to reformulate Puritan Calvinism to render it more harmonious toward the spiritual experiences of the Great Awakening.* In order to justify the results of the latter, Edwards set about to wrestle with freedom versus sovereignty. In his monumental work *Freedom of the Will* he introduced a subtle change into Calvinism which taught that man's role in salvation was negligible. Edwards conceived his doctrine of the inclined will, having borrowed heavily from John Locke's philosophy as an aid. In essence, Edwards said that God, in sovereign disposition, through the Holy Spirit's work, makes man's will able to respond to grace. In short, God inclines the will to render man able to respond to salvation. He sought middle ground between the "enthusiasts" of revival and Charles Chauncy,* who accused the revival of mindless emotion. Edwards agreed that "heat without light" was wrong, but one could not divorce truth from experience. He hoped to pacify the older Calvinists who spurned the revival and the opposite party of Arminian extremists.

This "New England Theology" of Edwards dominated conservative Congregational schools, such as Edwards's Yale, from about 1750 to the late 1800s, when German critical theology won the day. Later exponents were Jonathan Edwards, Jr. (1745-1801), Timothy Dwight,* Samuel Hopkins,* and Nathaniel Taylor.* The last-named seriously modified the doctrine of original sin. The movement was a gradual retreat from Calvinism in the face of greater emphasis on self-determination. ROBERT C. NEWMAN

NEW ENGLISH BIBLE, see BIBLE, ENGLISH VERSIONS

NEW HAVEN THEOLOGY. An American theological position associated with N.W. Taylor,* his students, and Yale Divinity School in New Haven, Connecticut. Sometimes known as "Taylorism," the theology of this school was a modified Calvinism used to provide an apologetic for the revivalism of the Second Great Awakening.* The New Haven Theology stood in contrast to the somewhat older theology of Samuel Hopkins* known as "Consistent Calvinism," the system stressing divine sovereignty, total human depravity and inability, and the idea of "disinterested benevolence"—the willingness to be damned for the glory of God. The New Haven Theology developed at a time when the Unitarian controversy was dividing many New England churches. Taylor and his followers attempted to use a rationalistic apologetic to defend Trinitarianism and to support experiential religious conversion. Taylor made a distinction between certainty and necessity: man sins inevitably and certainly, but not necessarily. Thus sin is voluntary. The reaction against Taylorism, or the New Haven Theology, led in 1834 to the formation of Hartford Theological Seminary, but Taylorism lent its support to a growing number of modifications in the older Puritan, Calvinistic theology. Gradually the "governmental" replaced the "satisfaction" theory of the Atonement: a universal atonement, and the idea of "limited atonement" and original sin came to be understood as moral or dispositional rather than imputational.

BIBLIOGRAPHY: F.H. Foster, *A Genetic History of the New England Theology* (1907); S.E. Mead, *Nathaniel William Taylor* (1942); H.S. Smith, *Changing Concepts of Original Sin* (1955); S.E. Ahlstrom, "Theology in America: A Historical Survey," *The Shaping of American Religion* (1961), pp. 231-321. DONALD M. LAKE

NEW JERUSALEM, CHURCH OF THE. (The New Church, commonly called "Swedenborgians.") A group organized in London in 1787 by followers of the theological teachings of Emanuel Swedenborg.* The organization and growth of the church is peculiar because the movement was started by books without the influence of any personal leadership. Swedenborg never preached a sermon and made no effort to gather followers about him, but he left his Latin works in twenty volumes to ministers and university librarians. These were translated and won disciples who were organized by Robert Hindmarsh, a Methodist. Ministers were ordained, other groups were started, and by 1789 the first general conference was held at their chapel in Great Eastcheap, London. By 1792 a Swedenborgian church was established in Baltimore, Maryland, and in 1817 the General Convention of the New Jerusalem met in Philadelphia. A division of the church (1897) resulted in a branch, the General Church of the New Jerusalem, with headquarters at Bryn Athyn, Pennsylvania. The emphasis in their worship is liturgical, concentrating on Jesus Christ with preaching based on the "inspired" parts of the Bible, twenty-nine books of the OT and five in the NT.

Baptism and the Lord's Supper are observed, and in addition to the usual Christian holidays, New Church Day (June 19) is observed. Worldwide membership is about 40,000 with 4,500 in Britain and 5,800 in the General Convention in America. The General Church has about 2,000 members and concentrates its activity at Byrn Athyn, where it supports an academy and a theological seminary. The General Convention seminaries are at Cambridge, Massachusetts, and Islington, London. The New Churches maintain an active missionary program and have had very successful work in Africa. There is a foundation in New York that distributes Swedenborg's writings, and the churches publish the monthly *New Church Messenger* and *Journal of the General Convention.*

BIBLIOGRAPHY: W. Wunsch, *An Outline of New Church Teaching* (1926); M. Block, *The New Church in the New World* (1932); H. Keller, *My Religion* (1964). ROBERT G. CLOUSE

NEW LICHTS, see AULD LICHTS

NEWLIGHTISM. New England Congregational preachers who supported the Great Awakening* of the 1740s with its emphasis on an instantaneous or sudden conversion experience and attendant emotional and mystical features became known as "New Lights"—because they sought to get their congregations "new-lighted" by the Spirit of God. A majority of these men were moderates (e.g., Jonathan Edwards*) and took an irenic approach, but many became radical separatists, vehemently critical of the established churches and advocates of the complete separation of church and state. Later, followers of B.W. Stone* became known as New Lights.

NEWMAN, FRANCIS WILLIAM (1805-1897). English scholar. Younger brother of John Cardinal Newman,* he was converted to the evangelical faith at fourteen, and in 1822 went to Oxford, where John was already an established don. Francis soon came to doubt the efficacy of infant baptism and so declined to take his M.A. degree. He associated with the early Brethren and joined A.N. Groves* in the mission to Baghdad; he was stoned by Muslims and just escaped martyrdom. Returning in England in 1833 to collect funds for the mission, he suffered intensely as a result of rumors of his "unsoundness," which culminated in an attack by J.N. Darby* and exclusion from Brethren circles. Spiritually isolated for many years, though earnestly longing for Christian fellowship, he eventually lost his faith and, though remaining a theist, became for a time England's foremost anti-Christian writer. Newman, who became professor of classics at University College, London, was a highly original thinker, an expert in many fields, including mathematics. His knowledge was encyclopedic, but his judgment was often warped, and he became renowned as a defender of lost causes. R.E.D. CLARK

NEWMAN, JOHN HENRY (1801-1890). Tractarian and cardinal. He was born into a family with Evangelical sympathies, and this was the strongest influence upon him until entering Oxford University in 1817. In 1822 he was elected to a fellowship at Oriel College, then a center of influence. He gradually relinquished Evangelicalism under the influence of R. Whateley,* who impressed upon him the divine appointment of the church, and Hawkins, who taught him to value tradition. E.B. Pusey,* J. Keble,* and above all R.H. Froude* took him further in High Church beliefs.

In 1828 he was appointed vicar of St. Mary's, the University Church in Oxford. The aim of Newman and his friends was to show that the Church of England was a *via media* between Protestantism and Romanism, a position based upon the teaching of the early "undivided church." His pulpit and the wider distribution of his sermons under the title of *Parochial and Plain Sermons* with the publication of *Tracts for the Times* provided the means for disseminating these views. The tracts came to an end when Tract 90—attempting a reconciliation between the Thirty-Nine Articles* and Romanism—came under widespread criticism. His researches into the early church had resulted in a book, *The Arians of the Fourth Century,* and also led to doubts about the Church of England which were raised again in 1839 while he was studying the Monophysite controversy. In 1843 he resigned St. Mary's and in 1845 was received into the Roman Church.

At first his career in the Roman Church seemed destined to be a succession of failures. Although the rectorship of the newly founded university in Dublin led to another book, *Idea of a University,* this scheme was to founder. The editorship of *The Rambler* was short-lived; a scheme to build a hostel for Catholics in Oxford in which Newman would have been warden was forbidden. But in 1864, in response to a personal attack upon him by Charles Kingsley,* Newman replied in an autobiographical sketch, *Apologia Pro Vita Sua.* This once again brought him into prominence. In 1870 he published the *Grammar of Assent* in defense of religious belief, and in 1879 he was made a cardinal.

His influence, within both the Church of England and the Roman Catholic Church, has been immense. In the former it is seen in the subsequent influence of the Anglo-Catholic tradition within the Church of England, and in Rome particularly in regard to theories about the development of doctrine (cf. his *Development of Christian Doctrine,* 1845).

See W. Ward, *The Life of John Henry, Cardinal Newman* (2 vols., 1912), and C.S. Dessain (ed.), *The Letters and Diaries of John Henry Newman* (1961).

PETER S. DAWES

NEW TESTAMENT. A translation of two Greek words which, better rendered "new covenant," occur in 1 Corinthians 11:25; 2 Corinthians 3:6; Hebrews 8:8, 9, 13. The term is based upon the prophecy of Jeremiah 31:31-34 that God would make a new covenant with his people. The apostolic church believed that He had done so in Christ. When the writings of the apostles and their companions were collected and put on a similar footing to the scriptures of Israel, it was natural for Christians to refer to "the old covenant" and "the new covenant." The NT canon was finally fixed to include twenty-seven books. The first four are the gospels, with details of the life, death and resurrection, and teaching of Jesus. Then there is the Acts of the Apostles, showing how after Pentecost the Gospel was taken from Jerusalem to Rome, followed by the epistles, letters of apostles and others to Christian congregations, and finally the apocalyptic Book of Revelation.

R.E. NIXON

NEW TESTAMENT CRITICISM. The application to the NT of techniques used by scholars in the study of ordinary literature in the attempt to determine the original wording of the various documents and to decide questions of date, authorship, literary composition, and the like. Although "criticism" has particularly negative connotations in some Christian circles—partly due to the hostility of some leading critics to orthodox theology and partly to an inadequate understanding of the nature of the biblical writings—the task of the NT critic is an essentially positive one. In the broadest sense a critic (from Gk. *krisis,* "judgment") is one who seeks to make intelligent judgments about fundamental questions which arise out of a serious study of the NT. Thus the distinction between "the critics" and "Bible-believing scholars" is a false one: anyone who studies the NT in any depth is by definition a biblical critic, for he must deal with the same necessary questions that others face.

For convenience, NT criticism may be divided into the areas of textual, linguistic, historical, literary, form, and redaction criticism.

Textual criticism seeks to ascertain the original wording of a book, particularly in the event this has been altered in the process of transmission (as in the case in all documents which have been copied by hand over the course of centuries). The discipline has been developed into a carefully scientific enterprise during the past two hundred years by men such as Griesbach,* Lachmann,* Tischendorf,* Tregelles,* Westcott,* and Hort,* von Soden, and a host of more recent scholars. Textual criticism is sometimes called "lower" criticism, since it represents the primary stage in the study of the NT and is therefore foundational for all subsequent work.

Linguistic criticism seeks to understand the nature of the words of a document. Here one is concerned with matters of Greek grammar and philology, idiomatic expressions and connotative overtones of words and phrases, precise relationships of words to one another in a particular context, the special vocabulary of an individual author, and so forth. In this area great advances have been made by scholars like Blass,* J.B. Lightfoot,* Westcott,* Deissmann,* G. Milligan,* J.H. Moulton,* A.T. Robertson, W. Bauer,* and the many contributors to the monumental *Theological Dictionary of the New Testament* (ed. Kittel-Friedrich, 1932ff.). The language of the NT is now identified as basically the ordinary Greek of the Eastern Roman Empire of the first century, but with a very strong Semitic flavor; this latter element is due both to the Aramaic mother-tongue of the earliest Christians (as also Jesus) and to the profound influence of the OT on the writers' vocabulary and style.

Historical criticism is the attempt to understand a document, a concept, or even a word in its historical setting. This is profoundly important, for a failure to grasp the historical background of a text may lead to grossly misleading interpretations. Thus one cannot understand Revelation without knowledge of the situation facing the church in Asia Minor toward the end of the first Christian century, or the fourfold prohibition of the Jerusalem Council (Acts 15:20, 29) apart from the problem of Jewish-Gentile relations in the early church, or the sayings and parables of Jesus independently of knowledge of the basic features of Semitic rhetoric and Palestinian Jewish customs. The greatest advances have been made in this area during the past two centuries of criticism.

Literary criticism—sometimes called "higher" criticism, because it builds on the results of textual or "lower" criticism—is concerned with questions about authorship, sources, composition, literary form, date, and place of writing, etc. As a comprehensive term it includes the whole scope of what goes under the rubric of "NT Introduction."

Redaction criticism represents the latest phase in the development of gospel criticism. It aims at understanding the special contribution of each evangelist, i.e., the way he (the redactor) has shaped the traditional material with which he works and how his approach differs from that of the other evangelists.

Form criticism—see separate article.

BIBLIOGRAPHY: S. Neill, *The Interpretation of the New Testament, 1861-1961* (1964); W.G. Kümmel, *Introduction to the New Testament* (ET 1966); G.E. Ladd, *The New Testament and Criticism* (1967); D. Guthrie, *New Testament Introduction* (rev. ed., 1971). W. WARD GASQUE

NEWTON, BENJAMIN WILLS (1807-1899). Early Plymouth Brethren* leader. Of Quaker stock, he had a distinguished academic career at Oxford where also he was influenced by J.N. Darby* who came on a visit. He began his ministry in Plymouth, and also traveled throughout the county preaching. About 1835 he was used in the conversion of his cousin S.P. Tregelles,* the textual critic, to whose researches he gave generous financial aid. Newton and Darby differed over prophetical interpretation and church order; in 1847 Newton was charged with heresy through some teaching on Christ's humanity, but he withdrew the doctrine. That year he left Plymouth and for many years ministered in a chapel in Bayswater, where he drew large congregations. His written ministry continued till he was nearly ninety. An austere man of Calvinist views and high personal honor, Newton influenced many leading ministers of his time. G.C.D. HOWLEY

NEWTON, SIR ISAAC (1642-1717). Scientist, theologian, and master of the Mint. Born of a Lincolnshire farming family, Newton early showed a mechanical bent. Converted as a student at Cambridge, his paramount aim was to understand Scripture. Science was a "garden" given him to cultivate; every discovery he made was, he believed, communicated to him by the Holy Spirit. Though an Anglican, he rejected infant baptism, believed that Scripture taught Arianism,* and held that all who believed simply in the love of God were entitled to Communion in church. His unorthodoxy was rarely suspected; he avoided controversy in religion as in science. Among his main interests were church history, chronology, alchemy, prophecy, mathematical science, and the relation of science to religion. In science he is remembered for the law of gravity, the infinitesimal calculus (with Leibnitz), the separation of white light into colors by the prism, Newton's rings, and his work as president of the Royal Society.* In earlier years he was lovable and generous and helped in the distribution of Bibles to the poor. In later years with the acquisition of power, his character seemed to deteriorate, and he could be singularly ungenerous to those (such as Whiston and Hooke) who ventured to disagree with him. At the Mint he was merciless to counterfeiters of coin. Newton was knighted by Queen Anne in 1705.

See biographies by J.W.N. Sullivan (1938), E.N. da C. Andrade (1954), and H. Sootin (1964).

R.E.D. CLARK

NEWTON, JOHN (1725-1807). Anglican clergyman and hymnwriter. Son of a merchant sea captain, he had an unsettled childhood and turbulent youth, including several periods of intense religious experience. He was forced to join the Royal Navy, tried to escape, was arrested in West Africa, and eventually became virtually the slave of a white slavetrader's black wife. She humiliated him, and he lived hungry and destitute for two years, involved in the slave trade. In 1747 he boarded a ship for England, but a violent storm in the North Atlantic nearly sank them. For Newton it was a moment of revelation, and he turned to God.

Nevertheless, further slave trading followed, but in 1755 he gave up the sea, and in 1764 became curate of Olney in Buckinghamshire. There, in a successful ministry of fifteen years, he befriended the poet William Cowper* and also became widely known. The two produced the *Olney Hymns,* of which a number are still in general use, including "Amazing grace," "How sweet the name of Jesus sounds," and "Glorious things of thee are spoken." In 1779 Newton moved to London, becoming vicar of St. Mary, Woolnoth. His influence was widely felt, especially in the evangelical world. Handel's *Messiah* had made an enormous impact on London, and Newton preached a famous series of sermons on the texts Handel had used as libretto. After one of these the young William Wilberforce* sought his counsel. In his latter years, Newton played a leading part in Wilberforce's political campaign which led to the abolition of the slave trade.

See B. Martin, *John Newton* (1950), and M.L. Loane, *Oxford and the Evangelical Succession* (1950). A. MORGAN DERHAM

NEW ZEALAND. Christianity was founded in New Zealand by nineteenth-century European missionaries and settlers, with a leaven of American influence. Anglican missions (1814) followed by Wesleyans (1822) and Roman Catholics (1838) made slow progress. Missionaries were frequently used by astute chiefs like Hongi (1777-1828) to further their political aims. Communication was difficult, and missionary lives were more persuasive than their preaching. They were often peacemakers in tribal wars, and freed slaves frequently assisted the work of conversion ahead of missionaries. The King Movement inspired by W. Tamihana (1802-66) combined Christian and Maori ideas, but aroused deep official suspicions. Bitter land wars and unjust confiscations gave the missions an irrevocable setback and inspired Hauhauism (c.1863) and Ringatu founded by Te Kooti (c.1830-93). By 1900 there were few Maori clergy, and even in the heavily Maori Waiapu Diocese no native synodsmen till 1900.

The healer, T.W. Ratana (1870-1939), inspired a significant independent church combining Maori and Christian religion, which had by 1931 linked with the Labour Party and played a vital role in improving the social lot of the Maori, as did those like Sir A. Ngata (1874-1950) educated at Te Aute College (1854). Ngata's important religious ideas had more influence on J.G. Laughton (1891-1965) and the Presbyterian Church than his own Anglican Church, which belatedly consecrated F.A. Bennett (1871-1950) as suffragan bishop to counter the impact of Ratana. Numerical European dominance hampered the development of indigenous Maori Christianity, though K. Ihaka (1921-) and R.H. Rangiihu (1912-) are significant leaders. After World War II, migrants from Samoa and the Cook Islands introduced a vigorous Polynesian Christianity, since 1969 largely Presbyterian.

European Christianity was dominated by Anglicans, Presbyterians, Roman Catholics, and Methodists. Outside Wakefield colonies like Presbyterian Otago (1848) or Anglican Canterbury (1850), there was no attempt to create an established church. Formerly national churches often found it difficult to adjust to being free churches. Even after the abolition of the provinces, regional loyalties remained strong and nationwide denominations developed slowly, especially among nonepiscopal churches. Presbyterians united in 1901 and Methodists in 1913. Shortage of clergy, isolation, and egalitarianism all contributed to greater lay participation than in Britain, notably among Anglicans, due to the Constitution of 1857, the first in the British Empire to reestablish synodal government.

Despite strong resistance to establishment, cooperation of churches and government has always been important in medical and welfare work and is increasing. Initially churches played a major educational role, but the 1877 Education Act established a free, secular, and compulsory primary system which effectively ended Protestant Schools and left religious instruction on a voluntary basis of doubtful legality till 1962. Led by Bishop P. Moran (1832-96), Roman Catholics established at great sacrifice a virtually complete school system. Apart from some notable secondary schools like Christ's College (1851), Protestants have worked within the state system. In Dunedin and Christchurch, churchmen were active in foundation of universities, though no faculty of theology emerged till 1945 at Otago. Theological colleges like St. John's College (1844), the Theological Hall (1876), Holy Cross College (1900), and halls of residence like Knox College (1909) have been the main Christian contribution to tertiary education. Scholars like J. Dickie (1875-1942), H. Ranston (1878-1971), J.A. Allan (1897-), and E.M. Blaiklock (1903-) have been of more than local importance. New Zealand culture has not been strongly influenced by Christianity, though writers like J.K. Baxter (1926-) are a sign of change.

Interdenominational cooperation has grown steadily from the Bible in Schools movement, through the first National Council of Churches in the Commonwealth (1941) to reunion negotiations which since 1964 have included Anglicans, Associated Churches of Christ, Congregationalists, Methodist, and Presbyterians. Protestant-Roman Catholic relations varied from bad to correct, but since 1945, the leveling influence of war and rapid population growth has strengthened previous tendencies to shed inherited and imported denominational differences. Vatican II and the establishment of a Joint Working Committee in 1967 led to important cooperative ventures like Inter-View '69 and an ecumenical faculty of theology at Otago University (1972).

Radicals like O.E. Burton (1893-) or L.G. Geering (1918-) have been rare, but R. Waddell (1850-1932), J. Gibb (1857- 1935), C. Julius (1847-1938), C. West-Watson (1877-1951), J.J. North (1871-1950), J. Liston (1881-), and P.B. McKeefry (1899-) have been distinguished national religious leaders, while A.A. Brash (1913-) and A.H. Johnston (1912-) have contributed to the international ecumenical movement. Sir W. Nash (1882-1967) and A.H. Nordmeyer (1901-) have made important Christian contributions to politics. New Zealand Christianity is sober, conservative, and practical, still very British, but increasingly aware of its Pacific and Asian brethren.

BIBLIOGRAPHY: J.J. Wilson, *The Church in New Zealand* (1910-26); J. Elder, *History of the Presbyterian Church of New Zealand* (1939); J.M. Henderson, *Ratana* (1963); J. Binney, *Legacy of Guilt* (1968); J.M.R. Owens, "Religious Disputation at Whangaroa 1823-7," *Journal of Polynesian Society* (1970); W.P. Morrell, *The Anglican Church in New Zealand* (1973); E.W. Hames, *Coming of Age* (1974); J.M.R. Owens, *Prophets in the Wilderness* (1974); P. Clark, *Hauhau* (1975); J.E. Worsfold, *History of the Charismatic Movements* (1975).

IAN BREWARD

NIAGARA CONFERENCES. Gatherings for Bible study at Niagara-on-the-Lake, Ontario, in the closing decades of the nineteenth century. These assemblies marked the beginning of the Bible Conference movement. The idea of the conferences probably originated in 1868 when eight men associated with the premillennial (or millenarian) periodical *Waymarks in the Wilderness* met informally in New York City. Other conferences followed, but early in the seventies several of the original group died, and the meetings were interrupted until younger men assumed leadership.

In 1875 another small group met near Chicago. Nathaniel West, James H. Brookes, W.J. Erdman, H.M. Parsons, and two other men agreed to meet the following summer. In July 1876, these six along with A.J. Gordon* and others met at Swampscott, Massachusetts, for fellowship and Bible study. They agreed to call their group "Believers' Meeting for Bible Study." This Swampscott meeting marked the birth of the Bible Conference Movement, for each year following these men and a growing company met for Bible study. From 1883 to 1897 the conference gathered at Niagara-on-the-Lake from which it received its name.

The conferences usually opened with a Wednesday evening prayer meeting. Then, for the next week the participants heard two Bible lessons each morning, two each afternoon, and another each evening. Topics studied during the Swampscott meetings were typical: "The Person and Work of the Holy Spirit," "How to Study the Bible," and "The Second Coming of Christ." Both the method of "Bible readings" and the topics of the conferences strongly suggest that the gatherings were a result of J.N. Darby's* travels in the United States and the influence of the Plymouth Brethren.*

Due to differences in the 1877 conference, Brookes drew up a fourteen-point doctrinal statement in 1878 which was officially adopted in 1890. The first article affirms that "the Holy Ghost gave the very words of the sacred writings," and the last article professes belief in "the premillennial advent" after "a fearful apostasy in the professing Christian body." Thus the confession reflects its background in the teachings of Darby and anticipates twentieth-century fundamentalism.

See C.N. Kraus, *Dispensationalism in America* (1958), and E.R. Sandeen, *The Roots of Fundamentalism* (1970). BRUCE L. SHELLEY

NICEA, COUNCIL OF (325). This was called by Emperor Constantine* to deal with Arianism,* which was threatening the unity of the Christian Church. The bishops assembled at Nicea (modern Isnik, in Turkey), a city of Bithynia close to Constantine's capital. According to tradition, the emperor formally opened the proceedings on 20 May. The council was hardly representative of the Western Church. Of some 300 bishops present, almost all were from the Eastern half of the empire. The Latin West seems to have been represented by four or five bishops, and two priests delegated by the bishop of Rome. One of the Western bishops, Hosius* of Cordova, presided over the council, probably because he was a confidant and respected friend of the emperor.

After an examination of the charges against Arius, the council sought a formula to express orthodoxy. A submission by Eusebius of Nicomedia* was rejected because of its blatant Arian teaching. Then Eusebius of Caesarea,* a moderate churchman, produced the baptismal creed of his church. This creed may have become the basis of the Nicene Creed,* but it is more likely that the creed of the council was a conflation from many sources, especially the baptismal creeds of the churches of Antioch and Jerusalem.

The main emphases of the Nicene Creed are: (1) the "sonship" of Christ is preferred to the Logos* concept; (2) the phrase is inserted that Christ is of the being *(ousia)* of the Father; (3) to the phrase "begotten" is added "not made," to deny the Arian contention that the Logos was "made"; (4) the Son is "one substance" *(homoousios)* with the Father—a momentous anti-Arian phrase; (5) to the words "became flesh" was added "and was made man"; (6) anti-Arian anathemas were appended to the creed.

The Nicene faith was received and signed by the majority of the bishops although not a few signed with hesitations. Arius and his friends were then anathematized along with two bishops who refused to accept the creed. Then, with a dangerous precedent, Constantine banished those anathematized to Illyricum.

Other matters dealt with by the council included the Melitian Schism* and the date of Easter.* The canons of the council are concerned with the problems of clerical discipline, heresy, and schism.

BIBLIOGRAPHY: A.E. Burn, *The Council of Nicaea* (1925); T.H. Bindley, *The Oecumenical Documents of the Faith* (rev. F.W. Green, 1950); G. Forell, *Understanding the Nicene Creed* (1965). G.L. CAREY

NICEA, SECOND COUNCIL OF (787). This seventh ecumenical* council was convoked to deal with the question of iconoclasm. In 730 Emperor Leo III* issued a decree forbidding the veneration of images or pictorial representations of Christ and the saints. Despite Jewish and early Christian beliefs, the practice of veneration of images had grown up gradually in both East and West, and by the eighth century it was well established throughout the empire; therefore Leo's decree met with fierce opposition. Both the patriarch of Constantinople and Pope Gregory III opposed the emperor, and Gregory held a synod in Rome in 731, where he excommunicated all who destroyed images. Leo's successor, Constantine V, nevertheless continued the iconoclastic policies and instituted a violent persecution of those who venerated images (see ICONOCLASTIC CONTROVERSY). Only when Constantine's widow Irene became regent for her minor son was there a change in imperial policy. In 786 she convoked a council in Constantinople to deal with the question, but it was broken up by iconoclastic soldiers. In the following year she reconvened the council at Nicea.

The council met in eight sessions over a month and was attended by over 300 prelates, mostly from the West, and included two legates sent by the pope. The position of the iconoclasts was condemned, and a statement was produced which declared that pictorial representations were lawful. They might receive "veneration" which honored the persons represented by the image, but not "adoration" which was due to God alone. In addition the council promulgated twenty-two disciplinary decrees. The decrees of the council on images were, however, not quickly accepted. Charlemagne rejected them at the Synod of Frankfurt (794), and in the West the council was not officially acknowledged as an ecumenical council until the late ninth century. In the East a number of emperors continued the iconoclastic policies until 843, when a local synod finally confirmed the decrees at Nicea.

RUDOLPH HEINZE

NICENE CREED. The Nicene Creed (N) was promulgated in 325 by the Council of Nicea* to defend the orthodox faith against the Arian heresy and to assert the consubstantiality of the Son with the Father. This relatively short creed was probably based upon creed(s) of Syro-Palestinian origin into which the Nicene emphases were interpolated.

But the term Nicene Creed is also used ambiguously of the creed used in the eucharistic worship of the church which is not only longer but different in many respects from N. The former is known as the Niceno-Constantinopolitan Creed (C). The hybrid title reflects the popular, although mistaken, view that at the second ecumenical council at Constantinople* (381) another creed was put forward which enlarged the Nicene formulary. A number of considerations, however, shed considerable doubt about the identity of C and its relationship to N: (1) no mention of a creed is made in the four canons of the Council of Constantinople or in the official letter to Theodosius.* The first appearance of C is, in fact, at the Council of Chalcedon* (451) where "the faith of the 150 fathers" was read out. There is, then, an absolute silence regarding a Constantinopolitan creed from 381 to 451; (2) a comparison of C with N shows that key formulae of the Nicene faith, such as the Son's participation in the "substance of the Father," are missing. Such omissions make it difficult to accept that it is a modified version of N; (3) the Creed of Jerusalem which Epiphanius incorporates in his tract *Ancoratus* (c.374) is practically identical to C.

Among the many suggestions for the solution of this mystery are: that C was used at the baptism and episcopal consecration of Nectarius*; that Cyril of Jerusalem presented the revised creed of Jerusalem at the Council as testimony to his orthodoxy; that the second ecumenical council reaffirmed N; and C, a creed of Syro-Palestinian origin, was embodied as an illustrative formula in its *tomos.*

The famous and divisive *Filioque* clause was added to C at the Third Council of Toledo in Spain in 589, but the Church of Rome continued to use the creed in its original form until the start of the eleventh century.

BIBLIOGRAPHY: F.J.A. Hort, *Two Dissertations* (1876); J. Kunze, *Das nicänisch-konstantinopolitanische Symbol* (1898); A.E. Burn, *An Introduction to the Creeds and to the Te Deum* (1899) and *The Nicene Creed* (1909); J.N.D. Kelly, *Early Christian Creeds* (1950); T.H. Bindley, *The Oecumenical Documents of the Faith* (4th ed. rev. by F.W. Green, 1950).

G.L. CAREY

NICEPHORUS (c.758-829). Patriarch of Constantinople. Like his father, he was an imperial secretary and strong defender of icons, a cause bringing the older man torture and banishment. In his secretarial post Nicephorus was commissioned to the Second Council of Nicea* (787), signing its promulgation sustaining the veneration of images. Shortly thereafter he withdrew to a monastery which he had founded on the Propontis, but without taking orders. Having returned to Constantinople as director of a home for the indigent, he was appointed by Emperor Nicephorus to succeed Tarasius as patriarch. His lay background, and his exoneration of a priest earlier deposed for countenancing the adulterous marriage of Emperor Constantine VI, evoked stiff opposition to him from the strong, renascent Studite Order.

With the revival of iconoclasm under Emperor Leo V, however, Nicephorus and Theodore the Studite joined forces against their mutual enemy. A majority of the clergy endorsed the policy of the emperor, who in 815 deposed and exiled Nicephorus to his monastery. Here he continued his polemic against the iconoclasts. *Apologeticus minor* (perhaps predating his exile), *Apologeticus major,* three *Antirhetikoi,* and several unedited writings constitute an apologetic corpus for icon veneration which is dogmatically definitive, apologetically thorough, and rich in its interpretation of patristic sources and preservation of imperial statements. His history of Byzantium from 602 to 769, *Historia syntomos,* acclaimed for objectivity and rhetorical style, wrestles with a theological explanation for the scourge of Islam. *Chronographia* is his chronology from Adam until 829.

JAMES DE JONG

NICEPHORUS GREGORAS (1295-c.1359). Byzantine historian. At the age of twenty he went to Constantinople where he studied under Patriarch Glykys and the Grand Logothete Theodore Metochites. He soon gained a reputation for his ability and intelligence. He was a prolific writer, and his works treat a variety of subjects such as theology, astronomy, hagiography, philosophy, grammar, and history. In addition, his large correspondence has also been edited and preserved. He is best known for his *Roman History,* a history of the period from 1204 to 1359 in thirty-seven books, which includes not only factual material but interesting insights on the political and cultural developments of this period. During the Hesychasm* controversy, surrounding the extreme mystical claims by Eastern monks at being able to arrive at a vision of the Uncreated Light of the Godhead, a claim which Gregoras opposed, he was confined to the monastery near Chora. Released in 1355, he died about four years later.

DONALD M. LAKE

NICETA (c. 335-414). He became bishop of Remesiana in the Balkans c. 370, but little is known about his life and work. Information about his work is gleaned from the writings of his contemporary and friend, Paulinus of Nola,* particularly in his poem commemorating Niceta's pilgrimage to Nola to visit the grave of St. Felix, in which is described Niceta's missionary success among the Goths, Scythians, and Dacians. Jerome also writes of the apostolic labors of Niceta, while Gennadius of Marseilles mentions that Niceta wrote at least six books, including instructions for baptismal candidates. There was also an important exposition on the Apostles' Creed, and a short work on the value of Psalm singing in which he makes some interesting remarks about people's attitude when singing in church: "Sing wisely, that is, understandingly, thinking of what you are singing . . . not savouring of the theatre. . . . do not show off. . . . our worship must be done as in God's sight, not to please men." Paulinus of Nola praises his ability as a hymnwriter, and some scholars attribute the *Te Deum** to Niceta and not to Ambrose. GORDON A. CATHERALL

NICETAS ACOMINATOS (d. after 1210). Byzantine historian and theologian. Born at Colossae into a rich family and educated at Constantinople, he eventually became governor of Philippopolis. When the members of the Third Crusade (1189) passed through his region, he became friendly with the emperor Frederick Barbarossa, and this connection allowed him to rise to prominence in the Eastern court of Constantinople. He remained here until its fall in 1204. Then he fled to Nicea. His two most famous writings are *Thēsauros Orthodoxias (A Treasury of Orthodoxy)* and *Chronikē Diēgēsis* (a history of Constantinople 1180-1206). The former is a valuable source of information for the decisions and councils of the Eastern Church from 1156 to 1166, while the latter has a valuable account of the taking of Constantinople by the Latins. PETER TOON

NICHOLAS I (the Great) (d.867). Pope from 858. He owed his election to Louis II. The Roman clergy wanted Anastasius, son of the bishop of Orte, whom Nicholas named as his secretary of state—an important choice in the subsequent schism with Photius* and the struggle with Hincmar of Reims.* During his pontificate the question of universal primacy of jurisdiction over the church was reopened. The diocese of LeMans about 850 forged a whole body of law by assigning to each decree a papal or conciliar decision as far back as the second century. These False Decretals* brought new detail to the defense of papal primacy as LeMans invoked the historical prestige of Rome. In two synods at Aix (860) and Aachen (860-62), Lothair II had his wife Theutberga repudiated. A synod at Metz (863) authorized his marriage to Waldrada. Hincmar supported Theutberga. Nicholas convened a synod at the Lateran which quashed the Metz decree and Aachen divorce. When in 862 Hincmar deposed Rothad II, bishop of Soissons, the pope ordered an examination which restored Rothad in 865. Nicholas now used the False Decretals. Together with his deposition of John, bishop of Ravenna, these events explain the papal claims to authority over Western sees.

Nicholas invited the Moravian Byzantine missionaries Cyril and Methodius* to come to Rome after 863. In March 862 Nicholas had ordered a trial of Photius, patriarch of Constantinople. This so infuriated the latter that he deposed the pope in 867. Nicholas died before learning of this action or the subsequent Photian schism which his own actions had precipitated. Rome's authority over the West increased, and so did its imperious attitude toward the brilliant Christian civilization at Constantinople. MARVIN W. ANDERSON

NICHOLAS V (1397-1455). Pope from 1447. Born Tommaso Parentucelli, son of a physician, he studied at Bologna and Florence. He became bishop of Bologna and negotiator to the Holy Roman Empire to secure the enforcement of the reforming decrees of the Council of Basle.* In 1447, on the death of Eugenius IV, he was elected pope. Nicholas then proceeded to secure the dissolution of the Council of Basle and the abdication of the antipope, Felix V (1449). In 1450 he proclaimed a Jubilee which attracted many pilgrims to Rome, thus strengthening papal prestige and finances. The pope's chief claim to fame is the encouragement he gave to the Renaissance in Rome. The city which had been neglected for over a century during the period of the Avignon* papacy and the Great Schism* was now made the center of a magnificent building program. Bridges, roads, palaces, churches, and fortresses were built. To decorate the buildings artists were brought in, including Fra Angelico.*

Nicholas's principal interest, however, was in books. Papal agents searched for rare manuscripts throughout Europe, copyists were employed to reproduce these finds, and many outstanding humanists were employed to translate them. Many of the ancient Greek authors—such as Herodotus, Thucydides, Homer, Polybius, and Strabo—and several of the Greek Fathers were translated into Latin in this effort. The pope left a large collection of manuscripts to be the foundation of the Vatican Library. The year 1453 which featured the twin blows of a plot on the pontiff's life and the fall of Constantinople led to his death. He claimed on his deathbed that the papal patronage of the Renaissance was necessary to increase Rome's reputation and thus insure its religious leadership. ROBERT G. CLOUSE

NICHOLAS, HENRY (c.1502-c.1580). Also Heinrich or Hendrik Niclaes, founder of the Family of Love or Familist* sect. Born at Münster, he was reared in a devout Christian home and as a boy demonstrated theological precocity and claimed to receive visions from God. The son of a merchant, he too became a businessman. Around 1540 he founded a religious movement which emphasized communality, mystical enlightenment, experiential holiness, the possibility of immanent union with God, the second coming of Christ, and a church comprising believers only. This last belief often led to Familists being misidentified as Anabaptists.* Nicholas spent much of his adult life in Amsterdam and Emden where he was constantly in trouble because of his aberrant religious views. However, his clandestine movement grew to substantial numbers in Germany, the Netherlands, and England. Nicholas nurtured the Familists with his many books and his pastoral travels, which probably included a trip to England in 1552-53. Having experienced consistent persecution, he died in Cologne around 1580.

ROBERT D. LINDER

NICHOLAS OF BASLE (fourteenth century). Proponent of the Free Spirit Movement. Nicholas won followers for the movement along the Rhine River from Constance to Cologne. Some of his adherents were burnt in Heidelberg and Cologne; among those executed was Martin of Mainz, a renegade monk. Nicholas claimed to be a new Christ and viewed himself as the sole source of authority and ordination; he could interpret the gospels better than the apostles; lacking his sanction, the church's hierarchy could perform no valid act. He called that person sinless who executed any one of his commands, including an order to murder or fornicate. Full submission to him would give a man his primal innocence. For his views Nicholas was apprehended and burnt in Vienna about 1395. JOHN GROH

NICHOLAS OF CUSA (1401-1464). German philosopher and cardinal. A native of Kues, he studied at Heidelberg and Padua, and became a doctor of canon law of Cologne in 1423. After lecturing at Cologne he became dean of St. Florin's, Coblenz (1431). At the Council of Basle* he wrote *De concordantia catholica,* which supported the superiority of the council over the pope. Annoyed with the proceedings of the council and its unconcern for Greek union, he supported Eugenius IV's move to Ferrara. In 1437-38 he was sent to Constantinople in the interests of reunion, and later served the papal cause in Germany (diets of Mainz, 1441; Frankfurt, 1442; Nuremberg, 1444) until settlement between pope and emperor came in the Concordat of Vienna (1448), by which time Nicholas V had made him cardinal.

Created bishop of Brixen (1450), he was appointed papal legate to Germany to preach the Jubilee indulgence, reform religious and diocesan clergy, and hold synods. He visited Vienna, Magdeburg, Haarlem, and Trier. Taking up his post in 1452, he worked hard at preaching, synods, and visitations until opposition from Duke Sigmund of Austria forced him to flee in 1457. He took refuge at Buchenstein in the Dolomites, where he wrote *De beryllo,* an essay on human knowledge; he resigned in 1458. Pius II appointed him vicar-general in 1459. With his father and sister he built a hospital in Kues which still exists and contains his library. He spent his final years as camerarius of the Sacred College in Rome. Having read the Fathers and contemplated mathematical, philosophical, and theological matters, he wrote *De docta ignorantia* (1440); *De Coniecturis* (1442); and four dialogues, *Idiota* (1450). The fall of Constantinople inspired his *De pace fidei.* More than 300 sermons and other writings remain. His non-Scholastic work was inspired by Augustine, Dionysius the Areopagite, Bonaventure, and Eckhart. Contributing to mathematics and astronomy, and responsible for the first geographical map of central Europe, he was also a legal historian, questioning the Donation of Constantine* and the Pseudo-Isidorian decrees.

BIBLIOGRAPHY: H. Bett, *Nicholas of Cusa* (1932); M. de Gandillac, Nikolaus von Cues (1953); E. Meuthen, *Die letzten Jahre des Nikolaus von Kues* (1958); P.E. Sigmund, *Nicholas of Cusa and Medieval Political Thought* (1963).

C.G. THORNE, JR.

NICHOLAS OF FLÜE (Bruder Klaus) (1417-1487). Hermit. Born near Sachseln, Switzerland, he pursued a military career in early life and worked as a deputy and judge. Influenced by the Friends of God,* he decided in 1467 to leave his wife and ten children and live a hermit's life. Visionary experiences influenced him to locate in Ranft near his birthplace. He prayed twelve hours daily and gave advice to his many visitors. Governmental officials also sought his counsel. He urged them to admit Fribourg and Soleure to the newly independent Swiss Confederation. He was canonized in 1947.

NICHOLAS OF HEREFORD (d. c.1420). Lollard writer. While a fellow of Queen's College, Oxford, and engaged in translating the Latin Bible into English, he became an ardent supporter of Wycliffe* and in 1382 began to preach Lollard doctrines. Condemned and excommunicated by the archbishop of Canterbury, he journeyed to Rome to petition against the sentence, but was imprisoned by the pope. Having escaped during a popular uprising in 1385, he returned to England and soon became the leader of the Lollard party in the west country, appealing especially to the common people by addresses and handbills. He was captured in 1391, was imprisoned at Nottingham, was tortured, then suddenly recanted, leaving his followers depressed and bewildered. Royal and episcopal favors were now heaped upon him. He was made chancellor and treasurer of Hereford Cathedral, but ended his life as a Carthusian monk at Coventry. He probably played an important part with John Purvey* in a revision of Wycliffe's English Bible.

IAN SELLERS

NICHOLAS OF LYRA (c.1265-1349). Franciscan scholar. Born in Lire (now Vieille-Lyre, Eure), he entered the Franciscan Order at Verneuil (c.1300). He studied theology at Paris, becoming a doctor (c.1308) and a regent master at the university. From 1319 he was provincial of his order in France, being present in that capacity at the general chapter of Pérouse (1321). In 1325 he was provincial of Burgundy. As executor of the estate of Jeanne of Burgundy, widow of Philip VI, he founded the college of Burgundy at Paris, where he died. The best-equipped biblical scholar of the Middle Ages, knowing Hebrew and acquainted with Jewish commentaries, notably Rashi, he was especially concerned to expound the literal sense of Scripture as against the current allegorical interpretation. He wrote two commentaries on the whole Bible, one being the first biblical commentary printed. He wrote a treatise on the Beatific

Vision, directed against Pope John XXII, and other works. J.G.G. NORMAN

NICHOLAS OF MYRA. Very little is known about Nicholas, from whom Santa Claus is derived, and of one of the most popular of all the saints in Christendom the prototype. He is reputed to have suffered imprisonment under the Roman emperor Diocletian,* only to be released to serve as one of the bishops at the Council of Nicea* (325), a supposition which is not supported by any of the records of Nicea. The patron saint of Russia as well as of sailors and children, he was believed to bring gifts to children on his feast day, 6 December. The first clear evidence of Nicholas is in the Church of St. Priscus and St. Nicholas built by Emperor Justinian at Constantinople. Veneration for this saint spread to the West after 1087, when his remains were claimed to have come into the possession of the people of Bari. DAVID C. STEINMETZ

NICHOLSON, WILLIAM PATTESON (1876-1959). Irish evangelist. After a wild youth spent at sea, he was converted in 1898, trained at the Bible Training Institute, Glasgow, and served as an evangelist with the Lanarkshire Christian Union. In 1914 he was ordained as an evangelist in the Presbyterian Church in the USA, and subsequently joined the staff of the Bible Institute of Los Angeles. His greatest work, however, was done in Ulster, where his preaching—vivid, uncompromising, and unconventional—brought about a significant spiritual awakening in the 1920s. Thousands were converted, and the moral tone of Belfast's dockland was raised. In 1926 he deputized for J. Stuart Holden as CICCU missioner in a united mission to Cambridge University (another missioner was William Temple). Nicholson's preaching created a sensation and led to numerous conversions. He campaigned also in Australasia and South Africa.

HAROLD H. ROWDON

NICODEMUS, GOSPEL OF see APOCRYPHAL NEW TESTAMENT

NICOLAI (Ivan Kasatkin) (c.1835-1912). Russian Orthodox missionary bishop to Japan. He offered as chaplain to the Russian Consulate at Hakodate, Hokkaido, became a monk, and took the name Nicolai. Arriving in Japan in 1861, he acquired a deep knowledge of the Chinese and Japanese languages. Christianity being a prohibited religion, he proceeded cautiously. Not until 1868 did he baptize his first three converts—a Samurai named Sawabe, a physician Sakai, and one Urano. Returning to Russia in 1869, he was responsible for the constitution of the Orthodox Mission. The penal laws against Christianity were abrogated (1873), so he moved to Tokyo where he built a cathedral. He was consecrated bishop (1880) and archbishop (1906). He encouraged the indigenous aspect of the church, selecting promising young men as catechists to evangelize their own people. The Russo-Japanese War (1904-5) severely tried the work, but at his death there were over 30,000 converts. J.G.G. NORMAN

NICOLAI, PHILIPP (1556-1608). Pastor and hymnwriter. Son of a Lutheran pastor, he was born at Mengeringhausen, Waldeck, studied at Erfurt and Wittenberg, and was appointed Lutheran preacher at Herdecke (1583). When Spanish troops invaded in 1586, the Mass was reintroduced and he resigned. After a pastorate at Niederwildungen, he became chief pastor at Altwildungen, then court preacher to Countess Margaretha of Waldeck (1588). He took part in the Sacramentarian Controversy and was forbidden to preach for a time (1592-93). In 1596 he became pastor at Unna, where he wrote his celebrated "*Watchet auf, ruft uns die Stimme*" ("Sleepers wake"), based on Matthew 25:1-13, during a terrible pestilence. In 1599 he wrote "*Wie schön leuchet de Morgenstern,*" marking the transition in German hymnody from the objective churchly period to the subjective experimental period. His last pastorate was St. Katherine's Church, Hamburg (1601-8). J.G.G. NORMAN

NICOLAITANS. Followers of one Nicolaus, they formed a sect in the early church at Ephesus and Pergamum and were condemned by John in Revelation 2:6, 15. They appear to have been a heretical group who retained the pagan practices of idolatry and immorality, which were contrary to Christian thought and conduct. Virtually nothing is known about these sectaries beyond John's references to them. Their works are hated and rejected, but not described, in the letter to Ephesus, while in Pergamum their teachings are held in like manner to those of Balaam (Num. 24:1-25; 31:16). Irenaeus asserts that this sect was founded by that Nicolaus who was the proselyte from Antioch, one of the seven appointed by the apostles in Jerusalem (Acts 6:5); but the weight of recent scholarship seems to be against this view. The sect disappeared after the second century, although in the Middle Ages the term was sometimes applied to married priests by those who were staunch supporters of clerical celibacy.

WATSON E. MILLS

NICOLE, PIERRE (1625-1695). French theologian. Native of Chartres, he took degrees in philosophy and theology at Paris, and taught at Port-Royal where began the close association with Antoine Arnauld.* Sometimes using a pseudonym, as with his Latin version of Pascal's *Lettres provinciales* (1658) signed "William Wendrock," he was a moderate Jansenist despite his extreme rejection of mysticism. He wrote also *Traité de l'oraison* (1679) and (his last work) *Réfutation des principales erreurs des quiétistes* (1695), prompted by Bossuet. Nevertheless he was keen to defend Jansenism,* opposing the Jesuits. Other volumes include a defense of transubstantiation which attacks the Calvinist as well as the general Protestant position, and his posthumous publication *Traité de la grâce générale* (2 vols., 1715) which revealed his absolute abandonment of Jansenism. C.G. THORNE, JR.

NICOLL, WILLIAM ROBERTSON (1851-1923). Religious journalist. Son of a Free Church of Scotland minister, he was educated at Aberdeen and subsequently entered the ministry, serving charges at Dufftown (1874-77) and Kelso (1877-85). Ill-health caused his demission. Moving to London, he edited *The Expositor* (1885-1923) and *The British Weekly* (1886-1923). He wrote also for the secular press and published many books—scholarly, expository, devotional, and literary. A friend of Lloyd George, he supported World War I, doing much to overcome Nonconformist pacifist inclinations. He was a formidable opponent of Erastianism and the Roman Catholic theory of "tactical succession" of bishops. He was knighted by Edward VII in 1909.

R.E.D. CLARK

NICOMEDES. Early Christian martyr. The date and circumstances of his death are unknown. He is mentioned in the Roman Martyrology (the official record of the Roman Catholic Church) issued in 1584, but omitted from the Hieronymian Martyrology (compiled in the fifth century). Moreover, there is mention in the fifth century of a titular church at Rome dedicated to him. The Book of Common Prayer, listing his feast day as 1 June, spells his name "Nicomede."

NIEBUHR, HELMUT RICHARD (1894-1962). Neoorthodox theologian. Missouri-born Niebuhr, professor of Christian ethics at Yale University from 1931 to 1962, personified American Neoorthodoxy,* the product of social and intellectual travail between the two world wars. An Evangelical minister, he brought to Yale much experience as seminary professor, college president, pastor, and author, and continued there to struggle with historicism, a task reflected in his doctoral dissertation at Yale, "The Religious Philosophy of Ernst Troeltsch" (1924). His *Social Sources of Denominationalism* (1929) attacked the church's gullible acceptance of middle-class values and described the socioeconomic origins of sect, denomination, and church. Various factors led him away from his early liberalism, however, and in *The Kingdom of God in America* (1937) he hoped for the restoration of Reformation roots in American Christianity. *The Meaning of Revelation* (1941) and *Radical Monotheism and Western Culture* (1961) grappled with the nature of religious experience, and *Christ and Culture* (1951) further explored the nature of Christian association with the world. More scholarly than his elder brother Reinhold,* he attempted to explore the relationship of faith and civilization by combining belief in God's sovereignty with modern scholarship, in order to effect a creative tension between the church and society.

DARREL BIGHAM

NIEBUHR, REINHOLD (1893-1971). Neoorthodox theologian. An Evangelical (now United Church of Christ) pastor who left a Detroit industrial parish in 1928 to begin thirty-two years on the faculty of Union Theological Seminary, New York, he struggled throughout his life with the question he raised in lectures at Edinburgh in 1939: "Man has always been his most vexing problem. How shall he think of himself?" Niebuhr thought of man as both nature and spirit, neither damned nor perfectible, but capable of transcending himself. A polemicist as well as a scholar, he contributed hundreds of articles to magazines and journals, was active in the creation of the National Council of Churches, New York's Liberal Party, and Americans for Democratic Action, and wrote seventeen major books, including *Moral Man and Immoral Society* (1932), *The Nature and Destiny of Man* (1941 and 1943), *Faith and History* (1949), and *Christian Realism and Social Problems* (1953). Thoroughly American, he was as critical of Karl Barth's* bibliolatry and aloofness from society as he was of doctrinaire reformers, although he believed that God and man were radically separate and that society needed fundamental changes.

Always sensitive to problems of church and nation, he chose the ministry in 1915 after theological studies at Eden Theological Seminary and Yale Divinity School because he wanted to work in society rather than toward an advanced theological degree. His pastorate taught him the meaning of modern technocracy, and for a time he was a member of the Socialist Party. He ran for Congress as a Socialist in 1930. The New Deal and imminent world war, however, cured him of socialism and pacifism, and in 1941 he founded *Christianity and Crisis* to bring realism into American Christianity's view of world ills. After the war he helped to create Americans for Democratic Action to keep Communists out of liberal Democratic affairs. A dynamo who slowed slightly after a 1952 heart attack, he opposed those who deluded the American people. Critical of reformers who ignored human self-glorification, and impatient with theologians who were pessimistic about man's chances of self-improvement, he united pragmatism and Christian orthodoxy to effect a theology that accepted God's sovereignty and encouraged men to reform institutions. A major figure in recent American Protestantism, he brought a needed sense of tragedy into American progress and influenced many secular thinkers.

BIBLIOGRAPHY: D.B. Robertson, *Reinhold Niebuhr's Works* (1954); C.W. Kegley and R.W. Bretall (eds.), *Reinhold Niebuhr: His Religious, Social, and Political Thought* (1956); J. Bingham, *Courage to Change: An Introduction to the Life and Thought of Reinhold Niebuhr* (1961); D.B. Meyer, *The Protestant Search for Political Realism, 1919-1941* (1961); N.A. Scott, Jr., *Reinhold Niebuhr* (1963).

DARREL BIGHAM

NIELSEN, FREDRIK KRISTIAN (1846-1907). Danish bishop and theologian. In his youth he was for some time affected by neo-rationalism, but later he found his permanent position in the conservative and ecumenically minded tradition of N.F.S. Grundtvig.* In his views of church order and in his practical ecclesiastical work and policy, he was strongly influenced by the Anglican Church. In 1877 he became professor of church history at Copenhagen University and there was characterized more by a staggering amount of knowledge and great narrative skill than by a sense of historical and doctrinal coherence and development. He became bishop of Aalborg in 1900 and was transferred to the see of Aarhus in 1905. His writings include a history of the Roman

Church in the nineteenth century (ET 1906), and he edited a Scandinavian dictionary of the church. N.O. RASMUSSEN

NIEMÖLLER, MARTIN (1892-). German theologian. Born in Lipstadt, Westphalia, son of a pastor, he received his education at Elberfeld Gymnasium and the University of Münster. During World War I he served as a submarine commander in the German Navy. He was ordained to the ministry in 1924.

In 1931 Niemöller became pastor of the fashionable church in Berlin-Dahlem. He soon joined with others such as Bonhoeffer* and Hildebrandt to oppose Hitler's Nazi rule and use of the Evangelical Church. This led him to help form and become the president of the *Pfarrer-Notbund* (Pastor's Emergency League) in 1933 which voiced its opposition to the Nazi anti-Jewish laws. He made a brave protest to Hitler in person, but was unsuccessful. As a further move of protest he became active in *Die Bekennende Kirche* (Confessing Church*). After several years of increasing restrictions and difficulties, he was arrested and imprisoned first at Sachsenhausen and then at Dachau concentration camps (1937-45).

After the war he served as president of the *Kirchliches Aussenamt* (Office for External Relations of the Evangelical Church in Germany) from 1945 to 1956, and president of the Evangelical Church in Hesse and Nassau from 1947 to 1964. In 1961 he was elected one of the presidents of the World Council of Churches, and served until 1968. JOHN P. DEVER

NIETZSCHE, FRIEDRICH (1844-1900). Philosopher and philologist. Born in Röcher, Prussia, son of a Lutheran minister, he showed early brilliance. Before passing his final examination he was appointed an associate professor of classical philology at the University of Basle on the recommendation of F.W. Ritschl. He resigned from this post in 1870, volunteering as a medical orderly in the Franco-Prussian war. Due to ill-health, he returned to the university the same year, finally retiring on a small pension in 1879. He went insane in January 1889. He had been "awakened" by the work of Charles Darwin* and what he took to be the nihilistic implications of evolutionary theory. Nietzsche attacked Christian dogma (e.g., in *The Antichrist,* 1895), but more especially he attacked the prevalent idea that Christian ethics could survive the overthrow of the Christian view of man which he believed the work of Darwin had brought about. "Supernature" is not something that men have in virtue of their creation in the divine image, it is a goal for the future. The "superman," capable of self-mastery, must go "beyond good and evil," beyond the values of a defunct Christianity. Several factors make Nietzsche's thought hard to grasp: his aphoristic, wide-ranging, immensely fertile work; misrepresentations of (for example) the superman idea, by fascists; and difficulties over the authenticity of several of his writings. PAUL HELM

NIGERIA, see WEST AFRICA

NIKON (1605-1681). Patriarch of Moscow. He had a monastic education and became a monk after the early deaths of his three children, separating from his wife, who became a nun. On a visit to Moscow in 1643 he became friendly with Czar Alexis, who subsequently promoted him to the patriarchate (1652-60) and allowed him to exercise considerable power: he acted as regent in the czar's frequent absences from Moscow. Nikon quickly reformed the Russian liturgy, bringing it into conformity with Greek and Ukrainian practice, thus aiding the czar politically in the absorption of the Ukraine (1654-67). Nikon's reforms were made permanent, but much opposition was aroused, the Old Believers* continuing to observe the traditional forms of worship. Feelings were so strong that both Nikon and the czar were variously identified with Antichrist, and bloodshed more than once ensued. Nikon was deposed in 1660 and exiled to a remote monastery. He was pardoned shortly before his death by the next czar, Fedor III. A man of immense energy and influence, though lacking in tact, Nikon is reckoned the greatest of the Russian patriarchs. R.E.D. CLARK

NINETEENTH AMENDMENT (1920). This Constitutional Amendment instituted nationwide woman suffrage, prohibiting the United States or state governments from denying or abridging citizens' right to vote "on account of sex." It culminated the national suffragist movement begun in 1848 by the Women's Rights Convention at Seneca Falls, New York. Supported by organized labor and prohibitionists, the National Association for Woman Suffrage of Elizabeth Cady Stanton and Susan B. Anthony and the American Woman Suffrage Association of Henry Ward Beecher,* Lucy Stone, and Julia Ward Howe* crusaded after 1869, merging in 1890 as the National American Woman Suffrage Association. Carrie C. Catt, Anna H. Shaw, and Alice Paul pressed the Amendment through Congress (1919) exactly as written by Susan Anthony when first submitted in 1878. Voting patterns remained substantially unchanged after passage. D.E. PITZER

NINETY-FIVE THESES, THE (1517). The generally accepted view that this episode marks the first public declaration of Reformation principles requires some modification. On 31 October (or 1 November) 1517, Martin Luther, angered at deceptions practiced on the common people by Tetzel's* sale of indulgences* at Jüterborg and Zerbst near Wittenberg, and agitated by the spiritual crisis through which he was then passing, nailed Ninety-Five Theses upon Indulgences to the door of the castle church, as a preliminary to a disputation which was never in fact held. The theses, which are heavily theological though shot through with outbursts of outrage and anguish, are really quite conservative in character. Justification by faith is not mentioned therein, nor were they intended to force a breach with the papacy, but merely to direct the pope's attention to a particular scandal in the confident expectation that it would be suppressed.

Briefly, the theses affirm that penance implies repentance, not priestly confession; mortification of the flesh is a useless exercise unless accompanied by inward repentance; the merits of Christ alone avail for the forgiveness of sins, penances and works prescribed by the Church having validity only insofar as they proclaim and confirm this divine pardon; the real "treasure of the Church" is the gospel of the grace of God in Jesus Christ. Though Luther would thus have strengthened the authority of the church by placing it upon a proper basis, the papal authorities were in no mood to distinguish attacks on ecclesiastical abuses from attacks on the church itself, and Luther's action led directly to the Curia's proceeding against him on the grounds of suspected heresy in June 1518. IAN SELLERS

NINIAN (c.400). British saint. According to Bede he was the son of a British chieftain, trained for the church in Rome. He settled in SW Scotland at Whithorn where he built a white church, Candida Casa, named in honor of Martin of Tours.* There has been considerable controversy over Ninian's work recently, as some Scottish historians have maintained that the Picts of Scotland were Christianized by Columba (521-97) while others, using the evidence of place-names and similar materials, believe that Ninian was largely responsible for the conversion of the Picts of E Scotland.

NITRIAN DESERT. Site of fourth-century Christian monasticism, known also as Nitrian Valley and Desert of Scete. A shallow desert valley about 50 miles south of Alexandria, it extends diagonally across the NE tip of the Libyan desert for some thirty miles. The name derives from "natron," the sodium carbonate found in its numerous lakes. The founder of its monastic settlements was reputed to be Amun (c.320), one of the Tall Brothers. The colony of hermits, living in *lauras* (i.e., clusters of windowless cells), numbered about 5,000 in the time of Macarius the Egyptian. According to Cassian, there were four churches, each with a presbyter. The colony became a center of learning; there Origenism* flourished, the *Apothegmata Patrum** began to be recorded, and Evagrius Ponticus* wrote. Barbarian attacks in the fifth century brought it low, but four ninth-century Coptic monasteries remain.

J.G.G. NORMAN

NITSCHMANN, DAVID (1696-1772). Moravian missionary and bishop. Born in Moravia and a carpenter by trade, he joined the Herrnhut* community as an evangelist. In 1732 he accompanied J.L. Dober* on the first Moravian Brethren* mission to the Negro slaves on St. Thomas, Virgin Islands, but returned to Germany after a few months. In 1735 he was the first Moravian to be consecrated bishop, thus establishing the principle of historical succession among the Brethren. Immediately after this he led a group of sixteen missionaries to Georgia and made a deep impression on John Wesley,* who was on the same ship. Constantly active as a bishop, he made at least fifty sea voyages before his death in Bethlehem, Pennsylvania, a settlement he had founded.

RICHARD V. PIERARD

NITZSCH, KARL IMMANUEL (1787-1868). German Lutheran theologian. Son of a general superintendent of the Lutheran Church and professor at Wittenberg, he was educated at Schulpforta and Wittenberg. He became a privatdozent in Wittenberg in 1810 and assistant preacher at the Castle Church in 1811. In 1822 he was called to a chair in theology at the University of Bonn and in 1847 succeeded P.K. Marheineke as professor of theology at the University of Berlin. As provost of the *Nicolaikirche* (1854) and a member of the supreme church council of the Prussian Evangelical Church, Nitzsch was a vigorous and articulate supporter of the Evangelical Union. As a theologian he represented the ecclesiastical-pietistic wing of the *Vermittlungstheologie,* a movement which sought to mediate between the culture of the early nineteenth century and the tradition of historical Christianity. He stressed the immediacy of religious feeling as the basis of religious knowledge, uniting elements of Schleiermacher's* theology with classical Protestant dogmatics. His chief works include the *System der christlichen Lehre* (1829) and his *System der praktischen Theologie* (1847-67). He was a co-founder of three theological journals: *Theologische Studien and Kritiken* (1828), the *Bonner Monatschrift,* and the *Deutsche Zeitschrift für christliche Wissenschaft und christliches Leben* (1850). DAVID C. STEINMETZ

NOBILI, ROBERT DE, see DE NOBILI

NOLASCANS, see MERCEDARIANS

NOETUS. He died about the end of the second century, and was considered heresiarch by the early patristic writers, but no direct evidence of his thought has survived. Even the polemic which recalls him is minimal, restricted to the cataloging of heresies which can be reconstructed for Hippolytus,* and to the few fragments of the latter's detailed refutation which survive. Noetus is identified with the position that so related the Father with the Son that the Father was considered to have suffered in the crucifixion. Sometimes called Patripassian Monarchianism this view Noetus held in common with Praxeas,* Victor I,* and Sabellius (see SABELLIANISM). Noetus, whose date must precede the work of Hippolytus, was thought to have come to Rome from Smyrna.

CLYDE CURRY SMITH

NOMINALISM (Lat. *nominalis,* "belonging to a name"). The theory of knowledge which insists that universals are created by reason. Essences have no independent reality of their own, but are only names or mere vocal utterances. Reality attributed only to particulars or individual things. This epistemological position is in opposition to Platonic realism, which insisted that universal essences existed. The nominalistic pattern of thought was first introduced by Porphyry* (233-304) who attributed it to Aristotle.

In the Middle Ages, nominalism emerged as a strong reaction to realism and was championed by Roscellinus* of Compiègne (c. 1050-1125). Abelard* gave these early disputes prominence when he tried (somewhat successfully) to find a middle position. His efforts led to a moratorium until William of Ockham* (1280-1349) revived nominalism and established the first frankly nominalistic system (ontological nominalism). He taught that terms are mental realities and are universals only insofar as they can stand for many. The terms themselves are like any reality, singular and unique. This universality is purely functional and does not refer to a common essence possessed by many things outside the mind. Reality was a collection of absolute singulars and, therefore, could not give evidence or provide scientific support for God's existence. This is clearly anti-Thomistic. Faith became the grounds of belief. The Ockhamist School eventually faded away, but nominalistic thought has continued to have its effect on philosophical thought. The teachings of John Locke,* David Hume,* and certain branches of logical positivism incorporate some of the thought of the Ockhamists. JOHN P. DEVER

NOMMENSEN, LUDWIG INGWER (1834-1918). German missionary to Sumatra. Born on Nordstrand and apprenticed to a schoolmaster, he entered the Rhine Mission's school at Barmen in 1857. Four years later he was sent out to the mission's new field of Sumatra and soon proved to be a man of indomitable resolution and faith. His ministry among the Bataks proceeded slowly until a number of chiefs were converted, and then Nommensen was overwhelmed by a great people's movement (103,500 Christians in 1911) and undreamed-of problems of church organization. He decided it should be a Batak, not a Western, church: for example, as much as possible of the traditional culture would be retained; Bataks were to be trained for the ministry; and congregations would establish systems of lay elders. Actually, the church organization was patriarchal, and the missionaries held all the positions of influence and authority until German control ended in 1940. RICHARD V. PIERARD

NONCONFORMITY. A term used generally to describe the position of those who do not conform to the doctrine and practices of an established church. The word "nonconformist" was first used in the penal acts following the Restoration (1660), to describe those who left the Church of England rather than submit to the Act of Uniformity* (1662). The earlier term "Dissenters" was superseded by "Nonconformist," and this to a certain extent by "Free churchmen." In England at present, nonconforming Protestants are composed largely of the Methodists, the Baptists, the Congregationalists, and the Presbyterians, but in Scotland the Church of Scotland is presbyterian, and the Episcopal Church in Scotland is therefore nonconformist. HOWARD SAINSBURY

NONJURORS. A title used to describe those members (usually clergy) of the Church of England and Episcopal Church of Scotland who refused to take the Oath of Allegiance to William and Mary after the Revolution of 1688. Six English bishops (and all the Scottish bishops) together with about 400 English clergy felt bound in 1689 by virtue of their oath to James II* to refuse the oath and were deposed in 1690. They set in motion a kind of High Church, episcopalian nonconformity, separate from the national church but claiming to be the true, historical church of England. The English prelates deprived were Archbishop W. Sancroft* and bishops T. Ken* (Bath and Wells), J. Lake (Chichester), F. Turner (Ely), T. White (Peterborough), and W. Lloyd (Norwich). With the permission of James II, three of these took part in the consecration of two bishops in secret—Thomas Wagstaffe (Ipswich) and George Hickes (Thetford). The latter consecrated three more.

In 1714 the ranks of Nonjurors were enlarged by the accession of those who did not swear allegiance to George I. Though agreeing in their opposition, the Nonjurors were divided among themselves on secondary matters. They disagreed as to the lawfulness of worshiping in parish churches and, after the death of Hickes, on whether or not they should use a new liturgy (based on early Christian liturgies and the 1549 Book of Common Prayer) written by a group of their own men, or continue to use the 1662 BCP. In the latter controversy the participants were called Usagers and Non-Usagers, but their quarrel was settled in 1732 in favor of the use of the new form. The four "usages" were the mixed chalice, the oblatory prayer, the offering of the elements to the Father, and the prayer for the descent of the Holy Spirit on the elements.

Apart from their belief in the Divine Rights of Kings* and the doctrine of nonresistance to rightful authority, they held a high doctrine of the historical episcopate and of liturgical worship. They admired the Eastern Orthodox Churches and even made efforts to unite with them. Their movement, however, did not last much more than a century, the last congregation dying out in 1805. Not a few able laymen were attracted to the Nonjurors—e.g., Henry Dodwell, the Camden professor of history at Oxford. Because of their high doctrine of the ministry and emphasis on liturgy, they are usually classed with the seventeenth-century Caroline and the nineteenth-century Tractarian divines.

See J.H. Overton, *The Nonjurors: their lives, principles and writings* (1902), and H. Broxap, *The Later Nonjurors* (1924). PETER TOON

NORBERT (c.1080-1134). Founder of the Premonstratensians.* He was born of a noble family at Xanten, Germany. After a period as a member of Henry V's imperial court and a canon of Xanten, he renounced this official type of clerical life in 1115. He gave away his goods, received ordination, and went about preaching in poverty. In 1120 Norbert founded a community at Prémontré, which grew into the Premonstratensian Order. He preached in Germany, France, and Belgium, where his fame was responsible for his summons to Antwerp to win back its inhabitants from the heretic Tanchelm in 1124. In 1126, the year

Honorius II confirmed his order, Norbert became archbishop of Magdeburg and won great respect as a clerical reformer. He traveled to Rome with the emperor Lothair in 1132-33 and successfully supported Innocent II against the antipope Anacletus II. In 1133 Lothair made him chancellor of Italy. Norbert died at Magdeburg, and was canonized by Gregory XIII in 1582. PETER TOON

NORIS, HENRI (1631-1704). Roman Catholic scholar and cardinal. Born in Verona, Italy, of English descent, he studied with the Jesuits at Rimini but joined the Augustinian Hermits* there in 1646. Lecturing in theology and ecclesiastical history at Pesaro, Perugia, Florence, Padua, and Pisa (where he held the chair, 1674-92), he was then appointed by Innocent XII as the first custodian of the Vatican Library, consultor to the Holy Office (1694), and cardinal (1695). *Historia Pelagiana* and *Vindiciae Augustinianae* (1673) were his chief works, containing his interpretation of Augustine's soteriology and defending his doctrine of grace, which he believed Baianism and Jansenism misunderstood. The Spanish Inquisition later accused him of heresy, but in 1748 Benedict XIV ordered his books to be removed from the Index. C.G. THORNE, JR.

NORTH AFRICA. Unlike Egypt and the Lebanon, North Africa had no remnants of ancient churches to keep alive some semblance of Christian testimony. When Islam swept across the area in the early seventh century, the church of Augustine and many early martyrs (see AFRICA, ROMAN) was swept away and Christianity for centuries was repressed, apart from the Roman Catholic presence in some Moroccan coastal stations and neighboring islands. It was not until the 1860s that Roman Catholic missionaries were concerned with much more than ministering to the European population, and not until 1908 was the mission raised to the status of a vicariate apostolic. Meanwhile, in Algeria and Tunisia during that same century there was extensive immigration from France, Spain, and Italy, and this predominantly Roman Catholic population helped also the work of missions (there had been sporadic work in Algiers and Tunis from the mid-seventeenth century). By 1866 the archbishopric of Algiers claimed to have 187 parishes and 273 secular priests; by 1884 its head was designated archbishop of Carthage and primate of Africa; by 1930 the Roman Catholic population was reported to have been 805,000. Converts among non-Christians were few, and the work was made harder, not only by the hostility of Islam, but by the discouragement of the French civil authorities. J.D. DOUGLAS

Less than a century has passed since Protestant missionaries entered North Africa. An English businessman, Edward Glenny, was the prime mover in the establishment of a mission to the Kabyles. These people, who live in the central mountains of Algeria, were fully Islamized and formed part of the larger group of Berber people who live across North Africa. The stated aim of the mission was to give them the gospel of Jesus Christ. The work eventually spread to Libya, Tunisia, and Morocco, and became known as the North Africa Mission.

Other work followed. Lilias Trotter* was responsible for the formation of the Algiers Mission Band, which made special efforts to reach the oases dwellers of the south. In the same year (1888) John Anderson founded the Southern Morocco Mission, seeking especially to evangelize the Berbers and Arabs of that large area. More recently these three missions have combined to form the North Africa Mission, with an international structure and headquarters in the south of France.

Over the same period the American Methodists, the Emmanuel Mission, Mennonite Mission, Sahara Desert Mission, Southern Baptist Convention, Gospel Missionary Union, Bible Churchman's Missionary Society, and the Christian Brethren have all worked in various parts of North Africa. Most of them, if not all, have suffered setbacks in the past decade, due to an aggressive nationalism which has been strongly Islamic and, generally speaking, unfavorable to the West.

Islam, both socially and religiously, has been fiercely resistant to the Gospel, and from the beginning faithful personal witness and testimony has been the pattern. Classes for all ages gradually became possible; such skills as carpet making, woodwork, pottery, embroidery, and knitting usually formed the popular appeal which enabled the missionaries to make their first friendly contact with the people. Large-scale medical work has never been allowed, but in Morocco and Algeria small but intensely busy clinics have been maintained, with many smaller ones in the interior. Often they have specialized in midwifery. There are two hospitals run by missionaries in the same two countries.

Arabic is the main language of North Africa, and the literary or Van Dyke Bible has been widely distributed. The NT and various portions have been published in the colloquials, and the whole Bible has been published in North African Arabic by the British and Foreign Bible Society, whose agents have long worked in North Africa. Bookshops, as selling bases and centers of testimony, have been established. Sometimes the work has been of a polemical nature, but there has been a more positive outreach to students.

Two world wars saw the emergence of Islamic nationalism and a resentful feeling that Christianity was the religion of the occupying power and that the latter and its religious emissaries should go. There have been no mass expulsions of missionaries, but many individuals and some complete groups have ceased to function. During the period of intense nationalism, the radio and Bible correspondence course ministry of the North Africa Mission based in Marseilles has proved a powerful means of gospel penetration, permitting thousands to study the basic facts of the Gospel and its implications in a way unknown before. The Muslim authorities are well aware of this and lose no opportunity by press and radio of warning the masses against the Christian impact. Earlier there were scattered converts, often without contact with other Christians and rarely able to form worshiping groups. These combined ministries have resulted in the establishment of many small groups, often meeting informally and frequently facing much opposition, all over North Africa.

The criticism has been made that though there were converts here and there over the area, the missionaries of different evangelical societies had rarely been able to bring them together to form witnessing, functioning churches. It should be said, however, that there have been various emphases in the whole approach to the building of the church. The Methodists have been church-conscious from the beginning, and through their hostels and training centers have sought to provide a trained, recognized leadership of national churches. Evangelical missions generally tend nevertheless to be individualistic, with their policies reflecting the patterns of the countries from which the missionaries have come. North Africa presents a background of ideological conflict and spiritual inquiry, and it is against this that the Christian is called to fulfill his mandate to preach the Word of reconcilation, and to plan afresh in strength the true church of Jesus Christ.

BIBLIOGRAPHY: J. Rutherford and E.H. Glenny, *The Gospel in North Africa* (1900); R. Kerr, *Morocco after Twenty-Five Years* (1912); A. Philippe, *Missions des Pères Blancs en Tunisie, Algérie, Kabylie, Sahara* (1931); A. Pons, *La Nouvelle Église d'Afrique ou le Catholicisme en Algérie, en Tunisie et au Maroc depuis 1830* (n.d.); B.A.F. Pigott, *I. Lilias Trotter* (n.d.); F.R. Steele, "Tolerance and Truth," *Cross and Crescent* (June 1963); R. Stewart, "The New Algeria," *North Africa* (September/October 1963); K.S. Latourette, *A History of the Expansion of Christianity*, vol. 6 (rep. 1970), pp.9-20. ROBERT I. BROWN

NORTH INDIA, CHURCH OF. A union of six denominations inaugurated on 29 November 1970 at Nagpur. The six were: Anglicans (the Church of India), with an estimated membership of 280,000; the United Church of Northern India, 230,000; Baptists, 110,000; Methodists of the British and Australian Conferences, 20,000; Church of the Brethren, 18,000; Disciples of Christ, 16,000. These figures were greatly modified by provisional statistics of the CNI, giving total membership as only 569,546, with 230,959 communicants. Initially nineteen dioceses were formed, stretching from Assam in the north to Nandyal in Andhra Pradesh (which is more "South" India than "North"). There were seventeen bishops, eight of them former Anglican bishops, and 917 presbyters. The synod comprised all the bishops plus equal lay and clerical representation from dioceses.

Efforts for union had begun in 1929; one of the bodies, the United Church of Northern India, was itself a union of Presbyterians and Congregationalists in 1924. A late development which greatly altered the complexion of the united church was the decision not to join by the Methodist Church in Southern Asia, numbering about 600,000. Other churches, such as the Lutherans, remained outside the union, and the churches of NE India (where the UCNI had half its membership) were yet to form a separate union. Although limited, the CNI union was nevertheless an important landmark in Indian church history and was generally regarded as a stage to still further union in making an All-India Church. In two respects the CNI union differed from that of the Church of South India* (1947) and had significance for other parts of the world:

(1) The CSI began with a "mixed" ministry, some ministers having been ordained by bishops and some not; and congregations could insist on having only an episcopally ordained minister celebrate Holy Communion. Thirty years were given for this problem to sort itself; all new ordinations were to be by bishops. Meanwhile the CSI had limited recognition by some other churches. The CNI avoided such a "mixed" ministry by a "representative act of unification of the ministry" at the inauguration of the church. Three CNI ministers, one a bishop, had hands laid on them by ten other ministers. The ten were (a) six representatives of the uniting churches, and (b) four ministers from outside the CNI area, including two bishops in the historic episcopate. The ten said, "May (God) continue in you his gifts and, in accordance with his will, may he bestow on you grace, commission and authority for your ministry, whether as a presbyter or as a bishop. . . ." The ten laid hands in silence on the three. After this, the three laid hands in turn on representative ministers of the uniting churches, including the aforementioned six, using the same words. At later services in the dioceses, the remainder of the ministry was "unified." It was a rite which seemed to satisfy all around. An Anglican statement had recognized that "it is on the human level legitimate to place different interpretations upon what God does in the act." And differ they did.

(2) The CNI permitted both infant baptism and believer's baptism (the latter having been practiced by Baptists, Disciples, and Church of the Brethren). Where there was no infant baptism, there was to be a service of infant dedication, and believer's baptism was to be followed by a service admitting to communicant membership, and including laying on of hands, which paralleled the Confirmation service for those baptized as infants. The mode of baptism could be immersion, affusion, or sprinkling.

See also INDIA; and official publications *Plan of Church Union in North India and Pakistan* (4th ed., 1965) and *Forward to Union: The Church of North India* (1968). ROBERT J. MC MAHON

NORWAY. Norwegians came under Christian influence through the contacts of Vikings with Christian countries and through missionary efforts from Denmark* and England. The first Christian kings (such as Olav*) promoted the Christianization of the country. An archiepiscopal see was established in Trondheim in 1153. The Reformation was introduced in 1537 by command of the king of Denmark and Norway. The seventeenth century was characterized by Lutheran orthodoxy. Pietism* came shortly after 1700 and left a lasting influence upon Norwegian church life. Confirmation was introduced by royal command in 1736, and was made compulsory.

The Pietistic "Explanation" to Luther's shorter catechism written by Erik Pontoppidan in 1737 was used in Norway for more than 150 years.

The era of Enlightenment lasted in Norway from 1750 to 1820. The theology professors of the newly established university of Oslo (1811) and the awakening by H.N. Hauge* heralded a period of richer spiritual life. In 1842 lay preaching became lawful, and in 1845 a "Dissenters' Law" for the first time gave citizens opportunity to cancel membership in the state church and to organize free churches. Free churches, however, never became strong in Norway; Christian believers chose to stay and make their influence felt within the state church. The Norwegian Missionary Society was founded in 1842, later followed by several other organizations for foreign missions. The lay movement initiated by Hauge entered a stage of organization in the second half of the nineteenth century. The Pietistic theology of Professor Gisle Johnson (d.1894), who greatly influenced both clergy and laity, formed the "Luther Foundation" (1868) which in 1891 was reorganized under the name of Norwegian Home Mission Society.

New revival movements toward the end of the nineteenth century led to the foundation of a new type of free organization which took a critical attitude toward the state church, emphasizing their independence in relation to the clergy. The greatest of these new organizations was the Norwegian Lutheran Mission, whose spiritual leader was the lay preacher Ludvig Hope (d.1954). To a great degree church life in Norway is characterized by free organizations for home mission and foreign mission. Christian believers worship in "prayer houses," listening to lay preachers, and some also celebrate the Lord's Supper there. The same believers may or may not go to the parish church to worship there. This lay activity, which runs parallel to the activity of the established church, has been of vital importance for the missionary activity of Norwegians.

In the twentieth century, liberal theology has caused severe struggle, which led to the formation of the conservative Free Faculty of Theology *(Menighetsfakultetet)* where the majority of the pastors are educated (see HALLESBY). During the years of German occupation the church resisted the attempts to Nazify the schools, etc. (see BERGGRAV). By the middle of the twentieth century the life and thinking of the people were greatly influenced by modern secularism. By 1966 there were 1,078 pastors in the state church and 1,688 lay preachers and staff workers in the free organizations. There are ten dioceses in the country. By 1960 the free churches had a total of 134,551 members. Of the population of Norway (3,867,000 in 1970), some 96 percent belong to the state (national) church.

BIBLIOGRAPHY: T.B. Wilson, *History of the Church and State in Norway from the Tenth to the Sixteenth Century* (1903); K. Gjerset, *History of the Norwegian People* (2nd ed., 1915); B. Høye and T.M. Ager, *The Fight of the Norwegian Church against Nazism* (1943); K. Larsen, *A History of Norway* (1948); J.M. Shaw, *Pulpit under the Sky: A Life of Hans Nielsen Hauge* (1955); T.K. Derry, *A Short History of Norway* (1957); E. Molland, *Church Life in Norway 1800-1950* (1957) and "Lutheranism in Norway," *The Encyclopedia of the Lutheran Church*, vol. III (1965).

CARL FR. WISLOFF

NOTKER (c.940-1008). Bishop of Liège. Nephew of Emperor Otto I, Notker was born in Swabia. After entering monastic life at St. Gall in Switzerland he was named imperial chaplain in Italy in 969. Appointed bishop of Liège in Flanders in 972, he was an effective administrator. He played an important role in establishing the famous Liège schools, had a broad influence on monastic reform and educational affairs, and served in governmental posts under Otto III and Henry II.

NOVALIS (1772-1801). Pseudonym of Friedrich Leopold Freiherr von Hardenberg, lyric poet and early German Romantic. Coming from a Pietistic background, he came under the influence of Goethe, Schiller, and Fichte at Jena, and became with Schlegel a spokesman for early German Romanticism. In *Die Christenheit oder Europa* (1799) he attacked both the Reformation and the Enlightenment, the first because it was responsible for what Novalis regarded as the fragmented character of present culture in comparison to the Middle Ages, the second because of its worship of reason. Novalis's own religion was that of a Romantic mystic with no clear distinction between finite and infinite, nor between immanence and transcendence. In his view poetry is an attempt to display the infinite, the source of the meaning of the universe, which cannot be conceptualized, only hinted at. PAUL HELM

NOVATIANISM. Novatian of Rome is noteworthy for two reasons. In the first place he was the "antipope" of the "Puritan" party in the church. Secondly, he gave to the Western Church the first full-length treatment of the Trinity. Perhaps disappointed by the elevation of Cornelius* as pope (251), Novatian joined those who demanded that the Christians who had apostatized during the Decian persecutions (249-50) should not be welcomed back into the fellowship of the church. Novatian's group formed themselves into their own party, under strict discipline. This separatist movement continued for many centuries. Their orthodoxy was never in doubt: Acacius,* one of their leading bishops during the Arian controversy, strongly repudiated Arianism. The Council of Nicea, which set out terms for the reception of the Novatians back into the church, demanded no change of doctrine. Novatian's work on the Trinity was strongly Trinitarian in character and maintained the full deity of Christ, though inclining to a form of the "kenotic" theory. Men such as Cornelius were not unnaturally critical of Novatian, who was nonetheless a vigorous champion of true Christology. Novatian died as a martyr during the persecutions under Valerian. H.D. MC DONALD

NOVATUS (third century). Presbyter of Carthage. An implacable enemy of Cyprian, he opposed his election as bishop and intrigued constantly against him. In the problem of the Lapsed after the Decian persecution, he attacked Cyprian's deprecation of the *libelli pacis* and championed the right of "confessors" to grant them. As leader of an anti-Cyprian faction he gained control of relief funds by appointing Felicissimus, a wealthy layman, as a deacon. In Rome he supported Novatian as bishop in opposition to Cornelius. Cyprian supported Cornelius, and when a delegation of Novatianists led by Novatus came to Carthage, the anti-Cyprian group appointed Novatian's delegate, Maximus, rival bishop of Carthage. J.G.G. NORMAN

NOVENA (Lat. *novem*, "nine"). A Roman Catholic practice intended to encourage devotion and piety by means of nine successive days of prayer, public or private, to obtain special favors and grace. Unlike the more festal octave, and though recommended by the church, the novena has no place in the liturgy. The scriptural prototype is seen in the nine days of apostolic waiting in Jerusalem prior to the descent of the Holy Spirit in Acts 2. Although the observance is practiced as a preparation of the soul for some event of spiritual significance, it is primarily associated with a period of mourning. Since the early nineteenth century, novenas have been enriched by the granting of indulgences.
H. CROSBY ENGLIZIAN

NOVICE. The term for a candidate (male or female) for admission to a religious order during the probationary period which customarily follows the postulancy, prior to the profession of any vows. Throughout the noviciate, established by the Council of Trent as not less than a full and continuous year, the novice lives according to the rule of the order, under the direction of the novice-master, and separate from the rest of the community, although he enjoys its privileges, immunities, and indulgences. He is free to leave, as the order is free to dismiss him, and therefore may not renounce his property.

NOYES, JOHN HUMPHREY (1811-1886). Religious and social reformer. Born in Brattleboro, Vermont, he graduated from Dartmouth College (1830), and after conversion, studied at Andover Theological Seminary and Yale Divinity School. Developing perfectionist and adventist views contrary to Calvinism, he was ousted from Yale and the ministry when he pronounced himself sinless in 1834. He established two communes—at Putney, Vermont (1840-48), and Oneida, New York (1848-81)—to practice and propagate his ideas of perfectionism, biblical communism, complex marriage, male continence, population control, mutual criticism, and education. Under public pressure he emigrated in 1876 to Niagara Falls, Ontario, where he died. His classic *History of American Socialisms* (1870) was based on wide acquaintance with contemporary communitarians. D.E. PITZER

NUELSEN, JOHN LOUIS (1867-1946). Methodist bishop. Born of American parents in Zurich, he was educated at Drew Theological Seminary, at Halle, and at Berlin. An ordained Methodist, he held pastorates in Missouri and Minnesota, lectured in ancient languages and exegetical theology, and was made bishop in 1908. From 1912 he was in charge of the Methodist Church's work in Europe with Zurich as his headquarters. In 1936 he organized the first Methodist Central Conference of Germany at Frankfurt. A member of the Central Bureau for Relief of Evangelical Churches of Europe, he was also known for his preaching. He wrote widely on Methodism and biblical criticism and was an editor of major theological works. C.G. THORNE, JR.

NUN. The term for a woman who has professed vows of poverty, chastity, and obedience—precisely applied only to women living in strictly cloistered communities, and less accurately used for religious women engaged in service in the world, properly termed "sisters." The three great monastic leaders, Pachomius, Basil, and Benedict, each established a foundation for women under his sister, and Augustine prepared an influential letter of direction for a community of women. The rigorous seclusion of religious women imposed by the Eastern Church had no counterpart in the West until the bull *Periculoso* of Boniface VIII, confirmed at the Council of Trent. By its terms, nuns may neither leave their cloisters nor receive outsiders, including females. MARY E. ROGERS

NUNC DIMITTIS. The song of Simeon (Luke 2:29-32), which derives its name from the first two words in the Latin version. Simeon was an aged Israelite who cherished the messianic expectation and was under the inspiration of the Spirit. When he saw Joseph and Mary bring Jesus into the Temple for the rites of purification forty days after his birth, he took hold of the baby Jesus and uttered the words of his song. This consists of (1) a statement that he can now die in peace because he has at last been granted what he was promised—a view of the Messiah, and (2) a description of the salvation which the Messiah brings—light for the Gentiles and glory for Israel. This is appropriate to its setting and it is unnecessary to read into it a Christian theology of the Gentile mission and then to dismiss it as unauthentic. It has been used as an evening canticle in the worship of the Eastern and Western churches since at least the fourth century. R.E. NIXON

NUNCIO AND LEGATE, PAPAL. A nuncio is an official permanent papal representative from the Holy See to both the state and the church of a given area. Usually a titular bishop or archbishop, he relates to the Holy See through the Cardinal Secretary of State. As papal envoy to the state, he has duties diplomatic in character, not unlike an ambassador in international relations; his ambassadorial status was recognized by the Congress of Vienna (1815). As papal envoy to the area church, he has duties ecclesiastical, like those of an apostolic delegate in areas where there is no nuncio.

"Legate" was the usual designation of a papal representative during the era of the seventeenth century and earlier. Three or four types are at times distinguishable: Legate *nati,* the principal resident bishop, who also held certain special authority from the pope, duties partially retained by a primate today; Legates *missi,* sent by the Pope on *ad hoc* missions; and Legate *a latere,* the highest rank of special papal envoy, today a title usually reserved for ceremonial functions. A fourth type, the *nuncii et collectores,* were financial officials charged with gathering papal funds. The modern nuncio gradually replaced and absorbed duties drawn from all such earlier legates as the structures of the papacy, the area churches, and the states became more distinct.

See R.A. Graham, *Vatican Diplomacy, A Study of Church and State on the International Plane* (1959), and I. Cardinale, *Le Saint-Siège et la diplomatie; aperçu historique, juridique, et pratique de la diplomatie pontificale* (1962).

C.T. MC INTIRE

NUREMBERG DECLARATION. A German Old Catholic* theological statement. It was drafted by fourteen German Catholic professors, including Johannes Friedrich,* Ignaz von Döllinger,* and F.H. Reusch,* at a meeting in Nuremberg in 1870 which J.F. von Schulte of Prague had called to protest the decrees of the Vatican Council. It declared: (1) Vatican I* was not a true ecumenical council because it was neither free nor morally unanimous; (2) chapters three and four of the dogmatic constitution *Pastor Aeternus* (which defined the primacy of the pope in the church and asserted that papal statements when made *ex cathedra* are infallible) were not dogma because they had not been universally believed and taught; (3) papal infallibility would stir up conflicts between church and state, Catholics and non-Catholics; and (4) an unfettered general council should meet in Germany. The thirty-three academic and clerical signatories of the manifesto constituted the nucleus of the Old Catholic movement.

RICHARD V. PIERARD

NUTTALL, ENOS (1842-1916). First archbishop of the West Indies. English-born, he went to Jamaica first in 1862 as a Methodist probationer, but in 1866 was ordained in the Anglican Church. He first appears as a man with sensible suggestions in the complex negotiations surrounding disestablishment in 1870. Ten years later he was elected bishop of Jamaica. He became archbishop in 1893. His interests cover the range of contemporary problems. He regarded the British Empire as, on the whole, a good thing, advocated an institutional connection with Canterbury, recruited clergy first in England, but in 1893 founded the first diocesan theological college in Jamaica. He advocated a modified system of public education, including secondary schools, with religious instruction at all levels. He pressed also for agricultural education. Eminently Victorian, he regarded Jamaican revivalism with suspicion, but blamed the church for its existence. He pioneered in the founding of a church nursing home in 1893. A competent man, convinced of the virtues of British culture, but without racial inhibitions, he was primarily concerned with individual rather than social problems.

GEOFFREY JOHNSTON

NYE, PHILIP (1596-1672). Congregational minister and theologian. Born in Sussex and educated at Oxford, where he graduated, he entered the ministry and by 1630 was at St. Michael's, Cornhill, but fled to Holland in 1633 to escape Laud's rigid ecclesiastical policies. Back in England by 1640, he became minister of a Congregational church in Kimbolton, Huntingdonshire. In June 1643 he was appointed to the Westminster Assembly,* there taking a leading part among the "dissenting brethren." Eventually he worked harmoniously with the Presbyterians and helped shape the Confession of Faith. With Stephen Marshall* in 1643 he established a working agreement between Scots and parliamentary forces, strongly promoting the Solemn League and Covenant.* He was very active in Cromwell's protectorate, was a leader at the Savoy Declaration* of 1658, was deprived at the Restoration, and during his latter years ministered privately among dissenting churches.

BRIAN G. ARMSTRONG

NYGREN, ANDERS THEODOR SAMUEL (1890-). Swedish theologian. Born in Göteborg, he studied at Lund and was ordained in the Church of Sweden in 1912. He was deputy pastor in Ölmevalla (1914-20) before returning to Lund as instructor in the philosophy of religion until 1924, when he became professor of systematic theology. He was active also in Lutheran world affairs, and in the formation of the World Council of Churches, having been a delegate at the conferences held at Lausanne (1927), Oxford and Edinburgh (1937), Amsterdam (1948), and Lund (1952). He was president of the Lutheran World Federation (1947-52) and was chairman of the WCC's Faith and Order Commission on Christ and the Church (1953-63). He served as bishop of Lund from 1949 until his retirement in 1958. His writings include *Agape and Eros* (1930; ET 1953), *Commentary on Romans* (1941), *Christ and His Church* (1955; ET 1957), and *Meaning and Method* (1972).

CLYDE CURRY SMITH

OAK, SYNOD OF THE (403). When Theophilus,* patriarch of Antioch, was summoned to Constantinople by imperial order to answer charges of persecution against a group of pro-Origenist monks, he formed alliances with those, including Empress Eudoxia, who disliked John Chrysostom's* reforming zeal. Helped by John's political innocence, Theophilus maneuvered the trial away from Constantinople to a suburb of Chalcedon called "The Oak." There, with the help of some thirty-six anti-Chrysostom bishops, he was able to turn it into a trial of John on forty-six indictments ranging from living like Cyclops to insulting the imperial majesty. John would not appear before judges who were his declared enemies and was therefore condemned and was exiled by Emperor Arcadius. He was recalled within a few days, however, partly because of the disquiet of the people of Constantinople, partly because of Eudoxia's disturbed conscience.

C. PETER WILLIAMS

OATES, TITUS (1648-1705). A leader of the 1678 Popish Plot.* Born at Oakham, Rutland, son of a Baptist minister of checkered career, he received Anglican ordination in 1673 without an academic degree, was converted to Catholicism but expelled from their schools in Spain, and returned to London with alleged information of a plot to assassinate Charles II and place the Roman Catholic duke of York on the throne. With Israel Tonge he produced documentary "proof" which caused a great stir and for a time made him a hero. The plot was discounted, Oates was convicted of perjury in 1685, and died in obscurity.

OATHS. An oath is an appeal to something held sacred as support for the truthfulness or sincerity of a statement or vow. It is held by many ethicists that swearing, whatever the circumstances and conditions, is not desirable but an evil necessity of the present age—a really Christian morality would require only a simple "yes, yes" and "no, no" (Matt. 5:33-37). Reformers and certain Protestant communities have held that only when oaths issue from the lower egotistical affections and impulses of human nature are they sinful. If they are pronounced for the sake of high ethical interests, then they are valid. These would interpret Matthew 5:33-37 as a prohibition against the frivolous and promiscuous oaths practiced in the everyday life of the Jews in the time of Christ (see Exod. 20:7; 1 Chron. 12:19; Matt. 23:16-22). The canon laws of the Roman Catholic Church and the Anglican Church have taken their cue from Jeremiah 4:2 which states that an oath should be given in truth, in judgment, and in righteousness. All these moral theologians would deem an oath before God as binding under grave punishment. Throughout the history of the church there have been those who have interpreted Matthew 5:33-37 as totally forbidding all oaths—Waldensians, Quakers, Jehovah's Witnesses, etc. Some governments have acknowledged the rights of these groups to refuse to take oaths.

BIBLIOGRAPHY: W. Lockhart, *On Oaths* (1882); C. Ford, *On Oaths* (1903); H. Silving, "The Oath," *Yale Law Journal* 68 (June/July 1959), pp. 1329-90, 1527-77. JOHN P. DEVER

OBERAMMERGAU. The location in upper Bavaria of the noted passion play. Such religious dramas were common in Bavaria, and legend holds that the Oberammergau villagers were spared from a plague in 1633 and in gratitude vowed to reenact the passion of Christ every ten years. From 1680 it was held on the decennial year, but was canceled in the war years 1870 and 1940. Originally it was staged in the church, then in the churchyard, and in 1830 the present site, a special theater, was occupied. Hitler admired the play for its alleged anti-Semitic qualities, and it was rewritten for the tricentennial performance in 1934 to make Jesus and the disciples appear as Aryan heroes. The text in current use, with some minor changes, was written by J.A. Daisenberger (1799-1883) in 1860. Performed by 700 villagers, the play lasts more than seven hours and is a lucrative community enterprise. In 1970 there were nearly 100 performances before a total audience of 500,000. RICHARD V. PIERARD

OBERLIN, JEAN FRÉDÉRIC (1740-1826). Alsatian Lutheran minister and philanthropist. Born in Alsace, he studied theology at Strasbourg and in 1769 became a pastor in Waldersbach. He soon gained distinction as a community servant and social reformer through his endeavors in the formation of schools, building roads and bridges, encouraging better agricultural techniques, and establishment of factories, stores, and savings-and-loan associations. Although deeply pious and devoted to his parishioners, he had an ecumenical outlook which embraced both Catholics and Calvinists. He welcomed the French Revolution and saw in republican ideals the earthly realization of the spirit of Christianity. He commanded the respect of the various French regimes for his philanthropies, while his work was admired and imitated elsewhere. Those contemporaries who were dissatisfied with rationalism in the church regarded his love for Christ and deep mystical devotion combined with a desire to promote the welfare of mankind as a symbol of hope.

RICHARD V. PIERARD

OBERLIN THEOLOGY. The work of C.G. Finney* and Asa Mahan (1799-1889), this was a moderate form of Christian perfectionism.* In *Scripture Doctrine of Perfection* (1839), Mahan, the first president of Oberlin College, wrote that the Christian might eventually attain unbroken peace and not come into condemnation. Finney, the Presbyterian evangelist who in 1835 began a second career as Congregationalist and professor of theology at Oberlin, indicated in *Lectures on Systematic Theology* (1846) that he had gone far beyond N.W. Taylor* and brought liberal Calvinism close to Methodist perfectionism. To him, God was benevolent and man capable of growing toward perfection, although not absolute perfection, and thus society was perfectible. At the little Congregationalist college in Ohio, Mahan and Finney trained professional evangelists and stimulated zeal for social reform, but their approach rested on faith in individual conversion as the key to social justice. This was transcended by the organicism of Horace Bushnell* and his supporters.

DARREL BIGHAM

OBLATE (Lat. *oblatus*, "one who has offered himself"). A term used in different historical periods with different connotations, but always with reference to monasteries. In the contemporary Roman Catholic Church it describes the member of a specified religious community, e.g., the Oblates of St. Charles Borromeo founded in 1857 by H.E. Manning, archbishop of Westminster. In medieval and modern times it referred either to children placed in a monastery by their parents in order to learn from the monks (cf. Benedictine Rule, chap. 49), or to those who shared in the common life of a monastery without taking the vows.

OBOOKIAH (Opukahaiah), HENRY (1792-1818). Hawaiian Christian who inspired American missionary interest in what were then the Sandwich Islands. Born on the island of Hawaii, at twelve years of age he saw his parents slain in a local war and himself taken prisoner. Later he found refuge with an uncle, a priest, who trained Obookiah for the same occupation. Discontented, however, the youth managed to leave the island for America in the ship of a Captain Brintnall, arriving in New York in 1809. Aided by some college students, he learned English, became a devoted Christian, and planned to return to Hawaii with the Gospel. He inspired the sending of the first missionary party, but himself died of typhus before he could return with them.

HAROLD R. COOK

OBRECHT, JACOB (1450-1505). Netherlandish composer. The first great composer to come from what is now Holland, he is the first important composer to leave compositions with Dutch text. He was organist at various centers in the Netherlands and had Erasmus as one of his choir boys in Utrecht. He is distinguished especially for his Masses and was one of the first to have such works appear in print at the press of Petrucci in Venice. He showed great individuality in handling the *cantus firmi* (preexistent melodies) in his Masses, and was one of the experimenters of his day.

O'BRYAN, WILLIAM, see BIBLE CHRISTIANS

OCCAM, see WILLIAM OF OCKHAM

OCCOM, SAMSON (1723-1792). Best-known American Indian preacher of the eighteenth century. Born in Mohegan, Connecticut, he was converted with his mother during the Great Awakening* in 1740. He studied theology with Eleazar Wheelock,* who began the Indian school that later developed into Dartmouth College. Occom served as teacher and minister to the Montauk Indians of Long Island from 1749 to 1764. Ordained in 1759, he also served as missionary to the Oneidas, 1761-63. From 1765 to 1768 he was in Great Britain, where he succeeded in raising £10,000 for Wheelock's school. He then returned to his preaching and missionary work with the Oneidas. He established the Indian town of Brotherstown, New York, in 1784, and published an Indian hymnal.

HAROLD R. COOK

OCHINO, BERNARDINO (1487-1564). Italian Reformer. From Siena, he entered the Franciscan Observants (c.1504) and later, desiring a yet stricter rule, the Capuchins* (1534), of whom he was elected vicar-general (1538-42). A popular penitential preacher, in 1536 he met Juan de Valdés* and his circle. He became convinced that ecclesiastical mediation could not gain salvation for man. Thus there arose a conflict between his convictions and his vocation which climaxed when he criticized the Inquisition* in a sermon at Venice (1542). Summoned to Rome, he escaped to Geneva where he was cordially received by Calvin. Licensed to preach, he ministered to the Italians of Geneva, 1542-45, and published several works including his *Apologhi.*

In 1545 Ochino settled at Augsburg, where he became minister of the Italian church. When the city fell to the imperial forces, he escaped to Basle and Strasbourg before finding refuge in England (1547-53). Thomas Cranmer received him kindly and secured for him a prebend in the church of Canterbury and a royal pension. Ochino in addition to writing preached to the Italians in London. His main works of this period were *The Usurped Primacy of the Bishop of Rome* and the *Labyrinth,* an attack on the Calvinist teaching of predestination. When Queen Mary came to the throne, he returned to the Continent, becoming minister of the Italian congregation in Zurich. In 1563 he issued the *Thirty Dialogues,* which treated the doctrine of the Trinity and monogamy in a free way, thus causing his expulsion. Moving next to Poland, he preached at Cracow for a time, but was forced to leave and finally settled in Slavkov, Moravia, where he died of the plague.

See C. Benrath, *Bernardino Ochino* (ET 1876), and R. Bainton, *Bernardino Ochino* (1940).

ROBERT G. CLOUSE

OCKEGHEM, JOHANNES (c.1420-c.1495). Musical composer. This great composer belongs to the second generation of highly talented musicians, born near what is now the French-Belgian border, who rapidly advanced the art of counterpoint until it became the expressive medium of the great Renaissance choral legacy. Unlike many of his great countrymen he did not go to Italy, and spent much of his career at the French court. Allowing for the loss of a number of his works, his output was surprisingly small. We have only eleven complete Masses by him, and a small number of motets and secular chansons. His work is noted in particular for his ability to write a "seamless web" of polyphonic voices interrupted by a minimum of cadences points. He was one of the first to write in five voice-parts. He enjoyed an extraordinary degree of esteem by his fellow musicians, and a number of *déplorations* were written upon his death. An exceptionally beautiful one was set to music by Josquin Desprez,* and there is also one with words by Erasmus,* who has also left witness to Ockeghem's worth as a man.
J.B. MAC MILLAN

OCKHAM, see WILLIAM OF OCKHAM

ODENSE, DIET OF (1527). A meeting in Denmark* of the Catholic majority and the Lutheran minority, marking a decisive turning point in the history of the Reformation in that country. At his coronation in 1523 Frederik I had promised to protect and preserve the Catholic Church and to oppose and suppress the Lutheran heresy. When the Reformation grew stronger, however, and spread throughout his kingdom, he followed a policy of tolerance and even extended his personal protection to Hans Tausen* and other Lutheran preachers. When at Odense the Catholics demanded that the king fulfill his obligations, withdraw protection from the Lutherans, and deliver them up to clerical jurisdiction, the king stated the principle that "the king's power and authority is good for life and property, not for the souls." The norm of the church is the Gospel, according to Scripture. When interpretations of the Gospel differ, it is not for the king to pass sentence or use coercion in matters of faith. The preaching of the Gospel must therefore be free, until an ecumenical council has finally decided which interpretation of the Gospel is the true one. This principle of religious toleration formed the legal foundation for Frederik's policy of coexistence of a national Catholic Church alongside free Lutheran congregations, until the final accomplishment of the Reformation in Denmark in 1536.
N.O. RASMUSSEN

ODILIA (Othilia) (d. c.720). Abbess and patroness of Alsace. She is reputed to have been the blind daughter of a Frankish nobleman, Adalric, and to have founded a nunnery at his castle of Hohenburg (Odilienberg) in the Vosges Mountains after recovering her sight. Her shrine in the abbey there became a great place of pilgrimage, especially for those with eye diseases. Famous people such as Charlemagne also visited it. In modern times it has enjoyed a revival as a place of pilgrimage.

ODILO (c.962-1049). Fifth abbot of Cluny. He became acting abbot in 994, three years after he entered the monastery, and was abbot for fifty years from 999. A man of outstanding ability, he was responsible for a significant building program at Cluny, a tremendous expansion of daughter houses, and more centralized control of the daughter houses by the main monastery at Cluny. In addition he was an active leader in diplomatic affairs, both as a counselor to Emperor Henry II* and as a mediator between Emperor Conrad II and the king of France, Robert II, in 1025. He introduced the commemoration of All Souls Day (2 November) at Cluny, which was later extended to the whole church, and was responsible for the extension of the Truce of God* to S France and Italy. Known for his concern for the poor, he sold treasures of the monastery to aid the poor during the severe famine of 1033. Odilo was canonized in 1063.
RUDOLPH HEINZE

ODO (879-942). Abbot of Cluny. Son of Lord Abbo I of Déols, he was educated at the court of William, duke of Aquitaine, and later in Paris under Remigius of Auxerre.* When nineteen years old he became canon of St.-Martin of Tours. Three years later he joined the Cluniac* community at Baume, where Abbot Berno placed him in charge of the monastery school. Odo succeeded Berno in 924, and became abbot of Cluny in 927. Under his leadership Cluny received a papal privilege from John XI which facilitated the spread of Cluniac reforms in numerous French and Italian houses. Several times he acted as Vatican negotiator in Italian political disputes. His alleged works include *Occupatio,* an extended poetic meditation on redemptive history; *Collationes,* three collections of moral essays; *Vita S. Geraldi Auriliacensis comitas* and *Vita Gregorii Turonensis episcopi,* two biographies; *Moralia in Job;* a number of musical pieces dedicated to St. Martin; and several sermons.
JAMES DE JONG

OECOLAMPADIUS (1482-1531). German Reformer. A native of the Palatinate whose original name was Hussgen or Hauschein, he became the leading Protestant Reformer of Basle. A brilliant philologist in Latin, Greek, and Hebrew, he began his university studies in law at Bologna, but transferred to theology at Heidelberg. Employed for a time as a tutor, he later secured a prebend at Weinberg (1503-12). Further studies at Tübingen and Stuttgart led to his contact with Reuchlin* and Melanchthon.* In 1515 he was called as minister to Basle, where he met Erasmus* and assisted him in the publication of the Greek NT. Later (1518) Oecolampadius became a pastor at Augsburg, where he was deeply affected by Luther's teachings. Then the pressure of his work caused him to enter a monastery (1520), only to withdraw after a short while.

In 1522 he accepted the position of court chaplain to Franz von Sickingen,* but returned to Basle, becoming lecturer on Holy Scripture at the university in 1523. Later he became a minister in the city, and his influence through lectures and sermons led to the establishment of the Reformed Church. He also promoted the evangelical cause throughout Switzerland by his writings and participation in disputations such as those held at Baden (1526) and Bern (1528). He attended the Colloquy at Marburg* (1529) where he defended the eucharistic teaching of his friend Ulrich Zwingli.* When the city council of Basle ordered the removal of images from the church and abolished the Mass, he supervised the work. He also introduced the ban and organized a board of ministers and laymen to see that discipline was executed.

See G. Rupp, *Patterns of Reformation* (1969), pp. 1-46. ROBERT G. CLOUSE

OECUMENICAL, see ECUMENICAL

OECUMENIUS (early sixth century). Rhetor and philosopher, not to be confused with the tenth-century bishop of Tricca in Thessaly with the same name. An adherent and contemporary of Severus* of Antioch, Oecumenius wrote the oldest extant Greek commentary on the Apocalypse, discovered early in the twentieth century. The work in twelve parts approached the biblical text historically. A commentary on the same book by Archbishop Andrew of Caesarea, written sometime between 563 and 614, referred to Oecumenius's work repeatedly. Other writings attributed to him came from different pens, including a commentary on Acts and expositions of the General and Pauline epistles, as well as a commentary on the Apocalypse edited by Donatus of Verona in 1532. JOHN GROH

OENGUS (ninth century). Irish monk. Member of the monastic community at Tallaght, near Dublin, he is frequently referred to as Oengus the Culdee, a term that indicates his simple faith and earnest devotion to Christ. He is best known as the author of a litany which throws light on the influence of the Eastern Syrian Church on the monastic life of the Celtic Church.* It commemorates the fact that large numbers of scholars from the Middle East sought refuge in Ireland in the eighth and ninth centuries. They built round towers for their protection and introduced a study of Hebrew and Greek in the monastic settlement. Oengus gives lists of bishops and pilgrims who lived in groups of seven. The oldest copy of his work, about 1200, is in the Franciscan Convent on Merchant's Quay, Dublin. In addition he wrote *The Martyrology of Tallaght* and *Saltair na Rann,* a poem on OT history. ADAM LOUGHRIDGE

OIKONOMOS, CONSTANTINE (1780-1857). Greek scholar and theologian. A keen patriot, he was active in politics as well as ecclesiastical affairs, and strongly opposed Western influences in Greek life which had assumed increasing importance in the eighteenth and nineteenth centuries. He wrote works on philology and the history of literature, but his most notable achievement was a massive four-volume study of the Septuagint in over 3,700 pages, published in Athens, 1844-49, entitled *Peri tōn ho hermēneutōn.* He believed the Septuagint text was canonical and inspired rather than the Massoretic text, an idea rarely held. Even so, it contained much useful information, e.g., concerning the difference in chronology between the Greek and Hebrew texts. J.G.G. NORMAN

OLAV (995-1030). King and national hero saint of Norway. As a Viking chieftain he fought for Ethelred II in England, and for Richard in Normandy. He was baptized in Rouen. In 1015 he became king of Norway. In his zeal for the Christian religion he made considerable use of violence. His harsh rule caused discontent, and his chieftains in cooperation with the Anglo-Danish king Canute tured him out of the country. In 1030 he made an attempt to regain his kingdom, but died in the battle of Stiklestad. Very soon his death was regarded as a martyr's end, and he was invoked as a saint. Over his shrine was built the cathedral of Trondheim which was a place of pilgrimage. The legendary account of Olav has the usual medieval shape of a saintly king, formed in accordance with Augustine's conception of the "righteous king" *(rex iustus).*

CARL FR. WISLOFF

OLCOTT, HENRY STEEL, see THEOSOPHY

OLD BELIEVERS. Dissident groups of Russian Orthodox, in schism (*raskol,* hence called also "Raskolniki") since 1666, when Nikon, patriarch of Moscow, introduced reforms of ritual in line with Orthodox practice elsewhere, but declared heretical at a council of 1551. The dissidents adhered to the older practices calling themselves *staroveri,* adherents of the old rite, misrendered "Old Believers." Avvakum,* their leader, went into exile from 1664; others were exiled, and fierce persecution led many to flee to remote parts, e.g., Siberia and Karelia. Monks, peasants, Cossacks, and townfolk were among them. They regarded the official apostasy and their persecution as works of Antichrist. They were brutally treated and many suicides—e.g., self-immolation of whole groups by fire—are known, to escape their certain fate.

Persecution lasted until the end of Peter the Great's reign; Peter III and Catherine II were more tolerant, but Nicholas I (1825-55) renewed efforts of coercion. Penal laws persisted until 1903 when alleviations were introduced. Their lot since the Revolution has been that of all religious communities, yet some of their remoter strongholds still succeed in retaining a vigorous life. There are two main groups: the *popovtsy,* i.e., with priests, were first supplied with priests leaving the Orthodox Church, but a hierarchy was established in 1846 by Ambrose of Bosnia at Bielo-Krinitz in the Austrian Empire. The second group, *pezpopovtsy,* i.e., without priests, evolved new ways of organization in the remoter places, sometimes with extravagances. They are both said to maintain a movingly sincere manifestation of traditional Russian Christian piety.

BIBLIOGRAPHY: F.C. Conybeare, *Russian Dissenters* (1921); P. Pascal, *Avvakum et les débuts du Raskol* (1938); R. Janin, *Églises orientales et rites orientaux* (1955), pp. 196-201; W. Kolarz, *Religion in the Soviet Union* (1961), chap. IV; M. Bordeaux, *Opium of the People* (1965), pp. 27-31; R.O. Crummey, *The Old Believers and the World of Antichrist* (1970). J.N. BIRDSALL

OLDCASTLE, SIR JOHN (Lord Cobham) (c.1378-1417). English Lollard.* Born on the Welsh border, he served as a soldier, represented Herefordshire in Parliament (1404), then through his second wife's title became a baron in 1409. In the following year, however, he incurred the displeasure of Thomas Arundel,* archbishop of Canterbury, for Lollard sympathies. The warning evidently went unheeded, for in 1413 Oldcastle was summoned before convocation, and handed over to the civil arm for trial, despite his friendship with Henry V. Among other things, he rejected transubstantiation and confession, denounced the pope as Antichrist, and denied the hierarchy's right to dictate what a man should believe. Sentenced to death, he somehow escaped, and with others conspired to seize the king and to organize a Lollard assembly. The plot came to nothing, many of the Lollards were captured, but Oldcastle evaded custody until taken late in 1417. He was "hung and burnt hung." J.D. DOUGLAS

OLD CATHOLICS. A movement in German-speaking Europe, especially Bavaria, which rejected the dogma of Papal Infallibility declared by the Vatican Council* (1870), and organized the Old Catholic Churches, in Germany, Switzerland, and Austria, not in communion with Rome. The movement was motivated by Febronianism,* and then Jansenism,* when it associated with the already established Church of Utrecht by a common adoption of the Declaration of Utrecht* (1889). The First International Old Catholic Congress convened at Cologne (1890).

J.J.I. von Döllinger, excommunicated from Rome, led the initial break with Rome, joined by other scholars, including Johann Friedrich* and Johann von Schulte. A congress of 300 convened at Munich, September 1871, and began organization of a separate church. Their aim was to perpetuate true Catholicism, claiming the Vatican decrees and other modern enactments of the Roman Church had in fact created a new church; hence the name "Old Catholics." Meanwhile Dö llinger declined further participation. In 1873 they elected Joseph Reinkens* first bishop, with see at Bonn; he was consecrated by the bishop of Deventer of the Dutch Church of Utrecht. The Swiss elected a bishop in 1876, with see at Bern. The Bismarckian *Kulturkampf** supported the Old Catholics in a play against Rome, by making available subsidies and some church buildings.

Adoption of the Declaration of Utrecht effected a bond with the Church of Utrecht, which had broken communion with Rome (1724) over Jansenism, and gave the Old Catholics a definite doctrinal tradition. The Declaration accepted as valid "the faith of the primitive Church" as specified by the first seven ecumenical councils, before the schism of 1054. The bishop of Rome was acknowledged as historically *primus inter pares.* It rejected: the Vatican decrees (1870); the dogma of Immaculate Conception* of Mary (1854); certain encyclicals including *Unigenitus* (1714) against Jansenism; the Syllabus of Errors* (1864); the decrees of the Council of Trent* except as they concur with the primitive church; and transubstantiation.* In church government and liturgy also they differed from Rome. Bishops were elected by synods; clergy and laity enjoyed equality in synods, councils, and courts; priests were elected by parishs; clergy may marry; liturgies were in the vernacular; auricular confession was not obligtory.

An international Old Catholic congress has met regularly since 1890, with the archbishop of Utrecht as president. Some other autonomous churches have affiliated, including the Polish National Catholic Church, founded in Scranton, Pennsylvania (1897), and the Yugoslav Old Catholic Church (1924). The Philippine Independent Church (Aglipay) established sacramental communion with Old Catholics in 1965.

From the start, Anglicans have been close to Old Catholics. The bishops of Ely and Lincoln sent communications to the Munich Congress (1871) and attended the second at Cologne (1872). Anglicans participated in an international conference of theologians, convened at Bonn by Old Catholics, to discuss reunion of churches outside Rome (1874). Old Catholics recognized Anglican ordinations (1925) and achieved full intercommunion with the Church of England (1932) and most other Anglican churches thereafter. Old Catholics number (in 1957) approximately 350,000 mainly in Germany, Holland, Switzerland, Austria, Poland, and North America.

BIBLIOGRAPHY: J.J.I. von Döllinger, *Über die Wiedervereinigung der Kirchen* (1872); J.F. von Schulte, *Der Altkatholizismus* (1887); C.B. Moss, *The Old Catholic Movement, Its Origin and History* (2nd ed., 1964). C.T. MC INTIRE

OLDHAM, JOSEPH HOULDSWORTH (1874-1969). Ecumenical pioneer. Son of an army officer, he was educated at Edinburgh Academy and Trinity College, Oxford. A warm-hearted evangelical, he became secretary of the Student Christian Movement* in 1896, then secretary to the World Missionary Conference which paved the way for the famous Edinburgh Conference* of 1910 and of its continuation committee from 1910 to 1921. He was joint secretary of the International Missionary Council* from 1921 to 1938, in all these early ecumenical gatherings working as secretary in close accord with J.R. Mott* as chairman. He founded and edited the *International Review of Missions* from 1931 to 1938. In 1937 he organized the Universal Christian Council for Life and Work Conference at Oxford,* from which emerged the striking phrase "Let the Church be the Church." He edited the *Christian Newsletter* from 1939 to 1945, but his later years were mainly devoted to improving the educational and social standards of the native African peoples. IAN SELLERS

OLD LATIN VERSIONS. Translations of the Bible in Latin, antedating the revisory work of Jerome.* Such translation was undertaken for the church in North Africa, Gaul, and N Italy from the second century onward. In time, the work of Jerome ousted the older versions, with the result that the Old Latin is not widely attested in manuscripts. The latest list enumerates 453, among which the largest group (41) represents MSS containing the gospels. Any other part of Scripture has far less attestation in this form. Quotations in Christian writings fill in the resultant lacunae. For some books we have no evidence.

Two main types of text are distinguished, named "African" and "European"; by this it is meant that the attestation of such-and-such a form is derived from quotations in African or European writings, not that particular forms originated in these regions. The distinction is on the basis of "rendering the same basic Greek by different Latin words." A gradual process of revision is to be seen by which original errors of translation, regional forms of speech, and textual differences are corrected and brought into line with Greek norms. Especially in the earliest strata many Greek words were, however, retained, producing at length such familiar words as "angel," "deacon," and "synagogue" in W European tongues. The antecedents of the translations were as a rule Greek (Septuagint in OT), but there may have been recourse to the Hebrew even in the Old Latin period, and in the NT some links with Aramaic traditions or Syriac texts have been discerned. For our knowledge of the Old Latin, the work of Pierre Sabatier* has long been a standard source. Many further advances have led to the establishment of the *Vetus Latina Institut,* Beuron, Germany, and to the work in progress there. We now possess definitive collections of the material for Genesis, the Catholic epistles, Ephesians, Philippians, and Colossians.

J.N. BIRDSALL

OLD ORDER RIVER BRETHREN, see BRETHREN IN CHRIST

OLD ROMAN CREED. As early as the end of the second century, this archetype of the Apostles' Creed* was the official creed of the Church of Rome. The earliest extant copy of the creed is found in the Hippolytus's (d.235) *Apostolic Tradition,* although the creed is first referred to about 150 in Tertullian's *De praescriptione.* The first writer of the West to give the text with a commentary was Rufinus* in the fourth century. The creed may also be found in Codex E[a] (Codex Laudianus [35] of the Bodleian Library at Oxford) which dates from the late sixth or early seventh centuries. The creed no doubt grew out of the confession of Peter (Matt. 16:16), which furnished its nucleus. WATSON E. MILLS

OLD SYRIAC VERSIONS OF THE NEW TESTAMENT. These are represented by two incomplete texts of the four gospels: (1) a manuscript discovered by William Cureton and subsequently published by him in 1858; (2) a palimpsest found in 1892 by Mrs. A.S. Lewis in St. Catherine's Monastery, Mt. Sinai. The manuscripts date from the fourth or fifth century, but their texts are as old as the second or third century, the Sinaitic being the earlier. Differences between the texts show that we have two recensions of the Old Syriac gospels. "Western" readings may be attributed either to their Greek *vorlage* or to the influence of Tatian's *Diatessaron.* There is also some evidence of an Old Syriac gospel text of the Caesarean type. Traces of Palestinian dialect in the Sinaitic recension suggest that the authors were converts from Judaism; the Curetonian text has largely been purged of Palestinian elements and has been revised according to a later Greek text. Burkitt's opinion that the Old Syriac version originated in Edessa is rejected by Kahle in favor of Adiabene. Ephraim Syrus's commentary on Acts, extant only in an Armenian translation, is evidently based on a text which differs from the Peshitta and which may be identifiable as an Old Syriac recension.

BIBLIOGRAPHY: W. Cureton, *Remains of a Very Ancient Recension of the Four Gospels in Syriac, hitherto unknown in Europe* (1858); F.C. Burkitt, *Evangelion da-Mepharreshe: the Curetonian Version of the Four Gospels, with the Readings of the Sinai Palimpsest and the early Syriac Patristic evidence* (2 vols., 1904); A. Hjelt, *Syrus Sinaiticus* (1930); P.E. Kahle, *The Cairo Geniza* (2nd ed., 1959). ROBERT P. GORDON

OLD TESTAMENT. "Do you understand what you are reading?" (Acts 8:30). This question, raised early in the church's history, has been the key issue in OT study. Dispute over the contents of the OT canon was not sparked until the Reformation. The need for vernacular translations was acknowledged in the early centuries when missionaries turned the Greek* and Hebrew* into Syriac,* Coptic,* Latin,* Ethiopic,* Georgian,* Armenian,* Gothic,* and a host of other tongues. Only during the Middle Ages was such translation proscribed. Then John Wycliffe's* work presaged an era which has seen parts of the Bible rendered into some 2,000 languages. Higher criticism,* with its queries into date, background, composition, and authorship, has been prominent only in the past two centuries. Not "which books nor what languages nor who wrote them," but "how do we interpret them" is the continuing question.

(1) *From the apostles to Augustine.*

The Ethiopian eunuch asked Philip about the meaning of Isaiah 53; "Then Philip began with that very passage of Scripture and told him the good news about Jesus" (Acts 8:35). Following their Lord's example ("These are the Scriptures that testify about me," John 5:39) the apostles began to read their Scriptures through eyes opened to deeper meanings.

In the OT they found prophecies which Jesus fulfilled, types which He completed, shadows of which He is the substance. Without questioning the factual character of the earlier events, they saw in them pointers to the fullness of time. What God had done in limited fashion in rescuing Israel from Egypt and settling her in the Promised Land, He was now doing for the church in lavish measure through Jesus, the Messiah.

(a) *Alexandria.* Almost immediately the post-apostolic generation found in the OT, not only pointers, but full charts of the future and detailed guides to holiness. Clement of Rome* (fl.c.90-100) used lives of individual Israelites as examples of waywardness and righteousness, and Rahab's scarlet cord as a picture of Jesus' blood.

The roots of the allegorizing technique adopted and exported by scholars like Clement of Alexandria* (155-c.220) and Origen* (c.185-c.254) were at least three: (i) rabbinic exegesis which allows a text great elasticity; (ii) OT studies of Philo,* who allegorized the Scriptures to find in them eternal verities akin to those of Greek philosophy; (iii) Plato's* teachings which separated what we know with our senses from eternal, real truths.

Planted in Christian soil, these roots were nourished by three convictions: (i) every word in Scripture was divinely inspired and, therefore, charged with hidden meaning; (ii) the OT everywhere spoke of Christ; (iii) where Christ was mentioned, His body, the church, was also in view.

Attempting to snatch the OT from his Jewish countrymen to protect it against the assaults of Marcion's* (second century) followers, Origen (whose comparisons of Greek and Hebrew versions made him a pioneer in textual criticism) devoted his monumental talents to the task of allegorizing it. Where the literal sense seemed absurd, unworthy, or immoral, he virtually disowned it in favor of the spiritual. As man is body, soul, and spirit, so Scripture has three senses: literal, moral, spiritual. The greatest of these is spiritual. The width of the gap between Origen's use of allegory* and the apostles' use of typology* is seen in the fact that Origen regularly allegorized the *New* Testament as well as the Old, setting a pattern that persisted through the Middle Ages.

(b) *Antioch.* Here Jewish tradition had interpreted the OT quite literally, and the influence of Plato and Philo was not dominant. Like Diodore of Tarsus* (d. c.392), the men of Antioch (including the gifted preacher, John Chrysostom,* d.407) distinguished between "theory" *(theoria)* and allegory, "theory" being the true meaning of the text. The Antiochenes, especially Theodore of Mopsuestia* (c.350-428), correctly sensed the importance of history as a foundation for theology and severely chided Origen for undermining this foundation: if Adam is not Adam, created and fallen, how can Christ be Last Adam, redeeming mankind from a fall that in allegorical exegesis did not really take place? When the Reformers sought to find a way back to apostolic methods of interpreting the OT, they took the road that passed near Antioch.

(c) *The West.* By allegorizing, Origen had sought to repel Marcion's attack on the OT as an unworthy Jewish book. Irenaeus* (fl.c.175-c.195), by contrast, answered Marcion by a system of typology in which the NT recapitulated and fulfilled the Old: Christ is the new Adam; Mary, the new Eve; the Cross, the new tree in the garden. Had the church-at-large adopted Irenaeus's scheme of progressive revelation, of continuity yet contrast between the two Testaments, the whole history of her understanding of the faith would have been altered for the better.

Tertullian* (c.160-c.220) tried to counter the Marcionites, Valentinians, and other heretics by depriving them of their right to the Bible, which he viewed as the legal possession of churches that remained true to apostolic teachings. He also gave to the church final authority in interpretation, anticipating the pattern of authority in Roman Catholicism. While not enthusiastic about allegories, Tertullian did not hesitate to use them in the service of orthodoxy's struggle with heresy or philosophy.

Jerome's* (c.345-c.419) knowledge of Hebrew (which set him apart from most of the Fathers), his familiarity with the geography of Palestine, his recognition of a Palestinian canon not containing the Apocrypha, his massive accomplishments in translation and revision of earlier Latin versions, his substantial commentaries on OT books—all indicate that, as philologist and biblical scholar, he had no peer for a thousand years. Although with Origen Jerome speaks of three senses of interpretation, most of the time he deals with only two—literal and spiritual. His insistence that the spiritual must be based on the literal, and his attention to the Hebrew text, gradually widened the gulf between him and Origen who, however, was not without his Western disciples like Hilary of Poitiers* and Ambrose.*

Augustine* (354-430) combined in his exegesis and preaching the approaches of a number of his predecessors. From Tyconius* the Donatist he borrowed rules for interpreting Scripture, especially OT prophecies, which had been carefully catalogued as to whether they referred to Christ or to His church or to both. Like Tertullian he stressed the church's authority in canon and interpretation. Holding that the OT was to be read both literally and figuratively, he frequently followed Ambrose's use of allegory as a weapon to resist the Manichaeans who derided the OT anthropomorphisms of God and accounts of patriarchal immoralities.

(2) *From Gregory the Great to Nicholas of Lyra*

(a) *Gregory I* and his successors.* Gregory's (d.604) commentaries on Job, Ezekiel, Kings, and parts of the gospels became a channel through which the knowledge and methods of the Fathers (especially allegorical interpretation) were conveyed to the Middle Ages, stripped of the polemics which had dominated much of patristic exegesis.

During the following centuries, the monasteries of Europe and England were beehives of literary activity, almost all of which consisted of fanciful applications or topical collections of patristic exegesis. Accomplishing more than this were Isidore of Seville* (d.636)—who compiled handbooks explaining difficult passages, proper names, dates, etc., based on material from Jerome—and the Venerable Bede* (d.735)—who sought to interpret figures of speech.

Charlemagne's* educational reforms (late eighth century) resulted in a number of commentaries, e.g., by Alcuin* (d.804) and Rabanus Maurus* (c.776-856). "The common characteristic in all this use of Scripture in the period from the seventh to the eleventh century was the constant link between the Bible and prayer, both public and private" (*Cambridge History of the Bible,* vol. 2, p. 188).

(b) *Monasticism.* Bernard of Clairvaux,* a Cistercian (1090-1153), wrote sermons on the Song of Songs and magnificent hymns which breathe with his mystical understanding of Scripture. Questions of history, chronology, and philology were largely beyond the monks' reach. Following Origen, they sought the NT's saving message of Christ in the OT, without clear agreement as to the number of senses a passage or word had: frequently four (literal or historical, typical, moral or allegorical, analogical or eschatological); sometimes three (historical, typical, moral).

(c) *Scholasticism.** The Schoolmen taught from copies of the Scripture with explanatory comments (usually from the Fathers) in the margins or between the lines. The best of these *glosses* (used, e.g., by Peter Lombard,* c.1095-1169) had been produced by Anselm* (d.1117) and Ralph of Laon. Hugh of St.-Victor (1096-1141), an abbey in Paris, gave great impetus to the teaching of the whole Bible by placing it at the heart of the curriculum and developing guidelines for exegesis, a work nobly advanced by his pupil, Andrew. A secular teacher, Peter Comestor* (whose title indicates that he had "eaten" the Scriptures, d.c.1180), produced his *Historia Scholastica* (a summary of sacred history), which joined the *gloss* as a required textbook.

Some of the later doctors (e.g., Bonaventure,* 1221-74; Thomas Aquinas,* 1224-74; and Albertus Magnus,* 1193-1280), though largely devoted to the relationship between Aristotle's* philosophy and their theology, did produce commentaries, particularly on the wisdom literature, whose ethics fascinated them. More than his fellows, Thomas Aquinas labored to bring system to exegesis by clarifying the ways in which the senses of Scripture were to be defined and discovered.

If Thomas's preference of the literal sense in proving doctrines (despite his frequent use of allegory) left the door slightly ajar for later reformers, it was Nicholas of Lyra* (c.1265-1349) who opened it wide. A Franciscan, Nicholas produced a commentary on the whole Bible which paved the way for hermeneutical reform by stressing the literal meaning of the text. Ironically, it was Jewish commentators like Ibn Ezra (c.1055-1135), Rashi (1040-1105), and the Kimchis (Joseph, c.1105-70; Moses, d. c.1190; David, c.1160-1235) who pointed Nicholas away from the fourfold formula which had hobbled medieval exegesis. Attempting to rescue their faith from Christians bent only on allegorizing the law and finding Christ in the prophets, these Spanish Jews moved away from kabbalistic and Talmudic exegesis and sought the plain sense of the Hebrew words and sentences. The allegorical method had developed in Alexandria to save the OT from its Jewishness. Now it was Jews who provided the means to save the OT from the church for the church.

(3) *The Reformers*

(a) *Luther.* Martin Luther's* (1483-1546) vernacular translation opened the Bible to the common people, established the order in which books of the OT have been generally published since (history, devotion, prophecy), separated the Apocrypha* from the canonical books, and paved the way for the critical investigation of Scripture. More important, his exegesis of the OT, especially Genesis and Psalms, moved away (notably after 1525) from the fourfold medieval pattern and emphasized the history of God's dealings with His people. Luther stressed the unity, the diversity, and the clarity of the Scriptures: unity, because the one God spoke the Word of Christ in both Testaments; diversity, because of the sharp antithesis between Law (both ceremonial and moral) and Gospel; clarity, because the plain truth of justification could be read by all, even in parts of the OT. Books that did not ring with the Gospel or sound the stern warnings of Law—e.g., Esther, Ezra-Nehemiah—he relegated to secondary importance without removing them from the Bible.

(b) *Calvin.* Like Luther, John Calvin* (1509-64) stressed the plain meaning of the OT although he played down the disjunction between Law and Gospel and (in the later editions of his *Institutes*) stressed the unity of the two Testaments by viewing the *moral* law (not the ceremonial) as an eternal statement of divine demands and by finding one major covenant of grace from the Fall through the Incarnation. Calvin's commentaries on the OT are models of scholarly exegesis, honoring every part of Scripture, while refusing to gloss over critical problems. His strong attachment to the OT was encouraged both by his desire to find Christian guidelines for Geneva's government and by his opposition to Anabaptists,* who treated the OT as a Jewish book with only limited Christian value. Calvin's doctrine of the *internal testimony of the Holy Spirit* as the chief argument for Scripture's authority sprang from his desire to refute the Roman Catholic claim that Scripture's authority is derived from the church.

(4) *From the Reformers to Schleiermacher*

After the Reformation,* theological positions both Catholic and Protestant hardened into systems in which matters of inspiration, authority, and canon were precisely determined.

(a) *The Council of Trent.** The council (8 April 1546) decreed that (i) unwritten traditions of the church be received with an authority equal to the Bible's; (ii) the Apocrypha have canonical status; (iii) the Vulgate* be the official version for public teaching or preaching; (iv) a definitive edition of the Vulgate be produced to settle the questions of textual variants; (v) all biblical books, in view of their divine inspiration, be accorded equal authority; (vi) the church have final say in interpretation.

(b) *Eastern Churches.* In 1672 a synod in Jerusalem defined the OT to include the Apocrypha, while forbidding unqualified lay people to read certain OT books.

(c) *The Reformed Creeds* (e.g., French, 1559; Anglican Thirty-Nine Articles, 1562; Belgic, 1566; Second Helvetic, 1566; Westminster, 1648). These set up their counter positions: (i) a fixed canon, excluding the Apocrypha (the reading of which was still permitted); (ii) the Spirit's inner witness, not ecclesiastical verdict, as the basis of Scripture's authority.

Lutherans who, in the Formula of Concord (1577), had affirmed Scripture's sole authority against the Catholic claims for tradition gradually moved closer to Calvinistic views of the Spirit's inner testimony and of the unity of Testaments, as Anglicans had already done.

(d) *Seventeenth-Century Orthodoxy.* Theological positions here were so minutely defined that the results have been labeled "Protestant Scholasticism." Dogmaticians like Johann Gerhard* (1582-1637), J.A. Quenstedt (1617-88), Leon Hü tter (1563-1616), J.B. Carpzov (1607-57), and Abraham Calovius* (1612-86) used the OT as a collection of proof-texts to oppose groups like the Socinians who valued the OT historically but rejected its doctrine. Meanwhile, free churches, especially in England, found comfort in the prophets' indictments of Israel's political and religious establishments. "The Old Testament influence . . . helped to make the Anglo-Saxon mentality different from any other Western European mentality" (E.G. Kraeling, p. 42).

(e) *Reactions to Orthodoxy.* Both textual ("lower") and historical ("higher") criticism began to chip away at the orthodox opinions that the inspiration of Scripture was a process of divine dictation. J. Cocceius* (1603-69) derived a scheme of progressive revelation from the OT and questioned its normative role for dogmatics. Philip Spener* (1635-1705) and J.A. Bengel* (1687-1752) and other Pietists sought to turn orthodoxy's attention from polemic debate to the teaching of the Scriptures which alone could produce true love and holiness.

Rejecting the Bible's revelatory claims, rationalistic philosophers like Thomas Hobbes* (1588-1679) and Benedict de Spinoza* (1632-77) sought to build systems of truth and morality on reason alone. Spinoza, though a Jew, rejected the doctrine of Israel's election. G.E. Lessing* (1729-81) further undercut the OT's uniqueness by distinguishing between the eternal truths of Christianity and their temporal contingent foundation in history, and by propounding that through reason God had educated other peoples even more than the Jews.

Though Immanuel Kant* (1724-1804) refuted the rationalists, he continued their attack on the OT, branding its laws as inferior to conscience because they were imposed from without. In academic circles, the philosophers won the day. Intellectually the church is still recovering from their victory.

(f) *Friedrich Schleiermacher** (1768-1834). He gave little more importance to the OT than to Greek wisdom, which he also saw as part of the preparation for Christianity. Defining true religion as the "feeling of absolute dependence on God," Schleiermacher treasured only parts of the OT that spoke of this new relationship (e.g., New Covenant in Jer. 31:31-34). For the first time since Marcion, a persuasive spokesman of the church encouraged her to close her first Scriptures.

(5) *From Wellhausen to the Present*

In the vast range of OT higher critics stretching back to J. Astruc* (1684-1766) and J.G. Eichhorn* (1752-1827), the highest peak is Julius Wellhausen* (1844-1918). Applying evolutionary canons stemming from Hegel* and Darwin,* Wellhausen reconstructed Israel's religious history as a naturalistic process that gradually matured from primitive animism and polytheism to the ethical monotheism of the prophets. The patriarchs had little historical basis; the Mosaic legislation was largely a product of the postexilic period; the prophets, not Moses, were the true pioneers; the Pentateuch, compiled from four main sources, was assembled only after the Exile.

Immediately a rash of critical studies broke out in the universities of Great Britain, Germany, and America, some of which led to sharp confrontation with churchly authorities. Scurrying to the aid of laymen and clergy who felt that higher criticism had snatched more than half the Bible from their hands came a flock of critical scholars who sought to point out the abiding relevance of the OT, e.g., G.A. Smith,* *The Higher Criticism and the Preaching of the Old Testament* (1901); A.S. Peake,* *The Bible, its Origin, its Significance and its Abiding Worth* (1913); H. Gunkel,* *What Remains of the Old Testament?* (1928).

Among the conservatives who sought to dull the impact of higher criticism by attacking its naturalistic presuppositions and by exposing weaknesses in its methods were William Henry Green (*The Higher Criticism of the Pentateuch,* 1895) and James Orr* (*The Problem of the Old Testament,* 1906).

In the decades since World War I some factors have combined to increase Christian understanding of the Old Testament:

(a) *Archeology* has confirmed the Bible's historical character and illuminated the setting and context of life in Bible times—cf. W.F. Albright, *Archeology and the Religion of Israel* (1942); G.E. Wright, *Biblical Archaeology* (1957); D. Winton Thomas (ed.), *Archeology and Old Testament Study* (1967). John Bright (*A History of Israel,* 1959) and Martin Noth (*The History of Israel,* 1958) represent respectively the more confident and the more skeptical approaches toward OT archeology, particularly in the period before David.

(b) *Philology* has opened a window through which we can look at the thought processes, value systems, and religious beliefs of Israel's neighbors, e.g., the Egyptians, the Sumerians, Assyrians and Babylonians, the Syrians (especially in Ugarit), the Moabites, the Hurrians and the Hittites, and the Arameans. As no other generation in history, ours has the tools and skills to confront antiquity on its own terms.

While pioneers in comparative studies were impressed by the *parallels* between Israel's religious practice and thought and her neighbors', their successors have appreciated the *contrasts* which point to the uniqueness of Israel's revealed, historical, convenantal, moral faith—cf. G.E. Wright's *The Old Testament Against Its Environment* (1950) and W.F. Albright's *From the Stone Age to Christianity* (1940).

(c) *The behavioral sciences*—anthropology, sociology, psychology—have provided entrée to the sociopolitical systems and to the inner life of Israel. Customs and relationships puzzling to our Western modernity are beginning to become clear—cf. J. Pedersen's *Israel: Its Life and Culture* (2 vols., 1926, 1940) and R. DeVaux's *Ancient Israel* (1961).

(d) *The study of literary forms* (e.g., by H. Gunkel, S. Mowinckel,* C. Westermann, B. Childs) has pointed out the varieties of literary genre and the emotional connotations that they brought to the Israelite who heard them from priest or prophet: hymn, complaint, instruction; proverb, fable, parable, riddle; work song, love song, battle cry; defendant's plea, plaintiff's argument, judge's verdict, king's commission to a messenger.

(e) In *theology,* for the first time since Calvin, a major theologian, Karl Barth,* stressed the unity of the Testaments and the way in which the OT is a unique witness to revelation as it anticipates history's pivotal revelatory event—the Incarnation. Conservatives have, with good reason, criticized Barth's view of Scripture, which restricts revelation to divine encounter rather than extending it also to the inspired record of that encounter, the biblical texts; but they should be grateful for his emphasis on the creating, redeeming God of the OT in an era when that picture of God was much maligned.

Heartened by these recent insights and sobered by two world wars and a great depression, a distinguished company of scholars have tackled the massive task of writing OT theologies, where most of their predecessors had shied away from the very term "theology," daring to speak only of Israel's "religion." Writers who have been keenly conscious of the OT's contribution to the Christian faith include W. Vischer *(The Witness of the Old Testament to Christ),* O. Procksch *(Theology of the Old Testament),* Th. Vriezen *(An Outline of Old Testament Theology),* H.H. Rowley *(The Unity of the Theology),* H.H. Rowley *(The Unity of the Bible),* G.A.F. Knight *(A Christian Theology of the Old Testament),* F.F. Bruce *(This is That: The New Testament Development of Some Old Testament Themes),* and especially G. von Rad *(Old Testament Theology).*

The debate continues. The teachers of the church struggle to examine the ties that unite and the uniqueness that separates the Testaments. Final resolution is probably out of reach. But the most promising approaches seem to combine a commitment of the reality of God's revelation in His words and deeds to Israel, a progress in that revelation through which more and more of God's ways and will are understood by His people, and a consummation of that revelation in the Incarnation of Christ, God's final Word to man. History and theology—event and the meaning of the event—must both be taken with full seriousness. We can be hopeful in the midst of the struggle: the Lord of the Testaments and of the church does not await scholarly consensus to get His speaking done.

BIBLIOGRAPHY: F.W. Farrar, *History of Interpretation,* Bampton Lectures (1886); B. Smalley, *The Study of the Bible in the Middle Ages* (1952); R.E. Brown, *The Sensus Plenior of Sacred Scripture* (1955); E.G. Kraeling, *The Old Testament Since the Reformation* (1955); R.P.C. Hanson, *Allegory and Event* (1959); A.S. Wood, *Luther's Principles of Biblical Interpretation* (1960); A.D.R. Polman, *The Word of God According to Saint Augustine* (ET 1961); R.M. Grant, *A Short History of the Interpretation of the Bible* (rev. 1963); J. Bright, *The Authority of the Old Testament* (1967); H. Bornkamm, *Luther and the Old Testament* (ET 1969); G.W.H. Lampe (ed.), *The Cambridge History of the Bible,* vol. II: *The West from the Fathers to the Reformation* (1969); J.S. Preus, *From Shadow to Promise* (1969); P.R. Ackroyd and C.F. Evans (eds.), *The Cambridge History of the Bible,* vol. I: *From the Beginnings to Jerome* (1970). DAVID A. HUBBARD

OLEVIANUS, KASPAR (1536-1587). Reformed theologian. Born at Trèves (Trier), he studied at Paris, Orléans, and Bourges, where he accepted Reformation ideas. The drowning of a friend impelled him to become a preacher. He studied theology in Geneva, Zurich, and Lausanne, becoming acquainted with Farel, Calvin, Peter Martyr, Beza, and Bullinger. He returned to Trèves to teach in the Latin school (1559), but his fervent preaching led to imprisonment. Invited by Elector Frederick III to Heidelberg, he became pastor of St. Peter's Church and helped to reconstruct the church on Reformed lines. With Zacharias Ursinus,* he drafted the final revision of the Heidelberg Catechism. He was involved in an "Arian" controversy and voted for the death penalty for the "blasphemers." Banished during a Lutheran reaction under Louis VI, he went to Berleberg, Wittgenstein, and to Hernorn, Nassau, where he established a complete Presbyterian organization. He also wrote NT commentaries.

J.G.G. NORMAN

OLIER, JEAN-JACQUES (1608-1657). Founder of the Society of Priests of St.-Sulpice. Born in Paris, he studied theology at the Sorbonne. His spiritual renewal came between 1630 and 1632, and was due in part to the influence of Vincent de Paul* and a pilgrimage to Loreto. In 1641 he founded a seminary for priests at Vaugirard. As parish priest of St.-Sulpice in Paris (1642-52) he sought to reform the area with the help of the seminary which moved there in 1642. The seminary grew in size and fame, and in 1657 Olier's missionary zeal led him to send priests to Montreal. Probably his best-known book was the *Catéchisme chrétien pour la vie intérieure* (1656), but all his writings were valued and influenced the spiritual renewal of many clergy. PETER TOON

OLIVETAN (c.1506-1538). Protestant Reformer and cousin of John Calvin. His real name was Pierre Robert, and like his cousin he was a native of Noyon in Picardy and a student successively at Paris and Orléans, where he earned the nickname "Olivetanus" because he burned the midnight oil. An early French Protestant, he fled from Orléans to Strasbourg in 1528 following his evangelical conversion. Beginning in November 1531, he preached briefly at Neuchâtel, leaving the following year for Piedmont where he contacted the Waldensians.* Olivetan taught in Geneva from 1533 to 1535, but resigned to return to Italy, where he died. He is chiefly remembered for two things: his religious influence on young Calvin, and his translation of the Bible into French. Olivetan was one of several important sources which influenced his cousin toward evangelical Christianity. Further, his French Bible, originally prepared for the Waldensians of Piedmont, was the version used by the first-generation Calvinist Reformers as they preached the Gospel in France. Published in Neuchâtel in 1535, it contained a preface by Calvin in which for the first time he gave a public confession of his biblical faith.

ROBERT D. LINDER

OLIVI, PETRUS JOANNIS (c.1248-98). Augustinian philosopher and leader of Spiritual Franciscans. Born at Sérignan, he entered the Franciscan Order at Béziers (c.1260). He studied at Paris and possibly Oxford. Accused of heresy at Strasbourg (1282), he established his orthodoxy at Montpellier (1287) and Paris (1292). He became lector at the convents of Nîmes, Florence, and Montpellier. He was greatly venerated by the Spiritual Franciscans after his death. His commentary on Revelation, *Postilla super Apocalypsim,* with its overtones of Joachitism and identification of the papacy with Antichrist, was condemned by John XXII (1326). Anthropologically, he distinguished between ontological and conceptual orders, concluding that man is composed of matter and spirit, and that there is plurality of forms in corporal beings. Concerned to keep body and soul separate, he divorced the intellectual soul from the vegetative and sensitive souls, making the latter alone the body's form. This idea was condemned by the Council of Vienne* (1311-12).

J.G.G. NORMAN

OLOPUN, see ALOPEN

OMAN, JOHN WOOD (1860-1939). Presbyterian theologian. Born in Orkney, and educated at Edinburgh and Heidelberg, he became minister at Alnwick, Northumberland, after which (1907) he joined the staff of Westminster College, Cambridge, where he was later principal (1925-35). He developed an early interest in Schleiermacher,* who held that religious experience is self-authenticating. Like Schleiermacher, Oman stressed feeling, defining religion as the direct feeling of the "supernatural." He did not mean the miraculous, but simply a wider environment than physical nature. This view is similar to that of R. Otto, who emphasized the feeling of the "awesome holy." Of his thirteen books the best known is *The Natural and the Supernatural* (1931). His obscurities, real or apparent, his inconsistencies, and his overindulgence in generalities do not make for easy reading, but sympathizers with German philosophy claim to plumb new depths of meaning in his carefully construed sentences.

R.E.D. CLARK

ONTOLOGICAL ARGUMENT. First formulated by Anselm* (c.1033-1109) while abbot of Bec, and later by Descartes* (1596-1650), this is an *a priori* argument for the existence of God, based on the idea that the concept of God as the most perfect being requires His existence, since a god that existed only *in intellectu* would be less than perfect (would lack the perfection of existence), and so could not be God. Another version of the argument which Norman Malcolm finds in Anselm's *Proslogion* involves the predication of necessary existence of God: God is the sort of being who does not just happen to exist, but exists necessarily. As Kant showed, the first version of the argument involves the unacceptable premise that "existence" functions as a predicate, while the second version, given prominence by contemporary philosophers such as Malcolm and Charles Hartshorne, involves a confusion between logical and ontological senses of "necessity." God may be said to be ontologically necessary—i.e., His existence is not contingent upon other states of affairs—but the second version requires the idea of the nonexistence of God to be self-contradictory, which it obviously is not. The ontological argument has aroused considerable interest in modern analytic philosophers, partly because of the many independent conceptual issues it raises about perfection, necessity, existence, and so forth.

PAUL HELM

ONTOLOGISM (*ontos,* "being"; *logos,* "science"). A speculative system of philosophy which (in a theistic form) maintains that we know God immediately as the primary and natural object of human cognitive powers. The intuition of God is the first act of our intellectual knowledge. This system claims descent from Plato* and Augustine* and was advocated by Malebranche* in the seventeenth century on the basis of his Occasionalism, which maintained that finite things have no efficient causality of their own and that our sensations and ideas are caused not by body or mind but produced by God, the universal Cause. Vincenzo Gioberti* and Antonio Rosmini* were prominent exponents. The term itself first appears in Gioberti's *Introduzione allo studio della filosofia* (1840). Attacked on the grounds that our idea of God is analogical, not direct, and that such teaching leads to pantheism, seven propositions of the ontologists were condemned by the Holy Office in 1861. In 1887 forty propositions from the works of Rosmini were condemned by the Vatican Council.

HOWARD SAINSBURY

OOSTERZEE, JAN JAKOB VAN (1817-1882). Dutch Reformed minister and theologian. Born in Rotterdam, he was educated at the University of Utrecht and by 1862 had served churches at Eemnes-Binnen, Alkmaar, and the chief pulpit in Rotterdam. Well known as pulpit orator and evangelical leader, his piety and scholarship returned him to Utrecht to teach practical theology, which position he kept until his death. He rejected naturalistic, critical views and vigorously maintained biblical and supernatural presuppositions toward theological science. His writing career began in 1845 with the editorship of a theological journal. His published works included *The Image of Christ as Presented in Scripture* (1874); commentaries on Luke, the Pastoral epistles, Philemon, and James; and *Praktische Theologie* (1878), which covered the full range of his pastoral and tutorial prowess.

H. CROSBY ENGLIZIAN

OPEN BRETHREN, see PLYMOUTH BRETHREN

OPHITES. Gnostic sect. Origen, and Celsus through Origen, provide what is principally known about them: the Ophites' diagrammatic approach to the total cosmos, having at its highest level the Kingdom of God with the Son as inner core; and their liturgy, which is a symbolic rite of passage through the stages thus diagrammed. Theologically one sees further Gnosticization of the Apostle John's celestial geography, with corresponding denial of the reality of history and/or time.

OPTATUS (fourth century). Bishop of Milevis in Numidia and author of an untidy work known as *On the Schism of the Donatists* or *Against Parmenian the Donatist* (bishop of Carthage). Books 1-6 were written about 367 and revised when book 7 (incomplete) was added about 385. An appendix compiled between 330 and 347 comprises ten documents important for Donatist origins. A Christmas sermon also survives. Optatus's historical and theological rebuttal of Parmenian's lost *Against the Church of "Traditores"* advances arguments later developed by Augustine. The "endowments" *(dotes)* of the true church include catholicity (worldwide extension), unity (communion with the *cathedra Petri* of Rome and the bishops), and the holiness of God's gifts, especially the sacraments, not of a membership embracing good and evil alike. Baptism sanctifies by virtue of divine grace, not the minister's standing. Optatus's church belongs unashamedly to the Christian empire and approves its measures against schismatics. D.F. WRIGHT

ORANGE, COUNCILS OF. Two synods held at Orange (Arausio) in S France (Vaucluse) in 441 and 529. Hilary of Arles* presided at the first, with sixteen bishops attending, where thirty canons were issued on disciplinary matters. Caesarius of Arles* presided at the second, championing Augustinian prevenient grace in the struggle against Semi-Pelagianism.* In a contest with Felix IV he triumphed, submitting to thirteen bishops in Orange a declaration on grace and free will, in the form of *capitula,* which they signed and which Boniface II approved in 531. These upheld much of Augustine's* doctrine on grace as the corrective to the views of John Cassian* and Faustus of Riez,* stressing the need for grace and condemning predestination of man to evil. Theories, not persons, were condemned. The 529 council ended the Semi-Pelagian controversy in S Gaul. C.G. THORNE, JR.

ORANGEMEN. Members of the Orange Order, an organization of Protestants, originating in Ireland,* with lodges also in England, Scotland, Australia, New Zealand, Canada, and Africa. The order derives its name from its attachment to the memory and achievements of William III,* Prince of Orange, whose victories over the Roman Catholic forces of James II, particularly at the Battle of the Boyne in 1690, are commemorated in annual processions.

The Orange Order was founded in 1795, at a time when there was considerable guerrilla warfare between Roman Catholic bodies such as the "Ribbonmen" and the "Defenders," and Protestant organizations such as the "Peep o' Day Boys," particularly in County Armagh. After a pitched battle at a place called "the Diamond," in which the Roman Catholics were defeated, the surviving Protestant victors formed an organization in Loughgall to pledge themselves to mutual protection and in defense of Protestantism. The movement grew rapidly both in Ireland and England in the early years of the nineteenth century, but its influence in England at that time waned when it was suspected of political intrigues involving the duke of Cumberland, uncle of Queen Victoria. Beginning as a largely defensive organization, with a decisively evangelical Protestant constitution, the Orange Order, with the passing of the years and particularly in opposition to Roman Catholic sympathies with Irish Republican politics, became itself increasingly a political movement and has wielded considerable influence on the political situation in Northern Ireland. HUGH J. BLAIR

ORATORIANS. The name of two associations of secular priests:

(1) *Italian Oratory,* founded by Philip Neri* in Rome out of an informal association of priests (1564). He was the first religious leader to add social and artistic aspects to devotional exercises. Palestrina,* one of his penitents, composed music for his external brotherhood, or "little oratory," thus giving the name to the art-form "oratorio." Formally approved in 1575, they spread through Italy, France, and Spain. J.H. Newman* introduced them to England at Old Oscott, 1847. They live in community without vows, supported by private means.

(2) *French Oratory,* founded by Pierre de Bérulle* at Paris (1611), approved as the *Oratoire de Jésus-Christ* (1613). Though inspired by the Italian Oratory, it is a separate institute, a centralized organization governed by a superior-general. Its principal activity was training priests in seminaries. During 1672-1733 it was dominated by Jansenism.* Dissolved in 1790, it was reestablished (1852) by L.P. Pététot and A.J.A. Gratry.* J.G.G. NORMAN

ORATORIO, see MUSIC, CHRISTIAN

ORATORY (Lat. *oratorium,* "place of prayer"). From its more general original meaning, the term is now associated in Roman Catholicism with a place intended for divine services, especially celebration of the Mass, primarily in connection with certain designated persons. Modern Roman Catholicism recognizes three types of oratory:

(1) A *public* oratory, chiefly for private individuals or a community, but open to all worshipers at least when divine services are held.

(2) A *semi-public* oratory is designed for the use of a particular community which may, however, at its pleasure admit or exclude others.

(3) A *private* or *domestic* oratory (the adjectives were previously distinguishable but are now

virtually interchangeable) is set up in a private home for the use of a family or individual. In such instances they are a reversion to historical usage —i.e., places of prayer which are no substitute for the local churches and are used only exceptionably for Mass. J.D. DOUGLAS

ORDEALS. A form of trial whereby, according to God's will and judgment, the guilt or innocence of an accused person was determined by some feat of physical endurance. The practice predates the biblical era and was almost universal in scope. The ordeal by water is mentioned in the Code of Hammurabi (nos. 2, 132) and the bitter-water ordeal is recorded in Numbers 5:11-31. The ordeal was commonly used in medieval Europe and held the favor of the church until 1215 (Fourth Lateran Council), when the clergy was forbidden to take part in the practice. Up to that time the ordeals were usually preceded by Mass. There were many canonists who had opposed it before 1215, but they had been unable to have it abolished. The ordeal took many forms: ordeals by poison, by water, by hot iron, by fire, and by combat. The ordeal by poison is used particularly by the peoples of West Africa. Europeans often required the accused to carry a ball of hot iron in his hand for a certain distance or to plunge his arm to the wrist or elbow into a caldron of boiling water. Festering on the third day proved guilt. JOHN P. DEVER

ORDINATION. The separation and commissioning of particular persons by the church for the work of the Christian ministry; but the outward calling by the congregation should correspond to and be consequent on the inward calling of the Holy Spirit. There is general consent that essential to the form of ordination are prayer and the laying on of hands, in accordance with what seems to have been the practice of the apostolic church. It is true that in the accounts of the calling of the Twelve (Mark 3:13ff.) and the commissioning of the Seventy (Luke 10:1ff.) it is not said that Christ prayed and laid hands on them; but this does not rule out the possibility that He did so. On the other hand, it might be concluded that a direct dominical commissioning rendered these acts unnecessary.

There is specific mention of prayer and the laying on of hands in connection with the appointment of the Seven (Acts 6:6) and the setting aside of Barnabas and Saul for the work of evangelism (Acts 13:3, where fasting is added). It is important to notice that the Seven were men "full of the Spirit and wisdom" as a prerequisite to, not a result of, their appointment, and that Barnabas and Saul were set apart by command of the Holy Spirit and were "sent on their way by the Holy Spirit" (Acts 13:2, 4; cf.6:3). Paul himself insists that it was Jesus Christ who appointed him to the ministry and that his apostleship came to him, not through men, but from God and by the will of God (see 1 Cor. 1:1; 2 Cor. 1:1; Gal. 1:1; Eph. 1:1; Col. 1:1; 1 Tim. 1:12; 2 Tim. 1:1). To the same effect Christ assures His apostles that it was not they who had chosen Him, but He who had chosen and appointed them (John 15:16); and Paul admonishes the Ephesian elders to take heed to all the flock over which the Holy Spirit had made them bishops or overseers (Acts 20:28). In the Pastoral epistles, Paul refers to the charism which was given to Timothy by prophetic utterance when the elders laid their hands on him (1 Tim. 4:14; cf. 5:22; 2 Tim. 1:6) and instructs Titus to appoint elders in every town of Crete (Titus 1:5). Within the biblical perspective, then, ordination is primarily an act of God's calling and appointment, and only secondarily an act of the church, which by prayer seeks to know and follow the will of God. PHILIP EDGCUMBE HUGHES

ORGAN. Traditionally known as "the King of Instruments," the organ today stands as a remarkable example of artistry and craftmanship. The early Christians associated it with their Roman persecutors; it is in fact probable that Nero's famous performance during the burning of Rome was actually on a water-organ rather than the proverbial fiddle. By the Middle Ages, however, organs were evidently in regular use in churches, a detail we know from pictures and manuscripts of the time. Since then it has been the instrument *par excellence* of the church and for the praise and adoration of Almighty God. One reason for this is the enduring quality of organ tone. On any other instrument the tone is born, fades away, and dies. But organ sound endures as long as the power is there; the subtle suggestion of a timeless eternity is felt. Another reason is the ensemble effect of the organ: the many voices and colors of the organ give a full, rich sound impossible of achievement in any other way. This has always been so with the pipe organ; now at last it is beginning to be true also of the best of the electronic organs. No longer need the latter be considered a poor substitute for pipes; it has now, in at least a few fine examples, graduated to the position of being a musical instrument worthy of acceptance on its own terms.

The literature of the organ spans many centuries. The clean, bright sound of a well-voiced Diapason or Principal Chorus is ideal for the performance of contrapuntal music. The rich, more romantic sound of the massed strings gives life to music of the Romantic Era. The many varieties of flute, and the solo and chorus reed stops, enrich the tonal palette of the instrument. In considering specific repertoire for the instrument, the name of J.S. Bach* naturally comes first to mind. But there were towering giants both before and after this great man. Pachelbel, Buxtehude,* Sweelinck,* and many others are among the predecessors of Bach whose music is heard frequently today. Since Bach, the list is long and constantly growing. Franck, Mendelssohn,* Liszt,* Schumann, and Reger are a few of the names of great stature that come to mind.

Today composers are writing music of the future for the organ as well as writing in the older forms and styles. How enduring the "music of the future" will be remains to be seen—certainly it is in a language foreign to many music-lovers. Yet without this experimentation in new forms and styles the art of music would stagnate; all of us must be grateful to the pioneers.

One criticism frequently directed at the organ is that it is by definition a mechanical instrument. Because of this, some maintain it is impossible to project any very deep sense of emotion through organ music. This is quite untrue. Although the performer is not as close to his tone as is, for example, the violinist, the truly artistic organist who masters the technical demands of his instrument can transcend these difficulties and project not only the spirit of the music but, if he possesses the Holy Spirit, also the Spirit of God. This is essentially the aim of all Christian music: to be a vehicle for the Holy Spirit. The organ, because of its traditional affiliation with the church and more especially because of its tonal qualities, is ideally suited to this noble purpose. ROBERT ELMORE

ORGANIC ARTICLES (1802). A French law unilaterally amending and implementing the Concordat of 1801,* to which it was attached without papal agreement. On grounds of concern for "public tranquility" (Concordat, Article 1), Napoleon hereby succeeded in seizing total control of the French Church, tightly centralized under the state. The law contained seventy-seven articles arranged under four main titles. Title 1 required all papal communications and papal representatives, but also decrees of general councils and any French national or diocesan synods, to receive state approval to be valid. The last three titles provided for a total reorganization of the French Church and seminary structures to conform to secular statist interests, including mandatory acceptance of the four Gallican Articles* (1682) in the seminaries, and minute regulation of worship and salaries. Although condemned by Pius VII and later popes, the Organic Articles remained law, with some parts disused, until 1905.

C.T. MC INTIRE

ORIGEN (Origenes Adamantius) (c.185-c.254). Alexandrian theologian. Most of the information about his life is found in the sixth book of Eusebius's* *Ecclesiastical History;* a panegyric by Gregory Thaumaturgus,* a disciple of Origen; Jerome's* *Of Illustrious Men;* and in the fragment of an apology written by Eusebius and Pamphilus.* Origen, born in Egypt and raised by Christian parents, studied under Clement* in the Catechetical School in Alexandria. During the persecution of Septimus Severus in 202, his father, Leonidas, was captured and martyred. Origen's wish to die with his father was prevented when his mother hid his clothes. He was able to continue his study after his father's death because of the generosity of a wealthy widow. He became, and remained for twenty-eight years, the head of the Catechetical School while pursuing an ascetic and extremely pious life. In his early manhood he apparently took the passage of Matthew 19:12 literally (cf. KJV) and castrated himself.

While in charge of the school in Alexandria he became famous, and according to Eusebius, thousands came to hear him, including many prominent pagans such as the mother of the emperor Alexander Severus. A wealthy convert allegedly hired secretaries to copy down his lectures and then published them. He studied with the father of Neoplatonic thought, Ammonius Saccas*; he traveled to Rome and heard Hippolytus.* During the persecution of Caracalla in 215, Origen went to Palestine where he was invited to preach by the bishops of Caesarea and Aelia. Bishop Demetrius* of Alexandria was displeased with this invitation to preach, since Origen was still a layman. Consequently Demetrius recalled Origen to Alexandria; Origen then devoted himself to writing. In 230 he was back in Palestine and ordained a priest by the same two bishops who had invited him to come the first time. This time Demetrius declared Origen deposed as a priest, deprived him of his teaching post in Alexandria, and exiled him. The deposition was generally not recognized outside Egypt.

Origen then established a school in Caesarea which became famous. He continued to preach and write. The persecution of Decius in 250 caught him, and he was put in chains and tortured, suffered the experience of the iron collar, was placed in stocks and confined to a dungeon. He did not survive long after this ordeal was over and he was released.

Origen was one of the Greek Fathers of the church. He is considered to be one of the first textual critics of the Bible; one of the first to set forth a systematic statement of the Faith; and one of the first Bible commentators. He was an effective apologist. From his own example and from some of his writings one can find some of the early principles that spawned the monastic movement. The number of his works varies from the 6,000 reported by Epiphanius* to 2,000 reported by Pamphilus to 800 reported by Jerome. Most of the works are lost although fragments survive and some have been discovered recently. His most famous works include the *Hexapla*—an edition of the OT in Hebrew, Greek, Greek versions of Aquila, Symmachus, the Septuagint, and Theodotion,* all arranged in six columns. *De Principiis* is another of his important works, being one of the first systematic theologies. Book 1 of *De Principiis* deals with the Heavenly Hierarchy of the Father, Word, and Spirit and their relation to earthly beings; book 2 deals with the material world including the place of man, his fall, and his redemption; book 3 deals with the freedom of the will in its struggle with the forces of good and evil; book 4 deals with biblical hermenuetics and the literal, moral, and allegorical interpretation of Scripture.

On Prayer was written later in his life and discusses prayer in general and the Lord's Prayer specifically. Here Origen argues prayer is not a petition, but a participation in God's life. *Contra Celsus* demonstrates that Origen could argue against his opponents using their philosophical grounds to prove the contrary of their position. He was accused of Subordinationism* by Jerome and Epiphanius and condemned by some synods, such as the Synod of Constantinople of 543. Perhaps one should not take these rejections too seriously, since he was the first of the systematic theologians and a seminal thinker.

BIBLIOGRAPHY: A. Roberts and J. Donaldson (eds.), *The Ante-Nicene Fathers,* vol. 4 (1951); R.P.C. Hanson, *Origen's Doctrine of Tradition* (1954); H.U. von Balthasar, *Parole et mystère chez*

Origène (1957); H. Crouzel, *Origène et la philosophie* (1962); G.W. Butterworth (ed.), *On First Principles* (1966); C.V. Harris, *Origen of Alexandria's Interpretation of the Teacher's Function* ... (1966); C. Bigg, *The Christian Platonists of Alexandria* (rep. 1970); R. Farina, *Bibliografia origeniana, 1960-1970)* (1971).

ROBERT SCHNUCKER

ORIGENISM. Doctrines attributed to Origen which later became controversial. Methodius of Olympus* (d.311) rejected most of Origen's speculations, especially his concept of human preexistence and the temporary character of the body, in *Symposium* and *De Resurrectione.* Origen was defended by Eusebius of Caesarea* and Pamphilus.* Eustathius* of Antioch (d.336) was another prominent anti-Origenist, but most of his arguments were based on Methodius.

The Nitrian monks of Egypt were strongly Origenist, and one, John, became bishop of Jerusalem. His Origenism was attacked by Epiphanius* of Salamis (c.310-c.403), who included Origenism among heresies enumerated in his *On Heresies.* Epiphanius persuaded Jerome,* previously an ardent defender of Origen, to become a violent opponent. John, however, refused to condemn Origenism. Jerome attacked Origen's doctrines of the resurrection body, the condition of souls, the devil's ultimate repentance, and the Trinity. Rufinus* of Aquileia, Jerome's friend, continued to support Origen and published a Latin translation of *De Principiis* (c.397), in which he injudiciously named Jerome's former allegiance to Origen, which Jerome furiously rebutted. Siricius* of Rome supported Rufinus, but his successor Anastasius condemned him. Meantime, the Council of Alexandria (400) had condemned Origenism, and Theophilus* of Alexandria expelled the Origenist monks, "the Tall Brothers," who found refuge with Chrysostom* in Constantinople.

The fierce disputes of Origenist and orthodox monks for the possession of the monasteries of St. Saba in Palestine led to the emperor Justinian's* famous letter to Mennas of Constantinople in which the "errors" of Origen were anathematized, including the preexistence of souls, the Incarnation, the resurrection body, and Restorationism.* A synod at Constantinople (543) issued an edict giving effect to this condemnation. The Origenists themselves were in two parties: the *Protoktistae,* who regarded the soul of Christ as not equal to other souls but divine; and the *Isochristi,* who held that at the final restoration all souls would become like Christ's. The Protoktists made common cause with the orthodox after renouncing the doctrine of preexistence. The fifth general council at Constantinople (553) listed Origen among ancient heretics. All bishops submitted except Alexander of Abila, who was deposed.

For bibliography see under ORIGEN.

J.G.G. NORMAN

ORIGINAL SECESSION CHURCH. More properly "The Synod of United Original Seceders," this was constituted in Scotland in 1842. It was a union of various groups which were heirs of the Secession of 1733 from the Church of Scotland. The term "church" was not officially used, as they considered themselves in secession from the national church and not as a separate church. They rejoined the Church of Scotland in 1956. The original secession had taken place because Ebenezer Erskine* and others had felt inhibited from protesting effectively against abuses in the Church of Scotland, particularly patronage. This meant that ministers were presented to parishes by patrons instead of being elected by the congregations as the seceders demanded. Later the Seceders divided into "Burghers"* and "Antiburghers" over the rightfulness of taking the Burgess Oath professing the "true religion," and then into "Auld Lichts"* and "New Lichts" over the interpretation of the clauses in the Westminster Confession* regarding the civil magistrate.

HENRY R. SEFTON

ORIGINAL SIN. There are in the OT several acknowledgments and confessions of the universal sinfulness of man (Gen. 8:21; 1 Kings 8:46; Ps. 130:3), of resignation or penitence that man is born into an inevitable sinful condition (Ps. 51:5; Job 15:14; 25:4), and of the impossibility that for any man it can be otherwise (Job 4:17; 14:4). These texts are all expressions of the fact of original sin. Man seems to inherit this bondage from his birth. He seems to sin involuntarily and inveterately, and yet feels responsible for so doing. Before he wakens up to the seriousness of his struggle with evil, the battle is already lost. Sin not only develops within him, it also envelopes him. He finds his community life warped and inhuman in its ideals and even in the best zeal it can muster for its own reformation or revolution.

The Jewish rabbis were aware of such problems in a limited way, and attributed the universal sinfulness of man to the fall of the sons of God described in Genesis 6:1-4. Thereby, they believed, man has developed an evil impulse or imagination in the soul, exerting the strongest pressure toward sin (Gen. 6:5; 8:21). It was Paul who first linked up the phenomenon of original sin with the story of the fall in Genesis 3. In Paul's thought, the fall was modeled on salvation. Thinking back from Christ to Adam, he saw that as mankind was somehow totally involved and thus saved in Christ, so also mankind was somehow totally involved in the fall of Adam, the true significance of Adam being revealed in Christ. Paul (Rom. 5:12-21) is thus original in his clear explanation of the universality of sin, but brought to light what was already implied by the OT writer. The sins of all men are thus the unfolding of the original sin of Adam.

Jesus Himself does not formulate such a doctrine. Its full expression had to wait until the revelation of His cross. Yet He speaks and acts often in such a way as to imply it. All men are lost (Mark 2:17; Luke 19:10) and need forgiveness. They are all "evil" (Matt. 7:11; Luke 11:4). The things of man's nature are contrary to God (Mark 7:23). Paul, continuing the rabbinical doctrine of an evil impulse inborn in man, speaks of sin as seated in the "flesh" which through sin has become a principle radically antagonistic to God, lying behind all man's activity apart from the grace of God. James speaks of an overwhelmingly powerful "desire" or "lust" within man (James 1:13, 14; Gal. 5:16-24).

The doctrine of original sin has been the center of much theological controversy. Tertullian* coined the term "concupiscence" for man's inborn evil desire. Augustine* insisted that the phrase translated often "because all sinned" (Rom. 5:12) implied that all men subsisted in Adam when he sinned, and that his sin was their sin. We are condemned, not by our willing assent, but by our ancestry. Moreover, ever since the Fall, man has lost all capacity to obey the will of God. In concupiscence itself we have original sin. Pelagius denied the doctrine of Augustine, insisting that God did not demand more than man can render, that we are born without virtue or vice and are free to choose the goal of our lives. We can live sinless lives. He believed that sin consists of bad acts rather than of bad dispositions, and is caused by bad example and deficient education (see PELAGIANISM).

During the Middle Ages, original sin tended to be defined as the absence of original righteousness—the privation of supernatural grace through the Fall—rather than as concupiscence. The latter was interpreted as the temptation to fleshly lusts which could become the material of sin if assented to—a weakness which could be helped, or even healed, by works. Among the Nominalists the will was regarded as possessing a *synteresis* —a tiny motion toward God which could be trained into the love of God. The Reformers, following Luther,* returned to the Augustinian teaching, insisting that concupiscence affected the whole of human life, including the intellect and will, inclining man to evil in everything he does.

As to the transmission of sin, theologians like Origen* have thought in terms of a personal pretemporal fall for each individual. Tertullian regarded the whole race as seminally present in Adam, the corruption being passed on through propagation. Augustine recognized the importance of the latter view, but regarded the passing on of sin as the punishment of Adam's sin. Adam's sin was imputed to his posterity. When Covenant Theology* became popular in the seventeenth century, the imputation of Adam's guilt to his successors was emphasized as well as the transmission of corruption.

BIBLIOGRAPHY: F.R. Tennant, *The Origin and Propagation of Sin* (1906); E.J. Bicknell, *The Christian Idea of Sin and Original Sin* (1922); N.P. Williams, *The Ideas of the Fall and Original Sin* (1927); C.R. Smith, *The Biblical Doctrine of Sin* (1953); H. Haag, *Is Original Sin Scriptural?* (1969). RONALD S. WALLACE

OROSIUS, PAULUS (early fifth century). Historian and presbyter. In 414 he fled Vandal invasions of his home in Spain, coming to Augustine of Hippo, young but already a presbyter. He provided Augustine with a treatise against the errors of both Priscillianists and Origenists. As trusted messenger, Orosius was sent by Augustine to Jerome* in Palestine, to assist in the indictment of the Pelagianists, but in 415 the Council of Diospolis upheld Pelagius. On Augustine's advice, Orosius composed a Christian world history in seven books, the philosophy of which, designed to explain Alaric's sack of Rome in 410, appears in its title: *Adversum Paganos.* The turning points in history which delimit his books became normative for subsequent authors. His material from 378 takes on value since his sources have not survived, as does his occasional citation of otherwise lost portions of Tacitus and the epitome of Livy. After the final date he deals with (417), Orosius himself disappears. CLYDE CURRY SMITH

ORR, JAMES (1844-1913). Scottish theologian. Born in Glasgow, he graduated from the university there and studied theology under the United Presbyterian Church.* He was minister of East Bank Church, Hawick (1874-91), taught church history in the UP theological college (1891-1901), and thereafter was professor of apologetics and theology in the United Free Church* college in Glasgow (having been one of the promoters of the union between his church and the Free Church). He was appreciated as a lecturer and popularizer of evangelical truth on both sides of the Atlantic. Writing in the heyday of liberal Protestantism, Orr contended for historic evangelicalism from the standpoint of "modified Calvinism," both by his exposition and defense of key doctrines (*The Resurrection of Jesus, God's Image in Man,* 1905; *The Problem of the Old Testament,* 1906; *Revelation and Inspiration,* 1919) and by his analysis of the dominant Ritschlian theology. His chief work in apologetics is *A Christian View of God and the World* (1893). He contributed also to *The Fundamentals** (1909-15). PAUL HELM

ORTHODOX CHURCHES, see EASTERN ORTHODOX CHURCHES

ORTHODOX PRESBYTERIAN CHURCH (formerly known as the "Presbyterian Church of America"). It was founded in 1936 after a long struggle within the Presbyterian Church in the USA between theological conservatives who sought to conform the denomination to its doctrinal constitution, the Westminster Confession* of Faith, and their opponents who were willing to tolerate theological "modernism."* The conservative group was led by J. Gresham Machen* who in 1929 left his professorship at Princeton Theological Seminary to found Westminster Theological Seminary in Philadelphia, and in 1933 founded "The Independent Board for Presbyterian Foreign Missions." This latter act brought the suspension of Machen and several others from the ministry, which in turn precipitated the schism. The Orthodox Presbyterian Church, a relatively small group, emphasizes strongly the infallibility of Scripture and faithfulness to traditional Presbyterian doctrine. The general assembly in 1975 approved a merger with the Reformed Presbyterian Church, Evangelical Synod, but the latter denomination voted against union.

GEORGE MARSDEN

ORTHODOXY (Gr. *orthos,* "right," "true"; *doxa,* "opinion"). The closest NT concept is "truth" and correct belief; the NT writers declare an orthodoxy revealed and governed by the Spirit of God. A determined attack has been mounted in modern times by critic after critic, attempting to prove that the early church was divided into many camps, and the original "message of Jesus" was either altered or utterly lost (see, e.g., TÜBINGEN SCHOOL). While it would be untenable to hold that *all* of the universal church will be agreed on all points, the standard of orthodoxy—the Word of God set forth in the first Christian century—continually demands that all else be measured by it. KEITH J. HARDMAN

ORTLIEB OF STRASSBURG. Early thirteenth-century leader of the *Ortlibarii,* an ascetic, heretical sect known also as the Ortlibenses and Ortlibians. He taught abstention from material things and guidance by one's inner spirit. Identifying pope and church with the apocalyptic harlot, his followers rejected ecclesiastical authority and necessity of the sacraments. Heretical on the doctrines of creation, resurrection, incarnation, and the Trinity, they were condemned by Innocent III. Their historical obscurity is attested by historians associating them with such varying groups as Waldensians, Cathari, Brethren of the Free Spirit, and Amalricians. The movement was seemingly short-lived and is best described in the so-called *Passau Anonymous.* JAMES DE JONG

OSIANDER, ANDREAS (1498-1552). German Reformer. Born at Gunzenhausen, near Nuremberg, he studied at the University of Ingolstadt and was ordained priest in 1520. He revised the Vulgate on the basis of the Hebrew text. As a reformer in Nuremberg he promoted the distribution of the Lord's Supper under both kinds (bread and wine.) With Lazarus Spengler he furthered the Lutheran movement in Nuremberg in its doctrinal and liturgical forms. As a participant in the Colloquy of Marburg* (1529) he sided with Luther and Melanchthon against Zwingli and Oecolampadius regarding the Lord's Supper. He attended also the Diet of Augsburg (1530). He frequently faulted Melanchthon, particularly after the signing of the Leipzig Interim (1548).

George von Brandenburg requested him to conduct the visitation in Brandenburg. The twenty-three articles of the Schwabacher Visitation of 1528 were expanded by him and Schleufner. The Brandenburg-Nuremberg Church Orders, which contained the sermons on Luther's Catechism, were translated into Latin by Jonas and became the so-called Cranmer's Catechism (1548). After Osiander's appointment as pastor and professor at Königsberg (1549), he attacked Melanchthon on forensic justification, setting forth that in justification the new believers become partakers of the divine nature. In the ensuing controversy Osiander received little support, and his views were repudiated in Article III of the Formula of Concord* (1577). Osiander is noteworthy too because he wrote the anonymous preface to the first edition of Nicholaus Copernicus's *De Revolutionibus Orbium Caelestium.*

See E. Hirsch, *Die Theologie des Andreas Osiander und ihre geschichtlichen Voraussetzungen* (1919). CARL S. MEYER

OSIUS, see HOSIUS

OSWALD (c.605-642). King of Northumbria. He was the son of Ethelfrith. Fleeing to Scotland after his father's death in 616, he became a Christian through the work of the monks on Iona.* On the death of King Edwin in 633, he returned to Northumbria. After prayer, he gained a victory over the British king Cadwalla at Heavenfield, near Hexham. He was determined to establish the Christian faith in his kingdom, and sent to Iona for missionaries. The first sent was unsuccessful, but the second was Aidan,* who took up his see on Lindisfarne. Oswald worked with him and sometimes accompanied him on missionary journeys, acting as his interpreter. He was killed in a battle with the heathen king Penda of Mercia at Maserfield (now identified as Oswestry: "Oswald's Tree or Cross"). He was a fine example of a leader influencing his people, and his relationship with Aidan set a primitive pattern of church-state cooperation. R.E. NIXON

OSWALD (d.992). Archbishop of York. Of Danish birth, he spent some time in the Benedictine monastery of Fleury in France. In 961 he succeeded Dunstan* as bishop of Worcester. Prominent in the introduction of the Benedictine way of life, he gradually reformed the church in his diocese, especially tightening up discipline among the clergy. He established or refounded a number of monasteries, including that at Ramsey in Huntingdonshire. In 972 he was appointed to York, continuing to hold the see of Worcester in plurality. With Dunstan (archbishop of Canterbury) and Ethelwold (bishop of Winchester) he played a great part in the revival of monasticism in England in the period before the Norman conquest. R.E. NIXON

OTHILIA, see ODILIA

OTTERBEIN, PHILIP WILLIAM (1726-1813). Co-founder of the Church of the United Brethren in Christ.* He was born in Prussia, son of a Reformed minister, studied at Herborn, and was ordained to the ministry of the Reformed Church. In 1752 he responded to an appeal from Michael Schlatter to do missionary work in America. While serving as pastor to the German Reformed congregation of Lancaster, Pennsylvania, he struggled earnestly in his attempts to preach a vital message, but felt inadequate in explaining how one might know the assurance of salvation. A moving spiritual experience brought him into a more complete awareness of salvation in Christ, and he began to preach with more power and effectiveness the need for every person to experience repentance and a new birth. He organized prayer meetings, trained laymen for evangelistic work, and worked closely with ministers of other denominations, including Methodist bishop Francis Asbury,* in whose consecration service he participated. Bishop and co-founder with Martin

Boehm* of the United Brethren in Christ (1800), he served as minister of the German Evangelical Reformed Church in Baltimore from 1774 until his death. HARRY SKILTON

OTTLEY, ROBERT LAWRENCE (1856-1953). Anglican theologian. Born at Richmond, Yorkshire, he was educated at Canterbury and Oxford, where he was successively vice-principal of Cuddesdon College, dean of divinity at Magdalen College, principal of Pusey House, and regius professor of pastoral theology. He contributed the article on Christian ethics to *Lux Mundi,** pleading for a recognition of personality in the sphere of economics. His book *The Doctrine of the Incarnation* (1896) is an essay in both historical theology and kenotic Christology, while his Bampton lectures, *Aspects of the Old Testament* (1897), followed by *The Hebrew Prophets* (1898) and *The Religion of Israel* (1905), introduced advanced critical ideas to a large section of the Anglican clergy. A liberal High Churchman and a severe critic of liberal Protestantism, Ottley devoted his later years to writings on Christian morals and ethics. His *Studies in the Confessions of St. Augustine* (1919) is a notable product of this period. IAN SELLERS

OTTO (c.1060-1139). Bishop of Bamberg. Born in Swabia of noble parentage, he was educated in monastic schools and served at the courts of several Polish nobles before entering the service of Emperor Henry IV. About 1101 he was appointed chancellor, and in 1102 bishop of Bamberg, but was not consecrated until 1106. He was involved in the Investiture Controversy* and suspended from office for a time. Within his diocese Otto was very active in building churches and monasteries—some twenty monasteries are said to have been founded or renewed by him. He was called on as arbiter in several disputes. As emissary to Pomerania he was eminently successful in establishing Christianity among the Slavs then inhabiting that region. In 1189 he was canonized by Clement III. CARL S. MEYER

OTTO I (the Great) (912-973). German king and emperor. A member of the Saxon dynasty, he succeeded his father, Henry I, in 936. By emphasizing his position as head of the Christians in Germany, Otto revived Charlemagne's alliance of church and state. He reasserted the old Carolingian rights over the appointment and control of ecclesiastical lords, established new bishoprics in frontier districts to support missionary activities, and patronized to a limited extent the cultural and scholarly endeavors of the church. Otto's policy was that of giving political power to the church in order to counterbalance the secular lords.

In foreign affairs he kept France weak and divided, defeated the Magyars at the Lechfeld in 955, pursued a Germanization policy east of the Elbe River, and extended his control over Italy. His coronation by John XII in 962 marked the foundation of the Holy Roman Empire, a union of the Roman imperial title with the German kingship which included only Germany and part of Italy. Although the Ottonian empire was based on an alliance of church and state, the two partners' interests were by no means identical. Otto's action in deposing and replacing the unpopular John set a clear precedent for imperial control of the papacy. RICHARD V. PIERARD

OTTO, RUDOLF (1869-1937). German theologian. Born in Hanover, he was educated at Erlangen and Göttingen, and taught theology at Göttingen (1907-14), Breslau (1914-17), and Marburg (1917-37). His book *Kantish-Friessche Religionsphilosophie* (1909; ET *The Philosophy of Religion,* 1931) reflected what became one of his chief interests. Otto's major work, however, was *Das Heilige* (1917; ET *The Idea of the Holy,* 1923), in which he stressed an approach largely neglected by liberal Protestantism and greatly influenced contemporary thinking. Not unconnected with this was his effort to deepen public worship in Lutheranism. He wrote extensively also on Hinduism after travel in the East.

J.D. DOUGLAS

OTTO OF FREISING (c.1111-1158). Bishop of Freising. Maternal grandson of Emperor Henry IV and son of Margrave Leopold III of Austria, Otto moved in imperial circles all his life. Supported by a benefice near Vienna, he studied in Paris, perhaps with Hugh of St.-Victor, Gilbert de la Porrée, and Abelard. He entered the Cistercian monastery of Morimund, became abbot there about 1137, and a year later was consecrated as bishop of Freising, Bavaria. In this capacity Otto instituted diocesan reforms, chartered several monasteries, stimulated study of Aristotle among German contemporaries, and joined Conrad III, a stepbrother, in the Second Crusade, 1147-48. His interpretive rather than factual chronicle of world history, *Historia de duabus civitatibus* (1143-46) modified Augustine's thesis in *Civitas Dei.* Otto began a contemporary history of Frederick Barbarossa, but died before its completion.

JAMES DE JONG

OUEN (c.609-684). Archbishop of Rouen. Known also as Owen, Dadon, and Audoin, he was born in Sancy near Soissons and educated at the abbey of St. Médard. He served the Merovingian monarchs Clothaire II and Dagobert I. As the latter's able chancellor he established the monastery of Rabais in 634. On Dagobert's death, Ouen was ordained and in 641 was consecrated as archbishop of Rouen. He built churches and monasteries, fostered theological education, and undertook evangelization in a diocese still infested with paganism and barbarism. Still the diplomat, he supported Ebroin, mayor of the palace, against the aristocracy, and negotiated peace between the Franks of Neustria and Austrasia. A life of Eligius, a scholarly friend at Dagobert's court, may be Ouen's work. JAMES DE JONG

OVERBECK, FRANZ CAMILLE (1837-1905). Theologian and church historian. Professor of church history at Jena (1864-70) and of NT and early Christian history at Basle (1870-97), he was an advocate of a totally secular interpretation of church history and particularly of early Christian origins. Although he was a member of a Protestant faculty of theology, he was a professed atheist from about 1870 and was an indefatigable critic of both orthodox and liberal theology. He was a close friend of Friedrich Nietzsche.* Among his most significant writings are a large commentary on Acts (1870); *Studien zur Geschichte der alten Kirche* (1875); and *Christentum und Kultur* (posthumous, 1919). W. WARD GASQUE

OWEN, JAMES (1654-1706). Dissenting Academy* tutor. Born in Wales of a Cavalier family, he was educated at Carmarthen and at Samuel Jones's academy at Brynllywarch. He became a Nonconformist preacher and in 1676 was fined for holding an illegal conventicle. In that year he became private chaplain to Mrs. Baker at Swinney near Oswestry, and had oversight of the Dissenting congregation in the town. Owen was a close friend of Philip Henry,* and both engaged in public debate with Bishop William Lloyd of St. Asaph in 1681. In 1690 Owen opened his academy, and moved it in 1700 to Shrewsbury when he became fellow-minister there to Francis Tallents. Owen had a wide reputation as a tutor who ruled his students strictly and insisted that their conversation should be in Latin. He was in everything a moderate, as his books, *Moderation a Virtue* (1703) and *Moderation still a Virtue* (1704), show. He was a Congregationalist with definite Presbyterian sympathies. In theology he favored Richard Baxter's* form of Calvinism. It was Owen who provided the material Edmund Calamy* used in his *Account* when describing the ministers ejected in Wales. He published books in Welsh, and composed hymns, one or two of which are still in use.

See C. Owen, *Some Account of the Life and Writings of James Owen* (1709).

R. TUDUR JONES

OWEN, JOHN (1616-1683). An advocate of the Congregational way and a Reformed theologian, he was educated at Queen's College, Oxford. Because of the Laudian innovations he left Oxford in 1637. His first parish was Fordham, Essex, to which he went in 1643. At this time he was a moderate Presbyterian, but the reading of a book by John Cotton convinced him of the biblical basis of the Congregational way. In his next parish, Coggeshall, he formed a gathered church. In the Civil War his sympathies were wholly with Parliament and he accompanied Cromwell in expeditions to Ireland and Scotland in 1649-51 as a chaplain. In 1651 Parliament appointed him dean of Christ Church, Oxford. A year later Cromwell made him a vice-chancellor, a post he held until 1657.

From 1651 to 1660 he devoted his energies to the production of "godly and learned" men, and to the reform of the statutes and ceremonies of the university. During this period he was also influential in national affairs. Apart from serving on many committees he was the chief architect of the Cromwellian State Church. He helped to compose the Savoy Declaration* of Faith and Order (1658), and he wrote important books against Arminian and Socinian views. Ejected from Christ Church in 1660, he settled temporarily at Stadhampton, the village of his youth, and gathered a church in his home. For the next twenty-three years he was an acknowledged leader of Protestant Nonconformity. He was pastor of a church in London, friend and guide to many ejected ministers, defender of the legal rights of Dissenters, expounder of the Congregational way, biblical commentator, and devotional writer. His books were treasured by Nonconformists and have constantly been reprinted. He was buried in Bunhill Fields, London.

See P. Toon (ed.), *Correspondence* (1970) and *Oxford Orations* (1971); *idem, God's Statesman* (1972). PETER TOON

OXFORD CONFERENCE (1937). Second conference of the Life and Work movement held at Oxford under the general title "Church, Community and State." There were 425 delegates representing most churches apart from the Roman Catholics and the German Evangelical (the latter under Nazi control). Younger churches had only twenty-nine representatives. It was directed by J.H. Oldham,* secretary of the International Missionary Council,* who had organized the Edinburgh Conference,* 1910. The chairman was John R. Mott.* Its watchword was "Let the Church be the Church," and statements were made on religious freedom, the criteria for a responsible economic order, and the Christian attitude to war. It went far beyond the first conference at Stockholm* (1925) in theological acumen and depth of social analysis, and it laid the groundwork for an ecumenical social ethic. It emphasized that the church must guard its moral and spiritual integrity and render a true critique of all social systems, especially those of Western civilization which it might be inclined to defend uncritically. Concurrently with the Faith and Order Conference (Edinburgh,* 1937) it proposed the formation of the World Council of Churches,* in which the concerns of both movements were integrated. A report was issued later that year under the title *The Churches Survey Their Task*.

J.G.G. NORMAN

OXFORD GROUP. Later called "Moral Re-Armament," the movement was instituted by Frank Buchman* as a moral and spiritual force to transform men and societies. The founder promoted spiritual conversion through the techniques of confession, surrender, guidance, and sharing. These techniques lead to the four absolute standards of life: absolute purity, absolute unselfishness, absolute honesty, and absolute love. The application of these principles through the Moral Re-Armament movement has brought Buchman decorations from France, Germany, Greece, Japan, the Republic of China, the Philippines, Thailand, and Iran for outstanding services rendered. Prime Minister Holyoake of New Zealand said, "He has done as much as any man of our time to unite the peoples of the world by cutting through the prejudices of color, class, and creed."

After a period of working out experimentally basic principles in the art of remaking men, Buchman became convinced that the backbone of Christianity was a set of absolute moral standards and that men needed the courage to pursue these. In 1921 he embarked on the task of raising an army of men and women to overcome the worldwide breakdown of morals. Using the method of spiritual house parties where discussions, meditations, testimonies, recreation, quiet hours, and public confessions were the order of the day, Buchman began to appeal to students at many of the great universities—Harvard, Princeton, Yale, Oxford, and Cambridge. In 1928 he took a team of Rhodes scholars to South Africa, and while they were there the South African press referred to them as "the Oxford Group." The label became the name of Buchman's movement for the next ten years. Late in 1938 a new worldwide program was launched under a new label of Moral Re-Armament.

The Moral Re-Armament movement has produced hundreds of thousands of supporters around the world. Greatest success has come in England, Switzerland, Germany, Africa, and the USA. Greatest appeal has been to the very wealthy, but many people with more modest incomes have participated. Buchman's success can be attributed in part to his ability to train others and his use of drama, musicals, and films to present his message. The "Up With People" program has been especially popular in the United States and Germany. The corporate headquarters and finance office are located in New York, but various centers are located throughout the USA and the world. Two international training centers are located in Caux, Switzerland, and Odawara, Japan. Although the movement has its devoted crusaders, there is no claim of denominational status; rather, the movement claims to be "an expeditionary force from all faiths and races. . . ."

Peter Howard* assumed the leadership of the worldwide operations upon the death of Buchman in 1961. Howard died in 1965. The organization is now administered by a group of directors.

See W.H. Clark, *The Oxford Group: Its History and Significance* (1951); B. Entwistle and J.M. Roots, *Moral Re-Armament: What Is It?* (1967).

JOHN P. DEVER

OXFORD MOVEMENT. The title given to the movement within the Church of England which opposed the growth of liberalism in the mid-nineteenth century. It had its roots in the High Church party of the seventeenth century. It was influenced by the Romantic Revival in its veneration of the medieval. Its leaders, such as J.H. Newman,* J. Keble,* and E.B. Pusey* (some of whom came from Evangelical families), were all members of Oriel College, Oxford, in the 1820s.

The movement began with an attack on the English government's bill to reduce the number of bishoprics of the Church of Ireland. Keble preached a sermon entitled "National Apostasy" in St. Mary the Virgin, Oxford, in 1833. He saw the state's action as an attack on the church and a direct disavowal of the sovereignty of God. In their efforts to revive the church, the leaders of the movement published the first of the *Tracts for the Times* (1833). The name Tractarianism* soon attached to the movement. In its resistance to liberalism in the church, the appointment of R.D. Hampden as regius professor of divinity in the University of Oxford was opposed unsuccessfully in 1836. The movement created wider hostility as it became clear that its teaching ran counter to the spirit of the Reformation. Despite the publication of Tracts against Roman Catholic teaching, the anti-Reformation tendency seemed to be confirmed by the publication of R.H. Froude's* *Remains* in 1838-39. The revelation of Froude's spiritual and ascetic practices and his attacks on the Reformers incited a Protestant reaction. In Oxford this took the form of raising a subscription for a memorial to the Oxford martyrs of the Reformation.

From 1840 onward, part of the movement led by Newman moved in a Roman Catholic direction. In 1841 he published Tract 90 in which he argued for a Roman Catholic interpretation of the Thirty-Nine Articles. While the other leaders of the movement approved the Tract, they were startled by the instant storm of opposition. From 1843 Newman began to withdraw from the leadership of the movement. The condemnation of W.G. Ward's *The Ideal of a Christian Church* in 1845 by the University of Oxford led to the reception of some members of the movement into the Roman Catholic Church. At the end of 1845 Newman himself became a convert.

The defection of Newman marked the end of the dominance of Oxford in the movement. While Pusey, an Oxford professor, remained prominent, the titles "Anglo-Catholic" and "Ritualist" marked a new phase in the movement. The main interest of the Oxford Movement was in a revival of a high doctrine of the church and its ministry. The revival of ceremonial that later attached to the movement stemmed from the Cambridge Camden Society. In pursuing its goal the movement adopted such practices as frequent Communion, confession, and the renewal of the monastic life, which in time greatly affected the character of the Church of England.

BIBLIOGRAPHY: J.H. Newman, *Apologia pro vita sua* (1864); R.W. Church, *The Oxford Movement* (1891); H.P. Liddon, *Life of E.B. Pusey* (4 vols., 1893-97); S.L. Ollard, *A Short History of the Oxford Movement* (1915); E.A. Knox, *The Tractarian Movement* (1933); C. Dawson, *The Spirit of the Oxford Movement* (1933); G. Faber, *Oxford Apostles* (rev. 1936); W.O. Chadwick (ed.), *The Mind of the Oxford Movement* (1961); E.R. Fairweather (ed.), *The Oxford Movement* (1964).

NOEL S. POLLARD

OXYRHYNCHUS PAPYRI. In 1897 and thereafter numerous fragments of Greek papyri were discovered at Oxyrhynchus, one of the chief cities of ancient Egypt, by two archaeologists, B.P. Grenfell and A.S. Hunt. The most important finds are three collections of the sayings of Jesus, Numbers 1, 654, and 655. The close similarity of these texts to the Coptic text of the *Gospel of Thomas**

suggests that the Oxyrhynchus papyri are fragments of a Greek version of this gospel.

OZANAM, ANTOINE FRÉDÉRIC (1813-1853). French Catholic literary historian; founder of the Society of St. Vincent de Paul (1833, 1835) to work among the poor. The society, formed at first by students of the Sorbonne in Paris and composed exclusively of nonclergy, was more than mere charity, but an instrument to implement genuine social regeneration according to Catholic faith. Ozanam published (1831) his rejection of the social principles of Saint-Simon,* and later worked out the beginnings of Catholic societal theory in lectures as professor of law in Lyons (1839). He was open to participation in the new democratic republic (1848), but without absorbing its secularist basis as did Lamennais,* his contemporary. With Lacordaire* he established a newspaper *Ère nouvelle* to promote his ideas; he wrote for Montalembert's *Correspondent.* By the time of his death, the society had nearly 3,000 chapters in Europe, Africa, the Near East, and North America. As a literary historian he won a professorship in foreign literature at the Sorbonne (1844), published creative studies of Dante (1839), early Franciscan poetry (1852), and a history of Christian civilization among the Franks (1849). His education included doctorates in law (1836) and literature (1839).

BIBLIOGRAPHY: G. Goyau, *Frédéric Ozanam* (1925); F. Méjecaze, *Ozanam et l'église catholique* (1932); E. Renner, *The Historical Thought of Frédéric Ozanam* (1959). C.T. MC INTIRE

P

PACHOMIUS (c.287-346). Egyptian pioneer of cenobitic monasticism. After a pagan upbringing at Latopolis in the Upper Thebaid, he was won to Christianity while a military conscript by the kindness of the Christians of Thebes, and after his release was baptized at Chenoboskion (Shenesit, near Nag Hammadi). For a few years he learned the solitary life under the anchorite Palamon before settling at the abandoned village of Tabennisi c.320 to fulfill his vision of an ascetic *koinōnia* patterned on the primitive Jerusalem community. By the time of his death thousands of monks in eleven monasteries within a radius of sixty miles to north and south along the Nile (not all of his foundation, some apparently antedating Tabennisi), including two for women (his sister Mary followed his example), obeyed his direction as superior of the congregation, observed his rudimentary rule, and gathered each Easter and August in general assembly at Pboou, his second settlement and headquarters from c.337.

Details of his life and evolving Pachomian monasticism remain contested while the value of various sources, especially Greek and Coptic *Lives*, is still debated. The Rule, which survives complete only in Jerome's Latin version, consists of four series of *Pracepta* and grew up as a collection of *ad hoc* regulations not only in Pachomius's lifetime but also under his successors Theodore (d.368) and Horsiesius (d.380). The moderate regime reflects the continuing appeal of the solitary ideal. The monasteries appear as self-sufficient, profitable agricultural colonies or manufacturing complexes, whose hierarchical organization perhaps bears a military stamp. Within the enclosure the monks were allotted according to their skills to houses of thirty-to-forty each which formed the basic unit of community life. Instruction in and memorizing of Scripture were prominent, two daily offices and a weekly Eucharist, were observed though Pachomius resisted ordained monks. His achievement, which through the Rule exerted a wide influence in East and West, still left to Basil the Great* a nobler realization of Christian ideals of community and service.

BIBLIOGRAPHY: Best edition of *Rule* in A. Boon and L.T. Lefort, *Pachomiana Latina* (1932). Other works in L.T. Lefort, *Oeuvres de S. Pachôme et de ses disciples* (1956). For *Lives* see J. Quasten, *Patrology* 3, pp. 154-59, and A. Veilleux, *La Liturgie dans le Cénobitisme Pachômien au Quatrième Siècle* (1968). See D.J. Chitty, *The Desert a City* (1966).

D.F. WRIGHT

PACIAN (c.310-c.390). Bishop of Barcelona. He was praised for his learning, sanctity, and pastoral zeal in Jerome's *De Viris Illustribus.* Three authentic works of Pacian are known: *De Baptismo,* a sermon to catechumens; *Contra Novatianos,* three letters to Symphronian, who propagated Novatiansim, in which Pacian defended the Catholic doctrine of forgiveness (in the first letter appears the famous epigram, "My name is Christian; my surname is Catholic"); and *Paraenesis ad Poenitentiam,* an earnest plea for penitence.

PACIFISM. Although this term is usually related to the renunciation of war by the individual or a nation, it admits of no single definition. Historic or biblical pacifists argue their case on biblical grounds, but modern pacifists seem to confront the problem more on moral and philosophical bases. Currently the term "pacifist" is regularly reserved for anyone who renounces all war, specifically wars fought with modern weapons.

Historic or biblical pacifists, such as Quakers and Mennonites, have based their beliefs on the words of Jesus in the Sermon on the Mount (Matt. 5–7). The NT message may not be quite as explicit as some pacifists advocate. The basic message is one of peace among men of good will (Luke 2:14) and brotherhood, but Christ warns that He had come to bring a sword, not peace (Matt. 10:34). It is clear that He used some physical force when driving the merchants from the Temple (John 2:14-16). To add to this ambivalence, Paul wrote: "If it is possible, as far as it depends on you, live at peace with everyone" (Rom. 12:18), but he also advocated the use of the sword in defense of the state (Rom. 13:4). After Constantine's rapprochment with Christianity, Christians served in the army, and in medieval times Aquinas upheld the "just war" idea. The Anabaptists of the Reformation generally renounced war on biblical grounds.

The first widespread peace movement emerged after the Congress of Vienna (1814-15) and was the forerunner of many more to come. Some were dedicated to the personal rejection of all war; others advocated a progressive extermination of war through education, arbitration, and international organization. The mid-nineteenth-century wars brought most of these movements to an abrupt halt. During the latter part of the century, the Russian author of *War and Peace,* Leo Tolstoy,* emerged as an advocate of pacifism.

A strong peace movement also rode the tides of political optimism into the early 1900s in both Europe and the United States. The American Peace Society claimed such members as Andrew Carnegie, J.R. Mott,* and W.J. Bryan.* World War

I soon brought the capitulation of the society; patriotism and the "just war" philosophy won the day. Great Britain was less hysterical in its reactions and eventually recognized the conscientious objections to war of thousands of citizens. The Fellowship of Reconciliation,* the world's largest modern peace organization, was organized in England (1914).

Peace movements reached an all time high in popularity between 1914 and 1939. In England in the latter year the Peace Pledge Union claimed more than 100,000 members pledged never again to support war. Reinhold Niebuhr* joined the pacifist movement, but later resigned and founded his own group, Christian Action, and actively professed a "just war" doctrine. The USA and Britain recognized the conscientious objector more readily during World War II. In the USA, 12,000 were assigned to civilian public service camps and 25,000 more served in the armed forces as noncombatants; 6,000 were in prison. Britain's figures showed nearly 24,000 in civilian work and more than 17,000 noncombatants.

After 1945 the peace churches emphasized world relief programs, and the Fellowship of Reconciliation reported worldwide organizational growth. The total postwar peace movement was, however, quantitatively smaller, but committed to total disarmament, nuclear pacifism, and nonviolence, more on philosophical than on biblical grounds.

BIBLIOGRAPHY: C.J. Cadoux, *The Early Christian Attitude to War* (1919); N. Thomas, *The Conscientious Objector in America* (1923); C.E. Raven, *War and the Christian* (1938); L. Richards, *Christian Pacifism After Two World Wars* (1948); D. Hayes, *Challenge of Conscience* (1949); G.F. Hershberger, *War, Peace and Nonresistance* (1953); M.L. King, Jr., *Stride Toward Freedom* (1958); G.F. Nuttall, *Christian Pacifism in History* (1958); G.H.C. MacGregor, *New Testament Basis of Pacifism* (1960); R.H. Bainton, *Christian Attitudes Toward War and Peace: A Historical Survey and Critical Re-Evaluation* (1960); P. Ramsey, *War and the Christian Conscience* (1961).

JOHN P. DEVER

PAGE, KIRBY (1890-1957). Social evangelist and author. Born in Texas, he was a Disciples of Christ minister who, after service with the YMCA in World War I and a brief term as pastor in New York City, turned to writing and lecturing to combat what he assumed were society's ills: war and capitalism. In addition to many tracts and articles, he wrote such books as *Christianity and Social Problems* (1921); *War—Its Causes, Consequences, and Cure* (1923); and *Jesus and Christianity* (1929). Pacifist and socialist, he represented many Protestants in the 1919-41 period, notably those who had been disillusioned by the Versailles Treaty and modern industrialism. The rise of Neoorthodoxy* and welfare capitalism in the 1930s, however, left Page with an ever-diminishing audience. DARREL BIGHAM

PAGNINUS, SANTES (1470-1536). Dominican scholar and disciple of Savonarola.* Born at Lucca, Italy, he went to Rome in 1516, was in Avignon from 1523 to 1526, and thereafter in Lyons. It was in the latter place that he finished his Latin version of the Bible made from the original languages. He compiled in 1529 a Hebrew lexicon *(Thesaurus linguae sanctae)* which was often reprinted. His Bible, republished in 1541 and 1564, was better then the Vulgate, but was not intended to supersede it. In 1542 Michael Servetus re-edited and used the Bible with notes to further his own views.

PAINE, THOMAS (1737-1809). Deistic writer and political propagandist. Born in Thetford, Norfolk, he had a monotonous life as corset maker, exciseman, teacher, and grocer until he sailed to America in 1774 with letters of introduction from B. Franklin.* In 1776 he published *Common Sense,* the 500,000 copies of which argued for a republic: "it must come to that some time or other." He was secretary to the foreign affairs commission in 1778-79. He returned to England in 1787, and 1792 was indicted for treason on publication of his book *The Rights of Man.* He escaped to France, where he had been made a citizen, and was elected to the Convention. *The Age of Reason* was published there (1794-96). This brought suspicion and imprisonment and roused British and colonial indignation with its Deistic arguments. He returned to America in 1802 and died there seven years later, having alienated most of his friends by his unpredictable allegiances.

C.G. THORNE, JR.

PAINTING, see ART, CHRISTIAN

PAKISTAN. In the first century, Greek traders used Barbarike, at the mouth of the Indus, for the export of Chinese silk and leather, Persian turquoise, and spikenard from Kashmir. The *Acts of Thomas,** written about 230, is clearly fiction, but may contain echoes of a tradition that Thomas followed this trading route, and it is possible that he preached at Taxila during the reign of Vindafarna (Gondaphoros), about A.D. 20-48. About 196, Bardaisan speaks of Christians among the Kushans, whose empire included the Punjab. Attendance records at synods of the (Nestorian) Church of the East between 410 and 775 show an organized church in Afghanistan, with a metropolitan at Herat, and seven bishops south and east as far as Kandahar, but give no such proof of bishops in West Pakistan. Cosmas Indicopleustes writes of Christians among "the rest of the Indians" in 525, after speaking of Malabar* and Kalyan. In 1321 Friar Jordan does not speak of Christians further north than Broach in Gujarat, but in 1430 Niccolo di Conti states that "the Nestorians are scattered all over India, as the Jews among us." Probably there were unorganized groups of Christians in West Pakistan, engaged in trade, but there is no evidence of an organized church, and no certain Christian remains.

Armenian traders, soldiers, and artisans settled in Lahore from 1601, built a church, and for a time had a bishop. Lahore had an Armenian Christian governor in the 1630s, and in 1735 Armenians were "the elite of the Mughal army." An Armenian founder made the famous Zam-Zammah ("Kim's gun"). After 1750 the community dwindled, and there is no trace today of the organized church. From at least 1714 there were Armenians in Dacca, and their beautiful church (1781) is still extant. They were pioneers in the jute trade, but after 1947 most of them left East Pakistan.

Before 1600, Jerome Xavier and other Jesuit missionaries followed the court of Akbar the Great when he moved to Lahore. A church was built and many converts made, but in 1632 Shah Jehan closed the church, and it cannot be traced today. Louis Francis began Carmelite work in Thatta, Sind, in 1618, built a church and monastery, and in four years baptized some converts. Augustinians followed in 1624, but with the waning of Portuguese power all missionaries were withdrawn by 1672, and no trace of churches or Christian communities has survived. In East Pakistan there were chaplains in Chittagong from 1534, but Jesuit missionary work there (1598-1602) had to be abandoned because of political opposition to the Portuguese. Evangelism carried out in the Dacca area by Antony, an enslaved Rajah's son liberated by a missionary, was followed up by Jesuit missionaries in 1678-84, and though the missionaries had to withdraw, and many converts reverted to Hinduism, a church was established which has remained to this day.

In more modern times William Carey* preached in Dinajpur District between 1794 and 1800. Baptist work was begun in Dinajpur in 1800 by an ex-Catholic; in Jessore and Khulna in 1812; and in Barisal in 1829. The Church Missionary Society began work in Kushtia District in 1821, and about 5,000 baptisms followed a severe famine. Other missions followed. Since 1947 there have been small tribal movements in the Garo Hills, Sylhet, and the Chittagong Hill Tracts. The typical East Pakistan Christian (now, of course, a citizen of Bangladesh) is a clerk, artisan, mechanic, or small trader. In 1961 there were about 100,000 Roman Catholics, 40,000 Baptists, and 10,000 Anglicans in that area.

In West Pakistan, American Presbyterians began work in Lahore in 1849, followed by the (Anglican) CMS (1851), American United Presbyterians (1855), and Church of Scotland (1857). Roman Catholic work was resumed in 1843 in Karachi and 1852 in Lahore. The events of 1857 led to the death of Thomas Hunter, the first Church of Scotland missionary, with his wife and son, in Sialkot. In the mid-1870s a mass movement began among low-caste people in the Punjab, which brought thousands into the church, but slowed down after 1915. Since 1947 there have been group movements, especially among the Kohlis in Sind, and the Marwaris in Bahawalpur. In 1961 there were 584,000 Christians in West Pakistan, of whom perhaps three-fifths were Protestants.

In 1970 Anglicans, Methodists, Lutherans, and some Presbyterians united to form the Church of Pakistan, which claimed a membership of 200,000 in both sections of the country. The other main Protestant denominations in West Pakistan are Presbyterian, Salvation Army, and Seventh-Day Adventist. The church is drawn mainly from the lower stratum of the population, and though it has an increasingly educated leadership, it plays little part in the political life of the country.

The former East Pakistan became at the end of 1971 the independent state of Bangladesh ("The Land of Bengal"), and this densely populated area with seventy-five million people presented a new challenge to Christian missions. There were only 200,000 professed Christians in the country, half of them Roman Catholic. There is a Muslim majority, a Hindu minority. There were fears that the new state would be influenced by India's hostile attitude toward missionary work, but there have been reports also of a growing interest in Christianity, especially among Hindus. Many missionary societies were heavily involved in the extensive relief operations mounted after the end of the nine-month struggle.

BIBLIOGRAPHY: A.J. Dain (ed.), *Mission Fields Today* (2nd ed., 1956); V. Stacey, *Focus on Pakistan* (1969); W.G. Young, "The Life and History of the Church in Pakistan," in *Al-Mushir* (May/June 1971). See also bibliography under MISSIONS, CHRISTIAN. WILLIAM G. YOUNG

PALESTINIAN SYRIAC TEXT OF THE NEW TESTAMENT. An Aramaic version used by Palestinian Christians and now preserved mainly in lectionaries; only the modified Estrangelo script may properly be called Syriac. It was begun about the beginning of the fifth century to meet the needs of Christians whose liturgy had been in Greek; alternatively, it may have originated with the secession of the Malkites from the Monophysite Syrian Church after the christological controversies of the fifth century. The version is independent of the Syriac translations identified with Edessa. Its underlying Greek text is of the Lucianic type, but it preserves many non-Byzantine variants; the influence of the Peshitta is often perceptible. ROBERT P. GORDON

PALESTRINA, GIOVANNI PIERLUIGI SANTI DA (1525-1594). Italian composer. Palestrina (the name is actually that of the composer's native town) went early in life to Rome as a choirboy, but returned to Palestrina as organist, where he married and where his talented older sons were born. Most of his life was spent in Rome, however, where he held several important church appointments. The tragic death of his older sons and wife through an epidemic seemed for a time to stifle his genius. After his remarriage he became involved in his second wife's fur business for a time, but soon returned to music and to his greatest achievements.

His role in the musical developments of the Counter-Reformation have been romanticized and misrepresented. There is in his music, however, a consistent blend of seriousness and serenity, and the adherence to a disciplined technique which undoubtedly contributed to making it the traditional ideal of true church style to later generations. The demythologizing which has accompanied modern musicological research has tended toward an underestimate of Palestrina's true worth. His often-performed *Sicut cervus* (Ps. 42:1,2) and his eight-part *Stabat Mater* are among his most representative works that are heard today. The short and popular *O Bone Jesu* is spurious.

Palestrina left over 250 motets, some of the most beautiful being on texts from the Song of Solomon. Lassus is considered his superior in this genre, but as a composer of Masses, Palestrina has no peer. Over one hundred of them survive. The little *Missa Brevis* and the majestic *Missa Papae Marcelli* are the best known, but are only two among many masterpieces. The latter was written before the Council of Trent and not as a demonstration piece for that august body, as was long believed. It is nevertheless true that Palestrina was one of the great musical voices of the Counter-Reformation. J.B. MAC MILLAN

PALEY, WILLIAM (1743-1805). Anglican scholar; archdeacon of Carlisle from 1780. Educated at Cambridge, he gained fame through his books, several of which had long-lasting influence as textbooks, especially in his own university, though their power to convince was eroded in the nineteenth century by scientific and philosophical developments. He was not an original or subtle thinker, but was an "unrivalled expositor of plain arguments." He claimed that his works formed a system; his thought drew upon and reflected the main elements of English theology as molded by eighteenth-century controversies. His *Natural Theology* (1802) sought to prove the being and goodness of God from the order of the world; the *Evidences of Christianity* (1794) argues on both internal and external grounds that Christianity is the true revelation of God; his *Principles of Moral and Political Philosophy* (1785), based on Cambridge lectures, is concerned, like several other works, with the duties resulting from natural and revealed religion. Here his utilitarianism anticipated Bentham, except that Paley retained a supernatural sanction. Paley took a lax view of subscription to the Thirty-Nine Articles,* perhaps he inclined to Unitarianism at certain points, and he was a conservative apologist for the Church of England and the British Constitution. Nevertheless the sincerity and strength of his faith has too often been underestimated.

See M.L. Clarke, *Paley: Evidences for the Man* (1974). HADDON WILLMER

PALLADIUS (c.363-425). Bishop and historian. Evidently a native of Galatia, he entered monastic life in Jerusalem at the age of twenty-three, and later went to Egypt to follow the ascetic life, first in Alexandria, then in the Nitrian Desert.* Returning to Palestine in 399 through illness, he was consecrated bishop of Helenopolis (400) by his friend John Chrysostom.* He appeared with the latter at the Synod of the Oak* (403) and in trying to defend Chrysostom's banishment was himself exiled to Egypt (406), where at Syene he almost certainly wrote the important *Dialogus de vita Sancti Joannis Chrysostomi.* He returned to his diocese when all opposition ceased (412), and five years later was transferred to the diocese of Aspuna, where he wrote *Historia Lausiaca* (419-20), a significant treatment of early monasticism. *Epistola de Indicis gentibus et de Bragmannibus,* which cannot, however, be confirmed as his, suggests a trip to India, and deals with the ascetical ideal. C.G. THORNE, JR.

PALLADIUS (fifth century). Missionary to Ireland. Native of Gaul and disciple of Germanus at Auxerre, he was sent as a delegate from Auxerre to Rome in 429. He made a good impression on Celestine and was commissioned by him to go to Ireland. The wording of the commission, recorded in the Chronicle of Prosper of Aquitaine, is significant: "Palladius, ordained by Pope Celestine, is sent as the first Bishop of the Scots who believe in Christ." The date was 431; the inference, that there were Christians in Ireland before the days of Patrick; the purpose, to combat the errors of Pelagius. Palladius had partial success on landing at Wicklow, and three churches were founded. But the hostility of Chief Nathi and the general unbelief of the people discouraged him and he left no enduring impression on Ireland or her church. He decided to return to Rome, but died in Britain on the way.

ADAM LOUGHRIDGE

PALLADIUS, PEDER (1503-1560). Danish bishop and Reformer. Seized by the Reformation movement about 1530, he pursued theological studies at Wittenberg (1531-37) where he became closely connected with Melanchthon and Bugenhagen. On the latter's recommendation Palladius was appointed the first Evangelical bishop of Zealand, the main diocese of Denmark. From 1538 he was also a professor of theology at the University of Copenhagen. As a man of solid theological learning, endowed with great administrative talents, and as a gifted and eloquent popular preacher, Palladius was highly qualified for the task of carrying out the Reformation in the local congregations. He worked at this with untiring energy and zeal; a very interesting and famous memorial of this is his *Book of Visitations,* with its many examples of blunt application of Evangelical teaching upon various life situations. Palladius was also the author of the 1556 Service Book and had a share in the translation of Christian III's Danish Bible (1550). N.O. RASMUSSEN

PALMER, BENJAMIN MORGAN (1818-1902). Presbyterian minister. Born in Charleston, South Carolina, he received degrees from the University of Georgia and Columbia Theological Seminary, served as pastor from 1841 to 1853, then became professor of ecclesiastical history and polity at Columbia Seminary. He preferred the pastorate, however, and in 1856 accepted a call to the First Presbyterian Church of New Orleans, where he remained until his death. Regarded as one of the great ministers of the Southern Presbyterian Church, he was the first Presbyterian moderator under the Confederacy in 1861, and defended slavery.

PALMER, WILLIAM (1803-1885). Theologian and ecclesiastical antiquary. A graduate of Trinity College, Dublin, and Magdalen, Oxford, he became a fellow of Worcester College, Oxford. In 1832 he wrote *Origines Liturgicae,* a learned treatise on the history of English liturgy. This brought him into touch with Keble,* Froude,* J.H. Newman,* and other Tractarian leaders. He was a rigid High Churchman and a doughty controversialist, who strongly opposed both Roman Catholicism and Dissent, and wrote against Cardinal Wiseman and others. His *Narrative of Events connected with the Publication of Tracts for the Times* precipitated the crisis which led to the secession of W.G. Ward and Newman from the Anglican Church. He was prebendary of Salisbury, 1849-58, and assumed the title of baronet on his father's death in 1865. J.G.G. NORMAN

PAMMACHIUS (d.410). Aristocrat and senator, he studied at Rome with Jerome.* When his wife died, he took monastic vows while remaining in the senate, and gave to the poor and the church. He helped Fabiola* establish a hostel for sea-travelers at Portus and left his home as a church (discovered in the Church of SS. John and Paul). Although he found Jerome's letter against Jovinian* too violent, he encouraged Bishop Anastasius (400) to condemn Origenism.* Jerome translated for him Origen's *Peri Archon* and dedicated to him such books as his commentaries on the Minor Prophets (406) and Daniel (407). He encouraged Numidians to return from Donatism* to the Catholic Church (401). He died in the siege of Rome by Alaric. G.T.D. ANGEL

PAMPHILUS (c.250-310). Christian martyr. Born in a wealthy home in Beirut, he pursued Hellenistic studies and engaged in Phoenician public affairs until his conversion to asceticism, philanthropy, and biblical studies. Tutored by Pierius at Alexandria, he moved to Palestinian Caesarea, where he founded a Christian school and restored the library of Origen, recopying biblical manuscripts, in particular the *Hexapla* edition of the Septuagint, and recovering the writings of Origen and others. His devoted pupil Eusebius of Caesarea* adopted the name "son of Pamphilus," wrote his life (not extant), and panegyrized him in the "Martyrs of Palestine." Arrested under Maximin Daza, he spent fifteen months in prison, working especially on five books of an *Apology of Origen,* to which Eusebius added a sixth. Pamphilus was beheaded early in 310.

G.T.D. ANGEL

PANTAENUS (d. c.190). First known head of the Catechetical School at Alexandria. He left no written indications of his own work or thought, but made a significant impact on the development of church theology. The direct succession from him provides the source of information. Clement's metaphor of the Sicilian bee makes probable his place of origin. His Christianity had taken him as missionary evangelist as far east as India. He is dated by the direct role he played as teacher of Clement (who succeeded him in 190) and of Alexander* (who studied under them both). Origen may have been too young to have been taught directly by him, but his martyred father was Pantaenus's contemporary and of like mind. Pantaenus's philosophy was Stoicism and his literary interests classical. Considering the exegetical writings of his successors, this would account not only for their strong emphasis upon a divine literature, but also for their allegorical methodology in interpreting the same. CLYDE CURRY SMITH

PANTALEON (d. c.305). Christian martyr. Nothing certain is known of him, but he is said to have been physician to Emperor Galerius of Nicomedia. Converted in early life through his mother, he fell away through the worldly court life, but was restored to faith through one Hermolaus. He suffered martyrdom under Diocletian's purge of Christians from the court. His cult has been very popular, especially in the East, and he is regarded as one of the patron saints of physicians.

PANTHEISM. The view that the universe is to be identified with God, i.e., that there is only one reality, alternately describable as "God" or "nature." Pantheism is clearly incompatible with the Christian view of creation and of the creature-Creator distinction so fundamental to Christian theology. In the tradition of Western culture, Spinoza* is the classic exponent of pantheism, but it is capable of a great diversity of expression, such as materialistic pantheism (d'Holbach), or psychological or mystical pantheism as in some Eastern religions. Pantheism is also an important strand in absolute idealism (e.g., Josiah Royce's *The Conception of God,* 1893) and in the Romanticism of Goethe* and Lessing.* There are interesting remarks of a pantheistic kind in the early philosophical jottings of Jonathan Edwards.* The term "pantheism" is first used in Toland's *Socinianism Truly Stated* (1705). PAUL HELM

PAPAL LEGATE, see NUNCIO AND LEGATE

PAPAL STATES. From 756 to 1870, certain civil territories in Italy acknowledged the pope as their temporal ruler. Constantine* probably gave the Lateran Palace to the Church of Rome after 321, when it could legally own property. By 600, gifts of large estates formed the Patrimony of Peter,* around which grew the legend that Constantine had donated these lands to Pope Sylvester I. A forged document called "The Donation of Constantine"* buttressed these claims until proven a forgery in the fifteenth century. Laurentius Valla in particular exposed the use of oriental language such as "satraps" to describe fourth-century Rome.

Until the eighth-century dispute over Iconoclasm, the largest papal estates were in Sicily. Even with the loss of these properties to the Byzantine emperor, the pope still controlled more land than any other person in Italy. The Isle of Capri, Gaeta, Tivoli, and other properties in Tuscany and about Ravenna and Genoa speak of wealth and trouble. Pope Gregory I* used his income for charitable purposes as well as food for Rome. From the sixth to eighth centuries the popes supported Byzantine authority at Ravenna against the Lombards.

Under Gregory II,* the Roman popes filled the vacuum left by Byzantine collapse in central Italy. The exarchate about Ravenna and the duchy of

Rome were central, especially the Pentapolis (Rimini, Pesaro, Fano, Sinigaglia, Ancona) and the fortress city of Perugia. When the Lombard king Liutprand* cut off Perugia in 738, Gregory II turned to Charles Martel.*

Pope Stephen II* (III) left Rome in 753 to travel to St.-Denis. There in January 754 he anointed Pepin,* giving him and his sons the title "Patrician of the Romans." Pepin promised in writing to give certain territories to the pope. Lands in central Italy perhaps stem from this last document of 754. In the summer of that year Pepin forced the Lombard king Aistulf to give up the Pentapolis and the exarchate. Pepin obtained such a deed for the pope, only to find necessary a second invasion and deed in 756. Pepin founded the States of the Church when in 756 he refused to return the territories wrested from the Lombards to the Byzantines. The pope was freed from foreign interference and entered into alliance with the West. In 781 Charlemagne* guaranteed to Pope Adrian I* the Pepin donations. The coronation of Charlemagne in 800 cemented this policy.

After this independence and Western policy the Papal States provoked a bitter quarrel among Roman nobility for control of this temporal authority and the papacy itself, and this in turn entailed internal turmoil for the church. Poverty of mind and political opportunism describe papal affairs until the sack of Rome in 1527. In modern times, papal administration or arbitration of direct control led one conqueror after another into central Italy, from Charles VIII of France and the emperor Charles V to Napoleon Bonaparte. The Congress of Vienna returned the confiscated properties of 1798, the Roman Republic, to the pope in 1815. The Austrians protected these properties until, first in 1846 under Massimo d'Azeglio and finally under Garibaldi from 1860, foreign influence ended. Italian troops of the *Risorgimento** entered Rome on 20 September 1870. On 13 May 1871, the Vatican, Lateran, and Castel Gandolfo were declared to be papal territory. Pius IX* refused to accept the Papal Guarantee, though under Mussolini Pius XI* signed the Lateran Pact on 11 February 1929. After 1944 this pact became part of the Republican Constitution.

MARVIN W. ANDERSON

PAPHNUTIUS (d. c.360). Bishop of Upper Thebaid, Egypt. A monk, and disciple of Antony,* he was badly mutilated during the persecution of Maximin Daza (305-13), and Constantine is said to have kissed his seared eye before the Council of Nicea (325). According to Socrates and Sozomen he dissuaded that council from enforcing celibacy on the clergy. He supported Athanasius* at the Synod of Tyre (335), walking out in protest at the unjust proceedings.

PAPIAS (c.60-c.130). Bishop of Hierapolis in Phrygia. Said to be "a man of primitive age, a hearer of John [the apostle], a companion of Polycarp" (Irenaeus). Fragments survive, chiefly in Irenaeus and Eusebius, of the *Exposition of Dominical Oracles* in five books (c.110), for which he collected unwritten traditions (without setting oral tradition above apostolic writings in principle) from the circles of "the elders [presbyters]," associates of the apostles, including Aristion and "the elder John" in Asia and the daughters of Philip the apostle (or evangelist?) in Hierapolis. Such traditions transmitted historical reminiscence (much-discussed accounts of the origins of the gospels of Mark and Matthew, and of John's gospel activity and martyrdom), miracle stories, noncanonical *pericopae* (a variant on John 7:53–8:11 found in the *Gospel according to the Hebrews* by Eusebius and perhaps Papias too), but chiefly beliefs of primitive Judeo-Christianity, including the millenarian enjoyment of a miraculously fruitful earth, the fall of angels commissioned to govern the world, and interpretations of early Genesis in terms of Christ and the church.

Eusebius, despising Papias's millenarianism, disparages his material and intelligence and argues (his *Church History* here contradicting his earlier *Chronicle*) that Papias heard only "the elder John," whom he distinguished from the apostle. Papias's *Exposition* influenced later writers like Victorinus of Pettau to an extent probably no longer demonstrable.

BIBLIOGRAPHY: For texts and translations, see APOSTLIC FATHERS. Further literature in J. Quasten, *Patrology* 1 (1950), and in ET by W.R. Schoedel (1967).

D.F. WRIGHT

PAPYROLOGY. The scientific study of papyrus, commonly narrowed to its use as a writing material and what is written thereon. There are NT references in 2 John 12; 3 John 13. Papyrology is a by-product of the archaeological recovery of antiquity. Ancient Upper Egypt has been the major source, since the dry desert climate provided the chief preservative factor in the survival of this material. Oddly, however, the initial finds were made at Herculaneum in 1752, when the charred remains of a library, sealed by the volcanic eruption of Vesuvius in A.D. 79, were uncovered. Beginning in 1815, bundles of papyri unearthed at Memphis and Thebes were acquired by museums in London, Paris, Turin, Vienna, and Leyden.

The winter of 1895-96 saw the first archaeological expedition undertaken specifically for the recovery of papyri—that of the Egyptian Exploration Fund in the north of the Fayum. Of importance to the history of Christianity, apart from the discovery of important examples of its major literary works, is the vast array of documentation illustrating nearly every facet of its inner life, as well as its conflict with the Roman state. The finding of fresh sources of *koinē* vocabulary and usage led not only to new understandings of the inherent meanings within the NT and other early Christian literature, but also to the obvious awareness that such literature could no longer be treated as an anomaly in the history of the Greek language. Papyrology had demonstrated that *koinē*, not academic Atticism, stood in the mainstream, more adequately representing the reality which was the ongoing, living speech of men.

CLYDE CURRY SMITH

PARABLES. The Greek word *parabolē* is used to translate the Hebrew *māshal*, which can refer to different kinds of sayings, including a byword or a proverb. The particular form of saying which biblical scholars refer to as a "parable" proper consists of a brief story involving some point of comparison with a situation in life to which the speaker wishes to draw the attention of his hearers. The usual purpose is that a judgment can be passed upon the situation through the exercise of the motions of indignation or ridicule. A classical example in the OT is the story of the ewe lamb told by Nathan to David (2 Sam. 12:1-10).

In the NT one of the most striking features about the ministry of Jesus is His teaching in parables. Christian commentators used to treat these as allegories, with every detail being supposed to have some significance. In reaction to that it was suggested by A. Jülicher that a parable had one point only. It seems wiser to allow that we must look for the main point, but that there may be subsidiary points of importance and that some parables may contain features of well-known symbolism which must, in a limited sense, be taken allegorically. The parables of Jesus are concerned with the nature and coming of the kingdom of God. C.H. Dodd has suggested that the whole of their original meaning was connected with the immediate situation of the ministry of Jesus, but there is no valid reason for denying some future reference in addition.

BIBLIOGRAPHY: C.H. Dodd, *The Parables of the Kingdom* (1935); J. Jeremias, *The Parables of Jesus* (ET 1954); A.M. Hunter, *Interpreting the Parables* (1960). R.E. NIXON

PARABOLANI. From the Greek meaning "to venture" or "to expose one's self," the name denotes members of a brotherhood which in the early church, first at Alexandria and then at Constantinople, nursed the sick and buried the dead. They risked their lives in their exposure to contagious diseases, and probably originated during an epidemic. They were also a kind of bodyguard for the bishop. Their number was never large: the Codex Theodosianus (416) restricted the enrollment to 500 in Alexandria, with a later increase to 600, while in Constantinople their number was reduced from 1,100 to 950, according to the Codex Justinianus. Chosen by the bishop and under his control, they probably had neither orders nor vows, although they were listed among the clergy and enjoyed those privileges. Their presence at public gatherings or in theaters was legally forbidden, but they did take part in public life. It appears they are not mentioned after Justinian's time. C.G. THORNE, JR.

PARACELSUS (1493-1541). Pseudonym of Philippus Aureolus Theophrastus Bombastus von Hohenheim, medical doctor, chemist, philosopher, and writer. Born in Switzerland, he later lived in Carinthia, a center of mining, smelting, and alchemy—interests that were to occupy him the remainder of his life. After study at various German universities, he completed his doctorate at Ferrara in 1515. For the next eleven years he traveled through Europe practicing medicine. Then in 1526-28 he was medical lecturer at the University of Basle, but after losing this post he wandered for the rest of his life. He was an innovator in the field of medicine, developing chemical urinalysis, a biochemical theory of digestion, chemical therapy, antisepsis of wounds, and new treatments for syphilis. His philosophy of alchemy and kabbalism grew from Neoplatonism, hermetic studies, and Gnosticism. Paracelsus's numerous books are mostly variants on the idea of man (the microcosm) in relation to nature (the macrocosm). A friend of Erasmus and Oecolampadius, he was accused by Erastus of the Dualist heresy, but he died a member of the Roman Church. ROBERT G. CLOUSE

PARAGRAPH BIBLES. In 1755 John Wesley published his NT in which he returned to the pre-Geneva Bible practice of paragraphing, as opposed to the arrangement in verses followed in the King James Version (Authorized Version) and other versions. The Religious Tract Society brought out an edition of the KJV (AV) in paragraphs in 1838, and there were two further editions in 1853. The Revised Version (1881) adopted the paragraph arrangement, and this is followed in most modern translations.

PARIS, MATTHEW, see MATTHEW PARIS

PARISH. Derived from the Greek *paroikia*, or "district," the term seems till about the fourth century to have corresponded to a whole diocese and only later to small subdivisions of the same. By the later Middle Ages the parish had emerged as a definite geographical district, its inhabitants restricted to a particular church to which they paid tithes and which had a single incumbent appointed either by the bishop, patron, or less usually, by the parishioners themselves. At the Reformation both Lutherans and Calvinists retained the parish system, the latter for administrative convenience only. In England the establishment of the parochial system has usually been attributed to Archbishop Theodore (seventh century) but its origins are now placed much earlier, even as far back as pre-Christian times.

From the Middle Ages onward the English parish became a unit of civil administration, the priest in his ecclesiastical duties being aided by constables, churchwardens, overseers of the poor, and elected vestries. This parochial pattern was changed only when the growth of the population led to the creation of new parishes by Acts of 1710, 1818, and 1824 and, more recently, through Orders in Council on the initiative of the Church Commissioners. In recent years, with the development of specialized and team ministries, the traditional parochial system has come under attack, but is often defended by Anglican Evangelicals who cherish the individual minister's freedom which it guarantees. The term was imported into the USA where, however, it is often applied in general to a Protestant minister's congregation or cure of souls without reference to geographical limitations. IAN SELLERS

PARK, EDWARDS AMASA (1808-1900). Congregational theologian. Professor of Christian theology at Andover Theological Seminary (1847-81), where he helped to found and edit *Biblio-*

theca Sacra, Park was preeminent among those who tried to reconcile the thinking of Jonathan Edwards and Jacksonian America. To Lyman Abbott, he was the last of those striving "to relieve Calvinism of the objections apparent in the dawn of the more ethical and humanistic spirit of our times." Unlike Charles Hodge, who fused Edwards to Scottish philosophy, Park allied divine sovereignty to "theology of the heart." His former student, Frank Hugh Foster, described him as an eclectic who by 1881 had become an anachronism. Gallant in defeat, ironically he has encouraged the rise of modern liberalism by humanizing the deity. DARREL BIGHAM

PARKER, HORATIO WILLIAM (1863-1919). American composer. The most distinguished composer of church music born in America up to that time, he studied in Munich with Rheinberger. He held various organ posts about New York and in Boston. He taught at Yale, where he had many distinguished pupils, including Charles Ives. He was also a founder of the American Guild of Organists. Many of his hymn tunes and anthems were widely used. His most ambitious work was his *Hora Novissima,* using a portion of the great poem by Bernard of Cluny. The success of this impressive score brought him commissions from several of the British choir festivals, and Cambridge University awarded him an honorary doctorate. Parker's work marked a movement away from the quartet anthems then popular to a style of greater liturgical propriety.
J.B. MAC MILLAN

PARKER, JOSEPH (1830-1902). English Congregational preacher. Born of humble, pious stock in Hexham, he was ordained in 1853 to the ministry in Banbury Congregational Church, though his formal education had ceased when he was sixteen. He moved to Cavendish Street Chapel, Manchester, in 1858, then went in 1869 to Poultry Chapel, London, which congregation built the City Temple, opened in 1874. There Parker ministered until his death, preaching twice a Sunday and every Thursday morning, and earning a reputation as one of the city's greatest pulpit masters, alongside Spurgeon and Liddon. His theology was the whole system of evangelical truth enshrined in the Apostles' Creed. With an impressive appearance, regal personality, commanding voice, impeccable diction, and histrionic manner, he preached authoritatively and appealingly. During 1885-92 he preached through the Bible, and these discourses were published in the twenty-five volumes of *The People's Bible.*
ARTHUR CLARKE

PARKER, MATTHEW (1504-1575). Archbishop of Canterbury from 1559. Born in Norwich and educated at Corpus Christi College, Cambridge, he became a fellow of his college and was ordained in 1527. He was probably attracted to the teaching of the Reformers by Thomas Bilney.* In 1535 he became chaplain to Anne Boleyn. He was appointed in 1544 as master of Corpus Christi College. As a reformer he courageously opposed the royal plan to seize the revenues of the chantries and colleges which appeared as a menace to the universities.

With the new freedom under Edward VI* Parker became a close friend of the continental Reformer Martin Bucer* while he worked in Cambridge. Parker was appointed dean of Lincoln Cathedral in 1552. During the reaction under Mary Tudor* he resigned his college post and was deprived of his preferments in 1554. For most of the reign he remained hidden in the house of a friend.

He enjoyed the years of quiet scholarly leisure and resisted Elizabeth I's* appointments as archbishop of Canterbury in 1559. Despite the controversy there is no doubt that he was properly consecrated. The service was performed by four bishops of Edward VI's reign, according to the ordinal attached to the 1552 Book of Common Prayer. Once the settlement of religion had been adopted by Parliament, Elizabeth expected Parker to enforce it. He consecrated and trained all the new bishops. Throughout the 1560s he struggled with the Puritans over vestments. He published the *Advertisements* of 1566 without royal support. He completed the Elizabethan Settlement* on the authority of the church alone. In 1572 the Admonition* controversy began. In this new phase of the struggle with the Puritans, Parker used J. Whitgift* as his chief agent. Parker was a good administrator, and despite powerful opponents he did much to form the character of the Elizabethan church. It was typical of him that he treated the deprived Marian bishops with tolerance and kindness. His scholarly interests continued to the end of his life, when he bequeathed his valuable collection of manuscripts and books to his college in Cambridge. In 1575 the fruit of his antiquarian researches was published in his *De Antiquitate Britannicae Ecclesiae....*

See modern *Lives* by W.M. Kennedy (1908) and V.J.K. Brook (1962); and E.D.W. Perry, *Under Four Tudors* (1940). NOEL S. POLLARD

PARKER, PETER (1804-1888). First medical missionary to China. Born in Framingham, Massachusetts, he studied both medicine and theology. The American Board sent him out in 1834, and in the following year he opened an eye hospital in Canton, the first Christian hospital in the Far East. In 1838 he helped organize the Medical Missionary Society in China and opened a hospital in Macao. After the first Opium War he was drawn more and more into diplomatic affairs, though continuing his medical missionary work. He helped negotiate the first treaty between China and the USA in 1844. The next year he left the mission and became secretary of the American legation. From 1855 to 1857 he was commissioner and minister plenipotentiary to China and helped revise the treaty. Ill-health forced him to spend his last thirty years in Washington, D.C., where he was elected regent of the Smithsonian Institution and interested himself in Christian enterprises such as the American Evangelical Alliance. HAROLD R. COOK

PARKER, THEODORE (1810-1860). Congregational clergyman. Born in Lexington, Massachusetts, he studied in Harvard Divinity School (1834-36) and was ordained in 1837 as pastor in not-too-distant West Roxbury. His sermon, "The Transient and Permanent in Christianity," in 1841 denied biblical authority and the deity of Christ. He became minister in 1845 of the Twenty-Eighth Congregational Society of Boston. He moved from Unitarianism to transcendental ideas that Christianity rested on universal truths gained by intuition transcending revelation or Christ. Religion was essentially morality growing out of moral oneness with God. This practical ideology led him to support prison reform, temperance, and abolition of slavery.

GEORGE MARSDEN

PAROUSIA, see SECOND COMING

PARTICULAR BAPTISTS. So called because of their belief in a particular atonement in which Christ died only for His elect people. Their origins in England can be traced to the adoption of believers' baptism by a group of Calvinistic London separatists in 1633. They retained the theological emphasis of the church from which they seceded and remained independent or congregational in polity. By 1660 there were 131 Particular Baptist churches in England. The first Particular Baptist Confession was published in 1644 and has been revised on numerous occasions, the last being in 1966. Apart from baptism, the theological emphasis of the confessions has always been Reformed. The Baptist Missionary Society was formed in 1792 by the Northamptonshire Association of Particular Baptist Churches at the call of William Carey,* so initiating the modern missionary movement. John Bunyan* and C.H. Spurgeon* were also associated with the Particular Baptists.

JAMES TAYLOR

PASCAL, BLAISE (1623-1662). Mathematical prodigy, physicist, religious thinker, inventor, and literary stylist. One of the great minds of Western intellectual history, he was born in Claremont in central France, where his father, a man of upper-class status, was a lawyer, magistrate, and tax commissioner of the area. When Pascal was three his mother died, and five years later the father moved with his three children to Paris, drawn there by the intellectual atmosphere which he cherished.

Instead of providing a tutor for his children, Étienne Pascal chose to educate them himself at home. History and science were taught through games, religion through reading the Bible. Geometry was to be the crowning study, withheld until Blaise was old enough to fully relish its beauty. But at age eleven he worked out on his own some of the basic Euclidian propositions. Building later on his mathematical knowledge, he was to create the theory of probability.

When the Pascal family moved to Rouen in 1640, where Étienne was to become tax collector, young Blaise observed the burdensome calculations which often kept his father up until two in the morning. Putting his remarkable mind to work to solve a practical problem, the son devised the first calculating machine, based on a series of rotating discs, a system that has been the basis of arithmetical machines up to modern times. In physics a notable discovery, known as Pascal's Law, states that pressure exerted on any part of an enclosed liquid is distributed equally to all parts of the liquid. This principle makes possible all modern hydraulic operations.

After his conversion in 1654 following a miraculous vision, Pascal set about preparing an *Apology for the Christian Religion.* The work was never finished, for Pascal died at the age of thirty-nine, leaving only a set of remarkable notes, later published as *Pensées.* The work is a classic of apologetics as well as literature. It undertakes to put the case for Christianity as against the rationalism of Descartes and the skepticism of Montaigne. For Pascal, God is to be known through Jesus Christ by an act of faith, itself given by God. Faith is not of reason; it is of the heart. Man's need for God becomes evident when he recognizes his misery apart from God. God is to be known by faith, but the evidences for validating Christianity are great: the prophecies, the miracles, the witness of history, the self-authentication of Scripture.

In 1657 Pascal's *Provincial Letters* appeared. This masterpiece of irony was directed against the Jesuits in defense of Jansenism,* a conservative reform movement within the Catholic Church which urged a return to the Augustinian emphasis upon grace alone as the basis of salvation.

BIBLIOGRAPHY: E. Boutroux, *Pascal* (ET 1902); H.F. Stewart, *The Holiness of Pascal* (1915) and *The Secret of Pascal* (1941); D.M. Eastwood, *The Revival of Pascal* (1936); E. Cailliet, *The Clue to Pascal* (1944); J. Mesnard, *Pascal, l'homme et l'oeuvre* (ET 1953).

PAUL M. BECHTEL

PASCHAL II (d.1118). Pope from 1099. An Italian monk, he lacked the worldly astuteness of his predecessor, Urban II.* He faced the unresolved problem of investiture, at first apparently successfully. In England the quarrel between Anselm and Henry I was settled in 1107 when Henry renounced investiture with the spiritual symbols, while retaining the right to receive homage for temporalities before consecration and to be present at episcopal elections. With Louis VI of France a similar agreement was reached, though he was satisfied with an oath of fealty rather than homage.

Paschal was not directly responsible for these settlements, but together with the First Crusade, they greatly enhanced the prestige of the papacy. In the empire he supported Henry V's rebellion against his father, only to find that he refused to surrender investiture. Paschal denounced him at several synods, but when Henry appeared in Rome for his coronation Paschal incredibly agreed to renounce the regalia—the secular rights and possessions of the church—if Henry would give up investiture. This produced incensed feelings in the church, and bishops and princes helped to create an uproar of riot proportions at Henry's coronation in 1111. Paschal's surrender was, however, unacceptable to the reforming conscience of the church, and by 1116 he was condemning his own concessions. Driven from Rome in 1117, he returned the following year and died. The fact that the regalia increasingly emerged during his reign as, in some sense, a royal right had important bearings on the concordat which was eventually agreed to.

See INVESTITURE CONTROVERSY.

C. PETER WILLIAMS

PASCHAL CANDLE, see CANDLES

PASCHAL CONTROVERSIES. These concerned the date for the celebration of Easter* and occurred from the second to the eighth centuries. From their earliest history, the Eastern and Western Churches used a different basis for determination of the date. The Eastern Church followed the custom of observing it on the day on which the Jews celebrated the Passover, i.e., the fourteenth day of the month of Nisan. This meant that it might be observed on any day of the week. The Western Church always observed Easter on a Sunday. It was not until the time of Charlemagne,* however, that the present custom of observing it on the first Sunday after the full moon on or next-after the vernal equinox was firmly established throughout the West.

Polycarp traced the Eastern custom back to the Apostle John; Eusebius, the Western custom back to Xystus, bishop of Rome early in the second century. Victor I,* later in the same century, attempted unsuccessfully to impose the Western custom on the church at large. In 325 the Council of Nicea* tried to stabilize the date of celebration, decreeing that it must be the first Sunday following the vernal equinox; but technical difficulties prevented a clear settlement of the issue. These difficulties stemmed from the use of different calendars.* Some churches followed the Jewish lunar calendar. Since this calendar was eleven days short, Easter could fall before the actual equinox, though 14 Nisan marked the full moon after the calendar equinox. Rome eventually fixed the equinox on 25 March; Alexandria, on 21 March, which was its correct date in the Julian calendar. But still other dates and methods of computation were used, especially in the Celtic churches. Though both Western and Eastern churches eventually resolved the technical difficulties and thus established a common practice within their respective jurisdictions, to this day a different method is followed in the East from that of the West, and the time of celebration can vary as much as five weeks.

MILLARD SCHERICH

PASCHASIUS RADBERTUS (c.785-860). Abbot and scholar. From the vicinity of Soissons, he entered the monastery of Corbie under the direction of Adalhard, its first abbot. Well versed in the Scriptures, Church Fathers, and Latin classics, he became an instructor of younger monks. His exemplary humility refused to allow him advancement beyond the order of deacon. Following the death of Abbot Isaac, however, he accepted the abbacy of Corbie, a post he renounced for unencumbered study about 853. Meanwhile he attended the synods of Paris (847) and Quiercy (849). He produced several biographies and dogmatic works—especially contributions in Mariology—and extensive commentaries on Matthew, Psalm 44, and Lamentations; yet *De corpore et sanquine Domini,* written in 831 and revised in 844, is his most famous work. His realistic interpretation of Christ's presence at the sacrament, graphically depicted as being in the Lord's same crucified and risen flesh, was sharply opposed by Ratramnus* and Rabanus Maurus* and later gave way to a subtler, Aristotelian explanation.

JAMES DE JONG

PASSAVANT, WILLIAM ALFRED (1821-1894). American Lutheran clergyman, editor, and philanthropist. Of Huguenot and German ancestry, he began his ministry in 1843 as a New Lutheran of Schmucker's school, but later espoused Old Lutheranism. He was, in 1867, one of the founders of the conservative General Council of the Evangelical Lutheran Church in North America. He lived for the last fifty years of his life in Pittsburgh, where he was pastor, until 1855, of the English Lutheran Church. To strengthen the missionary movement, in which he was deeply interested, he issued in 1848 the monthly *Missionary* which became a weekly in 1856 and was incorporated with the *Lutheran* of Philadelphia in 1861. He also edited, until his death, the *Workman* first published in 1881. He was largely responsible for introducing the order of deaconesses into the United States in 1849. He opened hospitals and orphan asylums in a number of American cities. He also founded the Chicago Lutheran Theological Seminary and Thiel College at Greenville, Pennsylvania.

JAMES TAYLOR

PASSIONISTS. Popular name for the "Congregation of Discalced Clerks of the Most Holy Cross and Passion of our Lord Jesus Christ," founded in 1725 by Paul of the Cross.* The first house or "retreat" was opened on Mt. Argentaro (1737). After 1840 the order expanded, founding houses in thirteen countries in Europe and America. In England they were the first religious since the Reformation to lead a strict community life and wear their habit in public. Emphasizing the contemplative life, they take a fourth vow to further the memory of Christ's passion in the faithful. Their chief activities are missions and retreats. Their black habit bears the emblem of a white heart inscribed *Jesu XPI Passio.*

The Passionist Nuns, founded by Paul with Faustina Gertrude (Mother Mary Crucifixa), were approved by Clement XIV in 1770. Strictly enclosed and contemplative, they take the fourth vow, practicing devotion to the passion. They have convents in Europe, America, and Japan.

J.G.G. NORMAN

PASSION PLAY, see OBERAMMERGAU

PASTOR, LUDWIG VON (1854-1928). Church historian. Born in Aachen of a Protestant father and Catholic mother, he had by 1874 decided to write a history of the popes. Early in his academic career he was befriended by the historian Johannes Jenssen; later the two would collaborate on some books. Pastor studied at Louvain, Bonn,

Berlin, Vienna; his Ph.D. was earned at the University of Grosz in 1878. He apparently played a role in the opening of the Vatican Archives to all scholars in 1883. He taught at the University of Innsbruck between 1881 and 1901; was director of the Austrian Historical Institute in Rome; and finally Austrian Ambassador to the Holy See in 1920. His major work was *The History of the Popes From the Close of the Middle Ages* (16 vols., 1886-1933; ET 40 vols., 1891-1953). This was based upon extensive research in the Vatican and in over 200 other European archives. He was of the opinion that only Roman Catholics could really understand and interpret papal history.

ROBERT SCHNUCKER

PASTORAL THEOLOGY. A practical application of the Scriptures to the relationship between a minister of the Gospel and the people for whose spiritual well-being he is responsible. It is theology because it deals with the things of God and His Word. It is pastoral, because it relates to a pastor and his people.

The basic precepts of pastoral theology are found in the NT. Paul addressed the elders of the Ephesian church at Miletus in words that showed his compassion and concern for the people and his own personal commitment to Christ and His service. This vital relationship between pastor and people is further emphasized in Paul's letters to Timothy and Titus. Details of organization and administration are given, but interwoven with them are solemn warnings and appeals that the minister's chief concern is the cultivation of mature Christian character in himself and his people.

The importance of this branch of theology is recognized by every denomination in the Christian Church, and every training college for ministers has a department of pastoral or practical theology. As a rule it embraces training in the art of preaching and the science of homiletics, but the main emphasis is on the character of the pastor and the care of souls. This care is exercised in different ways, such as the visitation of the homes for the discussion of spiritual problems, or personal interviews in the minister's home or in the office at the church. In recent years there has been a greater emphasis on psychology as an aid to pastoral care. Many of the larger churches have widened the scope of their pastoral concern by appointing chaplains for industry, schools, and colleges.

ADAM LOUGHRIDGE

PATARINES. A lay reform movement in N Italy during the late eleventh century, directed against clerical immorality. Centering in Milan, it was directed primarily at the archbishop and other simoniac priests, but also at the upper-class laity who had crept into the ranks of the religious by similar unethical means. Certain radical Patarines, as lay preachers, inveighed against these corrupt clergy, forbade the faithful to attend their ministrations, and by violent means removed refractory priests and bishops from their altars and their benefices. Papal encouragements angered German monarchs whose clergy had received only royal (lay) authority; the Reformation thus became an element in the bitter, centuries-long investiture* struggle between pope and emperor. By 1075, upon the excision of the more corrupt elements from the northern clergy, the movement soon disappeared. Patarine activity produced a strengthening of papal political authority in Lombardy, and served also to destroy an ecclesiastical network established upon simonaic practices.

H. CROSBY ENGLIZIAN

PATMOS. An island of the Sporades group in the Dodecanese, lying some thirty-seven miles WSW of Miletus on the coast of Asia Minor. It has an area of about twenty-two square miles, some eight miles long and six miles in its maximum width. It is of volcanic origin, with rocky slopes rising in three peaks of about 900 feet. Such islands of the Aegean Sea were used for political banishment, and the reference to Patmos in Revelation 1:9 suggests such a condition of exile for John the Seer. Tradition has it (Irenaeus, Eusebius, Jerome, and others) that the apostle was exiled there during the fourteenth year of Domitian's reign (A.D. 95) and was released during the reign of Nerva (A.D. 96), some eighteen months later. Much of the imagery of the Book of Revelation is about the sea. The island is close enough to Asia Minor to have kept John in touch with events there, as underscored in his letters to the seven churches (Rev. 2–3).

JAMES M. HOUSTON

PATON, JOHN GIBSON (1824-1907). Pioneer Presbyterian missionary in the New Hebrides. He was born in Kirkmahoe, near Dumfries in Scotland. He was educated at the University of Glasgow and studied theology at the divinity hall of the Reformed Presbyterian Church of Scotland. During 1847-57 he was a city missionary in Glasgow. At the end of this period he was ordained by his church as a missionary to the New Hebrides. He and his wife left Glasgow in 1858 for the island of Aneityum in the New Hebrides and subsequently became the pioneer missionaries on the island of Tanna. His wife died in childbirth in 1859. Paton was in almost daily danger of his life and was forced to leave the island in 1862. He became a traveling ambassador for the New Hebrides mission. In Scotland in 1864 he secured more recruits and remarried. In 1866 he moved to the island of Aniwa and saw the conversion of most of the islanders. After many years of hard labor on the islands, in the 1880s he made Melbourne in Australia his headquarters for work to support the mission. Until his death he traveled the world for the mission. At the Ecumenical Missionary Conference in 1900 at New York he was hailed as a great missionary leader. His autobiography (1889) published by his brother was an effective way of gaining support for the mission.

NOEL S. POLLARD

PATON, WILLIAM (1886-1943). Missionary organizer and writer. Born in England of Scottish parents, he became a minister of the Presbyterian Church of England after education at Pembroke College, Oxford, and Westminster College, Cambridge. From 1911 to 1921 he traveled Britain as missionary secretary of the Student Christian Movement,* then went to India for the YMCA. At the formation of the National Christian Council of India, Burma, and Ceylon he became general secretary for its seven formative years until brought back to be joint secretary of the International Missionary Council* for the rest of his life, based in Britain. He helped prepare the important conferences at Jerusalem (1928) and Madras (1938), and was an indefatigable worker toward making younger churches self-supporting and indigenously led. He took a considerable part in the formation of national Christian councils. A strong exponent of intermission cooperation and ecumenical relations, he was one of the architects of the World Council of Churches,* although he died before its inception. During World War II he did much for the Orphaned Missions Fund which helped the survival of missions and missionaries cut off from their home bases. Paton was editor of the *International Review of Missions* for sixteen years and an influential writer on missionary aims and methods. Among his better-known books are *Jesus Christ and the World's Religions* (1916) and *The Church and the New Order* (1941).

See M. Sinclair, *William Paton* (1949).

JOHN C. POLLOCK

PATRIARCH. The Septuagint translators of Chronicles had coined *patriarchēs* to define royal officers (always plural) variously expressed in Hebrew (1 Chron. 24:31; 27:22; 2 Chron. 19:8; 23: 20; 26:12); Hellenistic Judaism applied the concept to those special ancestral progenitors narratively identified in Genesis—Abraham (Heb. 7:4), with Isaac and Jacob (*4 Macc.* 7:19), the latter's twelve sons (Acts 7:8,9; cf. *Testaments of the Twelve Patriarchs*)—and to David (Acts 2:29). While pre-Nicene authors retained this usage, enlarging its significance by their prevailing typological exegesis, in the late fourth century Epiphanius* also indicated the word was being used for the hereditary chief office of Judaism, from which by analogy it was introduced for the highest ecclesiastical office within Christianity. Patristic citation thereafter employs "patriarch" for the sees of Old and New Rome, Jerusalem, Alexandria, and Antioch, though Socrates used it more widely to cover all imperial dioceses, and Gregory of Nazianzus for senior bishops *(presbuteroi episkopoi)* in general. CLYDE CURRY SMITH

PATRICK, SIMON (1626-1707). Bishop of Ely. Educated at Cambridge, he was influenced by John Smith, the Cambridge Platonist.* He was ordained a Presbyterian minister in 1648, but in 1654 was episcopally ordained in private. From 1662 he was vicar of St. Paul's, Covent Garden, remaining throughout the Plague. In 1687 he opposed James II's Declaration of Indulgence,* and took the oath of allegiance to William and Mary in 1689, when he was appointed bishop of Chichester. His translation to Ely came two years later. Patrick founded schools in London, helped found the Society for Promoting Christian Knowledge, and supported the Society for the Propagation of the Gospel. His voluminous writings include an allegory, *The Parable of the Pilgrim;* commentaries on the OT up to the Song of Solomon; devotional works; and polemical works against Roman Catholics and Nonconformists. JOYCE HORN

PATRICK OF IRELAND (c.390-c.461). His dates, origin, and career have long provoked controversy among historians. The only reliable sources of information are his own short writings: *The Confession* and *The Letter to the Christian Subjects of the Tyrant Coroticus,* often called erroneously *The Letter to Coroticus.* These have been supplemented by many medieval traditions which are largely valueless. Part of the difficulty is that in the medieval sources Patrick may have been confused with Palladius,* who was sent by Pope Celestine to Ireland in 431. Patrick's writings indicate no connection whatsoever with Rome. Linguistic and other considerations suggest that he received his theological training in Britain; the peculiarities of the vulgar Latin which he used point to a British background.

The dates of Patrick's life cannot be fixed with certainty. His father Calpurnius was a deacon and a Roman magistrate *(decurio),* son of Potitus, a presbyter. The place of his birth is defined in *The Confession* as Bonavem Taberniae, and later he speaks of his parents as living in Britain, and calls it his country. It seems highly probable that his birthplace was Old Kilpatrick, near the Scottish town of Dumbarton. At sixteen he was taken captive by marauders from Ireland, and became a slave in East Antrim, near a hill called Slemish, to a farmer called Milchu. His conversion dates from this period, when, as he says, "The Lord opened to me the sense of my unbelief that I might remember my sins and that I might return with my whole heart to the Lord my God." After six years he escaped from captivity and procured a passage, probably to Scotland. But he did not remain long at home. A night vision called him back to Ireland, and he returned about 432. His ministry and wanderings in Ireland for the next thirty years are obscure, although the subject of many legends; but the view may be accepted that he traveled throughout Ireland and that he had a considerable influence on the Irish chieftains of his day. He had special links with Tara, Croagh Patrick, and Armagh. There is no doubt that he broke the power of heathenism in Ireland and that his teaching was scriptural and evangelical, and that the church which he founded was independent of Rome. He was buried probably in Downpatrick. HUGH J. BLAIR

PATRIMONY OF ST. PETER, THE. This denotes the material wealth and possessions of the Chuch of Rome. Historically it refers to gifts of land given to the Holy See in 754 and 756 by Pepin* the Short. These gifts comprised what was later known as the Papal States.* Pepin's gift is significant in that it launched the temporal power of the bishop of Rome, and its correlating events were the beginning of Rome's claim to papal supremacy over the crowns of France and Germany. Papal temporal power over this vast part of Italy ended in 1870, when during the Italian *Risorgimento** King Victor Emmanuel took possession of Rome as the capital of free and united Italy. Papal economic power was restored, however, by Mussolini when he signed the 1929 Lateran Treaty.* This declared Rome a holy city, returned it to the spiritual domination of the Church of Rome, established the autonomous Vatican State, and granted the Holy See $90 million reparations for lands lost in 1870. These monies were invested in the economy of Italy and abroad.

Estimates in 1971 declared the Vatican to be the world's largest business corporation, and Nino Lo Bello puts the Vatican wealth in 1970 at $5.6 billion. In 1962 the Italian parliament investigated the Vatican's power over the Italian economy, and made the Roman Catholic Church liable for corporate taxes on its investments. The European Common Market ruled in December 1970 that the Vatican is ineligible for tax privileges inside the Common Market. Paul VI modernized the Vatican's financial administration in 1968 by establishing the prefecture of economic affairs of the Holy See, and under the prefecture he created the Administration for the Patrimony of the Holy See. This latter office is responsible for overseeing the Vatican's worldwide investments.

ROYAL L. PECK

PATRIPASSIANISM, see MONARCHIANISM; SABELLIANISM

PATRONAGE. The right of a patron to nominate for appointment to, and to help to administer, a benefice. It originates from Anglo-Saxon times when a landowner felt it a duty to build a church and provide a priest for those who lived on his estate. These were approved by the bishops, who themselves set up similar churches on their own estates. Because patronage was often attached to land and carried with it income, it became a central feature in struggles between church and state. After the dissolution of the monasteries there was a distribution of patronage. In the Church of England it is now held by archbishops and bishops, private individuals, universities and colleges, trusts, cathedral chapters, the Crown, incumbents of mother churches, and diocesan boards of patronage. When a vacancy arises, the patron obtains the views of the churchwardens and the parochial church councils, who have the right to state the needs and conditions of the parish and to resist any nominee whom they do not consider suitable for their requirements. The patron must receive the support of the bishop for his nominee. In England patronage is now increasingly coming under the control of the ecclesiastical authorities.

HOWARD SAINSBURY

PATTESON, JOHN COLERIDGE (1827-1871). First missionary bishop of the Church of England in Melanesia. Educated at Eton and at Balliol College, Oxford, he became a fellow of Merton College and held a curacy in Devon. He was persuaded by G.A. Selwyn,* bishop of New Zealand, to go out to Melanesia as a missionary in 1855. In 1856 he made his first journey to Melansia to encourage boys to return with him to study at the college Selwyn had set up first at Auckland and later on Norfolk Island. The training given by Patteson and his ability to acquire the many languages of the islands provided a strong basis for the mission. In 1861 he was consecrated bishop of Melanesia and traveled constantly, supporting his English and native workers. On the main island of Mota he saw the conversion of most of the population. His work was often made dangerous by the activities of white traders, known as "black birders," who forcibly took natives to labor in Australia. In September 1871, unaware of a recent outrage by these traders, he landed on the island of Nukapu and was speared to death.

NOEL S. POLLARD

PAUL III (1468-1549). Pope from 1534. Born Alessandro Farnese at Canino of an influential Italian family, he was educated at Rome and Florence where he received instruction by well-known humanists. Rising rapidly in the church he was made a cardinal by Alexander VI in 1493 and eventually became dean of the Sacred College. He led a scandalous moral life, fathering four illegitimate children, until his ordination in 1519, after which he became somewhat reformed. When Clement VII died in 1534, he was elected pope.

Although nepotism was a prominent feature of his pontificate and he was an enthusiastic participant in such Renaissance activities as the hunt and art patronage, he nevertheless was instrumental in setting reform trends in motion. He appointed several men to the cardinalate who were dedicated reformers and in 1536 formed a commission of nine distinguished churchmen to examine abuses in the church and report on the necessary steps for reformation. This report, the *Concilium de emendenda ecclesia* (1538), was criticized by Protestants for being superficial but eventually became the basis for much of the work of the Council of Trent.* It was during the reign of Paul that the Society of Jesus was recognized (1540) and several other reform orders given encouragement, including the Ursulines, the Barnabites, and the Theatines.

Both religious and secular rulers often frustrated many of the reform attempts of Paul. In 1536 he issued a bull which called for an ecumenical council at Mantua in 1537, but the actions of Protestant rulers and the duke of Mantua prevented the council from convening. Opposition between Francis I and Charles V blocked further attempts at a conciliar movement, but with the peace of Crespy (1544) Paul was able to convene the successful Council of Trent in 1545.

ROBERT G. CLOUSE

PAUL IV (1476-1559). Pope from 1555. Born Giovanni Pietro Caraffa of a noble Neapolitan family, he received a good education in the home of his uncle, Oliviero Cardinal Caraffa, who was able to secure an appointment in the Roman Curia for him. Beginning as a chamberlain under Alexander VI, he became bishop of Chiete (Teate) in 1506. He also served as a papal envoy in several instances: in 1513 he was sent to England by Leo X to collect Peter's Pence* from Henry VIII; he then went to Flanders (1515-17) and Spain (1517-20). It was perhaps this last mission which provoked an anti-Spanish feeling which was to affect the policies of his pontificate. Always active in reform movements within the church, Caraffa was a member of the Oratory of Divine Love from 1520 to 1527 and co-founder of the Theatines in 1524. In 1536 he was made a cardinal by Paul III and remained a staunch supporter of the Counter-Reformation, serving on reform commissions and reorganizing the Italian Inquisition.

Elected pope in 1555, he displayed a zeal for reform that produced some drastic measures. His anti-Spanish and anti-imperial policies along with his fervent promotion of the Inquisition,* especially against holders of high office, and the publication of the Index* of Prohibited Books lessened his popularity. Nepotism also marred part of his pontificate, although he expelled his nephews before the end of his reign. In his attempt to apply medieval concepts of papal power to sixteenth-century politics he was unable to stem the tide of Protestantism throughout N Europe.

ROBERT G. CLOUSE

PAUL V (1552-1621). Pope from 1605. Born Camillo Borghese in Rome, he studied canon law at Padua and Perugia and maintained a lifelong reputation as an able and strict canonist. He served in the Curia beginning about 1580 and rose rapidly to the top, being created cardinal in 1596, becoming vicar of Rome in 1603, and elected pope two years later as a compromise candidate. He supported educational reform, approved new reform orders, and sponsored improvements in the city of Rome.

But although he led a saintly life (except for nepotism), his pontificate was plagued with painful struggles: (1) the bitter Jesuit*-Dominican* dispute over grace, which he finally decided on the side of the Jesuits without condemning their opponents. His reaction was to ban all further discussions of the topic (1607); (2) the problem arising from the teachings of Galileo,* especially in the light of Bellarmine's* attacks. The Congregation of the Index decided this issue by condemning Galileo (1616); (3) a political struggle with Venice. Paul was forced to take action in Venice where in 1605 the "benefit of clergy"* provision was violated. The bitter struggle expanded into a pamphlet warfare on the old "Two Swords" issue, with Bellarmine siding with the papacy against the animated Paolo Sarpi.* Ultimately Paul placed an interdict on the city; the Venetians retaliated by expelling all groups supporting the pope. Matters were finally settled through the mediation of Henry IV of France; (4) a political struggle involving England. James I's* divine-right theories (especially the Oath of Allegiance, 1606) occasioned two papal briefs of condemnation. A fierce contention raged for many years; (5) finally, the outbreak of the Thirty Years' War. Paul sided with the Hapsburgs, but died shortly after the Battle of White Mountain (1620).

See C.P. Goujet, *Histoire du pontificat de Paul V* (2 vols., 1765). BRIAN G. ARMSTRONG

PAUL VI (1897-). Pope from 1963. Born to an upper-class family whose father was editor (1881-1912) of the daily *Il Cittadino di Brescia,* Giovanni Battista Montini attended a Jesuit institute near Brescia (1903-14) and because of ill-health lived at home while studying in the diocesan seminary. After ordination (1920) he went to Rome for graduate study at the Gregorian University and the University of Rome. After a brief visit to Warsaw, Montini in 1924 entered the papal secretariate of state. In 1924 he became chaplain at the University of Rome. In 1925 as national moderator of the Italian Catholic University Federation he opposed a similar, Fascist group.

Montini rose in Vatican esteem when Cardinal Pacelli appointed him to his staff (1937). The future Pope Pius XII* used Montini rather than appoint a new secretary of state (1944). In 1952 Montini and Domenico Tardini refused elevation to the cardinalate, remaining as dual prosecretaries of state. In 1954 the pope appointed Montini archbishop of Milan. During 1955-1963 he attempted to reach the workers. In 1963 he published *The Christian in the Material World,* a plea for social justice. "From evangelical poverty flows liberty of the spirit," Cardinal Montini wrote, in defense of the workers. Named a cardinal by Pope John XXIII* in 1958, he became John's successor and heir of Vatican II* on 30 June 1963.

Vatican II reconvened on 29 September 1963 for its second session, opened with a remarkable papal appeal for renewal and unity. At the end of that session Pope Paul proclaimed the *Constitution on the Liturgy.* The third session led to promulgation also of the *Constitution on the Church* and *Decree on Ecumenism.* At the end of Vatican II (1965) Paul promised a reorganization of the Curia.

Pope Paul VI traveled widely, making a historic visit to the Holy Land (1964) and the United Nations in New York (1965). From Bombay in 1964 to South America and the Philippines, he traveled more than any other pope. Pope Paul VI continued his social concern expressed while archbishop of Milan. Especially in Rome he continued as pastor of his huge diocese, visiting the ill and those in prison. The furor over the papal encyclical *Humanae Vitae* (1968) perhaps continued the erosion of papal authority.

BIBLIOGRAPHY: W.J. Wilson, *Paul, the Missionary Pope* (1968); M.E. Marty, "Self-criticism or self-demolition?" in *Frontier* 12 (1969), pp. 57-60; J.F. Andrews, *Paul VI: Critical Appraisals* (1970). MARVIN W. ANDERSON

PAUL, ACTS OF, see APOCRYPHAL NEW TESTAMENT

PAUL, APOCALYPSE OF. An apocryphal account of the apostle's journey in heavenly and infernal regions (cf. 2 Cor. 12:2ff.). Originally in Greek (not preserved in that form), it is best represented in the Latin tradition, and also known in Coptic, Syriac, Armenian, Georgian, and Slavonic versions, differing somewhat one from another. It evidently draws upon earlier works, such as the *Apocalypse of Peter;* consequently it is probably of fairly late composition. The document declares itself to have been hidden until 388, when its hiding-place was made known in a dream. Paul receives the divine commission and hears the complaint of creation against man, whose deeds are reported by angels. He sees the fate of righteous and wicked, and makes two visits to Paradise, meeting patriarchs and prophets on the first visit, and others on the second (doublet accounts), while Hell is visited and torments described in detail. The work is attested by Augustine, is used by some Latin poets, and influenced medieval imagery, including Dante's. J.N. BIRDSALL

PAULA (347-404). Roman matron and friend of Jerome; descended from the Gracchi and Scipio. The mother of five children, when aged thirty-three she dedicated herself to the ascetic life. In 385, with her daughter Eustochium, she followed Jerome* to Palestine, despite the pleas of her other children. After visiting the holy places and hermits in the Egyptian deserts, they settled at Bethlehem, where Paula founded three nunneries and a monastery. She gave away her remaining wealth and died in poverty.

PAULICIANS. Evangelical antihierarchical sect originating in the seventh century (possibly earlier) on Rome's eastern borders in Armenia, Mesopotamia and N Syria. Characteristic doctrines include: Adoptianist* Christology; rejection of mariolatry,* images, and hagiolatry; the authority of Scripture (especially esteeming Luke and Paul, and rejecting the OT, like Marcion); believers' baptism. Some, but not all, were dualists, though they repudiated Manichaeism.* The earliest reference to them occurs in 719, when John Otzin, catholicos of Armenia, warned against "obscene men who are called Paulicians." The name may be derived from their regard for Paul the Apostle, or from Paul of Samosata* (with whose teaching they had some affinity), or from an unknown Paul who learned the doctrine from his mother Callinike.

Their founder was probably Constantine-Sylvanus (c.640) of Mananali, a Manichaean village near Samosata, who labored at Cibossa for twenty-seven years before being stoned to death (c.684). His persecutor, Simeon, was himself converted and became Constantine's successor, only to be martyred (690). The sect was protected by Emperor Constantine Copronymus (741-775), himself probably a Paulician. Numbers increased greatly, especially under Sergius-Tychicus (801-35). Savage persecution under Empress Theodora (842-57), in which some 100,000 were martyred, developed into a war of extermination under Basil. Though victorious for a time under the leadership of Carbeas and Chrysocheir, with help from the Saracens, with whom they found refuge, after Chrysocheir's murder (873), the Paulicians were decimated and dispersed.

In 973 John Zimisces transported a great colony to Thrace, effectually introducing their thought to Europe. They continued to exist in scattered communities in Armenia, Asia Minor, and the Balkans, influential at least until the twelfth century, even spreading to Italy and France. Probably they developed into and amalgamated with sects like the Bogomiles,* Cathari,* and Albigenses.* The Crusaders found them everywhere in Syria and Palestine. Anabaptists in the sixteenth century had contact with apparent Paulicians. A colony holding their beliefs settled in Russian Armenia in 1828, bringing with them the manual of Paulician doctrine, *The Key of Truth.*

BIBLIOGRAPHY: K. Ter-Mkrttschian, *Die Paulicianer* (1893); F.C. Conybeare (ed.), *The Key of Truth* (1898); G.H. Williams, *The Radical Reformation* (1962). J.G.G. NORMAN

PAULINE EPISTLES, see EPISTLES, PAULINE

PAULINES, see BARNABITES

PAULINUS (c.730-802). Bishop of Aquileia. A Lombard born in Friuli, he excelled in juridical and theological scholarship. His reputation motivated Charlemagne to appoint him as master at the French court, where he became a firm friend of Alcuin,* and in 787 as bishop of Aquileia. Paulinus contended against Adoptianism at the synods of Regensburg (792), Frankfurt (794), and Cividale (796). At Frankfurt he wrote *Libellus Sacrosyllabus,* a refutation of the heresy, and he presided at Cividale, where the *Filioque* was adopted. Contact with the pagan Avars stimulated his significant contributions to medieval missionary theory and practice. Ecclesiastically he was a wise diplomat and progressive administrator, procuring the right of free episcopal elections. His writings include hymns or rhythms, poems, letters, and anti-Adoptianist pieces. JAMES DE JONG

PAULINUS OF NOLA (353/4-431). Bishop of Nola. A rich Aquitanian landowner who, under the influence of Martin of Tours* and Ambrose,* with the consent of his wife, renounced the world and settled in 395 in Campania by the shrine of Felix of Nola. He lived a life of great austerity and built hospitals for monks and poor in addition to water works for Nola. Ordained priest in 394, he became bishop of Nola in 409. As one of the foremost Christian Latin poets of the period, he wrote a long poem for each annual festival of St. Felix. Thirty of his poems and fifty of his letters to his friends (including Augustine) are preserved. Many of his letters contain valuable notes on ecclesiastical architecture. JAMES TAYLOR

PAULINUS OF TRIER (d.358). Anti-Arian bishop. Succeeding Maximin as bishop of Trier, he was a staunch opponent of Arianism* when Constantius induced other Western bishops at the Synod of Arles (353) to accept an Arian formula. Paulinus alone stood firm, and was banished to Phrygia where he died in exile.

PAULINUS OF YORK (c.584-644). Monk and missionary. Probably a Benedictine at St. Andrew's Monastery in Rome, he was part of the second contingent of missionaries sent in 601 by Gregory the Great to assist Augustine in Britain. He worked in Kent, where he was consecrated bishop in 625, then traveled north as the chaplain of Ethelburga, the Christian wife of Edwin of Northumbria. His efforts were rewarded in 627 with the baptism of Edwin at York, where he established his see. His preaching and baptizing continued until Edwin was killed at the Battle of Hatfield (632). Paulinus administered the vacant bishopric at Rochester until his death. Since he deserted York before receiving the pope's pallium, it is disputed whether he was York's first archbishop. JOHN GROH

PAUL OF CONSTANTINOPLE (d. c.351). Native of Thessalonica who became bishop of Constantinople about 335, he was opposed by the Eusebian party, and on nontheological grounds was exiled shortly afterward. He returned after the death of Constantine (337), but when Constantius settled in Constantinople, an Arian synod met and banished Paul and elected in his place Eusebius of Nicomedia* (338). On Eusebius's death (341), Paul returned once again. The Arians consecrated Macedonius and a mob riot developed between the opposing supporters in which much life was lost, including that of Hermogenes—the master of cavalry—sent by Constantius to deal with the situation. Paul was exiled, but returned probably in 344 and 346, when Constantius was under Western pressure. He was finally exiled about 351, and soon died—possibly strangled, but more probably of natural causes. He was a courageous upholder of orthodoxy against the increasingly heretical tendencies of the East.
C. PETER WILLIAMS

PAUL OF SAMOSATA (fl. 260-272). Bishop of Antioch. From Samosata on the Euphrates, he became bishop of Syrian Antioch c.260. His political role under Odenath and Zenobia of Palmyra—governor over tax collection—entitled him to a bodyguard, the position of judge, and a private council chamber. His critics charged him with encouraging adulation of himself and keeping *virgines subintroductae.* At local synods they failed to depose him in 264, but succeeded in 268, when they elected his successor, Domnus. Popular support retained Paul, and Antioch had two bishops until 272 when Aurelian recovered the city from the Palmyrene regime. Paulianists survived at least until the Council of Nicea (325).

The sources for his teaching are accepted with differing degrees of confidence by scholars. Apparently Paul distinguished the "heavenly Word" from the man Jesus, whereas his Origenist opponent Malchion* argued for a unity in Christ echoed later by Apollinarius.* Paul objected to attributing *ousia* to the Word, probably because two *ousiai,* of the Father and of the Word, would make a divided Godhead. Loofs interpreted Paul as an economic Trinitarian, foreshadowing Marcellus of Ancyra.* Fourth-century critics generally accused him of making Christ "a mere man" and stressing His human soul. Heretics later charged with his error include Marcellus of Ancyra, Theodore of Mopsuestia,* and Nestorius.*

See H. de Riedmatten, *Les Actes du Procès de Paul de Samosate* (1952). G.T.D. ANGEL

PAUL OF THEBES (d. c.340). Traditionally the first Christian hermit. Jerome's *Vita Pauli* is the sole authority for his life. Born in the Thebaid, during the Decian persecution (249-51) he fled to a cave near the Red Sea at the age of sixteen, where he lived a life of prayer and penitence reportedly for about 100 years. He is said to have been visited by Antony when he was 113; there is no confirmation of this visit in the *Vita Antonii,* and much is doubtful regarding his life. In later art Paul is represented with a palm-tree, from the leaves of which he wove a tunic, and two lions, which dug his grave.

PAUL OF THE CROSS (1694-1775). Founder of the Passionists.* Born Paolo Francesco Danei at Ovada, Italy, he led a life of austerity and became a hermit in 1720. In a forty-day retreat he drew up a Rule for a religious order to honor Christ's passion. Benedict XIII permitted him to receive novices (1725), and he was ordained (1727). He labored as a missionary at Monte Argentaro. The first Passionist "Retreat" was opened (1737), his Rule was approved by Benedict XIV (1746), and Paul was elected superior-general (1747). From 1769 he lived in Rome. He founded the Order of Passionist Nuns (1771), and in 1773 the Church of SS. John and Paul in Rome became the headquarters of the Passionists. By the time of his death he had established twelve monasteries in Italy. An eloquent preacher, especially on the Passion, he was renowned also as a miracle worker and spiritual director. J.G.G. NORMAN

PAUL THE APOSTLE. Paul, or—to use his Jewish name—Saul, was born in Tarsus and educated in Jerusalem under a leading Pharisaic rabbi, Gamaliel. After coming to prominence as a leading persecutor of the infant Christian Church, he experienced an abrupt *volte-face;* he was confronted by the risen Jesus in an experience which was for him both a conversion from his former zeal for Judaism and its law and also a call to redirect that zeal into being a missionary to the Gentiles. He spent three years in Arabia, presumably as a missionary, then returned via Jerusalem to his native town where he spent the better part of fourteen years in evangelism.

When he was summoned by Barnabas to Antioch to assist in his work there, he entered upon a new, well-documented stage in his career which was of decisive importance in the expansion of the church. Along with Barnabas he was sent out by the church at Antioch to do missionary work in Cyprus and Galatia. Then with other companions he worked successively in Macedonia (Philippi and Thessalonica), Greece (Corinth), and Asia (Ephesus), planting churches and providing both them and posterity with a series of writings which have fundamentally shaped Christian theology. His plans for a further mission in the west of the Roman Empire were hindered by his arrest and imprisonment, first in Jerusalem and Caesarea, and then in Rome. Whether he was ever released from imprisonment, and, if so, whether he ever reached Spain, are questions difficult to answer. The evidence suggests an affirmative answer to the former (cf. the hopes expressed in Phil. 2:24 and Philem. 22), but a negative one to the latter (the Pastorals imply work in the Aegean area). In any case, he concluded his life as a martyr—according to tradition—during the Neronian persecution.

Because of his considerable personal correspondence we know Paul better than any other character in the NT. Though stylo-statistical studies have sometimes suggested doubts about the authenticity of all but four or five of his letters, the methods used are themselves doubtful in their reasoning, particularly when their results clash with those of other, well-tried methods of literary and historical study. We may fairly confidently accept the Pauline authorship of the thirteen letters traditionally attributed to him, although there must remain some uncertainty whether a few of them (the Pastorals, and perhaps Ephesians) come directly from him (see EPISTLES, PAULINE). The picture which they give of Paul's career and theology agrees closely with that given in Acts, thus demonstrating (over against skeptical viewpoints) the substantial reliability of that source (see ACTS OF THE APOSTLES).

The picture that emerges is primarily that of a missionary. This was how Paul regarded himself. The key to the man is his sense of apostleship (Gal. 1:1; Rom. 1:1-6), of being called by the risen Christ (1 Cor. 9:1) to be His missionary to the Gentiles (Gal. 1:15-17; Rom. 15:15f.); through this mission he hoped to make his fellow Jews envious of the blessings brought by the Gospel and so lead them to faith in Christ (Rom. 11:13f.). In this missionary task Paul saw himself as *the* missionary to the Gentiles (Gal. 2:7f.), who could claim to have evangelized the whole of the Aegean area and Greece; by setting his sights on Spain he could look forward to having covered most of the northern half of the Roman Empire (Rom. 15:17-29).

There emerges incidentally a pattern of missionary preaching (1 Tim. 2:7) with the leader selecting strategic centers where he worked long enough to establish a self-propagating church and entrusted the detailed outworking of the campaign to his assistants, both local and itinerant like himself (e.g., Epaphras, Col. 1:7; cf. 2:1; Timothy and Titus). Paul also appears as a man with a tremendous pastoral concern for his churches, conscious of the responsibility of a father for their continued growth (2 Cor. 11:28; 1 Cor. 4:14-21).

Paul's basic theology rested firmly on that of the primitive church; he frequently is indebted to it for theological and ethical material. Throughout his career he was beset by opponents who were envious of his success or anxious to upset his work. His theology is thus very much shaped by polemics, and it owed its individual development to the exigencies of debate. Two main types of opponent may be distinguished.

On the one hand, there were Judaizers, men who insisted that in order to be saved Gentiles as well as Jews must keep the law of Moses, including circumcision and the observance of Jewish festivals and food regulations, and who forbade fellowship between Jewish Christians and uncircumcised Gentile Christians. Paul's refusal to tolerate such requirements nearly led to a break in fellowship with the Jewish Christians, but although the more extreme Jewish Christians may have kept to themselves, the leaders (Peter, James, and John) sided with Paul and accepted the principle of freedom for the Gentiles (Galatians; Phil. 3; the opponents of Paul in 2 Corinthians appear to have been Jewish Christians of a similar character).

On the other hand, there were Hellenistic Christians who held incipient Gnostic views. They postulated a sharp dualism of spirit and matter, holding that the latter was unsavable but that spiritual salvation was possible here and now for an elite group who possessed a higher "knowledge" not accessible to all Christians. This affected their attitude to bodily life, producing curious mixtures of asceticism and moral license. Such beliefs appear to have been held in Corinth (1 Corinthians reflects them; they are much less obvious in 2 Corinthians), Colossae, and the churches indirectly addressed in the Pastorals. It is not always easy to draw a firm line between Judaizing and Gnostic outlooks, since there were strong elements of syncretism in Diaspora Judaism.

The theology developed by Paul in this situation demonstrated an essentially Jewish background. It shows his constant indebtedness to the OT and to his rabbinical training as a disciple of the Hillelite teacher Gamaliel (Acts 22:3). Nevertheless, what he says is in conscious opposition to Jewish thinking. He strongly attacks the view that men can find favor with God by obedience to the Mosaic law, since on his view of it the effect of the law is actually to heighten human sinfulness (Rom. 7:7-12). Thus the advantage of the Jew is cut away at a stroke (Rom. 3:9). But although Gentiles may have the law in their hearts, they too have failed to keep it, and so the whole world is guilty before God (Rom. 3:22f.). It is a plight from which it can be rescued only through Christ, and only through a crucified Christ who takes on Himself the curse of human sin (2 Cor. 5:21; Gal. 3:13) and offers Himself obediently to God as a sacrifice for it (Rom. 3:24f.). Thus through the sheer grace of God (Rom. 5:8) men are reconciled to Him (2 Cor. 5:18-20) and redeemed from the power of sin to belong to Him (1 Cor. 6:19f.). On the human side the gift is to be received by faith alone, the attitude in which a man accepts humbly what Christ has done for him, instead of trying to please God (Rom. 3:27-31; Rom. 4; Gal. 2:15–3:9; Eph. 2:8). To describe this experience Paul developed the terminology of justification, a divine act tantamount to forgiveness (Rom. 4:4-8).

For Paul, however, being a Christian is more than having a new status before God. In a rich variety of ways he speaks of a new experience of God. Through faith in Christ, the Christian stands in a personal relationship to his new Lord. He can be said to have died with Christ to his old life of sin and to be alive with Christ (Rom. 6:1-11). Paul claims to "know" Christ in a relationship of close spiritual communion (Phil. 3:10). The very frequent phrase "in Christ" has often been understood in this "mystical" sense, but more probably it refers basically to having one's life determined by the "fact" of Christ. In any case, what is of supreme significance for Paul is the death and resurrection of Christ. He knows no other Christ than the crucified, risen, and returning Christ (1

Cor. 2:2; 15:3-5, 20-23), and almost lets His earthly life fade into insignificance.

At the same time the Christian life is characterized by the experience of the Spirit (variously called the Spirit of God and the Spirit of Christ, Rom 8:9), who comes and enters the life of the believer (Rom. 8:9-11). It seems that the practical difference between the presence of Christ (Rom. 8:10) and of the Spirit was minimal, although Paul was quite clear that they were distinct persons. Possession of the Spirit is the essential mark of the Christian; through the Spirit all the power of God leading to holiness and ultimate transformation to resurrection life is given to the believer.

Such thinking could lead Gnostics to a highly spiritual view of Christ, severed from historical reality, and suggest that salvation in all its fullness was already present (1 Cor. 4:8; 2 Tim. 2:18). Against such dangers Paul emphasized the historical reality of the crucifixion of Christ (1 Cor. 1:23), and the reality of temptation and suffering in the life of the Christian who is not yet perfect (Phil. 3:12) and looks forward to the coming of Christ (2 Cor. 4). And against the Gnostic depreciation of the body he stressed the hope of the resurrection of the body as the bearer of human personality (1 Cor. 15; 2 Cor. 5:1-10). He combated any suggestion that Christ was merely one divine power among many, and insisted that the fullness of deity was present in Him. He was preexistent (Phil. 2:5-11), the divine agent in creation (1 Cor. 8:6), the expression of divine love in redemption (Gal. 2:20), the very image of God (Col. 1:15-20; 2:9), the supreme Lord who is yet subordinate to His Father (1 Cor. 15:28).

None of the above should be understood in an individualistic manner. Paul's thought is basically corporate as he thinks of believers in the plural as sharers in a common salvation. Together they form the body of Christ (Rom. 12:4f.; 1 Cor. 12), subordinate to their Head (Col. 2:19); they are a temple indwelt by the Spirit (Eph. 2:20-22); they are the bride whom Christ loved and died to redeem (Eph. 5:25-33). As such, Christians exist to glorify and serve God (1 Thess. 2:12), and this they do, not only by worshiping Him, but also by a mutual love which leads them to fulfil God's commandments (Rom. 13:8-10) in every aspect of their family and social life (Eph. 4:17–6:9). Hence the church is a society of which all are at one and human divisions cease to divide (Gal. 3:28f.).

Paul's gospel was a message, to be preached. His basic activity was that of a preacher, leading men to salvation through what was not merely a human message but one empowered by the Spirit and thus itself the word of God (1 Thess. 1:5; 2:13). So great was Paul's emphasis on the word that he attached comparatively little significance to the sacraments; Christ sent him to preach, not to baptize (1 Cor. 1:17). The outward act of baptism was secondary in importance: what mattered was what it signified, cleansing from sin (1 Cor. 6:11) and union with Christ in His death and resurrection by faith (Rom. 6:1-11; Col. 2:12). Nor should we have heard about the Lord's Supper from him, divinely ordained rite though it also is, had it not been for disorders at Corinth (1 Cor. 11:23ff.); for Paul it was a means of proclaiming the Gospel, and of communion with Christ, although participation in it was no automatic guarantee of salvation.

This stress on preaching must not be misunderstood to mean that the message is everything, so that the historical Christ does not matter, and Christ becomes merely the content of a message challenging men to existential decision. This is the error of R. Bultmann,* who reduced Pauline theology to an anthropology of man's existence prior to faith and under faith. To think thus is to miss the essentially christological and theological orientation of Paul; it also misrepresents the place of the historical Jesus for Paul and the OT "history of salvation" that preceded Him. "If Christ has not been raised, our preaching is useless and so is your faith"; for Paul, theology derived from his meeting with the risen Christ, who was the historical Jesus and the eternal Son of God.

BIBLIOGRAPHY: C.A.A. Scott, *Christianity according to St. Paul* (1927, 1961); W.D. Davies, *Paul and Rabbinic Judaism* (1948, 1955); R. Bultmann, *Theology of the New Testament,* vol. I (1952); L. Cerfaux, *Christ in the Theology of Paul* (1959); B.M. Metzger, *Index to Periodical Literature on the Apostle Paul* (1960); J. Munck, *Paul and the Salvation of Mankind* (1960); E.E. Ellis, *Paul and His Recent Interpreters* (1961); H.J. Schoeps, *Paul: The Theology of the Apostle in the Light of Jewish Religious History* (1961); E.E. Ellis, "Paul" in *NBD* (1962): full bibliography to that date; C.K. Barrett, *From First Adam to Last: A Study in Pauline Theology* (1962); R.N. Longenecker, *Paul: Apostle of Liberty* (1964); D.E.H. Whiteley, *The Theology of St. Paul* (1964); A.M. Hunter, *The Gospel according to St. Paul* (1966); L. Cerfaux, *The Christian in the Theology of Paul* (1967); O. Cullmann, *Salvation in History* (1967); J.A. Fitzmyer, *Pauline Theology: A Brief Sketch* (1967); M.F. Wiles, *The Divine Apostle: The Interpretation of St. Paul's Epistles in the Early Church* (1967); G. Ogg, *The Chronology of the Life of Paul* (1968); B. Rigaux, *Letters of St. Paul: Modern Studies* (1968); H. Conzelmann, *An Outline of the Theology of the New Testament* (1969); J.C. Pollock, *The Apostle: A Life of Paul* (1969); E. Käsemann, *Perspectives on Paul* (1971); G. Bornkamm, *Paul* (1971); R.N. Longenecker, "Paul" in *ZPEB* (1975), IV, 624-65; J.W. Drane, *Paul: Libertine or Legalist?* (1975).

I. HOWARD MARSHALL

PAUL THE DEACON (c.720-800). Italian chronicler, known also as "Paulus Levita" and "Warnefridi." He was called "the Deacon" from 782 onward. After an excellent education he became the tutor of Adelperga, daughter of King Desiderius. Then he was a member for a brief period of the Benedictine monastery of St. Peter at Civate before settling at Monte Cassino Monastery. In 782 he visited Charlemagne and remained in Francia for four years, writing the history of the diocese of Metz. After his return to Monte Cassino he wrote the famous *Historia Gentis Langobardorum (History of the Lombard People),* which was translated and published at Philadelphia in 1907 by W.D. Foulke. He wrote also a history of Rome, *Historia Romana,* which continued the *Breviarium* of Eutropius, an exposition of the Rule of St. Benedict, and various poems, biographies, and liturgical pieces.

PETER TOON

PEAKE, ARTHUR SAMUEL (1865-1929). English scholar and writer. Born in Leek, Staffordshire, son of a Primitive Methodist* minister, he graduated at Oxford from a scholarship, became in 1889 lecturer at the new Mansfield College, and soon thereafter achieved the rarity for a Nonconformist of a theological fellowship at Merton College. He gave up these prospects to become tutor of the Primitive Methodist (later Hartley) College, Manchester, in 1892, and held that tutorship for the rest of his life, transforming the college and its reputation. In 1904 he became also the first Rylands professor of biblical exegesis at Manchester University, and the first dean of its theological faculty. His many books, though scholarly, were principally directed to a wide, rather than a learned, audience. Best known was the one-volume commentary on the Bible which he edited (1919) and which introduced thousands of students and laymen to biblical criticism.

A member of a church of the Revivalist tradition, he combined a gentle personality, warm evangelical piety, broadly traditional theology, and frank acceptance of literary criticism. This combination enabled him to procure wide acceptance for critical methods. He was perhaps the main British mediator of biblical scholarship in his day, and—though always a layman—the most widely known member of his church. Of ecumenical spirit, he worked hard for the union of the British Methodist churches, though he died before its consummation.

See J.T. Wilkinson (ed.), *Arthur Samuel Peake, 1865-1929* (1958).

A.F. WALLS

PEARSON, JOHN (1613-1686). Bishop of Chester. Educated at Eton and Cambridge, he was ordained in 1639 and became a prebendary of Salisbury in 1640. He served as chaplain to the Royalist forces in 1645, devoting himself later to theological study until the Restoration. With Peter Gunning in 1658 he debated with two Roman Catholics whether England or Rome was schismatic at the Reformation. He promoted the Polyglot Bible, and his weighty learning was demonstrated in the publication of his *Exposition of the Creed* (1659), a closely reasoned statement of the faith, with copious notes and references to early Christian Fathers. This classic work went through numerous editions, abridgments, and translations. Pearson became archdeacon of Surrey and master of Jesus College, Cambridge (1660), and held other academic posts till his appointment to Chester in 1673. He strongly supported the Restoration settlement at the Savoy Conference* of 1661, rejecting comprehensiveness in favour of uniformity. Convocation appointed him to superintend the translation of the Prayer Book into Latin. In 1672 he wrote *Vindiciae Epistolarum Sancti Ignatii,* an elaborate defense of the authenticity of letters ascribed to Ignatius of Antioch. Other works defended the Church of England against Roman Catholics and Nonconformists.

JOYCE HORN

PEASANTS' REVOLT (1524-25). German revolutionary mass movement. Unrest had spread among German peasants because the territorial princes ignored their customary rights and introduced new taxes. After vainly looking to a reformed and strengthened empire for justice, some engaged in sporadic violence (*Bundschuh* movement, 1502-17). Luther's theological views and attacks on greedy princes, merchants, and clergy, although misunderstood by the peasant leaders, helped to ignite the uprising. It broke out at Stühlingen (Black Forest) in June 1524 and spread rapidly over southern and central Germany until by late April 1525 some 300,000 peasants were under arms. Their program, *Twelve Articles of the Peasantry,* called for: congregational election of pastors, modification of tithes, abolition of serfdom, discontinuing enclosure of common lands, elimination of feudal dues, and reforming the administration of justice. Although some knights participated (Götz von Berlichingen and Florian Geyer), the peasants generally lacked capable direction and organization. Even Thomas Münzer* in Thuringia was a better preacher and agitator than military leader.

Luther's harsh, uncompromising opposition (*Against the Murderous and Thieving Hordes of Peasants*) and the combined forces of the Hessian, Saxon, and Brunswick princes led to Münzer's defeat at Frankenhausen on 15 May 1525. The Swabian League under Count Truchsess suppressed the movement in Swabia and Franconia, and within six weeks it had been brutally crushed almost everywhere. The revolt was extremely detrimental to the Reformation. Even though Luther soon moderated, the disillusioned peasants turned against him, and Lutheranism lost its popular appeal. It enabled princes to centralize their authority, including that over their churches. Catholics portrayed it as a divine judgment against Protestantism, thus discouraging further defections from Rome.

BIBLIOGRAPHY: W. Zimmermann, *Allgemeine Geschichte des grossen Bauerkrieges* (3 vols., 1841-43); F. Engels, *The Peasant War in Germany* (1850, 1956); E.B. Bax, *The Peasants' War in Germany* (1903); G. Franz, *Der deutsche Bauernkrieg* (1956).

ROBERT G. CLOUSE

PECK, JOHN MASON (1789-1858). Pioneer Baptist missionary in Missouri, Illinois, and Indiana. Born in Litchfield, Connecticut, he gained scant early education. He moved with his wife to New York and joined a Baptist church, where he was soon licensed to preach. Ordained in 1813, he ministered several years as a pastor before hearing the missionary challenge through Luther Rice.* In 1817 he and James E. Welch were appointed by the Foreign Mission Board to start work in the Mississippi Valley. Three years later the Board dropped the mission, but Peck stayed on; the Massachusetts Baptist Missionary Society assumed partial support. Besides itinerating widely, Peck founded Rock Spring Seminary, which became Shurtleff College; edited *The Pio-*

neer; helped start the American Baptist Home Mission Society; and for several years was connected with the Baptist Publication Society.

HAROLD R. COOK

PECKHAM, JOHN (c.1225-1292). Archbishop of Canterbury from 1279. Born at Patcham, Sussex, he joined the Franciscans about 1250 and studied at Oxford, then at Paris under Bonaventure. He returned to Oxford in 1270 and later became English provincial of the Franciscans. In 1276 he was summoned to Rome to be the first theological lecturer at the papal schools, but in 1279 Nicholas III appointed him archbishop of Canterbury, displacing the royal candidate, Robert Burnell. As archbishop, Peckham defended the papal position and sought to initiate church reforms. At a provincial synod at Reading (1279) he legislated against pluralities and other abuses. He tried to raise clerical standards, advanced the Dominicans and Franciscans, and used the Welsh War to bring the Welsh Church more closely under the control of Canterbury. His reforms brought him into conflict with Edward I and with many of the clergy. He was an able theologian and a poet, and wrote also on scientific subjects.

JOHN A. SIMPSON

PECOCK, REGINALD (c.1393-1461). Bishop of Chichester. Born in Wales, he was educated at Oriel College, Oxford, was made master of Whittington College, London, in 1431, bishop of St. Asaph in 1444, and of Chichester in 1450. An ardent but wayward defender of the church system of his day against the followers of John Wycliffe,* Pecock employed a doubtful apologetic which led as early as the 1430s to a suspicion of heresy, while prosecution was only narrowly avoided in 1447. In 1455 his *Repressor of Over Much Blaming of the Clergy,* while a useful guide to his Lollard opponents' standpoint, seems to exalt natural religion above Scripture, while the *Book of Faith* (1456) deals unsatisfactorily with the respective spheres of faith and reason. A year later Pecock's unorthodox defenses of "orthodoxy," his tampering with the Apostles' Creed, and his political attachment to the falling Lancastrian party led to a charge of heresy. He recanted and was confined in an abbey with a small pension for the rest of his life.

IAN SELLERS

PECULIAR PEOPLE. The term is a concept deeply embedded in Scripture and the structure of the primitive church, and appears in Tyndale's translation of the NT (1526). The name was applied to themselves by the Quakers of the seventeenth century and by the Tractarians (by way of disparagement) to the Evangelicals. More particularly, the Peculiar People, or Plumstead Peculiars, were a small sect of evangelical faith-healers founded by William Bridges in London in 1838. A humble and pious folk, they are largely confined to Kent and Essex. The Society of Dependents or Cokelers of West Sussex are an offshoot of the Peculiars, founded by John Sirgood.

PÉGUY, CHARLES PIERRE (1873-1914). French Catholic writer and poet who reflected the spiritual conflict between Catholic faith and the socialist ideal under the Third Republic. He abandoned his Catholic upbringing in *lycée,* and soon was drawn into the debate over Dreyfus (1894). The affair convinced him of the decay of traditional church and social ideals and pointed him toward involvement with the poor and those left out of society. He joined the socialist party under Jean Jaurès, but by 1900 found himself out of line because of his criticisms of the leaders and policies. The *Cahiers de la Quinzaine* (225 issues, 1900-1914), occasional articles, and essays which he published from the bookshop he had established became his forum. He began to suspect that socialism of Jaurès's sort was only half-good without the dimension of *mystique,* which he began to find in medieval Catholicism; indeed the modern world lacked this crucial dimension. By 1908 Péguy called himself a Catholic again. Whereas his early version of Jeanne d'Arc (1897) conceived salvation secularly, his new version (1910) displayed a sincerely deepened awareness of Catholic spirituality. *Eve* (1913) was his principal expression of his Christian faith. Between 1909 and 1913 he published a variety of essays and poetic *mystères* on medieval themes. He was killed in battle (1914). Only much later (1940s) was wide interest in his work aroused.

BIBLIOGRAPHY: H. Daniel-Rops, *Péguy* (rev. 1935); A. Rousseaux, *Le prophète Péguy* (2 vols., 1942-46); R. Rolland, *Péguy* (2 vols., 1945); A. Dru, *Péguy* (1956).

C.T. MC INTIRE

PELAGIA (d. c.311). Martyred in Antioch when only fifteen, she threw herself into the sea from a window to preserve her chastity when her home was surrounded by soldiers probably during the persecution by Diocletian. The name is also associated with a fourth-century recluse living in a grotto on the Mount of Olives practicing severe penances, and also with a virgin martyr of Tarsus who was reputed to have been burnt for refusing to become the mistress of the emperor.

PELAGIANISM. An ascetic movement with distinctive theological emphases, named after Pelagius, a Christian moralist. A well-educated Briton, trained in law, Pelagius was active in Rome c.383-409/10, teaching Christian perfection to aristocratic circles associated with Rufinus of Aquileia and Paulinus of Nola. He attacked Jerome's denigration of marriage, without accepting Jovinian's equation of marriage and virginity, and inveighed against the implications of Augustine's prayer (*Confessions* 10:29:40), "Give what you command, and command what you choose." His *Expositions of the Thirteen Epistles of Paul* drew on Augustine (e.g., *Free Will*), "Ambrosiaster," and Origen-Rufinus, but perhaps already criticized Augustine's opinions. He also wrote *The Hardening of Pharaoh's Heart; Faith in the Trinity* (anti-Arian and anti-Apollinarian); *Virginity;* and *The Law;* and ascetic manifestos to Demetrias and Celantia. Pelagius sought to be a catholic teacher, opposed especially to Manichaeism, which encouraged moral pessimism and fatalism and, like Jerome's extremism, discredited asceticism. He viewed the Church as the community of the (adult) baptized committed to perfectionist ideals, and magnified man's incorruptible created capacity for freedom from sin. Grace comprised this God-given ability, the illumination of instruction and example, and forgiveness of sins.

Marius Mercator says Pelagius was inspired by Rufinus "the Syrian,"* but Rufinus's influence is more evident in Celestius* and African Pelagianism. The oriental affinities of Pelagian ideas, e.g., in Theodore of Mopsuestia,* require further investigation.

The Gothic attack on Rome dispersed Pelagius's coterie; many passed via Sicily to Africa. The subsequent presence of Pelagians (including Celestius) in Sicily is attested by Augustine and by the writings (*Riches; Evil Teachers; The Possibility of Sinlessness; Chastity*, etc.) of an anonymous Sicilian, who inculcated a severe asceticism, denying salvation to the rich unless they renounced their wealth.

In Carthage, Celestius's views on infant baptism and original sin were condemned by churchmen traditionally sensitive on these issues. (It was 411, when Donatists* and Catholics convened in Carthage; similarities were discernible between Donatists and Pelagians, e.g., in ecclesiology.) Paulinus of Milan was the chief prosecutor.

Augustine's notice was first caught by Celestius's assertion that infants were baptized for sanctification, not forgiveness. He wrote *The Merits and Remission of Sins* in defense of original sin, but remained respectful toward Pelagius, with whom he shared friends and enemies, criticizing only extravagant ascetic claims. *Nature and Grace*, his reply to Pelagius's *Nature* (which depends on *The Sentences* of Sextus*), still refrains from attacking Pelagius by name (415).

Pelagius quickly left Africa for Palestine, welcomed by John of Jerusalem, Jerome's old opponent. Both Pelagius and Jerome wrote to Demetrias in 414, and Jerome, who with some justice viewed Pelagius as an Origenist, began his *Dialogues against the Pelagians.* In 415 the Spanish heresy-hunter Orosius brought news of Africa's excommunication of Celestius and possibly suspicions of the Syrian Rufinus, Jerome's presbyter. A Jerusalem synod cleared Pelagius and others of Orosius's accusations, but resolved on a reference to Rome, while Orosius's heavy-handedness furthered the concretion of "the Pelagian heresy" out of disparate elements. Pelagius was again acquitted at the Synod of Diospolis (415), but only after equivocating and disowning Celestian views. A raid on Jerome's monastery at Bethlehem was blamed on Pelagians (416).

Palestinian developments and Celestius's ordination at Ephesus cast Eastern aspersions on Africa orthodoxy, reasserted in conciliar condemnations of Pelagius's *Free Will* at Carthage and Milevis (416). Pope Innocent I* obligingly excommunicated Pelagius and Celestius (417), but Pope Zosimus* lifted the ban after appeals from Celestius in person and Pelagius in writing. Against an increasingly identifiable "Pelagianism," another Carthage council issued nine canons (418), denying salvation to unbaptized infants. Violence between Catholics and Pelagians in Rome resulted in their banishment by Honorius, which together with African pressure evoked a *volte-face* condemnation from Zosimus (418). Rome's rejection of Pelagian views did not endorse the full African position to whose defense Augustine progressively harnessed Catholic tradition.

The popes rejected further appeals, Boniface from Celestius (423), Sixtus III from Julian of Eclanum* (439), who led eighteen Italian bishops deposed for refusing to subscribe Zosimus's verdict. He prosecuted a polemical controversy with Augustine until the latter's death. The author of the *Predestinatus* was possibly an associate of Julian. (Another Italian Pelagian, Annianus, deacon of the unknown "Celeda," was the first identifiable Latin translator of Chrysostom.)

Pelagian ideas were propagated in Britain by men like Agricola, Bishop Severianus his father, and Bishop Fastidius, a likely author of parts of the Pelagian corpus. The theory that Pelagianism's success in Britain was that of a movement of social protest against unjust Roman rule has been severely criticized. Gallic clergy visited Britain to eradicate the heresy, notably Germanus of Auxerre in 429 (sent by Pope Celestine) and perhaps c.447. In Gaul, too, Pelagianism found supporters, such as Leporius.*

Barred from Palestine by a synod at Antioch (424), Pelagius disappeared, and probably died in Egypt. Julian and Celestius found refuge in Nestorius's Constantinople (429), but his inquiries on their behalf, Celestine's intervention and Marius Mercator's *Commonitory on the Name of Celestius* led to banishment by Theodosius II and condemnation with Nestorius at the Council of Ephesus (431). Rejection was now final in East and West. Later Pope Gelasius tried to flush out Pelagian pockets in Dalmatia and central Italy and wrote a refutation of Pelagianism.

Recent research has emphasized not only Pelagianism's diversity and relation to other controversies, e.g., over Origenism,* but also its preservation of features of primitive Christianity.

BIBLIOGRAPHY: F. Loofs in *Realencyclopädie für protestantische Theologie und Kirche* 15 (1904), pp. 747-774; A. Bruckner, *Quellen zur Geschichte des Pelagianischen Streits* (1906); G. de Plinval, *Pélage: Ses Écrits, Sa Vie et Sa Réforme* (1943), and *Essai sur le Style et la Langue de Pélage* (1947); T. Bohlin, *Die Theologie des Pelagius und ihre Genesis* (1957); R. Pirenne, *La Morale de Pélage* (1961); R.F. Evans, *Pelagius: Inquiries and Reappraisals* (1968); G. Greshake, *Gnade als Konkrete Freiheit: Eine Untersuchung zur Gnadenlehre des Pelagius* (1972); E. Teselle in *Augustinian Studies* 3 (1972), pp. 61-95; G. Bonner, *Augustine and Modern Research on Pelagianism* (1972); P. Brown, *Religion and Society in the Age of St. Augustine* (1972).

D.F. WRIGHT

PELOUBET, FRANCIS NATHAN (1831-1920). Writer of Sunday school literature. Born in New York City, he graduated from Williams College (1853) and Bangor (Maine) Theological Seminary, and was ordained (1857) as a Congregational minister. He was pastor of several churches in Massachusetts from 1857 to 1883. He produced between 1875 and 1920 the annual volumes of *Select Notes on the International Sunday School Lesson,* as well as writing Sunday school quarterlies from 1880 to 1919. He also revised several Bible dictionaries.

PEMBERTON, EBENEZER (1704-1777). American pastor. Born in Boston, son of a Congregational minister, he graduated from Harvard College (1721), and after serving as a chaplain was called to the pastorate of the Presbyterian Church in New York City (1727), having been ordained in Boston by a Congregationalist council. He formed close ties with J. Dickinson,* Aaron Burr, and John Pierson, all based in New Jersey. These four, of New England Puritan background, represented a powerful element in the formative stages of the Presbyterian Church: a group which resisted the doctrinal mentality of the Scotch-Irish and fought against slavish subscription even to the Westminster Confession.* Pemberton became a close friend of George Whitefield,* and with Dickinson was prominent in the founding of the College of New Jersey (1746). In 1754 he left the Presbyterian Church for a twenty-year ministry in the famous (Congregational) Old South Church, Boston. KEITH J. HARDMAN

PENANCE; PENITENCE (from Lat. *poena*). Regarded as a sacrament in the Eastern and Roman churches, it originated as a development of the idea of repentance which included not only an inward feeling of contrition, but also an outward act of self-abasement. Gradually the latter predominated and ultimately took the place of the former. The penitent was required to confess his guilt *(exomologesis),* and submit to discipline (e.g., exclusion from the Eucharist; committal to prayer, fasting, almsgiving), eventually to receive absolution and restoration. It probably arose from NT passages on discipline (e.g., Matt. 18:18; 1 Cor. 5:3-5; 1 Tim. 1:20; Titus 3:10). Early references to such discipline can be found in *1 Clement* 57 and Hermas, *Visions* 3.5.

After the irruption of Montanism,* penance was part of the regular discipline of the church. Prior to Novatianism,* the censures were short and administered simply, e.g., exclusion from participation in or in sight of the sacrament, exclusion from the church, for a few weeks. During the third to fifth centuries the time lengthened considerably. The practice was to permit only one penance after baptism, it was public and formal, and distinction was made between graver and lesser sins. The Novatianists refused to remit postbaptismal sins, and in reaction to them the penitential system tended to become more rigid and systematic (e.g., in the letters of Pacian to Symphronianus). Penance came to be regarded more as a penalty and less of a privilege. The Council of Elvira* (c.305) reveals the position at the end of the third century. The Penitential stages were developed, first laid down in the councils of Neocaesarea and Ancyra (314). Private penance eventually replaced public penance; the first reliable evidence for private penance as a sacrament was canon 2 of the Third Council of Toledo (589), which condemned it. In the East, after 1000 the pattern became (1) confession, (2) interrogation, (3) absolution, (4) assignment of penance. In the West, penance led to the growth of Indulgences,* whereby canonical penance for sin could be remitted by money payments.

At the Fourth Lateran Council (1215) private penance was made compulsory at least once a year. The Council of Trent* (1551) stated that the sacrament of penance was absolutely necessary for the forgiveness of postbaptismal sin, and consisted in confession, contrition, absolution, and satisfaction. (For the Protestant position see, e.g., Calvin's *Institutes.*)

BIBLIOGRAPHY: J.F. Bethune-Baker, *An Introduction to the Early History of Christian Doctrine* (1903); O.D. Watkins, *A History of Penance* (1920); R.C. Mortimer, *The Origins of Private Penance in the Western Church* (1939); J.N.D. Kelly, *Early Christian Doctrines* (1958).

J.G.G. NORMAN

PENINGTON, ISAAC (1616-1679). Quaker apologist. The eldest son of Sir Isaac Penington, lord mayor of London (1642-43) and a staunch Puritan. In 1654 he married Mary Springett, a widow; both had for some time been among the "Seekers," a group eventually amenable to Quakerism, and Isaac's "convincement" took place in London, 1658. He was imprisoned six times beginning in 1661 for refusing to take the Oath of Allegiance.* Several long imprisonments in the Reading and Aylesbury jails were borne with cheerful endurance, characteristic of his innocency of life universally recognized. As a leading Friend, he was a close acquaintance of George Fox.* Five children were born to him, one son, Edward (1667-1711), emigrating to Pennsylvania. His stepdaughter, Gulielma Springett, married William Penn.* KEITH J. HARDMAN

PENN, WILLIAM (1644-1718). English Quaker; founder of Pennsylvania. Born in London, he early became a Quaker and was expelled from Christ Church, Oxford, for Nonconformist views (1661). He traveled for a time in Europe, served briefly in the British navy, and studied law in London. In 1666 he went to Ireland to manage his father's property. He was imprisoned several times, using the occasions to write defenses of Quakerism. Finally freed in 1670, Penn made a missionary trip in Europe and married in 1672. He engaged in further missionary journeys, and by pen and pulpit advocated religious and political freedoms. He helped to send 800 Quakers to New Jersey (1677-78), and in 1681 secured a charter for Pennsylvania from Charles II because of a debt owed his father by the king. He later acquired the region of Delaware. Penn's finest accomplishment was the founding of Pennsylvania as a refuge for religious dissenters and freedom of expression. It was his "Holy Experiment," and his four "Frames of Government" (1682, 1683, 1696, 1701) and his fair and just treatment of the Indians set the pattern for Pennsylvania's colonial history. His friendship with James II cost him control of the colony from 1692 to 1694. In later years financial hardship plagued him, and for a time he was in a debtors' prison. In 1712 he almost completed transfer of the colony's control to the Crown when he became ill. His wife managed his affairs until his death in London. Among his writings are *The Great Case of Liberty of Conscience* (1670), *Christian Quaker and His Divine Testimony Vindicated* (1673), and *An Address to Protestants of All Persuasions* (1679).

BIBLIOGRAPHY: W.W. Comfort, *William Penn 1644-1718: A Tercentenary Estimate* (1944); E.B. Bronner, *William Penn's Holy Experiment: The Founding of Pennsylvania, 1681-1701* (1962); M.M. Dunn, *William Penn: Politics and Conscience* (1967). ROBERT C. NEWMAN

PENRY, JOHN (1559-1593). Elizabethan Puritan divine. Born in Wales, he graduated from Cambridge, afterward becoming an itinerant preacher in Wales, calling the attention of the government to the need for the propagation of the Gospel there, in *A Treatise addressed to the Queen and Parliament* (1587). Archbishop Whitgift ordered his arrest and the seizure of the copies of this book. Penry himself received a short prison sentence. Later he was directly associated with the production of the Marprelate Tracts* (1588) in which the prelates were severely attacked. Being suspected as their author (which has not yet been verified), he fled to Scotland where he was protected by sympathetic clergy. When the controversy over the tracts had died down, he returned to London and joined a society of separatists. He was soon recognized and arrested. After a trial he was executed. PETER TOON

PENTECOST. The name is derived from the Greek word for "fiftieth" *(pentēcostos)*, for it was seven weeks after Passover that the "Feast of Weeks" (Exod. 34:22; Deut. 16:10) or the "Feast of Harvest" (Exod. 23:16) was observed. It marked the end of the barley harvest and the beginning of the wheat harvest. It was one of the three occasions in the year on which male Israelites were to appear before the Lord (Deut. 16:16), but it was much less observed as an occasion of pilgrimage than the feasts of Passover and Tabernacles. Pentecost was regarded in later Judaism as the conclusion of the Passover rather than as a harvest festival. After the destruction of the Temple in A.D. 70 it was taken to commemorate the giving of the Law on Mount Sinai. Acts 2 records how the Holy Spirit was given to the first Christians on the day of Pentecost, which no doubt symbolized both the completion of the redemptive act of Good Friday and Easter and the beginning of the harvest of the nations. Pentecost was observed by the second century as a Christian feast, second only in importance to Easter. The name "Whitsunday" came to be attached to it because of its being a major occasion for baptisms, the baptisands being clothed in white. New proposals for the calendar of the churches in England refer (as does the Church of South India) to "Sundays after Pentecost" instead of "after Trinity." R.E. NIXON

PENTECOSTAL CHURCHES. A number of fundamentalist Protestant sects that emphasize Spirit baptism as an experience different from conversion and evidenced by speaking in tongues (Acts 2:1-13). They also teach the inspiration of the Bible, salvation by conversion and revival, instantaneous sanctification, divine healing; and claim to be a restoration of original Christianity. Early Pentecostal meetings were characterized by outbursts of ecstatic enthusiasm featuring healings, speaking in tongues, and motoric movements.

Pentecostalism began as an outgrowth of the Holiness Movement.* In 1901 a Bible school called Bethel College was started at Topeka, Kansas, by Charles F. Parham who, using no texbook but the Bible, drilled his students in Spirit baptism teaching. These pupils carried the message of the Spirit into Kansas, and when the school closed both teacher and students went throughout the South preaching Pentecostalism. Houston, Texas, became the next center of "Spirit baptism" when Parham and a local minister, W.F. Carothers, opened a school. One of their converts, William J. Seymour, brought the teaching to Los Angeles in 1906 where he founded the Apostolic Faith Gospel Mission on Azusa Street. Seymour, a black with only one eye, was described by one who attended his mission as being "meek, plain spoken and no orator," in short, not a very charismatic personality. Despite his unimpressive appearance, the results of Azusa revival attracted nationwide attention. Besides the many visitors, including ministers, who were influenced by the revival, publications were put out from this headquarters which caused the rapid growth of the movement. As other churches were started in different parts of the United States, the importance of Los Angeles decreased.

Pentecostalism became an international movement early in its history. One of the important leaders in spreading its teaching to Europe was Thomas Ball Barratt, a Cornishman who was a pastor of a Methodist church in Oslo, Norway. Barratt came to the United States in 1905 to solicit funds so that he might build a larger church. Some believe he visited the Azusa Street Mission, but even though that is debatable it is certain that he attended a Pentecostal meeting in New York City and experienced a Holy Spirit baptism and spoke in tongues. He returned to Norway and became an evangelist for Pentecostalism. People thronged to his meetings, and by 1916 he was able to found the Filadelfia Church, which became the largest dissenter body in Norway. Barratt was also influential in spreading Pentecostalism in Denmark and Sweden. In 1907 an Anglican clergyman, Alexander A. Boddy, after visiting Barratt, returned to England determined to promote a similar Pentecostal awakening there. Boddy wrote a pamphlet, *Pentecost for England*, which was widely distributed, and he invited Barratt to Britain (September 1907). Within a few months there was a Pentecostal revival in England.

Later the movement spread into Germany also. Pentecostalism also appeared in India, where it centered in Mukti and the orphanage of Pandita Ramabai. In 1909 groups from Mukti and missionaries from the United States and Great Britain had extended its teachings throughout India. In addition to W Europe and India, Pentecostalism also spread to Latin America, where it is claimed that eight out of every ten evangelical Protestants are Pentecostal. In Chile the very successful Pentecostal work began with the ministry of Willis C. Hoover, pastor of a Methodist Church in Valparaiso. He began holding charismatic meetings until his church became the "Azusa" of Chile. Later he was forced from the Methodist pastorate and started the Methodist Pentecostal Church, beginning a movement which has swollen to the present 500,000 Pentecostals in the country. Another area of rapid growth has been Brazil, where the charismatic revival began in 1910 with the establishment of the *Congregacioni Christiani* which has burgeoned to 1,400 congregations with nearly 500,000 communicants. Other Brazilian groups such as the Assemblies of God have used purely indigenous workers and scored equally impressive gains.

Early Pentecostals never desired new denominations, but rather felt they should call all Christians back to what they believed to be apostolic faith. Everywhere the work was to be under the guidance of the Holy Spirit, which in practice meant the control of visiting evangelists. But as their teaching was opposed by other groups, especially the Holiness churches, they began to organize denominations. Among the more important Pentecostal churches one could list the Assemblies of God, the Church of God in Christ, the Church of God (Tomlinson), the International Church of the Foursquare Gospel, and the United Pentecostal Church International.

The Assemblies of God* is the largest of these groups, with a total of 645,000 members. Founded in Hot Springs, Arkansas (1914), it maintains a denominational headquarters in Springfield, Missouri. The church has been very active in foreign missions and publication activities. In contrast with early Pentecostalism, the Assemblies give careful attention to the training of ministers. The church combines congregational and presbyterian forms of government and represents the most cultivated group in movement. Their meetings while emotional have departed from the ecstatic form of the early Pentecostal revival.

The Church of God in Christ is the largest and most influential black Pentecostal body. Though C.P. Jones made a notable contribution, C.H. Mason was the founder and the original leader of this church as a Pentecostal body. The church, organized like the Assemblies of God, publishes a periodical called the *Whole Truth* and had 425,000 members in 1964. Among the other Pentecostal churches using the title "Church of God," one of the more interesting is that over which A.J. Tomlinson* was the general overseer. Starting as a Holiness church in 1886, this Church of God turned Pentecostal and suffered many divisions. Fragmentation is typical of the Pentecostal movement as a whole, and Elmer Clark claims that after Tomlinson's death in 1907 the Church of God divided into more than two dozen organizations.

The International Church of the Foursquare Gospel, with its leader Aimee Semple McPherson,* is the best-known Pentecostal body. The church, organized in 1927, centers at the Angelus Temple located in Los Angeles. The colorful preaching of Mrs. McPherson started the movement, which continues today with 741 churches and over 90,000 members.

Recent developments have excited a lively interest in Pentecostalism. Its impressive growth while the major Protestant churches have been declining has caused concern in many circles. The fact that higher social classes are being attracted to its teachings—coupled with the building of attractive modern church buildings, accredited colleges (such as Oral Roberts University), orphanages, and other institutions—has also brought increasing public attention. In the post-World War II period a spate of new "independent" Pentecostal groups has appeared, including the New Order of the Latter Rain, Wings of Healing, the World Church, the Gospel Assemblies, and the Full Gospel Fellowship of Ministers and Churches, International. In addition to these, practically every major denomination, including the Episcopal, Roman Catholic, and Lutheran churches, now has its own charismatic element. The explosive growth of indigenous Pentecostal churches in Chile, Brazil, and South Africa has caused some to predict that the future center of Christianity will be in the southern hemisphere among non-Caucasian Pentecostals. The "Jesus people" have expressed interest in charismatic experiences, and the Pentecostal antiestablishment, egalitarian approach to women and blacks has made it attractive to a revolutionary age.

BIBLIOGRAPHY: E.T. Clark, *The Small Sects in America* (1949); K. Kendrick, *The Promise Fulfilled: A History of the Modern Pentecostal Movement* (1961); B.R. Wilson, *Sects and Society* (1961); N. Block-Hoell, *The Pentecostal Movement* (1964); J.T. Nichol, *Pentecostalism* (1966); V. Synan, *The Holiness-Pentecostal Movement in the United States* (1971); W.J. Hollenweger, *The Pentecostals: The Charismatic Movement in the Churches* (1972). ROBERT G. CLOUSE

PEPIN III (the Short) (714/15-768). King of the Franks. Son of Charles Martel,* he with his brother Carloman succeeded in 741 to the office of *major domus,* "mayor of the palace." When Carloman retired to a monastery in 747, Pepin received the mayoralty title. Considering himself king in all but name, he secured the sanction of Zacharias,* bishop of Rome, to the setting aside of Childeric III, last of the Merovingians, who in order to maintain the legal fiction of that dynasty was named king of all the Franks in 743 (in fact all power had devolved on the Carolingian line). Zacharias died, however, before he ratified the new dynasty. Boniface* and several other bishops consecrated Pepin at Soissons in 751; his anointing with his two sons, Charles and Carloman, by Pope Stephen, thus confirming the dynasty, took place in 754 at St.-Denis, whose monks had educated Pepin and where he was to die.

Though the more dramatic creation of the new "Roman Empire" awaited the coronation of Pepin's son Charles (Charlemagne*) in Rome in 800, Pepin III had in fact brought into being that interrelationship of W European political power and W Christian religious structure which reshaped and revitalized the remnants of Latin Roman civilization. That such occurred in the face of Arabic Islamic expansion about the Mediterranean and of Eastern Christianity's Iconoclastic Controversy* showed how Pepin actively involved himself in affairs of both church and state. He thus assisted Boniface in the ecclesiastical reform of the whole Frankish church, and assisted Chrodegang,* his relative, in a comparable monastic reform and, through defeat of the Lombards and the recovery of the Eastern-claimed exarchate of Ravenna, in creating by "donation" the Papal States (756).

See P. Laski, *The Kingdom of the Franks* (1971). CLYDE CURRY SMITH

PERFECTIONISM. The teaching that moral or religious perfection (in some cases sinlessness) is not only an ideal toward which to strive, but a goal attainable in this life. Within the Christian tradition, perfectionism has attempted to be faithful to certain neglected scriptural themes (cf. Matt. 5: 48; 1 Cor. 2:6; Eph. 4:13; Col. 1:28; 4:12; Heb. 6:1; 1 John 4:18, etc.). Most proponents have identified Christian perfection with "perfect love." Perfectionism in the early church reveals Gnostic and Platonic influence (cf. Clement of Alexandria*). Origen* developed perfectionism in the direction of ascetic and monastic renunciation of the world. This monastic ideal was dominant in the Middle Ages and remains a powerful force in Eastern Orthodoxy and Roman Catholicism. Jerome* revealed Pelagian influence, and Augustine's* teaching tended toward perfectionism though he drew back from such conclusions in controversy with Pelagianism.* Mysticism* influenced perfectionism and became intertwined with it in the Middle Ages. The Reformers were generally anti-perfectionistic, though perfectionism appeared among some forms of Anabaptism and to some extent in Arminius and his followers. In Anglicanism the perfectionism of William Law* and Jeremy Taylor* deeply influenced John Wesley,* whose teachings on the subject became the central concern of Methodism, the major advocate of Christian perfection. Wesley distinguished absolute perfection from Christian perfection, and defined the latter as freedom from sin only in the sense of "a voluntary transgression of a known law." For Wesley, perfection was received instantaneously by faith and confirmed by the witness of the Holy Spirit. In America a form of perfectionism was advocated by Asa Mahan and C.G. Finney* in the first half of the nineteenth century. At midcentury there emerged from Methodism the American Holiness Movement,* a more revivalistic and rigorist advocate of Wesley's perfectionism. From this developed the Church of the Nazarene, the Wesleyan Church, some forms of Pentecostalism, and other modern advocates of perfectionism.

BIBLIOGRAPHY: J. Wesley, *A Plain Account of Christian Perfection* (rep. 1921); B.B. Warfield, *Perfectionism* (2 vols., 1931-32); R.N. Flew, *The Idea of Perfection in Christian Theology* (1934; rep. 1968); W.E. Sangster, *The Path to Perfection* (1943); G.A. Turner, *The Vision Which Transforms: Is Christian Perfection Scriptural?* (1964);
DONALD W. DAYTON

PERKINS, JUSTIN (1805-1869). American Congregational missionary to Persia. Born in West Springfield, Massachusetts, and educated at Westfield Academy, Amherst College, and Andover Theological Seminary, he was ordained in 1833. He went to Persia under the American Board and there labored among Nestorian Christians. In 1835 he established his mission in Urumiah and founded several schools and a mission press from which issued many of his works, notably *Missionary Life in Persia* (1861). A noted scholar, Perkins translated the Bible into Syriac and was the first to reduce the Nestorian vernacular to writing. He acquired valuable Syriac manuscripts for European libraries.
ROBERT C. NEWMAN

PERKINS, WILLIAM (1558-1602). English Puritan scholar. Born in Marston Jabbet, Warwickshire, and educated at Christ's College, Cambridge, he was fellow there till 1595. He was thereafter lecturer at Great St. Andrews, Cambridge, till his death. A noted preacher and pastor, he influenced many undergraduates who later became Puritan leaders (W. Ames* was perhaps the best known). Though associated with the classical movement, he never publicly advocated a presbyterian polity, but was concerned for pastoral renewal and practical piety. He wrote many popular spiritual guides like *A golden chaine* (1590), which went through numerous editions in England and abroad, as far away as Hungary. A notable systematic theologian, Perkins had a rare capacity for popularization and presenting important issues without trivializing them. In addition to writing substantial treatises like *De Praedestinatione* (1597), which provoked Arminius to reply, Perkins was a prolific commentator on Scripture, a formidable patristic scholar, and polemicist on subjects ranging from Roman Catholicism to witchcraft and astrology. His writing on preaching, the role of the ministry, and collection of cases of conscience had considerable influence in the Church of England and the Netherlands. He was one of the founders of the tradition of English practical divinity which considerably influenced continental Pietism during the seventeenth century.

BIBLIOGRAPHY: T. Wood, *Five Pastorals* (1961); T.F. Merrill, *William Perkins* (1966); I. Breward, *The Work of William Perkins* (1970).
IAN BREWARD

PERPETUA (d.203). Young Carthaginian noble and martyr. Mother of a baby son, she was arrested with four fellow catechumens, all plausibly from one household, including Felicitas (perhaps a slave) who bore a daughter prematurely in prison. They were baptized and joined by their catechist Saturus, with whom four died in the amphitheater. The *Passion of Perpetua* incorporates accounts of their prison experiences, especially visions, by Perpetua and Saturus. Its compiler, probably not Tertullian, was Montanist like the martyrs, but both martyrs and *Passion* were revered in the African Church. The apocalyptic, movingly feminine *Passion* vividly reflects the eschatological conceptions, internal tensions, and liturgical customs of contemporary Carthaginian Christianity.

See Acts of the Martyrs; J. Quasten, *Patrology* 1 (1950), pp. 181-83; T.D. Barnes, *Tertullian* (1971).

D.F. WRIGHT

PERRY, CHARLES (1807-1891). First Church of England bishop of Melbourne, Australia. Educated at Harrow and Trinity College, Cambridge, where he was influenced by Charles Simeon,* he became a fellow of Trinity College. He gained pastoral experience by creating the new parish of St. Paul's, Cambridge, where he was vicar from 1842 to 1847. In 1847 he was chosen as first bishop of the new colony of Victoria. On his arrival in Melbourne there were only three colonial chaplains. During the gold rushes of the 1850s he found great difficulty in supplying clergy or churches for the vast increases of population. Perry's Evangelical convictions actuated him to involve the laity in the government of his diocese. His conference of 1851 was the first of its kind in Australia. He resigned from his diocese in 1876 and became a canon of Llandaff Cathedral. He played an important part in many Evangelical societies and in the foundation of Evangelical theological colleges in Oxford and Cambridge. He published many tracts and sermons against rationalism and ritualism, and he was a speaker at the church congresses from 1874 to 1888.

NOEL S. POLLARD

PERSECUTION. Persecution of Christians has occurred for a variety of reasons. Those persecuted have lost possessions, liberty, or life because they were considered dangerous or offensive. Responsibility for persecution has rested with individuals, mobs, the state, or the church.

In the early period, persecution was initiated partly by Roman emperors and partly by mobs. Reasons for the imperial opposition varied. Nero* (54-68) persecuted Christians to escape suspicion for burning Rome, Domitian* (81-96) out of fear of possible rivals and of Christian influence, Aurelius* (161-180) because he was sympathetic to Stoicism, Decius* (249-251) for the political threat of a growing Christian body, and Diocletian* (284-305) because he feared that Christians were disloyal and an impediment to a reorganization of the state.

The means of persecution similarly varied. Persecution was restricted to imprisonment and execution in Nero's time, but in Diocletian's an edict required that churches be demolished, Scriptures confiscated, clerics tortured, and Christian civil servants deprived of their citizenship and executed if unrepentant. Although mob violence was somewhat restrained during the reigns of Trajan (98-117), Hadrian* (117-38), and Pius (138-61), Christians nevertheless suffered. The legal grounds on which persecution was carried out are obscure. According to rumor, Christians were cannibals, atheists, and incestuous. No proof, however, was ever produced in substantiation. Apparently the *nomen upsum* of Christian became punishable without any other attendant vice, but the refusal to worship the emperor's *nomen* could easily be viewed as politically treasonous. A general law making Christianity a *religio illicita* was probably not in effect before the second century, contrary to what Tertullian* claimed. Galerius's edict of toleration in 311 ended this era of persecution.

With the growth of papalism in the Middle Ages, church and state became intermingled. Thus when Charlemagne* conquered the Saxons in the late eighth century, he forcibly Christianized them at the same time. In 1179 the Third Lateran Council ordered secular rulers to punish heretics, and in 1215 the Fourth Lateran Council gave the same charge to bishops. The notion that religious deviation was best settled by the sword was further confirmed by the Crusades against the Islamic people in Palestine in the twelfth and thirteenth centuries. Heretics, such as the Albigenses in S France and the Waldenses, were also the victims of crusades, many dying in the process.

At first the pursuit of heretics was disorganized, but in 1231 Gregory IX* established the Inquisition.* It reached its peak activity at the end of that century. In 1478 it was resuscitated in Spain, as it was later for the Counter-Reformation.* It survived officially until the nineteenth century, being suppressed in France, for example, in 1808, and in Spain in 1834. Convicted heretics who recanted were allowed to spend the rest of their lives in prison as penance, but those who persisted in their heresy were executed.

The martyrdom of John Hus* (1415) was an ominous reminder to the Protestant Reformers of the difficulties they could encounter. Many thousands are said to have died under Henry VIII.* In the reign of Mary Tudor,* Hugh Latimer,* Nicholas Ridley,* and Thomas Cranmer* were among those who perished. In Europe religious differences erupted in civil wars, principally in the Netherlands and France. Protestants also persecuted Catholics. In Lutheran lands such as Scandinavia, Catholics suffered greatly for their faith. Anabaptists* were also martyred by both Catholics and Protestants. Similarly the death penalty was passed on Michael Servetus* by both Catholics and Protestants, but the authorities in Geneva apprehended him and carried out the sentence.

During the Thirty Years' War (1618-48), persecution and intolerance did much to discredit Christian faith. The secular philosophy of the Enlightenment* resulted, and Christianity was immediately declared the enemy of the people. During the Revolution in France, the land was "dechristianized" and much suffering was incurred. In 1801, however, Napoleon ended this treatment of the church by entering into a concordat* with Rome.

In the nineteenth century, many countries sought to grant effective religious toleration to all. Despite this, harassment of the church still continued in some lands. In France, for example, it was done in the name of liberalism. In 1864 the pope countered by reasserting, in the name of God, his ancient theocratic claims over secular authority. But it was not until the twentieth century that carefully planned persecution was again undertaken by the state. Totalitarian regimes have ruthlessly and fanatically tried to eliminate Christianity. It is impossible to estimate the loss of Christian life under Nazism and Communism, but it would appear to be greater than at any other comparable period in history.

In summary, it may be said that persecution has occurred, first, for religious reasons as heresy or theological deviation has threatened the religious establishment which has then resorted to force as a means of self-protection; second, for political reasons when Christian actions or scruples were interpreted as disloyalty or treason; third, for racial or nationalistic reasons when, for example, the presence of Christians in the Roman Empire was felt to be the explanation of natural disasters and military defeats on the grounds that the nation's gods had thus been alienated; and finally, for ideological reasons as totalitarian regimes have found their philosophy opposed by Christian faith.

BIBLIOGRAPHY: P. Allard, *Histoire des persécutions* (5 vols., 1903-9); H.B. Workman, *Persecution in the Early Church: A Chapter in the History of Renunciation* (1923); A.C. Shannon, *The Popes and Heresy in the Thirteenth Century* (1949); L. Gussoni and A. Brunello, *The Silent Church: Facts and Documents concerning Religious Persecution behind the Iron Curtain* (1954); J. Lecler, *Toleration and the Reformation* (1960); H. Gregoire et al., *Les persécutions dans l'Empire romain* (1964); W.H.C. Frend, *Martyrdom and Persecution in the Early Church* (1965); D. Hare, *The Theme of Jewish Persecution of Christians in the Gospel according to St. Matthew* (1967); L.H. Canfield, *The Early Persecution of the Christians* (1968); F. Norwood, *Strangers and Exiles: A History of Religious Refugees* (1969); J.C. Pollock, *The Faith of Russian Evangelicals* (1969). DAVID F. WELLS

PERTH, FIVE ARTICLES OF (1618). A drastic innovation in Scottish Presbyterian ritual and worship passed by the general assembly under the direct coercion of James VI, who held that Episcopacy was more congenial to his policies. These Articles decreed kneeling at Communion, private Communion in cases of necessity, private baptism in similar cases, observance of the great annual festivals of the church, and confirmation by bishops. The Articles were abolished by the Covenanting assembly of 1638, but after the Restoration of Charles II in 1660 an even more stringent Episcopacy was forced upon Scotland. Such was the eventual reaction to the article on private Communion, nevertheless, that it was not until 1954 that the Church of Scotland general assembly revoked the Act of 1690 which, after the deposition of the Stuart dynasty, prohibited the private celebration of Communion. J.D. DOUGLAS

PESHITTA, see SYRIAC VERSIONS

PETAVIUS, DIONYSIUS (1583-1652). French Jesuit scholar. Born Denis Pétau, native of Orléans, he took his master's degree at sixteen and lectured in philosophy at Bourges (1603-5), became a Jesuit, and read theology at the Sorbonne. He then lectured in rhetoric in the Jesuit colleges at Reims (1609-12), La Flèche, and the *Collège de Clermont* in Paris (1618), taking the chair of dogmatic theology there in 1621 and continuing as librarian after retirement. The editions he produced included works of the fourth-century bishop Synesius (1612), three orations of Julian the Apostate (1614), and the complete works of Epiphanius of Constantia (1622). He attacked Calvinists and Jansenists; the latter had resented his *Dogmata theologia* (4 vols., 1644-50), unfinished at his death. He upheld frequent confession and Communion, found the patristic tradition not infallible, and believed in the progressive development of doctrine. C.G. THORNE, JR.

PETER, ACTS OF, see APOCRYPHAL NEW TESTAMENT

PETER, EPISTLES OF, see EPISTLES, GENERAL

PETER, GOSPEL OF, see APOCRYPHAL NEW TESTAMENT

PETER, PREACHING OF. A work, probably of the late second century, purporting to contain a collection of Peter's sermons written by the apostle himself. No copy of it has survived, but there are fairly extensive quotations in Clement of Alexandria* and Origen.* It seems to have been an orthodox book, which Origen considered might have been genuine, and to have originated in Egypt. It had an apologetic purpose to show the superiority of Christianity to Judaism and paganism. Christians find Christ to be the Law and the Word, and they worship God through Him with direct and perfect knowledge.

PETER CANISIUS, see CANISIUS, PETER

PETER CLAVER, see CLAVER, PETER

PETER COMESTOR (c.1100-c.1180). Scholar, exegete, and historian, hence known as "Peter the Eater" (of books). A priest at Notre Dame in Troyes, he was chapter dean there from 1145 to 1167. By 1160 he belonged to the chapter at Notre Dame in Paris; soon he became chancellor of the cathedral school (1164-68 and 1178-80) and teacher of theology (1164-68). Near the end of his life he retired to the abbey of St.-Victor in Paris as a canon regular (1169). With other Victorines he used allegory and etymology in his study of the Scriptures. His writings included 150 sermons; glosses and commentaries on Lombard's *Sentences;* commentaries on the gospels, Romans, 1 and 2 Corinthians; and *Sententiae de sacramentis.* His most influential work was *Historia scholastica* (1169-73), a Bible history written in the perspective of ecclesiastical and world history. The work earned him the title "master of histories" and was popular for several centuries.

JOHN GROH

PETER DAMIAN, see DAMIAN, PETER

PETER DE BRUYS (d. c.1131). Heretical preacher. Beginning in the insignificant French village of Bruys, he preached against the church of the day. During twenty years he gained a considerable influence in S France, and toward the end of his life he joined forces with Henry of Lausanne.* His teaching is known largely from the hostile abbot of Cluny, Peter the Venerable,* who points to five heretical doctrines. Peter taught that infant baptism was not valid, as only personal faith could bring salvation; that churches are unnecessary, as God hears according to the worthiness of the individual and not of the place; that the cross should not be an object of veneration but rather of execration, as it pointed to Christ's torture; that there is not a Real Presence in the Sacrament; and that sacrifices, prayers, and good works on behalf of the dead have no effect. Underlying this teaching is the belief that the Christian should be prepared to interpret the gospels even against the church (other Scripture he evidently regarded as inferior to the gospels), and a remarkable emphasis upon personal faith as the sole means of salvation.

His following was part of the widespread evangelical ferment of the period, but the iconoclasm of his followers and his own burning of crosses enraged the conservative devotions of the mob, and he was burnt at St. Gilles. Attempts to identify him with the Cathari* fail to do justice to his distinctive teaching which relates faith to salvation more clearly than almost anybody before Luther. His followers were known as "Petrobusians."

See J.C. Reagan, "Did the Petrobusians Teach Salvation by Faith Alone?" in *Journal of Religion* VII (1927), pp.81ff.; and W.L. Wakefield and A.P. Evans, *Heresies of the High Middle Ages* (1969), pp.118ff. C. PETER WILLIAMS

PETER GONZALEZ, see ELMO

PETER LOMBARD (c.1095-1169). Known as "the Master of the Sentences." Born in Lombardy, he was educated at Bologna and went to Paris, where by 1141 he had written a commentary on the Psalms and a gloss on Paul and had become a canon at Notre Dame. In 1159 he was elected bishop of Paris. His fame rests chiefly upon his *Book of Sentences (Libri Quatuor Sententiarum),* finished in 1157 or 1158. The book is basically a compilation with numerous citations to the Church Fathers and to near contemporaries such as Anselm of Laon,* Peter Abelard,* Hugh of St.-Victor,* the *Decretum* of Gratian,* the anonymous *Summa Sententiarum,* and the canons of Ivo of Chartres.* Lombard's great achievement was in organizing these materials into a sound, brief, objective summary of doctrine.

The work is divided into four books: (1) "On the mystery of the Trinity"; (2) "Concerning the creation and formation of corporal and spiritual things and many other items pertaining thereto"; (3) "Concerning the incarnation of the word and other matters pertaining thereto"; (4) "Concerning the sacraments and sacramental signs." He showed originality in arranging his texts, in using various currents of thought, in avoiding extremes between authoritarians and dialecticians, and in presenting the theology of the sacraments. He was one of the first to insist on the number seven as the proper group of sacraments, to distinguish them from sacramentals, and to state that they are not merely "visible signs of invisible grace" but also "the cause of the grace it signifies."

Lombard's work marked the culmination of a long tradition of theological pedagogy. By 1222 Alexander of Hales* had introduced it into his theological course as the standard text, and from here it passed into the curriculum of other universities in Europe to such an extent that all candidates in theology were required to comment on it as preparation for the doctoral degree.

The work continued to be used and commented upon (there were 180 commentaries written on it in England alone) until well into the seventeenth century, when it was finally replaced by the work of Thomas Aquinas* as amplified by Cajetan.* Despite the wide acceptance there were those who opposed Lombard's teachings both during and after his lifetime. Contemporaries such as Robert of Melun* criticized his apparent acceptance of Abelard's* teaching that, in Christ, God is not man, but has humanity. This understanding, called "christological nihilism," was condemned by Pope Alexander III* (1177). In the latter part of the twelfth century Lombard's Trinitarian teaching was opposed by Gilbert de la Porrée and Joachim of Fiore.* Efforts to have his work condemned were unsuccessful, and at the Fourth Lateran Council (1215) Joachimism was anathematized and Lombard was acknowledged as orthodox. There were still disputes, however, and from the thirteenth and fourteenth centuries there remain lists of "articles in which the Master of the Sentences is not commonly held by all."

BIBLIOGRAPHY: E.F. Rogers, *Peter Lombard and the Sacramental System* (1917); S.J. Curtis, "Peter Lombard, a Pioneer in Educational Methods," *Miscellanea Lombardiniana* (1957); P. Delhaye, *Pierre Lombard, sa vie, ses oeuvres, sa morale* (1961); also, there are many fine articles in the review *Pier Lombardo* which appeared between 1957 and 1962. ROBERT G. CLOUSE

PETER MARTYR (c.1205-1252). Dominican reformer. Born of Cathari* parents in Verona, he was converted to Romanism while a student in Bologna. He entered the Dominican Order in 1221, perhaps under Dominic's personal influence, and soon distinguished himself as an ardent, eloquent opponent of the Cathari. He won many back to Roman Catholicism in the cities of N and central Italy, organized cells of laymen to combat heresy, and successively held priorships of several Dominican houses. The papacy recognized his effectiveness and zeal by appointing him papal inquisitor in 1232 and again in 1251. En route from Como to Milan in 1252 he was assassinated by Cathari. The following year he was canonized by Innocent IV. His martyrdom became a favorite theme in late medieval art and was depicted by Titian's work in the Venetian church of SS Giovanni e Paulo. He was named a patron of the Holy See. JAMES DE JONG

PETER MARTYR (Pietro Martire Vermigli) (1491-1562). Protestant Reformer. Described by Beza* as "a phoenix sprung from the ashes of Savonarola," he was born in Florence and dedicated by his father to St. Peter Martyr* (d.1252). Educated by the Augustinian Order, he himself became a monk and later prior of the Neapolitan monastery of San Pietro-ad-aram. Influenced at first by Juan de Valdés,* he soon turned to the writings of Bucer* and Zwingli* and became a serious student of the Bible. B. Ochino* was his companion in his spiritual quest. Because of his unorthodox views he was transferred to Lucca as prior of San Frediano. Here his expository preaching attracted large crowds; and for the monastery he recruited such teachers as G.E. Tremellius and G. Zanchius.

Having fully adopted Protestant views, he fled Italy and took refuge first at Zurich, then at Basle. Then with the help of M. Bucer he became professor of theology at Strasbourg. During his five years here he married Catherine Dammartin, a former nun. With Ochino he answered T. Cranmer's* invitation to visit England. From the government he gained a pension of forty marks, and in 1548 he was appointed regius professor at Oxford. His preaching in Christ Church, where he was a canon, attracted much attention. He entered fully into English religion and had a part in the Prayer Book of 1552, the reform of canon law, and debates on Holy Communion. His views angered many Oxfordshire clergy.

After Mary's accession he was imprisoned, but with S. Gardiner's* help he was allowed to return to Strasbourg, where he regained his professorship. On account of his theology of Holy Communion, however, he felt it necessary to find a new home in Zurich. Here he enjoyed a happy relationship with Bullinger* and the English refugees who were there. He was appointed professor of Hebrew and had a small part in the preparation of the Second Helvetic* Confession.

BIBLIOGRAPHY: For his correspondence with England see *Zurich Letters* (1842-45). Biographies by J. Simler (Latin 1563; ET 1583) and C. Schmidt (1835). A specialized study is P. McNair, *Peter Martyr in Italy* (1967). For eucharistic views see J.C. McLelland, *Exposition of the Sacramental Theology of Peter Martyr* (1957).

PETER TOON

PETER MONGO (Peter Mongus; "Peter the Hoarse") (fifth century). Monophysite bishop of Alexandria, 477-90. Assisting the Alexandrian patriarch Dioscorus* as deacon at the Robber Synod of Ephesus* in 449, Peter reportedly collaborated with other intransigent Cyrillians in the death of Flavian, bishop of Constantinople. Theodore of Antinoe consecrated him bishop of Alexandria. In 482 he accepted the *Henotikon**; subsequently he was recognized as official patriarch in communion with Constantinople. In a delicate balancing act over Christology he claimed to regard Pope Leo I* and Chalcedon* with horror, but avoided terms that would startle officials. Initially the monks in his area opposed him for being too Chalcedonian, but eventually they yielded to him.

JOHN GROH

PETER NOLASCO (c.1182-1249 or 1256). Cofounder of the Order of Our Lady of Ransom (Mercedarians*). Born in Barcelona or Languedoc, probably of a merchant family, he lived a life later obscured by legend and falsified documents. Sometime between 1218 and 1234 (probably in the latter year) he established the Mercedarians on a constitution provided by a Dominican, Raymond of Penafort,* and with his assistance. In 1235 Gregory IX* approved the order under the Augustinian Rule, and King James of Aragon gave his help. The order worked in Spain and Africa to ransom Christians subjugated or terrorized by the Moors. Peter himself was imprisoned in Algiers. He died in Barcelona, and was canonized in 1628. His order now has about 1,000 members.

JOHN GROH

PETER OF ALCÁNTARA (1499-1562). Spanish ascetic. Nobly born Peter Garavita in Alcántara, he studied at Salamanca and joined the Franciscans in 1515. He was closely linked with the controversial discalced* movement within the order, and because of his association it spread to Italy, Mexico, East Indies, and Brazil. A statue in the Vatican acknowledges him to have been the restorer of the order. Known for the severity of his mortifications, he was honored in the autobiography of St. Teresa of Jesus, whom he encouraged latterly in her Carmelite reform. His authorship of the famous *Tratado de la oración y meditación* (1556) has been challenged (the book has gone through 175 editions and many translations). Canonized in 1669, he was made patron saint of Brazil in 1826 and named co-patron of Estremadura in 1962.

C.G. THORNE, JR.

PETER OF ALEXANDRIA (d.311). Martyr. He served as head of the Catechetical School at Alexandria, and succeeded Theonas as bishop about 300. He administered the church from prison during Diocletian's* persecution, and drew up rules for readmission of the Lapsed into the church in his Paschal letter (306), reflecting the milder attitude. The usurpation of his duties by Melitius of Lycopolis led to the Melitian Schism.* Peter excommunicated Melitius at a synod in 306, deciding that Melitian baptism was invalid. He returned after the edict of tolerance (311), but was beheaded on 24 November in Maximin's persecution. He wrote treatises against certain Origenist doctrines, e.g., the preexistence of the soul, the premundane Fall.

Another Peter of Alexandria (d.380), a presbyter who succeeded Athanasius* in 373, is noteworthy for his opposition to the efforts of Basil of Caesarea* to restore unity after the Arian controversy. He later intrigued against the appointment of Gregory of Nazianzus* as bishop of Constantinople.

J.G.G. NORMAN

PETER OF BLOIS (c.1130-c.1204). Author, ecclesiastic, and royal officer. Born at Blois of a noble Breton family, he was probably a student of Robert of Melun.* He studied law at Bologna, and theology and Scripture at Paris, gaining a reputation as a theologian. After acting as tutor and councillor for William II of Sicily (1167-69) and returning to France, he went to England to become archdeacon of Bath, and served also Henry II and Richard, archbishop of Canterbury, whom he represented unsuccessfully at Rome. After the death of Henry II he was secretary to Queen Eleanor (1191-95). His last years were bitter, especially when he was deprived of his offices. His writings include sermons in the allegorical style, commentaries on Scripture, attacks on the morals of the clergy, an appeal to the Third Crusade, and a diatribe against the Jews, among others. He is best known, however, for his letters, addressed to such notable contemporaries as Henry II, John of Salisbury,* Thomas à Becket,* and Innocent III.* Despite Peter's vanity, the letters are generally factual and full of historic interest. MARY E. ROGERS

PETER OF LAODICEA (c.seventh-eighth century). Virtually nothing is known of this Greek theologian who is reputed to have written commentaries on the gospels. Although various passages in ancient manuscripts have been attributed to him, the only certain work of his that is extant is his *Exposition of the Lord's Prayer,* printed as *Exposition in Orationem Domini* in J.P. Migne, *PG* LXXXVI (2), pp. 3321ff.

PETER OF TARANTAISE (d.1175). Archbishop of Tarantaise. Exceptional devotion as a Cistercian monk led to his being made superior of a new house at Taimé before he was thirty. In 1142 he unwillingly became archbishop and immediately began to reform his lax diocese. Unwelcome fame as a reformer and miracle-worker caused him in 1155 to try to return to monastic life disguised as a lay brother. After a year he was discovered and brought back to his see, where he continued to establish hospices for the poor, the sick, and for travelers. He also inaugurated the custom of distributing free bread and soup in the lean months before the harvest (May Bread). He strongly supported Alexander III,* who sent him on a mission to reconcile Louis VII of France and Henry II of England. He died on the return journey to France. C. PETER WILLIAMS

PETER'S PENCE. A tax sent from England to the pope in Rome avowedly with the purpose of helping poor English pilgrims resident in the *Schola Saxonum.* Its origin is not clear, but probably the custom was begun by King Offa of Mercia in 787 on the occasion of the visit of two papal legates when the archdiocese of Lichfield was being created. Supposedly one penny was raised on each hearth (i.e., house). After the Norman conquest the tax continued to go to Rome, but now, it appears, exclusively for the pope. By agreement, £199.6s.8d. was the figure sent from England. After the rejection of papal supremacy by Henry VIII* and his Parliament, Peter's Pence was abolished in 1534. PETER TOON

PETER THE APOSTLE. Peter was among the first disciples whom Jesus called to follow Him. By profession a Galilean fisherman, he was promised that Jesus would make him a fisher of men, a prediction amply fulfilled in subsequent events. He possessed natural gifts of leadership and appears both in the gospels and the Acts as the leader among the twelve apostles. He is consistently portrayed as a man of impulse who could rise to the heights or be plunged to the depths.

Three events in which he figured prominently are significant during the ministry of Jesus. It was he who made the first confession of faith in Jesus as Christ and Son of God at Caesarea Philippi (e.g., Matt. 16:13-20); it was this event which marked the turning-point in the ministry of Jesus. Closely following this, Peter was one of the three who witnessed the Transfiguration and the one who wanted to preserve the experience by constructing booths (e.g., Matt. 17:4). Yet this was the man who denied his Master three times during the trial of Jesus, an action which led to bitter remorse (e.g., Luke 22:54-62). His impulsive nature is vividly illustrated by his action in striking off the ear of the high priest's servant on the occasion of the arrest of Jesus (John 18:10). Peter did not lack courage, and his denial must be judged against this background. The restoration of Peter is described in one of the Resurrection narratives. It came while he was fishing (John 21), a striking parallel to the circumstances of the original call. The instruction to him to become a shepherd of the people of God, given three times, is further borne out by subsequent experiences.

At Pentecost (Acts 2) his sermon, delivered in the power of the Spirit, resulted in the conversion of about three thousand people. He is notable also as the first apostle through whom the first Gentile convert, Cornelius, was admitted into the church (Acts 10). It needed a special vision to persuade him to undertake this mission in a Gentile home, although he had difficulty later over the same problem when he incurred Paul's remonstration because he withdrew from having fellowship with Gentiles at Antioch (Gal. 2:11ff.). In spite of the fact that his missionary work was eclipsed by that of the Apostle Paul, he remained a highly respected leader of primitive Christianity. Since Acts contains no details of his activities after the Council of Jerusalem, it is impossible to be certain what those activities were, and much must be left to conjecture.

Certain traditions concerning Peter have been preserved. Papias* speaks of Mark as Peter's interpreter, and the conviction that he was the eyewitness behind the gospel of Mark had wide support until challenged by the presuppositions of the Form Critics in their interpretation of gospel traditions. Moreover the NT canon contains two epistles under the name of the apostle, which if accepted as authentic enable a fair assessment of his theological position to be made. He was strongly influenced by Pauline theology. But some scholars dispute that either epistle is authentic, although many are prepared to accept 1 Peter but not 2 Peter.

Tradition also associates the death of the apostle with Rome, and this is usually dated at approximately A.D. 68. The further claim of the Roman Catholic Church that Peter founded the church at Rome and was for twenty-five years its bishop is without support in the earliest testimony. The growing ecclesiastical reverence for Peter, however, is reflected in the great quantity of pseudo-Petrine literature which circulated during the second and third centuries. This included among others a gospel, acts, and apocalypse attributed to his name (see APOCRYPHAL NEW TESTAMENT). His influence seems to have been particularly strong among Gnostic and other heterodox groups. DONALD GUTHRIE

PETER THE FULLER (d.488). Monophysite theologian. Reputedly a monk of the convent of the Acoemetae,* where he practiced the trade of fuller (so Alexander of Cyprus, sixth-century monk). Expelled for Monophysitism,* he eventually accompanied Zeno the Isaurian (emperor from 474) to Antioch. There he joined the Apollinarians* and, supported by Zeno, supplanted Martyrius as bishop during his absence in Constantinople (470). Gennadius of Constantinople, however, obtained from Emperor Leo a decree for Peter's exile, commuted later to imprisonment in the convent of the Acoemetae (471). Through Emperor Basiliscus he regained the see (475), only to be deposed again and interned with the Messalians* (477). By assenting to Zeno's *Henoticon** (482), he again—and finally—became patriarch of Antioch. At a council he induced his bishops to assent to the *Henoticon.* According to Theodore Lector, he introduced the recitation of the Nicene Creed* at the Eucharist, the solemn blessing of the chrism,* and the commemoration of the *Theotokos** at every service.

Peter is chiefly remembered for his addition of the words (here italicized) to the Trisagion, viz. "Holy God, Holy Mighty, Holy Immortal, *who wast crucified for us,* have mercy upon us," i.e., Theopaschitism.* The formula became a test of Monophysitism. Despite being a dogmatically objectionable innovation, it ultimately was tolerated as a barrier against Nestorianism.* Pope Hormisdas* regarded it as heretical, but his successor, John II, agreed with Justinian in sanctioning the statement *"unim crucifixim esse ex sancta et consubstantiali Trinitate"* (533), approved by the Council of Constantinople (553).

See W.H.C. Frend, *The Rise of the Monophysite Movement* (1972). J.G.G. NORMAN

PETER THE HERMIT (c.1050-1115). Preacher of the First Crusade. Born near Amiens, he was an ascetic middle-aged hermit when Urban II* announced the crusade, and immediately he began to preach it with evangelical fervor. He gained a remarkable following, mainly among peasants moved by eschatological hopes and economic hardship. In 1096 he set out with some 20,000 people for the Holy Land. Motley and lacking in discipline, they alienated many of the areas through which they passed, and eventually in Asia Minor their lack of military sense led to a horrible massacre at Civitot, at the hands of the Turks. Being absent at the time, Peter survived, and his involvement with the crusade continued and, though he deserted at the siege of Antioch (1098), he was present when Jerusalem was taken (1099). On his return to Europe he founded the monastery of Neufmoutier. As legends proliferated, in less than a century he rather than Urban was popularly regarded as the instigator of the crusade.

See CRUSADES. C. PETER WILLIAMS

PETER THE VENERABLE (c.1092-1156). French abbot and scholar. Born at Auvergne of noble family, he was educated at the monastery of Sauxillanges of the congregation of Cluny, making his profession under its abbot, Hugh of Cluny,* in 1109. He was successively claustral prior at Vézelay, conventual prior at Domène, and finally chief abbot (1122) over 2,000 dependent houses across Europe. Peter effected financial and educational reforms, but could not halt general decline. His interest in studies at Cluny brought opposition from his close friend, Bernard of Clairvaux,* who wanted only prayer and manual work enforced. Peter supported Innocent II against the antipope Anacletus II (a Cluniac monk) and won reconciliation for Peter Abelard* after the Council of Sens (1140). But his attempts to divert the crusading spirit from deed to word failed. Peter traveled to Spain and England twice each, and often to Rome, but frequently withdrew to the hermitage for meditation and study. His sermons and poems show a careful knowledge of Scripture. He wrote treatises against Peter de Bruys,* the Jews, and the Saracens, and he was the first to have the Koran translated into Latin. Both Bernard of Cluny,* who often eclipsed him, and Frederick Barbarossa called him "venerable." C.G. THORNE, JR.

PETRARCH (1304-1374). Early Italian scholar, called "the Father of Humanism." He left the study of law to devote his time to the classics. The great figures of the Greco-Roman period became so real to him that he called Cicero his father and Virgil his brother. A very influential scholar, he wrote many Latin works, searched tirelessly for classical manuscripts, and edited many of them. Although he could not read them, he saved many writings in Greek also. He was the first to call medieval times "the Dark Ages," for he felt that a golden age was dawning when men would "be able to walk back into the pure radiance of the past." He fell in love with a beautiful lady named Laura, but because she was happily married he had to worship her from afar. The sonnets that he wrote to her had a great influence in Italian literature. His religious feelings are expressed in a work called *Secretum* (1352), where he attempts to reconcile piety with a love of the world.

ROBERT G. CLOUSE

PETRI, LAURENTIUS (1499-1573). Archbishop of Uppsala. Born at Örebro, he may have studied at the Carmelite monastery there. He was one of four royal scholars sent by Gustavus Vasa* to Wittenberg, and he helped in the translation of the 1526 Swedish NT. In 1531 he was, amid some protest, elected archbishop of Uppsala, a chapter still Catholic. Two Catholic bishops were, however, replaced by evangelicals in 1536. Petri contributed to the first complete Swedish Bible of 1541. He strove to protect church revenues from state expropriation. In 1558 he attacked Calvinism in his *A Little Instruction ... concerning the*

Eucharist, and sought to transform the Swedish Church after the manner of his beloved Melanchthon* at Wittenberg. Finally in Johann III he found royal company for his biblical, liturgical, and apostolic work. MARVIN W. ANDERSON

PETRI, OLAVUS (1493-1552). Swedish Reformer. Also "Olaus" or "Olaf." Born at Örebro, he (like his brother Laurentius) was educated at Wittenberg and witnessed there the posting of the Ninety-Five Theses.* His 1529 manual was the first vernacular service-book of the Reformation. At the 1527 Diet of Västeras an order for preaching of the pure Word of God ended the Roman ascendancy. Petri prepared a Swedish Mass in 1531, and a collection of songs. His Postils also determined Swedish religious instruction. In 1540 he retired after a failure to report a murder plot against the king. Between 1526 and 1531 he produced sixteen Swedish books (there had before been only eight in the vernacular). He adapted Luther's work, and by his death had transformed the Swedish ecclesiastical scene.

MARVIN W. ANDERSON

PFAFF FRAGMENTS. Published in 1713 by C.M. Pfaff, these four fragments were said to have been found in the Turin library and, moreover, according to Pfaff, they were written by Irenaeus.* These documents dealt with diverse subjects ranging from the Eucharist to a synopsis of the "true gnosis." Since their first appearance in Maffei's *Giornale de' letterati d'Italia,* they were suspected to be pseudonymous, although until the nineteenth century they were regularly quoted and discussed by scholars. In 1900 A. Harnack* showed that they were a fabrication of Pfaff himself. He built his case primarily upon the theology reflected in the documents (it was Pfaff's own Lutheran doctrine of the Eucharist) and certain linguistic considerations—e.g., he showed the dependence of these documents upon the Textus Receptus* and certain defective Greek editions of Irenaeus which were current and available to Pfaff in the early eighteenth century.

WATSON E. MILLS

PFLEIDERER, OTTO (1839-1908). German Protestant scholar. Born in Württemberg, he studied at Tübingen (1857-61) under F.C. Baur* and became an adherent of the Tübingen School.* He studied also in Britain, and in 1870 became chief pastor and superintendent at Jena, then professor ordinarius of theology. In 1875 he took the chair of theology at Berlin. Already he had begun a series of influential works that was to include *Der Paulinismus* (ET 1873), *The Development of Theology Since Kant* (1890), and *The Philosophy of Religion on the Basis of Its History* (ET 1886-88). He lectured in London and Oxford "On the Influence of the Apostle Paul," differing not only with the orthodox position on the consistency of Pauline and other NT theology, but also later with the Tübingen School which posited hostility between the Pauline, Petrine, and Johannine factions in the early church. Pfleiderer considered Pauline theology a very logical outworking of Christian teaching. KEITH J. HARDMAN

PFLUG, JULIUS (1499-1564). Bishop of Naumburg. Born at Cytra, he attended the universities of Leipzig and Bologna and pursued humanistic studies also at Padua. He was in the service of Duke George of Saxony at the Diet of Augsburg (1530), hopeful that Erasmus* and Melanchthon* could find a peaceful solution to the religious split caused by the Lutheran movement. Duke George conferred various benefices on him. Even after death of the duke he continued his efforts at healing the rift. Giving the chalice in the Eucharist to the laity and allowing priests to marry would, in his opinion, bring about a return to Rome. His election as bishop of Naumburg was nullified by Elector John Frederick* for a period of eight years. Pflug attended the Diet of Ratisbon in 1541, where he made suggestions for concessions to Cardinal Contarini.* In 1548 he counseled Charles V* regarding the Interims and participated in the negotiations with the Lutherans. He attended the Council of Trent* in 1551. His *Oratio de Ordiando republica Germaniae* (1562) is a plea for religious unity, the furtherance of political peace, and the strengthening of imperial power. CARL S. MEYER

PHARISEES. With the spiritual failure of the Hasmonean priest-kings, the staunch upholders of the Law divided. An influential minority withdrew from society, becoming known as the Essenes*; the majority remained in ordinary life, though separated from laxer Jews by their strict views on ritual purity and tithing—hence their name *Perushim* ("separated ones"). They opposed Alexander Jannai bitterly, and he avenged himself by crucifying about 800 of their leaders. His widow, Alexandra Salome (76 B.C.) entrusted the leadership of the country to her brother Shimon ben Shetah, a leading Pharisee. He enforced their views, but they lost their political role once the Romans took control and they learned under Herod that their power must be spiritual. Since the Sadducees,* their main rivals, were Temple-centered, they concentrated on the Synagogue, and through it won the support, though not necessarily imitation, of the people at large.

Their spiritual leaders were the Scribes, later called Rabbis, who continued and developed Ezra's principles. Most were ordinary "laymen," but they were never numerous. Josephus* estimates their number in Herod's reign as something over 6,000. They formed a closely knit order into which one had to be initiated and from which one could be expelled for nonconformity. Some points of controversy between them and the Sadducees have been preserved. In most they were defending the interests of townsmen, especially in Jerusalem, against the aristocracy and richer priests. When bitter controversy broke out in their order between the followers of Shammai and of Hillel (early first century A.D.), though the former represented the richer and stricter Pharisees, the victory almost always went to the latter's accommodation of the Law to the needs of the poor. The disasters of A.D. 66-72 broke the power of the Sadducees and the influence of Essenes and Zealots.* So the Pharisaic system became normative Judaism, more especially as it was merely a stricter version of the generally accepted Diaspora outlook (see JUDAISM).

Much controversy has waged recently over the picture the Pharisees painted of themselves and that implied by Jesus' criticism of them. The answer lies firstly in recognizing that the criticism came from Jesus rather than the disciples or Paul. Second, it has been established that "hypocrite" meant at the time an actor rather than a deceiving pretender.

BIBLIOGRAPHY: A.T. Robertson, *The Pharisees and Jesus* (1920); R.T. Herford, *The Pharisees* (1924); G.F. Moore, *Judaism in the First Centuries of the Christian Era* (3 vols., 1927, 1930); L. Finkelstein, *The Pharisees, The Sociological Background of their Faith* (2 vols., 1938); J. Jocz, *The Jewish People and Jesus Christ* (1949); H.L. Ellison, "Jesus and the Pharisees," *Journal of the Transactions of the Victoria Institute* LXXXV (1953); J. Parkes, *The Foundations of Judaism and Christianity* (1960). H.L. ELLISON

PHILADELPHIANS. Seventeenth-century English sect. John Pordage (1607-81), a Berkshire rector, impressed by mystical doctrines associated with J. Boehme,* gathered around him a group which in 1670 was formally named the Philadelphian Society for the Advancement of Piety and Divine Philosophy. A prominent member was Jane Lead (1623-1704), who from 1670 kept a diary entitled *A Fountain of Gardens* and subsequently published *The Heavenly Cloud* (1681) and *The Revelation of Revelations* (1683). The group was dependent on the personality first of Pordage (who had astrological interests), then of Mrs. Lead, and though Oxford graduate and surgeon Francis Lee (1661-1719) tried to extend the work, the sect did not long survive Mrs. Lead's death. J.D. DOUGLAS

PHILARET, DROZDOV (1782-1867). Metropolitan of Moscow. Son of a church cantor, he was educated at the Troitskii *laura,* near Moscow, became lecturer at its seminary in 1803, and took monastic vows in 1808. Ordained in 1809, he held the chair in philosophy in the seminary in St. Petersburg, also lecturing in theology in the Ecclesiastical Academy. Appointed to the Holy Synod (1818), he was made bishop of Jaroslav (1820), then archbishop (1821) and metropolitan (1826) of Moscow. His liberal episcopal career was restricted by the reactionary reign of Nicholas I (1825-55). The work of a gifted theologian, printed sermons rather than books preserve his thought, together with several volumes of letters which demonstrate his administrative judgments. Having early been exposed to and much appreciated Protestant thinking, he protested the Russian Church's insinuation of heresy, even declaring that their official pronouncements were only private opinions, doctrinal decisions being invalid so long as there were no administrative canons. With the liberal reforms of Alexander II he was honored by the production of a manifesto (1861) whereby the czar released the peasants from serfdom. C.G. THORNE, JR.

PHILARET, THEODORE NIKITICH ROMANOV (c.1553-1633). Russian patriarch. A respected soldier and diplomat under his cousin, Theodore I, the last czar of the House of Ruvik, he was later confined under Boris Godunov to the Antoniev monastery (1598-1605). When Godunov was overthrown by pseudo-Demetrius I, Philaret became metropolitan of Rostov (1605) and four years later was made patriarch of all Russia by the impostor pseudo-Demetrius II. Imprisoned by the Poles (1610-18), he was freed under the truce of Deulino, and in 1619 he was enthroned patriarch of Moscow, remaining virtually Russia's ruler until his death, even although his son Michael was the czar, a co-regency. Equalizing taxation, halting peasant migration off the land, and reorganizing the army, he established also theological projects such as a seminary in every diocese and the founding of a patriarchal library. C.G. THORNE, JR.

PHILASTER, see FILASTER

PHILEAS OF THMUIS (d.306). Martyr and first known bishop of Thmuis, Lower Egypt. Of noble birth and great wealth, he held important civil offices before his conversion. Imprisoned during Diocletian's* persecution with three other Egyptian bishops, he addressed a protest to Melitius of Lycopolis whose adherents had invaded their dioceses. Tried before the prefect, Culcianos, he was executed at Alexandria, in company with a Roman official, Philoromus. The *Acta* of his trial and a letter to his people written in prison survive.

PHILEMON, EPISTLE TO, see EPISTLES, PAULINE

PHILIP II (1527-1598). King of Spain from 1556. Only son of Emperor Charles V* and Isabella of Portugal, he was born in Valladolid and educated by clergy in Spain. He grew up grave, self-possessed, and distrustful, loved by his Spanish subjects, but not elsewhere. He married four times: (1) Maria of Portugal (1543), who died in childbirth (1546); (2) Mary I of England (1554)—a marriage of policy; (3) Elizabeth of Valois (1559); (4) Anne of Austria (1570).

On his father's abdication (1556), he became Europe's most powerful monarch, ruling Spain, Naples and Sicily, Milan, the Netherlands, Franche Comté, Mexico, and Peru. He governed his empire from his desk, possessed an unbounded power of work, and an absolute love of reading, annotating and drafting dispatches. He defeated the French at St. Quentin (1557) and assured Spain's ascendancy for a time. He defeated the Turks in the naval battle of Lepanto (1571). In 1580 he obtained the crown of Portugal and Brazil. He reactivated the Inquisition* in Spain, using it to establish his absolute power. Revolt, however, continued in the Netherlands, resulting in the independence of the Dutch republic (1579). He supported the Guises in France against Henry of Navarre, but his intrigues failed. His attempt to conquer England ended in hopeless disaster with the destruction of the Armada (1588).

He possessed great abilities, but lacked political wisdom. He crushed the chivalrous spirit of Spain, and destroyed its commerce by oppressive exactions and by bitter persecution of the industrious Moriscos (whom he expelled, 1570). He was bigoted, morose, and morbidly suspicious, though a tender husband and affectionate to his daughters. He encouraged art and built El Escorial. Under him, and supported by the Jesuits and the Inquisition, Spain became the intellectual, financial, and military spearhead of the Counter-Reformation.

BIBLIOGRAPHY: M. Hume, *Philip II of Spain* (1897); B.J. Kidd, *The Counter-Reformation* (1933); W.T. Walsh, *Philip II, King of Spain* (1938). J.G.G. NORMAN

PHILIP IV (the Fair) (1268-1314). King of France from 1285. He reigned when papal power was beginning its decline, and opposed the Roman Catholic Church's claim to temporal power. In 1296 he began his feud with Pope Boniface VIII over taxing the clergy, and this was renewed in 1301 in a dispute over Bishop Saisset of Pamiers, who had been accused of speaking against the king. In 1302 the latter summoned the States-General, which later approved his condemnation of the papal bull *Unam Sanctum.* Philip's agents captured Boniface and later humiliated him at Anagni. The king finally saw the papacy capitulate, elect Clement V,* and begin the Babylonian Captivity* of the church, namely, the removal of the papacy from Rome to Avignon. In 1307 Philip seized the riches of the church's Knights Templar* and arrested their grand master. At his insistence the order was abolished by Clement in 1312. ROBERT C. NEWMAN

PHILIP, GOSPEL OF, see APOCRYPHAL NEW TESTAMENT

PHILIP, JOHN (1775-1851). Scottish missionary to South Africa. Born at Kirkcaldy, he was a Congregational minister in Aberdeen (1804-19) before beginning his thirty years' work as resident director of the London Missionary Society in South Africa. He exercised strong personal control of all LMS work, making frequent tours and conducting a voluminous correspondence. He aimed to silence critics by improving the quality of missionary work. Eager for expansion, he assisted the Rhenish, Paris, and American Board missions to enter the field.

Philip played a controversial role in colonial politics. His vigorous campaign on behalf of the Hottentots prepared the ground for Ordinance 50 of 1828 which extended civil rights to colored people. He criticized the commando system on the Eastern Frontier, blamed the colonists for the Sixth Frontier War (1834-35), and vigorously opposed the proposed expulsion of the Xhosa from the Ciskei. His influence upon British philanthropists was sufficient to upset this policy, but he would have preferred the extension of British rule, without confiscation of land, to the unsatisfactory Treaty System which emerged. He likewise favored British protection for the Griqua, but obtained only treaties of friendship. He is often condemned as an ignorant negrophile. His information and judgments were sometimes faulty, but few men were better informed. His aggressive and intolerant manner did him harm, as did his unwillingness to admit mistakes and his unsympathetic attitude toward colonists. But these faults are outweighed by his passionate concern for justice and his acute understanding of the colony's true interests. D.G.L. CRAGG

PHILIP NERI (1515-1595). Founder of the Congregation of the Oratory. Born in Florence he was influenced in youth by Dominicans and Benedictines. In 1533 he went to Rome, where he earned his living as a tutor, wrote poetry, studied philosophy and theology. From 1538 he devoted himself to helping the city's sick and poor, and this resulted in the great Trinity hospital. There gathered around him a group which ministered to the needs of the many pilgrims who came to Rome. In 1551 he was priested, and was while living at the clergy house of San Girolamo that he began to establish the Oratory,* which work was later found also in Spain and France. His work did not go unchallenged, for his unconventional methods of talking about faith, his emphasis upon action, and his direct missionary methods horrified many. For him it was love and spiritual integrity, not physical austerity, that counted. It is reported that laughter was a word frequently linked with Philip Neri; this may well account for the success he had in his missionary work. He died in Rome, and was canonized in 1622.

GORDON A. CATHERALL

PHILIP OF HESSE (1504-1567). Landgrave of Hesse. The ablest of Luther's princely supporters, he was born at Marburg, his father dying when he was five. Following quarrels during his minority he assumed power in 1519. He soon proved himself a shrewd ruler and asserted his authority. He first met Luther in 1521 at the Diet of Worms,* but only after his marriage to Christina of Saxony in 1524 did he embrace Protestantism and encourage the Reformation in his state. He defended his new principles at the Diet of Speyer* (1526) and founded the University of Marburg in 1527. Suspecting that a League of Catholic princes was forming against him, he joined Saxony, Nuremberg, Strasbourg, and Uln in a secret understanding in 1529. About this time he became acquainted with Zwingli* and invited the Swiss Reformer to visit Germany to promote the unity of the Lutheran and Reformed churches.

In 1530 he formed with the elector of Saxony the Smalcald League* of Protestant powers for protection against the emperor. At first war with the empire was staved off by diplomacy, but fighting broke out in 1534, and the League was strengthened by the Concord of Wittenberg* in 1536. But the two Protestant confessions failed to agree, and this, together with Philip's bigamous marriage to Margarethe von der Saale in 1540, shattered the prospects of the League's political triumph throughout Germany. For a time Philip deserted his allies and made peace with the emperor, but within a few years he became aware once again of the dangers confronting Protestantism and formed a revived League which lead to the Smalcald War of 1546-47. Military defeat ensued, and Philip threw himself on the emperor's mercy in the interests of his state. Between 1547 and 1552 he was an imperial prisoner and was compelled to assent to the imposition of the Interim on Hesse, permitting Roman Catholic practices. On his release he renewed his efforts to bring about Lutheran-Calvinist unity, while at the same time he worked for a great Protestant federation and gave aid to the Huguenots.* Just before his death he gave a permanent organization to the Hessian Church by the great agenda of 1566-67.

IAN SELLERS

PHILIPPIANS, see EPISTLES, PAULINE

PHILIPPINES. When in 1521 Magellan landed in the central Philippines, he planted a wooden cross on a hill and so "took possession of the country in the name of Spain," while Father Pedro de Valderrama said the first Mass. But serious Catholic missionary work did not begin until 1565, when five Augustinian missionaries arrived with the conquering Spanish army.

Muslim missionaries, however, had been at work for two centuries, and had established sultanates in the southern islands of Mindanao and Sulu. What emerged was an "Islamized paganism," with Islam superimposed on indigenous animism, spiritism, and polytheism. Similarly the Spanish friars achieved often little more than "Christianized paganism." During the early years of the Philippines' colonial experience, Spain was discouraged by the small economic returns from the islands, but the church persuaded the state to remain because of the great potential of the islands for missionary work. Thus most of the colonizing was left to the friars, who in the early years stood against Spanish exploitation and did much for cultural development. But they had so much power that they soon became corrupt and exploitative. By the early seventeenth century the Augustinians had been joined by Franciscans, Jesuits, Dominicans, and Recollects. And within a few years they could claim that most of the population had been baptized.

Because relatively few secular colonizers were prepared to go to the Philippines, the church became an integral part of the colonial government. The friars thus became very wealthy, receiving generous expenses from the state, and taking tribute, fees, food, free labor, and vast areas of land from the Filipinos. From the beginning there was strong prejudice against ordaining Filipinos. Various popes insisted on the need for a national clergy, but the friars successfully used the threat of mass resignation. Eventually, after pressure from the Spanish throne, the first Filipino priest was ordained in 1702. By 1750 about one-quarter of all the parishes were controlled by national priests, a process accelerated when the Jesuits* were expelled in 1768. The consequences were disastrous. Unprepared and unsuitable men were ordained. Resultant scandals led the king to issue a decree in 1776, suspending the secularization of the parishes and allowing Filipino clergy to become only assistants to the friars. Thus by the end of the nineteenth century, less than one-sixth of all parishes were controlled by Filipinos—and these were small, poor and in distant areas.

There were various Filipino protests and revolts against the corruption of the Spanish friars in the seventeenth and eighteenth centuries, but it was the nineteenth century that really produced the freedom fighters of the Filipino Church. In 1841 Apolinario de la Cruz* became the first martyr, executed as a subversive. The revolution in 1868 and subsequent short-lived republic in Spain resulted in a brief spell of liberalism in the colony; but with the restoration of the monarchy in 1870, censorship was revived, the Filipinization of the church was reversed, and demand for political reform was declared treasonable and punishable by death. This seems to have included the demand for Filipino leadership of the church. In 1872 three priests—Jose Burgos,* Mariano Gomez,* and Jacinto Zamora—were executed for precisely this "crime."

But the friars could not reverse the clock. Filipinos had tasted liberalism. The Suez Canal had been opened in 1869, the telegraph had been invented, and new ideas were flowing quickly. A propaganda movement developed, with such inspiring writers as Jose Rizal* and Marcelo del Pilar daring to question traditional Catholic beliefs. They had been influenced by the rationalism and agnosticism of nineteenth-century Masonry, but were primarily opposed to foreign friars and their corruption, not to religion as such. Their writings paved the way in some measure for the arrival of Protestantism in 1899.

The revolution began in 1896; the Spanish-American war of 1898, when the U.S. Navy sank the Spanish fleet in Manila Bay, enabled the revolutionaries to proclaim the republic. But the Treaty of Paris, without consulting the Filipinos, ceded their country to the USA. The Filipinos revolted and were eventually suppressed, but such political upheavals inevitably affected the church. The 1898 revolutionary government expelled the friars, confiscated their lands, and appointed Gregorio Aglipay*—the only clerical member of the revolutionary congress—as head of the Philippine Church. He called an assembly of national clergy to set up a provisional government of the church until the pope would name Filipino bishops. Their request to Rome was ignored. Thus the Philippine Independent Church (PIC) was born, with Aglipay somewhat reluctantly taking the leadership. In 1902 the U.S. Congress paid the friars $7 million compensation for the loss of their lands, and the PIC severed its moorings from Rome. Carrying the torch of nationalism, it drew some two million former Roman Catholics into membership; but in 1906 the supreme court ruled that all the churches they were using should be returned to the Roman Church. This devastating blow seriously weakened the new denomination. Under the theological leadership of Isabelo de los Reyes* the PIC adopted a Unitarian, rationalistic stance, but after his death in 1938 it returned to a more Catholic position and entered (1961) into intercommunion with the Philippine Episcopal Church, with which it now shares a seminary.

The friars, meanwhile, who had vast financial power, invested it in establishing schools and colleges. At first American bishops replaced the Spanish ones, but in 1905 Jorge Barlin was consecrated first Filipino bishop. The religious orders, with the exception of the Jesuits, are still dominated by foreigners. In 1960 Rufinos J. Santos became the first Filipino cardinal.

The first Protestant missionary to settle was James B. Rodgers, an American Presbyterian, who arrived in 1899. He was quickly followed by others, representing most major denominations. They early agreed on a policy of mission polity and divided the country accordingly. The Episcopalians were unwilling to evangelize Roman Catholics, and went only to the Muslims in the south, the Chinese, the Caucasians, and the animistic tribes. The Seventh-Day Adventists, arriving in 1905, refused to observe any comity agreements. The only non-American arrival was the British and Foreign Bible Society. They had already begun translation into the vernacular from Europe in the 1880s, but much more was to be done: the Catholics had attempted no Scripture translation, and there are at least seventy languages.

There has been a strong ecumenical movement from the beginnings of Protestantism. The early founding of the Union Theological Seminary in 1907 led eventually to the forming of the United Church of Christ in the Philippines in 1948. But a stronger movement has been away from unity, aggravated by the militant nationalism of many Filipinos. Congregations often broke away from their too-foreign mother bodies. In 1909, for example, Nicolas Zamora, the first Filipino to be ordained by Protestants, led in the formation of a national independent Methodist Church. The picture is thus one of great fragmentation, seen notably in some seventy-five different Pentecostal groups.

The Philippines has produced its own cults, most significantly the *Iglesia ni Kristo,* or Church of Christ, founded in 1914 by Felix Manalo* and claiming a membership of several million. There are numerous smaller sects, like the Church of the Holy Savior, and a number of groups which venerate Jose Rizal as a "Second Christ."

BIBLIOGRAPHY: F.C. Laubach, *The People of the Philippines* (1925); G.F. Zaide, *Catholicism in the Philippines* (1937); D.E. Stevenson, *Christianity in the Philippines* (1955); J.L. Phelan, *The Hispanization of the Philippines* (1959); E.A. Hessel, *The Religious Thought of Jose Rizal: Its Context and Theological Significance* (1961); A.J. Sanders, *A Protestant View of the Iglesia Ni Cristo* (1962); R.L. Deats, *Nationalism and Christianity in the Philippines* (1967); D.J. Elwood, *Churches and Sects in the Philippines* (1968); G.H. Anderson (ed.), *Studies in Philippine Church History* (1969); P.G. Gowing, *Islands Under the Cross* (1969); A.L. Tuggy and R. Toliver, *Seeing the Church in the Philippines* (1972).

RICHARD DOWSETT

PHILIPS, DIRK (Dietrich) (1502-1568). Mennonite theologian. Son of a Dutch priest, he was well educated and able to use Latin, Greek, and some Hebrew, though it is not known where he studied. He left the Franciscans and converted to Anabaptism in 1533. At the wish of the brethren at Groningen, his elder brother Obbe ordained him an elder in 1534. With Obbe and Menno,* he was a firm opponent of Münsterite doctrines. He wrote extensively and systematically and was probably the leading theologian of the early Dutch and N German Mennonites. But largely because of his greater severity and rigidity, he was somewhat less influential than Menno Simons, and was partially responsible for schism within the Mennonite brotherhood.

KENNETH R. DAVIS

PHILIPS, OBBE (c.1500-1568). Netherlands Anabaptist leader. Brother of Dirk Philips,* he studied medicine. At Leeuwarden he witnessed the execution of the first Anabaptist martyr in the Netherlands, Sicke Freercks (1531). Drawn to Anabaptists influenced by Melchior Hofmann, he was baptized (1533), began to preach and baptize, and was forced to move first to Amsterdam, and then to Delft where he baptized and ordained David Joris (1534). He also baptized and ordained Menno Simons* (c.1536). He remained aloof from the revolutionary Anabaptists responsible for the Münster catastrophe, and for a time led the peaceful Anabaptists, so that they were often called "Obbenites" or "Obbites." Later he withdrew, and Menno Simons became leader of this group. After his death, Philips's *Confession* was published, in which he described his religious development.

J.G.G. NORMAN

PHILIP SIDETES (early fifth century). Historian. Native of Side in Pamphylia, he was ordained deacon in Constantinople by his friend John Chrysostom,* and was later three times unsuccessful in his candidacy for the patriarchate of Constantinople. Between 434 and 439 he wrote *Christian History,* in thirty-six books, treating world history from the Creation to about 426. The ecclesiastical historian Socrates (c.380-450) and later Photius have commented upon it, remarking unsympathetically about his Asiatic style, poor chronological order, and the mass of often unrelated information. Only fragments of the work remain and are valuable as a supplement to Eusebius. These deal with Papias* of Hierapolis, his alleged assertion that the Jews had martyred both John the Divine and James, and the Catechetical School of Alexandria. A refutation of Julian the Apostate's treatises against Christianity and other tracts are apparently no longer extant.

C.G. THORNE, JR.

PHILIP THE ARABIAN (Marcus Julius Philippus) (d.249). Roman emperor from 244. He succeeded after the army had murdered the boy emperor Gordian. He made peace with the Persian Shapur I, but had to deal with the serious invasions of Germanic peoples across the Danube. His troop commander in Illyrium, Decius,* was proclaimed emperor by his men, and in the ensuing battle at Verona, Philip was slain. The Christian Church later recalled Philip as the first Christian emperor, but there is some doubt about the accuracy of the description. Certainly he gave some benefits to the church, which was free of persecution during his reign. The tradition may have originated in the sharp contrast with the reign of Decius, when persecution was severe.

CLYDE CURRY SMITH

PHILO JUDAEUS (or Alexandrinus). Jewish writer. Living in the time of Christ, his blending of OT monotheism with Greek philosophy anticipated early Christian thought, notably on the "Word" in creation. Probably past middle age in A.D. 40, he was included then in a deputation to the Roman emperor after anti-Jewish rioting at Alexandria. Philo's extensive writings include Jewish apologetic, Pentateuchal criticism, and descriptions of Jewish monastic sects, e.g., the Essenes.* Philosophically he embraced the then fashionable amalgam of Stoicism and Platonism. He combined the abstractness of the philosophers' Supreme Being with the intensely personal, moral Yahweh. Similarly his concept of the divine "Word" (Logos*) unites Hebrew and Greek ideas—it represents the creative word of Genesis 1, the personalized "Wisdom" of Proverbs 8, the vehicle of God's activity (cf. Isa. 55:11), the World of Forms (see PLATONISM), and the immanent principle of natural-cum-moral law which Stoicism also called *logos.* Elsewhere Philo uses such terms as Son of God, Ideal Man, and Paraclete.

More Greek than Hebrew, but also anticipating Christianity, is his stress on spirit (*pneuma*) at the expense of body; in Philo this is related to asceticism and to mysticism, a type of religious ecstasy ("sober drunkenness") typical of the age (see also PLOTINUS). Philo links philosophy to the Pentateuch through allegory, a method widely used in the Greek world for giving relevance to mythology and early poetry. Much early Christian exposition of Scripture follows his lead—see e.g., Ambrose* and Origen.* His Jewish loyalties were only intensified by his contact with Greek culture, which he saw as offering favorable conditions for the propagation of Judaism, with its monotheism and strong practical morality, as a world-religion.

BIBLIOGRAPHY: I. Heinemann, *Philons griechische und jüdische Bildung* (1932); E.R. Goodenough, *An Introduction to Philo Judaeus* (1939); H. Lewy (ed.), *Philo* (1946); H.A. Wolfson, *Philo* (1947); C.H. Dodd, *Interpretation of the Fourth Gospel* (1953).

GORDON C. NEAL

PHILOSOPHY OF RELIGION. Since philosophy is a necessary activity of the human mind and religion an actual phenomenon of the human spirit, a philosophy of religion becomes an inescapable discipline. However much it may be emphasized with Bonaventura* that the heart makes the theologian, sooner or later head and heart must seek accord. It will not do for the religious man to be with himself at war. Besides, religion is more than a private monopoly of a privileged few; it is both a historical and universal phenomenon and as such must needs become the subject of inquiry and questioning. Philosophy, as man's reflection upon the existence of the world and the significance of human experiences, arose and could only do so in the context of a certain advanced state of civilized life. It is, therefore, the fruit of society's maturer age, not of its youthful springtime. Religion, on the other hand, is as old as man, so that there is cogency in the remark of Max Müller* that the true history of mankind is the history of religion. It is essentially a reality of human experience. Since, therefore, philosophy has been regarded historically as a reflection on experience in order to apprehend and understand its ultimate meaning, a philosophy of religion has been generally defined as a reflection on religious experience in an effort to discover its final ground.

In remote times there was no clear-cut distinction between religion, ethics, art, and other aspects of man's psychical life. It was when these began to assert their autonomy that many of the problems proper to a philosophy of religion emerged. A philosophy of religion arose consequently when religion came to be taken out of the domain of pure feelings or practical experience and became the object of reflective thought.

In Greek thought, however, the idea of a philosophy of religion did not arise, since hardly any distinction between religion and philosophy was recognized. In Judaism there was religion but little philosophy, although the great prophets did give meaning to the facts which underpinned their faith. It was within post-apostolic Christianity that the philosophical reflection on religious faith began. Yet here the result was not a philosophy of religion in the modern sense, since "religion" itself as a general fact of human experience was not the subject-matter of investigation. What first appeared was more properly a religious philosophy or a philosophical theology—the reflection upon a particular historical religion and that supremely from the perspective of an apologetic.

In Christianity the speculative movement was hastened by influence from without. The various Gnostic systems, for example, challenged the reflective Christian to consider the ultimate ground, meaning, and value of his faith. Thus the Greek Apologists, and the Alexandrian theologians, unashamedly used Greek philosophical concepts in the defense and propagation of the Gospel. The Middle Ages found the Schoolmen philosophizing in support of a foregone conclusion, the dogmas of the church. In the seventh century the English Deists began a search for the principles common to *all* religions and laid the foundation for a philosophy of religion in distinction from a religious philosophy. Kant* sought religion within the bounds of pure reason; and Schleiermacher* within pure feelings. Hegel* was the first to write a philosophy of religion in the modern sense. He saw all forms of religion as manifestations of the absolute religion in the process of becoming. He stressed that religion per se, and as a universal phenomenon, must be taken as the subject-matter of philosophical reflection.

The term "philosophy of religion" appeared in Germany for the first time at the close of the eighteenth century and occurs as a title in J.C.G. Schaumann's volume, *Philosophie der Religion* (1793), and J. Berger's *Geschichte der Religionsphilosophie* (1800). After Hegel, philosophies of religion separated into broad movements, speculative idealist, personal idealist, pragmatist, existentialist, empiricist, phenomenologist.

Consonant with the present dominant approach to the subject of philosophy which is concerned with the analysis of concepts and which operates on the basis of an empirical epistemology, contemporary philosophies of religion have been interested in the problem of theological language and with the empiricist challenge. In Roman Catholic circles, starting with the presupposition that reason is prior to faith, a philosophy of religion has often been equated with natural theology. With the Barthian school, in which reason is excluded as vitiated and depraved, and experience is regarded as an unsatisfactory basis for religious faith, a philosophy of religion is anathema. Christianity, it is here insisted, is not one of the religions, nor a particular manifestation of man's religious apriority. It is not some universal truth, nor some universal religious experience, but a definite fact which as such is opposed to every universal, be it religion or philosophy. But insofar as the Gospel becomes a fact of experience, it becomes at the same time a view on existence and a judgment about the world, and must at least meet challenges to these claims, and be prepared for inquiry into the nature, function, value, and truth of its religious experience, and the adequacy of its theistic faith as an expression of the nature of ultimate reality.

BIBLIOGRAPHY: A. Galloway, *The Philosophy of Religion* (1914); F. von Hügel, *Essays and Addresses* (1921); R. Otto, *The Philosophy of Religion* (1931); E. Brunner, *The Philosophy of Religion* (1937); E.S. Burtt, *Types of Religious Philosophy* (1938); J. Baillie, *Our Knowledge of God* (1939); E.L. Mascall, *He Who Is* (1943); I. Ramsey, *Religious Language* (1957); H.A. Wolfson, *Religious Philosophy* (1961); H.D. Lewis, *Philosophy of Religion* (1965); H.D. McDonald, *I and He* (1966); J. Collins, *The Emergence of a Philosophy of Religion* (1967); H.P. Owen, *Our Knowledge of God* (1969); S.M. Cahn (ed.), *Philosophy of Religion* (1970); D.Z. Philipps, *Faith and Philosophical Enquiry* (1970).

H.D. MC DONALD

PHILOSTORGIUS (c.368-430). Native of Borissus in Cappadocia, he wrote an ecclesiastical history from the Arian heresy (300) to Valentinian II's rise to the Western throne in 425. Like most histories emanating from the late empire, that of the Eunomian Philostorgius was inaccurate and patently biased. According to Photius (*Bibliotheca*, Cod. XI), whose ninth-century epitome is our chief source, Philostorgius's lost work was "less a history than an eulogy of the [Arian and Eunomian] heretics and a defamation against the orthodox." By the early fourteenth century only this epitome reached Nicephorus Callistus. Other fragments survive in the *Passion of Artemius*, an Arian martyr (c.362). Philostorgius's twelve books began respectively with the letters of his name, forming an acrostic. His work has value for its accounts of several important Arians and as a corrective to the partisan views of orthodox church historians. He knew considerable geography and astronomy and had an elegant style.

DANIEL C. SCAVONE

PHILOXENIAN VERSION, see SYRIAC VERSIONS

PHILOXENUS (c.450-523). Monophysite theologion. Born at Tahal, Persia, he was natively named Xenaya. He studied in Edessa when Ibas* was bishop, rejecting the current Nestorianism* of that school. He went to Antioch, where his ardent championship of the *Henoticon** led to his expulsion by the patriarch Kalanaion. Peter the Fuller* appointed him bishop of Hierapolis (Mabbug) in 485. One of the most learned Syrian theologians, he was spokesman for Monophysitism* in the Antioch patriarchate, his views approximating to those of Julian of Halicarnassus. His extensive writings include thirteen *Discourses on the Christian Life*, works on the Incarnation, many letters, and a Syriac version of the NT (508). He was exiled, first to Thrace (c.518), then to Gangra in Paphlagonia where he died violently.

J.G.G. NORMAN

PHILPOT, JOHN (1516-1555). Protestant martyr. Born at Compton in Hampshire and educated at Winchester and New College, Oxford, he qualified in law and then traveled on the Continent. He was nearly arrested by the Inquisition* for expounding heretical (Protestant) notions in controversy with a Franciscan friar. Returning to England, he became archdeacon of Winchester in the reign of Edward VI; but after attacking transubstantiation he fell foul of his bishop and his queen (Mary) and was imprisoned. After examination at Newgate, London, he was burned at Smithfield. He admired Calvin and translated some of his homilies. His extant works were printed by the Parker Society in 1842.

PETER TOON

PHOEBADIUS. First known bishop of Agen in S France. His birth and death dates are uncertain. He became bishop after 347, and was alive in 392. He wrote *Liber contra Arianos* to refute the Arian heresy. At the Council of Arminum* he at first refused to sign the Arian Confession, but was tricked into doing so by Valens. When he discovered the deception he protested strongly and cleared himself. He attended the Council of Valence in 372 and probably that of Saragossa in 380. He also penned *De Fide Orthodoxa contra Arianos* (which may have been written by Gregory of Elvira) and the *Libellus Fidei* which were thought to be part of the Orations of Gregory Nazianzus. He was known to Jerome and written to by Ambrose.

HAROLD LINDSELL

PHOTINUS (fourth century). Bishop of Sirmium. Pupil and former deacon of Marcellus of Ancyra,* learned and eloquent, he was condemned with Marcellus at the Council of Antioch (c.344) and again at a synod in Sirmium (347), but each time remained quietly in Sirmium supported by adherents. He was finally deposed from his see, which he had held for some seven years, and exiled at the Council of Sirmium in 351, after a theological disputation with Basil of Ancyra.* None of his writings have survived, but he seems to have developed Marcellus's doctrine (tending toward Sabellianism*) to an approximation with the views of Paul of Samosata.* Augustine said that Photinus denied Christ's preexistence, though allowing His birth of a virgin and endowment with superhuman excellence. His followers ("Photinians") were condemned by the Council of Constantinople in 381. J.G.G. NORMAN

PHOTIUS (c.820-c.895). Patriarch of Constantinople. Born into an aristocratic iconodule family which had suffered persecution, he attained learning renowned in his own day for its breadth and depth, and confirmed by modern research. He taught in the imperial university and also, as was customary, functioned as a civil servant and diplomat. In 855 he was a member of a diplomatic mission to the Arabs, concerned with exchange of prisoners. After his return, palace intrigues between Theodora and Michael III led to the deposition of Ignatius from the patriarchate in 858. Photius, still a layman, was elected his successor, receiving the ecclesiastical orders within the space of one week, a procedure not without precedent. He was consecrated by Gregory Asbestas, whom Ignatius had deposed.

The whole history of what followed is vitiated by much misrepresentation, about which there is still debate. Photius had been accepted only by way of compromise and, when dissension broke out, canonically deposed Ignatius by a synod. Papal legates, present for a council in 861 dealing primarily with the residual problems of Iconoclasm, reopened the case of Ignatius and confirmed his deposition. But Pope Nicholas I would not accept the action and asked for a further investigation of the Ignatian case. Behind this lay some claims to jurisdiction over Sicily, Calabria, and Illyricum. A synod in Rome in 863 condemned Photius and declared Ignatius patriarch.

Although in 865 reconciliation might have come about, it was prevented by the new problem of the Christianization of Bulgaria, where the khan Boris* upon his conversion had requested missionaries from the West. The differences between East and West in matters of practice came to the fore with the result that Photius in an encyclical letter condemned Latin practices and especially the added phrase *Filioque* in the Nicene Creed,* and a council at Constantinople in 867 deposed and excommunicated the pope. But in the same year, Byzantine politics led to the murder of Michael and the accession of Basil. Photius was deposed, Ignatius restored. Yet the schism with Rome continued, as Ignatius stood firm about the Bulgarian issue. After some years and with papal changes, reconciliation might once more have been effected, but in 877 Ignatius died and Basil reinstated Photius. The legates of John VIII in the synod of 879-80 acknowledged Photius and reversed the earlier condemnations.

Controversy broke out again for obscure reasons. On his accession Emperor Leo VI deposed Photius (886), and Pope Formosus may have excommunicated him in 892. Photius's last years are unrecorded; he died in exile in the last decade of the century. He was a complex character, sometimes high-handed, but he remained on friendly terms with Ignatius, and after his death canonized him. The monuments of his scholarship are the *Amphilochia,* dealing with doctrinal and exegetical questions; the *Bibliotheke,* in which his reading is recorded and from which knowledge of many lost works can be gained; the *Lexicon,* of which the full text has lately come to light; and the books against the Manichaeans, of which the authenticity has been doubted. Photius is recognized as a saint in the Orthodox Church.

BIBLIOGRAPHY: Edition: Migne, *PG,* pp. 101-4; J. Hergenroether, *Photius, patriarch von Konstantinopel* (3 vols., 1867-69); F. Dvornik, *The Photian Schism* (1948) and *The Patriarch Photius in the light of recent research* (Berichte zum XI. Internationalen Byzantinisten-Kongress, 1958); L. Politis, *Die Handschriftsammlung des Klosters Zavorda u. die neuaufgefundene Photios-Handschrift* (1961). J.N. BIRDSALL

PIARISTS. A Roman Catholic order which provided for the free education of the young, especially boys. The name is based on the last word of its formal title, *Regulares pauperes Matris Dei scholarum piarum,* used when recognized as an order by Gregory XV in 1621. Its founder was Jose Calasanze (Joseph Calasanctius), a Spanish nobleman who was ordained in Rome in 1593 after studying law and theology at Lerida and Alcalá. In 1597 he opened the first free elementary school in Europe to educate the children from the streets of Rome. By 1612 he and his helpers were looking after 1,200 children. Following Gregory's recognition of the group as an order, it gained in 1622 the privileges of the Mendicant orders (e.g., the right of members to work or beg for their living). By 1631 the order was working in Italy, Germany, Poland, Hungary, and other places. Its success aroused the jealousy of the Jesuits. From 1645 to 1698 its status was unstable. First reduced to an association or secular brotherhood in 1646, it was restored to an order in 1669 and then to its full mendicant privileges in 1698. It flourished in the seventeenth century in Europe, especially in Spain and its empire. Calasanze was canonized in 1767 by Clement XIII. The order is now governed by a general with four assistant generals. PETER TOON

PICO DELLA MIRANDOLA, GIOVANNI (1463-1494). Italian philosopher. Born count of Mirandola, he was fluent in Latin and Greek by the time he was sixteen, had studied at several universities, and was renowned for his remarkable memory. At the University of Padua (1480-82) he studied both Averroism* and the Kabbalah* the latter interest producing a current of Christian Kabbalism leading to J. Reuchlin* and the Hebrew studies of sixteenth-century Protestantism. After 1484 the Platonic Academy at Florence dominated his thought, but his major objective became the discovery of a unity underlying these various philosophical traditions, a unity that would accord also with Christianity. His major writings reflect both this objective and the humanist-Christian themes of the dignity of man and the incompatibility of astrology with human freedom. Though he became a follower of the Christian reformer Savonarola,* his last work, *De ente et uno*—unfinished—again sought to prove the essential unity of Plato and Aristotle and that all truth and knowledge are one. The depth and range of Pico's learning and vision have earned him the admiration of serious scholars.

KENNETH R. DAVIS

PIDGEON, GEORGE CAMPBELL (1872-1971). First moderator of the United Church of Canada.* Born in Quebec, he was educated at McGill University and Presbyterian College, and served as pastor of influential churches in Ontario for over fifty years, except for an interval teaching practical theology (1909-15). He was a key figure in the foundation of the United Church of Canada, being its first moderator and also the last moderator of the general assembly of the Presbyterian Church of Canada before the union. Pidgeon provided much of the spirit and direction for the new church in its early years. As scholar, preacher, and writer he was vitally concerned with the application of the Social Gospel* to the poor of the cities. ROBERT WILSON

PIERSON, A(RTHUR) T(APPAN) (1837-1911). Presbyterian minister and writer. Educated at Hamilton College and Union Seminary, New York, he held pastorates in New York State, in Detroit, Indianapolis, and Philadelphia. At Bethany Presbyterian, Philadelphia, an institutional church, he organized what grew into the city's First Penny Savings Bank. He wrote for many periodicals, was editor of *Missionary Review of the World,* and was a student of prophecy, missionary history, and comparative religion, as well as a dispensationalist. From 1891 to 1893 he was minister of Spurgeon's Tabernacle, London. A leader in the Bible Conference Movement and the Student Volunteer Movement,* he was also a consulting editor for the Scofield Bible, lecturer in Moody Bible Institute from 1893, and one of the few Americans to speak at Keswick. The Pierson Bible Institute, Seoul, Korea, was an outgrowth of his ministry. His many books deal with biblical criticism, missions, and devotion.

C.G. THORNE, JR.

PIETISM. A movement among Protestants in the seventeenth and eighteenth centuries which emphasized the necessity for good works and a holy life. It began in Germany shortly after the Thirty Years' War* (1618-48) when the churches had become entangled in confessional rigidity, and the time is often called the Age of Orthodoxy or the period of Protestant Scholasticism. The ideas of the Reformers had become so systemized and schematized that there was little comfort to be found in them.

The leader of the Pietist revival was Philipp Jakob Spener* (1635-1705). In 1674 he was invited to write an introduction to a new edition of sermons by Arndt. His work took the form of an independent tract prefixed to the book and entitled *Pia desideria (Pious Longings).* This manifesto of the Pietist Movement condemned the sins of the day and presented six requirements for reformation. These included a better knowledge of the Bible on the part of the people, the restoration of mutual Christian concern, an emphasis on good works, avoidance of controversy, better spiritual training for ministers, and a reformation of preaching to make it more fervent.

Spener's influence spread widely. Some praised and imitated him while others attacked him and even accused him of being a Jesuit. His prestige increased when he was called to be court preacher at Dresden (1686), and his teachings were taken up at Leipzig University where a group led by A.H. Francke* met for prayer and Bible study. When these men were expelled from Leipzig, Spener helped Francke secure an appointment at the University of Halle (1692).

The history of the Pietist Movement next revolves around Francke, who wrote the story of his activities in an account entitled *Pietas Hallensis: or a Public Demonstration of the Footsteps of a Divine Being Yet in the World, in an Historical Narration of the Orphan House and Other Charitable Institutions at Glaucha near Halle in Saxony* (1701; ET 1727). The work that he discusses grew out of his concern for the destitute and deprived people of Halle and its environs. A whole series of institutions were founded, including a school for the poor, an orphanage, a hospital, a widows' home, a teachers' training institute, a Bible school, book depot, and Bible house. Foreign missions were also emphasized, and in 1705 two young men, Bartholomaeus Ziegenbalg* and Heinrich Plütschau,* went to serve in India. From their activities a mission work was established which was directed by Francke until his death. Several other Pietists—such as Count von Zinzendorf* who created the Moravian Church from refugee fragments of Hussitism; J.A. Bengel; and the community that fostered the "Burleburg Bible"—deserve mention.

Historians disagree as to the nature of Pietism. Some feel it was essentially a revival of medieval monastic and mystical piety stimulated by contact with the Puritans. Others believe it represented progress in Lutheranism and looked forward to the modern world. One author finds in it a force that made for the rise of German nationalism. Whatever view one takes, Pietism has fostered a desire for holy living, biblical scholarship, and missions without which Protestantism would be much poorer.

BIBLIOGRAPHY: A. Ritschl, *Geschichte des Pietismus* (3 vols., 1880-86); K.S. Pinson, *Pietism as a Factor in the Rise of German Nationalism* (1934); P.J. Spener, *Pia Desideria* (ET 1964); M. Schmidt, *Das Zeitalter des Pietismus* (1965); F.E. Stoffler, *The Rise of Evangelical Pietism* (1965).

ROBERT G. CLOUSE

PIETROCOLA-ROSSETTI, TEODORICO (1825-1883). Italian poet and patriot. Born at Vasto (Abruzzi), he studied law at Naples University and joined Mazzini's organization, *La Giovane Italia.* Involved in the 1848 uprising, he had to flee and took refuge first at Leghorn, then Lyons,

Paris, and London where he was warmly received by the many exiles in the Mazzinian circle and by his cousin and poet Gabriel Rossetti, whose surname he joined to his. He earned his living by giving Italian lessons and was invited by one of his pupils to attend an Open Brethren meeting, which led to a conversion experience. He did not entirely abandon political activity; he wrote articles in journals, and two very important treatises, *La religione di Stato* and *Il problema religioso,* but Christ became the center of his life. He renounced political success and devoted his life to the preaching of the Gospel, which he considered the only remedy for the sad plight of Italy.

In 1857, after a most solemn commendation meeting, Rossetti left London for Alessandria (Piedmont) to begin, amid all kinds of difficulties and persecution, a successful work of evangelism. This included the teaching and preparing of young converts who gave evidence of real vocation to become evangelists in the newly formed congregations. These were called Free Italian churches, and in spite of a sad division in 1863 (see GAVAZZI and RISORGIMENTO), by 1870 they had grown to more than thirty communities, some large, scattered throughout Italy. Rossetti spent the last years of his life in Florence; his death occurred on a Sunday at the morning meeting after giving a message which, in the words of the hearers, "had led the congregation up to heaven." Generous, warmhearted, impulsive, but also deeply spiritual and highly intelligent, he was also a great lover of children. He wrote many valuable commentaries and hymns still sung today, and which are the best in Italian both for content and poetry. DAISY RONCO

PIGHI, ALBERT (c.1490-1542). Roman Catholic apologist. Born in Kampen, the Netherlands, he graduated from Louvain in 1509 and lived in Paris until he went to Rome in 1522. He is remembered as an opponent of Martin Bucer* and a defender of papal infallibility. In 1538 appeared his *Hierarchiae ecclesiasticae assertio;* three years later came the *Controversarium quibus nunc exagitatur Christi fides* ... which was widely used to answer the Protestant challenge. A posthumous work (1543), *Apologia A. Pighii ... adversus M. Bucer calumnias ...,* was used by Richard Smith in 1550 to attack Peter Martyr* and Bucer, and Pighi was often cited at the Council of Trent.* Peter Martyr chose to answer Pighi in his 1551 *Romans* lectures at Oxford and his 1558 printed commentary. MARVIN W. ANDERSON

PIKE, JAMES ALBERT (1913-1969). American bishop. Born in Oklahoma City, he was educated at the University of Southern California and Union Theological Seminary. He practiced and lectured on law (1936-42). At one time he had studied for the Roman Catholic priesthood, but in 1944 he was ordained deacon in the Episcopal Church; he was "priested" in 1946. He was chaplain and head of the religion department of Columbia University (1952-58) and bishop of California (1958-66). A noted champion of civil rights, planned parenthood, and social reform, he was also a controversial figure, often assailed by fellow churchmen for his rejection of several basic Christian doctrines, including the Virgin Birth. After his son's suicide in 1966 he explored Spiritualism, and in the same year resigned his bishop's office. He was found dead in the Judean wilderness during a Palestinian visit. His writings include *Beyond Anxiety* (1953); *A Time for Christian Candor* (1964); and *You and the New Morality* (1967). ROBERT C. NEWMAN

PILATE, ACTS OF, see APOCRYPHAL NEW TESTAMENT

PILATE, PONTIUS. Roman procurator or governor of Judea beween c. A.D. 26-36, appointed by Emperor Tiberius. Jesus's trial and crucifixion took place somewhere in the middle of Pilate's tenure. His attempt to evade responsibility here (Luke 23:1-25; John 18:28–19:22; and parallel passages), despite his recognition of Jesus' innocence of the Jews' charge of sedition, was caused by his fear of the high priest's power and by his difficult responsibility for the peace of Palestine. Pilate's headquarters were in Caesarea, and Herod Antipas' in Tiberias. The former usually came to Jerusalem with reinforcements at Passover to preserve order among the Jewish crowds, while the latter came to gain favor with his subjects, as the tetrarch of Galilee and Perea. A ruthless ruler, Pilate caused a massacre of Galileans (Luke 13:1), was removed for this, according to Eusebius, and committed suicide at Rome. He was tactless, hot-tempered, and often weak in ruling; to cover his weakness he often resorted to brutal acts. A fourth- or fifth-century book of the Pseudepigrapha, *The Acts of Pilate* (see APOCRYPHAL NEW TESTAMENT), tells of his ending as a Christian, but this is pure fancy. KEITH J. HARDMAN

PILGRIMAGE OF ETHERIA. Etheria was a late fourth-century or early fifth-century nun who made a pilgrimage to the Holy Land, Egypt, Asia Minor, and Constantinople. On the journey she identified various sites traditionally linked with OT events. She also visited a number of churches built on greatly revered places in and around Jerusalem itself—Mount Zion, Calvary, Bethlehem, Gethsemane, the Mount of Olives, and the site of the Finding of the True Cross (which discovery Etheria is one of the first to attribute to Helena*). Her account has preserved valuable information about the liturgical practices of the period. Egypt and Jerusalem both celebrated the Nativity on 6 January, today the feast of the Epiphany. Lent had its own liturgy; so had Holy Week, Easter, Pentecost, and the whole Paschal period. Early references to the various offices of the day are also included—the office before dawn, tierce, sext, none, vespers. At first the manuscript, discovered in 1884, was attributed to St. Silvia, but it was now generally accepted as the work of Etheria. L. FEEHAN

PILGRIMAGE OF GRACE (1536-37). A revolt that began primarily against the ecclesiastical policies of Henry VIII* of England. It had, however, many other causes and facets, for social discontent in the north of England where it took place was rife. Led by Robert Aske* and a number of the leading members of the northern aristocracy such as Sir Thomas Darcy, the rebels who came primarily from Lincolnshire, Yorkshire, and Durham demanded that Henry change both his councillors and his policies. In particular they demanded that the move toward Protestantism be arrested. Although the rebels professed great loyalty to him, Henry defeated them by dividing and then overwhelming them, after which in true Tudor style he exacted a bloody vengeance with over 200 executions. W.S. REID

PILGRIMAGES. Journeys undertaken for religious reasons, usually to shrines or holy places. The concept predates Christianity and is common to several religions, reaching its highest development in Islam.* Although the practice of pilgrimage barely appeared in the earliest Christian centuries, the desire to visit the actual scenes of Christ's earthly life was a natural one. Among the earliest well-known pilgrims to the Holy Land were Constantine* and his mother Helena,* both of whom built churches there. From the veneration of the sites of the Savior's earthly ministry, the concept of pilgrimage was extended to the sites of the martyrdoms of His witnesses, which quickly acquired miraculous associations.

As the cult of martyrs burgeoned, pilgrimages frequently became expeditions to collect relics. For Western Christians, the journey to the East was often out of the question, but Rome, a city associated with both Peter and Paul and many lesser saints, held an irresistible attraction, and pilgrimage was encouraged by the papacy, which eventually stipulated that every bishop must receive his pallium in Rome. Other popular shrines include those of St. Martin at Tours, St. James at Compostella, and St. Thomas at Canterbury. Charlemagne* indicated his concern for pilgrims by erecting a hostel in Jerusalem, with the cooperation of Haroun-al-Raschid.

As life became less precarious in the West, pilgrimages to the Holy Land became more frequent and more organized and included larger numbers of people. Pilgrimages thus formed an impetus to the Crusades, attacks on pilgrims helping to incite the papacy to initiate the movement. The church sought to aid and protect pilgrims by the creation of hostels and by ruling that the person of the pilgrim was inviolate, and that it was a pious act to assist him on his way. Indulgences* were promised to pilgrims to certain shrines, and became general with the crusading movement. This aspect was strengthened in 1300 with the establishment of the Year of Jubilee.* The denunciations of pilgrimages by the Reformers merely echoed the reservations expressed by John Chrysostom, Jerome, Augustine, Gregory of Nyssa, and Boniface, among others. In Protestant areas the movement declined, though the motif reappeared in the great classic, *Pilgrim's Progress.*

BIBLIOGRAPHY: *The Library of the Palestine Pilgrims' Text Society* (13 vols., 1888-97); J.J. Jusserand, *English Wayfaring Life in the Middle Ages* (ET 1892); P. Geyer (ed.), *Itinera Hiersolymitana Saeculi* IV—VIII (*CSEL* XXXIX, 1898); E.R. Barker, *Rome of the Pilgrims and Martyrs* (1913); B. Kötting, *Peregrinatio Religiosa* (1950); A.M. Besnard, *Le Pèlerinage chrétien* (1959); R. Oursel, *Les Pèlerins du moyen âge* (1963). MARY E. ROGERS

PILKINGTON, GEORGE LAWRENCE (1865-1897). Missionary to Africa. Born in Dublin into an old Protestant landlord family, he was educated at Uppingham and Cambridge. A promising classical scholar, he was converted in 1885. Although lacking theological training, he felt called to Christian service and went to Uganda under the Church Missionary Society in 1890. Because of his keen linguistic skills, he undertook the task of translating the Bible into the Luganda tongue. In his personal ministry he stressed the need for the baptism of the Holy Spirit. He advocated in his writings the principle of the self-supporting and propagating indigenous church, and contended most European missionaries should go to those areas where a strong, aggressive national church was active. By reinforcing the work already undertaken by native Christians, they would enhance the effects of the existing spiritual momentum. He was killed in a Sudanese militia uprising in Uganda. RICHARD V. PIERARD

PILKINGTON, JAMES (c.1520-1576). Bishop of Durham. He was born near Rivington in Lancashire, and became a fellow of St. John's College, Cambridge, in 1539 and president in 1550. Being a warm supporter of the Reformation, during the persecutions of Mary he had to flee to the Continent and lived in Zurich, Basle, Geneva, and Frankfurt. He returned to England in 1558 and was the first to sign the "Peaceable Letter" to the English Church at Geneva. He was appointed to the commission for the revision of the Book of Common Prayer.* In 1559 he became master of St. John's College and regius professor of divinity. The following year he was appointed bishop of Durham, the first Protestant to hold that see. He had a hand in the drawing up of the Thirty-Nine Articles* in 1562 and in the diocese pursued a vigorous Protestant policy. He took action against superstitious ornaments, but also allowed many of the buildings in the diocese to fall into ruin. In 1566 he gave a charter to Durham city. In the 1569 northern rebellion he had to flee. He wrote some OT commentaries and a treatise entitled "The Burning of Paul's Church in London by Lightning." R.E. NIXON

PIONIUS (d.250). Martyr. An elder of the Church in Smyrna, he was killed in the Decian persecution. The *Acta Pionii,* recording his trial and burning, are extant in the Greek and in two divergent Latin versions. They represent an almost contemporary document, probably incorporating materials from the martyr's own writing in prison, and known to Eusebius, who erroneously makes Pionius a contemporary of his townsman Polycarp.* The narrative bears impressive testimony to the martyr's character and culture, and gives valuable insight into the treatment of Christians. The surviving *Life of Polycarp* ascribed to Pionius is considered spurious, though accepted by C.J. Cadoux. Itself a legendary compilation, it was apparently intended to be part of a *Corpus Polycarpianum* fathered on the pious and literary Pionius, to which we may owe the preservation of the *Martyrdom of Polycarp.* COLIN HEMER

PIRCKHEIMER, WILLIBALD (1470-1530). German scholar. Born at Eickstätt, he studied at Pavia and Padua. In Nuremberg he promoted the study of the classics as his father, Johann, before him had done. Willibald's sisters were said to have been among the best educated women of Germany. Six of the seven entered religious orders; the most famous was Charitas, who brought about Willibald's return to Roman Catholicism after he had for a few years followed Martin Luther. Three of his five daughters became nuns. His *Bellum Helveticum seu Suitense,* published in 1610, earned him the title of "the German Xenophon." His *Oratio Apologetica,* written a year before his death, is a defense of his sister's convent and the Old Religion. Pirckheimer and Albrecht Dürer* were close friends. Pirckheimer's translations of classical writers and his original historical, scientific, and artistic studies put him into the front ranks of the German humanists.

CARL S. MEYER

PISA, COUNCIL OF (1409). Convoked by cardinals in an effort to end the Great Schism.* Eight of Gregory XII's* cardinals deserted him and joined the Avignonese cardinals to summon this general council in June 1409. The assembly was attended by approximately 500 members, representative of much of the Western Church. England, France, Portugal, Poland, Bohemia, and Sicily were represented, but there was no approval given by Scotland, Scandinavia, Hungary, Castile, Aragon, Ladislas of Naples, or the emperor Rupert in Germany. The council claimed authority and legitimacy on the basis of arguments developed by Cardinal P. d'Ailly* (who had deserted Benedict XIII*), F. Zabarella (created cardinal in 1411), and Jean Gerson.* They declared that the assembly—although not called together by a pope—fully expressed the unity of the church and had power to end the Schism. Peace was maintained because of the presence of Cardinal Cossa (later John XXIII*). The council deposed the existing popes (Gregory XII and Benedict XIII) as heretics and schismatics, and authorized the cardinals to elect a new pope. They elected the Greek cardinal, Peter Philargi, who became Alexander V.* The council also made an effort to deal with Wycliffism and the Bohemian movement.

There has been much criticism of this council. It is not recognized by the Roman Catholic Church as being ecumenical, because of the irregularity of its convocation. It did not end the Schism, which was now further complicated by having three rival popes. It was unable to enforce its decrees because it lacked sufficient support from the secular rulers in the Christian world. It is, however, agreed that this council prepared the way for the final healing of the Schism, which took place at the Council of Constance* in 1415.

JAMES TAYLOR

PISTIS SOPHIA. Gnostic writing in Sahidic. Probably a fourth- or fifth-century translation of a Greek original, a product of late Egyptian Gnosticism.* It has been variously ascribed to the Valentinians, the Ophites, and the Barbelo-Gnostics. It purports to be a revelation of esoteric mysteries made known by the risen Christ to the inner circle of disciples, showing how they may attain to the Light-world and escape the present mixed world which is doomed to destruction. The title is derived from the name of the heroine, "Pistis Sophia," a personification of Philosophy, who is delivered from *Authadēs* (= "self-will" or "arrogance") by Jesus during His ascension through the spheres and in conflict with the aeons. She is led from Chaos by a Power of Light sent by Jesus. Other unconnected sections are appended, including a Gnostic survey of hierarchies, aeons, and spheres, dialogues between Jesus and the disciples (especially with Mary), and discourses reflecting Gnostic sacramentalism. The work is miscellany of fantastic fragments loosely strung together, taking its name from one of the parts. In its own strange way it shows a real devotion to Christ. It exists in a single manuscript in the British Museum. J.G.G. NORMAN

PISTOIA, SYNOD OF (1786). Held under the presidency of Scipione de' Ricci,* and quietly manipulated by Leopold II, grand duke of Tuscany, this was a Jansenist attempt at ecclesiastical renewal. Its fifty-seven points of church reform included a deep desire for catechetical and liturgical revision, a de-emphasized episcopate, reorganization of clergy, just distribution of church goods, and purification of private and public piety. Initially a diocesan project, it attracted clergy and laymen from Europe as a whole. Doctrine was widely discussed, and the proceedings published in many languages. A proposed national council was thwarted, but the synod affirmed the views of Jansen (see JANSENISM), Arnauld,* Quesnel,* and Febronius (see FEBRONIANISM), adopted the Four Gallican Articles,* urged the vernacular, and decentralized authority throughout, thus enabling more popular participation in worship. The bull *Auctorem fidei* (1794) condemned many of the proposals, and Ricci recanted, but the theological concerns expressed through sources, writings, and seminary studies were to find positive development.

C.G. THORNE, JR.

PITRA, JEAN-BAPTISTE (1812-1889). Cardinal and patristic scholar. After studying at the seminary at Autun, he became a Benedictine and in 1843 prior of St. Germain in Paris. He cooperated here for a brief period with J.P. Migne,* the patrologist. Then he traveled widely in search of Latin and Greek MSS and to raise money for the abbey of Solesmes. The pope sent him on an ecumenical mission to Rome after which he became Pius IX's adviser on oriental matters. In 1869 he was appointed librarian at the Vatican, where he cataloged precious Greek codices. In 1879 he was made bishop of Frascati and in 1880 cardinal legate of Monte Cassino. He published important material on the patristic period as well as in canonical law. PETER TOON

PIUS I (d.154). Bishop of Rome from about 140. Information about his life and pontificate is at best sketchy. The Muratorian Fragment lists his brother as the author of the *Shepherd of Hermas*,* while the *Liber Pontificalis* indicates that he was born in Aquileia. The tradition that he was a martyr does not extend to the early period of the church.

PIUS II (1405-1464). Pope from 1458. Born Aeneas Sylvius Piccolomini near Siena, he was the most famous papal representative of Renaissance humanism. After studying at the University of Siena and at Florence, he became secretary to Domenico Capranica and accompanied him to the Council of Basle* (1431-35). Rising rapidly in the council's service due to his oratorical skill, he went on diplomatic missions to England and Scotland and became secretary to the conciliar pope, Felix V. Sensing the growing futility of the council, he entered the service of Frederick III* of Germany (1442). Personal suffering and his conviction that action must be taken against the Turks led him to adopt a more serious lifestyle. He submitted to Pope Eugenius IV (1445) and was ordained (1446) and made bishop (1447). After negotiating the Concordat of Aschaffenburg he was created a cardinal (1456).

Elected pope he preached a crusade against the Turks, who had recently captured Constantinople (1453). He convened a congress of Christendom at Mantua (1458) to formulate plans for the crusade, but received little support from the temporal princes for his project. When the Germans opposed his levy of a crusade tax by pointing to the sins of his youth, he issued his bull, *Execrabilis* (1460), condemning all appeals from the pope to an ecumenical council. In 1464 he personally led a crusade against the Turks, but was stricken with fever and died at Ancona, Italy.

Pius was a brilliant writer and produced a number of prose treatises in defense of conciliarism, poetry, history, fiction, and orations. Among his works are *Historii Frederili Imperatoris; Historica Bohemica, Cosmographiae in Asiae et Europae;* and *Miseriae Curialum*, a series of Latin poems that influenced fifteenth-century English satire. His *Commentaries* are a most valuable legacy as they consist of the only autobiography left by any pope.

BIBLIOGRAPHY: L. von Pastor, *The History of the Popes* (40 vols., 1891-1953); F.A. Cragg (tr.) and L.C. Gabel (ed.), *Memoirs of a Renaissance Pope* (1959); J.G. Rowe, "The Tragedy of Aeneas Sylvius Piccolomini (Pope Pius II)" in *Church History* 30 (1961), pp. 288-313; R.J. Mitchell, *The Laurels and the Tiara: Pope Pius II 1458-1464* (1963). ROBERT G. CLOUSE

PIUS IV (1499-1565). Pope from 1559; brought Council of Trent* to a successful conclusion. Gian Angelo de' Medici, son of a Milanese notary, studied law and medicine at Pavia, and entered papal service in 1527. He was made archbishop of Ragusa in 1545 and cardinal in 1549. He succeeded Paul IV as pope in 1559 and quickly ended nepotism in papal circles. No profound theologian, he was yet a skillful and amiable politician. He reassembled the Council of Trent in 1561. Negotiating privately with the emperor Ferdinand and the French and Spanish kings, he used the majority of Italian bishops in the council to maintain the power of the Curia. He conciliated Ferdinand by conceding Communion in both kinds to the laity in the Empire. The council was dissolved in 1563, thus completing the legal enactment of the Counter-Reformation.* Pius published a new Index* in 1564, and prepared an edition of the Roman Catechism.

J.G.G. NORMAN

PIUS V (1504-1572). Pope from 1566. Born Michele Ghislieri, he entered at the age of fourteen the Dominican Order wherein he later held the offices of master and prior, served the Inquisition in Milan, was appointed commissary general of the Inquisition in 1551, bishop of Nepi and Sutri in 1557, became a cardinal, and with the backing of Cardinal Borromeo* and the Ultramontane party, was elected pope. Fanatical and austere, Pius had as his ideal no less than the refashioning of the whole church on the model of his own household. To this end he vigorously enforced the recommendations of the Council of Trent, revised the Breviary and the Missal, and saw to the republication of the works of Aquinas.* He stamped out the Reformed faith in Italy while encouraging Spain to do likewise in the Netherlands, personally blessing the military campaigns of the duke of Alva. In France he acquired an ascendancy over Catherine de' Medici* and Charles IX, while in England his excommunication of Queen Elizabeth* led to divisions among and persecutions of the Catholic community. He helped promote the alliance of the Papal States,* Spain, and Venice which triumphed in the great naval battle at Lepanto in 1571. He was canonized in 1712.

IAN SELLERS

PIUS VI (1717-1799). Pope from 1775. Born of noble parents, he was educated by the Jesuits. In 1740 he went to Rome as secretary to Cardinal Ruffo and became secretary to Benedict XIV and a canon of St. Peter's in 1755, but was not ordained until 1758. In 1773 he was created a cardinal despite his opposition to the suppression of the Jesuit Order which took place the same year. As pope he contrived by delicate diplomatic efforts to secure the Jesuits' resettlement in Prussia and Russia. His first years were taken up with domestic concerns, but soon he was threatened by an outbreak of national church movements similar to Gallicanism.* In the Holy Roman Empire, Febronianism* spread rapidly with the encouragement of the archbishop-electors, though at the Ems Congress* of 1786 their aims were cleverly frustrated by the pope, and the movement soon came to an end. In Tuscany the grand duke Leopold adopted a similar course which reached its height at the Synod of Pistoia* in 1786. By 1790, however, a clerical reaction was well under way. Most seriously, Josephinism in the Hapsburg Empire led to the pope's journeying to Vienna in 1782 to plead with the reforming emperor, though the following year, after a threat of excommunication, Joseph returned the visit and Pius partially reasserted his authority. Finally when in 1786 Joseph extended his policies to the Spanish Netherlands, the devoutly Catholic inhabitants rose in a clerical-nationalist revolt, and the pope had the satisfaction of seeing princely reforming attempts defied by the people themselves. The French Revolution and the Civil Constitution of the Clergy (1790) led the pope to anathematize the revolutionaries and those clerics who accepted their reforms; this in turn led to the Papal States* being included in the first anti-French Coalition (1796), the invasion and occupation of a portion thereof, the humiliating Peace of Tolentino (1797), the seizure of Rome itself (1798), and the carrying off of the pope's person by the French and his untimely death in the depths of an Alpine winter. It is sometimes asserted that under Pius VI the papacy reached its nadir, but the successes of the papal counter-revolution were considerable and public sympathy for his sufferings strengthened the institution in popular esteem. IAN SELLERS

PIUS VII (1740-1823). Pope from 1800. Born Barnaba Chiaramonti and trained as a Benedictine, when he took the name Gregorio, he held the sees of Tivoli (1782) and Imola (1785) before election to the papacy. There he was immediately confronted by the self-aggrandizing demands of Napoleon, who, in order to secure total hold over France, wanted a new concordat. The new pope was conciliatory and conceived the principle of remaining immovable in exercising and defending his spiritual authority while seeking to accommodate the church to the new forms of society. The Concordat* (1801) did restore the Church in France, but as amended by the Organic Articles* (1802), left Napoleon in complete charge. Pius protested, but then approved a similar Concordat (1803) with Napoleonic northern Italy. Pius, still conciliatory, agreed to give the authority of his personal presence to Napoleon's self-coronation as emperor in Paris (December 1804). The emperor was now intent on full expansion of his regime across Europe, including the universal extension of his statist ecclesiastical system and the absorption of the States of the Church.* The latter process was completed in February 1808; the pope, maintaining constant protest, was bodily seized (1809) and deported to Savona near Genoa, and finally to Fontainebleau (1812) near Paris. Napoleon, who never successfully dominated the pope. released him and Pius reentered Rome (May 1814).

Cardinal Consalvi,* the talented papal secretary of state, induced the Congress of Vienna (1814-15) to effect a near-complete reestablishment of the Papal States. With Pius's support he reorganized the government amid resistance from both Sanfedisti and Carbonari. The church was effectively restored by concordats with Bavaria and Sardinia (1817), Naples and Russia (1818), Prussia (1821), and other Italian and German states. Pius restored the Jesuits (1814) and revitalized Catholic missions in Asia and Latin America. His singular piety and resistance to Napoleon provided an example of devotion for masses of the faithful.

BIBLIOGRAPHY: J.T. Ellis, *Cardinal Consalvi and Anglo-Papal Relations* (1942); A. Latreille, *L'église catholique et la révolution française* (2 vols., 1946-50); J. Leflon, *Pie VII* (1958); E.E.Y. Hales, *Revolution and Papacy, 1769-1846* (1966).

C.T. MC INTIRE

PIUS IX (1792-1878). Pope from 1846. Born Giovanni Maria Mastai-Ferretti, he was archbishop of Spoleto (1827) and bishop of Imola (1832) before his election as pope. He enjoyed the longest pontificate in history and consummated the spiritual renewal of the Roman Church in the nineteenth century. His central task was the identification and promotion of devoutly Catholic faith and practice in distinction from the many non-Christian and anti-Catholic philosophical or societal movements. Two world-historical events summarized his reign: the end of papal temporal power (1859-61, 1870), and the First Vatican Council (1869-70).

Pius IX experienced revolution firsthand in the Revolution of 1848-49 when he was forced by Mazzinians and Garibaldinians to flee Rome. French troops restored him (April 1850) and occupied Rome and its environs with only one interruption until 1870. Nevertheless, uprisings and Sardinian-Italian invasions terminated the temporal power over the States of the Church* (1859-61, 1870) after a thousand-year rule. He maintained a policy of nonrecognition *(non possemus)* of the Italian absorption of the Papal States, which he considered necessary to his spiritual independence.

Out of this experience, Pius identified the principles, including liberalism, democratism, rationalism, anticlericalism, which motivated anti-Catholic assaults, and condemned them in a series of addresses, excommunications, and encyclicals, notably the *Quanta Cura* and the appended Syllabus of Errors* (1864). These were consistent with his first encyclical *Qui pluribus* (1846); contrary to many liberal enthusiasts (1846-48), he never had a sympathy for secular liberalism.

Concomitantly he constructively promoted Ultramontane* renewal of his spiritual power by defining the Immaculate Conception* of Mary (1854), which encouraged wide popular Catholic revival among the faithful; by timely canonizations and papal jubilees; by urging extensive missionary work worldwide; by his own example of obvious piety and faith amid extreme adversity; and by convening the Vatican Council* (1869-70). Pius IX, supported by the mass popular revival, succeeded in centralizing the church in the papacy, especially in the promulgation of Papal Infallibility (1870), and undermining all attempts to continue Gallican and Febronian churches under the new motive of nationalism. He reestablished the hierarchies in England (1850) and the Netherlands (1853); secured favorable concordats with Russia (1847), Spain (1851), and Austria (1855); and established numerous new dioceses. In the process, Pius IX achieved for the church a remarkable independence from state domination.

Except for one diplomatic mission to Latin America (1823-25), he spent his whole life in the central Italian peninsula.

BIBLIOGRAPHY: T.A. Trollope, *The Story of the Life of Pius the Ninth* (2 vols., 1877); R. Aubert, *Le pontificat de Pie IX* (1952); E.E.Y. Hales, *Pio nono* (1954); P. Fernessole, *Pie IX* (2 vols., 1961-63); K.S. Latourette, *Christianity in a Revolutionary Age,* vol. 1 (1969), pp. 266ff.

C.T. MC INTIRE

PIUS X (1835-1914). Pope from 1903. Born Giuseppi Melchiorre Sarto, he became bishop of Mantua in 1884 and patriarch of Venice in 1893. He devoted his pontificate to the continuing spiritual purification of the Catholic Church. His struggle against Catholic modernism in the writings of Loisy,* Houtin, and others, led to its condemnation in the encyclical *Pieni l'animo* (1906), in a sixty-five point summary of its errors (1907), and finally in the requirement that clergy takes an oath against modernism (1910). He followed Leo XIII in recommending Thomism as the Catholic philosophy (1910) and founded the Pontifical Biblical Institute in Rome (1909). He showed some favor to Bernigni's League of St. Pius V, which professed Integralism, but condemned, graciously, Charles Maurras and *Action Française** (1926). To strengthen the faith of ordinary believers, threatened as he saw them by modernism, he promoted renewal in worship and personal devotion: he provided for enhancement of the church's music, including a revival of Gregorian chant (1904), a revision of the Breviary prayers (1911), new devotion to Mary (1904), more frequent Mass for the faithful (1905), an earlier First Communion for children, and better religious instruction (1905). He began the codification of canon law completed under Benedict XV.

The Catholic struggle with the secularist French Third Republic reached a new stage with the break in diplomatic relations (1904), and the French Law of Separation (1905), unilaterally terminating the Concordat of 1801*; among other things, the state confiscated further church properties and transformed its semi-favorable legal stance toward the church into active legal antagonism, Pius condemned the action in the encyclicals *Vehementer Nos* and *Gravissimo officii munere* (1906). He maintained, but softened, the papal *non possemus* policy against the Italian overthrow of papal temporal power, allowing limited Catholic participation in local elections (1905). *Pieni l'animo* (1906) also showed his favored model for politics: Catholic Action,* social action by nonclergy under effective hierarchical control, rather than independent Catholic movements advocated by the new Christian democracy. He was canonized in 1954.

BIBLIOGRAPHY: C. Ledré, *Pie X* (1952); P. Fernessole, *Pie X: Essai historique* (2 vols., 1952-53); V.A. Yzermans (ed.), *All things in Christ* (1954, his papal documents in ET); G. Dal-Gal, *Pius X: The life story of the Beatus* (1954).

C.T. MC INTIRE

PIUS XI (1857-1939). Pope from 1922. Born Ambrogio Damiano Achille Ratti, he earned three doctorates and before election as pope was archbishop of Milan. He significantly advanced Catholic formulation of the church's role in the secular post-World War I era. With the Lateran Treaties (1922), the long and complex question of the end of papal temporal power (1859-61, 1870) attained new status and apparently a conclusion: by mutual agreement between the papacy and Mussolini's Italy, the State of Vatican City was established, a Vatican-Italian concordat was signed, and Italy paid the Vatican a substantial indemnity. For Mussolini, the pact won initial Catholic support of his emerging totalitarian regime, while it gave the Vatican the independence it had sought.

The encyclical *Quadragesimo anno* (1931) developed Catholic societal principles, consistent with Leo XIII, along the lines of subsidiarity and corporativism: a pluralism of societal relationships arranged under the dogmatic and moral teachings of the church. His was a clear alternative to Communism or collectivist Socialism, individualism, and then Nazism. Mussolini eclectically tried to coopt some of the ideas into his Fascist system. Pius elsewhere condemned the principles of Fascism (1931), German Nazism (1937), and Soviet Communism (1937). Pius's advocacy of Catholic Action* (1922, 1928) provided a concrete way for nonclergy to join in social reconstruction in harmony with episcopal control and the church's teachings. He thought of it as "lay" sharing in the apostolic mission of the church. His encyclical *Divini illius magistri* (1929) defined the basis for Catholic school education and argued against exclusive state control of education.

He concluded many concordats, especially establishing the church in the new central E European states, like Poland (1925) and Romania (1927). He defined an accommodating role for the church in secularist France by resuming diplomatic relations (1921) and obtaining a new ecclesiastical agreement (1928). In missions he promoted indigenous episcopal leadership by naming six Chinese bishops (1926) and a Japanese bishop (1927). A number of World Eucharistic Congresses displayed the church's international character.

BIBLIOGRAPHY: *Sixteen encyclicals of His Holiness Pope Pius XI* (1938); R. Fontenelle, *His Holiness Pope Pius XI* (1938); R.J. Miller, *Forty years after: Pius XI and the social order* (1947).

C.T. MC INTIRE

PIUS XII (1876-1958). Pope from 1939. Ordained priest in 1899, Eugenio Pacelli had his first taste of life in the Vatican in 1901 when he entered the secretariat under Leo XIII. He held a succession of important posts before being created a cardinal and the papal secretary of state by Pius XI in 1930. He served as a papal legate to the Eucharistic Congress in Buenos Aires and at other important events in the USA and Europe. Never before had a secretary traveled outside Italy. Pius XI unofficially made it known that he wished his secretary to succeed him, and at the conclave following his death his wishes were realized and Pius XII appeared to give his blessings *Urbe et Orbi.*

According to his admirers he was a very able statesman, a great teacher, a custodian of sound doctrine, a champion of neutrality and internationalism, a follower of Pius XI's policy of concordatory relations, and a militant anti-Communist. To his detractors he was a gifted politician who skillfully adapted traditional curial practice to the circumstances of World War II and its aftermath. On internal church affairs it seemed at first that Pius XII would respond to the call for renewal. This is seen, for example, in the encyclicals of 1943—*Mystici Corporis* (which emphasized the church as mystical body of Christ) and *Divino Afflante Spiritu* (which gave the hope of a return to biblical studies in the church)—and that of 1947, *Mediator Dei*, on the Liturgy. But in 1950 he issued *Humani Generis*, which revoked some concessions made in the sphere of biblical studies, and thereby prepared the way for the proclamation of the dogma of the Assumption* of the Virgin Mary (1950), and the promulgation of the Marian Year for 1954.

His conservatism also was revealed in his project to discover the tomb of Peter beneath the Vatican and his zeal to canonize his predecessors. He elevated Pius X and Innocent I, revived the cause of the sanctification of the Blessed Innocent V and Gregory X, and initiated the cause of Pius IX. He died at his villa in Castel Gandolfo, and his death seems to have been the end of an epoch in the history of the papacy.

BIBLIOGRAPHY: O. Halecki, *Eugenio Pacelli: Pope of Peace* (1951); J.R. Mc Knight, *The Papacy: A New Appraisal* (1952); K. Burton, *Witness of the Light: The Life of Pope Pius XII* (1958); S. Friedlander, *Pius XII and the Third Reich* (1965); C. Falconi, *The Silence of Pius XII* (1970).

PETER TOON

PLAN OF UNION, THE (1801-1852). The scheme to prevent duplication of Presbyterian and Congregational work on the western American frontier fostered the development of Presbyterianism more than Congregationalism. John B. Smith, president of Union College in Schenectady, New York; Eliphalet Nott, a Congregational missionary; and the younger Jonathan Edwards all had a part in its adoption by the Presbyterian and Congregationalist denominations in 1801. The Plan united adherents of the two denominations in the West into congregations with local church government being that of the majority. The minister could be of either denomination. Larger disputes were to be resolved by the presbytery or association or a bidenominational council. Presbyterianism gained most adherents as the Plan developed from the Hudson River to Chicago. The Presbyterian general assembly ended cooperation in 1837, and the Plan was finally ended by its abrogation by a Congregational convention at Albany in October 1852.

EARLE E. CAIRNS

PLATINA, BARTOLOMEO (Bartolommeo de Sacchi) (1421-1481). Humanist scholar and historian. Born in Piadena, Italy, he was a soldier and tutor who traveled to Florence to study Greek. In 1462 he joined the entourage of Francesco Cardinal Gonzaga in Rome, then was appointed to the College of Abbreviators by Pius II in 1464. Soon he was arrested for refusing dismissal by Paul II as a member of Pomponius Laetus's academy and a reputed pagan. Imprisoned again in 1468 for heresy and for plotting against Paul's life, he was acquitted and released in 1469. Sixtus IV named him librarian at the Vatican in 1472. His *Lives of the Popes* (1479), the first systematic history of the popes—by no means a dispassionate one—portrayed Paul II as an opponent of the arts.

JOHN GROH

PLATONISM. It is customary to distinguish three groups among the intellectual heirs in antiquity of the Athenian philosopher Plato (428-348 B.C.), the first his immediate successors, then Middle Platonism (first century B.C. to second century A.D.), and finally the Neoplatonists (third century A.D. onward), who profoundly influenced Augustine of Hippo,* and are discussed further under their most important representatives Plotinus* and Porphyry.* But beyond these bounds Plato's ideas were widely adopted, sometimes adapted to fit other creeds (for some early Christian thinkers who show a debt to Plato, see articles on ATHENAGORAS, CLEMENT OF ALEXANDRIA, JUSTIN MARTYR, and ORIGEN). In the Middle Ages Plato's description of the physical universe in his *Timaeus* was known (and largely accepted) through Chalcidius's Latin translation. The Cambridge Platonists* were seventeenth-century scholars and preachers who used Platonist and Cartesian ideas against Hobbes's materialism.

(1) *Plato.* To Socrates's original but down-to-earth stress on man as moral being, Plato gave an other-worldly twist. To justify an absolute morality, he postulated an eternal, changeless, absolute World of Forms, above and apart from our changing, material universe but related to it as model to imperfect copy. Morals, science, art, etc., must be derived from these eternal principles (known only to the philosopher), not from experience. The body is inferior to the soul, which is immortal, subject to reincarnation, akin to the Forms, and capable of achieving fulfillment only after death. One Form (or Idea) corresponds to each group of things in this world which we call by a common name—e.g., dogs, beds, triangles—but also good, justice, beauty. Plato introduced to Greece the idea of an actively good (but not omnipotent) God and of creation (not *ex nihilo*, but out of preexisting chaos and using the Forms as blueprint); rejected the more immoral myths but not polytheism as such; and developed a version of the cosmological argument (*Laws* 893b ff., see AQUINAS).

(2) *Middle Platonism.* Plato's doctrines were revived from the first century B.C. more with religious than philosophical motives. The moral absolutes, the other-worldly values, the Creator remain, but each thinker makes his own significant adaptations. Influenced by Aristotle,* all postulate a remote, transcendent God (or Mind), incomprehensible except in momentary revelations, indescribable except by negatives, and active in creation only through intermediaries, e.g., a second Mind, the World-Soul, and mul-

tifarious deities, planets, spirits (Albinus). The independent Platonic Forms now become thoughts in the divine Mind. Theories about the cause of evil—matter itself (e.g., Numenius) or an evil soul within matter (e.g., Plutarch)—resemble Gnosticism.*

(3) See NEOPLATONISM.

BIBLIOGRAPHY: W.R. Inge, *The Platonic Tradition in English Religious Thought* (1926); G.M.A. Grube, *Plato's Thought* (1935); R.E. Witt, *Albinus and the History of Middle Platonism* (1937); P. Shorey, *Platonism Ancient and Modern* (1938); R. Kiblansky, *The Platonic Tradition during the Middle Ages* (1939); F. Solmsen, *Plato's Theology* (1942); A. Fox, *Plato for Pleasure* (1945); A.H. Armstrong, *An Introduction to Ancient Philosophy* (1947). GORDON C. NEAL

PLESSIS, JOSEPH OCTAVE (1762-1825). Roman Catholic archbishop of Quebec. Born near Montreal, the son of a blacksmith, he received a classical education at Montreal College and trained for the priesthood at the *Petit Séminaire* in Quebec. Completing these studies, he taught for a time at Montreal College, became secretary to Bishop Briand,* and was ordained a priest in 1786. He became a leader in resisting the predominance of the British over the French, and the British party attempted to hinder his rapid rise in the hierarchy. He was consecrated as bishop-coadjutor in 1801 and as bishop of Quebec in 1806. Enmity between the two groups increased at this time, but at the outbreak of the War of 1812 he urged the French to be loyal to Canada, thus winning the appreciation of the government. In 1814 he was granted a seat in the legislative council, and he used this to further the position of the Roman Catholic population, especially those of French extraction. He was consecrated archbishop of Quebec in 1818. KEITH J. HARDMAN

PLINY'S LETTER TO TRAJAN. Pliny was sent by Emperor Trajan about 112 to reorganize the affairs of the province of Bithynia. In one of his letters to the emperor, Pliny asked questions about the treatment of Christians and thereby supplied information about early Christian life and worship—e.g., he referred to services before dawn and late in the evening on one day (Sunday). Admitting that the "contagion" of Christianity had penetrated both cities and villages and that he had executed some resolute Christians, he asked Trajan how he should deal with those who were accused of Christianity. In reply, Trajan advised him not to search out Christians, but only to punish those accused and convicted.
PETER TOON

PLOTINUS (205-270). Originator of Neoplatonism.* A Greek-speaking Egyptian educated in Alexandria, he established a school in Rome in A.D. 245 which had a wide influence in antiquity and later notably on Augustine of Hippo* (for the later history of Neoplatonism, see PORPHYRY).

As in Middle Platonism, religious motives prevail, but as one of the more successful examples of philosophical monism, Plotinus's system was a useful ally against the dualism of Gnosticism.* He also influenced orthodox explanations of the Trinity, complementing Philo* in many ways, although direct dependence cannot be proved.

Plotinus's single ultimate being serves primarily as a focus for mystical meditation (religious ecstasy), but combines the transcendence of Aristotle's* Unmoved Mover with a creative role as sole cause of the universe, which however avoids the ambiguities attaching to divine Fire in Stoicism. He postulates distinct but inseparable levels ("hypostases") of being, the lower both proceeding from the higher and in turn aspiring toward it. At the top is the One (or the Good), pure Unity without any trace of duality, and therefore strictly nameless and engaged solely in self-contemplation, but producing as a necessary by-product (as the sun by being itself necessarily radiates light) the second hypostasis, the Divine Mind. This has two aspects: Intelligence and its object Truth (the Platonic Forms); it aspires towards Unity, but in doing so necessarily produces the third level, Soul (or Spirit). Soul has a higher aspect which contemplates Truth, and a lower on which the physical world, though itself eternal, depends for existence. This hierarchical "trinity" produces a universe in which neither evil nor matter is an independent principle, but is seen negatively as the point where, because of its distance from the One, creativity (inexplicably) fails.

The individual soul, like the cosmic, must aspire to Truth and avoid bodily indulgence. Because of an original "fall," our souls are subject to reincarnation until completely purified by asceticism, thought, and ecstasy (the soul transcending itself and achieving momentarily direct communion with the One). GORDON C. NEAL

PLUMSTEAD PECULIARS, see PECULIAR PEOPLE

PLUMTRE, EDWARD HAYES (1821-1891). Theological writer, classical scholar, and poet. Born in London, he was educated at University College, Oxford, and later became a fellow of Brasenose. Subsequently he held the chair of pastoral theology at King's College, London. In later years he held benefices in Kent and in 1881 became dean of Wells. He enjoyed a high reputation for classical learning. Apart from his translations of classical plays and books of poems written by himself, he helped in the translation of the Bible for the Revised Version, and contributed commentaries to various series. Theologically he was an eclectic; in his eschatology (*The Spirits in Prison,* 1884) he showed leanings toward the "larger hope." R.E.D. CLARK

PLUNKET, OLIVER (1629-1681). Primate of Ireland. Born in Loughcrew, Meath, he studied in Dublin and Rome, and was professor of theology in Propaganda College (1657-69). After twenty-five years in Rome he was consecrated at Ghent and returned to Ireland in 1670 as archbishop of Armagh and titular primate. An ultramontane who favored the Jesuits, he established in Dublin a school under their management. He labored unceasingly within and beyond his diocese, often barely housed or fed, and had a good relationship with Protestants. During the persecutions, though guiltless, he was hanged for treason by Charles II. His head remains in the Dominican convent at Drogheda, founded in 1722 by his grandniece Catherine. C.G. THORNE, JR.

PLÜTSCHAU, HEINRICH (1677-1747). Cofounder with B. Ziegenbalg* of the Halle-Danish Mission, the first Protestant mission in India. When the two were summoned from theological studies at Halle to become missionaries of King Frederick IV of Denmark, they at first believed that West Africa was the destination. They went instead in 1706 to Tranquebar, the tiny Danish settlement on the coast of Tamil-speaking India. Plütschau concentrated on the Portuguese-speaking congregation, and proved himself an able partner of the more renowned Ziegenbalg. He returned to Europe in 1711.

PLYMOUTH BRETHREN. Though originating in Dublin, they were so named because their first congregation was formed in Plymouth (1831). The beginnings were essentially informal, with many showing a desire to return to the simplicity of apostolic days and worship, and to break down the walls that divided Christians. The movement was a protest against the prevailing conditions of spiritual deadness, formalism, and sectarianism marking the earlier years of the nineteenth century.

Edward Cronin, medical student at Trinity College, Dublin, withdrew from church attendance for a time because he was refused Communion unless he entered into membership with one of the dissenting churches. This he regarded as a denial that "the church of God was one, and that all that believed were members of that one Body." Joined by a small group of like-minded persons, he met with them "for breaking of bread and prayer" in a private house. Others equally disenchanted with existing ecclesiastical conditions soon associated with them, including A.N. Groves,* John Vesey Parnell (afterward Lord Congleton), John Gifford Bellett, and J.N. Darby.* Their studies confirmed them in their belief that they could observe the Lord's Supper without an ordained clergyman. They broke bread simply, recognizing that the Lord, who was present, would guide by His Spirit as to audible participation in the gathering.

The unique character of the meetings created considerable interest and many more attended to inquire further and learn for themselves from Scripture. Numbers grew rapidly, and they used a hired room to accommodate the people. Darby was the outstanding teacher of the group. The gatherings were marked by deep devotion to Christ, zeal for evangelism, and a strong leaning toward prophetic studies. Groves, observing the dominance of Darby, warned him against strengthening the very elements of legalism from which they had withdrawn, but there is no evidence that Darby gave heed to Groves's words.

Among the many who came under Darby's influence were Francis Newman* and B.W. Newton.* The latter began a ministry in Plymouth with others, and the congregation grew to be large and influential. Darby and Newton had never agreed on prophetical interpretation, and on his return from the continent in 1845 Darby visited Plymouth, wishing to change some of the established customs of the church. Later in 1845 Darby initiated a breakaway; but in 1847/8 a more serious division occurred through attacks made against Newton's teaching on Christ's humanity. Though Newton withdrew his doctrine and never taught it again, the tension, discussions, and charges continued, leading to a major division affecting all the churches.

Two members of the Plymouth church applied for fellowship at Bethesda Chapel, Bristol, where George Müller* and Henry Craik were joint pastors. A few demanded that they should be refused as unsound because they had sat under Newton's teaching. The elders, however, claimed the right to examine such visitors for themselves to ascertain if they actually held false doctrine. As an autonomous church they believed they should settle their own affairs without being stampeded into some joint action of which they might disapprove. Darby pressed for division, and those who followed him broke off all relations with those who agreed with Bethesda. From that time Brethren became two distinct groups, the mainstream of the movement (Open Brethren) maintaining its original principles, while the Darbyist group (Exclusive Brethren) became increasingly centralized in government and separatist in relation to other Christians.

Exclusive Brethren have had several divisions among themselves, these coming to a head in recent times through the extremist teachings of James Taylor (d. 1970). There are other, smaller groups who are much less extreme, in no way associated with Taylorism and who repudiate its doctrines. Many of these are hardly distinguishable from Open Brethren.

During the 1880s a movement developed among Open Brethren that sought to reduce to more formal terms matters such as baptism, the Lord's Supper, and church government. These teachings, propagated through a journal named *Needed Truth,* led to a division in 1889, with the movement's taking the name of its magazine. Highly centralized in structure, it tended in a few years to produce small splinter groups and many of its members withdrew, some returning to Open Brethren though not always shedding "Needed Truth" principles. In this way some of the narrower principles they had imbibed were reintroduced into the Open Brethren churches. Gatherings under this influence still refuse other than their own members a place at the Lord's Table, and practice separation from other Christians. The vast majority of the Open Brethren, however, maintain their original "open" principles, mixing freely with other believers.

Their most distinctive gathering is the weekly breaking of bread, when there is freedom for brethren to lead in thanksgiving and prayer, or to participate in other ways, everything bearing in some way on the central purpose of the service, the remembrance of Christ in the Lord's Supper. While no order of clergymen is acknowledged, those who are gifted by God for the public ministry of the Word are gladly recognized, some being set apart for full-time service in evangelism or Bible teaching. Eschatology had prominence in earlier years, but from the first, differing views were held by leaders, and no one prophetic scheme has been imposed on the churches. Open Brethren practice believer's baptism, while the majority of Exclusive Brethren observe infant (or household) baptism.

Missionary concern marked Brethren from the first. Groves, his wife, and some friends journeyed to Baghdad, and later to India, in the cause of the Gospel. From that small beginning has grown a missionary outreach, with a missionary body of about 1,150. This work, known as Christian Missions to Many Lands, is represented by the magazines *Echoes of Service* (Bath, England) and *The Fields* (New York).

Brethren have firmly rejected the term "Plymouth," believing that a distinctive name placed a barrier between them and other Christians. Today, however, many feel that the term "Brethren" or "Christian Brethren" could be used without establishing denominational status or plunging them into sectarianism. Brethren have always exercised an influence among evangelical Christians out of all proportion to their numbers; they can be found in most parts of the world today.

BIBLIOGRAPHY: J.N. Darby, *Collected Works* (ed. W. Kelly, 32 vols., 1867-83) and *The Letters of J.N.D.* (1832-82); W.B. Neatby, *A History of the Plymouth Brethren* (1902); N.L. Noel, *The History of the Brethren* (1936); D.J. Beattie, *Brethren: The Story of a Great Recovery* (1940); H.A. Ironside, *A Historical Sketch of the Brethren Movement* (1942); H.H. Rowdon, *The Origins of the Brethren* (1967); F.R. Coad, *A History of the Brethren Movement* (1968). G.C.D. HOWLEY

PNEUMATOMACHI (Gr. = "fighters against the Spirit"). A fourth-century group which denied the deity of the Holy Spirit. Anticipated by the "Tropici" answered by Athanasius in his letters to Serapion of Thmuis, they came to the fore in 373 when Eustathius* of Sebaste became their leader after breaking his friendship with Basil of Caesarea. They were condemned by Damasus of Rome (374), and their doctrines were attacked by the Cappadocian Fathers and Didymus the Blind of Alexandria. The more moderate among them accepted the consubstantiality of the Son, but the more radical (led by Eustathius) regarded both Son and Spirit as "*like* in substance" or "*like* in all things" to the Father. Formally anathematized with other heresies at the Council of Constantinople in 381, the sect disappeared after 383, victims of the Theodosian antiheresy laws. Some early writers (Socrates, Sozomen, Jerome, Rufinus) regarded Macedonius of Constantinople as their founder, and they are sometimes called Macedonians; but Macedonius disappeared from sight after his deposition by the Arian Council of Constantinople in 360, and there is no known connection with the later sect, unless he worked out the theories in retirement. Possibly his followers amalgamated with the Pneumatomachi. Occasionally they were also called "Marathonians," after Marathonius of Nicomedia, another supporter of this teaching. J.G.G. NORMAN

POISSY, COLLOQUY OF (1561). An assembly of French Roman Catholic prelates and Reformed Protestant theologians, ministers and laymen, convened by the regent and queen mother, Catherine de' Medici.* Since the Council of Trent* had been adjourned in 1552 and was, moreover, clearly not following a policy of compromise with Protestantism, Catherine hoped to achieve religious peace and unity for France by a national program of reform, doctrinal and disciplinary, by calling a national council of the Gallican Church. The papacy prevented such a council by reconvening the Council of Trent. But Catherine went ahead, giving her assembly the designation "colloquy." As neither the Roman Catholic Church nor Calvinism was purely national, and as Catherine was also incapable of appreciating the depth of the doctrinal differences, this attempt at a political solution failed. Instead the Protestants gained an aura of royal acceptance, and religious passions were intensified, leading to open hostilities by 1562. KENNETH R. DAVIS

POLAND. To a degree hardly true of any other modern nation, the history of the church in Poland has been bound up with the history of the state. In 1966 the Poles celebrated the millennium of the baptism of Prince Meiszko, a symbol of Poland's entrance into the Catholic fold. For over a thousand years the population has been ardently Roman Catholic.

Before World War II the Jews comprised about a tenth of the population, but since the Hitlerian massacres their numbers are negligible. The non-Catholic, mainly Protestant, population is only about a tenth of the whole. Their cultural unity seems to have been strengthened, rather than weakened, by the Reformation, which they forcibly resisted, and by the more recent pressures, first by the Germans and later by the Russians. To break down their resistance the Soviets worked through a group of Catholics who were willing to support government policy: the Pax Association, led by Boleslaw Piasecki. Pax engaged in an enormous propaganda campaign through daily, weekly, and monthly publications, giving tacit approval to the repressive measures undertaken. It did not even protest the imprisonment of priests and bishops, including the Polish primate, Stefan Cardinal Wyszynski. In 1958 the cardinal had said, "The Church is against any form of large-scale state ownership.... I see no elements of humanism in socialism." In 1962 he berated government authorities as "enemies of God." Following his imprisonment he was no less adamant. As late as 11 June 1972 he stated that "it was the new Polish leadership rather than the Church that would have to make major compromises," insisting that "real (national) unity can be achieved only through the faith," that it is "neither comprehensive nor justified when attempts are made to destroy unity by leading workers from unity with Jesus Christ."

In December 1970, the party chief, Wladyslaw Gomulka, was replaced by Edward Gierek. Gierek, working through a group of Catholics conciliatory with the Soviets, known as *Znaks*, took the line that "it is now realized that the socialist system is not a temporary phenomenon, but at the same time it is also realized that Catholicism in Poland is not temporary either." Such a statement can only be understood as conciliatory, especially when it is coupled with accession by the government to the cardinal's major demand that full title be given to the church for nearly 7,000 buildings, most of them churches. In 1972 the cardinal began a campaign for the construction of several thousand new churches throughout Poland, fifty in Warsaw alone.

The Protestant Church is hardly a factor in national politics. Whether it will fare better under the present "normalization" of church-state relations is debatable. Government conciliation could open the way for greater acceptance of Soviet economic policies and to a lessening of resistance to the Protestant evangelical witness. There is some evidence that both of these effects are actually taking place.

See P. Fox, *The Reformation in Poland* (1924); and F. Siegmund-Schultze (ed), *Die evangelischen Kirchen in Polen* (1938): full bibliography.

MILLARD SCHERICH

POLE, REGINALD (1500-1558). Cardinal; archbishop of Canterbury from 1556. Son of the countess of Salisbury, niece of Edward IV, Pole studied at Oxford under Thomas Linacre and William Latimer, and received ecclesiastical advancement from Henry VIII* without being ordained. From 1521 he studied in Europe, corresponding with Thomas More* and Erasmus.* Made dean of Exeter in 1527, he refused the sees of both York and Winchester in 1530, and in order to avoid taking sides over the king's divorce went abroad in 1532. He was a friend of Gaspar Contarini* and knew Gian Pietro Caraffa (later Paul IV*). In 1536 Pole published *Pro Ecclesiasticae Unitatis Defensione* against Henry VIII's assumption of supremacy, and was appointed by Paul III* to a committee for the reform of church discipline.

After his ordination as deacon, the pope made him a cardinal and sent him as legate to persuade France and Spain to break with England. In 1540 an act of attainder was passed against him and his family in England, and his mother was executed. In 1542 he was among those appointed to preside at the Council of Trent,* and in 1549 he was nearly elected pope. On Mary's accession, Pole came to England as legate and absolved Parliament from schism. The day after Archbishop Cranmer* was burnt, and two days after being ordained priest, Pole was consecrated archbishop of Canterbury (1556). He supported Mary's persecution without taking an active part, but found the full restoration of Roman Catholicism impossible without the return of monastic property. He died twelve hours after Queen Mary.

See A.M. Quirini (ed.), *Collected Letters of Reginald Pole* (1744; rep. 1967); W. Schenk, *Reginald Pole, Cardinal of England* (1950).

JOYCE HORN

POLLARD, SAMUEL (1864-1915). Missionary to China. Born in Cornwall, son of a Methodist minister, he reached China under the Bible Christian* Mission in 1887 and was appointed to the SW region. He soon began the evangelization of the Miao tribe on the Kweichow-Yunnan border and spent twenty years among them. He reduced the language to writing and prepared literature in the script he devised. His educational efforts were greatly resented by those who gained from their workers' illiteracy, and on one occasion at least Pollard was severely beaten. Before he died of typhoid he witnessed a mass movement of the Miao into the Christian Church.

LESLIE T. LYALL

POLLOCK, ALGERNON JAMES (1864-1957). Plymouth Brethren* minister. Born in Newcastle-upon-Tyne of a distinguished legal family, he was a banker until in his late twenties he was called to the ministry. For the rest of his life he "lived by faith," without stipend and traveling the world as evangelist and teacher. A member of the Glanton Brethren (a branch of Plymouth but not Exclusive), his influence kept them from divisions. He studied Spiritualism and wrote articles, tracts, and books on devotion and prophecy (he was a dispensationalist), and for many years edited *The Gospel Messenger*.

POLYCARP (c.70-155/160). Bishop of Smyrna and martyr. He is depicted in the sources as a faithful pastor, champion of apostolic tradition, and pillar of catholic orthodoxy. The young Irenaeus heard him in Roman Asia describing his conversation with John the Apostle (the Elder?) and other eyewitnesses of Christ, and later prized his link with the primitive era through Polycarp, allegedly made bishop by apostles.

While en route to Rome, Ignatius was warmly received at Smyrna about 110 and afterward wrote from Troas both to the Smyrnaeans and to Polycarp, already their bishop. In a letter to the Roman presbyter, Florinus, another former disciple of Polycarp, Irenaeus mentions his letter-writing ministry, but only one to the Philippians is extant, two-thirds in Greek but complete in a Latin translation. P.N. Harrison has convinced majority opinion that it consists of two letters—one sent soon after Ignatius passed through Philippi, to accompany Ignatius's collected letters requested by the Philippians, and another written after they inquired about (Paul's teaching on) "righteousness." Few accept Harrison's date (c.135-37) for the second letter, which may be little later than the first. It illumines the development of the Philippian community (still without monepiscopacy), but is virtually a catena of quotations and echoes covering at least thirteen NT books and *1 Clement.* It warns against heresy, Docetism,* and avarice, which had corrupted a presbyter, Valens, and his wife. Polycarp visited Rome about 155, agreeing amicably to differ with Bishop Anicetus on the Quartodeciman* issue, and converting Valentinians and Marcionites. He clashed with Marcion* in person, at Rome or earlier in Asia.

The *Martyrdom of Polycarp,* a letter from the Smyrnaeans to the Church of Philomelium in Phrygia and "to all the Christian congregations in the world," is the earliest extant "acts" of a martyr, compiled by Marcianus (Marcion) within a year of the event from eyewitness accounts. Endless discussion of the dating in the appendix has yielded no consensus. The day was 23 February (22 if a leap year). Many argue for 154-60, others for the 160s or even 177. Precision would be invaluable because Polycarp professed to have been a Christian for eighty-six years, which probably indicates his age (and perhaps his infant baptism). The *Martyrdom* may have undergone interpolation or redaction. It presents Polycarp as an imitator of Christ, at times fancifully, but its historical value (e.g., on the embryonic martyr cult) and spiritual stature are unquestionable.

BIBLIOGRAPHY: See APOSTOLIC FATHERS for editions (especially Lightfoot) and English translations, and for *Martyrdom* see ACTS OF THE MARTYRS; ed. T. Camelot (*Sources Chrétiennes* 10, 4th ed., 1969), ET W.R. Schoedel (1967). See also C.P.S. Clarke, *St. Ignatius and St. Polycarp* (1930); P.N. Harrison, *Polycarp's Two Epistles to the Philippians* (1936); J. Quasten, *Patrology* 1 (1950), pp.76-82; P. Meinhold in Pauly-Wissowa-Kroll, *Realencyklopädie der klassichen Altertumswissenschaft* 21 (1952), 1662-93.

D.F. WRIGHT

POLYCHRONIUS (d. c.430). Bishop of Apamea in Syria and brother of Theodore of Mopsuestia.* One of the most prominent of the exegetes of the Antiochene School, he wrote commentaries on Job, Daniel, and Ezekiel of which only fragments have been preserved. Considered a heretic though never formally condemned, he criticized the Alexandrine method of biblical exegesis.

POLYCRATES (fl. c.190). Bishop of Ephesus. He was excommunicated for his leadership of Asia Minor Quartodecimanian* churchmen against Pope Victor's* (188-99) encyclical placing Easter uniformly on Sunday. Victor's strong stand against Quartodeciman tradition, which allowed a variable feast day after Jewish custom, was disapproved by Irenaeus and generally. Passover was not celebrated at all in Rome until Soter, Victor's predecessor, first celebrated Easter on Sunday. Nothing more is known about Polycrates.

POLYGLOT BIBLES. Bibles which print the text in several languages. One might point back for the origins, at least in intention, to the *Hexapla* of Origen in the third century, but from Origen to printing no further known experiments of like proportions were attempted, because of the difficulties involved in hand-copying. Origen's work did not completely disappear, and various bilingual fragments of Scripture portions also survive. In 1502 Francisco Jiménes* de Cisneros began a comprehensive edition of Scripture; his death in 1517, and the delay in obtaining papal sanction, postponed publication of his six-volumed "Complutensian Polyglot"* until 1522. Its OT presented a revised Hebrew Massoretic Text, the Lucianic version of the Greek text, and the Vulgate Latin. Its NT offered Greek and Latin. The sixth volume added dictionaries and a grammar, completing a pattern which the derivative polyglots were to follow.

From Antwerp under the patronage of Philip II* (hence *Biblia Regia*) and the editorship of Arias Montanus, a polyglot was printed by Christopher Plantin (8 vols., 1569-72). There were added to the OT—except for Daniel, Ezra-Nehemiah, and Chronicles—Targumim with Latin translation; to the NT, the Syriac with Latin rendering; to the helps extended, treatises of a philological and archaeological nature.

From Paris under the editorship of J. Morinus, G.M. LeJay republished an enlarged Antwerp polyglot (10 vols., 1629-45). The NT had both Syriac and Arabic versions; additional volumes contained the Samaritan Pentateuch with its Samaritan Targum, Gabriel Sionita's edition of the Peshitta, and the Arabic version of the OT—each with a Latin translation.

From London with public subscription, Brian Walton,* assisted by Thomas Hyde, edited the *Biblia Sacra Polyglotta* (6 vols., 1653-57), which included Ethiopic Psalms and Persian gospels, and had available for notes from collation the first uncial manuscript—that of *Alexandrinus,* received by James I* from Cyril Lucar*—an addition which stimulated manuscript collection and modern textual criticism. In 1699 as two supplementary volumes E. Castellus's *Lexicon heptaglotton,* a dictionary of Hebrew, Aramaic, Syriac, Samaritan, Ethiopic, and Arabic, to which separate Persian and a pioneering effort at comparative Semitics were appended. Other than these four, while polyglots are numerous, only parts of Scripture have received this kind of multilanguage treatment. CLYDE CURRY SMITH

PONTIFEX MAXIMUS. The term of highest office in the ancient religion of Rome, indicating the highest priest of the city and its cult, passed into the hands of Octavian in 12 B.C. Subsequent emperors, as is documented numismatically, bore the title until the termination of the Western half, by which time Rome itself had been abandoned as seat of government—though it had been refused by Gratian,* who is also noted for his supposed recognition of the Roman bishop's primacy. It was only natural, then, that the Christian bishop of the city should receive this designation among many others, just as he remained the ranking officer therein—a situation mockingly anticipated by Tertullian. A fascinating play with the term occurs in Erasmus's *Julius Excluded.*

CLYDE CURRY SMITH

PONTIUS PILATE, see PILATE

POOLE, MATTHEW (1624-1679). Biblical commentator. Born at York and educated at Emmanuel College, Cambridge, he became rector of St. Michael-le-Querne in London. In 1658 he devised and set in motion a scheme for the training of young men for the Christian ministry, but this came to an abrupt halt at the Restoration. Of strong Presbyterian sympathies, he was ejected in 1662, but he did little Nonconformist preaching afterward. Rather, his energy was devoted to study and the production of the *Synopsis*, a Latin compendium of textual commentary and interpretation (*Synopsis Criticorum aliorumque Sacrae Scripturae Interpretum*, 5 folio vols., 1669-76). Later he wrote *Annotations upon the Holy Bible* in English, a work that was completed by his friends after his death in Holland.

PETER TOON

POOR CLARES *(Les Clarisses).* The second Order of St. Francis,* founded by him and St. Clare* about 1213. Beginning at the church of St. Damien, the order spread rapidly through Italy and into France and Spain. Cardinal Ugolino (later Gregory IX*) placed Clare and her nuns temporarily under the Rule of St. Benedict, adding some very strict austerities, e.g., perpetual fasting, lying on boards, and almost complete silence. In 1224 Francis gave a written rule to Clare easing some of these restrictions. In 1247 and in 1253 further rules were sanctioned calling for complete poverty of the individual and the group. In 1263 Urban IV sanctioned a rule which was less severe and was followed by the majority, the Urbanists. The minority, who adhered to the stricter rules, were known as Clarisses. In 1436 the reform of St. Colette brought back many of the houses to the strict observance of the Rule of St. Francis. The two branches of the order are the Urbanists and the Colettines. The life of the Poor Clares is contemplative and most austere, including penance, manual work, and severe fasts. They wear a dark frieze habit, black veil, and cloth sandals on bare feet.

JAMES TAYLOR

POPE, THE (Lat. *papa* from Gr. *papas*, "father"). The supreme head of the Roman Catholic Church. Formerly denoting all Christian bishops, the title in the West has from the ninth century been appropriated exclusively by the bishop of Rome. In the Eastern Orthodox Church, however, it is still used for the patriarch of Alexandria, and the term is applied also to ordinary priests.

POPE JOAN, see JOAN, POPE

POPE, WILLIAM BURT (1822-1903). Wesleyan divine. Born in Nova Scotia and educated in England, he trained at the Wesleyan Theological Institution at Hoxton and was ordained in 1842. He traveled in several circuits and established a reputation as a linguist and translator of German antirationalist critics. From 1867 to 1886 he was a tutor at Didsbury Wesleyan College, Manchester. In 1875-76 he produced his greatest work, *A Compendium of Christian Theology* (3 vols.). This, while containing several specifically Wesleyan features, especially a very high doctrine of the ministry and an elaborate exposition of Christian holiness, is dedicated to what Pope called "the old doctrines of the Reformation." Impeccably orthodox and the most powerful of all Wesleyan essays in dogmatic theology, it undoubtedly held back the impact of destructively critical ideas on English Methodism for several decades. Pope died after a long and painful illness.

IAN SELLERS

POPISH PLOT. An alleged Jesuit conspiracy in England to assassinate Charles II* and to replace him by the Roman Catholic James, duke of York, sworn to by Titus Oates* in 1678. Oates, who had twice been expelled from Roman Catholic seminaries, fabricated the story for his own advancement, in association with a fanatical Jesuit called Israel Tonge. Their story was sufficiently credible to cause panic, and the ensuing witchhunt hurried some thirty-five suspects to the scaffold before Oates was discredited as liar and perjurer.

PORPHYRY (232-c.305). Neoplatonist writer. Born in Tyre and originally named Malchus, he was the pupil, successor, and editor of Plotinus.* A historian of philosophy and religion rather than an original thinker, he nevertheless initiated certain tendencies conspicuous in later Neoplatonism, e.g., in his own pupil Iamblichus (c.250-330), who taught mainly in Syria, and Proclus,* the most important of the Neoplatonists who took over Plato's Academy at Athens in the fifth century A.D.

Porphyry's work of fifteen books attacking Christianity has not survived, but was influential enough to merit suppression by the Council of Ephesus* in 431. We learn from Jerome that he rejected an early date for the Book of Daniel because it described second century B.C. events too accurately for prophecy. He evidently criticized the gospels for inconsistency, and although admiring Christ's teaching, felt that the apparent failure of His mission disproved His divinity. Porphyry himself wrote extensively on ethics, and preaches the cardinal virtues of "faith, truth, love (desire), hope." He may be credited with sparking the pedantic overelaboration, doctrinaire vegetarianism, and extravagant allegorizing which characterized much of later Neoplatonism.* Its interest in magic and divination should probably be blamed on Iamblichus.

See T. Whittaker, *The Neoplatonists* (1918), and W. Theiler, *Porphyrius und Augustine* (1933).

GORDON C. NEAL

PORTEUS, BEILBY (1731-1808). Bishop of London. Educated at Cambridge, he became rector of Hunton which had had no resident incumbent for thirty years, and there displayed that pastoral concern, marked by residence, visitation, and catechesis, that he was to urge as a bishop. In 1762 he became chaplain to Archbishop Secker* (whose biography he wrote), and in 1769 a royal chaplain. He was appointed bishop of Chester in 1776 and of London in 1787. As bishop he encouraged residence of incumbents, better stipends for curates, regular preaching, and higher standards of clerical duty. In the House of Lords he battled for public morality and vainly tried to hedge divorce legislation. Responsible as bishop of London for the overseas interests of the church, he opposed slavery, founded the "Christian Faith Society" for West Indian slaves, and proposed other forms of church mission. He was more sympathetic to the Evangelicals (and to Evangelical enterprises like the Church Missionary Society and the Bible Society) than were most high ecclesiastics, while remaining apart, an orthodox churchman in the Secker tradition. His *Collected Works* were published in 1811.

A.F. WALLS

PORTUGAL. The country secured freedom from Spanish rule in the twelfth century, when her independent political and ecclesiastical history can be said to begin. Though at first in bondage to the papacy, the nation under King Sanche I vigorously asserted its autonomy even against Innocent III,* the most powerful of popes. Disputes continued throughout the thirteenth century, the Friars and the widespread anti-Spanish sentiment helping the pope to maintain his authority. Under Prince Henry the Navigator, Portugal began in the later fifteenth century to build up her overseas empire, embarking on a policy of subjugation and conversion of the native peoples. During the first half of the seventeenth century she was again absorbed into Spain, but recovered her independence between 1640 and 1668, when the papacy was actively allied with Spain. Thereafter, as in Spain, the Catholic Church fell into a torpor from which it has never really recovered.

With the irruption of liberal and anticlerical movements into the country following the French Revolution, the history of Portugal closely resembles that of Spain. After the republican revolution of 1910 the church was disestablished and its power curtailed, but Dr. Salazar's New Constitution of 1933 and Concordat of 1940 restored harmonious relations between church and state. Since 1917 popular devotion in the country has been enormously strengthened by the cult of Our Lady of Fatima. This, which is now the hallmark of Portuguese Catholicism, began in a small town in the middle of the country in May 1917, when three poor children were alleged to have seen a vision of the Virgin on six occasions. The cult and the accompanying miraculous cures were at first frowned on by the church, but after 1930 were officially favored; Our Lady of Fatima is now identified, unofficially at least, as the queen of Portugal.

Portugal was immune to the Reformation, due to its isolation, the absence of pre-Reformation movements such as the Hussites, the complete hold of church and government over the common people, the feeling of national self-confidence in this particularly buoyant phase of expansion which discouraged expressions of dissent, and the Inquisition* (established 1536). Protestantism did not effectively reach Portugal till 1845 when meetings were commenced simultaneously in Lisbon and Oporto. Since then a large variety of missionary agencies has been active, and the progress of the Reformed faith has been similar to that in Spain, with English Methodism playing a more active role. The chief denominations are the Lusitanian Church (episcopal and Anglican in origin), the Evangelical Church (Congregational Presbyterian), the Baptists, Brethren, Methodists, and Pentecostalists who first arrived from Sweden in 1930. Persecution has never been as fierce as in Spain, and as a minority the Protestant population which now numbers about 33,000 is proportionately larger.

BIBLIOGRAPHY: F. de Almeida, *Historía de igreja em Portugal* (4 vols., 1910-22); J.C. Branner (tr.), *History of the Origin and Establishment of the Inquisition in Portugal* (1926); E. Moreira, *The Significance of Portugal: A Study of Evangelical Progress* (1933); C.E. Nowell, *A History of Portugal* (1952); H.V. Livermore, *A New History of Portugal* (1966).

IAN SELLERS

POSSIDIUS (d. c.440). Biographer and friend of Augustine.* He was a member of the monastery at Hippo until he was made bishop of Calama about 400. As such he was active in supporting Augustine's opposition to Donatism* and to the Pelagian* heresy. In addition to the biography he left a careful though incomplete list of Augustine's works.

POSTLAPSARIANISM, see SUPRALAPSARIANISM

POSTMILLENNIALISM. An optimistic type of theology which predicts a "golden age," a Christianized millennium of predominantly human achievement before the Second Advent and the subsequent eternal realm. The prophetic form of it is devout, the liberal form purely humanistic. An early exponent was Joachim of Fiore,* who in the twelfth century divided historical ages into an OT dispensation under the Father, a NT and early church one under the Son, and the Eternal Evangel (age of the Spirit) to begin in 1260. The modern term was popularized by the Unitarian freethinker Daniel Whitby,* later enjoyed immense vogue in Britain during the prosperous century from Waterloo to World War I, but was thereafter increasingly discredited by postwar realities, and even more so after World War II.

See also MILLENARIANISM.

ROY A. STEWART

POSTULANT. One who lives in a religious house under the supervision of the religious superior in a probationary period before entering formally into the novitiate. The postulant can leave at any time, but postulancy is not required for either valid entrance into the novitiate or for valid profession of vows subsequently. Postulancy ends when the candidate is admitted to the novitiate.

POTHINUS (Potheinos) (c.87-177). Martyr and first bishop of Lyons (Lugdunum). Born probably in Asia Minor and a disciple of Polycarp,* he is said to have introduced Christianity into S Gaul. In the persecution that broke out in Lyons in 177, described in the *Epistles of the Churches in Vienne and Lyons*, the ninety-year-old bishop was questioned by the governor and so badly treated that he died within two days.

POTTER, JOHN (c.1674-1747). Archbishop of Canterbury from 1737. Educated at University College, Oxford, he was elected a fellow of Lincoln College in 1694 and regius professor of divinity at Oxford in 1707. As a High Church Whig he was a safe appointment to the bishopric of Oxford in 1715, hut his translation to Canterbury in 1737 was particularly unexpected, since

Edmund Gibson, bishop of London and another High Church Whig, was the more popular candidate. Potter was an opponent of the Low Church party, particularly of Bishop Hoadly,* but his Whig political views ensured his good favor with the government. Among his works were *A Discourse on Church Government* (1707); *Archaeologica Graeca* (2 vols., 1697-99); and an edition of the works of Clement of Alexandria (1715).

JOHN A. SIMPSON

POWELL, VAVASOR (1617-1670). Welsh Puritan divine and activist. He was born in the hamlet of Knucklas in Radnorshire. He seems to have been educated at Oxford and served for some time as a schoolmaster at Clun, where he was converted under the influence of Walter Cradock's preaching and Richard Sibbes's *Bruised Reed.* By 1640 he was actively engaged in a vigorous preaching mission along the borders of Radnorshire and Brecknock, and as a result he came into conflict with the authorities.

With the outbreak of the Civil War he withdrew to London and in 1644 became Puritan vicar of Dartford, Kent. He participated in some of the military campaigns, but in 1646 he was authorized as a preacher by the Westminster Assembly* and named a preacher in N Wales by the Committee for Plundered Ministers. His status among the Puritan leaders is suggested by the fact that he preached before the lord mayor of London in December 1649 and before the House of Commons in the following February. He reached the zenith of his influence after his appointment as an approver under the Act for the Better Propagation of the Gospel in Wales, and he dedicated his uncommon energy to the task of making this Act successful. He saw it as a providential opportunity to make the Gospel known to the people of Wales, and in order to overcome the difficulty posed by a shortage of preachers, he devised the method of appointing itinerant preachers to serve fairly large areas of the country. His activities brought him the fierce hatred of his Anglican critics and accusations that he had misapplied the funds of the church in Wales to his own benefit. But the evidence points to his innocence. The Act was discontinued in 1653.

Like many other Puritans, Powell was a millenarian and believed in the early return of Jesus Christ to begin His personal reign upon earth. It was his millenarian activities that brought him into conflict with Cromwell's Protectorate, which he interpreted as a betrayal of Christ's sovereignty. He initiated a campaign in Wales against the Protectorate which culminated in the petition, "A Word for God" (November 1655).

With the collapse of the Puritan ascendancy, he became a marked man and was imprisoned in April 1660. Except for a period of eleven months in 1667-68 he spent the remainder of his life in jail. Some of the manuscripts that he wrote at this time are a moving proof of the way in which his fiery spirit continued to support the ideals of happier days while his concern for tolerance deepened. After his recapture in 1668 he was brought to trial and imprisoned in the Fleet, where he died. He was buried at Bunhill Fields. He was the author of some thirteen published works. His redoubtable character, tireless energy, and indomitable courage put him in the front rank of Welsh Puritans.

R. TUDUR JONES

PRAEDESTINATUS. A theological treatise, probably composed in Rome (c.432-40), against the Augustinian doctrine of predestination. Written from the standpoint of Semi-Pelagianism* (if not actually Pelagian), its bitter antipredestinarian views have been called "a cruel parody of Augustinianism." Of its three books, the first is a plagiarized reproduction of Augustine's *De Haeresibus,* the second purports to be written by a supporter of Augustine's doctrine, and the third is a refutation of the second. Edited originally by J. Sirmond in 1643, it was much discussed in the Jansenist controversies of the seventeenth and eighteenth centuries.

PRAEMUNIRE, STATUTES OF (Lat. *praemunir,* "to protect, to secure"). The first words of the writ *(praemunire facias)* used by Edward III in 1353 to protect the rights of the English crown against encroachments by the papacy give the basis for the title. This statute was revised in 1365 and followed in 1393 by the famous Statute of Praemunire, which was primarily aimed against Pope Boniface IX. The expression "praemunire" is now used to describe the statute, the offense against it, or the punishment under it. In essence the three statutes required that clergy were not to take to Rome matters that should be settled in England, and that papal bulls and excommunications were not to be promoted in England. As a result, appeals to the Vatican were diminished. The statute of 1393 was variously used in the sixteenth and seventeenth centuries by English monarchs to deal with Roman Catholics (e.g., by Henry VIII against Wolsey*). The Royal Marriages Act of 1772 is the last Act which subjects anyone to the penalties of praemunire.

PETER TOON

PRAEPOSITINUS OF CREMONA (c.1140-1210). Paris theologian. Born in Lombardy, he studied theology and canon law at Paris, then became a prebend of Mainz Cathedral. He held this post for about ten years (c.1194-1203), during which time he sought to convert the Cathari* to orthodox Catholicism. By 1206 he was chancellor of the University of Paris. His influential *Summa Theologica* contains the teaching he gave in this period and reveals that as yet Paris had not absorbed the Aristotelian philosophy that was beginning to be accepted in Europe. His other (certain) books include *Summa super Psalterium* and *Summa de Officiis.* Possibly he wrote also the *Summa contra Haereticos,* which is usually attributed to him.

PETER TOON

PRAETORIUS, MICHAEL (1572-1621). German composer. This prolific and influential Lutheran composer is known widely today for one tiny piece—"Lo, how a Rose e'er blooming" *(Es ist ein' Ros entsprungen).* It is taken from his monumental work *Musae Sioniae,* published in nine volumes and containing over 1,200 simple works covering the church year. He published several other collections of music for the Lutheran service. He was one of the first Germans to write in the new Venetian *concertato* style in up to twelve or fifteen parts, combining voices and instruments in glittering array. He applied this style to works based on the Lutheran chorales with fine effect. He also published his *Syntagma musicum* in three volumes (1615-19), which contains valuable information about performance practice and also about the instruments of the period, accompanied by accurate scale drawings. His section on the organ has been very influential in recent times on the movement of organ builders back to Baroque principles.

J.B. MAC MILLAN

PRAGMATIC SANCTION OF BOURGES (1438). This was issued by Charles VII of France following the National Synod at Bourges, which had close links with the Council of Basle* (1431-49). It was a statement of Gallicanist principles, contained in twenty-three articles, which effectively reduced the power of the papacy in France. Councils were superior to the pope, nomination of bishops and high ecclesiastical dignitaries was to be in the hands of the French king and princes, and French ecclesiastical affairs were to be settled in France. When Louis IX repealed the sanction in 1461, the parliament of Paris refused to endorse his action; however, in 1516 an agreement known as the Concordat of Bologna,* between Pope Leo X and Francis I, ended the sanction but preserved many Gallicanist principles.

PETER TOON

PRAGMATISM. An attempt to avoid Spencerian determinism and Hegelian metaphysics. It was born in late nineteenth-century America and stands as that nation's greatest contribution to Western philosophy. Combining empiricism with evolutionism's belief in an unfinished universe, it was given varied forms by its chief exponents—Charles Sanders Peirce (1839-1914), William James,* and John Dewey (1859-1952). Peirce's "pragmaticism," for example, was supposed to prevent the alleged mystical, individualistic approach of James. Dewey, in turn, created an "instrumentalism" to apply evolutionary science to democratic society's problems; he defined ideas as plans for action, and truth as that which best controlled the conditions and consequences of experience at any given moment. Often confused with expediency, pragmatism was an optimistic philosophical method which rejected dualism, stressed life's incompleteness and morality's relativity, and trusted the scientific method to cure every aspect of life. DARREL BIGHAM

PRAXEAS (fl. c.200). Modalist Monarchian. He is little known apart from Tertullian's treatise *Adversus Praxeam* (c.217) and has indeed been identified with Noetus and Epigonus, and even with Pope Callistus. He was said to have arrived in Rome toward the end of the second century from Asia where he had suffered imprisonment for his faith, and he may have gone later to Carthage. He was strongly anti-Montanist. He became leader of the so-called Patripassian* Monarchians—i.e., those concerned to maintain the unity of the Godhead even to the point of declaring that God the Father suffered. As Tertullian put it, "He drove out prophecy and introduced heresy: he put to flight the Paraclete and crucified the Father." Praxeas conceived of Father and Son as one identical Person, the Word's having no independent existence. Consequently it was the Father who entered the Virgin's womb, thus becoming, so to speak, His own Son who suffered, died, and rose again. J.G.G. NORMAN

PRAYER. When God approached men in OT times, they conversed with Him, confessed their sins, gave Him adoration and thanks, and asked things for themselves and others. Prayer is this conversation with God which arises out of communion with Him. Prayer has many elements. It can be a sacrifice of thanksgiving and adoration (e.g., Ps. 50:14,23; 107:22). It often begins with confession (e.g., Ps. 51), but it is essentially the element of asking—petition, supplication, and intercession—which constitutes prayer as prayer.

This asking arises out of the depths of human need. Prayer is a pouring out of the heart to God (1 Sam. 1:15) in all its moods and in the needs of the concrete human situation (Exod. 17:8ff.; 1 Chron. 5:20; Luke 22:44; Phil. 4:6). But prayer should arise more out of the promise and challenge of God's Word than out of the urgency of human need. Though the Word has its own power to go forth and fulfill itself, its fulfillment can be facilitated and hastened by such prayer as "Your kingdom come." Prayer is therefore an asking directed by the Word of God. The Lord's Prayer,* moreover, teaches that the first concern to be expressed in prayer should be for the hallowing of God's name. If prayer is thus directed by the Word, it is saved from triviality and self-centeredness (John 15:7).

Prayer is inspired by man's confidence that God has already drawn near to hear before men are there to speak (Ps. 27:8; 139:1-6). In the OT it is inspired by complete confidence in the faithfulness and power of God, man's Rock and Stronghold (Deut. 33:29; 2 Sam. 23:3; Ps. 46:1), who is appealed to in His omnipotence and eternity (Ps. 124:8). It is significant that in nearly all the prayers of Jesus, God is addressed as "Father." Paul's prayers are inspired by a filial trust in God's fatherhood (Rom. 8:15; Eph. 3:14). The presence and Word of God can arouse in men such boldness in prayer that it can take the form of importunate argument (Gen. 18:22-33; Exod. 32:11; Luke 18:1-7), or of an entreating of favor (1 Sam. 13:14; Ps. 119:58). Prayer can indeed be a striving with God, as Jacob strove with the angel (Hos. 12:3,4).

In the NT, prayer is in the name of Christ. This means prayer in union with the Christ who at the right hand of God continually makes intercession. Since prayer takes place in Christ and in the Spirit, the individual even though alone is praying in and with the community and is encouraged to say "*our* Father." True prayer thus tends to become communal (cf. Matt. 18:19,20; Acts 2:1, etc.), though it can nevertheless remain intensely private (Matt. 6:5,6). It is noteworthy that in the Psalms and the great prophetic writings the liturgical community prayer can perfectly express the personal needs and longings of the solitary individual also.

Prayer is essentially something within the heart and arising from the heart. The command to "pray without ceasing" suggests that prayer is a continuous attitude of heart (1 Thess. 5:17; cf. Luke 18:1). There can be "ejaculatory" prayers uttered when men are engaged in quite other than devotional tasks (Neh. 2:4,5). Yet prayer often demands expression in outward attitude. Various postures and gestures are described as the accompaniments of prayer—prostration, kneeling, standing, stretching out hands (Gen. 18:2; 2 Chron. 7:3; Ps. 28:2, etc.). Prayer also tends always to clothe itself in language. It can thus vary from a charismatic utterance in "tongues" to a carefully composed choral hymn. It tends to like music. At the best it remains simple (Matt. 6:8). To pray is a duty, and the prayer aspect of the Christian life remains the more healthy if times are set aside for prayer and the prayer life is brought under discipline (Ps. 5:3; Dan. 6:10; Acts 3:1). Jesus was familiar with the traditional prayers of Israel and made them His own. He prayed before meals and attended community worship. At times He spent whole nights in prayer—especially at times of decision and crisis.

BIBLIOGRAPHY: J. Hastings (ed.), *The Christian Doctrine of Prayer* (1915); F. Heiler, *Prayer* (1932); F.L. Fisher, *Prayer in the New Testament* (1964); J.G.S.S. Thomson, *The Praying Christ* (1965); R. Simpson, *The Interpretation of Prayer in the Early Church* (1965); J. Ellul, *Prayer and the Modern Man* (1970); I.T. Ramsey, *Understanding Prayer* (1970). RONALD S. WALLACE

PRAYERS FOR THE DEAD. The earliest Christian Father to refer to the practice of praying for the departed, Tertullian,* admits also that there is no direct biblical authority for doing so. Third-century inscriptions indicate the kinds of petitions made in these prayers, usually a simple and general request for the dead person to be with God or to know the forgiveness of sins. It is possible that such prayers arose out of the confused ideas over the consequences of postbaptismal sin, which caused much debate in the church of Tertullian's time. One suggested solution to this problem was the idea of a purgatorial discipline after death, which was discussed at Alexandria in the early third century and spread in the West through the powerful advocacy of Augustine* and Gregory the Great.* Meanwhile, at Jerusalem in the mid-fourth century the Eucharist came to be regarded as a propitiatory sacrifice which could be offered on behalf of both the living and the dead. Consequently intercessions for the departed came to be inserted in the *anaphora,* or canon of the Mass. In the Roman Church, a Mass offered specifically for a dead person is called a "requiem," although since the early Middle Ages the dead have also been remembered in the daily Mass.

In England, Cranmer's* second Prayer Book (1552) abolished all prayers for the dead; but a thanksgiving for the faithful departed was added to the intercessions in 1662. In modern times pastoral needs in the Church of England, where many non-churchgoers are given Christian burial, have led to the consideration of a form of prayer to include the unfaithful departed, and this has been included as an option in the Series 3 Orders for Holy Communion and Funerals. There remains a tension in Christian thought between the best way of expressing the biblical truth of an unbreakable fellowship of believers in Christ, and a sub-Christian desire to provide for, and communicate directly with, the spirits of the dead.

See *Prayer and the Departed:* A Report of the Archbishops' Commission on Christian Doctrine (1971). JOHN TILLER

PREACHERS, ORDER OF, see DOMINICANS

PREACHING, see HOMILETICS

PREACHING OF PETER, see PETER, PREACHING OF

PREBEND; PREBENDARY. In most English medieval cathedrals and collegiate churches, endowments were divided into separate portions in order to support members of the chapter. Each portion was known as a "prebend," because it supplied *(praebere)* a living to its holder, who became known as a "prebendary." The territorial names of many cathedral prebendal stalls indicate that the revenue for the particular stall came from cathedral lands in that area. In cathedrals of the "Old Foundation," the names prebend and prebendary have been retained, though in most cases the office is only honorary, the income having been transferred to the Ecclesiastical Commissioners. In cathedrals of the "New Foundation," there are canons, not prebendaries, though that title was used until the nineteenth century.

JOHN A. SIMPSON

PRECENTOR. Broadly, the official responsible for the singing of a church choir or congregation. In most English cathedrals of the "Old Foundation" (and many on the Continent), he ranks in the chapter after the dean. He has the oversight of the choral service and the choristers, but in practice customarily delegates his duties to a deputy (the succentor). In "New Foundation" cathedrals, however, the precentor is merely a minor canon or chaplain. In Scottish Presbyterian churches the precentor was a layman who literally led the singing, his function made necessary by the absence of any musical instrument. The office survives in some Highland and island congregations, and at the annual general assembly of the Church of Scotland a precentor leads the worship services. J.D. DOUGLAS

PRECISIAN. One who is rigidly precise or punctilious in the observance of moral and religious rules. It was a term in the sixteenth and seventeenth centuries applied by critics and opponents to those who were otherwise styled "Puritans";

the abuse intended referred to both strict morals and "purity of worship."

PREDESTINATION. In theology, predestination refers to the predetermination by God of the individual's ultimate destiny. Controversies regarding predestination have centered on the apparent contradiction between such predetermination and man's free will. (As noted below, they are thus analogous to controversies regarding man's freedom in a universe seemingly determined by scientific law.) The doctrine of predestination is associated particularly with Christianity, but also occurs elsewhere; in Islam,* e.g., during its scholastic period, the orthodox position was strongly predestinarian, but some theologians stressed free will (the Mutazalites). In the first centuries of the Christian Church, predestination was not an issue. Theological energy was taken up with definitions of the Trinity and arguments regarding the nature of Christ. In the Orthodox Churches this has remained the case (with minor exceptions: notably Cyril Lucar in the 1500s).

In the Western Churches the issue was raised (as imperial rule in the West tottered) by Pelagius, who taught that man had the freedom to accept or reject God. This was countered by the great theologian Augustine,* who held that man's will was enslaved by sin, that grace was needed to choose for God, and that this grace was given to those whom God had predestined. The Augustinian position was upheld by the Synod of Orange (529)—but by this time the barbarian invasions were in full swing, and there was little talent or time for theology. An aftermath to the Pelagian controversy occurred in the age of Charlemagne*: the monk Gottschalk* held (apparently) that God actively willed the nonelect to be damned, a position which was rejected (Synod of Quiercy, 849).

The medieval revival of learning, from around 1050, produced schools and universities in abundance. Theology was held to be the "queen of the sciences," the key to the understanding of reality. The task of the Schoolmen, or Scholastics, was to reconcile Christianity with the newly rediscovered heritage of classical philosophy; in a sense, to harmonize reason and faith. By the late 1200s, after Peter Lombard,* Bonaventura,* Albertus Magnus,* Thomas Aquinas,* and a host of other scholars, the task seemed completed, briefly, with several Scholastic systems, differing in detail, available. Predestination generally was handled in the context of God seen as Supreme Intellect, who predestined on the basis of His foreseeing the choice the individual would make (for God, all temporal things are "present"; He is outside time).

But this "solution" was soon attacked. With Duns Scotus, and especially with William of Ockham* and his followers in the 1300s, God was seen as Sovereign Will, and the problem of predestination shifted. How can man's choice be free, if foreseen? How can God be called fully sovereign, if He is bound to follow a future which is already determined? How can God bind His will in advance? The tangled controversies which followed seemed to raise insoluble questions. There were reactions: Thomas Bradwardine* revived a rigid Augustinian view, stressing divine predestination as basic to an ordered universe. From a different vantage point, John Wycliffe* and John Hus* stressed election as a key theological concept and viewed the church as the community of the elect, those already saved, rather than the source of desperately needed aids to salvation.

The Protestant Reformers followed this emphasis also. Luther,* Zwingli,* Calvin*—all held to predestination, the true church as made up of the elect, the enslavement of the will (e.g., Luther against Erasmus*), the need for unconditioned grace to enable a choice for God. Yet this strongly Augustinian approach did not escape criticism. In the Lutheran churches the fierce "synergistic" controversy of the later 1500s resulted from Melanchthon's* attempt to save some role for the human will. Similar debates arose in Calvinism over the teachings of Arminius (condemned at the Synod of Dort,* 1618-19). Scholastic refinements in Protestant theology brought further disagreements: the Calvinist quarrels between sublapsarian* and supralapsarian* theologians, the controversy in the Huguenot* churches over Amyraut's teachings, and the like. As Protestant Scholasticism declined during the 1700s and Pietism arose to regain a "heart-felt" religion, the question arose in Methodism: John Wesley favored "Arminianism,"* George Whitefield* a "Calvinist Methodism."*

In postmedieval Catholicism the issue flared up several times. The Council of Trent,* though avoiding a definitive stand, leaned toward a Semi-Pelagian* position. The teachings of Luis de Molina (d.1600) aroused much controversy, with Jesuits tending to support, and Dominicans oppose, his complex attempt to give man's will a role in the process of salvation. Around the same time, Baius* at Louvain, followed by Cornelius Jansen, returned to a rigid Augustinian. Jansenism* produced a notable controversy (Port-Royal, Pascal's* defense of Jansenism, etc.) which finally produced a minor schism (the Old Catholic Church* of Utrecht, from 1713).

As interest in traditional theological argument faded into the background in the 1800s, the problem appeared in other areas. If the universe is determined by scientific law, how can man have free will? If man's actions are not in a sense determined by such law, how is any political science or economic science or science of history possible? If heredity and environment determine actions, how can the courts punish a man (for doing what was predetermined)? And the like. In the twentieth century, Karl Barth's* revival of a Scholastic Calvinism has raised again the theological issue of predestination. Barth attempts to cut through previous controversy by stressing God's election of man in Christ (which perhaps involves some sort of universalism).

In summary, predestination and the debates about it deal with a recurring problem, whether in theology or in other fields: the relation between man's freedom and a universe which seems in some sense determined.

BIBLIOGRAPHY: J.B. Mozley, *A Treatise on the Augustinian Doctrine of Predestination* (1855); L. Boettner, *The Reformed Doctrine of Predestination* (1932); M. Luther, *On the Bondage of the Will* (ed. J.I. Packer and O.R. Johnston, 1957); P. Maury, *Predestination* (1960); H.G. Hageman, *Predestination* (1963). DIRK JELLEMA

PREMILLENNIALISM. The view which asserts that Christ will come a second time before the 1,000 years of His millennial rule, upholds a general chiliastic theology of Millennialism, and places the rapture of saints, the first resurrection, the tribulation, and Second Advent before the Millennium in prophetic time sequence, with the brief release of bound Satan, the second resurrection, and Last Judgment afterward. This view was held by early Church Fathers until Origen,* Eusebius,* and Augustine* modified it, and it has been revived in the modern era by J.N. Darby,* W.E. Blackstone,* and C.I. Scofield,* among others.

See also MILLENARIANISM.

PREMONSTRATENSIANS. Norbert* founded the first community of Premonstratensians, or "White Canons," at Premontré, near Laon, in 1120. The Premonstratensians adhered strictly to the Augustine Rule, but Norbert, a friend of the Cistercian monk, Bernard of Clairvaux,* also adopted certain monastic features and the Cistercian federal organization. Premontré was the order's mother-house, and its abbot was the order's abbot-general. Houses were arranged into regional and national *circaria,* which in turn constituted the "Grand Congregation." The Premonstratensians undertook an "apostolic" role of parochial work and preaching, and their order rapidly spread throughout Christendom. Their missionaries played an important part in conversion and colonization east of the Elbe. At first they admitted women to double monasteries, but this practice had ceased by 1200. The Reformation and the French Revolution seriously affected the order's size. PETER TOON

PRESBYTER (Gr. *presbuteros,* "an older person"). In the Septuagint the term was applied to "the elders" of Israel who carried out various governing and administrative functions in both civil and ecclesiastical government. In the NT they were those mature Christian men who were appointed to supervise the work of the church (Acts 14:23). In this respect they fulfilled an official function as bishops *(episkopos)*—(cf. Acts 20:17, 28; Titus 1:5-9; 1 Pet. 5:1-4), although the relationship between the use of the two terms is not always clear. Gradually one elder, probably the teaching member of the group (1 Tim. 5:17) assumed presidency to become in the second century the "bishop," with special powers and privileges. During the Middle Ages the term "presbyter" was shortened to "priest," while the presbyter-bishop assumed a superior position, often becoming a feudal lord. Although neither the Lutherans nor the Anabaptists stressed the presbyterate, John Calvin and his followers did, believing that there were in the NT four orders: pastors, doctors (teachers), deacons, and presbyters (elders), the last being primarily responsible for discipline—the admission of new members and the supervision of the individual and corporate lives of the congregation. The minister or pastor was a teaching elder. This view is still held by most Reformed and Presbyterian churches.

W.S. REID

PRESBYTERIAN CHURCH IN CANADA. Canadian Presbyterianism began in Nova Scotia just prior to the American Revolution and subsequently followed the patterns of migration from the United States and the United Kingdom. As with other Canadian denominations, American influence did not last long, and it was soon the arrivals from Scotland and Ireland who formed the backbone of the church.

The Secessionist bodies, who had followed the Erskines out of the Church of Scotland in the eighteenth century, and who would be known as the United Presbyterians in the nineteenth, were the first to see the growing population of Canada as a missionary responsibility. The Church of Scotland was much slower, but as the Evangelical or "popular" party gained the ascendancy, the Glasgow Colonial Society was formed in 1825 to encourage and support ministers in coming to Canada. Robert Burns of Paisley, later of Knox Church and Knox College, Toronto, poured his energy and vision into this movement, so that the arrivals of the 1820s and 1830s from Highlands and Lowlands, were often greeted by a congenial Evangelical ministry.

The Disruption* of 1843 in Scotland, with the formation of the Free Church,* was followed by a similar sympathy movement both in the Maritimes and in the province of Canada, today known as Ontario and Quebec. The Free Church Movement in Canada displayed remarkable vitality in the fields of home and overseas missions and theological education. A new country, however, could not long afford division, and in 1861 the "Free Church" and the United Presbyterians joined, and in 1875 the Church of Scotland amalgamated to form a national Presbyterian body of some 88,000 communicants.

The new church was absorbed in the opening of the West, under the dynamic leadership of James Robertson, while overseas missionaries such as Jonathan Goforth* made Canadian Presbyterianism known worldwide. The church continued to grow rapidly and threw itself wholeheartedly into the job of ameliorating the social problems of the day. Preoccupied with wresting a living from the country, and full of turn-of-the-century optimism, many paid little or no attention to the changes in thought that were taking place. Evolution, idealistic philosophy, and biblical criticism were creating a new mood. As the older Calvinism ebbed, the movement for union among Methodists, Congregationalists, and Presbyterians flowed. Church union was consummated in 1925, with 40 percent of the 380,000 Presbyterian members outside. Most of the large middle party had entered union while the continuing Presbyterians were an interesting and sometimes irreconcilable combination of traditionalism and evangelicalism.

Soon faced with the challenges of economic depression and war, it was amazing that the reconstruction of the Presbyterian Church proceeded as effectively as it did. The situation was complicated by the fact that there were strongly liberal elements in the two theological colleges. This emphasis was increasingly challenged by the charismatic W.W. Bryden of Knox College, who was virtually a Barthian before Barth. Today there is again an increasing evangelical movement in the denomination of almost 200,000 members, and a census constituency of over 800,000.

See W. Gregg, *Short History of the Presbyterian Church in the Dominion of Canada* (1892); and N. Smith, A. Farris, and H.K. Markell, *A Short History of the Presbyterian Church in Canada* (1967). IAN S. RENNIE

PRESBYTERIAN CHURCH IN THE U.S. Popularly known as the Southern Presbyterian Church, it shares a common history and heritage with the larger nationwide Presbyterian Church. The southern presbyteries comprised more than a third of the Old School branch of the Presbyterian Church in the USA, and did not renounce their connection with it until the Old School assembly, meeting in Philadelphia (May 1861), adopted resolutions pledging the church's support to the Federal Union, even although most of the Southern states had already seceded and civil war had begun. The first assembly of the Presbyterian Church in the Confederate States met in Augusta, Georgia (4 December), and was organized by commissioners from forty-seven Southern presbyteries. The United Synod of the South, comprising twenty-one presbyteries, was formed in 1858 by those who broke with New School Presbyterians.

The whole South suffered severely during the war. After the war the present name was adopted. Some of the early growth—slow during Reconstruction but substantial and steady thereafter—came by union of the southern branches of the New School and Old School in 1864. In 1969 there were some 4,000 churches, 4,593 ministers, and nearly a million communicants, primarily in urban areas, contributing a total of nearly $134 million. The assembly holds membership in the World Alliance of Reformed Churches,* the National and World Councils of Churches, and the Consultation on Church Union. Union with the Reformed Church in America was defeated by that body in 1969. Union with the northern and United Presbyterians was defeated in 1954; another plan of union with the northern body is now being studied in draft form.

For the greater part of its history the denomination has been somewhat homogeneous and committed to Calvinistic orthodoxy, strict subscription to the Westminster Standards, scriptural authority and inerrancy, *jure divino* Presbyterianism, and the exclusively spiritual mission of the church. Increasing internal tension developed after 1935, as the leadership has tended to modify its position on these and other theological, social, and ecumenical issues. Recent years have witnessed a proliferation of dissenting conservative groups who pledge to maintain a continuing Southern Church despite efforts by the leadership to carry the denomination into church unions and to draft a new confession of faith. In late 1973 some members withdrew from the denomination to form a new, separate body, now called the Presbyterian Church of America.

BIBLIOGRAPHY: T.C. Johnson, *History of the Southern Presbyterian Church* (1894); R.C. Reed, *History of the Presbyterian Churches of the World* (1905); H.A. White, *Southern Presbyterian Leaders* (1911); J.M. Wells, *Southern Presbyterian Worthies* (1936); T.W. Street, *The Story of Southern Presbyterians* (1960); M.H. Smith, *Studies in Southern Presbyterian Theology* (1962); E.T. Thompson, *Presbyterians in the South* (2 vols., 1963-72). ALBERT H. FREUNDT, JR.

PRESBYTERIANISM. The term derives from the word "presbyter."* Its reference is primarily to a church which is governed by presbyters, usually elected by the people of a congregation or of a group of congregations. Traditionally it is the general title given to the English-speaking Reformed or Calvinistic churches coming out of the Reformation and their daughter churches in many different lands.

Presbyterians trace their concept of church government back to the OT synagogue which was governed and directed by a group of "elders." Calvin held that since the NT church used the same form of organization, this is the structural pattern that the contemporary church too should follow in order to be as close to the NT as possible. This was in accord with his idea that the NT church provided the permanent example not only of the succeeding generations' beliefs but also of their ecclesiastical organization. Calvin did recognize, however, that other forms might be adopted, although he believed that the presbyterial was that closest to NT example.

According to Calvin, the NT church had four different offices: pastor, doctor or teacher, deacon, and presbyter or elder. The pastor was the preacher and the counselor of the Christians; the doctor taught in a more formal way than the pastor and might also hold the position of a theological professor. The deacon was primarily responsible for the material needs of the church and of the members, while the elders were those who had the oversight of the spiritual needs and the lives of the congregation. The pastors and doctors were usually elected and approved by the pastors and elders of other congregations, while the deacons and elders were elected by individual congregations on the advice of the existing consistory or session, made up of elders and sometimes deacons.

During the Middle Ages the NT organization had been radically changed with the establishment of a hierarchical organization consisting of priests, bishops, and pope, with many intermediate officials. While the Lutherans for convenience in administration had retained bishops or superintendents, the Genevan Reformer brought in a new and different pattern in seeking to reinstitute what he considered to be the proper NT form of church organization. Although he did not establish a completely presbyterial system as it came to exist later, he laid the foundation. The fact that his original structure dealt only with the four churches in Geneva meant that it would be different from the French and Scottish plans to devise an organization for a national church, covering a much larger area and a greater number of people.

The beginnings of an English-speaking church organized on the presbyterial basis took place in Geneva in the congregation of Marian exiles (1555-58) under the Scottish preacher, John Knox.* Unwilling to accept the Anglican *Book of Common Prayer* or the direction of the exiled bishops, they found it necessary to leave the congregation established in Frankfurt and move to Geneva where they could worship in accordance with their conscientious beliefs. In Geneva they set up a congregation ruled by elders and led by two elected pastors: John Knox and Christopher Goodman. They also adopted a confession of faith, an order of worship, and a form of discipline which followed the teachings of Calvin.

With the accession of Elizabeth* to the throne of England, the English Protestant exiles returned home, but Knox because of his earlier attack upon the idea of a woman's ruling a country was not permitted to go to England. Consequently he returned directly to Scotland, where the Reformation was beginning to come out into the open. Under his leadership the Protestant forces succeeded in having Parliament adopt a Reformed Confession (see SCOTS CONFESSION) in August 1560, but it did not accept the *Book of Discipline** submitted to it somewhat later, which would have established a Reformed structure of church order. The Reformed church, however, which was now established at least doctrinally, organized itself along lines that under Andrew Melville* in the latter part of the century became fully presbyterian with a hierarchy of courts extending from the local session through presbytery and synod to the national general assembly.

In England the failure of the Genevan refugees to establish a presbyterial system was due largely to Queen Elizabeth and her advisers, who disliked the popular aspects of the presbyterian form of government, favoring instead an episcopal organization that left the ultimate authority over the church in the hands of the civil authorities. Although Thomas Cartwright,* trained in Geneva under Calvin's successor Theodore Beza,* led a strong campaign to bring about a more radical reform of the Church of England, he was unsuccessful. The same was true in the seventeenth century when the Presbyterians in Parliament attempted to set up a uniform presbyterian system throughout the British Isles. The Independents* under Oliver Cromwell* and then the Anglicans under Charles II* prevented this. Only in Scotland did Presbyterianism gain the day, and only after much suffering, particularly during the Anglican persecution of the Covenanters* (1665-88). Not until 1692 was Presbyterianism finally established, although after the Union of the Parliaments of 1707 various modifications were made in the establishment by the British Parliament. These in turn led to a number of divisions within the Church of Scotland.*

With the colonial expansion of Britain during the eighteenth century, Scots and Scotch-Irish from Ulster carried with them their presbyterian form of government, doctrine, and worship to the empire, resulting in the establishment of large Presbyterian churches overseas. Consequently one finds churches of presbyterian structure and belief scattered across the globe. Although some may have in some ways modified their doctrinal views and even their form of government, fundamental presbyterian characteristics still remain.

The primary presupposition of Presbyterianism is that the risen Christ is the only head of the church. He rules His people by His Word and Spirit, directing believers as a whole. Thus there is no idea of a special elite group which has received through direct revelation or by the laying on of hands extraordinary powers or authority. Those who govern the church are chosen by all the church members, who recognize that God has given them gifts and abilities to teach and to direct the church in its life upon earth. The foundation of the church's structure is the session of the local congregation, which is elected by all communicant members and is led by the minister or "teaching elder," also known as the "moderator." The minister is chosen and called by the congregation, but is inducted into his charge by the presbytery, which is composed of the minister and the "representative" elder of each congregation within the presbytery's geographical bounds. This body has the oversight with extensive powers over all the congregations under its jurisdiction. It in turn is responsible to the synod, which is made up of representatives either appointed by a number of presbyteries or directly by the various sessions. With increased ease of communication in many churches, synods are increasingly recognized as being of no real importance, particularly since presbyteries now usually deal directly with the general assembly, which is made up of equal numbers of ministers and elders who are presbyterial representatives. The assembly, the highest court in any Presbyterian church, has final authority in all matters legislative or judicial, but in most cases a change in doctrine, government, or worship must be referred back to presbyteries under a Barrier Act* for ratification by a majority of those courts. In this way every major change must be considered and approved at the most general level of the church.

Although each Presbyterian church has its own particular standards of faith, government, and worship, the first complete statement of the Presbyterian position came from the Westminster Assembly* of Divines (1643-49), which prepared a Confession of Faith, two catechisms, a Directory of Worship, and a Form of Government, on the instructions of the English Parliament. This English body later rejected the Westminster symbols, but they were adopted by the Scottish Parliament and Church, and have since been accepted as the base upon which all other Presbyterian structures have been erected.

BIBLIOGRAPHY: A.H. Drysdale, *History of the Presbyterians in England* (1889); W.T. Latimer,

History of the Irish Presbyterians (1902); W.M. Macphail, *The Presbyterian Church* (1908); J.N. Ogilvie, *The Presbyterian Churches of Christendom* (1925); J. Moffatt, *The Presbyterian Church* (1928); J.L. Ainslie, *Doctrines of Ministerial Order in the Reformed Churches of the Sixteenth and Seventeenth Centuries* (1940); J.T. McNeill, *The History and Character of Calvinism* (1954); R.S. Louden, *The True Face of the Kirk* (1963); J.T. Cox (ed.), *Practice and Procedure in the Church of Scotland* (5th ed., 1964). W.S. REID

PRESBYTERIAN WORLD ALLIANCE, see WORLD ALLIANCE OF REFORMED CHURCHES

PRESBYTERY. The central legislative and judicial body in presbyterian polity. It is composed of equal numbers of ruling elders and ministers (teaching elders) from each congregation within its geographical bounds. It exercises episcopal oversight over all the congregations under its jurisdiction, ordains candidates for the ministry, inducts ministers into congregational charges, acts as an appeal court from decisions of sessions, and transmits petitions and overtures to the provincial or national general assembly. Changes in the constitution (faith, polity, and worship) of the church are usually referred back to presbyteries under the Barrier Act* for their approval before the general assembly takes final action. The moderator or chairman is usually elected, although attempts have been made to have bishops as permanent chairmen. The classis in the Reformed churches corresponds to the presbytery in Presbyterian churches.

See also PRESBYTERIANISM. W.S. REID

PRICHARD, RHYS (1579?-1644). Welsh clergyman and poet. Details of his family background and early education are uncertain, but he graduated from Jesus College, Oxford. He was ordained and given the living of Wytham in Essex (1602), but in August of the same year he was presented the vicarage of Llandovery and in 1613 added to it the rectorship of Llanedi—both in Carmarthenshire. In 1614 he was made a canon at Brecon Collegiate Church and in 1626 chancellor of St. Davids. He composed a substantial body of moral and religious poetry in a popular idiom, and it is symptomatic of their flavor that they were (posthumously) published by the Puritan Stephen Hughes between 1659 and 1681, the 1681 volume being a complete collection of his poetry bearing the title *Canwyll y Cymru.* The simplicity and directness of these poems, and their similarity to the folk poetry popular among the uneducated, commended them to a large public, and they soon became an integral part of the Welsh religious tradition. R. TUDUR JONES

PRIEST. The institution of priesthood is found in virtually all the great religions, usually in connection with some kind of sacrifice. The term "priest" either alone or in combination with "high" or "chief" occurs over 700 times in the OT and over 80 times in the New. Etymologically the English term "priest" is a contraction of *presbuteros,* which itself is rendered regularly in English as "elder." "Priest" renders *hiereus,* which never refers to a Christian minister in the entire NT, though in the gospels and Acts it usually refers to Jewish priests. In the OT, the pre-Mosaic order of the priesthood was patriarchal. Later a more formal priesthood appears to have emerged. Moses consecrated Aaron and his three sons (Exod. 28: 1). Next, the tribe of Levi was set aside and consecrated to the Lord (Exod. 32:26-29) and given charge of the services in the tent of meeting while only the sons of Aaron exercised the function of the priesthood. Thus the Book of Deuteronomy, which reflects the period of the monarch, refers to the levitical priests (Deut. 18:1).

In postexilic times the priesthood was divided into three orders: (1) the high priest; (2) the ordinary priest; and (3) the Levites. In theory, the members of all three orders were descendants of Levi, one of the twelve sons of Jacob. Thus the priesthood proper was confined to those Levites who were descendants of Aaron, one of Levi's grandsons and sometimes known as Aaronites. Those who could not claim kinship with Aaron became a lower order whose task it was to minister to the Aaronites (Ezek. 44:14).

In the NT, the idea of Christ as the culmination of the high priesthood (a mediator between God and man) finds expression in the Book of Hebrews, the only NT book with a specifically Jewish name. Here Christ is a High Priest (Heb. 5:10) and through His sacrifice He is able to reconcile man to God, which was the purpose symbolized by the older Jewish sacrifices of sheep and goats. But He Himself became the victim (Rev. 13:8) as well as the intercessor. In the Christian Church the "priesthood" did not emerge as a function until well after the apostolic period. Apart from a questionable reference in Ignatius, the term does not appear to have been applied to Christian ministers until c.200. Perhaps by the early fifth century, the priest had accrued the authority to administer the sacraments, and thus the way opened for the doctrine of the priesthood which would reach full flower in the medieval period. The Reformers, in general, rejected the concept of the priesthood because it had come to be seen in connection with the Mass.

The present use of the term is not limited to the Roman Catholic Church. This is perhaps due to a rediscovery of the relationship of the priesthood to Christ, rather than merely to church authority.

BIBLIOGRAPHY: E.R. Fairweather and R.F. Hettinger, *Episcopacy and Reunion* (1952); E.O. James, *The Nature and Function of Priesthood* (1955); C.C. Eastwood, *The Royal Priesthood of the Faithful* (1963); A.G. Hebert, *Apostle and Bishop* (1963). WATSON E. MILLS

PRIESTLEY, JOSEPH (1733-1804). A Nonconformist minister famous for his work in the chemistry of gases, and who also published his ideas on philosophy, religion, education, and political theory. Born in Yorkshire of a strict Calvinist family, he rejected the religion of the established churches, became a Dissenting minister, and by the age of twenty-seven was a teacher of classics and literature at a Dissenting academy in Warrington. While there he wrote *A Chart of Biogra-*

phy (1765) and *A New Chart of History* (1769) which earned him a doctor of law degree from the University of Edinburgh and a fellowship of the Royal Society.* In 1767 he became the minister of a congregation at Mill Hill, Leeds, which shared his views. Although before this he had rejected the doctrine of the Atonement and of the Trinity, now he took the final step to Unitarianism* and argued that Christ was only a man. He subsequently wrote works such as *History of Early Opinions Concerning Jesus Christ* (1786) to try to demonstrate that Unitarianism was the teaching of the early church. By 1780 he was living in Birmingham, where he was active in the Lunar Society. When the French Revolution broke out, Priestley supported it, and because of this, a Birmingham mob broke into his house and destroyed his belongings. Discouraged by this turn of events, he went to America (1794), settling in Northumberland, Pennsylvania, where he spent the last ten years of his life.

BIBLIOGRAPHY: J.T. Rutt (ed.), *The Theological and Miscellaneous Works of Joseph Priestley* (25 vols. in 26, 1817-32); T.E. Thorpe, *Joseph Priestley* (1906); E.F. Smith, *Priestley in America, 1794-1804* (1920); R.E. Crook, *A Bibliography of Joseph Priestley 1733-1804* (1966).

ROBERT G. CLOUSE

PRIMASIUS, (sixth century). Bishop of Hadrumetum, North Africa. He strongly supported pope Vigilius in the "Three Chapters"* controversy. He is chiefly remembered for his commentary on Revelation, in which he drew extensively on Tyconius and Augustine, written before 543-44 (when it was mentioned by Cassiodorus) and before he was embroiled in the controversy aroused in Africa by the "Three Chapters." Textually the commentary is important because it preserves almost completely the African Latin text of the Apocalypse.

PRIMATE. In an episcopally structured hierarchy, the title of one accorded first position or primacy for purposes of administration. It does not appear that the term was used before the seventh century. In the English Church, by ruling of Pope Innocent VI (1354), the archbishop of York was designated "Primate of England," the archbishop of Canterbury "Primate of All England."

PRIMITIVE METHODIST CHURCH. Formed in 1811 by the amalgamation of the Camp Meeting Methodists, led by Hugh Bourne,* and the Clowesites, followers of William Clowes.* These groups in turn derived from a revivalist movement which also produced the Independent Methodists in 1806. Bourne, a Staffordshire millwright, was impressed by the camp meetings* introduced into England by the revivalist preacher Lorenzo Dow, who had found them to be a great success in America and Canada. Bourne planned to hold such a gathering at Norton-on-Moors in 1807. In preparation for it members of the Harriseahead class meeting arranged for a "day's praying on Mow Cop"—a nearby promontory. This was where Primitive Methodism was born, although to call it a camp meeting is something of a misnomer.

Meanwhile, however, the Wesleyan Conference in Liverpool had passed a resolution condemning camp meetings as "highly improper" and "likely to be productive of considerable mischief." Bourne felt it right nevertheless to proceed with his plans for the Norton camp meeting, and as a result was expelled from the Burslem Quarterly Meeting. Thus in 1810 the Camp Meeting Methodists were formed. At the same time, Clowes had been similarly excommunicated, and in 1811 the two groups merged. In 1812 the name "Primitive Methodist" was officially adopted, the title being derived from the words of John Wesley* himself. The appointment of James Crawfoot as a traveling preacher is generally regarded as inaugurating the Primitive Methodist ministry. A set of rules was confirmed in 1814.

Growth was slow at first, but a revival in the Midlands led to a period of remarkable extension from 1819 to 1824. After weathering a crisis when lack of discipline threatened disintegration, the position was consolidated by the time a deed poll was executed in 1829. In 1843 the removal of the bookroom to London and the reorganization of the General Missionary Committee marked a transition toward connectionalism. Colonial missions were begun in the same year. There were already three Primitive Methodist conferences in America and a growing cause in Canada which in 1884 brought 8,000 members into the United Methodist Church of the Dominion. By 1901 a further development was recognized in the revised edition of the *Consolidated Minutes* with the replacement of the term "Connection" by "Church." In 1932 the Primitive Methodist Church joined with the Wesleyans and United Methodists to constitute the Methodist Church, bringing 222,021 members.

BIBLIOGRAPHY: H. Bourne, *History of Primitive Methodism* (1825); H.B. Kendall, *The Origin and History of the Primitive Methodist Church* (2 vols., 1905); J. Ritson, *The Romance of Primitive Methodism* (1910); A. Wilkes and J. Lovett, *Mow Cop and the Camp Meeting Movement* (1947); W.E. Farndale, *The Secret of Mow Cop* (1950); J.T. Willinson, *William Clowes* (1951) and *Hugh Bourne* (1952).

A. SKEVINGTON WOOD

PRIMUS. The title given to the presiding bishop of the Episcopal Church in Scotland. Elected by the college of bishops and addressed as "Most Reverend," he is nevertheless not given metropolitical powers. His function resembles that of the moderator of the general assembly of the Church of Scotland—i.e., he presides at his church's meetings—but unlike his Presbyterian counterpart he is not elected for one year only. The office is not attached to any one diocese.

PRINCE, H.J., see AGAPEMONISM

PRIOR. Head or deputy head of a monastery. During the early Middle Ages the word was used in a vague sense and could be applied to secular officials. Under Benedictine influence the title "prior" (or "claustral prior") came to denote the monk who ranked next to the abbot, deputizing for him and generally concerned with discipline. With the formation of the Cluniac Order in the tenth century there appeared the "conventual" prior who ruled as head of the monastery. The Canons Regular (Augustinians), Carthusians, Carmelites, Servites, and Dominicans so used the title. There is also the "obedientary" or "simple" prior, the ruler of a dependent priory.

PRIORESS. Head or deputy head of certain houses of nuns. In general the term corresponds to that of "Prior" in the equivalent male order—i.e., the "claustral prioress," assisting the abbess in the government of an abbey; the prioress of a dependent house; and the prioress of an independent or "conventual" priory. The terminology is preserved into modern times by Benedictines, Cistercians, Dominican nuns, Poor Clares, Carmelite nuns, various congregations of the Canonesses Regular, and the Brigittines.

PRIORY. Religious house presided over by a prior or prioress. In certain orders, especially those following the Augustinian Rule, the priory is the normal unit. The Benedictine and allied orders distinguish between "conventual" (self-governing) and "obedientary" (dependencies of abbeys) priories. In medieval England there was also the "cathedral" priory, where the bishop's chapter was constituted by an independent Benedictine priory.

PRISCA (Priscilla). The wife of Aquila the tentmaker (Acts 18:2). The best readings of the Pauline references give the form "Prisca"; Luke in Acts 18:2,18,26 uses the diminutive variant "Priscilla." Both writers give particular prominence to the wife, and this has been thought to show that she was of higher social standing or of greater importance in the church than her husband. Paul's association with the couple began in Corinth when they had come from Italy after Claudius's expulsion of Jews from Rome in A.D. 49. They were later his fellow-workers in Ephesus, where they instructed Apollos, and in Rome. Harnack* ingeniously urged the attribution to them of the epistle to the Hebrews. The name Priscilla is well attested, chiefly later and in Asia Minor, where its frequency may be due to the influence of Montanism.* COLIN HEMER

PRISCILLIAN (d.385). Heretical bishop of Avila in "Hispania Tarraconensis." Of noble birth, rich, learned, pious, ascetic, and eloquent, he was seemingly influenced by Gnostic doctrines brought to Spain by an Egyptian named Marcus. In the eyes of the orthodox he was soon judged a heretic; he caused real problems to the church, since his views and influence were widely spread. Soon he had many followers (Priscillianists) who included a few bishops. Eight of the canons of the Council of Saragossa (380) were directed against them. They retaliated by consecrating Priscillian bishop of Avila.

In 381 the church and empire combined to force the Priscillianists, now accused of teaching Manichaeism, into exile in France. From this point began a complicated series of appeals by Priscillian and other leaders to ecclesiastical and secular judges, to Pope Damasus,* to Ambrose* of Milan, and to Macedonius, Gratian's master of the offices, who finally restored them. However, when Maximus came to power he agreed to a trial of Priscillian and Instantius* at a synod at Bordeaux. Instantius was condemned and later banished, but Priscillian unwisely appealed to the emperor at Trier. By the latter's order he was tried by the prefect Evodius and found guilty of using "magic arts" (associated with Gnosticism*). The emperor decided he should be put to death, and so with six others he was beheaded at Trier in 385, the first people to suffer death as heretics in the history of Christianity. The teaching of Priscillian was not thereby stopped, and his body and those of the six others were conveyed to Spain and given funerals worthy of martyrs. Priscillianism was condemned at the Council of Toledo* in 400 and was still flourishing in 447.

Modern scholarship is divided on the question of whether Priscillian was a heretic or merely an eccentric enthusiast. His doctrine is known only through the statements of others, since manuscripts attributed to him are probably not genuinely his. PETER TOON

PROBABILIORISM (Lat. *probabilior*, "more likely"). A term used in Roman Catholic theology when the question was asked, What ought a Christian to do when duties seem to clash? What if Christian freedom suggests course A, but legalistic considerations point to course B? Two answers were given: (1) arguments in favor of A and B should be examined, and that course followed which seems more probably correct (hence Probabiliorism); (2) provided the nonlegalistic line of conduct can be defended, even as less probable than the legal, it may be followed (Probabilism*).

Acute controversy on the issue broke out in 1656 when Blaise Pascal,* the Probabiliorist Jansenist, in his *Lettres provinciales* lashed out at the Probabilist Jesuits for their hypocrisy, stressing that their view permitted a man to do anything, however wicked, which he could successfully rationalize. Probabiliorism was triumphant in the ensuing battle, but Probabilism was revived later (with some safeguards) by A. Liguori,* and is now official Roman Catholic teaching. R.E.D. CLARK

PROBABILISM. The view that, if there is doubt about the rightness of any course of action, that action is to be preferred which is probably right, even though the action which is *more* probably right is also the action that is in accord with the law. ("Probably" is not used in a statistical sense: a probable opinion is the opinion of some *doctor gravis et probus.*) The view, which is one of several approaches that have been adopted in areas of moral uncertainty, was unknown before the end of the sixteenth century. It was formulated, or perhaps first given prominence, by Bartholemew Medina (1527-81) and rapidly gained hold.

The rationale of the view lies in the need for casuists in the Roman Catholic Church to balance the verdicts of conflicting church authorities.

Probabilism was developed as a casuistical device by the Jesuits, as against Probabiliorism,* the view that the libertarian course of conduct is to be preferred only if it is more probable than the view that is in accordance with the law. The presence of many contradictory moral authorities in the Roman Church created endless possibilities of going against the legally accepted view by this approach.

Other competing views, besides Probabiliorism, are Rigorism, which holds that only if the less safe opinion is most probable can it be practiced, and Laxism,* which maintains that an only slightly probable opinion could be followed with good conscience.

Probabilism was one of the matters strongly controverted by the Jansenists* in their battle with the Jesuits in the seventeenth century. The Jansenists took a Probabilioristic or Rigoristic view. Pascal made Probabilism, and the ethos in which it flourished, notorious in his *Provincial Letters* (1656). Pope Alexander VII* condemned the teaching in 1665 after continued protests from the Sorbonne, and in 1700 the Assembly of the clergy of France forbade it to be taught.

PAUL HELM

PROBATIONER. In Presbyterianism the term is used of one who, having completed his theological course, has received authority to preach the Word, but has not yet been by ordination given authority to administer the sacraments. Other denominations, notably Methodists and Baptists, make use of the term with substantially the same meaning.

PROBST, FERDINAND (1816-1899). German liturgical scholar. Educated at Tübingen under Karl Hefele, he afterward concentrated on liturgical studies. He was ordained into the Roman Catholic Church in 1840, becoming parish priest at Pfärrich in 1843, professor of pastoral theology at Breslau in 1864, and finally dean of Breslau Cathedral in 1896. Despite the rigid conservatism of his views, some of his conclusions were very speculative, and hence were treated with suspicion by other scholars, in particular his belief that the Liturgy of Book VII of the fourth-century *Apostolic Constitutions** was the universal Christian eucharistic rite from the earliest days. Probst was a prolific writer, though few if any of his works were translated into English.

JOHN A. SIMPSON

PROCESSION OF THE SPIRIT. The distinguishing of the Holy Spirit from the Father and Son in the Godhead. Implied in the NT (e.g., John 15:26), it was asserted in the Niceno-Constantinopolitan Creed and first developed at length by the Cappadocian Fathers.* It signifies, as against the Macedonians, that the Spirit is not a "creature," but that His being is eternally derived from the one fount of deity.

According to the doctrine of *Double Procession,* the Spirit proceeds from both Father and Son. In the Eastern Church, Didymus the Blind,* Epiphanius,* and Cyril of Alexandria* ascribed the origin of the Spirit to both Father and Son, without actually using the term "procession" of the Spirit. This was expressly denied by Theodore of Mopsuestia* and Theodoret.* The Western Church added the formula *Filioque* (Lat. = "and the Son") to the clause in the Niceno-Constantinopolitan Creed, "the Holy Ghost which proceedeth from the Father." This arbitrary interpolation was first introduced at the Third Council of Toledo* (589). Western Fathers who supported the doctrine include Hilary of Poitiers,* Jerome,* Ambrose,* and especially Augustine* in *De Trinitate,* but Pope Leo III* (795-816) refused to authorize the use of the clause. Controversy flared when Photius* of Constantinople asserted it was contrary to the teaching of the Fathers (c.866). The interpolation received official sanction in Rome by Benedict VIII (1017) and the East-West schism followed (1054). Mark of Ephesus repeated Photius's assertion at the Council of Florence* (1439), when there was agreement on a compromise statement, but it came to nothing, as have all subsequent negotiations.

J.G.G. NORMAN

PROCLUS (d.446). Patriarch of Constantinople. Formerly secretary to and ordained priest by Atticus, patriarch of Constantinople, he was consecrated archbishop of Cyzicus in 426, though opposition kept him out of his see. A renowned preacher, he delivered a sermon on the *Theotokos** (428) before Archbishop Nestorius, which appears to have precipitated the Nestorian* controversy, although his personal involvement was minor. In 434 he became patriarch of Constantinople, proved himself a moderate supporter of orthodoxy, and gained popularity by translating John Chrysostom's body there in 438. His writings are mostly sermons and letters, sometimes attacking Jewish beliefs and morals. His *Tomus ad Armenios de fide,* discussing Christ's two natures, while addressed to the Armenians, points to the errors of Theodore of Mopsuestia (he did not name him, however). A letter containing the famous formula "*unum de Trinitate secundum carnem crucifixum,*" which became the center of the Theopaschite* controversy, is mistakenly ascribed to him.

C.G. THORNE, JR.

PROCOPIUS OF CAESAREA (d.565). Byzantine historian. Born in Palestine, he went to Constantinople at the accession of Justinian (527) and was appointed *consiliarius* (legal secretary) to the military commander, Belisarius (505-565), whom he accompanied on campaigns for the next fifteen years—details of which served as the basis of his history. After 542 his duties included that of senator in Constantinople, and probably that of city prefect from 562. His *Wars* in seven books are divided in sequence among Persian, Vandal, and Gothic campaigns, with an eighth (550-53) updating all fronts; the whole series showed a valuable objectivity. He produced also a panegyrized account of Justinian's architectural achievements in seven volumes, and *Anekdota* (commonly called "Secret History") which is a character assassination of the chief figures of the time based on unprovable accusations and at a level of scurrilous gossip.

CLYDE CURRY SMITH

PROCOPIUS OF GAZA (c.475-538). Man of letters and Christian exegete. He lived in Constantinople during the reign of Justin I (518-27), wrote poetry in the pagan tradition, and was one of the foremost members of the "School of Gaza," a group of Christian exegetes. His commentaries on the Octateuch, Isaiah, Kings, and Chronicles were constructed out of extensive extracts from the older commentators such as Philo, Basil, Theodoret, and Cyril of Alexandria. This type of commentary became known as a catena,* or chain, and achieved considerable popularity.

PROMOTER FIDEI, see DEVIL'S ADVOCATE

PROPAGANDA (Sacred Congregation for the Propagation of the Faith). The organization responsible for the direction and administration of Roman Catholic missionary activity. It was created in 1622 by Gregory XV to combat the lack of unity in missionary endeavors among the various religious orders and to weaken the firm control that Spain and Portugal exercised over missionary enterprises through the right of patronage. Under its first secretary, Francesco Ingoli, Propaganda gathered a wealth of data on missionary activities and, from this, formulated basic principles to govern later work. It laid down standards for the training of missionaries and founded a seminary *(Collegium Urbanum)*, promoted the development of an indigenous clergy, and urged the preservation of the cultural traits of non-Western peoples. The Chinese Rites Controversy,* however, severely limited the extension of the principle of cultural adaption in mission work. It set up a press in 1626 and by 1800 was printing material in forty-four African and Asian languages. Propaganda had been given supreme and exclusive authority in every mission region, but this was substantially limited by Pius X in 1908. In recent years the congregation has done much to bring about the creation of indigenous hierarchies in mission lands.

See R.H.S. Song, *The Sacred Congregation for the Propagation of the Faith* (1961).

RICHARD V. PIERARD

PROPHECY (Heb. *nebuah;* Gr. *prophēteia*). The word used in the OT to describe the message of men who spoke to the people of Israel under the inspiration of the Spirit of God. This message, often introduced with a "thus saith the Lord," came sometimes in the form of a commandment; sometimes as a promise of deliverance either in the immediate or more remote future; sometimes as a word of judgment and condemnation; and sometimes as a word of admonition, a lamentation, hymn of praise, or the like. In the written form in which OT prophecy has been preserved for us, these various types of oral utterance have been worked together either by the prophet himself or by later editors to convey his total message.

In declaring his message—be it one of reproof and admonition of the wicked, or one of comfort and consolation for the righteous—the prophet is always conscious of the fact that he speaks as God's mouthpiece, that his word is the word of Him who is the Lord of history, who declares "the end from the beginning" and makes His counsel to stand (Isa. 46:10). It is for this reason that the prophet's message embraces the future ("foretelling" as well as "forthtelling") and that the people know that what does not come to pass is not spoken of the Lord (Deut. 18:22). (The ultimate test of true prophecy, however, is not the outcome of a prediction, but faithfulness to the relevation given through Moses—Deut. 13.)

Much has been said about the nature of prophetic inspiration in an effort to understand how the word of the Lord came to certain men in a unique way; but it is doubtful that we can say more than that the prophets were personally aware of a mighty divine influence which came over them and gave the words that they used the ultimate authority of a message from God Himself. (Hence the New English Bible correctly concludes many prophetic oracles with: "This is the very word of the Lord.")

Whereas the spoken word is the primary form of prophecy, the prophets sometimes embodied the word in symbolic acts to convey their message. Isaiah walked naked and barefoot (Isa. 20); Jeremiah shattered a potter's vessel (Jer. 19); Ezekiel dug through a wall (Ezek. 12); and Ahijah rent his coat, giving ten pieces to Jeroboam (1 Kings 11:29f.). One might assimilate these symbolic acts to the visible word of the sacraments in the NT.

Prophecy, in the NT, means much the same things as in the Old, only now the word of consolation and admonition, as well as the word of prediction, relates to Christ and His kingdom and to His coming triumph over the powers of evil in the world. (In this sense the Apocalypse as a whole is called a word of prophecy—Rev. 1:3; 22:18,19.) In contrast with the OT, every Christian, in a sense, is a prophet, since the Spirit of Christ has come upon all flesh "and they shall prophesy." In the NT church, however, there seems to have been a special group called "prophets" who were next to the apostles in rank, though ministering to a single congregation ("in the church God has appointed first of all apostles, second prophets, third teachers"—1 Cor. 12:28). It would seem from this passage that Paul gave great weight to the gift of prophecy as uniquely edifying to the church. It was a gift used in the worshiping congregation; it involved the knowledge and communication of spiritual truths, resting upon inspiration; it was uttered, not in ecstatic, but rational, speech; and, in contrast with teaching, which was bound to the "tradition," prophecy had the character of revelation (see 1 Cor. 12–14 *passim*). However the "spirits" are to be "discerned" (1 Cor. 14:29), the criterion being the "analogy of the faith" (Rom. 12:6), that is, the faith which was once for all delivered to the saints by the apostles, who were the ultimate authoritative teachers in the church.

The exact content of these prophecies is difficult to ascertain, but evidently they consisted of utterances given by a sudden impulse in the form of a lofty discourse and in praise of the divine goodness and wisdom—utterances which illuminated the truth in a way that could not be achieved by reason alone. (Note the juxtaposition of "the gift of prophecy" and the understanding of "all mysteries and all knowledge" in 1 Corinthians 13:2.)

With death of the apostles, who had no successors, gradually those with the gift of prophecy also disappeared, so that from the third century onward, of the original triad of apostles, prophets, and teachers, there remained only the teachers. In the *Didache** the church is admonished to respect apostles and prophets, if they are true prophets, and instructions are given concerning them which would indicate that they were wandering teachers who might settle with a given congregation for a certain length of time.

The author of the *Didache,* Justin Martyr,* and Eusebius all discuss the problem of false prophecy. With the rise of Montanism* in the second century claiming new prophetic insights which did not correspond with the tradition received from the apostles, the church began to distinguish such prophecies from the true prophecies contained in Scripture. From this time on, the prophetic gift appears here and there, but increasingly it gives place to teaching. By the time of Hippolytus (235) and Origen (250), the word "prophecy" is limited to the prophetic portions of Scripture. In the place of the prophet one finds the teacher, specifically the catechist and apologist, who oppose all false doctrine and seek to support their exposition of true doctrine by appealing to the authoritative word of Scripture.

Enthusiasts have arisen on the fanatical fringe of the church down through the ages, claiming prophetic inspiration. Their followers, however, like those of Theudas and Judas, have, for the most part, come to nought. In our own day, Pentecostal assemblies have revived what they believe to be the NT "gift of prophecy." It differs from tongues in that it is given in intelligible speech, but unlike conventional preaching, it is unprepared and spontaneous. It is not given to fanaticism.

BIBLIOGRAPHY: A.B. Davidson, *Old Testament Prophecy* (1903); H.H. Rowley (ed.), *Studies in Old Testament Prophecy* (1946); H.A. Guy, *New Testament Prophecy, Its Origins and Significance* (1947).

PAUL KING JEWETT

PROPITIATION. This word is the correlative of "wrath" and can only be understood with reference to it. In theology it applies to the turning aside of divine wrath against sinful man. It is uncongenial to much modern thought; it is thought pagan and therefore to imply a sub-Christian view of God. In Scripture, however, it is related not only to the holiness of God (as in 1 John 2:2, noting the earlier context) but also to His love (as in 1 John 4:10). Paradoxically the God who is propitiated also lovingly provides the propitiation. The propitiation is Christ Crucified, and through His work God can be righteous and yet also (in grace) justify him who had faith in Jesus (Rom. 3:24-26). This NT idea has as its background the OT doctrine of sacrifice. There too, sacrificial blood is not only offered to God, but provided by Him (Lev. 17:11). The term is often replaced in modern translations by some weaker expression such as "expiation," but Leon Morris (*The Atonement in New Testament Teaching,* 1955, pp. 125-85) has demonstrated the inadequacy of such translation of the Greek and Hebrew words concerned. The use of the word "propitiation" serves to safeguard the penal element in the Atonement.*

G.W. GROGAN

PROSELYTE. This technical concept, coined by the Septuagint to translate the Hebrew "sojourner," came also to describe those "who had come over" *(prosēlutos)* religiously from the Hellenistic environment (cf. Matt. 23:15; Acts 2:11, etc.). In the case of males the process entailed circumcision. Proselytism was a result of active efforts of a missionary enterprise which persisted throughout the Talmudic period in spite of the changing relations with the Roman state, and over against the emergence of Christianity which carried on the activity even more vigorously while simplifying the requirements. Subsequent usage makes the terminology synonymous with "conversion," which in the case of Judaism, under Christian domination, was often prevented, seldom pursued, but never impossible.

CLYDE CURRY SMITH

PROSPER OF AQUITAINE (c.390-c.463). A scholar whose background is unknown save that he had a classical education, was learned in theology, was married, and was part of a monastic community in Marseilles at the outbreak of the Semi-Pelagian controversy* (426), which he opposed. Together with a friend, Hilary, he wrote to Augustine* in Africa (428) concerning the opposition to his theology of grace and predestination, especially among the disciples of John Cassian,* to which Augustine's reply was the *De praedestinatione sanctorum* and *De dono perseverantiae.* In 431 he went to Rome to gain Celestine I's* support for Augustine's doctrines, then published several works in their defense, with attacks on Vincent of Lérins* *(Pro Augustino responsiones)* and Cassian *(Contra collatorem),* including the *Capitula Caelestiana* which went to the bishops of Gaul as part of a papal letter. While initially in agreement, he finally rejected Augustine's position *(De vocatione omnium gentium),* believing God willed to save all men.

As secretary to Leo I* after 440, he aided him with correspondence and theological writings against the Nestorians.* His own writings were of various forms: *De ingratis,* a poem of 1,000-plus hexameters on grace; probably *Poema conjugis ad uxorem* in sixteen anacreontic verses and fifty-three distichs; a series of epigrams including those against Semi-Pelagians and *Epitaphium Nestorianae et Pelagianae haereseos;* and *Psalmorum a C ad CL expositio* after the Council of Ephesus.* *Epitoma chronicorum,* a synthesis of the chronicles of Jerome,* Sulpicius Severus,* and Orosius,* reflecting also his own time (433-55), was edited and augmented by Cassiodorus* and Paul the Deacon.*

C.G. THORNE, JR.

PROTASIUS, see GERVASIUS AND PROTASIUS

PROTESTANT DISSENTING DEPUTIES. Since 1732 these have consisted of two members (chosen annually) from each congregation of the Presbyterian, Congregationalist, and Baptist denominations in and within twelve miles of the City of London. From this large body, which has usually met only once or twice a year, the Committee of Twenty-one has been chosen in order by all legal means to lead the fight for the obtaining of full civil rights for Protestant Dissenters. The origin of the deputies is usually traced to a general meeting of Dissenters in 1732 at the Meeting House in Silver Street, London. Here the necessity of political representation to further the liberties enjoyed by Dissenters was emphasized —especially the need to gain freedom from the corporation and Test Acts (of Charles II's reign) which were still in force and were the basis for preventing Dissenters from playing full part in local government, etc. Through their committee the deputies were an effective force in eighteenth- and nineteenth-century politics; by their influence Nonconformists* gained the right to be buried in churchyards, to register their children in civil registers of births, and to enter the universities of Oxford and Cambridge without offense to their consciences. Also they saw the repeal of the Corporation and Test Acts.

For full details see B.L. Manning, *The Protestant Dissenting Deputies* (1952). The manuscript records of the Deputies are located in the Guildhall Library, London. In the twentieth century the work of the Deputies has virtually ceased, since the original objectives have all been gained.

PETER TOON

PROTESTANT EPISCOPAL CHURCH IN THE USA. The first Anglican services on North American shores took place during Martin Frobisher's Hudson Bay expedition in 1578, when Chaplain Wolfall preached and administered the sacrament. In June 1579, in the course of Sir Francis Drake's voyage along the west coast, a similar service was held near San Francisco, at which Drake's chaplain, Francis Fletcher, officiated. Various attempts to establish colonies on North American territory during this period were, however, unsuccessful. In 1607 a small band of colonists succeeded in settling at Jamestown, Virginia. There they built the first Anglican church in America, and public worship was regularly conducted by their chaplain, Robert Hunt. By 1624 Anglicanism was firmly established in Virginia.

The colonial clergy and parishes were under the jurisdiction of the bishop of London. In America there was prolonged and sometimes fierce opposition to the appointment of bishops (though not generally to the Prayer Book and its worship) on the part of non-Anglicans, in which some Anglicans shared, particularly in the South. Many of the early settlers had left England in order to escape from Laudian intolerance and the combined might of church and state, and to win for themselves freedom and independence, ecclesiastical as well as civil, in the New World. They feared that the appointment of bishops would mean the extension across the Atlantic of the lordly prelacy and royal dictation from which they had fled—hence to the present day the unremitting American insistence on complete separation of church and state. Because of this, Anglicanism suffered heavily during the American Revolution. Many ministers went over to the English side.

On 14 November 1784, Samuel Seabury* was consecrated the first bishop of the Protestant Episcopal Church in Connecticut and Rhode Island, by bishops of the Episcopal Church in Scotland. A general convention held in Philadelphia in 1789 drafted a constitution and canons and the revised Prayer Book for the PEC. The election of its bishops and the government of the church were organized along democratic lines. The church's highest council is its general convention, which meets ordinarily every three years, and its highest officer is the presiding bishop, who is elected by the general convention. The church has a membership of 3½ million members (baptized persons), of whom over 2¼ million are communicants. There are some 11,000 clergy.

The general convention's approval of the ordination of women in 1977 spawned the schismatic Anglican Church in North America amid considerable controversy.

BIBLIOGRAPHY: J.S.M. Anderson, *History of the Church of England in the Colonies and Foreign Dependencies* (3 vols., 1845); W.S. Perry, *The History of the American Episcopal Church* (2 vols., 1885); A.L. Cross, *The Anglican Episcopate and the American Colonies* (1902); W.W. Manross, *A History of the American Episcopal Church* (2nd ed., 1950); G. MacL. Brydon, *Virginia's Mother Church and the Political Conditions under Which It Grew* (2 vols., 1952); E.A. White and J.A. Dykman, *Annotated Constitution and Canons* (1952); C. Bridenbaugh, *Mitre and Sceptre* (1962); W.A. Clebsch (ed.), *Journals of the Protestant Episcopal Church in the Confederate States of America* (1962); R.W. Albright, *A History of the Protestant Episcopal Church* (1964).

PHILIP EDGCUMBE HUGHES

PROTESTANTISM. The name came from the Protestation of the German princes and cities at the Diet of Speyer* in 1529. The verb *protestari* from which the adjective "Protestant" is derived means not simply "to protest" in the sense of "to raise an objection," but also "to avow or witness or confess." Protestants believed they were confessing the primitive faith of the early church, which had been obscured by the later innovations of medieval Catholicism. More specifically, they regarded their message as a recovery of Pauline theology. Their main points were these:

(1) *Scripture and tradition.* There is a single source of revelation from which the Christian Church draws its teaching, and that source is Holy Scripture. Every doctrine which the church wishes to teach is found in Holy Scripture of necessity. This does not mean that the Protestant appeal to Scripture alone *(sola scriptura)* implies a total rejection of the tradition of the church; on the contrary, tradition is highly respected as an aid for the understanding of Holy Scripture. The wisdom of the past is not rejected, but neither is it looked upon as a second source of revelation. The Protestant Christian attempts to understand Scripture with the assistance of all those who have

labored on it before him. Nevertheless it is Scripture itself and not the exegetical traditions of the church which is the final norm of Christian doctrine.

(2) *Justification by faith.* In Catholic thought, justification is a gracious release of power which makes the Christian actually righteous. God considers the Christian to be righteous to the extent that he is conformed to the will of God and purged from the guilt of sin. That means that the Catholic expects at death to go to purgatory, where the satisfaction that he owes for his sins will be expiated. He cannot hope to enter heaven until this process is completed. The merits of Christ, which gained the sacraments for him, must be supplemented by his own merits, earned in cooperation with sacramental grace. Protestants had a quite different view. Righteousness is not a human property; it is not something which a man possesses. When a man trusts the Gospel, the good news of God's love in Christ, God pronounces him righteous, not because he already is, but because he possesses in faith the righteousness of another, the righteousness of Christ. All ideas of human merit are excluded from this understanding of justification.

(3) *Certitude of salvation.* While the Catholic Christian can have objective certitude of salvation—i.e., confidence that all the elect will be saved, that the sacraments of the church are reliable and do confer grace—he cannot have subjective certitude of salvation—i.e., confidence that he himself is elect and will finally be saved. At most he can have conjectural certitude, based on the reliability of the promises of God and the observable signs of his own growth in grace. Protestant Christians do not seek for certitude of salvation by examination of conscience or by an attempt to measure their own growth in grace. Certitude is based on the Word of God, which stands outside the self and which may even contradict the self's religious experiences. Luther* maintained that the Christian is like an invalid in hospital who has begun to get well. Looking at himself, the patient can only conclude that he is as ill as when he was admitted to hospital, but he clings to the word of the physician and trusts it. His comfort and certitude are found outside himself in the word of another. The same holds true for the Christian. He grounds his faith, not in the present state of his recovery, but on the Word of the divine Physician alone. He seeks his certitude and righteousness, not in himself, but in the absolutely naked Word of God.

(4) *Sacraments.* There is for Protestants only one means of grace: the Word. But this takes many forms; Scripture, preaching, pastoral conversation, and the sacraments. The sacraments are a visible Word of God. They do not offer the church something which it does not have when it trusts the Word of God in Scripture and proclamation, but they offer the church another mode or form of participation in that Word. Protestants accept as sacraments only those two sacraments for which there is NT warrant for believing they were established by Jesus Christ Himself: baptism* and the Lord's Supper.* Penance is rejected, or subsumed under baptism; repentance is a remembrance and reaffirmation of baptism. While there are very important differences between the various Protestant churches on the meaning of the Lord's Supper, there is fairly unanimous agreement: (a) that the Lord's Supper is not a sacrifice; (b) that there is no transubstantiation* of the elements into the body and blood of Christ—though Christ is in some sense really present, if not in the elements, at least in His body the church; (c) that living faith is important for participation in the benefits offered to the church in the Eucharist; and (d) that the service of Holy Communion* is a visible proclamation of the Gospel.

(5) *The Church.* The church is created by the gifts of God: His calling, election, Word, sacraments, and gifts of faith and love. Protestant churches, though they lack the juridical structure and hierarchy of the Roman Catholic Church, do not lack any of the elements essential to the existence of the church of Jesus Christ. Though election is hidden and faith is invisible, the church of Jesus Christ can be recognized by the signs of the proclamation of the Gospel and the proper administration of the sacraments of baptism and the Lord's Supper.

(6) *Priesthood of all believers.* The Protestant idea of the priesthood of all believers refers principally to the common right of all Christian brethren to hear the confession of sin. Luther was not opposed to confession; he was opposed to making it a clerical monopoly. All Christian brethren may hear confession, may be bearers to each other of God's Word of judgment and grace. To be such a priest is to be Christ to the neighbor, but this in no way, of course, supersedes the right of every man to have direct access to God through Christ, which needs no human intermediary.

(7) *Order and ministry.* Since every Christian is a priest, there is no spiritual difference between pastor and people, only a difference of function in the body of Christ. The Protestant minister bears an office. He may have gifts which differ from those of the layman whom he serves as pastor—but not necessarily. He does not bear an indelible sacramental character which sets him apart from laymen. He has been ordained to do publicly what all Christians have been commissioned through baptism to do privately: to bear witness to Jesus Christ. There is no question of higher and lower, but solely of order and function.

While Protestantism is a historical phenomenon which cannot be understood simply in terms of the theological convictions of the first generation of Protestant Reformers, these theological motifs have nevertheless remained in Protestantism, with greater or lessened intensity, throughout its history.

See also REFORMATION; LUTHERANISM; CALVINISM; PURITANISM; FUNDAMENTALISM; EVANGELICALISM; and entries under individual Protestant churches.

BIBLIOGRAPHY: J.S. Whale, *The Protestant Tradition* (1955); A.S. Wood, *The Inextinguishable Blaze* (1960); K. Heim, *The Nature of Protestantism* (1963); F.F. Bruce, *Tradition Old and New* (1970); D.C. Steinmetz, *Reformers in the Wings* (1971).

DAVID C. STEINMETZ

PROTESTERS. The name given to that section of the Scottish Covenanters* which regarded as criminal any dealings with Charles II* (who had been crowned on the first day of 1651 in Scotland) and which protested against the reinstatement of those formerly hostile to the Covenant. They opposed the majority of the clergy (see RESOLUTIONERS), accusing the latter of putting loyalty to the king above the rights of Christ in His church.

PROTEVANGELIUM. see JAMES, BOOK OF

PROVENCHER, JOSEPH NORBERT (1787-1853). Roman Catholic bishop in Canada. He was educated at the Nicolet Seminary in Montreal and ordained in 1811. In 1818 he was sent to Fort Douglas (Winnipeg) to minister to the people of the Red River, and two years later became coadjutor to the bishop of Quebec for the northwest. He was thus responsible for Roman Catholic policy in the west, and sought to weld together a new nation of Métis,* French, and Germans in order to preserve the French culture and Roman Catholic religion in the west. Provencher labored over thirty years among the Métis, Indians, and Eskimos of the northwest. He became bishop in 1847 of the new diocese of the Northwest, later named St. Boniface. ROBERT WILSON

PROVIDENCE, see GOD

PROVINCE. The limits of an archbishop's* or a metropolitan's* jurisdiction. The word signifies also the territorial division of certain Roman Catholic orders wherein area chiefs are known as "provincials."

PROVISORS, STATUTES OF. This name is given to several Acts of Parliament which strove to check the practice of papal "provision" or nomination (principally of foreigners) to vacant benefices over the heads of the rightful patrons. The first Statute of Provisors (1351) stated this principle broadly and provided for the expulsion of intruders. The first Statute of Praemunire* (1353) imposed penalties for controverting the same in foreign courts. The Second Statute of Praemunire (1365) confirmed these laws which, however, Edward III with the support of the papacy chose to ignore. The anger of Parliament and laity in general broke out more violently against Richard II, who breached the statutes even more flagrantly than his grandfather, and led to the Second Statute of Provisors (1390) and the Third Statute of Praemunire (1393), which enforced the previous regulations more stringently. Despite this the practice of provision continued down to the Reformation and was revived under Queen Mary (Tudor). Catholic historians have explained away these acts by arguing that they were not clashes between pope and people but between royal and papal administrations, and that papal provision was positively beneficial in comparison with the native variety. But the evidence is clear: papal provision of nonresident aliens was a scandal and contributed directly to the Reformation.

IAN SELLERS

PROVOOST, SAMUEL (1742-1815). First Protestant Episcopal bishop of New York. He graduated from King's College (Columbia), New York, and later went to England for study at Cambridge. He was ordained deacon in 1766 and priest shortly afterward. After resigning an assistant pastorate because he felt his pro-Whig politics offended Tory parishioners, he served in the Revolution as chaplain of the Continental Congress. In 1786 he was consecrated in England as a bishop, without the customary oath of allegiance to the king. He aided the formation of the new Episcopal Church, but was forced to resign in 1801 because of poor health.

PROVOST. This ecclesiastical title was used in pre-Reformation England to signify the head of some collegiate churches. It is still used in the Church of England to denote the head of the cathedral chapter in "New Foundations." In Scotland it was earlier used in a similar sense, and still is in the Episcopal Church.

PRUDENTIUS (Galindo) (d.861). Bishop of Troyes. He left his native Spain as a youth, probably due to the Saracen persecution, and went to the Frankish Empire, changing his name, Galindo, to Prudentius. Educated at the Palatine School, he was chaplain at the court of Louis the Pious before becoming bishop of Troyes about 845. He supported the Augustinian position in the controversy on predestination between Hincmar* and Gottschalk,* opposing the former first in an epistle and then in *De Praedestinatione contra Johannem Scotum* (852), clearly denying the general salvation of all men. At the Synod of Quiercy* (853), however, he apparently subscribed to the four anti-Augustinian propositions, either out of reverence for Hincmar or fear of Charles the Bald—only to negate that with *Epistola tractoria* (c.856), addressed to Venilo, archbishop of Sens, upholding his former Augustinian position even more strictly. He continued the *Annales Bertiniani* for the years 835-61, valuable for Frankish history, and wrote *Vita Sanctae Maurae Virginis*.

C.G. THORNE, JR.

PRUDENTIUS CLEMENS, AURELIUS (348-c.410). Christian Latin poet. Born in Spain, he is best known for the hymn "Of the Father's love begotten." Lawyer and civil servant, he published his extensive collection of poems at the age of fifty-seven. He applied classical Latin verse-forms to Christian teaching, and although the educational and moral outweighs the imaginative and lyrical, his poetry has considerable artistic merit. Plain, but concise and effective in description, he writes with great fluency in an exceptional variety of meters (still quantitative, not accentual). He extends the allegorical method of biblical interpretation (e.g., see AMBROSE, ORIGEN) to nature, society, and wherever he sees the possibility of pointing a Christian moral.

In the *Psychomachia* (915 lines), an extended allegory, the virtues battle with the vices for possession of the soul. Three slightly longer poems defend orthodoxy against pagans and heretics, combining poetry and doctrine in a manner recalling Lucretius's exposition of Epicureanism.* Twelve hymns and fourteen poems on Christian martyrs exceed the usual length of classical lyric. But typical of his instructional aims are forty-eight four-line biblical "snapshots" *(Dittochaei)*—e.g., No. 30, "Christ is baptised":

> The Baptist plunges men in water, his diet
> Locust and honey, his clothing camel-skins.
> When Christ bathed with those bathers, heaven's Spirit declared
> That this bather absolves every bather's sins.

BIBLIOGRAPHY: A.S. Walpole, *Early Latin Hymns* (1922); J. Bergman (ed.), *Prudentius* (*CSEL* vol. 61, 1926); B.M. Peebles, *The Poet Prudentius* (1951); J.E. Raby, *A History of Christian Latin Poetry* (2nd ed., 1953); R. Herzog, *Die allegorische Dichtkunst des Prudentius,* Zetemata 42 (1966). GORDON C. NEAL

PRYNNE, WILLIAM (1600-1669). Puritan pamphleteer. Born in Somerset and educated at Bath Grammar School and Oriel College, Oxford, he was called to the bar in 1628. In Lincoln's Inn he was influenced by the preaching of John Preston. He developed a strong opposition to Arminianism,* the attacking of which was the subject of his early books. Also he sought to reform the morals of his age, his lengthy *Histriomastix* (1632) being an exposure of the immorality of stage plays. For his outspoken criticisms, which were resented at court, he was imprisoned, fined, and pilloried, losing both his ears. Nothing, however, could stop the flow of pamphlets from his pen.

When the Long Parliament met in 1640, he was soon released from prison and restored to his membership of Lincoln's Inn. He defended the parliamentary cause and attacked prelacy, being especially active in preparing the case against Archbishop Laud.* From 1645 he turned his attention to defending Erastian principles and attacking Independency* and, later, the Commonwealth government. Remaining fairly quiet in the Protectorate of Oliver Cromwell,* he returned to active political activity in 1659 and helped to restore Charles II* to the throne. He was a member of the Convention and Cavalier parliaments and therein argued the case of the Presbyterians and of the need to comprehend them within the Church of England. His books and pamphlets total about 200.

See W.L. Lamont, *Marginal Prynne* (1966).

PETER TOON

PSELLUS, (CONSTANTINE) MICHAEL (1018-c.1078). Byzantine scholar and statesman. Born in Nicomedia, he lived through fourteen Byzantine administrations and provided character sketches of all the rulers from Basil II (d.1025) to Michael VII (abdicated 1078) in his *Chronographia.* The greatest detail covers Constantine IX (1042-55). In contrast with other writers over the same period, he avoided the context of universal history; thus his work became court annals weak in foreign affairs and without even the variety of natural disasters typical of Byzantine chroniclers. Psellus studied law at Constantinople and turned also to philosophy, including Plato, Aristotle, and the Neoplatonists. Under Michael IV (1034-41) he held an imperial appointment as judge at Philadelphia; under Constantine he was head of the faculty of philosophy at the newly founded Constantinople University. His preference for Aristotle was against the Platonic tendency of the church of the time.

The schism of Eastern and Western churches of 1054 left no apparent mark for his history, but Constantine's turning against him at that date and his withdrawal to the monastery of Olympus in Mysia (whence comes his monastic name Michael) can hardly be unrelated. In 1057 he was recalled by Theodora, and under Isaac Comnenus (1057-59) and his successors he held the office of imperial secretary. With the accession of his imperial pupil Michael VII (1071-78) he became prime minister. He was responsible for the dismissal of Michael Cerularius as patriarch in 1058, yet he made a laudatory oration at the latter's funeral in 1059. He was clearly a man of his times, a fact borne out by his ability to accommodate himself and be of service with increasing influence.

CLYDE CURRY SMITH

PSEUDEPIGRAPHA. The term is the technical designation of a large collection of Jewish writings not included in the OT canon, ranging from 200 B.C. to A.D. 200, some of which contain Christian additions. Written in Hebrew, Aramaic, and Greek, they include apocalypses, legendary histories, psalms, and wisdom literature. The fact that some of them are ascribed to Adam, Enoch, Moses, Isaiah, and Ezra caused them to be known in Protestant circles as "Pseudepigrapha." In the Roman Catholic Church they are called "apocryphal," which is to be distinguished from those other writings known by Protestants as the "Apocrypha," since the latter are known to Roman Catholics as "deuterocanonical." The term "Pseudepigrapha" is unsatisfactory in that it fastens attention on a feature of the literature which is not of major importance. In any event, the material contains many works which are anonymous. The rabbinical designation, "outside books," is used by some scholars, but this also gives rise to difficulties since other such writings, particularly the Dead Sea Scrolls,* would properly come under this category. In view of its usage over the years it is probably wisest to retain the term "Pseudepigrapha," on the understanding that it is being used in a technical sense without any judgment being expressed as to the nature of the contents.

There is no agreed order in the arrangement of the material. Generally it is classified according to its Palestinian (written in Hebrew or Aramaic) or Hellenistic (Greek) origin, and dated as follows.

(1) *Palestinian.*
- 1 Enoch, or Ethiopic Enoch (165-80 B.C.)
- Book of Jubilees
- Testaments of the Twelve Patriarchs (140-110 B.C.)
- Psalms of Solomon (70-40 B.C.)
- Testament of Job (first century B.C.)
- Assumption of Moses (A.D. 7-28)
- Lives of the Prophets (first century A.D.)
- Martyrdom of Isaiah (A.D. 1-50)

Testament of Abraham (A.D. 1-50)
Apocalypse of Abraham (A.D. 70-100)
Apocalypse of Baruch, or 2 Baruch (A.D. 50-100)
Life of Adam and Eve, or the Apocalypse of Moses (A.D. 86-110)

(2) *Hellenistic.*
Letter of Aristeas (200 B.C.-A.D. 33)
Sibylline Oracles (fifteen books, three of which are missing; written over six centuries, some by Christian authors; Book III dates 150-120 B.C., IV c. A.D. 80, V before A.D. 130)
3 Maccabees (toward end of first century B.C.)
4 Maccabees (toward end of first century B.C.
2 Enoch, otherwise known as Slavonic Enoch or the Book of the Secrets of Enoch (A.D. 1-50)
3 Baruch, or the Apocalypse of Baruch (A.D. 100-174)

The largest and most influential of the surviving Pseudepigrapha is *1 Enoch.* The story in Genesis 5:18-24 gave rise to the belief that Enoch was taken into heaven and shown the secrets of God. *1 Enoch* purports to describe Enoch's experiences. It contains visions, cosmology, angelogy, demonology, eschatology, the vindication of the Jews, the heavenly Jerusalem, the tree of life, eternal rewards and punishments. One is reminded of the Book of Revelation. Of particular interest to Christian readers is the description of the Messiah in the middle section of the book, chapters 37-71 which are known as the "Parables of Similitudes." He is called "the Elect One, the Son of Man." He is a preexistent heavenly being, who dwells close to God and pronounces judgment upon men and angels (46:1ff.).

By contrast, the *Psalms of Solomon* reflect the more staid Pharisaic background. Modeled on the canonical Psalter, these eighteen psalms contain such traditional themes as sorrow and consolation, exhortation and praise, human injustice and divine mercy, punishment for the wicked and rewards for the righteous. The nation is depicted as divided into two groups, the righteous (almost entirely the Pharisees to whom the author belongs) and the sinners (the Sadducees). Jerusalem has been captured and the Temple plundered by an arrogant Gentile invader (probably Pompey). The description of the Messiah in Psalms 17 and 18 is especially interesting: he is of the house of David, and will cleanse Jerusalem, gathering together the people of God and ruling both Jews and Gentiles.

Significant for its information of the origin on the Septuagint* is the *Letter of Aristeas.* The author claims to be an official of Ptolemy Philadelphus, whom, he claims, instigated what transpired to be a miraculous translation of the Scriptures into Greek. The legendary nature of the work detracts from its historical value, but the writing remains an interesting witness to the fusion of Jewish and Greek traditions in the Dispersion.

An ambitious example of Hellenistic Judaism is the *Sibylline Oracles.* Assuming the role of the great pagan prophetess Sibyl, the different authors of the Jewish Oracles sought on the one hand to keep alive the ancient hope of Israel and on the other hand to demonstrate to sophisticated Greeks that Zion easily vies with Hellas for the heart and mind of humanity. A few of the Oracles are either of Christian origin or edited by Christian writers.

Although the literary genre of the Pseudepigrapha differs greatly, it all has one common aim: to keep alive the faith of the Jews by offering a theology for the times.

The prophets of the OT had interpreted historical vicissitudes as the righteous judgment of God upon Israel for her sins. They promised forgiveness and restoration if Israel repented and obeyed God. Now a new interpretation seemed necessary. The centuries had passed, but Israel still suffered. It could not be reasserted that the nation was unfaithful, since the law was venerated and obeyed. Consequently faith was sorely tested. Hence the writers of the Pseudepigrapha, like those of the Apocrypha, sought to defend God and help Israel.

For the most part, they used the apocalyptic interpretation of history. They sought to explain the misfortunes of the nation by saying that God had consigned the world to evil. Deliverance would come not in this age but in the age to come. Since the world was irretrievably wicked, the most that the righteous could hope to do was to preserve their souls. This would be achieved by devotion to the law of God. At the great judgment day all will be called to give an account, the dead as well as the living. Thus those who grieve over the eclipse of righteousness can take comfort.

The message of the Pseudepigrapha made life tolerable for the Jews. It reaffirmed belief in the sovereignty of God and in His care for His dispossessed people. The description of the Messiah in *1 Enoch* on the one hand and in the *Psalms of Solomon* on the other hand is evidence of the increasingly important role which this individual was expected to play in the great drama of the last days. But it is the doctrine of God in these writings which was most influential. God was now conceived of on a grand scale. Elaborate time schemes represent His rule as extending backward to Creation and forward to the age to come. It was this which gave history its unity and meaning. It also helped the Jews grow cosmopolitan in outlook. Israel was encouraged to see herself, not as an inferior downtrodden race, but as playing the central role in the panorama of world events. Prophetic insights were thus taken over and developed to meet the needs of the new situation.

The part played in this development by extraneous influence is often emphasized. Evidence of the Zoroastrian idea of a struggle between good and evil is clear enough in these writings. But the extent to which imported concepts were reshaped in terms of the traditional faith of Israel is more remarkable than is sometimes allowed. Thus the dualistic principle of Persian religion was strongly subordinated to the monotheistic one inherent in the Israelite tradition. Influenced as they were by ideas foreign to their biblical heritage, the pseudepigraphal writers sought to remain true to this heritage, while developing and enriching it in ways which would answer the questions of the times in which they lived. One may say that the theology of these writers is essentially a theology of hope.

The value of the Pseudepigrapha for Jews and Christians alike is considerable. No serious student can pass from a study of the OT to the study of either the NT or rabbinical Judaism without considering these writings. Along with the Dead Sea Scrolls they form an indispensable background for understanding the developments which took place after the OT was written. The discovery of fragments of these writings in the caves of Qumran serves to emphasize their importance.

BIBLIOGRAPHY: R.H. Charles (ed.), *The Apocrypha and Pseudepigrapha of the Old Testament* (2 vols., 1913); C.C. Torrey, *The Apocryphal Literature* (1945); D.S. Russell, *Between the Testaments* (1960).

R.J. MC KELVEY

PSILANTHROPISM (from Gr. = "a mere man"). The doctrine that Christ was only "man" and not "truly God and truly man" in one person. According to *The Little Labyrinth,* quoted by Eusebius, it was taught by Theodotus the Cobbler, the second-century Adoptianist Monarchian. Christ had the status of "a mere man" whom the Spirit inspired. The Ebionites,* Artemas,* and Paul of Samosata* all held similar views. The single word *psilanthrōpos* apparently occurs for the first time in Anastasius of Sinai, while the noun was probably of nineteenth-century origin, coined by S.T. Coleridge. "Humanitarianism" is also used to describe the doctrine.

PSYCHOLOGY OF RELIGION. An attempt to apply scientific methods to the study of the facts of the religious consciousness. Interest in and concern for "religious affections" has always been present in the Christian Church, but the Romantic Movement, coming in the wake of Kant's* denial of the possibility of the knowledge of God, led to a concern for "religion" considered as a postulation or projection of the existence of God (for a variety of reasons) and for a delineation of those states of consciousness that ought properly to be called "religious." The views of Schleiermacher* and Feuerbach are of importance here.

Besides offering phenomenological accounts of religious experiences such as conversion, sanctification, and mystical experience, psychology of religion has been concerned also to offer explanations of such phenomena in causal terms correlating, e.g., conversion with factors such as age, sex, various personality traits, family upbringing, and so on. William James in his *Varieties of Religious Experience* is notable for his antireductionism, but usually the thrust of such explanations has been to show that religious experiences are *nothing but* compensatory devices, wish-fulfillment (Freud), etc. Another line of inquiry, more sympathetic to religion, attempts to isolate those features that all and only genuine religious experiences have—a feeling of absolute dependence (Schleiermacher), an awareness of the numinous (R. Otto*). But it may be that such attempts are doomed to frustration if it is denied (as it usually has been in post-Kantian Protestantism) that religious experiences have characteristic *objects.* Work in the psychology of religion has also been stimulated by comparative religion.*

Although students of religious psychology have unquestionably turned up much interesting data, and offered engaging hypotheses as explanations of religious phenomena, serious psychological explanation has often been vitiated by naturalistic and anti-Christian ideological assumptions.

OONAGH MC DONALD

PUFENDORF, SAMUEL (1632-1694). German philosopher. Born at Dorf-Chemnitz, son of a pastor, he became professor at Heidelberg (1661), at Lund (1670), historiographer to the court of Sweden (1677), and privy councillor to the elector of Brandenburg (1687). The first German professor of natural and international law, he elaborated, in a notable essay written in 1672, on the ideas of Grotius,* basing natural law on the instinctual responses of society and stressing its independence of revelation. Theology he treated as a type of mathematics, incurring the opposition of the orthodox theologians of Jena and Leipzig. In an essay on the relationship of the Christian religion to civil society (1687) he advocated the theory known as Collegialism,* asserting the voluntary nature of the church and criticizing the "territorialism"* of previous theorists. He also wrote historical accounts of the European kingdoms, of the papacy, Lutheranism, and of the Prussian royal house.

IAN SELLERS

PULCHERIA (399-453). Eastern empress from 450. Daughter of Arcadius, East Roman emperor from 395 to 408, she was made regent at fifteen for her younger brother Theodosius II* by the Constantinopolitan senate (414). Under the *de facto* rule of this pious saint of the Greek Church, the court assumed a charitable and ascetic character. Pulcheria arranged the marriage of Theodosius to Athenais, daughter of Leontius, a pagan philosopher of Athens (421). Assuming the name Eudocia, Athenais became a Christian and personal rival to Pulcheria. They differed in the Monophysite (Eutychian) and Nestorian controversies, Pulcheria espousing orthodoxy in both. We have Cyril* of Alexandria's letter to both women condemning the views of Nestorius, patriarch of Constantinople from 428 until the Council of Ephesus (431).

In 438 Pulcheria healed the thirty-year religio-political schism in Constantinople by returning John Chrysostom's bones to be interred in the Church of the Apostles there. Her temporary eclipse and departure from court about 440 resulted from the conspiracy of Eudocia with the eunuch prime minister Chrysaphius. Pulcheria returned in 450, Eudocia having retired to Jerusalem estranged from her husband, and upon Theodosius's death became empress and nominal wife of Marcian, now Eastern emperor (until 457). The Council of Chalcedon* (451), which condemned Eutychianism* and Nestorianism,* was convened at her order. Pulcheria founded three churches to Mary and left her possessions to the poor. DANIEL C. SCAVONE

PULE, HERMANO, see DE LA CRUZ, APOLINARIO

PULLEN, ROBERT (d. c.1146). English theologian and "Sentence writer." Born in England probably about 1080, he studied at Paris under William of Champeaux and Abelard.* By 1133 he was teaching Scripture at Oxford and was also archdeacon of Rochester, but with the troubles following the death of Henry I (1135), he returned to Paris to teach logic and theology, and among his pupils was John of Salisbury. Innocent II, influenced by Bernard of Clairvaux, summoned him to Rome, and in 1144 Pullen was made a cardinal. In 1145 he became chancellor of the Holy Roman Church. In his "Sentences" he tried to unify theological contradictions by the dialectical and Aristotelian methods.

JOHN A. SIMPSON

PUNSHON, WILLIAM MORLEY (1824-1881). Methodist minister. Born, educated, and ordained (1849) in England, he served there for eighteen years, then went to the Metropolitan Church in Toronto in 1868. He returned to England in 1873. A man of "moving eloquence," he had a vision of Canadian Methodism united from Atlantic to Pacific, and was the prime mover behind the negotiations for the unification which led in 1874 to the emergence of the Methodist Church of Canada.

PURCELL, HENRY (1659-1695). English composer. The Purcell family produced several musicians active in the Restoration period of English history. Henry is considered the only great and original genius between Elizabethan times and the renaissance of music that occurred in England with Elgar and Vaughan Williams at the end of the nineteenth century. During that period, except for relatively minor figures, foreign-born composers dominated the British musical scene. Purcell was organist of Westminister Abbey from 1679 until his death. He was also active in the Chapel Royal until its virtual dissolution with the advent of James II. His anthems were written for that institution, many of them with exacting solos for the famous bass, John Gostling. Purcell wrote much secular music throughout his short life, but his best anthems still rank high among his works. The simpler ones have been most performed in more recent times. Best known is his beautiful "Rejoice in the Lord alway," normally rendered without the elaborate string interludes ("symphonies") he wrote for the chapel of Charles II.

J.B. MAC MILLAN

PURDIE, JAMES EUSTACE (1880-). Canadian clergyman and educator. Born in Charlottetown, Prince Edward Island, he graduated from Wycliffe Theological College, Toronto, was ordained in 1907, then ministered successively in several Anglican parishes in E and W Canada and in the USA. He was noted for his evangelistic fervor and his deeper-life and healing ministries. While rector of St. James', Saskatoon, in 1919, he received a Pentecostal spiritual experience including "speaking in tongues." This brought contact with leaders of the emerging Pentecostal assemblies of Canada; subsequently he became principal of the first Pentecostal Bible college in Canada, which post in Winnipeg he retained until partial retirement in 1950. His theological materials formed the basic curriculum for other Pentecostal colleges at home and overseas. Other writings include a catechism, *Concerning the Faith* (1951), which was translated into several languages, and a booklet, *What We Believe* (1954). He deserves much of the credit for the stability and basic orthodoxy of much of the early twentieth-century charismatic revival in Canada.

KENNETH R. DAVIS

PURGATORY. Roman Catholic theology maintains that while eternal punishment and the guilt of moral sin is absolved by the sacrament of penance, the requirement of satisfaction and temporal as opposed to eternal punishment is not. If appropriate satisfaction has not been made for sins committed and absolved in life, then satisfaction must be made after death. Purgatory is not hell, since all the souls in purgatory are on their way to the heavenly Jerusalem, though it is a place of temporal punishment. The Catholic Church has usually appealed to texts such as *2 Maccabees* 12:39-45; Matthew 12:31ff.; 1 Corinthians 3:11-15 to support the idea of an intermediate place between heaven and hell, where the unfinished business of earth is settled.

The notion of purgatory appears fairly early in the writings of the Greek Fathers. Clement of Alexandria* near the end of the second century alludes to the sanctification of deathbed penitents by purifying fire in the next life. Even when the Greek Fathers do not talk about purgatory, they do advocate prayers and eucharistic services on behalf of the dead. The Latin Fathers echo these sentiments. Augustine,* for example, teaches purification through suffering in the afterlife. The medieval doctors systematized and developed the patristic heritage, teaching that the smallest pain in purgatory is greater than the greatest pain on earth, though the souls in purgatory are comforted by the knowledge that they are among the saved and are aided by the prayers and Masses offered for them by the church. The doctrine was developed and popularized by Gregory the Great,* and Thomas Aquinas* gave the idea greater elaboration.

The Greek Church had difficulty with the final form of the Latin doctrine of purgatory, rejecting the notion of atonement through suffering and the idea of material fire. The Greeks and Latins, however, were able to agree at Florence* (1439) that there is such a place as purgatory, and that prayers for the dead* are both useful and appropriate. In the West, purgatory was questioned and categorically denied by the Protestant Reformers, but was reaffirmed at the Council of Trent.*

See B. Bartmann, *Purgatory* (1936), and H. Berkhof, *Well-Founded Hope* (1969).

DAVID C. STEINMETZ

PURITANS; PURITANISM. Initially a movement within the English Church during the reign of Elizabeth I,* whose general aim was to implement a full Calvinistic reformation in England, Puritanism later also became a way of life, an interpretation of the Christian pilgrimage in terms of an emphasis upon personal regeneration and sanctification, household prayers, and strict morality.

The Bible, interpreted in the spirit of the early continental Reformers (e.g., Bullinger* and Beza*), was held by Puritans to be the only valid source from which doctrine, liturgy, church polity, and personal religion should be constructed. The spread of biblical theology was seen as the only way to halt the advance of Antichrist (Roman Catholicism). Bible reading in the homes from the annotated Geneva Bible* was encouraged. So also was regular biblical preaching from parish pulpits and weekly catechizing of parishioners in their homes. Various schemes were put forward and executed to train more preaching ministers (e.g., the founding of Emmanuel College, Cambridge).

The history of Puritanism may be divided into three periods: (1) from the accession of Queen Elizabeth to the crushing of the Presbyterian movement by her in 1593; (2) from 1593 to the calling of the Long Parliament in 1640; and (3) from 1640 to the restoration of Charles II* in 1660.

From 1559 to 1593 the governing classes became Protestant, the House of Commons created a Protestant National Church, and the queen decided in favor of a traditional diocesan episcopate rather than a Reformed episcopate favored by some of her subjects. Returning from exile in the Rhineland and Switzerland from 1558 onward, convinced Protestants had great hope for the Elizabethan church, but they were disappointed with the Settlement of Religion (1559) and what followed it, since they felt that too many relics of Roman Catholicism were preserved. They and their friends in Parliament pressed for further reformation according to the Word of God and the example of the best Reformed churches. Some people called them "Puritans," since they wanted to purify the church of all ceremonies, vestments, and customs inherited from the medieval church. Certain clerical Puritans also wanted to reform the polity of the church along presbyterian lines, but Elizabeth would have none of this.

After James I* (VI of Scotland) made it clear at the Hampton Court Conference* (1604) that he did not intend to make any important changes in the church, Puritans—especially ministers—faced real problems. Many compromised to the extent that they gave a minimum conformity and then used the parish as a center of evangelism by means of preaching and catechizing. Others became lecturers and preached on market days and other agreed times, being financially supported by voluntary gifts, not tithes. Yet others became Separatists* and of these some went to Holland and New England (e.g., the Pilgrim Fathers). After 1630 there was a large exodus of Puritans to Massachusetts, where they sought to create a purified Church of England, as an example to the homeland.

In 1640 Puritans were united in their desire to purify the national church and remove prelacy. Thus they were the religious force behind Parliament in the civil wars. They preached and fought for the opportunity to create a godly nation before the last days of the age dawned. However, the atmosphere of freedom that war brought led to open divisions in the Puritan movement. With the execution of Charles I and the advent of the Commonwealth and Protectorate, the Puritans became divided and opposed to each other. Cromwell in his liberally conceived national church sought to unite them, but it was not possible. Henceforth there were Presbyterians, Congregationalists, Baptists, Quakers, and other groups; the Clarendon Code* of the Cavalier Parliament ensured that the former Puritans remained outside the church. Thus Nonconformity was born. The Puritan spirit continued in various ways—e.g., in the emphasis on practical divinity and on sabbatarianism—but the Puritan ideal of the Reformed nation and church was gone forever.

BIBLIOGRAPHY: P. Miller, *Orthodoxy in Massachusetts* (1933); W. Haller, *The Rise of Puritanism* (1938); C. Hill, *Society and Puritanism* (1964); P. Collinson, *The Elizabethan Puritan Movement* (1967); P. Toon, *Puritans and Calvinism* (1973).

PETER TOON

PURVEY, JOHN (c.1353-c.1428). Colleague of John Wycliffe.* Said to have been a native of Lathbury, Buckinghamshire, and possibly educated at Oxford, though he was never referred to as a graduate, he was ordained in 1377. He was closely associated with Wycliffe at Lutterworth, and to him is attributed the revision into vernacular idiom of the verbatim translation of the Vulgate into English by Wycliffe and Nicholas of Hereford.* This revision was more intelligible and less pedantic and was probably completed in Bristol between 1388 and 1395. In 1387 Purvey was forbidden by the bishop of Worcester to itinerate in his diocese. He was imprisoned at Saltwood, Archbishop Arundel's castle, and in March 1401—just before the passing of the act *De Heretico Comburendo* and at the end of the week in which the first Lollard martyr, William Sawtrey,* was burned—he recanted. He was inducted to the vicarage of West Hythe, Kent, but resigned in 1403 because he could not conscientiously abandon his Lollard convictions. He was imprisoned in 1421 by Archbishop Chicheley, and was alive in 1427, but nothing further is known of him. Presumably he continued to disseminate Lollard doctrines as circumstances allowed.

J.G.G. NORMAN

PUSEY, E(DWARD) B(OUVERIE) (1800-1882). Leader of the Oxford Movement.* Educated at Eton and Christ Church, Oxford, where he graduated in 1819, he became fellow of Oriel College in 1822. During 1825-27 he studied biblical criticism in Germany, and in the process acquired a good knowledge of oriental languages. Although the book he published on his return was understood to sympathize with the rationalism he had studied in Germany, Pusey denied this, and all his subsequent biblical work was strongly conservative; the best known is his commentary on the Minor Prophets and on Daniel. In 1828 he was appointed regius professor of Hebrew. As a member of Oriel he was already acquainted with Keble* and Newman,* and when the latter began the *Tracts for the Times* in 1833, Pusey also contributed. Two of them, one on baptism and the other on the Eucharist, were much longer than the previous tracts. Pusey was instrumental also in the publication of the Oxford *Library of the Fathers.*

Outside his academic studies he further contributed to the Oxford Movement by his opposition to Dr. Hampden's appointment as professor of theology, and by his support of Newman in the storm over the publication of Tract 90. In 1843 he was inhibited as a university preacher because of a sermon on the Eucharist. When in 1845 Newman seceded to the Roman Church, Pusey became the best-known figure in the Church of England. The very fact of his staying in the Church of England retained many others who might otherwise have left also. He founded sisterhoods, encouraged private confession, and supported the revival of ritualism, though he himself kept to a very simple ceremonial.

His own desire for reunion with the Roman Church led to his publishing in three stages his *Eirenicon,* which met with a disappointing response, especially after Vatican I* in 1870. In later years, however, he was more occupied with combating the growing strength of liberalism represented in Oxford by Benjamin Jowett,* in publications such as *Essays and Reviews* (1860), and in proposals to truncate or omit the Athanasian Creed.* A man of great personal devotion, his private life was haunted by tragedy: his wife, to whom he was devoted, died after eleven years of marriage, and all but one of his children predeceased him. After his death Pusey House in Oxford was founded as a center for theological study, and this contains his library.

See Lives by H.P. Liddon (4 vols., 1893: full list of Pusey's works) and G.L. Prestige (1933).

PETER S. DAWES

PUSEY, PHILIP EDWARD (1830-1880). English scholar. Only son of E.B. Pusey,* he was both deaf and crippled. He graduated from Oxford in 1854, and when his physical defects prevented his ordination, he resolved to devote his life to helping his father. This he did by preparing a critical edition of the Peshitta and the works of Cyril of Alexandria. Despite his ill-health, he pursued his studies with exemplary thoroughness, visiting libraries throughout Europe and the Near East.

PUSEYISM, see TRACTARIANISM

QUADRAGESIMA. The forty days of Lent. The term has been used also to denote the first Sunday in Lent.

QUADRATUS (second century). This name heads a list of apologists from the early second century, with the few references placing him within the administrative period of Emperor Hadrian* (117-38). These specifically reflect the impact of the incipient Christian Church as a result of imperial policy and rescripts against Christianity by Hadrian and his predecessor Trajan. The occasion for Quadratus's *Apology* may have been Hadrian's visit to Athens in the winter of 124/5. Other than a brief fragment quoted by Eusebius, this writing is assumed lost, though some suggest it is preserved in the *Epistle to Diognetus,** with the Eusebian fragment fitted into the lacuna. The Quadratus who became bishop of Athens subsequent to the martyrdom of Publius could be the apologist. CLYDE CURRY SMITH

QUAKERS, see FRIENDS, SOCIETY OF

QUARLES, FRANCIS (1592-1644). English poet. Born near Romford, he was educated at Cambridge, studied law at Lincoln's Inn, and held official positions in royal and episcopal service. In 1639 he became city chronologer of London. He supported the king in the Civil War and as a result suffered the loss by plunder of his manuscript collection. His works include *A Feast for Worms* (1620), *Sion's Elegies* (1625) and *Divine Emblems* (1635). Such fame as he has rests upon this last work. The emblems were drawn from two continental collections, Herman Hugo's *Pia Desideria* (1624) and Philippe de Mallery's *Typus Mundi* (1627). They form a version of the Bible moralized, in which Quarles includes a good deal of that paradoxical mode now often associated with the Metaphysical poets.

ARTHUR POLLARD

QUARRIER, WILLIAM (1829-1903). Founder of the Orphan Homes of Scotland. Born into a humble Greenock home, fatherless at three, starving Glasgow slum-dweller at five, he was working the following year for one shilling a week and became a journeyman shoemaker at the early age of twelve. He probably had no formal schooling. Converted at seventeen, he never forgot the plight of children such as he had known. So the work began: a shoeblack brigade, a news(paper) brigade, a parcels brigade, an orphanage in Glasgow, the sending of residents to new lives in Canada. In 1878 the first cottage homes were opened at Bridge of Weir, in Dr. Barnardo's presence. Then came the first tuberculosis sanatorium in Scotland. Quarrier made no appeals, had no collectors, bazaars, or entertainments for money-raising purposes, relying on God's supply. When Quarrier died, having arranged for the country's first (and only) colony for epileptics, 1,526 children were in care at Bridge of Weir, with kindred work in Glasgow, Argyllshire, and Canada.

J.D. DOUGLAS

QUARTODECIMANISM. An early church practice, especially in Asia Minor, celebrating Easter on 14 Nisan *(die quarta decima),* the day of the Jewish Passover. About 155, Polycarp,* bishop of Smyrna, tried unsuccessfully to persuade Pope Anicetus to adopt Quartodeciman practice. Pope Victor* was determined that Quartodecimans should join the rest of Christendom in observing Sunday as the day of resurrection. There seemed legitimate objection to observing the chief Christian feast on the same day as the Jewish Passover. Irenaeus* of Lyons protested against Victor's efforts, and the Quartodecimans later separated and continued as a sect into the fifth century. It has been argued that the two traditions were not concerned to commemorate the same event, but rather were complementary festivals, both rooted in the Israelite calendar. Quartodecimans did not mourn Christ's death; only with the Franciscans* did Western Christians begin to dwell on the Passion. Ultimately the debate turned on differing theological interpretation; the Sunday *Pasch* triumphed, subsuming all that was commemorated on 14 Nisan.

See also PASCHAL CONTROVERSIES.

C.G. THORNE, JR.

QUATTRO CORONATI (Lat. = "four crowned ones"). Martyrs commemorated in the Western Church on 8 November. Roman tradition lists them as four brothers (Severus, Severian, Carpophorus, Victorinus) who held offices of trust in Rome and were beaten to death publicly and buried on the Lavican Way at the beginning of the Diocletian persecution about 304. They are confused with five Pannonian stonemasons (Nicostratus, Claudius, Symphorian, Castorius, Simplicius), who, refusing to sacrifice to the gods, were enclosed in leaden boxes and drowned. Modern hagiographies have been unable to distinguish the two groups. An ancient basilica on the Celian Hill, Rome, is dedicated to them. They are the patrons of stonemasons. J.G.G. NORMAN

QUEBEC ACT (1774). An Act of the British Parliament which superseded the Royal Proclamation of 1763, this was the constitution of the colony of Quebec. The Act extended the boundaries to the west and south so that they approached the earlier French limits of the colony. The Roman Catholic Church was officially recognized and permitted to collect its accustomed dues. The colony was given French civil law, but English criminal law remained in force. The Act marked the abandonment of the policy of assimilation or Anglicization and was resented by the English in Quebec and the other British North American colonies.

QUENTIN, HENRI (1872-1935). Biblical scholar. Born at St. Thierry, France, he studied at Reims and entered the Benedictine abbey of Maredsous, Belgium (1894-95). Transferred to Solesmes, France (1897), he was ordained in 1902. Pius X called him to Rome in 1907 to serve on the Commission for Revision of the Vulgate. He superintended the photographing of most of the important Vulgate MSS and was editor-in-chief of the Pentateuch. He became a member of the Pontifical Roman Academy of Archaeology (1923), and Pius XI appointed him to the historical section of the Congregation of Sacred Rites* (1930). He had a substantial part in the creation of the new Sacred Heart liturgy (1928). He was made first abbot of the Abbey of St. Girolamo, Rome, in 1933. J.G.G. NORMAN

QUESNEL, PASQUIER (1634-1719). French Jansenist* theologian. Educated by the Jesuits, he studied philosophy and theology at the Sorbonne and joined the Congregation of the Paris Oratory in 1657, subsequently becoming director. In 1672 he produced his *Réflexions morales,* a reprint of the NT with moral comments on every verse. The bishop of Châlons-sur-Marne commended it, and it was generally well received, but as new editions were prepared, it was alleged to be increasingly rigorist and Jansenist in tone. It was praised by the archbishop of Paris, but Quesnel's work was fiercely attacked, especially by the Jesuits, and he moved to Brussels to escape harassment. Louis XIV, convinced that Jansenism was a public danger, had Quesnel's arrest in Brussels engineered by Philip V of Spain, but he escaped to Protestant Holland where he continued to defend his views. In 1708, after fourteen years of discussion, the *Réflexions morales* was condemned by Pope Clement XI. Quesnel's reply led to further bitter controversy, complicated by the Gallican issue, and in 1713 the bull *Unigenitus* condemned 101 propositions, which amounted to the entire theological, ascetic, and moral doctrine of Jansenism. Quesnel died in Amsterdam without retracting his beliefs. HOWARD SAINSBURY

QUIERCY, SYNODS OF. Several such assemblies were held in the ninth century. The first met in 838, when Florus,* a supporter of Agobard,* archbishop of Lyons, alleged that parts of Amalar of Metz's* book on liturgical ritual were heretical. At Quiercy, Amalar's interpretation of the ceremonies of the Mass was condemned. A second synod of Quiercy met in 849 to condemn the views on double predestination allegedly held by the Augustinian theologian Gottschalk,* a monk of Fulda. The local archbishop, Rabanus Maurus,* attacked Gottschalk's position, and the Council of Mainz (848) condemned Gottschalk. Through the efforts of Hincmar* of Reims the Synod of Quiercy confirmed this condemnation in 849, and Gottschalk was scourged, defrocked, and imprisoned. Hincmar discovered, however, that other noted scholars—Ratramnus* of Corbie, Lupus of Ferriers, and Prudentius* of Troyes—broadly supported Gottschalk's position. The latter's opponents, who themselves held Semi-Augustinian views, argued that he had made God the author of sin, but in fact his position is based on "prescience" and not on "pre-ordaining."

At Charles the Bald's suggestion, Hincmar held a third synod at Quiercy in an effort to sort out the problem. Four propositions were passed: predestination to glory was accepted, and the reprobate, from not being helped, would go to hell through their own free choice; grace restores man's ability to do good; God desires to save all men; and Christ suffered for all men. These propositions were not universally accepted, and further synods subsequently met, including two at Quiercy (857-58), in an effort to achieve a solution based on the moderate Augustinianism of the Council of Orange* (829). L. FEEHAN

QUIETISM. This system of spirituality spread rapidly in Christendom in the later seventeenth century, and is best understood as an introverted and mystical reaction to the dogmatism and oppressions of the Thirty Years' War. It had three leading advocates—Fénelon,* Molinos,* and Madame Guyon*—and many lesser supporters. Most of these were persons of intense spirituality who suffered constraint or persecution, especially when their movement was condemned by Innocent XI* in the bull *Coelestis Pastor* (1687).

Quietism is basically an exaggeration of the orthodox doctrine of interior quiet, and of elements found in the medieval mystics—indeed the term is first encountered in the fourteenth century. It teaches firstly that the human soul's highest attainment is passive contemplation of the divine. This passivity is deliberately stressed: there is no recall of the medieval mystics' belief that contemplation implies a "busy rest" and calls for energetic human response to God's outgoing love, nor, as with Quietism, had it been generally held that the intellect as well as the will and emotions must be renounced in the quest for spiritual union, or that the Christian soul ultimately loses itself in the boundlessness of infinity. Secondly, Quietism insists that the soul surrenders to God in one decisive act after which it enjoys, despite all temptations, irrefragable union with the divine (cf. the Reformed doctrine of final perseverance). Lastly, the doctrine of pure or disinterested love, found especially in the popular and less informed ranks of Quietists, teaches that the renunciation of self and of desire is reached only by disregarding thoughts of heaven and hell and all external distractions, including spiritual exercises and the ordinances of the church. The result is a state of "mystic death," a dehumanization of man and a vague pantheism which is closer to Buddhism than to Christianity.

The antiecclesiastical implications of Quietism were at once seized upon by Rome, which condemned the movement as a logical outcome of the Reformation: the Quietists, however, derived far more from John of the Cross* and other Counter-Reformation mystics than from any Protestant source.

BIBLIOGRAPHY: P. Pourrat, *Christian Spirituality*, vol. IV (ET 1922); M. Petrocchi, *Le quietesmo italiano* (1948); R. A. Knox, *Enthusiasm* (1950); M. Bendiscioli, *Der Quietismus zwischen Häresie und Orthodoxie* (1964). IAN SELLERS

QUIMBY, PHINEAS PARKHURST (1802-1866). Founder of mental healing in America. Born in Lebanon, New Hampshire, he abandoned clockmaking to become a mesmerist after hearing the lectures of Charles Poyen in 1838. By 1847 he forsook mesmerism for mental healing, establishing settled practice in Portland, Maine (1859). He felt all disease arose in the mind, that it came mostly from erroneously attributing illness to physical causes. He held that God, or Wisdom, constituted all reality, and that matter was either an illusion or a manifestation of God. Quimby's philosophy became the foundation for Christian Science*—his term for properly understanding the relation between the divine and the human. Based on his treatment and manuscripts, his patient and disciple, M. B. Eddy,* practiced his methods after his death, and by 1875 established Christian Science as a distinct religion. By 1900 Quimbyites Warren Evans and Julius Dresser constructed the New Thought movement from his ideas. D.E. PITZER

QUIÑONES, FRANCISCO DE (1480-1540). Spanish reforming cardinal. Born at Léon, he entered the Franciscan Order in 1498 and originated the mission of the "Twelve Apostles" to Mexico in 1523. He was subsequently for five years minister-general of his order, became cardinal in 1527, and bishop of Coria, 1531-33. He negotiated with Charles V on behalf of Pope Clement VII* and prepared the treaty of Barcelona in 1529. He defended the interests of Catherine of Aragon in the matter of Henry VIII's* divorce. As representative of the Catholic reformation movement, he was asked by Clement to compile a new Breviary, which was published in 1535. He reduced to a minimum readings from the lives of the saints, and eliminated all elements of a choral nature. He also introduced the recital of all the Psalms through the week and the reading of nearly all the Bible through the year. It became a best seller until proscribed in 1558, and greatly influenced Thomas Cranmer* in his preparation of the Book of Common Prayer.* J.G.G. NORMAN

QUINQUAGESIMA. The period which begins with the Sunday immediately preceding Lent, and ends on Easter Day. The term is applied also to the first week of this period, and also to the Sunday before Lent.

QUIRINIUS PUBLIUS SULPICIUS (d. A.D. 21). Roman imperial legate. The fixed events in his career, based on Tacitus and the interpretation of supporting Latin inscriptions, include being in 12 B.C. consul with Marcus Valerius Messalla Barbatus, in A.D. 2 adviser in the East to Gaius Caesar (the emperor's grandson), and in A.D. 6 the legate of Syria, succeeding Lucius Volusius Saturninus with a commission to make a tax census of the newly incorporated procuratorial Judea (cf. Luke 2:1, 2). There is identified for him also a campaign against a desert tribe while he was proconsul of Crete and Cyrene (c.15 B.C.) and the governance of Galatia some years later with a victory over the Homonadenses. To the gospel allusion the events of Acts 5:37 are compounded by the use of Josephus, whereby Quirinius's name and office have become the main obstacle in computing the year of Jesus' birth, if Herod the Great (37-4 B.C.) is involved. CLYDE CURRY SMITH

QUMRAN, see DEAD SEA SCROLLS

QUR'AN (Koran). The holy book of Islam,* containing 114 chapters or *surās*, varying greatly in content. As well as historical material, it includes doctrinal and legal argument, exhortation, warnings and eschatological teachings. According to Muslim tradition, the *Qur'ān* exists in heaven on a preserved tablet; its contents were revealed to Prophet Muhammad by Archangel Gabriel. There is evidence that Muhammad was familiar with at least parts of the Old and New Testaments, and the *Qur'ān* contains references to Adam, Noah, Abraham, Moses, David—and Jesus. Jesus is called "the son of Mary, the apostle of God" (4:156, etc.), and "Messiah"; but "the Messiah the son of Mary is only a prophet" (5:79)—since the radical monotheism of Islam does not permit any attribution of sonship to Jesus or any other. In Muslim tradition the *Qur'ān* is held to be divinely inspired and therefore infallible. There have of course been disputes over interpretation, but the basic authority of the *Qur'ān* is never contested. Consequently historical criticism has made very little headway among Muslims. E.J. SHARPE

RABANUS MAURUS (c.776-856). Archbishop of Mainz. Born of a noble family at Mainz and educated at Fulda where he was made deacon in 801, he went the following year to Tours. There he studied under Alcuin,* from whom he received the name "Maurus," referring to St. Maur, disciple of St. Benedict, in recognition of his scholastic abilities. He became master of the monastery school at Fulda, one of the most influential in Europe, where Walafrid Strabo* and Otfrid of Weissenburg were pupils. Ordained in 814, he was abbot of Fulda from 822 to 842, advancing its intellectual, spiritual, and temporal welfare, erecting buildings, collecting manuscripts and art, and engaging in writing. In the struggle between Louis the Pious and his sons he supported Louis and then Lothair I, but with the defeat by Louis the German (840) he fled the monastery, returning briefly before retiring to nearby Petersberg for prayer and study. In 847 he became archbishop of Mainz, where he instructed clergy and laity, combated social disorders, and defended sound doctrine. He held three provincial synods: on ecclesiastical discipline (847), on Gottschalk* and his doctrine of predestination (848), and on the rights and disciplines of the church (852).

His writings are immense in subject and number: a study on grammar, a collection of homilies for the church year, two penitentials, a martyrology, and Latin poetry. A manual for monks and clerics in three books *(De institutione clericorum)* dealt with, e.g., sacraments, public prayer, and fasts, and relied much on Augustine, Gregory the Great, and Isidore. He also wrote many commentaries on Scripture. Learned in Scripture and patristics, he was not an original thinker; his writings are largely compilations, more important for their place in the Carolingian Renaissance* than for themselves. Not canonized but honored as a saint, he was acclaimed *praeceptor Germaniae.*

C.G. THORNE, JR.

RABAUT, PAUL (1718-1794). French Huguenot leader. Born into a Protestant family, he decided at the age of sixteen to accompany and help the itinerant preacher Jean Bétrine. The four years' experience thus gained served him well when he began in 1738 his work as a pastor and an opponent of repressive legislation. Except for a brief period in a seminary at Lausanne in 1740, he was associated with the church in Nîmes all his life. In 1756 he was voted president of the national synod of the Huguenot* Church, which at this period was in difficulties due to the revocation of the Edict of Nantes.* Though encountering many problems both from the state and from his own church, he fought on and did much to rehabilitate the Protestants of France. With Antoine Court* he is to be regarded as of major importance in the history of eighteenth-century French Calvinism. One significant success in which he and his sons shared was the passing of the Edict of Toleration in 1787.

PETER TOON

RABBI. This Greek word is derived from a Hebrew title used to honor the Jewish religious teachers, dating from the first century before Christ. Jesus is addressed by His disciples as "rabbi," because the outward form of His life appeared like that of a Jewish teacher of the law. But the gospel writers interpret the term so as to underscore His lordship (cf. Mark 9:5; Luke 9:33), thus indicating that early Christians thought of Him as much more than a rabbi. The term does not occur in early Christian literature subsequent to the closing of the canon. As the recognized title given to teachers of the law in the Jewish community, it is passed on from teacher to pupil by ordination, and qualifies one who has the proper training to function as preacher, teacher, and pastor in the Jewish synagogue.

PAUL K. JEWETT

RABBULA (c.350-435). Syrian theologian. Born near Aleppo of a Christian mother, he entered the civil service and became a prefect. He was converted about 400 and became a monk. In 411 he was elected bishop of Edessa and devoted himself to church reform, strongly opposing Jewish, pagan, and Gnostic influences in Syria. He supported Cyril* of Alexandria at the Council of Ephesus* in 431, and attacked Nestorianism,* especially the writings of Theodore of Mopsuestia.* Many scholars hold that he made the Peshitta version of the Syriac NT, issuing it to supersede the unorthodox Tatian's *Diatessaron.** He translated Cyril's *De Recto Fide* into Syriac. He is credited with having written some of the hymns in the Jacobite liturgy.

J.G.G. NORMAN

RACISM. According to Pierre van den Berghe, this is any set of beliefs that organic, genetically transmitted differences (whether real or imagined) between human groups are intrinsically associated with the presence or absence of certain socially relevant abilities or characteristics, and these differences form a legitimate basis for invidious distinctions between groups socially defined as races. It is not so much the presence of objective physical differences between groups that creates races, but the recognition of such differences as socially significant. "Race," which emphasizes physical appearance, should be distinguished from "caste"—a hereditary social group limited to persons of the same rank, occupation, or economic position—and "ethnic group"—one sharing a common and distinctive culture.

Race or color prejudice far antedates the period of European expansion. In Greek, Latin, Persian, and Sanskrit the words for "black" had negative connotations (bad, wicked, dismal, unlucky) while those for white were more favorable. This symbolism of color with its association of moral qualities was taken over by Christianity, and religious language from the beginning was full of dark deeds and fair promises, black thoughts and white angels. Thus the metaphors of light and darkness came to be applied to human conduct. The devil and those who scourged Christ were often portrayed as being black. Bede* asserted that the Ethiopian eunuch's skin was changed after his baptism by Philip (Acts 8) so he would no longer have to wear the badge of evil. Teresa of Avila* on successfully resisting temptation had a vision of a Negro boy enraged at his frustration. James I* of England in his *Demonologie* asserted a black man presided over witches' covens.

There were several historical factors in the development of the Western variety of racism. The capitalist exploitation of non-European peoples, particularly the institution of Negro slavery in the New World, fostered a complex ideology of paternalism and racism in which the black was seen as inferior, childish, and needful of civilization. The egalitarian and libertarian ideals of the Enlightenment* and the French and American revolutions led to a dichotomy between civilized and savage peoples, wherein these ideals applied only to the white, civilized ones. Extremely important was Darwinism, which regarded races as permanent, specieslike divisions possessing differential hereditary capacities for achieving civilization. Social Darwinists held that due to the processes of natural selection whites were far ahead of other races in the struggle for power. Now most scholars hold that qualitative differences between races are cultural rather than genetic in origin.

Although racism in the United States conflicts with the nation's democratic ethos, it has been institutionalized in the structure and culture of society. Neighborhood segregation remains basic even though nearly all legal supports to discrimination have been removed. The most extreme form is the apartheid policy of South Africa and Rhodesia which prescribes concrete measures for the total separation of the races—Whites, Coloureds, Indians, and Blacks—and the dominance of the white minority. The racial problem in Great Britain represents possibly more an expression of a general xenophobia—a resentment against immigrants from the West Indies, Pakistan, and India and desire to limit their numbers—than race prejudice as such. The Northern Ireland conflict is more ethnic, cultural, and religious than racial in nature, in spite of popular views to the contrary. The violent excesses of National Socialism in Germany transformed anti-Semitism in modern Europe from a policy of ethnic and religious discrimination to out-and-out racism.

Race prejudice also prevails in such non-Western areas as Brazil, China, India, and Africa. Varieties of black racism include the "Negritude" concept developed by writers in Francophone West Africa which emphasizes a sense of power and identity, and some African zionist and messianic movements, in which a black prophet seeks a new Jerusalem on earth for his chosen people or excludes white men from heaven. The pastoral Tutsi in Burundi rule over the agricultural Hutu in a racist manner, while the open animosity against Indians in East Africa possesses strong racial overtones. Both the Black Muslim movement, with its policy of territorial separation, and black anti-Semitism are expressions of black racism in the USA.

Although greatly at variance with the principles of biblical Christianity, white Christian racism is especially pervasive. The existence of separate races and the institution of slavery viewed as divinely ordained, and scriptural prooftexts (such as the "curse of Ham," slaveholding by the patriarchs, and instructions concerning the behavior of slaves and masters) are utilized to justify Negro subordination. Others argue that Christians should devote their attention to "spiritual" matters, and regard race prejudice as merely a "secular" concern.

In the USA, after the colonial era, blacks were segregated within the churches by the early nineteenth century and soon were pressured to form separate congregations. The Negro church became their fundamental social institution, serving both as a refuge from the hostile white society and as the training school for black leadership. Although some white Christians were involved in the abolitionist campaign, their acceptance of racism was evidenced by their consistent support of segregationist laws and practices in both the North and South in the post-Civil War decades.

The civil rights movement, initiated by the National Association for the Advancement of Colored People (founded 1910), gained momentum after World War II and reflected a strong Christian dimension, e.g., Martin Luther King, Jr.* and the Southern Christian Leadership Conference.* Although most of the early initiatives came from theological liberals, there were some noteworthy actions by American evangelicals on the racial front. These included Billy Graham's decision in 1953 to desegregate his meetings, the initiation of a distinctively evangelical civil rights publication *The Other Side* in 1965, the Tom Skinner Crusades, and the formation of urban ministries in such cities as Boston, Philadelphia, and Chicago. Nevertheless, there is still strong resistance to racial equality in many quarters of conservative and evangelical Christianity in America, and this has been accentuated by reactions to the new militancy of the "black power" movement and concomitant campaigns.

BIBLIOGRAPHY: J.O. Buswell III, *Slavery, Segregation and Scripture* (1964); K. Haselden, *The Racial Problem in Christian Perspective* (1964); G.K. Hunter, *Othello and Colour Prejudice* (1967); P.L. van den Berghe, *Race and Racism* (1967); M. Banton, *Race Relations* (1967); P. Mason, *Patterns of Dominance* (1970) and *Race Relations* (1970); C. Salley and R. Behm, *Your God*

Is Too White (1970); C. Bolt, *Victorian Attitudes to Race* (1971); O. Edwards, "Christian Racism" in *The Cross and the Flag* (ed. R.G. Clouse, R.D. Linder, and R.V. Pierard, 1972).

RICHARD V. PIERARD

RACOVIAN CATECHISM. One of the earliest confessions of modern Unitarianism,* it appeared in 1605 just after the death of Faustus Socinus. It is a very clear expression of Socinian* theology. Beginning with the question, "What is the Christian religion?" it gives as the answer: "The Christian religion is the way revealed by God for securing eternal life." The catechism clearly indicates that Unitarianism of the Socinian type held that both the OT and NT are inspired documents. But reason alone apprehends their spiritual truths. The truths of revelation are superior to reason, but are never contrary to its dictates. Christ is more than man, but not truly God, for if he had been truly divine he could not have died.

C. GREGG SINGER

RADBERTUS, PASCHASIUS, see PASCHASIUS

RADEWYNS, FLORENTIUS, see FLORENTIUS RADEWIJNS

RADICAL REFORMATION. The term covers congeries of movements flourishing from the 1520s which were often initially indebted to the "magisterial reformation" of, e.g., Zwingli* and Luther,* but wished to push changes farther and on different bases. Especially in its earliest stages, the radical reformation was a continuation of medieval movements of lay piety, heresy, and social protest. It drew many recruits from peasants and lower orders of townspeople and craftsmen, who were then in economic difficulties.

A radical movement appeared early in Wittenberg, where Luther's colleague A.B. von Carlstadt* introduced a vernacular Mass, abandoned vestments and the use of images, adopted a thoroughly lay interpretation of the priesthood of all believers, and was drawn to a mystical Quietism. Prophets claiming direct inspiration from God influenced T. Münzer* who, combining genuine compassion for the poor with apocalyptic fanaticism, was killed in the Peasants' War.

In Zwingli's Zurich, radicals such as C. Grebel* appeared in 1525-26, questioning infant baptism, and those who survived magisterial persecution were dispersed widely. A lively movement in the Tyrol was begun by G. Blaurock* and J. Hutter (see HUTTERITES). In Moravia a moderate movement under the pacific conservative spiritualist B. Hubmaier and M. Sattler (d.1527) lost the support of rulers after the more socially radical J. Hut (c.1490-1527) arrived. The movement, reorganized more conservatively by Hutter, issued in the Hutterite communities.

Other radicals traveled down the Rhine; Strasbourg was the scene of lively debates, 1528-34, involving C. Schwenkfeld (see SCHWENKFELDERS), S. Franck,* M. Servetus,* and M. Hofmann. The last inspired a strong movement (Melchiorites) in the Netherlands, long a hotbed of religious deviance, which became more militantly apocalyptic under the leadership of J. Matthijs and John of Leyden* and took over the Lutheran town of Münster (1535-36), set up a millennial community, restoring Paradise and Old Israel in anticipation of the universal rule of Christ; the practice of polygamy and the violent charismatic rule of the saints produced terror, and Catholic and Protestant forces combined to restore order after a long siege. Münster ensured that the term "Anabaptist"* symbolized, for over a century, social disorder and immorality; yet it was the last purgative outburst of visionary fanaticism in the radical movement. The future lay with more pacific Anabaptist church movements, like the Hutterites, the Mennonites,* and with a less orthodox, more rationalist wing in the Socinians.* Through such movements the radical reformation has continued to play an influential part in the history of Christianity.

The radical reformation was spasmodic, turbulent, and fragmented through persecution, travel, partisan strife, and theological debate. Its principles cannot therefore be characterized without allowing for large exceptions and paradoxes, but a sketch must be attempted here. Radical reformation rested on the thorough separation of the church and the world, of believer and unbeliever. It was not denied that sinners might be in the church; but no theoretical or practical concession was made to the presence of the unregenerate. The church was to be consistently defined as the company of true believers and disciples, one with the suffering and/or exalted Christ, and so not essentially conditioned by the world or the flesh. They had little patience with Luther's Augustinian theology of election and bondage of the will and the consequent acceptance of the hiddenness of the true church. Believers' baptism and the ban (excommunication), the distinctive life of the church, gathered in brotherly love and separate from the world (abstention from secular office, bearing the sword, oaths), and the experience of martyrdom as the climax of the practical imitation of Christ were all fruits of a quest for the visibility of the true church.

This stress on visibility went with a cultivation of inward spirituality, which occasionally made outward forms unnecessary (Schwenkfeld, Socinus), but more commonly was treated as their basis. This deeply felt spiritual unity with Christ sometimes produced unorthodox christologies, and was the basis for the spiritualist exegesis of Scripture, since it was held that spiritual experience not academic learning was the key to Scripture—in part this was a lay attack on professionals. Generally the movement did not bypass Scripture, but claimed to read it spiritually, and often this meant literally, to the point of challenging the reason or order of the world.

The movement was constantly inspired by the biblical concept of the people of God; sometimes this was understood more in terms of the warlike saints of some parts of the OT, or of Christ's suffering pacific people.

Not all the radical groups were initially Anabaptist, but the practice became increasingly important and widespread. Rebaptism was significant because it not only effected a visible church of confessing believers and was a restoration of primitive Christian practice; but implied also a break with the partnership of church and civil order, essential to Christendom, in which paedobaptism ensured that all citizens could be treated as Christians. Reformers like Zwingli who were occasionally attracted by the biblical arguments for believers' baptism resisted the radicals so harshly because they wished to reform, not to destroy, the traditional order of Christendom.

See G.H. Williams, *The Radical Reformation* (1962). HADDON WILLMER

RADICAL THEOLOGY, see DEATH OF GOD SCHOOL

RAIKES, ROBERT (1735-1811). Promoter of Sunday schools. Born in Gloucester, he succeeded his father as publisher of the *Gloucester Journal*, which business enabled him to maintain his interest in neglected children. After a meeting with Thomas Stock (1749-1803), who had started a Sunday school in Ashbury, Berkshire, he set up one in his own parish in 1780 which met a glaring need. The idea caught fire, and schools sprang up in other places. Despite popular opinion, however, he is not the movement's founder, and never claimed to be. By 1786 some 200,000 children were being taught in England, and a London society for establishing Sunday schools had been organized (1785) by William Fox. They spread into Wales (1789) through Thomas Charles* of Bala, and to Scotland, Ireland, and America. John Wesley encouraged them, and Adam Smith praised their cultivating good manners. They taught children to read and write along with giving Bible instruction. An interview Raikes had with Queen Charlotte led to Mrs. Trimmer's starting schools visited by George III. At first teachers were paid, later they volunteered. In 1803 a Sunday School Union was founded. Although Raikes was accused of excessive vanity, his immense benevolence cannot be disputed.

See also EDUCATION, CHRISTIAN.

C.G. THORNE, JR.

RAINOLDS (Reynolds), JOHN (1549-1607). Moderate Puritan theologian. Born at Pinhoe in Devon, he was educated at Corpus Christi College, Oxford, where after graduation he became a tutor and lecturer in Greek. One of his students was Richard Hooker,* Rainolds's lectures brought him fame in the university, but after some controversy in his college, he moved in 1586 to Queen's College; he also held at this time a lectureship "for the confutation of popish tenets." When Queen Elizabeth visited the university in 1592, it is said, she advised him to proceed in religion with moderation. In December 1593 he was made dean of Lincoln in order to facilitate his subsequent promotion to the presidency of Corpus Christi. He was well known as a moderate Puritan and a convinced Calvinist. He used the Book of Common Prayer, but objected to certain ceremonies—e.g., the churching of women and use of the sign of the cross. At the Hampton Court Conference* in 1604 he represented the Puritan interest, and afterward took a major part in the translation of the Bible (King James Version). PETER TOON

RAINY, ROBERT (1826-1906). Scottish minister and scholar. Born in Glasgow, he graduated from the university there, studied theology at New College, Edinburgh, and in 1851 became minister of the Free Church of Scotland* at Huntly. In 1854 he transferred to the Free High Church, Edinburgh, and after eight years assumed the chair of church history at his former college, a post he held for the next forty-four years and with which from 1874 he combined the college principalship. Not naturally a controversialist, he found himself nevertheless drawn into great issues of the time. He opposed public lotteries and the Boer War, and was much criticized for his (reluctant) acquiescence in the deposition of W.R. Smith (1881).

Reunion of the Scottish Presbyterian churches was a burning cause with him, and he saw disestablishment of the Church of Scotland as the only expedient way. This brought him into correspondence with a not unsympathetic Prime Minister Gladstone, who in 1895 called him "unquestionably the greatest of living Scotsmen." He led the Free Church into union with the United Presbyterian Church* in 1900, and his own third moderatorial term was over the United Free Church* assembly. His *Three Lectures on the Church of Scotland* (1872) completely refuted Dean A.P. Stanley's* astonishing denigration of Covenanters* and Seceders. Rainy's other works included *The Bible and Criticism* (1878). His funeral, delayed because he had died in Australia, was reported to have been the greatest spectacle Edinburgh had seen since that of Thomas Chalmers in 1847. J.D. DOUGLAS

RAMABAI, PANDITA (1858-1922). Indian Christian reformer. Daughter of a Brahman who had a remote hilltop ashram, she was born near Mangalore, lost both her parents during pilgrimage in South India in 1874, and with her brother wandered on, making a living by reciting the Hindu scriptures. In Calcutta Ramabai's prowess in Sanskrit so impressed pandits that she was designated "Pandita," a name retained even after she was a Christian. Her drift from orthodoxy was seen in her association with a reform movement, by her speeches in favor of female emancipation, and then by her breaking caste in marrying a man of lower caste who died two years later (1882).

Meanwhile, in Bengal Ramabai had met Christians and discovered the Bible. She went to Poona, organizing women's societies for reform and pursuing her new interest in Christianity in which she was helped by father Nehemiah Gore, an Anglican who was a convert from her own caste. The Wantage Sisters helped her go to Britain for education in 1883, and she and her young daughter were baptized at Wantage. In 1886 she went to America, studying kindergarten methods and getting support for a scheme to educate high-caste widows. A Ramabai Association guaranteed finance for ten years, and her book on *The High-Caste Hindu Woman* (1887) paid for Marathi textbooks and stirred concern in the West. Her boarding school opened at Bombay (1889) and later moved to Poona. Support from eminent Hindu reformers withered when some of her child widows were attracted to Christianity. Ramabai's evangelical conversion in 1891 made her an ardent evangelist as well as social worker, and baptisms followed.

Land purchased at Kedgaon, near Poona, became the scene of Ramabai's greatest work when she rescued hundreds of girls and women after the famines of 1896-97 and established a Mukti ("Salvation") Mission which grew to a community of over 1,300. Mukti was in 1905 a notable center of revival. During her last eighteen years Ramabai made a simplified Marathi translation of the Bible. After her death Mukti Mission continued, having a strong link with the Christian and Missionary Alliance, but as a separate organization. In 1971 there was a community of 649.

See Ramabai's *Testimony* (1917), and biographies by N. Macnicol (1926) and P. Sengupta (1970). ROBERT J. MC MAHON

RAMSAY, SIR WILLIAM MITCHELL (1851-1939). Classical scholar and archaeologist. Born in Glasgow, he studied at the universities of Aberdeen, Oxford, and (briefly) Göttingen. During 1880-90 and 1900-14 he engaged in extensive exploration among the antiquities of W Turkey. He was the first professor of classical art and archaeology at Oxford (1885-86) and then professor of humanity (i.e., Latin) at Aberdeen (1886-1911). His chief significance lies in his contribution to classical archaeology and geography. His *The Historical Geography of Asia Minor* (1890) lays the foundation for all future work in this area, and his *Cities and Bishoprics of Phrygia* (1895 and 1897), *Asianic Elements in Greek Civilization* (1927), and *The Social Basis of Roman Power in Asia Minor* (1941) are still standard works.

He also made important contributions to the study of the NT. His early work, *The Church in the Roman Empire before A.D. 170* (1893), set the tone of his future writings, but it is his *St. Paul the Traveller and Roman Citizen* (1895) which is best known. Though he had earlier accepted the conclusions of radical German scholarship concerning the historicity of Acts (see F.C. BAUR), his study led him to an increasingly high estimate of Luke as a historian. His work on the NT served to fill in the historical background for the life of Paul, establish the so-called South Galatian destination of the epistle to the Galatians, and commend the historical reliability of the Lucan writings to scholars. His principal writings were a series of articles in Hastings' *Dictionary of the Bible* (1898-1904), *A Historical Commentary on St. Paul's Epistle to the Galatians* (1899), *Letters to the Seven Churches of Asia* (1904), and *The Cities of St. Paul* (1907).

See W.H. Buckler and W.M. Calder (eds.), *Anatolian Studies presented to Sir William Mitchell Ramsay* (1923); and W.W. Gasque, *Sir William M. Ramsay: Archaeologist and New Testament Scholar* (1966): with bibliography and indexes. W. WARD GASQUE

RAMSEY, ARTHUR MICHAEL (1904-). Archbishop of Canterbury from 1961 to 1974. Educated at Repton and Magdalene College, Cambridge, where he read classics, and then theology in which he gained a first class degree, he was ordained in 1928 and served a curacy at St. Nicholas', Liverpool. From 1930 until 1936 he was subwarden of Lincoln Theological College, where he wrote his first major theological work, *The Gospel and the Catholic Church.* After parochial work in Boston (Lincolnshire) and Cambridge, he was appointed professor of divinity at Durham in 1940. In 1945 he published *The Resurrection of Christ,* and in 1949 *The Glory of God and the Transfiguration of Christ*—two profound biblical-theological works. By this time he was recognized as one of the outstanding modern Anglican theologians.

In 1950 he was appointed regius professor of divinity at Cambridge. Two years later he succeeded to the see of Durham, and was translated to York in 1956 and to Canterbury in 1961, continuing his theological writing alongside all his episcopal duties. An Anglo-Catholic himself, he showed sympathy for and understanding of other traditions, and committed himself to work for church unity. In 1964 he paid a visit to Pope Paul VI, and he created strong contacts with the Orthodox as well as the Free churches at home. In preaching and writing Ramsey has emphasized the importance of personal spirituality; has taken an uncompromising stand for social justice; and has proved himself a stable leader.

He retired upon his seventieth birthday in 1974. JOHN A. SIMPSON

RAMUS, PETER (1515-1572). French humanist. From Picardy, he was educated at the *Collège de France* and led an anti-Aristotelian movement. By 1551 he had become a professor at his alma mater and in 1561 was converted to Protestantism. After spending the years 1568-71 in Germany to escape persecution, he returned to France in 1571 and was killed in the massacre of St. Bartholomew's Day.* For a time Ramus engaged in an effort to establish congregational government in the French Calvinist churches (1568-71), but it is as a reformer of Aristotelian logic that he is best known. He believed that the concepts and abstractions of the human mind draw their validity, not from temporal or expedient constructs, but from eternal truth in the mind of God. Hence man can develop a methodology for inferring such universals from his experience and then relating them to infinity. Absolute truth then becomes available through the careful analysis of human perception. In practice the Ramist felt that facts could be analyzed in a series of sucessive dichotomies. These analyses were often arranged in a diagram which became a hallmark of the Ramist method. Ramism was especially popular among the Puritans in England and New England.

ROBERT G. CLOUSE

RANCÉ, A.J. LE B. DE, see DE RANCE

RANDALL, BENJAMIN (1749-1808). Founder and organizer of Freewill Baptists. Born in New

Castle, New Hampshire, he early went to sea with his sailor father, and became a sailmaker. Three sermons by George Whitefield* and the shock of Whitefield's death converted him in 1770. He became a Congregationalist, but immersionist Arminian views led him in 1776 to become a Baptist. He joined in 1778 an Arminian Baptist church in New Durham, New Hampshire, where by 1780 he was ordained and organized a Free Baptist Church. He drew up the covenant which became the basis for the later Freewill Baptist Church. His ardent evangelism won so many that by 1783 he organized a quarterly and in 1792 a yearly meeting of his group. When he died, there were about 6,000 adherents. EARLE E. CAIRNS

RANKE, LEOPOLD VON (1795-1886). Lutheran historian. Born in Saxony, he became professor at the University of Berlin (1825-71), and is most significant in church history for his *History of the Popes* (2 vols., 8 editions, 1834-85). Although mainly on the sixteenth and seventeenth centuries, in the last editions it covered the period from the Reformation to the Vatican Council (1869-70). He wrote many other works, especially separate histories of Germany (5 vols., 1839-47), France (5 vols., 1852-61), and England (7 vols., 1859-68), mainly of the Reformation period. He is rightly credited with transforming historiography from polemic to systematic history. This he did through fresh archival work in Italy and Germany and a commitment to write history as a task distinct from apologetics; he tried to be fair to Catholics and the popes. Protestants commonly charged him with favor to Catholics, while Gregory XIV added his work to the Index (1841).

Ranke's histories reflected a distinctive perspective: a Lutheran and Prussian concentration on political and diplomatic affairs led by great figures as the core of history; his treatment of the popes, for example, was largely political. He stood against revolution and secularistic progress, and for the stabilizing hand of the established authorities as rooted in national traditions. He believed in God's Providence in history. The Prussian state named him official historian (1841) and granted him the aristocratic title *von* (1865) for his efforts. Secular positivists in England wrongly attributed to him a belief in their notion of historical objectivity.

See T.H. von Laue, *Leopold Ranke. The Formative Years* (1950), and P. Geyl, *From Ranke to Toynbee* (1952). C.T. MC INTIRE

RANTERS. An epithet given to an antinomian movement during the time of the English Commonwealth (mid-seventeenth century). They were part of the effort of the time to restore primitive, apostolic Christianity. This involved a repudiation of the Church of England in its established form and much greater emphasis on individual thought and action. The individualism of the movement tended to produce a great variety of groups around prominent leaders so that England appeared to "swarm with sects." The Ranters, so far as they can be differentiated from this general ferment, showed two marked characteristics: they were pantheistic and antinomian. Joseph Salmon and Jacob Bauthumley represent two characteristic examples of Ranter leaders. Bauthumley wrote *The Light and Dark Sides of God* (1650) which develops an extreme doctrine of the Inner Light. Salmon authored a strange tract recounting his experience with God and teaching an extreme pantheism. Contemporary writers agreed in the opinion that the Ranters led morally disordered lives and that they considered themselves above the usual distinction of right and wrong. George Fox* wrote against them and converted many to Quakerism. Richard Baxter* also denounced the Ranters. Many of them were severely punished for their immoral and blasphemous acts; thus the movement was suppressed.

See R.M. Jones, *Studies in Mystical Religion* (1909), pp. 467-81, and N. Cohn, *The Pursuit of the Millennium* (1970), pp. 287-330.

ROBERT G. CLOUSE

RAPHAEL, SANZIO (1483-1520). Renaissance painter. He studied first under his father, and later under Perugia in his native town of Urbino. At Florence he became famous under the tutelage of Leonardo da Vinci* and Michelangelo.* From the former he learned the softness and the sweetness, from the latter the strength and the drama, and from them both the depth of composition and the pyramidal figured masses that characterized virtually all his later works. He continued his study in Rome under these masters, with whom he became linked as symbols of the Renaissance, and in Rome he died. In the main, Raphael's subjects were religious or philosophical and were intentionally symbolic, even allegorical. Thus in his *School of Athens,* Plato is pictured as a grave old man, pointing upward to the heavenly font of Forms; Aristotle, a vigorous and youthful figure, points downward, whence his truth came. Though Raphael is best known for his Madonnas, his subjects include the whole of the life of Christ. So great has been his influence that the image of Bible personages to our day is the image literally made colorful by the genial and ever-popular Raphael. MILLARD SCHERICH

RAPHAEL THE ARCHANGEL. One of the seven archangels. The name is not found in the canonical writings, but appears in the OT Apocrypha and Pseudepigrapha. In *Tobit* 12:15, Raphael described himself as "one of the seven holy angels, which present the prayers of the saints." In *Enoch* 10:7 he is described as having healed the earth when it was defiled by the fallen angels; this tradition is based on the meaning of the name Raphael, i.e., "God has healed" (Hebrew). Raphael became canonized in both Eastern and Western churches. Since the sixteenth century he has often figured as the patron of travelers.

RAPP, J.G., see HARMONY SOCIETY

RASHDALL, HASTINGS (1858-1924). Anglican moral philosopher and theologian. Educated at Harrow and New College, Oxford, he taught philosophy at Lampeter, Durham, and Oxford. An advanced liberal and moderate High Churchman, he strove to revive the inspiration of the Cambridge Platonists* in contemporary Anglicanism, and deplored the current immanentist theology as well as Inge's* mysticism and Ritschl's* historicism. To many he seemed to deny the validity of religious experience altogether and to profess a cold moralism. A notable book, *The Theory of Good and Evil* (1907), tried to work out a harmony between Utilitarian and Idealist ethics and adumbrates Rashdall's curious doctrine of a limited God, for God as Personality must be less than the Absolute which contains him. Vice-president of the Modern Churchman's Union from its foundation in 1898, Rashdall delivered the Bampton Lectures on *The Idea of the Atonement* (1915) which have been called a brilliant restatement of the Abelardian or Exemplarist theory; in fact they combine distorted criticism of Paul, Augustine, and the substitutionary view with a wayward interpretation of Scripture and a doctrine of the Atonement far more rationalistic than Abelard ever professed. IAN SELLERS

RASKOLNIKI, see OLD BELIEVERS

RAS SHAMRA TABLETS. The cuneiform documents discovered by archaeologists (from 1929 onward) at the tell of Ras Shamra in N Syria. The site was a major Canaanite city of the third and second millennia B.C., named Ugarit; the city was destroyed in the early twelfth century B.C. by invaders akin to the Philistines. The work of excavation, and of the publication and translation of the many documents found, still continues. The majority of the tablets are in the Ugaritic and Akkadian languages; the former was unknown prior to the discoveries, but although the script (alphabetic cuneiform) was completely new to scholarship, the language itself is NW Semitic, and a very close relative of early Hebrew. Ugaritic lexicography is therefore being increasingly used in the elucidation of biblical Hebrew. For OT study in general, the most valuable Ras Shamra documents are the mythological texts, which have greatly increased our knowledge of Canaanite religious beliefs and practices. D.F. PAYNE

RASTAFARIANS. A Jamaican movement, originating in the early 1930s, placing the political pan-Africanism of Marcus Garvey (1887-1940) in a mythological and messianic setting. Emperor Haile Selassie (Ras Tafari, crowned 1930) is the only true God; Ethiopia is the only heaven. While biblical texts (e.g., Rev. 5:5) are used in support, and modified Methodist and Sankey hymns regularly sung, Rastafarians denounce Christian preachers as false prophets, since the whites have perverted the Scriptures to hide the black identity of Adam, Israel, and Jesus. God's (i.e., Ras Tafari's) parousia to accomplish Black deliverance is imminent; meanwhile political activism is pointless. Rastafarians are predominantly male, bearded, and unkempt, from deprived urban areas with high unemployment. *Ganja* (marijuana) is greatly valued, and its use ascribed to Solomon. The movement seems to have gained no tangible results from a tour of African states in 1961 or the more recent visit of Ethiopian Emperor Haile Selassie to Jamaica. A.F. WALLS

RATHERIUS (c.890-974). Bishop and polemicist. Born in Liège of a noble family, his career was filled with polemics. He served sporadically as bishop of Verona (three terms between 931 and 968) and of Liège (953-955). His first stay in Verona ended with imprisonment in Como, from which he escaped in 939; he returned to Lobbes in 944 and to Italy in 946, but became a prisoner of Berengarius until he regained his see in 946. Forced to flee again in 948, he unsuccessfully joined forces with Ludolph of Saxony to reclaim his position. He taught briefly at the cathedral school in Cologne before he was named bishop of Liège. Forced to Lobbes from Verona in 968, he had to retire to the monastery in Aulne, Flanders, where he died. JOHN GROH

RATISBON (Regensburg), COLLOQUY OF (1541). This marked the high point of Charles V's efforts to reconcile the Roman Catholics and Lutherans; it came after the colloquies of Hagenau* and Worms.* Here Melanchthon,* Bucer,* and Pistorius were the spokesmen for the Protestants; for the Roman Catholics, Pflug,* Eck,* and Gropper.* Gropper and Bucer were largely responsible for the *Regensburger Buch,* which contained twenty-three doctrinal articles. Not much difficulty was experienced in agreeing on the first four articles: Man before the Fall; Free Will; the Cause of Sin; Original Sin. The fifth article, concerning Justification, did not state either Luther's doctrine or the later Tridentine position clearly. Gaspar Cardinal Contarini* endorsed it. In Rome and Wittenberg (by Martin Luther) it was rejected. There was no agreement at Ratisbon on the formulation of the article concerning the church. Melanchthon upheld the Lutheran view of the Lord's Supper. No agreement was reached on other doctrines. The Colloquy of Worms did not succeed in healing the breach between Roman Catholicism and Protestantism.

CARL S. MEYER

RATRAMNUS (d. c.868). Early medieval theologian. Little is known of his life. His importance rests on his books and his involvement in a number of theological controversies. His most famous work, *De Corpore et Sanguine Domini* was written in reaction to a tract on the sacrament written by his former teacher, Radbertus,* which taught an excessively realistic doctrine. Ratramnus emphasized a more symbolic interpretation which denied the identity of Christ's sacramental and historical body. His book was condemned in 1050 as medieval theology moved in the direction of defining the doctrine of transubstantiation.* During the Reformation some of the Reformers cited Ratramnus as a precedent for their teaching, and the book was placed on the Index* in 1559, where it remained until 1900. In 850 Ratramnus wrote *De Praedestinatione* in support of Gottschalk's position on double predestination.* His last book, *Contra Graecorum Opposita,* was written at the urging of Nicholas I* and provided a defense of the Latin Church against the attacks of the Eastern Church. He pleaded for unity, but maintained that the Spirit proceeds from the Father and the Son *(Filioque)*, and he held to the primacy of Rome. RUDOLPH HEINZE

RAUCH, CHRISTIAN HEINRICH (1718-1763). First Moravian missionary to the American Indians. Born in Anhalt, he became a missionary under the Moravians and arrived in New York in 1740. He soon made contact with a group of Mohicans, who accepted his offer to serve as a teacher among them. Rauch located in the village of Shekomeko, and after months of severe hardships baptized several converts in the presence of Count Zinzendorf.* By 1743 the Shekomeko mission prospered, and a chapel was built, but growing white settler opposition resulted in the expulsion of the Moravians and termination of the work by 1746. Rauch then served charges in Pennsylvania and North Carolina for ten years before going to Jamaica, where he spent his last years working among black people.

RICHARD V. PIERARD

RAUSCHENBUSCH, WALTER (1861-1918). Baptist minister and educator. Born in Rochester, New York, of German immigrant parents, he was early educated in Germany and graduated from Rochester University (1884) and Seminary (1886). His pastorate (1886-97) at the Second German Baptist Church in New York City led him to see the plight not only of immigrants but all classes of society who were socially and economically disadvantaged. He taught NT (1897-1902) and church history (1902-17) at his former seminary, and soon gained a national reputation for strong views on social change. He distinguished his brand of (Christian) socialism from the doctrinaire socialism of Marxism. Democracy to Rauschenbusch was both an economic and a political ideal, but he was not overly optimistic about the realization of these goals in history. There is a more sobering view of sin in his theology. His commitment to social Christianity gained him the title "Father of the Social Gospel in America." His works include *Prayers of the Social Awakening* (1910), *Christianizing the Social Order* (1912), and *A Theology for the Social Gospel* (1917). He was ostracized for his German ancestry in the years preceding World War I.

DONALD M. LAKE

RAYMOND NONNATUS (1204-1240). Patron saint of midwives. Very little is known about him except what has survived in a sixteenth-century manuscript. He was born in Portello, Spain, and apparently delivered by Caesarean section after his mother's death in childbirth—hence his name "*non natus*" ("not born"). He was a member of the Order of Our Lady of Ransom which ransomed Christian prisoners on the Barbary coast of North Africa. The Muslims at times subjected him to gross ill-treatment. He persevered in his work, and on one occasion even surrendered himself as surety for the ransom of others. After his return to Spain, Gregory IX made him a cardinal (1239). He died at Cerdagne at the start of a journey to Rome. L. FEEHAN

RAYMOND OF PENAFORT (c.1175-c.1275). Spanish canonist. Born near Barcelona where he studied and taught rhetoric and logic at the cathedral school until 1210, he then left for Bologna, where in 1216 he was appointed professor of law. In 1222 he joined the Dominicans and returned to Spain, where with Peter Nolasco* he helped to found the Mercedarians,* an order dedicated to redeeming Christian captives. He was appointed confessor, chaplain, and grand penitentiary to Gregory IX, who commissioned him to organize the papal decretals, a prodigious task he completed in four years (1234). His *Summa de poenitentia*, a handbook on penance, exercised a decisive influence on subsequent practice. In 1236 Raymond again returned to Spain where he became general of his order, and in this role reissued the constitutions of the order in definitive form. Resigning from the generalship, and concerned with the conversion of Jews and Muslims, he set up schools for the study of Hebrew and Arabic, and commissioned Thomas Aquinas to write his *Summa contra Gentiles.* He was canonized by Clement VIII in 1601.

MARY E. ROGERS

RAYMOND OF SEBONDE (fl.1434-36). Spanish physician, philosopher, and theologian. Latterly regius professor of theology at the University of Toulouse, it was there he wrote his major work, *Liber naturae sive creaturarum,* originally written in Spanish, a landmark in the development of natural theology, though Raymond never employs the term. Influenced by Raymond Lull,* and opposing the position of such thinkers as William of Ockham* that faith and reason, theology and philosophy are irreconcilable, Raymond asserted that the book of nature and the Bible are concordant divine revelations, the one general and the other specific. He claimed to find rational, extrabiblical proof for the basic Christian doctrines, particularly through self-knowledge, since man is the image of God, and made it a practice in his work to cite neither Scripture nor other authorities. The Latin translation was printed repeatedly under the title *Theologia naturalis* from about 1484, most later editions omitting the *Prologus,* placed on the Index* in 1595. Montaigne popularized it in a French translation (1569).

MARY E. ROGERS

REBMANN, JOHANNES (1819-1876). German missionary to East Africa.* Born in Württemberg and trained at Basle, he was sent to East Africa by the Church Missionary Society in 1846. With J.L. Krapf* he established the Rabia Mpia mission among the Nyika. He undertook several exploratory journeys, was the first European to see Kilimanjaro (1848), and prepared a map which helped inspire the Burton-Speke expedition (1857). His linguistic studies in three vernaculars laid sound foundations for future workers. After 1855 he was alone at Rabai except for a brief period. In 1875 he returned to Württemberg, blind and broken in health. His death coincided with the effective entry of the CMS into East Africa. D.G.L. CRAGG

RECARED (sixth century). Visigothic king who ruled in Spain from 586 to 601. The younger of two sons, he succeeded his father Leovigild in 586 and set to work at once in crushing several Arian revolts. The Lusitanian peninsula in Portugal was officially Arian until he converted to Roman Catholicism in 586, although the Arian kings had been extraordinarily tolerant. In 589 the Third Council of Toledo proclaimed Catholicism as the kingdom's official religion. Thereafter little was heard of Arianism*; the king ordered Arian books to be burned, and no Gothic text from Spain has survived. With religious barriers removed, the Visigothic invaders and the native Hispano-Romans, who continued to follow Roman law, could assimilate. JOHN GROH

RÉCOLLETS. A reformed branch of the Franciscan* Order, they were asked by Samuel de Champlain* to give aid in the evangelization of the Indians in Canada. In 1615 three fathers and a lay brother arrived in Quebec to begin their devoted and self-sacrificing labors among the Algonquin tribes whose nomadic way of life soon made the fathers despair of their conversion. They decided therefore to seek out more settled tribes and traveled west to labor, with some success, among the Hurons. Financial problems in 1623 forced them to ask the Jesuits for assistance, but cooperation between the two orders was difficult. The *Récollets* encouraged intermingling of Indians and French and cooperation with the Huguenots* in civil matters; the Jesuits held opposite views. The English in 1629 and again in 1760 put an end to the *Récollets'* labors, but in each case they returned to continue their Canadian ministries.
ROBERT WILSON

RECTOR. In the Church of England, an incumbent who has charge and care of a parish and, as distinct from a vicar, receives the full amount of the tithe rent therefrom. In the Episcopal churches of Scotland and the USA, the term is used more loosely of the generality of parish priests. In the Roman Catholic Church, a rector is an ecclesiastic in charge of a congregation, an important mission, a college or religious house, especially a Jesuit seminary. In seventeenth-century literature the term is often applied to God Himself as ruler of mankind.

RECUSANTS. The name given to those in England and Wales and Ireland who refused to obey the Act of Uniformity* (1559) which required all subjects to acknowledge Queen Elizabeth as the Supreme Governor of the church and to attend services conducted according to the Prayer Book. The term was derived from the Latin verb *recusare,* "to refuse." Although at first a fine of one shilling was levied against all those who disobeyed the law, because of threats of Roman Catholic invasions, plots against the queen's life and efforts, particularly in the north of England, to stir up rebellion, the penalties were greatly increased, culminating in death for treason. Numerous civil disabilities were also laid upon recusants, particularly in Ireland, reaching their climax after the "Glorious Revolution of 1688." Not until the passing of the Catholic Emancipation Act of 1829 were most of the restrictions removed throughout Great Britain. A Roman Catholic still cannot become king or queen by the terms of the coronation oath. W.S. REID

REDACTION CRITICISM, see SYNOPTIC GOSPELS, NEW TESTAMENT CRITICISM

REDEMPTION. The idea comes from legal and religious transactions in OT life. The firstborn male child was regarded as uniquely owed to God, unless he was redeemed by some kind of sacrifice. The Hebrew root word used for this kind of redemption was *padah* (Exod. 13:13; Num. 18:25ff.). By payment of a price of redemption a man could save his own life from being forfeited if his ox killed another man (Exod. 21: 30). Within family life, redemption (root *ga'al*) was a process by which, if a man had forfeited property or had himself fallen into slavery, what was lost could be brought back to its true ownership or liberty through repurchase (Lev. 25:25; Ruth 4:4-6). The redeemer was in this case the next of kin who thus protected and upheld the rights of the unfortunate relative. Sometimes the obligation to redeem a kinsman in this way meant avenging him of wrongs committed against him. The essential purpose of redemption was therefore deliverance from loss or bondage.

When God is described in the OT as bringing about the redemption of Israel, the phrase is to be interpreted by such analogies, especially by that of the kinsman-redeemer. In this role God effects the redemption of His people from Egypt (Deut. 9:26) and from Babylon (Isa. 43:1). Indeed, God is given the name of the "Redeemer." The redemptive action of God in these great historical events is regarded as a sign that His redeeming hand can extend in the same way over sin, evil powers, and even death (Isa. 33:22f.; Ps. 130:8). It is recognized too that His redemptive activity is also exercised toward the individual when he is involved in a helpless struggle with life's varied ills (2 Sam. 4:9, 10; Ps. 34:22).

Redemption, in the NT, describes an aspect of salvation. The use of the word reminds us that Christ has come to free man from the control of every alien power, from all the tyrannies that oppress the individual and cast a blight upon his life, and from all iniquity (Titus 2:14; cf. Rom. 7). It includes the redemption of the body (Rom. 8:23; cf. Phil. 3:4). Though the price of this redemption has been fully paid, and the immediate fruits of this payment are fully enjoyed in reconciliation with God, the full enjoyment of the ultimate fruits of redemption must wait until the Second Coming (Eph. 4:30; Rom. 8:23).

Jesus regarded His healing of the sick, casting out of devils, and raising of the dead as signs of the redemptive aspect of His work. He described Himself as the Son of Man who came "to give his life a ransom *[lutron]* for many" (Matt. 20:28; Mark 10:45). In this description he seems to have had in mind the suffering servant of Isaiah 53, whose life was yielded up as a vicarious sacrifice to bring healing and liberty to many. The suggestion is that Jesus' own sufferings and death as a substitutionary sacrifice pay the cost of man's redemption. The OT recognizes that any act of redemption carried out with what any mere man has to offer, even though he gives his best, is totally inadequate to meet the deepest human need (Ps. 49:7,8). But God is willing to lavish what He has, regardless of cost (Isa. 43:3ff.), to win His people back to Himself. It is in this context that the NT speaks of the blood of Christ, i.e., the offering of His obedient life poured out in death, as the cost of our redemption (1 Pet. 1:18, 19; Rom. 3:24, 25; Heb. 9:14).

BIBLIOGRAPHY: H.A.A. Kennedy, *St. Paul's Conception of the Last Things* (1904); H.E.W. Turner, *The Patristic Doctrine of Redemption* (1952); L.J. Sherril, *Guilt and Redemption* (1957); Leon Morris, *The Apostolic Preaching of the Cross* (2nd ed., 1960).

RONALD S. WALLACE

REDEMPTORISTS. A name commonly given to the "Congregation of the Most Holy Redeemer," a community of priests and lay brothers, founded by Alphonsus Maria di Liguori* at Scala, Italy, in 1732 for mission work among the poor. It has steadfastly refused to engage in purely educational activities. Its purpose is the sanctification of members through the imitation of Christ and through preaching. The order received papal approval in 1749, and a community of nuns (the Redemptorines, also founded by Alphonsus) was approved in 1750. Under Clement Hofbauer* the Redemptorists moved across the Alps into N Europe, and they entered the USA in 1832 and England in 1843. They are governed by a "Rector Major" who holds office for life and resides in Rome. J.G.G. NORMAN

REES, THOMAS (1815-1885). Welsh religious historian. He was born at Penpontbren, Carmarthenshire. His only formal education was three months of elementary school. He became a coal miner at Aberdare in 1835, but soon gave it up and opened a school. He was ordained in 1836 minister at the Congregational church at Craig-y-fargod, Merthyr Tydfil, and supplemented the stipend of ten shillings a month which he received from the twelve members of the church by opening a shop. This proved a failure, and he suffered a week's imprisonment as a debtor. His fortunes revived, however, and his subsequent career as a minister was a distinguished one. He ministered at Aberdare, Siloa (Llanelli), Beaufort, and finally at Ebenezer, Swansea (1861-85). He was twice elected to the chair of the Union of Welsh Independents and, in 1885, to the chair of the Congregational Union of England and Wales, but died before taking up his duties.

He was a prolific writer, but his most substantial contribution was in historical studies. He is best known to English readers for his *History of Protestant Nonconformity in Wales* (1861, extended in the 1883 edition). He was also coauthor with John Thomas of Liverpool (1821-92) of a four-volume history of the Welsh Congregational churches (to which Thomas added a fifth volume in 1891). When it is recalled that Thomas Rees was virtually self-educated, his work as a historian is outstanding. He had a gift for discovering manuscript sources at a time when their significance for historical study was not generally appreciated. Unfortunately he was cavalier in his use of them. He would abbreviate or omit passages without warning as the whim took him, and his strong prejudices in favor of Congregationalism and moderate Calvinism often tempted him to interpret evidence in a tendentious fashion. But when all his faults are admitted, his historical writing is still of great use and interest to the student of modern Welsh religious history.

Rees possessed a winsome personality and was in great demand as a preacher whose sermons never failed to move the hearts of the large congregations that loved to hear him. His hymns still find a place in the collections of the various churches. R. TUDUR JONES

REES, THOMAS BONNER (1911-1970). English evangelist. Born in Blackburn and converted in his teens, he immediately engaged in active evangelism and subsequently became youth organizer and lay worker at St. Nicholas' (Anglican) Church, Sevenoaks. He also joined with the Church Pastoral Aid Society in arranging camps for London slum boys. After leaving Sevenoaks he conducted many united missions in Northern Ireland, where thousands were converted. For several years he served on the staff of the Scripture Union. After World War II he conducted fifty-four mass rallies in the Royal Albert Hall, London, and campaigns in a number of major British cities. He went to America more than fifty times for campaigns, Bible conferences, and church retreats, and he was a speaker at the famous conventions at Keswick and Portstewart. He founded successively three conference centers, and the work at Hildenborough Hall is continued by his son Justyn as a memorial to his work. Among Tom Rees's written works was *Breakthrough*, a complete handbook on home evangelism. J.D. DOUGLAS

REES, WILLIAM (1802-1883). Welsh Congregational minister, author, and social leader, better known by his pen name, "Gwilym Hiraethog." Born at Llansannan, Denbighshire, Rees obtained his only formal education in a few terms at the local school. He spent his early years as a farm laborer and shepherd. This did not prevent him from acquiring a considerable amount of the literary culture that was available to him in the Welsh language. He became a Congregational minister and served at Mostyn (1831-37); Denbigh (1837-43); Tabernacle, Liverpool (1843-53); and Salem, Liverpool, until his retirement in 1875. Rees provides the most vivid example of the way in which evangelical Christianity in its Calvinistic form inspired cultural and political activity of a radical kind in nineteenth-century Wales. He was one of the most powerful preachers of the midcentury and no mean theologian when it is recalled that he had no formal training in its study. His little catechism (*Y Cyfarwyddwr*, 1833) proved popular and influenced the minds of many young people. He wrote a treatise on natural and revealed religion in 1841, most of which is devoted to an exposition of the divine authority of Scripture.

Like many Victorians, Rees was a man of boundless energy. He published a vast amount of poetry—his hymn *"Dyma gariad fel y moroedd"* became a kind of "signature-tune" during the 1904-5 Welsh Revival—but much the greater part of his work has not survived the test of time. His prose works are now of more interest, and in this field again he was a prolific author. His most influential work was done as an editor. He edited the newspaper *Yr Amserau* ("The Times") from 1843 to 1852, and in articles cast in the form of an old countryman's letters he introduced the main themes of radical politics to Welsh readers. He was also one of the founding fathers of modern Welsh nationalism. He corresponded with the Italian patriot Mazzini, and a Hungarian deputation visited him to express gratitude for the support he had given Kossuth. William Rees was one of the key figures in linking the evangelical churches of Wales with what later became Liberalism, but unlike that of some of his contemporaries, his theological and religious enthusiasm was not overshadowed by social interests.

See E. Rees, *Memoir of William Rees* (1915).

R. TUDUR JONES

REEVE, JOHN, see MUGGLETONIANS

REFORMATION, THE. A broad term used to denote a religious movement in Western Christendom which arose about 1500 and culminated around the mid-seventeenth century, with direct antecedents going back to the fourteenth century. Although conditioned by political, economic, social, and intellectual factors, the course of events and the writings of the Reformers themselves reveal that it was above all else a religious revival which had as its goal Christian renewal.

The Reformation occurred against a vast backdrop of unrest and change in Europe. Politically the most salient feature of the era was the emergence of national states which challenged the old order, including traditional papal prerogatives and the medieval concept of higher loyalties. In the economic realm it was a time of mounting discontent among the exploited peasantry as well as a period of revival of trade, the return of the money economy, and the growth of cities. These developments brought into existence a virile new socioeconomic class, the bourgeoisie. This upset tidy medieval social arrangements and led to increasing political tensions because of the rising expectations of the middle class. Further, beginning in the fourteenth century, the Renaissance* produced a new era of cultural achievement and expression, as well as widespread intellectual unrest. Moreover, a high moral sentiment, the desire for a restoration of past greatness, and growing racial and ethnic pride are common themes in the pre-Reformation literature of discontent.

But most important of all was the troubled state of the Western Church on the eve of the Reformation. It was an age of decline for a church faced with persistent heresy (e.g., the Waldenses* in the Alps, the Lollards* in England, and the Hussites [see HUS, JAN] in Bohemia); an outburst of popular piety (e.g., the flowering of German mysticism and the preaching of Savonarola* in Florence); a loss of papal credibility resulting from the years of "Babylonian Captivity"* in Avignon, the Great Schism* which followed, and a secularized Renaissance papacy; widespread clerical ignorance and abuse; and the unrelenting insistence of the Christian humanists that the church be reformed. Thus the seeds for Reformation in the sixteenth century were nurtured in the fallow soil of discontent at nearly every level of human existence.

The actual beginning of the Protestant Reformation in 1517 in Germany was a combination of the confluence of events with a man of dynamic personality, considerable talent, and deep religious concerns. In the coming of the Reformation, Martin Luther* was the catalytic individual, and the sale of indulgences* near his parish at Wittenberg the precipitating event. Convinced that it was time to challenge the perversion of the doctrine of indulgences and the papal authority which made such abuses possible, Luther drafted his Ninety-Five Theses for debate among theologians. At the time, he had no thought of disrupting the church or starting a new religious movement. Rather, his concern flowed from his desire to reform the church and his conviction that it had departed from its apostolic foundations. Luther, in a desperate search for personal peace with God, had found it, not in the sacraments or the works of merit prescribed by the church, but in Jesus Christ. Luther had recovered, he maintained, NT Christianity with its prescription of salvation by grace through faith in Christ, and not by works of righteousness *(sola fide).*

Thus Luther never espoused a radical rupture with the church's immediate past or its abolition as an institution as did the Anabaptists,* but only reform based on apostolic principles. Nevertheless, a combination of papal inability to comprehend the nature and intensity of the religious issues raised by Luther and of the temper of the times led to a breach between Rome and the German priest. After several debates over papal authority, and after attempts to reconcile and then coerce Luther, the fundamental differences between Roman and Lutheran Christianity became increasingly clear. The Reformation spread with the preaching of justification* through faith in Christ, and Luther's doctrinal position developed more fully to include biblical authority in the place of the teaching church as mediated by the pope, the priesthood of the believer, and two rather than seven sacraments. After his confrontation with the emperor and church authorities at Worms* in 1521 in which Luther refused to recant his views, the rupture was complete.

By 1529, when the imperial diet convened at Speyer,* six German princes and the representatives of fourteen upper German cities embraced the name "Protestant" ("protesting" the emperor's attempts to suppress Luther) and identified themselves as adherents of the Reformation. Luther's own town of Wittenberg in Saxony became the center of the movement which had by the time of the Reformer's death spread to every German-speaking land. By midcentury the

Lutheran Church had taken form and become the dominant faith of much of Germany and most of Scandinavia. It also had by this time made a significant impact on the religious life of the remainder of Europe.

Ulrich Zwingli* of Zurich was one of those touched by the Lutheran Reformation. Guided partly by Luther and partly by his own biblical insights, Zwingli introduced the Reformation in his native canton. Gradually the movement spread westward through German Switzerland, finally reaching the French cantons, where John Calvin* became its leader. Calvinism* became the most important expression of the Reformation, historically speaking, and by the middle of the century Geneva replaced Wittenberg as the main center of the Protestant world. In the last half of the sixteenth century Calvinism became the driving force of the Reformation, especially in Switzerland, W Germany, France,* the Netherlands and Scotland, and to a lesser extent in England, E Germany, Hungary,* and Poland.*

Calvinism triumphed in Scotland largely because of the work of John Knox,* who was the guiding spirit behind the Scots Confession* adopted by the Parliament of that land in 1560. Knox, impressed by Calvin's example, established the church of Scotland* on Presbyterian polity and Calvinist theology while at the same time linking the fortunes of the Reformation as closely as possible with growing Scottish national feeling against the Roman Catholic Mary, Queen of Scots.*

In England the fortunes of Calvinism were more varied. The development of the English Reformation was uneven compared with the reform movement in other countries. Beginning as an act of state in 1534 when Henry VIII* severed connections with Rome and assumed the title of Supreme Head of the Church, the Reformation soon became a genuine attempt to restore the ancient Christian faith to England. Building on the work of John Wycliffe* and his Lollards, the English Christian humanists, and imported Lutheran and Calvinist ideas, the Church of England* began to take shape. More thoroughgoing reform during the reign of Edward VI* was erased by Mary Tudor* when she tried to restore the English Church to the Roman fold. Her attempt failed, and under Elizabeth I* the Church of England once and for all became non-Roman, but not entirely Protestant. Rather, it developed as a *via media* between the former Roman faith on one hand and Protestant Calvinism on the other.

Often overlooked as an important part of the Reformation are the Radical Reformers. These advocates of the Radical Reformation* appeared early in the sixteenth century, represented the left wing of the movement away from Rome, and emphasized "restitution" rather than "reformation." Anabaptists and other Radicals to their left wanted to abolish all the accumulated practices, traditions, and ceremonies of the medieval Catholic Church and instead build a restored church entirely on NT principles. The majority of Anabaptists felt the true church was local, autonomous, governed by democratic polity, and composed only of heartfelt believers who had been baptized after their confession of faith in Jesus Christ.

Although there were a few notable exceptions, such as the fanatical millenarians of Münster (1534-35) and the unitarian Socinians,* most Radicals did not participate in politics and were pacifists. They were also the first in modern times to call for full religious liberty and separation of church and state. After 1535, Menno Simons* a former Roman priest, emerged as the primary leader of the Anabaptists in the Low Countries.* Gradually mainstream Anabaptists became known as "Mennonites."* Other enduring expressions of the Radical Reformation include the Hutterites* and Schwenkfelders,* while the Baptists* and Quakers* of the next century grew out of related and similar foundational principles and impulses.

Finally, the Catholic Reformation represented an attempt to renew the established church from within, both in reaction to the Protestant threat and in response to certain internal developments. The Oratory of Divine Love (founded 1517), a reformed papacy, the establishment of the Jesuits* (1540), the reforming Council of Trent* (1545-63), the Roman Inquisition, and Spanish mysticism were all expressions of this growing emphasis on reform, renewal, and retrenchment within the Roman Church (see COUNTER-REFORMATION).

Unlike the Renaissance, the Reformation directly affected nearly every European and forced almost everyone to make a choice between the old and the new. As it did, the Reformation movement profoundly changed the course of Western civilization and touched every facet of human existence. The modern, pluralistic, culturally fragmented Western World, for better or for worse, is largely the child of this tumultuous and significant movement.

BIBLIOGRAPHY: R.H. Bainton, *Here I Stand: a Life of Martin Luther* (1950) and *The Reformation of the Sixteenth Century* (1952); G.H. Williams, *The Radical Reformation* (1962); G.R. Elton, *Reformation Europe, 1517-1559* (1963); P. Janelle, *The Catholic Reformation* (1963); A.G. Dickens, *The English Reformation* (1964) and *Reformation and Society in Sixteenth-Century Europe* (1966); J.D. Douglas, *Light in the North* (1964); J.T. MacNeill, *The History and Character of Calvinism* (rev. ed., 1967); J.H.M. Salmon, *The French Wars of Religion: How Important Were Religious Factors?* (1967); J. Atkinson, *Martin Luther and the Birth of Protestantism* (1968); W.S. Reid (ed.), *The Reformation: Revival or Revolution?* (1968); R.M. Kingdon and R.D. Linder (eds.), *Calvin and Calvinism: Sources of Democracy?* (1970); H.J. Grimm, *The Reformation Era, 1500-1650* (3rd ed., 1973). ROBERT D. LINDER

REFORMED CHURCHES. Those ecclesiastical bodies or denominations which hold to the system of doctrine and government or polity as set out by John Calvin* in the sixteenth century, and expressed in various "Reformed" confessions. They include the Calvinistic churches deriving from Europe which are specifically known as "Reformed," as well as the "Presbyterian" churches of the English-speaking world. Some Baptist churches claim to be Reformed in doctrine, and some Anglicans also hold that the Church of England is "Reformed," a position denied by the Anglo-Catholics. In 1875 the Alliance of Reformed Churches was formed and still exists; the Reformed Ecumenical Synod was later established, made up of churches not in the Alliance.

W.S. REID

REFORMED CHURCH IN AMERICA. Known also as the Dutch Reformed Church,* this was one of the continental groups transplanted to the United States. It came with the original Dutch settlers to New Amsterdam, but the first congregation was not organized by Jonas Michaelius until 1628. During the seventeenth century it received much strength from the flow of Dutch to the colony, but it remained dependent on the church in the Netherlands for its ministers. The change to English rule in 1664 had little effect on the church, and the harmony between the church in the colonies and the classis of Amsterdam prevailed until after 1700.

With the arrival of T.J. Frelinghuysen* in 1720 and the Great Awakening,* tension developed between opponents and proponents of the revival. Frelinghuysen and his group felt that the church in the colonies should not be dependent on the Netherlands and should train its own ministers. As a result, the classis of Amsterdam finally agreed to the creation of a subordinate assembly, or *coetus,* which in 1770 obtained a charter for and organized Queen's College (now Rutgers) which after the revolution emerged as a college for the training of Christian ministers. Hope College in Michigan was chartered in 1866. In 1794 a general synod based on the Constitution of 1792 was formed from the various particular synods. During the nineteenth century the church expanded westward to Illinois, Michigan, and the Pacific coast and Canada. In 1867 it became known as the Reformed Church in America. It is Calvinistic in theology and Presbyterian in government.

C. GREGG SINGER

REFORMED EPISCOPAL CHURCH. This schism from the Protestant Episcopal Church* in America was founded by G.D. Cummins,* its first presiding bishop, in 1873, because of deep dissatisfaction with the rising ritualism and sacerdotalism among American Episcopalians. Their bishops trace their consecration from Canterbury, yet are not a third order in addition to deacons and presbyters, but rather first among equals. Its government involves much lay participation at all levels. Clergy from other groups are received without reordination, and laity without reconfirmation. Their *Book of Common Prayer* of 1874 is based on the Prayer Book of 1785, the colonial church's revision of the BCP under Bishop William White. Their doctrine of ministry is Reformed in that the minister is a minister of Word and sacrament rather than a mediator. Absolution, baptismal regeneration, and transubstantiation are not held, and the Prayer Book contains their own 1873 Declaration of Principles and Articles of Religion. It is an American church, divided into the Synod of New York and Philadelphia, the Synod of Chicago, and the Jurisdiction of the South, each with its own bishop. A few parishes exist in Canada and England (Free Church of England*). The total American membership is a little over 7,000. Its theological seminary is located in Philadelphia.

C.G. THORNE, JR.

REFORMED PRESBYTERIAN CHURCH. A body which claims unbroken descent from the Scottish Covenanters* who contended for a pure Gospel, a simple form of worship, national righteousness, and civil and religious freedom. The name indicates that in the matter of church government it is Presbyterian, and in doctrine and practice adheres to the attainments of Scotland's Second Reformation (1638-49). After 1690 the Reformed Presbyterian Church in Scotland consisted of the remnant societies that had adhered to Richard Cameron* and James Renwick.* They had no minister until 1706, and from then until 1743 only John MacMillan served their needs. A second minister enabled them to form a presbytery in 1743, and an increase in numbers led to the forming of a synod in 1811. Ministers were trained at a theological hall in Stirling and Paisley. Dissension on the question of the parliamentary elective franchise split the church in 1863, and the majority who favored leaving the matter an open question joined the Free Church of Scotland* in 1876. A small denomination of five congregations still survives and adheres to the Covenanting position.

The origin of the church in Ireland is related to the coming of Scottish settlers to Ulster in the early seventeenth century. The majority of them approved of the Presbyterian form of church government and signed the Solemn League and Covenant* in 1644. At the Revolution Settlement of 1690 a small minority adhered to Covenanting principles and became the Reformed Presbyterian Church of Ireland. They had no minister after the death of David Houston in 1696 and were dependent on visits from Scottish ministers until William Martin was ordained in 1757. In 1763 a presbytery was formed, and rapid growth after 1800 led to the formation of a synod in 1811. For many years their ministers were educated in Scotland, but a theological hall was established in 1854 and still operates with four part-time professors.

The two churches in Scotland and Ireland united to begin mission work in Syria in 1871. After almost a century in the Middle East, the mission field has since 1963 been Ethiopia. Close links are maintained with the Reformed Presbyterian Church of North America, organized by settlers from Scotland and Ulster in the mid-eighteenth century. Despite divisions and secessions the church still maintains a witness with about seventy congregations. It supports a theological seminary at Pittsburgh, a liberal arts college at Beaver Falls, and active missions in Cyprus and Japan.

The church accepts the continuing obligation of the Scottish Covenants. In worship the Psalms only are sung without instrumental accompaniment. Members do not participate actively in politics and do not join societies which require an oath of secrecy at initiation. The doctrinal standards are the Westminster Confession of Faith* and the Larger and Shorter Catechisms. Total membership with adherents is estimated at 20,000. ADAM LOUGHRIDGE

REGENERATION. A supernatural work of the Holy Spirit in the individual heart in which a new and holy spiritual life is imparted. The actual term occurs only twice in the NT (Matt. 19:28; Titus 3:5 KJV), and in the first of these it refers not to the individual but to the universe. The idea represented by the word is frequent, however (e.g., John 1:12ff.; 3:1-10; Gal. 4:23,29; James 1:15-18; 1 Peter 1:3,23; 1 John 2:29). Moreover, the ideas of spiritual resurrection, new creation, and circumcision of the heart are further figurative expressions conveying the same idea of an act of God involving a decisive break with the past. The language used brings out its supernatural character. There has been controversy as to the relation of regeneration to baptism and to conversion. Evangelicals see baptism as the sign and seal of regeneration while Roman Catholics see it also as conveying the regenerating grace it signifies. Calvinists see regeneration as the cause of conversion (repentance and faith) while Arminians see conversion as the cause of regeneration.

G.W. GROGAN

REGENSBURG, CONFERENCE OF, see RATISBON

REGINALD OF PIPERNO (c.1230-c.1290). Theologian. Born in Italy, he joined the Dominican Order at Naples and around 1260 became the confessor of Thomas Aquinas* as well as his constant companion. In 1272 he took a teaching post in Naples, succeeding Thomas in the chair of philosophy in 1274. He zealously collected Thomas's works, including the four *Opuscula* which he recorded from his master's lectures.

REGULAR (Lat. *regula,* "rule"). General term for members of the clergy who are bound by the vows of religion and live in community, following a "rule," in distinction from the "secular" clergy, i.e., priests living in the world. Monastic "rules" go back as far as Pachomius* (c.305); one of the most influential was that of Benedict of Nursia* (529).

REICHSBISCHOF. Protestant church office in Nazi Germany. Following Hitler's accession to power, many churchmen favored uniting the twenty-eight separate provincial churches into one with a national *(Reich)* bishop as head. The pro-Nazi German-Christians* supported the candidacy of Ludwig Müller (1883-1945) for the post, but he lost the election in 1933 to Friedrich von Bodelschwingh,* director of the Bethel charitable foundation in Westphalia. The German-Christians, with Nazi support, forced his resignation, and Müller was elected later that year at a national synod in Wittenberg. The Nazis soon abandoned the German-Christians, and with the creation of the Ministry of Church Affairs in 1935 the reichbishop was deprived of any effective authority. RICHARD V. PIERARD

REIMARUS, HERMANN SAMUEL (1694-1768). German scholar. Born in Hamburg, he was educated at Jena, taught philosophy at Wittenberg, and in England came under the influence of the English Deists. He was rector of the *Hochschule* at Wismar (1723-27) before becoming professor of Hebrew and oriental languages in the Gymnasium Johanneum in Hamburg. He published works on Dio Cassius, logic, and the instincts of animals, and worked from 1744 on a comprehensive study which remains as a whole only in manuscript (perhaps because of its controversial content), though segments of its 4,000 pages were published piecemeal and anonymously after his death under the title *Wolfenbüttel Fragments.** One published fragment, "The Object of Jesus and His Disciples," provided a perspective on Jesus which was to revolutionize the image of Him in modern theology, and to become the point of departure for A. Schweitzer's* *Quest of the Historical Jesus.* Though Reimarus's historical reconstruction was too influenced by Deistic rationalism concerned to eliminate the miraculous, he was one of the few in the history of Christian thought to identify correctly that Jesus had proclaimed the nearness of the messianic age to the Jews. Reimarus's efforts, were premature, since his generation was not prepared to break its traditional conception of the preaching of Jesus for a consideration of an imminent eschatological expectation.

See C.H. Talbert (ed.), *Reimarus: Fragments* (1970). CLYDE CURRY SMITH

REIMS NEW TESTAMENT, see DOUAI-REIMS BIBLE

REINKENS, JOSEPH HUBERT (1821-1896). Old Catholic* bishop. Born at Burtscheid, near Aix-la-Chapelle, he was professor of church history (1850) and rector (1865) at Breslau University. He opposed the Vatican Council* definition of papal infallibility (1870), joining J.J.I. von Döllinger* in the Nuremberg Declaration. Excommunicated, he was elected the first bishop of the German Old Catholics at Cologne (1873), with his see at Bonn. Consecrated by Bishop Hermann Heykamp of Deventer of the Old Catholic Church of Utrecht, he in turn consecrated Edward Herzog as the first Swiss Old Catholic bishop (1876). He took a prominent part in the Bonn* Reunion Conferences (1874-75) and devoted the rest of his life to the Old Catholic cause. He visited England (1881) and defended the validity of Anglican orders against the Dutch Old Catholics (1894). He wrote a treatise on Cyprian and the unity of the church (1875). J.G.G. NORMAN

REITZENSTEIN, RICHARD (1861-1931). German historian of religions and classical philologist. Beginning in 1889 he held several posts in German universities; from 1914 he taught at the University of Göttingen. After spending some years studying philology, he developed an interest in applying its methodology to the study of ancient religions, particularly the origins of Christianity. Interested in the Hermetic* literature, he attempted in *Poimandres* (1904) to show that the religion devoted to Hermes Trismegistus had some influence on early Christianity. Broadening his attempt to show Christian dependence on Gnosticism and mystery-religions, he published in 1910 *Die hellenistischen Mysterienreligionen*, and in 1916 *Historia Monachorum und Historia Lausiaca*. Such attempts did not win much scholarly acceptance. KEITH J. HARDMAN

RELICS. According to Roman Catholics, the material remains of a saint and any other objects which have had contact with him. These are to be venerated on the grounds that the bodies of the saints, now with Christ, were once living members of the church and temples of the Holy Ghost, destined to be raised to eternal life and glorification. Since the Godhead makes them the occasion for miracles, they are to be venerated by the faithful for in this way God bestows many gifts on men. Biblical justification for this cult is sought in Acts 19:12, where healing power is shown to be in handkerchiefs which had been in touch with Paul's body. OT references are also cited (e.g., 2 Kings 2:14; 13:21, where miracles are said to have occurred through Elijah's mantle and Elisha's bones).

The earliest classical instance of the veneration of relics is said to be found in a letter written by the inhabitants of Smyrna about 156, describing the death of Polycarp* in which they said, "We took up his bones, which are more valuable than precious stones and finer than refined gold, and laid them in a suitable place, where the Lord will permit us to gather ourselves together, as we are able, in gladness and joy, and to celebrate the birthday of his martyrdom." The cult spread rapidly in both East and West, and the increasing demand for relics which arose led in the East to the translating and dismemberment of the bodies of the saints. In 1084 the Council of Constantinople approved the veneration of relics for the Eastern Church, although it has always been overshadowed by the widespread use of icons. At Rome the cult was associated with the prayer services held in the catacombs (burial places for Christians), and from the fourth century the Eucharist was celebrated over the tombs of the martyrs. Unlike the East where no repugnance was felt over dismemberment, in Rome the Theodosian Code expressly forbade the translation, division, or dismemberment of the remains of the martyrs. The practice was, however, introduced in the West in the seventh and eighth centuries. The Council of Nicea* (787) decreed that no church should be consecrated without relics. The Crusades* gave a special impetus to the cult since relics, often spurious, were brought back in abundance from Palestine to Europe.

The cult has often been associated with superstitious practices, inevitable perhaps because of the influx of converted pagans into the church. Veneration of relics is said to be a primitive instinct of man and has been associated with many non-Christian religions such as Buddhism. The very nature of the cult lends itself to abuses, and with religious centers eager to be known as possessors of some unusually startling relic, fabrications were inevitable. The ecclesiastical authorities have made some efforts to secure the faithful against deception. Canon law forbids relics to be venerated which have not been authorized by a cardinal or bishop, and the sale of genuine relics as well as fabrications or distributions of false ones is punished by excommunication. It is still true, however, that many of the ancient relics exhibited for veneration in the great sanctuaries of Christendom are open to grave suspicion.

S. TOON

RELIEF CHURCH. Formed in Scotland in 1761 by Thomas Gillespie* and two other ministers, "for Christians oppressed in their Church privileges." The reference is to those who had been compelled to accept a minister appointed by the patron, contrary to their wishes. Gillespie had been deposed by the general assembly of the Church of Scotland for his support of the town council, kirk session, and members of the parish of Inverkeithing who protested against the patron's choice of minister there. In 1766 it is reported that the Relief Church and the Associate Synods drew as many as 100,000 to their places of worship. In 1847 there was a union with the United Secession Church* to form the United Presbyterian Church.* J.W. MEIKLEJOHN

RELIGIOUS CONGREGATIONS, see CONGREGATIONS

RELIGIOUS DRAMA, see DRAMA, CHRISTIAN

RELIGIOUS EDUCATION, see EDUCATION, CHRISTIAN

RELIGIOUS MIMICRY, see MIMICRY

RELIQUARY. A box, shrine, or casket used to keep or display a relic.* In the early Middle Ages the supernatural power of the unseen world was viewed as accessible to men in relics. Appropriately encased, they were carried by armies and noblemen and hidden in royal crowns and necklaces, which served as reliquaries. Charlemagne's throne, built to the specifications of Solomon's, was filled with cavities for the deposit of relics. Treaties were signed and oaths taken on reliquaries. St. Louis built Saint-Chapelle in Paris as a reliquary to house the crown of thorns purchased from Baldwin II of Constantinople after he used the item as collateral to secure a Venetian loan. Pilgrims were drawn to Canterbury, Compostela in Spain, Regensburg, Alt-Ötting in Bavaria, and other places by reliquaries and their contents.

JOHN GROH

REMBRANDT VAN RIJN (1606-1669). Dutch painter. Nurtured in the humanistic culture of Holland's "golden age," at Leyden University, he broke young into the fashionable art patronage circle with his group portrait, *The Anatomy Lesson of Dr. Tulp* (1632). From the time of his marriage to Saskia till her early death (1634-42), Rembrandt flourished by painting life-size portraits, biblical story topics, and fantastic landscapes in the fashionable Baroque manner; touches of oriental exotica and his own genial, somewhat darker chiaroscuro gave the vigorous, framebursting action or penetrating character study an extra dimension of aplomb—e.g., the *éclat* of the wall-size masterpiece, *Night Watch* (1642).

Thereafter the paintings grew with a darkling mysteriousness, so that the golden light spots did not dramatize external features and events so much as illumine covert meanings, depths deeper than meets the eye. That warm black, which Picasso has said he covets to have painted, leaves behind mythological fantasies and rules out mere sense perception; the darkness of Rembrandt's later portraits reveals the creatural truth of how men can suffer misfortune while still believing in the presence of God. The stern look of authority determined to be just, the pensive tenderness of an ennobling sadness, and the ache of life's troubled happiness come through again and again in the enormously richly colored, almost sketching, heavy brush strokes or thick impasto scraped away by Rembrandt's palette knife; see, for example, *Man With a Gilt Helmet* (1652), *Bathsheba After the Bath* (1654), *Saul and David* (c.1660). Other late works like the glorious *Flayed Ox* (1655, in the Louvre), the famous *Syndics of the Cloth Guild* (1662), and the unforgettable *(Jewish) Bridal Couple* (c.1665) show a dimension of insight not surpassed in the history of painting.

The truth rediscovered by the Reformation—that men in our sin-filled world can still disclose the handiwork of God, and that vocation, God's calling man to a task, is an intangible but everlasting reality, and that Grace can be *inside* daily human life, sanctifying matters as normal and pervasive as work and marriage—Rembrandt softly and masterfully paints for men with eyes to see.

CALVIN SEERVELD

REMIGIUS (Remi) (c.438-c.533). Archbishop of Reims and "Apostle of the Franks." Son of a count of Laon, he was proclaimed archbishop of Reims at age twenty-two. In 496 he had his greatest achievement—the baptism of Clovis,* king of the Franks, and 3,000 of his subjects after the battle of Tolbiac. Various legends are associated with the relationship of Remigius and Clovis, e.g., that the bishop conferred on the king the power of "touching for the king's evil." By his untiring efforts he also founded bishoprics at Arras, Cambrai, Laon, Térouanne, and Tournai, and sent missionaries to the Arians in Burgundy. His remains were transferred to the Abbey of St. Remi by Leo IX in 1049.

PETER TOON

REMIGIUS OF AUXERRE (c.841-908). Leader in the later Carolingian Renaissance.* He was educated at the Benedictine monastery of St. Germain at Auxerre, where he later taught. He taught also at Reims and made a significant contribution to the revival of classical learning which had been begun during the reign of Charlemagne. He wrote glosses and commentaries on a number of Latin authors, including Virgil and Terence. In addition, he wrote a commentary on Boethius, *De Consolatione Philosophiae,* and homilies on the gospel of Matthew.

REMONSTRANTS; REMONSTRANCE (1610). The Remonstrants were a revisionist group in Dutch Calvinism, associated with the controversies leading to the Synod of Dort* (1618-19), and there condemned. They were followers of Arminius (Hermandszoon), whose teachings at the theological faculty at Leyden aroused extensive controversy. After Arminius's death in 1609, Uytenbogaert took the lead in drawing up the Remonstrance of 1610, directed to the Estates of the province of Holland (where Leyden was located), and presenting the positions of the Arminian* party. The Remonstrance sets forth five points, all dealing with Arminius's attempt to soften the orthodox Calvinist idea of predestination and save something of man's free will. It holds that the decree of predestination is not absolute, but conditioned on man's response; that the offer of salvation is directed to all men, and all men in principle can be saved; that man can exercise his free will properly only after receiving grace; but, that this grace can be accepted or denied; thus, believers can fall from grace.

Acceptance of these points would have meant revising the Belgic Confession* and the Heidelberg Catechism,* generally accepted as doctrinal standards by the Dutch Calvinist churches. The Remonstrance provoked the Contra-Remonstrance of 1611, setting forth the orthodox position; to the Contra-Remonstrants it seemed as though a Semi-Pelagian* position was clearly proposed, and thus the assurance of salvation taken away. The controversy became mixed with political issues; the Remonstrants were supported by the powerful Oldenbarneveldt, but opposed by the *stadhouder* Maurice of Orange. The Estates-General issued edicts forbidding further controversy; these were ignored. By 1618 the political struggle was ended with the imprisonment of Oldenbarneveldt, and the Estates called the Synod of Dort to settle the religious issue. Deprived of their chief political supporter, the Remonstrants were helpless, and the synod speedily declared their teachings erroneous. Remonstrant minsters, some 200, were ousted from their pulpits, and many exiled for disturbing the peace. Uytenbogaert and Episcopius* established the Remonstrant Brotherhood, starting with the ousted ministers, and at the death of Maurice in 1625 the Remonstrants were again tolerated. A seminary was founded at Amsterdam (1630), with Episcopius as its leading figure, becoming steadily more "liberal" as time went on, under the able theologians Courcelles (d.1659), Limborch (d.1712), Leclerc* (d.1736), and Wettstein* (d.1754). During the 1700s the Remonstrants declined in numbers, losing members to Socinianism and Deism. During the later 1800s, however, as many found the Dutch Reformed Church too orthodox for their taste, they enjoyed a modest growth. The seminary was moved to Leyden in 1873. Present membership is over 25,000.

See A.W. Harrison, *Arminianism* (1937), and C.O. Bangs, *Arminius* (1971).

DIRK JELLEMA

RENAISSANCE, THE. A term used by historians to describe a special period of European history, roughly the fourteenth, fifteenth, and sixteenth centuries. Etymologically the word itself is French for "rebirth," meaning in general a revival of culture, although most historians today tend to see it more as an age of movements and accelerated transition rather than one of sharp departure from the medieval past.

The Renaissance began in Italy with a renewal of interest in the study of the classics known as "humanism." Thus, intellectually the Renaissance was a period of intense study of both the form and content of classical texts. Petrarch* (d.1374) is generally thought of as the "first humanist," followed by a host of other brilliant men of letters: Giovanni Boccaccio (d.1375), Lorenzo Valla* (d.1457), and Giovanni Pico della Mirandola* (d.1494), to name a few. The focus of this classical revival in Italy was more on man and his relation to the present material world than on God and the world to come, as had been true in the medieval past. But above all, the Italian Renaissance was a time of supreme cultural achievement. It was a period studded with geniuses and men of influence: Leon Battista Alberti (d.1472), Leonardo da Vinci* (d.1519), Raphael* (d.1520), Niccolò Machiavelli (d.1527), Michelangelo* (d.1564), and Benvenuto Cellini (d.1571), for example.

As the Renaissance moved north of the Alps in the fifteenth and early sixteenth centuries, it became more religious in tone and emphasis. The majority of northern humanists were more interested in the Christian classics (e.g., the NT and the Fathers) than in pagan texts. They also were concerned with reforming the church according to apostolic principles. Because of their desire to apply humanism to the question of reform, these northern scholars generally are called "Christian humanists." Among their number were John Colet* (d.1519), Johannes Reuchlin* (d.1522), Thomas More* (d.1535), Jacques Lefèvre d'Étaples* (d.1536), and the great Erasmus* (d.1536).

Although the question of the exact relationship between the Renaissance and the Protestant Reformation is still debated, it is clear that the former movement affected the course of Christian history in several important ways. First, Renaissance attitudes, values, and practices penetrated the Roman hierarchy in this period. By the time Martin Luther* (d.1546) drafted his Ninety-Five Theses in 1517, the papal chair had a long history of occupants insensitive to the spiritual needs of the faithful and more interested in real estate than reform, more concerned with politics than piety. Second, the Christian humanists' sharp criticisms of clerical abuse, and their call for reform, added to the growing unrest in Western Christendom. The old saw that "Erasmus laid the egg that Luther hatched" contains a great deal of truth. Third, after 1517 many younger humanists turned Protestant, for example, Ulrich Zwingli* (d.1531), Philip Melanchthon* (d.1560), John Calvin* (d.1564), and Theodore Beza* (d.1605).

ROBERT D. LINDER

RENAN, JOSEPH ERNEST (1823-1892). French humanist historian of religion and oriental philologist, who unsettled both Catholics and Protestants with *La Vie de Jésus* (1863). Using the new German textual and philological criticism with a rationalist skeptic's assumptions, he depicted Jesus as a truly remarkable itinerant preacher, but certainly not the Son of God. His portrait came at the right historic moment for him and achieved immense popularity among enlarging skeptical readership. The work expanded into *Histoire de origines du christianisme* (7 vols., 1863-81), with further studies of the apostles, Paul, Antichrist, the early church, and the end of the ancient world. He wrote *Histoire du peuple d'Israël* (5 vols., 1887-93) with the same perspective. Renan was raised Catholic and prepared by seminary study to enter the priesthood. This he abandoned (1845) along with his Catholic faith. Thereafter he went on numerous archaeological digs in the Near East and became professor of Hebrew at the *Collège de France* (1862) until removed because of the furor over his *Jésus.* He was reinstated (1870) and appointed director of the college (1879) under the secularist Third Republic.

C.T. MC INTIRE

RENAUDOT, EUSÈBE (1646-1720). Roman Catholic scholar. Although trained for the Society of Jesus, he became a secular priest. He was particularly noted for his knowledge of oriental languages, becoming one of the principal advisers of Louis XIV's minister Colbert on matters relating to the East. His writings were mainly devoted to showing the relationship between the Orthodox and the Roman churches, as he sought to prove the "continuity of the faith" in opposition to Protestant contentions that the Roman Catholic Church had perverted the faith. His most important works were *De la Perpetuité de la foi de l'église sur les sacrements et autres points* (1713) and *Historia patriarchum Alexandrinorum* (1713).

W.S. REID

RENQVIST, HENRIK (1789-1866). Finnish pastor. While still at school he experienced a personal revival and was greatly influenced by a book on conversion by Arthur Dent, the English Puritan. During Renqvist's hard inner struggles he came into contact with the "Prayer Movement," and his own conviction became clear. He liked especially the emphasis on "repentance" and "conversion" in the Pietistic movement called the "Prayers." Another important influence on him was the Scot, John Paterson, who underlined the work of the Bible Society (founded in Britain in 1804). Thereafter Renqvist was especially interested in the printing and spreading of inexpensive Bibles and Christian literature. He started writing and translating Christian books—some sixty titles altogether. In his own religious outlook he was a mystic. He stressed in his books the necessity of spiritual exercise, especially praying on one's knees. He was a pioneer in temperance work and missions in Finland.

STIG-OLOF FERNSTROM

RENUNCIATION OF THE DEVIL. From early times a formal renunciation of the devil accompanied the Christian's confession of faith at baptism. In the *Apostolic Tradition* of Hippolytus (c.215) the form is: "I renounce thee, Satan, and all thy service and all thy works" (xxi.9). Later in the West, the renunciation took the form of response to an interrogation. At Jerusalem in the fourth century, the renunciation was recited facing west, and likewise the baptismal creed facing east, and the turning thus became a symbol of conversion. This practice spread widely in the church, and a relic of it remains in turning east for the creed.

RENWICK, JAMES (1662-1688). Last of the Scottish Covenanter* martyrs. Born at Moniaive, Dumfriesshire, the only son of poor and God-fearing parents whose several daughters had died in infancy, he showed early signs of piety, though his sensitive mind was at times clouded with doubt. While a student at Edinburgh University, he heard Donald Cargill* preach and saw his martyrdom. Cargill's influence led him to associate with the Covenanting societies that had adhered to Richard Cameron.* They sent him to Holland to train for the gospel ministry. He was ordained there and returned to Scotland through Dublin in the summer of 1683. He began a passionate four-year ministry at Darmead, preaching in all weathers and at all hours. In one year he baptized 600 children. Though anxious to refrain from violence, he could not allow the enthronement of James II* to pass without a solemn Protestation that denounced the king as a murderer and idolater, and an enemy of true religion. He was arrested in Edinburgh condemned to die, and executed at the Grassmarket. Letters and sermons are extant that give valuable information about his character and the times in which he lived.

ADAM LOUGHRIDGE

REORDINATION. Strictly speaking, the term is a misnomer since in practice it applies only to those who have received ordination which is regarded as "invalid"—that is, as no ordination at all—and who consequently are in need of ordination* rather than reordination. In the early church the appearance of heretical or partially heretical sects and schismatic groups forced the issue into prominence. Unanimity of judgment is not to be found, however. Cyprian* (third century), for instance, insisting that outside the Catholic Church there could be no salvation, regarded every action of separated bodies as null and void: their baptism was no baptism and their orders were no orders, and acceptance was to be gained only by (re)baptism and (re)ordination in the unity of the Catholic Church, which is reality were first, not second, baptism and ordination.

The Council of Nicea* (325) represents a more moderate position which attempts a distinction between, on the one hand, heretics whose sacraments and orders were judged invalid, and on the other hand, schismatics who otherwise were orthodox in their articles of belief and whose sacraments and orders were judged acceptable in the event of their wishing to end their schism. The position of greatest tolerance is represented by Augustine* (d.430), in whose view rebaptism and reordination were unnecessary for the reconciling and acceptance of those who hitherto had been divorced from the Catholic Church, since Christ—not man or the church—is the source of grace and validity.

The Cyprianic position regarding orders (but not baptism) has been that of the Roman Catholic Church since the Reformation. Nonpapal orders have been dismissed as invalid and to all intents and purposes nonexistent. It is an attitude, too, that has become widely accommodated in the Anglican Communion,* though it is contrary to the teaching and practice of classical Anglicanism. The insistence, whether explicit or implicit, on the necessity of episcopal ordination (and therefore, despite pious talk regarding the blessing of the Holy Spirit that has attended nonepiscopal ministries, on the invalidity or insufficiency of such ministries) has proved a stultifying factor in attempts to achieve reunion between Anglican and nonepiscopal churches, especially when "catholic" Anglicans demand a method of reconciliation which can be interpreted by them as a reordination of ministers who lack episcopal orders. In view of the nonrecognition of Anglican orders by Rome, such manipulations have a distinctly Gilbertian flavor.

In the situation now prevailing, when there is so much disunity and fragmentation, the Nicene principle—according to which adherence to the apostolic faith is the criterion of the genuineness of a minister's calling and ordination, and the ground of reunion without question of reordination—is the principle which should consistently be applied.

PHILIP EDGCUMBE HUGHES

REORGANIZED CHURCH, see MORMONISM

REPENTANCE. The English translation of the Greek *metanoia,* signifying "a change of mind," and which is often used in the Septuagint to translate the Hebrew *nacham.* So defined, repentance might appear to be purely intellectual. In fact this is not the case, for the biblical writers were strongly aware of the unity of human personality. To change the mind was to change the attitude and so, at least in principle, to change the actions and even the whole way of life. It is an important element in biblical preaching (Jer. 25:1-7; Mark 1:15; 6:12; Luke 1:16f.; Acts 2:38, etc.). An OT passage which does not use the word well expresses the meaning of it (Prov. 28:13). Repentance is one aspect of conversion, the other being faith. They are two aspects of the one experience in which a man turns from sin to Christ. Initial repentance should lead to habitual renunciation of sin. It should not be confused with "penance,"* which appears as a translation of *metanoia* in the Roman Catholic Douai Version. Penance is understood in modern English as the performance of ecclesiastically prescribed acts to make satisfaction for postbaptismal sin, and this has no place in NT Christianity.

G.W. GROGAN

REPROBATION (Gr. *adokimos,* "rejection after test"). In the NT it is normally applied to man's sinful condition and implies that judgment will fall on the man thus described (Rom. 1:28; 2 Tim. 3:8; Heb. 6:8). In 1 Corinthians 9:27 it has sometimes been understood to refer rather to disqualification from Christian service, while in 2 Corinthians 13:5ff. Paul's readers are told to test themselves concerning their Christian standing. A study of these passages will reveal that the cause of such rejection by God is represented as man's sin. In none of them is this related to the eternal counsels of God. The idea that it is so related has sometimes been seen in 1 Peter 2:8; Jude 4 (where the actual word *adokimos* is not in fact used), although what is in view in these passages may simply be the fact that the men referred to were the subject of OT prophecy. Romans 9:1-29 deals with the specific problem of the apparent rejection of Israel by God. Note that although Paul speaks of God as the personal Author of salvation (v.23), he uses an impersonal form of expression when he speaks of damnation (v.22). In chapter 10 he goes on to show that in fact God's rejection of them is due to their rejection of Him. G.W. GROGAN

REQUIEM. A musical Mass for the dead, celebrated in the Roman Catholic Church. It receives its name from the first word of the Introit. *Requiem aeternam dona eis Domine* ("Give them eternal rest, O Lord"). This introit is also often recited by itself as a prayer for the departed soul. Although the requiem was intended to be sung only at funerals, at anniversary memorials for the departed, and on All Souls' Day (1 November), due to its limited use it became a concert form. The opening sections of the requiem uses traditional Gregorian chant melodies, and by the eighteenth century orchestral accompaniment had gained importance. Guiseppe Verdi's Requiem in memory of Italian novelist A. Manzoni is probably the best known. ROYAL L. PECK

REREDOS. A medieval Anglo-French word meaning "screen," this denotes the decoration found behind and above the altar in the chancel of a church, and usually under the east window. In its earliest manifestation it took the form of Christian paintings on the wall, but developments in the Middle Ages led to the production of elaborate rich silk or jeweled metal screens. Other forms which became common were carved wooden panels and sculptured stone or alabaster figures and symbols. Though used widely in Episcopalian churches of all types, evangelical Anglicans tend to dislike their usage.

RESCISSORY ACT (1661). Passed by the Scottish Parliament after the Restoration of Charles II,* this rescinded without distinction all the statutes passed since 1633. This erased much that was worthy in Presbyterian legislation and was the prelude to the reestablishment of Episcopacy and the persecution of Covenanters.* Even some of the nobility argued this Act, but they were overruled.

RESERVED SACRAMENT. The practice of keeping the bread (and sometimes the wine) consecrated at the Eucharist for the purpose of Communion, especially for the sick. Justin Martyr mentions the custom of sending a portion of the elements to those absent. Tertullian spoke of reservation on fast-days, and of the practice of home-communion. Reservation by private persons in their own homes was common at least until the late fourth century, surviving among hermits till the thirteenth and fourteenth centuries. From the fourth century, however, it was normally kept in churches. The custom was liable to abuse and was prohibited by the Council of Saragossa (380) and by a fourth-century Armenian canon (apart from sick-communion). There is no trace of reservation for the purpose of adoration before the development of the doctrines of transubstantiation* and concomitance. Luther and the Reformers rejected reservation. In Anglicanism, the first Prayer Book (1549) provided for reservation for the sick. This was dropped in the second Prayer Book (1552), and the 1661 Prayer Book ordered the elements remaining after the service to be consumed. By the nineteenth century the practice among Anglicans died out largely, though the Scottish Episcopal Church retained it, and it has now been widely restored. J.G.G. NORMAN

RESOLUTIONERS. In December 1650 the Estates in Scotland ordered the commission of the general assembly to decide whether it was lawful to reinstate those formerly purged from the army by the 1649 Act of Classes, which had excluded from civil and military posts all who were hostile to the National Covenant* and the Solemn League and Covenant.* The commission agreed that it was lawful to reinstate all but a small minority, and this decision was known as the first Public Resolution. In March 1651 the commission was asked about the legality of admitting to the Committee of Estates those who, formerly debarred, had now renounced their anti-Covenanting attitude. The commission recommended the admission of all save a few "pryme actors against the State." This was the second Public Resolution. Those who upheld these decisions were known as Resolutioners, and they were mostly moderate Presbyterians and moderate Royalists. Those who disagreed were called "Protesters."*

J.D. DOUGLAS

RESTORATION, THE (1660). A title used by British historians to describe the return of Charles II of the House of Stuart to the throne of England. He landed at Dover on 25 May after an exile on the Continent. On 8 May he had been proclaimed king by Parliament, but previous to this and since the execution of his father in 1649 he had been *persona non grata* in London. His return to the throne was followed by the restoration of the old character of the national church, with prelates and Prayer Book. For the Scots, who had crowned him ten years earlier, the Restoration meant the recovery of national independence.

RESTORATIONISM. The doctrine of universal salvation, or universalism, also known as "Apocatastasis." It claims that all free moral creatures—men, angels, devils—will ultimately be saved. If there is a hell, it is purgative only. Some have held that it is taught in the NT (e.g., C.H. Dodd on Romans 11). The doctrine seems present in Clement of Alexandria* and clearly appears in Origen,* who hoped that the devil would be finally redeemed, though even Origen's universalism needs to be qualified by his insistence on the eternity of man's freedom. Gregory of Nyssa* supported restorationism, but it was attacked by Augustine* and formally condemned in the first anathema against Origenism* at the Council of Constantinople (543). It was held by John Scotus Erigena (ninth century), by mystics like Eckhart and Tauler, by Johannes Denck and some Anabaptists, and by several Moravians and Pietists. In modern times exponents include Friedrich Schleiermacher, Erskine of Linlathen, F.D. Maurice, Ethelbert Stauffer, and many others in all sections of Christendom. J.G.G. NORMAN

RESURRECTION OF CHRIST. The belief that Jesus Christ had really died upon the cross and had been raised by God to life in a new sphere is central to the NT and was constitutive of the Christian Church. Its importance for the historical and theological assessment of the truth of Christianity today is generally recognized as being paramount. There was little teaching about resurrection in the OT. After death, men were thought to go to Sheol, the place of the departed, to an unsatisfying sort of existence (Ps. 88). But it was realized that God was there (Ps. 139:8) and there was therefore hope of deliverance (Ps. 16: 10). Specific ideas of resurrection are found in Isaiah 26:19 and Daniel 12:2, where there is a connection with the thought of judgment.

There are also instances of life being restored to dead children by Elijah and Elisha (1 Kings 17:17-23; 2 Kings 4:32-36), but they resume the same sort of life that they had before. The instances of the raising Jairus's daughter and the son of the widow of Nain in the gospels (Mark 5:35-43; Luke 7:11-17) belong to the same category, though raising the dead is quoted by Jesus as being one of His messianic works (Matt. 11:2-6; Luke 7:18-23). Even the raising of Lazarus, who had been in the tomb for four days, was a restoration of him to the life that he had before (John 11:1-44). It is in a sense a dramatic foretaste of the resurrection of Jesus, but it has a different nature. By the time of the ministry of Jesus there had been some development in Jewish thinking beyond the OT, but Pharisees* and Sadducees* were divided about the doctrine of resurrection (Mark 12:18-27; Acts 23:6-8).

Jesus' own teaching about resurrection was largely concerned with predictions that He Himself would rise from the dead (Mark 8:31; 9:31; 10:34, etc.). The use of the phrase "after three days" located it as a definite action in the sphere of history. Many passages in the synoptic gospels about judgment may be taken to assume the idea of resurrection, but only in Luke 14:14 is recompense at the resurrection made explicit. The only discussion of resurrection occurs in the answer to the trap question of the Sadducees (Mark 12:18-27). John's gospel records teaching about the resurrection of life and the resurrection of judgment (John 5:28f.), and Mary of Bethany voices a belief in the resurrection at the last day (John 11:24). Jesus, however, claims to be the Resurrection and the Life, operating in the present as well as in the future (John 11:25f.).

The vague hopes and dim foreshadowings that had gone before are replaced in the apostolic proclamation by the certainty of resurrection because of the certainty of the resurrection of Christ. The qualification for apostleship was to have been a witness of the resurrection (Acts 1: 22), and the fact that Jesus had been raised from the dead was at the center of the preaching of the apostles, whether by Peter at Jerusalem (Acts 2: 29-32) or by Paul at Athens (Acts 17:30-32). Paul states that the point of first importance in his teaching was the death and resurrection of Jesus, with its attestation (1 Cor. 15:3-11). It is of course to the gospels that we must go for a fuller account of the resurrection.

There is no description of how it happened in any of the canonical gospels. They all agree in referring to two things—the empty tomb and the appearances to the disciples, though in the case of Mark, the original ending of which may have been lost, the latter is predicted and not described in the authentic text (Matt. 28; Mark 16:1-8; Luke 24; John 20–21). The accounts differ in detail as to the number of women who went to the tomb, and concerning the inclusion of a reference to an angel (Matthew), a young man (Mark), or two men (Luke) to give them a message about the resurrection. Mark predicts and Matthew describes an appearance in Galilee, while Luke records appearances in Jerusalem and John includes appearances first in Jerusalem and then in Galilee. It is possible to make some sort of harmony of the accounts, but this is speculative and it is more important to see the basic agreement in essentials, together with an uncontrived variety of presentation which does much to suggest experience rather than propaganda. The narrative as we have it in Mark gives a vivid impression of the resurrection representing the end, for it is the breaking in of the world to come. Matthew stresses that the permanent presence of the risen Christ is connected with the world-mission of the church, and Luke likewise sees the resurrection as a key piece of the framework of redemption history which is to be continued through the gift of the Holy Spirit. In John the resurrection brings everything to a climax (John 20), but the appendix to the gospel is a reminder that there is work to be done before the final consummation (John 21).

The consequences of the resurrection for the present and the future are worked out in the epistles, particularly in Paul's magnificent exposition in 1 Corinthians 15. It is a guarantee of the efficacy of the Atonement and the sure pledge of the resurrection of believers in the future. This must be distinguished from Greek ideas of the immortality of the soul because it implies that the present physical body is the seed of the future "spiritual body." It is a clear proof of the reality of the new order of creation and the ultimate victory of God in Christ. Elsewhere the Christian is said to have gone through an experience similar to Christ's, symbolized in baptism (Rom. 6:1-11; Col. 2:12), and this means the ability to have heavenly aspirations (Col. 3:1) and to enjoy the power of the risen Christ (Phil. 3:10).

From the earliest times there have been those who have been unwilling or unable to believe in the resurrection of Christ. Alternative theories have included the suggestions that He never really died, that the women went to the wrong tomb, or that either His friends or His enemies stole the body. None of these suggestions will bear critical examination or account for the extraordinary psychological and moral change induced into so many different people and with such lasting effects in the face of those who had every reason to discredit the belief if they could. Nor will it do to believe in some sort of spiritual presence of the risen Christ without the raising of His body from the tomb. History, theology, and experience combine to show that "the glorious fact is that Christ *did* rise from the dead" (1 Cor. 15:20, Phillips).

BIBLIOGRAPHY: B.F. Westcott, *The Gospel of the Resurrection* (1866); W. Milligan, *The Resurrection of Our Lord* (1881); W.J. Sparrow Simpson, *The Resurrection and Modern Thought* (1911); P. Gardner-Smith, *The Narratives of the Resurrection* (1926); A.M. Ramsey, *The Resurrection of Christ* (2nd ed., 1961); O. Cullmann, *Immortality of the Soul or Resurrection of the Dead?* (ET 1958); W. Kunneth, *The Theology of the Resurrection* (ET 1965); C.F.D. Moule (ed.), *The Significance of the Message of the Resurrection for Faith in Jesus Christ* (1968). R.E. NIXON

RETREAT. This signifies, in general, those periods of time specifically set apart for spiritual contemplation, religious devotions, and inward renewal. Participants usually seclude themselves from their normal occupations. The practice is not uniquely Christian, but Christians have drawn inspiration from the examples of Jesus praying in the desert (Matt. 4), and from the apostles tarrying in the upper room before Pentecost (Acts 1: 13,14). Ignatius Loyola fully developed a complete, practical method of retreat in his volume *Exercitia spiritualia*, approved by Paul III in 1548. Taking inspiration from the Jesuits, Roman Catholic retreat houses have sprung up in phenomenal numbers around the world. Canon law obligates all priests to attend one retreat every three years and members of all religious orders to attend at least one each year. Retreats are not, however, limited to the Roman Catholic Church. They were practiced in the Oxford Movement* and adopted formally by the Church of England in 1856. Keith Miller, director of Laity Lodge in Texas, has given the retreat method of spiritual renewal great impetus in the USA since 1961.

ROYAL L. PECK

RETZ, CARDINAL DE (1614-1679). Archbishop of Paris. After taking part in abortive plots against Richelieu,* he devoted himself to an ecclesiastical career and was made coadjutor and successor to his uncle, the archbishop of Paris. In Notre Dame he became a popular preacher and encourager of political pamphleteers; when de Conde opposed the king, de Retz supported the court party. In 1651 he was made cardinal, which aggravated the enmity between him and the all-powerful Mazarin. Mazarin had him imprisoned in 1652 in Vincennes, from where he escaped to Rome. Upon the death of his uncle de Retz made legal claim to the see of Paris, instructing his clergy by letters which were publicly burned. On the death of Mazarin, Louis XIV made it known that de Retz would be unwelcome in Paris. He therefore resigned, compensated with the abbey of St. Denis, the revenues of which were greater than those of the see of Paris. He exercised greater influence in Rome than the French ambassador, and took part in the elections of Alexander VII, Clement IX, and Clement X; he also mediated in the struggle between Louis XIV and Rome. He traveled in Germany and Holland on his own behalf and in support of the restoration of the Stuarts in England. He was a church politician rather than a churchman.

GORDON A. CATHERALL

REUBLIN, WILHELM (c.1482-c.1559). Anabaptist* reformer. Born at Rottenburg, he studied at Freiburg, receiving there a clerical consecration. After parish work in Tübingen and Griessen, he became people's priest in Basle (1521). Crowds listened to his Scripture expositions, and supported him when the bishop complained about his attacks on the Mass. He went to Zurich, and was the first Swiss priest to marry (1523). With others he was expelled after the 1525 Disputation. During his subsequent wanderings, he won B. Hubmaier and the city of Waldshut to the Anabaptist cause, debated on baptism with Capito* in Strasbourg, called Michael Sattler to Horb, was whipped from Esslingen (1528), and denounced as a false prophet by fellow Anabaptists in Moravia (1531). By 1535 he had withdrawn from Anabaptism, and latterly lived in Znaim, Zurich, and Basle. J.G.G. NORMAN

REUCHLIN, JOHANNES (1455-1522). German humanist. Born in Pforzheim, he studied under the Brethren of the Common Life* in Schlettstadt; he attended the University of Paris and studied law at Paris, Freiburg, Basle, and Orléans. He served as legal adviser to the duke of Württemberg. Erasmus hailed him as "the triple-tongued" Reuchlin for his expert knowledge of Latin, Greek, and Hebrew. He was the outstanding Hebraist among the humanists of the early sixteenth century. His *Rudimenta Hebraica* was the authoritative Hebrew grammar of the period. Reuchlin steeped himself in Greek philosophy and became a proponent of Pythagorean philosophy. His study of Hebrew writings involved him deeply in cabalistic speculations. He wrote *On the Wonder-Working Word* and *On the Cabalistic Art.*

Reuchlin became involved in a controversy with the theologians of Cologne, especially Johannes Pfefferkorn and Jakob Hoogstraten. In *A Mirror for the Jews* (1506) Pfefferkorn argued that all Hebrew books should be confiscated. In 1509 Maximillian I issued a decree ordering the Jews to turn in their books. Reuchlin, when consulted, replied that the books should not be destroyed, saying that only those which were openly blasphemous might be burned after they had been condemned according to proper legal procedure. He defended the use of Jewish works on philosophy and science, the Talmud, the Kabbalah, biblical manuscripts, prayer books, and hymns in Hebrew. Hoogstraten cited Reuchlin to appear before his Court of Inquisition on the charge of heresy. Through John von der Wyck, Reuchlin won an appeal to Rome. The controversy evoked not only *A Mirror for the Jews* by Pfefferkorn, but also Reuchlin's famous reply *Augenspiegel (A Mirror for the Eyes).* The humanist world rallied to the side of Reuchlin; *Letters of Famous Men* (1515) contained their testimonies. Cortus Rubeanus (the first edition) and Ulrich von Hutten (the second edition) produced the scathing satires, *Letters of Obscure Men.*

Although he recommended Philip Melanchthon,* his grandnephew, to Frederick the Wise as instructor at Wittenberg, Reuchlin did not join the Lutheran cause. However, he hindered the burning of Luther's books in Ingolstadt. Leo X finally condemned Reuchlin's writings (1520). Nevertheless, he ranks as the outstanding German humanist of the first years of the sixteenth century, a promoter of Greek and Hebrew scholarship.

See L. Geiger, *Johann Reuchlin: Sein Leben und Seine Werke* (1871), and L.W. Spitz, *The Religious Renaissance of the German Humanists* (1963). CARL S. MEYER

REUNION. The theological premise upon which any concept of reunion is founded is that the visible church of God on earth is properly, or at its best, or archetypally, a single entity—a body, or an organism. The concept will be opaque wherever Christians hold either that the only proper unity of the church is invisible (and eschatological) and thus indivisible, or that the given oneness of the visible church (e.g., in and by baptism) is totally indivisible. On either of these views the church *is* a union which cannot be broken and thus does not admit of reuniting. The former view is characteristic of classical independency or Brethrenism, the latter of twentieth-century ecumenists rationalizing their failures.

In 1 Corinthians Paul seems to indicate that, while it was theologically unthinkable and virtually immoral for division to enter into the body, yet it was in fact not only possible but actually happening. It was the harmony and true oneness of the local "body" which was endangered, and Paul does not allocate blame separately to either party, but simply appeals to them to belong and live with each other, acknowledging no "party leaders" lower than Christ.

The post-apostolic church, in various ways, had to make more sophisticated appeals than this. The existing institution had prior claims to be "the church" and those who separated themselves had to return to that unity. The true church was identifiable to Ignatius, Irenaeus, Cyprian, and their contemporaries as those who held the true faith and were joined to a true bishop; the difficulty was that in disputes there were no higher courts of appeal to decide who had the true faith and which bishops were true ones. The true church had in the last resort to be self-authenticating to its members, and if more than one organization was thus self-authenticating, then sheer weight of numbers or the secular power (as with Augustine and the Donatists) had to settle it.

In the early centuries this concept of the single universal body of the church meant that reunion had to come through individual seceders or schismatics returning into union. What it could not mean was that two equal and mutually respecting denominations merged, with each making its contribution to the future. A schismatic returning had to renounce all that he had known and done while in schism; he reunited by submitting. Augustine of Hippo* made one small step toward the Donatists* when he afforded recognition to baptisms and orders given in schism, but this was novel—and perhaps ultimately misleading also in terms of Augustine's own ecclesiology. It gave some first shadow of respectability to reunion in that it allowed certain ecclesial marks to the Donatists.

The major schism of the church in the centuries following was that between East and West, formalized in 1054, but probably inevitable from long before that. At this point the theological thinking of each part about the other became obscure. There was an uncertain claim (coming down to the present day) on each side that it alone was the true church to which submission was required from the other. But it *was* uncertain, and each side retained at least some concept that the schism needed to be "healed" by something less than total surrender by the other. Attempts were occasionally made toward reconciliation, the Council of Florence* (1438-39) being the most notable.

The Reformation* brought new ecclesiologies in its train, and new concepts of the unity of the church in consequence. The Reformers took the view that Rome had left the truth, to which they clung, and therefore the guilt of schism and the duty to "return" lay with the pope. They differed as to whether Rome still exhibited the marks of the Christian Church at all, and they were far more concerned to protect themselves from her than to promote reunion with her. Their view of their own churches was broadly *cuius regio eius religio,* and this left them with no constitutional duties to each other except a fraternal strengthening of each other's hands. Finally they showed various degrees of tolerance to dissenters who arose within their parishes or districts, but the tolerance was rarely more than grudging, and in England the persecution of the Puritans* (as under Whitgift* and Bancroft*) became almost fanatical. There never existed two or more bodies which could face each other with mutual understanding and sympathy, and seek to unite with each other. The Hampton Court* and Savoy* conferences (1604 and 1661) from this standpoint were parodies of conferences—however they were occasioned, they quickly became mere instruments of triumphalist policy by the ruling (i.e., Anglican or Episcopalian) party. A greater possibility emerged in 1689 when the two parties had worked together in the overthrow of James II,* but the upshot was not a comprehensive united church (as many had hoped), but a wider tolerance for a multiplicity of churches.

A rare and interesting variant on these moves was found in the early part of the eighteenth century when Archbishop Wake* corresponded for some time (1717-19) with Du Pin, a Roman Catholic theologian of the Sorbonne. The end in view was an Anglicanizing of the French Church away from the new Ultramontanism* until the Churches of England and France could meet in a fraternal unity. More bizarre was the temporary flirtation of the English Nonjurors* with Eastern Orthodoxy.

The eighteenth century in England itself saw the Evangelical Revival. This led to the building of Methodist chapels alongside parish churches throughout the land, and fixed a deep division between the two—a division which was cultural and sociological as well as theological, a prime instance of the part "nontheological" factors play in the keeping of Christians separate from each other. The problem was now exactly that to which 1 Corinthians had been addressed, that of rival denominations cheek-by-jowl with each other. No attempt was made at reunion once John Wesley* had died, and the Oxford Movement* added the impression of a strong *theological* division to justify the already existing separation. England was a very powerful force in the nineteenth-century missionary expansion, so that the English divisions reproduced themselves around the world, with the further complications added of Lutheranism, Presbyterianism, and Moravianism from the Continent.

In the twentieth century the ecumenical movement has spawned or strengthened moves to corporate reunion throughout the world. Mention can be made of Scottish Presbyterian unions in 1900 and 1929, English Methodist union in 1932, and Presbyterian-Congregationalist union in 1972, and the formation of the United Church of Canada.* Anglicans were involved in the Church of South India* in 1947 and in the Churches of North India* and Pakistan* in 1970, in company with six other denominations. Many other unions have been projected but a number not implemented, including schemes in the USA, between Anglicans and Methodists in England (finally defeated in 1972), in Nigeria, East Africa, Australia, and New Zealand.

In all these moves neither the Roman Catholic Church nor the evangelical independent churches have been involved, due to their ecclesiologies and traditions. A stage is now arriving where Roman Catholics can and do become very sympathetic "observers" in schemes between other churches, but they have no mandate to negotiate toward local or national unions. There has at the same time arisen among those most in favor of reunion an impatience, or at least an ennui, concerning the whole concept of *schemes* for reunion. There is a proneness to trace the rise of antiinstitutional Christianity to the preoccupation with institutions which has marked the schemes of recent years. Thus in the early 1970s the specific schemes for reunion which have been projected but not yet implemented lie somewhat under a cloud, with the question arising as to whether institutional Christianity has not more important items on its agenda.

BIBLIOGRAPHY: S.L. Ollard, *Reunion* (1919); A.C. Headlam, *The Doctrine of the Church and Christian Reunion* (1920); G.K.A. Bell (ed.), *Documents on Christian Unity* (3 vols., 1924-48): covers 1920-48; K.D. Mackenzie (ed.), *Union of Christendom* (1938); J.D. Murch, *Cooperation Without Compromise* (1956); S.C. Neill, *The Church and Christian Union* (1968). See also under ECUMENICAL MOVEMENT and WORLD COUNCIL OF CHURCHES.

COLIN BUCHANAN

REUSCH, FRANZ HEINRICH (1825-1900). Old Catholic* theologian. Born in Westphalia, he studied at Paderborn, Bonn, Tübingen, and Munich, was ordained priest in 1849, and taught OT exegesis at Bonn from 1854 (professor from 1861). Close friend of J.J.I. von Döllinger,* he strongly opposed the Vatican Council* infallibility decrees (1870), being present at the Nuremberg Declaration.* He was excommunicated in 1872, served an Old Catholic church in Bonn, was Bishop Reinkens's* vicar-general, and in 1873 became rector of Bonn University. He took a leading part in organizing the Old Catholic Church, and arranging the Bonn* Reunion Conferences. Disagreeing with the Old Catholic abolition of clerical celibacy in 1878, he retired into lay communion. He wrote many books on OT subjects and modern ecclesiastical history, some with von Döllinger, notably a history of post-Tridentine moral theology (1899).

J.G.G. NORMAN

REUSS, EDWARD (1804-1891). Biblical scholar. A native of Strasbourg, he spent most of his life there. He taught at the Protestant seminary from 1834 (professor of NT from 1836 and of OT from 1864), and also at the state theological faculty from 1838. When the seminary was incorporated into the theological faculty of the new German university in 1872, Reuss was the first dean. Reuss was a keen advocate of the historico-critical approach to the Bible. His opposition with H.J. Holtzmann* to Baur's tendency criticism led to the latter's decline.

While recognizing a Jewish element in the early church, he considered Baur's theory of a conflict with Hellenism exaggerated and untenable. As early as 1834 he advocated a late date for the *Grundschrift,* later called the "Priestly Document," behind the Pentateuch. His views paved the way for the later hypotheses of Graf, Kuenen, and Wellhausen. He founded the *Theologische Gesellschaft* (1828) and was a regular contributor to Colani's *Revue de Théologie et de Philosophie.* His many writings brought the ideas of German critical theology before the French public. He was an editor of the standard edition of Calvin's works in the *Corpus Reformatorum* (59 vols., 1863-1900). Reuss produced a new French translation and commentary on the Bible (16 vols., 1874-81). Reuss's writings include *Geschichte der heiligen Schriften Neuen Testaments* (1842), *Histoire de la théologie chrétienne au siècle apostolique* (2 vols., 1854), *Geschichte der heiligen Schriften Alten Testaments* (1881), and *Das Alte Testament* (7 parts in 6 vols., 1892-94).

COLIN BROWN

RÉVEIL, LE. Literally, "The Awakening," it was an evangelical revival which began in French-speaking Switzerland in the early nineteenth century and spread to France and the Netherlands by 1825. It shook the state churches of Geneva and Vaud, spawned free churches in those two cantons, deeply touched the French and Dutch Reformed communities, and complemented contemporary revivals in the British Isles and the United States. *Le Réveil* was basically a reaction against the rationalism and materialism which the Enlightenment* had brought to the established churches of the Continent. Its leaders reemphasized historic Reformed doctrines, especially biblical authority, the sovereignty of God, the lost condition of man, justification by faith in Christ, and the necessity of personal conversion.

The revival began as early as 1810 with the formation of a society of "friends" in Geneva which began to study the Scriptures in search of spiritual renewal. A number of outside influences stimulated the awakening, including visits from Scottish evangelicals Robert Haldane* (d. 1842) and Henry Drummond* (d.1860). However, the movement was not a mere reflection of British revivalism, but developed its own distinctive character under the able leadership of César Malan* (d.1864), François Gaussen* (d.1863), and Merle d'Aubigné* (d.1872) in Geneva; Alexandre Vinet* (d.1847) in Vaud; Felix Neff (d.1829), Henri Pyt (d.1835), Adolphe (d.1856) and Frédéric (d.1863) Monod* in France; and Willem Bilderdijk* (d.1831), Isaak da Costa* (d.1860), and Guillaume Groen van Prinsterer (d.1876) in the Netherlands. By the end of the century, *Le Réveil* had won over a majority of the Venerable Company of Pastors of Geneva, permeated the French Reformed Church, and rejuvenated hundreds of Dutch congregations. ROBERT D. LINDER

REVELATION. A central concept in modern theological discussion, concerned as it has been with questions about how God may be known, about religious authority, and (more recently) about language. The idea of God *making Himself known* in acts of redemption and judgment and in prophetic, interpretative utterances pervades the Bible. It is not so much *a* biblical idea, as it is *the* biblical idea. Man is lost in his sinful ignorance unless and until God discloses Himself. Revelation is therefore gracious. Thus, to think of revelation as being essentially man's response to God, or as his insight into the ways of God, though a dominant viewpoint in much modern theology, is profoundly unbiblical.

God's revelation in history must be understood against the background of general or natural revelation. The Bible itself teaches that God reveals Himself to men generally, in nature, in history, and in their moral consciousness (Ps. 19; Rom. 1). However, such general revelation is rendered ineffectual by sin. Thus God's special revelation is *redemptive* in character. He reveals Himself, not generally—to all men—but specially—to His own chosen people. It was God's eternal purpose thus to display His glory by revealing His provision of redemption for a rebellious creation.

In the history of Christian thought about revelation, various emphases have been placed on the two aspects of "general" and "special" revelation. For Aquinas, general revelation (and with it the idea of natural theology) is very prominent. Man is able, by the use of his reason, to come to a rudimentary knowledge of God on the basis of God's natural revelation. For Calvin, natural theology is not prominent. A man knows God (and in knowing God knows himself) in Scripture, and true knowledge is conditioned upon inner spiritual illumination. For Karl Barth, revelation is understood in activistic terms—God makes Himself known *now,* as Scripture is read and preached and received in faith. So Scripture is not revelation, but the vehicle of it.

Special revelation concerns the activity of God in human affairs. It is therefore historical, and it proceeds in stages. The mode of revelation is thus suited to the epoch and the stage of redemptive history, but it has its culmination in "the fact of Christ." God has spoken, finally, in His Son, in His teaching, His work of atonement, and in the interpretative apostolic testimony. The OT revelation prepares for Christ. Christ does not repudiate the OT, He fulfills it.

The historical character of revelation makes clear its uniqueness. It is not a mere republication of the truths of natural religion, nor is it to be thought of in the same terms as the "revelations" of mysticism or the occult. Further, by calling revelation "historical," stress is laid on the *actuality* of the events recorded in Scripture. The events are not simply projections of the religious consciousness onto history. The testimony of the Christian Church is that God revealed Himself in human history, and now, in Scripture—in the very words and propositions of Scripture—God reveals Himself.

BIBLIOGRAPHY: J. Orr, *Revelation and Inspiration* (1910); P.K. Jewett, *Emil Brunner's Concept of Revelation* (1954); H. Bavinck, *The Philosophy of Revelation* (1954); G.C. Berkouwer, *General Revelation* (1955); J. Baillie, *The Idea of Revelation in the Light of Recent Discussion* (1956); C.F.H. Henry (ed.), *Revelation and the Bible* (1959); B. Ramm, *Special Revelation and the Bible* (1961); H.D. McDonald, *Ideas of Revelation* (1962). H.D. MC DONALD

REVELATION, BOOK OF. The last book of the Bible is generally regarded as a typical apocalypse. Apocalyptic literature* flourished in the last two centuries B.C. and the first of our era. It is marked by a lavish use of symbolism, often of a bizarre kind, by a pessimism as to the outcome of man's best efforts, and by a conviction that God is in supreme control. He will deliver His people out of their current trouble and will bring in the end of the world when His kingdom will be set up. Revelation is like the apocalypses in many respects (indeed the name of this book in Greek, *apokalypsis,* gives its name to this class of literature). But it differs in not being pseudonymous (most apocalypses are fathered onto some great figure of the past like Moses or Enoch, whereas our author gives his name as John). And it is specifically called a prophecy more than once (1:3; 22:7,10,18,19). It differs also in its more stringent demand for repentance where the apocalypses set out simply to comfort God's people (there is comfort, of course, in Revelation, but there is also the other note). This is not to say that there are not apocalyptic features in this book. There are, especially in the symbolism. But we must not simply view it as no more than another apocalypse.

It has been interpreted very variously throughout church history. The preterists see it as referring to events of the day under its symbols, with the author conveying to the church the message that God would help them in due course. The historicists hold that the book sets out the whole of human history in a panoramic view; unfortunately, most see it in terms of the history of W Europe—and even so there is no agreement as to what events are referred to. The futurists think the book a prophecy of events at the last day; as such it has no relevance to any generation of Christians except the last. Idealists do not see events at all; they think the author has imaginatively set forth with his vivid symbolism some important ideas for the Christian as he lives out his faith.

Probably elements from more than one view are needed. We cannot hold that the book was meaningless to its first readers, so it must refer in some way to current needs. But we should not (with the preterists) confine it to the first century. There is surely more than that. It seems that the best view is that which sees it as setting forth a theology of power. It sets out the great principles which we may observe in God's moral government of the world. There are references to contemporary happenings which illustrate the point, but the book goes far beyond that and pictures those principles as active to the end of time. Indeed, they will be especially operative at the end, and John's conviction that God will one day bring this present system to an end dominates the whole.

BIBLIOGRAPHY: H.B. Swete, *The Apocalypse of St. John* (1907); R.H. Charles, *A Critical and Exegetical Commentary on the Revelation of St. John* (1920); M. Kiddle, *The Revelation of St. John* (1940); W. Hendriksen, *More than Conquerors* (1956); G.B. Caird, *The Revelation of St. John the Divine* (1966); L. Morris, *The Revelation of St. John* (1969). LEON MORRIS

REVISED VERSION, see BIBLE, ENGLISH VERSIONS

REVISED STANDARD VERSION, see BIBLE, ENGLISH VERSIONS

REVIVALISM. A spontaneous spiritual awakening by the Holy Spirit among professing Christians in the churches, which results in deepened religious experience, holy living, evangelism and missions, the founding of educational and philanthropic institutions, and social reform. Revival should not be confused with evangelism, which is a result of revival. Revival has been linked with the Anabaptists,* Puritans,* and Pietists* and has occurred mainly in Protestantism since the Reformation.

As a determinative influence in America, revivalism may be dated from the Great Awakening* which began after 1720 when T.J. Frelinghuysen* came from European centers of Pietism to pastor four New Jersey Dutch Reformed churches. Under his influence, revivals were held by the Presbyterian Gilbert Tennent,* whose father William* had founded the "Log College," which produced many revivalists. In 1734, revivals broke out in New England under the preaching of Jonathan Edwards,* and by 1740 when George Whitefield* arrived bearing the spirit of the Wesleyan revival in England, the Awakening was widespread in America. Many were swept into the churches, controversy raged, denominations were split, and humanitarian efforts and the spirit of democracy were strengthened.

The Second Awakening occurred mainly among middle- and upper-class Anglicans in England after 1790 and in America in college awakenings, such as that in Yale under Timothy Dwight* in 1802 and with great emotional and physical manifestations in the Western America frontier camp meetings.* This latter technique, first used by Presbyterians in Kentucky, was developed by Baptists and especially the Methodists. Charles G. Finney's* urban revival meetings of the 1830s has been considered a later flowering of the Second Great Awakening.*

A lay interdenominational revival which developed through noonday prayer meetings in New York in 1857 led to more than half a million coming into the churches. A revival in 1863-64 in the Confederate Army brought 150,000 soldiers a vital Christian experience. Similar awakenings occurred in Britain and on the Continent.

After the Civil War, professional, planned, urban mass evangelistic meetings held in public auditoriums by men such as D.L. Moody* and R.A. Torrey* and in the twentieth century by Billy Sunday* and Billy Graham* replaced the earlier spontaneous, rural, pastoral congregational awakenings with the exception of the 1904 Welsh revival, which stimulated world-wide revival. Since 1904 there has been no general revival in Western industrial societies, but there have been regional revivals in less advanced societies in Korea, East Africa, and Ethiopia.

BIBLIOGRAPHY: W.W. Sweet, *Revivalism in America* (1944); T.L. Smith, *Revivalism and Social Reform* (1957); B. Weisberger, *They Gathered at the River* (1958); W.G. McLoughlin, *Modern Revivalism: Charles Grandison Finney to Billy Graham* (1959). The many books by J. Edwin Orr, such as *The Eager Feet, Fervent Prayer,* and *The Flaming Tongue,* present details from sources on revivals. DONALD W. DAYTON

REYES Y FLORENTINA, ISABELO DE LOS (1864-1938). Filipino journalist, radical, and amateur theologian. For writing revolutionary, antifriar propaganda he was exiled to Spain. There he attempted a translation of the Scriptures into Tagalog (Filipino). On his release he was commissioned by Aglipay* to negotiate with Rome for the rights of Filipino clergy (1899-1901). Having failed to obtain their request, he returned to the Philippines* and founded the Philippine Independent Church, which he persuaded Aglipay to lead. As the PIC's theologian, he led it increasingly to a Unitarian position. He was deeply involved in politics and labor disputes all his life. In 1936, after a political disagreement with Aglipay, he retracted all his writings and was reconciled to the Roman Catholic Church.

RICHARD DOWSETT

REYNOLDS, EDWARD (1599-1676). Bishop of Norwich. Educated at Merton College, Oxford, he was a moderate Anglican with Puritan sympathies. Member of the Westminster Assembly* in 1643, he served on the committee of twenty-two which examined ministers presented by parishes, and took the Solemn League and Covenant* in 1644. From 1645 to 1662 he was vicar of St. Lawrence Jewry and in 1647 one of the parliamentary visitors to Oxford University. He was vice-chancellor of Oxford in 1648, and dean of Christ Church, 1648-50, and again in 1659, when he refused to subscribe the Engagement and was ejected. At the Restoration he conformed, and in June 1660 drew up proposals for reconciling Episcopalians and Presbyterians. In 1661 he made similar efforts at the Savoy Conference* and was made bishop of Norwich, where he treated dissenters with moderation. The numerous sermons and short religious works which he published remained popular until the nineteenth century.

JOYCE HORN

REYNOLDS, JOHN, see RAINOLDS, JOHN

RHEIMS NEW TESTAMENT, see DOUAI-REIMS

RHENANUS, BEATUS (1485-1547). German humanist. Born in Alsace, son of a butcher, he studied at Paris (1503-7) when Lefèvre d'Étaples* was teaching Aristotelian philosophy there. At Basle (1511-26) he took an active share in the publishing activities of John Froben, returning to Schlettstadt (his birthplace) to pursue his studies. He wrote a biography of Geiler von Kaysersberg* (1510), published editions of the classics and the Fathers, especially Tertullian (1521), and his studies of German antiquities, *Rerum Germanicarum libri tres* (1531). While at Basle he became a close friend of Erasmus, and published his works and a biography in nine volumes (1540-41). He was also a friend of Luther and Martin Bucer, introducing Bucer to the works of Erasmus. At first, like Erasmus, Rhenanus favored the Reformation, but later reacted against it. He died at Strasbourg.

J.G.G. NORMAN

RHENIUS, KARL (1790-1838). Missionary to South India. From a Prussian officer's family, he attended Jänicke's* mission school in Berlin and went to India under the Church Missionary Society in 1814. After serving briefly in Madras, he moved to Tinnevelly where he proved to be a competent scholar, effective teacher-catechist, and outstanding organizer. He prepared a Tamil grammar and NT, and introduced a system linking the village church and school where the schoolmaster had responsibility for worship and religious instruction. The work was self-supporting and self-propagating, and so many lower-class people were converted that the British authorities feared social unrest. Having never received Anglican orders, Rhenius ordained Indian ministers himself, but the CMS forbade the practice. In 1835 he challenged this and was discharged. Because many Tinnevelly congregations remained loyal to him, he formed a separate church, but the schism was healed after his death.

RICHARD V. PIERARD

RHODES, KNIGHTS OF, see HOSPITALLERS

RHODO (second century). Anti-Gnostic apologist. All that is known of him is recorded in Eusebius, *Ecclesiastical History* (5.13). Born in Asia Minor, a disciple of Tatian in Rome, he wrote against the Marcionites. In particular, he had a confrontation with Apelles, an aged disciple of Marcion* in Rome. He also wrote a commentary on the *Hexaemeron.*

RICASOLI, BETTINO (1809-1880). Italian politician and patriot. He worked for the unity of Italy, was minister of Tuscany, and later, after Cavour's death (1861), was prime minister of Italy. In touch with the leaders of Tuscan evangelism and with Swiss Protestant circles, he was greatly influenced by them and hoped for a reform of the Roman Catholic Church. He never, however, severed his connection with the latter. As a minister he tried in vain to solve the problem of the relationship between the Italian state and the Vatican.

RICCI, MATTEO (1552-1610). Missionary to China. An Italian Jesuit, he reached Macao in 1582 in response to an appeal from Alexander Valignano in that city. He at once set about mastering the spoken and written Mandarin dialect of China. In 1583 he entered China proper at the invitation of the magistrate of Chao-ching and there translated the Ten Commandments while enduring much opposition from the people. But gradually his famous map of the world, his clocks, his books, and his mathematical instruments made an impression upon the learned. In 1594 Ricci moved to Shao-chow where he adopted the dress and the etiquette of the Chinese literati. In 1599 he set up a base in Nanking and was introduced to the learned society of that city, where he also instructed Paul Hsü Kuang-ch'i, the father of the mission at Shanghai. In 1600 Ricci set out for Peking, but reached the capital in 1601 only after imprisonment in Tientsin.

Once Ricci arrived in Peking he never left it, and he soon won the esteem of the learned and of the emperor for his scholarship and knowledge of Chinese culture. By his sympathetic approach to Confucian culture, which he did not regard as inconsistent with the Christian faith, and by his approval of the Confucian Rites ceremonies he started what became known as the Chinese Rites Controversy.* He witnessed many conversions, including some among the highest court officials. He died in Peking and was buried in the Tartar City in land granted by the emperor.

LESLIE T. LYALL

RICCI, SCIPIONE DE' (1741-1810). Bishop of Pistoia-Prato. A Florentine, he was influenced by Jansenism* in the Roman College, and while a student at Pisa encountered Gallicanism* and a rigid Augustinianism. Ordained in 1766, he became vice-general of the archdiocese of Florence (1775-80), and bishop of Pistoia-Prato in 1780, through Leopold I of Tuscany. He introduced severe disciplinary, liturgical, and doctrinal changes and enabled Jansenist writings to be published widely. In 1783 he founded a theological college at Prato, where lecturers were sympathetic with his views. He presided over the Synod of Pistoia* (1786), where his schismatic propositions brought his downfall. He resigned his diocese in 1791 and spent his final years in confinement in Rignana. His zeal for reform lacked proper training and culture, making his innovations dangerous and subject to political pressures.

C.G. THORNE, JR.

RICE, LUTHER (1783-1836). American missionary to India and promoter of missionary interest among Baptist churches. Born in Northboro, Massachusetts, he was one of the group of students who sparked the formation of the first American foreign mission society. When the American Board appointed its first four missionary couples in 1812, Rice's name was added a few days before sailing, on condition that he secure his own support. Ordained a Congregational minister, after reaching India he experienced the same change of mind about baptism as did Adoniram Judson,* and became a Baptist. To stimulate missionary interest among Baptists, they decided he should return temporarily to the USA. He helped start the Baptist foreign mission society and was so effective in gaining support for the mission that he never returned to the Orient.

HAROLD R. COOK

RICHARD I (1157-1199). King of England from 1189. Known as *Coeur-de-lion* ("Lionheart"), he was a leader of the Third Crusade, together with Philip Augustus of France and Emperor Frederick Barbarossa. The latter died en route, and when Richard and Philip arrived by sea (1191), they found Guy of Lusignan besieging Acre and helped in its capture and the massacre of its inhabitants. Richard and Philip quarreled incessantly. They failed to capture Jerusalem, but Richard had a series of victories and made a treaty (1192) by which the Christians held the coastal cities as far south as Jaffa and were granted access to the Holy Places. Returning home, Richard was captured and held for ransom by Emperor Henry VI. He died in battle in France.

J.G.G. NORMAN

RICHARD, TIMOTHY (1845-1919). Missionary to China. Born in Wales, he went to China with the Chinese Evangelization Society in 1870. Later he joined the Baptist Missionary Society. Like Ricci,* he planned to evangelize China through influencing the devoutly religious and the intelligentsia by adapting Christianity to Chinese culture and by education and literature. He hoped to see all phases of China's life transformed. During the 1877 famine in Shansi, Richard lectured to officials and scholars. In 1891 he became secretary of the newly formed Christian Literature Society and started an ambitious publications program. He also became adviser to the liberal reformers, some of whom were friendly toward Christianity. Finally he succeeded in forming a university in Taiyuan, the Shansi capital, with Boxer indemnity money, and remained in charge for ten years.

LESLIE T. LYALL

RICHARD OF CHICHESTER (1197-1253). Bishop of Chichester. Born at Droitwich, he studied at Oxford and Paris before becoming chancellor of Oxford University in 1235. But soon afterward he was called away to be chancellor to Edmund (of Abingdon) at Canterbury. When Edmund went into exile, Richard accompanied him, remaining with him until his death in 1242. After being ordained priest in France, Richard returned to England, where he was elected bishop of Chichester against the wishes of Henry III. He was a good diocesan administrator as well as a reforming cleric, with no sympathy for simony and other abuses. He died in the Maison Dieu at Dover, the day after consecrating a new church in honor of Edmund. He was canonized by Urban IV in 1262, and his shrine was in Chichester Cathedral until destroyed in 1538 during the reign of Henry VIII.

PETER TOON

RICHARD OF MIDDLETON (d. c.1300). Philosopher. His birthplace, "Mediavilla," is of unknown location, some claiming a French, others an English or Scottish site. He joined the Franciscans, and the early 1280s found him a theologian, philosopher, and preacher in Paris. In 1283 he was appointed to a commission investigating the views of P.J. Olivi.* By 1295 he had completed his commentary on the four volumes of Peter Lombard's* *Sentences.* At several points he favored Thomistic over traditional Augustinian positions, e.g., Thomas's theory of knowledge. He rejected Anselm's ontological argument. Although the close of his life is obscure, he served as tutor in Naples for the sons of Charles II of Sicily and may have accompanied them to Barcelona when they were taken hostage. A number of other works have been attributed to him, perhaps erroneously. He certainly wrote *45 Quaestiones disputatae* and *Quodlibeta.* The latter reflects his interest in hypnosis and telepathy.

JAMES DE JONG

RICHARD OF ST.-VICTOR (d.1173). Scholar and mystic. Born in Scotland, at an early age he entered the abbey of St.-Victor in Paris, becoming superior (1159) and prior (1162). Learned in Scripture and the Latin Fathers and given to theological questions and contemporary problems, he—like his esteemed master, Hugh of St.-Victor*—possessed great grammatical, dialectical, and rational ability. He shared more with Augustine, Anselm of Canterbury (using his *rationes necessariae* method), and Bonaventura than with Abelard, Peter Lombard, and Gilbert de la Porrée. Contemplation he practiced and understood, breaking it down into six stages; and his

Benjamin minor and *Benjamin maior* demonstrate these processes personally. Each stage corresponds to the progressive categories of knowledge—*imaginatio* and *ratio* to *intelligentia* —and has its own object from the visible to the spiritual and invisible. In *De Trinitate* he tried to understand the personal nature of God, which led to analyzing supreme goodness. This teaching was not followed later, although his mystical theology had influence through Bonaventura and the Franciscan school. He adumbrated Aquinas, but also believed more in speculative reasoning to unravel doctrines like the Trinity. He also wrote *Liber de Verbo incarnato; De statu interioris hominis; De Emmanuele;* and *Adnotationes mysticae in psalmos.*

See A.M. Éthier, *Le 'De Trinitate' de Richard de Saint-Victor* (1939), and G. Dumeige, *Richard de Saint-Victor et l'idée chrétienne de l'amour* (1952). C.G. THORNE, JR.

RICHELIEU, ARMAND-JEAN DUPLESSIS, DUC DE (1585-1642). French cardinal and statesman. Born in Paris of a noble family, he was trained for the army at the *Collège de Navarre* (Paris). The family needed money, however, and instructed him to seek ordination to become bishop of Luçon—a see at their disposal. In 1606 he was consecrated bishop, and spent nearly ten years building up the parish. In 1616 he managed to insinuate himself into political life, being appointed secretary of state. Until 1624 he served intermittently as Marie de Medici's principal adviser; with her assistance he was created cardinal in 1622. In 1624 he was made head of the royal council, and for the next eighteen years virtually ruled France, although not in full control as first minister until the "Day of Dupes" (1630).

Richelieu was ambitious and ruthless, but not so much for personal gain as for the state. Above all, he labored to make France great. He supported Gallicanism* and pursued an anti-Hapsburg foreign policy designed to centralize power in the king. A centralized state required destruction of the political and military power of the Huguenots* and the nobility. The Huguenots were taken care of at the Siege of La Rochelle (1628), but the struggle with the nobility was protracted and bitter. He systematically nullified their privileges by demolishing their fortresses (1626), by forbidding dueling (1627), by creating officials (called *intendants*) directly responsible to the king, and by attempting a national army. Recalcitrant nobles were mercilessly executed. Internationally he opposed the Hapsburgs by building a navy and by bringing France into the Thirty Years' War* on the "Protestant" side. France emerged from the conflict the dominant power in Europe, but Richelieu died before seeing this lifelong dream fulfilled. A patron of the arts and literature, he founded the French Academy (1635), supported playwrights (e.g., Corneille), and built great edifices (e.g., the Palais Royal and Sorbonne Chapel). He wrote widely, but not brilliantly.

BIBLIOGRAPHY: *Lettres* ... (8 vols., 1853-77); *Mémoires* ... (10 vols., 1908-31): critically edited for the *Société de l'Histoire de France; Testament politiques* (1947): best critical ed. Biographical studies in English include those by C.V. Wedgwood (1954); O. Ranum (1963); and D.P. O'Connell (1968). BRIAN G. ARMSTRONG

RICHTER, JULIUS (1862-1940). German missiologist. After serving fifteen years as a pastor in Germany, he became professionally interested in the science and history of missions. He presided over various missionary committees and organizations, and succeeded Gustav Warneck as editor of the *Allgemeine Missions-Zeitschrift* in 1911. In 1920 he was appointed to the first chair of missions at the University of Berlin. His wide travel and extensive personal connections made him one of the major ecumenical personalities of his time. Author of thirty books and innumerable essays on the development of missions and the church on all continents, he devoted special attention to the historiography of missions. His most significant work was a five-volume history of Protestant missions, *Allgemeine evangelische Missionsgeschichte* (1906-32).

RICHARD V. PIERARD

RIDLEY, NICHOLAS (c.1500-1555). Reformer and bishop of London. Born near Haltwhistle in Northumberland, he went to Pembroke College, Cambridge, in 1518. In 1524 he became a fellow of Pembroke and in 1527 went to the Sorbonne and Louvain, where he may have witnessed some of the Reformation controversies. He came back to Pembroke in 1530 where he spent much of his time reading the Scriptures and learning them by heart. In 1537 he was appointed chaplain to Cranmer* and the following year vicar of Herne, Kent, in Cranmer's diocese. In 1540 he became a chaplain to the king, and master of Pembroke. In 1547 he was consecrated bishop of Rochester and in 1550 was translated to London.

Ridley's long involvement in academic life was to stand him in good stead for the brief period of his episcopate. He seems to have been won round toward reformed views of the Eucharist through the study of *De Corpore et Sanguini Domini,* the work of a ninth-century monk, Ratramnus* or Bertram, who was refuting transubstantiation.* He had previously thought transubstantiation to be a primitive doctrine. From 1545 Ridley was convinced of the error of transubstantiation, and the following year he persuaded Cranmer, who in his turn persuaded Latimer.* His influence was recognized by Brooks, who said at his trial, "Latimer leaneth to Cranmer, Cranmer to Ridley, and Ridley to the singularity of his own wit." Ridley helped compile the Book of Common Prayer* of 1549 and its revision in 1552, in which his eucharistic theology was given clearer liturgical expression. He was prominent in carrying through reforms in both his dioceses and when in London took the lead in the removal of stone altars and the substitution of wooden Communion tables. He was active in preaching on social questions and promoted the foundation of schools and hospitals.

He was to have returned to his native see of Durham, but on the death of Edward VI* he supported the attempt to put Lady Jane Grey on the throne, and when that failed he was deprived and imprisoned. In 1554 he was taken with Cranmer and Latimer to Oxford, where they had to engage in various disputations. He stood firm by his views and, after burning of the Reformers had begun in 1555, Ridley and Latimer were sentenced to die at the stake. As the fires were lit Latimer cried out, "Be of good comfort, Master Ridley, and play the man. We shall this day light such a candle by God's grace in England as I trust shall never be put out!"

BIBLIOGRAPHY: N. Ridley, *Works* (1843); *The Acts and Monuments of John Foxe* (ed. J. Pratt, 1870); *A Brief Declaration of the Lord's Supper* (ed. H.C.G. Moule, 1895); J.G. Ridley, *Nicholas Ridley* (1957). R.E. NIXON

RIENZO, COLA DI (1313-1354). Italian leader. Son of a Roman innkeeper, he was a man of genius, but exhibited also signs of madness. Partly through his very considerable knowledge of the classical Latin authors, Rienzo became obsessed with the restoration of Rome to its former greatness. Petrarch's* laureation at the Capitol in 1341 intensified this obsession. In 1343 Rienzo was sent to Avignon to seek Clement VI's* return to Rome and the consequent end of the existing misrule by the leading Roman noble families—the Colonni, Orsini, and Savelli families. The mission failed, but Rienzo gained Clement's favor and was made a notary of the *Camera Urbana.*

In 1347, by accident or design, popular revolution broke out in Rome and Rienzo was swept into power. He was proclaimed "Tribune" and given wide-ranging powers. His government consciously looked back to ancient Rome. At first his rule was enlightened. He organized a Civic Guard and deprived the ruling families of their powers. Gradually his pretensions grew. He called on the pope to return to Rome. He called on the rival claimants to the imperial throne to look to him for judgment. He tried to convene meetings of all the Italian governments to formulate a common Italian policy under his leadership. The pope decided that Rienzo's intentions were a threat to papal power and turned against him in September 1347; the Colonni revolted, unsuccessfully, against him in November.

Meanwhile Rienzo's arrogance and luxurious life were alienating popular support. He was excommunicated for heresy, and when the Colonni again revolted, the Romans would no longer support him. He therefore abdicated and fled from Rome in December 1347 to the Franciscan Spirituals in the Abruzzi where he remained for two years. Then he went to Prague to urge Emperor Charles IV to be a "real" Roman emperor and bring peace, harmony, and justice to the world. Charles imprisoned Rienzo for two years. Then he handed him over to Clement VI, who sentenced him to death. Clement soon died and his successor Innocent IV,* hoping to use Rienzo in his Italian schemes, released him. In 1350 the Romans had again revolted, but Rienzo, sent to Italy in 1354 with Cardinal Albornoz, was successfully reestablished in control, this time as "Senator." His cruelty and luxurious life again alienated the Romans, and they slew him at the Capitol. L. FEEHAN

RILEY, WILLIAM BELL (1861-1947). Baptist minister and educator. Born in Indiana, he graduated from Hanover College (1885) and the Southern Baptist Theological Seminary (1888). He was ordained to the Baptist ministry in Kentucky and was pastor of several Indiana churches; Calvary Baptist Church, Chicago (1893-97); and First Baptist, Minneapolis (1897-1942). Riley founded and was president of Northwestern Bible Training School (1902) and Northwestern Evangelical Seminary (1935) and founded Northwestern College (1944). He strongly opposed theological liberalism and evolution. He was president of the Minnesota Baptist Convention (1944-45). The forty-volume *Bible of the Expositor and the Evangelist* (1924-38) is one of his major writings. ROBERT C. NEWMAN

RIMINI, SYNOD OF, see ARIMINUM

RINCKART (Rinkart), MARTIN (1586-1649). Born in Saxony, a cooper's son, he was educated at Eilenburg and Leipzig, where he graduated in theology. A good musician, he became a Lutheran cantor, then deacon at Eisleben. In 1617 he became archdeacon at Eilenburg, a walled town of refuge during the Thirty Years' War* which suffered from famine and pestilence. For some time he was the only clergyman there and in 1637, it is said, he buried nearly 5,000 people, including his wife. He also dissuaded a Swedish commander from imposing on the town an excessive tribute. "Now thank we all our God" (Winkworth*), a grace for his children based on Ecclesiasticus 50:22-24 and the *Gloria Patria,* became a thanksgiving for the Peace of Westphalia. Rinckart was also a poet and dramatist. JOHN S. ANDREWS

RISORGIMENTO. The period in Italian history between 1815 (the Congress of Vienna) and 1870 (the liberation of Rome). After the defeat of Napoleon, Italy (a merely geographical expression) was divided into various small states: the Kingdom of Sardinia under a Savoy king; the Lombard Venetian kingdom directly governed by Austria; the duchies of Parma, Modena, and Lucca, and the grand duchy of Tuscany; the Papal States*; and the kingdom of the two Sicilies under a Bourbon king.

Most of the Italian people accepted the situation, but some of the more enlightened had high ideals of independence, constitutional liberty, and unity and so formed secret societies. Their activity led to various uprisings in which temporary successes were followed by repression and the reinstatement by Austrian and French forces of the old regimes. Patriots were executed, imprisoned, or exiled. Gioberti* for a time considered a confederation of Italian states under Pius IX* as the best solution. That pope by his liberal attitude had raised great hopes in the minds of the patriots, but dashed them later by a complete *volte-face.* Finally, however, the wars of independence completed the unity of Italy (except for Trento and Trieste) when in 1870 Italian troops marched on Rome, causing the pope to withdraw into the Vatican.

Many reforms were badly needed: a constitutional regime, equality of citizens, and that separation of civil and religious powers so essential for religious liberty. It is important to see how the Roman Catholic Church and other religious movements were connected with the great political and national developments. The papacy, and Gregory XVI* (1831-46) in particular, held a position of marked conservatism, opposing any liberal and progressive movement. In the Papal States, bishops and the Inquisition tribunal had unlimited power, and there was no participation of the laity in government. Manzoni* and Rosmini* pointed out the importance of a closer association between church and people, suggesting even a renewed liturgy. Lambruschini* urged liberalism in all its aspects, but this was condemned by Pius IX both in 1864 and in the dogma of papal infallibility (1870). The laws of *Guarentigie*, establishing the relationship between the Vatican and the newly formed Italian state, were refused by Pius, who remained in voluntary exile in the Vatican. Roman Catholics were ordered not to vote—and thus to withdraw from political life.

Not to be overlooked is the contribution of the various "evangelical" movements to social and educational reform and the cause of religious liberty. Though small in numbers, they made their presence felt by their missionary zeal and by the opposition they met everywhere. Various factors contributed to the birth and growth of evangelism in nineteenth-century Italy: the Waldenses* in their valleys with a well-established church, the activity of Reformed foreign churches in most Italian cities, the contacts established with European Protestantism by many political and religious exiles, the development of philanthropic enterprises financed by rich Protestant foreigners. A great impact was made also by publication of S. Sismondi's *Histoire des republiques italiennes dans le Moyen Age* (1818), which attributed to the Roman Catholic Church and its ethics the decadence of Italian customs. Finally there was the great influence of *Le Réveil.**

The Gospel spread at first especially in Tuscany, where there was a certain religious tolerance under Grand Duke Leopold II. Of great importance was a meeting held in Florence in 1844, in the house of a Swiss Christian, Charles Cremieux, at which leading reformers such as Lambruschini, Guicciardini,* and Montanelli were among those present. All considered religious reform an essential part of the national movement, but while some thought this would be better achieved from within the Roman Church, others wished to break away from it. The failure of the 1848-49 uprisings and the suppression of reformers brought the cause of Italian liberty to international notice, and in Britain particularly public opinion rose against the reactionary government which was stifling liberty of conscience.

The religious exiles in Geneva, Malta, and London formed evangelical communities. In London, Ferretti* published a journal aimed at spreading the Gospel among Italians and exposing the errors of Roman Catholicism. In Piedmont, however, the constitution had not been revoked, and the Waldensians had obtained permission to build a church in Turin (1851) while other communities were founded in Genoa and Nice. Desanctis* and Mazzarella* collaborated for a time with the Waldensians. In 1854 Guicciardini settled at Nice; in 1857 his collaborator Pietrocola-Rossetti* settled in Alessandria, a small Piedmontese town. An independent evangelical movement was thus started, and gradually communities of believers were formed in many places in Piedmont and thereafter throughout Italy. The communities formed were called Free Italian Churches, and it was the intention of the founders that the movement should be neither Roman Catholic nor Protestant, but simply Christian.

In the 1860s a division occurred within the Free Italian Churches. A. Gavazzi* had joined the movement hoping to overthrow the papacy and unite Italians into one national Protestant church, but his sermons were mainly political and violently anti-Roman Catholic. He founded a new Free Italian Church which at the turn of the century was absorbed by the Methodist mission. Most of the communities of the original movement are still in existence, with many more added to them, under the name of Christian Brethren Churches *(Chiesa cristiana dei fratelli).*

With the gradual liberation of Italy, foreign missions also increased their efforts, and contributed to the spread of the Gospel. Between 1860 and 1870 three Baptist missions (two British and one American) and two Methodist missions (one Episcopalian and one Wesleyan) were founded. Because of social conditions and appalling illiteracy, side by side with all missionary endeavor, whether Italian or foreign, has gone the founding of schools, orphanages, and hospitals.

Lastly, mention should be made of the attempts to found an autonomous Roman Catholic Church by those who opposed the dogma of papal infallibility. A first attempt in 1872 failed, but a second lasted from 1885 to 1900 when the founder of the movement, Count Campello, returned to orthodox Roman Catholicism and his collaborator Ugo Janni joined the Waldensian Church.

DAISY RONCO

RITES, CONGREGATION OF SACRED. A department of the Curia responsible for the liturgy of the Latin Rite and the canonization of saints. It was created on 22 January 1588, when Sixtus V issued his famous bull *Immensa aeterni Dei,* reorganizing the church's central government into fifteen commissions of cardinals. These commissions replaced the ancient use of consistories in regulating church affairs. The Congregation is presently composed of some twenty cardinals and oversees its two areas of responsibility by means of three subsections dealing respectively with theological questions concerning sainthood, the commission for liturgical emendations, and historical questions concerning both liturgy and sainthood. Following the will of Vatican Council II, Paul VI de-Italianized and liberalized this and the other Congregations of the Curia, issuing his radical Apostolic Constitution *Regimini Ecclesiae Universae.*

ROYAL L. PECK

RITSCHL, ALBRECHT (1822-1889). German Protestant theologian. Born in Berlin, son of an Evangelical bishop, he studied at various German universities, and was thereafter professor at Bonn (1852-64) and Göttingen (1864-89). He began his career as a disciple of F.C. Baur* and defended Baur's thesis of a radical conflict between Petrine Judaism and Pauline Hellenism, but the second edition of *Die Entstehung der altkatholischen Kirche* (1857) broke with the theory. Ritschl's chief works were *The Christian Doctrine of Justification and Reconciliation* (ET 3 vols., 1870-84); *Die christliche Vollkommenheit* (1874); *Geschichte des Pietismus* (3 vols., 1880-86); *Theologie und Metaphysik* (1881); and *Gesammelte Aufsätze* (2 vols., 1893-96). His brief *Instruction in the Christian Religion* (ET 1901) provides a compact survey of his views.

Ritschl's thought was characterized by his rejection of metaphysics. This was expressed in his opposition to speculative reinterpretations of Christianity in terms of Hegelian idealism, and also in his caution about doctrines which went beyond verifiable history and immediate Christian experience. In this he was influenced by H. Lotze. It led to the celebrated distinction between judgments of fact and judgments of value *(Werturteile).* Thus the divinity of Christ is an expression of the revelational value of the church's faith based on Christian experience; it is not a matter of objective demonstration. Mysteries may be recognized, but they transcend knowledge, and hence nothing more may be said about them.

Ritschl believed in the uniqueness of Christ who was the historical author of the church's communion with God and fellowship among its members. He saw Christianity as an ellipse with two foci: the kingdom of God, and personal redemption or justification. Christ's vocation was to found the kingdom of God among men and be the bearer of God's ethical lordship over them. It is man's vocation to practice his civil calling morally and to serve the kingdom of God. The latter was understood in moral terms as the goal of God's plan for man living in mutual love. Through justification man is put in a position to realize it by the removal of his consciousness of guilt which issues in his reconciliation to the will of God.

He rejected the concept of the penal wrath of God. Christ's death was not a propitiation of just judgment, but the result of His uttermost loyalty to His vocation. Christ's object was to bring men into the same fellowship with God by sharing His own consciousness of Sonship which He preserved to the end. For Ritschl, religion was always social, and the individual can experience the effects which proceed from Christ only in connexion with the community founded by Him. Among the many scholars influenced by Ritschl were A. Harnack,* W. Herrmann, N. Söderblom,* and J. Kaftan.*

BIBLIOGRAPHY: O. Ritschl, *Albrecht Ritschls Leben* (2 vols., 1892-96); J. Orr, *The Ritschlian Theology and the Evangelical Faith* (1897); A.E. Garvie, *The Ritschlian Theology* (2nd ed., 1902); H.R. Mackintosh, *Types of Modern Theology* (1937), pp. 138-80; K. Barth, *From Rousseau to Ritschl* (1959), pp. 390-97; P. Wrzecionko, *Die philosophischen Wurzeln der Theologie Albrecht Ritschls* (1964); P. Hefner, *Faith and the Vitalities of History; A Theological Study Based on the Work of Albrecht Ritschl* (1966); R. Schäfer, *Ritschl: Grundlinien eines fast verschollenen Systems* (1968).

COLIN BROWN

RIVER BRETHREN, see BRETHREN IN CHRIST

RIZAL, JOSE (1861-1896). Filipino physician and political writer. He studied in various European universities, where he was influenced by Masonry and theological and political liberalism. His parents were turned off their family lands by Dominican friars. He wrote much propaganda against the tyrannical Spanish colonial rule in the Philippines* and against the corruption of the friars. Notable examples of this were *Noli Me Tangere* (1887) and *El Filibusterismo* (1891). He was also strongly critical of much traditional Roman dogma which he regarded as superstition. He worked for reform in the Philippines, but was opposed to the use of violence. He was nevertheless branded as a revolutionary, and first exiled and later shot by a Spanish firing squad. His death was used by revolutionary leaders to arouse Filipino fury. Formerly a saint of the Philippine Independent Church, he is venerated as the Philippines' national hero. Roman Catholic historians dubiously claim that he recanted and returned to the Roman fold before he was executed. Various *Rizalista* cults now worship him as a "Second Christ," and even look for his return.

RICHARD DOWSETT

ROBBER SYNOD, see EPHESUS, ROBBER SYNOD OF

ROBERT OF ABRISSEL (c.1055-1117). Itinerant preacher and founder of the Order of Fontevrault. He studied at Paris under Anselm of Laon and taught at Anger between 1085 and 1090. He then took up life as a hermit and attracted a number of disciples. He founded a monastic community at La Roe, and in 1096 Urban II visited him and urged him to become an itinerant preacher. He spent the remainder of his life preaching, and was particularly noted for his work among the poor and prostitutes. He gained a reputation for his saintly life, and a number of miracles were attributed to him. In 1100 he built a monastery at Fontevrault and in 1116 drew up a constitution for the new order.

RUDOLPH HEINZE

ROBERT OF JUMIÈGES (d. c.1055). Archbishop of Canterbury, 1051-52. Prior of St. Ouen, Rouen, and then abbot of Jumièges, he befriended in Normandy the exiled English royal claimant, Edward the Confessor,* and went to England with Edward when the latter became king in 1042. Robert established himself as one of the king's closest advisers and the leader of the Norman group at court. He was bishop of London (1044-51) and archbishop of Canterbury (1051-52), and he might have been responsible for modeling Edward's Westminster Abbey on the abbey church at Jumièges. It has been suggested that while Robert was on his way to receive his pallium from the pope in 1051 he also visited William of Normandy to tell him that Edward was recognizing him as his heir. In 1051 Robert's political rival, Godwin, the king's father-in-law and the leader of the English party at court, rebelled and was exiled, only to return in 1052, when Robert fled to Rome. He died at Jumièges a few years later. PETER TOON

ROBERT OF MELUN (d.1167). English scholar and bishop. Student and successor of Abelard* at St. Geneviève, he taught John of Salisbury* and Thomas à Becket* at Melun, and may have held a chair of theology at St.-Victor. After teaching in France for forty years he was summoned home about 1160 by Thomas à Becket, who hoped for his support, and was consecrated bishop of Hereford in 1163. Robert at first cast his weight on Henry II's side and tried to moderate Becket's rigidity, but he also deterred Henry from any violence against the archbishop, and was one of the mediators sent by Becket to request the king's permission to leave England. Robert later veered to Becket's side and was intending to obey the archbishop's summons to join him on the Continent when he was prevented by royal authority, and he died soon afterward.

In his thinking, Robert was a moderate realist, as evidenced in his *Sententiae,* his best-known work, in which his position provided a transition from the nominalism of Abelard to the realism of the school of St.-Victor. In this work, which may well have influenced Thomas Aquinas,* Robert did not hesitate to counter Bernard of Clairvaux and to side in part with Abelard in the controversy on the Trinity. His other major works were his *Quaestiones de divina pagina,* a work in the style of Abelard's *Sic et non,* which went beyond its model in providing resolutions for the issues, and his *Quaestiones de epistolis Pauli,* which established him as the founder of a school of commentators on Paul. In all his works, Robert exhibited a sturdy intellectual independence.

MARY E. ROGERS

ROBERT OF MOLESME (c.1027-1111). Abbot of Molesme in Burgundy. Of noble birth, he entered the abbey of Moutier-la-Celle when fifteen years old, and later became prior. He spent a brief period at St.-Michel-de-Tonnerre, but here, as at his first abbey, he was unable to promote the observance of a stricter interpretation of the Benedictine Rule. In response to the call of some hermits living in the forest of Colan he was able to join them in founding a monastery at Molesme on strict principles. This establishment flourished, but after some disagreements Robert and some monks left to found a new monastery at Cîteaux (of which Bernard of Clairvaux was to be a member in 1113). Robert, however, agreed to return to Molesme, which under his long rule became a famous Benedictine house. PETER TOON

ROBERTS, EVAN JOHN (1878-1951). Welsh revivalist. Born in Glamorgan, the ninth of the fourteen children of Henry Roberts, pitman, and his wife Hannah, his education at the parish school ended when at twelve years of age he accompanied his father to the coalmine. In 1902 he was apprenticed as a blacksmith, but was accepted as a candidate for the ministry by the Calvinistic Methodist* Church in 1904, and entered a preparatory school at Newcastle Emlyn. Even as a young man he was a remarkable character. For eleven years he devoted himself to intense intercession for an outpouring of the Holy Spirit. He was also granted visions and vivid experiences of the divine presence.

By 1904 there were indications in many parts of Wales that a revival was about to happen, and in that year while attending a meeting to deepen the spiritual life at Blaenannerch, Roberts underwent a profound experience of being anointed by the Holy Spirit. He returned home to Loughor and began to hold prayer meetings at his home church, Moriah. On successive nights these meetings drew ever larger crowds, and within a matter of weeks the revival had swept across Glamorganshire with tremendous power. The most significant feature of the revival was its concentration on the gift of the Holy Spirit; the meetings, even when Evan Roberts was present, were conducted with complete spontaneity. People were urged to pray, testify, confess, or sing as the Spirit moved them. Soon Roberts and a group of young friends began to make revival tours, first in Glamorgan, then Liverpool, Anglesey, and finally Caernarvonshire (November 1904–January 1906).

Yet the revival was by no means limited to places visited by Roberts: it was a national phenomenon and it was calculated that it led to some 100,000 conversions. By any standards it was a mighty movement of the Spirit, and since it was followed in great detail by the press, it had worldwide publicity. But physically the revival broke Evan Roberts. He retired from public life and went to live in Leicester. He returned to Wales about 1925 and died some twenty-five years later at Cardiff.

See E. Evans, *The Welsh Revival of 1904,* (1969). R. TUDUR JONES

ROBERTSON, FREDERICK WILLIAM (1816-1853). Anglican preacher. Educated at Edinburgh and Oxford universities, he became incumbent of Trinity Chapel, Brighton, where in the six years before his death he established a reputation as preacher equal to that of any man in the nineteenth century, Newman not excluded. Robertson was a Celt with all the fire of his race, but he was also apparently subject to fits of depression. He began as an Evangelical, but the practice of some of those professing this view of Christianity during his first curacy at Cheltenham did much to drive him from any sympathy with them. Nonetheless, he brought Evangelical passion to his beliefs and his preaching.

Robertson's fame coincided with the rise of Christian Socialism* under F.D. Maurice,* and his views were closely allied with those of the movement. He was, indeed, variously accused of being a socialist and a rationalist, and he certainly had a social gospel to preach, but he was not a socialist in the secular political sense of the term. Nor was he a rationalist. He would perhaps have been considered to approach the modernism of the early twentieth century, with his moral equations of biblical phenomena, characters, and events. He sought to interpret in the light of man's actual experience; the Virgin, for instance, epitomized the adoration of womanly purity. This is what points to Robertson's strength. He was a great psychological preacher, understanding the motivations of those characters who formed his subjects and relating these to the motivations of his hearers. His style—and all he left is notes—is simple, direct, forceful, a mirror of the feverish energy and wholehearted dedication which he brought to his task. His *Sermons* were published in 1906.

See also H. Henson, *Robertson of Brighton*, (1916). ARTHUR POLLARD

ROBINSON, HENRY WHEELER (1872-1945). English Baptist scholar. Born in Northampton, he was educated at Edinburgh University and Mansfield College, Oxford, then studied at continental universities. He held Baptist pastorates at Pitlochry in Scotland and Coventry before becoming tutor in Rawdon Baptist College in 1906. Thereafter for fourteen years he showed the teaching skill and administrative gifts which he was to display most notably when he was appointed principal of Regent's Park College, a post he held until his retirement in 1942. Known widely in scholarly circles beyond his own denomination, he served also for seven years from 1934 as reader in biblical criticism at Oxford. His many works include *The Religious Ideas of the Old Testament* (1913; rev. 1956), *The Christian Experience of the Holy Spirit* (1928), and *Redemption and Revelation* (1938). J.D. DOUGLAS

ROBINSON, JOHN (c.1575-1625). English Separatist and pastor of the Pilgrim Fathers. He was born in England, though early records of Robinson's life are scanty. He may have graduated at Cambridge in 1598. He was for a time curate in a Norwich church. He imbibed Puritan views and joined a Separatist* group at Scrooby Manor of which he became pastor. He fled to Holland with this group to avoid persecution, settling in Leyden by 1609. He was ordained as the pastor of their newly formed church in May 1609. For a time he was a student at the University of Leyden. He held a Separatist view of Communion, sided with the Calvinists in the Arminian* controversy, and advocated congregational church government. He urged emigration, and a small part of his church sailed from Holland, stopping first at England in 1620. The *Mayflower* landed at Plymouth, Massachusetts, on 11 November 1620. Robinson never emigrated, choosing rather to remain as pastor of the major portion of the church, which stayed in Holland. Though he was never to arrive in New England, his influence on the Plymouth Separatists was profound, due to his teaching before the voyage, his tracts and letters to the flock, and his guidance of Brewster as their spiritual adviser in New England. He is perhaps best known for his farewell sermon to the Pilgrims at Leyden on 21 July 1620. ROBERT C. NEWMAN

ROCH (c.1350-c.1378). Late medieval miracle-worker. Little trustworthy information is available on his life, despite early biographies which unfortunately are filled with legend and chronological errors. Born in Montpellier, France, he went to Italy where he became famous for his ability to bring about miraculous cures of the plague by prayer and the sign of the cross. After his death he was widely venerated in France and Italy, and a cult of St. Roch developed. He became the saint to invoke against the plague, and Luther mentioned him in this context in his denunciation of medieval saint worship in his Large Catechism. Roch's relics were taken to Venice in 1485, where a shrine was erected in his honor. RUDOLPH HEINZE

ROCK, JOHANN FRIEDRICH (1678-1749). German religious leader. Born at Oberwälden, Württemberg, of poor parents, he became a harnessmaker, but in 1707 underwent a mystical experience, was "seized with inspiration," and became a prophet or "vessel" of the ecstatic communities of the True Inspired, closely related to, if not descended from, earlier French prophetic movements. He led his sect together with Eberhard Ludwig Gruber till 1728, and thereafter alone. A powerful teacher and leader, he lost many followers through personal quarrels, including that with Zinzendorf* in 1734 over Rock's rejection of preaching and the sacraments, through defections to the Herrnhutters and allied movements, and through emigration to America. After Rock's death his mystical communities declined rapidly, despite a temporary revival in Germany and America in the early nineteenth century. IAN SELLERS

RODEHEAVER, HOMER ALVAN (1880-1955). Song evangelist and publisher. Born in Hocking County, Ohio, he learned the cornet during boyhood in Newcomb, Tennessee. He studied music, becoming a trombonist, at Ohio Wesleyan University between 1896 and 1904. As music director for evangelists W.E. Biederwolf* and Billy Sunday* he developed group singing techniques to a fine art, effectively using lively, nondoctrinal gospel songs such as "Brighten the corner." His Rodeheaver Publishers of Sacred Music (1910-35) and Rodeheaver Hall-Mack Company (since 1935) became the world's largest gospel music publishers. "Rody" wrote many songs and after 1920 conducted the Rodeheaver Sacred Music Conference annually at Winona Lake, Indiana. In 1950 he founded Rodeheaver Boys' Ranch, Palatka, Florida. D.E. PITZER

ROGERS, JOHN (c.1500-1555). Protestant martyr. Born near Birmingham, he was educated at Pembroke Hall, Cambridge, was from 1532 rector of a London church, then in 1534 became chaplain to the English merchants in Antwerp, where he assisted in smuggling forbidden books into England. Here he met William Tyndale,* then engaged in his translation of the OT, and embraced the Reformed faith, After Tyndale's martyrdom he used his manuscripts together with the already published translation of Miles Coverdale* to produce his influential "Matthew's Bible"* in 1537. Rogers's own share in the work was largely confined to the prefaces and marginal notes. Matthew's Bible was to be a major inspiration in all those translations which led up to the Authorized Version of 1611. Rogers now married and removed to a pastorate at Wittenberg, where he studied Melanchthon's* writings, some of which he later translated. He returned to England in 1548, was presented to two crown livings in London in 1550, and a year later was made prebendary and divinity lecturer at St. Paul's Cathedral. His advanced political and religious views had already caused him trouble under Northumberland's protectorate. With the advent of Queen Mary he was arrested and put into Newgate Gaol with J. Hooper* and others. After cruel sufferings he was burnt at the stake in 1555 in the presence of his wife and children—the first Protestant martyr of the new reign. IAN SELLERS

ROLLE OF HAMPOLE, RICHARD (c.1295-1349). Born in Thornton Dale, Yorkshire, he studied at Oxford, leaving at nineteen to live as a hermit. Moving from place to place, and preaching, he spent his last years at Hampole near a convent of Cistercian nuns who were under his spiritual guidance. He left seven treatises in Latin, a dozen commentaries on and translations of books of Scripture in Latin and English, several letters, and a number of English lyrics. Standing within the late medieval mystical tradition, he left as his two major works *Incendium Amoris* and *Melum Contemplativorum.* Read until the Reformation, he directed his writing not to the cloistered community but to those in the world, expressly using the vernacular and simple language. He does not reckon with the higher degrees of the mystical life, not having experienced that, nor did he know contemporary German mystics and the Dionysian influence there. While at Oxford he rebelled against the Scholastic teaching, which explains the simplicity and lack of formal learning in his work, despite the seeming influence of Augustine, Hugh of St.-Victor, and Bonaventura. He comes down equally on piety and learning as well as worldliness and fashion, and views the hermit as not inferior to the prelate or monk. Later English mystics knew his work, and he did influence the Lollards.

See F.M.M. Comper, *The Life of Richard Rolle* (1928), and H.E. Allen, *English Writings of Richard Rolle Hermit of Hampole* (1931).

C.G. THORNE, JR.

ROMAINE, WILLIAM (1714-1795). Evangelical Anglican. Born in Hartlepool and educated at Houghton-le-Spring Grammar School, Durham, he graduated from Christ Church, Oxford, was ordained in the Church of England, and served curacies at Lewtrenchard, Devon, and Banstead, Surrey. He was chaplain to Daniel Lambert in his year of office as lord mayor of London in 1741. One of his absorbing interests was the Hebrew language, and he spent much time preparing a new edition of the Hebrew *Dictionary* of Marios de Calasio (d.1620). Moving to London in 1748, he became lecturer at St. George's, Billingsgate, and later at St. Dunstan-in-the-West. By this time he was a convinced Evangelical and the friend of George Whitefield, the countess of Huntingdon, and others. Excelling as a preacher, he attracted large crowds in London and on preaching tours in the country. He wished to see the Gospel penetrate the whole church and nation, and organized days of prayer to this end. In 1766 he was appointed rector of St. Anne's Blackfriars, and remained there until his death. He also continued his lectureship at St. Dunstan's. His message was a warm, Calvinistic evangelicalism. His most famous work is the trilogy, *The Life, Walk, and Triumph of Faith* (1771-94). A new edition with biography (ed. P. Toon) was published in 1970.

PETER TOON

ROMAN AFRICA, see AFRICA, ROMAN

ROMAN CATHOLICISM. In the past the unchanging character of Rome made it comparatively straightforward to describe Catholicism. The shattering developments of the twentieth century have made that task much more complex. The constantly changing situation means a wide variety of opinions, making a comprehensive survey almost impossible, especially within the restricted confines of a short article. It is, however, possible to detect two major groupings within Rome, and while admitting that the dividing lines are not always clearly drawn, it seems legitimate to speak of "traditional" Catholicism and "the new Catholicism" and to survey the whole from these two aspects.

Traditional Catholicism. The dogmatic formulation may be found in the decrees of the Council of Trent,* the Creed of Pope Pius IV,* the decrees of Vatican I* and II,* papal utterances claiming infallibility, and the body of Roman canon law.* Alongside these there is the liturgy, and behind them the hierarchically organized church.

It is in fact the doctrine of the church which is fundamental to an understanding of traditional Catholicism. Developed across the centuries, it bears the marks of the Middle Ages when the imperial background lent weight to the concept of the church as an imperium with a resultant stress on ecclesiastical structure, on the hierarchy, and on the claims of the pope as absolute monarch. These claims reached their zenith in the bull of Boniface VIII,* the *Unam Sanctam* issued in 1302 with its affirmation: *"extra ecclesiam nulla salus"* ("outside the church there is no salvation"). Submission to the authority of the pope was written into the terms of salvation.

The stormy days of the Reformation* with the subsequent attempt by the Counter-Reformation movement to regain ground led to a continued rigidity and an emphasis on structure and organization. This trend was accelerated dramatically at the First Vatican Council in 1870 when the dogma of papal infallibility was promulgated—the concept of the imperium had reached its climax.

But there is another element in the traditional doctrine of the church which has its roots in the Pauline analogy of the church as the body of Christ. Paul, however, does not make the disastrous mistake which has vitiated so much Roman

Catholic thinking on this issue—he does not so overstress the unity of the head and members that he ignores the distinction. Christ always remains the head with the strong overtones of authority implicit in that headship. Rome on the contrary has so emphasized the unity of head and members as virtually to identify them. Hence the emergence of the concept of the church as the extension of the Incarnation. Just as Jesus of Nazareth was the Son of God incarnate, so, it is claimed, the church is Christ continuing incarnate in the world. The two streams of thought—the church as imperium outside which is no salvation, and as the continuing incarnation—flowed together in the encyclical of Pius XII,* the *Mystici Corporis* issued in 1943. The body of Christ is firmly equated with the hierarchical Roman Catholic Church in which Christ speaks and works as He did in Nazareth and at Calvary.

The traditional line continued at Vatican II. Pope Paul VI in a speech declared the church to be "Christ's continuation and extension." His speech was echoed in the Vatican II decree on the church. While there is a new stress on the biblical idea of the people of God, the old conception still stands. "We must not think of the church as two substances but a single complex reality, the compound of a human and a divine element ... the nature taken by the divine Word serves as the living organ of salvation in a union with Him which is indissoluble" (I:8). Hence it is stated that the church "is incapable of being at fault in belief" (II:12). The church, it must be stressed, is still assessed by Vatican II in traditional terms: "this church ... has its existence in the Catholic Church under the government of Peter's successor and the Bishops in communion with him" (I:8).

This fundamental concept colors the interpretation of Christ's messianic office. As prophet He still speaks within His church with the same infallibility as in the days of His earthly ministry. As priest He still offers Himself on the altar in the Mass as really as He did on the cross. As king He exercises His royal authority through His appointed agency the hierarchy of the church. The consequence of all this is quite clear: the church deprives herself of the divinely given corrective to error, for she equates the Word of God in Scripture and the Word of God in the church. She becomes herself the standard of truth. Such a church, says a critical Catholic, Hans Küng, turns itself into a revelation.

Allied to this basic doctrine is Rome's sacramental interpretation of Christianity. The sacraments which are ministered by the church are channels by which the grace of God flows to the recipient. There are seven sacraments—baptism, confirmation, the Mass, holy orders, penance, matrimony, extreme unction. While baptism may be administered *in extremis* by a layman, the normal administration of this sacrament and the essential administration of the others are by a priest or a bishop. In a valid sacrament three conditions are required—there must be the correct matter (e.g., water in baptism), the right form (e.g., words of consecration), and the true intention (the one who ministers must intend what the church purposes). Granted that the recipient places no obstacle in the way, the sacraments work *ex opere operato* ("by virtue of the performance of the work").

In baptism, original sin is cleansed and original righteousness restored by the infusion of grace. Behind this teaching lies the Roman view of man. Man's original righteousness is viewed as a *donum superadditum* ("an additional gift"). The Fall meant the loss of that gift, but not the impairment of his essential integrity. So original sin is dealt with by the replacement of the gift. This infusion of sanctifying grace is deepened in confirmation and is sustained by the Mass and by the regular ministry of the sacrament of penance in the confessional.

The focal point of traditional Roman Catholic worship is the Mass. The dogma which lies behind the Mass is that of transubstantiation.* Promulgated at the Fourth Lateran Council in 1215 and reaffirmed with great vigor by Pope Paul in the *Mysterium Fidei* issued during Vatican II, this dogma asserts that after the words of consecration, the substance of the bread and wine become actually and really the body and blood, the soul and divinity of Christ. With transubstantiation accepted, and with the conception of the priest as himself another Christ (see the encyclical of Pius XI* *Ad Catholici Sacerdotii*), the Mass is viewed as the sacrifice of Christ. Although now offered in an unbloody manner, yet the sacrifice is the same as that at Calvary. It is a "propitiatory" sacrifice to meet the judgment of God, and an "impetratory" one to invoke and to ask for specific blessings. The comment on this whole position is the strong emphasis of the epistle to the Hebrews on the finished work of the once-for-all sacrifice and on Christ as the one, only, and all-sufficient high priest.

Vatican II in its decree on the priestly life continues the traditional teaching. Priests are "given the power of sacred Order, to offer sacrifice, forgive sin, and in the name of Christ publicly to exercise the office of priesthood in the community of the faithful" (chap. 1, para. 2). This fails to face the fact that in the NT the term for a sacrificing priest *(hiereus)* is never applied to a Christian minister; he is an elder, a pastor, a bishop, but never a priest.

The confessional has always played a key role in traditional Catholicism with the sacrament of penance* as its basis. Sins are classified as mortal, which deprive the soul of sanctifying grace, and venial, which are not so serious. Repentance* is either contrition (i.e., true sorrow for sin) or attrition (i.e., sorrow for sin for a lesser motive, such as fear of punishment). Venial sin may be dealt with by attrition, moral sin either by contrition or by attrition plus recourse to confession—though it should be noted that contrition implies an intention to go to confession. Guilt requires not only absolution but reparation to be offered to divine justice, hence the imposition of penance and hence also the practice of indulgences,* in which the benefits of the alleged heavenly treasury of merit* may be set to the sinner's account either at his own request or at his friends' request after his death and during his detention in purgatory.* Comment on this must be sought in the NT stress on the sufficiency of Christ's sacrifice and the complete satisfaction of God's justice.

Mary. A prominent feature in Roman Catholic worship is the cult of Mary.* Consideration of this provides an appropriate transition to a description of the new Catholicism, for "Mary" provides a common ground on which very varied viewpoints come together so that Vatican II theologians as radical as Schillebeeckx of Holland can be as conservative as any traditionalist in dealing with Mary—witness his book, *Mary the Mother of the Redemption.* Development of the cult stems from the period when the Constantinian settlement brought an influx of pagan ideas into the churches—the mother goddess of the Mediterranean world, the female goddess with such titles as "Star of the sea" and with such roles as "Our lady" of this or that city, the mother and child motif of the Horus and Isis cult in Egypt. All these were to be reflected in Marian developments. The centuries-old debate between Dominicans and Franciscans was finally settled by Pius IX* in 1854 with the promulgation in the bull *Ineffabilis Deus* of the dogma of the Immaculate Conception* of Mary—that Mary was conceived without sin. It marked the triumph of the Franciscan theory of Duns Scotus* that in Mary we see "redemption by exemption," i.e., her preservation from the stain of original sin. The cult was carried further in 1950 by the dogmatic pronouncement of Pius XII* in the Apostolic Constitution *Munificentissimus* of the bodily Assumption* of Mary. Vatican II in spite of the reservations of some liberals went further still in pronouncing Mary the mother of the church and in giving a tacit recognition to Marian devotions.

The cult of Mary is significant theologically in that, as Sebastian Bullock pointed out *(Roman Catholicism),* it embodies the Roman Catholic view of human merit and of human cooperation in the work of salvation which are exemplified supremely in Mary. It is significant emotionally in that it shows how such a cult can exercise so firm a hold. It is significant in the contemporary situation as it indicates the difficulty of a liberal Catholic in breaking from his traditional past. In reply to the cult, one must point to the total silence of the NT on the roles and honors accorded by Rome to Mary and to such telling evidence against them as Mark 3:33-35; Luke 1:47; 2:49; 11:27, 28. He offered Himself (1 Tim. 2:5; Heb. 9:14; John 10:18)—it was not Mary who offered Him.

The New Catholicism. The prelude to the present developments in Rome may be traced to the modernist movement at the turn of the century. Pius X* acted firmly and condemned the movement in the decree *Lamentabili* and in the encyclical *Pascendi* of 1907, and in 1910 the antimodernist oath was imposed on the clergy. The reaction against modernism continued as far as the pontificate of Pius XII,* whose encyclical *Humani Generis* in 1950 firmly asserted the teaching authority of the papal office.

It was Pope John XXIII* who opened the door to the progressives both by convening the Vatican Council, which was to give them an invaluable forum, and also by introducing a distinction, vitally significant to them, when in his speech at the opening of the council he declared that "the substance of the ancient doctrine of the deposit of faith is one thing, and the way in which it is presented is another." Pope Paul VI* tried to stem the flood. His assertion of traditional eucharistic dogmas and devotions in the *Mysterium Fidei* was a counterblast to the reforms proposed in the Constitution on the Sacred Liturgy. His Credo delivered at the time of the Uppsala* meeting of the World Council of Churches was a further call for a return to traditional dogma. But the tide was running too swiftly and continues to surge forward, so that Hans Küng, for example, in his *Infallible* can tear papal infallibility to shreds with the vigor of an old-style Protestant controversialist and substitute for the dogma of infallibility his own emphasis on the indefectibility of the church. At the same time transubstantiation is being replaced by trans-signification; clerical celibacy* is in the arena of debate; cherished positions and devotions are questioned, and in many cases abandoned.

It is difficult to generalize about a movement which is itself in state of flux—today's description can be quickly dated. Then again, the variety of men involved means a variety of opinions. However, one can detect certain governing aims in the movement. There is an attempt to be biblical. There is a strong ecumenical emphasis. There is a firm desire to remain within Rome and to work for reform.

Undoubtedly there is a new biblical emphasis, not only in the realm of theological studies, but also at the popular level of encouraging Bible reading. There is an attempt to get away from the Council of Trent's* insistence on two sources of revelation, "Scripture and Tradition," and to make tradition the living voice of the church commenting on Scripture, although Vatican II retained the traditional position. The Dogmatic Constitution on Divine Revelation firmly welds together in a unity "sacred tradition, holy Scripture and the Church's magisterium" (II:10). The biblical emphasis of the progressives is also affected by their acceptance of the theories of contemporary radical biblical criticism with its "demythologizing" approach. There is also in some quarters a measure of agreement with the radical Protestant's rejection of the whole conception of propositional truth—existential experience rather than divine revelation becomes the criterion.

The ecumenical interest has moved the "new" Catholics away from the old view of Protestants as heretics. The latter have become "separated brethren," unity with whom being the aim. But it is not the ultimate aim, for the progressives look beyond the churches of the Reformation and Greek Orthodoxy to Judaism, to the great non-Christian religions, and even to atheism as they increasingly emphasize the ultimate synthesis which is Catholicism's goal. Behind this far-reaching vision is the widespread acceptance of some form of universalism. Behind that again is the incarnational theology which sees Christ sanctifying everything by His coming and redeeming all men by His death. The old idea of a latent faith is introduced, allied to the acceptance of "baptism by desire." The way is open, not only to acknowledge Protestants as "brothers by baptism," but to see in Muslims, Hindus, and even atheists those who by exercising "implicit" faith are in the "hidden" church in contrast with those whose explicit faith and sacramental initiation make them members of the explicit church.

Their attempt to remain loyal to Rome meets the countercharges of the conservatives that they are the old modernists in a new guise. The liberal reply has been developed at length by Hans Küng. It is that the earlier dogmatic statements are accepted, but words must be interpreted and a sixteenth-century creed must be understood in its own setting. The resultant interpretation of a "new Catholic" differs markedly from that of a traditionalist—hence the tension and at times the acrimonious debate.

Catholic Pentecostalism. This movement emerged in the autumn of 1966 among faculty members at Duquesne University, Pittsburgh, Pennsylvania. Influenced by Protestant Pentecostalism, they yet remained firm in their attachment to Rome; and within six years they were claiming 50,000 adherents. The early leaders of the movement in the USA were mainly laymen. They viewed the movement as heaven's answer to Pope John's prayer for a renewal of the wonders of Pentecost, hence their avoidance of what they considered some of the excessive emotionalism of Protestant Pentecostalism, hence too their encouragement of their followers to remain within Rome.

There is a warm stress in the movement on personal faith in Christ. Prayer, both personal and corporate, is a prominent feature. Bible study is encourged—as of course it is in the wider field of the new Catholicism. Witness to the outsider is emphasized. All these are elements linking the movement with Protestant evangelicals, but there are other features which root the movement firmly in Roman Catholicism, and still others which point in a radical direction.

The sacramentalism of traditional Catholicism is strongly taught. Leaders in the American movement (like Kevin and Dorothy Ranaghan) or in the English wing (like Simon Tugwell) expound the idea of "Baptism in the Spirit" as an explicit realization and manifestation of the divine life received in a hidden way, and *ex opere operato,* in baptism in water. Speaking in tongues is seen as one step in the process by which one is permeated by the divine life imparted in baptism—it involves handing over "one little bit of our body to God." The result, it is claimed, is a deeper appreciation of the Mass (with the traditional background of transubstantiation) and of the confessional.

The cult of Mary is particularly prominent. The Ranaghans quote testimonies of a new devotion to Mary and of the use of the Rosary. They found special significance in a tongue-speaking session where the initial "Hail Mary"—spoken in Greek, recognized as such by a participant and interpreted—led to a Marian emphasis which to their added delight proved to be a prelude to the next day—one of the major Marian festivals of the year.

One is forced to ask how it can be claimed that the Spirit of truth can lead into such doctrines as transubstantiation and the eucharistic sacrifice, and how the Spirit whose ministry is to glorify Christ could lead men to derogate from that glory by devotion to Mary. But the question comes even more insistently when it is discovered that the universalism already noticed in the new Catholicism appears, and the Pentecostal experience is linked with Zen, Transcendental Meditation, and even Marxism as steps toward the ultimate Catholic synthesis of human experience.

The ultimate criterion in the movement turns out to be experience. Scripture and tradition may be invoked, but ultimately it is the intensity of experience which authenticates the doctrine. The old position was that "doctrine precedes exegesis"; now, however, we read Scripture through the eyes of experience. This approach is more akin to existentialism than to the biblical stress on the mind—"faith comes by hearing," not by a blind leap in the dark.

BIBLIOGRAPHY: By Roman Catholics: R.A. Knox, *The Belief of Catholics* (1927); T. Corbishley, *Roman Catholicism* (1950); L. Ott, *Fundamentals of Catholic Dogma* (1962); H. Küng, *Justification: The Doctrine of Karl Barth and a Catholic Reflection* (1964), *Structures of the Church* (1965), *The Church* (1967), and *Infallible? An Inquiry* (1971); K. Rahner, *The Teaching of the Catholic Church* (1966); W.M. Abbott (ed.), *The Documents of Vatican II* (1966); K. and D. Ranaghan, *Catholic Pentecostals* (1969); S. Tugwell, *Did you receive the Spirit?* (1971).

By others: G. Salmon, *The Infallibility of the Church* (abridged, 1953); G. Miegge, *The Virgin Mary* (1955); G.C. Berkouwer, *The Conflict with Rome* (1958) and *The Second Vatican Council and the New Catholicism* (1965); A.F. Carillo de Albornoz, *Roman Catholicism and Religious Liberty* (1959); V. Subilia, *The Problem of Catholicism* (1964); D.F. Wells, *Revolution in Rome* (1972); H.M. Carson, *Dawn or Twilight—A Study of Contemporary Roman Catholicism* (1976).

H.M. CARSON

ROMANIA. The origins of Christianity in the area comprising modern Romania are uncertain. The Roman occupation of the region (Dacia) ended in A.D. 275. There may have been Christians among the Roman legions and colonists. By the fourth century, Christian communities had grown up in the Dacian regions as missionaries from centers on the right bank of the Danube carried on an expanding ministry. Shortly after this period, the waves of barbarian invasions began, and not until the medieval epoch can one speak of a full-fledged Christian church in Romania. With the establishment of the two principalities of *Moldavia* and *Wallachia,* organized Christianity began the thread which leads to the present. The metropolitanate of Ungro-Vlahia was founded in 1359, that of Moldavia in 1401; their recognition by the patriarchate of Constantinople signaled an acceptance of important religious and political developments rather than their inception.

With the fall of Constantinople in 1453, Romanian princes, notably Stefan the Great (1457-1504) and Neagoe Basarab (1512-21), became for a time the heirs of Byzantium and secular leaders of the entire Orthodox Church. They led campaigns against the infidel Turks, assumed patronage over Athos* monasteries, sponsored cultural advances, and instituted a remarkable series of monastic foundations whose architectural and artistic brilliance can still be seen. At the same time began an influx of Greek churchmen and culture. Their intellectual and theological influence led to a Romanian-Greek-Slavonic synthesis of merit, but their moral impact was not always as salutary.

The apogee of the Romanian Orthodox Church came in the seventeenth century. Powerful, cultured rulers coincided with notable church metropolitans. Under their aegis, theological synods were held, prolific presses established, monastic reforms enacted. Church leaders such as Varlaam (1632-53) and Dosofteiu (1671-86) of Moldavia, and Stefan (1648-68), Teodosie (1668-1708), and Antim (1708-16) of Wallachia carried on a wide program of cultural and religious activities, with key achievements being the establishment of the Romanian language in the liturgy and the publication of the Bucharest Bible in 1688, the first complete translation into Romanian.

In the next century, however, the Romanian Church fell increasingly under Greek domination as Phanariot princes replaced native rulers. More and more monasteries became Greek fiefs, and the general level of religious life declined. The language remained Romanian, however, and important reforms were carried out on lower levels. In the latter part of the century, Eugumen Paisie sparked a reform in Moldavia, and this spread into the nineteenth century under Metropolitans Iacob Stimati (1792-1803) and Veniamin Costache (1803-42). Both were men of the Enlightenment,* possessors of vast erudition and literary skill. The union of the two principalities in 1859 produced various attempts at unifying the church also. These schemes failed, though the Romanian Church became officially autocephalous in 1885. The secularization of the Greek-controlled dedicated monasteries in 1863 was an important event. The nineteenth century saw also the challenge of "new" denominations which transcended ethnic lines. Most important of these were the Baptists, who spread eastward from Hungary* beginning in 1870. At the same time Roman Catholicism became strong enough in Romania to form the archbishopric of Bucharest (1883).

Transylvania, the third Romanian principality, fell under Hungarian domination in the eleventh century, becoming subsequently the target of strenuous Catholic missionary efforts under the sponsorship of the Crown. These met with little success among the Romanian population. In the sixteenth century, Hungary's defeat by the Turks coincided with the beginnings of the Reformation,* and the German (Saxon) population of Transylvania was rapidly Lutheranized under the Reformer Johannes Honterus. The Magyar nobility, on the other hand, slowly became Calvinists (and a few, Unitarians) through a complex of events. The ethnic Romanian population remained staunchly Orthodox. The Reformation in Transylvania produced the first printed books in Romanian as well as a general religious and cultural revival among all the national groups of the area; its influence on the other side of the Carpathians remained largely cultural.

The brief unification of all three Romanian principalities under Michael the Brave (1600-1601) saw the first stable Orthodox hierarchy in Transylvania. The restoration of the now-Calvinist Magyar nobility forced the new church to accept many points of Reformed doctrine, though the great metropolitan, Sava Brancovici (1656-80), gained a brief reduction of Calvinist influence. In 1691 Transylvania came under (Hapsburg) Catholic rule again, and Catholicism (with Calvinism, Lutheranism, and Unitarianism) was declared one of the four "received religions." Jesuit efforts were unsuccessful, but in 1698 the Transylvania Greek Catholic Church (Uniate) was born through the adherence of the Orthodox metropolitan Anastasiu. In exchange for the suzerainty of the pope and the doctrines of purgatory and the *Filioque,* the Romanians kept their dogma and liturgy intact and were granted equal rights with the "received" clergy. The great majority of Orthodox clergy and believers refused to accept the new church, but the Transylvania Orthodox Church disappeared for half a century anyway.

The Transylvania Uniate Church, paradoxically, became the rallying center of the great Romanian national revival of the eighteenth century. Educational and spiritual reforms were effected, a vote obtained in the Diet, and the right established to build Romanian schools and churches. In 1758 the Orthodox hierarchy of Transylvania was reconstituted and included more than 80 per cent of the Romanian population. Reform began under Episcop Vasile Moga of Sibiu (1811-46) and his successor Andreiu Saguna (1848-73), the greatest modern Romanian churchman, who refounded a Transylvanian metropolitanate free of Serbian control in 1864 and reorganized and democratized church structure.

The union of all three Romanian principalities resulting from World War I led to a unification of the three Orthodox hierarchies. In 1925 Miron Cristea was named the first patriarch of Romania. The Orthodox and Uniate cults were designated as the "national" cults, comprising respectively 72.6 percent and 7.9 percent of the population (the two were merged after 1948). The Catholics (7 percent), Calvinists (3.9 percent), Lutherans (2.2 percent), and Unitarians (0.4 percent) continued their previous status as "received" cults, while the Baptists (0.3 percent) and other Protestant groups were classified as "tolerated" cults.

BIBLIOGRAPHY: Most of the literature is in Romanian, but see M. Beza, *The Rumanian Church* (1943); and *The Rumanian Orthodox Church* (ed. by the Bible and Orthodox Missionary Institute, Bucharest, 1962). See also bibliography under EASTERN ORTHODOX CHURCHES.

PAUL E. MICHELSON

ROMANOS (d.556). Greek hymnwriter, known as "Melodus." A Syrian by birth, he served as deacon in the Church of the Resurrection at Beirut, then went to Constantinople under the patronage of Patriarch Anastasius I. There he wrote over 1,000 hymns of which only some eighty survive. They are dramatic and vivid and have at least twenty-four strophes. Their subjects include the OT and NT and the church year. The most famous is the Christmas Day hymn, "On this day the Virgin gave birth to the Transcendent One." This was regularly sung on Christmas Eve in the imperial palace until the twelfth century.

ROMANS, see EPISTLES, PAULINE

ROME. The oldest settlement was made seventeen miles NE upstream from the Tiber mouth on the Tyrrhenian coast of the Italian peninsula, where a cluster of hills provided a natural location for village farming complexes. The river valley itself was too swampy and unhealthy for original occupation, but provided a burial ground both for cremation (Villanovan; northern) and inhumation (Picene; southern) practices from the ninth to the sixth centuries B.C., the Tiber serving as a rough distinguishing boundary. At the site of Rome, however, their overlapping complicates the use of burial practice for identifying its particular settlers. The late first-century annalist Tacitus summarized the most ancient history of Rome thus: "In the beginning Rome was ruled by kings." His predecessor Titus Livius set the traditional date of foundation at 21 April 753 B.C.

With the passing of the kings the Republic came into being, late enough in time (509 B.C.) for its chroniclers not to resist the achievement before them of the democratic governments of the Greek city-states, especially Magna Graecia and Sicily. The defining event which initiated actual Roman achievement was the sacking of the city in 387 B.C. by marauding Gauls. Thereafter Rome by sword wielded, colonies founded, and roads constructed brought all Italy south of the Po into a political confederation. For eight centuries the city was to stand inviolate, confirming in history her own mythos, but by no means immune from internal wranglings. Gradually the exploiting of party politics and the city's inhabitants transformed the republican empire into the imperial state. Yet the architectural impact of this process, with corresponding levels of "full" employment throughout the Roman world as well as in the city, was remarkable. The enormous profits from political murders and confiscations were turned into building programs that gave external symbolization to the revolution. Pompey brought Rome its first stone theatre, the Campus Martius; Caesar built a great new basilica in the republican forum, and to its north began a new Forum which was to serve as prototype for his imperial successors.

By the time of Augustus the population seems to have been well over one million. The period from Augustus to Diocletian and the next great reform of state, a span of three full centuries, may be divided into two phases. The first lists those sixteen emperors beginning with Tiberius and ending with Commodus, through the first two centuries A.D., of whom eight died natural deaths, ten had reigns of ten years, and three (Tiberius, Hadrian,* Antoninus Pius*) more than twenty. While unrest beset succession with the suicide of Nero,* the whole period has been termed *Pax Romana*, reflecting the stability of the internal economy and the transfer of power as well as of the external peace.

It was during this phase that Christianity came into being. Individual Christians, presumably undifferentiated from Jews, were in Rome already in the reign of Claudius, as his edict forcing Jews to leave the city affected Priscilla and Aquila. Yet Paul's* letter to the Romans, written not later than A.D. 56, reflects a continuing church there which he had not yet visited. Acts* terminates with Paul's arrival and initial work in Rome. The tradition, as cited by Irenaeus and Eusebius, denotes by unbroken list the succession from Paul and Peter,* beginning with Linus,* but records and dates are scanty before the Latin-speaking Victor* at the end of the second century. Apart from Paul and Peter, five names are given for the first century, nine for the second, and seventeen (including two antipopes) until the reign of Diocletian.* Of these the two dissidents, Hippolytus* and Novatian,* are best known, though to the first-century Clement* is ascribed a near-canonical epistle.

Those decisions of Roman policy made about Christianity are also a reflection of the currents constituting the other side of the *Pax Romana;* even Augustus had not finally solved the larger human problem inherited from the Republic. Under at least six of the rulers of the first phase of the succession, Christianity was beset by some governmental opposition, with increasing severity as the Flavian century wore on. Such interaction between church and state inspired the apologetic series of Christian literature.

The second phase of the succession from the Severi to Diocletian involves no less than twenty-eight imperial claimants, of whom only one (Septimus Severus) died a natural death, and only one other (Alexander Severus) reigned more than ten years. Nevertheless only four of the names (if one excludes Diocletian) are identified as persecutors. In this phase the church as institution emerged with properties, some at least above the subterranean level of the catacombed cemeteries—all of which lay outside the walls of the city (see CATACOMBS). But of the estimated 450 older churches now within urban Rome, not one survives as a pre-Constantinian structure in whole or part, though it may be definitely asserted that such "house-churches" existed in this phase, as both the interrogation of martyrs and imperial rescripts of restitution bear witness. That pre-Constantinian imperial public buildings were later converted into churches of the post-Constantinian state only reflects the subsequent changed relationship of church and state.

Diocletian's reorganization of the Empire, though attacked by Lactantius* as a result of the emperor's renewed assault upon the highly organized church, was a genuine effort to salvage the state by terminating military anarchy and recognizing East-West division. That the main beneficiary in time to come was the church reflects the debt owed to him. While Rome remained capital of the Western portion, the division posed the problem of retaining unity for a state with two centers. Constantine* was committed to Diocletian's genuine reform program, and at his death there emerged the dual state with two capitals after his creation of the "New Rome" in the East. Constantinople* was built from the ground up as a Christian version of the imperial capital.

While Diocletian and Constantine might restore the internal stability by economic measures and political reorganization, nothing could alter the developing situation beyond the frontiers of the fourth-century empire. The increased Germanic pressures on the Rhine-Danube line were matched beyond the Euphrates by the resurgence of a new Sassanid Persia. Rome itself as capital became less advantageous, and once it had proved vulnerable to Visigothic assault in A.D. 410 it was not long before all vestige of imperial government was withdrawn to safer terrain less distant from the fronts. That the sacking of the city after 800 years proved critical for a state newly allied to Christianity is borne out by Augustine's *City of God.* But that the city, now grossly reduced in dignity and size—estimates suggest a fourth-century population as low as half a million —should become instead the Christian Rome is but a reflection of its bishop being the sole remaining officer of rank within, when first the imperial government was transferred north, and then in 476 officially ceased to exist.

The symbolism of Rome's change of hands is best illustrated in its bishop, Leo, assuming the defunct title *pontifex maximus.* The Christian Church was thus successor to the ancient religion which was the city, and in such terms the primacy of the Church of Rome came to be declared. Thereby also the city was prepared to survive, though barely, the sequence of besieging disasters that followed: in 455 at Visigothic hands; in 537-38, 546, and 549 in the wars between Ostrogoths and Byzantines, both equally destructive of the city. The latter events saw the city become at one point uninhabited and her aqueducts destroyed (her water supply unrenewed until the sixteenth century). The Saracen threat of 846 saw Leo IV enwall the Vatican, where the circus of Gaius and gardens of Nero had given way to Constantine's basilica dedicated to Peter—the center of a Christian Rome.

BIBLIOGRAPHY: S. Dill, *Roman Society in the Last Century of the Western Empire* (1899; rep. 1958); R. Syme, *The Roman Revolution* (1939); C.N. Cochrane, *Christianity and Classical Culture: A Study of Thought and Action from Augustus to Augustine* (1944); A.H.M. Jones, *The Later Roman Empire, 284-602* (2 vols., 1964); A. Alföldi, *Early Rome and the Latins* (1965); M. Sharp, *A Guide to the Churches of Rome* (1966); G. Barraclough, *The Medieval Papacy* (1968); R.M. Grant, *Augustus to Constantine: The Thrust of the Christian Movement into the Roman World* (1970). CLYDE CURRY SMITH

ROSARY. In Roman Catholicism a Marian prayer composed of fifteen decades of "Hail Marys," with an "Our Father" preceding and a "Glory be to the Father" following each decade. The recitation is accompanied by meditation upon the fifteen mysteries pertaining to the joys, the sufferings, and the glories of Christ and the Virgin. In Romanism it is a very characteristic aspect of devotion to the Madonna. Existence of the Rosary is attested to by the twelfth and thirteenth centuries among the monks and nuns of the Cistercian* and Mendicant* orders. Later, use of the chaplet (endless string of beads) and the meditations were added to the vocal recitations. Pius V established the formula of the Rosary as it is known today. The feast of the Rosary (7 October) commemorates the victory of Charles VI over the Turks at Lepanto in 1571. The victory was attributed to the intervention of the Virgin. ROYAL L. PECK

ROSCELLINUS (c.1050-1125). Regarded as the founder of medieval Nominalism.* He was probably born at Compiègne and studied at Soissons and Reims. He then taught at Compiègne, and in 1902 was accused of Tritheism at the Synod of Soissons. Although he denied the charge, the synod ordered him to recant and he complied. He then went to England, where he came into conflict with Anselm and returned to France. He became a canon at Bayeaux and taught at Loches, where he again became involved in controversy. He was once again teaching his original position on the Trinity, and Peter Abelard,* one of his students, attacked the position and defended the unity of God in the Trinity. A letter defending himself is the only one of Roscellinus's writings that has survived; therefore his philosophical and theological teachings are known primarily from descriptions of his enemies: Anselm,* Abelard, and John of Salisbury.* His reasoning applied to theology implied that the three persons of the Trinity were three separate Gods. A Nominalist, he does appear, however, to have tried to preserve the unity of the Trinity by maintaining that the three persons had one will and power.

RUDOLPH HEINZE

ROSE, HUGH JAMES (1795-1838). Pre-Tractarian High Churchman. Born at Little Horstead, Norfolk, and educated at Uckfield and Cambridge, he was ordained priest in 1819, served in several parishes, was professor of divinity at Durham (1833-34) and principal of King's College, London (1836-38). Rose was a NT scholar and a member of the group of early nineteenth-century High Churchmen known as the "Clapton Sect" or "Hackney Phalanx." In 1824 he traveled in Germany and wrote a book exposing the dangerous trends of German radical criticism, for which he was, surprisingly, criticized by Pusey. In 1832 he founded the *British Magazine,* in which Newman's and Keble's "Lyra Apostolica" were first published. A meeting at Rose's Hadleigh Rectory in 1833 is rightly said to mark the beginning of the Tractarian* Movement. Though critical of some of the latter's tendencies, he managed to hold the Oxford and Clapton schools together until his death. IAN SELLERS

ROSENMÜLLER, ERNST FRIEDRICH KARL (1768-1835). German biblical scholar. Born in Hessberg, son of an Evangelical Lutheran pastor, he was educated at Leipzig and taught there from 1792, becoming professor of oriental languages in 1813. He prepared in sixteen parts *Scholia* for the OT (1788-1817), drawing together the insights from the writings of rabbis, Church Fathers, and medieval and Reformation scholars. He published also (1823-31) a handbook to the natural history of the biblical world. Living on the verge of the rediscovery of the ancient Near East, he contributed significantly to the accumulation of knowledge and interpretation which preceded that event, placing it within the restored framework of a concern for the literal meaning of the text. CLYDE CURRY SMITH

ROSE OF LIMA (1586-1617). Peruvian ascetic. Baptized "Isabel de Flores," she later took the name "Rosa de Santa Maria" for the special place of the Virgin Mary in her life. Catherine of Siena* was her model. Enduring mortifications from childhood and forbidden to leave her home, she joined the Third Order of St. Dominic at twenty. Living in a hermitage in the garden, she had an infirmary in the house to care for the destitute children in Peru. Widely known for her mystical gifts and admired by all classes in Lima, she was noticed but untouched by the Inquisition. She prophesied her death exactly, and has been proclaimed patron saint of Peru, all America, the Indies, and the Philippines. She was canonized by Clement X in 1671. C.G. THORNE, JR.

ROSICRUCIANS. One must distinguish between Rosicrucian ideas and Rosicrucian societies. The name derives from Christian Rosenkreuz (Rosy-cross), who is probably an allegorical figure (c.1378-1484). Between 1614 and 1616 four pamphlets described his travels in the East and his initiation into occult secrets. A Lutheran pastor, J.V. Andreae,* may have been the author of one or more. The pamphlets hinted at a Rosicrucian Fraternity with supernormal powers, but no such society can be traced, until in the eighteenth century several Rosicrucian groups were formed in Germany, Russia, and Poland. They were closely associated with Freemasonry,* and Masonic lodges still have an optional degree, established in 1845, known as the Rose Croix of Heredom, which includes the candidate's symbolic death and resurrection. In Britain this rite is Trinitarian and includes the reading of Isaiah 53. The American rite is given a wholly pagan interpretation.

The Societas Rosicruciana in Anglia was founded by Wentworth Little in 1865. Although not recognized by Masonry, this society limits its eight degrees to Masons only. Early in the present century H. Spencer Lewis founded the Ancient Mystical Order Rosae Crucis (AMORC), which now is centered in California and advertises postal courses in good-class periodicals. Its publications include the "lost" *Book of Jashar,* an obvious forgery compiled by Jacob Ilive in 1751, and also a book of the *Secret Teachings of Jesus.* Its chief rival is the group associated with the late Max Heindel, also based in California.

Although there are secrets for initiates only, the indications are that they are gnostic-theosophical. An AMORC brochure suggests that it concentrates on developing psychic powers. Max Heindel has publications on the Rosicrucian Cosmo-Conception, dealing with the world's past evolution under the guidance of great creative hierarchies, and including lost civilizations of Lemuria and Atlantis. Reincarnations provide fresh experiences, rather than the working out of *karma,* as Theosophy holds. The cross is not the symbol of Christ's atonement, but represents the human body. In the rose at the center there is a chaste and pure vital fluid to overcome the passion-filled blood of the human race.

BIBLIOGRAPHY: M. Heindel, *The Rosicrucian Cosmo-Conception* (1909); A.E. Waite, *The Brotherhood of the Rosy Cross* (1924); H.S. Lewis, *Rosicrucian Questions and Answers* (1932); J.K. Van Baalen, *The Chaos of Cults* (1956); F. King, *Ritual Magic in England* (1970).

J. STAFFORD WRIGHT

ROSMINI, ANTONIO (1797-1855). Also "Rosmini-Serbati." Italian philosopher and founder of the order later known as the *Suore della Providenza* ("Sisters of Providence"). Born in Rovereto Trentino, he was ordained priest in 1821. Despite early conservative views, he became the author of many significant progressive works dealing with various aspects of philosophy, politics, law, economics, and natural science. His treatise *Delle cinque piaghe della Santa Chiesa,* which he wrote after the election of Pius IX in 1846, denounces the insufficient education of the clergy, the divisions among bishops, and the riches of the church, and advocates its return to apostolic poverty and freedom. He laments, moreover, the division between clergy and lay people who are the vital part of the church, and proposes a far greater participation in worship and other church matters. With Gioberti* he hoped for a confederation of Italian states under the leadership of the pope, whose temporal power he considered essential to guarantee his freedom. DAISY RONCO

ROSMINIANS. The Institute of Charity founded by Antonio Rosmini-Serbati,* the Italian philosopher, at the instigation of Maddalena Canossa (1828). It was formally approved by Gregory XVI (1839). Members profess the three religious vows and live retired in prayer and study until called by the pope or by a particular need to some kind of external work such as teaching, preaching, missions, and literature. There are two grades: the presbyters (who take a fourth vow of obedience to the pope) and the coadjutors. There is no distinctive habit, only the cassock. In 1832 the congregation became associated with the Sisters of Providence, founded by one of Rosmini's disciples. The Rosminians went to England in 1835, introducing there the clerical (or Roman) collar and other innovations. There are houses also in Eire and the USA. The central house of the Institute is at St. John at the Latin Gate, Rome.

J.G.G. NORMAN

ROSSETTI, CHRISTINA GEORGINA (1830-1894). Anglican poetess. Daughter of a Dante scholar and sister of two Pre-Raphaelites, she was educated at home and later helped her mother run a school. A High Anglican, she rejected two suitors on religious grounds. These disappointments and ill-health, which led to a secluded life, intensified her religious temperament. She published several books in prose for the Society for Promoting Christian Knowledge. Her books of verse, beginning with *Goblin Market, and Other Poems* (1862) and ending with *New Poems* (1896), are rich in devotional feeling. Her sonnet sequences, *Monna Innominata* and *Later Life,* exalt divine over human love. Some of her poems are sung as hymns, e.g., "In the bleak midwinter"; "Love came down at Christmas"; and "None other Lamb, none other Name."

JOHN S. ANDREWS

ROSSETTI, TEODORICO PIETROCOLA, see PIETROCOLA-ROSSETTI

ROTHMAN, BERNT (c.1495-1535). German Anabaptist* leader. A priest of obscure origins and a powerful preacher, he was the chief evangelical Reformer of his native town of Münster. He introduced a Lutheran-style Reformation into Münster in 1532-33 in spite of vigorous official opposition. However, the course of reform in Münster took an unexpected twist when in May 1533 Rothman became an Anabaptist. Events took yet another turn in early 1534 when large numbers of Melchiorities, including the charismatic Dutchmen Jan Matthijs (d.1534) and Jan van Leyden (d.1535), flocked into the city. Rothman then changed his religious orientation once more and accepted Matthijs's radical Melchiorite millennialism. In February 1534 the radicals took over Münster in order to make ready for "the kingdom of God" to be established there shortly at Christ's second coming. First Matthijs and then van Leyden became the dictator of a theocratic state in which both communism and polygamy were introduced. Rothman seems to have been swept along with the tide and served as state preacher. When the ill-fated Münster kingdom fell to a besieging force in June 1535, Rothman reportedly died in the fighting. He left behind several books written at stages in his theological pilgrimage from Rome to radical millenarianism, including his *Restitution* (1534). ROBERT D. LINDER

ROUAULT, GEORGES (1871-1958). French painter. Born in Paris, he helped for a time on the restoration of the stained glass windows in Chartres Cathedral. He never worked from the Renaissance ideal of pleasing proportion, but reached back, in a fully modern artist's way, to a medieval awareness of suffering, death, and painterly decoration. His apprenticeship in stained glass carried over into his studies of prostitutes, judges, and clowns, where glowing colors of hopeful rage are closed in black-bordered compartments so that the ugliness of sin, the brutality of corruption, the bitterness of human misery come through firmly with a horrified and compassionate understanding. Rouault's black India ink drawings begun around World War I made into prints, *Miserere (et Guerre),* are a moving series of black and white penitential tears at the ruinousness of war and human inhumanity; the stark, controlled sprawl of lines have a Hebraic roughness, solitary grief muted by dignity. Many of his Christ figures have a mustard-olive, pasty, or bloodied character; but later in life red, green, and yellow friendlier colors come gently and serenely more to the fore, as in the joyous, smiling *Sarah* (1956), an earnest of the resurrection. But there is always something Jansenist* about Rouault, both sensuous and austere, disquieting and soaring, earthy and mystical together, kindred in spirit to the music of his French compatriot Oliver Messiaen.

CALVIN SEERVELD

ROUSSEAU, JEAN-JACQUES (1712-1778). French writer and philosopher. Born in Geneva and raised a Calvinist, he converted to Catholicism in 1728. A former Pietist and his benefactress for the next decade, the French Madame de Warens contributed to the formation of his religious outlook, a Deism tempered by Quietist sentimentalism. After immersing himself in contemporary philosophical literature, Rousseau went to Paris in 1742. Denis Diderot* introduced him to the circle of the *philosophes,* and he contributed several articles on music to the *Encyclopedia.* Around 1746 he took Thérèse Levasseur, a servant girl, as his common-law wife. They had five children, all of whom he placed in an orphanage. Rousseau made his literary debut in 1750 with a prize-winning essay submitted to the Dijon academy. This *Discourse on the Sciences and Arts* contended that public morals and character were corrupted by the progress of knowledge and art. Virtue could be found only in simplicity, when man lives close to nature. After returning to Geneva and the Calvinist faith, he produced his "second discourse" in 1755, an unsuccessful contest entry. It argued that inequality among men was the result of organized society. The natural man was free and happy, but class divisions and despotism arose from land ownership and laws.

Settling in Montmorency in 1756, Rousseau composed three of the eighteenth century's most influential books. In *Julie, or the New Héloïse* (1760), a passionate love story, he popularized the idea of irresistible love and the beauty of nature, and propounded a natural religion necessary for morality. In *Emile* (1762), a treatise on education in novel form, he set forth a pedagogical scheme to protect man, inherently good, from the corrupting influences of society. The program's religious aspect was summarized in the chapter on the vicar from Savoy, an unfrocked priest who advocated a sentimental Deism. This included a belief in the existence of God (whose law is written in the conscience) and the immortality of the soul, but no eternal punishment. *The Social Contract* (1762) contained Rousseau's concept of a just state. The free man voluntarily surrenders his will to the community and submits to its laws which are based on the general will of the people. Particularly significant is his idea of the civil religion, a civic faith necessary for a government's stability. Although the regime fixes its doctrines and they are binding on all citizens, other religions are permitted, if they do not claim absolute truth.

After his works were officially condemned in 1762, he spent the next eight years wandering. In 1765 he produced the *Confessions,* a curious autobiographical mixture of vanity and self-accusation, and a number of other literary works in his last years upon his return to Paris.

Rousseau's emphasis upon irrationality, subjectivism, and sensualism made him the forerunner of both Romanticism and modern totalitarianism, while his elevation of the individual above society contributed to individualism and democratic thought. By substituting a sentimental faith for revealed religion and by removing Christian doctrines from their supernatural context, he paved the way for humanistic liberalism.

BIBLIOGRAPHY: *Oeuvres complètes* (13 vols., 1874-87); *Correspondance complète* (12 vols., 1965-70); I. Babbitt, *Rousseau and Romanticism* (1947); J.H. Broome, *Rousseau, A Study of His Thought* (1963); J. Guéhenno, *Jean-Jacques Rousseau* (2 vols., 1966); W.H. Blanchard, *Rousseau and the Spirit of Revolt: A Psychological Study* (1967); M. Einaudi, *The Early Rousseau* (1967); R.D. Masters, *The Political Philosophy of Rousseau* (1968); L.G. Crocker, *Jean-Jacques Rousseau* (1968); J.N. Sklar, *Men and Citizens: A Study of Rousseau's Social Theory* (1969).

RICHARD V. PIERARD

ROUTH, MARTIN JOSEPH (1755-1854). English patristic scholar. Most of his life he was associated with Magdalen College, Oxford, of which he became president in 1791, an office held until his death. He was made deacon in 1777, but not priested until 1810. Routh was a patristic scholar much revered by the Tractarians.* When Samuel Seabury* sought episcopal succession for the American Church, Routh advised him to approach the Scottish Episcopal Church. Descended from a niece of Archbishop Laud, theologically he linked the outlook of the Nonjurors* and Caroline Divines with the Oxford Movement,* to which he gave support. He opposed the Hampden appointment, and also the censuring of Tract 90. JOHN A. SIMPSON

ROWLAND, DANIEL (1713-1790). Welsh Methodist leader. With Howel Harris* he has the distinction of being co-founder of Welsh Calvinistic Methodism.* He was born at Pantybeudy, Cardiganshire, the son of Daniel Rowland, parish priest of Nantgwnlle and Llangeitho. Information about his youth and education is scanty. He was ordained deacon (1734) and priest (1735) and served as curate to his brother, John, at the parishes served by their father. He was converted under the ministry of Griffith Jones of Llanddowror and began a preaching ministry of great power in the neighborhood of Llangeitho. He soon extended his labors beyond his own parishes and made contact with Howel Harris in 1737. Like Harris he began to found "societies" where his converts could be established in their newfound faith.

For a number of complex reasons, the Calvinistic Methodists in Wales split into two groups in 1752, Harris leading one group and Rowland the other. Ten years passed before reconciliation was effected. Meanwhile Llangeitho became the national center of Welsh Methodism since Harris had withdrawn from public work for a time. People traveled from all parts of Wales to hear Rowland preach and to receive Holy Communion at his hands. His position in the church was an anomalous one. When his brother was drowned in 1760, the authorities passed over Daniel Rowland and gave the living to his son, to whom he now became curate. But he was finally dispossessed in 1763 and continued his ministry in the "New Church" that had been built for him at Llangeitho. Rowland was both a hymnwriter and an author, but above all he was a preacher whose influence extended all over Wales. His sweetness of spirit and the magnetism of his delivery kept congregations spellbound—sometimes for hours on end—as he expounded the Gospel to them at Llangeitho. R. TUDUR JONES

ROWLEY, HAROLD HENRY (1890-1969). English Semitist and biblical scholar. Educated at Bristol and Oxford universities, he became successively minister of the United Church, Wells, Somerset (1917-22); professor in Shantung Christian University, China (1924-29); lecturer in Hebrew in University College, Cardiff (1930-34); professor of Hebrew in University College, Bangor, North Wales (1935-45), and in Manchester University (1945-59). In addition to his many published contributions to scholarship (among which his studies of the linguistic and critical problems of the Book of Daniel take preeminent place), which were characterized by judicial wisdom and comprehensive and exact bibliographical knowledge, he excelled himself at the end of World War II by his efforts to restore international correspondence and cooperation among OT scholars. In the postwar years, as foreign secretary of the (British) Society for Old Testament Study, he gained worldwide esteem immense and well-merited. F.F. BRUCE

ROWNTREE, JOSEPH (1801-1859). Quaker social reformer. Born in Yorkshire, he left school at thirteen, later becoming a grocer and member of the Merchants' Company, with a lifelong interest in education. He was a founder of the York Quarterly Meeting Boys' and Girls' Schools (1828, 1830) at Bootham and The Mount, and of the Friends' School at Rawdon (1832) for children of a different class. A founding trustee of the Flounders' Institute, Ackworth, for training teachers, he with Samuel Tuke helped establish the Friends' Educational Society (1837) and served on the committee of the Friends' Retreat for the insane at York. An alderman from 1853, he declined the mayorship of York in 1858 on conscience. He inaugurated several schemes for municipal reform, wrote pamphlets on colonial slavery and on education, and helped reform the marriage regulations of the Society of Friends, so that marriage to a non-Quaker would no longer mean disownment. C.G. THORNE, JR.

ROYAL SOCIETY. On 15 July 1662 the young society was incorporated by a royal charter; further charters, extending its privileges, followed in

1663 and 1669. The origin of the society was in the meetings for scientific experiment and study held in Wadham College, Oxford, during the Commonwealth and Protectorate. At the Restoration the meetings were transferred to London. During the seventeenth century virtually all the members were orthodox Christians. Since the presidency of Isaac Newton (1703-27) the relationship with orthodox Christianity has diminished. The Royal Society is the national academy for science for Great Britain, but it is not controlled by the government. Over the years, however, it has advised governments on many matters (e.g., change of calendar in 1751) and conducted surveys for governments. Its headquarters are in Somerset House, London, where its massive library and historical museum are located.

PETER TOON

RUBENS, PETER PAUL (1577-1640). Flemish painter. He glorifies whatever he paints with a passionate, red-blooded, lusty life of adventure. The mock heroic trappings of classical mythology, the Baroque bravura of superabundance, brilliant color, intense light, swirling lines of movement: the Rubens style of grand turbulence exalts man, woman, flower, and tree into being bigger than life. Nothing ordinary is conceivable in a Rubens canvas. His women figures have the weight and power of Michelangelo bodies softened by warm, Titian tints, all reconceived and heightened by a buxom exuberance peculiar to the Flemish master. Motifs of imaginary reality, historical occurrence, and an Italian Renaissance sense of the timeless past swim together richly in Rubens's composition and cast the spell of an empyreal Nature. The spirit of the paintings, no matter it be Christ on the cross, is one of Titanic desire, unchristened naïveté, and pure human nobility.

CALVIN SEERVELD

RUFINUS "THE SYRIAN" (late fourth/early fifth century). Palestinian presbyter, author of a *Treatise on the Faith* (*Liber de Fide,* first published 1650). He was probably the presbyter Jerome* sent from Bethlehem to Milan in 399 on a legal mission. His *Treatise,* written c.400 in Rome, like Jerome condemns Traducianism* and especially Origenism.* Marius Mercator* depicts him as the Syrian inspirer of Pelagius, but he more clearly influenced Celestius,* who appealed to his denial of transmission of sin at his trial in Carthage in 411, and similarly taught that infants were baptized not for forgiveness but to inherit the kingdom. Unlike Celestius, he held that Adam, though born mortal, would not have died had he remained sinless (like some persons in the OT). The *Treatise,* possibly implicitly critical of Augustine's published opinions, was attacked in the latter's *The Merits and Remission of Sins.* Rufinus perhaps also wrote a markedly anti-Origenist *Pamphlet on the Faith (Libellus de Fide),* a defense against his alleged errors, which included a form of perfectionism. His Syrian affiliations, though not explicit, appear reasonably identifiable. He probably died before the Pelagian* storm broke. D.F. WRIGHT

RUFINUS TYRANNIUS (345-410). Presbyter and scholar. Born near Aquileia in N Italy, he was at school in Rome when he began a long association with Jerome.* He went to Egypt, suffered in the persecution which followed the death of Athanasius (373), settled in Alexandria for eight years, studied Scripture and Origen under Didymus the Blind* and Gregory of Nazianzus,* and visited the Desert Fathers. In 381 he founded a monastery in Jerusalem. His many translations of Greek theological works into Latin aided Western asceticism and theology. The works included Basil's Rule for the monks at Pinetum, Gregory of Nazianzus, and Eusebius with additions, as well as his own commentary on the Apostles' Creed with the earliest Latin text of its fourth-century form. His rendering of Origen's *De Principiis,* simultaneously vindicating its orthodoxy, led to damaging strife with Jerome. C.G. THORNE, JR.

RUINART, THIERRY (1657-1709). French Benedictine scholar. Born at Reims, he studied in his native city before entering the Benedictine abbey of St.-Remi in 1674. Eventually he went to the monastery of St.-Germain-des-Prés, the great center of Maurist* learning. Here he became a pupil and friend of Jean Mabillon, with whom he collaborated in producing volumes VIII and IX of *Acta sanctorum ordinis sancti Benedicti* (1701). Ruinart's quiet life in the monastery enabled him to pursue his academic aims, and among other books he wrote *A Life of Jean Mabillon, A Life of Pope Urban II,* and *An Apology for the Mission of St. Maurus.*

RUMANIA, see ROMANIA

RUOTSALAINEN, PAAVO (1777-1852). Finnish Pietist leader. Son of a farmer, he already in his youth experienced a personal revival in a Pietistic spirit. His spiritual discernment developed only later through a blacksmith counselor who told him: "Lacking one thing, you lack everything—the inner knowledge of Christ." This sentence has become a motto for the Pietistic Movement. He appears as one of the greatest leaders in Finland, succeeding in uniting two branches of the Pietistic revival there, and developing what is now the biggest and perhaps the most significant movement in Finland which still influences large groups of people. In their regular annual summer meetings outdoors some 20,000-30,000 people gather. Ruotsalainen has influenced the Christian life in Finland more than anyone else in that country's modern church history.

STIG-OLOF FERNSTROM

RUPERT OF DEUTZ (c.1075-1129). Medieval theologian and exegete. Born in Germany, he early entered the Benedictine monastery of St. Laurence at Liège. He wrote and taught at Liège and Siegburg before being appointed abbot of Deutz near Cologne about 1120. He was strongly opposed to simony, to strict predestinarianism, and to the introduction of logic into theology. Possessing no formal theological system, he defended the mystical theology traditionally held by the Benedictines, especially against the dialectic methods of Anselm of Laon* and William of Champeaux.* He was a noted scriptural exegete, writing several commentaries and other works, in which he interpreted the Bible literally, allegorically, and morally. He developed an Augustinian theology of history. His imprecise language has led to conflicting interpretations of his doctrine of the Eucharist.

ALBERT H. FREUNDT, JR.

RURAL DEAN. Each diocese in the Church of England is divided into deaneries. The rural dean is the bishop's deputy in the deanery. He is sometimes elected by the clergy, but more often chosen by the bishop. The office in England dates back to the eleventh century, although its modern form largely stems from the nineteenth century. His chief duty is to preside over the chapter (the clergy in the deanery) and at rural deanery conferences of clergy and laity. With synodical government he is a joint chairman of the deanery conference (clergy and lay), and in some dioceses new titles are being introduced for the office since particularly in large towns the title "rural" is inapposite. Other functions include an annual inspection and seeing that provision is made for services in parishes where for whatever reason there is no minister. PETER S. DAWES

RUSKIN, JOHN (1819-1900). Victorian author and critic. Born into a wealthy Evangelical family, he trained early for the ministry. His memorization of large portions of the Bible affected his opinions and tastes permanently. His father, a wine merchant, took his son on continental tours and introduced him to beautiful landscapes, architecture, and art, which inspired in him a profound love of beauty. While at Oxford (1836-40) he won the Newdigate Prize for poetry and gave up his ministerial ambitions. The first period of his career was devoted to problems of art. *Modern Painters* (2 vols., 1843-46) was begun as a defense of the painter J.M.W. Turner; it accepted the principle that art is based on national and individual integrity and morality. Ruskin also defended the work of the Pre-Raphaelites. His interest in architecture is seen in *The Seven Lamps of Architecture* (1849) and *The Stones of Venice* (1851-53), the latter maintaining that the Gothic architecture of Venice reflected national and domestic virtue, while Venetian Renaissance mirrored corruption.

Despite recurring mental illness during his last years, accentuated in part by an indifferent public and unhappy personal life, Ruskin wrote without much study in the area of political economy: *The Political Economy of Art* (1857), *Unto This Last* (1860), and *Munera Pulveris* (1862-67), which attacked the ugliness of industrial England. Other works include *Sesame and Lilies* (1865), *The Crown of Wild Olive* (1866), six volumes of his lectures on art delivered while he was Slade professor of fine arts at Oxford (1869-79, 1883-84), *Fors Clavigera* (letters to workmen), and his autobiography to *Praeterita* (1871-74).

His works were to influence the British Labour Party and such writers as William Morris, George Bernard Shaw, and D.H. Lawrence.

BIBLIOGRAPHY: E.T. Cook and A. Wedderburn, *The Works of John Ruskin* (39 vols., 1903-12); E.T. Cook, *The Life of Ruskin* (2 vols., 1911); J. Evans, *The Lamp of Beauty: Writings on Art by John Ruskin* (1958); J.D. Rosenburg, *The Darkening Glass* (1961). ERWIN RUDOLPH

RUSSELL, CHARLES TAZE (1852-1916). Founder of Jehovah's Witnesses.* As a young man he built up a chain of drapery shops in Allegheny, Pennsylvania. He reacted against doctrines of hell and was attracted by date-fixing for the Second Coming, which he estimated first as 1874, then as 1914. He caricatured the Christian doctrine of the Trinity as "three gods in one person" and held that Christ was the first created being. In 1879 he launched a magazine, *Zion's Watchtower and Herald of Christ's Presence.* In 1884 he set up Zion's Watchtower Tract Society in Pittsburgh, and this publishing house produced Russell's six volumes of *Studies in the Scriptures* (1886-1904). These set out what has remained basically the Jehovah's Witnesses' theology. A seventh volume on the Book of Revelation was completed by others in 1917 after Russell's death, and Arius and Russell are called two of the angels of the Seven Churches.

Russell's wife left him in 1897 and obtained a legal separation in 1906. Neither party asked for divorce. In 1911 Russell advertised so-called "Miracle Wheat" in his magazine, to be sold in aid of the society's funds. The *Brooklyn Eagle* challenged his claims; Russell sued for libel and lost. He lost also a libel case against the Rev. J.J. Ross, who attacked his doctrines and scholarship. During the hearing Russell committed perjury by asserting under oath that he knew the Greek alphabet, whereas he could not name the letters when he was shown them in court.

He traveled widely and was an able writer and speaker. His sermons were syndicated in some 1,500 newspapers. He produced a photo-drama of the Bible story, with colored slides, films, and synchronized gramophone records. It lasted eight hours (four parts of two hours each) and included twelve talks by Russell.

See also JEHOVAH'S WITNESSES; RUTHERFORD, J.F. J. STAFFORD WRIGHT

RUSSIA. Although Christianity was introduced into Russian lands in the first century A.D. and had some success, it was not lasting. Very little penetration of Christianity took place in the next 700 years until in the ninth century the Church of St. Elias was established at Kiev, where there was a minority of Christians. In 867 the first Russian metropolitanate was set up by the patriarch of Constantinople and was probably located at Tmutorakan.

The first Kievan ruler to accept Christianity is usually considered to be Olga, who was regent between 945 and 964. The traditional account states that she was converted and baptized in Constantinople, refused an imperial offer of marriage, and returned to Kiev where her faith had little influence upon either her son or her people. Because of problems in reconciling the Byzantine records with the Russian *Chronicle*, many historians believe Olga was probably converted in Kiev around 955 and went to Constantinople in 957 to plead for autonomy for the Russian Church. Because she did not gain what she sought, the pagan party remained in power in Kiev under her son Sviatoslav.

Vladimir,* the grandson of Olga who reigned from 978 to 1015, was the man responsible for the Christianizing of Russia. After a strong pagan revival, the story is told of Vladimir's sending out missions to study the religions of Judaism, Islam, Roman Christianity, and Greek Christianity. His acceptence of Greek Christianity was supposedly on the basis of the beauty of its worship. It could be argued, however, that political and economic ties with Constantinople were advantageous to the Kievan state at this time. Included in the agreement of conversion for Vladimir was his marriage to Anna, the sister of the Byzantine emperor. In 988 Christianity was proclaimed the official faith of the realm, and baptism was ordered for Vladimir's subjects. The upper classes and those in the cities accepted the faith, but only slowly did it penetrate the lower classes and the countryside, which remained pagan until the fourteenth and fifteenth centuries. Orthodoxy became the state religion and remained so until 1917. The church was under the leadership of the metropolitan of Kiev until the fourteenth century, when the leadership changed to Moscow. The Russian Church also remained subject to Constantinople during the Kievan period, and the metropolitans chosen for the Russians were usually Greek.

Monasticism played an important role in early Russian Christianity. The most significant monastery was the Monastery of the Caves in Kiev. It was founded in the eleventh century by Antony and then reorganized by his successor, Theodosius, who emphasized poverty and humility.

The Mongol invasions of Kievan Russia started in 1237, and by 1240 Russia was under the Tartar yoke. Christianity survived, and the church helped to keep alive Russian national consciousness. Three men, all canonized by the church, are significant in the Mongol period. Alexander Nevsky, the victor over the Swedes and Teutonic Knights in the 1240s, is credited with saving the church from the papacy. Stephen of Perm in the fourteenth century became a missionary to the Zyrian tribes, and Sergius of Radonezh established the Monastery of the Holy Trinity, which became the greatest religious house in the land. Sergius encouraged resistance to the Mongols in the fourteenth century and helped the advance of Moscow by inspiring colonist monks to go into the forest regions.

With the rise of Moscow, Peter, the metropolitan from 1308 to 1326, moved the seat of the church there. After the fall of Constantinople to the Turks in 1453, Moscow increasingly advanced the claim of being the "third Rome." Ivan III's marriage to Sophia, niece of the last Byzantine emperor, helped to enhance that claim. After 1448 a council of Russian bishops elected the metropolitan. The Mongols were finally defeated in 1480, and a powerful Russian state developed under Ivan IV.

At a church council in 1503 a dispute between Nilus of Sora and Joseph, abbot of Volokalamsk, resulted in the formation of two groups. The Possessors, Joseph's followers, emphasized the social obligations of monasticism in caring for the sick and poor. The Non-Possessors, Nilus's followers, insisted that almsgiving is the duty of the laity, while a monk's major work is prayer and detachment from the world. The Possessors supported the ideal of the "third Rome" and believed in a close alliance between church and state. Their victory eventually led to a great subservience of the church to state, especially during the reign of Ivan IV. In 1589 the head of the Russian Church was elevated from the rank of metropolitan to patriarch.

Following the "Time of Troubles," the election of Michael Romanov as ruler in 1613 initiated a dynasty that was to reign until World War I. Church reforms were started by Abbot Dionysius, Philaret,* and Avvakum.* Then in 1652, during the reign of Czar Alexis, the new patriarch, Nikon,* attempted to reform the Russian Church by bringing it in line with the ideas of the four ancient patriarchates. There was great opposition to the reforms, especially among those of the Josephite traditions who eventually formed a separate sect known as Old Believers.* They were the conservatives opposing an official church which they thought had carried reform too far. Old Believers still remain in Russia and are divided into the *Popovtsy* who have kept the priesthood and the *Bezpopovtsy* who have no priests. The church accepted Nikon's reforms, but he was deposed and exiled.

Under Peter the Great, no new patriarch was appointed when Adrian died in 1700. In 1721 Peter abolished the patriarchate and set up a Holy Synod composed of twelve members. Its members were nominated by the Czar, and thus the church became a department of state. This system of church government continued until 1917. The synodal period is often described as a period of decline and of Westernization, but others would say true Orthodox life continued, and in the nineteenth century there was a revival in the Russian Church. New enthusiasm for missionary work occurred. The religious renewal began at Mt. Athos, where a monk named Paissy laid emphasis upon continual prayer and obedience to an elder, or *starets.* This was the age of the *starets,* and the greatest of them was Seraphim of Sarov. He was followed by the elders of Optino. In theology Russia broke with the West, and Aleksei Khomyakov,* leader of the Slavophile circle, became the first original theologian of the Russian Church.

In 1917, after the abdication of Nicholas II, an all-Russian church council met in Moscow and began a program of church reform which eventually restored the patriarchate and elected Tikhon* to that office in November. After the Bolshevik Revolution, the Soviet government decreed the separation of "church from state, and school from church." Tikhon was arrested following his criticism of Communist policies. The "Living Church" was organized with Communist recognition and was used by the regime for political purposes. It convened a council which unfrocked Tikhon and abolished the patriarchate. Tikhon was released after promising not to oppose Soviet rule. Following Tikhon's death in 1925, Peter was named head of the church, but was exiled the next year. The League of the Militant Godless was formed in 1925 and grew rapidly, claiming a membership of five million in 1932. When Sergei succeeded Peter, the Soviet regime refused to recognize him until 1943.

Stalin attacked the church directly, closing churches and monasteries. The Soviet constitution of 1936 guaranteed freedom of religious *worship* but at the same time granted freedom of antireligious *propaganda.* With the coming of World War II the regime eased its antireligious campaigns and got the church to cooperate with the war effort. In May 1944 Sergei died, and Metropolitan Aleksei of Leningrad was elected patriarch early in 1945. No statistics are available as to the number of Christian believers still in Russia and as to the number involved in persecutions and purges.

Besides the Orthodox Church in Russia, other forms of Christianity have existed. Roman Catholicism entered Russia mainly from Poland* and was propagated by Jesuits and Dominicans. Protestantism entered from Germany in the sixteenth century and from France and Holland in the eighteenth century. Out of the Protestant groups and some of the dissenting sects native to Russia have come strong evangelical groups. There is some evidence to indicate that such evangelical groups have grown in numbers and religious zeal even under Communist restrictions.

See also EASTERN ORTHODOX CHURCHES.

BIBLIOGRAPHY: G. Vernadsky, *Kievan Russia* (1948); J.S. Curtiss, *The Russian Church and the Soviet State, 1917-1950* (1953); M.T. Florinsky, *Russia* (2 vols., 1953); *The Russian Primary Chronicle, Laurentian Text* (tr. and ed. by S.H. Cross and O.P. Sherbowitz, 1953); F.C. Conybeare, *Russian Dissenters* (1962); T. Ware, *The Orthodox Church* (1963); G.P. Fedotov, *The Russian Religious Mind* (2 vols., 1966).

BARBARA L. FAULKNER

RUSSIAN ORTHODOX CHURCH, see EASTERN ORTHODOX CHURCHES

RUTHENIAN CHURCHES. Name given to Uniate Churches* found mostly in Polish Galicia, Czechoslovakia, and Hungary, with colonies in North America. The name is simply a Latinized form of "Russian." Sometimes they are known as "Ukrainians," and occasionally as White Russians and Slovaks. Their ancestors, converts of Vladimir,* were part of the Russian Church under the metropolitan of Kiev until his expulsion after the Union of Florence (1443). Pope Pius II appointed a Roman Catholic metropolitan of Kiev (1485) who was permitted by Casimir IV of Poland to exercise jurisdiction over the eight eparchies of the province under the control of Poland and Lithuania. In the sixteenth century they reverted to Orthodoxy, but in 1595 the metropolitan of Kiev, with the bishops of Vladimir, Lutsk, Pololsk, Pinsk, and Kholn, sought communion with Rome, which was achieved by the Union of Brest-Litovsk (1595-96). They were joined by the bishops of Przemysl (1694) and Lvov (1700). Despite a decree of Urban VIII (1624), during the seventeenth century most of the nobility and landowners in Poland adopted the Latin Rite.

After the partition of Poland (1795), most of the Ruthenians (except in Galicia) passed under Russian control and were gradually suppressed in favor of Orthodoxy. In the Kholn district (ceded by Austria in 1815), they survived till c.1875. Byzantines were still illegal, so most of the survivors passed to the Latin Rite. The Ruthenians of Galicia under the sovereignty of Austria were separated politically from their metropolitan, so Lvov was constituted an archbishopric (1807) to look after the interests of Lvov and Przemysl. They enjoyed religious toleration during the nineteenth century, but political troubles in E Europe have since engendered ill-feeling between the Latin Poles and the Byzantine Ruthenians.

A Ruthenian college was founded in Rome by Leo XIII (1897) which since 1904 has been controlled by Ruthenian Basilian monks. There is a strong monastic element, especially fostered by the austere Studites (founded c.1900). The Ruthenian liturgy is based on the Byzantine Rite with certain modifications adopted from Rome.

The Podcarpathian Ruthenians are another Ruthenian community who were granted a separate jurisdiction by the setting-up of the eparchy of Mukachevo, subject to the primacy of Hungary by Clement XIV (1771). The eparchy was created to settle the dispute between the Ruthenian metropolitan north of the Carpathians and the settlement, dating from the fourteenth century, of Little and White Russians south of the Carpathians, who had been brought into communion with Rome by the Union of Uzhgorod (1646).

There are considerable Ruthenian communities in the USA, Canada, Brazil, and Argentina. The Ruthenians are the largest Uniate group, numbering about 4,500,000. Since 1946, those in the Ukraine have been separated from the Roman Catholics and aggregated to the Russian Orthodox Church.

See D. Attwater, *The Catholic Eastern Churches* (1935) and *The Christian Churches of the East* (2 vols., 1961-62). See also under RUSSIA and EASTERN ORTHODOX CHURCHES.

J.G.G. NORMAN

RUTHERFORD, JOSEPH FRANKLIN (1869-1942). Successor of C.T. Russell* as head of the Jehovah's Witnesses.* As a lawyer he had served as a special judge in Missouri. In June 1918 he and six others were sentenced to twenty years' imprisonment for propaganda against military service, but the sentence was quashed on appeal after nine months in prison. He wrote some twenty-two books and many booklets and laid the foundations of the directed dogmatic studies that characterize Jehovah's Witnesses. Russell had taught that Jesus Christ would return in 1914 or shortly afterward. Rutherford rallied his disappointed followers by discovering in 1921 that Christ had in fact returned invisibly in 1914 and had begun to purge his spiritual temple in 1918. Rutherford also "received" the new name for the Watchtower followers in 1931, i.e., "Jehovah's Witnesses." When Armageddon was delayed and Witnesses numbered more than the expected 144,000, he found a way of including a second class in God's future blessings. He maintained also that the "higher powers" of Romans 13:1 were Jehovah and Jesus Christ, and not earthly rulers—an interpretation which has been rejected by current Jehovah's Witnesses' publications. J. STAFFORD WRIGHT

RUTHERFORD, MARK. Pen name of William Hale White (1831-1913), English novelist. Born and educated at Bedford, he studied for the Congregational ministry, but was expelled for doctrinal deviation. He subsequently enjoyed a successful career in the civil service. Two of his novels describe his own spiritual history, *The Autobiography of Mark Rutherford* (1881) and *Mark Rutherford's Deliverance* (1885). He wrote also four others, *The Revolution in Tanner's Lane* (1887), *Miriam's Schooling* (1890), *Catherine Furze* (1894), and *Clara Hopgood* (1896). Rutherford's novels are poorly constructed and rather colorless in tone, but they convey a sympathetic and realistic view of the world of the petty bourgeoisie who made up the Dissenting congregations with which he dealt. He manages to bring out particularly well the impact in the mid- and late-nineteenth century of the new scientific ideas and higher criticism of the Bible on the fundamentalist attitudes of such people. *The Autobiography* deals with his own emancipation from Calvinism.

ARTHUR POLLARD

RUTHERFORD, SAMUEL (1600-1661). Scottish pastor and theologian. He was born of farming stock at Nisbet in Roxburghshire, and gave evidence of grace and of spiritual insight in boyhood; his mind was always sensitive to spiritual impressions. He entered Edinburgh University in 1617, graduated M.A. in 1621, and two years later after a competitive examination was appointed professor of Latin language and literature in the university. Some unpleasantness in his relations with his colleagues that may have been connected with his marriage led him to resign his office and study theology. He was ordained at Anwoth in Kirkcudbrightshire in 1627 and exercised a fruitful ministry there until 1636, when he was deposed from office for Nonconformity and ordered to be confined to prison at Aberdeen during the king's pleasure. From there came his famous *Letters* to former parishioners and friends at Anwoth. These 365 letters are classics in the field of devotional literature. Released from prison in 1638, he returned to Anwoth for eighteen months before being appointed professor of divinity at St. Andrews.

In 1643 he went to London as one of the Scottish commissioners to the Westminster Assembly* of Divines. His insight and devotion contributed significantly to the Confession and Catechisms. Two children died during his four years absence in London, and his experience of this sorrow enlarged his compassion for the sorrowing. During his time in London he was an industrious student and a prolific writer, largely on matters of church polity. His monumental work was *Lex Rex, or The Law and The Prince; a Dispute for the Just Prerogatives of King and People.* It dealt more with political science than theology and is still regarded as a classic on constitutional government. The Revolution Settlement of 1690 embodied the principles of *Lex Rex.*

In 1647 Rutherford was appointed principal of St. Mary's at St. Andrews, and later, rector of the university. He was preeminent in Scotland as a scholar and leader. He was well known on the Continent and in 1648 and 1651 declined appointments to Dutch universities. The Restoration of Charles II in 1660 put him in great peril. He was removed from office, but died on 29 March 1661 before the full fury of the storm of persecution broke. ADAM LOUGHRIDGE

RUYSBROECK, JAN VAN (1293-1381). Flemish mystic. Born near Brussels, and ordained a priest in 1317, he was for nearly three decades vicar at St. Gudele in Brussels. In 1344 he retired to the nearby wooded valley of Groenendaal, where after some years he founded an Augustinian monastery (1350). At Groenendaal he did most of his writing, and acted as spiritual adviser. Tauler,* Groote,* and many others came to him for advice on the life of the spirit. Ruysbroeck was one of the great fourteenth-century mystics. Written in powerful and often exalted Flemish prose, his writings helped shape the language. He wrote in the tradition of Augustine, Bernard, and perhaps especially Richard of St.-Victor.* The aim of mysticism is the union of the spirit with its Creator, by way of three stages: the active life, the inner life, and the final vision of God. His best-known work is *Die Chierheit der gheestelijke Brulocht* (tr. as *The Spiritual Espousals,* 1952). He opposed abuses in the church (writing during the "Babylonian Captivity" of the church of Avignon) and pantheistic versions of mysticism. He was beatified in 1908, and his feast is observed locally. His works were translated into Latin and influenced later mystics such as John of the Cross. A complete modern edition is available (4 vols., ed. J. Van Mierlo et al., 1944-48).

DIRK JELLEMA

RYCAUT (Ricaut), SIR PAUL (1628-1700). English traveler, diplomat, and writer. Born at Aylesford, Kent, he was educated at Trinity College, Cambridge, where he graduated in 1650. After traveling in Europe, Asia, and Africa, he spent eight years as secretary to H. Finch, earl of Winchelsea, ambassador extraordinary to Turkey. *The Present State of the Ottoman Empire, in three books*... (1670) resulted, a widely circulated history. In 1666 he became a member of the Royal Society and published in the *Philosophical Transactions* a treatise on sable mice. In 1667 he began a twelve-year consulate at Smyrna for the Levant Company. From this experience he wrote *The Present State of the Greek and Armenian Churches*... (1679). He was knighted in 1685 by James II, and served William III as resident in Hamburg and the Hanse towns where he remained until shortly before his death.

BRIAN G. ARMSTRONG

RYERSON, ADOLPHUS EGERTON (1803-1882). Methodist leader and educationalist. Born and educated near London, Ontario, he was called to the Methodist ministry and became a successful saddlebag preacher and missionary to the Credit River Indians. In 1829 he became the first editor of the influential *Christian Guardian* and secretary of the Wesleyan Missionary Society. An early advocate of the secularization of the clergy reserves* and other political reforms, he refused to support W.L. Mackenzie's rebellion in 1837. Vitally concerned about education, he helped found the Upper Canada Academy, which became Victoria College in 1841 with himself as principal. The educational system of Ontario after 1870 was largely based upon his 1846 *Report* which he wrote while superintendent of common schools. Probably the most influential Canadian Methodist of his time, he was from 1874 to 1878 the first president of the general conference of the Methodist Church of Canada.

ROBERT WILSON

RYLE, HERBERT EDWARD (1856-1925). Anglican bishop and preacher. Son of Bishop J.C. Ryle,* his strong Evangelical home background gave him a deep personal faith and an evangelistic outlook, although he moved away from his father's theological views. After Eton and a brilliant academic career at Cambridge, he won distinction as a moderate and cautious OT higher critic. He became principal of Lampeter College in 1886, Hulsean professor of divinity at Cambridge in 1888, and president of Queens' College, Cambridge, in 1896. Thereafter he was successively bishop of Exeter (1900) and Winchester (1903). In 1911 weak health obliged him to accept the less onerous deanery of Westminster, and his fame rests on his preaching and leadership there in World War I. His gospel sermons were called "wonderfully simple and simply wonderful." He did much to establish the (Church of England) church assembly.

G.C.B. DAVIES

RYLE, JOHN CHARLES (1816-1900). Bishop of Liverpool. Born at Macclesfield and educated at Eton and Christ Church, Oxford, the son of a wealthy banker, he was destined for a career in politics. A fine athlete, he rowed and played cricket for Oxford, and also took a first class degree in Modern Greats, but declined offers of a college fellowship. He was spiritually awakened in 1838 on hearing Ephesians 2 read in church, and was ordained by Bishop Sumner at Winchester in 1842. Country livings followed at Helmingham and Stradbrooke in Suffolk, until at the age of sixty-four he was appointed in 1880 at Disraeli's recommendation as first bishop of Liverpool.

Ryle was a prolific writer, the author of numerous tracts and books, of which *Knots Untied* is probably the best known. His leadership of the Evangelicals was sound and sensible, persuading them not to isolate themselves from the mainstream of church life by boycotting church congresses, and so leave Anglo-Catholics alone to put forward their views. In his diocese he exercised a vigorous and straightforward preaching ministry, and was a faithful pastor to his clergy, taking particular care over ordination retreats. He formed a clergy pension fund, built over forty churches, and proved an able administrator. A commanding presence and fearless advocacy of his principles were combined with a kind and understanding attitude in his personal relationships, while vast numbers of working men attended his special meetings. His strength of character was shown in that, despite strong criticism, he declared it his policy to put first the raising of clergy stipends rather than commence the building of a cathedral.

See M. Smout and P. Toon, *John Charles Ryle: Evangelical Bishop* (1976).

G.C.B. DAVIES

S

SABAS (439-532). Founder of the Order of Sabaites. Born in Cappadocia, he entered a monastery at the age of eight. In 457 he went to Jerusalem and lived as a hermit monk in the desert. He became renowned for his holiness, and founded a number of monasteries. He was ordained into the priesthood in 490, and in the later years of his life was an active defender of the orthodox faith against Origenism* and Monophysitism.*

SABATIER, LOUIS AUGUSTE (1839-1901). French Protestant scholar. Of Huguenot stock, he was brought up in the early nineteenth-century Protestant revival and became a leading exponent of liberal Protestantism in France. His work in the Protestant faculty of theology in Strasbourg (1868-70) was cut short for political reasons, but eventually in 1877 he helped to refound the faculty in Paris. From 1886 he taught also in the nonsectarian religious studies department of the *École des Hautes Études* of the Sorbonne. His theology was evolved in relation to his wide interests in modern cultural problems, evidenced in prolific regular writings on literature and politics. His view that concepts in religion could be no more than symbols undermined the traditional authority of dogma. He held that the proper method of theology was the historical and psychological study of religious phenomena, which at once relativized dogma, seen as changing historical forms, and revealed faith as the enjoyment of God's gift of spiritual life, which was the unchanging essence of religion. His presentation of this approach has a christological center, reminiscent of Schleiermacher.* His chief works were *Outlines of a Philosophy of Religion* (1897) and *The Religions of Authority and the Religion of the Spirit* (1903). HADDON WILLMER

SABATIER, PAUL (1859-1928). French Calvinist scholar and pastor. After studying at Besançon and Lille, he enrolled at the Protestant faculty of the University of Paris where his brother, Auguste Sabatier,* and Ernst Renan* were among his teachers. After serving from 1885 to 1889 as vicar of the Protestant church in Strasbourg, he was expelled from Germany. Returning to France, he was a pastor from 1889 to 1894, resigning to devote himself to a life of scholarship. His historical interests caused him to travel to Assisi in Italy where he studied the life of Francis and the Franciscan Order. Later he became professor of Protestant theology at Strasbourg (1919) and continued teaching there until his death. His *Life of St. Francis of Assisi* (1893, ET 1894) was an immediate success and went into forty editions within his lifetime. The biography shows a sympathetic understanding of Francis, but Sabatier has been accused of molding him after the image of a nineteenth-century liberal. In addition, he studied and published early Franciscan sources and documents such as the *Actus Beati Francisci et Sociorum Ejus* (1902) and the *Speculum Perfectionis* (1898).

Sabatier also became involved (1904-14) in the modernist movement within the Roman Catholic Church, writing *An Open Letter to His Eminence Cardinal Gibbons* (1908) and delivering the Jowett Lectures on *Modernism* (1908). When World War I broke out, he wrote a defense of the spiritual ideals of the allies, *A Frenchman's Thoughts on the War* (1915) and served as interim minister for pastors who were in the armed forces. He made monumental contributions to Franciscan scholarship, and his *Franciscan Studies* (1932) reveal a much deeper understanding and sympathy for the medieval religious outlook than his earlier work. RICHARD V. PIERARD

SABATIER, PIERRE (1683-1742). French biblical scholar. Born in Poitiers, he studied in the monastery of St.-Germain-des-Prés, the great center of Maurist* learning, under Thierry Ruinart.* After the latter's death he made his life's work the search for the pre-Vulgate Latin text of the Bible. Though not completed at his death, his virtually exhaustive collection of manuscript material was published in 1743 at Reims as *Bibliorum Sacrorum Latinae Versiones Antiquae.* This was the first work of its kind and is still of great value today. Known as a deeply pious man, he removed to Reims in the latter part of his life because of accusations that he was a Jansenist.*

SABBATARIANISM. In its developed form, Sabbatarianism demands a strictly religious use of Sunday which transfers the rest of the Jewish Sabbath to the Christian Sunday. Some communities like Seventh-Day Adventists* regard the rest of Christendom as seriously in error because Sunday has replaced the literal Sabbath, but increasingly, strict observance of Sunday is declining even among churches of Anglo-Saxon origin, where it reached its most striking development. Resting on the conviction that the Fourth Commandment is part of the perpetual moral law, Sabbatarianism has led not only to ecclesiastical censures, but also to strict civil law against work and recreation on Sundays. Though there were signs of popular strictness in keeping Sunday in the early and medieval church, civil and canonical requirements were based on tradition and utility rather than natural law.

Saints' days were observed far more strictly than Sundays and during the Reformation of the sixteenth century this legalism was strongly attacked by Reformers without insisting on similarly strict observance of Sunday. Reformers like Beza* and Zanchius emphasized that the Fourth Commandment was natural, universal, and moral. In continental churches this did not lead to Sabbatarianism, but in England and Scotland this doctrine combined with strongly antipapal attitudes and local needs to produce a strictness of Sunday observance which was unique. Originally a feature of Puritanism it gained wide support, and Commonwealth legislation was consolidated in 1677. Future legislation in 1781 and 1871 closed further loopholes, and similar legislation was found in many colonies and parts of North America. There were fierce nineteenth-century battles over Sunday trains, opening of museums, libraries, and limited recreational facilities, but groups like the Lord's Day Observance Society (1831) have had decreasing success since 1945. Reaction against legalistic and joyless Sabbatarianism has been practical rather than theological, and the theological issues involved have largely been ignored by modern Protestantism, except by K. Barth.*

BIBLIOGRAPHY: R. Cox, *Literature of the Sabbath Question* (1835); P. Schaff, *The Anglo-American Sabbath* (1863); W. Whitaker, *Sunday in Tudor and Stuart Times* (1933) and *The Eighteenth Century English Sunday* (1940); M. Levy, *Der Sabbath in England* (1933); P. Collinson, "The origins of English Sabbatarianism" in *Studies in Church History* I (1964); C. Hill, *Society and Puritanism in Pre-Revolutionary England* (1964); W. Rordorf, *Sunday* (1968); R.D. Brackenridge, "The Sabbath war of 1865-6," *Records of Scottish Church History Society* 16:1 (1966).

IAN BREWARD

SABBATH. Denoting the seventh day of the Jewish week, the name derives from a Hebrew word meaning "to desist" (cf. Gen. 2:2, God desisted from His work of creation). It was a day to be set apart for God, no work for purely human profit being done on it, the lighting of a fire being especially prohibited (Exod. 35:3). The stress is on avoiding normal work, not obtaining rest, though the latter is included. No real traces of the Sabbath can be found outside Israel. None of the many efforts to derive it from Mesopotamian religion or from the phases of the moon carry conviction. Exilic and postexilic Judaism* considered that its institution was part of the Mosaic lawgiving (Ezek. 20:12; Neh. 9:14). The widely held Christian belief that it was part of a primitive revelation seems to be a false interpretation of Genesis 2:3; had it been, it is difficult to believe that no traces would have survived outside Israel.

Amos 8:5b shows that the Sabbath was generally enforced in the preexilian period, though Jeremiah 17:19-23 and Ezekiel 20:16,21 suggest that a good deal of laxity was shown. A more rigorous attitude is attributed to Nehemiah (13:15-22), and Sabbath observance soon became an outstanding characteristic of Judaism. In 168 B.C. Antiochus Epiphanes's representative took advantage of the Sabbath to capture Jerusalem (*2 Macc.* 5:25); a little later, a large group allowed itself to be massacred rather than defend itself on the Sabbath (*1 Macc.* 2:29-38). As a result, self-defense on the Sabbath was permitted. The enumeration of thirty-nine main classes of work prohibited on the Sabbath probably antedates the NT period, though many of their subdivisions are later. The Rabbis insisted on the joyous nature of the Sabbath and so, in contrast to Qumran and the later Karaites, they were concerned to make its observance no burden. Not merely worship and the study of the Torah, but also promotion of family life and the joyful use of God-given food were enjoined.

It is clear that Jesus, His disciples, and the Jewish Christians observed the Sabbath. The clash between Jesus and the Pharisees on the subject was confined to a narrow area. When Jesus' disciples plucked the ears of corn (Mark 2:23-28), there was a conflict of duties, for they were enjoined to avoid hunger on the Sabbath. Jesus' "fault" was His claim to have the right to legislate (Mark 2:28). The remaining incidents are connected with miracles of healing. The Rabbis permitted medical care on the Sabbath only where there was a risk to life. Jesus laid down the principle that acts which glorified God were not a breach of the Sabbath (cf. Heb. 4:9f.).

For the Sabbath in the primitive church, see SUNDAY.

See M. Friedländer, *The Jewish Religion* (1921), and Tractate *Shabbath* in Mishnah (tr. H. Danby, 1933). See also bibliography under JUDAISM.

H.L. ELLISON

SABELLIANISM. Another name for Modalistic Monarchianism* or Patripassianism. This was an influential theological movement at the beginning of the third century A.D. It seems to have originated in Asia Minor. Noetus* of Smyrna taught Patripassian views; his disciple Epigonus brought the teaching to Rome, where through Praxeas* and Sabellius it gained a strong foothold. Sabellius, whose name is given to the movement, was active in Rome during the early third century. Tertullian* in North Africa vigorously opposed Praxeas, as did Hippolytus* at Rome. Motives for the struggle may not be unmixed. However, while Bishop Zephyrinus* at Rome fought Montanism* (which Tertullian favored) and Zephyrinus and his successor Callistus* engaged in a bitter power struggle with Hippolytus, the theological implications of Sabellianism on the orthodox side were serious. A modern form of Sabellianism is Unitarianism.*

Little is known about Noetus, Praxeas, and Sabellius except through the writings of Tertullian *(Adversus Praxean)* and Hippolytus *(Refutation, Contra Noetum)* and other secondary sources. Sabellianism was an attempt to solve the problem of how to accept the deity of Christ and also maintain the unity of God. The Sabellians achieved this at the expense of a trinity of persons in the Godhead. They reduced the status of the persons to modes or manifestations of the one

God. The term is frequently coupled with the word "monarchy" to denote the primacy of God as the Father. The Son and Holy Spirit are thus revelatory and apparently temporal modes of God the Father's self-revelation. Tertullian sneered that Praxeas had put the Holy Spirit to flight and crucified the Father. If God the Father became incarnate, then He also suffered (Patripassianism).

See also MONARCHIANISM (for bibliography); SUBORDINATIONISM; INCARNATION; TRINITY.

SAMUEL J. MIKOLASKI

SACCAS, see AMMONIUS SACCAS

SACRAMENT. A religious rite variously regarded as a channel or as a sign of grace. In all its work and witness, the NT church gave a prior place to preaching (see HOMILETICS) through which, it recognized, Christ gave men fellowship with Himself and participation in the power of His death and resurrection. It recognized also, however, that Christ meant the Word to be accompanied in this unique ministry by baptism and the Lord's Supper, and it gave to both these ordinances also a special place in its life. It was always characteristic of God's approach to man in the OT that when He wanted to speak to men and have communion with them, He not only used words, but also gave signs along with the Word. For example, He used dreams and visions, symbolic objects and miracles, in addition to speaking. These sometimes illustrated and drew attention to what He had to say, or were sometimes simply signs that He was really present there and then in the saying of it. It was characteristic of Jesus that in His own ministry on earth He not only preached, but He added miracles and other signs to help to effect and to draw attention to what His Word proclaimed. Miraculous signs as seals of the Word continued within the early church for only a short time, and it was accepted that baptism* and the Lord's Supper* were to continue as the settled permanent signs attached to the Word.

In later days these two ordinances were called "sacraments." This word is the Latin for the Greek *mysterion*, which in the NT denotes the divine plan of salvation hidden in past ages, but now brought to light in the preaching of the Word. This mystery proclaimed in the Word was fully realized in the God-man Himself, in His person and work, and is now being realized in the union of the individual to Christ by faith. The revelation of this fulfilled mystery will be consummated in the last day (cf., e.g., Eph. 3:3-6; 1 Tim. 3:16; Col. 1:27; 1 Cor. 15:51,52). In the thinking of the ancient Catholic Church there was only one sacrament or mystery—that of Christ Himself—but baptism and the Lord's Supper were called "mysteries" or "sacraments" because they enabled men to participate in this sacramental union of God and man, through the atoning death and resurrection of Christ. The sacraments were regarded as effecting within the church nothing more than the Word itself effected when it was received by faith. They too required the same faith.

It was Augustine* who first gave the general definition of "a sacrament" which later became traditional—i.e., an outward and temporal sign of an inward and enduring grace. This general definition led later to the incorporation of other sacramental ceremonies into the life of the church. A sacramental theology developed in which ultimately seven sacraments were regarded as containing and causing grace. The church came to regard itself as a sacramental institution dispensing a special grace for every important occasion in life. The sacraments were regarded as effective *ex opere operato*, provided the recipient placed no obstacle in the way of their reception. The Reformers used the Augustinian definition of a sacrament and restricted their number to the two for which they believed they had Christ's command and promise. They insisted that the sacraments were given to serve the Word of God and were effective only when received by faith within a personal relationship with Christ.

BIBLIOGRAPHY: O.C. Quick, *The Christian Sacraments* (1927); J.K. Mozley, *The Gospel Sacraments* (1933); R.S. Wallace, *Calvin's Doctrine of the Word and Sacrament* (1953); B. Leeming, *Principles of Sacramental Theology* (1956); D.M. Baillie, *The Theology of the Sacraments* (1957); L. Bouyer, *Word, Church and Sacraments in Protestantism and Catholicism* (1961).

RONALD S. WALLACE

SADDUCEES. We first meet them in Josephus's account of John Hyrcanus (135-104 B.C.). Our information about them is meager and derived exclusively from hostile sources. Clearly they consisted mainly of the most influential priestly and aristocratic families. Normally their name is interpreted as "descendants of Zadok," i.e., David's high priest. They may have preserved some of the Hellenistic views adopted by many priests in the time of Antiochus Epiphanes, but they are presented essentially as the preservers of ancient priestly traditions.

In most conflicts with the Pharisees* they were clearly defending an older view. The greater popularity of the Pharisees among the people was mainly due to their trying to interpret the law of Moses, with all their strictness, with the needs of the poor in mind. Many of the Sadducean leaders were murdered by the Zealots* during the revolt against Rome, as real or suspected collaborators; the destruction of the Temple deprived their survivors of their position of religious significance. They disappeared, and the Pharisees saw to it that they left no traces behind them. Though they respected the prophetic books, they denied normative value to them. Hence they denied the resurrection as unprovable from the Pentateuch. They also maintained the concept of complete freedom of the will, and rejected scribal traditions. The denial of angels and spirits (Acts 23:8) was presumably as media of revelation.

BIBLIOGRAPHY: E. Schürer, *A History of the Jewish People in the Time of Jesus Christ* (ET 5 vols., 1886-90); J.W. Lightley, *Jewish Sects and Parties in the Time of Jesus* (1925), pp. 11-78; H.L. Strack and P. Billerbeck, *Kommentar zum Neuen Testament aus Talmud und Midrash*, IV (1928), pp. 339-52. Most works dealing with the

Pharisees include a treatment of the Sadducees.
H.L. ELLISON

SADOLETO, JACOPO (1477-1547). Cardinal, humanist, and biblical scholar. Born at Modena, he joined Oliviero Cardinal Carrafa as a minor poet in 1498. As a member of Clement VIII's Curia from 1524 to 1527, Sadoleto emerged as an exegete. He left Rome after its sacking in 1527, and from the diocese of Carpentras in France published the *De Laudibus philosophiae* (1538) and a controversial commentary on *Romans* (1535). Reginald Pole* warned Sadoleto not to neglect theology and was greatly impressed by him; Sadoleto responded that Pole was his guide. "The book of the Gospels contains the entire way and knowledge of our salvation," he held. John Calvin* replied to Sadoleto's famous letter to the Genevans (1539), deploring Sadoleto's emphasis on the safety of one's soul in contrast to God's glory. Whether for his Catholic friends like Pole or Protestant foes like Calvin, Sadoleto never ceased to work for reform. A great mass of his diocesan work at Carpentras remains unexamined, while his presence on the papal reform commission of 1536 guarantees his place as an irenic bishop in a polemical age.
MARVIN W. ANDERSON

SAHIDIC, see EGYPTIAN VERSIONS

SAILER, JOHANN MICHAEL (1751-1832). Jesuit scholar. Born in upper Bavaria, he entered the Society of Jesus in 1770 and was ordained to the priesthood in 1775. Five years later he became professor of dogmatics at Ingolstadt, and other academic appointments followed at Dillingen (1784) and Ingolstadt (1799), which latter university was in 1800 removed to Landshut. As the mentor of prospective priests, Sailer influenced a circle of evangelicals which included Martin Boos,* Johannes Gossner,* and Baron von Wessenberg.* Sailer's appointment to the bishopric of Augsburg was rejected by the Curia in 1819. After serving as vicar-general for four years he was appointed in 1829 to the see of Regensburg.
WAYNE DETZLER

SAINT-CYRAN, ABBÉ DE (1581-1643). Jansenist* theologian. Born Jean Duvergier de Hauranne, he was a pupil of Justus Lipsius at the Jesuit College, Louvain, and a fellow-student and close friend of Cornelius Jansen at Paris (1604-10) and Bayonne (1611-17). He became secretary to Bishop de la Rocheposay (1617) and abbot of Saint-Cyran (1620), thereafter living mainly in Paris. Attracted to the writings of Augustine,* he sought with Jansen to combat the Jesuits' moral laxity and to reform Catholicism on Augustinian lines in the hope of defeating Protestantism with its own weapons. From 1623 he was closely associated with the Arnauld family and the Cistercian convent of Port Royal, near Paris. As spiritual counselor at Port Royal from 1633, he exerted immense influence, making it the center of Jansenism in France. He devised a lay community, *les Solitaires de Port Royal,* which numbered among its adherents some of the most distinguished scholars of the age. Jansen wrote him a long series of letters from Louvain between 1617 and 1635. Cardinal Richelieu* considered him dangerous and had him incarcerated in the donjon at Vincennes from 1638 till Richelieu's death (1643). Here he wrote *Letters chrétiennes et spirituelles* (pub. 1645). Later Jansenists regarded him as a martyr.
J.G.G. NORMAN

SAINTS. According to Roman Catholic doctrine, saints are those now in heaven because of their exemplary lives, who can make intercession with God for the living as well as for those in purgatory.* The practice of the veneration of saints claims biblical foundation, e.g., Genesis 18:16-31; Matthew 19:28; Hebrews 12:1; Revelation 6:9f.; see also Paul's doctrine of Christ's mystical body, with all members as "fellow-citizens with the saints, and of the household of God" (Eph. 2:19). The nonbiblical sources for the practice range through Christian history, beginning in the pre-Nicene period with the *Odes of Solomon** and the *Martyrium Polycarpi* (c.156). Origen* was probably the first of the Fathers to permit the cult of martyrs a theological claim, and Cyril* and Chrysostom* made the distinction between those commemorated at the Eucharist and the ordinary dead.

As devotion to the saints grew, idolatry arose, to counter which theologians tried to make clear the difference between worshiping God and honoring the saints: the Greek terms *latreia* and *douleia* respectively. By the Carolingian period, popular devotion was flowering with pilgrimages,* greater concern with relics,* naming of patrons, and even making feasts civil festivals. The first formal canonization occurred in 993 with Ulrich.* Increasingly the "lives" were publicized, more stereotype than fact, which only added abuses. The great task of revising these lives, begun by Lipomani, Surius, and Baronius,* was critically done by Jean Bollandus (1596-1665) in his *Acta Sanctorum,* which provided a model. Leo I,* Gregory I,* and John of Damascus* furthered the theology of the practice, while liturgical developments reflected popular devotion and patristic teaching. Iconography grew accordingly, and even some angels were elevated (Raphael,* Gabriel,* Michael*).

A saint, by description not title, could be designated unofficially without beatification* or canonization.* These many practices brought protest from within, at the councils of Avignon (1209) and the Fourth Lateran (1215), as well as without (Cathari,* Waldenses*). The fiercest objections came at the Reformation from Zwinglians and Calvinists. The Council of Trent* approved the practice, but encouraged moderation, with Robert Bellarmine's* principles governing much of the present teaching. The modern practice is governed by canon law,* distinguishing between the worship of God and the honoring of Mary, the saints, and angels.

The Eastern Churches are similar to Rome in their attitude on the subject. With Protestant refusal to venerate the saints (since sanctity is potentially the province of all who enjoy salvation) their legends have diminished. The literary interest of John Milton* and some of his contemporaries kept them alive, but the Enlightenment,* especially Voltaire* and the Encyclopedists,* only reinforced the Protestant view. In the Anglican Communion the practice was revived with the Oxford Movement* despite early Tractarian misgivings.

BIBLIOGRAPHY: H. Delehaye, *Sanctus* (1927) and *Les Origines du culte des martyrs* (1933); R. Aigrain, *L'Hagiographie* (1953); H. Roeder, *Saints and Their Attributes* (1955); J. Douillet, *What Is a Saint?* (1958). C.G. THORNE, JR.

SAINT-SIMON, CLAUDE HENRI DE ROUVROY (1760-1825). French social philosopher. Although he came from a noble family in Paris, he was convinced by the French Revolution and industrialization that the end of the Catholic and aristocratic era had come. After an undistinguished military career and life as a titled profligate, he abandoned his titles of nobility in support of the Revolution. He engaged in intensive reflection (from 1797), culminating in a brief period (1814-25) in which he formulated his new societal and cultural ideas. Especially important were the periodicals *L'Industrie* (from 1816), and its successor *L'Organisateur* (1819-20). He considered the new *industriels* as the hope for the future, for they could administer society and arrange production for the good of all, whose life would be devoted to productive work of all kinds, for which they would receive just reward. The plan rested on the conviction that science—first physics, later biology—could point the way to social reconstruction. Positive science and industry, he believed, had replaced medieval Christianity and feudalism, as historical evolution had progressed. In *Nouvelle Christianisme* (1825) he argued that the ethics of Christianity could be useful to insure the solidarity of the new society, but it had to be stripped of its dependence on elements he considered metaphysical, supernatural, and dogmatic. Auguste Comte, a younger collaborator of Saint-Simon from 1817-24, later developed positivism as an implication of his teacher's scientific treatment of social organization. A group of Saint-Simonians, including Barthélemy Enfantin, developed his incipient biological organicism more romantically into a "utopian" socialism.

See F.E. Manuel, *The New World of Henri Saint-Simon* (1956). C.T. MC INTIRE

SAINT-SULPICE, SOCIETY OF, see SULPICIANS

SAINT-VALLIER, JEAN BAPTISTE DE LA CROIX DE CHEVIERES (1653-1727). Second bishop of Quebec. Ordained at the age of twenty-two, the French-born Saint-Vallier became one of Louis XIV's chaplains, then was nominated by Laval to succeed him in Quebec. He arrived in Canada in 1685 as vicar-general of the diocese and in 1688 was consecrated bishop. Zealous, pious, dictatorial, and undiplomatic, he was constantly in conflict with the civil authorities and other ecclesiastics over his efforts to maintain a high moral tone in colonial life by enforced asceticism. Constantly on the move, he faithfully administered his vast diocese and demonstrated his concern for the welfare of the poor by building several hospitals and setting up a rudimentary welfare service. Requests for his recall were often sent to the king, but it was the English who obliged his enemies by capturing him in 1704, so that he was absent from his diocese for thirteen years. ROBERT WILSON

SAKER, ALFRED (1814-1880). Missionary to Cameroon. Born at Borough Green, Kent, he was an engineer who joined the Baptist Mission which left Jamaica to work among liberated slaves in Fernando Po (1843). In Cameroon (1845) he founded Bethel station in Douala country, giving himself to preaching, teaching, and translation. When the Spanish authorities forbade Protestant worship on Fernando Po (1858), Saker with ninety families founded a settlement at Victoria on Cameroon mainland. He introduced crafts and building work and completed the Douala Bible (1872). With Sir Richard Burton he made the first ascent of Cameroons Mountain (13,352 feet). Retiring to England in failing health in 1876, he encouraged the founding of the Baptist Missionary Society Congo Mission in his closing years.

J.G.G. NORMAN

SALESBURY, WILLIAM (1520?-1584?). Welsh NT translator. He sprang from a family that had gained money and social prestige since the fourteenth century. He was born at Llansannan, but spent most of his life at Plas Isa in Denbighshire. No more detail is available about his university career except that he was educated at Oxford. After a period at the Inns of Court he became a lawyer in the service (possibly) of the lord chancellor. But he seems to have retired early to Wales to devote himself to his scholarly pursuits.

He dedicated himself to the task of providing the Welsh people with the Scriptures in their own language. After trying his hand as an author by publishing seven books, he began the great work of his life by publishing *Kynniver llith a ban* ("all the lessons and articles") in 1551, a Welsh translation of the lessons for Holy Communion according to the 1549 Book of Common Prayer. Salesbury was a typical man of the Renaissance in his scholarship and linguistic expertise, and it is apparent that in translating into Welsh he made use of Erasmus's Greek Testament, Luther's German, the Vulgate, Tyndale's NT, and the Great Bible. But he was always selective in his choice of translation. With the accession of Elizabeth I,* it became imperative to complete the work of translation. The Act of 1563 commanded the Welsh bishops together with the bishop of Hereford to have a Welsh translation of the Bible and of the Book of Common Prayer available by 1 March 1567. Salesbury was invited by Bishop Richard Davies to cooperate in the work of translation.

In May 1567 the Welsh Book of Common Prayer was published, followed by the NT in October. It is now agreed that the Common Prayer was very largely the work of Salesbury. While Bishop Davies translated 1 Timothy, Hebrew, James, 1 and 2 Peter, and Thomas Huet, dean of

St. David's, translated Revelation, the remainder was Salesbury's work. By now the Geneva Bible had been published as well as Theodore Beza's great work on the NT (1565). They influenced Salesbury to be more meticulous about the details of the Greek text and to adopt a literal translation very frequently rather than an idiomatic one—the common emphasis among Calvinistic translators. All in all, Salesbury's work was a very fine achievement, and although his NT is marred by the adoption of idiosyncratic ideas about Welsh orthography, it is nevertheless the basis of all subsequent translations. It was Salesbury who fused the spiritual energy of the Protestant Reformation and the enthusiasm of the Renaissance into the context of Welsh national life.

See I. Thomas, *William Salesbury and his Testament* (1967); G. Williams, "The achievement of William Salesbury," *Transactions of the Denbighshire Historical Society.* R. TUDUR JONES

SALESIANS (The Salesian Society of St. John Bosco). Founded in Turin in 1841 by Giovanni Bosco,* it is the third largest Roman Catholic order. In 1970 it had 20,423 members throughout the world and supported 1,533 institutions. Don Bosco envisioned reaching young men spiritually by means of song festivals (a daily average of 576 festivals are sponsored). Soon the work spread to operating orphanages, day schools, evening schools, savings banks, sports associations, asylums for poor students, and agricultural colonies. The order maintains 140 mission stations in Asia, Africa, and South America and cares for nearly half-a-million orphans. Pius IX gave the order apostolic approval in 1868. ROYAL L. PECK

SALMASIUS, CLAUDIUS (Claude de Saumaise) (1588-1653). French Protestant scholar. Born in Semur, Burgundy, he studied philosophy at Paris (1604-6) under Isaac Casaubon,* through whom he was converted to Calvinism. While studying jurisprudence in Heidelberg he discovered the manuscript of the Palatine anthology. He edited two previously unprinted fourteenth-century tracts against papal supremacy by Nilus of Salonica and the monk Barlaam (1608). After publishing classical works, including an edition of Solinus's *Polyhistor* (1629), he became professor at Leyden (1632), succeeding J.J. Scaliger. Here he remained apart from a visit to the court of Queen Christina of Sweden (1650-51). He defended the compatibility of usury with Christianity in *De usuris liba* (1638) and *De modo usurarum* (1639). His *Defensio regia pro Carolo I* (1649), an accusation of regicide against the English people, provoked a counterblast from John Milton, *Pro populo anglicano defensio* (1651). J.G.G. NORMAN

SALMON, GEORGE (1819-1904). Anglican divine. Born at Cork, Ireland, son of a Protestant linen merchant, he had a brilliant career at Trinity College, Dublin, and this led to a lifetime in the college as fellow (1841), professor (1866), and finally provost from 1888. He had been ordained in the Church of Ireland in 1845. He pursued two separate academic disciplines. Internationally recognized as a mathematician, he was also widely known for his theological writings. A strong Protestant, he cooperated with Archbishop Whately* in *Cautions for the Times* (1853), an answer to the Tractarians.* His widely read *Infallibility of the Church* (1889) was a brilliant, trenchant exposition of Roman claims which he answered with clarity, learning, and humor. His *Introduction to the New Testament* (1885) punctured several extravagantly liberal theories concerning Christian origins. He later questioned successfully many of the less happy hazards of Hort's Greek NT text. An able administrator, Salmon's financial acumen aided the Church of Ireland after the shock of disestablishment.

JOHN C. POLLOCK

SALTMARSH, JOHN (1612?-1647). Anglican writer and controversialist. As rector of Heslerton he was a keen conformist, but changed his views and resigned in 1643, becoming rector of Brasted in 1645 and, according to Thomas Fuller, "a violent oppressor of bishops and ceremonies." He was a prolific pamphleteer in favor of greater latitude in church government, with a gentle, quaint, controversial style. In 1646 he became an army chaplain and "prophesied" to Fairfax at Windsor that "the army had departed from God." He is chiefly memorable for the controversy with the Westminster divine Thomas Gataker over Saltmarsh's sermon *Free Grace* (1645), which Gataker regarded as dangerously lacking in emphasis on Christian responsibility. Saltmarsh with others was labeled "Antinomian" because of insistence that Christians have no responsibility to keep the law of God because they are God's children and so not "under law"—probably an unwise exaggeration of the biblical teaching of justification through the free grace of God, but not necessarily antinomian. Saltmarsh's views were further expounded in *Sparkles of Glory* (1647).

PAUL HELM

SALVATION. In the Bible the word may mean deliverance by God from almost any kind of evil, whether temporal and material or spiritual—defeat in battle (Exod. 15:2), trouble (Ps. 34:6), enemies (2 Sam. 3:10), exile (Ps. 106:47), death (Ps. 6:4), sin (Ezek. 36:29). It does not necessarily have a theological connotation. At first the Israelites thought of salvation primarily as deliverance in a material sense and as a national thing, but as their sense of moral evil deepened, salvation acquired a profound ethical meaning, and it gradually was seen to include Gentiles as well as Jews (Isa. 49:5,-6; 55:1-5). With the unfolding of the messianic idea, it came to be used of their deliverance from sin to be brought in with the Messianic Age. Among the Israelites, salvation was acquired through a sincere observance of the law, moral and ceremonial. The ritual sacrifices could not of themselves bring pardon of sin, for they were merely typical of the Lamb of God who was to die for the sins of the world (Isa. 53).

In the teaching of Jesus, salvation usually denotes deliverance from sin, to be experienced now, although its complete fulfillment is eschatological. He taught that salvation is only through Him, the incarnate Son of God (John 3:16). In the apostolic age salvation is through the death of Christ (Eph. 2:13-18) and includes all the redemptive blessings which believers have in Christ, chief of which are conversion, regeneration, justification, adoption, sanctification, and glorification. It is God's solution to the whole problem of sin, in all its aspects. It brings deliverance not only from the guilt of sin, but also from its power, and ultimately from its presence. Although provided through Christ's sufferings, death and resurrection, salvation becomes realizable in experience through the Holy Spirit, on the condition of faith. Its effects will someday embrace the whole universe. The curse will be removed from nature, and all history will find its consummation and completion in Christ (Rom. 8:21,22; Eph. 1:10).

BIBLIOGRAPHY: G.B. Stevens, *The Christian Doctrine of Salvation* (1905); E.F. Kevan, *Salvation* (1963); E.M.B. Green, *The Meaning of Salvation* (1965); O. Cullmann, *Salvation in History* (1967). STEVEN BARABAS

SALVATION ARMY. Founded by William Booth* as the "Christian Mission" in East London in 1865, the Salvation Army first took that name in 1878. It was an essentially evangelical movement, biblically orientated, theologically conservative. Its basis of belief includes the divine inspiration of the Bible, the doctrine of the Trinity, the salvation of believers "by faith through grace," the "immortality of the soul," the resurrection of the body, the final judgment. The doctrinal distinctives of the Army include an Arminian emphasis on free will and a "holiness" experience which can be subsequent to conversion—this is traceable to William Booth's Methodist origins—and the nonobservance of the sacraments of baptism and the Lord's Supper.

By 1879 William Booth "commanded" eighty-one stations, manned by 127 full-time evangelists, with another 1,000 voluntary speakers holding 75,000 services a year. Fifty-one new stations were opened in 1878; in that same year the first brass band featured in an Army event, and soon Salvationist words were being set to secular song tunes and bands were springing up everywhere. In 1880 standard uniforms were adopted, preceded in 1878 by the first volume of *Orders and Regulations for the Salvation Army.* Also in 1880 came the first overseas advance, into the USA; in 1882, Canada was "invaded," followed by India. By 1884 the Army had more than 900 corps, over 260 of them outside Britain, and headquarters in the city of London. This in spite of fierce opposition, at times leading to serious rioting.

The Army was inevitably led to operate on a broader front, faced with the appalling social needs of Victorian London and other great cities. A sensational case in which William Booth's son Bramwell was involved in exposing the white slave traffic forced the Army into prominence. Within five years, thirteen homes for girls in need of care and protection had been set up in the United Kingdom, and a further seventeen overseas. The first Prison Gate home for discharged prisoners was opened in Melbourne in 1883. In 1887 Booth, seeing homeless men sleeping rough on London Bridge, decided to do something practical. After intense research he set out the facts in a best-selling book, *In Darkest England—and the Way Out,* appealing for a fighting fund of £100,000. Cheap food depots, an unofficial employment exchange, a missing persons bureau, night shelters, a farm colony, soup kitchens, leper colonies, woodyards in the USA, home industries in India, hospitals, schools, and even a lifeboat for the fishermen of Norway—these marked successive stages in the Army's massive program of social action. Permeating it all was the basic concern for personal salvation which had been the motivation of its beginnings.

World War I gave many opportunities for service which broke down further barriers of prejudice and misunderstanding, and helped the Army to its present general acceptance both as an agency of goodwill and compassion, and as a member of the world family of Christian churches. Today the Army is at work in 74 countries, and numbers altogether some 2 million members. There are 25,039 full-time officers. Typical statistics for a recent year (1970) include: 20 million low-cost meals served, 10 million hostel beds provided, 7,035 missing persons traced, 19,722 unmarried mothers cared for. The Salvation Army has 40 general and specialist hospitals, with some 148,000 in-patients in a recent year.

See R. Sandall, *The History of the Salvation Army* (3 vols., 1947-55). A. MORGAN DERHAM

SALVIAN (c.400-c.470). Christian writer. Of a noble Christian family and probably born near Cologne, he had married early a young woman (Palladia) of pagan background, with whom by mutual consent the ascetic life had been chosen, to the vexation of her family. Separating according to agreement, he joined the community first at Lérins, later at Marseilles as presbyter. Apart from nine known letters, Gennadius provides what little else is known of the man. Under Augustinian influence Salvian composed (after 439) in eight books a treatise *de Gubernatione Dei,* falling back upon the fundamental biblical perspective of greater moral judgment upon the elect, while providing a vivid picture of the corrupt bureaucratic administration and its socioeconomic practices. As agent of the divine retribution, the "barbarian" peoples served rather than contravened Providence because of the greater perversity of the Romans in their sexual laxity, addiction to the games, and oppression of the poor. A work in four books also survives, concerning the willing of possessions to the church rather than to familial heirs. CLYDE CURRY SMITH

SAMARITANS. The name "Samaria" came to be used for that central section of Palestine coinciding approximately with the tribal portions of Ephraim and Western Manasseh. Its inhabitants, an amalgam of those Israelites not deported by the Assyrians and the various foreign elements introduced by them (2 Kings 17:24; Ezra 4:2,10), were known as Samaritans. At first their religion was syncretistic, but by the time of Zerubbabel the heathen elements seem to have been eliminated. Disappointed in their hope of sharing in the Jerusalem temple and worship, except on Judean terms, they built a temple on Mt. Gerizim in the fifth or fourth century B.C., served by a Zadokite priest from Jerusalem, whose descendants still function among them. This temple was destroyed about 107 B.C. by John Hyrcanus, rebuilt in A.D. 135, and finally destroyed in 484 for political rather than Christian motives. The Passover lambs are still sacrificed annually on Mt. Gerizim.

The chief features separating them from Rabbinic Jews is that they accept only the Pentateuch as canonical, their interpretation of it being stricter than that of the Talmud, and they consider Mt. Gerizim to be the site chosen by God for the Temple. Though Rabbinic tradition calls them "Cutheans," it also recognizes their right to share in the worship of Israel, if they abandon their special principles. They remained a prosperous, closely knit community until they began to decline owing to Muslim persecution. By 1955 they numbered about 250 in Nablus and seventy near Tel Aviv, but since then they have shown signs of increase.

BIBLIOGRAPHY: J.A. Montgomery, *The Samaritans* (1907); M. Gaster, *The Samaritans* (1923); J.W. Lightley, *Jewish Sects and Parties in the Time of Jesus* (1925), pp. 179-265; M. Simon, *Jewish Religious Conflicts* (1950), pp. 17-25.

H.L. ELLISON

SAMSON OF ARBIL (d.123). First known martyr east of the Roman Empire. Deacon to Paqida, the first bishop of Arbil (104-114) in what is now N Iraq, he himself succeeded to the diocese when in 120 he was consecrated by the bishop of Bait Zabdi. Samson carried out successful evangelism among surrounding villages whose inhabitants followed a debased form of fire-worship involving annual child-sacrifice. Arrested in 123 by nobles and Magians, he was tortured and decapitated.

SANCROFT, WILLIAM (1617-1693). Archbishop of Canterbury from 1678 till 1689. Graduate and fellow of Emmanuel College, Cambridge, he was ejected in 1651 and went abroad for some years, studying in Padua, and returned to England at the Restoration. He became chaplain to Bishop Cosin and was involved with him at the Savoy Conference* and afterward on the work of revising the Prayer Book. He also became chaplain to Charles II, and a prebendary of Durham in 1662. In the same year he became master of Emmanuel and founded a new college chapel. After a brief period as dean of York (1664), he became dean of St. Paul's and was closely associated with Wren in rebuilding that cathedral after the Great Fire. He became archbishop of Canterbury in 1678 and in that capacity crowned the Roman Catholic James II* in 1685, having revised the service to omit Communion. He refused to serve on James's ecclesiastical commission or to read his declaration of indulgence.* He was imprisoned in the Tower with six other bishops for his opposition. He supported actively the intervention of William of Orange, but refused to recognize him as lawful king, for which he was suspended in August 1689 and then deprived. With others of like mind he founded the schismatic body of Nonjurors*, but spent his last years in retirement.

JOHN TILLER

SANCTIFICATION. The separation of one's entire being from all that is polluting and impure, and a renunciation of the sins toward which the desires of the flesh lead. Both the Greek *hagiasmos* and the biblical doctrine of sanctification disallow any idea of progressively becoming holy. What God has once made holy in election and redemption is always thereafter holy, and there can be no degrees in the state of absolute holiness. Our moral progress is not a growth into holiness out of a state of comparative unholiness, but a growth *in* holiness effected by a supernatural act of God. The Christian can and is expected to cooperate by the proper use of the means God makes available (Rom. 12:1,2), but it is God who does the all-important work (Gal. 5:16-25). Sanctification involves the mortification of the old man (Col. 3:8-10; Rom. 6:6), and the giving of vitality to the new man, created in Christ unto good works (Rom. 6:11-23).

See PERFECTIONISM; HOLINESS CHURCHES; KESWICK CONVENTION. KEITH J. HARDMAN

SANCTIS, LUIGI DE, see DESANCTIS

SANCTUARY, RIGHT OF. Sanctuaries or asylums, wherein people could find temporary protection, existed among many different peoples—e.g., the cities of refuge of the Jews. Christian sanctuaries, first recognized by Roman law in 399, were given further legal rights in 419 and 431. Justinian, however, in a statute of 535, limited the privilege to persons not guilty of the grosser crimes. Canon law later allowed sanctuary for a prescribed time to persons guilty of crimes of violence so that terms of compensation might be worked out and agreed. In English common law a person accused of felony could have sanctuary within a church, but once there had to decide between submitting to a trial or confessing the crime and leaving the country. If no decision was made after forty days, he was starved into submission. In the sixteenth century the right of sanctuary was severely curtailed throughout Europe. Henry VIII limited it to seven cities of refuge—Wells, Westminster, Northampton, Manchester (later Chester), York, Derby, and Launceston. In 1623 James I abolished sanctuaries in cases of crime; in 1697 the law was tightened by the "Escape from Prison Act," and in 1723 a further act completed the work of extinction. In Scotland sanctuary was abolished at the Reformation, but certain debtors were accorded sanctuary around Holyroodhouse, Edinburgh, until about 1700. In Europe the practice lingered on until the time of the French Revolution.

See further, J.C. Cox, *The Sanctuaries and Sanctuary Seekers of Medieval England* (1911).

PETER TOON

SANDAY, WILLIAM (1843-1920). Biblical scholar. He was a fellow of Trinity College, Oxford, from 1866 until 1869 and then a country clergyman in three different parishes before becoming principal of Hatfield Hall, Durham, in 1876. In 1882 he returned to Oxford as Dean Ireland's professor of the exegesis of Holy Scripture and was Lady Margaret professor of divinity from 1895 until 1919. He was one of the original fellows of the British Academy. A patient and thorough scholar who embraced modernist convictions from 1912, he published a large number of books. Most of his work was in the field of the gospels, as is shown by his earlier works *The Authorship and Historical Character of the Fourth Gospel* (1872) and *The Gospels in the Second Century* (1876). His Bampton Lectures were published with the title *Inspiration* (1893). The work for which he is best remembered, however, is the commentary on *Romans,* written with A.C. Headlam for the *International Critical Commentary* in 1895 because he felt that a professor of exegesis should do some exegetical work. After his appointment to the Lady Margaret chair he concentrated on the life of Christ. His works included *Outlines of the Life of Christ* (1905), *The Criticism of the Fourth Gospel* (1908), *The Life of Christ in Recent Research* (1907), *Christologies Ancient and Modern* (1910), and *Personality in Christ and in Ourselves* (1911). In conjunction with his seminar he also published *Oxford Studies in the Synoptic Problem* (1911). R.E. NIXON

SANDEMANIANS. A body of Bible-loving Christians founded by John Glas* (hence the alternative name "Glasites") which flourished from 1725 until about 1900. Robert Sandeman (1718-71), son-in-law of Glas, came to the fore after the publication of James Hervey's *Theron and Aspasio* (1755), a Calvinist evangelical work which Sandeman attacked on the ground that it made faith a work of man which earns salvation. Sandeman held that bare assent to the work of Christ is alone necessary. After the controversy many new churches were founded; numerous Inghamite churches in Yorkshire joined the movement after 1759, and the London church in the Barbican area, of which Michael Faraday* was a member, was founded in 1760. Sandeman left England in 1764 to found churches in the USA, where the group survived until 1890. Sandemanians upheld the views of Glas and Sandeman: infant baptism and foot-washing were practiced; churches were organized with several coequal presbyters; and agreement (not a majority vote) was deemed essential. Excommunication was practiced. The sect was exclusive, and intermarriage was usual. It was the butt of much ill-informed criticism. Conditions of membership were strict (the church could control the use of members' private money), and membership low, although attendance at worship was large, at least in London.

Sandeman's works, which compared with those of Glas are repetitive and of low intellectual level, include *Some Thoughts on Christianity* (1762) and *Discourses on Passages in Scripture* (the 1857 edition of which contains a biography of Sandeman). R.E.D. CLARK

SANDERS, NICHOLAS (c.1530-1581). Roman Catholic scholar and controversialist. Born in Surrey and educated at Winchester and Oxford, he became a lecturer in canon law at Oxford (1551) until the accession of Elizabeth in 1558 made him leave England for the Continent. By 1561 he had been ordained priest at Rome, and he served as theologian at the Council of Trent* in the entourage of S. Cardinal Hosius.* Sanders's abilities were further recognized in 1565 with his appointment as professor of theology at Louvain, and there followed seven years in which he produced a number of works upholding the claims of Rome against the Anglican Church. He was summoned to Rome in 1572 to advise Gregory XIII* on English affairs, and worked and schemed for a military invasion of England by Catholic armies, with mixed results. He went to Ireland in 1579 in pursuance of his plans, but died there two years later. His works include *The Rock of the Church* (1567), *De Visibili Monarchiae Ecclesiae* (1571), and an uncompleted volume which was translated into English and published in 1877 as *The Rise and Growth of the Anglican Schism.* J.D. DOUGLAS

SANDYS, EDWIN (1516?-1588). Archbishop of York. After a brilliant academic career at St. John's College, Cambridge, he was chosen in 1547 as master of Catherine Hall and two years later became a canon of Peterborough. In 1553 he was appointed vice-chancellor of Cambridge. When Mary became queen he was arrested for his support of Lady Jane Grey's cause, but he escaped to Strasbourg where he enjoyed the friendship of Peter Martyr.* After Elizabeth's accession he became successively bishop of Worcester (1559), bishop of London (1570), and archbishop of York (1575). He helped to translate the Bishops' Bible, and was firmly committed to maintaining the Protestant character of the Church of England. He did, however, somewhat modify his earlier pronounced Puritanism after he became a bishop. PETER TOON

SANGSTER, WILLIAM EDWYN ROBERT (1900-1960). Methodist preacher and scholar. Born in London, he was educated at Shoreditch Secondary School and Richmond College, Surrey. He served in World War I, was ordained into the Methodist ministry (1926), and ministered at Bognor Regis, Colwyn Bay, Liverpool, Scarborough, and Leeds. He was for sixteen years minister of Westminster Central Hall, London, where his passionate, scholarly preaching drew large crowds. He was president of the London Free Church Federation (1944-46) and president of the Methodist Conference (1950). He was Cato Lecturer (1954) and member of the Senate of London University (1944-56). He became secretary of the home mission department of the Methodist Church. His doctoral thesis was published as *The Path to Perfection* (1943), a study of John Wesley's doctrine. His many other books include *The Pure in Heart* and several on preaching. J.G.G. NORMAN

SANHEDRIN (Gr. *synedrion,* "a council"). The term was used by the Rabbis both for the supreme council and court of the Jews in Jerusalem of seventy-one members and for the lesser tribunals of twenty-three. The rules governing them are found in tractates *Sanhedrin* and *Makkot* of the Mishnah, but it is certain that in their present formation they are later than the fall of Jerusalem, and many probably represent an ideal. The Sanhedrin in this form probably began with the election of Simon to the high-priesthood (142 B.C.); it became an ideal when John Hyrcanus broke with the Pharisees. It may have been reintroduced under Alexandra Salome (76 B.C.), but it certainly ceased to exist from the time of Herod. In the NT the Sanhedrin was a body dominated by the high priest and aristocratic Sadducees. Büchler's idea that there were two sanhedrins, one political and the other religious, has little to commend it. H.L. ELLISON

SANKEY, IRA DAVID (1840-1908). Singing evangelist and associate of D.L. Moody.* Born in Lawrence County, Pennsylvania, he early acquired musical talent. After service in the Union Army during the Civil War, he returned to Newcastle, Pennsylvania. Singing soon became his chief interest. He often sang at Sunday school conventions. In 1870 as a delegate to the international convention of the YMCA at Indianapolis, he impressed D.L. Moody, who persuaded Sankey to join him in his evangelistic work in Chicago. This meeting inseparably linked the two for the next quarter-century. He assisted Moody in a remarkable series of meetings in the British Isles (1873-75), during which their popular *Sankey and Moody Hymn Book* was published. On their return to the USA they were national figures. Sankey's baritone voice was not exceptional but, accompanying himself on a small reed organ, he could pour great feeling into the simplest hymn. BRUCE L. SHELLEY

SANQUHAR DECLARATION (1680). Delivered at the Burgh Cross of Sanquhar in S Scotland by Richard Cameron* and a band of Covenanters,* this was an audacious but significant piece of defiance against Charles II. The Declaration directs its anathemas against him as the chief author of the persecution against the Covenanters, and does so, moreover, for his civil as well as ecclesiastical tyranny. It warns against the implications of the (Roman Catholic) duke of York as heir to the throne, and is the first Covenanting statement in which allegiance to Charles II was renounced because of his claim to supremacy over the church. Ostensibly a futile gesture, representing only a tiny minority in 1680, it was to represent the mind of Great Britain as a whole nine years later, when it was the basis for the Revolution Settlement. J.D. DOUGLAS

SAPHIR, ADOLPH (1831-1891). Presbyterian minister. Born in Budapest, Hungary, the son of a Jewish merchant, he and the rest of his family were converted to Christianity by the Jewish mission of the Church of Scotland. Deciding to enter the ministry, he studied at the Free Church College, Edinburgh, and the universities of Aberdeen and Glasgow (M.A., 1854). He served as a missionary to the Jews in Hamburg in 1854, as minister of English Presbyterian churches at South Shields, Greenwich, Notting Hill, and Belgrave, London. Throughout his lifetime he maintained a great interest in the conversion of Jews and other non-Christians in Europe, serving in many capacities in various missionary agencies. His writings include *Christ and the Scriptures,* (1864), *Expository Lectures on the Epistle to the Hebrews* (1874-76), and *The Divine Unity of Scripture* (1892). KEITH J. HARDMAN

SARABAITES. A widely spread pre-Benedictine class of ascetics who lived either privately or in small groups in or near cities, with neither superior nor definite rule, keeping the result of their manual labors. Jerome speaks of them under the name of "Remoboth," and John Cassian notes their existence in Egypt and elsewhere, both references being adverse. The name, of which the original meaning is undetermined, later signified degenerate monks.

SARACENS. A word used by medieval Christians to describe the infidels and Muslims during the period of the Crusades.* The term appears on ancient inscriptions and seems to have applied originally only to a single tribe in the Sinai area. Later the meaning was expanded to refer to nomads in general, and after the Arab conquests of the seventh century the Byzantine empire called all Muslims Saracens. Western Christians adopted the term during the Crusades, and it was common to call all Muslims Saracens until the fall of Constantinople in 1453.

SARAVIA, HADRIAN (1531-1613). Anglican scholar. Born in Hesdin, Artois, to a Spanish Protestant, he was a drafter of the Belgic Confession and a refugee in Guernsey and England before returning to a chair of theology in Leyden (1582). Unsuccessful in persuading Elizabeth I of England to intervene more actively in the Netherlands, and finding his position untenable, Saravia returned to England in 1587 and became rector of Tattenhill the following year. Friendship with men like L. Andrewes* and R. Hooker* profoundly modified his Calvinism. Though firmly upholding the supremacy of Scripture, he insisted that no doctrinal changes would be introduced against the witness of the Fathers. He engaged in vigorous controversy with T. Beza* over divine-right Presbyterianism, argued powerfully for episcopacy, and attacked Calvin's views on Christ's descent into hell and predestination. Saravia received considerable preferment and was one of the OT translators of the Authorized Version as well as one of the first Protestant advocates of foreign missions. IAN BREWARD

SARDICA, COUNCIL OF (342). A council summoned by the emperors Constans and Constantius at the request of Pope Julius to settle the orthodoxy of Athanasius,* Marcellus of Ancyra,* and Asclepiades of Gaza, deposed at the Council of Tyre (335). Though it was intended to be an ecumenical council, the seventy-six Eastern bishops—including Acacius of Caesarea, Basil of Ancyra, Maris of Chalcedon—and the Western bishops Ursacius of Singidunum and Valens of Mursa refused to take part because Athanasius was accepted as a proper council member. Nearly 300 Western bishops met under the presidency of Hosius* of Cordova and Protogenes of Sardica. They confirmed the restoration of Athanasius, acquitted Marcellus of heresy, and restored Asclepiades. They deposed Acacius, Basil, Gregory of Alexandria, Ursacius, and Valens, among others, as Arians. They also passed disciplinary canons, among which canons 3, 4, and 5 constituted the bishop of Rome as a court of appeal for accused bishops in certain circumstances. They set the date of Easter for the following fifty years. They also promulgated in general terms a formula of faith declaring that the "hypostasis" of the Father and Son was one, taking the word in the sense of nature or substance.

The Eastern bishops withdrew to Philippopolis, where they subscribed the Fourth Creed of Antioch (341). They explained in more detail their reasons for deposing Athanasius and Marcellus and issued their own list of condemnations, which included Julius, Hosius, Protogenes, and Maximin of Treves. The Council of Sardica began a schismatic process leading straight to the East/West separation of 1054. Sardica is the modern Sofia. J.G.G. NORMAN

SÄRKILAX, PETER (Pietari Särkilahti). (d.1529). Finnish Reformer. At the time of the Reformation, Finland was politically joined with Sweden. The first impulses towards a reformation of the Church of Finland came, however, not from Sweden but through a Finnish scholar. Särkilax studied from 1516 at the University of Rostock and thus was in Germany when the Reformation began. Returning to Finland in 1523, he was seized with the spirit of the Wittenberg movement, acting boldly and openly in favor of it as a member of the diocesan board of Åbo (Turku). During his time in Germany he married, and so was one of the first Lutheran pastors to break the celibacy regulations. In his work he was constantly stressing the necessity of a pure doctrine free from papal heresy. The most important result of his activity was that Mikael Agricola,* the chief figure in the Finnish Reformation, during his youth was deeply influenced by Särkilax.

STIG-OLOF FERNSTROM

SARPI, PAOLO (1552-1623). Servite* theologian. He attended a school for nobility in Venice, coming under the Servite tutelage of Gian Maria Capella. At eighteen the bishop of Mantua made Sarpi a reader in canon law. After studying Greek, Hebrew, philosophy, law, natural science, and history, he was elected provincial of the Servite Order. While in Rome during 1580 and between 1585 and 1588 he met Bellarmine* and other influential men. Meanwhile in 1578 Sarpi took a doctorate at Padua and mingled in Venetian society. He was a close friend of French Protestants, many of whom he met in Venice. While yet in his twenties, he was accused by the Inquisition in Milan for denying that Genesis 1 taught the Trinity. In 1601 the papal nuncio reported his identity with the errors taught possibly by Morosini. The pope blamed Sarpi in 1606 for dissensions in Venice, using this as one excuse for the famous interdict of 1606. Sarpi prepared the Venetian defense for the Senate's formal reply; his defense of Venetian liberties stands behind his famous *Istoria del Concilio Tridentino,* a critical account of Trent's* attempt to reform Catholicism (ET 1619). MARVIN W. ANDERSON

SARTRE, J.P., see EXISTENTIALISM

SATAN, see DEVIL

SATURNINUS (second century). Gnostic. In the catalogs of earliest Gnostics and their systems, found in Justin Martyr and Irenaeus and repeated in Eusebius, Saturninus is the third named in the chain stemming from Simon Magus through Menander, and issuing in the further developments of Basilides, the Ophites, and Valentinus. The movement was already established in the late first century, and Saturninus can be dated no later than the first decades of the second. Like Menander he taught at Antioch, his teaching falling into the kind of pattern outlined in the *Apocryphon of John,* which originated in Greek but has been recovered from Coptic papyri. The material also shows an affinity to the kind of development stemming from the gospel of John, which found its earliest commentators in those same Gnostic circles (see GNOSTICISM).

CLYDE CURRY SMITH

SATYRUS (d. c.375). Elder brother of Ambrose,* whose household and property he managed. Ambrose preached his funeral oration *(De Excessu Fratris),* paying warm tribute to one who passed on an "undivided patrimony ... neither distributed nor diminished but preserved" with Ambrose and his sister as "stewards, not heirs."

SAUL OF TARSUS, see PAUL THE APOSTLE

SAVA (Sabas) (c.1175-c.1235). Patron saint of Serbia. The third son of the Serbian monarch Stephen Nemanya, he retired in 1191 to the monastery of Mt. Athos. Five years later his father abdicated and joined him. Sava and his father founded the monastery of Khilandari which became a center of Serbian culture, and Sava remained there until 1208 when he returned to become active in political affairs. He was the first archbishop of an autonomous Serbian Church, and was responsible for the organization of the Serbian Church and the building of many churches. He died in Bulgaria while returning from a trip to Palestine.

SAVONAROLA, GIROLAMO (1452-1498). Italian reformer. Born in Ferrara and destined at first for a career in medicine, he joined the Dominicans (1474) and served in several N Italian cities. Although at first unsuccessful as a preacher, he achieved oratorical confidence and fame through a series of sermons on the Apocalypse preached at Brescia (1486). In 1490 he settled in Florence, where he preached at the Medici foundation of San Marco, calling for repentance on the part of the city's leaders and pleading the cause of the poor and oppressed. Elected prior of San Marco and invited to preach in the cathedral, he grew in influence. By 1494 he predicted a flood of divine judgment would be unleashed on Florence.

When the French king, Charles VIII, invaded Italy, it seemed that God's wrath had struck. Twice Savonarola persuaded the king not to sack the city, and finally Charles left without having done any deliberate damage. At the approach of the French the Medici "boss" of the city, Piero, had left, and then with the removal of Charles it seemed to Savonarola that divine grace had intervened in behalf of Florence. He announced that a "golden age" had come and the city would soon have temporal and spiritual power over all Italy. He encouraged the establishment of a republican government similar to that of Venice. Under this administration he held the city in moral tension and initiated tax reform, aided the poor, reformed the courts, changing Florence from a lax, corrupt, pleasure-loving city into an ascetic, monastic-type community. This was done through the use of censorship and violent methods—for example, during the carnival of 1496 he inspired the "burning of the vanities" when the people made a great bonfire of their gambling equipment, cosmetics, false hair, and lewd books.

Savonarola also denounced Alexander VI* and the corrupt papal court. The pope, unhappy because of Florence's alliance with the French and the preaching of the "meddlesome friar," excommunicated Savonarola and threatened to place the city under an interdict if he was permitted to preach again. Although Savonarola denied the validity of the ban since, as he put it, Alexander was the representative of Satan not Christ, the people of Florence were frightened. Moreover, some of the wealthy citizens were impatient with the friar's ideas. The Franciscans arranged an ordeal in which one of their number and a follower of Savonarola would march through a fire. When the flames were kindled for the test, an argument broke out between the two groups and a sudden rainstorm quenched the fire. This incident helped to discredit Savonarola, who was tried for heresy, found guilty, and executed.

BIBLIOGRAPHY: M. de la Bedoyère, *The Meddlesome Friar and the Wayward Pope* (1958); R. Ridolfi, *The Life of Girolamo Savonarola* (1959); D. Weinstein, *Savonarola and Florence, Prophecy and Patriotism in the Renaissance* (1970).

ROBERT G. CLOUSE

SAVOY CONFERENCE (1661). An official conference of twelve Anglican bishops and twelve Puritan (mostly Presbyterian) divines, with nine coadjutors for each side, called after the Restoration of Charles II* and episcopacy (1660) to settle differences concerning the Book of Common Prayer.* The Puritans were determined to revise the sections which had caused so much agitation in the past—sections calling for the wearing of the surplice, kneeling at Communion, making the sign of the cross at baptism, bowing at the name of Jesus, and so on. The design was, of course, that the Puritans would gain concessions allowing them to serve in good conscience in the established church. Richard Baxter* and Edmund Calamy* led the Puritans, Gilbert Sheldon* and Accepted Frewen the Anglicans. The bishops took a defensive and unyielding position, rejecting Baxter's alternate liturgy. The Royalist sentiment in Parliament supported their hard line. No substantial changes resulted, and the 1662 Act of Uniformity* deprived more than 2,000 Puritans of their livings. BRIAN G. ARMSTRONG

SAVOY DECLARATION (1658). The first and basic English Congregational (or Independent) statement of doctrine and church polity. It was the product of a semi-official meeting of Congregational churchmen at the Savoy Palace in London. The reasons for it, its composition, and its daily agenda are imperfectly known. Probably about 200 representatives attended, with the majority probably laymen. Most of the leading Congregational ministers, however, were present, including Philip Nye,* Thomas Goodwin,* John Owen,* William Bridge,* William Greenhill,* and Joseph Caryl. Philip Nye was most likely the moderator. In a remarkably short time and with remarkable unanimity the group drafted and approved three documents which make up the Declaration—a preface, a confession of faith, and a church polity. The preface is notable only for its verbosity and tolerant spirit. Except for the chapters on church government, the confession is essentially the same as the Westminster Confession*—not surprisingly, since all who worked on it were at Westminster, excepting John Owen. The church polity section is brief and clear, establishing complete autonomy for local congregations under the headship of Christ. The confessional section became more or less standard in New England Congregationalism.

BRIAN G. ARMSTRONG

SAWTREY, WILLIAM (d.1401). Lollard* martyr. A priest in Norfolk in 1399, he was summoned before the bishop of Norwich and charged with heresies, which he supposedly abjured. In 1401, while he was attached to St. Osyth's, London, his heretical teaching came to the attention of Archbishop Arundel.* Sawtrey was the first to be tried before the newly passed statute *De Haeretico Comburendo,* appearing before convocation at St. Paul's. The charges were that he refused to adore the cross save as a symbol; maintained that priests should omit repetition of the hours for more important duties such as preaching, holding that money spent on pilgrimages should instead be distributed to the poor; and believed that bread after consecration was essentially unchanged. He appealed to king and Parliament on the basis of the NT and Augustine. The archbishop tried to force a change of heart, but Sawtrey held his ground. He was condemned as a relapsed heretic, in view of his former abjuration, degraded from priest to doorkeeper, and stripped of every clerical function and vestment including his tonsure. Finally he was burnt in chains at Smithfield. C.G. THORNE, JR.

SAXON CONFESSION *(Confessio Saxonica)* **(1551).** Protestant confession of faith drawn up by Philip Melanchthon* at the emperor's request for the Council of Trent.* It appeared first in Latin at

Basle, 1552. It followed the main lines of the Augsburg Confession* (1530), but was less conciliatory, there remaining no hope of Protestant-Roman Catholic reconciliation. The Scriptures as understood by the ancient church in the ecumenical creeds were declared the only and unalterable foundation of faith. The distinctive Christian doctrines were elaborated around the two articles in the Apostles' Creed* on the forgiveness of sins and the church. The former was held to exclude merit and justification by works; the latter to prove the church to be spiritual though visible communion of believers in Christ. The sacramental character of the eucharistic gifts was asserted to be confined to their use in the service. Unlike the Augsburg Confession, only the theologians signed, among them Melanchthon, J. Bugenhagen, and Georg Major. It was presented to the Council of Trent in 1552, together with the Württemberg Confession.* J.G.G. NORMAN

SAYBROOK PLATFORM (1708). A confessional document produced by the Congregational churches of Connecticut, it contained a confession (a reaffirmation of the Savoy Confession) and articles (a fifteen-point statement of ecclesiastical polity). The articles provided for the establishment of consociations in each county, with powers of oversight of the local congregations; ministerial associations in each county, with powers to examine ministerial candidates on doctrine and morals; and a general annual association, with undefined responsibility but composed of delegates from each consociation. Participation in the general association was voluntary. The Saybrook Platform replaced the Cambridge Platform* of 1648 as the most important confessional document of New England. DONALD M. LAKE

SAYCE, ARCHIBALD HENRY (1845-1933). Assyriologist. Born at Shirehampton, son of a vicar, he was educated at Queen's College, Oxford, where he became a fellow in 1869. Ordained and unmarried, he pursued a life of leisurely scholarship—in part a reflection of that fragile appearance which took him almost every winter from 1879 to 1908 to his "houseboat on the Nile." He became deputy professor of comparative philosophy in 1876, and first professor of Assyriology in England from 1891 until retirement in 1919. He was a member of the OT revision company (1874-84). No literalist, he nonetheless became—with his sharp wit and realization of the importance of discoveries—a sagacious opponent of a rampant higher criticism too preoccupied with its own theories of literary formation to look again at the empirical evidence. He held that Wellhausen's Pentateuchal literary divisions were as foolish as those of Homer by F.A. Wolf, since both were based on the same error that writing had not antedated the fifth century B.C. His *Reminiscences* (1923) were full of insight into the early history of, and men responsible for, the rediscovery of the ancient Orient. CLYDE CURRY SMITH

SAYERS, DOROTHY LEIGH (1893-1957). English writer. Born in Oxford, she graduated at the university there and then embarked on a teaching career. In 1923 she published her first of a long series of detective novels which were to make her probably the most popular mystery writer in England. During World War II she lived in Essex and was a member of the group that included C.S. Lewis, Charles Williams, J.R.R. Tolkien, and Owen Barfield. By nature and by preference she was a scholar and an expert on the Middle Ages. Her translation of the *Divine Comedy* is one of the finest, with unexcelled notes illuminating the Christian meanings of the poem. *The Man Born to Be King* (1941) is a series of radio plays on the life of Christ which displays her fine insights and her substantial endowments as a dramatist. In *The Mind of the Maker* she is at her best as a lay apologist for Christian doctrine, especially the doctrine of the Trinity. PAUL M. BECHTEL

SAYINGS OF JESUS. This term is used to denote either the extracanonical sayings of Jesus ("Agrapha") or collected sayings of Jesus ("Logia") supposedly in "Q," the *Gospel of Thomas,** and other writings. The Oxyrhynchus* discoveries (papyri I, 654, 655 and 840) are examples of collected sayings of Jesus. Recent scholarship indicates that the Oxyrhynchus collections are based upon the *Gospel of Thomas.*

SCALIGER, JOSEPH JUSTUS (1540-1609). Huguenot* scholar. Of Italian descent, he was born in Agen in S France, son of a renowned humanist, J.C. Scaliger, of whom he was companion and student in Latin from age fourteen. On his father's death in 1558 he went to Paris where he taught himself Greek before he could pursue the lectures of Adrianus Turnebus. In 1563 Scaliger, by then a Huguenot, was introduced to the nobleman Louis Chasteigner, with whose family he traveled and lived from time to time over the next thirty years while he pursued private research under their patronage. He escaped the St. Bartholomew's Day* Massacre by retreating to Geneva (1572-74), where he lectured until his return to France. His interest in textual criticism was enlarged by his study of ancient astronomy, and he it was who initiated the modern science of chronology. In 1593 he went to Leyden as professor of classics and there continued his studies toward his *Thesaurus Temporum* (1606), which had at its base his recovery of the *Chronicle* of Eusebius. Among the disciples he influenced were Daniel Heinsius (1580-1655) and especially Hugo Grotius.* CLYDE CURRY SMITH

SCHAFF, PHILIP (1819-1893). Theologian, church historian, and pioneer ecumenist. Born in Switzerland, son of a carpenter, he gained his education through scholarships. After graduating from the gymnasium at Stuttgart, he entered Tübingen University where he studied under F.C. Baur. Later he attended Halle and Berlin universities, where he studied with F.A.G. Tholuck, E.W. Hengstenberg, and J.A.W. Neander. A brilliant student, he was invited in 1844 to become professor of church history and biblical literature in the theological seminary of the German Reformed Church at Mercersburg, Pennsylvania. *The Principle of Protestantism* (1844; new ed.,

1964), his inaugural address, viewed the history of the Christian Church as a divine development leading to a merger of Protestantism and Roman Catholicism into a renewed evangelical Catholicism. This address caused him to be cited for heresy, but he was later exonerated. In the following years he and John Nevin* shaped the Mercersburg Theology.* From 1870 until his death he was professor at Union Theological Seminary, New York. Beginning in 1866, he was also active in the cause of Christian unity through working in the Evangelical Alliance.* A prolific writer, he published *A History of the Christian Church* (7 vols., 1858-92) and *The Creeds of Christendom* (3 vols., 1877). He edited the translated editions of Lange's *Commentary* (1864-80) and the Schaff-Herzog *Encyclopedia of Religious Knowledge* (1884) and helped to prepare the Revised Version of the Bible. In 1888 he founded the American Society of Church History and served as its first president.

See D.S. Schaff, *The Life of Philip Schaff* (1897). ROBERT SCHNUCKER

SCHALL, JOHANN ADAM (1591-1666). Jesuit missionary to China. Born in Cologne, he became a Jesuit, and in 1619 arrived in Macao. In 1630 he succeeded Ricci* in Peking, and by prophesying an eclipse gained an immediate reputation and a place on the board of astronomy engaged in reforming the calendar. He also became chaplain of a chapel within the imperial palace, having won the esteem of the first Manchu emperor. Many palace eunuchs were converted. After again successfully predicting an eclipse in 1645, Schall was appointed president of the board of astronomers, but he suffered through his support of Ricci in the Chinese Rites Controversy.* Moreover Confucian literati, envious of him, falsely accused him of planning a Portuguese invasion. An amnesty saved him from a cruel death, and eventually he died naturally in Peking. LESLIE T. LYALL

SCHEEBEN, MATTHIAS JOSEPH (1835-1888). German Roman Catholic theologian. After studies at the Gregorian University, Rome, he became professor of dogma at the Cologne seminary from 1860 until his death. His vast writings combined patristic and Scholastic learning with speculative thought. Although written in a style which was sometimes obscure, they have exerted considerable influence. Scheeben was regarded as the greatest speculative Catholic theologian of his day and a pioneer of Neoscholasticism. He was a great opponent of rationalism, and a leader of conservative Catholic theology. He took a vigorous part in the controversies attending the First Vatican Council* (1869-70), was a staunch defender of papal infallibility, and an opponent of Döllinger* and the Old Catholicism. His writings include *Nature and Grace* (1861; ET 1954); *Die Mysterien des Christentums* (1865); and the magisterial *Handbuch der katholischen Dogmatik*, I-III (1873-87; 4th volume added later by L. Atzberger). *A Manual of Catholic Theology based on Scheeben's "Dogmatik"* was published in English in 1890. Extracts were published as *Mariology* (1946). A *Festschrift* celebrating the centenary of his birth was published in Rome in 1935. COLIN BROWN

SCHEEL, OTTO (1873-1954). German church historian. Born in N Schleswig, Scheel studied historical theology at Halle and Kiel and was appointed professor of church history at Tübingen in 1906. For many years he was chairman of the *Verein für Reformationsgeschichte.* His most significant work, *Martin Luther: Von Katholizismus zur Reformation* (1914-17), was not merely an analysis of Luther's youth, but also a study of education in the period. Scheel left it uncompleted when he accepted a new chair of Schleswig-Holstein local history at Kiel in 1924. He became interested in his homeland, especially the influence of Pietism and the social history of the peasantry, and he was active in organizations that promoted cultural identity among N Schleswigers.

RICHARD V. PIERARD

SCHEFFLER, JOHANN (Angelus Silesius) (1624-1677). Polish hymnwriter. Son of a Lutheran Polish nobleman, he graduated in medicine and about 1649 became physician to the duke of Württemberg-Oels. His interest in mysticism, aroused by the Dutch followers of Boehme,* made him clash with Lutheran leaders. He resigned his post in 1652 and returned to Breslau, where in 1653 he became a zealous Roman Catholic. In 1671 he entered a monastery. His hymns, mostly uncontroversial and written while he was a Lutheran, were collected in *Heilige Seelenlust* (1657), e.g., "Thee will I love, my strength, my tower" (tr. J. Wesley); "O God, of good th'unfathomed sea" (Wesley); and "O Love, who formedst me to wear" (Winkworth). His religious epigrams were collected in his *Cherubinischer Wandersmann* (2nd ed., 1675). JOHN S. ANDREWS

SCHEIDT, SAMUEL (1587-1654). German composer. Most of his creative life was spent as organist of the *Moritzkirche* in Halle. He studied organ in Amsterdam with Sweelinck and ranks as one of the first great masters of the organ-chorale (chorale prelude). He is perhaps the first composer of such pieces whose works have more than an antiquarian interest to the church organist today. His *Tabulatura nova* (1624) is the first great collection of organ music in Germany to use staff notation exclusively instead of letter tablature. He also wrote many choral concertatos based on the Lutheran chorales. Later he published the *Görlitz Tabulaturbuch* (1650) containing 100 chorales in four-part harmony, ostensibly for the purpose of accompanying congregational singing. It is the first known book of its kind. He is in the first generation of great N German organists that leads up to Buxtehude and so to Bach.

J.B. MAC MILLAN

SCHEIN, JOHANN HERMANN (1586-1630). German composer. With his two contemporaries, Schütz and Scheidt, he stands out as one of the great Lutheran musicians of the early Baroque era. He was one of a number of distinguished names who acted as cantor of the *Thomasschule* in Leipzig, the position held by J.S. Bach a century later. Along with Scheidt, he contributed largely to the development of the Lutheran chorale concertato, one of the forerunners of the church cantata of Bach's time. He applied the Italian style of the early Baroque to such works as these, breaking up the chorale melody into florid passages and employing dramatic declamatory sections. He was probably the first to publish a chorale book with just the melodies and the continuo bass, which later became the norm in such collections. Schein also holds an honored place in the development of the instrumental suite.

J.B. MAC MILLAN

SCHELLING, FRIEDRICH WILHELM JOSEPH VON (1775-1854). German idealist philosopher. Son of a Württemberg pastor, he was trained at Tübingen, and from 1798 to 1803 held a professorship at Jena, the center of German Romanticism. He became friends with its leading figures—Friedrich and August Schlegel, Fichte, Hegel, and Goethe. Schelling's transcendental idealism, with its emphasis on the importance of the individual, the value of art, antirationalism, organicism, and vitalism, was the epitome of German Romantic philosophy. He later taught at Würzburg, Erlangen, and Munich and finally went to Berlin in 1841. His thought underwent considerable change and development. Starting from subjective idealism, he gradually worked out a philosophy of nature where the pure object and subject were integrated into an absolute unity of spirit and nature in God, the divine essence which can only be apprehended by will. The root of existence is God, the ungrounded, the eternal nothing, and only he is reality. Finite things are unreal and can only exist in removal from the Absolute, who then creates his own counterpart which is freedom. Schelling later became dissatisfied with this logical pantheism, stressed ideas as a route to ultimate reality, and even tried to reconcile Christianity with his philosophy. This emphasis resembles that of modern existentialism and has resulted in a renewed interest in Schelling.

RICHARD V. PIERARD

SCHERESCHEWSKY, SAMUEL ISAAC (1831-1906). Missionary and translator. Born at Taurrogen in Russian Lithuania, of Jewish parents, he graduated from Breslau University. Through reading the NT he became a Christian. Going to the USA, he was baptized in a Baptist church (1854), studied in a Presbyterian seminary (1855-58), joined the Protestant Episcopal Church, and studied further at the General Theological Seminary, New York (1858-59). He served as a missionary in Shanghai (1859) and Peking (1863-75); gifted in languages, he collaborated with others in translating the Prayer Book and NT into Mandarin, and undertook the OT by himself. He became bishop of Shanghai in 1877, but four years later was struck with paralysis. Despite his incapacity, one of his later achievements was the translation of the whole Bible into Wen-li, typing with one finger.

J.G.G. NORMAN

SCHISM (Gr. *schisma*). Ecclesiastical term for division in or separation from a church, distinguished from heresy in that the separation involved is not basically doctrinal. It may not entail loss of orders, i.e., schismatic ordination and administration of sacraments are valid. In this technical sense the word first occurs in Irenaeus.* Cyprian* discussed the relation of the church to schism, and he condemned schismatics for endangering men's souls, regarding them as worse than apostates and their baptism as worthless. Augustine* took a similar view, but did not regard schismatic sacraments as invalid. In time obedience to the Roman pontiff became the test of catholicity. Outstanding schisms include the Novatianist* and Donatist* churches, the "Great Schism"* between the Greek and Latin churches (finally established in 1054), the Avignon* schism (1378-1417), and the schism between the Church of England and Rome (since 1570).

J.G.G. NORMAN

SCHLATTER, ADOLF VON (1852-1938). Swiss NT scholar. Born in St. Gall, he studied theology at Basle and Tübingen. His later essay on J.T. Beck* suggests a primary influence on his thought. After a pastorate in Switzerland he taught at Bern (1880-88) before becoming NT professor successively at Griefswald (1888), Berlin (1893), and Tübingen (1898) where he remained until retirement in 1922. He was allied with no school, ecumenical in outlook, and concerned to mediate between liberals and Pietists. With A.H. Cremer* he edited from 1897 the *Beiträge zur Förderung Christlicher Theologie,* to which he frequently contributed and on whose origins and importance for his own theological work he bore witness. His theological writing from *Der Glaube im Neuen Testament* (1885) to his mature theology of the NT in two volumes—*Die Geschichte des Christus* (1921) and *Die Theologie der Apostel* (1922)—puts the emphasis on the importance of Jesus, finding anchorage in the facts of faith rather than in speculative thought. Schlatter stressed that both theology and history must not forget God, and he wrote histories both of Israel from Alexander to Hadrian (1901) and of the early church (1926; ET 1955). His specific studies on NT books displayed similar independence; he was one of the few to break with the trend of his times to continue support for the priority of Matthew.

CLYDE CURRY SMITH

SCHLATTER, MICHAEL (1716-1790). Organizer of the German Reformed Church in the Thirteen Colonies. Born in St. Gall, Switzerland, and educated at the University of Gelmstadt, he was ordained in 1739 as a minister in the German Reformed Church. In 1746 he was sent to America to organize and supervise the scattered German Reformed congregations, and did so ably. In 1751, on a trip to Europe to secure ministers for the colonial churches, one of those recruited was Philip Otterbein.* Schlatter served as superintendent of schools in Philadelphia (1754-56) and as a British army chaplain (1756-59). He then became a pastor in Philadelphia. His ardent patriotism led to loss of property and imprisonment in the Revolutionary War.

EARLE E. CAIRNS

SCHLEIERMACHER, FRIEDRICH DANIEL ERNST (1768-1834). German theologian. Of Silesian Pietistic background, he in his student days rejected the narrowness of Pietism,* but in later life regarded himself as a Pietist, only of a higher order. He studied at Halle and was ordained in 1794. In 1796 he became minister of the Charité Hospital, Berlin, and in this period he was drawn into the circle of Romantic writers who constituted the intellectual *avant-garde* of the time. Returning as a professor to Halle (1804), his next years were overshadowed by the Napoleonic Wars and the revival of German nationalism. From 1809 Schleiermacher was minister of the *Dreifaltigkeitskirche* in Berlin. The closing of the University of Halle led to the foundation of a new university at Berlin. Schleiermacher played a leading part in its foundation and in the establishment of its theological faculty, in which he was a professor. He was the first dean of the faculty, and for a time rector of the university. He was instrumental in bringing Hegel* to Berlin, though their relationship became strained through differences of outlook. Schleiermacher was a great supporter of German national unity and a leading advocate of the scheme to unite the Lutheran and Reformed churches in Prussia. He preached regularly and wrote voluminously. His writings embraced systematic theology, hermeneutics, philosophy, translations of Plato, and ten volumes of sermons.

Schleiermacher's first major work, *On Religion: Speeches to Its Cultured Despisers* (1799; ET 1894), is sometimes regarded as a theological expression of Romanticism. As an apology for Christianity in the post-Enlightenment world, it was neither a restatement of biblical orthodoxy nor a refurbishing of enlightened moralistic religion. It defined religion as "sense and taste for the infinite" and sought to show that life without religion is incomplete.

In many ways the work anticipated Schleiermacher's fuller statement of belief, *The Christian Faith* (1821; 2nd ed., 1830-31; ET 1928). Schleiermacher sought to avoid the alternatives of an orthodoxy based on revealed truth and a natural theology based on abstract speculation. He adopted a positive approach to religion based on a descriptive analysis of religious experience. He sought to analyze the essential elements in Christian experience and show how they were related to the main articles of Christian faith. The basis of religion is neither activity nor knowledge, but something which underlies them both: the continuum of feeling or awareness which we call self-consciousness. The common factor of religious experience is the feeling or sense of absolute dependence. This concept became not only the key to understanding religion, but the criterion for assessing the teaching of the past and the means of reinterpreting Christianity for modern man.

Sin is seen as essentially a wrongful desire for independence. The orthodox two-natures doctrine of Christ is replaced by the picture of a man in whom dependence was complete. It was his profound experience of God through his sense of dependence that constitutes an existence of God in him. On account of it, Jesus is able to mediate a new redemptive awareness of God to humanity. The same approach was elaborated in the posthumously published lectures on *Einleitung ins Neue Testament* (1845) and *Das Leben Jesu* (1864). Although he was ostensibly empirical, the question may be asked whether Schleiermacher was empirical enough: whether his concept of religion was not in fact too narrowly defined and used in an arbitrary and unrealistic manner. The result is that Christian doctrine is forced into the straitjacket of a preconceived system.

Schleiermacher's influence extended far beyond his disciples who made up the school of Mediating Theology *(Vermittlungstheologie)* in the mid-nineteenth century. To Karl Barth* he epitomized the liberal approach to religion which dwelt upon man rather than God. He has found renewed following among twentieth-century radicals. In many respects his method, his view of God and of man, and his Christology anticipated those of Paul Tillich* and J.A.T. Robinson.

BIBLIOGRAPHY: T.N. Rice, *Schleiermacher Bibliography* (1966). See also *Theology and Church* (1962), pp. 136-216; R.R. Niebuhr, *Schleiermacher on Christ and Religion* (1965); *SJT* 21, 3 (1968) and *Journal for Theology and the Church* 7 (1970) for articles on his relevance today; S. Sykes, *Friedrich Schleiermacher* (1971); M. Redeker, *Schleiermacher: Life and Thought* (1973). COLIN BROWN

SCHMALKALDIC, see SMALCALD

SCHMIDT, GEORG (1709-1785). Missionary to South Africa. He joined the Moravian Brethren* at Herrnhut in 1727, was imprisoned by Catholic authorities, and recanted after six years to obtain his freedom. As a punishment for this "weakness" he was sent alone to the Cape as first missionary to the Hottentots. Arriving in 1737, he settled at Baviaanskloof (later Genadendal), some eighty miles from Cape Town, where he gathered a small community of interested Hottentots. In 1742 Zinzendorf* ordained him by letter, and he baptized five converts. This action was resented by the local Dutch Reformed clergy who already doubted Moravian orthodoxy. Schmidt was told to discontinue baptisms pending the decision of the Amsterdam Classis. Lonely and depressed, he left for Holland in 1744, hoping to remove obstacles to his work and to return. This was not permitted, and only in 1792 was Moravian work resumed. After 1752 Schmidt lived at Niesky, a Moravian settlement, where he died. D.G.L. CRAGG

SCHMIEDEL, PAUL WILHELM (1851-1935). NT scholar. Professor of NT exegesis at the University of Zurich from 1893 to 1923, he tended toward a very radical criticism of the NT. He is best known for his famous thesis concerning the nine "pillar-passages" foundational for "a truly scientific life of Jesus" (his words), which was set forth in the lengthy article on the "Gospels" which appeared in the *Encyclopedia Biblica* in 1901. He also penned the article on "Acts" in the same encyclopedia (1899), in which he took a very poor view of the historicity of Acts. In a study of the fourth gospel (1906; ET 1908) he strongly contrasted John and the synoptics as sources for knowledge of Jesus. His *Die Person Jesu im Streit der Meinungen der Gegenwart* (1906) was the center of much controversy in his day.

W. WARD GASQUE

SCHMUCKER, SAMUEL SIMON (1799-1873). Lutheran clergyman. Dominant figure of American Lutheranism in his day, he sought to Americanize the eastern churches, organized as the General Synod in 1821. A founder of the Lutheran Theological Seminary in Gettysburg, Pennsylvania (1825), the first of its professors, and the creator of Pennsylvania (now Gettysburg) College (1832), he asserted that Lutheran Pietism was the best weapon against rationalism. Lutheranism, however, had to be freed from German culture and "her former lifeless and distracted condition." Accordingly he published *Elements of Popular Theology* (1834), the first English-language Lutheran theology in America, and declared in *A Fraternal Appeal* (1838) that Lutheranism could thrive as ally of New School evangelicalism. An organizer of the American branch of the European-based Evangelical Alliance* (1846), he attempted to have Lutherans adopt his "Definite Synodical Platform" (1855), but this prompted critics, notably C.P. Krauth,* to form the General Council to resist attempts at ecumenism.

DARREL BIGHAM

SCHOLARIUS, GEORGE, see GEORGE SCHOLARIUS

SCHOLASTICISM. The theology and philosophy taught in the medieval schools from the eleventh to fourteenth centuries, and revived in later periods such as in the late sixteenth and seventeenth and nineteenth and twentieth centuries. It features the application of Aristotelian categories to the Christian revelation and attempts to reconcile reason and faith, philosophy and revelation. As a theological method it is associated with organized textbook theology and the thesis method.

The movement appeared in the cultural unity created by Christianity in Carolingian times. Despite the breaking up of Charlemagne's empire, the Carolingian intellectual tradition continued in the monasteries such as those emanating from Cluny and Cîteaux and the Franciscan movement. Scholasticism was taught in the monastic and cathedral schools, and with the founding of the universities (c.1200) such as Oxford, Pisa, Bologna, and Salerno the tradition was given great impetus. The rise of humanism in the fifteenth century and the Reformation* of the sixteenth helped to destroy the medieval synthesis that made Scholasticism possible. Later attempts to revive it have never been so successful as the medieval effort, since the knowledge explosion and the fact that no one system has a monopoly on truth have altered the condition that produced and fostered its growth.

John Scotus Erigena,* who used a system of education based on Greek thought and relied on the use of reason in studying revealed data, is sometimes called the first Scholastic, but that honor seems rather to belong to Anselm of Canterbury,* who asserted that faith should precede understanding but understanding could in turn deepen faith through reason. One of the best known and most creative figures of early Scholasticism was Peter Abelard,* who in his revolt against tradition and his insistence upon the right of the philosopher to use his own reason did much to shape medieval thought. In *Sic et Non (Yes and No)* he demonstrated that tradition and authority were insufficient in themselves by making a list of questions—such as, Is God omnipotent? Do we sin without willing it? Is God a substance? Is faith based upon reason?—and then quoting authorities on both sides of the question. Although Abelard left these questions unresolved, Peter Lombard* and Thomas Aquinas* used the same method and supplied answers. These scholars wrote a type of formal treatise, a *summa,* which dealt exhaustively with a given subject. In works like these a reader could find the distinctions necessary for a complete, logical analysis of all the arguments for and against each proposition that made up the subject. The Bible was the basis for Scholastic theology, and the Church Fathers were referred to when there was an especially difficult passage to deal with. The use of Aristotelian logic by men like Abelard led to controversies such as the problem of universals. Early medieval thought was Platonic, insisting on the reality of ideas such as Soul, Honor, Tree, or Chair. However, during the twelfth century the Nominalists, who believed that reality consisted of individual items, challenged the Realists. Abelard worked out a compromise idea, called Conceptualism, that ideas are real in the human mind.

By the early thirteenth century, Scholastics were caught in a new wave of thought as they were forced to cope with the influx of a vast philosophic and scientific literature, including the advanced work of Aristotle* translated from Arabic and Greek. For the first time they confronted a world-system which relied completely on reason and operated without reference to the Christian God. Such ideas as the prime mover, eternal motion, a denial of providence and creation, uncertainty about immorality and the soul, and a morality based on reason alone caused much anguish for many medieval scholars. Some such as Siger of Brabant* followed the work of a Muslim commentator, Averroes (d. 1036), and advocated a theory of double truth, i.e., that there is one truth in human reason—Aristotle—and another in religion—the Christian revelation. Others rejected Aristotle completely, and in 1215 and 1231 decrees were passed by the University of Paris and the papacy prohibiting the study of some of his works. The problem of Aristotle, however, was solved not by censorship but by intellectual debate.

In the end, it was not Siger's ideas that won out, but the Rational Scholasticism of the Dominicans.* The greatest of these men was Thomas Aquinas, whose method represents the ultimate Scholastic refinement of the organization of knowledge. Even more important than his method, however, was Thomas's use of reason. He believed that reason can tell what God cannot be, and one can assume that what is left is something like what He is. There is, he felt, no contradiction between faith and reason as long as rational inquiry is properly conducted. When the two do conflict, faith is to correct reason. If faith leads to a conclusion that defies human reason, the incompatibility exists because of a failure of rationality. Thomas accepted the Aristotelian view that the universe is orderly, but this is due to the reflection of order of the divine mind in the universe. In harmonizing Aristotelian metaphysics with Christianity, it is the Aristotelian elements that must be fitted into the Christian system. Critics have pointed out that this is done by taking particular statements from Aristotle out of context and placing them into a new Thomistic context.

Thomas fashioned a view of the universe in which a being was good to the extent that it resembled God. Since God is a simple, incorporeal, purely spiritual being, that which is most spiritual is closest to God and that which is most material is farthest away from Him. This permits the construction of a Great Chain of Being consisting of God, angels, men, animals in order of their intelligence, plants, and inanimate objects. The moral plan of the universe is rational, hence reason as well as revelation can tell men what to do since God wills nothing arbitrary, but everything according to man's needs. Good and evil are objective realities. For example, God forbids theft and adultery because they disrupt society and make people unhappy. Aristotle had stated that the happiness of man consists in realizing his true nature, and Aquinas agreed, adding that man's true nature is union with God.

The position of Aquinas was attacked by scholars like William of Ockham* who advocated a type of double truth, but a more effective countertrend came from the Mystical Scholasticism of Bonaventure.* A professor of theology at Oxford and governor-general of the Franciscans,* he taught that rational knowledge of God is impossible, because God is different from man in quality as well as quantity. Thus knowledge of God can only be equivocal, hazy, and analogous. One may prepare for an understanding of God by separating himself from the world and by looking for reflections or shadows of God in objects. Then a man may advance to finding God within himself and experiencing His presence through grace. Finally, God's being is infused within the soul.

Another variety of Scholasticism, that based upon Empiricism, helped to prepare the way for modern science. Franciscans were the leaders in this movement as in Mystical Scholasticism, and two of their great scholars at Oxford, Robert Grosseteste* and Roger Bacon,* studied optics and the behavior of light, investigating perspective and the properties of prisms, rainbows, and mirrors. These scholars emphasized three principles of science now taken for granted: first, they believed in a scientific cosmology, a view of the world consistent with the observations of the senses; a second contribution was their emphasis on experimentation; and a third approach was the use of measurement and of quantitative concepts in the explanation of the world.

More modern Scholasticism is generally associated with a method of systematically applying reason to revealed knowledge. The Roman Catholic Church in the Counter-Reformation* used Scholasticism; in fact, at the Council of Trent* in the sixteenth century the works of Aquinas* lay open on the high altar along with the Bible as works of reference. A nineteenth-century revival of Scholasticism was inspired by the papal encyclical *Aeterni Patris,* which declared Thomism eternally valid.

Following the first generation of Reformers, despite Luther's condemnation of Aristotle and the Schoolmen, "Protestant Scholasticism" developed. A struggle within Lutheranism after the founder's death between the Gnesio-Lutherans and the Philippists was brought to a close by the Formula of Concord* (1577). This developed into a system of doctrine resembling Scholasticism and led to the late seventeenth-century growth of Pietism* as a challenge to orthodoxy. The Calvinists attempted to use the *Institutes* as the same sort of dogmatic statement as the Formula, but this led to a long series of disputes and the growth of the Arminian* movement. The major statement of Reformed Scholasticism is the decrees of the Synod of Dort* (1618-19). The contemporary heirs of Protestant Scholasticism are groups such as the American fundamentalists. The Scholastic emphasis of the Roman Catholic Church has been modified during the twentieth century, especially since the meeting of Vatican II.*

BIBLIOGRAPHY: F.C. Copleston, *Aquinas* (1955) and *A History of Philosophy* (1962-63); M. DeWulf, *An Introduction to Scholastic Philosophy* (1956); G. Leff, *Medieval Thought—Saint Augustine to Ockham* (1958); J. Pieper, *Scholasticism* (1960); J. Dillenberger, *Protestant Thought and Natural Science* (1960); D. Knowles, *The Evolution of Medieval Thought* (1962); R.W. Southern, *The Making of the Middle Ages* (1962); R. Scharlemann, *Thomas Aquinas and John Gerhard: Theological Controversy and Construction in Medieval and Protestant Scholasticism* (1964).

ROBERT G. CLOUSE

SCHOOLMEN. Teachers of philosophy and theology at the medieval European universities ("Schools"). They were the exponents of Scholasticism,* which developed after the Dark Ages, and which was concerned to systematize theology and to justify the claims of theology to reason. They discussed the Aristotelian logic and applied it to the doctrines of the church. Their chief method of teaching was known as "Dialectic"—i.e., investigating the truth of opinions by logical discussion. Prominent earlier Schoolmen include Anselm* and Abelard,* and after the thirteenth century many of the greatest came from the Mendicant Orders*—e.g., Thomas Aquinas* the Dominican, and Duns Scotus* the Franciscan.

J.G.G. NORMAN

SCHÜRER, EMIL (1844-1910). German Protestant scholar. Following studies at the universities of Erlangen, Berlin, Heidelberg, and Leipzig, he taught successively at Leipzig (1869-78), Giessen (1878-90), Kiel (1890-95), and Göttingen (1895-1910). His life-work was to establish the study of Judaism of the late intertestamental and early

Christian periods on a firm historical basis. Begun as a student's handbook of NT history (1874), his famous *History of the Jewish People in the Time of Jesus* (3 vols., 1886-90; ET in 5 vols., 1890-91) represents the fruit of his labors and, though dated, remains a standard work. Other influential publications included a study of the preaching of Jesus in relation to the OT and to Judaism (1882) and a monograph on the messianic consciousness of Jesus (1903). He was the founder of the journal, *Theologische Literaturzeitung,* which he edited from 1876 to 1880 and from 1888 to 1910.

W. WARD GASQUE

SCHÜTZ, HEINRICH (1585-1672). German composer. Unlike his great Lutheran contemporaries, Schein and Scheidt, Schütz (who was also known as Henricus Sagittarius) left no purely instrumental works. He can be confidently asserted to be the greatest Lutheran composer before J.S. Bach, who was born just 100 years later. He was sent by his patron, the landgrave of Cassel-Hesse, in 1609 to Venice to study with Giovanni Gabrieli.* There he remained until after his teacher's death in 1612, and there his first collection of works, a set of magnificent Italian madrigals, was published. These revealed him to be already an outstanding master.

Apart from a single opera now lost, the first by a German, Schütz's subsequent output was exclusively sacred. In his full maturity, this remarkable man returned to Venice to study with Gabrieli's successor, the outstanding Monteverdi.* Schütz was *Kapellmeister* at the court in Dresden for fifty-five years, except for an interim when the disruption of the Thirty Years' War forced him to find sanctuary in Copenhagen. His many modest but beautiful *Kleine geistliche Konzerte* (settings for solo voices and organ of texts from the Psalms) were the product of these years when few performers were available. Earlier he had written many works in the monumental style of the Venetians, his *Symphoniae sacrae.* He attained unprecedented refinement in the musical setting of the German language. A later collection, *Geistliche Chormusik,* consisted of magnificent scriptural motets, in which he produced a remarkable fusion of Renaissance and Baroque elements. One of these is his exquisite five-part setting of John 3:16. Rarely did Schütz employ any of the traditional Lutheran chorale melodies, but a few of them did appear among his own in his simple settings of Becker's metrical psalms. His oratorio, *The Story of Christmas,* the *Seven Words from the Cross,* and his three Passions (those of Matthew, Luke, and John) which adopt an extraordinary, severe, and economical style are works of his old age.

See H.J. Moser, *Heinrich "chuetz: His Life and Work* (tr. C.F. Pfatteicher, 1959).

J.B. MAC MILLAN

SCHWABACH, ARTICLES OF. Lutheran confessional document written in 1529. Probably composed by Luther on the basis of his *Confession Concerning the Last Supper of Christ* (1528) prior to the Colloquy at Marburg,* and shortened to serve as the fifteen Marburg articles, the text was reworked by Luther, Melanchthon, and Jonas (among others) after the Marburg Colloquy (3 October 1529) and submitted to the elector of Saxony and the margrave of Brandenburg-Anspach at Schwabach on 16 October 1529.

Whatever the history of these articles prior to their acceptance by the princes at Schwabach (evidence on this matter having been the subject of some dispute), the seventeen articles in their final form became the basis for the first part of the Augsburg Confession* (1530) and the test of admission to the Lutheran League of the North German States. The articles were directed against Catholics, Zwinglians, and Anabaptists, and affirmed the main lines of the Lutheran understanding of the Eucharist. The tenth article, for example, asserted "that in the bread and wine the body and blood of Christ are truly present, according to the word of Christ."

Walther Köhler maintained that the Schwabach Articles were not narrowly Lutheran, but should be understood as a union formula which had as its goal the reconciliation of Wittenberg and Strasbourg and the exclusion of the Swiss alone. In this view the Schwabach Articles were not simply the basis for the Augsburg Confession, but also the forerunner of the Wittenberg Concord* (1536). The text is in the *Corpus Reformatorum* 26 (1857), cols. 151-60.

DAVID C. STEINMETZ

SCHWARTZ, CHRISTIAN FRIEDRICH (1726-1760). Missionary to India. Born at Sonnenberg, Prussia, son of a master baker, he was educated at Halle University, the Pietist center. There he encountered Benjamin Schultze, a former missionary who had extended the work of Ziegenbalg* of Tranquebar (and completed the latter's Tamil translation of the Bible), and this led to the call to India. Having learnt Tamil even before sailing, Schwartz arrived in India in 1750 and spent his first years at the Danish-Halle Mission in Tranquebar. In 1760 he paid a notable visit to Ceylon. During his travels out from Tranquebar he opened up work at Trichinopoly, and in 1767 was appointed chaplain to the British there. He was therefore one of the remarkable succession of Germans who built up "English" missions in South India.

From 1772 his work moved to the kingdom of Tanjore, at the invitation of the rajah, who showed his estimate of Schwartz by wishing to appoint him guardian of the heir to the throne. The British in turn used Schwartz as emissary to their enemy in Mysore, Hyder Ali, who equally trusted the missionary; and for a period he was virtually prime minister of Tanjore. All of these political duties never deflected him from his primary calling as missionary. At Tinnevelly in the far south he appointed the catechist Sattianaden, and thus had a share in building what became a famous church. Often regarded as the greatest of the eighteenth-century German Protestant missionaries in South India, Schwartz died at Tanjore.

ROBERT J. MC MAHON

SCHWARTZ, EDUARD (1858-1940). German scholar. Born in Kiel, he studied at Göttingen, Bonn, Berlin, and Greifswald, showing great ability in ancient languages. From 1887 he held chairs in six universities, including Strasbourg and Göttingen, before settling at Berlin in 1919. Early in his career he turned his attention to the early history and literature of the church, and published editions of Eusebius's *Ecclesiastical History* (with T. Mommsen) and Tatian's *Oration to the Greeks*. Later followed studies on Athanasius, Nestorius, Theodosius the Deacon, and Constantine and the church. His greatest work was the publication (interrupted but not stopped by World War I) of the *Acta Conciliorum Oecumenicorum*, an edition of the Greek councils. In it were critical editions of the "decrees" of the councils of Ephesus (431) and Chalcedon (451). A friend of the English scholar C.H. Turner, he saw through the press for him the seventh fascicule of Turner's *Ecclesiae Occidentalis Monumenta* in 1939. PETER TOON

SCHWEITZER, ALBERT (1875-1965). German theologian, medical missionary, and musician. Born in Alsace, he gave himself to his own study till he was thirty, achieving much in theology and music (he became an expert on J.S. Bach and organs). From 1905 he studied medicine, and in 1913, loosely associated with the Paris Missionary Society, went to Gabon to found a hospital at Lambaréné. Except for interruptions in and after World War I, and for money-raising lecture and recital tours, he gave the rest of his life to developing the hospital on idiosyncratic lines, incurring criticism sometimes for old-fashioned paternalism, sometimes for his slowness to make Western standards normative for his people.

As a theologian Schweitzer was an heir of the nineteenth-century German Protestant tradition of historical and critical theology at the time of its high prosperity. Yet his work is part of that self-questioning and loss of confidence that beset it already in the 1890s. In his study of Jesus (cf. especially *The Quest of the Historical Jesus*, 1906; ET 1909), he believed he had discovered the real Jesus from the gospels by historical means, thus bringing the century-long quest to a successful conclusion. The common assertion that he ended the quest by showing that it is impossible is false, but he argued that the historical Jesus was so different from the figure beloved by the Christian humanism of liberal Protestantism that the latter position could claim no foundation in historical fact.

He maintained that Jesus was dominated throughout His career by the world-negating expectation of the imminent coming of God's kingdom as that was understood in contemporary Jewish apocalyptic, and that Jesus finally tried to force its coming by seeking His death. This view ("consistent" or "thoroughgoing" eschatology) meant that the teaching of Jesus, with its radical demand, was to be seen as appropriate to the situation in which there was only a short time left to the world ("Interim Ethic"). It implied too that Jesus' life was centrally directed by His mistaken expectation. But Schweitzer argued paradoxically that Jesus' saving achievement was partly to destroy, in His death, the eschatology by which He had lived and so to free men from it. Thus Schweitzer revealed how much of a liberal Protestant he remained. He followed the same method in his interpretation of Paul, treating Jewish eschatology as the distinctive basis of Pauline mysticism and sacramentalism. He has had considerable influence in NT scholarship, though few would now hold his extreme and distinctive positions.

Schweitzer was uncertain about traditional Christian dogma; for instance, he veered between theism and pantheism. But his own life shows how seriously he believed that Jesus' call to discipleship could still be heard in a way that determined the whole of life. Ethics was in fact a predominant concern of his. He thought philosophical ethics had failed by becoming too remote from life; in Africa he discovered and developed the ethical principle of "reverence for life" as an answer to this problem. Like many Germans of his generation, in the shadow of Schopenhauer and Nietzsche, he was dismally assured of the decay of civilization, and he became well known for his thought on its restoration.

In his last years Schweitzer was much honored, notably in the award of the Nobel Peace Prize in 1952. He had shown himself to be something of a polymath and a modern St. Francis together.

BIBLIOGRAPHY: A. Schweitzer, *Aus meiner Kindheit und Jugendzeit* (1924: ET *Memoirs of Childhood and Youth*, 1925), *Aus meinem Leben und Denken* (1931; ET *My Life and Thought*, 1933), and *Afrikanische Geschichten* (1938; ET *From My African Note-Book*, 1938); E.N. Mozley, *The Theology of A. Schweitzer* (1950); G. Seaver, *A. Schweitzer: the Man and His Mind* (5th ed., 1955); N. Cousins, *Dr. Schweitzer of Lamberéné* (1960); G. McKnight, *Verdict on Schweitzer* (1964). HADDON WILLMER

SCHWENKFELDERS. Those who follow the teachings of the aristocratic German diplomat and lay theologian, Kaspar von Ossig Schwenkfeld (1489-1561). Early acquaintance with Andreas Carlstadt* and Thomas Münzer* led him to adopt many of the principles of the Reformation,* but he had definite convictions of his own concerning the Lord's Supper, Christology, and church discipline, and these led him into successive conflicts with Luther,* Zwingli,* the Catholics, and Bucer.* He was forced to leave Silesia (1529) and Strasbourg (1534). The event that precipitated the 1540 Lutheran anathema was the publishing of Schwenkfeld's most characteristic doctrine—the deification of the humanity of Christ—in the *Grosse Confession* of 1540. In it he expressed his belief that all creatures are external to God, and God is external to all creatures. Therefore Christ's relationship to God must be entirely unique, and this uniqueness comes because He was "begotten" and not "created." God is the Father of Christ's humanity and deity. Christ's flesh stood in a very special relationship with God. This led to Schwenkfeld's being branded as a religious outlaw in 1540 by a convention of Evangelical theologians led by Melanchthon.*

With his followers he withdrew from the Lutheran Church after 1540 and established a community of worshipers who were originally called

"Confessors of the Glory of Christ." Following a middle way between the great ecclesiastical and religious parties of their day, their congregations grew most readily in Silesia and Swabia, in the towns their founder had visited, and in Prussia. The movement flourished in the vicinity of Goldberg until 1720 when an adverse tract caused Emperor Charles VI to dispatch a Jesuit coercive mission against them. Some escaped by emigrating into Saxony and, being denied tolerance, proceeded to Holland, England, and finally by 1734 to E Pennsylvania. The colony in Silesia was restored by Frederick the Great in 1742 and existed until 1826. A small group of about 2,500 still exists in Pennsylvania and is very similar to Quakers in practice and belief.

BIBLIOGRAPHY: O. Kadelbach, *Ausführliche Geschichte Schwenkfeldts und der Schwenkfeldtianer* (1861); H.W. Kriebel, *The Schwenkfelders in Pennsylvania* (1904); *Corpus Schwenkfeldianorum* (13 vols., 1907-37). JOHN P. DEVER

SCIENTOLOGY. Founded by Lafayette Ronald Hubbard (b.1911 in Nebraska), millionaire explorer and retired U.S. naval officer, the Church of Scientology of California teaches what it calls "an applied religious philosophy" aiming at "spiritual recovery and the increase of individual ability." The methods of Scientology (originally called "Dianetics") are claimed to be "technological," and make extensive use of apparatus, elaborate types of classification, and an involved—frequently cryptic—quasi-scientific language in expounding the stages by which the individual may progress, under close guidance, toward total self-determination. Its basic concept is "survival," and Scientology teaches the techniques by which the individual may "survive" most effectively.

Although passing reference is sometimes made to, for instance, "Supreme Being, the ultimate Creator, and God, when so meant" (the implications of the last clause are obscure), Scientology maintains that "the human mind and inventions of the human mind are capable of resolving any and all problems which can be sensed, measured, or experienced directly or indirectly." Man, then, is to be his own savior. Scientology further denies the existence of absolute good and absolute evil in the world of matter, energy, space, and time, and holds that "that which is good for an organism may be defined as that which promotes the survival of that organism." Use of the word "religious" to describe the methods, techniques, and aims of Scientology seems therefore entirely arbitrary, since Scientology is in no way concerned with any power or powers external to man.

Its authoritarian teachings and techniques have aroused much controversy, not least in Britain. In 1968, foreign Scientologists were prohibited from entering Britain, and an official inquiry was instituted into the movement's activities. It has nevertheless continued to expand and in 1968 claimed two to three million followers. Its international headquarters are near London.

Scientology is a type of "mind-cure" movement, more sophisticated than those known in the West for more than a century, and extremely wealthy. It is notable only as an example of the sort of mental and spiritual panacea so often resorted to in times of stress by those who are strangers to living faith. E.J. SHARPE

SCILLITAN MARTYRS. A group from Scilli in Numidia (location unknown) beheaded at Carthage on 17 July 180 by the proconsul Saturninus, the first persecuting governor in Roman Africa, according to Tertullian. The account of their trial is the earliest literary evidence of African Christianity and the oldest dated Christian document in Latin. Saturninus's sentence lists seven men and five women, but apparently only three of each were tried on this occasion (the others perhaps previously, if they are not interpolated). Their names suggest they belonged to the noncitizen classes. They had with them "the books" (gospels?) and Paul's epistles, indubitably in Latin. Their simple steadfastness verged on provocative defiance ("I do not recognize the empire of this world"). D.F. WRIGHT

SCOFIELD, C(YRUS) I(NGERSON) (1843-1921). American biblical scholar. Born in Michigan and raised in Tennessee, Scofield won the Confederate Cross of Honor while serving with Lee's army. Turning to law, he was admitted to the Kansas bar in 1869. After two years as U.S. attorney, he practiced law in St. Louis. After his conversion he was ordained to serve a small Congregational church in Dallas (1882-95). At Moody's request he took over the Moody Church in East Northfield, Massachusetts (1895-1902). He returned to the Dallas church (1902-7) before taking up Bible conference work at home and in the British Isles, and founding the Central American Mission. The year 1909 saw the publication of his dispensational, premillennial Bible, which he edited with the financial assistance of prominent businessmen. EARLE E. CAIRNS

SCOPES TRIAL (1925). So named after the defendant, John T. Scopes, a young Tennessee high school science teacher, who was charged with teaching biological evolution contrary to a recently enacted state law. The trial was, however, transformed into a sensationally publicized national contest pitting William Jennings Bryan*—a famous agrarian politician and champion of fundamentalism, literal interpretation of Scripture, and antievolution—against Clarence Darrow—a leading criminal lawyer, representing Scopes and modern skepticism. Although Scopes was found guilty and fined $100, the trial (reinforced by the death of Bryan a few days after its conclusion) helped to discredit fundamentalism in the public mind. Perhaps as a result, the strength of fundamentalism in major American churches declined very sharply after 1925. GEORGE MARSDEN

SCOTISM, see DUNS SCOTUS

SCOTLAND. Although there is evidence to show that Christianity first came to Scotland during the Roman occupation, we have little information concerning it at that time. The first Christian missionary of whom we have any knowledge is Ninian,* who about 400 set up a church in the southwest at Whithorn, whence he carried on missionary work in the interior and up the east coast. In the sixth century other missionaries, mainly from Ireland, came to Scotland, the most famous being Columba,* who founded the monastery on the island of Iona,* off the west coast of Argyll in 563. The Celtic Church,* resulting from the work of these men, was different from that of Rome in the form of the tonsure, the date of Easter, and in not accepting the authority of the bishop of Rome. Endowed with strong missionary zeal but rather unorganized, they carried Christianity as far south as the River Thames, but in 663/4 at the Synod of Whitby* the king of Northumbria, Oswy, accepted the Roman supremacy, with the result that Celtic Christianity withdrew into Scotland and Ireland, eventually conforming to Roman practices and submitting to Roman authority.

The spread of Roman control in Scotland was brought to its culmination in the reign of Malcolm Canmore (1057-93), through the influence of his Anglo-Saxon queen, Margaret. Under her influence and that of their sons, monasticism spread rapidly through the country, and the church was gradually organized on a diocesan basis. As Scotland had no primate, however, the archbishop of York sought to bring the church under his control. To this the Scots objected, with the result that in 1225 Pope Honorius III granted the bishops the right to hold a council without a metropolitan. It was largely this anti-English attitude that made most of the clergy take a stand on the side of Scottish nationalism during the Wars of Independence (1296-1328) under Wallace and Bruce. They did so even against the orders of the pope, whom they defied for some twenty-five years. Even though they later submitted once again to papal authority, their relations with Rome were often strained, as indicated in their support of conciliarism and their violent opposition to the papal creation of the archbishopric of St. Andrews in 1472. Despite formal submission to Rome, therefore, the Scottish Church always adopted a rather independent attitude toward the central administration.

Although the Scottish Church during the Middle Ages produced a number of important scholars and ecclesiastical leaders, by 1400 it was probably one of the most decadent churches in Europe. This was one reason why the new doctrines of the English "heretic" John Wycliffe* made headway in certain parts, particularly the southwest. Real reform began, however, only with the activities of the followers of Martin Luther* and later of John Calvin.* Under the influence of Patrick Hamilton,* George Wishart* (both martyrs for their faith), John Knox,* Andrew Melville,* and others supported primarily by the gentry and the burgesses of the towns, a radical Calvinistic reformation took place. A Reformed church was established that has since exercised a wide influence both within and without Scotland. The Scots Confession* and the First and Second Books of Discipline* have formed the basis of most of the English language formulae of doctrine and church polity in use even today.

Since the sixteenth century there has been constant conflict in Scotland over the question of Christianity. Naturally the Roman Catholic minority has striven to maintain itself against the Reformed church. At the same time, there have been divisions to right and to left of the established church. The Scottish Episcopal Church has sought to maintain a position very similar to that of the Church of England, although not subordinate to Canterbury. On the other hand, the question of patronage by the landowners was a constant irritant that has caused a number of schisms, the most important and largest being that of the Disruption* of 1843 which led to the formation of the Free Church of Scotland.* At the same time, "moderatism" which grew out of the rationalistic Enlightenment of the eighteenth century has likewise caused many problems in all the Scottish churches. Nevertheless, with the disappearance of the Scottish Parliament in the Union of 1707, the Church of Scotland* became the only popular representative of the Scottish people. While it no longer holds that position, the Scottish churches probably still represent the people more effectively and fully than any existing political body, which is quite appropriate since Christianity has had a stronger influence upon Scotland than upon most nations.

BIBLIOGRAPHY: G. Grub, *An Ecclesiastical History of Scotland (to 1861)* (4 vols., 1861); J. Cunningham, *Church History of Scotland* (1882); A. Bellesheim, *History of the Catholic Church of Scotland* (1887); J. Dowden, *The Mediaeval Church in Scotland* (1910); A.R. MacEwen, *A History of the Church in Scotland (to 1560)* (1913, 1918); J.R. Fleming, *A History of the Church in Scotland 1843-1929* (2 vols., 1927, 1933); A.B. Scott, *The Rise and Relations of the Church of Scotland* (1932); J.A. Duke, *History of the Church of Scotland to the Reformation* (1937); W.D. Simpson, *Saint Ninian and the Origins of the Christian Church in Scotland* (1940); J. Knox, *History of the Reformation* (ed. W.C. Dickinson, 1949); W.C. Dickinson and G. Donaldson, *Source Book of Scottish History* (3 vols., 1950-54); J.H.S. Burleigh, *A Church History of Scotland* (1960).

W.S. REID

SCOTLAND, CHURCH OF. Since the Reformation in 1560, Scotland's national church has been Presbyterian, except for two periods of modified episcopacy enforced by the Stuart kings. John Knox* is generally regarded as the founder of the modern Church of Scotland; his successor, Andrew Melville,* has been described as the "father of Presbyterianism."* After the Stuarts were deposed, William III* reestablished Presbyterianism as the national form of church government, and successive monarchs since the 1707 England-Scotland parliamentary union have sworn at their coronation to maintain this polity. Controversy arose when the early nineteenth-century Moderates* were accused of neglecting the Reformation principle which gave the people a voice in the election of ministers. Battle was joined over the vexed question of patronage, to maintain which system the civil power was appealed to in a number of notorious cases. This led to the 1843 Disruption,* when more than one-third of the ministers formed the Free Church of Scotland.*

The breach was largely healed finally in 1929 (see UNITED FREE CHURCH OF SCOTLAND), and all but some 50,000 of Scottish Presbyterians are now in membership of the Church of Scotland—i.e., about 1.1 million—constituting a larger proportion of population than any other Protestant church in the English-speaking world. National and free of state control (the present position regulated by the 1921 Act of Parliament), the Church of Scotland is organized in twelve provincial synods and sixty-four presbyteries, in each of which as well as in the general assembly there is equal representation of ministers and laity. Appeal to the assembly is open to any member of the church. Since 1694 the sovereign's Lord High Commissioner* has attended each general assembly, but his presence is not necessary for the transacting of the church's business. Ministers are elected by individual congregations, subject to formal ratification by the local presbytery, and all ministers have equal status.

The church's courts are courts of the realm, to carry out the decisions of which the assistance of the civil courts can be enlisted if necessary. The church's four theological colleges are virtually also the divinity faculties of the four ancient Scottish universities, thus maintaining the Scottish tradition which demands of its ministry a high educational standard. The Church of Scotland professes the evangelical faith, and bases its doctrine on Holy Scripture. The Westminster Confession* is still officially regarded as its subordinate standard. Two sacraments are celebrated—baptism, normally of infants and as part of morning worship; and the Lord's Supper, generally celebrated quarterly or semiannually, but there is now a tendency, particularly in city parishes, toward more frequent celebration. In some parts of the more conservative Highlands and in the Western Islands, on the other hand, this sacrament is an annual occasion, extending from the Fast Day service on Thursday right through to Thanksgiving on Monday.

The Church of Scotland, in addition to its four presbyteries in England, Europe, and the Near East, maintains work in twenty-one overseas mission fields.

See also general article on SCOTLAND.

BIBLIOGRAPHY: J.T. Cox, *Practice and Procedure in the Church of Scotland* (6th ed., 1976); J.H.S. Burleigh, *A Church History of Scotland* (1960); R.S. Louden, *The True Face of the Kirk* (1963); *The Church of Scotland Year-Book* (annually).

J.D. DOUGLAS

SCOTS CONFESSION. The Scots Confession was prepared in four days in August 1560 and submitted to the Scottish Parliament which ratified it with very little opposition. The individuals responsible for its preparation were John Winram, subprior of St. Andrews; John Spottiswoode, later superintendent of Lothian; John Willock, later superintendent of Glasgow and the West; John Douglas, rector of St. Andrews; John Row, minister of Perth; and John Knox,* minister of St. Giles's, Edinburgh—the "six Johns." The dominant figure in the preparation was undoubtedly Knox, who had already been involved in the formulation of a number of confessions on the Continent—in Frankfurt, in Geneva, and perhaps also in France.

The confession's theology is Calvinistic, although the document itself is by no means merely a copy of a statement by Calvin or some other continental leader. It would seem that Knox and his collaborators took into account the thinking and statements of a number of Reformers. The "marks of the church" were taken over from Valerian Poullain's *Liturgia Sacra* used in the French church in Frankfurt, while some of the other elements show clearly the influence of the French Confession prepared and adopted at the First National Synod of the Reformed Church in Paris in 1559. Although adopted by Parliament in 1560, Queen Mary, still in France, refused to ratify the decision, with the result that it did not become the official confession until 1567, when Parliament reenacted it after her deposition. It remained the confession of the Scottish Reformed Church until the adoption of the Westminster Confession* of Faith in 1647.

W.S. REID

SCOTT, THOMAS (1747-1821). Biblical commentator. Born at Braytoft, Lincolnshire, son of a grazier, Scott was employed for nine years in menial farm work which permanently ruined his health. Driven from home by his father's cruelty, he was ordained deacon in 1772 by the bishop of Lincoln, held a number of curacies, and in 1781 succeeded John Newton at Olney. From 1785 to 1801 he was chaplain at the Lock Hospital and lecturer at St. Mildred's, London, and from 1801 to 1821 rector of Aston Sandford, Buckinghamshire, where he helped to train missionaries for the Church Missionary Society. Apart from his theological works (5 vols., 1805-8), he is chiefly famous for *Force of Truth* (1779), his spiritual autobiography which recounts his development from early Unitarianism to the adoption of evangelical Calvinism, under the influence of Newton, and for his *Commentary on the Bible* which appeared in weekly numbers between 1788 and 1792. Scott's exegesis, though occasionally wooden, is remarkable for its candor, and as an exercise in experimental Christianity, eliciting the meaning of each passage of Scripture for the author's own soul. The commentary became immediately popular, but brought its author no financial rewards—and charges of Arminianism from extreme Calvinist critics.

IAN SELLERS

SCOUGAL, HENRY (1650-1678). Devotional writer. Born at Leuchars, Fife, son of Patrick Scougal, bishop of Aberdeen, he attended King's College, Aberdeen, and became a college tutor on graduating in 1668. In 1672 he was ordained and inducted to the parish of Auchterless, Aberdeenshire, but after a year returned to King's as professor of divinity. He died of consumption five years later. As a teacher he shared the aims of Archbishop Robert Leighton* to make sure that his lectures recommended holiness of life as well as orthodoxy of theology. Scougal is famous for the classic *The Life of God in the Soul of Man.*

SCRIPTURE UNION. Founded in England in 1867 as the "Children's Special Service Mission," Scripture Union is now an international, interdenominational evangelical youth and Bible-reading movement, with offices or representatives in seventy countries, mostly former or present British Commonwealth territories. Its international structure is regionalized; the international secretariate is in Switzerland, and there are some 220 full-time executive staff around the world. Basic activities are children's evangelism and youth work, especially through schools' groups. In most countries there is a related program of vacation camps and house parties which serve both for evangelism and training.

Scripture Union Bible-reading aids for personal use are issued in graded series for all ages and levels of Christian experience and are used by more than one million readers throughout the world. Scripture Union, especially in Britain, publishes books and booklets: a complete range of graded Sunday school lesson aids; story books for children and youth; training literature and discussion group material, as well as soundstrips and other audiovisual material. Bookshops are operated in many of the major cities of the British Commonwealth. Control of the movement is through national and state committees, mostly made up of laymen. A. MORGAN DERHAM

SCRIVENER, FREDERICK HENRY AMBROSE (1813-1891). NT scholar. Born in Southwark and educated at Falmouth School and Trinity College, Cambridge, he taught for ten years at King's School, Sherborne, was headmaster of Falmouth School (1846-56), and prebendary of Exeter (1874-91). Throughout his entire working life he studied NT texts; he published the texts of twenty manuscripts and listed all known manuscripts. He was an ardent supporter of the Textus Receptus, the Byzantine text (substantially that of the Codex Bezae) used in the King James Version, in opposition to the Westcott and Hort text used in the Revised Version—in which view he failed to convince other scholars. He also devised a method of classifying ancient manuscripts.

R.E.D. CLARK

SCROGGIE, WILLIAM GRAHAM (1877-1958). Baptist minister, Bible expositor, and author. Educated in Exeter, Malvern, and Bath, he studied for the ministry at Spurgeon's College, London. After early pastorates in London, Halifax, and Sunderland, he began in 1913 his most influential ministry at Charlotte Chapel, Edinburgh, remaining there till 1933. There his notable expositions of Scripture drew large audiences from many denominations, and in 1927 the degree of doctor of divinity was conferred upon him by the university. From 1933 to 1937 he exercised a traveling ministry in South Africa, Australia, New Zealand, North America, and the British Isles. He was pastor of the Metropolitan Tabernacle, London (1938-44), and from 1948 to 1952 he was lecturer in English Bible at Spurgeon's College. He wrote many books, mostly expository studies. J.G.G. NORMAN

SCROPE, RICHARD LE (c.1360-1405). Archbishop of York. Of noble family, after ordination (1377) he became chancellor of Cambridge University (1378). In Rome (1382) he was made auditor of the Curia. Appointed bishop of Coventry and Lichfield (1386), at Richard II's request he was translated to the archbishopric of York (1398). He acquiesced in Richard's abdication (1399) and assisted at Henry IV's enthronement, but growing disillusioned with the new ruler, he took up arms with Northumberland and Bardolf. He composed a manifesto demanding justice, security, and lighter taxation, and gathered an army of discontented citizens. The earl of Westmorland, pretending agreement with the reforms, tricked him into disbanding his followers. He was popularly venerated in N England as St. Richard Scrope. J.G.G. NORMAN

SCUDDER, IDA SOPHIA (1870-1960). Missionary doctor and founder of the Christian Medical College at Vellore in South India. Born in Ranipet in then Madras Presidency where her father, Dr. John Scudder, was in the North Arcot mission of the (Dutch) Reformed Church of America, she was at first determined not to be a missionary. She was on only a short-term commitment to India, really to be near her sick mother, when in 1893 she received a "call" which was not the least remarkable feature of a remarkable life. Three different men came to her door, asking her to come and attend to their respective wives in childbirth; they refused the services of any male doctor, such as her father, preferring rather that their wives die. Ida could do nothing, and the wives died; but the experience sent her home to America to study medicine and commit her life to service for India and its women. She returned as a doctor in 1900 and opened a hospital at Vellore. Later she began a nursing school and, in 1918, the medical college for women. This involved a great fund-raising campaign in America. A great new hospital arose and, at a separate site, the college, which later (1947) accepted men students as well, affiliated with the full Madras University medical course (1950) and was one of the outstanding interdenominational Christian institutions of Asia.

ROBERT J. MC MAHON

SCUDDER, JOHN (1793-1855). American Dutch Reformed missionary to India. Graduate of the College of New Jersey (1811) and the New York College of Physicians and Surgeons (1813), he was turned to missions in the reading of a tract. In 1819 he and his wife left for Ceylon under the American Board of Commissioners for Foreign Missions. Following ordination in 1821, Scudder founded a hospital and several schools in Ceylon, later establishing a printing press and mission at Madras, India. Failing health caused him to go to South Africa, where he died. Scudder printed many tracts in Tamil and English. Seven of his sons became medical missionaries and pastors in India. ROBERT C. NEWMAN

SEABURY, SAMUEL (1729-1796). First bishop of the Protestant Episcopal Church (USA). Born in Groton, Connecticut, he graduated from Yale

(1748), then studied theology and medicine in Edinburgh. He was ordained in 1753, served as a missionary in New Brunswick (1754-56), and thereafter was rector in Jamaica, Long Island (1757-66), where he also practiced medicine and taught school. He was rector in St. Peter's, Westchester, thereafter for eight years. His Tory sympathies, evidenced in pamphlets he wrote, brought brief imprisonment during the American Revolution, after which he went over to the British side and served as a hospital and later regimental chaplain. He was chosen bishop by Connecticut Episcopal clergymen in 1783 and consecrated in 1784 at Aberdeen by Scottish bishops. He served as rector of St. James's Church, New London, Connecticut (1785-96), and became presiding bishop of the new Protestant Episcopal Church in 1789 as well as being bishop of Connecticut and Rhode Island. His death came while making parish calls.

EARLE E. CAIRNS

SE-BAPTISTS, see SMYTH, JOHN

SEBASTE, FORTY MARTYRS OF. Members of the so-called Thundering Legion who were left naked on the ice of a frozen lake at Sebaste, Lesser Armenia, about 320, and who froze to death in sight of baths of hot water set on the bank to tempt them to apostatize. The place of one who gave way was taken by one of the pagan guards who was moved to conversion by what he saw. The martyrs' ashes were recovered by the empress Pulcheria and were venerated in the East. The incident was recorded by Basil of Caesarea and Gregory of Nyssa.

SEBASTIAN. Nothing is known of him historically, other than the mention in *Depositio martyrum* that he is buried in the cemetery *in catacumbas* (catacomb on the Appian Way). Ambrose affirms that Sebastian was from Milan and perished in Rome during the persecution of Diocletian. Many legends are connected with his name, and artists have depicted him pierced with many arrows. In 1967 Roman Catholic scholar Lancelot Sheppard refuted this Renaissance concept as false.

SECESSION CHURCH, see ORIGINAL SECESSION CHURCH; ERSKINE, EBENEZER

SECKER, THOMAS (1693-1768). Archbishop of Canterbury from 1758. A distinguished ecclesiastical career which included a royal chaplaincy, the sees of Bristol and Oxford, and the deanery of St. Paul's culminated in his appointment to Canterbury. Secker was a conformist, readily accepting established dogma and doctrines. His theology was of that rational, ethical brand usually associated with his predecessor, John Tillotson,* but it went somewhat beyond this in his doubts about the sufficiency of a prudential morality. His characteristic note, however, is moderation, so much in contrast with his contemporary Wesley; and his style matches his tone—dignified, sane, but rather colorless.

ARTHUR POLLARD

SECOND COMING. The teaching that Christ will return in glory to judge the living and the dead and to terminate history. The main Christian traditions, while holding that Christ's words predict the certainty of a final judgment and the replacing of the present order by the eternal state, yet have opposed speculation as to the exact manner and time of the coming. Many believers, however, have advocated a more detailed plan of the event and believe that there will be a reign of Christ on earth for a long period, the Millennium, before the last judgment. They feel that the coming of Christ will be followed by a binding of Satan and the resurrection of the saints, who will join Him in a temporal kingdom when He reigns on earth. At this time the prophetic statements of the OT which indicate that there will be a kingdom of peace, plenty, and righteousness are to be fulfilled (such as Isa. 11).

The early church holding this premillennial view looked for the imminent return of Christ as witnessed by the writings of Papias, Irenaeus, Justin Martyr, Tertullian, Hippolytus, Methodus, Commodianus, and Lactantius. When the Christian Church was given a favored status under Emperor Constantine (fourth century), the Millennium was reinterpreted. This amillennial view was presented most clearly in the work of Augustine, who taught that the 1,000-year period is no literal piece of history; it is a symbolic number coextensive with the history of the church on earth between the resurrection of Christ and His return. Hence there is to be no millenial reign of Christ either before or after His second coming. Throughout the Middle Ages the Augustinian view held sway except for the teaching of isolated individuals such as Joachim of Fiore.*

In the late sixteenth and seventeenth centuries premillennial teaching about the second coming of Christ was emphasized once again by men such as J.H. Alsted* and Joseph Mede,* and it encouraged many of the participants in the English Civil Wars of the 1640s. At the close of the seventeenth century the development of postmillennial views and the growth of Enlightenment thought led to a decline of premillennialism. Daniel Whitby* and other postmillennialists believed that the return of Christ would not occur until the kingdom of God had been established by the church in human history. Thus Christ will triumph through the church and after this golden age will return to raise the dead, judge the world, and inaugurate the eternal order.

During the nineteenth century there was a revival of premillennialism which has continued to the present. A new element, Dispensationalism, added through the Plymouth Brethren* movement, held that the references to the Second Coming in Scripture are primarily concerned with the fate of restored Israel in the last days and not with the church. This interpretation of the Second Coming has become prominent among evangelicals in the twentieth century through the work of men such as W.E. Blackstone,* C.H. MacIntosh, Harry Ironside,* A.C. Gaebelein,* C.I. Scofield* (and the Scofield Reference Bible), John F. Walvoord, and most recently Hal Lindsey.

BIBLIOGRAPHY: J.A. Seiss, *The Last Times, or Thoughts on Momentous Themes* (1878); D.H. Kromminga, *The Millennium in the Church* (1945); L.E. Froom, *The Prophetic Faith of Our Fathers* (4 vols., 1946-54); E.L. Tuveson, *Millennium and Utopia* (1949); G.E. Ladd, *Crucial Questions About the Kingdom of God* (1952); L. Boettner, *The Millennium* (1958); C.B. Bass, *Backgrounds to Dispensationalism* (1960).

ROBERT G. CLOUSE

SECOND GREAT AWAKENING. This second "national revival" in the United States (c.1787-1825) served as a corrective to the spiritual declension that set in during and following the revolutionary period. Deism and skepticism were popular among the educated, especially the students. The rigorous life on the rapidly expanding frontier without benefit of church and society was demoralizing.

The revival in the East was centered in the colleges and towns along the coast. Hampden-Sydney College experienced revival during 1787, and the movement spread to Washington College. At Yale under the preaching of Timothy Dwight* revival came in 1802. Amherst, Dartmouth, and Williams colleges became part of the movement. Through the influence of students and preachers the revival spread. The eastern phase was characterized by orderliness and restraint.

The revival in the West was filled with religious excitement and emotional outbursts. It apparently began in 1797 in the three Presbyterian churches James McGready* pastored in Logan County, Kentucky, which climaxed in a large outdoor Communion service during the summer of 1800. Barton Stone* carried the revival to Cane Ridge, Kentucky, where a year later (1801) a larger interdenominational six-day camp meeting* was held with 10,000 to 20,000 in attendance from as far away as Ohio. This technique was widely used later by the Methodists. The revival, accompanied by unusual physical phenomena, spread throughout the western frontier, largely among the Presbyterians, Methodists, and Baptists.

Significant church growth, improvement of morals and national life, check to the spread of Deism, schism and emergence of new religious groups such as the Cumberland Presbyterians* and Disciples,* home and foreign missionary outreach, abolition and social reform movements, introduction of the camp meeting, and influence upon great men like Archibald Alexander,* Adoniram Judson,* and Samuel J. Mills—these were all results of this Awakening.

HOWARD A. WHALEY

SECRÉTAN, CHARLES (1815-1895). Swiss Protestant philosopher and theologian. Born in Lausanne, he studied at Munich under F.W.J. Schelling and Franz Baader, who led him into a speculative, mystical view of religion. He held professorships at Lausanne (1839-45, being dismissed for political reasons), Neuchâtel (1850-66), and again Lausanne (from 1866). With A.R. Vinet* he was a leader of liberal Swiss Protestant thought. He increasingly emphasized, under Kant's* influence, the moral significance of faith and the importance of freedom, adopting a Scotist view of deity as "absolute indetermination." His chief book was *La philosophie de la liberté* (1849-79).

J.G.G. NORMAN

SECTARY. A term applied to some Protestant dissenters in England during the seventeenth and eighteenth centuries. Translations of the works of Luther and other Reformers used it to describe Anabaptists* and others of the Radical Reformation.* In modern times it is occasionally used in a derogatory sense about overzealous adherents of different opinions.

SECULAR CLERGY. A term used since the twelfth century to distinguish priests living in the world (Lat. *saeculum*) from the "regular clergy" who were members of religious orders living according to "rule" (Lat. *regula*). They are not bound by vows, but owe canonical obedience to their bishops, and may possess property. According to canon law they must be celibate. They take precedence over regular clergy of equal rank.

SECULARISM. A term invented by G.J. Holyoake (1817-1906) to indicate a way of life which leaves out of consideration God, revelation, heaven, and hell, but bases morality on that which will enhance the public good. Originally no denial of religious belief was implied, but as a result of Charles Bradlaugh's (1833-91) later powerful advocacy of secularism, the term increasingly became associated with atheism.* In later usage it often means the view that religion should not be taught in schools.

SEDULIUS SCOT(T)US (fl. 848-858). Poet and scholar. During ten years at Liège, which he established as an important center of his native Irish culture, Sedulius gained a reputation for versatility if not profundity. His poetry, whether religious, bucolic, mock epic, or occasional, is unique in its time for variety of meter. He compiled unoriginal *collectanea* on Paul's epistles and Matthew; that on grammar exhibits wide reading, especially Cicero. His theologically oriented *de rectoribus Christianis,* on Christian government, was perhaps a *Fürstenspiegel* for his pupil, a son of Lothair I, Carolingian emperor. Based on patristic, especially Augustinian, authority, it held that kings have primacy over religious leaders.

DANIEL C. SCAVONE

SEEKERS. A small seventeenth-century sect of Independents* which was heir to a Quietistic tendency on the Continent first noted by Sebastian Franck* in his *Chronica* (1531) and embraced by Dirck Coornheert,* a Dutch theologian. The visible church, with its "notions" (i.e., doctrines), organization, and ceremonies, was repudiated, while the true believer waits and "seeks" for the church of apostolic power which God will establish. The English "Seekers" during the Cromwellian period developed these ideas. The word "Seekers" as the name of a sect first appears in *Truth's Champion* (1617), probably written by John Murton. Seeker ideas had been taught by

Bartholomew Legate (c.1575-1612), an English cloth merchant trading with Holland, who was burnt at Smithfield for Arianism.* The Seekers were earnest, peaceable, spiritually minded people, and appear to have had large meetings in N England and in Bristol. From 1652 they were almost entirely absorbed by the Quakers.* Other names for them are "Legatine-Arians" and "Scattered Flock." J.G.G. NORMAN

SEELEY, SIR JOHN ROBERT (1834-1895). English historian. Born in London, son of a publisher and author, he was educated at the City of London School and Christ's College, Cambridge, where he read classics. He then returned to teach at his old school and later became successively professor of Latin at University College, London (1863), and professor of modern history at Cambridge (1869), where he succeeded Charles Kingsley.* He wrote voluminously on history and politics, especially on the life of Napoleon and the foundation of the British Empire. He is best known, however, for his brilliantly written (originally anonymous) *Ecce Homo* (1865) which tells the story of Jesus and His subsequent influence on the morals of the world. As it dealt only with the human side of the story, the book was construed as an attack on Christianity and gave rise to much controversy. Seeley followed it up with *Natural Religion,* a less successful book in which he tried to show that religion can subsist in the absence of supernaturalism. R.E.D. CLARK

SEGNERI, PAOLO (1624-1694). Jesuit preacher. Born at Nettuno, he studied at the Roman College, entered the Jesuit Order (1637), and was ordained (1653). He was a pupil of Sforza Pallavicino, whose sacred oratory much influenced him and led to the study of classical and patristic oratory. His preaching ministry, sometimes accompanied by self-flagellation and penitential processions, began in cathedral pulpits, and from 1661 until 1692 he conducted many popular missions and Lenten series. Innocent XII called him to Rome as his preacher (1692), making him theologian of the Sacred Penitentiary. Likened to Savonarola,* he did much to reform pulpit oratory. He also wrote works against Quietism* and Probabiliorism.* C.G. THORNE, JR.

SELDEN, JOHN (1584-1654). Jurist and Orientalist. Born in Sussex, he was educated at Hart Hall, Oxford (1602-4), and Clifford Inn, London, before being called to the bar in 1612 and practicing in the Temple. His interests were wider than law, however, and included history, oriental studies, and Judaism. The book for which he is most famous is *The History of Tythes* (1617) in which his contemporaries believed he argued that tithes were not by divine ordinance. Nevertheless their collection was legal. Thus, when he launched his political career in 1621 he already had made a name for himself. In Parliament he continued to gain fame as a defender of the common law and the ancient "liberties" of Englishmen. As a result he was imprisoned in 1629 for two years. He became prominent again in the Long Parliament, and he sat in the Westminster Assembly,* where his learning and Erastianism* were a double embarrassment to many divines. In 1645 he declined the mastership of Trinity Hall, Cambridge, and spent the last years of his life quietly pursuing his literary work. He had a vast memory and a large collection of books, most of which are still in the Bodleian Library, Oxford.

PETER TOON

SELEUCIA, COUNCIL OF (359). This assembly of the Eastern Church was called by Emperor Constantius to settle the Arian* controversy. Constantius had endorsed a creedal formulation which was intended to be a compromise between the contending parties in this controversy. It conceded to the Semi-Arians the view that Christ comes before all time the Son who is like, but not the same as, the Father. Despite the emperor's wishes, the council stood firm on this issue, whereupon Constantius instructed the council to send ten delegates who were persuaded to accept a compromise creed like that accepted by the Council of Arminum* (Rimini) in the West. This compromise was later reaffirmed by the Council of Constantinople in 381. C. GREGG SINGER

SELWYN, GEORGE AUGUSTUS (1809-1878). First Church of England bishop of New Zealand. Educated at Eton and St. John's College, Cambridge, he became a fellow of his college and was ordained in 1833. He gained parish experience in Windsor and was a private tutor at Eton. He was consecrated bishop of New Zealand in 1841 and reached Auckland, which later became his see city, in 1842. He traveled widely throughout New Zealand. Through an error of latitude in his letters patent he claimed that his diocese took in much of the Pacific. From 1848 he made frequent journies to Melanesia. He founded a college in New Zealand to train young men from the islands for the proposed mission. In 1861 he was instrumental in making Melanesia a separate diocese under J.C. Patteson.* The impetus for this missionary program came from the 1850 conference of the Australasian bishops under W.G. Broughton* in Sydney. Selwyn also planned the division of his diocese in the years following his visit to England in 1854. In 1857 a constitution for the church in New Zealand was drawn up. By 1859, when the first general synod met, there were four other bishoprics.

Selwyn was a pioneer in the concept of the independence of the Church of England in the colonies. Despite differences of churchmanship from the missionaries of the Church Missionary Society, Selwyn supported their evangelization of the Maoris even in the difficult days of the Maori wars. In 1867, while present at the first Lambeth Conference,* Selwyn was appointed bishop of Lichfield and did much to pioneer industrial chaplaincy work there until his death.

NOEL S. POLLARD

SEMI-PELAGIANISM. The designation since the seventeenth century of a largely monastic reaction against Augustine's* developed anti-Pelagianism, better called "Semi-Augustinianism." In 426/7 monks at Hadrumetum in Byzacena (Susa in Tunisia) were alarmed that Augustine's *Ep.* 194 on predestination apparently undermined free will and hence monastic and missionary endeavor. After disappointing inquiries, a deputation visited Augustine, who produced *Grace and Free Will* and, when this seemed to invalidate moral correction, *Rebuke and Grace* (427). To Vitalis,* a Carthaginian monk who c.427 affirmed that the unaided will performed the initial act of faith, Augustine's *Ep.* 217 stressed the necessary preparation of the will by prevenient grace.

Widespread anxieties arose in S Gaul in the monasteries of Lérins (founded c.410 by Honoratus) and Marseilles. John Cassian's *Conferences* (428/9) argue implicitly against Augustine's works for Hadrumetum, that the beginning of the good will is man's doing, but grace supervenes immediately thereafter. Such criticism, shared by others like Helladius (Euladius?), bishop of Arles, were reported to Augustine by two lay Marseilles monks, Hilary and Prosper of Aquitaine* (429). Augustine's response was *Predestination of the Saints* and *The Gift of Perseverance* (ET by M.A. Lesousky, 1956).

After Augustine's death Prosper became his stalwart champion, replying *seriatim* to questions or objections raised by two Genoese presbyters, anonymous Gallic critics and probably Vincent of Lérins* (cf. too his *Commonitory*). Prosper sought Roman backing, but Pope Celestine I* generally praised Augustine and condemned sophistic innovations, and Sixtus III and Leo I* were no more explicit. In Rome Prosper concentrated more on opposing Semi-Pelagianism than defending Augustinianism (attacked there c.450 by Arnobius Junior,* an African monk).

Semi-Pelagian beliefs remained dominant in Gaul. Faustus of Riez,* formerly abbot of Lérins, forced his priest Lucidus to retract apparently Augustinian views condemned by councils at Arles (472/3) and Lyons (474), and wrote *The Grace of God and Free Will* (473/5), which verged at points on Neo-Pelagianism. Gelasius I* criticized bishops who tolerated attacks on Jerome* and Augustine, and called for a confession of faith from Gennadius and Honoratus of Marseilles (496). The writings of Julianus Pomerius, an African presbyter of Arles, reveal continuing agitation.

At Constantinople in 519, Scythian monks led by John Maxentius inquired through Possessor, an African bishop then in Constantinople, about the orthodoxy of Faustus of Riez. Pope Hormisdas's* reply referred them to the Scriptures, councils, and Fathers, notably the later Augustine. The Scythians consulted refugee African bishops in Sardinia, where a synod condemned Faustus (523); and Fulgentius* of Ruspe wrote a lost refutation, which influenced Caesarius of Arles, trained at Lérins but suspected in Gaul for his Augustinian convictions. Under him the Second Council of Orange* (529) condemned Pelagian and Semi-Pelagian opinions and endorsed a moderate Augustinianism, in terms drawn largely from Prosper's extracts from Augustine submitted earlier to Pope Felix IV (probably from a preliminary synod at Valence) and amended by him. Confirmation by Pope Boniface II (531) made Orange the basis of medieval teaching on grace.

BIBLIOGRAPHY: L. Loofs in *Realencyklopädie für protestantische Theologie und Kirche* (3rd ed.), 18 (1906), pp.192-203; E. Amann in *Dictionnaire de Théologie Catholique* 14 (1941), cols. 1796-1850; G. de Plinval in *Histoire de l'Église* (ed. A. Fliche and V. Martin), 4 (1948); pp. 397-419; J. Chéné in *Recherches de Science Religieuse* 35 (1948), pp.566-88, and 43 (1955), pp.321-41, and *L'Année Théologique August* 13 (1953), pp.56-109; N.K. Chadwick, *Poetry and Letters in Early Christian Gaul* (1955); on Prosper's Augustinianism, R. Lorenz in *Zeitschrift für Kirchengeschichte* 73 (1962), pp.217-52, and his authorship of the Pseudo-Augustine's *Hypomnesticon Against the Pelagians and Celestians* (ed. of J.E. Chisholm, vol. I, 1967). D.F. WRIGHT

SEMLER, JOHANN SALOMO (1725-1791). German biblical scholar. Born in Thuringia, son of a Pietistic pastor, he entered the University of Halle and came under the influence of the rationalist J.S. Baumgarten whose assistant he became. He was professor at Coburg and Altdorf before returning to Halle in 1752 as professor of theology. When Baumgarten died five years later, Semler became head of the theological faculty. Semler pioneered in biblical and church historical criticism, investigating the origins of NT books in a manner unacceptable to Lutheran orthodoxy, and developing a threefold classification of Greek manuscripts which enabled textual criticism to go beyond quantity to quality based on age and geographical origin (Alexandrian, Oriental, or Occidental). He produced a major work on the canon (4 vols., 1771-75) in which a historical view of its development was first introduced. But when he replied to the fragments of H.S. Reimarus's* work published by Lessing, it appeared that he was moving away from his rationalistic stance.

CLYDE CURRY SMITH

SENS, COUNCIL OF (1141). This most important of the many councils held in the French town was called to hear the charges of heresy launched against Abelard* by Bernard of Clairvaux.* In 1121 Abelard had been forced to burn his own works, particularly his treatise on the Trinity, on the ground that he was advocating tritheism. Abelard was later charged by William of St. Thierry* with thirteen errors concerning the doctrines of the Trinity, the person of Christ, the Holy Spirit, and the grace of God in the redemption of man. Although Abelard had personally accused Bernard of instituting novel practices at Clairvaux, Bernard nevertheless visited Abelard in a vain effort to persuade him to retract his errors. Abelard refused to retract. Bernard came to Sens determined to have Abelard condemned for his heresy, and formally presented charges against him. Abelard refused to defend himself and appealed the case to Innocent II, apparently expecting to receive a more friendly hearing by the pope. But Innocent declared Abelard a heretic and imposed on him the penalty of perpetual silence and banishment. C. GREGG SINGER

SEPARATISTS. During the latter part of the reign of Elizabeth I of England, a small number of people took the doctrines of Puritanism to their logical conclusion and separated themselves from the "impure" national church to form small gathered churches. Though never more than several hundred in numbers, they were hunted down and severely punished by the agents of Elizabeth and James I, as well as being strongly criticized by the Puritan preachers. Famous names connected with this movement were Robert Browne,* John Smyth,* and John Robinson.* Some of these men became martyrs. Separatism, often termed "Brownism," was illegal until in the Commonwealth and Protectorate (1649-59) it became widespread and acceptable. Under the Clarendon Code* after 1660 it became illegal again.

See also INDEPENDENCY. PETER TOON

SEPTUAGINT. The conventional name of the earliest translation of the OT into Greek (from *Septuaginta,* Lat. "seventy" = LXX). In the *Letter of Aristeas* (second century B.C.), it is alleged that seventy-two Jewish translators sent from Jerusalem produced the version for Ptolemy II (Philadelphos) for his library. This cannot be accepted as historically true, but it contains reliable indications, namely that the version is of third-century B.C. origin, was made in Egypt by companies of translators, but was a product of community needs rather than imperial request. The prologue to the Greek *Ecclesiasticus* shows that the whole canon was in Greek by 132 B.C., but it is clear on internal grounds that the translation was first made of the Pentateuch, and that Prophets and Writings were rendered later, in that order. Compared with the Hebrew OT there are a number of additional books and portions thereof in this corpus.

While modern scholarship appears to accept the concept of an "original text" (Urtext), it is clear from the materials available in Dead Sea fragments, quotations in NT, early Fathers, and others (e.g., Josephus), that a number of distinct local texts arose, in part by correction of paraphrastic renderings, in part for theological reasons, in part from different Hebrew bases. In the early Christian centuries, stimulated by controversy, several new translations were undertaken by Jews, giving a more literal rendering of a now standard text, close to the Massoretic (viz., by Aquila, Theodotion, Symmachus), sometimes on the basis of the earlier local texts. In the early third century, Origen,* largely for controversial reasons, set out in six columns *(Hexapla)* the Hebrew text, a transliteration, the three Jewish translations, and the LXX with an apparatus of signs showing the divergences of the latter from the Hebrew in omission and addition. Ironically this work complicated the textual problem further by the contamination of other texts from this column.

In Antioch, in the late third or early fourth century, Lucian* produced a recension to some degree closer to the Hebrew. A third recension mentioned by Jerome, that of Hesychius, has proved impossible to identify. Modern research has identified in some books the R-recension transmitted in the catenae. Modern scholarship, based on some fine eighteenth-century antecedents, took its rise from Paul de Lagarde* and his pupil Alfred Rahlfs. Lagarde sought the Urtext; Rahlfs concentrated more upon the recensions. Their work has more recently been continued by M. Margolis and P. Katz (P. Walters). Two major publications are in progress: the larger *Cambridge Septuagint,* named "Brooke-MacLean" after its first editors; and the *Goettingen Septuagint,* whose first editor was J. Ziegler. The establishment of both the Urtext and the recensional forms is of great significance for the history of Hellenistic Judaism, early Christianity (for which the LXX was inspired Scripture), and the history of the Greek language.

BIBLIOGRAPHY: H.B. Swete, *An Introduction to the Old Testament in Greek* (1914); B.J. Roberts, *The Old Testament text and versions* (1951); P. Katz, "Septuagintal studies in the mid-century," *The Background of the New Testament and its eschatology* (ed. Davies and Daube, 1956), chap. 10; S. Jellicoe, *The Septuagint and modern study* (1968); S. Talmon, "The Old Testament text," *CHB* I (1970), chap. 7; P. Walters, *The Text of the Septuagint* (1972). J.N. BIRDSALL

SERAPION (d.211). Bishop of Antioch. He remains almost unknown despite identification by Eusebius as eighth in the succession of bishops of Antioch. Eusebius quotes from a letter, conjointly signed by other bishops, sent by Serapion to Caricus and Pontius, persons unknown, as a supporting cover for a writing of Claudius Apollinarius,* bishop of Hierapolis, against the Montanists.* Otherwise Eusebius knows only a few other epistolary compositions, including one to the church of Rhossus in Cilicia against the *Gospel of Peter,* the Docetic imprint of which had led to Serapion's rejection of it.

SERAPION OF THMUIS (d. c.362). Bishop of Thmuis. Superior of a colony of monks and friend of Antony, he was consecrated bishop of Thmuis, Lower Egypt, before 339. He was sent by Athanasius with four other Egyptian bishops in 356 to the court of Emperor Constantius II to refute Arian charges. Later he was ousted from his see by an Arian, Ptolemaius (359). He reported to Athanasius concerning some Egyptian Christians who held that the Holy Spirit was merely a "creature," doubtless forerunners of the Pneumatomachi.* Athanasius wrote him four letters, *Ep. ad Serapionem,* which constitute the first formal statement of the deity of the Spirit (c.359). Serapion wrote a treatise *Against the Manichaeans* (pub. 1931), a lost work on the titles of the Psalms, and letters (three of which are extant) addressed to Bishop Eudoxius, the monks of Alexandria, and some disciples of Antony. He was also author of a sacramentary, the *Euchologion.*

J.G.G. NORMAN

SERBIAN ORTHODOX CHURCH, see EASTERN ORTHODOX CHURCHES

SERGIUS (d.638). Patriarch of Constantinople. Born in Syria, he quickly gained preferment in the church and was consecrated patriarch in 610. Emperor Heraclius looked to him for advice in ecclesiastical matters, and since a major problem was that of the Monophysites,* Sergius sought a way to reconcile them to the rest of the church. He approved the formula "one mode of activity" *(energeia)* in Christ, which had been agreed to by the emperor and Monophysite leaders. Later this doctrine was modified to state that there was only one will *(mia thelēsis)* in Christ. In the *Ecthesis,* written by Sergius and issued by the emperor in 638, Monothelitism* was formally propagated. It was accepted by two synods at Constantinople in 638 and 639, but the West rejected it and ultimately so did the East at the Council of Constantinople in 681. Apart from his fame as the propagator of Monothelitism, Sergius is traditionally regarded as the author of the famous Greek hymn, the "*Akathistos,*" in honor of the Virgin Mary, sung during Lent. PETER TOON

SERGIUS (c.1314-c.1392). Russian saint. Born at Rostov, he became a monk in 1336 and built a chapel in honor of the Trinity in the forest of Radonezh which later became the famous monastery of Troitskaya Laura. He gained a reputation for miracles and visions, and was active in the political life of Russia. Several times he intervened to stop civil wars between the Russian princes, and he encouraged Dmitri Donsloi's resistance to the Mongols in 1380. In 1387 he refused to accept the patriarchate of Moscow, and he continued as a humble monk for the remainder of his life. He was canonized in the mid-fifteenth century and is considered one of the most important Russian saints.

RUDOLPH HEINZE

SERGIUS I (d.701). Pope from 687. A native of Palermo of Syrian descent, he was elected pope after a contest between two other candidates. He consecrated Willibrord, bishop of the Frisians (695), and ordered Wilfrid of York restored to his see. His refusal to recognize as authoritative the Trullan Synod which favored Eastern custom over Latin led the emperor to seek his arrest Byzantine power was, however, weak in Italy at the time, and Sergius was able to defy him.

SERIPANDO GIROLAMO (1493-1563). Archbishop of Salerno. Born at Troja (Apulia) of noble parentage, he lost his parents when very young and at the age of fourteen entered the house of the Augustinian Order at Viterbo. He studied Greek, Hebrew, theology, and philosophy. In 1515 he began to lecture at Siena, moving in 1517 to Bologna to be a full professor. Subsequently he became vicar-general (1532) and then superior-general (1539) of his order. He attended the Council of Trent* during 1546-47 where he showed great concern in the debates to preserve the doctrine of the purity of the text of Scripture and to make known his somewhat unorthodox views on original sin and justification. After acting as legate to Pope Paul III he refused the offer of the bishopric of Aquila. In 1551 he also resigned as superior-general and withdrew for two years to the quietness of a small convent, from where he emerged in 1553 to go on a mission for the city of Naples to Emperor Charles V. This complete, he became archbishop of Salerno. Pius IV made him a cardinal and second legate of the Holy See at the Council of Trent. Upon the death of his colleague, Cardinal Gonzaga, Seripando became president of the council. He was a prolific writer and controversialist. He wrote commentaries on Paul's letters to Galatia and Rome (1569, 1601) and a book on prayer (1670). PETER TOON

SERVETUS, MICHAEL (1511-1553). Anti-Trinitarian theologian and physician. Born in Spain of a pious family, he studied the biblical languages as well as mathematics, philosophy, theology, and law at the universities of Zaragoza and Toulouse, then went as a secretary to Charles V's confessor. He left the imperial court for Basle, then for Strasbourg where he met Martin Bucer* and possibly some of the Anabaptist leaders. These contacts stimulated his early radical theological ideas.

Servetus came to believe that in order to convert the Moors and Jews, the Christian teaching of the Trinity would have to be reinterpreted. He decided that the most serious error involved in Trinitarian doctrine was the belief in the eternal existence of the Son. He expressed his ideas in several books (1531-32), which led to attacks on his work by orthodox theologians. To avoid further trouble he adopted a disguise and began a second career as a physician. After studying at Lyons, he published a description of the pulmonary circulation of the blood and worked on geography and astrology. He also worked for a time in Vienne as physician to the archbishop and returned to his study of theology. Repeating his earlier attacks on the doctrine of the Trinity, he also rejected infant baptism, while proclaiming a christocentric pantheism developed from Neoplatonic, Franciscan, and Cabalistic elements.

In answer to Calvin's *Institutes of the Christian Religion* he wrote *Restitutio Christianismi* (1553). He was arrested and condemned by the Inquisition in Vienne, but escaped and went to Geneva. Again he was arrested and condemned, and this time he was burnt. His execution provoked a controversy over toleration of religious differences.

See E.M. Wilbur, *A History of Unitarianism* (2 vols., 1945-52), and R.H. Bainton, *Hunted Heretic: The Life and Death of Michael Servetus* (1953).

ROBERT G. CLOUSE

SERVITES. The Order of the Servants of the Blessed Virgin Mary was founded in 1240 by seven influential Florentines, who had already withdrawn from the world in order to enter the service of Mary. The Servites adopted a black habit and the Rule of St. Augustine, modified to some extent by provisions taken from the constitutions of the Dominicans. Servite piety centers on the Sorrowful Virgin. While the Second Order of Servite nuns is principally a contemplative religious community, Servites also engaged in activity in the world. Servite missionaries had reached India by the second half of the thirteenth century, and Servite nuns of the Third Order (founded in 1306) devoted themselves to the relief of the sick and the poor and to the training of the young. The order received official sanction for its work from Pope Benedict XI in 1304.

DAVID C. STEINMETZ

SEVEN SACRAMENTS, see SACRAMENT

SEVENTH-DAY ADVENTISTS. A religious denomination that grew from the work of William Miller* (d.1849), who began to preach that the end of the world was at hand, that a fiery conflagration would usher in the new heaven and the new earth, and that the date for this would be sometime between 21 March 1843 and 21 March 1844. The movement weathered its first difficulty when the deadline passed; another date was set: 22 October 1844. A general apocalyptic fervor aided the growth of the group, and it soon had between 50,000 and 100,000 adherents. The new date passed, and the early Millerite fervor was largely diminished. A few, however, continued to believe that the end was near. One of these, Hiram Edson, saw a vision of Christ entering the second compartment of heaven. This proved to Edson that Miller's prophetic calculations were correct, though the event foretold was not the Second Advent, but the opening of an investigative judgment in heaven to determine who among the dead are worthy of resurrection.

Other adventists believed that the Second Coming* had been hindered by their failure to maintain the biblical law of keeping the seventh day as the Sabbath. Sabbath-keeping was also confirmed by visions, especially those of Ellen G. White* (d.1915), whose importance to the movement cannot be overstressed. Although possessing only a third grade education, she wrote 45 major books and over 4,000 articles. One of her works, *Steps to Christ,* sold more than 5 million copies and appeared in 85 languages. The early Adventists were found chiefly in the New England states, but by 1855 their westward expansion was marked by the establishment of a headquarters in Battle Creek, Michigan. The denomination was organized in 1863, and by 1874 their first missionary, J.N. Andrews, was sent out.

Seventh-day Adventists today believe that the only prophetic texts awaiting fulfillment concern their church and its ministry. When the Gospel message has been proclaimed throughout the world and the church has grown to its predetermined size, then the end of the age will come. At that time the righteous dead will be raised and together with the righteous living will be taken to heaven, where they will spend the Millennium. While believers enjoy heavenly bliss, Satan will be left on earth for 1,000 years. At the end of this period, Christ will descend with His saints, destroy the wicked with fire, and create a new earth with the New Jerusalem as its center. Adventists also teach soul sleep, free will, the deity of Christ, and believer's baptism by immersion. Holy Communion is observed four times a year, preceded by a foot-washing service. The Sabbath, from sundown Friday until sundown Saturday, is scrupulously observed. Despite the relatively low economic status of its membership, insistence on tithing has led to the church's being among the leading American churches in per capita giving.

Adventists operate parochial schools from primary through university level. They also insist upon the proper care of the body, abstaining from foods forbidden in the OT such as pork, ham, and shellfish; do not smoke or drink; and conduct an extensive medical program with hospitals and clinics centering at Loma Linda University in California. The church also opposes secret societies, card playing, gambling, and the use of jewelry and cosmetics. "Worldly entertainments" such as motion pictures, television, the theater, and dancing are also avoided by them.

The church operates a sizable publishing industry. Their leading paper, *Review and Herald,* is one of the oldest continuously published religious periodicals in America. Adventists have a congregational government which is tied to a series of local and national conferences. The denomination's activities are centralized in the General Conference of Seventh-day Adventists at Washington, D.C. Sessions of this group consisting of delegates elected from the local conferences meet every four years. Between sessions of the world conference, business is conducted by an executive committee. About four-fifths of the church's 2.5 million members live outside North America.

BIBLIOGRAPHY: F.D. Nichol, *The Midnight Cry* (1944); L.E. Froom, *The Prophetic Faith of Our Fathers* (4 vols., 1946-54); A.W. Spalding, *Origin and History of Seventh-day Adventists* (4 vols., 1961-62); A.A. Hoekema, *The Four Major Cults* (1963); D.F. Neufeld (ed.), *Seventh-day Adventist Encyclopedia* (1966); B. Wilson, *Religious Sects, A Sociological Study* (1970); E.S. Gaustad (ed.), *The Rise of Adventism* (1974).

ROBERT G. CLOUSE

SEVERIAN (d. c.408). Opponent of John Chrysostom.* Bishop of Gabala, Syria, who came to Constantinople about 401, Severian was well received by Chrysostom, but attempted to undermine the latter's authority during his absence in Asia. At the Synod of the Oak* (403), under Theophilus's influence, he acted as accuser and judge of Chrysostom. Again he accused him (404) and finally effected Chrysostom's exile to Cucusus, demanding in 407 his removal to a severer place of exile. He wrote several homilies, including six on the *Hexaemeron.*

SEVERINUS (Severin) (d.482). "Apostle of Austria." An Eastern monk of Latin origin, after Attila's death (453) he came to Noricum Riponse, then occupied by barbarian invaders. For thirty years he evangelized the lands around Comagene and Astura (modern Stockerau and Haineburg), founding monasteries at Boiotro (near Passau) and Faviana. He established a kind of theocracy in Noricum when the district was left unprotected by Rome, caring especially for the poor. He won the friendship and respect of all, including that of Odoacer, the barbarian leader. He died at Faviana, and his body was taken to Lucullanum, near

Naples (488), where his companion Eugippus wrote his life (511). J.G.G. NORMAN

SEVERUS (OF ANTIOCH) (c.460-538). Leading Monophysite* theologian. Born in Pisidia, he studied in Beirut and Alexandria and showed a keen interest in theology at an early age, although he was not baptized until 488. Soon he became a monk in Egypt, then took up residence in Constantinople to represent Monophysitism. After visiting the court there from 508 to 511, he was consecrated patriarch of Antioch in 512. With his accession to power, Monophysites came into full control of Antioch, and soon after his consecration he condemned the Council of Chalcedon* and the Tome of Leo.* He encountered great difficulty with bishops and clergy hostile to his position, and perhaps used violence to retain his position.

Exiled from Antioch in 518 when Justinian I became emperor, he sought refuge in Alexandria where the Monophysite patriarch Timothy welcomed him. He visited Constantinople several times, including in 536 when Monophysites and Chalcedonians were engaged in religious debate there. In Alexandria he firmly opposed Bishop Julian of Halicarnassus, another Monophysite in exile. His theology was widely accepted in Syria, and he retained his position of primacy among the Monophysites until his death. He rejected attempts to separate Christ's two natures and to gloss over his humanity, but was probably closer to the Chalcedonians than his writings indicate. Syriac translations remain of his Greek treatises (including *Philalethes*), sermons, liturgical writings, and 4,000 letters. JOHN GROH

SEVERUS, GABRIEL, see GABRIEL SEVERUS

SEVERUS, SULPICIUS (c.362-early fifth century). Hagiographer. Born of a noble family in Aquitaine, he received that classical education in rhetoric which prepared him for the law. He partially followed the pattern of his friend Paulinus of Nola,* (among whose letters there remains one to Severus) into a semi-monastic life under the guidance of Martin,* bishop of Tours, who ordained him presbyter. Severus began a "life" of Martin and thereby contributed to the development of Christian hagiography. While this work shows itinerary and contacts through Gaul of his time, the form is almost devoid of general historical data—a situation partly remedied by Severus's *Chronicle* in two books based on biblical and classical materials, extended to 403, devised as a textbook for the Christian reader. Whether Severus died before or after the sacking of Rome (410) is not known. CLYDE CURRY SMITH

SEXTUS, SENTENCES OF. A collection of 451 Greek religious and ethical maxims compiled in the late second/early third century by an unknown Christian, largely from Pythagorean material partly Christianized. The compilation reflects a Hellenizing or paganizing of Christianity. Origen cited it as a Christian work and attested its popularity, and about A.D. 400 Rufinus* of Aquileia translated it into Latin under the title *The Ring (Anulus)*, recording the "traditional" ascription to Pope Xystus II (d.258). This version was widely read, e.g., by Pelagius, but condemned as Pythagorean paganism by Jerome the anti-Origenist (he earlier quoted it with approval). The *Sentences* were expanded and translated also into Syriac, Armenian, and Georgian. They appealed chiefly to ascetic and monastic circles committed to Christian perfection. D.F. WRIGHT

SHAFTESBURY, ANTHONY ASHLEY COOPER, Seventh Earl of (1801-1885). Evangelical social reformer. He was educated at Harrow and Christ Church, Oxford, then entered Parliament in 1826. He was a Tory, though his growing concern with social issues—particularly his desire to improve working-class conditions which had been created by the Industrial Revolution—made him more independent politically. In 1828 he became a member of the Metropolitan Lunacy Commission and began his work for the mentally ill. In 1845 he persuaded Parliament to establish a permanent Lunacy Commission for the whole country, and he was its chairman until he died. From 1833 to 1847 his main political concern was with the factory question, which after a long battle resulted in the Ten-Hours Act (1847), though the question continued to occupy his attention until the Factory Act (1874).

From 1840 he gave his support to other social questions. He championed the cause of the women and children working in mines and collieries, and secured the setting up of a Royal Commission of Inquiry into children's employment in general. It was not, however, until 1864 and 1867 that parliamentary acts regulated child and female labor, and not until 1875 did the Climbing Boys Act protect children used as chimney sweeps. He also promoted legislation to protect milliners and dressmakers. From 1859 Shaftesbury devoted more of his time to direct social work in connection with the slums, the Ragged School Union (of which he was chairman), and his own schemes of industrial schools and training ships. He was the leading evangelical in the mid-century, and strongly opposed Ritualism and Rationalism. He supported Catholic Emancipation (1829). As Lord Palmerston's stepson-in-law, he advised him on ecclesiastical appointments during his premiership. He was president of the British and Foreign Bible Society, and closely associated with the London City Mission, Church Missionary Society, YMCA, and Church Pastoral-Aid Society.

See E. Hodder, *Life and Work of the seventh Earl of Shaftesbury* (3 vols., 1886), and G.F.A. Best, *Shaftesbury* (1964). JOHN A. SIMPSON

SHAKERS. The common name of the celibate and communistic "United Society of Believers in Christ's Second Appearing," originating in the Quaker revivals of mid-eighteenth-century England. Mother Ann Lee* (d.1784) is generally considered the founder of the movement. Persecution, limited success, and a direct revelation led Mother Ann and seven followers to emigrate to New York in 1774. In 1787 the first Shaker settlement was established at New Lebanon, New York, which became the main base for the society's missionary enterprise in America. By the time Ann died, in spite of bitter opposition, there were growing numbers of Shakers in New York, Massachusetts, and Connecticut. The movement spread in the wake of revivalism* on the American frontier throughout the first part of the nineteenth century. Shaker communities were established as far west as Kentucky and Indiana. The society reached its zenith in the decade before the American Civil War when there were some 6,000 members in eighteen different settlements. After the war, the order began a steady decline which has lasted to the present. In 1900 there were fewer than 1,000 members; by 1970 only a handful of Shakers remained in three small communities.

Shaker celibacy rested on a form of dualism imposed on the society by Ann Lee, who came to regard sexual intercourse as the cardinal sin. According to Ann, God was both male and female, as was Christ who appeared in Jesus as the male principle. In Mother Ann the female principle of Christ was manifested and in her the promise of the Second Coming was fulfilled. From that time on, for believers, the two sexes were to be equal but separate. The Millennium started with the official foundation of the United Society in 1787. A Shaker's faith began with confession of sin and included celibacy, common property, separation from the world, uniformity of dress, the healing gift, and unstructured freedom of expression in worship which often involved dancing, marching, laughing, barking, singing, and shaking. Today the Shakers are largely remembered for their unique form of worship, their legendary thrift and industry, and their high quality craftmanship, especially in the making of furniture.

ROBERT D. LINDER

SHARP, JAMES (1613-1679). Archbishop of St. Andrews. Born in Banff, he graduated from King's College, Aberdeen, was appointed professor of philosophy at St. Andrews in 1643, and in 1649 became minister of nearby Crail. His ambition, Episcopal sympathies, and persistent reports of a scandalous private life made him an object of suspicion in more thoroughgoing Covenanting circles, but prior to the Restoration of Charles II he was sent to London as the representative of the Resolutioners,* to ensure that the Kirk's lawful privileges were maintained. Cromwell had labeled Sharp an atheist, and Charles II regarded him as "one of the worst of men." Nevertheless the king was to use him for his own ends.

Sharp betrayed the moderate Presbyterians whose emissary he was, and when Episcopacy was restored in Scotland he was Charles's choice as archbishop of St. Andrews. As primate his ruthless persecution of the Covenanters* involved deceit and treachery deplored even by Sir Thomas Dalziel, the king's Scottish military commander whom the Scots called "the Muscovy Brute." An unsuccessful attempt was made on Sharp's life in 1668, but a second was more successful when in 1679 his coach was ambushed on Magus Moor, near St. Andrews, by a group of nine zealots. His murder was swiftly repudiated by the Covenanting leaders, but it precipitated the second major rebellion of Charles's reign. An astonishing inscription in Latin is still decipherable in Holy Trinity Church, St. Andrews, referring to Sharp as a "most holy martyr" and "an example of piety."

J.D. DOUGLAS

SHAW, WILLIAM (1798-1872). Wesleyan Methodist missionary to South Africa. Born in Glasgow, he accompanied the 1820 settlers to South Africa. Although officially chaplain to one party, he established Methodism throughout Albany and used this settler church as a base for advance beyond the frontier. Between 1823 and 1830 he planted six missions in the Ciskei and Transkei, others being added later. He also supervised work north of the Orange River from 1838 and in Natal from 1842. Shaw's high standing among the settlers and friendship with several African chiefs gave him unique breadth of outlook and sympathy. Most colonial governors respected his advice. In 1856 he returned to England. His proposal for a South African conference was premature, and he did not return to the field. He was elected president of conference in 1865, a fitting tribute to the father of South African Methodism.

D.G.L. CRAGG

SHEDD, W(ILLIAM) G(REENOUGH) T(HAYER) (1820-1894). American theologian. Born in Massachusetts and educated at the University of Vermont and Andover Theological Seminary, he gained pastoral experience in both Congregational and Presbyterian churches, but the major portion of his life was spent first as professor of English literature at the University of Vermont, then in teaching in the theological seminaries at Auburn, Andover, and Union (New York). His theology was strongly conservative and Calvinistic. His *Dogmatic Theology* (3 vols., 1888-94) is a clear statement of Westminster Presbyterian Calvinism. He defended orthodoxy in controversy with his colleague at Union, C.A. Briggs,* whose OT critical views led eventually to his dismissal from the Presbyterian Church and the withdrawal of Union from the Presbyterian denomination. Shedd's other major works are *Lectures on the Philosophy of History* (1856); *History of Christian Doctrine* (1863); *Homiletics and Pastoral Theology* (1869).

DONALD M. LAKE

SHEEN, FULTON JOHN (1895-). Roman Catholic bishop and broadcaster. Born on a farm at El Paso, Illinois, he graduated from St. Viator College, Catholic University of America, and the University of Louvain (Ph.D., 1923). He was ordained in 1919, taught philosophy at Catholic University (1926-50), and then until 1966 was national director of the Society for Propagation of the Faith. From 1930 to 1952 his broadcasts of the "Catholic Hour" were heard around the world, while he was also in the same period preacher at St. Patrick's Cathedral, New York. His telecast "Life is Worth Living" was seen by an estimated 30 million each week from 1951 to 1957. He was made auxiliary bishop of New York in 1951, and bishop of Rochester in 1967. He has written many books on philosophical, devotional, and anti-Communist topics. EARLE E. CAIRNS

SHELDON, CHARLES MONROE (1857-1946). Clergyman and writer. Born in Wellsville, New York, he was educated at Phillips Academy, Brown University, and Andover Theological Seminary, and received Congregational ordination in 1886. He held pastorates in Vermont and Kansas, but is best known as author of many books. *In His Steps* (1896) became a religious best seller because it poignantly challenged Christians to base their behavior on the answer to the question "What would Jesus do?" and because, never copyrighted, numerous publishers printed it. Although many thought its sales second only to the Bible's, it is generally conceded that six million might be a realistic estimate of copies sold. He became editor-in-chief of *Christian Herald* (1920-25) and contributing editor thereafter.

D.E. PITZER

SHELDON, GILBERT (1598-1677). Archbishop of Canterbury from 1663. Educated at Trinity College, Oxford, he was elected a fellow of All Souls in 1622, the year of his ordination. In 1626 he became warden of All Souls and actively supported William Laud's* university reforms. During the Civil War he was often with Charles I,* and consequently in 1648 was ejected from his wardenship and imprisoned for a time. Under the Commonwealth he lived quietly in the Midlands. At the Restoration he became bishop of London, but during Juxon's* archiepiscopate he exercised the real power. In 1663 he succeeded to Canterbury. A strong Laudian, he supported the severe measures against dissenters, and the restoration of the Anglican Church to its pre-Civil War position was due mainly to him. He negotiated with Clarendon the arrangement whereby Convocation no longer taxed the clergy. Chancellor of Oxford University (1667-79), he built the Sheldonian Theatre at his own cost. JOHN A. SIMPSON

SHEMBE, ISAIAH (c.1870-1935). Founder of the nativistic ama-Nazaretha in South Africa. He grew up in a heathen environment in Natal, but in response to revelations he began to preach and heal. He was baptized and ordained by a minister of the African Native Baptist Church, but later formed his own organization to which he applied all OT references to the Nazirites. His headquarters were fixed at Ekuphakameni ("the high place") near Durban, but he traveled throughout Natal and acquired immense influence among the Zulu. He composed many remarkable hymns *(Izihlabelelo zama Nazaretha).* Biblical influences mingled with traditional Zulu beliefs and practices in his teaching and ritual. He claimed to be the Zulu Messiah, and is revered as such by his followers who believe he has risen. The leadership was inherited by his son, J.G. Shembe.

D.G.L. CRAGG

SHENOUTE (d.451). Abbot and Sahidic scholar. About 388 he succeeded his uncle, Pgol, who about 350 had established an important cenobitic community, and White Monastery, based on the Rule of Pachomius,* near Sohag, some fifty miles downstream from Nag Hammadi.* Several accounts of his life remain, the oldest by his pupil and successor Besa, which describe his severe temperament and organizational abilities. Extant also is a considerable volume of expository epistles and sermons, nearly half unpublished. Shenoute was present at the Council of Ephesus* (431) —he dates his ministry from some forty-three years earlier—and influential on the deliberations of Chalcedon (451), having aided the church against both Gnosticism* and Monophysitism.*

CLYDE CURRY SMITH

SHEPHERD, ROBERT HENRY WISHART (1888-1971). Missionary to South Africa. Born near Dundee, Scotland, he was ordained in 1918 and went to South Africa in 1920 as a missionary of the United Free Church of Scotland.* After seven years in Tembuland he moved to Lovedale as chaplain (1927-42) and principal (1942-55). On his retirement Lovedale passed into government control. He was a gifted writer and edited the *South African Outlook* (1932-63). As director of Lovedale Press and literary secretary of the Christian Council of South Africa he encouraged African authorship and promoted Christian literature. He was elected moderator of the general assembly of the Church of Scotland in 1959, and was a member of the Monckton Commission which visited Central Africa in 1960.

D.G.L. CRAGG

SHEPHERD OF HERMAS, see HERMAS

SHEPPARD, HUGH RICHARD LAWRIE ("Dick") (1880-1937). Anglican clergyman and pacifist. After schooling at Marlborough he entered Trinity Hall, Cambridge, and thereafter Cuddesdon for theological training. He was ordained in 1907 and held various curacies in the London diocese before his appointment as vicar of St. Martin's-in-the-Fields in 1914. He had a deep concern to reach people and made St. Martin's one of the best-known churches in Britain. With the invention of broadcasting, he was one of the first to appreciate and use its immense possibilities for Christian influence. He believed in the necessity of ecclesiastical reform and identified himself with the Life and Liberty Movement in its early stages, though he was dissatisfied with the reform achieved by the Enabling Act of 1919. Because of ill-health he resigned St. Martin's in 1926, but held the deanery of Canterbury from 1929 to 1931, and was a canon of St. Paul's for a year from 1934. In his last years he was an ardent pacifist, forming the Peace Pledge Union in 1936. JOHN A. SIMPSON

SHIELDS, ALEXANDER (1660-1700). Scottish Covenanter.* Born at Haughhead on the Scottish Borders, he grew up in a Covenanting atmosphere and was probably familiar with conventicles from boyhood. Educated in Holland and committed to the Covenanting cause, he suffered frequent imprisonments in London, Edinburgh, and on the Bass Rock. After the martyrdom of James Renwick* (1688) he became leader of the

Covenanting minority. At the Revolution Settlement he left the Covenanting societies to join the newly reestablished Church of Scotland (Presbyterian). In a short busy life he was successively chaplain to the Cameronian Regiment in Scotland and Flanders, minister in London and St. Andrews, and chaplain to the ill-fated Scots expedition to Darien, where he died. Despite his many activities and travels he wrote *A Hind Let Loose*, a unique work which expounded a philosophy of the rights of man, social, political, and spiritual, and exposed the folly of the Stuart claim to the Divine Right of Kings.* ADAM LOUGHRIDGE

SHIELDS, THOMAS TODHUNTER (1873-1955). Baptist minister. Born and educated in Bristol, England, he emigrated to Ontario where he preached in various towns from 1897. In 1910 he began his forty-year ministry at Jarvis Street Baptist Church, Toronto. A militant fundamentalist, he founded and edited in 1922 the *Gospel Witness,* which he used to attack modernism and McMaster University (of which he was a one-time board member). The virulence of his popular and controversial sermons drew a vote of censure against him by the Baptist Convention of Ontario and Quebec. He therefore founded the Ontario-Quebec Association of Regular Baptist Churches, which in turn was soon split by his anti-dispensational views, and the Independent Baptist Fellowship was formed. He was president of the Baptist Bible Union of North America from 1923 to 1930, and for a brief period in 1927 was the acting president of Des Moines University. Jarvis Street remained the heart of his movement, for it was the home of Toronto Bible Seminary, of which he became president in 1927, in addition to being the center of his pulpit ministry.

ROBERT WILSON

SHOEMAKER, SAMUEL MOOR (1893-1963). Episcopal clergyman and writer. Born in Baltimore, Maryland, he studied at Princeton University and was ordained priest after graduation from Union Theological Seminary in 1921. (Between Princeton and seminary he had served for two years as a YMCA secretary in China.) From 1925 he was rector of Calvary Episcopal Church, New York. Attracted by Frank Buchman's Moral Re-Armament* Movement, he later left it, but without loss of his evangelical enthusiasm. As a popular lecturer, counselor, and radio speaker, Shoemaker inspired laymen of all walks to put warm Christian faith into practice in their everyday lives, and to evangelize on a personal basis. He assisted also the founders of Alcoholics Anonymous in the formulation of their useful "Twelve Steps." His numerous books included *Realizing Religion* (1921); *Religion That Works* (1928); *Twice Born Ministers* (1929); *How You Can Help Other People* (1946); and *By the Power of God* (1954). ALBERT H. FREUNDT, JR.

SHORT, AUGUSTUS (1802-1883). First Church of England bishop of Adelaide. Educated at Westminster and Christ Church, Oxford, where he became a tutor in 1829 and censor in 1833, he was influenced by the Tractarians* and wrote a defense of Tract 90. In 1835 he accepted the living of Ravensthorpe, and in 1847 became bishop of Adelaide, Australia. On his arrival he found five clergy, and by 1851 all state aid was discontinued. As a result of the 1850 conference of all the Australian bishops held in Sydney, and of legal advice gained in England in 1853, Short decided to organize his diocese as a voluntary body in 1855. In 1856 the creation of the diocese of Perth relieved him of his oversight of Western Australia. In 1881 he retired through ill-health. He died in Eastbourne, England.

NOEL S. POLLARD

SICARD (1160-1215). Bishop of Cremona. He studied in Bologna and taught canon law and theology at Paris, thereafter teaching in the cathedral school of Mainz before becoming bishop of Cremona in 1185. He won the city's independence from Frederick I, defended it against Brescia and Milan, and planned its fortifications. In 1202-5 he assisted the papal mission of Peter of Capua in Armenia and Constantinople. His chief works are *Mitrale* (1200), on liturgy; *Chronica universalis* (1213), a history from Creation, especially valuable for the Fourth Crusade; and *Summa decretorum,* his Paris lectures reworked at Mainz following only the form of Gratian's *Decretum.* C.G. THORNE, JR.

SICILIAN VESPERS. The massacre of the French in Sicily (30 March 1282), signaled by the tolling of the bell for Vespers. Three to four thousand died. It marked the end of the ambitions and intrigues of Charles of Anjou and Pope Martin IV,* and led eventually to Sicilian independence at the treaty of Caltabellota in 1302.

SICKINGEN, FRANZ VON (1481-1523). Perhaps the most colorful among the German knights, an anachronism at the beginning of the Modern Era, von Sickingen fought for Emperor Maximilian I and supported Charles V, but lost his life in the Knight's War, or Sickingen Feud (1522-23). The decline of feudalism, the rise of national states, the importance of the burgher class because of the growth of industry and commerce, and even the impact of humanism were not understood by him, although he contended for social reform. He was a religious independent, ready to support Martin Luther. He offered Luther a haven in 1520, if he should have to leave Saxony. He gathered some troops together outside Worms at the time of the Diet (1521), but a question remains if he would have used them to protect Luther had he been hired as a mercenary to side against him. Von Sickingen was disappointed in Luther's leadership, because Luther disclaimed the use of the sword for the spread of the Gospel. Luther dedicated his book *On Confession* (1522) to him, and Oecolampadius served as his chaplain during 1522. As a freebooter he earned the animosity of his neighbors. In 1522 von Sickingen attacked Trier, whereupon Richard von Greiffenklau, archbishop of Trier, the elector of the Palatinate, and Philip of Hesse* combined against him. They defeated him at

Landstuhl, and von Sickingen fell, mortally wounded. CARL S. MEYER

SIDETES, PHILIP, see PHILIP SIDETES

SIDONIUS APOLLINARIS, G. SOLLIUS (c.431-c.482). Gallic Latin poet and bishop of Clermont. Born into a noble family of Auvergne and well educated, Sidonius wrote panegyrics to the emperors Avitus, his father-in-law (456), Majorian (458), and Anthemius (468), which won for him his statue erected in Trajan's Forum and two terms as prefect of Rome (456 and 468). Reluctantly accepting the see of Clermont though not a cleric (468), he was naturally weak in Scripture and dogma but always faithful to his religious duties. From the same time he ceased adding to his total of twenty-four secular poems *(carmina)* and began his flow of 147 epistles, valuable sources for the history of his era, especially the resistance of Auvergne (469-75), which he led, and its ignominious surrender to Euric, Visigothic king, by Emperor Nepos. His writing was graceful, but jejune in content, reflecting the life of an urbane Roman gentleman, despite the barbarians' encroachments. Sidonius influenced the Middle Ages in his epistolary and panegyric styles and in his late classical rhetorical mannerism.
DANIEL C. SCAVONE

SIEUR DE MONTS (Pierre Du Guas de Monts) (1560?-1611). French colonizer in Canada. He belonged to the *gens de robe* of his day. Having fought in the cause of Henry IV* of France, carrying his Calvinist convictions with great pride, he nonetheless, in the interest of trade, agreed to have Indians in New France instructed in the dogmas of the Church of Rome. To exploit his trading monopoly in furs which extended over 40-46 degrees north latitude, he organized a trading company in 1604. He used the ablest and the worst elements of French society to colonize Acadia. De Monts distinguished himself as trader, explorer, and the first lieutenant-governor in Acadia. He is credited with having been the founder of the first permanent settlement in New France. Some accounts give his dates as 1558-1628. EDWARD J. FURCHA

SI-GAN-FU STONE. An inscription in Chinese and Syriac discovered in 1625 by Jesuit missionaries. The text believed to be dated between 779 and 781 describes the fortunes of the (Nestorian) Christians since 635 when the emperor T'ai-tsung received Alopen.* Under his successor Kao-tsung, permission was granted to create monasteries in several regions, some monks being supported by the emperor. A bishopric was established in 650. Later in that century there were conflicts with Buddhists and persecutions, but in the early decades of the eighth century Catholicos Selibhazekha became metropolitan of China. Many of the Christians seem from their names to have been Syrian or Persian, but there is evidence for some native names in the inscription. The stone set up by a synod of 779 contains also a confession of the Christian faith: this is noteworthy by its considerable adaptation to the Chinese idiom and religious-philosophical terminology. It may indicate that Christianity had lost some distinctive features in its eastward journey, or that Christians hesitated to speak openly of their mysteries.
J.N. BIRDSALL

SIGEBERT OF GEMBLOUX (c.1030-1112). Medieval chronicler. Born probably near Gembloux, he was educated at the monastery there. He taught for twenty years in Metz and returned in 1070 to Gembloux, where he spent the remainder of his life. He was a highly respected teacher and an extremely productive writer. He also became involved in the Investiture Controversy* and wrote three tracts in support of the imperial position. His most important works were written in the last years of his life. His *Chronica* is a history of events from 381 to 1111, and *De Viris Illustribus* is a collection of ecclesiastical biographies. He also was the author of a number of saints' lives, and a history of the abbots of Gembloux.
RUDOLPH HEINZE

SIGER OF BRABANT (c.1235-c.1282). Radical Aristotelian philosopher. A canon at St. Martin's, Liège, he later taught philosophy in Paris (c.1266-76). He expounded a heterodox Aristotelianism while professing Christianity, and was attacked by Bonaventura* and Thomas Aquinas.* Thirteen errors taken from his teaching were condemned by the bishop of Paris (1270). This was ineffective and, summoned to appear before an inquisitor in 1276, Siger fled to Italy. The so-called Great Condemnation followed, when 219 propositions were condemned by the bishop, Étienne Tempier (1277). Siger retired to Orvieto where he was reportedly stabbed to death by an insane cleric. Siger was leader of a group inaugurating purely rational teaching, unconcerned with Christian dogma. His chief source was Aristotle*; secondary sources included Proclus,* Avicenna,* Averroes,* Albertus Magnus, and Aquinas. Typical teachings were: the First Being is the immediate cause of a single creature; all other creatures derive indirectly from God by a progressive emanation; the created world is necessary and eternal, and every species of being (e.g., man) is eternal; there is only one intellectual soul for mankind, and consequently one will; this unique soul is eternal, but human individuals are not immortal; human will is a passive potency moved by the intellect.
See SCHOLASTICISM. J.G.G. NORMAN

SIGISMUND (1361-1437). Holy Roman Emperor. Second son of Emperor Charles IV, Sigismund inherited the mark of Brandenburg on his father's death in 1378. For six years thereafter he studied at the Hungarian court. In 1385 he married Maria, daughter of King Louis of Hungary and Poland, and in 1387 he succeeded his father-in-law as king of Hungary. Domestic Hungarian problems, Turkish attacks, and intrigues in his bid for succession in Germany and Bohemia weakened his rule. In 1410 he was elected German king, or "king of the Romans." Solution of the Great Schism being in the best imperial interest, he pressured John XXIII* to convoke the Council of Constance* (1414-18), and his international travels and appeals during the sessions were instrumental in restoring a unified papacy. He guaranteed John Hus* safe passage to the council, where the Reformer was martyred. The Hussite Wars in Bohemia (c.1420-36) erupted after Sigismund succeeded Wenceslas as king of Bohemia in 1419 and pledged to prosecute heresy. Vexed by yet another Ottoman advance on Hungary, Sigismund was unable to consolidate his power in Germany. Although Pope Eugene IV crowned him Holy Roman Emperor in 1433, Sigismund died without having achieved his goal of unifying Christendom against Islamic advance.

JAMES DE JONG

SIGN OF THE CROSS. The act done by both clergy and people of reproducing the shape of the cross of Christ with the general idea of recalling that from Christ's death the grace of God flows to the church. In services of baptism and confirmation, from early times the priest or bishop has made the sign of the cross as part of the giving of God's blessing to the candidates. Also, in worship the priest gives God's blessing while making the sign of the cross in front of himself. Ordinary worshipers are taught to make the sign of the cross as they enter churches and during certain services. The traditional way is to take the right hand from forehead to the center of the chest, then from shoulder to shoulder, and then back to the center of the chest. In the West, the crossing of the chest is from left to right, and in the East from right to left.

PETER TOON

SIMEON, CHARLES (1759-1836). Evangelical leader. Educated at Eton and King's College, Cambridge, on entering the latter he discovered that attendance at Holy Communion was compulsory. His preparation for taking the sacrament was the main factor in his subsequent conversion. His own adoption of Evangelical views was fostered by his friendship with Henry and John Venn.* Appointed vicar of Holy Trinity in Cambridge in 1782, he ministered there until his death. He overcame early opposition mainly through his pastoral care, and while firmly attached to the Church of England, he became the center of evangelicalism in Cambridge. He had immense influence with undergraduates both from the pulpit and in small groups. As well as encouraging the British and Foreign Bible Society, he helped to found the Church Missionary Society and the London Jews Society (later the Church Mission to Jews), and his curate, Henry Martyn,* as a chaplain of the East India Company became one of India's best-known pioneer missionaries. He established the Simeon Trust (which still exists) which purchased livings for Evangelicals. His own sermon outlines, *Horae Homileticae*, were collected and published in twenty-one volumes in 1840.

PETER S. DAWES

SIMEON OF THESSALONICA (d.1429). Greek Orthodox archbishop and theologian. Little is known of his life; his fame rests primarily on his chief work, *Dialogue Against all Heresies and on the One Faith.* The first part of the book was devoted to a discussion of the Trinity and Christology, while the second section dealt with the liturgy and the sacraments. The work included polemics against Jews, Bogomiles,* and Muslims, and questioned the primacy of the papacy. Simeon was willing to grant the pope considerable authority if he upheld the true faith, but since the papacy had added the word *Filioque* to the Creed, Simeon maintained it had forfeited its primacy. He also wrote a number of shorter works dealing with the Nicene Creed* and doctrinal questions.

RUDOLPH HEINZE

SIMEON THE STYLITE (c.390-459). Pillar ascetic. Born in Cilicia, son of a shepherd, he moved to Antioch where as a teenager he became an anchorite. For twenty years he lived in various monasteries in N Syria. About 423 he started to live on a pillar at Telanissus (Dair Sem'an). For thirty-six years he lived in great austerity on a platform at the top of the pillar, the height of which was gradually increased until it reached sixty feet from the ground. Thousands came to see him and hear his preaching, with the result that his influence was extensive. After his death a monastery and sanctuary were built on the site of the pillar. A famous disciple of his was Daniel the Stylite.

PETER TOON

SIMON. Apostle. One of the Twelve, he is called by Matthew and Mark "the Cananaean" (Matt. 10:4; Mark 3:18 RSV), and by Luke "the Zealot" (Luke 6:15; Acts 1:13). The latter is a translation of the Aramaic underlying the former. Little is known about him, but his designation suggests that at some stage he had been associated with the active opposition to Roman rule which was characteristic of the Zealot* party. Some early Christian writers identified him with Simeon, son of Clopas, said by Hegesippus to have succeeded James as the head of the Jerusalem church.

SIMON MAGUS. A sorcerer mentioned in Acts (8:9-24) as combining the practice of magic with a Hellenistic-Jewish syncretism. He claimed to be a divine emanation: "that power of God which is called Great" (RSV). For a time he professed Christian faith, but was later condemned by Peter for seeking to obtain spiritual powers by the payment of money (hence the term "simony"). The passage in Acts is the only mention of Simon in the NT, though he features in later Christian literature. Justin Martyr says Simon was a native of Gitta in Samaria and was widely acclaimed as a god; he came to Rome during the reign of Claudius (A.D. 41-54). Irenaeus, Hippolytus, and Epiphanius consider him the prototype of the later Gnostic heresies and describe the alleged doctrines of his followers; it is uncertain how reliable their accounts are. Simon features in the later pseudo-Clementine literature (third/fourth century) as the opponent of Peter, but the material is clearly fictitious. It has been suggested that the heresiarch mentioned by the Fathers and Simon of Acts 8 are two different people, the former living in the second century.

W. WARD GASQUE

SIMON OF SUDBURY (d.1381). Archbishop of Canterbury from 1375. Born at Sudbury, he studied at Paris, afterward becoming chaplain to Innocent IV who in 1361 appointed him bishop of London. He was soon active in politics, and it was of advantage to the Lancastrian party when Gregory XIII appointed him archbishop of Canterbury in 1375. He was lenient toward the Wycliffites, and when he had to try John Wycliffe* (1378) he dismissed him with an injunction to silence. He was, however, stern in denouncing clerical abuses, particularly that of nonresidence. In 1380 he became chancellor, and his poll tax made him unpopular. In the revolt of 1381 his lands were spoiled and with Richard II he took refuge in the Tower of London. Though he resigned the chancellorship, the mob beheaded him.

JOHN A. SIMPSON

SIMON PETER, see PETER THE APOSTLE

SIMONY. The term is derived from Simon Magus (Acts 8:18-24) who attempted to purchase from the Apostles Peter and John the gift of conferring the Holy Spirit by the laying on of hands. Throughout Christian history it has assumed sophisticated and nuanced definitions in both civil and ecclesiastical jurisprudence. Essentially, however, simony refers to the deliberate conferment or acquisition of anything spiritual or sacred for remuneration, monetary or otherwise. An unsolicited gratuity, therefore, is not generally considered simony. Its classic manifestation was the medieval traffic in indulgences* and sale of clerical preferments. The legal recognition of Christianity under Constantine and the church's subsequent rise to wealth and power insured the appearance of simony as a problem of some magnitude. In its several forms it was condemned by such councils as Chalcedon (451), Third Lateran (1179), Trent (1545ff.) and by such leaders as Gregory I and Thomas Aquinas. Aquinas contributed significantly to its treatment in Roman Catholic canon law. Both Wycliffe and Hus polemicized against it in major writings. Wherever discovered, simony always requires ecclesiastical restitution and in extreme forms may result in deposition from office or even in excommunication. In some countries the offense is punishable under civil statutes.

JAMES DE JONG

SIMPLICIANUS (d.400). Bishop of Milan. He was first heard of in Rome (c.350-60), where he was instrumental in the conversion of Victorinus, a Platonist professor of rhetoric. When his pupil Ambrose became bishop of Milan (373), he removed there to prepare him for baptism and ordination. He helped the seeking Augustine by recounting the story of Victorinus's conversion and encouraging him in his reading of the Platonists; some of Augustine's early treatises were addressed to him. He succeeded Ambrose as bishop in 397.

SIMPLICIUS (d.483). Pope from 468. His pontificate saw the final collapse of the Western Empire when Odoacer assumed title as king of Italy in 476. Unlike his famous predecessor Leo I, Simplicius exercised very little influence over the political developments in Italy and was unable to maintain that high degree of leadership which made the papacy impressive from 440 to 461 in the midst of political upheaval. His influence was at best minimal, and he seems to have been the captive of the events of the day.

SIMPSON, A.B., see CHRISTIAN AND MISSIONARY ALLIANCE

SIMPSON, SIR JAMES YOUNG (1811-1870). Discoverer of the anaesthetic effect of chloroform. Son of a Scottish village baker, Simpson qualified in medicine at Edinburgh University and in 1840 was elected to the chair of medicine and midwifery. Ether was used in an operation in Edinburgh in 1846 and Simpson, experimenting on himself, looked for more suitable compounds for application, especially in midwifery—and in 1847 discovered the effect of chloroform. Violent controversy ensued which died out after 1853, when Queen Victoria was given chloroform at the birth of Prince Leopold. In his later days Simpson's fame was worldwide.

As a student Simpson showed little interest in religion; after his marriage in 1839, however, he joined the kirk. Duns believes he was converted about this time. He made good use of his detailed knowledge of the Bible, imbibed from his deeply religious home background, in his *Religious Objections to the Employment of Anaesthesia* (1848). In this he effectively argues that God Himself used anaesthesia to prevent pain (Gen. 2:21), that "sorrow" in the curse of Genesis 3:16, 17 does not mean "pain" but "labor," and that in any case the curse was not immutable (otherwise it would be sinful to pull up thorns and thistles or to use a tractor).

BIBLIOGRAPHY: J. Duns, *Memoir of Sir James Y. Simpson* (1873); and J.A. Shepherd, *Simpson and Syme of Edinburgh* (1969).

R.E.D. CLARK

SIMPSON, JAMES YOUNG (1873-1934). Scottish natural scientist. He studied at Edinburgh University where he became involved in the evangelical work of the day (e.g., the Moody mission). Strongly influenced by Henry Drummond,* who persuaded him to take up science as a Christian work, Simpson became professor of natural science at Edinburgh. He wrote five books on science and religion, the best known being *Landmarks in the Struggle between Science and Religion* (1925) in which he effectively demolished A.D. White's *History of the Warfare of Science with Theology* (1896). Where White saw a struggle between religion and science, Simpson saw one between each older generation of scientists and the new generation about to succeed it. His own forebear and great-uncle, Sir James, had strongly opposed Joseph Lister, inventor of antiseptic surgery.

R.E.D. CLARK

SIN. When the Bible seeks to describe the deepest of all the problems of human existence, it speaks in terms of sin, its consequences, and its guilt. It is sin done by man, and/or consented to by man, which creates the situation which requires the atonement with its infinite cost to God in the death of His Son. Sin involves man in many forms of personal failing within himself—loss of integrity, self-centeredness, and failure to measure up to the external standards and laws which even he himself sets for his achievement. Though such aspects of sin are taken note of in the OT, sin in itself is regarded as an attitude of hatred and mistrust toward God, of senseless pride before God. It is calculated to make impossible any true personal relationship to God (Exod. 20:5; Deut. 5:9; Rom. 5:10). The NT regards the true nature of sin as having been revealed in the attitude and response of men to the truth, love, and challenge of God in Christ (Acts 3:14,15,19). The true nature of Paul's own sin, for example, was brought home to him most forcibly when he found himself persecuting Christ and His church (1 Cor. 15:9; 1 Tim. 1:15; Gal. 1:23; Phil. 3:6).

Sin manifests itself in various ways. Some sins are regarded as more serious than others. The OT separates sins done through haste or weakness from sins done "with a high hand" (Num. 15:28-31). Jesus was more severe in His condemnation of certain types of pride and hypocrisy than in His condemnation of some other human failures. He spoke of some failures as meriting "few," some "many," stripes (Luke 12:47,48). Yet though acts are regarded as sinful, sin is much more than a series of overt misdeeds. The badness which makes sin sin is deeply entwined in the depth of man's personality. Sin belongs to the man and dwells in the man (Rom. 7:20,23) rather than in the act. A Christian has to ask pardon for what he is as well as for what he does. Paul regards sin as a power which can not only dwell in a man, but can possess and reign over him (Rom. 5:12; 6:6, 14). The whole of man's being which is infected with sin ranging from sensuality to pride Paul calls "the flesh." Even man's goodness can become tragically perverted by the evil will of the flesh.

Sin is regarded in the Bible as the result of man's being left free to choose to trust and love God. His refusal to do so is absurd, but its consequences are infinitely tragic. It is denied that God has any responsibility or complicity in this refusal. God foresaw sin, even risked the occurrence of sin. Sin is not essential to His creation. Nor does He hold sin in being as He holds good things in being. Sin can only be a perversion of what is good. God can cause the sinner's activity, but not His sin. When sin occurs it seems to fall in such a way under His sovereign power and purpose that it can seem to have been ordained. But God is in no sense its author.

BIBLIOGRAPHY: J. Muller, *The Christian Doctrine of Sin* (1868); W.E. Orchard, *Modern Theories of Sin* (1910); F.R. Tennant, *The Concept of Sin* (1912); E.J. Bicknell, *The Christian Idea of Sin and Original Sin* (1922); C.R. Smith, *The Biblical Conception of Sin* (1953); S. Porubcan, *Sin in the Old Testament* (1963). RONALD S. WALLACE

SIRICIUS (c.334-399). Pope from 384. Roman-born, he was much influenced by Ambrose,* but had his own decided views about the exalted status of his office and responsibilities. This was made clear in a letter to Bishop Himerius of Tarragona, regarded as the first papal decretal. During his pontificate the Melitian Schism* was ended, Priscillianism opposed, extremism of various kinds disapproved of, and the new basilica of St. Paul dedicated.

SIRMIUM, COUNCIL OF (357). Technically in the jurisdictional sphere of the Western Empire, the city was the venue chosen by the emperor Constantius, an Arian sympathizer, to bring the Western bishops into Arian line. It began with the expurgation of the Greek concepts *ousia* ("substance"), *homoousios,* and *homoiousios,* on the ground that they were nonscriptural and above human understanding, substituting a baptismal Trinity and a subordinated, begotten Son. Hilary of Poitiers* who reports upon the events and gives the Latin text of the creed, identifies it as the "Blasphemy of Sirmium."

SIRMOND, JACQUES (1559-1651). Jesuit scholar. Born at Riom, Auvergne, France, he studied at the Jesuit college in Billom, and became a member of the order in 1576. He taught literature at Pont-à-Mousson and Paris (1581-90) and was secretary to the Jesuit general, C. Aquaviva, in Rome (1590-1608). He returned to Paris, becoming rector of the *Collège de Clermont* (1617) and confessor to Louis XIII (1637-43). One of the greatest scholars of the age, from 1610 he published many works, especially editions of the Fathers—e.g., Fulgentius, Paschasius Radbertus, Theodoret, Eusebius of Caesarea, and Rufinus. He also distinguished between Dionysius the Areopagite and Pseudo-Dionysius. J.G.G. NORMAN

SIX ARTICLES, THE (1539). The Act of Six Articles was pushed through Parliament by Henry VIII of England and came to be known as "the whip of six strings" because noncompliance with it (classed as a felony) was punishable by death and confiscation of property. Protestants complained that the "Reformation goes backwards" in England because this was merely a restatement of some basic tenets of Romanism under the auspices of the English state church which Henry now headed. The English Church had become independent of Rome, but it did not change its theology under Henry. The Six Articles taught transubstantiation,* auricular confession to a priest, celibacy of the clergy, and Communion in one kind (bread only need be given to laymen).

HOWARD F. VOS

SIXTUS II (d.258). Pope 257-58. He resumed relations with the churches of Africa and Asia Minor, as well as with Cyprian, so ending a rupture over the validity of baptism by heretics. The latter he held valid, as had his predecessor Stephen I. He was nevertheless tolerant toward the rebaptism policy of the Eastern churches, probably influenced by Dionysius of Alexandria. Among the most highly venerated of the early martyrs (he was beheaded under Valerian while conducting services in the cemetery of Praetextatus), his name was in the Roman calendar of the mid-fourth century, and remains in the Canon of the

Mass. There is no proof for his having written *Ad Novatianum* or composed, even edited, the "Pythagorean Sentences of Sextus" translated by Rufinus of Aquileia. C.G. THORNE, JR.

SIXTUS IV (1414-1484). Pope from 1471. Born Francesco della Rovere, he entered the Franciscan Order early, studied at the universities of Padua and Bologna, and became general of his order in 1464. Further promotion soon came: cardinal in 1467, pope four years later. His international forays and plans involving Turks, French, and Russians met with little success, but he did establish the University of Copenhagen in 1475. Thereafter he concentrated on Italian politics, becoming embroiled in strife, nepotism, and conspiracy to an astonishing extent. Nevertheless he condemned abuses in the Spanish Inquisition, championed the Mendicant Orders, was a patron of the arts, built the Sistine Chapel, and enriched the Vatican Library. J.D. DOUGLAS

SIXTUS V (1521-1590). Pope from 1585. Born at Grottamare, Felice Peretti was educated by the Franciscans of Montalto and took the habit at the age of twelve. Ordained priest in 1547, he soon became a noted preacher and friend of Loyola and Philip Neri. In 1560 he was appointed Consultor of the Inquisition, becoming general of his order and bishop of St. Agata in 1566. In 1570 he was created cardinal by Pius X and was bishop of Fermo from 1571 to 1577, but during the pontificate of Gregory XIII (1572-85) was kept deliberately in the background. In 1585 he was chosen as pope.

At once the somewhat uncritical scholar and book collector showed himself an energetic reformer, suppressing brigandage in the Papal States by frequent executions, reforming the Curia and the cardinalate, over which he asserted papal authority, and placing the finances of Rome on a sounder footing by increased taxation and the elimination of graft. With the help of the architect Fontana he erected massive structures in Rome itself in the then fashionable Rococo style: the Lateran Palace, the Via Sistina, the Vatican Library, and the restored aqueduct renamed after him the Acqua Felice are all his creation. In foreign policy he was less able, trying with some vacillation to restore a balance of power among the Catholic powers and restrain Philip of Spain's ambitions at the expense of France. The revision of the text of the Vulgate known as the Sistine was also begun during this pontificate.

IAN SELLERS

SLEIDANUS, JOHANNES (1506-1556). Annalist of the Reformation. Born in Schleiden, near Aachen, he studied classics at Liège and Cologne, and law at Paris and Orléans. He entered the service of Cardinal du Bellay and represented Francis I in diplomatic negotiations with the Smalcald League* (1537). Dismissed for his Protestant opinions, he settled at Strasbourg (1542). As Sleidanus was accustomed to copying all papers bearing upon the Reformation to which he had access, Martin Bucer* persuaded Philip of Hesse to appoint him historian of the Reformation (1544). The first volume was finished in 1545. He used diplomatic visits to England and Marburg to collect materials. When war interrupted his work, at Cranmer's intercession he was granted a pension from Edward VI of England. He represented Strasbourg and a group of imperial cities at the Council of Trent (1551). Appointed professor of law at Strasbourg, he finished his great work, entitled *De Statu Religionis et Republicae Carolo V Caesare Commentarii* (1555). He died in poverty. His book remains the most valuable contemporary history of Reformation times, containing the largest collection of documents. Because of its impartiality, however, it pleased neither Protestant nor Catholic. J.G.G. NORMAN

SLESSOR, MARY (1848-1915). Missionary to West Africa. Born in Aberdeen, Scotland, and brought up in Dundee, she came of a very poor family, but her mother was a devout Christian deeply interested in the United Presbyterian Church's Calabar Mission. Mary was converted in her teens, and after experience of youth work in the Dundee slums she sailed for Nigeria in 1876 and worked there almost continuously until her death, first in the Okoyong area and then at Itu among the Ibo people. She fought against witchcraft, drunkenness, twin-killing, and other cruel customs. She believed in "the daily mixing with the people" to break down suspicion and fear. She acquired great skill in the languages and had an almost uncanny insight into the African mind.

Such was her influence over men that even the most savage and powerful chiefs made her a trusted arbiter in disputes of all kinds. She was instrumental in establishing trade between the coast and inland areas to their benefit, and in beginning the Hope Waddell Institution to train Africans in useful trades and to carry out medical work. She became the first woman vice-consul in the British Empire when British rule was established in the area. She had an unusual combination of qualities, humor and seriousness, roughness and tenderness, vision and practicality. These with a cool nerve and disregard for personal health and comfort helped to make her a powerful influence for Christianity. As a result of her work under God the Ibo people became more Christian than tribes in other parts of Nigeria.

See W.P. Livingstone, *Mary Slessor of Calabar, the White Queen* (1916). J.W. MEIKLEJOHN

SMALCALD ARTICLES (1537). Drawn up by Martin Luther, the Smalcald Articles were subscribed to by the leading Lutheran theologians as their response to the invitation of Paul III to the council called to be held in Mantua in 1537. To them is appended the "Treatise of the Power and Primacy of the Pope," written by Philip Melanchthon.* They were endorsed by the princes and estates, although not formally accepted by the Smalcald League.* They belong to the Lutheran Confessions, or Symbols.

The Smalcald Articles consist of three parts, after an introduction. The first part briefly reaffirms the ancient creeds. The second part deals with Christology, the Mass, chapters and cloisters, and the papacy. Purgatory, pilgrimages, monastic life, relics, indulgences, and the invocation of saints are condemned. The pope is branded as the "very Antichrist" and the apostle of the devil. Fifteen points of doctrine are singled out in the third part for special treatment: sin, the Law, repentance, the Gospel, baptism, the sacrament of the altar, the Keys, confession, excommunication, ordination and the Call, the marriage of priests, the church, how one is justified before God and of good works, monastic vows, and human traditions. Extended treatment is given to the "False Repentance of the Papists."

In Melanchthon's "Of the Power and Primacy of the Pope" the claim that the pope rules by divine right is repudiated by citations from the Scriptures and the Fathers. The arguments of the Roman Catholics are countered. The power and jurisdiction of the bishops are taken up separately. Among those signing "the Augsburg Confession and the Apology" and the Articles at Smalcald in 1537 were Martin Bucer, Ambrose Blaurer, Paul Fagius of Strasbourg, and the Scot, John Aepinus, superintendent of Hamburg.

See T.A. Tappert (ed.), *The Book of Concord: The Confessions of the Lutheran Church* (1959).

CARL S. MEYER

SMALCALD LEAGUE. A league of Lutheran princes formed to protect their religious interests. In 1525 the Roman Catholic princes had come together to form the League of Dessau. Early in the following year Elector John the Constant of Saxony and Landgrave Philip of Hesse made an alliance, which other princes joined. After the Diet of Augsburg the formal organization of the Smalcald League took place (December 1530). It was purely a defensive league. Strasbourg, Ulm, Constance, Reutlingen, and other cities joined in. In 1535, at the Diet of Smalcald, it was agreed that new members would have "to provide for such teaching and preaching as was in harmony with the Word of God and the pure teaching of our [Augsburg] Confession." Henry VIII of England found this stipulation one barrier, which prevented him from joining the League in 1536 and 1538. Attempts to reconcile the Lutherans and Roman Catholics failed at Ratisbon in 1541. In 1546, at the meeting with the emperor, no agreement was reached. The Smalcald War culminated in the defeat of the Smalcald League in the battle of Muehlberg and the Wittenberg Capitulation (1547). Mortiz of Saxony joined the emperor—for which he received the electorate—and thus insured the defeat of the Smalcald League. In the Peace of Augsburg* (1555), by the principle *cuius regio, eius religio* the Lutheran princes secured the right to regulate the religious affairs of their territories. CARL S. MEYER

SMART, CHRISTOPHER (1722-1771). Religious poet. He was born at Shipbourne, Kent, and educated at Cambridge, where he gained the Seaton prize for religious poetry on several occasions. He moved to London and worked as a hack journalist, and eventually he became insane. During his confinement he wrote *A Song to David* (1763) and *Jubilate Agno,* a work not published until the 1930s. He also produced a version of the Psalms and wrote several hymns. Though some of these were specifically written "for the fasts and festivals of the Church of England," his is a highly individualistic, even "enthusiastic" faith. He approaches questions of belief through the imagination, and hence his is visionary poetry. Thus *A Song to David* is a sustained, but also highly patterned, paean on the abundant and beneficent creativeness of God. David was the chosen of God, and Smart saw himself in the same capacity—an instrument of praise who must be pure for the task he had to fulfill. *Jubilate Agno* is even more complex in its pattern than *A Song to David* and owes much to Smart's knowledge of Hebrew.

ARTHUR POLLARD

SMART, PETER (1569-1652). Puritan divine. Educated at Westminster School and Christ Church, Oxford, he was ordained in 1595 and became master of Durham Grammar School in 1598 and, later, prebend of Durham Cathedral. Under the rule of the High Church bishop Neile (1617-27) he absented himself from the services and in 1628 published a sermon directed against John Cosin (a fellow prebend) and the "ritualistic" services. He was suspended and fined, and his sermon was burned. Refusing to pay the fine, he was imprisoned in 1631 and remained in custody until released by the Long Parliament in 1641. His further publications—e.g., *Catalogue of Superstitious Innovations* (1642) and *A Short Treatise of Altars* (1643)—reveal that imprisonment had not changed his views, which were similar to those of the sermon, *The Vanity and Downfall of Popish Ceremonies* (1628). After receiving several sequestered benefices, he died at Baxter Wood, near Durham. PETER TOON

SMITH, EDWIN WILLIAM (1876-1957). Missionary, writer, and anthropologist. Born of missionary parents in Aliwal North, South Africa, he entered the Primitive Methodist* ministry. Between 1902 and 1915 he worked among the Ila people of Zambia, producing a grammar, dictionary, and NT translation. He served the British and Foreign Bible Society (1916-39), latterly as editorial (translations) superintendent, and also taught in various American Negro colleges. Although he had little academic training, he achieved international recognition as an anthropologist, notably for his works, *The Ila-Speaking People of Northern Rhodesia* (with A.M. Dale) and *The Golden Stool.* He was elected president of the Royal Anthropological Institute (1933-35) and received various distinctions. He published numerous books on topics related to African religion and missions. D.G.L. CRAGG

SMITH, ELI (1801-1857). American Congregational missionary and Orientalist. Born at Northford, Connecticut, he graduated from Yale (1821) and Andover Seminary (1826). Ordained the latter year, he engaged in mission work under the American Board of Commissioners for Foreign Missions in Malta and Syria. In 1830, with H.G.O. Dwight, he explored Asia Minor, Armenia, Georgia, and Persia. The two published a book dealing with their trip which led to the

establishment of the American mission among Nestorian Christians. In 1838 he and Edward Robinson explored Sinai, Palestine, and S Syria. The last ten years of Smith's life were spent translating the Bible into Arabic. He died at Beirut.
ROBERT C. NEWMAN

SMITH, SIR GEORGE ADAM (1856-1942). OT scholar. Born in India, where his father was editor of the *Calcutta Review*, he read arts and theology at Edinburgh and pursued further studies at German universities and in Cairo. Following a brief period as tutor at the Free Church College, Aberdeen, after the suspension of W.R. Smith, he became minister of Queen's Cross Free Church in the city (1882-92). He saw the issues involved in Smith's trial and set himself "to reconcile the outlook of an advanced scientific scholar with the spirit of devout reverence" (his wife's words). While professor of OT at the Free Church College, Glasgow (1892-1909), he campaigned for proper labor conditions and also undertook four lecture tours in America. His Yale lectures, *Modern Criticism and the Preaching of the Old Testament* (1901) threatened a heresy trial in Scotland. From 1909 until retirement in 1935 he was principal of Aberdeen University. He became a knight, Fellow of the British Academy, moderator of his church's assembly, and a royal chaplain in Scotland. His many publications include the incomparable *Historical Geography of the Holy Land* (1894; rev. 1931).
C.G. THORNE, JR.

SMITH, HANNA WHITALL (1832-1911). Quaker author. Member of a pious Quaker family in Philadelphia, she is best known as the writer of *The Christian's Secret of a Happy Life* (1875) and as speaker with her husband, Robert Pearsall Smith, at interdenominational "Higher Christian Life" meetings in America and England. Undergoing early years of depression and skepticism, she was converted in 1858 under Plymouth Brethren influence at the same time as her husband, a Presbyterian layman. In 1867 she entered into a new experience of faith in Christ based upon Romans 6:6. She now testified to a life of spiritual victory and rest through complete commitment to Christ. Her husband, skeptical at first, later testified to the same experience and joined her as leader of Christian assemblies devoted to the study of biblical teaching on the life of victory in Christ. In 1872 they moved to England because of her husband's declining health. For two years they experienced phenomenal success in interdenominational meetings devoted to biblical exposition of their newfound religious experience. The movement thus initiated led to the founding of the Keswick Convention* in 1874, where annual sessions are still devoted to a consideration of biblical teaching on the higher life for the believing Christian.
S. RICHEY KAMM

SMITH, HENRY PRESERVED (1847-1927). OT scholar. Born in Troy, Ohio, of Puritan descent, he was educated at Amherst College, Lane Theological Seminary, and the University of Berlin. He taught at Lane from 1874 (apart from a year at Leipzig, 1876-77), was ordained to the Presbyterian ministry, and until about 1882 companied with conservatives. But already his mind was changing, for in that year an article on Wellhausen in the *Presbyterian Review* carried his observation that textual corruption of the Bible implied noninfallibility. It was not, however, until he defended C.A. Briggs* that he too was tried for heresy by the Cincinnati presbytery (1892), suspended from the ministry, denied appeal by the general assembly (1894), and forced to give up his Lane professorship and his home in Cincinnati. Subsequently he taught at Amherst (1898-1907), Meadville (Pennsylvania) Theological School (1907-13), and Union Theological Seminary (1913-25). His books included *The Religion of Israel* (1914), *Essays in Biblical Interpretation* (1921), and his autobiographical *Heretic's Defense* (1926).
CLYDE CURRY SMITH

SMITH, JOHN TAYLOR (1860-1937). Bishop and chaplain-general. He was converted at the age of eleven, educated at Kendall Grammar School and St. John's Hall, Highbury, and after five years' curacy in Norwood, sailed in 1891 to Sierra Leone as canon missioner, becoming bishop in 1897. An honorary chaplain to Queen Victoria, he was appointed chaplain-general in time for World War I, retiring in 1925. "Everybody's Bishop" was jovial, rotund, and saintly, a constant helper of the Children's Special Service Mission and the Keswick Convention.* Early rising for prayer, reading, and physical exercise was his lifelong custom. Daily he passed on to his friends his celebrated "Best Thought," gleaned in his morning devotions. He believed in, and relied upon, God's minute ordering of his life. He died aboard ship in the Mediterranean and was buried at sea.
ARTHUR CLARKE

SMITH, JOSEPH (1805-1844). Mormon* prophet and founder of the Church of Jesus Christ of Latter-day Saints. Born into a shiftless, frontier family in Vermont, he moved in 1816 to Palmyra, New York. Here he experienced conversion during a religious revival four years later. Subsequently he claimed to have received a direct revelation from God engraved upon golden plates. The contents he translated and published as the *Book of Mormon* (1830) with the assistance of Sidney Rigdon, a former Campbellite minister. *A Book of Commandments* (1833), authored by Smith and later republished as *Doctrine and Covenants* (1835), provided the basis of Mormon theology. Smith's practice of polygamy, for which he claimed divine approval in a revelation announced in 1843, and his continued mismanagement of community affairs led to his downfall. Non-Mormon Illinois neighbors arrested Joseph and his brother and confined them in the nearby Carthage jail, where they were murdered by a mob in 1844.
S. RICHEY KAMM

SMITH, RODNEY ("Gipsy") (1860-1947). English evangelist. Born in a tent near Epping Forest, he was the son of gipsies who traveled in East Anglia. He was greatly affected by his mother's death from smallpox. Soon after, his father was converted and began to hold services. Rodney was himself converted in 1876 in Cambridge, and in 1877 joined William Booth* in his "Christian Mission," serving as a captain in the Salvation Army until 1882. In 1889 he went to America on an evangelistic tour, after which he joined the Manchester Wesleyan Mission. Following a world preaching tour (1897-1912) he was missioner for the National Free Church Council. He served with the YMCA during World War I, and George VI made him a member of the Order of the British Empire. His preaching was characterized by "the wooing note" and constantly revealed his love of nature and the Bible. He sang simple gospel solos.

J.G.G. NORMAN

SMITH, SYDNEY (1771-1845). English clergyman, writer, and wit. Educated at Winchester and Oxford, he was ordained in 1796. In 1802 he helped to found the *Edinburgh Review* and contributed to it for twenty-five years. Bigotry and tyranny, hypocrisy and cruelty were his particular targets. Particularly noteworthy were *The Letters of Peter Plymley* on Catholic emancipation. His own prejudices, however, led him to unjustified attacks on William Carey* and other early missionaries abroad as well as evangelicals at home whom he described as "fanatics ... in one general conspiracy against common sense and rational orthodoxy Christianity." A fine talker and wit, he was prevented from receiving a preferment by his outspoken views, except to a canonry first at Bristol and then at St. Paul's, London, which at least delivered him from his vicarage in the rural countryside which he loathed.

PETER S. DAWES

SMITH, WILLIAM ROBERTSON (1846-1894). Scottish Old Testament scholar. A son of the manse, he was educated in Scotland and Germany and in 1870 appointed professor of Oriental languages and Old Testament exegesis at the Free Church College, Aberdeen. Seven years later he was suspended after contributing to the *Encyclopaedia Britannica* articles that allegedly undermined belief in the inspiration of Scripture. The charges were dropped in 1880, but renewed attacks on him, coupled with an uncompromising spirit that was the despair of his friends, led to dismissal from the college in 1881. He subsequently became editor-in-chief of the *Encyclopaedia Britannica* and professor of Arabic at Cambridge (1883-94), where he also served for a time as chief librarian. His books included *The Old Testament in the Jewish Church* (1881), *The Prophets of Israel* (1882), and *Lectures on the Religion of the Semites* (1889).

J.D. DOUGLAS

SMYTH, JOHN (c.1565-1612). Father of English General Baptists. Born in E England, he studied theology at Cambridge, becoming fellow of Christ's College. His tutor was Francis Johnson, later Separatist pastor. Ordained by the bishop of Lincoln, Smyth became lecturer at Lincoln Cathedral (1600), only to be dismissed for "personal preaching" (1602). He served as pastor of a Brownist congregation in Gainsborough, but to escape persecution they went to Amsterdam (1607). He did not join existing Separatist groups because of differences of view on the church, but formed a new congregation. Among its members was Thomas Helwys,* a friend from Lincoln days. Smyth had come to a new understanding of the church as a company of believers, and of the necessity of believer's baptism. In 1608 he baptized himself (which led to his being called "Se-Baptist"), then Helwys and the others on confession of faith. For a meeting-house they acquired a bakehouse belonging to a Mennonite, Jan Munter. Becoming distrustful of his self-baptism, Smyth made overtures to the Waterlander Mennonite congregation. Helwys and about ten others held back, having misgivings about the Hofmannite Christology of the Mennonites. Smyth died before he could be received by the Waterlanders. His last book was a plea for full liberty of conscience in religion.

See W.T. Whitley (ed.), *The Works of John Smyth* (1915).

J.G.G. NORMAN

SMYTH-PIGOTT, J.H., see AGAPEMONISM

SOBORNOST. From the Russian *sobor*, "gathering" or "synod," corresponding somewhat to the Greek *koinōnia*, with no proper English equivalent, this term in the Russian Orthodox Church denotes an ineffable unity. Viewing the Roman Church's unity as superficially based on exterior authority and Protestantism's individuality as inherently disruptive, the Orthodox claim a *sobornaja*, "oneness," among clergy and laity, for example through the liturgy. This highly personal-corporate approach they account as ecumenically viable ultimately (cf. the work of the Society of SS. Alban and Sergius).

SOCIAL GOSPEL. No clear definition can be given to the term, nor can its first use be dated, but the ideas it represents were those of the nineteenth century. The variants center on the theme of the relationship of the Christian gospel to the improvement of the social environment, and all represent a response to the Protestant doctrine that the salvation of the individual is the center of the Christian gospel, and that only through a prior reformation of the individual will society be improved through the fruit of the Spirit in the lives of individuals. That orthodox belief led some to engage in active social work as a means of Christian witness, the most notable being the Salvation Army,* and so to provide one, less common, interpretation of the term "Social Gospel."

Its more common use can be traced to the response to a number of beliefs held by different groups in the nineteenth century. First, some Christians virtually advocated a position which minimized the importance of all things material, regarding them as passing phenomena, and so had no interest in their amelioration—the "pie in the sky" attitude so lampooned by their critics. Second, and more widespread, was a fear that participation in works of social improvement would lead to evangelistic activities being swamped in social work. To avoid the danger, the form of social work adopted was narrowed to include care for the individual but to exclude concern with forms of institutional change, especially if they involved work of a political nature. It was understandable that some reacted to these two beliefs by stressing the need for institutional and environmental changes apart from the changes in the individual. The reaction led to the adoption of a position which could easily lead to a denial of the need for change in the individual.

The move from orthodoxy was encouraged still further by a third influence of the nineteenth century, the ideas of the biological and social scientists. An optimistic doctrine of man and of his improvement in society was accepted by many in the eighteenth century, but the optimism of the psychological basis of such theories was increased by the biological studies of the nineteenth. In the Western world they could be related to the economic progress of the time, and so it became easy to adopt a position which was contrary to the orthodox Christian view, and which held that society may be improved by institutional change. Man was to be perfected through change in society. ROY H. CAMPBELL

SOCIALISM, CHRISTIAN, see CHRISTIAN SOCIALISM

SOCIETY OF FRIENDS, see FRIENDS, SOCIETY OF

SOCIETY OF ST. JOHN THE EVANGELIST. Popularly known as the "Cowley Fathers," this is the oldest men's religious community in the Church of England. In 1850 R.M. Benson was appointed vicar of Cowley, and in 1866 took the vows with three others. The society has worked in India, and now works in the USA, Canada, and Africa. The members are in demand as conductors of missions and retreats and as spiritual directors to other communities. In recent years they have played an increasing part in ecumenical affairs.

SOCINIANISM. A rationalist movement that grew from the thought of Lelio Sozzini (1525-62) and his nephew Fausto (1539-1604) which became one of the forerunners of modern Unitarianism.* Lelio, a Sienese lawyer, was led by his attempt to restore primitive Christianity to denounce the "idolatry of Rome." The opposition this provoked forced him to wander through Switzerland, France, England, Holland, Germany, Austria, Bohemia, and Poland. He died at Zurich. Fausto, influenced by Italian humanism and his liberal uncle, also left his native land, settling in Basle. In 1578 he moved to Poland, where he spent the remainder of his life organizing a church of his persuasion. The most important of his writings is *De Jesu Christu Servatore* (1578).

Socinianism taught: a rationalist interpretation of Scripture with an emphasis on the early part of the OT and the NT; an acceptance of Jesus as the revelation of God but nevertheless solely a man; nonresistance; the separation of church and state; and the doctrine of the death of the soul with the body except for selective resurrection of those who persevered in obeying Jesus' commandments.

Fausto's work in Poland with the Minor (Reformed) Church led him to revise the Catechism of Racov (1574). This document, published in 1605 as the "Racovian Catechism," became the most famous expression of Socinianism. The Minor Church, centered in a communitarian settlement at Racov, NE of Cracow, propagated its teachings through an academy (at one time having more than 1,000 students enrolled) and a printing operation that published books and pamphlets in many languages. In addition to this center there were about 300 churches in Poland, including among their leadership such men as Andreas Wiszowaty (d.1678), Socino's grandson, and Samuel Przypkowski (d.1670). These churches attracted a number of converts from German Protestantism who moved to Poland.

In 1638, responding to the Counter-Reformation, the parliament of Poland closed the school at Racov and destroyed its buildings. The publishing house was also forced out of business, and churches were suppressed. A number of ministers were banished while others left voluntarily. In 1658, when the parliament passed the death penalty for adherents to the Racovian confession, there was a mass migration of Socinians to Hungary (Transylvania), Germany (Silesia and Prussia), England, and the Netherlands.

During the seventeenth and eighteenth centuries, Socinian influences in England can be traced in the opinions of Latitudinarianism* and Arians in the Church of England, in the Cambridge Platonists,* in the views of philosophers and scientists such as Isaac Newton* and John Locke,* and in the ideas of early Unitarians such as Stephen Nye—whose *History of Unitarianism, Commonly Called Socinianism* touched off the Trinitarian controversy in the Church of England in 1687.

BIBLIOGRAPHY: E.M. Wilbur, *A History of Unitarianism: Socinianism and Its Antecedents* (1945) and *A History of Unitarianism: In Transylvania, England and America* (1952); A.J. McLachlan, *Socinianism in 17th Century England* (1951); G.H. Williams, *The Radical Reformation* (1962). ROBERT G. CLOUSE

SOCIOLOGY OF RELIGION. The sociologist of religion is interested in the vast range of differences and similarities in beliefs and practices. Therefore his definition of religion cannot be the specific definition of any one religious cult. The only possible alternative is a functional definition which concentrates upon the function of religion and ignores the basic faith presuppositions. This is acceptable only if the sociologist realizes what he has done in the construction of his definition. Milton Yinger, taking this approach, defines religion as "a system of beliefs and practices by means of which a group of people struggle with the ultimate problems of human life."

If one defines sociology as the study of man in society and accepts the position that a complete analysis of human action requires the study of social, cultural, and personality facts, then sociology of religion may be defined as the scientific study of the reciprocal influences of religion and society, culture, and personality.

Sociology of religion is a very young science and as such is experiencing the growing pains of youth. Much interest is being shown, but little scientific theory is being developed, yet it takes time and competent, interested researchers to develop sound theory. At times sociological inquiry into religion has been central to the most important work being done in sociology; this was true at the turn of the century during the "golden era" of Ernst Troeltsch,* Max Weber, and Emil Durkheim. However, during the period between World Wars I and II, little attention was given to this field of endeavor. Since World War II there has been a resurgence of interest; even more recently hope has been rekindled by the establishment of dialogue between the three principal types of persons addressing themselves to the problem: college and university professors, seminary professors, and religious researchers. The right questions are beginning to be asked at both empirical and theoretical levels.

The earliest scientific approaches to a study of religion were highly influenced by August Comte's hierarchy of intellectual disciplines. Theological thinking was primitive and completely outmoded in the positive society. Religion would survive only in the form of liturgy based on science rather than revelation. In America two different groups developed. One viewed religion as a type of cultural lag and was antireligious in nature. A second group, of whom Walter Rauschenbusch* is a striking example, saw the possibilities for a "Social Gospel." Sociology could form the basis of a religious, social reform. Many of the early American sociologists were recruited from the ranks of the Protestant clergy. However, this "cult of progress" phenomenon vanished under the onslaught of two world wars and a world depression.

The development of functional sociology began to have its effects upon the sociological study of religion in the early twentieth century. In Germany this was marked by the works of Weber, Troeltsch, and Georg Simmel. The French parallels to these social theorists were Durkheim and Robert Will. In the USA, the attack upon the positivistic views of religion came from the field of anthropology. Bronislaw Malinowski found that primitive man was forced to seek answers to the unknown in order to adjust to his cultural environment. Therefore religion had a functional value in enabling man to meet the problems of life, and religious ritual strengthened the moral beliefs and social cohesion essential to community life.

The work of Talcott Parsons represents an attempt to systemize the ideas of the structure-function approach, and it allowed the American sociologists to make a break with the Comtian emphasis on religion as a passing phenomenon. Parsons prepared the way for an understanding of socioreligious organization, while detailed sociological analysis of relevant phenomena was carried on by William I. Thomas, Florian Znaniecki, M.E. Gaddis, Arthur E. Holt, Samuel Kincheloe, W. Lloyd Warner, and others. Systematic treatments with broad perspective have been undertaken more recently by Pitirim Sorokin and Joachim Wach.

The sociological theorists who have been most concerned with religion are usually identified as functionalists. In the struggle to relate religion to society, culture, and personality they have been able to isolate at least six basic functions of religion.

(1) Religion provides support, consolation, and reconciliation. In the face of uncertainty, men need support; the pain of disappointment demands consolation; alienation from the goals and norms of society gives rise to a longing for reconciliation.

(2) Through cultic practices and formal worship, religion offers a transcendental relationship which provides security in the midst of contingency.

(3) The norms and values of established society are sacralized by religion. This allows the group goals to be maintained over individual wishes.

(4) In contradiction to the above, religion can also provide a prophetic function as it provides the standards of value in terms of which institutionalized norms may be critically examined and found seriously wanting.

(5) Religion performs important identity functions. It is within the realm of religion that many are able to find the answers to who they are and what they are.

(6) Finally, there is a relationship between religion and maturation. Religion sacralizes norms and ends which support the expectations for each age level.

The functionalist is not without his critics, and even he is willing to admit that this approach has its weaknesses. Yinger points out that functional analysis is sometimes used to "prove" the ultimate validity of some specific practice or belief. This empirical proof of a nonempirical proposition is impossible.

The field of sociology of religion branches out into several specific areas of interest: Religion and the Economic Order, Religion and Family, Religion and Social Stratification, Religious Leadership and Authority, Religion and Conflict, Religious Attitudes, Typology of Religious Institutions, etc. The growing edge of the discipline is interested in religious organization, leadership, and authority. The sociological theories of organization, bureaucracy, and role-playing are being incorporated into the study of religious organization, leadership, and authority.

The immediate future holds much promise for the work on religion more akin to social psychology than sociology. Here the relationships between religious values and other kinds of values in the culture will be explored. Recent developments in methodology such as small-group research, interviewing techniques, survey research, and conceptual analysis have paved the way for this type of study.

BIBLIOGRAPHY: E. Troeltsch, *The Social Teachings of the Christian Churches* (2 vols., 1931); J. Wach, *Sociology of Religion* (1944); E.K. Nottingham, *Religion and Society* (1954); T.F. Hoult, *The Sociology of Religion* (1958); G. Simmel, *Sociology of Religion* (ET 1959); P. Benson, *Religion in Contemporary Culture* (1960); G. Vernon, *Sociology of Religion* (1962); O.R. Whitley, *Religious Behavior: Where Sociology and Religion Meet* (1964); T. O'Dea, *The Sociology of Religion* (1966); J.M. Yinger, *The Scientific Study of Religion* (1970). JOHN P. DEVER

SOCRATES OF ATHENS (469-399 B.C.). Greek philosopher and one of the great stimulating forces in Western philosophy. He died by drinking hemlock following his condemnation for "not believing in the gods the state believes in, and introducing different new divine powers; and also for corrupting the young." Socrates's background was in Sophism, his philosophical method that of a "midwife"—the raising of questions, the creation of a climate of doubt about accepted truths by a process of cross-examination ("dialectic"), combined with high intellectual standards and a sense of moral purpose. His magnetic personal influence, indifference to wealth, and freedom from ambition made the conservative elements suspicious and led to the political charge. Plato's *Crito* and *Phaedo* offer an intriguing portrayal of Socrates's last days. Despite his enormous influence, there is no agreement on the contribution of Socrates. Part of the difficulty is that he wrote nothing on philosophy. At one extreme there is the view that the Socrates of Plato's *Dialogues* is historical, at the other that nothing is known about him. Superficial parallels have led some to make rather fanciful comparisons between Socrates and Jesus Christ. PAUL HELM

SÖDERBLOM, NATHAN (1866-1931). Archbishop of Uppsala and ecumenist. Born at Trönö , son of a Pietistic pastor, he was educated at the University of Uppsala and soon proved himself to be a brilliant scholar, not least in esoteric byways of Orientalism. Ordained to the Lutheran ministry in 1893, he was successively chaplain in Paris, professor and pastor in Uppsala, and professor at Leipzig before his unexpected election as archbishop and primate of Sweden in 1914. Despite his father's misgivings, Söderblom never forgot his upbringing, told his daughter that he could not live one day "without unceasing prayer," and once electrified an American dinner audience by interrupting his address to deliver himself of all five verses of "There were ninety and nine." He participated in dialogue with Lambeth, early ecumenical stirrings in South India, talks with the Orthodox, and tentative advances toward Rome, and he organized the 1925 Stockholm Conference.* His concept of ecumenicity, according to his biographer, "was to emphasize the importance for the movement of its being nurtured by worship and prayer rather than [its being] an alliance of secretaries." His ecumenical work was recognized in 1930 when he was awarded the Nobel Peace Prize. Among his writings were *Christian Fellowship* (1923) and the substance of his Gifford Lectures at Edinburgh in 1931, published two years later as *The Living God: Basal Forms of Personal Religion.*

See studies by C.J. Curtis (1966 and 1967) and B. Sundkler (1969). J.D. DOUGLAS

SOGA, TIYO (c.1829-1871). First African ordained minister in South Africa. Son of a Christian mother and a polygamous councillor of Chief Ngqika, he entered Lovedale Seminary in 1844 and was sent to Scotland when war closed the institution in 1846. He returned to South Africa in 1848 as a catechist, but went back to Scotland for theological training (1851-56). As a minister of the United Presbyterian Church he served at Mgwali (1857) and Tutura (1868), proving himself a fine preacher and a faithful pastor. His work on the revision of the Xhosa Bible and a translation of *Pilgrim's Progress* (1866) revealed great literary ability. Ill-health prevented the full expression of his gifts, and he died at the early age of forty-two. D.G.L. CRAGG

SOISSONS, COUNCIL OF (1121). This was called to deal with the writings of Peter Abelard,* particularly with the heretical tendencies in his *Sic et Non.* The charge of heresy resulted from Abelard's insistence on applying reason or logic to the doctrine of the Trinity. How heretical this work was is not entirely clear, but it gave the impression that the Fathers of the Church were in great disagreement over important doctrines. The root of the problem lay in his insistence that unless we understand we cannot believe. The council forced him to retract his position, and ordered his book to be burned.

SOLAFIDIANISM. The doctrine that eternal salvation is had only through faith by grace in the work of Jesus Christ. This work of Christ is His obedience to the will of His Father and emptying or humbling of Himself. He died to reconcile man to God, overcoming sin, death, the devil, and hell as Victor and as the sacrifice and propitiation for all the sins of all men of all times, thereby obtaining the forgiveness of their sins. The acceptance and seal of this death by God is attested by Christ's resurrection and session. Justification of the sinner, therefore, is through faith, not as a work, but as a gift of God. Solafidianism opposes synergism in any form, whether it be Pelagianism* or Semi-Pelagianism* or any variations of these. It is not antinomian, but insists that good works are not necessary for salvation; they are the fruits of faith. The propounders of Solafidianism cite Scripture passages such as Galatians 2:16; 3: 11; Ephesians 2:8; and Romans 4:5. *Sola fide,* "faith alone," was one of the principles of the Reformers of the sixteenth century, together with *sola gratia,* and *sola Scriptura;* sometimes the three are brought together into one phrase, *solus Christus.* CARL S. MEYER

SOLEMN LEAGUE AND COVENANT (1643). Drawn up by Alexander Henderson,* this was approved by the general assembly of the Church of Scotland and transmitted to the English Parliament for ratification. After slight changes, the English legislators publicly swore to maintain the provisions of the League, jointly signing it along with the members of the Westminster Assembly* of Divines. The document was for a religious alliance between England and Scotland, rather than either a military or civil agreement. It guaranteed the maintenance of the Reformed Church of Scotland (Presbyterian) and promised to reform the churches of England and Ireland according to the Scriptures and the examples of the "best Reformed churches," without identifying the latter. It also promised loyalty to both Parliament and Crown. Thus it prepared the way for the establishment of Presbyterianism in England and Ireland in place of the churches of England and Ireland as then constituted. As a consequence of its acceptance by the English Parliament, Scottish commissioners attended the Westminster Assembly, taking in it a prominent and influential part.

The reason for the signing of this bond of mutual agreement was that in the spring and summer of 1643, the parliamentary forces fighting Charles I were in serious difficulties. Thus Parliament wished to have the assistance of the effective Scottish army led by experienced soldiers such as the Leslie brothers. Some of the English leaders were opposed to the League, but had to agree for military reasons. Charles II later signed the League in 1650, but never kept it. W.S. REID

SOLOMON, ODES OF. A series of hymns, forty-two in number, preserved as a whole (Ode 2 is completely missing) only in Syriac, which version appends the eighteen Psalms of Solomon. Five are known in the Coptic Gnostic *Pistis Sophia.* One has recently come to light in Greek. Many scholars consider Greek the original language, others argue for Syriac origin. They were known only by reference until 1905, when J.R. Harris discovered them. They present in allusive language an early Syriac Christianity akin to John's gospel and Ignatius's letters, and with much common to the *Acts of Thomas.** While most are general praises of God, with an emphasis upon mystical communion with Him, some more-developed theological themes are found: a Logos Christology akin to Jewish Wisdom thought; the notion of a miraculous birth, without pains of labor for His mother; the passion of Christ described with much reminiscence of the Psalms; the resurrection linked to the notion of the "harrowing of hell." The specific date and milieu of the Odes has not been determined with precision. Within the church a setting as baptismal hymns has been proposed, while recently a hinterground in the Jewish circles of Qumran has been investigated, but not certainly established. J.N. BIRDSALL

SOLOVIEV, VLADIMIR SERGEEVICH (1853-1900). Russian theologian and philosopher. The son of a Russian historian, he was graduated from the University of Moscow in 1873 and was appointed a fellow in the faculty of philosophy. After research in London and Egypt he returned to Moscow in 1876. In 1877 he moved to the University of St. Petersburg. Dostoevsky* and Leo Tolstoy* were among those present at his lectures on "Godmanhood." He was forced into retirement in 1881 after he advocated mercy for the assassins of Alexander II. He then devoted his life to his writings.

Soloviev was deeply influenced by German idealistic philosophy and Gnostic mysticism. The nucleus of his religious and philosophical system was his doctrine of Godmanhood, by which he meant the union of humanity and divinity through identification of man with Christ, the Incarnate Word. Included in this is the concept of "positive total unity," his synthesis of religion, philosophy, and science. His Sophiology which originated in his early mystical experiences is identified with Eternal Womanhood or Divine Wisdom. He advocated the reunion of the Eastern and Western churches and the establishment of a universal theocracy. Because of his connection with Rome, he is sometimes called "the Russian Newman." In 1896 he made his profession as a Byzantine rite Catholic. Among his works are *Lectures Concerning Godmanhood* (1878), *The History and the Future of Theocracy* (1886), *La Russie et l'Église Universelle* (1889), and *Three Conversations* (1899-1900).

See N. Zernov, *Three Russian Prophets* (1944), and E. Munzer, *Solovyev, Prophet of Russian-Western Unity* (1956). BARBARA L. FAULKNER

SOMASCHI, see JEROME EMILIANI

SON OF MAN. The commonest title used by Jesus to refer to Himself. In the OT it is often found in Ezekiel as the name by which God calls the prophet. It is sometimes found as a parallel to "man" (e.g., Pss. 8:4; 80:17), and in Daniel 7 it has an apocalyptic significance. This last idea is expanded in the *Similitudes of Enoch,* where there is particular emphasis on his role in judgment. The term is found frequently in the gospels as a self-designation of Jesus. Outside the gospels it is found only in the NT, in Acts 7:56; Revelation 1:13; 14:14. This makes it clear that it was not a term in general use after the Resurrection, when clearer messianic and divine titles could be used. There can therefore be little doubt, despite the objection of some modern scholars, that it was a term used of Himself by Jesus during His ministry which has been faithfully retained by the Evangelists.

Recently there has been a good deal of discussion about the source and meaning of the term as it is used in the gospels. It is possible that the varied OT usage and that of *Enoch* had some influence. The meaning of the term seems to fall into three main categories. First, there are passages where the term is a periphrasis meaning simply "I" (Matt. 8:20; 11:19) or may possibly mean "man" (Mark 2:10,28). Secondly, there are those sayings which refer to the suffering and death of the Son of Man followed by the resurrection (Mark 8:31; 9:31; 10:33f., etc.). This has been closely linked with the idea of Jesus as the suffering Servant of the Lord, and has sometimes been held to have something of a corporate significance, including the disciples with Jesus. Thirdly, there are references to the Son of Man as the exalted Lord who will come again in glory as judge of the world (Mark 8:38; 13:26; 14:62, etc.). In John the emphasis is on the Son of Man's being the ladder between heaven and earth (John 1:51), and His descending and ascending again (John 3:13f.; 6:62f.; 8:28). Paul's doctrine of the Second Adam (Rom. 5:12ff.; 1Cor. 15) may be linked with the idea of the Son of Man.

BIBLIOGRAPHY: A.J.B. Higgins, *Jesus and the Son of Man* (1964); F.H. Borsch, *The Son of Man in Myth and History* (ET 1967); M.D. Hooker, *The Son of Man in Mark* (1967). R.E. NIXON

SOPHRONIUS, PATRIARCH (c.560-638). Patriarch of Jerusalem. Born in Damascus, he is probably to be identified with Sophronius "the Sophist." He was a monk in Egypt (c.580), then in the Jordan area, and from 619 in Theodosius Monastery, Jerusalem. He opposed the doctrinal compromise concocted by Emperor Heraclius, Cyrus of Alexandria, and Sergius I of Constantinople which endeavored to reconcile the Monophysites* by a Monothelite formula. When elected patriarch of Jerusalem (634), he issued the customary encyclical which included his own statement of faith, holding that the Chalcedonian doctrine of two natures necessarily implied two wills. Shortly after, Jerusalem was captured by the Saracens under Caliph Omar (637), hastening his death. He wrote sermons and poems, and lives of Cyrus and John, Alexandrian martyrs.

J.G.G. NORMAN

SOTERIOLOGY, see SALVATION

SOTO, DOMINIC DE (1494-1560). Spanish theologian. Born in Segovia, he was a sacristan before reading logic and philosophy at Alcalá and theology at Paris. He taught briefly, spent some time in a Benedictine abbey, then made his profession in the Dominican Order in Burgos (1525). After teaching for seven years in the house of studies in Segovia, he went to Salamanca (1532) to take his order's university chair in theology. Charles V, to whom he was later confessor, appointed him imperial theologian to the Council of Trent.* Refusing the see of Segovia, he was elected prior at Salamanca in 1550, and two years later took the first chair of theology there. His writings include *De institia et iure* (1556), *Deliberatio in causa pauperum* (1547), and commentaries on Romans (1550) and Aristotle (1544).

C.G. THORNE, JR.

SOUL. Usage of this term in the history of the Christian Church has been affected by three main factors: the biblical data, particularly about man's creation in the image of God, and the resurrection of the body; dualistic philosophies, both ancient Greek and modern Cartesian and post-Cartesian versions; and, through Thomas Aquinas,* the influence of Aristotle.* "Soul" has thus often been a theory-laden term and even in popular usage has reflected the theories of metaphysics as well as the data of revelation.

According to Greek dualism—e.g., Plato's teaching in the *Phaedo*—the human person consists in an immortal soul, which has had preexistence, enclosed in a mortal body. This view is clearly incompatible with God's creation of man, "a living soul." Moreover, the influence of this Greek dualism has often had an unfortunate effect on Christian ethics—e.g., in the idea that man is a composite of "higher" and "lower" elements ("soul" and "body"), and hence that bodily activities and desires as such are unworthy of the true Christian (see MANICHAEISM). In Cartesian dualism the soul is an "incorporeal substance" located in a corporeal substance, the body. The chief philosophical difficulty is in accounting for an interaction between the two substances. Besides the philosophical difficulties there are other features of Cartesianism that are unwelcome from a Christian standpoint—e.g., the Cartesian idea of free will, and the view that the soul is *in principle* inaccessible to scientific investigation. Cartesian dualism has been heavily attacked in Gilbert Ryle's *Concept of Mind* (1949).

Aristotle's monistic view that the soul is the form of the body may come nearer to the biblical view in making intelligible the idea of bodily resurrection, but it seems to entail the noncontinuity of the individual as an individual after death, and before resurrection. The many conceptual problems in this area are partly due to the apparently untechnical and often opaque character of biblical terminology—e.g., Paul's phrase "a spiritual body" (1 Cor. 15:44). Perhaps some of the difficulties can be at least minimized by thinking of "soul" and "body" not as names for two substances, but as referring to two levels at which the human person can be regarded.

PAUL HELM

SOUTER, ALEXANDER (1873-1949). Scottish NT and patristic scholar. Educated at Aberdeen and Cambridge universities (where he was respectively influenced by W.M. Ramsay and J.E.B. Mayor), Souter was Yates professor of NT Greek and exegesis at Mansfield College, Oxford (1903-11), then regius professor of humanity, Aberdeen University (1911-37). He is widely known for his trilogy of handbooks to the study of the Greek NT: the *Oxford Greek Testament* (1910; rev. ed., 1947), in which he provided a select *apparatus criticus* for the text presumed to underlie the Revised Version of 1881; *The Text and Canon of the New Testament* (1913; posthumous revision, 1954); and *A Pocket Lexicon to the Greek New Testament* (1916).

But his most important work was done in the study of the Latin Fathers, especially the commentators on the Pauline epistles—a field where two great interests, the Latin language and Paul, converged; cf. his Princeton Stone Lectures, *The Earliest Latin Commentaries on the Epistles of St. Paul* (1927). Two outstanding achievements in this field were (1) his demonstration in *A Study of Ambrosiaster* (1905) that the author of the pseudo-Ambrosian commentaries on Paul (called "Ambrosiaster"* since Erasmus's time) was identical with the author of the pseudo-Augustinian *Quaestiones Veteris et Noui Testamenti* (which he edited for the Vienna *Corpus Scriptorum Ecclesiasticorum Latinorum,* 1908), and (2) his recovery and publication of the original text of Pelagius's expositions of the Pauline epistles (2 vols., 1922, 1926; with the pseudo-Jerome interpolations, 1931). His *Glossary of Later Latin to A.D. 600* (1949) presents the quintessence of over half a century's research in Latin lexicography.

See Memoir by R.J. Getty in *Aberdeen University Review* 33 (1949-50), pp. 117-24.

F.F. BRUCE

SOUTH, ROBERT (1634-1716). Anglican clergyman. Born at Hackney, he was educated at Oxford, where in 1660 he became public orator. He accepted a royal chaplaincy, but declined further preferment. South's sermons are among the classics of English divinity. He himself prescribed clarity, simplicity, and fervor as the necessary ingredients of a good sermon. He was extremely conscious of his preaching mode and reacted against the differing extravagances both of the Puritans and of a man like Jeremy Taylor. His prose is fluent and skillfully modulated, and he has a penchant for memorable statement—a quality aided by his very ready, epigrammatic wit and his occasional acerbity. ARTHUR POLLARD

SOUTHCOTT, JOANNA (1750-1814). Self-styled prophet. Born in Devon, daughter of a farmer, she was naturally religious, albeit eccentric, from an early age. Originally an Anglican, she became a Methodist in 1791. In 1792 she began to write a book of prophecies, the first of over sixty publications all "equally incoherent in thought and grammar." In 1802, for a charge, she began to issue seals to the faithful, including one to Mary Bateman, the notorious murderess. In 1805 she had a chapel in London, and in 1813 announced that—though unmarried—she was to bear a child called "Shiloh." She died of unknown causes. She left a box which was to be opened only in the presence of twenty-four bishops. In 1927, when it was opened in the presence of one bishop, it was found to contain only trivia. Rival followers, some still believed extant, claimed that it was the wrong box and that the real one still awaited opening. PETER S. DAWES

SOUTHERN AFRICA. Although Roman Catholic missionaries entered Mozambique and E Rhodesia in the sixteenth century, the Dutch settlement at the Cape was more significant for the development of Christianity in Southern Africa. For over a century only the Dutch Reformed Church was permitted, and it remains the spiritual home of most Afrikaners. The evangelization of the indigenous Hottentots and imported slaves received little attention until the late eighteenth century when Dii. van Lier and Vos awakened missionary interest among the Dutch colonists. This led eventually to a strong Dutch Reformed Mission Church among the Cape Colored people. After 1857, missions to Africans were also undertaken.

The first missionary at the Cape was the Moravian, Georg Schmidt,* who returned to Europe in 1744 after seven frustrating years. The Moravians resumed operations in 1792, and their work at Genadendal won universal respect. In 1799 the London Missionary Society entered the field, first among the Hottentots and subsequently among the Griqua and Tswana outside the colony. The concern of its missionaries, notably J.T. Vanderkemp* and John Philip,* for the rights of indigenous people made the LMS highly unpopular with the colonists. Methodism arrived with British soldiers at Cape Town and British settlers on the Eastern Frontier, and spread among black and white throughout South Africa. Although Anglicanism had official status after the British occupation of the Cape in 1806, its development awaited the arrival of Robert Gray, first bishop of Cape Town, in 1848. Roman Catholicism became well established after the arrival of its first bishop in 1838; its expansion owed much to the Oblates of Mary Immaculate and other missionary orders.

A healthy climate and relatively safe conditions attracted numerous societies. They included the Glasgow Mission (Ciskei, 1821), the Rhenish Mission (Cape and South-West Africa, 1829), the Paris Evangelical Mission (Lesotho, 1834), the American Board (Natal, 1835), and the South Africa General Mission (1889). The Berlin Mission (Cape, 1834) and the Hermannsburg Mission (Natal, 1854) found their most important fields in the Transvaal, where continentals were preferred to "meddling Anglo-Saxons."

Church life. In the eighteenth and nineteenth centuries, the Afrikaner *trekboer* ("wandering stock farmer") led an isolated existence. The paterfamilias conducted family devotions and occasionally took his clan to the distant town for *nagmaal* ("Communion"). Conservative Calvinism was the norm, and theological liberalism received short shrift when it appeared in the 1860s. British settlers transplanted the religious patterns of the homeland. In Anglican circles the tension between High and Low Churchmen led to a rift between the predominant Church of the Province and the small conservative-evangelical Church of England in South Africa—a division which has persisted to this day.

Missionaries gave devoted service as evangelists, translators, educators, administrators, and friends of their people. Unfortunately they often passed on denominational rivalries and paid insufficient attention to the indigenization of Christianity. During the nineteenth century the mission station was central to missionary strategy. In the Cape Colony it gave the Hottentots an alternative to vagrancy and farm labor. Among the Africans it both stemmed from and emphasized the rift between Christian and tribal society. As an interim measure it was perhaps inevitable, but Christianity only took deep root in African society when the Church was carried outside the mission station by African evangelists. In course of time the major mission stations became important educational centers, with the Scots at Lovedale setting the pace. Elementary schools were established in most congregations, and Christianity became almost synonymous with education. Provincial governments gradually assumed financial responsibility, but relied upon missionary managers until the Bantu Education Act (1954) asserted full government control and applied *apartheid* to educational policy.

Church development. South African churches began to emerge during the nineteenth century. The British occupation severed the links of the Cape Dutch Reformed Church with Holland, and in 1843 it was freed from government control. Legal and political considerations led to the establishment of separate synods in Natal, the Orange Free State, and Transvaal, but a general synod was formed in 1963. This does not include the *Hervormde* and *Gereformeerde* (Dopper) churches which originated in the Transvaal.

In 1853 the LMS began to withdraw financial support from its missions in the Cape Colony. The independent Coloured congregations eventually joined white Congregationalists in the Congregational Union of South Africa (1877). The Church of the Province of South Africa was constituted in 1870 as an independent member of the Anglican Communion, and the Methodists formed a South African conference in 1883. These so-called English denominations were (and remain) multiracial. By contrast the Dutch Reformed Church has established daughter churches for the Cape Coloured, African, and Indian communities. During the twentieth century, many overseas missions have formed self-governing local churches, some of which have sought a federal or organic relationship with white coreligionists.

A significant feature of the past eighty years is the growth of an estimated 3,000 independent African churches. These range from orthodox denominations, like the Presbyterian Church of Africa, to nativistic sects in which Christian elements have merged with traditional beliefs and customs. The complex causes of this movement include the rejection of white control; the search for significance and status in a small, personalized group; and the reaction of unsophisticated people to detribalization and urbanization.

Ecumenical movement. The ecumenical movement has made limited progress. There have been several confessional unions, and the Anglican, Congregational, Methodist, and Presbyterian churches have set up a church unity commission. The general missionary conferences (1904-32) promoted understanding and fellowship. They made way for the Christian Council (now South Africa Council of Churches) which is growing in importance and enjoys some cooperation with the Roman Catholic Church. The Dutch Reformed Church has, however, withdrawn from ecumenical contact, largely for political reasons, while black activists question the reality and value of racial partnership in the church. Many theologically conservative groups regard ecumenism with suspicion.

Christianity and politics have always been interrelated in South Africa. The DRC played a major role in the development of Afrikaner nationalism and has given its blessing to apartheid. On the other hand, there is a long-standing tradition of missionary support for black interests, but its influence is severely limited by the apathy of white churchmen and the concentration of political power in white hands. The Christian response to the developing political situation in Southern Africa will be fundamental to the future standing of the church in this region.

Rhodesia. Attention was drawn to Central Africa by the travels and writings of David Livingstone.* The LMS established a mission at Inyati in 1859, but general Christian penetration of Rhodesia followed its occupation by the Chartered Company in 1890. The pattern of evangelism and education was similar to that which evolved in South Africa, and many of the missions involved were the same. The Christian Council of Rhodesia promotes common action, and several of its member churches, together with the Roman Catholic hierarchy, have confronted (1968-72) the Smith government on aspects of its racial policy. Many Rhodesian missions are still heavily dependent upon overseas assistance.

Malawi. The attempt of the Universities' Mission to Central Africa to enter Malawi in 1861 ended in disaster. Not so the Scottish missions, which arrived in 1875 and exercised widespread influence from their headquarters at Livingstonia (Free Church) and Blantyre (Church of Scotland). The Dutch Reformed Church also entered this field in 1896 and joined the Scots in the Church of Central Africa Presbyterians in 1926. The UMCA undertook work among Muslims in the 1890s.

Zambia. Zambia was entered from several directions. F. Coillard* came from Lesotho to found the Barotse Mission in 1886, closely followed by the Primitive Methodists. The LMS entered Bembaland from Tanganyika; the Presbyterians, Dutch Reformed, and Anglicans came from Malawi; and the Wesleyan Methodists from Rhodesia. The development of the Copper Belt after 1925 led to the spontaneous formation of an African Union Church, and to united action by several missions. In 1965 the United Church of Zambia brought together churches in the Congregational, Methodist, and Presbyterian traditions, but it did not embrace considerable bodies such as the Plymouth Brethren, Reformed, and Anglicans.

In the 1920s and 1930s the government suppressed nativistic prophet movements which were blamed, perhaps unduly, upon the Watchtower Movement which has a large following. Another nativistic movement, the Lumpa Church of Alice Lenshina, was involved in violent conflict with the Zambian government after independence.

Mozambique. The sixteenth-century Roman Catholic mission faded away. During the late nineteenth century, work was resumed throughout Central Africa under the control of the White Fathers* and, in Rhodesia, the Jesuits. In Portuguese East Africa the missionaries had the support of a Catholic government which restricted Protestant missions, of which the most important were the Swiss Mission and the American Methodists. Since Vatican Council II, Roman Catholic pressure on Protestants had eased, and the White Fathers have recently clashed with the government on political issues.

BIBLIOGRAPHY: G.B. Scholtz, *Die Geskiedenis van die Nederduitse Hervormde of Gereformeerde Kerk* (2 vols., n.d.); J. Whiteside, *The History of the Wesleyan Methodist Church of South Africa* (1906); W.P. Livingstone, *Laws of Livingstonia* (1921); G.G. Findlay and W.W. Holdsworth, *The History of the Wesleyan Methodist Missionary Society*, vol. IV (1922); C.P. Groves, *The Planting of Christianity in Africa* (4 vols., 1949-58); J.N. Hanekom, *Die Liberale Rigting in Suid Afrika* (1951); G.B.A. Gerdener, *Recent Developments in the South African Mission Field* (1958); O. Chadwick, *Mackenzie's Grave* (1959); W.E. Brown, *The Catholic Church in South Africa* (1960); B.G. Sundkler, *Bantu Prophets in South Africa* (2nd ed., 1961); J. Taylor and D. Lehmann, *Christians in the Copperbelt* (1961); P.B. Hinchliff, *The Anglican Church in South Africa* (1963) and *The Church in South Africa* (1968); J. du Plessis, *A History of Christian Missions in South Africa* (rep. 1965); A. Ive, *The Church of England in South Africa* (1966); B. Kruger, *The Pear Tree Blossoms—the History of the Moravian Church in South Africa, 1737-1869* (1966); P. Bolink, *Towards Church Union in Zambia* (1967); E. Strassberger, *The Rhenish Missionary Society in South Africa, 1830-1950* (1968); D. R. Briggs and J. Wing, *The Harvest and the Hope—the Story of Congregationalism in Southern Africa* (1970); J. Sales, *The Planting of the Churches in South Africa* (1971); R.H.W. Shepherd, *Lovedale, South Africa, 1824-1955* (1971). D.G.L. CRAGG

SOUTHERN BAPTIST CONVENTION. The largest non-Roman Catholic religious body in America. It is a voluntary organization of Baptist churches contributing financially to the programs of its various agencies. According to the constitution of the SBC, its purpose is to organize Baptist churches "... for promotion of Christian missions at home and abroad and any other objects such as Christian education, benevolent enterprises, and social services which it may deem proper and advisable for the furtherance of the kingdom of God." The Convention neither claims nor exercises authority over any other Baptist body, whether local church, association, or state organization.

Baptists in the southern USA trace their origins to New England, and the Southern Baptist Convention to the older nationwide Triennial Convention. Baptists became the largest denomination in the South largely because of the great revival of 1755-75 and the work of Shubal Stearns (d.1784), a man of remarkable natural gifts and a profound sense of mission. A spinoff of the Great Awakening in New England, under Stearns's able leadership Baptist missionaries won thousands of people to Christ on the southern frontier in this period. In 1814 the Baptist churches of the South joined with those of other parts of the country to form the Triennial Convention, a nationwide Baptist fellowship organized to promote missions, education, and youth work.

However, the Triennial Convention faltered in the pre-Civil War period and the southern churches withdrew to form their own organization. Three factors led to this breakdown of national Baptist life. First, there was a basic difference between North and South over the associational principle, Southerners tending to place more emphasis on the role of the association. Second, many Baptist leaders in the South objected to the neglect of their region by the Baptist Home Mission Society, one of the agencies of the Triennial Convention. Third, foremost, and the precipitating cause of the split was the question of slavery. This divisive problem cut deep into American life because it was at once a political, economic, social, moral, and religious issue. When Baptist abolitionists insisted that slaveowners be ineligible to serve as missionaries, and some Northerners even suggested that their brethren in the South withdraw from the Triennial Convention, Southern Baptists met in Augusta, Georgia, in May 1845 to discuss the matter. Churches from eight states and the District of Columbia sent 293 representatives who voted to form the new Southern Baptist Convention.

The growth of the fledgling SBC was slow but steady in the period 1845-61. However, it suffered badly as a result of the Civil War (1861-65) and the subsequent withdrawal of its black members to form Negro Baptist churches. In the latter half of the nineteenth century it also was rent by quarrels between antimissionary Calvinists and evangelistically oriented general Baptists (the latter won), and between Landmark exclusivists and ecclesiastically unstructured elements (which was indecisive). In the early twentieth century the Convention emerged from these disputations strongly committed to evangelism and largely untouched by the Fundamentalist–Modernist controversy then raging in most of the other major denominations. Through it all, the SBC grew from 4,126 churches with 351,951 members in 9 states in 1845, to 34,441 churches with 11,826,463 members in all 50 states in 1971.

Southern Baptists hold the same basic beliefs as most other Baptist groups throughout the world: (1) the Bible as the sole norm for faith and practice in the Christian life; (2) a regenerate church membership safeguarded by baptism of believers only, and that by immersion; (3) autonomous local churches with Christ as the Head and democratic polity; and (4) religious liberty buttressed by the institutional separation of church and state. Above all, Southern Baptists are known as "people of the Book," meaning the Bible. Although rejecting binding creedal statements, messengers to the 1925 meeting of the SBC adopted a general confession known as "The Baptist Faith and Message" intended as a consensus document "for the general instruction and guidance of our own people" but "having no authority over the conscience." Individual Southern Baptists have the reputation of being devout, aggressively evangelistic, relatively strict Christians whose lives are oriented about the Bible and the fellowship of the local church.

The main features of Southern Baptist life include a heavy emphasis on evangelism and missions, a tradition of ministry to the common people, and a deep independent streak. The SBC's interest in evangelism is reflected in its steady growth and missions expenditures of $160,546,250 in 1971. Southern Baptists also claim a long tradition of eloquent evangelistic preachers, which includes more recently George W. Truett (d.1944), Robert G. Lee (b.1886), W.A. Criswell (b.1909), and Billy Graham* (b.1918). Further, many frontier observers recorded of the Baptist minister that "the common people heard him gladly." This is still largely true of the SBC, although there is some evidence of increasing class consciousness and a growing ecclesiastical inferiority complex in a few of the more affluent churches in the South. As far as independence is concerned, any annual gathering of the Convention clearly demonstrates that it is one of Southern Baptists' most cherished prerogatives.

In addition to normal growing pains, a number of serious problems face today's Southern Baptist Convention including how to deal with Landmarkism (especially the issues of "open Communion" and alien immersion); how to fulfill its new role as a national rather than a regional organization; how to decentralize its national agencies in accordance with historic Baptist principles without destroying their effectiveness or their contributions to Southern Baptist life; and how to shift from a fundamentally rural-oriented to a basically urban-oriented strategy for preaching the Gospel.

The Convention accomplishes its work through its Executive Committee, four general boards (Foreign Mission, Home Mission, Sunday School, and Annuity), six seminaries, seven commissions (American Baptist Seminary, Brotherhood, Christian Life, Education, Historical, Radio and Television, and Stewardship), two standing committees (Denominational Calendar and Public Affairs), and three associated organizations (Women's Missionary Union, Baptist World Alliance, and American Bible Society). Along with a number of other Baptist bodies, it also supports the Baptist Joint Committee on Public Affairs, Washington, D.C., which acts as both a watchdog and an informational clearinghouse for political matters of concern to Baptists.

BIBLIOGRAPHY: W.W. Barnes, *The Southern Baptist Convention, 1854-1953* (1954); *Encyclopedia of Southern Baptists* (2 vols., 1958); W.L. Lumpkin, *Baptist Foundations in the South* (1961); R.G. Torbet, *A History of Baptists* (rev. ed., 1963); S.S. Hill, Jr. and R.G. Torbet, *Baptists North and South* (1964); H. Wamble, "Landmarkism: Doctrinaire Ecclesiology Among Baptists," *Church History* XXXIII (December 1964), pp. 429-447; O.K. and M.M. Armstrong, *The Indomitable Baptists* (1967); W.R. Estep, Jr., *Baptists and Christian Unity* (1967); R.B. Spain, *At Ease in Zion: A Social History of Southern Baptists, 1865-1900* (1967). ROBERT D. LINDER

SOUTHERN CHRISTIAN LEADERSHIP CONFERENCE. Established by Martin Luther King* in 1957 to coordinate local nonviolent, direct-action movements emerging in the South, its goals are to win "full citizenship rights," including equality and integration of black people in American life. The SCLC's methods have combined the use of Ghandian nonviolent resistance and of the ballot box. Hence it encouraged not only voter-registration projects, but also mass demonstrations and civil disobedience. It also provides services to help black people obtain their rights. Despite its successes, it has been thwarted by the increasing decline of youthful support, primarily because of the alleged ineffectiveness of nonviolence, and the assassination of its founder (April 1968). The latter was succeeded as leader by Ralph D. Abernathy. DARREL BIGHAM

SOUTH INDIA, CHURCH OF. The result of a union of three churches effected 27 September 1947; noteworthy for being the first-ever union of episcopal and nonepiscopal bodies. The uniting churches were: the Anglican dioceses of Madras, Tinnevelly, Travancore and Cochin, and Dornakal (dioceses of the Church of India, Burma, and Ceylon); the South India province of the Methodist Church; and the South India United Church, which was originally formed by a union of Presbyterians and Congregationalists in 1908. The Church of South India had at its inauguration fourteen dioceses, of which one (Jaffna) was in Ceylon.

Negotiations toward union are generally reckoned to have begun at the 1919 Tranquebar Conference of Indian ministers from the Anglican and the South India United churches. Their published manifesto stressed that the existence of episcopacy in a united church need not call into question the "spiritual equality" of all members, nor imply any particular theory or doctrine about episcopacy. It was also proposed that a special service of "commissioning" by the laying on of hands by the bishops of the united church should give to each minister authority to officiate throughout the united church. An official joint committee of the two churches was appointed. Approaches were made to the Methodists, and they joined the committee in 1925. Then at Trichinopoly in 1926 the Anglicans suggested dropping "mutual commissioning," and the adopting of an "interim period" of the ministries growing together in a common episcopal frame. On this basis the first draft of a *Scheme of Union* was published in 1929.

From then until the seventh and final edition of the scheme in 1941, most interest and controversy centered upon the nature of the "interim period." Lambeth 1930 (see LAMBETH CONFERENCES) had given a cautious encouragement to the proposal, but in the latter stages the scheme provided that the thirty years' interim period would be terminated not by an exclusive rule requiring all ministers to be episcopally ordained, but by a review of the regulations which would not bind the united church to any particular course of action. This led to growing Anglo-Catholic opposition in England in the late 1930s and early 1940s, and in South India to a reconsideration of the old idea of "mutual commissioning." However, the four Anglican bishops in the area rescued the concept which had been in every draft of the scheme by announcing in 1946 that after union they would each without hesitation receive Communion from ministers not episcopally ordained. This was the breakthrough; the new church was inaugurated in 1947.

Anglo-Catholic hostility to the union continued; one Anglican society cut off all official grants to South India. In the Nandyal area, moreover, twenty Anglican clergy, with some 25,000 laity, refused to join the union and remained part of the Church of India, Pakistan, Burma, and Ceylon (ultimately joining the North India union in 1970). Lambeth 1948 reflected some of the suspicion from Catholic sources, and withheld approval. It and the succeeding Lambeth both recommended other areas to unite on the basis of "mutual commissioning" rather than by the South India procedure. Provinces of the Anglican Communion would not enter full communion with South India, as this would have involved accepting nonepiscopal ministers in principle in their own ministerial ranks.

In 1968, however, the Lambeth Conference and the Church of England Commission on Intercommunion (in its report *Intercommunion To-Day*) both recommended by majority votes that churches of the Anglican Communion should reexamine their relationship with the Church of South India with a view to entering full communion with it. Various churches and provinces have indicated their desire to do so, but none at the time of writing has altered its rule as to episcopal ordination.

The Church of South India now comprises roughly a million members with nearly a thousand presbyters. It has developed its own liturgical forms, and the South India Eucharistic Liturgy (1950) pioneered trends which have had an effect throughout the world. The church has striven to be free of dependence upon money from abroad, and has worked at being a genuinely Indian church. It has a small missionary work of its own in Thailand. It has also been engaged in far-reaching, further conversations with the Lutheran churches in South India and has always viewed its own unity as only partial, a challenge to continue a movement started as far back as 1908. It has made an international impact far greater than its size might be thought to warrant, both because its scheme of union has become a departure point in drafting elsewhere (e.g., the abortive Nigeria scheme wherein three drafts from 1957 to 1963 repeated the South India scheme verbatim for a very large part of its contents), and also because it has nurtured ecclesiastical statesmen like Bishop Lesslie Newbigin and, indirectly, Bishop Stephen Neill.

BIBLIOGRAPHY: J.E.L. Newbigin, *The Reunion of the Church: A Defence of the South India Scheme* (1948); B. Sundkler, *Church of South India: The Movement towards Union, 1900-47* (1954); R.D. Paul, *The First Decade* (1958); *The Book of Common Worship* (1963).

COLIN BUCHANAN

SOUTH SEAS. Christianity in this area began with Jesuit missionaries to the Marianas (1668-1769). There are tantalizing hints of Catholic influence in the Marquesas, Hawaii, and parts of Melanesia before rapid expansion of European contacts in the nineteenth century. Protestant missions began with the London Missionary Society in Tahiti and Tonga in 1797, at a time when many Polynesians were dissatisfied with traditional social and religious patterns and willing to undergo major changes to take advantage of the new horizons opened by Europeans. In Hawaii, Kamehameha I (1735-1819) made important political changes, and his wife Kaahumanu and son Liholiho overthrew traditional religion before American missionaries were admitted on trial (1820) after careful consultation with Tahitians about their experience. The Gilbertese similarly rejected their traditional culture for literacy and Christianity in 1868-69.

Enormous variety in language and culture, especially in Melanesia, made evangelism unusually difficult, and missionaries' attempts at individual conversion initially bore little fruit. Alongside considerable aversion to missionary teaching, difficulty of translating Christian concepts like sin often led to serious misunderstanding. When conversions came, they were often linked with the political aims of chiefs like Pomare II of Tahiti (c.1782-1821) and Cakobau of Fiji (1817-80), plus a desire for literacy as a key to European wealth, as in the Cook Islands (1821) or Samoa (1830). The resulting tribal movements created serious problems of education and pastoral care, and few missionaries realized the extent to which islanders were accepting Christianity on their own terms. On Tanna, European goods were obtained through the Queensland labor trade, and missionary preaching had no effect from 1842 till 1904, when the labor trade had ceased.

Early missionaries were often unprepared for island life. Of the ten pioneer artisans landed in Tonga, three were killed, one went native, and the remainder thankfully escaped to Australia. Deep dedication and a growing stock of experience led to an increasingly important religious impact. Allied with Islander readiness to travel for the Gospel, just as their ancestors had traveled for trade and war, this led to remarkable indigenous missionary effort, without parallel elsewhere in the nineteenth century. Great missionaries like J. Williams* (1796-1839) helped to inspire the pattern by leaving eight Tahitian teachers in Samoa (1830), but Tongans arrived in Fiji independently about 1823. Others reached parts of the New Hebrides by 1842 and the Loyalty Islands by 1841, thirteen years before European missionaries. Ta'unga from Rarotonga worked in New Caledonia and left a striking journal. Work in Melanesia was much more difficult because of language barriers and hostility to strangers, but pioneer evangelists like Soga and Marsden Manekalea laid the foundations of the church on Ysobel, and in 1925 Ini Kopuria founded an important missionary order—the Melanesian Brotherhood.

Protestant rivalries were sometimes sharp, but implicit or explicit comity agreements frequently led to Christian "kingdoms" like Tahiti and Tonga where missionaries exercised a considerable political role strikingly in contrast with their free church backgrounds. Their wishes to exclude undesirable European influences foundered on the resistance of traders and a combination of French imperialism and Roman Catholic missionary zeal (partly focused by P. Dillon, 1785-1847) which led to the annexation of Tahiti (1843) and New Caledonia (1853). Oceania was divided initially between the Picpus Fathers and the Marists, and though both were as anxious to convert Protestants as heathens, notable pioneering was done by P.M. Bataillon (1810-77) on Wallis and Futuna, which led to their conversion by 1849. Islanders utilized sectarianism to their own advantage, as in Fiji (1844), and by the end of the nineteenth century there were few groups without strong Catholic communities, organized in apostolic vicariates. The South Seas Evangelical Mission (1904) has also been used within the context of traditional tribal rivalries.

German annexation of Papua (1885) introduced yet another major denominational pattern into the Pacific. Although Australia assumed control of German territories in 1914, the work of the Neuendetteslau and Rhenish Missions (1886 and 1887) has resulted in a good-sized and well-organized Evangelical Lutheran Church, responsible since 1956 for its own affairs. The work of C. Keysser (1877-1961) and the success of his "tribal conversion" methods have profoundly affected missions on this huge island, where the Dutch began work in 1861, the London Missionary Society in 1871 with men like J. Chalmers,* S.M. McFarlane (1837-1911), and W.G. Lawes (1859-1907), not to mention Ruatoka of Mangaia and numerous other Polynesian pastors. Marists followed in 1885, Anglicans in 1891, Seventh-Day Adventists in 1914, and the Unevangelized Fields Mission in 1932, with expansion inland carefully controlled by the Australian government in order to ensure as far as possible missionaries' safety.

By the end of the nineteenth century a number of strong churches had emerged, and missionaries played a vital part in lessening the disastrous effects of European diseases, the labor traffic, and the brutal exploitation of some traders. Depopulation was heavy. In 1875, some 30,000–40,000 died from measles in Fiji alone, and in Hawaii population fell from c.300,000 in 1775 to 30,000 in 1900. Combined with the end of the labor trade and the needs of planters for workers, it led to the introduction of Asians, with particularly serious results in Fiji, where Indians now outnumber Fijians.

The relation of Christianity to traditional culture is still an unsolved problem, and in Melanesia has led to a number of movements with affinities to African independent churches. The Valaila Madness (1919), John Frum (1941 on), Marching Rule (1940s), and Federal Council (1951) are the most notable examples, and the administration and churches alike have found them very difficult to handle. The latter are still largely dominated by Europeans despite valiant official attempts at devolution of responsibility in line with developments elsewhere since 1945. In the Anglican diocese of Melanesia, for instance, local suffragan bishops were elected as late as 1963 and a house of laity was only introduced in 1958. The development of political independence in Western Samoa (1962), the Book Islands (1965), Nauru (1968), Fiji (1971); rapid moves toward responsible government in Papua-New Guinea and the Solomon Islands; and some devolution of responsibility in French territories have already had profound effects on the churches because of the unusually important place that they have in the community.

Migration and urbanization have already created serious social problems, and the development of secondary and tertiary education has necessitated radical changes in theological education inter-island and ecumenical cooperation, symbolized by Pacific Theological College in Suva (1966) and the Melanesian Council of Churches (1946), the Pacific Conference of Churches (1966), and the United Church of Papua-New Guinea and the Solomons (1968). Substantial pockets of paganism still exist in Melanesia and many village Christians are ill-equipped for dealing with their new political responsibilities. Most island economics cannot support welfare policies and rising economic expectations without substantial external aid, and few churches are as well placed as Lutherans where extensive plantations help to finance schools and hospitals.

European Christianity has helped to end cannibalism and tribal warfare, contributed some new economic skills, and significantly influenced family life, but the process has not been one-way colonialism. In many areas Christianity has been closely integrated into Island societies on local terms, and Christians can face their future with an enviable sense of community.

BIBLIOGRAPHY: C. Marau, *Story of a Melanesian Deacon* (1906); A.A. Koskinen, *Missionary Influence as a Political Factor in the Pacific Islands* (1953); C.R.H. Taylor, *A Pacific Bibliography* (1965); R. and M. Crocombe, *Works of Ta'unga* (1968); A.R. Tippett, *Solomon Islands Christianity* (1968); A. and S. Frerichs, *Anutu Conquers* (1969); R.P. Gilson, *Samoa* (1970). See also The Pacific Islands Yearbook; N. Rutherford, *Shirley Baker and the King of Tonga* (1971); F. Steinbauer, *Melanesische Cargo Kulte* (1971); A.R. Tippett, *People Movements in S. Polynesia* (1971); R. Jaspers, *Die Missionarische erschliessung Ozeanieus* (1972); R. Williams, *The United Church* (1972); S. Latu Kefu, *Church and State in Tonga* (1974); N. Threlfall, *A Hundred Years in the Islands* (1975). IAN BREWARD

SOUTHWELL, ROBERT (c.1561-1595). Roman Catholic poet. Born in Norfolk, he was educated at Douai, Paris, and Rome. He became a Jesuit and returned to England in 1586 as chaplain to the countess of Arundel. He was arrested in 1592 and executed three years later. His prose works include *St Mary Magdelene's Tears, A Short Rule of Good Life,* and *The Triumphs over Death,* but it is as a poet that he is now remembered, if at all. Even here he lives largely by a single poem, "The Burning Babe," which manifests early Metaphysical characteristics, tinged with that exotic sensuality that one associates with his fellow-Catholic Metaphysical, Crashaw. There is a note of intense adoration in his writing, which goes not unaptly with his daring comparisons that remind us of the poetic manner in which he was an early participant. ARTHUR POLLARD

SOWERBY, LEO (1895-1968). American composer. While his output of music has included orchestral and organ works, his name is inevitably connected with music for the church. He served much of his life as an Episcopal church organist and choirmaster, particularly at St. James' Church, Chicago. He might well be considered the leading figure in American church music at midcentury. In 1946 he was awarded the Pulitzer Prize for his *Canticle of the Sun.* His Passiontide cantata, "Forsaken of Man" (1940), is one of the finest works of its kind in recent years. Many of his anthems are on an ambitious scale, but he has also written a number of pieces of moderate difficulty and less pronounced modernity, such as "Love came down at Christmas." He served prominently on the Hymnal Commission of the

Episcopal Church and on the Joint Commission on Church Music. At the end of his life he was active in the school of church music associated with the cathedral in Washington, D.C.

J.B. MAC MILLAN

SOZOMEN (Salaminius Hermias Sozomenus) (fifth century). Church historian. A lawyer at Constantinople, but originally from Bethelia near Gaza in Palestine, he is named most commonly in the trilogy with Socrates and Theodoret as continuator of Eusebius's Ecclesiastical History, which as the *Tripartite History* was translated into late sixth-century Latin traditionally by Cassiodorus, perhaps by Epiphanius under the former's inspiration. Sozomen has been assumed to be the least independent of these initiating histories of the Christian Roman Empire, for large sections share directly without reference the parallel Socrates. Both have a concern to report the growth of monasticism as created by Christian Egypt. Yet he may have intended to revamp Socrates with fuller details both of the Western Empire and Church and of the Eastern Persian front with its Christian martyrs. The work in nine books specifies that it covered the period from the third consulate of the Caesars, Crispus and Constantine (324), to the seventeenth of Theodosius II (439), to whom it is dedicated before his death in 450. The conclusion is now missing, as is Sozomen's earlier epitome in two books of church history from the Ascension to the defeat of Licinius (323).

CLYDE CURRY SMITH

SOZZINI, see SOCINIANISM

SPACE EXPLORATION. The idea of space exploration can be traced back at least to the second century A.D. when the Greek satirist, Lucian, described an imaginary lunar voyage. Others have described space travel in fiction. In modern times the theoretical basis of space navigation was developed by R.H. Goddard in America, H.J. Oberth in Germany, and K.E. Tsiolkovsky in Russia. After World War II the early work on rockets carried out in Germany was continued in the USA and the USSR. In 1961 Yuri Gagarin was the first man in space, and the Russians triumphantly announced that he found no sign of God in the sky. In 1969, when Americans were the first men on the moon, Neil Armstrong and Edwin Aldrin read the Bible back to earth and partook together of the Lord's Supper on the surface of the moon. Space exploration and its kindred developments inevitably raise some questions of Christian concern: (1) Do they merit expenditure of so much money and resources when there is so much need on earth? (2) Do they bring nearer the concept of the superstate as prophesied (cf. Rev. 13, assuming the futurist interpretation)? (3) Has God made other intelligent creatures? Are they sinful? If so, by what means does He redeem them, since Christ died once only for sin? These questions are, of course, unanswerable in view of our ignorance. Man is beginning to realize, nevertheless, how vast is God's universe—so vast as to suggest that man's extraterrestrial investigation is circumscribed in God's comprehensive plan.

R.E.D. CLARK

SPAIN. According to tradition, Spain first received Christianity from St. Paul and St. James. Certainly by the third century, as Tertullian averred, a flourishing national church was in existence and a council was held at Elvira* in A.D. 300. The country was adversely affected by various heresies, and in the fifth century the Arian Visigoths overran the land, but their successors at the Third Council of Toledo* (589) accepted the Catholic faith. From 711 onward the country was taken over by the Muslim Moors who were at last checked by Charles Martel at Tours in 732. The Christian Church was now persecuted, but a reconquest was begun c.1000 and was finally completed with the absorption of Granada in 1492 and the union of the kingdoms of Aragon and Castile in 1494. Throughout this period French influence became everywhere predominant, the new religious orders arrived, and the ancient Spanish Mozarabic liturgy which had always been unpopular with the popes was suppressed.

In 1479 also the Inquisition* was introduced into Spain by Ferdinand and Isabella, the two "Catholic" monarchs, and a studied persecution of Marranos and Moriscos (Jews and Muslims) was begun; of the former alone 350,000 were accused of heresy, and 12,000 burnt. The Reformed faith was treated similarly, and in the sixteenth century —the age of Spain's greatest prosperity and expanding imperial power—the native church became the pope's most faithful ally against both France and England, and largely helped to shape the Counter-Reformation. St. Teresa,* St. John of the Cross,* and Ignatius Loyola* were notable figures of this period. Soon, however, Spanish power began to decline, the church during the seventeenth and eighteenth centuries became hopelessly intolerant and corrupt, and even drastic measures such as Charles III's expulsion of the Jesuits (1767) failed to halt the decline.

With the French occupation of Spain (1808) political liberalism and anticlericalism were introduced into the country, and these new forces confronted traditionalism in church and state and led to a century-and-a-half of civil strife. Socialism, anarchism, and various regional nationalisms latterly added weight to the forces of disruption. The Civil War of 1936-39 resulted in a victory for the Nationalists and the restoration of the Catholic Church, which had suffered severely during the period of the Republic (1931-36). Church and state were at one again, and a notable concordat was concluded in 1953, but of recent years individual church leaders have taken a more critical view of the Franco regime.

The Spanish Church today is divided into nine archbishoprics and sixty-one bishoprics. The secular clergy probably total 30,000 and the religious 45,000. Though Spain is regarded as the most devout of nations, probably only about 20 per cent of her people are practicing Catholics, while anticlericalism is widespread. The principal features of church life include an exaggerated cult of the Virgin Mary unparalleled elsewhere, the prominence of miracle-working relics, and elaborate and semi-pagan processions and pilgrimages connected especially with Holy Week.

Protestantism in Spain began with various pre-Reformation movements, especially those connected with Raymond Lull,* the thirteenth-century missionary to the Muslims; Alfonso de Madrigal, the expositor; and Pedro de Osuma, often called "the Spanish Hus." No doubt the sixteenth-century Reformation would have taken firmer root on Spanish soil but for the vigorous, if partial, reform of the church carried out by Jiménes de Cisneros* at the turn of the century consequent on the union of the Peninsular states. As it was, Protestantism made its appeal almost entirely to the privileged and educated classes and was confined to isolated families and individuals. Particularly noteworthy are Francisco de Enzinas, who translated the NT into Spanish and was for a time professor of Greek at Cambridge; Juan de Valdés,* who maintained his Protestant witness from Italy; Rodrigo de Valder, the "Spanish Wycliffe," who preached openly in Seville till he was imprisoned for life; and Archbishop Carranza of Toledo, primate of Spain who died after great suffering in 1576. No separatist churches were formed, however, and after 1530 the Inquisition vigorously suppressed all Reformed teaching; the first *auto-da-fé** was held in 1559.

The seventeenth and eighteenth centuries, a time of intellectual torpor and extreme social conservatism, saw the Reformed faith firmly excluded from Spanish soil. Not till the nineteenth century did Protestantism make a significant return to the country: in 1832 William Rule was preaching there, and in 1837 George Borrow,* as an agent of the International Bible Society, embarked on those adventures which he later described in *The Bible in Spain.* After the republican revolution of 1868, Protestantism could enter the country more freely, American, English, Irish, Swiss, and Swedish missionaries of various denominations being prominent in the field. Slowly native Spanish churches were built up: the Episcopal Reformed Church of Spain—whose first bishop was consecrated by three prelates of the Church of Ireland (to the anger of British High Churchmen) in 1894—which remains small, has recently developed liturgical interests and claims to preserve the native Mozarabic Rite in its purest form; the Spanish Evangelical Church (Congregational/Presbyterian and American in origin); the Baptists; and the Brethren.

With the revolution of 1931, Protestant activities could proceed unimpeded: churches were opened as well as schools, including *El Porvenir,* reputed to be the finest secondary school in Spain. The Nationalists, however, denounced the Protestants as being abettors of Republicanism, and after 1939 persecution began again. Only slowly has the position of this despised minority been improved, though evangelism and theological education have both been less restricted since the promulgation of the Organic Law of the State (1966). Even so, disabilities still attach to Protestants, in regard to marriages and burials, professional advancement, the printing and distribution of literature, and the position of the young Protestant conscript. In recent years Pentecostalism has been spreading to Spain in a manner not unlike its progress in the former Spanish American colonies. The total Protestant community in Spain is now about 43,000, which means that it has doubled during the last forty years.

BIBLIOGRAPHY: R.S. Alderson, *The Church in Caverns Hidden* (n.d.); G. Borrow, *The Bible in Spain* (1842); P.J. Hauben (ed.), *The Spanish Inquisition* (1869); C.R. Haines, *Christianity and Islam in Spain* (1889); F. Meyrick, *The Church in Spain* (1892); H.C. Lea, *The Moriscos in Spain* (1901) and *A History of the Inquisition in Spain* (4 vols., rep. 1967); E. Gill, *Europe and the Gospel* (1931); C.A. Garcia and K.G. Grubb, *Religion in the Republic of Spain* (1933); E.A. Peers, *Spain, the Church and the Orders* (1939); H.V. Livermore, *A History of Spain* (1958); H. Kamen, *The Spanish Inquisition* (1965); R.M. Smith, *Spain: A Modern History* (1965); *Journal of Religion* 35, pp. 242-51; *Theology Today* 16, pp. 338-44; *Ecumenical Review* 20, pp. 53-62.

IAN SELLERS

SPALATIN, GEORGE (1482-1545). German Reformer. Born "George Burkhard" at Spalt near Nuremberg, Spalatin (his Latinized name) attended the universities of Erfurt and Wittenberg. In 1505 he entered St. George's Monastery in Erfurt. In 1508 he became tutor to the future elector of Saxony, John Frederick. He served also as adviser and secretary to the elector Frederick the Wise. In 1525 he was made pastor at Altenburg, but continued to serve Elector John the Constant and John Frederick the Magnanimous. He was one of the earliest of Martin Luther's friends and promoters of this movement and greatly influenced Frederick the Wise in his tolerant and protective stance toward the Reformer. He carried on a voluminous correspondence throughout his active lifetime, only partially published. He outstandingly contributed to the promotion of education as a member of various church visitations. He helped to establish schools and libraries, actively trained pastors in pastoral theology, and helped in the training of young people. Spalatin's friendship for Luther during the last thirty years of his life (especially during the crucial years between 1517 and 1521) was of great significance in the work of Martin Luther. CARL S. MEYER

SPANGENBERG, AUGUST GOTTLIEB (1704-1792). Moravian missionary and church leader. Son of a Hanoverian court preacher, he first studied law and then theology and finally became a teacher in Halle. When he joined the Herrnhut* community in 1732, Zinzendorf* assigned him to conduct legal negotiations with various European colonial powers for permission to establish mission works abroad. He took personal charge of the Moravian group that settled in Georgia in 1735, and greeted and advised John Wesley who arrived there in February 1736. The next month, Spangenberg joined the Schwenkfelder colony in Pennsylvania as a simple farmer, and within a short time was preaching to the Indians. In 1744 he was placed in charge of the Moravian work in America and devoted himself intensely to Indian missions. Appointed Zinzendorf's successor in

1760, he returned to Herrnhut two years later and led the group until his death.

RICHARD V. PIERARD

SPEER, ROBERT ELLIOTT (1867-1947). Secretary of the Presbyterian Board of Foreign Missions, a post he held for forty-six years. Born in Huntingdon, Pennsylvania, he was one of the early Student Volunteers and traveled for a year for the movement. After becoming mission secretary, he made six extended visits to the fields, four to Asia and two to Latin America. A prolific writer, he authored sixty-seven books, many on missions. During World War I he was on the advisory committee on religious and moral activities of the army and navy, and was chairman of the General Wartime Committee of Churches. He chaired the Committee on Cooperation in Latin America from its inception until his retirement in 1937. He served as president of the Foreign Missions Conference of North America, as well as one term as president of the Federal Council of Churches. In 1927 he became the second layman to be elected moderator in the Presbyterian Church in the U.S.A.

HAROLD R. COOK

SPENCER, JOHN (1630-1693). Hebraist. Born in Kent, he was educated at King's School, Canterbury, then entered Corpus Christi College, Cambridge (1645), where he remained for the rest of his life. He was master of the college from 1667 until his death. In his *De Legibus Hebraeorum* (1685) he compared the laws and rites of the Jews with those of other Semitic nations, thus laying the foundations of comparative religion. His work was little heeded in his day, but was continued by Wellhausen and others of the Tübingen School* two centuries later.

SPENER, PHILIPP JAKOB (1635-1705). German Lutheran Pietist leader. Born in Rappoltsweiler, Alsace, and raised in a highly protective and deeply religious atmosphere (characterized by a mixture of Puritanism and Arndtian Pietistic mysticism), he studied theology at Strasbourg in 1651-59 under the strict Lutheran J.K. Dannhauer. During academic wanderings in 1659-62 to Basle, Geneva, Stuttgart, and Tübingen he came into contact with Reformed theology and Jean de Labadie, who preached repentance and regeneration. In 1663 he became a free preacher at Strasbourg, received a doctorate in theology in 1664, and served as pastor and senior of the ministerium in Frankfurt am Main (1666-85), where he emerged as the leader of the Pietist Movement. He was appointed court chaplain at Dresden in 1686, but his relations with the Saxon ruling family soon became strained, and in 1691 he accepted the invitation of the elector of Brandenburg to the pastorate of St. Nicholas Church in Berlin.

At Frankfurt he reformed religious instruction by preaching on whole books of the Bible, restoring the confirmation service, and setting aside days of fasting and prayer. He proclaimed the necessity of conversion and holy living, and in 1670 set up a conventicle *(collegia pietatis)* within the church where pastors and laymen met to study the Bible and pray together for mutual edification. He saw such conventicles as *ecclesiolae in ecclesia* which would aid the pastor in his spiritual duties and return the church to the spiritual level of the of the early Christian communities. Modeled upon similar bodies among the Reformed, the institution spread throughout Lutheran areas. In the tract *Pia Desideria (Pious Desires)*, published in 1675 as a preface to Johann Arndt's* *True Christianity*, Spener set forth the essence of his Pietistic doctrines—the central importance of Bible study, restoration of the priesthood of all believers, true faith expressed not in knowing but in deeds of love to one's neighbor, avoidance of theological disputation, emphasis upon spiritual life and devotional literature in the training of ministers, and preaching that should awaken in the hearers faith and its fruits.

As Spener's popularity spread, he became an increasingly controversial figure and his disciples were even expelled from Leipzig in 1690. Although little of his teaching was original (most had been expressed by Arndtian and Reformed Pietists before), his emphasis upon the new birth and exemplary life effectively undermined the position of scholastic orthodoxy and revitalized German Lutheranism.

BIBLIOGRAPHY: A.B. Ritschl, *Geschichte des Pietismus* (3 vols., 1880-86; rep. 1966); P.J. Spener, *Hauptschriften* (ed. Paul Grünberg, 1889) and *Pia Desideria* (ed. T.G. Tappert, 1964); P. Grünberg, *Philipp Jakob Spener* (3 vols., 1893-1906); K. Aland, *Spener-Studien* (1943); F. Stoeffler, *The Rise of Evangelical Pietism* (1965); J. Wallmann, *Philipp Jakob Spener und die Anfänge des Pietismus* (1970).

RICHARD V. PIERARD

SPEYER, DIETS OF. Speyer (Spires) on the Rhine in Bavaria was host to four meetings of the Diet, or parliament, of the Holy Roman Empire during the Reformation period. In each case religious and politico-military considerations became intertwined. Emperor Charles V wanted support of the princes of Germany in his struggle against the Frankish-Ottoman alliance, and he wanted suppression of Lutheranism as required by the Edict of Worms. Some leading princes (e.g., Frederick of Saxony) would not grant both, however. As Lutherans, they demanded relaxation of religious suppression as a price for military support. Some Catholic princes supported the Lutherans because they wanted a greater degree of freedom from imperial control. So at Speyer in 1526 the emperor was forced to accept the resolution of the diet: "Each one [prince] is to rule and act as he hopes to answer to God and his Imperial majesty." This opened the way for the spread of Lutheranism.

In 1529 Charles felt strong enough to demand that the diet of that year rescind the 1526 decision and ordered the rulers of Germany to enforce the Edict of Worms. Most complied but several, joined by fourteen free cities, drew up a strong protest to the emperor. Signatories came to be known as "Protestants," and all who eventually left the Catholic Church were given the same name. The military situation continued to be grave, and Charles needed the aid of Lutheran princes. At the third diet of Speyer in 1542, to get help against the Turks, and at the fourth in 1544, to get help against the French, he made concessions. Though he later tried to crush the Protestants by military force, he ultimately had to grant recognition to Lutheranism at the Peace of Augsburg* in 1555. HOWARD F. VOS

SPINOZA, BENEDICT (or "Baruch") **DE (1632-1677).** One of the foremost Rationalist philosophers. A Jew, he was expelled from the synagogue in 1656 for his unorthodox views. Most of his life was spent in Amsterdam, where he earned his living as a lens-grinder. Spinoza was an independent, original thinker whose work is difficult to interpret because of its self-contained character. Fundamental is the view that truth is formed of a system of interconnected deducible propositions. In his *Ethics,* geometrical-type reasoning was meant to lead to substantive ethical conclusions. This is only remotely plausible given Spinoza's rationalistic account of logical relations, and his failure to distinguish logical connections from causal connections.

Spinoza's view that there could only be one infinite substance entails pantheism, for any creation over against a creator would involve a limitation of the creator. With no concept of divine transcendence there can be no place for the idea of divine purpose; God's creativity is understood by Spinoza as nature's activity. Finite minds are modes of God's thought, bodies modes of God's extension. Spinoza has variously been described as a religious "God-intoxicated" thinker, and as irreligious, according to which side of the God-Nature equation has been stressed. His view that the Bible is written in the manner of unreflective irrational man was, historically, influential in the growth of rationalistic criticism of Scripture.

PAUL HELM

SPIRITISM. "Spiritism" and "Spiritualism" are terms, often used interchangeably, referring to the belief in communication between the living and the living spirits of the dead. History records periodic surges of interest in the spirit phenomenon, but as an organized religion it began at Hydesville, New York, in 1848. John D. Fox and his wife traced strange, tapping sounds in their home to the room of their teenage daughters, Margaret and Katie. After the continued experience of this and bedclothes pulled off the bed by invisible hands and chairs and tables removed from their places, the two girls devised a means of intelligent communication with the author of the noises, who would reply to questions with a number of raps. This widely advertised event set off a Spiritism revival in the United States that soon spread to England and Europe.

The breakdown of faith in the traditional, authoritarian doctrine in religion by the scientific revolution in the nineteenth century gave Spiritism an added boost. The number of adherents to the new faith grew rapidly, and at its heights the movement claimed over 10 million followers. Among the distinguished converts were Sir Arthur Conan Doyle, Oliver Lodge, and Alfred R. Wallace. Probably one of the most famous of the mediums was D.D. Home (1833-86), whose amazing feats and manifestations in the presence of leading scientists still stand without plausible explanation.

The Spiritism Movement produced many spirit-oriented phenomena: table-tipping, playing on musical instruments, levitation of various objects and even of the medium, appearance of objects in the atmosphere, spirit writing, rappings, mediumship, and materialization. Some mediums go into a trance and become a passive instrument for the spirit, while others report what they hear or see spirit forms say or do. The trance state is not generally practiced today. In the materialization séance a foggy, smokelike substance called "ectoplasm" (defined as "exteriorized protoplasm") is said to emanate from the body and mouth of the medium, forming an image. Usually materialized spirits do not give messages.

The Spiritualists draw heavily on the teachings of Emanuel Swedenborg,* Franz Mesmer, and Andrew Jackson Davis. Davis's book, entitled *Nature's Divine Revelations* (1847), stated the fundamentals of Spiritism. The doctrines of the Trinity and of the deity of Christ are rejected; they hold to the existence of an Infinite Intelligence expressed in the phenomena of nature, both physical and spiritual; true religion is the correct understanding and living in accordance with the Infinite Intelligence; personal identity continues after death; communication with the dead is possible; the highest morality is contained in the Golden Rule; the doorway to reformation is never closed against any human soul, here or hereafter; Christ was a medium; the teaching of God as love is central.

Spiritism is now organized on a basis similar to denominationalism. In the USA the main associations are the International General Assembly of Spiritualists, the National Spiritual Alliance of the U.S.A., and the National Spiritualist Association of Churches. The latter group is the orthodox body of American Spiritualism and the most prominent; they maintain a seminary for the training of their ministers. Regular services are held by the churches, and many of the standard religious rituals are observed—singing, praying, etc. This is combined with the practice of mediumship. Camps provide convenient centers for worship, instruction, and practice. Although the actual membership of the Spiritualist groups in the United States is much less than 200,000, the groups claimed in 1971 that there were over 1,000,000 believers.

As a result of the rise of Spiritism, the Society for Psychical Research was organized in 1882 in London, and similar societies were later organized in the United States. Psychical research, or "parapsychology" as it is now termed, has cast doubt on the proof of spirit survival by the evidence of mediums. Studies in extrasensory perception have shown that the mind can, at times, reach out beyond itself to acquire information which the senses and the reason could not obtain. It is quite possible that mediums possess such powers and simply attribute them to spirit origin. Much more research is needed in order to substantiate the claim.

BIBLIOGRAPHY: J.T. Stoddart, *The Case Against Spiritualism* (1909); J. Fox, "Spiritist Theologians," *Princeton Theological Review* (1920); R.B. Jones, *Spiritism in Bible Light* (1921); A.C. Doyle, *The History of Spiritualism* (2 vols., 1926); G.W. Butterworth, *Spiritualism and Religion* (1944); J.K. Van Baalen, *The Chaos of Cults* (1956): K.H. Porter, *Through a Glass Darkly* (1958); A.T. Schofield, *Modern Spiritism, Its Science and Religion* (1960); A. Blunsdon, *Popular Dictionary of Spiritualism* (1962).

JOHN P. DEVER

SPIRITUAL HEALING. Healing by spiritual and religious as distinct from scientific and medical means; also called "faith-healing," "divine healing." The means, as well as prayer, sometimes are sacramental, viz., unction, imposition of hands, Eucharist. It is a matter of dispute how far suggestion and psychological influences are involved—e.g., some regard cures at Lourdes* or Holywell as due at least partly to suggestion.

It was the oldest form of healing, and early medical practice grew up alongside it. It was known in Greece in the Dionysiac mysteries, and at the temple of Aesculapius, Epidaurus, and the healing gods (e.g., Apollo, Aesculapius, Zeus) were entitled *Soter.* The OT miracles of Elijah and Elisha foreshadowed the healing miracles of Jesus, which were a prominent feature of His ministry, and the apostolic healings in Acts.

During the patristic period the practice continued unbroken, the churches rivaling the pagan temples as places of healing. Tertullian wrote of Christ, "He reforms our birth by a new birth from heaven; He restores our flesh from all that afflicts it; He cleanses it when leprous, gives it new light when blind, new strength when paralyzed, when possessed by demons He exorcises it, when dead He raises it to life." Cyprian said that from the indwelling Spirit "is given power that is able to quench the virus of poisons for the healing of the sick, to purge out the stains of foolish souls by restored health." Irenaeus argued that the Gnostics, though they can produce miraculous effects, cannot perform works of healing like Christians, who "heal the sick by laying their hands upon them." Hermas said that those who know the sufferings of men, yet do not relieve those sufferings, commit great sin. A famous healer was Gregory Thaumaturgus* (c.213-c.270).

Though declining after the third century, the transition being marked by a growing veneration of relics,* the phenomenon recurred during the Middle Ages, healings being reported of Francis of Assisi,* Bridget,* Charles Borromeo,* Cuthbert,* Patrick,* among others. English monarchs of the eleventh through eighteenth centuries touched numerous people to cure them of "the King's Evil."* Spiritual healing proper reappeared among the Waldenses* and Bohemian Brethren.* Cures were recorded of Martin Luther and other Reformers.

In the seventeenth century, healings were effected by English Baptists (e.g., Hanserd Knollys, William Kiffin, Vavasor Powell); by Quakers (e.g., instances recorded in George Fox's *Journal);* and in other Puritan sects. A notable healer was Valentine Greatrakes (1629-83), who healed many in Ireland and England between 1662 and 1666 by laying on hands with prayer. In the eighteenth century John Wesley recounted several instances in his *Journal,* and the German Pietists also practiced healing. A famous nineteenth-century German healer was Prince Hohenlohe-Waldenburg-Schillingsfürst, canon of Grosswarden, and in Russia there was Father John of Kronstadt (1829-1909).

The Peculiar People* in sickness relied on oil, prayer, and nursing. Groups like the Irvingites and Mormons* practiced healing, and Christian Science* promoted it by teaching that pain and disease are an illusion. The Pentecostal churches* have always advocated healing; it came to the fore, however, in the campaigns of George Jeffreys between 1925 and 1935, and in America with Aimee Semple McPherson.* Recently there has been a revival among the older churches, e.g., "the Guild of Health" and "the Guild of St. Raphael" in Anglicanism, and in the Iona Community in the Church of Scotland. The Eastern Orthodox Churches have always retained a service of healing in their regular work. Their seventh sacrament is "Holy Unction" which, however, unlike the Roman practice in recent centuries, is the anointing with oil of the sick for *recovery,* in accordance with James 5:14f.

BIBLIOGRAPHY: P. Dearmer, *Body and Soul* (1909); J.M. Hickson, *Heal the Sick* (1924); G.G. Dawson, *Healing, Pagan and Christian* (1935); A.G. Ikin, *The Background of Spiritual Healing* (1937); E. Frost, *Christian Healing* (1940); L. Weatherhead, *Psychology, Religion and Healing* (1951); *The Church's Ministry of Healing* (report of the Church of England Archbishops' Commission on Divine Healing, 1958); B.E. Woods, *The Healing Ministry* (1961); M.T. Kelsey, *Healing and Christianity* (1974). J.G.G. NORMAN

SPITTA, KARL JOHANN PHILIPP (1801-1859). Lutheran hymnwriter. Born in Hanover and originally apprenticed to a watchmaker, he eventually graduated in theology at Göttingen (1824). As a result of his conversion about that time, he stopped writing secular verse, such as had appeared in his *Sangbüchlein der Liebe.* After four years as a tutor, he took Lutheran orders. He held clerical posts at Hamelin, Wechold, Wittingen, Peine, and finally Burgdorf. His *Psalter und Harfe* (two series, 1833-43) enjoyed in Germany a popularity comparable with that in Britain of *The Christian Year* by John Keble. The hymns were translated in full by R. Massie as *Lyra Domestica* (1860-64), and in part by others, notably Jane Borthwick* and Sarah Findlater.

JOHN S. ANDREWS

SPITTLER, CHRISTIAN FRIEDRICH (1782-1867). Swiss-German mission founder. A clerk from a German pastor's family, he became secretary of the Basle *Christentumsgesellschaft* in 1801. He helped to establish the Basel Mission Society in 1815, and during the 1820s sought to organize a work in Greece. His principal achievement was the St. Chrischona Pilgrim Mission which he founded in 1840. He opened a missionary training school for skilled craftsmen in a church near Basle, and its graduates first ministered among German immigrants in America. After 1846 the primary field became Palestine. Inspired by Samuel Gobat* in Jerusalem, Spittler attempted to extend the St. Chrischona work to Ethiopia, but his plan to establish a chain of mission stations along the Nile to link Jerusalem to Ethiopia came to naught. After his death St. Chrischona restricted itself to home missions until 1895 when it opened a China field in cooperation with Hudson Taylor.* RICHARD V. PIERARD

SPORTS, see BOOK OF SPORTS

SPURGEON, CHARLES HADDON (1834-1892). Baptist preacher. He was born in Kelvedon, Essex, with Dutch and Dissenting ancestry. His father and grandfather were Independent pastors. Early in 1850 he was converted in Artillery-street Primitive Methodist Chapel, Colchester, Essex, into which he came because of snowy weather. After baptism he became pastor of Waterbeach Baptist Chapel in 1851. In 1854 he was called to New Park Street Baptist Chapel, Southwark, London, which was soon filled to overflowing, necessitating the building of the Metropolitan Tabernacle in 1859.

In 1856 he married Susanna Thompson and also began the "Pastor's College" for training men "evidently called to preach the Gospel," which continues today as "Spurgeon's College." For fifteen years he bore the whole cost, after which the Tabernacle shared the burden. In 1865 he was one of the founders of the London Baptist Association, and in 1869 he established an orphanage at Stockwell, known now as "Spurgeon's Homes." Other charitable and religious organizations he founded and supported included Temperance and Clothing societies, a Pioneer Mission, and a Colportage Association.

He suffered periodic bouts of illness which sometimes kept him out of the pulpit. He preached at the Tabernacle for the last time on 7 June 1891 and died the following January at Mentone, S France. During his thirty-eight-year London ministry he had built up a congregation of 6,000 and added 14,692 members to the church.

During his early ministry he fought battles on two fronts, against hyper-Calvinism and Arminianism.* In 1864 he preached a sermon attacking the baptismal doctrine and practice of the Church of England, thus initiating the "Baptismal Regeneration Controversy." He accused the evangelical Anglicans of perjury in using the Prayer Book when they did not believe in baptismal regeneration. In the resulting furor he felt compelled to resign from the Evangelical Alliance for a time. In 1874 he was involved in a dispute on smoking. The "Downgrade Controversy" of 1887-89 arose out of his concern at the growth of radical teaching among Baptists. Several, including the Baptist Union secretary, pleaded with him to try to stop the trend. He made his protest, but was disregarded, so in October 1887 he withdrew with others from the Union. His resignation was accepted, and a motion of censure, never rescinded, was passed. The affair deeply grieved him and may have shortened his life, but he refused to form a new denomination.

Spurgeon was an evangelical Calvinist. He read widely and especially loved the seventeenth-century Puritans. A diverse author, he wrote biblical expositions, lectures to students, hymns, and the homely philosophy of "John Ploughman," among other works. Preeminently he was a preacher. His clear voice, his mastery of Anglo-Saxon, and his keen sense of humor, allied to a sure grasp of Scripture and a deep love for Christ, produced some of the noblest preaching of any age. His sermons have been printed and distributed throughout the world. Two popular works still widely used today are *Treasury of David* and *Morning and Evening*, the latter a compilation of devotional readings.

BIBLIOGRAPHY: G.J. Stevenson, *Pastor C.H. Spurgeon: His Life and Work to His Forty-Third Birthday* (1877); G.H. Pike, *The Life and Work of Charles Haddon Spurgeon* (1892-93); C.H. Spurgeon, *Autobiography* (4 vols., compiled by Mrs. Spurgeon and J.W. Harrald, 1897-1900, vols. I and II republished in 1962 as *C.H. Spurgeon, the Early Years*); W.Y. Fullerton, *C.H. Spurgeon* (1920); J.C. Carlile, *C.H. Spurgeon: An Interpretative Biography* (1933); H. Thielicke, *Encounter with Spurgeon* (1964); I.H. Murray, *The Forgotten Spurgeon* (1966); E.W. Bacon, *Spurgeon, Heir of the Puritans* (1967). J.G.G. NORMAN

SPURGEON, THOMAS (1856-1917). Baptist pastor. Twin son of C.H. Spurgeon,* he studied theology at the Pastors' ("Spurgeon's") College, and art and wood-engraving in London. In 1877 he visited Australia and Tasmania, and returning to Australasia two years later, he accepted a Baptist pastorate in Auckland in 1881, afterward being an evangelist for the New Zealand Baptist Union (1889-93). He then went to London as pastor of the Metropolitan Tabernacle, which he saw burned and rebuilt. He resigned for health reasons in 1908. He was also president of his old college, of the Colportage Association, and of Stockwell Orphanage. His son (Thomas Harold, 1891-1967) served for many years in Dublin as principal of the Irish Baptist College.

C.G. THORNE, JR.

SPYRIDON (Spiridion) (d. c.348). Bishop of Tremithius, Cyprus. According to tradition he was a simple shepherd who suffered in Diocletian's persecution and who, after becoming bishop, attended the Council of Nicea* (325). He certainly attended the Council of Sardica about 343. Many stories have gathered around him, recorded by Socrates, Rufinus, and Sozomen in their histories. He is reputed to have recited a statement of the Christian faith at the Council of Nicea to a pagan philosopher, Eulogius, who had previously refuted other Christian disputants, which led to Eulogius's conversion and baptism.

STAINER, SIR JOHN (1840-1901). English composer. Although his music is much out of favor with musicians today, he made a great contribution to his time. As the influential organist of St. Paul's Cathedral in London, he did much to raise the standards of choral performance by precept and example. He was a fine organist and teacher, and his manual on playing the organ attained a great circulation. As scholar and music historian, Stainer has a name respected still. His *Dufay and His Contemporaries* was one of the first important musicological works by a Briton.

His compositions were too facile and abounded in the clichés that have endeared them to countless churchgoers, while offending the taste of connoisseurs. "Grieve not the Holy Spirit," still in use with many choirs, illustrates his flair for dramatic expression. *The Crucifixion*, with all its musical banalities, provided a work easy enough for average choirs, while containing a timely message in a framework large enough to rank as an oratorio. The solo movements suffer from the sentimentalism of the age, but the recitative is dramatic and well fitted to the accentuation of the text. A number of the hymntunes, included after the example of a Bach Passion, have memorable melodies. The one really good ensemble piece, "God so loved the world," is a fine bit of vocal part-writing. With his musical gifts, had Stainer been reared in a more sophisticated tradition, his numerous anthems might well have had greater musical value.

J.B. MAC MILLAN

STALKER, JAMES (1848-1927). Scottish minister, scholar, and writer. Born in Crieff and educated at Edinburgh, Berlin, and Halle, he was ordained in the United Free Church* and ministered in Kirkcaldy and Glasgow before becoming professor of church history in the United Free Church College, Aberdeen (1902-26). He was well known as a visiting lecturer in various American colleges and seminaries, and the lucidity of his scholarship is seen in his many writings, among them his lives of Paul and (notably) Jesus (1891). Nevertheless it is as a preacher that he was best remembered. He had shared in the revival movement following the 1873 Moody and Sankey mission, and it had a lasting effect upon him. With evangelistic preaching he coupled also a marked social concern which "caused some douce hearers to become uneasy." He was fearless, untroubled by personal ambition (he declined both a principalship and the moderatorial chair), and encouraged every movement that carried the Gospel to the people.

J.D. DOUGLAS

STANFORD, SIR CHARLES VILLIERS (1852-1924). English composer. His name is linked almost inevitably with his distinguished contemporaries, Sir Hubert Parry and Sir Edward Elgar. The former, who was professor of music at Oxford and wrote the first authoritative life of Bach in English, made a contribution of some importance in the choral realm, but affected the course of church music less than Stanford. Elgar, who was the most original and distinctive composer of the three, distinguished himself with his oratorios and symphonic works. All three men helped to bring in a new era in British music, a veritable renaissance of native talent that led directly to the achievements of the present century.

Stanford was born in Ireland, received the best musical education available in England, and studied also in Germany. As well as teaching composition at the Royal College of Music, he was professor of music at Cambridge. A whole generation of significant figures were at some time his pupils—among them Tertius Noble, who became so influential in the USA, and R. Vaughan Williams.* Stanford wrote much music of every kind, but it is his church music that has lived. He was the first to apply to English anthems and cathedral services the technique of motivic development and sound formal construction found in the works of the great European masters from the classical era onward. His services in B-flat and C are the best known. The *Te Deum* and the *Magnificat* from the former, for example, found wide acceptance separately as anthems apart from their planned liturgical function, and illustrate well the important aspects of their composer's work. They have unity and musical logic, effective modulations, and the organ accompaniments possess a degree of independent interest without obscuring the text.

J.B. MAC MILLAN

STANISLAUS (1030-1079). Bishop of Cracow and martyr. Nobly born at Szczepanow, Poland, he was educated at the cathedral school of Gnesen and then at Paris. While Stanislaus was canon and preacher at Cracow, Alexander II appointed him bishop there in 1072. Opposing King Boleslaw II because of his long expedition against the Grand Duchy of Kiev (1069) and other scandalous conduct, Stanislaus excommunicated the king, only to be slain by him during Mass for treason. While the defeated king spent his remaining years penitent among Benedictines in Hungary, miracles and legends surrounded the martyred bishop, which cult extended to Lithuania and the Ukraine. He became the patron saint of Poland and was canonized by Innocent IV in 1253.

C.G. THORNE, JR.

STANLEY, ARTHUR PENRHYN (1815-1881). Dean of Westminster. Son of a rector who had private means and noble connections, he was educated at Rugby and Balliol, coming into touch with Pusey and the Tractarians.* He was elected a fellow of University College, Oxford, in 1839 and was ordained. In 1856 he became professor of ecclesiastical history at Oxford. Like the Tractarians, he desired more earnestness and order in the Church of England, which at this time suffered from lack of devotion, absenteeism, plurality, and indifference, but he was the antithesis of a sacramentalist. He was noted rather for Broad Church views, was strongly in favor of the state connection, and wanted a truly comprehensive Church of England. He would not have barred Unitarians, and he worked for the admission of Nonconformists to the universities. After he became dean of Westminster in 1863 he invited Keble, Liddon, and Pusey to preach; all refused, feeling that to do so would compromise them in view of Stanley's sympathy with German liberalism. Queen Victoria liked him, however, and would have made him a bishop but for Palmerston and Gladstone. He was a member of the NT revision committee of 1870 and one of those who wanted to remove the anathemas from the

Athanasian Creed in 1872. He opposed the disestablishment of the Church of Ireland. A widely traveled man, Stanley produced many books, including *Commentary on the Epistles to the Corinthians* (1855), *Lectures on the Eastern Church* (1861), and *Memorials of Westminster* (1868).

P.W. PETTY

STAPLETON, THOMAS (1535-1598). Roman Catholic apologist. Born in Sussex, he graduated from Oxford, but when Elizabeth became queen he left for Louvain and Paris. In 1569 he joined the English College at Douai, and in 1584 entered the Jesuit Order, subsequently leaving for health reasons. In 1590 he taught theology at Louvain. He was a skilled defender of controversial Roman Catholicism. Among his works is a translation of Bede's *History of the Church of England* (1565), a life of Thomas More* (1588), and a discourse appended to his translation of Staphylus's *Apologie ... of holy Scripture* (1565). In five million Latin and one million English words, Stapleton warned the English about the folly of abandoning Bede's faith. Like More, he was a persistent opponent of the Reformation.

MARVIN W. ANDERSON

STARETZ (Starets). A spiritual counselor in the Russian Church in the eighteenth and nineteenth centuries. Not surprisingly in view of the traditional aridity of church establishments, they were not professional appointees in the accepted sense. Originally those in this category were usually monks ministering to their fellow-monks, but gradually there was developed also a ministry to the laity, and the *startsy* were frequently the object of pilgrimages. Many heard of them for the first time in Dostoevsky's *The Brothers Karamazov.*

STATES OF THE CHURCH, see PAPAL STATES

STATIONS OF THE CROSS. A series of pictures or carvings depicting fourteen incidents in the last journey of Christ before His burial. They are usually placed around the walls of churches and used for popular devotions during Lent or Holy Week. The congregation goes from one picture to the next, led by the priest, recalling Christ's last hours before His death. The fourteen incidents are: (1) Pilate condemns Jesus to death; (2) Christ receives His cross; (3) Christ falls to the ground; (4) Christ meets His mother; (5) Simon of Cyrene takes the cross; (6) Christ's face is wiped by Veronica; (7) Christ falls a second time; (8) Christ tells the women of Jerusalem not to weep for Him; (9) Christ falls a third time; (10) Christ is stripped of His garments; (11) Christ is nailed to the cross; (12) Christ dies on the cross; (13) Christ's body is taken down from the cross; (14) Christ's body is placed in the tomb. The keeping of the Stations of the Cross was instituted by the Franciscans, but the form of service was not finally settled until the nineteenth century. Many Anglican churches as well as the Roman Catholic Church use this Lenten devotional service.

PETER TOON

STAUPITZ, JOHANNES VON (1460/69-1524). Roman Catholic scholar. Born in Motterwitz, Saxony, he studied at Leipzig and Cologne. He joined the Order of Hermits of St. Augustine and completed his studies at Tübingen (Th.D., 1500). He was prior in Munich and then professor of Bible at the University of Wittenberg. He became vicar-general of the Reformed Congregation of the Hermits of St. Augustine. Staupitz encouraged Martin Luther to study for his doctorate in theology. He tried to modify Luther's position in the first years of the Lutheran movement, but failed. In 1521 he received permission to join the Benedictines in Salzburg. In his theology Staupitz spoke of a covenant between God and man in which God set forth the terms, fulfilled them in Christ, and offered them to the elect unconditionally. He stressed the doctrine of election as central to soteriology. Grace in justification makes God pleasing to man, emphasizing the advent of Christ in grace.

CARL S. MEYER

STEBBINS, GEORGE COLES (1846-1945). American gospel hymnwriter. Born in East Carlton, New York, he studied music at Rochester, Chicago, and Boston. He was music director at Chicago's First Baptist Church from 1868, and in 1874 took up a similar post at Boston's Clarendon Street Church, and later at Tremont Temple. He was a lifelong acquaintance of D.L. Moody, Ira Sankey, P.P. Bliss, and D.W. Whittle. For nearly fifty years he led choirs, wrote music, and worked as an evangelistic music director. Stebbins produced over 1500 hymns, and was coauthor of several hymnbooks. Of his many popular gospel hymns one of the best known is "Take time to be holy."

ROBERT C. NEWMAN

STEINER, RUDOLF. see ANTHROPOSOPHY

STEPHEN (d. c. A.D. 36). Proto-martyr. As early Christianity emerged, according to Acts 6–7, the earliest response to the protest by Hellenists against discriminatory practices of the Hebrews saw the Twelve summoning the whole body of disciples and picking seven from the former group to perform the requisite ministry of service. Stephen headed the list. His vigorous ministry brought him into debate with the officials of the larger Judaism, from which the earliest Christian community was not yet distinguished. This led to his execution near Jerusalem (at least four sites are competitive) by stoning, in the presence of Saul as consenting bystander. Form-critical analysis of the summary of Stephen's theology (Acts 7:2-53) shows it stands in the tradition of "credo" passages wherein the proclamation of the activity of God is described by a recitation of that sacred history which gave structure to the OT narrative. Punning his name, Eusebius identifies him as first to win the crown *(stephanos)* reserved for martyrs; Sozomen's fragmented termination breaks off with the discovery of his relics presumably during the administration of Theodosius II (408-450), thereby intensifying that cult which keeps "the feast of Stephen" (26 December).

CLYDE CURRY SMITH

STEPHEN I (d.257). Pope from 254. A native Roman, he had a brief pontificate which saw several confrontations with Cyprian of Carthage* who held that "no one of us sets up to be bishop of bishops." Stephen restored two Spanish bishops, Basilides and Martialis, who had been deposed and replaced; refused to depose the bishop of Arles for Novatianism; and did not insist on rebaptism where the rite had been performed by heretics, so long as it had been done in the name of the Trinity. On the latter point Cyprian and eighty-seven of his colleagues differed, denying the validity of heretical baptism, at the Council of Carthage (256). J.D. DOUGLAS

STEPHEN II (III) (d.757). Pope from 752. A native Roman, he was chosen to succeed Zacharias as pope, in place of Stephen II who was elected but died before he could be consecrated. At his accession he appealed in vain to Constantinople for aid against a Lombard threat. After fruitless negotiations with the Lombard king, he secured protection from Pepin the Short (754). The papacy was now allied with the Franks rather than the Eastern emperor. Pepin compelled the Lombards to restore lands confiscated from the exarchate of Ravenna and from Rome. In a second campaign (756) Pepin forced the Lombard king to surrender to Stephen these and other territories, which were the foundation of the Papal States*; this "Donation of Pepin" established the papacy as a temporal power. ALBERT H. FREUNDT, JR.

STEPHEN III (IV) (d.772). Pope from 768. A Sicilian monk, he was elected in opposition to Constantine II and the antipope Philip, both the creatures of factions. Philip was forced to resign, Constantine was degraded and blinded, and terrible reprisals were inflicted upon their supporters. Charlemagne and Carloman, the Frankish rulers, sent several bishops to a Lateran council (769) that confirmed Stephen's election, excluded laymen from papal elections, decreed life imprisonment for Constantine and nullified his irregular election and ordinations, and condemned the iconoclastic synod of 754. Lack of unity among his Frankish allies forced Stephen to reach an agreement with the Lombard king that sacrificed the anti-Lombard party which had elected him. This weak and vacillating pope was regarded as a disgrace to the Roman church.

ALBERT H. FREUNDT., JR.

STEPHEN HARDING, see HARDING, STEPHEN

STEPHEN OF HUNGARY (c.975-1038). First king of Hungary. Baptized as a boy at the same time as his father, Duke Geza, by Adalbert of Prague,* he married Gisela, sister of the emperor. After succeeding to his father's dukedom he received a royal crown from Pope Sylvester II,* and in 1001 was made the first king of Hungary. He worked hard to convert his people to Christianity, founding both episcopal sees and monasteries. Unfortunately his last years were inglorious due to personal ill-health and quarrels about the succession to the throne. He was canonized in 1083, as was also his son Emeric (Imre) who was killed in a hunting accident.

STERN, HENRY AARON (1820-1885). Missionary to the Jews. Born to Jewish parents in Hesse-Cassel, he was educated at Frankfurt. At seventeen he began a commercial career in Hamburg, but became interested in Christianity, and he was baptized in London in 1840. He trained as a missionary in the College of the London Jews' Society. In 1844 he sailed for Baghdad and en route was ordained deacon by Bishop Alexander in Jerusalem. He worked among Jews and Muslims in Asia Minor and Persia until 1853, when he was transferred to Constantinople. In 1858-59 he went on missionary journeys to the Crimea and Arabia, and then joined J.M. Flad* in his work among the Falasha Jews of Ethiopia. Two years later, twenty-two were baptized, representing the firstfruits of the work. Later he incurred the hostility of the eccentric King Theodore, and he was imprisoned and tortured (1863-67), together with Flad and the consul and all other Europeans resident in the capital. Eventually Flad was released and brought the news to the British authorities, and an expeditionary force under Sir Robert Napier defeated Theodore, who committed suicide. Stern and the others were liberated and returned to England. The remaining years of his active ministry were spent in London, where he wrote and distributed literature among the Jews and became famous for his missionary sermons in Spitalfields and Whitechapel. Among his books were two on Ethiopia published in 1862 and 1868. J.G.G. NORMAN

STERNE, LAURENCE (1713-1768). English writer. He is rarely thought of as a theologian, even though he was for many years rector of Coxwold and prebendary of York. His fame rests on the whimsical, idiosyncratic, and even occasionally bawdy *Tristram Shandy* (1760-67), but in the same years he also published several volumes of the *Sermons of Mr Yorick.* Full though they are of unacknowledged borrowings from earlier writers, these sermons are as characteristically Sternean as anything he ever wrote. His biographer, Cross, described the best of them as "embryonic dramas," and so they are with their vivid account of incident, depiction of character and presentation of dialogue, and their unpredictable manipulation of language. In his preoccupation with conduct Sterne does not get much beyond eighteenth-century orthodoxy, but he is alone in presenting this with such original, versatile, and penetrating insights. Sterne's sermons are literature, because they are imaginative creations.

ARTHUR POLLARD

STERNHOLD, THOMAS (c.1500-1549). Versifier of Psalms. Educated at Cardinal College (Christ Church), Oxford, he became groom of the robes to Henry VIII, who left him a bequest of one hundred marks, and to Edward VI. He was a member of Parliament for Plymouth from 1545 to 1547. He composed metrical versions of the Psalms "for his own godly solace"—the precursor of the English Psalter. The first edition of nineteen Psalms appeared in 1547, dedicated to Edward VI, and a second edition of thirty-seven posthumously in 1549. In 1557 a third edition with seven further Psalms by John Hopkins, a Suffolk clergyman (d.1570), became well known as the "Sternhold and Hopkins" collection. Sternhold used mainly the ballad meter of "Chevy Chace," which helped to popularize psalm-singing in Elizabethan times. J.G.G. NORMAN

STERRY, PETER (d.1672). Chaplain to Oliver Cromwell.* Educated at Emmanuel College, Cambridge, he became a fellow there in 1636. During the 1640s he was a chaplain to Lady Brooke, a member of the Westminster Assembly* of Divines, and an occasional preacher before Parliament. In 1649 he became a regular preacher to the council of state of the Commonwealth, and also acted as a chaplain to the Lord Protector. After the latter's death he moved to Hackney where he taught some students in his home. Following the Restoration of Charles II, he was known as a Nonconformist and preached at conventicles. This period saw also some of his best literary work—e.g., *A Discourse of the Freedom of the Will,* published posthumously in 1675. He was a Calvinistic mystic, influenced by Neoplatonism and by such mystics as Jakob Boehme.*
PETER TOON

STIGMATA. Bodily wounds. This can be also a non-Christian phenomenon, but among Christians it dates back to later medieval times, despite references that have been made to Galatians 6:17. Stigmata were received on the hands, feet, side, shoulder, chest, or back and were reckoned to be a visible sign of participation in Christ's passion. Whether visible or invisible there is pain, sometimes accompanied by afflictions like lameness or blindness without logical causes, and nearly total abstinence from food and sleep. Stigmata are reported to resist treatment and to bleed periodically, especially during holy days and seasons, occur supernaturally and self-imposedly, and can represent evil as well as mystical contemplation. They spring from ecstasy, which can mean weakness; they can appear before or after a revelation, and to those of inferior piety and morality they are sometimes incomplete. Because of the connection with Christ's passion, there is even concern over their bodily position and shape. The Roman Catholic Church tends to be cautious over stigmata, and they have never been a reason for canonization. Francis of Assisi,* Catherine of Siena,* Teresa of Avila,* and Julian of Norwich* are well-known examples of those who have experienced stigmata, but the cases are numerous and involve especially women. The non-Roman Catholic traditions do not have this history, though they have known not dissimilar manifestations.
C.G. THORNE, JR.

STILLINGFLEET, EDWARD (1635-1699). Bishop of Worcester. Fellow of St. John's College, Cambridge, he gained rapid advancement through his great learning. He was preacher at Rolls Chapel and reader at the Temple Church, becoming a prebendary of St. Paul's in 1667 and dean in 1678. He wrote on the authority of Scripture and was an early advocate of the possibility of "comprehension" between Anglicans and Presbyterians. He also wrote antiquarian works, including *Origines Britannicae* in 1685 on the commencement of the British Church. He strongly defended the right of bishops to sit in the House of Lords, and was concerned with the reform of procedure of the consistory court. Frequently consulted by the bishops of his day, he was a close adviser of Archbishop Tillotson* on such controversial matters as the erroneous doctrines of Roman Catholics and Socinians. As one of the early Latitudinarians, he was appointed bishop of Worcester after the Revolution in 1689, where he proved himself an energetic and active pastor. G.C.B. DAVIES

STOCKHOLM CONFERENCE. An ecumenical gathering called the Universal Christian Conference on Life and Work which met in August 1925. Its moving spirit was Archbishop Söderblom.* Through World War I and afterward, Söderblom, a member of a neutral country, had sought to keep the ecumenical ideal alive. The conference dealt with the relationship between Christ and economics and industry, social and moral problems, and international relations and education. There were 600 delegates from thirty-seven countries, the latter figure surprisingly large since the war had left divisions which some thought could not be breached. There was some tension on the question of war guilt, but more concerning the question of whether the kingdom of God can or should be sought for here on earth. This was the first ecumenical conference when members were official delegates of churches and not just interested individuals. Particularly interesting was the presence of the Orthodox Church. Stockholm through its continuation committee may be seen as the initial step which led via Oxford* 1937 to the formation of the World Council of Churches* in 1948. PETER S. DAWES

STODDARD, SOLOMON (1643-1729). American Congregational pastor. Born in Massachusetts, he graduated from Harvard College and became its first librarian (1667). Ordained in 1672, he was pastor of the church at Northampton from 1672 until 1729. He was partially responsible for the formulation and defense of the Half-Way Covenant*; his apology for this practice was set out in his first major work, *Doctrine of Instituted Churches Explained and Proved from the Word of God* (1700). He moved further away from traditional Puritanism with his teaching that baptized persons should be admitted to the Lord's Supper even if they could not testify to a conversion experience. Stoddard argued that the Lord's Supper might act as a "converting influence." His other works include *An Appeal to the Learned* (1709), *Questions on the Conversion of the Indians* (1723), and *Safety in the Righteousness of Christ* (4th ed., 1792). Stoddard was the grandfather of Jonathan Edwards.*
DONALD M. LAKE

STOLBERG, FRIEDRICH LEOPOLD VON (1750-1819). Poet and diplomat. Born in Holstein, he was educated in a Lutheran Pietistic environment, thereafter reading law at Halle and at Göttingen, where he was a member of the *Hainbund,* a German poets' circle. Influenced by F.G. Klopstock* and himself a friend of Goethe,* he was envoy of the Protestant prince-bishop of Lübeck to the Danish court (1777), chief administrator at Eutin (1781), and Danish ambassador to Berlin (1789). His conversion to the Roman Catholic Church (1800) was widely noticed, and he resigned his posts. He wrote songs, plays, hymns, and odes, translated Homer, Plato, and Aeschylus, and produced *Geschichte der Religion Jesu Christi* (15 vols., 1806-18), which covered the period from OT times up to A.D. 430.

C.G. THORNE, JR.

STONE, BARTON WARREN (1772-1844). American frontier Presbyterian evangelist. Born in colonial Maryland, he crossed into Kentucky at the close of the American Revolution, with strong, pietistic reactions to war-induced vices. Essentially an Arminian revivalist, he broke with his Presbyterian heritage over unconditional election and limited atonement after the great Cane Ridge Meeting (1801), in which he was participant and recorder. He and five others set forth the *Last Will and Testament of the Springfield Presbytery* (1804), a declaration of biblical authority and the oneness of Christ's church. He organized the "Christian Church." His ecumenical outlook brought him into contact with many other "Christians" of "the Reformation of the Nineteenth Century," including especially Alexander Campbell,* with whose "Disciples" many of the Christians merged in 1832. Stone's *Address* (1814) and *Letters to Blythe* (1824) and the paper *The Christian Messenger* (1826) were molded by revivalism or the frontier experience—involvement in which moved him continually westward, despite successes in the Ohio Valley.

CLYDE CURRY SMITH

STONE, JOHN TIMOTHY (1868-1954). Presbyterian minister. Educated at Amherst College and Auburn Seminary, he held Presbyterian pastorates in New York State, Baltimore, and Fourth Presbyterian, Chicago (1909-30). His work in Chicago became the basis for the New Life Movement in the Presbyterian church nationally, while his own congregation showed remarkable growth. He served as chaplain in World War I. From 1928 to 1940 he was president of McCormick Seminary, Chicago, being made emeritus both there and at Fourth Church. He was moderator of his denomination's general assembly in 1913, and the author of devotional and practical books, notably *Winning Men* (1946).

C.G. THORNE, JR.

STONEHOUSE, NED BERNARD (1902-1962). NT scholar. Born at Grand Rapids, Michigan, he became a member of the Christian Reformed Church, and graduated in Arts from Calvin College in 1924. His theological studies were pursued first at Princeton, where in 1927 he received degrees in theology. Awarded the Alumni Fellowship in NT, he continued his researches at the University of Tübingen and the Free University of Amsterdam, which granted him its doctorate in 1929. Thereafter as instructor (1929), assistant professor (1930) and professor (1937) he served the NT department at Westminster Theological Seminary, Philadelphia, till his death. He was a gentle but penetrating scholar who gave particular attention to the synoptic gospels and to the Book of Revelation. Careful and prudent, he was tireless in his researches; his conclusions were widely respected because adequately based. His works include *The Witness of Matthew and Mark to Christ* (2nd ed., 1959); *The Witness of Luke to Christ* (1951); and *Origins of the Synoptic Gospels* (1963). He edited two posthumous volumes of the work of J. Gresham Machen* and wrote his biographical memoir (1954). Stonehouse was the founding editor of *The New International Commentary on the New Testament,* with which he was associated until his death.

PAUL WOOLLEY

STORCH, NICHOLAS, see ABECEDARIANS

STOWE, HARRIET ELIZABETH BEECHER (1811-1896). Abolitionist and author. Born in Litchfield, Connecticut, she studied and taught in Hartford. When Lyman Beecher,* her father, became president of Lane Theological Seminary, Cincinnati, in 1832, she went with him and married a professor, Calvin E. Stowe, in 1836. They sheltered fugitive slaves in their home until they moved in 1850 to Brunswick, Maine. *Uncle Tom's Cabin, or Life Among the Lowly* appeared in the magazine *National Era* in 1851-52, and as a book in the latter year. It evoked strong antislavery sentiment. For nearly thirty years she produced almost a book a year.

STRACHAN, JOHN (1778-1867). Educator and bishop. Born in Aberdeen, Scotland, he taught school while studying at the university there. In 1799 he emigrated to Canada and taught school in Kingston until 1803, when he was ordained in the Church of England and became curate of Cornwall. He was rector of St. James' Church, Toronto, from 1813 until 1867, during which period also he served for many years as a member of both the executive council and the legislative council of Upper Canada. He upheld the sole right of the Anglican Church to income from the Clergy Reserves.* From 1839 he was bishop of Toronto, and from 1827 the first president of King's College (University of Toronto) until he founded the University of Trinity College in 1851 and served as its first chancellor.

EARLE E. CAIRNS

STRACHAN, ROBERT KENNETH (1910-1965). Protestant missionary leader. Born in Buenos Aires, Argentina, son of Harry and Susan Strachan, British missionaries who later started the Latin America Evangelization Crusade (now Latin America Mission), he was educated in the United States, then joined his parents in the mission in Costa Rica in 1936. Following their death, he became general director in 1950. He developed a forward-looking team of workers in a multifaceted mission that came to include nationals in its membership. He was responsible for the beginning and the basic ideas of Evangelism-in-Depth.* His article on evangelism in the *International Review of Missions* in 1964 sparked a notable debate on that subject. In 1964-65 he was visiting professor of missions at Fuller Theological Seminary, where he helped establish the School of World Mission. His lectures there were published as *The Inescapable Calling.*

HAROLD R. COOK

STRANG, JAMES JESSE (1813-1856). Mormon* leader. Born of Baptist parents in Scipio, New York, he studied law and was admitted to the bar in 1836. He became interested in Mormonism through his wife's brother-in-law, Moses Smith, and was converted through Joseph and Hyrum Smith in 1844. When the former was killed, Strang claimed to be his successor, eventually forming the Mormon sect known by his name in St. James, Big Beaver Island, Lake Michigan, where he was crowned "king" in 1850. He was twice elected to the state legislature. He announced a revelation proclaiming plural marriage a divine institution (1850), himself taking four wives. He made many enemies and was finally assassinated.

J.G.G. NORMAN

STRAUSS, DAVID FRIEDRICH (1808-1874). German theologian. Born near Stuttgart, he studied under F.C. Baur* at Tübingen, where he obtained his doctorate and taught briefly. He achieved instant notoriety with his *Life of Jesus, Critically Examined* (2 vols., 1835-36; ET 3 vols., 1846). This destroyed for him all prospect of a career in theological teaching. The study consisted largely of a detailed examination of the events of the gospels, making extensive use of the concept of myth already known to German theology. Strauss admitted there was a basic historical framework behind the life of Jesus recorded in the gospels, but held that it had become so embellished and overlaid by pious reflection and fantasy that the life of Jesus had been mythically rewritten so as to make it repeat and fulfill the legends and prophecies of the OT. Thus the miracles of Jesus were virtually predetermined by popular expectation of how the Messiah should act. The true significance of Christianity is to be seen in the light of Hegelian philosophy. It is to be understood symbolically as the manifestation of the Absolute Spirit in man.

An enormous controversy followed. Strauss produced a sequel, *Christliche Glaubenslehre* (2 vols., 1840-41), arguing that biblical teaching cannot be harmonized with modern knowledge, and proposing a mixture of Platonic and Hegelian philosophy in its place. For the next twenty years he turned his back on theology, but returned with a study of *Hermann Samuel Reimarus* (1862) and a second life of Jesus (1864) which again ruled out the supernatural and miraculous and made considerable use of myth. But the Hegelianism of the first life was dropped in favor of the older rationalism of the Enlightenment.* A religion of humanity must supersede Christianity. Other writings include an attack on Schleiermacher* (1865) and a post-Darwinian statement of belief, *The Old and the New Faith* (1872; ET 1873).

In his day Strauss had more influence on freethinkers like George Eliot than on the mainstream of theology. His teaching on myth seems to have had little direct influence on Bultmann.*

See biographies by E. Zeller (1874) and T. Ziegler (2 vols., 1908); H. Harris, *David Friedrich Strauss and his Theology* (1973).

COLIN BROWN

STREETER, BURNETT HILLMAN (1874-1937). Biblical scholar. He was a member of Queen's College, Oxford, as student, fellow, and provost, during 1893-99 and 1905-37 (the intervening years being spent at Pembroke College, Oxford). In 1899 he was ordained deacon in the Church of England despite doubts about aspects of the Christian faith. A shy and retiring man, he nevertheless had great influence among students and won considerable respect from his colleagues. The liberal cast of his mind was shown by the fact that he was one of the seven contributors to *Foundations* (1912). In his latter years he came under the influence of Frank Buchman* and the Oxford Group,* and it was during his return from staying in Switzerland with members of the group that he was killed in an air crash. His works include *Reality: A New Correlation of Science and Religion* (1926) and *The Buddha and the Christ* (1926), but his most influential writings were on gospel criticism. His most famous work, still in wide use today, is *The Four Gospels: A Study of Origins* (1924). His origins were the "Four-Document Hypothesis," the "Proto-Luke Theory," and the demonstration of an early Caesarean text of the gospels.

R.E. NIXON

STRICT BAPTISTS. A group of Baptist churches, generally Calvinist in theology, which denies that saving faith is the duty of unbelievers. Saving faith is "not a legal duty, but the sovereign and gracious gift of God." Communion is restricted to "baptized believers" in the NT sense of those words, and the rule of life for believers is the Gospel, not the moral law. Early emphasis was placed on itinerant preaching, and recent efforts have failed to establish a more settled type of ministry. Strict Baptists withdrew their support, on theological grounds, from the Baptist Missionary Society, and founded in 1861 the Strict Baptist Mission. They maintain an aloofness from public affairs and show little interest in the social implications of the Christian faith.

JAMES TAYLOR

STRIGEL, VICTORINUS (1524-1569). German theologian. One of Melanchthon's* most distinguished associates, he was born in Kaufbeuren, Swabia, and after study at Freiburg and Wittenberg, taught at Magdeburg, Erfurt, and Jena. His biblical commentaries and historical and philological works showed him to be a wide-ranging scholar, in addition to being a notable theologian. Unfortunately he taught in territories wracked by bitter disputes among Lutherans on synergism and the Eucharist. Not willing to agree with definitions like the *Book of Confutation*, and sympathetic to Reformed theology, he found his position in Jena becoming more and more difficult because of his public disagreements with M. Flacius* and the inability of the duke of Saxony to reconcile the opposing factions. A move to Leipzig in 1562 brought a temporary remission, but by 1567 he was once again in difficulties and inhibited from teaching, because of suspected Calvinist views on the Lord's Supper. Appointment to a chair at Heidelberg in 1567 offered Strigel a congenial theological environment, but he died shortly after appointment. He was widely respected among Reformed theologians, including English scholars like W. Perkins.*

IAN BREWARD

STRONG, AUGUSTUS HOPKINS (1836-1921). American Baptist pastor and educator. Born in Rochester, New York, he graduated from Yale (1857) and Rochester Theological Seminary (1859) and studied at the University of Berlin. Ordained in 1861, he held pastorates in Massachusetts and Ohio. He was elected president of Rochester Theological Seminary (1872-1912) and was also professor of biblical theology. Although clearly conservative in theology, Strong was open to certain trends developing late in the nineteenth century, such as theistic evolution and German idealism. He was always active in the life of the American (Northern) Baptist denomination, serving as president of the American Baptist Missionary Union (1892-95) and first president of the Northern Baptist Convention (1905-10). He toured Baptist mission fields in 1916-17. His major writings include *Systematic Theology* (3 vols., 1886), *Philosophy and Religion* (1888), *The Great Poets and Their Theology* (1897), and *Christ in Creation and Ethical Monotheism* (1899).

DONALD M. LAKE

STRONG, PHILIP NIGEL WARRINGTON (1899-). Anglican primate in Australia. After service in World War I, he graduated from Cambridge in 1921 and was ordained the following year by Bishop Hensley Henson to serve in the diocese of Durham, in which he became in 1931 vicar of a Sunderland parish. Five years later he was consecrated as bishop of New Guinea, which in 1941 became the last land buffer between Australia and the Japanese forces. Bishop Strong, fifteen of his clergy, three laymen, and eighteen women missionaries chose to remain at their posts when the Japanese occupation came. The death in 1942 of eight missionaries and two Papuan Christians put a heavy burden on the heart and conscience of the bishop, but he devoted his energies unsparingly to the care of Papuan Christians and Australian soldiers alike, and was frequently exposed to great danger.

After the war Strong fostered the steady expansion of an increasingly indigenous church, and he was delighted in 1960 to consecrate a Papuan as assistant bishop: the first indigene anywhere in the Pacific to attain this office.

In 1962 Bishop Strong was elected archbishop of Brisbane, and in 1966 he became also primate of the Church of England in Australia, where his influence extended far beyond his own diocese, and was recognized when in 1970 he became a Knight Commander of the British Empire (KBE), a rare award for a churchman.

MARCUS LOANE

STROSSMAYER, JOSEPH GEORGE (1815-1905). Roman Catholic bishop. Born of German parents in Croatia, he was ordained to the priesthood in 1838 and nine years later became professor of canon law at Vienna. In 1850 he was elevated to the bishopric of Bosnien with its seat at Diakovár. At Vatican Council I* (1869-70) he opposed the promulgation of papal infallibility and then was the last bishop to publish the decrees of the council (December 1872). He persisted in maintaining relations with J.J.I. von Döllinger* and J.H. Reinkens* until October 1871. In spite of his German heritage, Strossmayer was an enthusiastic pan-Slavist, which brought him into conflict with Vienna. The pan-Slavic movement he advocated resulted in the formation of Yugoslavia after World War I. WAYNE DETZLER

STRYPE, JOHN (1643-1734). English church historian and biographer. Born at Houndsditch, London, he was educated at Cambridge and eventually became curate at Leyton, followed or accompanied from 1689 by a lectureship at Hackney and (1711) the sinecure of West Tarring, Sussex. A poor historian and worse stylist, he was a great and an unscrupulous collector of documents. His principal contribution is the publication of these sources in the valuable *Annals of the Reformation* ... (2nd ed., 4 vols., 1725-31) and *Ecclesiastical Memorials* (4 vols., 1721-33), and also in numerous appendices in biographies, e.g., of Archbishops Cranmer (1694), Grindal (1710), Parker (1711), and Whitgift (1718), and (especially worthwhile) of the Edwardian humanists Sir Thomas Smith (1689) and Sir John Cheke (1705).

BRIAN G. ARMSTRONG

STUBBS, JOHN (1543?-1591). Protestant controversialist. Born in Norfolk and educated at Trinity College, Cambridge, and Lincoln's Inn, he then returned to Norfolk to live on the family estate. He abhorred Roman Catholicism and viewed any compromise with it as dangerous. A proposed marriage between Queen Elizabeth and the Catholic Henry, duke of Anjou, was the occasion for his *Discoverie of a gaping gulf whereinto England is like to be swallowed* ... (1579). For this publication his publisher and his printer and he himself were arrested, and after much legal discussion, Stubbs and William Page, the publisher, had their right hands cut off (1679). Stubbs protested his loyalty to the queen, but was still sent back to prison, to be released a few months later. After his release he continued to defend Protestantism, and he also translated a book of Theodore Beza* on the Psalms. He was elected member of Parliament for Yarmouth in 1589 and died at Le Havre when on a visit to France. PETER TOON

STUBBS, WILLIAM (1825-1901). English historian and bishop. Born at Knaresborough, he passed from Ripon Grammar School to Christ Church, Oxford, in 1844. In 1848 he was elected a fellow of Trinity, and from 1850 to 1866 he was vicar of Navestock. Between 1864 and 1889 he produced his remarkable editions of English Medieval Chronicles, which made him the outstanding historian of his time, and the person who has laid the foundations of the modern approach to the study of medieval history. In 1866 he was appointed regius professor of modern history at Oxford, and in 1870 he published *Select Charters*—the history of the English constitution up to Edward I from original documents. This was followed (1873-78) by the *Constitutional History of England* (to 1485), a work which gave him a worldwide reputation. He became bishop of Chester in 1884 and was translated to Oxford in 1889. Theologically he was a High Churchman and called Pusey "the Master"; politically he was Conservative. JOHN A. SIMPSON

STUDD, C(HARLES) T(HOMAS) (1862-1931). Pioneer missionary. Third son of Edward Studd, a wealthy retired planter who was converted under D.L. Moody in 1877, he was himself converted in 1878. Educated at Eton and Trinity, Cambridge, he excelled at cricket and was in the England team in 1882. He volunteered for missionary service, and as one of a group of students known as the "Cambridge Seven" aroused much enthusiasm on missions in Edinburgh and elsewhere. He sailed for China with the China Inland Mission in 1885 and gave away his inherited fortune to Christian causes. Invalided home in 1894, he was working two years later among students in America, and the Student Volunteer Missionary Union was formed. He was pastor of the Union Church, Ootacumund, South India, in 1900-1906, until forced home again through illness. After a period of preaching around Britain, contrary to medical advice he sailed in 1910 for Africa, where he founded the "Heart of Africa Mission" in 1912, which later became the Worldwide Evangelization Crusade. He labored with Alfred Buxton and others in Central Africa until his death. J.G.G. NORMAN

STUDENT CHRISTIAN MOVEMENT. The British section of the World Student Christian Federation,* founded by John R. Mott* in 1895. Each national movement preserves its autonomy as a fellowship of students who "desire to understand the Christian faith and live the Christian life." The SCM was the product of several student movements in the latter part of the nineteenth century, of which the Student Volunteer Missionary Union was of special importance. This had begun in Cambridge in 1892 through the inspiration of the "Cambridge Seven." With a growing desire to include High Church and liberal representatives, the SCM gradually moved away from its Evangelical origins, till the Cambridge Inter-Collegiate Christian Union withdrew from membership of the movement on doctrinal grounds in 1910. Its desire for "comprehensiveness" included a determination to "shake itself free" from the conservative approach to the Bible.

In evolving the principle of being an "interdenominational" movement rather than "nondenominational," each denomination making its special contribution, the SCM paved the way for the modern ecumenical movement, many of whose leaders sprang from its ranks—J.H. Oldham,* William Temple,* N. Söderblom.* Through conferences and study groups it has sought to grapple with the relevance of Christianity to the contemporary world. Besides work in universities and among theological students, the movement operates also in schools. In 1929 the Student Christian Movement Press Ltd. was set up, primarily to supply students with literature at low cost, which has now grown into a substantial publishing house. Through the WSCF international contacts are maintained and encouraged.

See T. Tatlow, *The Story of the Student Christian Movement of Great Britain and Ireland* (1933). J.W. CHARLEY

STUDENT VOLUNTEER MOVEMENT. A movement dedicated to enlisting Christian college students for foreign missions. It originated in a summer Bible study conference in 1886 at Mount Hermon, Massachusetts, called by the YMCA and presided over by D.L. Moody. Students like Robert P. Wilder* aroused concern for missions, and before the conference ended one hundred had signified their intention of becoming missionaries. The following school year, Wilder and another toured the schools to stimulate more interest. In 1888 the movement organized formally, with John R. Mott,* one of the original hundred, as chairman, a position he held for thirty years. The often misunderstood motto became "The evangelization of the world in this generation." In 1891 came the first general convention in Cleveland, Ohio. After 1894 conventions were held quadrennially, reaching a peak at Des Moines, Iowa, in 1920, when 6,890 attended. Then came rapid decline. Before 1940 the SVM had lost its effectiveness. A brief resurgence after World War II was short-lived. In 1959 the organization was merged into the National Student Christian Federation which in 1966 became part of the University Christian Movement. In 1969 the UCM voted itself out of existence. During its lifetime the SVM saw more than 20,000 of its members become foreign missionaries.

HAROLD R. COOK

STUNDISTS. Russian evangelical sects tracing their origin to a group of Bible students in SW Russia about 1845. A certain Reformed pastor Bohnekämper conducted pietistic devotional hours (*Stunden*) for Russian peasants as well as German settlers. Under his son Karl the religious movement called "Stundism" arose about 1862, freed itself from all connection with the Reformed Church, and became purely Russian in character. Despite persecution by church and state, Stundism spread widely. The majority gradually linked up with Russian Baptists, being called "Stundo-Baptists" in distinction from the Baptist group originating from J.G. Oncken. The Stundists proper sought to maintain the original connection with the *Stundenhalter* among the Swabian colonists. J.G.G. NORMAN

STURM, JAKOB (1498-1553). Reformer and statesman. Born at Strasbourg, he was of a family that had given the city able magistrates for two centuries. He studied at Heidelberg and Freiburg with Capito and Eck. He joined the faculty at Freiburg in 1503 and later became a city councillor. He was appointed chief magistrate (*Stettmeister*) of Strasbourg in 1526, representing the city ninety-one times in the government of the empire. An early adherent of Reformation doctrine, he advocated an alliance of all German and Swiss evangelical groups, supported in this by Martin Bucer.* He was one of the original "Protestants" at the Diet of Speyer* (1529), took part in the Marburg Colloquy,* and presented the Tetrapolitan Confession* to the Augsburg Diet (1530). Through his influence Strasbourg joined the Smalcald League* (1531), but after the disasters of the Smalcaldic War had to sue for the pardon of his city from Charles V. He stood for liberty of conscience in church matters and was respected by all parties. J.G.G. NORMAN

STURM, JOHANNES (1507-1589). Protestant educationalist. Born at Schleiden (Sleida), he was educated at a school of the Brethren of the Common Life* and at Louvain University. Joining himself to the French humanists, he lectured on classics at Paris (1530-36). Bucer* influenced him to Protestantism, and he went to Strasbourg, actively furthering the Reformation. Conciliatory in spirit, he sought the reconciliation of all religious parties, including Protestants and Roman Catholics. He reorganized the educational system of Strasbourg, founding the gymnasium on a humanistic model (1538) with himself as rector. An academy followed in 1564. Generally regarded as the greatest educator of the Reformation, he was consulted by Calvin and by Thomas Platter of Basle. His ideas greatly influenced German education—and also the Jesuit educational system. He was expelled from Strasbourg in 1581 for his liberalism and extraconfessional sympathies, but was eventually permitted to return. His many writings include a life of Beatus Rhenanus.

J.G.G. NORMAN

STYLITE (Gr. *stulos,* "pillar"). An ascetic who lived permanently on the top of a natural or artificial pillar. Usually a kind of hut or platform was placed on the top of the stone in order to give protection from the weather. Food and basic requirements were normally provided by admiring disciples. Apart from the solemn duties of prayer and fasting, stylites were often gifted preachers and theologians, addressing the crowds which gathered at the foot of their pillars and pronouncing on current theological controversy. The traditional founder of this form of the religious life was Simeon the Stylite.* It was through his example that others sought to become hermits and stylites, so that in the Near East this form of asceticism was fairly common until the tenth century.

PETER TOON

SUÁREZ, FRANCISCO DE (1548-1617). Spanish Jesuit philosopher and theologian. Probably the greatest of the sixteenth- and seventeenth-century Scholastics, and perhaps even the greatest Jesuit theologian, he was born at Granada, studied canon law at Salamanca (1561-64), but after joining the Jesuits turned to theology and philosophy (1565-71). He was ordained in 1572 after he had already begun a lifelong career of teaching in Spanish universities (except for a five-year period at the Roman College from 1580). His last and longest appointment was as professor primarius at Coimbra (1597-1615).

Suárez published widely and made original contributions in legal and political theory, philosophy, and theology. His political and juridical doctrine is found primarily in *De legibus* (1612) and the polemical *Defensio fidei . . . adversus anglicanae sectae errores* (1613). With Francisco de Vitoria he helped lay the basis for international law, positing a natural community of nations whose relations are regulated by the "law of nations" *(jus gentium)*—a sort of natural/public law for the international community. His political doctrine is based on man's natural rights and on the idea that the people form the basis of political authority. He thus repudiated James I's* theory of divine right.* He held that the pope may depose for reasons of heresy (as a matter of Christian law), but may not violate a nation's natural rights.

His *Metaphysical Disputations* (1597?) combined Aristotelian and Thomistic logic with Scotist objections, producing a philosophical text widely used in Protestant and Catholic schools alike throughout the seventeenth century, and creating the system sometimes called "Suárezianism" or "Suarism." In theology, though basically a commentator on Aquinas's *Summa,* he contributed to the bitter Jesuit-Dominican dispute over the role and efficacy of grace on the side of the Jesuits, proposing a system known as "Congruism"—that God disposes an individual to salvation by giving congruent graces which by foreknowledge He sees will be useful in a given situation.

BIBLIOGRAPHY: *Opera* (23 vols., 1740-51) and with additional material (28 vols., 1856-78); L. Mahieu, *Francois Suarez, sa philosophie et les rapports qu'elle a avec sa théologie* (1921); J.H. Fichter, *Man of Spain* (1940); J. Mullavey, *Suárez on Human Freedom* (1950); B. Hamilton, *Political Thought in Sixteenth-Century Spain* (1963).

BRIAN G. ARMSTRONG

SUBLAPSARIANISM. This term refers to the position taken by one group of Calvinist theologians as the development of Calvinist scholasticism in the later sixteenth and the seventeenth centuries brought up the question of the precise purpose of predestination.* The opposing view was Supralapsarianism.* The Sublapsarian position held that God, in His decree of predestination, had as its object mankind-as-fallen. That is, loosely stated, that God created man with the possibility of the Fall, which happened, and then elected some men to salvation, leaving the rest in enmity with God. At issue is the logical order of the decrees, not the chronological (since God, as eternal, is outside time). The issue was thus notably difficult and abstruse, and Calvinist assemblies refused to make either position binding. Theologians concerned with working out a complete and logical dogmatic system tended toward Supralapsarianism. As the Sublapsarians pointed out, this position ran the danger of making God the author of sin. DIRK JELLEMA

SUBMERSION, see BAPTISM

SUBORDINATIONISM. An early, anti-Trinitarian, widely diffused sub-Christian Christology. One form of the doctrine concerned the origin of the preexistent Logos.* Most Christians rejected the Gnostic idea of intermediate beings, but that Christ is a divine being somewhat below the highest divine principle and that He derives His existence from it appealed to some, especially Origen.* Some see Subordinationist tendencies in Justin Martyr,* Irenaeus,* and Clement of Alexandria.* The fourth-century Arians (see ARIANISM) moved the christological issue back to the preincarnate origin of the Logos. Today, Jehovah's Witnesses* assign to Jesus Christ a preincarnate, derived existence.

Another form centered upon the man Jesus. He was a unique Galilean, perhaps sinless but still only a man, who was divinely endued (with the Christ) at his baptism for a special mission. The Ebionites,* Cerinthians, (see CERINTHUS) and Paul of Samosata* held similar views. The teachings condemned in 1 John are probably those of Cerinthus. The Trinitarian form of Subordinationism is "Dynamic Monarchianism." More recent Subordinationist Christologies are those of John Knox of New York and Norman Pittenger. The church has resolutely rejected christological reductionism in favor of the apostolic doctrine that Jesus Christ is the eternal Son of God made flesh.

See also INCARNATION; MONARCHIANISM; TRINITY. SAMUEL J. MIKOLASKI

SUBUNISTS. Name given to the party in Bohemia in the fifteenth century which defended the practice of Communion in one kind against the Utraquists* (Calixtines), according to the practice of the medieval church. The name derives from the Latin *sub una specie* ("under one kind"). With the Calixtines they received ecclesiastical recognition at the Prague Compacts (1433).

SUFFRAGAN. As the Latin etymology suggests, the concept is related to the vote and those processes involving its casting (*suffrag-*), with implications of favor or support. The late Roman Empire from the time of Constantine already had the problem of financial patronage for votes, and while often uprooted, the practice as often revived. The church likewise knew the practice; in the negative sense, the suffragan was that bishop whose vote could be counted upon by his metropolitan in synod. On the positive side, in the English Church from the thirteenth century the term could identify any auxiliary bishop who did not have the right of succession, in contrast to the *coadjutor*—though usage does not always bear out the distinction. CLYDE CURRY SMITH

SUICER, JOHANN KASPAR (1620-1684). Swiss Reformed theologian. Born at Frauenfeld and educated at the French academies of Montauban and Saumur, he taught Latin, Greek, and Hebrew at Zurich from 1644, and was professor of Greek at the *Collegium Carolinum* (1660-83). After publishing several works on Greek linguistics, he brought out his magnum opus, *Thesaurus Ecclesiasticus e Patribus Graecis Ordine Alphabetico* (2 vols., 1682). This work shows immense erudition and extensive reading of patristic literature, and is still indispensable for students of the vocabulary of the Greek Fathers and of Greek ecclesiastical institutions (2nd enlarged ed., 1728). His son, Johann Heinrich (1646-1705), succeeded him at the *Collegium Carolinum* in 1683. J.G.G. NORMAN

SUICIDE. The act of taking one's own life. Traditional Christian teaching has consistently regarded suicide as a crime, and in this it has the support of most religious and moral codes. Some societies have tolerated suicide, but even these have attempted to limit it to certain categories, of which the religious suicides of Japan and the courageous suicides of the Greco-Roman world are examples. Greek and Roman philosophers were divided on the subject. It was condemned by Plato, Aristotle, and Cicero, but regarded as a reasonable exercise of freedom by the Epicureans and Stoics, especially Seneca. The latter view found support in the work of Thomas More* and John Donne,* and also in that of Voltaire,* Montesquieu, and Hume.* More recently it has been defended as permissible or even virtuous on the grounds that a man's life is his own and that in the last resort he must be allowed to terminate it at his own discretion, or more narrowly, that suicide is justifiable in cases of extreme senility or painful and incurable disease.

The traditional Christian view has no direct support from the Bible, in which various cases of suicide are mentioned but without reference to a penalty and without condemnation of it—perhaps because the biblical emphasis is on the joyous acceptance of life as a gift from God. It was formulated by Augustine* and other early Fathers of the Church. Augustine condemned suicide on the grounds that it was self-murder; that it precluded any opportunity for repentance; and that it was a cowardly act. These views found expression in church law, by which suicide was denounced by a series of councils and was elaborated by Aquinas* and other Scholastics during the Middle Ages. Aquinas condemned suicide as being contrary to the natural law, to man's natural inclinations, and to a proper self-love. He held that man has no right to deprive society of his presence and activity or to reject the gift of life given him by God.

In the twentieth century, increasing attention has been paid to the psychopathology and sociology of suicide, and this has led to certain modifications of the Christian view. It now seems clear that suicide is probably only rarely a carefully premeditated act, but is more often the result of mental illness, of an overwhelming sense of failure or rejection, of loneliness, or of the loss of status and income during severe economic depression.

BIBLIOGRAPHY: In addition to Augustine's *City of God* I and Aquinas's *Summa Theologica* II-II, Q.65, art. 5, see G. Williams, *The Sanctity of Life and the Criminal Law* (1958); N. St. John-Stevas, *Death and the Law* (1961); S.E. Sprott, *The English Debate on Suicide from Donne to Hume* (1961). OONAGH MC DONALD

SULPICIANS. The Society of St.-Sulpice was organized in 1642 by J.J. Olier* in the parish of St.-Sulpice in Paris. Sulpicians are secular priests whose principal task is the theological education of parish clergy. As secular priests, Sulpicians take no special vows and are permitted to hold private property. Nevertheless they are expected to use their property in the service of Christ and to own it as though they did not have it. Theological training has never meant for the Sulpicians merely instruction in Scripture and dogmatics without spiritual formation and devotion to prayer and ascetism. To encourage the spiritual formation of their students, Sulpicians live a common life with them and share their spiritual exercises. In the past Sulpicians were strongly Thomist in their theology, though their spirituality followed the methods developed by their third superior-general, L. Tronson (1676-1700).

The society spread to Canada in 1657, even before it received papal sanction and its own constitutions. While suffering a partial eclipse during the French Revolution, it was restored during the reign of Napoleon. In 1791, under J.A. Emery, the order founded St. Mary's Seminary in Baltimore, the oldest Roman Catholic seminary in the USA, now a pontifical university. The society has numbered among its members many distinguished theologians, such as the French Church historian P. Pourrat and the American biblical scholar Raymond Brown.

See L. Bertrand, *Bibliothèque Sulpicienne, ou Histoire littéraire de la compagnie de Saint-Sulpice* (3 vols., 1900) and C.G. Herbermann, *The Sulpicians in the United States* (1916).

DAVID C. STEINMETZ

SULPICIUS SEVERUS, see SEVERUS, SULPICIUS

SUMMA (Lat. *summa*, "the totality"). A treatise giving a summary of the essence of a subject; in medieval times a compendium, used in the Schools as a textbook, of philosophy, theology, or canon law. The most famous *summae* are the *Summa Theologica* and *Summa contra Gentiles* of Thomas Aquinas,* but there were many others —e.g., the *Summa Creaturis* of Albertus Magnus.* Usually in the *summae* the subject under discussion is expounded by means of first stating, and then answering by means of the dialectical method, a series of questions. The *Summae* replaced the *Sentences* (e.g., those of Peter Lombard*) from the time of Aquinas onward.

SUMNER, CHARLES RICHARD (1790-1874). Bishop of Winchester. Brother of J.B. Sumner,* he was educated at Eton and Trinity College, Cambridge, ordained in 1814, and was made a royal chaplain by George IV who had a high regard for him. In 1826 he was consecrated bishop of Llandaff, which post he held in conjunction with the deanery of St. Paul's, and in 1827 was translated to Winchester. Although a convinced Evangelical, he voted for the 1829 Catholic Emancipation Bill (against the king's wishes) and later regretted it. In 1850 he strongly protested against the restoration of the Roman Catholic hierarchy. Sumner established new churches and poor schools in his diocese and improved the lot of agricultural laborers. Among his writings are the *Ministerial Character of Christ Practically Considered* (1824) and an edition of Milton's *De Doctrina Christiana.* He resigned his see in 1869. J.D. DOUGLAS

SUMNER, JOHN BIRD (1780-1862). Archbishop of Canterbury from 1848. Educated at Eton and King's College, Cambridge, he taught briefly at Eton, was ordained in 1803, became a canon of Durham (1820), and in 1828 was consecrated bishop of Chester, a see he held for twenty years before going to Canterbury. A strong Evangelical like his brother, C.R. Sumner,* he nevertheless like him voted for the Catholic Emancipation Bill in 1829. Later he opposed the Oxford Movement* and denied that baptismal regeneration was a fundamental doctrine of the Church of England. His published works include *Apostolical Preaching* (1815), *A Treatise on the Records of the Creation* (2 vols., 1816), and *The Evidence of Christianity* (1824). J.D. DOUGLAS

SUNDAR SINGH, SADHU (1889-c.1929). Indian Christian and mystic. The youngest of four children of wealthy Sikh parents in Rampur, North Punjab, he was deeply attached to his mother and was much distressed when she died in 1902. For a time he attended an American Presbyterian Mission school, but he was bitterly opposed to Christianity and publicly burned a copy of the gospels. Two days later he had a vision of Christ and was converted. Driven from home by his father, he became a preacher, wearing the saffron robe of a *Sadhu* ("holy man") in an endeavor to evangelize the Hindus. In 1905 he was baptized into the Church of England, but later refused to be restricted to a particular denomination. He traveled widely in Asia and visited the West, but was saddened by the love of comfort and luxury evident there. Despite ill-health, he persisted in evangelizing Tibet and disappeared there in 1929. J.G.G. NORMAN

SUNDAY. The primitive church in Palestine was almost entirely Jewish and as such continued Sabbath observance; it was a social necessity. In the Diaspora, Jewish Christians continued the practice as long as they preserved their Jewish identity, but Gentile Christians normally did not unless they accepted circumcision under the pressure of Judaizers. To be noted is that Sabbath observance was regarded as a specifically Jewish privilege and hence was not one of the Noachic commandments, the keeping of which was a prerequisite for social relationships between Jews and Gentiles, and so was not demanded by the council in Jerusalem (Acts 15:28f.). Paul mentions the Sabbath only once directly (Col. 2:16) and twice obliquely (Rom. 14:5f.; Gal. 4:10), thus showing how little practical importance the question had in his day. The social conditions of many Gentile converts, especially slaves, made Sabbath-keeping impossible, and provided a powerful motivation against its observance.

Most Jewish Christians continued to attend the synagogue until forced out—a process effectively carried through by the *birkat ha-minum* (about A.D. 90). This made the church's most natural time for the Lord's Supper Saturday evening, i.e., the beginning of Sunday, as seems to be the case in Acts 20:7. When under Trajan these evening gatherings apparently became illegal, the Supper was moved to the early Sunday morning. This move cut the last links with the Sabbath and made the connection with Christ's resurrection as a justification for specifically Sunday worship—something that must have been there from the first—virtually self-evident. The change from a natural tendency to a fixed rule will have been gradual but swift. The Quartodeciman* controversy shows that as late as the end of the second century it was not self-evident to all Christians that Easter had to fall on Sunday. 1 Corinthians 16:2 does not refer to a church gathering. While "the Lord's day" (Rev. 1:10) is probably Sunday (is Revelation from the time of Nero or Domitian?), it is unprovable. Sunday is called "the Lord's day of the Lord" in *Didache** 14 and "the Lord's day" by Ignatius.* For Justin Martyr* the Sunday service is standard. From then on, the term "the Lord's day" rapidly became the norm. Melito* of Sardis (d. c.190) wrote a thesis on *The Lord's Day.*

No evidence for the equating of Sabbath and Sunday is found before the end of the third century, but by that time there was an increasing stress on the true, i.e., spiritual, observance of the Sabbath, and it was, at least in theory, observed as a day of worship alongside Sunday. Emperor Constantine in 321 issued an edict requiring "rest on the venerable day of the sun" by the cessation of public works and the closing of the law courts, but agricultural labor was expressly excepted. From then on we find a growing stress on the necessity of Sunday rest, but the reason given is that men should be free to attend worship, not that Sunday is the "Christian Sabbath"—a phrase not found until the twelfth century. This stress on worship is the Roman Catholic and Orthodox position today.

The earlier Reformers—e.g., Luther, Zwingli, Calvin, Tyndale, Cranmer, Knox—insisted on the value of Sunday as a day of rest and worship, but refused to regard it as the Christian fulfillment of the Sabbath. In Britain, but not Europe, a rigorist reaction by the Puritans set in. Nicholas Bownde gave it classical expression in 1593 (see bibliography). Scotland adopted Sabbatarian legislation already in 1579; England followed suit after the Puritan triumph in the Civil War, and the legislation was only slightly relaxed under Charles II. Similar laws were enforced in most of the New England states in America. The Evangelical Revival and the growth of middle-class respectability led to a stiffening of these laws in Britain, and to the growth of Sabbatarian groups in Europe where, however, they never exercised much influence.

In Britain, especially in England, the observance of Sunday, both religiously and legally, has been steadily eroded since the mid-nineteenth century. The main influences have been the growth of the large towns; the steady rise of the working class, in large measure alienated from organized religion; the growth of public utilities demanding Sunday work for their functioning; radio and television; and two major world wars. The church is rapidly returning to the position it found itself in during the first two centuries as far as Sunday observance is concerned.

See also SABBATH; SABBATARIANISM.

BIBLIOGRAPHY: N. Bownde (or Bound), *Sabbathum Veteris et Novi Testamenti: or the True Doctrine of the Sabbath* (1593; enlarged 1606); J.A. Hessey, *Sunday, its Origin, History and Present Obligation* (1860); T. Zahn, *Geschichte des Sonntags vornehmlich in der alten Kirche* (1878); W. Rordorf, *Sunday* (1968).

H.L. ELLISON

SUNDAY, WILLIAM ASHLEY ("Billy") (1862-1935). American evangelist. A professional baseball player who worked between seasons as a fireman on the Chicago and Northwestern Railroad, he experienced an evangelical conversion through the Pacific Garden Mission (1886). After serving as assistant secretary of the religious department of Chicago YMCA (1891-93), he helped J. Wilbur Chapman* in mass evangelism for two years. From 1896 he worked independently, combining superb organization with sensational preaching. His campaigns, held in many American cities, were conducted in huge wooden tabernacles.

Because of the organizational methods, the impact of his campaigns was citywide. The "Sunday Party," a group of at least twenty experts, was responsible for such matters as advance planning, publicity, music, specialized work among businessmen, businesswomen, students, and so on. In addition to the central meetings, sectional meetings were held throughout the city, and "delegations" from various sectional interests were arranged. Thousands of church members were recruited to assist the running of the campaign, and the churches were closed for its duration. Sunday developed a preaching style which combined crude humor with florid rhetoric. He nevertheless played down emotionalism, and called for a down-to-earth commitment to Christ. Of the 100 millions who heard him preach, one million are said to have "hit the trail." Strongly fundamentalist in theology, he opposed evolution and advocated temperance. HAROLD H. ROWDON

SUNDAY SCHOOLS. The beginnings of an organized movement are usually dated from 1780 when Robert Raikes,* a Gloucester journalist, established a small school to care for local slum children who were neglected and illiterate. He wrote an article about his work which caught people's imagination and encouraged the setting up of Sunday schools throughout England. The first schools taught reading and writing as well as Scripture. The movement spread to the Continent and to America, where the First Day Society was established in Philadelphia in 1790. Raikes saw the culmination of his efforts when the Sunday School Union was founded in 1803 and received great support in evangelical circles. When general education became widespread in the nineteenth century, the Sunday schools concentrated increasingly on religious education. Most of the classes in Britain and the USA have been conducted by voluntary teachers with no special training for this type of work.

See also EDUCATION, CHRISTIAN.

J.D. DOUGLAS

SUPERINTENDENT. A literal translation of the Greek work *episkopos* usually rendered into English by "bishop." In the OT it was applied to the seventy delegated to assist Moses, and later to other officials during the period of the monarchy. In the NT this term is also employed, but usually is equated with "elder" (Acts 20:28; 1 Pet. 2:25). At the time of the Reformation the Lutheran churches in Germany and Scandinavia tended to use the term in place of "bishop," which had become distasteful to Protestants by virtue of its association with the great prelates of the Roman Catholic Church. After the Saxon visitation of 1527 they were instituted as state officials, but soon they came under the control of the consistory, a supervisory body with no very clearly defined powers.

The superintendent usually had authority over the clergy and congregations within his province, inducted pastors to their parishes, had general oversight of discipline, particularly excommunication, and acted as the church's administrative officer. In some areas there were general superintendents over other superintendents. In recent years the office has changed considerably. Attempts have been made to prove that the superintendents established in Scotland by the first *Book of Discipline* were of the same type, and while the title may have been taken over from the Lutheran churches, there were important differences, i.e., the office was temporary, the superintendents were directly responsible to the general assembly, and the superintendent could act in a number of instances only with the concurrence of the local ministers and church sessions. The office was finally abolished as the presbyteries became established.

W.S. REID

SUPRALAPSARIANISM. Particularly applied to the position taken by some Calvinist theologians as the development of Calvinist scholasticism after the mid-sixteenth century brought to the fore the knotty problem of the precise meaning of predestination.* Theologians such as Beza,* eager to have a fully worked out and internally consistent dogmatic, tended to be Supralapsarian (as against Sublapsarian*). At issue was the question of the logical order of God's actions in predestination (not the chronological, for God, as eternal, is outside time), and the problem was thus notably abstruse. The supra position held that God created mankind with the original idea in mind that some would be saved and some would not be; and then allowed the Fall, to bring about this intention. This seemed to imply, as the Sublapsarians pointed out, that God willed the Fall and was thus the author of sin. This the Supralapsarians denied, accusing their opponents of weakening God's sovereignty and indirectly holding to man's free will. Calvinist assemblies refused to support either position as binding. Thus, at the Synod of Dort,* Gomarus's attempt to have the supra position upheld failed; in the Swiss churches, later attempts to have the supra position condemned failed (cf. Formula of Consensus, 1675).

DIRK JELLEMA

SUPREMACY, ACT OF (1534). This piece of legislation passed by the English Parliament during the reign of Henry VIII* declared the king to be "the only supreme head on earth of the Church of England." Although it was repealed during the reign of the Roman Catholic queen Mary Tudor,* it was restored under Elizabeth I* in 1559, except that the reference now was to "supreme governor." This is the basis on which diocesan bishops are still appointed by the sovereign on the advice of the prime minister.

SUPREME COURT DECISIONS: RELIGIOUS LIBERTY. Religious liberty in the United States is a fundamental principle of the American constitutional system at both state and federal level. It is a tangible expression in law of the view laid down by Roger Williams,* Puritan "Seeker," that Christianity, being a life of personal faith, has no need of political support. This was expressed in part by various acts of religious toleration in the English colonies. The disestablishment of the Anglican Church in the State of Virginia (1786) laid the groundwork for the clause in the new Federal Constitution (1789) which forbade a religious test for public office.

Religious liberty is guaranteed by the First Amendment to the Federal Constitution which forbids Congress to legislate "respecting an establishment of religion, or prohibiting the free exercise thereof." This provision has been extended by the Supreme Court, chief arbiter of constitutional questions, to state governments under the provisions of the Fourteenth Amendment placing definite limitations upon state control of the civil rights of American citizens.

The "free exercise" clause received its first major test in *Reynolds* v. *United States* (1879). The court upheld a federal statute outlawing polygamy by denying its practice under constitutional guarantees of religious freedom. The court admitted that it could not pass upon the validity of religious beliefs because of "a wall of separation between Church and State" laid down by the First Amendment. It declared, however, that men's actions were subject to legal regulation in the interest of public welfare. Later the court upheld the power of Congress to annul the charter of the Mormon* Church for violation of federal law, and denied that Congress was creating "an establishment of religion" by barring the franchise to Mormons practicing polygamy.

Subsequent rulings involving religious liberty issues employed the public-welfare principle to justify a grant of federal funds to a nonsectarian hospital corporation, whose services were maintained by a Catholic sisterhood, and the purchase of textbooks by state funds for parochial school pupils.

More recently the "free exercise" clause has been invoked in cases involving Sunday closing laws, religious tests for public office, and the orthodoxy of religious tenets. The court continues to deny jurisdiction in matters of belief. It supports Sunday closing laws on public-welfare grounds, and protects individuals against loss of employment benefits because their religious convictions prohibit work on Sunday or Saturday.

To claim exemption from compulsory military service under the "free exercise" clause was long recognized by the court as legitimate, provided the applicant was a member of a church or sect whose tenets denied the morality of war. In 1970 the court modified this position by accepting "deeply held moral, ethical, or religious beliefs" as a basis for such exemption.

In *Cantwell* v. *Connecticut* (1940), the first of the Jehovah's Witnesses* cases, the First Amendment guarantees of religious liberty were extended to a state government under the restraining clauses of the Fourteenth Amendment. There followed a number of cases involving the right of this sect to propagate its views through the distribution and sale of handbills and religious books, and the right to claim exemption for their children from saluting the flag in the schools. The power of municipal corporations to tax vendors of such literature, the propriety of enforcing child labor laws in regulating the use of children in the distribution of religious propaganda, and objection to the verbal content of recorded messages became related issues.

The court early ruled that a state could require the flag salute as a demonstration of loyalty, but reversed its position in a succeeding opinion. In most instances the court considered the guarantees of the First Amendment, including religious liberty, as occupying a privileged position in American rights. Child labor laws, however, could be enforced, objectionable language was not protected, religious parades could be restricted by police regulations, and vendors could be limited when their presence in public or private premises became a recognizable public nuisance.

The court was unprepared for the controversy which developed over its opinion approving the busing of parochial school students at public expense. Justice Black, speaking for the majority in *Everson* v. *Board of Education* (1947), reaffirmed the "wall of separation" doctrine and denied that the use of public money for such purpose was "an establishment of religion." Rather, it was the use of tax power for the general welfare. Justice Jackson and others found this to be an important step toward breaking down the "wall of separation," even though the money was paid directly to the parents and not to the school. The court appeared to contradict itself the next year in *McCollum* v. *Board of Education* when it outlawed the use of public school facilities for "released time" (time devoted during the school day) religious instruction under church auspices. Such use, said the court, involved an "establishment of religion." Four years later it approved "released time" instruction when offered in other than school facilities.

Congressional approval of direct financial aid to private and public schools in 1965 alarmed those who feared that the establishment clause had been violated. Parochial aid statutes implementing these grants were adopted in New York, Pennsylvania, and Rhode Island. The first, requiring school boards to provide textbooks to all children at public expense, was approved by the court in 1968. The Pennsylvania statute, granting financial aid to parochial schools for instruction in nonreligious subjects, was invalidated in 1971, as was the Rhode Island attempt to supplement the salaries of parochial school teachers. The court declared that aid thus granted must have a secular legislative purpose, must neither advance or inhibit religion, and must not foster "an excessive government entanglement with religion." Champions of church and state separation were unhappy with a simultaneous decision approving construction grants to colleges and universities, regardless of their affiliation with religious bodies, under the Higher Education Facilities Act. The court justified its opinion on the ground that there was less danger of religious influence in higher education and that the facilities constructed would be conspicuously neutral.

Whatever controversy was engendered by the financial support issue appears minimal when compared with the public furor over decisions outlawing prayer and Bible reading in the public schools. In *Engel* v. *Vitale* (1962) the court held that the voluntary recitation of a prayer composed by representatives of the leading faiths and approved for use in the schools by the state was a step toward "an establishment of religion" and must be disallowed. Similarly, the court found that a Pennsylvania statute requiring the reading of ten Bible verses daily in the schools constituted state support of religion. Three successive efforts were made up to 1974 to secure congressional approval for an amendment to the First Amendment that would permit students to engage in voluntary prayer on school premises. All failed.

S. RICHEY KAMM

SURIN, JEAN JOSEPH (1600-1665). French mystic. He was born in Bordeaux and studied in the Jesuit college there, and at *Collège de Clermont*, Paris, entering the order in 1616. At Cardinal Richelieu's request he went to Loudun (1636) to be the exorcist in a community of Ursuline nuns diabolically possessed. As a result, he believed himself to be possessed, which devastating experience persisted for twenty years, finally advancing his own spirituality as a mystic. He was accused of Quietism and of overstressing the extraordinary in the mystical life. The Italian translation of his *Catéchisme spirituel* was put on the Index, though Bossuet defended his orthodoxy and Fénelon esteemed him. *Les Fondements de la vie spirituelle* (1667), *Dialogues spirituels* (1704-

9), and other works emphasize purification through suffering and self-denial.

C.G. THORNE, JR.

SURPLICE. A liturgical garment, medieval and monastic in origin. The term is a corruption of the Latin *superpelliceum,* meaning "to be worn over the pelisse or fur gown." In the unheated churches of N Europe it was necessary to wear a fur-lined gown, and the difficulty of getting the more primitive, tight-sleeved albe over this garment led to the evolution of the larger surplice. By the fourteenth century it was the essential choir-vestment everywhere, but it was never worn by the celebrant at the Eucharist. The 1552 English Prayer Book retained it as the only vestment, and under Elizabeth I the Puritans strongly objected to its use as "papistical." Parker's Advertisements (1566) and the 1604 Canons ordered its use for all church services. In the Revised Canons, it is one of the permitted forms of eucharistic vesture, and customary for all other services. The modern surplice is shorter and less full than the medieval.

JOHN A. SIMPSON

SUSO, HENRY, see HENRY SUSO

SUVERMERIAN. A term used to describe the eucharistic doctrine held by Archbishop Cranmer* between 1549 and 1555. Charles Smyth in his *Cranmer and the Reformation under Edward VI* (1926) reintroduced this term of a theology when in fact Saxon theologians latinized *Schwarmer* as *swermeros* to apply to theologians. More recently, Horton Davies has substituted the term "Virtualism" to link Cranmer with Bucer, Bullinger, Calvin, and Peter Martyr.

SVERDRUP, GEORG (1848-1907). American church leader. Of Norwegian origin, he studied theology in Oslo and at German universities. In 1874 he was called to America as a professor of theology at Augsburg Seminary, Minneapolis. Norwegian immigrants to America were mainly Lutherans, but by theological and ecclesiastical persuasion they were divided in several organized church bodies. By family tradition Sverdrup had a liberal political outlook, and he did not believe in an authoritarian church system such as the one he had known in Norway.* He had accepted the ideal of "living" Christianity, and in his new country he felt convinced that this would prosper only in the frame of a free church with lay preaching and independent congregations. He became therefore the champion of the free-church ideal. The Norwegian Lutheran Church was founded in 1897 and organized according to Sverdrup's ideas. A six-volume edition of his works was published in 1909-12.

CARL FR. WISLOFF

SWAINSON, CHARLES ANTHONY (1820-1887). Anglican theologian. Son of a Liverpool merchant, he was educated at Trinity College, Cambridge, from 1837. Later he became a fellow of Christ's College (1841), principal of Chichester Theological College (1854), Norrisian professor of divinity at Cambridge (1864), master of Christ's College (1881), and vice-chancellor of the university (1885). He was an authority on church creeds and liturgies; he traveled widely to view early manuscripts and published several works on the subject. His book, *The Greek Liturgies* (1884), is still used. Generous, unselfish, and devoted to the practical work of the ministry despite his high academic standing, he reckoned himself a disciple of Hooker* and the older English divines.

R.E.D. CLARK

SWEDEN. Politics and religion joined hands in early missionary enterprises; when Louis the Pious sought further domains, the missionary he chose was Anskar* (801-65). Failing in Denmark,* the latter turned to Sweden. King Bjorn gave permission to preach and build a church, the first in Scandinavia. Few Swedes responded, and the work faded for a time, but in the latter tenth century Christianity became established with bishops. Early in the eleventh century, King Olof Skotonung was baptized and established an archbishopric at Skara (1020). Later, under Svenkers (1130-55), paganism was overcome with the help of Cistercian monks from England and Germany, one of whom—Stephan—became first archbishop of Uppsala. Soon the Swedish bishops became subordinate to Rome.

For Sweden, as elsewhere, the Reformation was bound up with political reaction to foreign power —in this case, Denmark. Christian II attempted to subdue Sweden and murdered eighty of its leaders in 1520 ("The Stockholm Blood Bath"). The young Gustavus Vasa* gathered a peasant army, drove the Danes out, and was crowned king in 1528. All sympathizers with Christian fled, including Archbishop Tolle. Denuded of ecclesiastical leadership and finance, Vasa appropriated the possessions and revenues of the church.

Lutheranism came through Olavus Petri* (1480-1552), who taught at the cathedral school in Strangnas, becoming friendly with the archdeacon, Lars Anderson (1450-1552). Vasa met both men and invited them to Stockholm, Petri as preacher, Anderson as chancellor. Rome disapproved, whereupon Vasa applied for the consecration of four bishops-elect, adding that Sweden was unable to pay the customary annates. If the request were denied, the bishops would be consecrated by "Christ the only and highest pontiff." Two bishops who still protested their loyalty to Rome performed the consecration, thus maintaining the "apostolic succession." The rift with Rome was complete; 1527 saw the first national Protestant church established, in Sweden. Reforms included abolition of compulsory confessions, clerical celibacy, and preaching aided by a Swedish Bible as an important part of worship.

Lars Anderson, Olavus Petri, and his brother Lars Petri (1499-1573), later archbishop of Uppsala, played a significant part in the reforms. Olavus prepared a Swedish hymnal and a Swedish Mass, in all of which Luther was a prime influence. They saw that the principle of the Peace of Augsburg,* that rulers could determine religion, was reversed in Sweden, where faith was decided by the people. Restricting Vasa's power caused a rift between king and bishops, but his plan to restrict episcopal power was defeated by his death.

Thereafter there were many attempts to change church structure: Eric XIV introduced Calvinism; John III sought a rapprochement with Rome; Charles IX introduced Calvinism again. All failed, and the country remained Lutheran.

In 1638 commercial and evangelical interests in the Indians produced a Swedish colony in Delaware, and until 1791 the Church of Sweden continued to send clergy and finance. After 1815 and the humiliations of the Napoleonic Wars, there was a reaction against eighteenth-century rationalism, aided by Pietists and Moravians. Two theological trends appeared: Lund stressed the Church and became the center of High Church tendencies, Uppsala stressed a subjective and philosophical view of Christianity, discounting the Church. Furthermore, through the work of George Scott, an English Methodist, Bible and tract work developed; Sunday school and foreign missions were concentrated on, and the result was the Swedish Mission Covenant in 1878, bringing together most free churches.

With the religious revival, a church council, the *Kyrkomote*, came into being in 1863. Lay organizations and evangelism grew. In 1894 the World Student Christian Movement was inaugurated in Sweden. The Evangelical National Institute, formed in 1856, became a leading missionary movement. Nevertheless the effects of the nineteenth-century theological debates in Europe were evident; ideologies such as Marxism were being accepted; and noticeably in industry men were forsaking the faith. Having said that, it remains true that Protestantism was more vigorous in 1914 than in 1815.

Sweden made a distinguished contribution to the ecumenical movement through Nathan Söderblom.* Among those who reacted to nineteenth-century liberalism were the scholars Gustav Aulén* and Anders Nygren.*

BIBLIOGRAPHY: J. Wordsworth, *The National Church of Sweden* (1911); C.H. Robinson, *Anskar, Apostle of the North* (1921); H.M. Waddams, *The Swedish Church* (1946); B. Sundkler, *Nathan Söderblom: His Life and Work* (1969).

GORDON A. CATHERALL

SWEDENBORG, EMANUEL (1688-1772). Swedish scientist, philosopher, and theologian. Born in Stockholm, he was the son of a minister who was later appointed bishop of Skara. Emanuel was a keen student who studied the classics and Cartesian philosophy at Uppsala and became interested in mathematics and natural sciences. In 1709 he went abroad to study languages and mechanics at London, Oxford, Amsterdam, and Paris. After his return to Sweden (1715), he was appointed an assessor of the royal Board of Mines, a post he held until 1747 when he resigned to study the Scriptures. In 1719 he was made a noble by Queen Ulrika Eleanora and assumed the name "Swedenborg." He described his mining and engineering accomplishments in a large work, *Opera Philosophica et Minerazia* (3 vols., 1733), which explained the origin of the universe in a mechanical manner. Swedenborg abandoned the materialist view, however, and the next year he published *A Philosophical Argument on the Infinite and Final Cause of Creaton,* which followed the Neoplatonic teachings of the seventeenth-century Protestant mystics.

From this time he applied himself to discovering the nature of the soul and spirit by means of anatomical studies. For this purpose he studied at Paris, Venice, and Rome (1736-39) and published his results in the *Economy of the Animal Kingdom* (1739). Here he developed his doctrine of series and degrees which states that the soul must descend into matter by four degrees. Swedenborg experienced strange dreams and visions which increased in frequency after 1739. This led to a profound spiritual crisis (1743-45) relieved by a vision of Jesus Christ which he felt confirmed his interpretation of Christianity. Thereafter he spent the rest of his life expounding the ideas of the true Christian religion—in reality a Neoplatonic philosophy which admitted the historical Jesus Christ.

His first exegetical work was the *Heavenly Secrets* (8 vols., 1749-56), followed by many others, including *The True Christian Religion* (1771). In these works characterized by vivid descriptions of his experiences in the spiritual world, Swedenborg teaches the existence of spirits and angels, denies the Trinity and vicarious atonement, and describes God the invisible spaceless timeless One as manifesting himself on earth as Jesus Christ (the soul being the eternal Father, the body the son of Mary, and the Holy Spirit the action caused by the union of two). He held that man's spirit lives after death according to its earthly justification—good men gather in heaven, and the selfish seek their kind in hell. He also believed that the churches over the years had destroyed the original meaning of God's word and the Swedenborgian mission was to restore its primary sense. This message marked the transition in 1757 to a new age foretold in Scripture by the statements about the return of Messiah and the foundation of the New Jerusalem. Swedenborg did not try to win converts, however, but confined himself to publishing his revelations. His influence has been considerable, particularly on the Romantic movement and psychical science. In 1787 his religious followers organized into a group known as the New Church, or New Jerusalem Church.*

BIBLIOGRAPHY: J.J.G. Hyde, *A Bibliography of the Works of Emanuel Swedenborg, Original and Translated* (1906); M. Lamm, *Swedenborg* (1915); E. Benz, *Emanuel Swedenborg: Naturforscher und Seher* (1948); S. Toksvig, *Emanuel Swedenborg: Scientist and Mystic* (1948); C.S.L.O. Sigstedt, *The Swedenborg Epic* (1953); G. Trobridge, *Swedenborg: Life and Teaching* (4th ed., 1955). ROBERT G. CLOUSE

SWEELINCK, JAN PIETERSZON (1562-1621). Dutch composer. From about 1580 until his death he was organist of the *Oude Kerk* in Amsterdam. It is not clear that he ever went to Italy, but he was certainly greatly influenced by Zarlino, the great Italian theorist, for he left a theoretical work based on the Venetian's work. Sweelinck was much influenced by the English keyboard masters such as John Bull. In turn he was the chief figure behind the N German school of organists, a number of whom studied with him—notably Scheidt. He wrote many fine motets with Latin text, which show a growing sense for major and minor tonality rather than the older modes. Best known is his brilliant setting of the popular Christmas text "*Hodie Christus natus est*," a perennial favorite with cappella choirs. Some of his finest work is in his four volumes of motetlike settings of the Genevan Psalms, employing the original French texts. In view of the Calvinistic disapproval of polyphonic music, these were written not for church but for a *collegium musicum* of Dutch musical amateurs. Sweelinck was the last great composer to use extensively the Genevan Psalms.

J.B. MAC MILLAN

SWEET, WILLIAM WARREN (1881-1959). Methodist scholar. Born in Baldwin, Kansas, he was educated at Ohio Wesleyan University, Drew Theological Seminary, and Crozer Theological Seminary. After five years in the ministry he earned a Ph.D. at the University of Pennsylvania; his dissertation, published as *The Methodist Episcopal Church and the Civil War* (1912), initiated that prolific publication that was to make him "the dean of the historians of Christianity in America." He taught at Ohio Wesleyan (1911-13), DePauw University (1913-27), and the Divinity School of the University of Chicago (1927-46). His fundamental concern was to give such a reputation to church history that secular historians could no longer ignore its role. He influenced both the writing of general American history by calling attention to those often neglected "civilizing and cultural forces" of religion, and that of denominational histories by broadening individual examples to see their place within the total development of America and its peculiar form of Christianity. His many works include *Religion on the American Frontier* (1931-46), *The Story of Religion in America* (1930; 2nd rev. ed., 1950), and *Religion in Colonial America* (1942).

CLYDE CURRY SMITH

SWETE, HENRY BARCLAY (1835-1917). Anglican scholar. A clergyman's son, he was born in Bristol, educated at King's College, London, and at Cambridge, and after ordination served a number of parishes before becoming professor of pastoral theology in King's College, London (1882-90). He edited the Latin text of Theodore of Mopsuestia's commentary on Pauline epistles (1880-82) and issued *The Old Testament in Greek* (3 vols., 1887-91). In 1890 he became regius professor of divinity at Cambridge, from which post he retired in 1915. The Cambridge years were productive of various topics in liturgy and theology, and of the origination of the lexical project for patristic Greek (1906; published 1961-68) which emerged from his studies of various aspects of early Christian history. Apart from the Septuagint his Greek texts edited and annotated included the gospel of Mark (1898) and the Book of Revelation (1906). He was instrumental in founding the *Journal of Theological Studies* (1899), and the publication with J.H. Srawley of *The Cambridge Handbooks of Liturgical Study* (1910). He edited the *Gospel of Peter* from a newly discovered fragment (1893) and wrote major studies of the Holy Spirit in the NT (1909) and in the early church (1912).

CLYDE CURRY SMITH

SWIFT, JONATHAN (1667-1745). Irish satirist and clergyman. Born in Dublin, he was educated there at Trinity College. After a period as secretary to Sir William Temple, he took orders and became prebendary of Kilroot in Ireland. There he wrote *A Tale of a Tub* and *The Battle of the Books*, two of his wittiest satires. He received some ecclesiastical preferment, but his real interest in the first decade of the new century lay in London and politics, where he attached himself in particular to the Tories under Harley and Bolingbroke. After the Hanoverian succession in 1714 he withdrew from politics and spent the rest of his life as dean of St. Patrick's, Dublin, suffering periodically and ultimately continually from painful mental disorder. His greatest work is *Gulliver's Travels* with its ingenious but also very gloomy view of humanity. Swift had little faith in his fellowmen and immense capacity for showing it. Together these qualities make him one of the most powerful satirists in any literature. His prose is forthright and straightforward, and these same characteristics are to be found in his sermons with their short sentences, logical progression, and calm assurance.

ARTHUR POLLARD

SWISS GUARD. Formed in 1506 by Pope Julius II, it is entrusted with the personal protection of the pontiff. Members of the corps must be Swiss in nationality, Roman Catholic in faith, and eligible for service in the Swiss army. There are about sixty guards in all, and since the disbanding of three other corps of papal guards in 1970 they remain the only military order in the Vatican. The Guard retains its colorful Renaissance costumes, and although members are trained in the use of modern weapons they carry only the halberd and broadsword.

SWITHIN (Swithun) (d.862). Bishop of Winchester. Probably a secular clerk, he advised Egbert, king of Wessex, and taught his son Ethelwulf. Consecrated bishop on Ethelwulf's succession, he became one of his chief counselors. By his own wish his body was buried outside the cathedral's north wall. It was moved inside in 971, giving rise to the legend connecting forty days' rain with St. Swithin's Day (15 July). It was moved again in 1093. He was canonized by popular tradition, and his shrine was destroyed by Henry VIII.

SYLLABUS OF ERRORS. A list of eighty propositions condemning the doctrines of liberalism attached to the encyclical *Quanta Cura* issued by Pius IX* on 8 December 1864. The first impulse toward the drawing up of the Syllabus came from the Provincial Council of Spoleto in 1849 when Gioacchino Vincenza Pecci, bishop of Perugia (later Leo XIII*) requested a condemnation of modern errors. He wanted to bring together under the form of a constitution the chief errors of the time. Preparation of the Syllabus began in 1852 and continued over a period of twelve years. In 1860 O.P. Gerbet issued a Pastoral Instruction in which he listed eighty-five errors. This list, which became the basis of the Syllabus, was modified into sixty-one theses and was approved by an assembly of bishops at Rome in 1862. The final stage of preparation began with the appointment of a new commission by Pius IX which incorporated thirty of the approved sixty-one theses in its formulation of the eighty errors to be condemned. The wording of the errors was taken from the earlier official declarations of Pius IX. A reference was added to each of the eighty theses to indicate its content so as to determine the true meaning and theological value of the subjects treated.

The Syllabus was arranged under ten headings: Pantheism, Naturalism, and Absolute Rationalism; Moderate Rationalism; Indifferentism and False Tolerance in Religious matters; Socialism, Communism, Secret Societies, Bible Societies, and Liberal Clerical Associations; the Church and its Rights; the State and its Relation to the Church; Natural and Christian Ethics; Christian Marriage; Temporal Power of the Pope; and Modern Liberalism.

The pope's enemies saw in the Syllabus a formal rejection of modern culture and a declaration on the modern state. Belgians objected to it on the ground that it infringed constitutional rights. On 1 January 1865, publication of the Syllabus and encyclical was forbidden in France, although the prohibition was later withdrawn. In France and Germany it was seen as creating a cleavage between the church and the modern world. The Syllabus was a blow to liberalism. Roman Catholics saw the intellectual movement of the nineteenth century as a threat to the foundations of human and divine order in the world. They regarded the Syllabus as a necessary attempt to stem this tide which was undermining the influence of the Catholic Church on the life of nations and individuals. S. TOON

SYLVESTER II (Gerbert) (c.945-1003). Pope from 999. One of the important leaders of the intellectual revival of the late tenth century, he was born in Auvergne, was educated at the Benedictine monastery as Aurillac, and later studied in Spain. He became a teacher at the cathedral school at Reims and was noted for his knowledge of mathematics and natural science. He may have introduced Arabic numerals to W Europe, and he is credited with the invention of the pendulum clock. In addition, he was an avid and skilled letter writer, and his surviving letters are important historical sources. In 983 he aided Empress Theophano in securing the German crown for her son, Otto III. He also supported the election of Hugh Capet, the first Capetian king of France, in 996. In the following year he went to Otto III's court to become his teacher, and through Otto's influence he was made archbishop of Ravenna in 998, and pope in 999.

He was the first Frenchman to become pope, and he headed the church at a time when it was plagued by serious corruption. He attacked the outstanding evils of the day by fighting secular control of ecclesiastical appointments, simony, and nepotism. He was influential in the spread of Christianity in E Europe. He granted Poland its first archbishopric in 1000, and established an archbishopric and new bishoprics in Hungary. He was also skilled in philosophy, and wrote a philosophical tract, *De Rationali et de ratione uti.* A treatise on the Eucharist, *De Corpore et Sanguine Christi,* has also been attributed to him.

RUDOLPH HEINZE

SYLVESTRINES. A minor monastic order following a rather Benedictine rule, stressing poverty, founded by Sylvester Gozzolini (1177-1267) in 1231. Centered in Italy except for a few houses in Portugal, Brazil, and Ceylon (where they have had a large mission since 1855), they have also a convent of nuns. The habit is blue, the abbot-general resides in the mother-house in Rome, and the constitutions of 1690 obtain, following the short-lived union with the Vallumbrosans* (1662-80).

SYMMACHUS. This second-century Ebionite* is remembered primarily for his translation of the Hebrew OT into Greek. The translation was readable, but it lacked verbal accuracy and thus the extant fragments of his work have been of little value to textual critics, although Origen* did incorporate Symmachus's work into his *Hexapla.* He was particularly offended at the many anthropomorphic expressions in the OT and promptly removed them. He is referred to in Eusebius's *Ecclesiastical History.*

SYMMACHUS (d.514). Pope from 498. A double papal election followed when Pope Anastasius II died in 498. Division and schism marked the whole pontificate of Symmachus. Anastasius II had alienated many by accepting repentant schismatics. In 483 Pope Simplicius had bound his successors against the alienation of church property. At Anastasius's death each party elected a pope. One party decided on the intractable Laurentius and the other on Symmachus. There were riots and street battles as general disorder prevailed. Theodoric the Ostrogothic king intervened in behalf of Symmachus. Laurentius forced Symmachus to appear at Ravenna to clear his name. Theodoric chose a council of bishops who met at Rome in May 500. When Symmachus refused to appear, the bishops would not condemn him *in absentia.* Theodoric forced them to remain until the deadlock was resolved in October 501. The dating of Easter* was one point of dissension between Theodoric and Symmachus. The council broke up, giving "absolution by default." Laurentius reappeared in support of Theodoric. Symmachus called a synod which declared in November 502 that his see was beyond man's judgment. The decree of 483 was null and void. Not until Theodoric withdrew support of Laurentius in 507 was Symmachus undisturbed.

The Symmachan Forgeries appeared when Ennodius* of Milan argued that only God could judge the bishops of Rome. The popular works which support the decision of 501 purported to be the Acts of Sinuessa under Pope Marcellinus, the *Constitution* of Pope Sylvester I, the *Gesta* of Pope Liberius,* and certain acts clearing Pope Sixtus III. As incorporated in the *Liber Pontificalis,** these forgeries became history. The fourth-century *Liberian Catalogue** of Roman bishops was the primary authority for the first compiler of this papal biography written during the pontificate of Symmachus. Both sides wrote similar documents, evidence that papal schism could become a stimulus to propaganda.

MARVIN W. ANDERSON

SYNAGOGUE, see JUDAISM

SYNCELLUS. This Byzantine concept is a Greek-Latin hybrid: Greek *kella* from Latin *cella,* found in inscriptions from the second century A.D., identified a room or chamber especially for storage of wine, papyri, etc., whence it came to designate the "cell" of a monk. The chamber-mate, *synkellos,* known from the fifth century, was the ecclesiastic who shared living quarters with another. In the case the latter was a ranking prelate, the duties of the former were those of domestic chaplain, but it appears he functioned often as little more than ecclesiastical spy. Yet his association could make him an obvious choice for succession.

SYNCELLUS, GEORGE, see GEORGE SYNCELLUS

SYNCRETISM. The mixture of various systems of thought; a union of opposites on the basis of what they hold in common. Syncretism is to be distinguished from eclecticism* in that the latter results in a new system, whereas the elements in a syncretizing process still retain their old character. Syncretism can also be thought of as an attempt to distinguish between essential and non- or less-essential elements in a religion or philosophy. More specifically, the name is given to a movement in the Lutheran Church in the seventeenth century led by individuals such as Georg Calixtus* in the direction of interconfessional union coupled with a protest against what was regarded as dogmatic rigidity in Lutheran and Reformed churches. The interconfessional and, sometimes, interfaith emphasis of the modern ecumenical movement bespeaks the strong syncretist current in it.

PAUL HELM

SYNERGISM. The view that the will of man cooperates with the action of divine grace by having an independent part to play in conversion. As such, synergism is Semi-Pelagian* in character, denying the efficacy of divine grace and man's spiritual "death." Synergism came into prominence in second- and third-generation Lutheranism as a reaction against the strongly monergistic, Augustinian emphasis of Luther* himself ("Free will determined without grace has no power with respect to righteousness, but is necessarily involved in sin"). Melanchthon,* in his later period, taught the universality of divine grace and forbade further investigation into the divine and human factors in conversion. He spoke of "the Word, the Holy Spirit, and the will, not absolutely inert, but struggling against its own infirmity" as the "three concurrent causes of good action." The inadequacies of synergism reflect the methodological failure of attempting to analyze a theological issue in psychological terms.

PAUL HELM

SYNOD. An ecclesiastical deliberative and legislative assembly. In the Roman Catholic Church it is a gathering of priests and clergy of a diocese, called by the bishop to determine legislation for the diocese or to apply the canon law to particular situations. Benedict XIV ruled that a synod must always be considered as a convocation of the diocese, as opposed to a council, which is a convocation of all the bishops of the Catholic world. The first synod was probably held under Bishop Siricius in Rome in 387. Originally a synod differed from a council only in its finality. Vatican II gave rise to a nondiocesan synod in its teaching concerning the collegiality of bishops. In 1969 Paul VI opened the first biennial synod, composed of representative bishops elected by their respective episcopal conferences as delegates.

In Presbyterian churches, synod is a court of review immediately superior to the presbytery, and consists of all ministers and elders who are members of presbyteries. Synod in the Waldensian Church is the annual legislative assembly composed of clergy and laity.

ROYAL L. PECK

SYNOPTIC GOSPELS. The first three gospels, Matthew, Mark and Luke, are known as the Synoptic Gospels because they share a common outline of events as contrasted with the Fourth Gospel. Their basic similarity of structure, however, allows for considerable variation in the order of separate units of narrative. The similarity stretches not only to the considerable quantity of common material, but also to verbal agreements in much of the material. It is the combination of similarity and dissimilarity which constitutes the synoptic problem mentioned below.

These gospels are of utmost importance as sources of knowledge for the life of Christ. Since information from noncanonical sources is almost nonexistent, and since the character of the fourth gospel has caused many to question its validity as a historical source, the historian of Jesus Christ has relied almost exclusively on the synoptic gospels. From the Acts* and Epistles* only the barest outline of the historical Jesus is possible, while the secular evidence can do little more than establish the bare fact that he lived and died. Although the quest for the historical Jesus has itself been considered invalid by some modern schools of interpretation (notably form-criticism*), historians find it difficult to account for the subsequent emergence of the Christian Church unless some historical validity is granted to the records of his life and work. The synoptic gospels have therefore been in the forefront of the modern critical debate concerning Jesus Christ.

As documents which purport to be historical, the question of origin is of basic importance. It has engaged the attention of scholars throughout the period of criticism. The types of solution which have been proposed may be summarized in the following way.

(1) One theory is that all three synoptic gospels were composed from basically the same oral tradition and that the variations are the result of each author's choice regarding the material to be used. But most scholars do not see how so much common material in a fixed sequence could have been preserved by oral tradition. It is possible that insufficient weight has generally been given to the remarkable facility of memory possessed by the oriental mind and the Jewish religious practice of preserving the traditions of the elders by oral means.

(2) The most widely accepted theory is that Mark was the basic gospel, which both Matthew and Luke used as the core of their gospels. In addition to Mark, they both used another basic source ("Q") from which they derived their additional material. A modification of this proposes to restrict "Q" to material common to Matthew and Luke and postulates other sources for their special material ("M" and "L" respectively). Although this theory has wide support it is not without its problems. Many who accept some form of it do so for want of a better.

(3) A development of the source theory outlined above is the form-critical approach. This explains the origin of the sources by postulating that these were composed out of traditional material circulating in units, which can be classified according to the literary form in which they were preserved. This approach is therefore an attempt to describe the methods by which the oral tradition circulated. It is when its advocates assess historicity on the basis of "forms" that serious differences arise, some like Bultmann* taking a skeptical view which accepts very little as authentic history, and others treating the forms simply as literary units and declining to assess historical validity on this basis.

(4) A still further development from form-criticism is redaction-criticism, which sees the evangelists as theologians rather than as historians. There is clearly some advantage in emphasizing the personal contribution of the authors rather than thinking of the synoptic gospels as collections of isolated units. But there is a tendency to place theological interest over against historical data, with the result that many redaction-critics place little reliance on these gospels as sources of information regarding the historical Jesus. It must be noted that gospels written with a theological purpose need not be treated as unhistorical. Indeed a theological end would be better served by historically valid data than by data which was the product of imagination.

The three gospels possess individual characteristics which show their distinctive importance. Matthew's main theme is the messianic position of Jesus, who is seen as the fulfillment of OT predictions with special emphasis on His kingship. It is Matthew who records most of the attitude of Jesus toward the Law; it is clearly seen that He did not come to destroy the Law, but to fulfill it. Moreover He went beyond it to bring out the true spiritual implications of the Law. A characteristic expression of Jesus in this gospel is "Moses said, ... but I say." In some respects the teaching of Jesus, which in this gospel is arranged in five great blocks of material, appears in the role of a new Law for the Christian Church.

Mark's gospel places considerably more stress on the activity of Jesus than on His teaching. The dominant portrait of Jesus is as the Son of Man. Unlike Matthew's gospel there is no interest in the pre-ministry experience of Jesus, except a bare mention of the temptation and baptism. Mark launches his account immediately with a sequence of incidents in which he illustrates the relationships of Jesus with various groups with whom He mixed. The climax of the ministry of Jesus in his gospel, as also in Matthew and Luke, is found at Caesarea Philippi with Peter's confession, which marks the beginning of that period which led up to the cross and resurrection.

Luke's most distinctive feature is the concluding section of his gospel in which he incorporates in the form of an extended travel section from Galilee to Jerusalem much material which either does not occur at all in the other gospels or else occurs in a different context. Luke's portrait of Jesus is said to be less tragic than the others. But the passion narrative still occupies a major part of the gospel and shows the importance which Luke gave to the events surrounding the death of Christ. He includes more than the others on the theme of joy and on the activity of the Spirit. He also records more incidents which reflect the human interests of Jesus. Although each of the three gospels contains its own particular emphasis, it cannot be denied that they are alike in showing general agreement on the personality and purpose of Jesus.

It was not long before these gospels were received by the early church as part of their Scriptures. The need for authoritative books about the life and teaching of Jesus to be read alongside the OT arose at a very early stage. The fact that the synoptic gospels were known and used by the Gnostics of the early second century shows that they must have circulated long before that time in orthodox circles. In spite of a spate of pseudo-gospels, many showing dependence on the synoptic gospels, there was no serious consideration given to any other than these gospels and the gospel of John (see JOHN, GOSPEL OF). That true apostolic connection was the criteria used is evident from Tertullian's comment that, whereas Matthew and John were the work of apostles, Mark and Luke were written by pupils of apostles and could therefore be regarded as coming from Peter and Paul respectively. Although form-critics would discount this line of argument, there is no doubt that it played an important part in the final acceptance of the synoptic gospels and the gospel of John.

BIBLIOGRAPHY: B.H. Streeter, *The Four Gospels* (1924); W.L. Knox, *Sources of the Synoptic Tradition* (1953-57); L. Vaganay, *Le Problème synoptique* (1954); R. Bultmann, *The History of the Synoptic Tradition* (ET 1963); N.B. Stonehouse, *Origins of the Synoptic Gospels* (1963); D. Guthrie, *New Testament Introduction* (3rd ed., 1970).

DONALD GUTHRIE

SYRIAC. A Semitic language belonging to the East Aramaic group. Originally the dialect of Edessa, it was already in use there in pre-Christian times. With the increased stature of Edessa as a center of Christianity, the use of Edessan Syriac spread throughout Mesopotamia. As a literary language it is represented by texts from the second to the thirteenth centuries. By A.D. 800 Syriac had been replaced by Arabic as the language of Mesopotamia, but some Syriac-speaking communities survived, and Neo-Syriac dialects, chiefly of East Syriac, are still spoken in parts of Turkey and Iraq. The name "Karshuni" is given to Arabic texts written in Syriac script. Three distinct Syriac scripts, all of them cursive, were used. The oldest inscriptions and manuscripts are in Estrangelo; following upon the christological disputes and the division of the Syriac-speaking church in the fifth century, two other scripts were adopted by the respective traditions: Nestorian in the East, and Jacobite (or Serta) in the West. Separate systems of vowel notation were also evolved in the latter part of the first millennium A.D., the Jacobite being influenced by the Greek vowel symbols. Syriac literature is almost entirely Christian; much of the literary activity was centered upon the Edessan School of the Persians until its closure, by command of the emperor Zeno, in 489. In addition to the translation of Greek theological treatises, original Syriac works in prose and poetry were composed.

BIBLIOGRAPHY: C. Brockelmann, *Lexicon Syriacum* (1895); P.K. Hitti, *History of Syria* (1957); T.H. Robinson, *Paradigms and Exercises in Syriac Grammar* (1962); T. Nöldeke, *Kurzgefasste Syrische Grammatik* (rep. 1966); J.B. Segal, *Edessa "The Blessed City"* (1970).

ROBERT P. GORDON

SYRIAC VERSIONS OF THE BIBLE. There are three OT and five NT versions:

(1) *Old Testament.* (a) *Peshitta:* whether the Peshitta was originally a Jewish or a Christian translation is still a matter of debate. Some scholars stress the West Aramaic Targumic elements in the Pentateuch and argue that it must have originated among Jews who had a close connection with Palestine; later it will have been taken over by Christians. Kahle associated the translation of the Pentateuch with the province of Adiabene where the royal house embraced Judaism in the first century A.D. The Targumic influence has also been explained as the result of consultation of Jewish sources by Christian translators, probably converts from Judaism. Irrespective of its place of origin, the Peshitta was in use among Christians by the beginning of the third century. No uniform translation technique exists for other OT books: some are literal, some tend toward paraphrase; the degree of influence of the Septuagint and Targums varies greatly. As revision of the Peshitta proceeded, many Targumic features were eliminated and the text brought into greater conformity with the Septuagint.

(b) Some fragments of a *Palestinian Syriac text* of the OT have been preserved; the translation was made from the Septuagint sometime in the fifth century A.D. (see PALESTINIAN SYRIAC TEXT OF THE NEW TESTAMENT).

(c) *Syro-Hexaplar:* a Syriac translation of the fifth column of Origen's *Hexapla* made by Paul, bishop of Tella, in A.D. 616-17. It is an important witness to Origen's text of the Septuagint and reproduces the Hexaplaric signs; many readings from the minor Greek versions are noted. Reconstruction of the Septuagint *vorlage* is greatly helped by the slavishly literal nature of the translation.

(2) *New Testament.* (a) *Diatessaron:* a harmony of the four gospels composed by Tatian about A.D. 170 during his stay in the West; whether the original composition was in Greek or Syriac is uncertain. The version enjoyed immense popularity in the East for over two centuries and was often quoted by Syriac-speaking commentators; Ephraim Syrus wrote a commentary on it (see also DIATESSARON).

(b) *Old Syriac:* see separate entry.

(c) *Peshitta* version of the NT: The preparation of the standard text of the New Testament was completed sometime in the fifth century. By then the *Diatessaron* was in disfavor and the Old Syriac in obvious need of revision. The translation gives evidence of multiple authorship, but the name of Rabbula* is closely associated with the final phase of the standardization. The revision was made according to the regnant Byzantine Greek text and was sufficiently advanced to be retained by both sections of the Syrian Church after the division of 431. Manuscripts of the Peshitta go back to the fifth century.

(d) *Philoxenian and Harklean Versions:* in 508 Philoxenus, bishop of Mabbog, commissioned a new translation of the NT in which the Antilegomena were rendered for the first time. The version of Thomas of Harkel published in 616 was either a revision of the Philoxenian text or simply a reissue with marginalia added. The Harklean marginal readings for Acts are an important witness to the Western Text.

(e) *Palestinian Syriac Version* of the NT: see separate entry.

BIBLIOGRAPHY: B.J. Roberts, *The Old Testament Text and Versions* (1951); F.F. Bruce, *The Books and the Parchments* (3rd ed., 1963); B.M. Metzger, *The Text of the New Testament* (2nd ed., 1968).

ROBERT P. GORDON

SYRIAN CHURCHES. Christianity was established in Syria by the end of the second century; legend links it with Jesus Himself. Its gospel was the *Diatessaron** of Tatian. Its mode of life included a strong emphasis upon celibacy and asceticism. Its great teachers were Afrahat and Ephraem,* its centers Edessa and Nisibis. The doctrinal division of the church after the Council of Chalcedon* (451) left its mark very clearly upon Syrian Christianity, and the number of Syrian churches still witnesses to this. Already after the Council of Ephesus* (431) the Syrian churches of East Syria and Persia adhered to the teaching of Nestorius (see NESTORIANISM).

The Nestorian Church flourished, was tolerated under Islam, conducted missions in central Asia, and reached China in the seventh century. From the thirteenth century, Roman Catholic overtures were made, and some part became Roman Catholic. Both churches still survive, mainly in Iraq, after dreadful persecution in the early twentieth century. Most Christians in W Syria followed the Monophysite teaching after 451 (see MONOPHYSITISM); due to the organization by Jacob Baradaeus they prospered, being called Jacobites* after him. They flourished, adopted Arabic at length as their language, and are found in great numbers in the Near East. The links of some of them with Rome date from the seventeenth century. In South India (Malabar*) are Christians using Syriac in liturgy. These are no doubt the result of Jacobite or earlier missionary endeavor. After 451 the few who remained Orthodox were called Melkites*: these too survive in small numbers, some in communion with Rome. A small Syrian group adhered to the Monothelite* formula of reunion after it had been condemned in 680; centered around the shrine of St. Maro, they are called Maronites.* They became reunited with Rome in the twelfth and fifteenth centuries and are centered in Lebanon.

BIBLIOGRAPHY: F.C. Burkitt, *Early Eastern Christianity* (1904); D. Attwater, *The Christian Churches of the East* (2 vols., 2nd ed., 1961); B. Spuler, *Die morgenlaendischen Kirchen* (1964).

J.N. BIRDSALL

SYRIAN TEXT. One of the four different types of texts or text-families of the Greek NT distinguished by Westcott* and Hort.* A mixed text resulting from a revision made in the fourth century (cf. Lucianic Text*), it was regarded by Westcott and Hort as the furthest removed from the originals. It is best represented today by Codex Alexandrinus (in the gospels), the later uncial MSS, and the great mass of the minuscules. The Textus Receptus* is the latest form of the Syrian (=Byzantine=Koine) text.

SYRO-CHALDEANS, see CHALDEAN CHRISTIANS

SYRO-HEXAPLAR, see SYRIAC VERSIONS

SYRO-MALABAR CHURCH, see MALABAR CHRISTIANS

SYSTEMATIC THEOLOGY. In theology, faith seeks to interpret, understand, and unfold the wealth of the revelation of God with which it is confronted in His Word. Revelation,* as it is received by us, appears to consist of a multitude of events, saving facts, and truths which find their expression in a collection of separate theological doctrines. On the basis of the belief that God is one and intends to reveal Himself in the unity of His activity, theology seeks to present the whole range of its knowledge of revelation as one coherent, living whole.

Theological discussion has therefore always been undertaken in as orderly a manner as possible. Early attempts to seek to do justice to the unity of revelation were made, e.g., by John of Damascus* in *Fons Scientia,* and in the West by Peter Lombard* in his *Sententiae.* Aquinas* also tried to give unity to a theological system by absorbing it into what he regarded as a Christian philosophy. At the Reformation, Melanchthon* in his *Loci* tried to give a "system of doctrinal positions" drawn from the Word of God. Such theologies tended to take the order of discussion suggested in Holy Scripture, beginning with Creation and man's sin, then discussing the Law and the Gospel, and finishing with the Last Things.

More than any of his predecessors, Calvin* in his *Institutes* sought to do justice to the belief that the unity and rationality of the one God was reflected in His revelation. He sought to show how each doctrine was interconnected with the other and must be interpreted as part of the living whole. He tried to make Christ, rather than any one principle, the controlling center. In the seventeenth century, theology tended to relapse again into a discussion of a series of distinct doctrines, each of which apart from the whole could be justified in itself. Scripture, moreover, tended to be used atomistically in the support of individual propositions without reference to the whole of salvation history.

Schleiermacher* has been regarded as the first theologian ever to lay hold of a central theological principle, and in the light of it to build up a whole system in which each doctrine is carefully discussed in relation to the unity of the whole in the light of the controlling principle. In the nineteenth century, under pressure of the idea that revelation in itself had no inherent rationality, theologians, instead of using philosophy as a tool and medium for theological illustration, allowed their philosophy to subdue and refashion their theology, and to impose on it the pattern of contemporary thought in the production of systems that had often little relation to the Gospel or the Bible. Yet no one system can ever become finalized over against revelation and Scripture.

BIBLIOGRAPHY: C. Hodge, *Systematic Theology* (1872; rep. 3 vols., 1960); K. Barth, *Die Kirchliche Dogmatik* (1936; new ET in progress); T.F. Torrance, *Theological Science* (1970).

RONALD S. WALLACE

T

TABORITES. The most radical branch of the Hussite movement. Centered in S Bohemia where Hus had spent much of his exile between 1412 and 1414, the Taborites nevertheless did not really reflect his teachings. They were fundamentalists in the tradition of John Wycliffe and wished to confine doctrine to what was explicitly stated in the Bible. They rejected transubstantiation, purgatory, saints, relics, and the distinction between priests and laity. They were also militant millenarians who believed in an imminent second coming of Christ preceded by a period of turmoil. In addition, they represented the lower economic classes and were concerned about social and economic reforms. The movement became a mass movement in July 1419 when 40,000 people are said to have gathered on a hill to which the biblical name of Tabor was given. Under the brilliant military leadership of John Zizka they defeated the imperial crusades directed against them and were able to maintain a degree of unity with the more moderate party, the Calixtines.* This unity collapsed after Zizka died in 1424, and the Calixtines came to an agreement with Rome in 1433. The following year they combined with Catholic nobles to defeat the Taborites at the Battle of Lipany where their new leader, Procopius, was killed and the Taborite movement destroyed.

RUDOLPH HEINZE

TACHE, ALEXANDRE ANTONIN (1823-1894). Canadian Roman Catholic bishop. Born in Riviere du Loup, Quebec, and educated at the College of St. Hyacinthe and the Theological Seminary of Montreal, he became in 1844 a member of the Oblate Order, was ordained in 1845, and sent as a missionary to the Red River area of Manitoba. He became the second bishop of St. Boniface in 1853, and was consecrated archbishop and metropolitan there in 1871. He helped restore order during the 1869-70 Riel Rebellion by a too-generous promise of amnesty which involved him in later controversy. He was also a leader in the Manitoba separate-school struggle.

TACITUS, CORNELIUS (c.55-117). Secular Latin writer. Probably our most important single source for Roman history from Tiberius (A.D. 14-37) to Domitian (81-96), he provides independent confirmation for the NT at several points, and useful if biased background information about the more corrupt aspects of politics under the emperors. In recounting the persecutions of Nero* after the fire of A.D. 64 in Rome, Tacitus outlines the rise of Christianity ("a subversive cult," "atrocious practices," and more tellingly "its enmity against mankind") and mentions Christ's execution under Pilate, but his main target is Nero's perverted and pointless cruelty. Felix, procurator of Palestine in Acts 24 and brother of an influential freedman of the emperor Claudius, is condemned for serious misrule. Tacitus's account of the Jewish War of A.D. 66-70 (*Histories* 5.1ff.) supplements the fuller narrative of Josephus.*

GORDON C. NEAL

TAIT, ARCHIBALD CAMPBELL (1811-1882). Archbishop of Canterbury from 1868. Born of Scottish Presbyterian parents, he went from Glasgow University to Balliol College, Oxford, where he subscribed to the Thirty-Nine Articles* in accordance with the university statutes. In 1836 he took Anglican orders. He was one of the four tutors who publicly protested against Tract 90 in 1841. In 1842 he succeeded Thomas Arnold as headmaster of Rugby. Six years later an attack of rheumatic fever left him permanently weak and forced him to take lighter duty as dean of Carlisle. In 1856 Palmerston unexpectedly appointed the inexperienced Tait to the demanding and influential bishopric of London. He declined an invitation to move to York in 1862; and then in 1868 Disraeli, also against expectation, offered him Canterbury. The queen's influence was probably decisive.

By his strong leadership Tait restored the see of Canterbury to its position of preeminence in the Church of England. His chairmanship of the second Lambeth Conference in 1878 ensured its continuance as a regular meeting of Anglican bishops. Tait was a Broad Churchman in theology, though he opposed *Essays and Reviews.* He retained his Presbyterian distaste for ceremonial, and took vigorous steps to bring to order ritualistic clergymen like A.H. Mackonochie. He was the originator of the Public Worship Regulation Act of 1874. A firm believer in the establishment, Tait based his policies on his sense of the weight of opinion among the English people.

See R.T. Davidson and W. Benham, *Life of Archibald Campbell Tait* (2 vols., 1891); and P.T. Marsh, *The Victorian Church in Decline: Archbishop Tait and the Church of England, 1868-1882* (1969).

JOHN TILLER

TAIZÉ COMMUNITY. Founded in 1940 by the present prior, Roger Schutz, when he began to receive Jewish and other refugees into his home at Taizé in Burgundy. In 1942 the Gestapo forced him away, but in 1944 he returned with three brothers to begin the common life. By 1949 the monastic tradition took hold with the first seven brothers pledging themselves to celibacy, authority, and common property. More than seventy men from different Christian traditions from Europe and the Americas now make up the community, and a group of Franciscans and Eastern Orthodox monks have come to live with them. From 1968 Roman Catholics have joined, and the prior and a few brothers spend one month each year in Rome. Truly ecumenical, there are solid links with Rome, Constantinople, and the World Council of Churches (where brothers are on the staff). For social development, brothers go in small groups to many nations, notably Latin America and Africa, to live and work among varying situations ranging from scientific efforts to dishwashing. Wherever they are, brothers pray at appointed times thrice daily. Taizé has become a place of pilgrimage. Youth assemblies meet there regularly. C.G. THORNE, JR.

TALLIS, THOMAS (c.1505-1585). English composer. Considered the greatest composer in England before Byrd,* Tallis in his latter years was closely associated with the latter; they were jointly granted by Queen Elizabeth a twenty-one-year monopoly to print music in the realm. Tallis was organist of Waltham Abbey until the dissolution of 1540. Shortly thereafter he became a gentleman of the Royal Chapel, where he remained until his death. Probably much of his music has been lost, but the quantity which remains reveals him to be a refined and skillful craftsman. His Lamentations and office hymns are particularly beautiful. He wrote a *tour de force*—his motet, *Spem in alium,* for eight choirs of five parts each, which builds up to a climax in which all forty parts combine. Some of his English anthems are adapted from earlier Latin motets, but he wrote also much fine music for the English service, including the well-known "If ye love me." His responses, preces, and cathedral services have been much used. J.B. MAC MILLAN

TALMUD, see JUDAISM

TAMBARAM CONFERENCE (1938). Convened by the International Missionary Council,* the conference met near the Indian city of Madras during the closing days of 1938. It was notable for the impressive representation from the younger churches, whose problems were discussed and whose delegates brought a new dimension to the plea for Christian unity and the task of world mission. The conference, under the chairmanship of John R. Mott,* comprised 471 delegates from sixty-nine countries.

TARASIUS (c.730-806). Patriarch of Constantinople. Born in Constantinople, a well-educated layman and granduncle of Photius, he was secretary to the regent, Irene II, during the infancy of Constantine VI. At Irene's insistence he was elected patriarch, succeeding Paul in 784. He sought better relations with Rome, and persuaded Irene and Pope Hadrian I to call a council to condemn iconoclasm (see ICONOCLASTIC CONTROVERSY). An attempt to assemble it at Constantinople (786) failed through rioting inspired by iconoclasts. It met eventually in Nicea (787) under Tarasius's presidency as the seventh ecumenical council. Iconoclasm was condemned and the orthodox doctrine of image-veneration defined. Tarasius exercised leniency toward inconoclastic bishops who recanted, but was forced by rigorist monks to act against simony. In 795 he was attacked for his failure to condemn the adulterous second marriage of Constantine VI, though after Constantine's deposition he excommunicated the monk who performed the ceremony. In 802 he crowned as emperor Nicephorus, who had dethroned Irene. J.G.G. NORMAN

TARSICIUS (d. late third or early fourth century). Martyr. A poem by Damasus I (366-84) tells that Tarsicius was attacked by a pagan mob while carrying the sacrament, and accepted death for himself rather than defilement of the holy. Possibly a deacon who carried the Eucharist from the pope's Mass to the principal Roman churches as a sign of unity, he could have been an acolyte or layman appointed to bear the Eucharist to Christian prisoners.

TATE, NAHUM (1652-1715). Poet and dramatist. He was born in Dublin and educated there at Trinity College. Three of his children were killed and his house burned when he reported plans of a revolt to the government. He then left Ireland and settled in England, where he wrote poems and plays in which he adapted the work of earlier dramatists. With Nicholas Brady* he wrote a widely used metrical version of the Psalms (1696). The hymn "While shepherds watched" is attributed to him. Tate became poet laureate in 1692, but at the end of his life was very poor, and he died in debt.

TATIAN (c.110-172). Christian apologist and Gnostic. An Assyrian from Nisibitis on the Euphrates, he came to Rome about 150, was converted, and later became a pupil of Justin Martyr.* Like his master, he engaged in the defense of the faith against pagan misrepresentation. His "Address to the Greeks" (c.160) marks a retrograde step when compared with Justin's apologia. Unlike the latter's tolerant and courteous attitude to Greek learning and culture, Tatian had only mockery and contempt for pagan philosophy. After Justin's death he retired to Syria, where he became the founder of a group later called the Encratites.* Tatian's chief claim to fame is his *Diatessaron,* * used as a liturgical book in the Syrian Church until the fifth century. G.L. CAREY

TAULER, JOHN (Johann) (c.1300-1361). German mystic. Born in Strasbourg, he entered the Dominican Order of Preachers there about 1315 and came under Eckhart's* influence. Thomism was a part of him, but unlike Eckhart his ends were practical. He wrote only in German, never in Latin, and did not write learned works. His sermons were preached mainly to nuns. His was a simple, homely method, though obscurities do occur in the more mystical passages; dialogue was used by him for illustration to include his audience. His imagery, moreover, was local, derived from hunting, war, farming, trade, and natural history.

His was the kind of mysticism that was to dominate the fourteenth and fifteenth centuries, with its immense concern for everyone's spiritual health: no longer was mysticism only for the spiritual elite. Eckhart bridged Scholasticism and mysticism, and Tauler translated an academic approach to spirituality into a practical Christianity of high personal demands, designed for all men. His sermons demonstrate this where, with little biblical quotation and no personal testimony, the meaning of being a Christian is carefully unraveled. Many sermons have been ascribed to him due to his popularity, but only some are genuine. During the Black Death* period (1348) he devoted himself completely to the sick. His years in Basle (1338-43) found him a central figure in the Friends of God* *(Gottesfreunde).* He owed a great debt to the Waldensian layman Nicholas of Basle,* who advised him to stop preaching and meditate, which he did for two years amid his ministry with remarkable results. Luther read him profitably.

BIBLIOGRAPHY: C. Schmidt, *Johannes Tauler von Strassburg* (1841); S. Winkworth, *The History and Life of... John Tauler... with Twenty-five of His Sermons* (1857); D. Helander, *Johann Tauler als Prediger* (1923); J.M. Clark, *The Great German Mystics* (1949). C.G. THORNE, JR.

TAUSEN, HANS (1494-1561). Danish Reformer and bishop. In his youth a monk of the monastery of St. John of Antvorskov, he pursued theological studies in Rostock from 1516, Copenhagen (1521), Louvain (1522), and Wittenberg (1523). While studying in Germany, he became a fully convinced adherent of the Reformation. In 1525 he was recalled to Antvorskov and shortly afterward transferred to Viborg. Before long a large congregation gathered around his dauntless and popular preaching of evangelical truth. Tausen's superiors were alarmed and tried to stop him by expelling him from the monastery and from his order. He nevertheless continued preaching, and in 1526 King Frederik I put Tausen under his personal protection, so that he might continue his work of reformation unhindered by the clerical authorities. From 1529 he worked in Copenhagen, and here again his preaching gave rise to a vigorous reformation movement. After the official accomplishment of the Reformation in Denmark* in 1536, Tausen went on with his preaching ministry in Copenhagen, and from 1538 became also a lecturer in theology in Roskilde. In 1541 he was appointed bishop of Ribe. In this capacity he worked at the practical accomplishment of the Reformation within his diocese until his death. Tausen is the most outstanding figure among the pioneers of the Reformation in Denmark, not so much because of his originality as because of his frankness and lucidity. N.O. RASMUSSEN

TAVERNER'S BIBLE, see BIBLE, ENGLISH VERSIONS

TAYLOR, J(AMES) HUDSON (1832-1905). Missionary pioneer. Born in Yorkshire, the son of a Methodist chemist, he underwent at seventeen a deep conversion and soon felt a strong call to the almost closed empire of China.* He landed at Shanghai in 1854, after part-medical training, as agent of the short-lived Chinese Evangelization Society. The inefficiency of its home base threw him back on faith and prayer for his support, and a succession of providences caused him to sever connection. He made several evangelistic forays into the closed interior, and adopted Chinese dress. In 1858 he married Maria Dyer in Ningpo despite the opposition of other missionaries who viewed him as a "poor, unconnected Nobody." Invalided back to England, he bore a burden for inland China and the millions without Christ which grew even stronger. On the opening of the empire to Westerners, he could find no mission willing to back him, so he founded the interdenominational China Inland Mission (1865), asking God to send "24 willing, skillful laborers," two for each unreached province. They sailed in 1866. Maria died four years later.

Despite opposition from missionaries and mandarins, some internal dissension, and several riots, the CIM established itself as the "shock troops" of Protestant advance. Taylor's aim was to bring the Gospel to every creature: he was happy for others to reap where his pioneers had sown, although many CIM stations became permanent. By 1895 he led 641 missionaries, about half the entire Protestant force in China. His great spiritual qualities and the caliber of the CIM, together with his writings and world travels, gave him an influence far beyond China, and led to similar faith-missions being founded. Among his chief emphases were: identification with the people (e.g., all to wear Chinese dress), the direction of the mission to be on the field, not from the home base; dependence on God alone for supplies, with scrupulous efficiency in administration; the deepening of Christian life in the home churches as a sure means of encouraging missionary vocations. Taylor retired in 1901 and died four years later at Changsha, capital of the last province to open.

BIBLIOGRAPHY: H. and G. Taylor, *Hudson Taylor in Early Years* (1911) and *Hudson Taylor and the CIM* (1918; one-volume ed., 1965); J.C. Pollock, *Hudson Taylor and Maria* (1962).

JOHN C. POLLOCK

TAYLOR, JEREMY (1613-1667). Anglican bishop and writer. Born at Cambridge and educated there, he early attracted through his eloquence the attention of William Laud,* who nominated him as his chaplain and enabled him to be elected fellow of All Souls. He chose to be loyal to the king in the Civil War, but on the collapse of the royal cause he attached himself to Lord Carbery, remaining at his house in Carmarthenshire until the return of Charles II in 1660. His *Liberty of Prophesying* (1647) is a plea for toleration. It was followed by *The Life of Christ* (1649), *Holy Living* (1650) and *Holy Dying* (1651), and the manual of devotion, *The Golden Grove* (1655). After the Restoration his loyalty was recognized in his preferment to the see of Down and Connor, but his was a difficult task in an age when the toleration he so much valued was so little in fashion.

In addition to the works mentioned above, Taylor also published *XXVIII Sermons Preached at Golden Grove* (1651) and *XXV Sermons* (1653), as well as several scattered, individual sermons. He is the last practitioner of a "golden age" of preaching that began with Donne* and Andrewes.* Mason called him the "Shakespeare of divines." More accurately, Coleridge described him as the "Spenser of prose"—more accurately, because Taylor's prose compares in its color and elaborateness with the highly artificial work of the Elizabethan poet. The texture of his work consists in a wealth of biblical, patristic and classical allusion, woven in a rich and complex pattern. Basically Taylor's prose is still Ciceronian, but compared, say, with Hooker's at the end of the previous century, it has developed a certain mannered quality that is best described as baroque. This is revealed especially in his imagery, where he draws on his rich sensory, and especially his visual, imagination.

See *Works* (ed. R. Heber; rev. C.P. Eden, 1847-54); and C.J. Stranks, *The Life and Writings of Jeremy Taylor* (1952). ARTHUR POLLARD

TAYLOR, JOHN (1694-1761). Nonconformist* minister. Following a lengthy pastorate in Norwich, he was appointed in 1757 to the divinity chair of Warrington Academy. After reading Samuel Clarke's* *Scripture Doctrine of the Trinity,* Taylor adopted Arian views of the person of Christ, and in *The Scripture Doctrine of Original Sin* (1740), as well as in *Key to the Apostolic Writings,* he claimed that the orthodox Reformed view of the imputation of Adam's sin to his posterity lacked biblical support, and that Adam's sin had only natural, not moral, consequences. His views were fully answered by Jonathan Edwards* in *The Great Christian Doctrine of Original Sin Defended* (1758), but they continued to be influential in England and the USA. PAUL HELM

TAYLOR, NATHANIEL WILLIAM (1786-1858). American theologian and educator. Born in Connecticut, he graduated from Yale (1807) where he studied theology under Timothy Dwight.* After ordination and a pastorate at the First Church of New Haven (1812-22), he was appointed the first professor of theology at Yale Divinity School, where he remained the rest of his life. He tried to construct a consistent theology of revivalism suited to the Second Great Awakening.* This modified Calvinism was based on the earlier Edwardean revivalistic theology. Taylor's main thesis concerned the problem of moral depravity, and although he taught that sin was inevitable, each person was nevertheless responsible for his own moral choice—a position consistent with revivalistic preaching. His views created such controversy among Congregationalists that a more orthodox and Calvinistic seminary was formed at Hartford (1834). Taylor's works include *Practical Sermons, Lectures on the Moral Government of God* (1859), and *Essays and Lectures upon Select Topics in Revealed Religion* (1859). DONALD M. LAKE

TAYLOR, WILLIAM (1821-1902). American Methodist evangelist and missionary. Born in Vermont, he had not much early education, but in due course entered the Methodist ministry and was assigned to the tented city of San Francisco. His open-air meetings drew thousands of listeners and brought many conversions. Taylor organized the first Methodist church in the city. During 1856-61 he itinerated throughout North America; thereafter world travels took him to England, Australia, the West Indies, South America (three times), the Middle East, and South Africa. Elected missionary bishop for Africa in 1884, he systematized work particularly in the Congo and Liberia, advocating self-supporting missions wherever possible. His zeal and methods sometimes alarmed his home board, but his results were often impressive. He found time to write also; his works include *Seven Years' Street Preaching in San Francisco* (1856), *Christian Adventures in South Africa* (1867), and *Story of My Life* (1896). J.D. DOUGLAS

TEACHING OF THE TWELVE, see DIDACHE

TE DEUM. The name of an early Christian hymn written in Latin which comes from its first two words or, when its longer title is used, its first three words, *Te Deum Laudamus* ("We praise thee, O God"). There have been various traditions about its composition, the best known being that which ascribed it to Ambrose* and Augustine* at the latter's baptism. Nowadays it is generally accepted that the author was Niceta,* bishop of Remesiana about A.D. 400. The first nine verses are an ascription of praise, and the next twelve a confession of faith ending with prayer. The last eight verses are suffrages which became attached to it at an early date. The *Te Deum* has a central place in the morning worship of the Western Church. R.E. NIXON

TEILHARD DE CHARDIN, PIERRE (1881-1955). Roman Catholic Jesuit priest and paleontologist who advocated an evolutionary hypothesis that synthesized modern science and traditional Christian theology. Marie-Joseph-Pierre Teilhard de Chardin was born in Sarcenat, France, and after study at the Jesuit College of Mongré, Jersey and Hastings, he was ordained in 1911. His work at the Museum of Paris as a paleontologist was interrupted by the war, during which he served as a stretcher-bearer. He subsequently finished his doctoral thesis at the Sorbonne in 1922 and soon departed for China to serve as a consultant to the geological survey where he was associated with the discoveries of Pithecanthropus and Sinanthropus. He returned from China after World War II, but his teachings and ideas concerning cosmogenesis (that the world develops according to a law of increasing complexity and consciousness until the appearance of man) and Christogenesis (that the process converges in a rhythm of hypersocialization toward an Omega point) led his order to prohibit him from accepting a professorial chair at the *Collège de France* and from publishing *The Phenomenon of Man.* He moved to the USA and spent his last years working with the Wenner-Gren Foundation for Anthropological Research in New York.

Teilhard's ability as paleontologist was never questioned, and his more than 170 articles and technical papers declare his ability. However, his desire to incorporate the total knowledge of man into the understanding of the phenomenology of man led to controversy. His ideas bear some striking similarities to process philosophy.

See C. Cuénot, *Teilhard de Chardin* (1962), and C.E. Raven, *Teilhard de Chardin: Scientist and Seer* (1963). JOHN P. DEVER

TELEMACHUS (d.391). Eastern monk. Unknown but for his martyrdom (recorded by Theodoret), he rushed into the arena in 391 to separate the gladiators, only to be stoned to death by the exasperated spectators. As a result Emperor Honorius abolished gladiatorial combats soon after. Absence of this edict from the Theodosian Code has cast doubt on the historicity of the story, but Constantine's Edict already existed, and gladiatorial fights appear to end with Honorius; wild beast shows continued.

TELEOLOGICAL ARGUMENT. Otherwise known as the "argument from design," this has an analogical form, claiming that the purposiveness of the natural order requires the postulation of a designer, and that this designer is God. Probably originating in its modern form in William Derham's works, especially *Physico-Theology* (1713), the argument was popularized by rationalistic anti-Deists such as William Paley* and Joseph Butler.* It was effectively criticized by David Hume* in his *Dialogues Concerning Natural Religion* (1779), on the grounds that the evidence of design is ambiguous and establishes at best only a finite designer or designers, not God. Despite this criticism, the argument figured largely in popular Protestant apologetics in the nineteenth century and was given further currency by, e.g., the *Bridgewater Treatises* (1833-40). The impact of Darwinian evolutionary theory upon Protestantism is largely to be accounted for by the fact that that theory seemed to many to provide an alternative, naturalistic explanation of design.

PAUL HELM

TELESPHORUS (d. c.137). Martyr; bishop of Rome from about 127. According to the *Liber Pontificalis* he was a Greek. He apparently always observed Easter on Sundays, as against Quartodeciman* practice. Irenaeus* mentions him as the first Roman bishop to be martyred. His, in fact, is the only name known to us as a martyr of the emperor Hadrian's* reign.

TEMPERANCE. A nineteenth-century reform movement designed to encourage individuals to limit or abstain from the use of alcoholic beverages. It promised to improve the physical and moral well-being of individuals. Economic benefits, it was assumed, would follow.

Sporadic efforts of eighteenth-century ministers and humanitarians to encourage temperance were given medical support by Dr. Benjamin Rush's *An Inquiry into the Effects of Ardent Spirits on the Human Mind and Body* (1784). In 1826 the American Society for the Promotion of Temperance was formed in Boston under the dynamic leadership of Lyman Beecher.* Over 8,000 auxiliary societies were formed by 1835, representing about 1.5 million signers. Eleven weekly and monthly journals were published to promote temperance; millions of tracts were distributed. Songs and essays, plays and novels were written to dramatize the evils of strong drink. Timothy Shay Arthur's *Ten Nights in a Barroom* (1854) is the best known.

Division appeared within the American Society by 1836 over the issues of total abstinence from all alcoholic beverages, antislavery, and the legal control of the liquor traffic. The American Temperance Union, then formed, continued its campaign for total abstinence in the face of declining interest. Meanwhile the voluntary-abstinence movement gained new strength from the Washingtonians, a society of ex-drunkards formed in Baltimore on Washington's birthday, 1840. Members were enlisted through "experience meetings." John B. Gough became their leading evangelist and lecturer.

Many in the temperance movement now campaigned for government regulation of the sale and distribution of hard liquor and alcoholic beverages. Neal Dow, a Portland, Maine, merchant, headed the movement in that state. The legislature responded in 1846 by restricting the retail sale of intoxicating beverages. By 1851 it completely outlawed the liquor traffic. Within five years thirteen Northern states followed suit. Meanwhile the temperance movement gained strength in Britain, Ireland, and Canada. A World's Temperance Convention assembled in London, 1846, and in New York, 1853.

The temperance movement in America after the Civil War was characterized by a campaign for total abstinence on the part of individuals and the outlawing of liquor sales by local, state, and national governments. The Women's Christian Temperance Union, formed in 1874 under the leadership of Frances Willard,* spearheaded the drive for total abstinence. The National Prohibition Party (1869) entered local, state, and national political campaigns to insure the election of public officials committed to the abolition of the liquor traffic. Ably assisted by the Anti-Saloon League (1893), the prohibition forces secured congressional approval of wartime prohibition in 1919. A constitutional amendment outlawing the "manufacture, sale or transportation" of intoxicating liquors for beverage purposes was proposed by Congress in 1918 and approved as the Eighteenth Amendment the following year. The amendment was repealed in 1932 in spite of strenuous support from the Methodist Board of Prohibition and Morals and the two organizations previously mentioned.

BIBLIOGRAPHY: *The Cyclopedia of Temperance and Prohibition* (1891); E.H. Cherrington, *The Evolution of Prohibition in the United States* (1920); J.A. Krout, *The Origins of Prohibition* (1925); D.L. Colvin, *Prohibition in the United States* (1926). S. RICHEY KAMM

TEMPLARS. A military religious order founded c.1118 by Hugh des Payens, a knight from Burgundy, and Godfrey of St.-Omer, a knight from N. France. The original purpose of the Templars was to aid and protect pilgrims on their way to the Holy Land, thus acting in conjunction with the Knights of St. John, or Hospitallers,* who tended sick pilgrims in Jerusalem. These warrior-knights took the monastic vows of poverty, chastity, and obedience, and refrained from adopting many of the pompous rites and garbs prevalent among religious orders at the time. The official name of the order, "the Poor Fellow-Soldiers of Christ and the Temple of Solomon," is derived from their early state of poverty and from the portion of the king of Jerusalem's (Baldwin II) palace in which they lived, known as the Temple of Solomon.

After their modest beginnings the Templars grew rapidly during the twelfth century in purpose, size, and wealth. Their duties were expanded to include the defense of the Latin states of Palestine. In January 1128, at the Council of Troyes, a rule of order prepared by Bernard of Clairvaux* was accepted and the Knights were allowed to wear a white mantle, to which was added a red cross in later years. By the middle of the century, "commanderies" were established throughout Europe, governed by a hierarchy centering in the Holy Land headed by the grandmaster of the Temple of Jerusalem. Although knighthood was required of applicants during the early years of the order's existence, gradually three ranks were recognized in the membership: Knights, who joined for life; sergeants, consisting of wealthy bourgeois; and chaplains, who were priests, bound for life, to perform religious services for the Knights. As the Templars grew immensely wealthy due to gifts from royalty and pious donations, their influence spread to include financial and banking operations in Europe, and they even loaned money to the sultan of Damascus at one time.

With the fall of Jerusalem in 1187 and the expulsion of all Christians from the Holy Land after the fall of Acre in 1291, the Templars lost their crusading function. As a secret organization, independent of secular authority, they continued to acquire wealth and began to evoke hostility from European royal houses. In 1307 Philip IV* the Fair, king of a bankrupt France, moved against the Templars. By destroying this order Philip felt he could acquire much-needed land and money as well as striking indirectly at the authority of the papacy. In 1312 Pope Clement V,* pressured by Philip, issued the bull *Vox in excelso*, formally dissolving the order throughout Europe. In spite of protestations of innocence on charges of heresy, witchcraft, and sodomy, 120 Templars were executed in France, including Grand Master Jacques de Molay. Although Philip's debts to the disbanded order were eradicated and he seized much of their property, his desire to acquire all their wealth was not realized. Most of their possessions were transferred by the pope to the Hospitallers, with the exception of the Spanish and Portuguese branches, whose holdings went to other military religious orders.

BIBLIOGRAPHY: G.G. Addison, *The History of the Knights Templars, the Temple Church, and the Temple* (3rd ed., 1852); M. Dessubré, *Bibliographie de l'ordre des Templiers* (1928); M. Merville, *La vie des Templiers* (1951).

ROBERT G. CLOUSE

TEMPLES. Temples and holy places were integral to ancient society. Where the temple of Jerusalem differed from others was in the belief that God had chosen to dwell there and was in no way dependent upon it. He could forsake it and even destroy it (e.g., Jer. 7). In others words, the temple in the biblical tradition is part of the order of revelation and grace. The temple of Solomon (1 Kings 6–7) was built by Phoenician workmen, and its design had much in common with the temples of Egypt and Mesopotamia. But it soon inherited the theological traditions which had grown around the ancient tent of meeting, particularly when the ark was installed in it, and it did more than anything else to unite the tribes of Israel into a nation (cf. Deut. 12). Thus, in the exilic period the restoration and reunification of the scattered nation was symbolized in the hope of a new and glorious temple (Ezek. 40–48; cf. Isa. 54:11f; 60:13f.)

The temple which was built after the Exile seems to have had a rather pedestrian beginning, at least in the minds of contemporaries (Ezra 3:-12). At any rate, the desire for the glorification of the temple continued and grew in intensity (Mal. 3:1-4; cf. *Tobit.* 14:5; 2 *Macc.* 2:18). In apocalyptic thought, and subsequently in the rabbinic writings, the new temple is sometimes described in terms of a heavenly or supernatural temple descended to earth (*Enoch* 90:28f., etc.). Simultaneously another development, equally interesting, was taking place in the spiritualizing of the temple among the Jews of Qumran; here the company of believers is the "true temple."

The temple of Herod the Great was the third and last temple of Jerusalem. Herod had the existing temple considerably enlarged and magnificently embellished, but this temple met a sorry end in the war of A.D. 66-70. However, its destruction appears to have been less of a catastrophe than the destruction of its predecessor in 586 B.C. By this time the law and the synagogues, i.e., a nonsacrificial form of worship, had come to occupy a central place, with the result that Judaism could and did exist without a temple.

In the NT the Christian community is described as God's "new temple" (1 Cor. 3:16f.; 2 Cor. 6:16,17; Eph. 2:19-22; 1 Pet. 2:5; cf. 1 Cor. 6:19f.). Clearly our Lord's cleansing of the temple of Jerusalem was interpreted as more than an attempted reform of the cult. This is most apparent in the fourth gospel (John 2:19-22), where the saying of Jesus about the destruction of the temple is connected with the cleansing of the temple; but it is also present in the synoptic gospels (Mark 15:38; cf. 14:58). Whether the temple image is used to convey the unity and holiness of the church (1 and 2 Corinthians) or its inclusive character (Ephesians), the temple is always the temple of God and not (as in the image of the body) of Christ. Where Christ is mentioned, He is thought of as a part of the building (Eph. 2:19; 1 Pet. 2:4ff.; cf. 1 Cor. 3:11). The connection in 1 Peter 2:4ff. of the temple image with the image of Christ as the stone (cf. Mark 12:10; Acts 4:11) also points to the fact that the conception of the church as the "new temple" belongs to a very early tradition.

In Hebrews the concept of the heavenly temple is used, on the one hand, in a Platonic sense, to demonstrate the superiority of Christianity over Judaism (Heb. 8:5; 9:14), and, on the other hand, to depict the ongoing ministry of Christ (6:20; 8:1f; 10:21). In Revelation the heavenly temple forms the stage for the outworking of the divine drama in chapters 3–20 and is not in itself important. It is in chapter 21 that one finds the author's creative reinterpretation. There we read that the New Jerusalem has no temple (21:22). This bold thought can be taken to mean either that in the place where one would normally expect to find the sanctuary one finds God Himself (God and the Lamb), immediately accessible to all, or that the New Jerusalem is all temple. In any event, the meaning is the same.

See A. Parrot, *The Temple of Jerusalem* (1957), and R.J. McKelvey, *The New Temple* (1969).

R.J. MC KELVEY

TEMPLE, FREDERICK (1821-1902). Archbishop of Canterbury from 1896. Born in Greece where his father was a diplomat, he was educated largely by his widowed mother, to whom throughout his life he was closely attached. He entered Oxford on a Balliol scholarship and gained a first in classics and mathematics. In 1842 he was lecturer at Balliol and in 1846 was ordained. In 1850 he became principal of Kneller Hall, a training college for schoolmasters. In 1857 he was appointed headmaster of Rugby School where he gave vigorous leadership. His liberal sentiments had already found expression in his contribution to the celebrated volume *Essays and Reviews*, two of whose contributors were condemned by the church courts. Echoes of this controversy were found when in 1869 Temple was appointed bishop of Exeter and objections were made to his consecration. In Exeter he exhibited his lifelong interest in education, especially in the wake of the 1870 Education Act; another noteworthy field was in temperance reform.

He continued his interest in social questions after his appointment to the bishopric of London in 1885. In 1889 he helped to find a way of solving the dockers strike. In 1896 he was appointed archbishop of Canterbury, sold Addington Park, and with the proceeds bought the old Palace at Canterbury for his residence. He presided over the Queen Victoria Diamond Jubilee in 1897 and in the same year the Lambeth Conference.* He crowned Edward VII king. Throughout his time as primate he was involved in the ritual troubles. Straightforward and direct, often giving the impression of curtness, he was nevertheless a man of deep affection, as can be seen in his letters to his children, one of whom, William Temple,* also became archbishop of Canterbury.

See E.G. Sandford (ed.) *Memoirs of Archbishop Temple by Seven Friends* (2 vols., 1906).

PETER S. DAWES

TEMPLE, WILLIAM (1881-1944). Archbishop of Canterbury from 1942. Son of Frederick Temple,* he was educated at Rugby, then became an exhibitioner at Balliol, gaining a double first. In 1904 he was appointed a fellow of Queen's. In 1906 he was refused ordination by Bishop Paget of Oxford, who thought Temple was insufficiently certain concerning the doctrines of the Virgin Birth and resurrection. After further discussion (and with Paget's consent) he was ordained by Archbishop Davidson in 1908. In 1912 he became headmaster of Repton and contributed to the volume *Foundations.* He became in 1914 rector of St. James', Piccadilly, which he later resigned during the war years to become the secretary of the Mission of Repentance and Hope and later the leader of the Life and Liberty movement which resulted in the enabling act setting up church councils and the church assembly.

In 1921 he was appointed bishop of Manchester, and in 1929 archbishop of York. In 1942 he became archbishop of Canterbury. Temple combined a first-rate philosophical mind with theological acuteness and great social awareness. He was for many years president of the Workers Education Association. He chaired the Conference on Christian Politics, Economics and Citizenship in 1924, and his last book was entitled *Christianity and Social Order.* He was also in the forefront of ecumenical affairs and presided at the meeting which inaugurated the British Council of Churches. He had a great deal to do with commending the Church of South India* to the Lambeth Conference.*

His concern to express the Christian faith was done both in sermons, particularly in well-known missions to the universities, and in his books; particularly noteworthy among these are *Mens Creatrix, Christus Veritas,* and his Gifford Lectures *Nature, Man and God.* At the same time, his *Readings in St. John's Gospel* shows that his scholarship was matched by his devotion.

Toward the end of his life he himself noted the divergence between his own incarnational theology and that of the theology of redemption then coming into fashion. He was often said to be too much of a philosopher for the theologians, and too much of a theologian for philosophers. His sudden death in 1944 was widely felt as a serious loss to the worldwide church.

See F.A. Iremonger, *William Temple, His Life and Letters* (1948).

PETER S. DAWES

TEMPTATION OF CHRIST. All three synoptic gospels record that Jesus was tempted after His baptism and before He began His ministry. The account in Mark is very brief (Mark 1:12f.). Matthew and Luke give fuller accounts, the only substantial difference being the order of the last two temptations (Matt. 4:1-11; Luke 4:1-13). Matthew's account seems to work toward a psychological climax while Luke's appears to be more governed by geography. Jesus is tempted to turn stones into bread, to throw Himself off the pinnacle of the Temple so that God will rescue Him spectacularly, and to gain control of the world through worshiping Satan. Each of the temptations seems to be directed both to His personal relationship to His Father and to His mission on earth. "If you are the Son of God . . ." echoes the declaration made at His baptism and is heard again when His other period of intense testing in the passion reaches its climax on the cross (Matt. 27:40,43).

The temptations to be dissatisfied with God's provision, God's methods, and God Himself are all answered from the Book of Deuteronomy. This signifies Jesus' understanding that He was in His forty days in the wilderness recapitulating the experience of Israel in their forty years in the wilderness. Similar examples of Israel's temptations as also being experienced by Christians are found in 1 Corinthians 10:1-13. It seems as if there is also a parallel with the temptations of mankind as a whole as shown in Genesis 3 (cf. 1 John 2:16). Thus Jesus faces and overcomes representatively the temptations of all men (Heb. 2:14-18; 4:14-16, 5:7-10). The season of Lent, which was originally a preparation for the Easter baptism, in due course became a forty-day remembrance of the temptation of Jesus. R.E. NIXON

TEN ARTICLES, THE (1536). A summary of articles of faith of the "new" Church of England, these were adopted by Convocation to comply with the wishes of Henry VIII. The Bible, the three universal creeds, and acts of the first four councils were said to be authoritative. Baptism, the Lord's Supper, and penance were accepted sacraments. Transubstantiation in the Lord's Supper was not mentioned, but the real presence was asserted. Images could be used, but were not to be worshiped; but intercession might be made to the saints. Justification was more closely linked with faith, although works helped to justify. Prayers and Masses for the dead and purgatory were denied. These articles, which revealed Lutheran influence, were replaced in 1537 by another statement, the Bishops' Book.* EARLE E. CAIRNS

TEN COMMANDMENTS, THE, see COMMANDMENTS, THE TEN

TENISON, THOMAS (1636-1715). Archbishop of Canterbury from 1695. Ordained privately by Bishop Duppa in 1659, he became a fellow of his own Cambridge College (Corpus Christi) in 1662. As a don he attacked the views of Thomas Hobbes,* and as vicar of St. Andrew-the-Great won respect by his example during the plague. He became rector of St. Martin-in-the-Fields, London, in 1680. He preached the funeral sermon of Nell Gwynne (1687), attended Monmouth at his execution (1685), published *An Argument for Union* with the Dissenters (1683), and joined the seven bishops in their stand against James II. In favor after the Revolution for his Latitudinarian views and Whig sympathies, he became archdeacon of London in 1689, was consecrated bishop of Lincoln in 1692, and promoted to Canterbury three years later. As a member of the 1689 commission working on a comprehensive scheme, he compiled a list of all the Dissenters' scruples over the liturgy. The Tories greatly reduced his influence in Anne's reign, but he took active steps to secure the Hanoverian succession. He is also remembered for establishing the first public library in London (1684) and for his vigorous support for the foundation and early growth of the Society for the Propagation of the Gospel. John Evelyn approved of Tenison's stand for Christian morality in the loose-living days of Charles II.

See E. Carpenter, *Thomas Tenison* (1948).

JOHN TILLER

TENNENT, GILBERT (1703-1764). Presbyterian minister and revivalist. Eldest son of William Tennent,* he was born in Ireland, read theology under his father, and was licensed to preach in 1725. Next year he was ordained by Philadelphia Presbytery and became minister of the New Brunswick Presbyterian Church in New Jersey. Theodore Frelinghuysen* befriended him and guided him in his revivalistic ministry among the Scotch-Irish Presbyterians in what was the opening phase of the Great Awakening.* The New Brunswick Presbytery, founded by Tennent and graduates from his father's "Log College," became the center of Presbyterian revival and controversy in the middle colonies. Tennent accompanied George Whitefield* through the colonies in 1740-41.

His famous sermon, "The Danger of an Unconverted Ministry" (1740), was a broadside against the Philadelphia Synod's opposition to ordaining graduates of the Log College and was a factor in causing the Old Side-New Side schism in 1741. From 1743 until his death Tennent pastored the Second Presbyterian Church of Philadelphia. He and Samuel Davies* visited Britain (1753-55) to raise money for the new College of New Jersey (Princeton). HOWARD A. WHALEY

TENNENT, WILLIAM (1673-1746). Presbyterian minister and educator. Born in Ireland, he graduated from the University of Edinburgh, and in 1706 was ordained priest in the Church of Ireland. About 1717 he emigrated to Philadelphia, where he was admitted to the Presbyterian ministry. He held pastorates in Pennsylvania and New York before becoming minister at Neshaminy, Pennsylvania, where he stayed for the rest of his life. About 1735 he built a small log building on his property, where he educated his three younger sons for the ministry. Some fifteen other men were trained in what came to be derisively called the "Log College" by its detractors because the influential young men trained there were active revivalists who did not meet the educational requirements set by the Philadelphia Synod. In the ensuing controversy Tennent sided with the New Side Presbyterians. His sons Gilbert,* William Jr., John, and Charles and the "Log College men" bear eloquent testimony to Tennent's great contributions to Presbyterian history, especially to the Great Awakening.* HOWARD A. WHALEY

TEN YEARS' CONFLICT (1834-1843). A confrontation between Evangelicals and Moderates* in the Church of Scotland. Both parties believed in ecclesiastical establishment, but patronage was divisive—the presentation of a minister to a congregation upon the nomination of a patron, despite the opposition of the congregation. The

Moderates espoused this procedure. Rejected at the Reformation and in the Revolution Settlement, patronage was reintroduced in 1712. Making the minister dependent on the aristocracy, it tended to separate him from his congregation. The patron's social round was not always conducive to pastoral priorities. If the congregation showed any unwillingness to accept the minister, the civil magistrate would endorse the patron's choice. During the heyday of Moderatism the Church of Scotland lost one-sixth of its membership to secession groups bearing the Presbyterian name.

In 1834 the Evangelical party wanted to make the congregation's consent essential to the issuing of a call, and by passing the Veto Act gave the congregations the right to refuse the patron's nominee. That year the patron presented to the congregation of Auchterarder a nominee whom the congregation refused to accept; the presbytery upheld this decision. The nominee took action in the civil courts which declared his right to ordination and the emoluments of the church. After further clashes elsewhere between church and state, it became clear that the Church of Scotland was not free to govern itself. Evangelicals were faced with the choice of rescinding the Veto Act or leaving the church. Parliament would not move to help them. In 1843, therefore, the Disruption* took place, and the Ten Years' Conflict ended with the birth of the Free Church of Scotland.*

See R. Buchanan, *The Ten Years' Conflict* (2 vols., 1849). ARTHUR CLARKE

TERESA, MOTHER (Agnes Gonxha Bojaxhiu) (b. 1910). Missionary to India. Born in Yugoslavia, daughter of an Albanian grocer, she went to India in 1928 as a teacher under the Roman Catholic Church. Her heart soon went out to the poor of Calcutta, and after nursing training she moved into the slums. In 1948 she founded the Order of the Missionaries of Charity and organized schools and dispensaries. She became an Indian citizen, and adopted the sari as the habit of her order, which received canonical sanction from Pius XII in 1950. A leper colony was built, and the blind, crippled, aged, and dying were served. In 1963 the Indian government honored her. In 1964 Paul VI on his visit gave her a limousine which she promptly disposed of, giving the proceeds to aid her leper work. In 1971 she received the first Pope John XXIII Peace Prize. By the mid-seventies her order numbered some 700 nuns in 60 centers in Calcutta and 70 worldwide centers from Britain to Australia. J.D. DOUGLAS

TERESA OF AVILA (1515-1582). Carmelite reformer, mystic, and writer. Born into a good Spanish family at Avila and educated by Augustinian nuns, she entered the Carmelite Convent of the Incarnation there in 1533, at first suffering from serious illness which caused her withdrawal. Reentering, but purposeless, it was not until the 1550s that she sought the life of perfection, praying before a statue of the scourged Christ at the pillar, and soon to have divine locutions and heavenly visions. She knew ecstasy, and among her spiritual experiences was the mystical piercing of the heart by a spear of divine love. She wrote about this but never stressed it unduly, recognizing the dangers. Domingo Banez,* a Dominican priest, influenced her much in these years.

In middle life, with Peter of Alcántara* as her confessor, she founded a convent under the original (discalced) Carmelite Rule, St. Joseph's at Avila (1562). There she wrote her first work, *The Way of Perfection,* as instruction for her nuns. From 1567 she traveled in Spain, founding houses for both nuns and friars and receiving much assistance from John of the Cross.* She wrote *Life* (autobiography to 1562), *Book of Foundations* (account of her convents), and *The Interior Castle* which made her a doctor of the spiritual life, scientifically delineating the life of prayer from meditation through mystical marriage and noting intermediate stages. As the reformer of the Carmelite Order and a mystic, she proved in her distinguished life that great practical achievement and highest contemplation could coexist properly.

See V. Sackville-West, *The Eagle and the Dove* (1943), and E.A. Peers, *Mother of Carmel* (1945). C.G. THORNE, JR.

TERESA OF LISIEUX (1873-1897). Carmelite and devotional writer. Born Marie Françoise Thérèse Martin, she had a most trying childhood, including a grave illness. This led, however, at Christmas 1886 to a conversion experience that led to her monastic commitment. Religious from her earliest years, she now discovered what surrender and personal rebirth meant, and special permission was given her to enter the Carmelite convent in Lisieux at only fifteen. From 1893 she was acting novice-mistress and wrote *Little Way;* Benedict XV said it "contained the secret of sanctity for the entire world." Dying of tuberculosis, she wrote her autobiography, the wide circulation of which has led to her extensive cult. Canonized exceptionally early (1925), she has been named patroness of foreign missions and joined with Joan of Arc as patroness of France (1947). C.G. THORNE, JR.

TERMINISM. The teaching that there is a specific period of grace within which an individual can be converted. The view arose within the German Pietist Movement as a protest against what was regarded as the ineptitude of "deathbed" conversions. Originating as a response to an awkward pastoral situation, it soon became a point in speculative theology in which it was claimed that there is a time of grace for everyone, determined only by the will of God, beyond which it is unwarrantable to seek conversion. The issue is a good example of the speculative dogmatizing that has arisen out of acute practical problems. The chief controversialists were J.G. Böse (1662-1700), A. Rechenberg (1642-1721), and T. Ittig (1643-1710). It remained a general issue within Lutheran theology until about 1704. PAUL HELM

TERRITORIALISM. A theory of church government brought in with the Reformation but formulated later—e.g., by C. Thomasius (1655-1728)

and J.H. Böhmer (1749). The temporal ruler, by virtue of his office, possesses the right of regulating his country's ecclesiastical affairs and of banishing those disturbing the peace of the church. He himself is not subject to ecclesiastical discipline. Only one religion is permitted in a territory. It is summarized in the formula adopted at the Peace of Augsburg* (1555): *cuius regio, eius religio.* Opposite views are the "Collegialism" of H. Grotius,* S. Pufendorf,* and C.M. Pfaff, and the "Confessionalism" of the Anabaptist* tradition. J.G.G. NORMAN

TERSTEEGEN, GERHARD (1697-1769). German hymnwriter. Born in Westphalia, he received a classical education, after which he was apprenticed at fifteen to his brother, a Mülheim shopkeeper. After his conversion in the following year, he spent much time in prayer, fasting, and almsgiving. For some years he lived alone near Mülheim as a ribbon-weaver. He suffered great spiritual depression, but found assurance of faith in 1724. He left the Reformed Church, mainly because he would not take Communion with open sinners, but formed no sect. In 1725 he began to speak at prayer meetings in Mülheim and became known as a religious leader. He gave up weaving and was supported by gifts from friends. His cottage was known as a *Pilgerhütte,* a retreat for the *Stillen im Lande.* His influence spread throughout Germany, Holland, and Scandinavia, although temporarily restricted by a Prussian law against conventicles. He conducted an immense correspondence, translated works by mystics and Quietists, including Madame Guyon,* and delivered sermons which were published as *Geistliche Brosamen von des Herrn Tisch gefallen* (1769-73). Most of his 111 hymns were published in his *Geistliches Blumen-Gärtlein* (1729), which ran through many editions. The best known in English were both paraphrased by John Wesley*: "Thou hidden love of God" and "Lo, God is here!"

See also BEVAN, E.F. and BORTHWICK, J.L.

BIBLIOGRAPHY: H.E. Govan, *Life of Gerhard Tersteegen, with Selections from His Writings* (1898); W. Blankenagel, *Tersteegen als religiöser Erzieher* (1934); C.P. van Andel, *Gerhard Tersteegen* (1961). JOHN S. ANDREWS

TERTIARY. A lay person who belongs to the third order in a monastic order, after monks and nuns. Tertiaries entered the Franciscans* in the thirteenth century, the Augustinian Hermits* in 1400, the Dominicans* in 1405, the Servites* in 1424, and the Carmelites* in 1452. They were found also in other orders. Their order was established to permit lay people to participate in the orders' work. They were subject to the order's leadership in inner discipline, but external discipline rested with the bishop alone. Priests in the orders formed chapters of tertiaries who recited an office, underwent novitiate, were bound by discipline, but professed no vows. Tertiaries regular live in a convent. Only the Franciscan tertiaries regular have a separate Rule for themselves. JOHN GROH

TERTULLIAN (Quintus Septimius Florens Tertullianus) (c.160/70-c.215/20). African moralist, apologist, and theologian. Few details of his life are certain. Reared in the cultured paganism of Carthage, he imbibed a solid literary, rhetorical, and perhaps legal training. He possibly practiced as an advocate, but is not identifiable with the Roman jurist Tertullianus, though he probably visited Rome. After moderated immorality he became a Christian in unknown circumstances, plausibly influenced by the fortitude of martyrs. He married a Christian wife, and may have had children and become a widower, rejecting remarriage as a Montanist.* He was not a presbyter, but most likely a catechist or teacher. After espousing "the new prophecy," he left the catholic church c.206. Augustine* reported that he later abandoned Montanism and founded the Tertullianists, whose last remnant had rejoined the catholic church in Carthage in Augustine's lifetime. However, "Tertullianist" was probably the African name for Montanist.

Tertullian is known almost exclusively through his writings. His Greek works (on baptism, games, and shows, and on veiling of virgins, i.e., for local Greek-speaking Christians) have not survived, but thirty-one Latin works remain, the first significant corpus of Christian Latin literature. (Lost works included *Ecstasy, Paradise, Fate, The Hope of Believers, Flesh and Soul,* and *Against the Apellians.*) His writings span the period roughly from 196 to 212; their order and individual dating are often uncertain. They are nearly all controversial, revealing an initial preoccupation with apologetic and Christian mores, later partly displaced by refutation of heretics and Gnostics, though Montanism accentuated his ethical-ascetic thrust.

Several apologies, notably the *Apology* itself (c.197), highlight the legal and moral absurdities of persecution, while *To the Martyrs* (197) encourages Christian "athletes" in prison. The Stoic-indebted *Testimony of the Soul* (198) discerns the *anima naturaliter Christiana* in spontaneous ejaculations like "Good God!"

The *Prescription of Heretics* (203; ET S.L. Greenslade, 1956) refuses them appeal to the Scriptures, which rightfully belong only to churches with apostolic pedigrees. Gnostics* and Docetists* are the targets of several works c.204-7, especially *The Soul* (c.206; ed. J.H. Waszink, 1947), a lengthy learned rebuttal of Gnostic psychology. *Against Marcion* (207/8; ET E. Evans, 2 vols., 1972), twice rewritten and expanded, utilizes earlier lost refutations and constitutes an invaluable source. The Montanist *Against Praxeas* (c.210; ET Evans, 1948) is the most advanced exposition to date of Trinitarian doctrine.

From the first, Tertullian's practical works advocate disengagement from pagan society. *Idolatry* (196/7; ET Greenslade, 1956) blacklists numerous professions contaminated by paganism (cf. *The Soldier's Garland,* 208) while other treatises deal with *Women's Dress* (196-c.206) and *Shows* (196/7). His move into Montanism intensified this rigorism. After reluctantly condoning remarriage in *To His Wife* (c.200), he condemned it outright in *Monogamy* (c.210; both ET W.P. Le Saint, 1951). Once he tolerated flight from persecution, but later *Flight in Persecution* (c.208) outlawed any "unspiritual" avoidance of martyrdom. The catholic *Penitence* (c.200) permitted one postbaptismal penance, but when the bishop of Carthage extended it to adultery and fornication, the Montanist *Purity* (c.210; both ET Le Saint, 1959) reserved its authorization to the "pneumatics" who judged it never expedient.

Tertullian also produced the earliest exposition of the Lord's Prayer in *Prayer*, and the first extant treatise on *Baptism* (both c.200; ET Evans, 1953, 1964). They are homiletic-catechetical in form, like *Penitence* and *Patience* (c.200); other writings may have similar origin or structure.

Tertullian's sophistic brilliance and literary versatility, ruthless vigor as disputant and polemicist, fecundity in uttering memorable dicta, and fervent religious immediacy make him a captivating writer as well as a priceless mirror of early African Christianity. He influenced magisterically the ethos of the African Church and subsequent theology not only in the West, by providing terminology for classical Trinitarian and christological formulations and by advancing dogmatic development. He fostered juridically colored Latin interpretations of the work of Christ and relations between God and man, although facile assumptions have magnified the importance of his legal expertise. The chief non-Christian influences on his thought were Stoicism and the rhetorical tradition.

BIBLIOGRAPHY: I. Works: details in J. Quasten, *Patrology* 2 (1953), pp. 248-340; *Corpus Christianorum* 1, 2 (1954); T.D. Barnes, *Tertullian: A Historical and Literary Study* (1971); complete ET in *Ante-Nicene Christian Library* 7, 11, 15, 18 (1868-70).

II. Studies: P. Monceaux, *Histoire Littéraire de l'Afrique Chrétienne* 1 (1901); A. d'Alès, *La Théologie de Tertullien* (2nd ed., 1905); R.E. Roberts, *The Theology of Tertullian* (1924); J. Morgan, *The Importance of Tertullian in the Development of Christian Dogma* (1928); A. Beck, *Römisches Recht bei Tertullian und Cyprian* (1930); W. Bender, *Die Lehre über den Heiligen Geist bei Tertullian* (1961); R. Braun, *'Deus Christianorum': Recherches sur le Vocabulaire Doctrinal de Tertullien* (1962): J. Moingt, *Théologie Trinitaire de Tertullien* (4 vols., 1966-69); T.P. O'Malley, *Tertullien and the Bible* (1967); R.D. Sider, *Ancient Rhetoric and the Art of Tertullian* (1971); J.C. Fredouille, *Tertullien et la Conversion de la Culture Antique* (1972); G. Claesson, *Index Tertullianeus*, 3 vols. (1974-1975).

D.F. WRIGHT

TESTAMENT OF OUR LORD IN GALILEE. Known under this title from the earliest edited version (Ethiopic, in which it is combined with other writings), the *Epistula Apostolorum* is also known in Coptic, on which the best edition is based, and in a Latin fragment. In the form of a circular letter from the eleven apostles to the universal church, it deals with such supernatural questions as the Incarnation and Ascension, somewhat mythologically, but in an explicitly anti-Gnostic way, emphasizing the true flesh of Christ. First dated in the sixties of the second century and ascribed to Asia Minor, it has more recently been linked with Egypt in the early second century. Its antecedents are to be seen in Jewish Christianity, and the Dead Sea Scrolls* may cast some light on it. J.N. BIRDSALL

TESTAMENT OF OUR LORD JESUS CHRIST. An apocryphal work purporting (like, e.g., the *Gospel of Thomas**) to give post-resurrection words of Jesus to the disciples. An apocalypse of the end is followed by detailed prescriptions for church layout, services, ministry, and catechumenate. The prayers have a warm, devotional quality. The "church order" has clearly drawn on, *inter alia*, the lost *Apostolic Tradition* of Hippolytus,* but reflects a later period, with security for the church. The date is impossible to determine: probably in the fifth, perhaps the fourth century. Apollinarian or Monophysite ingredients are sometimes alleged but remain uncertain, though its later use in Monophysite churches may have affected the translated versions. The place of origin is also obscure: references in the Apocalypse suggest Asia Minor, the provenance Syria or Egypt.

The colophon of the published Syriac indicates its translation from Greek in A.D. 687. Coptic, Arabic, and Ethiopic versions exist, and a Latin arrangement of the Apocalypse. The direct attribution of these prescriptions and liturgies to the Risen Lord is the climax of a series of "apostolic" church orders, and perhaps reflects anxious championship of a particular local use.

A.F. WALLS

TETRAPOLITAN CONFESSION (1530). A Protestant confession of faith drawn up by Martin Bucer* and Wolfgang Capito* at the Diet of Augsburg, and presented by Jakob Sturm in the name of the cities of Strasbourg, Memmingen, Lindau, and Constance. Its purpose was to prevent a rupture in German Protestantism. It had Zwinglian affinities, but its doctrinal formulae were based on the Augsburg Confession, of which the compilers had obtained a copy. It was not generally accepted as was the Augsburg document, but it did become the symbolic formula of the four cities. With Bucer's "Greater Catechism" it was accepted by Strasbourg as binding on that city (1534), on the basis of which the magistrates decreed the banishment of persistent Anabaptists.*

J.G.G. NORMAN

TETZEL, JOHANN (c.1465-1519). Dominican friar. His one claim to fame is his hawking of indulgences* at Jüterbock, near Wittenberg, but just beyond the borders of Saxony. His exaggerated claims stirred Martin Luther's pastoral concerns (many Wittenbergers had bought indulgences from Tetzel) to such a degree that he formulated a set of theses for academic debate. These are the Ninety-Five Theses* of 31 October, 1517. Tetzel was subcommissary of the regions of Magdeburg and Halberstadt, and the indulgences were ostensibly meant for the building of St. Peter's Basilica in Rome, although the elector of Mainz profited in part from payment of debts to the Fuggers. CARL S. MEYER

TEUTONIC KNIGHTS, ORDER OF. German religious and military order. Founded by Lübeck and Bremen merchants during the siege of Acre in the Third Crusade and confirmed by Pope Clement III in 1199, it soon became an important order in Germany. After aiding Hungarian King Andrew II in repulsing the Cumans, the knights were invited by Polish Duke Conrad of Masovia to help crush the heathen Prussians. Frederick II named Grandmaster Hermann of Salza (1209-39) a prince of the empire in 1226 and authorized the acquisition of East Prussia. The Teutonic Order began the conquest in 1231, sponsored large-scale colonization by German peasants and merchants during the next century, and transferred its residence to the fortress Marienburg in 1309.

The order failed to unite its East Prussian and Livonian holdings and declined in the face of Polish, Lithuanian, and Russian resurgence. After the defeat at Tannenberg (1410) and a series of subsequent military disasters, the Treaty of Thorn (1466) reduced its territory to a portion of East Prussia. Grandmaster Albert of Hohenzollern (1490-1568) accepted Lutheranism in 1525 and secularized it as the duchy of Prussia under the suzerainty of the Polish king. Those remaining Catholic transferred the seat to Mergentheim in Franconia, and some participated in the Turkish wars. Although Napoleon dissolved the order in Germany in 1805, its existence continued in Austria with an archduke as grandmaster until 1918. It received a new Rule in 1929 establishing its strictly religious character, and the order now works chiefly in schools and hospitals.

RICHARD V. PIERARD

TEXTUS RECEPTUS. A Latin term meaning "the received text" of the NT. The discovery and collection of Greek manuscripts of the NT at the time of the Renaissance and the invention of printing led to a number of printed editions of the Greek NT in the sixteenth and seventeenth centuries. Erasmus* published his edition in 1516 and the "Complutensian Polyglot"* came out in 1522. These were used by R. Stephanus in producing his edition of 1550. There followed ten editions by Theodore Beza,* beginning in 1565. The Dutch brothers Elzevir drew especially upon Stephanus (only 287 variants are found between them) in their editions of 1624 and 1633. In the preface to the latter they use the phrase "*Textum ergo habes, nunc ab omnibus receptum*" ("You have the text which is now received by all"). This text was basically the Byzantine Text* appearing in most of the late manuscripts, and was assembled before the science of textual criticism had been developed. It underlies the Authorized Version (KJV).

R.E. NIXON

THADDAEUS. The name is found only in the lists of the twelve apostles recorded in the first two gospels (Matt. 10:3; Mark 3:18). In its place Luke has Judas the son (or brother) of James (Luke 6:16; Acts 1:13). The unpopularity of the name "Judas" because of the treachery of Iscariot may have led to this man's being known by another name. "Thaddaeus" is thought to have been derived from Aramaic meaning the "breast nipple." This might suggest that he was a character of almost feminine tenderness. Some Western manuscripts read "Lebbaeus" at Matthew 10:3. This is usually thought to be inauthentic and, as probably derived from Hebrew *leb,* "heart," may have been an explanation of the name "Thaddaeus." If Judas is the same as Thaddaeus, he is not likely to be the brother of our Lord or the author of the epistle of Jude, though he may be the same as "Judas (not Judas Iscariot)" of John 14:22. Jerome equates Thaddaeus, Lebbaeus, and Judas of James, and tells how he was sent on a mission to Abgar, king of Edessa. He was thought by Eusebius to be one of the seventy disciples sent out by Jesus (Luke 10:1).

R.E. NIXON

THAILAND. The story of Protestant Christian missionary work in this Buddhist land is one of repeated disappointments and frustration. Yet it has enjoyed the services of some outstanding missionaries. Karl Gutzlaff,* a German, was one of the first two to arrive in 1828. In less than three years he saw the Bible completely, though imperfectly, translated into Thai, and he produced a grammar and dictionary. But his wife and infant twin daughters died in 1831, and he himself had to leave the country, apparently in a dying condition. In all, sixty-one missionaries have died on this field.

The American Board took up the challenge in 1831 with David Abeel, but in 1849 it officially withdrew. The American Missionary Association, taking on the support of former American Board missionary Daniel Beach Bradley, had a longer history, but it came to be almost exclusively a one-man work. Physician and printer as well as preacher, no missionary made a more lasting impression on the country than Bradley. He was a good friend of King Mongkut. Yet he had few converts. After his death in 1873, the AMA also withdrew.

American Baptist work began with John Taylor Jones in 1833. He completed translating the NT in 1843 and baptized several Chinese, who comprise an important minority in the country. The Chinese Baptist Church organized in 1837 was the first Protestant church in the Far East. This mission's efforts were sporadic, and they ended in 1893.

The major continuing work is that of the American Presbyterians, begun in 1840. They had little success among the Thai of the south, but when Daniel McGilvary went to Chiengmai in the north in 1867 he began an important work with Laos and other hill tribes. He was largely responsible for the edict of toleration in 1878. In 1934 the Presbyterians formed the Church of Christ in Thailand with 8,713 members. In 1957 the mission dissolved and turned over all the work to this national church.

In 1929 the Christian and Missionary Alliance* entered neglected E Thailand from neighboring Cambodia. But the great influx of new missions came after World War II and the closing of China. Overseas Missionary Fellowship (the former China Inland Mission) began in 1951 primarily to reach hill tribes similar to those in China. It now has the largest number of missionaries. Work has extended also to the many Muslim Malays of the southern peninsula.

In spite of increased conversions, Christians still represent not more than 1 percent of the population. The Church of Christ in Thailand is a member of the World Council of Churches. It was host to the East Asia Christian Conference in 1949. It does not, however, represent many of the newer works.

BIBLIOGRAPHY: A.J. Brown, *One Hundred Years* (1936); E.A. Fridell, *Baptists in Thailand and the Philippines* (1956); I. Kuhn, *Ascent to the Tribes* (1956); K.E. Wells, *History of Protestant Work in Thailand* (1958); J.H. Hunter, *Beside All Waters* (c.1964); D.C. Lord, *Mo Bradley and Thailand* (c.1969).

HAROLD R. COOK

THANKSGIVING DAY. An annual holiday in the USA and Canada for expressing thanks to God for the harvest and other blessings. The original Thanksgiving of the Pilgrims was ordered by Governor Bradford after the first harvest in Plymouth Colony (1621). Special days were often appointed in Puritan New England for thanksgiving or fasting. Beginning in Connecticut (1649), the observance of an annual harvest festival spread throughout New England by the end of the eighteenth century George Washington proclaimed the first national Thanksgiving in 1789. With Lincoln's proclamation (1863) it became an annual observance. By an act of Congress (1941) Thanksgiving Day is the fourth Thursday of November. Although church services may be held, Thanksgiving is typically a family festival. The second Monday in October is observed in Canada.

ALBERT H. FREUNDT, JR.

THEANDRIC ACTS. A term coined by Dionysius the Pseudo-Areopagite (fifth century) to denote the characteristic activity of the God-Man. Christ "did not perform divine acts as God nor human acts as man, but as the God-Man he manifested a kind of new theandric activity." It was used by Monophysites and Monothelites (e.g., Severus* of Antioch) of the one nature and one will in Christ, substituting "one" for "new." Cyrus of Alexandria was won over to the formula by Sergius of Constantinople, including it in the Act of Union (633) by which the Theodosian Monophysites of Egypt were reconciled with the church. Its Monothelite use was condemned by Martin I* (Lateran Council, 649), who permitted it only to designate the union in Christ of the two distinct operations. Maximus the Confessor* and John of Damascus* used it in an orthodox sense.

J.G.G. NORMAN

THEATINES. An order founded early in the sixteenth century by Gaetano of Tiene, Bonifazio da Colle, Paolo Consiglieri, and Giovanni Pietro Caraffa (later Pope Paul IV) to combat heresy. Growing out of the work of the Oratory of Divine Love, this order made up of regular clergy sought to maintain the integrity of the Roman Catholic Church primarily by preaching, although in cooperation with the Inquisition* it became one of the principal instruments for suppressing the Reformation in Italy. The order spread across the Alps into France, Germany, Poland, and Spain. In some ways it provided a pattern for the organizational structure of the Society of Jesus.*

THEBAIC VERSION, see EGYPTIAN VERSIONS

THEISM. The term originated in England in the seventeenth century. It may refer either to a philosophical theory about the nature of God, or to a central, necessary aspect of historic Christianity. The latter is theistic in that it affirms that God is the creator of the universe (He is transcendent), and also that He sustains it, and by acts of particular providence and gracious redemption is redeeming it (He is immanent). It is possible to conceive of various versions of theism, e.g., dualistic or polytheistic, but the classic Christian view is monistic, hence the metaphysical problem of the origin of evil. Theism is usually contrasted with Deism,* the view that God is simply the transcendental "ground" of the universe, and with pantheism,* the belief that God is to be identified with the universe. Christian theism, although always recognizing the problem of referring to the transcendent God (see, e.g., Thomas Aquinas's doctrine of analogy and Calvin's view of "accommodation"), has come under heavy attack in modern times from antimetaphysical philosophies such as Logical Positivism and Existentialism,* and from those theologians who, like Karl Barth,* attempt to give an account of Christian theology in "christological" terms.

PAUL HELM

THEOBALD (d.1161). Archbishop of Canterbury from 1138. Norman-born, he became a monk at Bec and was later abbot. As archbishop he crowned Stephen in 1141, and his administrative capacity was shown during the politics of that reign. Appointed papal legate in 1150, having been bypassed earlier by Innocent II in favor of Henry of Blois, he worked for cooperation between church and state. In defiance of Stephen he attended the Council of Reims in 1148, whereupon he was exiled, which act led Eugenius III to put England under interdict. Again, in 1152, Theobald refused under papal orders to crown Stephen's son Eustace, and fled to Flanders. Having reconciled Stephen and Henry of Anjou in 1153, he crowned Henry II in 1154 and recommended Thomas Becket* as chancellor. John of Salisbury* was his own secretary and chief advisor during that time. He encouraged the study of canon law and opposed the monastic claims for exemption from diocesan ruling.

C.G. THORNE, JR.

THEOCRACY. The term was coined by Josephus to express the concept of a God-governed state. The Hebrew people expressed a distinct belief in this type of government, although it took various forms during their historical existence as a nation. At Sinai they became the "holy nation" of the Lord (Exod. 19:6) and soon established an amphictyony—a religious confederation of tribes pledged to the service of Yahweh their king. Later, during the period of the monarchy, the king became the representative of Yahweh's rule, "the

Lord's anointed" (Ps. 2:2; 20:6). Therefore God could use His prophets to dethrone even the king (1 Sam. 15:26; 16:1ff.). Still God remains the real King whom Isaiah sees as high and lifted up (Isa. 6:1). During the postexilic period, the mediation of Yahweh's spiritual rule was transferred to the priest, particularly the "high priest" (Hag. 2:2; Zech. 3:1). In the NT, the rule of God is more eschatological in nature, but this depends on one's interpretation of the Kingdom of God* concept. Islamic politics, Calvinism in Geneva, and Puritanism in New England are further examples of attempts at theocracy. It should be recognized that theocracy is always more idealistic than realistic and is an article of faith rather than a demonstrable system. JOHN P. DEVER

THEODICY. The question of the justification *(dikē)* of God Himself *(theos)* is raised as a response to the problem of actual human experience within a world in which fulfillment is qualified or shattered by premature death, mental or physical retardation, destructive social conditions including war, the accidents of natural or man-made catastrophe, or the terror of history itself. The term appears first in the title of a work published by G.W. Leibnitz* in 1710, and the ensuing discussion is shaped by the spirit of the so-called Enlightenment; but the problem is ancient, as illustrated by reference to the religio-secular wisdom literature of the Near East; Babylonian, Egyptian, and biblical. Presumably the problem rests ultimately in the effort to do justice to finite freedom in relation to divine creativity. Tillich* has rightly raised the observation that the appropriateness of the question of theodicy is maintained only with respect to the consideration of " 'my' creaturely existence," and that the correlate question cannot therefore be raised "with respect to persons other than the questioner," which observation is not far removed from that Reformation principle of the ultimate sovereignty and mystery of God upon which all notions of destiny and predestination rightly depend.
CLYDE CURRY SMITH

THEODORA I (c.500-548). Wife of the Byzantine emperor Justinian I.* After a notorious career as an actress, with adventures in Syria and North Africa, she became a Christian and married Justinian in 523, after he had persuaded his uncle Justin I to repeal the law forbidding the marriage of senators with actresses. Justinian and Theodora were crowned together in 527, and afterward she exercised a great influence over his policy. She took the side of the Monophysites (see MONOPHYSITISM), whom she had known in Egypt, and so caused the reactionary policy of Justinian which led to the Three Chapters Controversy.* She showed great bravery during the insurrection at Nika in 532, and she opened homes for prostitutes. There is a contemporary portrait of her in mosaic in the apse of the church of San Vitale, Ravenna. PETER TOON

THEODORE OF MOPSUESTIA (c.350-428). Antiochene exegete and theologian. Of wealthy Antiochene parentage, educated with John Chrysostom* under the eminent rhetorician and philosopher Libanius, like John he abandoned a secular career c.369 for the monastic school of Diodore* (of Tarsus). When marriage and the bar proved tempting, John persuaded him to persevere. He was ordained presbyter by Flavian c.383 and in 392 made bishop of Mopsuestia in Cilicia. During his lifetime his erudition and prolific literary versatility were renowned and his orthodoxy virtually unquestioned, but after the Council of Ephesus* (431) his standing became posthumously entangled with that of his condemned pupil Nestorius (see NESTORIANISM). Rabbula* of Edessa pressed the attack, Cyril* of Alexandria wrote *Against Diodore and Theodore,* and despite Chalcedon's* apparent favor, Theodore and his writings were anathematized in the first of the Three Chapters* by Justinian* (543/4) and the Second Council of Constantinople* (553), although defended by the West, especially Facundus* of Hermiane.

Scholars have judged the extracts from Theodore used by Leontius of Byzantium* to incriminate him so tendentious that modern expositions of his theology trust rather his undoubted works, several of which, mostly exegetical (see bibliography), have been recovered from manuscript catenae and translations, chiefly Syriac (whose reliability is sometimes questionable). With the *Commentary on the Minor Prophets* and numerous fragments—e.g., on Gen. 1–3 (he commented on most biblical books)—these reveal the most brilliant Antiochene exegete ("the Interpreter" of the Nestorian Churches) employing varied critical methods with remarkable insight (even rejecting the canonicity of some OT and NT books), though not wholly eschewing spiritual or typological meanings. His was the first attempt to place the Psalms historically. His lost *Against the Allegorists* adequately explains the Origenists' promotion of his condemnation.

Although only fragments (often uncertain) of his dogmatic-controversial works—mostly against Arius, Eunomius, and Apollinaris—are extant (except for a *Disputation with Macedonians* in defense of the Spirit's divinity in 392 at Anazarbus; Syriac ed. F. Nau, 1913), his creative contribution to Christology, especially in refuting Apollinarianism,* is increasingly acknowledged. If hindsight exposes his shortcomings, particularly in terminology, he also partly anticipated Chalcedon. The raw materials of Theodore's theology were more biblical and less philosophical than the Alexandrians'. It focused on immortality, achieved by a conjunction with God patterned on the divine-human conjunction in Christ and initiated through the sacraments. Thus his baptismal catecheses delivered at Antioch c.390 (his sole surviving practical works; lost are treatises on priesthood, monasticism, against magic, and his letters) reject a symbolic view of the Eucharist (yet interpret "daily bread" of normal food). They are invaluable commentaries on Antiochene baptismal, eucharistic, and penitential practices.

Theodore entertained Julian of Eclanum* c.421, wrote a lost anti-Augustinian *Against Defenders of Original Sin,* and was made the "father of Pelagianism"* by Marius Mercator.* His supposed teaching on the effects of Adam's sin and man's created mortality appears significantly similar, but again recently discovered works allegedly suggest otherwise. The Syrian affinities of Pelagian ideas are under renewed scrutiny (see RUFINUS "THE SYRIAN").

BIBLIOGRAPHY: L. Pirot, *L'Oeuvre Exégètique de Théodore de Mopsueste* (1913); F.J. Reine, *The Eucharistic Doctrine and Liturgy of the Mystagogical Catecheses of Theodore of Mopsuestia* (1942); E. Amann in *Dictionnaire de Théologie Catholique* (1946), pp. 235-79; F.A. Sullivan, *The Christology of Theodore of Mopsuestia* (1956); J. Quasten, *Patrology* 3 (1960), pp. 401-423; R.A. Greer, *Theodore of Mopsuestia: Exegete and Theologian* (1961); L. Abramowski in *Zeitschrift für Kirchengeschichte* 72 (1961), pp. 263-93; U. Wickert, *Studien zu den Pauluskommentaren Theodors von Mopsuestia* (1962); R.A. Norris, *Manhood and Christ: A Study in the Christology of Theodore of Mopsuestia* (1963); A. Grillmeier, *Christ in Christian Tradition* (2nd. ed., 1975).

D.F. WRIGHT

THEODORE OF RAITHU (sixth century). Monk and presbyter of a monastery in Raïthu, located on the SW coast of the Sinai peninsula. Some scholars claim that he is identical with Theodore of Pharan. The one work that can be attributed to him with certainty, *Proparaskuē (Praeparatio)*, was written between 537 and 553. It defended Chalcedonian Christology (see CHALCEDON) and the theology of Cyril* of Alexandria, and attacked two Monophysites, Julian of Halicarnassus and Severus* of Antioch. It also tried to counter the views of Manes, Nestorius (see NESTORIANISM), Eutyches*, Apollinarius, Paul of Samosata*, and Theodore of Mopsuestia.*

THEODORE OF STUDIUM (759-826). Byzantine abbot. Born in Constantinople, he became monk (787) and abbot (794) of Saccudion monastery in Bithynia. Having opposed Constantine VI's adulterous second marriage, he was briefly exiled. In 797 the community at Saccudion, vulnerable to Muslim attack, transferred to Studius in Constantinople, which became a center of monastic reform. After another two-year exile because of differences with Patriarch Nicephorus*, Theodore was again in disputation when in 814 the Iconoclastic Controversy* was revived, and he led the opposition to the Iconoclasts. Exiled again, he was subsequently allowed to return, but barred from reassuming his post as abbot. Theodore stoutly maintained the church's independence of the state, did not hesitate to protest when he saw compromise in his patriarchs, and even appealed to the popes of Rome. Apart from his defense of image worship he is known for his adaptation of the Rules of St. Basil which became the norm in the Eastern Church. The author of many works, he left much of particular historical value in nearly 600 letters. J.D. DOUGLAS

THEODORE OF TARSUS (c.602-690). Archbishop of Canterbury from 669. Educated at Tarsus and Athens, Theodore was among numerous Eastern Christian refugees from the Arab invasions, bringing to England a high standard of culture. He was appointed to the see of Canterbury by Pope Vitalian, recommended by Hadrian the African*, the monk to whom the post had been offered. After touring England, Theodore worked at establishing the primacy of Canterbury by convoking the first council of the entire English Church at Hertford* in 673 and regulating affairs of dioceses. With Hadrian and Benedict Biscop*, Theodore promoted conformity with Rome and sent to Pope Agatho a declaration of orthodoxy written at the synod of Hatfield (680). The school at Canterbury, enriched by Theodore's manuscripts, became a leading center of education for Roman law and Greek, whence Roman influence spread to Wearmouth and Jarrow monasteries in Northumbria, both founded by Benedict. None of Theodore's scholarly writings survives.

DANIEL C. SCAVONE

THEODORET (c.393-c.458). Bishop of Cyrrhus (Syria). Born in Antioch, he seems to have been early intended for the religious life, and duly entered a nearby monastery about 416. Seven years later he became bishop of Cyrrhus (an unwanted elevation), where he spent all but two years of the rest of his life, making for himself a reputation not only as theologian, historian, and controversialist, but as a faithful and diligent pastor. In the christological controversy sparked by Cyril* of Alexandria, Theodoret accepted neither of the extreme positions, but held that Christ had two natures, united in one person but not in essence. At the Council of Ephesus* (431) Theodoret protested against both Cyril's procedural opportunism and his doctrine, and afterward wrote a refutation of the anathemas directed by Cyril against Nestorius (see NESTORIANISM; THREE CHAPTERS CONTROVERSY; CONSTANTINOPLE, SECOND COUNCIL OF). Theodoret's continued opposition led to his deposition and exile at the Robber Synod of Ephesus* in 449, but he was restored at the Council of Chalcedon* in 451, though obliged to participate in condemning Nestorius and to accept *Theotokos** as the title of the Virgin Mary.*

As an exegete Theodoret was of the Antiochene* School, and this is reflected in his highly commended short commentaries on the Song of Solomon, the Prophets, Psalms, and the Pauline epistles. His other works included a collection of thirty biographies of monks, and a church history which continues the work of Eusebius down to 428. Of his letters 232 have survived.

J.D. DOUGLAS

THEODORE THE LECTOR (early sixth century). Church historian. He was a "reader" of the Hagia Sophia Church, Constantinople. He wrote two historical treatises. The first (c.520-30) was a *Tripartite History* compiled from the histories of Socrates, Sozomen*, and Theodoret.* The second was his own composition carrying the history from 438 to the accession of Justin I (518). The first has partially survived, but only fragments of the second remain.

THEODOSIUS I (c.346-395). Roman emperor from 379 (surnamed "the Great"). A Spaniard by birth, son of Count Theodosius who between 367 and 374 delivered Britain and Africa from barbarians, he distinguished himself in a campaign against the Sarmatians in 374. Appointed co-emperor by Gratian, he was given the task of restoring order in the Eastern provinces. Mostly by careful diplomacy with the Goths he did this. He was baptized in 380, gave up the use of the title *pontifex maximus,* and made it illegal to depart from the Nicene faith. In 381 he outlawed heretical churches and sects (e.g., Arians), and put their property at the disposal of the orthodox, and called the (First) Council of Constantinople.* In 390, while in a fit of temper, he ordered the punishment of the citizens of Thessalonica after a riot there; 7,000 were killed. Afterward, in response to the demands of Ambrose of Milan, he publicly acknowledged his guilt. PETER TOON

THEODOSIUS II (401-450). Eastern Roman emperor from 408. Grandson of Theodosius I, he was born in Constantinople and became ruler of the East after his father's death. He had to face difficulties caused by the invasion of Vandals in North Africa and by the accession of Attila, north of the Danube. In 425 he founded the University of Constantinople and three years later appointed Nestorius patriarch of the city. This led to the Nestorian controversy which he did not welcome (see NESTORIANISM). He called the Council of Ephesus* in 431, published the Theodosian (Law) Code in 438, and refortified Constantinople in 447. He had great respect for Simeon the Stylite,* from whom he accepted advice. He died after a fall from a horse. PETER TOON

THEODOTION (second century). Editor of a Greek version of the OT. Little is known of him; Irenaeus called him a Jewish proselyte, Jerome an Ebionite Christian, and Epiphanius a Marcionite. He was associated with Ephesus. His translation is a revision of the Septuagint from the Hebrew, in a style more readable than that of Aquila. Origen used it to fill lacunae in the Septuagint text, and the translation of Daniel entirely superseded that of the Septuagint. It is also especially valuable for the texts of Jeremiah and Job. Origen placed his text next after the Septuagint in his *Hexapla*; from this source considerable fragments survive.

THEODULF (c.750-821). Bishop of Orléans. A Goth by descent who fled his native Spain, he joined the court of Charlemagne. His intimate friendship with court personages, including Alcuin,* is engagingly depicted in his *Carmina,* a book of poems. The king appointed him bishop of Orléans by 798 and also to several neighboring abbacies. His administration produced a scholarly edition of the Vulgate,* established schools, reformed worship, and left architectural and artistic masterpieces, especially lavishly produced Bibles. *Versus contra judices* poetically describes his 798 mission to Visigothic France, and recommends legal reform. He accompanied Charlemagne* to Rome in 800 to adjudicate in charges against Leo III.* He pleaded against Adoptianism and for the *Filioque* in *De spiritu sancto,* and wrote a treatise on baptism. "All glory, laud and honor," a Palm Sunday hymn, is Theodulf's work. In 818 he was stripped of his benefices by Louis the Pious for alleged complicity in King Bernard of Italy's revolt. JAMES DE JONG

THEOGNOSTUS (d. c.282). Alexandrian priest and theologizer about whom little is known save through quotations in the writings of Photius, Athanasius, and Gregory of Nyssa. He authored a work titled *Hypotyposes* (in seven books), parts of which were hardly orthodox. In it he spoke of angels and devils having bodies, of Jesus Christ as a creature, and of the Holy Spirit in terms no more orthodox than Origen,* whose views he followed. As head of the Catechetical School he wielded some influence. Despite Theognostus's Origenistic tendencies and language, Athanasius* appealed to his writings in his struggle against Arianism.*

THEOLOGIA GERMANICA. An anonymous treatise probably originating in the late fourteenth century. It was written in the mystical tradition of the late Middle Ages associated with the names of Tauler* and Eckhart.* It emphasized humility, self-negation, and a mystical union with God. The work impressed Martin Luther, who thought he found in it precedents for his own theology, and he published an incomplete edition in 1516. In 1518 he published a complete edition with a preface in which he stated: "no book except the Bible and St. Augustine has come to my attention from which I have learned more about God, Christ, man and all things." The book does not in fact teach Luther's theology, but reveals the influence of a number of medieval theological traditions. It was placed on the Index* in 1621, although it contained nothing that made it antithetical to the doctrine of the Medieval Church. Since the Reformation it has appealed to a wide variety of different groups, including later German Pietists. Over 150 printed editions of the work have been identified since its first publication. An English translation by Susanna Winkworth with introduction and notes by W.R. Trask was published in 1949. RUDOLPH HEINZE

THEOPASCHITES. From a Greek word denoting those who teach that God suffered. The description was applied by contemporaries to that group of so-called Monophysites* who taught that when Jesus suffered on the cross it was in fact God who suffered; as their formula put it, "One of the Trinity suffered in the flesh and was crucified." This doctrine was publicly taught from 519 in Constantinople by certain Scythian monks, as well as by John Maxentius. Though their claim to orthodoxy was accepted by Emperor Justinian, it was disputed by the local patriarch and by the pope, with whom has gone the verdict of history.

THEOPHANY (Gr. *theos,* "God," *phainesthai,* "to appear"). A manifestation of God in some empirical form. The OT contains many such, but scholars are divided in their interpretations as to whether God Himself, or some divine agent, made the appearance. The problem revolves around the divine assertion, "God is spirit" (John 4:24), "No man has ever seen God" (John 1:18), and "You cannot see my face" (Exod. 33:20 RSV). But against this, see Genesis 32:30, "I have seen God face to face." There is no need to assume that in each case an angel or some other divine being appeared in the place of God, if we understand that it is the Father who remains spirit, and that the Son has manifested the Godhead empirically throughout history. This would be more correctly the concept of "Christophany." It would be a needed corrective to thinking that Christ assumed the likeness of the human form beginning at His conception in Mary's womb, however much His previous forms may have possessed superhuman powers. John 1:18 seems to hint at this interpretation. KEITH J. HARDMAN

THEOPHILANTHROPISTS. A Deistic sect founded in France at the Revolution during the rule of the Directory, with the object of establishing a religion completely free of dogma. Their chief patron was the director, L.M. La Réveillière-Lépeaux (1753-1824). In *Manuel des théophilanthropes,* J.B. Chemin-Dupontes set forth their creed as belief in God, virtue, and immortality, drawing his inspiration from Voltaire and Rousseau. In 1797 they met in Paris and were given by the director the use of Notre Dame cathedral and seventeen other Parisian churches. They appealed to few apart from some scientists, politicians, and artists, including Jacopus David. After the reestablishment of Catholicism by the Concordat of 1801* they lost ground, and Napoleon restored the churches to Roman Catholic worship (1802). Unsuccessful attempts were made to revive Theophilanthropism in the nineteenth century. J.G.G. NORMAN

THEOPHILUS (late second century). Christian apologist and bishop of Antioch. Of his works only his Apology, addressed to a pagan friend Autolycus, has survived. This work is in three books and seeks to show the superiority of the Christian revelation over pagan mythology. Although Eusebius called this apology "elementary," it cannot be denied that Theophilus's doctrine of the Godhead marks an important advance on his Christian predecessors. Proceeding from a theology influenced by Middle Platonism, he distinguished between two phases of the Logos: the *logos endiathetos* is the Logos innate in God, and the *logos prophorikos* is the Logos expressed from God for the purpose of creation. Theophilus is reticent concerning the person of Christ, but he clearly regarded him as the second Adam. However, there is no special emphasis on the redemptive work of Christ. The stress instead is upon the disobedience of the first Adam and the obedience of the second Adam by following whose example we may be saved. Theophilus was the first theologian to use the word Triad *(trias)* of the Godhead. G.L. CAREY

THEOPHILUS (d.412). Patriarch of Alexandria from 385. A learned and gifted man, he conducted his office well during the early years of his administration, but in his efforts to destroy paganism in Egypt he turned to violence and intrigue. Originally an admirer of Origen,* he turned against his theology and drove the Origenist monks from Egypt. When about fifty of these monks found a warmer welcome from Chrysostom* at Constantinople, he went to that city where he held a council of thirty bishops which drew up false charges against Chrysostom and deposed and banished him (403). His treatment of the venerable Chrysostom brought Theophilus into disrepute. C. GREGG SINGER

THEOPHYLACT (fl. 1070-1081). Archbishop of Achrida, and Byzantine exegete. Born in Euboea, he entered a monastery early in life. Quickly he showed great promise as a scholar and was chosen tutor to the young Prince Constantine, son of the emperor Michael VII (1071-78), to whom he dedicated a treatise "On the Education of Princes." About 1078 he became archbishop of Achrida (Ochrida), an uncivilized area. In his letters he often complained of the wickedness, ignorance, and bad manners of the Bulgars, who composed the majority of his flock. Nevertheless he was able to continue his literary work. He was a disciple of Michael Psellus* (c.1019-c.1078), the first professor of philosophy at the University of Constantinople, and this ensured his mastery of classical learning. He wrote commentaries on all the NT books except the Book of Revelation, as well as on several OT books. While his exegetical methods owed much to earlier Greek commentators, he was not himself lacking in lucidity of idea and expression. Apart from the commentaries, other extant works include homilies, letters, poems, and a conciliatory treatise concerning the Greak Schism.* The date of his death is not known, but he survived the accession of Alexius Comnenus in 1081. PETER TOON

THEOSOPHY. There are groups of Gnostic ideas that may be classified as theosophical, appearing in one form or another in such movements as Rosicrucianism,* occultism, advanced forms of Spiritualism,* and Anthroposophy.* They include reincarnation, the development of psychic and occult powers, the belief in *karma,* the influence of spirits of several grades, and enlightenment from great Masters, Adepts, or Mahatmas, who have completed their cycle of incarnations and yet have chosen to return once more to guide events from behind the scenes. God is generally an unknown God, or else God and the universe are treated in a pantheistic sense.

The term "Theosophy", however, is commonly identified with the Theosophical Society, founded in New York in 1875 by Colonel H.S. Olcott (1832-1907) and Mme. H.P. Blavatsky (1831-1891). They claimed to have been prompted by the latter's Himalayan Mahatmas. They were succeeded as leaders by Mrs. Annie Besant* (1847-1933), who was converted to the movement in 1889, after passing from a romantic Anglo-Catholicism into free thought under Bradlaugh. She was assisted for a time by C.W. Leadbeater, who later became bishop of the so-called Liberal Catholic Church.

At an early stage Olcott and Blavatsky moved to India, where Annie Besant also spent many years. Hence their ideas were much influenced by Hinduism* and Buddhism.* In 1912 Mrs. Besant and Leadbeater proclaimed a Hindu boy, Krishnamurti, as the reincarnation of the Supreme World Teacher. This led to the defection of Rudolf Steiner, who thereafter developed a more Western Anthroposophy. Krishnamurti repudiated his role.

The stated aims of the society are: "1. To form a nucleus of the Universal Brotherhood of Humanity, without distinction of race, creed, sex, caste or colour. 2. To encourage the study of comparative religion, philosophy, and science. 3. To investigate the unexplained laws of nature and the powers latent in man." The works of Annie Besant fill twenty-four columns in the British Museum catalog.

BIBLIOGRAPHY: H.P. Blavatsky, *The Secret Doctrine* (1888; abridged 1968); J.K. Van Baalen, *The Chaos of Cults* (1956); J. Symonds, *Madame Blavatsky* (1959); J.H. Gerstner, *The Theology of the Major Sects* (1960). J. STAFFORD WRIGHT

THEOTOKOS (Gr. *theotokos*, "God-bearer"). A title of Mary, the mother of Jesus. Favored by the Alexandrian School of theologians from the time of Origen* onward, this title helped to preserve for them the "Word-flesh" Christology. In the fifth century it was attacked by Nestorius who, wanting to emphasize the humanity of Jesus, proposed the compromise titles *Christotokos* ("Christ-bearer") or even *Theodochos* ("God-receiving"). *Theotokos*, however, had the militant support of Cyril* of Alexandria in his Paschal Letter of 429, and it was approved by the councils of Ephesus* (431) and Chalcedon* (451). The usual Latin equivalent was *Dei Genetrix* ("Mother of God"), not the literal *Deipara*. PETER TOON

THERAPEUTAE. An ancient sect of ascetics and recluses thought to have resided in the vicinity of Alexandria and Lake Mareotis, Egypt, in the first century A.D. The only account concerning them is in *De Vita Contemplativa*, attributed to the Jewish philosopher Philo.* There they are described as devoted to the study of the OT, particularly the Law and the Prophets, in the utmost solitude. From sunrise to sunset, the times for prayer, the OT was studied. Some of this sect ate only every second day, and others ate only once a week. On the Sabbath they met in worship and heard their oldest scholars discourse on their tenets. Their entire lives were regulated in the most austere manner, intended to promote piety.

Since only this one source remains for their existence and practices, scholars have had to conjecture concerning many matters. Previously some assumed they were Christians, following the error of Eusebius. Jerome, similarly confused, reckoned Philo to be an ecclestiastical writer of the Christians. In more modern times they were thought to be a branch of the Essenes,* but this was challenged by the church historian Adolf Harnack. While these two monastic groups do resemble each other especially in their strict discipline, other than their both being second- or first-century B.C. radical groups coming into being in that time when many new groups emerged, there was probably no direct connection.

KEITH J. HARDMAN

THERESA; THÉRÈSE, see TERESA

THESSALONIANS, EPISTLES TO THE, see EPISTLES, PAULINE

THESSALONICA (modern Salonika). Political capital and chief seaport of Macedonia, it is situated at the head of the largest gulf on the Aegean Sea. It has been a cosmopolitan and important commercial center ever since its founding (c.316 B.C.) by Cassander, who named it for his wife, the sister of Alexander the Great. When Macedonia was divided into four districts in 167 B.C., Thessalonica was made the capital of the second district which extended between the rivers Strymon and Axius. Then in 148 B.C. Macedonia was made a Roman province, and Thessalonica became the seat of the Roman administration and later was declared a free city. The Roman road, *Via Egnatia*, was built through the city and walls were built around the city. It was ruled by politarchs (Acts 17:6), which has been recently confirmed by the discovery of inscriptions. In Paul's day Thessalonica had about 200,000 inhabitants, with an important Jewish community, with many Gentile converts (Acts 17:4). The ease with which the Jews at Thessalonica could influence the civil authorities reveals their power (Acts 17:5) against Paul and Silas. The incident of their imprisonment and deliverance suggests that the politarchs were bent on justice and legal protection for Paul and his companions, for the inability of free cities to keep public order always raised the threat of Roman interference.

Because of its continued occupancy, little archaeological information of the Thessalonica of Paul's day has been excavated. The main street is still the alignment of the Egnatian Way. Until 1876, at the western extremity of the city stood the Roman arch built by the citizens in honor of Octavian and Antony, and known as the Vardar Gate. It is one of the inscriptions on it that makes mention of the city politarchs, a previously unknown word, yet accurately described in the narrative of Acts 17:6. It probably dates from the period 30 B.C. to A.D. 143. Because of its location, Thessalonica has remained an important city throughout the Christian era, and now it has a population of almost 400,000.

See E. Oberhummer, "Thessalonike," *Pauly-Wissowa*, Zweite Reihe, VI, 1 (1936), cols. 143-63. JAMES M. HOUSTON

THEUDAS. In his speech to the Sanhedrin about policy to be adopted toward the Christian movement, Gamaliel refers to Theudas who made an unsuccessful attempt at rebellion (Acts 5:36). He is said to have antedated Judas of Galilee who also failed in his uprising, made at the time of the census, which was in A.D. 6. Josephus refers to someone of that name who was a magician who promised to lead his followers through Jordan dry-shod. He was killed by the troops of the procurator Fadus (c. A.D. 46-48). It has often been argued that Luke has misread Josephus. This is most unlikely in view of Luke's known accuracy in other places, and it is much more probable that there were two men of this name who lived at different times and were both engaged in some sort of public disorder. R.E. NIXON

THIERRY OF CHARTRES (c.1100-c.1156). Scholastic philosopher and theologian. He was the younger brother of Bernard of Chartres, who was chancellor of Chartres from 1114 to 1119 and was a major figure in the humanist and Platonic tradition characteristic of the cathedral school at Chartres. Thierry followed the same emphasis. He taught at the cathedral school while his brother was chancellor, and in 1136 he became archdeacon of Dreux. He also taught in Paris where John of Salisbury* was one of his pupils. In 1141 he succeeded Gilbert de la Porrée as chancellor at Chartres. He attended the trial of Gilbert in Reims in 1148, where Gilbert was accused of a heretical position on the Trinity, and in the following year he attended the Diet of Frankfurt. Little is known of the remainder of his life, except that he probably spent his last years in a Cistercian monastery. His humanist emphasis is illustrated in his *Heptateuchon,* a manual of the seven liberal arts which provides an excellent description of the available knowledge in the period. He differed from his brother and Gilbert in his emphasis on scientific knowledge, and he was one of the first to promote Arabic science in the West. He wrote a commentary on creation, *De Sex Dierum Operibus,* which reveals his scientific interests and the influence of Platonic philosophy. In addition he wrote a commentary on Boethius's *De Trinitate* and Cicero's *De Inventione.*

RUDOLPH HEINZE

THIRTEEN ARTICLES (1538). A Latin manuscript entitled *A Book containing Divers Articles de Unitate Dei et Trinitate Personarum, de Peccato Originali, etc.* discovered among papers belonging to Archbishop Cranmer.* Possibly connected with discussions between conservative Lutheran divines invited to England in 1538 by Henry VIII* and an English committee of three bishops and four doctors, it may have been the basis of negotiation, or perhaps a record of doctrines actually agreed. The articles are closely akin to the Augsburg Confession.*

THIRTY-NINE ARTICLES. A doctrinal statement of the sixteenth century arising out of the controversies of the period, and defining the position of the Church of England* in relation to them. The Articles were not intended to be creedal, or a complete theological system. Their origin can be traced to the Ten Articles* of 1536, a compromise statement designed to establish Christian "quietness and unity" at a time of revolution, when the separation between church and state was just beginning. These Articles were followed in 1537 by the Bishops' Book (revised in 1543 as the King's Book), which expounded certain tenets of Christian doctrine, and dealt with the relationship between the Church of England and Rome. In 1539, when Thomas Cranmer's* influence was declining, the Six Articles* were brought in by Henry VIII* to check the growth of Reformed theology and practice. The year 1553 saw the publication under Edward VI* of the Forty-two Articles,* which were intended to avoid controversy and establish unity "in certain matters of religion"; they were largely the work of Cranmer and Nicholas Ridley.

The history of the Articles was interrupted by the reign of Mary Tudor* and began again under Elizabeth I.* Matthew Parker,* now archbishop of Canterbury, drew up as an interim measure his own profession of faith in the Eleven Articles of 1561. In 1563 Convocation* revised the Forty-Two Articles into thirty-nine—although Elizabeth struck out Article 29 (dealing with the "wicked, who do not eat the body of Christ") to placate the Romanists, and added an opening clause to Article 20, asserting the authority of the church to decree rites and ceremonies. The Convocation of 1571 restored Article 29, to give us the Thirty-Nine Articles as we now have them. Despite subsequent Prayer Book revision, the Articles have remained unchanged ever since.

The main reasons for enforcing the Articles at that time are set out by Matthew Parker in a letter to the queen dated 24 December 1566: (1) they are concerned with the advancement of true religion; (2) they are agreeable to God's Word; (3) they condemn doctrinal errors; (4) they establish unity. The Articles chiefly cover the Catholic and Reformed doctrines of Scripture, the triune God, salvation, and the church's sacraments and ministry. They should not be regarded as a compromise statement midway between Rome and Geneva, but as an answer to Roman and Anabaptist* extremities.

Assuming that the Thirty-Nine Articles have an interest which is more than historical, their positive function now, as well as in the past, may be conceived as fivefold: (1) to preserve the dogmatic order of the Anglican Church and Communion; (2) to exercise a purifying influence on liturgical and canonical action: (3) to test new teaching; (4) to provide a framework for continuing debate; (5) to maintain the challenge of a biblical and apostolic norm (G.W. Bromiley).

Clerical subscription to the Articles has been required since 1865. In the *Report* of the Archbishop's Commission (1968), the revision of the Articles, subscription to them and the formula of assent were discussed with an eye to Christian unity; but few positive proposals were made (see esp. pp. 38-45).

BIBLIOGRAPHY: W.R. Matthews, *The Thirty-Nine Articles* (1961): for a critique of the doctrine of the articles; H.E.W. Turner (ed.), *The Articles of the Church of England* (1964); R.T. Beckwith, "The Problem of Doctrinal Standards" in J.I. Packer (ed.), *All in Each Place* (1965), pp. 116-27; G.W. Bromiley, "The Purpose and Function of the Thirty-Nine Articles" in P.E. Hughes (ed.), *Churchmen Speak* (1966), pp. 82-87; D.B. Knox, *Thirty-Nine Articles* (1967); Report, *Subscription and Assent to the Thirty-Nine Articles* (1968).

STEPHEN S. SMALLEY

THIRTY YEARS' WAR (1618-48). This highly complex conflict in central Europe was three struggles telescoped into one—Protestants vs. Catholics in Germany, a civil war in the Holy Roman Empire between the emperor and estates, and an international contest between France and the Hapsburgs (Austrian and Spanish) for European hegemony in which other powers were implicated. Historians commonly divide the war into four periods:

(1) *Bohemian,* 1618-23. Hostilities began with the Bohemian revolt against the Hapsburgs (Defenestration of Prague, 1618). The Czechs deposed Emperor Ferdinand II* (1619-37) as king, replacing him with the Calvinist head of the Protestant Union, Frederick V, Elector Palatine. The Catholic League leader, Duke Maximilian of Bavaria, supplied the emperor with an army commanded by Count Tilly which crushed the rebellion in 1620 (Battle of White Mountain). In Bohemia a ruthless policy of reconversion, expulsion, and confiscation of Protestant property ensued. In 1521 Ferdinand gave Frederick's electoral title to Maximilian, Tilly overran the Palatinate, and Spain and Bavaria partitioned it.

(2) *Danish,* 1625-29. The controversial Albrecht von Wallenstein raised an imperial army for Ferdinand, while King Christian IV of Denmark entered the war with English subsidies. Wallenstein and Tilly subjugated N Germany in 1626-28, and Ferdinand concluded the Peace of Lübeck with Christian in 1629. He also issued an Edict of Restitution which ordered the restoration of church lands secularized since 1552.

(3) *Swedish,* 1630-35. Because of Catholic opposition to his vague nationalistic schemes, Wallenstein was dismissed soon after Swedish king Gustavus Adolphus* landed in Germany. Traditionally viewed as the Protestant savior of Germany, he has more recently come to represent the intervention of foreigners that caused the war to degenerate into a quest for power. The Protestant forces defeated the imperial army at Breitenfeld (1631), plundered Bavaria, and captured Prague. At Tilly's death in early 1632, Ferdinand recalled Wallenstein. Gustavus was killed at Lützen (November 1632), while Wallenstein was dismissed for privately negotiating with the enemy and was assassinated in 1634. At the Peace of Prague in 1635 a compromise was reached over the Edict of Restitution.

(4) *French,* 1635-48. Further devastation resulted as France, the ally of Sweden and the German Protestants, battled Austria, Spain, and Bavaria. After years of negotiation the Peace of Westphalia* was concluded at Münster and Osnabrück on 24 October 1648. The Franco-Spain and Baltic struggles continued outside Germany for another decade.

The peace marked the end of both the medieval papacy's political influence (Innocent X's* objections were ignored) and the medieval empire's significance. The recognition of Calvinism and designation of 1624 as the cut-off date for possession of ecclesiastical lands settled the German religious dispute. Princes could, if they wished, permit both faiths to exist in their territories. The Count Palatine was restored as the eighth elector. The independence of Holland and Switzerland was confirmed and autonomy granted to the 300 German entities.

BIBLIOGRAPHY: C.V. Wedgwood, *The Thirty Years War* (1938) and *Richelieu and the French Monarchy* (1949); F. Watson, *Wallenstein, Soldier under Saturn* (1938); B. Chudoba, *Spain and the Empire, 1519-1643* (1952); M. Roberts, *Gustavus Adolphus: A History of Sweden, 1611-1632* (2 vols., 1958); H. Holborn, *A History of Modern Germany,* vol. I (1959); T.K. Rabb, *The Thirty Years' War* (1964); F. Dickmann, *Der westfälische Friede* (1965); S.H. Steinberg, *The Thirty Years' War and the Conflict for European Hegemony, 1600-1660* (1966); G. Pagès, *The Thirty Years War, 1618-1648* (ET 1970).

ROBERT G. CLOUSE

THOLUCK, FRIEDRICH AUGUST GOTTREU (1799-1877). German Protestant theologian. Concentrating on the study of oriental languages at the universities of Breslau and Berlin, he was converted to Christ under Pietist influences and turned to the study of theology. After a brief period of teaching at Berlin, he was professor of theology at Halle for forty-nine years (from 1826), where he exerted a powerful influence on students and on the churches. A steadfast opponent of rationalism in biblical and theological studies, his important works include *Die Lehre von die Sünde und dem Versöhner* (1823), commentaries on Romans (1824), John (1827), the Sermon on the Mount (1833), Hebrews (1836), and the Psalms (1865), as well as a history of rationalism (1865). Through his connection with the revival movement and in his pastoral ministry among the students he did much to further the cause of believing scholarship in his day. W. WARD GASQUE

THOMAS. Apostle. The name apparently comes from an Aramaic word meaning "twin," but it is not certain whose twin he was. In the lists of the twelve, which are arranged in three groups of four, his name occurs in the second group, suggesting neither eminence nor obscurity (Matt. 10:2-4; Mark 3:16-19; Luke 6:14-16; Acts 1:13). He is most prominent in John's gospel and the Greek version "Didymus" is used three times (11:16; 20:24; 21:2). We find him here associated particularly with the death and resurrection of Jesus. He is prepared to go with Jesus to the tomb of Lazarus (John 11:16) even if it means death. He confesses himself to be ignorant of the meaning of Jesus when He talks about His departure (John 14:5), and he is unwilling to accept the account given him by the other disciples of the risen Jesus whose appearance he missed seeing (John 20:24f.). The climax of the fourth gospel comes when "Doubting Thomas" is given the evidence for which he asked, and in return ejaculates the supreme confession of faith, "My Lord and my God!" The last great beatitude is then pronounced on those who have not seen and yet believe. Thomas is named also in John 21:2. He was almost certainly active in missionary work in the East, possibly in Parthia (Eusebius), Persia (Jerome), or India (as the Mar Thoma Church believes). R.E. NIXON

THOMAS, ACTS OF. An aprocryphal account of the Apostle Thomas,* depicted as Christ's twin and recipient of His secret words. Thirteen wonderful deeds of the apostle are told, concluding with his martyrdom. Throughout the work, both in the symbolism of the stories and in explicit teaching, a Gnostic element of teaching is found, based on the myth of the soul sought by the Savior to be freed from this world and bodily bondage, and counselling asceticism. This aspect of the book is most strikingly seen in the two famous hymns, the Marriage Song, praising the "daughter of light" (probably heavenly wisdom), and the Song of the Pearl, or Hymn of the Soul, which depicts in terms of legendary adventure the Savior's quest for the soul. The teaching has close links with early Syriac Christianity, e.g., Bardaisan, and the book was taken over by the Manichaeans and bears some marks of their editing. It retained its appeal in orthodox circles, however, and in its Syriac version has undergone some accommodation to Catholic teaching. Its main transmission is in a Greek form, although it was composed in Syriac. A Syriac version is known, with Latin, Ethiopic, and Armenian. Among the Manichaeans,* it circulated in a five-book corpus with the *Acts* of Peter, John, Andrew, and Paul.

J.N. BIRDSALL

THOMAS, APOCALYPSE OF. In the so-called *Decretum Gelasianum,** a sixth-century list of books declared to be canonical and noncanonical, an item appears in the Apocryphal Books section entitled "Revelation which is ascribed to Thomas." Another attestation of this document is found in the *Chronicle* of Jerome of the Codex *Philippsianus No. 1829* in Berlin, in which the circumstances of the giving of this revelation to Thomas are described. Evidence shows that the book is fifth-century and tainted with a Manichaean flavor. It exists in two recensions, a longer and a shorter (ET of both versions is given in M.R. James, *The Apocryphal New Testament,* 1924); and there are critical comments on the two versions in *New Testament Apocrypha* II (ET ed. R. McL. Wilson, 1965, pp. 798ff.). The book based on the canonical Revelation describes the scenes presaging the events of the end-time, and does so according to a schema of seven days.

RALPH P. MARTIN

THOMAS, GOSPEL OF. An apocryphal work discovered in the Nag Hammadi* Coptic Gnostic library, in Codex II written c.350. Three third-century Oxyrhynchus* papyri (nos. 1, 654, 655) contain fragments of a Greek version, compiled c.140, of which the Coptic is a more gnosticized translation. It consists of some 120 "secret words" of "the living Jesus," many very close to synoptic parallels, others more obviously syncretistic, arranged according to no discernible pattern, except when connected by link-words. Its form *(Gattung)* is nearer the collection of (proverbial and parabolic) "words of the Wise" (cf. "Q" and *The Sentences of Sextus,* identified in fragments of Codex XII from Nag Hammadi) than the typically Gnostic post-resurrection "revelation." Though used by Naassenes* and Manichaeans,* it more plausibly reflects the encratism (return to unisex paradise, life of ascetic exile on earth) of early Christian Syria, its likely provenance, than Gnosticism proper. Its versions of (synoptic) sayings of Jesus preserve traces of a tradition more primitive than and independent of the canonical gospels, attested in other Jewish-Christian and Syriac sources (so especially G. Quispel in many studies). It may contain one or two authentic noncanonical (Agrapha) sayings of Jesus. It is not to be confused with the *Infancy Gospel of Thomas.*

BIBLIOGRAPHY: Coptic text with translation (ed. A. Guillaumont et al., 1959); R. McL. Wilson, *Studies in the Gospel of Thomas* (1960); H.E.W. Turner and H. Montefiore, *Thomas and the Evangelists* (1962); R. Kasser, *L'Évangile selon Thomas* (1961); full bibliography in D.M. Scholer, *Nag Hammadi Bibliography 1948-1969* (1971), pp. 136-65); J.E. Ménard, *L'Évangile selon Thomas* (1975).

D.F. WRIGHT

THOMAS, NORMAN MATTOON (1884-1968). Presbyterian clergyman and frequent presidential candidate. Born in Marion, Ohio, he studied at Princeton University and Union Theological Seminary. After ordination in 1911 he became pastor of East Harlem Church, and chairman of an American parish settlement house in New York City (1911-18). He then was secretary of the pacifist Fellowship of Reconciliation* and until 1921 edited its magazine, *The World Tomorrow,* which he had founded. He was co-director of the League for Industrial Democracy (1922-37). He demitted the ministry in 1931. Gradually assuming leadership of the Socialist Party, he ran unsuccessfully for several political offices, including the U.S. presidency (six times). He helped to forge his party's policies, giving "critical support" to the war effort in World War II and opposing Communism, Fascism, and social injustice. In later years he turned to the problem of international peace. He helped found the American Civil Liberties Union. Later writings include *A Socialist's Faith* (1951); *The Test of Freedom* (1954); *The Prerequisites for Peace* (1959); and *Socialism Re-examined* (1963).

ALBERT H. FREUNDT, JR.

THOMAS, OWEN (1812-91). Welsh Calvinistic Methodist minister and author. He was born at Holyhead, son of a stone-mason, and followed his father's craft when the family moved to Bangor in 1827. He entered Bala College in 1838 as a candidate for the Calvinistic Methodist* ministry and completed his studies at the University of Edinburgh. He was ordained in 1844. After serving several pastorates he moved in 1865 to Liverpool and spent the remainder of his ministry there. He received every honor that his church could bestow and was twice moderator of its general assembly (1868, 1888).

He was an outstanding figure in the life of the Welsh evangelical churches in the Victorian Age. His powerful preaching was characterized by seriousness and intensity. He was a learned theologian and possessed an exceptionally fine private library that showed the breadth of his interests in divinity and history. He was a firm defender of Calvinism in the moderate form that derived from the thinking of Dr. Edward Williams.

His literary output was extensive. Apart from a constant flow of articles in various journals and in the monumental Welsh encyclopedia known as *Y Gwyddoniadur,* he translated Thomas Watson's work on sanctification and Kitto's NT commentary. But his finest work was in the writing of the biographies. His biography of Henry Rees (1798-1869) is a late work, published in two volumes in 1890, and although accurate and imposing, it lacks the vigor of his earlier work. His greatest achievement was the huge biography of John Jones, Tal-sarn (1796-1857), published in 1874. Although John Jones was a preacher of quite exceptional influence, the book is more than his biography—it is the biography of Welsh evangelicalism in the first half of the nineteenth century. Its analysis of the development of theology in Wales during that age, and its description of Welsh preaching and its exponents, make it a basic document for the study of nineteenth-century Wales.

See biography by J.J. Roberts (1912).

R. TUDUR JONES

THOMAS, WILLIAM HENRY GRIFFITH (1861-1924). Anglican scholar and teacher. Born in Shropshire, he was educated at King's College, London; read theology at Christ Church, Oxford; and was ordained in 1886. After curacies and a ministry at St. Paul's Portman Square, London (1896-1905), he became principal of Wycliffe Hall, Oxford (1905-10). Removing to Canada, he was professor of OT at Wycliffe College, Toronto, from 1910, and active in Canadian Anglicanism. A dispensationalist, he was a founder of Dallas Theological Seminary and would have lectured there but for his death. He contributed weekly to *The Sunday School Times* and to the Toronto *Globe* and was the author of numerous books. A supporter of the Keswick Movement, he assisted the Victorious Life Testimony in America.

C.G. THORNE, JR.

THOMAS À KEMPIS (c.1380-1471). German mystic. Born in Kempen near Köln, his original name was "Hemerken" or "Hammerlein." He was educated in the school at Deventer run by the Brethren of the Common Life.* Later he entered the Augustinian Convent of Mt. Saint Agnes near Zwolle, which was a daughter house of Windesheim and of which his brother was prior. Thomas was ordained priest in 1413, became subprior in 1429, and spent his whole life in this house. He worked as a copyist and is said to have copied the whole Bible at least four times. As a director of the spiritual life he was much in demand; his methods and approach followed those of Gerard Groote* and Florentius Radewijns.* All his writings—letters, poems, homilies, etc.—are of a devotional nature, but his great fame rests primarily on *De imitatione Christi et contemptu omnium vanitatum mundi* (usually called in English *The Imitation of Christ*). This is a manual of devotion to help the soul achieve communion with God. What has made it acceptable to others than Roman Catholics is its supreme stress on Christ and fellowship with Him. It has gone through over 2,000 editions and printings. There is some doubt, however, that it was Thomas who wrote this book. From Bellarmine* in the seventeenth century until the present day, there have been those who denied his authorship.

For the complete works of Thomas, see the critical edition by M.J. Pohl (7 vols., 1902-22).

PETER TOON

THOMAS AQUINAS, see AQUINAS, THOMAS

THOMAS OF CELANO (c.1190-1260). Best known for his two biographies of Francis of Assisi,* he was founder of the Order of Friars Minor. Thomas joined the order c.1215 and was later sent to Germany. In 1228 the order's protector, Gregory IX,* appointed Thomas as the saint's biographer. The *Vita Prima* of 1229 remains the best single source for the life of Francis. At the 1244 general chapter, the then minister-general, Crescentius, asked all who knew stories about Francis to send them to Assisi. Thomas decided to incorporate these reminiscences of the saint's companions into a new life of Francis—the *Vita Secunda* (c.1246). In 1266 the order's general chapter directed that all previous lives of Francis be destroyed and that a recent biography by Bonaventure—the *Legenda Maior*—should become the official life. Fortunately copies of Thomas's two biographies escaped the destruction. He also wrote the *Tractatus de miraculis de Sancti Francisci* and the *Legend* of St. Clare.

PETER TOON

THOMAS OF JESUS (à Jesu) (1564-1627). Carmelite* leader and writer. Born in Spain, he took doctorates in law and theology at Salamanca. Entering the Discalced Carmelite novitiate at Granada because of the influence of Teresa of Avila's* autobiography (1585), he made his profession at Valladolid (1587) and became professor and vice rector of the College of Alcalá. He instituted an eremitical life in the order by establishing "deserts" at Bolarque (1593) and Las Batuecas (1599), was provincial of Castile (1597-1600), vicar, then prior, of Las Batuecas (1606), and prior of Zaragoza (1607). Called to Rome by Paul V in 1607, he met opposition there and left to found communities in Brussels (1610), Louvain (1611), Douai (1612), Cologne (1613), Lille (1616), and Marleine (1619), finally seeing the erection of the Belgian and German provinces. He wrote on mystical theology (*De Contemplatione Divina,* 1620; *Divinae Orationis Methodus,* 1623) and missions (*Stimulus Missionum,* 1610; *De Procuranda Salute omnium gentium,* 1613).

C.G. THORNE, JR.

THOMAS OF MARGA (ninth century). Nestorian* historian. In 832 he entered the Nestorian monastery of Beth-'Abhe, east of Mosul in Syria. After serving as chaplain and secretary to the Patriarch Abraham from 837 to 850, he was consecrated bishop of Marga. He is best known as the writer of the *Liber Superiorum (Book of Governors)* which is a history of his monastery and an important source for the early history of the Nestorian Church. The Syriac text of this with an English translation was published in London by E.A.W. Budge in 1893.

THOMISM. The term covers both the doctrines of Thomas Aquinas* and their development in the Roman Catholic Church.

THOMPSON, FRANCIS (1859-1907). English poet. Born at Preston and educated at Ushaw for the Roman Catholic priesthood, he was never ordained but chose instead to study medicine at Owens College, Manchester; again, however, he did not complete the course. Moving to London, he became an opium addict and was rescued from a vagrant life by Wilfred Meynell, who with his wife cared for Thompson for the rest of his life. He wrote on Shelley and St. Ignatius; but his poems are more important than his prose. These appeared in *Poems* (1893), *Sister Songs* (1895), and *New Poems* (1897). Thompson is a poet of the nineties with all the overwrought emotion and colorful expression of that decade. He is now remembered almost solely for "The Hound of Heaven," which appeared in the 1893 volume. It has the hectic intensity of the pursued fugitive that one finds in the work of other opium-afflicted poets, and like Herbert's poem "Love" it concludes in tender acceptance of the pursuing God:

> Ah, fondest, blindest, weakest,
> I am He Whom thou seekest!
> Thou dravest love from thee, who dravest Me.

See *Poems* (1937); and J.C. Reid, *Francis Thompson* (1959). ARTHUR POLLARD

THOMSON, WILLIAM (Lord Kelvin) (1824-1907). Physicist. Second son of a mathematician and author of textbooks, he was educated at home and at Peterhouse, Cambridge, and in 1846 was appointed professor of natural philosophy at Glasgow, holding the chair for fifty-three years and becoming the doyen of science. He made numerous discoveries: he was co-founder with J.P. Joule and others of the science of thermodynamics; he was the originator of the absolute (Kelvin) scale of temperature; founder of geophysics; inventor of numerous electrical instruments; pioneer of the first Atlantic cable (1858) and of electrical power transmission. In early life he was inspired by Faraday; his early researches originated in the desire to discover when God had created the world.

Thomson's character was exemplary: he was modest, never claimed priority in discovery, took special delight in praising the work of others, even the most junior, and treated assistants and students with the same deference as fellow professors. His niece, A.G. King, says she loved going to church with him ("there was something in his humble and quiet reverence which seemed to strengthen one's faith and bring one directly into the presence of God"). She tells also of his deep interest in and knowledge of the Bible. From about 1860 he was often involved in courteous controversy with geologists and materialistic evolutionists on topics relevant to the Christian faith.

See S.P. Thomson, *Life of Lord Kelvin,* (1910); and A.G. King, *Kelvin the Man,* (1924).

R.E.D. CLARK

THORN, CONFERENCE OF. Colloquy held in 1645, at Thorn in West Prussia, then under Polish protection, called by Wladislaw (Ladislaus) IV of Poland, who ruled over a religiously divided kingdom. The delegates included Catholic, Lutheran, and Reformed representatives. The controversial Lutheran ecumenicist Georg Calixt (Calixtus*) was present, as well as the irenic Moravian J.A. Comenius.* Hampered by Jesuit opposition and Lutheran internal quarreling, the conference resulted in little ecumenical progress. Perhaps surprisingly, the Lutheran and Calvinist delegates were able to agree that the (Polish) Consensus of Sandomir (1570), which combined Calvinist, Lutheran, and Moravian ideas, as well as the German "Saxon Confession"* (Revised Augsburg Confession) of 1551, agreed on scriptural essentials. The Calvinist "Declaration of Thorn" was composed at the conference and was adopted as a creedal standard by the Calvinist churches of Brandenburg. It is noteworthy for its explicit stress on continuity with the ancient church.

DIRK JELLEMA

THORNWELL, JAMES HENLEY (1812-1862). Presbyterian minister and scholar. Born in South Carolina and educated at South Carolina College, he engaged in pastoral work for a few years before becoming professor at that college which was later the University of South Carolina. He became president in 1851, but four years later moved to Columbia Theological Seminary as professor of systematic theology. He founded the *Southern Presbyterian Review* in 1847 and was moderator of his church's general assembly that same year. He helped to establish the Presbyterian Church in the Confederate States during the Civil War.

THORVALDSEN, A.B. (1770-1844). Danish sculptor. Living and working in Rome from 1797 to 1838, he was more attracted in his classicism by the deities of Greek mythology than by biblical figures. Nevertheless his most famous works are the biblical sculptures he made on request for the decoration of the cathedral of Copenhagen: the Christ figure, the twelve apostles, the kneeling angel with the font of baptism in his hands, and, on the frontal frieze of the cathedral, a more complex composition that represents John the Baptist preaching.

THREE CHAPTERS CONTROVERSY. The three chapters were three "subjects" condemned by Emperor Justinian* in an edict of 543/4. His purpose was to conciliate the powerful Monophysite* group and retain their allegiance to church and crown. The three chapters were (1) the writings of Theodore of Mopsuestia*; (2) the writings of Theodoret* of Cyrrhus against Cyril* of Alexandria and in defense of Nestorius; (3) the letter of Ibas* of Edessa to the Persian, Bishop Mari of Hardascir.

Although this edict did in fact undermine the authority and teaching of the Council of Chalcedon* (which had declared Theodoret and Ibas to be orthodox), the Eastern patriarchs accepted it without opposition. Pope Vigilius* rejected it at first, but after a visit to the emperor in Constantinople he endorsed it and made this public in his *Judicatum* of April 548, addressed to the patriarch of Constantinople. In the West, however, there was much opposition to the edict, and it found a leader in Bishop Facundus* of Hermiane.

When the Fifth Ecumenical Council met in May 553 at Constantinople, it decided in favor of the condemnation of the Chapters and against the efforts of Western bishops to have the condemnation withdrawn. A little later, Vigilius declared his submission to the decision of the council, but parts of the Western Church refused to follow the pope. The churches of N Italy, led by the clergy of Aquileia and Milan, broke off communion with Rome and caused a schism which lasted for half a century. PETER TOON

THYATIRA. A city in W Asia Minor, in the Hermus valley, a tributary of the Caicus. Situated on a fertile plain at about 330 feet, it had no significance as a stronghold, but its position on the imperial post road linking Italy-Greece-Asia Minor with Egypt, gave it commercial importance. Lydian in origin, Thyatira was refounded by Seleucus Nicator as an outpost against Macedonia, and then a Pergamene outpost. When the Pergamene kingdom became the Roman province of Asia in 133 B.C., it passed into control of Rome. It is probable there was a Jewish colony there (Acts 16:14), but it is not clear when Christianity reached Thyatira. One possibility is that Paul or one of his helpers ministered there from Ephesus (Acts 19:10). What is certain is that by about A.D. 95, when John addressed the church there, it was quite a strong church (Rev. 2:18-29). But little is known of Thyatira, and no archaeological excavations have been carried out. The ancient site is now occupied by the modern town of Akhisar (population about 30,000). JAMES M. HOUSTON

TIELE, CORNELIUS PETRUS (1830-1902). Dutch theologian. Born at Leyden, and educated in Amsterdam and at the seminary of the Remonstrant Brotherhood, he was pastor at Moordrecht and Rotterdam, and then professor at the Remonstrant seminary. He became professor of religious history, a chair specially founded for him, at Leyden University (1877-1901). During his professorship he exercised a great influence on the development of the study of comparative religion, especially in the Netherlands. He wrote many books, the best known being *Geschiedens van den Godsdienst tot aan der Heerschappij der Wereldgodsdiensten* (1876; ET *Outlines of the History of Religion*, 1877). He also gave the Gifford Lectures entitled *The Elements of the Science of Religion* (1897-99). These works came out of a vast knowledge of ancient languages and history and were widely useful because of their lucid and orderly arrangement. J.G.G. NORMAN

TIILILÄ, OSMO ANTERO (1904-1972). Finnish theologian. Ordained a Lutheran minister in 1926, he renounced the ministry in 1960 and later withdrew from membership in the Lutheran Church of Finland. This was done as a protest against the influence of liberal theology within the church without any disciplinary action taken against those who brought in the new teachings. Tiililä also wanted to point out that the social activities and duties of the church were stressed too much at the expense of the preaching of the Gospel. Member of the executive committee of the Lutheran World Federation and of its theological commission, and member of the Academy of Sciences in Finland, he also had a large literary output, including *Das Strafleiden Christi* (1941) and "A Hundred Years of Systematic Theology in Finland," in *Theologia Fennica* IV (1949). He edited the Finnish theological journal *Teologia ja kirkko.* STIG-OLOF FERNSTROM

TIKHON (Vasili Belavin) (1866-1925). Patriarch of Moscow. Born near Pskov, he trained for the priesthood in St. Petersburg and subsequently became bishop of Lublin before going to North America, where he held various posts (1899-1907) and was finally archbishop. He returned to Russia,* and after holding two more posts became in 1917 metropolitan of Moscow, then patriarch (the first holder of that post since 1700). When the Bolsheviks commandeered church land, withdrew church subsidies, decreed civil marriage only, and took over schools, Tikhon pronounced the anathema on the country's new rulers and their supporters. Four days later, church and state were officially separated, giving the signal for priests and congregations to be attacked.

During the famine in 1922 the government declared confiscation of all church treasures to relieve the hungry, though Tikhon had already called for all unconsecrated objects of value to be disposed of for this purpose. Priests who resisted the plundering were murdered or jailed. Tikhon himself was taken into custody, and released only because of the pressure of international opinion in 1923. During his imprisonment he concluded from Romans 13 that since the Soviet government was now the divinely sanctioned government, the church owed it secular obedience—an exegesis approved even by the Communists. His viewpoint was not shared by a section of his clergy, and his last years saw a division and the increased difficulty of a so-called Living Church Movement with more than a dash of socialism about it which the conservative Tikhon resisted.

J.D. DOUGLAS

TILAK, NARAYAN VAMAN (c.1862-1919). Marathi hymnwriter and poet. Born in Ratnagiri district of Maharashtra, he was a high-caste Chitpawan Brahman, but his father had forecast from a horoscope that the child would one day forsake Hinduism. Even before the break, his poetic gift and ardent religious spirit made him almost a youthful *sadhu.* On a train journey a foreigner gave him a NT, and thus began the road to Christianity (he had already been questioning Hindu orthodoxy). He was baptized at Byculla, Bombay, in 1895. Tilak found congenial scope for his gifts, and wise guidance, in the American Marathi Mission, and after ordination in 1904 worked as a preacher at Ahmadnagar. Writing was his great ministry: his hundreds of *bhajans* gave the Marathi Church a worthy Indian medium of praise in place of translated hymns in Western meter (though Tilak provided also worthy translations of his own). His greatest work was to be a life of Christ in verse, the *Christayan*, but only the first part had been completed by his death. In 1917 he left mission service in order to form a brotherhood of "the baptized and unbaptized disciples of Christ," with an ashram at Satara.

ROBERT J. MC MAHON

TILLEMONT, LOUIS SÉBASTIEN LE NAIN DE (1637-1698). Roman Catholic historian. Born in Paris and educated at Port-Royal under Pierre Nicole, he entered the seminary of Beauvais (1661), was ordained (1676), and collaborated with G. Hermant on the lives of SS. Athanasius, Basil, Gregory of Nazianzus and Ambrose, and with others in Paris on editions of patristic texts: Origen, Tertullian, Augustine. Returning to Port-Royal (1667), he was forced out by the persecution in 1679 and he retired to Tillemont, where he worked privately, as well as catechized children and helped the poor. He never accepted ecclesiastical office and even permitted his own work to be published under others' names. A member of the Jansenist sect, he took no part in the controversy. *Mémoires pour servir à l'histoire ecclésiastique des six premiers siècles* (16 vols., 1693-1712) established his reputation with its immense learning. Edward Gibbon spoke of its "inimitable exactitude" which led him through the rocky roads of Roman history with the surefootedness "of an Alpine mule." *Histoire des empereurs* (6 vols., 1690-1738) was intended as part of the *Mémoires,* but censorship demanded separate publication. He wrote a life of St. Louis which was published much later (6 vols., 1847-51).

C.G. THORNE, JR.

TILLICH, PAUL (1886-1965). Protestant theologian and philosopher. Born in Starzeddel, Germany, son of a Lutheran pastor, Tillich received schooling in the universities of Berlin, Tübingen, Halle, and Breslau. From the last he received his Ph.D. for a dissertation on Schelling.* After four years as chaplain in the army during World War I, he taught successively at Berlin, Marburg, Dresden, and Leipzig. In 1929, while professor of philosophy at the university of Frankfurt, he became involved in the Religious-Socialist movement. His opposition to Hitler and National Socialism led to his dismissal from the university in 1933; almost immediately he came to the USA, where he taught at Union Theological Seminary and Columbia University (1933-55), Harvard (1955-62), and the University of Chicago (1962-65).

The sources and philosophical foundations of Tillich's philosophical theology can be traced back to Platonism, medieval mysticism (Jakob Boehme), German Idealism (Schelling), and existentialism (Kierkegaard and Heidegger). His theological methodology, the "Method of Correlation," argues for a complementary role and relationship between philosophy and theology: philosophy poses the problems (questions) of ontology (the metaphysical structures of being or reality) with regard to the human situation, while theology provides the answers to these questions. In his three-volume *Systematic Theology* (1951-63) God is understood as the "Ground of Being" whom man knows as "ultimate concern." Existentially man derives his own being by "participation" in the "Ground of Being." "Ultimate concern" means the courage to affirm oneself ultimately in the face of non-being. The question of human existence also points to the issue of Christology, since Jesus Christ is the "New Being." In the sacrifice of Jesus upon the cross, He became "transparent" to the "Ground of Being" —i.e., the "New Being" or "The Christ." In the area of epistemology, Tillich has been one of the leading advocates of symbols or myths as signs that participate in the reality to which they point. Myth or symbol is therefore man's only way of grasping cognitively the meaning and structure of reality—God, who is the "Ground of Being."

The most serious charges against Tillich's theology are his dependence upon idealism which strongly implies pantheism and an impersonal deity, and his failure to grasp the *sola scriptura* principle of the Protestant tradition in which he stood. Other major works of his include *The Interpretation of History* (1936), *The Protestant Era* (1936), *The Courage to Be* (1952), *Love, Power and Justice* (1954), *Dynamics of Faith* (1957), and *Theology of Culture* (1959).

See also C.W. Kegley and R.W. Bretall (eds.), *The Theology of Paul Tillich* (1952), and W. Leibrecht, "Paul Tillich," *A Handbook of Christian Theologians* (ed. D.G. Peerman and M.E. Marty, 1965), pp. 485-500. DONALD M. LAKE

TILLOTSON, JOHN (1630-1694). Archbishop of Canterbury from 1691. He was born near Halifax, son of a Presbyterian, and educated at Cambridge. He accepted the Act of Uniformity* in 1661 and became an Anglican, thereafter securing speedy preferment to the deanery of Canterbury in 1672. He tried hard in troublous times to bring the Dissenters within a comprehensive Church of England. Tillotson was probably the most influential preacher in the history of the English sermon. He had as many imitators in the eighteenth century as Milton had among its poets. Yet his style was "plain and unaffected," to use his own phrase. He sought no heights, but both in language and structure his work is unsurpassed in its lucidity. As Bishop Burnet said in the funeral sermon, no man "knew better the art of preserving the majesty of things under a simplicity of words." ARTHUR POLLARD

TIMOTHY. A native of Lystra, son of a Greek and a Jewess (Acts 16:1; 2 Tim. 1:5), he was probably converted on Paul's first missionary journey and became the companion of Paul and Silas on the second. To avoid difficulties with the Jews, Paul had him circumcised (Acts 16:3). He was first sent to Thessalonica to encourage the church there (1 Thess. 3:2) and then to Macedonia and Corinth (1 Cor. 4:17). He went to Jerusalem with Paul, taking the collection (Acts 20:4f.), and was associated with Paul at the time of writing of the Philippian and Colossian epistles. Two of the Pastoral epistles are addressed to him, and he is shown there to have become Paul's representative at Ephesus (1 Tim. 1:3). At some stage he was imprisoned (Heb. 13:23). The general picture of his character, seen chiefly from the Corinthian letters and the Pastorals, is of an affectionate and loyal companion of Paul who lacked forcefulness of character and was self-conscious about his youthfulness. His role in his latter years, like that of Titus,* was that of "apostolic delegate." R.E. NIXON

TIMOTHY I (728-823). Patriarch of the East from 780. After study at Bashosh, he was appointed bishop of Bait Baghash near Arbil before 769 through the influence of his uncle. He obtained election as patriarch in 780 by unworthy methods, but was not accepted until 782. Despite this unpromising start, he was one of the greatest of the patriarchs of the East. On good terms with contemporary Caliphs, especially al-Mahdi (775-785) and Harun-al-Rashid (786-809), he shifted the patriarchal center from Ctesiphon to Baghdad and built a palace there. A firm and able administrator, he created at least six new metropolitan provinces—Damascus, Armenia, Rai, Dailam, Turkestan (Kashghar), and Tibet (Tangut). Widely read in both Church Fathers and Greek philosophy, he translated Aristotle's *Topics* into Arabic and arranged for the transcription of three copies of Origen's *Hexapla.* His famous *Dialogue with al-Mahdi* is a classic which has been a model for much Christian apologetic among Muslims since. It was not, however, effective as evangelism, and under Timothy there seems to have been no evangelization of Muslims.

In other spheres, Timothy was an outstanding missionary statesman, choosing the right men for missionary work and handling them sympathetically and imaginatively. Outstanding missionaries were the Arab Christian Shubhal-ishu', who died a martyr before 895; his successors Yab-alaha the scribe and Qardagh the bookbinder; and Elijah of Moqan.* Their spheres of labor were the heathen on the shores of the Caspian, and the Turks and Tartars eastward on the silk-route to China. Timothy wrote 200 letters, of which 59 are extant. His style is pithy, and his letters give an interesting picture of the conditions of his time. His correspondence extended even to India.

WILLIAM G. YOUNG

TIMOTHY, EPISTLES TO, see EPISTLES, PAULINE

TIMOTHY, PATRIARCH (d.517). Monophysite patriarch of Constantinople. Appointed by Emperor Anastasius I when the Chalcedonian patriarch Macedonius II was deposed (c.511-12), he was previously in charge of the cathedral ornaments of Constantinople. After some initial hesitation he became a leading exponent of Monophysitism.* In 512 he was involved in Severus's* attempt to introduce the Theopaschite* formula, "who was crucified for us," to the Trisagion, which provoked a riot. He formally condemned the Chalcedonian doctrine at a synod in 515, continuing to work with Severus, now bishop of Antioch. Theodore the Lector* ascribed to him the regular use of the Nicene Creed* in the liturgy of Constantinople; it had probably been introduced earlier at Antioch by Peter the Fuller.*

J.G.G. NORMAN

TIMOTHY AELURUS (d.477). Monophysite patriarch of Alexandria. His nickname *ailouros,* "weasel," was given by opponents because of his small stature. As presbyter he led the Alexandrian Monophysites. He became patriarch in 457 after Proterius had been lynched by the mob, but being unacceptable to the majority of bishops was banished by Leo I* in 460. In exile he wrote much to propagate Monophysitism.* He held that Christ is by nature God, not man; He became man only by *oikonomia* ("dispensation") thus His humanity is not His *phusis* ("nature"). However, he anathematized Eutyches* for holding that Christ's body was not of the same substance as other human bodies. Recalled by Basiliscus in 475, he died before another decree of banishment by Zeno could be carried out. Extant writings include three letters in a Syriac translation, and a collection of treatises and letters against the Council of Chalcedon* in Syriac and Armenian are attributed to him. He is venerated as a saint in the Coptic Church. J.G.G. NORMAN

TINDAL, MATTHEW (c.1656-1733). Deist writer. Son of a country clergyman, he was educated at Oxford and became a fellow of All Souls in 1678, a position he retained to the end of his life. He was subsequently a doctor of law. His High Church principles led him briefly to join the Roman Church under James II, but soon discerning "the absurdities of Popery," he returned to the Church of England. He began to advocate Erastian principles, first arousing a storm of opposition with *The Rights of the Christian Church asserted against the Romish and all other Priests who claim an Independent Power over it* (1706). The House of Commons had it burned by the common hangman in 1710. *Christianity as old as the Creation* (1730), hailed as "the Deists' Bible," was his mature and constructive attempt to show that true revealed religion is simply a republication of the religion of nature. It was answered by more than a hundred writers, but its translation into German extended Deistic influences to the Continent. J.W. CHARLEY

TINDAL, WILLIAM, see TYNDALE, WILLIAM

TINTORETTO, JACOPO (1518-1594). Italian painter. He worked in Venice most of his life. A sign supposedly hung on his studio wall read: "The drawing of Michelangelo, the color of Titian." This tells a great deal about Tintoretto's ambition and achievement. He reached success at the age of twenty-seven with his painting *Miracle of Saint Mark.* For the most part he was self-taught; Titian* threw him out of his studio after only a few weeks. Tintoretto's temperament was unusual. When the *Scoula de San Rocco* announced a competition for a painting of their patron saint, many famous painters were invited to compete. When the day came for the judging, the contestants arrived with their sketches—except for Tintoretto, who brought a huge finished painting. He excused himself by saying this was the way he worked.

Perhaps as a result, he became the club's official painter for nearly thirty years. In this place he attempted to marry the styles of Michelangelo* and Titian, but the result was wholly Tintoretto. His compositions exuded vitality with their dramatic lighting, brilliant color, and lithe bodies in unusual perspective. In his *Last Supper,* Tintoretto represents the moment when Jesus offered the bread and wine as the sacrificial body and blood of man's redemption. To illuminate the scene, Tintoretto bathed his canvas in a supernatural luminosity that emanates partly from the figure of Christ and partly from the flickering flames of the oil lamp, the smoke of which is formed into an angelic choir hovering around the head of Christ. He was perhaps the most skilled of the direct painters, painting onto the canvas without sketch or underpainting—a remarkable feat for any artist. ALVA STEFFLER

TISCHENDORF, LOBEGOTT FRIEDRICH KONSTANTIN VON (1815-1874). German Protestant theologian and textual critic. He studied at Leipzig, where he came under the influence of J.G.B. Winer,* the grammarian, who led him to combine a careful concern for Greek philology with a love for the sacred text. Though he was professor in the theological faculty at Leipzig for many years, he devoted himself to textual criticism and spent the greatest part of his time in the libraries of Europe and the Near East searching out unpublished ancient manuscripts. During his lifetime he published more manuscripts and critical editions of the Greek NT than any other scholar. His most famous discovery was the Codex Sinaiticus at St. Catherine's Monastery in Sinai, which he visited several times. Other important manuscripts which he edited include Codex Ephraemi rescriptus (1843-45), Codex Amiatinus (1850), and Codex Claromontanus (1852). His critical edition of the Greek Testament, the famous *editio octava critica maior* (2 vols., 1869-72), with its enormous critical apparatus, remains a basic reference tool for the NT scholar.

W. WARD GASQUE

TISSOT, JAMES JOSEPH JACQUES (1836-1902). French illustrator. Born at Nantes, he studied at the Academy of Fine Arts in Paris and fled to England when the Commune took over (1871). He gained a reputation as a portrait and genre painter, but finally became an illustrator (see his series of watercolors, *La femme à Paris,* and aquarelles on life in London). Returning to Paris, he experienced great attraction to Christ, arising from his work on *La femme qui chante dans l'église,* which took him regularly to church. This practice then inspired his *Christ appears to console two Unfortunates in a Ruin,* and OT illustrations later. After a year of study in Palestine, he spent ten years on several hundred aquarelles on gospel scenes. *Vie de Notre-Seigneur Jésus-Christ* (1896). C.G. THORNE, JR.

TITIAN (1477-1576). Italian painter. This widely heralded artist was unique in his versatility. He painted vast altarpieces alive with ecstasy and Christian tenderness, lusty alcove pictures, battle scenes with hundreds of people, and many portraits. His virtuosity made him the most sought-after painter of his time. Always in robust health, he lived to the age of ninety-nine. He was a good husband and father, a faithful friend, and a man of dignity. Early in his life, Titian's father recognized the outstanding talent in his son and sent him to Venice to learn painting. After some study in design and color with a mosaic-maker, Titian went to the studio of Giovanni Bellini. While there, he was more greatly influenced by Bellini's student, Giorgione, whom Titian admired and followed as a friend. Titian became famous with a series of paintings he and Giorgione did for a German warehouse.

Titian's paintings sparkled with a wide range of hues. In painting flesh, his method produced unusually natural results, looking as if the flesh would bleed if touched with a knife. His coloring became the model for later notable Venetians such as Tintoretto, and for the Baroque masters Rubens* and Velasquez. After Raphael's* death, Titian was much in demand. His style developed into an energetic painterly technique. In some of his later group portraits, quick slashing strokes endow the whole canvas with the spontaneity of a first sketch. His uncanny grasp of human nature is also observed in this freer technique. In *Christ Crowned with Thorns,* a masterpiece of Titian's old age, the shapes emerging from the semi-darkness now consist wholly of light and color; the shimmery surfaces have lost every trace of material solidity and seem aglow from within. Consequently the violent physical action of the moment has been suspended. What lingers in the mind is not the drama of the event, but a mood of deep religious serenity. ALVA STEFFLER

TITUS. A Gentile, he became one of Paul's missionary companions. He is mentioned only in Galatians, 2 Corinthians, and the Pastoral epistles. It is unlikely that he is the same as Titius Justus in Acts 18:7; W.M. Ramsay made the guess that he might not have been named in Acts because he was Luke's brother. On Paul's second visit to Jerusalem he took Titus with him, and the fact that he did not have to be circumcised was an important point of principle in the acceptance of the Gentiles. In 2 Corinthians he is shown as Paul's emissary to deal with strained relationships between the apostle and the church at Corinth. He seems to have had a more robust character than Timothy* (2 Cor. 7:14f.), and he was sent back to Corinth to supervise the taking of the collection. The epistle to Titus shows that he was sent by Paul to Crete to supervise the work there (Titus 1:5f.). He had to rejoin Paul at Nicopolis (Titus 3:12) and it was perhaps from there that he was sent to Dalmatia (2 Tim. 4:10). His role in his latter years seems to have been a roving one as a sort of "apostolic delegate." R.E. NIXON

TITUS, EPISTLE TO, see EPISTLES, PAULINE

TOLAND, JOHN (1670-1722). Deistic writer. Born of Roman Catholic parents in Ireland, he became a Protestant at the age of sixteen. His studies took him to Glasgow and Edinburgh, then to Leyden where he concentrated on ecclesiastical history. In *Christianity not Mysterious* (1696), clearly indebted to Locke,* he claimed to endorse all the essentials of Christianity, but stressed the primacy of reason and the subordinate role of revelation in merely supplying supplementary information. After revelation there was no "mystery" left. Nothing in Scripture was out of harmony with reason or above it.

Condemnation by the Irish Parliament compelled him to flee the country. A *Life of Milton* (1698) was thought to question the authenticity of the gospels, but on further elucidation he claimed reference only to apocryphal writings, of which he showed a remarkable awareness. His political works included *Anglia Libera* (1701), supporting the Hanoverian succession, which led to several visits to the court in Germany. In *Nazarenus* (1718) he anticipated F.C. Baur* in distinguishing Jewish and Gentile branches of Christianity. Finally he produced *Pantheisticon* (1720), a parody of Anglican liturgy, indicating the goal of his religious development. A facile rather than profound writer, he possessed a flair for expressing the latent feelings of the moment. Though his indiscretions compelled him to become a hack author, he was a notable exponent of Deism.

See L. Stephen, *English Thought in the Eighteenth Century,* I (1902), pp. 101-111; and G.R. Cragg, *From Puritanism to the Age of Reason* (1966), chap. 7. J.W. CHARLEY

TOLEDO, COUNCILS OF. Eighteen councils between 400 and 702 were held at Toledo in the Visigothic kingdom in Spain. Sometimes ten other councils held there from the eleventh to the sixteenth centuries are also included. The early councils were primarily assemblies of bishops called to deal with ecclesiastical affairs. The first (c.400) condemned Priscillianism,* and the second (527 or 531) dealt with the education of clerics and the obligation of celibacy. Considered one of the most important was the third council, summoned in 589 by King Recared* after his conversion from Arianism in 587. It recognized the orthodox creeds and established orthodox Christianity as the official state church in the Visigothic kingdom. After this the councils began to deal increasingly with political matters. The fifth (636) dealt with only one religious question, and the thirteenth (683) was both a political assembly and a church council concerned mostly with political matters. The canons of the councils are important sources for the study of the history of the Visigothic church, dealing with a variety of questions ranging from disciplinary decrees for the clergy through anti-Semitic legislation. The decrees for all but the last council have survived.

RUDOLPH HEINZE

TOLERATION. The question of the toleration of Christianity by Roman authorities did not become serious so long as the early church was regarded merely as a movement within Judaism. Christians were accorded the same peculiar status which Rome had granted the Jews. The fall of Jerusalem (A.D. 70), however, soon made it evident that the Church was distinct from Judaism, and Christianity soon became a *religio illicita.* For the next 240 years the church was subject to a series of persecutions, the last of which took place under Diocletian.*

The failure of persecution as a policy, and the precarious position in which Constantine* found himself, brought a drastic change in policy, and in 313 the emperor issued the Edict of Toleration which went beyond that which had been issued in 311 by Galerius.* Christianity thereby became a *religio licitor,* and Christians were not only free to profess their faith, but were also freed from the legal disabilities imposed on them by previous emperors. Theodosius (I)* the Great went even further in the policy of toleration when he issued the Edict of 380, making Christianity the official religion of the empire.

With the coming of the Reformation, the problem of toleration entered a new phase in which Protestants now were seeking toleration from Roman Catholic regimes, and in some cases, such as England and Germany, free-church groups were seeking toleration from those Protestant churches which had become established as the state church. Such toleration was not achieved in Germany until the 1648 Peace of Westphalia,* when Calvinism gained a recognition which it had not received in the 1555 Peace of Augsburg.* In England, dissenters from the Anglican Church received a new status with the passage of the 1689 Act of Toleration* which, however, did not cover Roman Catholics and Unitarians. The latter were granted toleration under George III, and a similar act passed in 1828 granted emancipation to Roman Catholics if they would change their attitude to the temporal supremacy of the papacy.

In colonial America, religious toleration came gradually, beginning in such colonies as Pennsylvania and Rhode Island before 1776, and coming to Virginia immediately after the War of Independence. The First Amendment to the Constitution proved to be an effective foundation for complete religious toleration.

"Toleration" is generally construed as the right to worship, often distinct from "freedom of religion," in which all religions have an equal base of civic rights. True religious liberty exacts no penalty of dissenters from an established church.

BIBLIOGRAPHY: J.H. Overton, *The Church in England,* vol. II (1897); N. Paulus, *Protestantismus und Toleranz im 16 Jahrhundert* (1911); W.K. Jordan, *The Development of Religious Toleration in England* (4 vols., 1932-40); A.P. Stokes and L. Pfeffer, *Church and State in Our United States* (rev. ed., 1962). C. GREGG SINGER

TOLERATION, EDICT OF, see MILAN, EDICT OF

TOLERATION ACT (1689). Passed by Parliament in England after the toppling of the Stuart dynasty, this provided a limited measure of relief to Nonconformists, apart from Unitarians. They were permitted to have their own places of worship (registration by the authorities was, however, mandatory), and to have their own pastors and teachers if such were willing to take certain oaths of loyalty and accept most of the established church's Thirty-Nine Articles.* The act did not remove social and political disabilities, and Nonconformists (like Roman Catholics) were still debarred from public office.

TOLKIEN, JOHN RONALD REUEL (1892-1973). English writer. Born in Bloemfontein, South Africa, where his father, who died in 1897, was a bank manager, Tolkien was sent to England for his education and graduated from Oxford in 1915. After service in World War I, he returned to Oxford as a professor of Anglo-Saxon and English literature. He has been a gifted and diversified writer of scholarly treatises, essays, novels, poems, and a play. *The Hobbit* (1937) and *The Lord of the Rings* (1954-55) are the works that have made Tolkien most widely known—the latter based on a series of myths of his own devising. A trilogy, the myths describe a cosmic war between good and evil in which the forces of evil are routed in a great struggle through courage and sacrifice. Peace and harmony are reestablished in the world order. Tolkien's work often reveals his enthusiasm for philology, legend, myth, and quests as a part of his fictional design. Like many medieval characters, Tolkien heroes are often engaged in perilous adventures that prove their moral strength. It is in this moral growth that there are often religious implications. Before he became a popular writer, Tolkien established his reputation as a serious scholar with critical studies of *Sir Gawain and the Green Knight* (1925) and *Beowulf: The Monsters and the Critics* (1936). He retired from active teaching in 1959.

PAUL M. BECHTEL

TOLSTOY, LEO (1828-1910). Russian novelist and social reformer. Born into a family of the ruling class, he knew comfort and social prestige in his youth on the family estate in Tula. After serving in the Crimean War, he returned home to write and study. In 1861 he freed his serfs. Tolstoy matured during the era when Russia was feeling the pressure for social reform. In the midst of his fame he experienced a mystical transformation and cast his lot with the peasants, adopting their dress and laboring in their trades. He rejected Russian orthodoxy and evolved his own form of faith, emphasizing as a central creed the nonresistance to evil. Disowning his title and his wealth, he turned over his property to his wife. In his later years he became embittered and left his home in company with his daughter. His best novels are *War and Peace* (1860), centered on the Napoleonic invasion of Russia; *Anna Karenina* (1877); *Kreutzer Sonata* (1890); *Resurrection* (1899); and an essay, *What is Art* (1899), which sets forth the conviction that good art is moral art. He stresses the conflict between reason and the natural desire to live without the restraints of social convention. He writes with a large comprehensiveness and is a master of analysis, characterization, and moral insights. PAUL M. BECHTEL

TOME OF LEO. A letter addressed by Leo I* to the Patriarch Flavian* of Constantinople in 449, containing a lucid and systematic exposition of the Catholic doctrine of the Incarnation. The occasion of the Tome was chiefly the appeal made to Leo by Eutyches* after he had been deposed for Monophysitism* by Flavian. The Tome, which upheld Flavian's condemnation of Eutyches, was endorsed by the Council of Chalcedon* two years later.

TOME OF ST. DAMASUS *(Fides Damasi)*. The collection of canons *(Post Concilium Nicaenum)* presented by Pope Damasus I* to bishops at a Roman synod which is variously dated between 369 and 382. It consists of a creed (a Latin rendering of Nicea, 325), to which are added twenty-four famous anathemas against heretics and schismatics, among them Sabellians, Eunomians, Macedonians, Melitius of Antioch, and Apollinarians. In the canons the Holy Spirit is said to be of one power and substance with the Father and the Son, and must be adored by all creatures, just as the Father and the Son. Damasus included the tome in his epistle *Per Filium* to Paulinus of Antioch. J.G.G. NORMAN

TOMLINSON, AMBROSE JESSUP (1865-1943). Church of God leader and revivalist. Born in a Quaker home near Westfield, Indiana, he was converted in 1892 and served as an American Bible Society colporteur in North Carolina. He joined the Pentecostal Church of God movement in Tennessee in 1896 and rapidly rose to leadership. He assumed the title of "general overseer" (1903-23) and established the headquarters of his Church of God at Cleveland, Tennessee (1908). He traveled widely and was especially successful in winning converts among poor Appalachian whites. His authoritarian rule, however, created friction, and factions had developed by 1917. After several secessions, his group took the name of "the Original Church of God." Two schisms after his death were headed by his sons, Homer A. and M.A. Tomlinson. Four large Tomlinson groups survive, and over forty religious groups can be traced to his movement.

ALBERT H. FREUNDT, JR.

TONGUES, see GLOSSOLALIA

TONSURE. The shaving of the hair at the top of the head of a priest or monk, by which he is distinguished from a layman. Traditionally the shaven area is said to represent the crown of thorns and to have been instituted by the apostles Paul and Peter. Probably the custom entered Christianity through the ascetics of the fourth and fifth centuries who perhaps adopted the custom from heathenism—e.g., the priest of Isis. In contemporary Roman Catholic practice the rite of entrance to the clerical life involves the cutting of five pieces of the candidate's hair, with full tonsure later (except in the USA, England, and some other countries); the tonsure, however, is much smaller than the earlier Roman or coronal tonsure which left only a fringe of hair around the head. Originally the Orthodox Church required the shaving of the whole head, but now it is usual for the hair to be cut short instead. The early British (Celtic) Church required a tonsure for which the hair was cut at the front and at the sides, leaving one half of the head with hair. PETER TOON

TOPLADY, AUGUSTUS MONTAGUE (1740-1778). Anglican hymnwriter. Educated at Westminster School and Trinity College, Dublin, he was converted through a Methodist lay preacher, took Anglican orders in 1762, and later became vicar of Broadhembury, Devon. In 1775 he assumed the pastorate of the French Calvinist chapel in London. He was a powerful preacher and a vigorous Calvinist, bitterly opposed to John Wesley.* He wrote the *Historic Proof of the Doctrinal Calvinism of the Church of England* (2 vols., 1774) and *The Church of England Vindicated from the Charge of Arminianism* (1769). His fame rests, however, on his hymns, e.g., "A debtor to mercy alone"; "A sovereign Protector I have"; "From whence this fear and unbelief?"; and especially "Rock of Ages" (appended to an article calculating the "National Debt" in terms of sin).
JOHN S. ANDREWS

TORAH, see JUDAISM

TORGAU ARTICLES, THE (1576). Twelve in number, these were adopted by the Lutheran theologians of Germany as "opinions as to how the dissensions prevailing among the theologians of the *Augsburg Confession* may, according to the Word of God, be agreed upon and settled in a Christian manner." The *Torgau Book* was submitted to the Lutheran princes (about twenty-five); most of the theologians in their territories approved it. Because some of them objected to its length, James Andreae* prepared an *Epitome.* James Andreae, Martin Chemnitz, and Nicholas Selneccer met at Cloister Bergen in March 1577. They met there again in May 1577, when they were joined by Andreas Musculus, Christopher Carnerus, and David Chytraeus. Here they completed the revision of the *Torgau Book*, known also as the *Belgic Book* or the "Solid Declaration" or the "Formula of Concord."* The Torgau Articles or the *Epitome* were also approved. The twelve articles deal with: (1) Original Sin; (2) Free Will; (3) the Righteousness of Faith Before God; (4) Good Works; (5) the Law and the Gospel; (6) the Third Use of the Law; (7) the Lord's Supper; (8) the Person of Christ; (9) the Descent of Christ into Hell; (10) Church Rites; (11) God's Eternal Foreknowledge (Predestination) and Election; (12) Other Factions (Heresies) and Sects.
CARL S. MEYER

TORQUEMADA, JUAN DE (1388-1468). Spanish theologian and cardinal. One of the leading defenders of papal authority against the conciliar theorists, he entered the Dominican Order in 1403 and attended the Council of Constance* in 1417. Afterward he finished his studies in Paris and taught in Spain, where he became prior of the Dominican house at Valladolid and Toledo. He attended the Council of Basle from 1432 to 1437 where he defended papal rights in a series of treatises. When the council deposed Eugenius IV, he led the faction which reconvened the council at Ferrara and was a leading figure in the negotiations with the Greek Orthodox Church which led to a short-lived decree of unity signed in 1439. He also wrote additional treatises defending papal primacy, and engaged in public debate with a leading conciliarist. For his services he was granted the title "Defender of the Faith" and was made cardinal in 1439. His major work, *Summa de Ecclesia* (1448), was a defense of the church both against heretics and conciliarists. In the last years of his life he was further rewarded by being appointed bishop of Palestrina in 1455 and of Sabina in 1463. RUDOLPH HEINZE

TORQUEMADA, TOMÁS DE (1420-1498). Spanish Grand Inquisitor. Born in Valladolid, he was the nephew of the cardinal and Dominican theologian Juan de Torquemada.* After graduating from the San Pablo Dominican convent he became the prior of Santa Cruz convent in Segovia (1452) and confessor to King Ferdinand V and Queen Isabella I in 1474. After Pope Sixtus IV* was prodded by the queen into making the Inquisition* a national institution in Spain (1483), Torquemada was given the power to organize the tribunals, and he did so with such effectiveness that the inquisitorial apparatus he set in motion lasted for three centuries. The regulations for the Spanish Inquisition were set down in Torquemada's *Ordinances* (1484). The victims of his persecution included Moors, Jews, Marranos (Jewish converts), Moriscos (Islamic converts), and other religious deviants from the Catholic norm. Since the Spanish Inquisition was not answerable to the papal Inquisition, complaints to the pope of Torquemada's excesses were usually unfruitful. Torture was used to extract evidence and confessions from prisoners. Some 2,000 executions occurred and numerous other methods of punishment were inflicted during his reign as Grand Inquisitor. His ruthless efficiency in enforcing his own austere religious character on those of other persuasions has left him with a reputation of cruelty and intolerance even for the times in which he lived.

See T. Hope, *Torquemada, Scourge of the Jews* (1939). ROBERT G. CLOUSE

TORREY, C(HARLES) C(UTLER) (1863-1956). Linguist who specialized in OT Aramaic, Apocrypha, and Pseudepigrapha of the OT, the Aramaic background of the NT and (later) Islam. He taught at Andover Seminary (1892-1900) and Yale University as professor of Semitic languages (1900-1932; emeritus 1932-56). Torrey was the first director of the American School of Oriental Research in Jerusalem (1900-1901). His writings include *The Translations Made from the Original Aramaic Gospels* (1912); *The Composition and Date of Acts* (1916); *The Four Gospels* (1933; rev. ed., 1947); *The Apocryphal Literature* (1945); and *Documents of the Primitive Church* (1946). His critical reconstructions of various OT books and theories regarding alleged Aramaic originals of large portions of the NT have found little acceptance among scholars. W. WARD GASQUE

TORREY, R(EUBEN) A(RCHER) (1856-1928). American evangelist and Bible scholar. A graduate of Yale College and Seminary, he also studied in German universities. He was ordained to the Congregational ministry in 1878, and became superintendent of the Congregational City Missionary Society of Minneapolis. He had a long association with D.L. Moody* and was the first superintendent of the Moody Bible Institute (1889-1908). He went on several world preaching tours between 1902 and 1921, visiting Britain and other European countries, Australia and New Zealand, and Asia. From 1912 to 1924 he was dean of the Bible Institute of Los Angeles and pastor of the Church of the Open Door in the same city from 1915 to 1924. He wrote numerous devotional and theological books; the most important were *What the Bible Teaches, How to Work for Christ,* and *The Person and Work of the Holy Spirit.* J.G.G. NORMAN

TOTAL DEPRAVITY. A position associated particularly with Calvin* and Calvinism,* although held by the Reformers generally, and associated with their return to Augustinian emphases. It concerns the definition of human nature. Medieval Scholasticism generally held that man was created with, and human nature consists of, both natural gifts (e.g., reason) and added or supranatural gifts (e.g., love of God), and that only the latter were lost at the Fall. "Total depravity," in contrast, holds that this distinction is false, that the totality of human nature was affected by the Fall, that man-as-totality turned against God at the Fall. Thus man cannot find truth about God through reason, nor turn to God with his will, but must be redeemed as a whole, a totality.
JAMES DE JONG

TOTAL IMMERSION, see BAPTISM

TOWARDS THE CONVERSION OF ENGLAND. In 1943, at the request of the church assembly, the then archbishop of Canterbury (William Temple*) set up a commission of nearly fifty people under the chairmanship of the bishop of Rochester to "... survey the whole problem of modern evangelism...." This report, entitled *Towards the Conversion of England,* was published in 1945. The report considered the Gospel itself, the need for laity to be fully involved in evangelism, the different needs of town and country, young and old, and the new opportunities in the postwar situation. The report, which had many practical suggestions, was widely acclaimed* (it was reprinted eight times in the first eight months), but never had any substantial effect. A recent writer, Roger Lloyd, has described it as a "damp squib."

Although the death of Archbishop Temple before the report was published may be one reason for its lack of influence, other reasons were the preoccupation of the Church of England with canon law revision, and the great and rapid theological and social changes unforeseen in the report. The report is confident in what the Gospel is; many in succeeding years were not. The report speaks of moral standards having reached a low point because of the war; there is no suggestion that they were to fall even further. The parish system and the 1662 Book of Common Prayer were still regarded as satisfactory; no radical reform of ministry or worship was contemplated. Furthermore, England was still seen as a Christian country to which "other nations look for leadership." Finally, the report—while making valuable suggestions in many different spheres—was never really a "plan" as the subtitle describes the report.
PETER S. DAWES

TRACT. Though this term is applied sometimes to the chant sung in Roman Catholic churches at Mass on certain penitential days, it more often refers to a type of propagandist literature larger than a handbill but shorter than a treatise, designed to promote spiritual or moral edification. Although the fondness of the Oxford reformers of the 1830s for this type of literature led to the term "Tractarians"* being applied to them, tracts and the colporteurs who distribute them are generally thought of as specifically Protestant. Thus the lesser writings of Wycliffe,* the Puritan *Marprelate Tracts** (1588), and the ephemeral literature of the Civil War period all fall into this category. The eighteenth-century Evangelicals relied heavily on tracts, and the famous Religious Tract Society was founded by George Burder and others in 1799, to be followed by numerous parallel societies in America, the continent of Europe, and the mission fields. Tracts are today used extensively by aggressive heretical sects.
IAN SELLERS

TRACTARIANISM. The name given to that stage of the Oxford Movement* when the *Tracts for the Times* were being issued. The first three *Tracts* were four-page leaflets published anonymously in 1833. In the first, J.H. Newman* sounded a clarion call to the clergy of the Church of England to exalt their office because of its "Apostolical Descent." The disciples of the Tract writers personally distributed the Tracts widely up and down the country vicarages. The Tracts began to change in character when E.B. Pusey* began to write in 1834. He produced longer theological documents, like Tract 18 on fasting which was signed with his initials. As the Tracts continued, they also included reprints of selections from the writings of the Caroline Divines,* with whom the Tractarians claimed affinity. The Tract writers included R.H. Froude,* R.I. Wilberforce, R.W. Church,* J.B. Mozley,* and I. Williams,* as well as the leaders of the Oxford Movement. Characteristic of their approach to religious teaching was the expression in two of the Tracts of "reserve in communicating religious knowledge." The Tracts came to a sudden end in 1841 with Tract 90 by J.H. Newman. His attempt to interpret the Thirty-Nine Articles in a "catholic" direction brought a storm of protest and forced the closure of the series. NOEL S. POLLARD

TRADITION. The Greek word *paradidōmi,* from which *paradosis* ("tradition") is derived, means "to hand something over." The NT employs the verb in a variety of ways (Matt. 11:27;

Acts 14:26; 1 Pet. 2:23), but the noun is reserved for teaching which has been handed down. Apostolic teaching—which included facts about Christ, their theological importance, and their ethical implications for Christian living—was described as tradition (1 Cor. 11:2; 2 Thess. 2:15). It had divine sanction (1 Cor. 11:23; Gal. 1:11-16) and, once committed to writing, was to be preserved by the church (Jude 3; 2 Tim. 1:13; Rom. 6:17). Jesus rejected tradition, but only in the sense of human accretion lacking divine sanction (Mark 7:3-9).

In the patristic period, the apostolic *paradosis* was usually distinguished from the church's *didaskalia*, or teaching, but a looser usage of *paradosis* is also discernible. Legends about the apostles, liturgical practices, biblical interpretation, and unwritten teachings, said to have come either openly or secretly from the apostles, were included under this rubric.

The Council of Trent* extended this practice, saying that revelational truth is to be found "in written books [Scripture] and unwritten traditions." Vatican I* also spoke of revelation being contained partly in the "written books" and partly in "unwritten traditions." More recently this unwritten *paradosis* has been identified, not so much with the Magisterium's teaching, as with the church's religious perception *(sensus fidei)*. Vatican II,* therefore, attempted to overcome the traditional polarization between Scripture and tradition by positing that there is only one source of revelation, not two. Scripture and tradition alike contain and reflect this one revelation, being its derivatives.

The Reformers distinguished between apostolic and post-apostolic tradition. The former they identified with divine revelation (cf. 1 Tim. 5:18; 2 Pet. 3:15; 1 Thess. 2:13), and the latter with human teaching. The latter was to be received only when it did not violate the former. The apostolic *paradosis* should be allowed to inform and structure Christian thought, providing an unchanging element of continuity through all ages.

BIBLIOGRAPHY: G.H. Tavard, *Holy Writ or Holy Church* (1960); J.P. Mackey, *The Modern Theology of Tradition* (1962); J.R. Geiselmann, *The Meaning of Tradition* (1966); Y.M.-J. Congar, *Tradition and Traditions* (1967); F.F. Bruce, *Tradition: Old and New* (1971). DAVID F. WELLS

TRADITIONALISM. An early nineteenth-century French Catholic response to European Enlightenment rationalism and the French Revolution, and a doctrine instrumental in forming ultramontanism.* Its main founder was Louis de Bonald (*La législation primitive*, 1802), who was joined by F.R. Lamennais* in his early years, and Joseph de Maistre.* It was a search for a principle of authority, necessitated by what was seen as a collapse of rationalism and social order. Individual human reason was incapable either of discerning metaphysical or moral truth by itself or of establishing right order in society. Instead, they argued, it was God who revealed the truths to men, through the first man, and embodied them in tradition—the transmission of God's original truth from generation to generation through the organic development of history, by instruction, and within the structures of authority, specifically the papacy for things spiritual and the rulers for things temporal. Belief was the certain apprehension and acceptance of this tradition.

The doctrine represented in Catholic philosophy a swing toward the pole of faith, denying that "natural reason" could attain such "natural truths" as God's existence, the principle of authority, and the moral law. Its contributions to Ultramontanism were accepted while its doctrines of faith and tradition were periodically condemned in mid-nineteenth-century papal encyclicals as blind faith. The decrees of Vatican I* and the reestablishment of Thomism* by Leo XIII* (1879) further destroyed the doctrine.

BIBLIOGRAPHY: E. Hocedez, *Histoire de la théologie au xix*[ième] *siècle* (2 vols., 1948, 1955); L. Foucher, *La philosophie catholique en France au XIX*[ième] *siècle avant la renaissance thomiste, 1800-1880* (1955); G. Boas, *French Philosophies of the Romantic Period* (1964); "Traditionalism" in *Sacramentum Mundi* VI (1970), pp. 274-75.

C.T. MC INTIRE

TRADITORS (Lat. *traditores*, "traitors"). A name given in Africa to Christians who saved their lives during Diocletian's* persecution by surrendering copies of the Scriptures. The Donatist* schism was partly caused by the Donatists' refusal to recognize Caecilian* of Carthage because he was consecrated bishop by an alleged traditor, Felix of Aptunga. Felix was later cleared of the charge. At the Council of Arles (314), persons consecrated by traditors were held to be duly recognized.

TRAHERNE, THOMAS (1637-74). English poet. He came from the Welsh borders, was educated at Brasenose College, Oxford, and was subsequently rector of Credenhill (Herefordshire). In 1673 he published *Roman Forgeries*, a criticism of the Roman Catholic Church based on extensive reading in early church history. *Christian Ethics* appeared in the year after his death, but the works by which he is best known were not printed until the present century: *Poems* (1903) and *Centuries of Meditation* (1908). Despite a certain limitation both of idea and of expression, Traherne is a fit member of the succession of Metaphysical poets. He shared Vaughan's delight in childhood and sense of the glory of the created universe. His meditative works also show his links with the Cambridge Platonists,* but his elevation into rhapsody on the divine love and wisdom takes him into a realm that borders on the mystical. He is a store of devotional wisdom, whose work has been not unaptly compared with that of Thomas à Kempis.* ARTHUR POLLARD

TRANSCENDENTALISTS. The American Unitarian Association was formed in 1825, but by 1836 the Boston churches were torn again as rebellious young preachers, members of the "Transcendental Club," abandoned Unitarianism,* feeling it to be complacent and sterile in its rationalism. Among the most prominent were R.W. Emerson,* George Ripley (1802-80), Theodore Parker,* James Freeman Clarke (1810-88),

Orestes Brownson (1803-76), and the layman Henry David Thoreau (1817-62). Much influenced by S.T. Coleridge's *Aids to Reflection,* they tried to introduce a strong note of mysticism and contemporary Romanticism's view of individual intuition, flashes and insights into truth, as the highest form of knowledge. Their thinking was basically syncretistic, viewed God as immanent in nature, rejected external authority, and saw every created thing as possessing deep religious meaning. For several years much of their writing appeared in the *Dial* (1840-44), and Emerson's "Divinity School Address" and Parker's sermon on "The Transient and Permanent in Christianity" were expositions of their thought, although they disagreed among themselves often in their intense individualism. KEITH J. HARDMAN

TRANSFIGURATION. The event in the life of Jesus when His appearance became radiant in the presence of Peter, James, and John (Matt. 17:1-9; Mark 9:2-10; Luke (9:28-36; 2 Peter 1:16-21). Three wonders accompanied the event: the transformation of Jesus' face and garments, the appearance of Elijah and Moses, and the voice of God speaking from a cloud. Tradition associates the event with Mt. Tabor, but a location in the foothills of Mt. Hermon provides a more likely setting. The gospel accounts of the event are laden with symbolic overtones: the revelation of Jesus' true nature (light), the fulfillment of the Law and the Prophets in Jesus' ministry (Moses and Elijah), the presence of God in the life of Jesus, and the divine commendation of His mission (cloud and voice). W. WARD GASQUE

TRANSUBSTANTIATION. The doctrine of the Eucharist maintained by the Roman Catholic Church and first defined by Radbertus, a Benedictine of Corbie in 831 on the basis of John 6. The influence of Greek views of substance and accident blurred the doctrine in the writings of Ockham* and Scotus,* while Biel* confessed that the miraculous presence of Christ was a mystery to be accepted only because of God's omnipotence. The Council of Trent* closed off other options by stating Christ is "truly, really and substantially contained in the sacrament under the appearance of sensible things.... By the consecration of the bread and wine a change is brought about of the whole substance of the body of Christ our Lord and of the whole substance of the wine into the body of his blood. This change ... is called transubstantiation." Vatican II* said of the Eucharist in terms of the body of believers, "Truly partaking of the body of the Lord in the breaking of the Eucharistic bread, we are taken up into communion with Him and with one another." And "no Christian community ... can be built up unless it has its basis and center in the celebration of the ... Eucharist." ROBERT B. IVES

TRANSYLVANIA, see ROMANIA

TRAPP, JOHN (1601-1669). English Bible commentator. Son of Nicholas Trapp of Kempsey, Worcestershire, he was born at Croome d'Abitot. Educated at Worcester Free School and Christ Church, Oxford, he became usher (1622) and headmaster (1624) of the Free School, Stratford-upon-Avon, and preacher at Luddington. He became vicar of Weston-on-Avon (1636). In the Civil War he sided with Parliament. Afterward he served as rector of Welford, Gloucestershire, but the former incumbent, Dr. Bowen, was reinstated in 1660, whereupon Trapp returned to Weston. He wrote commentaries on the whole Bible, furnishing an example of Calvinistic scholarship at its best, characterized by quaint humor and profound learning. The best-known edition is that of 1867-68. J.G.G. NORMAN

TRAPPISTS. Cistercian* monks of the reform instituted in 1664 by Armand Jean Le Bouthillier de Rancé,* abbot of La Trappe, a Cistercian abbey near Soligny, Normandy. One of the strictest orders, it emphasizes liturgical worship, demands absolute silence with no allowance for recreation, and imposes community life with a common dormitory. Meat, fish, and eggs are forbidden. The monks devote themselves to liturgical prayer and contemplation, theological study, and manual labor. Their habit is white with a black scapular and cowl. The order of nuns is called "Trappistines." The expulsion of monks during the French Revolution led to Trappist foundations in other parts of Europe and in China, Japan, and the USA. In 1817 they returned to La Trappe, and in 1898 they recovered possession of Cîteaux Abbey which had been secularized during the Revolution, and they declared it the mother church of the Reformed Cistercians. J.G.G. NORMAN

TRAVERS, WALTER (c.1548-1635). Puritan divine. Born in Nottingham, he matriculated at Christ's College, but studied at Trinity College, Cambridge, becoming a fellow in 1569. Whitgift's* new university statutes forced him out of Cambridge, and he traveled to Geneva where he became a close friend of T. Beza* and a convert to Presbyterian polity as the divinely ordained form of church government. His ideas on the subject were set out in *De Ecclesiasticae Disciplinae ... Explicatio* (1574). After a brief return to England in 1576, he became minister to the Merchant Adventurers in Antwerp, but tensions led to his resignation. His refusal to accept Anglican orders and his leadership in the classical movement made him unacceptable to Whitgift. He was passed over for the mastership of the Temple and finally inhibited from preaching. His last important post was the first provostship of Trinity College, Dublin, between 1594 and 1598, and for the rest of his life he lived in comparative obscurity. Though he played a vital role in editing the draft *Book of Discipline,* widely discussed by the classical movement, his final book, *Vindiciae Ecclesiae Anglicanae* (1630), suggested that he no longer so explicitly equated Christianity and Presbyterianism. His defense of Reformed theology and exposition of Presbyterian polity of this deeply learned man made him one of the most influential Elizabethan Puritans.

See S.J. Knox, *Walter Travers* (1962).

IAN BREWARD

TREGELLES, SAMUEL PRIDEAUX (1813-75). English NT textual critic. Brought up among the Society of Friends,* a fact which prevented him from pursuing a university career, he showed exceptional talent as a teenager by learning Greek, Hebrew, Aramaic, and Welsh while working at the same time in an iron works. His scholarly abilities were recognized by G.V. Wigram, who employed him to work on his famous *Englishman's Greek and Hebrew Concordances.* Quite independently he developed critical principles which paralleled those of Lachmann.* He traveled extensively across Europe for the purpose of systematically examining and collating all the then-known uncials and many of the more important minuscules; he was able to correct many erroneous citations by previous editors.

In *An Account of the Printed Text of the Greek New Testament* (1854), Tregelles surveyed previous work and laid down the principles for his own work. His Greek NT was published in six parts between 1857 and 1872. In addition to translating Gesenius's *Hebrew Lexicon* into English (1847), he authored many books on "Bible prophecy," including *The Man of Sin* (1840), *The Hope of Christ's Second Coming* (1864), and *The Prophetic Visions of Daniel* (1845), in which he defended what later came to be known as "posttribulational premillenialism." Associated with the Plymouth Brethren* in the early days of the movement (he was the brother-in-law of B.W. Newton*), he later worshiped with the Presbyterians and finally the Church of England.

See G.H. Fromow, *B.W. Newton and Dr. S.P. Tregelles* (n.d.); and B.M. Metzger, *The Text of the New Testament* (1964), pp. 127-28.

W. WARD GASQUE

TREMELLIUS, JOHN IMMANUEL (1510-1580). Italian Reformer and Semitic scholar. Born in Ferrara of Jewish parentage, he was educated at Padua and won to Christianity through Cardinal Pole in 1540. The following year, while Tremellius was teaching at Lucca, Peter Martyr Vermigli's influence led him to adopt Protestantism. He fled the Inquisition* (1542), journeying to Strasbourg, where he taught Hebrew in Johannes Sturm's school. In 1547, during the Smalcaldic War, he fled to England, and in 1549 became reader in Hebrew at Cambridge. At the accession of Mary Tudor he left England for the Continent, serving as tutor to the children of the duke of Zweibrücken (1555-59), as headmaster of the Hornbach gymnasium (1559-60), and professor of OT studies at Heidelberg (1561-77). He ended his career teaching Hebrew at Sedan, where he died. Tremellius is best known for his Latin translation of the Hebrew Scriptures (5 vols., 1575-79), long used as the most accurate Latin Bible. He also translated Calvin's Catechism into Hebrew and Greek (1551) and published Bucer's *Ephesians Commentary* from lectures he heard at Cambridge (1562), and an Aramaic and Syriac Grammar (1569).

BRIAN G. ARMSTRONG

TRENCH, RICHARD CHEVENIX (1807-1886). Archbishop of Dublin. Educated at Harrow and Cambridge, he was ordained in 1832. He was professor of divinity at King's College, London (1846-58), and dean of Westminster (1856-63) before going to Dublin. There he opposed unsuccessfully the disestablishment of the Irish Church and did much to settle the church after legislation had been passed. A High Churchman of saintly character, he had wide sympathies and versatile scholarship. His numerous works included *Notes on the Parables of our Lord* (1841), *Notes on the Miracles of our Lord* (1846), and *Lectures on Mediaeval Church History* (1877). He retired in 1884.

J. D. DOUGLAS

TRENT, COUNCIL OF (1545-63). By Roman Catholic reckoning the nineteenth ecumenical council, it was brought about by the continuing success of the Protestant Movement. The council was delayed by many problems. For example, although Charles V* was in favor of a council, he was opposed by Francis I of France. Charles envisioned the council as the means to reunite Christendom, while the papacy saw the council as the means to halt Protestantism.

Pope Paul III* summoned a council for Mantua in 1537, but it failed to meet. The council was transferred to Vicenza in 1538, but the indifference of Charles V and the Protestants resulted in the conference at the appearance of only a few churchmen. The failure of the conference at Ratisbon* spurred Paul III to try calling another council, this time for 1542, but the war between Charles V and Francis I prevented its meeting. After the peace of Crépy, the council was decreed by the bull *Laetare Hierusalem* issued 11 November 1544. The council was to meet in Trent on 15 March 1545 to settle the religious disputes brought about by the Protestants, to reform certain ecclesiastical abuses, and to begin a crusade against Islam. Due to another disagreement with Charles V over the purpose of the council, it was not convened until December of 1545. The council met in three stages: 1545-47, 1551-52, and 1562-63.

During the first stage, some basic ground rules were established that influenced the other two stages. First, the voting was to be by head rather than by nation; this gave the majority vote to the Italian representatives and thus to the papacy. Second, after some discussion with Charles V over which topics had precedence—reform or doctrine—it was decided to consider both concurrently. Those who could vote during the council included bishops, abbots, and generals of orders. Although the first stage included only thirty-four churchmen, the later meetings did include more. The majority of the participants came from Italy, Spain, France, and Germany. Although Protestants did attend some of the sessions, their impact on the council's deliberations was negligible.

During the first meeting of the council, the following important actions were taken: in session III, the Niceno-Constantinopolitan Creed was affirmed as the basis of faith; tradition and Scripture were declared to be equal sources of the faith: the canon of Scripture was fixed, and the Vulgate* was declared to be authentic for matters of faith; the Pelagian* and Protestant views of original sin were rejected; the critical problem of justification by faith was considered, with the affirmation made that man is inwardly justified by sanctifying grace and thus capable of good works only after his cooperation with the gratuitous divine assistance. In this same session, the Protestant position on the number of sacraments was rejected, and the council decreed there were seven, conferring grace *ex opere operato.* Apparently an epidemic broke out, and with political tension growing after some of the council left for Bologna, the pope suspended the council in 1547.

The second meeting of the council began in 1551, but the French were forbidden to attend on orders from Henry II; the Spanish representatives took a more independent stance with the support of Charles V; the Jesuits made their appearance in the form of Laynez* and Salmeron; and the Protestants made a brief appearance. A revolt of princes against Charles V plus internal friction brought an end to the meeting. The following actions were taken: the concept of the Eucharist* was carefully defined with the rejection of the Zwinglian and Lutheran positions; session XIV defined and affirmed the importance of auricular confession, the judicial character of absolution, the church's position on penance, and extreme unction; it also issued the reform decrees on discipline of the clergy.

With the third meeting of the council, called by Pius IV,* all hope of reconciliation with Protestantism was gone. One of the reasons for resummoning the council was apparently the fear of Pius IV that without the council France might become Calvinist. The council met in January of 1562, and 113 were present. The discussions were now more internal than previously. The struggle was not with Protestantism, but between pro- and antipapal forces. The latter usually were victorious, due partly to the consummate skill of Cardinal Morone.* The following actions were taken: a number of books were added to the Index in session XVIII; session XXI affirmed the belief that Christ is totally present in both species during the Eucharist, but only the bread was to be distributed to the laity; more precise definitions on the sacrificial aspects of the Mass* were arrived at. This was probably the second-most-important decision of Trent. A crisis arose over the appearance of the French bishops in November 1562 and their position on the question of clerical residency. The French and Spanish churchmen were opposed to the papal reform, but after ten months of adjournment due to this issue, a strict decree was promulgated. Other decrees were issued on matrimony, orders, founding of seminaries, and establishment of synods. The council ended 4 December 1563, and the decrees were confirmed 26 January 1564. Later in 1564 Pius IV issued a summary of the council's work called the "Tridentine Creed."

Although the council did not satisfy Protestants and some Catholics, it did provide the foundation for a revitalization of Catholicism through, for example, the Roman Catechism of 1566, the Revised Breviary of 1568, and the Missal of 1570. The council set the boundaries of Catholic belief, but did not always carefully and minutely define the details of belief, thus allowing some hope for further ecumenical efforts.

BIBLIOGRAPHY: Societas Goerresiana (ed.) *Concilium Tridentium* (1901ff.); H.J. Schroeder (tr.), *Canons and Decrees of the Council of Trent* (1941); L. Christiani, *L'Église à l'époque du Concile de Trente* (1948); G. Schreiber, *Das Weltkonzil von Trent* (1951); H. Jedin, *A History of the Council of Trent* (1957); J.A. O'Donohoe, *Tridentine Seminary Legislation* (1957); C.S. Sullivan, *The Formulation of the Tridentine Doctrine on Merit* (1959); M. Chemnitz, *Examination of the Council of Trent* (1971). ROBERT SCHNUCKER

TRINITARIANS. The "Order of the Most Holy Trinity," founded at Cerfroid, Meaux, in 1198 by John of Matha* and Felix of Valois, with approval from Innocent III. Known also as "Mathurins," they followed an austere form of the Augustinian Rule, wearing a white habit. Devoting themselves to redeeming Christian captives, they took a fourth vow to sacrifice their own liberty if necessary, using one-third of their revenues as ransoms. By the fifteenth century there were 800 houses as collecting centers and hospitals; they were particularly numerous in the British Isles. A reform movement, the "Barefooted Trinitarians," founded by Juan Bautista of the Immaculate Conception in Spain (1596), is the only surviving body, engaged in education, nursing, and ransoming of Negro slaves. Trinitarian nuns were affiliated from earliest times. The "Barefooted Trinitarian Sisters" date from 1612. J.G.G. NORMAN

TRINITY. The central tenet of the Christian faith is that God is one, personal, and triune. Trinitarian theology coheres with belief in the personal nature of God,* the Incarnation,* the Atonement,* the life in the Spirit, and the ultimate relation of redeemed men to God in Christ.

The Athanasian Creed* states: "We worship one God in Trinity, and Trinity in Unity; neither confounding the Persons, nor dividing the Substance." The truth that in the unity of God there is a trinity of persons can be known only by revelation,* but the truth is seen as neither irrational nor peripheral to faith. Trinitarian faith does not derive from the Church Fathers, but from the apostolic faith and teaching. The controversies of the first four centuries do not comprise attempts to impose alien Greek or other ideas upon Christianity, but attempts by the Fathers to assimilate adequately the empirical facts of the Christian revelation in an age which had neither categories nor language adequate to the new Christian realities.

The faith that the Father is God is held by all Christians. Monotheism* is deeply embedded in both Old and New Testaments, but the OT does contain important clues to plurality in God, including the *Sh'ma,* "Hear O Israel, Yahweh our Gods (Elohim) is Yahweh a unity;" the plural in Genesis 1:26 and 3:22; the three visitors to Abraham (Gen. 18:1-22); the captain of the Lord's hosts (Josh. 5:13-16); the Spirit of Yahweh passages (Gen. 1:2; Isa. 40:13); the triune invocation of the divine name (Isa. 6); and the striking words of Isaiah 48:16.

Historically, Trinitarian doctrine originated in the necessity Christians faced to distinguish Jesus from God, yet to identify Him with God. With the descent of the Holy Spirit at Pentecost, the empirical facts were all in hand for the subsequent formulation of the doctrine. Hence there is no hint of embarrassment in the NT to Jewish Christians due to Trinitarian theology. The doctrine is solidly embedded in the fabric of the NT (Matt. 28:19; 1 Cor. 12:3-6).

Through the Incarnation the first Christians learned to distinguish the Father and the Son while maintaining the faith that both are God. The Fatherhood of God was known in the OT. The unique NT teaching is that the Father and the Son are God, and that God is "the God and Father of our Lord Jesus Christ" (Rom. 15:6; 2 Cor. 1:3; Eph. 1:3; 1 Pet. 1:3).

Thus the doctrine of the Trinity is derived from the truth of the Incarnation and is to be tested by it. Jesus Christ is truly God the Son and distinctly God the Son (John 1:1,18; 20:28; Col. 2:9; Titus 2:13; Heb. 1:8,10). Subordinationism* and Adoptianism (Dynamic Monarchianism*) comprise two active, polemically minded erroneous alternatives. In the former the Son has a derived existence, in the latter he is only a man divinely energized for a mission. Neither of these alternatives adequately handles the empirical data of apostolic experience and witness. Their anti-Trinitarianism derives from a presupposition regarding the meaning of unity, rather than from the truth of the Incarnation.

While all Christians acknowledge the Holy Spirit* to be God, there remain two further levels of biblical understanding: the Holy Spirit is personal, and He is distinctly personal. Recent biblical studies are reluctant to make of the Spirit simply divine pervasive or invasive power. Some tend to identify Christ and the Spirit, although the Scriptures nowhere say that the Spirit is Christ. While some biblical passages do not demand a personal reading of the Spirit's reality, the controlling passages unambiguously declare the Spirit to be distinctly personal (Mark 3:22-30; Luke 12:12; John 14:26; 15:26; 16:7-15).

The most intractable problem faced by Christians has been how to conceive of the Trinity in unity. Traditional presuppositions that unity is simple and undifferentiated have forced many (including Subordinationists and Sabellians*) to jettison Trinitarian faith. However, if one sees unity as inclusive rather than exclusive, the problem is at least mitigated. If all approximations to unity are to be measured by a scale of degrees of absence of internal multiplicity, then Trinitarian theology and monotheism are irrevocably incompatible. But if the degree of unity is be be measured by the intensity of the unifying power in the life of the whole, then there is the prospect for at least partially comprehending the unity of the Godhead (cf. John 17:20-23) and other complex unities.

That God sent His Son to the cross and that God was in Christ is comprehensible on Trinitarian terms alone. Athanasius declared that only if Christ is truly God do we have contact with God in Him. Trinitarian faith in the NT enriched Christian experience. The Christian is said to be joined to the Trinitarian life of God through the redeeming work of Christ and the fellowship of the Spirit (Eph. 4:2-6).

BIBLIOGRAPHY: A.E.J. Rawlinson, *Essays on the Trinity and the Incarnation* (1928); G.L. Prestige, *God in Patristic Thought* (1936) and *Fathers and Heretics* (1940); L. Hodgson, *The Doctrine of the Trinity* (1955); S.J. Mikolaski, "The Triune God," in *Fundamentals of the Faith* (ed. C.F.H. Henry, 1969); K. Rahner, *The Trinity* (1970).

SAMUEL J. MIKOLASKI

TRITHEISM. Belief in three gods which denies the unity of substance in the Christian doctrine of the Trinity. Popular expressions of Trinitarian doctrine and some transactional soteriological theories can be tritheistic. Historical tritheism appeared in Monophysite circles c.550, associated with Johannes Askunages and Johannes Philiponus. Philoponus was an Alexandrian philosopher who opposed the Chalcedonian Christology, contending that Christ's was a single nature compounded of the divine and human, and that there are three divine substances *(ousiai)* in the Trinity. A speculative rather than a practical tritheism, it was opposed by John of Damascus in *de Fide orthodoxa.*

In medieval times the extreme Nominalism* of Roscellinus* of Compiègne and the exaggerated Realism of Gilbert de la Porrée led them into tritheistic positions which were condemned at the Councils of Soissons (1092) and Reims (1148) respectively. Gilbert's teaching influenced Joachim of Fiore,* who conceived the oneness of the three persons as a mere generic unity. Joachim's doctrine was condemned at the Fourth Lateran Council (1215), which defined clearly the numerical unity of the divine nature.

Anton Günther (1783-1863), opposing Hegelian pantheism, taught that the Absolute determined itself three times in a process of self-development. The divine substance is trebled, and the three substances attracted to one another through consciousness make a formal unity. This was condemned by Pius IX (1857). J.G.G. NORMAN

TROELTSCH, ERNST (1865-1923). Liberal German theologian, historian, and philosopher of religion. He taught mainly in Heidelberg and Berlin between 1894 and 1923, and served also for a time as minister of education in the government. Early associated with the "*religionsgeschichtliche Schule,*" he devoted himself to the problem raised for religion by the historical consciousness and method dominant in the West since the eighteenth century, and to relating Christianity to the cultural situation. He denied that dogmatic theology could have access to a super-historical absolute truth. The claims of all religions were depicted as relative to their total cultural settings, which both made them possible and limited them. His belief that Christianity could be shown to be the highest of religions weakened between *Die Absolutheit des Christentums und die Religionsgeschichte* (1902; ET *The Absoluteness of Christianity,* 1971) and *Der Historismus und seine Ueberwindung* (1924; ET *Christian Thought, Its*

History and Application, 1923). Yet he feared the skepticism of historical relativism and believed it could be overcome by living decisions, taken responsibly in the light of history. Thus Christianity was still the religion best suited to the Western world.

He was intensely concerned with social and political questions, and his interest in the possibilities and conditions for a fruitful contemporary relation between Christianity and civilization issued in *The Social Teaching of the Christian Church* (1931; first published in 1912 as *Die Soziallehren der christlichen Kirchen und Gruppen*), a work best but misleadingly known perhaps for its use of church, sect, and mysticism as three types of Christianity. By his classical explorations of the problem of historicism in relation to religion, Troeltsch could be regarded as marking an epoch in modern theology.

See R.H. Bainton, "Ernst Troeltsch—Thirty Years After," *Theology Today* 8 (April 1951), pp. 70-96. HADDON WILLMER

TROPHIMUS. A Gentile Christian who accompanied Paul to Jerusalem as a representative of the churches of Asia (Acts 20:4). He is identified as an Ephesian by the Western Text and in Acts 21:29. The supposition that Paul had brought him into the forbidden area of the Temple occasioned the riot which resulted in Paul's arrest and appearance before Felix and eventually his appeal to Caesar. 2 Timothy 4:20 refers to Trophimus as left behind sick at Miletus, ostensibly when Paul himself had been hurried under arrest to Rome with few trusted companions remaining.

The name is frequently recorded from Ephesus and district (e.g., British Museum Inscrs. 591; *Jahreshefte des österreichischen archäologischen Instituts* XLIV, 1959, Beiblatt, col. 369).

An early bishop of Arles named Trophimus is first mentioned in the fifth century. Tradition has attempted to identify him with the companion of Paul. COLIN HEMER

TROTTER, ISABELLA LILIAS (1853-1928). Missionary to North Africa. Daughter of a London businessman, she was privately educated, and converted under the ministry of Mr. and Mrs. Pearsall Smith. In 1876 she made the acquaintance of John Ruskin,* who admired her expert miniatures and exhorted her to devote her life to painting, but she determined to sail as a missionary to North Africa. She began her work in Algeria in 1888, making heroic and dangerous missionary journeys, securing converts among Arabs, French, Jews, and Negroes, and establishing preaching stations. Her Algiers Mission Band grew steadily from three to thirty full-time workers. Her translations of the NT into Algerian colloquial and her illustrated tracts for Muslim readers were greatly admired. She died while still on active service; her society is now incorporated in the North Africa Mission. IAN SELLERS

TRUCE OF GOD (Lat. *pax, treuga Dei*). A suspension of hostilities ordered by the Roman Catholic Church. The custom originated in France in the tenth century and was meant to lessen the impact on the lower orders of society of the incessant quarreling of the feudal nobles. Since the armistice was ordered by the church, it was traced back to the will of God and called "the truce of God." At first the period was from Saturday evening to Monday morning, but later holy days or seasons (e.g., Easter) were included. After the eleventh century the practice died out.

TRUMBULL, CHARLES GALLAUDET (1872-1941). Evangelical writer and journalist. Born in Hartford, Connecticut, he graduated from Yale and joined *The Sunday School Times,* founded by his father in 1893. He became editor in 1903 and later director (it ceased publication in 1967). He was for a long time a staff writer for the Toronto *Globe,* and he also wrote the weekly Sunday school lesson for several daily newspapers, including the Philadelphia *Evening Public Ledger.* His many interests included membership of the Victoria Institute, the Palestine Exploration Fund, and the Archaeological Institute of America. A supporter of missions, he wrote books that were evangelistic and prophetic.

C. G. THORNE, JR.

TÜBINGEN SCHOOL. In the early nineteenth century there was a Tübingen School of conservative theology led by J.C.F. Steudel at the University of Tübingen. But the Tübingen School commonly referred to is that headed by F.C. Baur,* who taught there from 1826 until his death in 1860. Baur's teaching was characterized by his anti-supernaturalistic attitude to history, tendency criticism in the interpretation of biblical writings, and the use of idealist philosophy in the interpretation of history. He saw a fundamental conflict between the Jewish church led by Peter and the Hellenistic Gentile church led by Paul. The degree in which NT books exhibited tendencies of this conflict determined their authenticity. Baur assigned most of them to the second century.

The organ of Baur's circle was the *Tübinger Theologische Jahrbücher* (1842-57), continued by A. Hilgenfeld* as the *Zeitschrift für wissenschaftliche Theologie* (1858-1914). Baur wrote in defense of the school in *An Herrn Dr. Karl Hase. Beantwortung des Sendschreibens "Die Tübinger Schule"* (1855) and *Die Tübinger Schule und ihre Stellung zur Gegenwart* (1859, rev. 1860). It is questionable, however, whether the school ever amounted to more than Baur and his immediate circle. Despite the attention Baur attracted, nineteenth-century German liberal theology tended to follow other paths. Of his disciples, A. Schwegler adopted Baur's approach in his *Geschichte des nachapostolischen Zeitalters* (2 vols., 1846), as did O. Pfleiderer.* But Baur's most famous pupils —D.F. Strauss,* E. Zeller,* and A. Ritschl*—developed their own approaches. After the death of his colleague F. Kern in 1842, Baur felt himself increasingly isolated within the Tübingen faculty and German academic theology.

BIBLIOGRAPHY: R.W. Mackay, *The Tübingen School* (1863); E. Zeller, "Die Tübinger Historische Schule" in *Vorträge und Abhandlungen geschichtlichen Inhalts* (1865), pp. 267-353; P.C.

Hodgson, *The Formation of Historical Theology: A Study of Ferdinand Christian Baur* (1966); H. Harris, *The Tübingen School* (1973).

COLIN BROWN

TUCKER, ALFRED ROBERT (1849-1914). Bishop of Eastern Equatorial Africa (1890-98) and of Uganda (1898-1911). He renounced an artistic career for the Anglican priesthood, and after seven years in English parishes he joined the Church Missionary Society and reached East Africa in 1890. His main achievement was to consolidate and extend the Anglican Church in Buganda and neighboring chiefdoms. He nurtured a locally supported African ministry, promoted educational and medical work, and traveled widely to supervise established missions and pioneer new work. His progressive views on church government aroused opposition among European missionaries who rejected full integration with the local church. Nevertheless, synodical government was accepted in 1909. Tucker campaigned for the British protectorate over Uganda (1894) and frequently championed the interests of its people. Ill-health compelled his resignation in 1911, at which time he became canon of Durham.

D.G.L. CRAGG

TUCKER, WILLIAM JEWETT (1839-1926). Congregational clergyman and educator. Educated at Dartmouth College and Andover Theological Seminary, he pastored churches in New Hampshire and New York City before teaching at Andover (1879-93). He was also associate editor of *Andover Review* (1884-93). Influenced by Horace Bushnell,* he asserted that only humane theology, or "progressive orthodoxy," would improve society. Because of this he founded in 1881 Andover House in Boston, patterned after Toynbee Hall. He was tried with others for heresy, but was acquitted. From 1893 until his retirement in 1909 Tucker was president of Dartmouth. Not a prolific writer, he best expressed his thinking in the autobiographical *My Generation* (1919), in which he praised the growth of the Puritan conscience in America. DARREL BIGHAM

TUCKNEY, ANTHONY (1599-1670). Cambridge Puritan. Born in Lincolnshire, he was educated at Emmanuel College, Cambridge, becoming a fellow and tutor of his college, which posts he held with distinction for nearly ten years. Three of his pupils were Benjamin Whichcote and Henry and William Pierrepont. In 1629 he became mayor's chaplain and in 1633 (in succession to John Cotton*) vicar of Boston, Lincolnshire. He served as a member of the Westminster Assembly* and on the drafting committee for the "Larger Catechism." In 1645 he became master of Emmanuel and eight years later, master of St. John's College. In 1654 he was appointed one of Cromwell's "Triers" and in 1655 regius professor of divinity. At the Restoration he lost his Cambridge posts and moved to London to live quietly in the parish of St. Mary Axe, but occasionally preached in private. Though nominated as a member of the Savoy Conference* (1661), he never attended. He published little, but did edit two books by Cotton. PETER TOON

TULCHAN BISHOPS. The name comes from an old Scots rural practice of stuffing a calf's skin with straw and leaving it beside the cow to induce it to give milk more freely. In 1572, during the infancy of James VI, through the influence of Regent Morton certain influential Protestant ministers were persuaded to agree to the retention in the kingdom of episcopal titles. The persons appointed to the vacant sees entered into an agreement with Morton and other powerful nobles to ensure that the incumbent should retain only a modest proportion of the episcopal emoluments and give the rest to their patrons. "The bishop," it was said, "had the title, but my lord had the milk." The general assembly denounced the practice, which did not long survive the end of Morton's regency (1578). J.D. DOUGLAS

TUNKERS, see BRETHREN IN CHRIST

TUNSTALL, CUTHBERT (1474-1559). Bishop of Durham. While studying at Oxford, Cambridge, and Padua, he made friends with William Warham,* Thomas More,* Erasmus,* and other foreign scholars. Warham made him his chancellor in 1511, and Henry VIII* sent him on various political missions, making him bishop of London in 1522 and of Durham in 1530. He at first opposed the royal supremacy, but later preached vigorously in its favor. This support of Tunstall was of crucial importance to Henry VIII because of the wide respect in which he was held. Tunstall remained Catholic in doctrine, however, defending auricular confession and striving to keep the Bishops' Book* of 1537 as Catholic as possible. Under Edward VI* he had second thoughts on royal supremacy, though he enforced the Act of Uniformity* in his diocese after voting against it in Parliament. He also voted against the abolition of chantries in 1547, and the act permitting priests to marry in 1549. In consequence he was imprisoned in his house in 1551, and in 1552 deprived of his bishopric. While in prison he wrote a defense of the Catholic view of the Mass, *De veritate corporis et sanguinis domini nostri Jesu Christi in eucharistia* (published in Paris in 1554). He was restored by Mary Tudor* in 1554, but took no part in persecution. On the accession of Elizabeth* he refused to take the Oath of Supremacy, or to help consecrate Matthew Parker* as archbishop of Canterbury. He was again deprived and kept in custody at Lambeth Palace until his death a few months later.

See register as bishop of Durham (ed. G. Hinde, 1951-52), and C. Sturge, *Cuthbert Tunstall: Churchman, Scholar, Statesman, Administrator* (1938).

JOYCE HORN

TURNER, CUTHBERT HAMILTON (1860-1930). Anglican historian of early Christianity, and NT scholar. He came to Oxford as a student in 1879 and there remained until his death, serving in various capacities as research scholar and lecturer, latterly as fellow of Magdalen College (1889-1930) and Dean Ireland's professor of exegesis (1920-30). His research centered in matters of chronology and textual criticism related to the Fathers and canon law. He failed to complete several major writing projects, but published in fascicles an extensive collection of documents relating to early Western canon law *Ecclesiae Occidentalis Monumenta Juris Antiquissima* (1899-1930) and a flood of learned essays—the more important being a classic study of NT chronology in Hastings's *Dictionary of the Bible* (vol. 1, 1898); "Greek Patristic Commentaries" in the same dictionary (extra vol., 1904); a general sketch of the development of early canon law for the *Cambridge Medieval History* (vol. 1, 1911); "Apostolic Succession" in *Essays on the Early History of the Church and the Ministry,* which he co-edited with H.B. Swete* (1918); and his inaugural lecture on *The Study of the New Testament, 1883 and 1920* (1920).

Two collections of his writings were published under the title *Studies in Early Christian History* (1912) and *Essays: Catholic and Apostolic* (posthumous, with a memoir by H.N. Bates, 1931). Turner was the first editor of *Journal of Theological Studies* (1899-1902), to which he contributed throughout his life, and a collaborator in the *Patristic Greek Lexicon* initiated by H.B. Swete and completed only recently. W. WARD GASQUE

TWELVE APOSTLES, GOSPEL OF THE, see APOCRYPHAL NEW TESTAMENT

TWELVE ARTICLES, THE (1525). A statement of the grievances of German peasants against their feudal lords, drawn up at Memmingen. The peasants wanted abolition of serfdom, rights to fish in streams, hunt game and cut wood, just rent, abolition of feudal death taxes as well as the right to appoint their pastors and control the amount of tithes. The last article asserted that their demands must be in conformity with Scripture and would be withdrawn if such were not the case. Luther agreed with these demands and urged the feudal lords to grant these articles, but when the peasants revolted against their feudal lords, Luther turned against the peasants and in a tract urged the lords to exterminate them to prevent anarchy.
EARLE E. CAIRNS

TWENTY-FIVE ARTICLES, see ARTICLES OF RELIGION

TYCONIUS (fl. c.370-c.390). Donatist lay theologian. For his "Catholic" views he was attacked by Parmenian, the Donatist bishop of Carthage, and excommunicated by a council at Carthage (?), but evidently did not become a Catholic. Two lost works, *On Internal War* and *Expositions of Various Causes,* discussed the Donatist–Catholic dispute. His *Book of Rules,* the first Latin essay in hermeneutics, presents seven keys to spiritual exegesis, which through Augustine's adoption (*On Christian Instruction* 3:30:42–37:56) had a wide influence. His spiritualizing *Commentary on the Apocalypse* may be recoverable from a partially extant Catholic recension and the later commentaries of Caesarius,* Primasius,* Bede,* and Beatus* of Liebana. Tyconius differed with Donatist theory in holding the church to be truly universal and "bipartite," an intermingling, inseparable until the end, of pure and impure, the "cities" of God and the devil. On this and other themes, including grace and faith, he decisively influenced Augustine; both were zealous Paulinists whose thought focused on the church.

BIBLIOGRAPHY: T. Hahn, *Tyconius—Studien* (1900); P. Monceaux, *Histoire littéraire de l'Afrique chrétienne* 5 (1920), pp. 165-219; W.H.C. Frend, *The Donatist Church* (1952), pp. 201-5, 316-19; G. Bonner, *St. Bede in the Tradition of Western Apocalyptic Commentary* (1966); R.A. Markus, *Saeculum: History and Society in the Theology of St. Augustine* (1970), pp. 115-22. *Comm. on Apoc.*, ed F. LoBue, *Texts and Studies* VII (1963), with bibliography. D.F. WRIGHT

TYE, CHRISTOPHER (c.1500-1573). English composer. For a time he was organist at Ely Cathedral, then sometime after the triumph of Protestantism he took orders in the Church of England, and died as rector of Doddington. A representative quantity of his Latin church music has survived and shows him to have been a superior composer. Most of his music with English text is adapted. His versification of fourteen chapters of Acts is inferior, but the music has yielded several much-used hymns. The short anthem, "O come ye servants of the Lord," is also an adaptation.

TYLER, BENNET (1783-1858). American Congregational theologian. Born at Middlebury, Connecticut, he graduated from Yale College in 1804. After studying theology, he was ordained in 1808. He became president of Dartmouth College in 1822, and pastor of the Second Church, Portland, Maine, in 1828. By this period the USA had entered the final phase of the Second Great Awakening,* which had begun in 1799. Charles G. Finney* and other evangelists interpreted popularly the doctrines of Nathaniel W. Taylor* and the "New Divinity" issuing from Yale. Orthodox Calvinists became increasingly alarmed, not only at the "new measures" used by revivalists, but more so at the concessions to Arminianism* in teaching human ability and choice. Open conflict broke out in 1828 when Taylor's address to the Connecticut Congregational clergy, *Concio ad Clerum,* dealt with the crucial issue of natural depravity and stated that men's depravity does not "consist in a sinful nature, which they have corrupted by being *one* with Adam, and by *acting in his act.*" This made it certain to the conservatives that the New Divinity departed from orthodox Calvinism at a most essential point.

Bennet Tyler, a Yale classmate of Taylor, in 1829 entered into discussions with him which continued for some years. The first results of the conflict was the founding of a Pastoral Union in 1833 and a Theological Institute at East Windsor (later Hartford Seminary) in 1834, with the express purpose of combating the New Haven Theology.* Tyler was called to assume the presidency and remained in this position until 1857. In his later years he entered into discussion with Horace Bushnell* on emerging theological issues. He published many sermons, articles, and books, on such subjects as the sufferings of Christ, the New Haven Theology, and New England revivals.

KEITH J. HARDMAN

TYNDALE, WILLIAM (c.1494-1536). English Reformer and Bible translator. Born in the west of England, he was educated at Magdalen Hall, Oxford, and afterward at Cambridge. He became tutor to Sir John Walsh's family, but seeing the ignorance of clergy and laity alike, he grew convinced that "it was impossible to establish the lay people in any truth, except the Scripture were plainly laid before their eyes in their mother tongue." From Cuthbert Tunstall,* bishop of London, he received no encouragement, so he left England, never to return. The printing of his first NT in English was begun in 1525 at Cologne, but a police raid stopped the work, and it had to be finished later that year at Worms. Tunstall, Thomas More,* and William Warham,* archbishop of Canterbury, attacked him relentlessly, and secret agents were sent to trap him as he moved around from his Antwerp base where sympathetic English merchants protected and helped him. Tyndale continued to revise his NT, though plagued by pirated versions and betrayed by his erstwhile helper George Joye who ran off another pirated version. Tyndale also embarked on the Pentateuch and left other OT translations uncompleted.

At the same time, he was writing OT commentaries, replying to More's longwinded attacks, writing NT expositions (1 John and Matt. 5–7), propounding justification by faith alone in *The Parable of the Wicked Mammon* (1528), penning a major constitutional and theological treatise in *The Obedience of a Christian Man* (1528), and dealing with Joye's speculations on the afterlife as well as the Anabaptist doctrine of soul-sleep (the subject of Calvin's first theological work). Tyndale's output was impressive, as the conditions under which he worked—a shipwreck and loss of manuscripts, secret agents after him, police raids on his printer, betrayal by friends—were daunting. His quality nevertheless remained high. He pioneered English Bible translations from the original languages. His style was lucid, crisp and concise, and above all appealed to ordinary people for its down-to-earth character. His literary work is now universally recognized. His theological works were often translations or rough paraphrases of Luther or Lutheran works, but there are traces from 1529 onward of the growing influence of Swiss Protestant theology. Arrested at Vilvorde near Brussels in 1535, he was finally strangled and burnt in the following year.

See *Works* (ed. H. Walter, 1848, and G.E. Duffield, 1964); and biographies by J.F. Mozley (1937) and C.S. Williams (1969).

G.E. DUFFIELD

TYPOLOGY, BIBLICAL. A system of biblical interpretation which features particularly in the NT treatment of the OT, and which has since been applied by some exegetes to the NT in turn. The term "typology" is derived from the Greek word *typos*, "pattern" or "figure" (cf. Rom. 5:14); the NT *antitypon* (cf. Heb. 9:24) means much the same thing. Typology and allegory* overlap to some extent, but typology is in general more historical in character; a "type" is accordingly an event, person, or object which by its very nature and significance prefigures or foreshadows some later event, person, or object. Thus Adam is explicitly described in Romans 5:14 RSV as "a type of the one who was to come" (i.e., of Christ), because of his special place in human history. The most fruitful OT source for NT typology was, however, the Passover and Exodus complex of events: cf. John 3:14; 6:31-35; 1 Corinthians 5:7; 10:1-5. Also used as types are Noah, Melchizedek, and Jonah. Galatians 4:22-31, though not dissimilar, is described by Paul as allegory. Typology began in the OT itself, as in the treatment of the Exodus theme in Isaiah 43:16ff.; 51:10f., and continued into postbiblical Jewish and Christian exegesis, especially at Alexandria.

BIBLIOGRAPHY: A.G. Hebert, *The Throne of David* (1941); G.W.H. Lampe and K.J. Woollcombe, *Essays in typology* (1957); J. Barr, "Typology and allegory" in *Old and New in Interpretation* (1966), pp. 103-48.

D.F. PAYNE

TYPOS (Gr. *tupos tes pisteos*, "model of faith"). An edict issued by Emperor Constans II in 647 or 648 superseding Heraclius's *Ecthesis.* With a view to securing theological peace, it forbade anyone to assert either Monothelite* or Dyothelete* beliefs, and required that teaching should be limited to the definitions of the first five ecumenical councils. It was drawn up by Paul, the Monothelite patriarch of Constantinople. At the Lateran Council (649), Martin I condemned both the *Ecthesis* and the *Typos,* and was banished.

TYRRELL, GEORGE (1861-1909). Roman Catholic modernist. Born in Dublin of an Anglican family, he studied briefly at Trinity College there before conversion to Roman Catholicism led to his entering the Jesuit Order in 1880. Ordained in 1891, he taught philosophy at Stonyhurst College (1894-96), then was called to his order's English headquarters in Farm Street, London. There he produced acceptable orthodox publications until 1899 and was remarkably successful in spiritual counseling. Among those he helped was the daughter of F. von Hügel,* who introduced Tyrrell to the works of French modernists.

An article on "Hell" in the *Weekly Register* (1899), in which be began overt questioning of Roman Catholic theology, led to his transfer to a provincial mission house, but Tyrrell maintained an active devotional life; and though superiors were uneasy about his tendencies, his books up to and including *Lex Orandi* (1903) were published with the Imprimatur.* His *Lex Credendi* (1906), with its oblique criticism of the church, was overshadowed that same year by a pseudonymous *Much Abused Letter* which led to dismissal from his order. Undeterred, Tyrrell wrote two letters to *The Times* in 1907, replying to Pius X's condemnation of modernism, and this led to his being refused the sacraments. In his publications he made the distinction between the "prison of theology" and the "liberty of faith."

When he died at forty-eight, having been plagued by ill-health all his life, he still considered himself a Catholic, but was unrepentant about his works. He was refused a Catholic burial, and lies in an Anglican churchyard in Sussex. Von Hügel, who ignored the virtual excommunication and continued to address Tyrrell as "Father," admitted his recklessness in correspondence, his bitterness, and his excessive reaction against extremism in others, but added: "if to be a saint is to be generous and heroic, to spend yourself for conscience and for souls, then T. is a saint."

J.D. DOUGLAS

TYRRELL, WILLIAM (1807-1879). First bishop of Newcastle, Australia. Educated at Charterhouse and St. John's College, Cambridge, he was ordained in 1833 and became incumbent of Beaulieu in Hampshire in 1839. Eight years later, when W.G. Broughton* divided the diocese of Australia, Tyrrell became bishop of the northern part of Eastern Australia based on the small town of Newcastle. He arrived in 1848 to find only eleven clergy to cover his vast diocese of over 20,000 square miles. In 1859 the diocese of Brisbane was separated from Newcastle, and in 1867 after a long delay the diocese of Grafton and Armidale took the northern part of New South Wales from Tyrrell's diocese. He was generous in supporting his clergy and building churches in his diocese. He died at Morpeth where he had lived, and left a large property as an endowment for the diocese.

NOEL S. POLLARD

UBERTINO OF CASALE (1259-c.1330). Franciscan Spiritual leader. Born at Casale, near Vercelli, Italy, he entered the Franciscan Order about 1272. He studied and lectured in Paris (1289-98), then returned to Italy as a preacher and teacher. Influenced by John of Parma* and P.J. Olivi,* ardent Spirituals, he became a leader of this party of strict Franciscans in Tuscany and Umbria. His principal work, *Arbor vitae crucifixae Jesu* (1305), written during enforced retirement, includes an account of Christ's life and suffering and a commentary on the Apocalypse following ideas of Joachim of Fiore,* with severe attacks upon the papacy and those Franciscans who abandoned rigid observance of the vow of poverty. For his part in the poverty controversy and failure to be reconciled to his order, John XXII* transferred him to the Benedictines (1317). Charged with heresy, he fled Avignon (1325). He preached against John XXII at Como in 1329; otherwise nothing is known of his last years.

ALBERT H. FREUNDT, JR.

UBIQUITARIANISM. This doctrine, derived by Luther from a variety of patristic and medieval sources, asserts that Christ is present in His human nature everywhere and at all times. The view was developed by the Reformer in two important works of 1527 and 1528 to uphold his belief in the Real Presence in the Eucharist.* Calvin and Melanchthon both recoiled from the doctrine, and Zwingli flatly denied it. It was hotly debated by Lutheran scholars of the Swabian and Saxon schools till the intellectual trends of the later seventeenth century caused christological argument to recede into the background.

UCHIMURA, KANZO (1861-1930). Founder of the Japanese non-church movement. The first son of a *samurai* (warrior-knight) family, he entered Sapporo Agricultural College, but within three months was won to Christianity by the zealous evangelism of students of Dr. W.S. Clark.* Under Uchimura's leadership the "Sapporo Band" became the first independent Japanese church in 1881. The years 1884-88 he spent mostly in New England, and on his return to Japan he became a prominent interpreter of Western culture and literature, and a fearless prophet of social righteousness, denouncing evils in society, government, churches, and missions. He is best known as founder and exponent of nondenominational Japanese Christianity, called *mukyokai* (non-church movement), and for his greatest legacy: a twenty-two-volumed Bible commentary—the fruit of his first love, the study of the Bible.

DAVID MICHELL

UDALL, JOHN (1560?-1592). Elizabethan Puritan divine, sometimes called "Uvedale." Educated at Christ's and Trinity College, Cambridge, where one of his friends was John Penry,* he served as a curate from 1584 at Kingston-on-Thames and was soon known as a Puritan. For the publication of his views he was summoned before the Court of High Commission at Lambeth, but was not severely punished. This was due in part to the help of aristocratic friends, such as the countess of Warwick. Still at Kingston, he continued to adopt a critical attitude toward the prelates and the church, and it was noted that he continued his friendship with Penry, who was implicated in the production of the *Marprelate Tracts.**

After being deprived of his living in 1588, Udall accepted an invitation to minister at Newcastle-on-Tyne. After a year there he was accused of complicity in the production of the *Marprelate Tracts* and summoned to appear before the Privy Council in London in January 1590. As a result he was put in prison until July, when he was tried at the Croydon Assizes, accused of the authorship of two anonymous tracts, *A Demonstration of the Truth of the Discipline*... (1588) and *The State of the Church of England* (1588). Later he was also tried at Southwark Assizes, found guilty of felony, and sentenced to death. Several eminent people, including Sir Walter Ralegh, tried to get him released. Eventually the governors of the Turkey Company offered to send him to Syria as a chaplain, but in June 1592 when all was set for his release he fell ill and died. Apart from his controversial writings, he did publish important works revealing his skill in Hebrew; the *Commentary on Lamentations* (1595), for example, was later highly prized by James I of England.

PETER TOON

UDALL, NICHOLAS (1505-1556). English scholar and dramatist. Going from Winchester College to Oxford in 1520, he gained a reputation for making Latin verses and for being favorably disposed to Lutheranism. In 1534 he was appointed headmaster of Eton, but was removed seven years later for "unnatural crime." After some time in jail he held benefices in the Church of England. He translated the first volume of Erasmus's NT into English which was published in 1548. He also wrote a tract against the rebels who opposed the first Edwardian Book of Common Prayer. When Mary Tudor came to the throne he reverted to Roman Catholicism, and later became head of Westminster School, but died shortly afterward. His best-known work is a play *Ralph Roister Doister,* probably written for presentation by the students at Eton. W.S. REID

UKRAINIAN ORTHODOX CHURCH, see EASTERN ORTHODOX CHURCHES

ULFILAS (Ulphilas) (c.311-c.381). Bishop of "the Christians in Gothia." Hard facts about him are scarce and amount to little more than that he translated a large part of the Bible from Greek into Gothic, involving him in the devising of a new alphabet, and that he took a leading part in the conversion of the Visigoths to a form of Arianism.* He is said to have come from north of the Danube; been consecrated (c.341) at the instigation of Eusebius of Nicomedia,* Arian bishop of Constantinople; obtained imperial permission to move his people into Roman territory south of the Danube to escape persistent persecution; and been conciliatory when Theodosius I* restored Nicene orthodoxy to the Roman Empire in 379. J.D. DOUGLAS

ULLMANN, KARL (1796-1865). German Lutheran theologian. He was deeply influenced by Schleiermacher* and Neander* and served as professor of theology in Heidelberg University from 1821, with the exception of a seven-year interlude at Halle (1829-36). Ullmann was the leading advocate of the theology of mediation *(Vermittlungstheologie),* and in contrast to rationalism he stressed the significance of salvation through Jesus. He also occupied several posts in the Baden church, including director of the consistory (1856-61). He tried to influence the Baden church in the direction of Pietism by introducing changes in the confession, order of worship, and catechism, and this led to his overthrow by liberals. In 1848 he assisted in founding the German Evangelical *Kirchentag.* RICHARD V. PIERARD

ULRICH (c.890-973). Bishop of Augsburg. His chief source of fame is that he is known as the first "saint" whose canonization was decreed by a pope (John XV), after an account of Ulrich's life and miracles had been submitted to the Lateran Synod in 993. A letter which circulated in the eleventh century against clerical celibacy and ascribed to him has been shown to be a forgery. The principal, albeit brief, source for his life is the biography by his contemporary, Gebhard, provost of Augsburg Cathedral. This reveals that Ulrich was firmly committed to the policy of Emperor Otto I,* and was a reforming bishop.

ULTRAMONTANISM. A movement of Catholic revival, especially after the French Revolution, which rediscovered and hoped to reimplement the unity and independence of the Roman Church under the papacy. This reassertion of Catholic Christian faith found inspiration in the pre-Renaissance Christian Commonwealth and the Catholic Reformation of the sixteenth and seventeenth centuries. The motifs of Romanticism, especially the new awareness of the medieval, were contributory. The movement, situated mainly in France but also in S Germany and England, wanted to terminate the power over the church of Enlightenment rationalism, especially as this was realized in the secularist state domination of the church since Louis XIV's time. This implied a renunciation of Gallicanism.*

The French Revolution was the decisive moment. Structurally Ultramontanism meant centralization of the church under absolute papal authority, coupled by independence of the Roman Church from state control and where possible subordination of the state and the rest of society to papal Catholic dogmatic and moral principles. The term, used derisively since the seventeenth century, implied attachment to Rome i.e., "beyond the mountains" (the Alps). The movement was shaped by the resistance of Pius VII* to Napoleon, by the support of subsequent popes, and public figures such as Joseph de Maistre* and Louis de Bonald, and Cardinal Manning* in England, by the reestablishment of the Jesuits* (1814), and the establishment of new orders, but especially by the singular devotion of countless parish clergy and faithful.

Because of its antirevolutionary role, Roman Catholicism was linked to hierarchical and legitimist societal ideals, the alliance of Throne and Altar. Pius IX* led the movement by his obvious Catholic devotion, exemplified by the dogma of the Immaculate Conception* of Mary (1854), by his resistance to the attacks on his temporal (political) power (1859-70), and especially by the dogma of papal infallibility* (1870). Vatican I* (1869-70) meant the official triumph of Ultramontanism as the church's stand and showed its view of the proper relation between pope and general council. The newfound spiritual power was the *sine qua non* of Leo XIII's* profound social encyclicals.

BIBLIOGRAPHY: E.E.Y. Hales, *Pio Nono* (1954) and *Papacy and Revolution, 1769-1846* (1966); K.S. Latourette, *Christianity in a Revolutionary Age,* I (1958); R. Aubert, *Le pontificat de Pie IX* (2nd ed., 1964). C.T. MC INTIRE

UNAMUNO, MIGUEL DE (1864-1937). Spanish scholar and writer. Born in Bilbao and educated at Madrid, he became professor of Greek at Salamanca, where he was later rector at different times and where he spent most of his working life. For his political opinions, which clashed with those of the ruling dictatorship, he was in 1924 exiled to the Canary Islands, spent some time in France, and did not return to Spain for six years. In due course, however, he became as much the critic of the socialists as of the monarchists.

He continually sent words to war in the service of ideas. "My painful duty," he said, "is to irritate people. We must sow in men the seeds of doubt, of distrust, of disquiet, even of despair." The world contained much that he disliked: parochialism, complacency, hypocrisy, dogmatism, and useless tradition. Although a religious man who called himself a Catholic and denounced twentieth-century materialism, he was regarded with alarm by the church and two of his books were placed on the Index.* He was described by some as a Catholic heart at war with the Protestant spirit. Out of his mystical and philosophical preoccupations and his mastery of some sixteen languages came a stream of publications: essays, tragedies, novels, religious poems, his uneasy spirit reflected in some of the titles—*The Tragic Sense of Life in Men and in Peoples* (ET 1926) and *The Agony of Christianity* (1928), with its reminder that Christ had not come to bring peace but a sword. It has been widely suggested that his *San Manuel Bueno, mártir* (1933; ET 1956), with its treatment of an unbelieving priest, expresses his own longing for immortality.

See F. Mayer, *L'Ontologie de Miguel de Unamuno* (1955), and J. Marías, *Miguel de Unamuno* (ET 1967).

J.D. DOUGLAS

UNCTION. A term for anointing with oil, used at baptism (and confirmation), ordination, coronations, consecrations of churches, on dead bodies, and on the sick. Called "chrism,"* it was linked with baptism almost universally from early times until the Reformation. It originally meant the consecration of Christians to the royal priesthood. It was also associated with the Spirit, especially when "laying on of hands" was dying out, signifying the gift of the Spirit at Christ's baptism.

For unction of the sick, the NT warrant was Mark 6:13; James 5:14f. Early references are many, and from the fifth century it is even more often cited, until Bede* could represent the rite as well established. The term "extreme unction" appeared first in Peter Lombard,* and from the twelfth century it was commonly regarded as a preparation for death, though in the Latin Rite it looks hopefully "for the health of body and soul." From the thirteenth century it was numbered among the seven sacraments.* Aquinas* expounded the medieval doctrine, and official Roman Catholic teaching was established at the Council of Trent.*

The Greek Orthodox Church called the rite *Euchelaion* ("oil of prayer"). When administered in full the ceremony is very long, involving seven priests. Primarily aimed at physical cure, it is also frequently received as a preparation for Communion, even by those not ill. Other Eastern Churches have the rite, though it has gone out of use among the Ethiopians and Nestorians.

In Anglican usage, the First Prayer Book (1549) permitted anointing if the sick desired it. Later versions omitted it, though the Nonjurors* restored it. The Scottish and American prayer books (1929) and others make provision for it.

See F.W. Puller, *The Anointing of the Sick in Scripture and Tradition* (1904).

J.G.G. NORMAN

UNDERHILL, EVELYN (1875-1941). English mystic. Daughter of a distinguished London barrister, she was educated at King's College, London, of which she was later made fellow. She was reared an Anglican, and her spiritual life began in 1907 while visiting a Franciscan convent. She professed conversion in 1911 with leanings toward the Roman Church. *Mysticism,* her first book, appeared that year, followed by *The Mystic Way* in 1913. Friedrich Von Hügel* was the divide in her life, and she was under his close direction for four years until he died in 1925—experiencing Christianity most personally and making a commitment to the Church of England. From 1911 her life was religious work: personal cases, social work, addresses, retreats, and books. Two volumes (*The Spiral Way* and *The Path of the Eternal Wisdom*) she published under the pseudonym "John Cordelier." She produced critical editions and translations of Ruysbroeck, Hilton, and other mystics and contributed to periodicals. *Practical Mysticism* (1915) and *The Essentials of Mysticism* (1920) were guidebooks. Her great work was *Worship* (1937) which includes the Orthodox Churches and their liturgy.

C.G. THORNE, JR.

UNDERWOOD, HORACE GRANT (1859-1916). Dutch Reformed missionary to Korea. Born in London, he emigrated to America in 1872 with his family and in 1881 graduated from New York University, and three years later from New Brunswick Theological Seminary. He studied also in France. He went to Korea in 1885 under the Presbyterian Board for Foreign Missions, and four years later married the queen's physician. He mastered the language, taught theology, helped establish a hospital and college, and also the Sai Mun An Church. A trusted confidant of the royal family, through his persistent efforts he brought more freedom for Christian missions. He published several volumes about Korea before his death at Atlantic City, New Jersey.

ROBERT C. NEWMAN

UNIAT(E) CHURCHES. Eastern Christian churches in communion with Rome who have retained their own liturgies, liturgical language, and ecclesiastical customs and rites—e.g., Communion in both kinds, baptism by immersion, marriage of the clergy. The name is derived from the Latin *unio* through the Polish *unia,* and was used in a derogatory sense by Russian and Greek Orthodox opponents of the Union of Brest-Litovsk (1595-96), when Byzantine Christians in the province of Kiev adhered to the Roman see. The term is now a common expression for all Roman Catholics of any Eastern rite. Its hostile flavor has not been lost, and it is not used in any Roman Catholic documents or by the groups so designated. The churches fall into several categories:

(1) *Antiochene Rite.* (a) The *Maronites** were the earliest Uniates, being Syrian Christians named from John Maron, patriarch of Antioch (eighth century), originally Monothelites* who renounced Monothelitism and united with Rome (1182). They use the Syriac "St. James" and other Anaphoroas,* and elect their own patriarch whose see town is Jebeil. (b) The *Syrian Uniates* are descended from the "Jacobites"* (Monophysites) who established relations with Rome in the sixteenth century, but who seem to have disappeared c.1700. The present church owes its existence to Mar Michael Garweh, a Roman Catholic who became archbishop of Aleppo (1783). (c) The *Malankarese Church* dates from 1930, resulting from an amalgamation of some Malabar Chris-

tians* with Jacobites in the seventeenth century, who negotiated for reunion with Rome in 1925.

(2) *Chaldean Rite.* (a) *Armenian Uniates.* There were always some Armenians who recognized Rome's authority, but during the Crusades (1198-1291) there was more formal contact under the patriarch of Cilicia at Beirut. From 1741 they had their own hierarchy and a patriarch at Constantinople. (b) *Chaldean Uniates.* Descended from the ancient Nestorians,* a group of which from Turkey and Persia were united with Rome in 1551, they had their first patriarch of Babylon at Mosul in George Hormuzd, appointed by Pope Pius VIII (1830). (c) The *Malabar Christians:* see separate article.

(3) *Alexandrine Rite.* (a) The *Coptic Uniates,* a small church numbering c.4,000 date from 1741 when Athanasius, Coptic bishop of Jerusalem, joined the Roman Catholics. (b) The *Ethiopian Uniates,* dating from 1839, live mostly in Eritrea and are governed by a vicar-apostolic. They observe the rites and canon law of the old church of Ethiopia.*

(4) *Byzantine Rite.* (a) The *Ruthenians** of E Galicia and Subcarpathian Russia date from the Union of Brest-Litovsk (1595-96). Since 1946 they have been aggregated to the Russian Orthodox Church. (b) The *Rumaics,* or Romanians of Transylvania, linked up with Rome under the archbishop of Abba Julia (1701). Since 1948 they have been aggregated to the Orthodox Church of Romania.* (c) There are also very small groups of *Hungarians* (reunited with Rome in 1595), *Yugoslavs* (1611), *Melchites** (1724), *Bulgars* (1860), *Greeks* (1860).

(5) The *Italo-Greek-Albanian* community of S Italy, which has never separated from Rome, are permitted to follow similar practices, under the bishop of Lungro.

BIBLIOGRAPHY: B.J. Kidd, *The Churches of Eastern Christendom* (1927); D. Attwater, *The Catholic Eastern Churches* (1935) and *The Christian Churches of the East* (2 vols., 1961-62).

J.G.G. NORMAN

UNIFORMITY, ACTS OF. These British parliamentary measures were four in number.

(1) *The Act of 1549.* This statute of Edward VI* commanded the use of the First Book of Common Prayer in English churches. Various penalties were imposed on clergy who failed to conform: a fine and imprisonment for the first offense, deprivation of one's living and imprisonment for the second, and life-imprisonment for the third. The act also declared that all services save in the universities and private devotions were to be in English.

(2) *The Act of 1552.* Another statute of Edward VI, but passed during the Protectorate of Northumberland, a time of growing political and religious conservatism, this enforced the use of the Revised Prayer Book and extended the penalties of the former act to include absence from church services and attendance at private conventicles. The acts of 1549 and 1552 were both repealed by the Catholic queen, Mary Tudor* in October 1553.

(3) *The Act of 1559.* This enforced Queen Elizabeth's* compromise religious settlement and regulated the ecclesiastical discipline of the Church of England for the next ninety years. It repealed all the legislation of Queen Mary which had restored Roman practices and commanded the use of a slightly modified edition of the 1552 Prayer Book. Penalties were again laid down, and ecclesiastical dress and ornaments were to be such as obtained in 1549, the queen as head of the church reserving for herself the privilege of introducing further needful ceremonies and rites—a provision to which the Puritans were later strongly to object.

(4) *The Act of 1662.* This was the most important of the laws restoring the Anglican establishment passed by the Cavalier Parliament of Charles II* following the Restoration, and the first of those acts of systematic repression known as the Clarendon Code.* It commanded universal adoption of a slightly revised form of the Elizabethan Prayer Book and received royal assent on 19 May. Before the ensuing St. Bartholomew's Day (24 August), all ministers had publicly to give their "unfeigned consent and assent" to the Book, and obtain episcopal ordination if not so ordained. A declaration of loyalty and repudiation of the National Covenant* had also to be taken. These provisions led to the "Great Ejection" of about 2,000 Presbyterian, Independent, and Baptist ministers, the final parting of the ways between Anglicans and Puritans, and the consequent birth of English Nonconformity. So far as Dissenters were concerned, the Act was made practically inoperative by the Toleration Act of William and Mary (1689), but it remained effective in regard to the Church of England, though it was later modified in several directions—most notably during the archiepiscopate of A.C. Tait.* Historically, Broad Churchmen have valued it as providing for unity based on comprehensiveness within the Established Church, and evangelicals have held to it as a safeguard for the Thirty-Nine Articles,* but High Churchmen, especially of the more extreme sort, have found it vexatious and restrictive.

IAN SELLERS

UNITARIANISM. A system of religious thought which rejects the doctrine of the Trinity and the deity of Christ, and seeks to show that genuinely religious community can be created without doctrinal conformity. It has evolved from emphasis on scriptural authority to a foundation on reason and experience. Unitarians believe in the goodness of human nature, criticize doctrines of the Fall, the Atonement, and eternal damnation, and require only openness to divine inspiration. In polity they are congregationalists.

As an organized movement Unitarianism, first in Poland and Hungary, dates from the Anabaptists* of the Reformation, but not until recently have there been Unitarian denominations. In the early church, implicit anti-Trinitarianism was expressed in Dynamic Monarchianism,* Arianism,* and Adoptianism,* and later in Paulician* circles. Severely limited by Nicene orthodoxy, it grew nonetheless in certain areas, notably Spain, until the condemnation of Felix of Urgel* by the

Frankish Church in A.D. 799. It was revived in the Reformation period and was most obvious among Socinians.* It spread particularly in Poland and Hungary in the sixteenth and seventeenth centuries, and later in England and America. Prominent anti-Trinitarian proponents were Girgio Blandrata, Francis David, Michael Servetus,* Fausto Sozzini, and John Biddle.* Servetus died at the stake for his views, but others fared better. In Poland the Piedmontese physician Blandrata dominated the early phases of the movement from 1558 until 1563. In 1565, Polish Unitarians were excluded from the synod of the Reformed Church, but under the Italian Sozzini from 1579 until 1604 they created their own synods as the "Minor Church" and issued the Unitarian Racovian Catechism* in 1605. They built a church college at Racow and had about 125 congregations, but after the death of Sozzini (1604) they lost influence. In 1638 Jesuits took over their college, and in 1658 Unitarians were expelled from Poland.

Meanwhile Blandrata had gone to Hungary as court physician (1563) and won his monarch John Sigismund to anti-Trinitarianism. David, made Unitarian bishop in 1568, had troubles after the king's death in 1570, partly because Blandrata was now retreating from Unitarianism, partly because David opposed prayer to Christ. Blandrata was instrumental in having him imprisoned, and in 1579 David died in the dungeon. Although harassed by the government, Unitarians created a common confession in 1638, and later were recognized.

English Unitarianism is traced to John Biddle, although no separate congregation existed until Theophilus Lindsey formed Essex Chapel, London, in 1774. A notable contemporary of Lindsey was Joseph Priestley,* the scientist, who ministered to Unitarian congregations in Leeds and later Birmingham before a mob, angry at his support of the French Republic, destroyed his chapel and his belongings. In 1794 he went to the United States and formed a church at Northumberland, Pennsylvania. English Unitarians were recognized by law in 1813; the British and Foreign Unitarian Association was formed in 1825; and in 1881 the national conference was created. One vehicle of Unitarian expression was the Robert Hibbert Fund, which sponsored the *Hibbert Journal* and the Hibbert Lectures.

The most successful Unitarian church body has been in the USA. Although prominent Americans like Thomas Jefferson* held anti-Trinitarian ideas, and although the first Unitarian congregation—King's Chapel, Boston—was formed out of the oldest Episcopal parish in America when the rector, James Freeman, ignored references in the Book of Common Prayer to the Trinity and the divinity of Christ (1785), American Unitarianism developed in the Congregational churches of E Massachusetts. Anti-Trinitarians won their first victory in 1805 when Henry Ware, a liberal, was appointed to the theological chair at Harvard. The Dedham decision by the Massachusetts Supreme Court in 1818 allowed the selection of unorthodox ministers by all voters in the parish, and many churches went over to Unitarianism. Prominent leader of the group was W.E. Channing* who in an 1819 sermon described the true church as creedless: "men made better, made holy, by His religion." Unitarians created a missionary and publication society, the American Unitarian Association, in 1825, but it was not until activities with the United States Sanitary Commission in the Civil War that they built a stronger national organization. Their first national conference was held in 1865.

In the early years much conflict existed over whether Jesus was divine or if the Bible was the Word of God, and that, combined with Unitarianism's alleged Rationalism, prompted R.W. Emerson* to break with the church in 1838 and to help form Transcendentalism.* Traditionally centered at Harvard Divinity School, American Unitarianism created other seminaries and preparatory schools, and in later years came much closer to Emerson's position. Its foreign work was carried out through the International Association for Liberal Christianity and Religious Freedom in Utrecht, Holland. Its growing national concern was manifest in 1940, not only in the creation of the Unitarian Service Committee and the United Unitarian Appeal, but also in its support of social justice movements. Unitarians were prominent, for example, in both the antislavery and civil rights movements of the USA. In 1961 it merged with Universalism* to form the Unitarian Universalist Association.

Unitarianism, briefly, has grown along rational, not biblical, lines. Begun by anti-Trinitarians, many of whom were otherwise orthodox, it has evolved into a creedless movement stressing the many forms of divine revelation and the inherent goodness of man. Many have held Unitarian ideas, therefore, without belonging to a Unitarian church. Unitarianism now stresses the religion of the Sermon on the Mount and the oneness of the human family.

BIBLIOGRAPHY: J.H. Allen, *An Historical Sketch of the Unitarian Movement since the Reformation* (1894) and *A History of the Unitarians and the Universalists in the United States* (1894); G.W. Cooke, *Unitarianism in America* (1902); H. Gow, *The Unitarians* (1928); H.J. McLachlan, *The Unitarian Movement in the Religious Life of England* (1934) and *Socinianism in Seventeenth-Century England* (1951); E. M. Wilbur, *History of Unitarianism* (1946); C. Wright, *The Beginnings of Unitarianism in America* (1955) and *The Liberal Christians: Essays on American Unitarian History* (1970); H.H. Cheetham, *Unitarianism and Universalism* (1962).

DARREL BIGHAM

UNITAS FRATRUM, see BOHEMIAN BRETHREN; MORAVIAN BRETHREN

UNITED BIBLE SOCIETIES, see BIBLE SOCIETIES

UNITED BRETHREN IN CHRIST. An American denomination, organized in 1800, which developed out of the activities of P.W. Otterbein* and Martin Boehm* among German settlements, chiefly in Pennsylvania, and out of conferences they held with other "United ministers." The new Methodist-type denomination was evangelical, Arminian, and perfectionist in doctrine, and episcopal in government. Otterbein and Boehm were the first bishops. When a new constitution was adopted in 1889, a group separated as the United Brethren in Christ (Old Constitution). In 1946 the parent body merged with the Evangelical Church, with which it had much in common, forming the Evangelical United Brethren.* The latter united with the Methodist Church in 1968 to create the United Methodist Church.*

ALBERT H. FREUNDT, JR.

UNITED CHURCH OF CANADA. Union negotiations between the Methodist Church and the Presbyterian Church in Canada were initiated in 1902 when the highest courts of these two prominent Protestant denominations agreed to initiate joint discussions. The union was consummated in solemn assembly in Toronto on 10 June 1925. It included the Congregationalist Churches in Canada and some 3,000 union churches, most of which had existed in the West for some time prior to this time. A large number of Presbyterian churches reorganized immediately to form the Presbyterian Church of Canada.* Nonetheless, the Union proved to be a successful one. It is delineated by a Basis of Union which seeks to incorporate the best traditions in Reformed theology and polity.

A presbyterial structure was developed which maintains a healthy balance between the various elements of the three uniting bodies in the nature and function of the various ecclesiastical courts. The general council, normally meeting every other year, is the chief legislative assembly. It is representative of the various lower courts and consists of an equal number of clergy and laymen. The ten conferences are in turn made up of Presbyteries in which local congregations have their membership. The basic structure of a congregation is the official board, which combines the session and the board of stewards. It is generally chaired by the minister, who is a member of the session. Items of concern and interest to the membership of the congregation are dealt with at the annual congregational meeting.

Ministers are ordained by the conference upon recommendation from colleges and presbyteries under whose oversight they trained for ministry. After an initial period of settlement every minister is eligible to be called to a charge, but he may also submit himself to the settlement committee of conference. Evangelism and social service of the United Church, the missionary outreach at home and abroad, and other important functions of the church are under the direction of boards and divisions, centralized in Toronto, but reproduced on conference and presbytery level throughout the church. Ecclesiastical affairs are laid down in *The Manual of the United Church.*

Experiments in restructuring courts of the church have introduced greater diversity and some amount of flexibility in recent years. Serious union negotiations have led to close cooperation with the Anglican Church in Canada and make the United Church one of the foremost agents of ecumenical concern on the Canadian scene.

The United Church is officially represented by the moderator, who is elected at the meetings of the general council for a two-year term. The secretary of general council plays an important administrative role. The United Church of Canada is in fraternal association with the World Alliance of Reformed Churches,* with Methodist world bodies, and with the World Council of Churches.* Its motto, *Ut Omnes Unum Sint,* seeks to express the goal of the church to act as an agent of unity among Christians in Canada. Several unions with other Christian communions have enriched the church in its history since 1925.

BIBLIOGRAPHY: C.E. Silcox, *Church Union in Canada: Its Causes and Consequences* (1933); K.J. Beaton, *Growing with the Years* (1949); G.C. Pidgeon, *The United Church of Canada* (1949); G.W. Mason, *The Legislative Struggle for Church Union* (1956).

EDWARD J. FURCHA

UNITED EMPIRE LOYALISTS. Lord Dorchester, governor-general of British North America, proposed in 1789 to honor all those who had by their act adhered to the unity of the empire. These emigrants to British North America who came during and immediately after the American Revolution and settled in the Niagara Peninsula and in Nova Scotia were inscribed on a list and entitled to distinguish themselves by affixing the letters *U.E.* to their names. Under Lord Simcoe, further emigrants from the thirteen colonies came to British North America, but these are not properly called United Empire Loyalists. It is generally estimated that one-third of the inhabitants of the thirteen colonies was against the Revolution. However, no exact numbers are known for those who migrated north. A conservative estimate places the number of emigrants at the time at about 50,000. The Loyalist migrations altered the distribution of French and British in favor of the British.

EDWARD J. FURCHA

UNITED EVANGELICAL CHURCH, see EVANGELICAL CHURCH (ALBRIGHT BRETHREN)

UNITED FREE CHURCH OF SCOTLAND. This body was formed in 1900 by the union of a majority of the Free Church of Scotland* and the United Presbyterian Church.* Unsuccessful efforts at union had been broken off in 1873 because of the strength of opposition in the Free Church. Thereafter both churches, which were opposed to the Church of Scotland,* were active in a disestablishment campaign aimed at the political parties. It failed, but it did something to bring the two groups together. Other factors were at work: revolt against hyper-Calvinism which found expression in doctrinal Declaratory Acts,* reverent biblical criticism, the spiritual awakening under Moody and Sankey, and changes in public worship by the introduction of instrumental music and hymns. In 1898 both churches accepted a modified formula to be signed by ministers and office-bearers by which they pledged their adherence to the fundamental doctrines and principles of the Church. The next two years were spent in fruitless efforts to bring in those who were to continue as the Free Church.

In 1929 the United Free Church united with the Church of Scotland to bring into being a church which had within its fold four-fifths of the churchgoing population of Scotland. As in 1900, however, a minority remained outside the union and called themselves the United Free Church (Continuing). With over ninety congregations and a membership of some 17,000, the denomination in recent years has entered into discussions about union with the Church of Scotland and several Scottish non-Presbyterian bodies.

J.W. MEIKLEJOHN

UNITED METHODIST CHURCH. One of the three Methodist Churches in England to participate in the union of 1932. In 1907 the Methodist New Connexion, the Bible Christians, and the United Methodist Free Churches came together in a single communion known as the United Methodist Church.

The Methodist New Connexion* was the first to secede from the parent body of Wesleyans. In 1797 Alexander Kilham* had unsuccessfully petitioned the conference to allow Methodists to receive Communion from their own preachers. He founded the "New Itinerancy" or "Methodist New Connexion" with equal representation of ministers and laymen in its oversight.

The Bible Christians* were so called because of their attachment to the Scriptures in drawing up the rules which governed their societies. Their Cornish founder, William O'Bryan, prefixed "Arminian" to the title as an indication of his theological standpoint. In his evangelistic zeal he launched out independently in a way which brought him under the discipline of the Stratton quarterly meeting. In 1815 he eventually established his own circuit at Week St. Mary, with James Thorne as his coadjutor. In 1829 O'Bryan himself left the connexion after a dispute over the leadership.

The United Methodist Free Churches date back to 1857, when the Wesleyan Reformers and the Wesleyan Methodist Association amalgamated. The Grand Central Association came into being in 1834 in protest against the growing authoritarianism of the Wesleyan Conference. In 1836 they were joined by the Protestant Methodists, formed in 1827 because of similar objections, and in successive years by the Arminian or Faith Methodist of the Midlands and some Independent Methodists from Wales and the North. Another dissident group known as Wesleyan Reformers had also seceded, and in 1857 most of them joined forces with the Wesleyan Methodist Association to comprise the United Methodist Free Churches, although a rump remained separate to establish the Wesleyan Reform Union in 1859.

BIBLIOGRAPHY: G. Eayrs in *A New History of Methodism* (ed. Townsend et al.), vol. I (1909), pp. 485-551; H. Smith, J.E. Swallow, and W. Treffry, *The Story of the United Methodist Church* (1932); O.A. Beckerlegge, *The United Methodist Free Churches* (1957) and *United Methodist Ministers and their Circuits* (1968); T. Shaw, *The Bible Christians* (1965).

A. SKEVINGTON WOOD

UNITED METHODIST CHURCH (USA), see METHODIST CHURCHES, AMERICAN

UNITED PRESBYTERIAN CHURCH (Scotland). A body formed in 1847 by the union of the United Secession Church* and the Relief Church,* representing respectively the main bodies of the first secession (1733) and the second secession (1761) from the Church of Scotland.* The union was not effected without difficulty, for the two churches differed considerably in nature. The former had a strictness of ecclesiastical discipline whereas the latter was more broadly evangelical and laxer in doctrine and discipline by the standards of the times. The United Presbyterian Church had 518 congregations and was largely confined to the cities and towns. In 1900 it united with the Free Church of Scotland* to form the United Free Church.* Among its contributions to the Christian cause was its zeal for missionary work in Calabar (now Nigeria) and India.

J.W. MEIKLEJOHN

UNITED PRESBYTERIAN CHURCH (USA). The largest of the American Presbyterian denominations following the merger in 1958 of the Presbyterian Church, USA and the United Presbyterian Church of North America. The roots of the denomination were in seventeenth-century immigration from Scotland and Ireland. The first essentially Presbyterian church was organized in 1629 in the Massachusetts Bay Colony with Samuel Skelton as pastor. Francis Makemie,* an early missionary, provided organizational structure for Presbyterians.

The first presbytery was formed in Philadelphia in 1706 by seven ministers representing ten congregations and 800 members, and with no official relationship to the church in Scotland. A synod of four presbyteries was organized in 1716. Immigration from Scotland and Northern Ireland and the Great Awakening* increased the numbers. The need to train new ministers led to the founding of academies, the most famous of which was William Tennent's* "Log College" in Neshaminy. It was replaced in 1747 by the College of New Jersey, later Princeton University.

Early tensions divided the church into three groups: the traditional and formal Scotch-Irish, the Log College men, and men trained in New England. The issues were: emotionalism in revivals, educational qualifications of ministers, polity, and itinerant ministers. George Whitefield's* preaching in 1739-40 aided the American groups. The basic struggle theologically was over latitude in interpreting the Westminster Confession,* a question resolved in 1729 by the Adopting Act which made the Westminster Confession and Catechisms official dogma. Disputes over revivalism and licensing untrained men to preach brought division into Old and New Side from 1741 to 1758, but the groups reunited in 1758.

In May 1789 the general assembly was formed in Philadelphia. John Witherspoon* presided. By 1900 the membership had increased to about one million members. This reflected nineteenth-century immigration patterns, revivals, and the growth and prosperity of the country. Frontiers were revival areas where men like James McGready* and Charles Finney* labored. The Plan of Union* for cooperation on the frontier with Congregationalists from 1801 to 1858 benefited Presbyterians most.

New growth led to the formation of Sunday schools and founding of foreign missions programs, but the revivals brought dissatisfaction from those who felt they sacrificed good theology and that the resultant church growth led to weakening the qualifications for ministers. These issues led to the secession of the Cumberland Presbyterians* in 1810 and brought hostility between the conservative Old School and the New School. Growth also led to the founding of seminaries to maintain the standards of the ministry—Princeton in 1812, and in the ensuing twenty years Western, Lane, and one in Chicago.

Division between Old School and New School lay partly in the influence on a growing number of ministers, such as Albert Barnes,* associated with the theology of Samuel Hopkins* and Nathaniel Taylor* and their weakening of the doctrine of original sin. It was affected also by slavery; the Plan of Union; polity differences; relation to voluntary societies such as the American Bible Society and the issue this raised of the nature of the church; and freedom in interpreting the Westminster Confession. These factors brought a split in the 1837 general assembly. The two groups finally reunited in 1870. Most of the Cumberland Presbyterians rejoined the main body in 1906.

The church in the twentieth century faced the effects of urban life with growing social concern, increased lay leadership, and a growing secularization of the clergy. The democratic character of the church made it increasingly difficult to discipline men or to obtain consensus on social and theological questions. Mass evangelism, with men like (non-Presbyterian) Billy Graham,* continued to add numbers to the church, as did the population growth and shift to the suburbs and the Western part of the country. By 1971, despite declines, some 13,000 ministers were serving nearly 8,700 churches with over 3.2 million members.

A strong missions program led by men such as R.E. Speer* declined in the latter part of the century because of conservative dissatisfaction with mission goals and because of the changing attitude toward missions in developing countries. Historical distinctives lost ground, paving the way for church mergers and federations in which E.C. Blake* took a major role, both as general secretary of the World Council of Churches and as proposer of the plan known as the Consultation on Church Union.

The conservative-liberal antithesis remained, focused in the fundamentalist-liberal controversy; among the names associated with it were H.E. Fosdick* and J.G. Machen.*

BIBLIOGRAPHY: R.E. Thompson, *A History of the Presbyterian Church in the United States* (1895); L.J. Trinterud, *The Forming of An American Tradition: A Reexamination of Colonial Presbyterianism* (1949); L.A. Loetscher, *The Broadening Church: A Study of Theological Issues in the Presbyterian Church since 1869* (1954); G.M. Marsden, *The Evangelical Mind and the New School Presbyterian Experience: A Case Study of Thought and Theology in Nineteenth Century America* (1970); E.R. Sandeen, *The Roots of Fundamentalism* (1970).

ROBERT B. IVES

UNITED REFORMED CHURCH. Britain's first union across denominational lines took place in 1972 when the Congregational Church in England and Wales merged with the Presbyterian Church of England. Negotiations had gone on since 1945. Less than 20 percent of the Congregationalists had exercised their right to vote and, on the other hand, some Congregationalists planned to opt out and continue as a Congregational association. In the much smaller Presbyterian Church, the two congregations in the Channel Islands declined to enter the new body and have been accepted into the Church of Scotland. The URC is divided into twelve provinces, each with a moderator, and sixty-one districts. Ministers number 1,100, and estimated membership is 200,000. Overseas work is carried out through the Council for World Mission (Congregational and Reformed).

J.D. DOUGLAS

UNITED SECESSION CHURCH. Formed in Scotland in 1820, it was a union of the New Light segments from Burghers* and Anti-burghers, which latter groups were children of the 1733 Secession under Ebenezer Erskine* and his brother Ralph. Though traditional Secession discipline was to some extent retained in the new body, the practice of renewing the National Covenant* and the Solemn League and Covenant* was abandoned. The "United Secession Synod" comprised some 280 congregations. In 1847 it united with the Relief Church* to form the United Presbyterian Church.*

UNITED STATES OF AMERICA. Christianity in the United States, like the Christian faith in other ages and lands, reveals the marks of time and space. The discovery of America and the birth of Protestantism were almost contemporaneous events. This fact helps explain the prevailing "protestant" character of American Christianity transplanted from the Old World. The major space factor is found in the fact that for almost three centuries of her history America and her churches were in continuous contact with frontier conditions and frontier needs. This combination of time and space shaped a unique type of Christianity, clearly distinguishable from the Christian faith in other ages and in other realms.

The English policy of private enterprise in establishing new colonies, and the westward spread of America's peoples, gave ample opportunity for the spread of *religious diversity* in the new land. From the settlement at Jamestown (1607) to the Civil War this denominational diversity was almost altogether within the Protestant, and chiefly the Puritan,* tradition. The second half of the nineteenth century saw much greater variety appear in the emergence of indigenous religious cults and large numbers of immigrants from Europe, most of whom had no part in the Puritan past.

The Founding Fathers of the new nation recognized this religious pluralism in the colonies and wrote into the First Amendment to the Constitution the *separation of church and state,* "the fair experiment," as Thomas Jefferson called it: no single church could stake a firm claim to a privileged status in the new nation. Thus the adoption of the First Amendment to the Constitution (1791) safeguarded the freedom for all by granting privileges to none.

The religious needs of the frontier coupled with this disestablishment policy of the new government forced the churches to employ new techniques, based on voluntaryism*, for winning people to the Christian faith. *Revivals* proved to be the highly successful means of planting vital Christianity across the continent. The Great Awakening,* the successful colonial crusade for souls, became the model for a series of spiritual awakenings spanning America's religious history.

Revivals consisted of appeals for conversions to Christ and His church, They were aimed at individuals. Thus the independent-minded pioneer found a kindred spirit in the leaders of the spiritual awakenings. This expression of *individualism* in religion, as well as politics, nurtured an almost endless assortment of voluntary societies aimed at bringing the kingdom of God to the American continent.

These characteristics of Christianity in the USA —religious pluralism, separation of church and state, revivalism,* and individualism—are (as well as social activism and ecumenism) the chief marks of America's religious uniqueness.

The history of Christianity on the American shores may be divided into four major periods: the Formative Years 1607-1776, the Frontier or National Years 1776-1860, the Critical Years 1860-1914, and the Post-Protestant Years 1914 to the present.

(1) *The Formative Years 1607-1776.* The Protestant Reformation* led to a host of national churches, sects, and dissenters. The refuge for many of those persecuted for conscience in Europe was colonial America. While politics, economics, and social advantage had their part in the early growth of the colonies, religion was responsible for the *founding* of more colonies than any other single factor. These colonies were English colonies, and the multiplicity of religious bodies within them was largely the result of a policy of toleration pursued by English authorities. The colonies were also commercial ventures. To be profitable they needed people to clear the forests and plant the fields. Thus colonial authorities promoted religious toleration in the New World as an inducement for persecuted peoples.

To this economic advantage we must add the growing religious diversity within England herself. Through the 1600s the British were struggling toward greater religious toleration at home. Many religious minorities, caught up in this struggle, chose the opportunities of the New World over the continued conflicts in their homeland. Although religious diversity early became a fact of life in the colonies, this multiplicity of sects was within an overarching unity. The vast majority of the religious groups stood within a common tradition, British in background and Puritan in theology. The first census in 1790 revealed this British predominance: 70 percent of the population was of English stock, and an additional 15 percent was of Scottish or Scotch-Irish descent. Even among the remaining non-British minorities—Germans, Dutch, French, Swedes—the Protestant background prevailed.

Once firmly rooted, this American Puritanism was naturally subject to change. The Great Awakening, in particular, gave Puritanism in the new land a decidedly evangelical character, even as it helped to create an American religious consensus. New England, where Puritanism first took root, is the best illustration of the difficulties encountered by Christians who attempted to maintain the traditional establishment idea. The first congregation in New England was the little Separatist group at Plymouth, planted in the New World by the 1620 landing of the *Mayflower.* Eight years later the much larger Puritan immigrations began in and around Boston. By securing a charter for the Massachusetts Bay Company and transporting it to the colony, the early Puritans were able to display to the whole world what a true church "after God's order" was like.

The first General Court of the colony (1631), composed of the governor and the freemen, linked the franchise with church membership. Five years later the court, in order to ensure religious uniformity, gave magistrates power over the churches. The clergy, however, through control of the franchise and through influence upon the magistrates, exerted considerable influence over public conduct. This functional alliance between magistrate and minister was the heart of the "holy commonwealth."

Dissent, however, was never far removed. Roger Williams,* who arrived in Boston in 1631, was among the first to challenge the Puritan "theocracy." He spread the idea that civil authority and spiritual authority should be separated. His persistence in this novelty led in 1635 to a sentence of banishment from the colony. By fleeing the colony in the middle of winter he was able, after securing land from the Indians, to settle at the present site of Providence, Rhode Island. After being joined shortly by others, he set up a new colony founded on the separation principle. Thus religious uniformity in New England was gravely threatened almost from the start. Other attempts at religious establishments—the Dutch in New Amsterdam and the Anglicans in the southern colonies—were even less successful than the Puritans in New England.

By the end of the century the shell of Puritan orthodoxy lingered in New England, but much of the spiritual vision of the first generation—a church of "visible saints"—had vanished. The recovery of a vital religious experience is the story of the Great Awakening.

(2) *The Frontier Years 1776-1860.* When winds of revolution filled the colonial air, many of the churches supported the cause of independence. The Congregationalists, the Presbyterians, and

the Baptists were almost universally in favor of it. Understandably the Episcopalian Church suffered most. While more Episcopalians signed the Declaration of Independence than any other colonial denomination, the royal governors and other colonial officials were usually Church of England men and many Loyalists were found within the church. Because of the religious diversity in the colonies, the founding documents of the new nation banned any religious test for public office and separated the spheres of state and church. Although a few state constitutions were slow in following the lead of the national documents, notably Massachusetts till 1833, most Christians regarded the "experiment" in religious freedom a wise course.

The Constitution had hardly been adopted when people began streaming westward. By 1860 states were rapidly forming west of the Mississippi. This movement of population continued until the entire continent had been peopled. The denominations most successful in moving with the people and in establishing churches in the new territories became, understandably, the largest bodies in the new nation. The Methodists, Baptists, and Presbyterians proved most adaptable to the frontier, while a fourth denomination, the Disciples of Christ, was born in the Ohio Valley through the preaching of Barton W. Stone* and Alexander Campbell.*

The technique widely used in reaching the unchurched masses was revivalism, a type of preaching that sought to make listeners vividly aware of their eternal destiny and of the importance of a thoroughgoing conversion to the Christian faith. A special type of revivalistic meeting also came into being by 1800, the "camp meeting."* These meetings were great outdoor gatherings for preaching that lasted several days. First employed by Presbyterians, they later became a characteristic Methodist technique for inducing excitement and conversions. The revivalistic spirit was in time tamed and channeled into voluntary societies. These were extra-church agencies, formed for specific purposes by individuals and unrelated structurally to the denominations. Societies were created for establishing Sunday schools, publishing literature, founding academies, and advancing a host of social reforms.

By midcentury this combination of revivals and social reforms had created an evangelical mood that minimized denominational differences before the greater cause of advancing Christ's kingdom throughout the youthful nation. By 1850 the Methodists were the largest denominational body, with a membership nearing 1.5 million, followed by the Baptists with about a million and then the Presbyterians with about a half-million. Major challenges to this evangelical consensus were not long in coming. The generation just prior to the Civil War was marked by controversy and division among the denominations. The Roman Catholic Church, whose roots in America ran back to the founding of Maryland and to the earlier Franciscan missions in the Southwest, received large numbers of immigrants, especially from Ireland. This sudden influx of Catholics aroused Protestant fears. In a similar way Lutherans, who had shown a willingness to adopt the cooperative spirit of the revivalistic denominations, were thrown into new internal tensions after 1830 with the arrival of many conservative Lutherans from Germany.

The greatest cause of controversy, however, was the national slavery issue. By 1830 a far-reaching agricultural revolution in the South made the region dependent upon slave labor. At the same time a radical abolitionist movement in the North contributed to the widening breach between the two sections of the country. Most of the denominations were torn apart by the diverging ideologies. The Presbyterians, owing to the presence of theological problems, divided first in 1837. In 1845 southern Methodists and Baptists split and formed denominations. The divisions among Presbyterians and Baptists have yet to be healed.

(3) *The Critical Years 1860-1914.* During Reconstruction, various church agencies poured money and men into the South to bring religion and education to the masses of Negroes just released from slavery. A number of independent Negro churches also gave expression to the newfound freedom. Baptist and Methodist churches made the greatest appeal to the Negro; by the end of the century most Negro Christians could be found in the National Baptist Convention, the African Methodist Episcopal Church, or the African Methodist Episcopal Zion Church (see AMERICAN NEGRO CHURCHES).

The years following 1860 also witnessed continued waves of immigrations destined to change markedly the religious face of the USA. Large numbers of Scandinavians, especially in the upper Midwest, led to the formation of independent Lutheran churches—generally along nationality lines—and made Lutherans the third-largest Protestant denomination. Other immigrations from E and S Europe after 1880 resulted in millions of additional Roman Catholics in the USA.

The critical nature of these years is most evident, however, in the turmoil created by the influx of new ideas relating to the Bible. German Idealism and the evolutionary theory, as set forth by Charles Darwin* in his *Origin of Species,* had serious implications for the traditional view of God and creation; and higher criticism, which tested the authenticity of the biblical writings by the same methods used in testing other ancient literature, appeared to undermine the foundations of traditional evangelical supernaturalism. Positive or negative attitudes toward these new views of the Bible threatened to divide the evangelical denominations and paved the way for the controversy involving "modernists" and "fundamentalists" after the turn of the century.

A related conflict swirled around the emerging social conscience within the churches. Evangelical denominations, chiefly Methodists and Baptists, which had once been identified with the poor, were rapidly becoming churches of the upper middle class. At the same time, the industrial and urban society that arose after the Civil War was attracting the attention of certain Protestant leaders who called for the application of the principles of Jesus to the new industrial-urban problems. This new concern was labeled the "Social Gospel."* It found its most persuasive advocate in Walter Rauschenbusch* and its specific goals stated in the Social Creed of the Federal Council of Churches. Nor could traditional revivalism escape these pronounced changes in American society. Revivals became mass, urban, professional, and organized movements through the ministry of D.L. Moody* and Ira Sankey.* Then, supported by Moody's great reputation, the Bible school* and Bible conference movements rallied many conservatives attempting to stem the tide of liberal views of Scripture. This wedding of revivalism and biblical conservatism fashioned the cradle of twentieth-century fundamentalism.*

(4) *The Post-Protestant Years 1914-1970.* These conflicting views of the Bible and plans of social actions resulted in the fundamentalist-modernist debate of the 1920s. This controversy produced a fundamentalism largely interdenominational in character and major Protestant denominations led by men more concerned with programs of action than theological soundness.

Scarcely had modernism tasted a measure of victory over fundamentalism in the traditional evangelical denominations than it was faced with a new theological challenge. The 1930s disclosed a deepening criticism of modernism's basic affirmations. The new mood was difficult to characterize, but was given an able American statement in the writings of Reinhold Niebuhr.* It was often popularly called "Neoorthodoxy."* This new theology reasserted the sovereignty of God and repudiated the notion that man has almost unlimited potential for good. Neoorthodoxy also rediscovered the "original sin" of man, not in the sense of an act of disobedience by a man named Adam, but in the sense of man's universal moral failure. Finally, the new theological mood stressed the central importance of the Bible and Christ as indispensable mediators of God's special revelation to man.

The years between the two world wars also witnessed the growth of the ecumenical spirit through interdenominational cooperation, organic reunion, and confederation.* The Student Volunteer Movement* and the formation of the United Presbyterian Church,* USA, illustrate the first two methods. The Federal Council of Churches of Christ in America, organized in 1908, was an early example of confederation. But the Federal Council gave way in 1950 to the more comprehensive National Council of Churches. In addition to the Federal Council, the new council embraced the Foreign Missions Conference of North America and the International Council of Religious Education and represented more than thirty denominations. Because of the long-standing differences with the Federal Council's liberal social orientation and doctrinal deficiencies, conservative evangelicals preferred to cooperate along the lines of voluntary agencies. The National Association of Evangelicals* (1942), the National Sunday School Association (1945), and the Evangelical Foreign Missions Association* (1945) were among the host of interdenominational agencies giving continued evidence of conservative cooperation.

The 1960s in the USA found the country filled with social unrest. Racial tensions and war-peace fevers were especially evident in the life of the churches. Churchmen were prominent in the public arena, demonstrating for racial justice or for peace in Vietnam. Against this background a "secular theology" arose which saw Christ as "a man for others" and the church's primary mission in terms of "humanizing" the social order. Conservatives, on the other hand, poured their energies into support of Billy Graham* Crusades or other evangelistic endeavors with the submerged hope that the world could be changed by the conversion of masses of individuals. These two emphases tended to polarize Christians into camps of "social activists" and "individual salvationists."

Late in the 1960s a rather unusual revival of fundamental Christianity erupted from the youth counterculture. The "Jesus Movement," as national magazines labeled it, was marked by remarkable conversions of former drug users, Bible study, some "speaking in tongues," and a lifestyle more in harmony with the earlier hippie culture than that of suburban churches.

BIBLIOGRAPHY: C.E. Olmstead, *History of Religion in the United States* (1960); W.S. Hudson, *Religion in America* (1965); S.E. Ahlstrom, *A Religious History of the American People* (1972).

BRUCE L. SHELLEY

UNITED ZION CHILDREN, see BRETHREN IN CHRIST

UNITY SCHOOL OF CHRISTIANITY. This began when in 1887 the wife of Charles Fillmore (1854-1948), a crippled real estate agent, was gradually healed of tuberculosis in an unusual manner. The Fillmores studied Christian Science* and New Thought, and from these studies emerged a new ideology to which Charles was converted by 1890. The Unity School is an independent institution of religious education that specifically denies sectarianism or denominationalism. Located at Unity Village, near Lee's Summit, Missouri, the school maintains a large printing and publishing house, a training school, a radio and television ministry, a personal prayer service (Silent Unity), and training and retreat facilities for students. The Silent Unity ministry answers over one-half million personal inquiries a year. No inquirer is asked to leave his own denomination. The Fillmores taught that God is Spirit or "Principle" and that Jesus was the perfect expression of the Divine Principle. Man is a trinity of spirit, soul, and body and receives his salvation through a series of reincarnations and regenerations of the body. All men will eventually become like Christ. Man overcomes want and illness through correct personal thought. The Unity School carries on a worldwide ministry through licensed teachers and ordained ministers who hold services in local "centers."

See C. Braden, *Spirits in Rebellion* (1963).

JOHN P. DEVER

UNIVERSALISM. An amalgam of several traditions including Gnosticism,* Anabaptism,* and mysticism.* Universalism was created in eighteenth-century America. The first congregation was organized in 1779, and the first creed was adopted in 1790. At first Universalists agreed on little beyond congregational polity and creedlessness, but the Winchester Platform (1803) was more specific in its stress on the perfectibility of man, the ultimate salvation of all men, the varied character of divine revelation, and the humanness of Christ. Prominent early leaders were John Murray,* Elhanan Winchester, and Hosea Ballou.* Many Universalists were also nonresistants. Disagreement over such matters as the reliability of the Scriptures, the nature of Christ, and the credibility of the Winchester document prompted the convention of 1899 to assert, among other things, unitarianism,* perfectionism,* and humanitarianism. By 1942 the group welcomed all humane men, Christian or not. It merged with Unitarians in May 1961, to form the Unitarian-Universalist Association. DARREL BIGHAM

UPPSALA ASSEMBLY (1968). Fourth assembly of the World Council of Churches.* Meeting 4-19 July in the Swedish university town, it had as its theme "Behold! I make all things new" and has been described as "the most document-laden Christian gathering in the last nine hundred years." The major issues were the gap between rich and poor nations, and the need to "humanize" the world. Elected as WCC presidents at the assembly were Hans Lilje, Ernest Payne, D.T. Niles, Alphaeus Hamilton Zulu, John Coventry Smith, and Patriarch German. The central committee chose as its chairman the Indian layman M.M. Thomas. Because of involvement in secular, economic, and social issues, the assembly contributed little that was fresh theologically. Of the 704 delegates, 3 per cent were from developing countries, 75 per cent ordained, and there were 15 Roman Catholic observers with nearly 200 other Roman Catholics present.

The main work proceeded in six study sections which met to discuss and amend 2,500-word drafts:

(1) "The Holy Spirit and the Catholicity of the Church." This revealed differences between the Protestant view of the Church as invisible and catholic, and the "Catholic" view of it as an unbroken, visible eucharistic unity; and emphasized the impossibility of intercommunion. It defined unity in terms of mankind and not of the church. Continuity comes in the future as well as the past, thus tradition relates to renewal. Catholicity is "the quality by which the church expresses the fulness, the integrity and the totality of life in Christ"—in economic, social, and political matters.

(2) "Renewal in Mission." This section found great controversy between evangelicals who viewed conversion and evangelism in terms of the individual, and those who saw conversion in terms of this world's societal structures and values; between those concerned with preaching and those concerned with dialogue. The report defined mission as "to serve suffering humanity and to aid the developing nations," to make "the world's agenda the church's business." Modern missions are not only in traditional areas, but in urban centers and among revolutionary movements where "Christian presence and witness are required." But there was no concern showed for man's spiritual hunger comparable to that expressed for physical hunger (pointed out one evangelical delegate). More delegates applied to join this section than any other.

(3) "World Economic and Social Development." Youth participants pressured this group to be more revolutionary. The group stressed the need for Christians to be more politically effective. It opposed the status quo, isolationism, violence in revolution, yet held that violent changes are "morally ambiguous." It was concerned to end discrimination, to cope with unemployment and underemployment, and with the questions posed by population growth and food shortages.

(4) "Toward Justice and Peace in International Affairs." In a time of political turmoil, "human rights cannot be safeguarded in a world of glaring inequalities and social conflicts." The discussion criticized U.S. involvement in Vietnam, inviting the comment by George McGovern, "no nation comes to this Assembly with clean hands." Christian peace is built on love of enemies, and reconciliation is based on the reconciling work of God in Christ; so Christians need to identify with the poor and oppressed in their struggle for justice. Opposition to war, particularly nuclear war and weapons control, desire to protect human rights, including support for the United Nations and protection of minority rights, and support for all efforts toward world peace including support for selective conscientious objection to war were stressed in the report.

(5) "Worship." A wide divergence of liturgical views were represented, with some finding joy in the old forms, as the Orthodox. Many Western churches wanted contemporary worship forms, evidenced in performance at the assembly of Sven Erik Back's "Mass of the Departure," Bo Nilsson's "Mass for Christian Unity," Sven Erik Johanson's "Cross in the Space Age," and Olov Hartman's "On That Day." The group opposed indiscriminate administration of baptism as a social custom and stressed the need for baptism to take place in the presence of the community. Desire was expressed for the Eucharist to be celebrated weekly, in new styles. "The eucharist shows the essential meaning of Christian worship, for the sacrament of the body and blood of Christ, shed for the remission of sins, is a communion meal in which Christians share in his life."

(6) "Toward New Styles of Living." This section, which had the highest proportion of women, laymen, and youth participants, defined "Styles" as the outward manifestations of inward convictions. The section disputed contextual ethics and moral principles, attempting to distinguish the ethics of the Gospel from cultural mores; but its members were unable to resolve the problem. They described a world divided along three lines—color, wealth, and knowledge—which bred specific problems: birth control, changes in traditional family patterns, chastity, and antinomian contextualism. They suggested that a Christian style of living is characterized by concern for suf-

fering of other people, struggle for social justice, and rules open to the Spirit.

Roman Catholics cooperated in many areas. Nine Roman Catholic theologians were added to the Faith and Order Commission, and Jesuit Father Robert Tucci gave an assembly address stressing that Roman Catholic membership in the World Council of Churches may soon come. A joint working group was established by the assembly to work out principles of cooperation. The Orthodox Churches remained the largest confessional group, with 140 delegates. The assembly established a new secretariat on racial equality and admitted four new denominations.

See N. Goodall (ed.), *The Uppsala Report 1968* (1968); and E.C. Blake in *A History of the Ecumenical Movement,* II, *1948-1968* (ed. H. Fey, 1970), pp. 411-445. ROBERT B. IVES

URBAN II (1042?-1099). Pope from 1088. Born in France, Odo of Lagery studied at Reims under Bruno, and entered the monastery at Cluny (c.1070) where he eventually became prior. Recommended to Gregory VII* by Abbot Hugh, he was called to Rome and in 1078 was made cardinal bishop of Ostia. He served also as legate to France and Germany. In 1088 he succeeded Victor III in the papacy as Urban II. The opposition, however, of the German king and emperor Henry IV,* the antipope Clement III (Guibert of Ravenna), and their supporters prevented him from settling permanently in Rome until 1093. He took possession of the Lateran Palace in the following year.

Adhering to the ideals professed by Gregory VII, Urban sought in a cautious and diplomatic manner to free the church from lay investiture and influence, and to strengthen the church internally by requiring priestly celibacy, eliminating simony, and healing the breach between Greek and Latin Christianity. All but the last of these objectives were given considerable attention at the Councils of Melfi (1089), Piacenza (1095), and Clermont* (1095). A serious effort to reunite Eastern and Western Christians was made at the Council of Bari (1098). In each area Urban achieved some measure of success. He is, however, probably most widely remembered for having proclaimed the First Crusade. In response to an appeal by the Eastern emperor Alexius I Comnenus to the Council of Piacenza for recruits for his own hard-pressed forces. Urban at the Council of Clermont called for the establishment of a papal army to rescue the Holy Places from Muslim hands. The Crusade not only achieved this goal, but also enhanced papal prestige and influence. Urban died, however, before news reached him of the capture of Jerusalem. His formal beatification was proclaimed in 1881.

BIBLIOGRAPHY: L. Paulot, *Un Pape français: Urbain II* (1903); H.K. Mann, *The Lives of the Popes in the Middle Ages,* vol. VII (1925-32); A. Becker, *Papst Urban II* (1964).

T.L. UNDERWOOD

URBAN V (1310-1370). Sixth pope from 1362 of the Avignonese line. Born Guillaume de Grimoard, of a noble family at Grisac, he was placed in the Benedictine monastery of Chirac as a child. He studied at Montpellier, Toulouse, Avignon, and Paris and taught canon law before becoming abbot of St.-Germain at Auxerre in 1352, and then of St.-Victor at Marseilles in 1361. His election as pope in 1362 resolved an impasse in the Curia. The fact that—though he had served on papal missions to Italy—unlike his predecessors he had no previous high administrative experience contributed both to his strengths and his weaknesses. His choice of the name "Urban" signaled his identification with both the initiator of the crusades and the original seat of the Holy See.

As pope he was a moderate reformer, living an exemplary monastic life in the papal palace, an overly generous patron and builder, especially devoted to educational projects. A methodical canon lawyer, he continued the policy of centralizing the ecclesiastical administration, and improved the papal palace at Avignon, adding gardens in good Benedictine fashion. His hopes of ending the conflict in the West and renewing the Crusade seemed in sight of fulfillment following the Peace of Bretigny and his humiliating peace with the Visconti, and encouraged his premature return to Rome in 1367. There he accepted the submission of Emperor John V Palaeologus and received Charles IV. The renewal of the Hundred Years' War forced his return to Avignon, where he died. Praised even by Petrarch, that captious critic of the Avignonese papacy, Urban is the only one of the Avignonese popes to have been beatified (1870). MARY E. ROGERS

URBAN VI (c.1318-1389). Pope from 1378. A native of Naples, he was archbishop of Acerenza (1363), then of Bari (1377), and a capable and irreproachable administrator to Gregory XI,* his predecessor, who ended the long Avignon residency of the papacy by returning to Rome in 1377. Four months after Urban's election, the French cardinals declared it invalid, claiming the Roman populace had forced them to choose an Italian. They were actually alienated by Urban's refusal to restore the papal court to Avignon, and by his tactlessness and evident intention to institute reform. In September they elected a new pope, Clement VII (Robert of Geneva), and thus initiated the Great Schism* which for thirty-nine years scandalized Christiandom, with two popes claiming to be the sole head of the church. Europe was almost equally divided, each pope's support based upon political expediency. Urban became involved in Italian quarrels. Deposing Queen Joanna of Naples, he gave her kingdom to Charles of Durazzo. He later placed Naples under interdict and had five cardinals killed for an alleged conspiracy with Charles to restrict Urban's authority. Urban died at Rome, possibly by poisoning. Despite good intentions, his pontificate was marked by anarchy and left the church confused and divided. ALBERT H. FREUNDT, JR.

URBAN VIII (1568-1644). Pope from 1623. This politically minded pope, classical scholar, and embellisher of the city of Rome was born into the wealthy Florentine Barberini family, and given the Christian name "Maffeo." Educated by the

Jesuits in Florence, he later studied at their Roman college and eventually took a doctorate of laws at Pisa (1589). He returned to Rome and held several offices in the church and Curia. He was legate and later nuncio to France (1601, 1604), was created cardinal (1606), made bishop of Spoleto (1608), legate of Bologna and prefect of Segnatura di Guistizia (1617), and finally was elected pope. Politically and militarily active, he spent great sums to establish an arms factory in Tivoli and various fortifications (including the Castel St. Angelo and Fort Urban), and even made the Vatican Library into an arsenal. He was jealous of the papal temporal power and pursued a balance-of-power policy consonant with that end. He reversed the pro-Hapsburg policy of his successor by fighting Spanish-imperial interests in Italy, and perhaps by supporting France and Sweden in the Thirty Years' War,* thus contributing to the dissolution of the Holy Roman Empire.

Urban was also reform-and mission-minded. He issued decrees on canonization, introduced reform into the Roman Breviary, enforced the Trent guidelines on episcopal residence, reduced the number of obligatory holy days, and approved new reform orders (including the Vincentians). He strongly supported missionary activity by opening the Far East to missionaries other than Jesuits and by founding Urban College for training missionaries (1627). His actions against heretics included the condemnation of Galileo* in 1632 (for the second time) and of Cornelius Jansen's *Augustinus* (1642). He spent great sums on projects to beautify and build up Rome. Works he sponsored include the Barberini Palace, the Fountain of the Triton, Bernini's refurbishing of St. Peter's Basilica, and the Vatican Seminary. Though his private life was above reproach, Urban was the last pope to practice nepotism widely.

BIBLIOGRAPHY: W.N. Weech, *Urban VIII* (1905); A. Leman, *Urban VIII et la rivalité de la France et de la maison d'Autriche de 1631 à 1635* (1920); G. Albion, *Charles I and the Court of Rome* (1935).

BRIAN G. ARMSTRONG

URSINUS, ZACHARIAS (1534-1583). German Reformer and theologian. Born at Breslau, he studied at Wittenberg, 1550-57. He is best known for his Heidelberg Catechism (1562), which exudes the spirit of Melanchthon* and agrees with the teaching of his favorite theologian, Peter Martyr* Vermigli. After a lengthy academic visit to Geneva in 1557, Ursinus taught at Breslau, where his 1559 thesis on the sacrament led to his dismissal. At Vermigli's request Elector Friedrich III appointed Ursinus to Heidelberg in 1561, where until 1568 he lectured on dogmatics. In 1570 the Palatinate adopted a church discipline at Ursinus's urgent request. He left Heidelberg in 1577 to continue with Zanchius at Neustadt. By their influence Ursinus and Zanchius made Melanchthon's *Loci Communes* standard reading for generations of Reformed pastors at Heidelberg.

MARVIN W. ANDERSON

URSULINES. The oldest women's teaching order of the Roman Catholic Church. Founded at Brescia (1535) by Angela Merici* and named after St. Ursula, patron of the foundress, it was a society of virgins dedicated to Christian education while living at home. Approved by Paul III (1544), regular community life and simple vows were introduced in 1572 at the instigation of Charles Borromeo.* On profession, members took a fourth vow to devote themselves to education. Paul V allowed the Ursulines of Paris solemn vows and strict enclosure (1612). Convents on these lines, following a modified Augustinian Rule, multiplied in France under Madeleine de Sainte Beuve, assisted by Madame Acarie, and Ann de Xainctongé. Temporarily halted during the French Revolution, growth again continued in the nineteenth century. In Quebec, Canada, convents were founded under Marie Guyard.* At a congress in Rome (1900), numerous convents belonging to different congregations united in the "Roman Union," members of which take simple perpetual vows. Many converts which have remained independent have solemn vows and papal enclosure. Their habit is black with long sleeves; the professed wear black veils, while novices and lay sisters wear white veils.

J.G.G. NORMAN

USSHER, JAMES (1581-1656). Irish archbishop and scholar. Born in Dublin and educated there (he was one of the first students of Trinity College), he was appointed in 1607 regius professor of divinity and chancellor of St. Patrick's Cathedral. An outstanding scholar, at the age of nineteen he engaged successfully in controversy with a learned Jesuit, Henry Fitzsimons. In 1615 Ussher showed his support of Calvinism when he shared in an attempt to introduce into the Irish Church a Calvinistic confession based on the Lambeth Articles* of 1595. In 1621 he became bishop of Meath and Clonmacnoise, and in 1625 archbishop of Armagh. During the years that followed, in which he made many visits to England, he did much to preserve the independence of the Irish Church and particularly its Calvinistic character. In 1640 he went to England for what he anticipated would be a short visit, but in 1641 rebellion broke out in Ireland and he was never able to return. He shared in ecclesiastical discussions thereafter. He was asked to be one of the commissioners at the Westminster Assembly* (1643), but declined. Despite his associations with Charles I, Ussher was indulgently treated by Cromwell, who accorded him a public burial in Westminster Abbey.

Among Ussher's many writings are a history of the Western Church from the sixth to the thirteenth centuries, *Discourse of the Religion currently Professed by the Irish and British* (1622), *Britannicarum Ecclesiarum Antiquitates* (1639), and *Annales Veteris et Novi Testamenti* (1650-54), which included his famous scheme of biblical chronology. Basing his dates on biblical genealogies, he concluded that the world was created in 4004 B.C. His scheme is generally discarded, but ranks as the first serious attempt to formulate a biblical chronology.

See R.B. Knox, *James Ussher* (1967).

HUGH J. BLAIR

USUARD, MARTYROLOGY OF. The most popular martyrology in medieval times. Usuard was so successful in obtaining relics of the saints for his order, the Benedictines, and for Charles the Bald that he was commissioned by Charles to draw up a martyrology. Taking into account the work of Ado, Bede, Florus, and the pseudo-Jerome, he completed his work in 875. The work proved very popular in monasteries throughout W Europe. In 1580 Gregory XIII ordered it to be revised and improved, and thus it became the basis for the *Martyrologium Romanum* (1583). The first critical edition of Usuard's work was prepared by Dom J. Bouillart and published at Paris in 1718.
PETER TOON

USURY. Although at present the term means an exorbitant or illegally high rate of interest on money, it originally meant any charge for the use of money. (Thus "usury" and the present meaning of "interest" would be identical.) In the OT the exacting of interest was forbidden in the case of Jewish debtors (Exod. 22:25; Deut. 23:19f.). Ancient Greek philosophers also condemned interest as unjust. Aristotle* taught that money is barren and that if two coins are put in a bag they will never reproduce; hence demanding the repayment of a larger sum than what is loaned not only violates justice, but also nature. The NT does not state anything explicit about the lending of money, but in the patristic age usury was condemned by most of the Church Fathers. During the Middle Ages the teachings of Aristotle were accepted by Scholastics like Aquinas. In the Decree of Gratian and at the Third Lateran Council (1179) usury was condemned, although Jews were allowed to engage in the practice by the Fourth Lateran Council (1215).

With the rise of capitalism, opposition to lending money for interest was gradually abandoned. Although some sixteenth-century Protestant Reformers like Luther, Zwingli, and Latimer condemned the practice, others such as Calvin and Beza justified it by distinguishing between consumption loans and those for production, insisting that in the latter case money was fertile or productive. Laws were passed in Geneva that allowed for a moderate rate of interest. Later (1571) England and Germany along with other continental lands followed suit, although in France interest was not legalized until 1789. Some scholars in the Roman Catholic Church began to shift their attitudes on usury during the sixteenth and seventeenth centuries, but as late as 1745 Pope Benedict XIV reaffirmed the Scholastic opposition to the charging of interest. By the nineteenth century, however, the Curia issued statements indicating that those who lend money at moderate rates of interest are "not to be disturbed," although the theoretical justification for the practice has never been settled in the Roman Church.

BIBLIOGRAPHY: R.H. Tawney, *Religion and the Rise of Capitalism* (1926); B.W. Dempsey, *Interest and Usury* (1943); J.T. Noonan, *The Scholastic Analysis of Usury* (1957); B.N. Nelson, *The Idea of Usury, From Tribal Brotherhood to Universal Otherhood* (2nd ed., 1969).
ROBERT G. CLOUSE

UTRAQUISTS. The more moderate group within the Hussite movement. Their name derives from the Latin *sub utraque specie* ("under both kinds") and refers to the bread and wine of the Communion. They believed that laymen should receive both elements, not merely the bread as the church decreed, and that certain clerical abuses should be eliminated. Their membership centered in Prague and consisted of many of the nobility and university professors of that city. They defeated the rival Hussite faction, the Taborites,* in the Battle of Lipany. Then they formulated a more moderate position and at the Council of Basle* were given an official status in the Roman Church. They remained the established church of Bohemia until the Thirty Years' War* (1620) resulted in the restoration of Catholicism.
ROBERT G. CLOUSE

UTRECHT, DECLARATION OF. A creedal summary issued in Utrecht in 1889, important in the history of the Old Catholic* movement. The original Old Catholic Church grew out of the Jansenist* controversies: Cornelius Steenhoven in 1724 was consecrated by a Catholic bishop (without papal approval) as archbishop of Utrecht, and the "Jansenist" Old Catholic Church continued as a small group thereafter, with some claim to valid orders. In the 1870s, when a larger secession from the Roman Church took place in protest against Vatican I's adoption of the doctrine of papal infallibility, they turned to Utrecht for ordination. J.H. Reinkens* was ordained bishop of the Old Catholic Church in Germany, and Edward Herzog similarly for Switzerland. The 1889 meeting in Utrecht thus brought together the bishops of the Old Catholic churches. The Declaration affirmed adherence to Catholicism, but rejection of Roman perversions of it, notably including the doctrine of papal infallibility. It was accepted as a doctrinal statement by the Old Catholic churches (including, since 1897, the Polish Old Catholic movement in the USA). DIRK JELLEMA

VAISON, COUNCILS OF. Two church councils (synods) were held at Vaison in SE France in the fifth and sixth centuries. The first (442) enacted ten canons, most of which were aimed at strengthening the power of bishops: e.g., presbyters should not have fellowship with those who were enemies of their bishop. The president was probably Nectarius of Vienne. The second was held in 529 and enacted five canons, one of which required regular prayers to be said for the bishop of Rome. Others related to the Mass, one of them requiring the repetition of the *Kyrie Eleison* ("Lord, have mercy").

VALDÉS, JUAN DE (c.1500-1541). Spanish humanist and reformer. He and his twin, Alfonso, were born at Cuenca, Castile. Educated at Alcalá University (1527), he was greatly influenced by Erasmus.* He accepted Luther's doctrine of justification, but remained a Catholic. He collaborated with Alfonso in two dialogues—*Mercury and Charon* and *Lactancio and the Archdeacon*—which criticizes the church, while he wrote his own *Dialogue on Christian Doctrine* which provoked a lawsuit and made it expedient for him to move to Italy (1531). He became chamberlain to Pope Clement VII (1533) and met Peter Carnesecchi and Ercole Cardinal Gonzaga. After Clement's death (1534) he lived in Naples, serving as Spanish imperial inspector of fortifications and becoming the spiritual adviser of Lady Giulia Gonzaga, the cardinal's sister, who presided over a group anxious for reform and spiritual revival. He wrote his *Christian Alphabet* (1536) based on discussions with Lady Giulia, and later wrote *110 Considerations,* commentaries, and translations of part of the Bible into Spanish. He paved the way for Protestant ideas by his emphasis on religious feeling and his disregard of ecclesiastical authority, so that after his death many of his friends, including Peter Martyr* Vermigli and Bernardino Ochino,* left the church. His followers were called "Valdesians." J.G.G. NORMAN

VALENCE, COUNCILS OF. Numerous church councils were held in this city on the Rhone River in the province of Dauphiné, France. Several are of particular significance. In 374 about twenty-four bishops convened and acted on four disciplinary issues. They ruled the ordination of digamists to be unacceptable; they defined the penance required of lapsed virgins and of idolatrous-then-rebaptized Christians; they considered the problem of clerics renouncing ordination on false pretexts. The last issue assumed concreteness in an extant conciliar letter regarding Acceptus, who refused the bishopric of Forum Iulii (Frejus).

About the time of the Council of Orange* (592), a council met in Valence to consider the doctrine of grace. Rejecting Pelagianism* and Semi-Pelagianism,* it favored the position of Caesarius of Arles,* also adopted by Pope Boniface II. A similar issue was treated by the Council of Valence called by Emperor Lothair in 855. Having investigated charges against the bishop of Valence, the council under the leading of Archbishop Remigius of Lyons refuted the position defended by Hincmar* of Reims and the Council of Quiercy* (853). It asserted double predestination in a manner guarding divine holiness and human responsibility, and it affirmed Christ's death for the elect only. Jansenists* later appealed the approval given these canons by Pope Nicholas I.*

Other councils were held in Valence in 585, 890, 1100, and 1209. In 1248 a so-called Council of Valence meeting at Montelimar anathematized Emperor Frederick II* and promoted the Inquisition.* JAMES DE JONG

VALENS (c.328-378). Co-emperor from 364. At the unexpected death of Julian,* the army had compromised on Jovian*; his equally sudden demise saw the choice fall to Valentinian (364), who a month later associated his younger brother Valens as co-augustus for the East. The brothers were from Christian peasant stock of Pannonia—Valentinian the more able militarily, Valens in finance. While both operated within Jovian's principle of broad religious toleration, Valens sought to deal with the internal dissension within Christianity which stemmed from the Arian* controversy. The synodical sessions since Nicea* (325) had in the East moved away from the *homoousion* position, and Valens undertook from 371 to suppress or remove from office remaining supporters. But the chief problems were at the frontiers. The Huns pushing behind the Goths had set the latter in motion. Valens received a petition for their resettlement in the empire, but before the matter could be negotiated they broke over the Danube. Without awaiting reinforcements coming from the West, Valens attempted to stop them at Adrianople, where in the ensuing disaster for the Roman legions he also met death. CLYDE CURRY SMITH

VALENS (late fourth century). Bishop of Mursa in Pannonia secunda. He was included with Auxentius (of Durostorum; later of Milan), Saturninus (of Arles), and Ursacius (of Singidunum) in the anathema of the Synod of Paris (361), with the latter alone in a letter to Athanasius* attempting reconciliation, and in the treatise of refutation by Hilary of Poitiers* *(Adversus Valentem et Ursacium)* who had actively championed their prosecution in preliminary synods and at Paris. This treatise of Hilary's, known to Jerome, has survived only in fragments. Valens was identified as an active proponent of Arianism* in the West, but his specific thought remains unclear. Since he is associated with Anomoeans* at Arles (353), but Homoeans* at Arminum (359), he may well have reflected those variations in compromising efforts under the impact of imperial pressures which sought to avoid *homoousion* extremes between the councils of Nicea* (325) and Constantinople* (381), by which date their generation of men had passed. CLYDE CURRY SMITH

VALENTINIAN III (419-455). Western Roman emperor from 425. He was the son of Constantius III and Galla Placidia. Throughout the long regency of his mother (425-450) and the *de facto* military authority of Aetius,* Valentinian remained dependent and degenerate. Though Attila's* famous march on Rome in 452 proved innocuous, the Western Empire declined greatly during his reign. Championing orthodoxy, Placidia's first acts in Valentinian's name in 425 were to restore church privileges revoked by the recently executed imperial usurper John, to exclude Jews and pagans from the bar and military rank, and to exile heretics and astrologers from the cities. Valentinian's *Novella* 17 of 445 endowed the bishop of Rome, then Leo I, with authority over provincial churches in the West, including Greece—an important advantage in the pope's rivalry with other patriarchs for universal supremacy. Valentinian was assassinated in 455.

DANIEL C. SCAVONE

VALENTINUS (second century). Prominent Gnostic, founder of the Valentinian sect. If one were to attempt to reconstruct the thought of Valentinus on the basis of the few fragments preserved in his earliest critics Irenaeus* of Alexandria, Clement, and Hippolytus,* he would appear but as another in that chain of Gnostics cataloged in their refutations, inseparable from his successor Ptolemaeus, who headed up the Valentinian school from about 160 and who was a sufficient systemizer to warrant more than fragmentary citation. Thus Irenaeus incorporated major passages of an extended work of the latter plus a portion of a commentary on John's gospel, and Epiphanius* quoted a most reasonable statement of the Valentinian position under the title "Letter to Flora."

Again, with quotations chiefly derived from the earliest opponents, Eusebius put together what was known of Valentinus within the framework of his historical chronology: Valentinus was understood to have arrived in Rome during the four-year bishopric of Hyginus (138-42), which he dated from the first year of Antoninus Pius. Tertullian's *Adversus Valentinianos,* built on his predecessors, likewise provides little fresh information about Valentinus himself other than the cryptic remark that Valentinus nearly became bishop of Rome—that is, he presumably failed to be elected, and thence withdrew from the community. Considering the low-key attack upon the person, it can only be assumed that Valentinus's thought was too close to Christian for comfort, originating from a strange but not impossible reading of John's gospel, and through it the synoptics, especially the teachings from Matthew. Certainly Alexandrian Christian commentaries on the NT are a response to Valentinian exegesis, and hardly distinguishable.

The discoveries of Coptic Gnostic papyri at Nag Hammadi* in Upper Egypt in 1945 have reopened the consideration of Valentinus, for among the texts were a series of writings which could very well be associated with him—particularly the one called the *Gospel of Truth,* which has been specifically named as his by Irenaeus. The text of this gospel is still close to revelatory in style; systematization had not yet set in. It is an announcement or declaration of what has not been known—namely, the name of the Father, possession of which enables the knower to penetrate that ignorance which has separated him and all creation from the Father. And Jesus the Christ in His work as Savior has functioned as the revealer of that name through a variety of modes laden with a language of abstract elements. The retention of belief in creation as the work of the Father makes this gospel an alternative to the contemporary Marcion,* but the notions are finally too esoteric for popular consumption, and the followers of Valentinus can only have been the learned.

BIBLIOGRAPHY: H.C. Puech et al., *The Jung Codex: Three Studies* (tr. and ed. F.L. Cross, 1955); R.M. Grant, *Gnosticism: A Sourcebook of Heretical Writings from the Early Christian Period* (1961), pp. 143-61, and *Gnosticism and Early Christianity* (rev. ed., 1966), chap. 5.

CLYDE CURRY SMITH

VALERIAN (Publius Licinius Valerienus) (third century). Emperor before October 253. Like the data for most of the Roman rulers of the troubled third century, the sources are less than adequate. Valerian would have been born near the beginning of the century, since at the time he became emperor, the son he associated with him, Gallienus (235-68), was already in his mid-thirties; Valerian must have been about sixty. He was from a noble family, and in the midst of the anarchical conditions of crisis from frontier assaults and internal disorder, he had already played important roles under preceding emperors. Under Decius (249-51), he was special finance officer, wherein he assisted in the efforts to compel by *libelli* revival of the state religion, which was the condition for the major persecution of the church. Under Gallus (251-53) he was military commander of troops from Raetia, but he was unable to prevent that emperor's death at the hands of soldiers in the conflict with the rival Aemilianus (253), who within three months met a similar fate when Valerian's troops made him the rival.

With his co-augustus he strove for stabilization, but the Goths and the Persians kept the fronts critical. Persecution of Christians remained severe under their joint administration; notable

martyrs (258) include Sixtus II* of Rome, his deacon Laurence* and Cyprian of Carthage.* But in the campaign on the Persian frontier, presumably by treachery, Valerian and his army were captured by Shapur I—a scene frequently cut in relief by the Persian. Neither the date of the disaster nor that of Valerian's subsequent death in captivity has been determined. But thereafter Gallienus brought a peace to the church which lasted for the next forty years, in spite of continuing internal revolt and strain on the frontiers.
CLYDE CURRY SMITH

VALERIAN (d. after 460). Bishop of Cemenelum. Among the minor ecclesiastical writers of the mid-fifth century is this bishop of a diocese of Alpes maritimae on the SE coast of Gaul, noted for homilies in the rhetorical style patterned after Seneca. He is associated in time, location, and thought with the developing monastic traditions of Lérins, just off that coast, which has been founded at the outset of the century and was by this time under abbot Faustus,* bishop of Riez, sending forth monk-bishops who were setting up daughter institutions in their episcopal cities, and from which was to spread during the sixth century the missionary enterprise to the non-Christian peoples of NW Europe. Valerian is identified as a kinsman of Eucherius* of Lyons. He attended the councils of Riez (439) and Vaison* (452), protesting with eighteen like-minded bishops on behalf of the primacy of Arles with its bishop Hilary* against Leo I* of Rome. Valerian's homilies provide the next major glimpse after Salvian of the movement and nature of the barbarian resettlement of Gaul, in the context of a strong emphasis upon the discipline of work. While not dogmatic in nature, Valerian represents a Semi-Pelagian* stance like those with whom he was associated. The rapid changes in the area reduced his see, and many of his homilies either disappeared or were ascribed to others, such as Eucherius or Petrus Chrysologus.
CLYDE CURRY SMITH

VALLA, LORENZO (1407-1457). Italian philologist and rhetorician, perhaps the most brilliant mind of the Renaissance. He was born in Rome, and after studying under Vittorino da Feltre he became teacher of Greek and Latin, wandering to Pavia, Milan, Genoa, Ferrara, Mantua, and Naples. The last decade of his life was spent at Rome working for Pope Nicholas V* in a position that gave him time for his literary activities. Valla's best-known book, *Elegances of the Latin Language* (1441), became a standard guide for humanists interested in precise expression and graceful style. Always a critical and independent thinker, he was led by his work into many controversies. His most famous book was the *Declamation Concerning the False Donation of Constantine* (1440) in which he demonstrated the spurious character of the document that allegedly proved that Constantine* had given central Italy over to papal control when he moved the Roman capital to the East. Valla demonstrated that the Donation* was an eighth-century forgery and thus could not be used to support papal claims to temporal power. Although his work aroused the ire of some churchmen, it had little practical importance, since the Renaissance popes did not base this claim to political power on the document. Valla's works, however, exerted a strong influence on Erasmus* and the Protestant Reformers.
ROBERT G. CLOUSE

VALLUMBROSAN ORDER. John Gualbert established this contemplative order about 1036 with a mother house in Vallombrosa near Florence. He incorporated many eremitic elements into the pattern of Benedict's Rule, including poverty, strict enclosure, and perpetual silence. Only the order's lay brothers, the *conversi*, worked the fields. The order still has a few followers in Italy, but the number of abbeys has fallen from a high of sixty to about eight cloisters. The mother house has been closed.

VALOIS, HENRI DE (Henricus Valesius) (1603-1676). French lawyer and classical scholar. Educated at the Jesuit college of Clermont and at Bourges, he renounced the bar in 1630 to devote himself to classical study. In 1650 he began to do research in early Greek church historians. Of special value was his Latin translation (1659-73) of Eusebius's *Ecclesiastical History*, which covers the period up to 324. Later Valois issued translations of the church histories of Socrates and Sozomen (1668), and Theodoret and Evagrius (1673).

VAN ALSTYNE, F.J., see CROSBY, FANNY

VANDERKEMP, JOHANNES THEODORUS (1747-1811). Missionary to South Africa. Born in Rotterdam, he revealed in his early career as army officer, doctor, and amateur philosopher a blend of intellectual ability, unconventionality, and stubborn independence. He was converted from Deism* in 1791 shortly after the drowning of his wife and daughter, and offered his services to the London Missionary Society in 1796. In 1799 he reached the Cape as leader of the pioneer LMS party. After an unsuccessful period among the Xhosa, he began work among the Hottentots and established a missionary institution at Bethelsdorp (1803). Conditions here were poor, and discipline weak. This drew criticism from the colonists, as did Vanderkemp's simple manner of life and marriage to a Malagasy slave. But his chief offense in their eyes was his defense of Hottentot interests in the face of widespread injustice.
D.G.L. CRAGG

VAN DYCK, ANTHONY (1599-1641). Flemish painter. Born to a bourgeois family in Antwerp, he was by seventeen already an independent artist, and two years later was admitted into the reputable Guild of Saint Luke as a master. In common with all the younger artists of Antwerp, Van Dyck was drawn into Rubens's* circle. His exact relationship with Rubens is difficult to define, although Rubens referred to him as a disciple. Van Dyck worked for him for two years, during which time the former carried on a sizable practice in portrait painting. In 1620 Van Dyck made his first visit to England, where he was briefly employed by James I. Two years later he went to Italy, where from his base in Genoa he visited the major cities. In 1632 he became court painter to Charles I, by whom he was knighted; yet his life was unsettled and unsatisfied. Van Dyck returned to Antwerp a number of times, each time hoping for an important religious commission.

His style passed through four phases which are generally labeled according to his place of activity at the time: thus his first and second Antwerp period, his Genoese and English periods. A fine example of his early Antwerp style is *Christ Crowned with Thorns,* painted when he was no more than twenty. In this and other works he brought an insistence on the down-to-earth realism of sacred events, as is evident in the bold, bare feet of the kneeling man who hands Christ the derisory scepter. Van Dyck interprets the scene at its most brutal level. Pathos is created by the relaxed, resigned body of Christ in comparison with the aggressive energy of His persecutors. The only sign of life in Christ is a raised index finger which reminds the onlooker that Christ suffered for mankind.

His output was large for a man who died at forty-two; yet there is no single great work by which he is remembered today. His compositions do not stand out against the mass of his portraits, and he lacks Rubens's robustness and fire. In England, however, he long remained the perfect example to which all portrait painters aspired.

ALVA STEFFLER

VANE, HENRY (1613-1662). Statesman and Puritan. Educated at Oxford and abroad, he was early converted to Puritanism. In search of religious liberty he went to New England in 1635, and in the following year became governor of Massachusetts. He became involved in doctrinal controversies, however, and returned to England in 1637. Through the influence of his father, Sir Henry Vane, he entered public life and in 1640 became a member of Parliament and was knighted. Although showing a religious tolerance not always evident in the Puritans he briefly lost the confidence of Charles I* for opposing episcopacy, and was a strong critic of William Laud.* He was one of the English representatives in drawing up the Solemn League and Covenant* with the Scots (1643), though he had some misgivings about Presbyterian attitudes. Vane did not approve of the execution of the king and took no part in the trial. He held office under Cromwell, but in 1653 differed from him and was later imprisoned for criticizing his regime. Imprisoned after the Restoration (1660), he was tried for treason and executed on Tower Hill.

J.D. DOUGLAS

VAN ESS, LEANDER (1772-1847). German biblical scholar. Born at Warburg/Westphalia, he entered the Benedictine Order (1790) and assumed the name "Leander" in place of his christened names "Johann Heinrich." His ordination to the priesthood came in 1796. Appointed professor of Catholic theology at Marburg in 1812, he remained there until 1822. In 1807 he collaborated with his cousin Karl in producing a German translation of the NT. The OT followed in 1822. Divergence from the Vulgate* led to Van Ess's version being placed on the Index in that same year. Nevertheless more than 500,000 copies of the NT were circulated with financial aid from the British and Foreign Bible Society.

WAYNE DETZLER

VAN EST, see ESTIUS

VAN EYCK, HUBERT (c.1366-1426) and JAN (c.1390-1441). Flemish painters. Hubert was called "second to none" on the inscription of the frame of the Ghent masterpiece, *The Adoration of the Lamb,* in the cathedral of St. Bavon. Apart from beginning this famous painting (which his brother finished in 1432) he is reckoned by many art historians to be the painter of *The Three Marys at the Sepulchre* in the Van Beuningen Collection, Rotterdam. The rest of his work is lost. Jan, however, is much better known. Between 1422 and 1424 he painted for John of Bavaria, bishop of Liège. In 1425 began his relationship with Philip the Good, duke of Burgundy. He was treated not as an artisan but as a friend, and he was also able to paint for wealthy Italians resident in the Netherlands. No painter has been more preoccupied with artifacts-e.g., the way a pin fits in a door hinge. This was probably because he pushed the problem of representation in art further than any other painter. The great emphasis on detail has the effect of spiritualizing the subject so that the division between secular and religious art is virtually done away with.

For a list of Jan's extant paintings and their whereabouts, see *The Van Eycks* (ed. R. Hughes, 1970).

PETER TOON

VAN MANEN, WILLEM CHRISTIAAN (1842-1905). Dutch theologian. Born at Nordeloos, he studied at Utrecht and became a preacher in the Dutch Reformed Church.* Recognized as an able theologian, he moved from a relatively orthodox position steadily toward an advanced higher critical viewpoint. In his early forties he was appointed professor at Leyden (1885) and became a brilliant exponent of radical higher criticism. He concluded that the Pauline epistles were sub-apostolic (in his *Paulus,* 3 vols., 1890-96) and extended this to most of the NT, dating it from the second century and seeing it as part of an effort to transform Judaism into a universal religion.

VÁSQUEZ, GABRIEL (1549-1604). Jesuit philosopher and theologian. Born near Belmonte, Spain (he was sometimes known as "Bellomontanus"), he read philosophy at Alcalá (1565-69) and became a Jesuit (1569). Lecturing in philosophy at Alcalá (1571-75), he also read theology there. He taught at Ocana and again at Alcalá, then for six years was at the Roman College until 1591, when he returned to Alcalá to succeed Francisco de Suarez* as professor of theology. His most important work is *Commentarii ac Disputationes* on Aquinas's *Summa* (8 vols., 1598-1615). He wrote a paraphrase and exposition of Paul's letters, and his *Disputationes metaphysicae* was formed out of assorted works by Murcia de la Llana (1617). Occasionally called "Augustine redivivus" and given to poverty, he often misdirected his learning; he labored heavily under Suarez's shadow and opposed his teaching.

C.G. THORNE, JR.

VATICAN I. Reckoned by Roman Catholics to be the twentieth ecumenical council, the First Vatican Council was convened by the papal bull *Aeterni Patris* on 29 June 1868. It sat from 8 December 1869 until 18 July 1870. The closure of the third session was precipitated by the withdrawal of French troops from Rome, due to the outbreak of war between France and Prussia, and by the occupation of the city by Italian troops.

In assessing the council it is necessary to take into account the political, cultural, and theological background. Politically the pope was still viewed as a temporal prince. Admittedly the Papal States had been taken over by the new Italy, and after 1860 only Rome remained. But across Europe there were grave suspicions that the concern for papal primacy and infallibility marked a reassertion of the old claims of the dominion of church over state. Culturally it was the period when Romanticism was in the ascendant and anti-intellectualism was strong—an opportune moment for a firmly traditional council.

Theologically the scene was set for a decisive step forward (though the minority was to consider it a long step backward). Already the pope had promulgated the dogma of the Immaculate Conception* in 1854. Ten years later he had issued the *Syllabus Errorum,** the reactionary trumpet blast against every manifestation of liberal or progressive thinking. The time had come in the judgement of Pius IX* and his supporters to put the coping stone on the edifice of papal absolutism gradually built up over a period of centuries via the forged pseudo-Isidorian decretals and the claims of such powerful medieval popes as Gregory VII,* Innocent III,* and Boniface VIII.*

The two theories which clashed in the Vatican debates carry the names Gallicanism* and Ultramontanism.* The Gallican theory went back via J.B. Bossuet (1627-1704) to the great conciliar theologians D'Ailly* and Gerson,* who at the time of the Council of Constance* asserted the supreme authority of a general council. Gallicanism did not go so far as the Protestant Reformers, but nevertheless it did reject the temporal claims of the papacy, subjected the pope to a general council, and denied that his decrees were beyond reform. The Ultramontane theory—this word meaning "across the mountains"—presented a traditional Italian and strongly papal position. Its classical exponent was the sixteenth-century cardinal Bellarmine,* and in the mid-nineteenth century there were advocates like the editor of the *Dublin Review* the English convert W.G. Ward, and the French editor of *L'Univers* ouis Veuillot,* who would have pushed the theory to its ultimate limits. In any event, Vatican I saw the decisive defeat of the Gallican theory and the triumph of Ultramontanism.

The key figure in the council was the pope himself. Pius IX succeeded to the papal throne in 1846. He began by favoring liberal ideas and Italian nationalism, but the revolution of 1848, his flight to Gaeta, his restoration to Rome by the French army in 1850—all these were a prelude to a thoroughly reactionary policy for the rest of his long reign (he died in 1878). Hans Küng* rather scathingly assessed him as being "without a trace of churchmanly or theological critical self-reflection." He himself made his own convictions clear in his famous reply to one dissident bishop—"Tradition? I am tradition."

The composition of the council helped the pope's ambitions. The 276 Italian bishops outnumbered the 265 from the rest of Europe. The 195 nondiocesan bishops were particularly dependent on the pope. Many of the bishops were theologically mediocre and so were open to the pressures of the majority party. The leaders of the latter were not unduly scrupulous about the methods used. Archbishop Manning* from England, exulting in the fact that he was the only "convert" at the council, and dedicated to the task of pushing through the decree, described the council as "a running fight" in which his group aimed "to watch and counteract" what the minority bishops were doing. By skillful intrigue he succeeded in packing the Deputation—the commission responsible for drawing up the doctrinal statement—so that every opponent was excluded.

The minority party included many outstanding leaders, such as the great historian Hefele,* two Austrian cardinals, Dupanloup* of France, Moriarty of Ireland, and many others. From outside the council they were supported by such notable figures as J.H. Newman* and Döllinger,* one of the foremost Roman Catholic theologians. They resisted the decree on various grounds—that it was unbiblical and unhistorical, that it denied the status of the bishops, that it made future councils redundant, that it was inopportune—this last objection coming from the Eastern Catholics from Orthodox lands and those from the Protestant nations.

But the resistance was in vain. On 13 July 1870 the placets ("ayes") numbered 471, 88 voted non-placet, 62 agreed in principle but not in detail, while 76 abstained. On 13 July 1870 the constitution *Pastor Aeternus* was passed by 533 placets to two non-placets. The rest abstained, many leaving Rome to avoid having to vote against the pope. Newman might query the validity of the decree because of the lack of unanimity, the Old Catholic* breakaway movement might claim many adherents, but the decree was a fact. Döllinger was excommunicated. The bishops who had fought so hard gradually capitulated. To the outsider they appeared as beaten men prepared to go against their conscience, but to a Roman Catholic commentator like Butler they appeared as good Catholics ready to sacrifice private judgment to the supreme authority of their church.

Vatican I dealt with other matters, including the promulgation of the Constitution on the Catholic Faith, but it is remembered chiefly for its formulation of the papal primacy and infallibility. Its dogmatic definitions present a major problem today to the New Catholicism.

BIBLIOGRAPHY: L. Jaeger, *The Ecumenical Council* (1961); C. Butler, *The Vatican Council* (1962); W.S. Kerr, *A Handbook on the Papacy*

(1962); H. Küng, *Infallible? An Inquiry* (1971).

H.M. CARSON

VATICAN II. The Second Vatican Council was the unexpected project of one whose election as pope was looked on as an interim appointment because of his age, but whose pontificate was to prove a landmark in the history of the Roman Catholic Church. Pope John XXIII* had been in office for only ninety days when on 25 January 1959 he declared his intention of convening the twenty-first ecumenical council. Although he did not live to see the council completed—he died during the preparations for the second session—the impact of his personality on the council deliberations was marked.

Pope John was himself a blend of the traditional Catholic approach and the new forward-looking attitude which was to be such a marked feature of the council. The impact of a kindly and genial personality, the obvious concern with people, the desire to let a breath of fresh air into the turgid atmosphere of the Vatican—all these should not lead us to the erroneous conclusion that in his doctrine he was a liberal, for in fact he was markedly conservative. Hence, while his liberal attitude opened the door for progressive thinkers, his own theological position was traditional. On the one hand, he gave the progressives a mandate for action with his often-quoted distinction between the unchanging affirmations of the faith and the changing representations (opening speech 11 October 1962). On the other hand, in his encyclical *Ad Petri Cathedram* he came down firmly on the side of the traditional doctrines of the Mass and of Mary, and made his appeal both to Scripture and tradition.* So too on the one side he changed dramatically the approach to those who were formerly heretics, but were now designated "separated brethren," yet at the same time he made it quite clear that reunion meant their return to the one true church and to the pope as center of unity.

This pattern is seen throughout the council's documents. While they are addressed to the twentieth century and while they bear evidence of the new movements of theological thought within Rome, they are at the same time in the mainstream of Catholic orthodoxy and in fact frequently reiterate their endorsement of both Vatican I* and Trent.* Where they seemed to go too far, as in the Constitution on the Sacred Liturgy and the *Lumen Gentium* (on the church), the pope replied—it was now Paul VI*—with counterbalancing statements of strongly traditional doctrine, the *Mysterium Fidei* on the Eucharist, and the appendix on papal prerogatives affixed to the decree on the church.

The council had four sessions. The first lasted from 11 October until 8 December 1962. The second session lasting from 29 September until 4 December 1963 produced the Constitution on the Sacred Liturgy and the Decree on the Instruments of Social Communication. The third session from 14 September until 21 November 1964 saw the promulgation of major documents on the church, on Ecumenism, and on the Eastern Catholic Churches. The fourth session from 14 September until 8 December 1965 produced a final quota of eleven documents. There were the decrees on the Bishops' Pastoral Office, on Priestly Formation, and on the Appropriate Renewal of the Religious Life; and the declarations on the Relationship of the Church to Non-Christian Religions and on Christian Education. These were promulgated 28 October and were followed on 18 November by the Dogmatic Constitution on Divine Revelation and the Decree on the Apostolate of the Laity. Finally, on 7 December 1965 came the last act of the council in the promulgation of four texts—the Pastoral Constitution on the Church in the Modern World, the Decree on the Ministry and Life of Priests, the Decree on the Church's Missionary Activity, and the Declaration on Religious Freedom.

Before turning to a consideration of some of these documents, some significant factors in the situation at Vatican II need to be kept in mind. This was the first Roman Catholic council at which there were non-Roman observers. Their presence strongly emphasized the ecumenical aims of the council and was clearly a factor in the debates, although of course they did not actually take part. Allied to this was the unprecedented blaze of publicity which made the council a major news item across the world, and which let not only the body of the Catholic faithful but outsiders as well see what was happening. A further significant element was the participation of the *periti*, the theological experts who were there as advisers to the council fathers. Their presence, particularly of those from N Europe and America, was to bring the ferment and turmoil of the current theological debate to the floor of St. Peter's. One final factor was the change in the papacy. Paul VI had been looked on as a liberal, but he clearly viewed with alarm the rapidity of the change developing within Rome, and the radical character of some of the proposals—hence his interventions. These were designed either to moderate the advances gained by the progressives, or (as in his personal intervention in the debate on religious toleration) to delay matters and to gain some measure of agreement in a situation where the depth of feeling between conservatives and liberals was leading to a dangerous polarization.

The Documents. First in order of importance, if not in order of promulgation, is the *Lumen Gentium* ("the light of the nations"), the decree on the church, the "De Ecclesia." It is the result of a drastic revision of the schema originally presented. This represented a traditional and polemical attitude and was replaced by a new document drafted between the first and second sessions, debated during the second and third, and promulgated at the end of this session on 21 November 1964. The basic conviction of the document is the traditional view of the church as the continuing incarnation of Christ. The analogy is quoted of the indissoluble union between the Word and His human nature in the Incarnation—the church is a like incarnation. "The Church exists in Christ as a sacrament or instrumental sign of intimate union with God and of unity for the whole human race." This church moreover is quite firmly stated to be the Roman Catholic Church, for although other Christian communities have marks of holiness, they cannot be accepted as being on the same level as the Roman Catholic Church.

It is true that the conception of the church as "the people of God" is brought into prominence, and the biblical testimony to the continuing people of God in the Old and New Testaments is expounded. But this biblical emphasis is vitiated in two ways. For one thing, the people of God in its fullness really means the Roman Catholic Church, and for another there is still the traditional distinction between people and priest. In the Bible *laos* embraces the whole people of God, but traditional Catholicism maintains a firm gap between the priesthood and the laity. Admittedly there is an attempt to accord priestly functions to the people of God as a whole. But in fact the old position of the priest remains the same—this is seen not only in the "De Ecclesia," but also in the Decree on the Priestly Ministry and Life. "There is an essential difference between the faithful's priesthood in common and the priesthood of the ministry or hierarchy, and not just a difference of degree" ("De Ecclesia"). Priests, defined in a levitical sense, are "given the power of sacred Order to offer sacrifice, forgive sin and in the name of Christ publicly to exercise the office of priesthood." By "a special sacrament . . . they are signed with a specific character and portray Christ the priest." "They reconcile sinners to God and the Church through the sacrament of penance. . . . They sacramentally offer the Sacrifice of Christ in a special way when they celebrate Mass" (Decree on Priestly Ministry.) They "share, at their own level of the ministry, the office of Christ, the sole mediator" ("De Ecclesia").

The church as defined above "is incapable of being at fault in belief." This involves a supernatural gift imparted to the people of God as a whole to recognize and to accept the authoritative teaching of the magisterium or teaching authority. This infallibility focused in the pope is diffused throughout the college of bishops. This area of teaching has, however, received a mixed reaction from theologians. One comment from a major commentary on the decree aptly sums up their reaction: "A dispersed episcopate which is infallible, but never quite knows when is a puzzling paradox."

The papal prerogatives are firmly stated in the decree with its stress on "the institution, the perpetuity, the power and the nature of the sacred primacy of the Roman Pontiff, and of his infallible magisterium." Papal definitions "stand in no need of approval of others and they admit of no appeal to another court." This very strong reaffirmation of papal infallibility was accompanied by a stress on the college of bishops for whom the claim is made that they "have a life-giving contact with the original apostles by a current of succession which goes back to the beginning." Their authority and infallibility are, however, hedged around by the proviso that they can never function without the head, viz., the pope. This aspect of the supreme, overriding, and unique authority of the pope was defined even more precisely in the "Nota" (the notes of explanation) appended to the "De Ecclesia."

Other traditional elements retained in the "De Ecclesia" are the belief in purgatory, prayers for the dead, and the invocation of the saints. Of particular significance is the section devoted to the position of Mary. That a special decree was not allocated to this subject was seen as some measure of victory for the progressive wing; but in fact the traditional dogmas are vigorously restated, and indeed a further dogmatic stage is reached. Mary is again declared to be immaculate in conception, perpetually virgin and sinless, sharing in the work of atonement, raised incorruptible to heaven where she reigns as queen; now she is also presented as the mother of the church. Far from minimizing or rejecting the excesses of the Marian cult, the council called for "a generous encouragement to the cult of the Blessed Virgin, especially to the liturgical cult."

The Dogmatic Constitution on Divine Revelation stands in clear lineal descent from the Council of Trent with its appeal to the two sources of revelation—Scripture and tradition. In spite of the new emphasis in Roman Catholic thinking and writing on the Bible, the Council still maintained the position that "both Scripture and Tradition should be accepted with equal sentiments of devotion and reverence." Both of them "form a single sacred deposit of the word of God." This conviction is linked with an acceptance of the infallible teaching authority of the church to produce the conclusion: "sacred Tradition, holy Scripture and the Church's magisterium are by God's most wise decree so closely connected and associated together that one does not subsist without the other two." The final authority in interpreting Scripture is thus the magisterium of the church.

The Decree on Ecumenism considers the possibilities of reunion from the standpoint of a firm insistence on the traditional claim of Rome to be the one true church. "Only through the Catholic Church of Christ, the universal aid to salvation, can the means of salvation be reached in all their fulness." The Eastern Churches have a special status because of their close approximation to Rome in doctrine, church order, and liturgy. The Anglican Communion also has particular mention because of her retention of Catholic traditions. The other churches while being defective from Rome's standpoint are still acknowledged to retain some elements of Catholic truth. What then is the basis for unity? It is the idea of "brothers by baptism." Baptism sets up a bond of unity which it is hoped will develop into a fully integrated unity and an ultimate return to the unity which Rome alone possesses.

Vatican II may be viewed either as a rearguard action by the conservatives, or as a transition to more radical developments in the future. The assessment may well be more an indication of the standpoint of the observer than of the essential position. Time will doubtless demonstrate which assessment is vindicated by subsequent events.

BIBLIOGRAPHY: Y. Congar, H. Küng, and D. O'Hanlon (eds.), *Council Speeches of Vatican II* (1964); G.C. Berkouwer, *The Second Vatican*

Council and the New Catholicism (1965); W.M. Abbott (ed.), *The Documents of Vatican II* (1966); J. Moorman, *Vatican Observed* (1967); K. McNamara (ed.), *Vatican II: The Constitution on the Church* (1968); A.M.J. Kloosterman, *Contemporary Catholicism* (1972). H.M. CARSON

VATICAN CITY, see PAPAL STATES

VAUGHAN, CHARLES JOHN (1816-1897). Dean of Llandaff, Wales. Educated under Thomas Arnold at Rugby, and at Trinity College, Cambridge, where he was elected fellow in 1839, Vaughan was ordained in 1841. After a short period in Leicester he became headmaster of Harrow, which he raised from some sixty ill-disciplined boys to a flourishing and well-organized school between 1844 and 1859. When vicar of Doncaster (1860-69) he began training young men for the ministry, who were known as "Vaughan's doves," continuing this work after he became master of the Temple (1869-94). Among his 450 pupils was the future archbishop, Randall Davidson. Vaughan was a good parish priest and an exceptionally fine preacher, his expository sermons being impressively delivered with strong conviction. He was hostile to High Church practices and to the contemporary German critical views of the Bible. When dean of Llandaff (1879-97) he refused several offers of preferment. His Nonconformist sympathies brought close links with the founding in 1883 of University College, Cardiff. He left strict instructions that no biography of him should be written. G.C.B. DAVIES

VAUGHAN, HENRY (1621-1695). Poet. From a Welsh family, he was educated at Jesus College, Oxford, fought briefly for the Royalists in the Civil War, and spent the rest of his life as a country doctor. He acknowledged the influence of George Herbert* in his religious poems, *Silex Scintillans* (1650, enlarged 1655), but in his experience he is more of a mystic than Herbert. In this regard he may owe something to the Hermetic interests of his twin brother, Thomas. Two of his emphases deserve particular mention. One is the link he saw between the nonhuman creatures and God in, by contrast with man, an unfallen intercourse; the other, his quasi-Platonic view of the prenatal existence of the soul with God. As a poet his work is uneven, and he is often unable to sustain the inspired level at which he opens a poem. ARTHUR POLLARD

VAUGHAN WILLIAMS, RALPH (1872-1958). English composer. Son of a clergyman in the west of England, he studied at Cambridge, at the Royal College of Music, and with Bruch in Berlin. For a time he was a church organist, but soon gave it up and devoted himself primarily to composition. He was noted as a lecturer, a folksong collector, and a choral conductor. He was a deep student of Elizabethan music, which was one of the contributing factors to his striking and original personal style. Like Schütz,* in his maturity he took time out to study with an acknowledged master, Maurice Ravel.

Although his personal views bordered on agnosticism, he was the only major twentieth-century composer for whom religious music has been a highly significant part of his total output throughout his career, from his early and unconventional oratorio, *The Holy City,* to his large Christmas cantata, *Hodie,* written at eighty-two. He was musical editor of *The English Hymnal* (1906, rev. 1933), to which he contributed such stirring original tunes as *Sine Nomine* and *King's Weston,* together with many adaptations of folk melodies, and many Welsh hymntunes previously little known outside of Wales. He also shared in editing *Songs of Praise* and *The Oxford Book of Carols.* Although his number of anthems is not great, these are fresh and new in approach. His festival setting for congregation, choir, organ, and full orchestra of "All hail the power of Jesus' name" is the most ambitious of all his pieces based on hymns. The unaccompanied Mass in G minor is an outstanding work. There are several compositions based on *Pilgrim's Progress,* culminating in the fullscale opera of 1951, which he called *A Morality.*

While not strictly church music, there are various other works that draw their inspiration from Christian themes: the *Mystical Songs* on poems by George Herbert, the *Tallis Fantasia* for string orchestra, the *Fantasia on Old 104th Psalm Tune* for piano, chorus, and orchestra are representative. It would not be too much to say that Vaughan Williams has been the greatest single force in Protestant music in the English-speaking world thus far in the present century. He is also the composer of nine symphonies, a variety of concertos, orchestral works, operas, and chamber music.

See M. Kennedy, *The Works of Ralph Vaughan Williams* (1964), and U.V. Williams, *R.V.W.: A Biography* of *Ralph Vaughan Williams* (1964). J.B. MAC MILLAN

VEDAST (Vaast) (d.539). Bishop of Arras. Born possibly in W France, he came into contact with Clovis, the first Christian king of the Franks. Vedast accompanied Clovis* to Reims for the latter's baptism after his legendary conversion in a battle against the Alemanni. On the way Vedast is reputed to have cured a blind beggar by prayer and the sign of the cross. This miracle is reported to have confirmed Clovis in the Christian faith. Vedast became bishop of Arras in 499 and spent the remainder of his life restoring Christianity to that area.

VENN, HENRY (1724-1797). Anglican clergyman. Descendant of a long line of clergymen, he was born at Barnes and educated at Cambridge, where he graduated in 1745. He became fellow of Queens' College in 1749, the year also of his ordination. He served various curacies until his appointment in 1759 as vicar of Huddersfield in Yorkshire. He found the town, says Bishop J.C. Ryle, "dark, ignorant, immoral," but when he left for health reasons twelve years later it was "shaken in the centre by the lever of the gospel." He then became vicar of Yelling in Huntingdonshire until just before his death. As a preacher he was esteemed highly by many, including the Countess of Huntingdon, William Cowper, and Charles Simeon. He was chosen to give the funeral orations for William Grimshaw and George Whitefield. Although the author of two volumes (*The Complete Duty of Man* and *Mistakes in Religion*), Venn is perhaps more highly regarded as a letter-writer. Living in a controversial age of Calvinism against Arminianism, he held a theological position evidently both scriptural and well-balanced in disputed points and also in pastoral counseling. J.D. DOUGLAS

VENN, HENRY (1796-1873). Anglican clergyman. Eldest child of John Venn,* he was born at Clapham Rectory. A fellow of Queens' College, Cambridge, he was presented by Wilberforce to the living of Drypool in 1827 and was incumbent of St. John's, Holloway, from 1834 to 1846. His greatest work was as secretary of the Church Missionary Society from 1841 until shortly before his death (for the first five years in conjunction with his pastorate). Venn saw in Matthew 28:19 the emergence of national churches, with marked national characteristics, throughout the world; missionary policy should thus aim at the "euthanasia of a mission" through the stimulation of "self-governing, self-supporting and self-propagating churches" (the "three-self" formula is attributed both to Venn and Rufus Anderson). His secretaryship witnessed a vast increase in "native clergy" and—sometimes against missionary opposition—in the responsibilities they carried, and the beginnings of the indigenization of the episcopate. His death saw his policies drastically modified, but never formally abandoned. A.F. WALLS

VENN, JOHN (1759-1813). The son of Henry Venn,* born shortly before the latter's institution at Huddersfield, he was educated at Cambridge, became rector of Little Dunham, Norfolk, in 1783, and by Charles Simeon's* instrumentality, became rector of Clapham in 1792. The Thorntons, Wilberforce, and Zachary Macaulay were among his parishioners, and he became virtually chaplain to the Clapham Sect* of Christian political activists. He took an active part in the formation of the Church Missionary Society, carried the main burden of its administration in its early years, and drafted some of the normative memoranda and reports. His health, always bad, required his resignation of these duties in 1808.
A.F. WALLS

VERBECK, GUIDO HERMAN FRIDOLIN (1830-1898). Dutch-American missionary to Japan. Reared in the Netherlands town of Zeist under Moravian influence, he studied engineering in Utrecht and at the same time developed linguistic, literary, and musical skills. Emigrating to the USA (1852), he studied at the Presbyterian Theological Seminary, Auburn, New York (1855-59), then was ordained by the Dutch Reformed Church and sailed to Japan as a missionary. He established a school in Nagasaki and taught English through the Bible. At government request he opened a school for Japanese interpreters, using the NT and the American Constitution as textbooks. Several of his students became prominent in national affairs. In 1869 he headed a school in Tokyo which became eventually the Japanese Imperial University. Later he was appointed official translator of foreign documents and treatises for the Japanese government. In 1879 he returned to missionary activity, teaching in the theological school, lecturing, and preparing literature for the Japanese church. S. RICHEY KAMM

VERBIEST, FERDINAND (1623-1688). Jesuit missionary to China.* A Dutchman trained in mathematics and astronomy, he became assistant to J.A. Schall* in Peking and unsuccessfully defended the Jesuits against the false accusers who would have had Schall executed. After the latter's death and a controversy over the Calendar, he was appointed, like Schall had been, president of the board of astronomy in succession to his discredited Chinese predecessor. Verbiest and other missionaries had close and friendly contacts with Emperor Kang Hsi (who studied mathematics under Verbiest's tuition), for whom they conducted all kinds of public works. Although the emperor turned a blind eye to the law that prohibited Chinese from becoming Christians, Verbiest was convinced that the future of Christianity lay in the hands of the Chinese clergy, and he founded a seminary in which they could be trained.
LESLIE T. LYALL

VESPERS. The evening Daily Office of the Western Church. It is preceded by None and followed by Compline.* Its structure and solemnity are similar to that of Lauds* (the morning Office), with four or five Psalms, a Bible-reading, a hymn, the *Magnificat,** and a collect. Often the altar is incensed during the chanting of the *Magnificat*. Appropriate changes and additions are made on holy days, ember days, etc. With Lauds it has the distinction of being the oldest of the seven daily offices, and its celebration in the late afternoon rather than at night may be traced back to the time of Benedict (c.500). The Anglican service of Evensong* is based on Vespers, but with additions, e.g., from Compline. PETER TOON

VESTIARIAN CONTROVERSY. The dispute in the English Church over clerical dress which began about 1550 and reached its peak in 1566. The controversy was in two parts. The first, principally involving John Hooper,* who had returned from exile in Switzerland, took place in 1550-51 in the reign of Edward VI.* Hooper refused to be consecrated bishop of Gloucester if he had to wear the surplice and rochet as required by the Prayer Book of 1559. Eventually he compromised, but only after a literary debate with N. Ridley* had begun, the advice of Jan à Lasco,* Martin Bucer,* and Peter Martyr* had been sought, and the Privy Council had acted. The second part came early in the reign of Queen Elizabeth I,* who restored vestments in her royal chapel in 1549. In 1560 the bishops required their clergy to wear a cope during Holy Communion and a surplice in other services. The "hotter" Protestants in Parliament and Convocation protested about this compulsion. The authorities, however, pressed on and in 1566 Archbishop M. Parker* issued his *Advertisements* which made certain vestments compulsory. Then followed a determined attempt, especially in London, to enforce conformity, and as a result some Puritan ministers were deprived of their livings.

A literary warfare broke out with many tracts (e.g., *A brief discourse against the outwarde apparell and Ministring Garments of the Popish Church*) attacking the compulsory use of vestments. For Parker the use of vestments belonged to those things termed *adiaphora,* but for the men who opposed him their use was a relic of popery. Though many Puritan ministers acquiesced, the opposition to vestments never left Puritanism, and it was very evident among the Elizabethan Separatists and later among the sects of the Puritan Revolution.

See J.H. Primus, *The Vestments Controversy* (1960). PETER TOON

VESTMENTS. The traditional eucharistic apparel of the Eastern and Western churches are, in origin, the dress worn by the Roman citizen in the first centuries A.D. In the primitive church there was no special ministerial garb for services, but from 400 to 800, while secular fashions changed, the clergy continued to wear in church the dress of earlier centuries. The albe, amice, chasuble, dalmatic, girdle, maniple, stole, and pallium were established liturgical vesture by the ninth century; and as the Middle Ages advanced, additions were made, and various symbolic, fanciful explanations were given to the different eucharistic garments, together with special prayers to be used when vesting. The albe was a development of the *tunica alba* of the Roman gentleman; the chasuble of the *paenula,* a cloak covering the body, sewn in front and put over the head; the maniple of the *mappula,* a handkerchief; and the stole of the *orarium,* a napkin. All these garments underwent modifications, particularly in size, in order to facilitate movement, and also by the addition of apparels and orphreys.

The 1549 Prayer Book allowed the use of a plain, white albe and a vestment or cope (a choir and processional vestment, dating back to the sixth century), but these were abandoned in the Second Prayer Book. The 1559 Ornaments Rubric may have allowed their use, but not until the nineteenth century was their use revived by the Anglo-Catholics. In the Revised Canons (1969) they are one of the permitted forms of eucharistic vesture, though with no particular doctrinal significance.

See H. Norris, *Church Vestments: Their Origin and Development* (1949), and C.E. Pocknee, *Liturgical Vesture* (1960). JOHN A. SIMPSON

VESTRY (Lat. *vestiarium,* "robing-room"). In England until the end of last century this was the place where ratepayers of a parish met to carry out parochial business, but the name came to be applied also to the collective body of parishioners. The vestry is still responsible for certain aspects of ecclesiastical business. In the USA, each Protestant Episcopal Church parish has a vestry with clearly delineated duties. In Scotland the word is used to describe the minister's retiring room where, in the case of small parishes, meetings of the kirk session are held.

VEUILLOT, LOUIS (1813-1883). French Roman Catholic writer. Born near Orléans, this self-taught son of a cooper returned to a living Catholic faith during a visit to Rome in 1839. Thereafter he devoted his exceptional journalistic skills to the defence of the Ultramontane* cause in France, chiefly as editor of *L'Univers* (1843-60 and 1867-74). He made it the most powerful Catholic journal of the time. His influence was especially great among the conservative lower rural clergy. He criticized Napoleon III's policy, particularly when it endangered the Papal States or other Catholic interests, and Napoleon suspended the journal from 1860 to 1867. His aggressiveness and hostility to any compromise with liberalism made him unpopular with more liberal Catholics, but he was generally protected from even episcopal critics like Dupanloup* by the support of Pope Pius IX.* Veuillot strove continually for a full restoration of Catholic order to France, thus reversing the French Revolution and its consequences. At first he opposed the Falloux Laws (1850) which once more allowed Catholic schools; he wanted nothing less than a totally Catholic educational system. He accepted the Laws only at the pope's persuasion. Residing in Rome during Vatican I,* he gave energetic support to the cause of the definition of papal infallibility. HADDON WILLMER

VIANNEY, J.B.M., see CURÉ D'ARS

VICAR (Lat. *vicarius,* "substitute" or "representative"). Ecclesiastically the term applies to Christ's earthly representatives. In the Roman Church it means the pope, who (as the "Vicar of Christ") claims universal jurisdiction from Christ's words to Peter (John 21:16ff.), and until the ninth century it referred also to emperors. Among Protestants, chiefly in the Anglican Communion, it can mean the parish priest, perpetual curate, or a minor cathedral official (vicar-choral). In the medieval office, the tithes went usually to a monastery in return for religious services, and when a secular priest was substituted for the religious he was called vicar also.

See too VICAR GENERAL and VICAR APOSTOLIC. C.G. THORNE, JR.

VICAR APOSTOLIC. Formerly one to whom the pope delegated responsibility in some distant outpost of his jurisdiction, the vicar apostolic is now more usually a titular bishop appointed to a territory either without bishops of its own, or where the latter are for some reason unable to carry out their episcopal functions.

VICAR GENERAL. A term known to have been applied to the pope, but more often used in the Church of Rome to describe the representative of a bishop, usually an archdeacon. In 1535 Henry VIII of England gave the title to Thomas Cromwell* as his representative in church matters. In the modern Church of England the title survives as a separate office only in one diocese; usually it is combined with the function of chancellor of the diocese.

VICELIN (c.1090-1154). Missionary bishop and apostle to the Wends. Born at Hameln, Germany, he studied at Paderborn, was a canon and teacher at Bremen, and a student at Laon, France. In 1126 he was ordained and sent by Bishop Adalbero of Bremen as a missionary to the Wagrian Wends. His little success depended upon rare intervals of peaceful relations between the pagans and neighboring Christian princes. After several disappointments, he founded a monastery of Augustinian Canons* at Neumünster in Holstein (1141) on the German-Wendish border, the center of his missionary attempts. Unfortunately the devastation of Wendish lands by Holsteiners and a crusade against the Wends in 1147 swept away the results of many years' labor and destroyed any possibility of peaceful conversion of the Wends. Vicelin was consecrated bishop of Oldenburg beyond the German frontier (1149) and became embroiled in an investiture conflict. He died at Neumünster two years after he became completely paralyzed. ALBERT H. FREUNDT, JR.

VICTOR (fifth century). Bible commentator. A presbyter of Antioch, he compiled a Greek commentary on Mark from studies in Matthew, Luke, and John by earlier writers such as Origen, Titus of Bostra, Theodore of Mopsuestia, Chrysostom, and Cyril of Alexandria. The earliest extant commentary on Mark, it was very popular in the East, and survived in more than fifty codices of the gospels. Victor wrote also a commentary on Jeremiah.

VICTOR (d.554). Bishop of Capua from 541. Very little is known of his life, and only fragments of his writings survive. Victor wrote a treatise on the date of Easter which was cited by Bede, a book on Noah's Ark, and a work on the Resurrection. Only fragments of these works are known, but his most important work, a Latin harmony of the gospels, was later translated into German and has survived intact. This work, the *Codex Fuldensis,* is based on the *Diatessaron** of Tatian but utilizes the Latin of the Vulgate.

VICTOR I (d.198). Pope from 189. Born in Africa of Latin stock, he was the first pope to bear a Latin name. The most significant feature of his ten years in office was his clash with the Quartodecimans* led by Polycrates,* bishop of Ephesus, with each side claiming apostolic authority for its dating of Easter. Victor seems to have regarded the matter just as much as a challenge to his authority, and to have seen it as a choice between Rome and Ephesus. He threatened with excommunication Polycrates and his fellow bishops from Asia Minor, and when they defied him went through the motions of carrying out the sentence. Irenaeus* considered the action too drastic, and though he disagreed with the Easterners he intervened, as did others, on their behalf. While it appears that the pope withdrew the sentence, the interlude did something to consolidate the position of the bishop of Rome. Nevertheless the matter was not finally disposed of until the Council of Nicea (325) when the Roman view was upheld. Victor acted with severity also in the case of others suspected of heresy. J.D. DOUGLAS

VICTORINUS (d. c.303). Earliest Latin exegete, and martyr under Diocletian. He was bishop of Pettau, near Vienne. Jerome, our chief source, lists commentaries by Victorinus on various biblical passages, but only that from Revelation is extant. Jerome criticized his style and repudiated the Millenarianist tendencies of this exegesis. Victorinus was influenced by Origen* especially, Papias* of Hierapolis, Irenaeus,* and Hippolytus.* Victorinus's belief in the reign of Christ on earth beginning with the resurrection of the just in the seventh millennium also appears in his partially extant *De fabrica mundi.* This latter is no longer considered to be part of his lost commentary on Genesis, but actually a treatise on the week of creation, in which symbolic emphasis is placed on the number seven. Probably because of his Millenariani views, Victorinus's writings were included among the apocryphal works by the *Decretum Gelasianum.** The treatise *Against All Heresies* attributed to him by Jerome is lost, but stylistic and other internal evidence suggests it is not identical with that treatise which was appended to Tertullian's *Prescription of Heretics.* DANIEL C. SCAVONE

VICTRICIUS (c.330-c.407). Bishop of Rouen. At the age of seventeen he enlisted in Roman military service, but renounced this when he was converted shortly afterward. He was flogged and sentenced to death, but following a miraculous delivery he later studied philosophy and theology, and was chosen bishop about 380 while still a layman. He founded many churches in rural areas, encouraged the monastic life, and brought many relics from Rome to the Rouen cathedral. He preached among the heathen in Gaul, Flanders, Hainault, and Brabant. When his orthodoxy was impugned, he went to Rome to defend himself. He met Emperor Honorius and Innocent I, who sent him a famous decretal relative to disciplinary matters. He died before 409, as a letter to Augustine from Paulinus does not name him in a list of living eminent bishops of Gaul.
HAROLD LINDSELL

VIENNE, COUNCIL OF (1311-12). Church council which is considered the fifteenth ecumenical council by the Roman Catholic Church. It was convoked by Clement V and was in session from 16 October 1311 until 6 May 1312. A major reason for calling it was the question of the Knights Templar.* The transfer of the papacy to Avignon in 1309 made the pope more subject to the influence of the king of France, Philip IV,* who desired the property of this wealthy crusading order. In 1307 he had ordered the arrest of the Templars in France and by torture had exacted confessions of heresy and immorality. He then brought pressure on Clement V to suppress the order—and his presence at the council forced the pope to meet the royal demands and to suppress the order. The council discussed also a new Crusade which, despite Philip's promise to undertake it, never materialized. The council tried to settle a dispute among the Franciscans,* condemned the Beguines and Beghards,* prescribed the teaching of Greek, Hebrew, and Chaldaic at universities to aid in missionary work, and issued a number of canons dealing with church reform. RUDOLPH HEINZE

VIGILANTIUS (b. c.370). Gallo-Roman opponent of excessive asceticism. Born in Calagurris, Aquitaine (Cazères), and brought up in an inn, he managed the estates of Sulpicius Severus,* acquired a considerable literary culture, and was ordained presbyter at Barcelona (395). He traveled in the East, but was repelled by the extreme asceticism there. Meeting Jerome* at Bethlehem, he returned to Gaul after a quarrel with him and published an attack on Jerome's asceticism and intolerance, labeling him an Origenist.* He declared that honor paid at martyrs' tombs was excessive, that hermit life was cowardice, and that presbyters should be married before ordination. Jerome replied in *Contra Vigilantium* (406), a work full of violent invective, from which most of our knowledge of Vigilantius comes. He nicknamed him "Dormitantius" (i.e., "Dormant" instead of "Vigilant") for his rejection of vigils.*

J.G.G. NORMAN

VIGILIUS (c.500-555). Pope from 537. Although his father, a Roman noble, served in Theodoric's Gothic administration, Vigilius's destiny lay with Byzantine political forces. When a Roman deacon, he was Boniface II's intended successor. In 532 the Roman clergy and senate blocked his consecration, and he was sent as Agapetus I's apocrisarius to Constantinople, where he became confidant of the Monophysite empress Theodora.* When the Goths abandoned Rome, Vigilius became the imperial candidate for the papacy. On the pretext of treason Silverius was deposed, tried, and exiled by the Byzantine general Belisarius, and Vigilius succeeded him in the spring of 537.

Acting contrary to his alleged commitment to Theodora, Vigilius demonstrated an effort to avoid secular manipulation of the papacy by refusing to exonerate and reinstate Anthimus, a deposed Monophysite patriarch, and by defending Chalcedonian orthodoxy in correspondence with Emperor Justinian* and Patriarch Menas.* Consistent with Western clerical conviction, Vigilius dissented from Justinian's 543/44 edict—an obvious concession to Monophysites—against the Three Chapters.* Forcibly brought to Constantinople under imperial orders, Vigilius was with difficulty persuaded to accept Justinian's position. He did so only after expressing reservations based on the Chalcedon decrees, stating his position in *Iudicatum* (548). Intense Western reaction to his concession moved him to return to his initial stance, which he set forth in *Constitutum.* When the Council of Constantinople* (553), over which Vigilius had refused to preside, confirmed the emperor, Vigilius again vacillated and under protracted pressure acquiesced in the council's decision. Allowed after a seven-year absence to return to Rome, he died on the way. His role in the Three Chapters Controversy has been cited as historical evidence against the Roman claim of papal infallibility in doctrinal matters.

JAMES DE JONG

VIGILIUS OF THAPSUS (fl. c.470-500). Bishop of Thapsus. After being summoned to appear before the Arian king Huneric at Carthage about 484, he fled to Constantinople, where he was well received. Here he wrote his *Libri quinque contra Eutychetem* ("Five Books Against Eutyches") in which he defended Chalcedonian orthodoxy and refuted Monophysitism.* He wrote other books against the developed Arianism* which still existed in the East, but only the *Dialogue* is extant. Not a few treatises attributed to him have now been shown to be by other authors.

VIGILS. Periods of prayer, services of worship, or times of fasting* held during the night or on the day before a church festival. Nocturnal services began very early in Christian history and were possibly influenced both by the example of Jesus praying at night and by His parable of the Ten Virgins with the arrival of the Bridegroom at midnight. Certainly by about A.D. 200, vigils were kept for a part of the night before the services at Easter and Pentecost. In the Western Church the vigils have gradually become daytime rather than nighttime activities. They are regarded as periods of fasting and preparation for major festivals. In the Roman Church a vigil of fasting is enjoined before Whitsunday and before the feasts of the Assumption, All Saints, and Christmas, while periods of devotional preparation are expected for others. The Church of England's Prayer Book enjoins sixteen vigils for sixteen holy or feast days.

PETER TOON

VILMAR, AUGUST FRIEDRICH CHRISTIAN (1800-1868). German Lutheran theologian. Vilmar's studies at Marburg (1818-20) led him to embrace Rationalism, a creed which he found increasingly less satisfying over the course of the next two decades. He began his career as a secondary school teacher in Rotenburg (1823), Hersfeld (1827), and Marburg (1833). Elected as a liberal candidate to the newly formed parliament of Hesse and appointed to a ministerial committee for religious education, he contributed significantly to the improvement of the quality of religious instruction and general education in Hesse.

His unhappiness with certain movements of his own time, and his studies in the Church Fathers and the Lutheran confessions, led him to break with the Rationalism and liberalism of his earlier years and to become a leader of the political and theological conservatives in Hesse. Refused confirmation as the successor to the general superintendent of Kassel, Vilmar was called instead to the chair of theology in Marburg (1855). His theological stance was reflected in his most famous book, *Die Theologie der Tatsachen wider die Theologie der Rhetorik* (1856). Vilmar contrasted the objective facts of salvation and their effect in human experience with theology that is nothing more than mere "talk" or "intellectual knowledge." These facts are embodied in the institutional church, which is identical with the body of Christ. Vilmar deemphasized the priesthood of all believers and stressed the mediation of salvation through the ministerial office. He regarded the institutional church as the last wall of defense against the disintegrating and disruptive forces at work in nineteenth-century Europe. Most of his theological words appeared posthumously, including *Die Augsburgische Konfession* (1870); *Die Lehre vom geistlichen Amt* (1870); *Christliche Kirchenzucht* (1872); *Pastoraltheologie* (1872); *Dogmatik* (1874); *Theologische Moral* (1871); *Collegium Biblicum* (6 vols., 1879-83); and *Predigten and geistliche Reden* (1876.)

See U. Asendorf, *Die europäische Krise und das Amt der Kirche. Voraussetzungen der Theologie von A.F.C. Vilmar* (1967); and Gerhard Müller, *Die Bedeutung August Vilmars für Theologie und Kirche* (1969). DAVID C. STEINMETZ

VIMONT, BARTHÉLEMY (1594-1667). French Jesuit missionary to Canada. He entered the Society of Jesus in 1613 and was ordained priest in 1626. In 1629 he came to North America as a missionary, where he served as chaplain to a settlement on Cape Breton Island, Canada, returning to France the following year. He arrived in Canada again in 1639 as third superior of the Jesuits in New France and served as *curé* of Notre Dame in Montreal. He returned to France nearly twenty years later and died there.

VINCENT, JOHN HEYL (1832-1920). Bishop of the Methodist Episcopal Church, and Sunday school educator. Born in Tuscaloosa, Alabama, he studied briefly in the Wesleyan Institute in New Jersey and was ordained elder in 1857. He held pastorates in Illinois, pioneered in Sunday school improvements such as uniform lessons (1872), and with Lewis Miller in 1874 started the world-famous Chautauqua* conferences in W New York State. The general conference of his church elected him bishop in 1888. He served in Switzerland from 1900 to 1904. *The Chautauqua Movement* (1886) is one of his more important books.

VINCENT DE PAUL (1581-1660). Founder of the Lazarists.* Born in Landes, France, of a peasant family, he studied humanities at Dax (1595-97) and theology at Toulouse (1604). He was for two years a slave in Tunisia after capture by pirates (1605-7). Following his conversion he was almoner to Queen Marquerite of Valois (1610); pastor of a congregation at Clichy (1612-26); and chaplain to the family of Philippe-Emmanuel de Gondi, general of the galleys, which included looking after their household staff and the peasants on their estates. Seeking God rather than benefices, he decided for a life of serving the poor; in 1617 he founded the first Confraternity of Charity. He was appointed superior of the Visitation convents in Paris (1622); principal of the *Collège des Bons-Enfants,* Paris (1924); superior of the Congregation of the Mission (Vincentians or Lazarists), and Daughters of Charity (1633). He started retreats for ordinands, organized the Tuesday Conferences for clergy, founded seminaries, established provincial relief during the Wars of Religion, ministered to Louis XIII, served on Louis XIV's Council of Conscience, actively opposed Jansenism,* and lived to see his work increase in and spread beyond France. Canonized in 1737, he was in 1885 named patron of all works of charity for which he is in any way the inspiration.

See studies by P. Coste (ET 3 vols., 1952), A. Dodin (French 1960), and M. Roche (1964).

C.G. THORNE, JR.

VINCENT FERRER (1350-1419). Dominican preacher. He was born at Valencia and entered the Order of Preachers in 1368. After studying at Tarragona and Barcelona and then teaching natural sciences and logic, he completed his studies at Toulouse in 1379 and became prior at Valencia. There he wrote his *De vita spirituali.* In 1384 he resigned in order to teach in the local cathedral school. His growth in holiness and intellect were parallel: "study followed prayer, and prayer study." Marvels accompanied his prayers. Keen to heal the Great Schism,* he supported Clement VII,* the Avignonese claimant, for whom he wrote *De moderno Ecclesiae schismate* to Pedro IV of Aragon. Repudiating Benedict XIII* for prolonging unnecessarily the travesty, he took no part in the Council of Constance which ended the Schism. Morally concerned, in 1399 he set out from Avignon to preach for the next twenty years across Europe. Known as "Angel of the Judgment," he also protected and converted the Jews.

C.G. THORNE, JR.

VINCENTIAN CANON. The test of religious truth laid down by Vincent of Lérins* (early fifth century) in *Adversus profanas omnium novitates haereticorum Commonitorium,* an attack on Augustine's predestination teaching. The criterion states "what has been believed everywhere, always and by all." Vincent maintained that the final ground of truth lies in Scripture; by this threefold test of universality, antiquity, and consent the church can differentiate between true and false traditions. The order of the tests should be noted: certain nineteenth-century English writers frequently misquoted the canon by putting "always" first.

VINCENT OF BEAUVAIS (c.1190-1264). Medieval French encyclopedist. Born in Beauvais, he studied at the University of Paris and entered the Dominican Order about 1220. He returned to Beauvais in 1228 and between 1247 and 1259 completed his major work, *Speculum Maius,* an effort to collect all the knowledge of his day in a great encyclopedia. The completed work consisted of eighty books and is the most extensive encyclopedia written during the Middle Ages. He was a close friend of Louis IX of France and became lector in the royal court about 1240. At the request of the king he wrote also a treatise on the education of princes.

VINCENT OF LÉRINS (d. before 450). Semi-Pelagian* presbyter in the monastery on the island of Lérins (now called St. Honorat off

Cannes). Under the pseudonym "Peregrinus," Vincent wrote two *Commonitoria* or "Notebooks" (434), of which only one has survived. With the growth of theological options in the West and with them growth in the number of questions, it was necessary to find a new standard by which the validity of a point of view might be assessed. Holy Scripture alone cannot resolve the controversies, since Scripture itself can be interpreted in different ways. All teaching must be assessed by the Catholic principle of tradition, the so-called Vincentian Canon.* While there may be growth in the understanding of the dogmatic deposit of faith, there may be no alteration in its content. Antiquity is in every case to be preferred to theological novelty or innovation. Vincent directed this principle against Augustine, whose doctrine of grace he regarded as a novelty. DAVID C. STEINMETZ

VINEGAR BIBLE. The popular name of a fine folio edition of the Bible, printed at Oxford in 1716-17 by John Baskett (d.1742), the king's printer. The headline of Luke 20 reads "The Parable of the Vinegar" instead of "The Parable of the Vineyard."

VINES, RICHARD (1600-1656). Puritan divine. Born in Leicestershire and educated at Magdalene College, Cambridge, he was ordained and taught in a grammar school at Hinckley before becoming rector of nearby Caldecote. Apart from these posts he lectured weekly at Nuneaton. When Civil War broke out, he went first to Coventry and then to London for safety. Here he was granted the sequestered rectory of St. Clement Danes. Appointed by Parliament to the Westminster Assembly,* he served on the drafting committee for the Confession of Faith. In 1644 he was made master of Pembroke Hall, Cambridge, where despite his other duties he did useful work by increasing student enrollment. He was not happy about the rise to prominence of the Independents,* and he solemnly opposed the execution of Charles I.* He refused the Engagement of Loyalty to the Commonwealth and was removed from the mastership of Pembroke. However, knowing his abilities, the parishioners of St. Lawrence Jewry, London, called him to be their minister. He accepted and later increased his influence by participating in regular lectures at St. Michael's, Cornhill. When Cromwell's first Parliament discussed in 1654 what were the fundamentals of the faith, Vines was one of the divines whose advice was sought. He was named also as a ministerial assistant to the ejectors in the London area in Cromwell's "National Church."

In theology he had views similar to R. Baxter.* He would have been happy with a modified episcopacy in the English Church, and on the Atonement he held Amyraldist* ideas. His publications are few, but his reputation with his contemporaries was such that by some he was called "the Luther of England." PETER TOON

VINET, ALEXANDRE RUDOLPHE (1797-1847). French-speaking Swiss theologian, often called the "Schleiermacher of French Protestantism." Born in Lausanne, he studied theology there, taught French at Basle for twenty years, then returned to Lausanne in 1837 as professor of practical theology. Ordained in 1819, he tended to deprecate traditional doctrines unless they had been confirmed by personal experience, and he put great stress on good conscience and right conduct. He advocated separation of church and state, wrote a book on the subject (ET 1843), and took a leading part in the founding in 1845 of the Free Church in the canton of Vaud. His other works included *Études sur Blaise Pascal* (1848). J.D. DOUGLAS

VIRET, PIERRE (1511-1571). Protestant Reformer. An important figure in the Calvinist Reformation, he acquired an interest in the study of the NT from Romain before attending the College of Montaigu at Paris (1518-31). There he joined the Reformers under the influence of G. Farel,* and became a preacher serving Reformed congregations at Payerne, Neuchâtel, and Lausanne. He joined Farel in Geneva (1534), where an attempt to poison him damaged his health. Later at Lausanne he was instrumental in founding the Reformed church and establishing a flourishing academy. His work here ended due to the opposition of Bern (1559). Viret was a trusted friend and correspondent of Calvin and was associated with him at the Lausanne disputation (1536) and in Geneva (1541-42; 1559-61). He transferred his activities to S France and presided over a Reformed national synod at Lyons (1563). Viret was an extremely effective preacher and a gifted writer. His major work is *Instruction chrestienne en la doctrine de la Loy et l'Évangile* (1564)—three volumes which contain a popular version of Calvinist teaching in dialogue form. ROBERT G. CLOUSE

VIRGILIUS OF SALZBURG (c.710-784). Early medieval Irish scholar, abbot, and bishop. Little is known of his early life. He began his career in Ireland where he served as abbot of the monastery of Aghaboe and gained a reputation for geographical knowledge. In 743 he went to the court of Pepin,* who later sent him to Bavaria to become bishop of Salzburg. Virgilius, however, refused consecration and administered only the temporal affairs of the diocese. He soon came into conflict with his archbishop, Boniface,* who disapproved of the arrangement. Boniface first directed Virgilius to rebaptize all who had been baptized by a priest who had utilized a grammatically incorrect version of the baptismal formula. Virgilius appealed to Pope Zacharias and was upheld. In 748 Boniface charged Virgilius with holding a heretical view on the spherical shape of the earth, and with making intrigue against him. No trial seems ever to have been held, and Virgilius was not condemned. In 767 he finally accepted consecration, and among the significant services he rendered the church was the conversion of the Alpine Slavs in 772. He was canonized in 1233 by Gregory IX. RUDOLPH HEINZE

VIRGIN BIRTH, THE. The two synoptic accounts of the birth of Jesus are complementary, yet they are evidently written to serve a different interest. Luke (1–2) stresses the personal intervention of God in the life of Mary, whose intimate feelings are recorded. In Matthew (1–2) the emphasis falls on the fulfillment of OT prophecy, with use made of the proof-text, Isaiah 7:14 (Septuagint). If Matthew follows an "apologetic" or evidential line, Luke's narratives raise questions of a metaphysical character (H. von Campenhausen prefers to call this "dogmatic"). Luke's record focuses on the relationship between the Son of God and Mary's child.

The remaining NT data are at best circumstantial and largely inferential, based on references to Jesus as "Mary's son" (Mark 6:3), as though the event of His birth lay under a cloud of suspicion, and to Paul's descriptions, such as Gal. 4:4: "born of a woman." Early patristic commentators looked to John 1:13 which was read as a singular: *"qui ex deo natus est."* It must, however, be acknowledged that the Virgin Birth does not belong to the public *kerygma* of the NT writers. Its esoteric character is best explained on the assumption that there were "secret traditions" (*disciplina arcani*) which were not divulged to the pagan public but reserved for believers once they had confessed the faith.

The two tracks of a "dogmatic" and an "apologetic" motif in this teaching are followed in the sub-apostolic age. Ignatius cites the teaching as part of the church's confessional statements; he is obviously intent on refuting docetism* by this assertion of the Lord's full humanity in His taking a human body. On the other side, Justin's apologetic writing against the Jew Trypho includes the Virgin Birth as an item fulfilled by the Messiah. The use of the teaching for theological purposes does not come until Irenaeus who, with his doctrine of recapitulation, needed to show that the last Adam as true man won back all that had been lost by the first Adam in paradise. Also as anti-Gnostic polemic it was needful to stress the taking of a human body against the idea that matter was inherently sinful. Once established in the church's dogmatic system, it was the dogmatic line which was pursued, especially when the mission to Israel petered out. By a strange irony of history, however, the dogma that began as an assertion of Jesus' full humanity was used to buttress an ascetic theology and practice, as the celibate state became highly valued.

BIBLIOGRAPHY: J.G. Machen, *The Virgin Birth of Christ* (1930); D. Edwards, *The Virgin Birth in History and Faith* (1943); H. von Campenhausen, *The Virgin Birth in the Theology of the Ancient Church* (ET 1964). RALPH P. MARTIN

VIRTUALISM. The doctrine that virtue from Christ is received in Holy Communion,* although the elements remain unchanged.

See also CALVIN, JOHN; TRANSUBSTANTIATION.

VISSER 'T HOOFT, WILLEM ADOLF (1900-). Dutch ecumenical leader. Born in Haarlem, he studied theology at Leyden and was successively secretary of the World Alliance of YMCAs, general secretary of the World Student Christian Federation,* and general secretary of the World Council of Churches* (not formally constituted as such until 1948). He was the author of numerous publications that distinguished him as one of the foremost ecumenical statesmen of modern times. A multilinguist with sound Dutch common sense, he was associated with many aspects of the WCC which raised doubts in evangelical circles. Nevertheless he never succumbed to the aridity often encountered in professional ecumenists, and just before his retirement in 1966, answering a question from *Christianity Today*, he expressed three convictions: "that it is the duty of every Christian to proclaim the divine lordship of Jesus Christ; that this Gospel is to be addressed to every man, whatever his religious or cultural background may be; that it is to be given in its purest form, that is, in accordance with the biblical witness and unmixed with extraneous or cultural elements."

J.D. DOUGLAS

VITALIAN (d.672). Pope from 657. Born at Segni, he was enthroned as successor of Eugenius I, and announcing his accession to the emperor Constans II, thereby instituted friendly relations between Rome and Constantinople during the early part of his reign and also during the Monothelite* controversy. Unsuccessful in attempts to exert jurisdiction over Maurus, bishop of Ravenna, who refused to appear in Rome, he had good relations with England since he consecrated Theodore* as archbishop of Canterbury in 668. Relations between East and West were not helped when Vitalian's name was removed from the Diptychs (list of people for whom prayers were said) in the church of Constantinople.

PETER TOON

VITORIA, FRANCISCO DE (c.1486-1546). Spanish Dominican theologian and philosopher. Born at Vitoria in the Basque country, he joined the Dominicans in 1504 and later studied at the University of Paris, where he encountered Nominalism and Renaissance humanism. From 1523 to 1526 he lectured in theology at Gregorian College, Valladolid, and from 1526 till his death held the prime chair of theology at Salamanca. Though he made original contributions in politico-legal theory, he himself published nothing, and his ideas are to be found in the summaries of his classroom lectures posthumously published. Most famous are *De Indis, De Iure Belli,* and *De Potestate Civili.* Following Aristotle,* he saw that man must live in an organized society, but argued for the independence, self-sufficiency, and sovereignty of a local state whose power lies in the body politic, and whose purpose is to promote the common good and protect its citizens. The sovereign local state is in turn part of an international society. In dealing with the governance of this international society Vitoria pioneered the ideas, later developed by Suárez,* which form the basis of international law. There is a natural "law of nations" (*jus gentium)* which must legally govern the international community. Building upon these politico-juridical ideas, Vitoria developed his most famous position—a staunch advocacy of the rights of the

New World Indians, and severe criticism of Spanish exploitation.

BIBLIOGRAPHY: Studies in Spanish by C. Barcí a Trelles (1928); L.G. Alonso Getino (1930); V. Beltrán de Heredia (1939); A. Truyol Serra (1946); and S. Lissarague (1947). ET of selections of his theological lectures was published by H.F. Wright (1917). See also H.F. Wright, *Vitoria and the State* (1932), and B. Hamilton, *Political Thought in Sixteenth-Century Spain* (1963).

BRIAN G. ARMSTRONG

VITRINGA, CAMPEGIUS (1659-1722). Protestant Orientalist and biblical exegete. Born at Leeuwarden, son of the recorder of the supreme council of Friesland, he learned Greek and Hebrew early, and studied philosophy and theology at Franeker and Leyden universities. From 1860 he taught at Franeker, refusing to leave for more prominent positions at Utrecht and Leyden, holding the chairs of oriental languages and of theology (1682) and becoming professor of sacred history (1693). His biblical exegesis from the orthodox Calvinist standpoint was marked by freshness and penetration. His chief work was a commentary on Isaiah (1714-20). Other writings include *De Synagoga Veteri* (1696; ET *The Synagogue and the Church,* 1842) and *Anakrisis Apocalypseos Ioannis Apostoli* (1705), a commentary on Revelation combining the Recapitulation and Chiliastic interpretations. This last was widely influential, and made Chiliasm popular among German Pietistic circles, despite the Augsburg and Helvetic confessions, which branded Millenarianism as a Judaistic heresy.

J.G.G. NORMAN

VITUS (d.303?). Martyr. He is said to have been born of pagan parents in Lucania, S Italy, and secretly brought up as a Christian by his nurse Crescentio and her husband Modestus. All three were martyred in Diocletian's* persecution. He is invoked against sudden death, hydrophobia, and the convulsive disorder known as chorea or "St. Vitus' Dance." His cult spread in the Middle Ages, especially among Germans and Slavs. He is sometimes regarded as the patron saint of comedians and actors.

VLADIMIR (956-1015). Prince of Kievan Russia responsible for the Christianization of Russia. He came to power about 980, after a civil war with his two brothers at a time when Russia was experiencing a pagan revival. The first Christian ruler of Russia was his grandmother, Olga (945-62), but her conversion does not seem to have made a significant impact on her subjects, and her son Sviatoslav did not adopt his mother's faith. According to legend, Vladimir investigated Judaism, Islam, and Western Christianity before adopting the Eastern Orthodox faith around 988. Two years later he proclaimed Christianity as the faith of his realm and ordered his subjects to be baptized. He was later canonized by the church and called "equal to the apostles" for his work in converting Russia. In addition, he is remembered as an able and successful ruler.

RUDOLPH HEINZE

VOETIUS, GISBERTUS (Gijsbert Voet) (1588-1676). Dutch Calvinist theologian. Born at Heusden near Utrecht, he studied at Leyden as the disputes between Gomarus* and Arminius* began the "Remonstrant controversy." As a minister and young theologian, he supported the Contra-Remonstrant party, which defended an orthodox and systematized Calvinism. He was a delegate to the Synod of Dort* (1618-19), which condemned the Remonstrants. As a mature scholar, skilled in oriental languages as well as theology, Voetius moved to Utrecht as a professor in 1634, and for three decades and more was known internationally as a defender of scholastic Calvinism. He defended vigorously the independence and importance of the church, attacking the idea (associated with the Remonstrants) that the state should oversee it and allow a wide range of doctrinal positions in it. He rejected state patronage, held that toleration of erroneous doctrines weakened both church and state, and viewed usury and related economic questions as matters for the church to judge. Holding that truth in religion and philosophy began with Scripture, he viewed with alarm the methodology of Descartes* and engaged in heated controversy with his Utrecht colleague Regius (De Roy) and with Descartes himself (1640s).

Voetius's personal religious life was devout, influenced by Puritan devotional writings, and Pietist. At first supporting the mystical pietism of Jean Labadie, he later denounced him for his disregard of the organized church. The most influential of his controversies was perhaps that with Cocceius* (De Cock), an able Dutch Calvinist theologian, who developed a system stressing a succession of divine-human covenants. This implied, as Voetius soon discerned, that OT regulations (e.g., on the Sabbath) would not apply under the "new covenant." The resultant polemics split the Calvinist church and theological faculties into contending factions for a generation after Voetius's death in 1676.

BIBLIOGRAPHY: *Politica Ecclesiastica* (4 vols., 1676; ed. selections F.L. Rutgers and P.C. Hoedemaker, 1885); *Selectae Disputationes* (5 vols., 1655; ed. A. Kuyper, 1887); A.C. Dyker, *Gijsbert Voet* (3 vols., 1897-1915).

DIRK JELLEMA

VOLTAIRE (François-Marie Arouet) (1694-1778). French Enlightenment philosopher and *littérateur.* He was an extremely versatile figure and the greatest formulator of the new Enlightenment* vision of secular and rationalist regeneration. Educated by the Jesuits, he was introduced early to Descartes,* Montaigne, and Pierre Bayle.* Exile in England (1726-29), where he learned from the English Deists* and especially Locke,* permanently transformed his worldview. His commitments were published in *Lettres philosophiques* (1734). Throughout most of his life he was out of favor in official France because of his biting and penetrating critique of the Establishment. He did serve Frederick II as philosopher-poet (1750-52), but his most productive years came under the patronage of Mme. de Chatelet in provincial Lorraine (1734-49), and his country estate at Ferney near Switzerland where as a kind of patriarch he implemented his social ideals (1758-78). Ferney became a model Enlightenment village of 1,200 people with a watch factory, a silk-stocking mill, and social contentment under paternalistic Voltaire. During his final year (1778) he was treated in Paris as virtually a living human deity. The devotees of the Revolution gave him a grand burial in the Pantheon (1792).

Voltaire's gifts were especially literary. In over twenty plays, beginning with *Oedipe* (1718), he forcefully presented his ideals; theater was one of his chief means of "evangelizing" the French social *élites* during fifty years. He was a master of wit and devastating ridicule. His satire *Candide* (1759) summarized his critique against prevailing notions of the ultimate goodness of evil. As historian he reversed traditional historiography with its emphasis on Divine Providence by recasting human history immanentistically, as in his *Essai sur les moeurs et l'ésprit des nations* (1769) and *Philosophie de l'histoire* (1765). His heroes were philosophers, scientists, and poets, not kings and generals.

He called himself a Theist,* but his god was, as Torrey puts it, "a vague impersonal being with no particular concern for the affairs of men." The organized Christian Church was an abomination to him. Christ he admired as a great man, and Christian ethics he considered correct insofar as they concurred with elements found in other religions. He had an intense commitment to humanistic justice, which he believed would be achieved by enlightened amelioration of society which he called "progress." His many articles in the *Dictionnaire Philosophique* summarized his religious and cultural ethic. He conducted an immense correspondence with kings, philosophers, poets, merchants, and ordinary people throughout the world—more than 20,000 letters to over 1,200 correspondents. Thirty-nine of his works were placed on the Index. A Voltairian outlook typified a large sector of the educated classes in the eighteenth and nineteenth centuries. He was the supreme example of the proud, self-sufficient humanist.

BIBLIOGRAPHY: *Oeuvres complètes* (ed. L. Moland, 52 vols., 1877-85); *Voltaire's correspondence* (ed. T. Besterman, over 100 vols., 1953ff.); R. Pomeau, *La religion de Voltaire* (1956); R. Waldinger, *Voltaire and reform in the light of the French Revolution* (1959); N.L. Torrey, *The Spirit of Voltaire* (1963); J.H. Brumfitt, *Voltaire, historian* (1970). C.T. MC INTIRE

VOLUNTARYISM. The belief that membership in a religious body should be free and uncoerced. It follows that such bodies should not be supported by the state but rather by voluntary contributions, and that all groups should stand equal before the law and independent of it and of each other. Voluntaryism is generally considered to be an American movement which grew out of the early insistence on the separation of church and state, and which by its belief in the legitimacy of a variety of religious tastes and ideas led to a pluralism of churches. Another effect of the movement was, on the one hand, to play down theology that tended to be a divisive issue and, on the other, to encourage an activism which emphasized the role of the laity and resulted in the founding of missions and the conducting of education and charitable work. Voluntaryism was a term used also by English Nonconformists, particularly in the nineteenth century, generally to distinguish their position in relation to the Church of England and more specifically to refer to their efforts to maintain an educational system which was church-directed and free from government aid and control. E. MORRIS SIDER

VOLUNTEERS OF AMERICA. An evangelical, social welfare organization founded in 1896 at New York City by Ballington Booth, son of Salvation Army leader William Booth, because his father refused to democratize the Army's administration. Retaining the quasi-military character of the Salvation Army,* the Volunteers hold nondenominational Protestant services and Bible classes, distribute Christian literature, comfort the aged in hospitals, and serve prisoners through the Volunteer Prisoners League. Dispensing social and material aid to over 2 million annually at nearly 600 service centers, they operate girls' homes, rehabilitation centers, summer camps, and day nurseries. Led by Maud Charlesworth Booth, Ballington's widow, from 1940 to 1948 and by their son Charles Brandon Booth after that, they now number about 33,000. D.E. PITZER

VON HARDENBERG, F.L.F., see NOVALIS

VON HÜGEL, FRIEDRICH (1852-1925). Roman Catholic philosopher and writer. Born at Florence, Italy, of an Austrian diplomat father and a Scottish mother, Friedrich went to England with his family when he was fifteen and stayed there for the rest of his life, apart from periods of travel. Baron of the Holy Roman Empire (an inherited title), student of many subjects, and master of seven languages, he never held an office in the Roman Catholic Church, and consorted indeed with its stormy petrels such as G. Tyrrell* and Loisy.* One of his major preoccupations was the relation of Christianity to history, but he is chiefly remembered as a guide and encourager of souls—and not always in an orthodox way. He was a Roman Catholic who did not believe in purgatory hereafter, a religious man who expressed horror that E.B. Pusey* read only religious books, and a mystic who walked the world with open eyes. He wrote copiously, his published works including a study of Catherine of Genoa (1908), *Essays and Address* (1921) and *The Reality of God* (1931). Not surprisingly, however, since he was one of the dying race of thoughtful correspondents, his *Selected Letters* (1928) have proved to be the most durable. His tombstone in an English country churchyard bears the simple inscription, "Whom have I in heaven but Thee?"

J.D. DOUGLAS

VORSTIUS, CONRADUS (Konrad von der Vorst). (1569-1622). Reformed theologian, a leader of the Remonstrant party in the Dutch Reformed Church.* Born in Cologne, he studied at Heidelberg, and then at Geneva under Beza.

Teaching at the academy at Steinfurt near Heidelberg, he was charged with Socinian* tendencies and had to clear himself before the Heidelberg theological faculty. Meanwhile, in the Netherlands the "Remonstrant controversy" was in full swing. Arminius,* at Leyden, tried to soften the orthodox Calvinist teachings on predestination. After Arminius's death (1609), Uytenbogaert and other followers issued the Remonstrance* of 1610. Vorstius was asked to take Arminius's place at Leyden, and accepted (1610). But a book on the nature and attributes of God aroused an immediate storm; the theological faculty at Heidelberg, and James I of England, an amateur theologian, were among those who felt that Vorstius followed Socinius and denied the divinity of Christ. Vorstius became involved in a series of bitter polemics with the Contra-Remonstrants,* and especially with the Frisian theologian Lubbertus.*

In 1612, though his salary at Leyden was continued, he was forbidden to teach; he retired to Gouda, where he translated some of Socinius's works. The Remonstrant controversy continued, mixed in with political factionalism, and grew more embittered. The Synod of Dort* (1618-19) upheld traditional Calvinist orthodoxy, condemned the Remonstrants, and arranged the exile of the Remonstrant leaders. Vorstius spent his last years in Holstein, dying at Tönningen.

DIRK JELLEMA

VOSS, GERARD JAN (1577-1649). Dutch church historian, associated with the Remonstrant or Arminian* controversy. Educated at Leyden, he served for several years as rector of the Latin school at Dort. There he developed his textbook on Latin grammar (published in 1618), which remained in use in the Netherlands for two centuries. A supporter of the Remonstrant ("liberal") party in the heated polemics over the revisionist ideas of Arminius and his followers, he was named regent of one of the colleges at the University of Leyden (1615). After the Synod of Dort* condemned the Remonstrants (1619), he was ousted from his post, but soon (1622) made professor of chronology and rhetoric, and later (1625) professor of Greek. In 1632, at fifty-five, with an established reputation, he was called to the new university at Amsterdam as professor of general and church history. Voss's mature field of study was the history of dogma, and he was one of the first to apply historical methodology to that sensitive field. His major work was his 1642 *Dissertationes tres de tribus symbolis* in which he demonstrated that the traditional attribution of the "Athanasian Creed"* to Athanasius* could not be correct. His collected works were published at Amsterdam after his death (6 vols., 1695-1701).

DIRK JELLEMA

VULGATE, THE. The Latin version of the Bible, derived from the work of Jerome,* declared by the Council of Trent* in 1546 the only authentic Latin text of the Scriptures. In 382 Pope Damasus* commissioned from Jerome a revision of the Latin gospels. He proceeded to revise the Psalms ("Roman" psalter) and other OT books, perhaps of his own accord. In 386 he settled in Bethlehem, where he engaged in patristic translation and commentary work: he revised more thoroughly the Old Latin psalter ("Gallican" psalter), and continued to revise further OT books on the basis of the Septuagint, but with reference to the Hebrew and Origen's* *Hexapla.* Little remains of this: Job and prefaces to Chronicles and works of Solomon. Increased knowledge of Hebrew and Aramaic made him convinced that a version on a Hebrew basis was necessary, and to this he applied himself from about 390; the end of 404 saw its completion. The different parts of the OT differ much in execution. For the NT, there are good reasons for denying to Jerome any part in the revision of any books other than the gospels.

The translation did not meet with acceptance at first, especially the OT. Even two centuries later we find some divergent opinions, and Old Latin biblical manuscripts continued to be copied for centuries in some areas. It was thus inevitable that by transcriptional error and by contamination Jerome's work should become obscured and in need of purification. The medieval process of edition, recension, and correction produced a complex history. The most significant names in this are as follows, although the part ascribed to some by earlier scholarship is not necessarily correct: Victor* of Capua, and Cassiodorus* (Italy, late sixth century); Peregrinus and Isidore of Seville* (Spain, fifth and seventh centuries); Ceolfrid (England, early eighth century); Theodulf* and Alcuin* (court of Charlemagne, ninth century); and a number of Cistercian, Dominican, and Franciscan scholars of the late Middle Ages. Although neither Greek nor Hebrew was widely known in the West, some knowledge was from time to time found, and this with the additional influence of exegetical interests and traditions left its mark on different recensions.

After the invention of printing, a number of impressions appeared, the Clementine Vulgate (1592-98) at length emerging as the authorized version. Modern philology has produced much advance in knowledge on both Old Latin and Vulgate. Two critical editions are outstanding. The NT was edited by a succession of British scholars between 1898 and 1954. Meanwhile, in 1908 Pius X commissioned an international group headed by Aidan Gasquet* to revise the Vulgate text of the whole Bible. Centered in the Benedictine Abbey of St. Jerome in Rome publication began in 1926, and in 1969 had reached the thirteenth volume, Isaiah. A hand edition edited by R. Weber appeared in 1969, containing the whole Bible with Apocrypha and some *spuria* such as the *Epistle to the Laodiceans.*

BIBLIOGRAPHY: S. Berger, *Histoire de la Vulgate* (1893); *Novum Testamentum Latine* (ed. J. Wordsworth et al., 1898-1954); *Biblia sacra iuxta latinam vulgatam versionem* (1926-); *Biblia sacra vulgata* (2 vols., 1969); H.J. Vogels, *Vulgatastudien* (1928); *Cambridge History of the Bible* (1963-70), vol. I, chap. 16; vol. II, chap. 5; vol. III, chap. 6; B. Fischer, *Bibelausgaben des fruehen Mittelalters* (1963), pp. 520-704.

J.N. BIRDSALL

WACE, HENRY (1836-1924). Dean of Canterbury. Born in London and educated at Rugby and Oxford, he was ordained in 1861 and for seven years from 1863 was curate of St. James's, Piccadilly, during which time he began regular contributions to *The Times.* He held a variety of other ecclesiastical positions in London (1872-1903), some of which overlapped: chaplain and preacher of Lincoln's Inn; professor of ecclesiastical history and principal, King's College; rector of St. Michael's, Cornhill; royal chaplain. In 1903 he was appointed dean of Canterbury, which post he held until his death. A strong exponent of Reformation principles, he never hesitated to make his views known in the church's assemblies. Himself a man of versatile scholarship, he rejected the claims of higher criticism. He wrote several books, including *The Gospel and Its Witnesses* (1883) and *The Bible and Modern Investigations,* but he is best known for his collaboration with William Smith in the *Dictionary of Christian Biography* (4 vols., 1880-86), and with Philip Schaff in the second series of *Nicene and Post-Nicene Fathers* (14 vols., 1890-1900). J.D. DOUGLAS

WAKE, WILLIAM (1657-1737). Archbishop of Canterbury from 1717. Educated at Oxford, he became chaplain to Lord Preston, the English ambassador to France, and went to Paris in 1682. He attained prominence through theological dialogues with Bossuet,* and met Gallicanism* which advocated the independence of the Roman Catholic Church in France from papal authority. On his return to England in 1685, he became successively preacher at Gray's Inn, rector of St. James, Westminster, dean of Exeter (1703), and bishop of Lincoln (1705) before going to Canterbury. In 1693 he published an *English Version of the Genuine Epistles of the Apostolical Fathers,* and in 1700 *The Principles of the Christian Religion* (commenting on the Catechism). In the convocation controversy he wrote *The Authority of Christian Princes over their Ecclesiastical Synods Asserted* (1697) and *The State of the Church and Clergy of England in their Convocations* (1703) in answer to Francis Atterbury's *Rights and Privileges of an English Convocation.* The archbishopric was a reward for his support of the Whigs and the Protestant succession during Anne's reign, but he opposed the government's bill to repeal the Occasional Conformity and Schism Acts in 1718. He engaged in discussions (1717-20) with French ecclesiastics on a projected union between the Anglican and Gallican churches, but these ended without result after the death of Louis Ellies du Pin. Wake regarded the Church of England as a *via media* between Rome and Geneva, but recommended changes in the Prayer Book to meet the scruples of Nonconformists.

See J.H. Lupton, *Archbishop Wake and the Project of Union (1717-20)* (1896); and N. Sykes, *William Wake, Archbishop of Canterbury, 1657-1737* (1957). JOYCE HORN

WALAFRID STRABO (c.808-849). Theologian and monk, named from Latin *strabus,* "squint-eyed." Trained at Reichenau, he studied under Rabanus Maurus* at Fulda and there befriended the famous Gottschalk.* Chaplain of the empress Judith after 829, he was named abbot of Reichenau in 838, but was unable to divorce himself from the political struggles between Charles the Bald and Charlemagne's successors. His scientific and theological writings showed him to be one of the second-generation scholars of the Carolingian Renaissance.* Cultural historians owe him a debt for studies of the day's liturgical and religious customs.

WALCH, JOHANN GEORG (1693-1775). Protestant theologian. Born at Jena and inclined to Pietism in his youth, he taught at Jena all his life, as professor of philosophy (1718), poetry (1721), and theology (1734). He attacked the ideas of Wolff in the name of Lutheran orthodoxy, though his heavy dependence on natural theology reveals his indebtedness to the Enlightenment.* He treated of Lutheranism in two massive works, one historical (5 vols., 1733-36), and one doctrinal (5 vols., 1730-39). He also edited Luther's works in twenty-four volumes (1740-52) with valuable introductions and critical apparatus. From 1754 he was ecclesiastical councillor for Saxe-Weimar. His sons Johann (1725-78) and Christian (1726-84) were also noted Lutheran theologians.

IAN SELLERS

WALDENSES. The key dates in Waldensian history are 1210—the Albigensian-Waldensian Crusade; 1532—the synod of Chanforans; and 1848—the Albertine Statute of Emancipation. Before 1210 much is legendary prehistory, but under the phrase the "first Reformation," the thirteenth to fifteenth centuries are currently studied by V. Vinay and the Czech A. Molnar for the mutual interplay of Waldensian, Hussite, Wycliffite, and Bohemian Brethren* ideas. The "second" (sixteenth-century) Reformation turned the Waldensians from a movement into a church, the Synod of Chanforans being the focal point. For 300 years before 1848 their history illustrates the tragic side of *Cuius regio, eius religio,* in an unbroken sequence of persecution, guerrilla war, exile, and return. Since 1848, it is a story of the cultural and evangelical penetration of the Italian nation and its overseas dependencies.

Notions of Waldensian origins have varied with the propaganda slant of the historian. The vestigial community in the Cottian Alps provided the early Protestants in their mainly theological or ecclesiological debates with a splendid riposte to the question, "Where was your Church before Luther?" just as in the earlier reformation, with its sociopolitical radicalism, did legends of fourth-century bishops who rejected the Constantinian church-state establishment, and of the ninth-century bishop Claudius who rejected Charlemagne's restatement of it. Radical Waldensians today prefer to derive their name from Peter Waldo and the Poor Men of Lyons in the eleventh century, recognizing a debt too, to Arnold of Brescia,* Peter de Bruys,* and Henry of Cluny.

The marks of the medieval Waldensians were: evangelical obedience to the Gospel, especially the Sermon on the Mount; a rigorist asceticism; a "donatist"* aversion to recognizing the ministry of unworthy-living priests; belief in visions, prophecies, spirit-possession, and Millenarianism; and a concern for social renewal. Though anti-Constantinian, anti-imagery, and anti-hierarchy, they tended to reject only Catholic practices that were clearly contra-Scripture. At times they attended Mass, being content that their own meetings were private ones after the style of religious societies. Their clergy *(Barbes)* itinerated. During the "first" (Hussite) Reformation, Waldensian influence spread all over Europe. We hear of communities with episcopal as well as presbyterian ministries. Possibly, however, "Waldensian"—especially on the pen of Catholic writers, and these are practically all the sources we have—was a portmanteau word for all varieties of un-Roman activity. This first Reformation was, of course, almost completely suppressed north of the Alps, and south of them survived only in such inaccessible parts as the Waldensian valleys and Calabria.

In the "second" Reformation Geneva and the Waldensians made early contact, and 1532 saw the latter accepting the pattern of a Reformed church and ministry that they have since retained. Their worship was now to be open and ordered, with no Mass, and they were to have regular Genevan Confession of Faith (a little light, perhaps, on anti-Constantinianism, but accepted). As Savoyards they were French-speaking (their Scriptures being Olivetan's* version), and under the Catholic House of Savoy they were to face 300 years of persecution, at first physical and violent (cf. Milton's 1655 sonnet), later civil and economic. These sufferings made them see themselves not just as a Reformation Church, but as an elect people with a God-given destiny—"the Israel of the Alps."

Their modern epoch began with the Statute of Emancipation of 1848 (celebrated on 17 February each year). Admired, assisted, and advised by Anglo-Saxon evangelicals (notably General Beckwith, who said to them, "Evangelize or perish"), they exploited fully Cavour's policy of "A Free Church in a Free State," and abandoning their French *patois,* they exchanged their ghetto image for that of an Israel in *diaspora*—a cultural and religious leaven at work all over Italy. By founding a theological college in 1855, first in Florence and later in Rome, they have come to terms with the highest values in Italian culture. From it Luzzi gave Italy an Italian Bible, and Giovanni Miegge a distinctive Barthian theology that supported intellectual, spiritual, and physical resistance to Fascism. Latterly, the *Facoltà Valdese* in Rome played a vital role in ecumenical contact with Roman Catholic theology, and V. Subilia is perhaps the acutest Protestant assessor of Vatican II. A daughter college in Buenos Aires (shared with Methodism) serves South American Protestantism.

The supreme court of the church, a synod meeting annually in Torre Pellice, elects a moderator (for seven-year terms) and an executive board *(Tavola).* Though only 20,000 communicant members strong, the church has schools, orphanages, homes for the aged, hospitals, and a publishing house (Claudiana). It supports missions in Africa as well as *Servizio Cristiano* in Riesi and Palermo, Sicily. In this last work, Tullio Vinay pioneers the sociopolitical involvement of radical Christianity.

All Italian Protestantism has been strongly influenced by the Waldensian ethos. No church union has taken place, but with the Italian Methodist Church, cooperation has reached the stage of mutual recognition of ministry and membership, and the regular holding of joint meetings of the two synods.

BIBLIOGRAPHY: J.A. Wylie, *History of the Waldenses* (1880); G.B. Watt, *The Waldenses in the New World* (1941) and *The Waldenses of Valdese* (1965); E. Comba, *Storia dei Valdesi* (4th ed., 1950); F. Junker, *Die Waldesner* (1970).

R. KISSACK

WALDENSTRÖM, PAUL PETER (1838-1917). Swedish theologian and churchman. Born at Lulea, he studied at Uppsala, then taught biblical languages and theology at Gävle, a seaport on the Gulf of Bothnia. Ordained in 1864 in the national church, he found its theological outlook depressing. Deeply interested in the revival movement, he stressed Scripture rather than creeds, and insisted that salvation came through a personal commitment to Christ. The Fall he held to have alienated man, not God. "God is love, and does not need to be reconciled, but a reconciliation which takes away the sin of the world is needed, and has been given in Christ." He resigned from the national church's ministry in 1882 and worked for the Evangelical National Association, a movement founded in 1856 for the reform of religion in Sweden, of which Waldenström became leader and editor of the movement's publication *Pietisten.* In 1878 he organized the Swedish Mission Covenant which, though technically still within the national church, adhered to congregational principles. Many of its members emigrated to the USA where they formed what is now the Evangelical Covenant Church. Waldenström was also a member of the Swedish *Riksdag.* His devotional writings were widely regarded as the best reading after the Bible. J.D. DOUGLAS

WALDO, PETER, see WALDENSES

WALES. Welsh Christianity traces its ancestry to the period of the Roman occupation although nothing is known about its first introduction to Britain. Three British bishops were present at the Council of Arles* (A.D. 314), and despite the crumbling of Roman power and the incursions of invaders, Christianity was able to survive. It was given a new unity and sense of purpose by Illtud and Dubricius between A.D. 500 and 547. Their work culminated in an upsurge of spiritual vigor in the period commonly labeled "the Age of Saints" when men of the caliber of Deiniol, Padarn, Cybi, Seiriol, Teilo, and David left an indelible imprint not only on the minds of the Welsh people but on their place-names. Welsh Christianity had by now developed traditions that it would not abandon, even at the behest of Augustine of Canterbury (603), and in consequence Welsh and English Christianity parted. The period of isolation in the history of the Welsh Church lasted until 750 when, belatedly, it accepted the Roman method of calculating Easter. Despite the tempestuous nature of European life in the following centuries, the Welsh Church was able to maintain its vigor and to withstand the challenge of barbarism.

By the time of the Norman Conquest, the church in Wales was virtually a national church. But changes were in the offing. By the middle of the twelfth century, the Welsh bishops had submitted to Canterbury. During the same period too the dioceses were defined and territorial parishes came into existence. In its internal life and administration the church began to follow the patterns of Western Christendom generally. The *clas*, the characteristic ecclesiastical unit of the Celtic Church, disappeared, and continental monasticism penetrated into Wales, with the Cistercian Order taking pride of place. At the same time, the English crown tightened its hold on the church and its revenues. After 1323, when the pope began to intervene in elections, there was a marked increase in the tendency to appoint foreigners to Welsh livings and offices. Inevitably there was growing frustrations among Welsh clergy as was demonstrated by the support they gave to the national insurrection under the leadership of Owain Glyn Dwr in 1400. Despite the ravages of that war of liberation and its failure, the church enjoyed a period of revival in the latter half of that century, in piety, discipline, and monastic vocations. But as in Western Europe generally, the beginning of the sixteenth century was also the time when decline set in. Although Wales was hardly touched by those spiritual and cultural forces which elsewhere made for reformation, its rather romantic attachment to the House of Tudor, its anticlericalism together with a gradual decay in spiritual seriousness, led it to acquiesce in the changes introduced by Henry VIII.*

The Protestant Reformation came to Wales by the same legal processes as it did in England. It was the Welsh language that constituted the main difference. The Book of Common Prayer* and the NT appeared in Welsh in 1567, translated mainly by Bishop Richard Davies* and William Salesbury.* In 1588 came William Morgan's* translation of the whole Welsh Bible—one of the most momentous events in Welsh history. Although a powerful group of Welsh Roman Catholic exiles kept alive the hope of reconverting Wales, the Reformation came to be accepted by the Welsh people. The hold of Protestantism on the Welsh people was greatly strengthened by Puritanism* and Methodism.* Under the leadership of men like Walter Cradock, Vavasor Powell,* Morgan Llwyd,* and John Miles,* Puritanism found support on a modest scale; Baptist, Congregationalist, Presbyterian, and Quaker congregations came into existence. They maintained their ground in the persecutions between 1660 and 1689, but suffered a period of stagnancy at the beginning of the eighteenth century.

In 1735 the Evangelical Revival started under the leadership of Howel Harris,* and he was soon joined by such men as Daniel Rowland* and William Williams,* Pantycelyn. By about 1780 this revival was developing into a massive folk-movement with far-reaching social and cultural effects. The Methodists themselves, hitherto a group within the Church of England, withdrew in 1811 to form the Calvinistic Methodist* Church of Wales. The older denominations shared in the new spiritual vigor and by midcentury Nonconformity* had become the dominant form of Welsh Christianity. It left no aspect of the life of the nation untouched, and by the Victorian Age it was the major force in education, culture, and politics. It was inevitable that the Church of England would have to be disestablished, and this occurred eventually in 1920. Welsh religious life throughout the nineteenth century continued to be revivified by religious revivals, the greatest of which occurred in 1859-60 and 1904-5.

But in the twentieth century Welsh Christianity ran into very great difficulties. After reaching a new zenith about 1908 it began to decline in its hold upon the public. The reasons for this are extremely complex. The decline of spirituality, the loss of a dynamic theology, the temptations of power, the intrusion of anti-Christian philosophies, World War I, the social distress that followed it—all these severely affected the churches. Yet by today there are real signs that this ancient Christian tradition is being revived in its faith.

BIBLIOGRAPHY: G.F. Nuttall, *The Welsh Saints, 1640-1660* (1960) and *Howel Harris, 1714-1773* (1965); G. Williams, *The Welsh Church from Conquest to Reformation* (1962); J.W. James, *A Church History of Wales* (n.d.).

R. TUDUR JONES

WALKER, THOMAS (1859-1912). Anglican missionary to South India. "Walker of Tinnevelly" (Tirunelveli) was born at Matlock Bath, Derbyshire; educated at St. John's College, Cambridge; ordained in 1882; and sent by the Church Missionary Society to Tinnevelly in 1885. Apart from his evangelism and his Bible teaching in the Tamil field, he was noted for preaching at conventions (on Keswick lines), particularly at hill stations, and he exerted considerable influence among the reformed section of the Syrian Orthodox Church (newly gathered into the Mar Thoma Church*) by preaching tours in Kerala at the Syrians' invitation, and especially at the Maramon Convention. He was associated closely also with the work of Amy Carmichael* of Dohnavur, who wrote his biography. ROBERT J. MC MAHON

WALKER, WILLISTON (1860-1922). American church historian. Born in Portland, Maine, he was educated at Amherst College, Hartford Seminary, and Leipzig University. In 1888-89 he lectured at Bryn Mawr College, then for twelve years was professor at Hartford Seminary before becoming professor of ecclesiastical history at Yale, where he remained till his death. A member of many learned societies, he wrote books ranging from *On the Increase of Royal Power under Philip Augustus* (1888) to American Congregationalism, biography of churchmen, *The Reformation* (1900), *French Trans-Geneva (1909)*, and his most noted, *A History of the Christian Church* (1918).

WALLACHIA, see ROMANIA

WALLOON CONFESSION, see BELGIC CONFESSION

WALTER, HUBERT, see HUBERT WALTER

WALTER OF ST.-VICTOR (d. c.1180). Prior of the canons regular of St. Augustine connected with the former abbey of St.-Victor in Paris. The house was founded between 1108 and 1110 by William of Champeaux,* teacher of Peter Abelard, and chartered in 1113. It swiftly became a home for a series of distinguished theologians, mystics, and biblical commentators. Walter, though not the most famous of the Victorines, gained a certain prominence with a treatise called *Against the Four Labyrinths of France,* which attacked Abelard,* Peter Lombard,* Peter of Poitiers, and Gilbert de la Porrée. Walter described the theology of these four masters as a danger to the church and castigated the dialectical method which they employed. His charges, however, did not seriously impede the growth in popularity of the dialectical method among his contemporaries.
DAVID C. STEINMETZ

WALTHER, CARL FERDINAND (1811-1878). American Lutheran theologian. Born in Langenchursdorf, Saxony, he attended the Gymnasium at Schneeberg and the University of Leipzig, was ordained into the Lutheran ministry in 1837, and made pastor in Braeunsdorf. While at Leipzig he had come under Pietistic influences, which combated the Rationalistic influences of his younger years. He was profoundly influenced by Martin Stephan, pastor in Dresden, and came into the *Erweckungsbegung,* the newly awakened Lutheran confessional movement. He joined the emigrants under Stephan, arriving in Missouri early in 1839. Here he was made pastor at Dresden and Johannisberg, Perry County, Missouri. He played an active role in the deposition of Martin Stephan, who had developed autocratic tendencies and was accused of moral turpitude. In 1841 Ferdinand succeeded his brother Otto Herman as pastor of the Saxon church in St. Louis. In 1849 he became professor of theology at Concordia Seminary, which was moved from Perry County to St. Louis in December of that year. Walther retained the post of chief pastor of the *Gesamtgemeinde* (the Lutheran congregations in St. Louis as they were newly established became part of the one-parish structure) while serving as professor and president of the seminary.

He was the first president of the Evangelical Lutheran Synod of Missouri, Ohio, and Other States (The Lutheran Church—Missouri Synod), founded in 1847, and the first president of the Evangelical Lutheran Synodical Conference of North America, founded in 1872, a federation of German and one Norwegian Lutheran synods which was strictly confessional. Walther participated in various theological controversies in his lifetime: the controversy on church and ministry with the so-called Buffalo Synod, with Wilhelm Loehe* and the Iowa Synod, a controversy with the German Methodists, a controversy within his own synod on Millennialism, and a controversy within the synodical conference on predestination and election. He was regarded as an outstanding preacher and the published output of his sermons is fourteen volumes.

His chief theological writings are on the question of church and ministry (mainly *Die Stimme unserer Kirche in der Frage von Kirche und Amt*), a manual on pastoral theology *(Americanisch-Lutherishe Pastoraltheologie),* and his lectures on Law and Gospel (*Die rechte Unterscheidung von Gesetz und Evangelium* translated as *The Proper Distinction between Law and Gospel*). Walther is generally regarded as the outstanding Lutheran theologian in America in the nineteenth century.

BIBLIOGRAPHY: L.W. Spitz, Sr., *The Life of Dr. C.F.W. Walther* (1961); C.S. Meyer (ed.), *Moving Frontiers: Readings in the History of The Lutheran Church—Missouri Synod* (1964) and (ed.) *Letters of C.F.W. Walther: A Selection* (1969).
CARL S. MEYER

WALTHER, JOHANN (1496-1570). German composer. A competent but not outstanding composer, he was closely associated with Martin Luther at Wittenberg. In 1524 they published in that city the famous *Geystlich Gesangk-Buchleyn (Little Sacred Songbook).* This was not a collection of hymns, but a choral collection in four and five vocal parts. The first edition contained forty-three pieces: five Latin motets, and the rest consisting of German *Lieder* (part-songs), of which twenty-three employed texts by Luther. It was the first Protestant book of its kind and went through many editions with changes of contents; it marked the beginning of the enormous literature of Lutheran choral music.
J.B. MAC MILLAN

WALTON, BRIAN (1600-1661). Bishop of Chester. Born in Yorkshire, he entered Magdalene College, Cambridge, in 1616, transferring later to

St. Peter's College. In later years he held numerous ecclesiastical posts. Around 1641 he aroused much opposition in controversy about city tithes in which, as a result of much legal research, he supported the rights of the clergy. After he had summoned his London parishioners for nonpayment of tithes, Parliament removed most of his preferments, and he was imprisoned for a short term. At the Restoration his benefices were returned, and the bishopric of Chester was conferred upon him. In the intervening years he worked on the vast English "Polyglot Bible" (6 vols., 1657), one of the first English works published by private subscription. This contained the entire Bible in nine languages. R.E.D. CLARK

WANGEMANN, HERMANN THEODOR (1818-1894). German mission executive. Influenced by the Pomeranian revival movement, he studied theology and in 1849 became director of a Lutheran teachers' college. Appointed director of the Berlin Mission in 1865, he held this post until his death. His leadership was so strong-willed and even authoritarian that the agency was popularly referred to as the "Wangemann Mission." His first love was the South African field, which he visited in 1867 and 1884-85, and he was not enthusiastic about the mission's move into China (1882) and German East Africa (1890). He built up a supportive constituency by organizing many local associations in E Germany, but he was seldom interested in cooperative efforts with other missions, either at home or abroad.

RICHARD V. PIERARD

WAR. This has been defined as any struggle between rival groups, featuring the use of arms or other means, which can be recognized as a legal conflict. Thus riots or individual acts of violence are excluded, but insurrections and armed rebellion would be included. The teachings of Christ if fully applied would rule out the use of violence (Matt. 5–7). However, since Christians live in a world where evil often predominates they have found it necessary to rationalize the use of force. In the process three major attitudes toward war have been articulated. Some, including most of the early churchmen, have adopted a *pacifist* position. Others have tried to formulate codes of *just war.* This attitude became prominent when Emperor Constantine made Christianity the most favored religion of the state and the barbarians invaded the Roman Empire. During the fourth and fifth centuries the church adopted from classical thought the teaching of the just war. As propounded by Augustine this type of conflict was to have as its goal the establishment of justice and the restoration of peace. It should be waged only under the authority of the ruler and be conducted justly (i.e., faith should be kept with the enemy, there should be no looting, massacres, or profaning of places of worship). Also, the clergy both secular and regular were not to participate in warfare.

The third major outlook toward war, the *crusading* ideal, became prominent during the Middle Ages. A crusade was a holy war fought under the auspices of the church for the sake of the ideal, the Christian faith. This represented a new way of thinking when compared with the just-war concept of justice conceived in terms of life and property. Since to the Crusades* the enemy was considered as a representative of evil, counsels of moderation toward the opposition tended to break down.

The pacifist, just war, and crusading views were well established by the close of the Middle Ages. With various adjustments they reappeared during the modern period of church history. At the time of the Reformation the Wars of Religion again forced Christians to articulate views of war. The Lutherans and Anglicans adopted the Roman Catholic position of the just war, the Reformed churches emphasized the crusade, and pacifism* was advocated by the Anabaptists* and the Quakers.* During the eighteenth and nineteenth centuries there was little Christian thought on the subject of war, but in the twentieth century, with the development of huge national armies and the advent of two world wars, the three historic Christian positions have been restated. The crusading ideal predominated in the major churches with regard to World War I, pacifism was prominent between the two wars, and the mood of the just war was present during World War II.

Today, under the charter of the United Nations, many authorities believe that warfare between nation states is illegal and that the only legal war is an international police action to prevent aggression or to punish aggressors. However, some nations have opposed the United Nations forces, claiming that their legal national interests were at stake. Despite the fact that the Roman Catholic Church and the major Protestant bodies have taught the principle of the just war, due to the development of rocketry and nuclear weapons since World War II many leaders of the major Christian groups have espoused pacifism.

BIBLIOGRAPHY: R. Bainton, *Christian Attitudes Toward War and Peace* (1960); Q. Wright, *A Study of War* (2nd ed., 1965); R.G. Clouse, "The Vietnam War in Christian Perspective" in *Protest and Politics, Christianity and Contemporary Affairs* (ed. R.G. Clouse, R.D. Linder, and R.V. Pierard, 1968), pp. 253-71; P. Ramsey, *The Just War* (1968); R.B. Potter, *War and Moral Discourse* (1970); R.G. Clouse, "The Christian, War and Militarism" in *The Cross and the Flag: Evangelical Christianity and Contemporary Affairs* (ed. R.G. Clouse, R.D. Linder, and R.V. Pierard, 1972), pp. 217-36. ROBERT G. CLOUSE

WARD, NATHANIEL (c.1578-1652). English Puritan clergyman and author. Born at Haverhill, Suffolk, he graduated from Cambridge, and after practicing law for a short time, turned to the ministry (1624-33). In 1634 he went to America because of persecution for Puritan views and became minister of a congregation at Ipswich. Although ill-health forced him to retire from the active ministry, his law degree from Cambridge made him extremely useful to New England's Massachusetts Bay Colony. He had a prominent part in writing the *Body of Liberties* (1641), the first codification of Massachusetts law. He is best remembered for his pseudonymous *The Simple*

Cobler of Aggawam (1647). This was a satirical rebuke of English Puritans for yielding to the pressures of toleration. He returned to England in 1647 to participate in the Civil War.

DONALD M. LAKE

WARFIELD, B(ENJAMIN) B(RECKINRIDGE) (1851-1921). American Presbyterian scholar. Born near Lexington, Kentucky, into an old American family, he prepared for college by private study, then took his Arts degrees at Princeton (1871, 1874). He traveled in Europe for a year, then became an editor of the *Farmer's Home Journal.* Later he trained for the ministry at Princeton Theological Seminary, studied at the University of Leipzig (1876-77), and became assistant minister at the First Presbyterian Church of Baltimore. In 1878 he became instructor of NT language and literature in Western Theological Seminary, Pittsburgh, holding the rank of professor from 1879 to 1887. In the latter year he became professor of didactic and polemical theology at Princeton Theological Seminary, where he succeeded A.A. Hodge.*

Warfield published a score of books on theological and biblical subjects, in addition to numerous pamphlets and addresses. An accomplished linguist of Hebrew, Greek, and modern tongues, he was at home in patristics, theology, and NT criticism. He was a committed Calvinist with a high regard for the Westminster Confession of Faith.* He held dogmatically to an inerrant Scripture, original sin, predestination, and a limited atonement. Among his writings are *An Introduction to the Textual Criticism of the New Testament, The Lord of Glory, The Plan of Salvation, The Acts and Pastoral Epistles,* and *Counterfeit Miracles.* After his death, collections of his articles were published in book form, entitled *Revelation and Inspiration, Studies in Tertullian and Augustine, Calvin and Calvinism, The Westminster Assembly and Its Works,* and *Perfectionism* (2 vols.). He fought a running battle with C.A. Briggs* and H.P. Smith* over biblical inerrancy, which he and Charles Hodge* defended vigorously. Some of Warfield's articles appeared in the *Presbyterian and Reformed Review* and in its successor, the *Princeton Theological Review,* both of which he edited. Perhaps no theologian of that age is as widely read and has had his books kept in print so long as Warfield.

HAROLD LINDSELL

WARHAM, WILLIAM (c.1450-1532). Archbishop of Canterbury from 1504. Educated in law at Oxford, he was sent by Henry VII on legal business to Rome and Antwerp (1490-91) and on a political mission to Flanders (1493). He became Master of the Rolls (1494), precentor of Hereford (1493), and archdeacon of Huntingdon (1496). He acted as envoy to Scotland, to Burgundy, and to the emperor Maximilian (1496-1502), negotiated treaties, and helped arrange the marriage of Arthur and Catherine of Aragon. Made bishop of London in 1502, he was translated to Canterbury in 1504, and appointed lord chancellor. In 1506 he arranged the marriage of Henry VII and Margaret of Savoy. He crowned Henry VIII* and Catherine of Aragon in 1509, and befriended Erasmus.

Wolsey* replaced Warham as lord chancellor in 1516, and from 1518 when Wolsey was made papal legate there was friction between them over precedence. Warham attended Henry VIII to France in 1520. Eight years later he was appointed counsel for Queen Catherine in the divorce proceedings, but was afraid to do anything to support her cause and was forced by Henry to advise Pope Clement VII* to annul the marriage. In 1531, when the clergy were obliged to recognize the king as supreme head of the church, Warham added the phrase, "so far as the law of Christ will allow." He protested ineffectively in 1532 against all parliamentary measures prejudicial to the pope's authority since 1529. He was a patron and benefactor of the New Learning, though entirely unsympathetic to Protestants.

See W. Hook, *Lives of the Archbishops of Canterbury* (1888).

JOYCE HORN

WARNECK, GUSTAV (1834-1910). Founder of the science of missiology. First a pastor and then an official in the Barmen (Rhine) Mission (1871-77), he was a biblicist and stood close to the *Heiligungsbewegung.* He founded in 1874 the important scholarly journal of German missions, the *Allgemeine Missionszeitschrift,* initiated in 1879 the practice of holding regular missions conferences in the German churches, assisted in founding the Evangelical *Bund* in 1885, and occupied the first university chair of missiology in Germany at Halle (1896-1908). In an 1888 paper he called for decennial general missionary conferences supported by a continuing central committee which would coordinate Protestant missionary activity, a vision that was finally realized with the Edinburgh Conference* of 1910 and the formation of the International Missionary Council* in 1921. Warneck's five-volume synthesis of mission theory, *Evangelische Missionslehre* (1892-1903), and his many historical works establish him as Germany's leading missiologist.

RICHARD V. PIERARD

WASHINGTON, BOOKER TALIAFERRO (1856-1915). Negro educator. Son of a slave mother and a white father, he was educated at Hampton Institute, where he came to believe that only vocational training produced income and virtue for blacks. Called in 1881 to organize Tuskegee Institute, an Alabama school for Negroes, he grew more convinced that manual training, unlike classical education, would prevent Negroes from learning egalitarian ideas and provide them with jobs—neither of which was offensive to whites. At the Atlanta Exposition of 1895 he further pleased whites by declaring that blacks were interested in hard work, not social advancement. Washington hoped that sobriety and perseverance would eventually prompt white recognition of human equality, but that also encouraged whites to consider blacks only as manual laborers, and projected artisan and yeoman status for blacks in an increasingly mechanized society. Whatever the merits of his ideas, he was the leading spokesman of blacks in his day.

DARREL BIGHAM

WATERLAND, DANIEL (1683-1740). Theologian. Born in Lincolnshire and educated at Lincoln School and Magdalene College, Cambridge, he became master of his college (1714-40) and vice-chancellor of the university (1715). Much of his life was involved in the Deist* controversy. His accurate scholarship and brilliant writing did much to restore Trinitarianism in England against Arians* and Deists. His scholarly work *Critical History of the Athanasian Creed* (1713) remained the standard work on the subject for 150 years. In *A Vindication of Christ's Divinity* (1719) he attacked Samuel Clarke* and Daniel Whitby.* Waterland placed little faith in the evidential value of mystical experience or philosophical argument; he sought always to base Christian faith on objective evidence.

R.E.D. CLARK

WATKINS, OWEN (1842-1915). Wesleyan Methodist missionary. Born near Manchester, he entered the ministry in 1863. In 1876 he was sent to Natal for health reasons, and in 1880 became first chairman of the Transvaal and Swaziland district of the Wesleyan Methodist Church. He and his colleagues traveled extensively contacting independent African evangelists, setting up stations, purchasing mission farms, and providing pastoral care for white communities, especially on the Witwatersrand gold fields. In 1891 he visited Mashonaland with the Rev. Isaac Shimmin and obtained land grants from Cecil Rhodes. He contracted fever while walking two hundred miles from Umtali to Beira, and was invalided home in 1892. His ambition to lead an advance into Central Africa was unfulfilled, and he finished his ministry in British circuits.

D.G.L. CRAGG

WATSON, RICHARD (1781-1833). Wesleyan minister. Apprenticed to a Lincoln joiner, he received his first conference appointment as a Methodist preacher when only sixteen (his name does not appear in conference minutes until later). At nineteen his first publication appeared. His inquiring mind—and his argumentativeness—brought him under suspicion of heresy in his circuit, and he resigned in 1801, becoming later a preacher with the Methodist New Connexion* and secretary of their conference. In 1807 shattered health induced his resignation. He became editor of a Liverpool newspaper. In 1812 he reentered the Wesleyan ministry and became, with Jabez Bunting,* one of its outstanding figures. For twelve years he had secretarial responsibility for Wesleyan missions, of which, as of the abolition of slavery, he was a leading advocate. His impressive *Christian Institutes* (1823) were the first major work of Methodist systematic theology.

A.F. WALLS

WATTS, ISAAC (1674-1748). Hymnwriter. He was born at Southampton, son of a Dissenting schoolmaster, and was educated at Stoke Newington Academy. After a few years as a private tutor he became pastor of the London nonconformist church at Mark Lane in 1702, a post he retained, despite recurrent ill-health, for the rest of his life.

He is now best known for his hymns, which first appeared in *Hymns and Spiritual Songs* in 1707 and ran through sixteen editions with numerous alterations in the author's lifetime. This work had been preceded by his collection of religious poems, *Horae Lyricae* (1706), and was followed, among others, by *Divine Songs* (1715) for children and *The Psalms of David Imitated in the Language of the New Testament* (1719). These last include two of his most famous pieces, "O God, our help in ages past" and "Jesus shall reign," based respectively on Psalms 90 and part of 72. To these should be added among his more celebrated pieces "When I survey the wondrous Cross" and "There is a land of pure delight." In his own day, and later, he was renowned as an educationalist, not least for *Logick* (1725) and *The Improvement of the Mind* (1741).

Watts stands at that point in Dissenting history which marks the transition from Calvinism to Unitarianism, and there is evidence in his work of the influence of this movement. But he never subscribed to those versions of Calvinism which espoused the doctrine of total depravity. The "remains" of reason after the Fall might be "ruinous," but Watts insisted on making the best use of them.

In his hymnwriting, despite his Christianizing the Psalms, there is an austere OT quality about his vision, especially of God as all-powerful Jehovah, and in his epic sweep of time and eternity. His simple measures and familiar images serve only to emphasize the majesty of this vision.

See A.P. Davis, *Isaac Watts* (1943), and J. Hoyles, *The Waning of the Renaissance* (1971).

ARTHUR POLLARD

WAYLAND, FRANCIS (1796-1865). Baptist pastor and educator. Born in New York City, he had intellectual gifts which enabled him to enter the sophomore class at Union College, from which he graduated in 1813. His immediate interest was medicine, but after completing a course of medical study he experienced a profound religious change which led him to Andover Theological Seminary for a year of study in 1816. From 1817 to 1821 he taught various subjects at Union College, then came a five-year pastorate at the First Baptist Society of Boston before he returned to Union College as professor of moral philosophy. In 1827 he was elected to the presidency of Brown University, which post he held with distinction until his retirement in 1855. He was author of a plan for free public schools in Rhode Island. He gained a national reputation for a printed sermon, "The Moral Dignity of the Missionary Enterprise," and sponsored prison reform, emancipation of slaves, and free trade. Baptist historians have hailed him as one of the strongest defenders of religious freedom and toleration. His many published works include *Domestic Slavery Considered as a Scriptural Institution* (1845).

DONALD M. LAKE

WEALTH, GOSPEL OF, see SOCIAL GOSPEL

WEBB, CLEMENT CHARLES JULIAN (1865-1954). Anglican religious philosopher. The son of Benjamin Webb, founder of the Cambridge Camden Society, he spent the whole of his academic life at Oxford. He was distinguished in two spheres. As a historian of both medieval and modern religious philosophy, he produced a critical edition of the works of John of Salisbury,* and two important studies: *Religious Thought in the Oxford Movement* (1928) and *Religious Thought in England from 1850* (1933). As a religious philosopher on his own account, Webb, working within the philosophical climate of late nineteenth-century Idealism, strove in a series of three volumes published between 1911 and 1920 to reconcile the idea of God as all-inclusive Absolute to that of God as Personality, apprehended in religious experience. His mediating position between the two schools of Absolute and Personal Idealism (he himself inclined to the latter) is not regarded as altogether satisfactory by modern scholars.

IAN SELLERS

WEBB-PEPLOE, HANMER WILLIAM (1837-1923). Anglican clergyman. Born in Herefordshire and educated at Marlborough and Cheltenham colleges and Pembroke College, Cambridge, he was ordained priest in the Church of England (1863), he served as curate of Weobley, chaplain of Weobley Union, then as vicar of King's Pyon with Birley, Herefordshire, and St. Paul's Onslow Square, London (1876-1919), and prebendary of St. Paul's Cathedral, London, from 1893. A champion gymnast, he suffered an injury while at Cambridge, putting him on his back for three years in which position he did all degree and ordination examinations. A leader of the Evangelical party and chief promoter of Keswick, he also addressed the Northfield Bible Conference (1895) and wrote several books, including *Christ and His Church* and *Calls to Holiness.* He was also a strong supporter of missions, particularly among the Waldensians.

C.G. THORNE, JR.

WEE FREES. A description often employed, usually by other Scottish Presbyterians and with a degree of affection, of the Free Church of Scotland.* The term may also be applied, perhaps even with greater accuracy, to those who, leaving the Free Church of Scotland, founded the Free Presbyterian Church* in 1893.

WEIGEL, VALENTIN (1533-1588). Lutheran mystical writer. Born at Naundorf, he studied at Leipzig and Wittenberg, and from 1567 was pastor at Zschopau, near Chemnitz. Suspected of holding impure doctrine (1572), he managed to clear himself, but his studies proceeded in an increasingly heterodox direction. In numerous writings, not printed till the following century, Weigel developed—under the supposed guidance of the inner light and from his readings in the Gnostics and medieval mystics—both an intensely subjective mysticism and a vast pantheistic system which left little room for Scripture, the church, or the means of grace, but all clothed misleadingly in Christian terminology stripped of its historic sense. Weigelianism became popular in the seventeenth century and exerted an influence on Boehme* and on the development of Rosicrucianism.*

IAN SELLERS

WEIL, SIMONE (1909-1943). French Jewish writer, social and political activist, and religious seeker. A graduate of the *École Normale Supérieure,* she employed her talents as a teacher for a time, then took a position of laborer in order to identify with the worker, and later joined the International Brigade against Franco in the Spanish Civil War. She was forced to flee France during World War II, but soon returned from the United States to London to work for the Free French government. Because of an empathetic, forced diet, death came at the age of thirty-four. Her agnostic and anticlerical position was weakened during the latter part of her life by a sincere attraction to Christianity. Until her death she was torn between the two positions. In her writings she makes clear her continued "waiting for God."

JOHN P. DEVER

WEISS, BERNHARD (1827-1918). German Protestant NT scholar. He taught at the universities of Königsberg (1852-63), Keil (1863-77), and Berlin (1877-1908). Demonstrating that criticism and positive evangelical theology were not mutually exclusive, Weiss is one of a long line of conservative German scholars who have not been given the recognition they deserve. In addition to an important handbook of NT theology (1868; ET 2 vols., 1882-83), an influential NT Introduction (1886; ET 2 vols., 1889), and a Life of Jesus (2 vols., 1882; ET 3 vols., 1883-84), he wrote commentaries in the famous Meyer series on Mark and Luke (6th-9th eds., 1878-1901), John (6th-9th eds., 1880-1902), Romans (6th-9th eds., 1881-1902), Mtthew (7th-9th eds., 1883-97), the Pastorals (5th-7th eds., 1885-1902), Hebrews (5th-6th eds., 1888-97), epistles of John (5th-6th eds., 1888-1900), and many other commentaries. He was a strong critic of the *Tendenzkritik* ("tendency criticism") of F.C. Baur* and the Tübingen scholars.

W. WARD GASQUE

WEISS, JOHANNES (1863-1914). German Protestant NT critic. Son of B. Weiss,* the famous conservative scholar, he was educated at the universities of Marburg, Berlin, Göttingen, and Breslau, and later taught NT at Göttingen (1888-95), Marburg (1895-1908), and Heidelberg (1908-14). With W. Bousset, H. Gunkel,* and R. Reitzenstein,* he represented the comparative religions *(religionsgeschictlich)* approach to the study of the Bible. His book *Die Predigt Jesu vom Reiche Gottes* (*Jesus' Proclamation of the Kingdom of God,* 1892), in which he interpreted the message of Jesus entirely in terms of futuristic or "consistent" *(konsequent)* eschatology, was his most influential work. Along with the similar works of Wrede and Schweitzer,* Weiss's work marked the end in Germany of the older liberal interpretation of Jesus and His message—which had interpreted the kingdom of God as an inward, spiritual experience, or as a system of ethics—and paved the way for the work of Bultmann* and his followers. In an article on the literary history of the NT which appeared in the first edition of the reference work *Die Religion in Geschichte und Gegenwart* (1912), Weiss expounded the principles of form-criticism which were later developed by M. Dibelius,* K.L. Schmidt, and Bultmann. His monumental history of early Christian history and literature, *Das Urchristentum* (1914; ET 1937) was completed and edited by R. Knopf and published posthumously. W. WARD GASQUE

WEIZSÄCKER, KARL HEINRICH VON (1822-99). German Protestant theologian. Successor to F.C. Baur* as professor of church history at Tübingen (1861ff.), he was the founder of the journal *Jahrbücher für deutsche Theologie,* which he edited from 1856. In his work on the history of the gospels (1864) he attempted a *rapprochement* between the radical criticism of Baur and more conservative criticism. His most influential work was *Die christliche Kirche in apostolische Zeitalter* (1886; ET *The Apostolic Age,* 2 vols., 1894-95), in which he insisted that one take a more positive view of the fourth gospel than had been common in critical circles. He also rejected the Baurian hypothesis of the Pauline-Petrine conflict in early Christianity. He was active in both ecclesiastical and academic life, holding offices from time to time in the church and in the university (being elected chancellor of the University of Tübingen in 1890). His translation of the NT into German was widely read and appreciated.
W. WARD GASQUE

WELD, THEODORE DWIGHT (1803-1895). American abolitionist. Born at Hampton, Connecticut, he grew up in W New York where he was profoundly influenced by Capt. Charles Stuart, principal of Utica Academy. Converted under C.G. Finney's* preaching, he spent two years preaching in the latter's "Holy Band," but by 1830 his concern had shifted to the abolition of slavery. He enlisted New York philanthropists Arthur and Lewis Tappan in the abolitionist cause in the financing of Lane Seminary, Cincinnati, where Weld and some of Finney's converts took up studies under the presidency of Lyman Beecher.* When abolitionist activities were banned, Weld and his followers transferred to Oberlin College. After 1835 Weld was employed by the American Anti-Slavery Society* he helped found. Using Finney's methods, Weld's agents spread the cause of abolition through writing and training of personnel. After 1836 he focused his energies on the society's publicity and lobbying in Washington, DC. His most influential writings were *The Bible Against Slavery* (1837) and *American Slavery As It Is* (1839), which provided the stimulus for H.B. Stowe's *Uncle Tom's Cabin* (1852). HOWARD A. WHALEY

WELLHAUSEN, JULIUS (1844-1918). German biblical critic. Born in Hameln, Westphalia, he studied at Göttingen, taught there for two years, then went as professor of OT to Greifswald (1872), where orthodox Lutherans were alarmed at the doubts he cast on the inspiration of Scripture; he resigned in 1882. He transferred to the teaching of oriental languages, first at Halle, then as professor at Marburg (1885) and Göttingen (1892). Building on the work of earlier scholars, he attracted widespread attention by suggesting that the basic document of the Pentateuch ("P") was the youngest rather than the oldest element, and that the development of OT religion became clearer if the Pentateuch were viewed as a composite document. His *History of Israel* (1878; ET 1883) gave him a place in biblical studies comparable, it was said, to that of Darwin in biology. Wellhausen also contributed significantly to Islamic and NT studies. J.D. DOUGLAS

WELSH BIBLE. The NT first appeared in Welsh in 1567, translated from the Greek mainly by William Salesbury,* who also collaborated with Richard Davies,* bishop of St. Davids, to produce the Welsh Prayer Book published the same year. Salesbury's NT was accurate and idiomatic, and served as the basis for the complete Bible published in 1588 by William Morgan,* bishop of St. Asaph, aided by Edmund Prys. A revision by Morgan's successor, Richard Parry (1560-1623), probably helped by John Davies (c.1567-1644), his chaplain, was published in 1620. This Bible, still in general use, used the language of the bards and was a formative influence on Welsh prose language. J.G.G. NORMAN

WENTWORTH, PETER (1530?-1597). Puritan leader. A man of property with considerable Puritan connections both by friendship and marriage, he was from 1571 to 1593 leader of the Puritan party in the House of Commons. Successively member for Barnstaple, Tregony, and Northampton, he delivered bold and (to the queen) impertinent speeches, demanding a revision of the Thirty-Nine Articles,* reforms in the church, a reduction of the powers of the higher clergy, and a clear settlement of the succession question as well as his championship of the privileges of the House of Commons. These led three times to his being committed to the Tower, where he eventually died. In the tone and content of his addresses he anticipates the next generation of Puritan leaders, especially Pym and Hampden.
IAN SELLERS

WESEL, JOHANN, see JOHN OF WESEL

WESLEY, CHARLES (1707-1788). The "sweet singer" of Methodism. Born at Epworth Rectory, Lincolnshire, the eighteenth child of Samuel and Susanna Wesley, Charles was at the age of nine sent to Westminster School. A distant Irish relative, Garret Wesley, wanted to adopt him and settle an inheritance on him. The offer was declined, and Charles went up to Christ Church, Oxford, in 1726. He was instrumental in forming the "Holy Club"* and in 1735 joined his brother John in an abortive mission to Georgia, acting as secretary to the governor, James Oglethorpe.

On his return to England he came under the influence of Peter Boehler,* the Moravian. Lying ill at the house of John Bray, he first read Luther on Galatians. On Whitsunday, 1738—three days before his brother—Charles experienced an evangelical conversion. "I now found myself at peace with God, and rejoiced in hope of loving Christ," he testified. He composed the birthday hymn, "Where shall my wondering soul begin."

He now threw himself into the work of evangelism. He began in the houses of friends, visited the prisons, and preached in the churches until the doors were closed against him. Eventually he took to the open air and became one of the most powerful of the field preachers in the revival. In 1749 he married Sarah (Sally) Gwynne, daughter of a Welsh magistrate, and made his home at the New Room in Bristol until he moved to London in 1771 where he supplied the City Road pulpit, among others.

He was the most gifted and most prolific of all English hymnwriters. Some 7,270 such compositions came from his pen—of varying quality, but including many of the very highest order. He gave expression to evangelical faith and experience in language at once biblical and lyrical.

See also HYMNS.

BIBLIOGRAPHY: T. Jackson, *The Life of the Rev. Charles Wesley* (2 vols., 1841); T. Jackson (ed.), *Journal* (2 vols., 1849); G. Osborn (ed.), *The Poetical Works of John and Charles Wesley* (13 vols., 1868-72); J.E. Rattenbury, *The Evangelical Doctrines of Charles Wesley's Hymns* (1941); F. Baker, *Charles Wesley as Revealed in his Letters* (1948); F.C. Gill, *Charles Wesley* (1964).

A. SKEVINGTON WOOD

WESLEY, JOHN (1703-1791). Founder of Methodism.* He was the fifteenth child of Epworth rector Samuel Wesley and his wife Susanna. Although John's father was a staunch High Churchman of the old school, both his grandparents were Puritan Nonconformists. Educated at Charterhouse and Christ Church, Oxford, John Wesley was elected in 1726 to a fellowship at Lincoln College in the same university. He had been ordained deacon the previous year and had preached his first sermon in South Leigh. On two separate occasions he served as his father's curate. In 1728 he was ordained priest by John Potter.

Returning to Oxford, he found that his brother Charles had gathered a few undergraduates, including George Whitefield,* into a society for spiritual improvement. The scope of what was nicknamed the "Holy Club"* was widened when John Wesley joined it and eventually took over the leadership. Its members met for prayer, the study of the Greek Testament, and self-examination. To their devotional exercises were added works of charitable relief.

In 1735 the Wesleys accepted an invitation from the Society for the Propagation of the Gospel to undertake a mission to the Indians and colonists in Georgia. The project proved a fiasco, and when he got back to England in 1738 Wesley wrote: "I went to America to convert the Indians; but, oh, who shall convert me?"

On the journey to America the Wesleys had met a company of twenty-one German Moravians whose simple faith had made a considerable impression on them. When, therefore, John Wesley was introduced in London to another Moravian, Peter Boehler,* he was predisposed to lean toward him. In the event, Boehler was to be the pedagogue to bring Wesley to Christ. As a result of conversations with Boehler, Wesley was "clearly convinced of unbelief, of the want of that faith whereby alone we are saved." On 24 May 1738, his heart was "strangely warmed" as he listened to a reading from Luther's preface to Romans at a meeting in Aldersgate Street. This experience made him an evangelist. "Then it pleased God," he declared, "to kindle a fire which I trust shall never be extinguished."

Shortly after his conversion, Wesley visited the Moravian settlement at Herrnhut and met Count Zinzendorf.* He returned to England and embarked on his life-work. His objective was clear. He set out "to reform the nation, particularly the Church, and to spread Scriptural holiness over the land." He declared that he had only "one point of view—to promote so far as I am able vital, practical religion; and by the grace of God to beget, preserve, and increase the life of God in the souls of men." Wesley knew himself to be an apostolic man, sent by God with an extraordinary commission to evangelize Great Britain.

In April 1739 he took to open-air preaching at the instigation of Whitefield. It was at Kingswood, Bristol, that he ventured on "this strange way of preaching in the fields," as he described it. But the most effective medium for reaching the masses had been discovered, and Wesley was to exploit it for the rest of his itinerancy. It gave him a flexibility which could have been acquired in no other way and brought him face to face with the common people who heard him gladly. The churches were increasingly reluctant to welcome him on account of his evangelical doctrine, and henceforward his preaching was largely extramural.

To conserve the gains of evangelism, Wesley formed societies in the wake of his missions. The organization of Methodism was thus a direct outcome of his success in preaching the Gospel. London, Bristol, and Newcastle-upon-Tyne represented the three points of a triangle so far as his itineration in England was concerned. He soon extended his journeys to include Ireland and Scotland. Wales was left to Howel Harris.* Although Wesley himself did not again visit North America, he sent preachers there and in 1784 ordained Thomas Coke* to superintend the work. Wesley's own account of his mission to the nation and beyond is contained in his now classic *Journal.* His other published writings consist of sermons, letters, expositions, treatises, tracts, translations, histories, and abridgments.

BIBLIOGRAPHY: *Explanatory Notes upon the New Testament* (1754); T. Jackson (ed.), *Works* (3rd ed., 14 vols., 1829-31); L. Tyerman, *The Life and Times of the Rev. John Wesley* (3 vols., 1870-71); J. Telford, *The Life of John Wesley* (1899); N. Curnock (ed.), *Journal* (8 vols., 1909-16); E.H. Sugden (ed.), *Sermons* (2 vols., 1921); J. Telford (ed.), *Letters* (8 vols., 1931); M.L. Edwards, *John Wesley and the Eighteenth Century* (1933); G.C. Cell, *The Rediscovery of John Wesley* (1935); M. Piette, *John Wesley in the Evolution of Protestantism* (ET 1937); T.W. Herbert, *John Wesley as Editor and Author* (1940); W.R. Cannon, *The Theol-*

ogy of John Wesley (1946); R.W. Burtner and R.E. Chiles (eds.), *A Compend of Wesley's Theology* (1954); W.L. Doughty, *John Wesley: Preacher* (1955); C.W. Williams, *John Wesley's Theology Today* (1960); V.H.H. Green, *The Young Mr. Wesley* (1961); M. Schmidt, *John Wesley: A Theological Biography,* vol. I (ET 1962); A.C. Outler (ed.), *John Wesley* (1964); A.S. Wood, *The Burning Heart* (1967). A. SKEVINGTON WOOD

WESLEY, SAMUEL SEBASTIAN (1810-1876). English composer. This influential cathedral organist and composer of anthems was the natural son of Samuel Wesley (1766-1837), the highly talented but somewhat unstable son of Charles Wesley.* Anglican prejudice against the Wesleys hampered him in his earlier years; he was successively organist at Hereford, Exeter, Winchester, and Gloucester cathedrals. Like Bach, he suffered from inadequate forces to carry out his ideals and was angered by official indifference. In 1849 he published a scathing monograph, "A Few Words on Cathedral Music and the Musical System of the Church, with a Plan of Reform" (rep. 1961). How much Wesley's influence was felt in later reform is open to argument. He did, however, write some fine anthems, certainly the best of his generation. "The Wilderness," with its effective choruses and beautiful bass solo is one of the classics of its genre. "Blessed be the God and Father" has still much appeal, and his Cathedral Service in E is a dignified and worthy example from a period when dullness and mediocrity were the rule. His well-known tune *Aurelia,* heard with "The Church's one foundation," was originally written for "Jerusalem, the golden." J.B. MAC MILLAN

WESLEYAN METHODISTS, see METHODIST CHURCHES

WESSEL, JOHANN (Johann Wessel of Gansfort) (1419-1489). Biblical humanist. Born at Groningen in the N Low Countries, and educated at Deventer under the Brethren of the Common Life,* he went on to Cologne, Louvain, and Paris. He taught at Heidelberg and then at Paris. An able scholar, he knew both Greek and Hebrew, which was unusual at the time. His pupils included Reuchlin and Agricola. At first a Thomist, Wessel turned to Augustinianism, and added Ockham's Nominalism. His attempts to combine Nominalism* and mysticism* earned him the nickname of "Master of Contradictions." Around 1474, in his mid-fifties, he returned to Groningen, where he directed a nuns' cloister, and talked of spiritual matters with a warm circle of friends (including David of Burgundy, bishop of Utrecht). He did not write extensively, exercising influence mostly through teaching.

In some ways he can be regarded as a forerunner of the Reformation (Luther, in 1521, edited some of his writings). He opposed superstition, clerical abuses, papal and conciliar infallibility. Man is forgiven because grace enables him to repent: justification is by faith (which must express itself as love), at least in a sense: Christ is love personified and lifts man to the divine likeness: Christ is present at the Eucharist, and transubstantiation does take place, but in a sense He is present only to believers. Wessel's theology was not notable for clarity. His writings were placed on the Index in the 1500s.

See E.W. Miller and J.W. Scudder, *Wessel Gansfort* (2 vols., 1917). DIRK JELLEMA

WESSENBERG, IGNAZ HEINRICH KARL VON (1774-1860). Radical Roman Catholic churchman. He was born in Dresden; his theological studies took him to Dillingen for lectures by Sailer, to Würzburg, and to Vienna. From Austria he went to Augsburg where he was recruited by Bishop Dalberg for the diocese of Constance. There he occupied posts as vicar-general (1802), priest (1812), and administrator (1814). Upon Dalberg's death in 1817 he was chosen unanimously by the cathedral chapter for the vacancy. While the grand duke of Baden supported his nomination, the Roman Curia refused to approve it. Von Wessenberg served as bishop-elect until 1827, when he retired to private life. The reasons for curial rejection lie in Von Wessenberg's liberal policies. These included the expansion of education for priests, more elementary schools, the conversion of monasteries into hospitals and schools, the suspension of clerical celibacy, permission for mixed marriage with Protestants, and the vernacular Mass. WAYNE DETZLER

WEST, JOHN (1775-1845). Church of England clergyman. He was born in Farnham, Sussex, England; some authorities give 1778 as the year of his birth. After studies at Oxford he was ordained priest in the Church of England in 1804, was appointed chaplain to the Hudson Bay Company in 1819, but did not arrive in the Red River Settlement until 1820. His three years under the auspices of the Church Missionary Society were filled with extensive travels on foot, by canoe, etc. A service of worship according to the rites of the Church of England, held in 1820 on the shores of the Hudson Bay, is the beginning of regular acts of worship in the area. West's letters, published by the CMS, and the substance of a journal of his missionary activities established his reputation as a pioneer churchman. During 1825-26 he traveled once again on Canadian soil on behalf of the New England Company to investigate the educational system in Nova Scotia and elsewhere. An account of the mission to the Mohawks was published in 1827. Little is known of the latter part of his life. EDWARD J. FURCHA

WEST AFRICA. This article covers countries south of the Sahara from Senegal to Congo People's Republic (the former French Congo). Countries are referred to by their current names.

Early Christian activity was mainly Roman Catholic and Portuguese. The kings of Portugal received from the papacy by the *ius patronatus* powers which included a commission to evangelize. A Wolof (Senegal) chief was baptized in 1489; Sao Tomé became a suffragan bishopric in 1539 and a diocesan bishopric in 1584. A Christian king ruled in Benin from 1550, and in Sierra Leone, King Farama III was converted in 1604.

A German Lutheran named Joachim Dannenfeldt was sent to the Gambia in 1654/55 as chaplain and missionary. W.J. Müller combined the same functions at Fort Frederiksborg (Ghana) from 1661 to 1669. The Anglican Thomas Thompson worked in Cape Coast (Ghana) from 1752 to 1756. His Ghanaian pupil Philip Quaque was ordained in 1765 as the first non-European Anglican priest. Scripture selections were published in Fante (Ghana) in 1764.

In the 1790s Sierra Leone became a home for freed slaves from England and Nova Scotia. Many of these were Christian. Early missionary work from 1795 by Baptists and others was unsuccessful, but more enduring work was begun in 1806 by German Lutheran missionaries of the (Anglican) Church Missionary Society (CMS), who were joined in 1811 by British Wesleyan Methodist missionaries. Sierra Leone long remained an important center for Christian expansion throughout West Africa.

Episcopalian ministers accompanied the first American settlers to Liberia in 1820-21. Baptist missionaries followed in 1822, and Methodists in 1833. In 1821 a British Methodist missionary arrived in the Gambia. In Ghana, the Basel Missionary Society began operations at Christiansborg (Accra) in 1828, and abortively at Kumasi in 1839. In 1834 a British sea captain found groups of Africans meeting for Bible study at Cape Coast, and addressed an appeal to British Methodists which they answered in the same year. Most of the early missionaries soon died or were invalided home, but there were notable exceptions, such as the British Methodist Thomas B. Freeman,* who lived in Ghana almost uninterruptedly from 1838 to 1890.

In the early 1840s, former slaves who had become Christians in Sierra Leone returned to their homes in Nigeria and elsewhere and appealed for missionaries. Anglican and Methodist work began in W Nigeria in 1841. British Baptists set up a mission in Fernando Po in 1841, and in Cameroun in 1844. Scottish Presbyterians began in 1846 to work around Calabar, where Mary Slessor* was to do outstanding work from 1876 to 1915. West Indian missionaries played a significant role in this area. American Presbyterians reached Gabon in 1842 and Rio Muni in 1864, later extending their activities into Cameroun. In 1866, Ga (Ghana) became the first West African language to have a translation of the whole Bible.

Throughout the nineteenth century, Roman Catholic missions were predominantly French. The Holy Ghost Fathers reached Gabon in 1844, followed by Roman Catholic sisters in 1849, by which time J.R. Bessieux of Gabon was responsible as vicar apostolic for Roman Catholic work throughout West Africa. This responsibility was divided in 1863. The *Société des Missions Africaines de Lyon* opened stations in Dahomey (1861), Nigeria (1868), Ghana (1879), and Liberia (1884).

African church leadership developed to a limited extent during the nineteenth century, and African lay initiative was often prominent. Three Senegalese Roman Catholic priests were ordained in 1840, but their number did not increase rapidly. The Anglican Samuel Ajayi Crowther* was in 1864 consecrated bishop "on the Niger," and in 1874 the Sierra Leonean Charles Taylor became chairman of the Sierra Leone Methodist district.

The training of African ministers formed part of a wider educational program, of which Fourah Bay College in Sierra Leone (founded 1827, affiliated to Durham University 1876) was an outstanding but not an isolated example. Christian education was generally practical as well as academic. The Holy Ghost Fathers developed handicrafts and agriculture, and the Basel Mission formed a Mission Trade Society in 1859.

Opposition to Christianity broke out violently from time to time. Baptists had to withdraw from Fernando Po in 1858, and Jesuits in 1870. All missionaries were expelled from Abeokuta in 1867. Baptist work in Douala (Cameroun) was destroyed in 1885.

With the Berlin Conference of 1884-85, the colonial "scramble for Africa" began in earnest, with mixed effects on the life of the churches. In 1886 the Basel Mission took over British Baptist work in Cameroun, and German Pallotines and Baptists entered the country four years later. French missions in British territories, and British missions in French, suffered certain disadvantages, and tended to co-opt, especially for educational work, missionaries of the same nationality as the colonial power. A painful reorganization, under missionary control, of Crowther's Niger Mission took place in 1890. Divisions within the church had occurred in Sierra Leone from the 1790s, but it was not until much later that the first important independent churches came into being (Native Baptist Church, 1888; United Native African Church, 1891—both in Nigeria).

During the period 1890-1945, the expansion of Christianity generally gained momentum, despite two world wars and the economic depression of 1929-31. Missions already active extended their work. In 1896 the Basel Mission finally established a center at Kumasi. Two years before, a Roman Catholic mission had entered the Central African Republic. In 1900 the CMS entered the largely Muslim area of N Nigeria; American Presbyterian activity increased in S Cameroun; N Ghana was entered by British Methodists in 1911.

New missions arrived, such as the Qua Iboe Mission from Northern Ireland (E Nigeria, 1897), the Sudan Interior Mission (N Nigeria, 1893), the Sudan United Mission (N Nigeria, 1904), the Africa Inland Mission (Chad, 1909), the Christian and Missionary Alliance (French-speaking Guinea, 1918), the Brethren Church of the United States (Brazzaville, 1918), the Assemblies of God (Upper Volta, 1921), and the World Evangelization Crusade (Ivory Coast, 1935). Many of these missions were conservative in theology and American in origin. New Roman Catholic orders entered the area (e.g., Sittard missionaries of the Heart of Jesus, Cameroun, 1911; Capuchins, Chad, 1929; St. Patrick's Society for Foreign Missions, Calabar, 1932).

Outstanding missionaries (the best-known being Albert Schweitzer,* Gabon, 1913-63) continued to take the lead in most churches, but Africans assumed increasing responsibility. The growth of Christianity was stimulated by a number of African prophets, such as Garrick Braid (Nigeria, from 1909), W.W. Harris (Ivory Coast, 1913-15), Sampson Opong (Ghana, c.1920), and Joseph Babalola, who in 1928 began in Nigeria a ministry which led to the founding of the Christ Apostolic Church and the Church of the Lord (Aladura) in Nigeria and Sierra Leone. Most of these movements led both to the extension of mission work and were one source independent churches now a major factor in West African Christainity.

The encyclical *Rerum Ecclesiae* (1926) crystallized Roman Catholic thinking on the indigenous priesthood. The first African Roman Catholic assistant bishop was consecrated in Sierra Leone in 1937.

World War I caused the repatriation or internment of German missionaries, particularly in Togo and Cameroun. The International Missionary Council* under its secretary J.H. Oldham* was largely responsible for organizing help for such "orphaned missions" when peace returned. Cooperation among missions, and between missions and government, increased, especially in education and medicine. An Eastern Nigerian Missionary Conference was held in 1911. Fourah Bay became an interdenominational college in 1919. Similar cooperative institutions were established at Kumasi, Lagos, Bunumbu (Sierra Leone), and elsewhere. The formulation of educational policy was advanced by a survey undertaken in 1920-21 by the Phelps-Stokes Commission, on the initiative of American missionary societies.

After 1945, churches developed (often before political independence) toward full autonomy under African leadership. An African was appointed chairman of the GHana Methodist district in 1948. The first African Anglican diocesan bishops since Crowther were consecrated in 1951, within an autonomous province. African Roman Catholic diocesan bishops were consecrated in Cameroun (1955), Dahomey (1957), and elsewhere; an African archbishop in Dahomey (1960) and the first West African cardinal (Paul Zougrana of Upper Volta) in 1965.

Interchurch cooperation became the norm in such areas of the churches' life as theological education, Christian literature, agriculture, and medical work. The All Africa Conference of Churches was inaugurated at Ibadan in 1958; the more conservative Association of Evangelicals of Africa and Madagascar was founded in 1966. No major church union, however, had taken place by 1971.

BIBLIOGRAPHY: K.S. Latourette, *A History of the Expansion of Christianity,* vols. 3-5 (1939-45); C.P. Groves, *The Planting of Christianity in Africa* (4 vols., 1948-58); A.F. Walls (ed.), *Bibliography of the Society for African Church History* (1967-); E. Dammann, *Das Christentum in Afrika* (1968). PAUL ELLINGWORTH

WEST INDIES. Christianity in the West Indies may be divided into two phases, "colonial" and "evangelical." The colonial church was that which came with the Europeans and was intended for them, although it could, and ultimately did, include Indians and Africans. Evangelical Christianity, dating from the eighteenth century, was a true mission, intended for everyone, but primarily the slaves.

European colonists of whatever nationality brought their faith with them and rapidly established churches in the islands they colonized. The Spanish had a fully developed hierarchy by 1522, but neither the British nor the French were so well organized. As first-comers, the Spanish had to cope with the Indians, and men like Las Casas carried on a running fight with the colonists over the humanity of the Indians, and therefore their right to decent treatment, winning significant improvements in the law, but less in practice. Officially the church came around to a constructive attitude; at a synod in 1622 elaborate regulations governing the treatment of Indians were promulgated. As African labor replaced Indian, steps were taken to incorporate the newcomers into Christianity as quickly as possible, although with instruction. The priesthood was largely provided by European regulars, although some Creole whites were included. The quality varied, and there were never enough, for the islands soon became a backwater, and traditional Christianity in some places was seriously altered by African imports, as can be seen in the syncretistic Voodoo cult in Haiti, perhaps the clearest example. Nevertheless the Catholic Church in Spanish and French islands has proved very resistant to Protestant infiltration.

The British made no attempt to convert the slaves, considering the project at best quixotic and at worst dangerous. The only serious attempt was that of the Society for the Propagation of the Gospel on its estates at Codrington, Barbados. British settlers accepted only as much of the church as they wanted, and that included neither evangelists nor episcopate. The real growth of the Anglican Church thus begins in 1825, with the arrival of bishops.

The earliest evangelicals were the Moravians, arriving in the Danish Virgins in 1732, and extending to a number of islands by 1800. Between 1786 and 1790 Thomas Coke* established the Methodists widely throughout the islands, and in 1782 a group of black American exiles opened a Baptist church in Jamaica. The black Baptists developed a number of African variations on evangelical Christianity, both in Trinidad and Jamaica, in the middle of the nineteenth century, but in the last hundred years this tradition has tended toward orthodox Pentecostalism.

Other missions entered early in the nineteenth century. Between 1800 and 1845 church growth was rapid, but those islands with a functioning establishment—Barbados (Anglican), most Windwards, and Trinidad (Catholic)—resisted the evangelicals. The Spanish and French islands were not attempted, except for a Methodist bridgehead in Haiti. By 1800 most evangelicals favored abolishing slavery, but pressure from the planters required them to keep their opinions to themselves. They were primarily interested in salvation from sin, and only when that object was threatened, as in Jamaica after the revolt of 1831, did even radical evangelicals like the Jamaica Baptists openly declare their views.

Emancipation came in the middle of a period of rapid growth and involved the churches heavily in education. Except for Trinidad and Guyana, which were still expanding ecclesiastically and economically, most churches stalled in the mid-forties, and advance was not resumed until after 1870. Although the church was largely white-led, the local ministry became more significant, educational and institutional development was marked, especially among the Anglicans who successfully adjusted to disestablishment after 1870. Serious missionary work was undertaken, both among West Indian migrants in Central America, and in Africa. Most significant of all was the growth of the church among the East Indian sugar workers, although the most important group in this field were the Canadian Presbyterians, who specialized in the problem after the arrival of John Morton in Trinidad in 1868. Their influence extended beyond Trinidad to Guyana, Grenada, St Lucia, and Jamaica. The Catholic Church shared in the general revival of Catholicism in the period, except in the Spanish islands, where its development was hindered by various political difficulties.

The twentieth century has been marked by the rapid growth of Pentecostalism and Holiness groups, mainly from the United States, which merged in Jamaica and Trinidad with the native tradition. The older churches have not grown much, but since World War II an indigenous theological movement has appeared, concentrating on the related problems of nationalism, identity, race, class and poverty.

BIBLIOGRAPHY: P. Duncan, *A Narrative of the Wesleyan Mission to Jamaica* (1849); G. Blyth, *Reminiscences of a Missionary Life* (1851); J. Buchner, *The Moravians in Jamaica* (1854); E.B. Underhill, *The West Indies, Their Social and Religious Condition* (1862) and *Life of James Mursell Phillippo* (1881); A. Caldecott, *The Church in the West Indies* (1898); J.B. Ellis, *The Diocese of Jamaica* (1913); G.G. Findley and W.W. Holdsworth, *The History of the Wesleyan Methodist Missionary Society*, vol. II (1921); J. Bennett, *Bondsman and Bishop* (1958); J.L. Gonzalez, *The Development of Christianity in the Latin Caribbean* (1969). GEOFFREY JOHNSTON

WESTCOTT, BROOKE FOSS (1825-1901). Bishop of Durham. He attended King Edward's School, Birmingham, where he was much influenced by the headmaster, James Prince Lee. In 1844 he went to Trinity College, Cambridge, where he became a fellow in 1849. His pupils included his old schoolfellows J.B. Lightfoot* and E.W. Benson,* and also F.J.A. Hort.* He was ordained in 1851 and the following year went to teach at Harrow School. In 1869 he was appointed a residentiary canon of Peterborough and the next year, at Lightfoot's instigation, was recalled to Cambridge as regius professor of divinity. He tidied up the courses and syllabi and himself lectured for the first three years on early church history and then for five years mainly on Christian doctrine. Thereafter he took books or selected passages of the NT. He was active in university administration and also in pastoral concern. He was prominent in the formation of the Cambridge Mission to Delhi and the founding of the Cambridge Clergy Training School (later known as "Westcott House").

In 1890, at the age of sixty-six, he was appointed to succeed Lightfoot as bishop of Durham. He did not have to face problems of reorganization similar to those facing Lightfoot, and he built upon his predecessor's work, particularly with the ordination candidates at Auckland Castle. He showed a deep concern for the social and industrial problems of the diocese and held conferences at Auckland for representatives of both sides of industry and of social work. He often addressed the miners, and in 1892 he helped to settle a coal strike.

Westcott published a considerable number of books, but he is best remembered for his work with Hort in establishing the text of the NT (1881) by making a scientific evaluation of the vast mass of manuscript evidence which had become available, and for his NT commentaries. It was intended that Lightfoot, Westcott, and Hort should between them write a complete commentary on the NT. Westcott was to undertake the Johannine literature and Hebrews, and he completed his share (apart from the Book of Revelation) with definitive volumes on John's gospel (1881), the epistles of John (1883), and the epistle to the Hebrews (1889). Westcott's knowledge of the patristic commentaries was unrivaled, and if at times he was oversubtle, his exegesis and exposition were always marked by great theological and spiritual depth. His theological position combined the learned historical conservatism of Lightfoot with the incarnational approach to social problems of F.D. Maurice,* whose works he avoided reading for fear of losing his originality.

See A. Westcott, *Life and Letters of Brooke Foss Westcott* (2 vols., 1903). R.E. NIXON

WESTERN TEXT. The name given by Westcott* and Hort* to a type of text of the Greek NT which had special affinities with the West. Its chief representatives are Codex Bezae (D) for the gospels and Acts, and Codex Claromontanus (Dp) for the Epistles (in both of which the text is written in Greek and Latin), the Old Latin version, and the Curetonian Syriac. Most of the Latin Fathers—including Marcion, Tatian, Justin, Irenaeus, Hippolytus, Tertullian, and Cyprian—made use of the Western form of the text in their quotations. The date of origin of the Western Text is thought consequently to have been as early as the middle of the second century. The characteristics which Westcott and Hort found in it included an apparent freedom to change things in order to bring out the meaning better. This might involve the omission or insertion of words, clauses, or even whole sentences. They also found a tendency to assimilate words and phrases found close to each other and, more seriously, through a process of harmonization to obliterate differences in similar or parallel passages.

This process of harmonization is of course most readily found in the gospels, where it was particularly easy through carelessness or particularly tempting in the interests of consistency to make passages conform to each other. It was in the writings of Luke, particularly in Acts, that the Western Text was found to diverge most from the other types. The Western Text may not have been a deliberate recension, as were the other text types. Westcott and Hort had a very low view of its value except in the passages which they called rather cumbersomely "Western non-interpolations," i.e., passages where the Neutral Text* had interpolations not found in the Western Text. Nowadays much less reliance is placed on the *a priori* likelihood of a particular text type being right or wrong, and readings from various sources are assessed on their merits.

BIBLIOGRAPHY: B.F. Westcott and F.J.A. Hort, *The New Testament in the Original Greek* (1881); A. Souter, *Text and Canon of the New Testament* (2nd ed., 1954); B.M. Metzger, *The Text of the New Testament, Its Transmission, Corruption and Restoration* (2nd ed., 1968). R.E. NIXON

WESTMINSTER ABBEY. According to legend, the abbey was founded in 616, and certainly existed by 785. Refounded by Edward the Confessor* as an abbey of Benedictine monks in 1050 with extensive property, its new abbey church was consecrated in 1065. The following year Edward was buried there, his canonization in 1161 subsequently attracting large numbers of pilgrims. Adjacent to the royal palace of Westminster, it held a central place in national life, with special privileges of sanctuary, and for three centuries the House of Commons met in its chapter house. Kings were traditionally crowned here, and from 1296 the Stone of Scone, the Scottish coronation stone, was incorporated into the throne. The present building was begun in 1245 and Henry VII's chapel was completed in 1519. In 1540 the abbey was dissolved and reorganized with a dean and twelve prebendaries. Some of the monastic buildings were used for the new Westminster Grammar School. Thomas Thirlby was bishop of Westminster in 1540-50, but the see was then suppressed. The west front and towers were designed by Christopher Wren* and Nicholas Hawksmoor and completed, 1740-50. From the eighteenth century the abbey has been the burial place for numerous national celebrities.

BIBLIOGRAPHY: J.P. Neale, *History and Antiquities of Westminster* (1818); A.P. Stanley, *Historic Memorials of Westminster Abbey* (1868); A. Fox, *Westminster Abbey* (1951). JOYCE HORN

WESTMINSTER ASSEMBLY (1643). During the English Civil War between Charles I and Parliament, the latter continued its program of reforms and declared its intention of establishing a church government that would be "more agreeable to God's Word and bring the Church of England into a nearer conformity with the Church of Scotland and other Reformed Churches abroad." To implement this design, Parliament convened "an Assembly of learned, godly and judicious Divines to consult and advise of such matters and things as should be proposed unto them...." The Assembly consisted of 121 divines with ten lords and twenty commoners as assessors with equal debating and voting rights. The Church of Scotland was asked to send commissioners and appointed four ministers and two elders. The Assembly was representative of very different viewpoints in matters of church government.

Sessions were held from 1 July 1643 to 22 February 1649. The Assembly was not a church court, and possessed no ecclesiastical authority. It was simply a council summoned by Parliament to give advice and guidance to the civil authorities for the promotion of unity and uniformity in the work of Reformation. Average daily attendance ranged between sixty and eighty members, though only about twenty took a leading part in all the debates. The Westminster Assembly was early associated with the Solemn League and Covenant,* approved the document, and was joined at its meeting place, St. Margaret's Church, Westminster, by both houses of Parliament for a formal swearing of the Covenant.

Main work of the Assembly was the preparation of the Westminster Confession* of Faith, the Larger and Shorter Catechisms, the Form of Church Government, and the Directory for Public Worship.* Rouse's metrical version of the Psalter was examined and approved for general use in the public worship of the church. The Westminster Standards were adopted by the Church of Scotland by a special act in 1647, and with minor adjustments became the subordinate standards of Presbyterian Churches throughout the English-speaking world. Some of these have in recent years relegated the standards to "historic document" status.

BIBLIOGRAPHY: A.F. Mitchell and J. Struthers, *Minutes of the Sessions of the Westminster Assembly of Divines* (1890); B.B. Warfield, *The Westminster Assembly and Its Work* (1931); S.W. Carruthers, *The Everyday Work of the Westminster Assembly* (1943). ADAM LOUGHRIDGE

WESTMINSTER CATECHISMS, see CATECHISMS

WESTMINSTER CONFESSION. One of the most influential creeds of Calvinism, a creedal standard for all Presbyterian churches, drawn up at Westminster (1643-46). The immediate background to the Confession lies in the tensions between Charles I* and his subjects, growing in large measure out of Charles's insistence on imposing Anglicanism. In an age when it seemed obvious that the state, concerned with the welfare of its citizens, was hence concerned with religious affairs, such a stance had political implications. The Puritans* felt that the creeds of the Church of England* must be revised, so that a pure religion would be taught and preached. The Scots, convinced Calvinists, resisted any attempt to remodel their creeds. In 1638 the historic National Covenant* affirmed this, and a Scots invasion of N England forced Charles to call Parliament into session. But it demanded far-reaching concessions from Charles which he refused, and by 1642 civil war had broken out.

In this context, as part of parliamentary efforts at reform, an assembly was called to meet at Westminster to formulate a creed suitable for the English and Scottish churches (1643). Civil strife continued as the assembly met. Dominated by Puritan Calvinists, with only a few Puritan "Independent" delegates, the assembly also included Scots Calvinists (from England, 121 clergy, 30 laymen; from Scotland, 4 clergy, 2 laymen; some 35 of the delegates did not appear due to the civil strife). Meeting for three years (1643-46), the delegates had little difficulty in agreeing on doctrine (two-thirds of the Confession), but the chapters on church and state took somewhat longer to draw up. The creed is a systematic exposition of orthodox Calvinism, in scholastic formulation. The sovereignty of God is stressed, and election to salvation emphasized. Questions disputed among Calvinists (notably Supralapsarianism) were avoided. Adopted in England and Scotland, the Confession stayed on as a creedal standard in the Presbyterian Church of Scotland.

In England, Cromwell rose to power, the king was executed (1649), and the Commonwealth set up. Resting on the power of the army, where Puritan "Independents" were strong, it granted religious toleration to all Protestants. Cromwell had to conquer Scotland by force (the Scots, though Calvinist, supported their royal family), and in Ireland, Catholicism was temporarily driven underground. The Commonwealth lasted but a decade. Charles II* became King (1660), and the Anglican Church again became the established church in England. Scotland retained its established Presbyterian Church.

See G.S. Hendry, *The Westminster Confession for Today* (1960). DIRK JELLEMA

WESTON, FRANK (1871-1924). Bishop of Zanzibar. Brought up as an Evangelical, he early became through school and university an extreme Anglo-Catholic. Graduating with a first in theology at Oxford, he served curacies in London before going to Africa under the auspices of Universities' Mission to Central Africa. Initially he was concerned in educational work and in the training of ordinands. In 1908 he was made bishop of Zanzibar, in which country he served until his death with zeal and a love for souls, but with all the strength and weaknesses of a nature akin to fanaticism.

He is best remembered for his opposition to the tentative scheme of reunion proposed at Kikuyu for the Protestant churches in East Africa.* In 1920, however, he was a strong supporter of the Appeal on Reunion issued by the Lambeth Conference, a support which was not wholly reechoed by all the Anglo-Catholic party. He was also implacably opposed to the liberal tendencies of his day, and when the bishop of Hereford (H.H. Henson*) made B.H. Streeter* a canon he excommunicated the former. It was this controversy which led him to write his best book *The One Christ* which, while opposing the prevalent kenotic theories, suggested a milder form which would preserve belief in our Lord as an infallible teacher. Like many of his contemporaries, he was from his undergraduate days interested in social affairs, an interest which bore fruit in his strong opposition to forced labor in East Africa.

See H.M. Smith, *Frank, Bishop of Zanzibar 1871-1924* (1926). PETER S. DAWES

WESTPHALIA, PEACE OF (1648). Collective term for the decisive treaties concluding the Thirty Years' War.* Discussions began in two Westphalian towns in 1643—at Münster with France, and at Osnabrück with Sweden. Little progress was made until in January 1648 Spain unexpectedly made peace with the Dutch (granting them *de iure* independence), whereupon the emperor negotiated settlements with France and Sweden. The settlements, based upon the principle of the sovereignty and independence of individual states, mark the practical end of the Holy Roman Empire and of the medieval age wherein religion and the concept of *respublica christiana* had dominated. They represented major triumphs for France and Sweden, opening the way for France to dominate completely European affairs for nearly two centuries. Germany was doomed to decentralized impotency for two centuries as some 343 separate sovereign states were confirmed within her borders. Religion was determined on the principle *cuius regio, eius religio* among Catholic, Lutheran, and Calvinist.

BRIAN G. ARMSTRONG

WETTSTEIN (Wetstein), JOHANN JAKOB (1693-1754). Swiss NT scholar. Born at Basle, he wrote a dissertation on variant NT readings (1713). He traveled in Switzerland, France, and England in search of manuscripts, meeting Richard Bentley (1716) who encouraged him in his textual studies. Made a deacon in Basle (1717), he became his father's assistant in St. Leonard's (1720). He devoted himself to NT study, and his rejection of the reading *theos* for *hos* in the Textus Receptus of 1 Timothy 3:16 led to his deposition in 1730 for alleged Socinianism.* Having answered his opponents (1732), he became professor of church history in the Remonstrants' college at Amsterdam (1733), remaining there till death. He published the prolegomena to his new edition of the Greek NT in 1730, the edition itself appearing in two volumes at Amsterdam, 1751-52. This was notable for its large collection of variants and the introduction of the *sigla* for denoting manuscripts still in common use.

J.G.G. NORMAN

WEYDEN, ROGIER VAN DER (1399-1464). Flemish painter. He was among the first to use oil paint, capitalizing upon its rich, brilliant color and shaded hues potential to give traditional biblical topics the appearance of veritable, three-dimensional reality and ordinary life settings simultaneously the luster of heightened, sanctified meanings. Rogier used less gold and worked with shadows on people's faces, complex folds in their robes, detailed attention to postural nuances, all of which focused sculpturally on the grief and sadness or interior emotions felt by the personages in the paintings. Rogier tuned the exacting, atmospherically real depiction of things pioneered by the van Eyck* brothers on a cosmic scale down to a disclosure, very warmly done, of human sensitivities. This double emphasis—oil detail and emotional intensity—of the Flemish master practically set the standard for painting on both sides of the Alps during the last half of the fifteenth century. CALVIN SEERVELD

WEYMOUTH NEW TESTAMENT. *The New Testament in Modern Speech,* subtitled "An idiomatic translation into everyday English from the text of the *Resultant Greek Testament,*" was the work of Richard Francis Weymouth (1822-1902). Weymouth was a fellow of University College, London, and at one time headmaster of Mill Hill School. He was a classical scholar, and his *Resultant Greek Testament* was an edition based on the greatest measure of agreement between the leading nineteenth-century editors. His translation was not intended to supplant the versions then in general use, but to act as a compressed running commentary on them. He expressed the hope that someday there might be a new translation of the Bible which would supersede the King James Version and the Revised Version. His own intention was to be free from doctrinal and ecclesiastical bias. He includes brief introductions to the various NT books and a fair number of linguistic footnotes. The translation was not published until 1903, the year after he died. In 1924 the fourth edition was published after the original had been revised by several scholars. R.E. NIXON

WHARTON, HENRY (1664-1695). Anglican scholar. Born at his father's rectory of Worstead in Norfolk and educated by him before entering Caius College, Cambridge, he studied diligently there under Isaac Newton and others. From 1686 he contributed very considerably—and without much acknowledgment—to W. Cave's *Historia Literaria.* Ordained under the canonical age because of his learning, Wharton worked first for T. Tenison,* producing treatises on celibacy and on the Scriptures, and for W. Sancroft.* In 1689 he took the oaths, separating from his Nonjuror* friends and ending his ecclesiastical hopes. The first two volumes of *Anglia Sacra* appeared in 1691—a masterly survey of English bishops and dioceses with a collection of relevant original texts. A third posthumous volume is unequal in quality. G.S.R. COX

WHARTON, PHILIP (Fourth Baron Wharton) (1613-1696). Philanthropist and friend to Nonconformist ministers. A member of both the Long Parliament and Westminster Assembly, as well as a soldier in the Civil War, he was a friend of Oliver Cromwell* but took little part in national affairs from 1649-60. After the Restoration and during the period of the persecution of Nonconformists (1660-89), he did what he could in Parliament to oppose repressive legislation—e.g., he opposed the second Conventicle Act* of 1670. At his home at Woburn he entertained Nonconformist leaders and helped them financially. By a deed made in 1662 he settled some of his lands at Healaugh, Yorkshire, upon trustees for 1,050 Bibles and Catechisms to be given to poor children.

PETER TOON

WHATELEY, RICHARD (1786-1863). Archbishop of Dublin. Born in London, he graduated from Oriel College, Oxford, in 1808 and was appointed a fellow in 1811. Fellow-students included Sir Robert Peel, John Keble, and John Henry Newman. From 1822 he spent four years as a parish minister on the borders of Norfolk, and for five years he served as president of St. Alban's Hall. In 1831 amid controversy he was appointed archbishop of Dublin. Though gifted with originality of thought, he was a poor preacher but showed his gifts as an essayist and satirist, a leader in education at a time when four university colleges were established in Ireland, and a stern disciplinarian who made many enemies. He was instrumental in establishing a chair of political economy in Trinity College, Dublin, and did much to raise the standard of theological education. He published about sixty volumes of essays and sermons. ADAM LOUGHRIDGE

WHEATON DECLARATION. A statement adopted by the Congress on the Church's Worldwide Mission at Wheaton (Illinois) College in 1966. Called by the Interdenominational Foreign Mission Association* and the Evangelical Foreign Missions Association,* 938 delegates from seventy-one countries representing over 250 groups registered for the conclave. Major study papers were written in advance on the relation of mission to ten specific problem areas: syncretism, neo-universalism, proselytism, neo-Romanism, church growth, foreign missions, evangelical unity, evaluating methods, social concern, and a hostile world. After extensive discussion by small groups, a final drafting committee, which drew upon their conclusions, prepared the Wheaton Declaration. It was adopted unanimously by the delegates as the collective opinion of the congress. The document dealt with the above issues, but clearly asserted biblical authority, proclamation of the Gospel, and social action as evangelicals.

RICHARD V. PIERARD

WHEELOCK, ELEAZAR (1711-1779). Congregational minister; founder and first president of Dartmouth. Graduating from Yale in 1733, he became two years later pastor of the Second or North Parish in Lebanon, Connecticut. While at Yale he was a member of a group similar to the "Holy Club"* at Oxford of which the Wesleys were members. During the first year of his pastorate a revival broke out in Wheelock's church under the influence of Jonathan Edwards'* ministry at Northampton, Massachusetts. During the Great Awakening,* Wheelock gave himself unstintingly to promotion of revival, preaching in Connecticut, Rhode Island, and Massachusetts. Though accused of encouraging radical separatists, he was a moderate, opposed to both the radicals and the Old Lights. Interested in converting and educating the Indians, he received from Col. Joshua More a gift of a house and schoolhouse at Lebanon to aid in his work. This became known as More's Charity School, which was later moved to Hanover, New Hampshire, under a 1769 charter and renamed Dartmouth College.

HOWARD F. VOS

WHICHCOTE, BENJAMIN (1609-1683). Cambridge Platonist.* Born in Shropshire and educated at Cambridge where he eventually became provost of King's College, he lost his post at the Restoration. He was incumbent of St. Lawrence Jewry, London, from 1668 till his death. Whichcote's main works are his *Discourses* and *Moral and Religious Aphorisms.* He was one of the leading members of the group of liberal divines known as the Cambridge Platonists. Indeed, his insistence on the text from Proverbs, "The spirit of man is the candle of the Lord" emphasizes the Platonists' belief in man's reason as the ultimate seat of authority in religion, while the importance for him of Paul's "For God's temple is sacred, and you are that temple" indicates the religious significance of moral behavior for this group. For Whichcote, God was "the Original of Man's being, the centre of his soul, his ultimate end."

ARTHUR POLLARD

WHISTON, WILLIAM (1667-1752). Church historian, mathematician, and translator. Son of the manse, he entered Clare Hall, Cambridge, in 1686, where he studied mathematics, became friendly with Newton, and was appointed a fellow in 1693. In 1698 he became vicar of Lowestoft, but in 1703 returned to Cambridge to succeed Sir Isaac Newton* as Lucasian professor of mathematics, on Newton's own recommendation. Unlike the cautious Newton, however, Whiston tactlessly vented his doubts in public, with the result that after four or five years of legal proceedings he was finally expelled from the university in 1710 on a charge of Arianism.* He suffered intensely as a result and thereafter lived in considerable poverty. Barred from the Anglican Communion, he at first held meetings in his home, but later (1747) joined the Baptists.

He regarded the Reformation as only half-completed: it showed the way back to the church of Augustine's time but not to the NT. He believed that when the works of the anti-Nicene fathers were translated, the way would be opened for a restoration of primitive Christianity, and to this end he worked unceasingly. He held that the miracle-gifts of the Holy Spirit were withdrawn from the church when the papists introduced alleged wonder-working relics, and he sought in his writings to establish the connection historically. Like Newton, he repudiated both the Athanasian Creed* and infant baptism. In his earlier days at least Whiston often seemed to write as Newton's mouthpiece; only on some aspects of prophecy did they differ sharply, but his intellectual level is set at a far lower level than Newton's. He wrote much on science and religion, but many of his ideas are unacceptable today. Of his piety and passionate desire to follow Christ whatever the consequences there can be no doubt; his Arianism, if mistaken, was based only on his understanding of Scripture. Today he is chiefly remembered as the translator of Josephus.*

See his *Memoirs of the Life and Writings of Mr. W ... written by himself* (1733). R.E.D. CLARK

WHITAKER, WILLIAM (1548-1595). Cambridge Puritan theologian. Born in Lancashire and educated in his hometown of Burnley and (with the aid of his uncle, Dean A. Nowell) at St. Paul's School, he then went to Trinity College, Cambridge, where he became a fellow. He excelled at Greek and translated the Book of Common Prayer,* as well as Nowell's *Larger Catechism,* into that language. In 1578 he became a canon of Norwich and two years later regius professor of divinity. Like other Puritans he was totally committed to Protestant principles in opposition to those of Roman Catholicism, and he heartily defended his views against those of R. Bellarmine* and T. Stapleton.* His Puritanism nearly prevented his being appointed master of St. John's in 1586 but once in this post, as also when he was master of Trinity (1593-95), he made sure that Calvinistic orthodoxy was the theology taught in college. He was one of the group of men responsible for the Lambeth Articles* (1595). Most of his twenty or so treatises were written in Latin and enjoyed a wide readership in Europe. PETER TOON

WHITBY, DANIEL (1638-1726). Anglican scholar. An erudite clergyman trained at Oxford, he engaged in several controversies, including an attack on Roman Catholicism, an attempt to gain concessions for Nonconformists so that they would join the Church of England, and a refutation of Calvinism. Among his thirty-nine published works the most famous is a *Paraphrase and Commentary on the New Testament* (2 vols., 1703). This work continued to be used throughout the eighteenth and nineteenth centuries. Its area of great significance was in popularizing Postmillennialism. Whitby held that the world would be converted to Christ, the Jews restored to the Holy Land, and pope and Turks defeated, after which the world would enjoy a time of universal peace, happiness, and righteousness for a thousand years. At the close of this millennium Christ would personally come to earth again and the last judgment would be held. This view was adopted by most of the leading eighteenth-century ministers and commentators. ROBERT G. CLOUSE

WHITBY, SYNOD OF (663/4). An important turning point in the history of the church in England. English Christianity in the seventh century had two main streams. One came from Rome via Augustine of Canterbury and Paulinus, and the other from the Celtic Church via Iona and Lindisfarne. There were a number of differences of ethos and of religious observance between these two streams, the most notable of the latter concerning the date on which Easter* was to be celebrated. The issue came to a head in 663 when King Oswy of Northumbria saw that in the following year he would be celebrating Easter when his wife, who had been brought up in Roman ways, would be observing Lent.

A synod was called at Streanshalch (Whitby) in Yorkshire, the site of Hilda's abbey. The delegates of the Celtic persuasion were King Oswy, who presided; Cedd,* bishop of the East Saxons; Hilda*; and Colman,* bishop of Lindisfarne. The Roman representatives included Oswy's son Alchfrith; Agilbert, bishop of Dorchester; Wilfrid,* abbot of Ripon; and James the Deacon. Colman argued that the Celtic tradition went back through Columba* and Polycarp* to John the Evangelist. Wilfrid pleaded the near-universality of the observance of a tradition going back to Peter and Paul. The king judged in favor of the Roman party on the grounds that he would rather be on good terms with "the keeper of heaven's gate" than with Columba. The decision caused some bitterness among the Celtic party and was influential in bringing England within the mainstream of Christendom (with its administrative advantages and theological dangers) for the next eight and three-quarter centuries. R.E. NIXON

WHITE, A.D., see SIMPSON, J.Y.

WHITE, ELLEN GOULD (1827-1915). Most prominent leader of the Seventh-day Adventist* Church. Born at Gorham, Maine, she received almost no formal education because of poor health. Her parents were devout Methodists, but in the 1840s embraced William Miller's* Advent preaching and were disbarred from the church. Miller's preaching and Mrs. White's testimony of her own revelations formed the beginning of the Seventh-day Adventist Church, which stresses a strong prophetic and eschatological note and health reform. She became the inspired leader and messenger, marrying Elder James White in 1846. In 1855 they moved to Battle Creek, Michigan, where the church headquarters became located. She spent some time in Europe and Australia after her husband's death in 1881. In 1903 the headquarters was moved to Washington, D.C. A woman of deep religious persuasion, she insisted she was not a leader but simply a divinely appointed messenger. Sixty-four of her works have appeared in print in English.

ROBERT C. NEWMAN

WHITE, FRANCIS (c.1564-1638). Bishop of Ely. Educated at Cambridge, he was ordained in 1588 and held livings at Broughton Astley, Leicestershire, and at St. Peter, Cornhill. In 1617 he published *The Orthodox Faith and Way to the Church* against a Roman Catholic treatise entitled *White dyed Black.* He was employed by James I* in 1622 in two disputes against the Jesuit John Fisher,* a report being later published as *The Fisher Catched in his owne Net.* He was made dean of Carlisle (1622) and bishop of that diocese, with some suspicion of simony (1629), and was translated to Norwich (1629) and to Ely (1631). White opposed the Puritan view of the Sabbath, in a conference with Theophilus Brabourne.* In 1635 he published his *Treatise of the Sabbath Day,* written at the command of Charles I* and dedicated to William Laud,* and in 1637 *A Brief Answer to a late Treatise of the Sabbath Day.*

JOYCE HORN

WHITE, JOHN (1866-1933). Wesleyan missionary. Born in Cumberland, he served in Southern Rhodesia (1894-1931), and as chairman of the district from 1903 supervised a growing work south and north of the Zambesi. He was the founder of Waddilove Institution and governor for two periods, a founder and chairman of the Southern Rhodesia Missionary Conference, and the translator of the Shona NT. White laid equal stress upon personal salvation and the social implications of the Gospel. Throughout his ministry he championed Shona interests, denouncing injustice, opposing discriminatory legislation, and earning much unpopularity among Europeans. His health was never robust and broke down completely in 1931. D.G.L. CRAGG

WHITE, WILLIAM (1748-1836). Organizer of the Protestant Episcopal Church* in the USA. Born into a wealthy Philadelphia family, he graduated in 1765 from the College of Philadelphia (University of Pennsylvania) and then studied theology. He was ordained priest in England in 1772 and served in Christ Church and St. Peter's in Philadelphia from 1772 until 1836. He was joint chaplain of the Continental Congress and its successor. He led in the formation of the Protestant Episcopal Church by drafting a constitution for a church free of the state in which laity were equally represented with the clergy, by writing a revised Book of Common Prayer and promoting conventions in 1785 and 1789 which created the Protestant Episcopal Church. He was elected bishop of Pennsylvania in 1786, was consecrated in London, and served in Philadelphia from 1787 till his death. He became presiding bishop in 1796. EARLE E. CAIRNS

WHITE FATHERS. The common name for the "Society of Missionaries of Africa," taken from their white cassocks and mantles. The society was founded in 1868 by Charles Cardinal Lavigerie (1825-92), archbishop of Algiers, to evangelize Africa. The Fathers are secular priests, together with lay brothers, who live in community, not taking the vows of regular religious communities, but bound by oath to lifelong work in African missions and to obedience to their superiors. They began their missions in Algeria and Tunisia. Early attempts to penetrate the Sahara were unsuccessful, so the missionaries withdrew from the desert to work in oases on the northern fringe, and to demonstrate the Gospel in loving action rather than by preaching. Later they entered Buganda, where they were very successful, subsequently going to Tanganyika, Nyasa, and Congo. They were also much concerned with the abolition of slavery, the improvement of agriculture, and the scientific exploration of Africa. J.G.G. NORMAN

WHITE, W.H., see MARK RUTHERFORD

WHITEFIELD, GEORGE (1714-1770). English preacher. Born at Gloucester, he was educated there and at Pembroke College, Oxford, where he associated with those who formed the "Holy Club"* and who would later be known as the first Methodists. There also he experienced an evangelical conversion. He was subsequently ordained, and his first sermon—in his native town—was of such fervor that a complaint was made to the bishop that he had driven fifteen people mad. He preached in several London churches, but quickly accepted an invitation from John and

Charles Wesley* to go to Georgia where, with the exception of a notable visit home, he remained from 1737 to 1741. The visit home included his first attempt at open-air preaching, in Bristol. He was to continue the practice to the end of his life, regularly delivering up to twenty sermons a week, covering vast distances that included fourteen visits to Scotland and, in those days of long and hazardous voyages, no less than seven journeys to America, where he died shortly after preaching his last sermon.

The association with Wesley in the early years quickly gave way to differences and even to bitter feud. This arose mainly from their opposed views of the availability of salvation, Wesley adopting the Arminian* interpretation and Whitefield the Calvinistic. As a result, the latter became closely associated with the work of the countess of Huntingdon,* and in his later years he opened several of the meetinghouses of her Connexion as well as the theological college at Trevecca in 1768.

In his kind Whitefield is supreme among preachers, sharing his eminence only with Latimer.* Others might be more learned, even more stylish, but none was more eloquent or more moving. J.C. Ryle has justly claimed, "No preacher has ever retained his hold on his hearers so entirely as he did for thirty-four years."

His theme is the basic evangelical message of man's irremediable sinfulness and Christ's effective salvation. Indeed, as we read them, there is a sameness about Whitefield's sermons that becomes rather tedious, an effect no doubt of too much preaching and too little preparation. Nevertheless this does not detract from their vividness. His vision of heaven and, more particularly, hell was too immediate for that, and his regard for the eternal welfare of the souls of each of his hearers too insistent. There is thus an intimate note in all his work, displaying itself in his earnestness and importunity. "My brethren, I beseech you" is a recurrent expression. Like open-air preachers before him, like the friars and like Latimer, his work abounds with vivid colloquial phrases and apt, familiar analogies. And none knew better than he how to use question and exclamation to produce a tense, dramatic atmosphere. He added antithesis, repetition, brevity, assertion to his range. Above all, he was, as contemporary record witnesses, a supreme actor, gifted in voice and gesture to pull out all the stops.

Others—Pope, Johnson, Fielding among them—criticized him. To William Cowper,* who thought and felt as he did but who in his timidity differed so much from the sometimes strident self-confidence of Whitefield, was left the task of tribute:

> He followed Paul—his zeal a kindred flame,
> His apostolic charity the same.

See L. Tyerman, *George Whitefield* (1876); and *Select Sermons* (1958). ARTHUR POLLARD

WHITELAW, THOMAS (1840-1917). Scottish Presbyterian minister and biblical scholar. Born in Perth, he was educated at St. Andrews and the United Presbyterian Theological Hall, Edinburgh, and was ordained in 1864. He held pastorates in Glasgow and Kilmarnock, refused a call to Australia, and was a special commissioner for the United Free Church* at the union of Australian Presbyterian churches in 1901. U.F. Church moderator in 1912-13, he traveled widely. He served on many key committees and wrote many books, including commentaries on Genesis, John, and Acts and a study of the divinity of Christ.

WHITGIFT, JOHN (c.1530-1604). Archbishop of Canterbury from 1583. He studied at Cambridge under Nicholas Ridley* and John Bradford* and adopted Reformed doctrinal views. In 1563 he became Lady Margaret professor of divinity at Cambridge, and in 1565 petitioned against the use of the surplice, but soon he became a convinced upholder of Anglican ritual and episcopal government. He became regius professor and master of Pembroke Hall (1567) and master of Trinity (1570). He secured the expulsion from Cambridge of Thomas Cartwright,* who had powerfully attacked episcopacy, and engaged in prolonged pamphlet controversy with him. In 1577 Whitgift became bishop of Worcester and vigorously enforced conformity by Puritans.* Recognizing the similarity of their aims, Elizabeth I* made him primate in 1583. With his private fortune he lived magnificently and had the queen's approval and friendship. Ministers were obliged to conform and increased powers secured for the Court of High Commission. Stringent control of printing by the bishops led to the secret publication of the *Marprelate Tracts,** whose instigators Whitgift punished most severely. In 1593 he secured the passing of an act banishing nonattenders at church, and some Nonconformists went to Holland. He tried to remedy the lack of precision in the Thirty-Nine Articles* by defining predestination more closely (and in an entirely Calvinist manner) in the Lambeth Articles* of 1595, but the indignant queen forced him to withdraw them. After attending Elizabeth on her deathbed, he was obliged by James I* to confer with Puritans at the Hampton Court Conference* of 1604, and died later that year.

BIBLIOGRAPHY: J. Strype, *Life and Acts of John Whitgift* (1822); *Works* (ed. J. Ayre, 1851-53); P.M. Dawley, *John Whitgift and the Reformation* (1955); V.J.K. Brook, *Whitgift and the English Church* (1957). JOYCE HORN

WHITMAN, MARCUS (1802-1847). Presbyterian medical missionary. Born at Rushville, New York, he qualified in medicine and for eight years practiced in Canada and New York. In 1835 the American Board of Commissioners for Foreign Missions sent him to study the possibility of Indian missions in the American Northwest. In 1836 his new bride Narcissa and a small missionary band accompanied him across the Rockies, founding a chain of mission stations in the Walla Walla River Valley. Encouraging at first, his work encountered multiplying obstacles. In the fall of 1842 he began a famous 3,000-mile horseback journey to Boston, convincing the Board to rescind its decision to close part of the work. He conferred with government officials in Washington and returned in 1843 with immigrants to the Oregon Territory. Due to a tragic misunderstanding, Cayuse Indians murdered him, his wife, and twelve others. ALBERT H. FREUNDT, JR.

WHITSUNDAY, see PENTECOST

WHITTIER, JOHN GREENLEAF (1807-1892). American Quaker poet and abolitionist. Born near Haverhill, Massachusetts, he had little formal education, but read extensively. His first book of poems was published in 1831. In 1833 he entered politics as an abolitionist, served in the Massachusetts legislature (1835), and was an important writer in the antislavery movement. He broke with W.L. Garrison by 1843, but continued a political activist, influencing the formation of the Republican Party and supporting Lincoln for the U.S. presidency. After the Civil War, poetry was his main interest and he became popular, especially on the rural New England themes that appear in his best-known work, *Snow-Bound* (1866). In old age he turned to religious verse, and his "Dear Lord and Father of mankind" and "Immortal Love forever full" are still widely used.

ALBERT H. FREUNDT, JR.

WHITTINGHAM, WILLIAM (c.1524-1579). Dean of Durham. After education at Oxford, he went to Orléans, Germany, and Geneva (1550-53), returning to England a convinced Protestant. As Mary Tudor* was now queen, he left after four months for Frankfurt, inviting other English exiles to gather there. He took a leading part in organizing the English congregation, but divisions arose between those content with Edward VI's* second Prayer Book, led by Richard Cox,* and those like John Knox* wanting a more thorough reformation. Knox was expelled in March 1555, and Whittingham in September followed him to Geneva, where he was successively elected elder, deacon, and minister. He was largely responsible for the translation of the Geneva (or "Breeches") Bible,* and when the other translators returned to England on the death of Mary in 1558, he remained in Geneva until it was printed in 1560. Whittingham also made metrical versions of the Ten Commandments and many psalms.

In 1563 he was made dean of Durham despite Elizabeth's dislike of his Puritanism. He held services twice daily, removed images, and improved the grammar and song schools. Proceedings were begun against him in 1566 for his refusal to wear the surplice, but Whittingham gave way. He incurred the hostility of Edwin Sandys, archbishop of York,* for resisting his attempt to visit the cathedral. Sandys questioned the validity of his ordination and attempted to deprive him, but Whittingham died before a decision was reached. His memorial inscription in Durham Cathedral stated that he married Catherine, sister of John Calvin, but this is erroneous.

See "A Brief Discourse of the Troubles at Frankfurt" in *The Works of John Knox* (ed. D. Laing, vol. IV, 1855); and contemporary life (ed. M.A. Everett Green) in *The Camden Miscellany*, VI (1871). JOYCE HORN

WHOLE DUTY OF MAN, THE. A work first published anonymously in London in 1658 and which has been attributed to various authors, although Archbishop Sterne of York seems one of the most likely candidates. Written from a Church of England and royalist point of view, it yet has much in common with the practical writings of Richard Baxter,* the contemporary Presbyterian divine. Seeking to deal with the ethical aspects of the Christian life, it was very popular in its own day and has been reprinted in numerous editions down to the present time.

WHYTE, ALEXANDER (1836-1921). Scottish minister, often described as the "last of the Puritans." Born in Kirriemuir, he was educated at King's College, Aberdeen, and at the Free Church of Scotland's* New College, Edinburgh. For four years from 1866 he was assistant minister of Free St. John's, Glasgow, then he was called to Edinburgh as colleague and successor to R.S. Candlish at Free St. George's. During nearly forty years there he established a reputation as a graphic and compelling preacher to an extent probably unparalleled even in a nation of preachers. In 1909 Whyte became principal of New College and taught NT literature there. He was moderator of the Free Church general assembly in 1898 and the author of a number of devotional books.

J.D. DOUGLAS

WICHERN, JOHANN HINRICH (1808-1881). German Protestant minister and founder of the Innere Mission.* Born in Hamburg, where he was influenced by the piety of the *Erweckungsbewegung*, Wichern studied theology at Göttingen under Lücke (1829-31) and at Berlin under Neander and Schleiermacher.* Returning to Hamburg as a candidate for ordination, Wichern was moved by the plight of underprivileged children in the poorer sections of that city. In 1833 he founded a school for neglected children in the village of Horn near Hamburg and christened it *Rauhes Haus* ("Rough House"). Under Wichern's strong leadership the school grew and expanded. In 1842 he established a training institute for his assistants and in 1844 began to publish a monthly periodical, *Die Fliegenden Blätter aus dem Rauhen Hause.* Children in the school were divided into families of approximately twelve, under the guidance of an overseer—generally a candidate for the ministry—and two assistants.

At the First Congress of the Evangelical Churches in Wittenberg in 1848, Wichern called for the coordination of all charitable activities in Germany through a single agency, the Innere Mission. While holding firmly to the confessions of the evangelical churches, Wichern wanted to bind the preaching of the Gospel to active social service and thus give practical expression to the Reformation principle of the priesthood of all believers. Wichern took part in prison reform in Prussia (1857) and in the establishment of the *Johannisstift* in Spandau (1858). He organized the *Felddiakonie* to minister to the wounded in the wars of 1864, 1866, and 1870-71. In 1872 he returned to the *Rauhes Haus*, and died in Hamburg.

See J.H. Wichern, *Gesämmelte Schriften* (ed. J.

Wichern and F. Mahling, 6 vols., 1901-8).
DAVID C. STEINMETZ

WICLIF, JOHN, see WYCLIFFE

WIED, HERMANN VON, see HERMANN VON WIED

WILBERFORCE, SAMUEL (1805-1873). Anglican bishop. Third son of William Wilberforce,* he was educated at Oxford, inherited his father's political skill, strong sense of mission, and charm, but earned the nickname "Soapy Sam." Brought up an evangelical, he was influenced strongly at Oxford by J.H. Newman* and by H.E. Manning,* to whom he was related by marriage. He broke with both on their conversion to Rome. Ordained in 1828, he spent ten years in parish work (more evangelical than Tractarian* in tone), and became successively bishop of Oxford (1845) and Winchester (1869), being known as a High Church bishop. Prime Minister Gladstone would have made him primate. Wilberforce was killed by a fall from his horse. He is remembered as the pioneer of the modern-style episcopate, organized chiefly for the pastoral care of the diocese; for pioneering the corporate training of ordinands (he founded Cuddesdon Theological College); and for the 1859 debate at Oxford on Darwin's theory, at which his intervention precipitated the later nineteenth-century conflict between science and religion. JOHN C. POLLOCK

WILBERFORCE, WILLIAM (1759-1833). Slave trade abolitionist. Born in Hull, where his house still stands as a museum to him, he was educated first at Hull Grammar School where he came under the influence of Joseph Milner,* the headmaster, and his brother Isaac. The latter used to lift the small boy onto the table so that the other scholars could hear him read with his beautiful voice. When he had been less than two years at the school, his father died; he went to live at Wimbledon with an aunt who was a staunch Methodist. His mother wanted to remove him from religious influences of this kind and brought him back to Yorkshire, where he went as a boarder to Pocklington School. At the age of fourteen he wrote a letter to a York paper about the evils of the slave trade. He largely wasted his time at St. John's College, Cambridge, but when he was twenty-five he met Isaac Milner* at Scarborough and invited him to come to Europe with him. He was converted through their conversation and study of the NT together on this trip. In 1780 he had been elected member of Parliament for Hull, after laying out a great deal of money on the election, and in 1784 he was again returned for his native city, but took instead the county seat for Yorkshire to which he had been chosen without a contest, and he was unopposed for that seat for twenty-three years. At this election James Boswell records how his smallness of stature was forgotten in the midst of his eloquence—"the shrimp grew and grew and became a whale."

Wilberforce became associated with the Clapham Sect,* a group of Evangelicals who were active in public life. Through his friendship with John Newton* and Thomas Clarkson* on one hand, William Pitt on the other, he was persuaded to put most of his energies into the abolition of the slave trade. The intellectual climate of the time was favorable to ideas of human liberty and happiness, and it was on the grounds of economics or national policy that slavery was defended. By a brilliant use of all the weapons available to them, he and his friends gradually undermined both the main grounds of defense, and in 1807 the slave trade was abolished. The complete abolition of slavery was not achieved until just before his death in 1833.

He was involved in many other good causes. In 1787 he founded a society for the reformation of manners, and ten years later published his *Practical View of the Prevailing Religious System of Professed Christians in the Higher and Middle Classes in this Country contrasted with Real Christianity,* which was a best seller for forty years. He and his friends sought to evangelize the upper classes as Wesley had the lower classes, and also to use their wealth and influence in a multitude of good causes. He helped in the formation of the Church Missionary Society (1799), and the British and Foreign Bible Society (1804).

BIBLIOGRAPHY: R. Coupland, *Wilberforce* (2nd ed., 1945); R. Furneaux, *William Wilberforce* (1974); R.T. Anstey, *The Atlantic Slave Trade and British Abolition 1760-1810* (1975).
R.E. NIXON

WILBUR, JOHN (1774-1856). Quaker preacher. A native of Rhode Island, descended from Samuel Wilbur, he taught in the public schools of Rhode Island, was recorded a minister of the Society of Friends in 1812, and became known as a rugged, effective speaker. On a preaching tour of the British Isles (1831-33) he zealously opposed the evangelical movement's entering Quakerism under the leadership of Joseph Gurney, Elizabeth Fry's brother. He published letters written to George Crosfield defending the old Quaker position on the Inner Light, and attacking "dangerous innovations" (1832). When Gurney preached in America (1837-38), Wilbur opposed him and was expelled (1843), becoming leader in 1845 of five hundred separatists known as "Wilburites" (officially the "New England Yearly Meeting of Friends"). Other groups supported him in New York, Ohio, and Philadelphia. He preached again in England during 1853-54. J.G.G. NORMAN

WILDER, ROBERT PARMELEE (1863-1938). Missionary to India and virtual founder of the Student Volunteer Movement.* Born in Kolhapur, India, he was studying and promoting missions at Princeton College when D.L. Moody* called for the summer Bible conference out of which the SVM came. For two years he visited colleges on behalf of the movement. In 1892, on his way to India, he founded the British Student Volunteer Missionary Union. In India he worked with students through the YMCA. When ill-health forced him to leave India in 1902, he spent fourteen years in Europe promoting the World Student Christian Federation.* From 1919 to 1927 he was general secretary of the SVM. Then for six years he was executive secretary of the Near East Christian Council, residing in Cairo.

HAROLD R. COOK

WILFRID (634-709). Bishop of York. Son of a Northumbrian nobleman, he was educated at Lindisfarne then went to study the Roman form of religious life at Canterbury. After visiting Rome with Benedict Biscop* in 654, he returned to England as abbot of Ripon. At the Synod of Whitby* (663/4) he was the chief and most vehement advocate of the Roman tradition concerning the dating of Easter. Shortly after the synod he was appointed bishop of Northumbria with his seat at York. He went to Gaul for consecration by Frankish bishops. Due to his dallying there for two years, King Oswy had Chad,* abbot of Lastingham, consecrated as bishop of York. On his return in 666 Wilfrid went to Ripon, but Theodore of Tarsus* (archbishop of Canterbury) had him installed at York in 669.

He was an ambitious and able man with a forceful personality. He fell foul of King Ecgfrith, and in 678 Theodore, concerned at Wilfrid's love of power, divided the diocese of York into four and appointed other bishops for the various sections. Wilfrid appealed to Theodore and then to Pope Agatho. The papal synod upheld his case, but on his return to England he was imprisoned by Ecgfrith. After his release he went to Sussex, where he had some success evangelizing the heathen South Saxons. After the death of Ecgfrith in 686 he was reconciled to Theodore and returned north as bishop of Ripon and abbot of Hexham. In 691 he was banished after a dispute with King Aldfrith. King Ethelred of Mercia invited him to become bishop of Leicester. In 703 the synod called by Archbishop Brihtwold at Austerfield in Yorkshire decreed that he should resign the see of York and retire to Ripon as a monk. He appealed again to Rome. His claims were upheld, but he agreed to the appointment of John of Beverley* as bishop of York and himself as bishop of Hexham.

He spent his last years in the monastery at Ripon till his death at the monastery at Oundle in Northamptonshire. His importance lies in the large part he played in the Romanization of the Celtic Church.

See J. Raine, *The Historians of the Church of York and its Archbishops* (ed. B. Colgrave, 1927), and B. Colgrave, *The Life of Bishop Wilfrid by Eddius Stephanus* (1927).

R.E. NIXON

WILKES, PAGET (1871-1934). Missionary to Japan.* Son of an Anglican clergyman, he was educated at Lincoln College, Oxford, was deeply influenced by the activities of the Oxford Inter-Collegiate Christian Union, and left to serve in Japan under the Church Missionary Society in a group led by the Rev. Barclay Buxton in 1897. Here Wilkes felt unduly restricted by orthodox Anglicanism and founded the interdenominational Japan Evangelistic Band, pledged to aggressive evangelism and the distinctive holiness doctrines associated with the Keswick Convention* movement. In 1903 the headquarters of the work was centered at Kobe, under the name of the "One by One Band." The whole of Wilkes's active life was spent in Japan, but his name became widely known through his books, of which the best known were *The Dynamic of Service* (1920), *The Dynamic of Faith* (1921), and *The Dynamic of Redemption* (1923). Though he seldom visited England, when he did come to Oxford or Cambridge he strongly urged students to offer for missionary service.

G.C.B. DAVIES

WILKINS, JOHN (1614-1672). Bishop of Chester. After studying at Oxford, he became vicar of Fawsley, Northamptonshire, in 1637, and subsequently became a private chaplain, devoting his leisure to scientific studies. His first work in 1638, *The Discovery of a World in the Moon,* sought to prove the moon habitable, and in 1640 he published *A Discourse concerning a New Planet,* showing the probability that the earth was a planet. From 1645 he promoted weekly meetings for scientific discussion in London, anticipating the Royal Society.* A supporter of the parliamentary side, he took the Solemn League and Covenant* and was made warden of Wadham College, Oxford, in 1648. Tolerant of Royalists, he gathered a distinguished group of scientists around him and held weekly meetings on the London pattern. He married Cromwell's sister in 1656 and became master of Trinity College, Cambridge, in 1659, but was deprived the following year. He accepted the Restoration settlement and after various preferments was made bishop of Chester (1668). In 1662 he became first secretary of the Royal Society, of which he was virtual founder. As bishop of Chester, he tried to promote the toleration and comprehension of dissenters. Besides mathematical works, he wrote *On the Principles and Duties of Natural Religion,* defending natural theology.

See P.A.W. Henderson, *The Life and Times of John Wilkins* (1910).

JOYCE HORN

WILKINSON, JEMIMA (1752-1819). Religious leader. Born in Cumberland, Rhode Island, her religious interest was roused about 1758 by George Whitefield's* sermons and by the meetings of the "New Light Baptists." In 1774 she was influenced by Ann Lee,* the founder in America of the "Shakers." Following fever, she claimed to have died and that her body was inhabited by the "Spirit of Life." Taking the name "Public Universal Friend," she held open-air meetings, led processions on horseback clad in a long robe over masculine attire, and established churches (1777-82). Her disciples' claim that she was Christ come again aroused hostility, forcing her to leave New England; she eventually established a colony "Jerusalem" in Yates County (1790). Internal disputes affected the movement, which disintegrated entirely after her death.

J.G.G. NORMAN

WILLAERT, ADRIAEN (c.1490-1562). Flemish composer. A pupil of Jean Mouton, an outstanding follower of Josquin Desprez, Willaert went to Italy early in his career, like so many of his countrymen. There he was active at Rome, Ferrara, and Milan, being closely associated with the Este family. In 1527 he was elected *Maestro di cappella* at St. Mark's in Venice. In his compositions he strove for the utmost perfection, and exemplified the humanistic desire for complete clarity of the text. He experimented with the expressive use of chromaticism. In Venice, partly because of the placement of his choral forces in the basilica, Willaert gave special attention to writing for multiple choirs, a feature that was further developed by his successors there. He is considered the founder of a distinct Venetian school. Zarlino, the greatest theorist of the time, was his pupil, and transmitted his master's ideas in his writings. Andrea Gabrieli also studied with him, carrying on his traditions at Venice, becoming a great organist also, and the teacher of his illustrious nephew Giovanni Gabrieli. Willaert was equally important as a composer of instrumental and vocal secular music.

J.B. MAC MILLAN

WILLAN, HEALEY (1880-1968). Musical composer. A native of England, he spent most of his professional life in Canada. He came to the prominent organist's position at St. Paul's Anglican Church, Toronto, in 1914, but soon transferred to the High Church atmosphere of St. Mary Magdalene's, where he served for the rest of his life. Here his *a cappella* gallery choir and chancel choir of men formed an attraction for connoisseurs of liturgical music for many years. Although he has written anthems not unlike Stanford, Wood, and Noble, his most characteristic works are the shorter, unaccompanied pieces, which he probably wrote for his own choir at St. Mary Magdalene's. A number of hymn anthems, based on traditional melodies such as "Sing to the Lord of Harvest," are in excellent taste and have wide appeal. *The Three Kings,* for six-part mixed chorus, has fine contrasts of choral timbre. *An Apostrophe of the Heavenly Host,* with a text from the Eastern liturgy coupled with the tune, *"Lasst uns erfreuen,"* is a more exacting work for eight-part chorus. Willan has also written larger works with orchestral accompaniment, and a quantity of useful hymn-preludes and other works for organ. Willan taught for many years at the conservatory in Toronto, and exerted a wide influence.

J.B. MAC MILLAN

WILLARD, FRANCES ELIZABETH CAROLINE (1839-1898). Educator, temperance leader, and suffragist. Born of Puritan ancestry at Churchville, New York, she grew up on the Wisconsin frontier. She attended Milwaukee Female College and graduated from Northwestern Female College, Evanston (1859), where she was converted, later becoming a Methodist. Never married, she taught at Pittsburgh and at Genesee before becoming president of Evanston College for Ladies (1871-74). She became president of the National Woman's Christian Temperance Union (1879) and of the World's Woman Christian Temperance Union (1891) and helped organize the Prohibition Party in 1882. A women's rights reformer, she was National Council of Women president. She wrote *Woman and Temperance* (1883) and *Glimpses of Fifty Years* (1889).

D.E. PITZER

WILLEHAD (d.789). Missionary to the Saxons. Born in Northumbria and educated for the priesthood at York, he went in 765 as a missionary to Friesland. His work was partly inspired by the example of the earlier Northumbrian, Willibrord,* the "Apostle of Frisia." At the direction of Charlemagne* he went about 780 to seek to evangelize the Saxons in Wigmodia (between the lower Weser and the Elbe), but his work was cut short by an insurrection in 782. After a visit to Rome he spent a few years at the monastery of Echternach (founded by Willibrord) copying ancient manuscripts. Then he returned to his earlier work, and in 787 became bishop of Bremen. He built for his diocese a cathedral dedicated to the Apostle Peter, which was opened in 789.

PETER TOON

WILLIAM III (1650-1702). King of England, Scotland, and Ireland. Born at The Hague, he was thoroughly educated at Leyden, raised a Calvinist (though religiously tolerant), and proved himself an iron-willed soldier-politician. From his birth until 1672 the Republican party dominated Dutch politics; when France invaded the United Provinces, the Republicans, who stood for appeasement, were overthrown and William appointed *stadholder* and captain-general for life. He managed the war defense successfully and dedicated his life to foiling Louis XIV's plans of European hegemony. To that end he married Mary, daughter of the future James II of England, in 1677. When the latter succeeded his brother Charles II and pursued pro-French and pro-Catholic policies, William was invited by English nobles to take the throne. He did; James fled. William and Mary were crowned as joint monarchs in 1689. William decisively influenced religious and political affairs, and though he was dead by 1702 after a fall from his horse, he brought his country to greatness in Europe by helping to seal the downfall of France in the War of the Spanish Succession.

BRIAN G. ARMSTRONG

WILLIAM OF AUVERGNE (c.1180-1249). Bishop of Paris. He studied in Paris and became a teacher first in the faculty of arts and then in the faculty of theology. In 1223 he became a canon of Notre-Dame, and in 1228 bishop of Paris. When suspicion of Aristotle* in ecclesiastical circles was rife, he tried to synthesize the traditional Augustinian method and doctrine with these new Neoplatonic, Arabian, and Aristotelian philosophical ideas, and so prepared the way for his more successful followers, Alexander of Hales,* Albertus Magnus,* and Thomas Aquinas.* In 1229 William, who is known also as "Guillaume d'Auvergne" or "Guillaume de Paris," was rebuked by Pope Gregory IX* for his laxity in dealing with problems in the university. Gregory expressed regret at William's election as bishop. His most important philosophical writings are *De Universo* and *De Anima* and, theologically, *De Virtutibus, De Sacramentis,* and *De Trinitate.*

HOWARD SAINSBURY

WILLIAM OF AUXERRE (1150-1231). Philosopher and theologian. After serving as archdeacon of Beauvais and then proctor of the University of

Paris at the Roman Curia under Honorius III (1216-27), he was appointed (1230) by Gregory IX as a member of a commission of three to correct the physical treatises of Aristotle in order to bring him into line with Christian thought and to make him acceptable at the University of Paris. William died, however, before he could complete his part in this important assignment. He was largely influenced by Augustine* and Anselm of Canterbury* in his theology, and to some degree by Hugh* and Richard of St.-Victor.* His most famous work is his *Summa Aurea* (1215-20), which generally follows the pattern of the *Sentences* of Peter Lombard,* but covers some issues not treated by the latter. C. GREGG SINGER

WILLIAM OF CHAMPEAUX (c.1070-1121). Medieval philosopher, theologian, and reforming bishop. He is best known for his controversy with Abelard on the question of universals. Reacting against Roscellinus* of Compiègne's Nominalist position of universals, William taught an extreme Realist doctrine. In 1100, while teaching at the cathedral school in Paris, Peter Abelard,* his pupil, attacked the position by illustrating the ludicrous logical results of this teaching. William left his post shortly afterward and retired to the Abbey of St.-Victor, where he modified his teachings and started a school of theology. He was consecrated bishop of Châlons-sur-Marne in 1113 and was responsible for a reform of the clergy in his diocese. Few of his writings have survived, and his views on universals must largely be based on the descriptions of Abelard.

RUDOLPH HEINZE

WILLIAM OF CONCHES (c.1080-c.1154). Norman philosopher. He was a disciple of Bernard of Chartres and himself taught at Chartres, where he sought to further classical learning and literature. John of Salisbury,* one of his pupils, considered him an accomplished grammarian. After 1140 he was attacked by opponents of classical studies and by William of St.-Thierry,* who detected in his writings the influence of Abelard's* heresies. Withdrawing from public teaching to the court of Geoffrey Plantagenet, he taught the future king of England, Henry II. He wrote commentaries on all the basic Platonic texts of the early Middle Ages. His treatises *Philosophia Mundi* and *Pragmaticon* reveal strongly Platonistic and Realistic tendencies. Leaning toward pantheism, he identified the Holy Spirit as the world soul. ALBERT H. FREUNDT, JR.

WILLIAM OF MALMESBURY (c.1090-c.1143). English historian and monk. Educated at Malmesbury Abbey in SW England, he became a monk there and helped to build up the library. He evidently could have become abbot in 1140, but relinquished it in favor of a colleague. Taking Bede* as his pattern, he set out to write English history in a popular form. His *Gesta regum Anglorum* (record of the kings of England from the end of the sixth century) was published about 1125, closely followed by *Gesta pontificum Anglorum* (which covered the English Church hierarchy over roughly the same period). His *Historia novella*, dealing with events after 1126, breaks off abruptly at the end of 1142. His chronology has been at points criticized, but William's work contains interesting anecdotes and perceptive comments and strictures, all presented in vivid and powerful style. He testified to the quality and discipline found in contemporary English monks. Many know his writing only for the striking and much-quoted passage in which he tells of the high moral motives that made men undertake the First Crusade (preached about the time he was born).

J.D. DOUGLAS

WILLIAM OF MOERBEKE (c.1215-1286). Philosopher and translator. He was born at Moerbeke, Belgium, and studied at Cologne. By 1260 he had served in the Dominican Order in Thebes and Nicea and, urged by Thomas Aquinas,* began to edit and translate ancient Greek authors. He translated much of Aristotle into Latin and became the most important and prolific translator of Greek in the thirteenth century. Translating Greek commentators, and writings of Proclus, Archimedes, Eutochius, Ptolemy, Hero, Galen, and Hippocrates, he through his work gave impetus to Neoplatonism* in the late Middle Ages. He was a chaplain and confessor to popes Clement IV and Gregory X. Zealous for reunion with the Greek Church, he participated in the Council of Lyons* (1274) and was appointed archbishop of Corinth (1278), where he resided until his death.

ALBERT H. FREUNDT, JR.

WILLIAM OF NORWICH (1132-1144). Supposed victim of a Jewish ritual murder. A pious tanner's apprentice in Norwich, William was—according to certain ecclesiastics, especially the prior, William Turbe—lured from his home in Holy Week and sacrificed by the Jews during their Passover celebrations. In fact he had probably died of a cataleptic fit and been buried prematurely by his parents. At first the civil authorities refused to believe this tale, which was the first accusation of ritual murder in English, and indeed European, history (there is no continental parallel till 1171), but William's reburial in Norwich Cathedral in 1151 aroused a wave of superstitious fanaticism. Visions and miracles were reported at the tomb, and the boy's relics were venerated as those of a saint and martyr till the Reformation. This ugly episode began a whole series of discoveries of boy saints and martyrs elsewhere, details of which are based suspiciously on the Norwich prototype. IAN SELLERS

WILLIAM OF OCKHAM (c.1280-c.1349). Medieval Scholastic theologian and philosopher. Born in Surrey, England, he entered the Franciscan Order about 1310 and studied at Oxford between 1318 and 1324. His ideas led to a summons to Avignon (1324) to answer charges of heresy. A dispute between Pope John XXII* and the Spiritual Franciscans was then at its height, and William identified himself with the Spirituals in opposition to John. In 1328 he left the city and went to the court of the emperor Louis of Bavaria. Excommunicated, William is supposed to have said to the emperor: "You defend me with your sword and I will defend you with my pen." From 1328 until his death he produced powerful defenses of the imperial theory against those who favored the pope. After Louis's death in 1347 William made an effort to be reconciled with his order and the church, but the outcome of this attempt is not known.

His writings fall into two groups associated with the two phases of his life. While working for Louis (1333-47) he wrote works about the relation of church and state such as *Dialogus Inter Magistrum et Discipulum, Octo Quaestiones Super Potestate ac Dignitate Papali,* and *Tractatus de Imperatorum et Pontificum Potestate.* The nonpolitical works that contain his contributions to philosophy and theology were written while he was at Avignon and Oxford (1317-28). These include lectures on Peter Lombard's* *Sentences,* an explanation of Aristotle's* *Physics,* commentaries and treatises on logic and natural science. His most important philosophical work is *Summa Logicae,* which he completed before he left Avignon.

Ockham criticized the accommodation of the philosophical system of Aristotle with Christian doctrine that had been fashioned by the thirteenth-century Schoolmen such as Thomas Aquinas.* This method had tried to achieve an accord between faith and reason by reinterpretation of the philosophical assumptions of Aristotle. Its purpose was to keep the philosophic system of Aristotelianism intact. Franciscan scholars from Bonaventure to Duns Scotus tried to argue for the Christian faith by destroying Aristotle's philosophy. All of these thirteenth-century systems, however, depended on the doctrine of Realism. Oakham rejected this teaching on the basis of a radical empiricism in which the base of knowledge is direct experience of individual things (Nominalism*). Involved in his explanation of reality is his view that "What can be done with fewer assumptions is done in vain with more" ("Ockham's Razor"). Called the *via moderna* as opposed to the *via antiqua* of Aquinas, Ockham's Nominalism was of great significance for science, since it suggested that natural phenomena could be investigated rationally. God to Ockham, however, was above all knowledge. He cannot be apprehended by reason, as the Thomists taught, or by illumination, as the Augustinians believed, but only by faith.

BIBLIOGRAPHY: E.A. Moody, *The Logic of William of Ockham* (1935); M.H. Carre, *Realists and Nominalists* (1946); P. Boehner, *Collected Articles on Ockham* (1956) and (ed.), *Ockham: Philosophical Writings* (1957); H. Shapiro, *Motion, Time and Place According to William of Ockham* (1957). ROBERT G. CLOUSE

WILLIAM OF ORANGE, see WILLIAM III

WILLIAM OF ST.-THIERRY (c.1085-1148). Scholastic philosopher. Born of a noble family, at Liège, he studied under Anselm of Laon* and entered the Benedictine Abbey of St. Nicasius of Reims (1113); he was later elected abbot of St.-Thierry, near Reims (1119). In the next three years he wrote *De Natura et dignitate amoris* and *De contemplando Deo.* He formed a lasting friendship with Bernard of Clairvaux,* but was refused admission there. In the conflict between Cluniacs and Cistercians (1120s) he urged Bernard to defend Cîteaux, resulting in Bernard's dedicating his *Apologia* (1124) and *De Gratia et libero arbitrio* (1128) to him, with a return favor of *De Sacramento altaris* (1128). Between 1128 and 1135 he wrote several treatises based on the Fathers: the Canticle of Canticles after Gregory the Great, Ambrose, and Origen; *De Natura Animae et corporis;* a study of Romans. He took part in the first general chapter of Benedictines of Reims province (1130), but resigned his abbacy in 1135 for a strictly contemplative life, joining the Cistercians at Signy. His writings increased: *Meditativae orationes* (1130-45) showing his inmost soul; *Speculum fidei* and *Aenigma fidei;* and an attack on Abelard. He also began, but did not complete, a life of Bernard. C.G. THORNE, JR.

WILLIAM OF TYRE (c.1130-c.1185). Archbishop of Tyre. Born in Jerusalem of a European merchant family, he returned to Europe c.1145 where for twenty years he pursued his studies of arts and theology in France, and civil and canon law at Bologna. Peter Lombard* and Hugh de Porta Ravennata were among his teachers. Ordained before 1161, he returned to Palestine in 1165, becoming archdeacon in 1167, and was consecrated archbishop of Tyre in 1175. Diplomatic missions took him to Constantinople and Rome in 1168-69. In 1170 he was appointed tutor to Baldwin,* son of Amaury the king of Jerusalem (1163-74), from whom he had been given earlier a stipend to write the official history of his reign, *Gesta Amaurici.*

At royal request, c.1170 he began his *Historia rerum in partibus transmarinis gestarum,* covering Crusade events from 1095 to 1184. This was translated into French in the thirteenth century and printed in Basle as early as 1549. A signal work of medieval historiography, it was the primary authority from 1127, where Fulcher of Chartes had stopped, and a contemporary chronicle from 1144. Though confusion exists in chronology, men and events are judged honestly in terms of religion, morality, and politics, even to comment on human physical and intellectual characteristics. He was also familiar with the works of Albert of Aachen, Fulcher of Chartes, and Balderic of Bourgueil, as with versions of the *Gesta Francorum.* A polyglot, he knew Latin, Greek, French, and Arabic. One of his lost works is the *Gesta orientalium principum,* on the Arabs, and another is on the Third Lateran Council. He became chancellor of the Latin Kingdom of Jerusalem in 1174, led the Jerusalem delegation to the Third Lateran Council (1179), and failed to procure the patriarchate of Jerusalem in 1183. He retired to his *Historia,* writing the Prologue in 1184 declaring his determined objectivity.

See A.C. Krey, "William of Tyre, The Making of an Historian in the Middle Ages," *Speculum* XVI (1941), pp. 149-66. C.G. THORNE, JR.

WILLIAM OF WYKEHAM (1324-1404). Bishop of Winchester. Born in humble circumstances, he was educated at Winchester and held various royal administrative posts before becoming chancellor in 1367. He received numerous ecclesiastical livings from 1357 although he was not priested until 1362. Five years later he was consecrated bishop of Winchester. Anticlerical agitation led to his dismissal from the chancellorship (1371), and John of Gaunt had him brought to trial (1376) to answer for his conduct while in office. Though found guilty on only one minor count, he was sentenced to forfeit the temporalities of his see and dismissed from court. On the accession of Richard II he was pardoned, and twelve years later resumed the chancellorship. His importance lies in his generous educational patronage; he endowed a college at Oxford (now New College) in 1379, and a school at Winchester in 1382. His school was the first independent and self-governing school in the country and became the pattern for Henry VI's foundation at Eton.

HOWARD SAINSBURY

WILLIAM (FITZHERBERT) OF YORK (d.1154). Archbishop of York. Of noble birth, he was a chaplain of King Stephen of England, and by 1114 had become treasurer and canon of York Cathedral. As Stephen's candidate he was elected archbishop in 1142. He was opposed, however, by the Yorkshire Cistercians, whose candidate was the strict Cistercian, Henry Murdac. They attributed his election to simony and royal pressure, and Theobald,* archbishop of Canterbury, refused William consecration. Both sides appealed to Rome. Despite the opposition of Bernard of Clairvaux* and his whole Cistercian Order, Innocent II cleared the way for William's consecration at Winchester (1143) by Henry, bishop of Winchester, who was the king's brother as well as papal legate. Complaints were renewed against William at the accession of a Cistercian to the papacy. Influenced by Bernard, Eugenius III suspended William from office in 1147. He was deposed by the Council of Reims (1147) after his supporters burned the Cistercian Fountains Abbey where Murdac was now abbot. Murdac was elected to York in his place, and William, who found refuge with his friend, the bishop of Winchester, devoted himself to prayer and study.

When Bernard, Eugenius, and Murdac died (1153), William's appeal for restoration was granted by Pope Anastasius IV (1154). William's death, one month after his return, was said to have been caused by poisoning. Considered a martyr and revered for his sanctity as well as for miracles alleged in connection with his return and after his death, William was canonized in 1226. The events of his career constitute a notorious example of twelfth-century ecclesiastical politics.

ALBERT H. FREUNDT, JR.

WILLIAMS, CHARLES (1886-1945). English writer. Born in London and educated at St. Albans and University College, London, he spent the greater part of his career in the service of the Oxford University Press. As a writer his range was wide, covering religious drama (*Thomas Cranmer of Canterbury* in 1936 followed Eliot's *Murder in the Cathedral* at the Canterbury Festival in the preceding year); Arthurian legend, best represented by *Taliesin through Logres* (1938); what he himself described as "metaphysical thrillers"; biography, criticism, and theology. The last includes *He Came Down from Heaven* (1937), *Descent of the Dove* (1939), and *The Forgiveness of Sins* (1942).

Williams in his youth had links with the group of "Rosicrucians,"* of which Yeats was at one time a member. To this association with the occult should be added the influence of the Christian mysticism of Evelyn Underhill.* The relationship of different spiritual states to various parts of the body, an idea central to the symbolism of *Taliesin,* owes something to the Kabbalah. The importance of Dante must also be noted in Williams's development, especially in the positive affirmations of his faith. Hence his stress on the Incarnation and on the active work of the Spirit in history *(Descent of the Dove)* and the life of society. For Williams there could indeed be a real *civitas Dei.*

He is perhaps best known as a novelist, but even here he appeals only to a special taste, prepared to accept his treatment of serious religious themes in a thriller mode and through stylized dialogue. His work in the drama also employs symbolism, but less subtly, and in any case his plays are altogether lesser achievements. In the end he will probably be remembered most for the difficult and Blake-like *Taliesin through Logres,* a poem which grapples with the religious significance of the Arthurian story in a series of epic odes.

ARTHUR POLLARD

WILLIAMS, DANIEL (1643?-1716). Presbyterian divine and founder of Dr. Williams' Library in London. Born near Wrexham in Wales, he became an itinerant preacher by the age of nineteen. Without formal academic training but nevertheless widely read, he accepted the post of chaplain to the countess of Meath in Ireland in 1665. From 1667 he was a Nonconformist pastor in Dublin. There he gained his largely theoretical admiration of Presbyterianism. Leaving Ireland in 1687 during the troubles, he moved to London where he became pastor of the Presbyterian congregation in Hand Alley, Bishopsgate. During the 1690s he was the acknowledged leader of the London Presbyterians, especially so after the death of Richard Baxter. His Calvinism was of the Amyraldian* variety, deemed "Neonomianism" by those with whom he engaged in bitter controversy. He was twice married, and it was money received from his wives that enabled him to be a philanthropist—hence the library.

PETER TOON

WILLIAMS, SIR GEORGE (1821-1905). Founder of the Young Men's Christian Association.* He was born at Dulverton, Somerset, the son of a farmer. Apprenticed to a draper in Bridgwater, he was converted through reading the work of C.G. Finney* and joined the local Congregational church. In 1841 he went to work in a London draper's, and later rose to be a partner in the firm. In 1844 a meeting of twelve young men in Williams's room is generally recognized as marking the founding of the London YMCA. Thereafter

the history of that international movement cannot be understood without reference to the tireless labors, practical wisdom, and catholic spirit of this businessman, evangelist, temperance advocate, and social reformer. He was knighted in 1894.
IAN SELLERS

WILLIAMS, ISAAC (1802-1865). Welsh Tractarian* poet and theologian. Born at Cymcynfelyn near Aberystwyth and educated at Harrow and Trinity College, Oxford, he was deeply influenced by John and Thomas Keble and by Richard Hurrell Froude. Ordained in 1829, he was tutor and dean of Trinity by 1833, and curate to J.H. Newman* at St. Mary's. He contributed verses to the *Lyra Apostolica* (1836) and wrote many other poems, including "The Cathedral" (1838) and "The Baptistery" (1842), and produced many translations of hymns from Greek and Latin, of which the best known is "Disposer Supreme." He was generally recognized to be the natural successor to John Keble* for the professorship of poetry at Oxford in 1841/2, but his Tract 80 on *Reserve in Communicating Religious Knowledge* aroused great alarm and antagonism in the Anglican Church and cost him the chair. He spent the rest of his life in semi-retirement, writing hymns, poetry, sermons, and devotional works.
HOWARD SAINSBURY

WILLIAMS, JOHN (1796-1839). Protestant missionary, known as "the Apostle of Polynesia." He and his wife were sent out by the London Missionary Society in 1817 to Eimeo, one of the Society Islands near Tahiti. In 1823 he discovered Rarotonga and founded a mission there. He later translated parts of the Bible and other books into Rarotongan. He was a born leader and a man of great zeal. A training school to augment the missionary force for carrying the Gospel to other islands was launched by Williams. He built a vessel, *The Messenger of Peace*, to be used in evangelizing the South Sea Islands. By 1834 no island of importance within 2,000 miles of Tahiti had been left unvisited. From 1834 to 1838 he returned to the British Isles to conduct an extensive speaking tour, familiarizing many with the evangelistic opportunities in the South Seas* and creating much enthusiasm. His *Narrative of Missionary Enterprises in the South Sea Islands* (1837) helped in this. Returning there on 20 November 1839, he landed at Dillon's Bay, Erromanga, in the New Hebrides, to be met by savages, killed, and eaten, in return for cruelties previously inflicted by British sailors. Thousands of converts mourned his martyrdom. A new burst of enthusiasm for missions was generated, and a succession of ships bearing the name *John Williams* was employed in evangelizing the area for many years.
KEITH J. HARDMAN

WILLIAMS, R.V., see VAUGHAN WILLIAMS

WILLIAMS, ROGER (1603?-1683). Founder of Rhode Island. Born in London, he so impressed the famed jurist Edward Coke with his ability at shorthand that Coke sent him to Pembroke College, Cambridge, where he graduated and went on until episcopally ordained in 1629. He became chaplain to William Masham's family, married, then in 1631 because of Separatist views migrated to Boston. There he refused a position as teacher because the church was not separated from the Church of England. He then traded with the Indians and was assistant pastor to Ralph Smith for house and land in Plymouth (1631-33). He became teacher in the Salem Church in 1633 until in 1635 his views brought his banishment. He fled to Providence in 1636, bought land from the Indians to found Rhode Island, and founded also a Baptist church. He went to England in 1642 and secured a charter for the colony. He engaged in a pamphlet war with John Cotton* and wrote *The Bloody Tenet of Persecution* (1644) and *The Bloody Tenet Yet More Bloody* (1652). During a second stay in England (1651-54) he was able to safeguard the colony's charter, and on his return became president of Rhode Island until 1657. His views that the church and state must be separate and that the state must not coerce the conscience of the individual are a treasured part of the Baptist heritage in America.
EARLE E. CAIRNS

WILLIAMS, SAMUEL WELLS (1812-1884). Early American missionary, diplomat, and authority on China. Specially trained as a printer, he was sent by the American Board to Canton in 1833. There he cooperated with Elijah Bridgman* in editing and printing *The Chinese Repository* and other literary works. Unsettled conditions forced them to move the press to Macao in 1835. During an extended furlough in the USA (1844-48), Williams produced his great two-volume *Middle Kingdom*, for many years the standard work on China. He also learned Japanese from some shipwrecked sailors and in 1837 was part of the expedition that tried to return them to Japan. After 1853 he became involved in diplomatic affairs, first as interpreter on Commodore Perry's visit to Japan, then for twenty years as secretary and interpreter to the American legation in China. He was professor of Chinese language and literature at Yale (1877-84), ninth president of the American Bible Society (1881-84), and president of the American Oriental Society.
HAROLD R. COOK

WILLIAMS, WILLIAM ("of Pantycelyn") (1717-1791). Welsh Methodist leader, author, and hymnwriter. Although known to English readers as nothing more than the author of "Guide me, O thou great Jehovah," Williams, whose work is almost entirely in Welsh, is the most significant literary exponent of the mind and spirit of the Evangelical Revival. He was born at Cefncoed, Llanfair-ar-y-byrn, Carmarthenshire, son of a ruling elder at the Congregational Church at Cefnarthen. Williams was educated at the Dissenting Academy at Llwyn-llwyd and it was then that he experienced evangelical conversion under the ministry of Howel Harris* when the latter was preaching in the churchyard of Talgarth. He joined the Church of England and was ordained deacon in 1740, but he was refused ordination as a priest in 1743. In the intervening years he served as a curate. After this unhappy experience, he devoted himself to the Methodist Revival as an itinerant preacher and the ablest conductor of the societies that were springing into existence.

Williams was the most prolific of Welsh Methodist authors. Some ninety titles were published under his name between 1744 and 1791. His greatest contribution was as a hymnwriter and poet. In these productions the power, the unction, and the passion of the Methodist Revival are fully expressed. In two long poems, *Golwg ar Deyrnas Crist* (*View of Christ's Kingdom,* 1756, 1764) and *Bywyd and Marwolaeth Theomemphus* (*Life and Death of Theomemphus,* 1764, 1781), he describes the objective christological framework of his thinking on the one hand, and on the other the spiritual trials and eventual triumph through grace of the redeemed sinner, Theomemphus. These two works provide the golden threads that run through his hymns, which were published from time to time in twenty-four publications. His masterly exploration of the inner life—its joys, fears, trials, and victories—was a unique contribution to evangelical literature. The other golden thread is his "praise of the Lamb" (to use Moravian terminology). Ann Griffiths* alone among Welsh hymnwriters comes close to him in the ability to celebrate successfully the grace and power of the Redeemer. Williams's hymns were not merely literary exercises, but were intended as a practical contribution to the work of the Revival; its history in Wales cannot be fully written without putting adequate emphasis on the role of these hymns in evoking and sustaining the revival spirit.

Williams's prose writings do not reach the same standard as his poetry and hymns, but they are of great interest to the student of the period. His *Ductor Nuptiarum* (1777), for example, is a treatise bearing upon sexual ethics that is far removed from the prudery that characterized Evangelicals of a later age. His handbook for conducting society-meetings (*Drws y Society Profiad,* 1777) throws valuable light on the principles that animated Methodist leaders at their class-meetings. The breadth of Williams's interests—the kind of enthusiasm that was later to flower in the foreign missionary enterprises—is demonstrated in his very substantial account of the religions of the world (*Pantheologia,* 1762-79).

Amid all these literary labors, Williams, like his colleagues, was constantly engaged in preaching tours throughout Wales. Of all the Methodist leaders, he was the ablest spiritual physician, with a rare gift for assisting people in their religious and psychological difficulties. His stature as a poet has continued to increase in the estimation of critics, and his influence continues among Welsh Christians, since large numbers of hymns are included in all the denominational hymnbooks.

BIBLIOGRAPHY: Biography by G.M. Roberts, *Y Pêr Ganiedydd;* his works are in process of being published in a definitive edition; S. Lewis, *Williams Pantycelyn* (1927), is a brilliant (though controversial) literary appraisal. There are a vast number of articles and monographs on his work.

R. TUDUR JONES

WILLIAMS, WILLIAM ("o'r Wern") (1781-1840). Welsh preacher. He was born in the parish of Llanfachreth, Merioneth, the son of a carpenter, a craft which he himself followed in early life. He was educated at Wrexham Academy and ordained into the Congregational ministry at Wern, near Wrexham, in October 1808. He moved to the Tabernacle, Crosshall Street, Liverpool, in 1836, but returned to Wern in 1839 and died there five months later. Tradition has it in Wales that Williams, John Elias,* and Christmas Evans* are the three greatest names among the nation's preachers. Unlike the two others, Williams's preaching style was quiet and persuasive rather than tempestuous. The very large congregations that gathered to hear him would listen in rapt silence to his carefully developed themes, and the stillness would be broken by an occasional sob. He was an early protagonist of the "Modern Calvinism" of Dr. Edward Williams, and in his sermons he demonstrated the philosophical cast of his mind by the care and lucidity with which he developed the doctrinal theses which were his chosen topics. His influence brought many congregations into existence in the Wrexham district, and throughout N Wales he brought a new evangelical and missionary enthusiasm into being just at a time when the great flood of the Methodist Revival was showing signs of ebbing. R. TUDUR JONES

WILLIBRORD (658-739). Anglo-Saxon missionary and "Apostle to the Frisians." Born in Northumbria, Willibrord was educated at the monastery of Ripon, near York, headed by Wilfrid.* In his twenties he went to Ireland, where he became enthusiastic about becoming a missionary; his mentor, the Anglo-Saxon monk Egbert* (active in trying to influence the Irish Church to join the Anglo-Saxon Church* in recognizing papal direction) suggested Frisia, where Wilfrid had already briefly preached in 677. Frankish power had just expanded northward, under Pepin of Heristal, to include the commercially important southern edge of Frisia (which at that time included the coastal regions from Schleswig south to Flanders). In 690 Willibrord and eleven companions sailed across the Channel to Frankish Frisia, were greeted by Pepin, and began missionary work. At thirty-seven Willibrord went to Rome in 695, to be made archbishop of a new Frankish church province to be centered at Utrecht. He established the famed monastery of Echternach, in Luxembourg, and supervised a vigorous mission effort, which gained much success in the area under Frankish control. The N Frisian counterattack after Pepin's death, led by the pagan Radbod, halted progress temporarily (715-19); but under Charles Martel the Franks regained S Frisia. Aided by Boniface,* Willibrord continued working with the Frisians, and much of the Frankish-controlled region was Christianized by the time of his death. Boniface carried on his work.

See G. Huelin, *St. Willibrord and His Society* (1960). DIRK JELLEMA

WILSON, DANIEL (1778-1858). Anglican bishop. Son of a rich London silk manufacturer, Wilson was intended for a business career, but experienced an evangelical conversion and decided to enter the ministry. He studied at St. Edmund Hall, Oxford, of which he was later vice-principal, and exercised several highly successful ministries, one of which was at St. Mary's, Islington (1824-32). Then through the influence of Charles Grant he was made bishop of Calcutta at the late age of fifty-four. He found the affairs of the diocese in confusion, but by unrelenting work and spiritual discipline restored order and established his influence over chaplains and missionaries alike. He waged war on the caste system, built Calcutta Cathedral and many new churches, secured in 1833 freedom of missionary activity from the control of the East India Company, and created two new sees (Madras and Bombay) which gave him metropolitan status over the whole of India. Vigorous, unbending, and determined to hold others to his own high standards, he helped the Anglican Church in India to grow in power, zeal, and esteem. He was in Serampore during the Mutiny, and died at Calcutta.

IAN SELLERS

WINCHELSEA, ROBERT OF (c.1240-1313). Archbishop of Canterbury from 1293/4. Born probably at Winchelsea, Sussex, he studied at Paris, where he became rector (1267), and at Oxford, where he was made chancellor (1288). In 1283 he was appointed archdeacon of Essex, and in 1293 was elected archbishop of Canterbury, but a vacancy in the papacy delayed his consecration until 1294. As champion of ecclesiastical rights and papal authority, he resisted Edward I's demands for clerical subsidies for war with France, but delayed for a year publishing Boniface VIII's bull *Clericis laicos* (1296) forbidding clergy to pay taxes to lay rulers. Accommodation was finally reached; but when Edward's vassal became Pope Clement V, Winchelsea was suspended (1306-8) until Edward's death. Recalled by Edward II, he was soon at odds again, joining the barons in their struggle with the king. Winchelsea was a capable administrator; he loved power, was jealous of his dignity, and was constantly involved in political and religious quarrels in which he used the power of excommunication.

ALBERT H. FREUNDT, JR.

WINDISCH, HANS (1881-1935). Biblical scholar. Born in Leipzig, he began his teaching career as a private tutor in his home city. In 1914 he became a professor of biblical literature in Leyden; in 1929 he moved to Kiel, and in 1935 to Halle. He was a leading member of the *Religionsgeschichtliche Schule* (History of Religion School, i.e., the use of comparative religious material in the interpretation of Christianity). His publications were either commentaries on books of the NT (e.g., on 2 Corinthians and Hebrews), expositions of parts of the Bible (e.g., *The Meaning of the Sermon on the Mount,* 1929), or studies of early Christian thought (e.g., *Baptism and Sin in Primitive Christianity,* 1908; and *Philo's piety and its significance for Christianity,* 1909).

PETER TOON

WINE, see COMMUNION, HOLY

WINEBRENNER, JOHN (1797-1860). German Reformed pastor. Born in Maryland, he studied at Dickinson College, Carlisle, Pennsylvania (1855-58). He was a pastor in the German Reformed Church from 1820 to 1825 at Harrisburg, Pennsylvania. At that time he conducted revivals among the Germans in the surrounding area and was censured for his evangelistic preaching. He withdrew from the German Reformed Church and in 1830 formed the General Eldership of the Church of God, stating his opposition to all creeds, forms, and nonbiblical names. The only creed was to be the Bible, since manmade creeds had led to sectarianism. The only cure for this was the restoration of the primitive apostolic faith and purity. By 1845 three elderships had been founded, and these were brought under the General Eldership of the Church of God in North America, the title still used. Arminian doctrines were held, and the ordinances of baptism, the Lord's Supper, and foot-washing became obligatory. Each local church has a council consisting of the pastor and of elders elected by the congregation. Originally in Maryland and Pennsylvania, the churches are now found elsewhere in the United States.

KEITH J. HARDMAN

WINER, JOHANN GEORG BENEDIKT (1789-1858). Protestant NT scholar. Born and educated in Leipzig, he became privatdocent in theology there. He was called to a chair at Erlangen in 1823, but returned to Leipzig as professor of theology (1832-58). He fixed the rules of grammatical interpretation of the NT, especially in his *Biblisches Realworterbuch* (1820). Of his many works, the most notable is his *Grammatik des neutestlichen Sprachidioms* (1821). W.F. Moulton* translated and edited the third edition of this work in 1870, and it is still invaluable within its limits, despite more recent advances in understanding Hellenistic Greek.

WINFRITH, see BONIFACE

WINKWORTH, CATHERINE (1829-1878). Translator of hymns. She lived near Manchester, England, until 1862, when she moved to Clifton. She was a pioneer in women's higher education, a founder of Clifton High School, and she prepared the ground for the establishment of the University College at Bristol. In 1853 she first met C.C.J. von Bunsen,* whose *Gesang- und Gebetbuch* (1833) she drew upon for her translations of German hymns. Over 300 were published in her *Lyra Germanica* (2 series, 1855-58) and *The Chorale Book for England* (1863; new ed., 1865), the latter having as music editors W.S. Bennett and O. Goldschmidt. In her biographical *Christian Singers of Germany* (1869) she included more than 100 translations, some having previously appeared. Among her many fine translations are those from Gerhardt,* Heermann,* Neander,* Nicolai,* Rinckart,* and Scheffler.*

JOHN S. ANDREWS

WINSLOW, EDWARD (1595-1655). American Pilgrim leader. Born at Droitwich, Worcestershire, of prosperous parentage, he received a private education and became associated with John Robinson's* Separatist movement in the Netherlands, where he married Elizabeth Barker in 1618. Winslow sailed with the *Mayflower*, landing at Plymouth in December 1620. He was employed as an Indian agent and envoy to England on several occasions. He was often Plymouth's assistant governor from 1624 to 1646, and governor in 1633, 1636, and 1644. He helped to found the "Society for the Propagation of the Gospel among the Indians in New England" (1649). Archbishop Laud* imprisoned him for four months while he was in England in 1634, for teaching and performing marriages as a layman. After serving under O. Cromwell* in a maritime dispute, he died on the way back to Plymouth in 1655 and was buried at sea. His published narrative of the years 1621-23 are an invaluable source to historians of early Plymouth.

ROBERT C. NEWMAN

WINTHROP, JOHN (1588-1649). English lawyer who became governor of Massachusetts. Born in Suffolk, he attended Trinity College, Cambridge, and then studied law at Gray's Inn. Elected governor of the migrating Puritans, he led the exodus of 1630 to Massachusetts where he served as governor many times. His principles, enunciated in *A Model of Christian Charity* (1630), focused the life of the colony on the covenant of God with His people, patterned after OT Israel, in which corporate obedience to God's law of love was the major condition determining whether God would bless or curse the enterprise. The civil magistrates accordingly attempted to maintain both moral purity and theological conformity, the latter principle leading to the exiling of Anne Hutchinson* and Roger Williams.* Winthrop helped to organize, and was the first president of, the New England Confederation of 1643.

GEORGE MARSDEN

WISEMAN, NICHOLAS PATRICK STEPHEN (1802-1865). English cardinal, first archbishop of Westminster (1850-65), and chiefly responsible for the reestablishment by Pius IX* of the Roman Catholic hierarchy in England (1850). Anglo-Irish by birth, rector of the English College, Rome (1828-40), he began promoting the restoration, based on hopes of Catholic revival and the conversion of England, prompted by news of the Oxford Movement.* He returned to England as president of Oscott College, Birmingham, and coadjutor bishop to the vicar apostolic of the Midlands (1840). In 1847 he undertook a special mission, in the name of the English bishops, to the new Pope Pius IX to urge restoration, which occurred in September 1850. Pius IX named him cardinal and first English Catholic primate. Wiseman's enthusiastic announcement of it (October 1850) awakened popular anti-Catholicism, abetted by Prime Minister Lord John Russell's alarm against "papal aggression." Wiseman promoted Ultramontane* principles and practices, established English branches of religious orders, and organized basic Catholic ministries among the newly immigrant Irish. Among many writings he published a scholarly study on Syriac (1827) and lectures on science and the Bible (1835), founded the *Dublin Review* (1835), and wrote a popular novel *Fabiola* (1854).

See biographies by W. Ward (2 vols., 1897), E. Gwynn (1950), and B. Fothergill (1963).

C.T. MC INTIRE

WISHART, GEORGE (c.1513-1546). Scottish Reformer and martyr. Born into a family with aristocratic connections, he evidently graduated from Aberdeen, having acquired a knowledge of NT Greek which he taught pupils at school in Montrose. Charged with heretical tendencies, he went first to England, then to the Continent, where he became acquainted with the first Helvetic Confession* which he was the first to translate into English. He returned to England about the end of 1542, spent a year teaching at Cambridge, then in 1543 or 1544 went back to Scotland and preached the Gospel, particularly in Angus. He based his exposition on the Ten Commandments, the Lord's Prayer, and the Apostles' Creed. In Dundee he ministered fearlessly to those sick and dying of the plague, and survived an attempt on his life. The converted priest John Knox* was found thereafter bearing a sword for his protection. Finally seized at Ormiston in East Lothian, he was taken to St. Andrews, and against the will of regent and people, at the instigation of David Cardinal Beaton* he was condemned to death and burnt at the stake.

J.D. DOUGLAS

WITCHCRAFT. The use of natural and/or supernatural powers to coerce or harm others in a way which arouses community concern. The boundaries between witchcraft and magic are indefinable, and both are closely related to prescientific cosmology, in which man is part of a world of spirit. Historically the demise of witchcraft in Europe is recent, and witch beliefs are still potent in Africa and Asia, where they have important social functions in relationships and as an explanation of misfortune. Yet witchcraft persists in the midst of technological and highly literate nations, and as recently as the mid-twentieth century has experienced a revival in which it takes the form of an organized and even institutionalized religion.

Christian attitudes to witchcraft have been shaped by the Bible, Roman law, and the folk customs of Europe. The early Christians believed that membership of Christ conferred immunity from demonic powers and that sin exposed one afresh. Witchcraft was associated with idolatry and a denial of love and truth, so that the sorcerer was the opposite of the saint. Penance was imposed for recourse to witches, and some penalties of Roman law were incorporated into canons. By the sixth century the witch was frequently regarded as a servant of Satan, but early medieval writers were skeptical about many of the alleged powers of witches, like aerial flight, following Augustine's argument that much witchcraft was based on illusion. Canonical sin and civil wrongs were distinguished, and though the medieval penitentials suggest the persistence of pagan witchcraft practices, there was no intellectual foundation for treating the witch as *the* menace to society. Eastern Churches developed no persecuting streak, but it is difficult to isolate the reasons for the European witch-craze.

The threat of dualist heresy, the stresses generated by Crusades, and development of the theology of Satan removed earlier ambiguities and skepticism and provided the intellectual basis for witch beliefs which were to lead to tragic persecutions in the fifteenth to seventeenth centuries. Alexander IV (1258) and John XXII (1320) permitted the Inquisition* to deal with witchcraft if it was associated with heresy. The influential *Malleus Maleficarum* (1487) further systematized witch beliefs and emphasized the need to rid society of witches. Though humanists and Reformers rejected some attributes of the medieval witch, they offered no challenge to the basic assumptions underlying witchcraft beliefs. G. Bruno* (1548-1600) showed the Renaissance fascination with the occult, and classical and biblical references to witchcraft were regarded as authoritative by Reformers.

Yet more is involved than the imposition of witch beliefs by a clerical elite. There is considerable evidence for persistent folk-beliefs, and witch-crazes which reached their peak between 1580 and 1650 also owe something to disease, deviant sexual behavior, personal and mass hysteria, blasphemous actions, hallucinogenic drugs, deceit, the search for social scapegoats, and the effect of village rivalries which were all readily interpreted in the framework of witch beliefs, which developed a momentum of their own. Torture undoubtedly led to gross exaggerations and confessions which the inquisitors like B. Carpzov (1595-1666) wanted to hear, but in England, where there was no judicial torture for witchcraft, many of the unpleasant details of continental trials still emerged. Fears of witches were also heightened by religious conflicts, social tensions, and suspicions of any strangers or extraordinary behavior. J. Weyer's (1516-68) attempts to explain that witches were only harmless old women (*De praestigiis daemonum,* 1563) was, for contemporaries, convincingly refuted by J. Bodin (1529-96) in *Démonomanie des sorciers* (1580).

Serious questions have recently been raised against the exclusively Protestant origins of witch beliefs in England, where the Essex trials were alleged to have close links with Puritan ideas about witchcraft imported from Europe. Translations of L. Daneau (1530-95) and L. Lavater (1527-86) and W. Perkins* (*Discourse on the damned art of witchcraft,* 1608) undoubtedly had some influence on the literate, but they did not erect the full-blown demonology of Europe, and magistrates were more influenced by legal precedents in sentencing than by theological considerations. Except for a brief period during the Civil War (1645-47) when M. Hopkins hunted out witches remorselessly, English witch accusations were rarely official in origin. Most Essex trials stemmed from tensions of village life, breakdown of mutual help for the needy, and the end of the ritual protections against witchcraft which had been provided by the Medieval Church. Recourse to law was provided by acts of 1563 and 1604, providing an important channel for the release of tensions over grievances and misfortune, until men began to apply other explanations.

Witchcraft beliefs retained considerable power in the latter part of the seventeenth century, as the Swedish trials (1668-77) and the Salem craze (1692) showed. In tolerant Holland, B. Bekker (1634-98) was disgraced for his denial of Satan in *Betoorverde Weereld* (1691), and another critic of witchcraft beliefs in Germany, C. Thomasius (1655-1728) was careful to assert his belief in both witches and a devil, but attacked witch trials (1701). Many great names were still to be found supporting the intellectual superstructure of witch beliefs.

The reasons for the decline of witch beliefs in Europe during the seventeenth to nineteenth centuries are obscure. More was involved than declining religious fanaticism and expanding rationality, for anthropological studies of modern witchcraft suggest that the replacement of witch beliefs is a long and complex process. A new cosmology, deepened insight into the theology of creation, decline in angelology and demonology, rejection of torture and witch hunts as a satisfactory legal procedure, growing religious skepticism—all contributed. Recourse to "cunning" men and women was still common, but witchcraft was no longer used as an overall explanation for the mysterious and misfortune, even though technology was not advanced enough to make magic superfluous. There were important regional differences in the process. Changes in legal procedure (abolition of the charge of *sorcellerie sabbatique* in 1672) were significant in France, while in England the emergence of a more individualistic social morality and public acceptance of responsibility for poor relief removed some of the tensions which inspired witchcraft accusations. In England there was also a growing reluctance to convict. The last trial was held in 1717 and the witchcraft laws were repealed in 1736, though in England and elsewhere there were extralegal acts of violence against suspected witches as late as the nineteenth century.

Witchcraft beliefs survived in small, esoteric groups in Europe, but they are still a serious practical problem in African churches where Christians remain close to the old cosmology. Between 1956 and 1964, Chikanga exercised enormous influence in Central and East Africa, and African Zionist churches show the power of the old beliefs.

BIBLIOGRAPHY: N. Paulus, *Hexewahn und Hexenprocess* (1910); H.C. Lea, *Materials towards a history of witchcraft* (1939); E. Delcambre, *Le Concept de la Sorcellerie dans Lorraine* (1949-51); J.C. Baroja, *The world of witches* (1964); C. Ginzburg, *I Benandanti* (1966); R. Mandrou, *Magistrats et Sorciers en France* (1968); H. Trevor-Roper, *Religion, the Reformation and Social Change* (1968); L. Mair, *Witchcraft* (1970); M. Douglas (ed.), *Witchcraft confessions and accusations* (1970); A. Macfarlane, *Witchcraft in Tudor and Stuart England* (1970); K. Thomas, *Religion and the Decline of Magic* (1971).

IAN BREWARD

WITHERSPOON, JOHN (1723-1794). President of Princeton University. Born at Gifford, East Lothian, a lineal descendant of John Knox, he graduated at Edinburgh in arts and divinity and was successively minister of the parishes of Beith (1745) and Paisley (1757). His *Ecclesiastical Characteristics* (1753) and *Serious Apology* (1764) attacked abuses and satirized the Moderates* as "paganized Christian divines." After refusing calls to Dublin and Rotterdam and the presidency of Princeton, he accepted the latter post (1768) and held it for twenty-five years. He improved finances, natural sciences, and languages, and himself lectured in divinity, moral philosophy, and eloquence. In American Presbyterianism he worked for union with Congregationalists and Dutch Reformed, favored a general assembly, and influenced Scots and Irish Presbyterians to support the Whigs. A member of several political assemblies and the Continental Congress (1776-79; 1780-83), he encouraged the Declaration of Independence and was the only cleric and educator to sign it. C.G. THORNE, JR.

WITTENBERG, CONCORD OF (1536). An agreement reached by Lutheran and Zwinglian theologians on the disputed doctrine of the Lord's Supper. Following a preliminary conference between Bucer* and Melanchthon* at Cassel in 1534, a large group of theologians including Luther gathered at Wittenberg. A doctrinal statement drawn up by Melanchthon, setting forth an essentially Lutheran doctrine (though not insisting on ubiquity), was accepted. Bucer admitted Luther's phrases, even the crucial statement proposed by Bugenhagen,* pastor of Wittenberg, that the body and blood are truly offered to the "unworthy," who receive it to their condemnation (understanding "unworthy," however, not as "unbelievers," as did Luther, but as "careless believers"). The reunion soon collapsed, largely through the refusal of the Swiss Zwinglians to accept the Concord. The S German pastors, however, used it as a bridge to cross over into Lutheranism. J.G.G. NORMAN

WOLFENBÜTTEL FRAGMENTS. A German Deist tract written by H.S. Reimarus,* after whose death the manuscript was given to G.E. Lessing* who published portions of it between 1774 and 1778 as "Fragments by an Anonymous Writer." It rejected the validity of biblical revelation and explained the origins of Christianity from a purely naturalistic standpoint. Jesus was merely a fervent mystic whose dream of a kingdom on earth was shattered on the cross. The apostles invented the fable of his resurrection to conceal his defeat.

WOLSEY, THOMAS (c.1475-1530). English cardinal. He was born at Ipswich and educated at Oxford. He became a fellow of Magdalen College in 1497 and was ordained priest in 1498. In 1503 he became chaplain to the governor of Calais and so began his public career. He served both Henry VII and Henry VIII as chaplain. Royal service speedily led to ecclesiastical preferment. Among the many rewards he received, in 1514 he was made bishop of Lincoln and archbishop of York. In the following years he added further bishoprics and other important appointments. From all of these he gained enormous profit. The year 1515 marked the zenith of his power, when he was made cardinal by the pope and lord chancellor by the king. When in 1518 he was made papal legate, he became supreme in both church and state under the king.

He was thoroughly an ecclesiastic, although he was immersed continuously in affairs of state. He was trained in Scholastic theology, although he was no theologian. He was not purely a secular figure and frequently said Mass. Although he was not zealous in the pursuit of heresy, he was an orthodox Catholic in his outlook. His religion was "probably highly conventional, but not purely formal."

Throughout his career the two authorities whom he served—pope and king—were in harmony. When that harmony was broken by the king's "divorce," Wolsey's career was shattered. His great house, Hampton Court, his college at Oxford, and most of his appointments and wealth were confiscated in 1530. In the days after his fall he made an attempt to fulfill his duties as archbishop of York for the first time. None of this saved him, and only his death on the way to London cheated the royal executioner. He is well described as "not creative or reflective," but an "uncomplicated activist" who was "a magnificent if extravagant manipulator of what was available."

See A.F. Pollard, *Wolsey* (1929); editions by G.R. Elton (1965) and A.G. Dickens (1966).

NOEL S. POLLARD

WOMEN IN THE CHURCH, PLACE OF. Although Christ was not married and none of the Twelve was a woman, women played a decisive role in His ministry, which was often directed to the needs of the female sex (cf. Matt. 9:20ff.; Mark 7:25ff.; Luke 10:38; John 4:7ff.). At the crucifixion "many women were there, watching from a distance. They had followed Jesus from Galilee to care for his needs" (Matt. 27:55). Mary Magdalene and the "other Mary" were first at the tomb on Easter morning. The Book of Acts calls special attention to the women who were in the upper room after the Ascension (Acts 1:14). Both men and women were baptized in the early church (Acts 8:12). Much attention was given to the important women who were attracted to the faith in this period of the church (Acts 13:50; 17:4,12).

Paul specifically excludes women from the official ministry of the Christian community, and requires that they neither practice the tongues gift in corporate worship or exercise places of leadership in the church (1 Cor. 14:26-36; 1 Tim. 2:8-15; 5:1-16). Perhaps more important is the fact of service rendered by women and the strong emphasis upon their full participation in the reality of the church as the body of Christ. In the former case, the important role of Priscilla and Aquila should be noted (cf. Acts 18:2,18,26; Rom. 16:3; 1 Cor. 16:19; 2 Tim. 4:19). In the extended list of believers to be greeted in the church at Rome, a large number were women (cf. Rom. 16:1-16). The specific mention of Phoebe as a "deaconess of the church at Cenchreae" clearly implies a recognized ministry of service if not of leadership. The early church never failed to note the purpose of the Incarnation as the Redeemer coming to "serve," not to be served (Mark 10:45). To speak of a "deaconess" ("one who waits on tables") is to pay the highest tribute to the role of women in the early church. Her full participation in the community is guaranteed also by the fact that sexual distinctions disappear in the body of Christ, and the marriage relationship derives its significance from the relationship of Christ and the church (Gal. 3:28; Col. 3:11; Eph. 5:21-33).

Specific mention should be made of the practice in the apostolic age of widows in the church. The ministerial class of deacons probably had its origin in the issue surrounding the neglect of the Greek-speaking (Hellenist) Jewish widows. The apostolic church seems to have assumed material responsibility for widows, but Paul warns of the abuses and dangers of this practice (1 Tim. 5:1-15; Acts 6:1-7).

Probably the two most distinguishing features of female relationships in the church in the postapostolic age are Syneisaktism and monasticism.* Syneisaktism (Gr. *suneisaktoi*; Lat. *subintroductae*) was a form of spiritual marriage appearing very early. The female partners were known as *agapētae* from the Greek term for "love" or "beloved." Some scholars have interpreted Paul's statement in 1 Corinthians 7:36-38 as referring to a spiritual marriage which under the stress of the sexual instinct has made legal marriage necessary; the former vow of spiritual marriage is then appropriately set aside. Syneisaktism grew out of the ascetic ideals popularized primarily in the East, and the strong emphasis in ecclesiastical circles upon the life of brotherly love. Evidence for this practice appears in the *Shepherd of Hermas*, and in the churches at Antioch under Paul of Samosata* and Constantinople under John Chrysostom.* At a later time, Irish monasticism was characterized by mixed communes which often led to charges of sexual abuses. A number of early church synods and councils condemned the practice, especially Nicea* (325), as well as church leaders such as Chrysostom and Cyprian.*

With the development of binding clerical celibacy* in the Middle Ages, it became customary for clergymen to have dwelling with them housekeepers whose relationship was functional as well as spiritual, if not sexual. From the same ascetic ideals developed a form of Christian monasticism among women. Macrina* is credited with being the founder of women's conventual life, especially in the Eastern Church. Paula,* one of Jerome's converts, took up the monastic calling and assisted Jerome throughout his life with his scholarly pursuits. Francis of Assisi* was equally successful in winning converts from the female section of the church, and established the Poor Clares.*

During the period of the Avignon papacy, Catherine of Siena* exercised a great influence in returning the papacy to Rome; in her is seen a kindred movement to monasticism—mysticism.* Noted mystics are Gertrude the Great,* Mechthild of Hackborn (d.1310), Mechthild of Magdeburg,* Bridget of Sweden.* The women of the Middle Ages always drew inspiration from the Virgin Mary, and stimulated the development of Mariology.* With the appearance of the Mendicant Orders of the thirteenth century, women began to play a more decisive role in education and social service, as did the monastic movement generally. Ordination and administration of the sacraments were never open to women throughout the postapostolic and medieval period.

With the Protestant Reformation came a general decline in the monastic movement, and Christian marriage took on a more positive appearance. Luther played matchmaker in arranging marriages between monks and nuns who were leaving convents and monasteries. Even within the Catholic Church, women such as Teresa of Avila* helped to bring about reform. The Sisters (or Daughters) of Charity carried on a commendable work among the poor and sick. Nevertheless women still played minor roles so far as ecclesiastical life was concerned, and general education of women, except in rare cases, did not begin until the nineteenth century.

During the second decade of the following century, Congregationalists and Baptists, particularly in England, were the first to admit women into the official ministry, as in the case of Agnes Maude Royden (originally an Anglican). But such women as Selina, countess of Huntingdon,* a patroness of evangelicalism, are also important. The Salvation Army* was a notable pioneer in giving opportunities for women to minister, and from the second half of the nineteenth century the role of missionary was open to women (in some cases the missionary movement was almost exclusively their responsibility). From 1950 most of the major denominations, apart from Roman Catholics and Orthodox, allowed women to be ordained, though their position in the Anglican Communion* is still to be precisely determined. The Protestant Episcopal Church in the USA approved the ordination of women in 1977.

In 1780 Robert Raikes* secured the services of four women to teach the catechism to children on Sunday. From these beginnings, the Sunday school* has become largely the responsibility of women in the church at almost every level. Not a few denominations, particularly in America, owe their origins to the leadership of women (see, for example, ANN LEE, MARY BAKER EDDY, and AIMEE SEMPLE MCPHERSON).

BIBLIOGRAPHY: E. Deen, *All the Women of the Bible* (1955); R.C. Prohl, *Woman in the Church* (1957); C.C. Ryrie, *The Place of Women in the Church* (1958); M.E. Thrall, *The Ordination of Women to the Priesthood* (1958); H. Lockyer, *The Women of the Bible* (1967); G. Harkness, *Women in Church and Society* (1972); D.M. Lake, "Woman," in *ZPEB* (1975). DONALD M. LAKE

WOODSWORTH, JAMES (d.1917). Canadian Methodist clergyman. He was born in Toronto and was ordained to the Methodist ministry in 1864. The same year he was sent to the Portage la Prairie circuit in the West, where he served for the rest of his life. For many years he was superintendent of Northwest missions for the Methodist

Church. Memories of his days in W Canada are recorded in the book, *Thirty Years in the Canadian North-West* (1917).

WOOLLEY, SIR CHARLES LEONARD (1880-1960). Archaeologist and author. His early life was spent in poor surroundings in Bethnal Green, London, and he was dependent on scholarships for education at St. John's School, Leatherhead, and at New College, Oxford, where he studied theology. His interest rapidly turned to biblical archaeology; for over forty years he was to excavate ancient sites, remaining a free lance throughout, although he received support from the British Museum and other institutions. In 1907 Woolley excavated in Nubia, and in 1914 (with T.E. Lawrence) he studied the ancient routes between Egypt and Palestine. He was taken prisoner (1916-18) by the Turks, and when released worked on the ancient Sumerian civilization. He is best known for his excavations in Ur of the Chaldees (1922-34), where he discovered the royal cemetery. Diggings at Atchana in the Hatay followed (1937-39; 1946-49). He had a remarkable flair for knowing where to dig.

R.E.D. CLARK

WOOLMAN, JOHN (1720-1772). American Quaker advocate of the abolition of slavery. Born in Northampton, New Jersey, he spent his youth on a farm. He always lived by the labor of his hands, principally as a tailor. Deeply pious, he became a recorded minister of the Society of Friends* in 1743 and traveled throughout the Thirteen Colonies. His *Journal* reveals a simple character lacking worldly or selfish motives, and conspicuous for an intense mystical piety. He preached against conscription and taxes for military supplies, Negro slavery, and ill-treatment of the Indians. His testimony ended in 1776 the practice by Quakers of the Philadelphia yearly meeting of owning slaves. Woolman died of smallpox on a visit to English Friends, and was buried at York. His main writings, *Some Considerations on the Keeping of Negroes* (1754) and his *Journal* (1774), greatly influenced nineteenth-century abolitionists.

IAN SELLERS

WOOLSTON, THOMAS (1669-1733). English Deist. Fellow of Sidney Sussex College, Cambridge, he wrote *Discourses on the Miracles of our Saviour* (1727-29), a blunt attack on the miracles of Christ in which Woolston claimed that they were allegories. He was a keen student of Origen* and a lover of allegorizing. As a result of publishing his theory he was sentenced to a year's imprisonment and a fine of £100. He died in prison, unable to pay the fine. There is reason to believe that Woolston may have been mentally unbalanced.

WORDSWORTH, CHRISTOPHER (1807-1885). Bishop of Lincoln. Son of a master of Trinity College, Cambridge, and a nephew of the poet, he had a brilliant career in classics and mathematics at Cambridge. He was headmaster of Harrow from 1836 to 1844, when he became canon of Westminster. He was Hulsean Lecturer in 1848-49. He then took a country living until 1869, when he was consecrated bishop of Lincoln. His many ecclesiastical works include *S. Hippolytus and the Church of Rome* (1853), a reply to Bunsen; *A Commentary on the Whole Bible* (1856-70); and *Church History to A.D. 451* (1881-83). In his *Holy Year* (1862) he provided hymns for every phase of every season. Some are still popular, eg., "Gracious Spirit, Holy Ghost"; "See the Conqueror mounts in triumph"; and "O Lord of heaven and earth and sea." A conservative High Churchman, he was involved in controversy with the Wesleyans in 1873.

JOHN S. ANDREWS

WORDSWORTH, JOHN (1843-1911). Bishop of Salisbury. Elder son of Christopher Wordsworth,* he first attained distinction as a Latin scholar. From 1878 he worked on a critical edition of the Vulgate text of the NT, amassing and collating a vast amount of manuscript material. He gave the Bampton Lectures of 1881 on comparative religion, entitled *The One Religion,* and became the first Oriel professor of the interpretation of Scripture in 1883. Made bishop of Salisbury in 1885, he was the close friend and adviser of Archbishop E.W. Benson.* He hoped for church reunion on the basis of episcopacy, maintained relations with Eastern Churches and especially studied the Swedish Church. His writings include *The Ministry of Grace* (1901), a history of Christian ministry.

JOYCE HORN

WORLD ALLIANCE OF REFORMED CHURCHES. The oldest international Protestant confessional body, the Alliance grew out of the cooperation engendered by the revivals and missionary movements of the nineteenth century. Professors J. McCosh of Princeton College and W.G. Blaikie of Edinburgh first mooted the possibility. Steps toward a meeting were initiated at the New York meeting of the Evangelical Alliance in 1873, and after extensive correspondence a meeting was held at the English Presbyterian College, London, in July 1875. The Alliance was the result. Its full title was "Alliance of the Reformed Churches throughout the world holding the Presbyterian system." Membership was open to any church organized on Presbyterian principles, which holds the supreme authority of the Scriptures of the Old and New Testaments in matters of faith and morals and whose creed is in harmony with the consensus of the Reformed Churches.

The Alliance has a confederate structure and since the first general council of 1877 has met regularly. There have been occasional constitutional changes like those of 1954, but the Alliance's role remains essentially consultative and advisory. An executive committee meets annually, and regional groupings of varying vitality exist in every part of the world. The Alliance has made major contributions to cooperation and understanding between Reformed Churches, and its significance has not been lessened by the growth of the World Council of Churches.* In addition to relief work, mutual theological consultation, and joint activities like editing Calvin's writings, the Alliance has played an invaluable role in dialogue with Rome since Vatican II. At the Nairobi General Council (1970), the Alliance merged with the International Congregational Council* and since 1963 has held conversations with Lutheran representatives in both Europe and North America. The Lüneburg Concord of September 1971 was a vital step in removing historic divisions between the two families and will have far-reaching ecumenical implications if taken seriously by member churches. The Alliance publishes a periodical called *The Reformed World.*

IAN BREWARD

WORLD CONGRESS ON EVANGELISM. This major global gathering devoted to fulfilling Christ's great commission to evangelize the earth was held in 1966 in West Berlin. Major ecumenical assemblies and conferences had been sponsored by the World Council of Churches* to discuss church unity, faith and order, and church and society concerns. The World Congress on Evangelism was a para-ecumenical effort inspired by the massive crusades of evangelist Billy Graham,* who served as honorary chairman. It was sponsored as a tenth anniversary project by the evangelical Protestant magazine *Christianity Today* whose founding editor, Carl F.H. Henry, was Congress chairman.

The congress drew participants dedicated to evangelism in more than 100 countries, most being nationals carrying evangelistic tasks in ecumenically aligned and independent denominations. Their identification within seventy-six church bodies inside and outside the conciliar movement constituted the Berlin Congress in some ways more ecumenical in scope than the World Council. Participants went back historically as far as the Mar Thoma Church* in India. Others came from young churches in Africa and Asia; youngest of all was the Auca church in Latin America sprung from the witness of five American missionary martyrs. The congress achieved a significant emphasis on evangelistic priorities and a correlation of theological and evangelistic concerns in a time when neo-Protestant reconstruction of both the doctrine and task of the church was displacing historic Christian commitments.

As the church moved into the last third of the twentieth century, Christianity represented only 28 percent of the world population and without new evangelistic vitality was doomed to become a diminishing remnant. Evangelistic momentum was slackened by Free World materialism and Communist-sphere atheism, by mass media emphasis on this-worldly concerns, and by the institutional church's preoccupation with sociopolitical issues.

The Berlin Congress achieved a significant correlation of theological and evangelistic concerns. Churchmen singled out as critically important target areas for contemporary engagement: the expanding great cities, the 20 million college and university students, the mass media, involvement of the laity, the world of computer technology, and the social dimensions of human life. Since the world population is expected to double by the year 2000, the importance of moving gospel witness into the space age and mass media age was evident. It was also noted that in the so-called silent world of 800 million illiterates, which revolutionary political forces are exploiting, evangelical Christianity has 40,000 Bible-teaching missionaries already familiar with the languages of people with an illiteracy problem.

The World Congress has stimulated subsequent regional and national conferences, including East Africa (Nairobi, 1968), Asia-South Pacific (Singapore, 1968), Latin America (1968), Eastern Europe (Novi Sad, Yugoslavia, 1969), United States (Minneapolis, 1969), Canada (Ottawa, 1970), and Western Europe (Amsterdam, 1971). In the USA, Key '73 signaled an attempt to coordinate evangelical energies at city and community levels in simultaneous and cooperative evangelism.

Asian and African interest in evangelism coincides with ecumenical missionary retrenchment abroad. Two-thirds of the world population now lives in Asia, where Christians who represent only 3 percent of the population have opened a coordinating office for Asian evangelism. A quarter of the human race, in mainland China, is sealed off from evangelism.

See *One Race, One Gospel, One Task* (Congress papers and reports, 2 vols., 1967).

CARL F.H. HENRY

WORLD COUNCIL OF CHURCHES. Founded in 1948, this is the main international agency of cooperation between the Christian churches. Its membership includes virtually all major autonomous Christian churches from both East and West, except the Roman Catholic Church and the most confessionally minded or separatist of evangelical bodies. Progressive steps in the formation and development of the WCC are symptomatic of the march of the ecumenical movement* itself in this century.

The point of origin is usually taken to be the Edinburgh (Missionary) Conference* in 1910. This was not specifically concerned with matters of Faith and Order, but with the cooperation of societies conducting missions to non-Christian peoples. The conference, however, led some to a vision of a united church, and this necessitated the facing of differences of belief through further forms of conference. Within weeks, movements were started which led to the formation of Faith and Order. World War I caused delay, and the first conference met at Lausanne* in 1927. The second was at Edinburgh* in 1937, from which came a proposal (which was accepted) made by the Life and Work Movement to form a "World Council of Churches."

Life and Work was a similar international agency, but its concern was the social program and political responsibilities of the churches. The stimulus came from Christian efforts toward peace in the decade 1910-20 (marked particularly by the World Alliance for Promoting International Friendship through the churches, and a conference it held on the eve of World War I). In 1919 a committee for Life and Work took on an existence independent of the World Alliance, and they convened a conference at Geneva in 1920 (which spent some energies on questions of war guilt). From there the movement went to its first Life and Work conference proper at Stockholm* in 1925; in the process the committee quoted approvingly the dictum "Doctrine Divides, but Service Unites."

At the second conference at Oxford* in 1937, proposals were received for the formation of a WCC, and were passed on with approval to the Faith and Order conference at Edinburgh that year. When both conferences had approved, a joint committee was appointed to bring the WCC into existence. In 1938 at Utrecht a provisional constitution was agreed, and a provisional committee of the "World Council of Churches in process of formation" set up, with headquarters at Geneva. World War II prevented the inaugural assembly from happening until 1948, when at Amsterdam* the delegates of 147 churches from 44 countries resolved that the formation of the WCC was now completed. The WCC has since held assemblies at Evanston* (Illinois) in 1954, New Delhi* in 1961, and Uppsala* in 1968, while its own departments, such as Faith and Order, continue their own conferences under the direction of the WCC itself and of its central committee.

The third strand of ecumenical work deriving most directly from the 1910 Edinburgh conference was the International Missionary Council.* This was formed in 1921 and was kept in touch with the formative stages of the WCC, without wishing to integrate with it, but it finally joined the WCC at New Delhi in 1961.

The basis of WCC membership, as amended at New Delhi, is: "The World Council of Churches is a fellowship of churches which confess the Lord Jesus Christ as God the Saviour according to the Scriptures and therefore seek to fulfil together their common calling to the glory of the one God, Father, Son and Holy Spirit."

National councils of churches (which used to be the delegating bodies to the International Missionary Council) do not send delegates to the WCC themselves, but are recognized by the WCC as "Associated Councils." The delegates are from member *churches,* and at Uppsala in 1968 there were 704 delegates from 235 member churches. Observers were also admitted at Uppsala, and the Roman Catholic Church, which until then had had no form of presence at all, participated in this way.

The year 1968 saw also a swing from the "Faith and Order" emphasis of the first two decades to a stronger "Life and Work" emphasis (often of a radical sort). A new leadership was also starting to emerge, and the long ecumenical careers of J.R. Mott,* J.H. Oldham,* and many others belonged to history. Actual progress toward union schemes was either being consummated or was in the doldrums. The needs of a torn world and the possibilities of a reformed Roman Catholicism predominate in the current thinking.

See R. Rouse and S.C. Neill (eds.), *A History of the Ecumenical Movement 1517-1948* (2nd ed., 1967); H.E. Fey (ed.), *The Ecumenical Advance: A History of the Ecumenical Movement Volume Two 1948-1968* (1970). COLIN BUCHANAN

WORLD EVANGELICAL FELLOWSHIP. This body was set up in 1951 at Woudschoten, Holland, from members of the former British-based World's Evangelical Alliance, founded in 1846 (see EVANGELICAL ALLIANCE). The European former members of the WEA did not join the new body until 1967. Membership of the WEF is at present open to national evangelical fellowships which represent a substantial proportion of the conservative evangelical community in their countries, though this is under review. Members must subscribe to a basis of faith similar to that of a number of other conservative evangelical organizations. In 1967 a WEF office was set up in Lausanne, but administration has since been transferred to London, in the care of the Evangelical Alliance there. The WEF has three associated programs: the theological assistance program; international Christian assistance, administered through Britain's Evangelical Alliance Relief Fund; and evangelistic and Bible ministries. A. MORGAN DERHAM

WORLD METHODIST COUNCIL, see METHODIST CHURCHES

WORLD STUDENT CHRISTIAN FEDERATION. Uniting forty autonomous Christian student groups, the Federation was founded in Sweden in 1895. Student leaders from Scandinavia, Europe, Great Britain, the United States, and nations receiving missionaries met under the leadership of John R. Mott,* then student secretary of the International Committee of the YMCA. For Mott and many others this was the culmination of a movement which had been drawing together such organizations for some years. Mott and others had been active in missionary-recruitment conferences previously, and growing enthusiasm in these endeavors was channeled in 1888 into a permanent organization, the Student Volunteer Movement,* which adopted as its motto "The evangelization of the world in this generation." The Federation reflected this evangelistic thrust in its stated purpose, "to lead students to accept the Christian faith in God—Father, Son, and Holy Spirit—according to the Scriptures, and to live together as true disciples of Jesus Christ." It is active today in organizing international conferences and projects in many areas, and in publishing. KEITH J. HARDMAN

WORMS, COLLOQUY OF (1540-41). The adjourned Colloquy of Hagenau* met in Worms in November 1540, although the talks did not get under way before January 1541. Johann Eck* was the spokesman for the Roman Catholics; Philip Melanchthon* for the Protestants. Agreement was reached on the doctrine of original sin. Because of the impending diet at Ratisbon,* however, the discussions were broken off, to be resumed there.

WORMS, DIET OF (1521). Charles V,* in accepting his election as emperor of the Holy Roman Empire of the German Nation, had pledged himself to call a diet as soon as possible. Moreover the Golden Bull of 1356 made a diet mandatory.

A pestilence in Nuremberg made it necessary to hold the diet elsewhere, and Charles chose Worms. In January the several estates of the empire started gathering in that city. The diet was confronted with the "Gravamina of the German Nation," the problem of what to do about Martin Luther,* and the problem of civic administration because of the many territories outside of the Holy Roman Empire held by Charles.

On 28 November 1520 Charles commanded Elector Frederick the Wise* of Saxony to bring Luther with him to the diet. Negotiations carried on by the papal legate caused Charles to rescind this request, and the *causa Lutheri* became a political question. Finally on 2 March, the emperor gave his consent to Luther's summons and promised safe conduct. Luther was required to appear because of his "teachings and books." The summons was handed on 26 March by Kaspar Sturm. Luther left Wittenberg on 2 April and arrived in Worms on the 16th. On 17 April he appeared before the diet and was asked whether the books present were his and whether he still subscribed to their contents. Luther's request (in his own words) "for time to think, in order to satisfactorily answer the question without violence to the divine Word and danger to my soul" was granted, and he was given one day for deliberation. On the 18th he answered (in part): "Unless I am convinced by the testimony of the Scriptures or by clear reason (for I do not trust either in the pope or in councils alone, since it is well known that they have often erred and contradicted themselves), I am bound by the Scriptures I have quoted, and my conscience is captive to the Word of God. I cannot and will not retract anything, since it is neither safe nor right to go against conscience. I cannot do otherwise, here I stand, may God help me, Amen." At least, so tradition has the last sentence.

On the 19th Charles notified the estates that he would defend the ancient faith against Luther. On the 24th Luther met with the archbishop of Trier and seven other princes or churchmen. At this meeting Luther continued to insist on the authority of Holy Scripture. The next day further attempts were made to deter him from his stand. Luther left Worms on the 26 April for his return trip to Wittenberg. En route he was kidnapped and taken to the Wartburg Castle. The Edict of Worms, dated 8 May 1521, declared Luther an "outlaw," together with his adherents. Other matters occupied the attention of the diet, especially the cause of justice (*Kammergericht*), but these have largely been forgotten because of Luther's appearance before the diet.

See *Luther's Works* (American ed.), XXXII (1958), pp.101-131; F. Reuter (ed.), *Der Reichstag zu Worms von 1521: Reichspolitik und Luthersache* (1971). CARL S. MEYER

WORSHIP, EARLY CHURCH. The church which meets us in the pages of the NT is a worshiping community of believing men and women. This is clear from the descriptions in the Acts of the Apostles (1:14; 2:42,46; 4:31; 5:12,42; 13:1-3; 20:7-12) and from the statements of Paul in his letters (notably 1 Cor. 10-14). By an application of the methods of form-criticism* to the NT epistles it is possible to gain a further access to the worshiping life of the Christian communities as liturgical passages (containing putative hymns, creeds, and confessions of faith) are exposed to view, and these sections are tentatively placed in a *Sitz im Leben* of the corporate life of the early Christians as they engaged in the worship of God.

From these data it is a reasonable deduction to conclude that, while Christian worship arose directly out of the matrix of the Jewish traditions in the Temple and synagogue, some distinctively new elements were added from the beginning. Many of the Jewish forms (such as the blessing of God as Creator and sustainer of life, the so-called *berakah*) were taken over and can be seen clearly in NT statements which open with "Blessed be God" (2 Cor. 1:3; Eph. 1:3; 1 Pet. 1:3 KJV), and in responses such as the familiar "Amen" (Cor. 14:16) are found. But these forms were filled with a new content which belongs to the new situation in the history of God's saving purposes for the world. Christians of the apostolic era were conscious of living in days of eschatological fulfillment which flowed from the Incarnation and redeeming achievement of Jesus of Nazareth in whom they recognized Israel's Messiah and the world's Savior. It was this conviction which stamped itself on their worship in every aspect and gave it a distinctiveness which is unique.

Tokens of that distinctively Christian pattern may be set down. Standing high in the list of features which marked out Christian worship from its antecedents in the OT and rabbinic Judaism and from the contemporary world of Greco-Roman religion is the christological reality of the risen Jesus whose promise to be with His people who assembled in His name (Matt. 18:20) was claimed and known. While there is a verbal parallel to this thought of worshipers meeting together in accord and being promised the divine presence (Pirqe Aboth 3:6 in the Mishnah, based on Mal. 3:16), the personal presence of the living Lord speaks of a dimension which is quite new. Several parts of what we may judge to be the structure of early Christian assemblies are explained only on the basis of Christ's coming to meet His people.

(1) Prayers were offered in His name (Acts 4:24; Eph. 5:20; Heb. 13:15) and hymns sung in His honor. The great christological declarations of the epistles (Phil. 2:6-11; Col. 1:15-20; 1 Tim. 3:16) most probably had their origin in gatherings for worship since they retain many of the poetic and hieratic features which enable us to classify them as hymnic in form and strophic in arrangement. Independent evidence from Pliny from a post-NT decade confirms this view of hymns offered to the cosmic Christ who as risen and exalted is now world Ruler. Ignatius appeals for church unity on the ground that a worship service is like a choir which sings hymns to Jesus Christ (*Ephesians 4*); and Tertullian alludes to the passage from Pliny in his description of Christian worship.

(2) The characteristic Christian liturgical act is the solemn meal of bread and wine which are taken "in remembrance of" Jesus Christ. Scholarly discussion has reached no consensus as to the

historical origin of this meal, whose rubric was taken from the last supper which Jesus held with His disciples before His death. Both the setting of the last meal and the precise transition by which the "last supper" became the "Lord's Supper" are subjects of debate. The simplest view is that Jesus' meal was set in the framework of the Jewish Passover meal, and that the followers of Jesus continued to observe a breaking-of-bread service in thanksgiving for what Jesus had accomplished as the counterpart of the Paschal sacrifice (1 Cor. 5:7,8) and the new life which they had come to share as the Israel of the new covenant. By the eschatological significance thus given to historical events, the Christian Communion* was practiced neither as a cult observance of a dead leader nor as a Christianized version of a Hellenistic mystery-religion, but it was a sacramental "re-presentation" of Jesus Christ, once crucified now alive, who came to greet His church and to extend His living presence to them. Only thus can we understand Paul's teaching in 1 Corinthians 10: 16; and fit the invocation *Maranatha*, "Come, O Lord!" (1 Cor. 16:22) into a meaningful pattern. See *Didache* 10 for a eucharistic setting which places in central position an invitation to the risen Lord to come to His people.

The history of the origin of the Lord's Day may be seen in this light. The shift from the Jewish holy day (Sabbath*) to the "eighth day" (Barnabas 15:8,9) of the new creation was made in recognition that Jesus "rose from the dead" and appeared to His own at a meal (so Luke 24; Acts 10:41). The thesis that the Lord's day was hailed as the first day of the week in remembrance of the appearance of the risen Lord in the context of the holy Supper is ably maintained by W. Rordorf, *Sunday* (ET 1968).

(3) It is not otherwise with the practice of baptism.* Again there are antecedent practices within Judaism,* both mainline and sectarian, which attest the influence of rites of washing as preparatory for a new spiritual experience. But the most decisive fact to leave its impress on Christian baptism was Jesus' use of the term to prepare for His impending death and vindication (Luke 12:50; 13:32,33). In Pauline teaching the roles of death-baptism are exchanged, so that whereas Jesus saw His death as a baptism, Paul describes the Christian's baptism as a death to sin (Rom. 6:4ff.; Col. 2:12). In both cases baptism was seen as an event which lacks meaning unless Christ is risen. It is not surprising therefore that baptism as initiation into the body of Christ by the Holy Spirit (1 Cor. 12:12,13) became a badge of Christian profession under the image of a "seal" (cf. Abercius's inscription, c. A.D. 182).

Attempts to find traces of a "church order" in the NT are not conspicuously successful, though parts of Ephesians and 1 John have been appealed to. The latter is supposed to contain allusions to confirmation and a confessional system. The early church is known, in its canonical literature, more for its spontaneity, improvised forms of worship (as in 1 Cor. 14:26f.), vivid awareness of the Spirit in His charismatic gifts (1 Cor. 12:4ff.; 14:1ff.; Rom. 12:6ff.) and rudimentary ministerial offices (Phil. 1:1; Eph. 4:11f.; 1 Tim. 3:1ff.; 5:17ff.; Heb. 13:17). Nothing resembling the Ignatian bishop's authority (see Ignatius, *Smyrnaeans* 8) is found, and there is no suggestion that worship is to be conducted only by a clerical elite. Signs of development in the direction of a standardizing of worship, both in structure and personnel, are seen in the *Didache** (c. A.D. 100) and *1 Clement* (A.D. 96) as well as Ignatius; and with these changes from spontaneous "congregationalism" of the earlier letters of Paul to a more ordered pattern with the use of rubrics for hymns and Scripture readings (in the later Pauline books) the ground is prepared for a decisive step toward a fixed form of worship in which the sacraments* play a vital role, church officers exercise full authority, and the Sunday* worship follows a set pattern (so, in Justin, A.D. 150).

BIBLIOGRAPHY: L. Duchesne, *Christian Worship, its Origin and Evolution* (1920); A.B. Macdonald, *Christian Worship in the Primitive Church* (1934); W.D. Maxwell, *An Outline of Christian Worship* (1936); J.H. Srawley, *The Early History of the Liturgy* (2nd ed., 1947); R.P. Martin, *Worship in the Early Church* (1964), pp.135ff. RALPH P. MARTIN

WORSHIP OF GOD, DIRECTORY FOR THE PUBLIC. Prepared by the Westminster Assembly* of Divines, this was adopted by the general assembly of the Church of Scotland and the English Parliament in 1645 as a replacement for the Church of England's *Book of Common Prayer.** Its purpose, as its title indicates, was to provide direction, rather than to lay down a liturgy that was to be followed exactly. The preface points out that the obligatory use of the Anglican Prayer Book had proven to be a detriment rather than a help to true worship, as it had insisted upon read prayers, had curtailed preaching, and had generally made worship a mechanical act. Therefore the Directory had been prepared in order to guide ministers in the conduct of services of worship, but not to provide a set form, since different circumstances might call for different orders and different actions. The principal interest of the authors was the stimulation of the spontaneous worship of God and the edification of the people. This work has formed the basis of most English-language orders of Presbyterian service drawn up since its publication. W.S. REID

WREN, CHRISTOPHER (1632-1723). Architect. Son of a clergyman and educated at Westminster School, he went up to Wadham College, Oxford, in 1650, where there was a scientific club (later the Royal Society*). He became a fellow of All Souls, where he developed some of his scientific interests. In 1657 he became professor of astronomy at Gresham College, London, and in 1661, Savillian professor of astronomy at Oxford. He held the latter post until 1673, but well before this date his major interest had moved from natural philosophy to architecture. In the 1660s he designed the new chapel at Pembroke College, Cambridge, and the Sheldonian Theatre at Oxford. His fame, however, is associated primarily with the City of London. After the Great Fire of 1666 he laid before Charles II plans for the restoration of the city. Soon afterward he became "surveyor general of the royal works." He designed St. Paul's Cathedral and some fifty-two churches and other important buildings in London. He died at the great age of ninety-one and was buried under the south aisle of the choir of St. Paul's.

PETER TOON

WULFSTAN (c.1010-1095). Bishop of Worcester. Ordained before 1038, he held various monastic offices in Worcester before his consecration in 1062—the only bishop, it was said, who obtained his see on spiritual grounds. Although he was a friend of King Harold—who said that he would go thirty miles out of his way to talk with him—he submitted to William at the Norman Conquest, and was the only Saxon bishop not replaced by a Norman. The invaders were appalled at the depths to which the English Church had sunk; Wulfstan was retained because of what William of Malmesbury calls his "simple goodness." With Archbishop Lanfranc* of Canterbury he strove against the slave trade till the practice was abandoned. Wulfstan was canonized by Innocent III in 1203.

J.D. DOUGLAS

WÜRTTEMBERG CONFESSION (1552). This document was one of a number of Interims prepared by Protestant theologians after the opening of the Council of Trent* as bases for ecumenical debate with the Roman ecclesiastics. Following on the Augsburg Confession* of 1548, Melanchthon's Leipzig Interim of 1549 and his Saxon Confession* of 1551, J. Brenz* at the request of Duke Christopher of Württemberg drew up this statement of belief containing thirty-five articles and reflecting the mind of the Protestant church in that state. Predominantly Lutheran, it contains some concessions to the Calvinists and a number to the Roman Catholics, especially in respect to the Real Presence, in which Brenz was inclined to pursue dogmatic definition more systematically than Luther. Though dispatched to the Protestant representatives at the council, the Confession was rendered abortive by the unexpected armed intervention of Elector Maurice, and all thoughts of such Catholic-Lutheran reconciliation were ended by the Settlement and Peace of Augsburg* (1555). The Confession influenced Archbishop Parker* and the Convocation of 1563, when the Edwardine Forty-Two Articles* were refashioned into the more conservative Thirty-Nine Articles,* particularly in relation to free will, justification, the canon of the OT, the Trinity, the Holy Spirit, and the Lord's Supper. A new edition of the *Confessio Virtembergica* was published in 1952, and it has featured in recent Lutheran-Catholic dialogue.

IAN SELLERS

WYCLIFFE, JOHN (c.1329-1384). English Reformer. A Yorkshireman who attended Oxford University, receiving the doctorate in theology (1372), he spent much of his life in association with the institution. By 1361 he was a lecturer at the university, but received his living from churches to which he was appointed rector. Wycliffe was a brilliant scholar, and master of the late Scholastic tradition. His talents were useful to John of Gaunt (duke of Lancaster), the son of Edward III, who summoned him to court (1376-78). Gaunt was the effective ruler of England from the death of his father until the emergence of Richard II from his minority (1381).

Wycliffe offended the church with his nationalist, pro-government views, among these being the idea that the civil government should seize the property of immoral clerics. Consequently a meeting was held at St. Paul's (1377) to which Wycliffe was called to answer for his ideas. The duke of Lancaster and the presiding bishop, William Courtenay, argued over their respective rights in the conduct of the session, and the meeting broke up without a word from Wycliffe. By 1377 the pope condemned Wycliffe's teaching in a series of bulls and warned the university to exclude him. Later (March 1378) Wycliffe appeared before the archbishop at Lambeth House, and even though an order from the government forbade his condemnation, he was told to stop spreading his views.

As long as Wycliffe's criticisms were limited to the wealth of the church and the civil power of the clergy, he kept many friends, both among the friars and the aristocracy. But when he attacked the doctrine of transubstantiation and taught a doctrine of the Real Presence (c.1380), he lost much of his support. There were also two other developments that hurt his cause, namely, the Great Schism* of 1378 which caused the English to form closer ties with the Roman Curia, and the Peasants' Revolt of 1381. Although he was not directly involved in the rebellion, his critics claimed that the disaster was implicit in his heresies.

This situation enabled Bishop Courtenay to force Wycliffe's followers from Oxford. Since he was ill, Wycliffe went to live at his parish of Lutterworth (1382). He died of a stroke (1384) and was buried in the church graveyard. In 1428, due to his heresy, Wycliffe's body was exhumed and burnt and the ashes were thrown into the Swift River.

Wycliffe was a prolific writer. Even during the last ten years of his life, when he was the focus of a sharp attack by the papacy and involved in several trials and hearings, he was so productive that even his enemies were amazed. During these years he completed a *Summa Theologica,* at least six other books, and numerous pamphlets. He instigated a translation of the Vulgate* into English (see BIBLE, ENGLISH VERSIONS), preached hundreds of sermons, continued to lecture at the university until his health failed, and counseled those involved in the "poor priest" movement. His earlier writings dealt with logical and metaphysical subjects. Later he turned to the problem of the relations between the church and the state. Some scholars believe that he was alienated from the papacy because he did not receive an important position; but it is just as likely that the Avignon* papacy caused his alienation.

He has been called "the Morning Star of the Reformation." Certainly his belief that the Bible was the only authoritative guide for faith and practice would substantiate this claim. In other ways he was a proto-Protestant. He denied transubstantiation, attacked the institution of the papacy, repudiated indulgences, and wished to have religious orders abolished. Wycliffe's teaching did not have much effect in England. His connection with the Lollard* movement is a matter of dispute. The persecution of his followers, especially by the act *De heretico comburendo* (1401), was effective. What failed in England was successful in Bohemia. Students from that land attended Oxford and took his teachings back to Prague. Through this means Jan Hus* and his followers adopted the ideas of Wycliffe and kept them alive until the Reformation era.

BIBLIOGRAPHY: H.B. Workman, *John Wyclif* (2 vols., 1926); J.H. Dahmus, *The Prosecution of John Wycliffe* (1952); K.B. McFarlane, *John Wycliffe and the Beginnings of English Non-Conformity* (1952); E.A. Block, *John Wyclif: Radical Dissenter* (1962); J. Stacey, *Wyclif and Reform* (1964). ROBERT G. CLOUSE

WYCLIFFE BIBLE TRANSLATORS. An organization dedicated to producing translations of the Bible in languages that heretofore had none. It came from the vision and initiative of L.L. Legters and W. Cameron Townsend. In 1934 they started Camp Wycliffe in Arkansas as a summer school in descriptive linguistics for pioneer missionaries. Work began among Mexican tribes in 1935. In 1942 they decided to organize formally. Two corporations were set up. The Summer Institution of Linguistics represents the scientific linguistic and cultural aspects, is non-sectarian, makes contracts with governments, and directs work on the fields. It conducts summer training programs in cooperation with state universities. Wycliffe Bible Translators represents the religious aspect, promotes the work among churches, secures financial support, and screens candidates. To qualify for service with SIL, candidates must also be accepted by WBT. Another organization, Jungle Aviation and Radio Service (JAARS), provides transportation and communications. Since 1944 operations have become worldwide with more than 1,800 active members.

HAROLD R. COOK

WYTTENBACH, THOMAS (1472-1526). Swiss Reformer. Born at Biel (Bienne), after study at Tübingen he lectured at Basle on the *Sentences*, and was influenced there by humanism and the new methods of biblical study. He lectured on the NT, especially Romans, and attacked indulgences in public several years before Luther. Zwingli* was among his pupils (1506) and said he learned from Wyttenbach that "the death of Christ alone is the price of the forgiveness of sins," and also claimed that he owed him his first serious contact with Scripture. He was people's priest at Biel from 1515, and from 1523 publicly supported the Reformation. His marriage in 1524 led to his deposition. After his death the reform of Biel was carried through by his successor, Jakob Würben.

J.G.G. NORMAN

XAVIER, FRANCIS, see FRANCIS XAVIER

XIMÉNEZ DE CISNEROS, F., see JIMENES

XYLOLATORS. From two Greek words meaning "worshipers of the wood," this was used as a term of reproach by the eighth-century Iconoclasts against the orthodox who reverenced both the symbol of their faith and representations of sacred persons and objects, which had led to the veneration of pictures, icons, and images. The Iconoclasts held that only the eucharistic bread was a true image of Christ; all other images signified a return to pagan idolatry.

YORK. City, county, and parliamentary borough and see town of the archbishopric of York. It was founded in A.D. 71 as Eboracum by the Romans as a headquarters for the ninth legion. It became the Roman military capital in Britain and was visited by the emperors Hadrian, Severus, and Constantine Chlorus (who died there). Constantine was proclaimed emperor at York. Mention is made of a bishop of York at the Council of Arles* in 314, but the Roman occupation ended soon after A.D. 400 and little is known about York or the church there till the appointment of Paulinus* as bishop in 625. Edwin, king of Northumbria, was baptized at York in 627 and founded a church there. Paulinus departed to Rochester in 633 after the defeat of Edwin by Cadwallon. For the next thirty years York came under the spiritual oversight of the bishops of Lindisfarne, who followed the Celtic customs of the church.

At the Synod of Whitby* in 663/4 King Oswy decided that the Northumbrian church should follow Roman customs and soon after, Wilfrid,* who was the leader of the Roman party, was appointed bishop of Northumbria with his see at York. He went to Gaul for consecration, and because of his delaying there, Chad* was appointed in his place. The see was restored to Wilfrid in 669, but in 678 Theodore,* archbishop of Canterbury, divided the diocese into four and appointed other bishops. In 735 under Egbert* the see was made into an archbishopric and a school was founded which included among its pupils the scholar Alcuin.* From the eleventh until the fourteenth centuries there was a struggle for precedence between the sees of Canterbury* and York. It was finally decided that the archbishop of Canterbury had precedence with the title "Primate of All England," while the archbishop of York was to be "Primate of England." Famous archbishops include Thomas Wolsey and William Temple. York Minster, dating largely from the thirteenth to the fifteenth centuries, stands on the site of Edwin's church, and the city has a unique collection of medieval parish churches still in use, as well as other ecclesiastical remains. R.E. NIXON

YORKERS, see BRETHREN IN CHRIST

YOUNG, BRIGHAM (1801-1877). Founder of the Mormon* settlement in Utah. Born in a worker's home in Whitingham, Vermont, he had little schooling, but great leadership ability. He joined Joseph Smith's* Mormons in 1832, led the group to Kirtland, Ohio, became an apostle in 1835, and chief of the Twelve Apostles in 1838. He led the Mormons from Independence, Missouri, to Nauvoo, Illinois, and thence to Salt Lake in 1847, and organized the state of Deseret. He became in 1850 governor of the Territory of Utah. He did most of the planning for the Mormon Temple in Salt Lake City and founded the University of Utah.

YOUNG, EDWARD (1683-1765). English poet. Born at Upham (Hampshire) and educated at Winchester and Oxford, he included among his works satires (*The Love of Fame,* 1725-28) and plays (*Busiris,* 1719), but his best-known work is *Night Thoughts* (1742-45). He had entered the church in 1727, becoming rector of Welwyn in 1730. His *Conjectures on Original Composition* (1759) place him among the early theorists of Romantic views of literature. Young thus belongs to two worlds, the Augustan and the Romantic. Likewise in matters religious, he is of the transition. *Night Thoughts* originates in real sorrow, in the bereavement of his wife and stepdaughter, but there is much in it of the rational Augustan theology with its moral and cosmological arguments. Meant as a poem of Christian triumph over death, *Night Thoughts* is often dismissed as a prolix and gloomy work. There is indeed too much argument, but there are also flashes of joy, particularly in the fourth *Night* with its celebration of Christ's victorious death and resurrection. These flashes show the intuitive response in its rare emergence from behind the repelling façade of reason. ARTHUR POLLARD

YOUNG, EDWARD JOSEPH (1907-1968). OT scholar. After graduating from Stanford University (1929) he continued his education by travel, especially in Spain and in the Near East. He trained in theology, first in Westminster Theological Seminary, Philadelphia, then under Albrecht Alt in Leipzig, before taking his Ph.D. (1943) at what is now Dropsie University, Philadelphia. He was instructor (1936-39), assistant professor (1939-46), and professor (from 1946) of OT at Westminster Theological Seminary. Young was especially outstanding for his linguistic brilliance, his gentleness and modesty, and his tenacious loyalty to the inerrant Scriptures. He conversed easily in many languages, including Russian and Arabic. Hebrew and other Semitic languages were his familiar companions. He presented his views with a genuine humility, and his deference to faculty colleagues was notable. Yet he would never surrender any demand made by the authoritative Scriptures, and he constantly adhered to that line. His major works include *An Introduction to the Old Testament* (1949); *The Prophecy of*

Daniel (1949); *My Servants the Prophets* (1952); *Thy Word is Truth* (1957); and the magisterial *Book of Isaiah* (3 vols., 1965ff.).

PAUL WOOLLEY

YOUNG, PATRICK (1584-1652). Biblical and patristic scholar. Born in Forfarshire, the fifth son of Sir Peter Young, he studied at St. Andrews (1603) and became librarian to the bishop of Chester. Incorporated at Oxford in 1605, he was ordained and made chaplain of All Souls. He became one of the most proficient Greek scholars of his time. He was librarian to Prince Henry, James I, and Charles I, assisting James in making a Latin translation of his works. He visited Paris in 1617, was made a burgess of Dundee in 1618, and became prebendary and treasurer of St. Paul's Cathedral, London, in 1621. From 1623 he was rector of Hayes, Middlesex, until deposed at the time of the Westminster Assembly (1647). In 1633 Young published from the recently arrived Codex Alexandrinus the *editio princeps* of the epistles of Clement. He published many other works until interrupted by the Civil War. After Charles I's execution he retired to his son-in-law's house in Essex.

J.G.G. NORMAN

YOUNG, ROBERT (1822-1888). Scottish theologian and Orientalist. Born in Edinburgh and apprenticed to a printer, he combined this work with bookselling from 1847, and spent much time in the study of languages. In 1856 he went to India as literary missionary and superintendent of the mission press at Surat, retaining his missionary interests when he returned in 1861. A Calvinist in theology, associated with the Free Church, and a man of meticulous mind, he is best known for his comprehensive *Analytical Concordance to the Holy Bible* (1879), which has gone through many editions.

YOUNG MEN'S CHRISTIAN ASSOCIATION. The YMCA, as an organized body of young men determined to win their fellows to a saving faith in Christ, appears to have had independent beginnings in several European countries, but its origin is traditionally ascribed to George Williams* and his meetings in London in 1844. These Bible classes and the Exeter Hall lectures that sprang from them were patronized by leading Evangelical laymen, including the Earl of Shaftesbury, and soon the movement had spread to France, Holland, the USA, and the British Empire. A series of international discussions where the presiding genius was that of Williams culminated in an important conference at Paris in 1855 which adopted the "Paris Basis" as the declaration of faith of the movement, and in 1878, enormously strengthened by the Second Evangelical Awakening, the World's Alliance of YMCAs set up a permanent executive at Geneva, the Central International Committee (CIC). In 1894 the jubilee of the movement was celebrated, appropriately enough in London.

These early decades were not untroubled, and Williams's primacy and single-minded purposefulness were often resented by British and foreign colleagues alike. Criticized at various times for being either too broad or too narrow, particularly in its prohibition of games and smoking, the YMCA gradually overcame prejudice and added recreational and relief work to its original evangelistic concern. In the world wars, with its symbol of the Red Triangle, it strove to provide especially for the needs of soldiers, the wounded and prisoners-of-war. Today with its elaborate organization of hostels, clubs, cafeterias, gymnasia, vocation training centers, and holiday homes, the YMCA has throughout the world about six million members. Students and young men living away from home or traveling abroad now chiefly avail themselves of the organization's resources. In the USA, where its principal strength now lies, education has been particularly stressed, and a number of degree-giving institutions are supported.

IAN SELLERS

YOUNG WOMEN'S CHRISTIAN ASSOCIATION. This was originally two separate organizations, both founded in Britain in 1855 by Emma Roberts and Lady Kinnaird. They united in 1877, and the first World Committee of YWCAs met in London in 1894, to be followed by the first World Conference in 1898. With the Blue Triangle as its symbol, it has had numerical growth and spread of its activities parallel to those of its brother organization. Internationalism and interdenominationalism remain the keynotes of both movements, and the original evangelistic inspiration is still to be found in them.

YOUTH FOR CHRIST, INTERNATIONAL. Set up in 1945, this organization engages in teenage evangelism, high school Bible clubs, and juvenile delinquency prevention programs. The first "Youth for Christ" rally was conducted by Paul Guiness in 1934 in Brantford, Ontario. From then on, but especially in 1943-44, Saturday night evangelistic youth rallies proliferated in large American cities. YFCI was founded in 1945 at Winona Lake, Indiana, and Torrey Johnson, leader of the Chicago rallies, was elected president. A conservative creed was adopted to help the organization keep "Geared to the Times and Anchored to the Rock." Through Saturday night rallies, "Teen Teams" sent overseas, and high school Bible clubs, youth has been evangelized at home and abroad. Youth Guidance programs help juvenile delinquents as well as prevent delinquency by counseling. Youth Guidance, *Campus Life* magazine, and high school Campus Life chapters are major thrusts in the seventies. The first world congress in 1948 in Switzerland demonstrated its spread from its present headquarters in Wheaton, Illinois.

EARLE E. CAIRNS

Z

ZABARELLA, FRANCESCO (1360-1417). Cardinal-deacon of Florence and distinguished canon lawyer. Trained in the church's legal traditions at Bologna, he taught canon law at Florence (1385-90) and at Padua (1390-1410), while serving simultaneously in the diplomatic corps of Padua and Venice. He was summoned to Rome by Boniface IX to aid in overcoming the Great Schism.* After participating in the Council of Pisa* (1409), he was created a cardinal by John XXIII* in 1411. His indebtedness to John did not prevent him from supporting the Council of Constance* (1414-18) and recommending the abdication of John as a step toward the reconciliation of the divided halves of Latin Christendom. His *Tractatus de schismate* (1402-8), based on the work of the decretists and decretalists, made proposals similar to those advocated by John of Paris,* an early conciliar theorist. The church was a corporation presided over by the pope. Papal power was of a derivative and limited kind, conferred on the pope by the members of the church. Should the pope fail to call a council in the crisis then rending the church, the right to summon a council would pass immediately to the cardinals and in the event of their inaction to the emperor, the representative of the whole people. Zabarella's defense of the supremacy of a council over a pope, and his untiring efforts at Constance, promoted the eventual healing of the schism, though he himself died before the council had completed its work.

BIBLIOGRAPHY: W. Ullmann, *The Origins of the Great Schism* (1948); E.F. Jacob, *Essays in the Conciliar Epoch* (1955); B. Tierney, *Foundations of the Conciliar Theory* (1955).

DAVID C. STEINMETZ

ZACHARIAS (d.752). Pope from 741; last of the Greek popes. A Greek from Calabria, he was the first pope to be elected without reference to imperial authority. He was noted for his charity, learning, and diplomacy, and for translating the *Dialogues* of Gregory the Great into Greek. He prevailed upon the Lombard king to abandon an attack upon Ravenna and to return four cities to the Roman duchies. A truce of twenty years was concluded. He wrote the Byzantine emperor Constantine V Copronymos in opposition to iconoclasm. He supported Boniface,* the "apostle to the Germans," whose mission everywhere extended papal authority. Zacharias had Boniface consecrate Pepin III* ("the Short") as king of the Franks, replacing the weak Merovingian line, creating the Carolingian-papal alliance, and establishing a precedent to papal claims to the right to make and depose kings. In turn, Pepin recognized the pope as head of the Papal States,* another precedent of lasting significance. Zacharias called synods in 743 and 745.

ALBERT H. FREUNDT, JR.

ZACHARIAS SCHOLASTICUS (c.465-after 536). Church historian and bishop of Mytilene. Also known as "Zacharias Rhetor," he was educated at Gaza, and later studied philosophy and law. In 492 he went to Constantinople to practice law. His two names "Scholasticus" ("advocate") and "Rhetor" ("pleader") refer to his legal career. Later in life he took up an ecclesiastical career, and although little is known of his activities, it is clear that by 536 he was bishop of Mytilene. Nothing is known of his later years, and even the date of his death cannot be established. Zacharias is best known for his church history, which was probably written before 515. It is a valuable source for events in Egypt and Palestine from 450 to 491. Among his other surviving works are biographies of Severus of Antioch and Peter the Iberian. He also wrote two polemical works, *Disputatio de Mundi Officio* and *Disputatio Contra Manichaeos.*

RUDOLPH HEINZE

ZAHN, THEODOR (1838-1933). German Lutheran biblical and patristic scholar. He taught at the universities of Göttingen (1868-77), Kiel (1877-78), Erlangen (1878-88 and 1892-1909), and Leipzig (1888-92). Though he was one of the greatest scholars of his day, he did not have the impact that he might have had, due possibly to his defense of orthodox theology in a day when this was far from popular and also to a tendency toward eccentricity in his exegesis. He was the author of many important monographs and commentaries, including twelve volumes on the canon of the NT (*Geschichte des neutestamentlichen Kanons,* 2 vols., 1888-92; *Forschungen zur Geschichte des neutestamentlichen Kanons,* 10 vols., 1881-1920), seven commentaries (Matthew, 1903; Luke, 1913; John, 1908; Acts, 2 vols., 1919-21; Romans, 1910; Galatians, 1905; Revelation, 2 vols., 1924-26), and a massive introduction to the NT (2 vols., 1897-99; ET 3 vols., 1909). In addition he edited (with Harnack and von Gebhardt) the works of the Apostolic Fathers (ed. major, 1875-78; ed. minor, 1877).

W. WARD GASQUE

ZAIRE (formerly Belgian Congo). H.M. Stanley's explorations in the basin of the Congo brought the first Protestant missionaries in 1878. The first two societies were the Livingstone Inland Mission and the British Baptists. The former was a branch of the Regions Beyond Missionary Union, and when the parent body found itself over-extended, the Congo mission was handed over to the American Baptists.

Other missions followed until they totaled some forty-six societies, a high percentage of them American. Two thirds of these societies entered after World War I. Yet in this vast region of more than 900,000 square miles there was little overlapping. Partly this was because of the Congo Protestant Council. The council began as a general conference of missionaries in 1902, when there were only eight missions and about 200 missionaries in the country. It developed into the authoritative voice for Protestantism in the Congo and a most effective channel of comity and cooperation. Today this intermission organization has been succeeded by an interchurch organization under Congolese leadership.

It was the missionaries who revealed the atrocities committed during the time of King Leopold's personal rule. This obliged the Belgian government to step in and take control in 1908. The Roman Catholic government showed a distinct favoritism toward Roman Catholic missions, which had followed the Protestants into the field, until after World War II, when a more liberal party came into power. For the next fifteen years all missions were officially on the same basis, enjoying considerable support from the government. This included subsidies for educational programs that met government standards. The government preferred to leave primary education in the hands of the missions, Roman Catholic or Protestant, with freedom to teach religion.

Medical work also has played an important role in the Congo, with most missions involved to some degree. A recent example of inter-mission cooperation is the medical center at Nyankunde, with doctors of four missions trying to meet the needs of the NE region.

By the time of independence in 1960, roughly half of the population were professed Christians, Roman Catholics outnumbering Protestants about two to one. Congo has been the scene of various native "prophet" movements, especially since 1960. The largest in all Africa is one that began with the late Simon Kimbangu in 1921 and now counts about half a million adherents.

Independence brought chaos to much of the country. In the next few years many missionaries had to be evacuated, some two or three times. A number were killed, both Catholics and Protestants, along with uncounted numbers of Congolese Christians. The church in Congo came out of these trials more self-reliant than before. The missionaries have returned, but now in a new relationship to the church, as assistants in the work.

BIBLIOGRAPHY: A.R. Stonelake, *Congo, Past and Present* (1937); G.W. Carpenter, *Highways for God in Congo* (1952); C.P. Groves, *The Planting of Christianity in Africa,* vols. III and IV (1955, 1958); R.M. Slade, *English-speaking Missions in the Congo Independent States* (1959); J.T. Bayly, *Congo Crisis* (1961). HAROLD R. COOK

ZAMORA, JACINTO, see GOMEZ, MARIANO

ZAMORA, NICOLAS (1875-1914). Filipino pastor. Grandnephew of the martyred priest Jacinto Zamora,* he was born into a family that was already studying a smuggled Bible. His father was exiled for this crime, but Nicolas continued Bible study secretly as a student of arts and then law. As a soldier in the revolutionary army, he translated his Spanish Bible into dialect and read it to the men. When the Methodists began evangelistic meetings in Manila in 1899, he was invited to speak because the Spanish interpreter failed to arrive. He attracted crowds. He was ordained deacon in the Methodist Episcopal Church in 1900, the first Protestant Filipino clergyman. After a few months of seminary study in Shanghai, he worked widely and effectively as an itinerant evangelist. From 1904 he became pastor of the largest Protestant church in Manila, where he normally preached in the local dialect.

In 1906 he was asked to quiet some nationalistic Methodists who wanted independent Filipino Methodism. Zamora eventually came to agree with them, and objected to the church's being foreign-financed and dominated by paternalistic missionaries. On 28 February 1909, after a question of church discipline had been raised by a foreign Methodist bishop, he announced the formation of *La Iglesia Evangelica Metodista en las Islas Filipinas.* Many of the older Tagalog Filipinos joined the new denomination. Zamora died in a cholera outbreak in 1914 while he was general superintendent of the new church.

RICHARD DOWSETT

ZEALOTS. The nationalistic Jewish partisans of the first century A.D., particularly those active in the Jewish War (66-70). The word "zealot" is Greek *zelotes,* and in the NT and the writings of Josephus represents the Hebrew *qanna'* or its Aramaic equivalent *qan'ana;* one of Jesus' disciples was a "zealot" or "Cananaean" (cf. Mark 3:18; Luke 6:15 RSV). The term could signify a religious enthusiast in general, then a militant nationalist, and finally a member of the religio-political party, the "fourth philosophy" (Josephus) of first-century Judaism,* which took a prominent part in the struggle against Rome.

The chief spiritual antecedents of the Zealot movement were the Maccabees, whose ardent religious zeal inspired them to take up the sword and wage a victorious campaign against their pagan Greek overlords in the second century B.C. The Maccabean ideal was not forgotten, and it revived after the Roman conquest of Palestine. The Zealot party as such was probably formed, or had its immediate origins, in the abortine revolt caused by the Roman census of A.D. 6; its first leader was Judas the Galilean (Acts 5:37), whose sons carried on the movement after his death. The last stand of the Zealots was at the fortress of Masada, captured by the Romans in A.D. 73.

BIBLIOGRAPHY: W.R. Farmer, *Maccabees, Zealots and Josephus* (1956); M. Simon, *Jewish sects at the time of Jesus* (ET 1967); F.F. Bruce, *New Testament History* (1969), pp. 88-95.

D.F. PAYNE

ZEISBERGER, DAVID (1721-1808). Moravian missionary to American Indians* for sixty-three years. Born in Zauchtenthal, Moravia, he was five when the family fled to Herrnhut* in Saxony. In 1736 the parents joined the Moravian colony in Georgia, David following two years later. In 1745, after moving to Pennsylvania, he began his missionary work with such acceptance that the Six Nations made him a *sachem* and "keeper of their archives." His greatest work, however, was with the Delawares and demonstrates the frustrations of Indian missions; persecuted in Pennsylvania, in 1772 they migrated to Ohio, where their settlements were destroyed in the Revolution. Particularly shocking was the massacre at Gnadenhütten in 1782 by colonial militia. Zeisberger accompanied his Indians to Michigan and finally to Canada, where he founded Fairfield. In 1798 Congress restored the Indians' land, so he returned to Ohio and built Goshen. HAROLD R. COOK

ZELLER, EDUARD (1814-1908). German Protestant theologian and philosopher. A student and later son-in-law of F.C. Baur,* Zeller first taught theology at Tübingen (1840-47), Bern (1847-49), and finally Marburg (1849-62); in 1862 he changed to philosophy, going first to Heidelberg (1862-72) and then Berlin (1872ff.). He was the founder of the journal *Theologische Jahrbücher* (1842-57), which served as an organ for publication of the views of the Tübingen School.* In a series of articles (1848-51) which were later published in book form under the title *The Acts of the Apostles according to its Contents and Origin Critically Investigated* (1854; ET, 2 vols., 1875-76), he attempted a detailed application and demonstration of the Tübingen theory of early Christian origins to Luke's second volume. Here he combined the *Tendenzkritik* (tendency-criticism) of Baur with the "mystical" approach of Strauss* and proceeded to call into question the essential historicity of Acts. Zeller also wrote a classic history of Greek philosophy (3 vols., 1845-52), a history of modern German philosophy (1873), and a biography of D.F. Strauss (1874). W. WARD GASQUE

ZENO. Little is known of this fourth-century bishop of Verona (c.362-75), except that he was an African by birth. His sermons, known as *Tractatus,* have certain affinities with the writings of Tertullian, although these did not come into circulation until the Middle Ages, and therefore were not known to Jerome.

ZEPHYRINUS (d.217). Bishop of Rome from 198. He succeeded Victor, but despite his nineteen-year pontificate little is known of his life. His severe critic, Hippolytus,* described him as being "lax" in the matter of enforcing the church's position against certain heresies—e.g., Sabellianism*—and characterized him as a man "without education." Zephyrinus did, however, excommunicate Theodotus, the "money changer" who defended Dynamic Monarchianism, since he had given support in large measure to two disciples of Noetus* (Epigonus and Cleomenes) who defended the Modalistic position. Tradition says Zephyrinus was martyred.

ZIEGENBALG, BARTHOLOMAEUS (1682-1719). Co-founder with H. Plütschau* of the first Protestant mission to India,* and the first to translate the NT into an Indian language. Born in Saxony, he was converted at sixteen, and was a theological student at Halle when a request came for missionaries to the overseas territories of Frederick IV of Denmark. After some hesitation about their official acceptance, Ziegenbalg and Plütschau arrived in 1706 at the Danish settlement of Tranquebar on the Coromandel coast. There they encountered incredible opposition from the authorities, apart from the predictable hostility of Hindus and Roman Catholics. As a result of one controversy Ziegenbalg was imprisoned for four months by the commandant. This first attempt to establish a Protestant mission to Indians was fought tooth and nail, but Ziegenbalg had the needed tenacity. The missionaries gathered Portuguese-speaking and Tamil congregations and built a church. The Tranquebar method became a pattern for later missions: schools, orphanage, Bible translation, printing, training of preachers, catechizing the young, and all aimed at personal conversion. Ziegenbalg produced the Tamil NT (1714) and translated the OT up to the Book of Ruth.

Annual letters of the missionaries, sent out from Halle and distributed in Britain, had great influence in stirring missionary interest (e.g., in the rectory where the Wesleys grew up). Ziegenbalg was received by the king and the archbishop of Canterbury when he visited Britain on his one home leave. "English" missions in the territories out from Tranquebar came to be financed from England although staffed by the German Lutherans of the Danish-Halle Mission. In the "home board" of the mission at Copenhagen, however, Ziegenbalg's institutional methods were severely criticized. and apparently unfairly, in his latter years. His writing on Hinduism was presumably an irrelevance to some in Europe, for it appeared in print only long after his death (one book in 1926). In every sense a pioneer, Ziegenbalg swam against the tide. He died at Tranquebar.

See H.M. Zorn, *Bartholomaeus Ziegenbalg* (in English, 1933). ROBERT J. MC MAHON

ZILLERTHAL EVANGELICALS. Protestant residents of Zillerthal, one of the Tyrol valleys, who seceded from the Roman Catholic Church and migrated to Prussia in the 1830s. Though Zillerthal was not officially Tyrolean until 1816, the people felt themselves to be Tyroleans even before the Protestant Reformation. For centuries, however, the valley had been under the archbishopric of Salzburg. A Baptist movement had been totally suppressed early in the seventeenth century, but a strong Lutheran group would not yield to the Catholic pressures. From certain other Austrian provinces Protestants had emigrated en masse in search of a land where religious toleration was practiced. Tyrol, however, had been traditionally more lenient and had avoided such mass migrations. But late in the seventeenth century two brothers named Stainer of Mairhofen preached the evangelical doctrine in the Ziller valley with such effect that the Catholic hierarchy was alarmed. Over a period of several generations they attempted by various means, including both teaching and harassment, to regain the people's loyalty. Their efforts never fully succeeded. When a Prussian court preacher named Strauss visited with them, he was so favorably impressed that he arranged for them to migrate to Prussia. In 1837 they set out in six wagons, arriving in Schmiedaberg in October of that year. They established a colony at Erdmannsdorf.

MILLARD SCHERICH

ZINZENDORF, NIKOLAUS LUDWIG, COUNT VON (1700-60). Founder of the Moravian Church. Born in Dresden to an Austrian noble family, he was the son of a high Saxon official who died during Zinzendorf's youth. He was raised by his maternal grandmother, a Pietist and close friend of Spener* and Francke,* and educated at the Halle *Pädagogium* (1710-16). A deeply religious youth, he became interested in foreign missions after meeting the Danish-Halle missionaries to India, but his family pressured him into a governmental career. In 1716-19 he studied law at Wittenberg, a center of orthodox Lutheranism, and he tried unsuccessfully to reconcile orthodoxy and Pietism.* While traveling in W Europe in 1719-20, he came into contact with Reformed theology, non-churchly groups, and Roman Catholicism which further broadened his understanding of Christianity. After entering the Saxon civil service in 1721, he sponsored religious assemblies in his Dresden home and purchased an estate at Berthelsdorf where in 1722 he invited a group of Bohemian Protestant refugees (Unitas Fratrum) to form a Christian community called "Herrnhut."* In 1727 he retired from government service to devote full time to the colony.

His religious thought matured during these years, and he broke with the Halle Pietists. He stressed "heart religion"—a deep mystical, spiritual, experiential faith—as well as Christian community, worldwide evangelism, and ecumenical relationships. He felt Francke's successors had become too rigid, while the Pietists questioned the validity of his conversion and criticized his extravagant mysticism, supposed heterodoxy, and utopian ideas of reunion with the Roman Catholic and Greek Orthodox churches. Orthodox Lutherans also attacked him, and in 1734 his beliefs were formally examined. He then became a theological candidate in Tübingen and in 1737 was ordained a bishop by the Berlin court preacher D.E. Jablonski,* which meant official recognition for Zinzendorf and his movement, although circumstances eventually forced the Moravians into a separate organization.

While visiting Copenhagen in 1731, a chance meeting with a West Indian Negro rekindled his interests in foreign missions. The first Moravian missionaries were sent to the Caribbean in 1734, and Zinzendorf himself visited St. Thomas in 1738-39. Expelled from Saxony in 1736, he settled in the Wetterau and traveled around Europe founding Moravian communities, the most significant being those in Holland and England. In 1741-43 he journeyed to America, where he labored in Indian missions and in building up the Moravian congregations. He attempted in vain to unify the German Lutheran churches in Pennsylvania, a task finally carried out by H.M. Mühlenberg* while the Moravian churches went their own way.

Zinzendorf returned to Herrnhut in 1747 and engaged in pastoral work there except for the five years (1749-50; 1751-55) he spent working with the congregation in England. His last years were marred by personal tragedy (death of his son and wife) and financial difficulties. His importance lies in the creation of a missionary, service-oriented, ecumenical free church based upon a common experience of salvation and mutual love, and the emphasis upon deep, emotional religious expression (especially in his hymns, prayers, poems, and "daily watch words") which infused new life into Protestant orthodoxy.

BIBLIOGRAPHY: J.R. Weinlick, *Count Zinzendorf* (1956); E. Beyreuther, *Zinzendorf und die sich allhier beisammen finden* (1959) and *Zinzendorf und die Christenheit, 1732-1760* (1961); A.J. Lewis, *Zinzendorf, the Ecumenical Pioneer* (1962); Zinzendorf's *Hauptschriften* (ed. E. Beyreuther, 7 vols., 1962-64); G.W. Forell, *Zinzendorf: Nine Public Lectures on Important Subjects in Religion* (1973). RICHARD V. PIERARD

ZONARAS, JOHANNES (twelfth century). Byzantine historian and canonist. At first the commander of the bodyguard of Alexius Comnenus, the emperor, Zonaras later became secretary of the chancery and then left Constantinople to become a monk on Hagia Glykeria. His most important work, a chronicle that extended to 1118, relied heavily on Dio Cassius's study. Other writings included a lexicon, several sermons, a commentary on Eastern synodal canons, and one of the canons of third- and fourth-century fathers.

ZOROASTRIANISM. The dominant religion of Persia for more than a millennium preceding the Mohammedan invasion (636), founded on the teachings of the prophet Zarathustra. (*Zoroaster* is the corrupt Greek form of the Iranian *Zarathustra.*) The historical personality is practically lost in the legendary promulgations of the followers. Although disputed, the most logical date of his birth is about the sixth or fifth century B.C. in Iran. At the age of thirty or a little later, Zarathustra had a life-changing religious experience in which he met *Ahura-Mazda* ("the Wise Lord"). This experience plus other revelations led him to become the prophet of a new, purified religion. Tradition says this new prophet was successful in converting King Vishtaspa, the ruler of E Iran, and found a powerful protector of the faith in Vishtaspa's son, Darius the Great. Zarathustra died at age seventy-seven.

He taught a new ethical religion that was firmly rooted in the old Iranian folk-religion. He bitterly attacked the cult of the gods of popular religion and promoted the worship of the one Spirit, *Ahura-Mazda* (later called *Ormazd.*) The good man joined the battle against *Angra Mainyu* (later called *Ahriman*), the chief agent of evil, in preparation for the final judgment involving the resurrection of the dead and the confinement of the wicked to the regions of torment. Each individual was to be judged according to his deeds. It is thought that this eschatology came to influence Jewish eschatology through exilic contact with the Persians. The scriptures, the *Avesta*, became the bases for the cultus which was administered by the priestly class known as the *Magi*. A major part of the worship is centered around the fire altar.

Zoroastrianism became the spawning grounds for other religions such as Mithraism* and Manichaeism,* but it came to a sudden end in Persia with the seventh-century Muslim conquest. It survives now in a small Parsi community in India located chiefly in and around Bombay. The followers have emphasized education and, therefore, hold many of the influential positions in Bombay.

BIBLIOGRAPHY: J.H. Moulton, *Early Zoroastrianism* (1913); M.N. Dhalls, *Zoroastrian Civilization* (1922) and *History of Zoroastrianism* (1938); E. Herzfeld, *Zoroaster and His World* (2 vols., 1947); J.J. Modi, *Religious Ceremonies and Customs of the Parsis* (2nd ed., 1954).

JOHN P. DEVER

ZOSIMUS (d.418). Pope from 417. Greek by birth, he inherited the controversy over Pelagianism* from Innocent I,* was initially deceived by the smooth-talking Celestius,* and summoned to Rome the African bishops, including Augustine, who had condemned Pelagius and Celestius. The Africans not only refused to comply, but mustered such support, including that of the emperor, for their viewpoint that Zosimus executed a *volte-face*. He issued a letter requiring Western bishops to endorse the condemnation. Those who refused to comply, among them Julian of Eclanum,* were deposed. Zosimus was thwarted also when he rashly tried to interfere in ecclesiastical affairs in Gaul. Several of his letters are extant.

J.D. DOUGLAS

ZOSIMUS (early sixth century). Byzantine historiographer. He provided a narrative of the Roman imperial state from its inception, skimming rapidly through its transition, but concerning events in detail from 270 until the city was sacked by Visigoths in 410. The account was heavily dependent especially upon the fifth-century authors Eunapius of Sardis and Olympiodorus of Egyptian Thebes, thereby preserving them. Consequently, considering also the nonpreservation of Books I-XIII (up to 351) of the major late Roman historian, Ammianus Marcellinus (whose Book XXXI terminates with 378), whom he does not cite, Zosimus becomes a principal source for the reigns of Constantine* and his sons, and again for Gratian* and his successors. Most importantly identified, however, not only as the last pagan historian, but as the first historian of Rome's fall, this proponent of the position that Rome's difficulties were due to the abandonment of the old gods and to the alliance with Christianity gave paradigm to Gibbon.

CLYDE CURRY SMITH

ZURBARAN, FRANCISCO DE (1598-1662). Spanish painter. Born at Fuente de Cantos, Estremadura, he painted directly from nature with pure colors, bluish tone, and careful shadow, his figures reflecting devotion and asceticism. Painting on the whole for monastic orders (1623-39), he was appointed about 1630 as painter to Philip IV who, according to legend, said: "Painter to the king, king of painters." The Madrid years (1634-39) made his style softer and lighter, later to return to earlier patterns, and influenced finally by Murillo.* His commissions were numerous: several of Bonaventura; *Apotheosis of St. Thomas Aquinas*; many for the Carthusians at Xeres; thirteen on Jerome for the Hieronymites at Guadalupe; a series of full-length portraits of Hieronymite monks; and great founders of religious orders from Elias to Loyola for the Capuchins at Castellon.

C.G. THORNE, JR.

ZURICH AGREEMENT *(Consensus Tigurinus).* In 1546 Bullinger* sent to Calvin* a work on the sacraments. Calvin criticized it frankly, and the ensuing correspondence resulted in the first draft of the Consensus in November 1548, consisting of twenty-four propositions, which Calvin prepared and Bullinger annotated. The two met in Zurich in May 1549 to draw up the twenty-six articles on the sacraments which were published in 1551. The Consensus was received by Zurich, Geneva, Neuchatel, Basle, St. Gall, Schaffhausen, and the Grisons. It stated that in the Lord's Supper we partake of Christ by the power of the Holy Spirit and the lifting of our souls to heaven. The sacramental grace, however, comes only to the elect.

ROBERT B. IVES

ZWEMER, SAMUEL MARINUS (1867-1952). "Apostle to Islam." Born in Michigan to a Dutch Reformed family, the thirteenth of fifteen children, he became a Student Volunteer when Robert Wilder* visited Hope College after the Mt. Hermon conference—and became a leader in the movement. At seminary he and James Cantine planned to start a mission in the world's most difficult field. The outcome was the Arabian Mission. Cantine went out in 1889, Zwemer in 1890. They concentrated on the Arabian Gulf area. In 1894 the Reformed Church in America assumed responsibility for the mission. In 1906 and 1911 Zwemer organized and chaired general conferences on Islam at Cairo and Lucknow. In 1911 he started *The Moslem* (now *Muslim*) *World*, and edited it for forty years. From 1913 to 1929 he made Cairo his center, worked with the Nile Mission Press, and traveled throughout the Islamic world, including India and NW China. From 1929 he had a ministry of teaching, first at Princeton Theological Seminary, later at Biblical Seminary in New York and the Missionary Training Institute at Nyack. He was also a gifted writer and authored some fifty books.

HAROLD R. COOK

ZWICKAU PROPHETS. Also known as "Storchites," Luther dubbed these three German radical reformers "the Zwickau prophets." Nicholas Storch, Thomas Drechsel, and Marcus Stübner, influenced by Taborite* and Waldensian* teachings, preached a radical biblicism which included rejection of infant baptism, denial of the need for a professional ministry and organized religion because all godly men were under the direct influence of the Spirit, special revelation through visions and dreams, the imminent return of Christ, and perhaps psychopannychism. Driven from the Saxon town of Zwickau where they originated and where they had influenced Thomas Münzer,* they visited Wittenberg in December 1521 during Luther's absence. Philip Melanchthon,* impressed with their biblical knowledge, gave them a hearing. However, their millennial enthusiasm and outspoken criticism of the Wittenberger's liturgy led to their expulsion in 1522. Little is known of their activities after this date except that they won a number of temporary converts including Gerhard Westerburg and Martin Cellarius. ROBERT D. LINDER

ZWILLING, GABRIEL (Didymus) (c.1487-1558). German Reformer. Born at Annaberg, he was an Augustinian monk and a colleague of Luther in the Wittenberg Reformation, together with Melanchthon* and Carlstadt.* During Luther's exile in the Wartburg (1521), Carlstadt and Zwilling gave a more radical turn to the Reformation, encouraged by the Zwickau Prophets* who joined forces with them. Zwilling was a fiery preacher and by October was denouncing the Mass and urging the abandonment of clerical vows. He gained a large following, especially in the Wittenberg Augustinian monastery, many monks renouncing their vows. Soon he was attacking images, and by December was leading in iconoclastic riot, encouraged by Carlstadt. The Wittenberg town council recalled Luther to restore order, which he did in March 1522. Luther recommended Zwilling to a pastorate in Zwickau; subsequently the patron dismissed him, despite protests from the people and Luther. In 1549 he spurned the Leipzig Interim of Duke Maurice, and suffered for it. He died in Torgau.

J.G.G. NORMAN

ZWINGLI, ULRICH (Huldrych) (1484-1531). Swiss Reformer. Son of a village magistrate in the Upper Toggenburg, he came from a family typical of that class of prosperous farmers who controlled the local government of the German Swiss cantons and looked to the church as the best means of improving their children's status. After attending the Latin school of Heinrich Wölflin (Lupulus) at Bern, Zwingli entered the University of Vienna where he made friends with Joachim von Watt (Vadianus), became aware of humanism, and was introduced to the *via antiqua* by Vellini (Camerinus). He completed his studies at Basle, where he absorbed the biblical interests of his teachers Thomas Wyttenbach* and Johann Ulrich Surgant, and formed a circle of friends, including Leo Jüd* and Glarean, which later brought him into direct contact with Erasmus.*

Erasmian humanism and his own experience as a chaplain with Swiss mercenaries in Italy led him to oppose the system of mercenary service. These views, which were expressed in terms of opposition to French influence in the Confederacy, and the fact that he was a papal pensioner led to his transfer from Glarus to the chaplaincy at Cloister Einsiedeln in 1516. At the end of 1518 he was called to be people's priest at the Zurich Great Church, largely because his views on the mercenary system were shared by an influential segment of the Zurich establishment. Despite opposition from some of the canons who feared him as a reform-minded Erasmian and accused him of immorality, his appointment was confirmed after he explained that his "immorality" was confined to contact with a known prostitute.

Between 1519 and 1525, when the Mass was abolished in the city, Zwingli advocated a practical program of reform in cooperation with the magistracy. His approach to the question of public worship and his view of the sacraments represented a far more radical break with past traditions than did the Lutheran reform movement. Indeed Luther had no profound influence upon Zwingli as he moved beyond Erasmus to form his own Augustinian-biblical theology within the environment of a Swiss city-state. Zwingli can be rightfully remembered as the first of the "Reformed" theologians. His own radical followers, led by Conrad Grebel* and Felix Manz,* endangered his alliance with the magistracy, whose support he believed was essential. After the Second Disputation in October 1523 they broke with him and in January 1525 formed a separate church, a conventicle, at Zollikon in which membership was symbolized by rebaptism. The rebaptizers (now called *Täufer*, then called Anabaptists*) were viewed as a threat to public order; the first of them were drowned in Lake Zurich with Zwingli's approval in 1527.

Zwingli's last years were marked by increasing political activity. He hoped both to open the entire Confederacy to the preaching of the Gospel and to create a European-wide anti-Hapsburg alliance. By 1528 the urban cantons Basle, Schaffhausen, and Bern, the most powerful of the Confederates, as well as Constance, had accepted Zwingli's reform program and had allied themselves with Zurich, but his hopes to extend the alliance to include the German Protestants led by Philip, landgrave of Hesse, were disappointed when he and Luther failed to reach agreement on the question of Christ's presence in the Eucharist (Marburg Colloquy,* 1529). This failure, with Bern's preoccupation with westward expansion, left the Swiss Protestants divided and exposed to a counteroffensive by the Forest Cantons which ended with Zwingli's death in the battle of Kappel and halted the expansion of the Reformation in German Switzerland.

Though frequently a member of one or another of the commissions set up by the Zurich government to find solutions for various domestic and diplomatic problems, Zwingli never held political office. His influence was the result of his ability and personal connections. He was able to exercise this influence in part because his own view of the church and his doctrine of election allowed the visible church in the world to be identified with civil society, a *corpus permixtum*, and left him free to grant the Christian magistrate the right to determine the external forms of the church's worship and life, and to govern the Christian Commonwealth in cooperation with the prophet who expounded the Scriptures for the spiritual well-being of the entire community. Zwingli's interpretation of the Eucharist* has been widely misunderstood, for during his last years he moved away from his earlier view which appears close to mere memorialism toward a doctrine of spiritual presence *(spiritualis manducatio)*.

BIBLIOGRAPHY: H. Zwingli, *Sämtliche Werke* (14 vols., ed. E. Egli and G. Finsler, 1905-69); S.M. Jackson (ed.), *The Latin Works of Huldreich Zwingli* (3 vols., 1912-29); O. Farner, *Huldrych Zwingli* (4 vols., 1943-60); G.W. Bromiley (ed.), *Zwingli und Bullinger* (1953); J. Courvoisier, *Zwingli: A Reformed Theologian* (1963); J.V. Pollet, *Huldrych Zwingli et la Réforme en Suisse* (1963); J. Rilliet, *Zwingli: Third Man of the Reformation* (1964); R.C. Walton, *Zwingli's Theocracy* (1967); M. Haas, *Huldrych Zwingli* (1969).

ROBERT C. WALTON